Compound class	Amine	Aldehyde	Ketone	Carboxylic acid
General formula	RNH_2, R_2NH, R_3N (R = alkyl, aryl)	R—CH=O (R = alkyl, aryl, H)	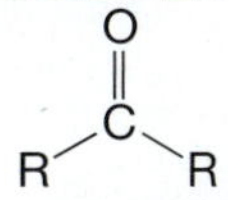 (R = alkyl, aryl)	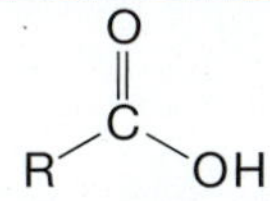 (R = alkyl, aryl)
Functional group	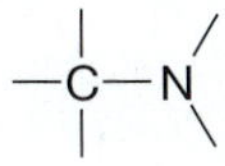	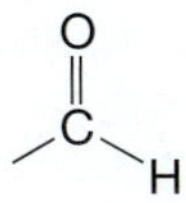	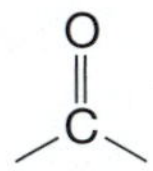	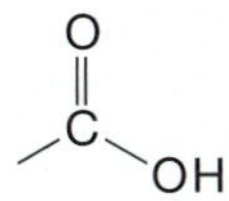
Specific example	$H_3C—CH_2—NH_2$	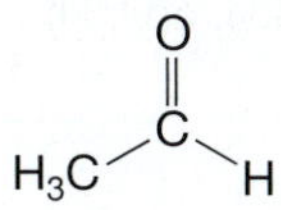	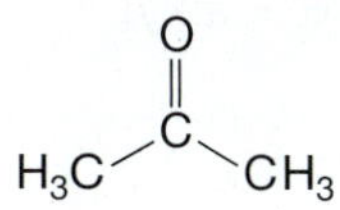	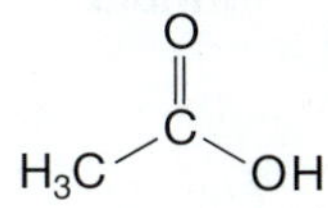
Common name	ethylamine	acetaldehyde	acetone	acetic acid
Substitutive name	ethanamine	ethanal	2-propanone	ethanoic acid

Compound class	Carboxylic Acid Derivatives				
	Ester	Amide	Anhydride	Acid chloride	Nitrile
General formula	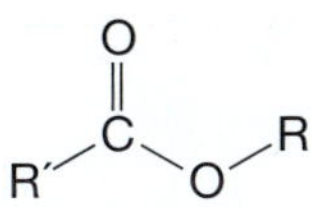 (R′ = alkyl, aryl, H) (R = alkyl, aryl)	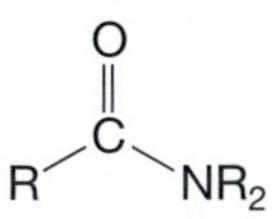 (R = alkyl, aryl, H)	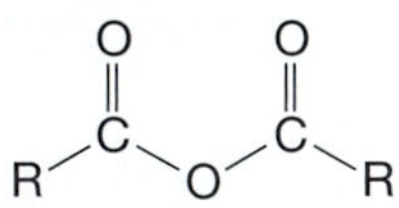 (R = alkyl, aryl, H)	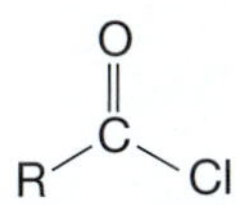 (R = alkyl, aryl, H)	R—C≡N (R = alkyl, aryl)
Functional group	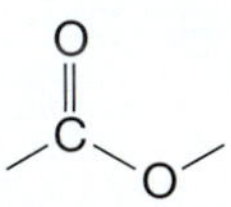	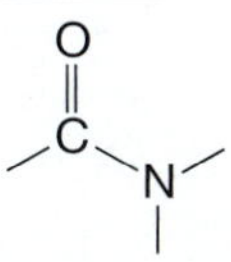	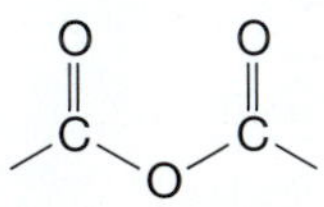	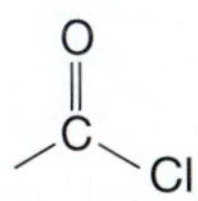	—C≡N
Specific example	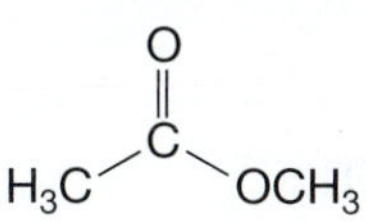	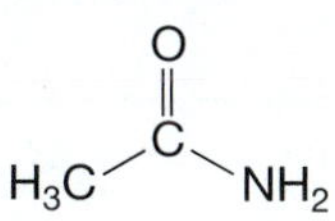	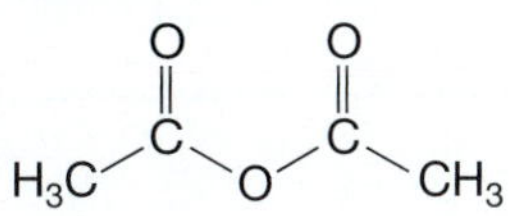	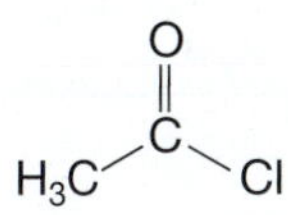	$H_3C—C≡N$
Common name	methyl acetate	acetamide	acetic anhydride	acetyl chloride	acetonitrile
Substitutive name	methyl ethanoate	ethanamide	ethanoic anhydride	ethanoyl chloride	ethanenitrile

ORGANIC CHEMISTRY

Seventh Edition

Marc Loudon
PURDUE UNIVERSITY

Jim Parise
UNIVERSITY OF NOTRE DAME

macmillan international
HIGHER EDUCATION

To Judy and my family—56 years of good chemistry
—ML

To Kathryn, Ariana, and Emilia
—JP

Senior Vice President, STEM: Daryl Fox
Program Managers: Beth Cole, Lori Stover, Jeff Howard
Marketing Manager: Maureen Rachford
Executive Content Development Manager, STEM: Debbie Hardin
Development Editor: John S. Murdzek
Executive Project Manager, Content, STEM: Katrina Mangold
Editorial Project Manager: Karen Misler
Director of Content, Physical Sciences: Heather Southerland
Senior Media Editor: Stacy Benson
Editorial Assistants: George Hajjar, Kaylin Fussell
Marketing Assistant: Morgan Psiuk
Director of Content Management Enhancement: Tracey Kuehn
Senior Managing Editor: Lisa Kinne
Senior Content Project Manager: Kerry O'Shaughnessy
Senior Workflow Manager: Paul Rohloff
Director of Design, Content Management: Diana Blume
Design Services Manager: Natasha Wolfe
Cover Designer: Laura de Grasse
Interior Designer: Patrice Sheridan
Art Manager: Matthew McAdams
Illustrations: Marc Loudon
Chapter Opener Art: Quade Paul
Director of Digital Production: Keri deManigold
Media Project Manager: Brian Nobile
Executive Permissions Editor: Cecilia Varas
Photo Researcher: Richard Fox, Lumina Datamatics, Inc.
Production Supervisor: Robert Cherry
International Contact: Sarah Oughton
Composition: Lumina Datamatics, Inc.
Printing and Binding: Transcontinental

ISBN-13: 978-1-319-37683-3
ISBN-10: 1-319-37683-5

Printed in Canada

1 2 3 4 5 6 25 24 23 22 21 20

Macmillan Learning
One New York Plaza
Suite 4600
New York, NY 10004-1562
www.macmillanlearning.com

BRIEF CONTENTS

CONTENTS

A green square ■ refers to sections containing "Chemical Biology Topics," and a blue dot ● refers to sections containing "Chemistry in the Real World" topics. These topics are also identified with margin icons within the text.

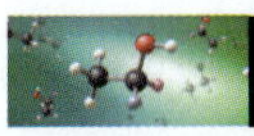

Dear Instructor,

Research in chemical education shows that many students who have trouble in organic chemistry are trying to memorize their way through the subject. One of the keys to students' success is to provide them with help in relating one part of the subject to the next—to help them see how various reactions that seem very different are tied together by certain fundamentals. An overarching goal of our text is to help students achieve a *relational understanding of organic chemistry*, which we have addressed in the following ways.

- **An Acid–Base Framework Is a Key to Understanding Mechanisms**

Although we have organized *Organic Chemistry,* Seventh Edition, by functional group, we have used mechanistic reasoning to help students understand the "why" and "how" of reactions. Mechanisms start with acid–base reactions and the curved-arrow notation and build from there.

- **Tiered Topic Development Reinforces Important Ideas**

Students will see many concepts introduced in a relatively simple way, revisited next with an added layer of complexity, and later reviewed again at a greater level of sophistication.

- **Everyday Analogies Help Students to Construct Their Own Knowledge**

We provide common analogies from everyday experience so that students can relate a new idea to something they already know.

- **An Increased Focus on Biological Applications Motivates Students Interested in the Allied Health Sciences**

Many organic chemistry classes are populated largely by premedical students, prepharmacy students, and other students interested in the life sciences. To that end, we have provided sections that stress the chemistry of biological processes and its relationship to reactions that the student has studied in the same chapter.

- **The Textbook Is a Flexible Resource**

Whether instructors want to take a biological orientation or teach a more traditional, synthetically oriented course, they will find everything they need in this text.

- **Solving Problems Is an Essential Component of the Learning Process**

We have provided a rich collection of more than 2000 problems across chapters: "Focused Problems" and "Skills Objectives" are mostly single-concept problems; "Integrated Problems" require students to draw on their knowledge of multiple sections within the chapter as well as multiple chapters and even earlier courses; "Study Problems" are worked examples.

ORGANIZATIONAL CHANGES IN THE SEVENTH EDITION

A new Chapter 10 covers free-radical reactions, main-group organometallic reactions, and carbenes, topics that were removed from Chapters 5 and 9 of the previous edition. We eliminated Chapter 14 on alkynes and redistributed this material in Chapters 4, 5, and 10. We provided a more unified treatment of acids and bases in Chapter 3. Finally, chapter summaries have been moved to the *Study Guide and Solutions Manual* to make room for more problems.

Supporting students and instructors without sacrificing the rigor or sophistication of organic chemistry

Available for the first time in **Achieve,** the highly respected ***Organic Chemistry,* Seventh Edition,** is rooted in scholarly tradition and informed by chemical education. ***Organic Chemistry*** helps students achieve a relational understanding of organic chemistry. Loudon and Parise emphasize how key organic chemistry concepts relate to one another through functional group organization and mechanistic reasoning to help students understand the "why" of reactions.

Overview of key features

The seventh edition has been rigorously updated and continues to be a trusted reference for instructors, with the most up-to-date research and the highest standard of accuracy. Hallmark features, including an elegant writing style, masterful problems, and biological applications, are retained and enhanced by Achieve, Macmillan's new online learning system.

NEW TEXT FEATURES AND KEY CHANGES:

- **New skills objectives** for each chapter guide students in the print text and align with Achieve
- **New study problems** throughout the text with **mirrored questions in Achieve**
- **New animated lessons** in the Achieve e-book to support author Marc Loudon's precise artwork
- A focus on **biological applications**
- **Extensive set of problems, written by the authors**

Achieve helps educators and students throughout the full flexible range of instruction and offers resources to support the learning of core concepts, visualization, problem-solving, and assessment. Powerful analytics and instructor support resources in Achieve pair with exceptional organic chemistry content to provide an unrivaled learning experience.

FEATURES OF ACHIEVE:

- A **design guided by learning science research.** Co-designed through extensive collaboration and testing by both students and faculty, including two levels of Institutional Review Board approval for every study of Achieve
- **An interactive e-book** with embedded multimedia and features for highlighting, note-taking, and accessibility support
- A **flexible suite of resources** to support learning core concepts, visualization, problem-solving, and assessment
- A **detailed gradebook with insights for just-in-time teaching** and reporting on student and full class achievement by learning objective
- Easy **integration and gradebook sync with iClicker** classroom engagement solutions
- Simple **integration with your campus LMS** and **availability through Inclusive Access** programs

NEW MEDIA AND ASSESSMENT FEATURES IN ACHIEVE FOR *ORGANIC CHEMISTRY*, SEVENTH EDITION:

- **Skills objectives** with accompanying problems at the end of each chapter ensure that students master the key concepts.
- **Animated lessons** in Achieve provide practice with visualization skills. Animated lessons are embedded within the e-book.
- **LearningCurve adaptive quizzing** serves up low Bloom's level problems to solidify students' recognition of nomenclature and core concepts. The game-like learning tool guides students by providing extra practice where they need it.

Achieve

Achieve is the culmination of years of development work put toward creating the most powerful online learning tool for organic chemistry students. It houses all of our renowned assessments, multimedia assets, e-books, and instructor resources in a powerful new platform.

Achieve supports educators and students throughout the full range of instruction, including assets suitable for pre-class preparation, in-class active learning, and post-class study and assessment. The pairing of a powerful new platform with outstanding organic chemistry content provides an unrivaled learning experience.

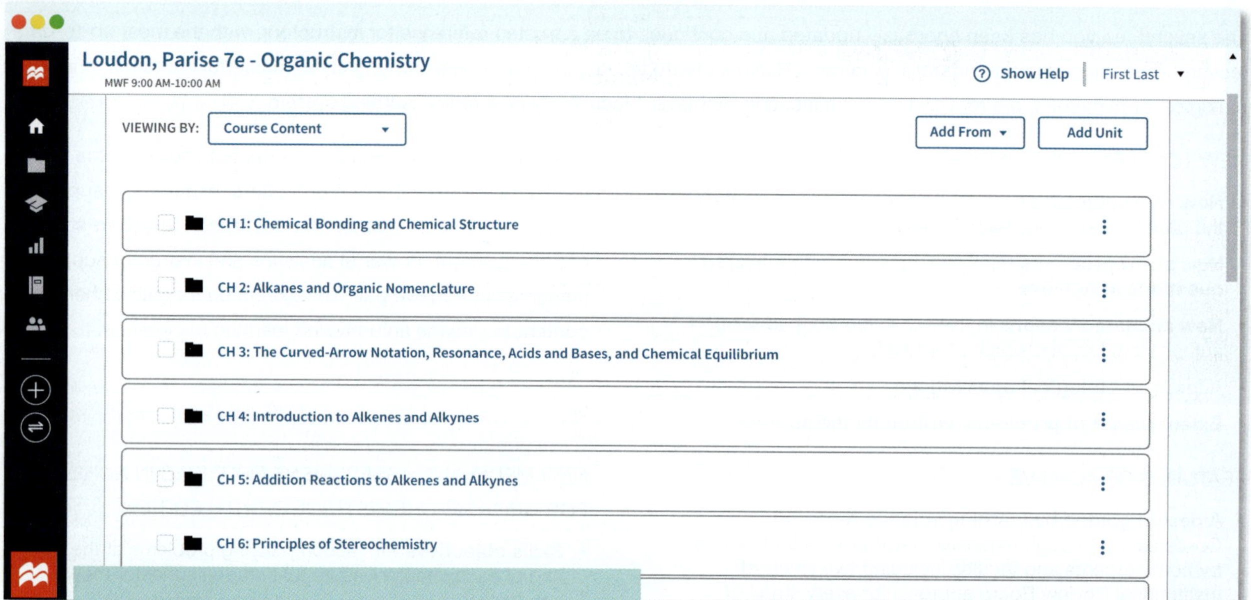

Highlights:

- **A design guided by learning science research.** Co-designed through extensive collaboration and testing by both students and faculty including two levels of Institutional Review Board approval for every study of Achieve.
- **A learning path of powerful content** including pre-class, in-class, and post-class activities and assessments. A detailed gradebook with insights for just-in-time teaching and reporting on student achievement by learning objective.
- **Easy integration and gradebook sync** with iClicker classroom engagement solutions.
- **Simple integration** with your campus LMK and availability through **Inclusive Access** programs.

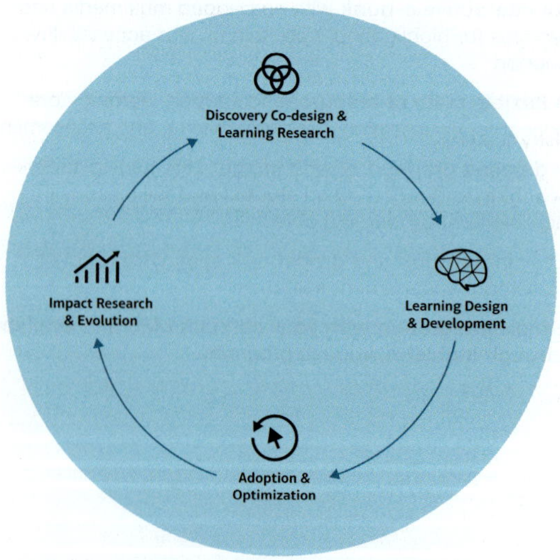

For more information or to sign up for a demonstration of Achieve, contact your local Macmillan representative or visit **macmillanlearning.com/achieve**

Engage Every Student

Achieve supports flexible instruction and engages student learning. This intuitive platform includes content for pre-class preparation, in-class active learning, and post-class engagement and assessment, providing students with an unparalleled environment and resources for learning!

LearningCurve's game-like quizzing motivates each student to engage with the course content, and reporting tools help teachers get a handle on what their class needs.

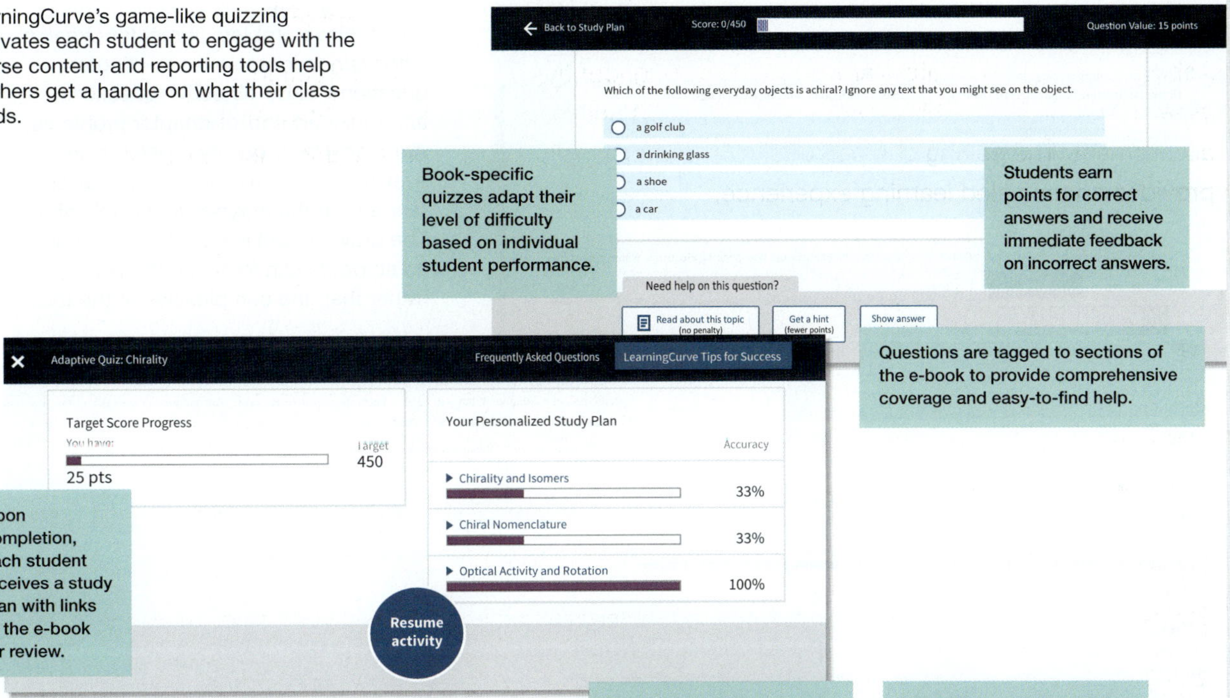

Book-specific quizzes adapt their level of difficulty based on individual student performance.

Students earn points for correct answers and receive immediate feedback on incorrect answers.

Questions are tagged to sections of the e-book to provide comprehensive coverage and easy-to-find help.

Upon completion, each student receives a study plan with links to the e-book for review.

Quiz adapts to each student's needs based on performance, automatically providing more questions on topics where the student is struggling.

Performance is measured by Learning Objectives and can also be tracked to give instructors actionable information about topics that need extra emphasis.

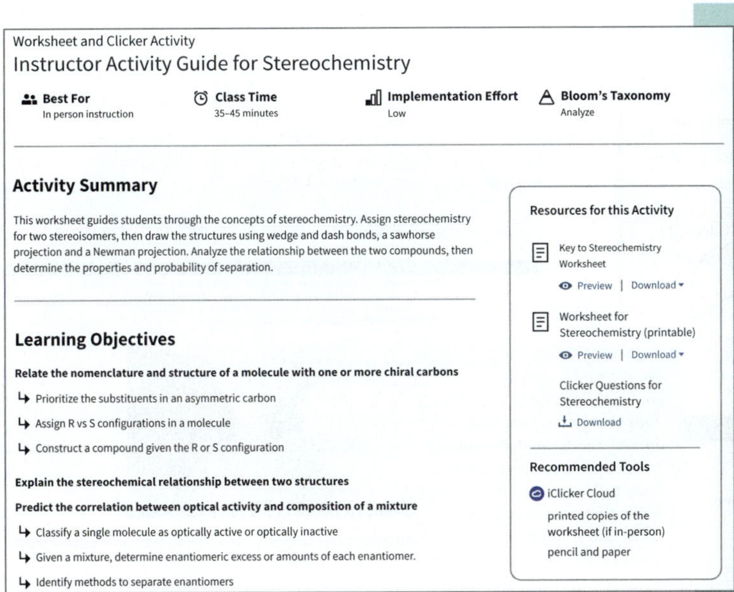

In-Class Activities

Active learning exercises provide students with hands-on exploration and help correct common misconceptions.

The activities have been vetted in the classroom and serve as a great off-the-shelf option for incorporating new methods of active learning in your classroom.

The activities contain everything you need, including an instructor's guide, presentation decks, and student handouts.

iClicker

Achieve includes gradebook integration with iClicker, allowing for a seamless pairing of the best classroom response system with the best in-class content.

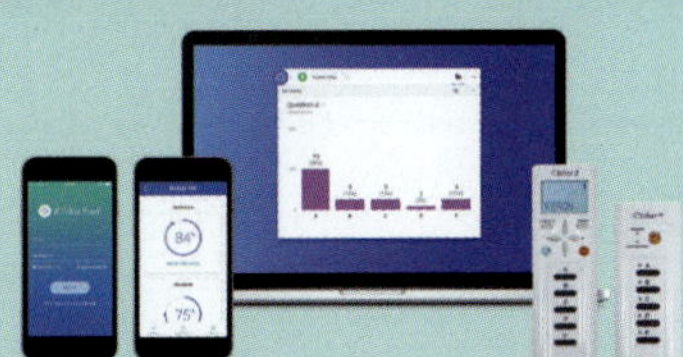

A Foundation for Problem Solving

The Achieve problem library for *Organic Chemistry* mirrors select in-chapter questions, skills-objective questions, and multipart end-of-chapter problems. Additional high-quality problems are available in the full library with a diverse range of problem types and levels of rigor. The drawing tool is easy to use and intuitive, so students can focus on the concepts rather than the complexities of the tool.

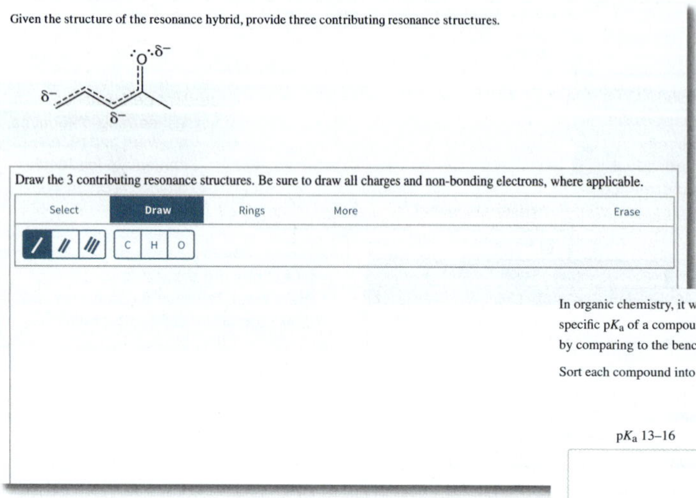

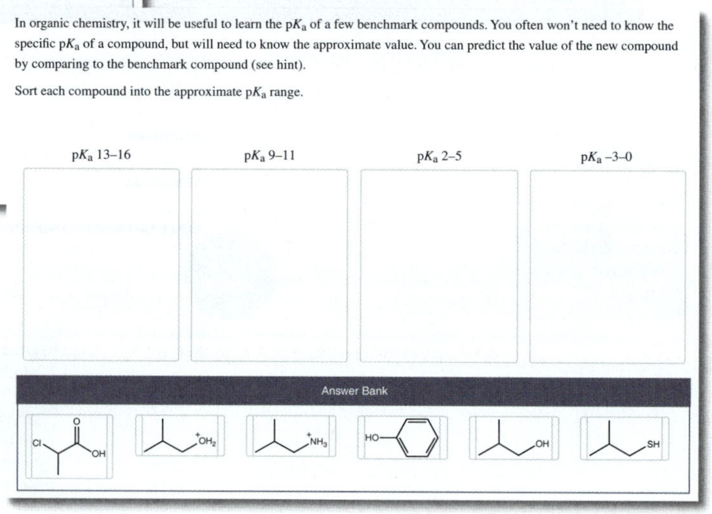

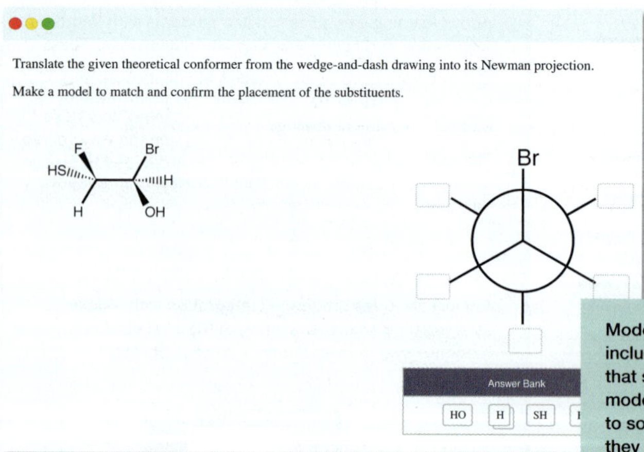

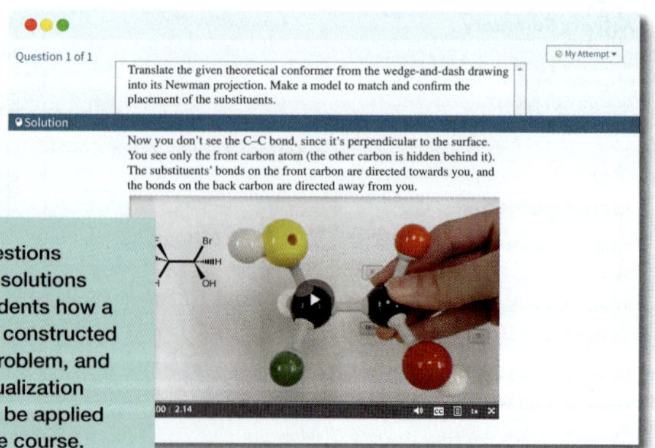

Model Kit Questions include video solutions that show students how a model can be constructed to solve the problem, and they build visualization skills that can be applied throughout the course.

Powerful analytics, viewable in an elegant dashboard, offer instructors a window into student progress. Achieve gives you the insight to address students' weaknesses and misconceptions before they struggle on a test.

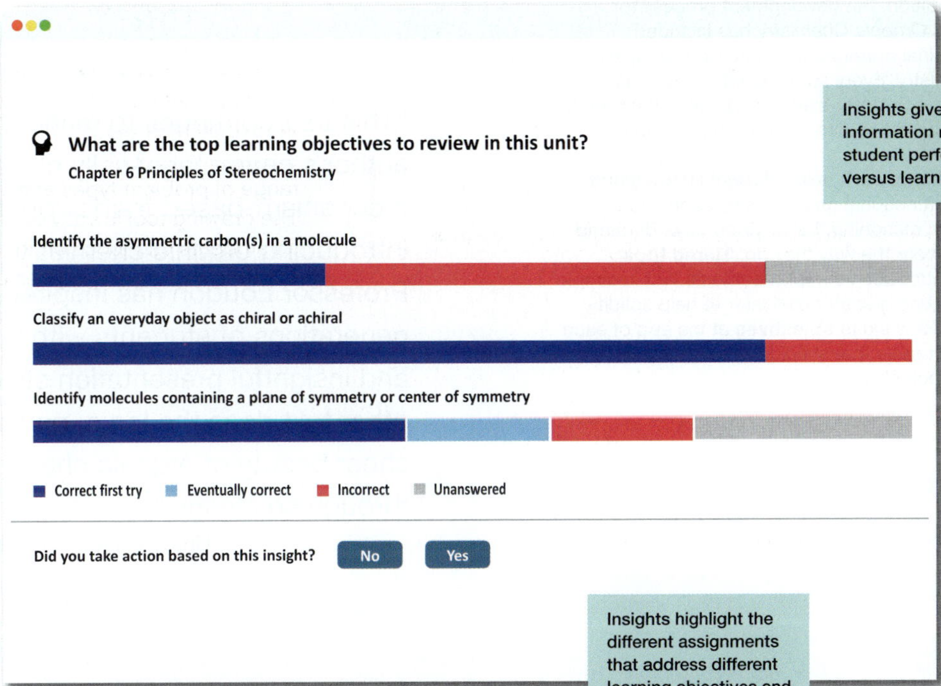

Insights give instructors information regarding student performance versus learning objectives.

Insights highlight the different assignments that address different learning objectives and show visualizations of student performance.

Reports provide powerful views of student performance and can be viewed in a number of ways.

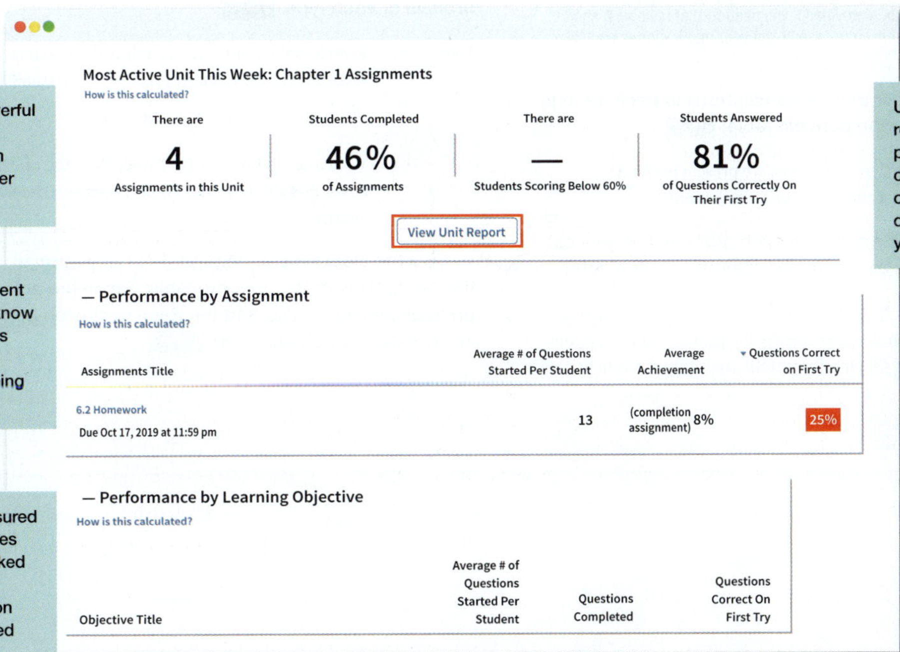

Unit and subunit reports give quick performance overviews based on the way you've decided to group your assignments.

At-a-glance assignment level details let you know how far your students have gone and how quickly they're grasping the material.

Performance is measured by Learning Objectives and can also be tracked to give instructors actionable information about topics that need extra emphasis.

Scholarly Tradition and An Organization Designed for Understanding

Since the first edition, the development process for Loudon/Parise's *Organic Chemistry* has included researching original or review literature of each topic in organic chemistry. Every subsequent edition has emphasized this **scholarly tradition**, as demonstrated by reaction examples taken directly from the literature.

An **acid–base framework** helps students understand mechanisms. A functional group organization, along with mechanistic reasoning, helps students understand why reactions occur the way they do. **Tiered topic development** introduces complex subjects first in simple terms and goes into greater detail later to help solidify understanding. New **skills objectives** at the end of each chapter highlight skills that should have been developed after studying each chapter.

> "This text continues to refine the author's characteristically elegant, mechanism-based framework for introducing organic chemistry. Professor Loudon has inspired several generations of students with his clear and insightful presentation style. In no other text does the logic, power, and sheer beauty of organic chemistry shine through so clearly."
>
> —Bruce Ganem, *Cornell University*

SKILLS OBJECTIVES WITH PROBLEMS

"Skills Objectives" are given at the end of each chapter. These describe skills that you should develop as you study each chapter. (The section number is provided if you need further review.) A short problem is provided in each case to test your application of each skill.

- **Give the number of valence electrons in each atom in the "A" groups of the periodic table. 1.2A**

1.38 How many valence electrons are present in the Al (aluminum) atom? In the I (iodine) atom?

- **Show the ion formed from each atom by the gain or loss of electrons, assuming the operation of the octet rule. 1.2A**

1.39 What ion should be formed by the gain or loss of electrons from the Mg (magnesium) atom according to the octet rule?

- **List the atomic orbitals in order of increasing energy up through 3d, and show the number of equivalent orbitals of each type. 1.2C**

1.40 (a) Which of the following orbitals is *not* possible according to the quantum theory of the atom?

4d 5s 2p 3p 2d 6p

(b) Which orbital in part (a) [other than the one that is not possible] has the highest energy? The lowest energy?

- **Give the electronic configuration of any atom in the "A" groups of the periodic table. Apply the aufbau principle, Hund's rules, and the Pauli exclusion principle in deriving this configuration. 1.2D**

Emphasis on Student Learning and Motivation with Varied and Creative Problems

Organic Chemistry motivates the large life science audience with a **focus on biological applications** and uses **analogies** to tie discussions of chemical principles to everyday experiences so that students can relate a new idea to something they are already familiar with. The **extensive problem sets are all written by the authors**, pulled from primary literature, and provide opportunity for realistic problem solving through varied levels of rigor. **New study problems** throughout the text will be mirrored in the Achieve problem library for *Organic Chemistry.*

B. Hydride Reduction in Biology

Chemical Biology Topic

In Sec. 11.8 we introduced NAD^+ as an important biological oxidizing agent. (The structure of NAD^+ is shown in Display 11.68, Sec. 11.8.) When NAD^+ is used for an oxidation, its reduced form, NADH, is formed. NADH is "nature's hydride reagent." It acts as a biological reducing agent in the same sense that $LiAlH_4$ and $NaBH_4$ act as laboratory reducing agents, as we illustrate in this section.

One of the most important biological functions of NADH is *carbonyl reduction.* An example of this reaction is the conversion of pyruvate into (*S*)-lactate under anaerobic conditions in muscle tissue. The pro-(*R*) hydride of NADH (*red*) is transferred stereospecifically.

Both NAD^+ and NADH exist in phosphorylated forms, $NADP^+$ and NADPH; the position of phosphorylation is also shown in Display 11.68, Sec. 11.8. These forms differ in their cellular location and the general role that they play in cellular biochemistry. You will learn these details if you study biochemistry. However, from a chemical standpoint, they function in the same way.

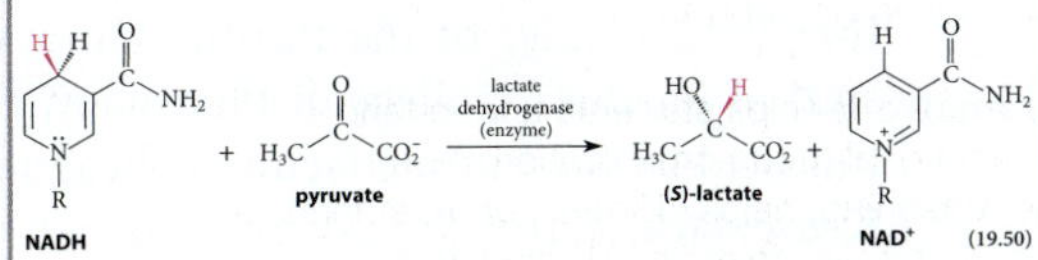

(19.50)

(The lactate produced in this reaction is the source of the burning in muscle that occurs after very intense exercise; the conditions of oxygen deprivation promote this reaction as a way of generating NAD^+ required in the metabolism of glucose.) This reaction is catalyzed by the enzyme *lactate dehydrogenase.* Although the enzyme is commonly called a "dehydrogenase," metabolically the reaction normally proceeds in the reductive direction as shown in Eq. 19.50. [The enzyme is "officially" named (*S*)-lactate:pyruvate oxidoreductase; it catalyzes the reaction in both directions, as it must by the principle of microscopic reversibility.]

In the reduction of pyruvate to lactate, the ketone carbonyl group is activated by hydrogen bonding from a protonated histidine residue of the enzyme. In this respect, the reaction resembles sodium borohydride reduction, in which the carbonyl group is activated by hydrogen-bond donation from a protic solvent (Eq. 19.46, Sec. 19.9A).

Study Problem 19.6

Outline two Wittig alkene syntheses of 2-methyl-1-hexene. Is one synthesis preferred over the other? Why?

Solution The analysis in Eq. 19.96 suggests that the "right-hand" part of the alkene can be derived from the ketone 2-hexanone:

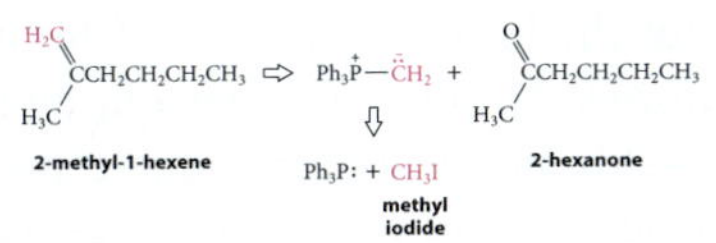

(19.98a)

Another possibility, however, is that the "left-hand" part of the alkene is derived from formaldehyde:

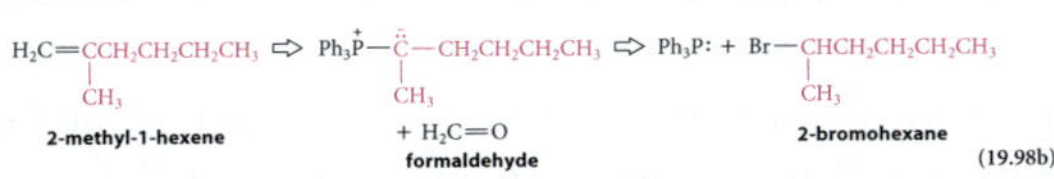

(19.98b)

Although both syntheses seem reasonable, the one in Eq. 19.98a requires an S_N2 reaction of triphenylphosphine with a methyl halide, whereas the one in Eq. 19.98b requires an S_N2 reaction of triphenylphosphine with a secondary alkyl halide. The first reaction is preferred because methyl halides are much more reactive than secondary alkyl halides (Sec. 9.4C).

The outlined synthesis is as follows:

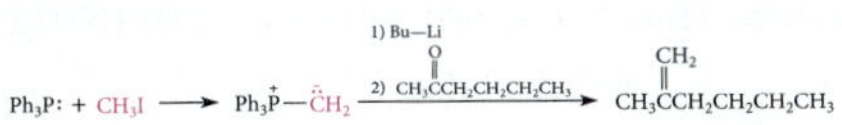

"From over thirty years in the classroom, Marc Loudon knows how students learn. His problems are the work of a master teacher; they are original, range in difficulty levels, and are all based on real examples from the literature....This book is a symphony of great ideas and will open students' eyes to the wonders of modern organic chemistry."

—John Grutzner, *Purdue University*

ACKNOWLEDGMENTS

The electronic resources of the Purdue Library have streamlined the research process for this text; this was essential for an author living 900 miles south of Purdue. We thank the faculty of the Purdue Library System Vicki Killion and David Zwicky, and the IT staff of Purdue's College of Pharmacy, especially Mark Sharp, for their able assistance. We are indebted to our Purdue faculty colleagues John Grutzner, George Bodner, David Nichols, Casey Krusemark, Carol Post, Chris Rochet, and Rodolfo Pinal, and Notre Dame colleagues Paul Helquist, Xavier Creary, and Seth Brown for valuable advice, assistance, and suggestions. We are indebted to our reviewers, particularly Animesh Aditya of Kennesaw State University and Lisa Bonner of Eckerd College, who were with us from beginning to end. Our development editor John Murdzek provided important review and suggestions along the way. We appreciate very much the work of Macmillan professionals Beth Cole, Lori Stover, and Jeff Howard, our editors; Kerry O'Shaughnessy, project manager; Matt McAdams, art coordinator; Cecilia Varas, permissions editor; and Karen Misler, schedule coordinator. We would also like to thank the many students and faculty from all over the country who made suggestions, offered comments, and reported errors in the previous edition. We welcome correspondence with the students and faculty using this edition. We can be reached by email at *loudonm@purdue.edu* and *jparise@nd.edu*.

ML could not have completed this project without the love and support of Judith and his family and friends, for which he is grateful beyond words. As time marches on, he is increasingly aware of the debt that he owes to his mentors—his many undergraduate professors at LSU; his mentor, the late Donald S. Noyce at Cal–Berkeley; his lab partner Charles A. Brown at Cal–Berkeley; and his biochemistry mentor, the late Daniel E. Koshland, Jr., also at Berkeley.

JP would like to thank his wife, Kathryn, and his family for their support and encouragement. He is indebted to his own educators, and his current and former colleagues, at Oswego State, Duke, and Notre Dame. He is especially grateful to work with and learn from his coauthor ML, and for his trust and guidance.

Our wish is that the students who use this text will see the amazing diversity and beauty of science through their study of organic chemistry, and that they will benefit from using this book as much as we have enjoyed writing it.

MARC LOUDON
October 2020
New Orleans, Louisiana

JIM PARISE
October 2020
Notre Dame, Indiana

REVIEWERS AND CONSULTANTS

The author and publisher wish to acknowledge with gratitude the extensive support received from the organic chemistry community in the development of this edition. Our reviewers and consultants are listed below; some people served in multiple roles.

Textbook Reviewers

Donald Aue, *University of California, Santa Barbara*
Peter Bell, *Tarleton State University*
Michael Berg, *Virginia Polytechnic Institute and State University*
Kenneth Darrell Berlin, *Oklahoma State University*
Thomas M. Bertolini, *University of Southern California*
Marco Bonizzoni, *The University of Alabama*
Maggie Bump, *Virginia Polytechnic Institute and State University*
Corey Causey, *University of North Florida*
J. Michael Chong, *University of Waterloo*
Elizabeth A. Clizbe, *SUNY Buffalo*
Cynthia Dowd, *George Washington University*
Brian Esselman, *University of Wisconsin, Madison*
Markus Etzkorn, *University of North Carolina, Charlotte*
Gregory Friestad, *University of Iowa*
Barnabas Gikonyo, *SUNY Genesco*
Nicholas Hill, *University of Wisconsin, Madison*
Alex John, *California Polytech State University, Pomona*
Arif Karim, *Austin Community College*
Monique Koppel, *University of Maryland*
Joseph Lauher, *Stony Brook University*
Jung-Jae Lee, *University of Colorado, Denver*
Philip Lukeman, *St John's University*
Kristen Mascall, *Brandeis University*
Deborah L Mead, *Seminole State College, Sanford*
Ognjen Miljanic, *University of Houston*
Christopher Nicholson, *Marian University, Indianapolis*
Hector Palencia, *University Nebraska, Kearney*
Christopher Petrie, *Eastern Florida State College*
Joshua G. Pierce, *North Carolina State University*
Allan Pinhas, *University of Cincinnati*
Jeffrey Raker, *University of South Florida*
Bala Ramjee, *Old Dominion University*
Jacqueline Richardson, *University of Colorado at Boulder*
Thomas R. Ruttledge, *Cornell University, Ithaca*
Alan Sellinger, *Colorado School of Mines*
Kevin Shaughnessy, *The University of Alabama*
Brian M. Stoltz, *California Institute of Technology*
Marcus Tius, *University of Hawaii*
Ross Weatherman, *Rose-Hulman Institute of Technology*

Media Reviewers

John Carran, *Queens University, Kingston*
Alice Cherestes, *McGill University, Macdonald Campus*
Wheeler Conover, *Southeast Kentucky Community and Technical College*
Sean Hickey, *Wayne State University*
David Hunt, *The College of New Jersey*
Joseph Lauher, *Stony Brook University*
Taylor Mach, *Concordia University Saint Paul*
Nasri Nesnas, *Florida Institute of Technology*
Hector Palencia, *University of Nebraska, Kearney*
Paul Sampson, *Kent State University*
Elijah St. Germain, *Florida Atlantic University*
Pritha Subramanian, *Kent State University*
Anil Waghe, *Plymouth State University*
Hua Zhao, *University of Northern Colorado*
Eugene Zubarev, *Rice University*

Sixth Edition Reviewers

Animesh V. Aditya, *Purdue University*
Igor Alabugin, *Florida State University*
John Bartmess, *University of Tennessee*
Peter Beak, *University of Illinois*
Jason Belitsky, *Oberlin College*
Thomas Berke, *Brookdale, The County College of Monmouth*
Daniel Bernier, *Riverside Community College*
Michael Best, *University of Tennessee*
Caitlin Binder, *California State University, Monterey Bay*
Dan Blanchard, *Kutztown University of Pennsylvania*
Lisa Bonner, *Eckerd College*
Paul Bonvallet, *College of Wooster*
Ned B. Bowden, *The University of Iowa*
Stephen G. Boyes, *Colorado School of Mines*
David Brown, *Davidson College*
Rebecca Broyer, *University of Southern California*
Paul Carlier, *Virginia Tech*
David Cartrette, *South Dakota State University*
Allen Clauss, *University of Wisconsin*
Geoffrey W. Coates, *Cornell University*
Bryan Cowen, *University of Denver*
Michael Danahy, *Bowdoin College*
William Daub, *Harvey Mudd College*
William Dichtel, *Cornell University*

Sally Dixon, *University of Southampton*
Andrew Duncan, *Willamette University*
Jason Dunham, *Ball State University*
Mark Elliott, *Cardiff University*
Brian Esselman, *University of Wisconsin-Madison*
Ed Fenlon, *Franklin & Marshall College*
Marcia France, *Washington and Lee University*
Lee Friedman, *University of Maryland*
Bruce Ganem, *Cornell University*
Sarah Goh, *Williams College*
Bobbie Grey, *Riverside City College*
Nicholas J. Hill, *University of Wisconsin-Madison*
John Hoberg, *University of Wyoming*
Joseph Houck, *University of Maryland*
Lyle Isaacs, *University of Maryland*
Madeleine Joullie, *University of Pennsylvania*
Sarah Kirk, *Willamette University*
Riz Klausmeyer, *Baylor University*
Brian Long, *University of Tennessee*
Leonard MacGillivray, *University of Iowa*
Charles Marth, *Western Carolina University*
Dan Mattern, *University of Mississippi*
James McKee, *University of the Sciences in Philadelphia*
John Medley, *Centre College*
Kevin P. C. Minbiole, *Villanova University*
Timothy Minger, *Mesa Community College*
Michael P. Montague-Smith, *University of Maryland*
Michael Nee, *Oberlin College*
Donna Nelson, *University of Oklahoma*
James Nowick, *University of California, Irvine*
Kimberly Pacheco, *University of Northern Colorado*
Laura Parmentier, *Beloit College*
Joshua Pierce, *North Carolina State University*
David P. Richardson, *Williams College*
Robert E. Sammelson, *Ball State University*
Paul Sampson, *Kent State University*
Nicole L. Snyder, *Davidson College*
Gary Spessard, *University of Oregon*
Brian M. Stoltz, *California Institute of Technology*
Scott Stoudt, *Coe College*
Jennifer Swift, *Georgetown University*
Eric Tillman, *Bucknell University*
Mark M. Turnbull, *Clark University*
David A. Vosburg, *Harvey Mudd College*
Ross Weatherman, *Rose-Hulman Institute of Technology*
Carolyn Kraebel Weinreb, *Saint Anselm College*
Travis Williams, *University of Southern California*
Jimmy Wu, *Dartmouth College*
Hubert Yin, *University of Colorado*

ABOUT THE AUTHORS

Judith L. Loudon

MARC LOUDON received his B.S. (magna cum laude) in chemistry from Louisiana State University in Baton Rouge and his Ph.D. in organic chemistry in 1968 from the University of California, Berkeley, where he worked with Professor Donald S. Noyce. After a two-year postdoctoral fellowship with Professor Daniel E. Koshland, Jr., in Biochemistry at Berkeley, Dr. Loudon joined the Department of Chemistry at Cornell University, where he taught organic chemistry to both preprofessional students and science majors. In 1977, he joined the Department of Medicinal Chemistry and Pharmacognosy (now the Department of Medicinal Chemistry and Molecular Pharmacology) in the College of Pharmacy at Purdue University, where he taught organic chemistry to prepharmacy students for 38 years. Dr. Loudon served as the Associate Dean for Research and Graduate Programs for the College of Pharmacy, Nursing, and Health Sciences from 1988 to 2007. He has received numerous teaching awards: the Clark Award of the College of Arts and Sciences at Cornell (1976); the Heine Award of the School of Pharmacy at Purdue (1980 and 1985); Purdue's Class of 1922 Helping Students Learn Award (1988); the Charles B. Murphy Award that recognizes the best teachers at Purdue (1999); and the Indiana "Professor of the Year" Award of the Carnegie Foundation (2000). Dr. Loudon was named the Gustav E. Cwalina Distinguished Professor in 1996, one of the first three faculty members to be recognized by Purdue as distinguished professors for teaching and teaching scholarship. He was inducted into Purdue's Teaching Academy in 1997, and he was listed in Purdue's permanent "Book of Great Teachers" in 1999.

In collaboration with Professor George Bodner and, more recently, Professor Animesh Aditya, Dr. Loudon has developed and implemented collaborative-learning methods for large organic chemistry classes from 1993 to the present. He was part of the HHMI-sponsored NEXUS project from 2010–2014, which resulted in the complete redesign of the first organic chemistry course to make it more relevant to the interests of students in prehealth profession programs. Dr. Loudon was named Distinguished Professor Emeritus following his retirement from Purdue in 2015. Dr. Loudon is the author of numerous research articles, co-author of an in-house laboratory manual, and co-author of the *Study Guide and Solutions Manual* for this text. The first edition of this text was published in 1984.

Dr. Loudon and his wife, Judith, who celebrated their 56th anniversary in 2020, live in New Orleans and have two grown sons and four grandchildren ranging in age from 11 to 25. Dr. Loudon is an accomplished pianist and organist who performs professionally. Dr. Loudon and his wife enjoy concerts, theatre, tennis, and travel.

Barbara Johnston, University of Notre Dame

JIM PARISE received his B.S. in chemistry from the State University of New York at Oswego in 2000 and his Ph.D. in organic chemistry in 2007 from Duke University, where he worked with Professor Eric Toone. After a postdoctoral fellowship with Professor David Lawrence at the University of North Carolina at Chapel Hill, Dr. Parise joined the Department of Chemistry at Duke University, where he taught organic chemistry and coordinated the organic chemistry laboratory courses. In 2011, he joined the faculty in the Department of Chemistry and Biochemistry at the University of Notre Dame. As a teaching professor at Notre Dame, he has received multiple awards for teaching excellence, including the Thomas P. Madden Award for Exceptional Teaching of First Year Students (2015) and the Rev. Edmund P. Joyce Award for Excellence in Undergraduate Teaching (2019).

Dr. Parise teaches primarily Organic Chemistry courses to prehealth (medicine, veterinary, and dental) students and chemistry and biochemistry majors. His scholarly interests include development of more effective teaching methods for large lecture courses and improving organic chemistry pedagogy in both the classroom and the laboratory. He created a training and mentoring program for new instructors and has developed auxiliary courses aimed at supplementing student preparation for introductory organic courses. He has also incorporated pedagogical technology into large lecture halls to facilitate active learning through the use of teaching tools like iPads, Surface Pros, and the Lightboard. In addition to teaching general and organic chemistry, he has also taught various laboratories and a course in the chemistry of fermentation and distillation. Dr. Parise has authored and co-authored laboratory manuals and a peer-reviewed book chapter on writing in the laboratory, and he is co-author of the *Study Guide and Solutions Manual* for this text.

Outside of teaching and writing, he enjoys spending time with his family, traveling, running, and beer making.

1 CHEMICAL BONDING AND CHEMICAL STRUCTURE

1.1 INTRODUCTION

A. What Is Organic Chemistry?

Organic chemistry is the branch of science that deals generally with compounds of carbon. The number of organic compounds is not known for sure, but there are *millions*—perhaps 10 million—known organic compounds. The number of *possible* organic compounds is essentially infinite. But all of these have one thing in common: they contain carbon. The large number of organic compounds results from carbon's singular ability to combine with other carbons to form long chains (as you will learn).

B. How Is Organic Chemistry Useful?

This is the structure of the anticancer drug imatinib (marketed as Gleevec).

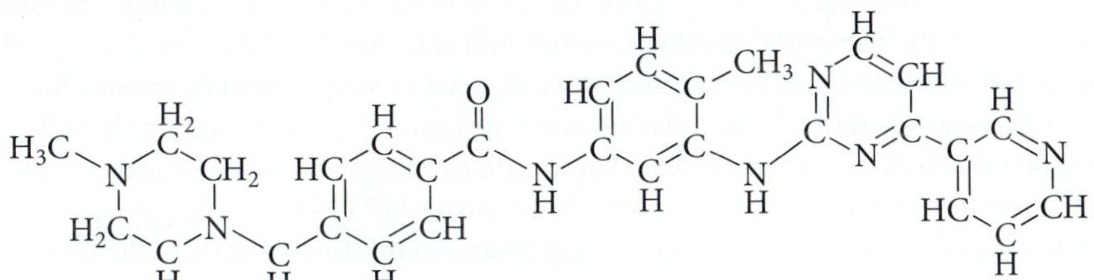

imatinib (Gleevec)

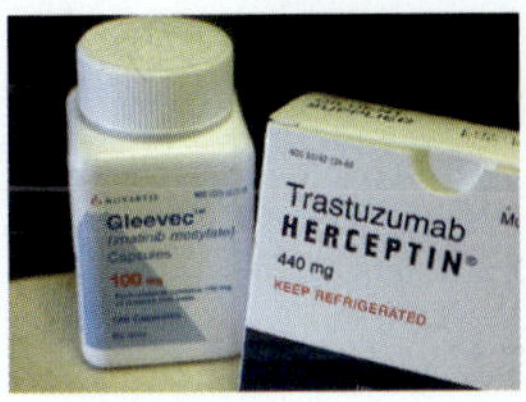

Carlos Chavez/Getty Images

Chances are, you might not know how to interpret this structure, but you will learn. Imatinib illustrates the utility of organic chemistry. Prior to 2001, a medical diagnosis of chronic myelogenous leukemia (CML), a relatively uncommon cancer of the blood and bone marrow, was a death sentence. However, an oncologist, Brian Drucker, and a biochemist, Nicholas Lydon, using results on the genetic basis of CML, were able to screen a number of organic compounds for their ability to inhibit a key enzyme, *bcr–abl*. This means that they found a way to block the progression of CML. Working with organic chemists, they produced analogs of their successful compounds (compounds of similar structure), ultimately landing on imatinib as their drug of choice. Clinical trials, conducted with physician Charles Sawyers, led the Food and Drug Administration (FDA) to approve imatinib in 2001. Imatinib cured CML in most cases, and it has proved useful in the treatment of other cancers as well. The ability to rapidly prepare and characterize a large number of organic compounds was crucial in the development of this drug, and this required a knowledge of organic chemistry. In the larger picture, the alliance of organic chemistry, molecular biology, and medicine is clearly the way of the future in the development of effective new drugs, because *most drugs are organic compounds.*

By the time you have worked your way through this text, you will certainly understand the structure and chemical properties of imatinib and many other important molecules. Should you want to participate in the excitement of drug discovery, you will be prepared for advanced work that can set you on that road. If you are headed for a career as a practicing health-care professional, you will be prepared to understand the chemical basis of biochemistry, which is fundamental to all life sciences.

Organic chemistry is not only useful in medicine. Many valuable materials come from organic chemistry. Things as diverse as textiles, body armor, artificial sweeteners, sports equipment, and computers are materials, or are based on materials, that come from organic chemistry.

Apart from its practical utility, organic chemistry is an intellectual discipline that has both theoretical and experimental aspects. You can use the study of organic chemistry to develop and apply basic skills in problem solving and, at the same time, to learn a subject of immense practical value. Whether your goal is to be a professional chemist, to remain in the mainstream of a health profession, or to be a well-informed citizen in a technological age, you will find value in the study of organic chemistry.

In this text we have several objectives. We present the "nuts and bolts"—the naming, classification, structure, and properties of organic compounds. We also cover the principal reactions and the syntheses of organic molecules. But, more than this, we develop underlying principles that allow us to understand, and sometimes to predict, reactions *rather than simply memorizing them.* We'll bring some order to the rather daunting array of 10 million organic molecules, their reactions, and their properties. Along the way, we'll continue to highlight some of the important applications of organic chemistry in medicine, industry, and other areas.

C. The Emergence of Organic Chemistry

Although the applications of organic chemistry, as we have seen, are not restricted to the life sciences, the name *organic* certainly implies a connection to living things. In fact, the emergence of organic chemistry as a science was closely associated with the evolution of the life sciences.

As early as the sixteenth century, scholars seem to have had some realization that the phenomenon of life has chemical attributes. Theophrastus Bombastus von Hohenheim, a Swiss physician and alchemist (ca. 1493–1541) better known as Paracelsus, sought to deal with medicine in terms of its "elements" mercury, sulfur, and salt. An ailing person was thought to be deficient in one of these elements and therefore in need of supplementation with the missing substance. Paracelsus was said to have effected some dramatic "cures" based on this idea.

By the eighteenth century, chemists were beginning to recognize the chemical aspects of life processes in a modern sense. Antoine-Laurent de Lavoisier (1743–1794) recognized the similarity of respiration to combustion in the uptake of oxygen and expiration of carbon dioxide.

At about the same time, it was found that certain compounds are associated with living systems and that these compounds generally contain carbon. They were thought to have arisen from, or to be a consequence of, a "vital force" responsible for the life process. Swedish chemist Jöns Jacob Berzelius (1779–1848) applied the term *organic* to substances isolated from living things. Somehow, the fact that these chemical substances were organic in nature was thought to put them beyond the scope of the experimentalist. The logic of the time seems to have been that life is not understandable; organic compounds spring from life; therefore, organic compounds are not understandable.

The barrier between organic (living) and inorganic (nonliving) chemistry began to crumble in 1828 because of a serendipitous (accidental) discovery by Friedrich Wöhler (1800–1882), a German analyst originally trained in medicine. When Wöhler heated ammonium cyanate, an inorganic compound, he isolated urea, a known urinary excretion product of mammals.

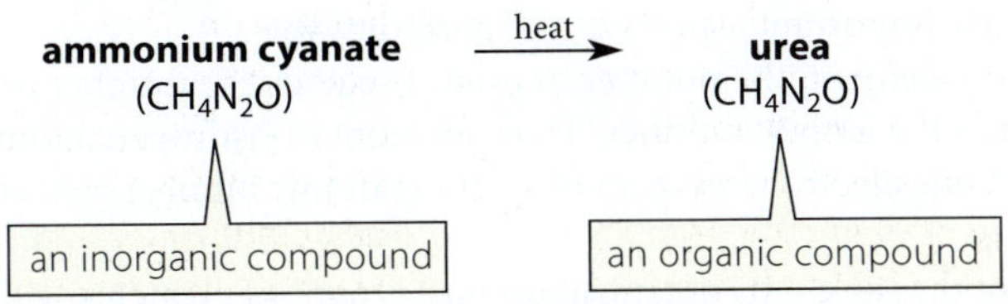

(1.1)

Wöhler recognized that he had synthesized this biological material "without the use of kidneys, nor an animal, be it man or dog." Not long thereafter followed the synthesis of other organic compounds: acetic acid by Hermann Kolbe in 1845 and acetylene and methane by Marcellin Berthelot in the period 1856–1863. Although "vitalism" was not so much a widely accepted formal theory as an intuitive idea that something might be special and beyond human grasp about the chemistry of living things, Wöhler did not identify his urea synthesis with the demise of the vitalistic idea; rather, his work signaled the start of a period in which the synthesis of so-called organic compounds was no longer regarded as something outside the province of laboratory investigation. Organic chemists now investigate not only molecules of biological importance, but also intriguing molecules with bizarre structures—molecules of purely theoretical interest. Thus, organic chemistry deals with compounds of carbon regardless of their origin. Wöhler seems to have anticipated these developments when he wrote to his mentor Berzelius, "Organic chemistry appears to be like a primeval tropical forest, full of the most remarkable things."

Now that we've seen how organic chemistry originated, let's begin our study of organic chemistry at the beginning, with a review of atomic and molecular structure. You may be familiar with much of the material in this chapter from your previous chemistry courses. If so, check yourself by working the "Focused Problems" (the in-text problems), and feel free to move quickly through these early sections.

1.2 ELECTRONS IN ATOMS

A. The Arrangement of Atoms in the Periodic Table

Chemistry happens because of the behavior of electrons in atoms and molecules. The basis of this behavior is the arrangement of electrons within atoms suggested by the periodic table. Because this arrangement is very important, let's first review the organization of the periodic table. (See the page facing the inside back cover for the complete table.) The elements in blue in the abbreviated table shown in **Fig. 1.1** are of greatest importance in organic chemistry. You should learn their atomic numbers and relative positions.

We here use the older group numbering for the periodic table, which assigns the group numbers 1A–8A to the main-group elements. The newer numbering gives groups 1A and 2A the numbers 1 and 2, and groups 3A to 8A the numbers 13–18. The transition element groups, which are numbered 1B–8B in the older system, receive the numbers 3–12. (Both numbering systems are shown in the periodic table on the page facing the inside rear cover.) We use the older system for main-group elements because it gives directly the number of valence electrons in the neutral element and because it relates more easily to the calculation of formal charge (Sec. 1.4B).

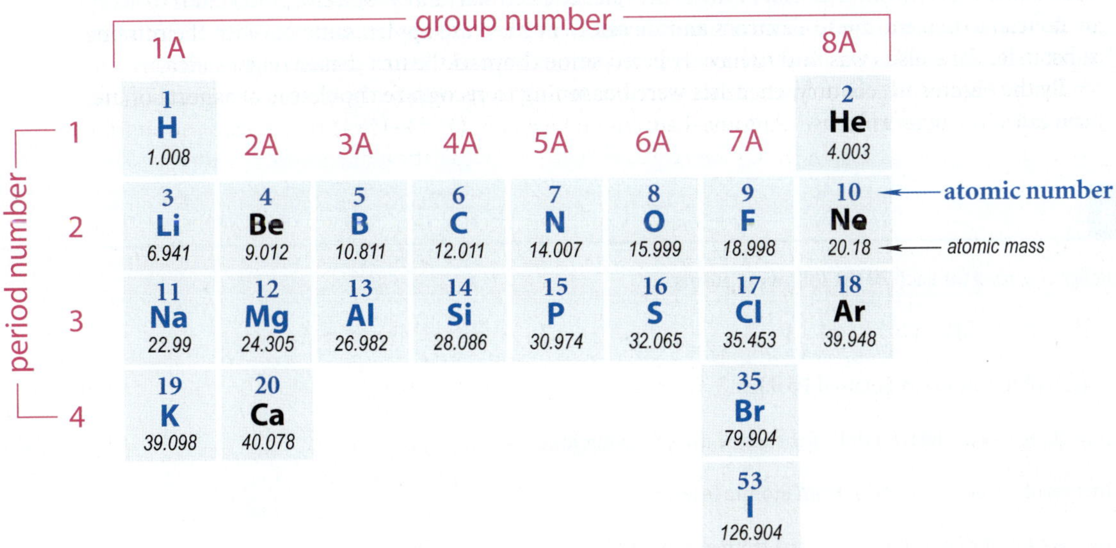

FIGURE 1.1 An abbreviated periodic table. Atoms that are especially important in organic chemistry are shown in blue.

Here are some particularly important aspects of the periodic table.

In a neutral atom of an element, the number of protons equals the number of electrons, and the number of each equals the atomic number. Thus, an atom of hydrogen (atomic number = 1) contains one proton and one electron; an atom of carbon (atomic number = 6) contains six protons and six electrons.

The periodic aspect of the table—its organization into groups of elements with similar chemical properties—led to the idea that electrons reside in layers, or shells, about the nucleus. *Each shell of electrons corresponds to a period (horizontal row) of the periodic table.* The outermost shell of electrons in an atom is called its **valence shell**, and the electrons in this shell are called **valence electrons.** *The number of valence electrons for any neutral atom in an A group of the periodic table (except helium) equals its group number.* Thus, lithium, sodium, and potassium (group 1A) have one valence electron, whereas carbon (group 4A) has four, the halogens (group 7A) have seven, and the noble gases (group 8A, except helium) have eight. Helium has two valence electrons.

Atoms are often represented with the element abbreviation along with dots corresponding to their valence electrons.

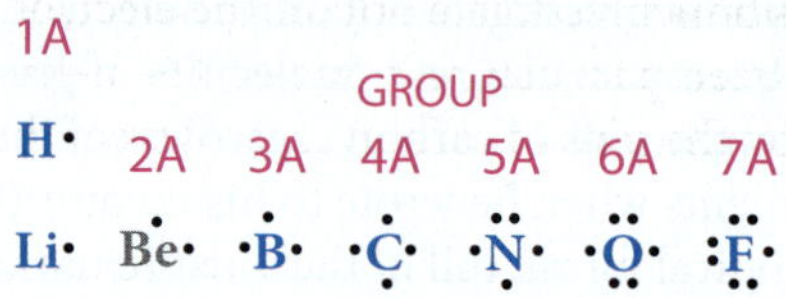

(1.2)

To draw dot structures for the elements, put single dots on each of the four sides of the atom symbol, in any order, to a total of 4; then pair the remaining electrons, also in any order.

Atoms can gain or lose electrons to form charged species called **ions.** When an atom gains an electron, it takes on a negative charge and is called an **anion.** When an atom loses an electron, it takes on a positive charge and is called a **cation.** To draw the dot structures of ions, follow the same convention used for the dot structures of neutral atoms:

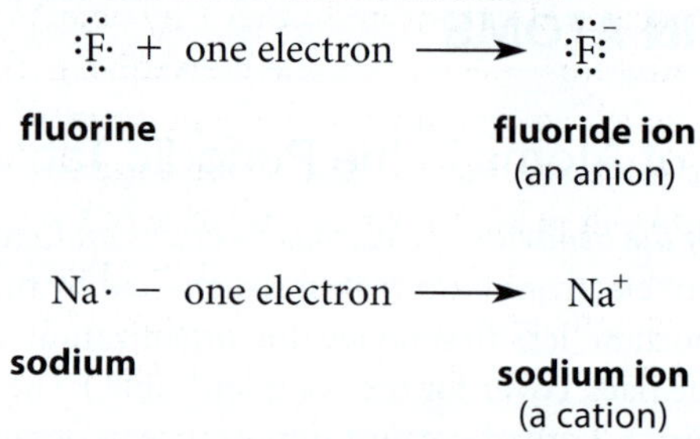

(1.3)

Walther Kossel (1888–1956), a German physicist, noted in 1916 that when atoms form ions, they tend to gain or lose valence electrons so as to have the same number of electrons as the noble gas of closest atomic number. Thus, sodium, with one valence electron (and 11 total electrons), tends to lose an electron to become Na^+, the sodium ion, which has the same number of electrons (10) as the nearest noble gas (neon). Fluorine, with seven valence electrons (and 9 total electrons) tends to accept an electron to become the 10-electron fluoride ion, F^-, which also has the same number of electrons as neon. Because fully occupied valence shells in periods beyond the first contain eight electrons, the tendency of atoms to gain or lose valence electrons to form ions with the noble-gas configuration has been called the **octet rule.**

Focused Problems

1.1 Give the number of *valence electrons* for each of the following atoms.

(a) B (b) S (c) P (d) Ca (e) I

1.2 Draw dot structures for each of the atoms in Focused Problem 1.1.

1.3 Draw a symbol, including charges and electron dots, for each of the following ions:

(a) The ion formed by loss of a valence electron from atomic bromine.

(b) The ion formed by the gain of two valence electrons from atomic sulfur.

B. Quantum Theory and Atomic Structure

To understand the electronic structure of atoms, and, ultimately, the chemical bond, we need to call upon a theory called *quantum mechanics.* Quantum mechanics deals in detail with, among other things, the behavior of electrons in atoms and molecules. Although the theory involves some sophisticated mathematics, we need not explore the mathematical detail to appreciate some general conclusions of the theory. The starting point for quantum mechanics is the idea that *small particles such as electrons also have the character of waves.* How did this idea evolve?

As the twentieth century opened, certain things about the behavior of electrons could not be explained by conventional theories. There seemed to be no doubt that the electron was a particle; after all, both its charge and mass had been measured. However, electrons could also be diffracted like light, and diffraction phenomena were associated with waves, not particles. The traditional views of the physical world treated particles and waves as unrelated phenomena. In the mid-1920s, this mode of thinking was changed by the advent of quantum mechanics. This theory holds that in the submicroscopic world of the electron and other small particles, there is no real distinction between particles and waves. The behavior of small particles such as the electron can be described by the physics of waves. In other words, matter can be regarded as a *wave-particle duality.*

How does this wave-particle duality require us to alter our thinking about the electron? In our everyday lives, we're accustomed to a deterministic world. That is, the position of any familiar object can be measured precisely, and its velocity can be determined, for all practical purposes, to any desired degree of accuracy. For example, we can point to a baseball resting on a table and state with confidence, "That ball is at rest (its velocity is zero), and it is located exactly 1 foot from the edge of the table." Nothing in our experience indicates that we couldn't make similar measurements for an electron. The problem is that humans, chemistry books, and baseballs are of a certain scale. Electrons and other tiny objects are of a much smaller scale. A central principle of quantum mechanics, the **Heisenberg uncertainty principle**, states that *the accuracy with which we can determine the position and velocity of a particle is inherently limited.* For "large" objects, such as basketballs, organic chemistry textbooks, and even molecules, the uncertainty in position is small relative to the size of the object and is inconsequential. But for very small objects such as electrons, the uncertainty is significant. As a result, the position of an electron becomes fuzzy. According to the Heisenberg uncertainty principle, we are limited to stating the *probability* that an electron is occupying a certain region of space.

In summary:

1. Electrons have wavelike properties.

2. The exact position of an electron cannot be specified; only the probability that it occupies a certain region of space can be specified.

C. Atomic Orbitals: Relative Energies

In an earlier model of the atom, electrons were thought to circle the nucleus in well-defined orbits, much as Earth circles the Sun. Quantum theory replaced the *orbit* with the *orbital*, which, despite the similar name, is something quite different. An **atomic orbital** describes the wave properties of an electron in an atom.

Orbitals have two properties that are especially important for us. First, each orbital is characterized by a discrete *energy*. That is, when an electron occupies an orbital, the electron has a precise total energy. It can't have a higher energy or a lower energy; the energy is *quantized.* The existence of certain energies and not others is characteristic of a wave. Second, each orbital occupies a certain region of space around the nucleus; the electron in an orbital is confined within that space. In other words, orbitals have *shape.*

Let's first consider the organization of orbitals within an atom and their energies. First, orbitals are arranged in principal energy levels 1, 2, 3, ... The integer used to designate an energy level is called the **principal quantum number**. The higher the principal quantum number, the greater the energy. Within each principal level are sublevels, labeled s, p, and d,

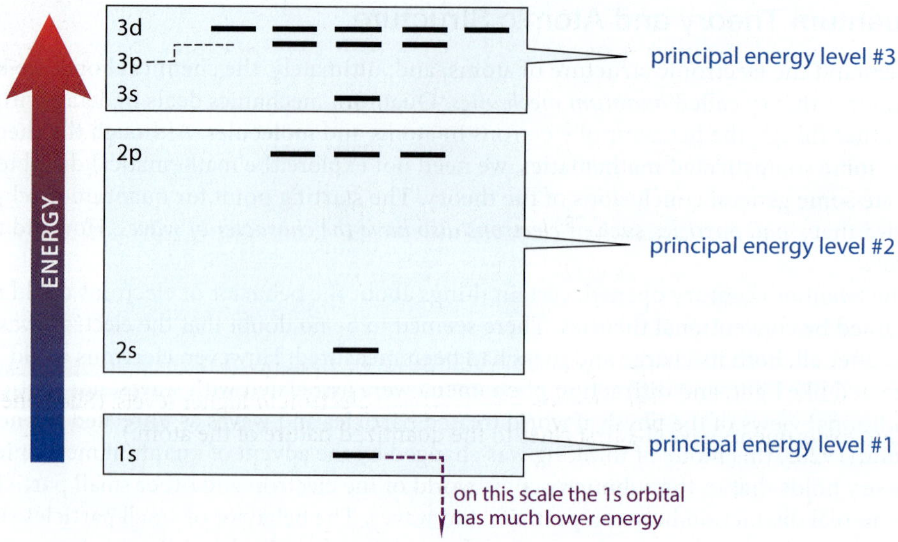

FIGURE 1.2 The relative energies of different orbitals shown for the first three principal energy levels. The energy spacing for the orbitals varies from atom to atom, but the order is the same. In some atoms with electrons in energy level 4s, this energy level is lower than the 3d level, but this will not affect the discussion in this section.

There are many more principal energy levels (a total of seven corresponding to the seven periods of the existing periodic table), and the higher energy levels have additional sublevels. However, we won't need to be concerned with these in organic chemistry.

also in order of increasing energy. A diagram of the atomic orbitals and their relative energies is given in **Fig. 1.2**. *This diagram applies to any atom.* (The only difference between atoms is the precise difference in energy between the levels.) Notice that principal energy level 1 has only one orbital, the 1s orbital. Principal energy level 2 contains four orbitals: one 2s orbital and three 2p orbitals. The 2p orbitals are **degenerate**; this means that they have identical energies. Principal energy level 3 contains three types of orbitals: a 3s orbital, three degenerate 3p orbitals, and five degenerate 3d orbitals. Regardless of the principal level number, there is only one s orbital, three p orbitals (in levels 2 and above), five d orbitals (in levels 3 and above), and so on.

D. The Electronic Structure of Atoms

The electrons in atoms reside in the atomic orbitals we've just described, and the way these electrons fill the orbitals follows very specific rules. The first is the **aufbau principle** (German *aufbau* = buildup), which stipulates that the electrons fill the atomic orbitals from lower to higher energy. Thus, the single electron of the hydrogen atom populates the 1s orbital of the hydrogen atom because it is the orbital of lowest energy. The electronic configuration of a hydrogen atom is sometimes represented as $1s^1$, which means that there is one electron in a 1s orbital.

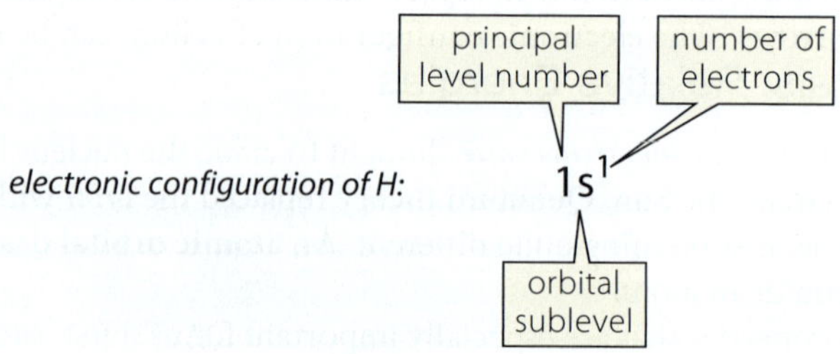

(1.4)

An electron in the hydrogen atom has available to it all of the orbitals shown in Fig. 1.2, but only the orbital of lowest energy is normally occupied with an electron. This "normal" state of affairs is called the **ground state** (lowest energy state) of the hydrogen atom. However, if a hydrogen atom is subjected to the exact amount of energy (say, from light) required to increase the energy of the electron to a state with a higher principal quantum number (for example, level 2), the electron absorbs energy and instantaneously assumes the new, more energetic, wave motion characteristic of the orbital of principal energy level 2.

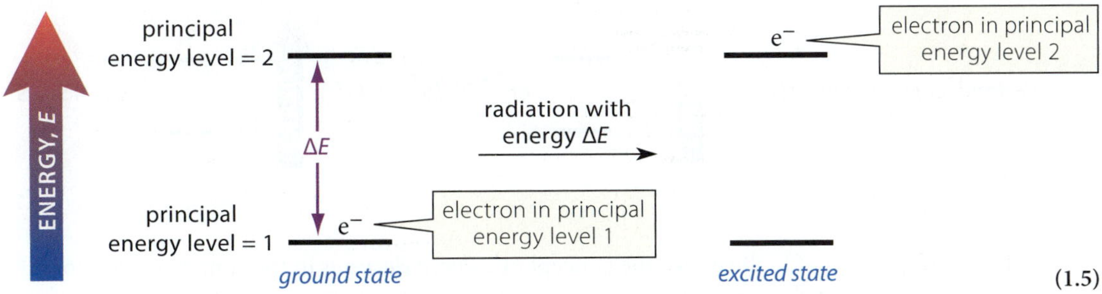

(1.5)

The resulting state of the hydrogen atom is an **excited state** (a state of higher energy). It is also possible to use higher energies to promote the hydrogen electron to higher levels. (Such energy-absorption experiments gave the first clues to the quantized nature of the atom.)

An Analogy for Energy Absorption by a Wave

Alenavlad/Shutterstock

If you have ever blown across the opening of a flute (or a soda-pop bottle, which is a less sophisticated example of the same thing), you know that the flute can produce only a certain pitch (if you don't change its length by pressing a key). If you blow harder, the pitch does not rise, but only becomes louder. If you blow hard enough, however, the sound suddenly jumps to a note of higher pitch. The pitch is quantized; only certain sound frequencies (pitches) are allowed. Such phenomena are observed because sound is a wave motion of the air in the flute or bottle, and only certain pitches can exist in a cavity of given dimensions without canceling themselves out. The progressively higher pitches you hear as you blow harder (called *overtones* of the lowest pitch) are analogous to the progressively higher energy states (orbitals) of the electron in the atom. Just as each overtone in the flute or bottle is described by a wavefunction with higher "quantum number," each orbital of higher energy is described by a wavefunction of higher principal quantum number n.

Helium contains two electrons in a 1s orbital, but now we must take into account a magnetic property of electrons called **spin**. Two values of the spin property are possible, which we'll call "up" and "down." (You'll sometimes see spins described with the numbers $+\frac{1}{2}$ for the "up" spin and $-\frac{1}{2}$ for the "down" spin.) The **Pauli exclusion principle** states that *when electrons occupy the same orbital, they must have opposite spin.* Because only two values of electron spin are possible, *each orbital can contain only two electrons.* In representing electron spin symbolically, we typically use up or down "half-arrows." The electronic configuration of helium can be represented in the following way:

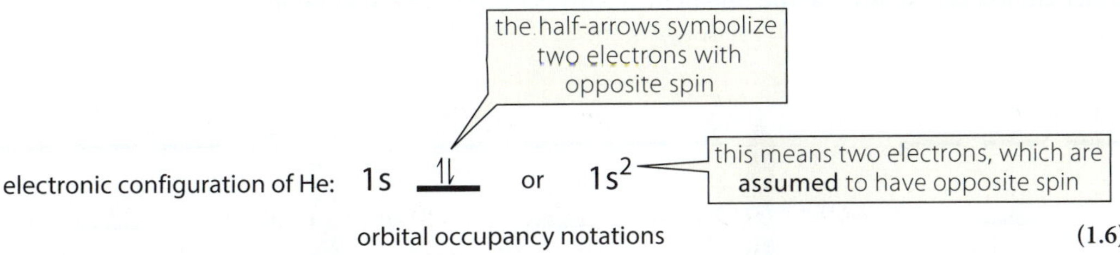

(1.6)

A lithium (Li) atom has three electrons. The aufbau principle requires that the 1s orbital is be filled first. Because the 1s orbital can only contain two electrons, lithium has two electrons in the 1s orbital, and the third electron must go into the orbital of next highest energy, the 2s orbital.

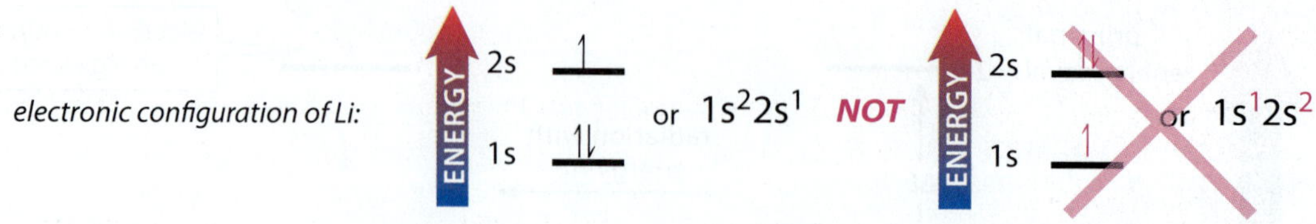

(1.7)

The configuration of Li illustrates the concept of *valence electrons* in the context of electronic structure. *The valence electrons are the electrons in the highest principal energy level.* For lithium, this is the lone electron in energy level 2—that is, the 2s electron. The other electrons are called **core electrons**. The valence electrons are typically involved in the chemistry of an atom, whereas the core electrons are not.

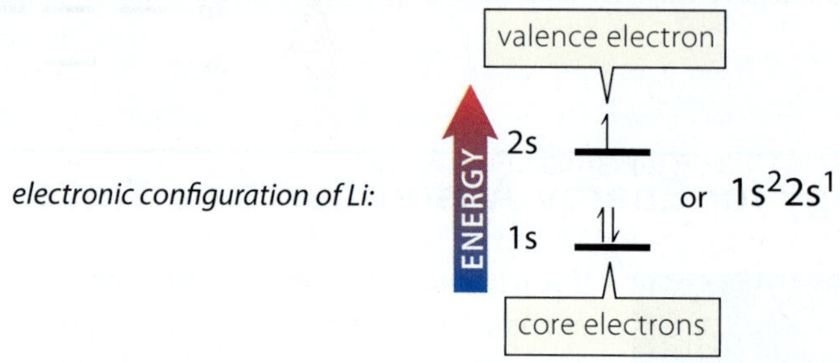

(1.8)

Carbon has six electrons. The first two electrons, the core electrons, fill the 1s orbital with opposite spin. The remaining four electrons are in quantum level 2, and are therefore valence electrons. The first two of these electrons fill the 2s orbitals with opposite spin, but the remaining two electrons go into *different* 2p orbitals as unpaired electrons *with the same spin.*

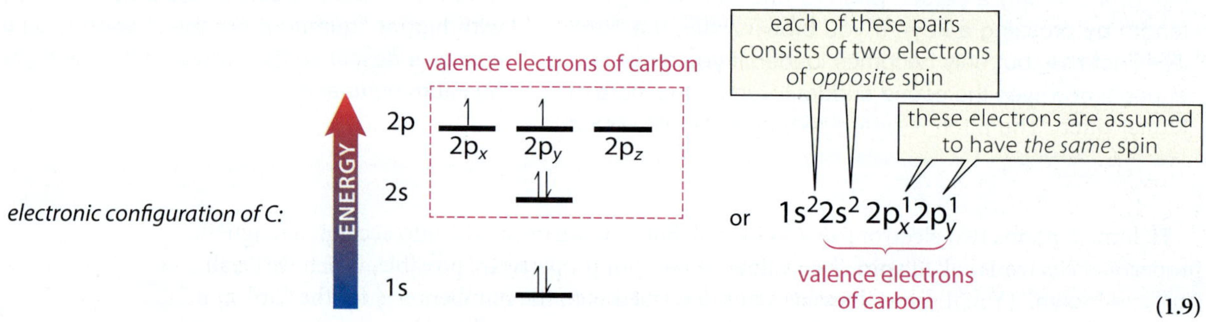

(1.9)

(The three 2p orbitals are differentiated by the labels $2p_x$, $2p_y$, and $2p_z$—we'll see subsequently that this notation is connected to their spatial properties.) This electron distribution is a consequence of physical principles called **Hund's rules**. One of Hund's rules states that in filling orbitals of equivalent energy, electrons will populate empty orbitals before they are paired. The other Hund's rule states that the unpaired electrons must have the same spin.

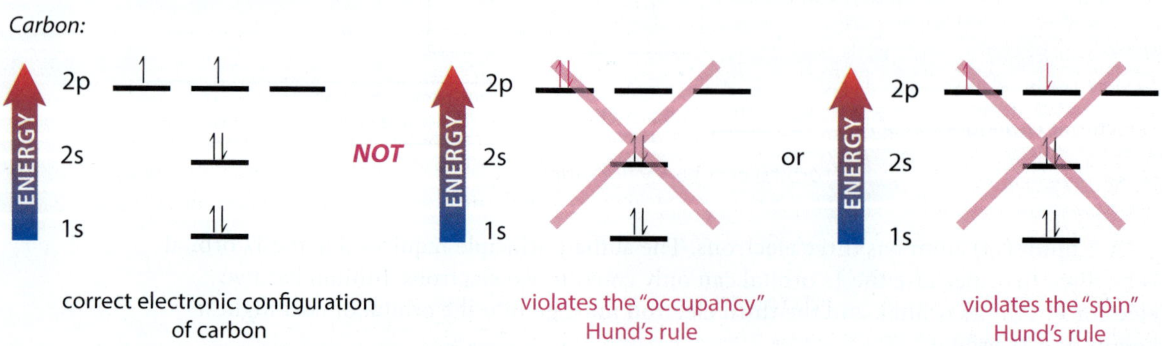

(1.10)

Study Problem 1.1

Describe the electronic configuration of the sulfur atom. Identify the valence electrons, the core electrons, and the valence orbitals.

Solution Because sulfur has an atomic number of 16, a neutral sulfur atom has 16 electrons. Following the aufbau principle, the first two electrons occupy the 1s orbital with opposite spins. The next two, again with opposite spins, occupy the 2s orbital. The next six occupy the three 2p orbitals, with each 2p orbital containing two electrons of opposite spin. The next two electrons go into the 3s orbital with paired spins. The remaining four electrons are distributed among the three 3p orbitals. Taking Hund's rules into account, the first three of these electrons are placed, unpaired and with identical spin, into the three equivalent 3p orbitals: $3p_x$, $3p_y$, and $3p_z$. The one remaining electron is then placed, with opposite spin, into the $3p_x$ orbital. To summarize:

sulfur, S: $(1s)^2(2s)^2(2p)^6(3s)^2(3p_x)^2(3p_y)^1(3p_z)^1$

or

$[Ne](3s)^2(3p_x)^2(3p_y)^1(3p_z)^1$

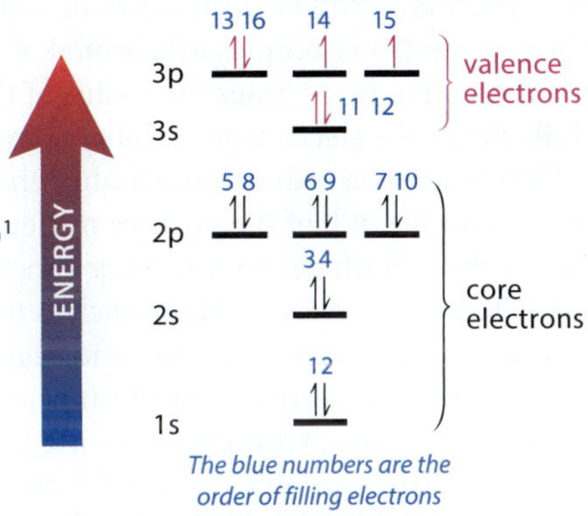

As shown in the diagram, the 3s and 3p electrons are the valence electrons of sulfur; the 3s and 3p orbitals are the valence orbitals. The other electrons are the core electrons.

Typically, the core electrons completely fill principal quantum levels. You can abbreviate the core electrons with the element symbol of the corresponding noble gas. For example, the symbol [Ne] can be substituted for the filled $n = 1$ and $n = 2$ levels, because these levels are also filled in the noble gas neon. This notation is also illustrated in the preceding diagram.

Focused Problems

1.4 Give the electronic configurations of the following atoms and ions. Show the configurations in the two ways illustrated in Study Problem 1.1. Identify the valence electrons, the valence orbitals, and the core electrons for each.

(a) the nitrogen atom (N)
(b) the oxygen atom (O)
(c) the chloride ion (Cl^-)
(d) the phosphorus atom (P)
(e) the magnesium atom (Mg)
(f) the Mg^{2+} ion

1.5 What similarities do you see in the electron configurations of nitrogen and phosphorus in Focused Problem 1.4a and 1.4d? In view of their positions in the periodic table, why is this similarity expected?

E. Atomic Orbitals: Spatial Properties

One of the most important aspects of atomic structure for organic chemistry is that *each orbital is characterized by a three-dimensional region of space in which the electron is most likely to exist.* That is, orbitals have *spatial* characteristics.

1. The *size* of an orbital is described mainly by its principal quantum number *n*: the larger *n* is, the greater the region of space occupied by the corresponding orbital. Orbitals with the same quantum number (for example, 2s and 2p) have roughly the same size.

Atomic-scale distances in this text are expressed in *picometers*, abbreviated pm. A picometer is 10^{-12} meter, or 100 ångstroms (Å). If you prefer distances in Å, divide the distance in pm by 100.

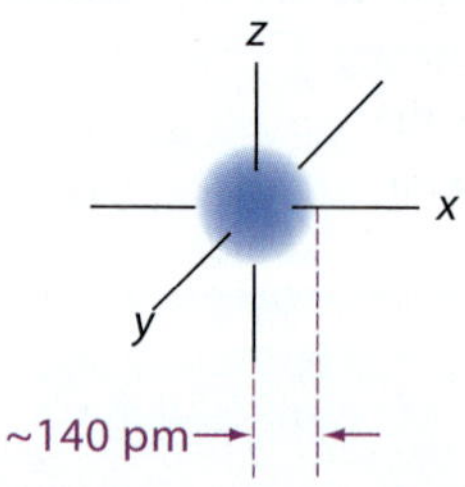

FIGURE 1.3 A 1s orbital. Most (about 90%) of the electron density lies within a sphere with a radius of about 140 pm.

Because orbitals are actually mathematical functions of three spatial dimensions, it would take a fourth dimension to plot the value of the orbital (or the electron probability) at each point in space. (See Further Exploration 1.1 for an additional discussion of electron probability.) Because we are limited to three spatial dimensions, Fig. 1.3 and the other orbital pictures presented subsequently show each orbital as a geometric figure that encloses about 90% of the electron probability.

FURTHER EXPLORATION 1.1
Electron Density Distribution in Orbitals

"Further Explorations" are short sections in the *Study Guide and Solutions Manual* that provide greater depth on the given topic.

2. The *shape* of an orbital is described by the orbital sublevel (s, p, d, etc.).

3. Multiple orbitals of a given type—for example, the three 2p orbitals—have the same shape and energy and differ only in their *direction*.

Let's see how detailed descriptions of several orbitals illustrate these characteristics.

When an electron occupies a 1s orbital, it is most likely to be found in a sphere surrounding the atomic nucleus (**Fig. 1.3**). According to the uncertainty principle (Sec. 1.2B), *we can't pinpoint the exact position of the electron*; locating the electron is a matter of probability. The mathematics of quantum theory indicates that the probability is about 90% that an electron in a 1s orbital will be found within a sphere centered on the nucleus with a radius of about 140 pm. This "90% probability level" is taken as the approximate size of an orbital. Thus, we can depict an electron in a 1s orbital as a *diffuse sphere of electron density*, most of which is within about 140 pm of the nucleus.

When an electron occupies a 2s orbital, it also occupies a sphere, but the sphere is considerably larger—more than three times the radius of the 1s orbital (**Fig. 1.4a**). The cross section in **Fig. 1.4b** shows the electron probability along a "slice" through the orbital in the plane of the page. The greater size is due to the higher energy of the electron, which can escape to greater distances from the pull of the positive nucleus.

The 2s orbital illustrates another very important spatial aspect of orbitals: a *node*. You may be familiar with a simple wave motions, such as the wave in a vibrating string or waves in a pool of water. If so, you know that waves have *peaks* and *troughs*, regions where the waves are at their maximum and minimum heights, respectively. We illustrate this idea with a simple wavefunction of one dimension—in this case, a sine wave. We are going to use the Greek letter *psi* (ψ) in general for wavefunctions. As you know from trigonometry, the wavefunction $\psi = \sin x$ has a positive sign at its peak and a negative sign at its trough (**Fig. 1.5a**). Because ψ is continuous, it must have a zero value somewhere between the peak and the trough. **Nodes** in a wavefunction of one dimension, such as a sine wave, are *points* at which the wave is zero. In a wavefunction of two dimensions, nodes are *lines* (**Fig. 1.5b**); the wave has a zero value along a nodal line. In the wavefunction of a three-dimensional wave, such as a 2s orbital, nodes are *surfaces*. The wave has a zero value everywhere on a nodal surface. The 2s orbital thus consists of a *wave peak*, a *spherical node*, and a *wave trough*. In other words, the 2s orbital is a sphere of electron density within a larger sphere, with the two spheres separated by a spherical node. Our convention in this text is to show wave peaks in blue and wave troughs in green.

! Two cautions: In the sine wave shown in Fig. 1.5a, the mathematical sign of the sine function in the "peak" regions is positive (+), whereas it is negative (–) in the "trough" regions. The mathematical sign of the 2s wavefunction shown in Fig. 1.4 is also positive (+) in the "peaks" and negative (–) in the "troughs." *Don't confuse this sign with the sign of the electron charge in the orbital*; electrons have negative charge.

Some students confuse the inner peak of electron density in the 2s orbital with the 1s orbital, because they are about the same size. The two orbitals are completely distinct; the inner part of the 2s orbital and the 1s orbital simply occupy roughly the same space. If an atom has electrons in both a 1s and a 2s orbital, they stay out of each other's way.

Nodes are a hallmark of increased orbital energy. Consider once again our sound-wave analogy. When you blow into a flute (or soda bottle) and the pitch of the sound jumps an octave, the new sound wave has an additional node. *Orbitals with a principal quantum number n have n – 1 nodes.* Thus, a 1s orbital has 0 nodes, a 2s orbital has 1 node, and a 3s orbital has 2 nodes. The higher the energy of an s orbital, the more nodes it has. You can also see the effect of wave energy on nodes if you create a wave in a rope attached to a stationary object (such as a wall) by shaking one end up and down. If you shake the rope faster (add more energy), you introduce a node into the "rope wave."

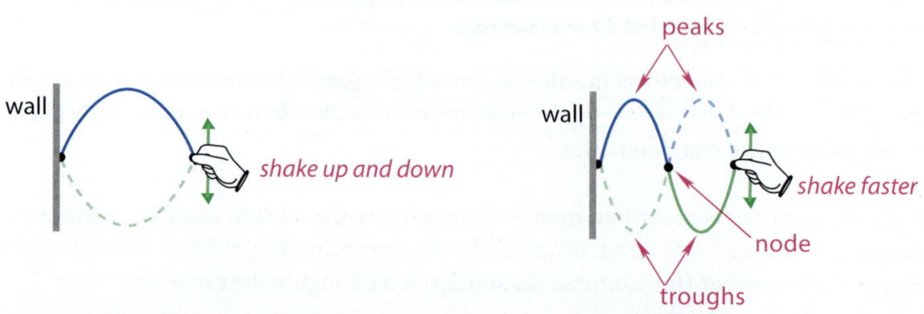

(1.11)

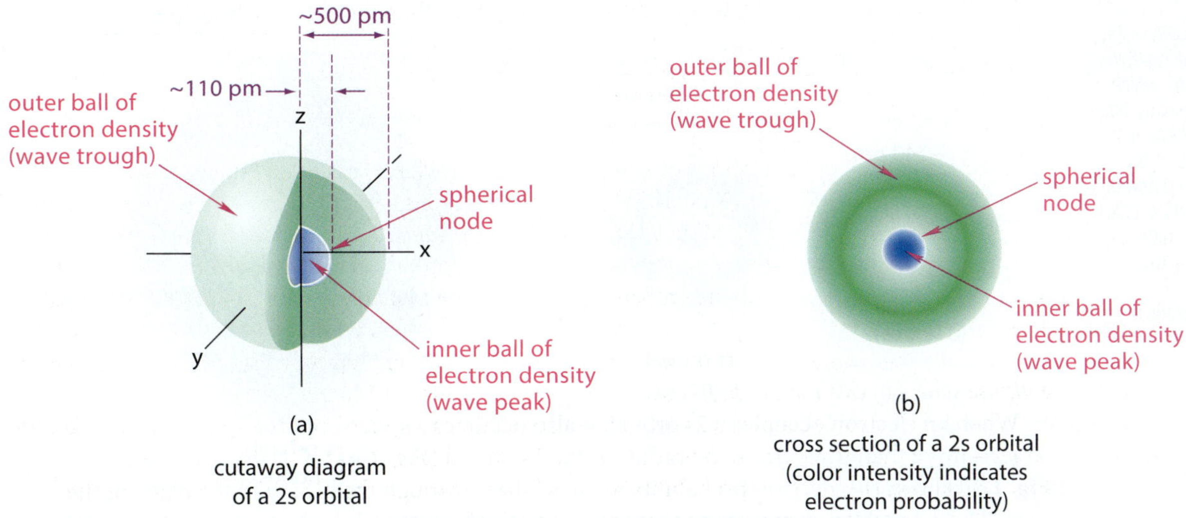

FIGURE 1.4 (a) A cutaway view of a 2s orbital showing the peak (blue) and trough (green) of electron density. This orbital can be described as a sphere within a sphere, with the two spheres separated by an infinitely thin spherical node. A 2s orbital is almost three times as large as a 1s orbital (see Fig. 1.3). (b) A cross section through the 2s orbital showing the electron probability.

If the electron wave is zero at the node, then how does the electron cross the node? The answer is that the electron *is* a wave, and the node is part of its wave motion, just as the node is part of the wave in our vibrating rope. The electron is not analogous to the rope; it is analogous to the *wave* in the rope.

A 3s orbital is also spherical and consists of *three* concentric spheres: a wave peak, a node, a wave trough, a node, and another wave peak. The 3s orbital is much larger than the 2s, and it has an additional node. The relative sizes of the 1s, 2s, and 3s orbitals are contrasted in **Fig. 1.6**.

We examine next the 2p orbitals (**Fig. 1.7**), which are especially important in organic chemistry. Comparison of the 2s and 2p orbitals illustrates how the orbital type (s, p, d) governs the *shape* of an orbital. Whereas all s orbitals are spheres, all p orbitals have dumbbell shapes and are directed in space (that is, they lie along a particular axis). One lobe of the 2p orbital corresponds to a wave peak, and the other corresponds to a wave trough; the electron density, or probability of finding the electron, is identical in corresponding parts of each lobe. The two lobes are *parts of the same orbital.* The node in the 2p orbital, which passes through the nucleus and separates the two lobes, is a plane. The size of the 2p orbital, like that of other orbitals, is governed by its principal quantum number; it extends about the same distance from the nucleus as a 2s orbital.

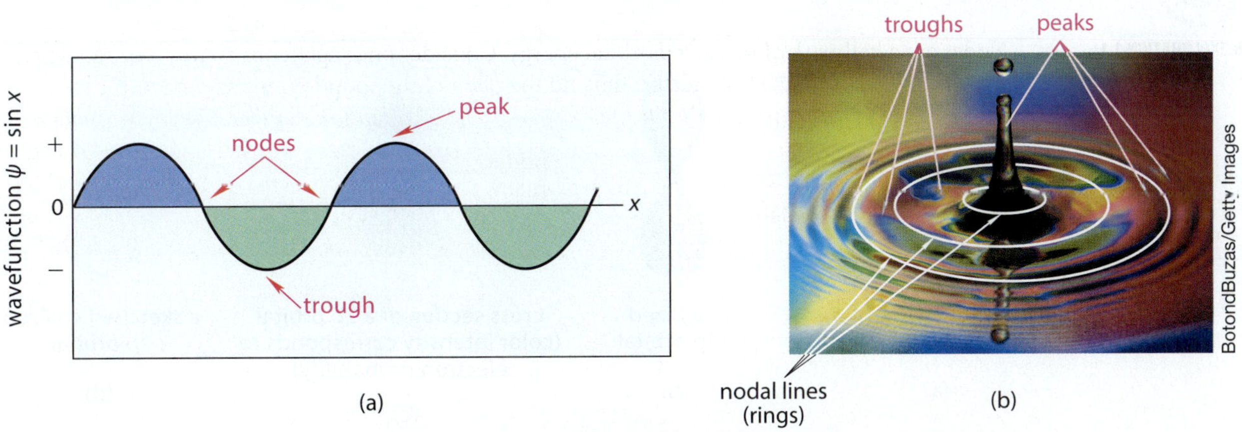

FIGURE 1.5 (a) An ordinary sine wave, which is a wavefunction of one dimension, showing peaks, troughs, and nodes. In this wave the nodes are points. (b) A planar wave in a pool of water. In this case, the nodes are lines (circles), shown in white.

The orbital sizes shown in this section are based on orbitals in the hydrogen atom, which can be calculated exactly. For atoms with many electrons, the orbitals of a given type can have somewhat different sizes. That is why orbital sizes are given as approximate values. Nevertheless, the general conclusions about orbital shapes and sizes are the same.

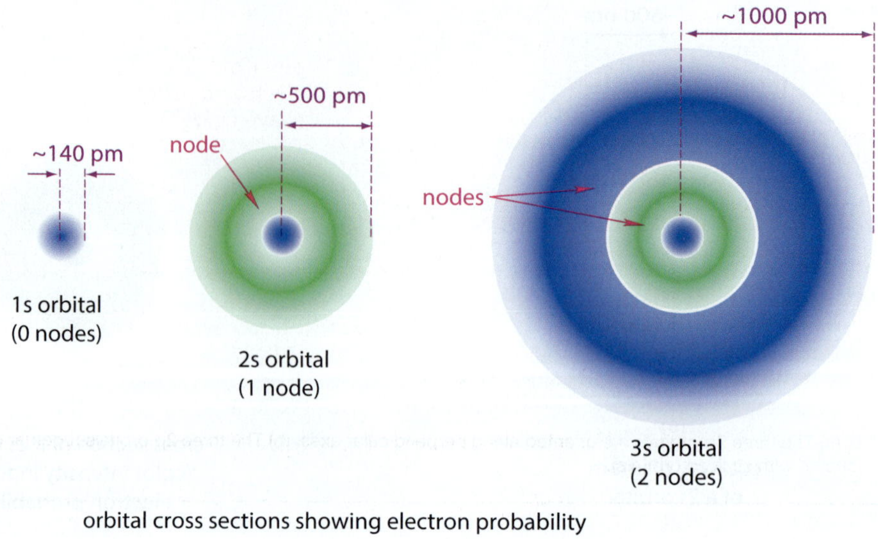

FIGURE 1.6 Cross sections of the 1s, 2s, and 3s orbitals. All three are spheres with spherical nodes. Notice, though, how the size and number of nodes increase with increasing principal quantum number.

Fig. 1.7b shows a more stylized way of drawing a 2p orbital as a "double teardrop" that is used frequently in texts (including this one); you can also sketch a 2p orbital simply as shown in **Fig. 1.7d**. Although easier to draw, a stylized orbital indicates clearly the directional character of a 2p orbital.

Recall from Fig. 1.2 that there are three 2p orbitals, which are identical except for their orientation in space. The axes of these orbitals are mutually perpendicular—that is, they are oriented along the imaginary *x*-, *y*-, and *z*-axes. The direction of a 2p orbital is sometimes indicated with a subscript. For example, if a 2p orbital is oriented along the *x*-axis, the orbital is called a $2p_x$ orbital. The three 2p orbitals are shown in **Fig. 1.8a**. The appearance of the orbitals together on a single atom is shown (with stylized orbitals) in **Fig. 1.8b**.

The 3p orbital (**Fig. 1.9**) reinforces the ideas of orbital size and orbital nodes. First, notice the greater size of this orbital, which is a consequence of its greater principal quantum number and its greater energy. The 90% probability level for this orbital is more than 900 pm from the nucleus, which is about twice the size of a 2p orbital (see Fig. 1.7). Next, notice the *shape* of the 3p orbital. It is generally lobe-shaped, and it consists of four regions separated by nodes. The two inner regions resemble the lobes of a 2p orbital. The outer regions, however, are large and diffuse, and they resemble mushroom caps. Finally, notice the number and the character of the nodes. A 3p orbital contains two nodes, as we expect for an orbital with principal quantum number 3. One node is a plane through the nucleus, much like the node of a 2p orbital. The other is a spherical node, shown in **Fig. 1.9b**, that separates the inner part of each lobe from the larger outer part.

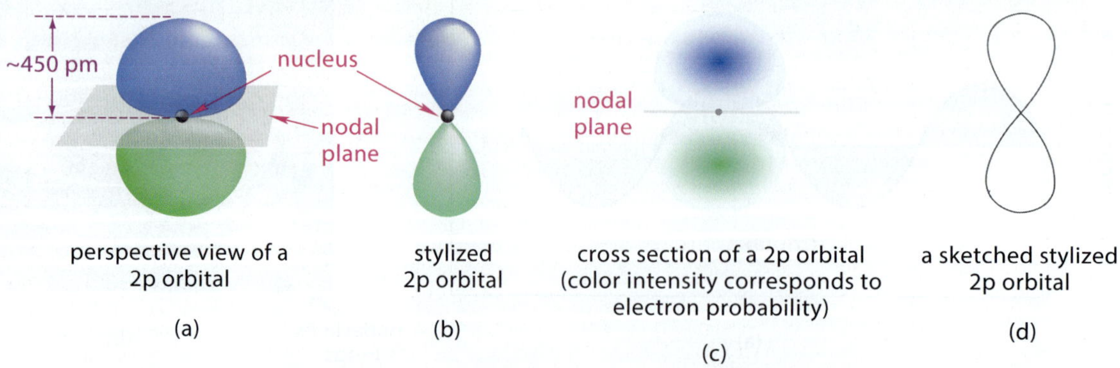

FIGURE 1.7 (a) Actual size of a 2p orbital. (b) A stylized "cartoon," or "double teardrop," version of a 2p orbital. (c) A cross section of a 2p orbital in which color intensity indicates electron probability. The electron probability is greatest near the center of each lobe and drops off in all directions. (d) A 2p orbital as it is frequently sketched.

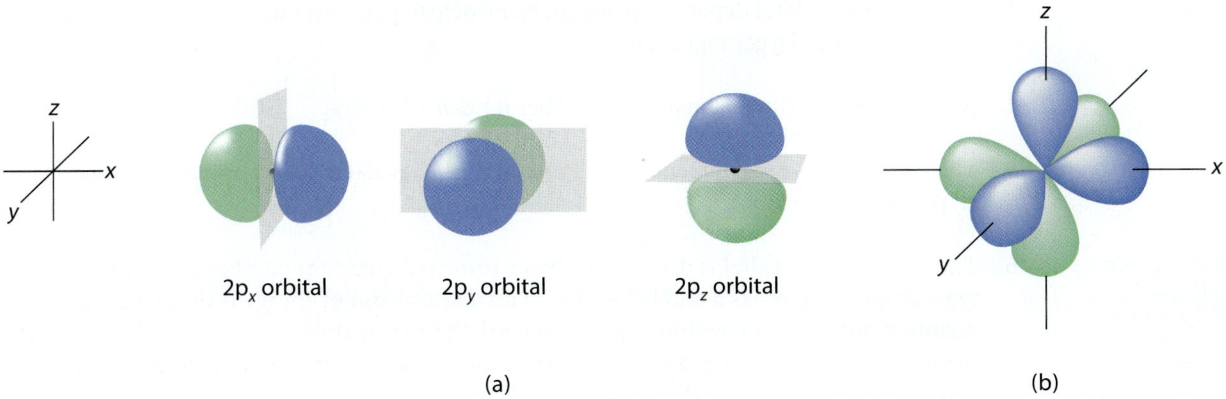

FIGURE 1.8 (a) The three 2p orbitals are oriented along perpendicular axes. (b) The three 2p orbitals together on a single atom (shown with stylized orbitals).

The presence of peaks, troughs, and nodes in orbitals is important for understanding the molecular orbital theory of bonding, which is discussed in Sec. 1.8. However, the actual electron probability in an orbital is proportional to the *square* of the wavefunction. This means that the electron probability in any orbital is positive everywhere, as it must be to have physical meaning.

Let's summarize what we've learned about orbitals:

1. Orbitals describe the wave character of electrons and are described by wavefunctions. When an electron occupies an orbital, its motion has the wave properties (including the energy and spatial properties) of the orbital.

2. In all orbitals above quantum level 1, orbitals consist of wave peaks (regions in which the wavefunction of the orbital is positive) and wave troughs (regions in which the wavefunction is negative) separated by nodes (surfaces at which the wavefunction is zero).

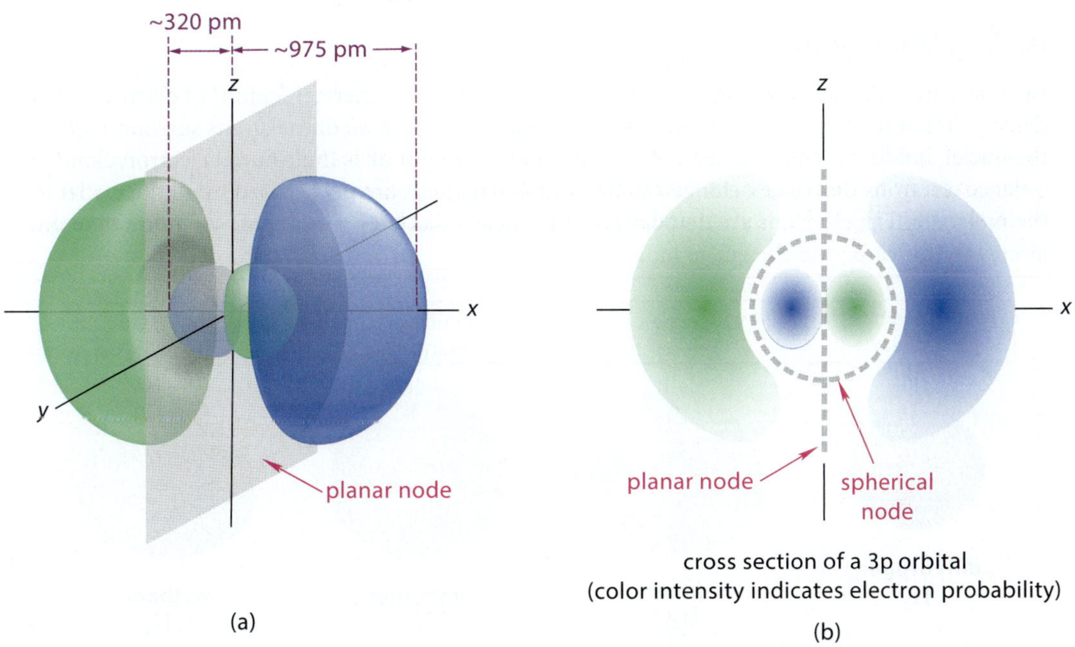

FIGURE 1.9 (a) Perspective representation of a 3p orbital; only the planar node is shown. There are three mutually perpendicular 3p orbitals. Notice that a 3p orbital is much larger than a 2p orbital (compare with Fig. 1.7). Most of the electron density lies in the large outer lobes. (b) Cross section of a 3p orbital showing both the planar and the spherical nodes.

3. The size of an orbital depends mostly on its principal quantum number n; orbitals with greater n occupy larger regions of space.

4. An orbital with principal quantum number n has $n - 1$ nodes.

5. The shape of an orbital depends on its orbital type (s, p, d, etc.) All s orbitals are spherical; all p orbitals have lobes.

6. The energy of an orbital is determined by its principal quantum number and the orbital type. In quantum levels 3 and below, the main determinant of energy is the principal quantum number, and, within a quantum level, the energy order is s < p < d. The increase in energy with principal quantum number explains the orbital size increase: more energetic electrons can stray farther from the positive nucleus.

7. Each orbital type except s has a number of degenerate (equal-energy) orbitals. These degenerate orbitals differ in their spatial orientation. The most important example for organic chemistry is the three p orbitals of identical energy in each quantum level; these orbitals differ solely in their spatial orientation along three perpendicular axes. Their orientation is sometimes indicated with a subscript (for example, $2p_x$, $2p_y$, $2p_z$).

You may recall from general chemistry that 4s orbitals are filled before 3d orbitals in some elements. The lower energy of 4s electrons, despite their higher principal quantum number, won't concern us in organic chemistry until we discuss transition-metal chemistry in Sec. 18.6.

Focused Problems

1.6 Sketch a plot of the wavefunction $\psi = \sin nx$ for the domain $0 \le x \le \pi$ for $n = 1$, 2, and 3. What is the relationship between the "quantum number" n and the number of nodes in the wavefunction?

1.7 Use the trends in orbital shapes you've just learned to describe the general features of a 4s orbital, including the size relative to a 3s orbital and the number and type of nodes.

1.3 COVALENT BONDS AND LEWIS STRUCTURES

A. Covalent Compounds

In atoms, all of the electrons surround the nucleus, creating a spherical "cloud" of electron density. In covalent molecules, all of the valence electrons from *all* of the atoms surround *all* of the nuclei, holding them together. (Molecules can be thought of as multinuclei electron clouds.) Valence electrons that once belonged to the individual atoms are now shared by all the nuclei in the molecule. The electrons are **delocalized**; this means that they are spread over more than one atom—in this case, over the entire molecule.

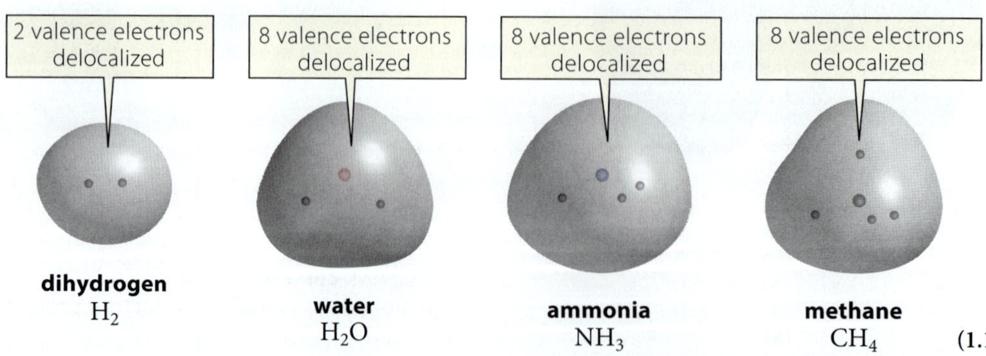

(1.12)

The origin of these bonding descriptions is *molecular orbital theory*, a theory of bonding that is discussed in Sec. 1.8. Although these depictions of molecules are realistic, they are cumbersome to deal with on a practical level. It is often more convenient, and perhaps more intuitive, to imagine that atoms are connected to each other by discrete bonds consisting of two electrons that are shared between the bonded nuclei. Bonding occurs because the electrons in the bond are simultaneously attracted to the two bonded nuclei.

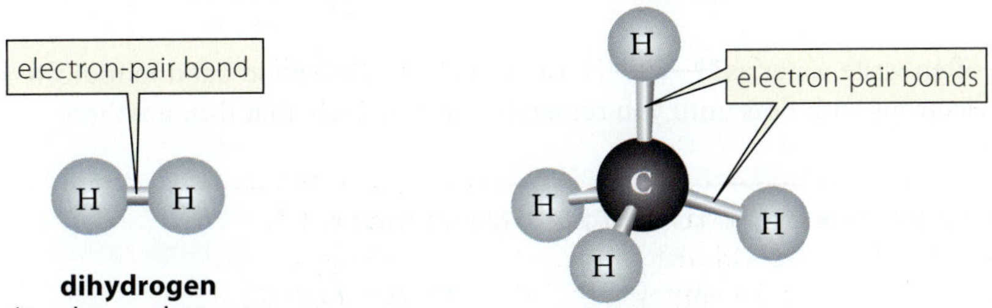

dihydrogen
(2 valence electrons in an electron-pair bond)

methane
(8 valence electrons: 4 × 2 in electron-pair bonds) (1.13)

These two-electron bonds are called **covalent bonds**. This model of covalent bonding is called **valence bond theory**.

B. Lewis Structures and Covalent Bonds

The electrons in covalent bonds originate from the valence electrons of one or both atoms. The simplest example of a covalent bond is the bond between the two hydrogen atoms in the hydrogen molecule, H_2. Conceptually, the bond can be envisioned to come from the pairing of the valence electrons of two hydrogen atoms:

$$H\cdot + \cdot H \longrightarrow H{:}H \qquad (1.14)$$

Any of the symbols ":," "··," and "—" can be used to denote a shared electron pair. *A shared electron pair is the essence of the covalent bond.* Molecular structures that use this notation for the electron-pair bond are called **Lewis structures**.

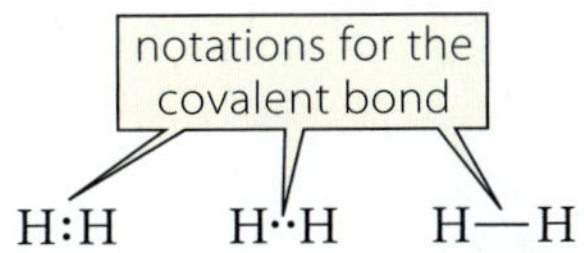

equivalent Lewis structures for the hydrogen molecule (1.15)

We can form methane (CH_4) conceptually by pairing each of the four carbon valence electrons with a hydrogen valence electron to make four C—H electron-pair bonds. (We explore the orbitals involved in the formation of these bonds in Sec. 1.7.)

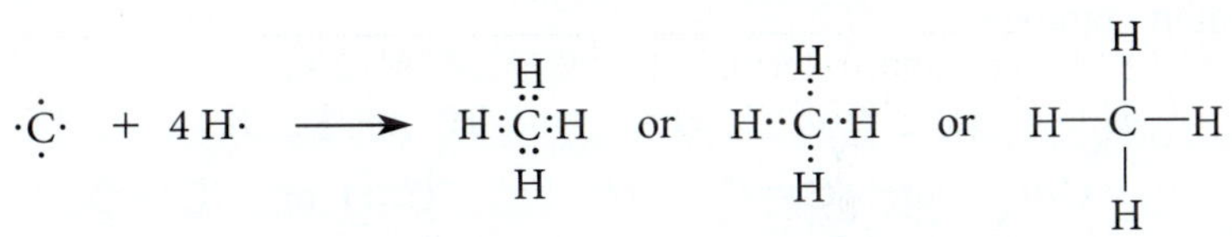

equivalent Lewis structures for methane (CH_4) (1.16)

In the previous examples, all valence electrons of the bonded atoms are shared as bonding electrons. However, in some covalent compounds, such as water (H_2O), some valence electrons remain unshared on one or both atoms—on oxygen, in the case of water. In the water molecule, oxygen has six valence electrons. Two of these electrons combine with electrons from two hydrogens to make two O—H covalent bonds; four of the oxygen valence electrons are left over. These are represented in the Lewis structure of water as electron pairs on the oxygen. Unshared valence electrons in Lewis structures are often depicted as paired dots and referred to as **nonbonding electrons, lone pairs,** or **unshared pairs**.

Lewis structures are named for Gilbert Newton Lewis (1875–1946), an American physical chemist, who proposed the shared-electron model of the covalent bond. From 1912 until the end of his life, Lewis was Professor of Chemistry at the University of California, Berkeley. Lewis' original paper used only electron dots for bonds in his structures. Some people reserve the name *Lewis structures* for these types of structures and refer to structures in which lines are used for electron-pair bonds as *bond-line structures.* We adopt the practice used by many chemists today and refer to both types of structures as Lewis structures. Any structure in which at least one bond is indicated by a line or by a pair of electron dots is classified as a Lewis structure in this text.

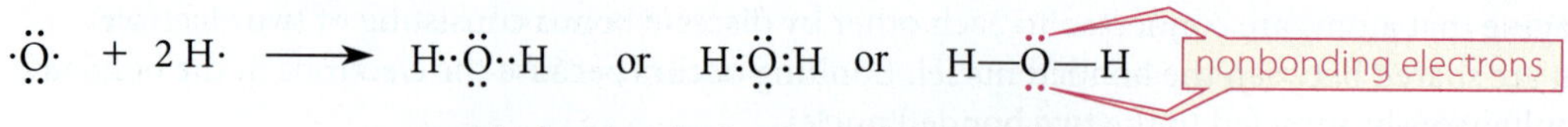

equivalent Lewis structures for water (H_2O) **(1.17)**

Although we often write water as H—O—H, or even H_2O, it is a good habit to indicate all nonbonding electrons with dots until you remember instinctively that they are there.

Be sure to write nonbonding electrons with dots; do NOT use lines:

Such lines can be confused with minus charges, so reserve the symbol "—" for covalent bonds.

The foregoing examples illustrate an important point: *The sum of all bonding and nonbonding valence electrons around each atom in many stable covalent compounds is eight (two for the hydrogen atom).* This is the **octet rule** for covalent bonding. When we count for the octet, each of the bonded atoms is credited with the *total* number of electrons in their common bond. (An analogy is the way that two married partners might express their finances. If each has a certain net worth before the marriage, then, when they are married, their total wealth is credited to each partner.) Because the octet rule will prove to be very important in understanding chemical reactivity, let's look at a few examples in more detail.

In the structure of methane, four shared pairs surround the carbon atom—eight shared electrons, an octet. (Each hydrogen shares two electrons.) Similarly, the oxygen of the water molecule has four shared electrons and two unshared pairs for a total of eight, and again the hydrogens have two shared electrons.

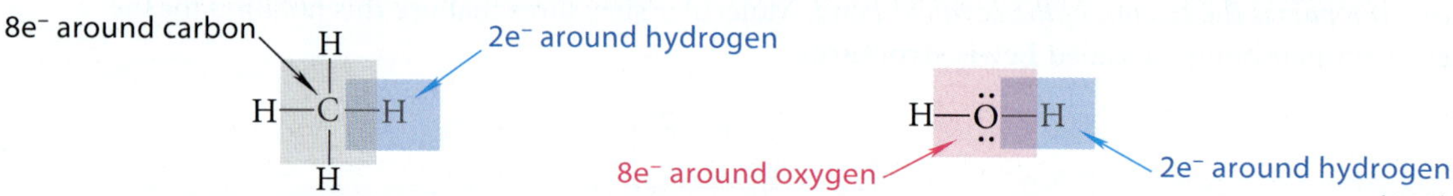

(1.18)

In methane, the carbon is credited with *both* electrons of each C—H bond, and the hydrogen is credited with the same two electrons. In water, the oxygen is credited with both electrons of each O—H bond, and the hydrogen is credited with the same two electrons. Notice, though, that the *nonbonding* electrons on oxygen are credited only to the oxygen, because they are not involved in a chemical bond.

Two atoms in covalent compounds may be connected by more than one covalent bond. Here are some common examples:

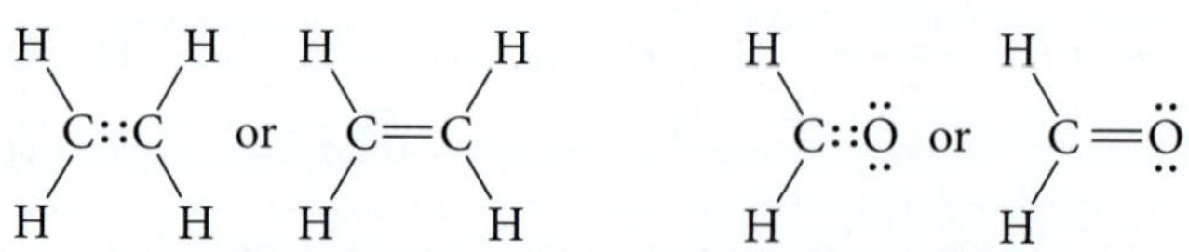

ethylene **formaldehyde**

H—C⋮⋮C—H or H—C:::C—H or H—C≡C—H

acetylene **(1.19)**

Ethylene and formaldehyde each contain a **double bond**—a bond consisting of two electron pairs. Acetylene contains a **triple bond**—a bond involving three electron pairs. In counting for the octet, all of the electrons in a double or triple bond are credited to both bonded atoms. Therefore, in these cases as well, every atom has an octet (or a "duet" for hydrogen). Again, *two different atoms can each claim all of the electrons within their common bond when counting for the octet rule.*

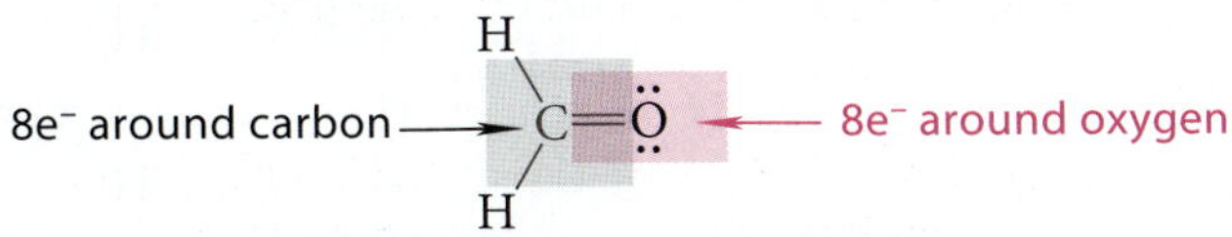

(1.20)

Carbon claims the four electrons of the two C—H bonds plus the four electrons in the carbon–oxygen double bond for a total of eight. Oxygen also claims the four electrons in the carbon–oxygen double bond; but *only* oxygen claims its four nonbonding electrons, also for a total of eight.

C. Common Covalent Bonding Arrangements: Building Lewis Structures

Some of the elements that we see most commonly in organic chemistry have a limited number of standard bonding arrangements. These arrangements are based on their number of valence electrons, which is determined, in turn, by their group number. (Refer to the abbreviated periodic table in Fig. 1.1, if necessary, to see the group numbers.) In the most common bonding arrangements, nonbonding electrons are paired. The rule for bonding to a *neutral* atom is as follows:

$$\text{bonds} + \text{nonbonding electrons} = \text{the group number} = \text{number of valence electrons} \qquad (1.21)$$

The reason for this rule is that it takes *one* valence electron to form a covalent (that is, electron-pair) bond with another atom; the valence electrons not involved in covalent bonds are nonbonding electrons.

Hydrogen (group 1A) typically forms one bond to other atoms, giving it a duet of electrons:

H— (1.22a)

Carbon (group 4A) tends to form four covalent bonds. There are four ways that carbon can form four covalent bonds:

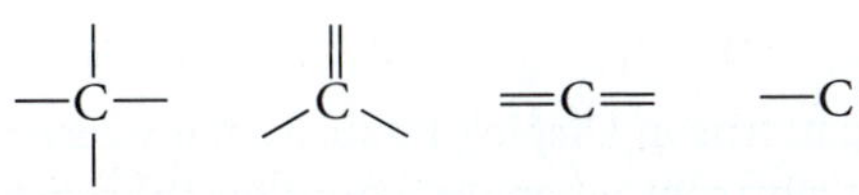

(1.22b)

Nitrogen (group 5A) typically forms three covalent bonds and has a nonbonding electron pair:

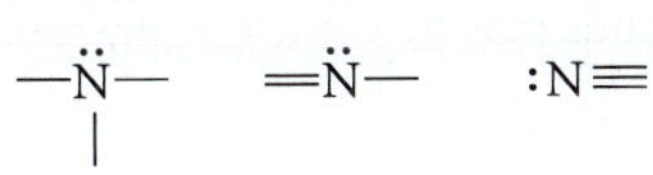

(1.22c)

Oxygen (group 6A) tends to form two covalent bonds and has two nonbonding electron pairs:

(1.22d)

The halogens (F, Cl, Br, I, group 7A) tend to form one bond and have three nonbonding electron pairs:

(1.22e)

In all of the cases just shown, the octet rule is obeyed. Verify for yourself that adding up *all* of the electrons (two for each bond and two for each nonbonding electron pair) around any atom in these bonding arrangements gives eight electrons. For example, for the two bonding patterns listed for oxygen:

counting for the octet on oxygen:

(1.22f)

You will find it helpful to learn these bonding patterns so that you can recognize them quickly. These bonding arrangements apply to atoms that are electrically neutral (that is, uncharged). Atoms in molecules are electronically neutral if the sum of the *bonds* (not *bonding electrons)* and unshared electrons matches the atom's group number. (Atoms can bear charges, and when this happens the bonding patterns change. We discuss these situations in Sec. 1.4B.)

Atoms in groups 1A, 2A, and 3A—atoms to the left of carbon in the periodic table—tend to form a number of bonds equal to their group number. Because these atoms lack enough electrons to fill out their octet with nonbonding electrons, their bonding patterns don't conform to the octet rule. For example, boron trifluoride (BF_3) has three single bonds to fluorine for a total of six electrons. These compounds tend to seek additional electrons to complete their octets through chemical reactions, as we explain in Sec. 3.1.

A few other atoms will show up less often in our study of organic chemistry, but they are also important: silicon, phosphorus, and sulfur. (Phosphorus and sulfur are particularly important in biological chemistry.) These atoms sit in period 3, below carbon, nitrogen, and oxygen, respectively, in the periodic table. This placement suggests that they share some of the same bonding patterns with other atoms in the same column.

Silicon, like carbon, tends to form four single bonds. Unlike carbon, it rarely forms multiple bonds to the same atom.

(1.23a)

Phosphorus, like nitrogen, can form three single bonds with a nonbonding electron pair left over. Sulfur, like oxygen, can form two single bonds, and occasionally a double bond, with two nonbonding electron pairs.

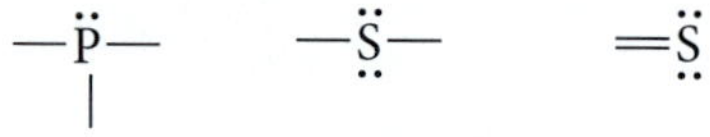

(1.23b)

In addition to the bonding patterns in Display 1.23b, in many cases phosphorus and sulfur are surrounded by more than eight electrons when we "count for the octet." Because these atoms still follow the "bonding rule" in Eq. 1.21, this means that they have more bonds and fewer unshared electron pairs than they would have in an octet structure. It is sometimes said that these elements can undergo **octet expansion.** Here are some examples of phosphorus and sulfur compounds that have expanded octets:

We discuss octet expansion in more detail in Secs. 3.3C and 11.10A.

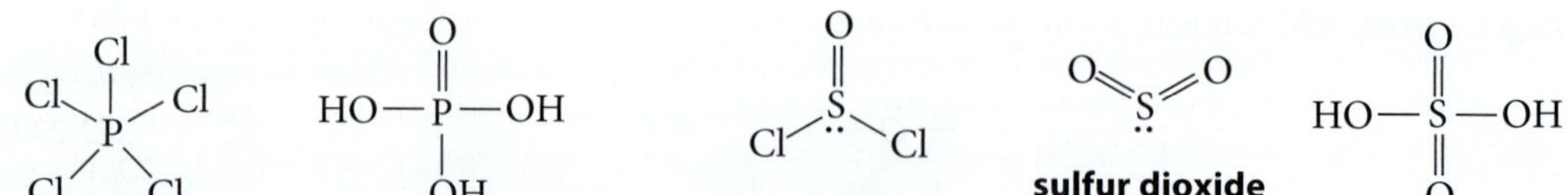

phosphorus pentachloride (pentachlorophosphorane) **phosphoric acid** **thionyl chloride** **sulfur dioxide** **sulfuric acid**

(1.24)

Compounds such as these with expanded octets are sometimes called **hypervalent** compounds. If we count for the octet in the hypervalent phosphorus compounds, we find that phosphorus is surrounded by 10 electrons, two more than the octet. Sulfur, on the other hand, is surrounded by 10 electrons in the thionyl chloride and sulfur dioxide molecules, and by 12 electrons in the sulfuric acid molecule. Nevertheless, the “bonding rule” in Eq. 1.21 still holds for all of these compounds; there are simply more bonds and fewer nonbonding electron pairs.

Building Lewis Structures In some cases you will be given a molecular formula and be asked to propose structures for covalent compounds. For all but the simplest formulas, many valid structures are usually possible.

If you know the order in which the atoms are connected, start writing the Lewis structure by placing single bonds between all atoms. If some atoms are short of an octet, then try using multiple bonds (double bonds or triple bonds) to provide the octet, and add nonbonding electron pairs. Use the common bonding patterns for the elements that we just discussed.

If you don't know how the atoms are connected, you'll have to experiment by trying different atomic connections. More than one correct structure may be possible. Then, follow the recommendations in the previous paragraph.

In either situation, sum up all of the valence electrons of the atoms involved, and make sure you account for all of them in the final structure (subtracting one for each positive charge and adding one for each negative charge). Make sure an octet of electrons is around each atom (except hydrogen, which should share two electrons).

Study Problems 1.2 and 1.3 illustrate this strategy.

Study Problem 1.2

Draw a Lewis structure for the covalent compound methylamine, CH_5N. Assume that the octet rule is obeyed for all elements (duet for hydrogen). Show all nonbonding electron pairs.

Solution First, sum the valence electrons: C(4) + 5H(5) + N(5) = 14. The formula doesn't tell us anything about the way the atoms are connected. Hydrogen can form only one bond, so the C and N must be connected:

C—N

Next, we need to add hydrogens. Putting three on carbon gives the carbon an octet. Putting the remaining two on nitrogen gives nitrogen six electrons, two electrons short of an octet.

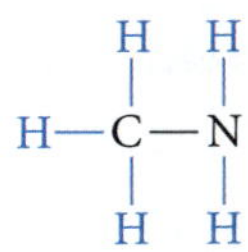

We've used 12 of the 14 electrons. Because all of the atoms are used up, finish the structure by giving a nonbonding electron pair to nitrogen, completing its octet. (Recall that a typical bonding pattern for nitrogen is three bonds and a nonbonding electron pair.)

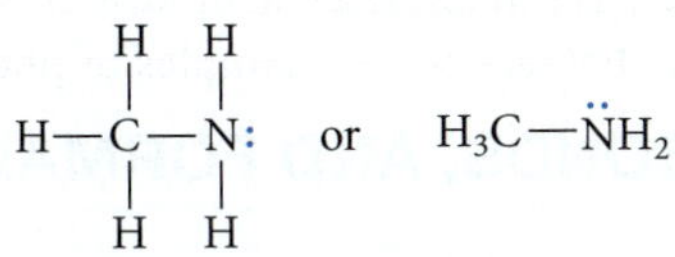

Study Problem 1.3

Draw a Lewis structure for the covalent compound acetaldehyde, C_2H_4O, which contains a carbon–carbon single bond. Assume that the octet rule is obeyed for elements other than hydrogen. Show all nonbonding electron pairs.

Solution First, count the available valence electrons: 2C(8) + O(6) + 4H(4) = 18. Because hydrogen can form only one bond, and because we are given that there is a carbon–carbon single bond, the following partial structure is required:

C—C—O

We know that carbon tends to form four bonds and oxygen tends to form two. Of the following two structures, *B* is ruled out because the problem states that there is a carbon–carbon single bond:

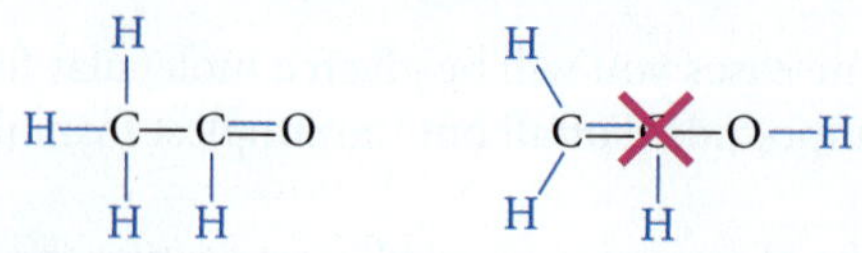

A *B*

We've used 14 of the 18 valence electrons. Recognizing the common bonding pattern of oxygen—two bonds and two nonbonding electron pairs—we complete the structure with octets for both carbon and oxygen and all 18 valence electrons.

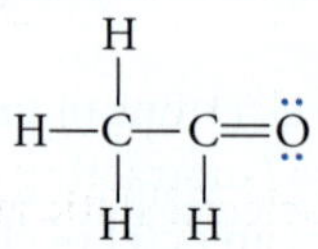

acetaldehyde

Focused Problems

1.8 Write Lewis structures for compounds with the following formulas. Assume that the molecules and all atoms in the molecules are uncharged and that the octet rule is obeyed. Show any nonbonding electron pairs.

(a) H_3CCN (the atoms are connected in the order shown)

(b) C_2H_4

(c) C_3H_6O, containing a carbon–oxygen double bond. Two answers are possible.

(d) $C_2H_4O_2$, a carboxylic acid

1.9 Draw two Lewis structures for each of the following molecules. In each example, one structure follows the octet rule, and the other structure involves octet expansion.

(a) phosphorous acid, H_3PO_3

(b) dimethylsulfoxide (DMSO), $(CH_3)_2SO$

1.4 IONS, IONIC BONDS, AND FORMAL CHARGE

A. The Ionic Bond

In Sec. 1.2A you learned that electrically neutral atoms can gain or lose electrons to form ions. A cation and an anion can come together to form an **ionic compound**, or **salt**. The bonding in such compounds is *not* covalent; the ions retain their charge. The oppositely charged ions in an ionic compound are held together by a through-space interaction called an **electrostatic attraction**. The **electrostatic law** of physics describes this attraction as a potential energy *E*:

$$E = k\frac{q_1 q_2}{r_{12}} \tag{1.25}$$

where *k* is a proportionality constant, q_1 and q_2 are the magnitudes of the charges (including their signs), and r_{12} is the distance between them. When q_1 and q_2 have opposite signs, *E* is negative. A negative energy by convention is *stabilizing*, so that when two ions of opposite charge come

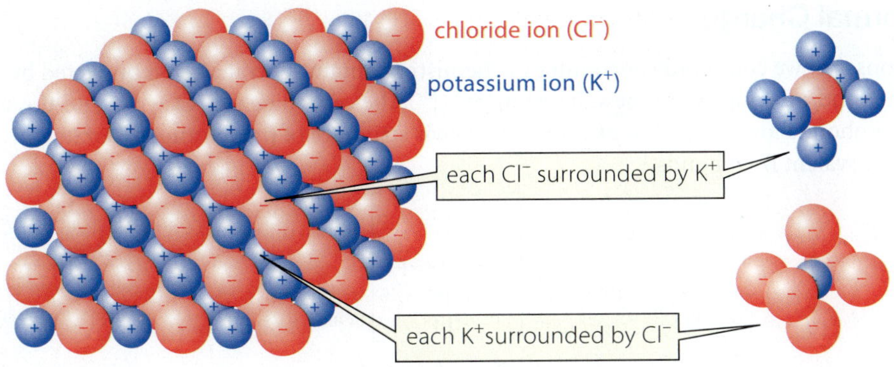

FIGURE 1.10 The structure of a crystal of potassium chloride (KCl). The attraction between the oppositely charged ions is an electrostatic attraction. Each potassium ion is surrounded by chloride ions, and each chloride ion is surrounded by potassium ions.

together, the resulting compound is stable. The electrostatic attraction between ions of opposite charge in an ionic compound is called an **ionic bond**.

Potassium chloride, KCl, is an example of ionic compound formed between a metal cation and a halide anion. The structure of crystalline KCl is shown in **Fig. 1.10**. In the KCl structure, each positive ion is surrounded by negative ions, and each negative ion is surrounded by positive ions. This example illustrates that the ionic bond is *not* directional like a covalent bond. Instead, a positive ion has the same attraction for each of its neighboring negative ions, and a negative ion has the same attraction for each of its neighboring positive ions. Ionic attraction is the same in all directions.

When we write the structures of ionic compounds, we generally write a minimal structure containing the smallest possible number of each ion to give an overall neutral compound. To illustrate, we can write potassium chloride in the following ways:

a line would indicate electron sharing in a covalent bond (Sec. 1.3), not ions

K^+ Cl^-	KCl	but *NOT*	K—Cl (crossed out)
charges written explicitly	charges assumed		

(1.26)

When an ionic compound such as KCl dissolves in water, it dissociates into free ions (each surrounded by water). (We discuss this process further in Sec. 8.6F.) In solution, each potassium ion moves around in solution more or less independently of each chloride ion. The conduction of electricity by KCl solutions shows that the ions are present. Thus, the ionic bond is broken when KCl dissolves in water. This behavior stands in marked contrast to the behavior of covalent compounds, which (barring chemical reactions with the solvent) remain as intact compounds when dissolved.

To summarize, here is our understanding of the ionic bond:

1. It is an electrostatic attraction between ions.
2. It is the same in all directions—that is, it has no preferred orientation in space.
3. It is broken when an ionic compound dissolves in water.

Focused Problem

1.10 Which ionic bond is stronger: the ionic bond between calcium ions (Ca^{2+}) and chloride ions (Cl^-) in crystalline calcium chloride ($CaCl_2$), or the ionic bond between potassium ions (K^+) and chloride ions (Cl^-) in crystalline KCl? Explain.

B. Formal Charge

Many ions that we commonly encounter in chemistry contain several atoms connected by covalent bonds. Examples are the sulfate ion $[SO_4]^{2-}$, the carbonate ion $[CO_3]^{2-}$, and the tetrafluoroborate ion $[BF_4]^-$. For example, the Lewis structure of the tetrafluoroborate ion contains covalent B—Fbonds.

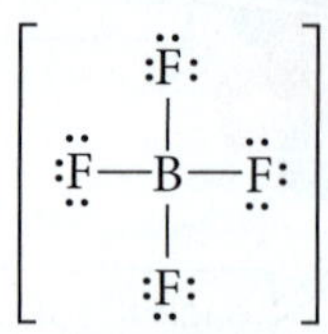

tetrafluoroborate ion (1.27)

Because the ion bears a negative charge, one or more of the atoms within the ion must be charged. But which ones? In reality, the charge is shared by all of the atoms, as we explain further in Sec. 1.9C. However, chemists have adopted a useful and important procedure for electronic bookkeeping that assigns a charge to specific atoms. The charge assigned to an atom with this procedure is called its **formal charge**. The sum of the formal charges on the individual atoms must equal the total charge on the ion.

Computation of formal charge on an atom involves dividing the total number of valence electrons between the atom and its bonding partners. Each atom receives all of its unshared electrons and *half* of its bonding electrons. To assign a formal charge to an atom, then, use the following procedure:

1. Write down the group number of the atom from its column heading in the periodic table. This is equal to the number of valence electrons in the neutral atom.
2. Determine the valence electron count for the atom by adding the number of unshared valence electrons on the atom to the number of covalent bonds to the atom. Counting the covalent bonds in effect adds half the bonding electrons—one electron for each bond.
3. Subtract the valence electron count from the group number. The result is the formal charge.

This procedure is illustrated for the tetrafluoroborate ion in Study Problem 1.4.

Study Problem 1.4

Assign a formal charge to each of the atoms in the tetrafluoroborate ion, $[BF_4]^-$, which has the structure shown in Display 1.27. Verify that the sum of formal charges equals the overall –1 charge of the ion.

Solution Apply the procedure outlined earlier to each atom in turn. We start with fluorine.

Start with the group number of fluorine:		7
Valence electron count:		
Nonbonding electrons	6	
Covalent bonds	1	
Subtract the total valence electron count:		–7
Compute the formal charge:		0

Because all fluorine atoms in $[BF_4]^-$ have the same electron count, they all have the same formal charge—0. It follows that boron must then bear the formal negative charge. Let's calculate it to be sure.

Start with the group number of boron:		3
Valence electron count:		
Nonbonding electrons	0	
Covalent bonds	4	

Subtract the total valence electron count: –4
Compute the formal charge: –1
The sum of formal charges equals the overall charge on the ion.

Because the formal charge on boron is –1, we can write the Lewis structure of $[BF_4]^-$ with the formal charge on the boron:

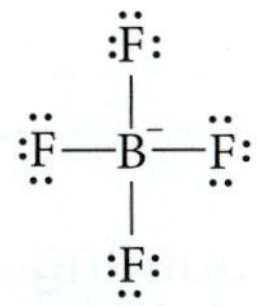

tetrafluoroborate ion

When indicating charge on a structure, show the formal charges on each atom, or show the formal charge on the ion as a whole, *but do not show both.*

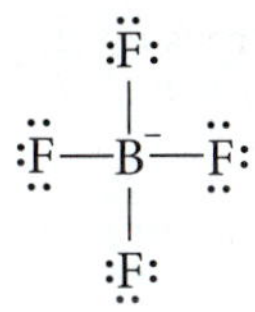

formal charge

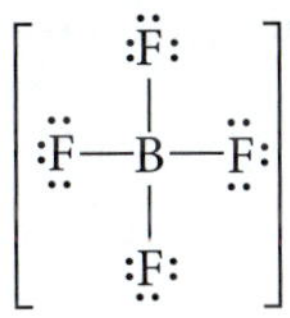

overall charge

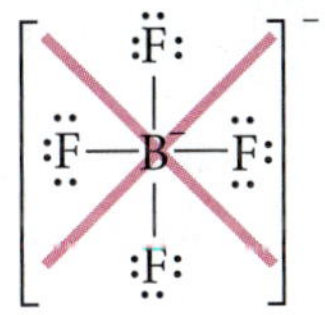

don't show both

Although you can do a formal-charge calculation every time "on the fly," it will save you time and effort if you learn the common bonding patterns and charges for the small handful of elements that are most common in organic chemistry. For example, an oxygen with three bonds and one nonbonding electron pair always has a +1 formal charge. (Verify this.)

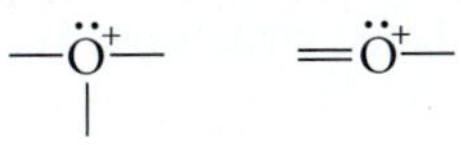

(1.28a)

An oxygen with one bond and three nonbonding electron pairs always has a –1 formal charge. (Verify this, too.)

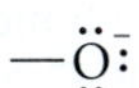

(1.28b)

Other common situations are explored in Focused Problem 1.11.

Focused Problems

1.11 Here are some formal-charge situations you'll encounter frequently. Give the formal charge, and draw a Lewis structure that illustrates the case. Which structures conform to the octet rule?

(a) Carbon has three bonds and one pair of nonbonding electrons.

(b) Carbon has three single bonds and no unshared electrons.

(c) Nitrogen has four single bonds and no unshared electrons.

(d) Nitrogen has two bonds and two pair of nonbonding electrons.

1.12 Assign a formal charge to each atom in the following structures. In each case, compute the formal charge on the entire structure.

(a) 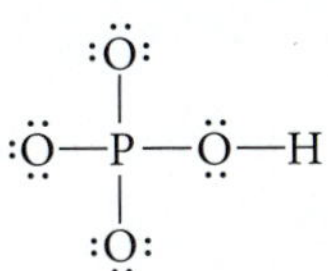

(b) 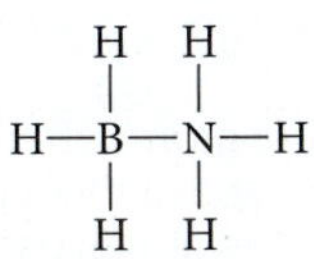

(c) 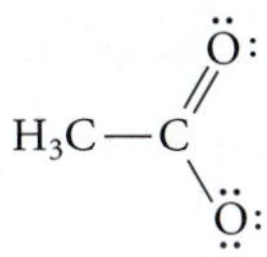

(d) 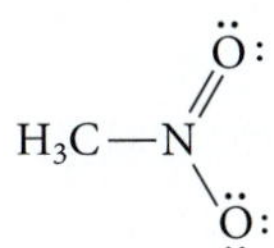

1.13 Add appropriate nonbonding electron pairs (if required) to the following structures given the charges (or absence of charges) provided.

(a) 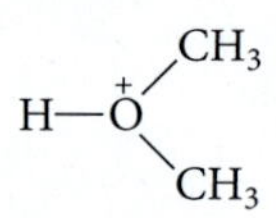

(b) 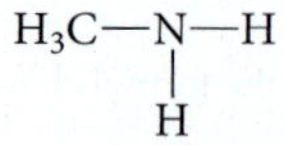

(c) 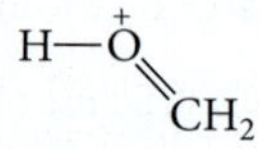

(d) 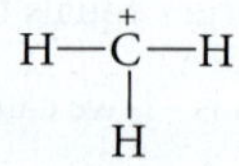

(e) 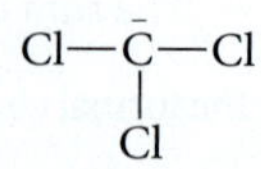

C. Two Types of Electron Counting

As you've seen, electrons can be counted in two different ways for different purposes:

1. To know whether an atom has a complete octet, count all nonbonding valence electrons and *all bonding electrons.*

2. To determine formal charge, count all nonbonding valence electrons and *half of the bonding electrons.*

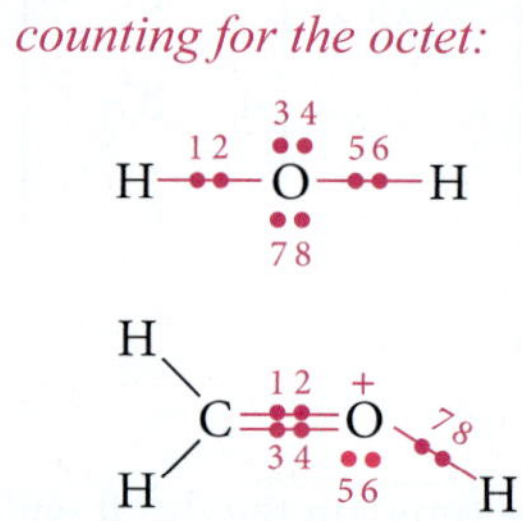

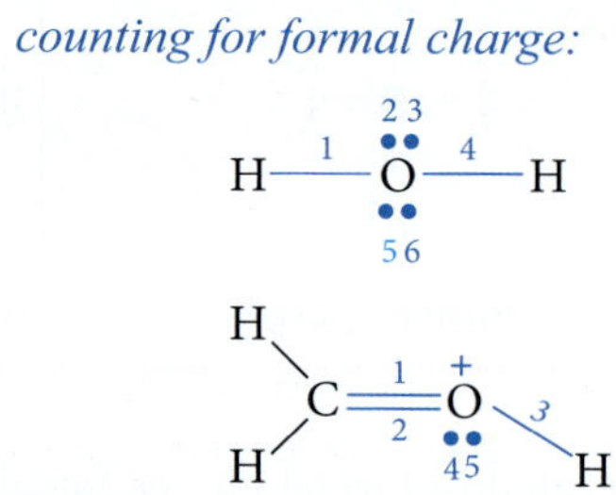

(1.29)

D. Rules for Writing Lewis Structures

We started learning how to write Lewis structures for neutral compounds in Sec. 1.3C. We've now added the possibility of charged atoms to Lewis structures. Let's now summarize the rules for writing Lewis structures:

1. Hydrogen can share no more than two electrons.

2. When you count for the octet, the sum of all bonding and nonbonding electrons for atoms in the second period of the periodic table (B, C, N, O, F) is never greater than eight (octet rule). These atoms may have fewer than eight electrons, but not more.

3. In some cases, atoms beyond the second period of the periodic table may have more than eight electrons (P, S, etc.).

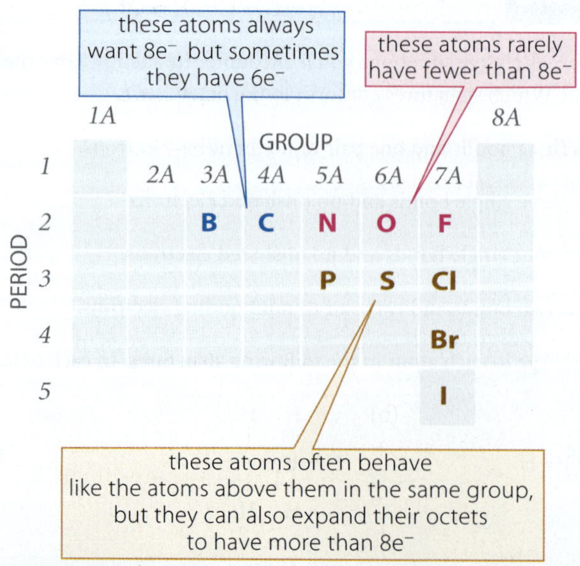

(1.30)

4. The formal charge is computed by the procedure given in Sec. 1.3B; if it is not zero, it is indicated with a plus or minus sign on the appropriate atoms.

5. Sum up all of the valence electrons of the atoms involved, and make sure you account for all of them in the final structure (subtracting one for each positive charge and adding one for each negative charge). Make sure an octet of electrons is around each atom (except hydrogen, which should share two electrons).

Study Problem 1.5

Draw a Lewis structure for the covalent compound acetonitrile *N*-oxide, H_3CCNO. (The atoms are connected as shown in the structure.)

Solution The total number of valence electrons is 3H(3) + 2C(8) + N(5) + O(6) = 22. Because the structure in the problem tells us how the atoms are connected, start by connecting all atoms with single bonds. These bonds account for 12 valence electrons.

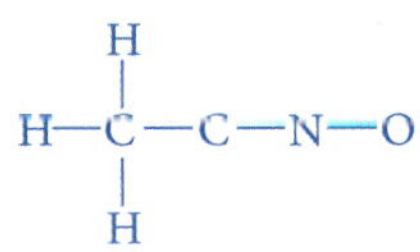

The first carbon has an octet. Give the second carbon an octet by writing multiple bonds between C and N (this also gives N an octet). These bonds account for 4 more valence electrons, or a total of 16 so far.

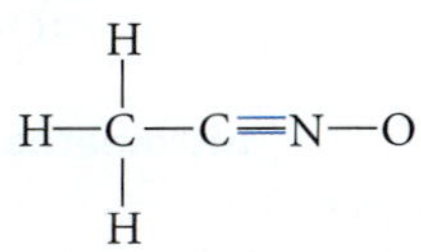

Give the oxygen an octet with electron pairs. This accounts for all 22 valence electrons.

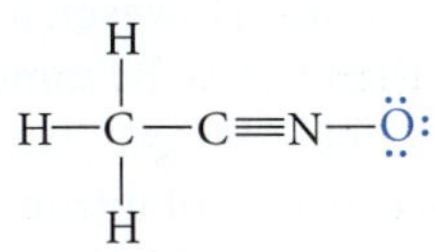

Add formal charges. If you learned to recognize common formal-charge patterns, you can do this by inspection to give a correct structure.

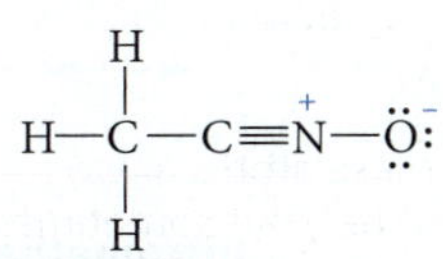

acetonitrile-*N*-oxide

You could have put double bonds between the carbon and nitrogen and between the nitrogen and oxygen, in which case you would have generated a different structure:

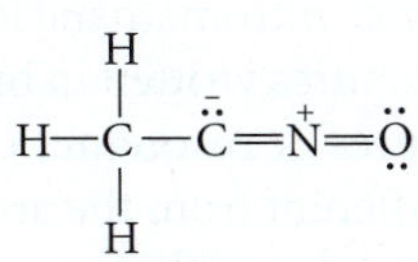

acetonitrile-*N*-oxide
(another structure)

This is also a correct structure! As this example illustrates, sometimes more than one correctly written Lewis structure is possible with the atoms connected in the same way that differ only in the placement of bonds, unshared pairs, and charges. We consider such situations in the next section.

Focused Problems

1.14 Draw a Lewis structure for each of the following species. Show all nonbonding electron pairs and the formal charges, if any. Assume that bonding follows the octet rule in all cases.

(a) $HCCl_3$ (b) $[NH_4]^+$ (c) $[C_2H_5O]^-$

(d) $[C_2H_8N]^+$ (Two structures are possible in which atoms are connected differently; draw both of them.)

1.15 Draw two possible Lewis structures for diazomethane (CH_2N_2). (*Hint:* Although the molecule has an overall charge of 0, both structures contain formal charges.)

1.5 INTRODUCTION TO RESONANCE

Some compounds are not accurately described by a single Lewis structure. Consider, for example, the structure of nitromethane, $H_3C—NO_2$.

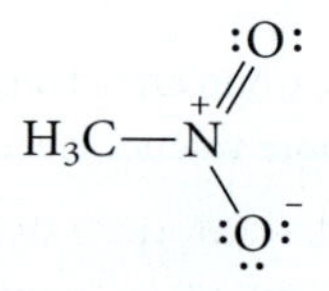

nitromethane (1.31)

This Lewis structure shows an N—O single bond and an N=O double bond. In our study of molecular geometry (in the next section, 1.6B), we explain that double bonds are shorter than single bonds between the same two atoms. However, it is found experimentally that the two nitrogen–oxygen bonds of nitromethane have the same length, and this length is intermediate between the lengths of single and double nitrogen–oxygen bonds found in other molecules. We can convey this idea by writing the structure of nitromethane as two different structures in brackets:

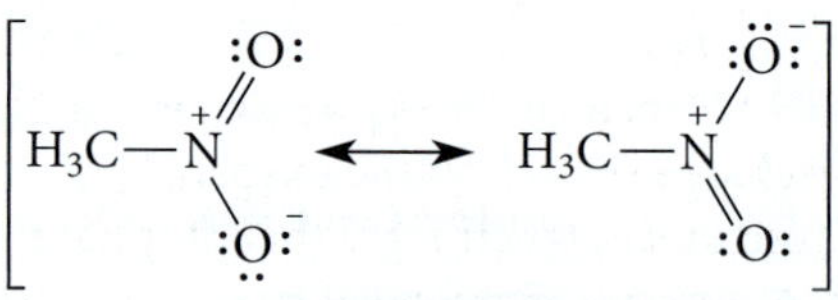

nitromethane
a *resonance hybrid;*
the individual structures are *resonance contributors* (1.32)

The double-headed arrow (⟷) means that nitromethane has a *single structure* that is the "average" of the individual structures; nitromethane is said to be a **resonance hybrid** of these two structures. The two individual structures written in brackets and connected by the double-headed arrow are called **resonance structures** or **resonance contributors**. Be sure to understand that the double-headed arrow, ⟷, is different from the arrows used in chemical equilibria, ⇌. The two resonance contributors for nitromethane are *not* rapidly interconverting and they are *not* in equilibrium. Atoms do *not* move in resonance structures. Resonance structures are alternative representations of *one molecule*. Resonance structures are necessary because of the inadequacy of a single Lewis structure to represent nitromethane accurately.

The two resonance structures in Eq. 1.32 represent *fictitious* molecules, whereas nitromethane is a *real* molecule. Because nitromethane can't be described accurately with a single Lewis structure, we must describe it as the hybrid of two fictitious structures.

When two resonance structures are equivalent, as they are for nitromethane, they are equally important in describing the molecule. (Two structures are *equivalent* when they have the same atoms connected in exactly the same way. For example, each nitromethane structure has an N—O double bond, an N—O single bond in which the nitrogen has a negative charge, and a positively charged N.) We can think of nitromethane as an average of the structures in Eq. 1.32. That is, each oxygen bears half of a negative charge, and each nitrogen–oxygen bond is neither a single bond nor a double bond, but a bond halfway in between. This hybrid character can sometimes be conveyed by a single structure in which dashed lines are used to represent partial bonds:

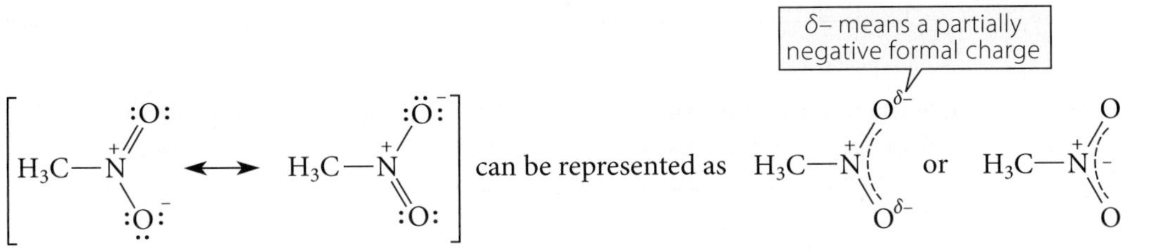

resonance structures for nitromethane

hybrid structures for nitromethane

(1.33)

In the first hybrid structure, a δ notation is used to represent partial charges on the oxygens. In the second structure, a single minus charge is written next to the dashed line, and it is understood that the charge is distributed between the two oxygens.

If two resonance structures are not equivalent, then the molecule they represent is a weighted average of the two. That is, one of the structures is more important than the other in describing the molecule. Such is the case, for example, with the methoxymethyl cation:

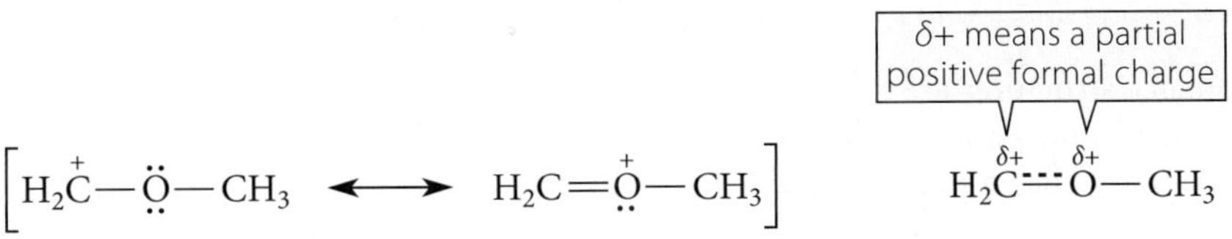

methoxymethyl cation

hybrid structure (1.34)

It turns out that the resonance contributor on the right is a better description of this cation because all atoms have complete octets. (We describe how to evaluate the importance of resonance structures in Sec. 3.3C.) Hence, the C—O bond has significant double-bond character, and most of the formal positive charge resides on the oxygen.

A very important aspect of resonance structures is that they have implications for the stability of the molecule they represent. *A molecule represented by resonance structures is more stable than any of its fictional resonance contributors would be.* For example, the actual molecule nitromethane is more stable than either one of the fictional molecules described by the contributing resonance structures in Display 1.32. Consequently, nitromethane is said to be a **resonance-stabilized** molecule, as is the methoxymethyl cation.

Why are compounds that have resonance structures more stable than their fictional contributors? Consider the two resonance structures for nitromethane. In either one, the double bond is *localized*—that is, confined to one place in the molecule. In the resonance hybrid, this bond, and therefore the electrons involved in the bond, are **delocalized**. *The delocalizing of electrons generally results in increased stability (lower energy),* because the electrons have more space over which to circulate, and repulsions between them are reduced. Delocalizing electrons also cause delocalization of the negative charge.

The connection between resonance and stability is considered in greater depth in Sec. 15.6.

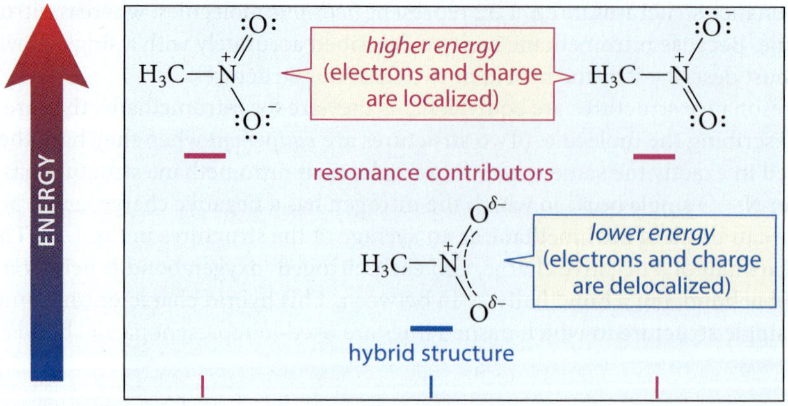

(1.35)

How do we know when to use resonance structures, how to draw them, or how to assess their relative importance? In Sec. 3.3A, we present a technique for deriving resonance structures, and in Sec. 3.3C, we return to a more detailed study of the other aspects of resonance. In the meantime, we'll draw resonance structures for you and tell you when they're important. Just try to remember the following points:

1. Resonance structures are used for compounds that are not adequately described by a single Lewis structure.
2. Resonance structures are enclosed in brackets and connected by double-headed arrows ($\longleftrightarrow$).
3. Resonance structures are sometimes summarized as a single hybrid structure in which delocalized bonds are shown as dashes and delocalized charges as partial charges using the delta ($\delta+$, $\delta-$) notation.
4. Resonance structures are not in equilibrium—that is, the molecule they describe does *not* have one resonance structure part of the time and the other resonance structure part of the time; rather, it has *a single structure.*
5. Although the positions of electrons differ between resonance structures, the nuclei of atoms do not move.
6. The structure of a molecule is the weighted average of its resonance structures. When resonance structures are equivalent, they are equally important descriptions of the molecule.
7. Resonance hybrids are more stable than any of the fictional contributors. Molecules described by resonance structures are said to be resonance-stabilized.

Focused Problems

1.16 The compound *benzene* has only one type of carbon–carbon bond, and this bond is intermediate in length between a single bond and a double bond. Draw a resonance structure for benzene which, taken with the following structure, accounts for the carbon–carbon bond length in benzene.

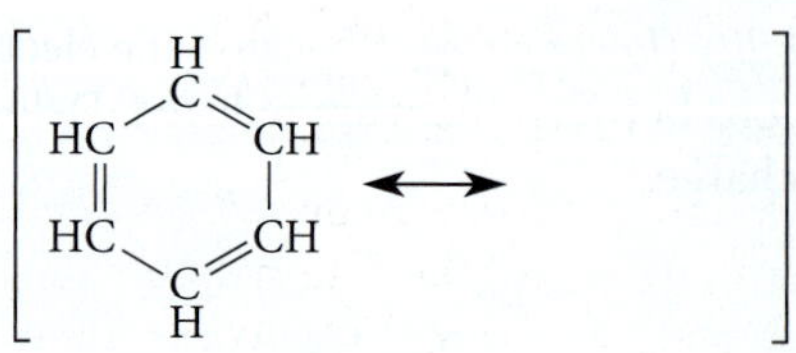

benzene

1.17 (a) Draw a resonance structure for the *allyl anion* that shows, along with the following structure, that the two CH_2 carbons are equivalent.

$$\left[H_2C{=}CH{-}\overset{-}{\ddot{C}}H_2 \longleftrightarrow \qquad\qquad\qquad \right]$$

allyl anion

(b) According to the resonance structures, how much negative charge is on each carbon?

(c) Draw a single hybrid structure for the allyl anion that shows delocalized bonds as dashed bonds and the charges as partial negative charges.

1.6 STRUCTURES OF COVALENT COMPOUNDS

Covalent bonds are directed in space. In other words, covalent molecules have specific structures. We know the **structure** of a molecule containing covalent bonds when we know its *atomic connectivity* and its *molecular geometry*. **Atomic connectivity**, or **constitution**, is the specification of how atoms in a molecule are connected. For example, we specify the atomic connectivity within the water molecule when we say that two hydrogens are bonded to an oxygen with single bonds. **Molecular geometry**, or **molecular configuration**, is the specification of how the atoms are situated in space.

Chemists learned about atomic connectivity before they learned about molecular geometry. The concept of covalent compounds as geometrical objects emerged in the latter part of the nineteenth century on the basis of indirect chemical and physical evidence. Until the early part of the twentieth century, however, no one knew whether these concepts had any physical reality, because scientists had no techniques for viewing molecules at the atomic level. Since then, a variety of physical techniques have emerged for determining the structures of covalent compounds. In this section, our goal is to survey the relatively small number of shapes that are typically encountered in organic chemistry. We present a few principles that will allow you to predict molecular geometry in a general way if you know the atomic connectivity within a molecule.

It's important for you to learn about molecular geometry because it is a key concept in organic chemistry that can affect chemical reactivity. It also lies at the heart of molecular biology.

Molecular structure comprises three elements:

1. Bond length
2. Bond angle
3. Dihedral angle

We define each of these terms, give examples, and learn some principles by which these elements of structure can be predicted.

A. Methods for Determining Molecular Geometry

Among the greatest developments of chemical physics in the early twentieth century were the discoveries of ways to deduce the structures of molecules. Such techniques include nuclear magnetic resonance (NMR) spectroscopy, infrared spectroscopy (IR) spectroscopy, ultraviolet (UV)–visible spectroscopy, and mass spectrometry, which we consider in Chapters 13–16. As important as these techniques are, they are used primarily to provide information about atomic connectivity. Other physical methods, however, permit the determination of molecular structures that are complete in every detail. Most complete structures today come from three sources: X-ray crystallography, electron diffraction, and microwave spectroscopy.

The arrangement of atoms in the crystalline solid state can be determined by *X-ray crystallography*. This technique, invented in 1915 and subsequently revolutionized by the availability of high-speed computers, uses the fact that X-rays are diffracted from the atoms of a crystal in precise patterns that can be mathematically deciphered to give a molecular structure. In 1930, *electron diffraction* was developed. With this technique, the diffraction of electrons by molecules of gaseous substances can be interpreted in terms of the arrangements of atoms in molecules. Following the development of radar in World War II came *microwave spectroscopy*, in which the absorption of microwave radiation by molecules in the gas phase provides detailed structural information.

Most of the spatial details of molecular structure in this book are derived from gas-phase methods: electron diffraction and microwave spectroscopy. For molecules that are not readily studied in the gas phase, X-ray crystallography is the most important source of structural information. No methods of comparable precision exist for molecules in solution, a fact that is unfortunate because most chemical reactions take place in solution. The consistency of gas-phase and crystal structures suggests, however, that molecular structures in solution probably differ little from those of molecules in the solid or gaseous state.

B. Bond Length

The structure of a covalently bonded diatomic molecule, such as Br_2, HCl, and CO, is completely defined by its **bond length**, the distance between the centers of the bonded nuclei. Bond length is measured in picometers (pm), where 1 pm = 10^{-12} m. It is also frequently reported in angstroms (Å), where 100 pm = 10^{-8} cm = 1 Å.

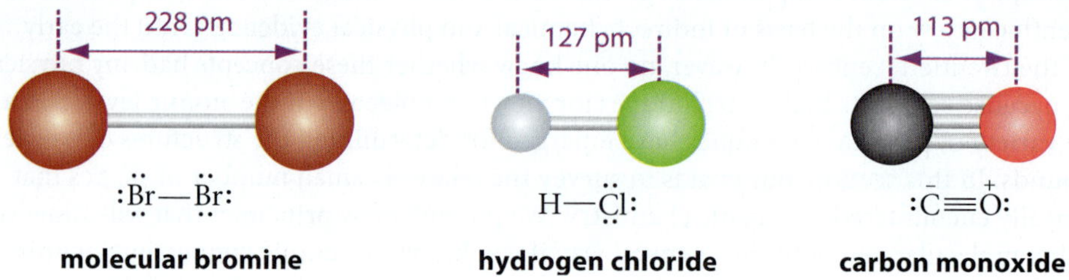

(1.36)

The following three generalizations can be made about bond length, *in order of importance:*

1. Within a group of the periodic table, bond lengths increase as the period number increases.

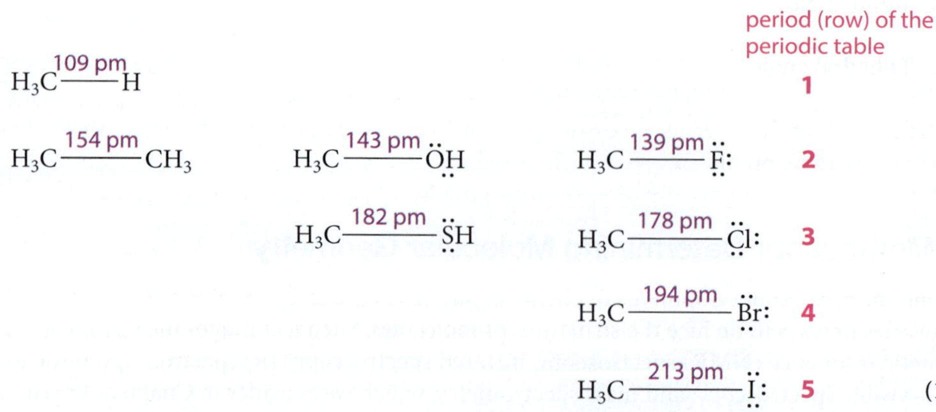

(1.37)

This trend is a reflection of the sizes of the bonded atoms. Although we often represent molecules as ball-and-stick models (spheres connected by bonds, as above), we could also visualize the atoms in a bond as spheres that are just touching. In this representation, the radius of each bonded atom is called its **covalent radius**. The covalent radius is one measure of atomic size. For example, the covalent radius of a hydrogen atom in H_2 is 37 pm.

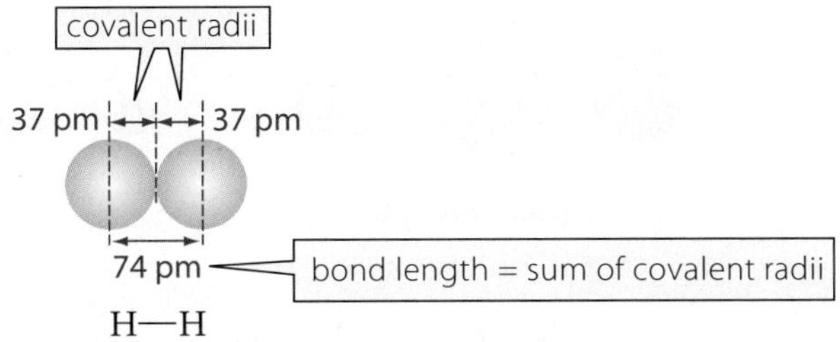

dihydrogen (H_2) (1.38)

The sum of the covalent radii of bonded atoms is the bond length. It follows, then, that bonds involving larger atoms are longer.

2. *Bond lengths decrease with increasing bond order.* **Bond order** describes the number of covalent bonds shared by two atoms. A C—C bond has a bond order of 1, a C═C bond has a bond order of 2, and a C≡C bond has a bond order of 3. In other words, triple bonds are shorter than double bonds to the same two atoms, which are shorter than single bonds.

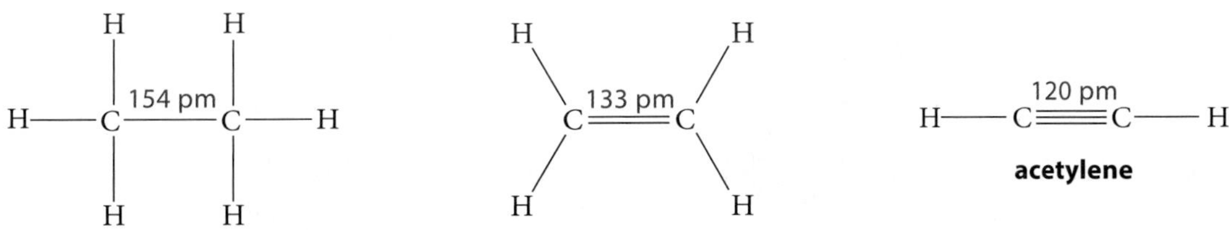

ethane **ethylene** **acetylene** (1.39)

3. Bonds decrease in length toward higher atomic number (to the right) along a row (period) of the periodic table. Like the increase between periods (generalization 1), this is an effect of atomic size, which tends to decrease as you move from left to right in a row of the periodic table. However, this effect is much less significant than the increases between atoms in successive periods.

methane **ammonia** **water** **hydrogen fluoride** (1.40)

C. Bond Angle

When a molecule has more than two atoms, a complete structure requires knowledge of the **bond angle**, the angle between each pair of bonds to the same atom.

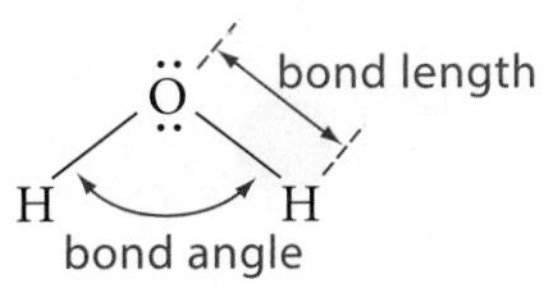

(1.41)

You can predict approximate bond angles for almost all organic molecules by considering the number of atoms and nonbonding electron pairs around an atom.

The central principle is that groups around a central atom should have maximum separation. When an atom is surrounded by *two* atoms or groups of atoms, and no additional nonbonding electrons are present, the maximum separation of the bonds requires a bond angle of 180°, and the atoms have a **linear** geometry or shape.

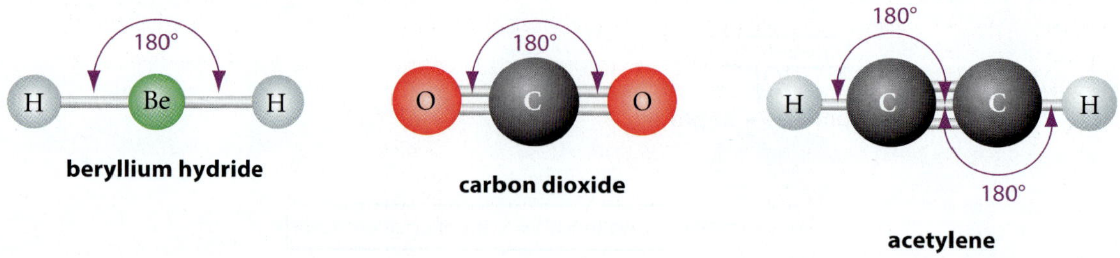

(1.42)

Notice that *double bonds and triple bonds are considered as one bond* for purposes of predicting the bond angle. In carbon dioxide, the carbon is surrounded by two oxygens, and the oxygens and their double bonds are as far apart as possible. In acetylene, each carbon is surrounded by two groups: a hydrogen and another carbon, even though the bond between the carbons is a triple bond.

If the central atom has one or two pairs of nonbonding electrons, these electrons take up space and squeeze the two bonds closer together. The bond angle becomes considerably less than 180°—typically less than 120°, the exact value depending on the molecule. These molecules or ions have a **bent** shape.

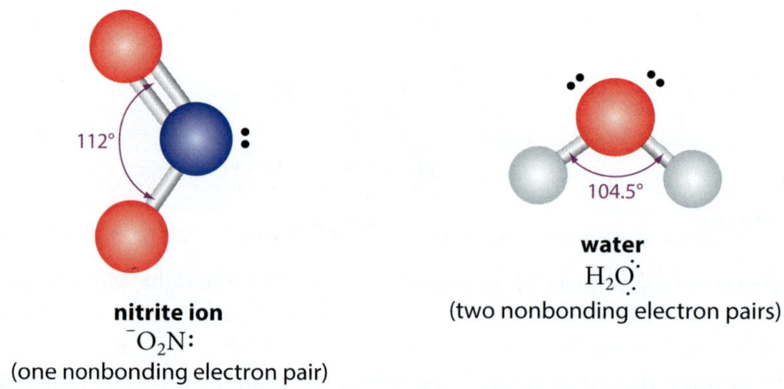

(1.43)

(Don't be concerned with the exact bond angles in these examples. What you should notice is that the shape of the molecule is *bent*.)

When *three* atoms and no nonbonding electrons surround a central atom, the bonds are as far apart as possible when all bonds lie in the same plane with bond angles of 120°.

As the structure of ethylene illustrates, there may be minor deviations from the ideal 120° bond angles when the three atoms surrounding the central atom are different.

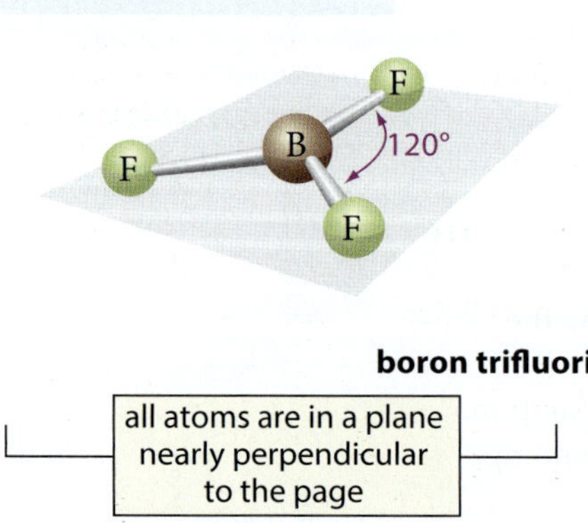
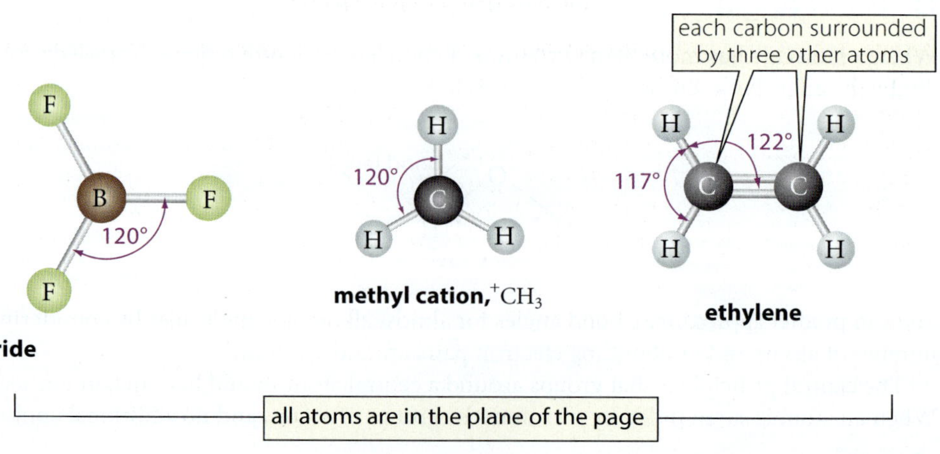

(1.44)

In these cases, the geometry at each central atom is called **trigonal-planar** geometry, because all atoms lie in the same plane.

If, in addition to three atoms (or groups of atoms), the central atom has a pair of nonbonding electrons, then two geometries are possible. If the electrons are *localized* (that is, confined to the central atom), these electrons take up space and squeeze the other bonds closer together, just as they do when there are two bonds. These bond angles are somewhat smaller than the 120° bond angles of trigonal-planar atoms, and the molecule has a "puckered" shape.

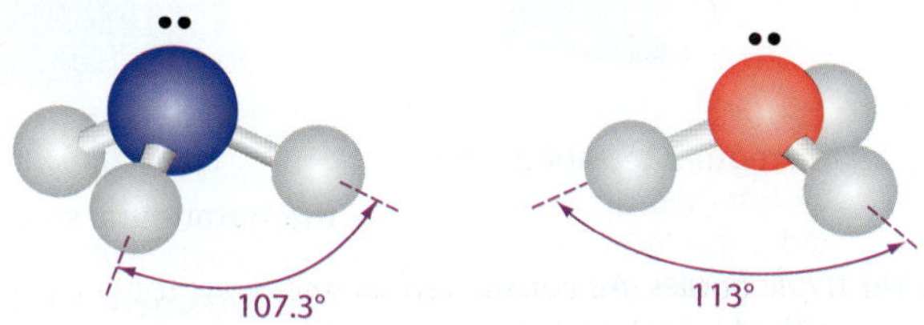

ammonia :NH_3 **hydronium ion** $H_3\overset{+}{O}$: (1.45)

This geometry is called **trigonal-pyramidal** geometry. The reason for this name is that a geometrical figure superimposed on such a structure with each atom at a vertex is a **trigonal pyramid**—that is, a pyramid with the atom bearing the nonbonding electron pair at the apex and the three attached atoms at the corners of a trigonal base. (The legs of a camera tripod form the three vertical edges of a trigonal pyramid.)

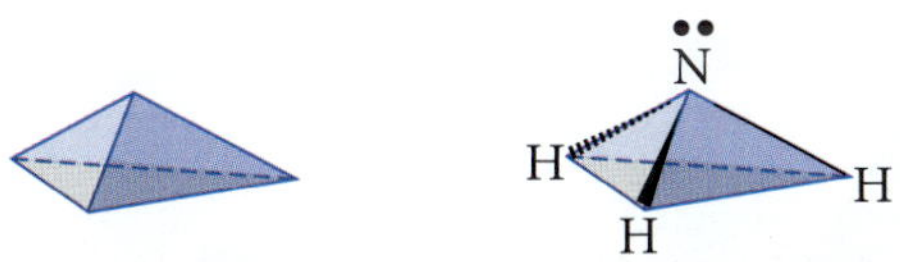

a trigonal pyramid trigonal-pyramidal molecular geometry (1.46)

If the electrons on the central atom can be *delocalized by resonance*, the atom has *trigonal-planar geometry*. Consider, for example, the nitrogen atom in formamide. The electrons on the nitrogen can be delocalized to the neighboring carbon by resonance.

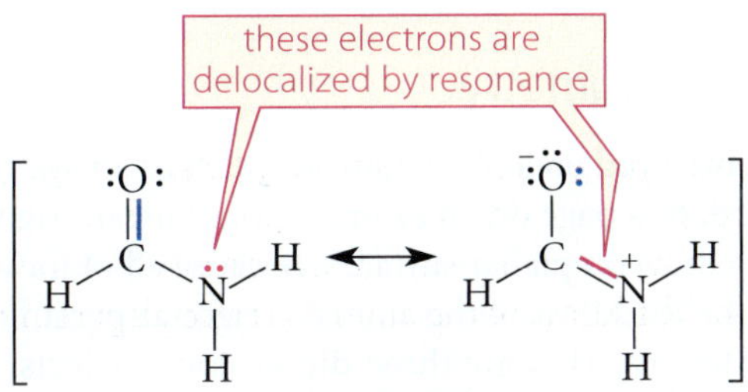

formamide
the nitrogen is trigonal planar (1.47)

Because the resonance structures, when averaged together, suggest that the nonbonding electrons on the nitrogen are really shared with other atoms, we say that the electrons are delocalized. The second resonance structure shows the nitrogen bonded to three other atoms, which suggests *trigonal-planar* geometry. In fact, this prediction is confirmed by the actual bond angles around nitrogen in formamide, which are close to 120°.

When four groups are bonded to a central atom, the bonds are farthest apart when the central atom has **tetrahedral** geometry. In this geometry, the four bound groups lie at the vertices of a **tetrahedron**.

We explain how to predict when electrons are delocalized by resonance in Sec. 3.3B. For now, all you need to understand is that when such resonance occurs, the shape around nitrogen is trigonal planar.

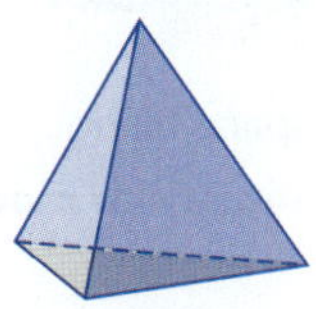

a tetrahedron (1.48)

Methane, CH_4, has tetrahedral geometry. The central atom is the carbon and the four groups are the hydrogens. The C—H bonds of methane are as far apart as possible when the hydrogens lie at the vertices of a *regular tetrahedron*. The tetrahedral shape of methane requires a bond angle of 109.5°.

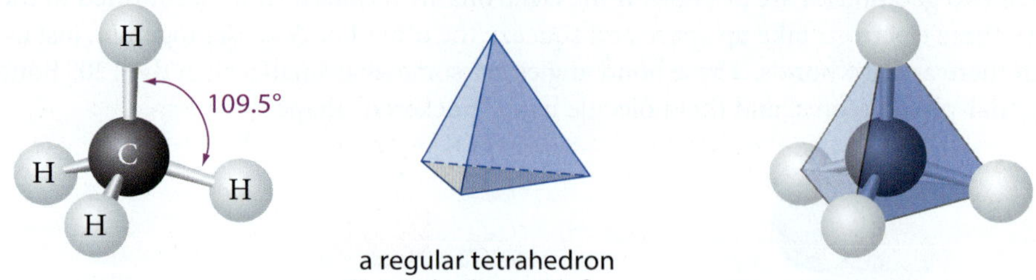

methane

a regular tetrahedron
(A tetrahedron with four sides that are equilateral triangles. All edges have equal lengths.)

the tetrahedral shape of methane
(The carbon is inside the tetrahedron and the hydrogens are at the vertices.)

(1.49)

Minor deviations from the exact tetrahedral bond angle of 109.5° can occur when different atoms surround the central atom, but the general tetrahedral shape exists regardless.

Let's contrast the tetrahedral and trigonal-pyramidal shapes. A tetrahedral molecule has one central atom *surrounded* by four other atoms or groups at the vertices of a tetrahedron, whereas a trigonal-pyramidal molecule has *all four atoms* at the vertices of a shorter (or "squashed") tetrahedron, which we call a "trigonal pyramid."

a regular tetrahedron

tetrahedral geometry

a trigonal pyramid
(a "squashed" tetrahedron)

trigonal-pyramidal geometry

(1.50)

D. Depicting Molecular Geometry

Molecular geometry is often portrayed on a planar surface such as a page of this text, your instructor's slides or blackboard, or a page on an examination. Linear, trigonal-planar, and bent structures are relatively easy to draw on planar surfaces, because all of these objects are one- or two-dimensional (neglecting the thickness of the atoms). Trigonal-pyramidal and tetrahedral structures are trickier to draw because they are three-dimensional objects. The goal of this section is to define conventions for drawing three-dimensional structures on a two-dimensional surface.

First, become familiar with the tetrahedral shape by building a **molecular model** of methane using a tetrahedral carbon and four hydrogens. *Almost all beginning students require models, at least initially, to visualize the three-dimensional aspects of organic chemistry.* Here is a model of methane constructed with the most common model types.

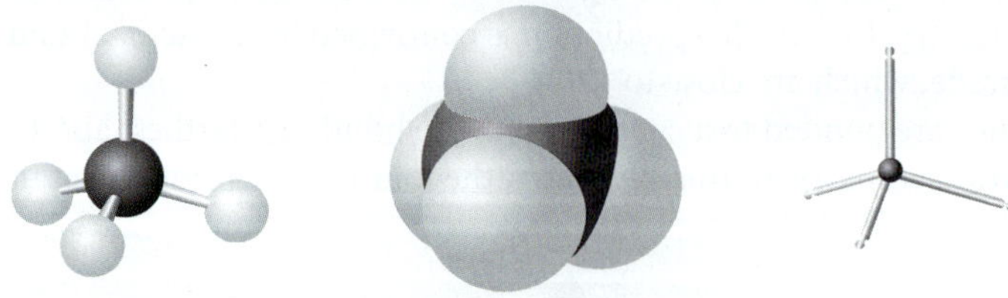

ball-and-stick

space-filling

framework

types of molecular models

(1.51)

In this text, we use *ball-and-stick* models to visualize the directionality of chemical bonds, and we use *space-filling models* to see the consequences of atomic and molecular volumes. A *framework* model emphasizes the network of bonds, with nuclei assumed to be at the positions at which the bonds intersect or terminate.

Orient your model of methane so that it matches the one above. The convention for representing tetrahedral geometry is to use **line-and-wedge** structures. In a line-and-wedge structure, bonds shown with regular lines are assumed to be in the plane of the page, *solid wedge bonds* project toward you in front of the page, and *dashed wedge bonds* project away from you into the page.

Relatively inexpensive student molecular model sets of the ball-and-stick or framework type are commercially available. You will find that the use of models will be valuable in understanding the spatial aspects of organic chemistry.

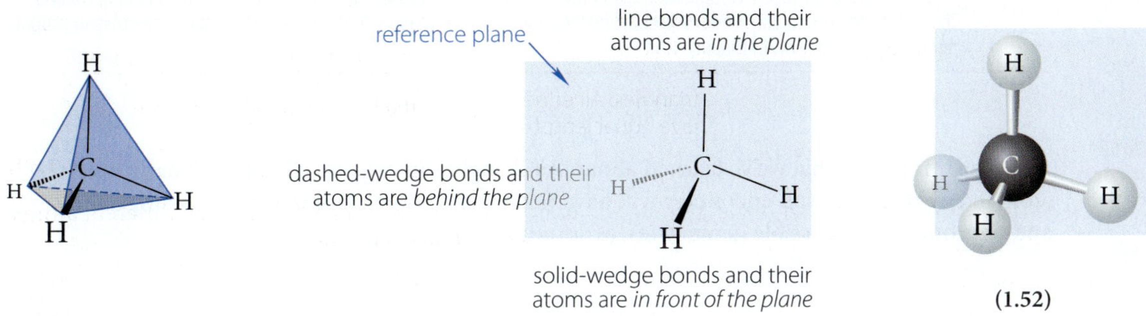

(1.52)

Focused Problem 1.19 asks you to turn a tetrahedral model in different ways and draw the resulting line-and-wedge structure. It is important that you practice this skill until you feel comfortable with it.

Because trigonal-pyramidal structures are "squashed tetrahedra," the same line-and-wedge convention is used to portray these structures as well.

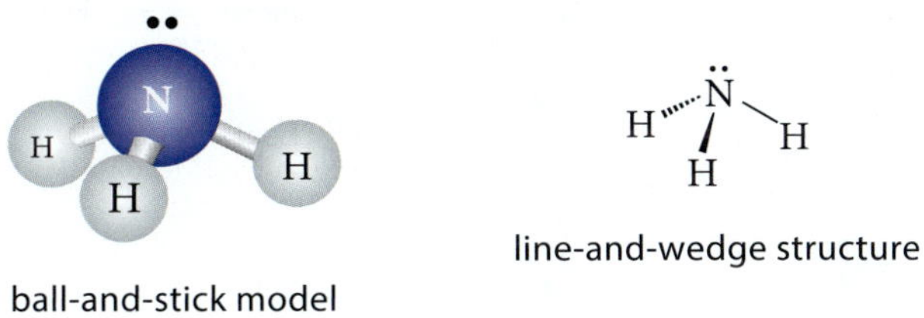

ammonia (1.53)

Although trigonal-planar structures can be drawn with all bonds in the plane of the paper, they can also be drawn with line-and-wedge structures if we want to view them from another perspective.

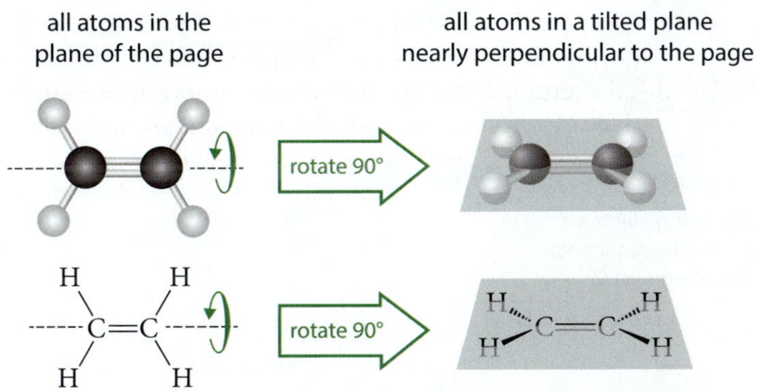

(1.54)

This perspective is very useful in discussing orbitals (Sec. 1.7) as well as the spatial aspects of reactions that occur at the carbon–carbon double bond.

Don't be confused by drawing or typesetting conventions that seem to ignore what you've just learned about molecular geometry. For example, you might see methane written as

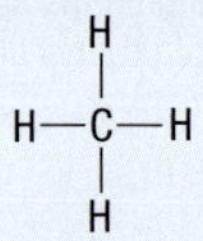

even though you know methane is tetrahedral and can be drawn as a line-and-wedge structure. This structure shows *connectivity* but not *molecular geometry.* It is perfectly acceptable to use "connectivity-only" structures when molecular geometry is not an issue because they are simpler and can be drawn more rapidly. Only when molecular geometry is under discussion do you need to use line-and-wedge structures.

A common error made by students is to use wedges and dashed wedges for in-plane structures:

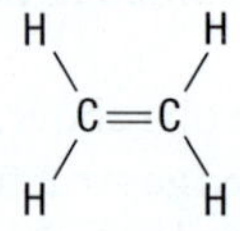

correct representation of an in-plane structure

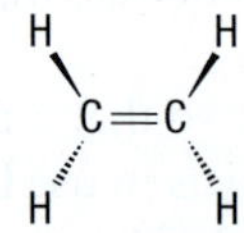

incorrect representation of an in-plane structure

This is incorrect because wedges are used only for bonds that are *emerging from the plane* or *receding behind the plane.* Be sure to use wedges and dashed wedges only for these situations, and draw the molecules in proper perspective, as shown in Display 1.54.

Let's summarize what we've learned about bond angles. In organic chemistry, nearly every atom in any molecule you encounter will have one of these seven atom/electron arrangements and thus one of five possible geometries, as summarized in Table 1.1.

TABLE 1.1 A Table of Common Molecular Shapes Encountered in Organic Chemistry

A = central atom X_n = number of atoms or groups bound to central atom
E_n = number of nonbonding electron pairs on central atom

Abbreviation	Electron arrangement	Geometry	Shape	Examples
AX_2E_0		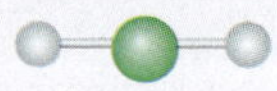	linear	$H-C\equiv CH$ $\ddot{O}=C=\ddot{O}$
AX_2E_1	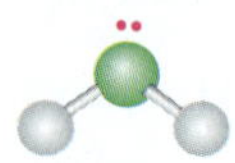	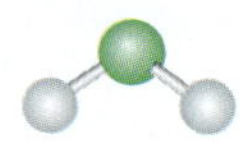	bent	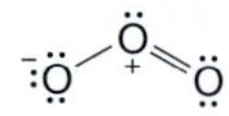
AX_2E_2	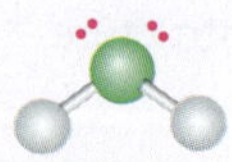	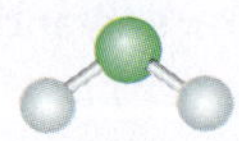	bent	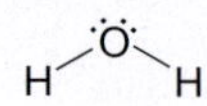
AX_3E_0	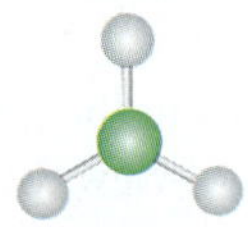	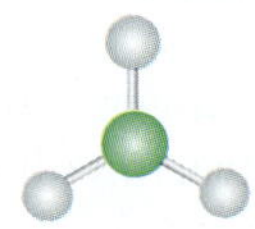	trigonal-planar	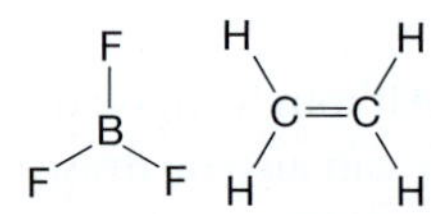
AX_3E_1	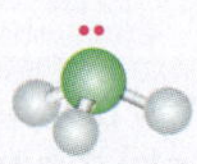 (nonbonding electrons localized)		trigonal-pyramidal	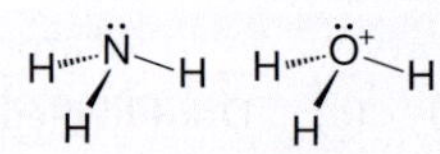
AX_3E_1	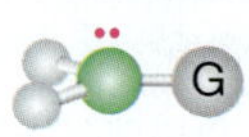 (nonbonding electrons delocalized by resonance into the group G)		trigonal-planar	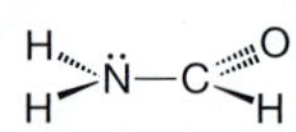
AX_4E_0	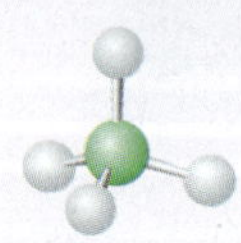	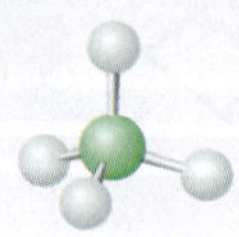	tetrahedral	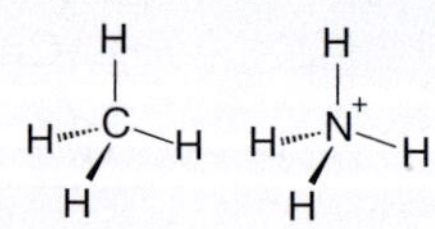

Study Problem 1.6

Estimate each bond angle in the following molecule, and order the bonds according to length, beginning with the shortest.

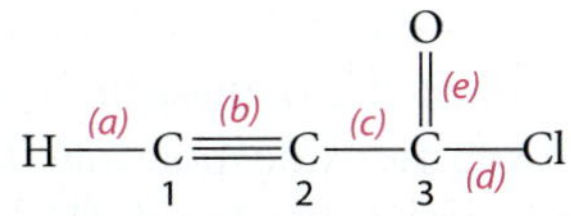

Solution Because carbon-1 is bound to two groups (H and C), its geometry is linear. Similarly, carbon-2 also has linear geometry. The remaining carbon (carbon-3) is bound to three groups (C, O, and Cl); therefore, it has approximately trigonal planar geometry.

To arrange the bonds in order of length, recall the order of importance of the bond-length rules. The major influence on length is the row in the periodic table from which the bonded atoms are taken. Hence, the H—C bond is shorter than all carbon–carbon or carbon–oxygen bonds, which are shorter than the C—Cl bond. The next major effect is the bond order. Hence, the C≡C bond is shorter than the C—O bond, which is shorter than the C═C bond. Putting these conclusions together, the required order of bond lengths is

$$(a) < (b) < (e) < (c) < (d)$$

Focused Problems

1.18 Predict the approximate molecular geometry in each of the following molecules or ions.

(a) $[BH_4]^-$ (b) $H_2C{=}\ddot{\underset{\cdot\cdot}{O}}$ **formaldehyde** (c) $H_3C{-}C{\equiv}N{:}$ **acetonitrile** (d) $H{-}\ddot{N}{-}H$ with Cl on N **chloramine** (e) $H{-}\ddot{C}{-}H$ **methylene**

1.19 Many line-and-wedge structures are possible for any given molecule, depending on how we view the molecule. Draw line-and-wedge structures for each of the following ball-and-stick models. The white balls are hydrogen (H) and the green balls are another atom, such as chlorine (Cl).

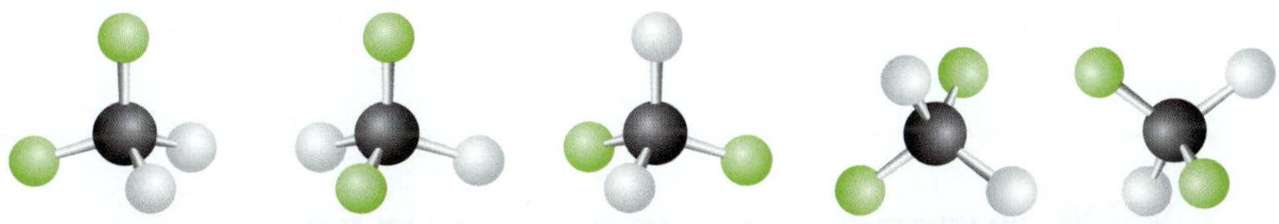

A *B* *C* *D* *E*

1.20 (a) Draw line-and-wedge structures for molecules (a) and (d) in Focused Problem 1.18.

(b) Draw a line-and-wedge structure for molecule (b) in Focused Problem 1.18 in which the H—C—H plane is perpendicular to the page.

1.21 Estimate each of the bond angles, and rank the bonds in order of increasing length in the following molecule. [*Hint:* Bond (*f*) is slightly longer than bond (*d*); bonds (*g*) are slightly longer than bonds (*a*).]

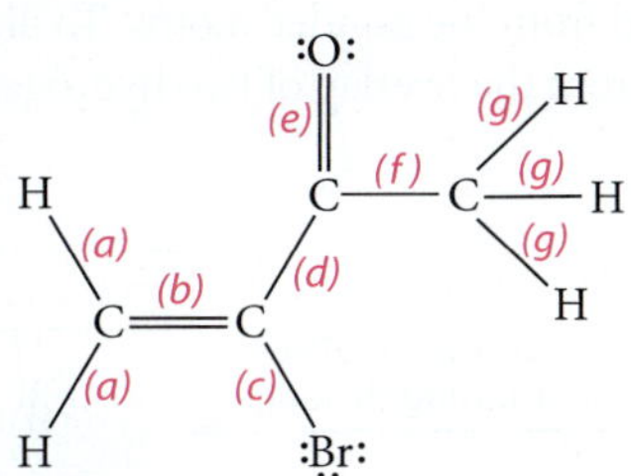

E. Dihedral Angle

To completely describe the shapes of molecules that are more complex than the ones just discussed, we need to specify not only the bond lengths and bond angles, but also the spatial relationship of the bonds on *adjacent* atoms.

To illustrate this problem, consider the molecule hydrogen peroxide, H_2O_2. Both O—O—H bond angles are 96.5°. However, knowledge of these bond angles is insufficient to describe completely the shape of the hydrogen peroxide molecule. To understand why, imagine two planes, each containing one of the oxygens and its two bonds (Display 1.55). To completely describe the structure of hydrogen peroxide, we need to know the angle between these two planes. This angle is called the **dihedral angle** or **torsion angle**. Three possibilities for the dihedral angle are shown here.

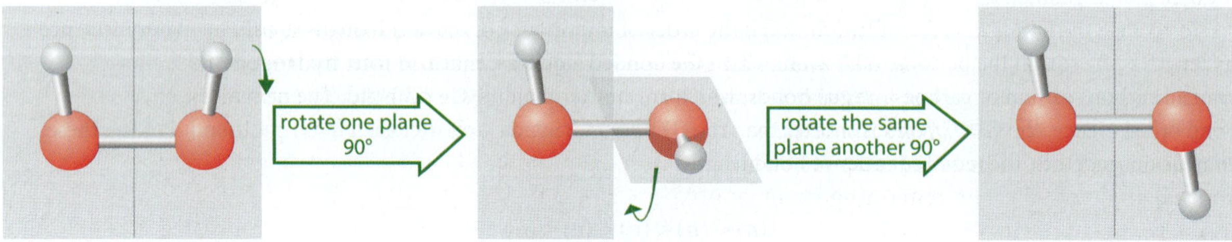

dihedral angle = 0° dihedral angle = 90° dihedral angle = 180°

(1.55)

You can also visualize these dihedral angles using a model of hydrogen peroxide by holding one O—H bond fixed and rotating the remaining oxygen and its bonded hydrogen about the O—O bond. (The actual dihedral angle in hydrogen peroxide is addressed in Problem 1.66.) Molecules containing many bonds typically contain many dihedral angles to be specified. We explain some of the principles that will allow you to predict dihedral angles in Chapter 2.

In summary, dihedral angles are required to describe a structure completely when a molecule contains adjacent atoms that are bonded by other groups.

Focused Problem

1.22 Which of the following ions and molecules require one or more dihedral angles to specify their geometry completely?

$H_3C—\overset{+}{N}H_3$ (*A*) $\overset{-}{P}F_6$ (*B*) $H\ddot{O}—\ddot{O}\!:^-$ (*C*) $H_3C—\ddot{O}—CH_3$ (*D*)

1.7 THEORIES OF THE COVALENT BOND: BOND ORBITALS AND HYBRID ORBITALS

A. Orbital Overlap

We've learned that a covalent bond consists of an electron pair that is shared between two bonded atoms. In this section, we consider in more detail how that "electron sharing" occurs.

Because electrons in atoms reside in atomic orbitals, covalent bonds result from the overlap of valence orbitals derived from the bonded atoms. To illustrate, consider the dihydrogen molecule, H_2. This molecule results from the overlap of two hydrogen 1s orbitals, each containing one electron.

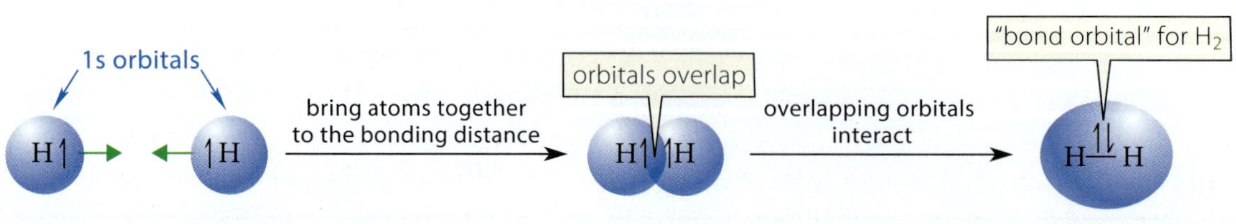

H—H σ-bond

(1.56)

In this model, the bond consists of two electrons in a **bond orbital**, a merged orbital of the molecule resulting from the overlap of two atomic orbitals. This bond orbital is oriented along the line connecting the two hydrogen nuclei and is cylindrically symmetrical about that line. In other words, the bond orbital looks the same no matter how it is rotated about the line connecting the two atoms. Bonds of this type are called **sigma bonds** or **σ bonds**.

Sigma is the Greek letter equivalent of *s*. Sigma is used because the simplest bond of this type—the "head-to-head" bond between two hydrogens—involves overlap of two s orbitals. (Not all sigma bonds involve s orbitals.)

B. sp^3 Hybridization

If we try to construct a bond-orbital model for methane, we would use the four valence orbitals of carbon—the 2s, $2p_x$, $2p_y$, and $2p_z$ orbitals—and let each of them overlap with a 1s orbital of hydrogen. With a total of eight valence electrons—four from carbon, four from the hydrogens—this process would create four electron-pair C—H bonds. However, this arrangement presents a problem of *molecular geometry*. We know that the four hydrogens in methane are at the corners of a regular tetrahedron, the hydrogens are completely equivalent, and the H—C—H bond angles in methane are 109.5°. In contrast, the bond-orbital picture just described would have three 2p–1s bonds at bond angles of 90°, because 2p orbitals have a mutual angle of 90°. The remaining bond would be a 2s—1s bond, occupying the remaining space around carbon.

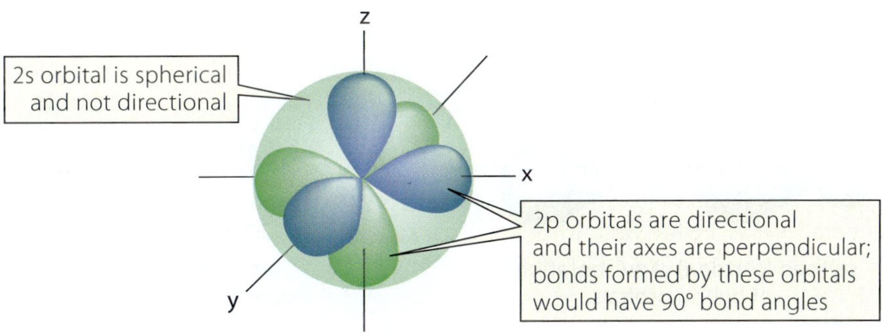

carbon valence orbitals

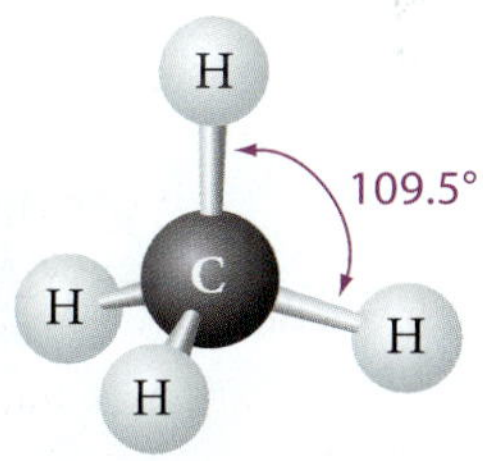

The use of pure 2s and 2p orbitals for C—H bonds is inconsistent with known methane bond angles.

(1.57)

This problem was solved in 1928 by Linus Pauling (1901–1994), a professor at the California Institute of Technology, who was awarded the 1954 Nobel Prize in Chemistry for his work on chemical bonding. Pauling's theory started with the premise that the *valence orbitals of the carbon in methane are different from the orbitals in atomic carbon*. However, the orbitals of carbon in methane can be derived simply from those of atomic carbon. For carbon in methane, we imagine that the 2s orbital and the three 2p orbitals are mixed to give four new equivalent orbitals, each with a character intermediate between pure s and pure p. It's as if we mixed a blue dog and three yellow cats and ended up with four *identical* yellowish-green animals, each of which is three-fourths cat and one-fourth dog. This mixing process applied to orbitals is called **hybridization**, and the new orbitals are called **hybrid orbitals**. Because each of the new hybrid carbon orbitals is one part s and three parts p, it could be called an $s_{0.25}\,p_{0.75}$ hybrid orbital. Four of these orbitals account for one 2s and three 2p orbitals. Conventionally, a hybrid orbital type is notated as sp^n. If f_s = fraction s character and f_p = fraction p character, n is given by

$$n = f_p / f_s \tag{1.58}$$

(Remember that by definition, $f_s + f_p = 1$.) So, for 0.25 s character and 0.75 p character, $n = 0.75/0.25 = 3$. Therefore, each hybrid orbital is an **sp^3 orbital** (pronounced "s-p-three," not "s-p-cubed"). The six carbon electrons in the hybrid-orbital model are distributed between one 1s orbital (the core electrons) and four equivalent sp^3 hybrid orbitals in quantum level 2. This process can be summarized as follows:

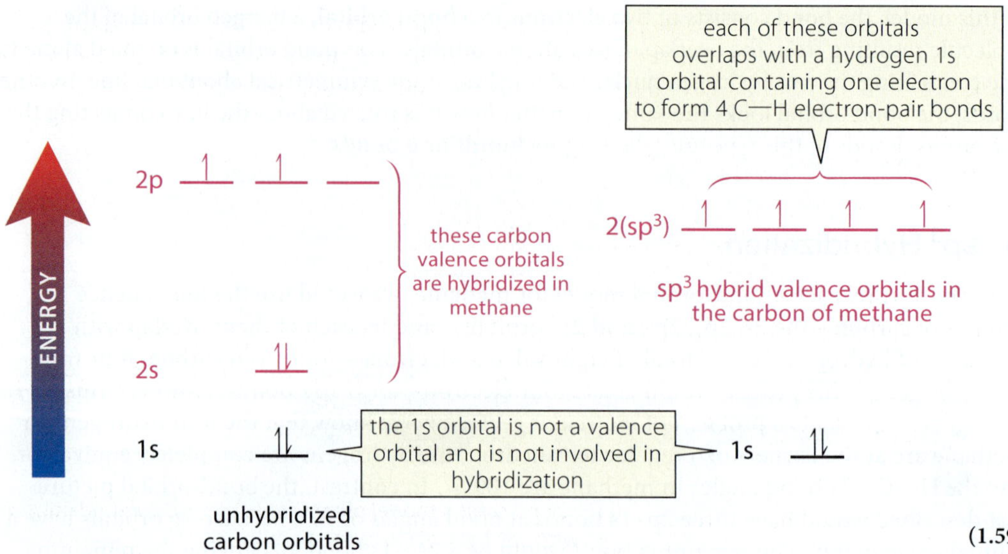

(1.59)

An sp^3 orbital, like a 2p orbital, has lobes. However, mixing in some 2s character makes one of the lobes smaller than the other.

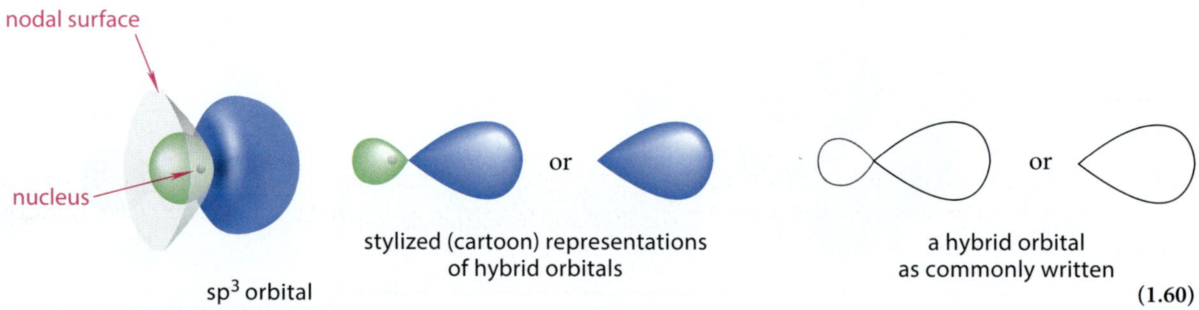

(1.60)

As Display 1.60 shows, we often sketch the hybrid orbitals as teardrops, which emphasizes their directional character.

Because there are *four equivalent sp^3 orbitals*, they are distributed spatially about the carbon nucleus in the same way that we would distribute any other four spatially equivalent objects: they are directed to the corners of a regular tetrahedron. The formation and spatial distribution of these hybrid orbitals are summarized in **Fig. 1.11**.

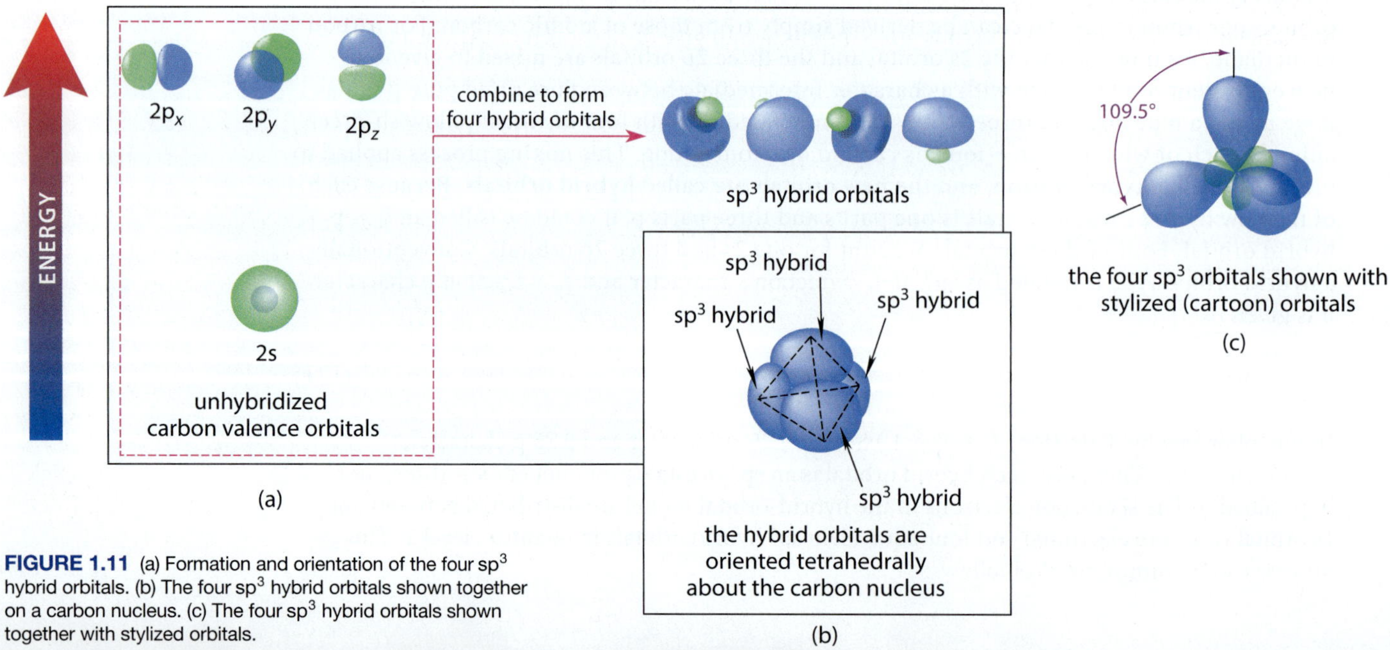

FIGURE 1.11 (a) Formation and orientation of the four sp^3 hybrid orbitals. (b) The four sp^3 hybrid orbitals shown together on a carbon nucleus. (c) The four sp^3 hybrid orbitals shown together with stylized orbitals.

Overlap of a hydrogen 1s orbital containing one electron with each of the sp^3 orbitals of carbon gives the four C—H bond orbitals of methane with tetrahedral molecular geometry. (The smaller rear lobes of the sp^3 orbitals in the second structure are omitted for clarity.)

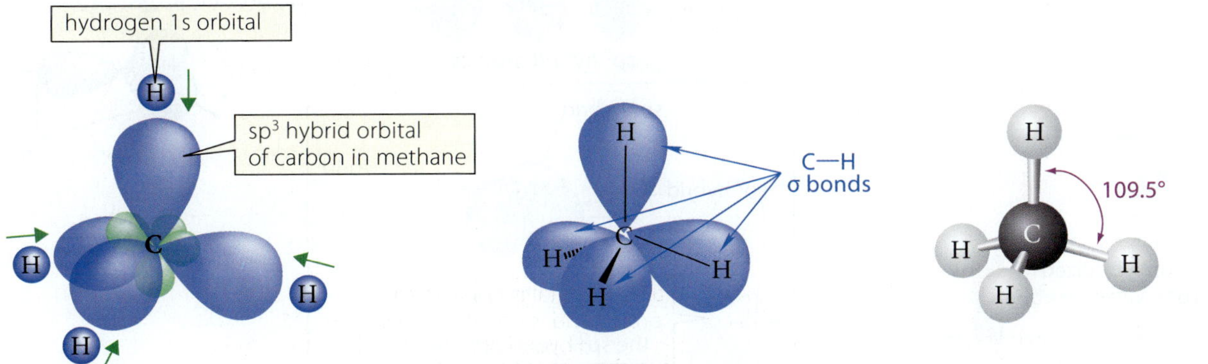

bond orbital model of methane with sp^3-hybridized carbon orbital (small rear lobes not shown)

ball-and-stick model of methane

(1.61)

Tetrahedral geometry is associated with sp^3 hybridization of the central atom. This is true whether the central atom is carbon or any other atom.

Focused Problem

1.23 Following steps like the ones used for the hybrid-orbital model of methane, construct a hybrid-orbital model of the ammonium ion, $[NH_4]^+$. (*Hint:* Start with an N^+ ion, form the hybrid orbitals, and let each hybrid orbital interact with a hydrogen atom.)

C. sp^2 Hybridization

Trigonal-planar central atoms utilize a different orbital hybridization, in which one of the 2p orbitals of carbon is left out of the hybridization. This type of hybridization accounts for the geometry of each carbon of ethylene, $H_2C{=}CH_2$. For each of these carbons, imagine that three valence orbitals of carbon, the 2s, $2p_x$, and $2p_y$ orbitals, are mixed, or hybridized, to form three equivalent hybrid orbitals; each of these hybrid orbitals has one-third 2s character and two-thirds 2p character. This means that, in the sp^n notation (Eq. 1.58), the hybrid orbitals are **sp^2 orbitals**. The remaining $2p_z$ orbital is not included in the hybridization and remains unchanged. The four valence electrons are redistributed to give one electron in each orbital.

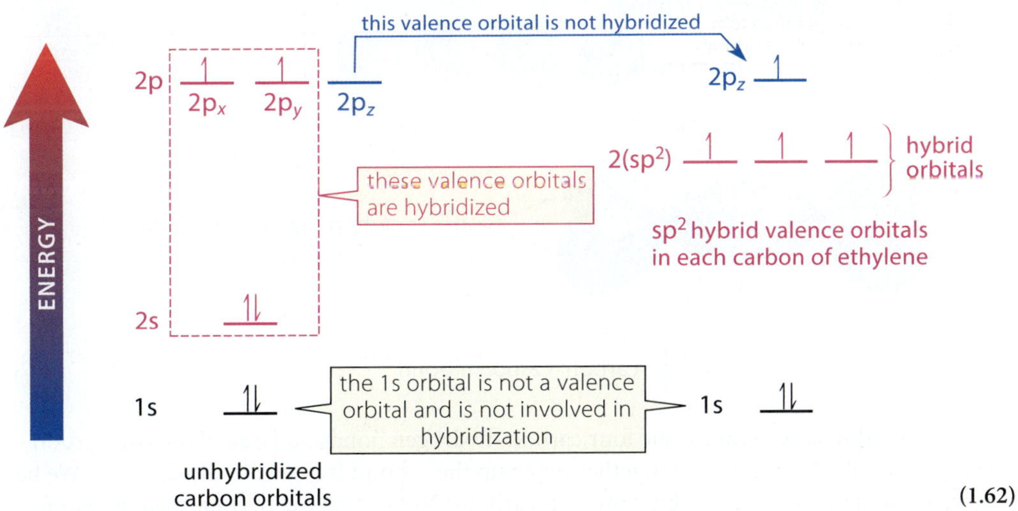

(1.62)

This electron distribution may seem like a violation of the aufbau principle, because the hybrid orbitals and the leftover 2p orbitals have different energies. However, we are not describing the electron distribution in an isolated carbon atom, but rather in a hybridized carbon atom within a molecule such as ethylene. The strengths of the bonds formed makes up for any energy cost of hybridization.

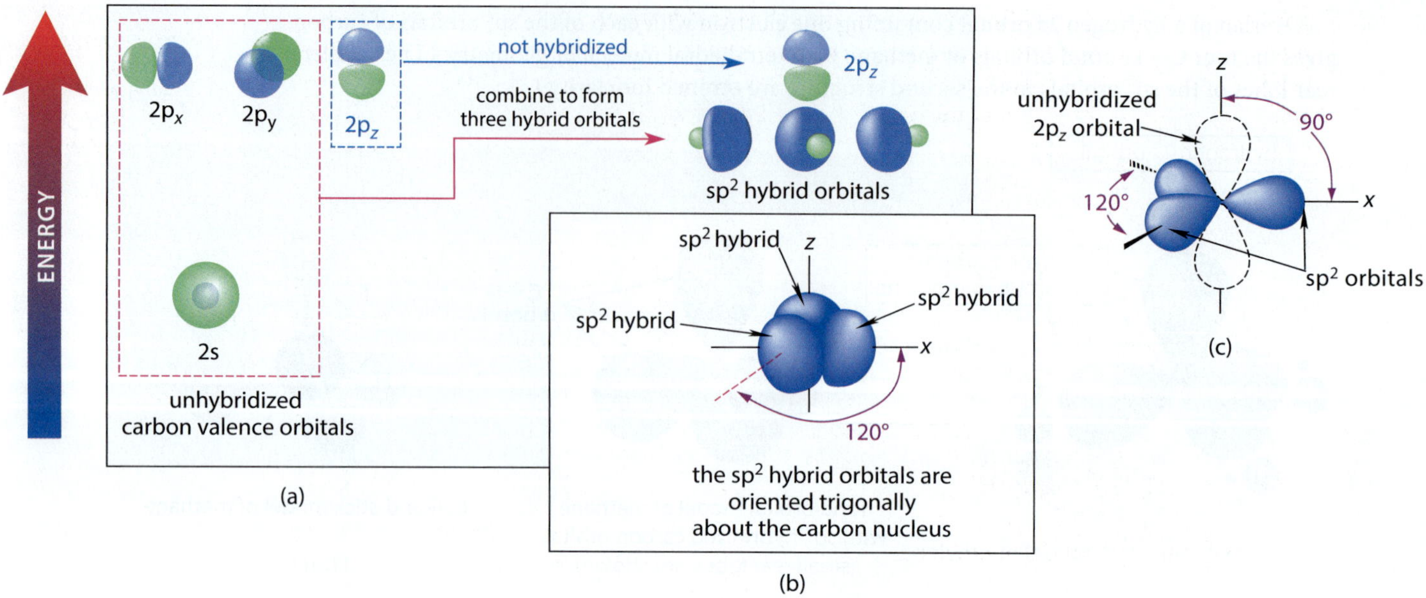

FIGURE 1.12 (a) Formation and orientation of the three sp^2 hybrid orbitals. (b) The three sp^2 hybrid orbitals shown together on a carbon nucleus. (c) The three sp^2 hybrid orbitals shown together in stylized form along with the unhybridized 2p orbital.

Because the $2p_x$ and $2p_y$ orbitals are used for hybridization, the axes of the three equivalent sp^2 hybrid orbitals lie in the *xy* plane, and they are oriented at the maximum angular separation of 120° for three equivalent objects. Because the axis of the "leftover" (unhybridized) $2p_z$ orbital is the *z*-axis, this orbital is perpendicular to the plane containing the sp^2 hybrid orbitals. The sp^2 hybridization of a carbon atom is summarized in **Fig. 1.12**.

The shape of the sp^2 hybrid orbital is very similar to that of the sp^3 hybrid orbital: it is lobe-shaped with large and small lobes.

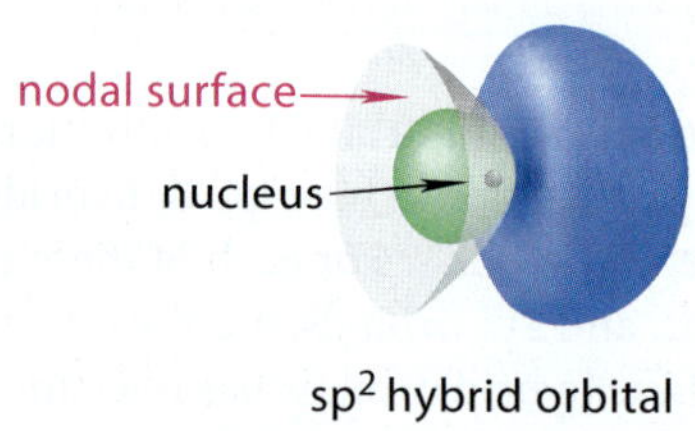

(1.63)

Conceptually, ethylene can be formed in the hybrid-orbital model by the bonding of two sp^2-hybridized carbon atoms with an sp^2–sp^2 carbon–carbon σ bond and four hydrogen atoms with sp^2–1s carbon–hydrogen σ bonds.

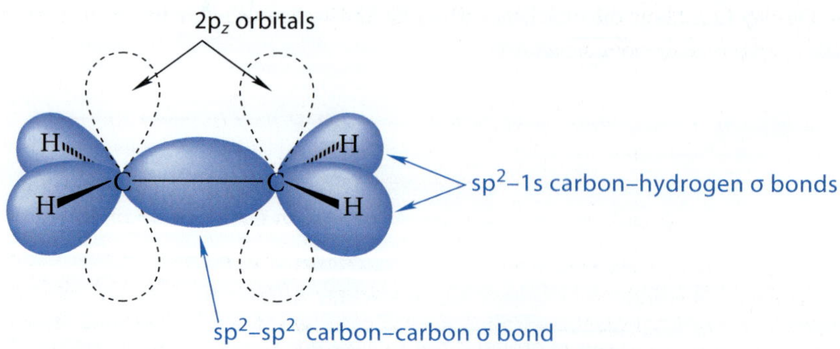

(1.64)

These bond orbitals account for the four carbon–hydrogen bonds and one of the two carbon–carbon bonds of ethylene, which together make up the σ-bond framework of ethylene. (We have not yet accounted for the $2p_z$ orbital on each carbon.) Notice that the trigonal-planar geometry of each carbon of ethylene is a direct consequence of the way its sp^2 orbitals are directed in space.

Once again, we see that hybridization and molecular geometry are related. *Whenever a main-group atom has trigonal-planar geometry, its hybridization is* sp^2.

The two $2p_z$ orbitals not used in σ-bond formation (dashed lines in Display 1.64) overlap *side to side* to form the second bond of the double bond, called a **π (pi) bond**.

> The symbol π (pi) is used because π is the Greek equivalent of the letter p and because the π bond is formed by the overlap of p orbitals.

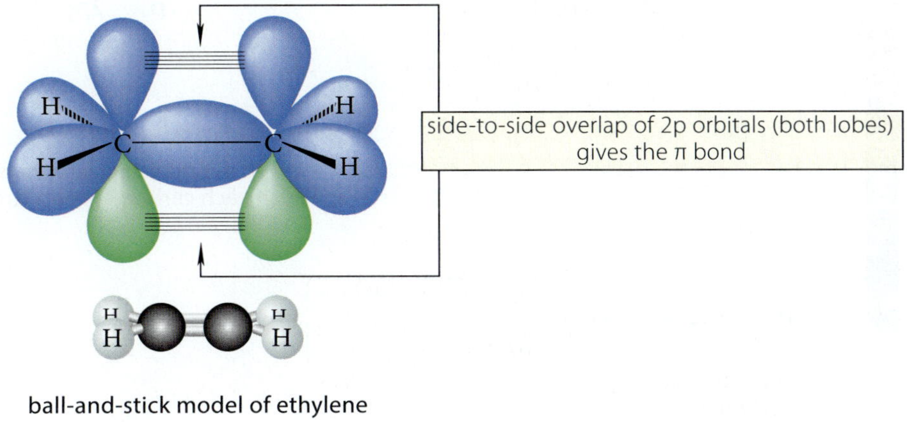

(1.65)

The π bond has electron density both above and below the plane of the ethylene molecule, but it has no electron density in the plane of the molecule.

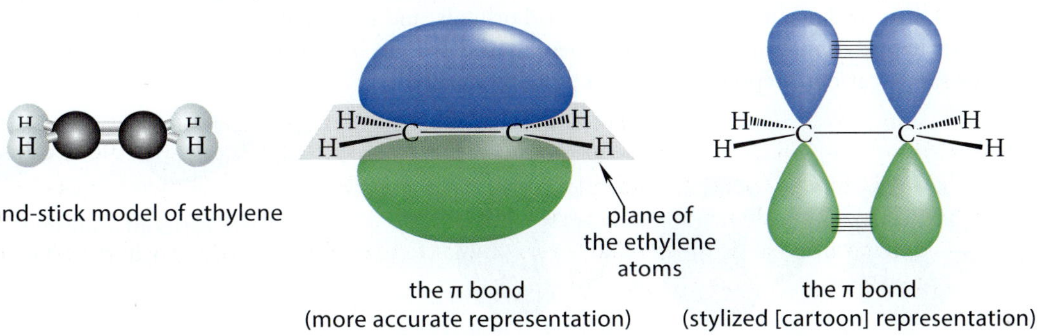

(1.66)

The π bond is *one bond with two lobes*, just as a 2p orbital is one orbital with two lobes. There are two distinct types of carbon–carbon bond in ethylene: a σ bond, with most of its electron density relatively concentrated between the carbon atoms, and a π bond, with most of its electron density concentrated above and below the plane of the ethylene molecule.

Focused Problem

1.24 Provide a hybrid-orbital description for the boron atom in trifluoroborane (boron trifluoride, BF_3) similar to the one shown for carbon in Display 1.62. Show the molecule with a line-and-wedge structure in a plane perpendicular to the page, and sketch the unhybridized 2p orbital on your structure.

D. sp Hybridization

Central atoms with linear geometry (that is, atoms with no nonbonding electrons bound to two groups) utilize a different orbital hybridization, in which two of the 2p orbitals of carbon are left out of the hybridization. This type of hybridization accounts for the geometry of each carbon of acetylene, HC≡CH. For each of these carbons, imagine that two valence orbitals of carbon, the 2s and $2p_x$ orbitals, are mixed, or hybridized, to form two equivalent hybrid orbitals. Each of these hybrid orbitals has 50% 2s character and 50% 2p character. Therefore, in the sp^n notation, $n = 1$ (Eq. 1.58), and these hybrid orbitals are therefore called **sp hybrid orbitals**. The remaining two 2p orbitals—the $2p_y$ and the $2p_z$ orbitals—are unaffected by the hybridization. The four valence electrons are redistributed to give one electron in each hybrid orbital and one electron in each of the two leftover 2p orbitals.

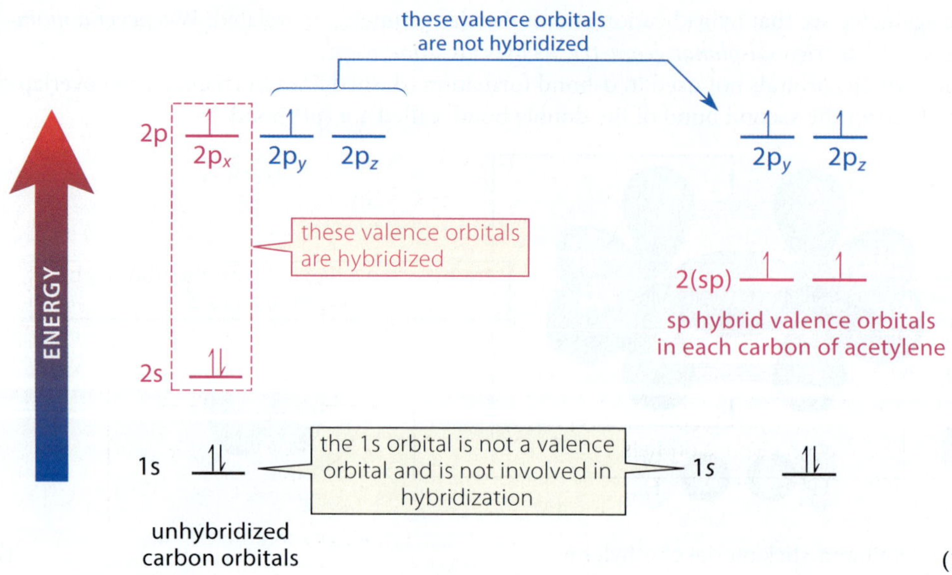

(1.67)

Like sp^3 and sp^2 hybrid orbitals, sp hybrid orbitals are lobe-shaped. The large lobe of the sp hybrid orbital, because it contains more s character, is somewhat more spherical (that is, "spread out") than the large lobe of the other hybrid orbitals, and the small lobe is smaller than that of the other hybrid orbitals. However, all of these hybrid orbitals have a similar appearance, and we use the same "cartoon" representations for all of them.

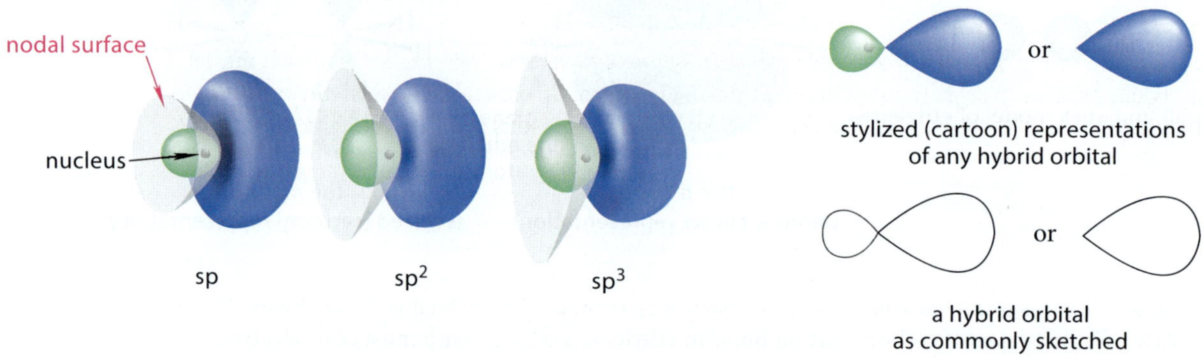

(1.68)

Because only the 2s and $2p_x$ orbitals are used for hybridization, these sp orbitals both lie on the x-axis, and they are oriented at the maximum angular separation of 180°. Because the axes of the "leftover" (unhybridized) $2p_y$ and $2p_z$ orbitals are directed along the y- and z-axes, these orbitals are perpendicular to each other and to the axis containing the sp hybrid orbitals. The sp hybridization of a carbon atom is summarized in **Fig. 1.13**.

Acetylene results from the combination of two sp-hybridized carbon atoms and two hydrogen atoms. One bond between the carbon atoms is an sp–sp σ bond resulting from the overlap of two sp hybrid orbitals, each containing one electron. The remaining sp orbital on each carbon overlaps with a hydrogen 1s orbital to form a carbon–hydrogen 1s–sp σ bond. The linear geometry of acetylene results from the 180° orientation of the sp orbitals on each carbon. Once again, *hybridization and geometry are correlated.*

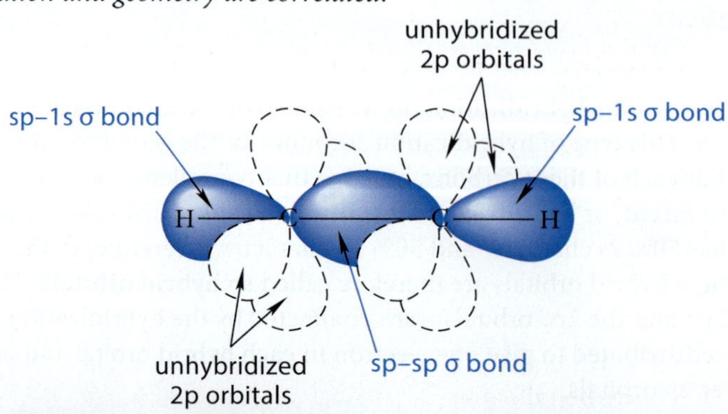

(1.69)

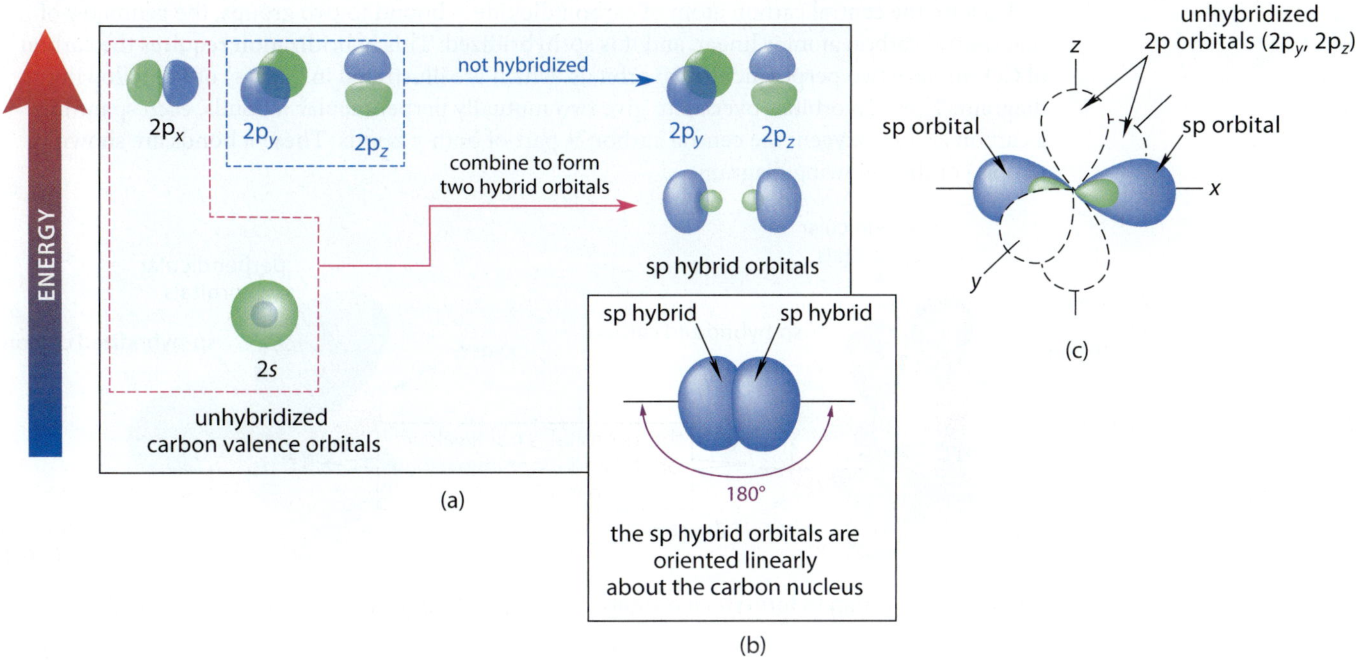

FIGURE 1.13 (a) Formation and orientation of the two sp hybrid orbitals. (b) The two sp hybrid orbitals shown together on a carbon nucleus. (c) The two sp hybrid orbitals shown together in stylized form along with the unhybridized $2p_y$ and $2p_z$ orbitals.

The leftover 2p orbitals on each carbon overlap side-to-side to form π bonds. Because each carbon of acetylene has two 2p orbitals, two π bonds are formed. Like the 2p orbitals from which they are formed, they are mutually perpendicular. The two π bonds that result from this overlap are as follows:

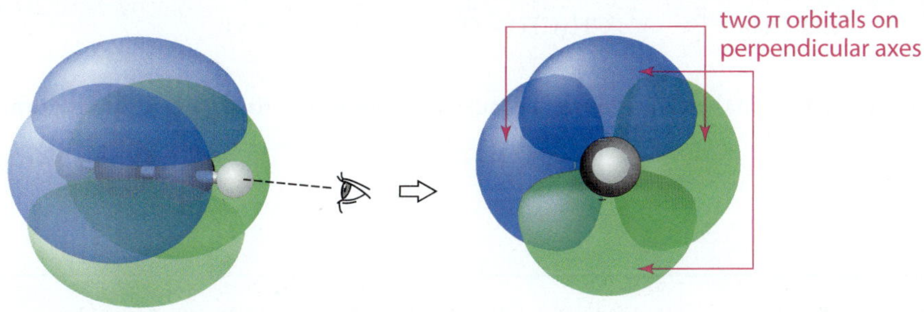

(1.70)

The acetylene molecule is literally surrounded by π electrons. The total electron density from all of the π electrons taken together forms a cylinder, or "doughnut," of electron density about the axis of the molecule.

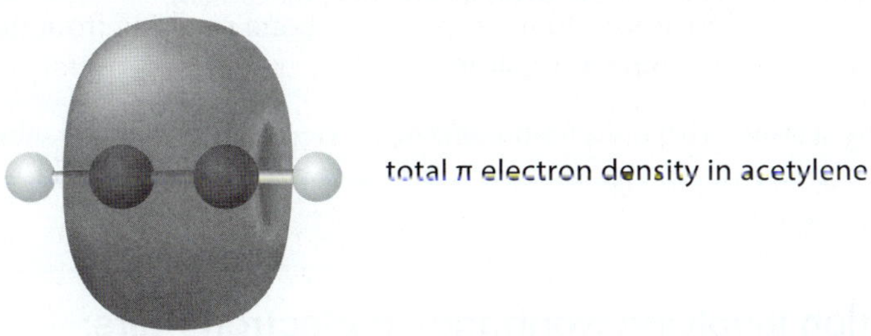

(1.71)

Although the triple bond is the bond most commonly associated with sp hybridization, this hybridization is also involved when one carbon (or other atom) takes part in two double bonds. One of the simplest examples of this type of bonding is the carbon dioxide molecule, CO_2.

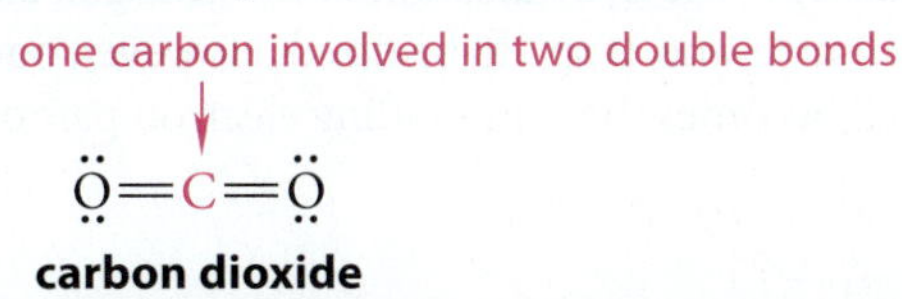

(1.72)

Because the central carbon atom of carbon dioxide is bound to two groups, the geometry of that central carbon atom is linear, and it is sp-hybridized. This hybridization requires the carbon of CO_2 to have two perpendicular 2p orbitals, which are illustrated in part (a) of the following diagram. These 2p orbitals overlap to give two mutually perpendicular π bonds, each spanning a carbon and an oxygen; the central carbon is part of both π bonds. These π bonds are shown in part (b) of the following diagram.

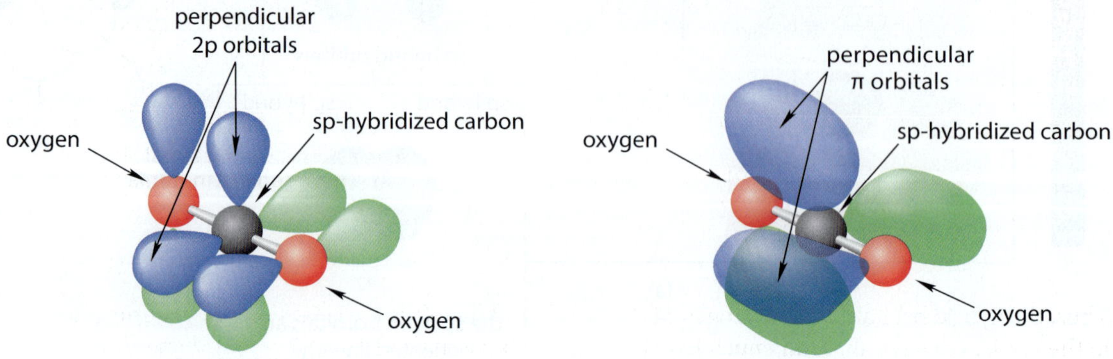

(a) the 2p orbitals that overlap to form the π orbitals (b) the two perpendicular π orbitals

carbon dioxide (1.73)

One other common case of sp hybridization occurs with *terminal atoms.* A **terminal atom** is an atom that is bonded to only one other atom; it doesn't matter whether the bond is single, double, or triple. The terminal atom in each of the following cases is sp-hybridized:

terminal atom; sp-hybridized	terminal atom; sp-hybridized	terminal atom; sp-hybridized
$CH_3CH_2—\ddot{\underset{\cdot\cdot}{F}}:$	$CH_3CH=\ddot{O}$:	$CH_3C≡N$:

(1.74)

The geometry at a terminal atom is linear, because it is connected to only one other atom.

Focused Problems

1.25 Beryllium dihydride (BeH_2) is a linear molecule. Create a hybrid-orbital description of BeH_2. Describe the Be—H bonds, and sketch an orbital picture of the molecule, including the unused 2p orbitals.

1.26 (a) What is the hybridization of the central carbon of *allene* ($H_2C=C=CH_2$)?

(b) On a Lewis structure of allene, sketch the arrangement of the 2p orbitals on all of the carbons.

(c) Sketch the arrangement of the π orbitals on a Lewis structure of allene.

(d) Each H—C—H group in allene defines a plane. What is the relationship of one H—C—H plane to the other?

E. Hybridization Involving Nonbonding Electron Pairs; Fractional s and p Character

The hybrid-orbital model can be used to describe bonding in molecules containing nonbonding electron pairs. Consider, for example, the hybridization of the nitrogen atom in ammonia, $:NH_3$. The nitrogen, like the carbon in methane, is sp^3-hybridized. However, one of the sp^3 hybrid orbitals contains an electron pair, which becomes the nonbonding electron pair on the nitrogen of ammonia.

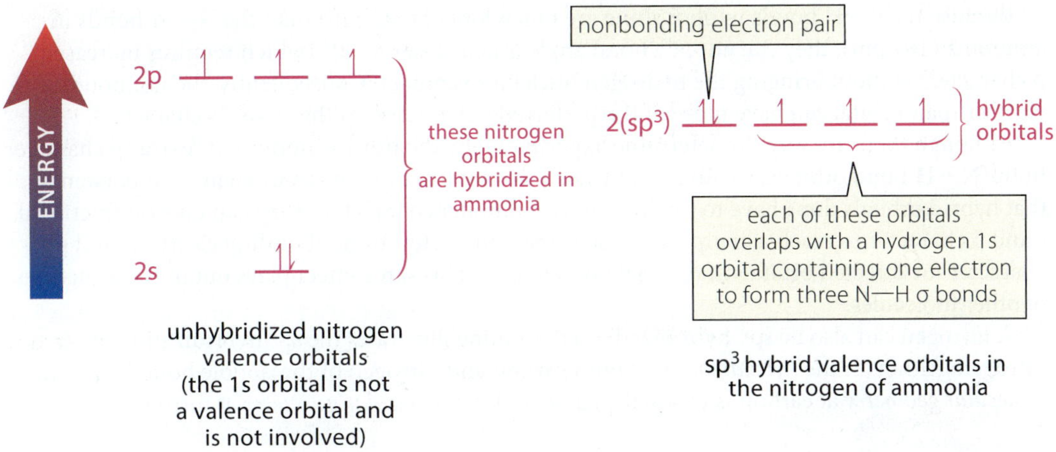

(1.75)

Thus, the hybrid orbitals in the nitrogen of ammonia, and therefore the N—H bonds, are oriented to the vertices of a tetrahedron, much like the hybrid orbitals and C—H bonds of methane. One of the hybrid orbitals, however, houses the nonbonding electron pair.

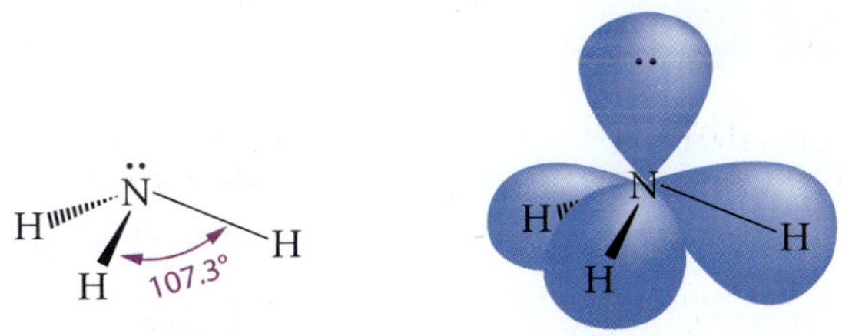

line-and-wedge Lewis structure of ammonia

orbital picture

(1.76)

Even though the *orbital structures* of methane and ammonia are similar, their *molecular geometries* are different, because molecular geometry describes the positions of the *nuclei.* Methane is tetrahedral, whereas ammonia is trigonal-pyramidal.

Although it's perfectly acceptable to say that the hybridization of the nitrogen in ammonia is sp^3, it's actually only *approximately* sp^3. We know this because the H—N—H bond angle is somewhat smaller than the tetrahedral value of 109.5° required for exact sp^3 hybridization. The reason for this difference lies in *the preference of nonbonding electron pairs for orbitals with as much s character as possible*. This preference is due simply to the lower energy of 2s orbitals relative to 2p orbitals. There is no need for directional character in this orbital, because these electrons are not involved in a chemical bond. Therefore, this "nonbonding-pair orbital" assumes as much 2s character as possible. Because the total amount of 2s and 2p character on an atom is fixed, the hybrid orbitals on nitrogen involved in bonding to the hydrogens must then have less 2s character and more 2p character. If the lone-pair orbital were completely 2s, then the nitrogen orbitals involved in the N—H bonds would be pure 2p and the H—N—H bond angles would be 90°. However, a 90° bond angle would squeeze the hydrogen nuclei too close together; they would "bump into" each other. The ammonia bond angles, then, are a compromise between the preference of a nonbonding electron pair for a 2s orbital and the need to separate the hydrogen nuclei so that they are clear of each other.

It is worth comparing the bond angles in phosphine, $:PH_3$, with those in ammonia. In phosphine, the H—P—H bond angles are 93.4°, much smaller than the H—N—H bond angles of ammonia (107.3°).

FURTHER EXPLORATION 1.2
Fractional Hybridization

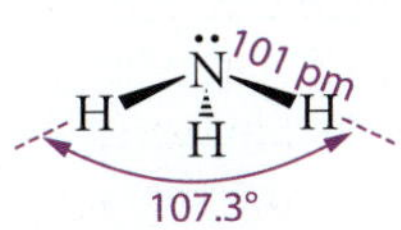

ammonia

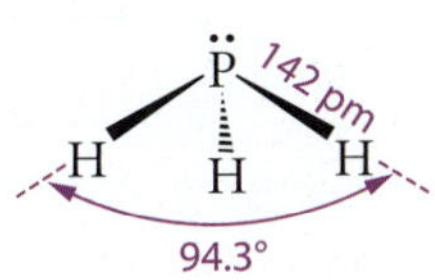

phosphine

(1.77)

Because the P—H bonds in phosphine are much longer (142 pm) than the N—H bonds in ammonia (101 pm), they can adopt a bond angle much closer to 90° (which requires increased p character) without bringing the hydrogen nuclei into contact. Consequently, the nonbonding electron-pair orbital can have more of the preferred s character (in this case, 3s character).

Although there are ways to determine experimentally the precise amount of "extra" p character in the N—H bond orbitals, we don't really need this information. It is sufficient to understand that hybrid orbitals don't have to involve integral numbers of orbitals; they can take on fractional s and p character to reach a compromise between s character for nonbonding electrons and p character for directed orbitals at optimal bond angles. This same effect plays out in the structures of other molecules.

A nitrogen can also be sp^2-hybridized. Methanimine illustrates the sp^2 hybridization of nitrogen. Although the *hybridization* of both carbon and nitrogen of the double bond is sp^2, the *molecular geometry* at carbon is trigonal-planar, and the *molecular geometry* at nitrogen is bent.

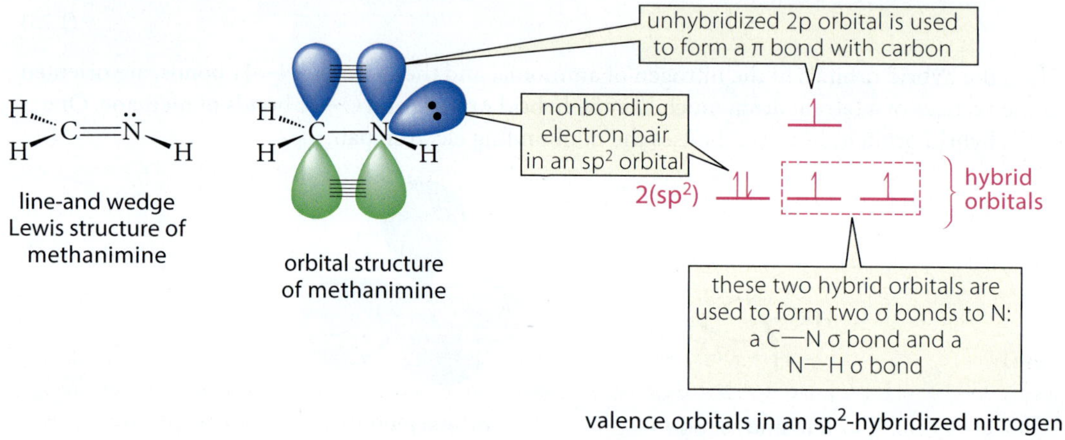

(1.78)

Because the nitrogen is sp^2-hybridized, the nonbonding electrons are in an sp^2 hybrid orbital, and the axis of this orbital is coplanar with the bonded carbon and hydrogen.

In true sp^2 hybridization, the C—N—H bond angle would be 120°. However, the nonbonding electron pair prefers to be in an orbital with more 2s character. As a result, the C—N and N—H bond orbitals have somewhat more 2p character, and the C—N—H bond angle is compressed. (It is about 116°.) This structure illustrates again that *orbital hybridization dictates the molecular geometry*.

As discussed in Sec. 1.6C, the nitrogen in formamide is trigonal-planar, and therefore it is also sp^2-hybridized. We justified its molecular geometry with a resonance structure that is trigonal-planar around the nitrogen. To form a π bond with carbon, the unshared electron pair of the nitrogen must occupy a 2p orbital. This means that the hybridization of nitrogen *must* be sp^2.

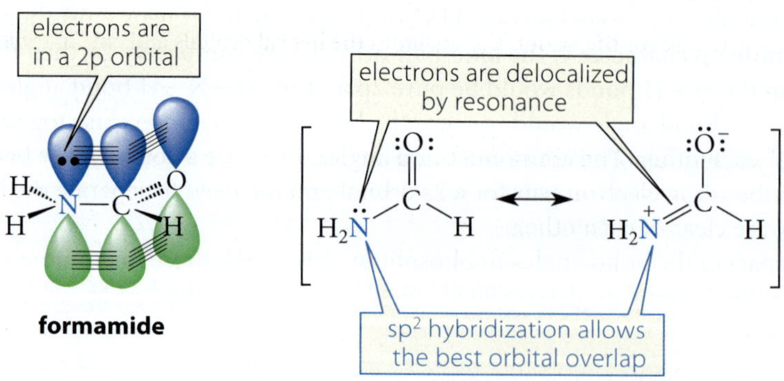

(1.79)

Remember that atoms do *not* move between resonance structures. This means that atoms cannot change geometry between resonance structures. The nitrogen is *not* sp^3-hybridized in one structure and sp^2-hybridized in the other. The nitrogen adopts the geometry and hybridization that allows the formation of the double bond, and this geometry must the same in both structures.

The oxygen atom in water (H_2O) is a situation in which there are *two* nonbonding electron pairs. We could envision two models for the hybridization of this oxygen that would predict a bent geometry. In one, the oxygen is sp^3-hybridized with the two nonbonding electron pairs in equivalent sp^3 orbitals. (This is sometimes called the "rabbit-ears" model.) In a second model, the oxygen is sp^2-hybridized. One unshared electron pair is in a 2p orbital, whereas the other is in an sp^2 hybrid orbital.

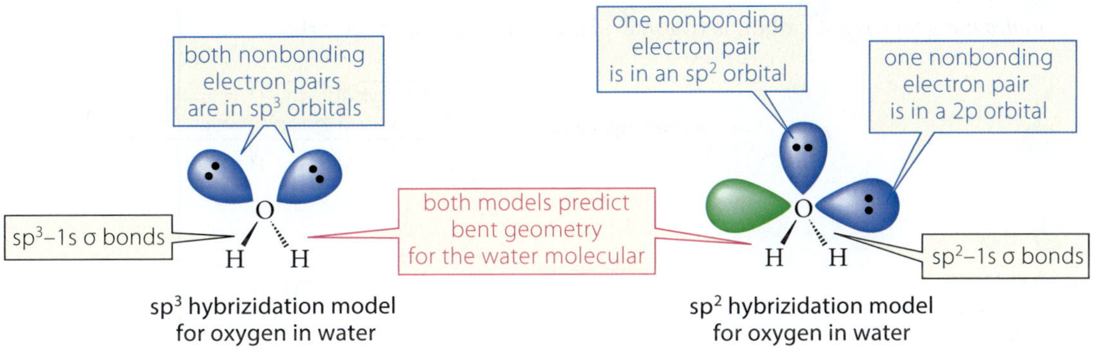

(1.80)

Research has shown that aspects of the electronic structure of water and other oxygen-containing molecules (which we haven't discussed) are best accommodated by an sp^2-hybridized oxygen. This is the general hybridization for an atom with two single bonds and two nonbonding electron pairs.

At the sp^2-hybridized oxygen of the water molecule, the H—O—H bond angle should be 120°. However, the unshared electron pair in the sp^2 orbital prefers to have as much 2s character as possible. This means that the O—H bond orbitals will have more 2p character, and the H—O—H bond angle will be compressed to the point that the two hydrogen atoms are not quite touching. In fact, the actual bond angle in H_2O is 104.5°.

Focused Problems

1.27 In acidic solution, formaldehyde adds a proton to give a cation $H_2C{=}\ddot{\overset{+}{O}}{-}H$.

(a) Construct a hybrid-orbital diagram for the oxygen atom in this cation. (*Hint:* Start with an oxygen cation.)

(b) What are the hybridizations of the carbon and the oxygen in this cation?

(c) Sketch a structure for this cation that includes the hybrid orbitals and any 2p orbitals.

(d) What is the molecular geometry at the carbon and the oxygen in this cation?

1.28 (a) What is the hybridization of the nitrogen in acetonitrile?

$$H_3C{-}C{\equiv}N{:}$$

acetonitrile

(b) In what type of orbital does the nonbonding electron pair reside?

(c) Sketch the molecule, showing the orbitals around the triply bonded carbon and nitrogen.

TABLE 1.2 A Table of Common Molecular Shapes and Hybridizations Encountered in Organic Chemistry

A = central atom X_n = number of atoms or groups bound to central atom
E_n = number of unshared electron pairs on central atom

Abbreviation	Electron arrangement	Geometry	Shape	Examples	Hybridization
AX_2E_0			linear	H—C≡CH O—C=O	sp
AX_2E_1	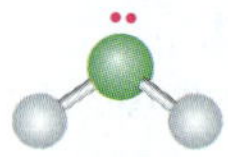	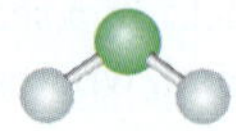	bent	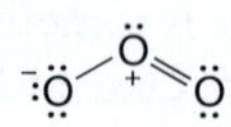	
AX_2E_2	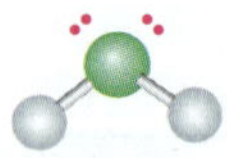	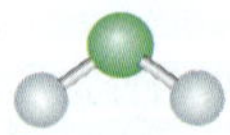	bent	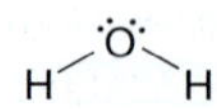	
AX_3E_0	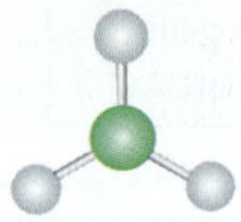	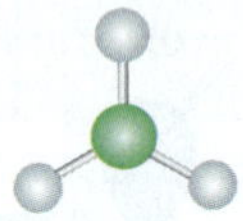	trigonal-planar	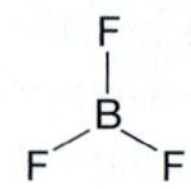 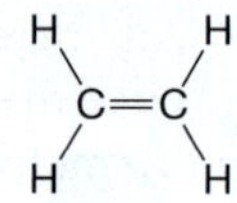	sp^2
AX_3E_1	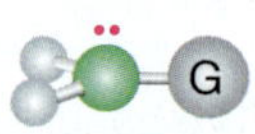 (nonbonding electrons delocalized by resonance into the group G)		trigonal-planar	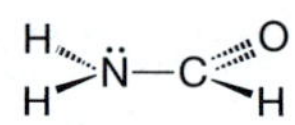	
AX_3E_1	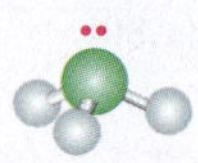 (nonbonding electrons localized)		trigonal-pyramidal	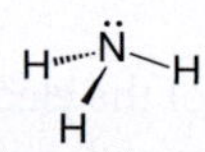 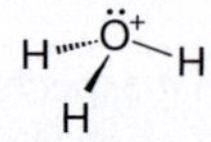	sp^3
AX_4E_0	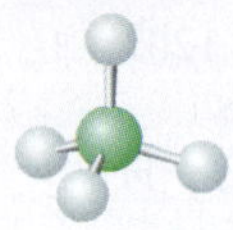	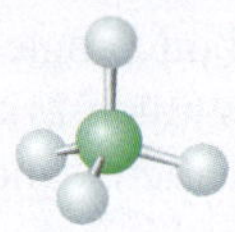	tetrahedral	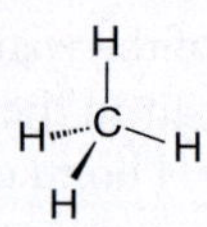 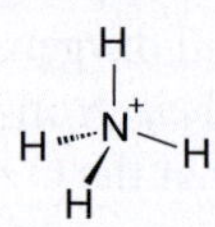	

F. Hybrid Orbitals and Molecular Geometry: Summary

We've seen that hybridization determines the arrangement of bonds and orbitals containing unshared electron pairs around a central atom. Because molecular geometry is the arrangement of nuclei about a central atom, and nuclei are attached by bonds, *hybridization dictates molecular geometry*. The connection between hybridization and geometry is summarized in **Table 1.2**. This table shows the "idealized" hybridizations (sp^3, sp^2, sp), which are all you really need to know to determine general molecular shape. Remember that bond angles are affected by the incorporation of fractional hybridization: an increase in s character for orbitals containing nonbonding electron pairs, and a compensating increase in p character for the bonding orbitals in the same molecules, results in smaller bond angles than the idealized value.

1.8 THEORIES OF THE COVALENT BOND: MOLECULAR ORBITAL THEORY

Hybrid-orbital theory will serve our needs in most of the situations that we cover in this text. The attractiveness of this theory is that atoms are connected by shared electron pairs in bond orbitals. We can associate each bond orbital with a bond in a Lewis structure or in a molecular model.

In certain situations, however, we'll need to dig a little deeper into bonding theory. For this purpose, we'll call upon the most accurate physical model of chemical bonding, called *molecular orbital theory.* In **molecular orbital theory** (abbreviated as **MO theory**), the electrons in molecules reside in orbitals of the entire molecule (called **molecular orbitals**, or **MOs**) rather than in localized bond orbitals. As in hybrid-orbital theory, molecules are still held together by the attractions of valence electrons for positive nuclei; however, the molecular orbitals in which valence electrons are housed in most cases span several (or all) atoms. This means that the electrons in molecules are not confined to electron-pair bonds but are *delocalized* across many atoms. In this section, we illustrate molecular orbital theory with three cases: hydrogen, helium, and methane.

A. Molecular Orbitals of Dihydrogen and (the Impossible) Dihelium

Molecular orbitals are formed from the valence orbitals of the constituent atomic orbitals when atoms come together to form molecules. The first principle of MO theory is that if all bonded atoms have *n* valence atomic orbitals, then *n* molecular orbitals are formed. Let's illustrate with the simplest case: the formation of dihydrogen from two hydrogen atoms. Each hydrogen atom provides a 1s valence orbital; therefore, the number of atomic orbitals involved in forming the hydrogen molecule = 2. When the two hydrogen atoms come to a bonding distance, the *two* 1s orbitals interact to form *two* molecular orbitals. Formation of these MOs follows a mathematical process described by quantum physics, but we won't need any math. To form the first MO, we simply *add* the two 1s wavefunctions at the bonding distance. When we add two wave peaks, the waves reinforce at the points of overlap—that is, we get a larger wave peak. The result of this wave addition is the first molecular orbital of the dihydrogen molecule. This is called a **bonding molecular orbital**, because when electrons occupy this orbital, there is enhanced electron density (that is, "electron cement") between the two nuclei. It is no coincidence that this looks exactly like the "bond orbital" for H_2 in Display 1.56.

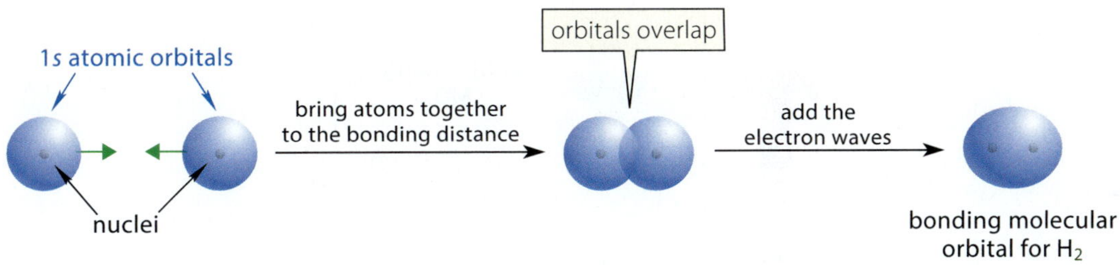

(1.81)

To get the second molecular orbital of dihydrogen, we *subtract* the two 1s electron waves in the region of overlap. This is mathematically the same as converting one wave peak to a wave trough and then adding. When a peak and a trough are added, wave cancellation occurs in the overlap region, and we get a *node*.

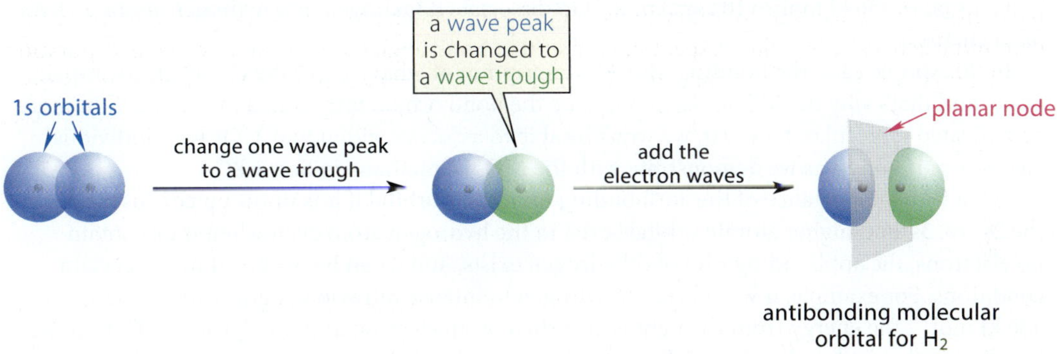

(1.82)

This MO is called an **antibonding molecular orbital** because, when an electron populates this orbital, there is no electron density (that is, no "electron cement" and no bonding) between the two nuclei.

Now let's arrange these molecular orbitals in order of increasing energy. As in atomic orbitals, a larger number of nodes indicates higher energy. Thus, the bonding molecular orbital has lower energy, and the antibonding molecular orbital has higher energy. Relative to the isolated hydrogen 1s orbitals, the bonding MO has a lower energy by 218 kJ mol^{-1} (52.1 kcal mol^{-1}), and the antibonding MO has a higher energy by about the same amount.

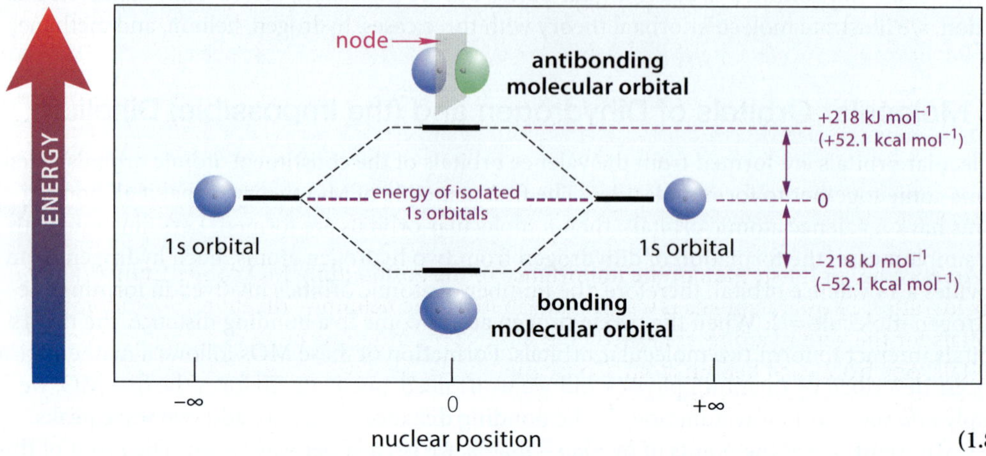

(1.83)

Finally, we add available electrons to the MOs using the same rules (aufbau principle, Pauli principle, Hund's rules) that we use for populating atomic orbitals. In the dihydrogen molecule, there are two electrons, so both go into the bonding MO with opposite spins. The antibonding MO is unoccupied.

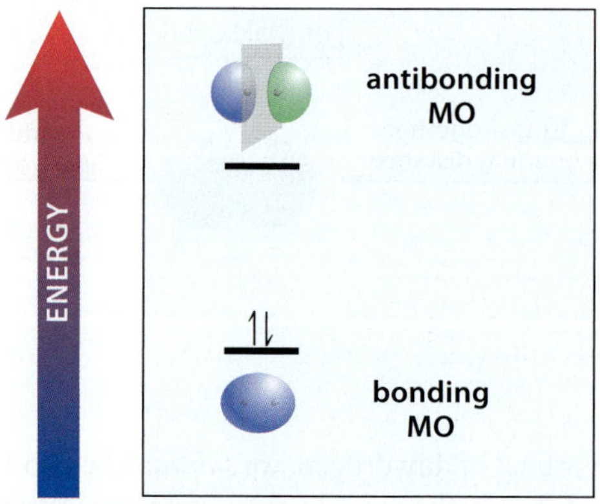

(1.84)

The stability of the hydrogen molecule is due to population of the bonding MO with two electrons. Each electron contributes 218 kJ mol^{-1} (52.1 kcal mol^{-1}) of stability to the dihydrogen molecule—that is, it takes 436 kJ mol^{-1} (104 kcal mol^{-1}) to dissociate dihydrogen into hydrogen atoms. (That's a *lot* of energy!)

In this simple case, the bonding MO is a description of what we call the H—H electron-pair bond, and that's why the MO looks exactly like the bond orbital introduced in Sec. 1.7A. In more complicated molecules, however, we won't be able to associate individual MOs with individual electron-pair bonds, as we demonstrate with the MOs of methane in Sec. 1.8C.

What is the importance of the antibonding molecular orbital if it is unoccupied? Just as the 2s, 2p, 3s, and higher atomic orbitals exist in the hydrogen atom even when they contain no electrons, the antibonding MO of dihydrogen exists, and it can be occupied under certain conditions. For example, if we subject dihydrogen to intense ultraviolet light with an energy of 436 kJ mol^{-1}, the energy from the light is absorbed by an electron and it is "promoted" from the bonding to the antibonding MO. When that happens, the high energy of the antibonding electron cancels the low energy of the bonding electron, and the "excited" dihydrogen molecule has no net stabilization. As a result, the molecule dissociates into two hydrogen atoms. (It is well known in chemistry that a good way to produce individual atoms from diatomic molecules is to irradiate them with light of the appropriate energy, a process called *photodissociation*.)

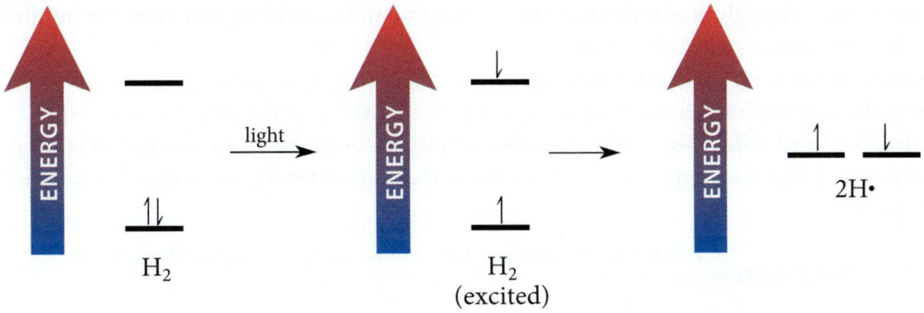

(1.85)

An even better illustration of the importance of the antibonding MO comes from a consideration of what happens if we try to form diatomic helium—"dihelium." The atomic orbitals for the helium atoms are exactly the same as the atomic orbitals of hydrogen: two 1s orbitals, one from each helium atom. Consequently, we get exactly the same two MOs when we mix the two atomic orbitals. When we populate these two MOs with the *four* available electrons (two from each helium atom), however, we find that there are two electrons in each MO. In this case, the high energy of the electrons in the antibonding MO *exactly cancels* the low energy of the electrons in the bonding MO. Therefore, dihelium has no reason to exist—and it doesn't! Helium (and the other noble gases) are monoatomic.

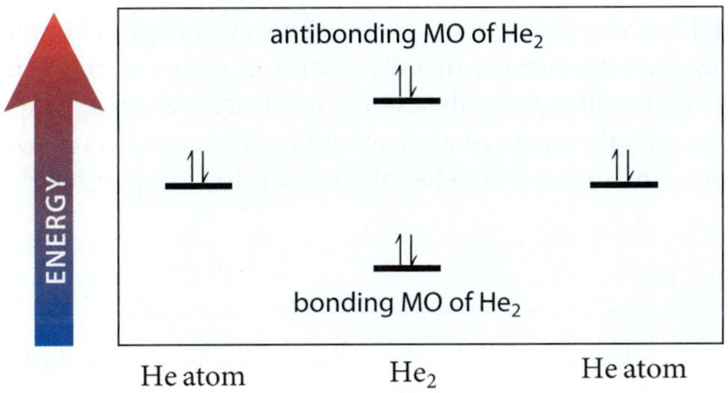

(1.86)

Focused Problems

1.29 Draw molecular orbital energy diagrams for each of the following species like the one shown in Display 1.83. Indicate which should exist as diatomic species and which should dissociate into two monoatomic fragments. Explain.

(a) He_2^+ ion (b) H_2^- ion (c) H_2^{2-} (d) H_2^+ ion

1.30 The bond dissociation energy of H_2 is 436 kJ mol^{-1} (104 kcal mol^{-1}). In other words, it takes 436 kJ mol^{-1} (104 kcal mol^{-1}) of energy to dissociate H_2 into hydrogen atoms. Estimate the bond dissociation energy of the H_2^+ ion, and explain your reasoning.

B. Molecular Orbital Theory and the Lewis Structure of H_2

Let's now relate the quantum-mechanical description of H_2 to the concept of the Lewis electron-pair bond. In the Lewis structure of H_2, the bond is represented by an electron pair shared between the two nuclei. In the quantum-mechanical description, the bond is the result of the presence of two electrons in a bonding molecular orbital and the resulting electron density between the two nuclei. Both electrons are attracted to each nucleus, and these electrons thus serve as the "cement" that holds the nuclei together. For H_2, therefore, the Lewis electron-pair bond is equivalent to the quantum-mechanical idea of a bonding molecular orbital occupied by a pair of electrons. The Lewis picture places the electrons squarely between the nuclei. Quantum theory says that although the electrons have a high probability of being between the bonded nuclei, they can also occupy other regions of space.

When we are dealing with a diatomic molecule, the formula for bond order in MO theory reinforces the connection between an occupied bonding MO and the electron-pair bond. Recall that the *bond order* (Sec. 1.6B) describes whether a bond between two nuclei is single (bond order = 1), double (bond order = 2), or triple (bond order = 3). The formula for bond order in MO theory is

$$\text{bond order} = \frac{\text{electrons in bonding MOs} - \text{electrons in antibonding MOs}}{2} \tag{1.87}$$

The number 2 in the denominator represents the fact that a full covalent bond requires two electrons. In the H_2 molecule, for example, there are two electrons in the bonding MO and no electrons in the antibonding MO; therefore, the bond order is $(2-0)/2 = 1$, in which case the covalent bond in dihydrogen is a single electron-pair bond.

Molecular orbital theory shows, however, that a chemical bond need not be an electron pair. For example, H_2^+ (the hydrogen molecule cation, which we might represent in the Lewis sense as $H\dot{}^+H$) is a stable species in the gas phase (see Focused Problem 1.30). It is not as stable as the hydrogen molecule itself because the ion has only one electron in the bonding molecular orbital, rather than the two found in a neutral hydrogen molecule. The hydrogen molecule anion, H_2^-, discussed in the previous section, might be considered to have a three-electron bond consisting of two bonding electrons and one antibonding electron. The electron in the antibonding orbital is also shared by the two nuclei, but shared in a way that reduces the energetic advantage of bonding. (H_2^- is not as stable as H_2.) This example demonstrates that the sharing of electrons between nuclei in some cases does *not* contribute to bonding. Nevertheless, the most stable arrangement of electrons in the dihydrogen molecular orbitals occurs when the bonding MO contains two electrons and the antibonding MO is empty—in other words, when there is an electron-pair bond.

Focused Problem

1.31 Referring to your solution to Focused Problem 1.29d, calculate the bond order of the covalent bond in the hydrogen molecule cation, H_2^+. How does this result bear on the answer to Focused Problem 1.30?

C. Molecular Orbitals of Methane

The molecular orbitals of methane illustrate well the difference between the molecular-orbital and hybrid-orbital descriptions of bonding. The molecular orbitals of methane come from mixing the four valence orbitals (2s and three 2p) of carbon and four 1s orbitals from hydrogen (one from each hydrogen). Therefore, we mix *eight* atomic orbitals, and we obtain *eight* molecular orbitals for methane. We then populate these molecular orbitals with eight valence electrons (four from carbon, one from each of four hydrogens).

Formation of these MOs also involves additions and subtractions of atomic orbitals; however, the process is not as simple as formation of the dihydrogen MOs, so we just consider the result, shown in **Fig. 1.14**. Methane has four bonding MOs and four antibonding MOs. There is one bonding MO of lowest energy, which has no nodes, and three degenerate bonding MOs of higher energy, each with one node. (Recall from the discussion of atomic orbitals that more nodes in an orbital means that it has higher energy.) The energies of the bonding MOs look a lot like the energy diagram of a carbon atom, and for good reason. The MO of lowest energy comes from mixing a 1s orbital from each of the four hydrogens with the 2s orbital of carbon. The three degenerate bonding MOs come from mixing each of the three degenerate 2p orbitals of carbon with the 1s orbitals of all four hydrogens. These three MOs, like the 2p orbitals from which they are derived, have one node, and they differ in their orientation. Notice from the orbital diagrams

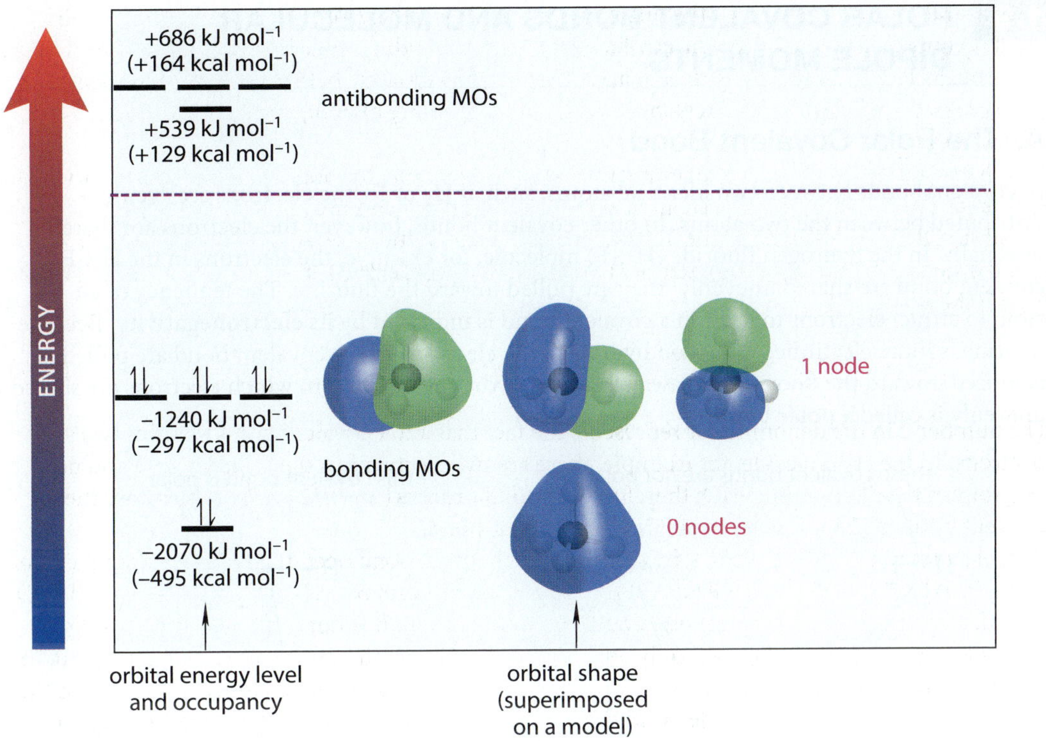

FIGURE 1.14 An energy level diagram for the eight molecular orbitals of methane. The shapes of the MOs are shown only for the bonding MOs. A ball-and-stick model of methane is superimposed on the MOs for reference. The orbital energies are computed; they can vary with the different types of MO calculation.

that *all of the MOs encompass several atoms, so that we can't associate a given MO with a particular bond of methane in the bond-orbital model.*

When we superimpose the electron densities from all of the occupied MOs, we get the following picture of electron distribution in methane: a "blob" of electron density with tetrahedral symmetry and imbedded nuclei. Notice the absence of discrete bonds in the electron density.

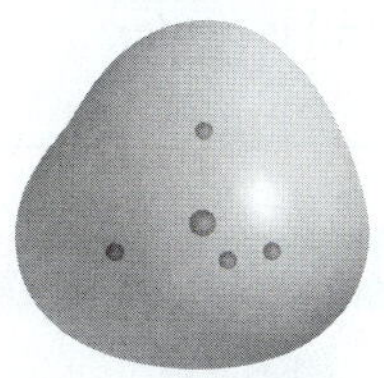

electron density of methane derived from molecular orbital theory with the positions of the carbon and hydrogen nuclei shown (1.88)

The MO bonding model has the greatest accuracy of all of the bonding models in common use in organic chemistry. However, because MO descriptions have no discrete bonds (except for simple diatomic cases like dihydrogen), MO bonding descriptions in most cases are difficult to relate to our conventional Lewis structures. We use hybrid-orbital descriptions most of the time because they are more intuitive. When necessary, however, we resort to MO descriptions for specific reasons.

It may bother you that we can use different models of bonding for different purposes. The ultimate test of a model is whether it predicts experimental results. The hybrid-orbital model does a good job in most cases; but when it doesn't, we have to use MO theory. An analogy is the prediction of hurricanes. There are some fairly simple computer models that give the general direction of storms. In fact, when you see a storm developing over the Azores and moving westward, it doesn't take a computer model to tell you to start paying attention if you live in the coastal United States, Mexico, or the Caribbean. Later, relatively unsophisticated computer models tell us what general regions should be on the lookout for an approaching storm. As we need more detailed information, scientists resort to more sophisticated models that can pinpoint more exactly where a storm might make landfall. Each model, with its own level of sophistication, has its uses.

Focused Problem

1.32 (a) If you were to construct the molecular orbitals of water (H_2O) from the valence orbitals of its atoms, how many MOs would you expect to obtain?

(b) When we add electrons to these MOs, how many of them would be fully occupied?

(c) How many nodes would the occupied MO of lowest energy have? Sketch this MO.

1.9 POLAR COVALENT BONDS AND MOLECULAR DIPOLE MOMENTS

A. The Polar Covalent Bond

In covalent bonds between two identical atoms, such as H_2 or F_2, the electrons are evenly distributed between the two atoms. In other covalent bonds, however, the electrons are shared unequally. In the hydrogen fluoride (H—F) molecule, for example, the electrons in the H—F covalent bond are shared unevenly: they are pulled toward the fluorine. The tendency of an atom to attract electrons to itself in a covalent bond is indicated by its **electronegativity**. Because fluorine is more electronegative than hydrogen, the electrons in the covalent bond are pulled, or *polarized*, toward the fluorine and away from the hydrogen. A bond in which electrons are shared unevenly is called a **polar bond.**

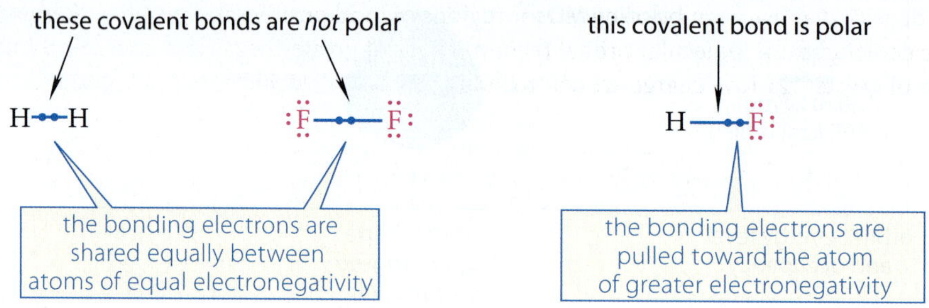

Figure 1.15 shows the electronegativities of the elements most commonly encountered in organic chemistry. Knowing these values exactly isn't as important as knowing their *relative* values and noticing the trend: *electronegativity values increase going to the right and up the table.* Atoms with very low electronegativity (colored blue in **Figure 1.15**) are said to be *electropositive.*

You can diagnose whether a bond is polar by considering the electronegativities of the bonded atoms. If there is a significant difference between the electronegativities of the bonded atoms (roughly speaking, a different color in Figure 1.15), then the bond is polar. If there is little or no difference in electronegativities, then the bond is *apolar.*

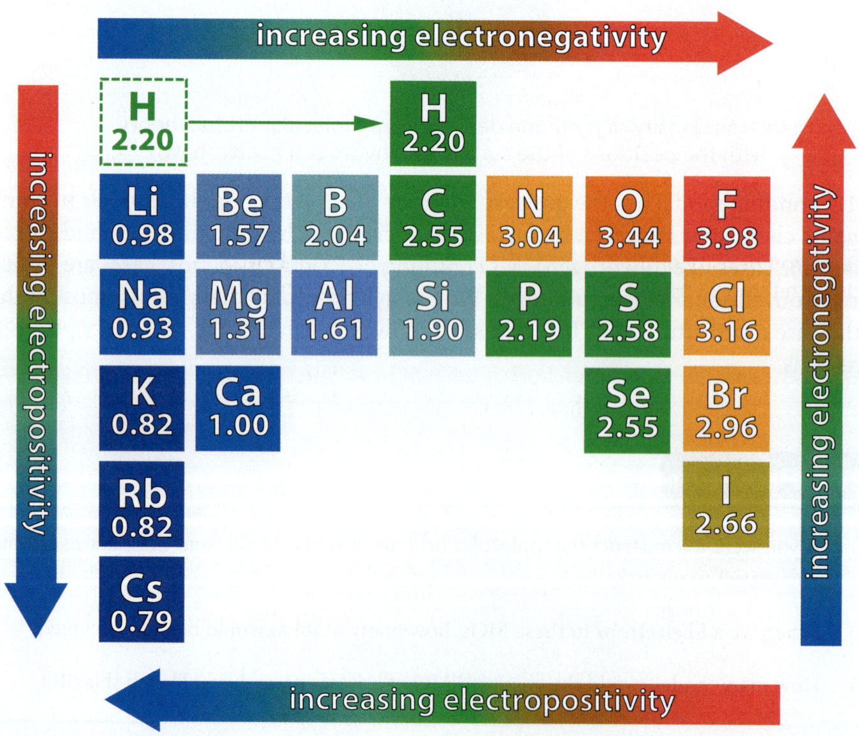

FIGURE 1.15 Electronegativities of elements commonly encountered in organic chemistry. Although hydrogen is in group 1A, its electronegativity places it closer to carbon.

In a polar bond, the excess of electrons at one end and the deficiency of electrons at the other creates partial charges on the bonded atoms. In H—F, because the bonding electrons are polarized toward the fluorine, the fluorine has a partial negative charge, and the hydrogen has a partial positive charge of equal magnitude. Sometimes we indicate these partial charges, and therefore the polarity of the bond, with the same lowercase italic Greek letter delta, δ, that we used in showing partial charges in resonance hybrids (see Display 1.33):

hydrogen is "partially positive" → $\overset{\delta+}{\text{H}}$—$\overset{\delta-}{\text{F}}$ ← fluorine is "partially negative"

Another more graphical way that we'll use to show polarities is the **electrostatic potential map**, or **EPM**. An EPM of a molecule starts with a map of the total electron density. This is a picture of the spatial distribution of the electrons in the molecule that comes from molecular orbital theory (Sec. 1.8B), which was illustrated for the methane molecule in Display 1.88. Whether you studied MO theory or not, think of this as a picture of "where the electrons are." The EPM is a map of total electron density that has been color-coded for regions of local positive and negative charge. (EPMs are also calculated from molecular orbital theory.) Areas of greater negative charge are colored red, and areas of greater positive charge are colored blue. Areas of neutrality are colored green.

The EPM is a "potential map" because it represents the interaction of a test positive charge with the molecule at various points. When the test positive charge encounters negative charge in the molecule, an attractive potential energy occurs; this is color-coded red. When the test positive charge encounters positive charge, a repulsive potential energy results, and this is color-coded blue.

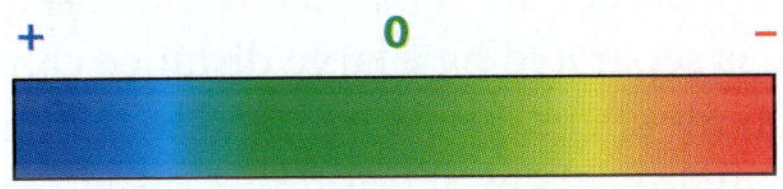

color scale for EPMs (1.89)

For example, the EPM of H—F starts with the total electron density of the molecule calculated from MO theory. On this is superimposed the color-coded EPM. The EPM of H—F shows the red region over the F (which has a $\delta-$ charge) and the blue region over the H (which has a $\delta+$ charge), as we expect from the greater electronegativity of F versus H. The EPM is superimposed on a ball-and-stick model for reference.

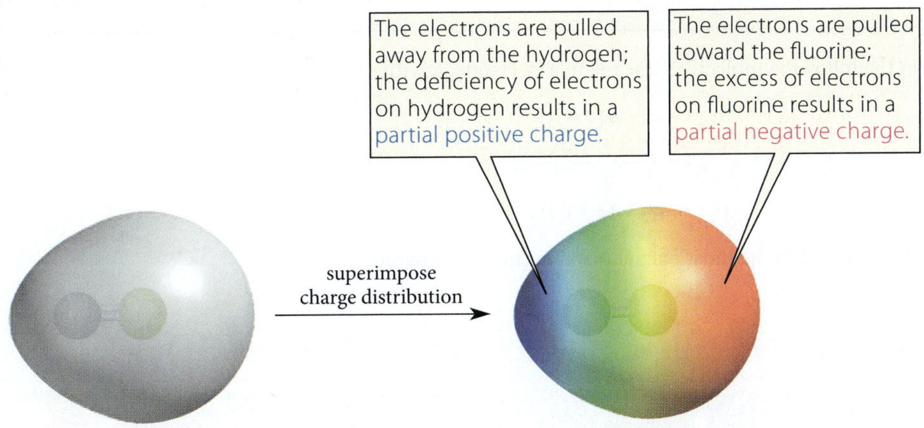

electron density of hydrogen fluoride (H—F)

EPM of hydrogen fluoride (H—F) (1.90)

In contrast, the EPMs of dihydrogen and difluorine shows the same color on both atoms because the two atoms in both molecules share the electrons equally. The green color indicates that neither atom bears a net charge.

If you study electricity in physics, you may find that physicists use the convention that the dipole vector is oriented from the negative to the positive charge. In the physics convention, then, the dipole moment vector of H—F points from F to H. Chemists, on the other hand, view the uneven distribution of *electrons* as the source of the molecular dipole moment, so they orient the dipole moment vector in the direction of excess electrons—from positive to negative. Be sure to understand that the two conventions do not differ in the location of the partial charges, but only in which charge is considered the head and which is considered the tail of the dipole moment vector. Either convention can be used in vector calculations as long as it is used consistently.

EPM of dihydrogen (H—H)

EPM of difluorine (F—F) (1.91)

It is useful to have some intuition about the significance of the debye in terms of charge separation. Full (+) and (–) charges separated by the bond length of HF would give a dipole moment of 4.4 D. Because the actual dipole moment is 1.84 D, the charge separation in HF is 1.84/4.4 = 0.42. In other words, there is –0.42 charge on the fluorine and +0.42 charge on the hydrogen. For more on this topic, see Further Exploration 1.3 in the Study Guide and Solutions Manual.

The electron distribution in a compound containing covalent bonds is measured by the **dipole moment**, which is abbreviated with the Greek letter $\boldsymbol{\mu}$ (mu). The dipole moment, which can be measured experimentally, is a vector quantity defined by the following equation:

$$\boldsymbol{\mu} = q\boldsymbol{r} \tag{1.92}$$

In this equation, q is the magnitude of the separated charge and $\boldsymbol{r}$ is a vector from the site of the positive charge to the site of the negative charge. For a simple molecule like HF, the magnitude of the vector $\boldsymbol{r}$ is merely the length of the HF bond, and it is oriented from H (the positive end of the dipole) to F (the negative end). The directions of $\boldsymbol{\mu}$ and $\boldsymbol{r}$ are the same—from the positive to the negative end of the dipole. As a result, the dipole moment vector for the HF molecule is oriented along the H—F bond from the H to the F:

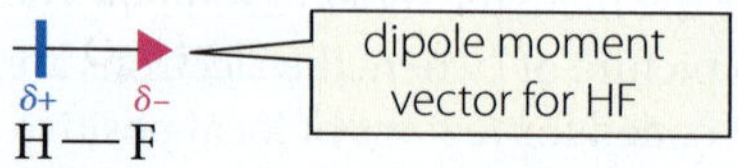

(1.93)

The magnitude of the dipole moment is affected not only by the amount of charge that is separated (q) but also by how far the charges are separated (r). Consequently, a molecule in which a relatively small amount of charge is separated by a large distance can have a dipole moment as great as one in which a large amount of charge is separated by a small distance. Nevertheless, relative electronegativities play the major role in determining dipole moments in most of the organic molecules we consider in this text.

FURTHER EXPLORATION 1.3
Dipole Moments and Charge Separation

The dipole moment is commonly given in derived units called debyes, abbreviated D, and named for the physical chemist Peter Debye (1884–1966), who won the 1936 Nobel Prize in Chemistry. For example, the HF molecule has a dipole moment of 1.84 D, whereas dihydrogen (H_2) and difluorine (F_2), which have uniform electron distributions, have dipole moments of zero.

Focused Problems

1.33 What is the most polar bond in the following molecule? Use relative electronegativities as your guide.

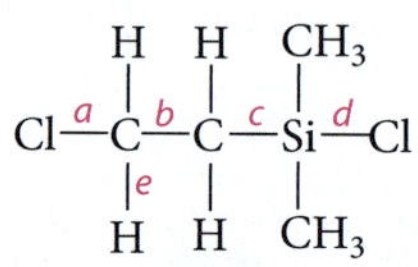

1.34 Which carbon in the following molecule has the most partial positive charge? Explain.

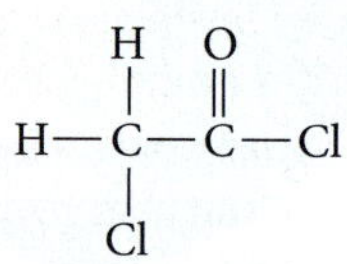

B. Polar Molecules

The dipole moment vector for an individual bond is called a **bond dipole** or **bond dipole moment**. For a diatomic molecule such as HF, the *bond dipole* is the same as the dipole moment. For molecules with more than one polar bond, the dipole moment of the molecule is the *vector sum* of its bond dipoles. That is, not only the magnitudes but also the directions of the component bond dipoles are important in determining the overall molecular dipole moment. The vectorial nature of bond dipoles can be illustrated with the carbon dioxide molecule, CO_2:

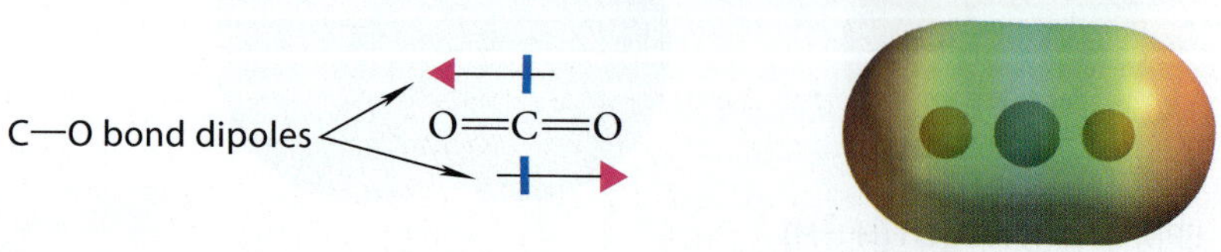

EPM of carbon dioxide **(1.94)**

Because the CO_2 molecule is linear, the C=O bond dipoles are oriented in opposite directions. Because they have equal magnitudes, they exactly cancel. (Two vectors of equal magnitude oriented in opposite directions always cancel.) *Consequently, CO_2 is a nonpolar molecule, even though it has polar bonds.*

In contrast, if a molecule contains several bond dipoles that do *not* cancel, the various bond dipoles add vectorially to give the overall resultant dipole moment. In the water molecule, for example, which has a bond angle of 104.5°, the O—H bond dipoles add vectorially to give a resultant dipole moment of 1.84 D, which bisects the bond angle. The EPM of water shows the charge distribution suggested by the dipole vectors—a concentration of negative charge on oxygen and positive charge on hydrogen.

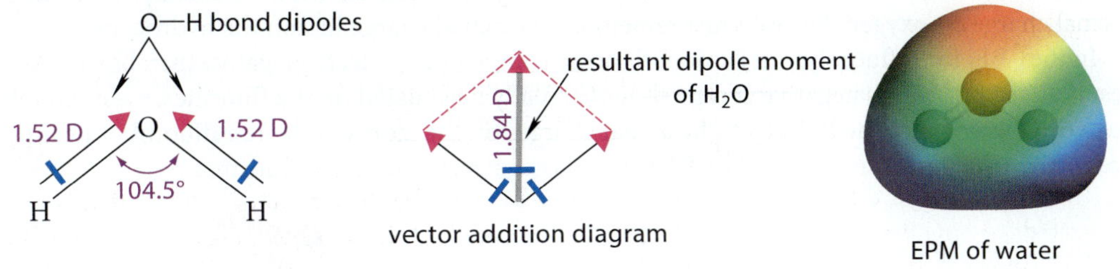

(1.95)

STUDY GUIDE LINK 1.1
A Refresher on Vector Addition

"Study Guide Links" are sections in the *Study Guide and Solutions Manual* that provide further assistance in understanding the listed topic.

Focused Problem

1.35 Two of the following molecules have an overall dipole moment of zero, and the other two have significant dipole moments. Identify the two molecules with zero dipole moments, and explain how you know.

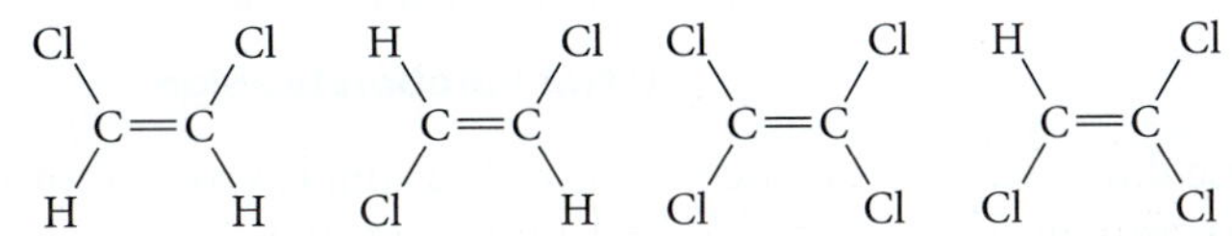

A *B* *C* *D*

C. Formal Charge versus Actual Charge

The location of charge in a molecule can have significant effects on chemical reactivity and physical properties. For this reason, it's important to keep in mind that formal charge (Sec. 1.4B) is only a bookkeeping device for dividing up the overall charge of a molecule or ion among its atoms. In some cases, formal charge corresponds to the actual charge on an atom in a molecule. For example, the negative charge on the hydroxide ion, ^-OH, is actually centered on the oxygen, because oxygen is much more electronegative than hydrogen. The EPM is very red (negative) around the oxygen.

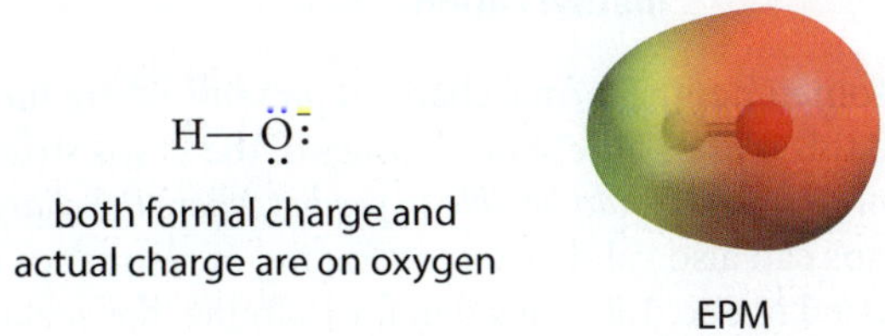

hydroxide ion (1.96)

In other cases, however, the formal charge does *not* correspond to the actual charge. In the Lewis structure of the hydronium ion, for example, oxygen has a formal charge of +1. However, because oxygen is more electronegative than hydrogen, the positive charge is actually distributed across the three hydrogens. This is reflected in the EPM, which is very blue (positive) around the hydrogens.

EPMs in this text are scaled so that relative charges in different parts of the molecule are emphasized. A certain color is not associated with the same absolute value of charge in different molecules. For example, the "red" oxygens in the hydroxide ion and the hydronium ion have different amounts of negative charge. The red color means that the oxygens in both molecules are substantially negative relative to the rest of the molecule.

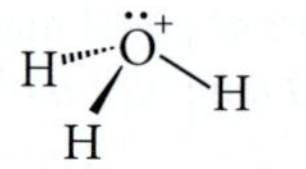

formal charge is on oxygen

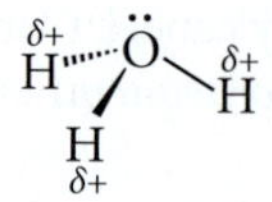

actual charge is shared by the hydrogens

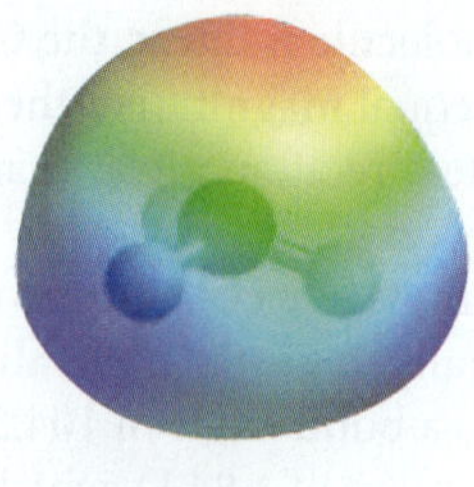

EPM

hydronium ion (1.97)

The sum of the $\delta+$ charges on the hydrogens equals 1, so the net charge on the hydronium ion is the same regardless of where we write the actual charge. For convenience, we usually write the formal charge on oxygen, but we must remember the actual charge lies on the hydrogens.

In the $^{-}BF_4$ (tetrafluoroborate) anion, fluorine is much more electronegative than boron. As a result, most of the ion's negative charge should actually be situated on the fluorines, even though we write the "–" near the B. In fact, the actual charges on the atoms of the tetrafluoroborate ion are in accord with this intuition. The EPM is red (negative) around the fluorines.

The charges in $^{-}BF_4$ and other cases like this are calculated with a quantum chemistry computer program. The exact magnitude of the charges depends on the type of calculation, so the precise numbers have some uncertainty. You are not expected to derive or memorize these charges. The important point is to understand the relative locations of the different charges.

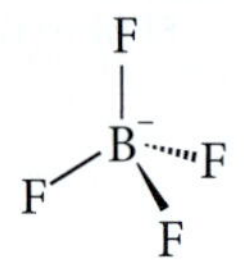

formal charge is on boron

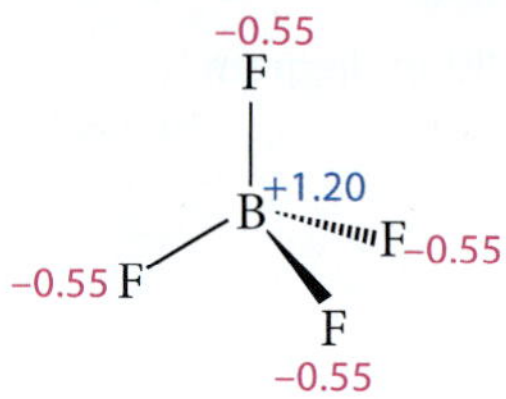

actual charges in the tetrafluoroborate anion show that the negative charge is shared by the fluorines

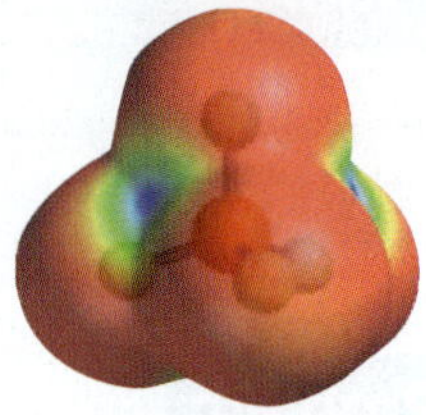

EPM

tetrafluoroborate anion (1.98)

Because the Lewis structure doesn't provide a simple way of showing this distribution, we assign the charge to the boron by the formal-charge rules.

In the methyl cation, the formal charge is on carbon, but the actual charge is distributed between the carbon and the hydrogens. This is reasonable because carbon and hydrogen have similar electronegativities. Nevertheless, the EPM is bluest (most positive) around the carbon, as the formal charge suggests.

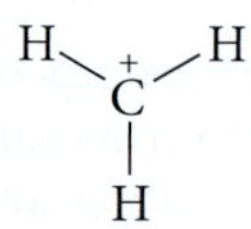

formal charge is on carbon

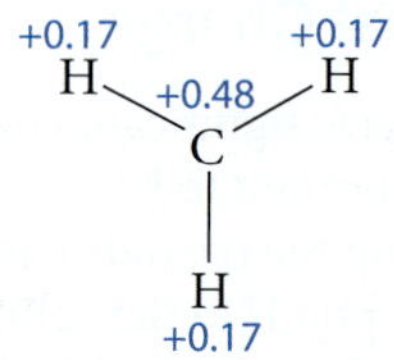

actual charge is shared by both carbon and hydrogen

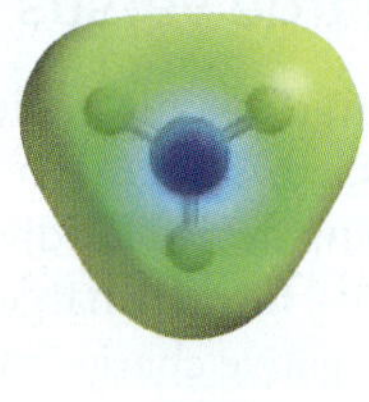

EPM

methyl cation (1.99)

It may seem strange that sometimes the formal charge represents where the actual charge resides, and sometimes it doesn't, but this is one of the shortcomings of the Lewis structure model. *Using relative electronegativities, however, you should be able to predict where the charges actually exist.*

Electrostatic potential maps can also validate our averaging of resonance structures to generate a resonance hybrid, as discussed in Sec. 1.5. Consider, for example, the acetate ion, which is represented by two resonance structures:

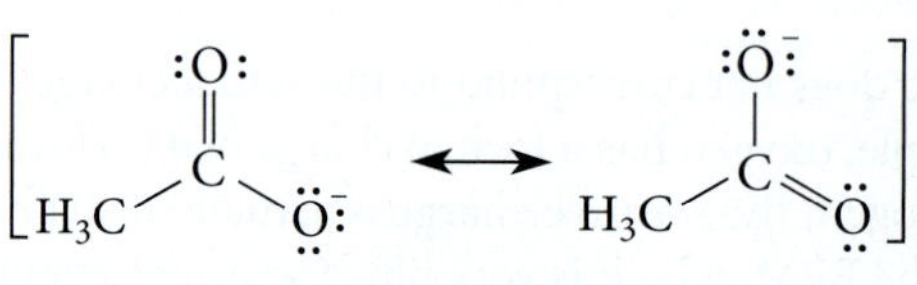

acetate ion
resonance structures

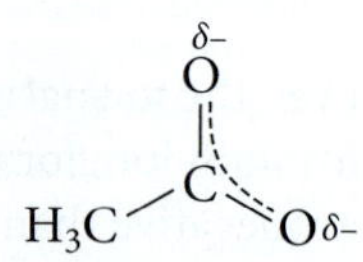

hybrid structure (1.100)

Each resonance contributor suggests that one oxygen bears a negative charge and the other is neutral. However, the *resonance hybrid* indicates that each oxygen bears half of a negative charge. The EPM of the acetate ion confirms this idea: the oxygens are equally red—one isn't redder than the other.

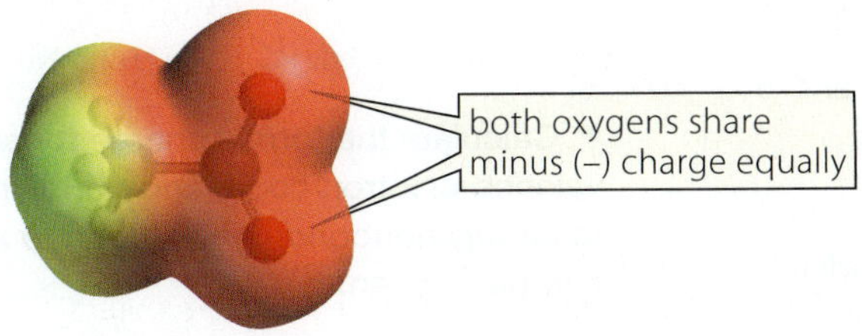

acetate ion EPM (1.101)

Focused Problems

1.36 For which of the following ions does the formal charge give a fairly accurate picture of where the charge really is? Explain in each case.

(a) $\overset{+}{N}H_4$ (b) $\overset{-}{:\ddot{N}}H_2$ (c) $H_3C{-}\overset{+}{\ddot{O}}{-}H$ (with H below O)

1.37 Explain why carbon monoxide (CO, structure below) has a very small dipole moment even though it contains full (+) and (–) formal charges. (*Hint:* Consider the relative electronegativities of C and O.)

$$:\overset{-}{C}{\equiv}\overset{+}{O}:$$

carbon monoxide

CHAPTER SUMMARY *For a summary of the chapter, see Chapter 1 in the Study Guide and Solutions Manual.*

SKILLS OBJECTIVES WITH PROBLEMS

"Skills Objectives" are given at the end of each chapter. These describe skills that you should develop as you study each chapter. (The section number is provided if you need further review.) A short problem is provided in each case to test your application of each skill.

- **Give the number of valence electrons in each atom in the "A" groups of the periodic table. 1.2A**

1.38 How many valence electrons are present in the Al (aluminum) atom? In the I (iodine) atom?

- **Show the ion formed from each atom by the gain or loss of electrons, assuming the operation of the octet rule. 1.2A**

1.39 What ion should be formed by the gain or loss of electrons from the Mg (magnesium) atom according to the octet rule?

- **List the atomic orbitals in order of increasing energy up through 3d, and show the number of equivalent orbitals of each type. 1.2C**

1.40 (a) Which of the following orbitals is *not* possible according to the quantum theory of the atom?

4d 5s 2p 3p 2d 6p

(b) Which orbital in part (a) [other than the one that is not possible] has the highest energy? The lowest energy?

- **Give the electronic configuration of any atom in the "A" groups of the periodic table. Apply the aufbau principle, Hund's rules, and the Pauli exclusion principle in deriving this configuration. 1.2D**

1.41 Give the electronic configuration of the bromine (Br) atom. Show it both as an electron-populated orbital diagram (as in Study Problem 1.1) and as a list of orbitals with superscripts indicating the number of electrons. Identify the valence orbitals, the valence electrons, and the core electrons.

- **Sketch the shapes, and show the nodes, of s and p orbitals through principal quantum level 3. 1.2E**

1.42 Which orbital has (a) a planar node? (b) Which has two spherical nodes (c) Which has no nodes? (d) Which has a spherical and a planar node? Sketch cross sections of the orbitals in (a) and (b), and show the nodes as lines or circles.

- **Given the Lewis structure of a covalent compound (including nonbonding electrons), determine whether any atom in the compound has an electronic octet. Indicate whether any atom has expanded its octet. 1.3B**

1.43 Which atom(s) in the following molecule have an electronic octet? Which (if any) has expanded its octet? Which has fewer than an octet?

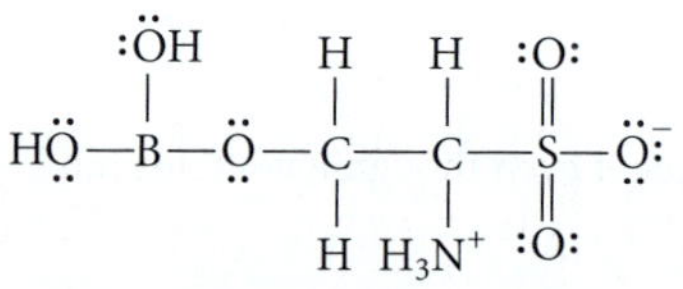

- **Know the common bonding patterns for both neutral and charged states of the following atoms: H, O, N, C, halogens, B, P, and Si. 1.3C**

1.44 Of the following three structures, one does not follow common bonding patterns for atoms or ions. Which is it and why?

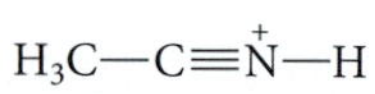

A

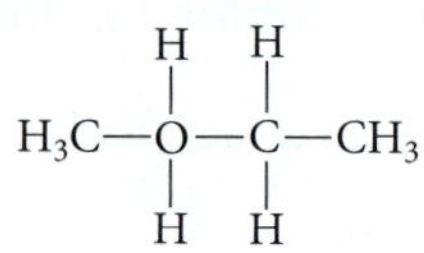

B

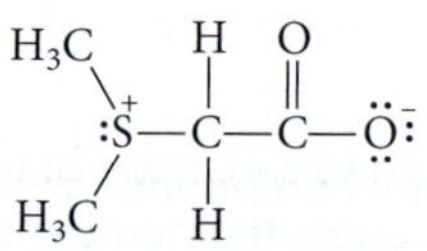

C

- **Be able to propose one or more valid Lewis structures for a molecule, given its molecular formula and atomic connectivity. Show all formal charges (if any) and nonbonding electrons. 1.3C**

1.45 Propose a Lewis structure that conforms to the octet rule for each of the following molecules. Show all formal charges and nonbonding electron pairs. For compound (b), give two structures: one that contains no formal charges and another that contains two formal charges.

(a) Azide ion, N_3^-

(b) Isocyanic acid (atom connectivity = HNCO)

(c) C_4H_{10} (Give two possible structures.)

(d) Ketene (C_2H_2O) (*Hint:* No ring!)

- **Calculate the formal charge for any atom, given its valence-electron count and bonding pattern; be able to supply nonbonding electrons to a structure in which charge is given. 1.4B**

1.46 (a) Give the formal charge on each atom (if any) and the overall charge (if any) in each of the following structures. All unshared valence electrons are shown.

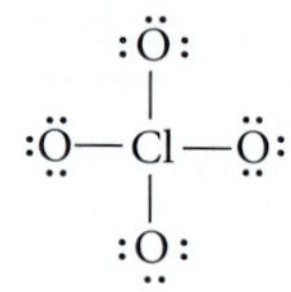

perchlorate

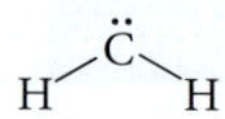

methylene

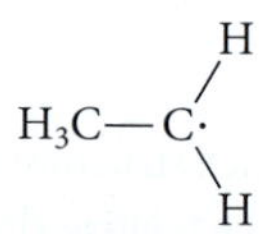

ethyl radical

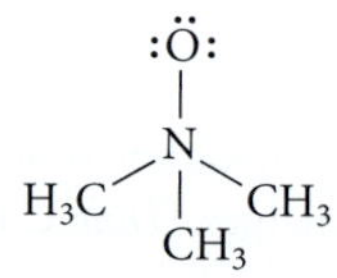

trimethylamine oxide

(b) Supply the nonbonding electrons (if any) in each of the following structures. All charges are shown.

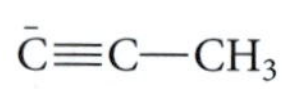

propyn-1-ide anion

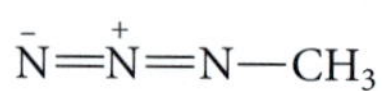

methyl azide

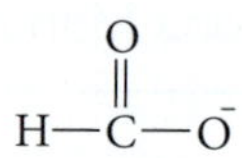

formate anion

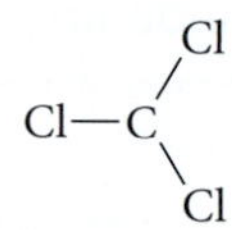

trichloromethyl

- **Use relative electronegativities to determine whether sites of formal charge correspond to sites of actual charge in a molecule or ion. 1.9C**

1.47 For each of the following atoms, explain whether the formal charge and actual charge are concentrated on the same atoms.

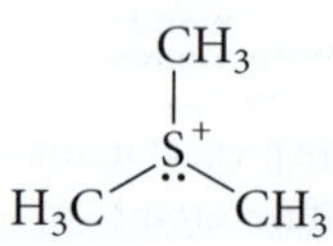

trimethylsulfonium ion

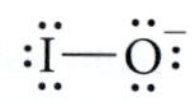

hypoiodite ion

- **Determine whether the resonance structures for a given molecule are equivalent.** **1.5**

1.48 In each case that follows, determine whether the resonance structures for the molecule are equivalent. If not, explain why not.

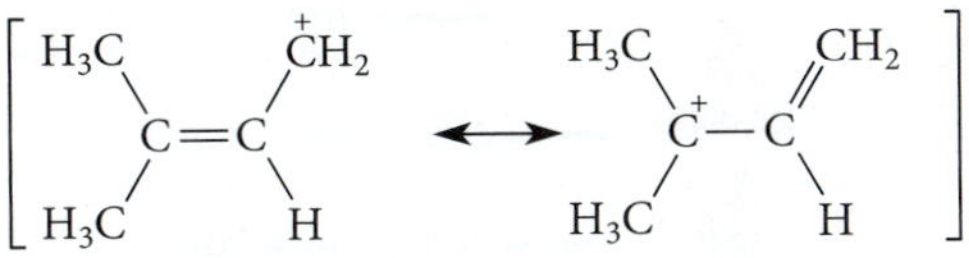

A

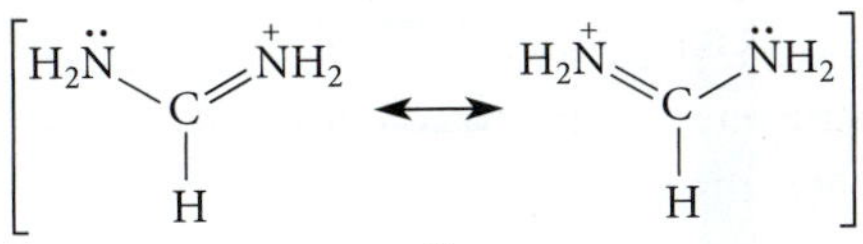

B

- **Determine the structural properties (charge on each atom, types of bonds to each atom) of a molecule that has two or more equivalent resonance structures.** **1.5**

1.49 Consider the resonance structures for the carbonate ion.

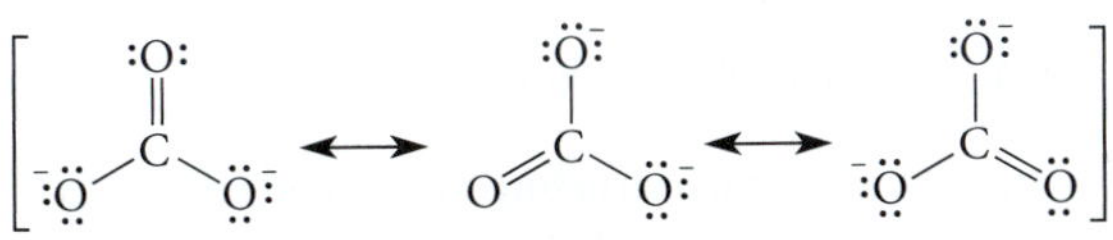

carbonate ion

(a) How much negative charge is on each atom of the carbonate ion?

(b) What is the bond order of each carbon–oxygen bond?

- **Given the resonance structures for a molecule, draw a hybrid structure using dashed lines and partial charges.** **1.5**

1.50 Given the resonance structures for ozone, draw a hybrid structure that uses dashed lines and partial charges.

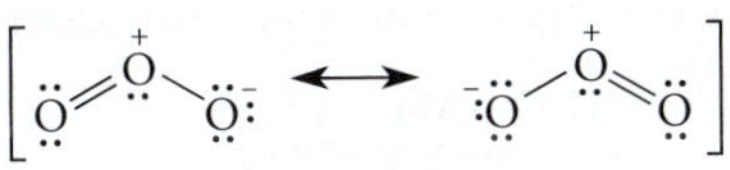

ozone

- **Know and apply the three generalizations in bond length to determine the relative lengths of covalent bonds in a molecule.** **1.6B**

1.51 Arrange the labeled bonds in the following structure in order of increasing length. State any points of ambiguity and explain.

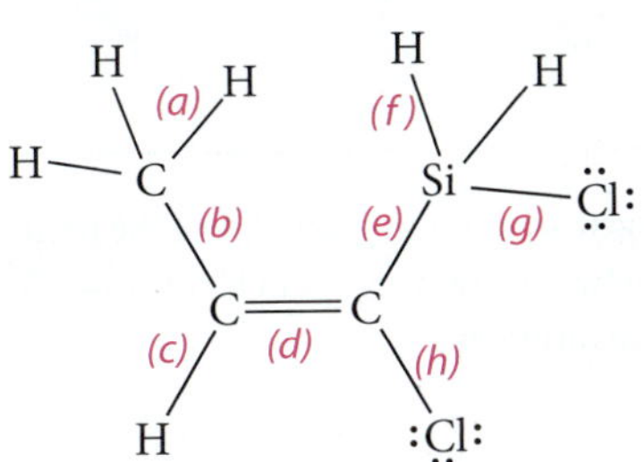

- **Know the common molecular geometries commonly encountered in organic chemistry. Determine the shape (geometry) of any molecule about each of its atoms that are bonded to two or more groups, including cases in which nonbonding electrons are present.** **1.6C, 1,7E**

1.52 Determine the molecular shape at each labeled atom of the following molecule. Identify the longest and shortest bond.

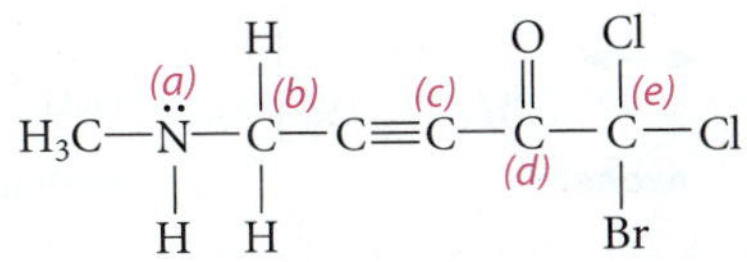

- **Sketch three-dimensional structures of any molecule (except linear molecules) using the line-and-wedge convention.** **1.6D**

1.53 Using a line-and-wedge structure, sketch the three-dimensional structure of the tetrachloroaluminate anion ($^{-}AlCl_4$).

- **Given a three-dimensional structure, determine the dihedral angle between bonds on adjacent atoms.** **1.6E**

1.54 (a) Give the hybridization of the asterisked (*) carbon and oxygen of methyl formate.

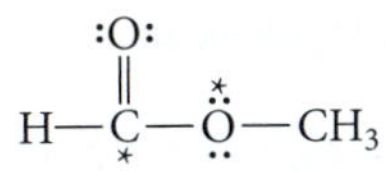

methyl formate

(b) Give the geometry at each of the asterisked atoms. (*Hint:* All of the atoms except for the CH_3 hydrogens are in the same plane.)

(c) One dihedral angle in methyl formate is the angle between the plane containing the O═C—O bond and the plane containing the C—O—C bonds. (Notice that one bond is common to both.) Sketch the two structures of methyl formate: one in which this dihedral angle is 0° and the other in which it is 180°. (Use the hint in part b.)

- **Construct hybrid-orbital bonding models for the central atom(s) of any compound, given its geometry and nonbonding electrons. Indicate what adjustments in hybridization and geometry are caused by the presence of nonbonding electrons.** **1.7B,C,D,E,F**

1.55 Construct hybrid-orbital models, and give the geometry, for each of the following molecules. How do the nonbonding electrons affect the hybridization of the sulfur in hydrogen sulfide?

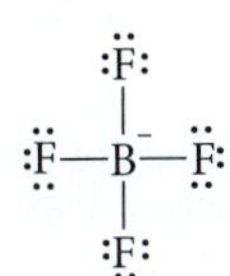

tetrafluoroborate

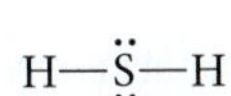

hydrogen sulfide

1.56 Give the ideal hybridization (that is, sp^3, sp^2, sp) of each atom indicated by letter in each of the following four structures. Indicate what changes in hybridization (and the resulting bond angle) would be caused by the presence of the nonbonding electrons on oxygen (a) in ozone, sulfur (b) in methanethiol, and nitrogen (f) in the *N*-methylamide anion.

ozone

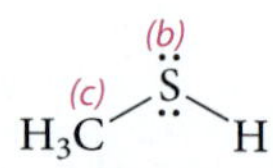

methanethiol

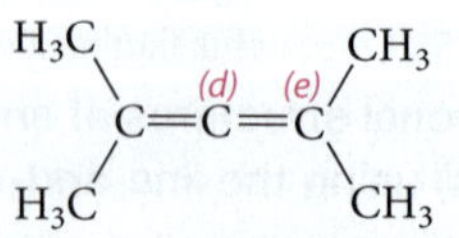

2,4-dimethyl-2,3-pentadiene

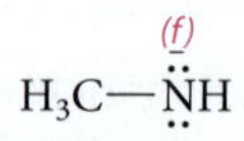

***N*-methylamide anion**

- **Construct a molecular-orbital bonding model for a situation in which two atoms come together with the overlap of two atomic orbitals. 1.8A**

1.57 Bring two 2s orbitals together to a bonding distance. (The wave troughs will overlap at this distance.) Use the addition/subtraction method to form bonding and antibonding molecular orbitals, and show the electron occupancy diagram for the Li_2 molecule.

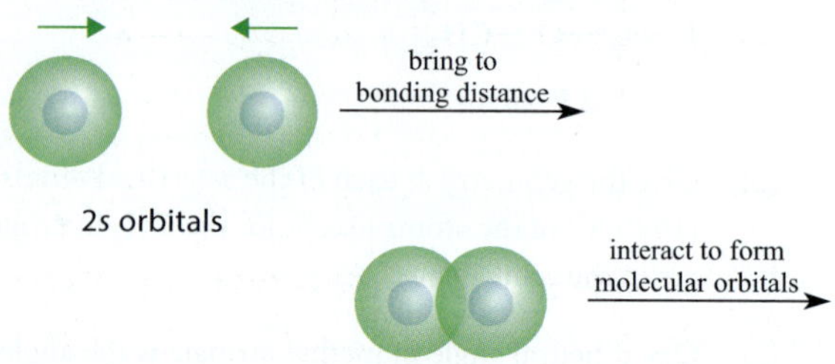

- **For any molecule, determine the total number of MOs formed when all valence orbitals of every atom interact. 1.8C**

1.58 (a) How many molecular orbitals are formed when the valence orbitals of three hydrogens and one carbon interact to form the methyl cation, $^+CH_3$?

(b) How many of these MOs are filled in the methyl cation?

- **Given the molecular orbitals of a molecule, fill them with available electrons using the aufbau principle, Hund's rules, and the Pauli principle. Calculate the MO bond order for diatomic molecules. 1.8A,B,C**

1.59 The molecular orbitals of the dinitrogen molecule (N_2) are shown in **Fig. P1.59**, in order of increasing energy. The asterisked MOs are antibonding; the others are bonding.

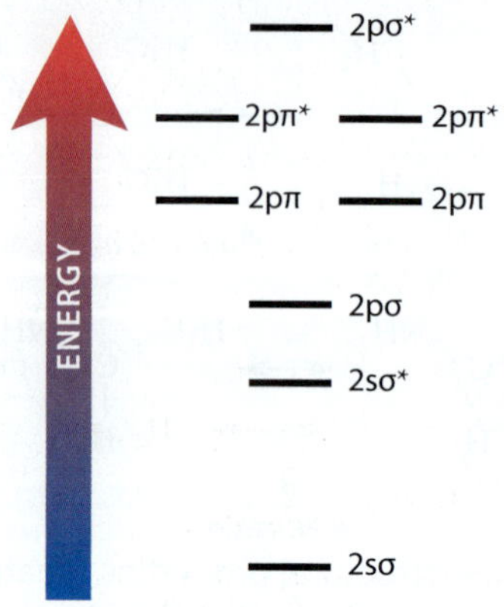

FIGURE P1.59 A molecular orbital energy diagram for dinitrogen. [The same diagram applies to dioxygen (Problem 1.70).]

(a) Fill in the MOs with the available electrons.

(b) Calculate the MO bond order for N_2.

(c) Draw a Lewis structure for N_2, and relate it to the bond order calculation you did in part (b). (*Hint:* Electron pairs not involved in bonds are unshared pairs.)

- **Determine whether a given covalent bond is polar. Show the bond dipole of a polar bond as a vector, and show the partial charges on the bonded atoms. From the polarity of its covalent bonds and its geometry, determine whether a covalent compound is polar. 1.9A,B**

1.60 (a) Draw appropriate bond dipoles for the carbon–magnesium bonds in dimethylmagnesium, and show partial charges on the appropriate atoms.

$H_3C—Mg—CH_3$

dimethylmagnesium

(b) What is the geometry of dimethylmagnesium? Explain.

(c) What conclusion can you draw about the dipole moment of dimethylmagnesium?

INTEGRATED PROBLEMS

"Integrated Problems" at the end of each chapter require you to integrate what you have learned from several sections of the chapter and from other courses, including mathematics.

1.61 Sulfur and oxygen are both in group 6A of the periodic table. Explain why the bond angles in H_2O (104.5°) and H_2S (92°) are so different.

1.62 The following three compounds have identical atomic compositions, but their atoms are connected differently. Provide a Lewis structure for each compound. The atoms are shown in the order that they are connected. Be sure to provide formal charges if appropriate.

(a) methyl isocyanate: H_3CNCO

(b) methyl cyanate: H_3COCN

(c) methyl fulminate: H_3CONC

1.63 The allyl cation can be represented by the following resonance structures.

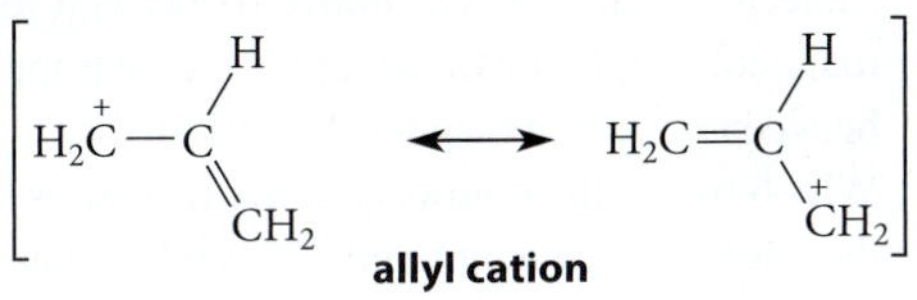

allyl cation

(a) What is the bond order of each carbon–carbon bond in the allyl cation?

(b) How much positive charge resides on each carbon of the allyl cation?

(c) What is the hybridization and molecular geometry at each carbon of the allyl cation? Use your answer to sketch a geometrically accurate structure.

(d) Draw a hybrid structure that shows electron delocalization with dashed lines and partial charges with the δ notation.

(e) Although the preceding structures are reasonable descriptions of the allyl cation, the following cation *cannot* be described by analogous resonance structures. Explain why the structure on the right is *not* a reasonable resonance structure.

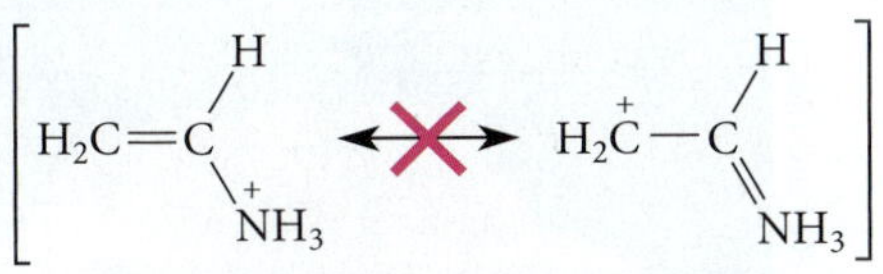

1.64 A well-known chemist, Havno Szents, has heard you claim that the water molecule has a bent geometry. Dr. Szents is unconvinced by your arguments and continues to propose that water is a linear molecule. He demands that you debate the issue with him before a distinguished academy. You must therefore come up with experimental data that will prove to an objective body of scientists that water indeed has a bent geometry. Explain why the dipole moment of water, 1.84 D, could be used to support your case.

1.65 Use your knowledge of vectors to explain why, even though the C—Cl bond dipole is large, the dipole moment of carbon tetrachloride, CCl_4, is zero. (*Hint:* Take the resultant of any two C—Cl bond dipoles; then take the resultant of the other two. Now add the two resultants to get the dipole moment of the molecule. Use models!)

1.66 Three possible dihedral angles for H_2O_2 (0°, 90°, and 180°) are shown in Display 1.55, Sec. 1.6E.

(a) Assume that the H_2O_2 molecule exists predominantly in one of these arrangements. Which of the dihedral angles can be ruled out by the fact that H_2O_2 has a large dipole moment (2.13 D)? Explain.

(b) The bond dipole moment of the O—H bond is tabulated as 1.52 D. Use this fact and the overall dipole moment of H_2O_2 in part (a) to decide on the preferred dihedral angles in H_2O_2. Take the H—O—O bond angle to be the known value (96.5°). (*Hint:* Apply the law of cosines.)

1.67 Given the dipole moment of water (1.84 D) and the H—O—H bond angle (104.45°), use a vector analysis to justify the statement in Problem 1.66(b) that the bond dipole moment of the O—H bond is 1.52 D.

1.68 Consider two 2p orbitals, one on each of two atoms, oriented head-to-head as follows:

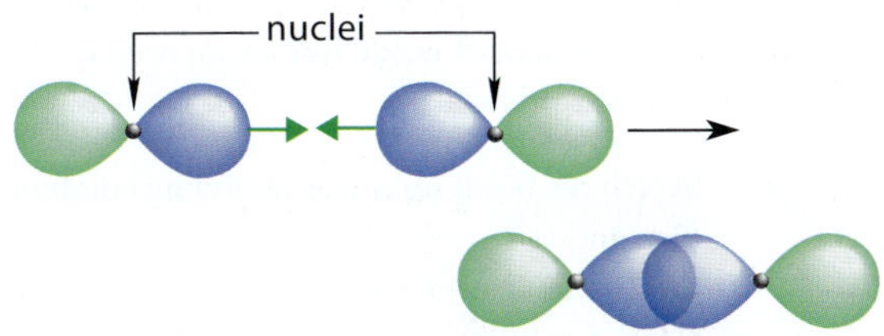

Imagine bringing the nuclei closer together until the two wave peaks (the blue lobes) of the orbitals just overlap, as shown. A new system of molecular orbitals is formed by this overlap.

(a) Sketch the resulting bonding and antibonding molecular orbitals.

(b) Identify the nodes in each molecular orbital.

(c) Construct an orbital interaction energy diagram for molecular orbital formation. That is, on an energy scale, show the two 2p orbitals on either side, and show the molecular orbitals in the middle, as in Display 1.83, Sec. 1.8A.

(d) If two electrons occupy the bonding molecular orbital, is the resulting bond a σ bond? Explain.

1.69 Consider two 2p orbitals, one on each of two different atoms, oriented side-to-side. Imagine bringing the nuclei closer together so that overlap occurs as shown in the figure. This overlap results in a system of molecular orbitals.

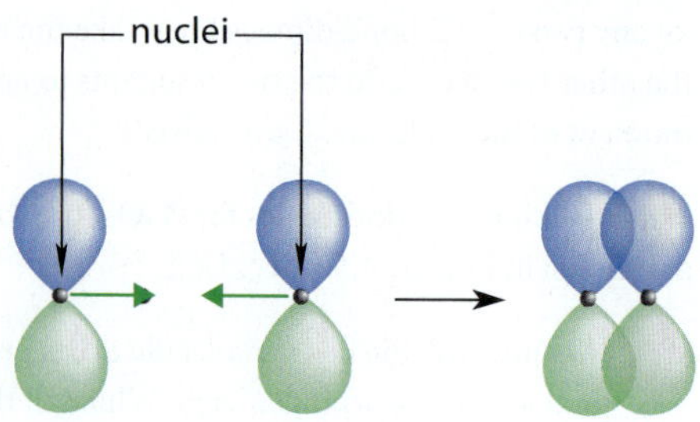

(a) Sketch the resulting bonding and antibonding molecular orbitals.

(b) Identify the node(s) in each.

(c) Construct an orbital interaction energy diagram for molecular orbital formation.

(d) When two electrons occupy the bonding molecular orbital, is the resulting bond a σ bond? Explain.

(The bonding MO is identical to the bond orbital shown in Display 1.66, Sec. 1.7C.)

1.70 In this problem, we construct the molecular orbitals of the dioxygen molecule (O_2). Imagine bringing two oxygen atoms to within a bonding distance along the x-axis (the horizontal axis in the plane of the page), and imagine the molecular orbitals that would form from overlap of the valence atomic orbitals.

(a) Sketch the bonding and antibonding combinations (2sσ and 2sσ*, respectively) that are formed when the 2s atomic orbitals interact. Show the nodes. (*Hint:* See Problem 1.57; if you worked that problem, you've already worked this one.)

(b) Sketch the bonding and antibonding combinations (2pσ and 2pσ*) that result when the $2p_x$ atomic orbitals overlap. (The x-axis is the horizontal axis in the plane of the page.) Show the nodes. (*Hint:* See Problem 1.68.)

(c) Sketch the bonding and antibonding combinations ($2p\pi_y$ and $2p\pi_y^*$) that result when the $2p_y$ atomic orbitals overlap. (Assume that the y-axis is the vertical axis in the plane of the page.) Show the nodes. (*Hint:* See Problem 1.69.)

(d) Show that the bonding and antibonding combinations ($2p\pi_z$ and $2p\pi_z^*$) that result when the $2p_z$ atomic orbitals overlap are identical to the $2p\pi_y$ and $2p\pi_y^*$ MOs in (c), but oriented at 90°. (Assume that the z-axis is the axis perpendicular to the plane of the page.)

(e) The relative energies of the MOs you constructed are the same as those for dinitrogen, shown in Fig. P1.59. Add the available valence electrons from two oxygen atoms to the orbitals, paying particular attention to Hund's rules.

(f) Molecules that have net (nonzero) electron spin are magnetic. Explain why liquid O_2 can be trapped between poles of a magnet (**Fig. P1.70**).

(g) Which one of the following Lewis structures best describes the covalent bond(s) in O_2? Explain. (*Hint:* Calculate the bond order from Eq. 1.87, Sec. 1.8B.)

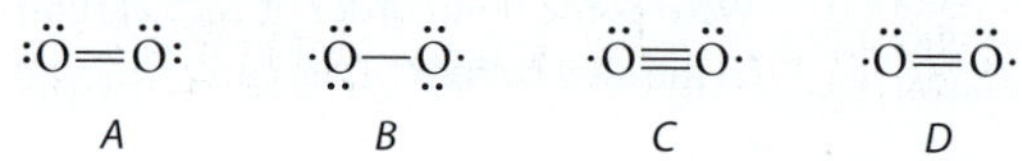

FIGURE P1.70 Liquid oxygen is trapped between poles of a magnet because it has a net electron spin.

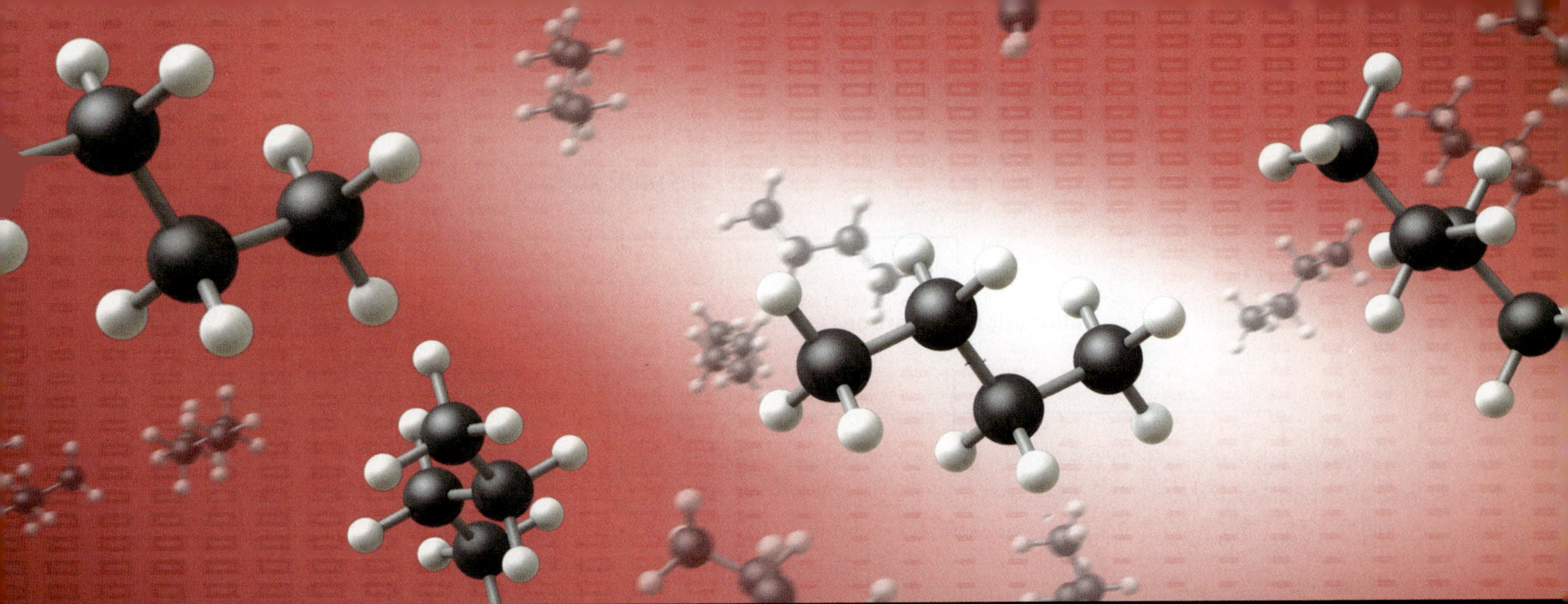

2 ALKANES AND ORGANIC NOMENCLATURE

Organic compounds all contain carbon, but they can also contain a wide variety of other elements. Before we can appreciate such chemical diversity, however, we have to begin with the simplest organic compounds, the *hydrocarbons.* **Hydrocarbons** are compounds that contain only the elements carbon and hydrogen. After summarizing the different types of hydrocarbons, this chapter considers primarily one hydrocarbon class: the alkanes.

2.1 CLASSIFICATION OF HYDROCARBONS

Hydrocarbons are divided into two broad classes: *aliphatic hydrocarbons* and *aromatic hydrocarbons.* The **aliphatic hydrocarbons** consist of three hydrocarbon families: *alkanes*, *alkenes*, and *alkynes.* We begin our study of aliphatic hydrocarbons with **alkanes**, hydrocarbons that contain only single bonds. Later we consider **alkenes**, hydrocarbons that contain carbon–carbon double bonds, and **alkynes**, hydrocarbons that contain carbon–carbon triple bonds. The last hydrocarbons we study are **aromatic hydrocarbons**, which include benzene and its substituted derivatives. The classification of hydrocarbons is summarized in **Fig. 2.1**.

Alkanes are sometimes called **paraffins,** alkenes are sometimes called **olefins,** and alkynes are sometimes called **acetylenes.** These older names are frequently used in the chemical industry.

For our study of hydrocarbons (and chemical structures in general), we need to understand some important concepts that are used in describing chemical structures. To fully describe a chemical structure, you need to know four things about it ("the four Cs"):

1. **Composition.** What elements do the compound contain? This information can be conveyed in a **molecular formula** like CH_4 or H_2O.

2. **Constitution.** How are the elements bonded together? This information can, in a few cases, be inferred from molecular formulas, but it is often more clearly conveyed with a Lewis structure. Another term for constitution is **connectivity**—how the atoms of the compound are connected.

3. **Configuration.** How are the elements arranged in three-dimensional space? We considered configuration in Chapter 1 when we explored various molecular geometries: tetrahedral,

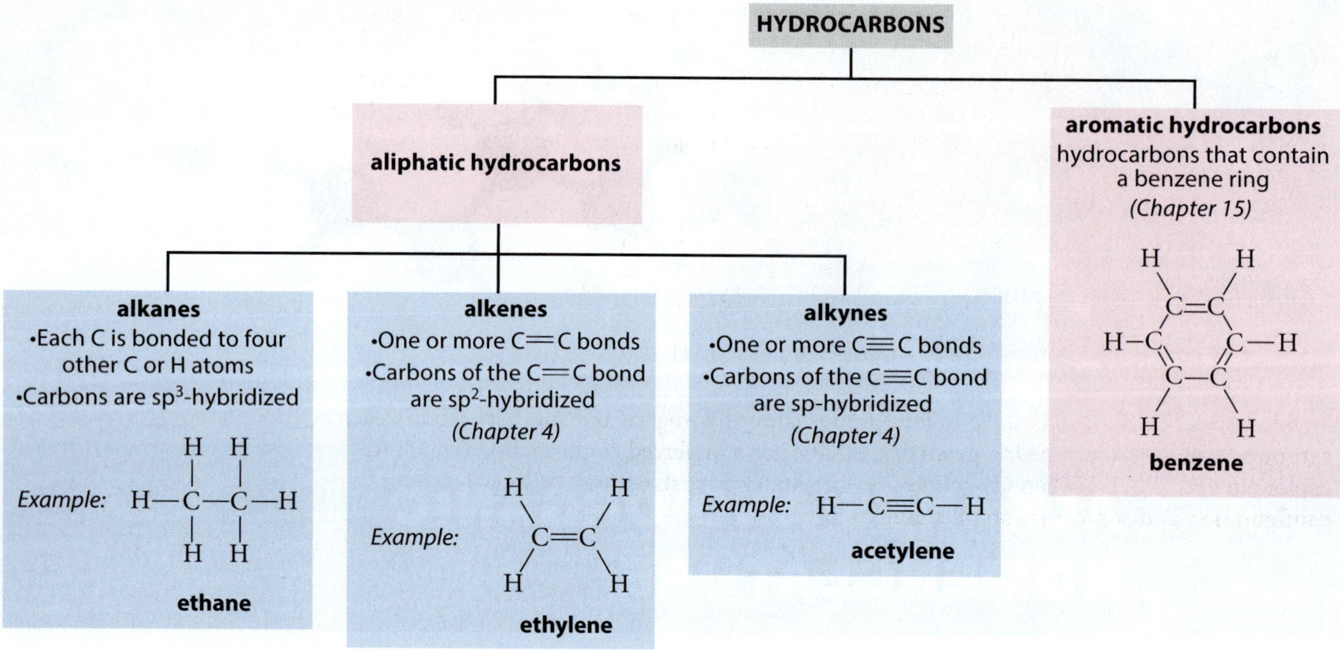

FIGURE 2.1 Classifications of hydrocarbons with a simple example of each class.

bent, or linear. This information can be conveyed with three-dimensional drawings (line-and-wedge structures), physical molecular models, or computer-generated images.

4. **Conformation.** What is the spatial relationship between bonds on adjacent atoms? This requires knowledge of *dihedral angles* (Sec. 1.6E), which we explore in more detail in this chapter. We introduce the *Newman projection,* a straightforward method of viewing dihedral angles.

2.2 UNBRANCHED ALKANES

A. Methane and Ethane

In Chapter 1, you learned about methane, the simplest alkane. The *composition* of methane is given by its molecular formula, CH_4, which shows us what atoms are present. Its *constitution*—how its atoms are connected—is given by its Lewis structure.

CH_4

formula of methane

composition: shows what elements are present and their amounts

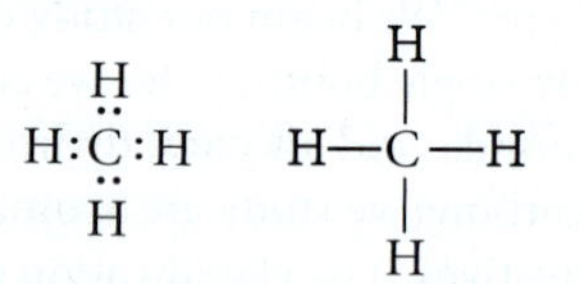

Lewis structures of methane

constitution: shows connectivity (2.1)

You also learned that in methane, the hydrogens occupy the corners of a regular tetrahedron. This is the *configuration* of methane. Two ways to depict the methane configuration are with molecular models and with line-and-wedge structures:

ball-and-stick line-and-wedge

configuration of methane

configuration: shows the three-dimensional arrangement of atoms (2.2)

Methane does not have different conformations. In Sec. 2.5, we use other alkanes, ethane and butane, to describe molecular conformations.

Instead of being bonded only to hydrogens, a carbon atom can also be bonded to a second carbon with enough hydrogens to complete its octet. The resulting compound is *ethane*, which has the formula (composition) C_2H_6. The constitution of ethane is indicated by its Lewis structures:

Lewis structures of ethane:

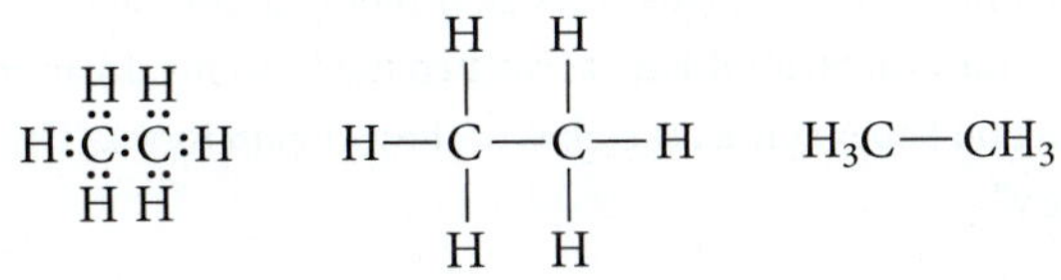

(2.3)

Because each carbon is bonded to four groups (three hydrogens and another carbon), each carbon in ethane has tetrahedral geometry. Ethane has a preferred conformation (which we explore in Sec. 2.5A). We can use models, or line-and-wedge structures, to show both the configuration and conformation of ethane.

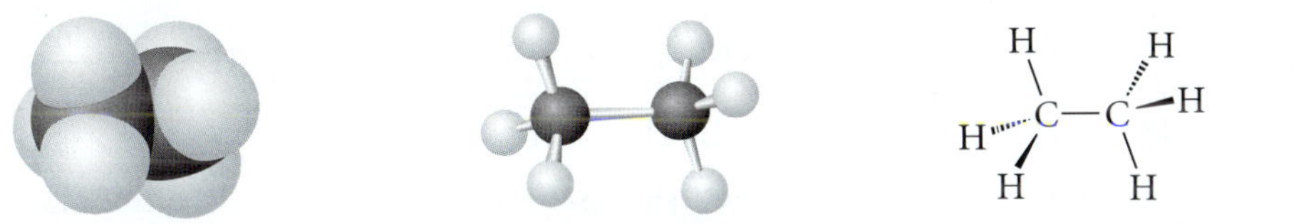

space-filling model | ball-and-stick model | line-and-wedge Lewis structure

ethane structures that show configuration and conformation (2.4)

In ethane, the bond between the two carbon atoms is longer than a C—H bond. Like the C—H bonds, it is a covalent bond. In terms of hybrid orbitals, the carbon–carbon bond in ethane consists of two electrons in a bond that is formed by the overlap of two sp^3 hybrid orbitals, one from each carbon. Thus, the carbon–carbon bond in ethane is an sp^3–sp^3 σ bond. The C—H bonds in ethane are like those of methane. They consist of covalent bonds, each of which is formed by the overlap of a carbon sp^3 orbital with a hydrogen 1*s* orbital—that is, they are sp^3–1*s* σ bonds. Both the H—C—C and H—C—H bond angles in ethane are essentially tetrahedral because each carbon bears four groups. (The overlapping atomic orbitals are shown as dashed lines for the two labeled bonds.)

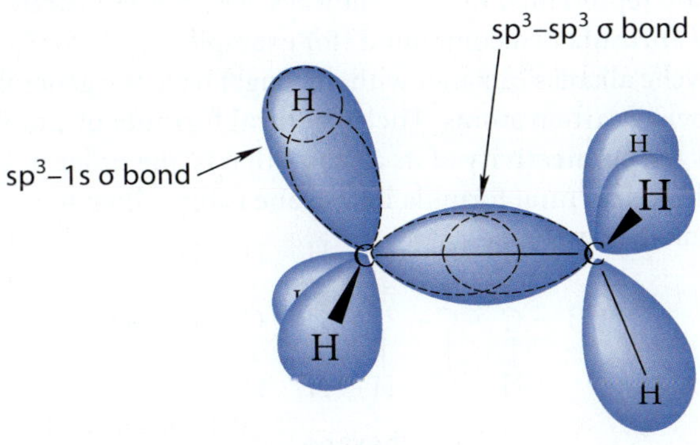

(2.5)

B. Other Unbranched Alkanes

We can go on to envision other alkanes in which any number of carbons are connected with sp^3–sp^3 σ bonds to form chains of carbons with their associated hydrogen atoms. Indeed, the ability of a carbon to form stable bonds to other carbons is what gives rise to the tremendous number of known organic compounds. The idea of carbon chains, a revolutionary one in the early days of chemistry, was developed independently by the German chemist August Kekulé (1829–1896) and the Scotsman Archibald Scott Couper (1831–1892) in about 1858. Kekulé's account of his inspiration for this idea is amusing.

> During my stay in London I resided for a considerable time in Clapham Road in the neighborhood of Clapham Common. … One fine summer evening I was returning by the last bus, "outside" as usual, through the deserted streets of the city that are at other times so full of life. I fell into a reverie, and lo, the atoms were gamboling before my eyes. Whenever, hitherto, these diminutive beings had appeared to me they had always been in motion. Now, however, I saw how, frequently, two smaller atoms united to form a pair. . . . I saw how the larger ones formed a chain, dragging the smaller ones after them but only at the ends of the chain. . . . The cry of the conductor, "Clapham Road," awakened me from my dreaming, but I spent a part of the night putting on paper at least sketches of these dream forms. This was the origin of the "Structure Theory."

Carbon chains take many forms in the alkanes; they may be branched, unbranched, or cyclic.

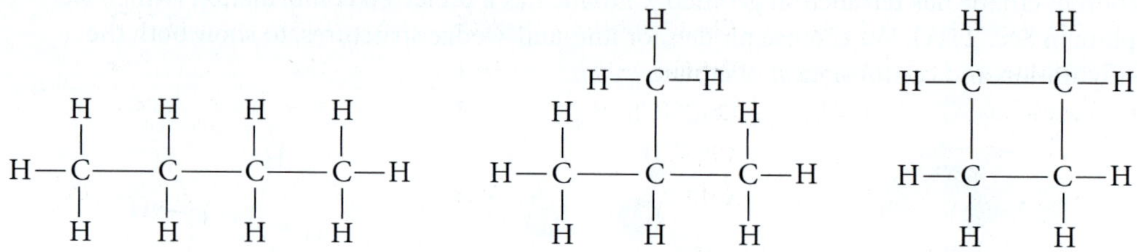

butane (an unbranched alkane) **methylpropane (isobutane)** (a branched alkane) **cyclobutane** (a cyclic alkane)

(2.6)

In an **unbranched** carbon chain, carbons within the chain are bonded to only two other carbons and two hydrogens. Unbranched alkanes are discussed subsequently in this section. In a **branched** carbon chain, an interior carbon can be bonded to three or four other carbons and correspondingly fewer hydrogens. (We meet many examples of branched carbon chains beginning in in Sec. 2.4A.) In a **cyclic alkane**, or **cycloalkane**, the carbon chains form closed rings. (We meet our first cyclic alkanes in Sec. 2.3.)

Alkanes with unbranched carbon chains are sometimes called **normal alkanes**, or ***n*-alkanes**. This terminology is not sanctioned in official nomenclature and is falling into disuse.

You should learn the names of the first 12 unbranched alkanes because they are the basis for naming many other organic compounds. The names methane, ethane, propane, and butane have their origins in the early history of organic chemistry, but the names of the higher alkanes are derived from the corresponding Greek numerical names: pentane (*pent* = five), hexane (*hex* = six), and so on.

Organic molecules are represented in different ways, which we illustrate using the alkane hexane. The **molecular formula** of a compound (for example, C_6H_{14} for hexane) gives its atomic composition. All noncyclic alkanes (alkanes without rings) have the general formula C_nH_{2n+2}, in which n is the number of carbon atoms. The **structural formula** of a molecule is its Lewis structure, which shows the **connectivity** of its atoms—that is, the order in which its atoms are connected. For example, a structural formula for hexane is the following:

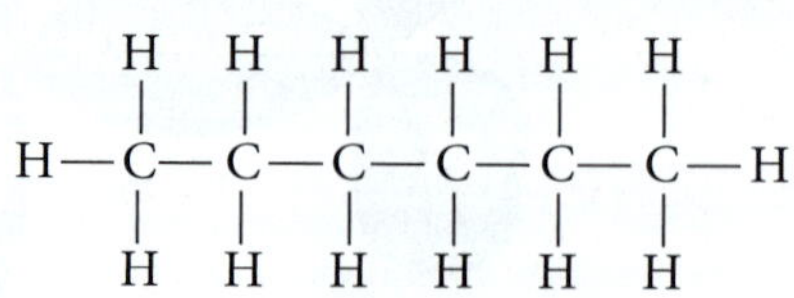

hexane (2.7)

Writing each hydrogen atom in this way is very time-consuming, and a simpler representation of this molecule, called a **condensed structural formula**, conveys the same information.

$$H_3C—CH_2—CH_2—CH_2—CH_2—CH_3$$

hexane (2.8)

In such a structure, the hydrogen atoms are understood to be connected to carbon atoms with single bonds, and the bonds shown explicitly are *bonds between carbon atoms.* Sometimes even these bonds are omitted, so that hexane can also be written $CH_3CH_2CH_2CH_2CH_2CH_3$.

TABLE 2.1 The Unbranched Alkanes

Compound name	Molecular formula	Condensed structural formula	Melting point (°C)	Boiling point (°C)	Density* (g mL^{-1})
methane	CH_4	CH_4	−182.5	−161.7	—
ethane	C_2H_6	CH_3CH_3	−183.3	−88.6	—
propane	C_3H_8	$CH_3CH_2CH_3$	−187.7	−42.1	—
butane	C_4H_{10}	$CH_3(CH_2)_2CH_3$	−138.3	−0.5	—
pentane	C_5H_{12}	$CH_3(CH_2)_3CH_3$	−129.8	36.1	0.6262
hexane	C_6H_{14}	$CH_3(CH_2)_4CH_3$	−95.3	68.7	0.6603
heptane	C_7H_{16}	$CH_3(CH_2)_5CH_3$	−90.6	98.4	0.6837
octane	C_8H_{18}	$CH_3(CH_2)_6CH_3$	−56.8	125.7	0.7026
nonane	C_9H_{20}	$CH_3(CH_2)_7CH_3$	−53.5	150.8	0.7177
decane	$C_{10}H_{22}$	$CH_3(CH_2)_8CH_3$	−29.7	174.0	0.7299
undecane	$C_{11}H_{24}$	$CH_3(CH_2)_9CH_3$	−25.6	195.8	0.7402
dodecane	$C_{12}H_{26}$	$CH_3(CH_2)_{10}CH_3$	−9.6	216.3	0.7487
eicosane	$C_{20}H_{42}$	$CH_3(CH_2)_{18}CH_3$	+36.8	343.0	(solid at 20 °C)

*The densities tabulated in this text are of the liquids at 20 °C unless otherwise noted.

The structural formula may be further abbreviated as shown in the third column of **Table 2.1**. In this type of formula, for example, $(CH_2)_4$ means $—CH_2CH_2CH_2CH_2—$. Hexane can thus be written $CH_3(CH_2)_4CH_3$.

$$CH_3CH_2CH_2CH_2CH_2CH_3 \qquad CH_3(CH_2)_4CH_3$$

two other representations of hexane (2.9)

The most efficient type of structure, and one that we'll use a great deal, is called a **skeletal structure.** In skeletal structures, we write the carbon backbone as a zigzag line without writing the C's and H's explicitly. In a skeletal structure, it is assumed that *at every vertex* and *at the end of every line* there is a carbon with enough hydrogens to fill out the carbon valence of four bonds. Skeletal structures can be used for compounds containing three or more carbons.

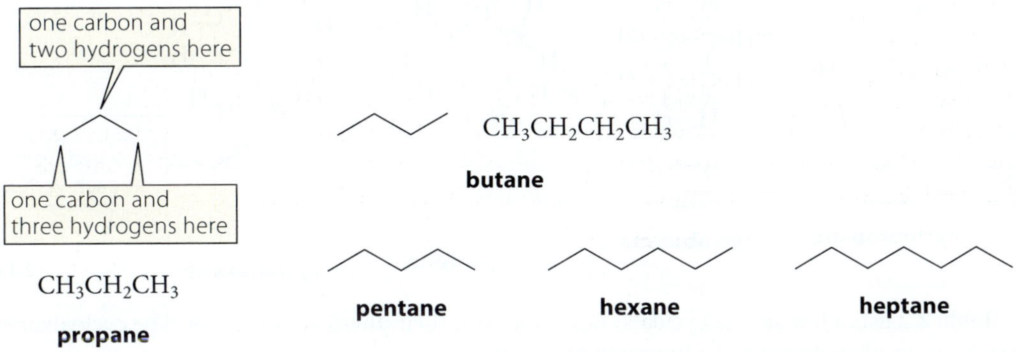

(2.10)

Despite their simplicity, skeletal structures can convey all four pieces of information needed to describe the structure of an alkane: composition (atoms present), constitution (how atoms are bonded), configuration (all alkane carbons are assumed to be tetrahedral, even though the hydrogens are not shown explicitly), and conformation (the spatial arrangement of bonds on different atoms, which discussed in Sec. 2.5).

The unbranched alkanes constitute a series in which successive members differ from one another by one CH_2 group (**methylene group**) in the carbon chain. A series of compounds that differ by the addition of methylene groups is called a **homologous series**. Thus, the unbranched alkanes constitute one homologous series. Generally, physical properties within a homologous series vary in a regular way. An examination of Table 2.1, for example, reveals that the boiling points and densities of the unbranched alkanes vary regularly with increasing number of carbon atoms. This variation can be useful for quickly estimating the properties of a member of the series whose properties are not known.

The French chemist Charles Gerhardt (1816–1856) made an important chemical observation in 1845 about members of homologous series. His observation still has significant implications for learning organic chemistry. He wrote, "These (related) substances undergo reactions according to the same equations, and it is only necessary to know the reactions of one in order to predict the reactions of the others." What Gerhardt was saying is that we can study the chemical reactions of propane with the confidence that ethane, butane, or dodecane will undergo analogous reactions.

Focused Problems

2.1 (a) How many hydrogen atoms are in the unbranched alkane with 18 carbon atoms?

(b) Is there an unbranched alkane containing exactly 23 hydrogen atoms? Is so, give its structural formula. If not, explain why not.

2.2 Give the skeletal formula of *tridecane*, the unbranched alkane with the formula $C_{13}H_{28}$.

2.3 (a) How many carbons and hydrogens are in the following unbranched alkane? Name it.

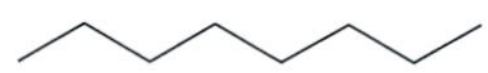

(b) Draw the expanded structure showing all of the bonds.

2.3 CYCLOALKANES

Some alkanes contain carbon chains in closed loops, or rings; these are called **cycloalkanes**. Cycloalkanes are named by adding the prefix *cyclo* to the name of the alkane.

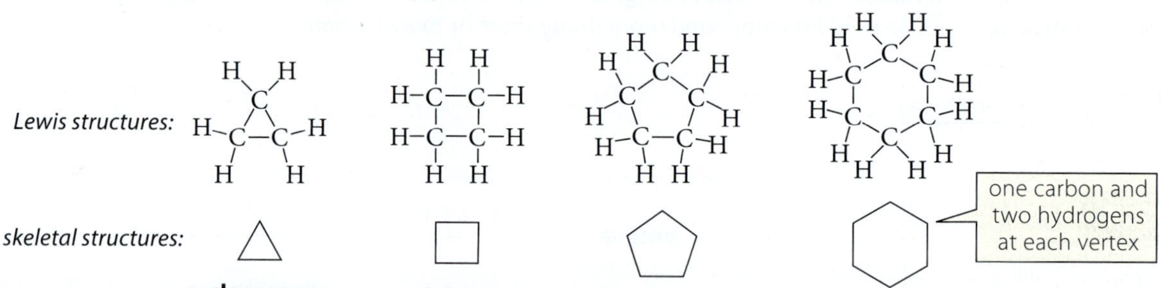

cyclopropane **cyclobutane** **cyclopentane** **cyclohexane** (2.11)

Table 2.2 lists a few simple cycloalkanes along with their physical properties. The cycloalkanes constitute another example of a homologous series.

Carbon has a tetrahedral configuration in the cycloalkanes, just as it does in the alkanes. For this reason, the carbon skeletons of cycloalkanes (except for cyclopropane) are not planar. (Building a model of cyclohexane will convince you of this point.) We study the conformations of cycloalkanes in Chapter 7. For now, just remember that the planar skeletal structures of cycloalkanes convey no information about their conformations.

The general formula for an alkane containing a single ring has two fewer hydrogens than that of the noncyclic alkane with the same number of carbon atoms. For example, cyclohexane has the formula C_6H_{12}, whereas hexane has the formula C_6H_{14}. The general formula for cycloalkanes with one ring is C_nH_{2n}. For each additional ring, a cycloalkane has two fewer hydrogens than the noncyclic alkane with the same number of carbons. For example, decane has 22 hydrogens

TABLE 2.2 Some Cycloalkanes and Their Physical Properties

Compound	Boiling point (°C)	Melting point (°C)	Density (g mL^{-1})
cyclopropane	–32.7	–127.6	—
cyclobutane	12.5	–50.0	—
cyclopentane	49.3	–93.9	0.7457
cyclohexane	80.7	6.6	0.7786
cycloheptane	118.5	–12.0	0.8098
cyclooctane	150.0	14.3	0.8340

($C_{10}H_{22}$). Cyclodecane ($C_{10}H_{20}$), with one ring, has two fewer hydrogens. Decalin ($C_{10}H_{18}$), with two rings, has four fewer hydrogens than decane.

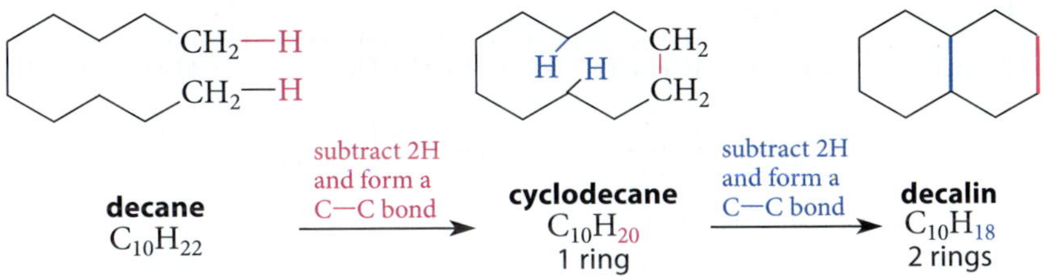

(2.12)

Focused Problems

2.4 How many hydrogens are in a cycloalkane of *n* carbons containing (a) two rings? (b) three rings? (c) *m* rings?

2.5 (a) How many rings does a cycloalkane have if its formula is C_8H_{10}? C_7H_{12}? Explain.

(b) Count the carbons in the following cycloalkane. Then determine the number of hydrogens without counting them explicitly.

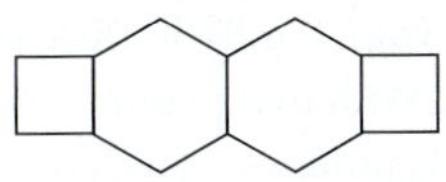

2.4 CONSTITUTIONAL ISOMERS AND NOMENCLATURE

A. Isomers

When a carbon atom in an alkane is bonded to more than two other carbon atoms, a *branch* in the carbon chain occurs at that position. The smallest branched alkane has four carbon atoms. As a result, there are two four-carbon alkanes; one is *butane*, and the other is *isobutane*.

Lewis structures: $H_3C—CH_2—CH_2—CH_3$

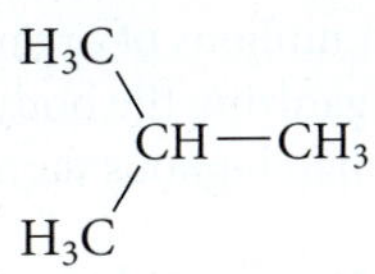

skeletal structures:

butane
bp –0.5 °C

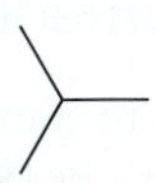

isobutane
bp –11.7 °C

These are different compounds with different properties. For example, the boiling point of butane is –0.5 °C, whereas that of isobutane is –11.7 °C. Yet both have the same molecular formula, C_4H_{10}. Different compounds that have the same molecular formula are said to be **isomers** or **isomeric compounds**.

There are different types of isomers. Isomers that differ in their *constitution*—that is, in the connectivity of their atoms—such as butane and isobutane, are called **constitutional isomers** or **structural isomers**. Recall (Secs. 1.6, 2.5) that *connectivity* is the order in which the atoms of the molecule are bonded. The atomic connectivities of butane and isobutane differ because in isobutane a carbon is attached to three other carbons, whereas in butane no carbon is attached to more than two other carbons.

Study Problem 2.1

Which of the following four structures represent constitutional isomers of the same molecule, and which one is neither isomeric nor identical to the others? Explain your answers.

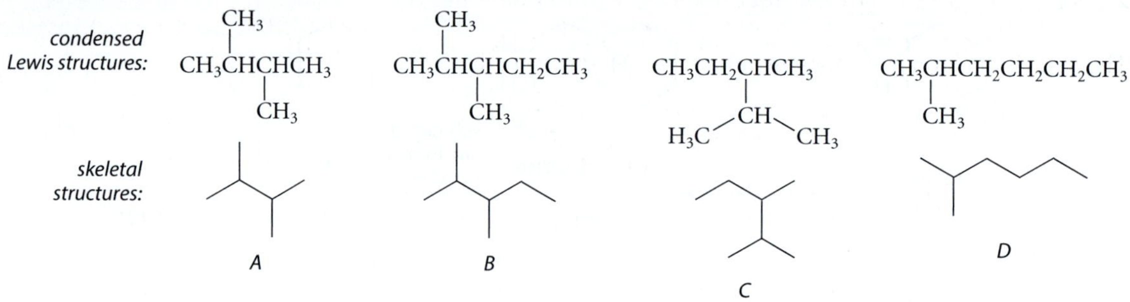

Solution Compounds must have the same molecular formula to be either identical or isomeric. Structure *A* has a different molecular formula (C_6H_{14}) from the other structures (C_7H_{16}), and hence structure *A* represents a molecule that is neither identical nor isomeric to the others. To solve the rest of the problem, we must understand that *Lewis structures show connectivity only.* They do not represent the actual shapes of molecules unless we start adding spatial elements such as wedges and dashed wedges. This means that *we can draw a given structure many different ways.* Have you ever heard the old spiritual, "The foot-bone's connected to the ankle-bone"? That's a song about connectivity of the typical human body. If the description fits you, its validity doesn't change whether you are sitting down, standing up, standing on your head, or doing yoga. Similarly, the connectivity of a molecule doesn't change whether it is drawn forward, backward, or upside-down. With that in mind, let's trace the connectivity of each of these structures. Consider structures *B* and *C*. Each has two CH_3 groups connected to a CH, and that CH is connected to another CH, which in turn is connected to both a CH_3 group and a CH_2CH_3 group. In *B*, this connectivity pattern starts on the left; in *C*, it starts on the bottom. But it's the same in both. Because both structures have identical connectivities, they represent the same molecule.

Structures *D* and *B* (or *D* and *C*) have the same molecular formula C_7H_{16}; but, as you should verify, their connectivities are different, so they are constitutional isomers.

Butane and isobutane are the only constitutional isomers with the formula C_4H_{10}. However, more constitutional isomers are possible for alkanes with more carbon atoms. There are nine constitutional isomers of the heptanes (C_7H_{16}), 75 constitutional isomers of the decanes ($C_{10}H_{22}$), and 366,319 constitutional isomers of the eicosanes ($C_{20}H_{42}$)! These few examples demonstrate that millions of organic compounds are known and millions more are conceivable. It follows that organizing the body of chemical knowledge requires a system of nomenclature that can provide an unambiguous name for each compound.

B. Organic Nomenclature

An organized effort to standardize organic nomenclature dates from proposals made at a conference in Geneva in 1892. From those proposals the International Union of Pure and Applied Chemistry (IUPAC), a professional association of chemists, developed and sanctioned several accepted systems of nomenclature. The most widely applied system in use today is called **substitutive nomenclature.**

The IUPAC rules for the nomenclature of alkanes form the basis for the substitutive nomenclature of most other compound classes. Hence, it is important to learn these rules and be able to apply them.

C. Substitutive Nomenclature of Alkanes

Alkanes are named by applying the following 10 rules *in order*. This means that if one rule doesn't unambiguously determine the name of a compound of interest, we proceed down the list *in order* until we find a rule that does.

1. *The unbranched alkanes are named according to the number of carbons, as shown in Table 2.1.*

2. *For alkanes containing branched carbon chains, determine the principal chain.*

The **principal chain** is the longest continuous carbon chain in the molecule. To illustrate:

Lewis structure:

skeletal structure:

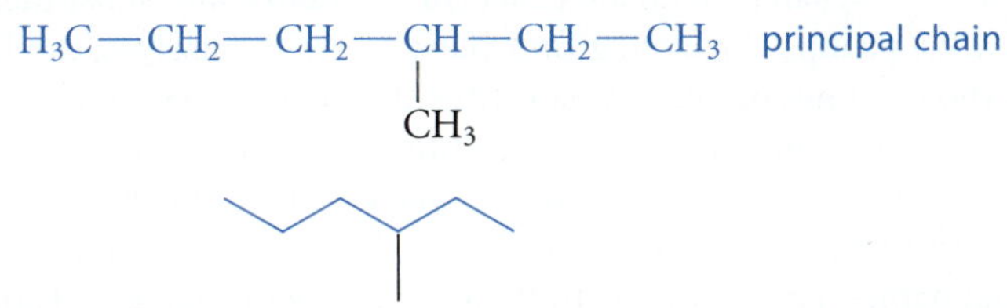

When identifying the principal chain, take into account that *the structure of a given molecule may be drawn in several different ways* (Study Problem 2.2). Thus, the following structures represent the *same molecule*, with the principal chain shown in blue:

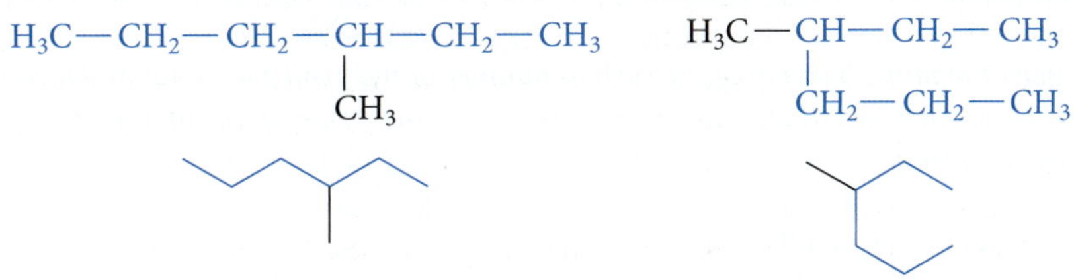

(Be sure to verify that these structures have identical connectivities and thus represent the same molecule.)

3. *If two or more chains within a structure have the same length, choose as the principal chain the one with the greater number of branches.*

The following structure is an example of such a situation:

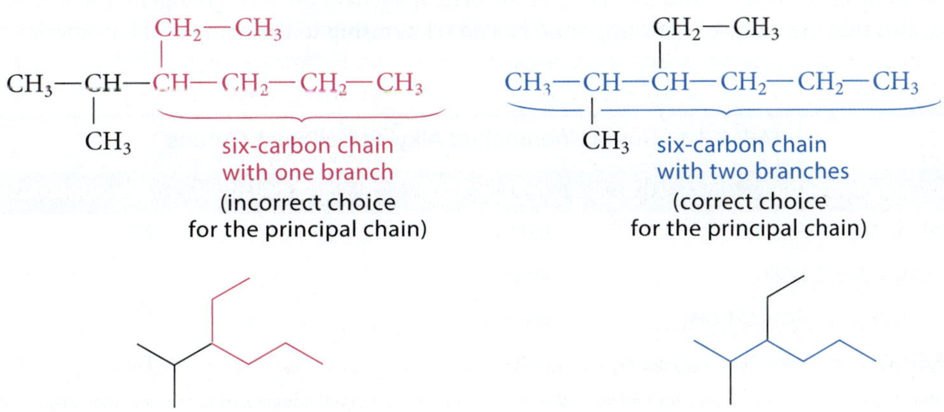

The correct choice of principal chain is the one on the right, because it has two branches; the choice on the left has only one. (It makes no difference that the branch on the left is larger or that it has additional branching within itself.)

4. *Number the carbons of the principal chain consecutively from one end to the other in the direction that gives the lower number to the first branching point.*

In the following structure, the carbons of the principal chain are numbered consecutively from one end to give the lower number to the carbon at the —CH_3 branch.

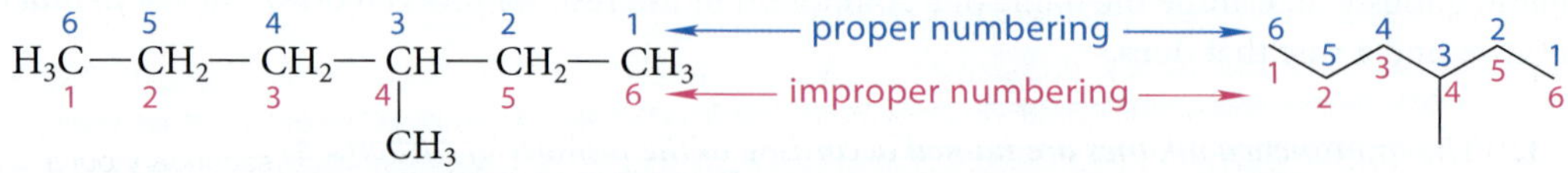

5. *Name each branch, and identify the carbon number of the principal chain at which it occurs.*

In the previous example, the branching group is a —CH_3 group. This group, called a *methyl group*, is located at carbon-3 of the principal chain.

STUDY GUIDE LINK 2.1
Nomenclature of Simple Branched Compounds

Branching groups in general are called **substituents**, and substituents derived from alkanes are called **alkyl groups**. An alkyl group, like alkanes themselves, can have any number of carbons. **Table 2.3** shows some simple alkyl substituents, their names, and the abbreviations commonly used for them. The name of the alkyl substituent is derived from the unbranched alkane with the same number of carbons by dropping the final *e* and adding *yl.*

Alkyl substituents themselves may be branched. The most common branched alkyl groups have special names, given along with their skeletal equivalents and common abbreviations in **Table 2.4**. These should be learned because they are encountered frequently. Notice that the "iso" prefix is used for substituents containing two methyl groups at the end of a carbon chain. Notice, too, the difference between an isobutyl group and a *sec*-butyl group; these two groups are frequently confused.

6. *Construct the name by writing the carbon number of the principal chain at which the substituent occurs, a hyphen, the name of the branch, and the name of the alkane corresponding to the principal chain.*

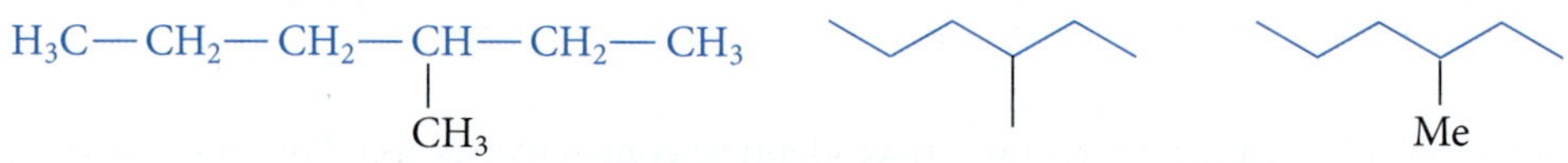

name: **3-methylhexane**
- **3-methyl**: number and name of the alkyl substituent
- **hexane**: name of the principal chain

The name of the branch and the name of the principal chain are written together as one word. Notice also that the name of the compound has *no* relationship to the name of the isomeric

TABLE 2.3 Some Unbranched Alkyl Substituent Groups

Substituent structure	Written name	Skeletal structure*	Text abbreviation
$-CH_3$ *or* CH_3- *or* H_3C-	methyl		Me
CH_3CH_2- *or* $-CH_2CH_3$	ethyl		Et
$CH_3CH_2CH_2-$ *or* $-CH_2CH_2CH_3$	propyl	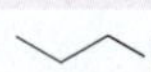	Pr
$CH_3CH_2CH_2CH_2-$ *or* $-CH_2CH_2CH_2CH_3$	butyl		Bu

*In the skeletal structures the bracket } refers to the point of attachment of the substituent to the principal chain.

TABLE 2.4 Some Branched Alkyl Substituent Groups

Lewis structure	Skeletal structure*	Condensed structure	Text abbreviation	Written name	Pronounced name
H_3C CH— H_3C	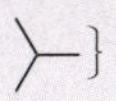	$(CH_3)_2CH$—	*i*-Pr or *i*Pr	isopropyl	isopropyl
H_3C $CHCH_2$— H_3C	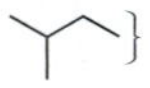	$(CH_3)_2CHCH_2$—	*i*-Bu or *i*Bu	isobutyl	isobutyl
CH_3CH_2CH— CH_3	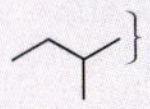	—	*s*-Bu or *s*Bu	*sec*-butyl	secondary butyl or "sec-butyl"
CH_3 H_3C—C— CH_3	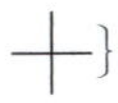	$(CH_3)_3C$—	*t*-Bu or *t*Bu	*tert*-butyl (or *t*-butyl)	tertiary butyl or "tert-butyl"
CH_3 H_3C—C—CH_2— CH_3	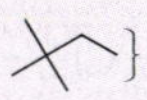	$(CH_3)_3CCH_2$—	—	neopentyl	neopentyl

*In the skeletal structures, the bracket } indicates the point of attachment to the main chain.

unbranched alkane—that is, the preceding compound is *a constitutional isomer of heptane* because it has seven carbon atoms, but it is named as a *derivative of hexane*, because its principal chain contains six carbon atoms.

Study Problem 2.2

Name the following compound, and give the name of the unbranched alkane of which it is a constitutional isomer.

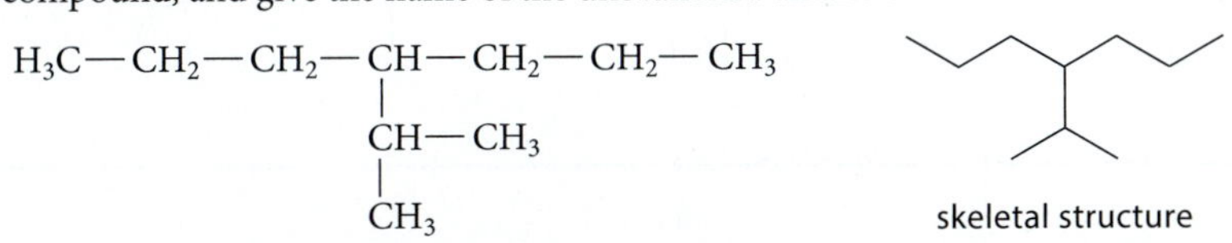

skeletal structure

Solution Because the principal chain has seven carbons, the compound is named as a substituted heptane. The branch is at carbon-4, and the substituent group at this branch is

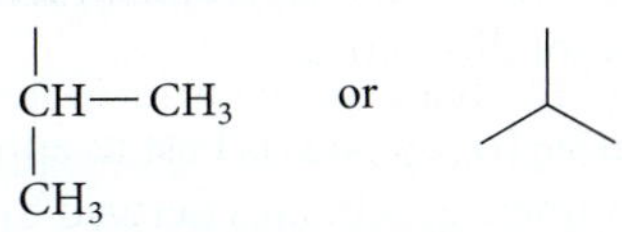

Table 2.4 shows that this is an *isopropyl group*. Thus, the name of the compound is 4-isopropylheptane:

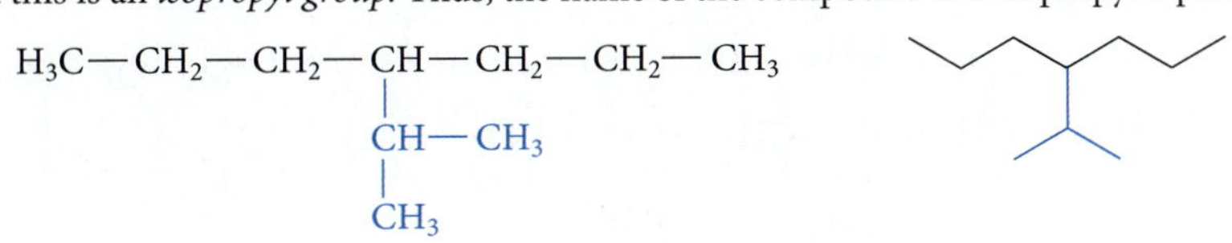

4-isopropylheptane

alkyl group name from Table 2.4

Because this compound has the molecular formula $C_{10}H_{22}$, it is a constitutional isomer of the unbranched alkane *decane*.

7. *If the principal chain contains multiple substituent groups, each substituent receives its own number. The prefixes di, tri, tetra, and so on, are used to indicate the number of identical substituents.*

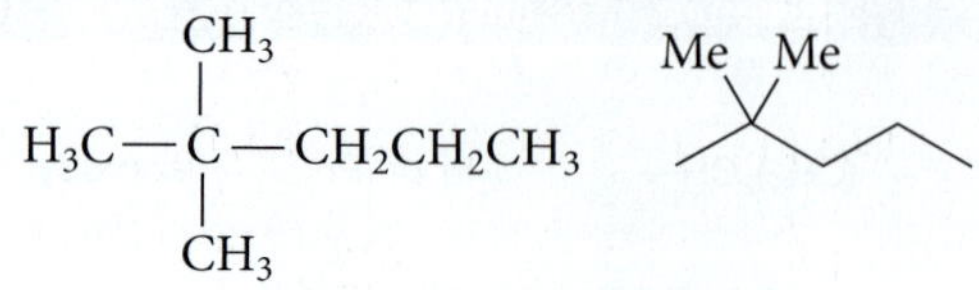

2,2-dimethylpentane

two methyl substituents

shows that both methyl branches are at carbon-2 of the principal chain

Study Problem 2.3

Which two of the following structures represent the same compound? Name the compound.

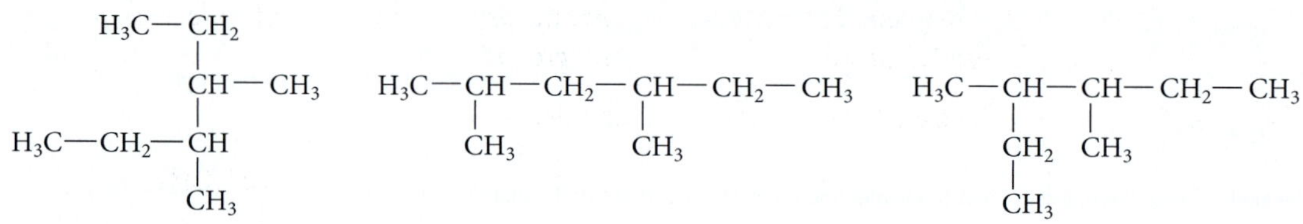

A *B* *C*

Solution The connectivities of both *A* and *C* are the same: [CH_3, CH_2, (CH connected to CH_3), (CH connected to CH_3), CH_2, CH_3]. The compound represented by these structures has six carbons in its principal chain and is therefore named as a hexane. There are methyl branches at carbons 3 and 4. Hence the name is 3,4-dimethylhexane. (You should name compound *B* after you study the next rule.)

Two acceptable skeletal structures are the following. (Others are possible.)

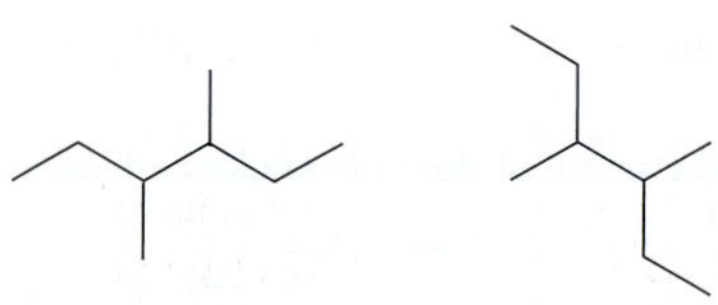

8. *If substituent groups occur at more than one carbon of the principal chain, alternative numbering schemes are compared number by number, and the one is chosen that gives the smaller number* at the first point of difference.

This nomenclature rule can be tricky, so it is best to approach it systematically. To apply this rule, write the two possible numbering schemes derived by numbering from either end of the chain. In the following example, the two schemes are 3,3,5- and 3,5,5-.

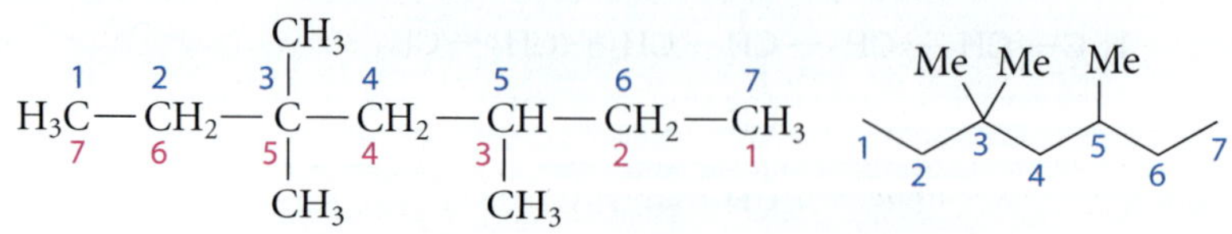

possible names:
3,3,5-trimethylheptane (correct)
3,5,5-trimethylheptane (incorrect)

A decision between the two numbering schemes is made by a *pairwise comparison* of the number sets (3,3,5) and (3,5,5).

How to do a pairwise comparison:

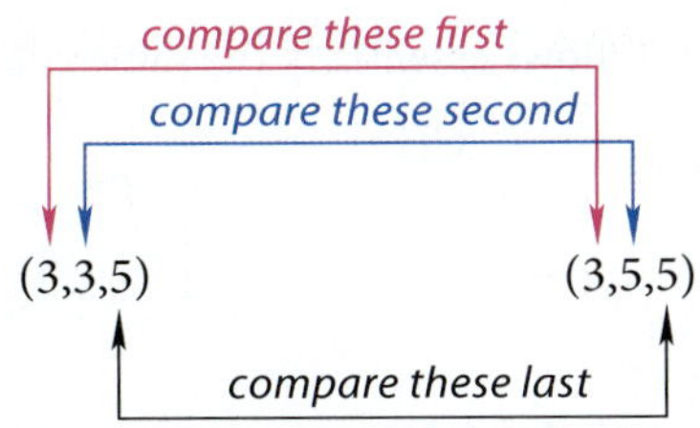

Because the *first point of difference* in these sets occurs at the second pair—3 versus 5—the decision is made at this point, and the first scheme is chosen, because 3 is smaller than 5. If there are differences in the remaining numbers, they are ignored. The sum of the numbers is also irrelevant. Finally, it makes no difference whether the names of the substituents are the same or different; only the numerical locations are used.

The next rule deals with the order in which substituents are listed, or "cited," in the name. Don't confuse the citation order of a substituent with its numerical prefix; they aren't necessarily the same.

9. *Substituent groups are cited in alphabetical order in the name regardless of their location in the principal chain. The numerical prefixes di, tri, etc., as well as the hyphenated prefixes tert- and sec-, are ignored in alphabetizing, but the prefixes iso, neo, and cyclo* (Sec. 2.4D) *are considered in alphabetizing substituent groups.*

The following compounds illustrate the application of this rule:

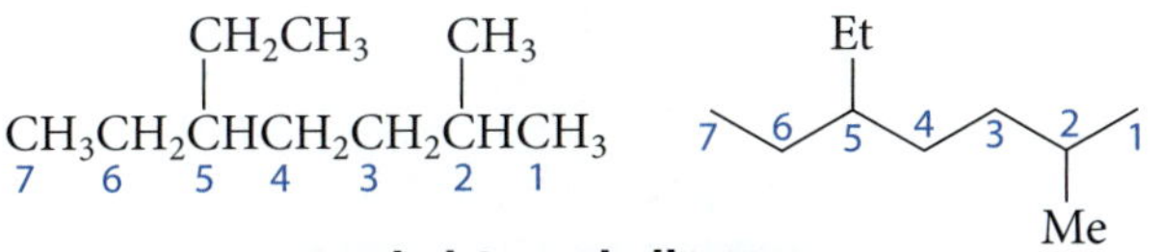

5-ethyl-2-methylheptane
(*ethyl* is cited before *methyl* even though it has a higher number)

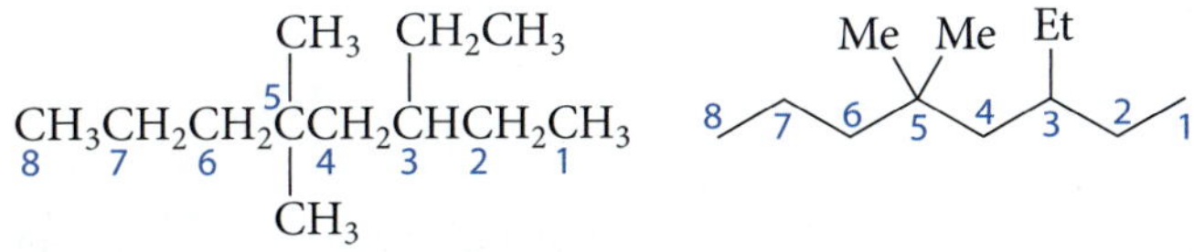

3-ethyl-5,5-dimethyloctane
(*dimethyl* begins with the letter *m* for purposes of citation)

10. *If the numbering of different groups is unresolved by the other rules, the first-cited group receives the lowest number.*

In the following compound, rules 1–9 do not dictate a choice between the names 3-ethyl-5-methylheptane and 5-ethyl-3-methylheptane. Because the ethyl group is cited first in the name (rule 9), it receives the lower number, by rule 10.

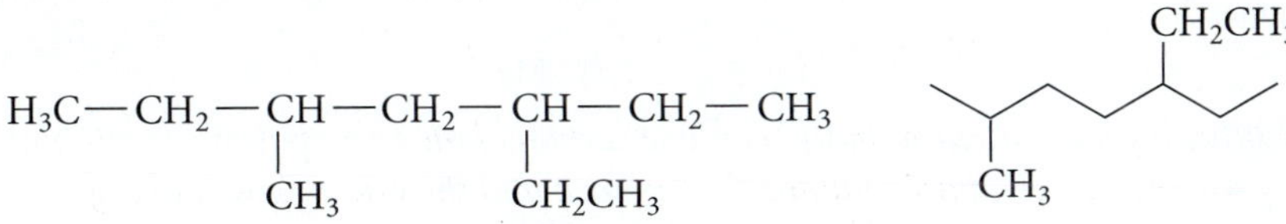

3-ethyl-5-methylheptane

Some situations of greater complexity are not covered by these 10 rules; however, these rules will suffice for most cases.

Focused Problem

2.6 Name the following compounds.

(a) 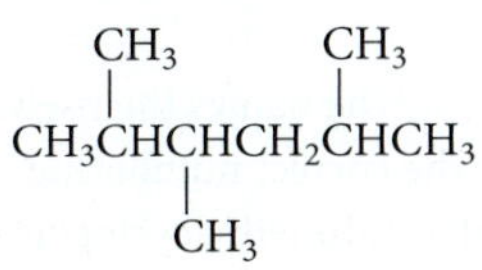

(b) Compound *B* in Study Problem 2.3

(c) 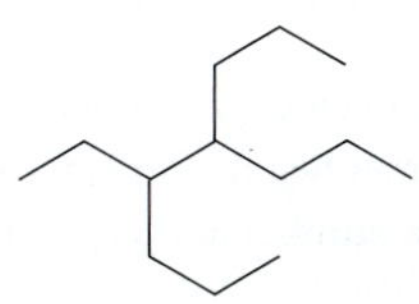

(d) 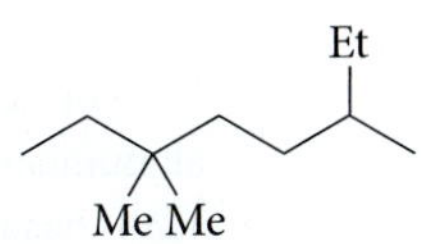

D. Nomenclature of Cycloalkanes

The nomenclature of cycloalkanes follows essentially the same rules used for noncyclic alkanes.

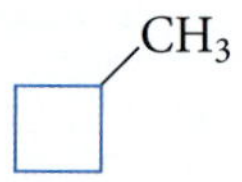

methylcyclobutane

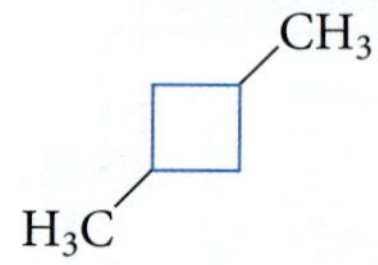

1,3-dimethylcyclobutane

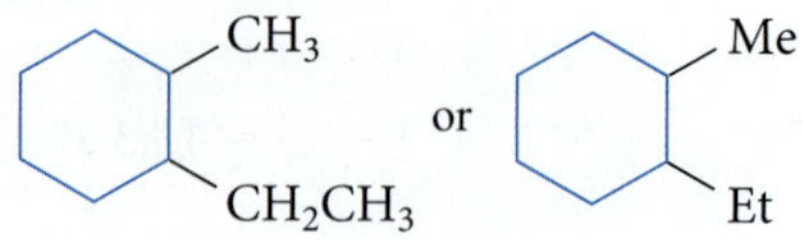

1-ethyl-2-methylcyclohexane
(Note the alphabetical citation, rule 9.)

The numerical prefix 1- is not necessary for cycloalkanes with only one substituent. Thus, the first compound is methylcyclobutane, not 1-methylcyclobutane. Two or more substituents, however, must be numbered to indicate their relative positions. The lowest number, usually 1, is assigned in accordance with the usual rules.

Most of the cyclic compounds used as examples in this text, like those in the preceding examples, involve rings with small alkyl branches. In such cases, the ring is treated as the principal chain. However, when a noncyclic carbon chain contains more carbons than an attached ring, the noncyclic chain is treated as the principal chain and the ring is treated as the substituent. The names of ring substituents are formed in the usual way: replace the *ane* of the hydrocarbon name with *yl*. For example, a cyclopropane ring as a substituent is named *cyclopropyl*; a cyclopentane ring used as a substituent is named *cyclopentyl*.

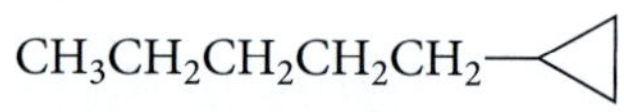

1-cyclopropylpentane
(not pentylcyclopropane)

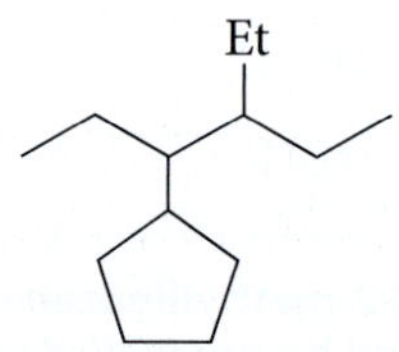

3-cyclopentyl-4-ethylhexane

In the second example, notice that alphabetical citation determines the numbering of the principal chain (rule 10). The prefix *cyclo* is considered in alphabetizing, as noted in rule 9.

Study Problem 2.4

Name the following compound.

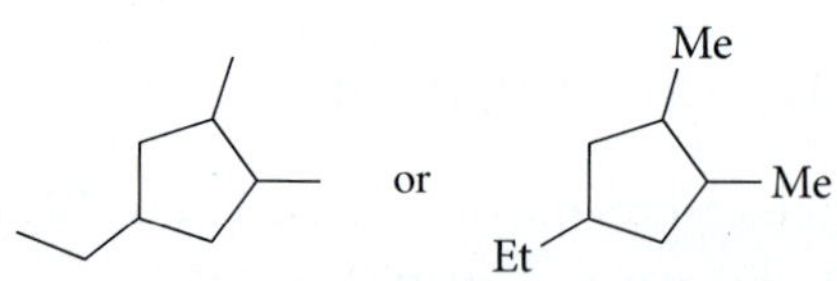

Solution This problem, in addition to illustrating the nomenclature of cyclic alkanes, is a good illustration of rule 8 for nomenclature, the "first point of difference" rule (Sec. 2.4C). The compound is a cyclopentane with two methyl substituents and one ethyl substituent. If we number the ring carbons consecutively, the following numbering schemes (and corresponding names) are possible, depending on which carbon is designated as carbon-1:

1,2,4-	4-ethyl-1,2-dimethylcyclopentane
1,3,4-	1-ethyl-3,4-dimethylcyclopentane
1,3,5-	3-ethyl-1,5-dimethylcyclopentane

The correct name is decided by nomenclature rule 8 using the numbering schemes (*not* the names themselves). Because all numbering schemes begin with 1, the second number must be used to decide on the correct numbering. The scheme 1,2,4- has the lowest number at this point. Consequently, the correct name is 4-ethyl-1,2-dimethylcyclopentane.

Study Problem 2.5

Draw a skeletal structure of *tert*-butylcyclohexane.

Solution The real question in this problem is how to represent a *tert*-butyl group with a skeletal structure. The branched carbon in this group has four other bonds, three of which go to CH_3 groups. Therefore:

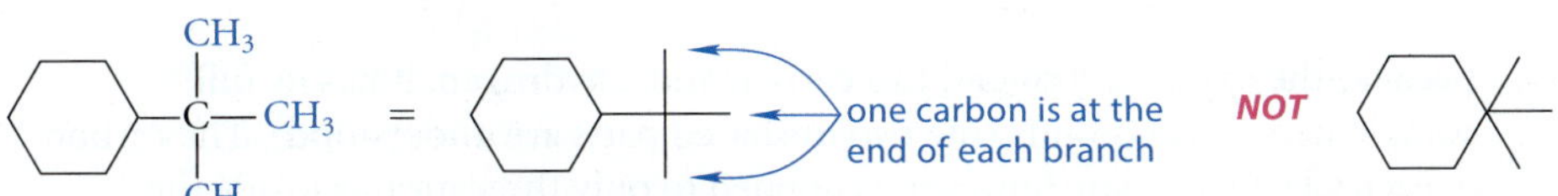

***tert*-butylcyclohexane**
(skeletal structure)

this structure has a carbon with 5 bonds; a bond to the *tert*-butyl group is missing

Focused Problems

2.7 Give the structure of each compound.

(a) A skeletal structure of ethylcyclopentane, in which a text abbreviation is used for the ethyl group

(b) A skeletal structure of 2,6-diethyl-1,1-dimethylcyclohexane

2.8 Give an IUPAC substitutive name for each of the following compounds.

(a)

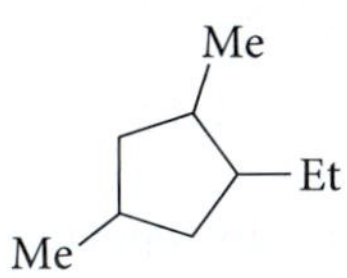

(b)

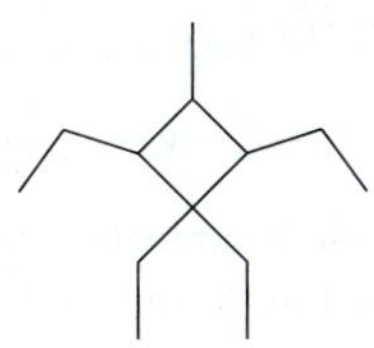

(c)

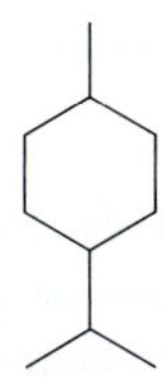

***p*-menthane**
(smells like fennel)

E. Highly Condensed Structures

When space is at a premium, parentheses are sometimes used to form highly condensed structures that can be written on one line, as in the following example.

$(CH_3)_4C$ or $C(CH_3)_4$ means

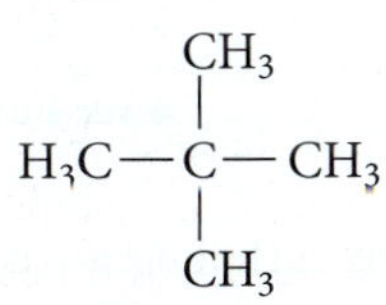

(2.13a)

When such structures are complex, it is sometimes not immediately obvious, particularly to the beginner, which atom inside the parentheses is connected to the atom outside the parentheses, but a little analysis will generally solve the problem. Usually the structure is drawn so that one of the parentheses intervenes between the atoms that are connected (except for attached hydrogens). However, if in doubt, look for the atom within the parentheses that is missing its usual number of bonds. When the group inside the parentheses is CH_3, as in the previous example, the carbon has only three bonds (to the H's). Hence, it must be bonded to the atom outside the parentheses. Consider as another example the CH_2OH groups in the following structure.

Research in student learning strategies has shown that student success in organic chemistry is *highly correlated* with whether a student takes the time to *write out* intermediate steps in a problem. Such steps in many cases involve writing structures and partial structures. Students may be tempted to skip such steps because they see their professors working things out quickly in their heads and perhaps feel that they are expected to do the same. Professors can do this because they have years of experience. Most of them probably gained their expertise through step-by-step problem solving. In some cases, the temptation to skip steps may be a consequence of time pressure. If you are tempted in this direction, remember that a step-by-step approach applied to relatively few problems is a better expenditure of time than rushing through many problems. Study Problem 2.6 illustrates a step-by-step approach to a nomenclature problem.

$(CH_3)_2CH—CH(CH_2OH)_2$ means

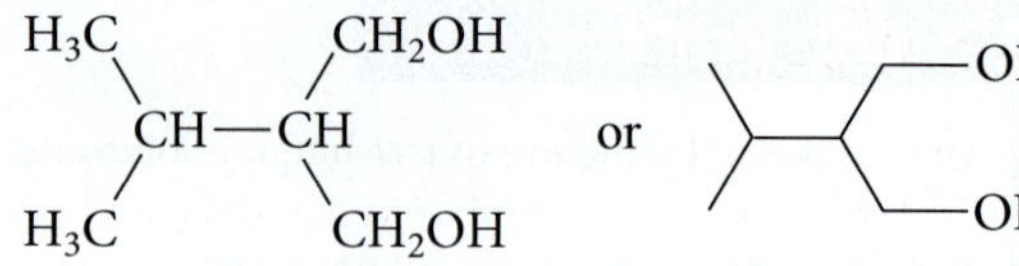

(2.13b)

Because the oxygen is bonded to a carbon and a hydrogen, it has its full complement of two bonds (the two unshared pairs are understood). The carbon of each CH_2OH group, however, is bonded to only three groups inside the parentheses (two H's and the O); hence, it is the atom that is connected to the carbon outside the parentheses.

If the meaning of a condensed structure is not immediately clear, *write it out in a less condensed form.* If you will take the time to do this in a few cases, it should not be long before the interpretation of condensed structures becomes more routine.

Study Problem 2.6

Write the Lewis structure of 4-*sec*-butyl-5-ethyl-3-methyloctane. Then draw a skeletal structure.

Solution Don't try to write out the structure immediately; rather, take a systematic, stepwise approach involving intermediate structures. First, write the principal chain. Because the name ends in octane, the principal chain contains eight carbons. Draw the principal chain without its hydrogen atoms:

C—C—C—C—C—C—C—C

Next, number the chain from either end, and attach the branches indicated in the name at the appropriate positions: a *sec*-butyl group at carbon-4, an ethyl group at carbon-5, and a methyl group at carbon-3. (Use Table 2.4 to learn or relearn the structure of a *sec*-butyl group, if necessary.)

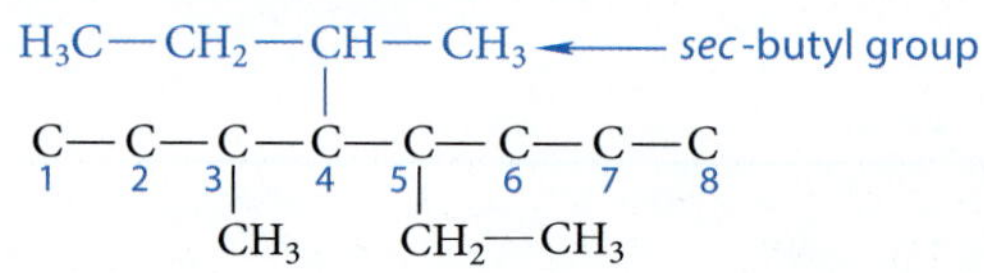

Fill in the missing hydrogens to obtain the complete Lewis structure. Each carbon should have a total of four bonds.

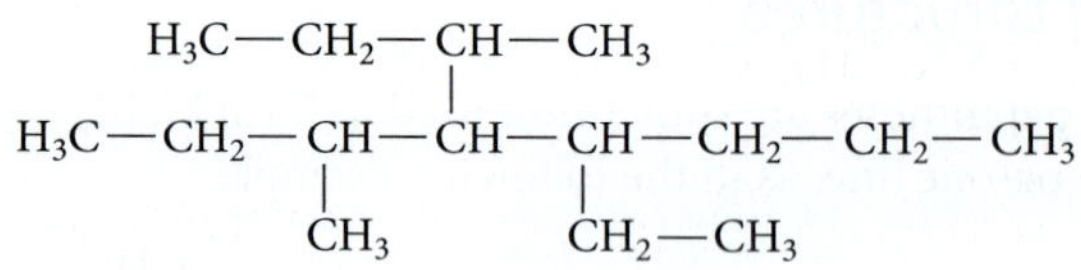

4-*sec*-butyl-5-ethyl-3-methyloctane
(Lewis structure)

Drop the explicitly written carbons and hydrogens, and connect all of the bonds to derive the skeletal structure:

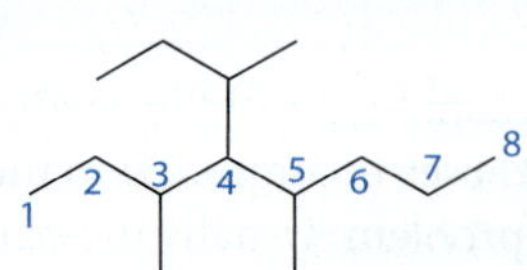

4-*sec*-butyl-5-ethyl-3-methyloctane
(skeletal structure)

Nomenclature and Chemical Indexing

The world's chemical knowledge is housed in the **chemical literature**, which is the collection of books, journals, patents, technical reports, and reviews that constitute the published record of chemical research. To find out what, if anything, is known about an organic compound of interest, we have to search the entire chemical literature. To carry out such a search, organic chemists rely on two major indexes. One is *Chemical Abstracts*, published by the Chemical Abstracts Service of the American Chemical Society, which has been the major index of the entire chemical literature since 1907. The second index is the *Beilstein Handbook of Organic Chemistry*, known to all chemists simply as *Beilstein*, which has published detailed information on organic compounds since 1881. Initially, a search of these indexes was a laborious manual process that could require hours or days in the library. Today, however, both *Chemical Abstracts* and *Beilstein* have efficient search engines (called *SciFinder* and *Reaxys*, respectively) that enable chemists to search for chemical information from a personal computer. Nomenclature plays a key role in locating chemical compounds, particularly in *Chemical Abstracts*, but it is also possible to search for a compound of interest by submitting its structure. A search of *Chemical Abstracts* yields a short summary, called an *abstract*, of every article that references the compound of interest, along with a detailed reference to each article. *Beilstein* yields not only the appropriate references but also detailed summaries of compound properties. You can also search for chemical reactions in both indexes.

Focused Problem

2.9 Draw a Lewis structure for $(CH_3CH_2CH_2)_2CHCH(CH_2CH_3)_2$ in which all carbon–carbon bonds are shown explicitly. Then draw a skeletal structure, and name the compound.

F. Classification of Carbon Substitution

When we begin our study of chemical reactions, it will be important to recognize different types of carbon substitution in branched compounds. A carbon is said to be **primary**, **secondary**, **tertiary**, or **quaternary** when it is bonded to one, two, three, or four other carbons, respectively.

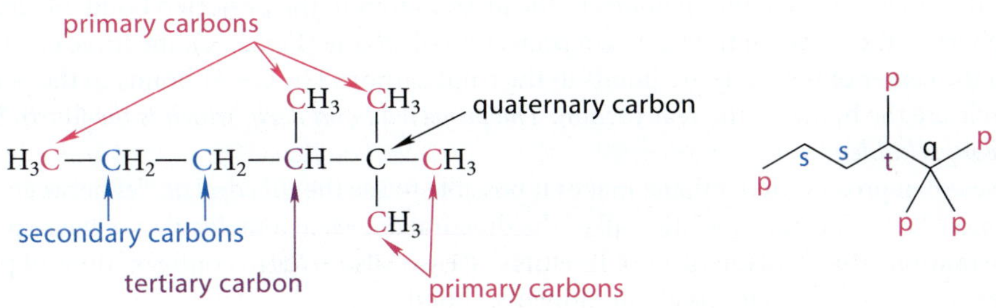

(2.14)

Likewise, the hydrogens bonded to each type of carbon are called primary, secondary, or tertiary hydrogens, respectively.

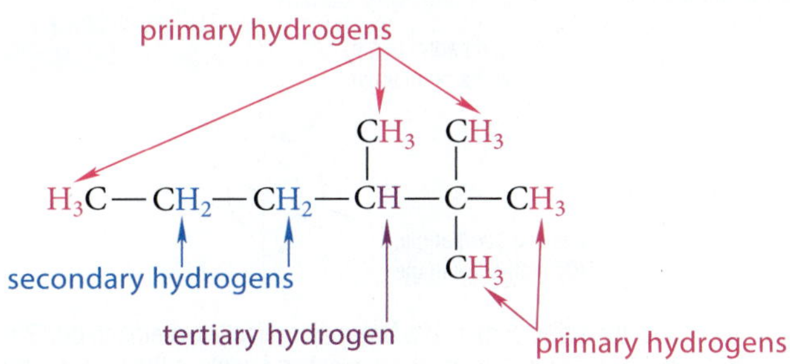

(2.15)

Focused Problems

2.10 (a) Draw a Lewis structure and a skeletal structure of 4-isopropyl-2,4,5-trimethylheptane.

(b) On the skeletal structure, identify the primary, secondary, tertiary, and quaternary carbons.

(c) On the Lewis structure, identify the primary, secondary, and tertiary hydrogens.

2.11 (a) Draw a skeletal structure of 3-isopropyl-1,1-dimethylcyclohexane.

(b) Identify the primary, secondary, tertiary, and quaternary carbons.

(c) Identify the primary, secondary, and tertiary hydrogens.

2.5 CONFORMATIONS OF ALKANES

In Sec. 1.6E, we learned that understanding the structures of many molecules requires that we specify not only their bond lengths and bond angles but also their dihedral angles. There we defined the dihedral angle as the angle between two intersecting planes. In this section, we describe a method to view dihedral angles easily. Then, we use the simple alkanes ethane and butane to develop some widely applicable simple principles that will allow us to predict the dihedral angles in more complex molecules. *You will find that the use of molecular models throughout this section will be very helpful.*

A. Conformations of Ethane

To specify the dihedral angles in ethane, we must define the relationship between the C—H bonds on one carbon and those on the other. A convenient way to do this is to view the molecule in a *Newman projection*, devised by Melvin S. Newman (1908–1993), who was a professor of chemistry at The Ohio State University. A **Newman projection** is a type of planar projection along *one* bond, which we'll call the *projected bond*. For example, suppose we wish to view the ethane molecule in a Newman projection along the carbon–carbon bond, as shown in **Fig. 2.2**. In this projection, the carbon–carbon bond is the projected bond. To draw a Newman projection, start with a circle. The remaining bonds to the *nearer* atom in the projected bond are drawn to the center of the circle. The remaining bonds to the *farther* atom in the projected bond are drawn to the periphery of the circle. In the Newman projection of ethane (Fig. 2.2c), the three C—H bonds drawn to the center of the circle are bonds to the front carbon. The C—H bonds to the periphery of the circle are the bonds to the rear carbon. *The projected bond itself, which is the fourth bond to each carbon, is hidden.*

The Newman projection of ethane makes it possible to see the *dihedral angles* between its C—H bonds. When we have specified all of the dihedral angles in a molecule, we have specified its **conformation**—the location of all of its atoms. We can also refer to conformations of parts of molecules, such as conformations about individual bonds.

Dihedral angles are angles between the bonds on *adjacent* carbons in a Newman projection. *Bond angles*, on the other hand, are the angles between bonds on the *same* carbon.

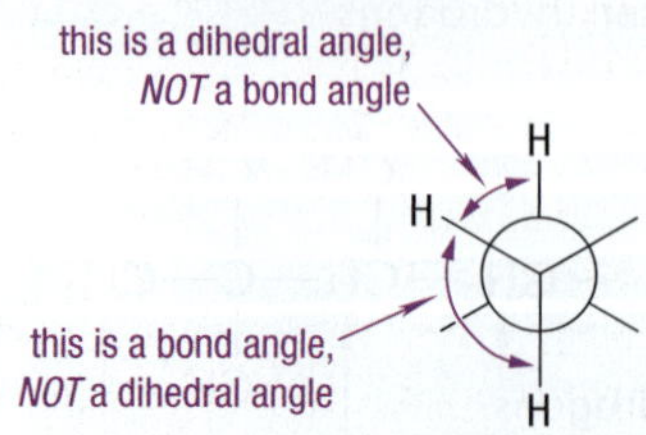

The angle between bonds on the same carbon in a Newman projection seems to be 120° because it is projected onto the plane of the page. As we know, though, the actual bond angle is the tetrahedral angle of 109.5°.

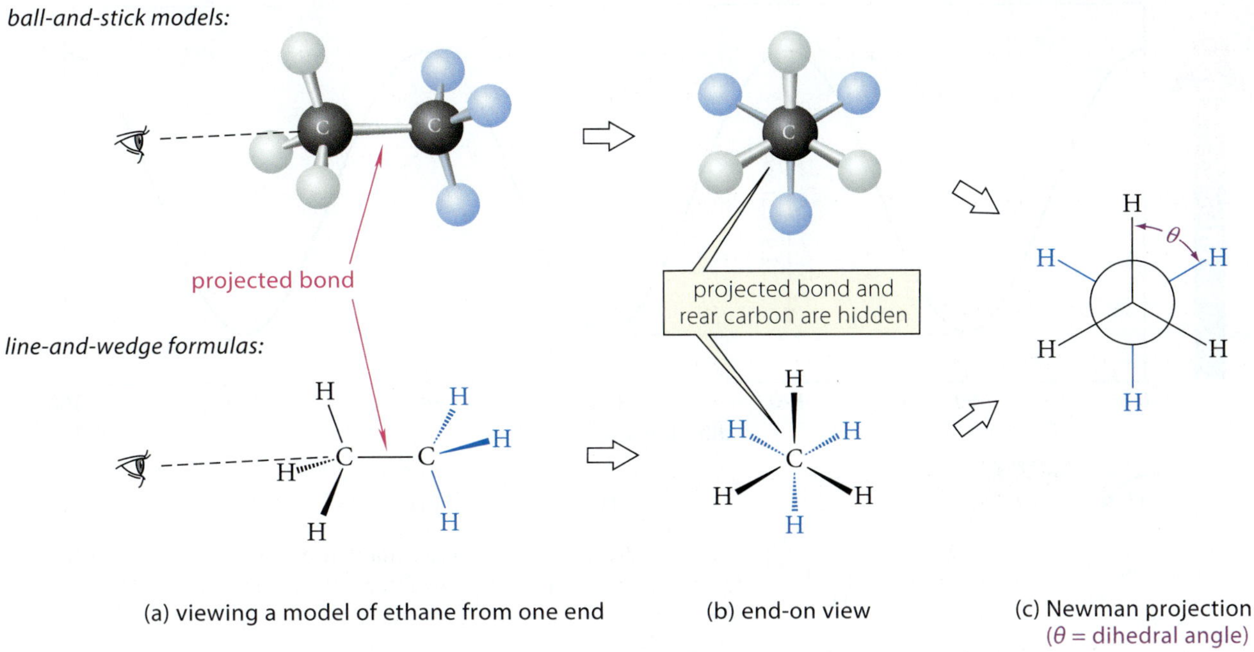

FIGURE 2.2 How to derive a Newman projection for ethane using ball-and-stick models (*top*) and line-and-wedge formulas (*bottom*). (The hydrogens and C—H bonds farthest from the observer are shown in blue.) First view the ethane molecule from the end of the bond you wish to project, as in part (a). The resulting end-on view is shown in part (b). This is represented as a Newman projection (c) in the plane of the page. In the Newman projection, the bonds drawn to the center of the circle are attached to the carbon closer to the observer; the bonds drawn to the periphery of the circle (*blue*) are attached to the carbon farther from the observer. The projected bond (the carbon–carbon bond) is hidden.

Two limiting possibilities for the conformation of ethane can be seen from its Newman projections; these are called the *staggered conformation* and the *eclipsed conformation.*

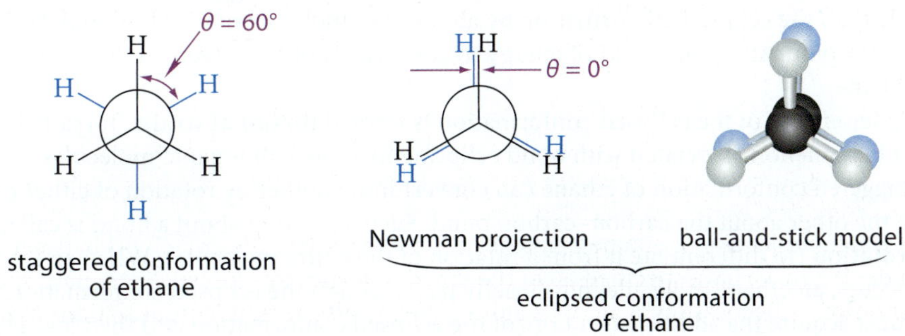

In the **staggered conformation**, a C—H bond of one carbon bisects the angle between two C—H bonds of the other. The smallest dihedral angle in the staggered conformation is 60°. (The other dihedral angles are 180° and 300°.) In the **eclipsed conformation**, the C—H bonds on the respective carbons are superimposed in the Newman projection. The smallest dihedral angle is 0°. (The other dihedral angles are 120° and 240°.) Although conformations intermediate between the staggered and eclipsed conformations are possible, these two conformations will prove to be of central importance.

Which is the preferred conformation of ethane? The energies of the ethane conformations can be described by a plot of relative energy versus dihedral angle, which is shown in **Fig. 2.3**. In this figure, the dihedral angle is the angle between the bonds to the colored hydrogens on the different carbons. To see the relationships in Fig. 2.3, build a model of ethane, hold either carbon fixed, and turn the other carbon about the C—C bond. It doesn't matter which carbon you choose to rotate; in Fig. 2.3, we've arbitrarily chosen to rotate the front carbon. As the angle of rotation changes,

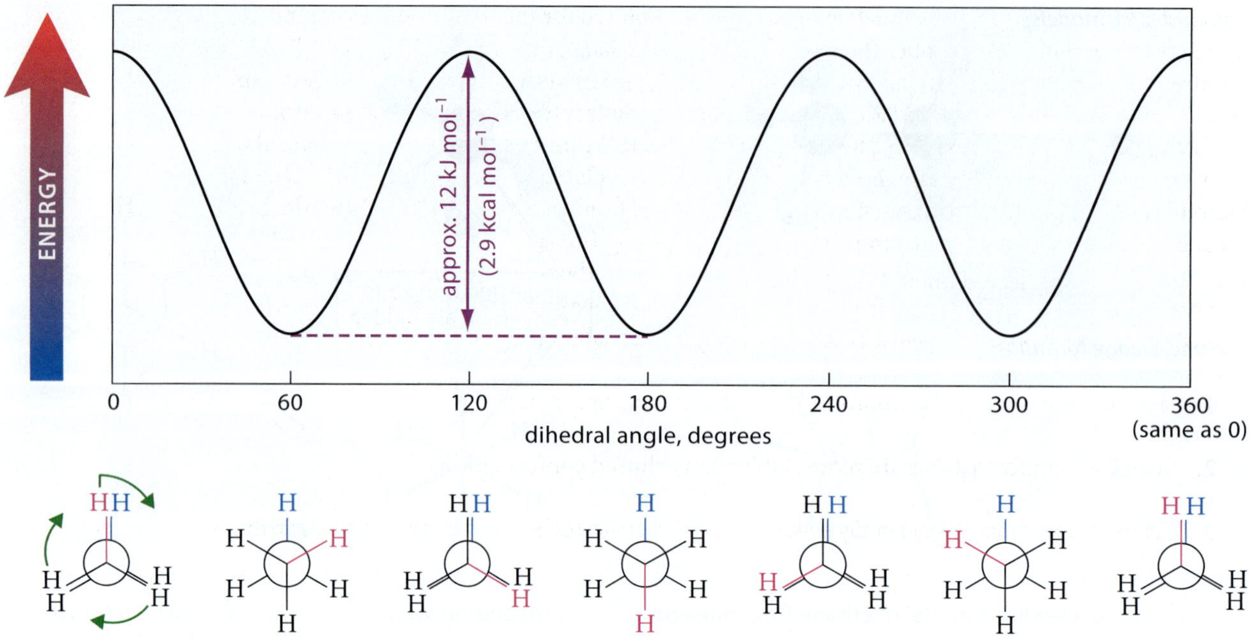

FIGURE 2.3 Variation of energy with a dihedral angle about the carbon–carbon bond of ethane. In this diagram, the rear carbon is held fixed and the front carbon is rotated, as shown by the green arrows. The dihedral angle plotted is the angle between the bonds to the red and blue hydrogens. The staggered conformations are at the energy minima, whereas the eclipsed conformations are at the energy maxima.

the model passes alternately through three identical staggered and three identical eclipsed conformations. As Fig. 2.3 shows, identical conformations have identical energies. Moreover, the eclipsed conformation is characterized by an energy *maximum*, whereas the staggered conformation is characterized by an energy *minimum. The staggered conformation is therefore the more stable conformation of ethane.* The graph shows that the staggered conformation is more stable than the eclipsed conformation by about 12 kJ mol^{-1} (about 2.9 kcal mol^{-1}). This means that it would take about 12 kJ of energy to convert 1 mol of staggered ethane into 1 mol of eclipsed ethane.

Torsional strain has also been called *eclipsing strain* and *Pitzer strain*, after Kenneth S. Pitzer (1914–1997), a professor of chemistry at the University of California, Berkeley, who studied internal rotation. The reasons for the relative stability of the staggered conformation have been debated for years. For example, an early explanation was that the electrons in eclipsed bonds, being closer than electrons in staggered bonds, experience a repulsion that increases the energy of the staggered conformation. More recently, molecular orbital theory has shown that the staggered conformation is more stable than the eclipsed conformation because of favorable orbital interactions in the staggered conformation.

The higher energy of the eclipsed conformation is termed **torsional strain**. Torsional strain is a general phenomenon associated with bond eclipsing in almost all organic molecules.

One staggered conformation of ethane can convert into another by rotation of either carbon relative to the other about the carbon–carbon bond. Such a rotation about a bond is called an **internal rotation** (to differentiate it from a rotation of the entire molecule). When an internal rotation occurs, an ethane molecule must briefly pass through the eclipsed conformation. To do so, it must acquire the additional energy of the eclipsed conformation and then lose it again. What is the source of this energy?

At temperatures above absolute zero, molecules are in constant motion and therefore have kinetic energy. Heat is a manifestation of this kinetic energy. In a sample of ethane, the molecules move about in a random manner, much as people might mill about in a large crowd. These moving molecules frequently collide, and molecules can gain or lose energy in such collisions. (An analogy is the collision of a bat with a ball; some of the kinetic energy of the bat is lost to the ball.) When an ethane molecule gains sufficient energy from a collision, it can undergo internal rotation, passing through the more energetic eclipsed conformation into another staggered conformation. Whether a given ethane molecule acquires sufficient energy to undergo an internal rotation is strictly a matter of *probability* (random chance). However, an internal rotation is *more probable* at higher temperature because warmer molecules have greater kinetic energy.

The probability that ethane undergoes internal rotation is reflected in its *rate* of rotation—how many times per second the molecule converts from one staggered conformation into another.

This rate is determined by how much energy must be acquired for the rotation to occur: the more energy required, the smaller the rate. In the case of ethane, 12 kJ mol^{-1} (2.9 kcal mol^{-1}) is required. This amount of energy is small enough that the internal rotation of ethane is very rapid even at very low temperatures. At 25 °C, a typical ethane molecule undergoes a rotation from one staggered conformation to another at a rate of about 10^{11} times per second! This means that the interconversion between staggered conformations takes place about once every 10^{-11} second. Despite this short lifetime for any one staggered conformation, an ethane molecule spends most of its time in its staggered conformations, passing only transiently through its eclipsed conformations. Thus, an internal rotation is best characterized not as a continuous spinning but as a constant succession of jumps from one staggered conformation to another.

Let's summarize what we have learned by studying internal rotation in ethane:

1. Internal rotation about carbon–carbon single bonds occurs rapidly.
2. Staggered conformations are more stable than eclipsed conformations.
3. Eclipsed conformations generally exist only as transient states between staggered conformations.

B. Conformations of Butane

Butane contains two distinguishable types of carbon–carbon bonds: the two terminal C—C bonds (*blue*) and the central C—C bond (*red*).

$$H_3C—CH_2—CH_2—CH_3$$

butane
two types of C—C bonds

Let's consider internal rotation about the central C—C bond. Although this rotation is a bit more complex than the ethane case, examination of this rotation leads to important new insights about molecular conformation. As with ethane, we use Newman projections, as shown in **Fig. 2.4**.

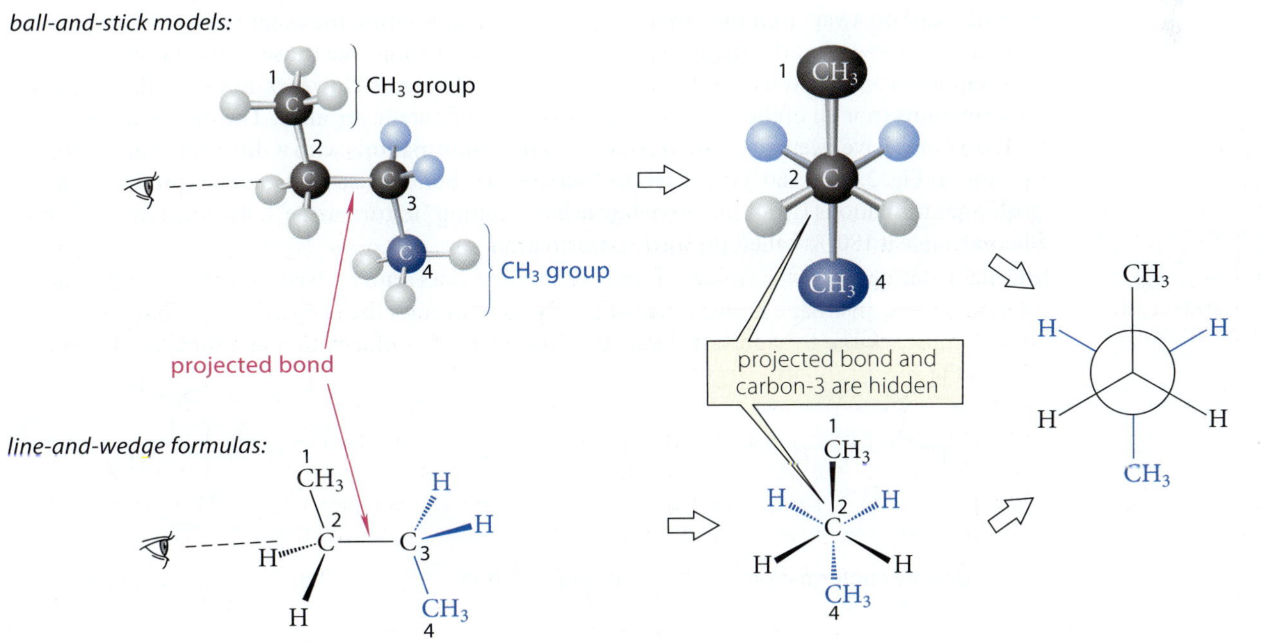

(a) viewing a model of butane from one end of the central carbon–carbon bond

(b) end-on view

(c) Newman projection

FIGURE 2.4 How to derive the Newman projection of the central carbon–carbon bond in butane using ball-and-stick models (*top*) and line-and-wedge formulas (*bottom*). *It is important to follow this description with your models.* The bonds and groups on the rear carbon of the projected bond are shown in blue. (Only one of the butane conformations is shown.)

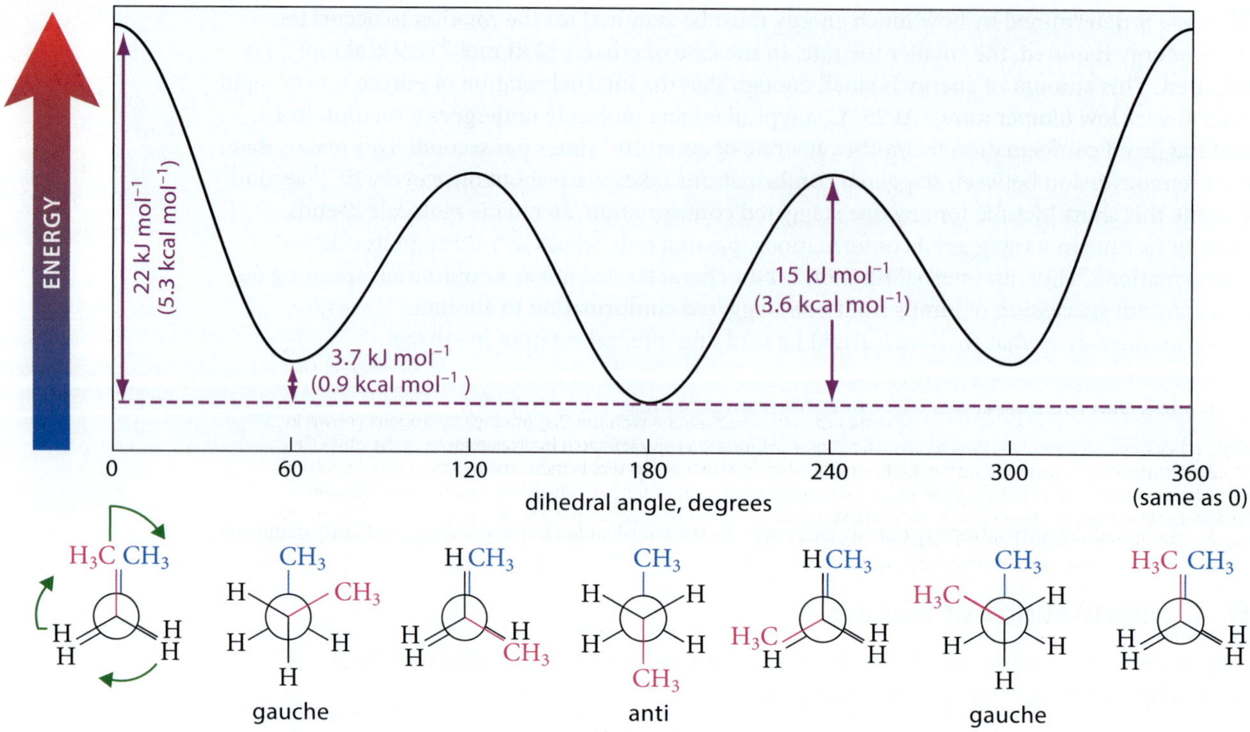

FIGURE 2.5 Variation of energy with a dihedral angle about the central carbon–carbon bond of butane. In this diagram, the rear carbon is held fixed and the front carbon is rotated, as shown by the green arrows. The dihedral angle plotted is the one between the bonds to the two CH_3 groups.

Remember again that the projected bond—the central C—C bond in this case—is hidden in the Newman projection.

The graph of energy as a function of dihedral angle in butane is given in **Fig. 2.5**. As with Fig. 2.3, the various rotational possibilities are generated with a model by holding either carbon fixed (the carbon away from the observer in Fig. 2.5) and rotating the other one.

Figure 2.5 shows that the staggered conformations of butane, like those of ethane, are at energy minima and are thus the stable conformations of butane. However, not all of the staggered conformations (nor all of the eclipsed conformations) of butane are alike. The different staggered conformations have been given special names. The conformations with a dihedral angle of 60° and 300° in Fig. 2.5 (or ±60°) between the two C—CH_3 bonds are called **gauche** (pronounced "gōsh") conformations (from the French *gauchir,* meaning "to turn aside"); the form in which the dihedral angle is 180° is called the **anti** conformation.

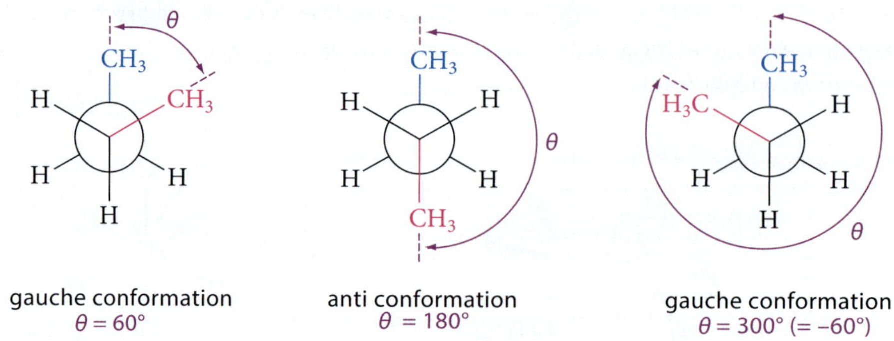

The relationship between bonds also can be described with the terms *gauche* and *anti.* Two bonds that have a dihedral relationship of ±60° are said to be **gauche bonds**. Two bonds that have a dihedral relationship of 180° are said to be **anti bonds**. These terms refer to bonds on *adjacent* carbons.

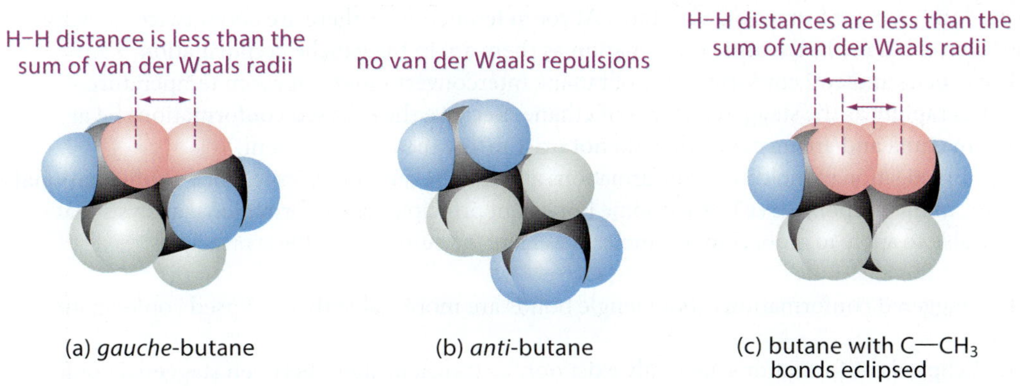

FIGURE 2.6 Space-filling models of different butane conformations with the CH_3 (methyl) hydrogens shown in color. (a) *Gauche*-butane. A hydrogen atom from one CH_3 group is so close to a hydrogen atom of the other CH_3 group that these hydrogens, shown in pink, violate each other's van der Waals radii. The resulting van der Waals repulsions cause *gauche*-butane to have a higher energy than *anti*-butane, in which this interaction is absent. (b) *Anti*-butane. This conformation is most stable because it contains no van der Waals repulsions. (c) Butane with the C—CH_3 bonds eclipsed. In this conformation, van der Waals repulsions between the hydrogens of the two CH_3 groups (*pink*) are even greater than they are in *gauche*-butane.

Figure 2.5 shows that the gauche and anti conformations of butane have different energies. The anti conformation is more stable by 3.7 kJ mol^{-1} (0.9 kcal mol^{-1}). The gauche conformation is less stable because the CH_3 groups are very close together—so close that the hydrogens on the two groups occupy each other's space. You can see this with the aid of the space-filling model in **Fig. 2.6a**.

This problem can be discussed more precisely in terms of atomic size. You learned about *covalent radius* in Sec. 1.6B as an indication of the size of *bonded* atoms. The measure of a *nonbonded* atom's effective size is its **van der Waals radius.** Van der Waals radii are significantly larger than covalent radii. Energy is required to force two *nonbonded* atoms together more closely than the sum of their van der Waals radii. The reason is that repulsions, called **van der Waals repulsions,** occur between the valence-electron clouds of the two atoms as they are forced together. Because the van der Waals radius of a hydrogen atom is about 120 pm, forcing the centers of two nonbonded hydrogens to be closer than twice this distance requires energy. Furthermore, the more the two hydrogens are pushed together, the more energy is required. Thus, to attain the gauche conformation, butane must acquire more energy caused by van der Waals repulsions between hydrogens. In other words, *gauche-butane is destabilized by van der Waals repulsions between nonbonded hydrogens on the two CH_3 groups.* Such van der Waals repulsions are absent in *anti*-butane (see **Fig. 2.6b**). Therefore, *anti*-butane is more stable than *gauche*-butane.

As with ethane, the eclipsed conformations of butane are destabilized by torsional strain. But, in the conformation in which the two C—CH_3 bonds are eclipsed, the major source of instability is van der Waals repulsions between the methyl hydrogens (**Fig. 2.6c**), which are forced to be even closer than they are in the gauche conformation. This is the most unstable of the eclipsed conformations (0° in Fig. 2.5).

FURTHER EXPLORATION 2.1 Atomic Radii and van der Waals Repulsion

Don't confuse *torsional strain* and *van der Waals repulsions. Torsional strain* is the additional energy of eclipsed conformations. *Van der Waals repulsions* are caused by atoms violating each other's van der Waals radii. These two sources of instability are different; but, because both can occur together, they can be confused. The energies of the gauche conformations of butane are raised only by van der Waals repulsions. The energies of the eclipsed conformations of ethane are raised only by torsional strain. The energies of the eclipsed conformations of butane are raised by *both* torsional strain and van der Waals repulsions.

It is important to understand the relative energies of the butane conformations because, when different stable conformations are in equilibrium, *the most stable conformation—the conformation of lowest energy—is present in the greatest amount.* Thus, the anti conformation of butane is the predominant conformation of butane. At room temperature, there are about twice as many molecules of butane in the anti conformation as there are in the gauche conformation.

The gauche and anti conformations of butane interconvert rapidly at room temperature—almost as rapidly as the staggered forms of ethane. Because the eclipsed conformations lie at energy maxima and are unstable, they do not exist to any measurable extent.

The investigation of molecular conformations and their relative energies is called **conformational analysis**. In this section, we've learned some important principles of conformational analysis that we'll be able to apply to more complex molecules. Here is a summary of these principles:

1. Staggered conformations about single bonds are more stable than eclipsed conformations.

2. Eclipsed conformations generally exist only as transient states between staggered conformations.

3. If staggered conformations have different energies, the most stable conformation is present in the greatest amount. The greater the difference in stability, the greater is the disparity of the amounts present.

4. Van der Waals repulsions (repulsions between nonbonded atoms) occur when atoms are forced to be closer than the sum of their van der Waals radii. Van der Waals repulsions raise the energy of any conformation (or molecule) in which they are present.

5. Van der Waals repulsions are greater between large groups than they are between smaller groups. Van der Waals repulsions between two large groups are greater when the groups are eclipsed than when the groups are staggered.

6. Rotation about C—C single bonds in most cases is so rapid that it is hard to imagine separating conformations except at very low temperature. The energy difference between a staggered conformation and its adjacent eclipsed conformation in an energy diagram determines the rate at which staggered conformations interconvert: the greater the energy difference, the lower the rate of interconversion.

Study Problem 2.7

Draw a Newman projection for the anti conformation about the C3–C4 bond of 2-methylhexane, viewing the bond so that C3 is nearest the observer.

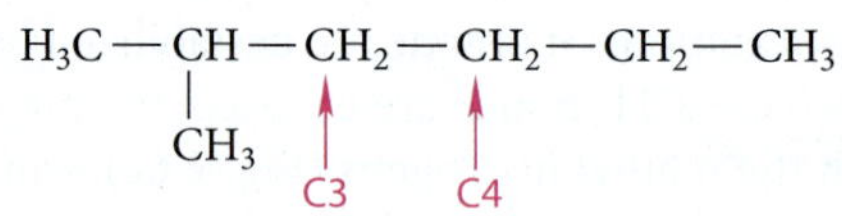

2-methylhexane

Solution First draw a "blank" Newman projection to represent the projected bond. Remember that the projected bond itself (the C3–C4 bond) is invisible in the projection. Either template below can be used.

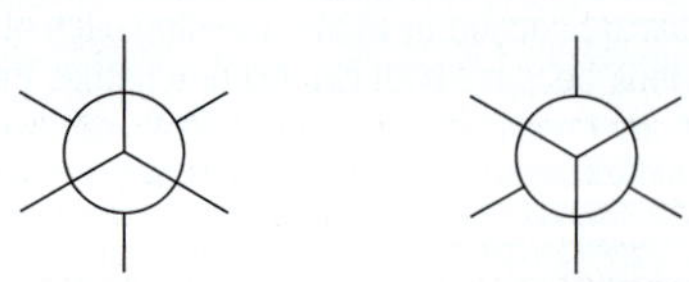

We arbitrarily pick the template on the left. In the view dictated by the problem, the front carbon is C3. Identify the three groups attached to C3 with bonds other than the projected bond. These groups are H, H, and the $-\underset{\displaystyle CH_3}{\underset{|}{CH}}-CH_3$ group. Put these on the *front* carbon of the Newman projection. *It doesn't matter which bonds go to which groups as long*

as all groups are on the front carbon. Because we are not examining the bonds within the large group, we can condense this group to —$CH(CH_3)_2$ or even —*i*Pr.

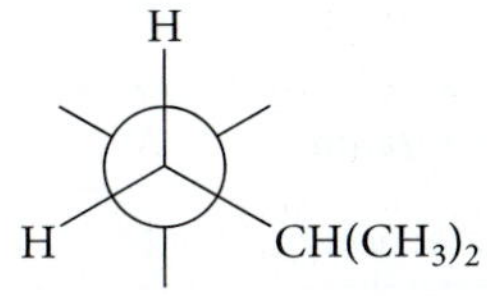

We then identify the groups attached to the back carbon (C4) by bonds *other than* the projected bond. These groups are H, H, and CH_3CH_2—. Now it *does* matter where we put these groups, because we are asked for the anti conformation. The CH_3CH_2— group must be placed anti to the —$CH(CH_3)_2$ group. Remember that "anti" means a dihedral angle of 180°.

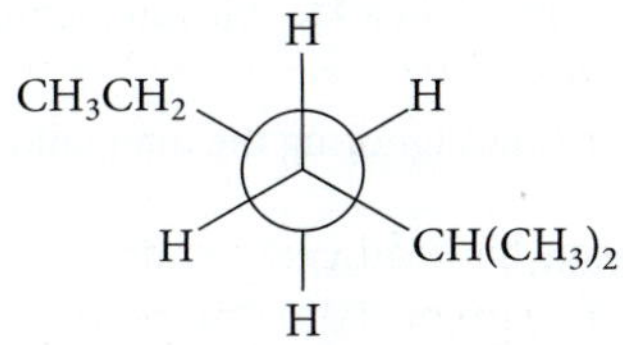

2-methylhexane
anti conformation about the C3–C4 bond

Remember that Newman projections are used to examine conformations about *a particular bond.* If we want to examine the conformations about several different bonds, we must draw a different set of Newman projections for each bond.

Study Problem 2.8 shows how to construct a conformational energy diagram.

Study Problem 2.8

Show all staggered and eclipsed conformations of 2,3,4-trimethylpentane about the C3–C4 bond, with C4 nearest the observer. Rank these conformations in order of increasing energy. Then use your rankings to construct a conformational energy diagram for internal rotation about this bond. Which conformation(s) are present in the greatest concentration at equilibrium?

Solution The first task is to draw the structure of the compound and identify the projected bond.

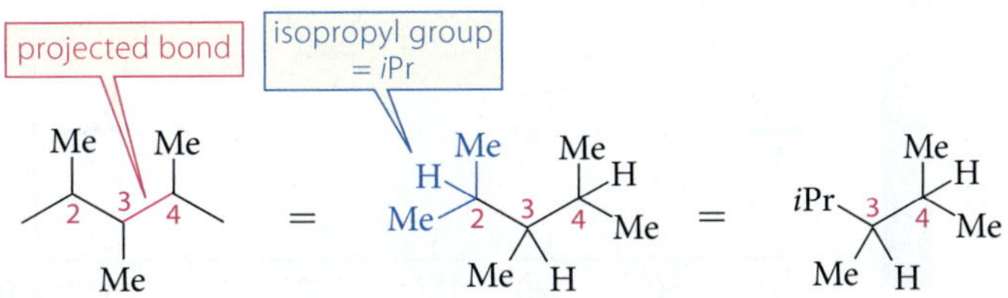

2,3,4-trimethylpentane

(The hydrogens are shown, and an isopropyl group is abbreviated, for convenience in constructing the Newman projections. If you build a model you can use a single colored ball for the isopropyl group.) Start with any one of the conformations, and then derive the others by successive 60° rotations of either C3 or C4. (We rotate the front carbon [C4]; the bonds and substituents to this carbon are colored blue.)

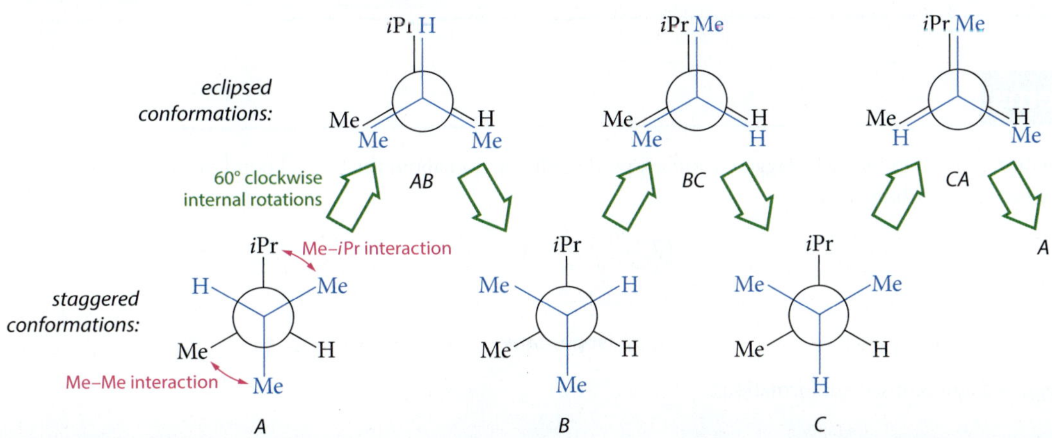

To rank the energies, first consider the staggered conformations. List the gauche interactions in each conformation. (We've identified two of these in the Newman projections in the previous diagram; you should identify the others; ignore interactions involving hydrogens.)

Conformation:	*A*	*B*	*C*
Me–Me gauche interactions:	1	2	1
Me–*i*Pr gauche interactions:	1	1	2
Relative energy:	low	medium	high

A methyl–*i*Pr gauche interaction has higher energy than a Me–Me interaction, because *i*Pr is larger than methyl. Therefore, in terms of energy, $A < B < C$.

Next, do the same for the eclipsed conformations, ignoring the interactions involving single hydrogen atoms:

Conformation:	*AB*	*BC*	*CA*
Me–Me eclipsed interactions:	1	1	0
Me–*i*Pr eclipsed interactions:	0	1	1
Relative energy:	low	high	medium

Assume that eclipsed conformations have greater energies than staggered conformations, not only because of torsional strain, but also because, when large groups are eclipsed, they are forced to a closer approach than when they are gauche. We can then rank the conformations by relative energy:

$$A < B < C \ll AB < CA < BC$$

On a graph of potential energy versus angle of internal rotation, mark the relative energies and connect them with a smooth curve. (You are not expected to provide a numerical value for the relative energies; therefore, the heights of the relative energy marks are arbitrary, as long as they are in the correct order.)

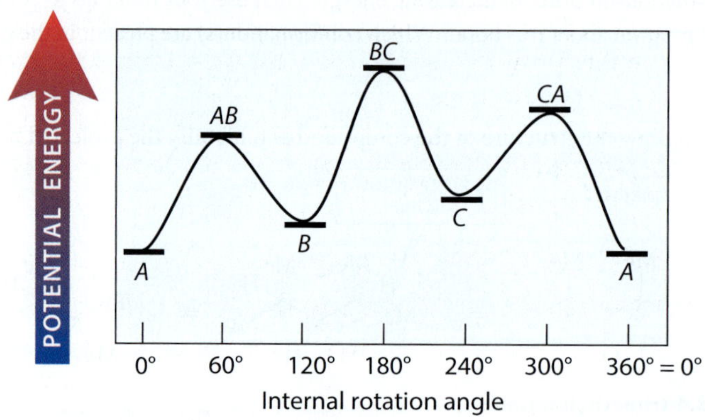

Conformation *A* is present in the greatest amount, conformation *B* in next greatest amount, and conformation *C* in the least amount. (The exact percentages depend on the energy differences.) The eclipsed conformations do not have any significant lifetime.

Focused Problems

2.12 (a) Draw a Newman projection for each staggered and eclipsed conformation about the C2–C3 bond of isopentane, a compound containing a branched carbon chain.

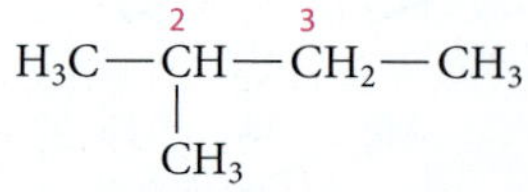

isopentane

Show all staggered and eclipsed conformations.

(b) Sketch a curve of potential energy versus dihedral angle for isopentane, similar to that of butane in Fig. 2.5. Label each energy maximum and minimum with one of the conformations you drew in part (a).

(c) Which of the conformations you drew in part (a) are likely to be present in the greatest amount in a sample of isopentane? Explain.

2.13 Repeat the analysis in Focused Problem 2.12 for either of the two terminal C—C bonds of butane.

C. Depicting Conformations

The previous section showed that Newman projections can be used to represent specific conformations. This section describes some other ways of drawing specific conformations of molecules. At the end of this section, you will know three ways of drawing specific conformations:

1. Newman projections (already learned)
2. Line-and-wedge structures
3. Sawhorse projections

These last two methods are essentially ways of showing perspective in our structures, and we want to be able to do this systematically without attending art school! We illustrate each of these using the anti conformation of butane with which you are already familiar. *You should approach the material that follows with models in hand.*

Line-and-wedge structures (sometimes called **dash–wedge structures**) are extensions of the technique introduced in Sec. 1.6D for drawing the bonds to tetrahedral carbon. As before, we use lines to represent bonds in the plane of the page, solid wedges to represent bonds emerging in front of the page, and dashed wedges to represent bonds receding behind the page. Now, however, we explain how to do this for two or more carbons simultaneously. To draw a line-and-wedge structure, we draw the bond of interest and two of the adjacent bonds (in this case, the carbon–CH_3 bonds) in the plane of the page as if we're looking at the structure "side-on." If you look at a model this way, you'll see that the hydrogens behind the page are obscured by the hydrogens in front (view *A* in Display 2.16).

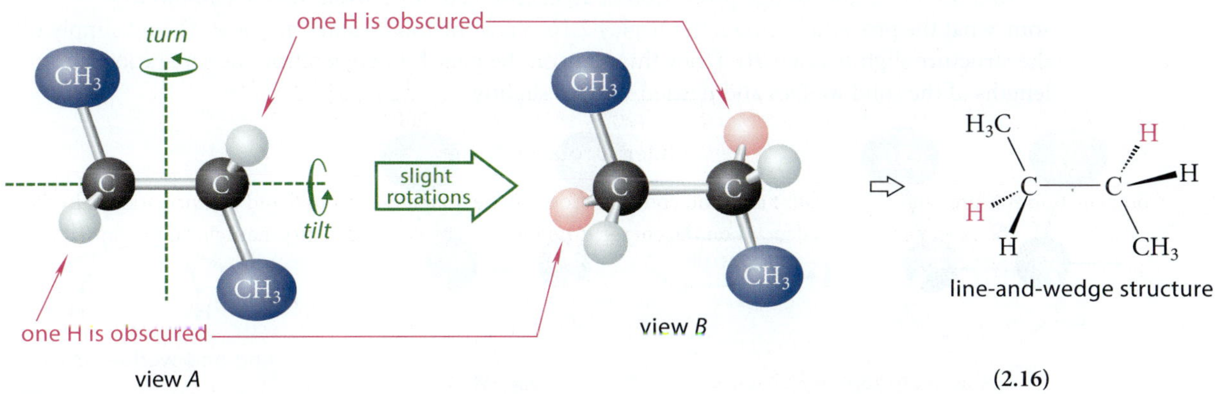

(2.16)

We then tilt and turn the model slightly as shown by the green arrows to give view *B*, and then draw this model using dashed wedges for bonds receding behind the page, solid wedges for bonds emerging from the page, and ordinary lines for bonds in the page. The "tilt and turn" about either axis can be in whichever direction brings all of the groups clearly into view, but for consistency, we'll turn the model as shown in Display 2.16 in most cases. The hydrogens that are obscured in view *A* are colored pink in view *B*. Notice the relationships between the solid- wedged and dashed-wedge bonds. In the view shown in Display 2.16, for example, the dashed-wedge (receding) bonds are above the solid-wedged (emerging) bonds on the page.

Viewing the line-and-wedge structure (or the corresponding model) end-on and projecting this view (flattening it) onto the page gives the familiar Newman projection.

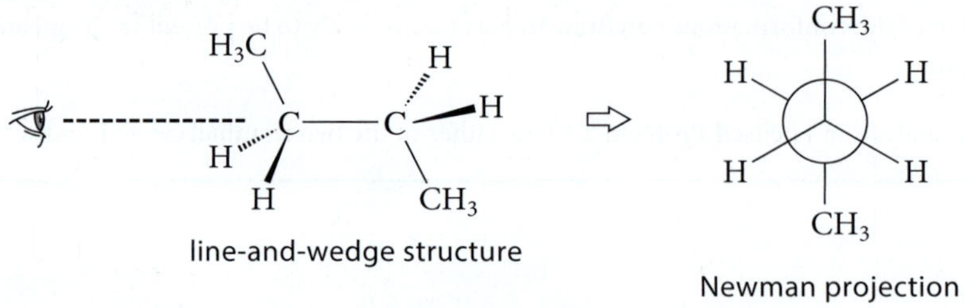

(2.17)

This procedure was also shown in more detail in Fig. 2.2 (Sec. 2.5A).

A **sawhorse projection** is a view of the projected bond as if we were looking at the model from above and to the right. We'll illustrate sawhorse projections with the same *anti*-butane we've been using. To draw a sawhorse projection, follow these steps:

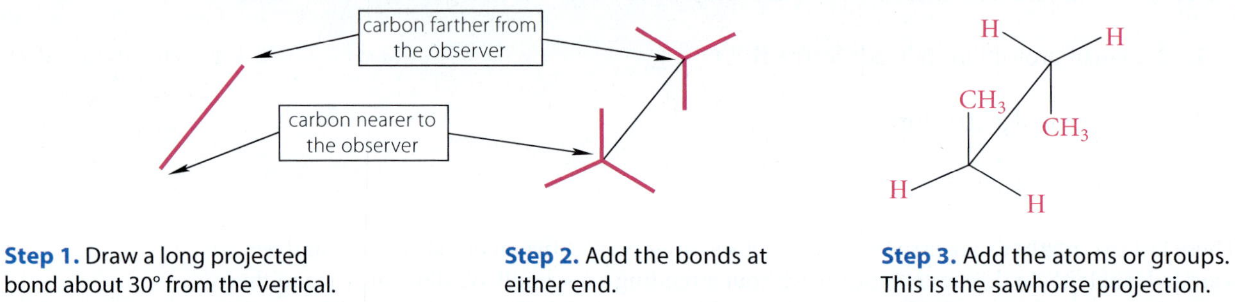

(2.18)

Some people like to draw sawhorse projections with a wedged projected bond to emphasize the perspective, but this perspective is understood whether or not the bond is wedged.

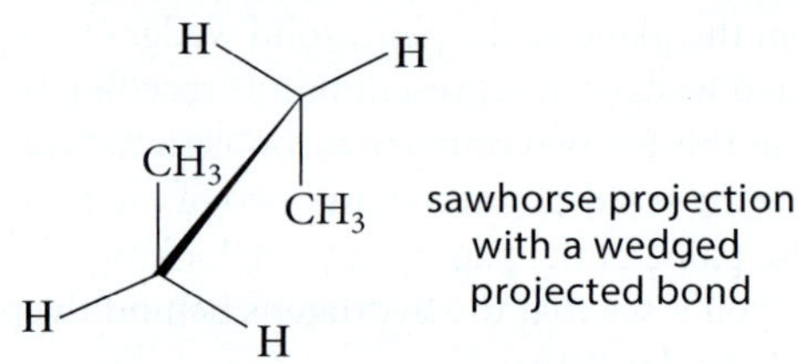

To draw a line-and-wedge projection of an eclipsed conformation, we have to modify somewhat the procedure shown in Display 2.16. View the model side-on (view *A*), and simply tilt the structure slightly (view *B*). Draw this view on the page by exaggerating the separations and lengths of the solid wedges and dashed wedges slightly.

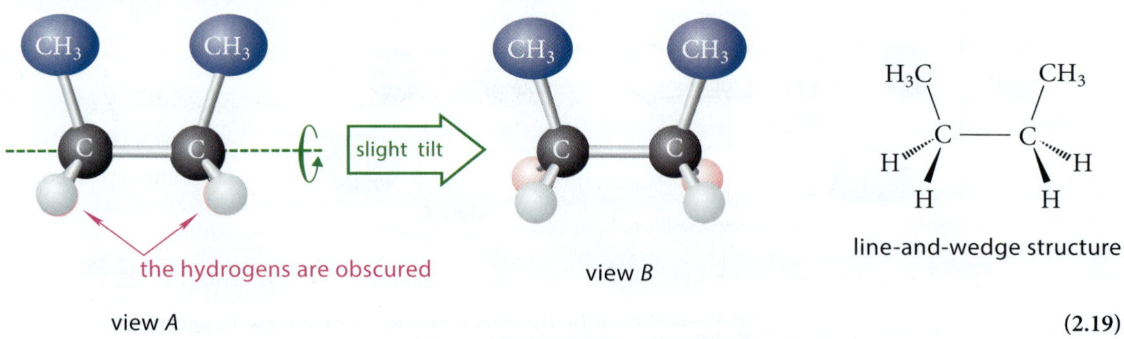

(2.19)

The procedures for drawing sawhorse and Newman projections of eclipsed conformations are the same as for staggered conformations.

Typically, we focus on one particular bond (the projected bond) when drawing conformations, but it is possible to draw line-and-wedge structures for the conformations about several bonds simultaneously, *provided that all of the projected bonds can be drawn in the plane of the page*. This is the case in the following all-anti conformation of hexane:

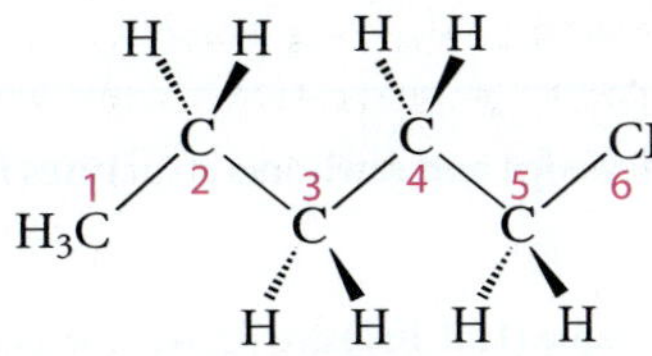

hexane
anti conformations about
bonds C2–C3, C3–C4, and C4–C5 (2.20)

All of the carbon–carbon bonds are in the plane of the page—that is, all of them are represented by lines. This allows the hydrogens to be represented with dashed and solid wedges.

We can also depict the result of an internal rotation of the C4–C5 bond to give a gauche conformation. In this case, one of the C5–H bonds is in the plane of the page.

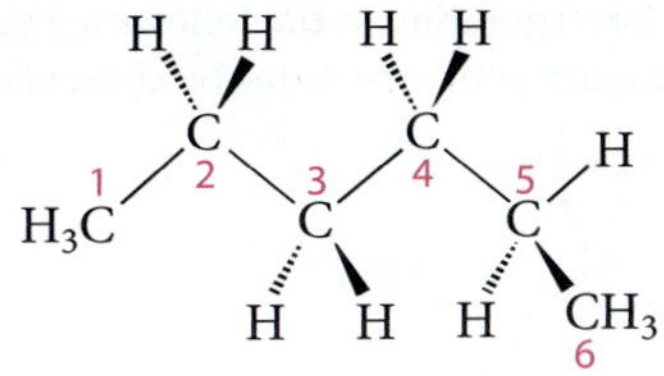

hexane
anti conformations about
bonds C2–C3 and C3–C4,
and gauche conformation about C4–C5 (2.21)

(We could depict a similar rotation about C2–C3.)

What we shouldn't do in such a line-and-wedge structure, however, is to show the rotation about C3–C4, because the C4–C5 bond would then become a dashed or solid wedge, and some of the bonds attached to C5 would also be wedges that are relative to a different plane. In other words, *you should avoid wedged bonds emanating from other wedged bonds.* (If we eliminated the lines and wedges at C5 by summarizing the C5–C6 group as "Et" or "C_2H_5," then we could rotate about C3–C4.) When constructing a multibond line-and-wedge structure, *every carbon atom in the backbone should have two bonds that are lines (in the plane of the page), one dashed-wedge bond, and one solid-wedge bond.*

Two additional rules: (1) When drawing line-and-wedge structures, the in-plane (line) bonds to a carbon should join at a bent (approximately tetrahedral) angle. (2) Wedged bonds that come from the same carbon should both emanate from the *convex side* of the junction. Only if these rules are followed do we obtain a meaningful perspective for the tetrahedral geometry of carbon. Here are some incorrect and correct line-wedge structures.

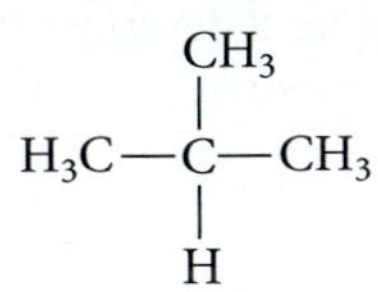

isobutane
(Lewis structure
not showing conformation)

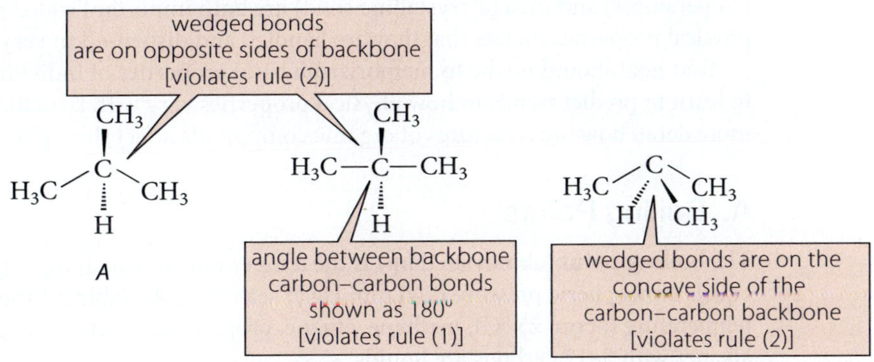

A B C

incorrect line-and-wedge structures

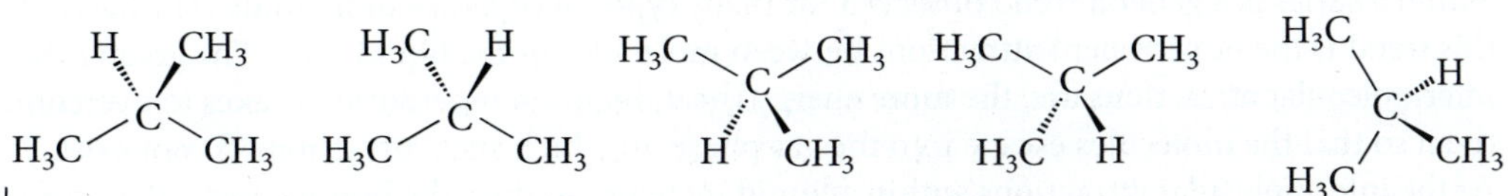

correct line-and-wedge structures

Focused Problems

2.14 Use the procedures described in this section to draw the line-and-wedge and sawhorse structures for both gauche conformations of butane about the C2–C3 bond. Don't hesitate to use models.

2.15 Which of the following line-and-wedge structures of dibromomethane (H_2CBr_2) are drawn correctly, and which are drawn incorrectly? Explain.

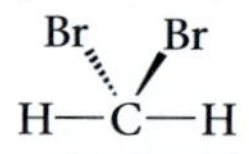

A

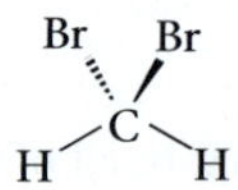

B

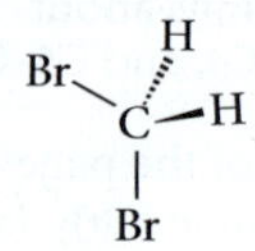

C

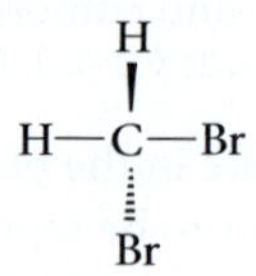

D

2.16 (a) Using models when necessary, draw a Newman projection corresponding to the following line-and-wedge structure. Project the C2–C3 bond, with carbon-2 nearer the observer, and simplify your structures by abbreviating large groups.

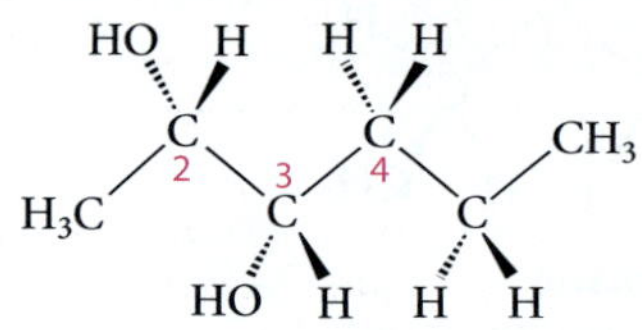

(b) In your Newman projection, rotate carbon-2 clockwise by 120° about the C2–C3 bond, and draw a line-and-wedge structure for the resulting conformation.

(c) In your Newman projection from part (a), rotate carbon-2 by 180° about the C2–C3 bond, and draw a line-and-wedge structure for the resulting conformation.

2.6 PHYSICAL PROPERTIES OF ALKANES

Each time we come to a new family of organic compounds, we examine the trends in their boiling points, melting points, densities, and solubilities. These physical properties of an organic compound are important because they determine the conditions under which the compound is handled and used. For example, the form in which a drug is manufactured and dispensed is affected by its physical properties. In commercial agriculture, ammonia (a gas at ordinary temperatures) and urea (a crystalline solid) are both important sources of nitrogen, but their physical properties dictate that they are handled and dispensed in very different ways.

Your goal should *not* be to memorize physical properties of individual compounds, but rather to learn to predict trends in how physical properties vary with structure. Chapter 8 describes in more detail how the structures of organic compounds affect their physical properties.

A. Boiling Points

The **boiling point** (abbreviated **bp**) is the temperature at which the vapor pressure of a substance equals atmospheric pressure (760 mm Hg at sea level). As Table 2.1 (Sec. 2.2B) shows, at room temperature (about 25 °C), methane, ethane, propane, and butane are gases. The unbranched alkanes with 5–17 carbons are liquids.

A regular increase in the boiling points of the unbranched alkanes occurs with increasing number of carbons (**Fig. 2.7**). The regular increase in boiling point of 20–30 °C per carbon atom within a series is a general trend observed for many types of organic compounds. The basis of this trend is the noncovalent attractions between molecules in the liquid state. The greater these intermolecular attractions are, the more energy (heat, higher temperature) it takes to overcome them so that the molecules escape into the gas phase, in which such attractions do not exist. As the intermolecular attractions within a liquid increase, so does the boiling point. Because boiling points increase with molecular size, it follows that noncovalent intermolecular attractions increase with molecular size. Now, it is important to emphasize that these intermolecular attractions are *not* covalent; there are no covalent bonds between alkane molecules in the liquid

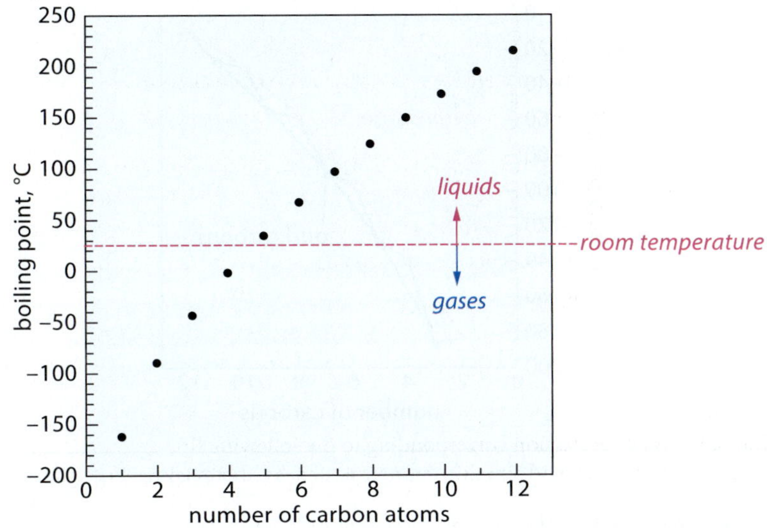

FIGURE 2.7 Boiling points of some unbranched alkanes plotted against the number of carbon atoms. Notice the steady increase in boiling point with the size of the alkane, which is in the range of 20–30 °C per carbon atom. The alkanes with fewer than five carbons are gases at room temperature.

state. Furthermore, intermolecular attractions within a liquid have nothing to do with the strengths of the covalent bonds within the molecules themselves.

A number of chemical and biological phenomena depend on noncovalent attractions. The physical basis of these noncovalent attractions is considered in Sec. 8.5.

B. Melting Points

The **melting point** (abbreviated **mp**) of a substance is the temperature above which it is transformed spontaneously and completely from the solid to the liquid state. The most typical solid state is the crystal, which consists of molecules in regular symmetrical arrays. The melting point of crystalline compounds is an especially important physical property in organic chemistry because it is used both to identify organic compounds and to assess their purity. Melting points are usually depressed, or lowered, by impurities. Moreover, the melting range (the range of temperature over which a substance melts) is usually quite narrow for a pure substance, and it is substantially broadened by impurities.

The melting point reflects the difference in the stabilizing noncovalent intermolecular forces in the liquid and solid state. The higher the melting point, the more stable is the crystal structure relative to the liquid state. However, intermolecular attractions are generally stronger in the crystal solid state than in the liquid; so, trends in melting point in many cases reflect trends in intermolecular interactions in the crystal state. Although most alkanes are liquids or gases at room temperature and have relatively low melting points, their melting points nevertheless illustrate trends that are observed in the melting points of other types of organic compounds.

One such trend is that melting points tend to increase with the number of carbons (**Fig. 2.8**). Another trend is that the melting points of unbranched alkanes with an even number of carbon atoms lie on a separate, higher curve from those of the alkanes with an odd number of carbons. This reflects the more effective packing of the even-carbon alkanes in the crystalline solid state. In other words, the odd-carbon alkane molecules do not "fit together" as well in the crystal as the even-carbon alkanes. Similar alternation of melting points is observed in other homologous series.

Branched-chain hydrocarbons tend to have lower melting points than linear ones. However, when a highly branched alkane has a substantial symmetry, its melting point is typically relatively high. For example, the melting point of the very symmetrical molecule neopentane, −16.8 °C, is considerably higher than that of the less symmetrical pentane, −129.8 °C. Compare also the melting points of the compact and symmetrical molecule cyclohexane, +6.6 °C, and the extended and less symmetrical hexane, −95.3 °C.

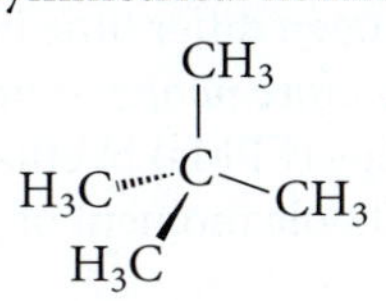

neopentane
mp −16.8 °C

pentane
mp −129.8 °C

cyclohexane
mp +6.6 °C

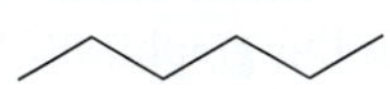

hexane
mp −95.3 °C

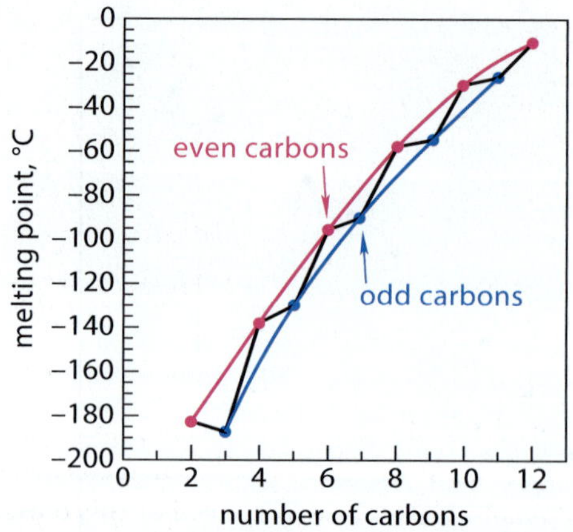

FIGURE 2.8 A plot of melting points of the unbranched alkanes against the number of carbon atoms. Notice the general increase of melting point with molecular size. Notice, too, that the alkanes with an even number of carbons (*red*) lie on a different curve from the alkanes with an odd number of carbons (*blue*). This trend is observed in a number of different types of organic compounds.

The reasons for the effect of branching and symmetry on melting point depend on the type of compound. These effects are discussed in Sec. 8.5D.

In summary, melting points show the following general trends:

1. Melting points tend to increase with increasing molecular mass within a series.
2. In many cases, highly symmetrical molecules have unusually high melting points.
3. A sawtooth pattern of melting point behavior (see Fig. 2.8) is observed within many homologous series.

Focused Problems

2.17 Arrange the following molecules in order of increasing boiling point.

(a) 2-methylheptane, pentane, 2-methylhexane

(b) toluene, benzene, and ethylbenzene

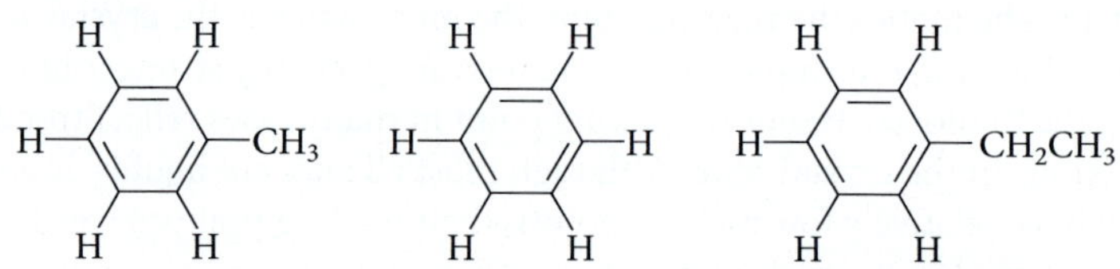

toluene **benzene** **ethylbenzene**

2.18 In each part, match the melting point with the compound.

(a) −109 °C, −56 °C, + 100.7 °C: octane, 2-methylheptane, and 2,2,3,3-tetramethylbutane

(b) −93 °C, + 5.5 °C: benzene and toluene (structures in Focused Problem 2.17)

C. Other Physical Properties

Among the other significant physical properties of organic compounds are dipole moments, solubilities, and densities. A molecule's dipole moment (Sec. 1.9A) determines its *polarity*, which, in turn, affects its physical properties. Because carbon and hydrogen differ little in their electronegativities, alkanes have negligible dipole moments and are therefore *nonpolar molecules*. We can see this graphically by comparing the electrostatic potential maps (EPMs) of ethane, with a dipole moment of zero, and fluoromethane, a polar molecule with a dipole moment of 1.82 D.

EPM of ethane EPM of fluoromethane ($H_3C—F$) (2.22)

Solubilities are important in determining which solvents can be used to form solutions; most reactions are carried out in solution. Water solubility is particularly important for several reasons. For one thing, water is the solvent in biological systems. For this reason, water solubility is a crucial factor in the activity of drugs and other biologically important compounds. There has also been increasing interest in the use of water as a solvent for large-scale chemical processes as part of an effort to control environmental pollution by organic solvents ("green chemistry"). The water solubility of the compounds to be used in a water-based chemical process is crucial. (We'll deal in greater depth with the important question of solubility and solvents in Sec. 8.6.) The alkanes are, for all practical purposes, insoluble in water—thus the saying, "Oil and water don't mix." (Alkanes are a major constituent of crude oil.)

The density of a compound is another property, like boiling point or melting point, that determines how the compound is handled. For example, whether a water-insoluble compound is more or less dense than water determines whether it will appear as a lower or upper layer when added to water. Alkanes have considerably lower densities than water. For this reason, a mixture of an alkane and water will separate into two distinct layers with the less dense alkane layer on top. An oil slick is an example of this behavior (**Fig. 2.9**).

Focused Problem

2.19 Gasoline consists mostly of alkanes. Explain why water is not usually very effective in extinguishing a gasoline fire.

Stan Honda/AFP/Getty Images

FIGURE 2.9 An oil spill in flood waters is contained by plastic tubes at a Texas refinery in the aftermath of Hurricane Rita in 2005. The inset shows how the lower density and insolubility of the oil allow it to be trapped by the physical barrier.

2.7 COMBUSTION

A. Combustion of Alkanes

Alkanes are among the least reactive types of organic compounds. They do not react with common acids or bases, nor do they react with common oxidizing or reducing agents.

Alkanes do, however, share one type of reactivity with many other types of organic compounds: they are flammable. This means that they react rapidly with oxygen to give carbon dioxide and water, provided that the reaction is initiated by a suitable heat source, such as a flame or the spark from a spark plug. This reaction is called **combustion**. An example is the combustion of methane, the major alkane in natural gas:

$$CH_4 + 2O_2 \longrightarrow CO_2 + 2H_2O \qquad (2.23)$$

This reaction illustrates *complete combustion*: combustion in which carbon dioxide and water are the only products. Under conditions of oxygen deficiency, incomplete combustion may also occur with the formation of such byproducts as carbon monoxide, CO. Carbon monoxide is a deadly poison because it bonds to, and displaces oxygen from, hemoglobin, the protein in red blood cells that transports oxygen to tissues. It is also colorless and odorless, and is therefore difficult to detect without special equipment.

The fact that we can carry a container of gasoline in the open air without its going up in flames shows that simple mixing of alkanes and oxygen does not initiate combustion. Once a spark is applied, however, the combustion reaction proceeds vigorously.

Among organic compounds, alkanes are one of the best chemical sources of energy because they liberate large amounts of energy on combustion. Their high energy and their ready availability account for their importance as fuels for both transportation and heating. For example, 1 mol (114 g, or about 1/3 cup) of liquid 2,2,4-trimethylpentane, a major component of automotive gasoline, liberates 5461 kJ (1305 kcal) when it undergoes complete combustion.

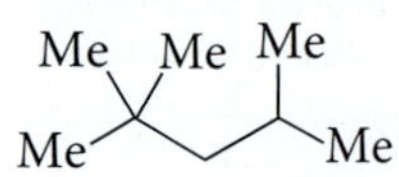

$$+ \frac{25}{2}O_2 \longrightarrow 8\,CO_2 + 9\,H_2O$$

2,2,4-trimethylpentane (2.24)

This is a *large* amount of energy; about 1/3 cup of this alkane liberates enough energy on combustion to propel a 3000-pound car at conventional mileage for about half a mile, or (assuming we could capture and use all of it) to convert almost 3 *gallons* of water at 0 °C to steam at 100 °C!

Two problems with the combustion of alkanes are (1) the efficiency with which energy can be recovered from the reaction as work, and (2) the products of the reaction, specifically carbon dioxide. The efficiency of the typical automotive engine is roughly 20–25%. This is not likely to increase significantly for gasoline engines. Other, more efficient, ways of powering motor vehicles involve the generation and use of electricity in various ways, in some cases from renewable fuels, and these are under active development. Burning hydrocarbons for heating is very efficient, but the problem of CO_2 generation remains.

As Eq. 2.23 illustrates, every carbon atom of a hydrocarbon combines with two atoms of oxygen to generate a molar equivalent of carbon dioxide, and every pair of hydrogens combines with one oxygen atom to generate a molar equivalent of water. The atmosphere can hold a relatively small amount of water, and when that is exceeded, water returns to Earth as rain or snow. However, natural processes of removing carbon dioxide from the atmosphere are limited. After eons in which the CO_2 content of the atmosphere remained relatively constant at about 290 parts per million (ppm), the amount of CO_2 in Earth's atmosphere began to rise dramatically with the advent of the industrial age. The CO_2 level now exceeds 400 ppm, an increase of more than 40% (**Fig. 2.10**). Most of this increase has taken place since the 1980s. Because so much of it is produced, carbon dioxide is the most significant of several compounds known to be *greenhouse gases*, atmospheric compounds that act as a heat-reflective blanket over Earth. Overwhelming evidence suggests that the temperature of Earth is being increased by the effect of greenhouse gases; this phenomenon is known as *global warming*. Scientists believe that global warming is

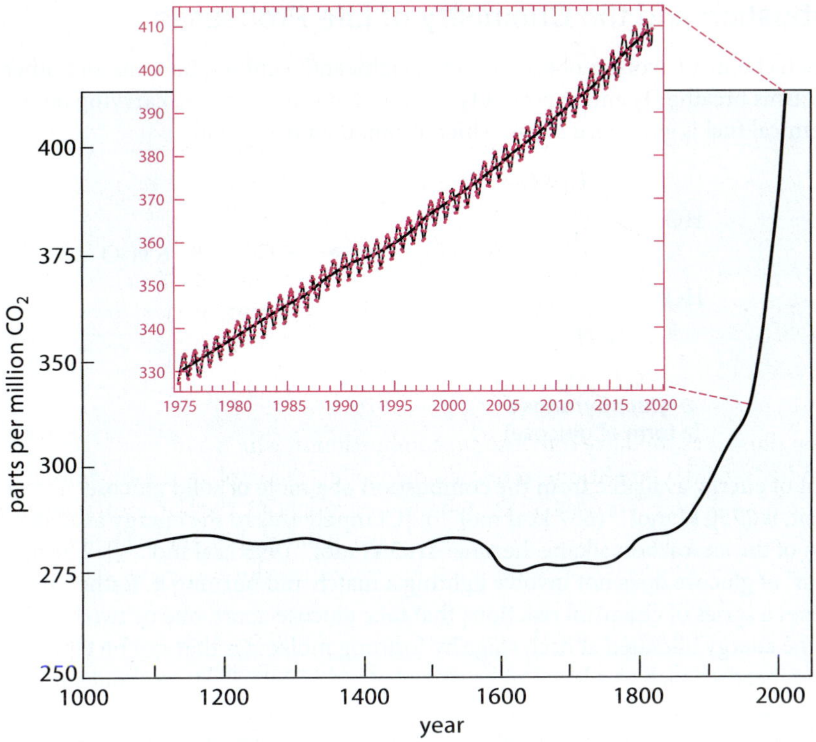

FIGURE 2.10 Atmospheric CO_2 levels for the past 1000 years. The data prior to 1958 were obtained from air bubbles trapped in dated ice core samples. More recent data were obtained from air sampling towers on Mauna Loa, Hawaii, by the Scripps Institution of Oceanography (1958–1974) and the National Oceanic and Atmospheric Administration (NOAA, 1974–present). The inset shows data obtained since 1975 in more detail. These data show the seasonal fluctuations normally observed in CO_2 levels. Notice the continuous rise in CO_2 levels since the nineteenth century.

beginning to have significant adverse environmental consequences, such as an increase in the intensity of hurricanes, the rapid receding of glaciers, and the extinction of animal and plant species at an increased rate. Global warming models predict that the ocean levels will rise as polar ice undergoes melting, and the resulting coastal flooding will likely displace hundreds of millions of people. As a result of these concerns, along with concerns about the political instability of some oil-producing regions of the world, the development of alternative fuels has become increasingly urgent. Ideally, the goal is to produce cheap and abundant fuels that will not, on combustion, increase the net CO_2 content of the atmosphere.

Combustion finds a minor but important use as an analytical tool for the determination of molecular formulas. In this type of analysis, the mass of CO_2 produced in the combustion of a known mass of an organic compound is used to calculate the amount of carbon in the sample. Similarly, the mass of H_2O produced is used to calculate the amount of hydrogen in the sample. (Procedures have been developed for the combustion analysis of other elements.) Combustion analysis is illustrated in Problem 2.55.

Focused Problems

2.20 Give a general balanced reaction for the following.

(a) The complete combustion of an alkane (formula C_nH_{2n+2})

(b) The complete combustion of a cycloalkane containing one ring (formula C_nH_{2n})

2.21 Calculate the number of pounds of CO_2 released into the atmosphere when 15 gallons of gasoline is burned in an automobile engine. Assume complete combustion. Also assume that gasoline is a mixture of octane isomers and that the density of gasoline is 0.692 g mL^{-1}. (This assumption ignores about 10 volume percent of oxygenated additives.) Useful conversion factors: 1 gallon = 3.785 L, 1 kg = 2.204 lb.

Chemical Biology Topic

B. Combustion and the Chemistry of Life Processes

As the French chemist Lavoisier observed in the eighteenth century, humans and other aerobic living organisms breathe O_2 and expire CO_2, and in that sense they are carrying out combustion. The biochemical fuel is glucose, a sugar, which is obtained from foods:

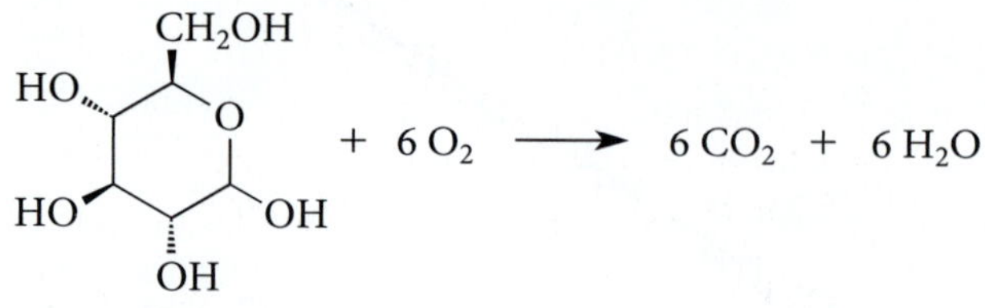

$C_6H_{12}O_6$
D-glucopyranose
(a form of glucose) (2.25)

The amount of energy available from the combustion of a mole of solid glucose, if it were released solely as heat, is 2750 kJ mol^{-1} (657 kcal mol^{-1}). [Compare this to the energy available from the combustion of the six-carbon alkane, hexane: 4163 kJ mol^{-1} (995 kcal mol^{-1})]. The biological "combustion" of glucose does not involve lighting a match and burning it. Rather, the living organism uses a series of chemical reactions that take glucose apart, one or two bonds at a time, and stores the energy liberated at each stage by forming molecules that can be tapped as energy sources when needed, such as adenosine triphosphate (ATP). We'll learn about some of these processes in this text, and, if you study biochemistry, you'll get a more thorough overview. The human body recovers the energy from glucose "combustion" with efficiencies that vary from 40–60%, depending on conditions. Given that human metabolism is 2–3 times as efficient as an automotive engine, more energy is recovered from the "combustion" of a mole of glucose than an internal-combustion engine recovers from a mole of 2,2,4-trimethylpentane!

2.8 FUNCTIONAL GROUPS, COMPOUND CLASSES, AND THE "R" NOTATION

A. Functional Groups and Compound Classes

Alkanes are the conceptual "rootstock" of organic chemistry. Replacing C—H bonds of alkanes gives the many functional groups of organic chemistry. A **functional group** is a characteristically bonded group of atoms that has about the same chemical reactivity whenever it occurs in a variety of compounds. Compounds that contain the same functional group constitute a **compound class**. Consider the following examples:

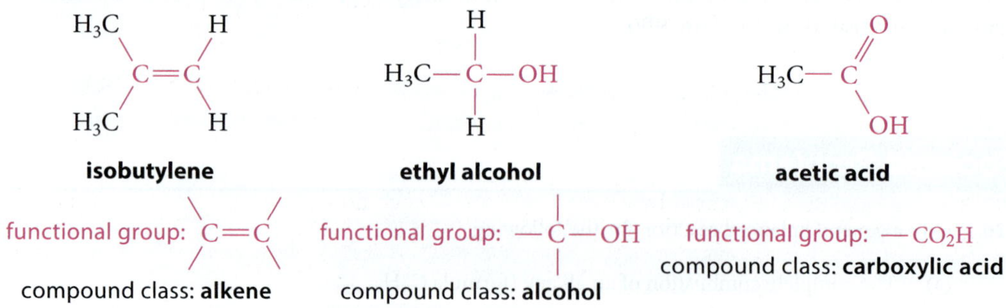

(2.26)

For example, the functional group that is characteristic of the alkene compound class is the carbon–carbon double bond. Most alkenes undergo the same types of reactions, and these reactions occur at or near the double bond. Similarly, all compounds in the alcohol compound class contain an —OH group bonded to the carbon atom of an alkyl group. The characteristic reactions of alcohols occur at the —OH group or the directly attached carbon, and this functional group undergoes the same general chemical transformations regardless of the structure of the remainder of the molecule. Needless to say, some compounds can contain more than one functional group. Such compounds belong to more than one compound class.

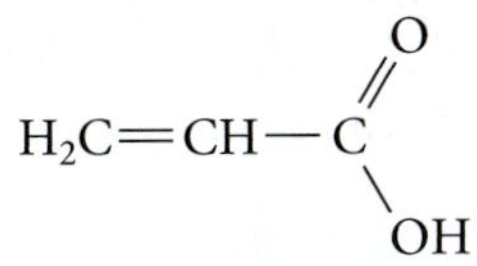

acrylic acid
contains both C=C and CO_2H functional groups and is thus both an alkene and a carboxylic acid

The organization of this text is centered for the most part on the common functional groups and corresponding compound classes. Although we describe in detail each major functional group in subsequent chapters, you should learn to recognize the common functional groups and compound classes now. These are shown on the pages inside the front cover.

B. "R" Notation

Sometimes we'll want to use a general structure to represent an entire class of compounds. In such a case, we can use the **R notation**, in which an R is used to represent all *alkyl groups* (Sec. 2.4C). For example, R—Cl can be used to represent an alkyl chloride.

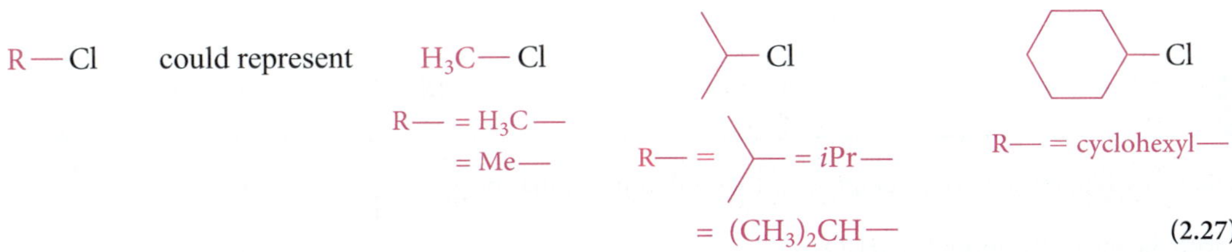

(2.27)

Just as alkyl groups such as methyl, ethyl, and isopropyl are substituent groups derived from alkanes, **aryl groups** are substituent groups derived from benzene and its derivatives. The simplest aryl group is the **phenyl group**, abbreviated Ph—, which is derived from the hydrocarbon benzene. Each ring carbon of an aryl group not joined to another group bears a hydrogen atom that is not shown. (This is the usual convention for skeletal structures; see Sec. 2.2B.)

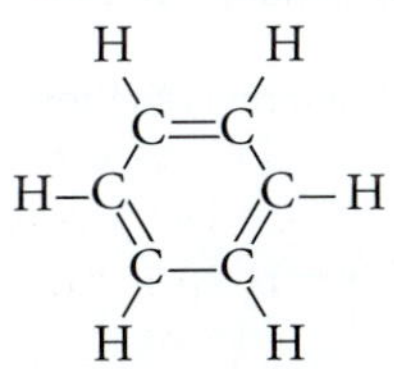

benzene

skeletal structure of benzene

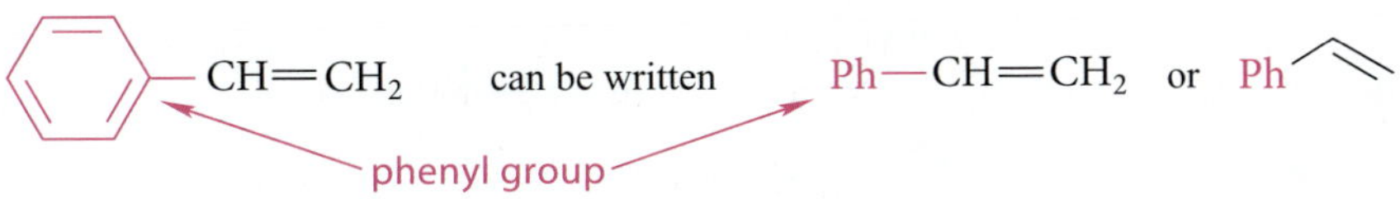

(2.28)

Other aryl groups are designated by Ar—. Thus, Ar—OH could refer to any one of the following compounds, or to many others.

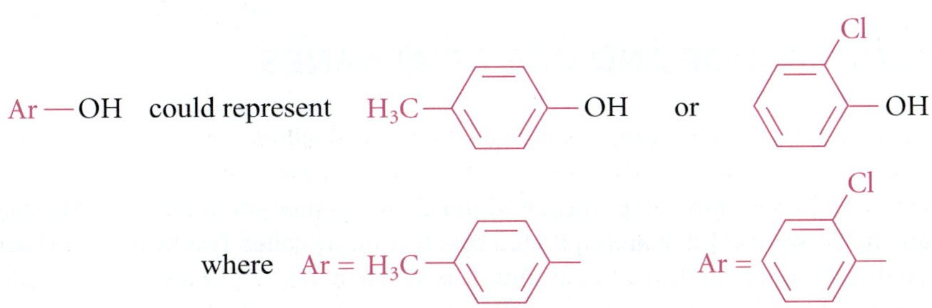

(2.29)

If a structure contains multiple R groups, it is understood that the R groups can be either the same or different.

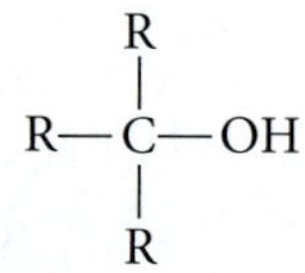

If you want to specify that all R groups are the same (a situation that is not common), you should make that point clear with an explanation (such as, "all R groups the same"). On the other hand, if you want to be more specific about differences among the R groups, you can use primes or superscript numbers with the R notation:

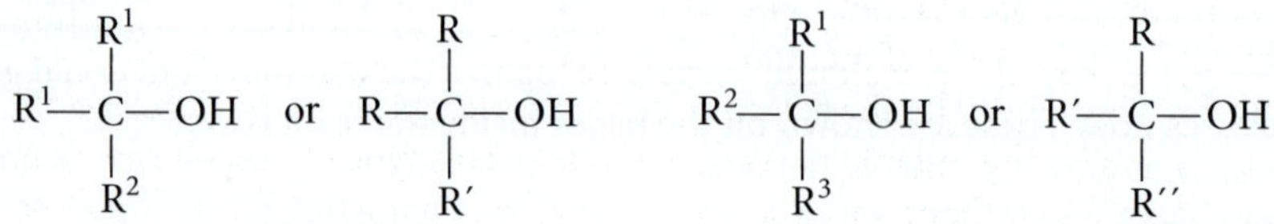

two R groups are the same, one is different

all three R groups are different

Although you will not study benzene and its derivatives until Chapter 15, before then you will see many examples in which phenyl and aryl groups are used as substituent groups.

Focused Problems

2.22 Draw a structural formula for each of the following compounds. (Several formulas may be possible in each case.)

(a) A carboxylic acid with the molecular formula $C_2H_4O_2$

(b) An alcohol with the molecular formula $C_5H_{10}O$

2.23 A certain compound was found to have the molecular formula $C_5H_{12}O_2$. To which of the following compound classes could the compound belong? Give one example for each positive answer, and explain any negative responses.

an amide an ether a carboxylic acid a phenol an alcohol an ester

2.24 Which of the following structures with the R notation (if any) would be a suitable abbreviation for the specific structure in each case?

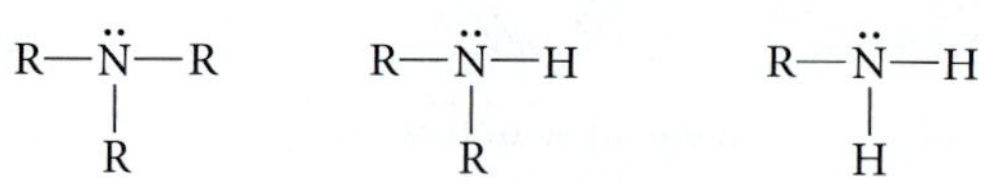

A *B* *C*

(a) 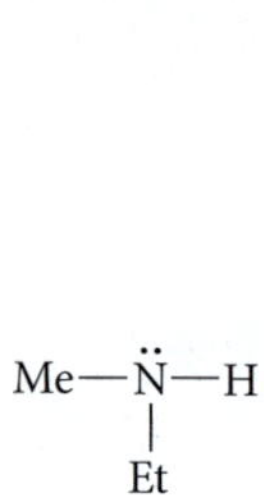 (b) 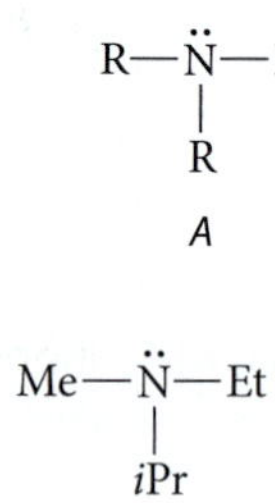 (c) 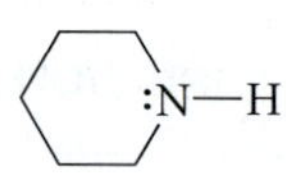 (d) 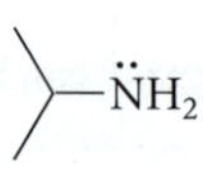

Chemistry in the Real World

2.9 OCCURRENCE AND USE OF ALKANES

Most alkanes come from **petroleum**, or crude oil. (The word *petroleum* comes from the ancient Greek word for "rock" (*petra*) and the Latin word for "oil" (*oleum*); thus, "oil from rocks.") Petroleum is a dark, viscous mixture composed mostly of alkanes and aromatic hydrocarbons (benzene and its derivatives) that are separated by a technique called **fractional distillation.** In fractional distillation, a mixture of compounds is slowly boiled; the vapor is then collected, cooled, and recondensed to a liquid. Because the compounds with the lowest boiling points vaporize most readily, the condensate from a fractional distillation is enriched in the more volatile components of the mixture. As distillation continues, components with progressively higher boiling points appear in the condensate. A student who takes an organic chemistry laboratory course will almost certainly become acquainted with this technique on a laboratory scale. Industrial fractional distillations are carried out on a large scale in fractionating towers that are several stories tall (**Fig. 2.11**). The typical fractions obtained from distillation of petroleum are shown in **Fig. 2.12**.

Jenna Wagner/Getty Images

FIGURE 2.11 Fractionating towers such as these are used in the chemical industry to separate mixtures of compounds on the basis of their boiling points.

Another important alkane source is natural gas, which is mostly methane. Natural gas comes from gas wells of various types. Horizontal hydraulic fracturing ("fracking") is used to release natural gas trapped in rock formations. This technique requires the use of large amounts of water and chemicals. Although it has contributed significantly to domestic natural gas supplies, fracking is controversial because of its environmental impact.

Significant biological sources of methane also exist that are exploited commercially. For example, methane is produced by the action of certain anaerobic bacteria (bacteria that function without oxygen) on decaying organic matter (**Fig. 2.13**). This type of process, for example, produces "marsh gas," as methane was known before it was characterized by chemists. This same biological process can be used for the production of methane from animal and human waste. Methane produced this way is becoming practical as a local source of power (see Fig. 2.13). The methane is burned to produce heat that is converted into electricity. Although this process generates carbon dioxide, the source of the carbon is the food eaten by the humans and animals,

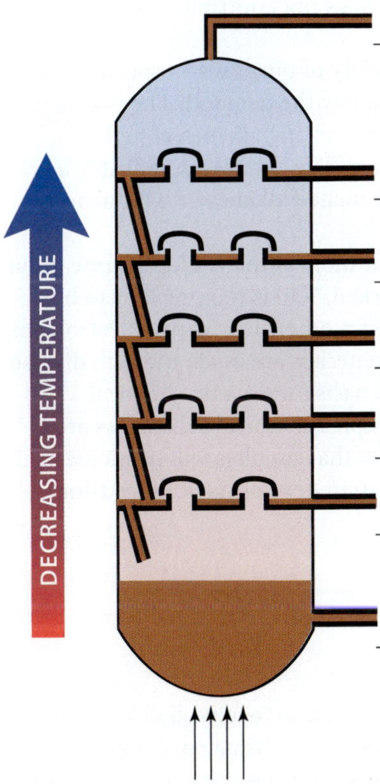

Fraction name	Fraction temperature	Number of carbons	Typical use
petroleum gas	up to 20 °C	C_1–C_4	bottled gas
gasoline	20–70 °C	C_5–C_{10}	automobile fuel
naphtha	70–120 °C	C_8–C_{12}	chemical feedstock
kerosene	120–240 °C	C_{10}–C_{16}	jet fuel, paraffin
light fuel oil	240–320 °C	C_{15}–C_{20}	diesel fuel
heavy oil	320–500 °C	C_{21}–C_{28}	lubricants, heating oil
asphalt, tar	above 500 °C	$>C_{28}$	road surfacing

FIGURE 2.12 Schematic view of an industrial fractionating column. The temperature of the column decreases from bottom to top. Crude oil is introduced at the bottom and heated. As the vapors rise, they cool and condense to liquids. Fractions of progressively lower boiling points are collected from bottom to top of the column. The figure shows typical fractions, their boiling points, the number of carbons in the compounds collected, and the typical uses of each fraction after further processing.

DR M.ROHDE, GBF/Science Source

Jim West/Science Source

FIGURE 2.13 Manure fermenters at a large midwestern dairy farm are used to produce methane. Electricity produced from burning the methane is fed into the power grid. The power produced can in some cases be sufficient to support a large fraction of the farm's power requirements. Such fermentations are carried out biologically by methanogens (methane-producing bacteria). An example of a methanogen is shown in the inset.

and the carbon in that food comes from atmospheric carbon dioxide by photosynthesis. In other words, the CO_2 produced in this process is "recycled CO_2" and does not contribute to a net increase in atmospheric CO_2.

Alkanes of low molecular mass are in great demand for a variety of purposes—especially as motor fuels—and alkanes available directly from wells do not satisfy the demand. The petroleum industry has developed methods (called *catalytic cracking*) for converting alkanes of high molecular mass into alkanes and alkenes of lower molecular mass. The petroleum industry has also developed processes (called *reforming*) for converting unbranched alkanes into branched ones, which have superior ignition properties as motor fuels.

Typically, motor fuels, fuel oils, and aviation fuels account for most of the world's hydrocarbon consumption. An oil minister from the Middle East once remarked, "Oil is too precious to burn." He was undoubtedly referring to the important uses for petroleum other than as fuels. Petroleum is the principal source of *carbon*, from which organic starting materials are made for such diverse products as plastics and pharmaceuticals. Petroleum is thus the basis for organic chemical *feedstocks*—the basic organic compounds from which more complex chemical substances are fabricated. However, the volatility of oil prices and the possibility that supplies will be increasingly constrained in the future have increased the interest and research in development of feedstocks from other sources such as plant-based compounds.

Alkanes as Motor Fuels; Fuel Additives

BARRI/Shutterstock

Alkanes vary significantly in their quality as motor fuels. Branched alkanes are better motor fuels than unbranched ones. The quality of a motor fuel relates to its rate of ignition in an internal combustion engine. Premature ignition results in "engine knock," a condition that indicates poor engine performance. Severe engine knock can result in significant engine damage. The octane number is a measure of the quality of a motor fuel: the higher the octane number, the better the fuel. The octane number is the number you see associated with each grade of gasoline on the gasoline pump. Octane numbers of 100 and 0 are assigned to 2,2,4-trimethylpentane (called "isooctane" in the petroleum industry) and heptane, respectively. Mixtures of the two compounds are used to define octane numbers between 0 and 100. For example, a fuel that performs as well as a 1:1 mixture of 2,2,4-trimethylpentane and heptane has an octane number of 50. The motor fuels used in modern automobiles have octane numbers in the 87–95 range. The octane number of octane itself, the unbranched isomer of 2,2,4-trimethylpentane, is –19; it is inferior even to heptane in engine knock tests.

Various additives can be used to improve the octane number of motor fuels. In the past, tetraethyllead, $(CH_3CH_2)_4Pb$, was used extensively for this purpose, but concerns over atmospheric lead pollution and the advent of catalytic converters (which are damaged by lead) resulted in a phase-out of tetraethyllead over the period 1976–1986 in the United States and in the European Union by 2000. This was followed by the use of methyl *tert*-butyl ether [MTBE, $(CH_3)_3C—O—CH_3$] as the major antiknock gasoline additive. After a meteoric rise in MTBE production, this compound became an environmental problem in the mid-1990s when several communities in the United States discovered it leaking from storage vessels into groundwater. Because MTBE has shown some carcinogenic (cancer-causing) activity in laboratory animals, many cities and states have enacted a phase-out of MTBE usage as a gasoline additive. Ethanol (ethyl alcohol, CH_3CH_2OH) can be used as a substitute for MTBE, and ethanol can be produced by the fermentation of sugars in corn. Political action by corn-farming interests in the United States has been successful not only in substituting ethanol for MTBE as an antiknock additive, but also in the partial replacement of the hydrocarbons in gasoline by ethanol as a fuel in its own right. Ethanol production by the fermentation of sugars in corn has been subsidized. The demand for fuel ethanol at one time was so great that the price of corn escalated sharply, and the demand for corn for ethanol production had a noticeable impact on the price of foods that depend on corn as an animal food (for example, milk, chicken, and beef). In rapidly developing Asian markets and in some European markets, however, which are not influenced by ethanol subsidies, MTBE continues to be used as an antiknock additive. A number of other oxygen-containing compounds can also be used for this purpose, but none of them can compete economically with MTBE and ethanol. MTBE, ethanol, and other oxygen-containing additives are collectively referred to in the fuel industry as *oxygenates*.

CHAPTER SUMMARY *For a summary of the chapter, see Chapter 2 in the* Study Guide and Solutions Manual.

SKILLS OBJECTIVES WITH PROBLEMS

- **Classify any hydrocarbon as an alkane, alkene, alkyne, aromatic hydrocarbon, or a combination of these. 2.1**

2.25 Classify each of the following hydrocarbons as an alkane, alkene, alkyne, or aromatic hydrocarbon.

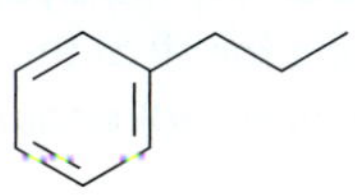

A

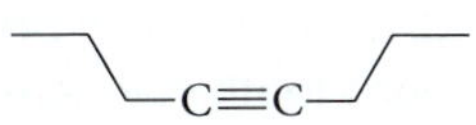

B

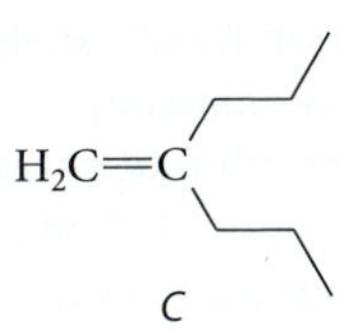

C

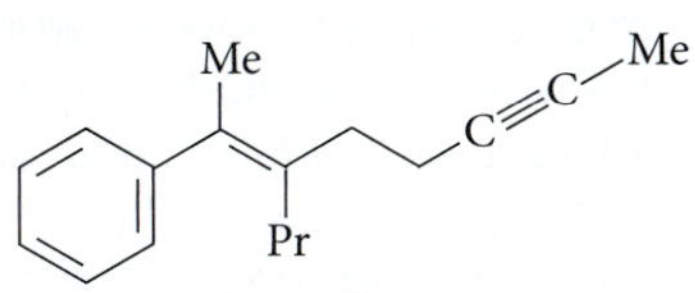

D

- **Define the terms *composition*, *constitution*, *configuration*, and *conformation*, and be able to illustrate these concepts for a given molecule. 2.1**

2.26 For butane, give (a) the composition, (b) the constitution, (c) the configuration, and (d) the possible conformation(s).

- **Know the names and structures of the first 12 unbranched alkanes. 2.2B**

2.27 Without consulting tables, give condensed Lewis structures, skeletal structures, and names of the unbranched alkanes (a) with 10 carbons, (b) with 5 carbons.

- **Given the number of carbons in a noncyclic alkane, give the number of hydrogens. Given the number of hydrogens, give the number of carbons. Given the number of rings and the number of carbons in a cycloalkane, give the number of hydrogens. 2.2B, 2.3**

2.28 (a) How many hydrogens are there in a noncyclic alkane that contains 15 carbons?

(b) How many carbons are there in a noncyclic alkane that contains 22 hydrogens?

(c) A cyclic hydrocarbon has no double or triple bonds, and it has 30 carbons and four rings. How many hydrogens does it have?

- **Be able to determine systematically whether two structures represent identical compounds, constitutional isomers, or neither. 2.4A**

2.29 Given skeletal structures *A–D*, what is the relationship between each pair? Choose between *identical, constitutional isomers,* and *not isomers.*

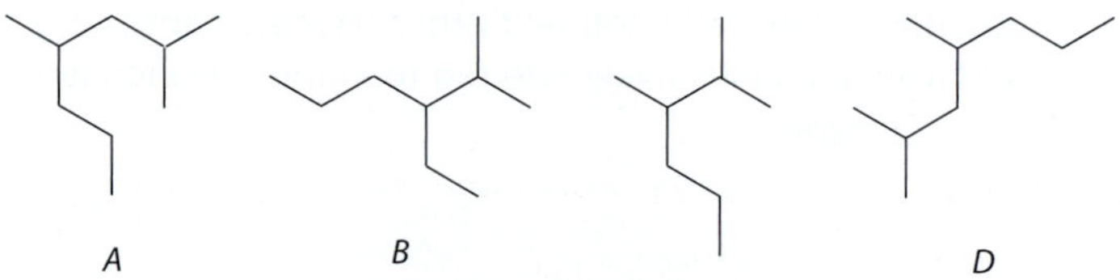

A *B* *C* *D*

- **Know the names and structures of the common substituent groups. TABLES 2.3, 2.4**

2.30 A chemist has come to you with three alkanes and gives you the following names, which, although incorrect, unambiguously define structures. Draw a skeletal structure corresponding to each name, and then give the correct name for each compound.

(a) Isopropylethane

(b) 2-*tert*-Butylpropane

(c) 2-*sec*-Butylbutane

- **Apply the 10 nomenclature rules to give the IUPAC substitutive name for any alkane or cycloalkane. Given the name of an alkane or cycloalkane, draw its Lewis structure and its skeletal structure. 2.4C, 2.4D**

2.31 Name each of the following compounds using IUPAC substitutive nomenclature.

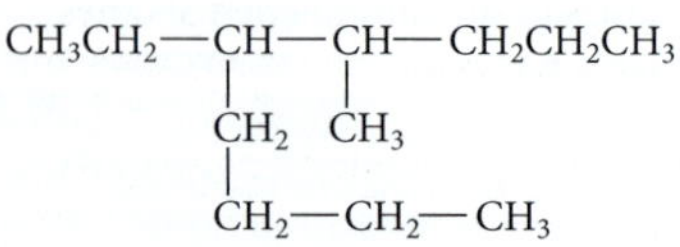

A

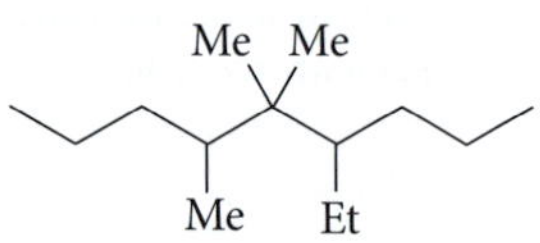

B

2.32 Draw structures of the following compounds.

(a) 4-Isobutyl-2,5-dimethylheptane (Lewis structure)

(b) 5-Neopentyldecane (skeletal structure)

(c) 1-Cyclopropyl-2,4-dimethylcyclohexane (skeletal structure)

(d) 5-Ethyl-3,3-dimethyloctane (Lewis structure with substituent abbreviations)

- **Redraw a highly condensed structure as both a Lewis structure and a skeletal structure. 2.4E**

2.33 Draw both a Lewis structure and a skeletal structure corresponding to the following condensed structure. Your structures should show all carbon–carbon bonds explicitly.

$$[(CH_3CH_2)_2CH]_3CH$$

- **Given a Newman projection, construct a line-and-wedge structure that shows the configuration of the two carbons in the projected bond. 2.5C**

2.34 Convert each of the following Newman projections into a line-and-wedge structure viewed from the direction shown by the eyeball. *Use models if necessary.*

(a)

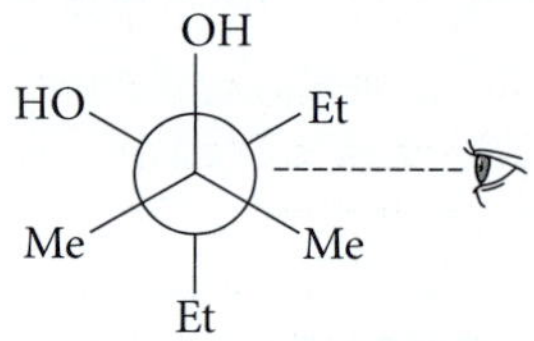

(b)

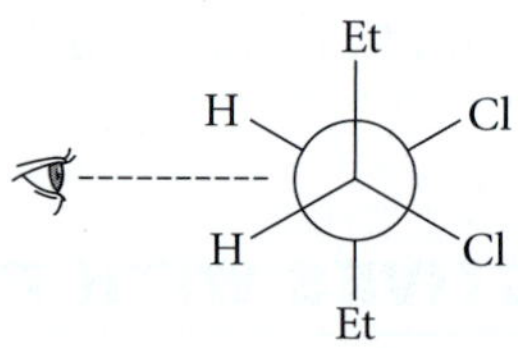

(c)

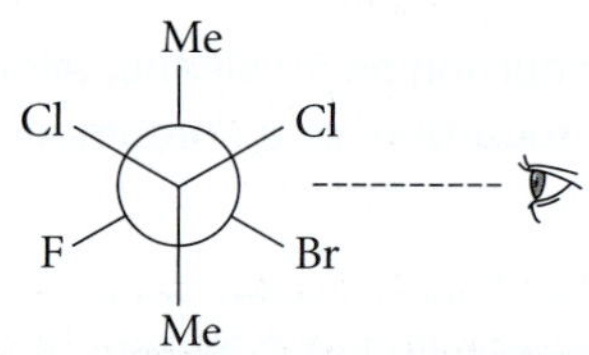

- **Given a line-and-wedge structure of any noncyclic compound, draw Newman projections for all of the conformations (both staggered and eclipsed) about any bond. 2.5B, 2.5C**

- **Given the Newman projections for the conformations of a compound about a carbon–carbon bond, analyze the relative magnitude of gauche and eclipsing interactions, and sketch an energy graph that shows the relative energies of the conformations. Tell which conformations are present in the greatest amount. 2.5B, 2.5C**

2.35 (a) Draw Newman projections of the three staggered conformations about the C2–C3 bond of 2,3,3,4-tetramethylpentane, with carbon-2 nearer the observer. (Use group abbreviations where appropriate.)

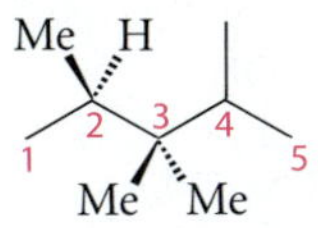

2,3,3,4-tetramethylpentane

(b) Rank the three staggered conformations in energy, lowest first. Explain your choices. (Some conformations may have identical energies.)

(c) Draw Newman projections about the same bond for the three eclipsed conformations.

(d) Rank the three eclipsed conformations in energy, lowest first. Explain your choices.

(e) Sketch a graph of energy versus angle of rotation for internal rotation about the C2–C3 bond.

(f) Which conformation(s) of 2,3,3,4-tetramethylpentane are present in the greatest amount in a sample this alkane?

• Classify any carbon in an alkane as primary, secondary, tertiary, or quaternary. Classify any hydrogen in an alkane as primary, secondary, or tertiary. 2.4F

2.36 (a) Classify each of the carbons in *tert*-butylcyclopentane as primary, secondary, tertiary, or quaternary.

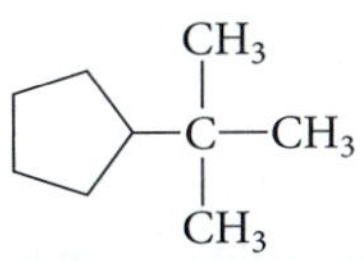

***tert*-butylcyclopentane**

(b) Classify each of the hydrogens in *tert*-butylcyclopentane as primary, secondary, or tertiary.

• Given the boiling point of any alkane, estimate the boiling point of another alkane in the same homologous series. 2.6A

2.37 The boiling point of 2-methyloctane is 149 °C. Estimate the boiling point of (a) 2-methylheptane and (b) 2-methyldecane.

• Apply symmetry criteria to determine which of several isomers has the highest melting point. 2.6B

2.38 Match each of the following constitutional isomers with their melting points, and explain your choices.

Isomers: undecane (see Table 2.1, Sec. 2.2B), 2-methyldecane, 2,2,3,3,4,4-hexamethylpentane

Melting points: –45 °C, –26 °C, + 65 °C

• Provide a general characterization of alkanes in terms of the following physical properties: water solubility, density, physical state at room temperature, and dipole moment. 2.6

2.39 Which of the following descriptions applies to a sample of (a) octane and (b) propane? Assume a temperature of 25 °C.

1. When added to a beaker of water, forms a separate layer on the bottom of the water
2. When added to a beaker of water, forms a separate layer on the top of the water
3. When added to a beaker of water, dissolves to form a solution
4. Is a gas at room temperature
5. Is a solid at room temperature
6. Is a liquid at room temperature
7. Is a very polar molecule
8. Is a nonpolar molecule

• Write a balanced equation for the complete combustion of any alkane or cycloalkane. 2.7A

2.40 (a) Which produces fewer moles of water as a combustion product: 1 mol of hexane or 1 mol of cyclohexane?

(b) Contrast the number of moles of carbon dioxide produced by the combustion of the two hydrocarbons in part (a).

• Learn the structures of the common organic functional groups so that you can identify the functional groups present in organic structures. 2.8A

2.41 To which compound class does each of the following compounds belong?

(a)

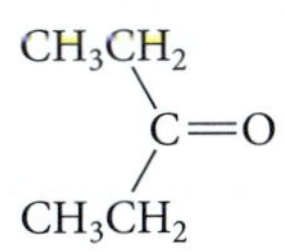

(b)

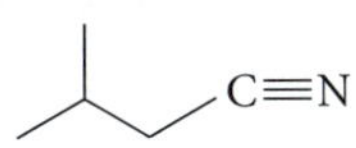

(c) 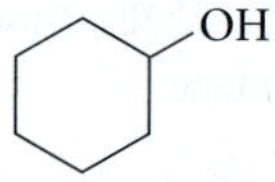

(d) 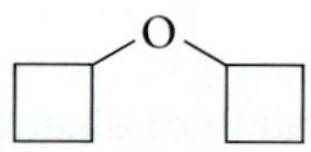

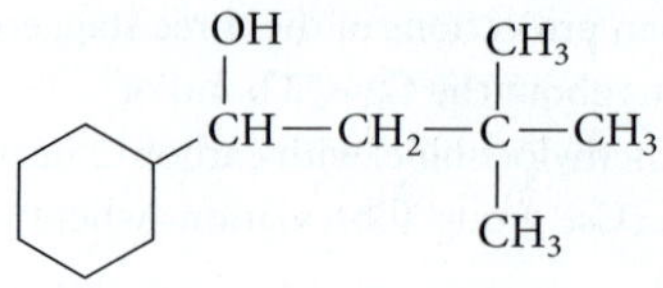

- **Represent the groups around any carbon in an organic compound with an appropriate R notation. Interpret the meaning of the R notation in specific cases. 2.8B**

2.42 Which of the structures *A–D* would be appropriate abbreviations for the following structure?

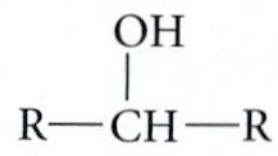

A

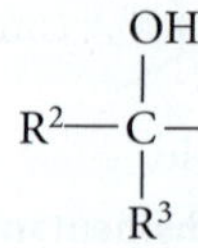

B

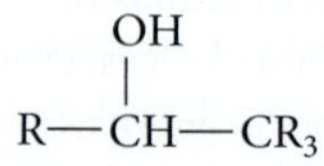

C

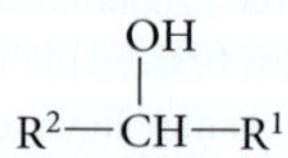

D

INTEGRATED PROBLEMS

2.43 Given the boiling point of the first compound in each set, estimate the boiling point of the second.

(a) 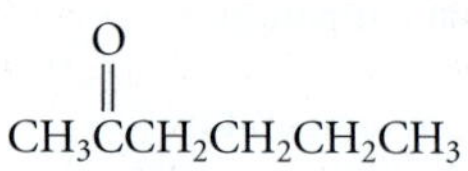 (bp 128 °C)

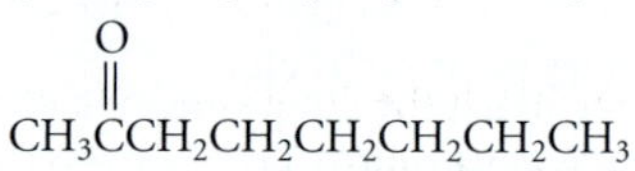

(b) 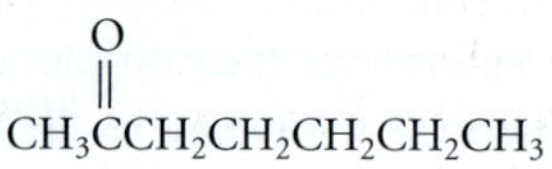 (bp 152 °C)

2.44 Draw their Lewis structures, and give the names of all isomers of octane that meet each of the following criteria.

(a) Five carbons in their principal chains

(b) Six carbons in their principal chains

2.45 Draw the structure of an alkane or cycloalkane that meets each of the following criteria.

(a) A compound that has more than three carbons and only primary hydrogens

(b) A compound that has five carbons and only secondary hydrogens

(c) A compound that has only tertiary hydrogens

2.46 Name each of the following compounds using IUPAC substitutive nomenclature.

(a) 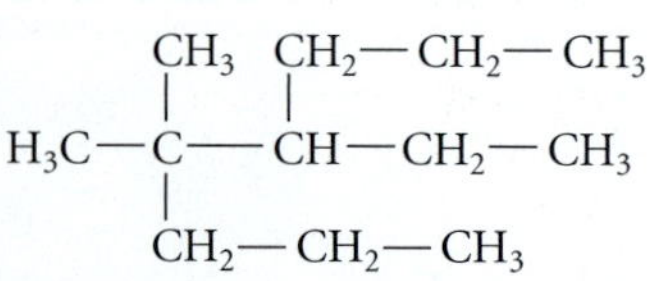

(b) 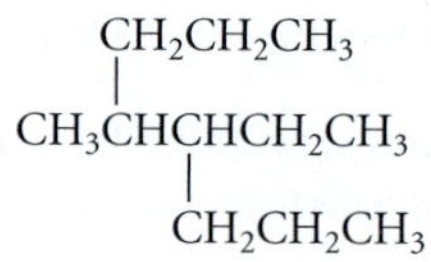

(c) 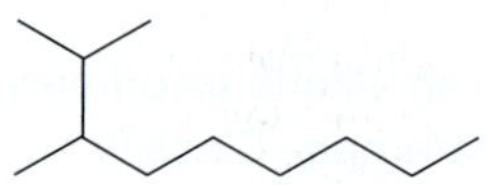

(d) 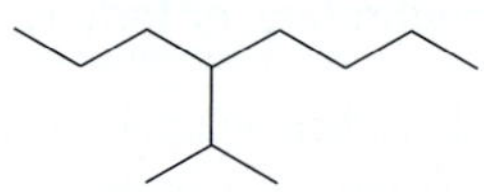

(e) 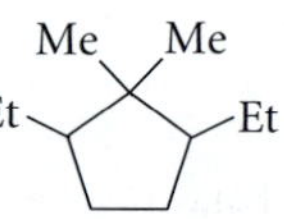

2.47 Draw structures that correspond to the following names.

(a) 3-Ethyl-4-isopropyl-5-methylheptane (a structure in which all carbons and hydrogens are shown)

(b) 2,3,5-Trimethyl-4-propylheptane (skeletal structure)

(c) 5-*sec*-Butyl-6-*tert*-butyl-2,2-dimethylnonane (skeletal structure)

2.48 The following labels were found on bottles of liquid hydrocarbons. Although each name defines a structure unambiguously, some are not correct IUPAC substitutive names. Give the correct name for any compounds that are named incorrectly.

(a) 2-Ethyl-2,4,6-trimethylheptane

(b) 5-Neopentyldecane

(c) 4-Ethyl-1,2-diisopropylcyclohexane

(d) 3-Butyl-2,2-dimethylhexane

2.49 Although compounds are indexed by their IUPAC substitutive names, sometimes chemists give whimsical names to compounds that they discover. Assist these four chemists by providing IUPAC substitutive names for their compounds.

(a) Chemist (and distance runner) Minnie Miles prepared the following compound that she named "marathane."

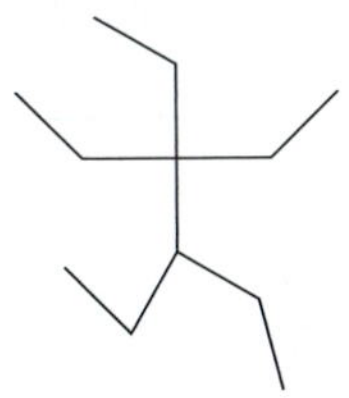

(b) Chemist Val Losipede isolated an alkane with the following skeletal structure from asphalt scrapings following a bicycle race and named it "tourdefrançane."

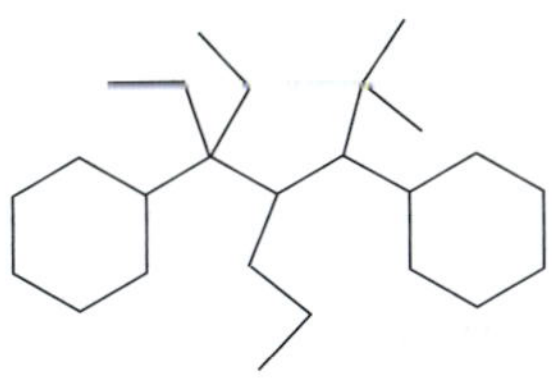

(c) Chemist Slim Pickins isolated a compound with the following structure from the floor of a henhouse and dubbed it "pullane" (*pullus*, Latin for chick).

(d) Yolanda Plane, a chemist and amateur pilot, prepared the following compound, which she called *puh-lane*. (*Hint:* An 11-carbon alkane is called *undecane*.)

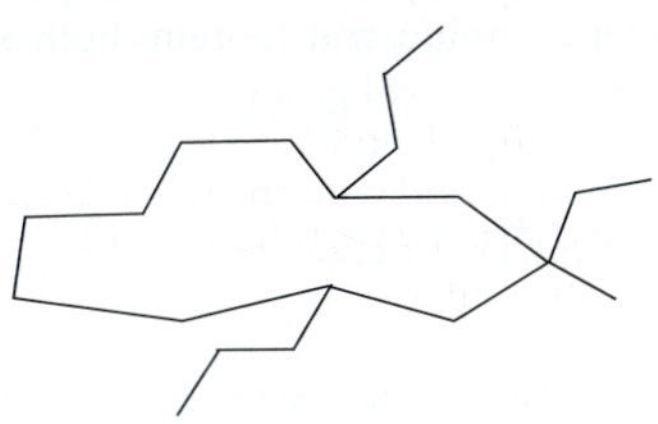

2.50 The anti conformation of 1,2-dichloroethane, $Cl{-}CH_2{-}CH_2{-}Cl$, is 4.81 kJ mol^{-1} (1.15 kcal mol^{-1}) more stable than the gauche conformation. The energies of the two eclipsed conformations (measured relative to the energy of the gauche conformation) for carbon–carbon bond rotation are 21.5 kJ mol^{-1} (5.14 kcal mol^{-1}) and 38.9 kJ mol^{-1} (9.30 kcal mol^{-1}).

(a) Sketch a graph of potential energy versus dihedral angle about the carbon–carbon bond. Show the energy differences on your graph, and label each minimum and maximum with the appropriate conformation of 1,2-dichloroethane.

(b) Which conformation of 1,2-dichloroethane is present in the greatest amount? Explain.

2.51 Which of the following compounds should have the larger difference between staggered and eclipsed conformations for rotation about the bond that is shown? Explain your reasoning carefully.

$(CH_3)_3C{-}C(CH_3)_3$ (*A*) $\qquad (CH_3)_3Si{-}Si(CH_3)_3$ (*B*)

2.52 From what you learned in Sec. 1.6B about the relative lengths of C—C and C—O bonds, predict which of the following compounds should have the larger energy difference between gauche and anti conformations about the indicated bond. Explain.

$CH_3O{-}CH_2CH_3$ (*A*) $\qquad CH_3CH_2{-}CH_2CH_3$ (*B*)

2.53 (a) What value is expected for the dipole moment of the anti conformation of 1,2-dibromoethane, $Br{-}CH_2{-}CH_2{-}Br$? Explain.

(b) The dipole moment μ of any compound that undergoes internal rotation can be expressed as a weighted average of the dipole moments of each of its conformations by the following equation:

$$\mu = \mu_1 N_1 + \mu_2 N_2 + \mu_3 N_3$$

in which μ_i is the dipole moment of conformation i, and N_i is the mole fraction of conformation i. (The mole fraction of any conformation i is the number of moles of i divided by the total moles of all conformations.) There are about 82 mole percent of anti conformation and about 9 mole percent of each gauche conformation present at equilibrium in 1,2-dibromoethane, and the observed dipole moment μ of 1,2-dibromoethane is 1.0 D. Using the preceding equation and the answer to part (a), calculate the dipole moment of a gauche conformation of 1,2-dibromoethane.

2.54 Carv and Di Oxhide drive their family automobile about 12,000 miles per year with an average mileage of about 25 miles per gallon of gasoline. What is the "carbon footprint" (pounds of CO_2 released into the atmosphere) of the Oxhide family car over one year? Ignoring the oxygenates present, take the density of gasoline as 0.692 g mL^{-1}. (Useful conversion factors: 1 gallon = 3.785 L, 1 kg = 2.204 lb.)

2.55 This problem illustrates how combustion can be used to determine the molecular formula of an unknown compound. A compound *X* (8.00 mg) with a molecular mass of 100 undergoes combustion in a stream of oxygen to give 24.60 mg of CO_2 and 11.51 mg of H_2O.

(a) Calculate the mass of carbon and hydrogen in *X*.

(b) How many moles of H are present in *X* per mole of C? Express this as a formula C_1H_y.

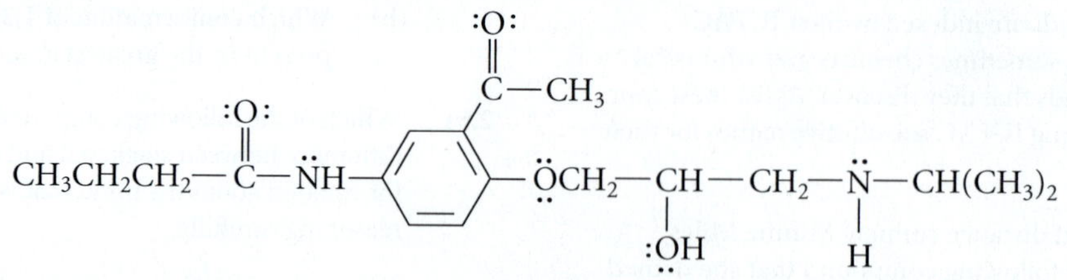

acebutolol

FIGURE P2.56

(c) Multiply this formula by successive integers until the amount of H—that is, y—is also an integer. This is the *empirical formula* of X—the smallest whole number ratio of the elements.

(d) Calculate the molecular mass of your empirical formula. If it matches the molecular mass of compound X, then the empirical formula is also the molecular formula. If not, multiply the molecular mass of your empirical formula by successive integers until it matches the molecular mass of X. When you have achieved a match, you have the molecular formula.

2.56 Organic compounds can contain multiple functional groups. Identify all of the functional groups present in acebutolol (**Fig. P2.56**), a drug used to treat high blood pressure. Name the compound class to which each functional group belongs.

2.57 The α-amino acids are the building blocks of proteins. Most have the following general structure:

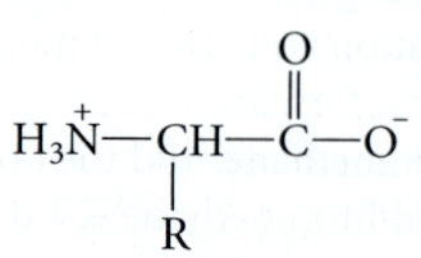

general structure of the α-amino acids

These amino acids differ only in their side chains —R. What functional groups are present in the side chains of each of the following amino acids?

(a)

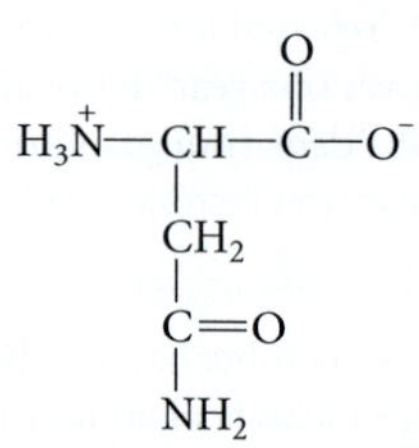

asparagine

(b)

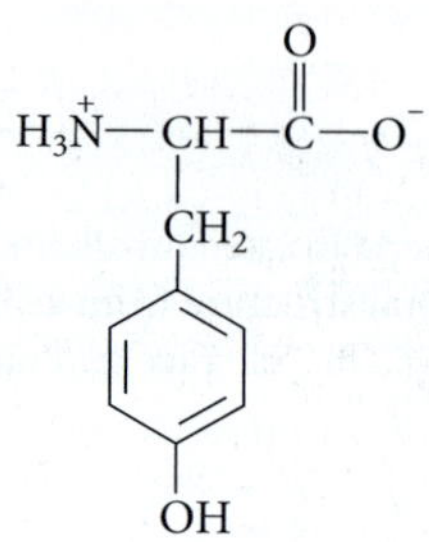

tyrosine

(c)

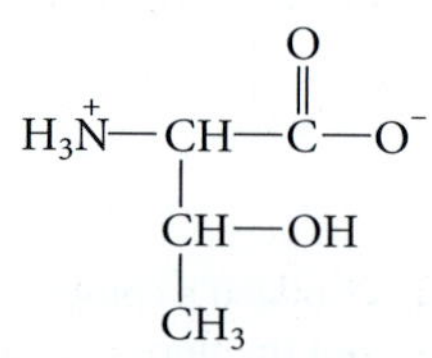

threonine

2.58 (a) Two amides are constitutional isomers and have the formula C_4H_9NO, and each contains an isopropyl group as part of its structure. Give structures for these two isomeric amides.

(b) Draw the structure of two other amides with the formula C_4H_9NO that do not contain isopropyl groups.

(c) Draw the structure of a compound X that is a constitutional isomer of the amides in parts (a) and (b), but is not an amide, and contains both an amine and an alcohol functional group.

(d) Could a compound with the formula C_4H_9NO contain a nitrile functional group? Explain.

3 THE CURVED-ARROW NOTATION, RESONANCE, ACIDS AND BASES, AND CHEMICAL EQUILIBRIUM

Chapter 3 concentrates on four topics that are essential for your preparation to study organic chemistry. The first is *acid–base reactions*, a topic that you probably studied in previous chemistry courses. Acid–base reactions are worth special attention for several reasons. First, many organic reactions are themselves acid–base reactions or are close analogs of common inorganic acid–base reactions with which you are familiar. This means that if you understand the principles behind simple acid–base reactions, you also understand the principles behind the analogous organic reactions. Second, acid–base reactions provide simple examples that can be used to illustrate some ideas that will prove useful in more complicated reactions. Finally, understanding acid–base chemistry is crucial to understanding biochemistry.

The second topic is an introduction to the *curved-arrow notation*, a powerful device to help you follow, understand, and even predict organic reactions. We use acid–base reactions as important examples of the curved-arrow notation.

We then take a deeper look at *resonance*, a topic that was introduced in Sec. 1.5. We apply the curved-arrow notation to the derivation of resonance structures, and we explain how to determine which resonance structures are most important in describing a resonance-stabilized compound.

Lastly, we consider the principles of chemical equilibrium and free energy, topics you may have also studied in general chemistry. We then apply these principles to acid–base reactions.

3.1 LEWIS ACID–BASE ASSOCIATION REACTIONS

A. Electron-Deficient Compounds

In Sec. 1.3B, you learned that covalent bonding in many cases conforms to the *octet rule*, which says that the sum of the bonding and nonbonding valence electrons surrounding a given atom equals eight (two for hydrogen). The octet rule (or "duet" rule in the case of hydrogen) holds without exception for covalently bonded atoms from the first and second periods of the periodic table. Although the electronic octet can be exceeded when atoms from period 3 and higher are involved in covalent bonds, the rule is often obeyed for main-group elements in these periods as well.

The octet rule stipulates the *maximum* number of electrons, but it is possible for an atom to have *fewer* than an octet of electrons. In particular, some compounds contain atoms that are *short of an octet by one or more electron pairs.* Such species are called **electron-deficient compounds.** Here are two examples of electron-deficient compounds:

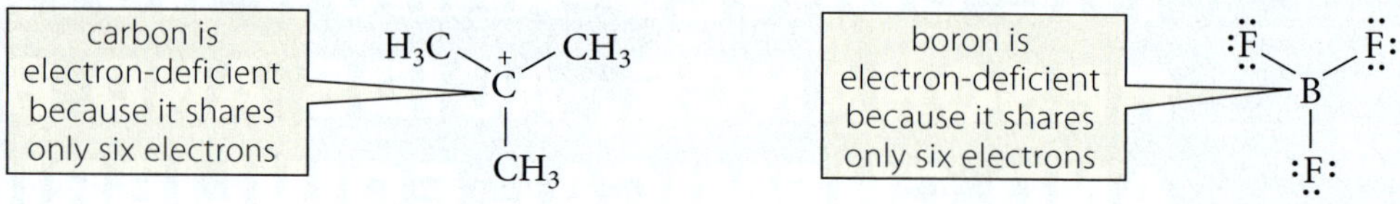

***tert*-butyl cation** **boron trifluoride**

(3.1)

The term *electron-deficient* does *not* mean "positively charged." An electron-deficient atom *may be* positively charged, as in the *tert*-butyl cation, above; but, as boron trifluoride illustrates, it may be neutral. The criterion for electron deficiency is an electron count short of an octet by one or more electron pairs.

B. Reactions of Electron-Deficient Compounds with Lewis Bases

Electron-deficient compounds have a tendency to undergo chemical reactions that complete their valence-shell octets. In such reactions, an electron-deficient compound reacts with a species that has one or more nonbonding valence electron pairs. An example of such a reaction is the association of boron trifluoride and fluoride ion:

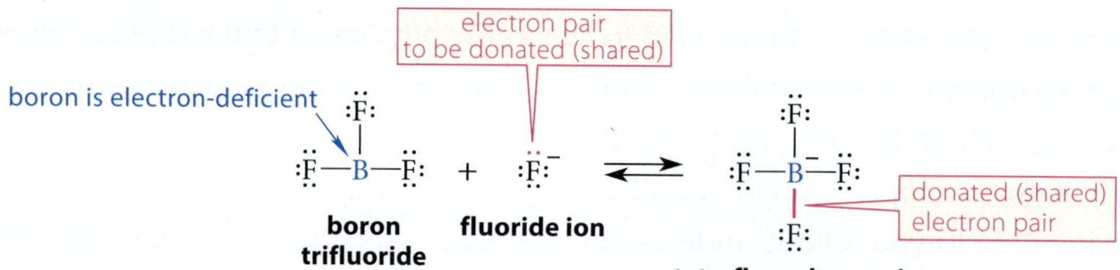

(3.2)

In such reactions, the electron-deficient compound acts as a *Lewis acid.* A **Lewis acid** is a species that accepts an electron pair to form a new bond in a chemical reaction. Boron trifluoride is the Lewis acid in Eq. 3.2 because it accepts an electron pair from the fluoride ion to form a new B—F bond in the product, tetrafluoroborate anion. The species that donates the electron pair to a Lewis acid to form a new bond is called a **Lewis base.** Fluoride ion is the Lewis base in Eq. 3.2. When the Lewis base donates an electron pair, it is not giving it away completely; rather, it is donating it to be shared in a covalent bond. (An analogy would be your "donation" of your tandem bicycle to a friend with the understanding that you will ride it together.) When an electron-deficient Lewis acid and a Lewis base combine to give a single product, as in this example, the reaction is called a **Lewis acid–base association reaction.**

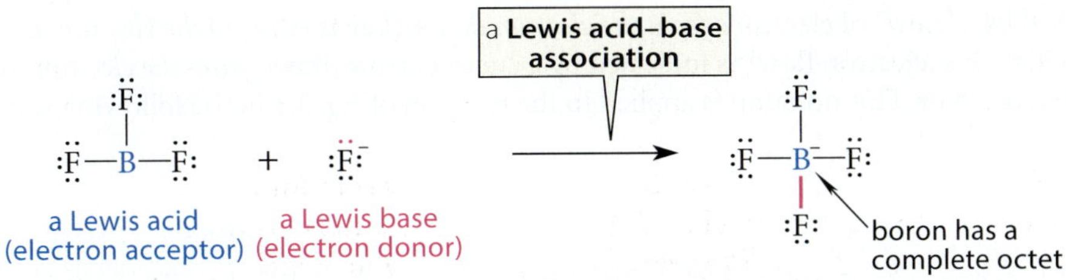

(3.3)

Remember that there are two types of electron counting (Sec. 1.4C). When we count for formal charge, the boron has a net gain of one electron, which gives it a –1 charge. But when we count for the octet, boron has gained two electrons. As a result of this association reaction, each atom in the tetrafluoroborate ion has a complete octet. In fact, *completion of the octet on boron provides the major driving force for this reaction.*

The reverse of a Lewis acid–base reaction is a **Lewis acid–base dissociation**. Therefore, the dissociation of fluoride ion from $^{-}BF_4$ to give BF_3 and F^-—that is, the reverse reaction in Eq. 3.2—is an example of a Lewis acid–base dissociation.

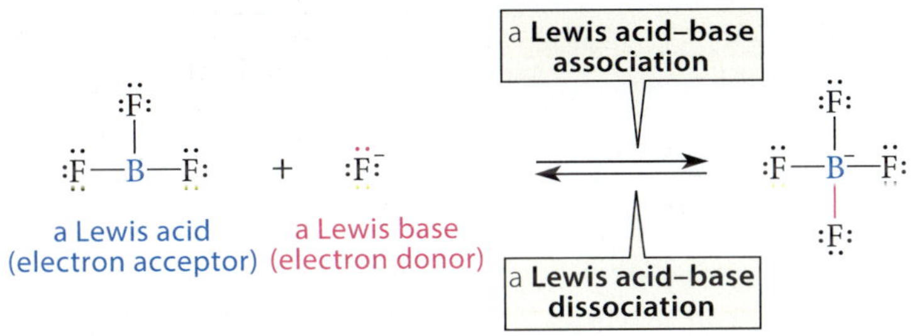

(3.4)

Study Problem 3.1

Which of the following compounds can react with the Lewis base Cl^- in a Lewis acid–base association reaction?

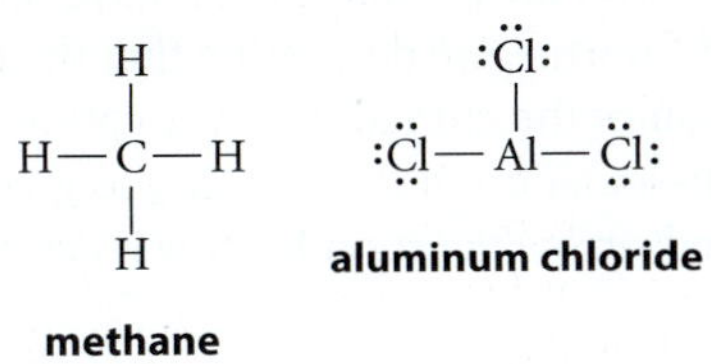

Solution For a compound to react as a Lewis acid in an association reaction, it must be able to accept an electron pair from the Lewis base Cl^-. In aluminum chloride, the aluminum is short of an octet by one pair. Hence, aluminum chloride is an electron-deficient compound and can readily accept an electron pair from chloride ion in an association reaction, as follows:

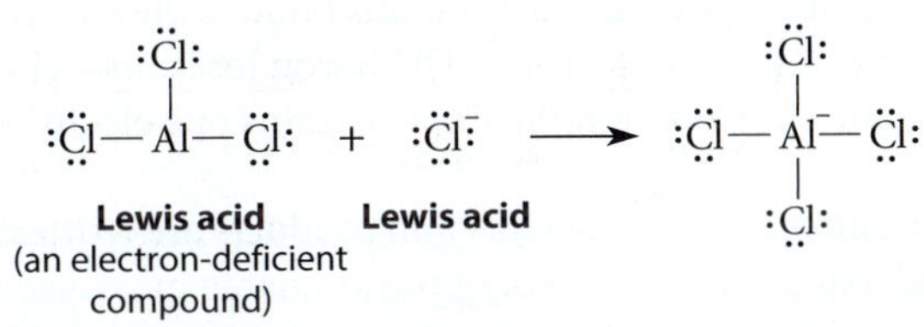

In contrast, every atom in methane has the nearest noble-gas number of electrons (carbon has eight, hydrogen has two). Hence, methane is *not* electron-deficient and *cannot* undergo a Lewis acid–base association reaction.

C. The Curved-Arrow Notation for Lewis Acid–Base Association and Dissociation Reaction

Organic chemists have developed a symbolic device for keeping track of electron pairs in chemical reactions; this device is called the **curved-arrow notation**. As this notation is applied to the reactions of Lewis bases with electron-deficient Lewis acids, the formation of a chemical bond is described by a "flow" of electrons *from the electron donor* (Lewis base) *to the electron acceptor* (Lewis acid). This "electron flow" is indicated by a curved arrow drawn *from the electron source to the electron acceptor.* This notation is applied to the reaction of Eq. 3.3 in the following way:

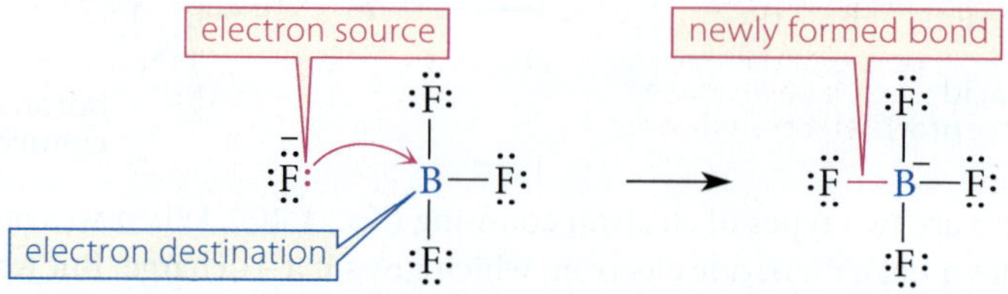

(3.5)

The red curved arrow indicates that a nonbonding electron pair on the fluoride ion becomes the shared electron pair in the newly formed bond of $^{-}BF_4$.

The arrow might seem to suggest that both electrons are "jumping" from one atom to the other, but that is not what the curved arrow symbolizes. The fluorine is *not* transferring both electrons; rather, it is sharing them in a bond with boron, as discussed in Sec. 3.1A.

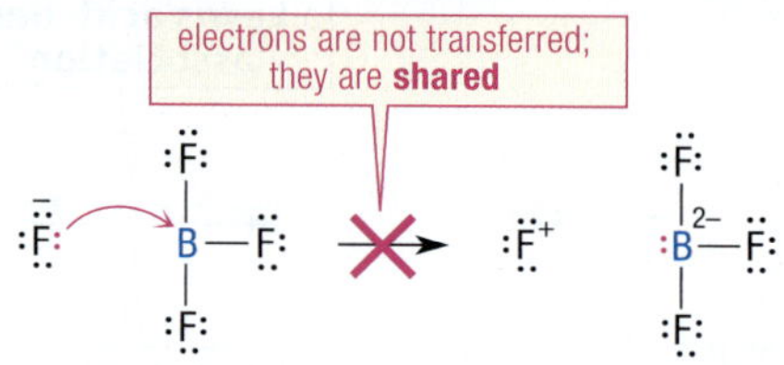

The correct application of the curved-arrow notation also involves computing and properly assigning the formal charge to the products. For each reaction involving the curved-arrow notation, *the algebraic sum of the charges on the reactants must equal the algebraic sum of the charges on the products.* In other words, *total charge is conserved.* Thus, in Eq. 3.5, the reactants have a net charge of -1; therefore, the products must have the same net charge. By calculating the formal charge on boron and fluorine, we determine that the formal charge must reside on boron.

To illustrate the application of the curved-arrow notation to a Lewis acid–base dissociation reaction, let's consider the dissociation of the ion $^{-}BF_4$ to give BF_3 and F^-; this reaction is the reverse of Eq. 3.5. The curved-arrow notation for this reaction is as follows:

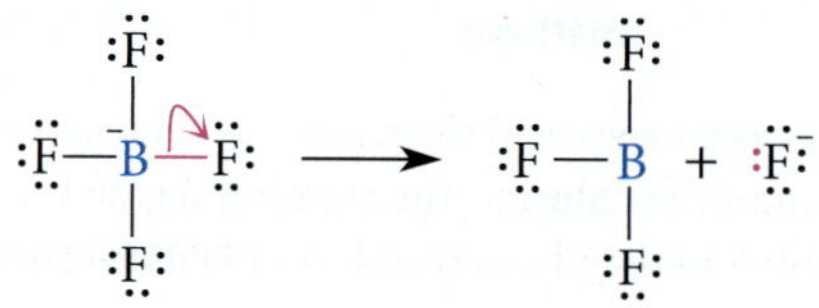

(3.6)

Because the B—F bond breaks in this reaction, this bond is the source of the electron pair that is transferred to a fluorine to give fluoride ion. The boron loses both electrons when we count for the octet; but it loses one electron (and the fluorine gains one electron) when we count for formal charge.

An important convention: When reactants and products are written with an overall net charge, as in Eq. 3.6, it is understood that an ion of opposite charge must also be present for overall electrical neutrality. For example, in this equation, a positive ion must be present; this ion could be Na^+, K^+, or some other positive ion that is "along for the ride" and *does not have an explicit role when we write the reaction.* Such an ion is called a **spectator ion**. If we include a spectator ion in this equation, it would be present on both sides of the equation. Typically, spectator ions are omitted.

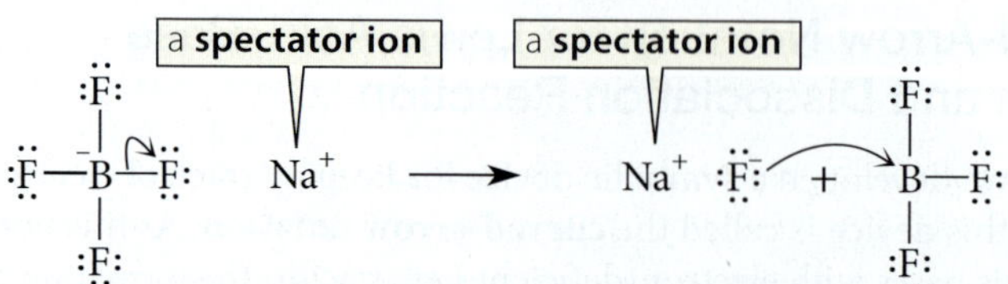

(3.7)

Study Problem 3.2

Sodium methoxide and aluminum trichloride react in a Lewis acid–base association reaction. Propose a curved-arrow notation for this reaction, and give the product. Then show the Lewis acid–base dissociation reaction along with the curved-arrow notation. Be sure to keep track of formal charge.

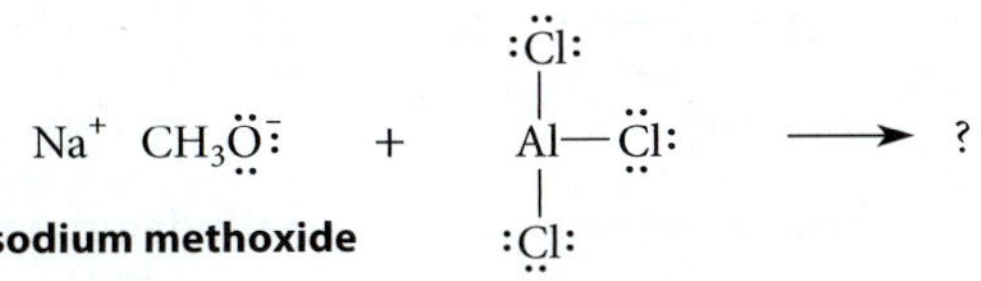

sodium methoxide

aluminum trichloride

Solution A Lewis acid–base association reaction requires an electron-deficient atom as the Lewis acid. The aluminum (Al) is such an atom because it is surrounded by only six electrons. The oxygen of sodium methoxide ion, with its nonbonding electron pairs, can serve as the Lewis base. Draw a curved arrow from the Lewis base to the Lewis acid and then complete the reaction as indicated by the curved arrows. Be sure to add the formal charges. To summarize:

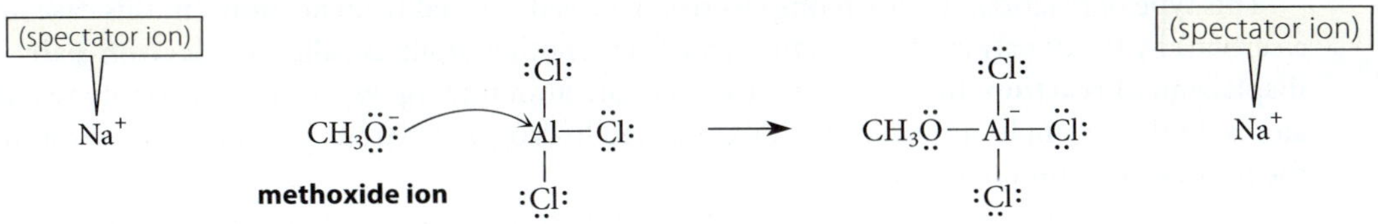

methoxide ion

aluminum trichloride

The reverse reaction shows the detachment of the methoxide ion with its electron pair. The curved-arrow notation shows the transformation of a bonding electron pair to a nonbonding electron pair on the oxygen.

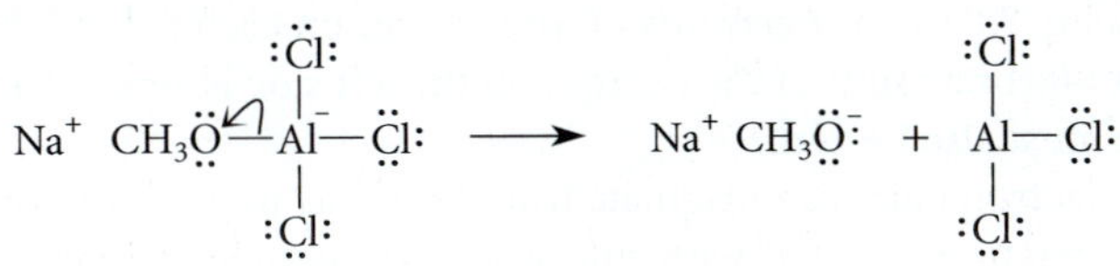

Focused Problem

3.1 Use the curved-arrow notation to derive a structure for the product of each of the following Lewis acid–base association reactions; be sure to assign formal charges. Then, on the structure of the reaction product, show the curved-arrow notation for the reverse reaction.

(a)

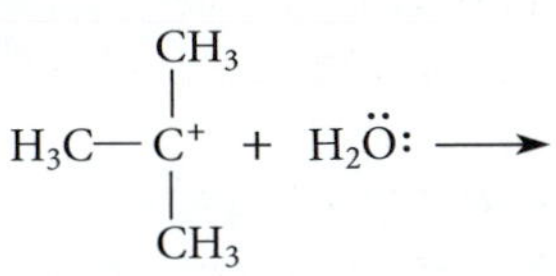

(b)

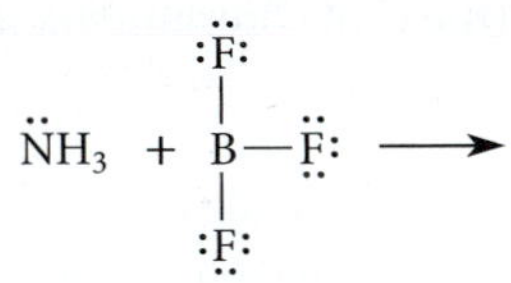

3.2 ELECTRON-PAIR DISPLACEMENT REACTIONS

A. Donation of Electrons to Atoms That Are Not Electron-Deficient

In some reactions, an electron pair is donated to an atom that is *not* electron-deficient. When this happens, another electron pair must simultaneously depart from the receiving atom so that the octet rule is not violated. The following reaction is an example of such a process.

When drawing curved arrows, the curvature doesn't matter. What matters is that *curved arrows always start at the electron source and point to the destination atom.*

both notations are valid:

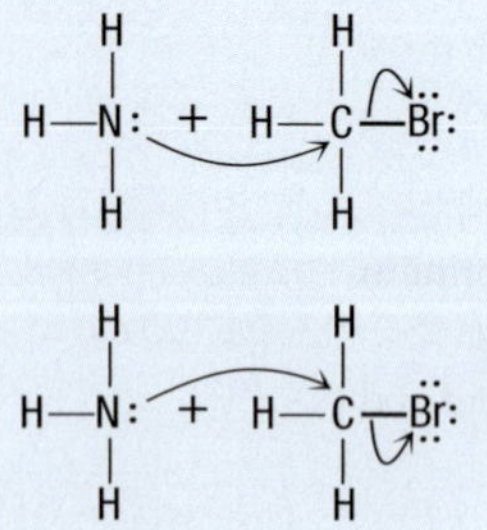

For a summary of the rules for using curved arrows, see Study Guide Link 3.1.

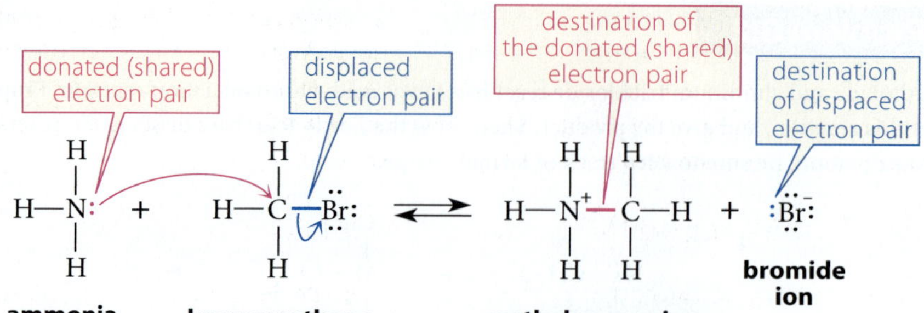

(3.8)

In this reaction, the carbon of bromomethane receives and shares an electron pair from the nitrogen of ammonia. As a result, this nitrogen becomes bonded to the carbon to give the methylammonium ion, and the electron pair in the C—Br bond of bromomethane becomes an additional nonbonding electron pair in the bromide ion. If this electron pair had not departed, carbon would have ended up with 10 electrons, which is more than is allowed by the octet rule.

This type of reaction, in which one electron pair is displaced from an atom (in this case, from a carbon) by the donation of an electron pair from another atom, is called an **electron-pair displacement reaction**. In many such reactions, an atom undergoes a transfer between two other atoms. In this example, a carbon (with its attached hydrogens) is transferred from the bromine to the nitrogen of ammonia.

The curved-arrow notation is particularly useful for following electron-pair displacement reactions. This usage is illustrated in Eq. 3.8. In this case, *two* arrows are required, one for the donated electron pair and one for the displaced electron pair. Notice again that each curved arrow originates at the *source* of electrons—in this case, a nonbonding electron pair—and terminates at the *destination* of the electron pair.

Notice also in Eq. 3.8 the *conservation of total charge* on each side of the equation, as discussed in Sec. 3.1C. The algebraic sum of the charges on the left side is zero; hence, the sum of all charges on the right side must also be zero.

The donated electron pairs can originate from bonds as well as nonbonding pairs. This is illustrated by the reaction of $^{-}AlH_4$ with chloromethane to give methane, AlH_3, and chloride ion:

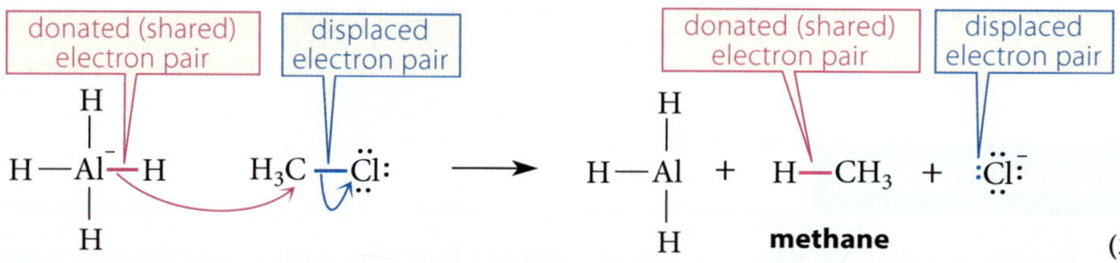

(3.9)

In this notation, the bond corresponding to the donated electrons is "hinged" at the transferred atom (the H of the Al—H bond); it swings away from the aluminum and toward the atom that receives the electrons (the C of chloromethane).

Study Problem 3.3

Give the curved-arrow notation for the following reaction.

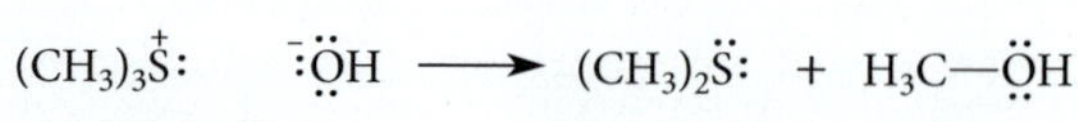

Solution In this reaction, a nonbonding electron pair from the oxygen forms a bond to one of the methyl carbons of the dimethylsulfonium ion. The carbon–sulfur bond is broken, because in the product, the sulfur is bound to only two carbons. Therefore, a carbon atom (along with its three hydrogens) is transferred from the sulfur to the oxygen. Because this is an electron-pair displacement reaction, two curved arrows are required. Remember that a curved arrow is drawn from the *source* of an electron pair to its *destination.* The *source* of the donated electron pair is the ^{-}OH ion.

The *destination* of the donated electron pair is the carbon atom. Hence, one curved arrow goes from an electron pair of the $^-$OH (any one of the three pairs) to the carbon atom. Because carbon can have only eight electrons, it must lose a pair of electrons to the sulfur, which gains an electron pair in the reaction. Hence, the source of this electron pair is the C—S bond; its destination is the sulfur. Notice that a carbon–sulfur bond *must be shown explicitly* to apply the curved-arrow notation. The curved-arrow notation for this reaction is as follows:

(Remember that curved arrows can be drawn in many different ways as long as the head and tail of the arrow are in the right places.) In this reaction, a methyl group is transferred from sulfur to oxygen.

Study Problem 3.3 shows how to write the curved-arrow notation for a completed reaction. Study Problem 3.4 shows how to complete a reaction for which the curved-arrow notation is given.

Study Problem 3.4

Given the following two reactants and the curved-arrow notation for their reaction, draw the structure of the product.

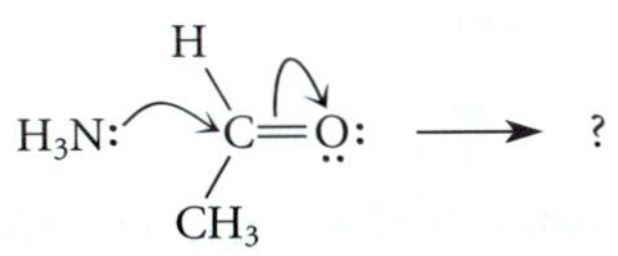

Solution The bonds or nonbonding electron pairs at the tails of the arrows are the ones that will not be in the same place in the product. The heads of the arrows point to the places at which new bonds or nonbonding electron pairs exist in the product. Use the following steps to draw the product.

Step 1. Redraw all atoms just as they were in the reactants:

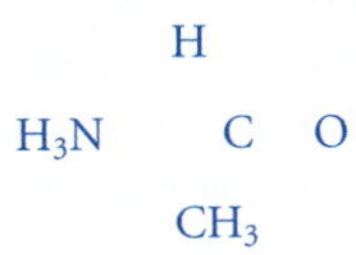

Step 2. Put in the bonds and electron pairs that do not change:

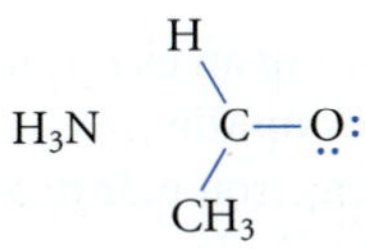

Step 3. Draw the new bonds and nonbonding electron pairs indicated by the curved-arrow notation:

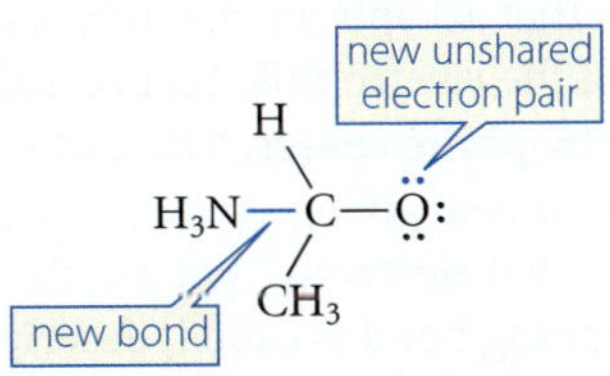

Step 4. Complete the formal charges to give the product. The *algebraic sum* of the formal charges in the reactants and products must be the same—zero in this case.

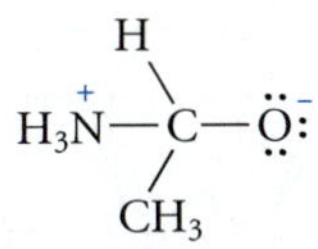

Study Guide Link 3.1
Rules for Use of the Curved-Arrow Notation

As you learn to use the curved-arrow notation, you will find the additional assistance in Study Guide Link 3.1 to be very useful.

Focused Problem

3.2 For each of the following cases, give the product(s) of the transformation indicated by the curved-arrow notation.

(a) 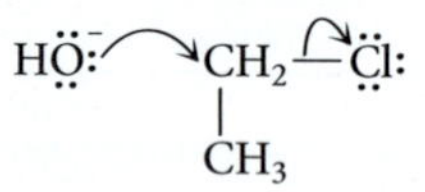

(b) 

(c) 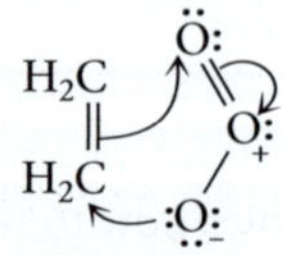

B. Nucleophiles, Electrophiles, and Leaving Groups

In this section we develop a terminology that is widely used for classifying the components of an electron-pair displacement reaction. Let's return to the reaction (3.8) that we used to introduce electron-pair displacements:

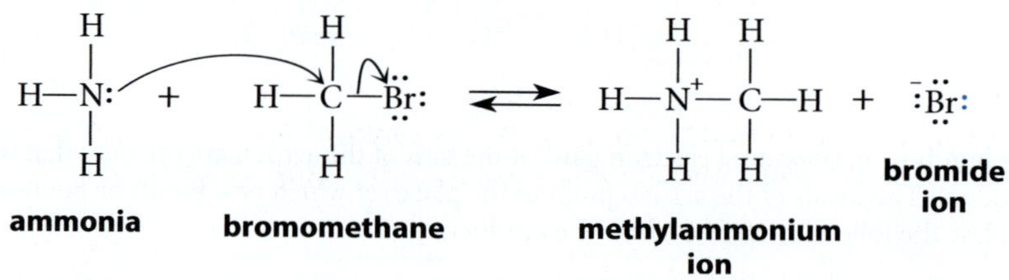

(3.10)

Let's first think about the left side of this reaction from a Lewis acid–base perspective. The ammonia is a Lewis base; it is donating a pair of electrons. The carbon is accepting this electron pair and seems to be a Lewis acid. However, it is also donating a bonding electron pair to the bromine, and it might also be considered simultaneously to be a Lewis base. The bromine is accepting this bonding electron pair, and it might be considered to be a Lewis acid. This example shows that Lewis acid–base terminology is not very useful for describing uniquely the roles of each "actor" in this reaction.

The terms used for the components of an electron-pair displacement reaction are *nucleophile, electrophile,* and *leaving group.* A **nucleophile** (from the Greek word *philos,* meaning "nucleus-loving") is a species that donates an electron pair to form a new bond. In Eq. 3.10, ammonia is the nucleophile. The atom that actually donates the electron pair is called the **nucleophilic atom** or **nucleophilic center**. Nitrogen is the nucleophilic center in ammonia. An **electrophile** ("electron-loving") is a species that accepts an electron pair from the nucleophile. Bromomethane is the electrophile. The atom of the electrophile that actually accepts the electron pair is called the **electrophilic atom** or **electrophilic center**. The carbon of bromomethane is the electrophilic center. (In this case, the carbon is also giving up electrons, and this behavior may seem inconsistent for an atom that "loves electrons," but put this point aside for now.) The group that accepts electrons from the breaking bond is called, descriptively enough, a **leaving group**. The bromine is the leaving group; it becomes the bromide ion after accepting an electron pair from the breaking bond.

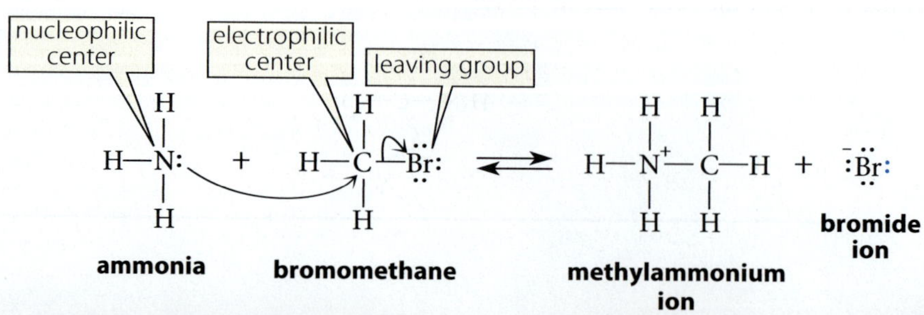

(3.11)

In the reverse reaction, the roles of the nucleophile and the leaving group are reversed, and the electrophilic center is the same.

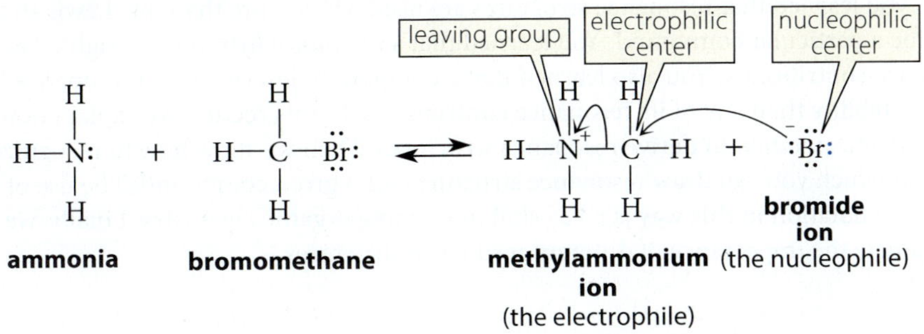

(3.12)

The same terminology can be applied to Lewis acid–base associations and dissociations. Using Eq. 3.5 as an example,

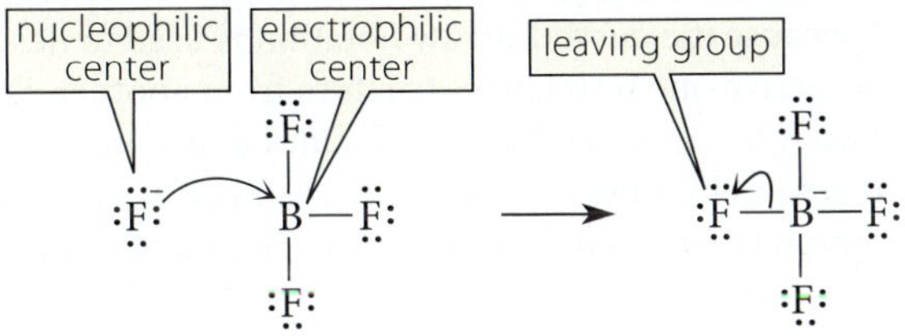

(3.13)

In the association (forward) direction of a Lewis acid–base association, there is a nucleophile and an electrophile, but no leaving group. In the dissociation (reverse) direction, there is a leaving group but no nucleophile.

We've already noted that in some electron-pair displacement reactions, the nucleophilic electron pair can originate from a bond rather than a nonbonding electron pair.

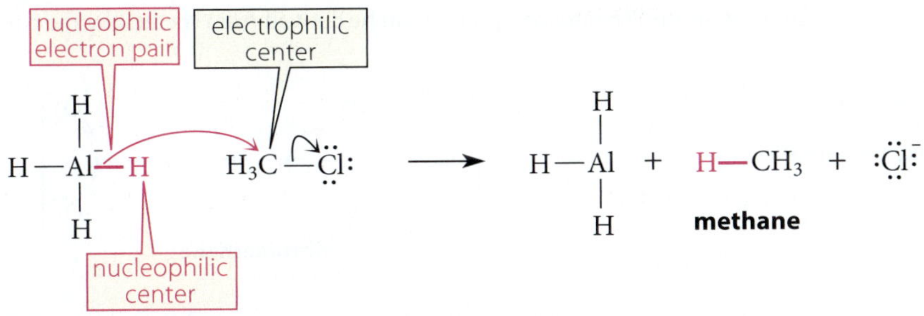

(3.14)

In this case, the nucleophilic center is the hydrogen with its pair of *bonding* electrons.

Focused Problems

3.3 Consider the reaction analyzed in Study Problem 3.3, reproduced here. Identify the nucleophilic center, the electrophilic center, and the leaving group in the forward direction. (Don't hesitate to draw out the bonds between the sulfurs and each methyl group, if necessary.)

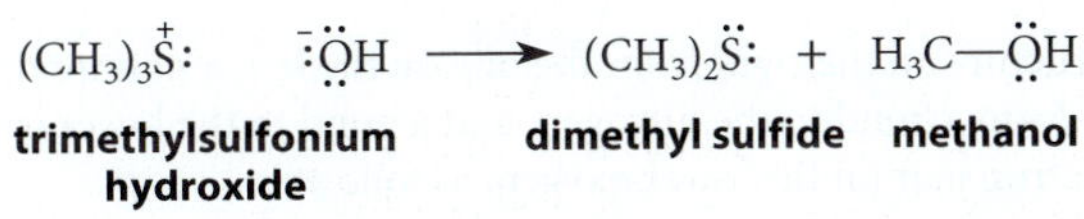

3.4 (a) Use the curved-arrow notation to complete the following Lewis acid–base association reaction.

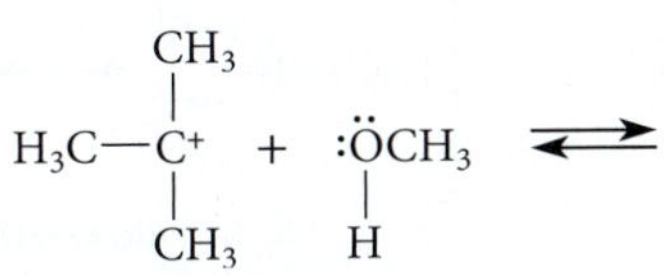

(b) After you have completed the reaction, give the curved-arrow notation for the reverse direction.

(c) Identify the nucleophilic center, the electrophilic center, and the leaving group in both forward and reverse directions of the reaction in part (a).

3.3 RESONANCE: A DEEPER LOOK

In Sec. 1.5 you learned that resonance structures are used when more than one Lewis structure can describe a particular compound. You learned that a *resonance hybrid* is a weighted average of the resonance contributors. You also learned that a compound described by a resonance hybrid has greater stability than any of its resonance contributors. In this section we explain how to use the *curved-arrow notation* to derive resonance structures. Then we show how to recognize the situations in which you can draw resonance structures for a given compound. The use of the curved-arrow notation in this way is a key skill in learning organic chemistry. Finally, we describe how to evaluate the importance of different resonance structures.

A. Use of the Curved-Arrow Notation to Derive Resonance Structures

Because resonance structures differ in the positions of electrons—nonbonding electron pairs and electron-pair bonds—and because the curved-arrow notation is used to describe electron flow, this notation can be used to derive one resonance structure from another. Study Problem 3.5 illustrates this point with two resonance-stabilized molecules that were used as examples in Sec. 1.5. The use of curved arrows in these two examples is similar to the use of curved arrows in the Lewis acid–base association reactions and the electron-pair displacement reactions in Sec. 3.1 and 3.2, respectively.

Study Problem 3.5

In each of the following sets, show how the second resonance structure can be derived from the first by the curved-arrow notation.

(a) 

methoxymethyl cation

(b) 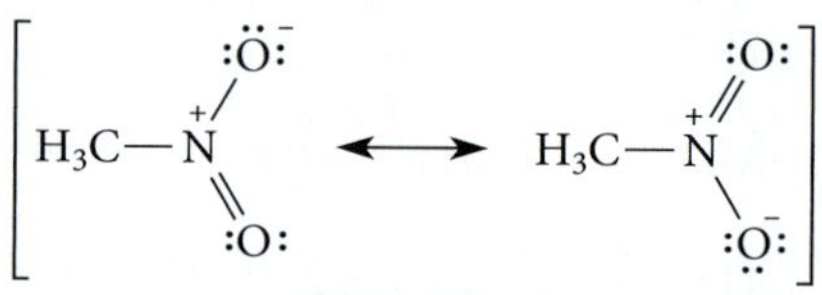

nitromethane

Solution

(a) In the structure on the left, the positively charged carbon is electron-deficient. The structure on the right is derived by the donation of a nonbonding electron pair from the oxygen to this carbon.

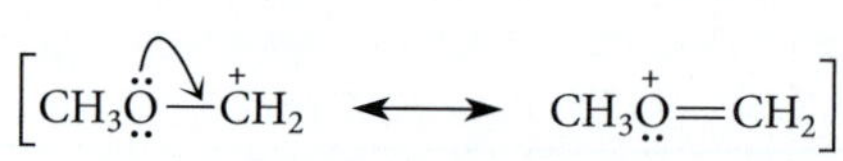

methoxymethyl cation

This transformation resembles a Lewis acid–base association reaction, and the same curved-arrow notation is used: a single curved arrow showing the donation of the nonbonding pair of electrons to the electron-deficient carbon.

(b) To derive the structure on the right from the one on the left, a nonbonding electron pair on the upper oxygen must be used to form a bond to the nitrogen, and a bond to the lower oxygen must be used to form a nonbonding electron pair on the lower oxygen, as follows:

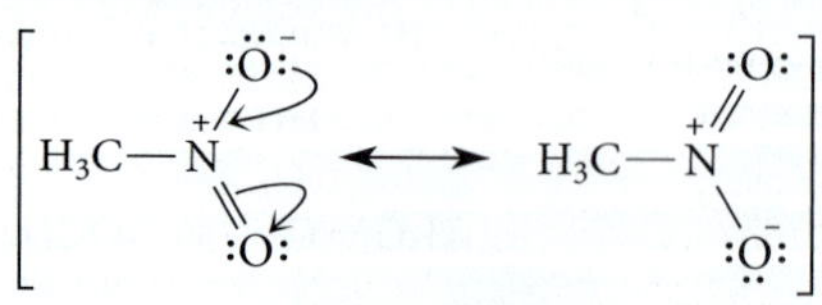

nitromethane

Two arrows are required because the formation of the new bond requires the *displacement* of another. So, we use the curved-arrow notation for electron-pair displacements.

In both of the preceding examples, the curved-arrow notation is applied in the left-to-right direction. This notation can be applied to either structure to derive the other. Therefore, for part (a) in the right-to-left direction, the curved-arrow notation is as follows:

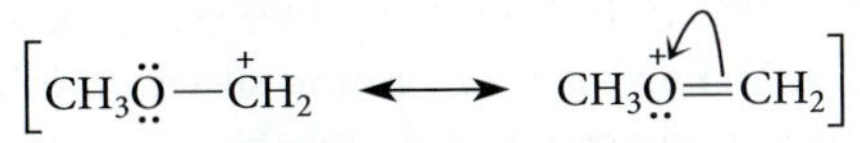

methoxymethyl cation

You should draw the curved arrow for part (b) in the right-to-left direction.

When applied to *reactions* (Secs. 3.1 and 3.2), the curved-arrow notation shows electron *movement* as reactants are converted into products. The implication is that, as electrons move, atoms move as well. When the curved-arrow notation is applied in exactly the same way to resonance structures, however, it describes electron *distribution* rather than actual electron movement. We'll *pretend* that electrons move between resonance structures, but we must be aware that a resonance hybrid has *a single* electron distribution and *a single* molecular structure described by a weighted average of the electron distributions and structures of the individual contributors. When an electron pair "moves" within resonance structures, we say that it is **delocalized.** Any charges that "move" as a result of the movement of electrons are also delocalized. *In general, delocalizing electrons and charges stabilizes (lowers the energy of) a molecule or ion.*

B. When (and How) to Draw Resonance Structures

Some common structural patterns should alert you to the possibility of drawing resonance structures. We consider each in turn with an example.

1. *Two atoms with significantly different electronegativities are connected by a pi (π) bond.*
   A carbonyl group (C=O group) is a common example of such a situation. In this case, the π electrons can be delocalized onto the atom of greater electronegativity (in this case, the oxygen).

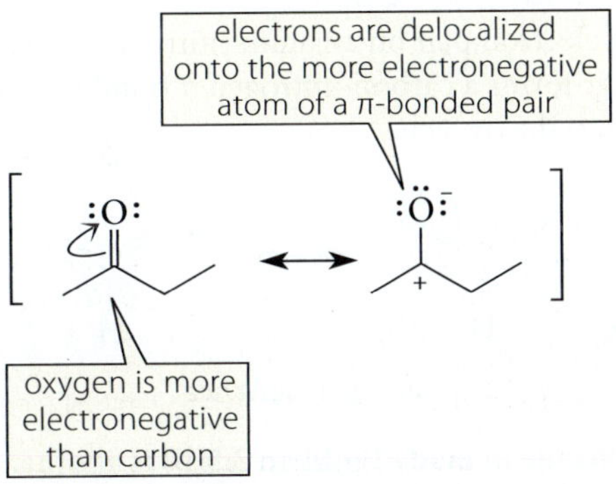

(3.15)

2. *An atom with a nonbonding electron pair is connected to an electron-deficient atom.*
   This situation is the reverse of the one in situation 1. Study Problem 3.5(a) is another example of this situation.

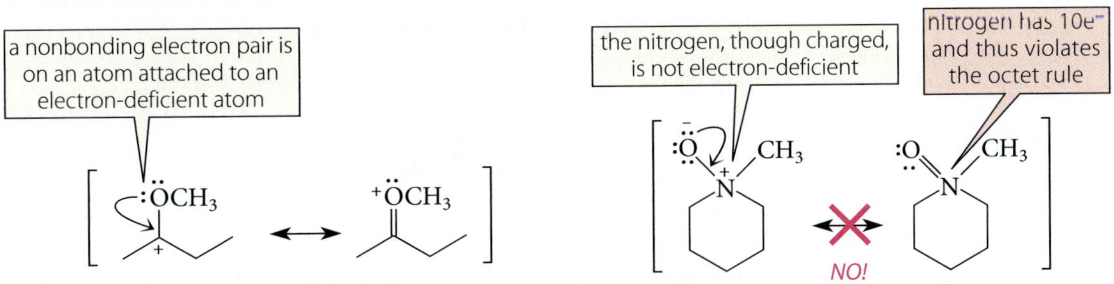

(3.16)

As the second example illustrates, be careful not to violate the octet rule for second-period atoms (B, C, N, O, F).

3. *An electron-deficient atom is in next to a π bond.*

The following example illustrates a common situation in which an electron-deficient, positively charged carbon is adjacent to a double bond.

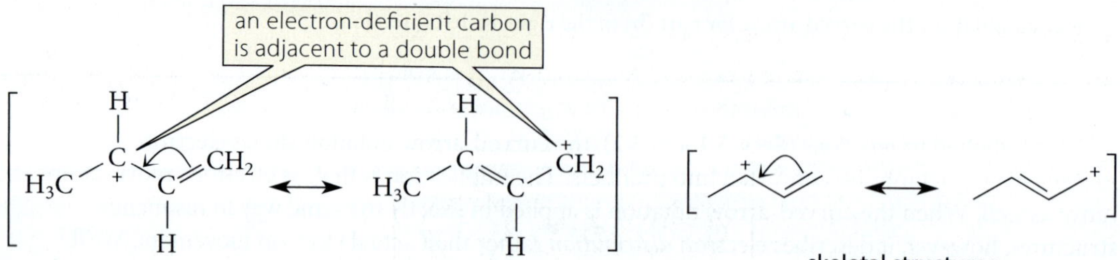

Lewis structures

skeletal structures

a resonance-stabilized cation (3.17)

Do not confuse this situation with one in which a cation is actually *on* a double-bonded carbon. Such cations are *not* resonance-stabilized.

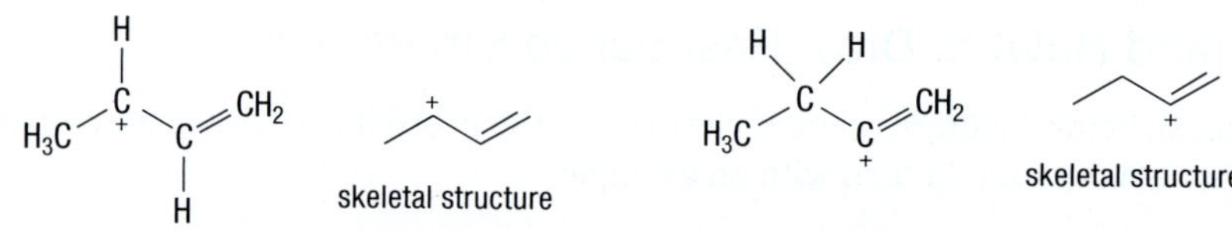

this cation is resonance-stabilized

this cation is *not* resonance-stabilized

4. *An atom with a nonbonding electron pair is next to a π bond.*

An *amide* contains an electron pair on an atom (nitrogen) adjacent to a carbon–oxygen π bond. The electron pair forms a carbon–nitrogen π bond that displaces the electron pair in the existing π bond onto the oxygen.

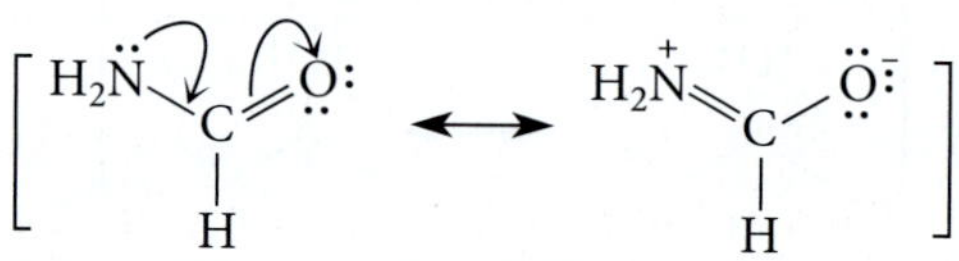

amide resonance (3.18)

The nitromethane molecule in Study Problem 3.5(b) is another example of this situation.

Again, distinguish carefully between an electron pair that is *adjacent* to a π bond and one that is participating in the π bond. An electron pair that is on an atom of a double bond cannot be delocalized by resonance.

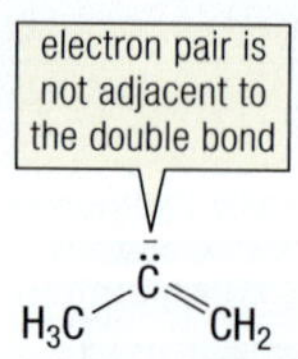

Because triple bonds involve π bonds, they can also be involved in resonance situations like those shown in examples 3 and 4.

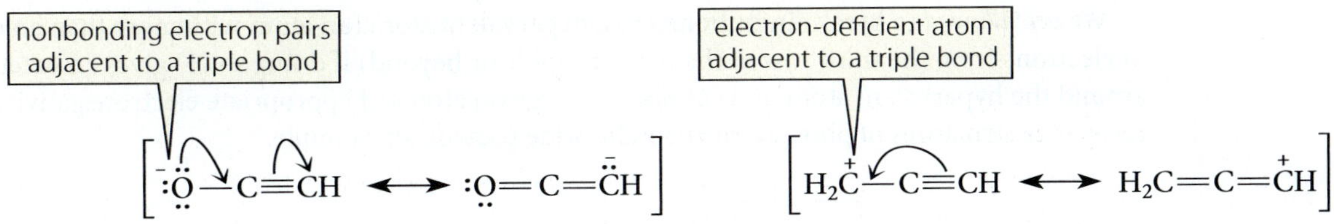

(3.19)

5. *Three alternating single and double bonds occur within a six-membered ring.*
Benzene and its derivatives are the archetypal examples of this situation. In this case, the resonance structures are derived when each double bond displaces the adjacent one in a cyclical fashion.

Benzene is the most commonly occurring case of cyclic electron delocalization; we introduce other aspects of this type of resonance in Chapter 15.

benzene (3.20)

Some molecules have more than two resonance structures. These structures are derived by successive application of some or all of the five structural patterns just illustrated.

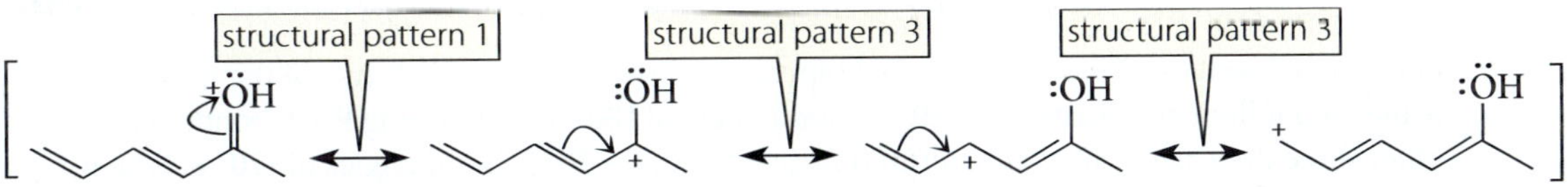

A *B* *C* *D*

(3.21)

The structural patterns cited apply to curved arrows for the structures in the left-to-right direction.

The curved arrows from individual structural patterns can be combined into single, more complex, curved-arrow notations. For example, we can show the conversion of structure *A* directly into structure *D* by combining all of the individual pairs of structures into one diagram:

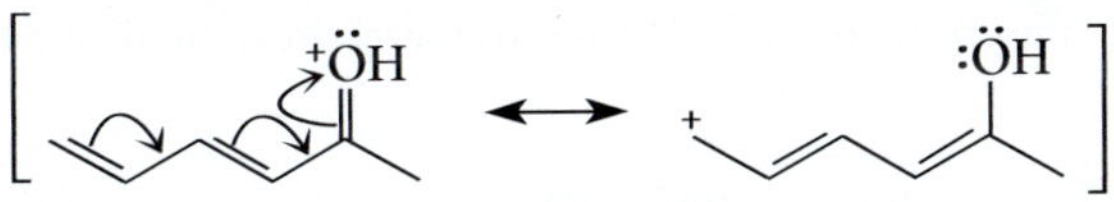

A *D* (3.22)

Conversely, if we were given the *A* to *D* conversion above, we could break out the notation into individual pairs of structures.

Drawing resonance structures can create or obliterate formal charges. Be sure to include the proper formal charges in each resonance structure. As a check on formal charge, *always make sure that the overall charge of each contributing resonance structure is the same.* This point is illustrated in the last two examples, in which the formal charge is +1 in each structure.

Don't break or form single bonds between two atoms of the first or second period when drawing resonance structures.

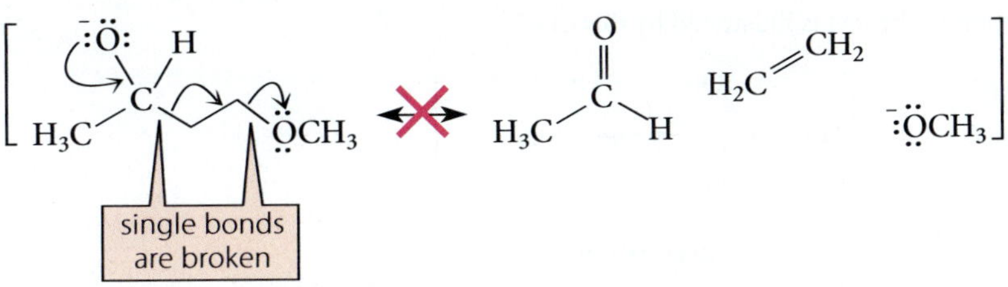

(3.23)

This could well represent a *reaction* in which atoms as well as electrons move, but it cannot be a resonance structure.

We *are allowed* to break single bonds to a hypervalent atom (an atom with more than an octet of electrons from the third row of the periodic table or beyond) if doing so (1) gives an octet around the hypervalent atom and (2) places charge on atoms of appropriate electronegativity. The resonance structures of phosphorus pentafluoride provide an example.

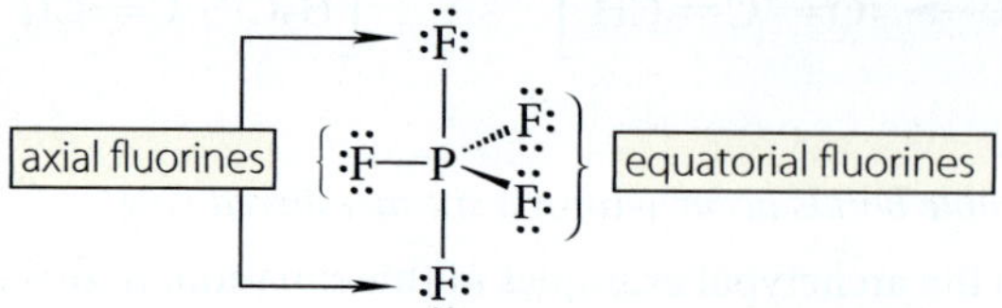

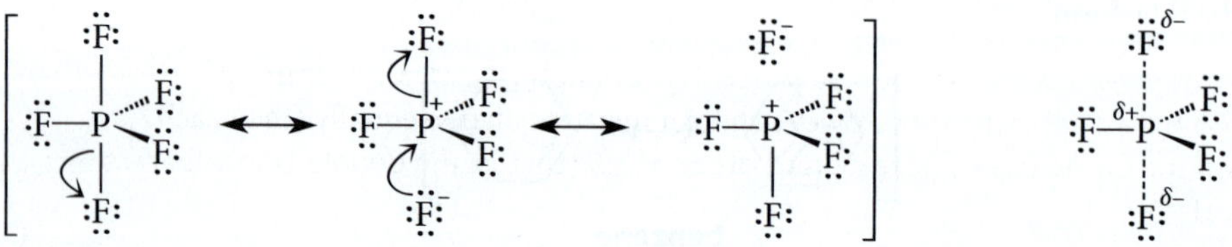

A B C

resonance structures of phosphorus pentafluoride

hybrid structure of phosphorus pentafluoride

(3.24)

Phosphorus pentafluoride has a trigonal bipyramidal structure in which two of the fluorines (called *axial* fluorines) have an F—P—F bond angle of 180° and very long P—F bonds. The other three fluorines (called *equatorial* fluorines) have F—P—F bond angles of 120° and lie in a plane containing the phosphorus perpendicular to the axial F—P—F bonds. Structure *A* has a hypervalent phosphorus. In structures *B* and *C*, the phosphorus has an octet and a positive charge. The compensating negative charge is shared by the two axial fluorines equally; fluorine is very electronegative. (Molecular orbital theory has confirmed this bonding picture.)

Focused Problems

3.5 In each case, provide the curved arrows that convert the structure on the left into the one on the right; then provide the curved arrows that convert the structure on the right into the one on the left. Cite the structural situation by number from the cases in this section.

(a) 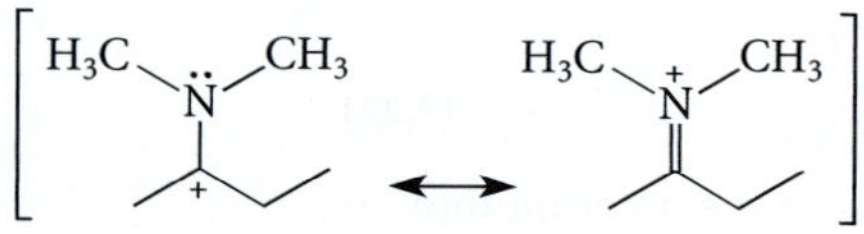

(b) 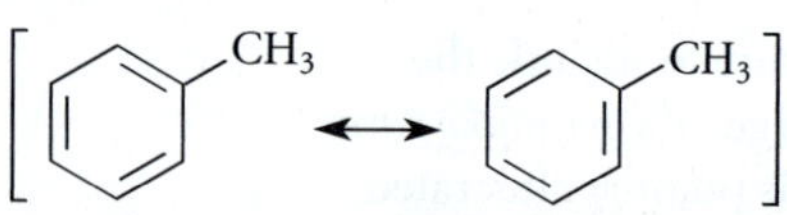

(c) 

(d) 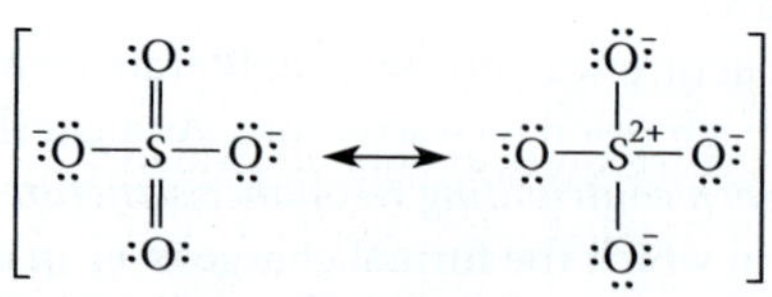

sulfate ion

3.6 (a) Derive a resonance structure for acetate ion using the curved-arrow notation in which the negative charge is delocalized to the other oxygen. Which situation in the text is illustrated by this case?

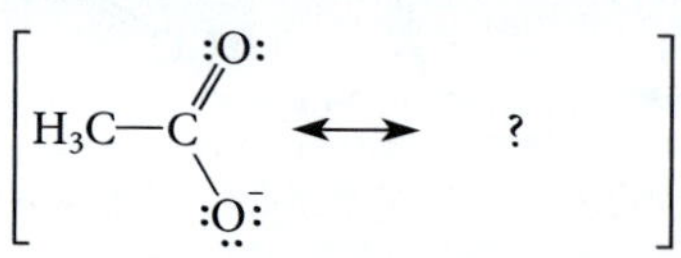

acetate ion

(b) Derive a resonance structure of acetonitrile-*N*-oxide using the curved-arrow notation in which negative charge is delocalized to a carbon atom.

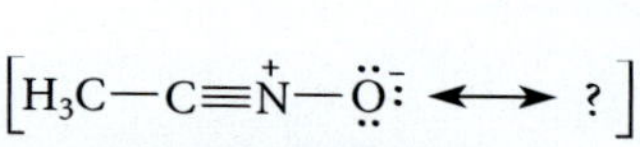

acetonitrile-*N*-oxide

3.7 Draw the resonance structure indicated by the curved arrows.

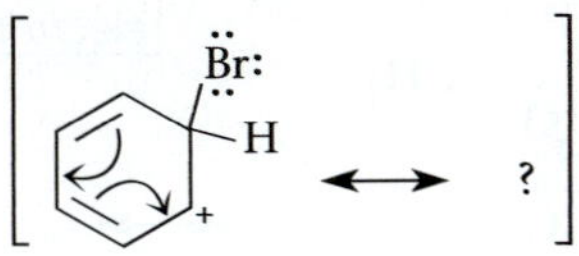

C. The Relative Importance of Resonance Structures

In Sec. 1.5, you learned how to determine whether two resonance structures are equivalent. When all resonance structures are equivalent, they are equally important in describing the molecule, and the hybrid is an average of the structures. For example, the two structures of nitromethane are equivalent; therefore, the two oxygens share a negative charge equally (they each have $-\frac{1}{2}$ charge); and the C—O bonds have a bond order of 1.5 (that is, the double-bond character is shared equally.)

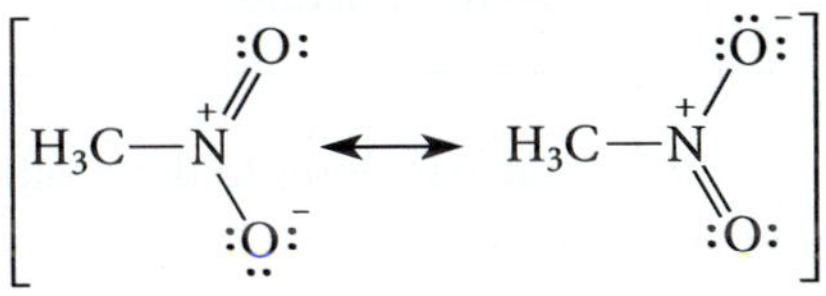

nitromethane (3.25)

Remember that the resonance hybrid (the actual molecule) is more stable than either of the fictional resonance contributors.

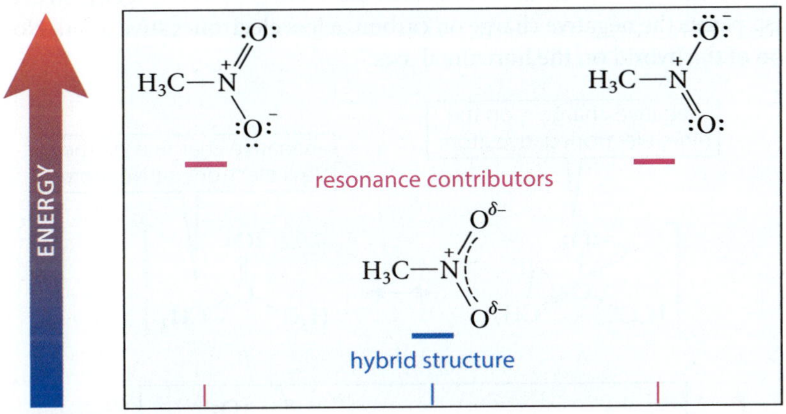

(3.26)

The hybrid is shown on the horizontal axis midway between the two fictitious structures because it reflects the two structures equally.

What if the resonance contributors are not equivalent? In that case we consider the relative energies of the individual resonance contributors as if they were real molecules. *The resonance structures of lowest energy are the most important structures in describing the resonance hybrid.* The problem of determining the most important resonance structure, then, devolves into a problem of determining the most stable resonance contributor. We haven't learned all of principles for evaluating the relative energies of molecules; some of these principles are presented in later chapters. However, there are a few important rules that we can address now. These are listed *in order of decreasing importance.*

1. *Structures with complete octets are much more important than structures with electron-deficient atoms.*

   Of the following resonance structures, structure *A* is more stable, and therefore more important, than structure *B*. In structure *A*, every atom has an octet; in structure *B*, the carbon is electron-deficient.

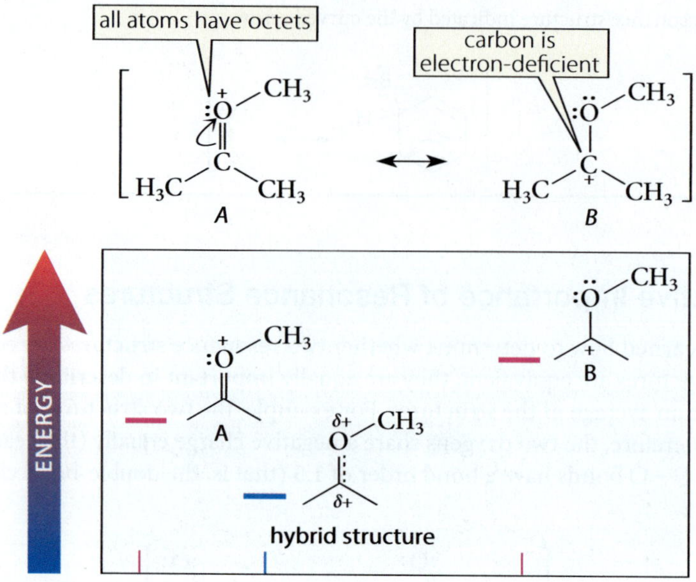

(3.27)

The hybrid is shown on the horizontal axis closer to the position of *A* (farther to the left) because structure *A* is more important.

2. *If structures contain charges, the more important structures have negative charges on atoms of greater electronegativity and positive charges on atoms of lesser electronegativity.*

   In the following ion, structure *A* is more important (more stable) than structure *B* because the negative charge is on the atom of greater electronegativity (oxygen). Structure *B*, in contrast, places the negative charge on carbon, a less electronegative atom. Notice the position of the hybrid on the horizontal axis.

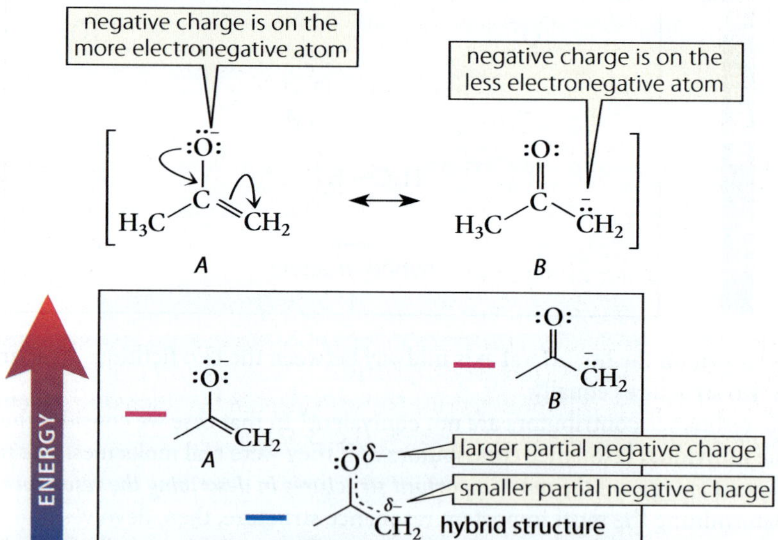

(3.28)

3. *Neutral structures and structures that delocalize charge are more important than structures in which opposite charges are separated.*

   According to the electrostatic law (Eq. 1.25, Sec. 1.4A), separating two charges of opposite sign requires energy. (It's somewhat like pulling two magnets apart.) In the following example, both structures have octets on every atom, but structure *B* involves charge separation, which raises its energy.

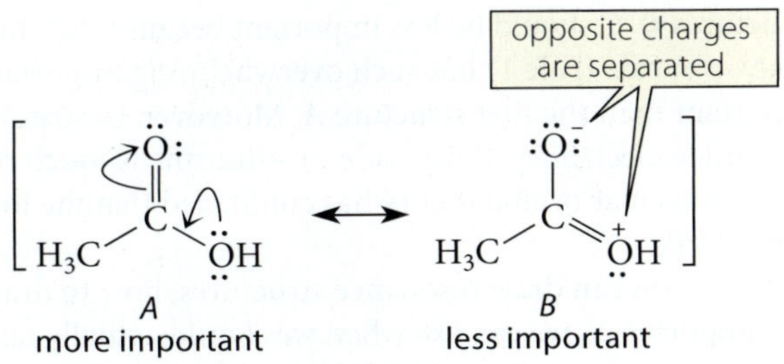

acetic acid (3.29)

In contrast, both of the following structures are equally important because positive charge is *delocalized*.

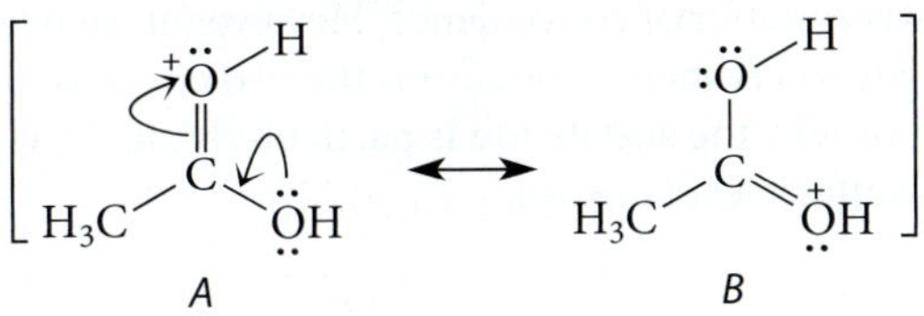

protonated acetic acid (3.30)

Don't confuse *charge separation* with *charge delocalization*. In charge *delocalization*, the *same* charge is delocalized onto multiple atoms (as in Eq. 3.30), which stabilizes the molecule or ion. In charge *separation*, opposite charges are created where there was none to start with, as in Eq. 3.29.

Sometimes more than one of the preceding rules will apply; when that happens, the most important rule determines the most important structure. For example, consider the case discussed under rule 1 (Display 3.27).

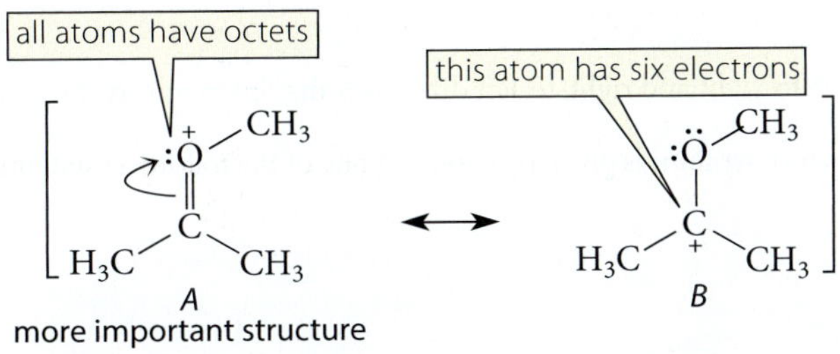

(3.31)

Although structure *A* violates rule 2, the most important rule is rule 1—that all atoms have octets. Therefore, structure *A* is the more important structure.

The overriding importance of the octet rule is also evident in the resonance structures of hypervalent compounds, as in the structure of PF_5 (Eq. 3.32).

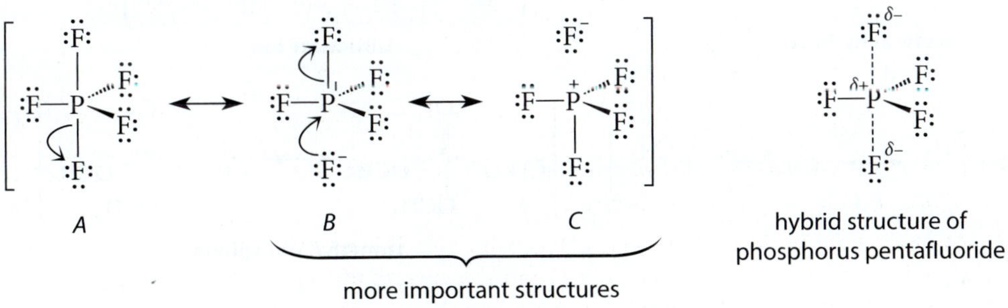

resonance structures of phosphorus pentafluoride (3.32)

Chemists use the hypervalent neutral structure for convenience because of tradition, and because they don't have to draw in the charges. Furthermore, the International Union of Pure and Applied Chemistry (IUPAC) nomenclature of hypervalent compounds is based on the neutral structure. So, it's acceptable to draw the neutral, "octet-expanded" structure, but we will need to keep in mind the greater importance of the charge-separated structures when we consider reactivity.

Rule 3 suggests that structures *B–C* should be less important because they involve charge separation. However, the octet rule (rule 1) has such overwhelming importance that these structures are more important than the first structure *A*. Moreover, in structures *B–C*, the charges reside on atoms of appropriate electronegativity (rule 2)—fluorine is much more electronegative than phosphorus. In fact, molecular orbital theory has confirmed that the ionic structures are the most accurate depictions of PF_5.

Now that you know when you can draw resonance structures, how to draw them, and how to evaluate their relative importance, we can ask when we should actually take the time and effort to draw out resonance structures. We certainly do *not* draw resonance structures at every opportunity; we normally draw single resonance contributors. However, when discussing a question of structure or reactivity that requires resonance structures for an explanation, then we must draw out the resonance structures.

For example, when discussing the chemistry of the acetate ion, we typically will use one of its resonance contributors as a matter of convenience. However, if we need to explain why the negative charge in the acetate ion is shared equally on the two oxygens, why the two C—O bonds have the same bond order, or why the acetate ion is particularly stable, we then need to invoke resonance structures for a suitable explanation.

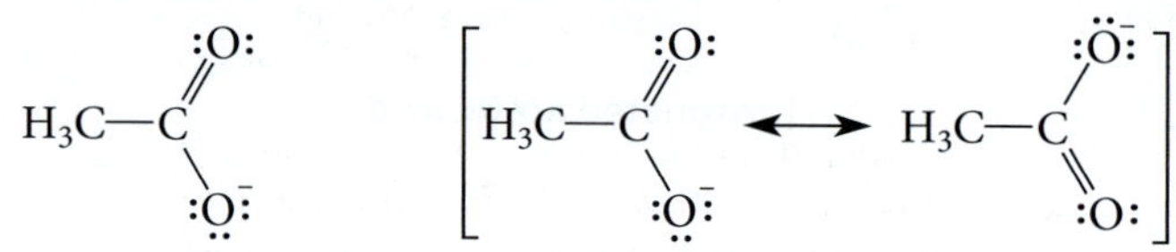

acetate ion in ordinary usage

resonance structures of acetate ion

(3.33)

Focused Problem

3.8 In each of the cases *A–D*, do the following:

(a) Provide the curved arrows in both the left-to-right and right-to-left directions that interconvert the two structures given.

(b) Indicate the relationship of the two resonance structures given by choosing one of the following statements about the *second* structure.

(1) Equally important structure

(2) More important structure

(3) Less important structure

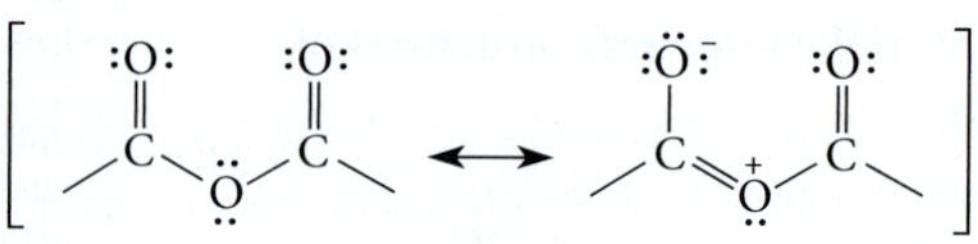

acetic anhydride

A

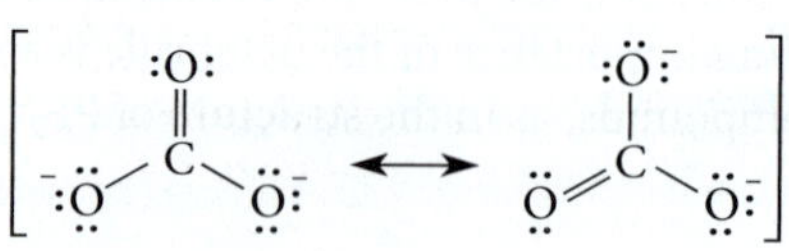

carbonate ion

B

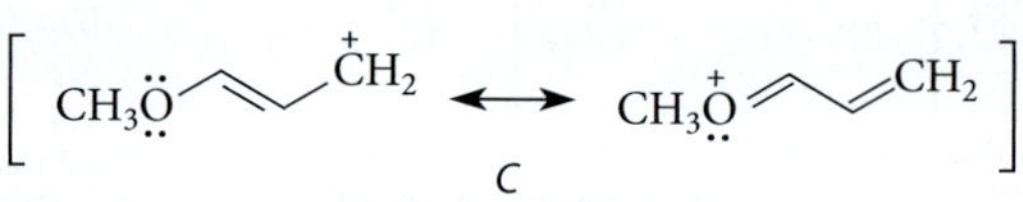

C

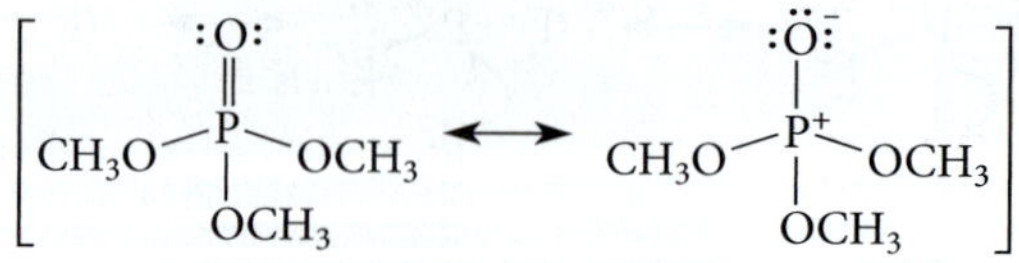

trimethyl phosphate

D

3.4 BRØNSTED–LOWRY ACIDS AND BASES

A. Definition of Brønsted Acids and Bases

Although less general than the Lewis concept, the *Brønsted–Lowry acid–base concept* provides another way of thinking about acids and bases that is very important and useful in organic chemistry. A species that donates a proton in a chemical reaction is called a **Brønsted acid**; a species that accepts a proton in a chemical reaction is a **Brønsted base**.

The reaction of ammonium ion with hydroxide ion is an example of a Brønsted acid–base reaction.

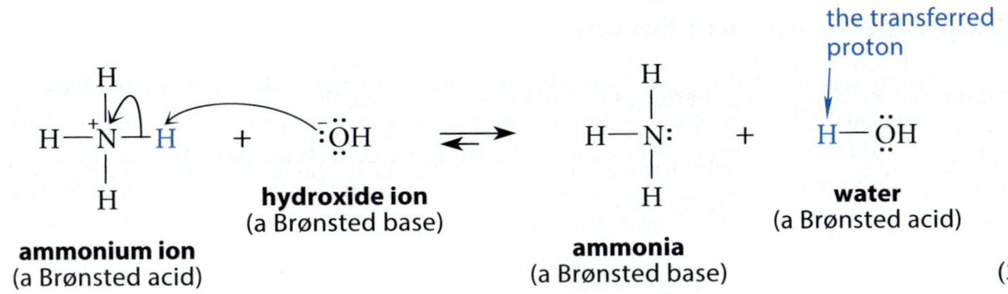

(3.34a)

On the left side of this equation, the ammonium ion is acting as a Brønsted acid and the hydroxide ion is acting as a Brønsted base; looking at the equation from right to left, however, water is acting as a Brønsted acid, and ammonia as a Brønsted base.

The "classical" definition of a Brønsted acid–base reaction given in the first paragraph of this section focuses on the movement of a proton. But in organic chemistry, we are always going to focus on the *movement of electrons*. As Eq. 3.34a illustrates, *any Brønsted acid–base reaction can be described with the curved-arrow notation for electron-pair displacement reactions.* A **Brønsted acid–base reaction** is simply a *special case* of an electron-pair displacement reaction in which the electrophilic center is a proton. It's the action of the electrons that causes the net transfer of a proton from the Brønsted acid to the Brønsted base. When the electrophilic center is a proton, the electron donor is called a **Brønsted base** rather than a nucleophile. A **Brønsted acid** is the species that provides a proton to the base.

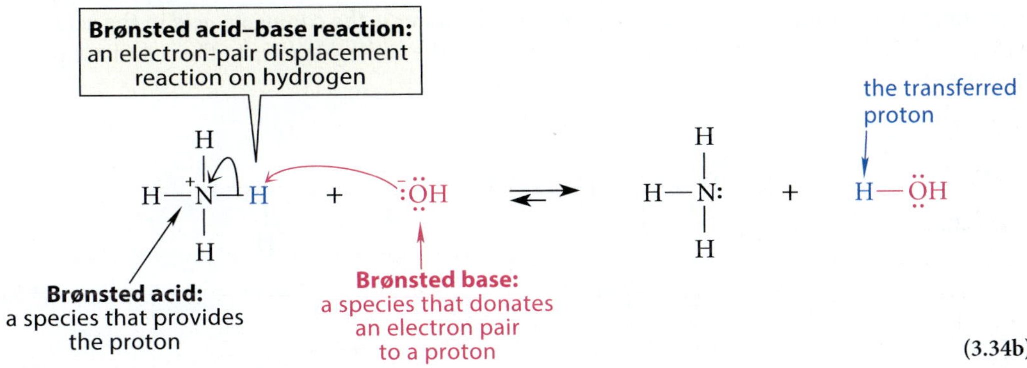

(3.34b)

An important caution about notation: Because the traditional definition of Brønsted acids involves "proton transfer," you may sometimes be tempted to write a curved-arrow notation incorrectly in the following way:

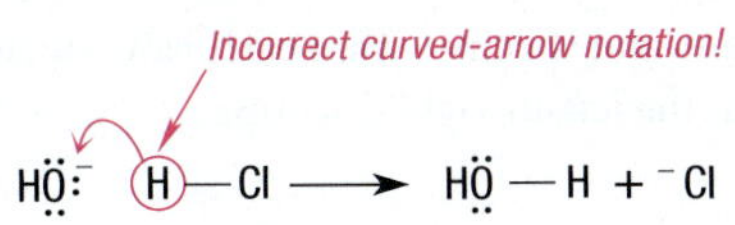

This is incorrect because it shows the movement of the proton rather than the flow of electron pairs. Someone accustomed to using the curved-arrow notation correctly would take this to imply the transfer of H^- to ^-OH, an impossible reaction! The correct use of the curved-arrow notation shows the flow of *electron pairs*:

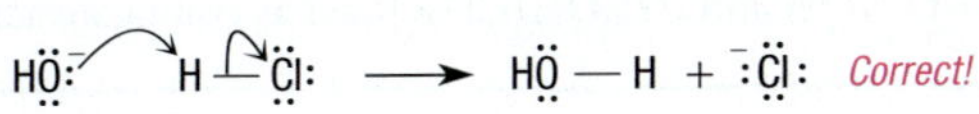

Focused Problem

3.9 (a) Identify the Brønsted acid and the Brønsted base on each side of the following reaction.

$$H_3C{-}\ddot{\underset{..}{O}}{:}^{-} + H{-}\ddot{\underset{..}{S}}{-}CH_3 \rightleftharpoons H_3C{-}\ddot{\underset{..}{O}}{-}H + {}^{-}{:}\ddot{\underset{..}{S}}{-}CH_3$$

(b) Show the curved-arrow notation for this reaction in both the forward and the reverse direction.

B. Conjugate Acids and Bases

When a Brønsted acid loses a proton, its **conjugate base** is formed; when a Brønsted base gains a proton, its **conjugate acid** is formed. When a Brønsted acid loses a proton, it becomes a Brønsted base; this acid and the resulting base constitute a **conjugate acid–base pair**. In any Brønsted acid–base reaction there are two conjugate acid–base pairs. Hence, in Eq. 3.34b, $^{+}NH_4$ and NH_3 are one conjugate acid–base pair, and H_2O and ^{-}OH are the other.

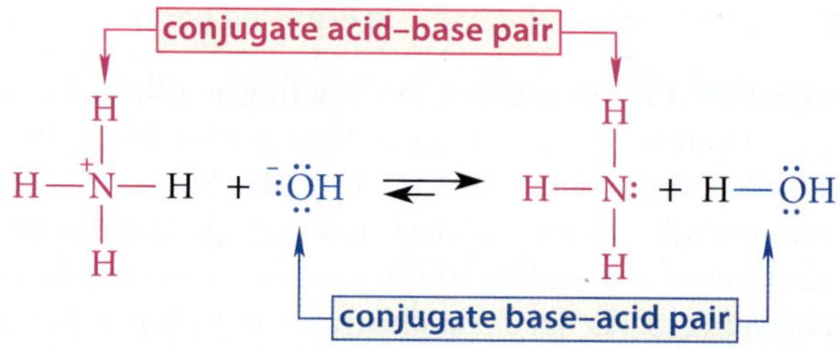

(3.35)

Notice that the conjugate acid–base relationship is *across the equilibrium arrows*. For example, $^{+}NH_4$ and NH_3 are a conjugate acid–base pair, but $^{+}NH_4$ and ^{-}OH are *not* a conjugate acid–base pair.

The identification of a compound as an acid or a base depends on how it behaves in a specific chemical reaction. Water, for example, can act as either an acid or a base. Compounds that can act as either acids or bases are called **amphoteric compounds**. Water is the archetypal example of an amphoteric compound. In Eq. 3.35, for example, water is the conjugate *acid* in the acid–base pair $H_2O/^{-}OH$; in the following reaction, water is the conjugate *base* in the acid–base pair H_3O^{+}/H_2O:

STUDY GUIDE LINK 3.2
Identification of Acids and Bases

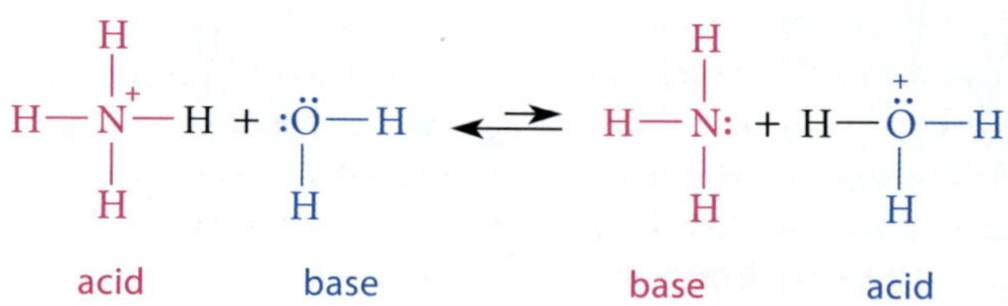

(3.36)

Focused Problems

3.10 In the following reactions, label the conjugate acid–base pairs and specify within each pair which is the acid and which is the base. Then draw the curved-arrow notation for these reactions in the left-to-right direction.

(a) $\ddot{N}H_3 + {}^{-}{:}\ddot{\underset{..}{O}}H \rightleftharpoons {}^{-}{:}\ddot{N}H_2 + H_2\ddot{\underset{..}{O}}{:}$

(b) $\ddot{N}H_3 + \ddot{N}H_3 \rightleftharpoons {}^{-}{:}\ddot{N}H_2 + \overset{+}{N}H_4$

3.11 Write a Brønsted acid–base reaction in which $H_2\ddot{\underset{..}{O}}/{}^{-}{:}\ddot{\underset{..}{O}}H$ and $CH_3\ddot{\underset{..}{O}}H/CH_3\ddot{\underset{..}{O}}{:}^{-}$ act as conjugate acid–base pairs.

Sections 3.2B and 3.4A are about analyzing the roles of the various species in reactions. You will find that most of the reactions you will study can be analyzed in terms of these roles, and hence an understanding of these sections will help you to understand and even predict reactions. The first step in this understanding is to apply the definitions you've learned in the analysis of reactions. Study Problem 3.6 illustrates such an analysis.

Study Problem 3.6

What follows is a series of acid–base reactions that represent the individual steps in a known organic transformation, the replacement of —Br by —OH at a carbon bearing three alkyl groups. Considering only the forward direction, classify each reaction as a Brønsted acid–base reaction or a Lewis acid–base association/dissociation. Classify each labeled species (or a group within each species) with one of the following terms: *Brønsted base, Brønsted acid, nucleophile, nucleophilic center, electrophile, electrophilic center*, and/or *leaving group.* For Brønsted acid–base reactions, show the conjugate acid–base pairs.

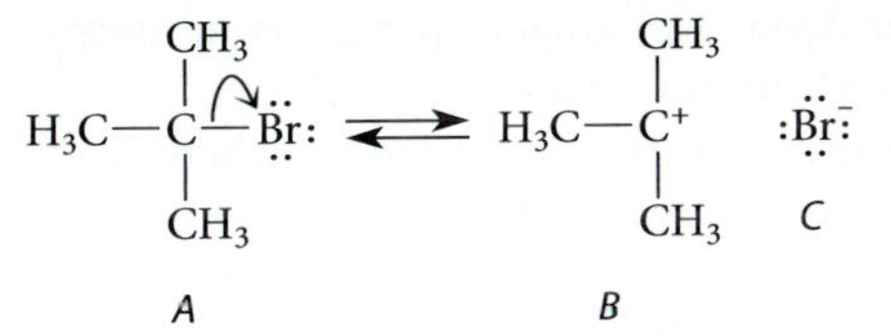

(3.37a)

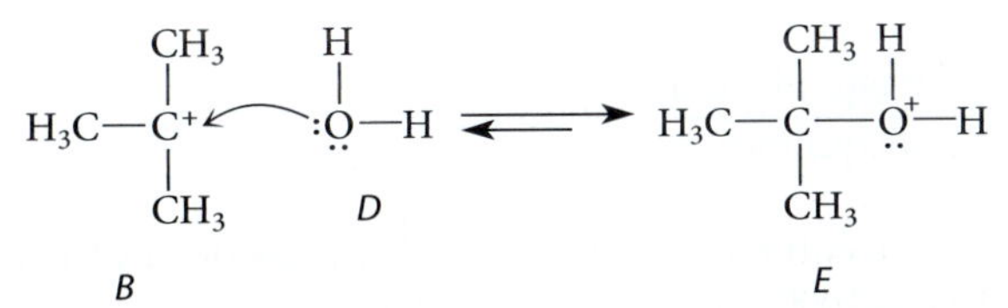

(3.37b)

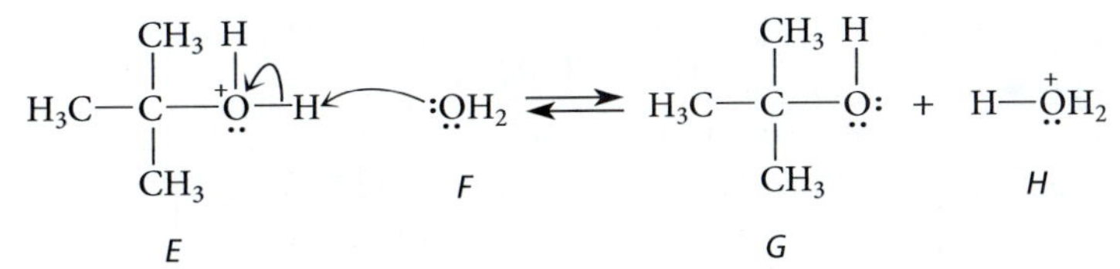

(3.37c)

Solution Classify each reaction first, and then analyze the role of each species. Reaction 3.37a is a Lewis acid–base dissociation. (Notice that a single curved arrow describes the dissociation.) In compound *A*, Br is the leaving group.

Reaction 3.37b is a Lewis acid–base association reaction. Cation *B* is an electrophile, and the electron-deficient carbon is the electrophilic center. Water molecule *D* is a nucleophile, and its oxygen is the nucleophilic center.

Reaction 3.37c is a Brønsted acid–base reaction. Ion *E* and compound *G* constitute a conjugate Brønsted acid–base pair, and compound *F* and compound *H* are a conjugate Brønsted base–acid pair. The water molecule *F* is a Brønsted base. The proton of *E* that receives an electron pair from water is an electrophilic center. The part of *E* that becomes *G* (everything but the proton) is a leaving group.

Focused Problems

3.12 Work Study Problem 3.6 for the *reverse* of each reaction 3.37a–c.

3.13 In each of the following processes, complete the reaction using the curved arrow given, classify the process as a Brønsted acid–base reaction or a Lewis acid–base association/dissociation, and label each species (or part of each species) with one of the following terms: *Brønsted base, Brønsted acid, nucleophile, nucleophilic center, electrophile, electrophilic center,* or *leaving group.* In each part, once you complete the forward reaction, draw the curved arrow(s) for the reverse reaction and do the same exercise for it as well.

(a) 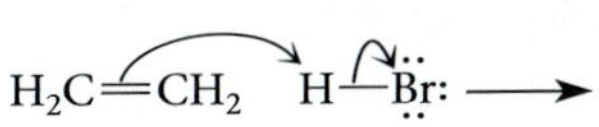

(b) 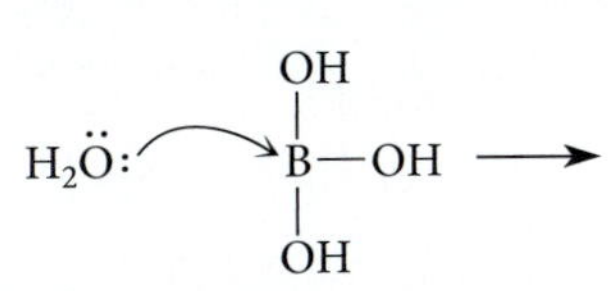

3.5 CHEMICAL EQUILIBRIUM AND FREE ENERGY

Our next step in studying acids and bases is to consider the relative strengths of Brønsted acids and bases. Understanding this topic requires that you understand the principles of chemical equilibrium and free energy, which are reviewed in this section. If you don't need this review, you can skip to Sec. 3.6.

A. The Equilibrium Constant

Imagine a scenario in which a sample of compound A can react to form compound B. Once the reaction is over, it has achieved **equilibrium**, in which the concentrations [A] and [B] no longer change. There are three general situations. After some period of time, either (1) there is more B than A at equilibrium—that is, $[B]/[A] > 1$; (2) there is more A than B—that is, $[B]/[A] < 1$; or (3) A and B are present in equal amounts—that is, $[B]/[A] = 1$. Equilibrium is *dynamic*—that is, at equilibrium, individual molecules of A or B may still interconvert into one another, but the rates of interconversion are equal so that the *concentrations remain constant.*

If you study physical chemistry, you will learn that the bracketed quantities in an equilibrium expression are dimensionless quantities called *activities* rather than concentrations. If the solution is very dilute, activities and concentrations in many cases have the same values. We are assuming that the "dilute solution approximation" is valid.

Suppose that at equilibrium, a lot of A has converted into B, so that there is 1000 times as much B as A. This ratio of 1000 B to 1 A is commonly represented as an *equilibrium constant,* or K_{eq}. The **equilibrium constant** for any reaction is the ratio of product concentrations *at equilibrium* to the reactant concentrations *at equilibrium.*

$$K_{eq} = \frac{[\text{products at equilibrium}]}{[\text{reactants at equilibrium}]} \quad (\text{concentrations are in moles L}^{-1}) \tag{3.38}$$

In this reaction, the equilibrium concentration of the product, [B], divided by the equilibrium concentration of the reactant, [A], is 1000, so K_{eq} is equal to 1×10^3.

$$A \rightleftharpoons B \qquad K_{eq} = \frac{[B]}{[A]} = \frac{1000}{1} = 10^3 \quad \text{means}$$

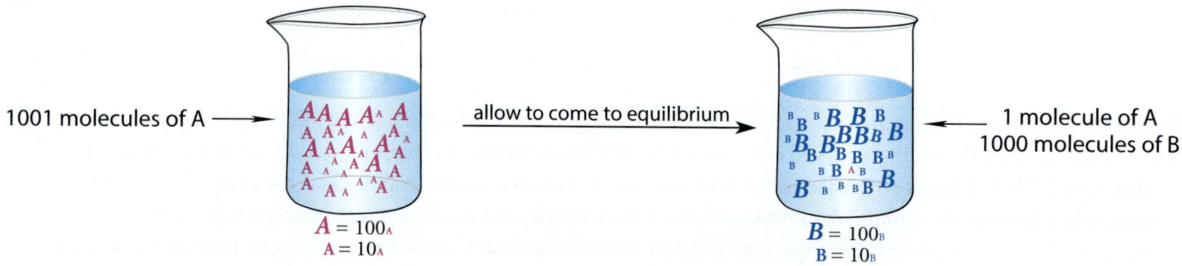

A sample containing A converts mostly into B at equilibrium when $K_{eq} > 1$. (3.39a)

Whenever K_{eq} is greater than 1, it means that there is more product than reactant at equilibrium. In these situations, we often say that "the equilibrium favors the products" or "the equilibrium lies to the right" (or *toward* B, in this example).

Sometimes "unbalanced" equilibrium arrows are used when one side of an equilibrium is strongly favored.

$$A \rightleftharpoons B \qquad \text{B is "favored" at equilibrium} \tag{3.39b}$$

In a different situation, suppose that at equilibrium only a little of A has converted into B, and that there is 1000 times as much A as B. In this reaction, [B] divided by [A] is 0.001, so K_{eq} is equal to 1×10^{-3}.

$$\mathrm{A} \rightleftharpoons \mathrm{B} \qquad K_{eq} = \frac{[\mathrm{B}]}{[\mathrm{A}]} = \frac{1}{1000} = 10^{-3} \quad \text{means}$$

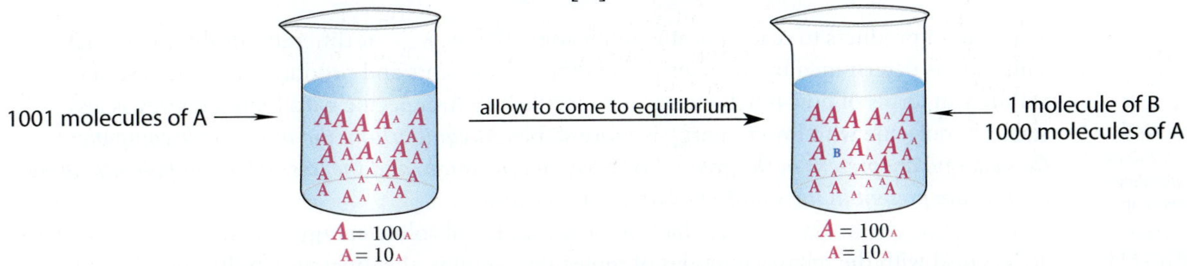

A sample containing A converts into very little B at equilibrium when $K_{eq} < 1$.

(3.40a)

Whenever K_{eq} is less than 1, it means that there is more reactant than product. In these situations, we often say that "the equilibrium favors the reactants" or "the equilibrium lies to the left" (or *toward* A, in this example).

$$\mathrm{A} \longleftarrow\!\rightharpoonup \mathrm{B} \qquad \text{A is "favored" at equilibrium} \tag{3.40b}$$

Finally, imagine that at equilibrium, half of A has converted into B. In this reaction, because there are equal amounts of A and B at equilibrium, or [A] = [B], K_{eq} is equal to 1.

$$\mathrm{A} \rightleftharpoons \mathrm{B} \qquad K_{eq} = \frac{[\mathrm{B}]}{[\mathrm{A}]} = \frac{500}{500} = 1 \quad \text{means}$$

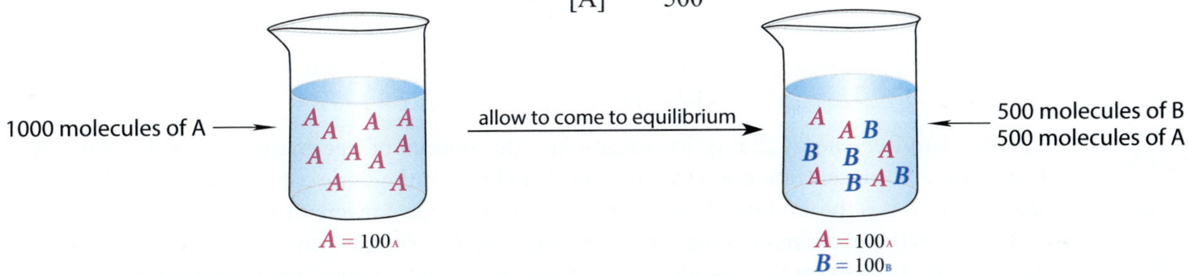

A sample containing A contains equal amounts of A and B at equilibrium when $K_{eq} = 1$.

(3.41)

Whenever K_{eq} is equal to 1, it means that the concentrations of reactant and product are equal at equilibrium and that "neither side is favored at equilibrium."

In summary, K_{eq} is a *ratio* that equals the *relative* concentrations of products and reactants at equilibrium.

Focused Problems

3.14 A reaction has $K_{eq} = [\mathrm{B}]/[\mathrm{A}] = 5.7 \times 10^3$. For every 1000 molecules of B present at equilibrium, how many molecules of A are present?

3.15 A reaction has $K_{eq} = [\mathrm{D}]/[\mathrm{C}] = 7.0 \times 10^{-3}$. For every 1000 molecules of D present at equilibrium, how many molecules of C are present?

B. Standard Free Energy: The Relationship Between Equilibrium and Relative Stability

The relationship between equilibrium concentrations and relative stability was noted when we discussed the conformations of butane (Sec. 2.5B): the most stable conformation is present in greatest amount in a sample of butane. In this section we consider this relationship in more detail.

The ratio of products to reactants at equilibrium—that is, K_{eq}—is different for different reactions. This ratio is determined by the *relative stabilities* of reactants and products. We can describe **stability** in terms of how much energy a molecule has. A molecule with higher energy is less stable; a molecule with lower energy is more stable. *At equilibrium, the more stable components (less energy) are present in the greatest relative concentration, and the less stable components (more energy) are present in the smallest relative concentration.*

In organic chemistry we never have to deal with the absolute energy of a molecule. Instead, we always deal with the relative energies of molecules—that is, the *difference* in the energies of two molecules. Energy can be expressed in different ways—potential energy, kinetic energy, enthalpy, free energy. We don't need to be concerned with the differences between these forms of energy here. What we need to understand is that the energy related directly to chemical equilibrium is the **standard free energy change,** $\Delta G°$. The Greek letter *delta* (Δ) means a *difference* of some quantity, in this case, standard free energy—specifically, the difference between the final state and the initial state of a process.

The superscript ° means that the relative free energy is measured under *standard conditions,* which typically means a pressure of 100 kPa (or 1 atm), and, if we are dealing with solutions, 1 *M* (mole per liter) concentration.

For a reaction, the standard free-energy change that occurs when 1 mol of reactant A is converted into 1 mol of product B is the difference between the standard free energies of the products and reactants under standard conditions:

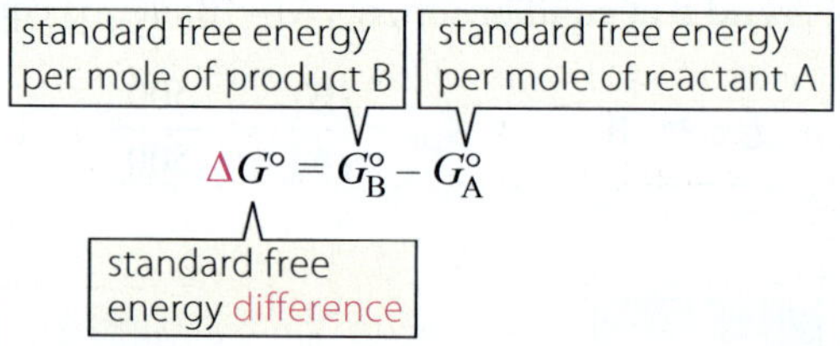

$$\Delta G° = G°_B - G°_A \tag{3.42}$$

The standard free-energy change for a reaction is the amount of free energy required to convert 1 mol of A at a concentration of 1 mol L^{-1} into 1 mol of B at the same concentration. (If B is more stable than A, it is the amount of energy released in the same conversion.)

Be sure to understand that 1 mol L^{-1} is *not* the equilibrium concentration of A or B; the relative equilibrium concentrations are given in the scenarios. By measuring the *actual* equilibrium concentrations, we learn how much energy would be released or required for the fictional conversion of 1 mol of A to 1 mol of B, each at 1 *M* concentration. The energy for conversion of 1 mol of reactants to 1 mol of a less stable product in principle could be provided by a battery; or, if the products are more stable, the energy released could be used to charge a battery. Usually, though, such a conversion is hypothetical; it is the standard by which free energy is measured.

Compounds that are more stable—have less standard free energy—are favored in a chemical equilibrium. As an example, consider the first scenario, in which most of compound A is converted into B (Display 3.39a). Because the equilibrium concentration of B is much higher than the equilibrium concentration of A, *B must be more stable than A*. The following free-energy diagram summarizes this situation.

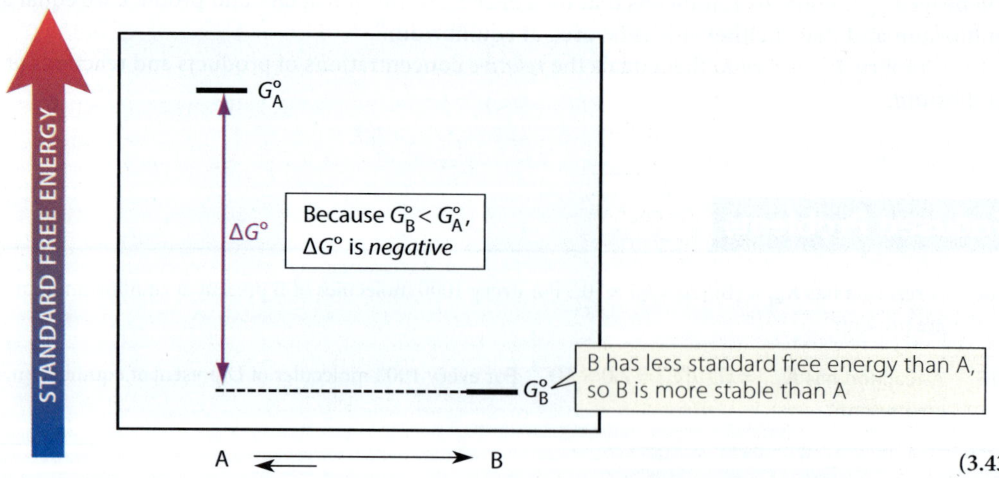

(3.43)

Without determining the standard free energies of A or B individually, we can say that the *difference* between the two energies is a negative value (Display 3.43) because B is favored in the equilibrium. Because the product B has less energy than the reactant A, an amount of energy $\Delta G°$ would be *released* if we were to convert 1 mol L^{-1} of A to 1 mol L^{-1} of B.

In the second scenario, very little B is present at equilibrium (Display 3.40a). In this case, A is more stable than B.

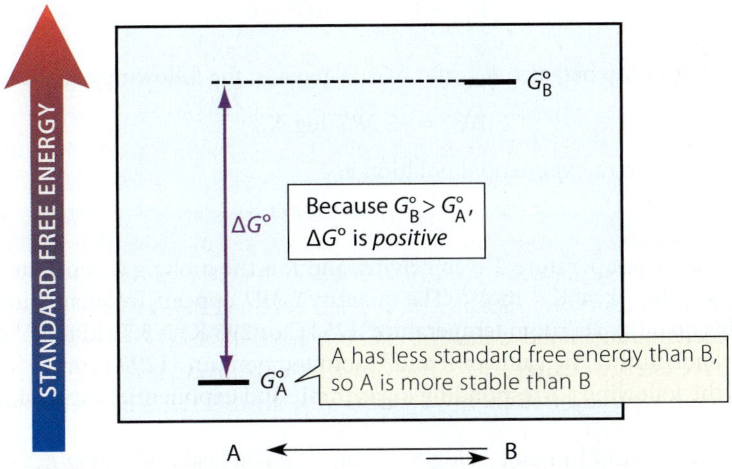

(3.44)

The *difference* between the two energies is a positive value (Display 3.44) because the product has more energy that the reactant. This means that an amount of energy ΔG° would have to be *added* to convert 1 mol of A into 1 mol of B, each at 1 *M* concentration.

When more than two components are in equilibrium, the expression for the equilibrium constant is altered by multiplying all of the reactant concentrations in the denominator and by multiplying all of the product concentrations in the numerator, as in the following examples:

$$\mathrm{C + D \rightleftharpoons E} \qquad K_{eq} = \frac{[\mathrm{E}]}{[\mathrm{C}][\mathrm{D}]}$$

$$\mathrm{F + G \rightleftharpoons H + I} \qquad K_{eq} = \frac{[\mathrm{H}][\mathrm{I}]}{[\mathrm{F}][\mathrm{G}]}$$

$$\mathrm{J + J \rightleftharpoons K} \qquad K_{eq} = \frac{[\mathrm{K}]}{[\mathrm{J}]^2}$$

(3.45)

The corresponding standard free energies are added. For example, the ΔG° for a reaction $\mathrm{A + B \rightleftharpoons C + D}$ is

$$\Delta G^\circ = \overbrace{(G_C^\circ + G_D^\circ)}^{\text{standard free energy of the products}} - \overbrace{(G_A^\circ + G_B^\circ)}^{\text{standard free energy of the reactants}}$$

standard free energy change for the reaction (ΔG°)

(3.46)

Focused Problem

3.16 Sketch the energy diagram for the equilibrium $\mathrm{A \rightleftharpoons B}$ in which A and B are present in equal amounts at equilibrium. What is ΔG° for this reaction?

You learned in the previous discussion that when K_{eq} for a reaction is greater than 1, ΔG° for the reaction is negative; when K_{eq} for a reaction is less than 1, ΔG° for the reaction is positive; and (from Focused Problem 3.16) when K_{eq} for a reaction is exactly 1, ΔG° for the reaction is 0.

$$K_{eq} > 1 \qquad \Delta G^\circ < 0$$
$$K_{eq} < 1 \qquad \Delta G^\circ > 0$$
$$K_{eq} = 1 \qquad \Delta G^\circ = 0$$

The precise relationship between K_{eq} and ΔG° is given in the following equation.

$$\Delta G^\circ = -2.3RT \log K_{eq} \tag{3.47a}$$

An equivalent expression in exponential notation is

$$K_{eq} = 10^{(-\Delta G^\circ/2.3RT)} \tag{3.47b}$$

In these equations, the temperature T is in kelvins, and R is the molar gas constant, 8.314×10^{-3} kJ K^{-1} mol^{-1} or 1.987×10^{-3} kcal K^{-1} mol^{-1}. The quantity $2.3RT$ appears frequently in energy calculations. This quantity at "room temperature" (25 °C or 298 K) is 5.71 kJ mol^{-1} or 1.36 kcal mol^{-1}. Substituting the value of R and "room temperature" (298 K or 25 °C) into these equations gives the following corresponding logarithmic and exponential expressions:

$$\Delta G^\circ = -(5.71 \text{ kJ mol}^{-1}) \log K_{eq} \quad \text{or} \quad -(1.36 \text{ kcal mol}^{-1}) \log K_{eq} \tag{3.48a}$$

$$K_{eq} = 10^{(-\Delta G^\circ/5.71 \text{ kJ mol}^{-1})} = 10^{-0.175(\Delta G^\circ)} \quad \text{or} \quad 10^{(-\Delta G^\circ/1.36 \text{ kcal mol}^{-1})} = 10^{-0.735(\Delta G^\circ)} \tag{3.48b}$$

Study Problem 3.7

The energy difference between *anti*-butane and either one of the *gauche*-butane conformations is 3.7 kJ mol^{-1} (see Fig. 2.5, Sec. 2.5B). Treating this difference as a standard free energy, calculate the amounts of *gauche*- and *anti*-butane present in equilibrium in 1 mol of butane at 25 °C. (Remember that there are two gauche conformations.)

Solution ΔG° for the reaction *gauche*-butane $\rightleftharpoons$ *anti*-butane $= -3.7$ kJ mol^{-1}; this free energy is negative because the product is more stable than the reactant. Using Eq. 3.48b,

$$K_{eq} = 10^{(-\Delta G^\circ/5.71 \text{ kJ mol}^{-1})} = 4.45$$

Because there are two different *gauche*-butanes, in 1 mol of butane we have

$$[\textit{anti}\text{-butane}] + 2[\textit{gauche}\text{-butane}] = 1 \text{ mol}$$

The value of K_{eq} means that $[\textit{anti}\text{-butane}]/[\textit{gauche}\text{-butane}] = 4.45$. Using this value in the previous equation, we have

$$4.45[\textit{gauche}\text{-butane}] + 2[\textit{gauche}\text{-butane}] = 6.45[\textit{gauche}\text{-butane}] = 1 \text{ mol}$$

Therefore, each $[\textit{gauche}\text{-butane}] = 1/6.45 = 0.16$ mol, and the amount of both *gauche*-butanes together are 0.32 mol. Therefore, the amount of *anti*-butane is $1.00 - 0.32 = 0.68$ mol. In other words, butane is about two-thirds *anti*-butane and one-third *gauche*-butane.

Equation 3.48b shows that *small* changes in ΔG° lead to *large* changes in K_{eq}, or for every 5.71 kJ mol^{-1} difference in energy between products and reactants, the equilibrium constant changes by an *order of magnitude* (which means "a factor of 10 times"). **Table 3.1** illustrates this relationship numerically.

TABLE 3.1 The Relationship among Standard Free-Energy Changes, Equilibrium Constants, and Relative Equilibrium Concentrations at 25 °C (298 K)

$\Delta G^\circ = -2.3RT \log K_{eq}$ or $K_{eq} = 10^{-\Delta G^\circ/2.3RT}$			
ΔG° (kJ mol^{-1})	ΔG° (kcal mol^{-1})	K_{eq}	[Products]:[reactants]
+34.2	+8.4	0.000001	1:1000000
+28.5	+7.0	0.00001	1:100000
+22.8	+5.6	0.0001	1:10000
+17.1	+4.2	0.001	1:1000
+11.4	+2.8	0.01	1:100
+5.71	+1.4	0.1	1:10
0	0.0	1	1:1
−5.71	−1.4	10	10:1
−11.4	−2.8	100	100:1
−17.1	−4.2	1000	1000:1
−22.8	−5.6	10000	10000:1
−28.5	−7.0	100000	100000:1
−34.2	−8.4	1000000	1000000:1

Let's summarize the important points of this section:

1. Chemical equilibrium favors the species with the lowest standard free energy.

2. The more two compounds in equilibrium differ in standard free energy, the greater their difference in concentration at equilibrium. Each 5.71 kJ mol^{-1} (or 1.36 kcal mol^{-1}) increment in ΔG° affects the equilibrium concentration ratio by a factor of 10.

Focused Problems

3.17 Are reactions with the following properties favorable (equilibrium lies toward the products), unfavorable (equilibrium lies toward the reactants), or neither?

(a) $K_{eq} = 5.7 \times 10^{-3}$

(b) $\Delta G^\circ = 1.2$ kJ mol^{-1} (0.29 kcal mol^{-1})

(c) $K_{eq} = 1.1 \times 10^5$

(d) $\Delta G^\circ = -21.96$ kJ mol^{-1} (−5.25 kcal mol^{-1})

3.18 In each of the parts of Focused Problem 3.17, (1) which has more energy, the products or the reactants, and (2) which is more stable, the products or the reactants?

3.19 (a) A reaction has a standard free-energy change of −11.4 kJ mol^{-1} (−2.72 kcal mol^{-1}). Without making a calculation, estimate the equilibrium constant for the reaction at 25 °C.

(b) Calculate the equilibrium constant for the reaction in part (a).

(c) Without calculating, estimate the standard free-energy difference between starting materials and products for a reaction that has an equilibrium constant of 305.

3.20 (a) A reaction $A + B \rightleftharpoons C$ has a standard free-energy change of −2.93 kJ mol^{-1} (−0.700 kcal mol^{-1}) at 25 °C. What are the concentrations of A, B, and C *at equilibrium* if, at the beginning of the reaction, their concentrations are 0.1 *M*, 0.2 *M*, and 0 *M*, respectively?

(b) Without making a calculation, tell in a qualitative sense how you would expect your answer for part (a) to change if the reaction has instead a standard free-energy change of $+2.93$ kJ mol^{-1} ($+0.700$ kcal mol^{-1}).

3.21 Complete each of the following statements with a number. Assume that the temperature is 25 °C (298 K).

(a) Two reactions have equilibrium constants that differ by a factor of 10. Their standard free energies differ by ____________ kJ mol^{-1} (or ____________ kcal mol^{-1}).

(b) For every 1 kJ mol^{-1} in standard free energy that two reactions differ, their equilibrium constants differ by a factor of ____________. For every 1 kcal mol^{-1} in standard free energy that two reactions differ, their equilibrium constants differ by a factor of ____________.

3.6 STRENGTHS OF BRØNSTED ACIDS AND BASES

A. The Acid Dissociation Constant K_a and the pK_a

Acid and base strength is a concept that can be widely used to understand a number of organic reactions.

The strength of a Brønsted acid is determined by how well it transfers a proton to a standard Brønsted base. The standard base traditionally used for comparison is water. The transfer of a proton from a general acid, HA, to water is indicated by the following equilibrium:

$$\mathrm{HA} + \mathrm{H_2O} \rightleftharpoons \mathrm{H_3O^+} + \mathrm{A{:}^-} \qquad (3.49)$$

It is well established that the proton in aqueous solution (the "hydrated proton") has more than one water of hydration. You may have seen the hydrated proton written as H^+(aq). We notate the hydrated proton as H_3O^+ because, in the curved-arrow notation, when we need to show water as a Brønsted base, this is the simplest possible structure of its conjugate acid. The fact that this is not a rigorously correct notation for the hydrated proton will not cause us any problems as long as we are consistent in its use.

The farther to the right the equilibrium for this reaction lies, the stronger is the acid.

The equilibrium constant for this reaction is given by

$$K_{eq} = \frac{[\mathrm{H_3O^+}][\mathrm{A{:}^-}]}{[\mathrm{HA}][\mathrm{H_2O}]} \qquad (3.50)$$

(The quantities in brackets are molar concentrations at equilibrium.) Because water is the solvent (and is therefore present in large excess), *we treat the water concentration as constant.* Rather than use the concentration of water (55.6 *M*), we let $[\mathrm{H_2O}] = 1$. This convention is called the **unit activity standard state of water**. When we adopt this convention, we call the resulting equilibrium constant expression the **acid dissociation constant**, abbreviated K_a.

$$K_a = \frac{[\mathrm{H_3\ddot{O}^+}][\mathrm{A{:}^-}]}{[\mathrm{HA}]} \qquad (3.51)$$

(numerator: products; denominator: reactants)

Each acid has its own unique dissociation constant that depends on the structure of the acid, something we explore in Sec. 3.7. The more the acid dissociates, the more H_3O^+ ions are formed when the acid is dissolved in water at a given concentration. The larger the ratio of numerator to denominator, the larger is K_a. Thus, *the strength of a Brønsted acid is measured by the magnitude of its dissociation constant.*

Because the dissociation constants (K_a values) of different Brønsted acids cover a range of many powers of 10, it is convenient to express acid strength in a logarithmic manner. Using "p" to mean "negative logarithm of," we can convert K_a values into pK_a values as follows:

$$-\log K_a = \mathrm{p}K_a \qquad (3.52)$$

This allows us to express K_a values, which can be very large or very small numbers, as simpler pK_a values, as shown in the following examples:

K_a	K_a (in exponential notation)	pK_a
0.0000001	10^{-7}	7
0.0001	10^{-4}	4
0.00326	3.26×10^{-3}	2.49
10	10^{1}	–1
100000	10^{5}	-5

For example, if an acid has a dissociation constant $K_a = 10^{-5}$, its $pK_a = 5$. An acid with K_a equal to 10^{-3} is 100 times stronger; its $pK_a = 3$. This example shows that *stronger acids have smaller* pK_a values.

The pK_a of water merits special attention. The autoionization of water—that is, one water molecule protonating another—is shown by the following transformation.

$$H_2O + H_2O \rightleftharpoons H_3O^+ + {}^-OH \qquad (3.53)$$

The equilibrium constant expression for this transformation is

$$K_{eq} = \frac{[H_3O^+][^-OH]}{[H_2O][H_2O]} \qquad (3.54)$$

Applying the *unit activity standard state of water* ($[H_2O] = 1$), the dissociation constant of water is then

$$K_a = [H_3O^+][^-OH] \qquad (3.55)$$

The K_a of water is 10^{-14}. The corresponding pK_a of water (often called pK_w) is then 14. From the balanced equation for the dissociation of water, the concentrations of hydronium ions and hydroxide ions in pure water are equal and therefore must be 10^{-7} *M* each.

The pK_a of water is often listed as 15.7. This number comes from the use of 55.6 *M* as the concentration of one of the water molecules in Eq. 3.54, which is a different convention than the unit activity convention used with other aqueous acids.

The dissociation constant is a special type of equilibrium constant that is used for acid dissociation. Although the K_a is sufficient to describe the favorability of a reaction and thus the strength of an acid, we often use the pK_a because, as described earlier, pK_a numbers are easier to work with. K_a values like 6.6×10^4 and 9.9×10^{-16} become pK_a values –4.8 and 15, respectively. Again, note the "negative logarithm" relationship of K_a and pK_a. *Stronger acids,* which have *large* K_a values, have *small* pK_a values, whereas *weaker acids,* with *small* K_a values, have *large* pK_a values.

Stronger acid	Smaller pK_a	Larger K_a
Weaker acid	Larger pK_a	Smaller K_a

Here is an important convention: When we discuss acidities of individual compounds, we'll often associate a pK_a value with an individual proton in a molecule.

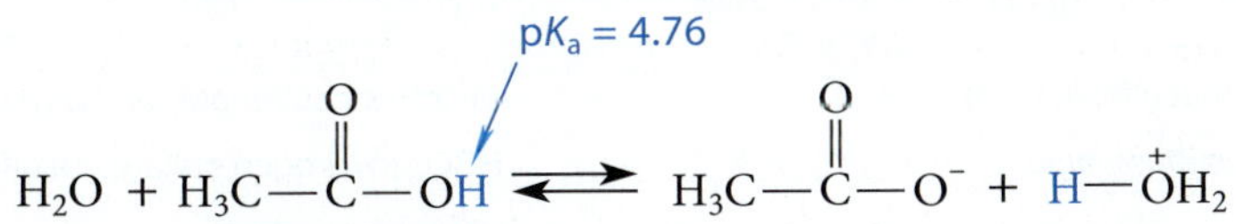

acetic acid (3.56)

Rigorously, molecules, not individual protons, have pK_a values. The acetic acid molecule is an acid with $pK_a = 4.76$ ($K_a = 1.74 \times 10^{-5}$), and the notation in Eq. 3.56 means that the dissociation of acetic acid occurs by loss of the labeled proton.

The pK_a values of some common acids are listed in Table 3.2 in order of increasing acidity (decreasing pK_a). For now, the purpose of the table is to give you a sense of the large range of acidities that we can encounter in common compounds. Don't try to memorize this table. We'll refer back to it in later chapters to point out trends that you'll be asked to learn at the appropriate time.

Two aspects of **Table 3.2** are worth special emphasis. First, many of the pK_a values in Table 3.2 are estimates or approximations. One reason for the estimates is that the exact pK_a values depend on the details of the structure abbreviated by "R" in the table. Another reason for the estimates of the pK_a values of very strong and very weak acids is that these values cannot be measured directly in water, because H_3O^+ is the strongest acid that can exist in water, and ^-OH is the strongest base that can exist in water. However, chemists have devised ways of estimating these pK_a values. Such estimates are the basis for the acidities of strong acids such as HCl and very weak acids such as NH_3 in Table 3.2. These approximate pK_a values will suffice for many of our applications.

The second aspect of Table 3.2 is that these pK_a values are for aqueous (water) solutions. Compounds generally have different pK_a values in solvents other than water. Because much organic chemistry is carried out in nonaqueous solvents, we'll note these pK_a values when we need to.

TABLE 3.2 Relative Strengths of Some Acids and Bases in Water

(GREATER ACIDITY OF CONJUGATE ACID ↓ ; GREATER BASICITY OF CONJUGATE BASE ↑)

Conjugate acid	pK_a	Conjugate base
$\ddot{N}H_3$ (ammonia)	~35†	$^-:\ddot{N}H_2$ (amide)
$R\ddot{O}H$ (alcohol)	15–19*	$R\ddot{O}:^-$ (alkoxide)
$H\ddot{O}H$ (water)	14.0	$H\ddot{O}:^-$ (hydroxide)
HPO_4^{2-} (hydrogen phosphate)	12.3	PO_4^{3-} (phosphate)
$R\ddot{S}H$ (thiol)	10–12*	$R\ddot{S}:^-$ (thiolate)
$R_3\overset{+}{N}H$ (trialkylammonium ion)	9–11*	$R_3N:$ (trialkylamine)
$\overset{+}{N}H_4$ (ammonium ion)	9.25	$H_3N:$ (ammonia)
HCN (hydrocyanic acid)	9.40	$^-:CN$ (cyanide)
$H_2PO_4^-$ (dihydrogen phosphate)	7.21	HPO_4^{2-} (hydrogen phosphate)
$H\ddot{S}H$ (hydrosulfuric acid)	7.0	$H\ddot{S}:^-$ (hydrosulfide)
R—C(=O:)—$\ddot{O}$H (carboxylic acid)	4–5*	R—C(=O:)—$\ddot{O}:^-$ (carboxylate)
$H\ddot{F}:$ (hydrofluoric acid)	3.2	$:\ddot{F}:^-$ (fluoride)
H_3PO_4 (phosphoric acid)	2.2	$H_2PO_4^-$ (dihydrogen phosphate)
$H_3\ddot{O}^+$ (hydronium ion)	0.0	$H_2\ddot{O}$ (water)
HNO_3 (nitric acid)	–1.3	NO_3^- (nitrate)
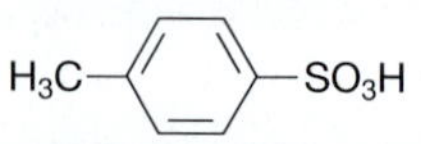 (p-toluenesulfonic acid)	–2.8†	 (p-toluenesulfonate, or "tosylate")
H_2SO_4 (sulfuric acid)	–3†	HSO_4^- (hydrogen sulfate, bisulfate)
$H\ddot{C}l:$ (hydrochloric acid)	–6 to –7†	$:\ddot{C}l:^-$ (chloride)
$H\ddot{B}r:$ (hydrobromic acid)	–8 to –9.5†	$:\ddot{B}r:^-$ (bromide)
$H\ddot{I}:$ (hydroiodic acid)	–9.5 to –10†	$:\ddot{I}:^-$ (iodide)
$HClO_4$ (perchloric acid)	–10†	ClO_4^- (perchlorate)

†Estimated value.
*Exact value depends on the structure of R.

Focused Problems

3.22 Chloroacetic acid is stronger than formic acid, which is stronger than propionic acid.

(a) Match the acids with their pK_a values: 4.87, 3.75, 2.85.

(b) Calculate the dissociation constants K_a for each acid.

3.23 (a) Write the equation for H_3O^+ acting as an acid toward the base H_2O. Be sure to write both reactants and products.

(b) Write the equilibrium constant expression. Use this to show that the pK_a of H_3O^+ is 0, as shown in Table 3.2.

B. Strengths of Brønsted Bases

The strength of a Brønsted base is directly related to the pK_a of its conjugate acid. Thus, the base strength of fluoride ion is indicated by the pK_a of its conjugate acid, HF; the base strength of ammonia is indicated by the pK_a of its conjugate acid, the ammonium ion, $^+NH_4$. That is, when we say that a base is weak, we are also saying that its conjugate acid is strong; or, if a base is strong, its conjugate acid is weak. Thus, it is easy to tell which of two bases is stronger by looking at the pK_a values of their conjugate acids: *the stronger base has the conjugate acid with the greater* (*or less negative*) pK_a. For example, ^-CN, the conjugate base of HCN, is a weaker base than ^-OH, the conjugate base of water, because the pK_a of HCN is less than that of water.

Focused Problem

3.24 (a) Sodium salts of different carboxylic acids are added to water, and an equilibrium is established for each as shown in the following equation. For which compound does the equilibrium lie farthest to the right? (Na^+ is a spectator ion.)

$$R{-}CO_2^-\ Na^+ + H_2O \rightleftharpoons R{-}CO_2H + Na^+\ ^-OH$$

sodium acetate (R = CH_3) | **acetic acid** (R = CH_3, pK_a = 4.76)
sodium formate (R = H) | **formic acid** (R = H, pK_a = 3.75)
sodium chloroacetate (R = $ClCH_2$) | **chloroacetic acid** (R = $ClCH_2$, pK_a = 2.85)

(b) Which sodium carboxylate (the compound on the left) in part (a), when dissolved in water at 0.1 *M* concentration, gives a solution with the *lowest* pH? How do you know?

C. Equilibria in Acid–Base Reactions

When a Brønsted acid and base react, we can tell immediately whether the equilibrium lies to the right or left by comparing the pK_a values of the two acids involved. *The equilibrium in the reaction of an acid and a base always favors the side with the weaker acid and weaker base.* For example, in the following acid–base reaction, the equilibrium lies well to the right, because H_2O is the weaker acid and ^-CN is the weaker base.

$$\underset{\substack{\text{p}K_a = 9.4\\ \text{(stronger acid)}}}{HCN} + \overset{\substack{\text{(stronger}\\ \text{base)}}}{^-OH} \rightleftharpoons \overset{\substack{\text{(weaker}\\ \text{base)}}}{^-CN} + \underset{\substack{\text{p}K_a = 14.0\\ \text{(weaker acid)}}}{H_2O} \qquad (3.57)$$

We'll frequently find it useful to estimate the equilibrium constants of acid–base reactions. The equilibrium constant for an acid–base reaction can be calculated in a straightforward way from the pK_a values of the two acids involved. To do this calculation, subtract the pK_a of the acid on the left side of the equation from the pK_a of the acid on the right and take the antilog of the resulting number. That is, for an acid–base reaction

$$HA + B^- \rightleftharpoons A^- + HB \qquad (3.58)$$

in which the pK_a of HA is pK_{HA} and the pK_a of HB is pK_{HB}, the equilibrium constant can be calculated by

$$\log K_{eq} = pK_{HB} - pK_{HA} \qquad (3.59a)$$

or

$$K_{eq} = 10^{(pK_{HB} - pK_{HA})} \qquad (3.59b)$$

This procedure is illustrated for the reaction in Eq. 3.57 in Study Problem 3.8, and it is justified in Problem 3.70 at the end of the chapter.

Study Problem 3.8

Calculate the equilibrium constant for the reaction of HCN with hydroxide ion (see Eq. 3.57).

Solution First identify the acids on each side of the equation. The acid on the left is HCN because it loses a proton to give cyanide ($^-$CN), and the acid on the right is H_2O because it loses a proton to give hydroxide ($^-$OH). Before doing any calculation, ask whether the equilibrium should lie to the left or right. Remember that the *stronger acid and stronger base are always on one side of the equation*, and the weaker acid and weaker base are on the other side. *The equilibrium always favors the weaker acid and weaker base.* This means that the right side of Eq. 3.57 is favored and, therefore, that the equilibrium constant in the left-to-right direction is > 1. This provides a quick check on whether your calculation is reasonable. Next, apply Eq. 3.59a. Subtracting the pK_a of the acid on the left of Eq. 3.57 (HCN) from the one on the right (H_2O) gives the logarithm of the desired equilibrium constant K_{eq}. (The relevant pK_a values come from Table 3.2.)

$$\log K_{eq} = 14.0 - 9.4 = 4.6$$

The equilibrium constant for this reaction is the antilog of this number:

$$K_{eq} = 10^{4.6} = 3.9 \times 10^4$$

This large number means that the equilibrium of Eq. 3.57 lies *far to the right*. That is, if we dissolve HCN in an equimolar solution of NaOH, a reaction occurs to give a solution in which there is *much* more $^-$CN than either $^-$OH or HCN. Exactly *how much* of each species is present could be determined by a detailed calculation using the equilibrium-constant expression, but in a case like this, such a calculation is unnecessary. The equilibrium constant is so large that, even with water in large excess as the solvent, the reaction lies far to the right. This also means that if we dissolve NaCN in water, only a minuscule amount of $^-$CN reacts with the H_2O to give $^-$OH and HCN. Typically, when K_{eq} is $\geq 10^2$, the reaction is said to lie "completely to the right"; when K_{eq} is $< 10^{-2}$, the reaction is said to lie "completely to the left."

Focused Problem

3.25 Using the pK_a values in Table 3.2, calculate the equilibrium constant for each of the following reactions.

(a) NH_3 acting as a base toward the acid HCN

(b) F^- acting as a base toward the acid HCN

Sometimes students confuse acid strength and base strength when they encounter an *amphoteric compound* (see Sec. 3.4B). Water presents this sort of problem. According to the definitions just developed, the *base strength* of water is indicated by the pK_a of its *conjugate acid*, H_3O^+, whereas the *acid strength of water* (or the base strength of its conjugate base hydroxide) is indicated by the pK_a of H_2O itself. These two quantities refer to very different reactions of water:

Water acting as a base:

$$H_2O + AH \rightleftharpoons H_3O^+ + A^{-} \qquad pK_a = 0 \tag{3.60a}$$

Water acting as an acid:

$$B^{-} + H_2O \rightleftharpoons BH + {}^{-}OH \qquad pK_a = 14.0 \tag{3.60b}$$

Focused Problem

3.26 Write an equation for each of the following equilibria, and use Table 3.2 to identify the pK_a value associated with the acidic species in each equilibrium.

(a) Ammonia acting as a base toward the acid water

(b) Ammonia acting as an acid toward the base water

Which of these reactions has the larger K_{eq} and therefore is more important in an aqueous solution of ammonia?

D. Acid Dissociation and Free Energy

Because the dissociation constant K_a is an equilibrium constant, there is a corresponding **standard free energy of dissociation**, abbreviated ΔG_a°:

$$\Delta G_a^\circ = -2.3\ RT \log K_a \tag{3.61a}$$

Substituting pK_a for $-\log K_a$, we obtain

$$\Delta G_a^\circ = 2.3RT\ (pK_a) \tag{3.61b}$$

This very useful equation says that the standard free energy of ionization of an acid is directly proportional to the pK_a of the acid. At 25 °C (298 K), $2.3RT = 5.71$ kJ mol^{-1} (1.36 kcal mol^{-1}); therefore, at room temperature,

$$\Delta G_a^\circ \text{ (in kJ mol}^{-1}) = 5.71\ pK_a$$
$$\Delta G_a^\circ \text{ (in kcal mol}^{-1}) = 1.36\ pK_a \tag{3.61c}$$

Let's see how this equation applies to the dissociation of hydrogen fluoride (HF), a moderately weak acid:

$$H{-}F + H_2O \rightleftharpoons F^{-} + H_3O^+ \tag{3.62}$$

From Table 3.2, the pK_a of HF is 3.2. For HF, then, $\Delta G_a^\circ = (5.71)(3.2) = 18.3$ kJ mol^{-1}. This standard free energy of dissociation ΔG_a° is equal to the difference between the standard free energies of the ionization products (H_3O^+ and F^-) and the un-ionized acid (HF) in the presence of H_2O as a solvent. The standard free energy of the solvent (and reference base) water, because it is the same for all acids, is arbitrarily set to zero (that is, ignored). The following is an energy diagram for this ionization.

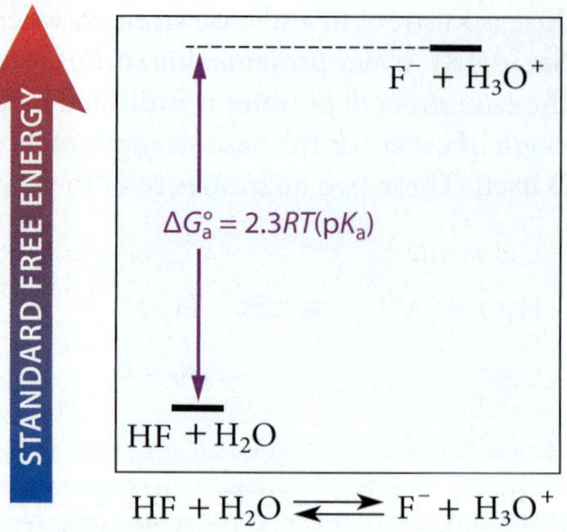

$$HF + H_2O \rightleftharpoons F^- + H_3O^+ \qquad (3.63)$$

We know that the standard free energy of the ionization products is higher than that of HF and H_2O because ΔG°_a is > 0.

The meaning of this standard free-energy change is that the products of the dissociation equilibrium, H_3O^+ and F^-, have 18.3 kJ mol^{-1} more free energy than the undissociated acid HF and solvent water—that is, the products are less stable than the reactants by 18.3 kJ mol^{-1} under standard conditions. Physically, this means that if we could somehow couple a free-energy source, such as a battery, to the HF ionization reaction, this battery would have to provide 18.3 kJ of energy to convert 1 mol per liter of HF *completely* into 1 mol per liter of hydrated protons and 1 mol per liter of fluoride ions.

Focused Problems

3.27 The pK_a of acetic acid is 4.76. What is its standard free energy of dissociation?

3.28 The standard free energy of dissociation of formic acid is 21.4 kJ mol^{-1} (5.11 kcal mol^{-1}). Is it a stronger or weaker acid than acetic acid?

Chemical Biology Topic

E. Dissociation States of Conjugate Acid–Base Pairs; The Henderson–Hasselbalch Equation

In many situations it is important to know the **dissociation state** of a conjugate acid–base pair—that is, whether it exists in the conjugate acid form, the conjugate base form, or a mixture of both. This knowledge is particularly important in biology and medicine for several reasons. For one, many biomolecules (such as enzymes) contain acidic and basic groups, and understanding the chemistry of these biomolecules requires that we know the dissociation states of these groups. Another important reason is that the biological activities of many drugs, including their uptake into cells, depend on their dissociation states. The goal of this section is to show how to determine the dissociation state of a conjugate acid–base pair in a straightforward way.

As a specific example, consider the dissociation equilibrium of a carboxylic acid:

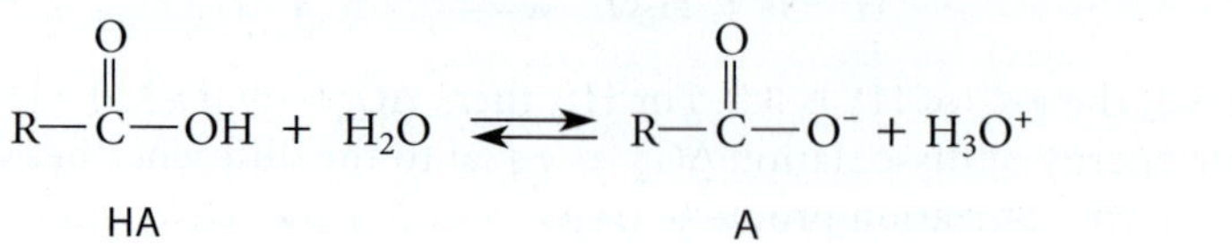

(3.64)

In this equilibrium, we'll refer to the conjugate acid as HA and its conjugate base as A, leaving the charge off of A for convenience.

Application of Le Chatelier's principle shows that in the presence of a very large H_3O^+ concentration—low pH—the equilibrium will favor the conjugate-acid form of the acid—that is, HA. Similarly, if the H_3O^+ concentration is very low—high pH—the equilibrium will favor the conjugate-base form of the acid—that is, A. Exactly what pH values are required to favor one form or the other and by how much? The answer depends on the equilibrium constant for the reaction, which in this case is the *dissociation constant*, K_a (see Sec. 3.6A and Eq. 3.51):

$$K_a = \frac{[H_3O^+][A]}{[HA]} \tag{3.65a}$$

Because $[H_3O^+]$ and K_a are usually cited as the logarithmic quantities pH and pK_a, we'll find it useful to put this equation in logarithmic form. Taking the logarithms of both sides of this equation, we have

$$\log K_a = \log [H_3O^+] + \log \frac{[A]}{[HA]} \tag{3.65b}$$

Adopting the customary definitions, $pH = -\log[H_3O^+]$ and $pK_a = -\log K_a$,

$$-pK_a = -pH + \log \frac{[A]}{[HA]} \tag{3.65c}$$

or, rearranging,

$$pH = pK_a + \log \frac{[A]}{[HA]} \tag{3.65d}$$

Equation 3.65d is known as the **Henderson–Hasselbalch equation**, and it is simply a logarithmic form of the expression for the dissociation constant (Eq. 3.65a).

Here is a very important point: the dissociation constant K_a is a property of the acid. We can't change this. However, the pH is a property of a solution that can be changed experimentally. Once we fix the pH experimentally, then Eq. 3.65d says that the ratio [A]/[HA] is fixed. Or, conversely, if we fix the ratio [A]/[HA] experimentally, we have then fixed the pH. Therefore, the dissociation state of an acid–base pair depends on the pH of the solution (which we can change) and the pK_a of the acid (which we *cannot* change).

Rearranging the Henderson–Hasselbalch equation to Eq. 3.65e shows that the dissociation state of the acid—the ratio [A] / [HA]—depends on the *difference* Δ between the pH of the solution and the pK_a of the acid.

$$\Delta = pH - pK_a = \log \frac{[A]}{[HA]} \tag{3.65e}$$

Let's say that an acid HA is dissolved in aqueous solution, and we want to fix the ratio [A]/[HA] at a certain value. There are two ways to fix this ratio. One way is to adjust the pH by adding ^-OH (for example, by adding NaOH or KOH); or, equivalently, if we dissolve the conjugate base A in solution, we can adjust the pH by adding H_3O^+ (for example, by adding HCl). The second way to fix the ratio [A]/[HA] corresponds to a situation that is particularly relevant to biology: we dissolve either the acid HA or its conjugate base A (or both) in a *large excess* of a buffer solution that has a fixed pH value. The "large excess" of buffer is necessary because we want any acid–base reaction of HA or A with the buffer, which would change the pH, to be negligible. In human cells, which are typically buffered at pH = 7.4 (called **physiological pH**) by the carbonate/bicarbonate/CO_2 buffer system, most acid–base pairs are present at a much smaller concentration than the buffer. Therefore, the dissociation state of the acid–base pair present in dilute solution in the cell depends on the relationship of its pK_a to the pH of the buffer solution.

Given an acid with a certain pK_a, how does its dissociation state depend on pH? We can obtain the answer from Eq. 3.65a or 3.65d. However, we can manipulate these equations to give us a particularly convenient way of looking at this problem. The total concentration of all forms of the acid is given by [HA] + [A]. The *fraction dissociated* f_A, then, is the ratio of the dissociated form A to the total:

$$f_A = \frac{[A]}{[A] + [HA]} \tag{3.66a}$$

Now solve Eq. 3.65a for [A] to obtain $[A] = K_a[HA]/[H_3O^+]$, substitute this result into Eq. 3.66a, and cancel the common factor [HA]:

$$f_A = \frac{K_a[HA]/[H_3O^+]}{(K_a[HA]/[H_3O^+]) + [HA]} = \frac{K_a}{K_a + [H_3O^+]} \tag{3.66b}$$

Using the definitions of pH and pK_a and the property of logarithms that $x = 10^{\log x}$, Eq. 3.66b can be rewritten as

$$f_A = \frac{10^{-pK_a}}{10^{-pK_a} + 10^{-pH}} \tag{3.66c}$$

In the same way, we can show that the *fraction undissociated acid* f_{HA} is given by Eq. 3.66d:

$$f_{HA} = \frac{[HA]}{[A] + [HA]} = \frac{[HA]}{(K_a[HA]/[H_3O^+]) + [HA]} = \frac{[H_3O^+]}{K_a + [H_3O^+]} = \frac{10^{-pH}}{10^{-pK_a} + 10^{-pH}} \tag{3.66d}$$

Convince yourself that the function in Eq. 3.66b goes from $f_A = 0$ at $[H_3O^+] >> K_a$ (that is, very low pH) to $f_A = 1$ at $[H_3O^+] << K_a$ (very high pH). Conversely, f_{HA} (Eq. 3.66d) goes from 1 to 0 at the same extremes. Prove to yourself also that $f_{HA} + f_A = 1$, which must be true by definition.

As we have seen—and as Eq. 3.65e shows—the dissociation state of an acid–base pair is a function of the *difference* between the pH of the solution and the pK_a of the conjugate acid. Letting pH − pK_a = Δ, you can show (Focused Problem 3.29) that f_A and f_{HA} can be expressed solely as a function of Δ:

$$f_A = \frac{1}{1 + 10^{-\Delta}} \tag{3.67a}$$

$$f_{HA} = \frac{1}{1 + 10^{\Delta}} \tag{3.67b}$$

If we plot these two functions, we get the S-shaped curves shown in **Fig. 3.1**. There are several things to notice about these graphs.

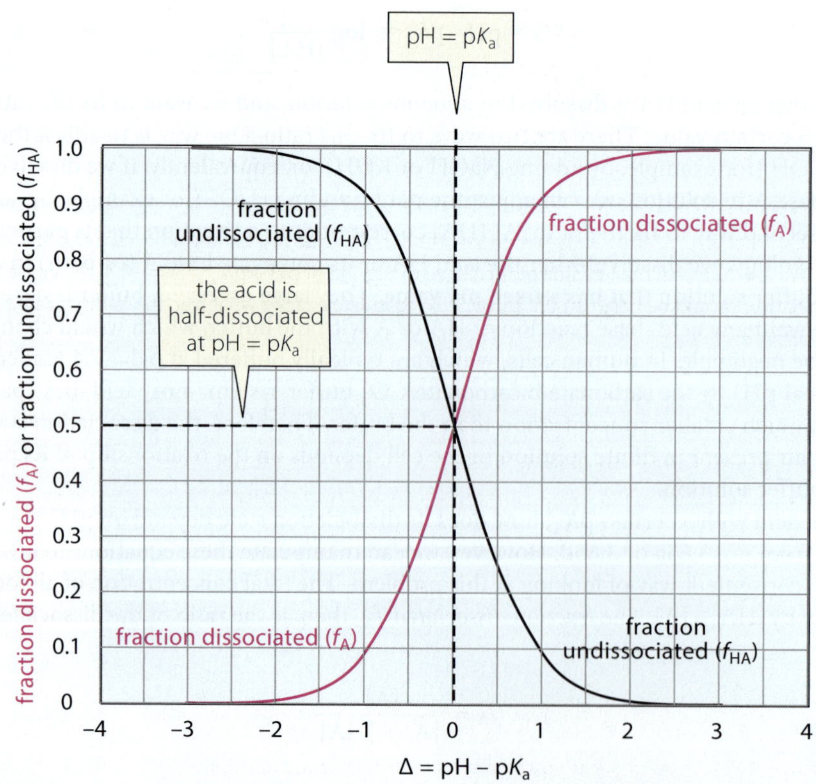

FIGURE 3.1. The fraction of an acid HA that is dissociated (red curve) or undissociated (black curve) as a function of the difference between the pH of the solution and the pK_a of the acid. On the left half of the curve, the pH is lower than the pK_a; on the right half, the pH is higher than the pK_a. The acid is half-dissociated when the pH and the pK_a are equal.

1. At pH = pK_a, the fraction dissociated = the fraction undissociated = 0.5. In other words, operationally, the pK_a is equal to the pH at which the acid–base pair is half-dissociated. Be sure you understand what this means. To say that an acid is "half-dissociated" does not mean that the proton is half removed from a given molecule. It means that in a population of A and HA molecules, their concentrations are equal; half of the molecules are in the A form, and half are in the HA form. Also, equilibrium is dynamic—that is, protons are rapidly jumping on and off of these molecules, but in such a way that the [A]/[HA] ratio is maintained at 1.0.

2. At pH values well below the pK_a—that is, at pH << pK_a—the acid is largely undissociated. When the pH is 1 unit lower than the pK_a, the acid is about 10% dissociated. When the pH is 2 units lower than the pK_a, the acid is about 1% dissociated.

3. At pH values well above the pK_a—that is, at pH >> pK_a—the acid is largely dissociated. When the pH is 1 unit higher than the pK_a, the acid is about 90% dissociated. When the pH is 2 units higher than the pK_a, the acid is about 99% dissociated.

Regarding points 2 and 3, we can see that an acid is never 0% dissociated at any pH, but as the pH is lowered, 0% dissociation is approached asymptotically. Likewise, an acid is never 100% dissociated at any pH, but, as the pH is raised, 100% dissociation is approached asymptotically. As a *practical* matter, we usually say that an acid is "completely undissociated" when the pH is 2 or more units *below* its pK_a, and "completely dissociated" when the pH is 2 or more units *above* its pK_a.

If you are handy with spreadsheets, you can reproduce Fig. 3.1 for yourself. Fill column 1 with closely spaced pH values from 1 to 13 (say, 0.1 unit apart). This can be filled in automatically after you enter the first few values. Fill column 2 with the pK_a, which will be the same in each cell. (Pick any value, say 4.5.) Program column 3 to calculate $[H_3O^+] = 10^{-pH}$, column 4 to calculate $K_a = 10^{-pK_a}$, and columns 5 and 6 to calculate f_A and f_{HA} from their formulas in Eqs. 3.66c and 3.66d. Then use the plotting function to plot columns 5 and 6 against pH (column 1) on the same set of axes. You can then change the pK_a in column 4 and see how all the numbers and the graphs change as a result.

Study Problem 3.9

A histidine residue (B), one of the functional groups in the structure of a certain enzyme, has a conjugate-acid pK_a = 7.8. What is the fraction of each form (HB and B) present at physiological pH (pH = 7.4)?

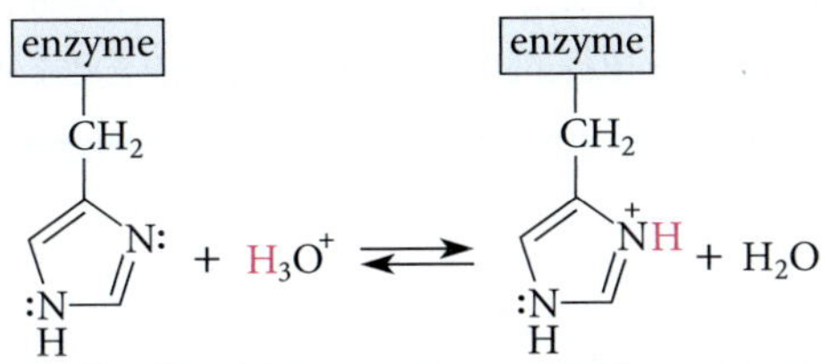

B HB (pK_a = 7.8)

Solution Before we do a calculation, let's think about what we should expect to find. Because physiological pH (pH = 7.4) is *below* the pK_a—that is, pH − pK_a < 0—we know that the histidine residue must be less than half-dissociated—that is, [HB] > [B]. We are, then, on the left side of the curve in Fig. 3.1 at pH − pK_a = Δ = −0.4; the curve indicates that the fraction dissociation should be between 0.2 and 0.3. To calculate exactly what fraction is dissociated, use Eq. 3.67a, with the dissociated form = B:

$$f_B = \frac{1}{1+10^{-(-0.4)}} = \frac{1}{1+10^{+0.4}} = \frac{1}{1+2.51} = \frac{1}{3.51} = 0.28$$

This shows that 28% of the enzyme molecules have histidine in the B form, and 1 −0.28 = 0.72, or 72%, of the enzyme molecules have histidine in the HB form. This calculation verifies our preliminary analysis.

Focused Problems

3.29 Let pH − pK_a = Δ. Starting with Eqs. 3.66c and 3.66d, derive Eqs. 3.67a and 3.67b.

3.30 Ibuprofen is a drug sold as a nonprescription anti-inflammatory medication.

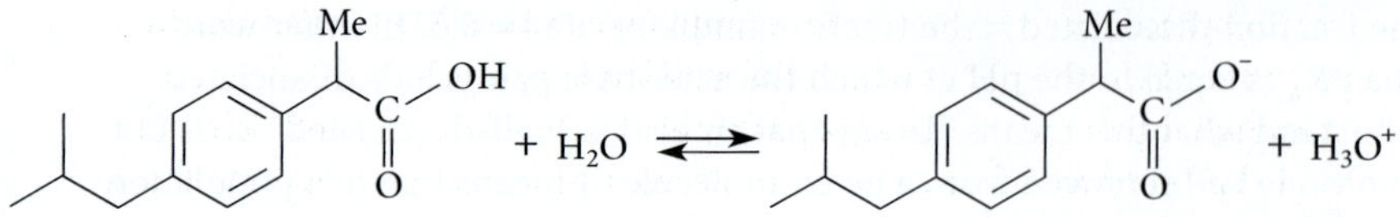

HA A

ibuprofen
($pK_a = 4.43$)

(a) What are the concentrations of ibuprofen and its conjugate base if 10^{-4} mol of ibuprofen is dissolved in an aqueous solution containing a large excess of a buffer at pH = 5.0?

(b) Ibuprofen is taken orally. What fraction of ibuprofen is dissociated in stomach acid? (Take the pH of stomach acid to be 2.0.)

(c) What is the dissociation state of ibuprofen in the bloodstream (pH = 7.4)?

3.31 Acetic acid, CH_3CO_2H, is a carboxylic acid with $pK_a = 4.76$. What is the fraction of acetic acid dissociated (f_A) if 0.1 mol of acetic acid is dissolved in pure water? (*Hint:* You'll have to calculate the pH of the solution first.)

3.32 Nicotine, a habit-forming compound found in tobacco, can undergo two successive acid–base reactions:

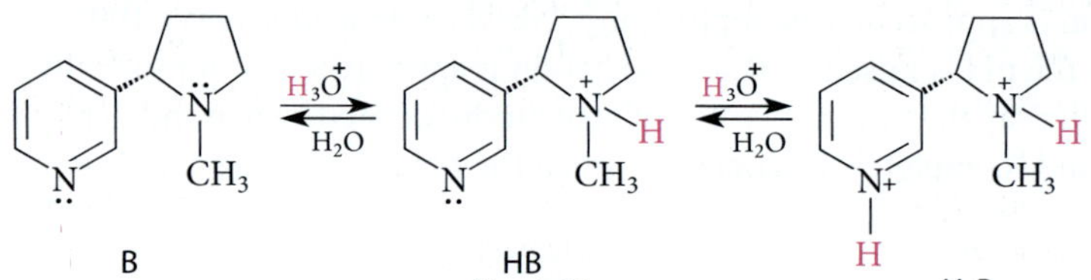

B — **nicotine**; HB — $pK_a = 8.02$; H_2B — $pK_a = 3.13$

(a) Using intuition gained from this section, and using the pK_a values shown in the equation, *sketch* on the same set of axes, *without performing any calculations*, plots of f_B, f_{HB}, and f_{H_2B} over the pH range 1 to 10.

(b) At what pH do you think f_{HB} is a maximum? Explain.

(c) What form of nicotine is present in greatest amount if a small amount of nicotine is dissolved in blood?

3.7 THE RELATIONSHIP OF STRUCTURE TO ACIDITY

In many situations you will need to estimate the relative acidities or basicities of two compounds. You can always go to a pK_a table to look up the acidities, but in many cases you won't need that sort of precision. Rather, you will be able to determine which compound is more acidic or basic, and in some cases by roughly how much, by analyzing their structures. The goal of this section is to help you learn to use the structures of acids to predict trends in their acidities. The reasoning you learn here can often be used to predict the effect of structure on other chemical properties. We consider several structural effects on acidity *in approximate order of importance.*

A. The Charge Effect

One of the most important effects of structure on acidity is the effect of charge on the atom bonded to the acidic hydrogen. For example, the pK_a of H_3O^+ is 0 and the pK_a of H_2O is 14.0. In these cases, oxygen gains an additional nonbonding electron pair as a result of dissociation. The positively charged oxygen in H_3O^+ is much more electronegative than the neutral oxygen in H_2O. Consequently, the positive oxygen attracts the electrons in the O—H bond more strongly than the neutral oxygen, pulls electrons away from the H, and causes the proton in H_3O^+ to dissociate more easily. This effect increases the acidity of H_3O^+ relative to the acidity of water.

Another effect that reduces the acidity of water relative to H_3O^+ is that the ionization of H—OH requires the separation of two charges: the positive charge on the proton and the negative charge on the ^-OH. Increasing the distance between two opposite charges requires energy. (This is like pulling two magnets apart.) Therefore, charge separation increases energy. (We discussed a similar idea in evaluating resonance structures; see Eq. 3.29 and associated discussion.)

Both effects—the electronegativity of oxygen in H_3O^+ and the separation of opposite charges required for the ionization of H_2O—operate in the same direction and conspire to increase the charge effect.

B. The Element Effect

Another important factor that determines the acidity of a Brønsted acid is *the identity of the atom to which the acidic hydrogen is attached.* Two trends in this effect are observed. The first trend comes from a comparison of atoms within the same column (group) of the periodic table.

$CH_3CH_2—O—H$	$CH_3CH_2—S—H$
ethanol	**ethanethiol**
(an alcohol)	(a thiol, or mercaptan)
$pK_a = 15.9$	$pK_a = 10.5$

These two compounds are structurally similar; the sole difference between them is the element (color) bonded to the acidic hydrogen. These elements come from the same group (group 6A) of the periodic table, yet the acidity of the thiol is almost a million times that of the alcohol. The variation in acidity with the element to which the acidic hydrogen is attached is called the **element effect**. The foregoing example illustrates the first important trend in the element effect:

1. *Brønsted acidity increases as the atom bonded to the acidic hydrogen has a greater atomic number (is farther down) within a column (group) of the periodic table.*

Another important example of the same trend is the relative acidities of the hydrogen halides, HF, HCl, HBr, and HI. HI is the strongest of these acids; HF is the weakest. (The relevant pK_a data are found in Table 3.2.) In water, HCl, HBr, and HI are fully ionized and appear to be equally strong; but in other, less basic, solvents, this acidity order is observed.

Let's now consider how acidities vary across a row (period) of the periodic table:

	$H—CH_3$	$H—NH_2$	$H—OH$	$H—F$
pK_a:	~55	~35	14.0	3.2

These data illustrate the second important trend in the element effect:

2. *Brønsted acidity increases as the atom bonded to the acidic hydrogen has a greater atomic number (is farther to the right) within a row (period) of the periodic table.*

Let's consider the origin of these two trends in the element effect. We can divide the ionization process of a typical acid H—A into three steps, as shown in Eqs. 3.68a–c. We are allowed to do this by the first law of thermodynamics, which states that the energy difference between two compounds doesn't depend on the pathway used to interconvert them, just as the height of a building doesn't depend on how one gets to the top to measure it.

Bond breaking $$H—A \longrightarrow H\cdot + A\cdot \qquad (3.68a)$$

Loss of an electron from H• $$H\cdot \longrightarrow H^+ + e^- \qquad (3.68b)$$

Electron transfer to A• $$e^- + A\cdot \longrightarrow A\colon^- \qquad (3.68c)$$

Sum: $$H—A \longrightarrow H^+ + A\colon^- \qquad (3.68d)$$

The first step (Eq. 3.68a) is the breaking of the H—A bond "in half," with one bonding electron going to each atom. The energy required for this step is called the **bond dissociation energy (BDE)**. The BDE is the direct measure of bond strength. When we compare different bonds, the greater the BDE, the stronger the bond. Because acid dissociation involves breaking the bond to hydrogen, smaller BDEs promote increased acidity.

The second step of acid dissociation (Eq. 3.68b) is loss of an electron from the hydrogen atom. The energy required for this step is the **ionization potential** of hydrogen. Because this is the same for all Brønsted acids, it does not enter into a comparison of different acids.

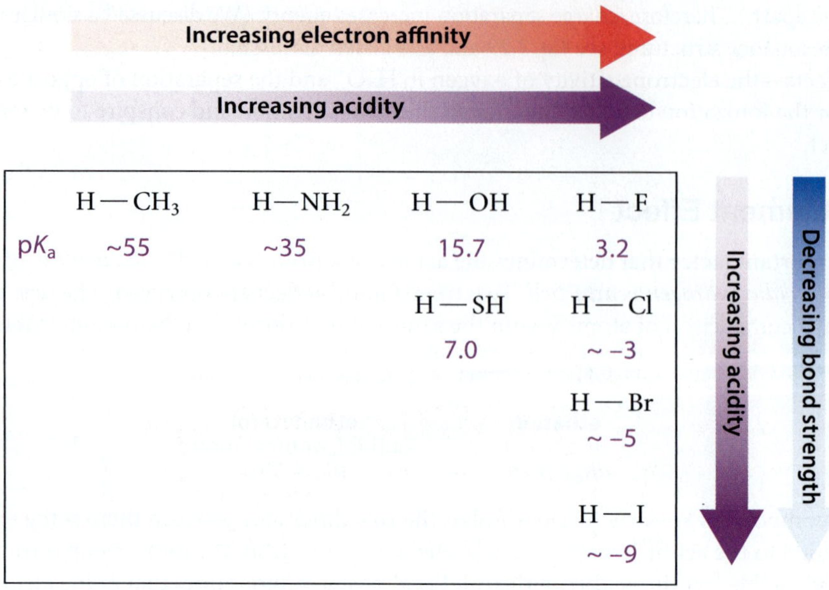

FIGURE 3.2 Sources of the element effect on Brønsted acidity. The acidities of a few compounds H—A are organized by the positions of the nonhydrogen element A in the periodic table. The purple arrows show the direction of increasing acidity within a row or column. The red and blue arrows show the major effect responsible for the trend in each direction.

The charge effect (Sec. 3.6A) can be thought of as an extreme example of the effect of electronegativity: The oxygen in H_3O^+ is much more electronegative than the oxygen in H_2O, and this difference is reflected in the lower pK_a of H_3O^+. In the charge effect, we are comparing the same atoms that differ in charge; in the element effect, we are comparing different atoms.

Bond dissociation energies, ionization potentials, and electron affinities are measured in the *gas phase.* However, we can't calculate actual solution acidities from these numbers because the *solvation of ions*—the stabilizing interaction of solvent molecules with ions—has a huge effect on solution acidities. (Ionic solvation is considered in Sec. 8.6F.) Nevertheless, the gas-phase numbers explain correctly the *relative* acidity orders in the element effect.

The third step of acid dissociation (Eq. 3.68c) is transfer of an electron to A• to form the anion. The energy required for this step is the **electron affinity** of A•. Larger electron affinities promote greater acidity. Electron affinities roughly correspond to electronegativities (see Fig. 1.15, Sec. 1.9A). This is reasonable because both are measures of electron attraction.

Figure 3.2 shows the acidities of a few compounds H—A, with the positions of acids corresponding to the positions of the elements A in the periodic table. The trends in acidity are shown with purple arrows. The *major factor* causing the trend is shown with the red or blue arrow. Within a group, or column, of the periodic table, bond strengths change significantly, but electron affinities change relatively little. Hence, *the major factor governing the acidity increase from top to bottom within a column of the periodic table is the weaker bonds.* Students are sometimes surprised that H—I is a much stronger acid than H—F. Even though fluorine is a much more potent "electron attractor" than iodine, the dominant effect governing acidity is bond strength: the H—I bond is much weaker than the H—F bond.

Across a row of the periodic table, bond strengths change relatively little, but electron affinities change significantly. Therefore, *the increase in acidity across a row of the periodic table is mainly controlled by the electron affinity (or electronegativity) of the elements to which the acidic hydrogen is bonded.*

Let's summarize what we've learned about the element effect:

1. The acidities of Brønsted acids H—A increase toward higher atomic number of atom A within a group of the periodic table. The main source of this increase is the decreasing strength of the H—A bond.

2. The acidities of Brønsted acids H—A increase toward higher atomic number of the atom A within a row of the periodic table. The main source of this increase is the increasing electron-attracting ability of the atom A.

Focused Problem

3.33 (a) Rank the following four acids in order of increasing acidity (decreasing pK_a):

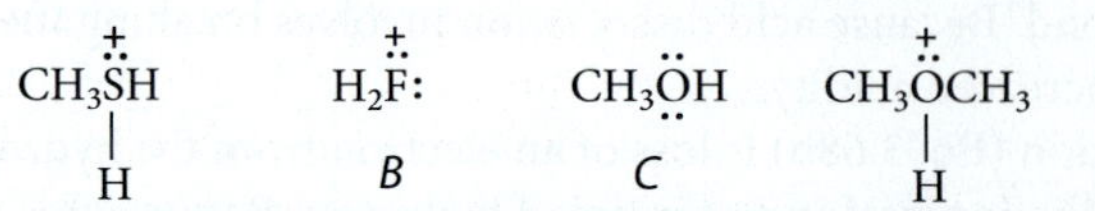

A *B* *C* *D*

(b) Rank the following three bases in order of increasing basicity. (*Hint:* Think about the acidity of their conjugate acids and the relationship between acid strength and base strength.)

$CH_3\ddot{\underset{..}{O}}:^-$ $CH_3\ddot{\underset{..}{O}}H$ $^-:\ddot{N}H_2$

A   *B*   *C*

C. The Hybridization Effect

Brønsted acidity depends strongly on the hybridization of the atom bonded to the acidic hydrogen. Specifically, an increasing amount of s character is associated with greater acidity. Consider, for example, the acidities of simple hydrocarbons.

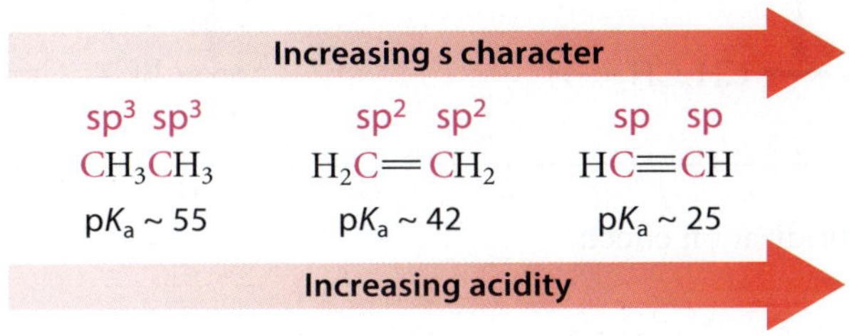

(3.69)

All of these hydrocarbons are *very* weak acids. Consequently, their conjugate bases are *very* strong bases.

According to this diagram, hydrogens attached to sp-hybridized carbons (50% s character) are more acidic than hydrogens attached to sp^2-hybridized carbons (33% s character), which are more acidic than hydrogens attached to sp^3-hybridized carbons (25% s character). You learned in the previous section that two factors control the acidity of an acid A—H: the strength of A—H bond (that is, its bond dissociation energy) and the electron-attracting ability of the group A. Let's see how these two factors contribute to the hybridization effect.

The bond strength of an A—H bond *increases* with increasing s character. If bond strength were the major determinant of acidity, then alkynes should be the *least acidic* of the hydrocarbons. This is not the observed result. Therefore, the other factor must be important: the electron-attracting character of the carbon. In fact, carbons *increase in electronegativity* in the hybridization order $sp^3 < sp^2 < sp$, and the hydrogens attached to these carbons therefore become more acidic in that order.

Why does carbon electronegativity increase in the order $sp^3 < sp^2 < sp$? A carbon with sp^2 hybridization has an electron in a 2p orbital. A carbon with sp hybridization has an electron in each of *two* 2p orbitals. Because 2p electrons are oriented along the *z*-axis, they are not very effective in screening the positive charge of the carbon nucleus in the *xy*-plane. Therefore, any bonding electrons in this plane are attracted more strongly by the carbon nucleus than sp^3 electrons would be. In other words, there is a significant bond dipole for C—H bonds in alkenes and alkynes.

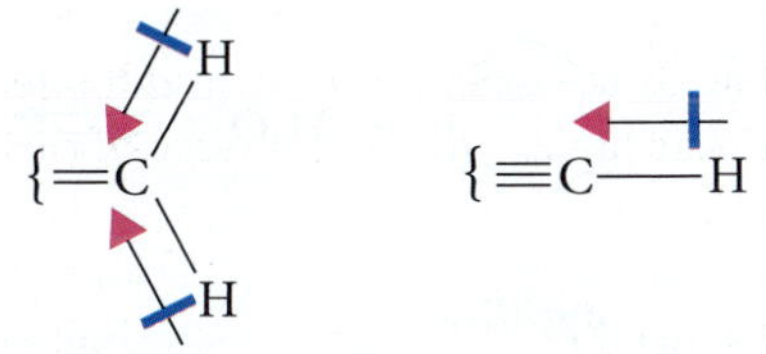

(3.70)

This means that carbons with sp^2 and sp hybridizations are electron-attracting—that is, electronegative; and sp-hybridized carbons are more electron-attracting than sp^2-hybridized carbons.

Another explanation of the hybridization effect comes from considering the hybridization of the orbital containing the nonbonding electron pair in the conjugate-base anion. *Nonbonding electron pairs are stabilized by orbitals of increasing s character.* (Recall that a 2s orbital has lower energy than a 2p orbital.) The conjugate base of an alkyne has a nonbonding electron pair in an sp orbital, which has 50% s character. The conjugate base of an alkane has a nonbonding electron pair in an sp^3 orbital, which has 25% s character. Therefore, the conjugate base of an alkene or alkyne is stabilized by the greater s character in the orbital containing its nonbonding electron pair, and the stabilization is greater for an alkyne. Stabilization of a base favors its formation from its conjugate acid, as we can see from an energy diagram:

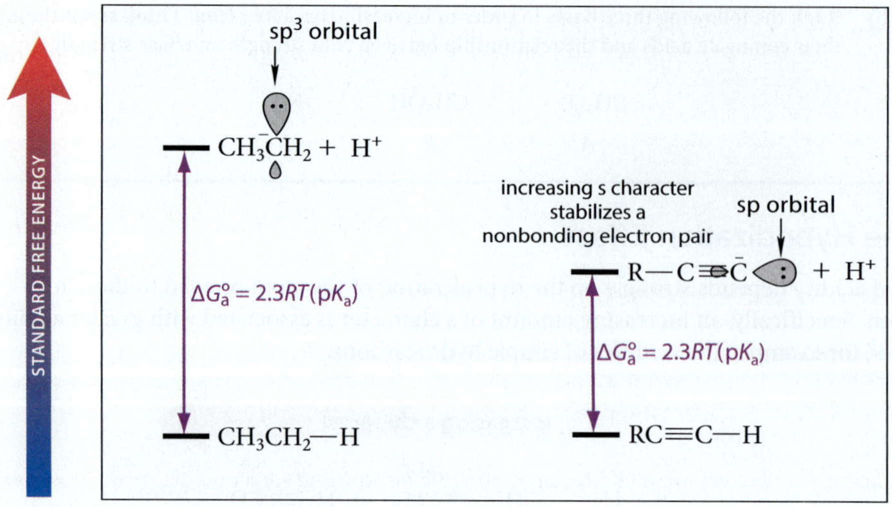

(3.71)

To summarize the hybridization effect:

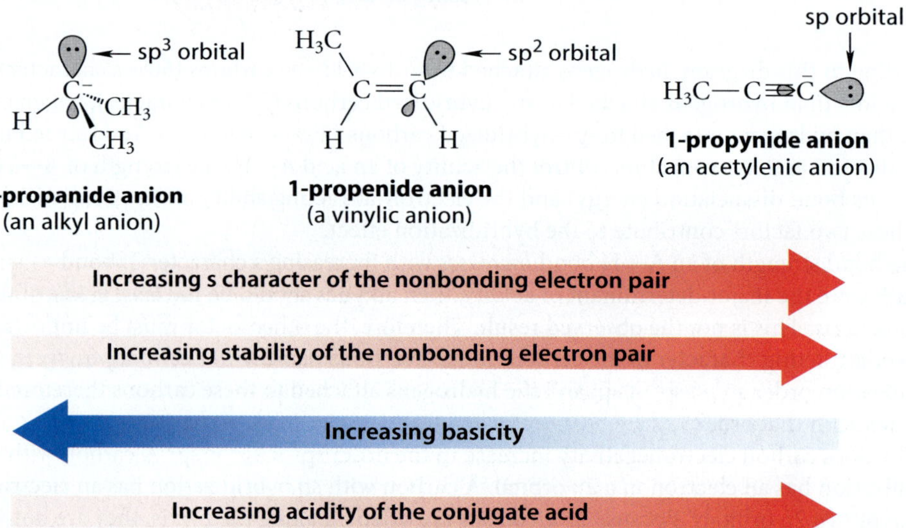

(3.72)

The hybridization effect is not restricted to carbon. The effect of nitrogen hybridization on the acidities of $^{+}N—H$ bonds is illustrated by the relative pK_a values of the conjugate acids of the two amines pyridine and piperidine:

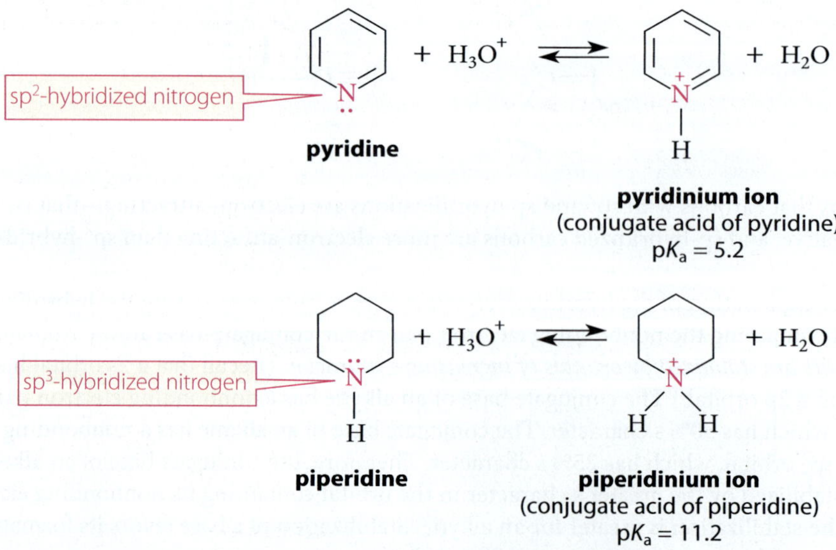

(3.73)

The conjugate acid of piperidine, with its sp^3-hybridized nitrogen, is less acidic than the conjugate acid of pyridine, which has an sp^2-hybridized nitrogen. Equivalently, piperidine is much more basic than pyridine.

Focused Problems

3.34 In each set, match each ion with the correct pK_a, and explain.

(a) $CH_3C{\equiv}\overset{+}{N}H$ $\quad CH_3CH_2\overset{+}{N}H_3$ $\quad$ p$K_a \approx -10$, p$K_a = 10.6$

(b) 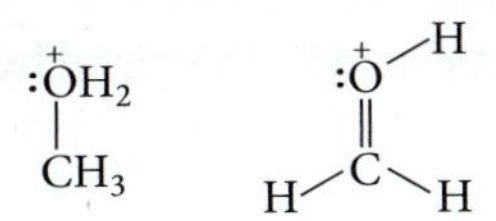

p$K_a \approx -6$, p$K_a \approx -2.5$

3.35 Nicotine is the compound in tobacco responsible for smoking addiction. The two nitrogens are basic. Which nitrogen is more basic and why?

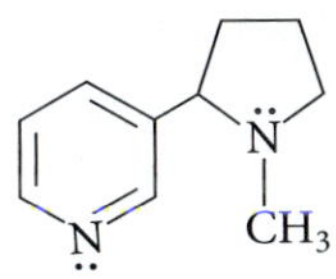

nicotine

D. The Resonance Effect

The acidity of a compound can be significantly affected if it or its conjugate base is resonance-stabilized. *Resonance stabilization of a base makes it less basic (or its conjugate acid more acidic). Resonance stabilization of an acid makes it less acidic (or its conjugate base more basic).*

To understand these resonance effects, we use the relationship of ΔG_a° to pK_a developed in Sec. 3.6D. Remember that resonance *lowers the energy* of a compound; it is a *stabilizing effect.* We draw a diagram showing the relative ΔG_a° of an acid HA and its conjugate base A; we omit the charge on A for convenience. In organic chemistry, we deal most of the time with weak acids (acids with p$K_a > 0$); therefore, the energy of HA is placed below the energy of A. (In other words, the dissociation is unfavorable.) The following energy diagram shows that if resonance lowers the energy of the conjugate base A, then it lowers the pK_a of A (makes HA more acidic, A less basic).

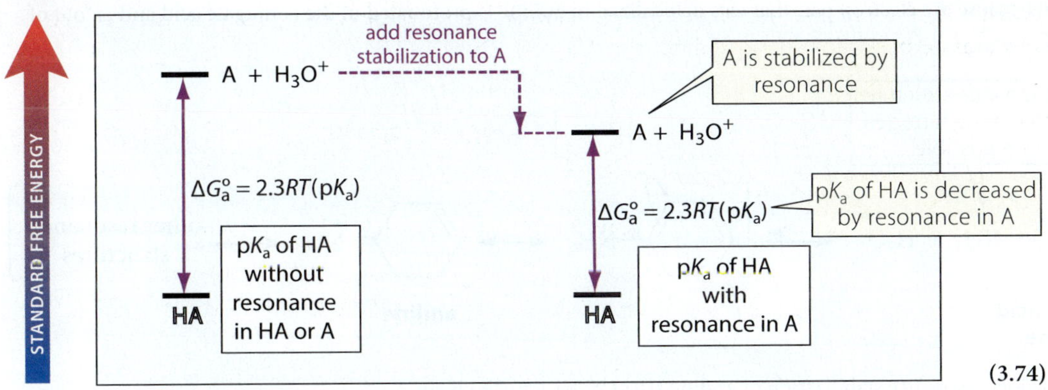

(3.74)

An illustration of this effect is the pK_a of a carboxylic acid such as acetic acid.

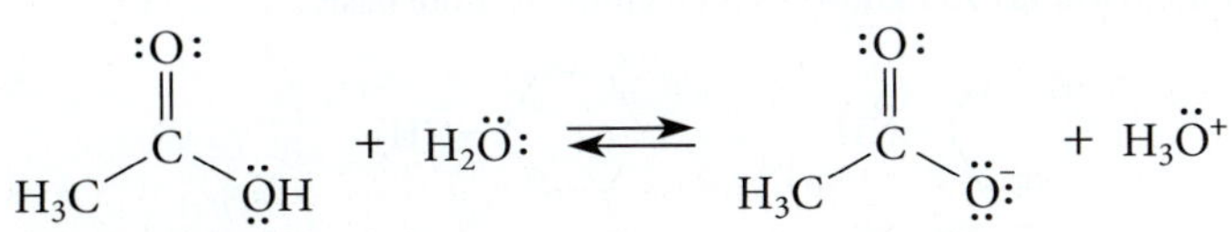

acetic acid p$K_a = 4.76$ $\qquad$ **acetate ion** (conjugate base of acetic acid) $\qquad$ (3.75)

The conjugate base acetate is resonance-stabilized:

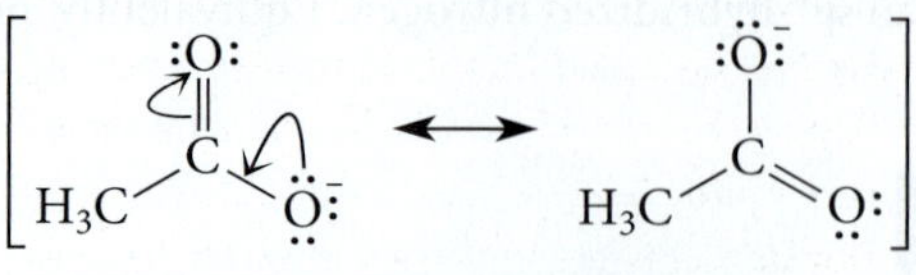

resonance structures of acetate (3.76)

As Display 3.74 shows, this stabilization of the conjugate base *enhances* the acidity of acetic acid. That is, acetic acid is more acidic than it would be without the resonance stabilization of its conjugate base.

The acid-strengthening effect of resonance on the pK_a of acetic acid is illustrated by the following comparison:

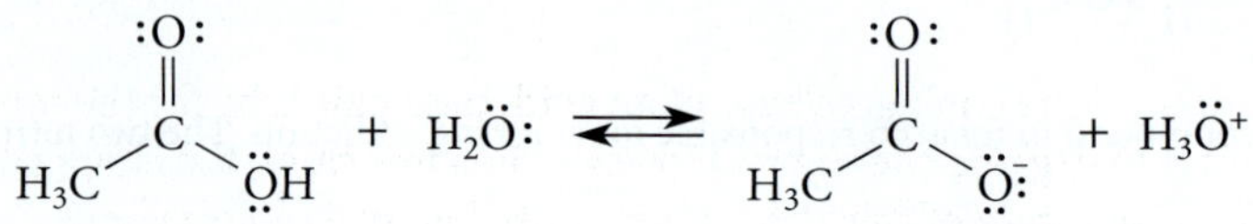

acetic acid
$pK_a = 4.76$

acetate ion
(conjugate base of acetic acid; resonance-stabilized))

The C=O group of acetic acid contributes an acidity-enhancing *polar effect* (discussed in Sec. 3.7E); so, the resonance effect is not the sole reason for the pK_a difference between the two compounds. Nevertheless, the resonance effect is a major contributor to this difference.

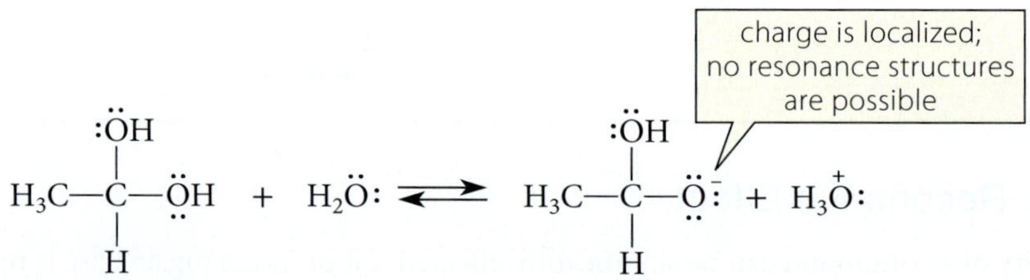

1,1-ethanediol
(acetaldehyde hydrate)
$pK_a = 13.5$ (3.77)

According to this comparison, acetic acid is almost a *billion* (10^9) times more acidic than the diol.

Study Problem 3.10

Aniline, an important molecule in the dye industry, is a resonance-stabilized amine base in which the nonbonding electron pair on the nitrogen is delocalized into the ring. The conjugate acid has no resonance structures involving the nitrogen because the electron pair that was delocalized in aniline is protonated in the conjugate acid and is "out of circulation."

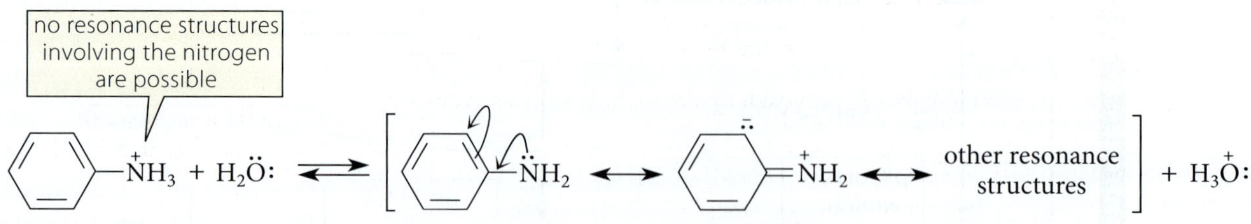

conjugate acid of aniline **aniline**

(a) How does resonance stabilization of aniline affect its basicity? Explain.

(b) One of the following amines has conjugate-acid $pK_a = 10.6$, and the other has conjugate-acid $pK_a = 4.6$. Which pK_a goes with which amine? How do you know? Which amine is more basic?

aniline **cyclohexylamine**

Solution

(a) We can use the diagram in Display 3.74 to predict the answer. Aniline is the conjugate base (A) in this figure, and its conjugate acid is HA. Because aniline is stabilized by resonance, the pK_a of its conjugate acid is reduced—that is, its acidity is *increased*. Saying that the acidity of the conjugate acid is increased is equivalent to saying that the basicity of the conjugate base aniline is therefore *decreased*.

(b) Neither cyclohexylamine nor its conjugate acid is stabilized by resonance. (Why?) That means that the basicity of cyclohexylamine is *not* affected by resonance. However, from part (a), the basicity of aniline *is* decreased by resonance. Therefore, we expect cyclohexylamine to be the stronger base. This means that its conjugate acid has the higher pK_a. We conclude, then, that the higher pK_a (10.6) belongs to the conjugate acid of cyclohexylamine.

The resonance effect on the conjugate-acid pK_a of aniline is smaller than 6 units, because the conjugate acid is stabilized by a *polar effect* of the phenyl group, discussed in the next section. This stabilization partially offsets the resonance effect. The actual resonance effect is closer to 3 pK_a units.

Resonance stabilization of the conjugate base of an acid–base pair is by far the most common situation you'll encounter in organic chemistry. However, in a few cases, the conjugate acid can be stabilized by resonance. (For a case of this type, see end-of-chapter Problem 3.74.)

Focused Problem

3.36 Phenol can ionize to give its conjugate base, phenoxide, as shown in the following equation.

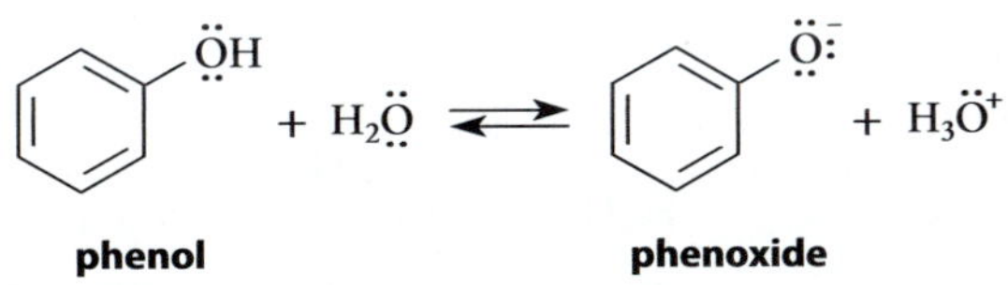

(a) Supply the curved arrows that lead to the resonance structures of phenoxides.

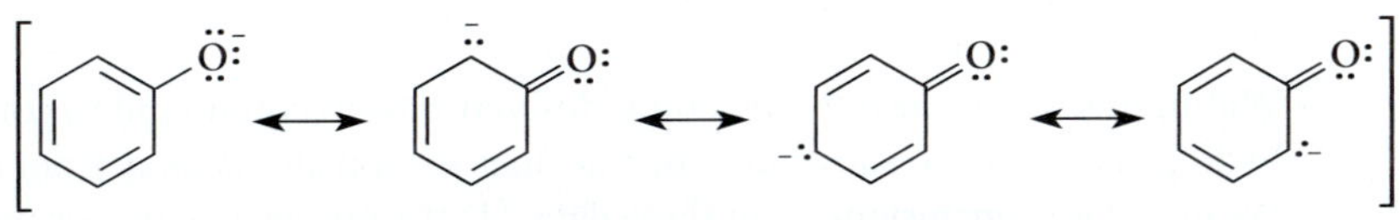

resonance structures for the phenoxide anion

(b) How does resonance in the phenoxide ion affect the pK_a of phenol? (Choose all that apply.)

1. Resonance in phenoxide increases the acidity of phenol.
2. Resonance in phenoxide raises the pK_a of phenol.
3. Resonance in phenoxide lowers the pK_a of phenol.
4. Resonance in phenoxide increases the basicity of phenoxide.
5. Resonance in phenoxide decreases the basicity of phenoxide.

(c) Use your answer to part (b) to decide which of the following compounds is the stronger acid (has the lower pK_a). Explain.

E. The Polar Effect

In this section we consider the effect on acidity by substituents that are remote from the acidic group. The acidity of a compound can be significantly affected if the compound has electronegative substituents near the acidic hydrogen. We use *carboxylic acids* as our case study.

Although carboxylic acids are classified as weak acids, they are more acidic than most other types of organic compounds. The typical carboxylic acid in aqueous solution undergoes a small degree of ionization to give its conjugate base, a *carboxylate ion*.

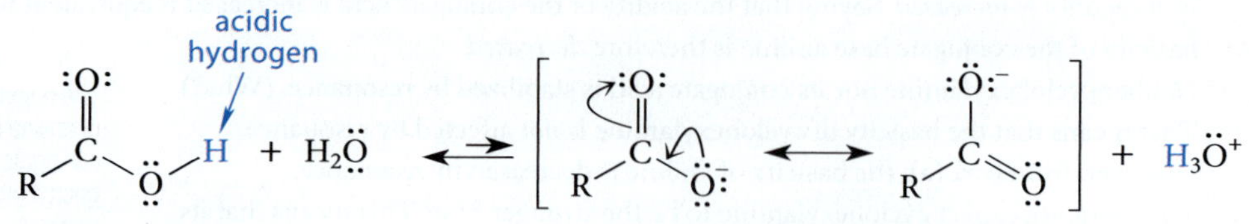

general structure of a carboxylic acid — general structure of a carboxylate ion (3.78)

As shown in Eq. 3.78, carboxylate ions are resonance-stabilized. For convenience, we'll sometimes use the following hybrid structure for carboxylate ions, which shows the sharing of double-bond character and negative charge with dashed lines and partial charges:

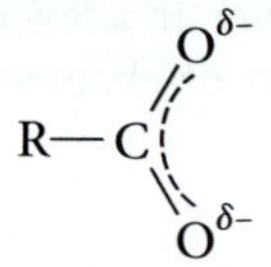

hybrid structure of a carboxylate ion (3.79)

Consider the trend in acidity indicated by the following data for acetic acid and some of its substituted derivatives:

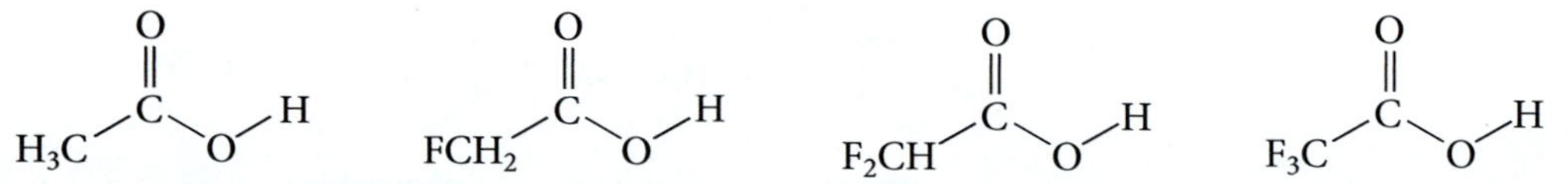

acetic acid pK_a = 4.76 — **fluoroacetic acid** pK_a = 2.66 — **difluoroacetic acid** pK_a = 1.24 — **trifluoroacetic acid** pK_a = 0.23 (3.80)

Within this series, the only structural difference from compound to compound is that hydrogens have been substituted by fluorines several atoms away from the acidic hydrogen. We draw two conclusions from these data: (1) the presence of one or more fluorine substituents strengthens the acid; and (2) the more fluorines are present, the greater is the acid-strengthening effect.

A similar effect is observed when other electronegative substituents are present:

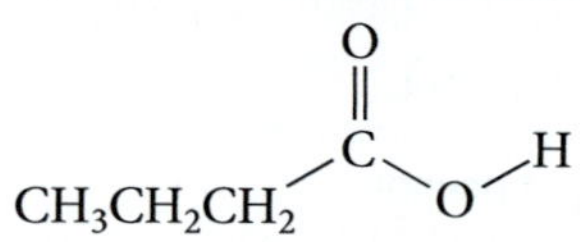

butanoic acid pK_a = 4.82

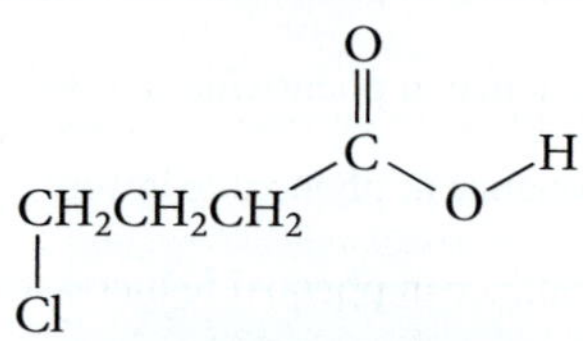

4-chlorobutanoic acid pK_a = 4.52

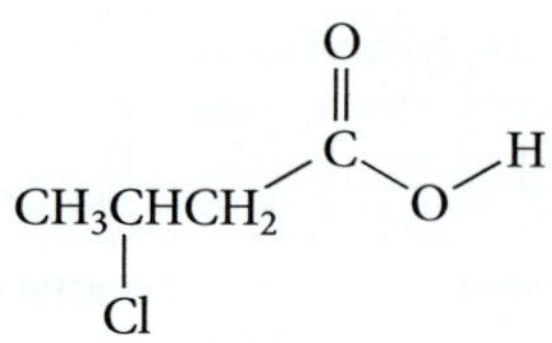

3-chlorobutanoic acid pK_a = 4.06

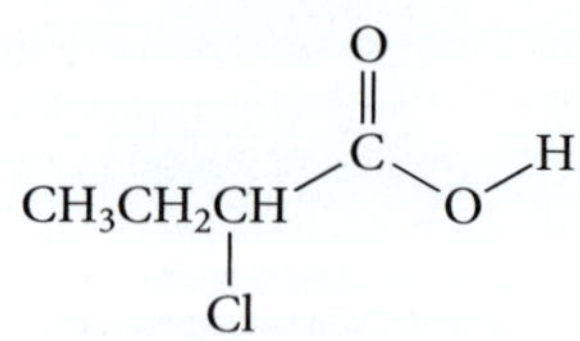

2-chlorobutanoic acid pK_a = 2.84 (3.81)

These data show that the closer the electronegative group is to the acidic hydrogen, the greater its effect on acidity. When the substituent is five or more atoms removed from the acidic hydrogen, its effect on acidity becomes negligible.

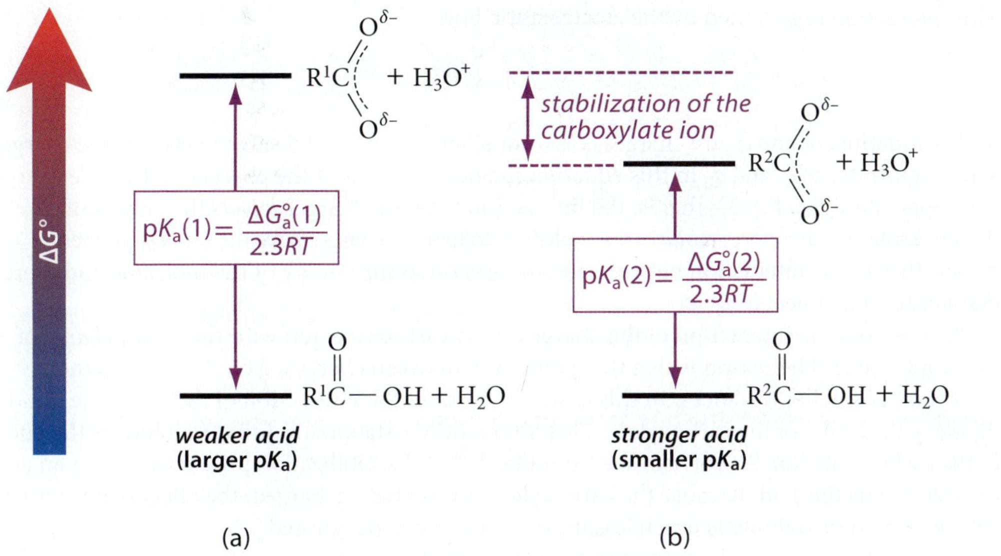

FIGURE 3.3 (a) The pK_a of an acid is proportional to the standard free-energy difference between an acid and its conjugate base. (b) Lowering the standard free energy of a conjugate base reduces the pK_a of the acid and increases its acidity. The two un-ionized carboxylic acids have been arbitrarily placed at the same standard free energy to focus attention on the relative free energies of the conjugate bases.

To understand these effects, we start with the standard free energy of the ionization process. Recall (Eq. 3.61b) that the standard free energy of ionization ΔG_a° is related to the dissociation constant K_a of an acid by the equation

$$\Delta G_a^\circ = 2.3RT\,(pK_a) \tag{3.82a}$$

Rearranging this equation, we have

$$pK_a = \frac{\Delta G_a^\circ}{2.3RT} \tag{3.82b}$$

Remember that the standard free energy of ionization for a carboxylic acid is equal to the *difference* between the products of ionization (the conjugate base and H_3O^+) and the carboxylic acid itself. This idea is diagrammed in **Fig. 3.3a**. Look at this part of the figure, and think about what would happen to the pK_a of a carboxylic acid if we did something to increase the relative stability (that is, lower the relative standard free energy) of its conjugate base. This is shown in **Fig. 3.3b**. Lowering the standard free energy of a conjugate base decreases ΔG° and, by Eq. 3.82b, also lowers the pK_a of the acid. In other words, *lowering the standard free energy of a conjugate base makes the conjugate acid more acidic.*

Electronegative substituent groups such as halogens increase the acidities of carboxylic acids by stabilizing their conjugate-base carboxylate ions. This stabilization originates in the polarity of the carbon–halogen bond—that is, its bond dipole. To visualize this idea, consider the electrostatic interaction (interaction between charges) of the negatively charged carboxylate oxygens in fluoroacetic acid with the nearby carbon–halogen bond dipole:

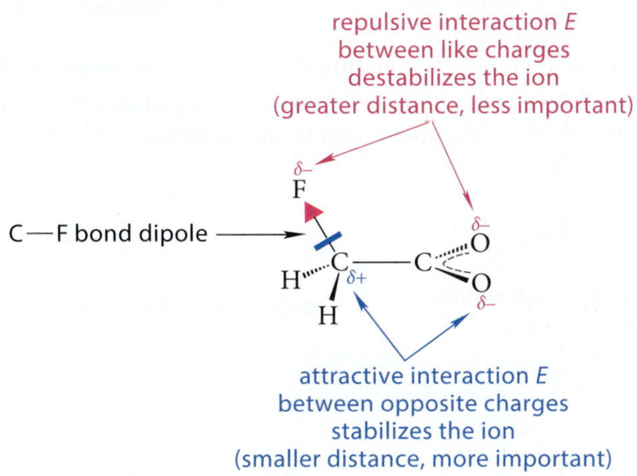

(3.83)

This interaction is governed by the **electrostatic law**:

$$E = k\frac{q_1 q_2}{r} \tag{3.84}$$

In this equation, q_1 and q_2 are charges, k is a constant, and r is the distance between the charges. The magnitudes of q_1 and q_2 in this equation include the signs of the charges. When the charges have opposite signs, $E < 0$—that is, the interaction between charges *lowers* the energy of the molecule and therefore *stabilizes* the molecule. When the charges have the same sign, then $E > 0$—that is, the interaction between the charges *raises* the energy of the molecule and therefore *destabilizes* the molecule.

We consider the interaction of the charge on the carboxylate ion with the *closest charge* of the bond dipole. The reason is that the interaction between charges, from Eq. 3.84, is more important at smaller distances. In this case, we consider the interaction of the $\delta+$ of the bond dipole with the $\delta-$ of the carboxylate. This interaction is stabilizing—that is, it lowers the energy of the carboxylate ion. As you can see from Fig. 3.3b, this stabilization lowers the pK_a of an acid, or strengthens the acid. Because the carboxylic acid itself is uncharged, the effect of a fluorine substituent on its stability is much less important and can be ignored.

When there are more electronegative substituents, the positive charge on the closest (carbon) end of the bond dipole is larger, and the electrostatic attraction is correspondingly greater:

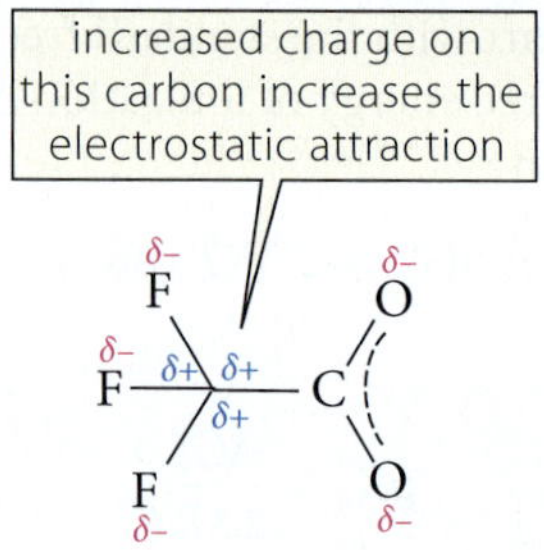

(3.85)

This accounts for the enhanced acidity of the acids with more than one fluorine substituent shown in Display 3.80.

The inverse relationship between the interaction energy E and r in Eq. 3.84 shows that the interaction energy between charges diminishes with distance. This accounts for the reduced effect of substituents as they are further removed from the carboxylate ion.

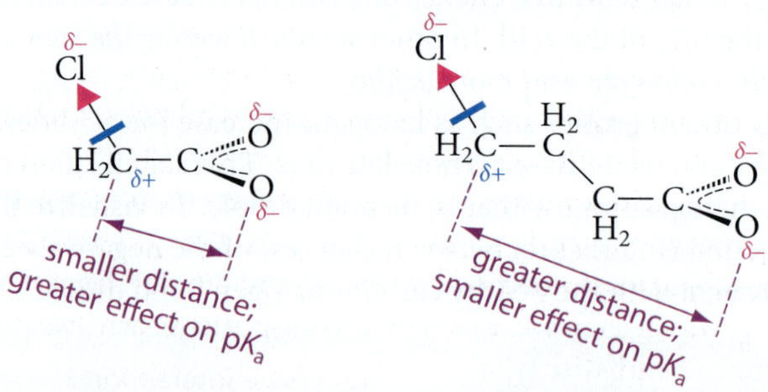

(3.86)

We analyze the electrostatic interaction within the *charged* component of the acid–base equilibrium. Therefore, if the conjugate acid is positively charged and the conjugate base is neutral, we analyze the electrostatic interaction in the *conjugate acid*. The following case illustrates such an analysis:

conjugate acid of ethanamine
pK_a = 10.6

conjugate acid of 2-fluoroethanamine
pK_a = 5.7

(3.87)

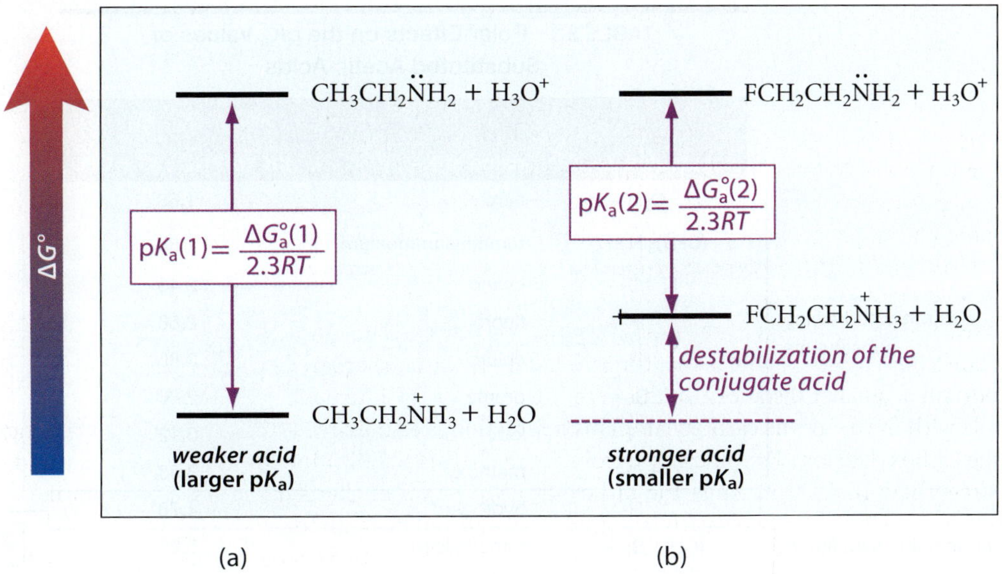

FIGURE 3.4 (a) The pK_a of an unsubstituted amine. (b) Electrostatic interaction between the positive charge of the conjugate acid and the bond dipole of the substituent raises the energy of the acid and reduces its pK_a.

In this case, the interaction at the smallest distance is the interaction of the positive charge of the acid with the positive end of the substituent bond dipole.

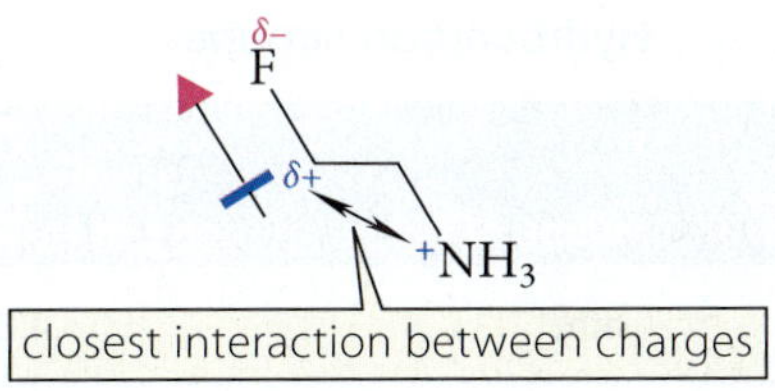

(3.88)

This interaction is a *destabilizing interaction*. Because the conjugate-acid component of an acid–base equilibrium is destabilized, the pK_a of the conjugate acid is reduced—*that is, its acidity is increased* (**Fig. 3.4**).

Even though the electrostatic interaction involves the conjugate base in one case (stabilizing the conjugate base) and the conjugate acid in the other (destabilizing the conjugate acid), an electronegative substituent is acid-strengthening in both cases. This effect of a "remote" substituent on the pK_a of an acid (or other chemical reaction) is called a **polar effect**. It is also sometimes called an **inductive effect**. Electronegative, acid-strengthening substituents are termed **electron-withdrawing substituents**, because, as their bond dipoles suggest, they pull electrons away from the carbon to which they are attached. Electropositive, acid-weakening substituents are called **electron-donating substituents**.

Table 3.3 shows the magnitude of the polar effect for substituted acetic acids. (X = H corresponds to acetic acid itself, which serves as our reference compound.) Notice that pK_a values decrease *roughly* as the electronegativity of the substituent increases (toward the top of the table). Notice also that the acid with the electropositive substituent [$(CH_3)_3Si$—] and the acid with the negatively charged substituent (—CO_2^-) have *higher* pK_a values than acetic acid.

The effect of hydrocarbon substituents, shown in **Table 3.4**, is also interesting. Substituents with sp^2 carbons (vinyl, phenyl) are more electronegative than the substituent with sp^3 carbons (ethyl), as we learned in 3.6B. The substituent with the sp carbons (ethynyl) is more electronegative still. In other words, phenyl, vinyl, and ethynyl are all *electron-withdrawing substituents* and are therefore acid-strengthening substituents.

The term *inductive effect* refers to an alternative mechanism for the polar effect in which the positive charge of the substituent bond dipole is transmitted through bonds rather than through space. The through-space mechanism discussed in this section is sometimes called a **field effect**. There has been considerable debate about the mechanism by which the polar effect operates. Regardless of what it is called, the result is the same.

TABLE 3.3 Polar Effects on the pK_a Values of Substituted Acetic Acids

	substituent X—	name of X—	pK_a of $H_2C(X)—C(=O)—OH$
electron-withdrawing substituents	O_2N—	nitro	1.68
	$(CH_3)_3\overset{+}{N}$—	trimethylammonium	1.83
	N≡C—	cyano	2.43
	F—	fluoro	2.66
	Cl—	chloro	2.86
	Br—	bromo	2.86
	I—	iodo	3.12
	CH_3O—	methoxy	3.53
	H—	hydrogen*	4.76*
electron-donating substituent	$(CH_3)_3Si$—	trimethylsilyl	5.22
negatively charged substituent	$^-O—C(=O)—$	carboxy	5.69

*acetic acid

Stronger acids ↑ / Weaker acids ↓

TABLE 3.4 pK_a Values of Acetic Acids Substituted with Hydrocarbon Groups

	substituent X—	name of X—	pK_a of $H_2C(X)—C(=O)—OH$
electron-donating substituent	CH_3CH_2—	ethyl	4.82
acetic acid, X = H	H—	hydrogen	4.76
electron-withdrawing substituents	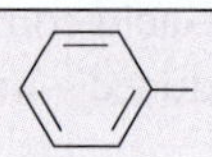	phenyl	4.35
	$H_2C═CH$—	vinyl	4.35
	HC≡C—	ethynyl	3.32

Focused Problem

3.37 In each of the following sets, choose the strongest acid, and explain your reasoning.

(a)  *A* 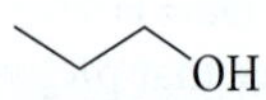 *B*

(b) 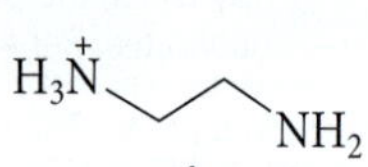 *A* 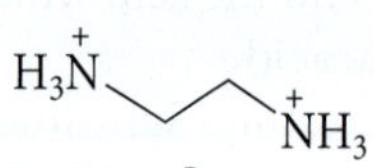 *B*

(c) 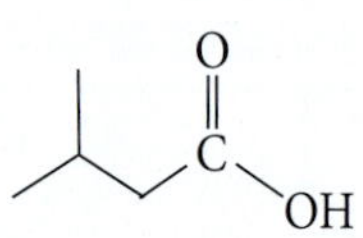 *A* 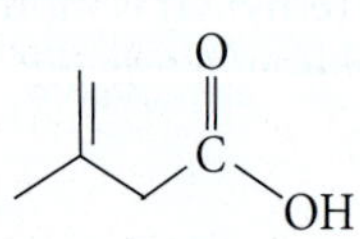 *B* 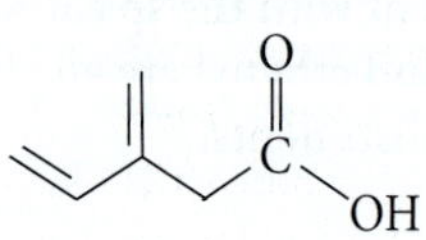 *C*

F. Summary of the Factors Affecting Acidity

Let's summarize the factors affecting acidity that we've learned about in this section:

1. The *charge effect* is the increase in acidity that results from increasing the positive charge on the atom bonded to the acidic hydrogen.

2. The *element effect* is the effect on acidity of changing the atom to which the acidic hydrogen is bonded. The element effect has its origins in the bond dissociation energies to acidic hydrogens and the electron affinities of the elements attached to the acidic hydrogens. Trends in acidity based on the element effect can be predicted from the relationship of the elements in the periodic table.

3. The *hybridization effect* is the effect of hybridization (sp^3, sp^2, sp) of the atom to which the acidic hydrogen is bonded. Acidity increases with fraction s character.

4. The *resonance effect* depends on whether the conjugate acid or the conjugate base is selectively resonance-stabilized. Stabilization of the conjugate base increases acidity, and stabilization of the conjugate acid decreases acidity. If both acid and base have resonance structures, the resonance effect is evaluated from the species with the more important resonance structures.

5. The *polar effect* is the effect on acidity that results from the interaction of the charged species in an acid–base equilibrium with the bond dipole (or charges) on substituents.

If you are asked to rank several compounds according to basicity, simply convert the compounds into their conjugate acids and rank the acidities of these acids. Then reverse the order for the relative basicities of the conjugate bases. Study Problem 3.11 illustrates this process.

Study Problem 3.11

Rank the following compounds in order of increasing basicity.

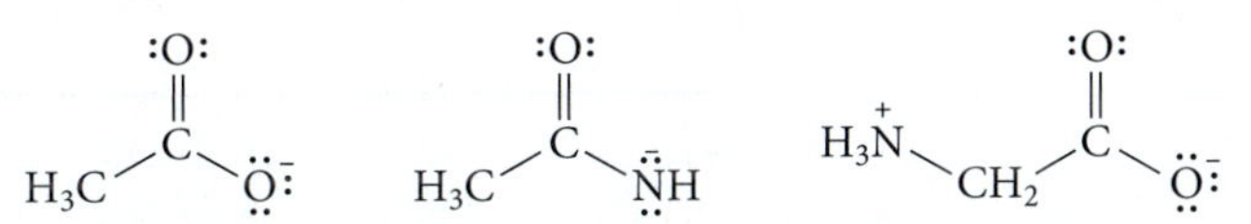

acetate ion *A* | **acetamide anion** *B* | **glycine** (an amino acid) *C*

Solution First, a problem in relative basicity is equivalent to a problem in relative acidity. If you can rank the acidities of the conjugate acids, you've solved the problem. The relevant conjugate acids are

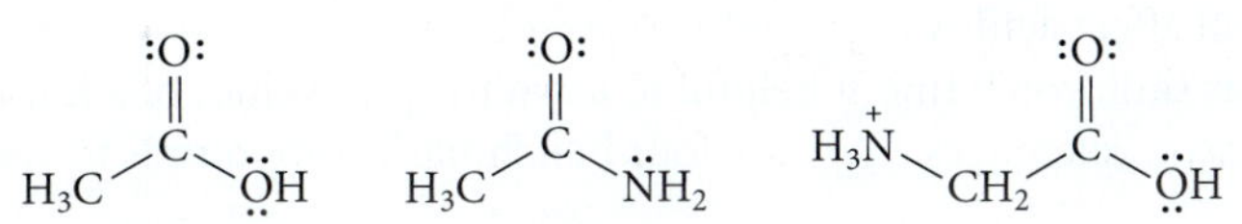

acetic acid *HA* | **acetamide** *HB* | **glycine** (conjugate acid) *HC*

Compound *HC* has two potentially acidic groups: the $H_3\overset{+}{N}$— (ammonium) group and the carboxylic acid group. However, the only basic group in *C* is the carboxylate group; the ammonium group cannot accept a proton because it is already bonded to four groups. Therefore, in terms of this problem, we can think of glycine as a carboxylic acid. In compounds *HA* and *HC*, the acidic hydrogen is bound to an oxygen, whereas in compound *HB*, the acidic hydrogen is bound to a nitrogen. The difference in acidities of *HB* and the other two compounds is therefore due primarily to the element effect along the first row of the periodic table, and this is the most important effect. This effect predicts that the O—H group should be more acidic than a comparably substituted N—H group, because oxygen is more electron-attracting than nitrogen. Thus, the acidities of both *HA* and *HC* are greater than the acidity of *HB*. The difference in the acidities of *HA* and *HC* is due to the polar effect of the $H_3\overset{+}{N}$— group in compound *HC* on the acidity of the nearby carboxylic acid group. The full positive charge on the nitrogen has a favorable interaction with the negatively charged carboxylate oxygen in the conjugate base *C*. As shown in Fig. 3.2, this interaction stabilizes the conjugate base and thus enhances the acidity of *HC*. Hence, the final order of acidity is *HC* > *HA* > *HB*. Because stronger acids have weaker conjugate bases, the basicity order of the conjugate bases is *C* < *A* < *B*. Our prediction is correct: The actual pK_a values are *HC*, 2.17; *HA*, 4.76; and *HB*, 15.

There is a resonance effect on the acidity of all three compounds: the conjugate bases of all three are stabilized by resonance. Because the resonance effect acts in the same way on all three compounds, it can be ignored.

Focused Problem

3.38 In each of the following sets, arrange the compounds in order of decreasing pK_a, and explain your reasoning.

(a)

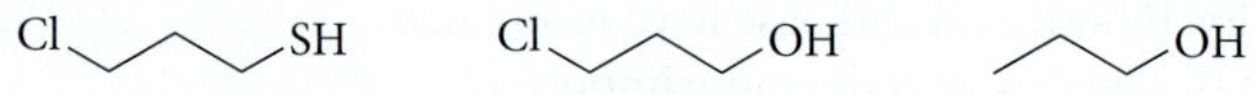

A *B* *C*

(b)

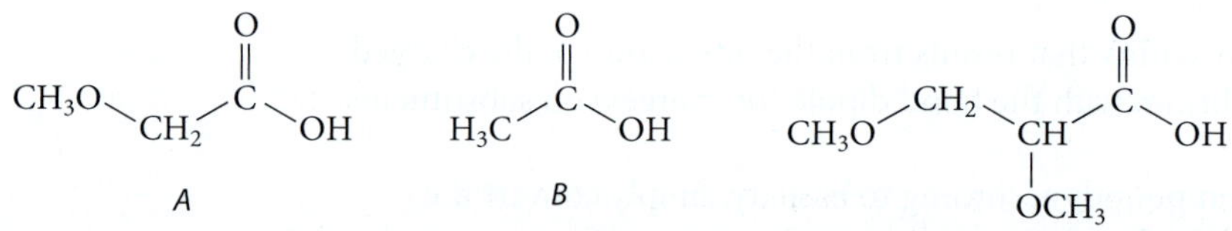

A *B* *C*

(c) (*Hint:* The charge effect is more important than the resonance effect.)

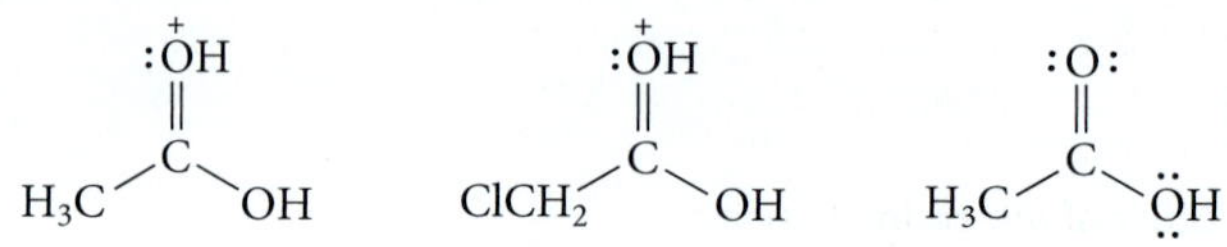

A *B* *C*

G. What pK_a Values Should You Learn?

In this section you've learned how certain elements of structure affect acidity and basicity. The whole point of learning these effects is that we don't have to memorize a large number of pK_a values. Knowing the pK_a of one compound, you can then predict *in a general way* how modifying its structure will affect acidity.

To apply this skill, you'll find it helpful to learn the pK_a values of a few *benchmark compounds* that you can use as references. Here are four benchmark compounds to start with:

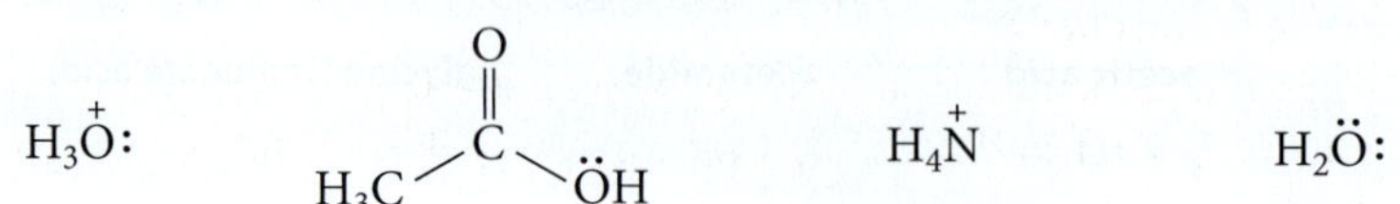

hydronium ion pK_a = 0 | **acetic acid** pK_a = 4.76 | **ammonium ion** pK_a = 9.25 | **water** pK_a = 14 (3.89)

Quite often, we won't need to know exact pK_a values, but only approximate ones (within a couple of pK_a units). A useful generalization for estimating pK_a values of certain organic compounds is that *substituting a hydrogen with an alkyl group* (methyl, ethyl, etc.) *has a relatively minor effect on* pK_a. Therefore, if we know the pK_a of water, we also know the *approximate* pK_a of an alcohol:

water pK_a = 14 **methanol (methyl alcohol)** pK_a = 15.5 (3.90)

Alcohols have pK_a values in the 15–19 range, and most of them are closer to 15.

If we know the pK_a of H_3O^+, then we also know the *approximate* pK_a values of protonated alcohols and ethers within a couple of pK_a units:

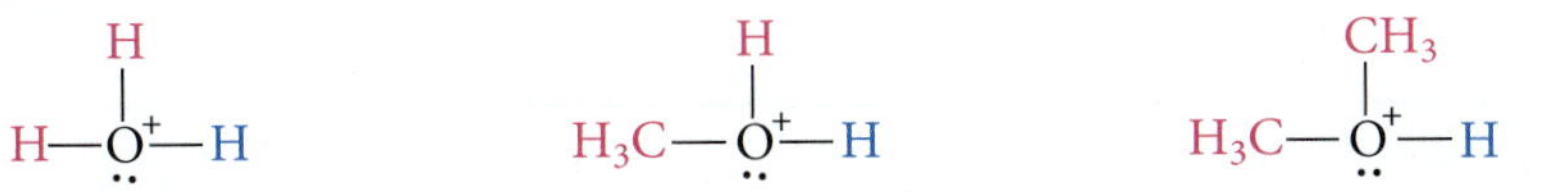

hydronium ion pK_a = 0 **protonated methanol** pK_a ≈ −2 **protonated dimethyl ether** pK_a ≈ −2.5 (3.91)

If we know the pK_a of ammonium ion, then we also know the approximate pK_a values of protonated amines:

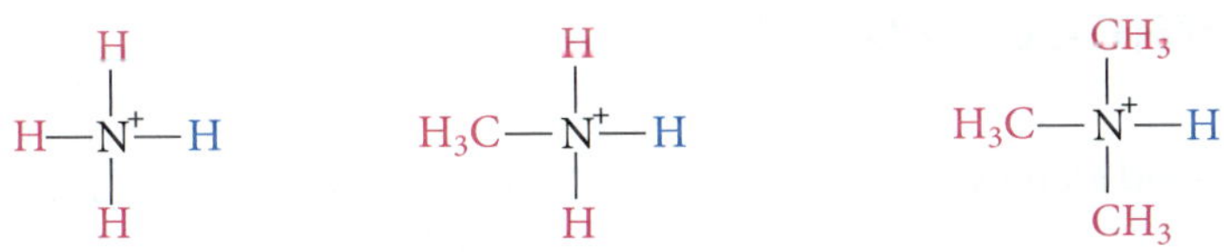

ammonium ion pK_a = 9.25 **methylammonium ion** pK_a = 10.6 **trimethylammonium ion** pK_a = 9.7 (3.92)

Protonated amines (the conjugate acids of amines) have typical pK_a values in the 9–11 range.

Finally, if we know the pK_a of acetic acid, then we also have a general idea of the pK_a of carboxylic acids with the hydrogen substituted by alkyl:

propanoic acid pK_a = 4.87 **acetic acid** pK_a = 4.76 (3.93)

Carboxylic acids have pK_a values typically in the 4–5 range.

The point is that we can go from the pK_a of our benchmark compounds to the *approximate* pK_a values of a large variety of organic compounds very simply. As discussed in the previous section, if we add polar substituents to our organic compounds, we know or can figure out the direction in which we have to adjust these pK_a values. If we introduce a group that can interact by resonance with the element bonded to the acidic hydrogen, we know how to reason the effect on pK_a. If we change the element bonded to the acidic hydrogen (for example, O to S), we should be able to forecast the direction of the pK_a change and, from the examples in this section, the approximate magnitude of the change.

One last point to remember is that very strong acids such as nitric acid (HNO_3), hydrochloric acid (HCl), sulfuric acid (H_2SO_4), and perchloric acid ($HClO_4$) all have negative pK_a values in water, which means they dissociate completely in water to H_3O^+ and their respective conjugate bases. Most hydrocarbons (such as methane, ethylene, and acetylene) and amines (such as ammonia and methylamine, CH_3NH_2) are very weak acids (very large, positive pK_a values), and they do not dissociate at all unless they are allowed to react with very strong bases. Despite their very low acidity, their conjugate bases can be formed under special conditions, and these prove to be useful reagents in organic chemistry.

Focused Problem

3.39 Place each compound into one of the following pK_a ranges: 13–16, 9–11, 3–5, –3 to 0.

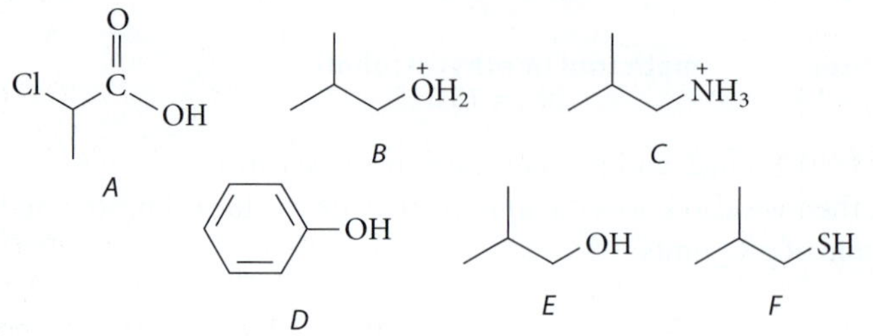

CHAPTER SUMMARY *For a summary of the chapter, see Chapter 3 in the Study Guide and Solutions Manual.*

SKILLS OBJECTIVES WITH PROBLEMS

- **Recognize electron-deficient compounds, and identify the electron-deficient atom. 3.1A**

3.40 Which of the following are electron-deficient compounds? Explain how you know.

(a) 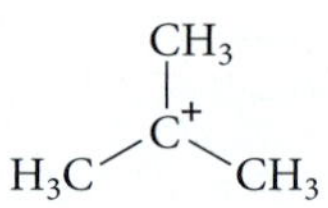

(b) 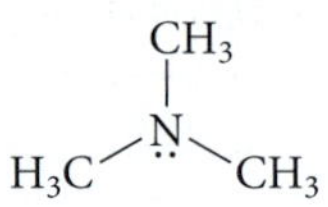

(c) 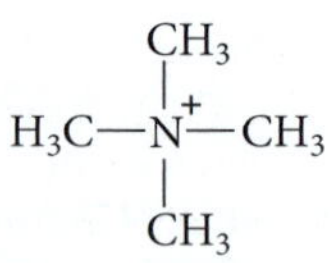

(d) 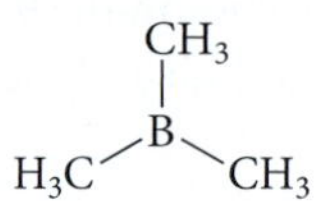

(e) 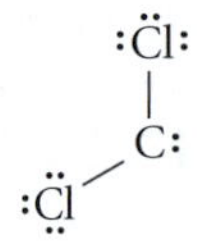

- **Show the reaction of an electron-deficient compound with a Lewis base using the curved-arrow notation. Identify the electrophile, the electrophilic center, the nucleophile, and the nucleophilic center. 3.1B,C**

3.41 Give the curved-arrow notation for, and predict the immediate product of, each of the following reactions. Each involves an electron-deficient Lewis acid and a Lewis base. Identify in each case the electrophilic center and the nucleophile.

(a) $H_3C-\ddot{\underset{..}{O}}-CH_3 + BF_3 \longrightarrow$

(b) 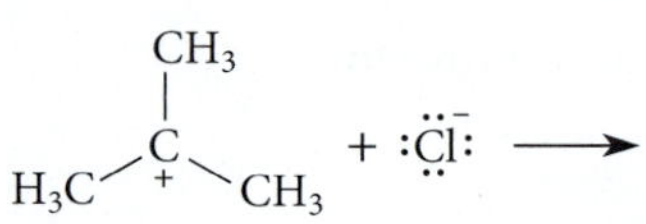

(c) 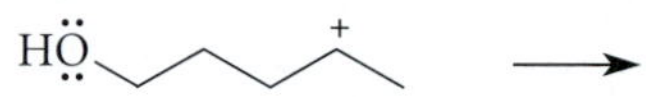

(*Hint:* This reaction forms a ring.)

(d) $(CH_3)_3B + :\overset{-}{C}\equiv\overset{+}{O}: \longrightarrow$

(e) 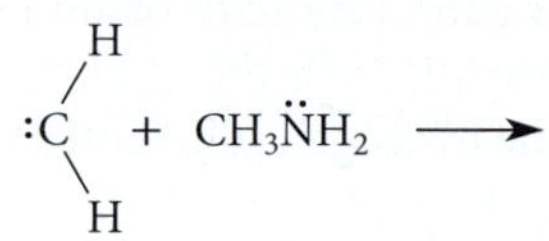

- **Given a Lewis acid–base dissociation, supply the curved-arrow notation. Identify the leaving group. 3.1C, 3.2B**

3.42 Supply the curved-arrow notation for the following reactions, and identify the leaving group.

(a) 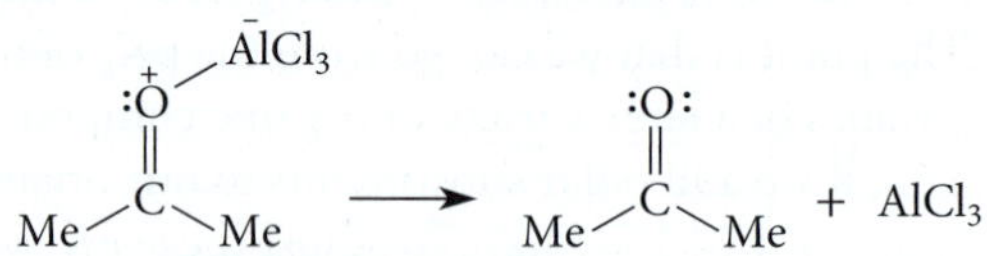

(b) 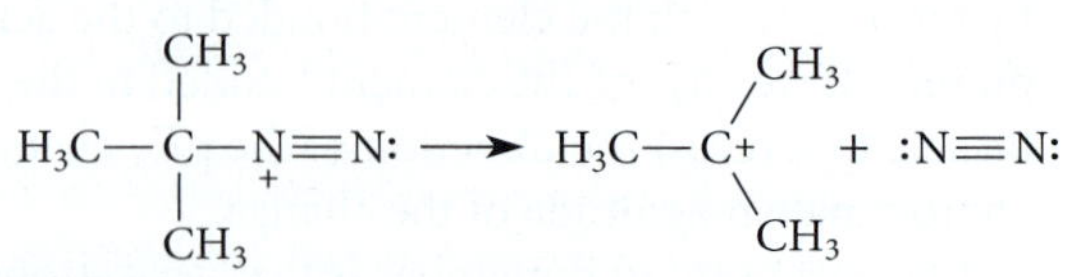

- **Supply the curved-arrow notation for an electron-pair displacement reaction. Identify the nucleophile, the nucleophilic center, the electrophile, the electrophilic center, and the leaving group. 3.2A,B**

3.43 Provide a curved-arrow notation for the following electron-pair displacement reactions. In each case, identify the nucleophile, the nucleophilic center, the electrophile, the electrophilic center, and the leaving group.

(a) 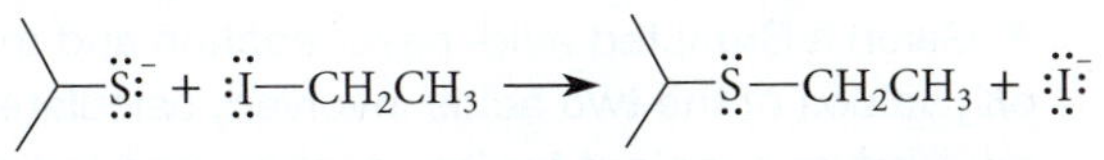

(b) 

- **Given the curved-arrow notation for a reaction, draw the structures of the products.** **3.2A,B**

3.44 Give the products of the following reactions.

(a) 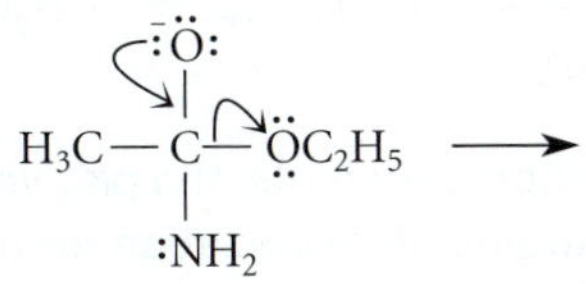

(b) 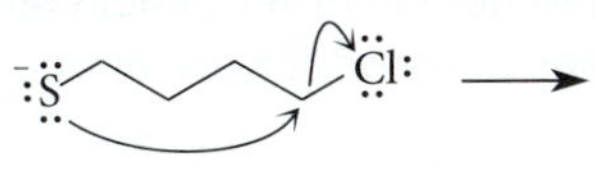

(*Hint:* This reaction forms a ring.)

(c) 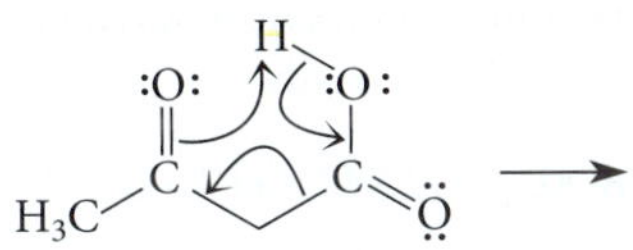 (One of the products is a gas.)

- **Recognize the common structural patterns for which resonance structures are appropriate. Derive the resonance structures using the curved-arrow notation, and represent the molecule as a hybrid of all of the important structures using dashed lines and partial charges.** **3.3**

3.45 Draw an important resonance structure for each of the following molecules. Each part represents one of the common situations described in Sec. 3.3. Depict the resonance hybrid as a single structure with dashed lines and partial charges.

(a) 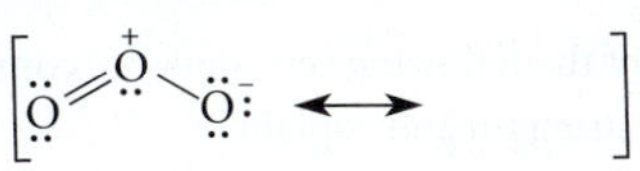

(b) 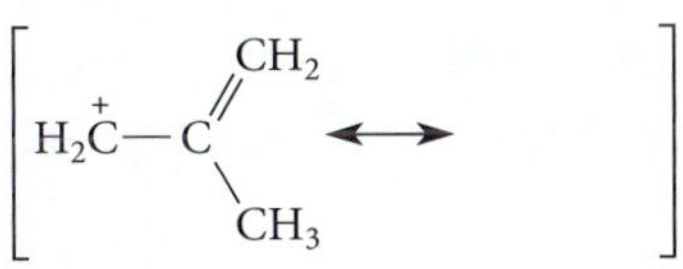

(c) 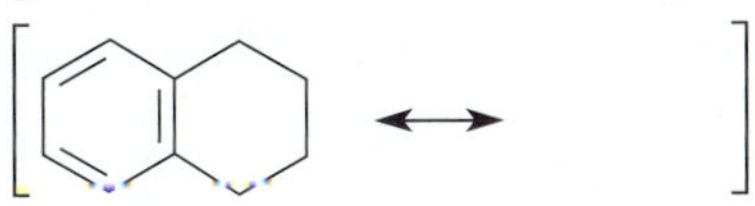

(d) 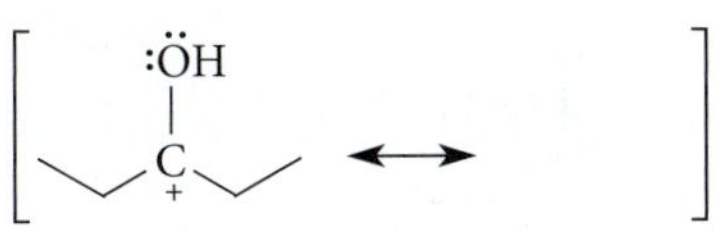

(e) 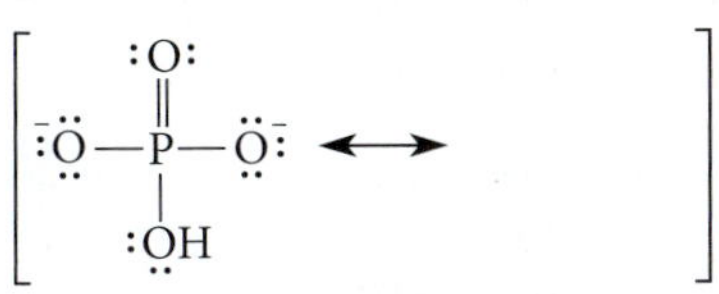

3.46 Given the following structures of resonance hybrids, provide the three contributing resonance structures for (a) and (b), and four structures for (c).

(a) 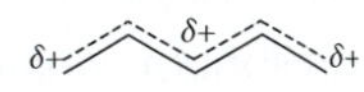

(b) 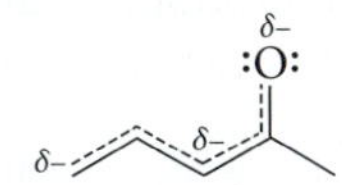

(c) 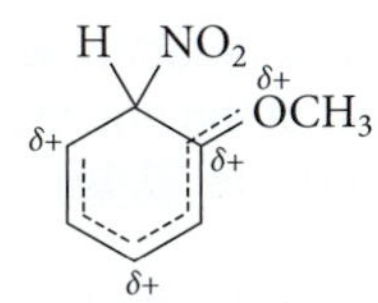

- **Within a group of resonance structures, determine their relative importance, and draw a diagram showing the relative energies of the resonance structures and the resonance hybrid on an energy scale.** **3.3C**

3.47 In each part, use curved arrows to show the conversion of the resonance structure on the left to the structure on the right. Draw a hybrid structure using dashed lines and partial charges. Determine the relative importance of the resonance contributors shown. Then, on an energy scale, indicate qualitatively the relative energies of each structure along with the energy of the resonance hybrid.

(a) 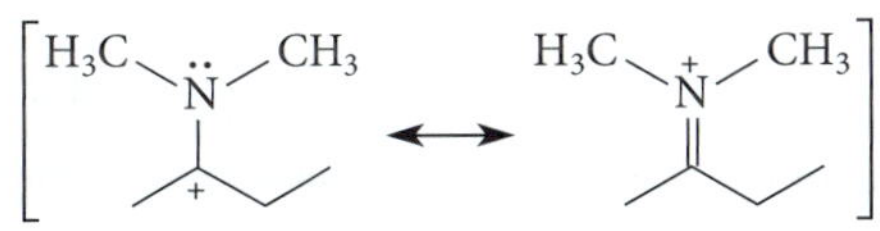

A *B*

(b) 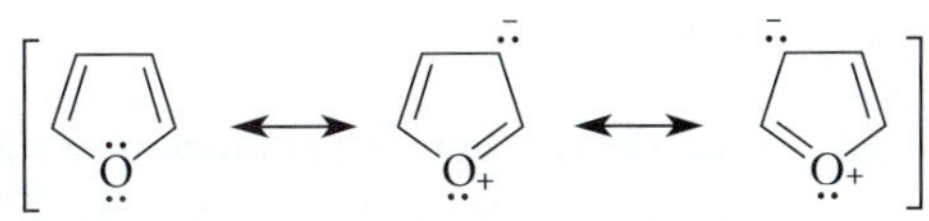

A *B* *C*

- **Recognize a Brønsted acid–base reaction, give its curved-arrow notation, and identify the Brønsted acid and the Brønsted base. Identify the conjugate acid–base pairs.** **3.4A,B**

3.48 Which of the following three reactions are Brønsted acid–base reactions? For those reactions, label the conjugate acid–base pairs. For all reactions, provide the curved-arrow notation. (All charges and nonbonding electron pairs are shown.)

(a) 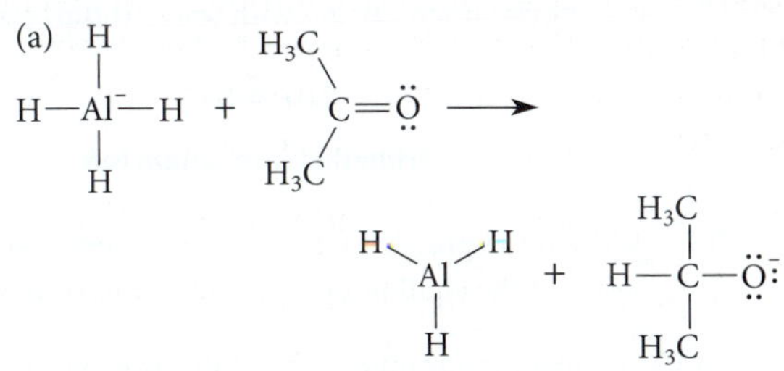

(b) 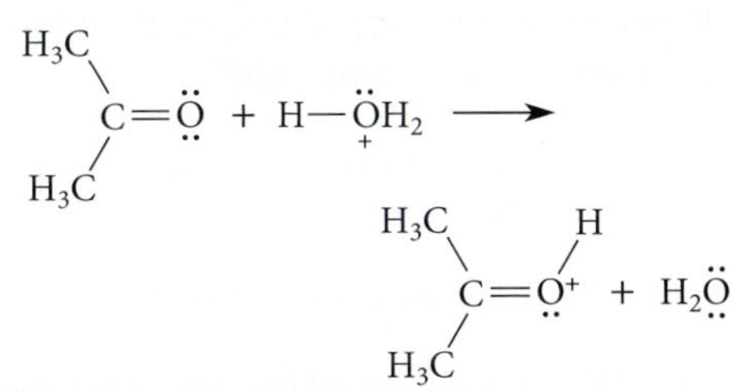

(c) 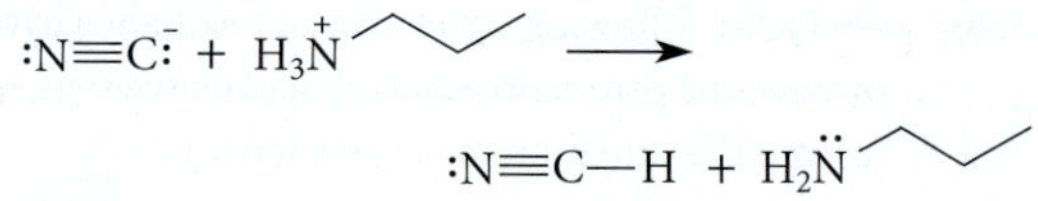

- **Given a reaction, write the equilibrium constant as a function of the equilibrium concentrations of all reactants and products. 3.5A**

3.49 For the reaction in Problem 3.48c, write the equilibrium constant expression in terms of the equilibrium concentrations of all species.

- **Given the standard free energy (ΔG°) for a reaction, calculate the equilibrium constant, and show the reactants and products on a standard free-energy scale. 3.5B**

3.50 The standard free-energy change for the following reaction (an example of *ester aminolysis*) is –28.5 kJ mol^{-1}.

$$\underset{\textbf{dimethylamine}}{Me_2NH} + \underset{\textbf{methyl acetate}}{Me-\overset{\overset{\large O}{\|}}{C}-OMe} \rightleftharpoons \underset{\textbf{\textit{N,N}-dimethylacetamide}}{Me-\overset{\overset{\large O}{\|}}{C}-NMe_2} + \underset{\textbf{methanol}}{HOMe}$$

(a) What is the equilibrium constant for this reaction at 25 °C?

(b) Show the reactants and products on a standard free-energy scale. (Show the free energies of the reactants together at one energy, and show the free energies of the products together at another energy.)

(c) If the reactants are mixed at 1 *M* concentration and allowed to come to equilibrium, what is the concentration of the products at equilibrium?

- **Write the expression for dissociation constant K_a in terms of the reactant and product concentrations in a Brønsted acid–base reaction. 3.6A**

3.51 (a) Write the chemical equation for acid dissociation of trimethylammonium ion with water as the base.

$$\underset{\textbf{trimethylammonium ion}}{H-\overset{+}{N}Me_3}$$

(b) Write the expression for the dissociation constant K_a in terms of the equilibrium concentrations of all species.

- **Given either the dissociation constant, the pK_a value, or the standard free energy of dissociation for a Brønsted acid, calculate the other two. 3.6D**

3.52 The pK_a for the dissociation of trimethylammonium ion (Problem 3.51) is 9.76.

(a) What is the dissociation constant?

(b) What is the standard free energy of dissociation at 25 °C?

- **Given a Brønsted acid–base reaction and the pK_a values of the two acids involved, calculate the equilibrium constant for the reaction, and indicate whether the reaction favors the right or left side of the equation. 3.6C**

3.53 Using Table 3.2 as well as the data given below, estimate the equilibrium constant for each of the following reactions.

(a) $(CH_3)_3N + H-CN \rightleftharpoons \underset{pK_a = 9.76}{(CH_3)_3\overset{+}{N}H} + {}^{-}CN$

(b) $\underset{pK_a = 10.5}{CH_3CH_2S-H} + {}^{-}OH \rightleftharpoons CH_3CH_2S^{-} + H_2O$

- **Given the dissociation constants, the pK_a values, or the standard free energies of dissociation for a series of Brønsted acids, rank them in order of acid strength; rank their conjugate bases in order of base strength. 3.6A,B,D**

3.54 Rank the acids *HA*, *HB*, and *HC* in order of increasing acid strength, and rank their conjugate bases *A*, *B*, and *C* in order of increasing base strength.

HA: $pK_a = -3.0$ *HB*: $\Delta G_a^\circ = 22.8$ kJ mol^{-1} *HC*: $K_a = 1.6 \times 10^{-7}$

- **Given the pH of a buffer solution and the pK_a of an acid, determine the percentage of conjugate acid and conjugate base that forms in solution. 3.6E**

3.55 Phenylacetic acid ($PhCH_2CO_2H$), pK_a = 4.31, was dissolved in a pH = 4.5 buffer at an initial concentration of 10^{-4} *M*. What is the concentration of phenylacetic acid and its conjugate base in the buffer solution?

- **Rank a series of Brønsted acids of unknown pK_a according to acid strength, and identify the structural features that determined your ranking. Rank a series of bases according to base strength. 3.7**

3.56 Within each of the following sets, rank the compounds in order of acid strength, and explain.

(a)

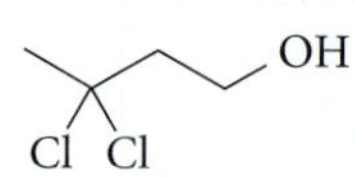

A

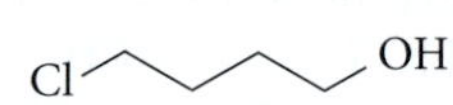

B

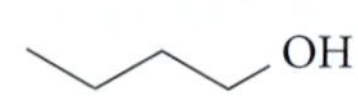

C

(b)

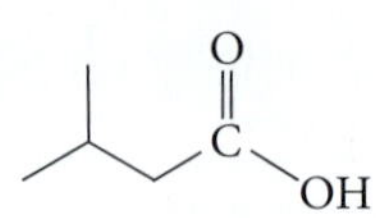

A

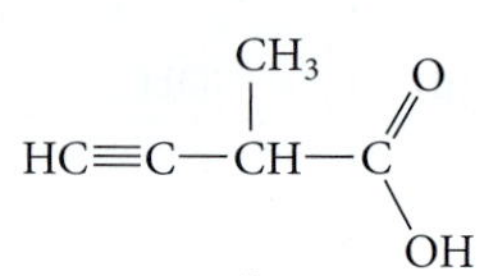

B

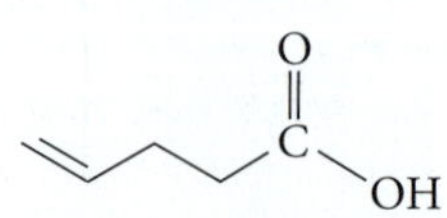

C

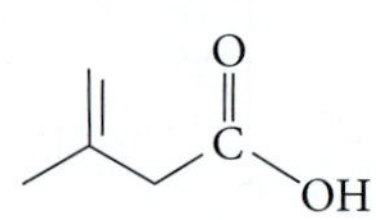

D

(c)

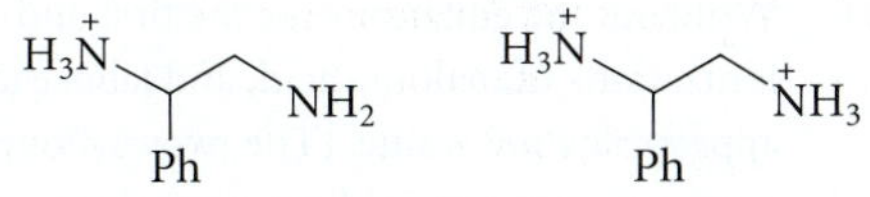

A *B*

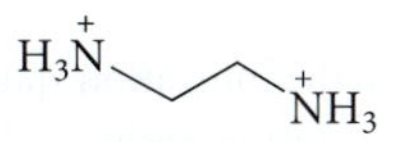

C

3.57 Within each set, rank the Brønsted bases in order of increasing basicity, and explain.

(a)

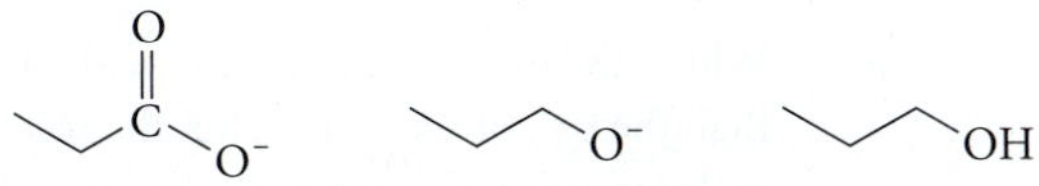

A *B* *C*

(b)

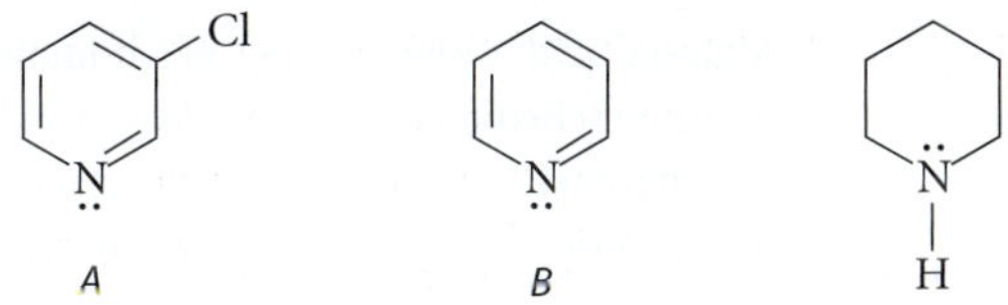

A *B* *C*

- **Learn the pK_a values of benchmark compounds, and use them to estimate pK_a values of other compounds where appropriate. 3.7G**

3.58 Arrange the following compounds in order of increasing acidity. (*Hint:* Two of the compounds have roughly the same acidity.)

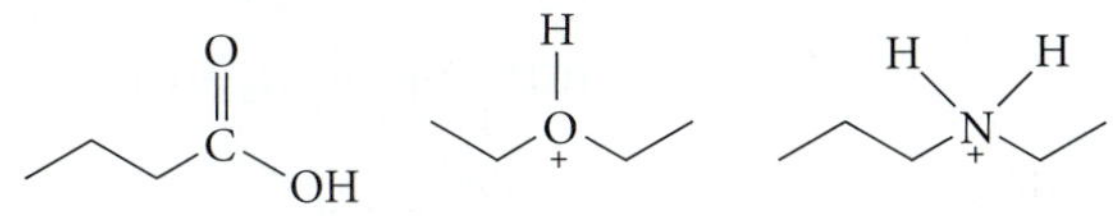

A *B* *C*

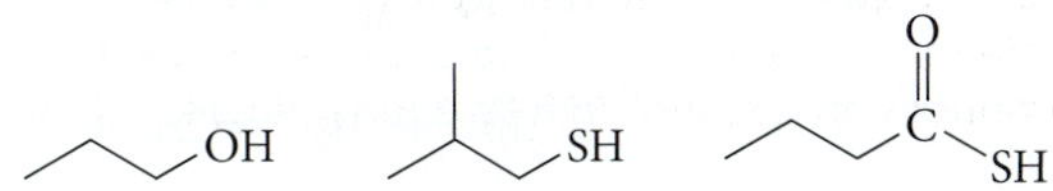

D *E* *F*

INTEGRATED PROBLEMS

3.59 Although we normally think of acetic acid as a weak acid, in very acidic solutions it can act as a weak base.

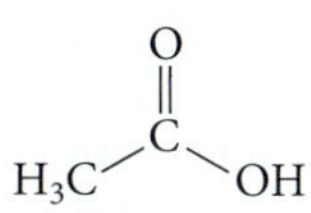

acetic acid

Which oxygen of acetic acid is the more basic one? (*Hint:* Draw both possible conjugate acids, and decide which of the two is more stabilized by resonance.)

3.60 Three examples of an *incorrect* application of the curved-arrow notation follow. In each case, indicate what is wrong with the notation, and then provide a corrected notation.

(a)

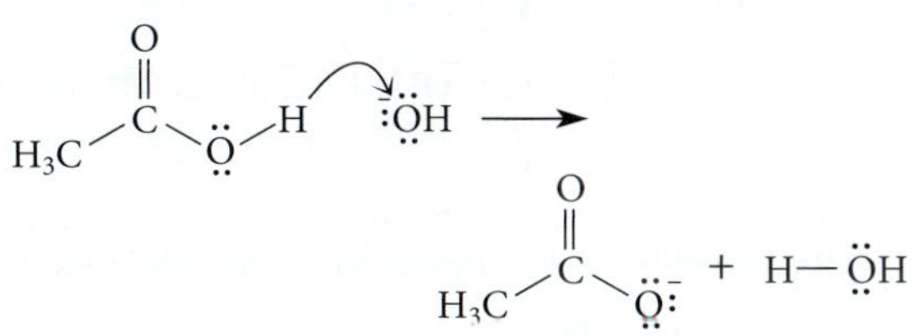

(b)

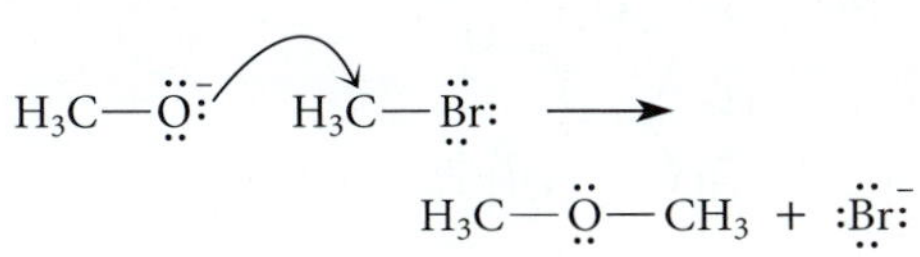

(c)

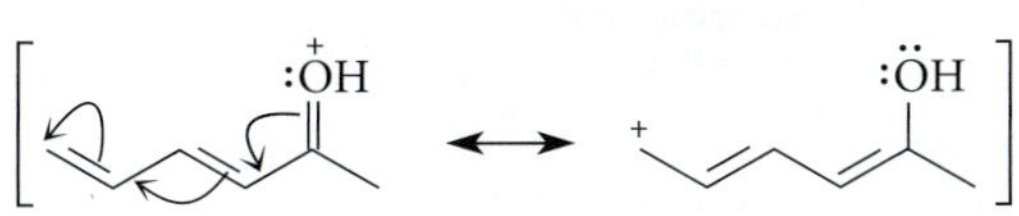

3.61 (a) Use the curved-arrow notation to derive two additional and equivalent resonance structures for the carbonate ion.

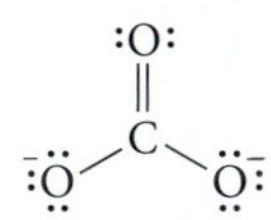

carbonate ion

(b) Use the curved-arrow notation to derive two additional resonance structures for naphthalene. Which, if any, of the three structures are equivalent?

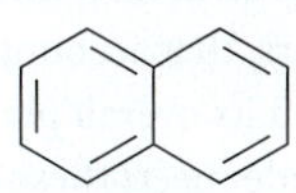

naphthalene

(c) Use the curved-arrow notation to derive five additional resonance structures for the 4-dimethylaminopyridinium ion. Rank the structures in importance, and give your reasoning. Which, if any, are equivalent?

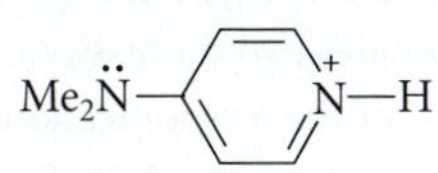

4-dimethylaminopyridinium ion

3.62 Provide a curved-arrow notation for each of the following reactions. Each reaction occurs as a single step. (*Hint:* Use more than two curved arrows.)

(a)

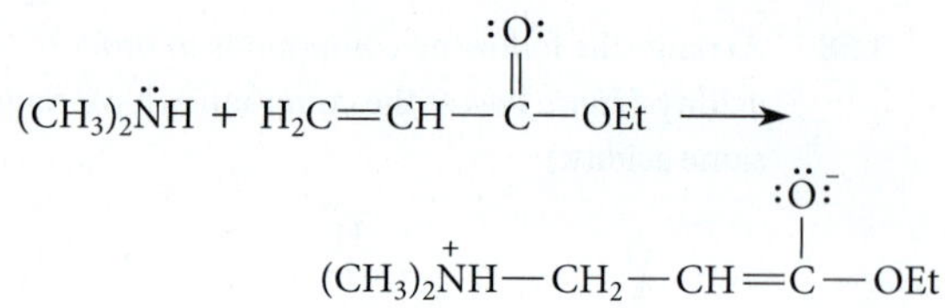

(b)

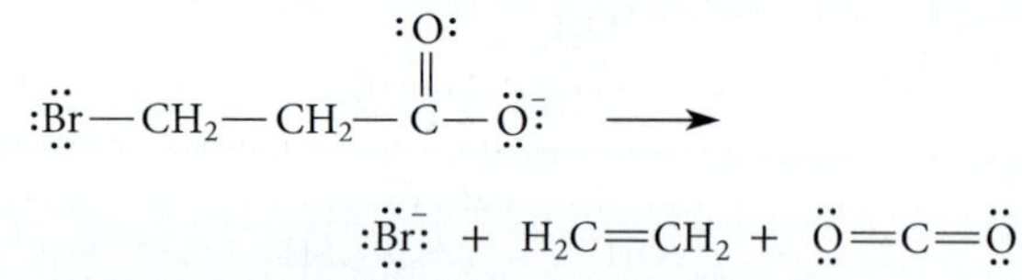

3.63 The standard free-energy difference between dimethylpropane and pentane is 6.86 kJ mol^{-1}; dimethylpropane is the more stable compound. If the two were present in an equilibrium mixture, what would be the percentage of each in the mixture at 25 °C?

3.64 A scientist at an Air Force research laboratory in California studies "highly energetic materials" (explosive materials) and, more significantly, lives to tell about it. In 2000, he and his equally adventurous collaborators determined the X-ray crystal structure of the N_5^+ cation; most salts of this cation are highly explosive. This is the first species ever isolated in modern times that contains more than three contiguously bonded nitrogen atoms. The crystal structure revealed a V-shape for the cation, as follows:

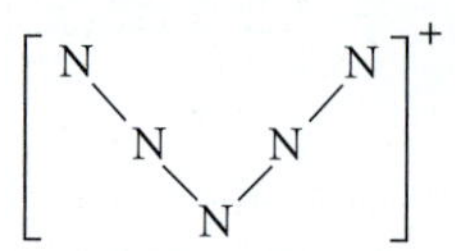

The lines do not indicate the types of bonds, but only the shape.

(a) Draw an acceptable Lewis structure, including nonbonding electron pairs, that accounts for the shape of the molecule and its overall plus charge. Explain why your molecule meets these criteria. Be sure to show in your structure the formal charge of every atom with nonzero formal charge.

(b) Using the curved-arrow notation, derive two additional resonance structures for this cation that meet the same criteria.

3.65 Malonic acid has two carboxylic acid groups and consequently undergoes two successive ionization reactions. The pK_a for the first ionization of malonic acid is 2.86; the pK_a for the second is 5.70. For comparison, the pK_a of acetic acid is 4.76.

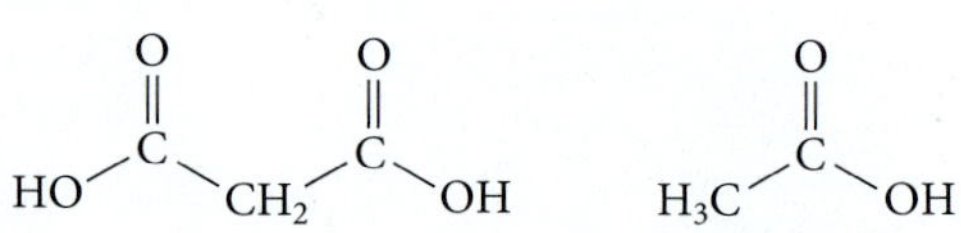

malonic acid **acetic acid**

(a) Write out the equations for the first and second ionizations of malonic acid, and label each with the appropriate pK_a value. (The two carboxylic acid groups in malonic acid are equivalent, so it doesn't matter which ionizes first.)

(b) Describe the ionization state of malonic acid if the pH of a solution of the acid is adjusted to 4.3.

(c) Use your answer to part (b) to determine the number of moles of NaOH per mole of malonic acid required to adjust the solution to a pH of 4.3.

(d) Why is the first pK_a of malonic acid much lower than the pK_a of acetic acid, but the second pK_a of malonic acid is much higher than the pK_a of acetic acid?

(e) Malonic acid is one member of a homologous series of unbranched dicarboxylic acids, so-called because they have two carboxylic acid groups. Compounds in this series have the following general structure.

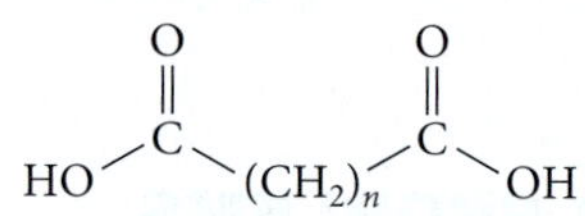

How would you expect the difference between the first and second pK_a values to change as n increases? Explain. (*Hint:* Look at the denominator of the electrostatic law, Eq. 3.84.)

3.66 Which of the following two reactions would have an equilibrium constant more favorable to the right? Explain your answer using a standard free-energy diagram. (*Hint:* Consider the interaction of the C—F bond dipoles with the positive charge in the product cation.)

(1)

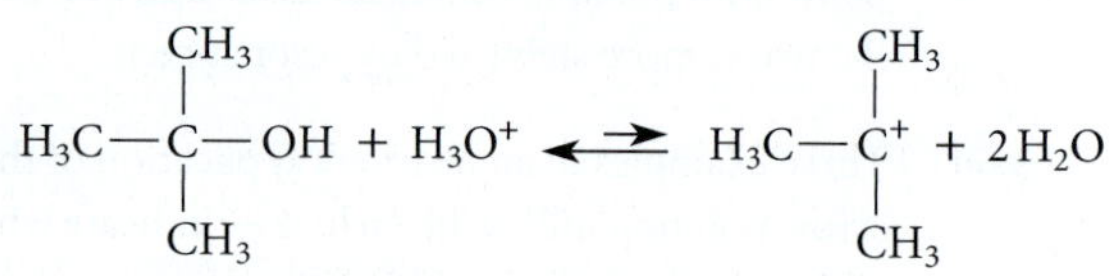

(2)

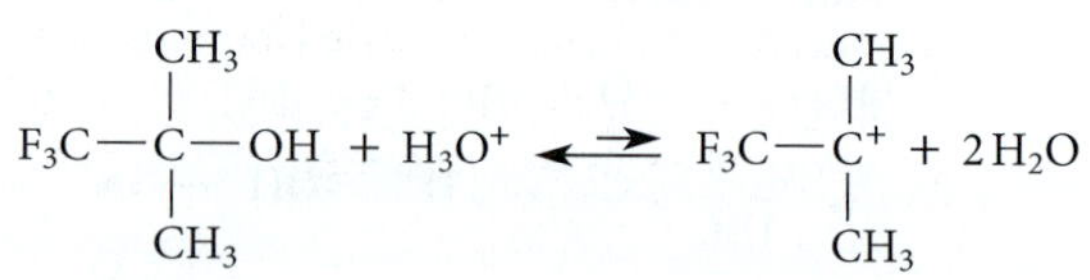

3.67 Ascorbic acid (vitamin C) has the following structure.

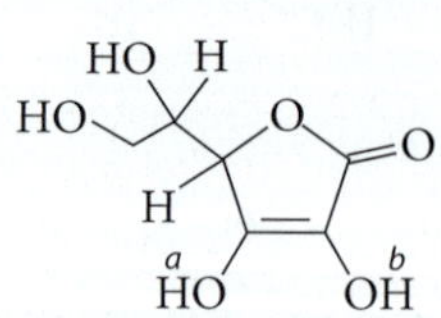

ascorbic acid
pK_a = 4.2, 11.6

Protons *a* and *b* are acidic, but proton *a* is the more acidic of the two.

(a) Draw the structure of ascorbic acid as it exists at physiological pH (pH = 7.4). Explain.

(b) Using the curved-arrow notation, derive a resonance structure for the structure you drew in part (a); the two structures together will show that negative charge is shared between two oxygens.

3.68 Arrange the compounds within each set in order of decreasing pK_a, and explain your reasoning.

(a) *Hint:* Use the benchmark pK_a values you learned in Sec. 3.7G.

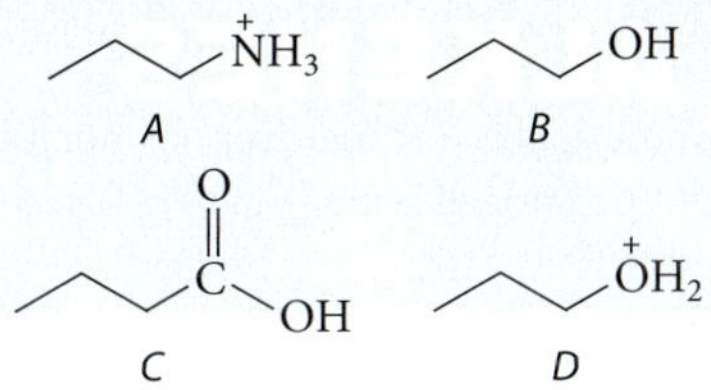

(b) $H—\ddot{A}s(CH_3)_2$ $H—\overset{+}{A}s(CH_3)_2$ (with H on As)

A   *B*

$H—\ddot{N}(CH_3)_2$ $H—\ddot{P}(CH_3)_2$

C   *D*

(c) $CH_3CH_2\ddot{O}H$ $(CH_3)_2\ddot{N}—CH_2CH_2\ddot{O}H$

A   *B*

$(CH_3)_3\overset{+}{N}—\ddot{O}H$

C

3.69 Kainic acid is a molecule from a red Japanese seaweed that binds to glutamate receptors in the brain and kills brain cells. Kainic acid undergoes three successive acid dissociations with pK_a values of 2.16, 4.57, and 10.42. Starting with the undissociated form shown here, give reactions for each of the dissociations, and label each with its pK_a. Explain your reasoning.

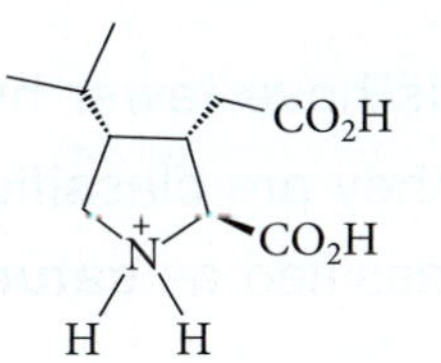

kainic acid

3.70 Derive Eq. 3.59b for calculating the equilibrium constant for the reaction between an acid HA and a base $B:^-$. (*Hint:* First show that $K_{eq} = K_{HA}/K_{HB}$, the ratio of dissociation constants for the two acids.)

3.71 For the general acid–base reaction,

$$HA + B\bar{:} \rightleftharpoons A\bar{:} + HB$$

derive the general relationship between $\Delta G°$ for the reaction and the ΔG_a° values for the individual acids.

3.72 (a) What is the standard free-energy change for the following acid–base reaction in the left-to-right direction at 25 °C?

$$(CH_3)_3N + H—CN \rightleftharpoons (CH_3)_3\overset{+}{N}H + {}^-CN$$

$pK_a = 9.40$ $pK_a = 9.76$

(b) How much of each compound in this reaction will be present at equilibrium if the initial concentrations of $(CH_3)_3N$ and H—CN are both zero and the salt $(CH_3)_3\overset{+}{N}H\ {}^-CN$ is present at a concentration of 0.1 *M*?

3.73 Trifluoroacetic acid ($F_3C—CO_2H$, $pK_a = 0.23$), its conjugate base, sodium trifluoroacetate ($F_3C—CO_2^-\ Na^+$), acetic acid ($H_3C—CO_2H$, pK_a 4.76), and its conjugate base, sodium acetate ($H_3C—CO_2^-\ Na^+$) are mixed together in water at the same initial concentration of 0.1 *M* each. When the solution comes to equilibrium, what are the concentrations of the four species? (*Hint:* Write out the acid–base reactions involved. You shouldn't need to do a complex calculation.)

3.74 (a) Which of the following two cations is predicted to be more acidic according to the hybridization effect?

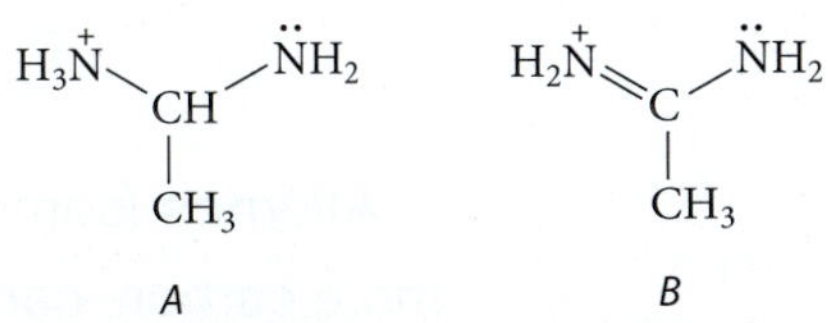

(b) Which of these two cations is predicted to be more acidic according to the resonance effect? Compound *B* is resonance-stabilized, and it is acting as an *acid.* Show your reasoning with a free-energy diagram.

(c) Given that the pK_a of cation *A* is about 9.2, and the pK_a of cation *B* is 12.4, which of the two effects seems to be more important in this comparison?

4 INTRODUCTION TO ALKENES AND ALKYNES

Structure and Reactivity

Alkenes are hydrocarbons that contain one or more carbon–carbon double bonds. Alkenes are sometimes called **olefins**, particularly in the chemical industry. *Ethylene* is the simplest alkene.

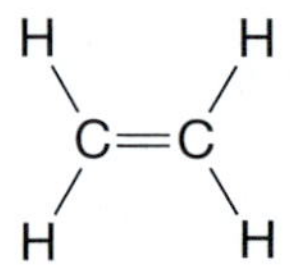 or $H_2C{=}CH_2$

ethylene
(substitutive name: **ethene**)

Alkynes (sometimes called **acetylenes**) are hydrocarbons that have one or more carbon–carbon triple bonds. *Acetylene* is the simplest alkyne.

$$H{-}C{\equiv}C{-}H \quad \text{or} \quad HC{\equiv}CH$$

acetylene
(substitutive name: **ethyne**)

Because compounds containing double or triple bonds have fewer hydrogens than alkanes with the same number of carbons, they are classified as **unsaturated hydrocarbons**. In contrast, alkanes are classified as **saturated hydrocarbons**.

This chapter covers the structure, bonding, nomenclature, and physical properties of alkenes and alkynes. Then, a few alkene and alkyne reactions are introduced, and the alkene reactions are used to introduce some physical principles that are important in understanding the reactivities of organic compounds in general.

4.1 STRUCTURE AND BONDING IN ALKENES AND ALKYNES

A. Structure of Alkenes and Alkynes

The double-bond geometry of ethylene is typical of the geometry of other alkenes. Ethylene follows the rules for predicting molecular geometry (Secs. 1.6B and C), which require each carbon of ethylene to have *trigonal planar* geometry—that is, all the atoms surrounding each carbon lie in the same plane with bond angles approximating 120°. The experimentally determined structure of ethylene agrees with these expectations and shows further that *ethylene is a planar molecule.* For alkenes in general, the carbons of a double bond and the atoms directly attached to them all lie in the same plane.

As expected from the rules for predicting molecular geometry, each carbon of acetylene has *linear* geometry—that is, the atoms bonded to each carbon of the triple bond have a bond angle of 180°. So, *acetylene is a linear molecule.*

Models of ethylene and acetylene are shown in **Fig. 4.1**. Comparisons of the structures of acetylene, ethylene, and ethane, and of propyne, propene, and propane, are given in **Fig. 4.2**. The carbon–carbon double bonds of ethylene and propene (133 pm) are shorter than the carbon–carbon single bonds of ethane and propane (154 pm). The carbon–carbon triple bonds of acetylene and propyne are shorter still (120 pm). These examples illustrate the relationship of bond length and bond order (Sec. 1.6B): triple bonds are shorter than double bonds, which are shorter than single bonds between the same atoms.

In Sec. 1.6D (Display 1.54), we showed two ways of drawing alkenes: in the plane of the page, and in a plane tilted nearly perpendicular to the page. Don't mix these two drawing methods. A common error is for students to draw the perspective structure with wedge and dashed-wedge bonds at incorrect angles:

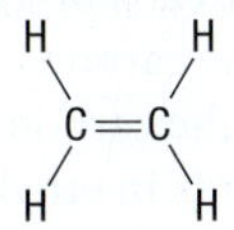

correct in-plane structure: H—C—H angle is about 120°

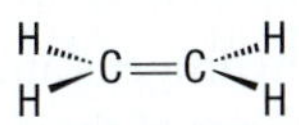

correct perspective structure: H—C—H angle is compressed

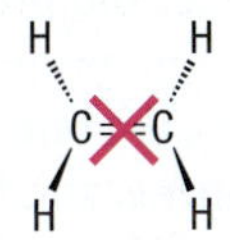

incorrect perspective structure: H—C—H angle is not compressed

If you intend to draw a perspective structure, make sure that the H—C—H angle is compressed to indicate proper perspective.

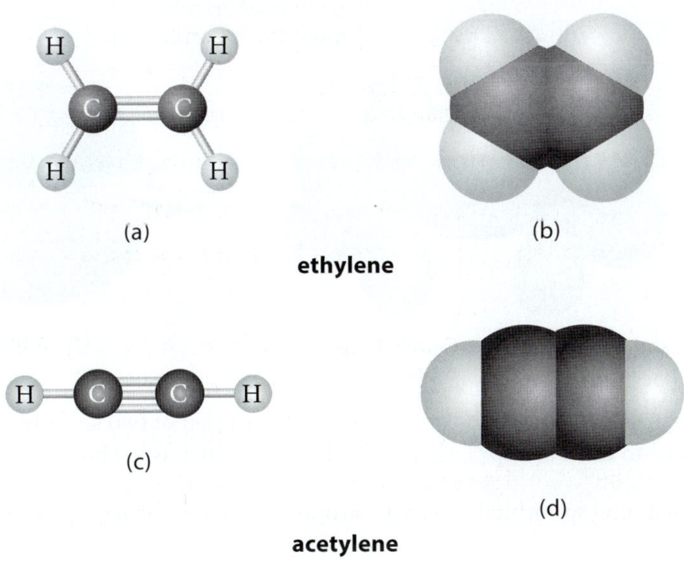

FIGURE 4.1 Models of ethylene and acetylene. (a) and (c) are ball-and-stick models. (b) and (d) are space-filling models. Ethylene is a planar molecule, whereas acetylene is a linear molecule.

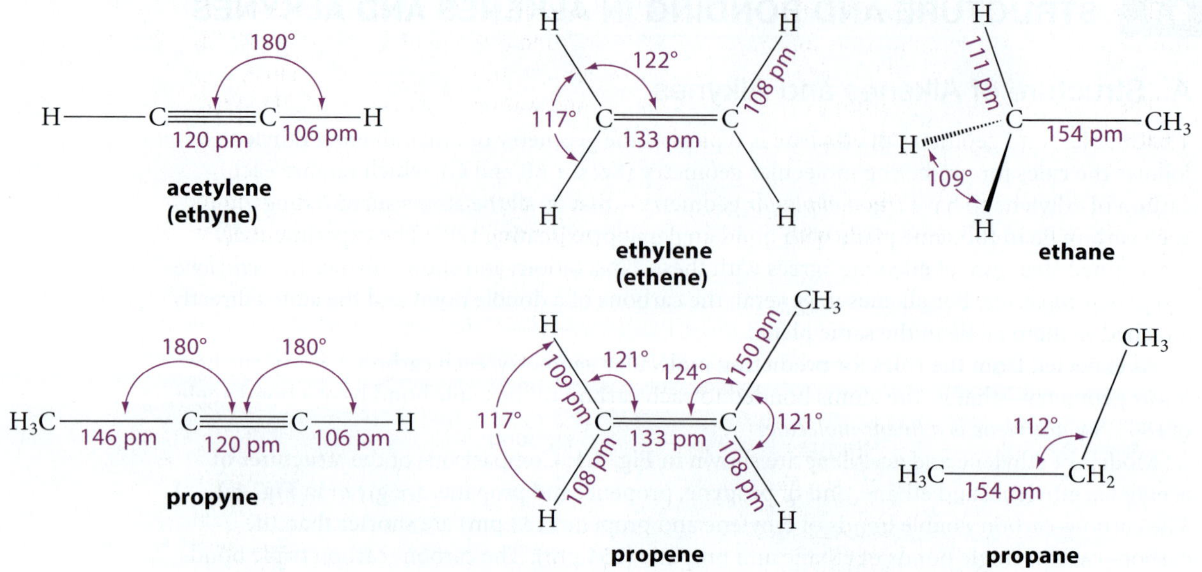

FIGURE 4.2 Structures of acetylene, ethylene, ethane, propyne, propene, and propane. Compare the linear geometry of acetylene (bond angles 180°), the trigonal planar geometry of ethylene (bond angles near 120°), and the tetrahedral geometry of ethane (bond angles near 109.5°). All carbon–carbon triple bonds are shorter than carbon–carbon double bonds, which are shorter than carbon–carbon single bonds. The carbon–carbon single bonds in propyne and propene, moreover, are somewhat shorter than the carbon–carbon bonds of propane.

Another feature of alkene and alkyne structure is apparent from a comparison of the structures of propyne, propene, and propane in Fig. 4.2. Notice that the carbon–carbon *single* bonds of propyne (146 pm) and propene (150 pm) are shorter than the carbon–carbon bonds of propane (154 pm). Likewise, the bonds to the hydrogens attached to the carbon of the triple bond in acetylene and propyne, and to the carbons of the double bonds in ethylene and propene, are shorter than the C—H bonds of propane. The shortening of all these bonds is a consequence of the particular way that carbon atoms are hybridized in alkenes.

B. Hybrid-Orbital Bonding Model in Alkenes and Alkynes

The sp^2 hybridization of carbons involved in a double bond was discussed in Sec. 1.7C. Let's summarize what we learned there by applying it to bonding in ethylene:

1. An sp^2-hybridized carbon has three sp^2 orbitals whose axes lie in a common plane (the *xy*-plane) at an angle of 120°, and one 2p orbital directed along the *z*-axis.

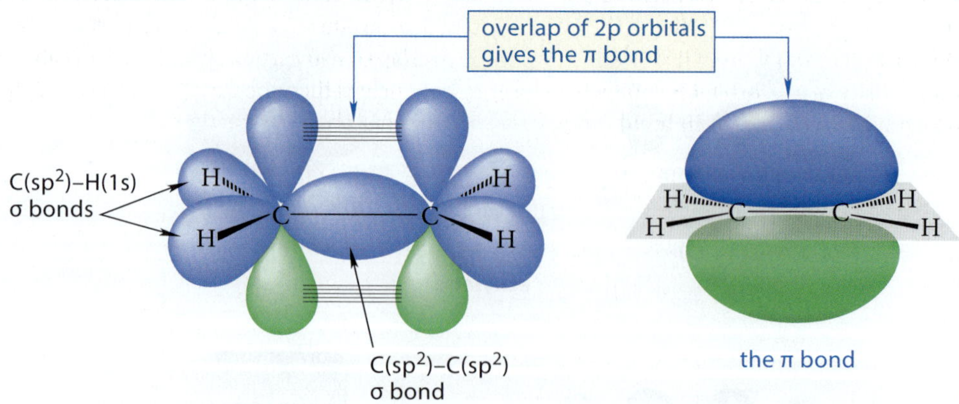

(4.1)

2. One of the carbon–carbon bonds results from the overlap of two sp^2 orbitals—one from each carbon—to form a $C(sp^2)–C(sp^2)$ single bond, which is a σ bond.

3. The two remaining sp^2 orbitals on each carbon are used to form $C(sp^2)–H(1s)$ σ bonds to hydrogens.

4. The 2p orbitals on each carbon overlap to form a π (pi) bond.

This bonding model shows why ethylene and other alkenes are planar around the π bond. If the two CH_2 groups were twisted away from coplanarity, the π bond could not form. In other words, the overlap of the 2p orbitals and the very existence of the π bond require ethylene to be planar.

The sp hybridization of carbons involved in a triple bond was discussed in Sec. 1.7D. Applying what we learned there to acetylene, we summarize:

1. Each sp-hybridized carbon has two sp orbitals oriented in opposite directions along a common axis (the x-axis), and two 2p orbitals directed along perpendicular axes (the y- and z-axes).

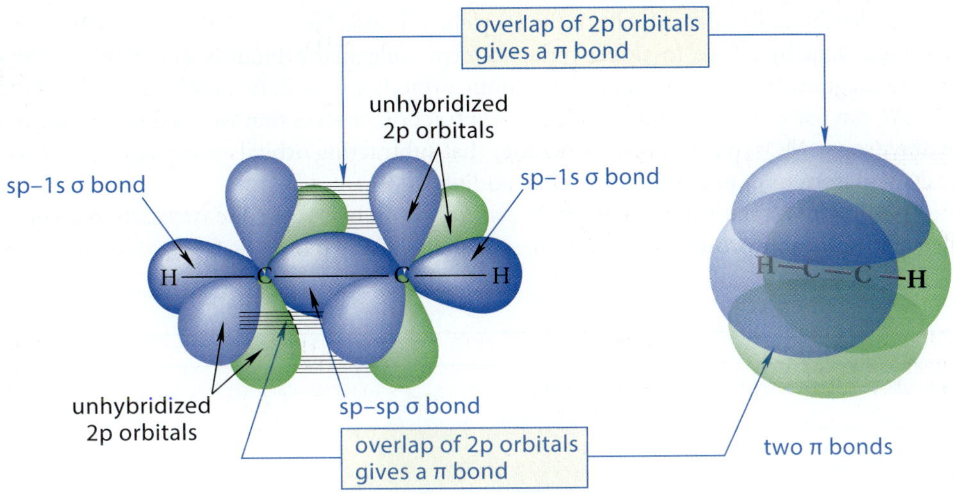

(4.2)

2. The carbon–carbon bond results from the overlap of two sp orbitals—one from each carbon—to form a C(sp)–C(sp) single bond, which is a σ bond.

3. The remaining sp orbital on each carbon is used to form a C(sp)–H(1s) σ bond to hydrogen.

4. The two 2p orbitals on each carbon overlap to form two π (pi) bonds.

Return to the structure of propene in Fig. 4.2, and notice that the carbon–carbon bond to the CH_3 group is shorter by about 4 pm than the carbon–carbon bonds of ethane and propane. The carbon–carbon bond to the CH_3 group in propyne is even shorter: about 8 pm shorter than the carbon–carbon bonds of ethane and propane. These small but real differences are general: single bonds to an sp-hybridized carbon are shorter than single bonds to an sp^2-hybridized carbon, which are shorter than single bonds to an sp^3-hybridized carbon. The carbon–carbon single bond of propyne, for example, is derived from the overlap of a carbon sp^3 orbital of the CH_3 group with a carbon sp orbital of the alkyne carbon. The carbon–carbon single bond of propene is derived from the overlap of a carbon sp^3 orbital of the CH_3 group with a carbon sp^2 orbital of the alkene carbon. A carbon–carbon bond of propane is derived from the overlap of two carbon sp^3 orbitals. Because the electron density of a 2s orbital is somewhat closer to the nucleus than the electron density of a 2p orbital, within bonds of a given bond order, *bonds with more s character are shorter.*

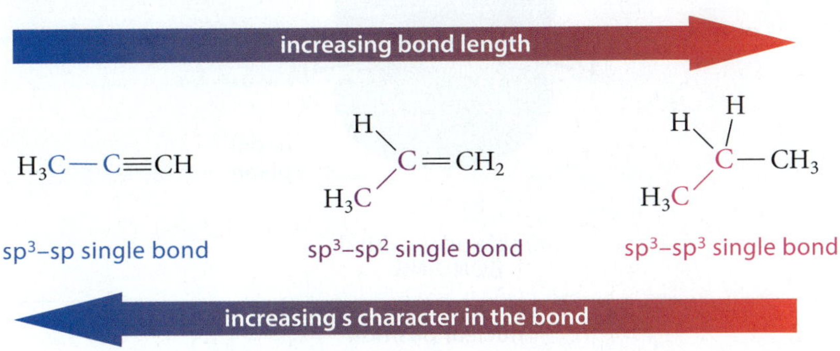

(4.3)

For exactly the same reason, C—H bonds show the same trend. As demonstrated in Fig. 4.2, The bond lengths of C—H bonds are in the order

$$\text{sp–1s C—H bonds} < \text{sp}^2\text{–1s C—H bonds} < \text{sp}^3\text{–1s C—H bonds.}$$

C. Molecular Orbital Description of the π Bond

Molecular orbital (MO) theory provides a richer description of the π bond that forms the basis for understanding ultraviolet spectroscopy (an important tool for molecular analysis; see Sec. 15.2), the properties of aromatic compounds such as benzene (Sec. 15.7B), and a class of reactions called pericyclic reactions (see Chapter 28). MO theory allows us to treat the π bond independently of the other MOs of the ethylene molecule. This is another relatively rare situation (as in dihydrogen, H_2; Sec. 1.8B) in which molecular orbitals are associated with a particular bond that we can draw in a Lewis structure.

We consider the molecular orbitals of the ethylene π bond. The interaction of two $2p_z$ orbitals of ethylene by a "side-to-side" overlap to form molecular orbitals is shown in an orbital interaction diagram (**Fig. 4.3**). Because two atomic orbitals are used, two molecular orbitals are formed. We can use exactly the same addition–subtraction process that we used for the formation of the dihydrogen MOs in Sec. 1.8A. Remember that subtracting orbitals is the same as reversing the peaks and troughs of one orbital and then adding.

The bonding molecular orbital that results from additive overlap of the two carbon 2p orbitals is called a π **molecular orbital**. This molecular orbital, like the p orbitals from which it is formed,

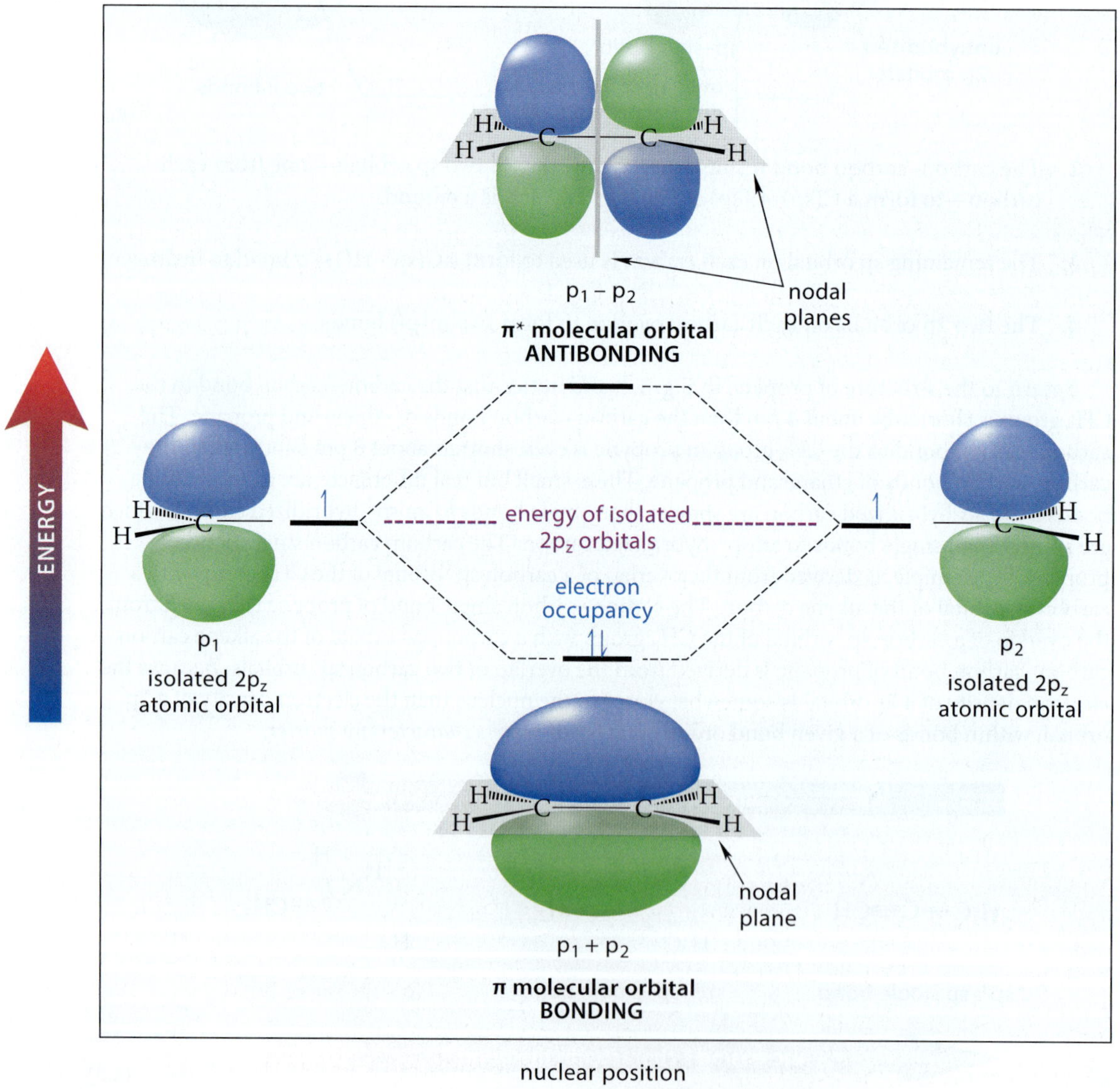

FIGURE 4.3 An orbital interaction diagram showing the overlap of 2p orbitals to form bonding and antibonding π molecular orbitals of ethylene. The π bond is formed when two electrons occupy the bonding π molecular orbital. Wave peaks and wave troughs are shown with different colors. The nodal planes are perpendicular to the page.

has a nodal plane (shown in Fig. 4.3); this plane coincides with the plane of the ethylene molecule. The antibonding molecular orbital, which results from subtractive overlap of the two carbon 2p orbitals, is called a **π* molecular orbital**. (The π* notation is pronounced "pi-star.") It has two nodes. One of these nodes is the plane of the molecule, and the other is a plane *between* the two carbons, perpendicular to the plane of the molecule. The bonding (π) molecular orbital lies at lower energy than the isolated 2p orbitals, whereas the antibonding (π*) molecular orbital, with its additional node, lies at higher energy. By the aufbau principle, the two 2p electrons (one from each carbon, with opposite spin) occupy the molecular orbital of lower energy—the π molecular orbital. The antibonding molecular orbital is unoccupied.

The filled π molecular orbital is the π bond. Unlike the σ bonds discussed in Sec. 1.7A, a π bond is not cylindrically symmetrical about the line connecting the two nuclei. The π bond has electron density both above and below the plane of the ethylene molecule, with a wave peak on one side of the molecule, a wave trough on the other, and a node in the plane of the molecule. This electron distribution is particularly evident from an electrostatic potential map (EPM) of ethylene, which shows the local negative charge associated with electron density above and below the molecule.

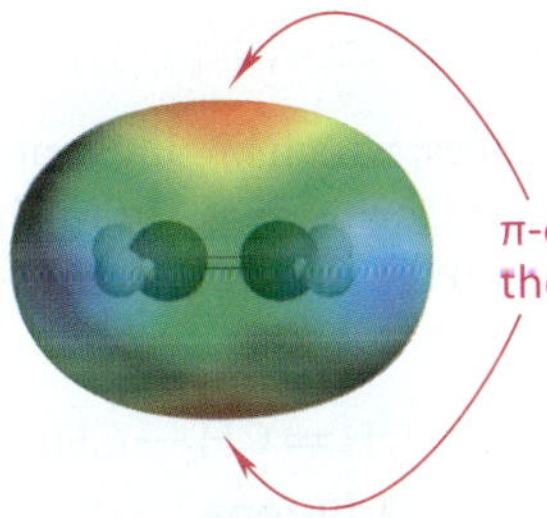

EPM of ethylene (4.4)

The π bond is *one bond* with two lobes, just as a 2p orbital is *one orbital* with two lobes. In this bonding picture, then, the ethylene double bond consists of two types of carbon–carbon bond: a σ bond, with most of its electron density relatively concentrated between the carbon atoms, and a π bond, with most of its electron density concentrated above and below the plane of the ethylene molecule.

Acetylene has two π bonds of equal energy. Therefore, acetylene has two degenerate bonding π molecular orbitals and two degenerate antibonding π* molecular orbitals. The EPM of acetylene shows a negative region associated with the two occupied bonding π molecular orbitals wrapping around the entire molecule:

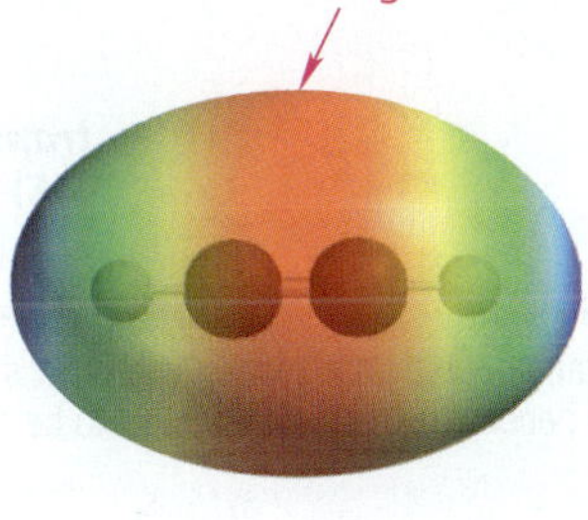

EPM of acetylene (4.5)

Contrast this EPM with the one for ethylene in Display 4.4.

An important aspect of the π electrons is their relative energy. The π bond is a weaker bond than typical carbon–carbon σ bonds because π overlap, which is "side-to-side," is inherently less effective than σ overlap, which is "head-to-head." It takes about 243 kJ mol^{-1} (58 kcal mol^{-1}) of energy to break a carbon–carbon π bond, whereas it takes much greater energy—about 377 kJ mol^{-1} (90 kcal mol^{-1})—to break the carbon–carbon σ bond of ethane. So, *π electrons generally have higher energy than σ electrons*, just as p electrons have higher energy than s electrons. A consequence of this higher energy is that π electrons are more easily removed than σ electrons. In fact, we'll find that electrophiles react preferentially with the π electrons in an alkene or alkyne because those electrons are most easily donated. In addition, because π electrons are on the periphery of the molecule, they are also more accessible to external reagents than σ electrons. *Many of the important reactions of alkenes and alkynes involve the electrons of the π bond, and many of these reactions involve the reaction of electrophiles with the π electrons.*

Focused Problem

4.1 Arrange the labeled bonds in the following molecule in order of increasing length, shortest first. Explain your reasoning.

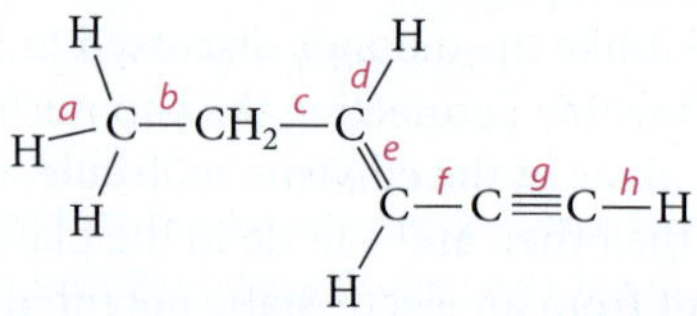

D. Double-Bond Stereoisomers

The bonding in alkenes has other interesting consequences, which are illustrated by the four-carbon alkenes, the butenes. The butenes exist in isomeric forms. First, in the butenes with unbranched carbon chains, the double bond may be located either at the end or in the middle of the carbon chain.

$$H_2C{=}CH{-}CH_2{-}CH_3 \qquad H_3C{-}CH{=}CH{-}CH_3$$

1-butene **2-butene**

Isomeric alkenes, such as these, that differ in the positions of their double bonds are further examples of *constitutional isomers* (Sec. 2.4A).

The structure of 2-butene illustrates another important type of isomerism. *There are two separable, distinct 2-butenes*, each with its own characteristic properties. One has a boiling point of 3.7 °C, whereas the other has a boiling point of 0.88 °C. In the compound with the higher boiling point, called *cis*-2-butene or (*Z*)-2-butene, the methyl groups are on the same side of the double bond. In the other 2-butene, called *trans*-2-butene or (*E*)-2-butene, the methyl groups are on opposite sides of the double bond.

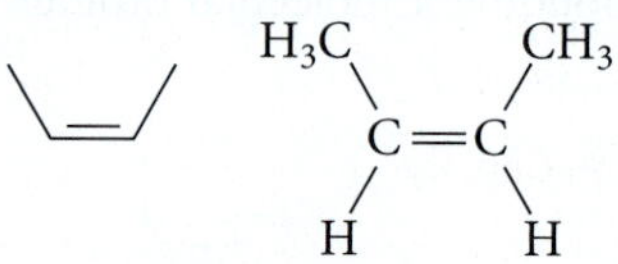

***cis*-2-butene**
(*Z*)-2-butene

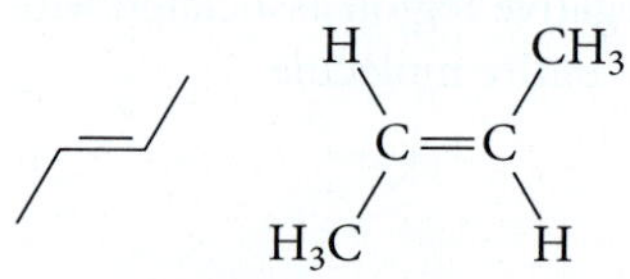

***trans*-2-butene**
(*E*)-2-butene

Be sure to differentiate between cis and trans double bonds when you draw skeletal structures. You don't have to draw the hydrogens (unless specifically necessary), but you do have to be sure that the bonds on either side of the double bond are oriented properly.

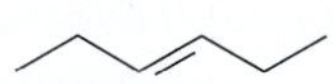

a trans alkene

a cis alkene

These isomers have identical atomic *connectivities* (CH_3 connected to CH, CH doubly bonded to CH, CH connected to CH_3). Despite their identical connectivities, *the two compounds differ in the way their constituent atoms are arranged in space*. Compounds with identical connectivities that differ in the spatial arrangement of their atoms are called **stereoisomers**. Therefore, *cis*- and *trans*-2-butene are stereoisomers. [The International Union of Pure and Applied Chemistry (IUPAC) has adopted the (*E*) and (*Z*) notation as a general way of naming cis and trans isomers. This notation is discussed in Sec. 4.2B.]

The interconversion of *cis*- and *trans*-2-butene requires a 180° internal rotation about the double bond—that is, a rotation of one carbon while holding the other carbon stationary.

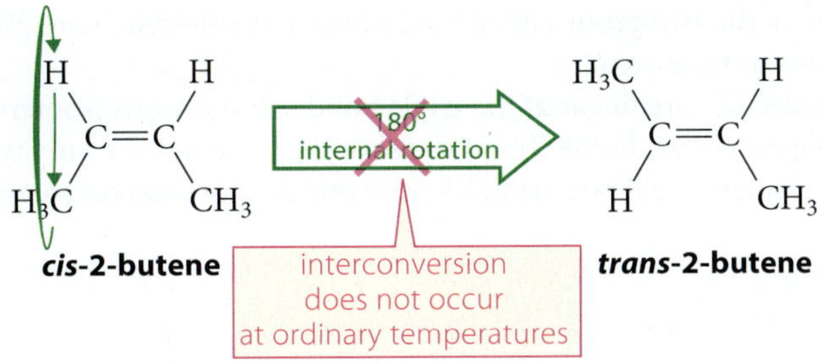

(4.6)

Because *cis*- and *trans*-2-butene do not interconvert, even at relatively high temperatures, this internal rotation must be very slow. For such an internal rotation to occur, the 2p orbitals on each carbon must be twisted away from coplanarity—that is, *the π bond must be broken.*

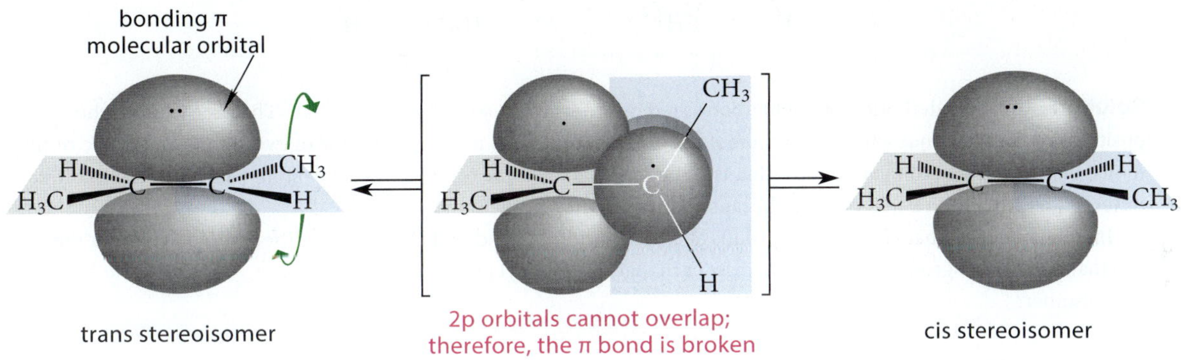

(4.7)

Because bonding is energetically favorable, breaking a bond is energetically costly. It takes *much* more energy to break the π bond than is available under normal conditions; therefore, the π bond in alkenes remains intact, and internal rotation about the double bond does not occur. In contrast, internal rotation about the carbon–carbon single bonds of ethane or butane can occur rapidly (Sec. 2.5) because no chemical bond is broken in the process.

Cis- and *trans*-2-butene are examples of *double-bond stereoisomers.* **Double-bond stereoisomers** (also called **cis–trans stereoisomers** or ***E,Z* stereoisomers**) are defined as compounds related by an internal rotation of 180° about the double bond. (We can always imagine such a rotation even though it does not occur at ordinary temperatures.) Another equivalent definition is that double-bond stereoisomers are different compounds related by exchange of the two groups at either carbon of a double bond.

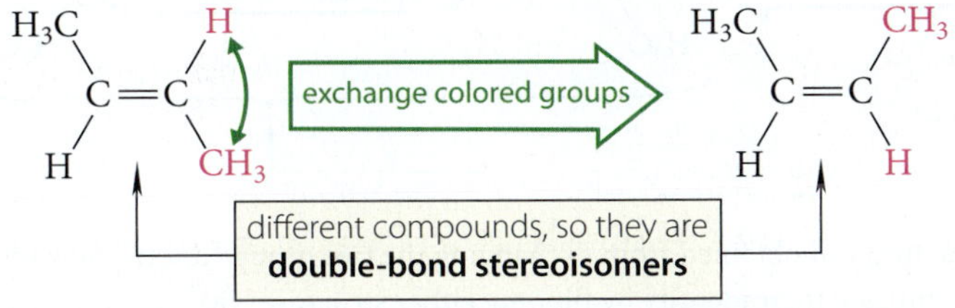

(4.8)

When an alkene can exist as double-bond stereoisomers, both carbons of the double bond are *stereocenters.* An atom is a **stereocenter** when the exchange of two bonded groups gives stereoisomers. (Other terms you might encounter that mean the same thing are **stereogenic atom** and **stereogenic center.**)

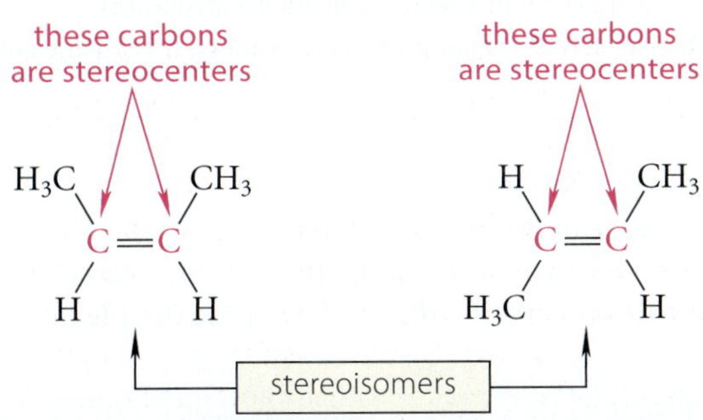

(4.9)

Because the exchange of the two groups at either carbon of the double bond gives stereoisomers, each of these carbons is a stereocenter.

Because alkyne molecules are linear at the triple bond, cis and trans isomers aren't possible.

You'll learn in Chapter 6 that double-bond stereoisomers are not the only type of stereoisomer. In every set of stereoisomers, we'll be able to identify one or more stereocenters.

Study Problem 4.1

Tell whether each of the following molecules has a double-bond stereoisomer. If so, identify its stereocenters.

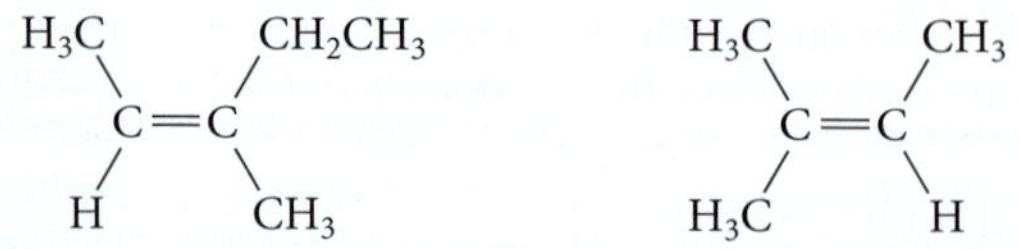

A *B*

Solution Apply the definition of double-bond stereoisomers as illustrated in Display 4.9. That is, exchange the positions of the two groups at *either* carbon of the double bond. This process gives one of two results: either the resulting molecule is congruent to the original—that is, superimposable on the original atom-for-atom—or it is different. If it's different, it can *only* be a stereoisomer, because its connectivity is the same.

In molecule *A*, exchanging the two groups at either carbon of the double bond gives different molecules. In the original, the methyl groups are trans; after the exchange, the methyl groups are cis. Therefore, *A* has a double-bond stereoisomer:

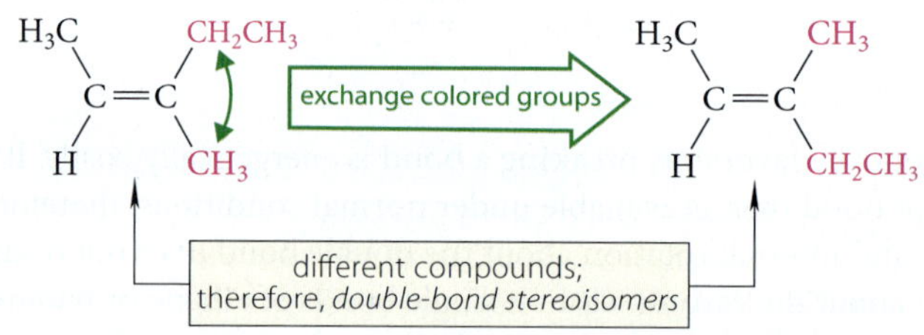

(You should verify that exchanging the two groups at the other carbon of the double bond gives the same result.) The two carbons of the double bond are both stereocenters.

In the case of structure *B*, exchanging the two groups at either carbon of the double bond gives back an identical molecule.

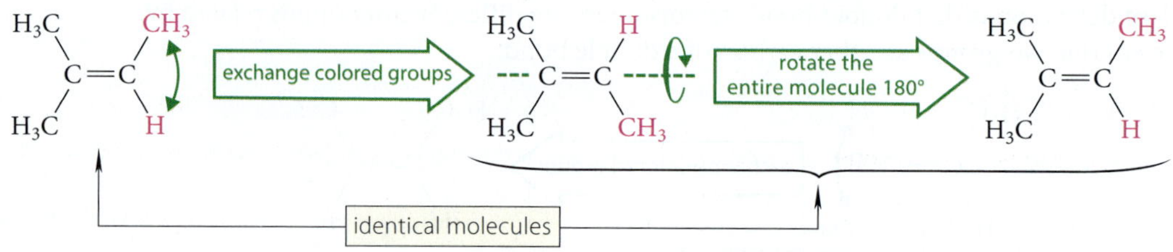

(4.10)

You may have found that the structure you obtained from exchanging the two groups doesn't *look* identical to the one on the left, but it is. You can demonstrate their identity by flipping either structure 180° about a horizontal axis (*green dashed line*)—in other words, by turning it over, as shown in the Display 4.10. But if you have difficulty seeing this, *you should build molecular models of both structures and convince yourself that the two can be superimposed atom-for-atom.* There is *no substitute for model building* when it comes to the spatial aspects of organic chemistry! After a little work with models on issues like this, you will develop the ability to see these relationships without models. Study Guide Link 4.1 offers more insights about how to achieve facility in relating alkene structures.

STUDY GUIDE LINK 4.1 Different Ways to Draw the Same Structure

Because exchanging two groups in Display 4.10 does *not* give stereoisomers, this alkene contains no stereocenters.

Study Problem 4.1 applies the definition of a stereocenter to the process of determining whether stereoisomers exist. Once you understand the idea of a stereocenter, you can understand that there is a shortcut for this process: If *either carbon* of the double bond has identical groups attached, then *neither carbon* is a stereocenter and the alkene *cannot* exist as double-bond stereoisomers.

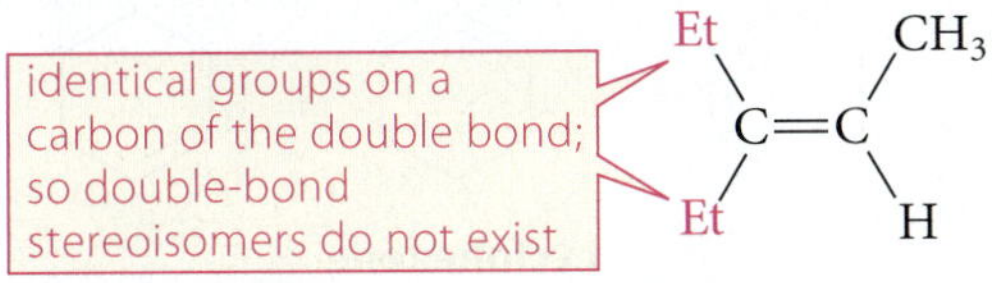

(4.11)

Focused Problem

4.2 Which of the following alkenes can exist as double-bond stereoisomers? Identify the stereocenters in each.

(a) $H_2C{=}CHCH_2CH_2CH_3$ **1-pentene**

(b) $CH_3CH_2{-}CH{=}CH{-}CH_2CH_3$ **3-hexene**

(c) $H_2C{=}CH{-}CH{=}CH{-}CH_3$ **1,3-pentadiene**

(d) 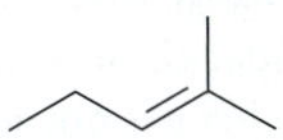 **2-methyl-2-pentene**

(e)  **cyclobutene**

[*Hint for part (e)*: Try to build a model of both stereoisomers, but don't break your models!]

4.2 NOMENCLATURE OF ALKENES AND ALKYNES

A. IUPAC Substitutive Nomenclature of Alkenes

The IUPAC substitutive nomenclature of alkenes is derived by modifying alkane nomenclature in a simple way. An unbranched alkene is named by replacing the *ane* suffix in the name of the corresponding alkane with the ending *ene*. The location of the double bond is specified with a number. The carbons are numbered consecutively from one end of the chain to the other, starting at the end that gives the double bond the lower number. The carbons of the double bond are numbered consecutively.

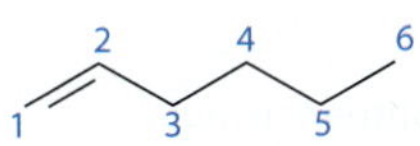

hexane + ene = hexene (with "ane" struck out)

6 5 4 3 2 1 incorrect numbering

1 2 3 4 5 6 correct numbering

$H_2C{=}CH{-}CH_2CH_2CH_2CH_3$

1-hexene

↑ position of double bond

The IUPAC recognizes an exception to this rule for the name of the simplest alkene, $H_2C{=}CH_2$, which is usually called ethylene rather than ethene. [*Chemical Abstracts* (Sec. 2.4E), however, uses the name ethene for substitutive nomenclature.]

When the name is unambiguous without a number, no number for the double bond or the substituent is used.

$$H_3C{-}\overset{\displaystyle CH_3}{\overset{|}{C}}{=}CH_2$$

methylpropene
(not 2-methylpropene; not 2-methyl-1-propene)

cyclohexene
(not 1-cyclohexene)

For example, contrast the nomenclature of cyclohexene and the noncyclic hexenes. Cyclohexene is the same compound no matter where you draw the double bond, so you don't need to number the position of the double bond; however, 1-hexene, 2-hexene, and 3-hexene are different compounds. (They are constitutional isomers.) For these compounds, the position of the double bond has to be specified.

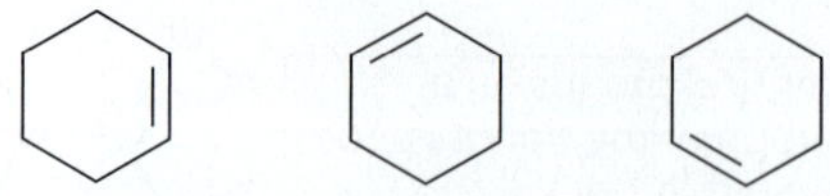

cyclohexene
(the same molecule drawn three different ways)

$H_2C{=}CHCH_2CH_2CH_2CH_3$ — **1-hexene**

$CH_3CH{=}CHCH_2CH_2CH_3$ — **2-hexene**

$CH_3CH_2CH{=}CHCH_2CH_3$ — **3-hexene**

different molecules

The names of alkenes with branched chains, like those of alkanes, are derived from their *principal chains*. In an alkene, the principal chain is defined as *the carbon chain containing the greatest number of double bonds*, even if this is *not* the longest chain. If more than one candidate for the principal chain have equal numbers of double bonds, the principal chain is the longest of these. The principal chain is numbered from the end that results in the lowest numbers for the carbons of the double bonds.

The principal-chain rule is modified when a molecule contains both double and triple bonds; this situation is covered in Sec. 4.2C.

When the alkene contains an alkyl substituent, the position of the double bond, not the position of the branch, determines the numbering of the chain. This is the main difference in the nomenclature of alkenes and alkanes. However, the position of the double bond is cited in the name *after* the name of the alkyl group. Study Problem 4.2 shows how these rules are implemented.

Study Problem 4.2

Name the following compound using IUPAC substitutive nomenclature.

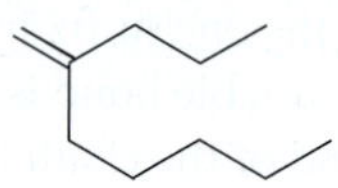

skeletal structure

$H_2C{=}C{-}CH_2CH_2CH_3$
$\quad\quad\ |$
$\quad\ CH_2CH_2CH_2CH_2CH_3$

condensed structure

Solution The principal chain is the longest continuous carbon chain containing *both carbons* of the double bond, as shown in color in the following structures.

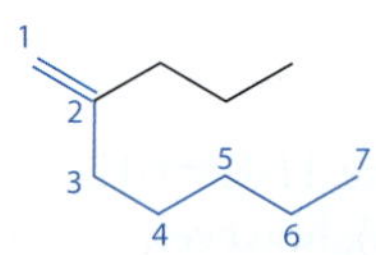

$\overset{1}{H_2C}{=}\overset{2}{C}{-}CH_2CH_2CH_3$
$\quad\quad\ |$
$\quad\ \underset{3}{CH_2}\underset{4}{CH_2}\underset{5}{CH_2}\underset{6}{CH_2}\underset{7}{CH_3}$ ← principal chain (longest chain containing the double bond; double bond receives the lowest number)

2-propyl-1-heptene
(2 = position of the substituent group; 1 = position of the double bond)

Notice in this case that the principal chain is *not* the longest continuous carbon chain in the molecule. The principal chain is numbered from the end that gives the double bond the lowest number—in this case, 1. The substituent group is a propyl group on carbon-2. Hence, the name of the compound is 2-propyl-1-heptene.

If a compound contains more than one double bond, the *ane* ending of the corresponding alkane is replaced by *adiene* (if there are two double bonds), *atriene* (if there are three double bonds), and so on.

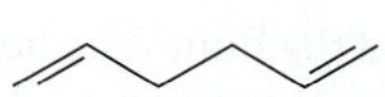

$H_2C{=}CHCH_2CH_2CH{=}CH_2$

1,5-hexadiene

Study Problem 4.3

Name the following compound:

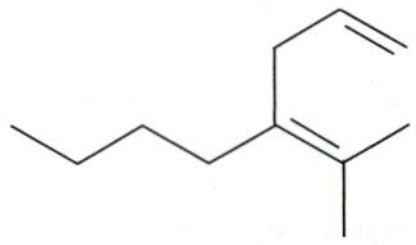

Solution The principal chain (blue in the following structure) is the chain containing the greatest number of double bonds.

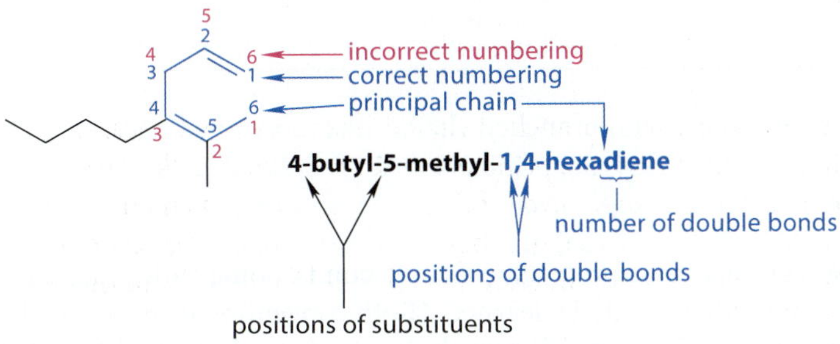

One possible numbering scheme (*blue*) gives the first-encountered carbons of the two double bonds the numbers 1 and 4, respectively; the other possible numbering scheme (*red*) gives the first-encountered carbons of the double bonds the numbers 2 and 5, respectively. We compare the two possible numbering schemes pairwise—that is, (1,4) versus (2,5). The lowest number at first point of difference (1 versus 2) determines the correct numbering. The compound is a 1,4-hexadiene, with a butyl branch at carbon-4 and a methyl branch at carbon-5. The correct name is 4-butyl-5-methyl-1,4-hexadiene.

If the name remains ambiguous after determining the correct numbers for the double bonds, then the principal chain is numbered so that the lowest numbers are given to the branches at the first point of difference.

Study Problem 4.4

Name the following compound:

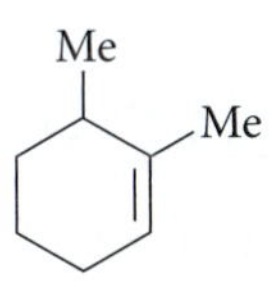

Solution Remember that Me = methyl. The carbons of the double bond can be numbered 1 and 2 in two ways:

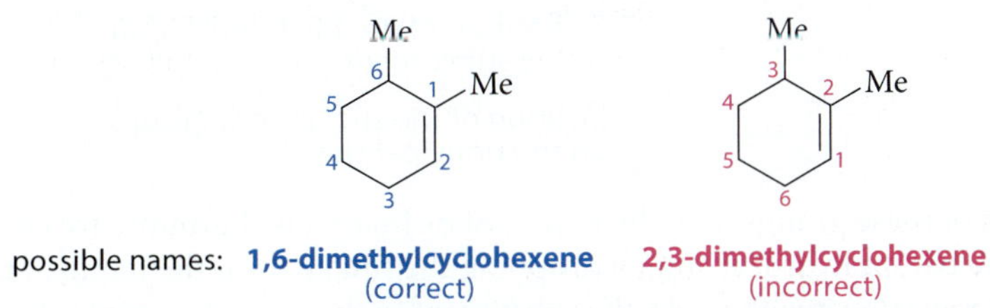

In this situation, choose the numbering scheme that gives the lowest number for the methyl substituents *at the first point of difference*. In comparing the substituent numbering schemes (1,6) with (2,3), the first point of difference occurs at the first number (1 versus 2). The (1,6) numbering scheme is correct because 1 is lower than 2. The number 1 for the double bond is not given explicitly in the name, because this is the only possible number. That is, when a double bond in a ring receives numerical priority, *its carbons must be numbered consecutively* with the numbers 1 and 2. That's why the following numbering scheme is incorrect. One carbon of the double bond has the number 1, but the other is not numbered consecutively.

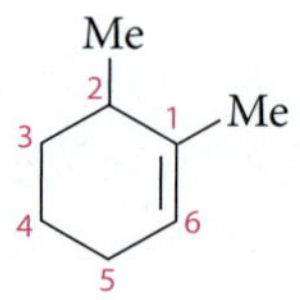

1,2-dimethylcyclohexene
(incorrect because carbons of double bond are not numbered consecutively)

Substituent groups may also contain double bonds. Some widely occurring groups of this type have special names that should be learned. (Both conventional and skeletal structures are shown. In these structures, the bracket indicates the point of attachment of the substituent group to the principal chain.)

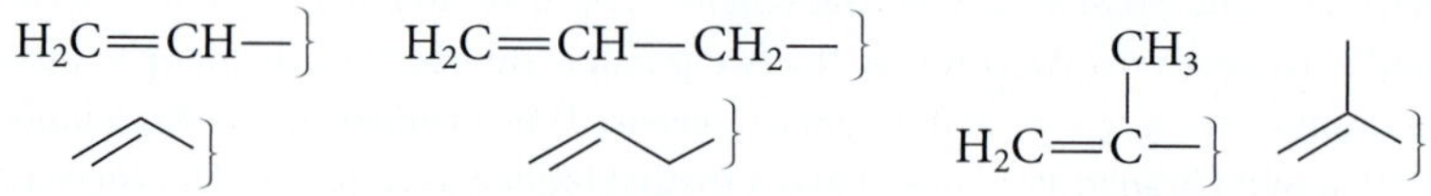

vinyl **allyl** **isopropenyl**

Here are some examples of structures containing these substituent groups. In each case, the ring is the principal chain because it has the greater number of carbons.

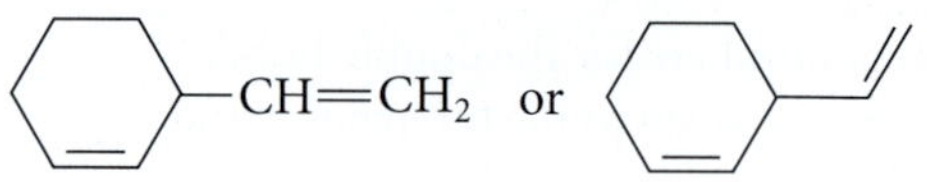

3-vinylcyclohexene

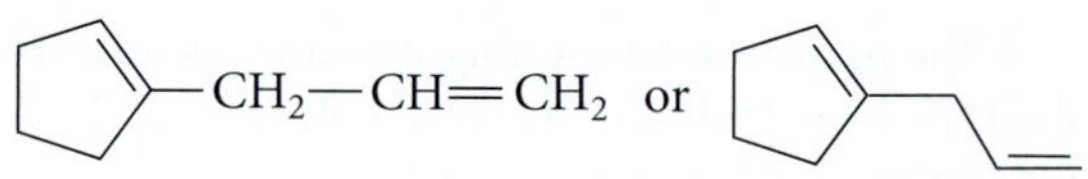

1-allylcyclopentene

Other substituent groups containing double bonds are numbered *from the point of attachment to the principal chain.*

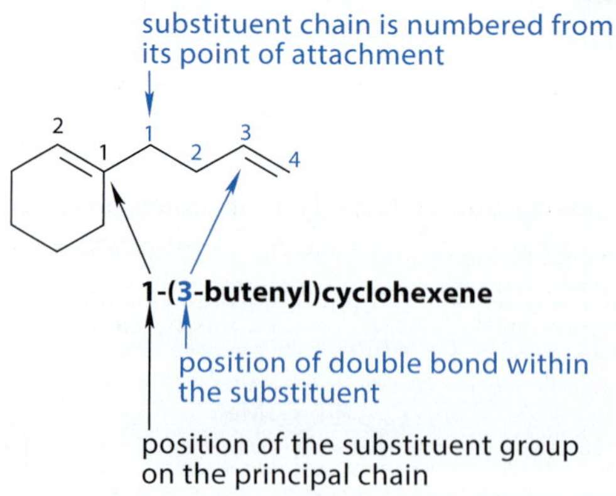

The names of these groups, like the names of ordinary alkyl groups, are constructed from the name of the parent hydrocarbon by dropping the final *e* from the name of the corresponding alkene and replacing it with *yl.* So, the substituent in the second example above is butene +/yl = butenyl. Parentheses are used to set off the names of substituents that contain internal numbering.

Finally, some alkenes have nonsystematic traditional names that are recognized by the IUPAC. These can be learned as they are encountered. Two examples are styrene and isoprene:

Ph—CH=CH_2 or 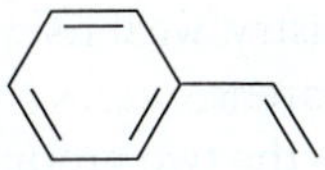

styrene

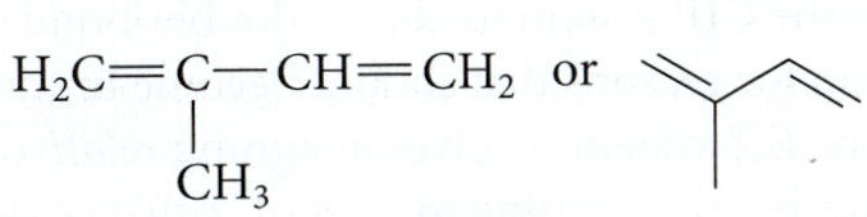

isoprene

(Recall from Sec. 2.8B that Ph refers to the phenyl group, which is a singly substituted benzene ring.)

1993 IUPAC Recommendations The nomenclature used in this text is based on the widely used 1979 IUPAC rules. In 1993, the IUPAC recommended an alteration in nomenclature that places the number of the double bond just before the *ene* suffix of the name. In this more recent system, 1-hexene is named hex-1-ene, and 2,4-hexadiene is named hexa-2,4-diene. This new system is being used by some chemists and not by others. For example, the official 2007 nomenclature guide of *Chemical Abstracts* continues to use the 1979 IUPAC system. In this text, we use the 1979 recommendations. However, conversion between old and new names is a simple matter of moving the numerical designation to a position prior to the final suffix in the name. Your use of either system is correct. We indicate the differences between the two systems when we introduce the nomenclature of other functional groups.

Focused Problems

4.3 Give the structure for each of the following.

(a) 1,3-Dimethylcyclohexene (b) 4-Methyl-1,3-hexadiene

(c) 1-Isopropenylcyclopentene (d) 5-(3-Pentenyl)-1,3,6,8-decatetraene

4.4 Name the following compounds.

(a) 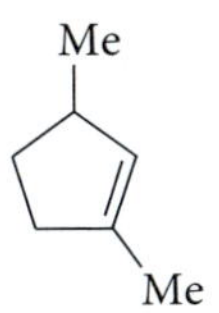

(b) $CH_3CH_2CH{=}CHCH_2CH_2CH_3$

(c) 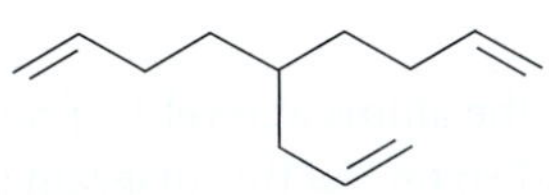

B. Nomenclature of Double-Bond Stereoisomers: The *E,Z* System

The cis and trans designations for double-bond stereoisomers are unambiguous when each carbon of a double bond has a single hydrogen, as in *cis*- and *trans*-2-butene. In some important situations, however, the use of the terms cis and trans is ambiguous. For example, is the following compound, a stereoisomer of 3-methyl-2-pentene, the cis or the trans stereoisomer?

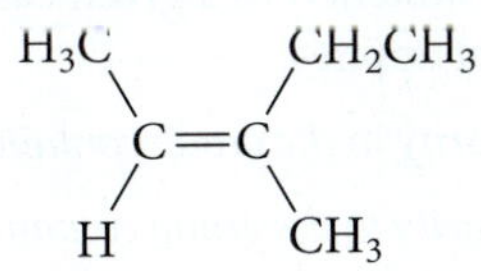

One person might decide that this compound is trans, because the two identical groups are on opposite sides of the double bond. Another might decide that it is cis, because the larger groups are on the same side of the double bond. Exactly this sort of ambiguity—and the use of both conventions simultaneously in the chemical literature—brought about the adoption of an unambiguous system for the nomenclature of stereoisomers. This system, first published in 1951, is part of a general system for the nomenclature of stereoisomers called the **Cahn–Ingold–Prelog (CIP) system**. When we

The Cahn–Ingold–Prelog system is named after its inventors, Robert S. Cahn (1899–1981), then editor of the *Journal of the Chemical Society*, the most prestigious British chemistry journal; Sir Christopher K. Ingold (1893–1970), a professor at University College, London, whose work played a very important part in the development of modern organic chemistry; and Vladimir Prelog (1906–1998), a professor at the Swiss Federal Institute of Technology, who received the 1975 Nobel Prize in Chemistry for his work in organic stereochemistry.

apply the CIP system to alkene double-bond stereochemistry, we'll refer to it simply as the ***E,Z* system** for reasons that should become immediately apparent.

The *E,Z* system involves assigning *relative priorities* to the two groups on each carbon of the double bond according to a set of *sequence rules* to be described here. We then compare the relative locations of these groups on each alkene carbon. If the groups of higher priority are on the same side of the double bond, the compound is said to have the *Z* configuration (*Z* from the German word *zusammen*, meaning "together"). If the groups of higher priority are on opposite sides of the double bond, the compound is said to have the *E* configuration (*E* from the German *entgegen*, meaning "across"). A convenient way to remember *E* and *Z* is that *ears* begins with *E*, and your *Ears* are on opposite sides of your head.

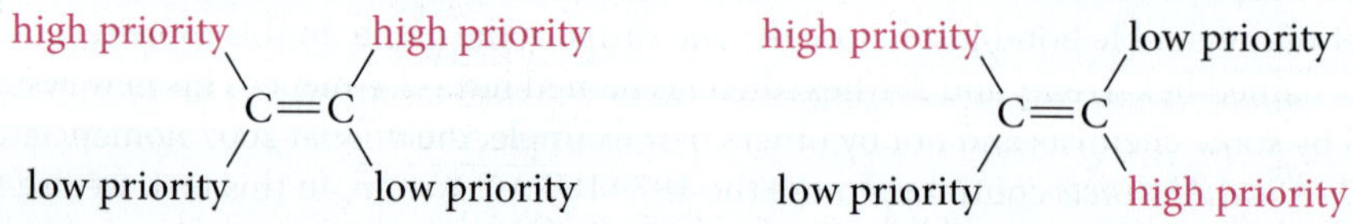

(*Z*) (*E*) **(4.12)**

For a compound with more than one double bond, the configuration of each double bond is specified separately.

The **sequence rules** used to assign relative priorities are the core of the CIP system. To apply these rules, we first organize the atoms in each of the two groups attached to a given carbon of the double bond. Level 1 consists of the atoms directly attached to the double bond. Level 2 consists of the atoms attached to the level-1 atoms, looking away from the double bond. Level 3 consists of the atoms attached to the level-2 atoms, and so on.

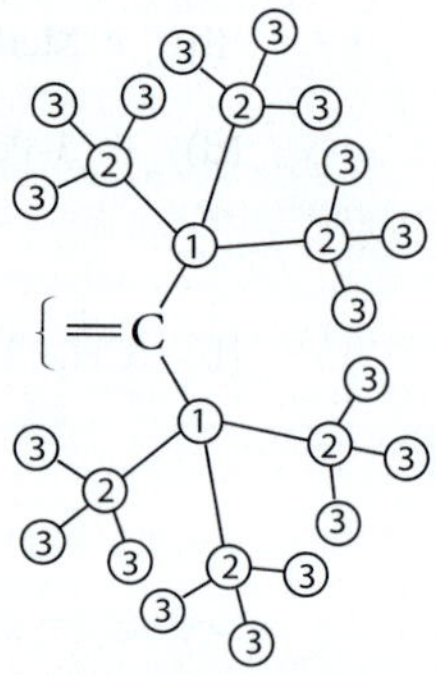

(4.13)

Application of the sequence rules starts with a comparison of the atoms at level 1. If no decision is possible, we proceed to the atoms at level 2, then level 3, and so on. As this diagram shows, the greater the level, the more possibilities there are for comparison at each level. The sequence rules specify both how to make the comparisons at each level and how to proceed to the next level if a decision is not possible.

To assign relative priorities, follow these steps and the accompanying rules *in order* until a difference is found. A decision must be made at the first point of difference. (Each step is illustrated with an example. Then, Study Problems 4.5–4.7 show how these steps are applied in three cases of increasing complexity.)

Step 1. Examine the atoms directly attached to a given carbon of the double bond (level 1), and then follow the first rule that applies.

Rule 1a *Assign higher priority to the group containing the atom of higher atomic number.*

Rule 1b *Assign higher priority to the group containing the isotope of higher atomic mass.*

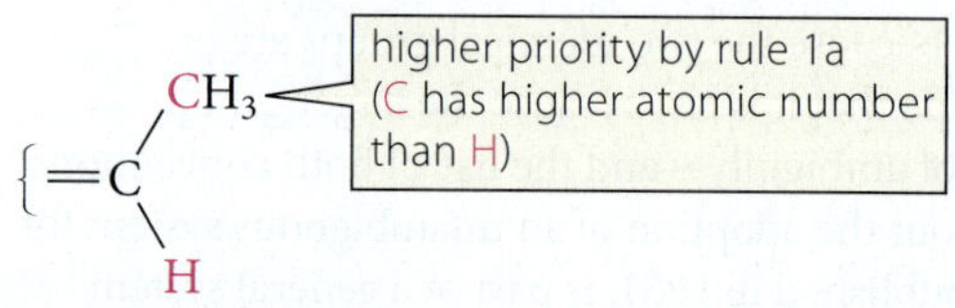

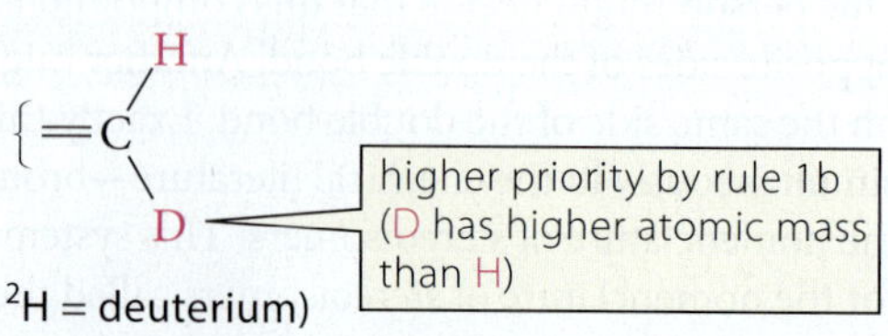

(D = ^{2}H = deuterium)

Step 2. If the level-1 atoms directly attached to the double bond are the same, then, working outward from the double bond, consider within each group the set of attached atoms—that is, the level-2 set. You'll have two level-2 sets—one for each group on the double bond. Apply rule 2 to each level-2 set.

Rule 2 *Arrange the atoms within each set in descending priority order, and make a pairwise comparison of the atoms in the two sets. The higher priority is assigned to the atom of higher atomic number (or atomic mass in the case of isotopes) at the first point of difference.*

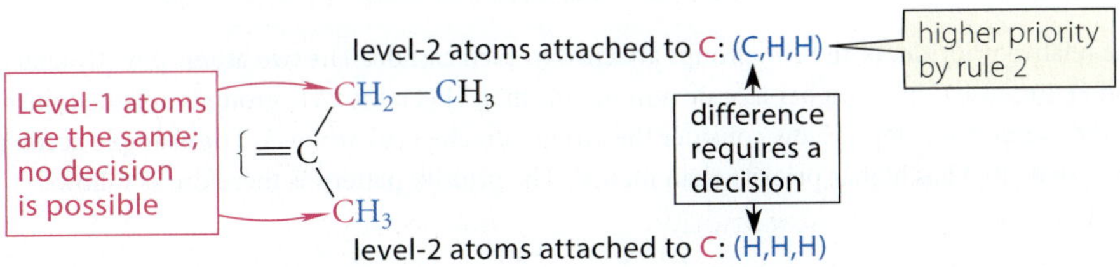

Step 3. If the level-2 sets in the two groups are identical, then, within each set, choose the atom of highest priority. Identify the level-3 set of atoms attached to it. Then compare the level-3 sets in each group by applying rule 2.

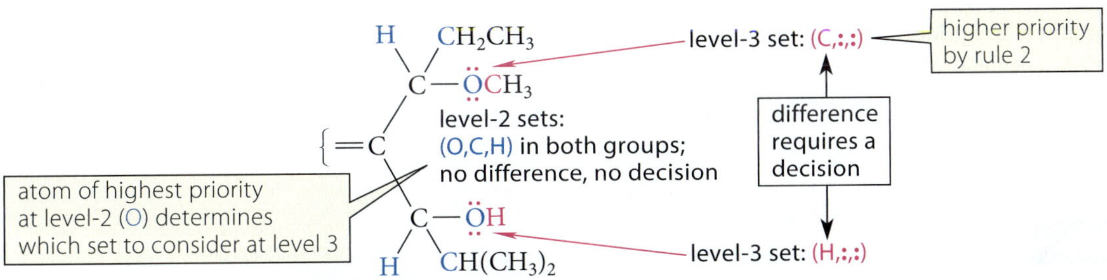

Step 3a. If no decision emerges at step 3, choose the atoms of next highest priority in the level-2 sets and repeat the process in step 3. Choose atoms of progressively lower priority in the level-2 sets until a difference is found.

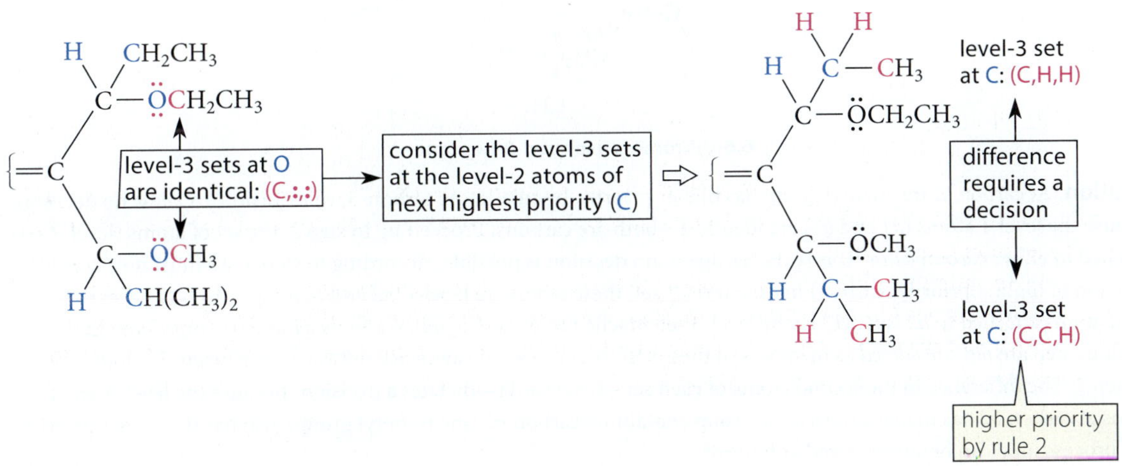

Notice that we consider *only the atom directly attached* to the oxygens (C in this example); we do not consider any atoms farther out.

Step 3b. If two or more atoms in any set are the same, decide on their relative priorities for step 3 by continuing to explore outward from each, and choose as the atom of higher priority the one that gives the path of higher priority. (Step 3b is illustrated in Study Problem 4.7.)

Step 3c. If no decision is possible, move away from the double bond within each group to atoms at the next level and repeat step 3. Continue this exploration, level-by-level, until the first difference is found.

Study Problem 4.5

What is the configuration of the following stereoisomer of 3-methyl-2-pentene? (The numbers are for reference in the solution.)

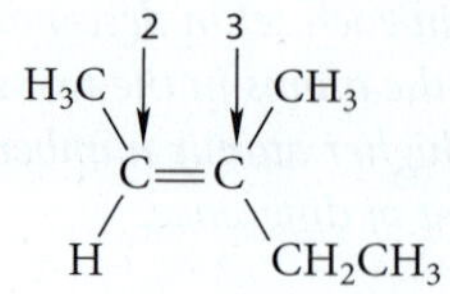

Solution Assign the relative priorities of the two groups attached to each carbon. The two atoms directly attached to carbon-2 are C and H. Because C has a higher atomic number (6) than H (1), the CH_3 group is assigned the higher priority, as shown in the example for step 1. Now consider the groups attached to carbon-3. The example accompanying step 2 and rule 2 shows that ethyl has higher priority than methyl. The priority pattern is therefore as follows:

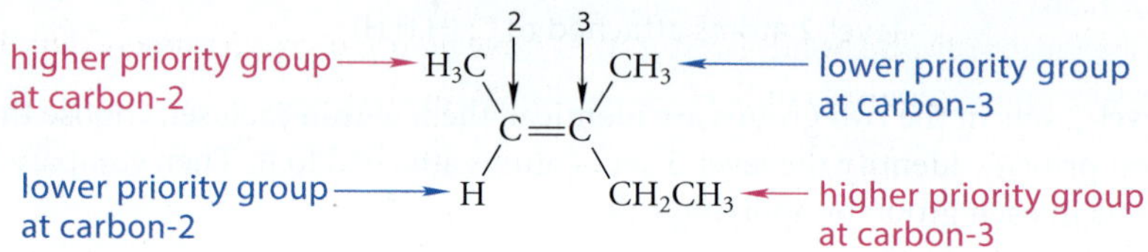

Because groups of like priority are on opposite sides of the double bond, this alkene is the *E* isomer; its complete name is (*E*)-3-methyl-2-pentene.

Study Problem 4.6

What is the configuration (*E* or *Z*) of the following alkene? (The numbers and letters are for reference in the solution.)

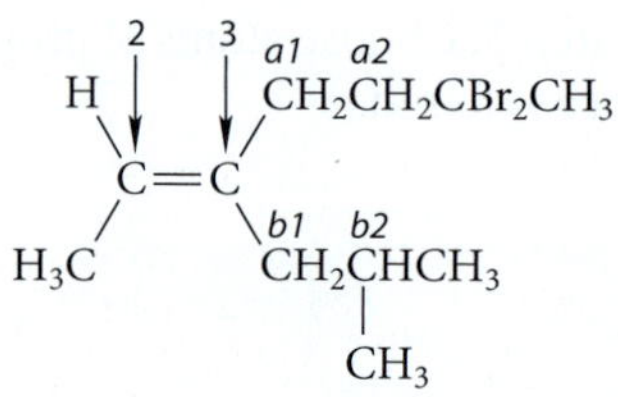

6,6-dibromo-3-isobutyl-2-heptene

Solution At carbon-2, the methyl group has higher priority, by rule 1a. At carbon-3, rules 1a and 1b allow no decision, because the level-1 atoms (*a1* and *b1*) are identical—both are carbons. Proceeding to step 2, the set of atoms (level-2 set) attached to either carbon *a1* or *b1* is (C,H,H); again, no decision is possible. According to step 3, we must now consider the atom of highest priority (carbon) in each level-2 set; these atoms are labeled *a2* and *b2*, respectively. The level-3 set of atoms attached to *a2* is (C,H,H); the level-3 set attached to *b2* is (C,C,H). Carbons *a1* and *b1* considered in the previous step are *not considered* as members of these sets, because we always work outward, away from the double bond, by step 2. The difference in the second atoms of each set—C versus H—dictates a decision. Because the level-3 set of atoms at carbon *b2* has higher priority, the group containing carbon *b2* (the isobutyl group) also has the higher priority. The process used can be summarized as follows:

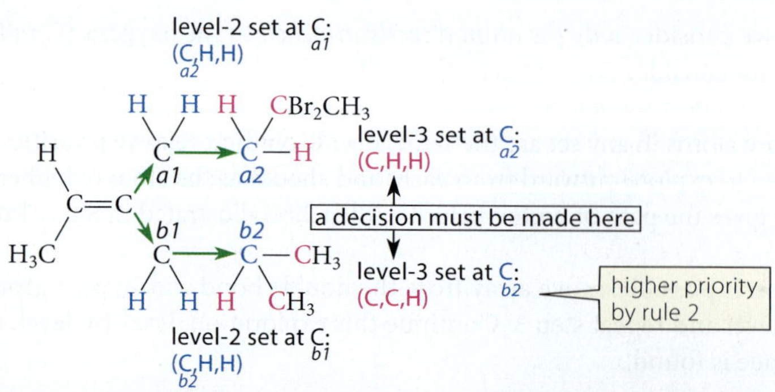

Although Br has a higher priority than H, the decision point occurs before we reach the Br (which is in a level-4 set). Therefore, the Br-versus-H comparison is irrelevant in this case.

Because the groups of like priority are on the same side of the double bond, this alkene has the *Z* configuration; the name is (*Z*)-6,6-dibromo-3-isobutyl-2-heptene.

Sometimes the groups to which we must assign priorities themselves contain double bonds. Double bonds are treated by a special convention, in which the double bond is rewritten as a single bond and the atoms at each end of the double bond are duplicated:

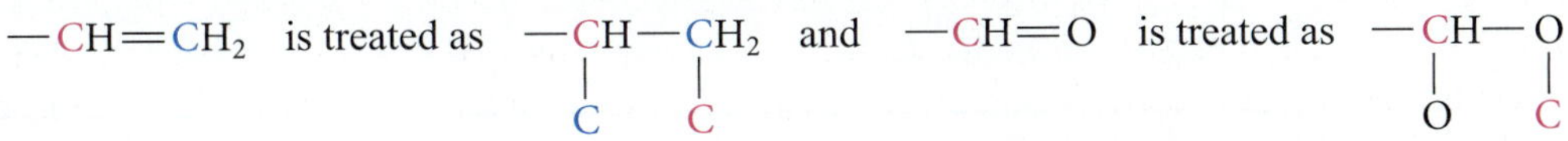

(4.14a)

The duplicated atoms bear only one bond—that is, they have no other groups attached to them. The treatment of triple bonds requires triplicating the atoms involved:

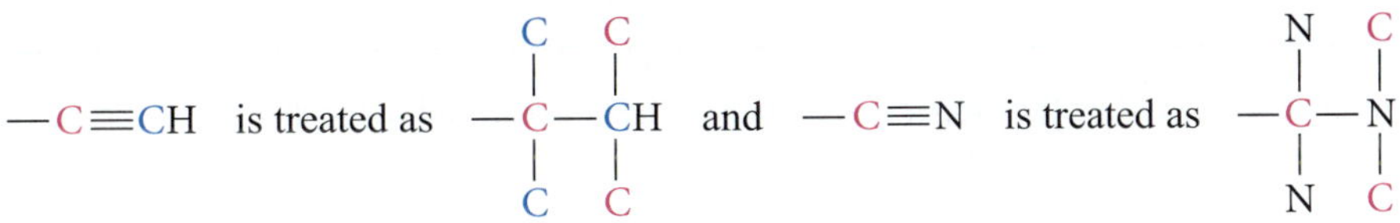

(4.14b)

The use of this convention is illustrated in Study Problem 4.7.

Study Problem 4.7

Give the IUPAC name of the following compounds, including the *E,Z* designation for the double-bond stereochemistry. (The carbon numbers are for reference in the solution.)

(a)

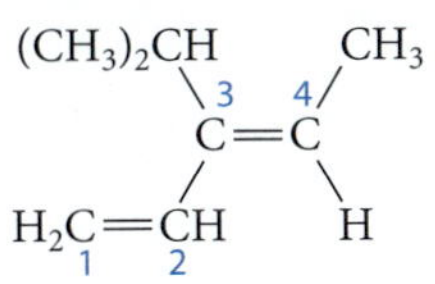

(b)

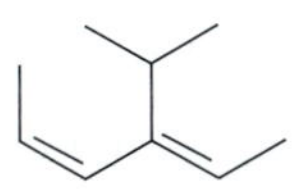

Solution (a) First, give the name without stereochemistry. By the nomenclature principles in Sec. 4.2A, the name is 3-isopropyl-1,3-pentadiene. Now assign stereochemistry. Carbons 1 and 2 are not stereocenters, but carbons 3 and 4 are. At carbon-4, the methyl group receives higher priority than H. The real challenge here is the relative priorities of the groups at carbon-3. We analyze the two groups as follows:

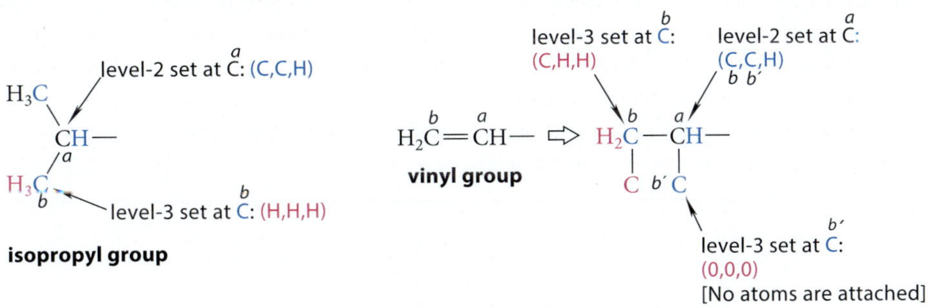

(The symbol ⇨ means "implies.") Notice that the carbons of the double bond are duplicated. The level-3 set (0,0,0) in the vinyl group indicates that nothing is attached to the duplicated carbon *b*′. We compare carbon *a* of the isopropyl group with carbon *a* of the vinyl group. These are the same. Furthermore, the attached (level-2) sets are the same: (C,C,H). We then consider the level-2 atom of highest priority in each group and examine the level-3 sets attached to this atom. In the isopropyl group, the level-2 atom of highest priority is either of the methyl carbons. In the vinyl group, we have to choose between carbon *b* and its duplicated image, carbon *b*′. Because both atoms are the same, we use step 3b.

According to step 3b, this choice is made by going out one level beyond carbons *b* and *b′*—in other words, within the vinyl group, we must compare the level-3 sets (C,H,H) for C^b with (0,0,0) for $C^{b'}$. Because any of the atoms attached to carbon *b* has a higher priority than "nothing" on carbon *b′*, carbon *b* represents the path of higher priority for vinyl. Now we are ready to compare the vinyl and isopropyl groups again. For isopropyl, the level-3 set at carbon *b* is (H,H,H), and for vinyl, the level-3 set at carbon *b* is (C,H,H). By rule 2, vinyl receives the higher priority. The name is therefore (*E*)-3-isopropyl-1,3-pentadiene.

You should be able to work part (b) using the same tactics. Try it. Be sure to convert the skeletal structure into a condensed structure if necessary. The name is (2*E*,4*Z*)-3-isopropyl-2,4-hexadiene. (The position of the isopropyl group determines the numbering of the double bonds, which is ambiguous otherwise.) This example also illustrates that when two or more double bonds require a stereochemical designation, the number of the double bond is included with the *E* or *Z*.

Focused Problems

4.5 Name each of the following compounds, including the proper designation of the double-bond stereochemistry.

(a) 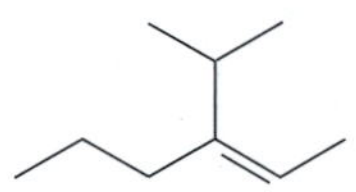

(b) 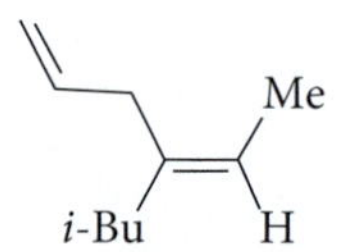

4.6 Give the structure of the following compounds.

(a) (*E*)-4-Allyl-1,5-octadiene (b) (2*E*,7*Z*)-5-[(*E*)-1-Propenyl]-2,7-nonadiene

Be sure to read Study Guide Link 4.2 if you have difficulty with this problem.

STUDY GUIDE LINK 4.2
Drawing Structures from Names

4.7 In each case, which group receives the higher priority?

(a) 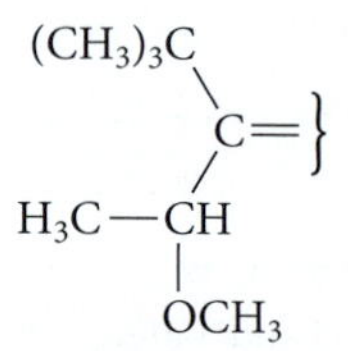

(b) 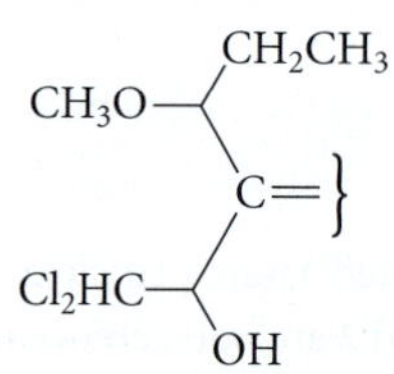

(c) 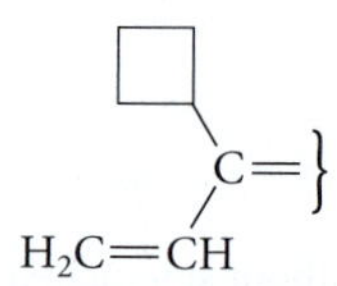

(d) 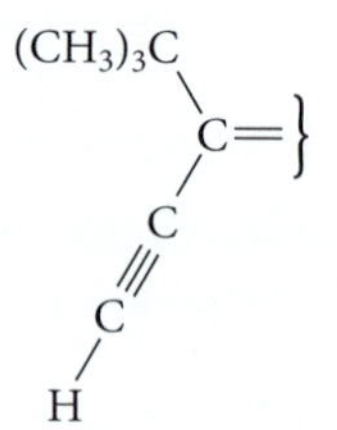

C. Nomenclature of Alkynes

There is a common nomenclature of alkynes in which simple alkynes are named as derivatives of the parent compound acetylene:

$H_3C—C{\equiv}C—H$	$H_3C—C{\equiv}C—CH_3$	$CH_3CH_2—C{\equiv}C—CH_3$
methylacetylene	**dimethylacetylene**	**ethylmethylacetylene**

The IUPAC substitutive nomenclature of alkynes is much like that of alkenes. The suffix *ane* in the name of the corresponding alkane is replaced by the suffix *yne*, and the triple bond is given the lowest possible number.

$H_3C{-}C{\equiv}C{-}H$ **propyne**

$CH_3CH_2CH_2CH_2{-}C{\equiv}C{-}CH_3$ **2-heptyne**

$H_3C{-}CH_2{-}C{\equiv}C{-}H$ **1-butyne**

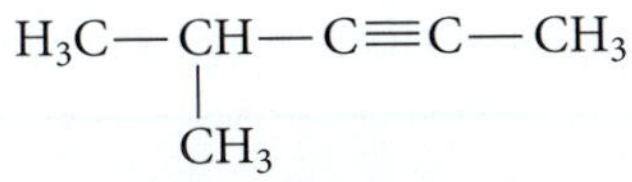

4-methyl-2-pentyne

$HC{\equiv}C{-}CH_2{-}CH_2{-}C{\equiv}C{-}CH_3$ **1,5-heptadiyne**

As with the stipulation of double-bond position in alkenes, the 1993 IUPAC recommendations place the number that indicates the triple-bond position prior to the final *yne* or *diyne* suffix. Thus, the examples here become hept-2-yne, but-1-yne, 4-methylpent-2-yne, and hepta-1,5-diyne.

Substituent groups that contain a triple bond (called *alkynyl groups*) are named by replacing the final *e* in the name of the corresponding alkyne with the suffix *yl*. (This is exactly analogous to the nomenclature of substituent groups containing double bonds; see Sec. 4.2A.) The alkynyl group is numbered from its point of attachment to the main chain:

$HC{\equiv}C{-}$ **ethynyl group** (ethyne + yl)

$HC{\equiv}C{-}CH_2{-}$ **2-propynyl group**

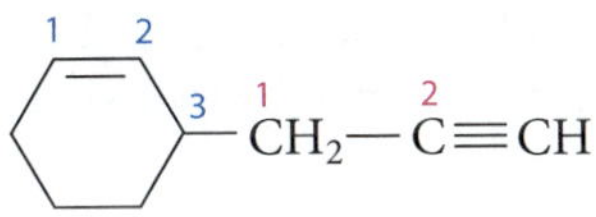

3-(2-propynyl)cyclohexene

position of the triple bond within the substituent

position of the 2-propynyl group on the ring

When a molecule contains both double and triple bonds, the principal chain is the carbon chain that has the greater number of *double and triple* bonds. Only if there is ambiguity does the number of double bonds take precedence.

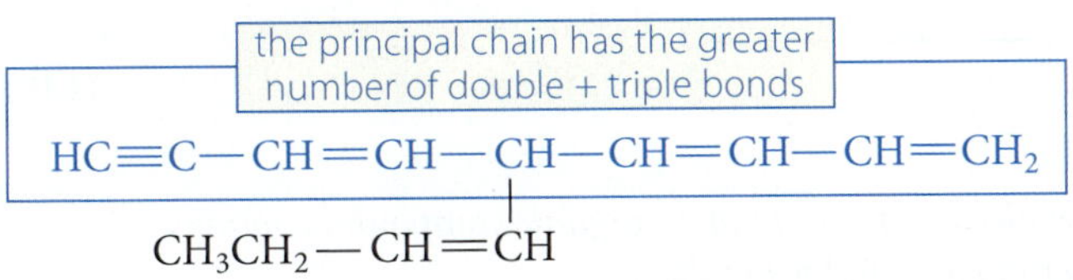

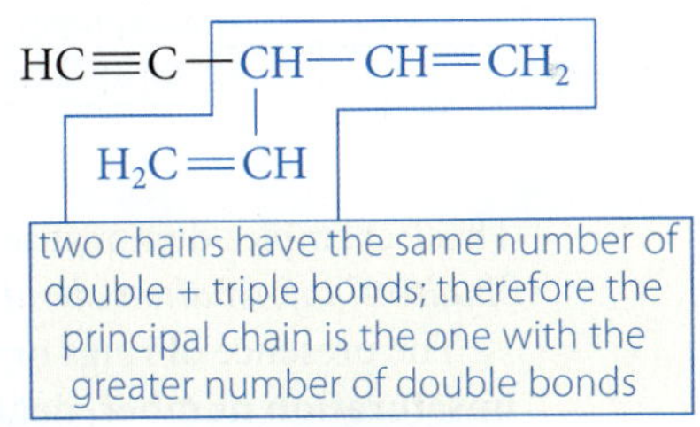

In numbering the principal chain, the double or triple bond that has the lower number at first point of difference receives numerical precedence. However, if this rule is ambiguous, a double bond receives numerical precedence over a triple bond.

$\overset{1}{H}C{\equiv}\overset{2}{C}{-}\overset{3}{C}H{=}\overset{4}{C}H\overset{5}{C}H_3$ **3-penten-1-yne**

$\overset{5}{C}H_3\overset{4}{C}{\equiv}\overset{3}{C}{-}\overset{2}{C}H{=}\overset{1}{C}H_2$ **1-penten-3-yne**

precedence is given to the bond that has lower number at first point of difference

$\overset{1}{H_2C}{=}\overset{2}{C}H\overset{3}{C}H_2\overset{4}{C}{\equiv}\overset{5}{C}H$ **1-penten-4-yne**

precedence is given to the double bond when numbering is ambiguous

Focused Problems

4.8 Draw a structure for each of the following alkynes.

(a) Isopropylacetylene (b) Cyclononyne (c) 4-Methyl-1-pentyne (d) 1-Ethynylcyclopentene

(e) 1,3-Hexadiyne

4.9 Provide the substitutive name for each of the following compounds. Also provide common names for (a) and (b).

(a) $CH_3CH_2CH_2CH_2C{\equiv}CH$

(b) $CH_3CH_2CH_2CH_2C{\equiv}CCH_2CH_2CH_2CH_3$

(c)

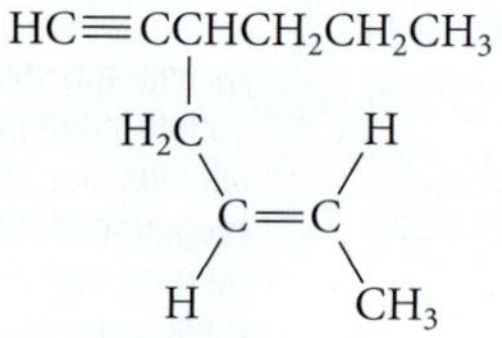

(d)

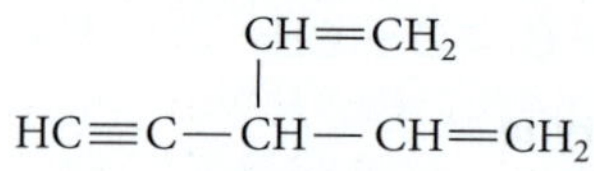

4.3 UNSATURATION NUMBER

In Sec. 2.3 we found that each ring in a cyclic hydrocarbon reduces the number of hydrogens in its formula by 2. Similarly, each π bond also reduces the number of hydrogens in its formula by 2. These two points are illustrated by four six-carbon hydrocarbons:

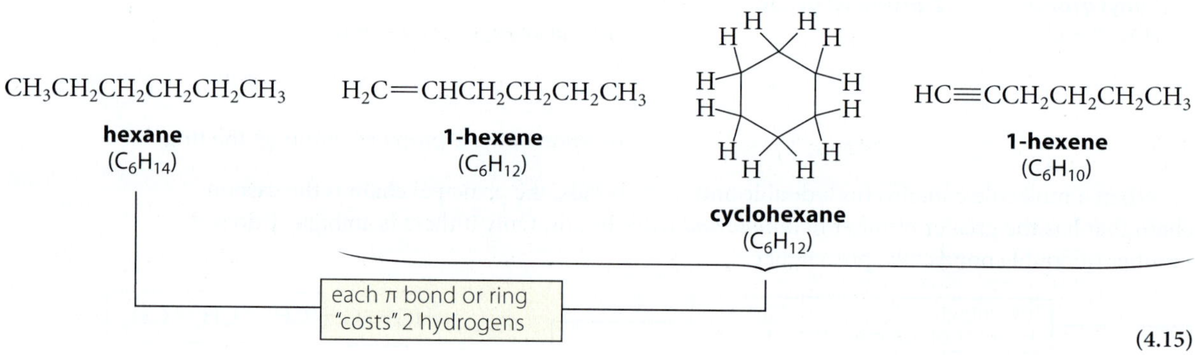

(4.15)

These examples demonstrate that *the molecular formula of an organic compound contains "built-in" information about the number of rings and π bonds.*

The presence of rings or π bonds within a molecule is indicated by a quantity U called the **unsaturation number, degree of unsaturation,** or **index of unsaturation.** *The unsaturation number of a molecule is equal to the total number of its rings and π bonds.* The unsaturation number of a hydrocarbon is readily calculated from its molecular formula as follows. The maximum number of hydrogens possible in a hydrocarbon with C carbon atoms is $2C + 2$. Because each ring or π bond reduces the number of hydrogens from this maximum by 2, the unsaturation number is equal to half the difference between the maximum number of hydrogens and the actual number H:

$$U = \frac{2C + 2 - H}{2} = \text{number of rings} + \pi \text{ bonds} \tag{4.16}$$

For example, cyclohexene, C_6H_{10}, has $U = [2(6) + 2 - 10]/2 = 2$. Cyclohexene has two degrees of unsaturation: one ring and one π bond. A triple bond (two π bonds) contributes two degrees of unsaturation. For example, 1-hexyne, $HC{\equiv}C{-}CH_2CH_2CH_2CH_3$, has the same formula as cyclohexene: C_6H_{10}.

How does the presence of other elements affect the unsaturation number calculation? From common examples (for instance, ethanol, C_2H_5OH) you can show that Eq. 4.16 remains valid when oxygen is present in an organic compound. Halogens are counted as if they were hydrogens, because halogens are monovalent, and each halogen reduces the number of hydrogens by 1. Therefore, if X represents the number of halogens,

$$U = \frac{2C + 2 - (H + X)}{2} \tag{4.17}$$

Another common element found in organic compounds is nitrogen. When nitrogen is present, the number of hydrogens in a saturated compound increases by 1 for each nitrogen. (For example, the saturated compound methylamine, $H_3C—NH_2$, has $2C + 3$ hydrogens.) Therefore, if N is the number of nitrogens, the formula for the unsaturation number becomes

$$U = \frac{2C + 2 + N - (H + X)}{2} \tag{4.18}$$

The unsaturation number is a valuable source of structural information about a compound of unknown structure. This point is illustrated in Focused Problem 4.12.

Focused Problems

4.10 Calculate the unsaturation number for each of the following formulas.

(a) $C_3H_4Cl_2$ (b) $C_5H_8N_2$ (c) $C_{20}H_{34}O_2$

4.11 What is the unsaturation number of each of the following compounds?

(a) Methylcyclohexane (b) 2,4,6-Octatriene

4.12 (a) Which of the following *cannot* be correct formula(s) for an organic compound? Explain.

$C_{10}H_{20}N_3$ (*A*) $C_{10}H_{20}N_2O_2$ (*B*) $C_{10}H_{27}N_3O_2$ (*C*) $C_{10}H_{16}O_2$ (*D*)

(b) Draw constitutional isomers of compounds with the formula C_4H_8O that contain (1) an alcohol functional group and (2) a ketone functional group.

4.4 PHYSICAL PROPERTIES OF ALKENES AND ALKYNES

Except for their melting points and dipole moments, many alkenes and alkynes differ little in their physical properties from the corresponding alkanes.

	$HC{\equiv}C(CH_2)_3CH_3$ **1-hexyne**	$H_2C{=}CH(CH_2)_3CH_3$ **1-hexene**	$CH_3(CH_2)_4CH_3$ **hexane**
boiling point	71–72 °C	63.4 °C	68.7 °C
melting point	−132 °C	−139.8 °C	−95.3 °C
density	0.710 g mL^{-1}	0.673 g mL^{-1}	0.660 g mL^{-1}
water solubility	negligible	negligible	negligible
dipole moment	0.87 D	0.46 D	0.085 D

Like alkanes, alkenes and alkynes are flammable, nonpolar compounds that are less dense than, and insoluble in, water. The alkenes of lower molecular weight are gases at room temperature.

As these examples illustrate, the dipole moments of some alkenes and alkynes, though small, are greater than those of the corresponding alkanes.

How can we account for the dipole moments of alkenes and alkynes? Recall from Secs. 3.7C and 3.7E that carbon electronegativity is in the order $sp^3 < sp^2 < sp$. In other words, carbon electronegativity increases with fraction s character. As a result, any sp^2–sp^3 or sp–sp^3 carbon–carbon bond has a small bond dipole (Sec. 1.9) in which the sp^3 carbon is the positive end of the dipole and the sp^2 or sp carbon is the negative end.

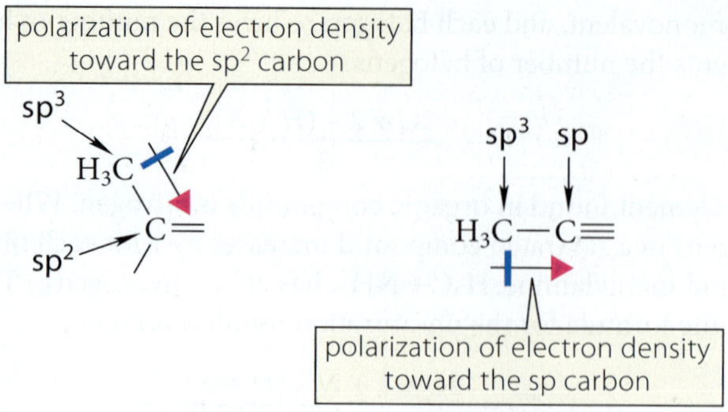

(4.19)

The dipole moment of *cis*-2-butene (0.25 D) illustrates the consequences of this bond dipole. The dipole moment of *cis*-2 butene is the vector sum of all of the H_3C—C and H—C bond dipoles. Although both types of bond dipole are oriented toward the alkene carbon, there is good evidence (Problem 4.51) that the bond dipole of the H_3C—C bond is greater. Therefore, *cis*-2-butene has a net dipole moment.

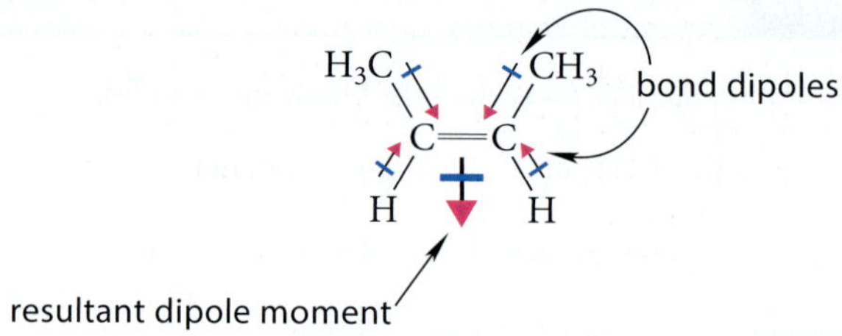

(4.20)

In summary: Bonds from alkyl groups to sp^2-hybridized and sp-hybridized carbons are polarized so that electrons are drawn away from alkyl groups toward the sp^2- or sp-hybridized carbon. This means that a carbon–carbon double bond or triple bond, when viewed as a substituent group, exerts an *electron-withdrawing polar effect*. The polar effect of a double bond is about 20% that of a chlorine, and the polar effect of a triple bond is about 75% that of a chlorine. (See Table 3.4, Sec. 3.7E.)

Focused Problem

4.13 Which compound in each set should have the larger dipole moment? Explain.

(a) *cis*-2-Butene or *trans*-2-butene (b) Propene or methylpropene (c) 1-Butyne or 2-butyne

4.5 RELATIVE STABILITIES OF ALKENE AND ALKYNE ISOMERS

When we ask which of two compounds is more stable, we are asking which compound has lower energy. However, energy can take different forms, and the energy we use to measure "relative stability" depends on the purpose we have in mind. We've learned that $\Delta G°$ for a reaction is the energy quantity related to the equilibrium constant, as we know from the equation $\Delta G° = -2.3RT \log K_{eq}$ (Sec. 3.5). Measuring the equilibrium constant is a good way to determine $\Delta G°$. However, if we are interested in the *total energy change* for a reaction, we use the **standard enthalpy change** for the reaction, $\Delta H°$. The $\Delta H°$ for a reaction approximates very closely the *total* energy difference between reactants and products, and it reflects the *relative stabilities of bonding arrangements in reactants and products*. The $\Delta G°$ and $\Delta H°$ for a reaction are related by the equation $\Delta G° = \Delta H° - T\Delta S°$, where $\Delta S°$ is the entropy change for the reaction and T is the absolute temperature. (For a structural interpretation of $\Delta S°$, see Further Exploration 4.1 in the Study Guide.) In other words, the $\Delta G°$ for a reaction differs from the total energy difference between reactants and products by an amount $-T\Delta S°$. In Sec. 4.5A we explain the conventions for presenting enthalpy data, and in Sec. 4.5B we use enthalpy data to investigate the relative stabilities of alkenes.

FURTHER EXPLORATION 4.1
Relationship between Free Energy and Enthalpy

A. Heats of Formation

The relative enthalpies of many organic compounds are available in standard tables as *heats of formation*. The standard **heat of formation** of a compound, abbreviated ΔH_f°, is the energy change that occurs when the compound is formed from its elements in their natural state at 1 atm pressure and 25 °C. Thus, the heat of formation of *trans*-2-butene is the ΔH° of the following reaction at 298 K (or other specified temperature):

$$4\,\mathrm{H_2}\,(\text{gas}) + 4\,\mathrm{C}\,(\text{solid}) \longrightarrow \textit{trans}\text{-2-butene}\,(\text{gas},\ \mathrm{C_4H_8}) \tag{4.21}$$

The sign conventions used in dealing with heats of reaction are the same as with free energies: the heat of any reaction is the *difference* between the enthalpies of the products and the reactants.

$$\Delta H^\circ(\text{reaction}) = H^\circ(\text{products}) - H^\circ(\text{reactants}) \tag{4.22}$$

A reaction in which heat is liberated is said to be an **exothermic reaction**, and one in which heat is absorbed is said to be an **endothermic reaction**. The ΔH° of an exothermic reaction, by Eq. 4.22, has a negative sign; the ΔH° of an endothermic reaction has a positive sign. The heat of formation of *trans*-2-butene (Eq. 4.21) is -11.2 kJ mol^{-1} (-2.67 kcal mol^{-1}); this means that heat is liberated in the formation of *trans*-2-butene from carbon and hydrogen, and that the alkene has lower energy than the 4 mol each of C and H_2 from which it is formed.

Heats of formation are used to determine the relative enthalpies of molecules—that is, which of two molecules has lower energy. Study Problem 4.8 illustrates how this is done.

Study Problem 4.8

Calculate the standard enthalpy difference between the cis and trans isomers of 2-butene. Specify which stereoisomer is more stable. The heats of formation are -7.0 kJ mol^{-1} for the cis isomer and -11.2 kJ mol^{-1} for the trans isomer (-1.67 and -2.67 kcal mol^{-1}, respectively).

Solution The enthalpy difference requested in the problem corresponds to the ΔH° of the following hypothetical reaction:

	cis-2-butene ⟶	*trans*-2-butene	
ΔH_f°	-7.0	-11.2	kJ mol^{-1}
	-1.7	-2.7	kcal mol^{-1}

To obtain the standard enthalpy difference, apply Eq. 4.22 *using the corresponding heats of formation in place of the H° values*. Thus, the ΔH_f° for the reactant, *cis*-2-butene, is subtracted from that of the product, *trans*-2-butene. The ΔH° for this reaction, then, is $-11.2 - (-7.0) = -4.2$ kJ mol^{-1} or $-2.7 - (-1.7) = -1.0$ kcal mol^{-1}. This means that *trans*-2-butene is more stable than *cis*-2-butene by 4.2 kJ mol^{-1} (1.0 kcal mol^{-1}).

Using heats of formation to calculate the standard enthalpy difference between two compounds (Study Problem 4.8) is analogous to measuring the relative heights of two objects by comparing their distances from a common reference, such as the ceiling. If a table top is 5 ft below the ceiling and an electrical outlet is 7 ft below the ceiling, then the table top is 2 ft above the outlet. The height of the ceiling can be taken arbitrarily as zero; its absolute height is irrelevant. When heats of formation are compared, the enthalpy reference point is the enthalpy of the elements in their "standard states," their normal states at 25 °C and 1 atm pressure; the enthalpies of formation of the elements in their standard states are arbitrarily taken to be zero.

The procedure used in Study Problem 4.8 is based on the fact that *chemical reactions and their associated energies can be added algebraically*. This principle is known as **Hess's law of constant heat summation**. Hess's law is a direct consequence of the first law of thermodynamics, which requires that the energy difference between two compounds doesn't depend on the path (or reactions) used to make the measurement. Thus, what we have done in Study Problem 4.8 is to add the two formation reactions and their associated enthalpies, one in the forward direction and the other in the reverse direction:

Equations:		ΔH° (kJ mol^{-1}):	ΔH° (kcal mol^{-1}):	
	$\cancel{4\mathrm{C}} + \cancel{4\mathrm{H_2}} \longrightarrow$ *trans*-2-butene	-11.2	-2.7	(4.23a)
	cis-2-butene $\longrightarrow \cancel{4\mathrm{C}} + \cancel{4\mathrm{H_2}}$	$+7.0$	$+1.7$	(4.23b)
Sum:	*cis*-2-butene $\longrightarrow$ *trans*-2-butene	-4.2	-1.0	(4.23c)

Heats of formation are not measured directly, because the formation reaction is not practical. Rather, heats of formation are determined by combining the enthalpies of other, more practical reactions, such as combustion (Sec. 2.7A) or catalytic hydrogenation (Sec. 4.9A) using Hess's law calculations. Heats of formation are the conventional way in which these various sources of enthalpy data are brought together and tabulated. (See Further Exploration 4.2 and Focused Problem 4.14.)

Because *cis*- and *trans*-2-butene are isomers, the elements from which they are formed are the same and *cancel in the comparison*, as shown Display 4.24.

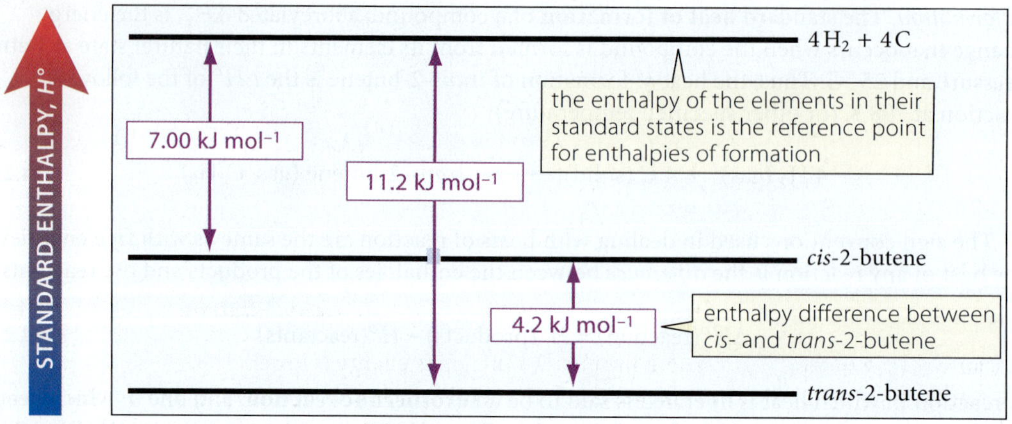

(4.24)

FURTHER EXPLORATION 4.2
Sources of Heats of Formation

Heats of formation can only be used to compare the relative energies of compounds that have the same molecular formulas—that is, isomers. Were we to compare the enthalpies of compounds that are not isomers, the two formation equations would have different quantities of carbon and hydrogen, and the sum would contain leftover C and H_2. This sum would not correspond to the direct comparison desired.

Focused Problems

4.14 (a) Calculate the enthalpy change for the hypothetical reaction 1-butene $\longrightarrow$ methylpropene. The heats of formation are 1-butene, –0.30 kJ mol^{-1} (–0.07 kcal mol^{-1}), and methylpropene, –17.3 kJ mol^{-1} (–4.13 kcal mol^{-1}).

(b) Which butene isomer in part (a) is more stable?

4.15 (a) If the standard enthalpy change for the reaction 2-ethyl-1-butene $\longrightarrow$ 1-hexene is +15.3 kJ mol^{-1} (+3.66 kcal mol^{-1}), and if the ΔH_f° for 1-hexene is –40.5 kJ mol^{-1} (–9.68 kcal mol^{-1}), what is the ΔH_f° of 2-ethyl-1-butene?

(b) Which isomer in part (a) is more stable?

4.16 The ΔH_f° of CO_2 is –393.51 kJ mol^{-1} (–94.05 kcal mol^{-1}), and the ΔH_f° of H_2O is –285.83 kJ mol^{-1} (–68.32 kcal mol^{-1}). Calculate the ΔH_f° of 1-heptene from its heat of combustion, –4693.1 kJ mol^{-1} (–1121.7 kcal mol^{-1}). (See Further Exploration 4.2.)

B. Relative Stabilities of Alkene Isomers

The heats of formation of alkenes can be used to determine how various structural features of alkenes affect their stabilities. We answer two questions using heats of formation. First, which is more stable: a cis alkene or its trans isomer? Second, how does the number of alkyl substituents at the double bond affect the stability of an alkene?

Study Problem 4.8 showed that *trans*-2-butene has a lower heat of formation than *cis*-2-butene by 4.2 kJ mol^{-1} (1.0 kJ mol^{-1}) (see Eq. 4.23c). In fact, almost all trans alkenes are more stable than their cis isomers. The reason is that, in a cis alkene, the larger groups are forced to occupy the same plane on the same side of the double bond. For example, a space-filling model of *cis*-2-butene shows that at least one hydrogen in each of the cis methyl groups is within a van der Waals radius of one or more hydrogens in the other. These hydrogens, shown in color in the following model, therefore experience *van der Waals repulsions* (Sec. 2.5B), which raise the energy of the molecule.

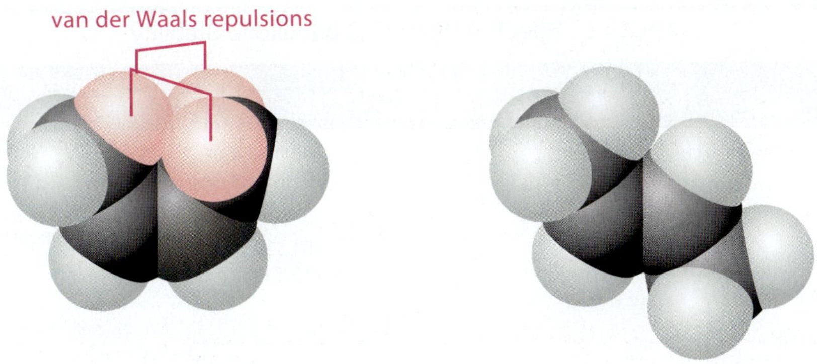

***cis*-2-butene** ***trans*-2-butene** (4.25)

Because these repulsions are absent in *trans*-2-butene, its energy is lower.

Another structural aspect of alkenes that has a considerable effect on stability is the number of alkyl groups *directly attached* to the carbons of the double bond. For example, we compare the heats of formation of two isomers in Displays 4.26a and 4.26b. The first has a single alkyl group directly attached to the double bond, whereas the second has two alkyl groups attached to the double bond.

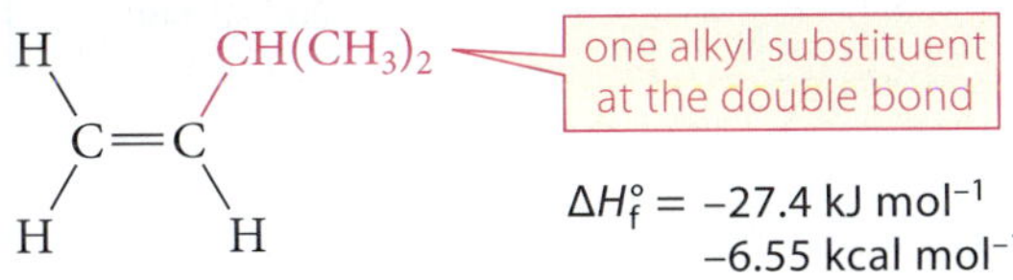

$\Delta H_f^\circ = -27.4$ kJ mol^{-1}
-6.55 kcal mol^{-1} (4.26a)

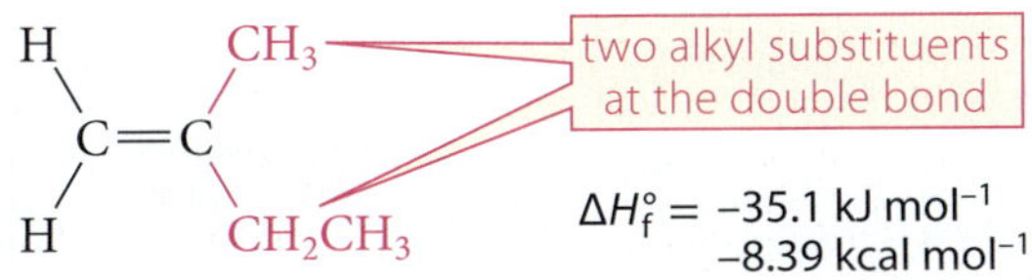

$\Delta H_f^\circ = -35.1$ kJ mol^{-1}
-8.39 kcal mol^{-1} (4.26b)

Because the isomer with two alkyl substituents at the carbon of the double bond has the more negative heat of formation, it is more stable. The data in **Table 4.1** for other isomeric pairs of alkenes show that this trend continues for increasing numbers of alkyl groups directly attached to the double bond. These data show that *an alkene is stabilized by alkyl substituents on the double bond.* When we compare the stability of alkene isomers, we find that *the alkene with the greatest number of alkyl substituents on the double bond is usually the most stable one.*

To a useful approximation, it is the *number* of alkyl groups on the double bond more than their *identities* that governs the stability of an alkene. In other words, a molecule with two smaller alkyl groups on the double bond is more stable than its isomer with one larger group on the double bond. The first two entries in Table 4.1 demonstrate this point: the second entry, (*E*)-2-hexene, with a methyl and a propyl group on the double bond, is more stable than the first entry, 1-hexene, which has a single butyl group on the double bond.

Why does alkyl-group substitution at the double bond stabilize alkenes? The answer lies in a phenomenon called **hyperconjugation**. In hyperconjugation, the electrons in an adjacent bond of an alkyl group overlap with the empty *antibonding* (π*) molecular orbital of the double bond:

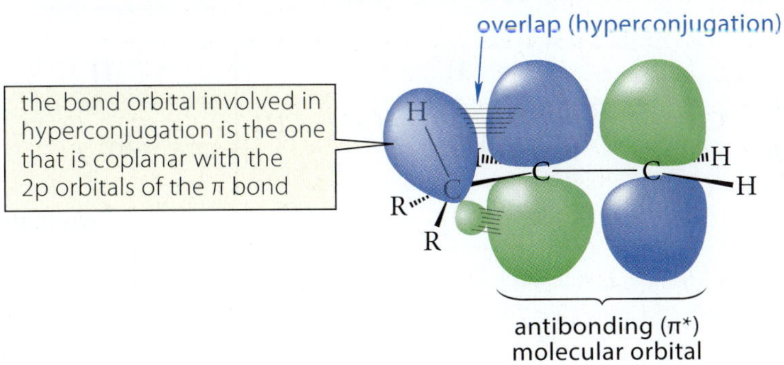

(4.27a)

TABLE 4.1 Effect of Branching on Alkene Stability

Alkene structure*	Number of alkyl groups on the double bond	ΔH_f°	Enthalpy difference
$H_2C{=}CH{-}CH_2CH_2CH_2CH_3$	1	–40.5 kJ mol^{-1} –9.68 kcal mol^{-1}	–10.6 kJ mol^{-1} –2.5 kcal mol^{-1}
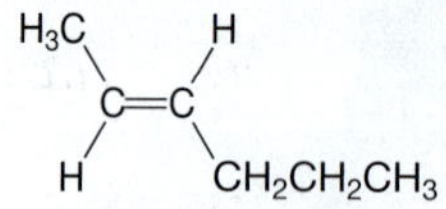	2	–51.1 kJ mol^{-1} –12.2 kcal mol^{-1}	
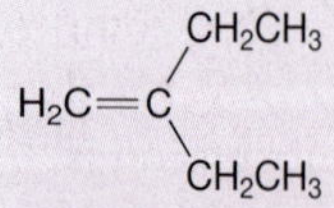	2	–55.8 kJ mol^{-1} –13.3 kcal mol^{-1}	–5.70 kJ mol^{-1} –1.4 kcal mol^{-1}
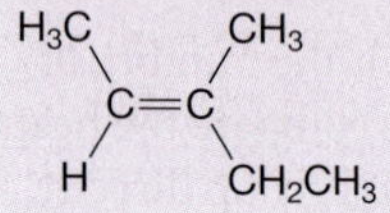	3	–61.5 kJ mol^{-1} –14.7 kcal mol^{-1}	
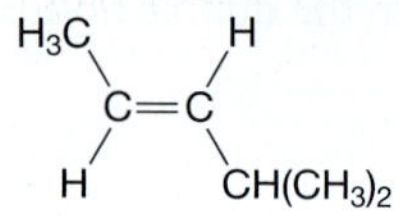	2	–60.1 kJ mol^{-1} –14.4 kcal mol^{-1}	–2.6 kJ mol^{-1} –0.6 kcal mol^{-1}
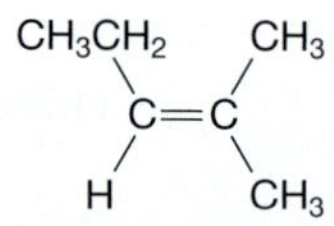	3	–62.7 kJ mol^{-1} –15.0 kcal mol^{-1}	
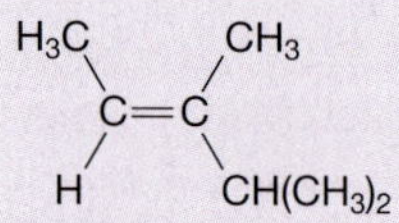	3	–88.4 kJ mol^{-1} –21.1 kcal mol^{-1}	–2.1 kJ mol^{-1} –0.5 kcal mol^{-1}
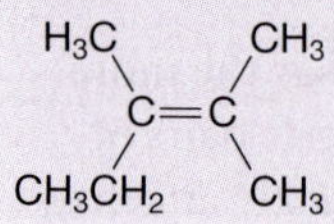	4	–90.5 kJ mol^{-1} –21.6 kcal mol^{-1}	

*In each comparison, the two compounds are equally branched; they differ only in whether the branch is at the double bond.

(Hyperconjugation involving a C—H bond of the alkyl group is shown here, but C—C bonds as well can be involved in hyperconjugation.) Because orbital overlap is involved, only one bond orbital in an alkyl group—the bond orbital that is coplanar with the 2p orbitals of the C═C bond—can participate in hyperconjugation at a time. This overlap results in additional bonding, which is a stabilizing effect.

Hyperconjugation can also be shown with resonance structures, but to do this we have to violate one of our rules for writing resonance structures—we have to break a σ bond. Electrons in the C—H bond are delocalized into the double bond:

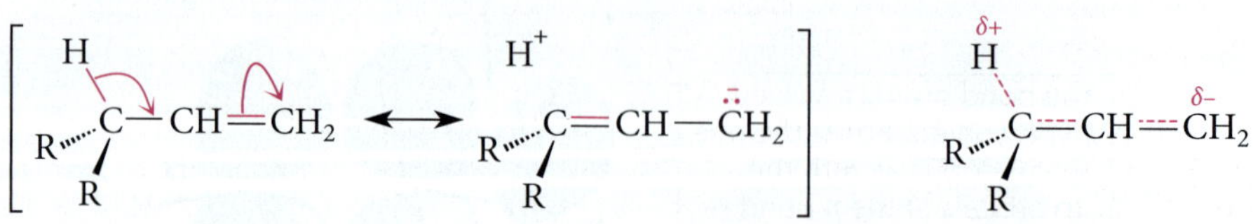

hybrid structure

(4.27b)

Carbon hyperconjugation is depicted in a similar way:

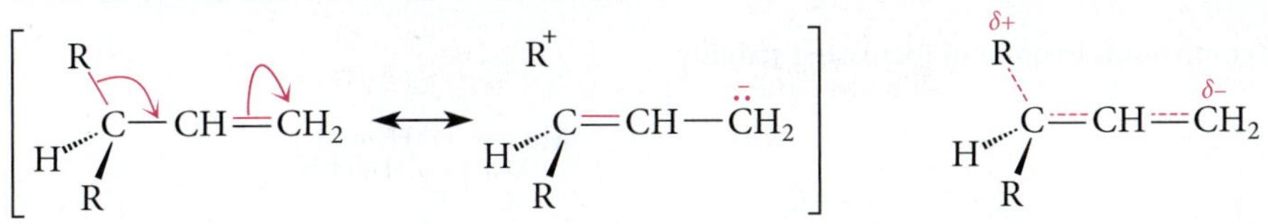

hybrid structure (4.27c)

The shared electrons are shown in color. Remember the meaning of resonance: the molecule is a single species that has some characteristics of both resonance structures. Therefore, the proton or R^+ in the right-hand structures haven't moved; they are still part of their molecules. Because the resonance structure *separates* charge, it is a relatively minor contributor, but its contribution is sufficient to have a measurable effect on energy.

We can draw analogous resonance structures for one of the C—H or C—C bonds in each alkyl substituent. In other words, each alkyl branch provides additional hyperconjugation and thus more stabilization. Consequently, alkyl substitution at alkene carbons stabilizes alkenes.

Heats of formation have given us considerable information about how alkene stabilities vary with structure. To summarize:

Increasing stability:

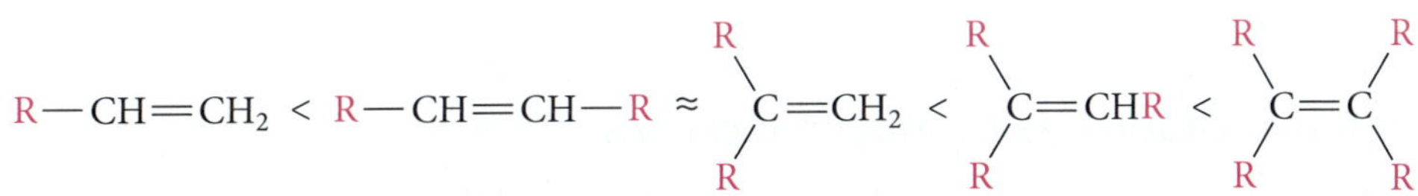

(4.28a)

and

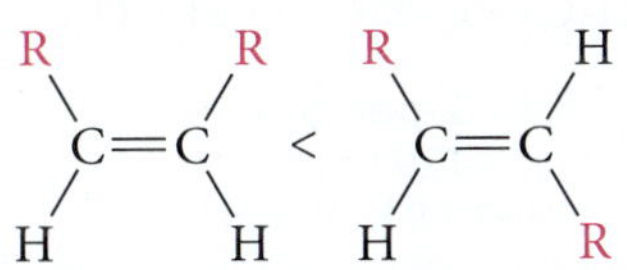

(4.28b)

Cyclic alkenes with small- and medium-sized rings are one exception to the greater stability of trans alkenes relative to their cis isomers. All cycloalkenes with six carbons or fewer exist solely as their cis isomers. If you attempt to build a model of *trans*-cyclohexene, for example, you'll see why: you can't bridge the two carbons attached to the double bond with only two other carbons.

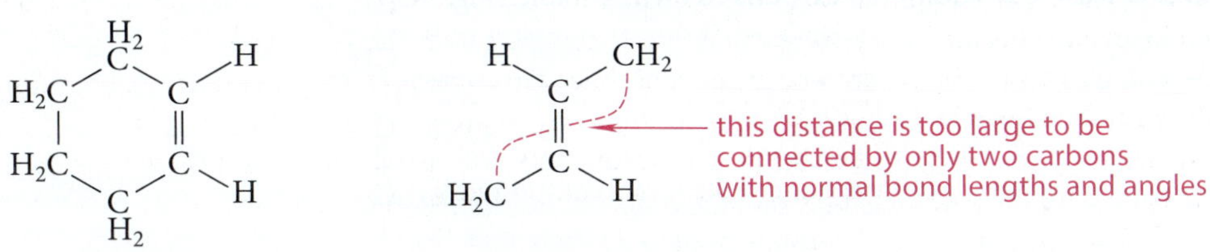

cyclohexene (a cis alkene) ***trans*-cyclohexene** (too unstable to exist) (4.29)

Trans-cyclooctene is the smallest trans cycloalkene that can exist at room temperature, and it is substantially (about 47.6 kJ mol^{-1}, 11.4 kcal mol^{-1}) less stable than its cis isomer. With a 10-membered ring (the cyclodecenes), the ring has become large enough that the trans isomer is more stable than the cis.

We justify breaking the rule for writing resonance structures, that a σ bond should not be broken, by the experimental demonstration with heats of formation that alkyl substitution has a measurable stabilizing effect on alkenes. Molecular orbital theory supports the proposition that there is a stabilizing interaction between bonding σ electrons and the antibonding π^* orbital. (See Further Exploration 4.3.) The only method we have to show this interaction with Lewis structures is to draw resonance structures.

FURTHER EXPLORATION 4.3
Molecular Orbital Description of Hyperconjugation

Focused Problems

4.17 Within each series, arrange the compounds in order of increasing stability:

(a) 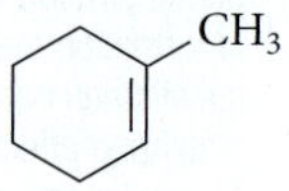

A

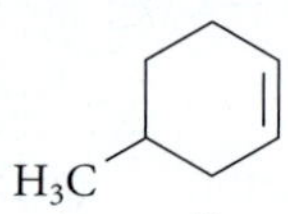

B

(b) 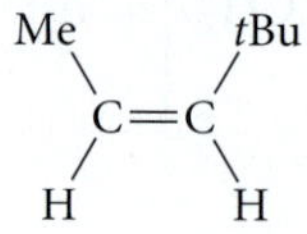

A

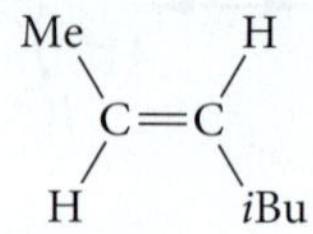

B

4.18 Alkenes can undergo the addition of hydrogen in the presence of certain catalysts. (This reaction is discussed in Sec. 4.10A.)

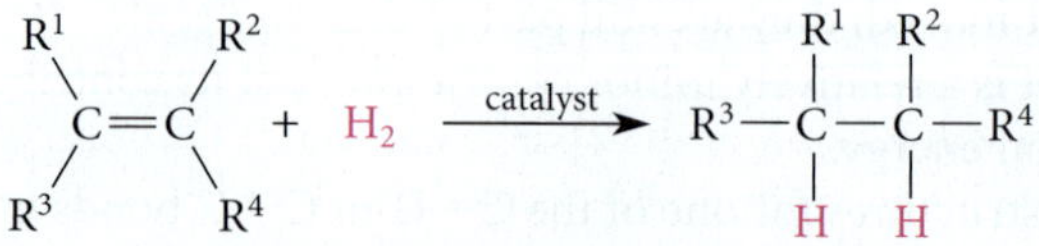

The $\Delta H°$ of this reaction, called the *enthalpy of hydrogenation*, can be measured very accurately and can serve as a source of heats of formation. Consider the following enthalpies of hydrogenation: (*E*)-3-hexene, –117.9 kJ mol^{-1} (–28.2 kcal mol^{-1}); (*Z*)-3-hexene, –121.6 kJ mol^{-1} (–29.1 kcal mol^{-1}). Calculate the heats of formation of these two alkenes, given that the $\Delta H_f°$ of hexane is –167.2 kJ mol^{-1} (–40.0 kcal mol^{-1}).

C. Relative Stabilities of Alkyne Isomers

Heats of formation show that alkynes, like alkenes, are also stabilized by alkyl substituents on the triple bond.

	$H—C{\equiv}C—CH_2CH_2CH_3$	$H_3C—C{\equiv}C—CH_2CH_3$	$H_2C{=}CH—CH_2—CH{=}CH_2$
	1-pentyne	**2-pentyne**	**1,4-pentadiene**
$\Delta H_f°$	+144 kJ mol^{-1} (+34.5 kcal mol^{-1})	+129 kJ mol^{-1} (+30.8 kcal mol^{-1})	+106 kJ mol^{-1} (+25.4 kcal mol^{-1})

The last example shows that dienes are more stable than isomeric alkynes. This means that *the sp hybridization state is inherently less stable than the sp^2 hybridization state.*

The smallest cyclic alkyne that can be isolated at room temperature is cyclooctyne. Cycloalkynes with smaller rings are unstable at room temperature, although there is evidence for the fleeting existence of cyclopentyne, cyclohexyne, and cycloheptyne. The reason for the instability of small-ring cycloalkynes is that the linear geometry of a four-carbon alkyne unit requires at least four additional carbons to form a stable ring.

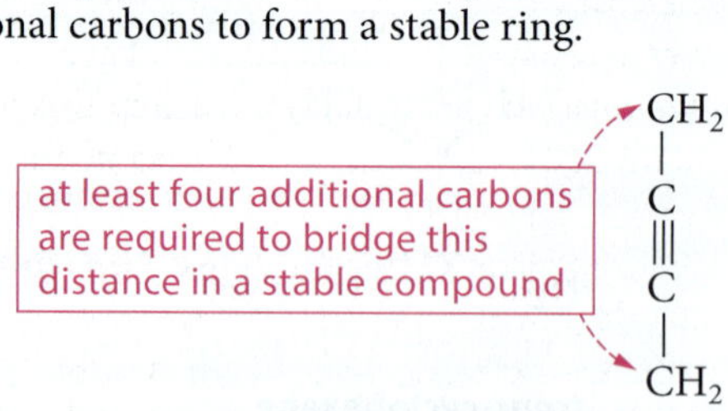

(4.30)

An attempt to build a model of a cycloalkyne with fewer than eight carbons should convince you that such compounds are very unstable.

Focused Problem

4.19 Which compound would produce more energy (heat) on combustion, 1-pentyne, 2-pentyne, or 1,4-pentadiene? (See their heats of formation given previously.)

4.6 ADDITION REACTIONS OF ALKENES AND ALKYNES

For the remainder of this chapter, as well as Chapter 5, we study examples of a characteristic reaction of alkenes and alkynes, which is **addition** to the carbon–carbon double bond and triple bond. The addition reaction can be represented generally for an alkene as follows:

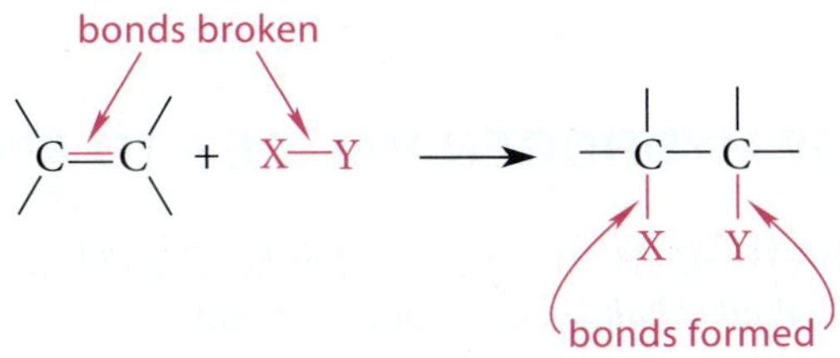

(4.31)

In an addition reaction, the carbon–carbon π bond of the alkene and the X—Y bond of the reagent are broken, and new C—X and C—Y bonds are formed.

Addition to a triple bond is conceptually similar. Because the product of the first addition has a double bond, a second addition can sometimes occur, and it could in principle occur in two different ways:

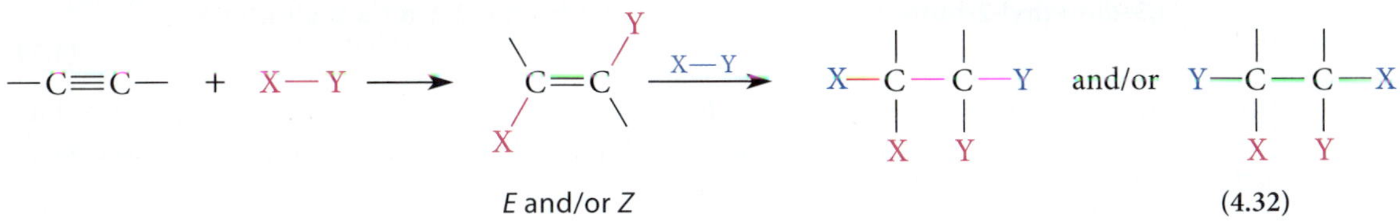

(4.32)

When an alkene is *unsymmetrical*—that is, when the two alkene carbons have different substituents—the addition can occur in two different ways when the atoms that add are different:

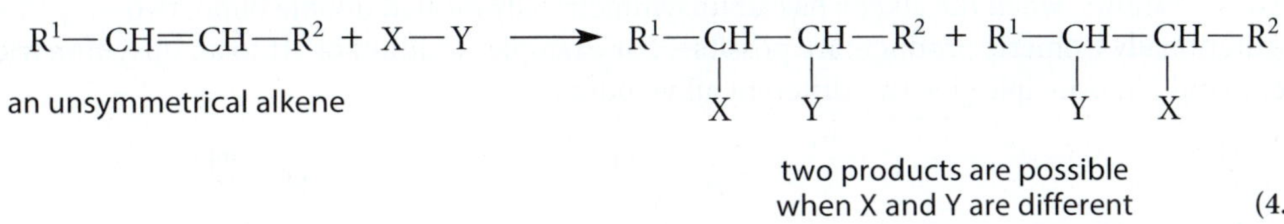

(4.33)

Unsymmetrical alkynes have the same issue, plus the possibility of a second addition shown in Eq. 4.32.

In the following section, we use the addition reaction of alkenes with hydrogen halides to establish the principles that determine how this reaction occurs in unsymmetrical alkenes—that is, whether one product is formed in significantly greater amount than the other. This discussion considers of the factors that determine the *relative rates* of competing reactions—that is, why is one reaction faster than another? To answer this question, we have to study the details of how this reaction occurs—the *reaction mechanism*. What we learn in this discussion will be generally useful in the discussions of other reaction mechanisms. The same issues are involved in addition to alkynes.

Having dealt in depth with hydrogen-halide addition, we then consider in subsequent sections two other addition reactions that deal with the concept of *catalysis* of organic reactions: catalytic hydrogenation of alkenes and alkynes, and hydration of alkenes. In hydration, we not only see a second example of catalysis, but we also put to use what we will have learned about reaction mechanisms in our discussion of hydrogen-halide addition.

Our discussion of alkene and alkyne addition reactions then continues in Chapter 5.

Focused Problem

4.20 Give the structure of the possible addition product(s) formed when

(a) Ethylene reacts with H—I

(b) Acetylene reacts with Br_2 (one equivalent)

(c) Ethylene reacts with BH_3 (*Hint:* Each of the B—H bonds undergoes an addition to one molecule of ethylene. That is, 3 mol of ethylene react with 1 mol of BH_3.)

(d) Propene reacts with H—Br (*Hint:* Two products can in principle be formed.)

4.7 ADDITION OF HYDROGEN HALIDES TO ALKENES

The hydrogen halides H—F, H—Cl, H—Br, and H—I undergo addition to carbon–carbon double bonds to give products called *alkyl halides*, compounds in which a halogen is bound to a saturated carbon atom:

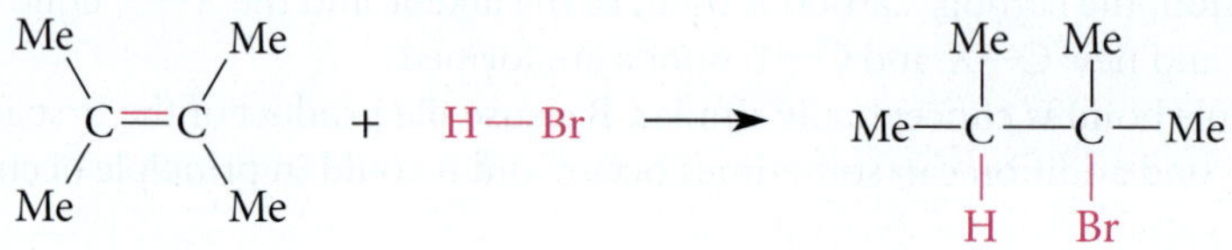

2,3-dimethyl-2-butene **2-bromo-2,3-dimethylbutane** (an alkyl halide) (4.34)

Although the addition of HF has been used for making alkyl fluorides, HF is extremely hazardous and is avoided whenever possible. Additions of HBr and HI are generally preferred to addition of HCl because additions of HBr and HI are faster.

A. Regioselectivity of Hydrogen Halide Addition

As Eq. 4.33 shows, when the alkene has an unsymmetrically located double bond, two constitutionally isomeric products are possible. For example, addition of HI to an unsymmetrical alkene could in principle give two different alkyl iodides.

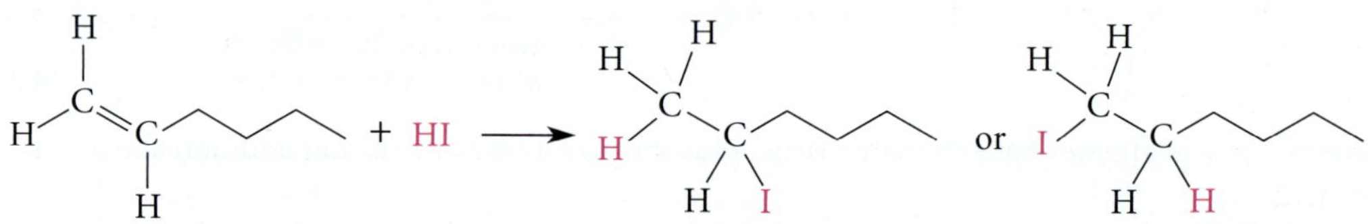

1-hexene **2-iodohexane** (observed) **1-iodohexane** (not observed) (4.35)

Only one of the two possible products is formed, however, from a 1-alkene in significant amount. Generally, *the main product is that isomer in which the halogen is bonded to the carbon of the double bond with the greater number of alkyl substituents, and the hydrogen is bonded to the carbon with the smaller number of alkyl substituents.*

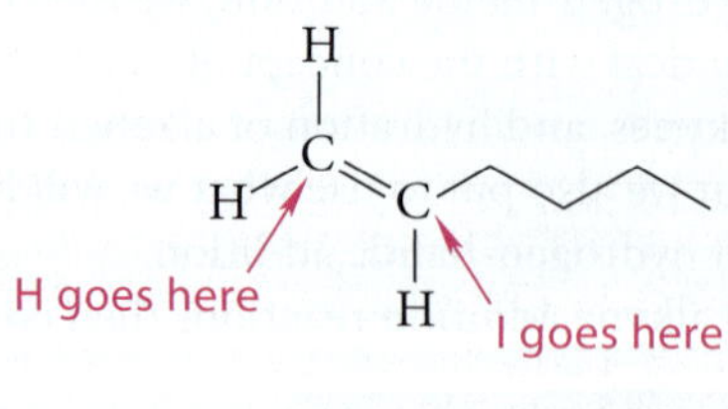

(Another way to think about this result is to apply the old saying, "Them that has, gets." That is, the carbon with more hydrogens gains yet another hydrogen in the reaction.) When the products of a reaction could consist of more than one constitutional isomer, and when one of the possible isomers is formed in excess over the other, the reaction is said to be a **regioselective reaction**. Hydrogen halide addition to alkenes is a *highly regioselective reaction* because addition of the hydrogen halide across the double bond gives only one of the two possible constitutionally isomeric addition products.

When the two carbons of the alkene double bond have equal numbers of alkyl substituents, little or no regioselectivity is observed in hydrogen halide addition, even if the alkyl groups are different.

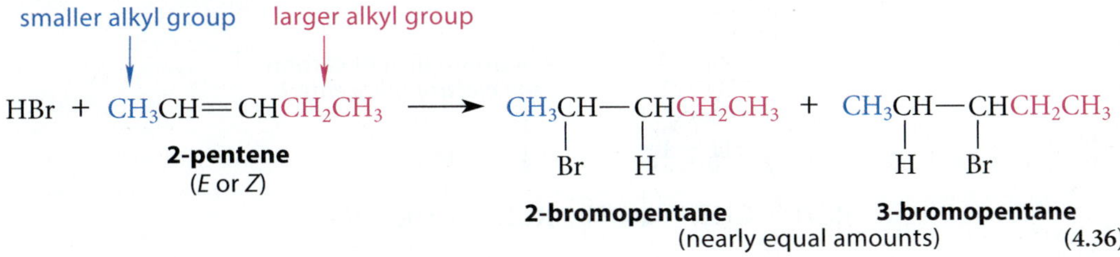

(4.36)

Markovnikov's Rule

S.Borisov/Shutterstock

In his doctoral dissertation of 1869, the Russian chemist Vladimir Markovnikov (1838–1904; also spelled Markownikoff) proposed a "rule" for regioselective addition of hydrogen halides to alkenes. This rule, which has since become known as **Markovnikov's rule**, was originally stated as follows: "The halogen of a hydrogen halide attaches itself to the carbon of the alkene bearing the lesser number of hydrogens and greater number of carbons."

Markovnikov's higher education was in political science, economics, and law. During required organic chemistry courses in the finance curriculum at the University of Kazan, he became infatuated with organic chemistry and eventually completed his now-famous doctoral dissertation. He was appointed to the chair of chemistry at the University of Moscow in 1873, where he was known not only for his chemistry, but also for his openness to students. He was forced to resign this position in 1893 because he would not sign an apology demanded of the faculty by a political official who had been insulted by a student. He was allowed, however, to continue working in the university for the duration of his life.

Focused Problem

4.21 Using the known regioselectivity of hydrogen halide addition to alkenes, predict the addition product that results from each of the following reactions:

(a) H—Cl with methylpropene (b) H—Br with 1-methylcyclohexene

B. Carbocation Intermediates in Hydrogen Halide Addition

For many years, the regioselectivity of hydrogen halide addition had only an *empirical* (experimental) basis. By exploring the underlying reasons for this regioselectivity, we set the stage to develop a broader understanding of not only this reaction but many others as well.

A modern understanding of the regioselectivity of hydrogen halide addition begins with the fact that the overall reaction actually occurs *in two successive steps*. Let's consider each of these in turn.

In the first step, the electron pair in the π bond of an alkene is donated to the proton of the hydrogen halide. The electrons of the π bond react rather than the electrons of σ bonds

The class nomenclature of species with a positively charged carbon atom has evolved. A species with an electron-deficient, positively charged, trivalent carbon was once called a **carbonium ion**. When chemists discovered $^{+}CH_5$ (which has a two-electron, three-center bond involving the carbon and two of the hydrogens), it was proposed that the class name *carbonium ion* be reserved for this type of hypervalent species, and that the class name **carbenium ion** be reserved for the species with an electron-deficient carbon. The IUPAC defines a *carbocation* as *any* species with a positively charged carbon and an even number of electrons. Thus, according to the IUPAC definition, both *carbonium ions* and *carbenium ions* are carbocations. In this text, we adopt the common practice of using the term *carbocation* to be synonymous with *carbenium ion*—a species with an electron-deficient carbon—because this is the type of carbon cation that we will almost always encounter as intermediates in organic reactions.

because π electrons have the highest energy (Sec. 4.1A). (They are also the most accessible electrons.) As a result, the carbon–carbon double bond is *protonated* on a carbon atom. The other carbon becomes positively charged and electron-deficient:

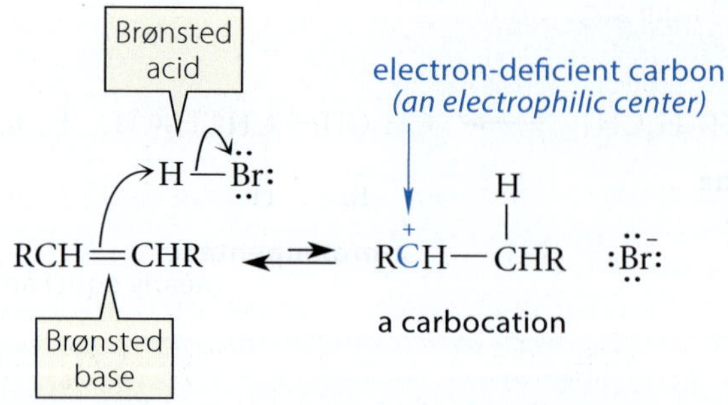

(4.37a)

A species with a positively charged carbon is called a **carbocation**, pronounced CAR-bo-CAT-ion. The carbocation formed in this reaction has an *electron-deficient carbon*. It is formed from the alkene in *a Brønsted acid–base reaction* (Sec. 3.4A) in which the π bond acts as a *Brønsted base* toward the *Brønsted acid* H—Br. The π bond is a *very weak* base. Nevertheless, it can be protonated to a small extent by a strong acid such as HBr.

The resulting carbocation is a powerful electron-deficient Lewis acid and is thus a potent electrophile. In the second step of hydrogen halide addition, the halide ion, which is a Lewis base, or nucleophile, reacts with the carbocation at its electron-deficient carbon atom:

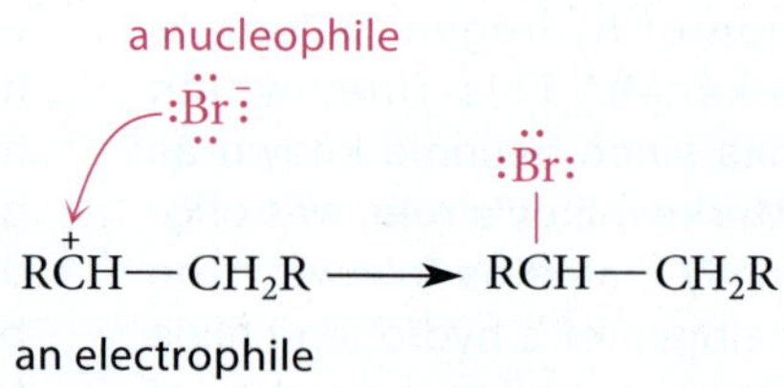

(4.37b)

This is a Lewis acid–base association reaction (Sec. 3.1B).

The carbocations involved in hydrogen halide addition to alkenes are examples of **reactive intermediates**, species that react so rapidly that they never accumulate in more than very low concentration. Most carbocations are too reactive to be isolated except under special circumstances. Therefore, carbocations cannot be isolated from the reactions of hydrogen halides and alkenes because they react instantaneously with halide ions.

The complete description of a reaction pathway, including any reactive intermediates such as carbocations, is called the **mechanism** of the reaction. To summarize the two steps in the mechanism of hydrogen halide addition to alkenes:

1. A carbon of the π bond is protonated (a Brønsted acid–base reaction).

2. A halide ion reacts with the resulting carbocation (a Lewis acid–base association reaction).

Now that we understand the mechanism of hydrogen halide addition to alkenes, let's see how the mechanism addresses the question of regioselectivity. When the double bond of an alkene is not located symmetrically within the molecule, protonation of the double bond can occur in two distinguishable ways to give two different carbocations. For example, protonation of methylpropene can give either the *tert*-butyl cation (Eq. 4.38a) or the isobutyl cation (Eq. 4.38b):

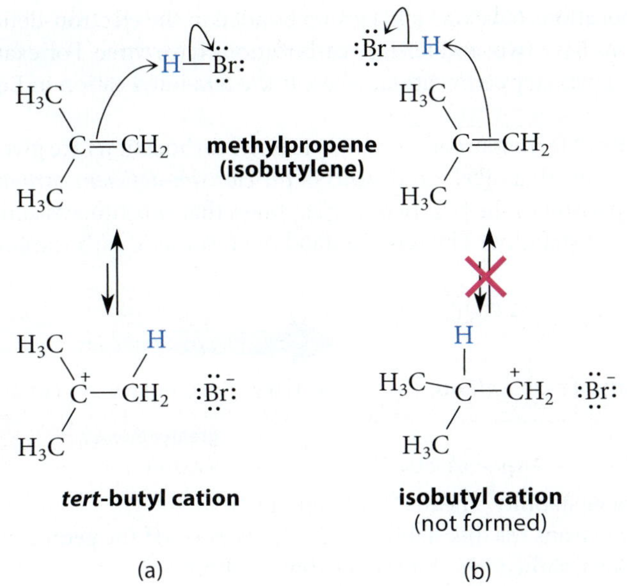

(4.38)

These two reactions are in *competition*—that is, one can only happen at the expense of the other because the two reactions compete for the same starting material. Only the *tert*-butyl cation is formed in this reaction. The *tert*-butyl cation is formed exclusively because reaction 4.38a is *much faster* than reaction 4.38b. Because the *tert*-butyl cation is the only carbocation formed, it is the only carbocation available to react with the bromide ion. Therefore, the only product of HBr addition to methylpropene is *tert*-butyl bromide.

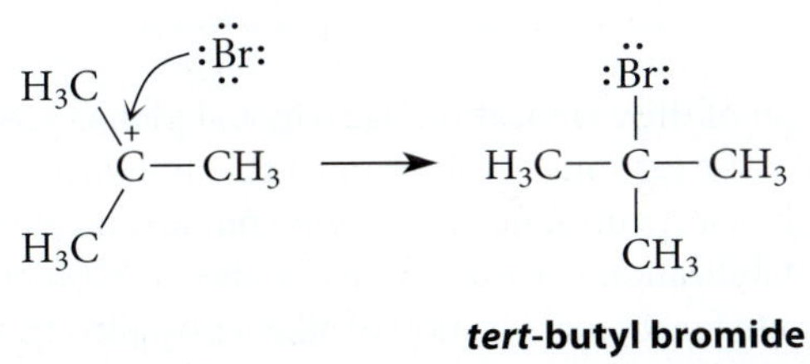

(4.39)

In this step, the bromide ion has become attached to the carbon of methylpropene bearing the greater number of alkyl groups. In other words, *the regioselectivity of hydrogen halide addition is due to the formation of only one of two possible carbocations.*

To understand why the *tert*-butyl cation is formed more rapidly than the isobutyl cation in HBr addition, we need to understand the factors that influence reaction rate. The relative stability of carbocations plays an important role in understanding the rate of HBr addition. A discussion of carbocation stability, then, is an essential prelude to a more general discussion of reaction rates.

C. Structure and Stability of Carbocations

Carbocations are classified by the degree of alkyl substitution at their electron-deficient carbon atoms.

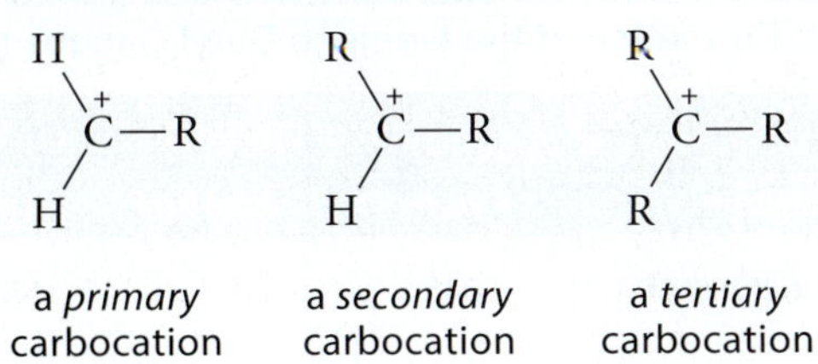

(4.40)

That is, primary carbocations have one alkyl group bonded to the electron-deficient carbon, secondary carbocations have two, and tertiary carbocations have three. For example, the isobutyl cation in Eq. 4.38b is a primary carbocation, whereas the *tert*-butyl cation in Eq. 4.38a is a tertiary carbocation.

The gas-phase heats of formation of the isomeric butyl carbocations are given in **Table 4.2.** The data in this table show that *alkyl substituents at the electron-deficient carbon strongly stabilize carbocations.* (A comparison of the first two entries shows that substituents at other carbons have a much smaller effect on stability.) The relative stability of isomeric carbocations is therefore as follows.

Stability of carbocations:

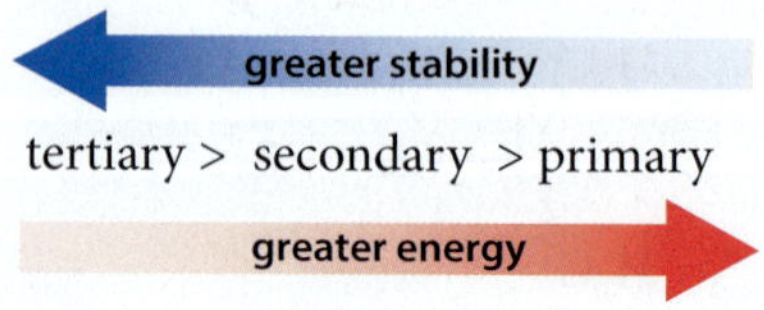

(4.41)

(Remember that "greater stability" means "lower energy.")

To understand the reasons for this stability order, consider first the geometry and electronic structure of carbocations, shown here for the *tert*-butyl cation:

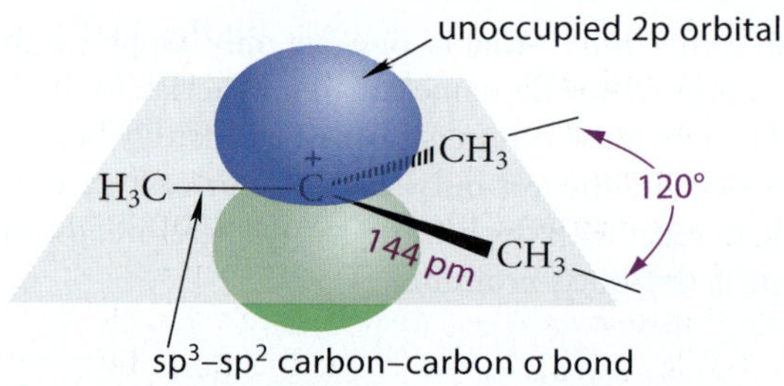

(4.42)

The electron-deficient carbon of the carbocation has trigonal planar geometry (Sec. 1.6C) and is therefore sp^2-hybridized, like the carbons involved in double bonds (Sec. 4.1A); in a carbocation, however, the 2p orbital on the electron-deficient carbon contains no electrons.

The explanation for the stabilization of carbocations by alkyl substituents is *hyperconjugation*, the same explanation that accounts for the stabilization of alkenes by alkyl substitution (Sec. 4.5B). In this case, hyperconjugation involves the overlap of bonding electrons from the adjacent σ bonds (in this case, the neighboring C—H bond) with the vacant 2p orbital of the carbocation:

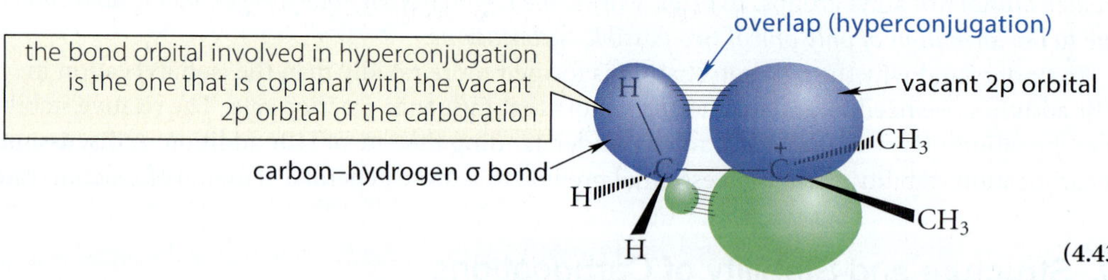

(4.43)

TABLE 4.2 Heats of Formation of the Isomeric Butyl Cations (Gas Phase, 25 °C)

Cation structure	Name	Heat of formation		Relative energy*	
		kJ mol^{-1}	kcal mol^{-1}	kJ mol^{-1}	kcal mol^{-1}
$CH_3CH_2CH_2\overset{+}{C}H_2$	butyl cation	845	202	155	37
$(CH_3)_2CH\overset{+}{C}H_2$	isobutyl cation	828	198	138	33
$CH_3\overset{+}{C}HCH_2CH_3$	*sec*-butyl cation	757	181	67	16
$(CH_3)_3\overset{+}{C}$	*tert*-butyl cation	690	165	(0)	(0)

*Energy difference between each carbocation and the more stable *tert*-butyl cation

The energetic advantage of hyperconjugation is that it involves additional bonding. That is, the electrons of the C—H bonds participate in bonding not only with the C and H, but also with the electron-deficient carbon. Additional bonding is a stabilizing effect.

As with hyperconjugation in alkenes, we can show this hyperconjugation with resonance structures:

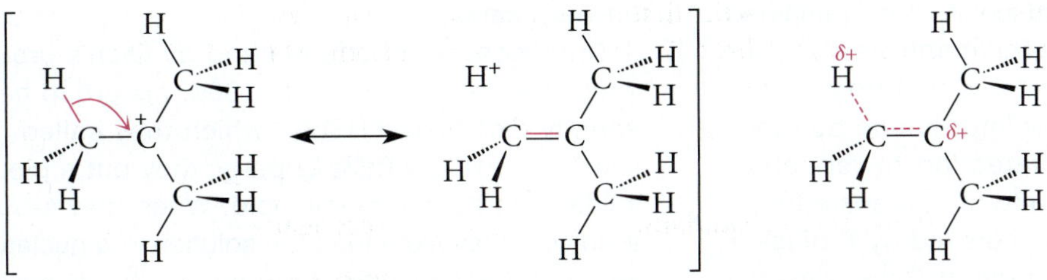

hybrid structure (4.44)

The shared electrons are shown in color. (Similar structures can be drawn for a C—H bond in each of the other methyl groups.) The double-bond character suggested by the resonance structure on the right is reflected in the lengths of the carbon–carbon bonds in the *tert*-butyl cation. These bonds are considerably shorter (144 pm) than the carbon–carbon single bond in propene (150 pm). For primary and secondary carbocations, fewer alkyl groups, and therefore fewer C—H bonds, are available to stabilize the carbocation by hyperconjugation. Therefore, primary and secondary carbocations are less stable than tertiary carbocations.

A comparison of this resonance structure with the one for hyperconjugation in alkenes (Display 4.27b) shows that both resonance structures violate the guideline that we generally don't break σ bonds when drawing resonance structures; the effects of alkyl substitution on both alkene and carbocation stability justifies this practice. For carbocations, however, resonance structures *delocalize* charge, whereas the structures for alkene hyperconjugation *separate* opposite charges. (See Sec. 3.3C, guideline 3.) Therefore, the resonance structures for carbocation hyperconjugation are relatively more important. As we can see by comparing the effects of alkyl substitution in alkenes (see Table 4.1) and carbocations (see Table 4.2), the effect of alkyl substitution on carbocations is more than 10 times as great.

When the electron-deficient carbon of a carbocation is substituted with alkyl groups larger than methyl, carbon hyperconjugation is as important for carbocations as it is for alkenes.

It is a C—H or C—C bond on an alkyl substituent attached to the electron-deficient carbon that is involved in hyperconjugation, *not* a bond to the electron-deficient carbon itself.

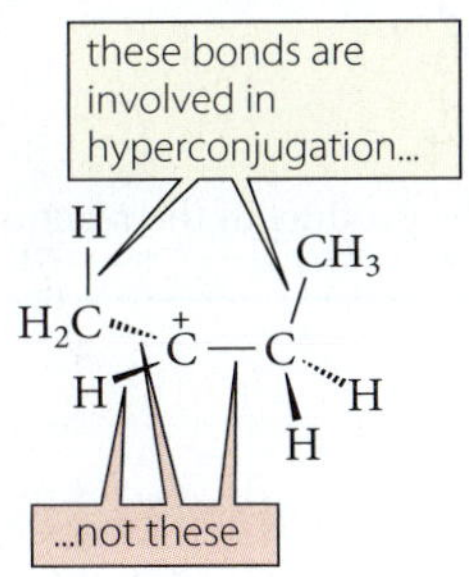

FURTHER EXPLORATION 4.3
Molecular Orbital Description of Hyperconjugation

Let's now bring together what you've learned about carbocation stability and the mechanism of hydrogen halide addition to alkenes. The addition occurs in two steps. In the first step, protonation of the alkene double bond occurs *at the carbon with the fewer alkyl substituents* so that *the more stable carbocation is formed*—that is, *the carbocation with the greater number of alkyl substituents at the electron-deficient carbon*. The reaction is completed when the halide ion, in a second step, reacts with the electron-deficient carbon.

An understanding of many organic reactions hinges on an understanding of the reactive intermediates involved. Carbocations are important reactive intermediates that occur not only in the mechanism of hydrogen halide addition, but also in the mechanisms of many other reactions. Therefore, your knowledge of carbocations will be put to use often.

George Olah, a Holiday Party, and a Nobel Prize

Lee Atherton/Superstock

Because carbocations are in most cases reactive intermediates that are too unstable to isolate, they remained hypothetical for many years after their existence was first postulated. However, their importance as reactive intermediates led to repeated, unsuccessful efforts to prepare them. In 1966–1967, a team of researchers led by Professor George A. Olah (1927–2017), then at Case Western Reserve University, figured out how to prepare a number of pure carbocation salts in solution and studied their properties. For example, they formed a solution of *tert*-butyl cation by protonation of methylpropene at –80 °C. As the acid, they used HF in the presence of the powerful Lewis acid SbF_5, which actually forms the very strong acid $H^{+}\,{}^{-}SbF_6$.

$$\underset{\textbf{methylpropene}}{(CH_3)_2C{=}CH_2} + H^{+}\,{}^{-}SbF_6 \xrightarrow[\text{ClSO}_2\text{F (solvent)}]{-80\,^\circ\text{C}} \underset{\textbf{\textit{tert}-butyl cation hexafluoroantimonate salt}}{(CH_3)_3C^{+}\,{}^{-}SbF_6} \quad (4.45)$$

The fluoride ion is so tightly held within the ${}^{-}SbF_6$ complex ion that it can't act as a nucleophile toward the *tert*-butyl cation.

The discovery of this method by Olah's group was somewhat serendipitous. In 1966, his group had only recently discovered $HSbF_6$, which they called "magic acid." Following a holiday party, they put a piece of a holiday candle into magic acid. When they saw that it dissolved, they examined the solution in a nuclear magnetic resonance (NMR) instrument. (We discuss NMR as a powerful method for determining structure in Chapter 14.) They saw unmistakable evidence for carbocations. This was the beginning of a productive series of investigations in which a number of carbocations were generated and examined structurally. Subsequently, Olah joined the faculty of the University of Southern California. He received the 1994 Nobel Prize in Chemistry for his work in carbocation chemistry.

Focused Problems

4.22 Classify the isomeric carbocations in each of the following parts as primary, secondary, or tertiary, and tell which is the most stable carbocation in each part and why.

(a)

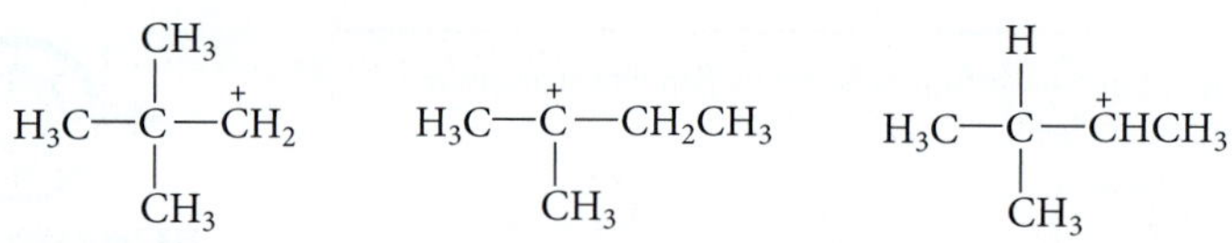

A B C

(b)

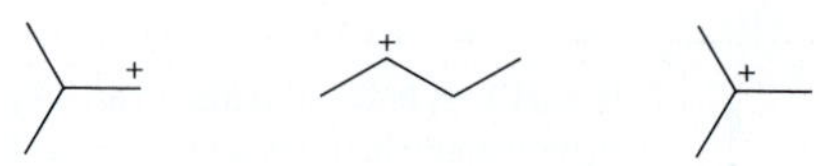

A B C

4.23 By writing the curved-arrow mechanism of the reaction, predict the product of the reaction of HBr with 2-methyl-1-pentene.

Study Problem 4.9

Draw the structure of an alkene that would give 2-bromopentane as the major (or sole) product of HBr addition. (The numbers are for reference in the solution.)

$$\text{an alkene} + \text{HBr} \longrightarrow \underset{\textbf{2-bromopentane}}{CH_3CH_2\overset{3}{C}H_2\overset{2}{C}H(Br)\overset{1}{C}H_3}$$

Solution The bromine of the product comes from the H—Br. However, there are many hydrogens in the product! Which ones were there to start with, and which one came from the H—Br? First, recognize that the carbon bearing the bromine must have originally been one carbon of the double bond. It then follows that the other carbon of the double bond must be an *adjacent* carbon (because two carbons involved in the same double bond must be adjacent). Use this fact to construct *all possible alkenes* that *might* be starting materials. Do this by "thinking in reverse": Remove the bromine and a hydrogen from *each adjacent carbon in turn.*

remove Br from carbon-2 and H from carbon-3 ⇨ $CH_3CH_2CH{=}CHCH_3$
2-pentene (cis or trans)

remove Br from carbon-2 and H from carbon-1 ⇨ $CH_3CH_2CH_2CH{=}CH_2$
1-pentene

(The symbol ⇨ means "implies as starting material.") Which of these is correct? Or are they both correct? You haven't finished the problem until you've mentally carried out the addition of HBr to *each compound.* Applying the known regioselectivity of HBr addition leads to the conclusion that the desired alkyl halide could be prepared as the major product from 1-pentene. However, both carbons of the double bond of 2-pentene bear the same number of alkyl groups. Equation 4.36 indicates that from this starting material we should expect not only the desired product, but also a second product, because both would involve secondary carbocations of similar stability:

$$\underset{\textbf{2-pentene}}{CH_3CH_2CH{=}CHCH_3} + HBr \longrightarrow \underset{\textbf{2-bromopentane}}{CH_3CH_2CH_2CH(Br)CH_3} + \underset{\textbf{3-bromopentane}}{CH_3CH_2CH(Br)CH_2CH_3}$$

Furthermore, the two products should be formed in nearly equal amounts. This means the yield of the desired compound would be relatively low and it would be difficult to separate from its isomer, which has almost the same boiling point. Consequently, 1-pentene is the only alkene that will give the desired alkyl halide as the *major* product (that is, the one formed almost exclusively).

In solving this type of problem, it isn't enough to identify potential starting materials. You must also determine whether they really will work, given the known characteristics—in this case, the regioselectivity—of the reaction.

Focused Problem

4.24 In each case, give *two different* alkene starting materials that would react with H—Br to give the compound shown as the major (or only) addition product.

(a) 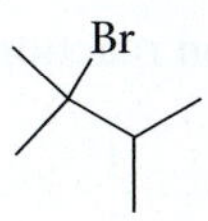

(b) 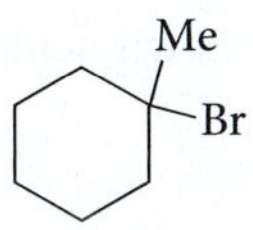

D. Carbocation Rearrangement in Hydrogen Halide Addition

In some cases, the addition of a hydrogen halide to an alkene gives an unusual product, as in the following example.

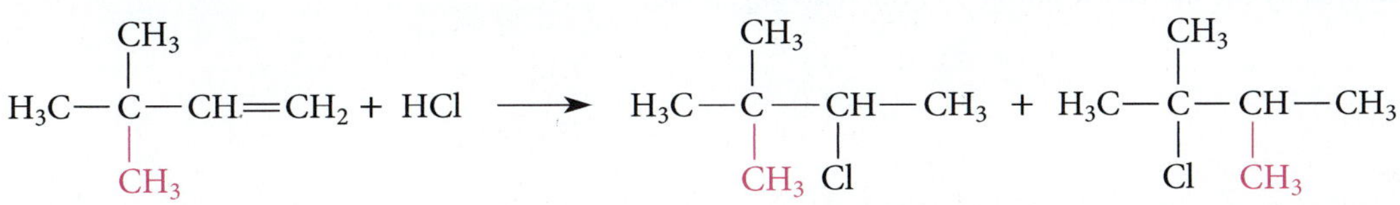

3,3-dimethyl-1-butene (17% of product) (83% of product)

(4.46)

The minor product is the result of ordinary regioselective addition of HCl across the double bond. The origin of the major product, however, is not obvious. Examination of the carbon skeleton of the major product shows that a *rearrangement* has occurred. In a **rearrangement**, a group from the starting material has moved to a different position in the product. In this case, a methyl group of the alkene (*color*) has changed positions. As a result, the carbons of the alkyl halide product are connected differently from the carbons of the alkene starting material. Although the rearrangement leading to the second product may seem strange at first sight, it is readily understood by considering the fate of the carbocation intermediate in the reaction.

The reaction begins like a normal addition of HCl—that is, by protonation of the double bond to yield the carbocation with the greater number of alkyl substituents at the electron-deficient carbon.

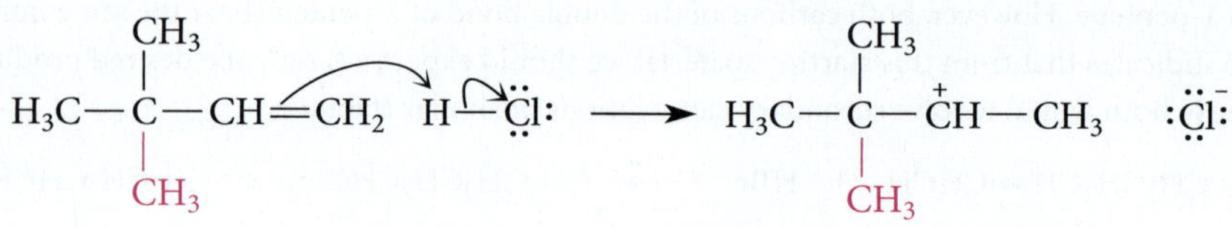

a secondary carbocation (4.47)

Reaction of this carbocation with Cl^- occurs, as expected, to yield the minor product of Eq. 4.46. However, the carbocation can also *rearrange*.

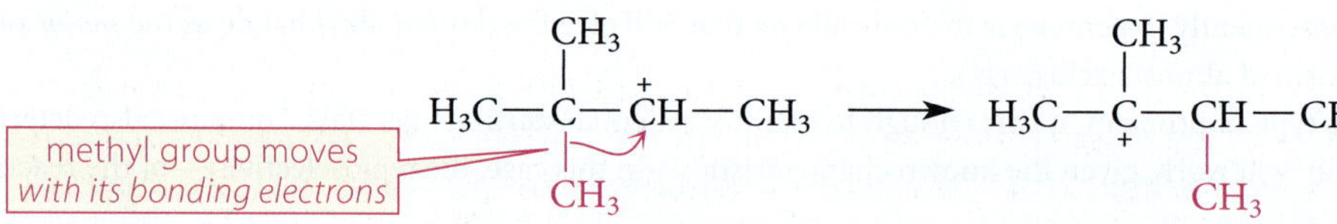

a secondary carbocation a tertiary carbocation (4.48a)

In this reaction, the methyl group moves *with its pair of bonding electrons* from the carbon *adjacent* to the electron-deficient carbon. The carbon from which this group departs, as a result, becomes electron-deficient and positively charged. That is, the rearrangement converts one carbocation into another. This is essentially a *Lewis acid–base reaction* in which the electron-deficient carbon is the Lewis acid and the migrating group *with its bonding electron pair* is the Lewis base. The reaction forms a new Lewis acid—the electron-deficient carbon of the rearranged carbocation.

The major product of Eq. 4.46 is formed by the Lewis acid–base association reaction of Cl^- with the new carbocation.

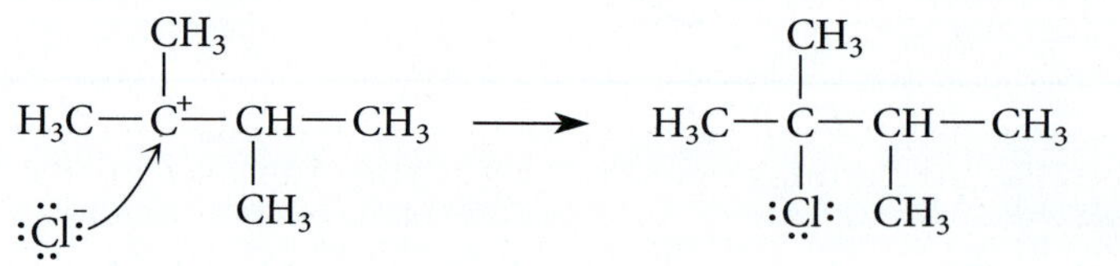

(4.48b)

Why does rearrangement of the carbocation occur? In the case of reaction 4.48a, a more stable tertiary carbocation is formed from a less stable secondary one. Therefore, *rearrangement is favored by the increased stability of the rearranged ion.*

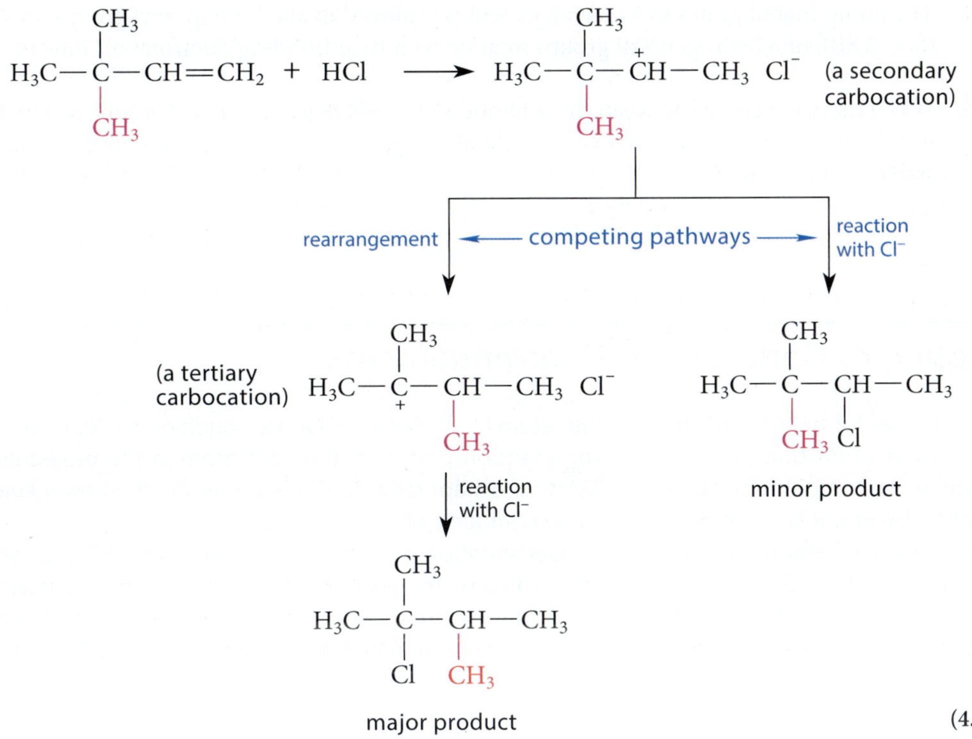

(4.49)

You've now learned two pathways by which carbocations can react. They can (1) react with a nucleophile and (2) rearrange to more stable carbocations. As Eq. 4.49 shows, the outcome of Eq. 4.46 reflects a competition between these two pathways. In any particular case, you cannot predict exactly how much of each different product will be obtained. Nevertheless, the reactions of carbocation intermediates show why both products are reasonable.

Carbocation rearrangements are not limited to the migrations of alkyl groups. In the following reaction, the major product is also derived from the rearrangement of a carbocation intermediate. This rearrangement involves a **hydride shift**, the migration of a hydrogen *with its two bonding electrons.*

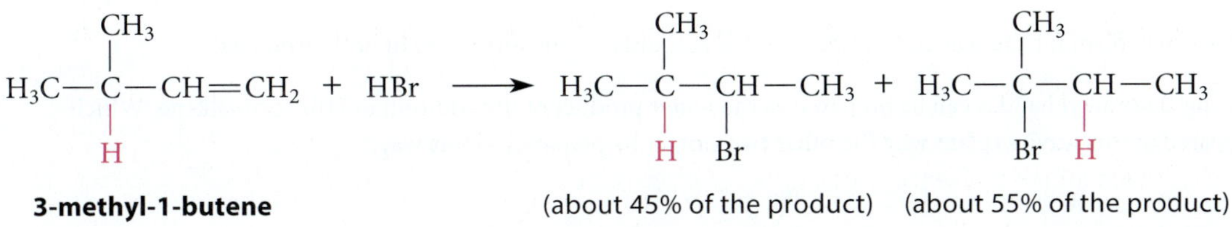

(4.50)

The hydride migrates instead of an alkyl group because the rearranged carbocation is tertiary and is more stable than the starting carbocation. Migration of an alkyl group would have given another secondary carbocation.

Keep in mind the following points about the rearrangement of carbocation intermediates, all of which are illustrated by the examples in this section:

1. A rearrangement almost always occurs when a more stable carbocation can result.

2. A rearrangement that would give a less stable carbocation generally doesn't occur.

3. The group that migrates in a carbocation rearrangement comes from a carbon *directly bonded* to the electron-deficient, positively charged carbon of the carbocation.

4. The group that migrates in a rearrangement is typically an alkyl group, aryl group (Sec. 2.8B), or a hydrogen. All groups migrate with their bonding electron pair intact.

5. When there is a choice between the migration of an alkyl group (or aryl group) or a hydrogen from a particular carbon, hydride migration typically occurs because it gives the more stable carbocation.

The First Description of Carbocation Rearrangements

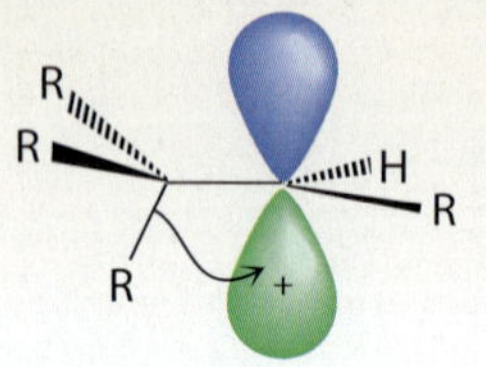

The first clear formulation of the involvement of carbocations in molecular rearrangements was proposed by Frank C. Whitmore (1887–1947) of Pennsylvania State University. (Such rearrangements were once called "Whitmore shifts.") Whitmore said that carbocation rearrangements result when "an atom in an electron-hungry condition seeks its missing electron pair from the next atom in the molecule." Whitmore's apt description emphasizes the Lewis acid–base character of the reaction.

Carbocation rearrangements are not just a laboratory curiosity; they occur extensively in living organisms, particularly in the biological pathways leading to certain cyclic compounds such as steroids (Sec. 17.6C).

Focused Problems

4.25 Which of the following carbocations is likely to rearrange? If rearrangement occurs, give the structures of the rearranged carbocations.

(a) 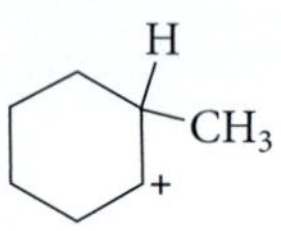

(b) 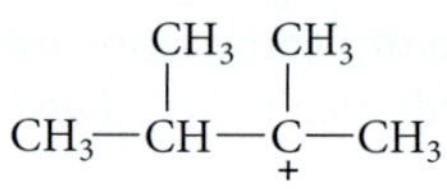

(c) 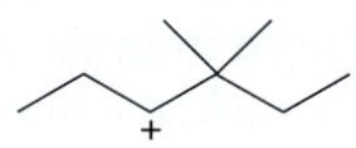

4.26 Draw the curved-arrow mechanism for the reaction in Eq. 4.50 that accounts for the formation of both products.

4.27 Only one of the following three alkyl halides can be prepared as the *major* product of the addition of HBr to an alkene. Which compound can be prepared in this way? Explain why the other two *cannot* be prepared in this way.

A

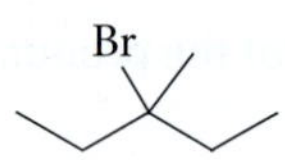

B

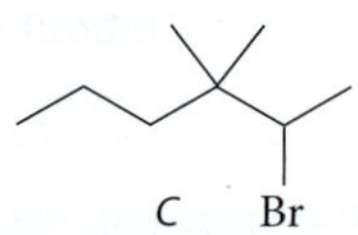

C

4.8 REACTION RATES

Whenever a reaction can give more than one possible product, two or more reactions are in competition. (You've already seen examples of competing reactions in hydrogen halide addition to alkenes.) One reaction predominates when it occurs *more rapidly* than other competing reactions. Understanding why some reactions occur in preference to others, then, is often a matter of understanding the *rates* of chemical reactions. The theoretical framework for discussing reaction rates is the subject of this section. Although we use hydrogen halide addition to alkenes as our example to develop the theory, the general concepts are used throughout this text.

A. The Transition State

The **rate** of a chemical reaction can be defined for our purposes as the number of reactant molecules converted into product in a given time. The theory of reaction rates used by many organic chemists postulates that as the reactants change into products, they pass through an unstable state of maximum free energy, called the **transition state**. The transition state has a higher energy than either the reactants or products and therefore represents an **energy barrier** to their interconversion. This energy barrier is shown graphically in a **reaction free-energy diagram** (**Fig. 4.4**). This is a diagram of the standard free energy of a reacting system as old bonds break and new ones form along the reaction pathway. In this diagram the progress of reactants to products is called the **reaction coordinate**. That is, the reactants define one end of the reaction coordinate, the products define the other, and the transition state is at the energy maximum somewhere in between. The energy barrier $\Delta G^{\circ\ddagger}$, called the **standard free energy of activation**, is equal to the difference between the standard free energies of the transition state and the reactants. (The double dagger, ‡, is the symbol used for transition states.) The size of the energy barrier $\Delta G^{\circ\ddagger}$ determines the rate of a reaction: *the higher the barrier, the slower the reaction*. Therefore, the reaction diagrammed in Fig. 4.4a is slower than the one in Fig. 4.4b because it has a higher energy barrier. In the same sense that relative free energies of reactants and products determine the equilibrium constant, the relative free energies of the transition state and the reactants determine the reaction rate.

Notice from Fig. 4.4 that *a reaction and its reverse have the same transition state*. An analogy is that if a certain mountain pass is the shortest way to get from one town to another, then the same mountain pass is the shortest way to make the reverse journey.

Because the transition state has central importance in determining the reaction rate, it would be useful to have a way of estimating its energy. Because it lies at an energy *maximum*, a transition state can't be isolated. However, the energy of a transition state, like the energy of any ordinary molecule, depends on its structure. So, what does the transition state look like? The power of transition-state theory is that we can visualize the transition state as a *structure*. To illustrate, consider the following Brønsted acid–base reaction.

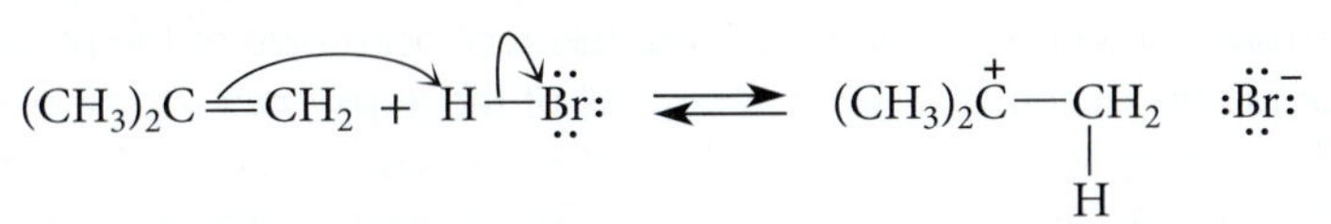

(4.51)

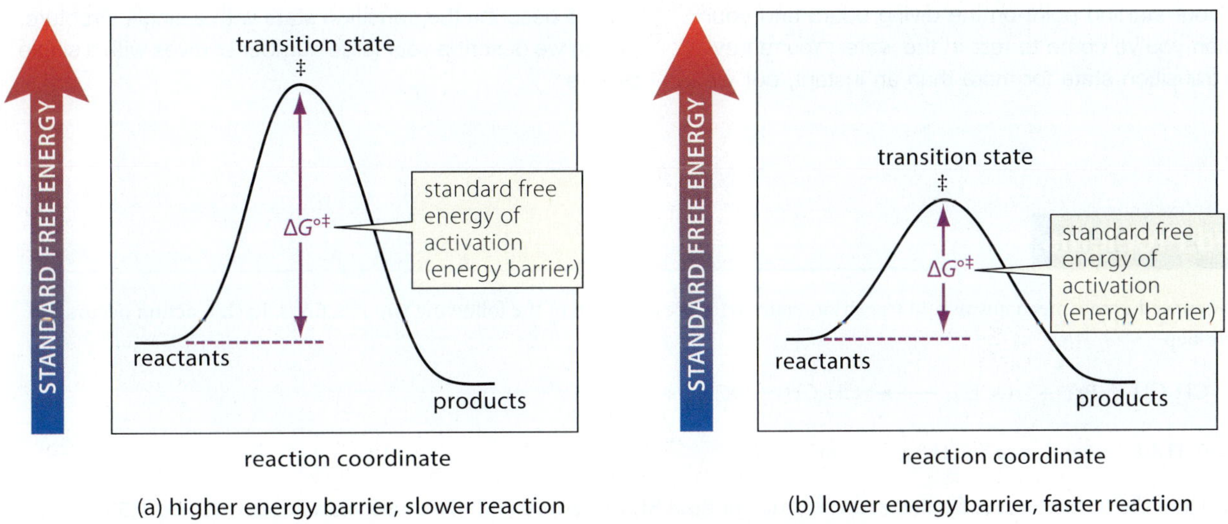

FIGURE 4.4 Reaction free-energy diagrams for two hypothetical reactions. The standard free energy of activation ($\Delta G^{\circ\ddagger}$), shown for the forward reaction, is the energy barrier that must be overcome for the reaction to occur. The reaction in part (a) is intrinsically slower because it has a larger $\Delta G^{\circ\ddagger}$ than the one in part (b).

This is the first step in the addition of HBr to an alkene; see Eq. 4.37a. In the transition state of this reaction, the H—Br bond and the carbon–carbon π bond are partially broken, the new C—H bond is partially formed, and the new charges are only partially established. We can represent this situation using dashed lines for partial bonds and using $\delta+$ and $\delta-$ for partial charges, as follows:

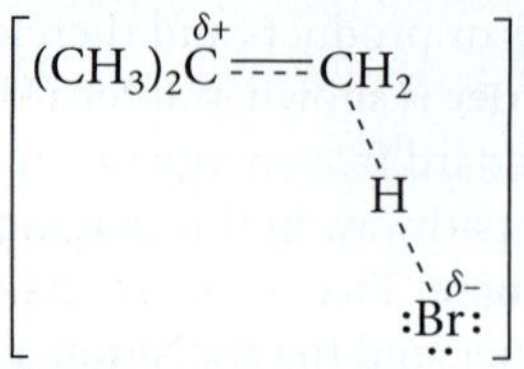

(4.52)

This shows the bonds breaking and forming. If we view this as an event frozen in time, we're looking at the structure of the transition state. The double dagger (‡) outside of the brackets indicates that the structure is a transition state.

If you can draw the curved-arrow notation for a reaction, you can draw a transition state. The curved arrows that we use to symbolize electron flow tell you which bonds are broken and where new bonds are formed. If you replace each of the bonds broken and bonds formed with a partial bond—that is, a dashed line—and replace charges in the reactant and product with partial charges, you will have drawn the transition state for the reaction.

An Analogy for the Transition State

Matthew Stockman/ Getty Images

Suppose you do a somersault off of a high diving board at the local pool. You have someone take your picture at the height of your jump with a very fast shutter. You can think of that picture as the transition state for your dive. It's the "in-between" state that defines the point of highest potential energy between your starting point on the diving board and your finish, when you've come to rest in the water. You're never at the transition state for more than an instant, but we don't have any problem conceptualizing how you would appear at that instant. Now imagine repeating your dive an Avogadro's number of times (you never get tired) and having a picture taken of each dive at the highest point. Now you average all of those pictures. Because each dive is a little different, the averaged picture of the transition state is a bit blurry. This averaged picture of your dive is more like what we are dealing with when we talk about the transition state for a mole of molecules, but chances are that the first picture you took is not far from the average. So, we describe the transition state with a single structure, just as we describe your large number of dives with a single picture.

Focused Problems

4.28 Draw curved-arrow mechanisms and transition-state structures for each of the following two reactions. Each reaction occurs as a single step.

(a) $CH_3CH_2{-}\ddot{\underset{..}{Br}}\colon + \colon\ddot{\underset{..}{O}}CH_3^{-} \longrightarrow CH_3CH_2{-}\ddot{\underset{..}{O}}CH_3 + \colon\ddot{\underset{..}{Br}}\colon^{-}$

(b) $(CH_3)_3C{-}\ddot{\underset{..}{Br}}\colon \longrightarrow (CH_3)_3C^{+} + \colon\ddot{\underset{..}{Br}}\colon^{-}$

4.29 (a) Draw the transition state for the reverse reaction of Eq. 4.51. Compare it with the transition state shown in Eq. 4.52.

(b) What general statement can you make about the transition-state structures for a reaction and its reverse?

B. The Energy Barrier

We've learned that the standard free energy of the transition state (relative to the standard free energy of reactants and products) defines the free-energy barrier $\Delta G^{\circ\ddagger}$ for the reaction. Let's learn a little more about the relationship between the size of this energy barrier and rate.

The relationship between rate and standard free energy of activation is an exponential one.

$$\text{rate} \propto e^{-\Delta G^{\circ\ddagger}/RT} = 10^{-\Delta G^{\circ\ddagger}/2.3RT} \tag{4.53}$$

where R = the gas constant (8.31×10^{-3} kJ K^{-1} mol^{-1}, or 1.99 kcal K^{-1} mol^{-1}) and T is the absolute temperature (K). (The sign $\propto$ means "is proportional to.") The negative sign in the exponent means that large values of $\Delta G^{\circ\ddagger}$—that is, large energy barriers—result in a smaller rate, as shown in Fig. 4.4. It follows that, if two reactions A and B have standard free energies of activation $\Delta G_A^{\circ\ddagger}$ and $\Delta G_B^{\circ\ddagger}$, respectively, then under standard conditions (all reactants at 1 M concentration), the relative rates of the two reactions are

$$\frac{\text{rate}_A}{\text{rate}_B} = \frac{10^{-\Delta G_A^{\circ\ddagger}/2.3RT}}{10^{-\Delta G_B^{\circ\ddagger}/2.3RT}} = 10^{(\Delta G_B^{\circ\ddagger} - \Delta G_A^{\circ\ddagger})/2.3RT} \tag{4.54a}$$

or

$$\log\left(\frac{\text{rate}_A}{\text{rate}_B}\right) = \frac{\Delta G_B^{\circ\ddagger} - \Delta G_A^{\circ\ddagger}}{2.3RT} \tag{4.54b}$$

These equations show that the rates of two reactions differ by a factor of 10 (that is, one log unit) for every increment of $2.3RT$ (5.7 kJ mol^{-1}, or 1.4 kcal mol^{-1}, at 298 K) difference in their standard free energies of activation. A factor of 10 in rate is roughly the difference in rate between a reaction that takes an hour and one that takes 10 hours. This means that reaction rates are *very sensitive* to their standard free energies of activation.

Where do molecules get enough energy to overcome the energy barrier? In general, molecules obtain this energy from thermal motions. The energy of a collection of molecules is characterized by a distribution (**Fig. 4.5a**), which is called a **Maxwell–Boltzmann distribution**. The rate of a reaction is directly related to the fraction of molecules that has enough energy to cross the energy barrier. This fraction is shown in Fig. 4.5a as a hatched area. The lower the barrier, the larger the

You may have learned about energy barriers if you studied reaction rates in general chemistry, and there you may have referred to the energy barrier by the name *activation energy* and the abbreviation E_a or E_{act}. The activation energy is very close to the $\Delta H^{\circ\ddagger}$ of the reaction (the *standard enthalpy difference*) rather than the standard free-energy difference between the transition state and starting materials. It is possible mathematically to relate transition-state theory to the theory of activation energy. (See Further Exploration 4.4.) For our discussion, however, the difference between the two theories is not conceptually important.

FURTHER EXPLORATION 4.4
Activation Energy

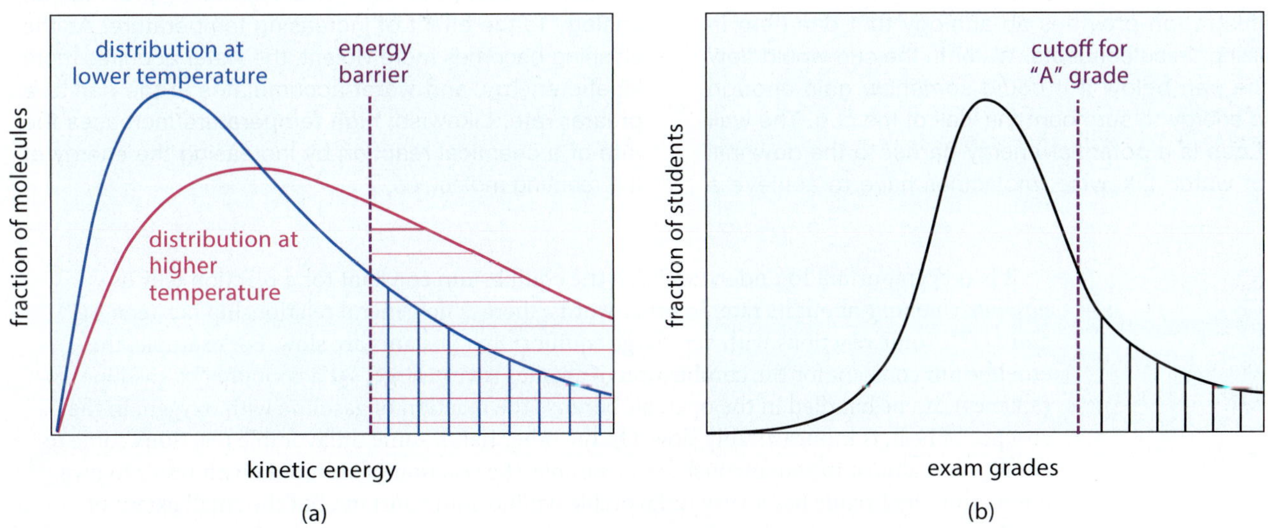

FIGURE 4.5 (a) A Maxwell–Boltzmann kinetic energy distribution at two different temperatures. (The right side of the distributions extend indefinitely and are cut off in the figure.) This is a plot of the number of molecules as a function of kinetic energy. The purple dashed line is the energy barrier for a reaction. The fraction of molecules with enough energy to cross the barrier is given by the hatched areas. At higher temperature, the Maxwell–Boltzmann distribution is skewed to higher energy, and the fraction of molecules with enough energy to cross the barrier is greater (*red hatched area*). (b) The results of an exam can also be characterized by a distribution, which is a plot of the number of students as a function of exam grade. The cutoff (*purple dashed line*) defines the part of the distribution that receives an "A" grade. The fraction of students receiving an "A" is equal to the hatched area under the curve.

hatched area will be and the greater the reaction rate will be. An analogy is the results of an exam by a distribution of grades (**Fig. 4.5b**). The fraction of students who receive an "A" on an exam depends on the cutoff imposed by the instructor; the lower the cutoff grade is, the more students receive an "A."

For a given reaction under a given set of conditions, we cannot control the size of the energy barrier; it is an intrinsic property of the reaction. Some reactions are intrinsically slow, and others are intrinsically fast. However, we can sometimes control the fraction of molecules with enough energy to cross the barrier. We can increase this fraction by *raising the temperature.* As shown in Fig. 4.5a, the Maxwell–Boltzmann distribution is skewed to higher energies at higher temperature, and, as a result, a greater fraction of molecules have the energy required to cross the barrier. In other words, reactions are faster at higher temperatures. Different reactions respond differently to temperature, although a *very rough* rule of thumb is that a reaction rate doubles for each 10 °C (or 10 K) increase in temperature.

Let's summarize. Two factors govern the intrinsic reaction rate:

1. The size of the energy barrier, or standard free energy of activation $\Delta G^{\circ\ddagger}$: reactions with smaller $\Delta G^{\circ\ddagger}$ are faster (see Fig. 4.4).

2. The temperature: reactions are faster at higher temperatures.

An Analogy for Energy Barriers

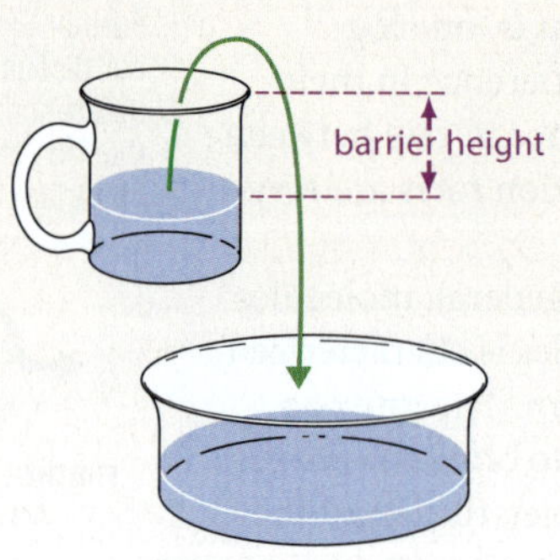

The illustration provides an analogy that can help in visualizing these concepts. Water in the cup would flow into the pan below if it could somehow gain enough kinetic energy to surmount the wall of the cup. The wall of the cup is a potential-energy barrier to the downhill flow of water. Likewise, molecules have to achieve a transient state of high energy—the transition state—to break stable chemical bonds and undergo reaction. An analogy to thermal motion is what happens if we shake the cup. If the cup is shallow (low energy barrier), the likelihood is high that the shaking will cause water to slosh over the sides of the cup and drop into the pan. This will occur at some characteristic rate—some number of milliliters per second. If the cup is very deep (high barrier), water is less likely to flow from cup to pan. Consequently, the rate at which water collects in the pan is smaller. Shaking the cup more vigorously provides an analogy to the effect of increasing temperature. As the sloshing becomes more violent, the water acquires more kinetic energy, and water accumulates in the pan at a greater rate. Likewise, high temperature increases the rate of a chemical reaction by increasing the energy of the reacting molecules.

It is very important to understand that the equilibrium constant for a reaction tells us *absolutely nothing* about its rate. In other words, there is no general relationship between ΔG° and $\Delta G^{\circ\ddagger}$. Some reactions with very large equilibrium constants are slow. For example, the equilibrium constant for the combustion of alkanes is very large; yet a container of gasoline (alkanes) can be handled in the open air because the reaction of gasoline with oxygen, in the absence of heat, is immeasurably slow. On the other hand, some unfavorable reactions come to equilibrium almost instantaneously. For example, the reaction of ammonia with water to give ammonium hydroxide has a very unfavorable equilibrium constant; but the small extent of reaction that does occur takes place very rapidly.

Focused Problems

4.30 (a) The standard free energy of activation of one reaction *A* is 90 kJ mol^{-1} (21.5 kcal mol^{-1}). The standard free energy of activation of another reaction *B* is 75 kJ mol^{-1} (17.9 kcal mol^{-1}). Which reaction is faster and by what factor? Assume a temperature of 298 K.

(b) Estimate the temperature increase of the slower reaction required for it to have a rate equal to that of the faster reaction.

4.31 The standard free energy of activation of a reaction *A* is 90 kJ mol^{-1} (21.5 kcal mol^{-1}) at 298 K. Reaction *B* is 1 million times faster than reaction *A* at the same temperature. The products of each reaction are 10 kJ mol^{-1} (2.4 kcal mol^{-1}) more stable than the reactants.

(a) What is the standard free energy of activation of reaction *B*?

(b) Draw reaction free-energy diagrams for the two reactions showing the two values of $\Delta G^{\circ\ddagger}$ to scale.

(c) What is the standard free energy of activation of the reverse reaction in each case?

C. Multistep Reactions and the Rate-Limiting Step

Many chemical reactions take place with the formation of reactive intermediates. Such reactions are called **multistep reactions**. We use this terminology because, when intermediates exist in a chemical reaction, then what we commonly express as one reaction is really a sequence of two or more reactions. For example, you've already learned that the addition of hydrogen halides to alkenes involves a carbocation intermediate. This means, for example, that the following addition of HBr to methylpropene

$$(CH_3)_2C{=}CH_2 + HBr \longrightarrow (CH_3)_3C{-}Br \qquad (4.55)$$

is a multistep reaction involving the following two steps:

$$(CH_3)_2C{=}CH_2 + HBr \rightleftharpoons (CH_3)_3C^+\ Br^- \qquad (4.56a)$$

$$(CH_3)_3C^+\ Br^- \longrightarrow (CH_3)_3C{-}Br \qquad (4.56b)$$

Each step of a multistep reaction has its own characteristic rate and therefore its own transition state. The energy changes in such a reaction can also be depicted in a reaction free-energy diagram. Such a diagram for the addition of HBr to methylpropene is shown in **Fig. 4.6.** Each free-energy maximum between reactants and products represents a transition state, and the minimum represents the carbocation intermediate.

Generally, the rate of a multistep reaction depends in detail on the rates of its various steps. However, it often happens that one step of a multistep reaction is considerably slower than any of

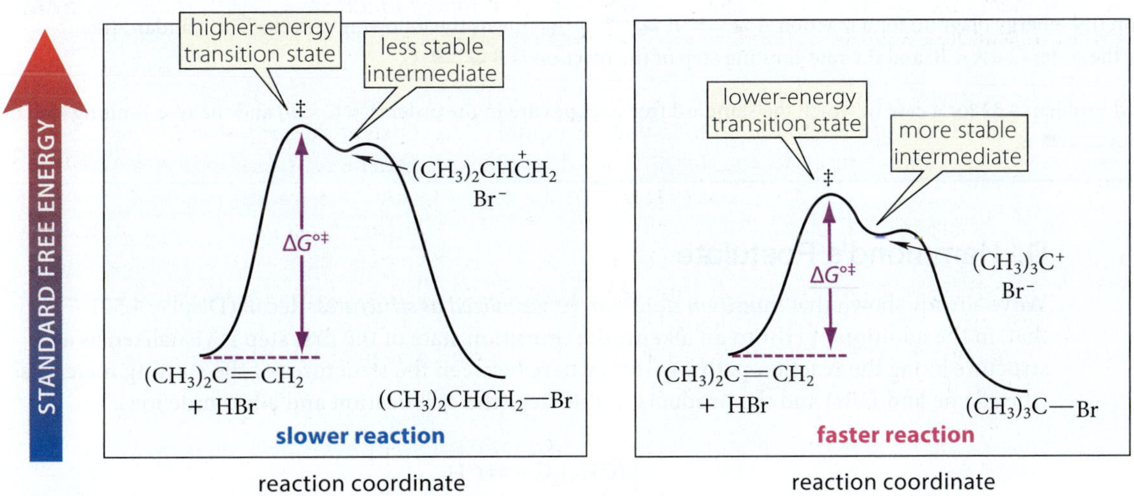

FIGURE 4.6 Reaction free-energy diagram for a multistep reaction. The rate-limiting step of a multistep reaction is the step with the transition state of highest standard free energy. In the addition of HBr to methylpropene, the rate-limiting step is the first step—protonation of the double bond to give the carbocation intermediate.

the others. This slowest step in a multistep chemical reaction is called the **rate-limiting step**, or **rate-determining step**, of the reaction. In such a case, *the rate of the overall reaction is equal to the rate of the rate-limiting step*. In terms of the reaction free-energy diagram in Fig. 4.6, *the rate-limiting step is the step with the transition state of highest free energy*. This diagram indicates that in the addition of HBr to methylpropene, the rate-limiting step is the first step of the reaction—the protonation of the alkene to give the carbocation. The overall rate of addition of HBr to methylpropene is equal simply to the rate of this first step.

The rate-limiting step of a reaction has a special importance. Anything that increases the rate of this step increases the overall reaction rate. Conversely, if a change in the reaction conditions (for example, a change in temperature) affects the rate of the reaction, it is the effect on the rate-limiting step that is being observed. Because the rate-limiting step of a reaction has special importance, its identification receives particular emphasis when we attempt to understand the mechanism of a reaction.

An Analogy for the Rate-Limiting Step

Philip Lange/Shutterstock

A rate-limiting step can be illustrated by a toll station on a freeway at rush hour. We can think of the passage of cars through a toll booth as a multi-step process: (1) entry of the cars into the toll area, (2) taking of the toll by the collector, and (3) exit of the cars from the toll area. Typically, paying the toll to the collector is the rate-limiting step in the passage of cars through the toll plaza. In other words, the rate of passage of cars through the toll plaza is the rate at which they pay their tolls. Cars can arrive more or less frequently, but as long as there is a line of cars, the rate of passage through the toll plaza is the same.

Installing automated change collectors generally increases the rate at which cars pass through a toll plaza. This strategy works because it increases the rate of the rate-limiting step. Increasing the speed limit at which cars can approach the toll plaza, on the other hand, would not increase the rate of passage through the toll plaza, because this change has no effect on the rate-limiting step.

Installing "E-ZPass," an electronic toll permit reader, increases the rate of toll payment even more. In fact, it's possible that with E-ZPass, toll payment is no longer the rate-limiting step. In this case the rate of passage through the toll plaza is limited by the first step, which is the rate at which cars enter the plaza. In this scenario, raising the speed limit of the approach would increase the rate, but increasing the speed of the E-ZPass monitor would have no effect.

Focused Problems

4.32 Draw a reaction free-energy diagram for a reaction $A \rightleftharpoons B \rightleftharpoons C$ that meets the following criteria: The standard free energies are in the order $C < A < B$, and the rate-limiting step of the reaction is $B \rightleftharpoons C$.

4.33 Repeat Focused Problem 4.32 for a case in which the standard free energies are in the order $A < C < B$, and the rate-limiting step of the reaction is $A \rightleftharpoons B$.

D. Hammond's Postulate

We've already shown that *transition states can be visualized as structures*. Recall (Display 4.52) that, in the addition of HBr to an alkene, the transition state of the first step is visualized as a structure along the reaction pathway somewhere between the structures of the starting materials (the alkene and HBr) and the products of this step (the carbocation and a bromide ion):

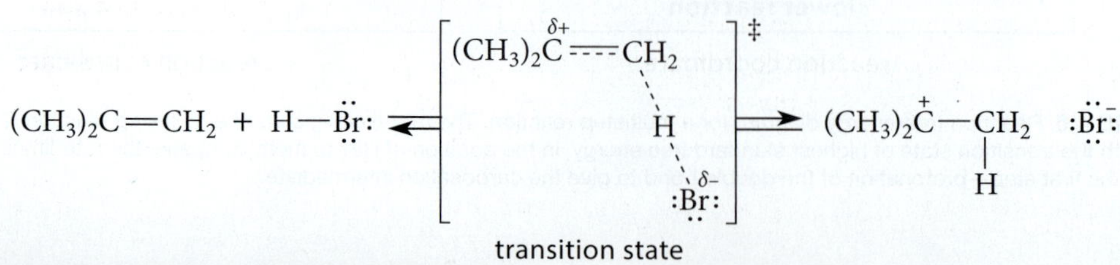

(4.57)

What makes this transition state so unstable? First, the bonds undergoing transition are neither fully broken nor fully formed. The unstable bonding situation is why the transition state lies at an energy maximum. But additionally, a significant contribution to the high energy of the transition state comes from the same factors that account for the high energy of the carbocation. One factor is that the carbocation has one bond fewer than HBr and the alkene. Because bonding releases energy, this fact alone means that the carbocation has a considerably higher energy than starting materials (or products). The other factor is the separation of positive and negative charge. The electrostatic law (Eq. 3.84, Sec. 3.7E) tells us that separation of opposite charges requires energy.

So, we've concluded that the energies of the transition state and the carbocation intermediate are similar, and that the structural elements that account for the high energy of the carbocation also account for most of the transition-state energy. In view of these similarities, the following approximation seems justified: *The structure and energy of the transition state in Eq. 4.57 can be approximated by the structure and energy of the carbocation intermediate.* This approximation can be generalized in an important statement called *Hammond's postulate*:

> **Hammond's Postulate**: *For a reaction in which an intermediate of relatively high energy is either formed from reactants of much lower energy or converted into products of much lower energy, the structure and energy of the transition state can be approximated by the structure and energy of the intermediate itself.*

Hammond's postulate is named for George S. Hammond (1921–2005), who first stated it and applied it to organic reactions in 1955 while he was a professor of chemistry at Iowa State University. (This is not Hammond's exact statement of his postulate, but it will prove to be the most useful version of it for us.)

The utility of Hammond's postulate in dealing with reaction rates can be demonstrated by showing how we could have used it along with a knowledge of carbocation stability to predict the regioselectivity of HBr addition to methylpropene. Recall (Sec. 4.8C) that the rate-limiting step in this reaction is the first step: protonation of the alkene by HBr to give a carbocation. As shown in Eqs. 4.38a and b, this protonation could occur in two different and competing ways. Protonation of the double bond at one carbon gives the *tert*-butyl cation as the unstable intermediate; protonation of the double bond at the other carbon gives the isobutyl cation. We apply Hammond's postulate by assuming that *the structures and energies of the transition states are approximated by the structures and energies of the unstable intermediates—the carbocations—themselves.*

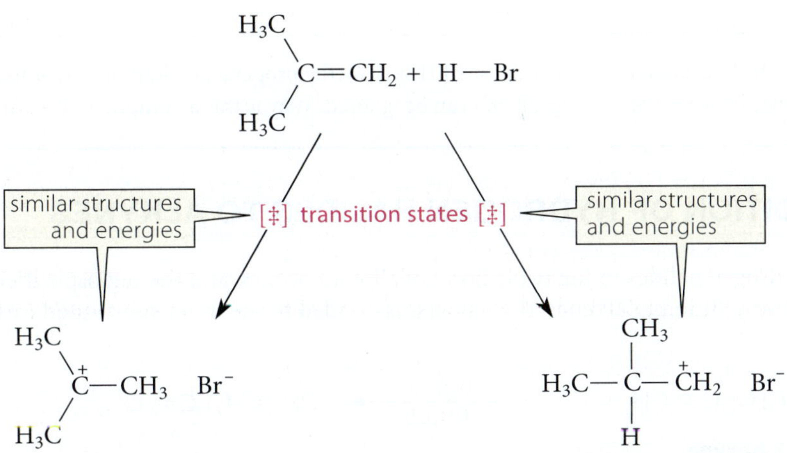

***tert*-butyl cation** (tertiary carbocation, more stable)

isobutyl cation (primary carbocation, much less stable) (4.58)

Because the tertiary carbocation is more stable, *the transition state leading to the tertiary carbocation should also be the transition state of lower energy*. As a result, protonation of methylpropene to give the tertiary carbocation has the transition state with the smaller free energy and is thus the faster of the two competing reactions. The following diagram summarizes the application of Hammond's postulate.

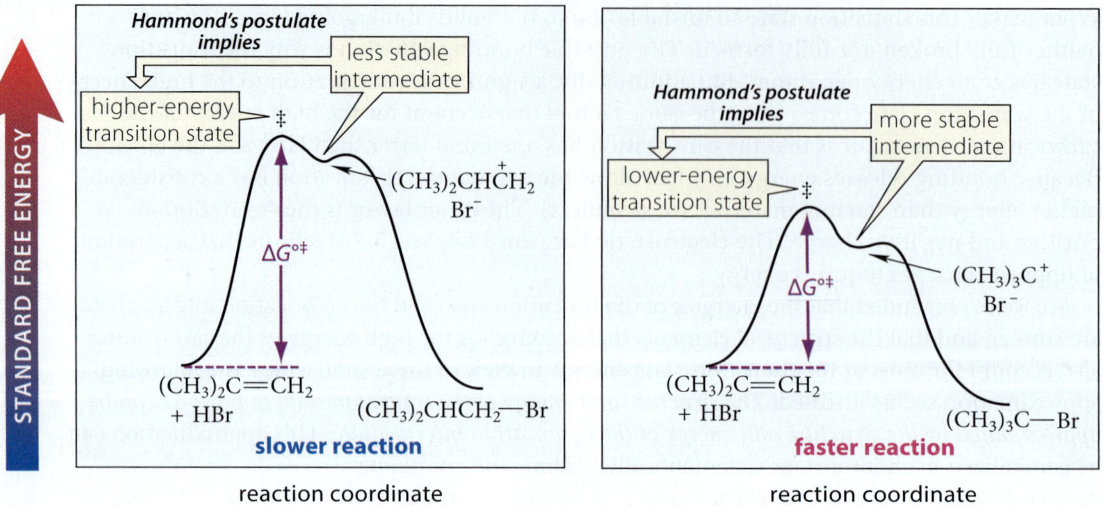

(4.59)

Addition of HBr to alkenes is regioselective because protonation of a double bond to give a tertiary carbocation has a transition state of lower energy than the transition state for protonation to give a primary carbocation. The stabilities of the carbocations themselves do not determine which reaction is faster; rather, *the relative free energies of the transition states for carbocation formation* determine the relative rates of the two processes. *It is Hammond's postulate that allows us to make the connection between carbocation energy and transition-state energy.*

We need Hammond's postulate because the structures of transition states are uncertain, whereas the structures of reactants, products, and reactive intermediates are known. Therefore, knowing that a transition state resembles a particular species (for example, a carbocation) helps us to make a good guess about the transition-state structure. In this text, we frequently analyze or predict reaction rates by considering the structures and stabilities of reactive intermediates such as carbocations. When we do this, we are assuming that the transition states and the corresponding reactive intermediates have similar structures and energies; in other words, we are invoking Hammond's postulate.

Focused Problem

4.34 Apply Hammond's postulate to decide which reaction is faster: addition of HBr to methylpropene or addition of HBr to *trans*-2-butene. Assume that the energy difference between the starting alkenes can be ignored. Why is this assumption necessary?

4.9 ADDITION OF HYDROGEN HALIDES TO ALKYNES

Addition of hydrogen halides to the triple bond of alkynes occurs, and the regiospecificity is the same as the same addition to alkenes: the halogen is bonded to the more substituted carbon of the triple bond.

$$\underset{\textbf{1-hexyne}}{CH_3(CH_2)_3C{\equiv}CH} + HBr \xrightarrow[CH_2Cl_2]{(Et)_4N^+\ Br^-} \underset{\substack{\textbf{2-bromo-1-hexene} \\ \text{(89\% yield)}}}{CH_3(CH_2)_3\underset{\displaystyle Br}{\underset{|}{C}}{=}CH_2} \qquad (4.60)$$

As with addition of HBr to alkenes, a carbocation intermediate is involved.

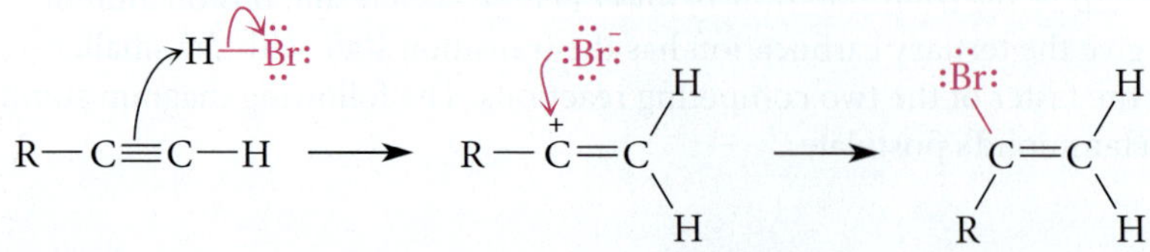

carbocation intermediate
(*a vinylic cation*)

(4.61)

This type of carbocation is called a **vinylic cation**. In a vinylic cation, the electron-deficient carbon is sp-hybridized, and it contains two 2p orbitals. One 2p orbital is empty, and the other is used to form the π bond to the other carbon of the double bond:

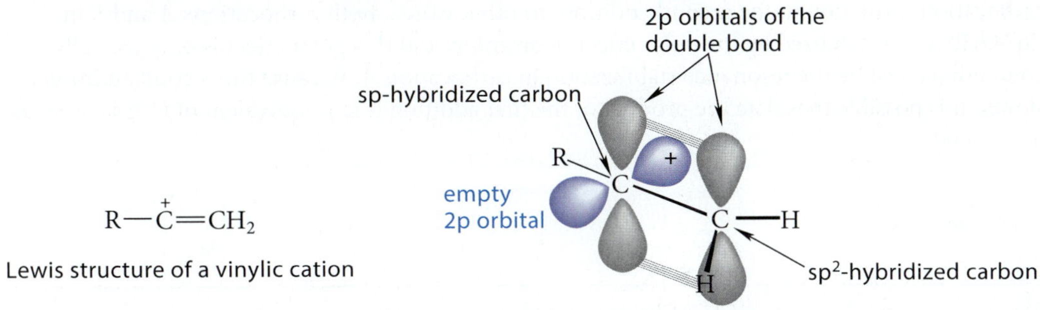

2p orbitals in the vinylic cation (4.62)

Because the axes of the 2p orbitals of the double bond are perpendicular to the axis of the empty 2p orbital, *there is no overlap between the two sets of orbitals.* In other words, this is *not* a resonance-stabilized carbocation. (Vinylic cations are among the least stable carbocations for reasons that we explore in Sec. 18.3. They can be formed from acetylenes because one of the carbons of the acetylene changes hybridization from the less stable sp hybridization to the more stable sp^2 hybridization.)

Like other carbocations, vinylic carbocations can undergo Lewis acid–base association reactions with nucleophiles. Such a reaction of the carbocation intermediate gives the addition product in Eq. 4.61.

Because addition to an alkyne gives a substituted alkene, a second addition can occur in many cases.

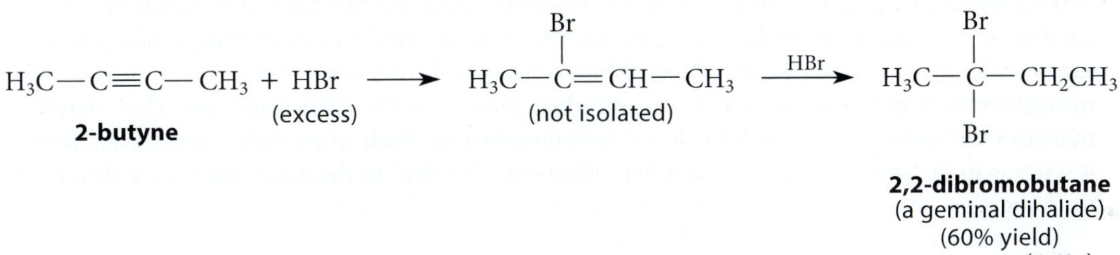

(4.63a)

This is one method for the preparation of **geminal dihalides**: compounds that have two halogens on the same carbon. (*Geminal* is from the Latin *geminus*, meaning "paired.")

The regioselectivity of this second addition reaction is determined by the relative stabilities of the two possible carbocation intermediates. One of the two possible carbocations (*A* in Eq. 4.63b) is stabilized by resonance. By Hammond's postulate (Sec. 4.8D), this carbocation is formed more rapidly.

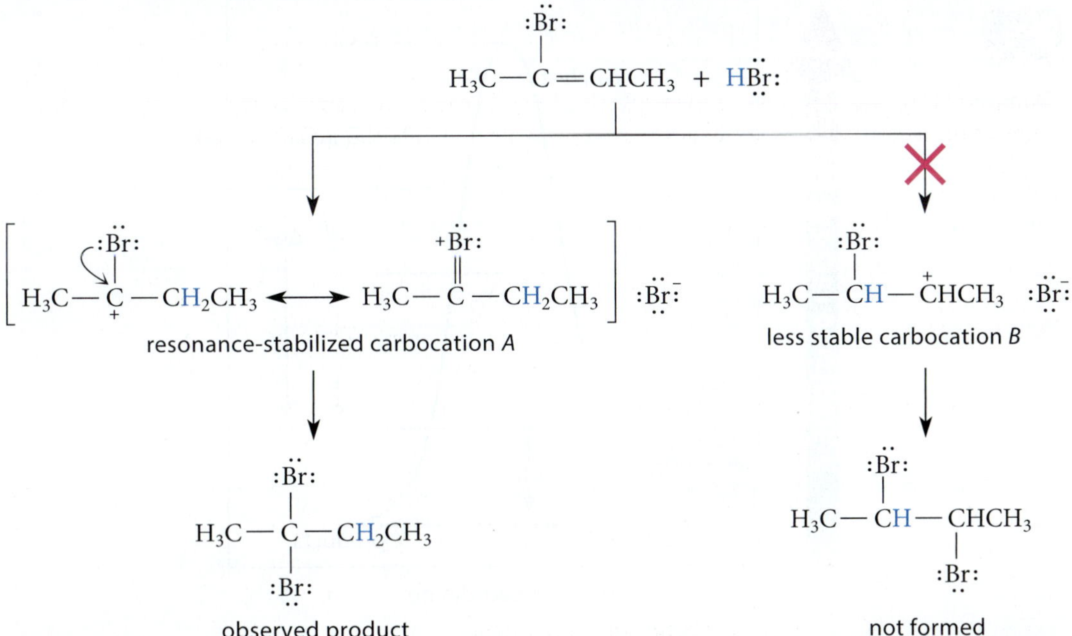

(4.63b)

In the addition of a hydrogen halide or a halogen to an alkyne, the second addition is usually slower than the first. This is because the halogen that enters the molecule in the first addition is an electron-withdrawing substituent and exerts a rate-retarding polar effect (Sec. 3.7E) on carbocation formation in the second addition. In other words, both carbocations *A* and *B* in Eq. 4.63b are destabilized by the polar effect of bromine, and this polar effect is only partially counterbalanced by the resonance stabilization in carbocation *A*. Because the second addition is slower, it is possible to isolate the product of the first addition if one equivalent of HBr is used, as in Eq. 4.60.

Focused Problem

4.35 (a) Give the structures of all the products that could form from the addition of one equivalent of HCl to 2-hexyne. Classify each pair as constitutional isomers or stereoisomers.

(b) Give the structure of the carbocation intermediate that accounts for each of the constitutional isomers formed in part (a).

4.10 CATALYSIS

Some reactions take place much more rapidly in the presence of certain substances that are themselves left unchanged by the reaction. A substance that increases the rate of a reaction without being consumed is called a **catalyst**. A practical example of a catalyst occurs in the catalytic converter on the modern automobile. The catalyst in the converter brings about the rapid oxidation (combustion) of unburned hydrocarbons, the conversion of nitrogen oxides into nitrogen and oxygen, and the conversion of carbon monoxide into carbon dioxide. The catalyst increases the rates of these reactions by many orders of magnitude; these reactions essentially do not occur in the absence of the catalyst. Despite its involvement in the reactions, the catalyst is left unchanged.

Here are some important points about catalysts:

1. A catalyst increases the reaction rate. This means that it lowers the standard free energy of activation for a reaction (**Fig. 4.7**).

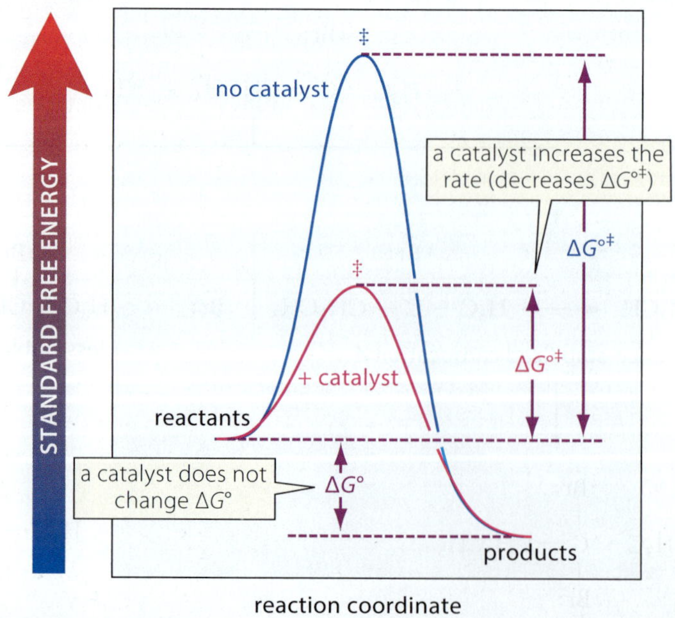

FIGURE 4.7 A reaction free-energy diagram comparing a hypothetical catalyzed reaction (*red curve*) to the uncatalyzed reaction (*blue curve*). Although both reactions are shown here as simple, one-step processes, a catalyzed reaction and the corresponding uncatalyzed reaction typically have different mechanisms and different numbers of reaction steps. The relative energies of reactants and products are unaffected by a catalyst.

2. A catalyst is not consumed. It may be consumed in one step of a catalyzed reaction, but if so, it is regenerated in a subsequent step.

   Points 1 and 2 imply that a catalyst that strongly accelerates a reaction can be used in very small amounts. Many expensive catalysts are practical for this reason.

3. A catalyst does not affect the energies of reactants and products. In other words, a catalyst does not affect the $\Delta G°$ of a reaction and consequently also does not affect the equilibrium constant (Fig. 4.7).

4. A catalyst accelerates both the forward and reverse of a reaction by the same factor.

The last point follows from the fact that, at equilibrium, the rates of a reaction and its reverse are equal. If a catalyst does not affect the equilibrium constant (point 3) but increases the reaction rate in one direction, equality of rates at equilibrium requires that the rate of the reverse reaction must be increased by the same factor.

When a catalyst and the reactants exist in separate phases, the catalyst is called a **heterogeneous catalyst**. The catalyst in the catalytic converter of an automobile is a heterogeneous catalyst because it is a solid and the reactants are gases. In other cases, a reaction in solution may be catalyzed by a soluble catalyst. A catalyst that is soluble in a reaction solution is called a **homogeneous catalyst**.

A large number of organic reactions are catalyzed. In this section, we introduce the idea of catalysis by considering three examples of catalyzed reactions. The first example, *catalytic hydrogenation*, is an important example of heterogeneous catalysis. The second example, *hydration*, is an example of homogeneous catalysis. The last example involves catalysis of a biological reaction by an enzyme.

Catalyst Poisons, the Catalytic Converter, and Leaded Gasoline

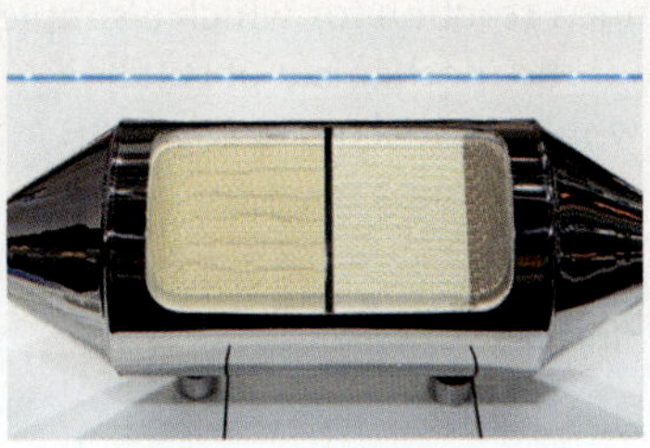

Zoonar GmbH/Alamy

Although in theory catalysts should function indefinitely, in practice many catalysts, particularly heterogeneous catalysts, slowly become less effective. It is as if they "wear out." One reason for this behavior is that they slowly absorb impurities, called catalyst poisons, from the surroundings; these impurities impede the functioning of the catalyst. An example of this phenomenon also occurs with the catalytic converter. Lead is a potent poison of the catalyst in a catalytic converter. This fact, as well as the desire to eliminate atmospheric lead pollution, are the major reasons why leaded gasoline is no longer used in automotive engines in the United States.

A. Catalytic Hydrogenation of Alkenes

When a solution of an alkene is stirred under an atmosphere of hydrogen, nothing happens. But if the same solution is stirred under hydrogen in the presence of a metal catalyst, the hydrogen is rapidly absorbed by the solution. The hydrogen is consumed because it undergoes an *addition* to the alkene double bond.

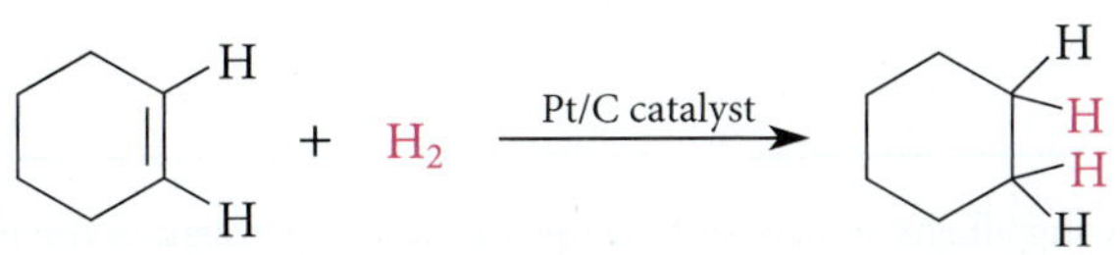

cyclohexene **cyclohexane** (4.64)

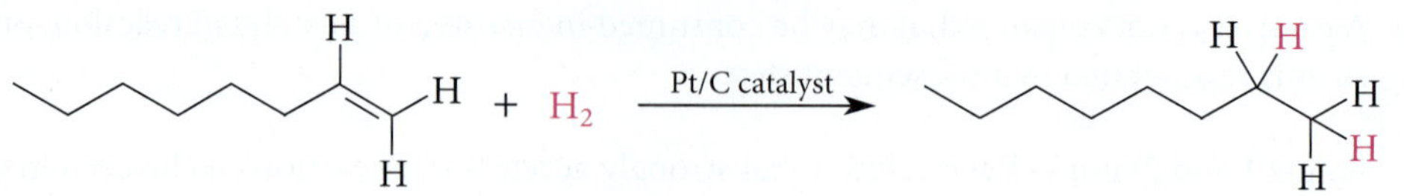

1-octene **octane** (4.65)

These reactions are examples of **catalytic hydrogenation**, the addition of H_2 to an alkene in the presence of a catalyst. Catalytic hydrogenation is one of the best ways to convert alkenes into alkanes. Catalytic hydrogenation is an important reaction in both industry and the laboratory. The inconvenience of using a special apparatus for the handling of a flammable gas (H_2) is more than offset by the great utility of the reaction.

In the preceding reactions, the catalyst is written over the reaction arrows. Pt/C is read as "Platinum supported on carbon" or simply "Platinum on carbon." This catalyst is finely divided platinum metal that has been precipitated, or "supported," on activated charcoal. This process produces very small catalyst particles and thus provides a high catalyst surface area. (Because the reaction occurs at the surface of the catalyst, the efficiency of the catalyst is improved by increasing its surface area.) A number of noble metals, such as platinum, palladium, and nickel, are useful as hydrogenation catalysts, and they are often used in conjunction with solid support materials such as alumina (Al_2O_3), barium sulfate ($BaSO_4$), or, as in the previous examples, activated carbon. Hydrogenation can be carried out at room temperature and pressure or, for especially difficult cases, at higher temperature and pressure in a "bomb" (a closed vessel designed to withstand high pressures).

Because hydrogenation catalysts are insoluble in the reaction solution, they are examples of *heterogeneous catalysts*. (Soluble hydrogenation catalysts are also known; Sec. 18.5E, Focused Problem 18.15.) Even though they involve relatively expensive noble metals, heterogeneous hydrogenation catalysts are practical because they can be reused. Furthermore, because they are exceedingly effective, they can be used in small amounts. For example, typical catalytic hydrogenation reactions can be run with reactant-to-catalyst molar ratios of 100 or more.

How do hydrogenation catalysts work? Research has shown that both the hydrogen and the alkene must be adsorbed on the surface of the catalyst for a reaction to occur. The catalyst is believed to form reactive metal–carbon and metal–hydrogen bonds that ultimately are broken to form the products and to regenerate the catalyst sites. Beyond this, the chemical details of catalytic hydrogenation are not thoroughly understood. This is not a reaction for which a simple curved-arrow mechanism can be written. Additional insight about the mechanism of catalytic hydrogenation comes from the stereochemistry of the reaction, which we discuss in Sec. 7.8E.

The benzene ring is inert to conditions under which normal double bonds react readily:

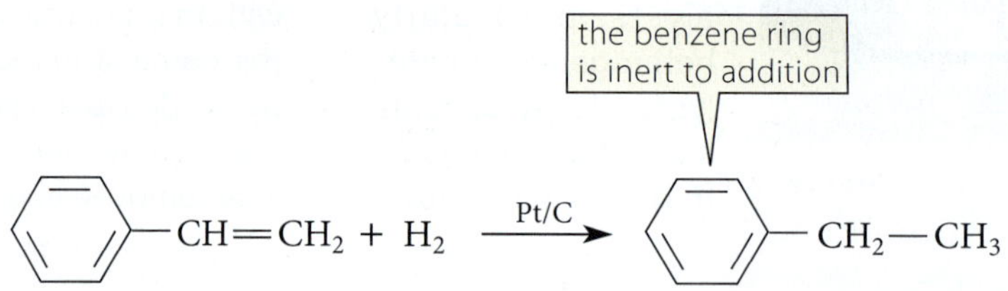

styrene **ethylbenzene** (4.66)

(Benzene rings can be hydrogenated, however, with certain catalysts under conditions of high temperature and pressure.) Many other alkene reactions also do not affect the "double bonds" of a benzene ring. The relative inertness of benzene rings toward the conditions of alkene reactions was one of the great puzzles of organic chemistry that was ultimately explained by the theory of aromaticity, which is introduced in Sec. 15.7.

Focused Problems

4.36 Give the product formed when each of the following alkenes reacts with a large excess of hydrogen in the presence of Pd/C.

(a) 1-Pentene (b) (*E*)-1,3-Hexadiene

4.37 (a) Give the structures of five alkenes, each with the formula C_6H_{12}, that would give hexane as the product of catalytic hydrogenation.

(b) How many alkenes containing one double bond can react with H_2 over a Pt/C catalyst to give methylcyclopentane? Give their structures. (*Hint:* For an approach to this type of problem, see Study Problem 4.9, Sec. 4.7C.)

B. Catalytic Hydrogenation of Alkynes

Alkynes, like alkenes, undergo catalytic hydrogenation. The first addition of hydrogen yields an alkene; a second addition of hydrogen gives an alkane.

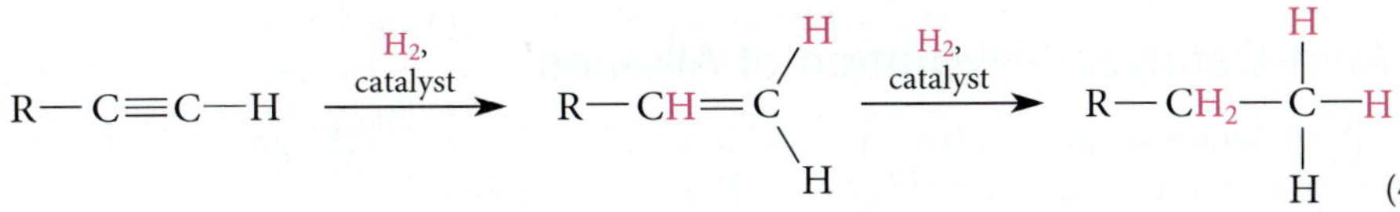

(4.67)

The utility of alkyne hydrogenation is enhanced considerably by the fact that hydrogenation of an alkyne stops at the alkene stage if the reaction mixture contains a **catalyst poison**: a compound that disrupts the action of a catalyst. Among the useful catalyst poisons are salts of Pb^{2+}, and certain nitrogen compounds, such as pyridine, quinoline, or other amines.

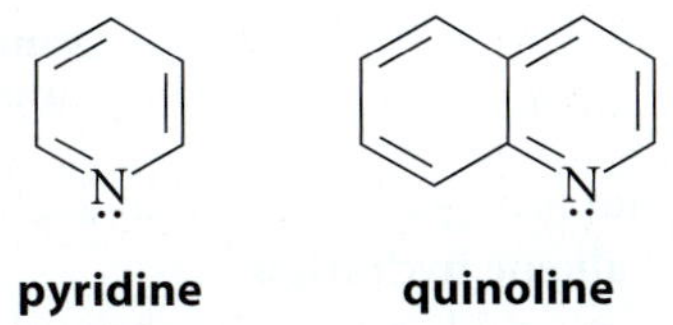

These compounds *selectively* block the hydrogenation of alkenes without preventing the hydrogenation of alkynes to alkenes. For example, a $Pd/CaCO_3$ catalyst can be washed with $Pb(OAc)_2$ to give a poisoned catalyst known as **Lindlar catalyst**. In the presence of Lindlar catalyst (or other poisons), an alkyne is hydrogenated to the corresponding alkene. Furthermore, if the alkyne is internal (that is, if it has a substituent on each carbon of the triple bond), the alkene product has cis stereochemistry.

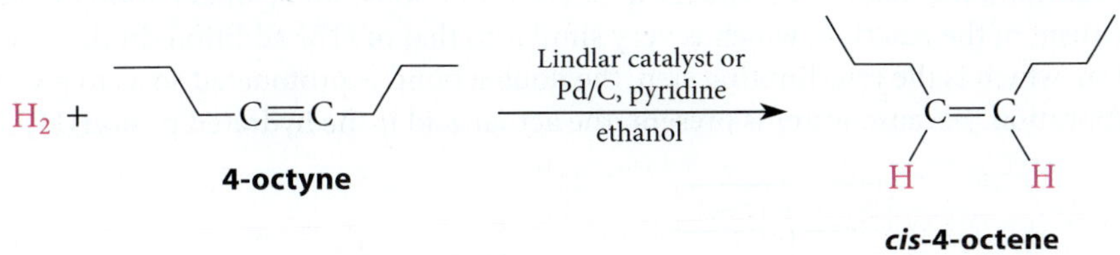

(4.68)

The formation of cis alkenes from alkynes by catalytic hydrogenation shows that the two hydrogen atoms must react with the alkyne from the same face (or side) of the alkyne molecule. The stereochemistry of addition reactions is discussed in more detail in Sec. 7.8.

As this example illustrates, *catalytic hydrogenation of alkynes is one of the best ways to prepare cis alkenes.* (The complementary preparation of trans alkenes from alkynes [by a method that does *not* involve catalytic hydrogenation] is discussed in Sec. 10.2.)

In the absence of a catalyst poison, two equivalents of H_2 are added to the triple bond.

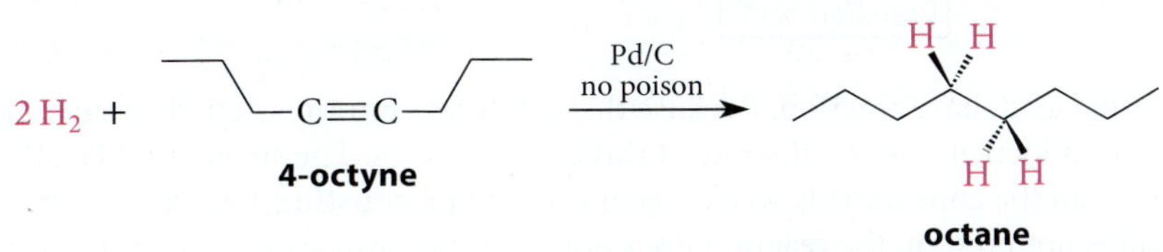

(4.69)

Therefore, the catalytic hydrogenation of alkynes can therefore be used to prepare either alkenes or alkanes by either including or omitting the catalyst poison. How catalyst poisons exert their inhibitory effect on the hydrogenation of alkenes is not well understood.

Focused Problem

4.38 Give the principal organic product formed in each of the following reactions.

(a) $CH_3(CH_2)_5C{\equiv}CH + H_2 \xrightarrow{\text{Lindlar catalyst}}$
1-octyne

(b) Same as part (a), but with no catalyst poison

(c) 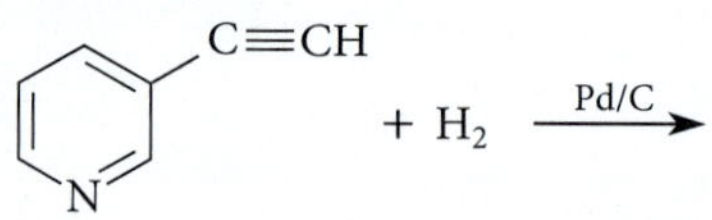

(d) $H_3C{-}C{\equiv}C{-}CH_2CH_3 + D_2 \xrightarrow{\text{Lindlar catalyst}}$
2-pentyne

C. Acid-Catalyzed Hydration of Alkenes

The alkene double bond undergoes reversible addition of water in the presence of moderately concentrated strong acids such as H_2SO_4, $HClO_4$, and HNO_3.

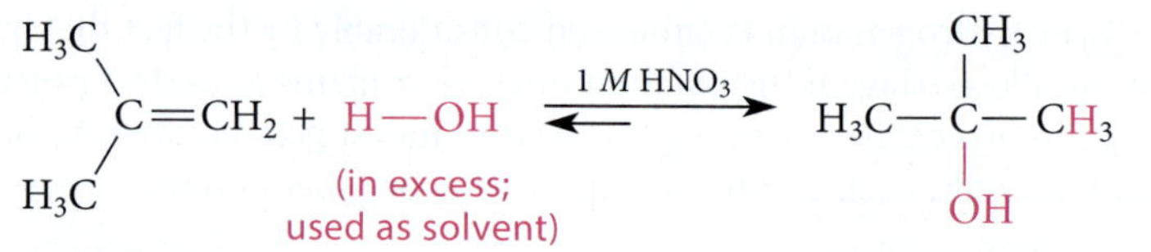

methylpropene (isobutylene) **2-methyl-2-propanol (*tert*-butyl alcohol)** (4.70)

Alkynes also undergo hydration. The hydration of alkynes is covered in Sec. 5.5.

The addition of the elements of water is in general called **hydration**. Hence, the addition of water to the alkene double bond is called **alkene hydration**.

Hydration does not occur at a measurable rate in the absence of an acid, and the acid is not consumed in the reaction. Hence, alkene hydration is an *acid-catalyzed reaction*. Because the catalyzing acid is soluble in the reaction solution, it is a *homogeneous catalyst*.

Acid-catalyzed hydration, like the addition of HBr, is *regioselective*. As in the addition of HBr, the hydrogen adds to the carbon of the double bond with the smaller number of alkyl substituents. The more electronegative partner of the H—OH bond, the OH group, like the Br in HBr addition, adds to the carbon of the double bond with the greater number of alkyl substituents.

In this reaction, the manner in which the catalyst functions can be understood by considering the mechanism of the reaction, which is very similar to that of HBr addition. In the first step of the reaction, which is the rate-limiting step, the double bond is protonated so as to give the more stable carbocation. Because water is present, the actual acid is the hydrated proton (H_3O^+).

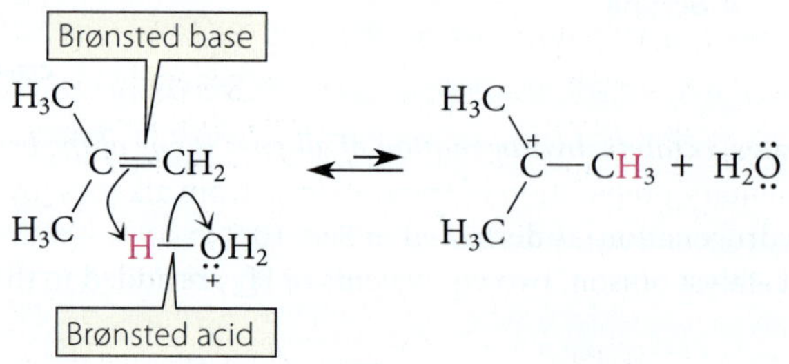

(4.71a)

This is a Brønsted acid–base reaction. Because this is the rate-limiting step, the rate of the hydration reaction increases when the rate of this step increases. The strong acid H_3O^+ is much more effective than the considerably weaker acid water in protonating a weak base (the alkene). If a strong acid is not present, the reaction does not occur because water alone is too weak an acid to protonate the alkene.

In the next step of the hydration reaction, the nucleophile water combines with the carbocation in a Lewis acid–base association reaction:

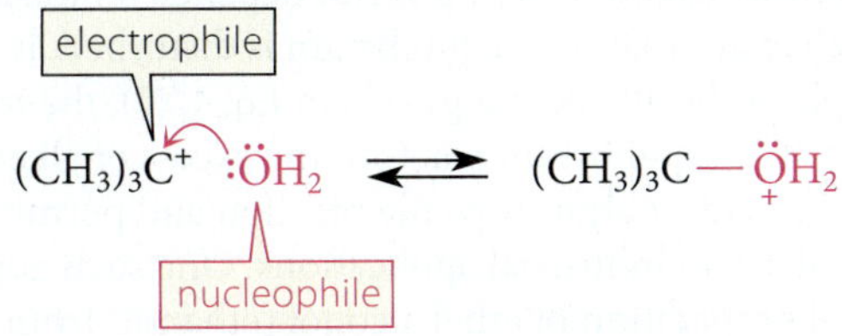

(4.71b)

Finally, a proton is lost to solvent in another Brønsted acid–base reaction to give the alcohol product and regenerate the catalyzing acid H_3O^+:

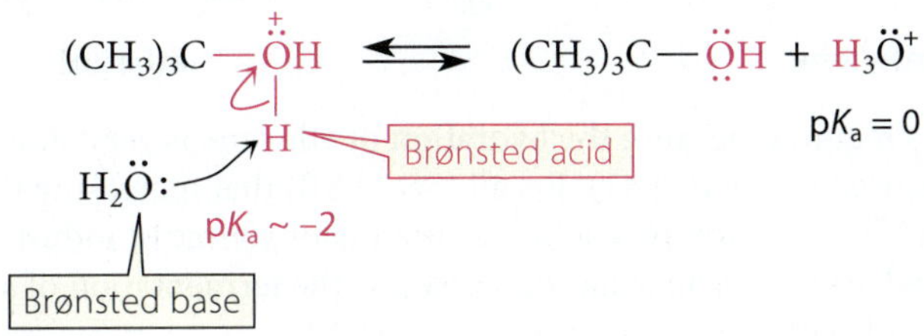

(4.71c)

Notice three things about this mechanism:

1. It consists entirely of Lewis acid–base association–dissociation and Brønsted acid–base reactions.

2. Although the proton consumed in Eq. 4.71a is not the same as the one produced in Eq. 4.71c, there is no *net* consumption of protons.

3. The nucleophile in Eq. 4.71b and the Brønsted base in Eq. 4.71c is water. Some students are tempted to use hydroxide ion in a situation like this because it is a stronger base:

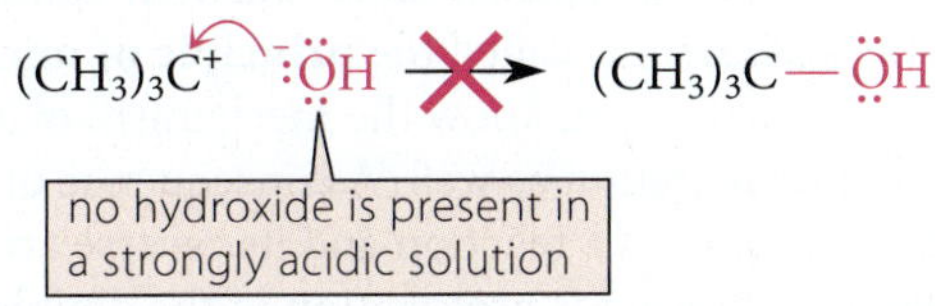

(4.72)

However, there is no hydroxide present in a 1 *M* nitric acid or sulfuric acid solution. Nor is hydroxide needed: the carbocation in Eq. 4.71b is such a strong electrophile that it reacts rapidly with water; and, as we can see from the pK_a values in Eq. 4.71c, the acid on the left of that equation is strong enough to donate a proton to the weak base water. We can generalize this result as follows:

Whenever H_3O^+ acts as an acid, its conjugate base H_2O acts as the base. (Read again about amphoteric compounds in Sec. 3.4B if this is unclear.) More generally, *acids and their conjugate bases always act in tandem in acid–base catalysis.If H_2O is the acid, then the base is ^-OH, and vice versa.*

Because the hydration reaction involves carbocation intermediates, some alkenes give rearranged hydration products.

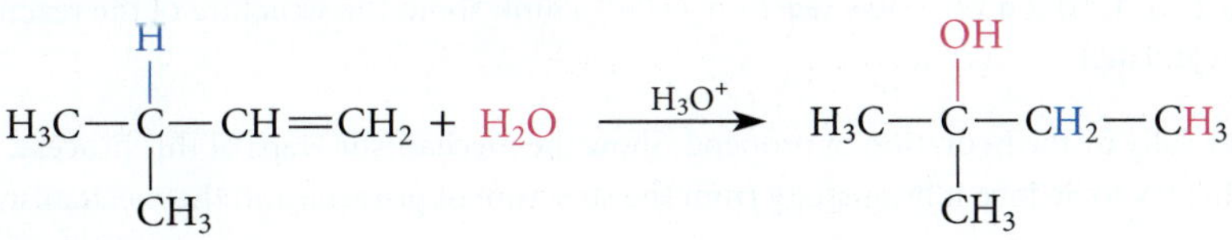

(4.73)

Focused Problem

4.39 Give the mechanism for the reaction in Eq. 4.73. Show each step of the mechanism separately with careful use of the curved-arrow notation. Explain why the rearrangement takes place.

The equilibrium constants for many alkene hydrations are close enough to unity that the hydration reaction can be run in reverse. The reverse of alkene hydration is called *alcohol dehydration*. The direction in which the reaction is run depends on the application of **Le Chatelier's principle**, which states that if an equilibrium is disturbed, it will react so as to offset the disturbance. For example, if the alkene is a gas (as in Eq. 4.70), the reaction vessel can be pressurized with the alkene. The equilibrium reacts to the excess of alkene by forming more alcohol. Neutralization of the acid catalyst stops the reaction and permits isolation of the alcohol. This strategy is used particularly in industrial applications. One such application of alkene hydration is the commercial preparation of ethyl alcohol (ethanol) from ethylene:

$$\underset{\textbf{ethylene}}{H_2C{=}CH_2} + H_2O \xrightarrow[300\ ^\circ C]{H_3PO_4\ \text{(adsorbed on a solid support)}} \underset{\textbf{ethanol}}{H_3C{-}CH_2{-}OH} \qquad (4.74)$$

A high temperature is required because the hydration of ethylene is very slow at ordinary temperatures (see Focused Problem 4.41). Recall (Sec. 4.8B) that increasing the temperature accelerates a reaction. This reaction was at one time a major source of industrial ethanol. Although it is still used, its importance has decreased as the fermentation of sugars from biomass (for example, corn) has become more prevalent (Sec. 11.13).

To run the hydration reaction in the reverse (dehydration) direction, the alkene is removed as it is formed, typically by distillation. (Alkenes have significantly lower boiling points than alcohols, as we further discuss in Sec. 8.5C.) The equilibrium responds by forming more alkene. Alcohol dehydration is more widely used than alkene hydration in the laboratory. We consider this reaction in Sec. 11.2.

Alkene hydration and alcohol dehydration illustrate two important points. First is one of the key points about catalysis: a catalyst accelerates the forward and reverse reactions of an equilibrium by the same factor. For example, because alkene hydration is acid-catalyzed, alcohol dehydration is acid-catalyzed as well. The second point is that alkene hydration and alcohol dehydration occur by the forward and reverse of the same mechanism. Generally, *if a reaction occurs by a certain mechanism, the reverse reaction under the same conditions occurs by the exact reverse of that mechanism.* This statement is called the **principle of microscopic reversibility.** Microscopic reversibility requires that, if you know the mechanism of alkene hydration, then you know the mechanism of alcohol dehydration as well. A consequence of microscopic reversibility is that the rate-limiting transition states of a reaction and its reverse are the same. For example, if the rate-limiting step of alkene hydration is protonation of the double bond to form the carbocation intermediate (Eq. 4.71a), then the rate-limiting step of alcohol dehydration is the reverse of the same equation—deprotonation of the carbocation to give the alkene.

Focused Problems

4.40 (a) Unlike the alcohol product in Eq. 4.70, the product in Eq. 4.73 does *not* come to equilibrium with the starting alkene. However, it *does* come to equilibrium with two other alkenes. What are their structures?

(b) Why isn't the alkene starting material in Eq. 4.73 part of the equilibrium mixture?

4.41 Explain why the hydration of ethylene (Eq. 4.74) is a very slow reaction. (*Hint:* Think about the structure of the reactive intermediate and apply Hammond's postulate.)

4.42 Isopropyl alcohol is produced commercially by the hydration of propene. Show the mechanistic steps of this process. If you do not know the structure of isopropyl alcohol, try to deduce it by analogy from the structure of propene and the mechanism of alkene hydration.

D. Enzyme Catalysis

Chemical Biology Topic

Catalysis is not limited to the laboratory or chemical industry. The biological processes of nature involve thousands of chemical reactions, most of which have their own unique naturally occurring catalysts. These biological catalysts are called **enzymes**. (Enzyme catalysis is discussed in Sec. 27.10.) Under physiological conditions, most important biological reactions would be too slow to be useful in the absence of their enzyme catalysts. Enzyme catalysts are important not only in nature; they are used both in industry and in the laboratory.

Many well-characterized enzymes are soluble in aqueous solution and hence are homogeneous catalysts. However, other enzymes are immobilized within biological substructures such as membranes and can be viewed as heterogeneous catalysts.

An example of an important enzyme-catalyzed addition to an alkene is the hydration of fumarate ion to malate ion.

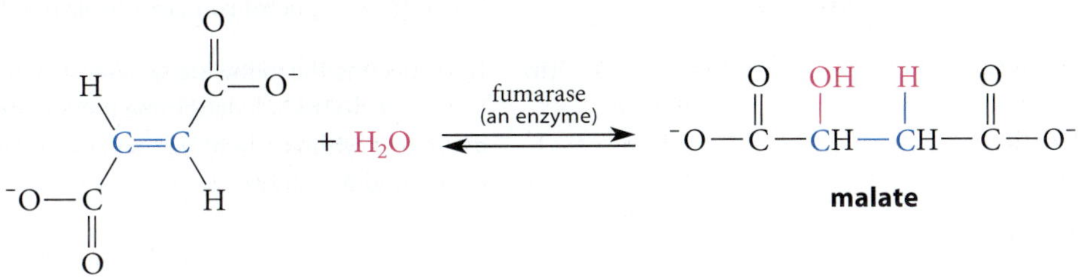

fumarate (4.75)

This reaction is catalyzed by the enzyme *fumarase*. It is one reaction in the Krebs cycle, or citric acid cycle, a series of reactions that plays a central role in the generation of energy in biological systems. Fumarase catalyzes only this reaction. The effectiveness of fumarase catalysis can be appreciated by the following comparison: At physiological pH and temperature (pH = 7, 37 °C), the enzyme-catalyzed reaction is about 10^9 (1 *billion*) times faster than the same reaction in the absence of enzyme. To put the catalytic effectiveness of fumarase in greater perspective, the enzyme-catalyzed reaction occurs in a fraction of a second. The uncatalyzed reaction requires *hundreds of thousands of years*—in other words, it doesn't occur!

The mechanism of catalysis by fumarase is discussed in detail in Sec. 22.9D.

CHAPTER SUMMARY *For a summary of the chapter, see Chapter 4 in the* Study Guide and Solutions Manual.

REACTION REVIEW *For a summary of reactions discussed in this chapter, see the* Reaction Review *section of Chapter 4 in the* Study Guide and Solutions Manual.

SKILLS OBJECTIVES WITH PROBLEMS

- **Describe the bonding in alkenes and alkynes using the hybrid-orbital model. Use the fraction s character in the component bond orbitals of single bonds to predict relative bond lengths. 4.1B**

4.43 Identify the hybrid orbitals involved in each labeled bond in the following structure.

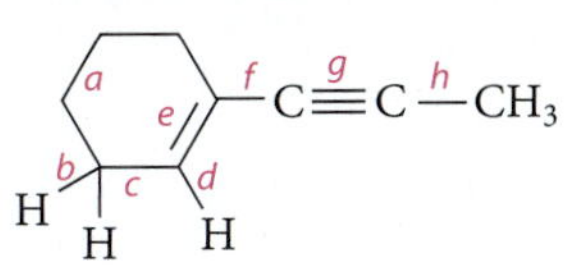

4.44 For the structure in Problem 4.43, rank the labeled bonds in order of increasing bond length, and explain your reasoning.

- **Construct the molecular orbitals for a system of two 2p orbitals overlapping "side-to-side." 4.1C**

4.45 (a) Sketch the bonding (π) and antibonding (π*) molecular orbitals for methylpropene.

(b) Sketch an energy diagram showing the relative energies of the (π) and antibonding (π*) molecular orbitals of acetylene.

- **Name alkenes and alkynes using the IUPAC system of substitutive nomenclature. Given the IUPAC name of an alkene or an alkyne, draw its structure. 4.2A,C**

4.46 Name the following compounds with IUPAC substitutive nomenclature.

(a) 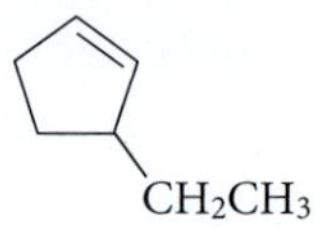

(b) 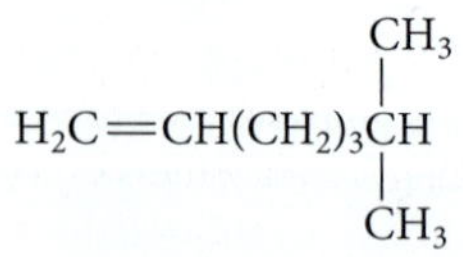

(c) 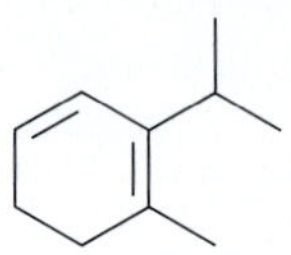

(d) HC≡C—CH=CH_2

(e) 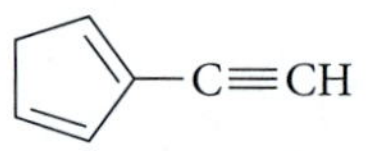

4.47 Draw a structure for each of the following compounds. (Ignore stereochemistry.)

(a) 2-Hexene

(b) 4-Methylcycloheptene

(c) 3,4-Diethyl-1,3,5-hexatriene

(d) 2,4-Nonadiyne

(e) 4-(3-Methyl-1-butynyl)cyclohexene

- **Determine whether an alkene has an *E*,*Z* stereoisomer. If so, identify the stereocenters. 4.1D**

4.48 Which of the following alkenes has an *E*,*Z* stereoisomer? Explain how you know.

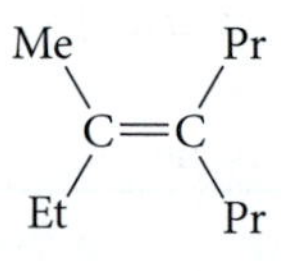

A

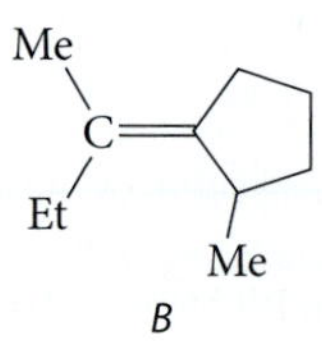

B

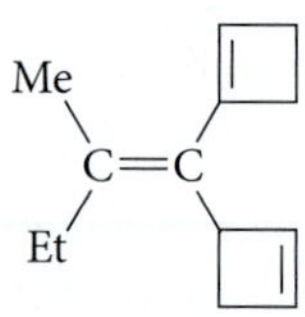

C

- **Determine whether a double bond in an alkene has the *E* or *Z* configuration by applying the Cahn–Ingold–Prelog sequence rules. 4.2B**

4.49 Specify the stereochemical configuration (*E* or *Z*) of each of the following alkenes. D = deuterium (^{2}H, the isotope of hydrogen with molecular mass = 2).

(a) 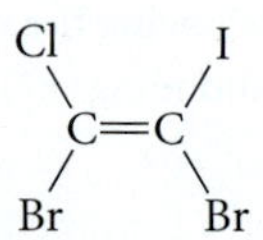

(b) 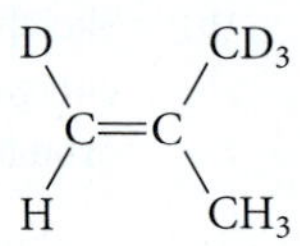

(c) 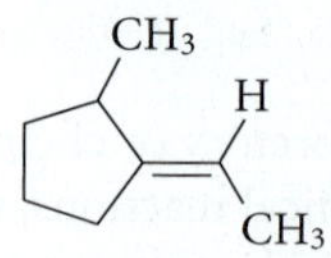

(d) 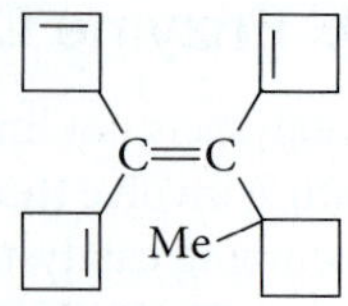

- **Use the unsaturation number (*U*) to determine the number of rings and π bonds in a compound from its molecular formula. Given a structure, determine the number of hydrogens without having to count them explicitly. 4.3**

4.50 (a) A compound *A* has molecular formula $C_7H_{13}ClN_2O$. How many rings and/or π bonds does it contain? Could the compound contain a triple bond?

(b) By inspecting the following skeletal structure of 17α-ethynylestradiol, determine the number of hydrogens without counting them explicitly. (The compound has 20 carbons.)

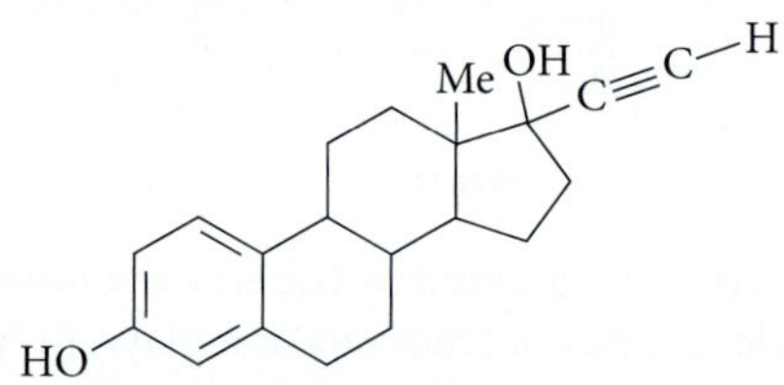

17α-ethynylestradiol
(a synthetic estrogen used in oral contraceptives)

(c) The drug molecule aspirin contains 8 hydrogens and 4 oxygens, and its unsaturation number is 6. How many carbons does it have?

- **Estimate the effect of the geometry around sp-hybridized and sp^2-hybridized carbons on molecular properties such as dipole moment. 4.4**

4.51 Consider the following compounds and their dipole moments:

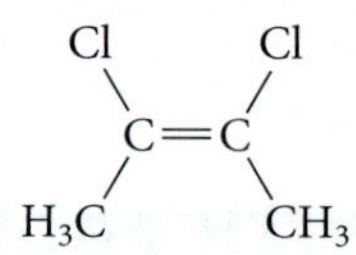

μ = 2.4 D

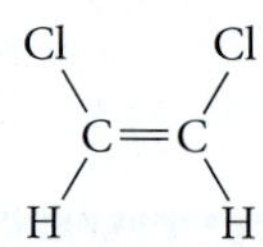

μ = 1.9 D

(a) According to these dipole moments, is a methyl group or a hydrogen more electron-donating toward a double bond? Explain.

(b) Which of the following compounds should have the greater dipole moment? Explain.

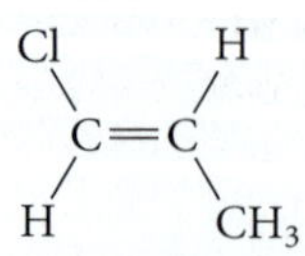

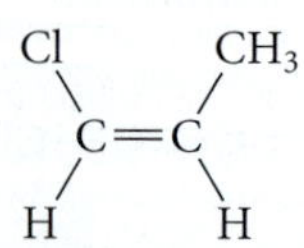

4.52 Rank the following compounds in order of increasing dipole moment.

$H_3C—C{\equiv}C—Cl$ (*A*) $Cl—C{\equiv}C—Cl$ (*B*) $H_3C—C{\equiv}C—H$ (*C*)

- **Rank alkene and alkyne isomers in order of relative stability. 4.5B,C**

4.53 In each set, rank the compounds in order of increasing stability (decreasing heat of formation). Explain your reasoning.

(a)

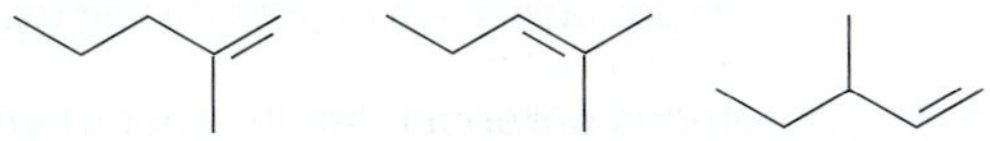

A *B* *C*

(b)

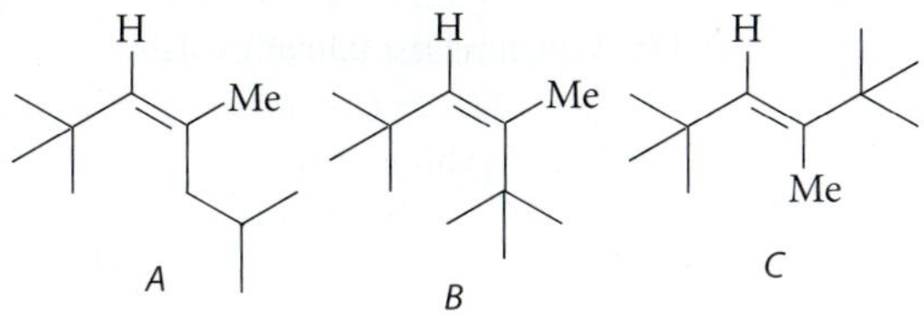

A *B* *C*

(c)

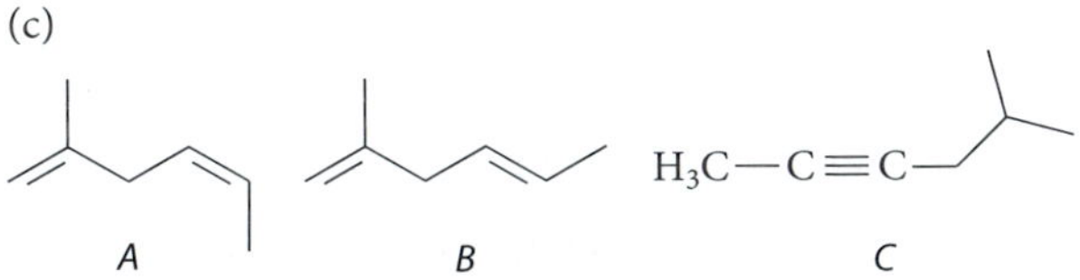

A *B* *C*

(d) 1-Hexyne and 3-hexyne

- **Use Hess's law calculations to calculate relative heats of formation. 4.5**

4.54 The $\Delta H°$ of hydrogenation is the heat liberated when a compound undergoes catalytic hydrogenation. Consider the $\Delta H°$ values for hydrogenation of the following three alkenes: 3-methyl-1-butene, –126.8 kJ mol^{-1} (–30.31 kcal mol^{-1}); 2-methyl-1-butene, –119.2 kJ mol^{-1} (–28.49 kcal mol^{-1}); and 2-methyl-2-butene, –112.6 kJ mol^{-1} (–26.91 kcal mol^{-1}).

(a) Draw an energy diagram in which the three alkenes are placed on the same energy scale along with their hydrogenation product, 2-methylbutane.

(b) Use these data to rank the three alkenes in order of stability, placing the most stable first. Explain how you reached your conclusion.

(c) By how much do the *heats of formation* of the three alkenes differ from each other? Explain.

(d) Explain why this stability order is expected from the structures of the three alkenes.

- **Write the general result of an addition reaction to an alkene or alkyne in which regioselectivity is *not* an issue. 4.6**

4.55 Write the structure of the product of an addition reaction to 4-octene with each of the following reagents. (Ignore stereochemistry, and assume that the bond shown is the one that breaks.)

(a) F—Cl

(b) Br—Br

(c) $H—BH_2$

(d) A general reagent X—Y

4.56 (a) Give the structures of all possible addition products when one equivalent of Br_2 adds to 3-hexyne, including stereoisomers.

(b) Give the structure(s) of all possible addition products when one equivalent of H—Cl adds to 3-hexyne, including stereoisomers.

(c) Give all possible addition products when two equivalents of Br_2 add to 3-hexyne.

- **For a given alkene or alkyne, give the product(s) of the following addition reactions (taking into account regioselectivity where appropriate): (a) hydrogen halide addition, (b) catalytic hydrogenation, and (c) hydration (for alkenes). 4.7A, 4.9, 4.10A–C**

4.57 For each part, draw the structure of the major (or only) product of an addition reaction to the following alkene with the reagents shown.

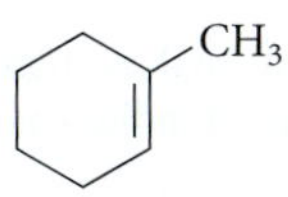

1-methylcyclohexene

(a) HBr

(b) H_2, catalyst

(c) H_2O, nitric acid (catalyst)

4.58 For each part, draw the structure of the major (or only) product of an addition reaction to each of the following alkynes with the reagents shown.

$CH_3CH_2C{\equiv}CCH_2CH_3$ — *A* — **3-hexyne**

$CH_3(CH_2)_3C{\equiv}CH$ — *B* — **1-hexyne**

(a) HCl (two equivalents)

(b) H_2, Pt/C

(c) H_2, Lindlar catalyst

- **Draw the structure(s) of the alkenes or alkynes that could give specified addition products of the following reactions: (a) hydrogen halide addition, (b) catalytic hydrogenation, and (c) hydration (for alkenes). 4.10A–C**

4.59 In each of the following cases, draw structures of *all* of the alkenes (if any) containing one double bond or alkynes containing one triple bond that will provide the given product as the *major* product of addition. (There may be no alkene that will give the product shown.)

(a)  from addition of HBr

(b) 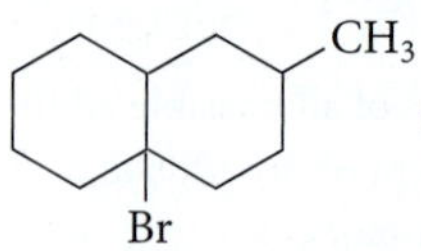 from addition of HBr

(c) 2,4-Dimethylpentane from catalytic hydrogenation; include both alkenes and alkynes, if any.

(d) 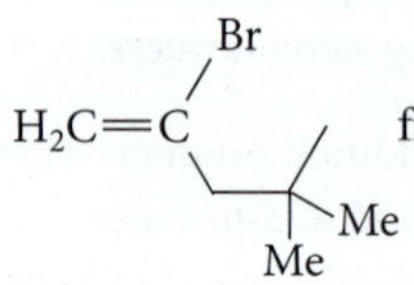 from HBr addition

(e) 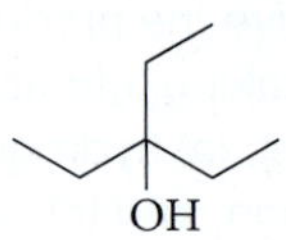 from acid-catalyzed hydration

(f) 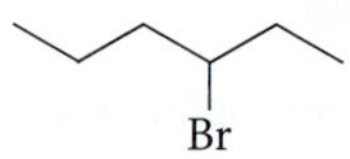 from addition of HBr

(g) 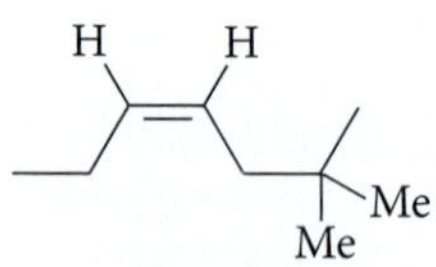 from catalytic hydrogenation with a Lindlar catalyst

• Draw a curved-arrow mechanism for (a) hydrogen halide addition and (b) acid-catalyzed hydration to a given alkene. Identify Brønsted acid–base reactions, electron-pair displacement reactions, and Lewis acid–base association and dissociation reactions that occur in the mechanism. 4.7A–C

4.60 Draw the curved-arrow mechanism for (a) HBr addition and (b) acid-catalyzed hydration of 2-methyl-2-butene. For each mechanism, identify steps that are Brønsted acid–base reactions, electron-pair displacement reactions, or Lewis acid–base association and dissociation reactions.

• Draw resonance structures and orbital diagrams that depict hyperconjugation in a carbocation. 4.7C

4.61 (a) Carbon–carbon bonds, like carbon–hydrogen bonds, can be involved in hyperconjugation. Circle the carbon–hydrogen bonds and the carbon–carbon bonds that could be involved in the stabilization of the following carbocation by hyperconjugation. (*Hint:* Expand the structure to show the appropriate bonds explicitly.)

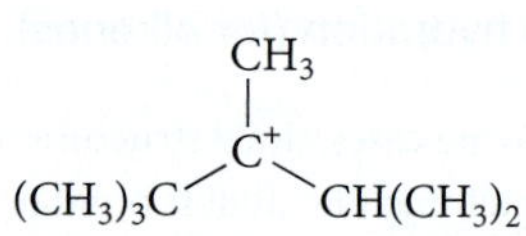

(b) Sketch the bond orbital and the empty 2p orbital of the carbocation, and show the overlap between them for one example of C—H bond hyperconjugation in this carbocation.

(c) Sketch the bond orbital and the empty 2p orbital of the carbocation, and show the overlap between them for one example of C—C bond hyperconjugation in this carbocation.

(d) Draw resonance structures that show the hyperconjugation in parts (b) and (c).

• Rank isomeric carbocations in order of stability. 4.7C

4.62 Which carbocation in each of the following groups is most stable? Which is least stable? Explain.

(a)

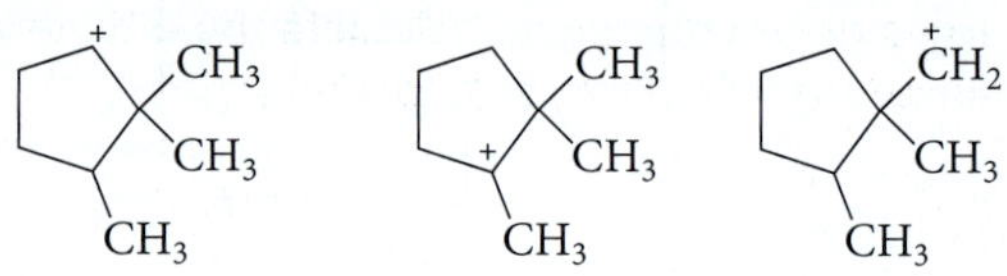

A *B* *C*

(b)

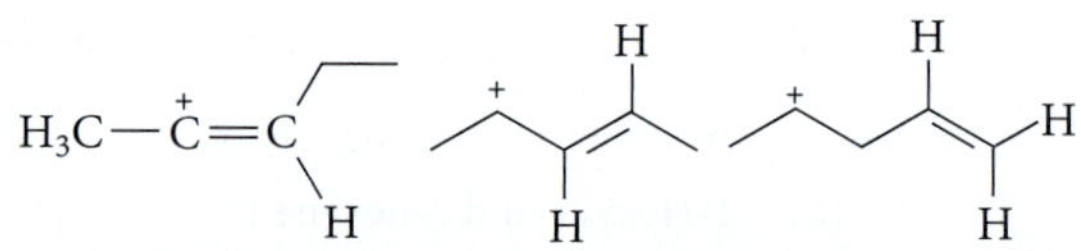

A *B* *C*

• Given the structure of a carbocation, indicate whether it is likely to rearrange, and give the structure of the rearranged carbocation. 4.7D

4.63 Of the three carbocations shown in Problem 4.62a, which one(s) is (are) most likely to rearrange? Give the structures of the rearranged carbocations. (*Hint:* In one case two rearranged carbocations are possible.)

• In a reaction coordinate-free energy diagram, identify the points corresponding to starting materials, reactive intermediates, and transition states. Identify the rate-limiting step, and show the standard free energy of activation diagrammatically.

4.64 A reaction $A \rightleftharpoons B \rightleftharpoons C \rightleftharpoons D$ has the following reaction free-energy diagram.

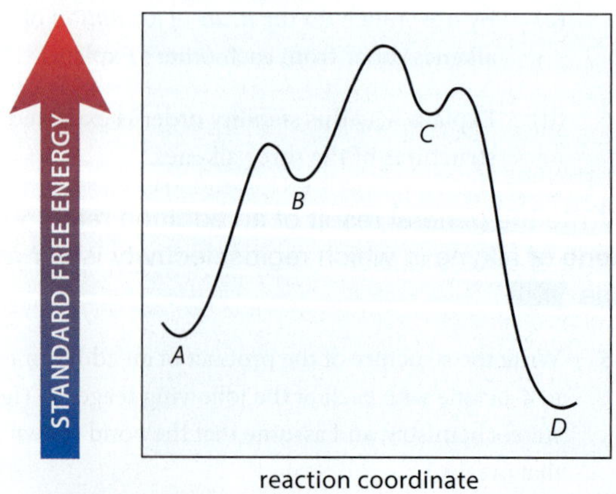

(a) Identify the points corresponding to starting materials (in the left-to-right direction), products, reactive intermediates, and transition states.

(b) Which compound is present in greatest amount when the reaction comes to equilibrium? In least amount?

(c) What is the rate-limiting step of the reaction?

(d) Using a vertical arrow, label the standard free energy of activation for the overall $A \longrightarrow D$ reaction.

(e) Which reaction of compound C is faster: $C \longrightarrow B$ or $C \longrightarrow D$? How do you know?

• Given the standard free energies of activation of two reactions, calculate their relative rates. Given the relative rates of two reactions, calculate the difference in their standard free energies of activation. 4.8

4.65 (a) A reaction $A \longrightarrow B$ has a standard free energy of activation of 81.5 kJ mol^{-1} (19.5 kcal mol^{-1}). A competing reaction $A \longrightarrow C$ has a standard free energy of activation of 87.2 kJ mol^{-1} (20.8 kcal mol^{-1}). Which reaction of compound A is faster and by what factor?

(b) A reaction $D \longrightarrow E$ is faster than a competing reaction of $D \longrightarrow F$ by a factor of 100. What is the difference in their standard free energies of activation?

• Given the curved-arrow mechanism for a reaction, sketch the structure of the transition state for each step using dashed lines and partial charges. 4.8A

4.66 (a) Give the product X expected when methylenecyclobutane undergoes acid-catalyzed hydration.

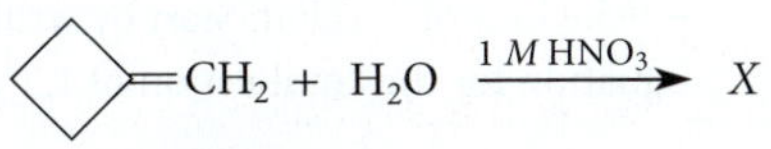

methylenecyclobutane

(b) The rate-limiting step is protonation of the double bond; use H_3O^+ as the acid catalyst. Draw the structure of the reactive intermediate formed in the rate-limiting step.

(c) Using dashed lines and partial charges, draw the transition state for the rate-limiting step.

(d) What is the rate-limiting step for dehydration of X [the reverse of the reaction shown in part (a)]?

• Apply Hammond's postulate to predict which of two closely related reactions is faster. 4.8D

4.67 Use Hammond's postulate to predict which of the following alkenes undergoes the faster addition of HBr. Explain your reasoning. (Ignore the energy difference between A and B.)

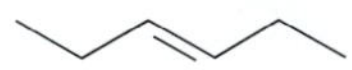

A

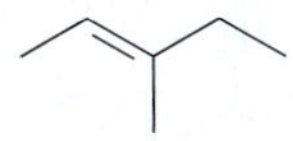

B

• Identify the effect of catalysis on the rate of a reaction. 4.10

4.68 Which of the following reaction free-energy diagrams for a reaction $A \longrightarrow B$ corresponds to a catalyzed reaction?

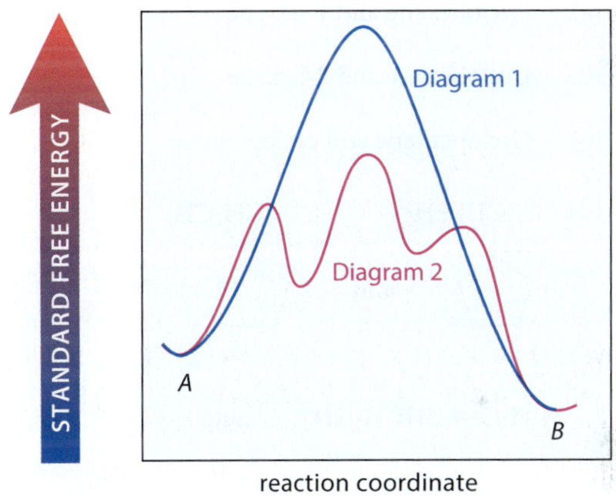

• Given the mechanism of a reaction, apply the principle of microscopic reversibility to determine the mechanism of the reverse reaction. 4.10C

4.69 (a) Draw a stepwise curved-arrow mechanism for the acid-catalyzed hydration of (E)-3-methyl-2-pentene (alkene B in Problem 4.67).

(b) Draw a stepwise curved-arrow mechanism for the reverse (dehydration) reaction.

INTEGRATED PROBLEMS

4.70 Give a structure for each of the following compounds.

(a) Cyclobutene

(b) 3-Methyl-1-octene

(c) 5,5-Dimethyl-1,3-cycloheptadiene

(d) 1-Vinylcyclohexene

(e) 4-Allyl-2-methyl-1-octen-5-yne

4.71 Name the following compounds, including the stereochemistry where appropriate.

(a) 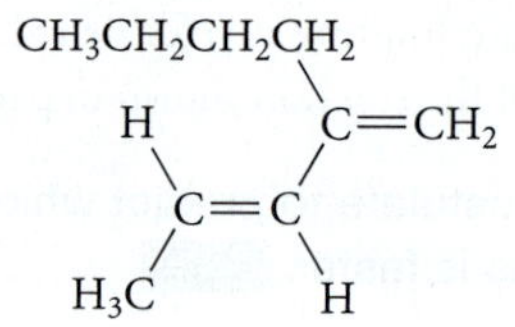

(b) 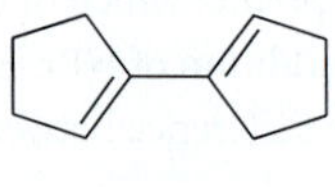

(c) 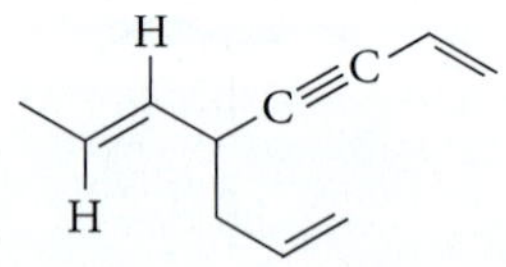

4.72 Classify the compounds within each of the following pairs as identical molecules (I), constitutional isomers (C), stereoisomers (S), or none of the above (N).

(a) Cyclohexene and 1-hexene

(b) Cyclohexene and 2-hexyne

(c) Cyclopentane and cyclopentene

(d) 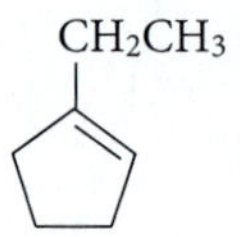 and 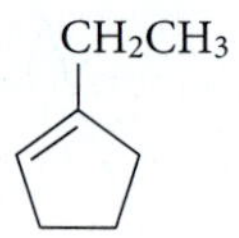

(e) $H_2C{=}CHCH_2CH_3$ and 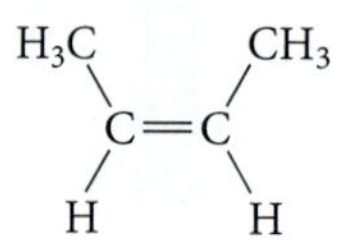

(f) 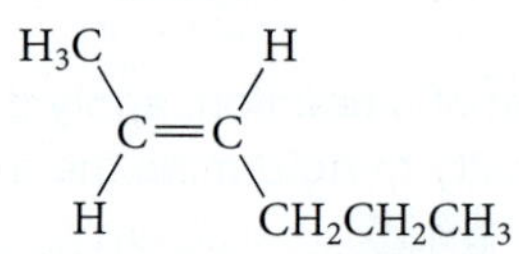 and 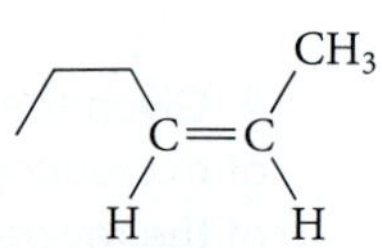

4.73 Give the structures of the two stereoisomeric alkenes with the molecular formula C_6H_{12} that react with HI to give the same single product and undergo catalytic hydrogenation to give hexane.

4.74 Give the structures of all the alkynes and alkenes (including double-bond stereoisomers) containing one double bond that would give propylcyclohexane as the product of catalytic hydrogenation.

4.75 (a) Draw the structure of a stable five-carbon alkyne that contains a ring.

(b) Draw the structure of a stable six-carbon trans alkene that contains a ring.

4.76 An alkene *X* with molecular formula C_7H_{12} adds HBr to give a single alkyl halide *Y* with molecular formula $C_7H_{13}Br$ and undergoes catalytic hydrogenation to give 1,1-dimethylcyclopentane. Draw the structures of *X* and *Y*. (For solution hints, see Study Guide Link 4.3.)

STUDY GUIDE LINK 4.3
Solving Structure Problems

4.77 You have been called in as a consultant for the firm Alcohols Unlimited, which wants to build a plant to produce 3-methyl-1-butanol, $(CH_3)_2CHCH_2CH_2OH$. The research director, Al Keyhall, has proposed that acid-catalyzed hydration of 3-methyl-1-butene can be used to prepare this compound. The company president, O. H. Gruppa, has asked you to evaluate this suggestion. Millions of dollars are on the line. What is your answer? Can 3-methyl-1-butanol be prepared in this way? Explain your answer.

4.78 A certain compound *A* is converted into a compound *B* in a reaction without reactive intermediates. The reaction has an equilibrium constant $K_{eq} = [B]/[A] = 150$ and, with the free energy of *A* as a reference point, a standard free energy of activation of 96 kJ mol^{-1} (23 kcal mol^{-1}).

(a) Draw a reaction free-energy diagram for this process, showing the relative free energies of *A*, *B*, and the transition state ‡ for the reaction.

(b) What is the standard free energy of activation for the reverse reaction $B \longrightarrow A$? How do you know?

4.79 The alkene 3,3-dimethyl-1-butene undergoes acid-catalyzed hydration with rearrangement. Use the mechanism of hydration and rearrangement to predict the structure of the hydration product of this alkene.

4.80 The heat of formation of (*E*)-1,3-pentadiene is 75.8 kJ mol^{-1} (18.1 kcal mol^{-1}), and that of 1,4-pentadiene is 106.3 kJ mol^{-1} (25.4 kcal mol^{-1}).

(a) Which alkene has the more stable arrangement of bonds? (We consider the reasons in Sec. 15.1.)

(b) Calculate the heat liberated when 1 mol of 1,3-pentadiene is burned. The heat of combustion of carbon is –393.5 kJ mol^{-1}, and that of H_2 is –285.8 kJ mol^{-1}. (*Hint:* Start by writing a balanced equation for the combustion of 1,3-pentadiene.)

4.81 Make a model of cycloheptene with the trans (or *E*) configuration at the double bond. Now make a model of *cis*-cycloheptene. By examining your models, determine which compound should have the greater heat of formation. Explain.

4.82 Supply the curved-arrow notation for the acid-catalyzed isomerization.

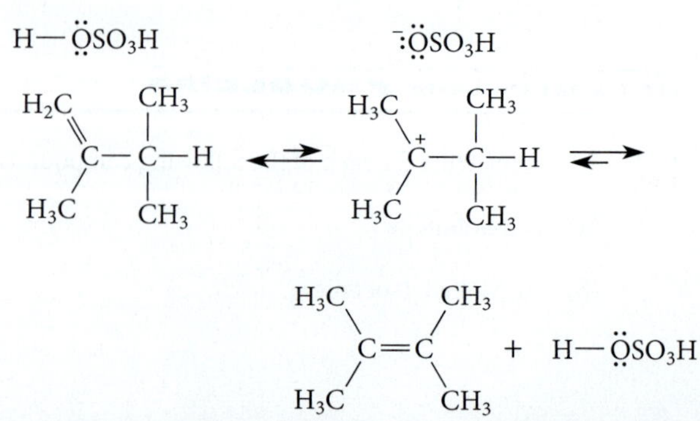

4.83 The industrial synthesis of methyl *tert*-butyl ether involves the treatment of methylpropene with methanol in the presence of an acid catalyst:

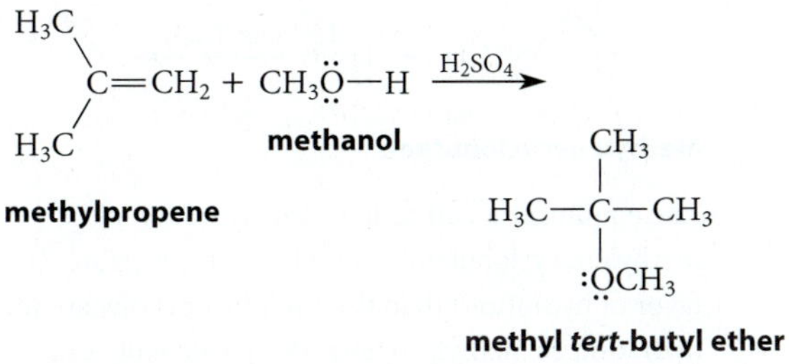

This ether has been used commercially as an antiknock gasoline additive. (It is banned in the United States but is used in parts of Europe.) Using the curved-arrow notation, propose a mechanism for this reaction. (*Hint:* Use the catalyst to protonate the double bond, and think of this reaction as an analogy to acid-catalyzed hydration.)

4.84 The curved-arrow notation can be used to understand seemingly new reactions as simple extensions of what you already know. This is the first step in developing an ability to use the notation to predict new reactions. Provide a curved-arrow mechanism for the following reaction.

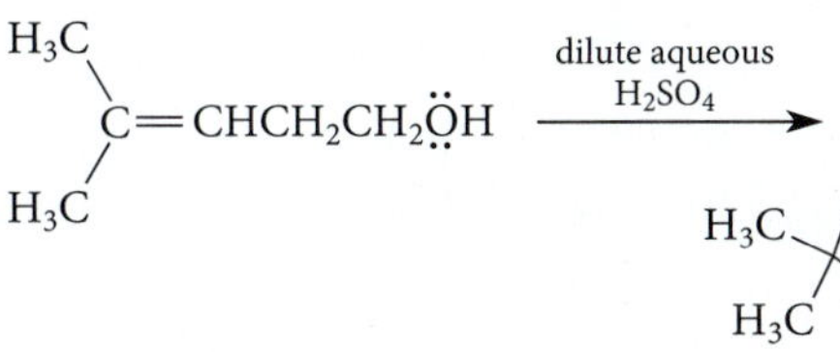

To do this, follow these steps:

1. Examine the reactants and products, and label corresponding atoms. If you're not sure, make a guess.

2. Describe what has happened to the functional groups in the starting material. In this case, focus on the double bond. Is this transformation similar in any way to a reaction you have seen before?

3. Make the connections you deduced in step 1 with a curved-arrow mechanism, trying to use steps that are similar to mechanistic steps you've seen in other reactions. Use separate structures for each step of the mechanism—that is, don't try to write several mechanistic steps using the same structure.

4. Use a Lewis acid–base association, Lewis acid–base dissociation, or Brønsted acid–base reaction for each step.

4.85 Using the curved-arrow notation, suggest a mechanism for the following reaction. [*Hints:* (1) Follow the problem-solving suggestions in Problem 4.84. (2) Use Hammond's postulate to decide which double bond should protonate first.]

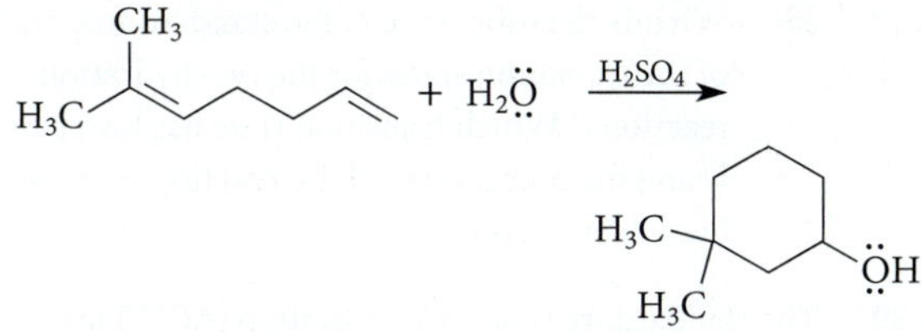

4.86 Which of the following two alkenes is the stronger Brønsted base? Rationalize your answer by showing the alkenes and their conjugate acids on a free-energy diagram. (Place the two alkenes at the same energy level for comparison—that is, ignore the difference in their energies.) (*Hint:* See the use of free-energy diagrams in Sec. 3.7D.)

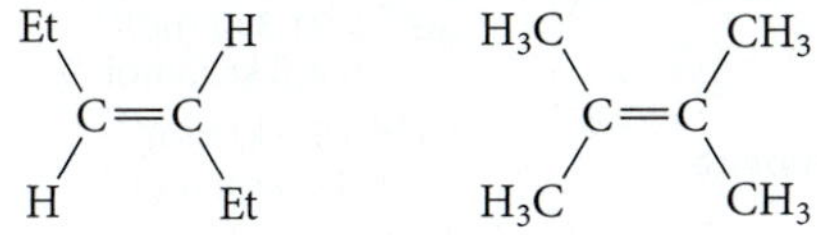

A *B*

4.87 The standard free energy of formation, ΔG_f°, is the free-energy change for the formation of a substance at 25 °C and 1 atm pressure from its elements in their natural states under the same conditions.

(a) Calculate the equilibrium constant for the interconversion of the following alkenes, given the standard free energy of formation of each. Indicate which compound is favored at equilibrium.

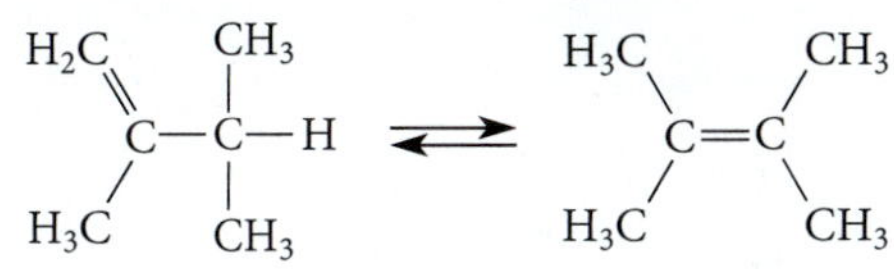

ΔG_f° 79.0 kJ mol^{-1} (18.9 kcal mol^{-1}) 75.9 kJ mol^{-1} (18.1 kcal mol^{-1})

(This is the same reaction for which Problem 4.82 asked for the curved-arrow mechanism.)

(b) What does the equilibrium constant tell us about the rate at which this interconversion takes place?

(c) Give a structural reason for the direction of the chemical equilibrium.

4.88 The *difference* in the standard free energies of formation for 1-butene and methylpropene is 13.4 kJ mol^{-1} (3.20 kcal mol^{-1}). (See Problem 4.87 for a definition of ΔG_f°.)

(a) Which compound is more stable? Why?

(b) The standard free energy of activation for the hydration of methylpropene is 22.8 kJ mol^{-1} (5.74 kcal mol^{-1}) less than that for the hydration of 1-butene. Which hydration reaction is faster?

(c) Draw reaction free-energy diagrams on the same scale for the hydration reactions of these two alkenes, showing the relative free energies of both starting materials and rate-determining transition states.

(d) What is the difference in the standard free energies of the transition states for the two hydration reactions? Which transition state has lower energy? Using the mechanism of the reaction, suggest why it has lower energy.

4.89 The standard free energy of activation ($\Delta G^{\circ\ddagger}$) for hydration of methylpropene to 2-methyl-2-propanol is 91.3 kJ mol^{-1} (21.8 kcal mol^{-1}). The standard free energy ΔG° for hydration of methylpropene is −5.6 kJ mol^{-1} (−1.3 kcal mol^{-1}).

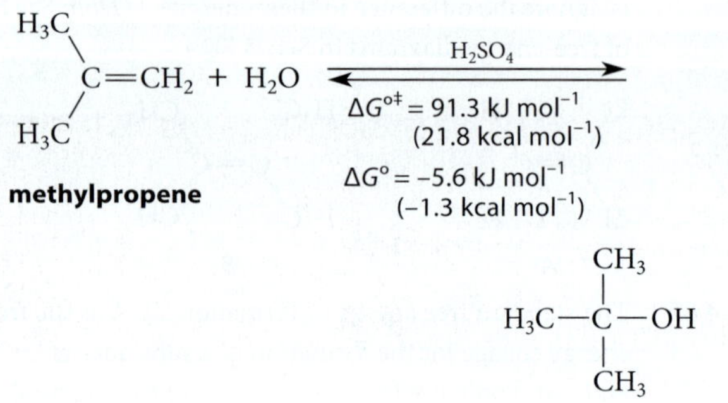

2-methyl-2-propanol

The rate of hydration of methylenecyclobutane to give an alcohol (compound *X*) is 0.6 times the rate of hydration of methylpropene.

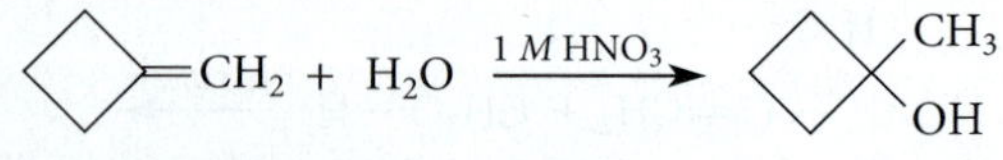

methylenecyclobutane *X*

The equilibrium constant for the hydration of methylenecyclobutane is about 250 times greater (in favor of hydration) than the equilibrium constant for the hydration of methylpropene. Which alcohol, *X* or 2-methyl-2-propanol, undergoes dehydration faster? How much faster is it? Explain.

5 ADDITION REACTIONS OF ALKENES AND ALKYNES

The most common reactions of alkenes and alkynes are *addition reactions*. In Chapter 4, we studied hydrogen halide addition and catalytic hydrogenation of alkenes and alkynes, and hydration of alkenes. We learned how the curved-arrow notation and the properties of reactive intermediates can be used to understand the regioselectivity of these additions. Here in Chapter 5, we survey some other addition reactions of alkenes and alkynes using the same approach.

5.1 AN OVERVIEW OF ELECTROPHILIC ADDITION REACTIONS

In subsequent sections, we present several more alkene and alkyne addition reactions in detail. First, though, we're going to look at four of these reactions in a more general way to see how they resemble two reactions we've already studied—namely, the addition of HBr (Sec. 4.7) and the addition of H_2O (Sec. 4.10C). Understanding these connections will help you to learn these reactions more easily.

Each of the four reactions is illustrated here with methylpropene, a prototypical alkene. First, convince yourself that each of these reactions is an addition. Then determine what has been added to the double bond. Notice where each group in the product comes from.

Bromine addition:

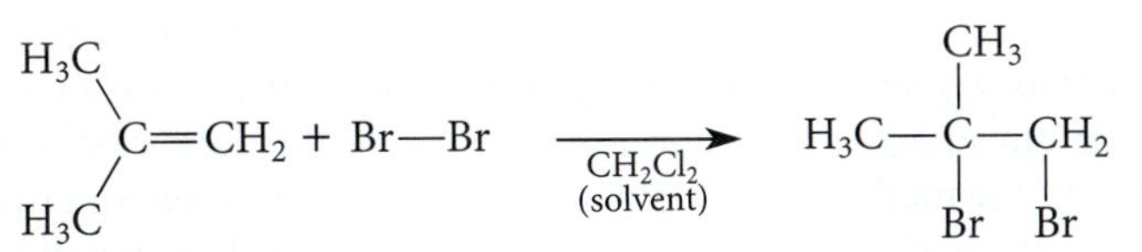

methylpropene (isobutylene) **1,2-dibromo-2-methylpropane** (5.1)

Oxymercuration:

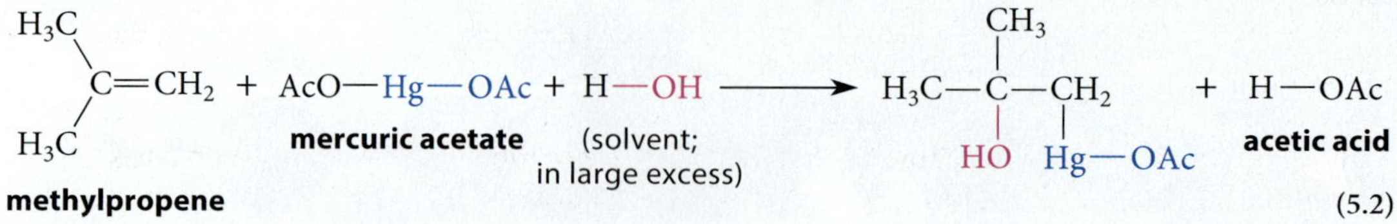

(5.2)

In this reaction, —OAc and AcO— are abbreviations for the *acetoxy group*:

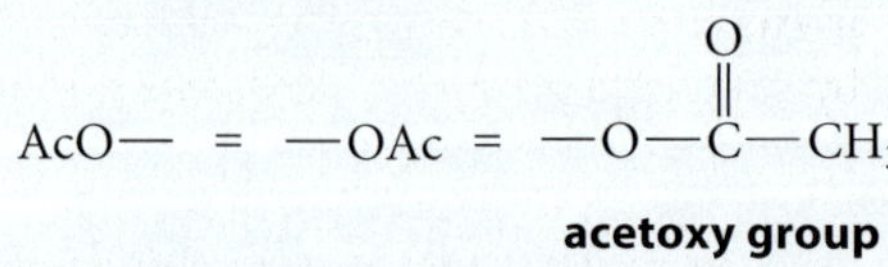

Hydroboration:

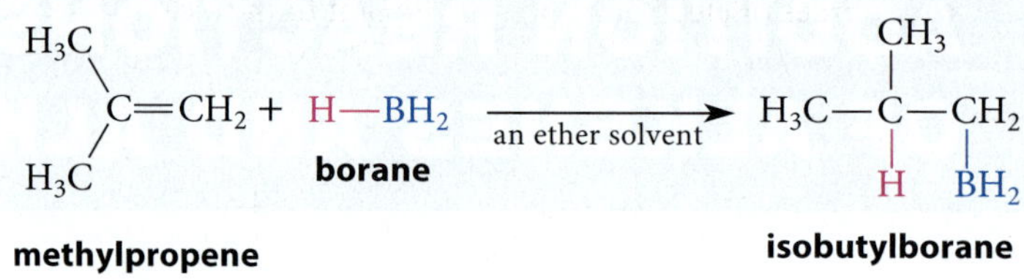

(5.3)

In this reaction, don't confuse the element *boron* (B) with *bromine* (Br).

Ozonolysis:

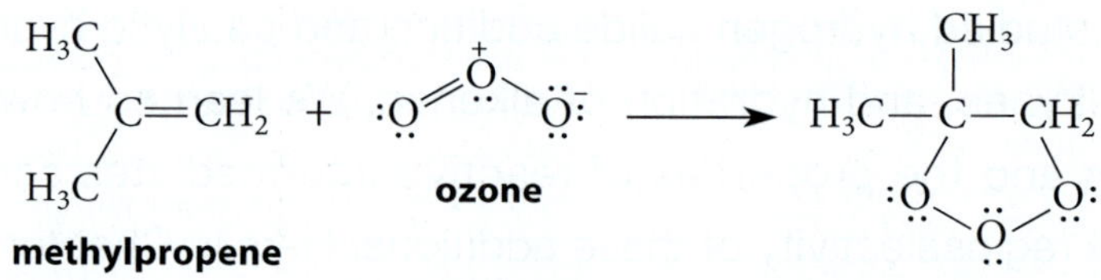

(5.4)

Ozonolysis is an example of a **cycloaddition**, which is an addition reaction that forms a ring.

Consider the reactions in which two different groups add to the double bond. If you focus on the relative electronegativities of the two groups, the outcome of each reaction is similar to the outcomes of HBr addition and hydration. In the additions of both HBr and H_2O, the hydrogen, which is the less electronegative group in both, adds to the CH_2 carbon of the double bond, and the more electronegative group (Br or OH) adds to the carbon with the methyl substituents.

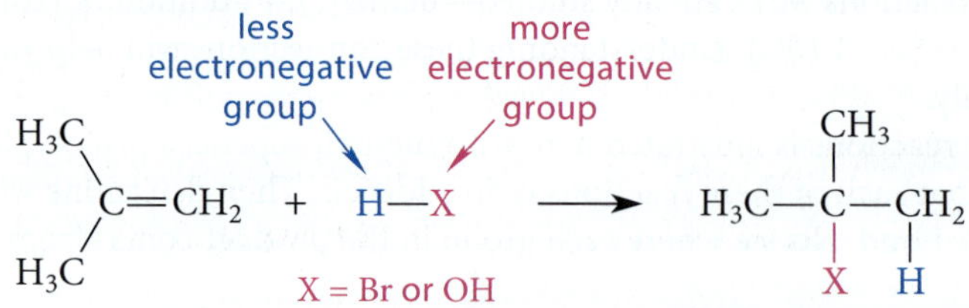

(5.5)

In oxymercuration (Eq. 5.2), Hg is a metal, which is much less electronegative than the oxygen of the OH group (2.00 for Hg vs. 3.44 for O; see Fig. 1.15, Sec. 1.9A). In this reaction, too, the less electronegative group adds to the carbon of the CH_2, and the more electronegative group adds to the carbon with the methyl substituents. The same pattern applies in hydroboration (Eq. 5.3), where the electronegativity of hydrogen (2.20) is *greater* than that of boron (2.04). In this case, boron (the *less* electronegative group) adds to the CH_2 carbon, and hydrogen (the *more* electronegative group) adds to the carbon with the methyl substituents.

To generalize these observations: In these addition reactions, when the two groups that add are different, *the carbon of the double bond with fewer alkyl substituents becomes bonded to the less electronegative group, and the carbon of the double bond with more alkyl substituents becomes bonded to the more electronegative group.* We can think of this statement as a "modified Markovnikov's rule" (Sec. 4.7A).

As we explain in the following sections, these reactions occur by different mechanisms, but two common threads run through all of the mechanisms:

1. The first step of each reaction is *the donation of an electron pair from the alkene π bond to an electrophilic center.* The alkene π bond *acts as a nucleophile* in this process. The electrophilic atom becomes bonded to the alkene carbon with *fewer alkyl substituents.*

2. A nucleophilic atom that was either part of the original reactant or present in solution completes the addition by *donating electrons to the alkene carbon with the greater number of alkyl substituents.*

Abbreviating the electrophile or electrophilic center as E (blue) and the nucleophile or nucleophilic center as X (red), this idea can be summarized as shown in Eq. 5.6. (The dashed curved arrows show the sources of electrons but aren't meant to indicate the actual reaction mechanism.)

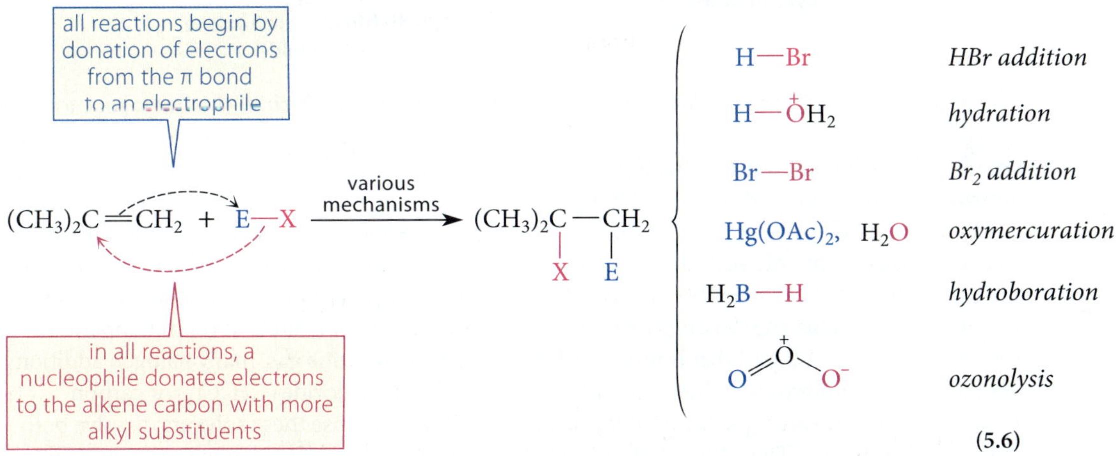

(5.6)

Review the mechanisms of HBr addition (Sec. 4.7B) and hydration (Sec. 4.10C), and notice how they fit this pattern.

Because of the similarity of these addition reactions and others like them, they are grouped as a class and referred to as *electrophilic additions.* An addition reaction is an **electrophilic addition** when it begins with the donation of an electron pair from a π bond to an electrophilic atom.

Focused Problem

5.1 (a) Iodine azide, I—N_3, adds to methylpropene in the following manner:

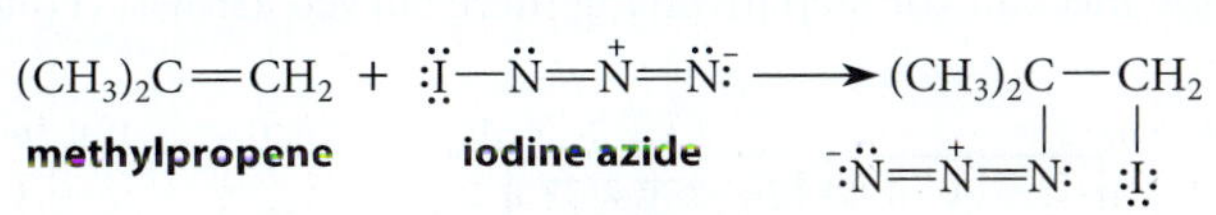

Which group or atom in iodine azide is the electrophilic group to which the π bond donates electrons? How do you know? Does this result fit the electronegativity pattern for electrophilic additions? Explain.

(b) Predict the product of the following electrophilic addition reaction, and explain your reasoning.

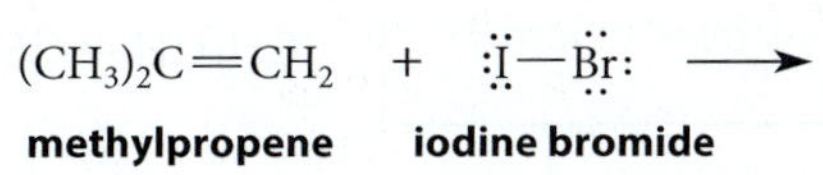

5.2 REACTIONS OF ALKENES WITH HALOGENS

A. Addition of Chlorine and Bromine

Halogens undergo addition to alkenes.

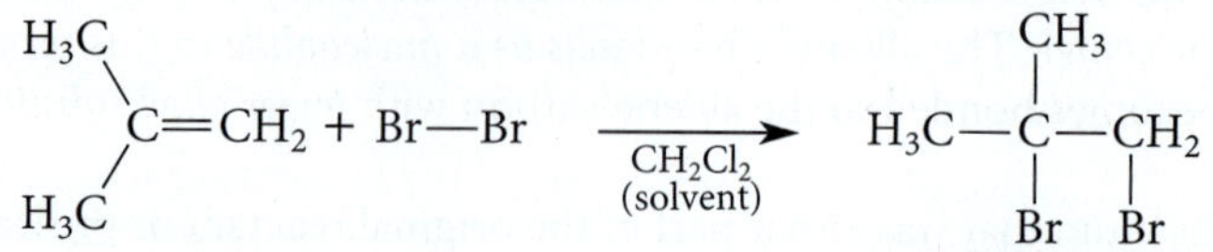

methylpropene (isobutylene) → **1,2-dibromo-2-methylpropane** (5.7)

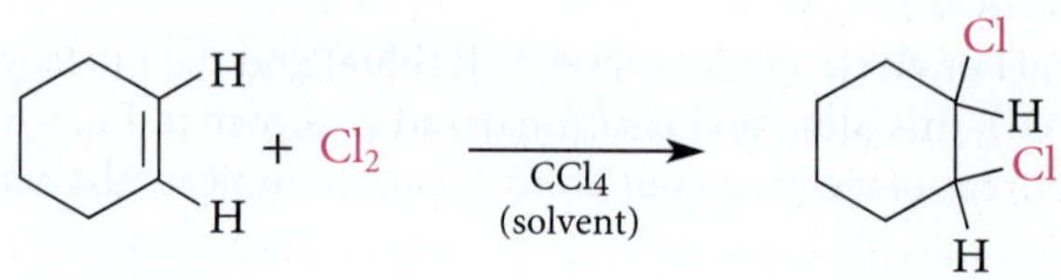

cyclohexene → **1,2-dichlorocyclohexane** (70% yield) (5.8)

The products of these reactions are examples of **vicinal dihalides**. *Vicinal* (Latin *vicinus*, for "neighborhood") means "on adjacent sites." Thus, vicinal dihalides are compounds with halogens on adjacent carbons. Contrast this type of dihalide with a *geminal dihalide*, in which two halogen atoms are on the *same* carbon (Sec. 4.9, Eq. 4.63a).

Bromine and chlorine are the two halogens used most frequently in halogen addition. Fluorine is so reactive that it not only adds to the double bond but also rapidly replaces all the hydrogens with fluorines, often with considerable violence. Iodine adds to alkenes at low temperature, but most diiodides are unstable and decompose to the corresponding alkenes and I_2 at room temperature. Because bromine is a liquid that is more easily handled than chlorine gas, many halogen additions are carried out with bromine. Inert solvents such as methylene chloride (CH_2Cl_2) or carbon tetrachloride (CCl_4) are typically used for halogen additions because these solvents dissolve both halogens and alkenes. The addition of bromine to most alkenes is so fast that when bromine is added dropwise to a solution of the alkene, the red bromine color disappears almost immediately.

Bromine addition can occur by a variety of mechanisms, depending on the solvent, the alkene, and the reaction conditions. One of the most common mechanisms involves a reactive intermediate called a *bromonium ion.*

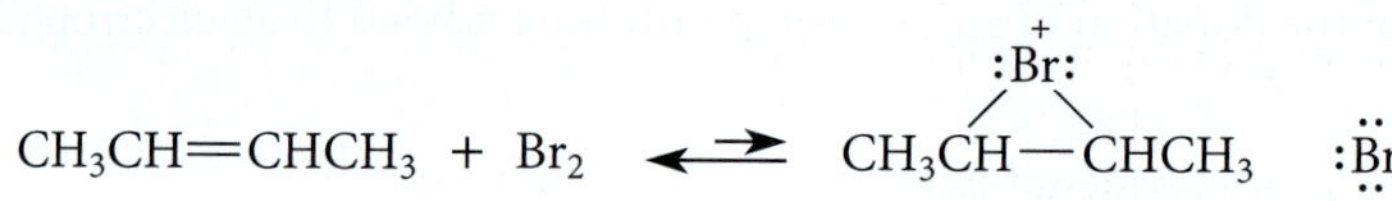

a bromonium ion (5.9)

A **bromonium ion** is a species that contains a bromine bonded to two adjacent carbon atoms; the bromine has an octet of electrons and a positive charge. Formation of the bromonium ion occurs in a single mechanistic step involving three curved arrows. (Follow the arrows below in order.)

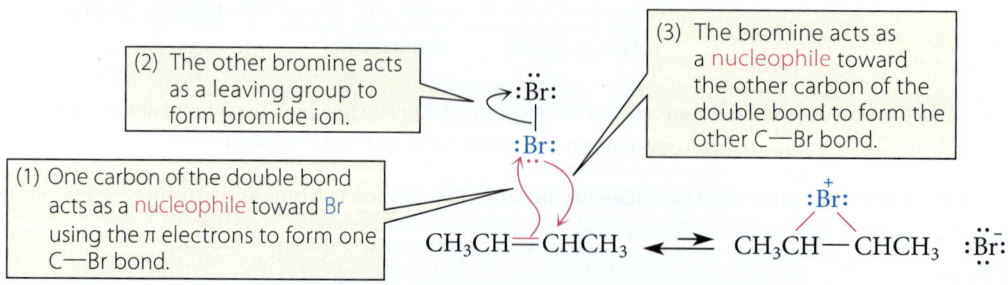

(5.10a)

Analogous cyclic ions form in chlorine and iodine addition.

Bromine addition is completed when the bromide ion donates an electron pair to either one of the ring carbons of the bromonium ion.

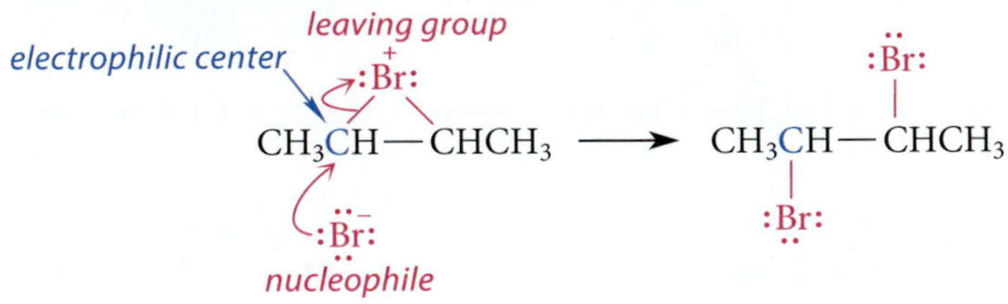

(5.10b)

This is another *electron-pair displacement reaction* (Sec. 3.2), in which the nucleophile is bromide ion, the electrophilic center is the carbon that accepts an electron pair from the nucleophile, and the leaving group is the bromine of the bromonium ion. (The leaving group doesn't actually leave the molecule, because it is tethered by another bond.) This reaction occurs because the positively charged bromine is very electronegative and readily accepts an electron pair. It also occurs because, as we explain in Chapter 7, three-membered rings are strained, so opening a three-membered ring releases considerable energy. (To better appreciate the strain, make a model of a three-membered ring.)

We might reasonably have written a mechanism for bromine addition that involves a carbocation intermediate, as we did for HBr addition:

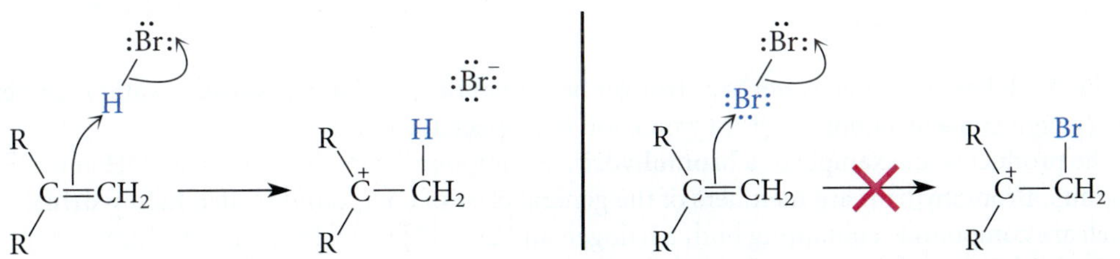

H—Br addition Br—Br addition (5.11)

How do we know that bromonium ions rather than carbocations are reactive intermediates in bromine addition? First, rearrangements are not typically observed in bromine addition. We have seen (Sec. 4.7D) that rearrangements are observed in HBr addition. Second, bromonium ions have been isolated under special circumstances, although, like carbocations, they are not stable enough to isolate under the usual reaction conditions. Finally, there is compelling experimental evidence for bromonium ions that we consider in Sec. 7.8C after we have developed some additional background in stereochemistry in Chapter 6.

Why does a bromonium ion form instead of a carbocation? In other words, why is a bromonium ion more stable than the corresponding carbocation? The reasons are that a bromonium ion has more covalent bonds than a carbocation, and every atom has an octet.

bromonium ion carbocation

B. Formation of Halohydrins

When alkenes react with bromine in an inert solvent such as CH_2Cl_2 or CCl_4, the only nucleophile available to react with the bromonium ion is the bromide ion (Eq. 5.10b). When other nucleophiles are present, they, too, can react with the bromonium ion to form products other than dibromides. A common situation of this type occurs when the solvent itself can act as a nucleophile. For example, when an alkene is treated with bromine in a solvent containing a large excess of water, a water molecule rather than bromide ion reacts with the bromonium ion, because water is present in much higher concentration than bromide ion:

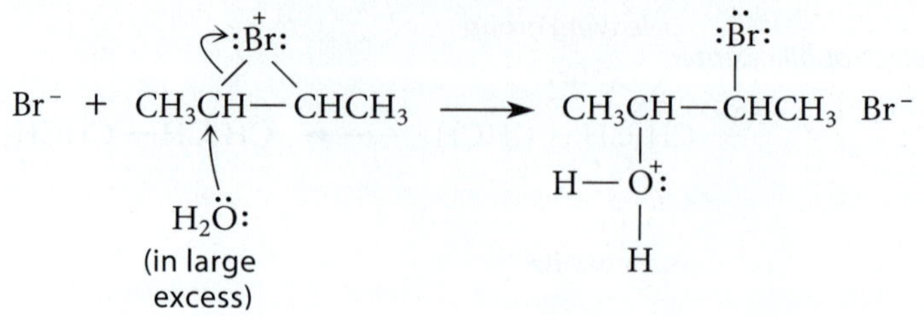

(5.12a)

The conjugate acid of an alcohol is very acidic—its acidity is comparable to that of H_3O^+. Therefore, the solvent H_2O can remove this acidic proton to give the product:

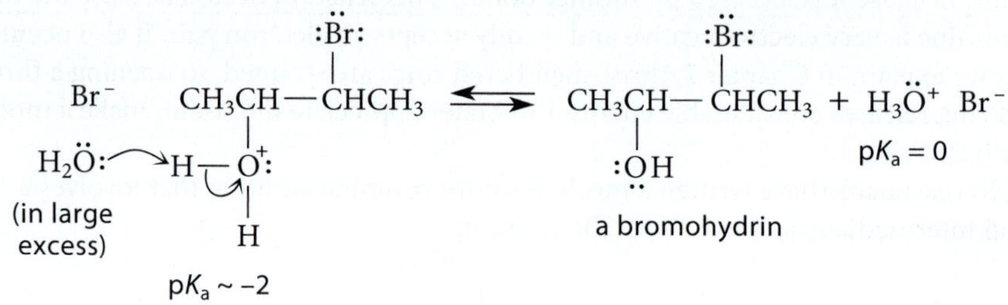

(5.12b)

(In this acid–base reaction, water is a stronger base than Br^-, and, as the solvent, water is present in very high concentration; so, the bromide ion is a "spectator ion.")

The product is an example of a **bromohydrin**: a compound containing both an OH and a Br group. Bromohydrins are members of the general class of compounds called **halohydrins**, which are compounds containing both a halogen and an OH group. In the most common type of halohydrin, the two groups occupy adjacent, or *vicinal*, positions.

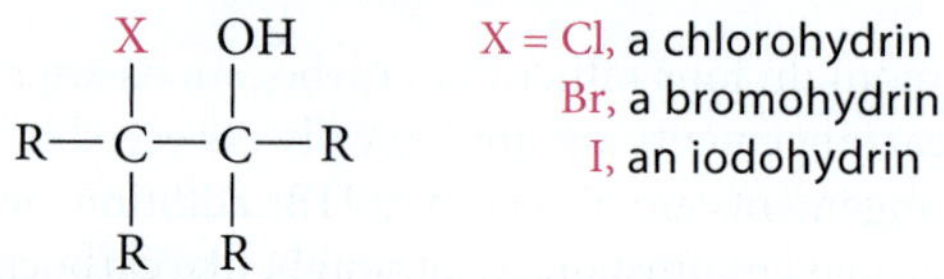

general structure
of a vicinal halohydrin
(R = alkyl, aryl, or H)

(5.13)

Although the products of I_2 addition are unstable (see Sec. 5.2A), iodohydrins can be prepared.

When the double bond of the alkene is positioned unsymmetrically, the reaction of water with the bromonium ion can give two possible products, each resulting from breakage of a different carbon–bromine bond. The reaction is highly regioselective, however, when one carbon of the alkene contains *two* alkyl substituents.

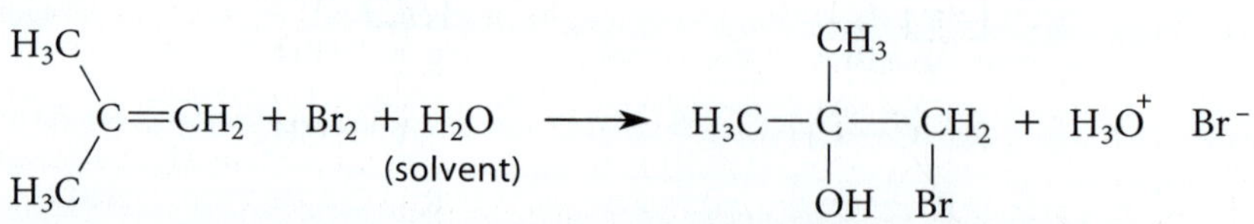

1-bromo-2-methyl-2-propanol
(77% yield)

(5.14)

The reason for this regioselectivity can be seen from the structure of the bromonium ion (**Fig. 5.1**). In this structure, about 90% of the positive charge resides on the tertiary carbon, and the bond between this carbon and the bromine is so long and weak that this species is essentially a carbocation containing a weak carbon–bromine interaction.

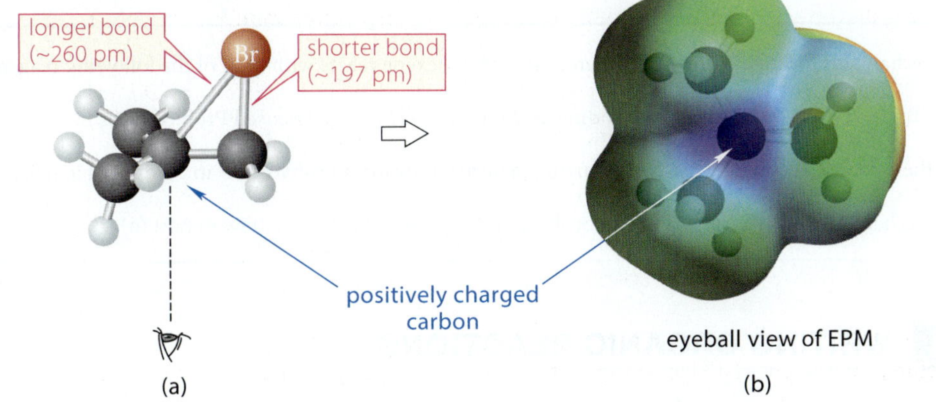

FIGURE 5.1 (a) A ball-and-stick model showing the structure of the bromonium ion involved in bromine addition to methylpropene. Notice the very long bond between the bromine and the tertiary carbon. (b) An electrostatic potential map (EPM) of the bromonium ion viewed from the direction of the eyeball in part (a). The concentration of positive charge on the tertiary carbon is indicated by the blue color. About 90% of the positive charge resides on this carbon.

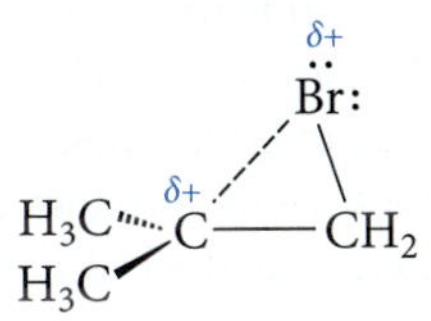

The weaker bond to the leaving group is broken, to give the observed regioselectivity.

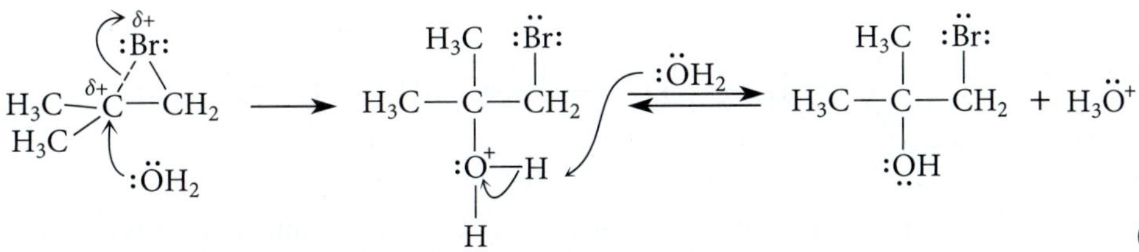

(5.15)

Study Problem 5.1

Which of the following chlorohydrins could be formed by addition of Cl_2 in water to an alkene? Explain.

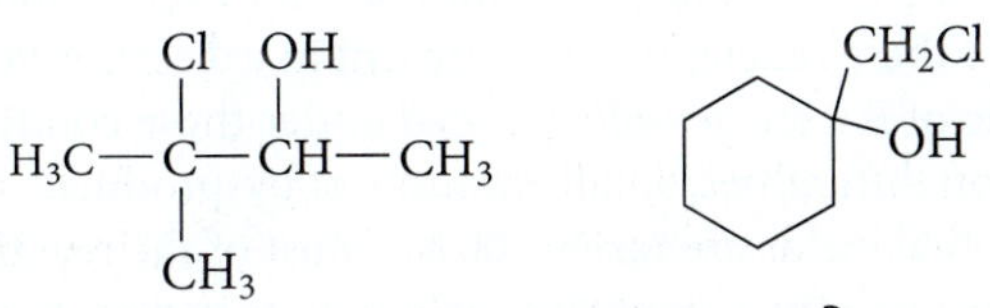

A *B*

Solution The mechanistic reasoning used in this section shows that the nucleophile (water) reacts with the carbon of the double bond that has more alkyl substituents. In compound *A*, the carbon bonded to the OH group has *fewer* alkyl substituents than the one bearing the Cl. Hence, this compound could *not* be formed in the reaction of Cl_2 and water with an alkene. Compound *B* could be formed by such a reaction, however, because the OH group is bonded to a carbon with more alkyl substituents than the Cl. Don't forget that the carbons of the ring are alkyl groups even though they are part of the ring structure.

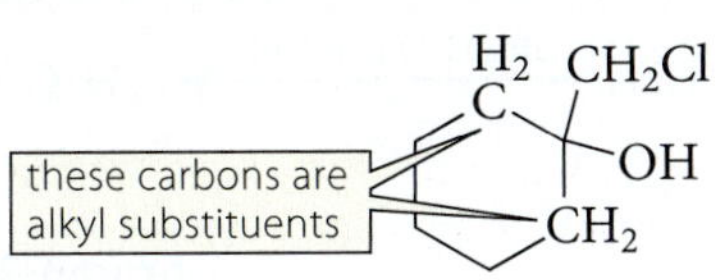

Focused Problems

5.2 Give the products, and the mechanisms for their formation, when 2-methyl-1-hexene reacts with each of the following reagents.

(a) Br_2 (b) Br_2 in H_2O (c) Iodine azide (I—N_3) (*Hint:* See Focused Problem 5.1a.)

5.3 (a) Draw the structure of the alkene that could be used as a starting material to form chlorohydrin *B* in Study Problem 5.1.

(b) Draw a curved-arrow mechanism for the formation of chlorohydrin *B* from the alkene you drew in part (a).

5.3 WRITING ORGANIC REACTIONS

As we continue with our study of organic reactions, confusion can be avoided if you learn a few widely adopted conventions for writing reactions.

The convention for writing solvents under the arrow and catalysts over the arrow is followed in this text, but in other sources conventions may differ.

The most thorough way to write a reaction is to use a complete, balanced equation. Equations 5.7 and 5.8 in Sec. 5.2A are examples of balanced equations. Other information, such as the reaction conditions, is sometimes included in equations. In Eq. 5.7, for example, the solvent is written under the arrow, even though the solvent is not an actual reactant. Also, catalysts are written over the arrow. In Eq. 5.16, for example, the H_3O^+ written over the arrow indicates that an acid catalyst is required (Sec. 4.9B).

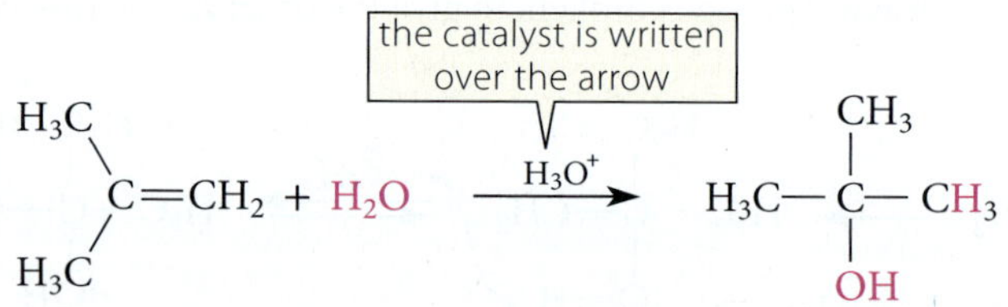

(5.16)

We can also tell reactants from catalysts because a catalyst is not consumed in the reaction.

Equation 5.8 includes a **percentage yield**, which is the percentage of the theoretical amount of product formed that has actually been isolated from the reaction mixture by a chemist in the laboratory. Although different chemists might obtain somewhat different yields in the same reaction, the percentage yield gives a rough idea of how free the reaction is from contaminating by-products and how easily the product can be isolated from the reaction mixture. Thus, a reaction $2A + B \longrightarrow 3C + D$ should give 3 mol of *C* for every 1 mol of *B* and 2 mol of *A* used (assuming that one of these reactants is not present in excess). A 90% yield of *C* means that 2.7 mol of *C* per mole of *B* were *actually isolated* under these conditions. The 10% loss may have been due to separation difficulties, small amounts of by-products, or other reasons. (Because of operational losses, actual yields are rarely 100%.) Most of the reactions given in this book are actual laboratory examples; the percentage yield figures included in many of these reactions are not meant to be learned, but are given simply to indicate how successful a reaction actually is in practice.

Here's a convention for writing reactions that you particularly need to understand. In many cases, organic chemists abbreviate reactions by showing only the *organic starting materials* and the *major organic product(s)*. The other reactants and conditions are written over the arrow. Thus, Eq. 5.14 will be written as

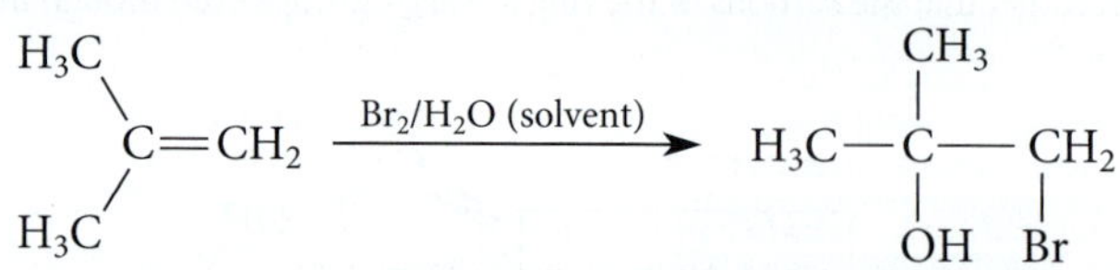

1-bromo-2-methyl-2-propanol
(77% yield)

(5.17)

This shorthand way of writing organic reactions is frequently used because it saves space and time. When equations are written this way, by-products are not given and, in many cases, the equation is not balanced. This shorthand can present ambiguities for the beginner (and sometimes for the experienced chemist as well!). Are the items written over the arrow reactants, catalysts, a solvent, or something else? In Eq. 5.17, we know that Br_2 is a reactant because it is consumed in the reaction. The H_2O solvent is also a reactant (because the product contains an OH group). The by-products H_3O^+ and Br^- are not shown.

To avoid such ambiguities, we present most reactions in this text initially in balanced form (when the balanced form is known). We also label catalysts and solvents at their first occurrence. This should help to clarify the roles of the various reaction components when abbreviated forms of the same reactions are used subsequently.

5.4 OXYMERCURATION–REDUCTION OF ALKENES

Although the hydration of alkenes (Sec. 4.10B) is used industrially for the preparation of particular alcohols, it is rarely used for the general laboratory preparation of alcohols because of the specialized conditions required. In Secs. 5.4 and 5.6, we present two reaction sequences that are especially useful in the laboratory for the conversion of alkenes into alcohols. These two processes, called *oxymercuration–reduction* and *hydroboration–oxidation*, bring about the *overall* addition of H and OH to a double bond. The two sequences are complementary, however, because they occur with opposite *regiospecificities*; in other words, they give different constitutional isomers. (Compare the positions of the —OH group in the products.)

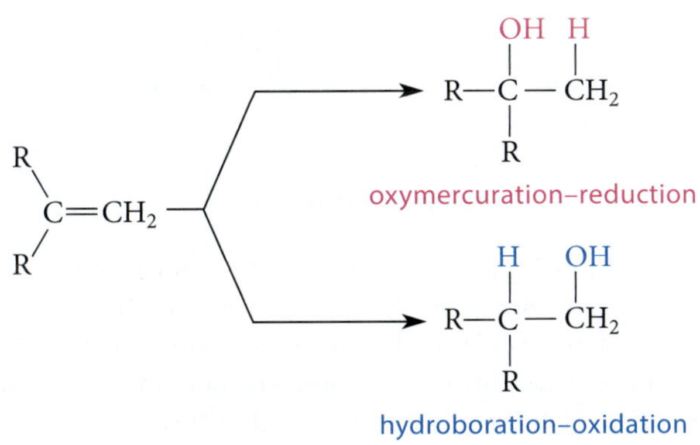

(5.18)

Each of these processes also occurs in two experimentally separate operations, which we consider in turn.

In Secs. 5.5 and 5.7, we also consider similar addition reactions of alkynes.

A. Oxymercuration of Alkenes

In **oxymercuration**, alkenes react with mercuric acetate, $Hg(OAc)_2$, in aqueous solution to give addition products in which an —HgOAc (acetoxymercuri) group and a —OH (hydroxy) group derived from water have added to the double bond.

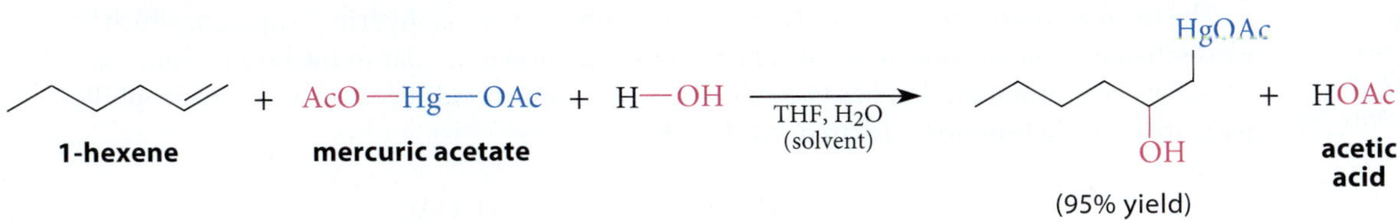

(5.19)

We discussed the —OAc (or AcO—) abbreviations when oxymercuration was introduced (Eq. 5.2, Sec. 5.1). Notice that the —HgOAc group goes to the carbon of the double bond with fewer alkyl substituents and the OH group to the carbon of the double bond with more alkyl substituents.

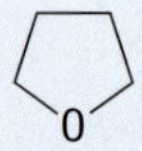

tetrahydrofuran, or **THF**

THF is an important solvent because it dissolves both water and many water-insoluble organic compounds. Its role in oxymercuration is to dissolve both the alkene and the aqueous mercuric acetate solution. (Recall that alkenes are insoluble in water alone; Sec. 4.4.) Water is required as both a reactant, as shown in Eq. 5.19, and as a solvent for the mercuric acetate.

The solvent (written under the arrow in Eq. 5.19) is a mixture of water and tetrahydrofuran (THF), a widely used ether.

In the mechanism of oxymercuration, the mercury, which is a Lewis acid, accepts *both* electrons of the alkene π bond to form a *mercurinium ion*, which has a **three-center, two-electron bond**. This is a bond in which two electrons are shared among three atoms—in this case, the two carbons of the double bond and the mercury.

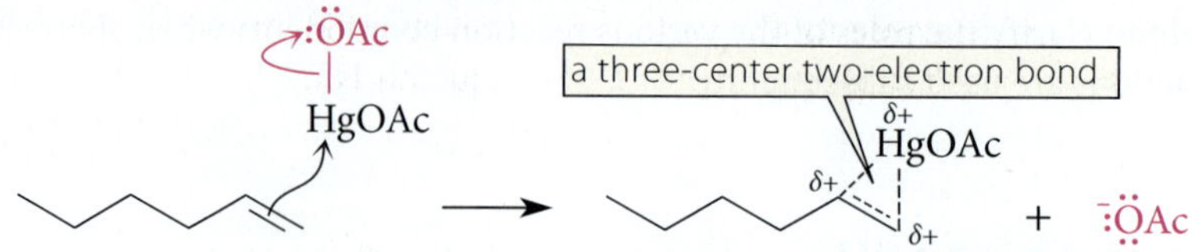

a mercurinium ion (5.20a)

We can represent this as a resonance structure in which there is a partial bond between the mercury and each carbon of the double bond:

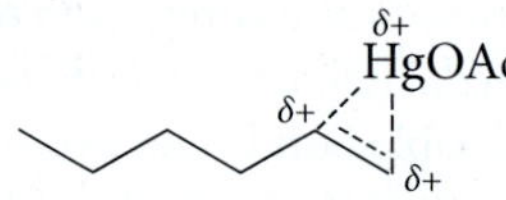

hybrid structure

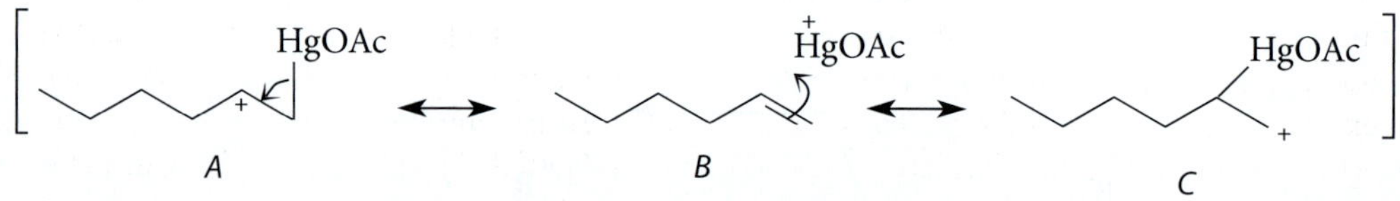

resonance structures (5.20b)

Resonance structure *A* is a secondary carbocation, whereas resonance structure *C* is a primary carbocation. As a result, resonance structure *A* is much more important; in other words, more positive charge and electron deficiency is on the carbon with the alkyl substituent. This accounts for the regioselectivity of the oxymercuration reaction—namely, the water in the solvent reacts as a nucleophile at the carbon that has the greater amount of positive charge.

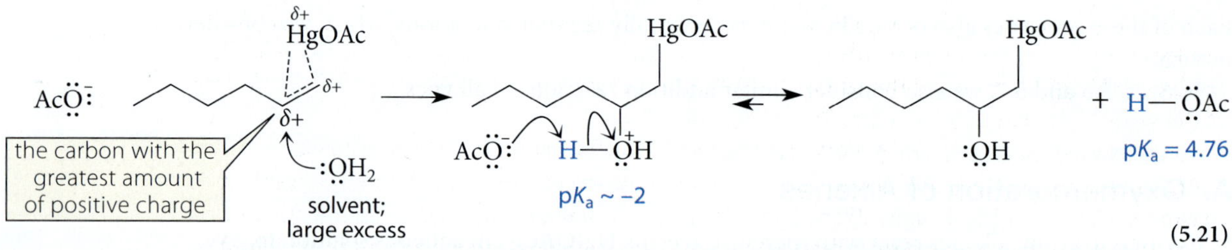

(5.21)

Nucleophiles other than water can be used as solvents in the oxymercuration reaction to prepare other types of compounds, as we show in Sec. 12.2B. (See also Problem 5.38.)

The final acid–base equilibrium lies well to the right, because acetate ion ($^{-}$OAc), the conjugate base of acetic acid, is a much stronger base than the alcohol product. (Recall Sec. 3.6C.)

The oxymercuration reaction bears a close resemblance to halohydrin formation, which we discussed in the previous section. The mercurinium ion is similar to the bromonium ion; however, the bromonium ion has two full carbon–bromine bonds because of the nucleophilic participation of a bromine unshared pair (Eq. 5.22).

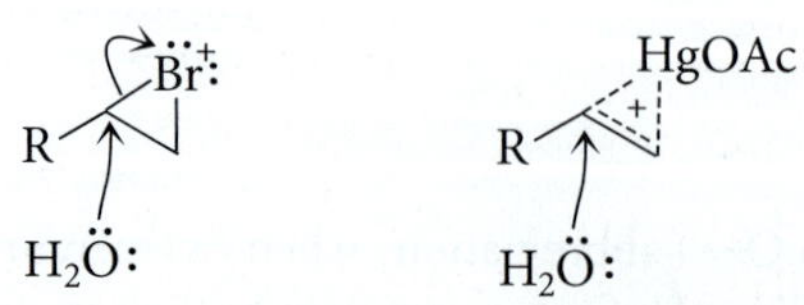

(5.22)

Both types of ion, however, react with solvent water.

Oxymercuration and halohydrin formation differ in their degree of regioselectivity. In oxymercuration, the reaction of water occurs almost exclusively at the carbon with more alkyl substituents, even if that carbon has only *one* alkyl substituent (as in Eq. 5.21). (In halohydrin formation, the reaction is highly regioselective *only* if one of the alkene carbons has *two* alkyl branches.) This difference is probably due to the greater amount of electron deficiency and positive charge in the mercurinium ion.

B. Conversion of Oxymercuration Adducts into Alcohols

Oxymercuration is useful because its products are easily converted into alcohols by treatment with sodium borohydride ($NaBH_4$) in the presence of aqueous NaOH.

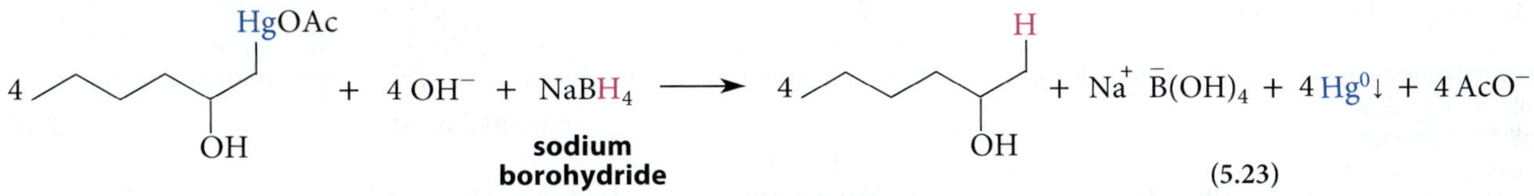

(5.23)

We won't study the mechanism of this reaction. The key thing to notice is its outcome: the carbon–mercury bond is replaced by a carbon–hydrogen bond (color in Eq. 5.23). In the lab, the oxymercuration adducts are not usually isolated, but are treated directly with a basic solution of $NaBH_4$ in the same reaction vessel.

The oxymercuration and $NaBH_4$ reactions, when used sequentially, are referred to collectively as the **oxymercuration–reduction** of an alkene. (The general classification of reactions as oxidations or reductions is discussed in Sec. 11.6.) The overall result of oxymercuration–reduction is the *net* addition of the elements of water (H and OH) to an alkene double bond in a highly *regioselective* manner: the OH group is added to the *more branched carbon of the double bond.* Here's the overall sequence applied to 1-hexene written in shorthand style. The numbers above the arrow mean that two steps are carried out *in sequence*—that is, *first,* the alkene is allowed to react with $Hg(OAc)_2$ and H_2O, and *then,* in a *separate step,* aqueous $NaBH_4$ and NaOH are added.

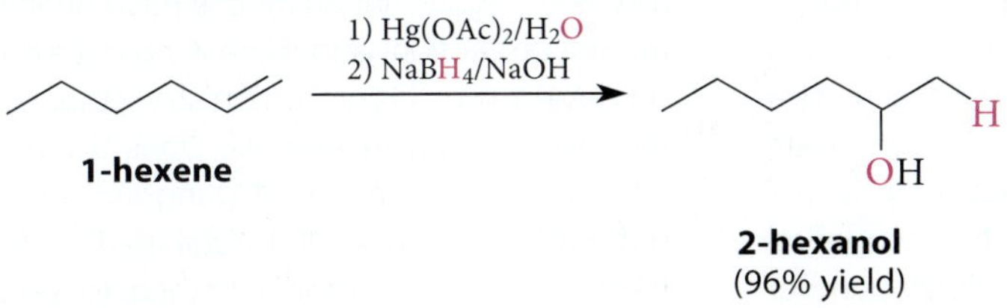

(5.24)

Writing consecutive reactions in this manner can save lots of time and space. However, if you use this shorthand, *be sure to number the reactions over the arrow.* If the numbers were left off, a reader might think that all of the reagents were added at once. Adding the reagents for both steps at the same time would *not* give the desired product!

Oxymercuration–reduction gives the same overall transformation as the acid-catalyzed hydration reaction (Sec. 4.10C). However, oxymercuration–reduction is much more convenient to run on a laboratory scale than alkene hydration, and it is free of rearrangements and other side reactions that are encountered in hydration, because carbocation intermediates are not involved in oxymercuration. (See the sidebar at the end of this section.) For example, the alkene in the following equation gives products derived from carbocation rearrangement when it undergoes hydration (Problem 4.79). However, no rearrangements are observed in oxymercuration–reduction:

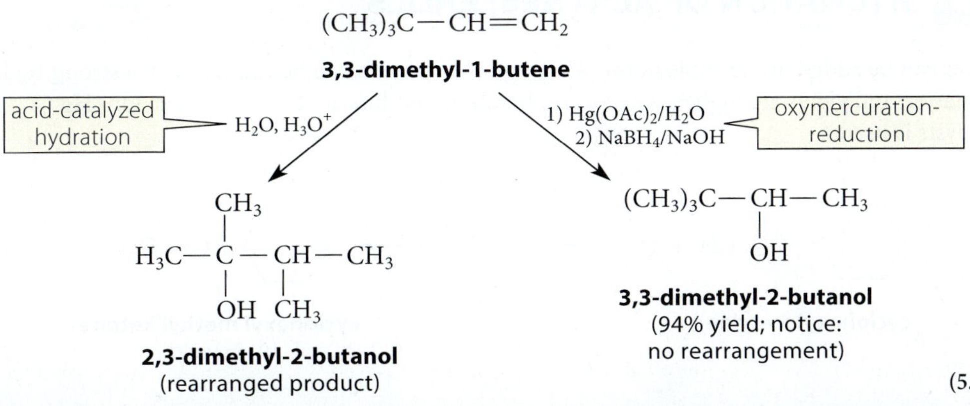

(5.25)

The absence of rearrangements is one reason that mercurinium ions, rather than carbocations, are thought to be the reactive intermediates in oxymercuration.

Focused Problems

5.4 Give the products expected when each of the following alkenes is subjected to oxymercuration–reduction.

(a) Cyclohexene (b) 2-Methyl-2-pentene

(c) *trans*-4-Methyl-2-pentene (d) *cis*-3-Hexene

5.5 What alkenes would give each of the following alcohols as the major (or only) product as a result of oxymercuration–reduction?

(a) 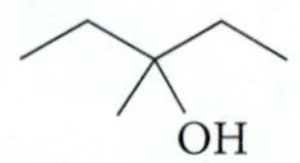

(three different alkenes, including stereoisomers)

(b) 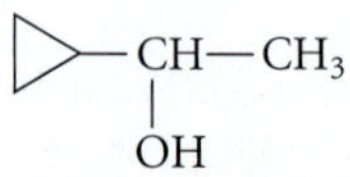

(one alkene)

Laboratory Use of Mercury and Other Toxic Reagents

MarcelClemens/Shutterstock

Mercury is a very toxic element because it can be converted in the environment into methylmercury, CH_3Hg^+, which can accumulate in the fatty tissues of animals such as fish. Ingestion of methylmercury leads to neurotoxicity. The oxymercuration–reduction reaction sequence illustrates the point that chemists use a significant number of highly toxic reagents. Three issues surround the toxicity of chemical reagents. The first is the *knowledge* that they are toxic. Chemists are provided with a *Safety Data Sheet* (SDS) that describes the known hazards of each reagent that they purchase, and most SDSs are readily available on the web. The second issue is the *safe handling* of toxic or dangerous reagents in the laboratory. Part of a scientist's laboratory training is to become familiar with common laboratory hazards and to learn how they can be avoided or confronted safely. For example, if you are taking a laboratory course, you undoubtedly are required to wear gloves and safety glasses. The third issue is *protection of the environment*. Significant advances have been made in environmental protection, and there is now a significant emphasis on developing "green" chemistry—that is, environmentally friendly chemical processes. Does this mean that a chemist must avoid the use of dangerous or environmentally unfriendly reagents? Not necessarily. The issue is to use them safely and to dispose of them (or better, recycle them) properly. To take oxymercuration–reduction as an example, the ultimate mercury-containing product of the reaction (Eq. 5.23) is metallic mercury. This can be collected and recycled. Although most chemists would avoid the use of mercury where possible—for example, replacing mercury-containing thermometers in laboratories—sometimes there simply is no alternative. Oxymercuration–reduction is such an effective reaction that it is attractive despite the inconvenience of proper mercury recycling.

5.5 HYDRATION OF ALKYNES; ENOLS

Water can be added to the triple bond. Although the reaction can be catalyzed by a strong acid, it is faster, and yields are higher, when a combination of dilute acid and mercuric ion (Hg^{2+}) catalysts is used.

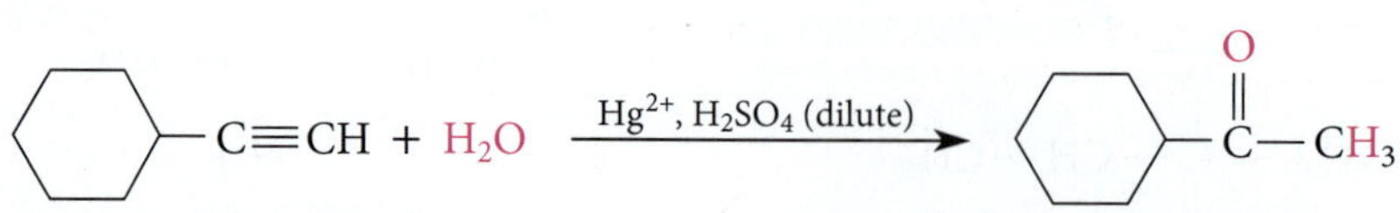

cyclohexylacetylene **cyclohexyl methyl ketone** (91% yield) (5.26)

The addition of water to a triple bond, like the corresponding addition to a double bond, is called **hydration**. The hydration of alkynes gives ketones (except in the case of acetylene itself, which gives an aldehyde; see Study Problem 5.2). Unlike alkene hydration, *alkyne hydration is irreversible.* (We'll see why shortly.)

Although the Hg^{2+}-catalyzed process is more useful synthetically, let's focus mechanistically on the acid-catalyzed hydration so that we can contrast this process with the hydration of alkenes (Sec. 4.10C). The hydration of alkynes and alkenes begins with exactly the same first step: protonation of a carbon–carbon π bond to give a *vinylic carbocation*:

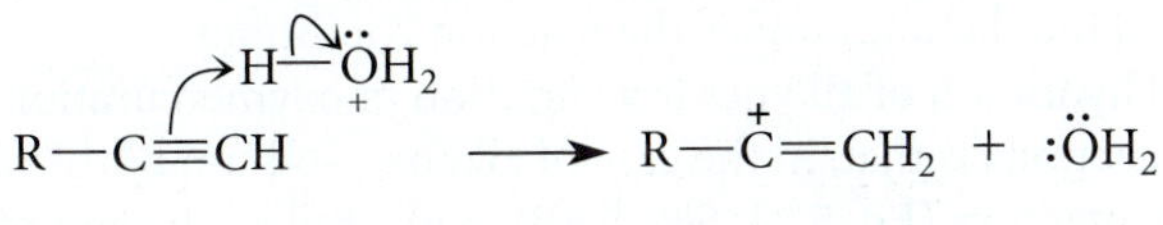

(5.27a)

This is the same type of carbocation that is involved in the addition of hydrogen halides to alkynes. (An orbital diagram of a vinylic cation is shown in Display 4.62, Sec. 4.9.) As in hydrogen halide addition, the carbocation is formed at the carbon with the alkyl branch (why?).

The vinylic cation undergoes a Lewis acid–base association reaction with a nucleophile, which in this case is water. A subsequent Brønsted acid–base reaction produces the product, which is a vinylic alcohol, or *enol.*

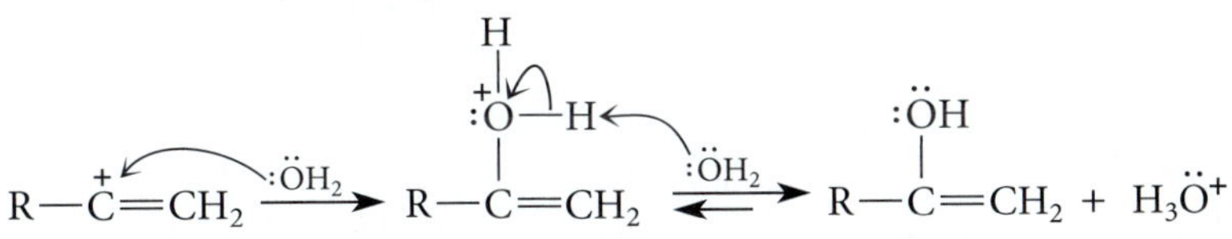

(5.27b)

An **enol** (pronounced ēn′-ôl) is a special type of alcohol in which the OH group is on a carbon of a double bond. *Most enols cannot be isolated because they are unstable and are rapidly converted into the corresponding aldehydes or ketones.*

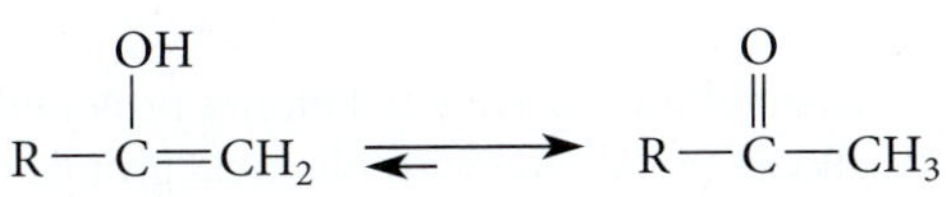

(5.27c)

Equation 5.27c also shows why alkyne hydration, unlike alkene hydration, is not reversible. The equilibrium between an aldehyde or ketone and its enol isomers *strongly* favors the aldehyde or ketone. The equilibrium concentrations of enols are in most cases minuscule—typically, one part in 10^8 or less. (The relationship among aldehydes, ketones, and their enol isomers is explored in more detail in Chapter 22.) The important point here is that because the equilibrium in Eq. 5.27c lies so far to the right, *the hydration of alkynes is effectively irreversible.*

What makes enols so unstable? Actually, nothing is particularly unfavorable about the structures of enols. They are unstable only because they can be rapidly converted into *even more stable* ketone isomers. Ketones are more stable than enols largely because the C═O bond of a ketone is considerably stronger than the C═C bond of an enol.

Let's see how the acid-catalyzed conversion of enols into ketones takes place. We've already learned (Sec. 4.9) that alkynes sometimes undergo two addition reactions because the first addition product contains a double bond, which can itself undergo an addition. We should not be surprised, then, that the enol reacts further, because it is an alkene "at heart" and its π bond can be protonated. Furthermore, the carbocation produced by protonation of the enol is *resonance-stabilized*:

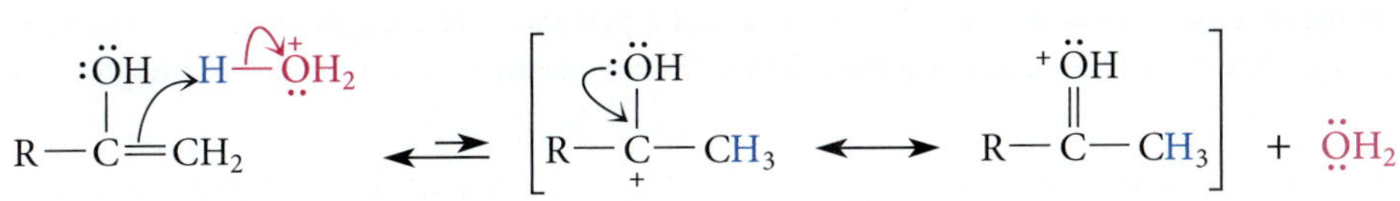

(5.28a)

The resonance structure on the right shows that this carbocation is the conjugate acid of a ketone. Loss of a proton gives the ketone product.

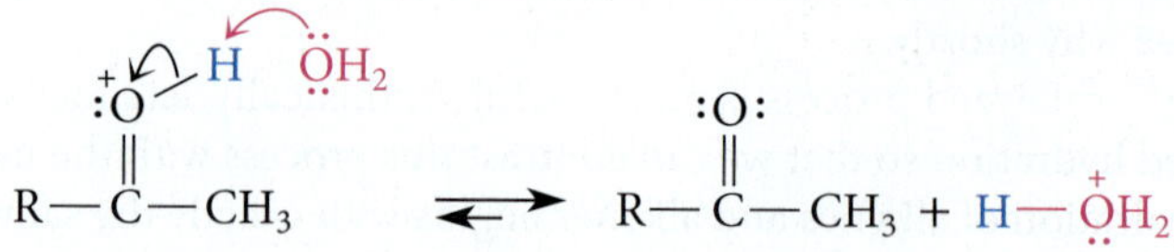

(5.28b)

In summary, then, alkynes undergo acid-catalyzed hydration to give enols, which are rapidly and irreversibly transformed into ketones under the reaction conditions.

The Hg^{2+}-catalyzed hydration of alkynes is a variation of oxymercuration (Sec. 5.4A). A difference in the two reactions is that, in the case of alkenes, sodium borohydride ($NaBH_4$) is required to remove the mercury (Eq. 5.23, Sec. 5.4B), and a full equivalent of Hg^{2+} is required. In the hydration of alkynes, however, the mercuric ion is a catalyst; it is regenerated by a reaction of the oxymercuration addition product with the acid co-catalyst.

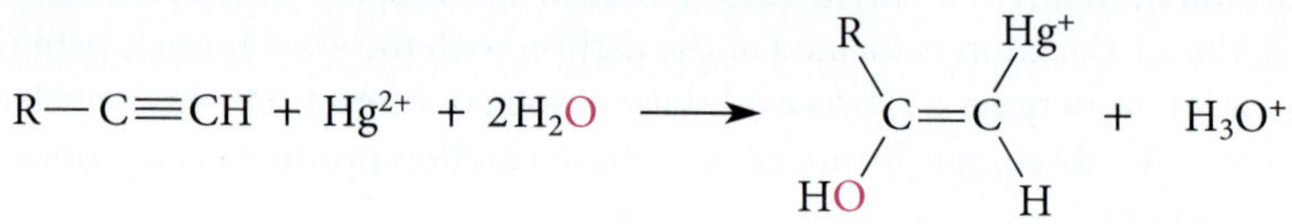

the oxymercuration addition product

(5.29a)

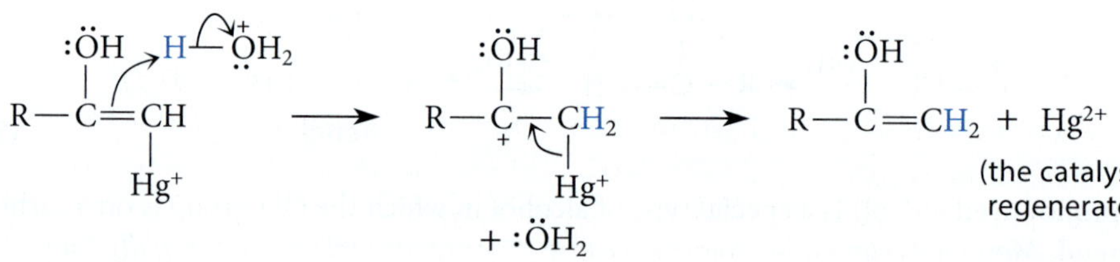

(5.29b)

Because mercuric ion is a catalyst in the hydration of alkynes, it can be used in relatively small amounts.

The hydration of alkynes is a useful way to prepare ketones provided that the starting material is a 1-alkyne or a symmetrical alkyne (an alkyne with identical groups on each end of the triple bond). This point is explored in Study Problem 5.2.

Study Problem 5.2

Which one of the following compounds could be prepared by the hydration of alkynes so that it is uncontaminated by constitutional isomers? Explain your answer.

(a) 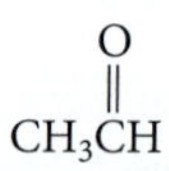

acetaldehyde

(b) 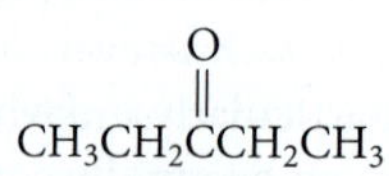

3-pentanone

Solution First, what alkyne starting materials, if any, would give the desired products? The equations in the text show that the two carbons of the triple bond in the starting material correspond within the product to the carbon of the C=O group and an adjacent carbon. Thus, for part (a), the only possible alkyne starting material is acetylene itself, HC≡CH. For part (b), the only possible alkyne starting material is 2-pentyne, $CH_3C{\equiv}CCH_2CH_3$.

Next, we need to show whether hydration of these alkynes gives *only* the products in the problem. Remember, a good synthesis gives relatively pure compounds. The hydration of acetylene indeed gives only acetaldehyde. (In fact, acetaldehyde is the only aldehyde that can be prepared by the hydration of an alkyne.) However, hydration of 2-pentyne gives a mixture consisting of comparable amounts of 2-pentanone and 3-pentanone, *because the carbons of 2-pentyne both have one alkyl substituent.* Thus, there is no reason that the reaction of water at either carbon should be strongly favored.

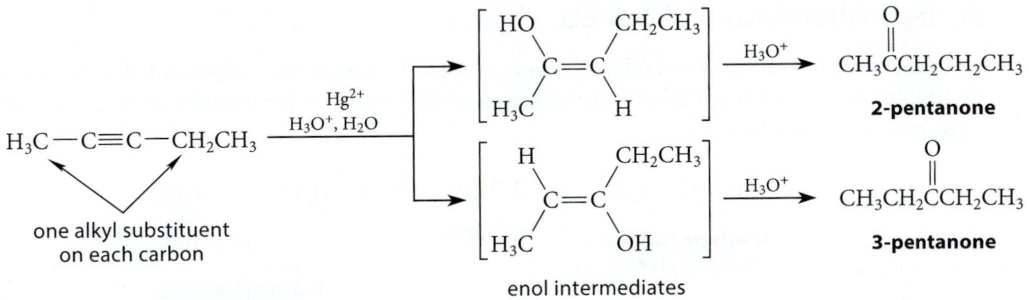

Hence, hydration would give a mixture of constitutional isomers that would have to be separated, and the yield of the desired product would be low. Consequently, hydration would *not* be a good way to prepare 3-pentanone. (However, 2-pentanone could be prepared by hydration of a different alkyne; see Focused Problem 5.6a.)

Focused Problems

5.6 From which alkyne could each of the following compounds be prepared by acid-catalyzed hydration?

(a) 

(b) 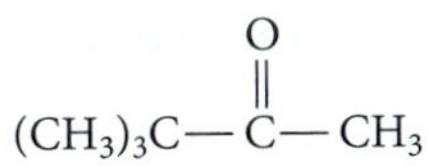

(c) 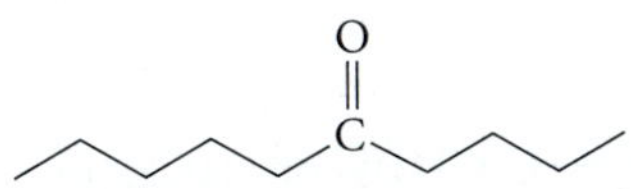

5.7 The hydration of an alkyne is *not* a reasonable preparative method for any of the following compounds. Explain why.

(a) $CH_3CH_2CH{=}O$

(b) 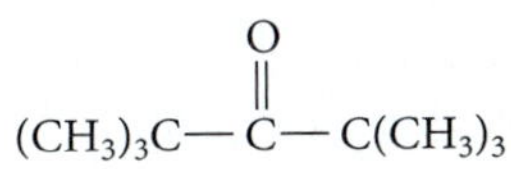

(c) 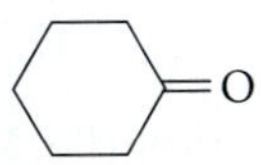

5.8 Which of the following compounds are enols? For each one that is an enol, draw the structures of the ketone into which it would be spontaneously converted.

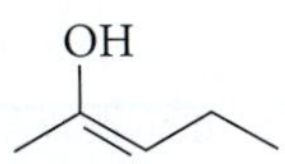

A

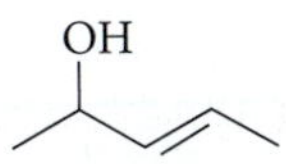

B

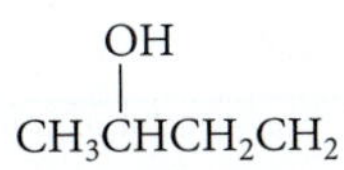

C

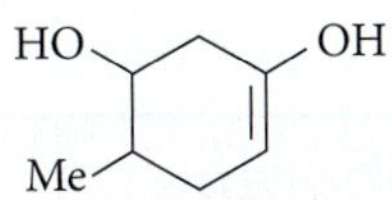

D

5.6 HYDROBORATION–OXIDATION OF ALKENES

In Sec. 5.4, we showed that oxymercuration–reduction brings about the addition of H and OH to a double bond so that the OH group adds to the carbon with more alkyl substituents. Suppose, though, that we want to add H and OH to a double bond so that the OH group adds to the carbon with *fewer* alkyl substituents. *Hydroboration–oxidation* is a method that can bring about this transformation. As with oxymercuration–reduction, hydroboration–oxidation involves two separate experimental steps, which we discuss in turn.

A. Hydroboration of Alkenes

Borane (BH_3) adds regioselectively to alkenes so that the boron becomes bonded to the carbon of the double bond with *fewer* alkyl substituents, and the hydrogen becomes bonded to the carbon with *more* alkyl substituents:

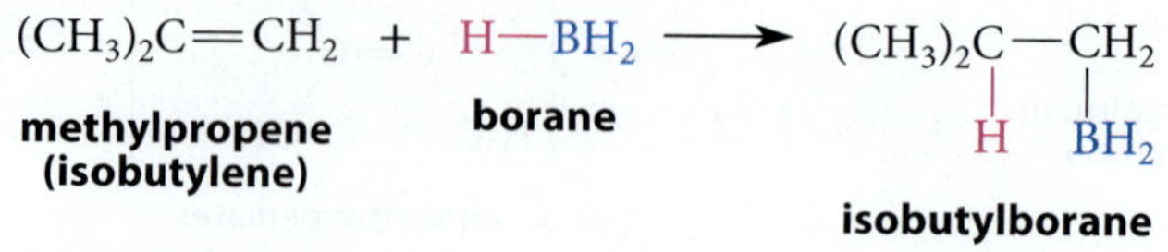

(5.30a)

Because borane has *three* B—H bonds, one borane molecule can add to three alkene molecules. The first of these additions to methylpropene is shown in Eq. 5.30a. The second and third additions are as follows:

Second addition:

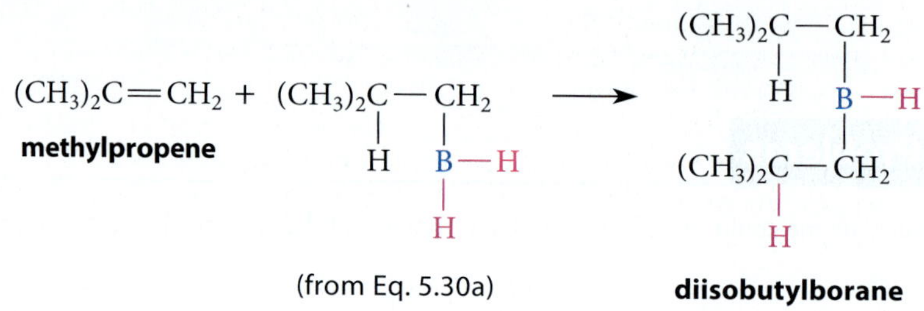

(5.30b)

Third addition:

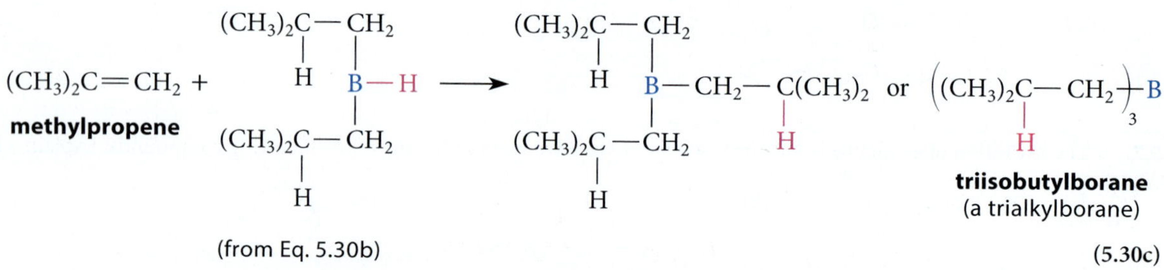

(5.30c)

An addition reaction of BH_3 is called **hydroboration**. The product of alkene hydroboration is a *trialkylborane*, such as the triisobutylborane shown in Eq. 5.30c.

Borane and Diborane

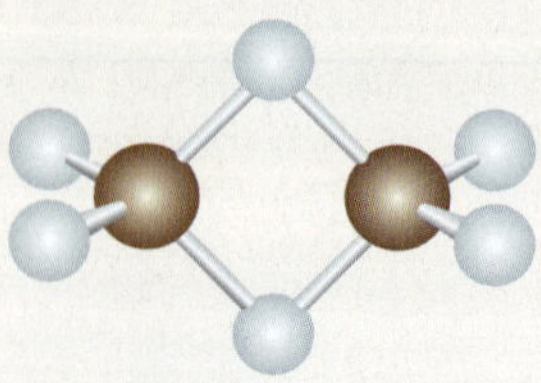

Borane actually exists as a toxic, colorless gas called *diborane*, which has the formula B_2H_6. Because borane is an electron-deficient Lewis acid, the boron has a strong tendency to acquire an additional electron pair. This tendency is satisfied by the formation of diborane, in which two hydrogens are shared between the two borons in unusual "half bonds." This bonding can be shown with resonance structures:

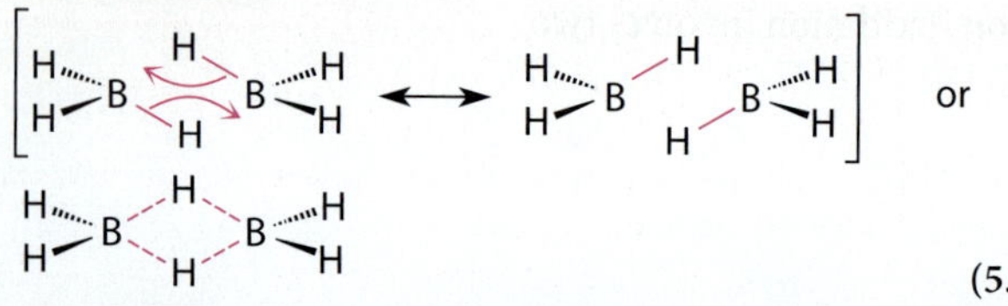

(5.31)

When dissolved in an ether solvent, diborane dissociates to form a borane–ether complex. Because ethers are Lewis bases, they can satisfy the electron deficiency at boron:

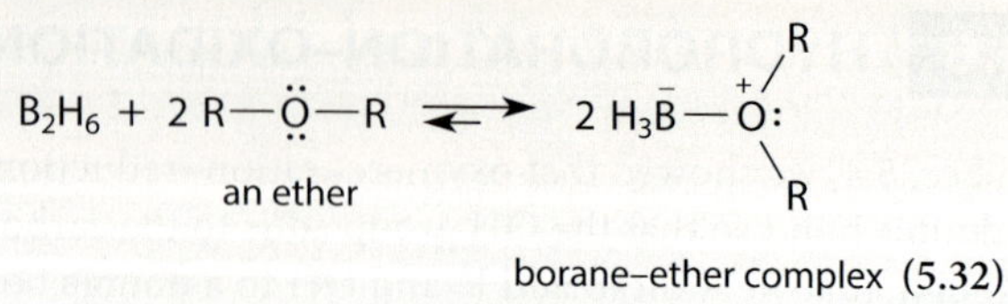

(5.32)

The following ethers are commonly used as solvents in the hydroboration reaction:

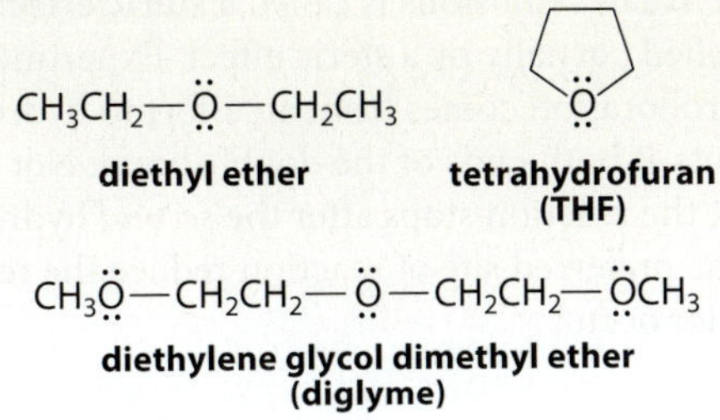

Borane–ether complexes are the actual reagents involved in hydroboration reactions. Although hydroboration was known as early as 1948, it was the use of ether solvents that was found to accelerate the reaction to the point that it was useful. For simplicity, we omit the ether solvent and use the formula BH_3 for borane.

Hydroboration is believed to occur in a single mechanistic step because carbocation rearrangements are not observed and because of stereochemical evidence we discuss in Sec. 7.8D.

(5.33)

A reaction like this one, which occurs in a single step without intermediates, is said to occur by a **concerted mechanism** because everything happens "in concert," or simultaneously. The regioselectivity of hydroboration is the result of two factors. First, some electron deficiency is built up on the tertiary carbon in the transition state of the reaction:

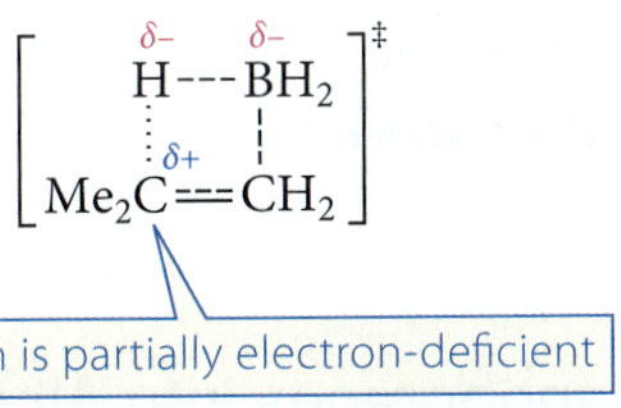

transition state for hydroboration (5.34a)

Just as alkyl substitution at the electron-deficient carbon stabilizes a carbocation, alkyl substitution at a *partially* electron-deficient carbon stabilizes a transition state. Thus, hydroboration occurs with the regioselectivity that places partial positive charge on the carbon with more alkyl substituents.

The second factor controlling the regioselectivity of hydroboration is that the groups around the boron avoid van der Waals repulsions with the alkyl substituents of the alkene when the boron bonds to the carbon with fewer substituents. This factor becomes particularly important in the second and third additions in the mechanism (Eqs. 5.30b and 5.30c), when one or two of the hydrogens of BH_3 have been replaced by alkyl groups. For example, suppose the boron were to form a bond to the more substituted carbon of methylpropene. Van der Waals repulsions would occur between the alkyl group on boron and the alkyl substituents in the alkene, which would be eclipsed in the transition state:

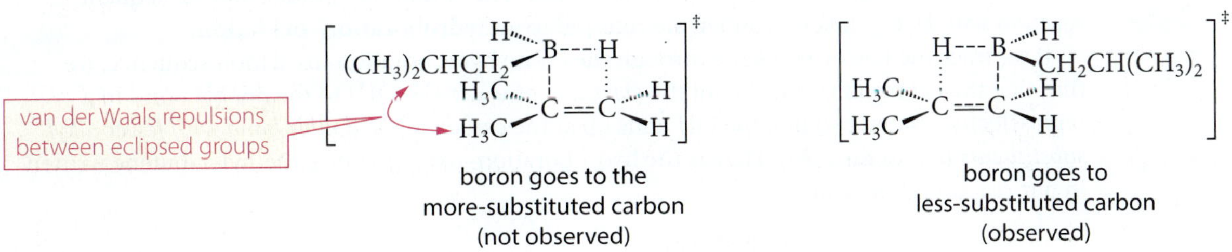

transition state for the second hydroboration

(5.34b)

These repulsions raise the energy of this transition state and cause this reaction to be slower. Therefore, addition of the boron at the less substituted carbon is faster and is the observed mode of reaction.

The control of a reaction outcome by van der Waals repulsions is called a **steric effect**. The regioselectivity of hydroboration, then, is controlled partially by a steric effect. Experimental evidence that steric effects are important in hydroboration comes from an effort to hydroborate 2-methyl-2-butene, which has methyl substituents at both ends of the double bond. Not only does boron end up on the less substituted carbon, but the reaction stops after the *second* hydroboration step, presumably because steric effects even at the preferred site of reaction reduce the reaction rate to the point that a third hydroboration cannot occur.

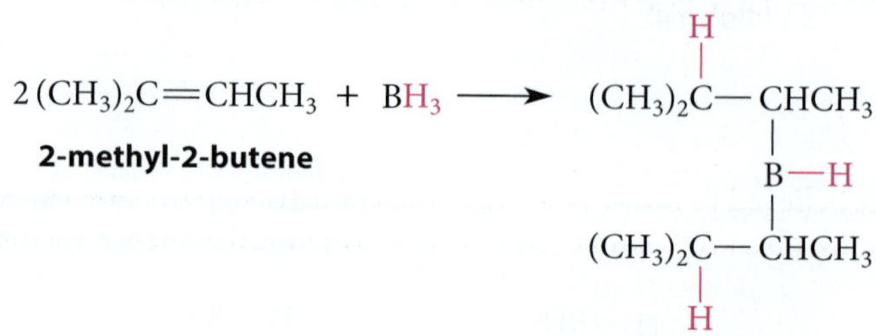

2-methyl-2-butene

"disiamylborane" (5.34c)

The names "disiamylborane" and "thexylborane" are nicknames that stand for "di-secondary-isoamylborane" and "tertiary-hexylborane," neither of which is an IUPAC name. Nevertheless, these nicknames have become ingrained in organoborane chemistry. Disiamylborane is used in the hydroboration of 1-alkynes (Sec. 5.7).

Hydroboration of 2,3-dimethyl-2-butene, which has two methyl branches at each of the alkene carbons, despite the electron-donating effect of the methyl branches, stops after only *one* round of hydroboration.

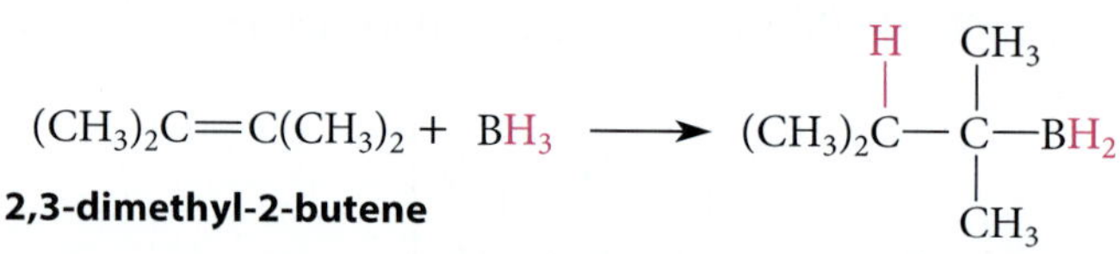

2,3-dimethyl-2-butene

"thexylborane" (5.34d)

B. Conversion of Organoboranes into Alcohols

The utility of hydroboration lies in the many reactions of organoboranes themselves. One of the most important reactions of organoboranes is their conversion into alcohols with hydrogen peroxide (H_2O_2) and aqueous NaOH.

$$\left((CH_3)_2CHCH_2\right)_3B + 3H_2O_2 + {}^{-}OH \longrightarrow 3(CH_3)_2CHCH_2{-}OH + \bar{B}(OH)_4 \quad (5.35)$$

triisobutylborane — **hydrogen peroxide** — **2-methyl-1-propanol (isobutyl alcohol)** (95% yield)

FURTHER EXPLORATION 5.1
Mechanism of Organoborane Oxidation

(The mechanistic details are given in Further Exploration 5.1 in the Study Guide and Solutions Manual.) The important thing to notice about this transformation is *the replacement of the boron by an* OH *in each alkyl group*. The oxygen of the OH group comes from the H_2O_2.

Oxidation and reduction reactions are covered extensively in Sec. 11.6.

Typically, the organoborane product of hydroboration is not isolated, but is treated directly with alkaline hydrogen peroxide to give an alcohol. The addition of borane and subsequent reaction with H_2O_2, taken together, are referred to as **hydroboration–oxidation**.

If we trace the fate of an alkene through the entire hydroboration–oxidation sequence, we find that the *net* result is addition of the elements of water (H, OH) to the double bond in a regioselective manner so that the OH ends up at the *carbon of the double bond with fewer alkyl substituents* (*red* in Eq. 5.36). Here is the hydroboration–oxidation of 2-methyl-1-butene written in our reaction shorthand.

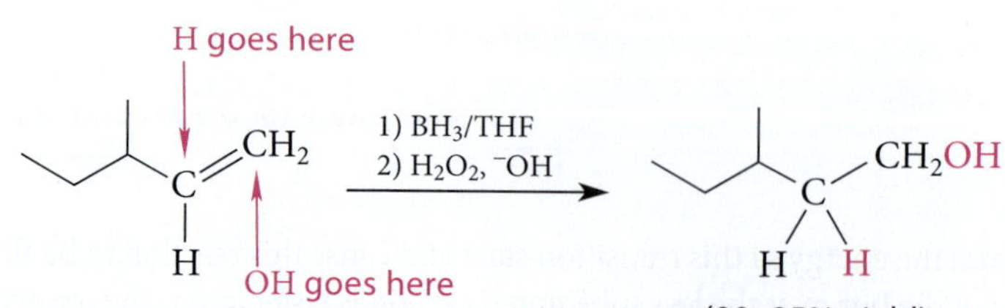

(90–95% yield) (5.36)

Notice again the numbered steps. Step 1 is the reaction of the alkene with borane. *After this step is complete*, a solution of hydrogen peroxide in aqueous NaOH is added in step 2.

Hydroboration–oxidation is an effective way to synthesize certain alcohols from alkenes. It is particularly useful to prepare alcohols of the general structure $R_2CH—CH_2—OH$ or $RCH_2—CH_2—OH$, as in Eqs. 5.35 and 5.36. Because carbocations are not involved in either the hydroboration or the oxidation reaction, the alcohol products are not contaminated by constitutional isomers arising from rearrangements.

The benzene ring does *not* react with BH_3, even though the ring apparently contains double bonds:

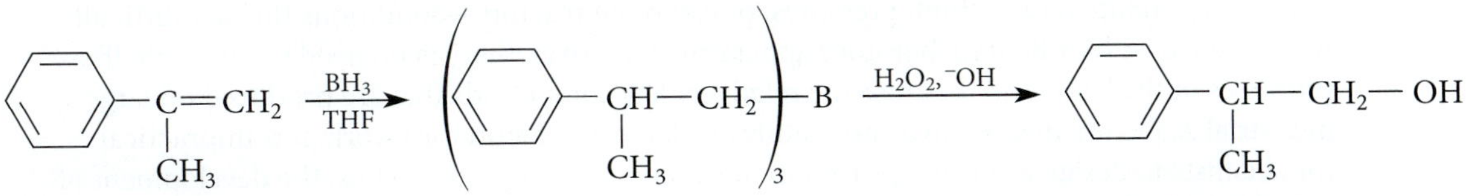

***α*-methylstyrene** **2-phenyl-1-propanol** (95% yield) (5.37)

We have seen a similar resistance of benzene rings to other addition reactions, such as catalytic hydrogenation (Sec. 4.10A). The reason benzene rings are resistant to addition reactions is discussed in Sec. 15.7. (You can safely ignore the benzene double bonds for all of the reactions in this chapter.)

H. C. Brown and Hydroboration

Editorial/Alamy

The Nobel Prize Medal

Hydroboration was discovered accidentally in 1955 at Purdue University by Professor Herbert C. Brown (1912–2004) and his colleagues. Brown quickly realized its significance and in subsequent years carried out research demonstrating the versatility of organoboranes as intermediates in organic synthesis. Brown called the chemistry of organoboranes "a vast unexplored continent." In 1979, his research was recognized with the Nobel Prize in Chemistry, which he shared with another organic chemist, Georg Wittig (Sec. 19.13).

Focused Problems

5.9 Give the product(s) expected from the hydroboration–oxidation of each of the following alkenes.

(a) Cyclohexene (b) 2-Methyl-2-pentene

(c) *trans*-4-Methyl-2-pentene (d) *cis*-3-Hexene

5.10 Contrast the answers for Focused Problem 5.9 with the answers for the corresponding parts of Focused Problem 5.4. For which alkenes are the alcohol products the same? For which are they different? Explain why the same alcohols are obtained in some cases and different ones are obtained in others.

5.11 For each of the following cases, provide the structure of an alkene that would give the alcohol as the major (or only) product of hydroboration–oxidation.

(a) 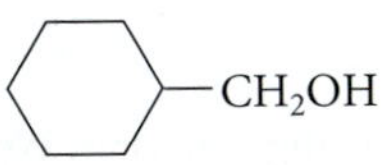

(b) 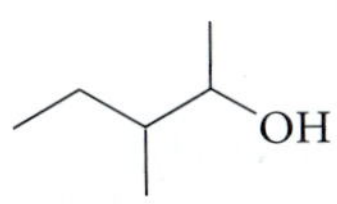

C. Comparison of Methods for the Synthesis of Alcohols from Alkenes

Let's now compare the different ways to prepare alcohols from alkenes. The *hydration of alkenes* is a useful industrial method for the preparation of a few alcohols, but it is a poor laboratory method (Sec. 4.10C). Indeed, many industrial methods for the preparation of organic compounds are not *general.* That is, an industrial method typically works well in the specific case for which it was designed, but it cannot necessarily be applied to other related cases. The reason is that the chemical industry has gone to great effort to work out conditions that are optimal for the preparation of *particular compounds* of commercial significance (such as a few simple alcohols) using reagents that are readily available and inexpensive. These processes in many cases require high temperatures, high pressures, or elaborate reactors—conditions that are difficult to reproduce with ordinary laboratory apparatus. Moreover, there is no need to duplicate these processes in the laboratory because the relatively few compounds that are produced on large industrial scales *are* inexpensive and readily available. For laboratory work, it is impractical for chemists to design a specific procedure for each new compound. Thus, the development of general methods that work with a wide variety of compounds is important. Because laboratory synthesis is generally carried out on a relatively small scale, the expense of reagents is less of a concern.

Hydroboration–oxidation and oxymercuration–reduction are both *general laboratory methods* for the preparation of alcohols from alkenes. That is, they can be applied successfully to a wide variety of alkene starting materials. A choice between the two methods for a particular alcohol usually hinges on the difference in their regioselectivities. As shown in Eq. 5.38, hydroboration–oxidation gives an alcohol in which the OH group has been added to the carbon of the double bond *with fewer alkyl substituents.* Oxymercuration–reduction gives an alcohol in which the OH group has been added to the carbon of the double bond *with the greater number of alkyl substituents.*

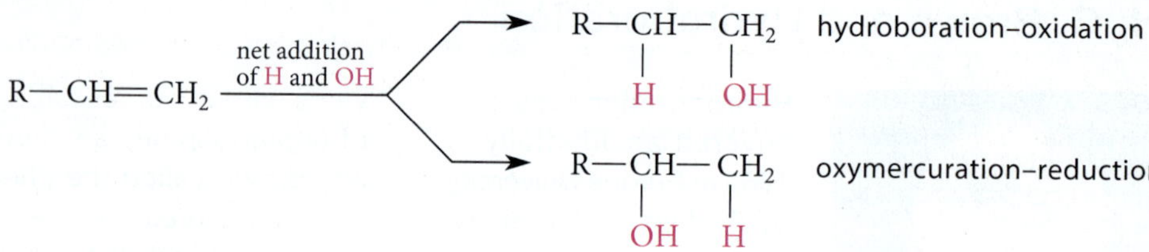

(5.38)

For alkenes that yield the same alcohol by either method, such as alcohols with symmetrically located double bonds, the choice between the two is in principle arbitrary.

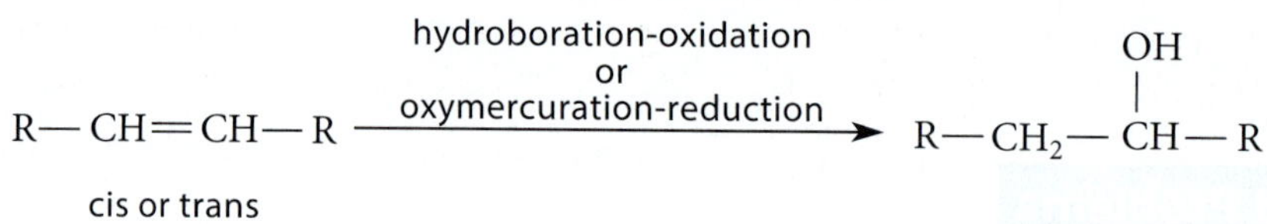

(5.39)

Study Problem 5.3

Which of the following alcohols could be prepared free of constitutional isomers by (a) hydroboration–oxidation, (b) oxymercuration–reduction, (c) either method, or (d) neither method? Explain your answers, and give the structure of the alkene starting material for the cases in which a satisfactory synthesis is possible.

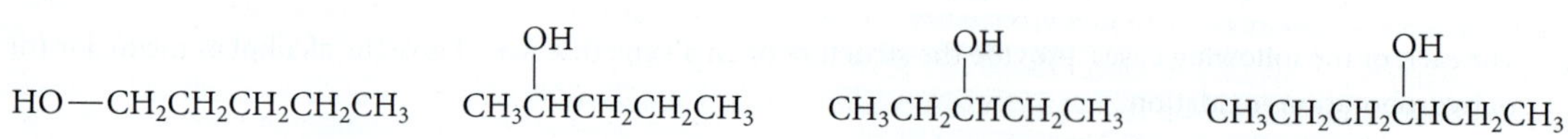

A *B* *C* *D*

Solution Use the following steps to solve this problem: (1) Draw the possible alkene starting materials. We should consider initially any alkene in which one carbon of the double bond is the one that bears the hydroxy group in the product. (2) Decide whether each reaction can be used on that starting material to give the desired alcohol and *only* that alcohol.

To prepare *A*, the only possible alkene starting material is 1-pentene.

$$\underset{\textbf{1-pentene}}{H_2C{=}CHCH_2CH_2CH_3} \xrightarrow{?} \underset{A}{HO{-}CH_2CH_2CH_2CH_2CH_3}$$

Hydroboration–oxidation has the correct regioselectivity to bring about this conversion, but oxymercuration–reduction does not.

To prepare *B*, the possible alkene starting materials are 1-pentene and (*E*)- or (*Z*)-2-pentene.

$$\underset{\textbf{1-pentene}}{H_2C{=}CHCH_2CH_2CH_3} \text{ or } \underset{\textbf{2-pentene}}{CH_3CH{=}CHCH_2CH_3} \xrightarrow{?} \underset{B}{CH_3\overset{OH}{\overset{|}{C}}HCH_2CH_2CH_3}$$

Oxymercuration–reduction has the correct regioselectivity for the conversion of 1-pentene to *B*. However, the 2-pentenes would yield a mixture of *B* and *C* with either method. Hence, neither of the 2-pentenes is a satisfactory starting material. The issue is not whether the 2-pentenes would react; *the issue is whether we would obtain the constitutional isomer we want or a mixture of constitutional isomers.* Remember also that the *number* of alkyl groups on each alkene carbon, not their *size*, determines regioselectivity.

The (*E*)- and (*Z*)-2-pentenes are the only potential starting materials for *C*.

$$\underset{C}{CH_3CH_2\overset{OH}{\overset{|}{C}}HCH_2CH_3} \Rightarrow \underbrace{CH_3CH{=}CHCH_2CH_3 \text{ or } CH_3CH_2CH{=}CHCH_3}_{\text{the same compound: }\textbf{2-pentene}}$$

As we just observed, neither method would yield a single compound with a 2-pentene as the starting alkene. Hence, *C* can*not* be prepared as a pure compound by *either* method. (There are other ways to make this alcohol.)

Finally, the possible alkene starting materials for alcohol *D* are (*E*)- or (*Z*)-2-hexene and (*E*)- or (*Z*)-3-hexene.

$$\underset{\textbf{2-hexene}}{CH_3CH_2CH_2CH{=}CHCH_3} \text{ or } \underset{\textbf{3-hexene}}{CH_3CH_2CH{=}CHCH_2CH_3} \xrightarrow{?} \underset{D}{CH_3CH_2CH_2\overset{OH}{\overset{|}{C}}HCH_2CH_3}$$

Because the double bond in 3-hexene is symmetrically situated, either method in principle would give alcohol *D* as the only product. Hence, either stereoisomer of 3-hexene is a satisfactory starting material. However, the reaction of a 2-hexene would, like the reaction of a 2-pentene, give a mixture of constitutional isomers by either method.

At this point, we have studied a number of new reactions, and so, this is a good time to ask, "What is the best technique for studying and learning reactions?" Staring at the page and highlighting the reactions in the text are *not* good methods. The best methods should be *active*—that is, they should require you to think about the reactions and mentally process them, and they should require you to *write* them. They should test not only your ability to complete reactions for which you are given the starting materials, but also to provide starting materials and appropriate conditions given a desired product, as in Study Problem 5.3. You should also focus on the reaction mechanisms, using them particularly to see the similarities and differences in related reactions. (Mechanisms should not be memorized by rote.) In Study Guide Link 5.1 of the *Study Guide and Solutions Manual,* we have outlined one method that seems to work fairly well. Be sure to read about this method and adopt it or some equivalent method. Then *work as many problems as you can*. If you apply this method as you go and not allow reactions to accumulate without learning them, you will find that you can master the large amount of material without "cramming" them right before an examination.

STUDY GUIDE LINK 5.1
How to Study Organic Reactions

Focused Problems

5.12 From what alkene and by which methods could you prepare each of the following alcohols essentially free of constitutional isomers?

(a) 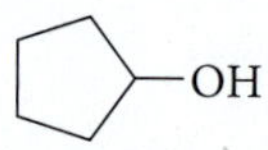

(b) 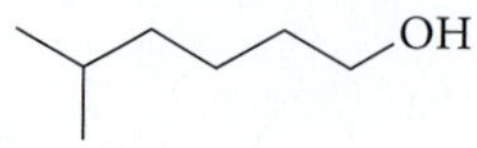

(c) $Et_3C—OH$ (*Hint:* Draw out the structure!)

5.13 Which of the following alkenes would yield the same alcohol from either oxymercuration–reduction or hydroboration–oxidation, and which would give different alcohols? Explain.

(a) *cis*-2-Butene (b) 1-Methylcyclohexene

5.7 HYDROBORATION–OXIDATION OF ALKYNES

Hydroboration of an alkyne gives the all-cis (all-*Z*) product, but as Eq. 5.40b shows, the stereochemistry of the addition becomes irrelevant at the next step. (The stereochemistry of hydroboration is discussed in detail in Sec. 7.8D.)

The hydroboration of alkynes is analogous to the same reaction of alkenes (Sec. 5.6A).

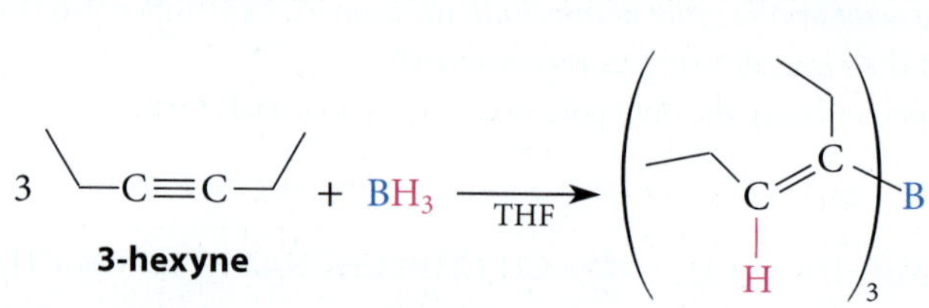

3-hexyne (5.40a)

As in the similar reaction of alkenes, oxidation of the organoborane with alkaline hydrogen peroxide yields the corresponding "alcohol," which in this case is an *enol.* As shown in Sec. 5.5 (Eq. 5.27c), enols react further to give the corresponding aldehydes or ketones.

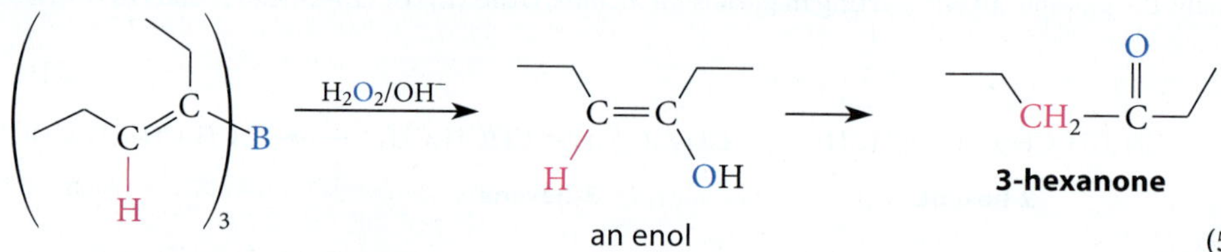

an enol **3-hexanone** (5.40b)

Because the organoborane product of Eq. 5.40a has a double bond, a second addition of BH_3 is in principle possible. However, the reaction conditions can be controlled so that only one addition takes place, as shown, provided that the alkyne is not a 1-alkyne.

If the alkyne is a 1-alkyne (that is, if it has a triple bond at the end of a carbon chain), a second addition of BH_3 cannot be prevented.

$$\underset{\text{a 1-alkyne}}{R—C\equiv CH} + BH_3 \longrightarrow \text{multiple addition reactions}$$

However, we learned in Sec. 5.6A that the hydroboration reaction is retarded by van der Waals repulsions if the hydroboration reagent contains large groups. We take advantage of this steric effect in the hydroboration of 1-alkynes by using an organoborane containing highly branched groups instead of BH_3 to avoid the second addition. An ideal reagent for this purpose is disiamylborane, represented with the skeletal structure shown in Display 5.41.

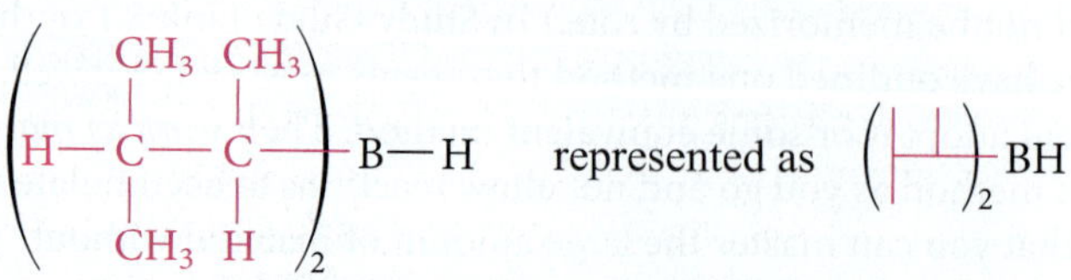

disiamylborane (5.41)

(The preparation of disiamylborane is shown in Eq. 5.34c, Sec. 5.6A.) The disiamylborane molecule is so large and highly branched that only one equivalent can react with a 1-alkyne; addition of a second molecule results in severe van der Waals repulsions in the product. In other words, van der Waals repulsions are used to advantage to prevent an *undesired* second addition from occurring:

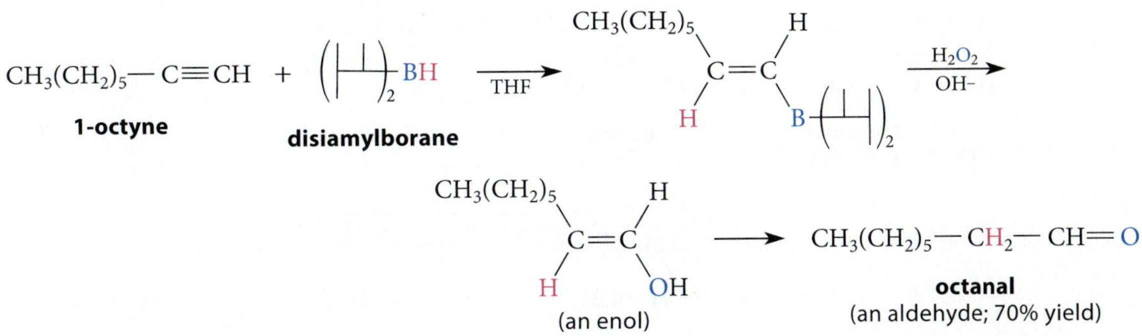

(5.42)

This example shows that the regioselectivity of alkyne hydroboration is similar to that observed in alkene hydroboration (Sec. 5.6A): boron adds to the unbranched carbon atom of the triple bond, and hydrogen adds to the branched carbon.

Because hydroboration–oxidation and Hg^{2+}-catalyzed hydration give different products when a 1-alkyne is used as the starting material (why?), these are *complementary* methods for the preparation of aldehydes and ketones in the same sense that hydroboration–oxidation and oxymercuration–reduction are complementary methods for the preparation of alcohols from alkenes.

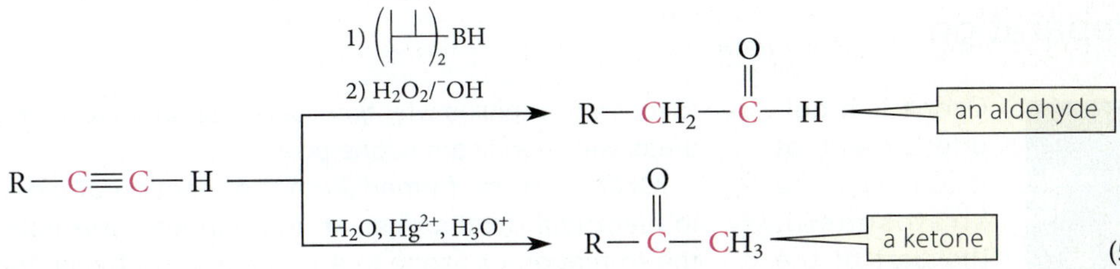

(5.43)

In summary: hydroboration–oxidation of a 1-alkyne gives an *aldehyde*; Hg^{2+}-catalyzed hydration of any 1-alkyne (other than acetylene itself) gives a *ketone* with a methyl group as one of the alkyl branches. Both hydroboration–oxidation and Hg^{2+}-catalyzed hydration of internal alkynes (alkynes in which the triple bond is not at the end of the carbon chain) give ketones.

Focused Problem

5.14 Compare the results of hydroboration–oxidation and Hg^{2+}-catalyzed hydration for the following two alkynes. For which is the result of the two reactions the same, and for which are the results different?

(a) Cyclohexylacetylene (b) 2-Butyne

5.8 OZONOLYSIS OF ALKENES

The reaction of ozone, O_3, with alkenes, like hydroboration–oxidation and oxymercuration–reduction, involves two chemically and experimentally distinct steps. In the first step, the alkene reacts with ozone to form an addition product called an *ozonide*. In the second step, the ozonide is treated with oxidizing or reducing agents to form various products. We consider each of these steps in turn.

A. Formation of Ozonides

The addition of ozone to alkenes occurs at low temperature and breaks the carbon–carbon π bond to give an unstable addition product. The spontaneous conversion of this addition product into an **ozonide** breaks the second carbon–carbon bond.

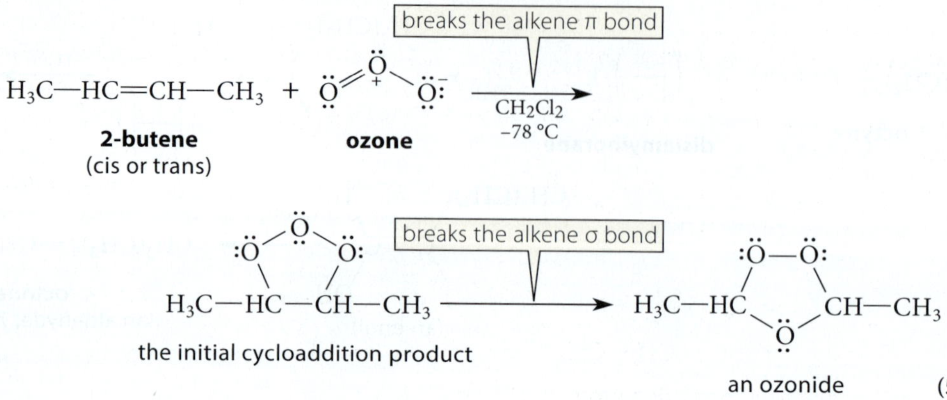

(5.44)

The reaction of an alkene with ozone to yield products of double-bond cleavage is called **ozonolysis**. (The suffix *-lysis* is used for describing bond-breaking processes. Examples are *hydrolysis*, "bond breaking by water"; *thermolysis*, "bond breaking by heat"; and *ozonolysis*, "bond breaking by ozone.")

Ozone and Its Preparation

Anettphoto/Shutterstock

Ozone is a colorless gas that is formed in the stratosphere, the part of the atmosphere that lies about 6–30 miles above Earth's surface, by the reaction of oxygen with short-wavelength ultraviolet radiation. Ozone shields Earth from longer-wavelength "UV-B" radiation by absorbing it. Although depletion of stratospheric ozone is a significant environmental concern, an increase in ozone near Earth's surface is also an environmental issue. This ozone, formed in complex reactions from nitrogen oxides and unburned hydrocarbons, is a significant contributor to smog. For example, the reaction of ozone with the double bonds in tire rubber (Sec. 15.5) explains why tires have a shorter life in urban areas with significant ozone pollution.

Ozone can be formed from the reaction of oxygen in electrical discharges. An atmospheric example is the formation of ozone in a thunderstorm by lightning (photo). The laboratory preparation of ozone involves a similar reaction:

$$3O_2 \xrightarrow{\text{electrical discharge}} 2O_3$$

Ozone is produced in the laboratory in a commercial apparatus called an *ozonator* either by passing oxygen through an electrical discharge or by treating it with ultraviolet radiation. Because ozone is unstable, it cannot be stored in gas cylinders and must be produced as it is needed.

Residential air purifiers that produce ozone are marketed commercially. Because ozone is a toxic gas and can react with a number of biomolecules, the wisdom of using these devices in the home is questionable.

The first step in ozonolysis is another addition reaction of the alkene π bond. The central oxygen of ozone is a positively charged electronegative atom, so it strongly attracts electrons. The curved-arrow notation shows that this oxygen can accept an electron pair when the other oxygen of the O=O bond accepts π electrons from the alkene.

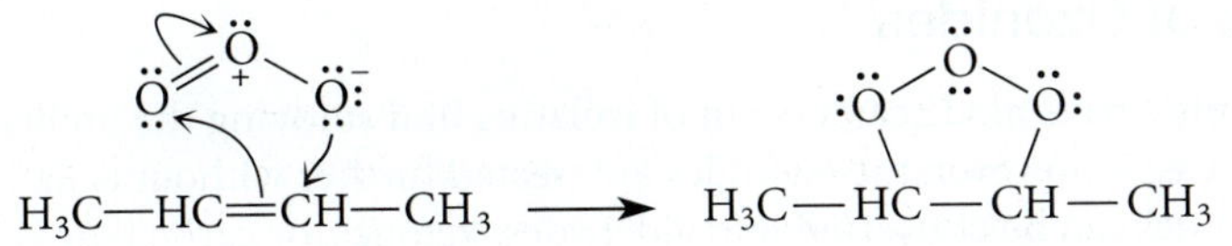

initial cycloaddition product (5.45)

This reaction results in the formation of a ring because the three oxygens of the ozone molecule remain intact. Additions that give rings are called *cycloadditions.* Furthermore, the cycloaddition of ozone occurs in a single step. Hence, this is another example, like hydroboration, of a *concerted mechanism.*

The initial cycloaddition product is unstable and spontaneously forms the ozonide. In this reaction, the remaining carbon–carbon bond of the alkene is broken.

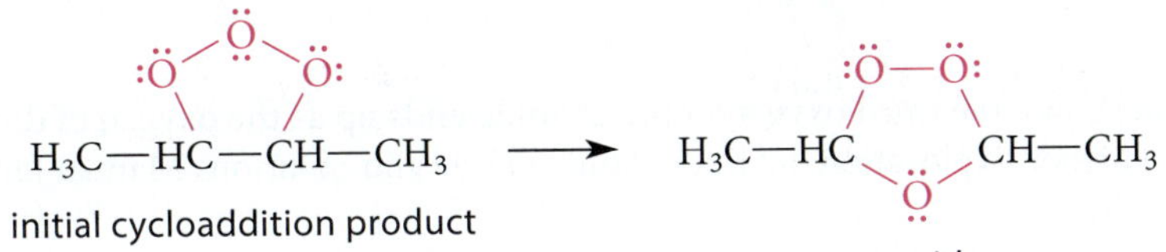

initial cycloaddition product

ozonide (5.46)

This process occurs, first, by a cyclic electron flow to form an aldehyde and an aldehyde oxide. An O—O bond (a very weak bond) is broken in the process.

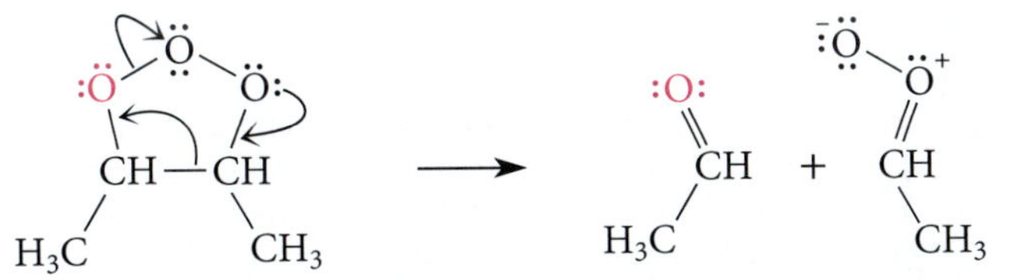

initial cycloaddition product | an aldehyde | an aldehyde oxide (5.47a)

The aldehyde flips over, and a second cycloaddition, much like ozonolysis itself—but this time, an addition to the C=O bond—completes the formation of the ozonide:

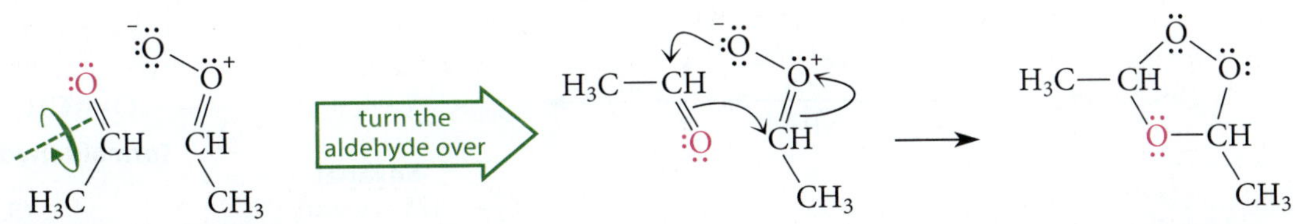

an aldehyde | an aldehyde oxide — products of Eq. 5.47a

an ozonide (5.47b)

Focused Problem

5.15 Ozonolysis of 2-pentene gives a mixture of the following *three* ozonides. Using the mechanism in Eqs. 5.47a and 5.47b, explain the origin of all three ozonides.

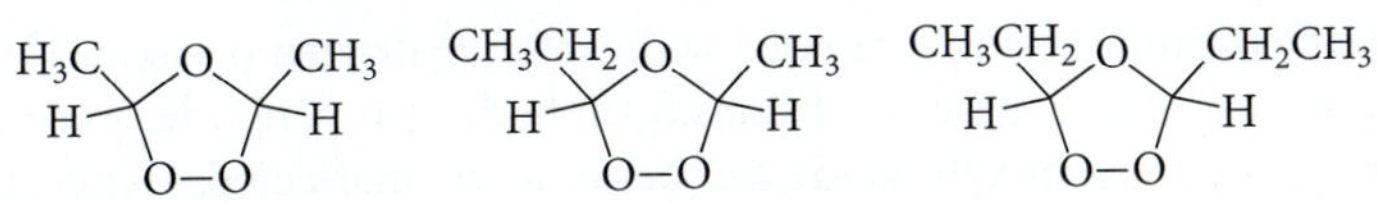

A B C

B. Reactions of Ozonides

A few daring chemists have made careers out of isolating and studying the highly explosive ozonides. In most cases, however, the ozonides are treated further without isolation to give other compounds. Ozonides can be converted into aldehydes, ketones, or carboxylic acids, depending on the structure of the alkene starting material and the reaction conditions. When the ozonide is treated with dimethyl sulfide, $(CH_3)_2S$, the ozonide is split:

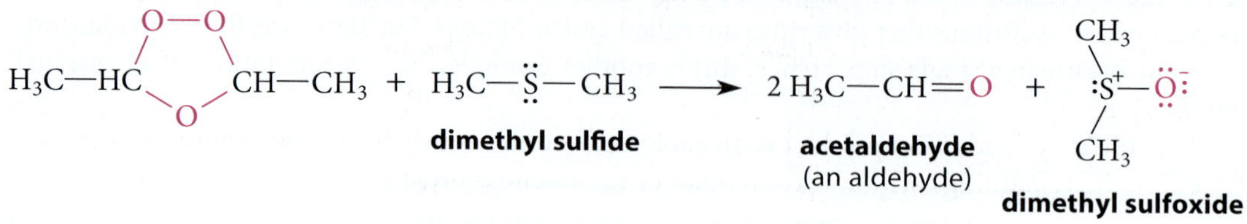

(5.48)

FURTHER EXPLORATION 5.2
Mechanism of Ozonide Conversion into Carbonyl Compounds

As this equation shows, the third oxygen of the ozonide ends up as the oxygen of dimethyl sulfoxide. (See Further Exploration 5.2 in the Study Guide and Solutions Manual for the mechanistic details.)

The net transformation resulting from the ozonolysis of an alkene followed by dimethyl sulfide treatment is the replacement of a C=C group by two C=O groups:

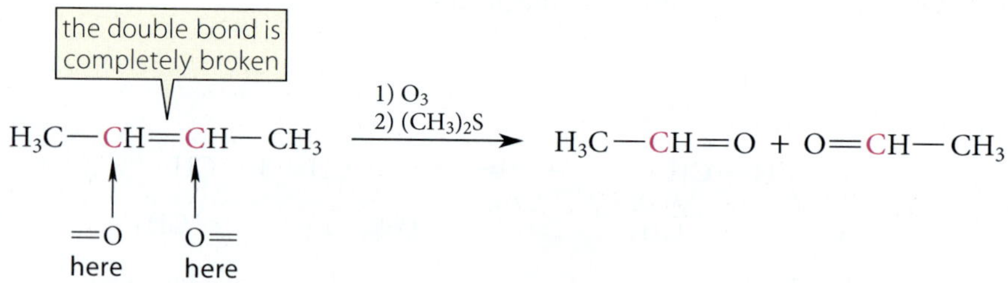

(5.49)

If the two ends of the double bond are identical, as in Eq. 5.49, then two equivalents of the same product are formed. If the two ends of the alkene are different, then a mixture of two different products is obtained:

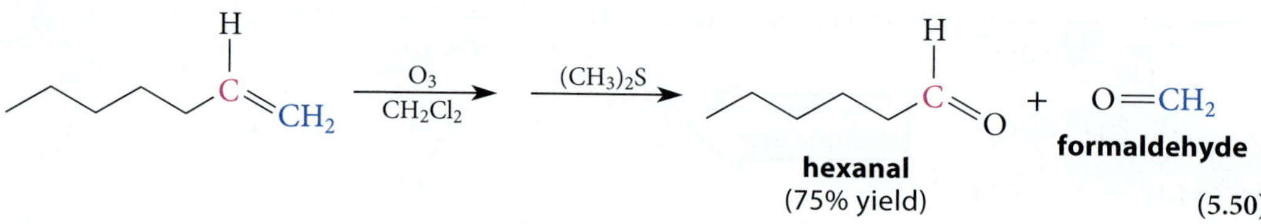

(5.50)

If a carbon of the double bond in the starting alkene bears a hydrogen, then an aldehyde is formed, as in Eqs. 5.49 and 5.50. If a carbon of the double bond bears *no* hydrogens, then a ketone is formed instead:

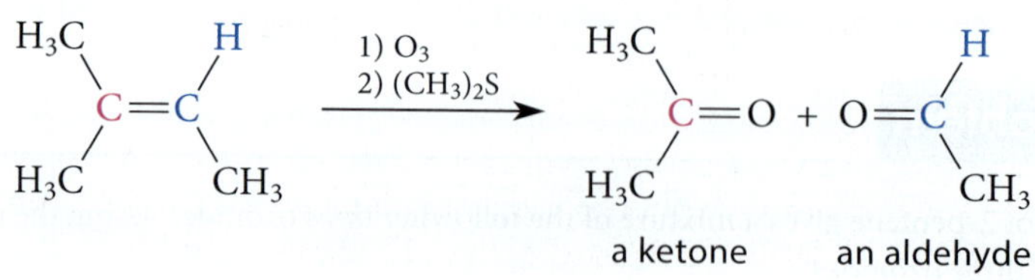

(5.51)

If the ozonide is simply treated with water, hydrogen peroxide (H_2O_2) is formed as a by-product. Under these conditions (or if hydrogen peroxide is added specifically), aldehydes are converted into carboxylic acids, but ketones are unaffected. Hence, the alkene in Eq. 5.51 would react as follows:

TABLE 5.1 Summary of Ozonolysis Results under Different Conditions

Alkene carbon	Conditions of ozonolysis: O_3, then $(CH_3)_2S$	Conditions of ozonolysis: O_3, then H_2O_2/H_2O
$R_2C{=}$	$R_2C{=}O$ **ketone**	$R_2C{=}O$ **ketone**
$R(H)C{=}$	$R(H)C{=}O$ **aldehyde**	$R(HO)C{=}O$ **carboxylic acid**
$H_2C{=}$	$H_2C{=}O$ **formaldehyde**	$H(HO)C{=}O$ **formic acid**

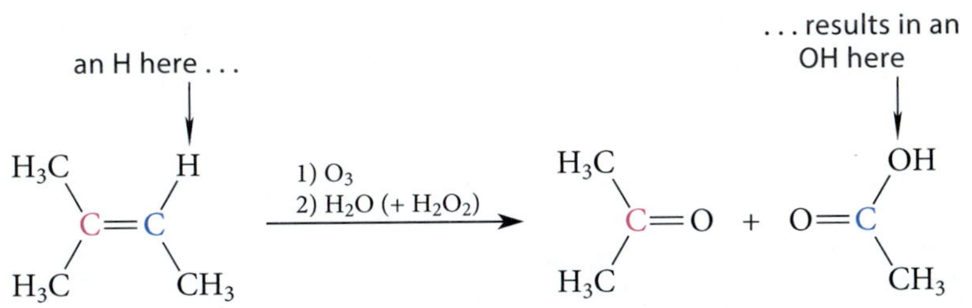

a ketone a carboxylic acid (5.52)

The different results obtained in ozonolysis are summarized in **Table 5.1**.

If the structures of its ozonolysis products are known, then the structure of an unknown alkene can be deduced. This idea is illustrated in Study Problem 5.4.

Study Problem 5.4

Alkene *X* of unknown structure gives the following products after treatment with ozone followed by aqueous H_2O_2:

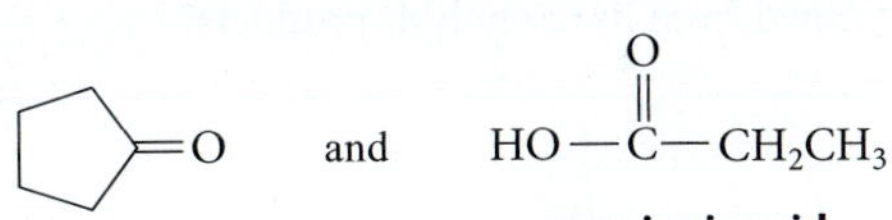

cyclopentanone **propionic acid**

What is the structure of *X*?

Solution The structure of the alkene can be deduced by *mentally reversing* the ozonolysis reaction. To do this, rewrite the C=O double bonds as "dangling" double bonds by dropping the oxygen. (The symbol ⇨ means "implies as the original structure.")

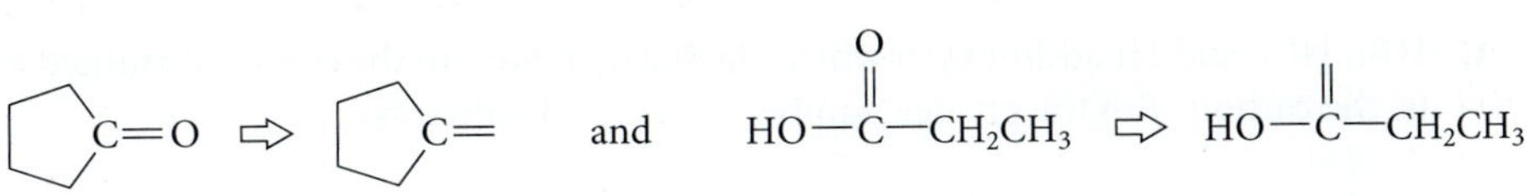

Next, replace the HO group with an H, because a carboxylic acid is formed only when a hydrogen is on the carbon of the double bond (see Table 5.1).

Finally, connect the dangling ends of the double bonds in the two partial structures to generate the structure of the alkene:

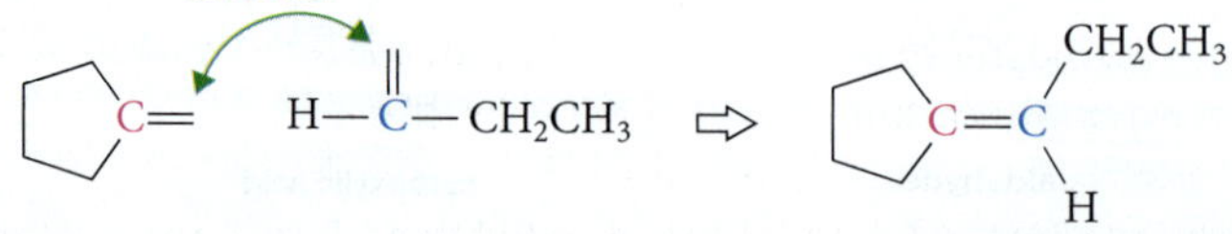

alkene structure

Focused Problems

5.16 Give the products (if any) expected from the treatment of each of the following compounds with ozone followed by dimethyl sulfide.

(a) 3-Methyl-2-pentene

(b) 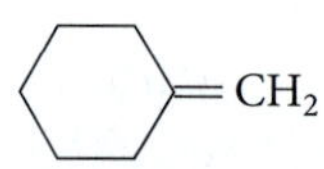

(c) Cyclooctene

(d) 2-Methylpentane

5.17 Give the products (if any) expected when the compounds in Focused Problem 5.16 are treated with ozone followed by aqueous hydrogen peroxide.

5.18 In each case, give the structure of an eight-carbon alkene that would yield each of the following compounds (and no others) after treatment with ozone followed by dimethyl sulfide.

(a) 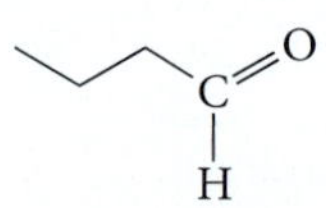

(b) 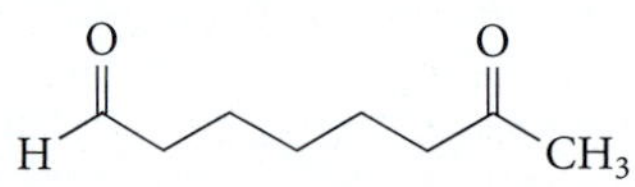

(c) 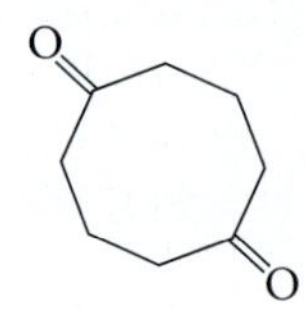

5.19 What aspect of alkene structure *cannot* be determined from its ozonolysis products?

5.9 ADDITIONS TO ALKENES AND ALKYNES: SUMMARY AND USE IN SYNTHESIS

A. Summary of Addition Reactions to Alkenes and Alkynes

We've now studied several examples of addition to alkenes in Chapters 4 and 5:

1. HBr, HCl, and HI addition, in which the halogen goes to the more substituted carbon (that is, the carbon with the greater number of alkyl substituents)

2. Acid-catalyzed addition of water (hydration), in which the OH goes to the more substituted carbon

3. Addition of Cl_2 and Br_2
4. Addition of Cl or Br and OH (as aqueous Cl_2 or Br_2), in which the OH goes to the more substituted carbon if it has two alkyl substituents
5. Oxymercuration–reduction, in which H and OH are added across the double bond so that the OH group goes to the more substituted carbon
6. Hydroboration–oxidation, in which the elements of H and OH are added across the double bond so that the OH group goes to the less substituted carbon
7. Ozonolysis, in which ozone undergoes a cycloaddition to the double bond to form an ozonide in which the carbon–carbon double bond is broken
8. Catalytic hydrogenation, in which a hydrogen is added to each carbon of the double bond

We've also studied several addition reactions of alkynes:

1. HBr, HCl, and HI addition, in which addition can be controlled to occur either once or twice to give a vinylic halide or a *geminal*-dihalide
2. Hg^{2+}-catalyzed hydration of alkynes, in which H and OH are added across the triple bond so that the OH group goes to the more substituted carbon; the resulting enols are spontaneously converted into ketones
3. Hydroboration–oxidation of alkynes to enols in which H and OH are added across the triple bond so that the OH group goes to the less substituted carbon; the resulting enols are spontaneously converted into aldehydes (if the alkyne was a 1-alkyne) or ketones
4. Catalytic hydrogenation, in which either one hydrogen is added to each carbon of the triple bond in the presence of a catalyst poison, or two hydrogens are added to each carbon of the triple bond if no poison is present

All of these reactions except catalytic hydrogenation are *electrophilic additions.* The mechanisms of electrophilic additions vary, but they all have in common *the reaction of the alkene π electrons with an electrophile.* The π electrons are the most reactive electrons in an alkene because the π bond is the weakest bond in an alkene. This is the case because the pi-type overlap ("side-to-side") of unhybridized 2p orbitals is not so strong as the sigma-type overlap ("head-to-head") of hybrid orbitals.

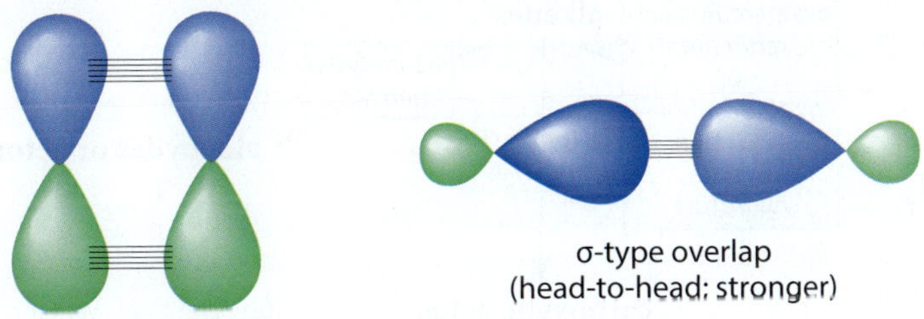

σ-type overlap
(head-to-head; stronger)

π-type overlap
(side-to-side; weaker)

(5.53)

In addition, the π electrons are the electrons that are most accessible to external electrophiles in an alkene or alkyne.

In some electrophilic additions, carbocations are formed as intermediates; in other cases, the reactive intermediates (bromonium ions, mercurinium ions) have carbocation-like resonance forms in which positive charge (and electron deficiency) is built up on the more branched carbon of the double bond. Even in hydroboration, a concerted reaction, the transition state has partial positive charge on the more substituted carbon.

The electrophilic character of these reactions accounts for their regioselectivity, which we summarize as a *modified Markovnikov's rule*: when the two groups that add are different, typically the more electronegative partner ends up on the more substituted carbon of the double bond. The preference for the more substituted carbon has its roots in the greater stability of carbocations (and transition states that resemble carbocations) on carbons that have the greater number of alkyl branches.

We discuss other electrophilic additions to alkenes in Secs. 12.3A and 12.6A. The same themes operate in these reactions. Because carbonyl (C=O groups) have π bonds, they, too, undergo electrophilic additions, which we cover in Chapter 19 and beyond.

Additions to alkenes can occur by other mechanisms. One of the most important of these is the free-radical addition of HBr, in which the Br ends up on the *less substituted* carbon of the double bond. This reaction provides a useful synthetic complement to "normal" HBr addition. Another important free-radical addition is polymerization, in which many small alkene molecules can be hooked up end-to-end to form very large molecules called *polymers* (see Sec. 5.10). This is one of the most important reactions of the chemical industry. All of these reactions involve a completely different type of reactive intermediate, called a *free radical.* Free-radical HBr addition and polymerization are presented in Secs. 10.1 and 10.3.

B. Using Alkene and Alkyne Addition Reactions in Synthesis

As you learn more reactions, one of the skills you should develop is the ability to see a molecule in terms of the ways that might be used to prepare it from other molecules. When we have a particular molecule in mind, the planning of ways that might be used to produce it from other molecules is called **organic synthesis.** As we progress throughout this text, we will develop increasingly sophisticated techniques for planning an organic synthesis.

The first technique is called **functional group transformation**, in which we learn to recognize how reactions can be used to transform one functional group into another. In Chapters 4 and 5, we have learned how to transform alkenes into various functional groups, as summarized in Display 5.54:

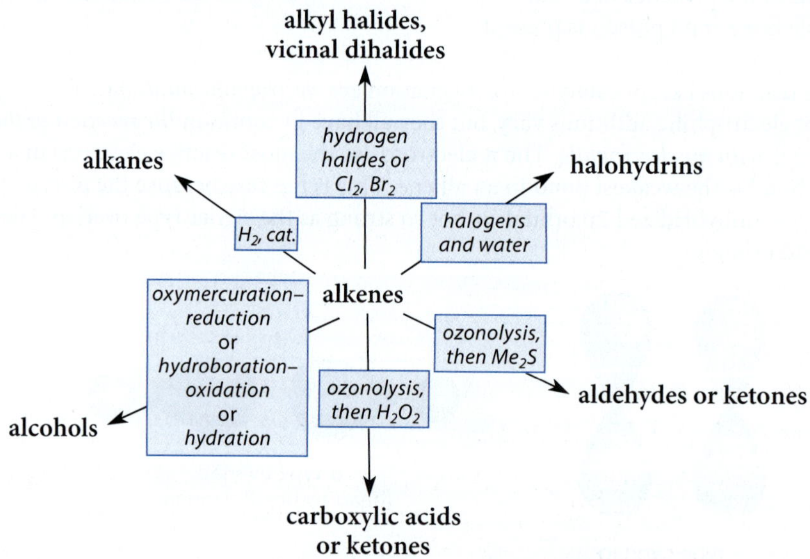

(5.54)

We have also learned how to transform alkynes into the several functional groups:

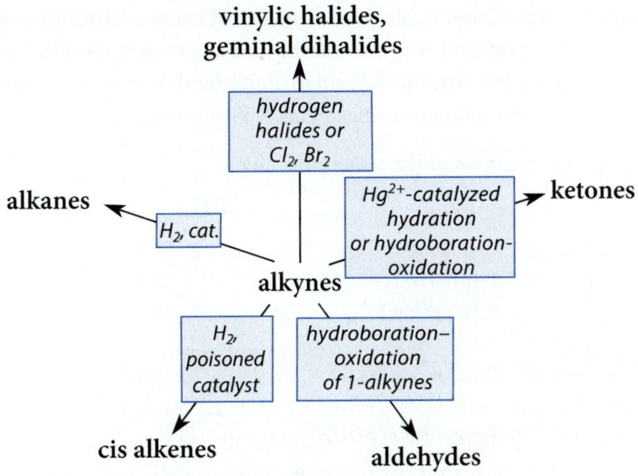

(5.55)

We refer to a molecule we wish to synthesize as a **target molecule**. The goal is to analyze the target molecule systematically in terms of particular starting materials and reactions:

1. To begin with, look at the functional group(s) in the target and ask whether you know any reactions that can produce this functional group. Make a list of the possible reactions. (Make and add to your own diagrams for each functional group as you progress through the text.)

2. Ask what starting materials might be used in each reaction to produce the target.

3. Finally, make sure that application of a reaction to a starting material produces only the correct constitutional isomer. For example, you can't prepare *any* alcohol from an alkene by hydroboration–oxidation, because the reaction occurs with a certain regioselectivity. The goal is to produce reasonably pure compounds that are free of constitutional isomers.

These ideas are illustrated in Study Problem 5.5.

Study Problem 5.5

Devise a synthesis of the following compound from an alkene with the same number of carbons.

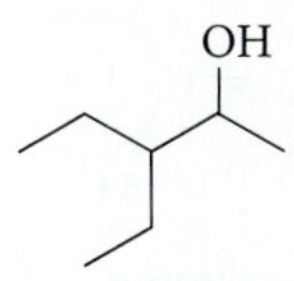

target molecule

Solution Apply each of the steps listed in the previous discussion.

Step 1: The functional group in the target molecule is an alcohol. We've studied three reactions for making alcohols from alkenes: acid-catalyzed hydration, oxymercuration–reduction, and hydroboration–oxidation.

Step 2: If we're going to use one of these reactions, the OH group *must* be at one of the carbons of the double bond in the alkene starting material. Two alkenes meet this criterion (the symbol ⇨ in this context means "implies as starting material"):

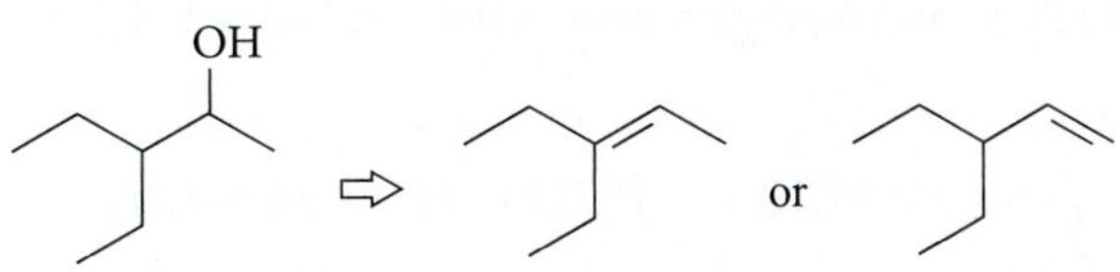

target molecule *A* *B*

Step 3: For *A* to serve as starting material, the OH group must be added at the *less substituted* carbon of the double bond. Hydroboration–oxidation could be used for this purpose. For *B* to be used as a starting alkene, the OH must be added to the *more substituted* carbon of the double bond. Would acid-catalyzed hydration work? In this case, the carbocation intermediate would rearrange; if we get any of the desired product, it would be obtained in a mixture with a rearranged alcohol. (Work through the mechanism to understand this point.) However, oxymercuration–reduction could be used, because carbocation rearrangements typically don't occur with this reaction.

Therefore, we've found two solutions to the synthesis of the target molecule.

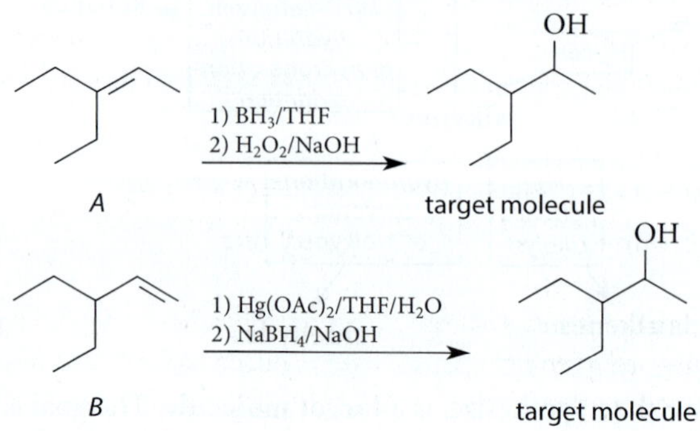

Which reaction we would actually use is in principle arbitrary. For example, we might choose hydroboration–oxidation because we could avoid the handling and disposal issues associated with mercury. However, we could use either reaction.

In a given synthesis problem, there may be one solution, there may be more than one—or there may be none.

A summary of reactions used in functional-group transformation is given in Appendix V. This appendix is organized by target functional group. All of the reactions in this text can be found in this appendix.

Focused Problem

5.20 Use the techniques described in this section to devise a synthesis for each of the following compounds from an alkyne.

(a) $(CH_3)_3CCH_2CH_2CH_2\overset{\displaystyle O}{\overset{\|}{C}}H$

5,5-dimethylhexanal

(b)

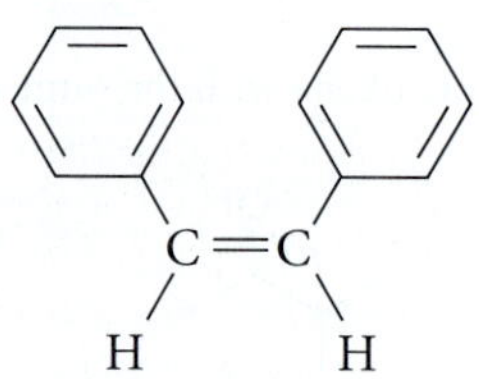

***cis*-stilbene**

Chemistry in the Real World

5.10 ALKENES IN THE CHEMICAL INDUSTRY

More ethylene is produced industrially than any other organic compound; it is the most widely used starting material for chemical production in the chemical industry. In the United States, it has ranked second in industrial production of all chemicals (behind sulfuric acid). About 330 billion pounds of ethylene are produced annually worldwide. Propene (known industrially by its older name *propylene*) is not far behind, with an annual world output of more than 225 billion pounds. Other important alkenes are styrene, 1,3-butadiene, and methylpropene (usually called isobutylene in the chemical industry).

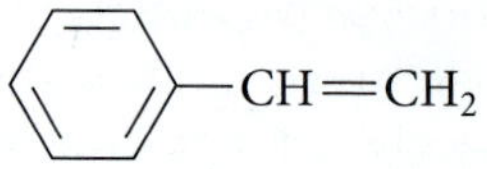

styrene

$H_2C{=}CH{-}CH{=}CH_2$

1,3-butadiene

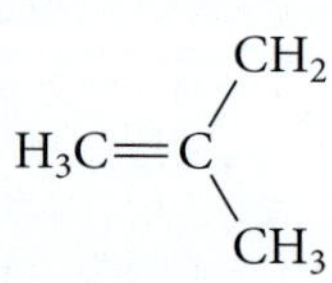

methylpropene (isobutylene)

Ethylene and propene are considered to be petroleum products, but they are not obtained directly from crude oil. Rather, they are produced industrially from alkanes in a process called **thermal cracking**. Cracking breaks larger alkanes into a mixture of dihydrogen, methane, and other small hydrocarbons, many of which are alkenes. In this process, a mixture of alkanes from the fractional distillation of petroleum (Sec. 2.9) is mixed with steam and heated in a furnace at 750–900 °C for a fraction of a second and is then quenched (rapidly cooled). The products of cracking are then separated. Specialized catalysts have also been developed that allow cracking to take place at lower temperatures ("cat-cracking").

In the United States, the hydrocarbon most often used to produce ethylene is ethane, which is a component of natural gas. In the cracking of ethane, ethylene and H_2 are formed at a very high temperature.

$$\underset{\textbf{ethane}}{H_3C-CH_3} \xrightarrow[900\ ^\circ C]{} \underset{\textbf{ethylene}}{H_2C{=}CH_2} + H_2 \tag{5.56}$$

Although petroleum is the major source of ethylene, there has been increasing interest in the production of "green" ethylene—ethylene from biological sources other than petroleum. The dehydration of ethanol produced by the fermentation of sugars in corn, sugarcane, and other sources of fermentable sugars is a promising source of "green" ethylene. (See Sec. 11.13.) Each ton of ethylene produced in this way captures 2.5 tons of atmospheric CO_2.

$$\text{plants} + CO_2 \xrightarrow{\text{photosynthesis}} \text{sugars} \xrightarrow{\text{fermentation}} \underset{\textbf{ethanol}}{CH_3CH_2OH} \xrightarrow[\text{heat}]{\text{acid catalyst}} \underset{\textbf{ethylene}}{H_2C{=}CH_2} + H_2O \tag{5.57}$$

Currently, this process cannot compete economically with ethylene produced from natural gas, but it may become more important in the future.

Thousands of alkene molecules can be strung together (like freight cars on a train) to make long molecules called *polymers.*

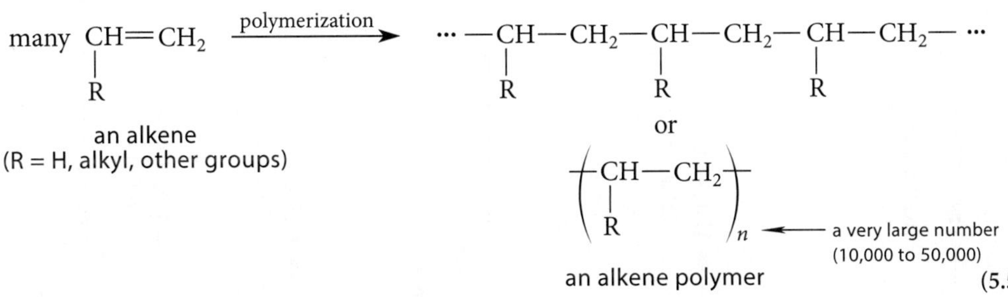

(5.58)

(Polymerization is discussed in Sec. 10.3, and the structures of some important polymers are shown in Table 10.3.) Polymerization is a major end use for ethylene, propene, and styrene, which give the polymers polyethylene, polypropylene, and polystyrene, respectively (Table 10.3). Different alkenes can be polymerized together to produce a *copolymer.* The diene 1,3-butadiene is polymerized together with styrene to produce a synthetic rubber copolymer, styrene–butadiene rubber (SBR), which is important in the manufacture of tires. (This process is discussed in Sec. 15.5.)

Ethylene is a starting material for the manufacture of ethylene glycol, HO—CH_2CH_2—OH, which is the main component of automotive antifreeze. Ethylene glycol is also a starting material for the production of polyesters (Eq. 21.105, Sec. 21.12A). Ethylene is also used, along with benzene, to produce styrene, which, as we noted previously, is another important alkene in the polymer industry. Propene is a key compound in the production of phenol, which is used in adhesives, and acetone, a commercially important solvent. In addition, propene is polymerized to give polypropylene, an important alkene polymer. Methylpropene (isobutylene) is used to prepare octane isomers that are important components of high-octane gasoline, and it is reacted with methanol (CH_3OH) to give a gasoline additive, methyl *tert*-butyl ether (MTBE; see Problem 4.83).

Some of these chemical interrelationships are shown in **Fig. 5.2**. This figure also shows how fundamental petroleum is to the chemical economy of much of the industrialized world.

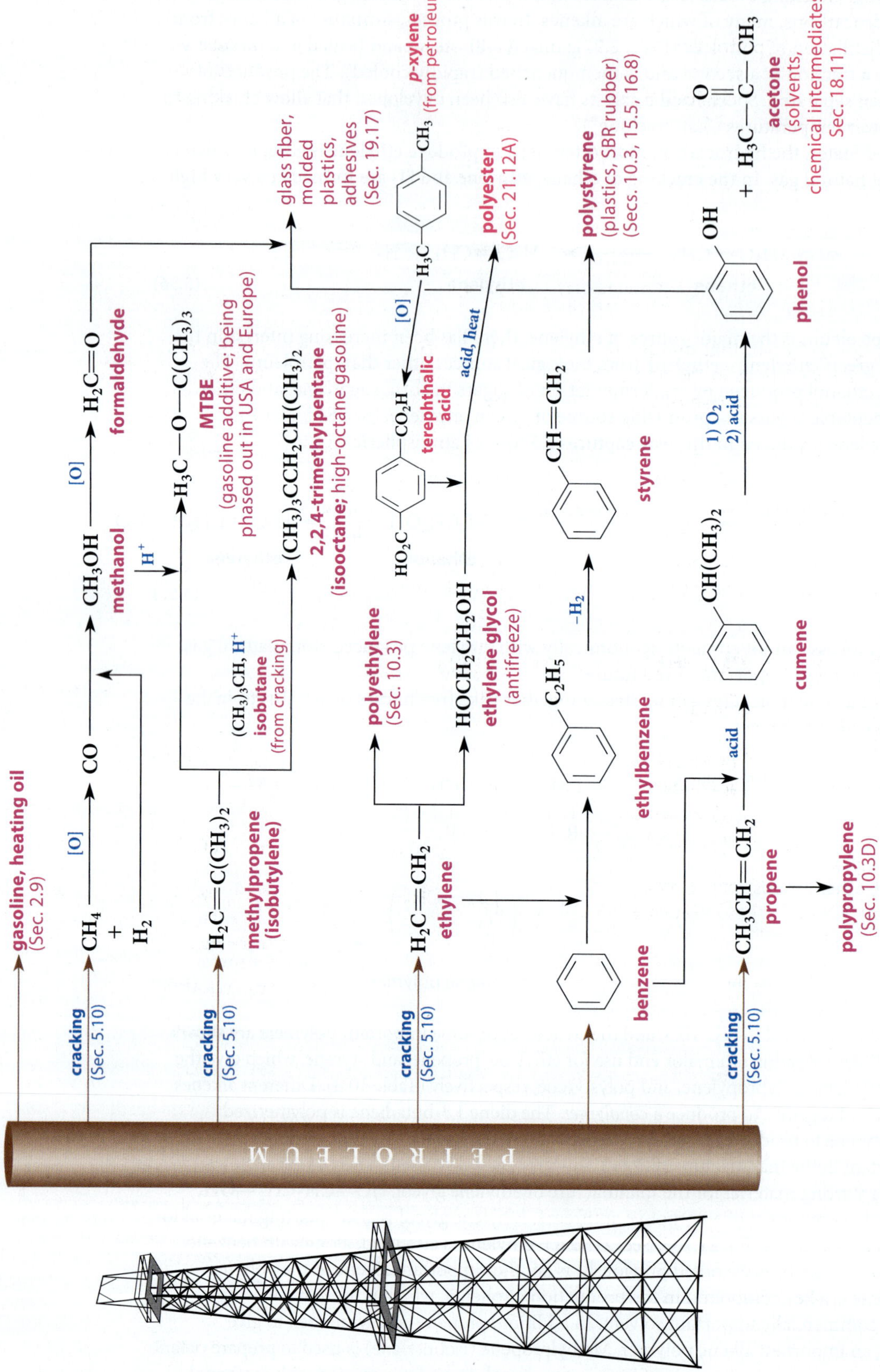

FIGURE 5.2 The chemical industry is an important part of the world economy that depends critically on the availability of petroleum. (In the reactions, **[O]** indicates oxidation.)

Ethylene as a Fruit Ripener

Denis and Yulia Pogostins/Shutterstock

An intriguing role of ethylene in nature has been put to commercial use. Ethylene produced by plants causes fruit to ripen. That is, ethylene is a *ripening hormone*. Plants produce ethylene by the degradation of a relatively rare amino acid:

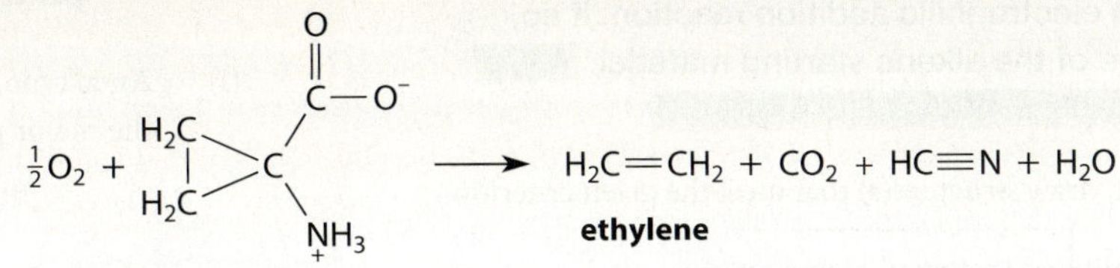

1-amino-1-cyclopropane-carboxylic acid

It is ethylene, for example, that brings green tomatoes to that peak of juicy redness in the home garden. Commercial growers have made use of this knowledge by picking and transporting fruit before it is ripe, and then ripening it "on location" with ethylene.

CHAPTER SUMMARY *For a summary of the chapter, see Chapter 5 in the* Study Guide and Solutions Manual.

REACTION REVIEW *For a summary of reactions discussed in this chapter, see the* Reaction Review *section of Chapter 5 in the* Study Guide and Solutions Manual.

SKILLS OBJECTIVES WITH PROBLEMS

(This includes a review of reactions in Chapter 4.)

- **Draw the structures of the products of the reactions from Chapters 4 and 5 when applied to a given alkene or alkyne, including regiochemistry where appropriate.**

4.7A; 4.9; 4.10A,B; 5.2A,B; 5.4A,B; 5.5; 5.6A,B; 5.7; 5.8

5.21 Give the principal organic products expected when 1-butene (or another compound listed) reacts with each of the following reagents.

(a) Br_2 in CH_2Cl_2 solvent

(b) O_3, −78 °C

(c) Product of (b) with $(CH_3)_2S$

(d) Product of (b) with H_2O_2

(e) O_2, flame

(f) HBr

(g) I_2, H_2O

(h) H_2, Pt/C

(i) BH_3 in tetrahydrofuran (THF)

(j) Product of (i) with NaOH, H_2O_2

(k) $Hg(OAc)_2$, H_2O

(l) Product of (k) with $NaBH_4$/NaOH

(m) HI

5.22 Repeat Problem 5.21 for 1-ethylcyclopentene.

5.23 Give the principal organic products expected when 1-pentyne (or another compound listed) reacts with each of the following reagents.

(a) HBr (1 equivalent)

(b) H_2 (large excess), Pt/C

(c) H_2, Lindlar catalyst

(d) H_2O, H_3O^+, Hg^{2+} (catalyst)

(e) Disiamylborane

(f) Product of (e) with NaOH, H_2O_2

5.24 Give the principal organic products expected when 2,7-dimethyl-4-octyne (or another compound listed) reacts with each of the following reagents.

(a) HBr (2 equivalents)

(b) H_2 (large excess), Pt/C

(c) H_2, Lindlar catalyst

(d) H_2O, H_3O^+, Hg^{2+} (catalyst)

(e) BH_3 in diethyl ether

(f) Product of (e) with NaOH, H_2O_2

• Given an organic compound, indicate whether it can be formed as the sole constitutional isomer from an alkene in a given electrophilic addition reaction. If so, give the structure of the alkene starting material. 4.7A; 4.9; 4.10A,B; 5.2A,B; 5.4A,B; 5.5; 5.6A,B; 5.7

5.25 In each part, draw structure(s) that meet the given criterion.

(a) Two alkenes that would yield only 1-methylcyclohexanol when treated with $Hg(OAc)_2$ in water, then $NaBH_4$/NaOH

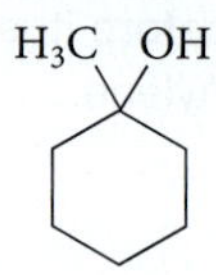

1-methylcyclohexanol

(b) Two stereoisomeric alkenes that would give 4-octanol as the major product of hydroboration followed by treatment with alkaline H_2O_2

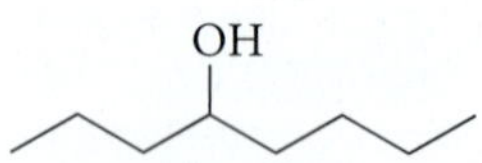

4-octanol

(c) A five-carbon alkene that would give the same alcohol as the major product of either hydroboration–oxidation or oxymercuration–reduction

(d) Four alkenes with the formula C_7H_{12} that would undergo catalytic hydrogenation to give methylcyclohexane

(e) The alkene that would give the following alcohol as the major product of hydroboration–oxidation

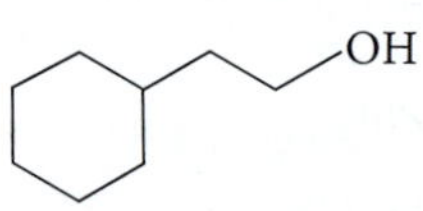

(f) The alkyne that would give the following ketone as the major product of Hg^{2+}-catalyzed hydration

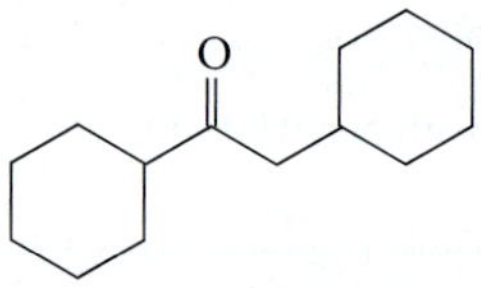

(g) The alkene that would, on treatment with concentrated HI, give the following iodoalkane as the only product

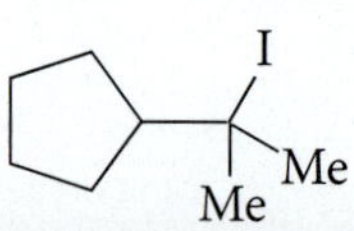

(h) Two alkenes that would give pentanal as the *only* product of ozonolysis followed by treatment with Me_2S

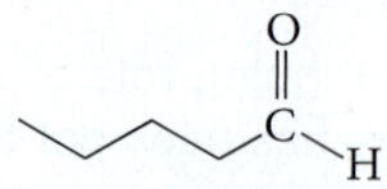

pentanal

(i) An alkyne that would give 3,3-dibromohexane as the major product of its reaction with excess HBr

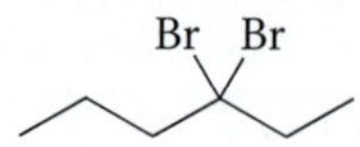

3,3-dibromohexane

(j) An alkyne that would give pentanal [structure in part (h)] as the only product of hydroboration with disiamylborane followed by alkaline H_2O_2

(k) A compound with the formula C_6H_{12} that would *not* undergo ozonolysis

• Draw the structure of the key reactive intermediate in each of the following alkene reactions: (a) the addition of H—X (X = Br, Cl, or I) 4.7B; (b) acid-catalyzed hydration 4.10C; (c) the reaction of a dihalogen in the presence of water 5.2A,B; and (d) oxymercuration 5.4A. Show how the intermediate in each case accounts for the regiochemistry of these reactions.

5.26 For parts (a)–(d), draw the structure of the principal organic product(s) and the structure of the reactive intermediate(s) involved for each of the addition reactions to the following alkene. Show how the intermediate accounts for the regioselectivity of each reaction.

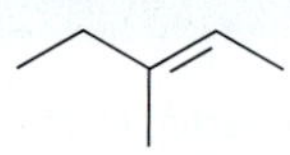

(a) Addition of HCl

(b) Acid-catalyzed hydration

(c) The reaction with I_2 in water

(d) Oxymercuration

• Draw the transition state for a hydroboration reaction of an alkene, including partial charges. Show how the structure of this transition state accounts for the regioselectivity of the hydroboration reaction. 5.6A

5.27 For hydroboration of the alkene in Problem 5.26, draw the structure of the addition product (for the first addition), and sketch the transition-state structure, including partial charges. Show how this transition state accounts for the regioselectivity of hydroboration.

- **Reconstruct the possible structures of an alkene from the structures of its ozonolysis products 5.8B.**

5.28 Draw the structure(s) of the compound(s) that meet each of the following criteria:

(a) An alkene with one double bond that would give the following compound as the *only* product after ozonolysis followed by H_2O_2

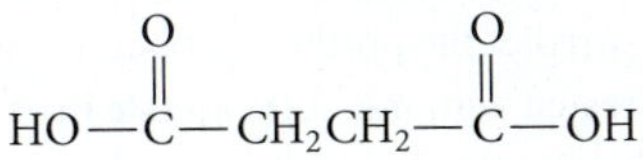

(b) Two alkenes that would give pentanal as the *only* product of ozonolysis followed by treatment with Me_2S

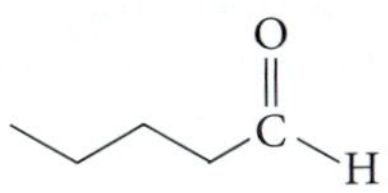

pentanal

(c) A compound with the formula C_5H_{10} that would *not* undergo ozonolysis

- **Devise syntheses of target molecules from alkene or alkyne starting materials. 5.9B**

5.29 Propose a synthesis of each of the following target molecules from an alkene or alkyne with the same number of carbons. Each synthesis should yield the target free of contaminating isomers.

(a)

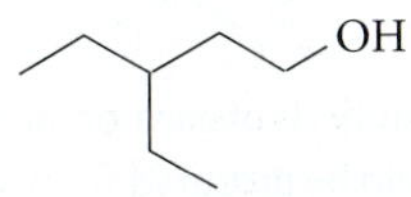

(b)

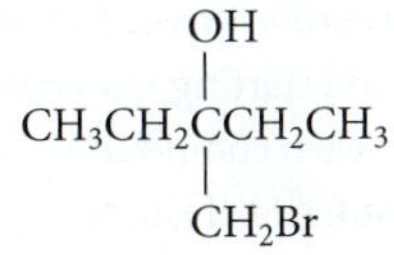

(c)

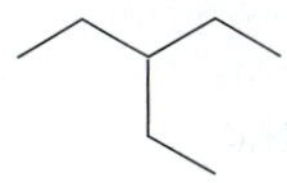

(d)

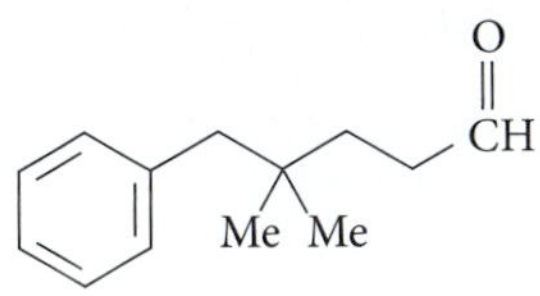

(e)

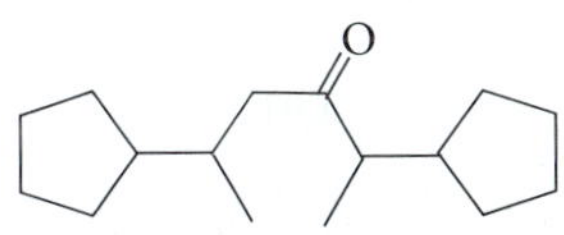

(f)

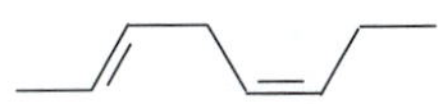

INTEGRATED PROBLEMS

5.30 Contrast the products formed when the following alkene undergoes (a) oxymercuration–reduction and (b) acid-catalyzed hydration. Explain the reason for the difference, if any.

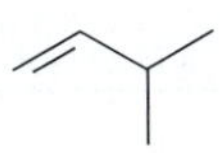

3-methyl-1-butene

5.31 Give the structures of both the reactive intermediate and the product in each of the following reactions.

(a)

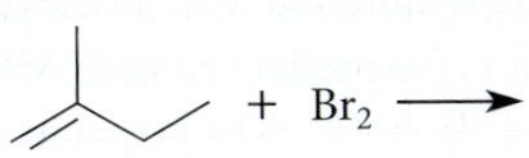

(b)

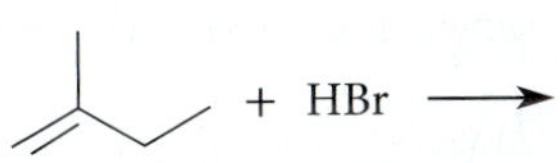

(c)

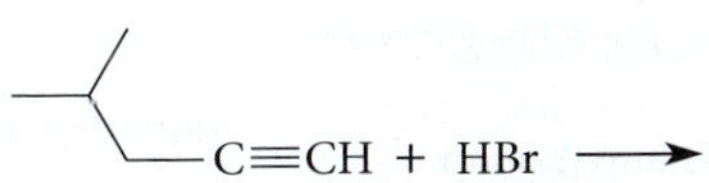

(d)

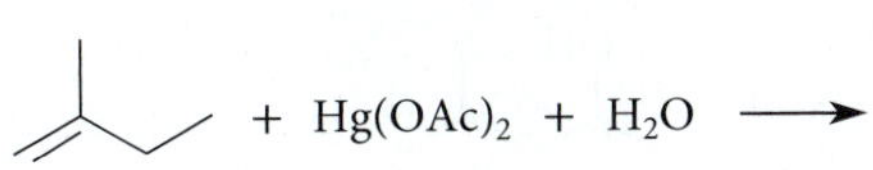

5.32 Show the products that form at each stage of the following reaction. Sketch the transition state for the last stage (the third addition).

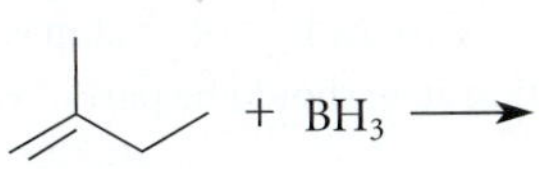

(excess)

5.33 Provide the structure of the missing starting material, product, or reagent in each of the following reactions.

(a)

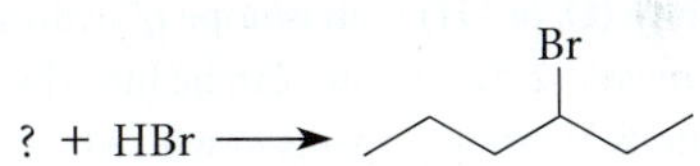

(the only product)

(b)

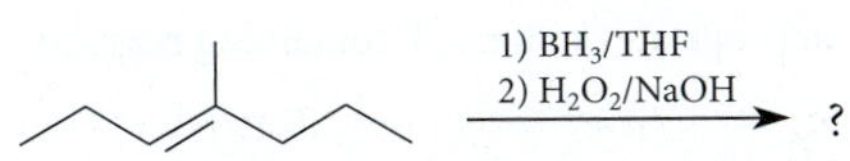

(c)

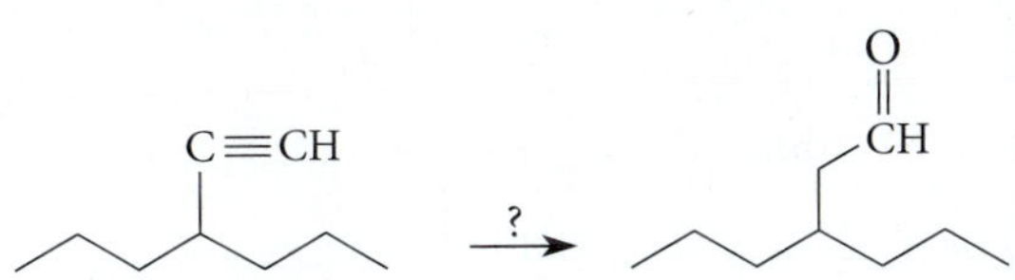

(d)

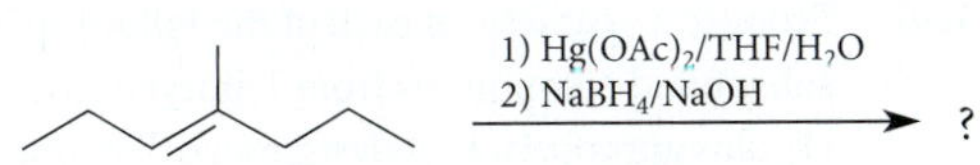

(e)

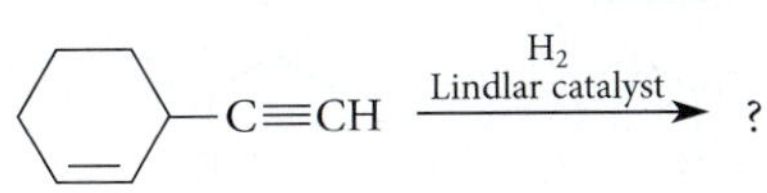

(f)

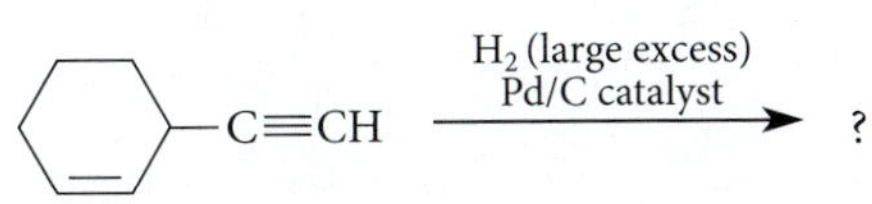

(g)

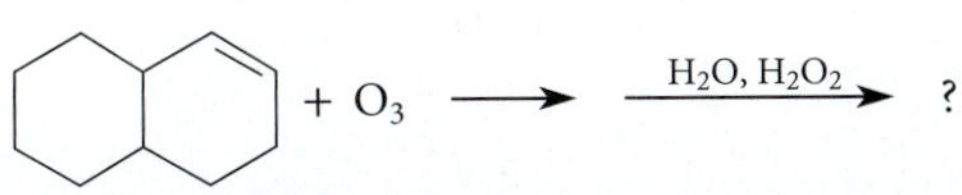

(h) (The formula of the missing starting material is $C_{10}H_{16}$.)

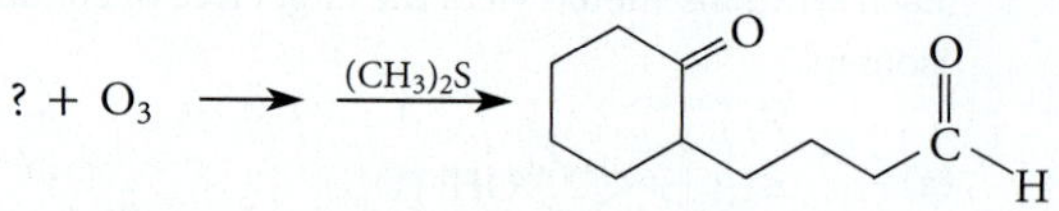

5.34 Outline a laboratory synthesis of each of the following compounds. Each should be prepared from an alkene *with the same number of carbon atoms* and any other reagents. The reactions and starting materials used should be chosen so that each compound is virtually uncontaminated by constitutional isomers.

(a) 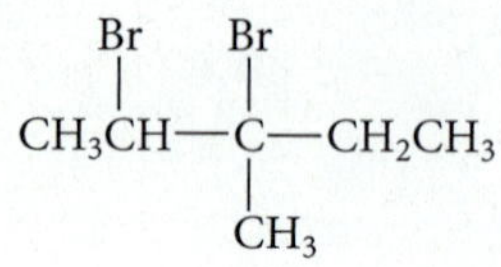

(b) 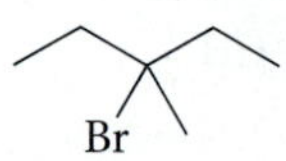

(c) 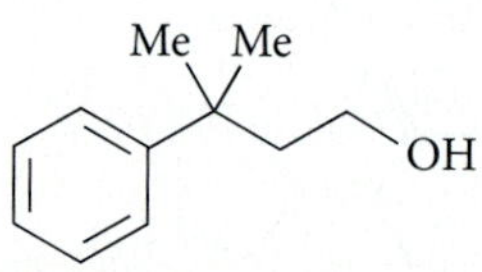

(d) 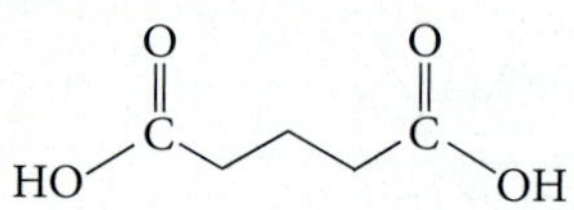

(e) 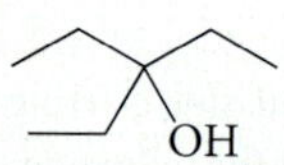

5.35 Deuterium (D, or ^{2}H) is an isotope of hydrogen with atomic mass = 2. Deuterium can be introduced into organic compounds by using reagents in which hydrogen has been replaced by deuterium. Outline preparations of both isotopically labeled compounds from the same alkene using appropriate deuterium-containing reagents.

(a) 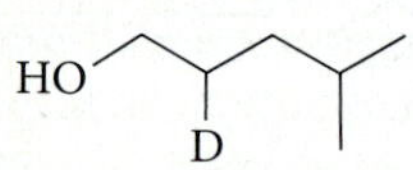

(b) 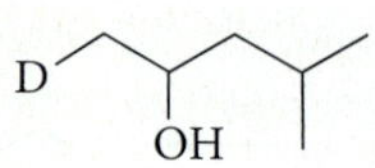

5.36 Propose a synthesis of each of the following isotopically substituted compounds from 1-butyne. You can use D_2 or H_2 plus appropriate catalysts; more than one reaction may be required.

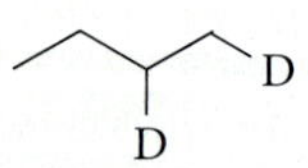

A

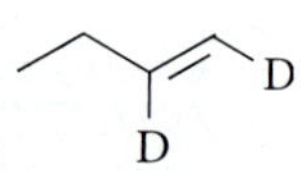

B

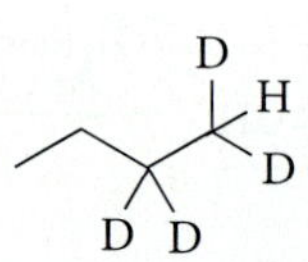

C

5.37 Using the mechanism of halogen addition to alkenes to guide you, predict the product(s) obtained when 2-methyl-1-butene is subjected to each of the following conditions. Explain your answers.

(a) Br_2 in H_2O

(b) I_2 in CH_3OH solvent

5.38 Using the mechanism of the oxymercuration reaction to guide you, predict the product(s) obtained when 1-hexene is treated with mercuric acetate in each of the following solvents and the resulting products are treated with $NaBH_4$/NaOH. Explain your answers, and tell what functional groups are present in each of the products. [Reaction (b) is called *alkoxymercuration,* and reaction (c) is called *acetoxymercuration.*]

(a) H_2O/THF

(b) $(CH_3)_2CH—\ddot{O}H$

isopropyl alcohol

(c) 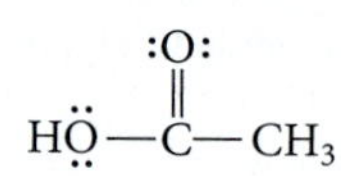

acetic acid

5.39 Complete each of the following alkyne reactions, and explain your reasoning.

(a) 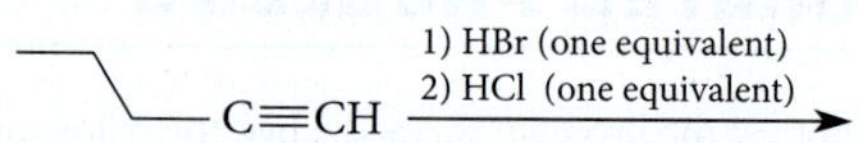

(b) (*Hint:* The products of this reaction are two isomeric compounds with the formula $C_6H_{11}BrO$.)

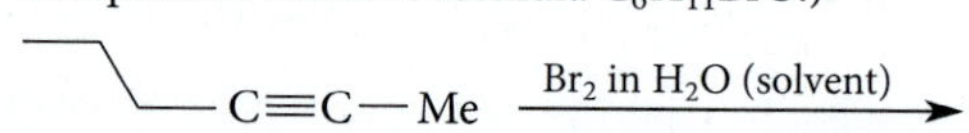

5.40 When (*E*)-4-methyl-2-pentene undergoes hydroboration with BH_3, followed by the usual oxidative treatment with alkaline H_2O_2, a product mixture consisting of 57% of alcohol *A* and 43% of alcohol *B* is obtained. When the same alkene undergoes hydroboration with disiamylborane, followed by oxidation, the product consists of 97% of alcohol *A* and 3% of alcohol *B*. (The preparation of disiamylborane is shown in Eq. 5.34c, Sec. 5.6A.)

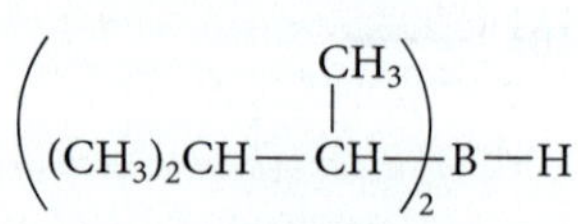

"disiamylborane"

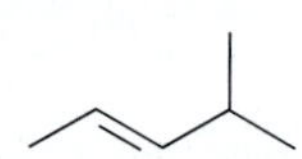

4-methyl-2-pentene

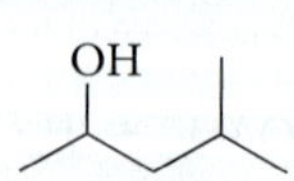

A

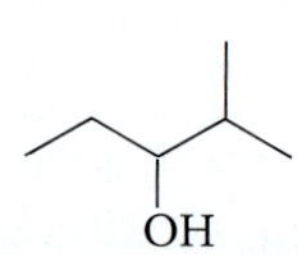

B

Account for the increase in the percentage of alcohol *A* when disiamylborane is used as the hydroboration reagent. A structure of the transition state should be part of your explanation.

5.41 A compound *A* with the molecular formula C_8H_{16} decolorized Br_2 in CH_2Cl_2. Catalytic hydrogenation of *A* gave octane. Treatment of compound *A* with O_3 followed by aqueous H_2O_2 yielded butanoic acid as the *sole* product.

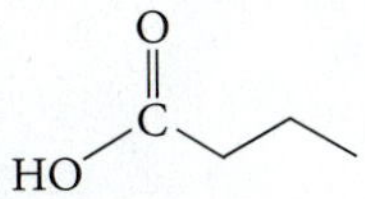

butanoic acid

What is the structure of compound *A*? What aspect of the structure of *A* is not determined by the data?

5.42 A compound *A* (C_6H_6) undergoes catalytic hydrogenation over Lindlar catalyst to give a compound *B*, which in turn undergoes ozonolysis followed by workup with aqueous H_2O_2 to yield succinic acid and two equivalents of formic acid. In the absence of a catalyst poison, catalytic hydrogenation of *A* gives hexane. Propose a structure for compound *A*.

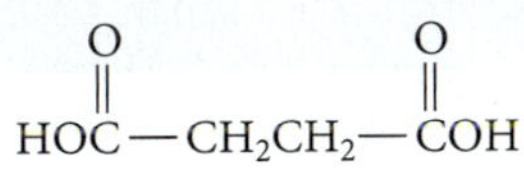

succinic acid

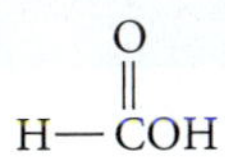

formic acid

5.43 In a laboratory a bottle was found containing a clear liquid *A*. The bottle was labeled, "$C_{10}H_{16}$, isolated from a lemon." Because of your skills in organic chemistry, you have been hired to identify this substance. Compound *A* decolorizes Br_2 in the inert solvent CH_2Cl_2. When *A* is hydrogenated over a catalyst, two equivalents of H_2 are consumed and the product is found to be 1-isopropyl-4-methylcyclohexane. Ozonolysis of *A* followed by treatment of the reaction mixture with H_2O_2 gives the following compound as a major product:

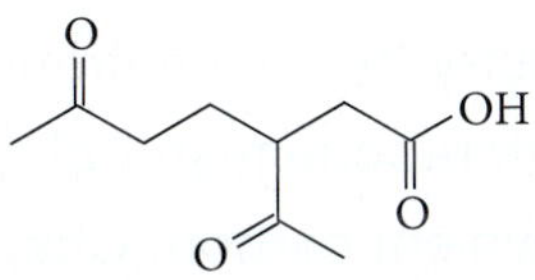

STUDY GUIDE LINK 5.2
Solving Structure Problems

Suggest a structure for *A*, and explain all observations. (See Study Guide Link 5.2 if you need help.) (*Hint:* The ozonolysis product has one less carbon than the formula on the bottle.)

5.44 When *trans*-3-hexene is subjected to ozonolysis in the presence of an *excess* of acetaldehyde containing the isotope ^{18}O, an ozonide is isolated that contains the isotope at one of the oxygens. Use the mechanism of ozonolysis to postulate a structure for the ozonide, including the position of the isotope.

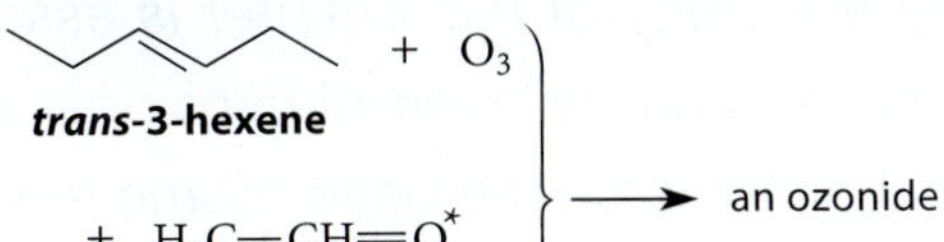

***trans*-3-hexene**

acetaldehyde-^{18}O
(*O = ^{18}O)

5.45 Provide a curved-arrow mechanism for each of the following reactions. Be sure to draw separate structures for each mechanistic step.

(a) (*Hint:* Number the carbon skeleton to assist in mapping the bonds breaking and forming in this reaction.)

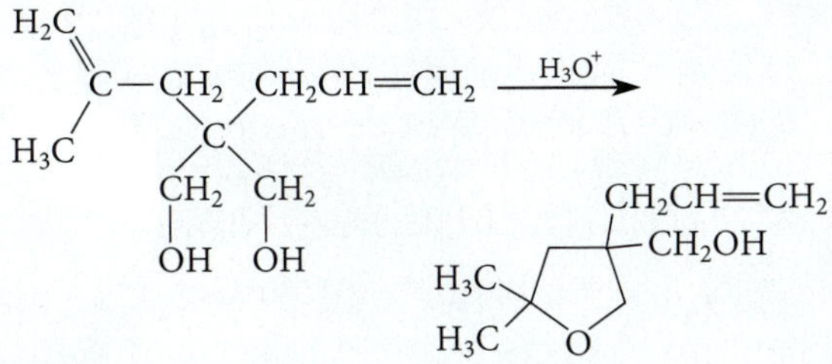

(b) (*Hint:* Start with the reaction of Br_2 at the double bond.)

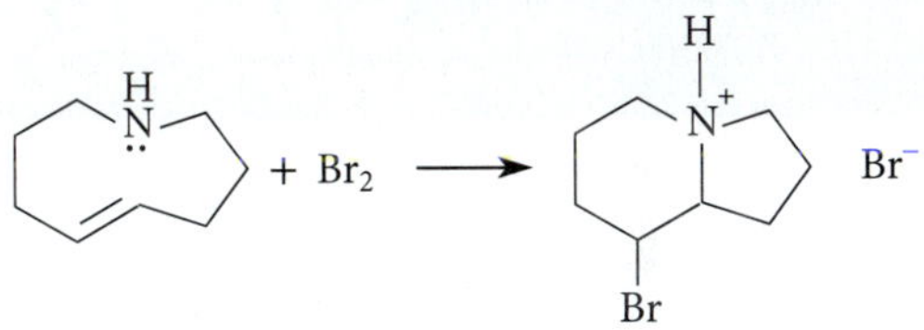

(c) (*Hint:* Sulfur is the electrophilic center in the reaction of sulfur dichloride with an alkene. Sulfur, like bromine, can form cyclic ions; in the case of sulfur, these are called *sulfonium ions.*)

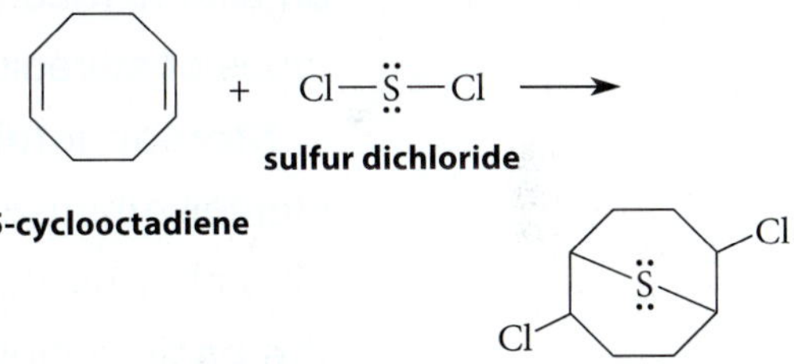

sulfur dichloride

1,5-cyclooctadiene

(d) For the following reaction, give the curved-arrow notation for only the reaction of the alkene with $Hg(OAc)_2$ and H_2O. Then show that the compounds that you obtain from this mechanistic reasoning can be converted into the observed products by the $NaBH_4$ reduction.

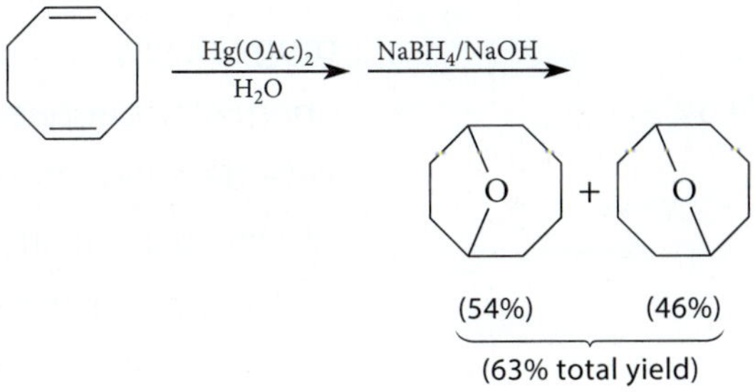

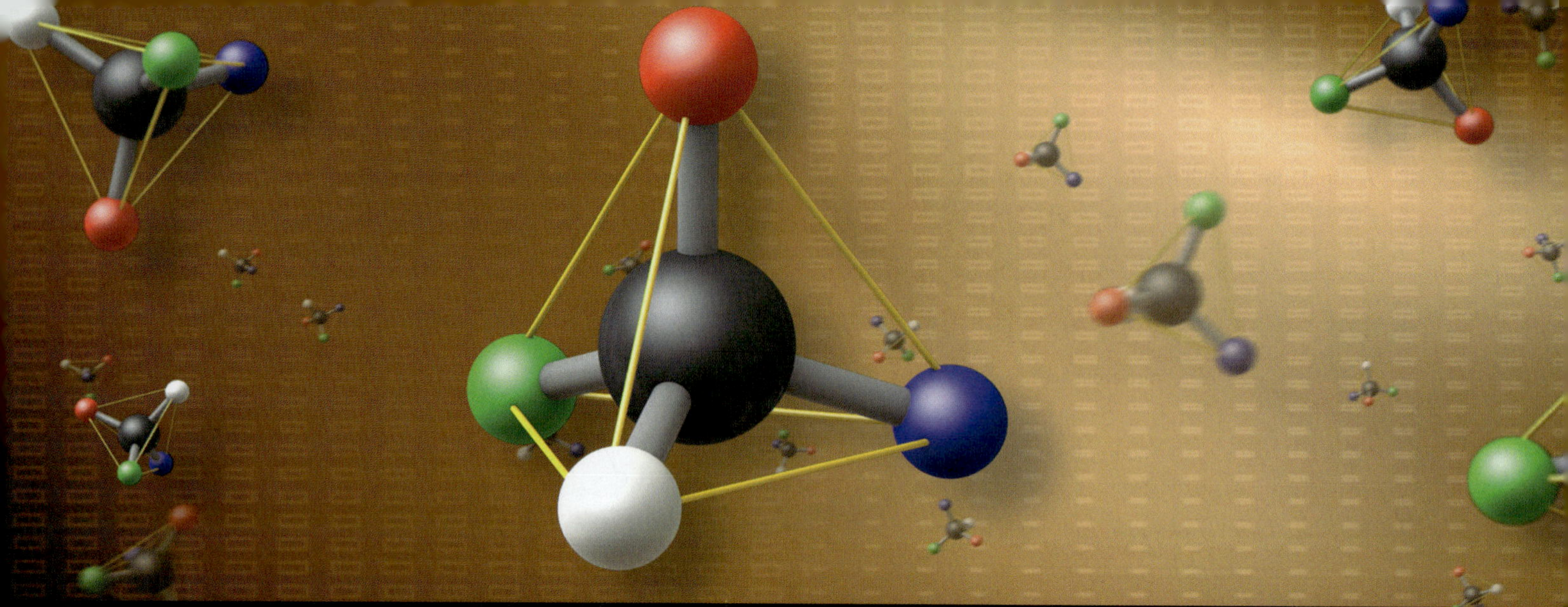

6 PRINCIPLES OF STEREOCHEMISTRY

This chapter and the one that follows deal with stereoisomers and their properties. **Stereoisomers** are compounds that have the same atomic connectivity but a different arrangement of atoms in space. Recall that *E* and *Z* isomers of an alkene (Sec. 4.1D) are stereoisomers. In this chapter we'll learn about other types of stereoisomers.

Stereochemistry is the study of stereoisomers and the chemical effects of stereoisomerism. A few ideas of stereochemistry were introduced in Sec. 4.1D. This chapter delves more generally into stereochemistry by concentrating on the basic definitions and principles. We'll see how stereochemistry played a key role in the determination of the geometry of tetravalent carbon. Chapter 7 continues the discussion of stereochemistry by considering both the stereochemical aspects of cyclic compounds and the application of stereochemical principles to chemical reactions.

Stereochemistry is important because many biological phenomena such as drug activity and enzyme catalysis depend on stereochemistry. It is important in chemistry because the stereochemical aspects of chemical reactions can provide detailed information on their mechanisms. Examples of these applications of stereochemistry will be developed in this and the following chapter.

The use of molecular models during the study of this chapter is essential. Models will help you develop the ability to visualize three-dimensional structures and will make the two-dimensional pictures on the page "come to life." If you use models now, your reliance on them will gradually decrease.

In this chapter you will also be using perspective structures of molecules, particularly line-and-wedge structures. The techniques for drawing and interpreting these types of structures were introduced in Secs. 1.6D and 2.5C. You should review this material as a preparation for this chapter.

6.1 ENANTIOMERS, CHIRALITY, AND SYMMETRY

A. Enantiomers and Chirality

Any molecule—indeed, any object—has a mirror image. Some molecules are **congruent** to their mirror images. This means that all atoms and bonds in a molecule can be simultaneously superimposed onto identical atoms and bonds in its mirror image. An example of such a molecule is ethanol, or ethyl alcohol, $H_3C—CH_2—OH$ (**Fig. 6.1**). Construct a model of ethanol and another model of its mirror image. For simplicity, use a single colored ball to represent the methyl group and a single ball of another color to represent the hydroxy (OH) group. Place the two central carbons side by side and align the methyl and hydroxy groups, as shown in Fig. 6.1. The hydrogens should then align as well. *The congruence of an ethanol molecule and its mirror image shows that they are identical.*

Some molecules, such as 2-butanol, are *not* congruent to their mirror images.

$$H_3C-\underset{*}{\overset{\overset{\large OH}{|}}{C}}H-CH_2CH_3 \quad \text{or} \quad H_3C-\underset{*}{\overset{\overset{\large OH}{|}}{C}}H-C_2H_5$$

2-butanol

Build a model of 2-butanol and a second model of its mirror image, and test them for congruence as shown in **Fig. 6.2**. If you align the carbon with the asterisk and any two of its attached groups, the other two groups do not align. Therefore, *a 2-butanol molecule and its mirror image are noncongruent and are therefore different molecules.* Because these two molecules have identical connectivities, by definition they are *stereoisomers*. Molecules that are noncongruent mirror images are called **enantiomers**. Therefore, the two 2-butanol stereoisomers are enantiomers.

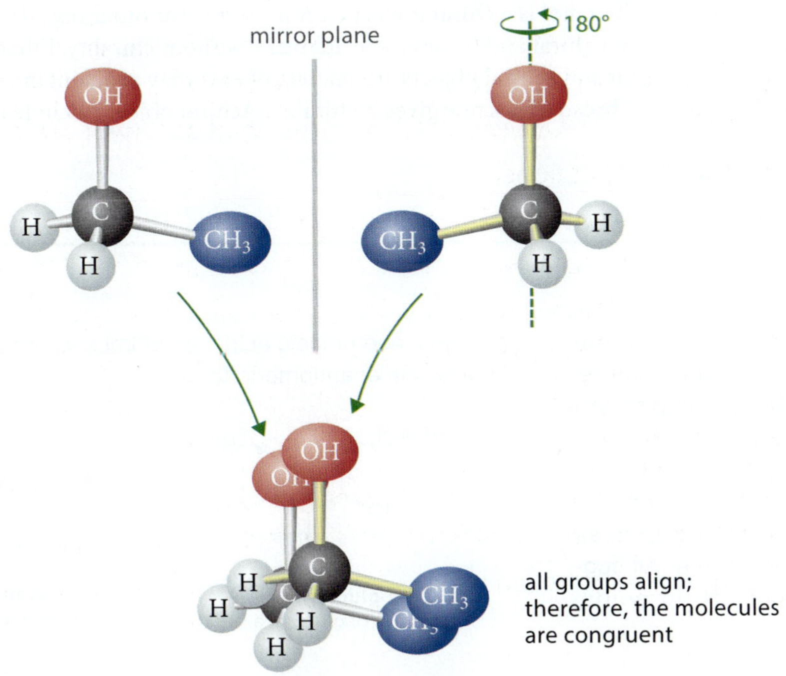

FIGURE 6.1 Testing mirror-image ethanol molecules for congruence. One mirror image is shown with yellow bonds to distinguish it from the other. Aligning the central carbons, the CH_3 groups, and the OH groups on the different molecules causes the hydrogens to align as well. To achieve this alignment, one of the molecules must be rotated in space.

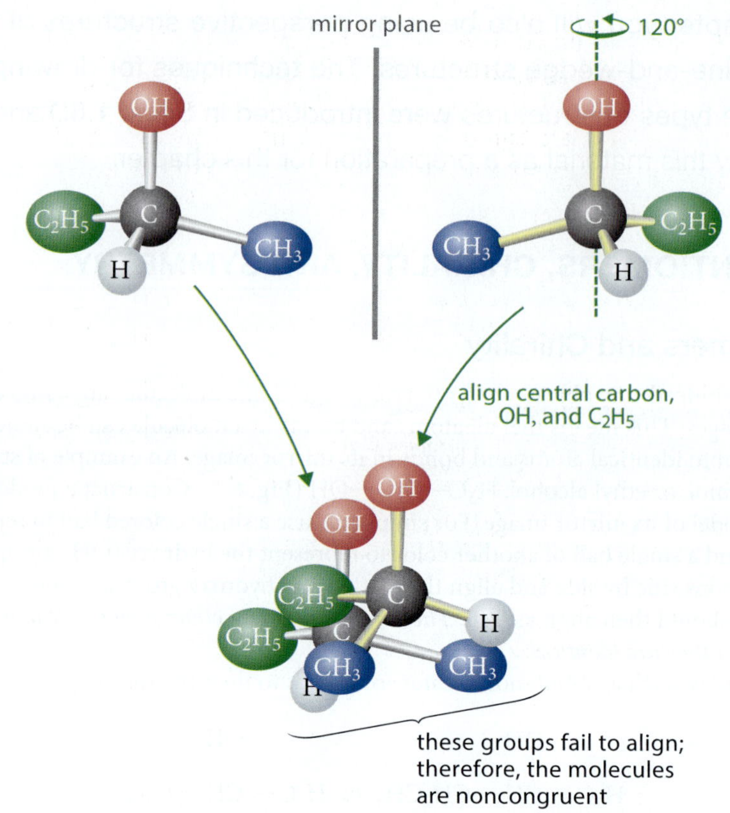

FIGURE 6.2 Testing mirror-image 2-butanol molecules for congruence. As in Fig. 6.1, the bonds of one mirror image are yellow. When the central carbon and any two of the groups attached to it (OH and C_2H_5 in this figure) are aligned, the remaining groups do *not* align.

Enantiomers must not only be mirror images; they must also be *noncongruent* mirror images. Therefore, ethanol (see Fig. 6.1) has no enantiomer because an ethanol molecule and its mirror image are congruent.

A molecule (or other object) that has an enantiomer (that is, a noncongruent mirror image) is said to be **chiral** (pronounced kī′rŭl); it possesses the property of **chirality**, or handedness. (*Chiral* comes from the Greek word for hand.) Enantiomeric molecules have the same relationship as the right and left hands—the relationship of an object and its noncongruent mirror image. Therefore, 2-butanol is a chiral molecule. A molecule (or other object) that has a *congruent* mirror image is not chiral and is said to be **achiral**—without chirality. Ethanol is an achiral molecule. Both chiral and achiral objects are matters of everyday acquaintance. A foot or a hand is chiral; the helical thread of a screw gives it chirality. Achiral objects include a ball and a soda straw.

Importance of Chirality

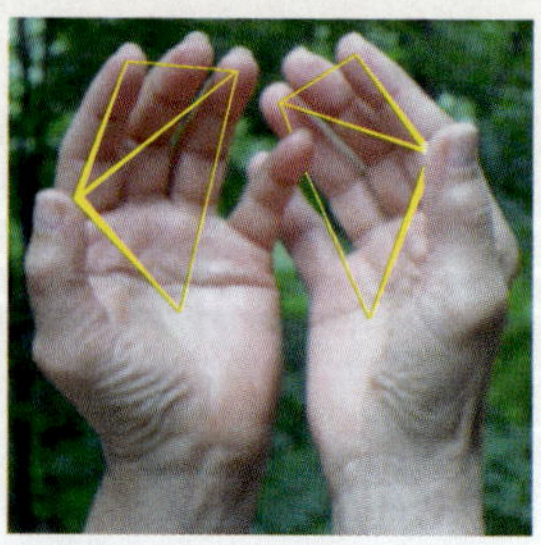

Chirality = handedness

Chiral molecules occur widely throughout nature. Glucose, an important sugar and energy source, is chiral; the enantiomer of naturally occurring glucose cannot be utilized as a food source. All sugars, most amino acids, proteins, and nucleic acids are chiral and occur naturally in only one enantiomeric form.

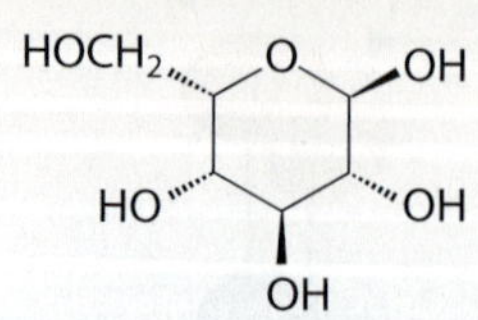

β-D-glucopyranose
(one form of glucose, a sugar)

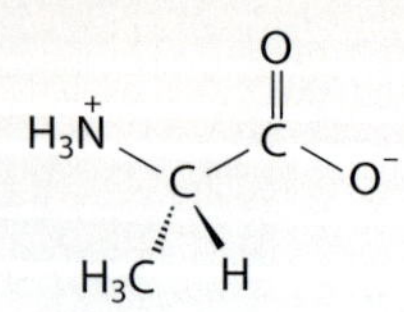

(*S*)-alanine
(an amino acid)

Chirality is important in medicine as well. More than half of the organic compounds used as drugs are chiral, and in most cases only one enantiomer has the desired physiological activity. Here are four of the many examples, with their active stereoisomers shown:

In rare cases, the inactive enantiomer is toxic (see the story of the drug thalidomide in Sec. 6.4B). The safety and effectiveness of synthetically prepared chiral drug molecules have been issues of concern for both pharmaceutical manufacturers and the U.S. Food and Drug Administration (FDA) for several decades.

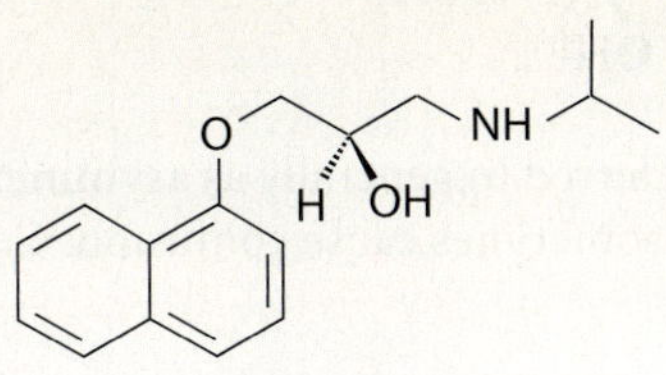

propranolol
(a β-blocker)

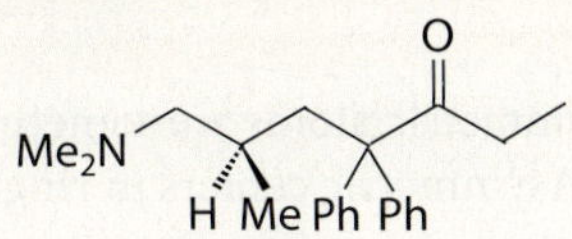

methadone
(treatment for opioid addiction)

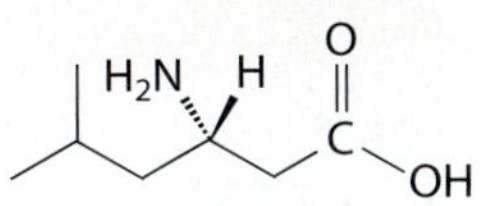

pregabalin (Lyrica)
(treatment for neuropathic pain, anxiety)

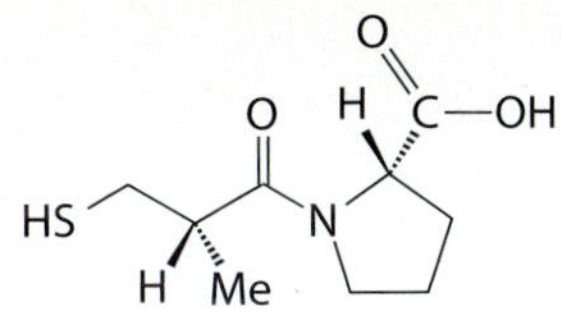

captopril
(treatment for hypertension)

B. Asymmetric Carbon and Stereocenters

Many chiral molecules contain one or more asymmetric carbon atoms. An **asymmetric carbon atom** is a carbon to which *four different groups* are bonded. Therefore, 2-butanol (see Fig. 6.2), a chiral molecule, contains an asymmetric carbon atom; this is the carbon that bears the four different groups —CH_3,—C_2H_5,—H, and —OH. In contrast, none of the carbons of ethanol, an achiral molecule, is asymmetric. *A molecule that contains only one asymmetric carbon is chiral.* No generalization can be made, however, for molecules with more than one asymmetric carbon. Although many molecules with two or more asymmetric carbons are indeed chiral, not all of them are (Sec. 6.7). Moreover, an asymmetric carbon atom (or other asymmetric atom) is not a *necessary* condition for chirality; some chiral molecules have no asymmetric carbons at all (Sec. 6.8). Despite these caveats, it is important to recognize asymmetric carbon atoms because so many chiral organic compounds contain them.

You may encounter other terms such as **chiral carbon** and **chiral center** that mean the same thing as *asymmetric carbon.*

Study Problem 6.1

Indicate whether 4-methyloctane has an asymmetric carbon. If so, identify it.

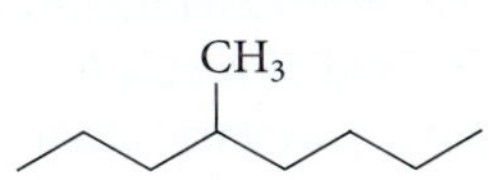

4-methyloctane

Solution The carbon marked with an asterisk (*) in the following structure has four different groups attached to it—a $CH_3CH_2CH_2$—(propyl) group, a —CH_3 (methyl) group, a H—, and a —$CH_2CH_2CH_2CH_3$ (butyl) group. Therefore, this is an asymmetric carbon.

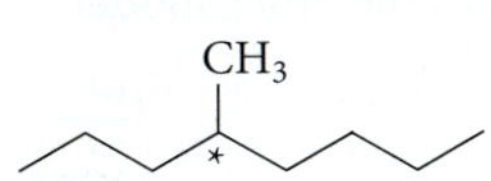

The propyl and butyl groups are *not* different at the point of attachment—both have CH_2 groups at that point, as well as at the next carbon removed. The difference is found at the ends of the groups. The important point is that *two groups are different even when the difference is remote from the carbon in question.*

Although carbon is the most common asymmetric atom in organic molecules, other atoms can be asymmetric as well. For example, the following chiral compound contains an asymmetric phosphorus.

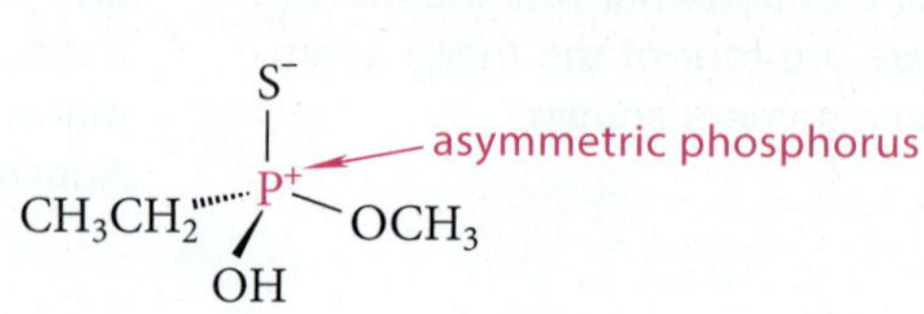

Asymmetric atoms are sometimes referred to generally as **asymmetric centers**.

Asymmetric centers in rings can sometimes cause confusion. Consider the following two cases:

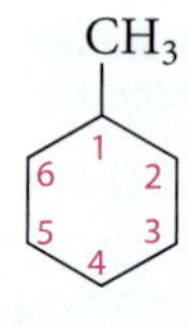

methylcyclohexane

A

Is carbon-1 in molecule *A* asymmetric? In the case of rings, *we regard each branch as a separate group, even though the groups are tied together*. We compare carbon-2 with carbon-6 and carbon-3 with carbon-5. When we reach carbon-4, which is common to both branches, we have found no difference. Therefore, the two ring branches at carbon-1 are identical, and carbon-1 is *not* asymmetric. Molecule *A* has no asymmetric carbons and is achiral.

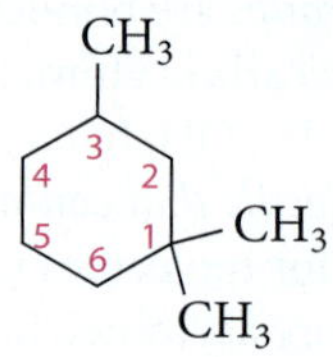

1,1,3-trimethylcyclohexane

B

Does molecule *B* contain any asymmetric carbons? Comparing carbons 2 and 4, there is no difference—there is a CH_2 group at each position. Comparing carbons 1 and 5, however, there is a difference. We conclude, then, that carbon-3 is an asymmetric carbon: it has four different groups attached, a —CH_3 group, a —H, and the two different parts of the ring. Carbon-1 is not asymmetric because it has two identical (methyl) groups attached. The remaining ring carbons each have two identical groups attached (H), so they aren't asymmetric carbons either. Molecule *B*, then, contains one asymmetric carbon—carbon-3—and molecule *B* is therefore chiral.

An asymmetric carbon (or other asymmetric atom) is another type of *stereocenter*, or *stereogenic atom*. Recall (Sec. 4.1D) that a **stereocenter** is an atom at which the interchange of two groups gives a stereoisomer. In Fig. 6.2, for example, interchanging the methyl and ethyl groups in one enantiomer of 2-butanol gives the other enantiomer. If this point is unclear from Fig. 6.2, *you should build two models to demonstrate this to yourself.* First construct a model of either enantiomer, and then construct a model of its mirror image. Then show that the interchange of *any* two groups on one model gives the other model.

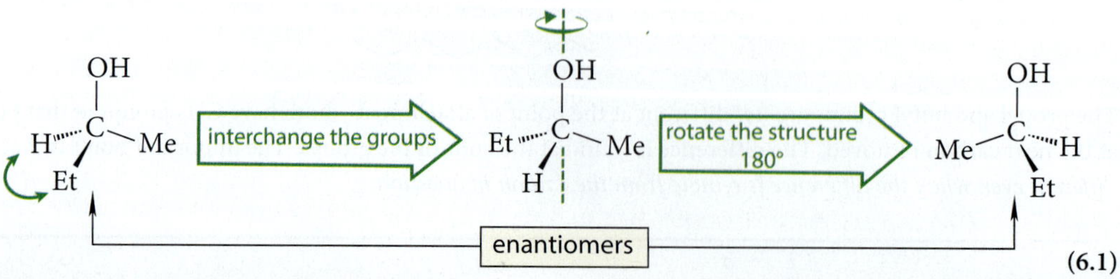

(6.1)

Not all carbon stereocenters are asymmetric carbons. Recall (Sec. 4.1D) that the carbons involved in the double bonds of *E* and *Z* isomers are also stereocenters. These carbons are not asymmetric carbons, though, because they are not connected to four different groups. In other words, the term *stereocenter* is not associated solely with chiral molecules. *All asymmetric atoms are stereocenters, but not all stereocenters are asymmetric atoms.*

STUDY GUIDE LINK 6.1
Stereocenters and Asymmetric Atoms

Focused Problem

6.1 Identify the asymmetric carbons, if any, in each of the following molecules.

(a) 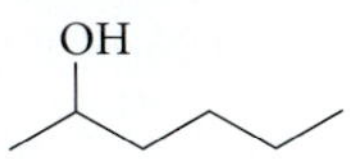

(b) 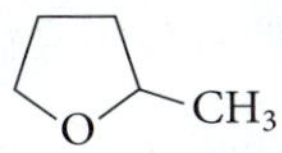

(c) 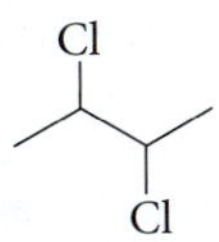

(d) 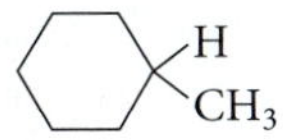

C. Chirality and Symmetry

What causes chirality? *Chiral molecules lack certain types of symmetry.* The symmetry of any object (including a molecule) can be described by certain **symmetry elements**, which are lines, points, or planes that relate equivalent parts of an object. A very important symmetry element is a **plane of symmetry**, sometimes called an **internal mirror plane**. This is a plane that divides an object into halves that are exact mirror images. For example, the mug in Display 6.2 has a plane of symmetry. Similarly, the ethanol molecule has a plane of symmetry, which, in the display, we are viewing end-on.

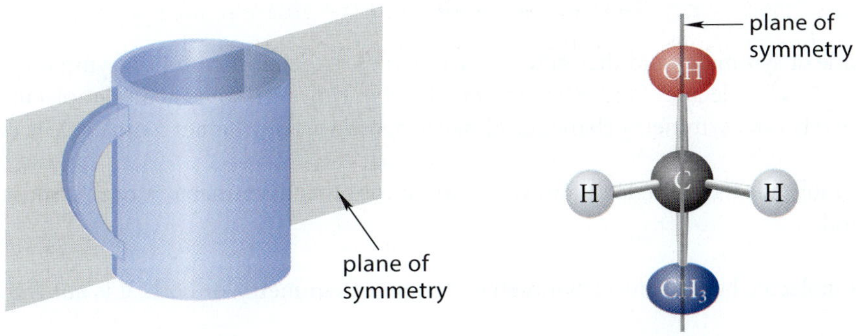

ethanol (6.2)

A molecule or other object that has a plane of symmetry is achiral. Therefore, the ethanol molecule and the mug are achiral. Chiral molecules and other chiral objects do *not* have planes of symmetry. The chiral molecule 2-butanol, analyzed in Fig. 6.2, has no plane of symmetry. A human hand, also a chiral object, has no plane of symmetry.

Another important symmetry element is the **center of symmetry**, sometimes also called a *point of symmetry.* A molecule has a center of symmetry if you can reproduce it by (1) forming its mirror image then (2) rotating this mirror image by 180° about an axis perpendicular to the mirror.

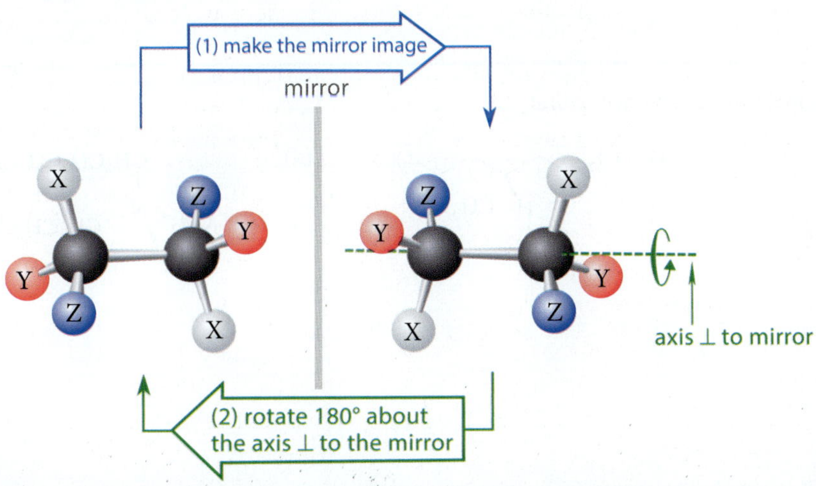

(6.3)

This description of symmetry is applied to a *particular conformation.* Different conformations of the same molecule may have different symmetries.

More descriptively, a center of symmetry is a point through which *any* line contacts exactly equivalent parts of the object at the same distance in both directions.

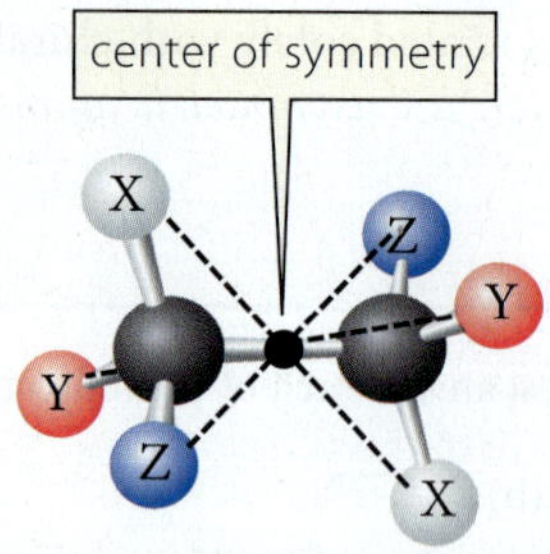

corresponding points on opposite sides are equidistant from a center of symmetry (6.4)

Some objects, such as a box, can have both a center and planes of symmetry.

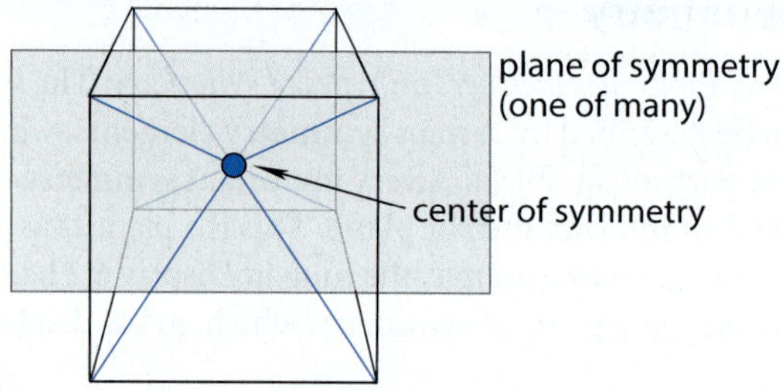

(6.5)

The plane of symmetry and the center of symmetry are the most common symmetry elements present in achiral molecules (and other achiral objects). However, some achiral molecules contain other, relatively rare, symmetry elements. How, then, can we tell whether a molecule is chiral?

1. If a molecule has a single asymmetric carbon (or other asymmetric atom), it must be chiral.
2. If a molecule has a plane of symmetry, a center of symmetry, or both, it is *not* chiral.
3. If neither (1) nor (2) applies, the most general way to assess chirality is to build two models or draw two perspective structures, one of the molecule and the other of its mirror image, and then test the two for congruence. If the two mirror images are congruent, the molecule is achiral; if not, the molecule is chiral.

At the end of Sec. 6.7, we expand this list to molecules with multiple asymmetric centers.

Focused Problems

6.2 State whether each of the following molecules is achiral or chiral.

(a) 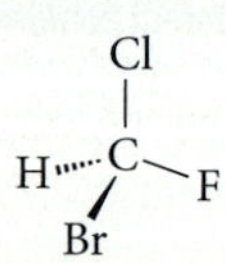

(b) 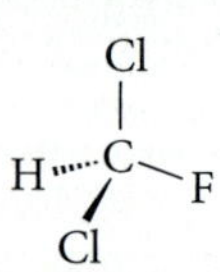

(c) 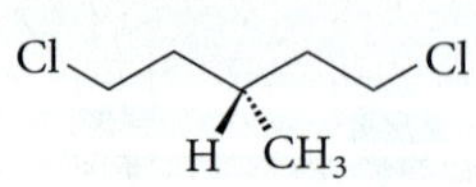

(d) 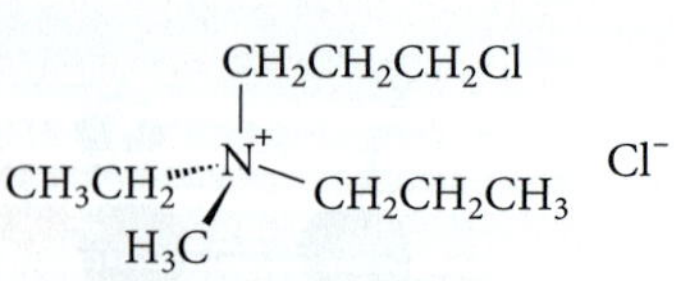

6.3 Ignoring specific markings, indicate whether the following objects are chiral or achiral. State any assumptions that you make.

(a) A shoe (b) A book (c) A person (d) A cone

(e) A pair of shoes (consider the pair as one object) (f) A pair of scissors

6.4 Show the planes and centers of symmetry (if any) in each of the following achiral molecules.

(a) Methane (b) Water (c) Ethylene (d) *trans*-2-Butene

(e) *cis*-2-Butene (f) The anti conformation of butane

6.2 NOMENCLATURE OF ENANTIOMERS

A. The *R,S* System

The existence of enantiomers poses a special problem in nomenclature. How do we indicate in the name of 2-butanol, for example, which enantiomer we have? The same Cahn–Ingold–Prelog priority rules used to assign *E* and *Z* configurations to alkene stereoisomers (Sec. 4.2B) can be applied to enantiomers. A **stereochemical configuration**, or spatial arrangement of atoms, at each asymmetric carbon in a molecule can be assigned using the following steps, illustrated for one enantiomer of 2-butanol.

The Cahn–Ingold–Prelog rules were first developed for asymmetric carbons and then later applied to double-bond stereoisomerism.

Step 1. Identify an asymmetric carbon and the four different groups bonded to it.

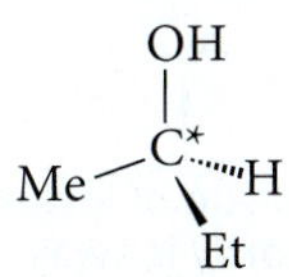

(6.6a)

Step 2. Assign priorities to the four different groups using the rules given in Sec. 4.2B. The convention used in this text is that the highest priority = 1 and the lowest priority = 4.

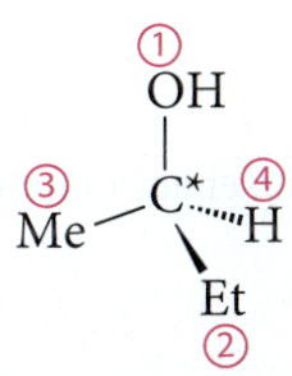

(6.6b)

Step 3. View the molecule along the bond *from* the asymmetric carbon *to* the group of lowest priority—that is, with the asymmetric carbon nearer and the lowest-priority group farther away. (This is essentially a Newman projection about this bond.) In this case, the hydrogen is the group of lowest priority, so we look down the bond from the C to the H.

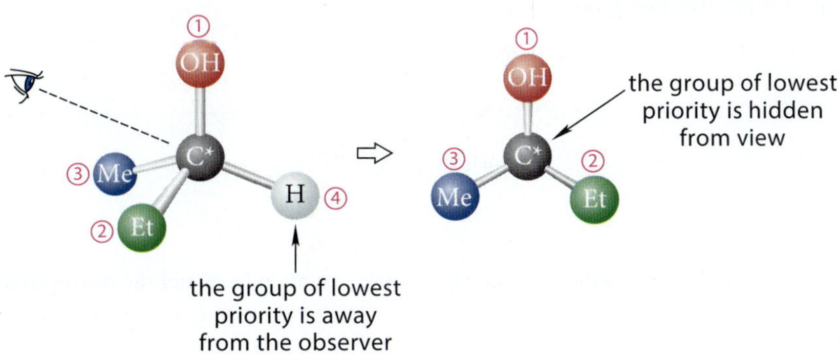

(6.6c)

Step 4. Determine whether the remaining priorities descend in the clockwise or counterclockwise direction. If the priorities descend in the clockwise direction, the asymmetric carbon is said to have the ***R* configuration** (*R* = Latin *rectus*).

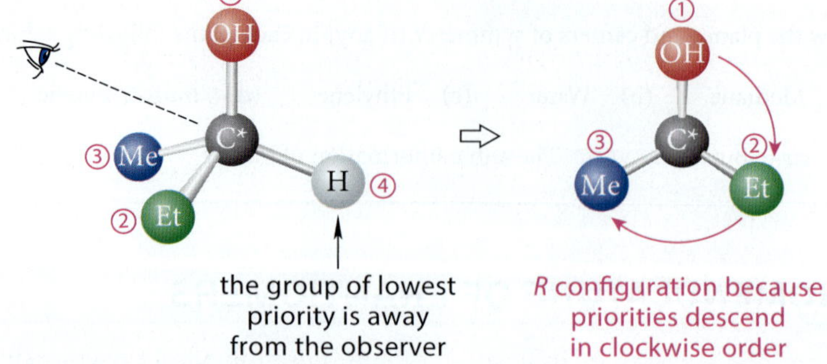

(6.6d)

If the priorities of these groups descend in the counterclockwise direction, as they do for the other enantiomer of 2-butanol, the asymmetric carbon is said to have the *S* **configuration** (*S* = Latin *sinister*).

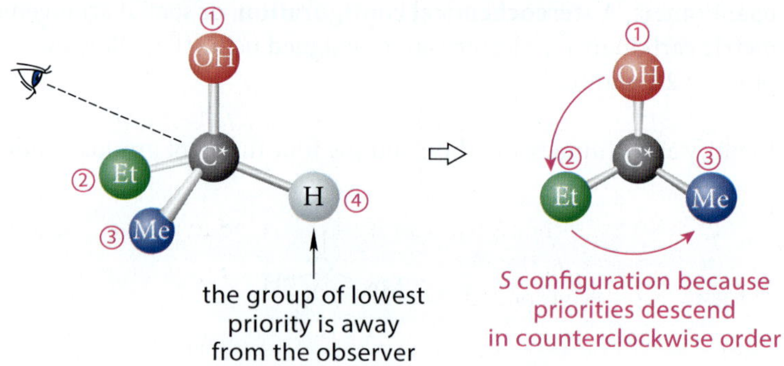

(6.6e)

Study Problem 6.2

Determine the stereochemical configuration of the following enantiomer of 3-chloro-1-pentene:

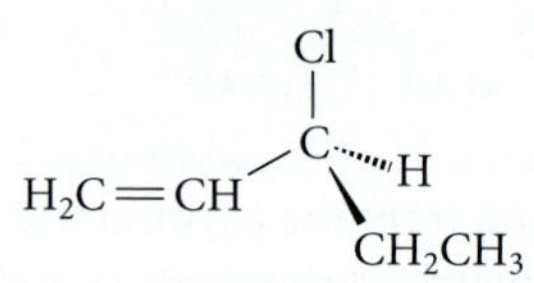

Solution Follow the steps in the text.

Step 1. Identify the asymmetric carbon. This is asterisked (*) in the structure shown in step 2.

Step 2. Assign priorities to the four groups.

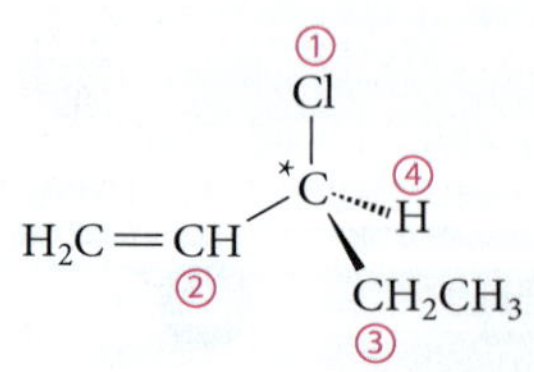

Step 3. View a projection of the molecule with the asymmetric carbon in front and the hydrogen in the rear.

Step 4. Apply the Cahn–Ingold–Prelog sequence rules from Sec. 4.2B to determine whether the priorities decrease from high to low in a clockwise or counterclockwise direction.

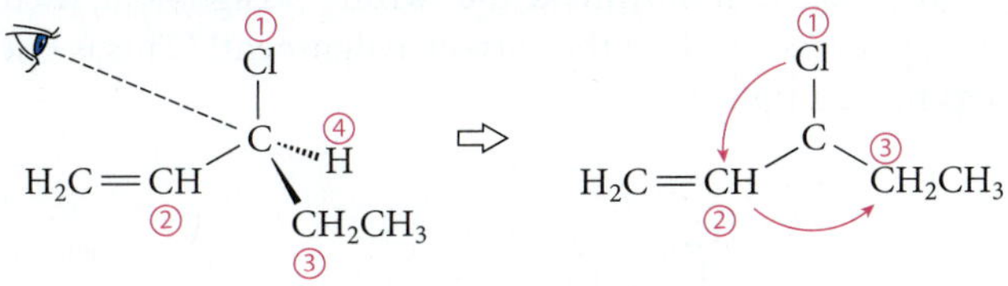

Because the priorities of the first three groups descend in a counterclockwise direction, this is the *S* enantiomer of 3-chloro-1-pentene.

A stereoisomer is named by indicating the configuration of each asymmetric carbon before the systematic name of the compound, as in the following examples:

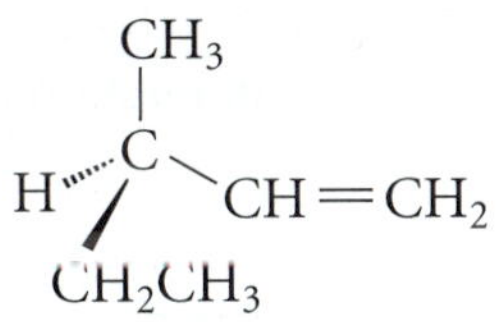

(*R*)-3-methyl-1-pentene

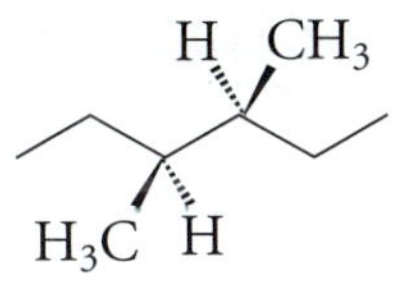

(3*S*,4*S*)-3,4-dimethylhexane

(6.7)

(Be sure to verify these and other *R*,*S* assignments you find in this chapter.) As the second example illustrates, numbers are used with the *R*,*S* designations when a molecule contains more than one asymmetric carbon.

An alternative way of applying the sequence rules, the “right-hand rule,” can be found in the *Journal of Chemical Education,* **1994**, 71(1), 20–23. This journal can be found in most chemistry libraries and online.

The *R*,*S* system is not the only system used for describing stereochemical configuration. The D,L system, which predates the *R*,*S* system, is still used in amino acid and carbohydrate chemistry (Chapters 24 and 27). With this exception, the *R*,*S* system has gained virtually complete acceptance.

Is *R* Right, or Is It Proper?

Choice of the letter *R* presented a problem for Cahn, Ingold, and Prelog, the scientists who devised the *R*,*S* system. The letter *S* stands for *sinister*, one of the Latin words for *left*. However, the Latin word for *right* (in the directional sense) is *dexter*, and unfortunately the letter *D* was already being used in another system of configuration (the D,L system). It was difficulties with the D,L system that led to the need for a new system, and the last thing anyone needed was a system that confused the two! Fortunately, Latin provided another word for *right*: the participle *rectus*. But this “right” does not indicate direction: it means *proper*, or *correct*. (The English word *rectify* comes from the same root.) Although the Latin wasn’t quite proper, it solved the problem! In passing, several chemists have noted that *R* and *S* are the first initials of Robert S. Cahn, one of the inventors of the *R*,*S* system (Sec. 4.2B). A coincidence? Perhaps.

B. Using the *R*,*S* System with Line-and-Wedge Structures

In this section we describe a few handy shortcuts for applying the *R*,*S* system to line-and-wedge structures. If the group of lowest priority is a dashed wedge, then you are essentially looking at it in the proper manner for assignment.

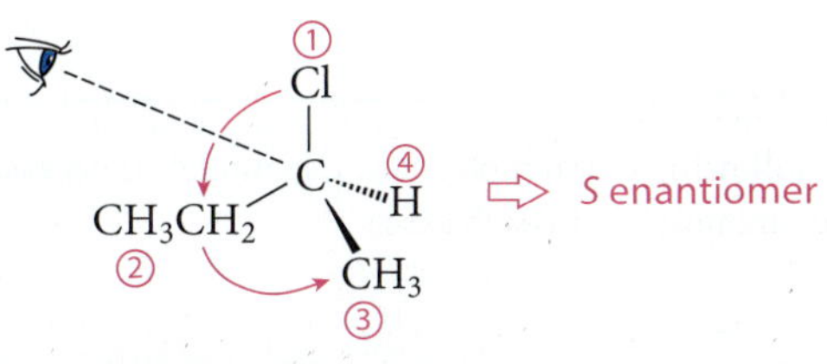

(6.8)

If the group of lowest priority is on a solid wedge (that is, pointing toward you), you are viewing the molecule from the wrong end of the bond to the lowest-priority atom. Because only two assignments are possible, simply make the "wrong" assignment; then, because you know your assignment is wrong, reverse it to get the correct assignment! (This is one of the few situations in which two wrongs make a right.)

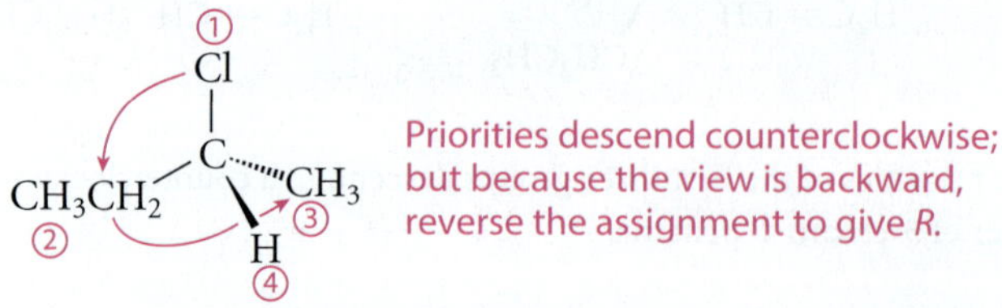

(6.9)

If the group of lowest priority is on neither a dashed wedge nor a solid wedge, then we can apply the definition of a stereocenter: *exchanging any two groups at a stereocenter produces the opposite configuration.* It follows that *two successive exchanges result in the original configuration.* We apply this "double switch" idea to the previous example:

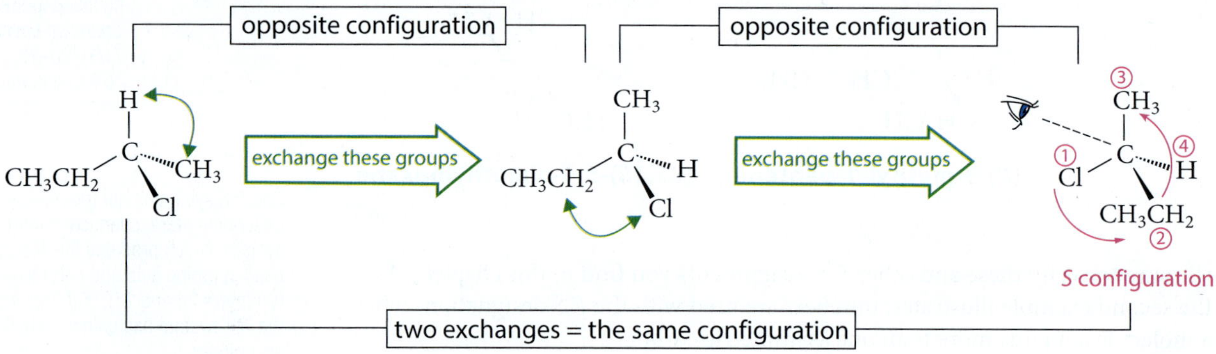

(6.10)

What if the molecule is very complicated? You don't need to make a complete model. Try this: Use a tetrahedral carbon from your model set and attach four different "balls" of different colors, and give priorities to the colors. To remember the priorities, you can make priorities decrease in alphabetical order of the colors—black = 1, blue = 2, green = 3 and red = 4—or you can use a yellow marking pen to put priority numbers on the four groups. Then consider each asymmetric carbon of the complicated structure in turn. For each, make sure the "priority balls" are oriented the same way as the group priorities in the complicated line-and-wedge structure. Then you can orient your simplified model to make the assignment. Repeat this for each asymmetric carbon.

The first exchange places the lowest-priority group (H) into the rear position, from which it is easy to determine configuration, but also gives the enantiomer; the second exchange inverts the configuration again to the original configuration. It makes no difference which groups are exchanged, or in which order; so, make whatever exchanges result in the simplest perspective.

Finally, you can imagine your "eyeball" in the proper orientation and draw the resulting projection:

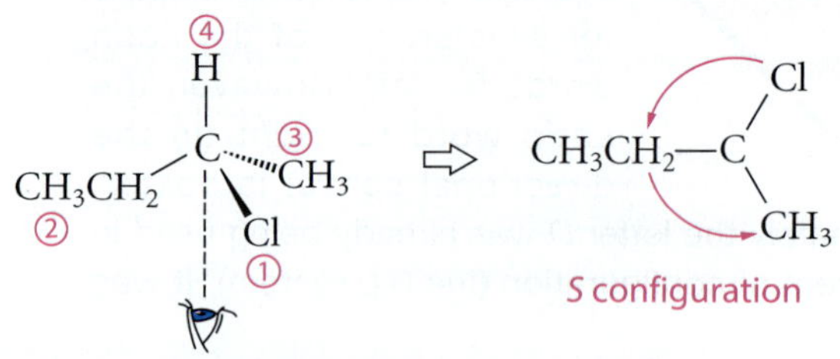

(6.11)

If you have trouble seeing the correct projection this way, try it a few times and check yourself with a model. With practice you'll get the hang of it.

Focused Problems

6.5 Draw line-and-wedge representations for each of the following chiral molecules. Use models if necessary. (D = deuterium = ^{2}H, a heavy isotope of hydrogen.) Several correct structures are possible in each case.

(a) 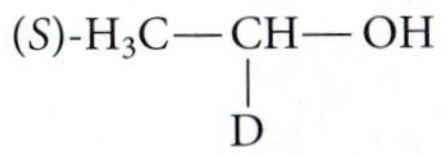

(b) (2*Z*,4*R*)-4-Methylhex-2-ene

(c) 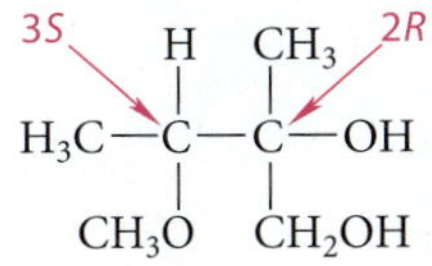

6.6 Indicate whether the asymmetric atom in each of the following compounds has the *R* or *S* configuration.

(a) 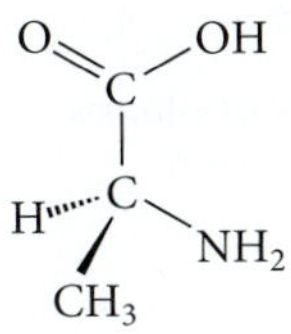

alanine

(b) 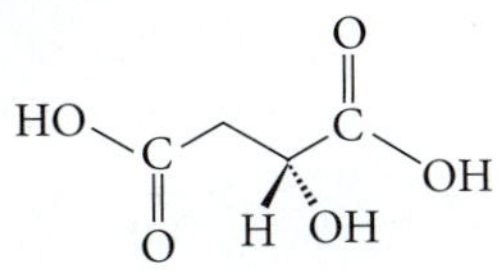

malic acid

(c) 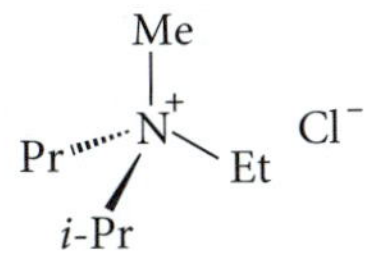

(d) 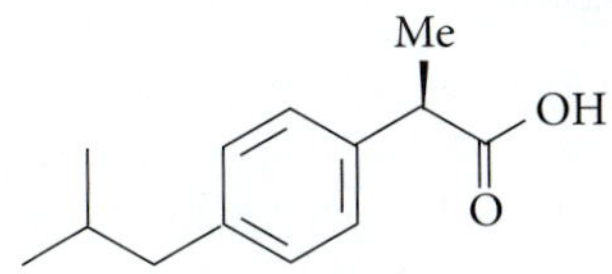

ibuprofen

6.7 Identify the asymmetric carbons in each of the four drug molecules shown in the sidebar, "Importance of Chirality," in Sec. 6.1A: propranolol, methadone, pregabalin, and captopril. Determine the *R*/*S* configuration of each asymmetric carbon.

6.3 PHYSICAL PROPERTIES OF ENANTIOMERS; OPTICAL ACTIVITY

Recall from Sec. 2.6 that organic compounds can be characterized by their physical properties. Two properties often used for this purpose are the melting point and the boiling point. However, *the melting points and boiling points of a pair of enantiomers are identical.* So, the boiling points of (*R*)- and (*S*)-2-butanol are the same—99.5 °C. Likewise, the melting points of (*R*)- and (*S*)-lactic acid are both 53 °C

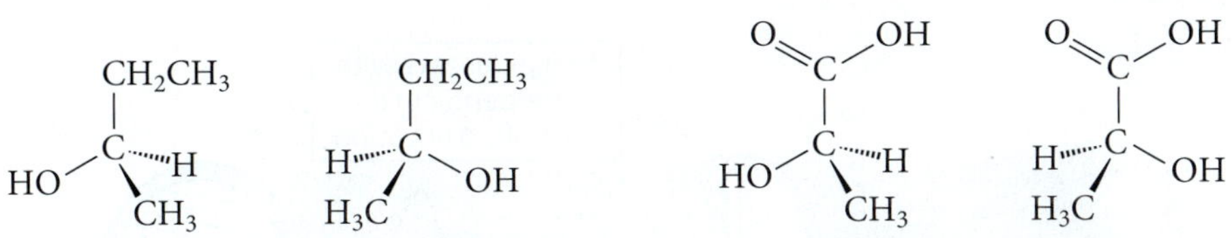

(*R*)-2-butanol **(*S*)-2-butanol**
both have bp = 99.5 °C

(*R*)-lactic acid **(*S*)-lactic acid**
both have mp = 53 °C (6.12)

A pair of enantiomers also have identical densities, indices of refraction, heats of formation, standard free energies, and many other properties.

If enantiomers have so many identical properties, how can we tell one enantiomer from the other? *A compound and its enantiomer can be distinguished by their effects on polarized light.* Understanding these phenomena requires an introduction to the properties of polarized light.

A. Polarized Light

Light is a wave motion that consists of oscillating electric and magnetic fields. The electric field of ordinary light oscillates in all planes, but it is possible to obtain light with an electric field that oscillates in only one plane. This kind of light is called **plane polarized light** or, simply, **polarized light** (**Fig. 6.3**).

Polarized light is obtained by passing ordinary light through a polarizer, such as a Nicol prism (a prism made of specially cut and joined calcium carbonate crystals). The orientation of the polarizer's axis of polarization determines the plane of the resulting polarized light. Analysis of polarized light hinges on the fact that if plane-polarized light is subjected to a second polarizer whose axis of polarization is perpendicular to that of the first, no light passes through the second polarizer (**Fig. 6.4a**). This same effect can be observed with two pairs of polarized sunglasses (**Fig. 6.4b**). When the lenses are oriented in the same direction, light will pass through. When the lenses are turned at right angles, their axes of polarization are crossed, no light is transmitted, and the lenses appear dark.

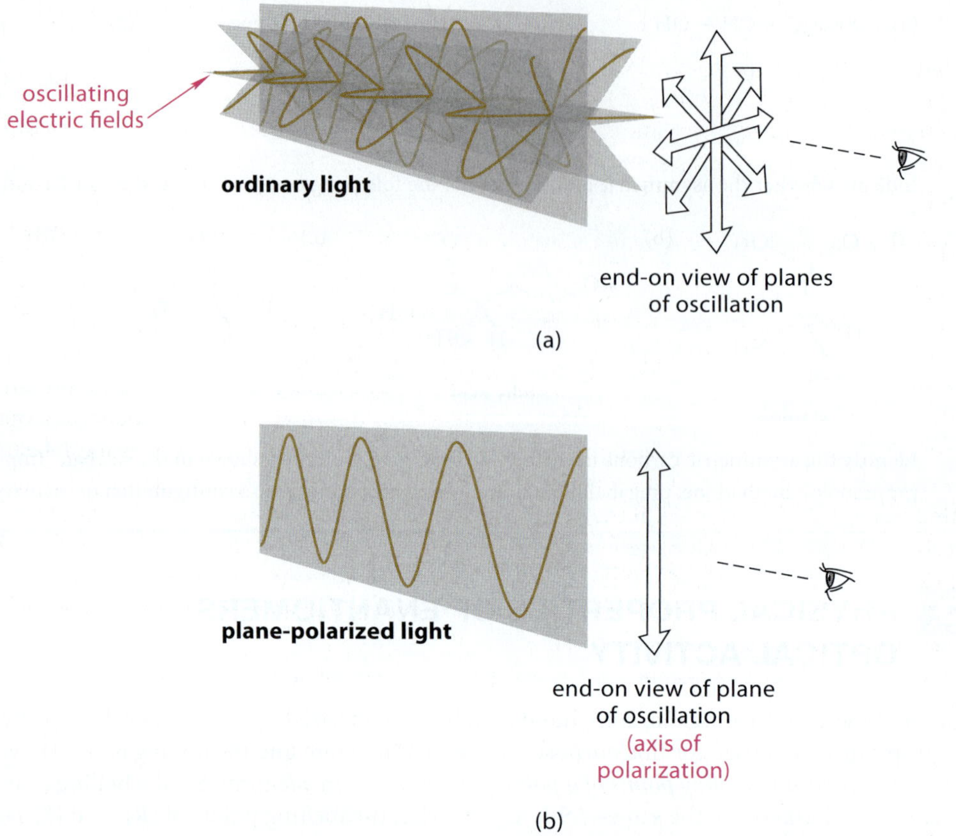

FIGURE 6.3 (a) Ordinary light has electric fields oscillating in all possible planes. (Only three planes of oscillation are shown.) (b) In plane-polarized light, the oscillating electric field is confined to a single plane, which defines the axis of polarization.

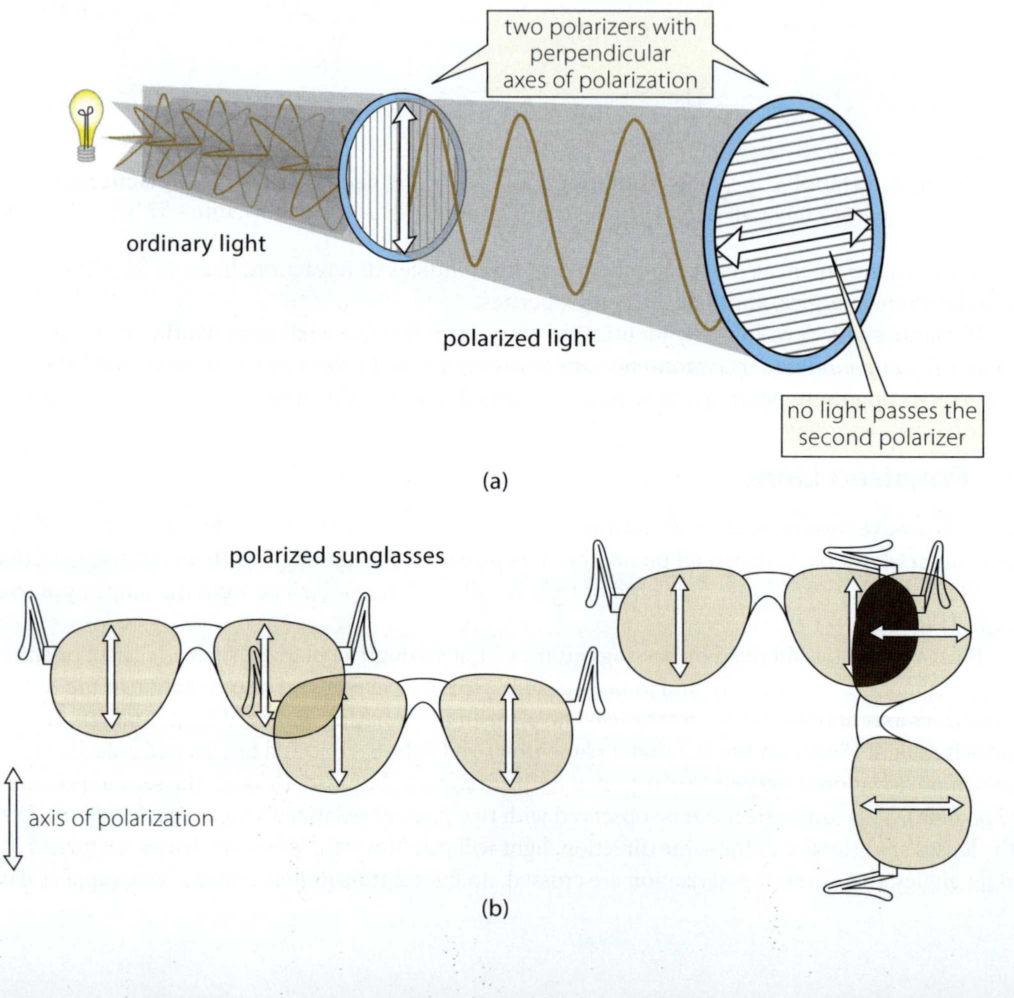

FIGURE 6.4 (a) If the polarization axes of two polarizers are at right angles, no light passes through the second polarizer. (b) The same phenomenon can be observed using two pairs of polarized sunglasses.

You can see the same effect with a tablet computer, which produces a polarized image. If the image disappears when you view your tablet with polarized sunglasses, turning your tablet 90° should restore the image.

B. Optical Activity

If plane-polarized light is passed through one enantiomer of a chiral substance (either the pure enantiomer or a solution of it), *the plane of polarization of the emergent light is rotated.* A substance that rotates the plane of polarized light is said to be **optically active**. *Individual enantiomers of chiral substances are optically active.*

Optical activity is measured with an instrument called a **polarimeter** (**Fig. 6.5**), which is basically the system of two polarizers shown in Fig. 6.4. The sample to be studied is placed in the light beam between the two polarizers. Because optical activity changes with the wavelength (color) of the light, monochromatic light—light of a single color—is used to measure optical activity. The yellow light from a sodium arc (the sodium D-line with a wavelength of 589.3 nm) is often used in this type of experiment. An optically inactive sample (such as air or solvent) is placed in the light beam. Light polarized by the first polarizer passes through the sample, and the analyzer is turned to establish a dark field. This setting of the analyzer defines the zero of optical rotation. Next, the sample whose optical activity is to be measured is placed in the light beam. The number of degrees α that the analyzer must be turned to reestablish the dark field is the **optical rotation** of the sample. If the sample rotates the plane of polarized light in the clockwise direction, the optical rotation is given a plus sign. Such a sample is said to be **dextrorotatory** (Latin *dexter*, meaning "right"). If the sample rotates the plane of polarized light in the counterclockwise direction, the optical rotation is given a minus sign, and the sample is said to be **levorotatory** (Latin *laevus*, meaning "left").

The optical rotation of a sample is the quantitative measure of its optical activity. The observed optical rotation α in degrees, is proportional to the number of optically active molecules present in the path through which the light beam passes. Therefore, α is proportional to both the

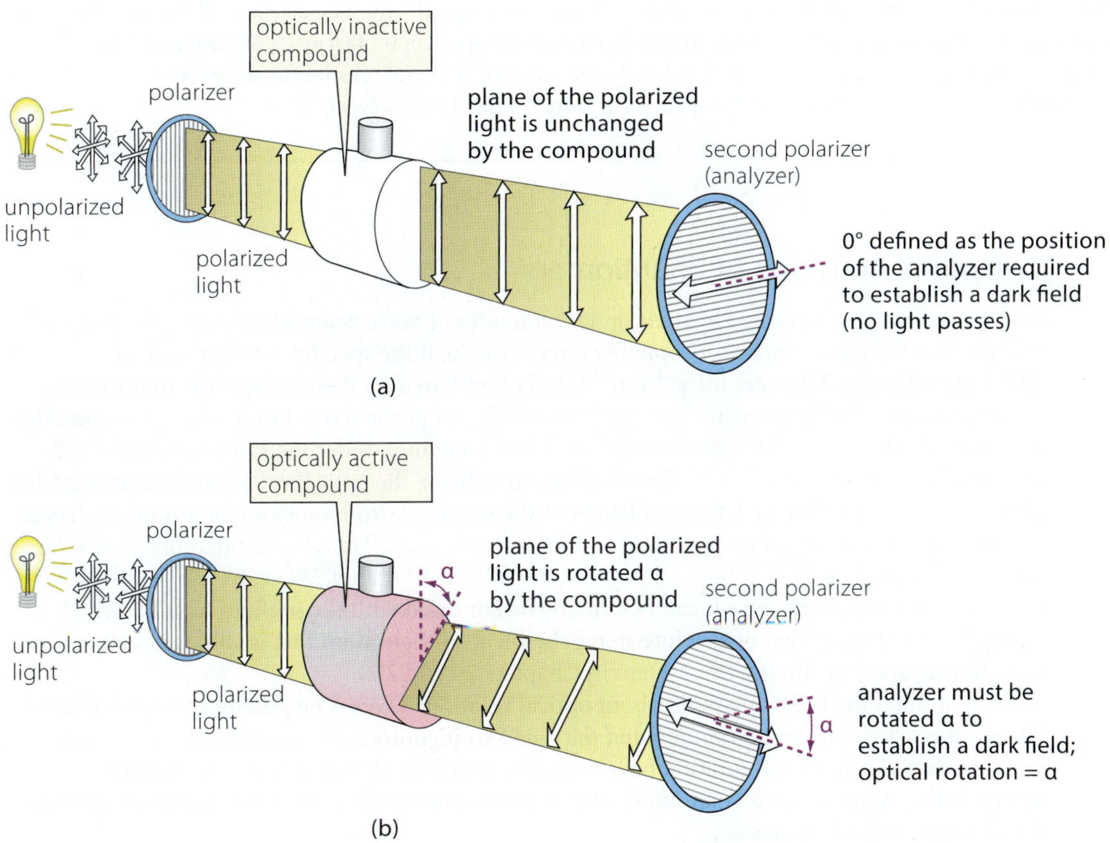

FIGURE 6.5 Determination of optical rotation in a simple polarimeter. (a) First, the reference condition of zero rotation is established as a dark field. (b) Next, the polarized light is passed through an optically active sample with observed rotation α. The analyzer is rotated to establish the dark-field condition again. The optical rotation α can be read from the calibrated scale on the analyzer.

One decimeter = 10 centimeters. The reason for using the decimeter as a unit of length is that the length of a typical sample container used in polarimeters is 1 dm. Specific rotation can, in some cases, have a nonlinear dependence on concentration—that is, Eq. 6.13 is not followed at all concentrations. One reason might be that the sample molecules undergo noncovalent association at higher concentration, and this association might be solvent-dependent. Yet specific rotations are often determined from Eq. 6.13 at a single concentration. It is important in practice, therefore, to report the conditions of solvent and concentration under which the specific rotation is determined. In this text, however, for simplicity, we assume the validity of Eq. 6.13.

concentration *c* of the optically active compound in the sample as well as the length *l* of the sample container:

$$\alpha = [\alpha]cl \quad \text{(Biot's law)} \qquad (6.13)$$

The constant of proportionality, [α], is called the **specific rotation**. *The specific rotation is the standard for reporting optical activity.* By convention, the concentration of the sample is expressed in grams per milliliter ($g\ mL^{-1}$), and the path length in decimeters (dm). (For a pure liquid, *c* is taken as the density.) Thus, the specific rotation is equal to the observed rotation at a concentration of $1\ g\ mL^{-1}$ and a path length of 1 dm.

Typically, the specific rotation [α] is determined as the slope of a plot of observed rotation α against the concentration *c*. Because the dimensions of the observed rotation are degrees, the dimensions of [α] are degrees $mL\ g^{-1}\ dm^{-1}$ (Often, specific rotations are reported simply in degrees, with the other units understood.) Because the specific rotation of any compound varies with wavelength, solvent, and temperature, [α] is conventionally reported with a subscript that indicates the wavelength of light used and a superscript that indicates the temperature. Therefore, a specific rotation reported as $[\alpha]_D^{20}$ has been determined at 20 °C using the light of a sodium D-line.

Study Problem 6.3

A sample of (*S*)-2-butanol has an observed rotation of +2.18° at 20 °C. The measurement was made with a 2.0 *M* solution of (*S*)-2-butanol in methanol solvent in a sample container that is 10 cm long. What is the specific rotation $[\alpha]_D^{20}$ of (*S*)-2-butanol in this solvent?

Solution To calculate the specific rotation, the sample concentration in $g\ mL^{-1}$ must be determined. Because the molecular mass of 2-butanol is 74.12, the 2.0 *M* solution contains $148.1\ g\ L^{-1}$, or $0.148\ g\ mL^{-1}$, of 2-butanol. This is the value of *c* used in Eq. 6.13. The value of *l* is 1 dm. Rearranging Eq. 6.13 and substituting these values, $[\alpha]_D^{20} = (+2.18°)/(0.148\ g\ mL^{-1})(1\ dm) = +14.7$ degrees $mL\ g^{-1}\ dm^{-1}$ in methanol solvent.

C. Optical Activities of Enantiomers

Enantiomers are distinguished by their optical activities because *enantiomers rotate the plane of polarized light by equal amounts in opposite directions.* So, if the specific rotation $[\alpha]_D^{20}$ of (*S*)-2-butanol is +14.7 degrees $mL\ g^{-1}\ dm^{-1}$ (Study Problem 6.3), then the specific rotation of (*R*)-2-butanol is −14.7 degrees $mL\ g^{-1}\ dm^{-1}$. Similarly, if a particular solution of (*S*)-2-butanol has an observed rotation of +3.5°, then a solution of (*R*)-2-butanol under the same conditions will have an observed rotation of −3.5°. Another way to indicate the optical rotation of a compound is with a lower-case prefix *d* or *l*, the first letters of the words *dextrorotatory* and *levorotatory*. These are sometimes used *instead of* the plus (+) and minus (−) signs. Thus, (+)-2-butanol can also be called *d*-2-butanol, and (−)-2-butanol can also be called *l*-2-butanol. We won't use this notation extensively in this text because it can potentially be confused with the prefixes D and L, which are used in an older system of absolute stereochemical configuration that is still used with amino acids and sugars. (We discuss this system in Chapters 24 and 27.)

Keep in mind the following point about optical rotation: *There is no general correspondence between the sign of the optical rotation and the R or S configuration of a compound.* Thus, some compounds with the *S* configuration have positive rotations, whereas others have negative rotations. For example, the *S* enantiomer of 2-butanol is dextrorotatory, whereas the *S* enantiomer of 1,2-butanediol is levorotatory.

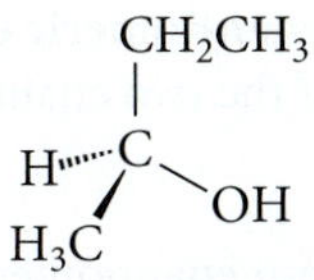

(S)-(+)-2-butanol
(d-2-butanol)
$[\alpha]_D^{20} = +14.7$ degrees mL g^{-1} dm^{-1}

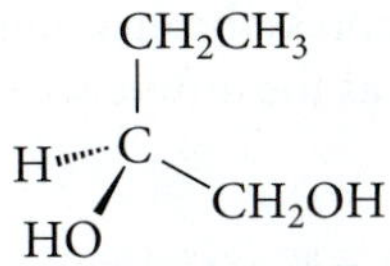

(S)-(−)-1,2-butanediol
(l-1,2-butanediol)
$[\alpha]_D^{20} = -15.4$ degrees mL g^{-1} dm^{-1} (6.14)

The only way to determine optical rotation is to measure it experimentally. For example, the name (*S*)-(+)-2-butanol implies that someone has measured and reported the optical rotation of the *S* enantiomer. We can then deduce that the *R* enantiomer should be (*R*)-(−)-2-butanol, because enantiomers have equal rotations of opposite signs.

Conversely, you can measure the optical rotation of a chiral substance without knowing its configuration. For example, let's imagine you have isolated a compound from a natural source (let's call it "newnol") and have found that it has a positive optical rotation. Your compound is therefore (+)-newnol, or *d*-newnol. But its sign of rotation does not allow you to deduce its stereochemical configuration. (The determination of absolute stereochemical configuration is discussed in Sec. 6.5.)

It would certainly be useful to be able to deduce the sign (and the magnitude) of the optical rotation from a structure. There are ways to do this, but these methods require complex quantum-chemical calculations.

Focused Problems

6.8 (a) The specific rotation of sucrose (table sugar) in water is +66.1 degrees mL g^{-1} dm^{-1}. What is the observed optical rotation in a 1 dm path of a sucrose solution prepared from 5 g of sucrose and enough water to form 100 mL of solution?

(b) Chemist Ree N. Ventdawil has said to you that he plans to synthesize the enantiomer of sucrose so that he can measure its specific rotation. You politely inform him that you already know the result. Explain how you can make this claim.

6.9 A sample of *S*-(−)-α-phenethylamine has an observed rotation of −9.2° at a concentration of 0.3 g mL^{-1} measured with the sodium D-line in a 1-dm measuring cell in chloroform at 25 °C. What is the specific rotation $[\alpha]_D^{25}$ of *S*-(−)-α-phenethylamine in chloroform?

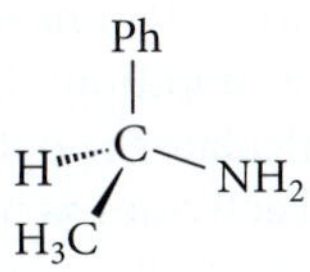

(S)-(−)-α-phenethylamine

6.4 MIXTURES OF ENANTIOMERS

A. Enantiomeric Ratio and Enantiomeric Excess

When one enantiomer of a chiral compound is uncontaminated by the other enantiomer, it is said to be **enantiomerically pure**. However, mixtures of enantiomers occur commonly. A 50:50 mixture of enantiomers is called a **racemic mixture** or **racemate**. (We discuss racemates further in Sec. 6.4B.) A mixture of enantiomers with proportions other than 50:50 is called a **scalemic mixture**. The enantiomeric composition of a mixture of enantiomers can be expressed in two ways. The simplest way is the **enantiomeric ratio (ER)**, which is the ratio of the major enantiomer to the moles of the minor enantiomer. For example, a scalemic mixture that is 80% *R* and 20% *S* has ER = 4, or ER = 80% *R*:20% *S*.

Another way to express the enantiomeric composition is the **enantiomeric excess**, abbreviated **EE**, which is defined as the difference between the percentages of the two enantiomers in the mixture:

$$\text{EE} = \%\text{ of the major enantiomer} - \%\text{ of the minor enantiomer} \tag{6.15}$$

For example, if a mixture contains 80% of the (+)-enantiomer and 20% of the (−)-enantiomer, the EE is 80% − 20% = 60%. Notice that EE is not the same as % purity. The use of enantiomeric excess comes from the effect of the contaminating enantiomer on the optical activity of the mixture. For example, if a mixture consists of 80% of the (+)-enantiomer and 20% of the (−)-enantiomer, then the optical activity of the mixture is 60% of the optical activity of the pure (+)-enantiomer. This is because the 20% of the (−)-enantiomer cancels the rotation of 20% of the (+)-enantiomer, leaving a net optical rotation of 60%. [Remember that the (+)- and (−)-enantiomers have equal rotations of opposite sign.]

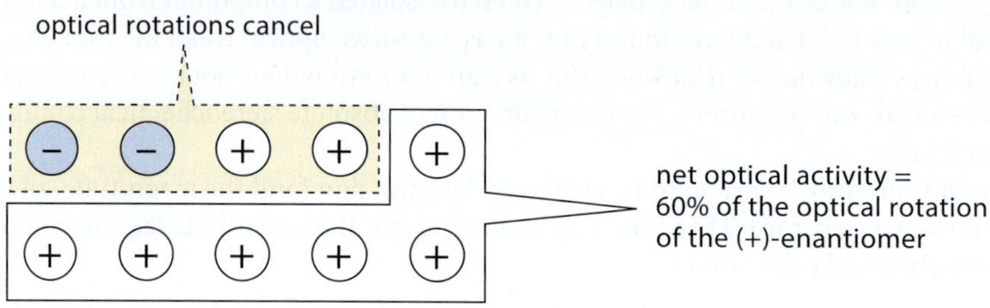

10 molecules: 20% (-)-enantiomer, 80% (+)-enantiomer (6.16)

From this example you can see that the optical activity of the mixture has a percentage optical activity of the pure (+)-enantiomer that equals the EE. It follows that if we know the optical activity of the *pure* major enantiomer, and if the two enantiomers are the only optically active substances present in a sample, then the enantiomeric excess in the sample can be determined from the specific rotation of the mixture:

$$\text{EE} = 100\% \times \frac{[\alpha]_{\text{mixture}}}{[\alpha]_{\text{pure}}} \tag{6.17}$$

EE and optical purity are the same only when Biot's law (Eq. 6.13) is valid—that is, when optical activity is proportional to concentration. When the optical activities of some compounds deviate from Biot's law at higher concentrations, enantiomeric excess has to be determined in some other way. We assume here the validity of Biot's law.

where $[\alpha]_{\text{mixture}}$ and $[\alpha]_{\text{pure}}$ are the specific rotations of the mixture and the pure major enantiomer determined under the same conditions. Conversely, if we know the optical activity of the mixture and the EE, then we can calculate the optical activity of the pure major enantiomer. When the EE is calculated from optical activities as in Eq. 6.17, it is sometimes also called **optical purity**.

Finally, the actual percentages of each enantiomer, and thus the ER, can be calculated from the EE. Remember that the percentages of the two enantiomers add to 100%.

$$\%\text{ minor enantiomer} = 100\% - \%\text{ major enantiomer} \tag{6.18}$$

Substituting this equation into Eq. 6.15, we have

$$\begin{aligned}\text{EE} &= \%\text{ major enantiomer} - (100\% - \%\text{ major enantiomer}) \\ &= 2(\%\text{ major enantiomer}) - 100\%\end{aligned} \tag{6.19a}$$

Solving for % major enantiomer, we find

$$\%\text{ major enantiomer} = (\text{EE} + 100\%)/2 \tag{6.19b}$$

If EE = 80%, for example, the % major enantiomer is 180%/2 = 90%. The remaining 10% is the minor enantiomer. The ER is 9, or 90%:10%.

Study Problem 6.4

Your lab partner tells you that a sample of (+)-2-butanol has an apparent specific rotation of +13.8 degrees mL g^{-1} dm^{-1} and is known to be a mixture of the two enantiomers with EE = 94%. (a) How much of each enantiomer is in the mixture? (b) What is the ER? (c) What is the specific rotation of enantiomerically pure (+)-2-butanol?

Solution

(a) The % of the major enantiomer (+)-2-butanol is calculated from Eq. 6.19b:

$$\%\ (+)\text{-enantiomer} = (94\% + 100\%)/2 = 97\%$$

Then there is (100 − 97)% = 3% of the minor enantiomer (−)-2-butanol in the mixture. Checking our work, this means that the EE has to be 97% − 3% = 94%, as given.

(b) The ER is 97% (+)-2-butanol:3% (−)-2-butanol, or 97/3 = 32.3.

(c) Rearranging Eq. 6.17, we calculate the specific rotation of pure (+)-2-butanol as follows:

$$[\alpha]_{\text{pure}} = 100\% \times \frac{[\alpha]_{\text{mixture}}}{\text{EE}} = \frac{100\% \times (13.8 \text{ degrees mL}^{-1}\text{ g}^{-1}\text{ dm}^{-1})}{94\%} = +14.7 \text{ degrees mL}^{-1}\text{ g}^{-1}\text{ dm}^{-1}$$

B. Racemates (Racemic Mixtures)

Recall from the previous section that a mixture containing equal amounts of two enantiomers is called a *racemate* or a *racemic mixture.* (In older literature, the term *racemic modification* was used.) A racemate is referred to by name in three ways. The racemate of 2-butanol, for example, can be called racemic 2-butanol, (±)-2-butanol, or *d*,*l*-2-butanol.

Racemates typically have physical properties that are different from those of the pure enantiomers. For example, the melting point of either enantiomer of lactic acid (Display 6.12, Sec. 6.3) is 53 °C, but the melting point of racemic lactic acid is 18 °C. The reason for the different melting points is that the crystal structures differ. Recall from Sec. 2.6B that the melting point largely reflects interactions between molecules in the crystalline solid. (Imagine packing a dozen left shoes—a "pure enantiomer"—in a box, and then imagine packing six right shoes and six left shoes—a "racemate"—in the same box. The interactions among the shoes—the way they touch each other—are different in the two cases.) The optical rotations of enantiomers and racemates are another example of differing physical properties. The optical rotation of any racemate is zero, because a racemate contains equal amounts of two enantiomers whose optical rotations of equal magnitude and opposite sign exactly cancel each other. In a racemate, the ER is 1.0 and the EE = 0.

FURTHER EXPLORATION 6.1
Terminology of Racemates

The process of forming a racemate from a pure enantiomer is called **racemization**. The simplest method of racemization is to mix equal amounts of enantiomers. As you will learn, racemization can also occur as a result of chemical reactions.

Because a pair of enantiomers have the same boiling points, melting points, and solubilities—exactly the properties that are usually exploited in designing separations—the separation of enantiomers poses a special problem. The separation of a pair of enantiomers, called an **enantiomeric resolution**, requires special methods that are discussed in Sec. 6.10.

Racemates in the Pharmaceutical Industry

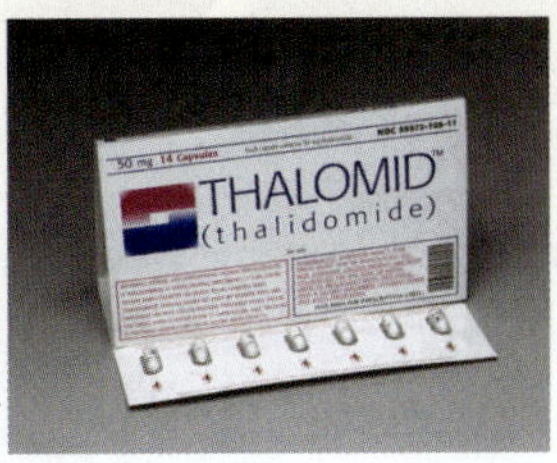

Science and Society/ Superstock

Over half of the pharmaceuticals sold commercially are chiral compounds. Drugs that come from natural sources (or drugs that are prepared from materials obtained from natural sources) have always been produced as pure enantiomers, because, in most cases, chiral compounds from nature occur as only one of the two possible enantiomers. (We'll explore this point in Sec. 7.7A.) Until relatively recently, however, most chiral drugs produced synthetically from *achiral* starting materials were produced and sold as racemates. The reason is that the separation of racemates into their optically pure enantiomeric components requires special procedures that add cost to the final product. (See Sec. 6.10.) The justification for selling the racemic form of a drug hinges on its lower cost and on the demonstration that the unwanted enantiomer is physiologically inactive, or at least that its side effects, if any, are tolerable. However, this is not always so.

The landmark case that dramatically demonstrated the potential pitfalls in marketing a racemic drug involved *thalidomide,* a compound first marketed as a sedative in Europe in 1958.

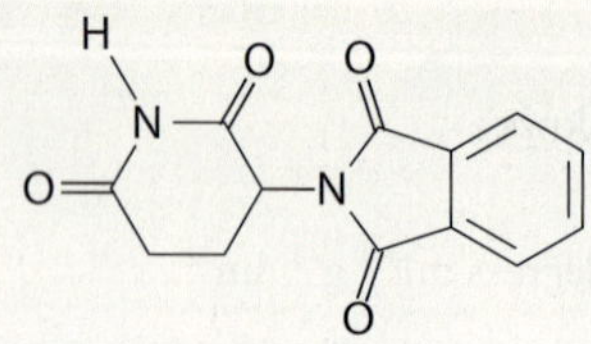

thalidomide

The (*R*)-(+)-enantiomer of thalidomide was found to have a higher sedative activity than the (*S*)-(−)-enantiomer, but, as was typical of the time, the drug was marketed as the racemate for economic reasons. This drug was taken by a number of pregnant women to relieve the symptoms of morning sickness. It turned out, tragically, that thalidomide is teratogenic—that is, it causes horrible birth defects, such as deformed limbs, when taken by women in early pregnancy. An estimated 12,000 children were born with thalidomide-induced birth defects, mostly in Europe and South America. The drug was never approved for use in the United States, although some was given to doctors and dispensed for "investigational use."

Although it is believed that only the (*S*)-(−)-enantiomer of thalidomide is teratogenic, it has been shown that either enantiomer is racemized in the bloodstream. So, it is likely that the teratogenic effects would have been observed with even the optically pure *R* enantiomer. Nevertheless, thalidomide illustrates the point that enantiomers can, in some cases, have greatly different biological activities.

A remarkable and happier postscript to the thalidomide story is evolving. One of the reasons that thalidomide is teratogenic is that it suppresses *angiogenesis* (the growth of blood vessels), which is essential for actively dividing cells. This effect is disastrous for a developing fetus, but is likely to be beneficial for some cancer patients, because the suppression of angiogenesis has been found in early trials to be effective in treating certain cancers. Thalidomide has been approved as part of a treatment for multiple myeloma and is also being used in the treatment of leprosy. Despite these potential benefits, it cannot be given to women who are pregnant or are likely to become pregnant.

The pharmaceutical industry, spurred in part by the U.S. Food and Drug Administration (FDA), has with increasing regularity developed synthetic chiral drugs as single enantiomers rather than racemates, thus ensuring that consumers will not have to contend with unanticipated side effects of therapeutically inactive stereoisomers.

Focused Problems

6.10 (a) Identify the asymmetric carbon of thalidomide.

(b) Draw a structure of the teratogenic *S* enantiomer of thalidomide using lines, a wedge, and a dashed wedge to indicate the stereochemistry of this carbon.

6.11 A 0.1 *M* solution of an enantiomerically pure chiral compound *D* has an observed rotation of +0.20° in a 1 dm sample container. The molecular mass of the compound is 150.

(a) What is the specific rotation of *D*?

(b) What is the observed rotation if this solution is mixed with an equal volume of a solution that is 0.1 *M* in *L*, the enantiomer of *D*?

(c) What is the observed rotation if the original solution of *D* is diluted with an equal volume of solvent?

(d) What is the specific rotation of *D* after the dilution described in part (c)?

(e) What is the specific rotation of *L*, the enantiomer of *D*, after the dilution described in part (c)?

(f) What is the observed rotation of 100 mL of a solution that contains 0.01 mol of *D* and 0.005 mol of *L*? (Assume a 1 dm path length.)

(g) What is the enantiomeric excess (EE) of *D* in the solution described in part (f)?

(h) What is the enantiomeric ratio (ER) in the solution described in part (f)?

6.12 A chemist has developed a new synthesis of ibuprofen and has reported that she has prepared the (*S*)-(+)-enantiomer in 90% EE, and that this material has a measured specific rotation of +51.7 degrees mL g^{-1} dm^{-1}.

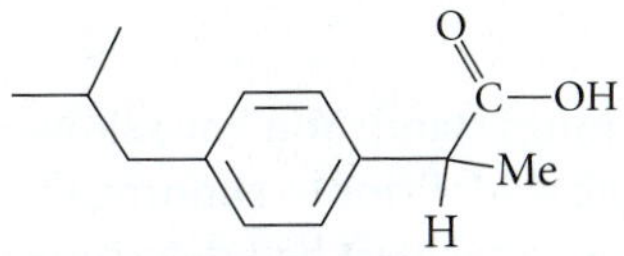

ibuprofen

(a) Draw a line-and-wedge formula of (*S*)-(+)-ibuprofen.

(b) What is the specific rotation of pure (*S*)-(+)-ibuprofen? Of pure (*R*)-(−)-ibuprofen?

(c) How much of each enantiomer is present in her sample?

6.13 What observed rotation is expected when a 1.5 *M* solution of (*R*)-2-butanol is mixed with an equal volume of a 0.75 *M* solution of racemic 2-butanol, and the resulting solution is analyzed in a sample container that is 1 dm long? The specific rotation of (*R*)-2-butanol is −14.7 degrees mL g^{-1} dm^{-1}.

6.5 STEREOCHEMICAL CORRELATION

You can't specify the configuration of a molecule with the *R,S* system until you know the actual three-dimensional arrangement of its atoms—that is, its **absolute configuration** or **absolute stereochemistry**. If you were the first person ever to synthesize one enantiomer of a chiral compound, how would you determine its absolute configuration? (Recall that the sign of optical rotation *cannot* be used to assign an *R* or *S* configuration; Sec. 6.3C.) One way is to use a variation of X-ray crystallography called *anomalous dispersion*. Although X-ray crystallography is more widely used than it once was, it still requires specialized expensive instrumentation that is not readily available in the average laboratory. The absolute configurations of most organic compounds are determined instead by using chemical reactions to correlate them with other compounds of known absolute configurations. This process is called **stereochemical correlation**.

To illustrate a stereochemical correlation, suppose you have obtained an optically active sample of the following alkene.

$$\mathrm{Ph{-}\underset{\displaystyle CH_3}{\underset{|}{CH}}{-}CH{=}CH_2}$$

$[\alpha]_D^{25} = -6.7$ degrees mL g^{-1} dm^{-1}

R or *S* configuration unknown

You've measured its optical activity experimentally and have determined that it is levorotatory—that is, it is the (−)-enantiomer. However, you don't know its absolute configuration—whether it is *R* or *S*. Remember, the optical rotation does *not* provide this information. You go to the chemical literature and you find that no one has ever determined the absolute configuration of either the (−)- or the (+)-alkene. How would you determine its absolute configuration? This compound is a liquid, so crystallization followed by X-ray crystallography is unrealistic.

Being an astute organic chemistry student, you know (Sec. 5.8B) that you can convert this alkene into a carboxylic acid by ozonolysis. This reaction breaks the double bond, but *it does not break any of the bonds to the asymmetric carbon*. You carry out this reaction on your alkene and obtain a sample of the carboxylic acid, which has the common name *hydratropic acid*. You find by direct measurement that your sample of hydratropic acid is optically active and dextrorotatory.

If you haven't studied ozonolysis yet, all you need to know about the reaction for this section is its result, shown in Eq. 6.20.

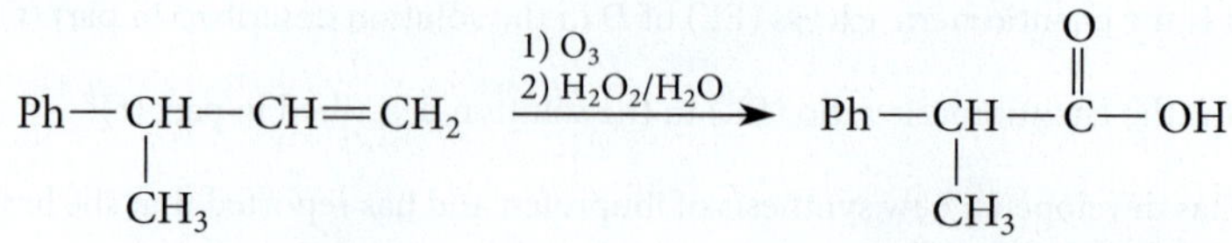

(–) optical rotation

hydratropic acid
(+) optical rotation (6.20)

Here is where things stand: You have shown that the (–)-alkene is converted by ozonolysis into (+)-hydratropic acid. Prior to running this reaction, you searched the chemical literature and found that someone in the past had determined the absolute configuration of (+)-hydratropic acid, perhaps by X-ray crystallography. (This search can be carried out rapidly by computer.) In this previous work, the (+)-enantiomer of hydrotropic acid was shown to have the *S* configuration. Therefore, you have converted the (–)-alkene of unknown configuration into (+)-hydrotropic acid, *known from earlier work* to have the *S* configuration:

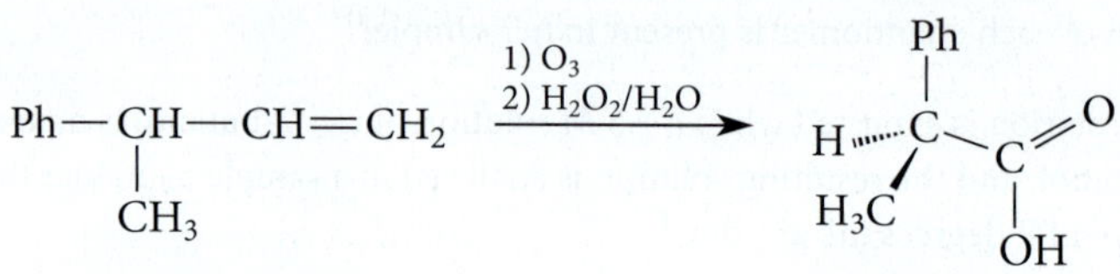

(–) optical rotation
unknown configuration

(*S*)-(+)-hydratropic acid
The *S* enantiomer is known to have (+) optical rotation from earlier work. (6.21)

You are now in a position to deduce the absolute configuration of your alkene, because *the corresponding groups in the two compounds must be in the same relative positions.* Remember, no bonds to the asymmetric carbon were broken.

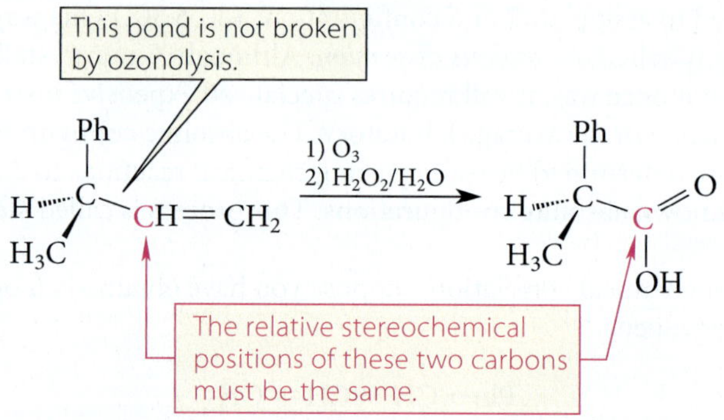

(6.22)

You now know the absolute configuration of the alkene—that is, you can build a three-dimensional model of it, as shown in Eq. 6.22. Assigning the configuration by the sequence rules (Sec. 6.2) shows that the alkene has the *R* configuration. You have now linked the optical rotation of the alkene to its configuration. [Note that the phenyl (Ph) group has higher priority than the alkene carbon in the *R*,*S* system.]

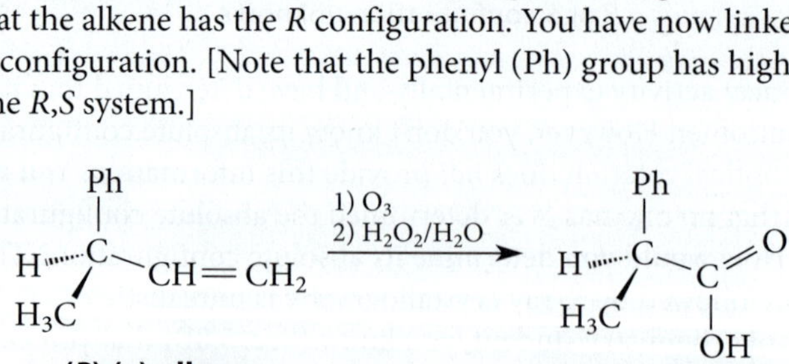

(*R*)-(–)-alkene
(configuration deduced from this correlation)

(*S*)-(+)-hydratropic acid
(configuration was previously known) (6.23)

Once this result is published in the chemical literature, others can use this assignment to carry out other correlations (Focused Problem 6.14). It also follows that the dextrorotatory alkene—the enantiomer of your starting alkene—must have the *S* configuration because enantiomers must have optical rotations of opposite signs. *Thus, a correlation carried out on either enantiomer establishes the configurations of both.*

Although the reactant and the product in this example have different *R,S* configurations, this relationship is not true in general. It is possible for the *R,S* configurations of the reactant and the product to be same. The result depends on the relative priorities of the groups at the asymmetric carbon and their relative positions in the three-dimensional model. If a reaction results in a change in the relative priorities of groups at the asymmetric carbon, as in this case, then the correlated structures will have different *R,S* designations. If relative priorities don't change, the correlated structures will have the same *R,S* designations. (See Focused Problem 6.14.)

The optical rotation of the two correlated compounds may have different signs, as in this example, or they may have the same sign. The relative signs of rotation must be determined by experiment.

To summarize: We can determine the absolute configuration of one compound by converting it into another compound whose absolute configuration is known. We could also take the same approach "in reverse" and deduce the configuration of a product from a starting material of known configuration. This approach of relating reactants and products is unambiguous when the bonds to the asymmetric atom(s) are unaffected by the reaction. (A reaction that breaks these bonds can also be used if the stereochemical outcome of such a reaction has previously been carefully established.)

Stereochemical correlation has been used to establish stereochemistry throughout the history of organic chemistry. One of the most spectacular examples was the determination of the stereochemistry of glucose and other sugars, which we show in Sec. 24.10.

Focused Problem

6.14 Use the known stereochemistry of the starting alkene (Eq. 6.23) to assign the stereochemical configuration of the product, which was found by experiment to be levorotatory.

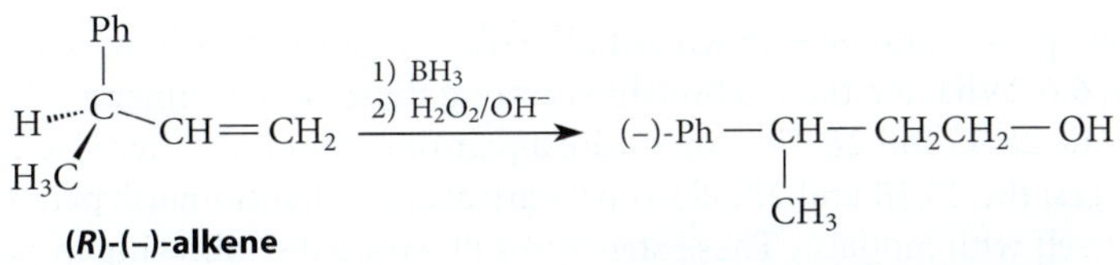

(*R*)-(−)-alkene

unknown configuration; the optical rotation is known from experimental measurement

6.6 DIASTEREOMERS

A. Stereoisomers That Are Not Enantiomers

Up to this point, our discussion has focused on molecules with only one asymmetric carbon. Now we consider molecules that have two or more asymmetric carbons. This situation is illustrated by 2,3-pentanediol, in which both carbons 2 and 3 are asymmetric.

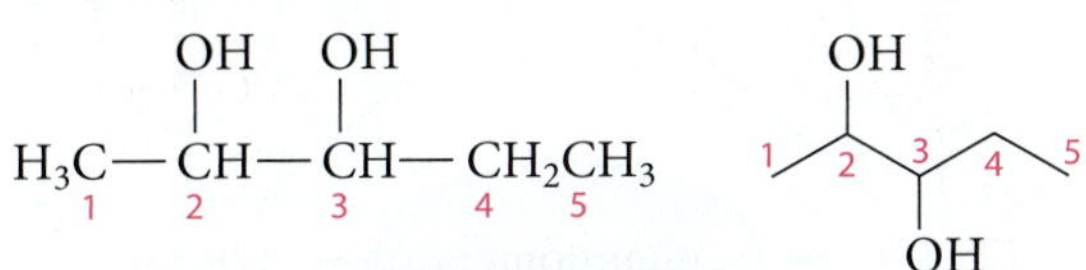

2,3-pentanediol

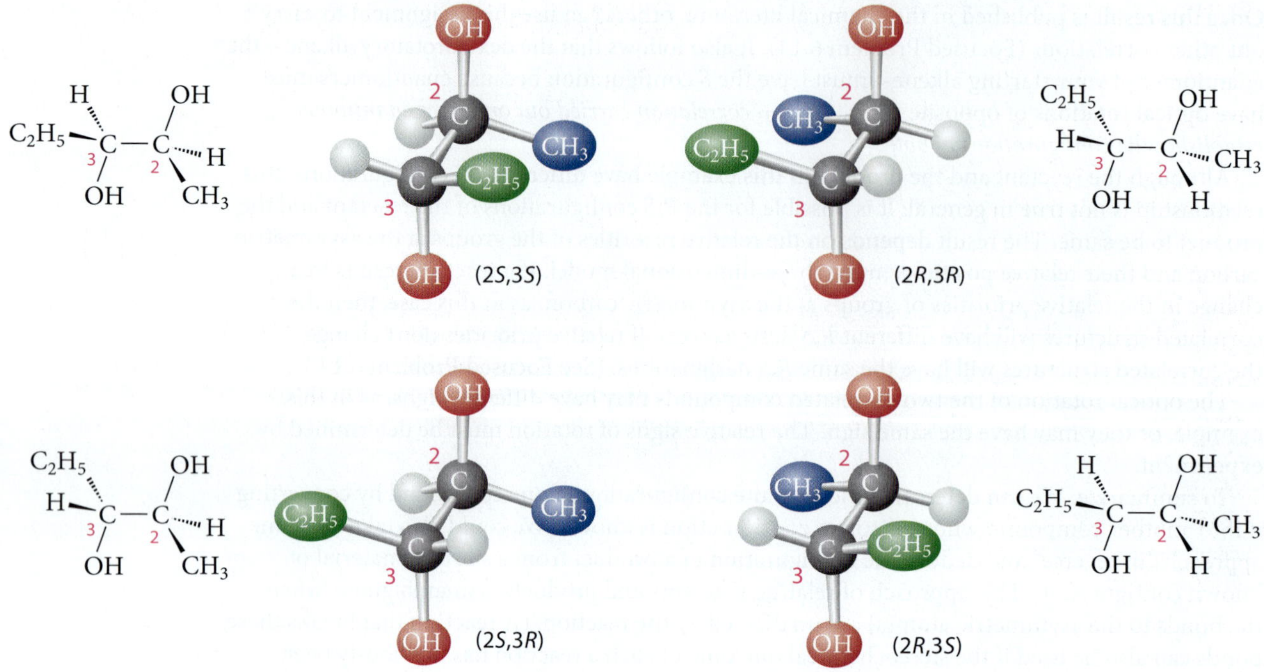

FIGURE 6.6 Stereoisomers of 2,3-pentanediol shown with both ball-and stick models and line-and-wedge structures. In each model, the small unlabeled atoms are hydrogens. This illustration uses a particular conformation of each stereoisomer, but the analysis in the text is equally valid for any other conformation. (Try it!)

Each asymmetric carbon might have the *R* or *S* configuration. With two possible configurations at each carbon, four stereoisomers are possible:

(2*S*,3*S*) (2*R*,3*R*)
(2*S*,3*R*) (2*R*,3*S*)

These possibilities are shown as both ball-and-stick models and line-and-wedge structures in **Fig. 6.6**. What are the relationships among these stereoisomers?

The 2*S*,3*S* and 2*R*,3*R* isomers are a pair of enantiomers because they are noncongruent mirror images; the 2*S*,3*R* and 2*R*,3*S* isomers are also an enantiomeric pair. (Demonstrate this point to yourself with models.) These structures illustrate the following important general point: *A pair of enantiomers must have opposite configurations at* every *asymmetric carbon.*

Because neither the 2*S*,3*S* and 2*S*,3*R* pair nor the 2*R*,3*R* and 2*R*,3*S* pair are enantiomers, they must have a different stereochemical relationship. Stereoisomers that are not enantiomers are called **diastereoisomers** or, more simply, **diastereomers**. Diastereomers are *not* mirror images. All of the relationships among the stereoisomeric 2,3-pentanediols are summarized in the following diagram:

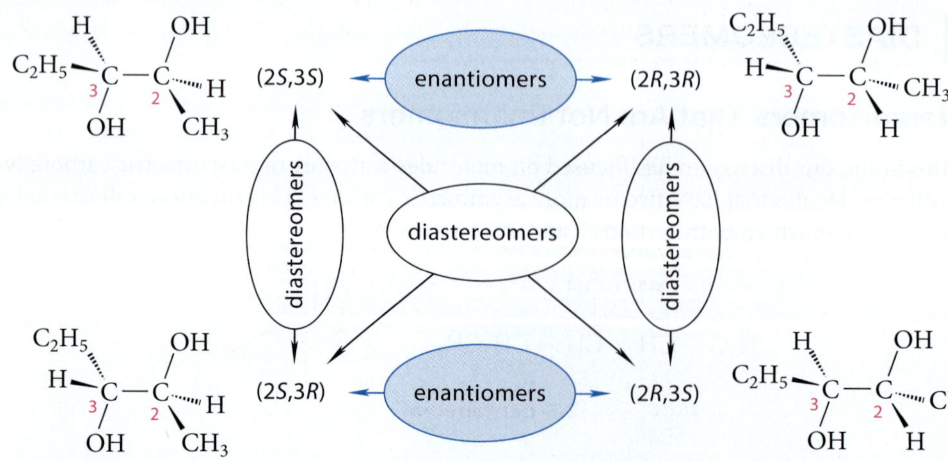

(6.24)

TABLE 6.1 Properties of Four Chiral Stereoisomers

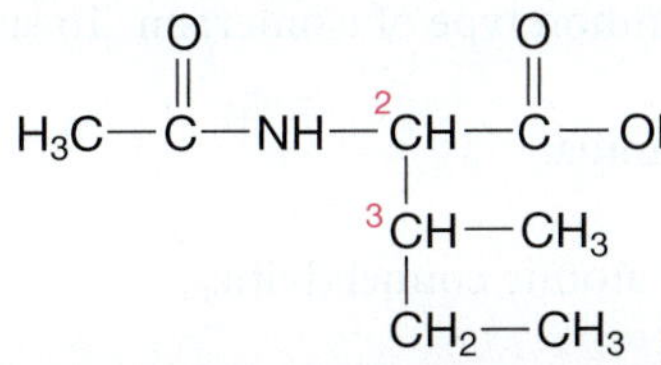

***N*-acetylisoleucine**

Configuration	Specific rotation $[\alpha]_D^{25}$ (ethanol), degrees mL g^{-1} dm^{-1}	Melting point, °C	Relationship	
(2*S*,3*S*)	+15	150–151	enantiomers	diastereomers
(2*R*,3*R*)	–15	150–151		
(2*S*,3*R*)	+21.5	155–156	enantiomers	
(2*R*,3*S*)	–21.5	155–156		
racemate of (2*S*,3*S*) and (2*R*,3*R*)	0	117–123		
racemate of (2*S*,3*R*) and (2*R*,3*S*)	0	165–166		

Diastereomers differ in all *of their physical properties*. Thus, diastereomers have different melting points, boiling points, heats of formation, and standard free energies. Because diastereomers differ in all of their physical properties, they can in principle be separated by conventional means, such as fractional distillation or crystallization. If diastereomers happen to be chiral, they can be expected to be optically active, but their specific rotations will have no relationship. These points are illustrated in **Table 6.1**, which gives some physical properties for four stereoisomers and their racemates for an amino acid derivative, *N*-acetyl isoleucine.

Let's contrast various aspects of enantiomers and diastereomers:

1. *Occurrence:* Enantiomers occur in pairs. If a molecule is chiral, it has one and only one enantiomer. There can be many diastereomeric relationships among stereoisomers that possess multiple asymmetric centers.

2. *Asymmetric centers:* No matter how many asymmetric centers a chiral molecule has, enantiomers differ in configuration at *every* asymmetric center; they cannot be mirror images otherwise. If two stereoisomers with asymmetric centers have the same configuration at one or more (*but not all*) asymmetric centers, they must be diastereomers.

3. *Chirality:* Each enantiomer of a pair is chiral and is therefore the noncongruent mirror image of the other. Some diastereomers are chiral (as is the 2,3-pentanediol example of this section), but some diastereomers are not chiral. (See Study Problem 6.5 for an example; another important example is discussed in Sec. 6.7.) Whether chiral or not, a pair of diastereomers *cannot* be mirror images, because two molecules that are mirror images must be either identical or enantiomers.

4. *Symmetry:* No chiral molecule can have either an internal plane of symmetry or a center of symmetry. Therefore, enantiomers must not have these symmetry elements. Any molecule with one or more of these symmetry elements is achiral.

There are other symmetry elements that can rule out chirality, but they are so unusual that we won't encounter them in this text.

5. *Optical activity:* All enantiomers are optically active; the optical activities of an enantiomeric pair have equal magnitudes and opposite signs. If diastereomers are chiral, they are optically active; but the optical activities of two diastereomers have no relationship in either magnitude or sign.

6. *Other physical properties:* A pair of enantiomers have identical physical properties—melting point, boiling point, heat of formation, standard free energy. Diastereomers have different physical properties.

B. Summary of Isomerism

You have now seen an example of every common type of isomerism. To summarize:

1. *Isomers* have the same molecular formula.

2. *Constitutional isomers* have different atomic connectivities.

3. *Stereoisomers* have identical atomic connectivities. There are *only two* types of stereoisomer:

   a. *Enantiomers* are noncongruent mirror images.

   b. *Diastereomers* are stereoisomers that are not enantiomers; furthermore, they are not mirror images.

The structural relationships among molecules are analyzed by working with *one pair of molecules at a time*. The flowchart in **Fig. 6.7** provides a systematic way to determine the isomeric relationship, if one exists, between two nonidentical molecules. Study Problem 6.5 illustrates the use of Fig. 6.7.

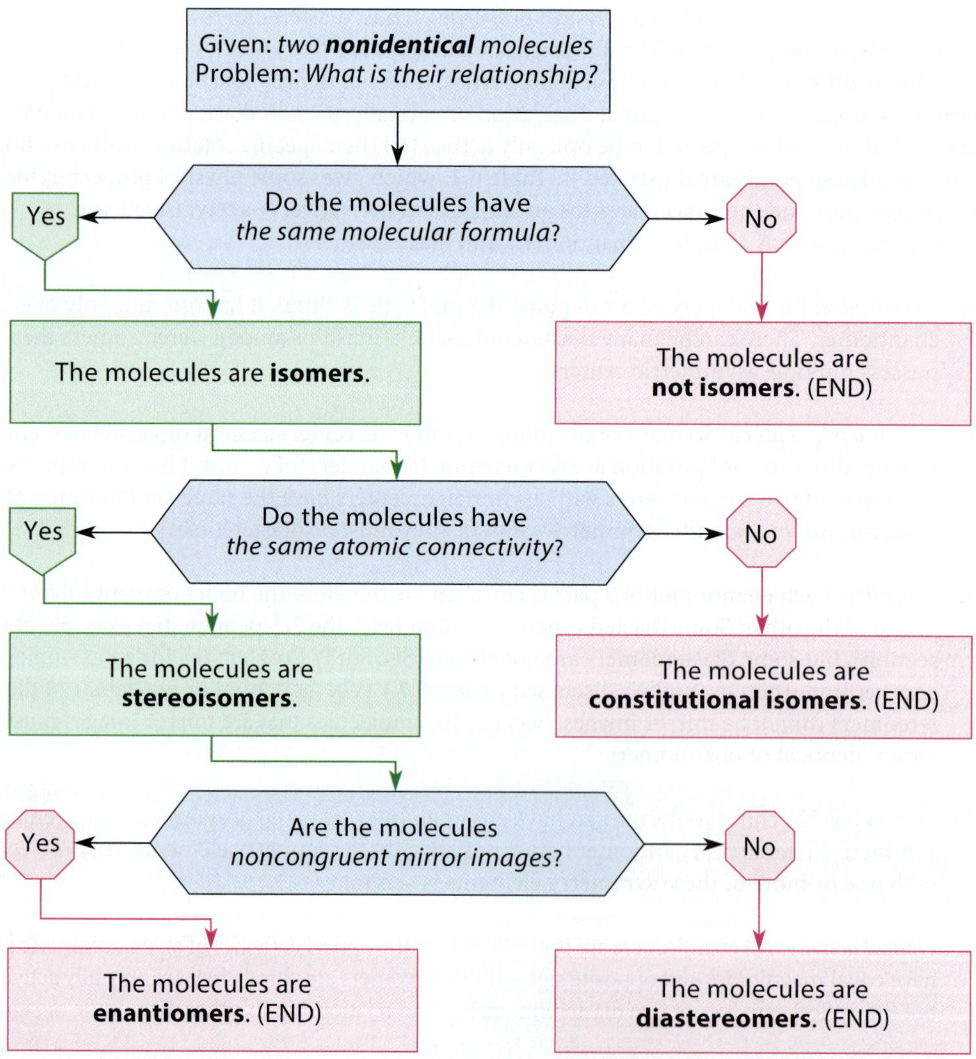

FIGURE 6.7 A systematic way to analyze the relationship between two nonidentical molecules. Given a pair of molecules, work from the top of the chart to the bottom, asking each question in order and following the appropriate branch. When you get to a red box labeled "END," the isomeric relationship is determined.

Study Problem 6.5

Determine the isomeric relationship between the following two molecules:

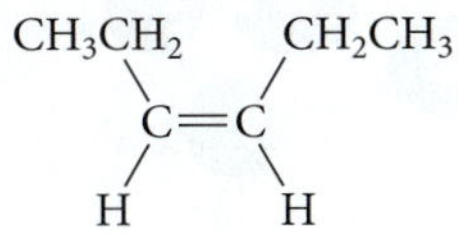
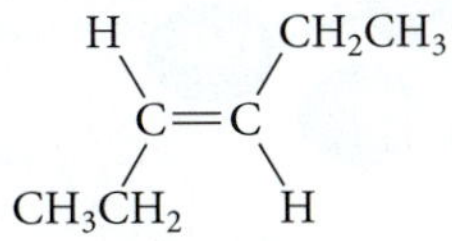

Solution Work from the top of Fig. 6.7, and answer each question in turn. These two molecules have the same molecular formula; therefore, they are isomers. They have the same atomic connectivity; therefore, they are stereoisomers. (In fact, they are the *E* and *Z* isomers of 3-hexene.) Because the molecules are not mirror images, they must be diastereomers. Thus, (*E*)- and (*Z*)-3-hexene are diastereomers.

We can see from Study Problem 6.5 that double-bond isomerism is actually one type of diastereomeric relationship. The fact that neither (*E*)- nor (*Z*)-3-hexene is chiral shows that some diastereomers are *not* chiral.

Focused Problem

6.15 Give the relationship between the molecules in each of the following pairs. Use E for enantiomers, D for diastereomers, C for constitutional isomers, I for identical, and N for none of these.

(a) 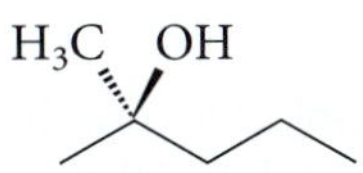 and 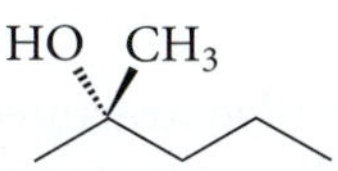

(b) 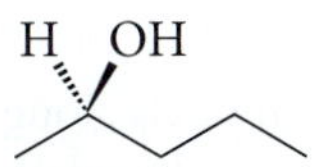 and 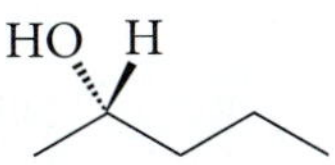

(c) 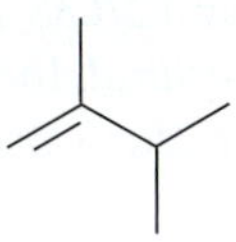 and 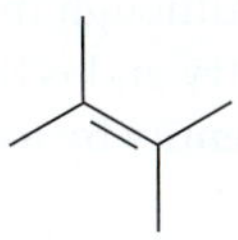

(d) 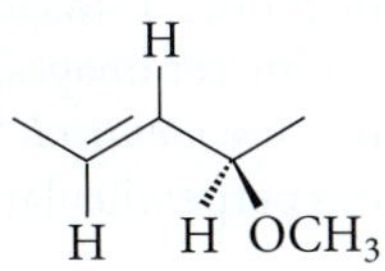 and 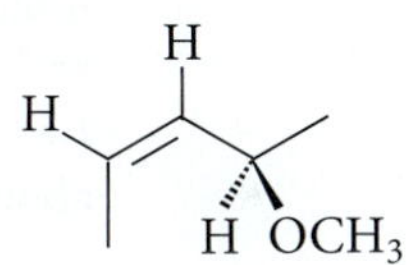

(e) 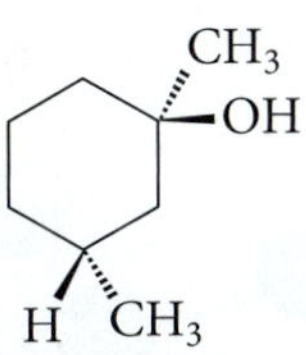 and 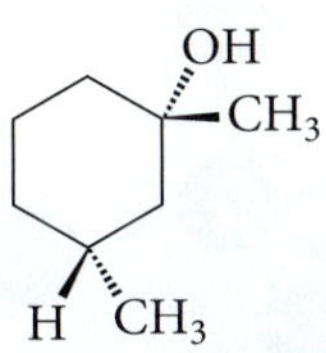

(f) 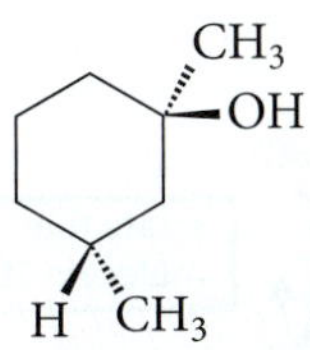 and 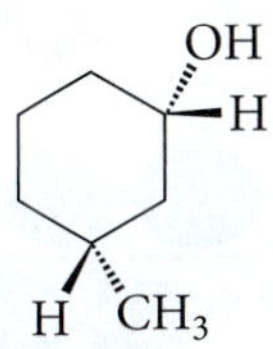

6.7 MESO COMPOUNDS

A. Achiral Compounds with Asymmetric Centers

Up to this point, each example of a molecule containing one or more asymmetric carbon atoms has been chiral. However, certain compounds containing two or more asymmetric carbons are achiral. 2,3-Butanediol provides an example:

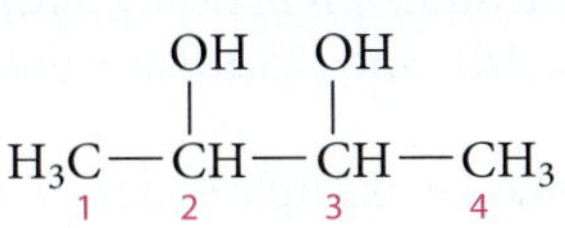
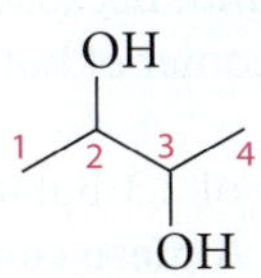

2,3-butanediol

As with 2,3-pentanediol in Sec. 6.6, four stereoisomers seem possible:

$$(2S,3S) \quad (2R,3R)$$
$$(2S,3R) \quad (2R,3S)$$

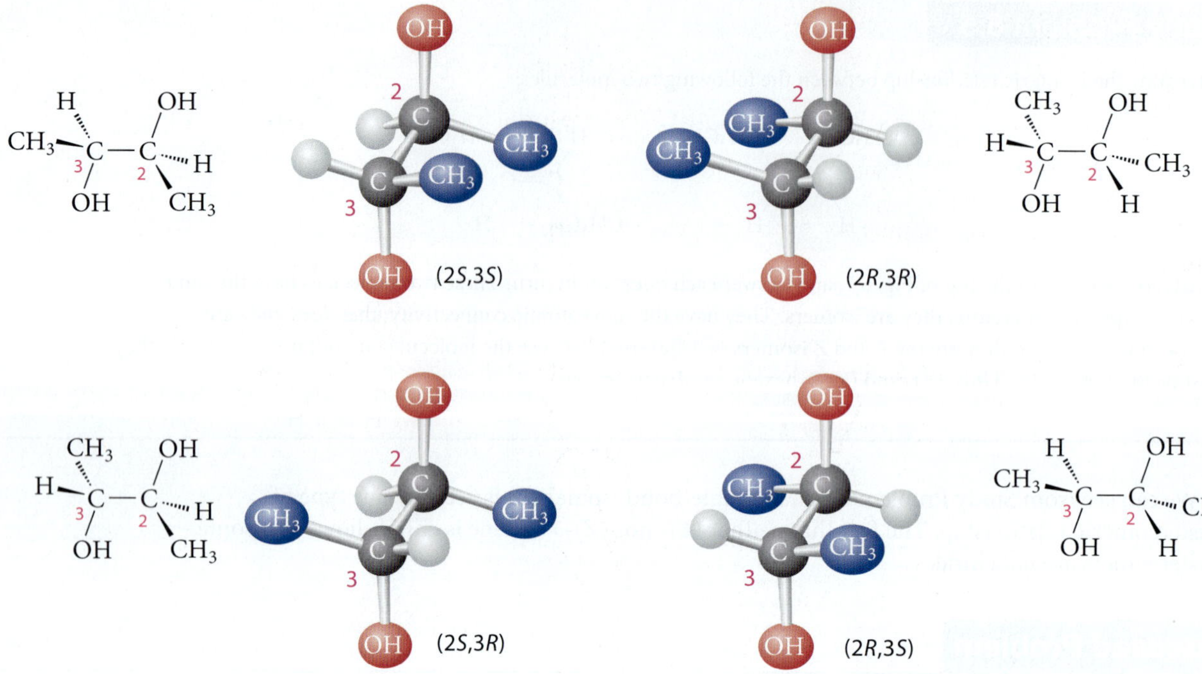

FIGURE 6.8 Stereoisomeric possibilities for 2,3-butanediol. As with Fig. 6.6, this illustration uses a particular conformation of each stereoisomer, and the small unlabeled atoms on each model are hydrogens.

Ball-and-stick models along with line-and-wedge structures of these molecules are shown in **Fig. 6.8**. Consider the relationships among these structures. The 2*S*,3*S* and 2*R*,3*R* structures are noncongruent mirror images, so they are enantiomers. Although the 2*S*,3*R* and 2*R*,3*S* structures are shown as mirror images, each has a center of symmetry and is therefore *achiral.* In fact, *these two structures are identical.* We can demonstrate their identity by rotating the 2*R*,3*S* structure 180° about an axis perpendicular to the C2—C3 bond:

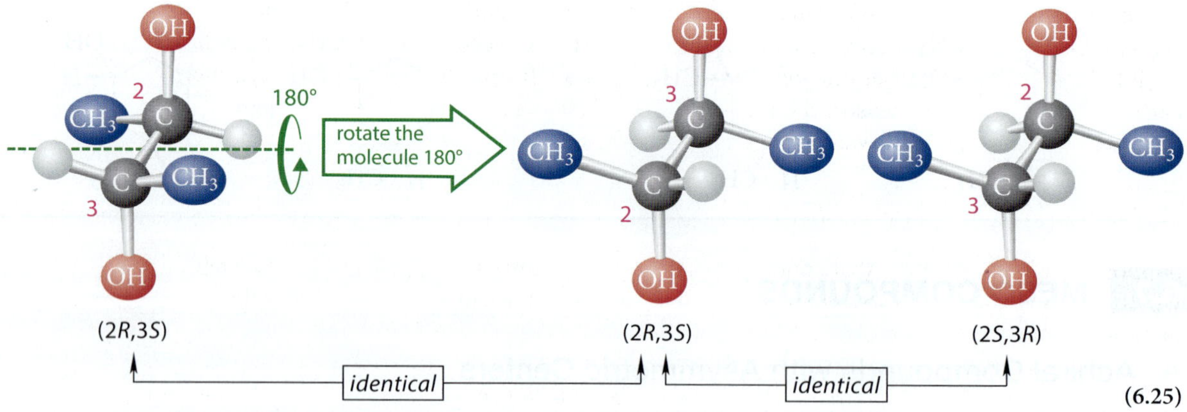

(6.25)

So, there are only *three* stereoisomers of 2,3-butanediol, not four as it seemed at first, because two of the possibilities are identical. Because the 2*R*,3*S*-stereoisomer of 2,3-butanediol is congruent to its mirror image, *it is achiral.* Because it is achiral, *it is optically inactive.* This stereoisomer is therefore an achiral diastereomer of both (2*R*,3*R*)- and (2*S*,3*S*)-2,3-butanediol, which, as we have seen, are chiral.

The achiral stereoisomer of 2,3-butanediol is an example of a *meso compound*, and it is called *meso*-2,3-butanediol. A **meso compound** is an achiral (and therefore optically inactive) compound that has chiral diastereomers. In virtually all of the examples we'll encounter, *a meso compound is an achiral compound that has at least two asymmetric centers.* Although there are a few unusual exceptions, we can use this statement as the operational definition of a meso compound. For example, *cis*- and *trans*-2-butene are stereoisomers, and they are achiral; but they are *not* meso compounds because neither has any asymmetric carbons.

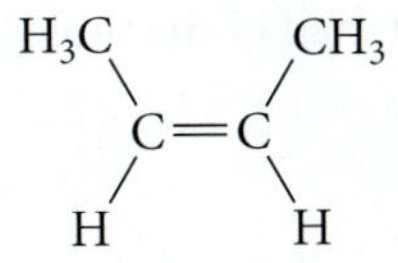
***cis*-2-butene**

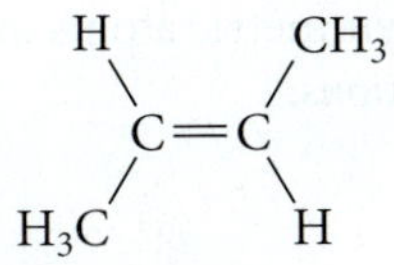
***trans*-2-butene**

Both compounds are achiral . . . BUT neither has chiral stereoisomers, and neither has asymmetric carbons. Therefore, these are *not* meso compounds.

The following diagram summarizes the relationships between the 2,3-butanediol stereoisomers.

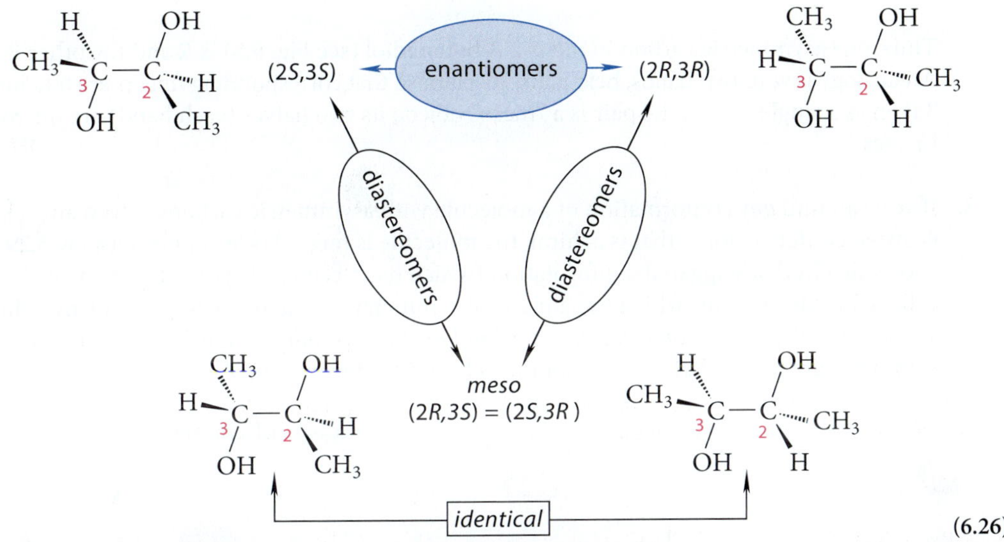

(6.26)

> **!** Notice carefully the difference between a meso compound and a racemate. Although both are optically inactive, a meso compound is a *single achiral compound*, but a racemate is a *mixture of chiral compounds*—specifically, an equimolar mixture of enantiomers.

The existence of meso compounds shows that *some achiral compounds have asymmetric carbons*. Therefore, the presence of asymmetric carbons in a molecule is an *insufficient* condition for it to be chiral, unless it has only *one* asymmetric carbon. If a molecule contains *n* asymmetric carbons, then it has 2^n stereoisomers unless there are meso compounds. If there are meso compounds, then there are fewer than 2^n stereoisomers.

B. Recognizing Meso Compounds

Suppose we have a structure that contains two or more asymmetric carbons. How can we tell whether it can exist as a meso stereoisomer? Here are a few shortcuts that can simplify the analysis:

1. *A meso compound is possible only when a molecule with two or more asymmetric atoms can be divided into halves that have the same connectivity.* (The word *meso* means "in the middle.")

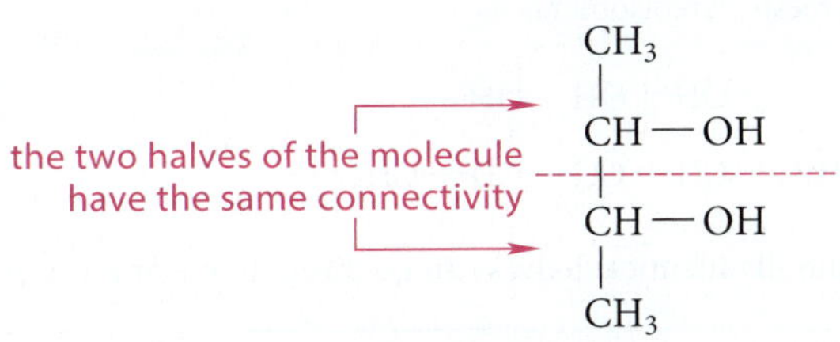

(6.27)

2. In a meso compound, the corresponding asymmetric atoms in each half of the molecule must have *opposite stereochemical configurations*:

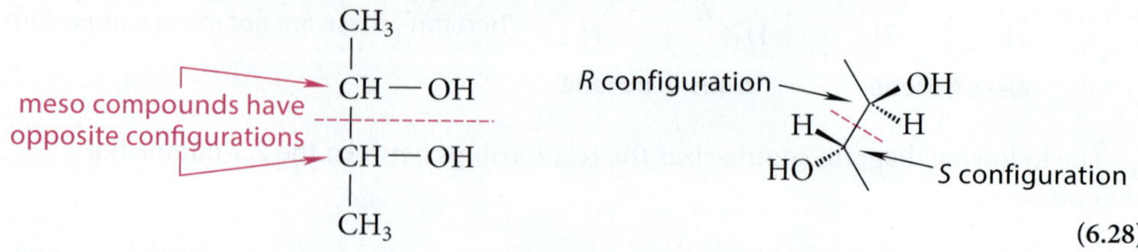

(6.28)

Thus, one asymmetric carbon in *meso*-2,3-butanediol (see Fig. 6.8) is *R* and the other is *S*. An analogy is your two hands, held palm-to-palm so that corresponding fingers are touching. Taken as a single object, the pair is a "meso" object; its two halves (each hand) are mirror images.

3. If you can find *any* conformation of a molecule with asymmetric carbons—even an eclipsed conformation—that is achiral, the molecule is meso. For example, Display 6.29 shows that both a staggered conformation (which has a center of symmetry) and an eclipsed conformation (which has a plane of symmetry) of *meso*-2,3-butanediol are achiral. *Recognition that either conformation is achiral means the molecule is achiral.* (Planes of symmetry are particularly easy to spot in eclipsed conformations.)

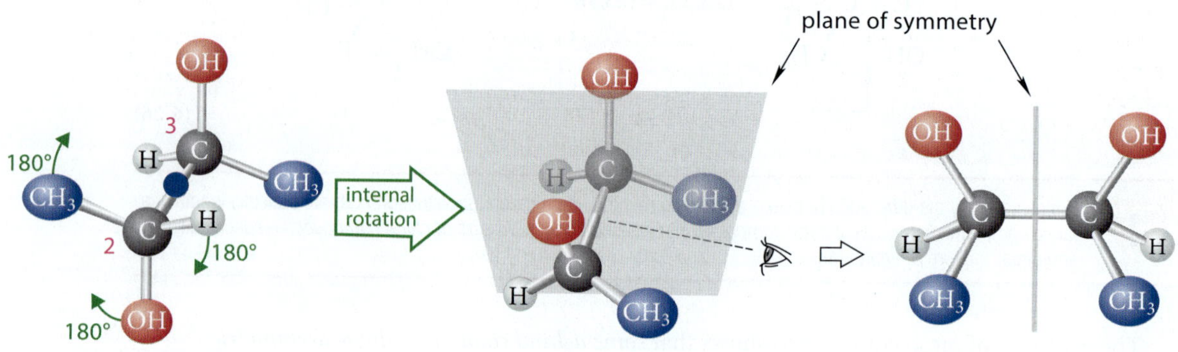

This conformation is achiral because it has a center of symmetry (●).

This conformation is achiral because it has a plane of symmetry.

(6.29)

Focused Problems

6.16 Tell whether each of the following molecules has a meso stereoisomer.

(a) 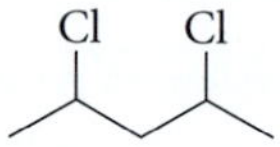

(b) 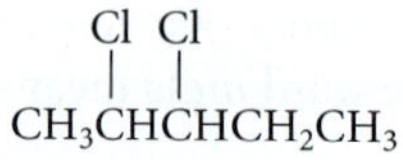

(c) *trans*-2-Hexene

6.17 Explain why the following compound has two meso stereoisomers.

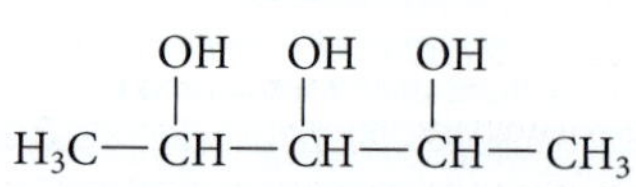

(*Hint:* The plane that divides the molecule into structurally identical halves can go through one or more atoms.)

C. Summary: Testing Molecules for Chirality

The points in the following list will allow you to assess the chirality of most compounds merely by inspecting their Lewis structures and knowing their *R/S* configurations. (This expands the list presented in Sec. 6.1C to include molecules with multiple asymmetric centers.)

1. If a molecule has a single asymmetric carbon (or other asymmetric atom), it must be chiral.
2. If a molecule has two or more asymmetric carbons, it is chiral unless it is a meso compound. (The previous section shows how to test for a meso compound.)
3. If a molecule has a plane of symmetry, a center of symmetry, or both, it is *not* chiral.
4. If you can't analyze for chirality by the previous three criteria, the most general way to assess chirality is to build two models or draw two perspective structures, one of the molecule and the other of its mirror image, and then test the two for congruence. If the two mirror images are congruent, the molecule is achiral; if not, the molecule is chiral.

6.8 CHIRAL MOLECULES WITHOUT ASYMMETRIC CENTERS

The existence of meso compounds demonstrates that the presence of asymmetric carbons is an insufficient condition for the chirality of a molecule. In this section you will learn that the presence of an asymmetric atom is also *unnecessary* for chirality. In other words, *some chiral molecules contain no asymmetric atoms.* An example is the following pair of enantiomeric 2,3-pentadiene molecules.

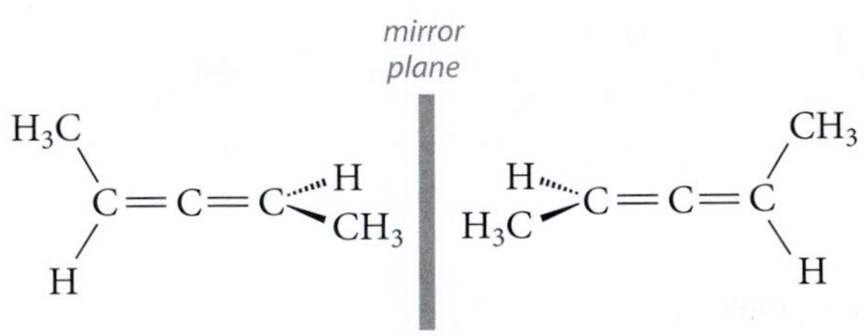

2,3-pentadiene (6.30)

These molecules have no asymmetric carbon; yet they are noncongruent mirror images and are therefore enantiomers. (If necessary, build models of them and convince yourself that this so.)

The chirality of 2,3-pentadiene is "twist" chirality—it is the same type of chirality as the thread of a screw. We are most familiar with right-handed screw threads—you tighten the screw by turning it to the right. However, left-handed screw threads are sometimes encountered—you tighten the screw by turning it to the left. If we view 2,3-pentadiene from one end in a Newman projection, in one enantiomer the methyl-group sequence turns to the right from the near carbon to the far carbon. In the other enantiomer, the methyl-group sequence turns to the left.

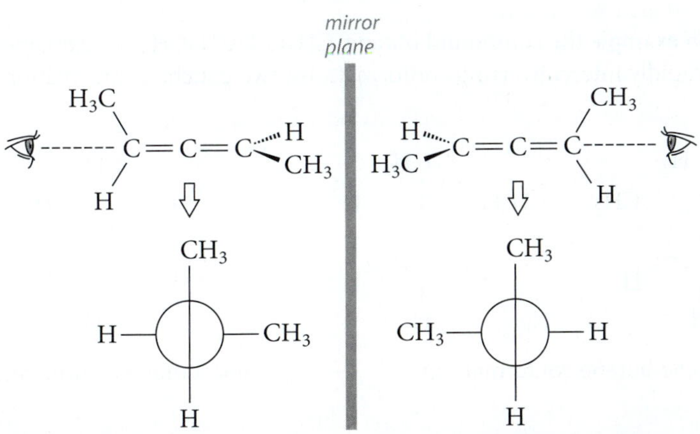

(6.31)

The twist in 2,3-pentadiene is due to the sp hybridization of the central carbon. This hybridization causes the two systems of π orbitals to be perpendicular. (This is like the carbon of CO_2; see Display 1.73, Sec. 1.7D.) Consequently, the planes of the three bonds to the two sp^2 carbons are also perpendicular.

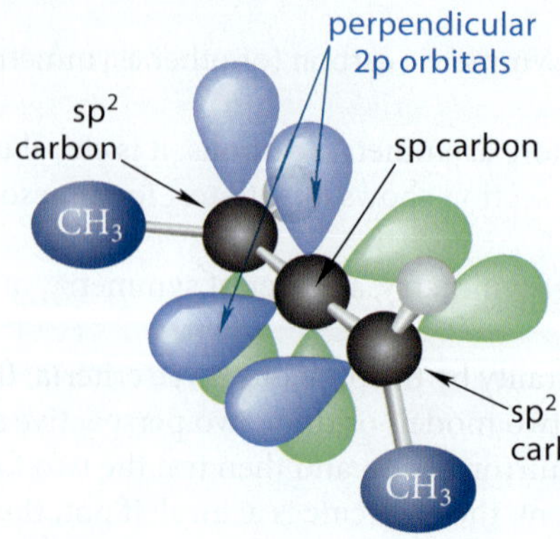

(6.32)

Compounds containing two carbon–carbon double bonds joined with a common sp-hybridized carbon are called **allenes**. We discuss allenes in more detail in Sec. 15.1C.

Focused Problem

6.18 (a) Indicate whether each molecule is chiral. If it is chiral, draw the structure of its enantiomer. (Assume that the rings are planar for purposes of this problem.)

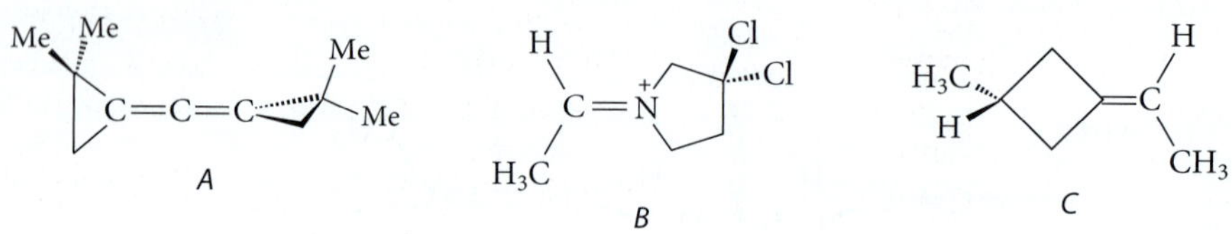

(b) Which atoms are stereocenters (if any) in each molecule?

(c) Which atoms (if any) are asymmetric centers in each molecule?

6.9 RAPIDLY INTERCONVERTING STEREOISOMERS

A. Stereoisomers Interconverted by Internal Rotations

Chirality is a *geometric* concept that can be rigorously applied only to static, rigid objects. In this section you'll learn to think about the chirality of molecules that are not static, but are changing rapidly.

Consider as an example the compound butane, $CH_3CH_2CH_2CH_3$. As explained in Sec. 2.5B, butane exists in rapidly interconverting conformations: two gauche conformations and one anti conformation.

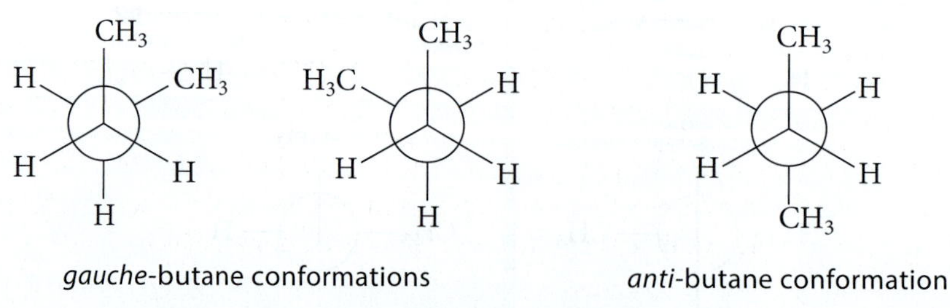

(6.33)

The anti conformation is achiral because it has a center of symmetry (verify this). However, *the two gauche conformations of butane are chiral*, and they are enantiomers, because they are noncongruent mirror images:

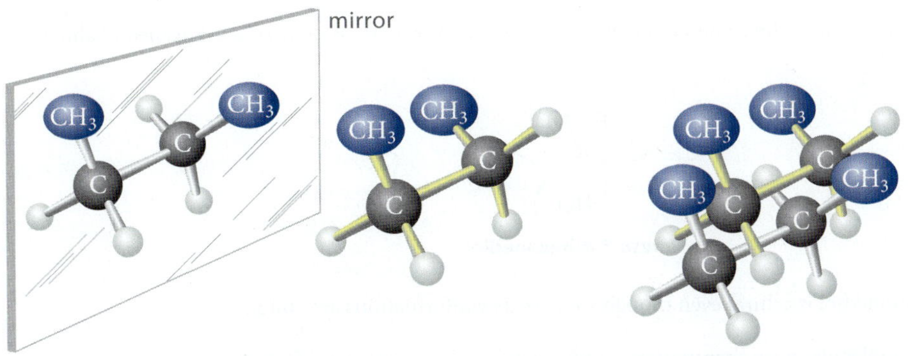

(a) mirror images (b) the two mirror images are not congruent (6.34)

Butane has *no* asymmetric carbons, so each *gauche*-butane conformation is another example of *a chiral molecule that has no asymmetric center*. The chirality of the gauche butanes is a "twist" chirality, which was introduced in Sec. 6.8.

The *gauche*-butanes are examples of **conformational enantiomers**: enantiomers that are interconverted by an internal rotation. Either of the *gauche*-butanes and *anti*-butane are **conformational diastereomers**: diastereomers that are interconverted by an internal rotation.

This analysis shows that the chirality of butane is actually changing rapidly over time. So, is butane chiral or achiral? To simplify situations like this, chemists agree to assess the chirality of butane by considering its *average* structure over time. The conformations of butane are interconverting so rapidly that *the two enantiomeric conformations are taken together as a single species* (like a pair of shoes). When we take this view, butane is achiral. Rigorously, butane is an achiral anti conformation plus a *racemic mixture* of the two gauche conformations, but we regard butane as an achiral molecule because we think of the racemic mixture as a single achiral object. Any analysis of butane in real time would not detect any evidence of chirality, such as optical activity. We couldn't separate the enantiomeric conformations; and even if we could, their optical activities would dissipate instantly as the conformations came to equilibrium.

Butane is not unique; many achiral compounds have enantiomeric chiral conformations, and we treat them just as we treat butane. You can apply the following principle to chiral conformations:

> When a molecule can exist as *rapidly interconverting enantiomeric conformations*, the molecule is considered to be achiral.

This principle is the reason that we can represent methyl groups or ethyl groups as "single balls" when we analyze models for chirality, as we did in Sec. 6.1A. If we draw out the hydrogens explicitly on the methyl or ethyl carbons, you might find chiral conformations; but these conformations are interconverting so rapidly with their enantiomeric conformations that we view their conformations in aggregate as one species. The methyl and ethyl groups can be viewed as rapidly spinning tops.

You can assume rapid rotation about single bonds unless you are told otherwise. This assumption allows you to use the principles given in the list in Sec. 6.7C (such as the presence of asymmetric carbons) to determine whether such a molecule is chiral or achiral without even having to examine individual conformations. If we identify a molecule as chiral using these principles, it *must* have at least one chiral conformation that is *not* in equilibrium with its enantiomer. If we identify a molecule as achiral using these principles, then every chiral conformation must be in rapid equilibrium with an enantiomeric conformation. In a meso compound, for example—despite the presence of asymmetric carbons—every chiral conformation must be in rapid equilibrium with an enantiomeric conformation.

How would butane behave on a *very cold* planet—so cold, that the rate of its internal rotations would be reduced to the point that its conformations did not interconvert? (It would be also interesting to meet the inhabitants of such a planet capable of appreciating this fact.) On this planet, there would be three distinct stereoisomers of butane: the two enantiomeric *gauche*-butanes, which would be optically active with equal and opposite optical rotations, and their diastereomer *anti*-butane, which would be achiral and would have different properties from the gauche stereoisomers. Life would become much more complicated for organic chemistry students on that planet when asked to assess the chirality of butane.

Chemists have actually been able to produce very unusual molecules that have such slow internal rotations that their individual conformations can be separately isolated and studied. When the conformations are enantiomeric, they are indeed optically active. (See Further Exploration 6.2.)

FURTHER EXPLORATION 6.2
Isolation of Conformational Enantiomers

Focused Problems

6.19 Taking the anti conformation of butane as an isolated structure, determine whether it has any stereocenters. If so, identify them.

6.20 (a) What are the stereochemical relationships among the three staggered conformations of *meso*-3,4-hexanediol about the C3–C4 bond?

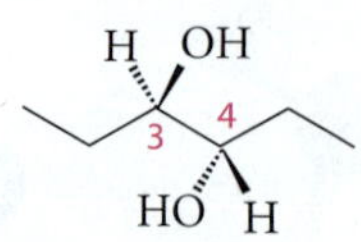

***meso*-3,4-hexanediol**

(b) Explain why *meso*-3,4-hexanediol is achiral even though some of its conformations are chiral.

6.21 Which of the following achiral molecules have enantiomeric conformations? Show one pair of enantiomeric conformations in each case.

(a) Propane

(b) 2,3-Dimethylbutane

(c) 2,2,3,3-Tetramethylbutane

B. Stereoisomers Interconverted by Inversion

Amine Inversion An interesting phenomenon occurs with amines, such as ethylmethylamine, in which the nitrogen is bonded to three different groups.

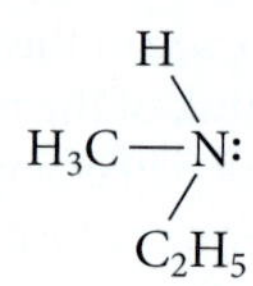

ethylmethylamine

Ethylmethylamine has trigonal pyramidal geometry. It seems to be a chiral molecule—it should exist as two enantiomers. The asymmetric atom is a nitrogen.

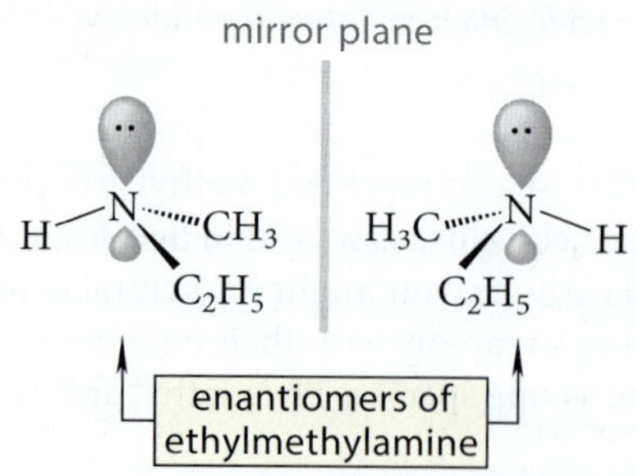

(6.35)

In fact, ethylmethylamine can exist as two enantiomers, but the enantiomers cannot be separated because they rapidly interconvert by a process called **amine inversion**, shown in **Fig. 6.9**. In this process, the larger lobe of the electron pair seems to push through the nucleus to emerge on the other side. (Imagine pulling an inflated balloon through a small hole.) *The molecule is not simply turning over; it is actually turning itself inside out!* This is something like an umbrella turning inside-out in the wind. This process occurs through a transition state in which the amine nitrogen becomes sp^2-hybridized and the nonbonding electron electron pair is in a 2p orbital. The energy barrier to the inversion is caused by the loss of s character in the nonbonding electron pair; a 2p orbital has higher energy than an sp^3 orbital. Figure 6.9b shows that amine inversion interconverts the

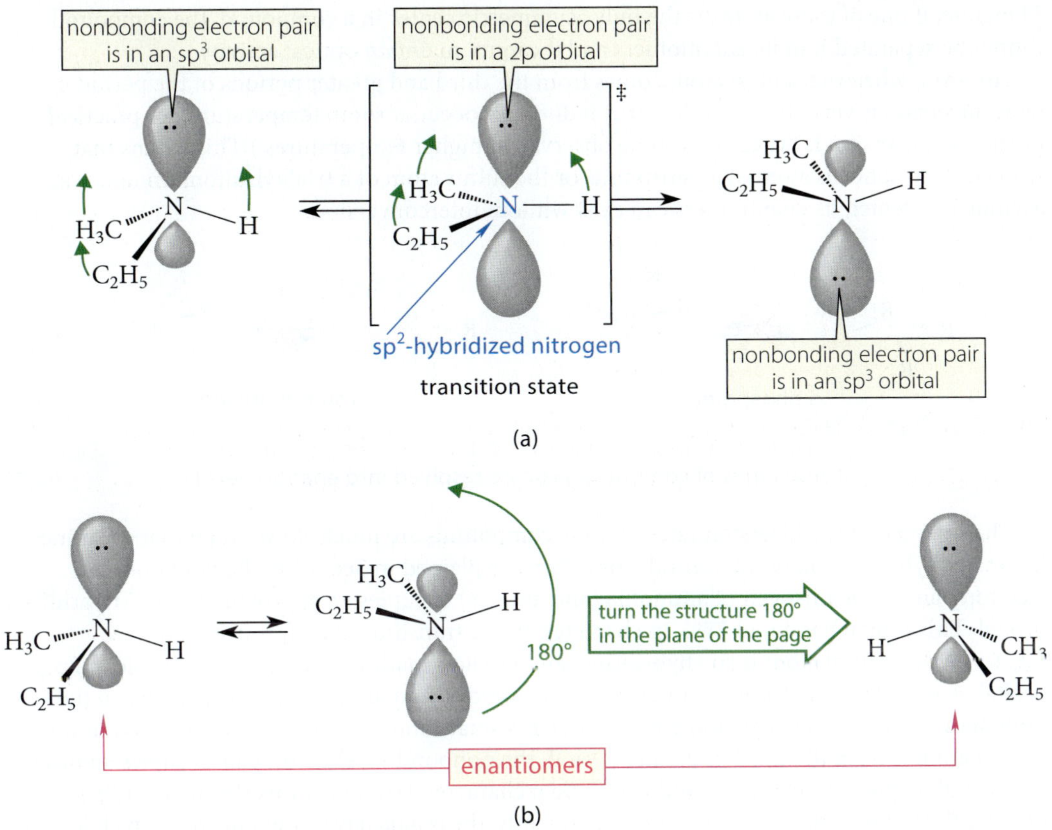

FIGURE 6.9 Inversion of amines. (a) As the inversion takes place, the large lobe of the nonbonding electron pair appears to push through the nitrogen to the other side. As this occurs, the three other groups move first into a plane containing the nitrogen, then to the other side (*green arrows*). Notice the change of hybridization that occurs in the transition state. (b) The mirror-image relationship of the inverted amines is shown by turning either molecule 180° in the plane of the page. Because the two mirror images are noncongruent, they are enantiomers.

enantiomeric forms of the amine. Because this process is rapid at room temperature, it is impossible to separate the enantiomers. Therefore, ethylmethylamine is a mixture of rapidly interconverting enantiomers. Amine inversion is yet another example of *racemization* (Sec. 6.4B).

For ordinary amines, inversion occurs at about 10^7 times per second.

Focused Problem

6.22 Assume that the following compound has the *S* configuration at its asymmetric carbon, as shown.

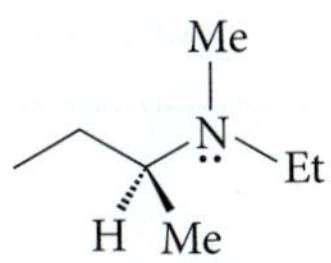

(a) What is the isomeric relationship between the two forms of this compound that are interconverted by amine inversion?

(b) Could this compound in principle be separated into enantiomers?

Inversion at Other Atoms Inversions can occur at other atoms. When the central atom comes from the second period of the periodic table, inversion is rapid, as it is with amines:

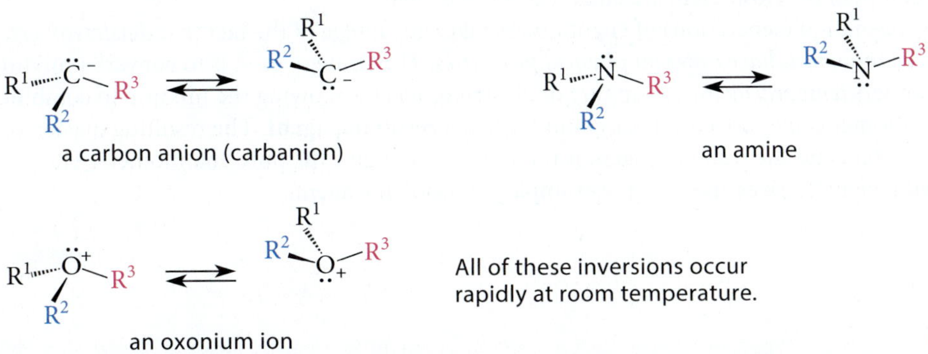

(6.36)

Therefore, if one of these atoms is the only asymmetric center in a compound, the compound cannot be separated into its enantiomers and it cannot maintain optical activity.

However, when the central atom comes from the third and greater periods of the periodic table, inversion is very slow—so slow that it does not occur at room temperature, for practical purposes. (Inversion is faster and can be observed at higher temperatures.) This means that when the phosphorus atom of a phosphine, or the sulfur atom of a trialkylsulfonium ion, is an asymmetric center, its enantiomers can exist without interconversion.

a phosphine

a sulfonium ion

Neither of these inversions occur at room temperature.
(These types of compounds can be resolved into enantiomers.) (6.37)

The reason that the inversion rates of these compounds are much slower than that of amines lies in the hybridization of the central atom. As we explained in Sec. 1.7E, the nonbonding electron pair on the nitrogen of ammonia (and amines) occupies an approximately sp^3-hybridized orbital. This orbital has about 75% 2p character. In the transition state for inversion (see Fig. 6.9a), the central atom is sp^2-hybridized, and the nonbonding electron pair occupies a 2p orbital. A relatively small amount of energy is required to add another 25% p character to the nonbonding pair, so the inversion energy barrier is small, and inversion is fast. However, if the central atom is from the third or greater period, the nonbonding electron pair occupies an orbital with a high degree of s character and very little p character. [We know this (from Sec. 1.7E) because the bond angle is close to 90°; consequently, the bonds have a high degree of p character, and the nonbonding electron pair then has most of the s character.] It takes significant energy to convert an electron pair in a 3s orbital to an electron pair in a 3p orbital. Therefore, the inversion barrier for these atoms is larger, and inversion is slower.

Focused Problem

6.23 Arsenic (As) is below nitrogen and phosphorus in group 5A of the periodic table. In an *arsine* (R_3As:), the R—As—R bond angles are about 92°. How would you expect the inversion rate of arsines to compare with that of amines and phosphines? Explain.

6.10 SEPARATION OF ENANTIOMERS (ENANTIOMERIC RESOLUTION)

As noted in Sec. 6.4B, the separation of two enantiomers (an **enantiomeric resolution**) poses a special problem. Because a pair of enantiomers have identical melting points, boiling points, and solubilities, we cannot exploit these properties for the separation of enantiomers as we might for other compounds. How, then, are enantiomers separated?

The resolution (separation) of enantiomers takes advantage of the fact that *diastereomers, unlike enantiomers, have different physical properties*. The strategy used is to convert a mixture of enantiomers *temporarily* into a mixture of diastereomers by allowing the mixture to combine with an enantiomerically pure chiral compound called a **resolving agent**. The resulting diastereomers are separated, and the resolving agent is then removed to give the pure enantiomers. The following vignette gives an everyday example of a resolving agent.

Analogy for a Resolving Agent

Photo Copyright © Marc Loudon

Suppose you are blindfolded and asked to sort 100 gloves into separate piles of right- and left-handed gloves. (Never mind how you got into this predicament!) The gloves are identical except that 50 are right-handed and 50 are left-handed. The mixture of gloves is a "racemate." How would you separate them? You can't do it by weight, by smell, or by any other simple physical property, because right- and left-handed gloves have the same properties. The way you do it is by trying each glove on your right (or left) hand. Your hand thus acts as an "enantiomerically pure" chiral resolving agent. A right-handed glove on your right hand generates a certain feeling (which we describe by saying "it fits"), whereas a left-handed glove on the right hand generates a totally different feeling. In fact, we could say that the right hand wearing the right-handed glove is the diastereomer of the right hand wearing the left-handed glove. The different sensations generated by the two situations are analogous to the different physical properties of diastereomers. After classifying each glove as "right" or "left" on the basis of this sensation, you then break the hand–glove interaction (you remove the glove) and put the glove in the appropriate pile. You have thus converted the diastereomer (the hand–glove combination) into the pure enantiomer (the glove) and the chiral resolving agent (the hand).

The principle used to separate enantiomers can be stated more formally as follows:

> **The principle of enantiomeric differentiation:** *The separation or differentiation of enantiomers requires that they interact with an enantiomerically pure chiral agent.*

This is an important general principle that we'll use often. In the hand–glove analogy, the "chiral agent" is the hand that "interacts with" (that is, tries on) the two gloves. When the "chiral agent" is a compound, the interaction results in a pair of diastereomers, which have different properties that can be used to separate or distinguish the two enantiomers.

Many techniques have been developed for enantiomeric resolution. In this section you'll learn about two examples: *chiral chromatography* and *diastereomeric salt formation*. In Sec. 8.7A, we discuss a third method that is widely used in the pharmaceutical industry, *selective crystallization*. Try to notice how the principle of enantiomeric differentiation is applied in each example.

A. Chiral Chromatography

Chromatography is an important technique for separating the components of mixtures. In a widely used version of this technique, called **column chromatography**, shown in **Fig. 6.10a**, a mixture of compounds is introduced onto the top of a cylindrical column of a finely powdered solid, called the **stationary phase (SP)**. The components of the mixture adsorb (bind) reversibly to the SP. This adsorption results from *noncovalent* attractions between the molecules of the SP and the molecules in solution to be separated. (Noncovalent interactions are discussed in Secs. 8.4–8.6.) The column is then eluted (washed) continuously with a solvent. If the components of the mixture adhere to the SP with sufficiently different affinities, the component with the smallest affinity for the SP emerges first from the column, and the components with progressively greater affinities for the SP emerge later. A graph of the concentrations of the compounds in the mixture against time (or volume of the solvent) is called a **chromatogram** (**Fig. 6.10b**).

The chromatographic separation of enantiomers is called **chiral chromatography**. In a widely used type of chiral chromatography, the stationary phase typically consists of microscopic porous glass beads to which an *enantiomerically pure chiral compound* has been *covalently* attached. This material serves as the *resolving agent*. This combination of glass beads and covalently attached resolving agent is called generally a **chiral stationary phase (CSP)**. For example, one of many CSPs that are available commercially is shown in Display 6.38. (Don't be at all concerned with the detailed structure; just notice that the pendant resolving agent is a *chiral* compound and that it is *enantiomerically pure*. As an analogy, think of it as a left hand hanging from the glass bead.)

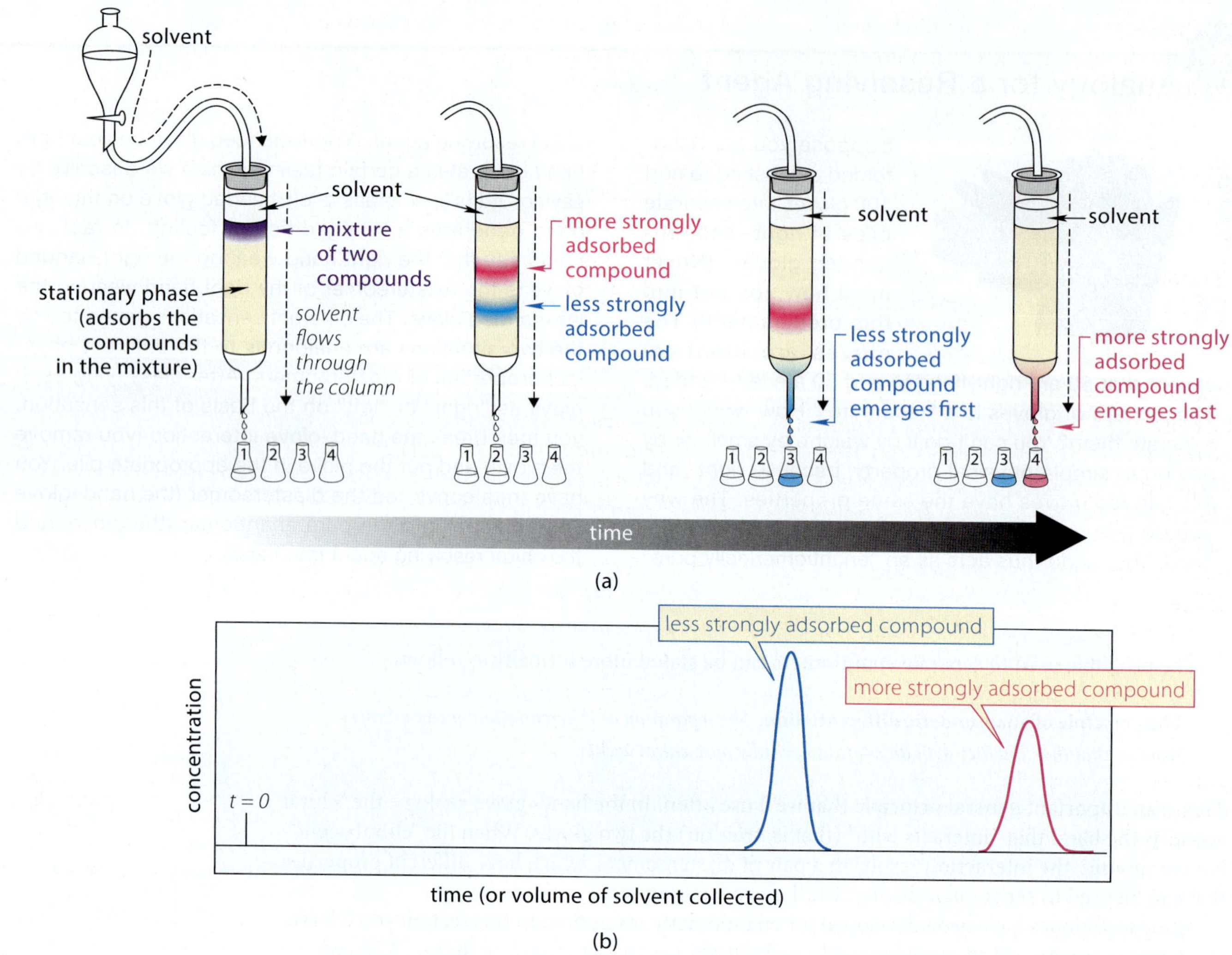

FIGURE 6.10 (a) In column chromatography, a mixture of two compounds is adsorbed reversibly onto the upper part of a column of a finely powdered solid (stationary phase). As solvent is passed through the column, the less strongly adsorbed component of the mixture moves through the column more rapidly and emerges first. The more strongly adsorbed component of the mixture is retained longer on the column and emerges later. (b) A plot of concentration of the mixture components vs. either time of elution or solvent volume. This is called a *chromatogram*. The area under each peak is proportional to the total amount of each component.

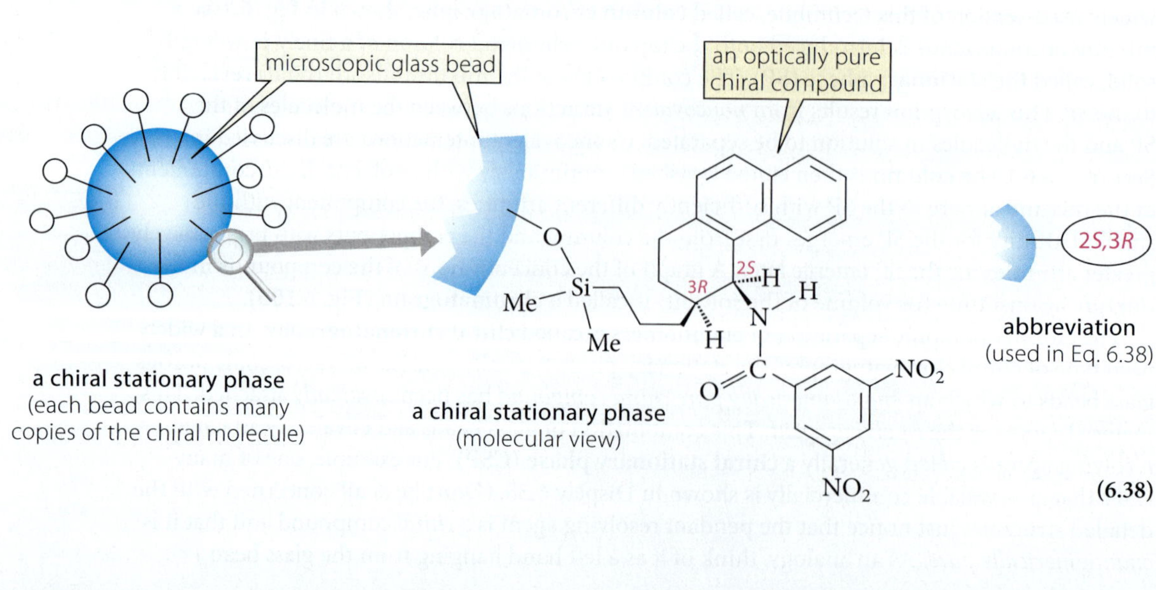

(6.38)

As Display 6.38 shows, each solid bead of the CSP contains many copies of the resolving agent. Now, if a mixture of enantiomers is passed through the CSP column, each of the two enantiomers forms a noncovalent complex with the immobilized resolving agent:

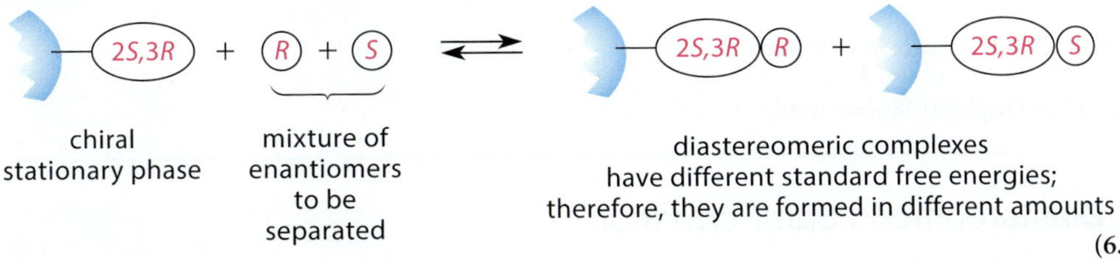

(6.39)

(We've used the abbreviation for the CSP in Display 6.39 as a specific example. Don't be concerned about the details of complex formation.) Because the resolving agent is enantiomerically pure, the two complexes differ in configuration at *only one* of their asymmetric carbons. In other words, the two complexes are *diastereomers*. In general, diastereomers have different free energies and different stabilities. For this reason, the *equilibrium constants for their formation differ*. This means that one of two enantiomers binds more tightly to the resolving agent than the other and, as a result, the concentrations of the two complexes are different. As solvent is passed through the column, the more strongly adsorbed enantiomer is retained longer on the column. Therefore, the solution of the less strongly binding enantiomer emerges first, and solution of the more strongly binding enantiomer next. **Figure 6.11** shows the chromatogram for the enantiomeric resolution of Nirvanol, a synthetic drug, by the CSP in Display 6.38.

Typically, chiral chromatography is used on a relatively small scale because the chiral stationary phases are fairly expensive. It is a superb method for the analysis of mixtures of enantiomers, and it is frequently used to determine enantiomeric ratio (ER) and enantiomeric excess (EE; Sec. 6.4A).

How do we know what CSP to use for a particular separation? We don't really know unless someone has done it before (as in this case). The selection of the CSP and the conditions to be used for a particular separation are sometimes matters of trial and error, but experience has led to some principles by which a resolving agent can usually be chosen rationally. The crucial point for us to notice is that the enantiomers are separated through their interaction with the CSP by *the temporary formation of diastereomers*. In terms of the analogy at the beginning of this section, the two enantiomers to be separated are the gloves, and the CSP is the hand.

Practical chiral chromatography was developed by William H. Pirkle in the mid-1970s while he was a professor of chemistry at the University of Illinois, and the CSPs he developed became known as *Pirkle columns*. (The CSP shown in Display 6.38 is an actual example of a Pirkle column.) A large variety of CSPs (including Pirkle columns) are available commercially.

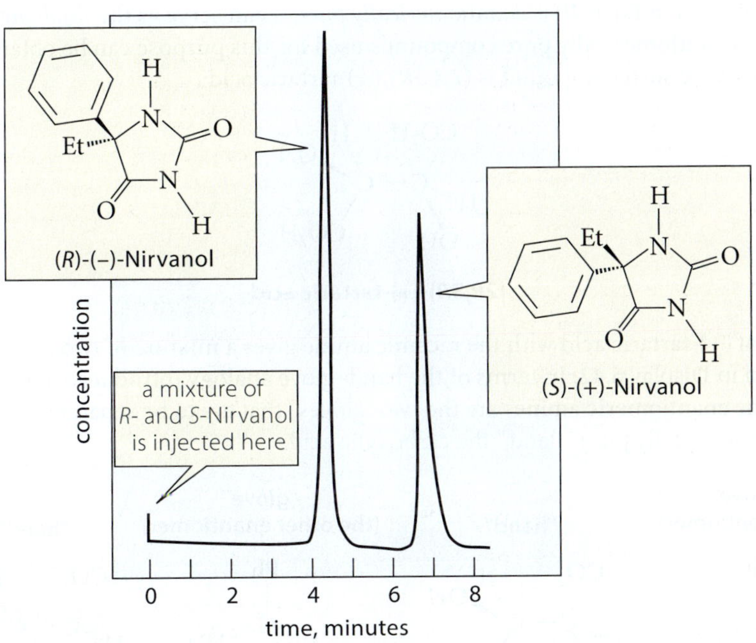

FIGURE 6.11 A chromatogram showing the enantiomeric resolution of Nirvanol, a synthetic anticonvulsant, on the chiral stationary phase in Display 6.38. (The elution solvent is an 80:20 hexane:isopropyl alcohol mixture.) Which enantiomer has the greater affinity for the chiral stationary phase?

Focused Problem

6.24 The enantiomeric resolution in Fig. 6.11 used the chiral stationary phase (CSP) in Display 6.38. How would the enantiomeric resolution in Fig. 6.11 be affected in each of the following situations?

(a) If the *enantiomer* of the CSP in Display 6.38 were used

(b) If the *racemate* of the CSP in Display 6.38 were used

B. Diastereomeric Salt Formation

Diastereomeric salt formation is a method used for the enantiomeric resolution of acidic or basic compounds. Particularly well suited for large-scale separations, this method is illustrated by the enantiomeric resolution of the racemate of α-phenethylamine:

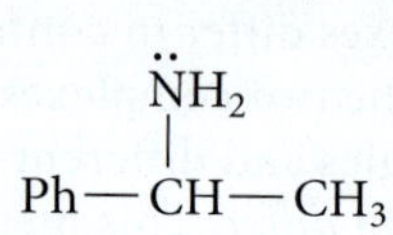

α-phenethylamine

Amines are derivatives of ammonia in which one or more hydrogen atoms have been replaced by organic groups. Diastereomeric salt formation involving amines takes advantage of the fact that amines, like ammonia, are bases; so, they react rapidly and quantitatively with carboxylic acids in a Brønsted acid–base reaction to form salts:

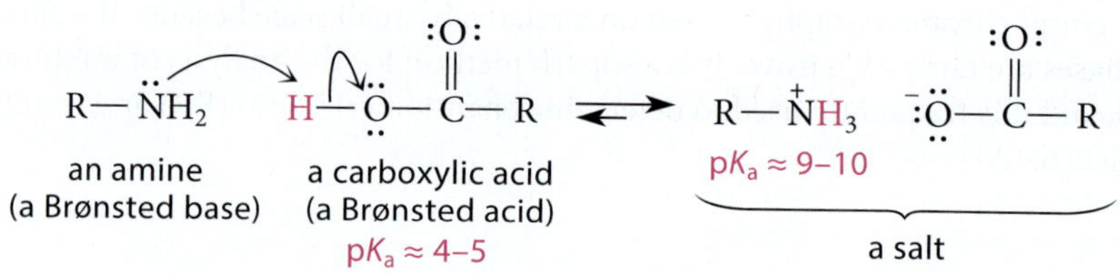

(6.40)

If the carboxylic acid in Eq. 6.40 is *enantiomerically pure*, it can serve as the *resolving agent.* In many cases, enantiomerically pure compounds used for this purpose can be obtained from natural sources. One such compound is (2*R*,3*R*)-(+)-tartaric acid:

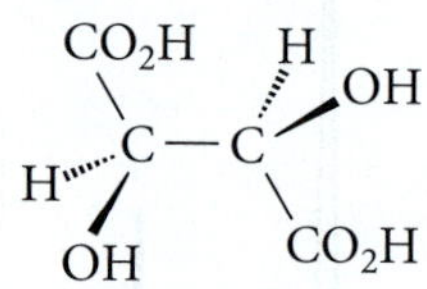

(2*R*,3*R*)-(+)-tartaric acid (6.41)

The reaction of (+)-tartaric acid with the racemic amine gives a mixture of two *diastereomeric* salts, as shown in Display 6.42. In terms of the hand–glove analogy introduced at the beginning of Sec. 6.10, the enantiomeric amines are the two "gloves" that are to be separated by interaction with an enantiomerically pure "hand," the carboxylic acid.

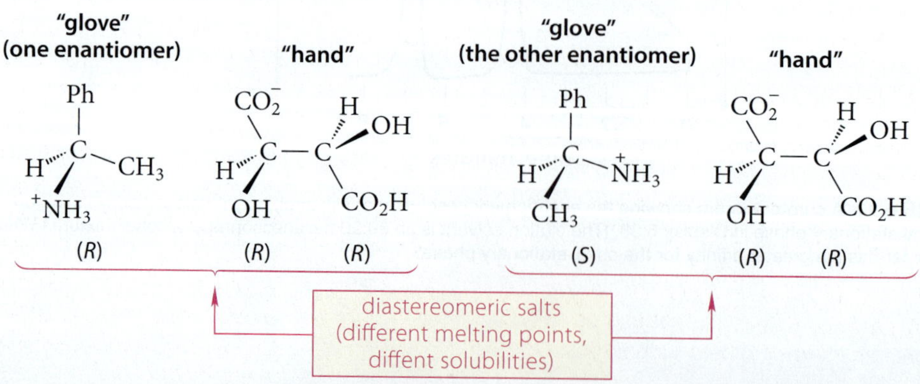

(6.42)

These salts are diastereomers because they differ in configuration at *only one* of their three asymmetric carbons. Because these salts are diastereomers, they have different physical properties. In this case, they have significantly different solubilities in methanol, a commonly used alcohol solvent. (The difference in solubilities was found by trying different solvents.) The (*S*,*R*,*R*)-diastereomer happens to be less soluble, and it crystallizes selectively from methanol, leaving the (*R*,*R*,*R*)-diastereomer in solution, from which it may be recovered. Once either pure diastereomer is in hand, the salt can be decomposed with aqueous base to liberate the water-insoluble, optically active amine, leaving the tartaric acid in solution as its conjugate-base dianion.

The exact conditions for selective crystallization of the salt are very specific, and you should not be concerned with learning these. The insolubility of the amine in water and the solubility of its conjugate acid are also crucial to the process. These conditions were determined by the application of solubility principles that we cover in Chapter 8. A certain amount of trial and error is also typically required for designing these conditions. The important points here are the use of a chiral resolving agent, the acid–base reaction used to form the salt, and the acid–base reaction used to liberate the amine.

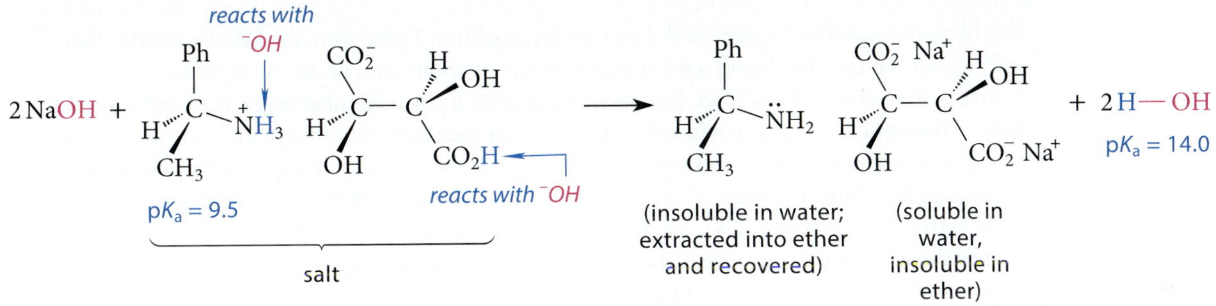

(6.43)

Salt formation is such a simple and convenient reaction that it is often used for the enantiomeric resolution of amines and carboxylic acids.

Focused Problem

6.25 Which of the following amines could in principle be used as a resolving agent for a racemic carboxylic acid?

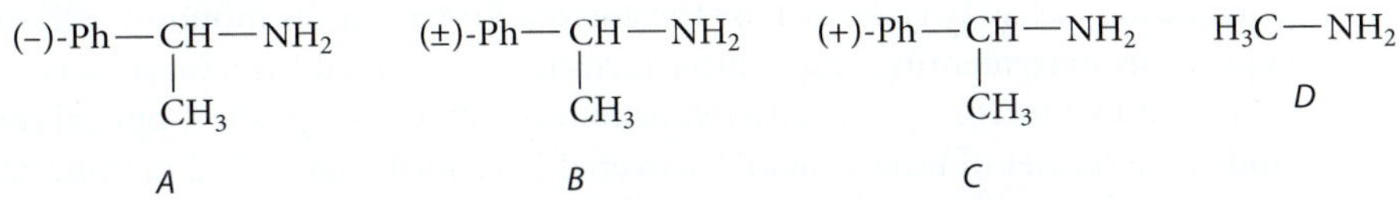

Chiral Recognition by Scent Receptors

In many cases, enantiomers have different odors. The enantiomers of carvone are one example.

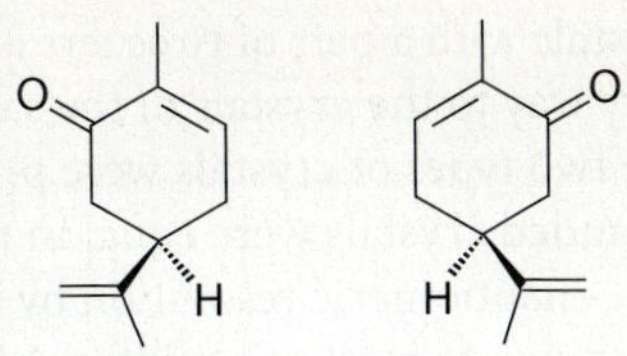

R-(−)-carvone (spearmint) **S-(+)-carvone** (caraway)

(*R*)-(−)-Carvone gives spearmint its familiar odor. [Naturally obtained (*R*)-(−)-carvone is used as a natural flavoring, and spearmint oil production is a $100 million industry in the United States.] Its enantiomer, (*S*)-(+)-carvone, is present in caraway seeds (actually, the fruit of the caraway plant), which give rye bread its characteristic odor.

The different odors of enantiomers provide a biological illustration of the principle of enantiomeric differentiation. Scent receptors are proteins, and they are enantiomerically pure, chiral molecules. (Humans have 347 scent receptor proteins.) Each scent receptor can therefore serve as a "chiral agent." Whether a carvone molecule interacts with one or (as is likely) several different scent receptors, the interactions of the two carvone enantiomers with any given scent receptor are diastereomeric. These diastereomeric interactions result in different neurological signals that the brain recognizes as different odors.

6.11 THE POSTULATION OF TETRAHEDRAL CARBON

Chemists recognized the tetrahedral configuration of a carbon with four bonds almost one-half century before physical methods confirmed the idea with direct evidence. This section shows how the phenomena of optical activity and chirality played key roles in this development, which was one of the most important chapters in the history of organic chemistry.

The first chemical substance in which optical activity was observed was quartz. When a quartz crystal is cut in a certain way and exposed to polarized light along a particular axis, the plane of polarization of the light is rotated. In 1815, the French chemist Jean-Baptiste Biot (1774–1862) showed that quartz exists as both levorotatory and dextrorotatory crystals. The Abbé René Just Haüy (1743–1822), a French crystallographer, had earlier shown that there are two kinds of quartz crystals, which are noncongruent mirror images. Sir John F. W. Herschel (1792–1871), a British musician and astronomer, found a correlation between these crystal forms and their optical activities: one of these forms of quartz is dextrorotatory and the other levorotatory. These were the key discoveries that clearly associated the chirality of a substance with the phenomenon of optical activity.

During the period 1815–1838, Biot examined several organic substances, both pure and in solution, for optical activity. He found that some (for example, oil of turpentine) show optical activity, and others do not. He recognized that, because optical activity can be displayed by compounds in solution, *it must be a property of the molecules themselves.* (The dependence of optical activity on concentration, Eq. 6.13, Sec. 6.3B, is sometimes called *Biot's law.*) What Biot did *not* have a chance to observe is that some organic molecules exist in both dextrorotatory and levorotatory forms. The reason Biot never made this observation is undoubtedly that many optically active compounds are obtained from natural sources as single enantiomers.

The first observation of the two enantiomeric forms of the same organic compound involved tartaric acid:

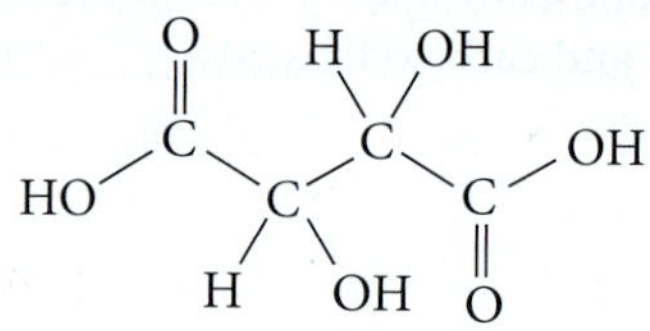

tartaric acid

This substance had been known by the ancient Romans as its monopotassium salt, *tartar*, which deposits from fermenting grape juice. Tartaric acid derived from tartar was one of the compounds examined by Biot for optical activity; he found that it has a positive optical rotation. Biot also studied an isomer of tartaric acid discovered in crude tartar, called *racemic acid* (*racemus*, Latin, "a bunch of grapes"), and he found it to be optically inactive. The exact structural relationship of (+)-tartaric acid and its isomer racemic acid remained obscure.

All of these observations were known to Louis Pasteur (1822–1895), a French chemist and biologist. One day in 1848 the young Pasteur was viewing crystals of the sodium ammonium double salts of (+)-tartaric acid and racemic acid under the microscope. Pasteur noted that the crystals of the salt derived from (+)-tartaric acid were "hemihedral" (chiral). He noted, too, that the racemic acid salt was not a single type of crystal, but was actually a mixture of hemihedral crystals: some crystals were "right-handed," like those in the corresponding salt of (+)-tartaric acid, and some were "left-handed" (**Fig. 6.12a**; thus the name "racemic mixture"). Pasteur meticulously separated the two types of crystals with a pair of tweezers and found that the right-handed crystals were identical in every way to the crystals of the salt of (+)-tartaric acid. When equally concentrated solutions of the two types of crystals were prepared, he found that the optical rotations of the left- and right-handed crystals were equal in magnitude, but opposite in sign. Pasteur had thus performed the first enantiomeric resolution by human hands. Racemic acid, then, was the first organic compound shown to exist as a mixture of enantiomers—that is, noncongruent mirror images. One of these mirror-image molecules was identical to (+)-tartaric acid, but the other was previously unknown. Pasteur's own words tell us what then took place:

> The announcement of the above facts naturally placed me in communication with Biot, who had doubts concerning their accuracy. Being charged with giving an account of them to the Academy,

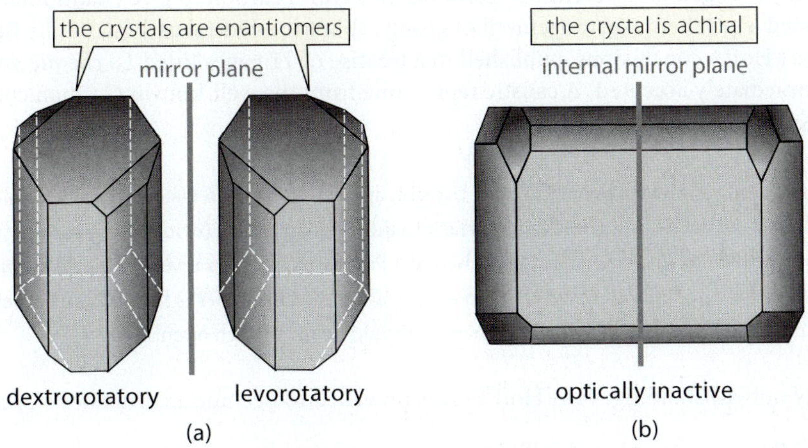

FIGURE 6.12 Diagrams of the crystals of the tartaric acid isomers that figured prominently in the history of stereochemistry. (a) The chiral crystals of sodium ammonium tartrate separated by Pasteur. (b) The achiral crystal of sodium ammonium racemate that crystallizes at a higher temperature.

> he made me come to him and repeat before his very eyes the decisive experiment. He handed over to me some racemic acid that he himself had studied with particular care, and that he found to be perfectly indifferent to polarized light. I prepared the double salt in his presence with soda and ammonia that he also desired to provide. The liquid was set aside for slow evaporation in one of his rooms. When it had furnished about thirty to forty grams of crystals, he asked me to call at the Collège de France in order to collect them and isolate, before his very eyes, by recognition of their crystallographic character, the right and left crystals, requesting me to state once more whether I really affirmed that the crystals that I should place at his right would really deviate [the plane of polarized light] to the right and the others to the left. This done, he told me that he would undertake the rest. He prepared the solutions with carefully measured quantities, and when ready to examine them in the polarizing apparatus, he once more invited me to come into his room. He first placed in the apparatus the more interesting solution, that which should deviate to the left [previously unknown]. Without even making a measurement, he saw by the tints of the images . . . in the analyzer that there was a strong deviation to the left. Then, very visibly affected, the illustrious old man took me by the arm and said, "My dear child, I have loved science so much all my life that this makes my heart throb!"

Pasteur's discovery of the two types of crystals of racemic acid was serendipitous (accidental). It is now known that the sodium ammonium salt of racemic acid forms separate right- and left-handed crystals only at temperatures below 26 °C. Had Pasteur's and Biot's laboratories been warmer, Pasteur would not have made the discovery. Above this temperature, this salt forms only one type of crystal: a "holohedral" (achiral) crystal of the racemate (**Fig. 6.12b**)! From his discovery, and from the work of Biot, which showed that optical activity is a *molecular* property, Pasteur recognized that some molecules could, like the quartz crystals, have an enantiomeric relationship, but he was never able to deduce a structural basis for this relationship.

Focused Problems

6.26 As described in the previous account, Pasteur discovered two stereoisomers of tartaric acid. Draw their structures [you cannot tell which is (+) and which is (−)]. Which stereoisomer of tartaric acid was yet to be discovered? (It was discovered in 1906.) What can you say about its optical activity?

6.27 Think of Pasteur's enantiomeric resolution of racemic acid in terms of the "resolving agent" idea discussed in Sec. 6.10. Pasteur's resolution involved a resolving agent. What was it?

In 1874, Jacobus Hendricus van't Hoff (1852–1911), a professor at the Veterinary College at Utrecht, The Netherlands, and Achille Le Bel (1847–1930), a French chemist, independently arrived at the idea that if a molecule contains a carbon atom bearing four different groups, these groups can be arranged in different ways around the central carbon to give enantiomers. Van't Hoff suggested a tetrahedral arrangement of groups about the central carbon, but Le Bel was less specific. Van't Hoff's conclusions, published in a treatise of 11 pages titled *La chemie dans l'espace*, were not immediately accepted. A caustic reply came from the well-known German chemist Hermann Kolbe:

> A Dr. van't Hoff of the Veterinary College, Utrecht, appears to have no taste for exact chemical research. Instead, he finds it a less arduous task to mount his Pegasus (evidently borrowed from the stables of the College) and soar to his chemical Parnassus, there to reveal in his *La chemie dans l'espace* how he finds atoms situated in universal space. This paper is fanciful nonsense! What times are these, that an unknown chemist should be given such attention!

Kolbe's reply notwithstanding, van't Hoff's ideas prevailed to become a cornerstone of organic chemistry.

How can the existence of enantiomers be used to deduce a tetrahedral arrangement of groups around carbon? Remember that these early chemists had none of our theories of chemical bonding. All they knew was that some molecules can exist in two noncongruent mirror-image forms. Let's examine some other possible carbon geometries to see the sort of reasoning that was used by van't Hoff and Le Bel. Consider a general molecule in which the carbon and its four groups lie in a single plane:

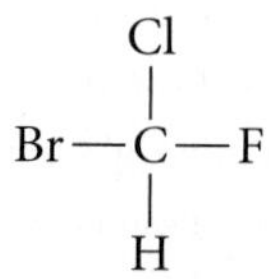

all atoms are in the same plane

Because the mirror image of such a *planar* molecule is congruent (show this!), enantiomeric forms are impossible. The existence of enantiomers thus *rules out* this planar geometry.

However, other conceivable nonplanar structures could exist as enantiomers. One structure has a square-pyramidal geometry:

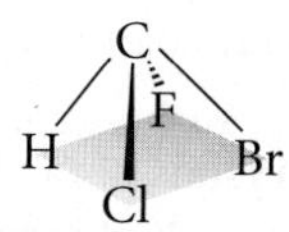

(6.44)

(Convince yourself that such a structure can have an enantiomer.) This geometry could not, however, account for other facts. Consider, for example, the compound dichloromethane (CH_2Cl_2). In the square-pyramidal geometry, two *diastereomers* would be known. In one, the chlorines are on opposite corners of the pyramid; in the other, the chlorines are on adjacent corners. (Why are these diastereomers?)

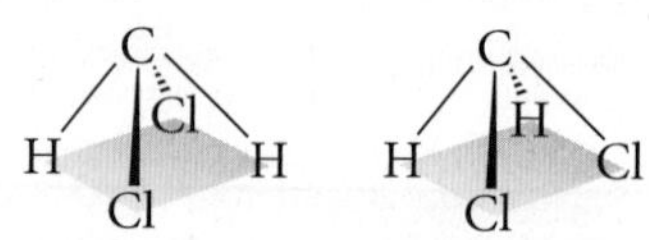

opposite adjacent

square-pyramidal dichloromethane molecules (6.45)

These molecules should be separable because diastereomers have different properties. In the entire history of chemistry, only one isomer of CH_2Cl_2, CH_2Br_2, or any similar molecule, has ever been found. Now, this is *negative* evidence. To take this evidence as conclusive would be like saying to the Wright brothers in 1902, "No one has ever seen an airplane fly; therefore, airplanes can't fly." Yet this evidence is certainly suggestive, and subsequent experiments by others (Problem 6.62) could only be interpreted in terms of tetrahedral carbon. Indeed, modern methods of structure determination have shown repeatedly that van't Hoff's original proposal—tetrahedral geometry—is correct.

CHAPTER SUMMARY *For a summary of the chapter, see Chapter 6 in the* Study Guide and Solutions Manual.

SKILLS OBJECTIVES WITH PROBLEMS

- **Identify stereocenters and asymmetric carbons within molecular structures, and know the difference between the two. 6.1B**

6.28 How many stereoisomers are there of 3,4-dimethyl-2-hexene?

(a) Show all of the carbon stereocenters in the structure of this compound.

(b) Show all of the asymmetric carbons in the structure of this compound.

- **Define the following terms: stereoisomers; enantiomers, chirality, diastereomers, racemate, scalemic mixture, meso compound. 4.1D; 6.1A,B,C; 6.4A,B; 6.6A; 6.7A**

6.29 Match each of the following terms with the statement that best describes it. (One statement has no match.) enantiomers (E), diastereomers (D), chirality (C), stereoisomers (S), racemate (R), scalemic mixture (SM), meso compound (M)

(a) Different compounds with the same molecular formula and the same connectivity

(b) A mixture of enantiomers with zero optical activity

(c) Compounds that are noncongruent mirror images

(d) A mixture of enantiomers with enantiomeric ratio not equal to 1.0

(e) Compounds with the same formula and the same connectivity that are not mirror images

(f) A compound that has at least two asymmetric centers and has no enantiomer

(g) A property possessed by all enantiomers

(h) A property possessed by all diastereomers

(i) A mixture of enantiomers with an enantiomeric excess of zero

(j) A mixture of enantiomers with optical activity not equal to zero

- **Determine whether a given physical property is the same for two stereoisomers. 6.3, 6.6**

6.30 Give the relationship between the given physical properties for each pair of stereoisomers, if a relationship exists.

(a) The melting points of enantiomers

(b) The melting points of diastereomers

(c) The optical activities of enantiomers

(d) The optical activities of two chiral diastereomers

(e) The optical activities of two different meso stereoisomers with the same connectivity.

(f) The melting points of two different meso stereoisomers with the same connectivity.

(g) The heats of formation of enantiomers

- **Apply the *R,S* system to determine the configuration of an asymmetric center. Given the *R* or *S* configurations of asymmetric centers in a compound, draw a correct line-and-wedge structure. 6.2**

6.31 Give the configuration of each asymmetric atom in the following compounds as *R* or *S*.

(a)

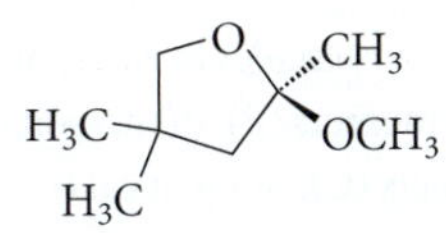

(b)

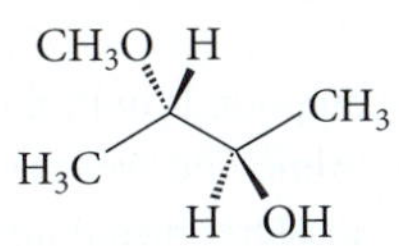

(c)

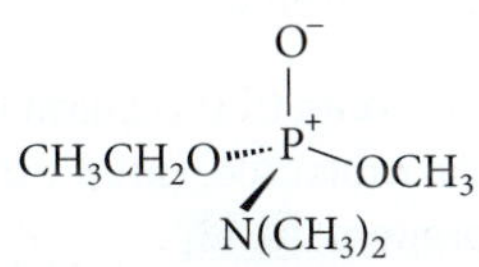

6.32 For each of the following compounds, draw a line-and-wedge structure in which all carbon–carbon bonds are in the plane of the page.

(a) (3*R*,4*R*)-3,4-Hexanediol

(b) *meso*-3,4-Hexanediol

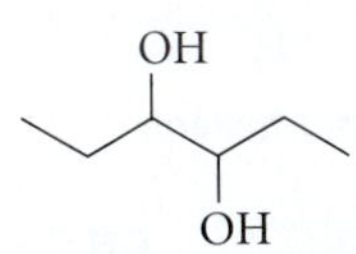

3,4-hexanediol

(c) (2*S*,3*R*,4*S*)-2,3,4-Hexanetriol

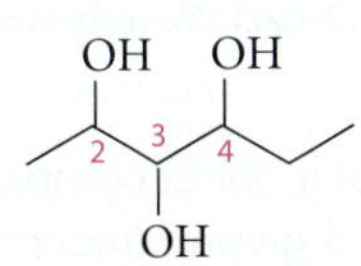

2,3,4-hexanetriol

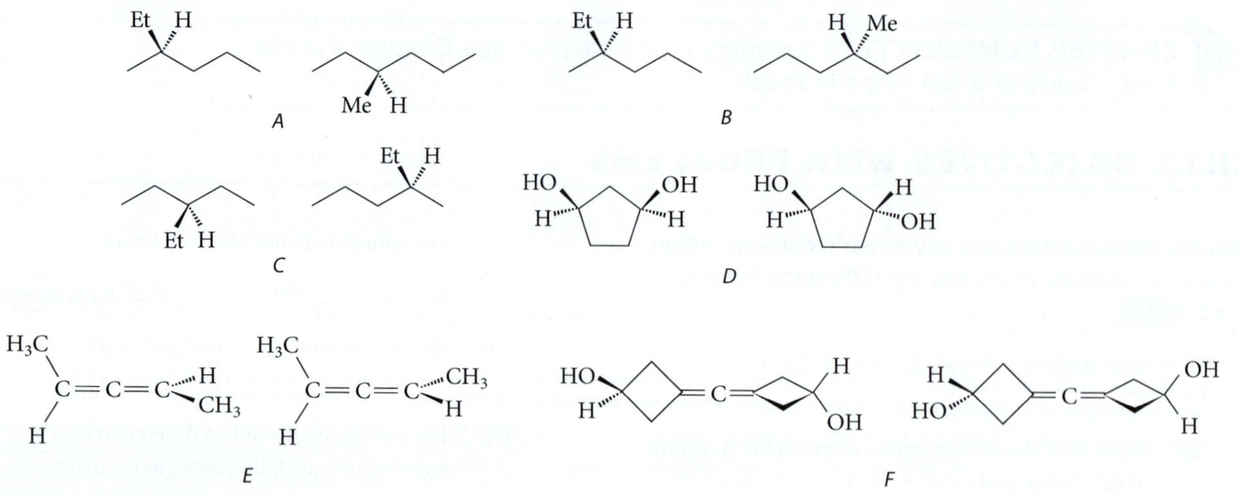

FIGURE P6.33

- Recognize from their structures whether any two isomers are enantiomers, diastereomers, or constitutional isomers. Given one structure, draw the structure of an enantiomer, a diastereomer, or a constitutional isomer, if one exists. **2.4A, 6.1, 6.6, 6.8**

6.33 (a) Indicate whether the molecules in each pair in **Fig. P6.33** are enantiomers (E), diastereomers (D), constitutional isomers (C), or identical (I). Explain your reasoning.

(b) For each of the two compounds in Problem 6.31, parts (a) and (b), draw line-and-wedge structures of an enantiomer and a diastereomer, if one exists. If one does not exist, explain why.

- Name two symmetry elements that a chiral molecule cannot have. For an achiral molecule, point out any of these elements that are present. **6.1C**

6.34 All of the following compounds are achiral. Show all of the symmetry elements in each that are responsible for their lack of chirality.

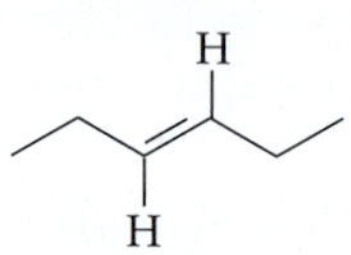

***trans*-3-hexene**

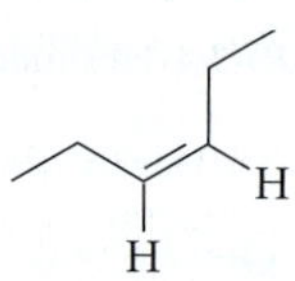

***cis*-3-hexene**

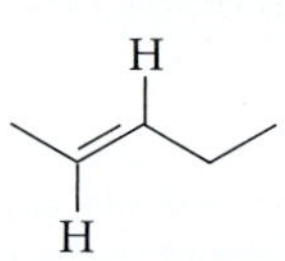

***trans*-2-pentene**

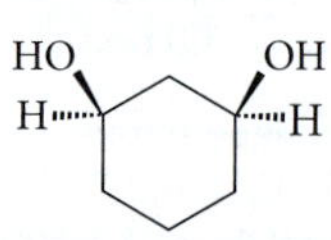

***meso*-1,3-cyclohexanediol**

- Given the specific rotation of a chiral compound, calculate its optical rotation at a given concentration. Given the optical rotation and concentration of a chiral compound, calculate its specific rotation. **6.3B**

6.35 The specific rotation of the *R* enantiomer of the following alkene is $[\alpha]_D^{25} = +79$ degrees mL g^{-1} dm^{-1}, and its molecular mass is 146.2.

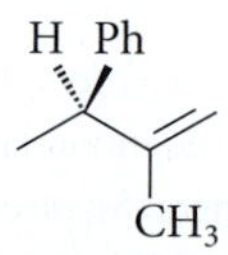

(a) What is the observed rotation of a 0.5 *M* solution of this compound in a 5 cm sample path?

(b) What is the observed rotation of a solution formed by mixing equal volumes of the solution from part (a) and a 0.25 *M* solution of the enantiomer of the same alkene?

6.36 Pure (*R*)-(−)-carvone, the compound responsible for the odor of spearmint, has an observed optical rotation of −58.7° in a 1 dm light path at 25 °C. The density of (*R*)-(−) -carvone is 0.962 g mL^{-1}. What is the specific rotation $[\alpha]_D^{25}$ of (*R*)-(−)-carvone?

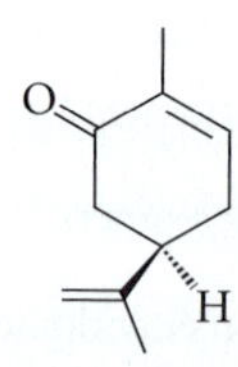

***R*-(−)-carvone**

- Given the specific rotation of a chiral compound and the specific rotation of a mixture of the compound and its enantiomer, calculate the enantiomeric excess (EE). Given the EE in a mixture of enantiomers and the specific rotation of either pure enantiomer, calculate the enantiomeric ratio (ER). Given the ER for a mixture of enantiomers, calculate the EE. **6.4A**

6.37 Enantiomerically pure (*R*)-(+)-2-methyl-1,2-butanediol has a specific rotation $[\alpha]_D^{20} = +9.3$ degrees mL g^{-1} dm^{-1} in chloroform solution.

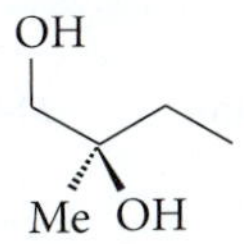

(*R*)-(+)-2-methyl-1,2-butanediol

(a) What is the EE of a mixture of (+)- and (−)-2-methyl-1,2-butanediol that has an apparent specific rotation of −6.3 degrees mL g^{-1} dm^{-1} under the same conditions?

(b) What percentage of each enantiomer is present in the mixture? What is the ER of the mixture?

(c) What is the EE of the major enantiomer in the solution formed in Problem 6.35, part (b)?

- **Given the structure of a compound, determine whether a meso stereoisomer exists. If one exists, draw a line-and-wedge structure. 6.7**

6.38 For each of the following structures, determine whether there is a meso stereoisomer. If there is one, draw a line-and-wedge structure. For any structure that does *not* have a meso stereoisomer, explain why.

(a) 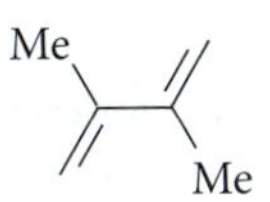

(b) 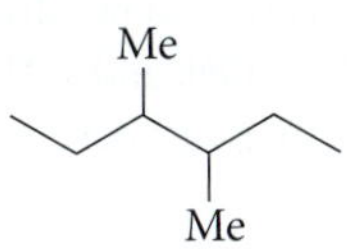

(c) 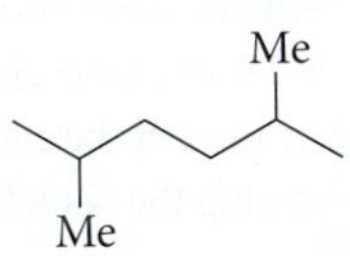

- **Interpret a stereochemical correlation experiment with correct absolute configurations. 6.5**

6.39 From the outcome of the following transformation, indicate whether the levorotatory enantiomer of the product has the *R* or *S* configuration. Draw a structure of the product that shows its absolute configuration. (*Hint:* The phenyl group has a higher priority than the vinyl group in the *R*,*S* system.)

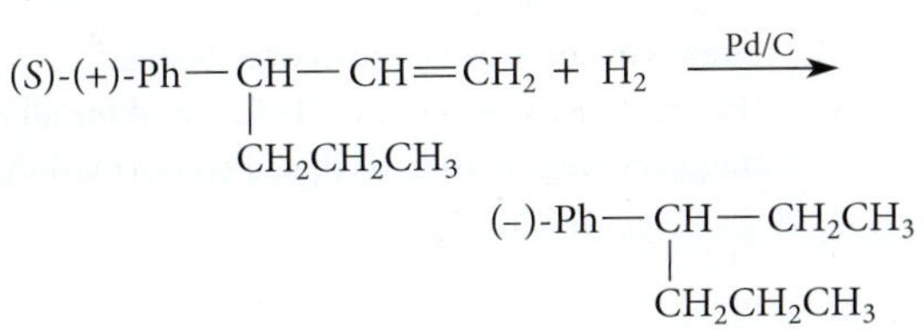

6.40 What is the absolute configuration of (+)-3-methylhexane if catalytic hydrogenation of (*S*)-(+)-3-methyl-1-hexene gives (−)-3-methylhexane?

- **Categorize conformations resulting from rapid rotation about single bonds in terms of their relationship (enantiomers, diastereomers, identical). 6.9A**

6.41 Construct sawhorse and Newman projections of the three staggered conformations of 2-methylbutane (isopentane) that result from rotation about the C2–C3 bond.

(a) Identify the conformations that are chiral.

(b) What is the stereochemical relationship of each conformation to each of the other two?

(c) Explain why 2-methylbutane is not a chiral compound, even though it has chiral conformations.

(d) Suppose each of the three staggered conformations of 2-methylbutane could be isolated and their heats of formation determined. Rank these conformations in order of increasing heat of formation (that is, smallest first). Explain your choice. Indicate whether the ΔH_f° values for any of the isomers are equal and why.

- **Draw structures resulting from inversion about a central atom, and give their relationship (enantiomers, diastereomers, identical). 6.9B**

6.42 (a) In each case, draw a line-and-wedge structure for the product of inversion at nitrogen or phosphorus.

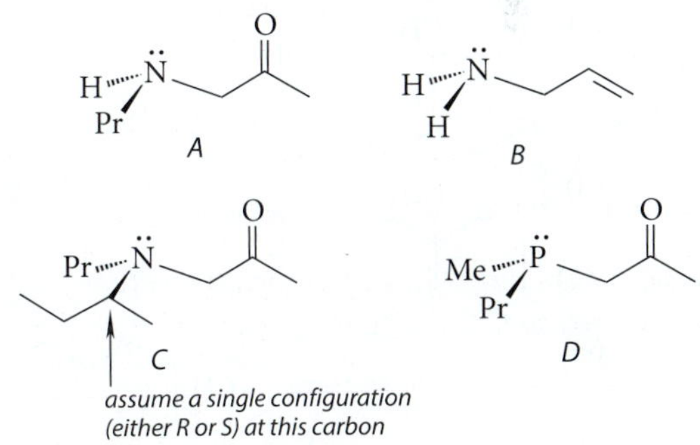

(b) In each case, give the stereochemical relationship between the two structures related by inversion at nitrogen or phosphorus.

(c) Indicate whether each compound could be optically active at room temperature.

• For a given resolution of enantiomers, show how the principle of enantiomeric differentiation is applied. 6.10

6.43 You have been given the task of resolving racemic ibuprofen into its enantiomers.

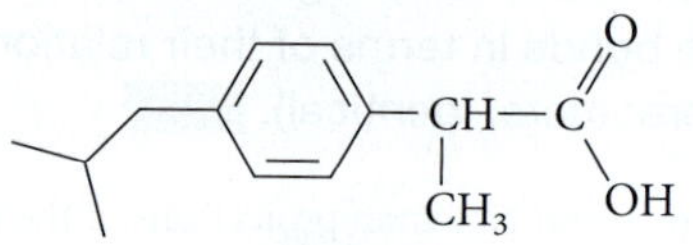

ibuprofen
(racemic)

(a) Because ibuprofen is a carboxylic acid, you have decided that it could probably be resolved by the diastereomeric salt method using a readily available amine. Which of the following amines could in principle be used to bring about this resolution? Explain.

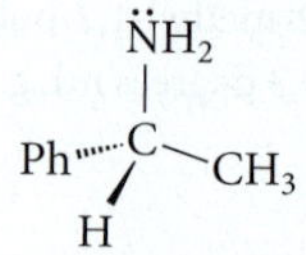

(*R*)-(+)-α-phenethylamine

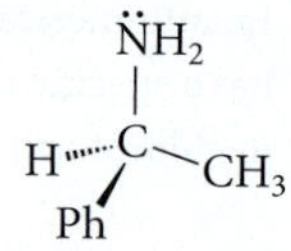

(*S*)-(−)-α-phenethylamine

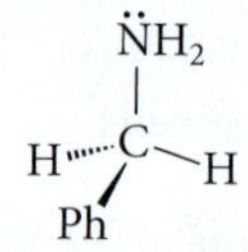

benzylamine

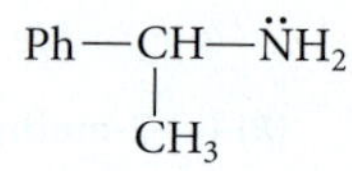

(±)-α-phenethylamine

(b) Choosing any one of the correct answers to part (a), give the structures of the two salts formed when racemic ibuprofen and the amine are mixed. Give the stereochemical relationship of these two salts. Why is it that one will crystallize preferentially from a solvent such as ethanol?

INTEGRATED PROBLEMS

6.44 Identify all of the asymmetric carbon atoms (if any) in each of the following structures.

(a)

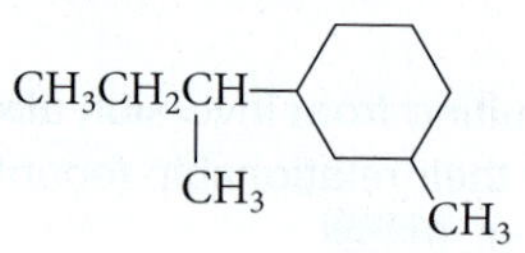

(b)

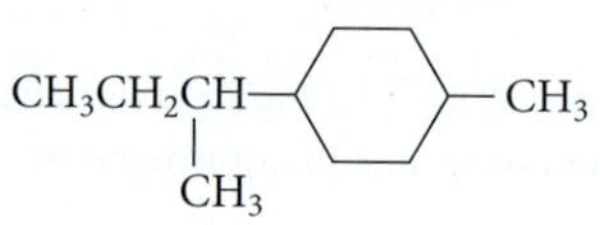

(c)

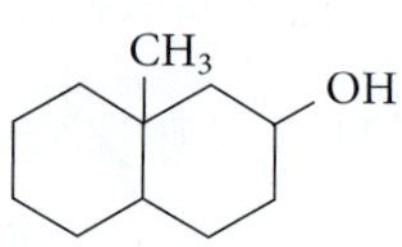

(d)

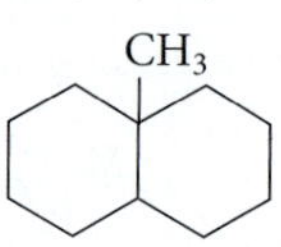

(e)

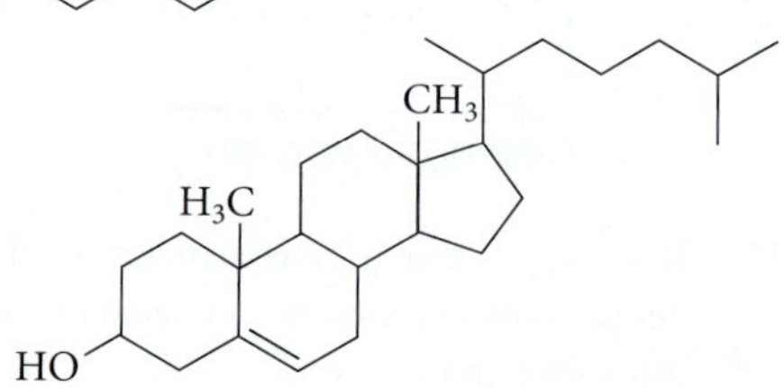

6.45 *Ephedrine* has been known in medicine since the Chinese isolated it from natural sources in about 2800 BCE. Its structure has been known since 1885. Ephedrine can be used as a bronchodilator (a compound that enlarges the air passages in the lungs), but it tends to increase blood pressure because it constricts blood vessels. *Pseudoephedrine* has the same effects, except that it is much less active in elevating blood pressure. Ephedrine is the (1*R*,2*S*)-stereoisomer of the following structure (Ph = phenyl). Pseudoephedrine is the (1*S*,2*S*)-stereoisomer of the same structure.

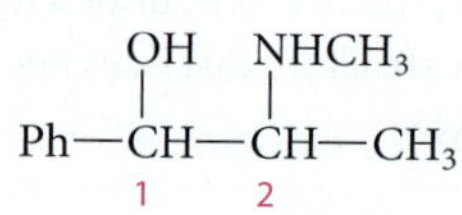

(a) Draw a line-and-wedge structure for each of these two stereoisomers in which the Ph, CH_3, and the two asymmetric carbons lie in the plane of the page. In this part and part (b), do not draw out the bonds within the $—CH_3$, —OH, and $—NHCH_3$ groups explicitly. (More than one correct structure is possible.)

(b) Draw a sawhorse projection and a Newman projection about the C1–C2 bond for each of the structures you drew in part (a). Let the carbon nearest the observer be the one bearing the —OH group.

(c) What is the relationship between these two compounds? Choose from *enantiomers*, *diastereomers*, *identical molecules*, and *constitutional isomers*. Explain how you know.

(d) Should the melting points of these two compounds be the same or different in principle?

(e) What, if anything, can you say about the optical activities of these two compounds?

6.46 (a) Draw sawhorse projections of ephedrine (Problem 6.45) about the C1–C2 bond for all three staggered and all three eclipsed conformations.

(b) Examine each conformation for chirality. How do the chiralities of these conformations relate to the overall chirality of ephedrine?

6.47 Draw the structure of the chiral alkane of lowest molecular mass not containing a ring. (No isotopes are allowed.)

6.48 Draw the structure of the chiral cyclic alkane of lowest molecular mass. (No isotopes are allowed.)

6.49 Indicate whether each of the following statements is true or false. If false, explain why.

(a) In some cases, constitutional isomers are chiral.

(b) In every case, a pair of enantiomers have a mirror-image relationship.

(c) Mirror-image molecules are in all cases enantiomers.

(d) If a compound has an enantiomer, it must be chiral.

(e) Every chiral compound has a diastereomer.

(f) If a compound has a diastereomer, it must be chiral.

(g) Every molecule containing one or more asymmetric carbons is chiral.

(h) Any molecule containing a stereocenter must be chiral.

(i) Any molecule with a stereocenter must have a stereoisomer.

(j) Some diastereomers have a mirror-image relationship.

(k) Some chiral compounds are optically inactive.

(l) Any chiral compound with a single asymmetric carbon must have a positive optical rotation if the compound has the *R* configuration.

(m) A structure is chiral if it has no plane of symmetry.

(n) All chiral molecules have no plane of symmetry.

(o) All asymmetric carbons are stereocenters.

(p) All stereocenters are asymmetric atoms.

6.50 Explain why compound *A* can be resolved into enantiomers but compound *B* cannot.

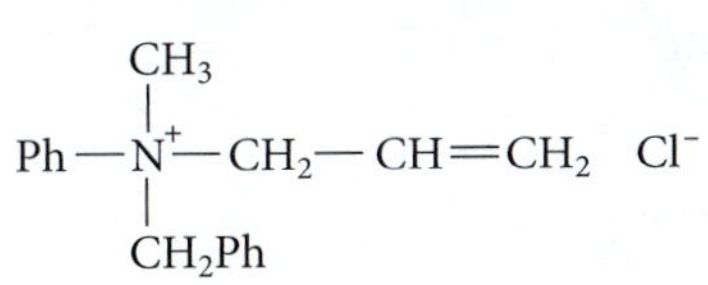

A

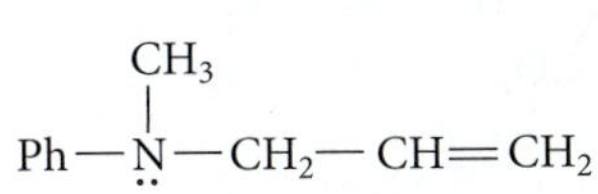

B

6.51 Which of the following compounds can in principle be resolved into enantiomers? Explain why or why not.

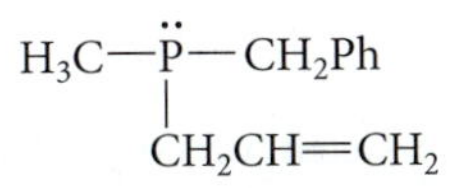

A

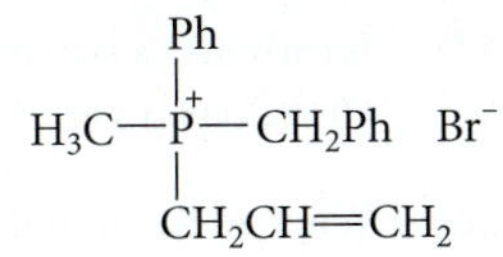

B

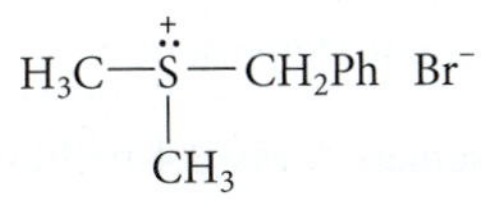

C

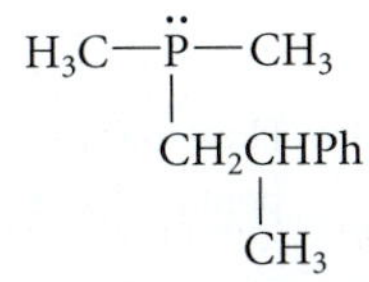

D

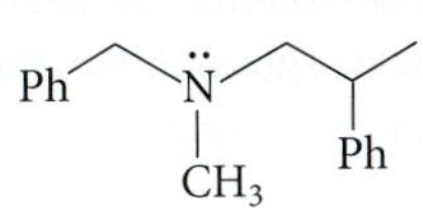

E

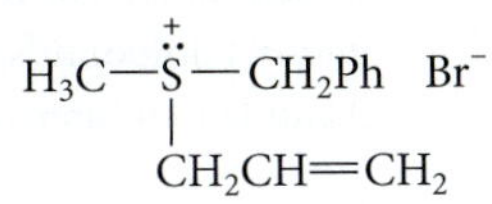

F

6.52 Omeprazole can be separated into two enantiomers that do not interconvert at room temperature.

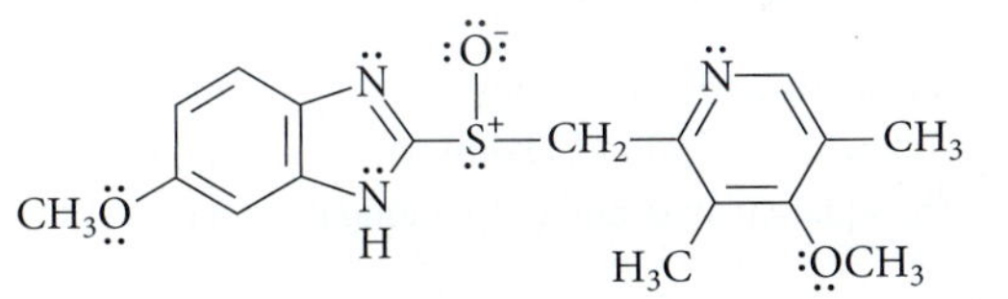

omeprazole

(a) What atom is the asymmetric center in omeprazole?

(b) Esomeprazole, the *S* enantiomer of omeprazole, is a drug used to control acid reflux. Redraw the structure of omeprazole to show it as the *S* enantiomer. (*Hint:* A nonbonding electron pair has lower priority than H.)

6.53 Explain why an optically inactive product is obtained when (−)-3-methyl-1-pentene undergoes catalytic hydrogenation.

6.54 (a) 2,3,4-Trichloropentane has *two* meso stereoisomers.

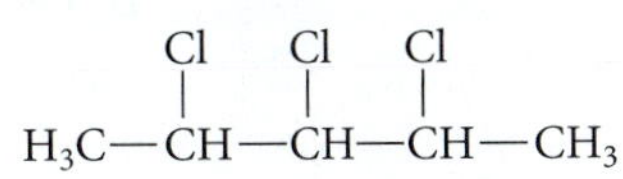

2,3,4-trichloropentane

Starting with the following template for each, complete line-and-wedge structures for the two meso stereoisomers.

(b) Show the symmetry element in each meso stereoisomer that makes the compound achiral. (*Hint:* A symmetry plane can go through atoms as well as bonds.)

(c) What is the relationship (enantiomers or diastereomers) between the two meso compounds?

(d) Is carbon-3 a stereocenter in these meso compounds? How do you know?

(e) What addition to the *R,S* system would you have to make to assign a configuration to carbon-3? Invent a rule, and then assign an *R* or *S* designation to each carbon in your two meso stereoisomers.

(f) How many stereoisomers does 2,3,4-trichloropentane have?

6.55 (a) Give the stereochemical relationship (enantiomers, diastereomers, or the same molecule) between each pair of compounds in the set shown in **Fig. P6.55**. Assume that internal rotation is rapid.

(b) Which, if any, compounds are meso? Explain.

(c) Which compounds should be optically active? Explain.

6.56 Which of the salts in **Fig. P6.56** should have identical solubilities in methanol? Explain.

6.57 Draw structures of the possible stereoisomers for the following compound, assuming, in turn, (a) tetrahedral, (b) square planar, and (c) pyramidal geometries at the carbon atom. For each of these geometries, what is the relationship of each stereoisomer with every other?

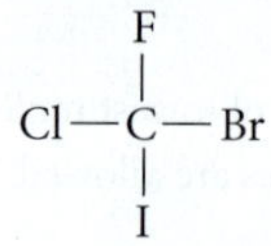

6.58 Two stereoisomers of the compound $(H_3N)_2PtCl_2$ with different physical properties are known. Show that this fact makes it possible to choose between the tetrahedral and square planar arrangements of these four groups around platinum.

6.59 In a structure containing a pentavalent phosphorus atom, the bonds to three of the groups bonded to phosphorus (called *equatorial* groups) lie in a plane containing the phosphorus atom (shaded in the following structure), and the bonds to the other two groups (called *axial* groups) are perpendicular to this plane:

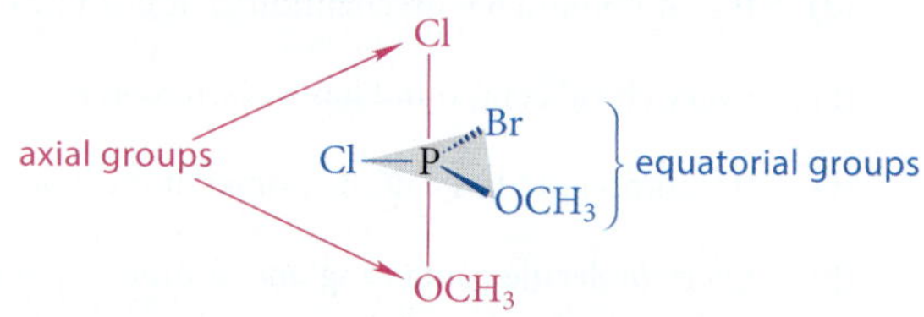

Is this compound chiral? Explain.

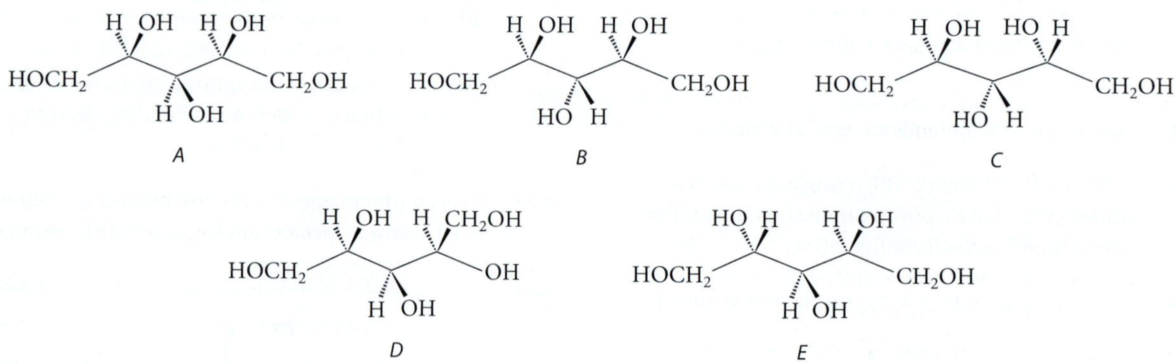

FIGURE P6.55

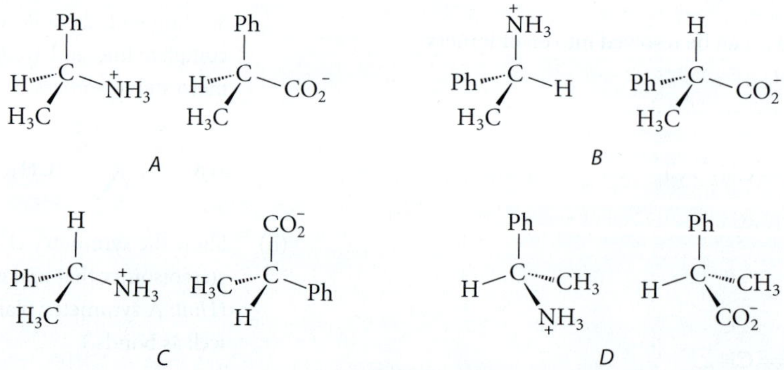

FIGURE P6.56

6.60 (*S*)-(+)-Aspartic acid is one of the naturally occurring α-amino acids.

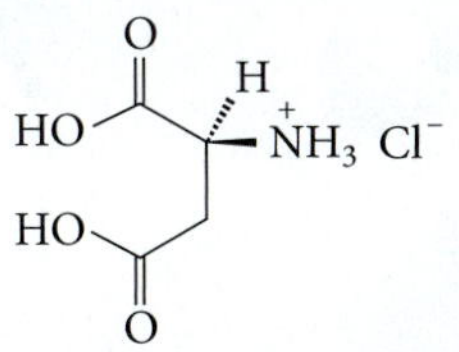

(*S*)-(+)-aspartic acid
(as it exists in HCl solution)

Over time in the environment or in aqueous solution, aspartic acid can undergo racemization very slowly (by a process that we won't consider here). The racemization of aspartic acid occurs at a known rate that can be used to date tissue samples in forensic investigations. The rate of racemization in skull samples follows the equation

$$\ln\left(\frac{1+r}{1-r}\right) = kt, \text{ where } k = 6.24 \times 10^{-4}\ \text{yr}^{-1} \text{ and } t = \text{age of the sample in years}$$

In this equation, r is the enantiomeric ratio (ER) of (*R*)-(−)- and (*S*)-(+)-aspartic acid.

The local coroner's office, knowing your expertise in organic chemistry, comes to you with an old skull sample that was excavated from a building site. You isolate partially racemized aspartic acid from the sample and find (by chiral chromatography) that the sample contains 6% of the (*R*)-(−)-enantiomer and 94% of the (*S*)-(+)-enantiomer.

(a) How old is the skull sample?

(b) What is the enantiomeric excess of the (*S*)-(+)-enantiomer in the sample?

(c) If (*S*)-(+)-aspartic acid has a specific rotation $[\alpha]_D^{20} = +24.5$ degrees mL g^{-1} dm^{-1} in 6 *M* aqueous HCl solution, what is the observed rotation of 100 mg of the forensic (partially racemized) sample in 10 mL of 6 *M* aqueous HCl?

6.61 The two most common forms of glucose, α-D-glucopyranose and β-D-glucopyranose, can be brought into equilibrium by dissolving them in water with a trace of an acid or base catalyst.

$$\alpha\text{-D-glucopyranose} \xrightleftharpoons{H_3O^+ \text{ or } ^-OH} \beta\text{-D-glucopyranose}$$

The specific rotation of the equilibrium mixture is +52.7 degrees mL g^{-1} dm^{-1}. The specific rotation of pure α-D-glucopyranose is +112 degrees mL g^{-1} dm^{-1}, and that of pure β-D-glucopyranose is +18.7 degrees mL g^{-1} dm^{-1}. What is the percentage of each form in the equilibrium mixture?

6.62 (a) In 1914, the chemist Emil Fischer carried out the following conversion in which optically active starting material was transformed into a product with an identical melting point and an optical rotation of equal magnitude and opposite sign. No bonds to the asymmetric carbon were broken in the process.

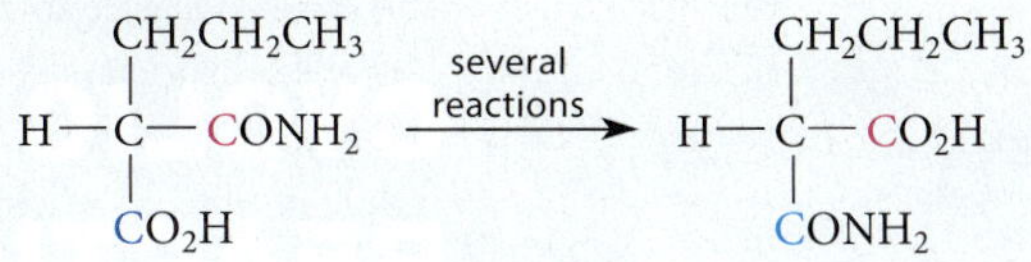

Show that this result is consistent with *either* tetrahedral or pyramidal geometry at the asymmetric carbon.

(b) Fischer also carried out the following pair of conversions. Again, no bonds to the asymmetric carbon were broken. Explain why this *pair* of conversions (but not either one alone) and the associated optical activities rule out pyramidal geometry at the asymmetric carbon, but are consistent with tetrahedral geometry.

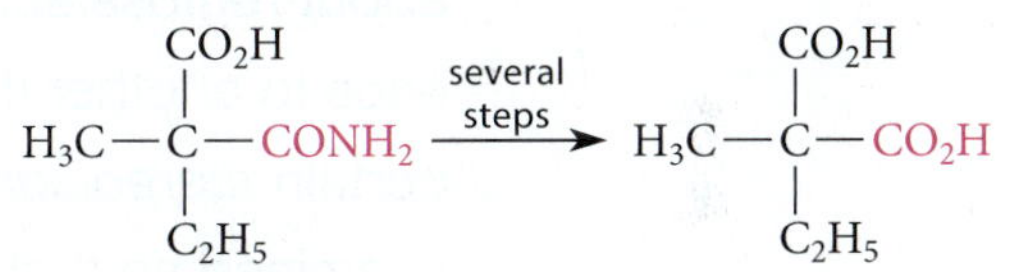

optically active — optically inactive

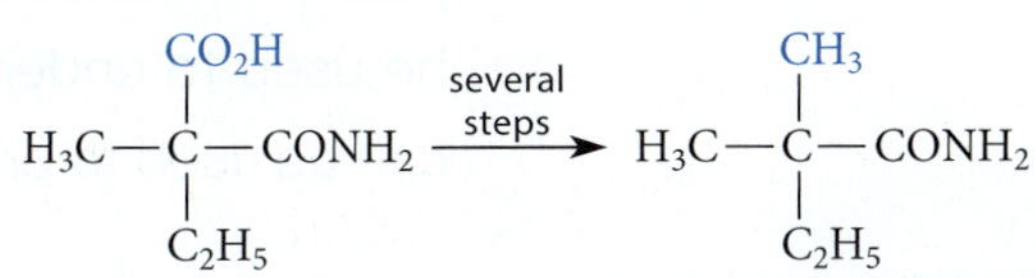

optically active — optically inactive

7 CYCLIC COMPOUNDS AND REACTION STEREOCHEMISTRY

Compounds with cyclic structures present some unique aspects of stereochemistry and conformation. This chapter deals with the stereochemical and conformational issues in cyclic compounds and their derivatives followed by a discussion of how stereochemistry enters into chemical reactions. We've already learned about *regioselective reactions*, which yield one *constitutional isomer* in preference to another (for example, HBr addition to alkenes). Many reactions also yield certain *stereoisomers* to the exclusion of others. Several such reactions will be examined so that we can understand some of the principles that govern the formation of stereoisomers. We also show how the stereochemistry of a reaction can be used to understand its mechanism, and how a known reaction mechanism can be used to predict the stereochemistry of the product.

7.1 RELATIVE STABILITIES OF THE MONOCYCLIC ALKANES

A compound that contains a single ring is called a **monocyclic compound**. The first few monocyclic alkanes are shown here.

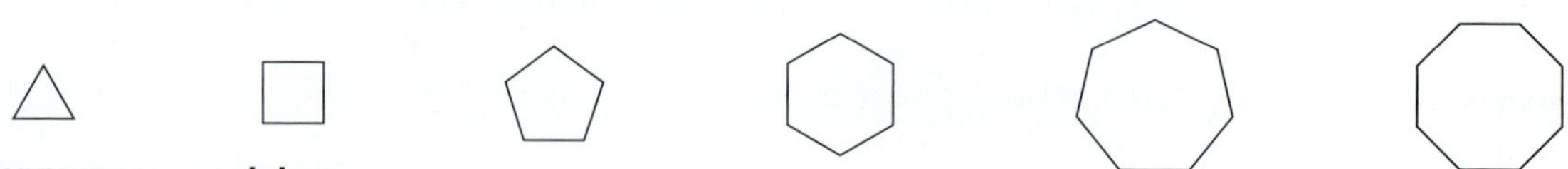

cyclopropane **cyclobutane** **cyclopentane** **cyclohexane** **cycloheptane** **cyclooctane**

(7.1)

For convenience, we represent these compounds in condensed structures with planar polygons. Except for cyclopropane, however, the carbon skeletons of these rings are not planar. The conformations of these compounds will occupy us for the first few sections of this chapter.

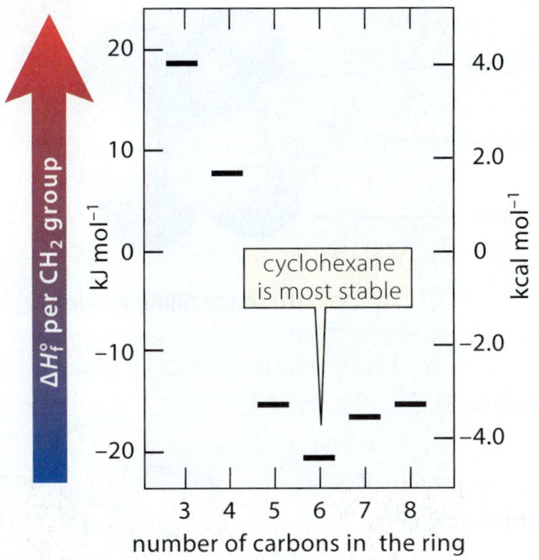

TABLE 7.1 Heats of Formation per CH_2 for Some Cycloalkanes

(n = number of carbon atoms)

		$\Delta H^\circ_f/n$	
n	Compound	kJ mol^{-1}	kcal mol^{-1}
3	cyclopropane	+17.8	+4.25
4	cyclobutane	+7.1	+1.7
5	cyclopentane	−15.4	−3.7
6	cyclohexane	−20.7	−4.95
7	cycloheptane	−16.9	−4.0
8	cyclooctane	−15.55	−3.7
9	cyclononane	−14.8	−3.5
10	cyclodecane	−15.4	−3.7
11	cycloundecane	−16.3	−3.9
12	cyclododecane	−19.2	−4.6
13	cyclotridecane	−18.95	−4.5
14	cyclotetradecane	−17.1	−4.1

The relative stabilities of the monocyclic alkanes give us some important clues about their conformations. These relative stabilities can be determined from their heats of formation, given in **Table 7.1** and shown graphically in the figure within the table. Although the monocyclic alkanes are not isomers, they have the same **empirical formula**, CH_2. This is the formula that gives the smallest whole-number proportions of the elements. When compounds have the same empirical formula, their heats of formation, and therefore their stabilities, can be compared on a *per carbon basis* by dividing the heat of formation of each compound by its number of carbons. The data in Table 7.1 show that, of the cycloalkanes with 14 or fewer carbons, cyclohexane has the lowest (that is, the most negative) heat of formation per CH_2. Therefore, *cyclohexane is the most stable of these cycloalkanes.*

Further insight into the stability of cyclohexane comes from a comparison of its stability with that of a typical *noncyclic* alkane. The heats of formation of pentane, hexane, and heptane are –146.5, –167.1, and –187.5 kJ mol^{-1} (–35.0, –39.9, and –44.8 kcal mol^{-1}), respectively. These data show that heats of formation, like other physical properties, change regularly within a homologous series; each CH_2 group contributes –20.7 kJ mol^{-1} (–4.95 kcal mol^{-1}) to the heat of formation. The data for cyclohexane in Table 7.1 show that a CH_2 group in cyclohexane makes exactly the same contribution to its heat of formation (–20.7 kJ mol^{-1} or –4.95 kcal mol^{-1}). This means that *cyclohexane has the same stability as a typical unbranched alkane.*

Cyclohexane is the most widely occurring ring in compounds of natural origin. Its prevalence, undoubtedly a consequence of its stability, makes it the most important of the cycloalkanes. Two questions emerge as we consider these data: (1) Why is cyclohexane so stable? (2) Why are the smaller rings so unstable? We consider the first question in Sec. 7.2 and the second in Sec. 7.5.

7.2 CONFORMATIONS OF CYCLOHEXANE

A. The Chair Conformation

Why is cyclohexane so stable? The stability data in Table 7.1 require that the bond angles in cyclohexane must be essentially the same as the bond angles in an alkane—very close to the ideal 109.5° tetrahedral angle. If the bond angles were significantly distorted from tetrahedral, we would expect to see a greater heat of formation. The carbons of cyclohexane, then, are sp^3-hybridized. Furthermore, cyclohexane must have a staggered conformation about each

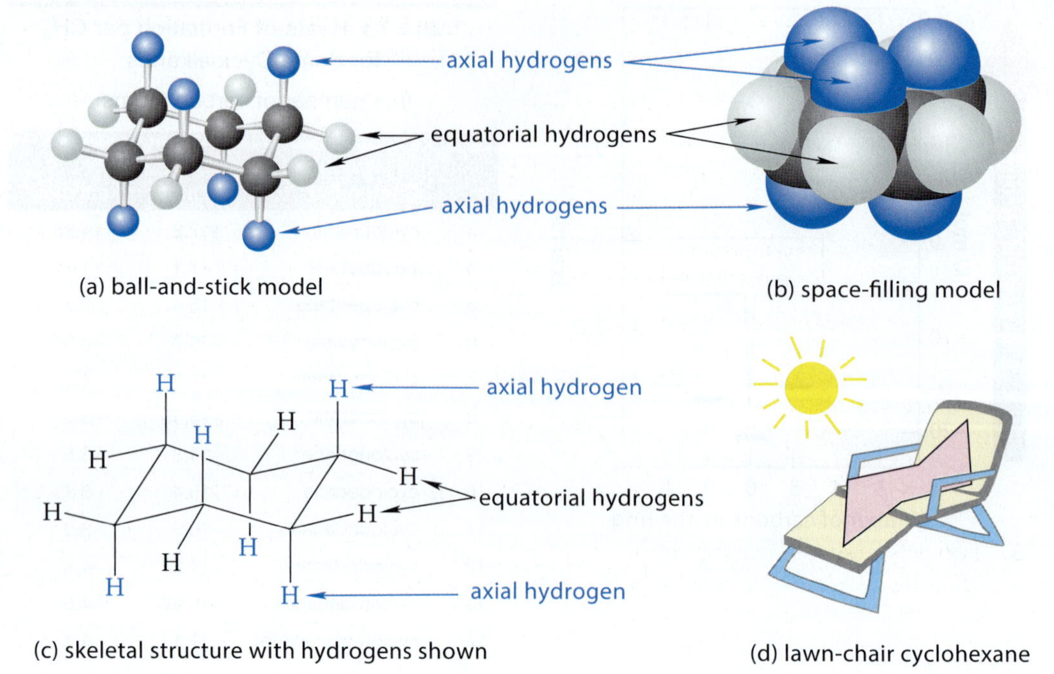

FIGURE 7.1 The chair conformation of cyclohexane. (a) A ball-and-stick model. (b) A space-filling model. (c) A skeletal structure. (d) Origin of the name "chair." The different types of hydrogens are color-coded. The axial hydrogens are blue in parts (a)–(c) of the figure. The equatorial hydrogens are gray in parts (a) and (b) and black in part (c).

carbon–carbon bond because, otherwise, eclipsing interactions (*torsional strain*; Sec. 2.5A) would also increase the heat of formation. These two geometrical constraints can only be met if the carbon skeleton of cyclohexane assumes a nonplanar, "puckered" conformation. This conformation, shown in **Fig. 7.1**, is called the **chair conformation** because of its resemblance to a lawn chair. If you have not already done so, construct a model of chair cyclohexane *now* and use it to follow the subsequent discussion. If your model kit allows, leave off the hydrogens; we'll deal first only with the carbon skeleton. If the carbons of your model set include the hydrogens "built in," then simply notice the positions of the carbons in the subsequent discussion; we'll deal with the hydrogens later.

Notice the following four points about the cyclohexane molecule and how to draw it:

1. To draw the cyclohexane ring, we use a "tilt-and-turn" technique similar to the one used for drawing line-and-wedge structures (Display 2.16, Sec. 2.5C).

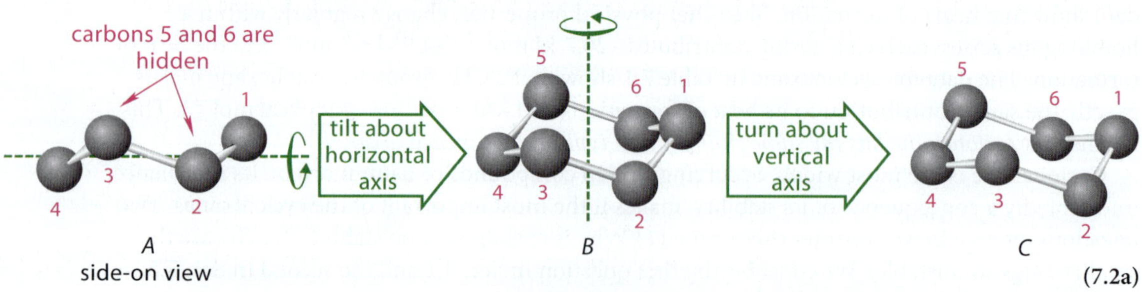

(7.2a)

First, we view the model side-on (view *A* in Display 7.2a). In this view, carbons 5 and 6 are obscured behind carbons 2 and 3. Then, we tilt the model about a horizontal axis to give view *B*. Finally, we turn the model slightly about a vertical axis to give the view used to draw the skeletal structure, as shown in Display 7.2b.

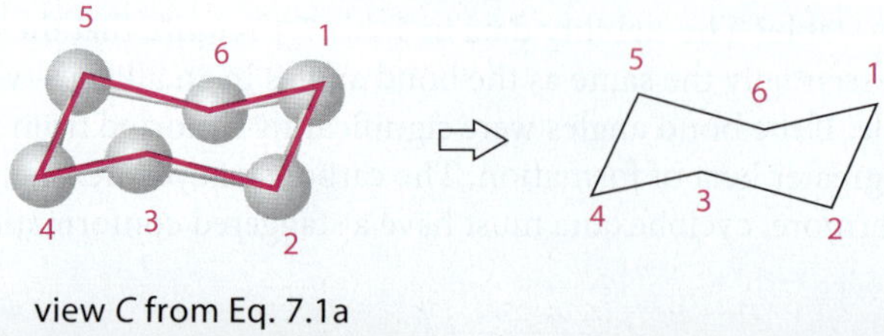

(7.2b)

If we imagine carbons 1 and 4 to be in the plane of the page, then carbons 2 and 3 are in front of the page, and carbons 5 and 6 are behind the page.

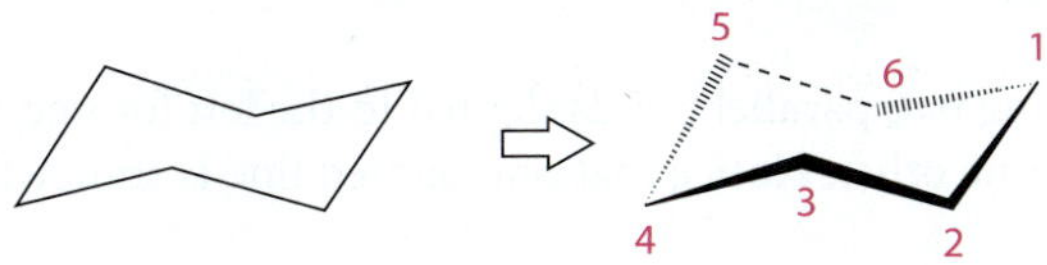

(7.2c)

Remembering that the lower part of the ring is in front of the page is essential to avoiding an optical illusion.

2. Bonds on opposite sides of the ring are parallel:

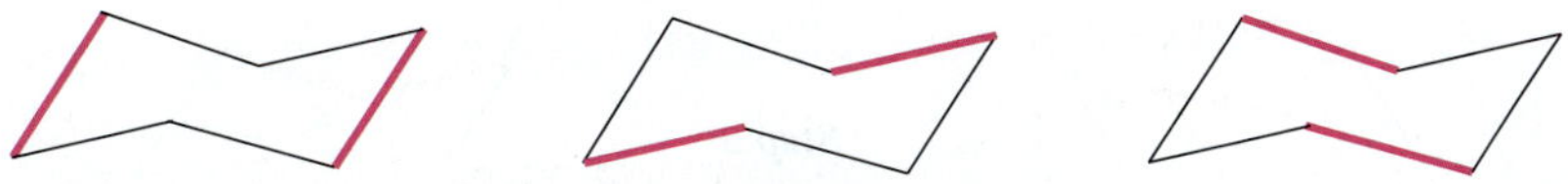

(7.2d)

3. Two perspectives are commonly used for cyclohexane rings. In one, the leftmost carbon is *below* the rightmost carbon; in the other, the leftmost carbon is *above* the rightmost carbon:

(7.2e)

These two perspectives are mirror images. As shown in Displays 7.2a–c, the perspective on the left is based on a view of the model from above and to the left. The perspective on the right is based on a view of the model from above and to the right. The tilt-and-turn procedure for producing this structure is the same except that the rotation is in the opposite direction.

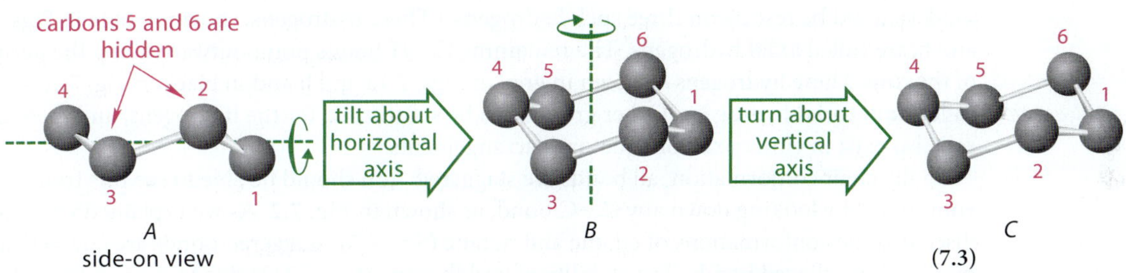

(7.3)

4. A rotation of either perspective by 120° regenerates the same perspective:

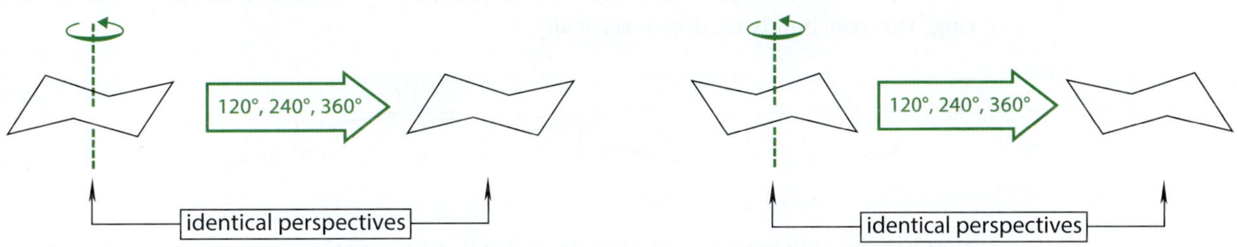

(7.4a)

However, rotation of either perspective by an odd multiple of 60° (that is, 60°, 180°, and so on), followed by the slight shift in viewing direction, gives the other perspective. *Be sure to verify this with models!*

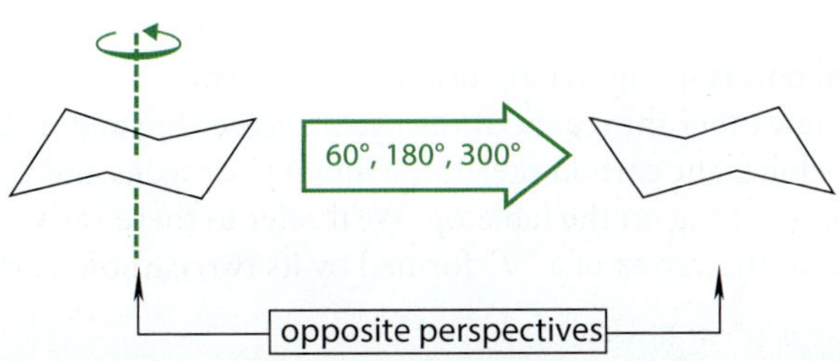

(7.4b)

It is important for you to be able to draw a cyclohexane chair conformation. Once you've examined the preceding points, practice drawing some cyclohexane rings in the two perspectives. Use the following three steps:

Step 1. Begin by drawing two parallel bonds slanted to the left for one perspective and slanted to the right for the other. Notice that one slanted line is somewhat lower than the other in each case.

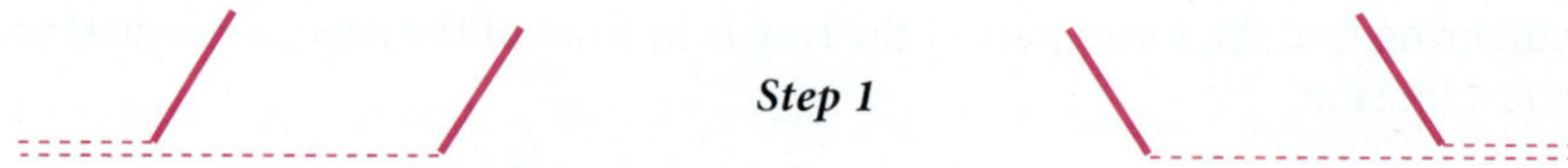

(7.5a)

Step 2. Connect the tops of the slanted bonds with two more bonds in a "V" arrangement.

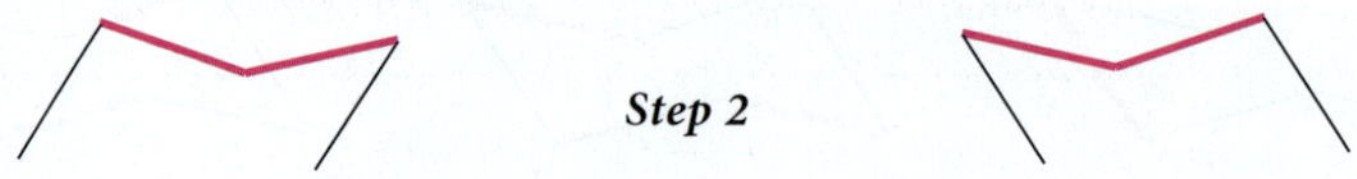

(7.5b)

Step 3. Connect the bottoms of the slanted bonds with the remaining two bonds in an inverted "V" arrangement. (Be sure to keep the appropriate bonds parallel, as shown in Display 7.2d.)

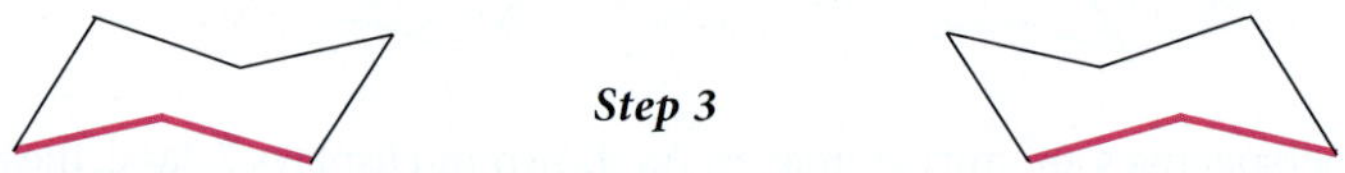

(7.5c)

Now, if you haven't already done so, add the hydrogens to your model. The hydrogens of cyclohexane are of two types. If you place your model of cyclohexane on a tabletop (you did build it, didn't you?), you'll find that six C—H bonds are perpendicular to the plane of the table. (Your model should be resting on three such hydrogens.) These hydrogens, shown in blue in Figs. 7.1a and b, are called **axial** hydrogens. The remaining C—H bonds point outward along the periphery of the ring. These hydrogens—shown in gray in Figs. 7.1a and b and in black in Fig. 7.1c—are called **equatorial** hydrogens. Other groups can be substituted for the hydrogens, and these groups can also exist in either axial or equatorial arrangements.

In the chair conformation, all bonds are staggered. You should be able to see this from your model by looking down any C—C bond, as shown in **Fig. 7.2**. As we explained when we discussed the conformations of ethane and butane (Sec. 2.5), staggered bonds are energetically preferred to eclipsed bonds. The stability of cyclohexane (Sec. 7.1) is due to the fact that all of its bonds can be staggered without compromising the tetrahedral carbon geometry.

Once you have mastered drawing the cyclohexane ring, it's time to add the C—H bonds to the ring. The axial bonds are drawn vertically.

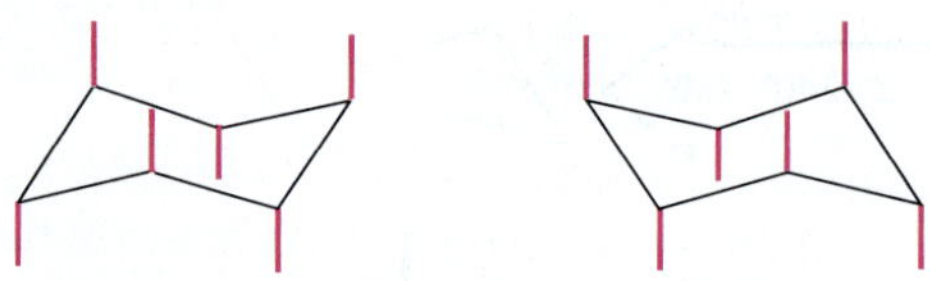

(7.6a)

When drawing the equatorial bonds, notice that pairs of them are parallel to pairs of nonadjacent ring bonds (red):

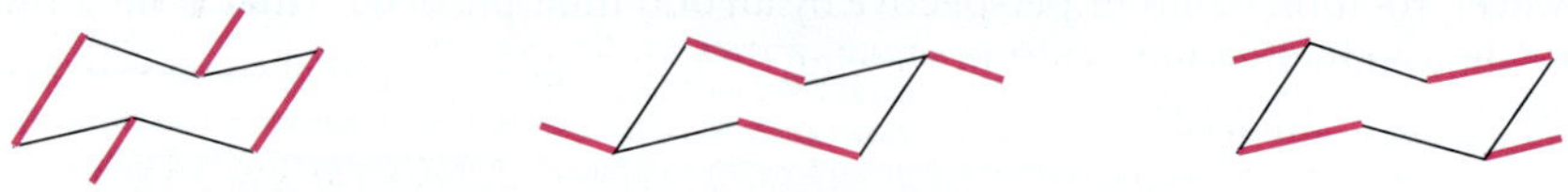

(7.6b)

(All of the equatorial bonds in Fig. 7.1 are oriented this way.)

You should notice a few other things about the cyclohexane ring and its bonds. First, if we make a model of the cyclohexane carbon skeleton *without* hydrogens and place it on a tabletop, then *every other* carbon is resting on the tabletop. We'll refer to these carbons as *down carbons*. Each of these carbons is at the vertex of a "V" formed by its two carbon–carbon bonds. The other

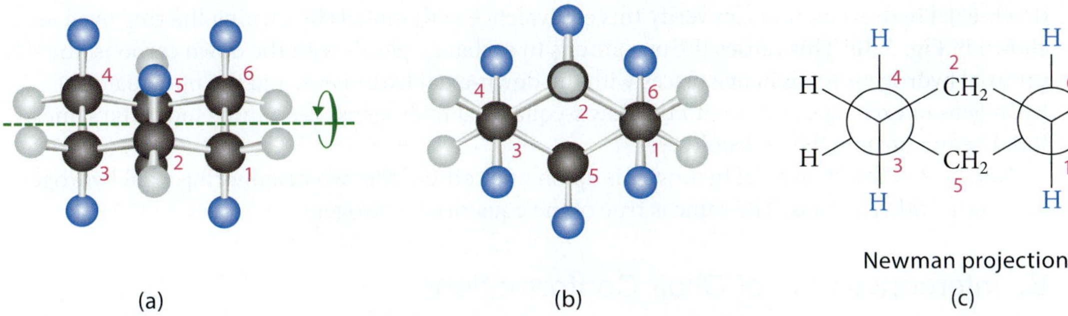

FIGURE 7.2 A demonstration that the bonds in chair cyclohexane are staggered. View a chair cyclohexane model as shown in (a); tilt the model slightly about the horizontal axis shown so that the model is viewed down the carbon–carbon bonds on opposite sides of the ring, as shown in (b). These bonds become the projected bonds in the Newman projection (c). The axial hydrogens are blue; the equatorial hydrogens are gray in (a) and (b), and black in (c).

three carbons lie in a plane *above* the tabletop. We'll refer to these carbons as *up carbons*. Each of these carbons is at the vertex of an inverted "V."

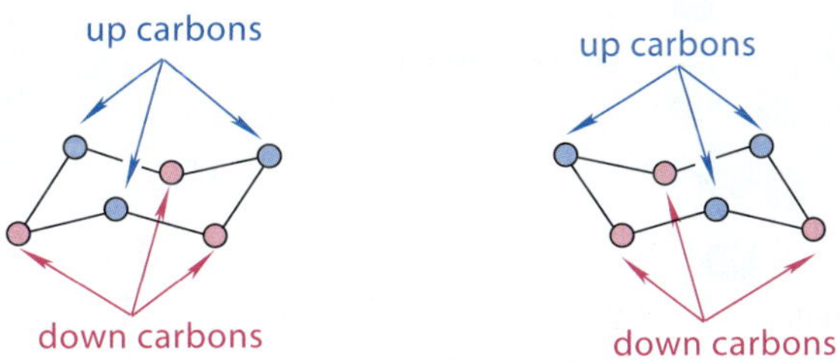

(7.7)

Notice that the three axial hydrogens on up carbons point up, and the three axial hydrogens on down carbons point down. In contrast, the three equatorial hydrogens on up carbons point down, and the three equatorial hydrogens on down carbons point up (**Fig. 7.3a**). *The up and down hydrogens of a given type are completely equivalent.* That is, the up-equatorial hydrogens are equivalent to the down-equatorial hydrogens, and the up-axial hydrogens are equivalent to the

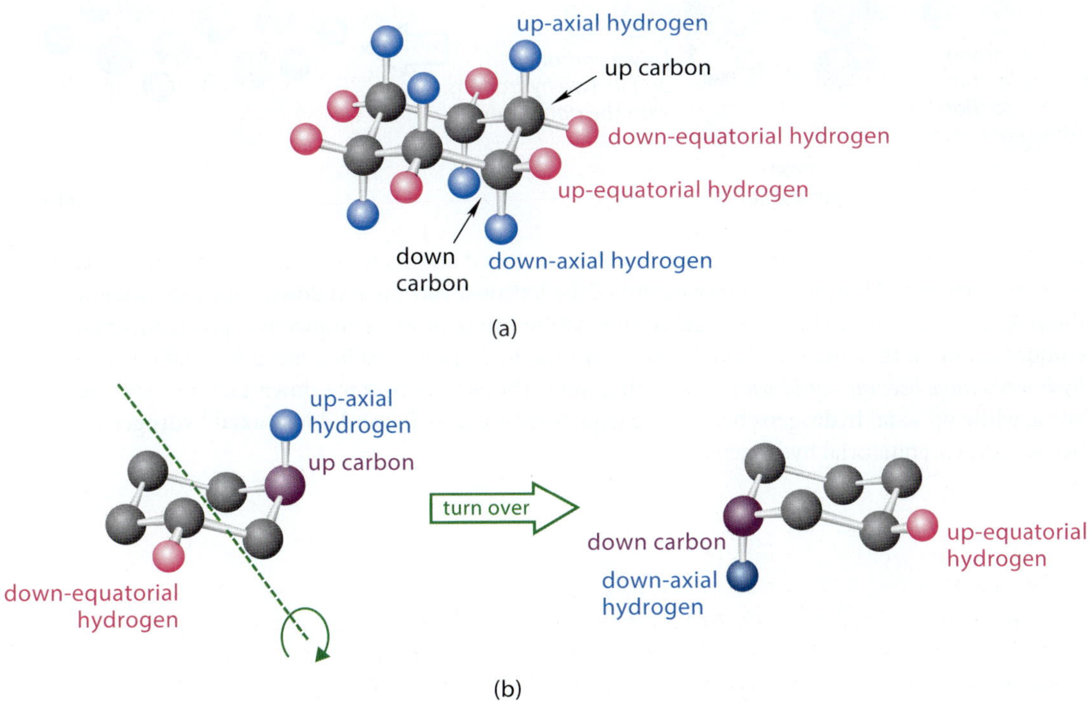

FIGURE 7.3 (a) Up- and down-equatorial and axial hydrogens. The up-axial hydrogens are on up carbons and the down-axial hydrogens are on down carbons. The opposite is true for equatorial hydrogens. (b) The up- and down-axial hydrogens are equivalent, and the up- and down-equatorial hydrogens are equivalent. This equivalence can be demonstrated by turning the ring upside down (*green arrow*). In doing so, the up carbons trade places with the down carbons, the up-axial hydrogens trade places with the down-axial hydrogens, and the up-equatorial hydrogens trade places with the down-equatorial hydrogens. (This part shows explicitly the fate of two hydrogens and the violet color shows the fate of one carbon.)

down-axial hydrogens. You can verify this equivalence with models by turning the ring over, as shown in **Fig. 7.3b**. This causes the up carbons to exchange places with the down carbons, the up-axial hydrogens to exchange places with the down-axial hydrogens, and the up-equatorial hydrogens to exchange places with the down-equatorial hydrogens. Everything looks the same as it did before turning the molecule over.

Notice, too, that if an axial hydrogen is up on one carbon, the two neighboring axial hydrogens are down, and vice versa. The same is true of the equatorial hydrogens.

B. Interconversion of Chair Conformations

Cycloalkanes, like noncyclic alkanes, undergo internal rotations (Sec. 2.5). However, because the carbon atoms are constrained within a ring, several internal rotations must occur at the same time. When a cyclohexane molecule undergoes internal rotations, *a change in the ring conformation occurs.* In this change, one chair conformation is converted into another, completely equivalent, chair conformation. Follow this four-step procedure for demonstrating this conformational change with a model. First, hold carbon-1—the rightmost carbon—so that it cannot move, and raise carbon-4 up as far as it will go. The result is a different conformation, called a **boat conformation**.

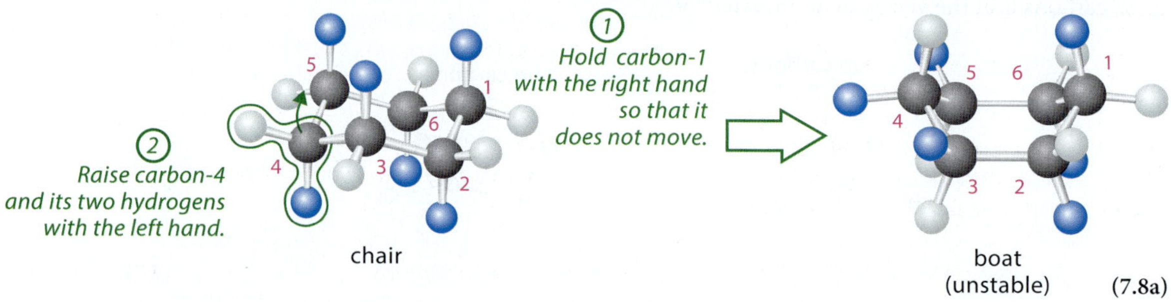

(7.8a)

(We consider the boat conformation in more detail in Sec. 7.2C.) Formation of the boat conformation involves *simultaneous internal rotations* about all carbon–carbon bonds except those to carbon-1. Now hold carbon-4 of the boat—the leftmost carbon—so it cannot move, and lower carbon-1 as far as it will go; the model returns to a chair conformation.

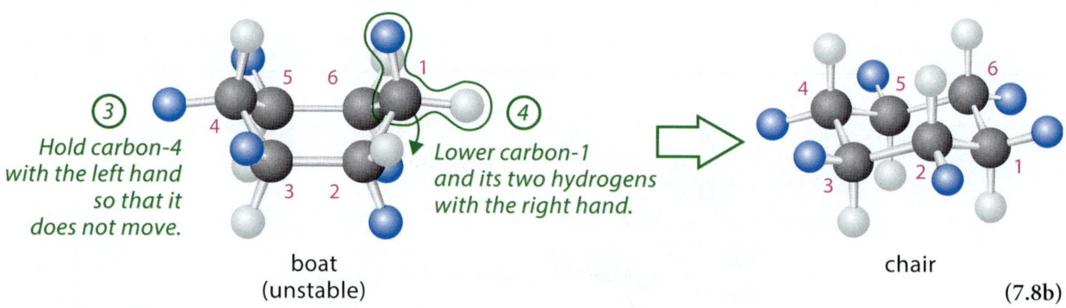

(7.8b)

In this case, simultaneous internal rotations have occurred about all carbon–carbon bonds except those to carbon-4. Thus, upward movement of the leftmost carbon and downward movement of the rightmost carbon change one chair conformation into another, completely equivalent, chair conformation. In this process, though, *the equatorial hydrogens have become axial, and the axial hydrogens have become equatorial.* In addition, up carbons have become down carbons, and vice versa, while up-axial hydrogens become up-equatorial hydrogens, and down-axial hydrogens become down-equatorial hydrogens.

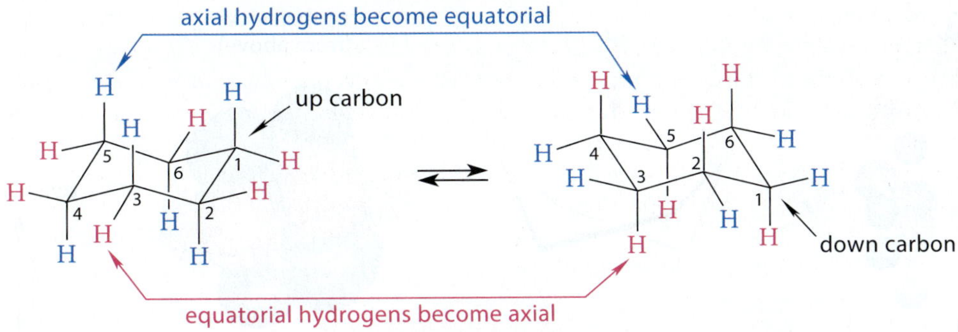

(7.9)

(Confirm these points with your model by using groups of different colors for the axial and equatorial hydrogens.)

The interconversion of two chair forms of cyclohexane is called a **chair interconversion**. It is sometimes also called a *chair flip*, although this terminology is somewhat misleading because it implies that the ring is simply turned over. As Eq. 7.9 shows by labeling the hydrogens, this is a series of coordinated *internal* rotations rather than an overall rotation of the entire molecule. The energy barrier for the chair interconversion is about 45 kJ mol^{-1} (11 kcal mol^{-1}). This barrier is low enough that the chair interconversion is rapid; it occurs about 10^5 times per second at room temperature.

To review: Although the axial hydrogens are stereochemically different from the equatorial hydrogens in any one chair conformation, the chair interconversion causes these hydrogens to change positions rapidly. Therefore, *averaged over time*, the axial and equatorial hydrogens of cyclohexane are equivalent and indistinguishable.

C. Boat and Twist-Boat Conformations

Displays 7.8a and 7.8b show a boat conformation of cyclohexane. Let's examine this conformation in more detail. The boat conformation is not a stable conformation of cyclohexane; it contains two sources of instability, both of which are shown in **Fig. 7.4**. One is that certain hydrogens (colored blue) are eclipsed. The eclipsing of these hydrogens is easier to see in an end-on view, which is essentially equivalent to a Newman projection:

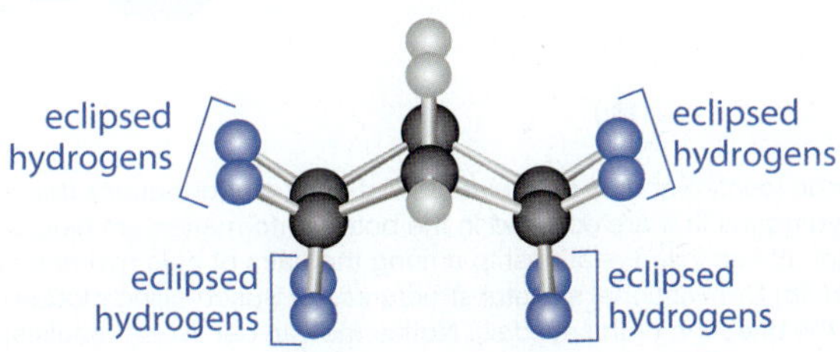

boat cyclohexane
(end-on)

(7.10)

The second source of instability is that the two hydrogens on the "bow" and "stern" of the boat, called *flagpole hydrogens*, experience modest van der Waals repulsion. (The flagpole hydrogens are colored red in Fig. 7.4.) For these reasons, the boat undergoes slight internal rotations that reduce both the eclipsing interactions and the flagpole van der Waals repulsions. The result is another stable conformation of cyclohexane called a **twist-boat conformation**. To see the conversion of a boat into a twist-boat, view a model of the boat conformation from above the flagpole hydrogens, as shown in Fig. 7.4b. Grasping the model by its flagpole hydrogens, nudge one flagpole hydrogen up and the other down to obtain a twist-boat conformation. As shown in Fig. 7.4, this motion can occur in either of two ways, so that two twist-boat conformations are accessible from any one boat conformation.

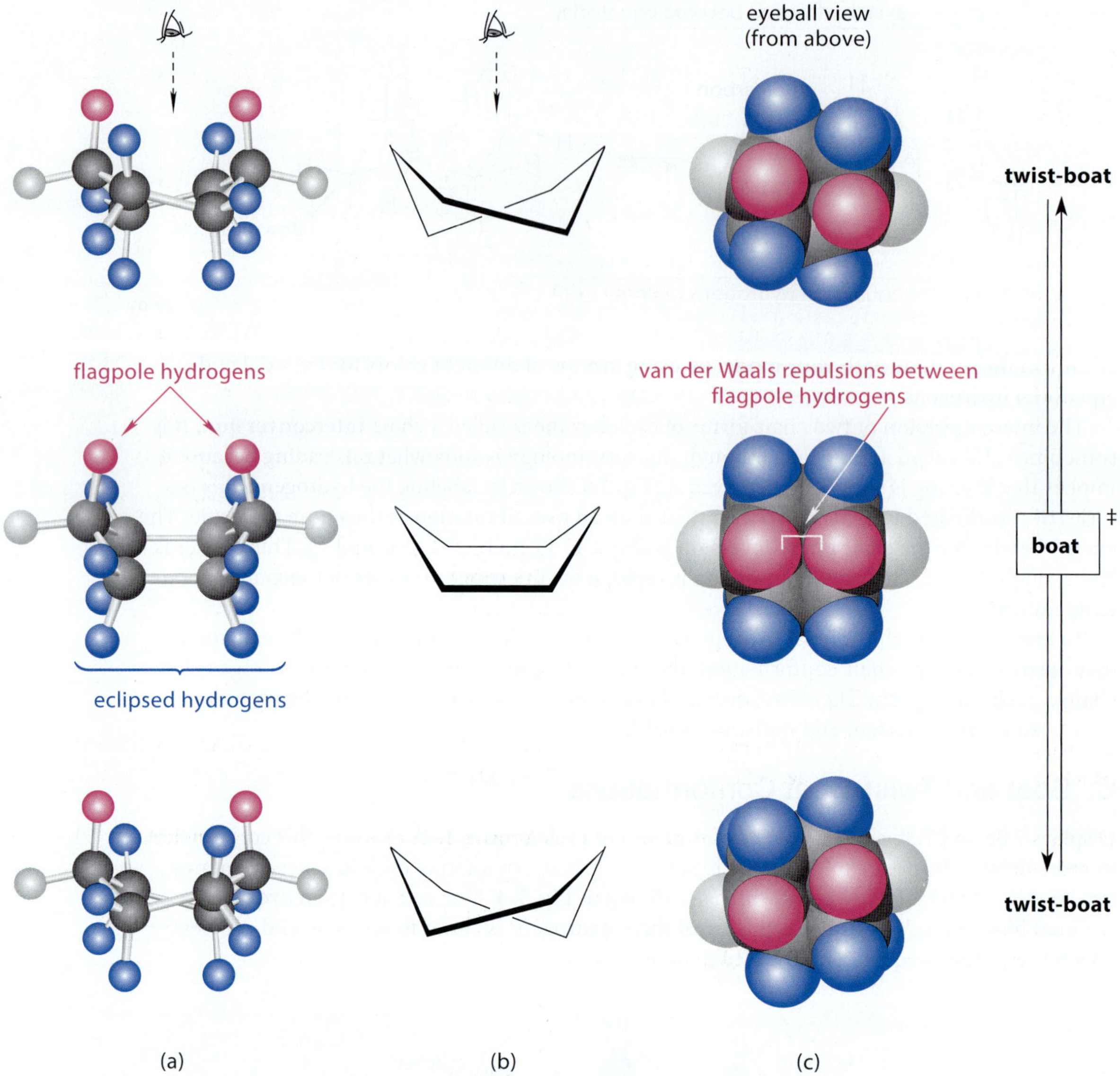

FIGURE 7.4 Boat cyclohexane (*center*) and its two related twist-boat conformations (*top* and *bottom*). The flagpole hydrogens are red, and the hydrogens that are eclipsed in the boat conformation are blue. (a) Ball-and-stick models. Notice in the boat conformation the eclipsed relationship among the pairs of blue hydrogens. This eclipsing is reduced in the twist-boat conformation. (b) Conventional skeletal structures. (c) Space-filling models viewed from above the flagpole hydrogens (from the direction of the eyeball). Notice the van der Waals repulsion between the flagpole hydrogens in the boat conformation. This unfavorable interaction is reduced in the twist-boat conformations because the flagpole hydrogens (*red*) are farther apart.

The relative enthalpies of the conformations of cyclohexane are shown in **Fig. 7.5**. Notice that the twist-boat conformation is an intermediate in the chair interconversion. Although the twist-boat conformation is at an energy minimum, it is less stable than the chair conformation by about 23 kJ mol^{-1} (5.5 kcal mol^{-1}) in standard enthalpy. The standard free-energy difference (14.4 kJ mol^{-1}, 3.44 kcal mol^{-1}) is also considerable. As Study Problem 7.1 illustrates, a sample of cyclohexane has very little twist-boat conformation present at equilibrium. The boat conformation itself is the *transition state* (Sec. 4.8A) for the interconversion of two twist-boat conformations.

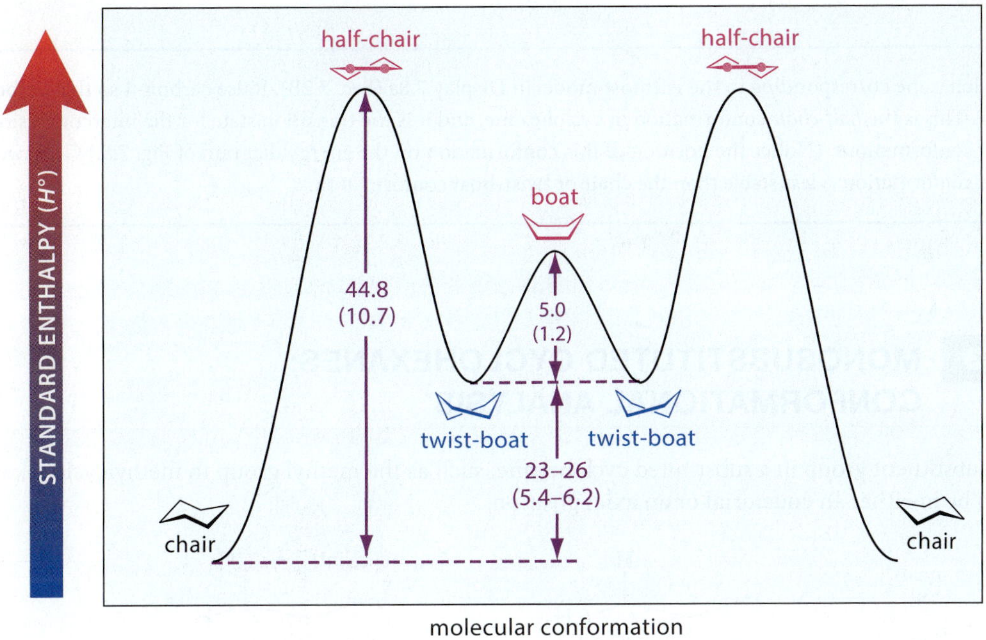

FIGURE 7.5 Relative enthalpies of cyclohexane conformations in kJ mol^{-1}. (The enthalpies in kcal mol^{-1} are in parentheses.) The inset shows the interconversion of twist-boat and boat conformations, which is much faster that the conversion of the twist-boat into either chair conformation. (The half-chair conformation is explained in Focused Problem 7.1 at the end of this section.)

Study Problem 7.1

Given that the twist-boat form is 14.4 kJ mol^{-1} (3.44 kcal mol^{-1}) higher in standard free energy than the chair form of cyclohexane, calculate the percentages of each form present in a sample of cyclohexane.

Solution We are interested in the equilibrium ratio of the two forms of cyclohexane—that is, the equilibrium constant for the equilibrium

$$\text{chair (C)} \rightleftharpoons \text{twist-boat (T)}$$

This equilibrium constant can be expressed as follows:

$$K_{eq} = \frac{[T]}{[C]}$$

The equilibrium constant is related to standard free energy by Eq. 3.47a (Sec. 3.5B):

$$\Delta G° = -2.3RT \log K_{eq}$$

or its rearranged form, Eq. 3.47b,

$$K_{eq} = 10^{-\Delta G°/2.3RT}$$

Applying this equation with energies in kilojoules per mole, $R = 8.31 \times 10^{-3}$ kJ mol^{-1} K^{-1}, and $T = 298$ K,

$$K_{eq} = \frac{[T]}{[C]} = 10^{-\Delta G°/2.3RT} = 10^{-14.4/5.71} = 10^{-2.52} = 3.01 \times 10^{-3}$$

Therefore, $[T] = (3.01 \times 10^{-3})[C]$. So, in 1 mol of cyclohexane, we have

$$1 = [C] + [T] = [C] + (3.01 \times 10^{-3})[C] = 1.00301[C]$$

Solving for [C],

$$[C] = 0.997$$

and, by difference,

$$[T] = 1.000 - [C] = 1.000 - 0.997 = 0.003$$

Therefore, cyclohexane contains 99.7% of the chair conformation and 0.3% of the twist-boat conformation at 25 °C.

Focused Problem

7.1 Make a model of chair cyclohexane corresponding to the leftmost model in Display 7.8a (Sec. 7.2B). Raise carbon-4 so that carbons 2–5 lie in a common plane. This is the *half-chair* conformation of cyclohexane, and it is the transition state for the interconversion of the chair and twist-boat conformations. (Notice the position of this conformation on the energy diagram of Fig. 7.5.) Give two reasons why the half-chair conformation is less stable than the chair or twist-boat conformation.

7.3 MONOSUBSTITUTED CYCLOHEXANES; CONFORMATIONAL ANALYSIS

A substituent group in a substituted cyclohexane, such as the methyl group in methylcyclohexane, can be in either an equatorial or an axial position.

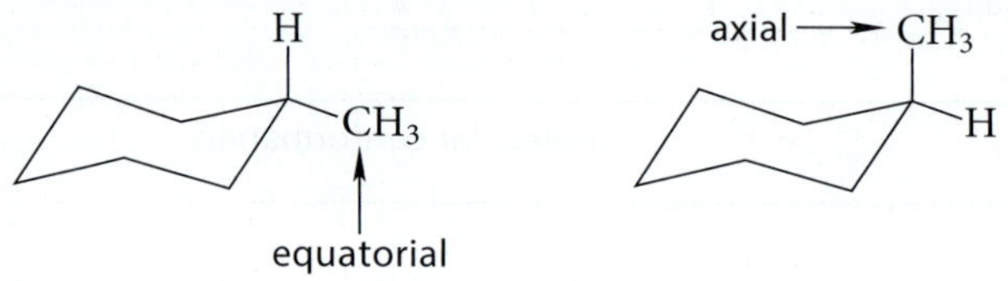

(7.11)

These two compounds are not identical, yet they have the same connectivity, so they are stereoisomers. Because they are not enantiomers, they must be diastereomers (Sec. 6.6). Like cyclohexane itself, substituted cyclohexanes such as methylcyclohexane also undergo the chair interconversion. As **Fig. 7.6** shows, axial methylcyclohexane and equatorial methylcyclohexane are interconverted by this process. In this interconversion, a down methyl remains down and an up methyl remains up. (Demonstrate this to yourself with models!) Because this process is rapid at room temperature, methylcyclohexane is a mixture of two *conformational diastereomers* (Sec. 6.9A). Because diastereomers have different energies, one form is more stable than the other.

Equatorial methylcyclohexane is more stable than axial methylcyclohexane. In fact, it is usually the case that *the equatorial conformation of a substituted cyclohexane is more stable than the axial conformation*. Why should this be so?

Examination of a space-filling model of axial methylcyclohexane (**Fig. 7.7**) shows that van der Waals repulsions occur between one of the methyl hydrogens and the two axial hydrogens on the same face of the ring. Such unfavorable interactions between axial groups are called generally **1,3-diaxial interactions**. These van der Waals repulsions destabilize the axial conformation relative to the equatorial conformation, in which such van der Waals repulsions are absent.

The energy (enthalpy) difference between axial and equatorial conformations of methylcyclohexanes is 7.4 kJ mol^{-1} (1.8 kcal mol^{-1}).

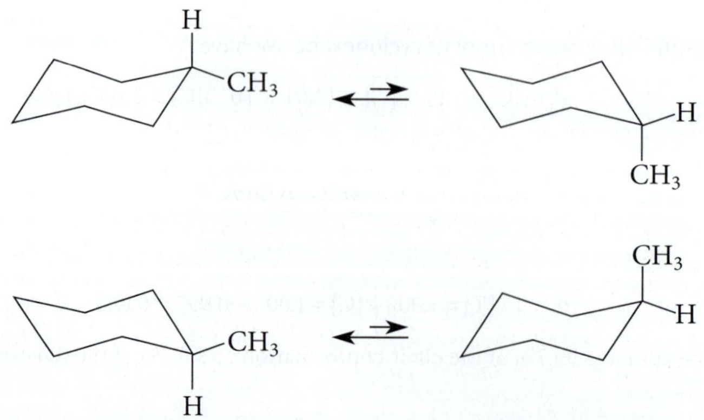

FIGURE 7.6 The chair interconversion results in equilibrium between equatorial (*left*) and axial (*right*) conformations of methylcyclohexane. The conversion is shown with two different ring perspectives. In this interconversion, a down methyl remains down and an up methyl remains up.

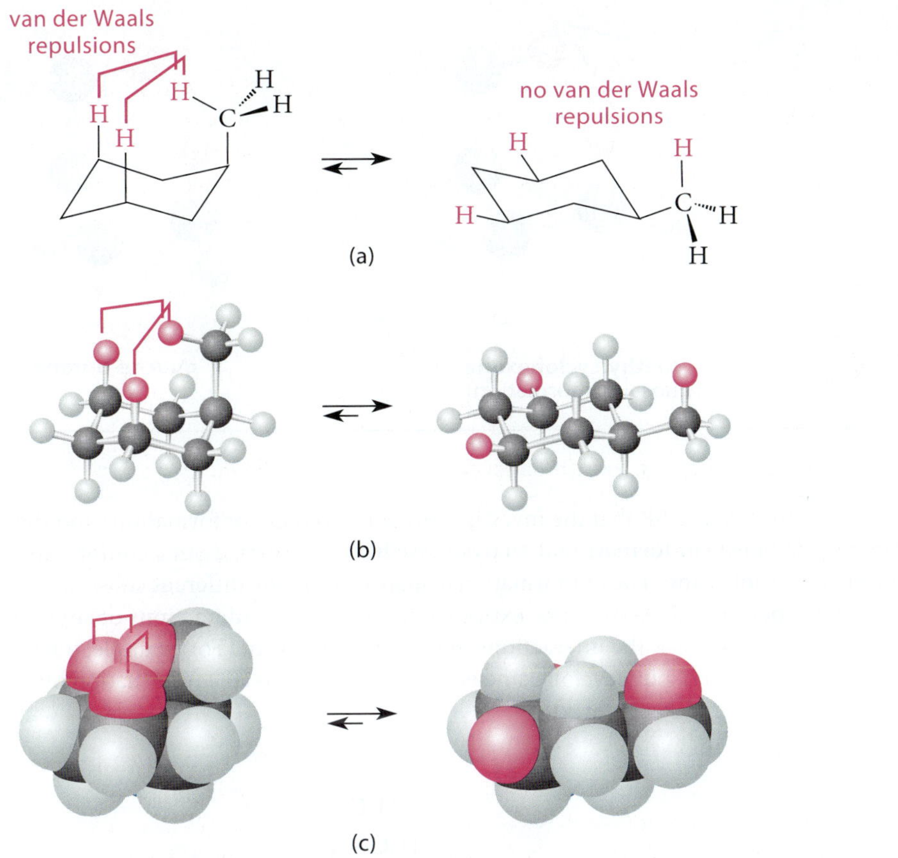

FIGURE 7.7 The equilibrium between axial and equatorial conformations of methylcyclohexane is shown with (a) Lewis structures, (b) ball-and-stick models, and (c) space-filling models. The hydrogens involved in 1,3-diaxial interactions in the axial conformation are shown in color, and the interactions themselves are indicated with red brackets.

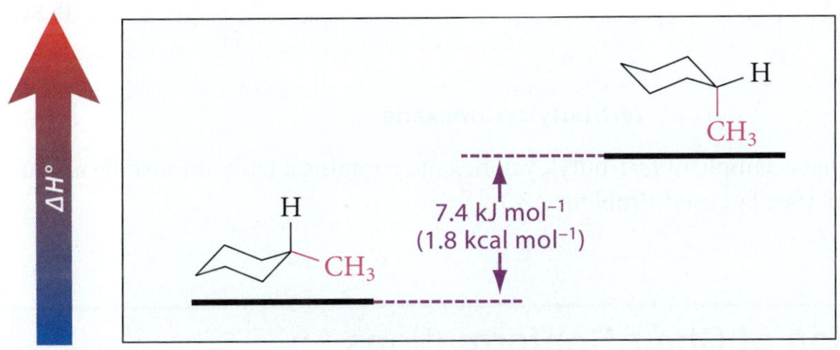

(7.12)

Because the axial conformation of methylcyclohexane has *two* 1,3-diaxial interactions, each interaction is responsible for one-half of the enthalpy difference, or 3.7 kJ mol^{-1} (0.9 kcal mol^{-1}). We'll find that we can use this value in predicting the relative energies of other methyl-substituted cyclohexanes. In other words, *each methyl–hydrogen 1,3-diaxial interaction in a cyclohexane derivative raises the enthalpy by 3.7 kJ mol^{-1} (0.9 kcal mol^{-1}).*

As shown in **Fig. 7.8**, the 1,3-diaxial interaction of a methyl group and a hydrogen in *axial*-methylcyclohexane looks a lot like the van der Waals interaction between methyl hydrogens in *gauche*-butane. The energy cost of this interaction in *gauche*-butane is 3.7 kJ mol^{-1} (see Fig. 2.5, Sec. 2.5B). Because *axial*-methylcyclohexane has *two* such interactions, the *gauche*-butane analogy would predict an energy cost of $2 \times 3.7 = 7.4$ kJ mol^{-1}. The actual value, 7.4 kJ mol^{-1}, is in excellent agreement with this prediction. For this reason, 1,3-diaxial methyl–hydrogen interactions in cyclohexane derivatives are sometimes called ***gauche*-butane interactions**.

The energy cost of placing a methyl group in the axial position of a cyclohexane ring is reflected in the relative amounts of *axial*- and *equatorial*- methylcyclohexanes present at equilibrium. As you will see when you work Focused Problem 7.2, methylcyclohexane contains very little of the axial conformation at equilibrium.

FIGURE 7.8 The relationship between the axial conformation of methylcyclohexane and *gauche*-butane. One *gauche*-butane part of methylcyclohexane is highlighted, and the corresponding van der Waals repulsion is shown with a red bracket. The second *gauche*-butane interaction in methylcyclohexane is shown with the blue bracket.

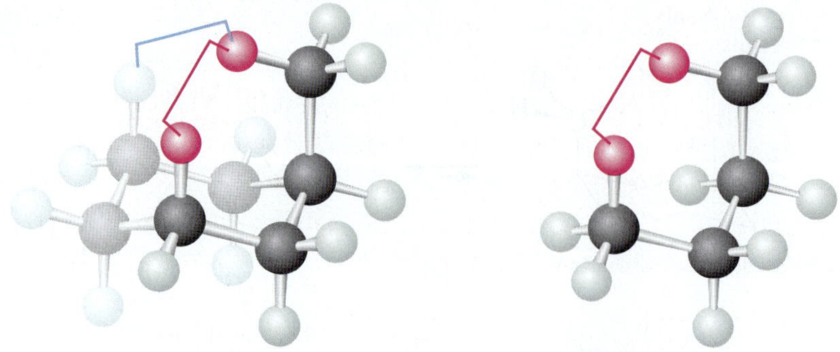

methylcyclohexane (axial conformation) **gauche-butane**

Recall from Sec. 2.5B that the investigation of molecular conformations and their relative energies is called **conformational analysis**. We have just carried out a conformational analysis of methylcyclohexane. The conformational analyses of many different substituted cyclohexanes have been performed. As might be expected, the 1,3-diaxial interactions of large substituent groups are greater than the interactions in methylcyclohexane. For example, the equatorial conformation of *tert*-butylcyclohexane is favored over the axial conformation by about 20 kJ mol^{-1} (about 5 kcal mol^{-1}).

In Eq. 7.13 there is so little of the conformation with an axial *tert*-butyl group that chemists say sometimes that the conformational equilibrium is "locked." This statement is somewhat misleading because it implies that the two conformations are not in equilibrium. The equilibrium indeed occurs rapidly, but it simply contains very little of the conformation in which the *tert*-butyl group is axial.

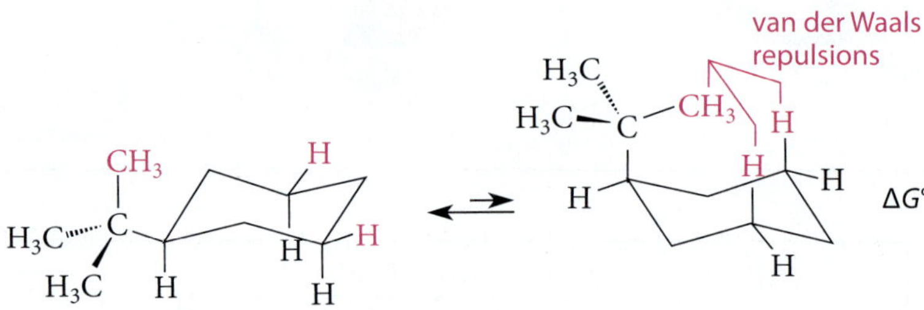

$\Delta G° = 20$ kJ mol^{-1} (5 kcal mol^{-1})

tert-butylcyclohexane (7.13)

This means that a sample of *tert*-butylcyclohexane contains a truly minuscule amount of the axial conformation. (See Focused Problem 7.3.)

Separation and Isolation of Chair Conformations

The two chair conformations of a monosubstituted cyclohexane are diastereomers. If these conformations could be separated, they would have different physical properties. In the late 1960s, C. Hackett Bushweller, then a graduate student in the laboratory of Professor Frederick Jensen at the University of California, Berkeley, cooled a solution of chlorocyclohexane in an inert solvent to –150 °C. Crystals suddenly appeared in the solution. He filtered the crystals at low temperature; subsequent investigations showed that he had selectively crystallized the equatorial form of chlorocyclohexane!

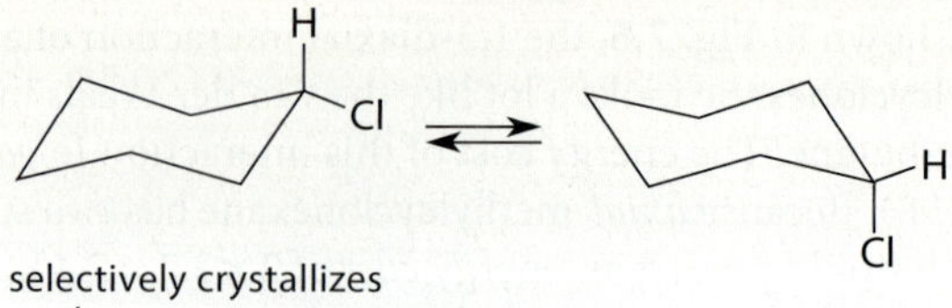

selectively crystallizes at low temperature

When the axial form remaining in solution was "heated" to –120 °C, the rate of the chair interconversion increased, and a mixture of conformations again resulted. Similar experiments have been carried out with other monosubstituted cyclohexanes.

Focused Problems

7.2 The $\Delta G°$ difference between the axial and equatorial conformations of methylcyclohexane (7.4 kJ mol^{-1}, 1.74 kcal mol^{-1}; see Display 7.12) is about the same as the $\Delta H°$ difference. Calculate the percentages of axial and equatorial conformations present in 1 mol of methylcyclohexane at 25 °C. (*Hint:* See Study Problem 7.1.)

7.3 Using the information in the previous problem and in Eq. 7.13, contrast the relative amounts of axial conformations in samples of methylcyclohexane and *tert*-butylcyclohexane.

7.4 (a) The axial conformation of fluorocyclohexane is 1.0 kJ mol^{-1} (0.25 kcal mol^{-1}) less stable than the equatorial conformation. What is the energy cost of a 1,3-diaxial interaction between hydrogen and fluorine?

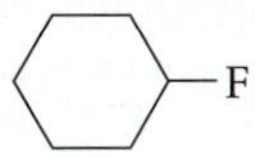

fluorocyclohexane

(b) Estimate the energy difference between the gauche and anti conformations of 1-fluoropropane.

$$H_3C—CH_2—CH_2—F$$

1-fluoropropane

7.4 DISUBSTITUTED CYCLOHEXANES

A. Cis–Trans Isomerism in Disubstituted Cyclohexanes

We use 1-chloro-2-methylcyclohexane as a case study for the stereochemical and conformational aspects of disubstituted cyclohexanes. First, we consider **planar ring structures**—structures in which the cyclohexane ring is drawn as a planar hexagon.

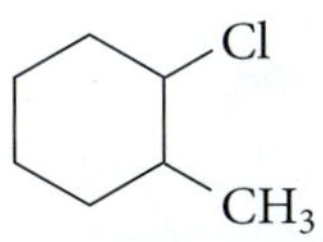

1-chloro-2-methylcyclohexane

The cyclohexane ring isn't planar, though, so a planar-ring structure is a *projection* of the ring into the plane of the page. In effect, it is an average of the two chair conformations. To indicate stereochemistry, we use the *line-and-wedge notation*: solid wedges to indicate "up" bonds—bonds to substituents in front of the page—and dashed wedges to indicate "down" bonds—bonds to substituents behind the page. 1-Chloro-2-methylcyclohexane has two asymmetric carbons and therefore four stereoisomers—two diastereomeric pairs of enantiomers. In one pair of enantiomers, the two substituents are both up or both down.

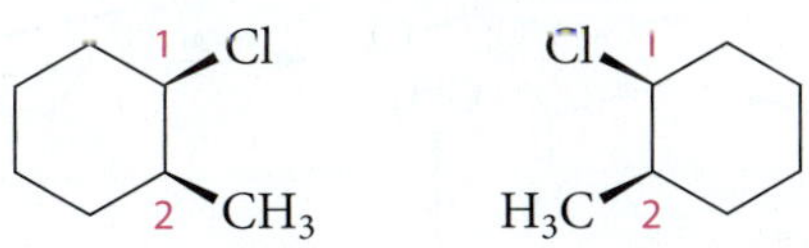

(1*R*,2*S*) (1*S*,2*R*)

***cis*-1-chloro-2-methylcyclohexane** (7.14)

We arbitrarily used the structures with both substituents up, but equivalent structures with both substituents down are equally valid. We can derive one from the other by rotating the structure 180° about the axis shown:

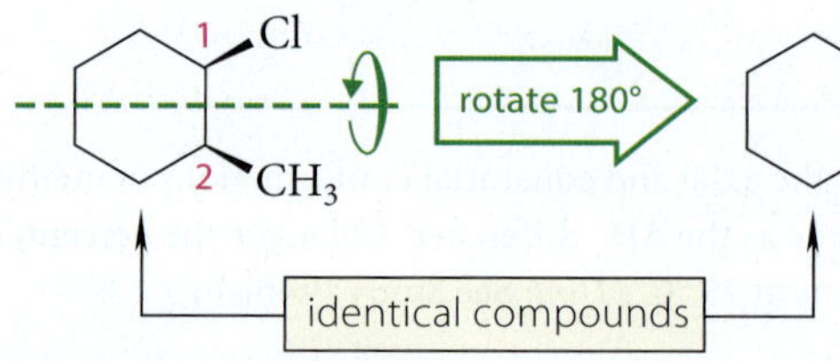

(7.15)

When both substituents have the same *relative* orientation—both up or both down—the substitution pattern is called **cis**. Therefore, the two compounds in Eq. 7.15 are enantiomers of *cis*-1-chloro-2-methylcyclohexane.

In the other pair of enantiomers of 1-chloro-2-methylcyclohexane, one substituent is up and the other down. When both substituents have different *relative* orientations—one up and the other down—the substitution pattern is called **trans**.

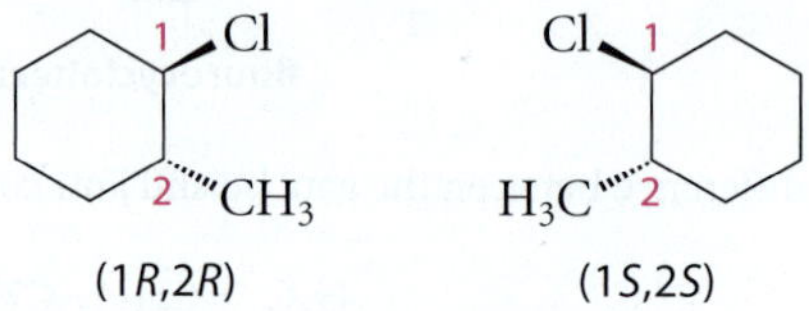

(1*R*,2*R*) (1*S*,2*S*)

***trans*-1-chloro-2-methylcyclohexane** (7.16)

The terms *cis* and *trans*, when used with cyclic compounds, refer to the *relative* orientations of substituents (that is, their up/down relationship) and *not* to their absolute (that is, *R*,*S*) configurations.

Next, we consider the relationship of the planar-ring structures to their chair conformations. Let's derive the chair conformations of (1*S*,2*S*)-1-chloro-2-methylcyclohexane. Start with two unsubstituted chairs, and then associate each of the substituted carbons in the planar structure with a carbon atom of the chairs. It is completely arbitrary where we start the numbering of the chair as long as we proceed in the same direction around the planar and chair forms. (Remember that the planar structure is a "bird's-eye" view—that is, viewed from above.) It is often easiest, although not essential, to let one substituted carbon of the chair be the rightmost or leftmost carbon.

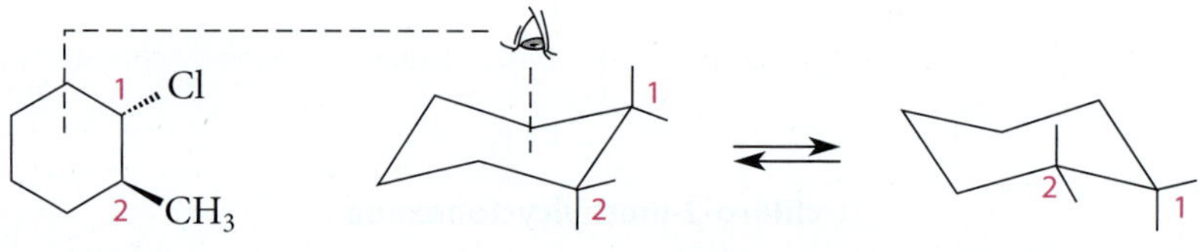

(1*S*,2*S*) (7.17)

The chlorine is a down substituent, so we put it in the down position at carbon-1 in each of the two chairs. In one chair form, the down position is equatorial, whereas in the other, it is axial. We then add the methyl substituent to the up position on each of the two chairs.

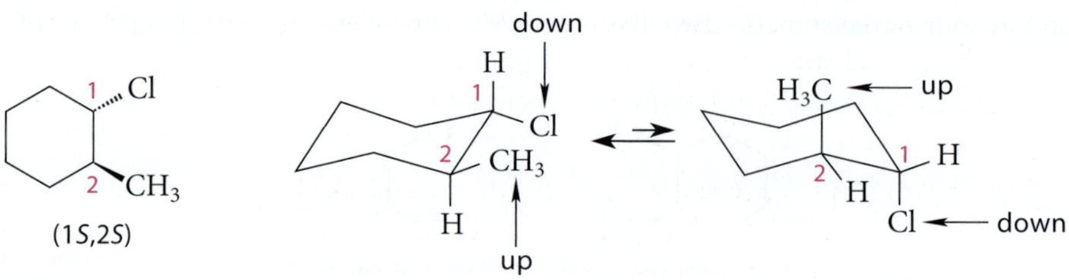

(1*S*,2*S*) (7.18)

It may look as if carbons 1 and 2 have shifted to the left in the second chair form. That's because the viewing perspective is different for the two chair forms (Display 7.3, Sec. 7.2A). This shift in perspective is the reason that it's helpful to use one of the substituted carbons as the rightmost or leftmost carbon. For example, carbon-1 is the rightmost carbon in either perspective in Display 7.18.

Notice two things: (1) A planar-ring structure contains information about configuration (*R*, *S*, cis, or trans), but no information about conformation. Each planar-ring structure has two chair conformations. Therefore, you can answer questions about chirality with planar-ring structures, but you can't answer questions about conformation—for example, whether a substituent is axial or equatorial. (2) The up and down substituent positions in the planar-ring structures carry over to the chair conformations, because the up/down positions are not affected by the chair interconversion—that is, in the chair interconversion, up-equatorial becomes up-axial, and down-equatorial becomes down-axial, as shown in Eq. 7.9.

The two chair conformations in this case are conformational diastereomers. They have different energies, and the diequatorial form is favored at equilibrium (why?), as shown by the unequal equilibrium arrows in Eq. 7.18. You should verify that *cis*-1-chloro-2-methylcyclohexane is also a mixture of conformational diastereomers.

Study Problem 7.2

(a) Show that *trans*-1,3-dimethylcyclohexane is chiral. (b) For the 1*R*,3*R* stereoisomer, draw the two chair conformations. (c) What is the relationship between the two chair conformations of this compound: identical, conformational enantiomers, or conformational diastereomers?

Solution

(a) The question about chirality is best answered with a planar-ring structure. Draw one of the stereoisomers, then draw its mirror image, and then test for congruence. Remember that *trans* means that the substituents have an up–down relationship—that is, one substituent is on a wedged bond, whereas the other is on a dashed-wedge bond.

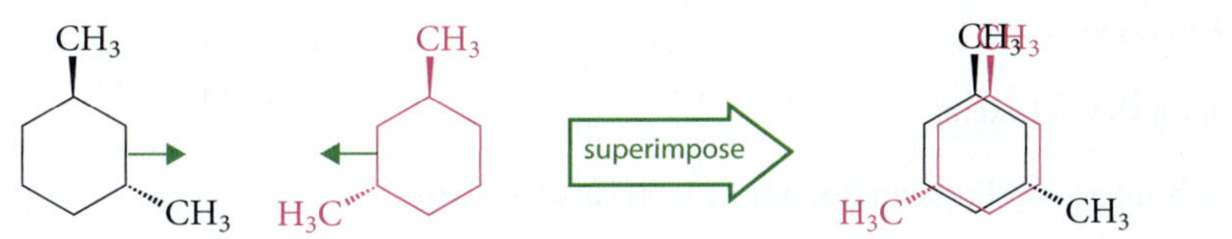

***trans*-1,3-dimethylcyclohexane** (mirror images)

demonstration of noncongruence; therefore, molecules are chiral and are enantiomers

Because its mirror images are noncongruent, *trans*-1,3-dimethylcyclohexane is chiral.

(b) Follow the procedure in the text for constructing the two chair conformations. (The carbons are numbered for reference.) For ease of reference, one of the substituted carbons is chosen as the rightmost carbon.

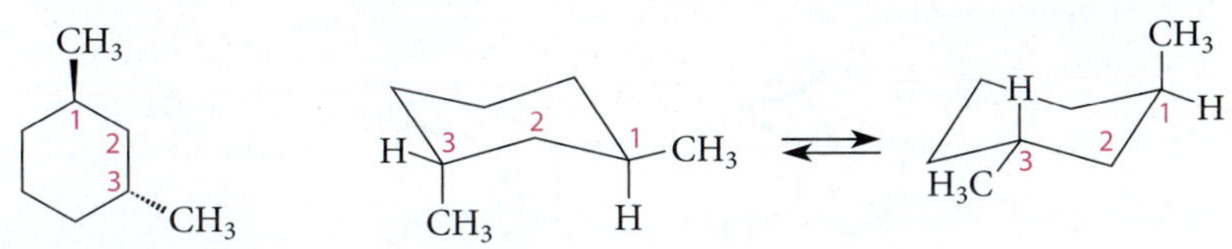

(1*R*,3*R*)-1,3-dimethyl-cyclohexane

(c) The two chair conformations are identical. Because they don't look identical, we should verify that they are; you should follow along with models. The goal is to rotate one of the chairs (without changing its conformation!) in various ways so as to superpose exactly one of its methyl-substituted carbons onto a methyl-substituted carbon of the other chair. The relationship of the two structures will then be easier to see. First, rotate the second chair structure 180° about the axis shown; then rotate the resulting structure 120° as shown about an axis through the ring.

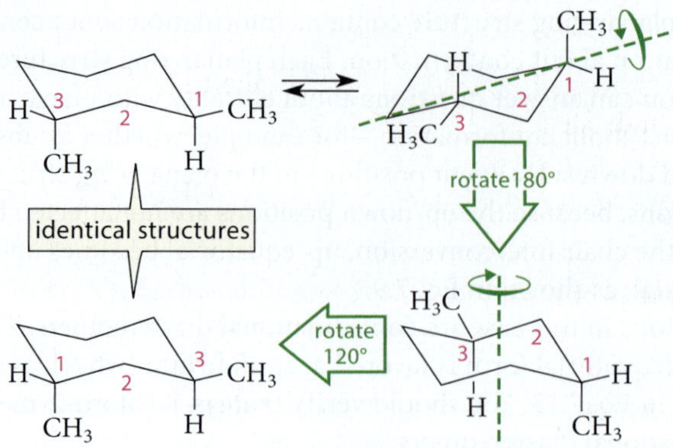

STUDY GUIDE LINK 7.1
Relating Cyclohexane Conformations

Indeed, the two chair conformations are identical. Carbon-1 in one chair is equivalent to carbon-3 in the other, and vice versa.

FURTHER EXPLORATION 7.1
Other Ways of Designating Relative Configuration

When a ring contains more than two substituents, cis–trans nomenclature is usually cumbersome. For such cases, other systems have been developed to designate relative configuration. (See Further Exploration 7.1.)

Focused Problems

7.5 For each of the following compounds, draw the two chair conformations that are in equilibrium.

(a) *cis*-1,3-Dimethylcyclohexane

(b) *trans*-1-Ethyl-4-isopropylcyclohexane

7.6 For each of the compounds in Focused Problem 7.5, draw a boat conformation.

7.7 Draw planar-ring structures for the following cyclic compounds.

(a) *cis*-1-Bromo-3-methylcyclohexane (both enantiomers)

(b) (1*R*,2*S*,3*R*)-2-Chloro-1-ethyl-3-methylcyclohexane

7.8 Draw the two chair conformations of each compound in Focused Problem 7.7.

7.9 Draw the planar-ring structure that results from each rotation of the following structure.

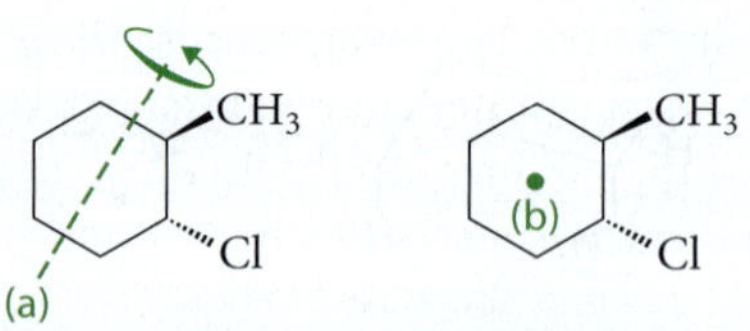

(a) 180° about axis (a)

(b) 120° counterclockwise about an axis (b) that runs through the center of the ring and perpendicular to the page

7.10 Redraw the following chair conformation after 180° rotations about each of the axes shown. Use models if necessary.

(a)

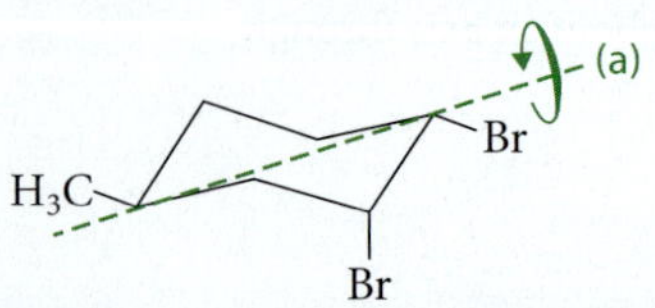

(b)

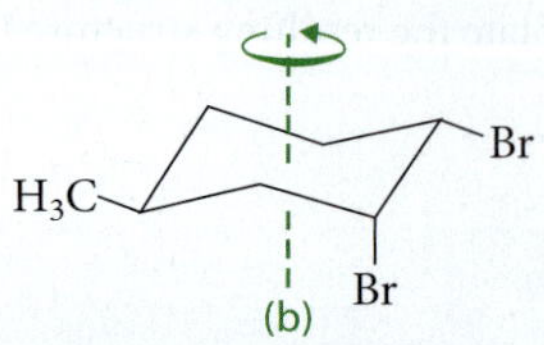

B. Cyclic Meso Compounds

The planar-ring structures of both *cis*-1,2-dimethylcyclohexane and *cis*-1,3-dimethylcyclohexane have a plane of symmetry, as indicated by the gray line:

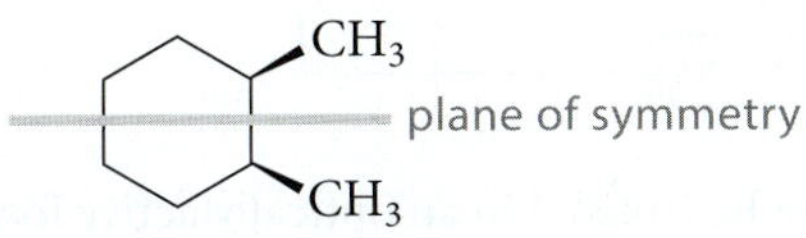

cis-1,2-dimethylcyclohexane

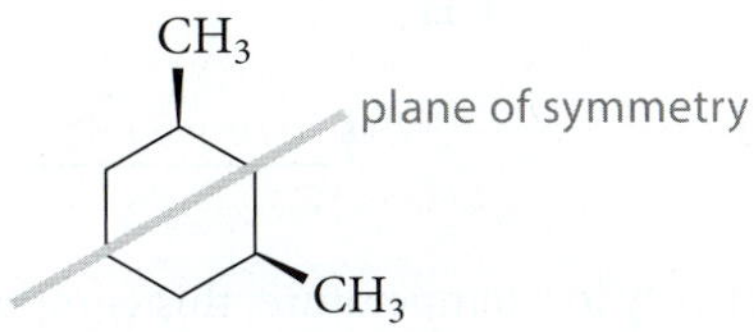

cis-1,3-dimethylcyclohexane (7.19)

The internal plane of symmetry shows that these compounds are achiral. Because these compounds have asymmetric carbons and have chiral stereoisomers (the trans isomers), they are examples of meso compounds.

An examination of their chair conformations shows, however, that the chair conformations of *cis*-1,2-dimethylcyclohexane are chiral, but *each conformation is the enantiomer of the other*:

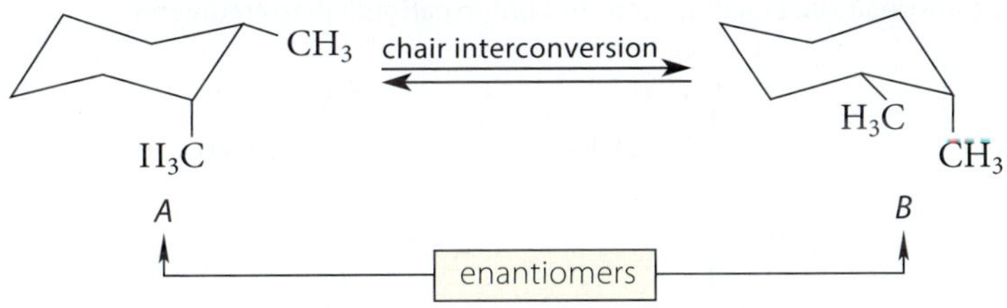

(7.20a)

From the way they are drawn, structures *A* and *B* may not look like enantiomers, but they are. This relationship can be verified by rotating structure *B* 120° about the vertical axis shown and comparing it with structure *A*. The two are noncongruent mirror images and therefore enantiomers.

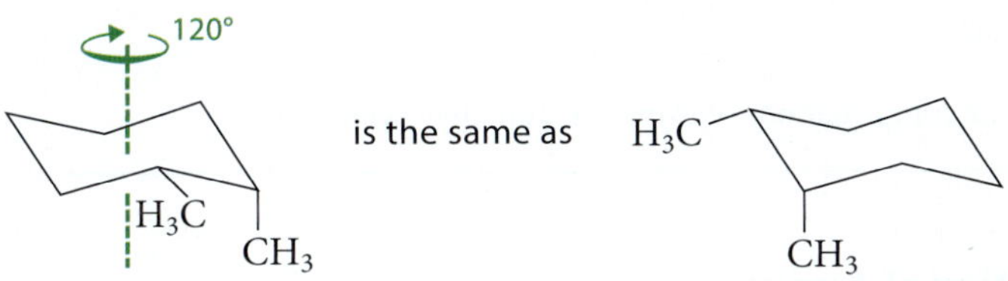

B enantiomer of structure *A* in Eq. 7.20a (7.20b)

In other words, *cis*-1,2-dimethylcyclohexane is a mixture of *conformational enantiomers* (Sec. 6.8A). If we could isolate the individual conformations (which might be possible at *very* low temperature), each would be chiral and optically active. However, because these conformations interconvert rapidly at ordinary temperatures, *cis*-1,2-dimethylcyclohexane cannot be isolated in optically active form. Over any realistic time interval, we can think of it as the time-averaged, planar-ring structure, which is meso. This same conclusion is true for any cis-1,2-disubstituted cyclohexane in which the substituents are identical. We discussed this situation in Sec. 6.9A, where we suggested that compounds with enantiomeric conformations in rapid equilibrium could be thought of as a single chiral object—like a pair of shoes—and are considered to be achiral.

Unlike the 1,2-dimethyl isomer, *cis*-1,3-dimethylcyclohexane consists of two chair conformations that are each true meso compounds, because *each one* has an internal plane of symmetry.

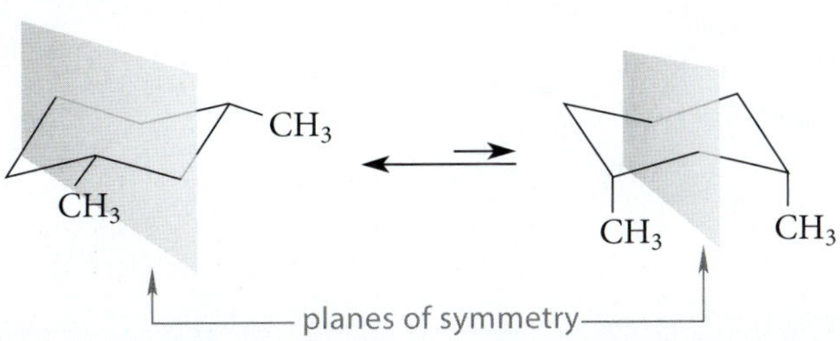

(7.21a)

These two conformations are *conformational diastereomers.*

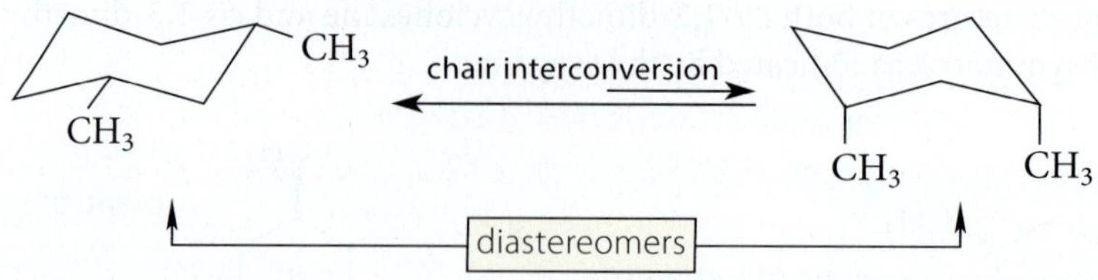

(7.21b)

Even at very low temperature, this compound cannot be isolated in an optically active form, because each conformation is achiral.

Focused Problems

7.11 Use their planar-ring structures to determine which of the following compounds are chiral. For any that are achiral, indicate whether their individual chair conformations are chiral. Tell how you know. For any achiral compounds, give the relationship of their two chair conformations (identical, conformational enantiomers, or conformational diastereomers).

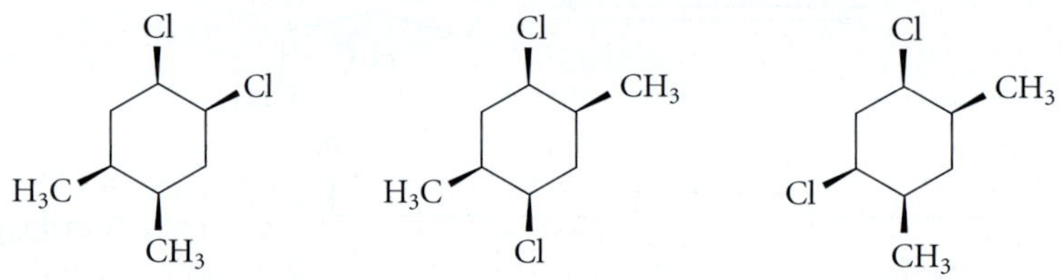

A *B* *C*

7.12 (a) Does *trans*-1,4-dimethylcyclohexane contain any asymmetric carbons? If so, identify them.

(b) Does *trans*-1,4-dimethylcyclohexane contain any stereocenters? If so, identify them.

(c) Is *trans*-1,4-dimethylcyclohexane chiral?

(d) What is the relationship of the two chair conformations of *trans*-1,4-dimethylcyclohexane?

C. Conformational Analysis

Disubstituted cyclohexanes, like monosubstituted cyclohexanes, can be subjected to conformational analysis. The relative stability of the two chair conformations is determined by comparing the 1,3-diaxial interactions (or *gauche*-butane interactions) in each conformation. Such an analysis is illustrated in Study Problem 7.3.

Study Problem 7.3

Determine the relative energies of the two chair conformations of *trans*-1,2-dimethylcyclohexane. Which conformation is more stable?

Solution The first step in solving any problem is to draw the structures of the species involved. The two chair conformations of *trans*-1,2-dimethylcyclohexane are as follows:

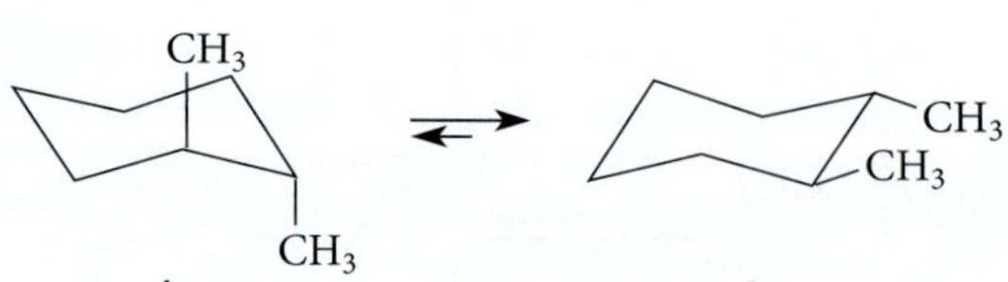

A *B*

Conformation *A* has the greater number of axial groups and should therefore be the less stable conformation—but by how much? Conformation *A* has four 1,3-diaxial methyl–hydrogen interactions (show these!), which contribute $4 \times 3.7 = 14.8$ kJ mol^{-1} ($4 \times 0.9 = 3.6$ kcal mol^{-1}) to its energy. What about *B*? You might be tempted to say that *B* has no unfavorable interactions because it has no axial groups, but in fact *B* does have one *gauche*-butane interaction—the one between the two methyl groups themselves, which have a dihedral angle between their carbon–carbon bonds of 60°, just as in *gauche*-butane. This interaction can be seen in a Newman projection of the bond between the carbons bearing the methyl groups:

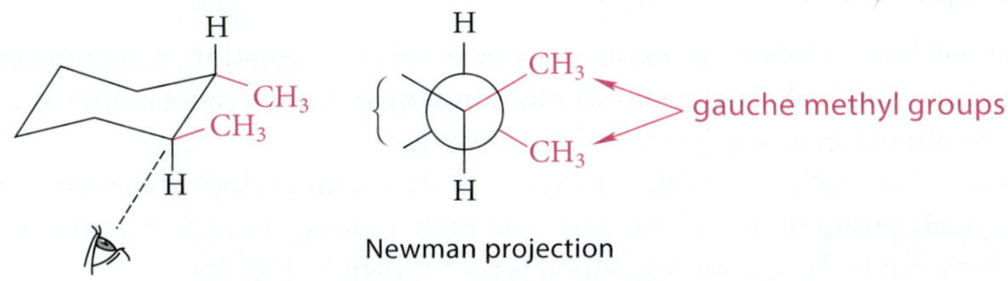

This *gauche*-butane interaction contributes 3.7 kJ mol^{-1} (0.9 kcal mol^{-1}) to the energy of conformation *B* (see Fig. 2.5, Sec. 2.5B). The relative energy of the two conformations is the difference between their methyl–hydrogen interactions: $14.8 - 3.7 = 11.1$ kJ mol^{-1} (or $3.6 - 0.9 = 2.7$ kcal mol^{-1}). The diaxial conformation *A* is less stable by this amount of energy.

When two groups on a substituted cyclohexane conflict in their preference for the equatorial position, the preferred conformation can usually be predicted from the relative conformational preferences of the two groups. Consider, for example, the chair interconversion of *cis*-1-*tert*-butyl-4-methylcyclohexane.

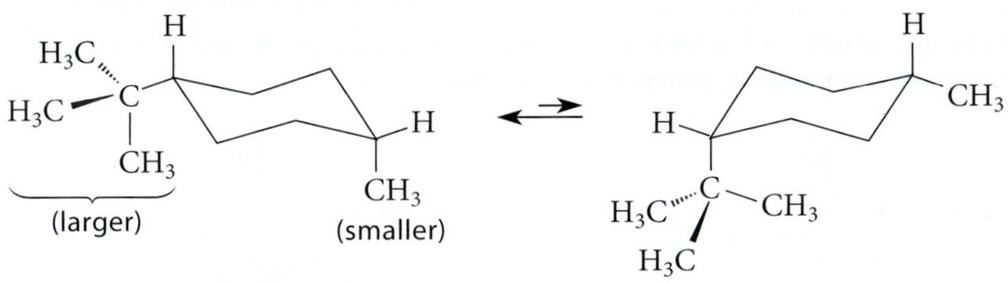

(7.22)

The *tert*-butyl group is so large that its van der Waals repulsions control the conformational equilibrium (see Eq. 7.13, Sec. 7.3). Therefore, the chair conformation in which the *tert*-butyl group assumes the equatorial position is overwhelmingly favored. The methyl group is thus forced into the axial position.

Focused Problems

7.13 (a) Calculate the energy difference between the two chair conformations of *trans*-1,4-dimethylcyclohexane.

(b) Calculate the energy difference between *cis*-1,4-dimethylcyclohexane and the more stable conformation of *trans*-1,4-dimethylcyclohexane.

7.14 Draw the more stable chair conformation for each of the following compounds.

(a) 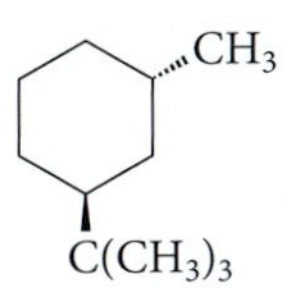

(b) 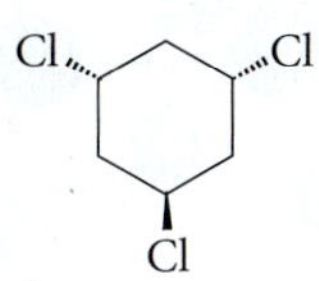

(c) 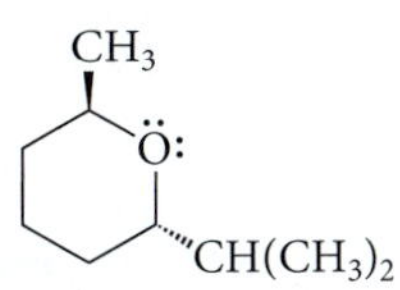

7.5 CYCLOPENTANE, CYCLOBUTANE, AND CYCLOPROPANE

Recall from Sec. 7.1 that cyclopentane, cyclobutane, and cyclopropane are all less stable than cyclohexane. In this section we consider the structures of these smaller cycloalkanes and the reasons for their greater energies.

A. Cyclopentane

Cyclopentane, like cyclohexane, exists in a puckered conformation, called the **envelope conformation** (**Fig. 7.9**). This conformation undergoes rapid conformational changes in which each carbon alternates as the "point" of the envelope.

The heats of formation in Table 7.1 (Sec. 7.1) show that cyclopentane has somewhat higher energy per CH_2 group than cyclohexane. The higher energy of cyclopentane is due mostly to eclipsing between hydrogen atoms, which is also shown in Fig. 7.9.

Substituted cyclopentanes also exist in envelope conformations, but the substituents adopt positions that minimize torsional strain (eclipsing) and van der Waals repulsions with neighboring groups. For example, in methylcyclopentane, the methyl group assumes an equatorial position at the point of the envelope.

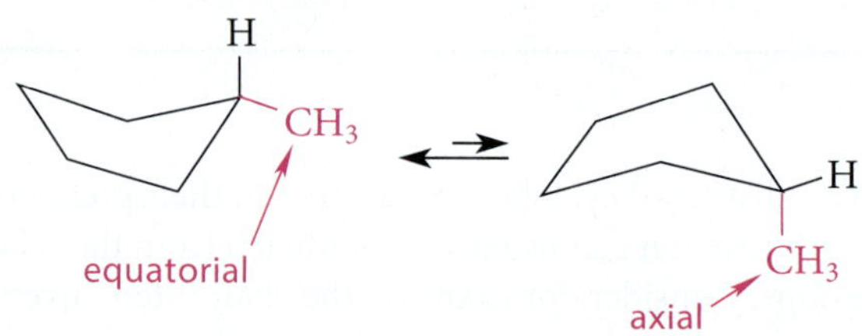

(7.23)

Conformations in which the methyl group is not on the point of the envelope are less important.

When a cyclopentane ring has two or more substituent groups, cis and trans relationships between the groups are possible, just as in cyclohexane.

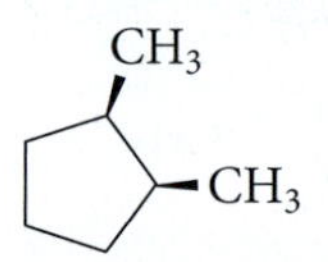

***cis*-1,2-dimethylcyclopentane**

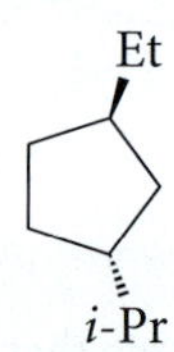

***trans*-1-ethyl-3-isopropylcyclopentane** (7.24)

B. Cyclobutane and Cyclopropane

The data in Table 7.1 (Sec. 7.1) show that cyclobutane and cyclopropane are the least stable of the monocyclic alkanes. In each compound, the angles between carbon–carbon bonds are constrained by the size of the ring to be much smaller than the optimum tetrahedral angle of 109.5°.

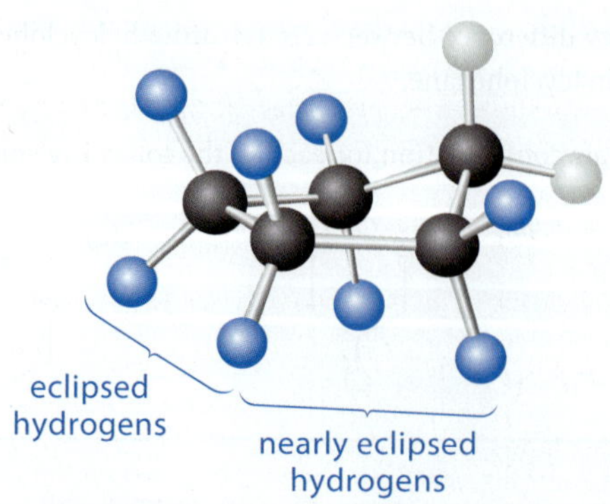

FIGURE 7.9 A ball-and-stick model of the envelope conformation of cyclopentane. The hydrogens shown in blue are either eclipsed or nearly eclipsed with the blue hydrogens on adjacent carbons.

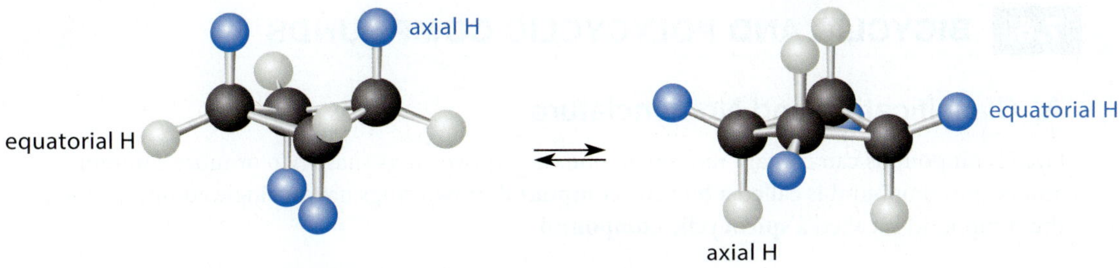

FIGURE 7.10 Cyclobutane exists in two equivalent puckered conformations in rapid equilibrium. As in cyclohexane, the equilibrium causes axial hydrogens (*blue on the left*) to become equatorial (*blue on the right*), and vice versa.

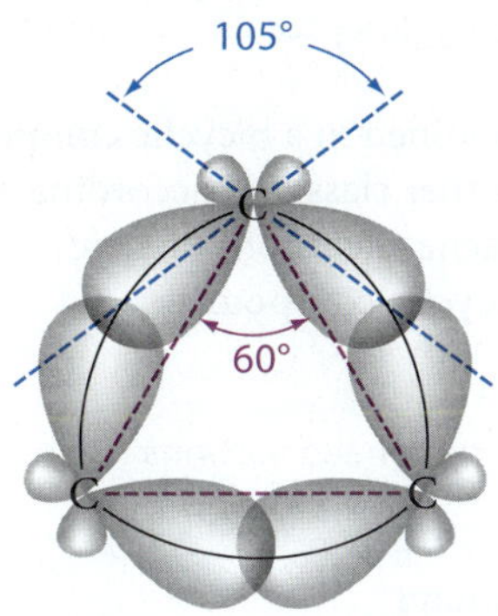

FIGURE 7.11 The orbitals that overlap to form each C—C bond in cyclopropane do not lie along the straight line between the carbon atoms. These carbon–carbon bonds (*black lines*) are sometimes called "bent" or "banana" bonds. The C—C—C angle is 60° (*purple dashed lines*), but the angles between the orbitals on each carbon are closer to 105° (*blue dashed lines*).

When the bond angles in a molecule deviate from their ideal values, the energy of the molecule is raised in the same sense that squeezing the handles of a hand exerciser increases the potential energy of the resisting spring. This excess energy, which is reflected in a greater heat of formation, is called **angle strain**. Angle strain contributes significantly to the high energies of both cyclobutane and cyclopropane.

Cyclobutane consists of two puckered conformations in rapid equilibrium (**Fig. 7.10**). This puckering avoids complete eclipsing between hydrogens.

Because three carbons define a plane, the carbon skeleton of cyclopropane is planar. Because cyclopropane cannot pucker, there is no way to relieve its angle strain or the eclipsing interactions between its hydrogens. As the data in Table 7.1 show, cyclopropane is the least stable of the cyclic alkanes. The carbon–carbon bonds of cyclopropane are bent in a banana shape around the periphery of the ring. Such "bent bonds" allow for angles between the carbon orbitals that are on the order of 105°, closer to the ideal tetrahedral value of 109.5° (**Fig. 7.11**). Although bent bonds reduce angle strain, they do so at a cost of less effective overlap between the carbon orbitals.

FURTHER EXPLORATION 7.2
Alkenelike Behavior of Cyclopropanes

Focused Problems

7.15 (a) The dipole moment of *trans*-1,3-dibromocyclobutane is 1.1 D. Explain why a nonzero dipole moment supports a puckered structure rather than a planar structure for this compound.

(b) Draw the more stable conformation of *trans*-1,2-dimethylcyclobutane.

7.16 Tell whether each of the following compounds is chiral, and give your reasoning.

(a) *cis*-1,2-Dimethylcyclopropane

(b) *trans*-1,2-Dimethylcyclopropane

7.6 BICYCLIC AND POLYCYCLIC COMPOUNDS

A. Classification and Nomenclature

Cyclic compounds can contain more than one ring. If two rings share two or more common atoms, the compound is called a **bicyclic compound**. If two rings have a single common atom, the compound is called a **spirocyclic compound**.

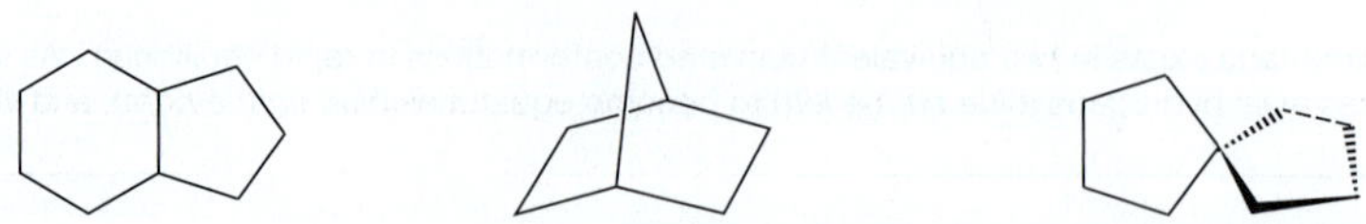

bicyclo[4.3.0]nonane **bicyclo[2.2.1]heptane** (bicyclic compounds) **spiro[4.4]nonane** (a spirocyclic compound) (7.25)

The atoms at which two rings are joined in a bicyclic compound are called **bridgehead carbons**. Bicyclic compounds are further classified according to the relationship of the bridgehead carbons. When the bridgehead carbons of a bicyclic compound are adjacent, the compound is classified as a **fused bicyclic compound**:

a fused bicyclic compound (7.26)

When the bridgehead carbons are *not* adjacent, the compound is classified as a **bridged bicyclic compound**:

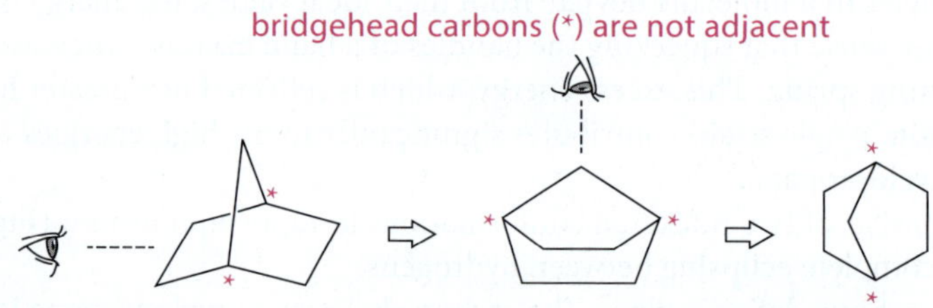

three views of a bridged bicyclic compound (7.27)

Many students are puzzled by these interesting structures. It is important to build models of them and relate these models to the structures that are drawn on paper.

The nomenclature of bicyclic hydrocarbons is best illustrated by example:

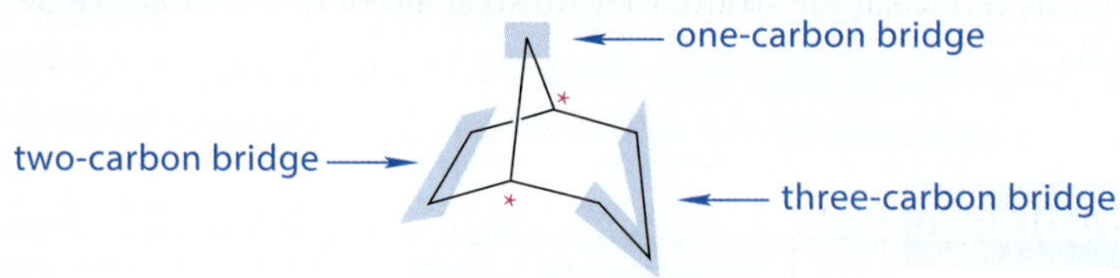

bicyclo[3.2.1]octane
(* = bridgehead carbons) (7.28)

This compound is named as a bicyclooctane because it is a bicyclic compound containing a total of eight carbon atoms. The numbers in brackets represent the number of carbon atoms in the respective bridges, in order of decreasing size.

Study Problem 7.4

Give the IUPAC name of the following compound. (Its common name is *decalin.*)

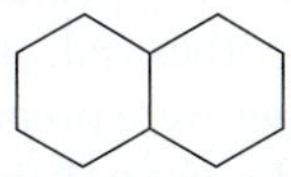

Solution The compound has two fused rings that contain a total of 10 carbons, so it is named as a bicyclodecane. Three bridges connect the bridgehead carbons: two contain four carbons, and *one contains zero carbons.* (The bond connecting the bridgehead carbons in a fused-ring system is considered to be a bridge with zero carbons.)

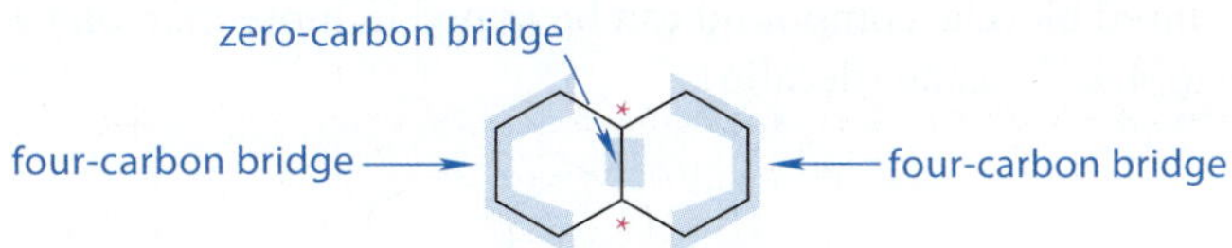

bicyclo[4.4.0]decane
(* = bridgehead carbons)

The compound is named bicyclo[4.4.0]decane.

Focused Problems

7.17 Name the following compounds, and tell whether each is a bridged or a fused bicyclic compound.

(a) 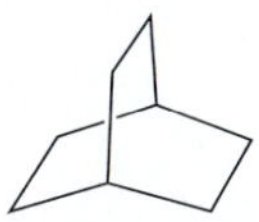 (b) 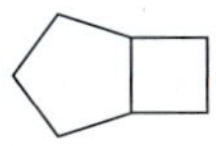

7.18 Without drawing their structures, tell which of the following compounds is a fused bicyclic compound and which is a bridged bicyclic compound, and explain how you know.

bicyclo[2.1.1]hexane (*A*) bicyclo[3.1.0]hexane (*B*)

Some organic compounds contain many rings joined at common atoms; these compounds are called **polycyclic compounds**. Among the more intriguing polycyclic compounds are those that have the shapes of regular geometric solids. Three of the more spectacular examples are cubane, dodecahedrane, and tetrahedrane.

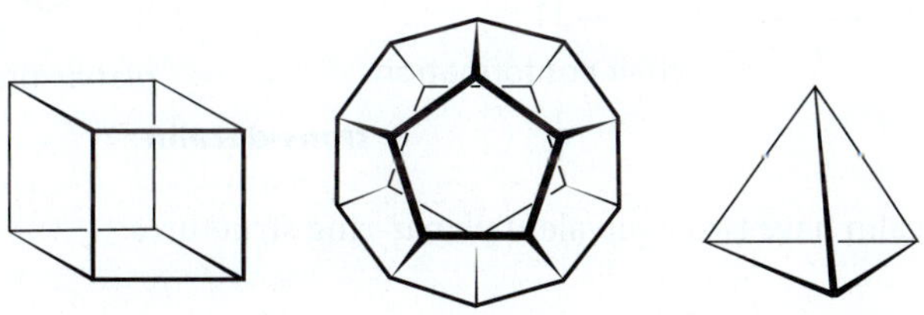

cubane **dodecahedrane** **tetrahedrane (tricyclobutane)** (7.29)

Cubane, which contains eight CH groups at the corners of a cube, was first synthesized in 1964 by Professor Philip Eaton and his associate Thomas W. Cole at the University of Chicago. Dodecahedrane, in which 20 CH groups occupy the corners of a dodecahedron, was synthesized in 1982 by a team of organic chemists led by Professor Leo Paquette of the Ohio State University. Tetrahedrane itself has not yet been synthesized, although derivatives containing *tert*-butyl and Me_3Si substituent groups at each corner were prepared in 1978, 2002, and 2011. Chemists tackle the syntheses of these unique molecules not only because they represent interesting problems in chemical bonding, but also because of the sheer challenges of the endeavors.

B. Cis and Trans Ring Fusion

Two rings in a fused bicyclic compound can be joined in more than one way. Consider, for example, bicyclo[4.4.0]decane (decalin).

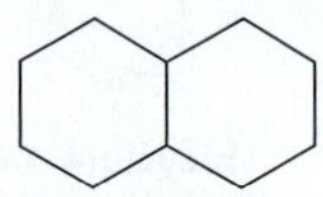

decalin
(bicyclo[4.4.0]decane) (7.30)

There are two stereoisomers of decalin. In *cis*-decalin, two CH_2 groups of ring *B* (circles) are cis substituents on ring *A*; likewise, two CH_2 groups of ring *A* (squares) are cis substituents on ring *B*. The cis ring fusion can be indicated in a planar-ring structure by showing the cis arrangements of the bridgehead hydrogens.

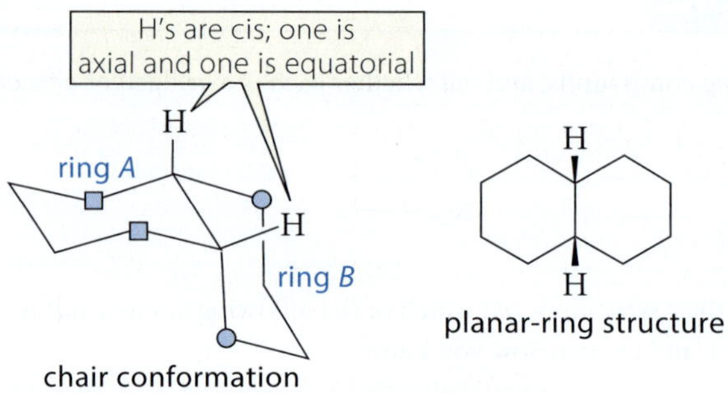

***cis*-decalin** (7.31a)

In *trans*-decalin, the CH_2 groups adjacent to the ring fusion are in a trans-diequatorial arrangement. The bridgehead hydrogens are trans-diaxial.

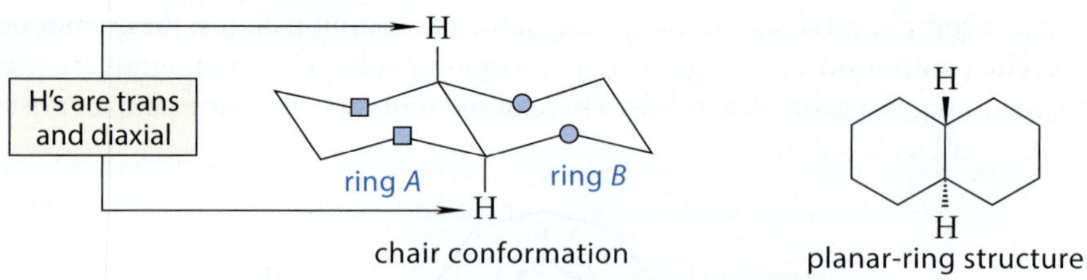

***trans*-decalin** (7.31b)

Both *cis*- and *trans*-decalin have two equivalent planar-ring structures:

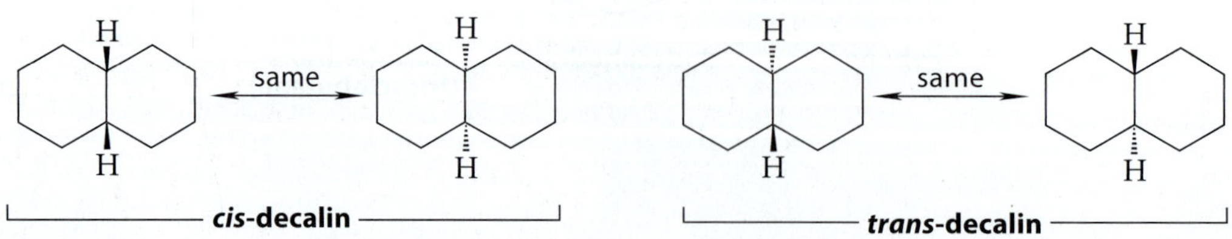

(7.31c)

Each cyclohexane ring in *cis*-decalin can undergo the chair interconversion. You should verify with models that when one ring changes its conformation, the other must change also. In *trans*-decalin, however, the six-membered rings can assume twist-boat conformations, but *they cannot change into their alternative chair conformations*. You should try the chair interconversion with a model of *trans*-decalin to verify this point for yourself. Focus on ring *B* of the *trans*-decalin structure in Eq. 7.31b. The two circles represent carbons that are in effect *equatorial* substituents on ring *A*. If ring *A* were to convert into the other chair conformation, these two carbons in ring *B* would have to assume *axial* positions, because, in the chair interconversion, equatorial groups become axial groups. When these two carbons are in axial positions, they are much farther apart than they are in equatorial positions; the distance between them is simply too great to be spanned easily by the remaining two carbons of ring *B*.

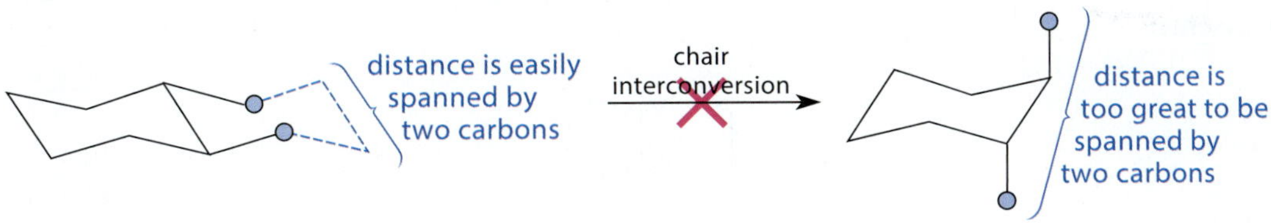

(7.32)

As a result, the chair interconversion introduces so much ring strain into ring *B* that the interconversion cannot occur. Exactly the same problem occurs with ring *A* when ring *B* undergoes the chair interconversion.

Focused Problem

7.19 How many 1,3-diaxial interactions occur in *cis*-decalin? In *trans*-decalin? Which compound has the lower energy and by how much? (*Hint:* Use your models, view the CH_2 groups attached to rings as if they were methyl groups, and don't count the same 1,3-diaxial interaction twice.)

Trans-decalin is more stable than *cis*-decalin (by 11.1 kJ mol^{-1} or 2.65 kcal mol^{-1}) because it has fewer 1,3-diaxial interactions (Focused Problem 7.19). Trans ring fusion, however, is not the more stable way of joining rings in all fused bicyclic molecules. In fact, if both of the rings are small, trans ring fusion is virtually impossible. For example, only the cis isomers of the following two compounds are known:

bicyclo[1.1.0]butane **bicyclo[3.1.0]hexane** (7.33)

Attempting to join two small rings with a trans ring junction introduces too much ring strain. The best way to see this is with models, using Focused Problem 7.20 as your guide.

Focused Problem

7.20 (a) Compare the difficulty of making models of the cis and trans isomers of bicyclo[3.1.0]hexane. (Don't break your models!) Which is easier to make? Why?

(b) Compare the difficulty of making models of *trans*-bicyclo[3.1.0]hexane and *trans*-bicyclo[5.3.0]decane. Which is easier to make? Explain.

In summary:

1. Two rings can in principle be fused in a cis or trans arrangement.
2. When the rings are small, only cis fusion is observed because trans fusion introduces too much ring strain.
3. In larger rings, both cis- and trans-fused isomers are well known, but the trans-fused ones are more stable because 1,3-diaxial interactions are minimized (as in the decalins).

Effects (2) and (3) are about equally balanced in the *hydrindanes* (bicyclo[4.3.0]nonanes); heats of combustion show that the trans isomer is only 4.46 kJ mol^{-1} (1.06 kcal mol^{-1}) more stable than the cis isomer.

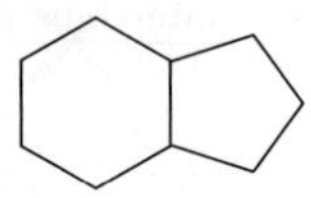

hydrindane
(bicyclo[4.3.0]nonane) (7.34)

C. Trans-Cycloalkenes and Bredt's Rule

Cyclohexene and other cycloalkenes with small rings have cis (or *Z*) stereochemistry at the double bond. In Sec. 4.5B, we noted that trans-cycloalkenes with six or fewer carbons have never been observed. Recall that the carbons attached to a trans double bond are so far apart that it is difficult to connect them with only two other carbon atoms. *Trans*-cyclooctene is the smallest trans-cycloalkene that can be isolated under ordinary conditions; however, it is 47.7 kJ mol^{-1} (11.4 kcal mol^{-1}) less stable than its cis isomer.

Closely related to the instability of trans-cycloalkenes is the instability of any small bridged bicyclic compound that has a double bond at a bridgehead atom. The following compound, for example, is very unstable and has never been isolated:

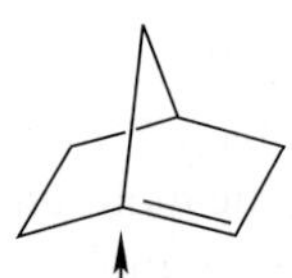

bridgehead

bicyclo[2.2.1]hept-1(2)-ene
(unknown) (7.35)

Julius Bredt (1855–1937) was a German chemist who, for the last 25 years of his career, was a professor and the director of the Organic Chemistry Laboratory at the Technische Hochschule of Aachen in Germany. In 1893, he proposed the correct (and then very unusual) bridged bicyclic structure of camphor, for which more than 30 incorrect structures had previously been proposed. His studies of bridged bicyclic compounds led him to formulate in 1913 the rule that bears his name.

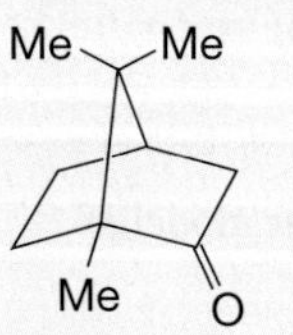

camphor

The instability of compounds with bridgehead double bonds has been generalized as **Bredt's rule**: *In a bicyclic compound, a bridgehead atom contained solely within small rings cannot be part of a double bond.* (A "small ring," for purposes of Bredt's rule, contains seven or fewer atoms within the ring.)

The basis of Bredt's rule is that double bonds at bridgehead carbons within small rings are twisted—that is, the atoms directly attached to such double bonds *cannot* lie in the same plane. To see this, try to construct a model of bicyclo[2.2.1]hept-1(2)-ene, the bicyclic alkene shown in Display 7.35. You will see that the bicyclic ring system cannot be completed without twisting the double bond. Twisting the double bond prevents overlap of the 2p orbitals required to form the π bond. This is similar to the double-bond twisting that would occur in *trans*-cyclohexene. Like the corresponding trans-cycloalkenes, bicyclic compounds containing bridgehead double bonds solely within small rings are too unstable to isolate. Bicyclic compounds that have bridgehead double bonds within larger rings, such as bicyclo[4.4.1]undec-1(2)-ene, are more stable and can be isolated.

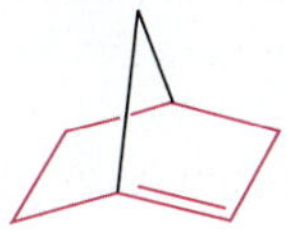

bicyclo[2.2.1]hept-1(2)-ene
The trans double bond is in a 6-membered ring; the compound is too unstable to isolate.

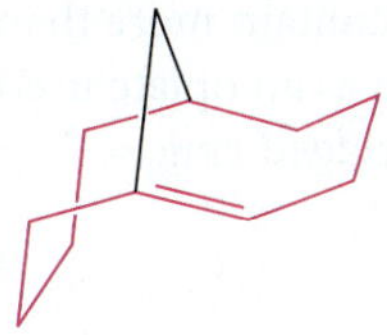

bicyclo[4.4.1]undec-1(2)-ene
The trans double bond is in a 10-membered ring; the compound is stable enough to isolate. (7.36)

Bredt's rule has implications for resonance structures, too. Resonance structures that place double bonds at a bridgehead in a small-ring bicyclic structure are not important. For example, Sec. 3.3B (Eq. 3.18) explained that amides have two important resonance structures.

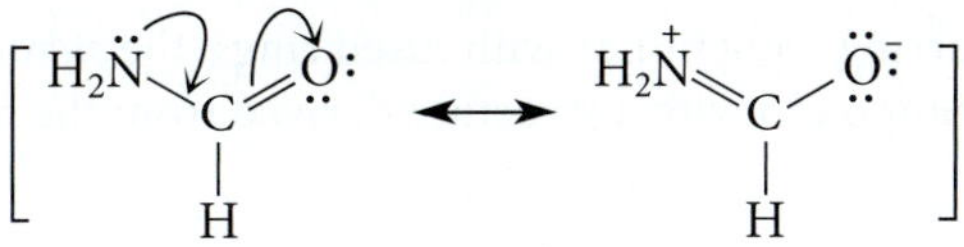

amide resonance (7.37a)

However, in the following bicyclic amide, resonance is not important because the resonance structure violates Bredt's rule.

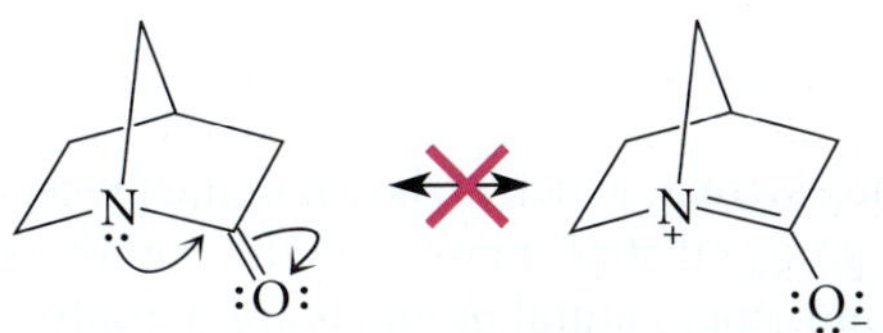

this resonance structure violates Bredt's rule (7.37b)

Remember that resonance implies orbital overlap. The basis of the Bredt's rule violation is that the bicyclic structure forces the sp^3 orbital of the nitrogen nonbonding electron pair and the carbonyl 2p orbitals to be perpendicular.

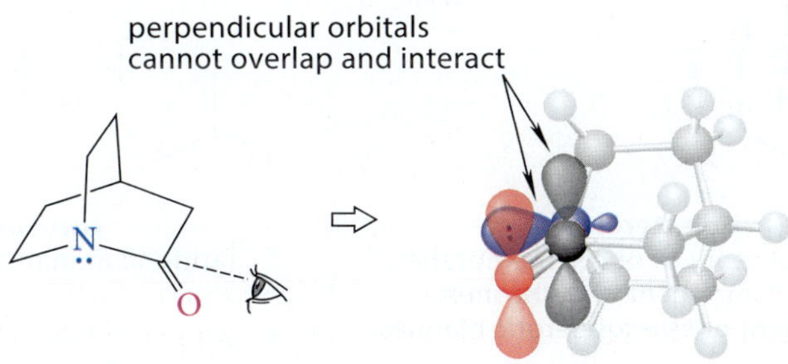

(7.37c)

Because perpendicular orbitals cannot overlap, resonance is impossible. In fact, this amide has been prepared, and it is very unstable. (See the sidebar in Sec. 21.5, "An Amide with a Twist.")

Focused Problem

7.21 Use models if necessary to help you decide which compound within each pair should have the greater heat of formation. Explain.

(a)

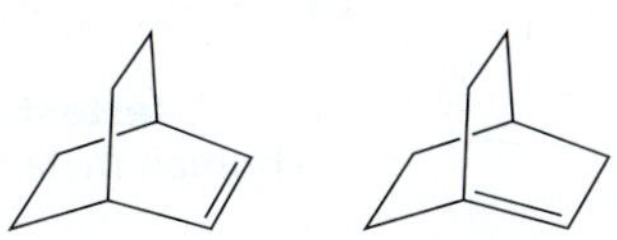

A *B*

(b)

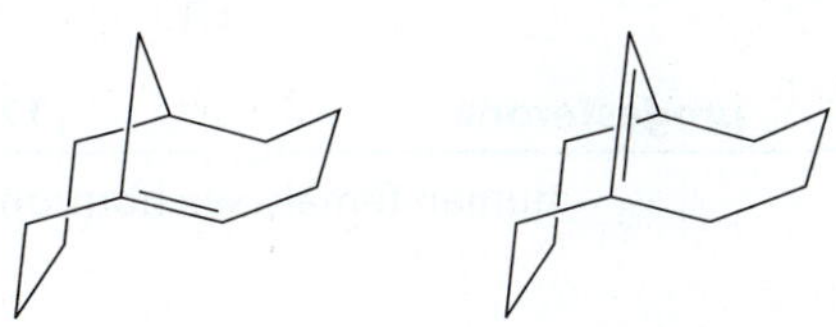

A *B*

Many compounds contain more than two rings. An example is the pentacyclic compound morphine, well known as an opiate used for both acute and chronic pain. Morphine has four fused rings and one bridged ring.

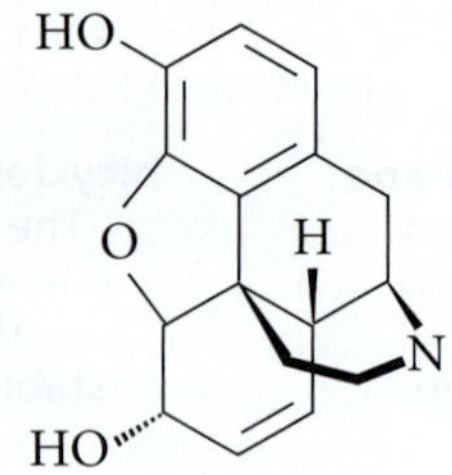

morphine (7.38)

D. Steroids

Of the many naturally occurring compounds with fused rings, the *steroids* are particularly important. A **steroid** is a compound with a structure derived from the following tetracyclic ring system:

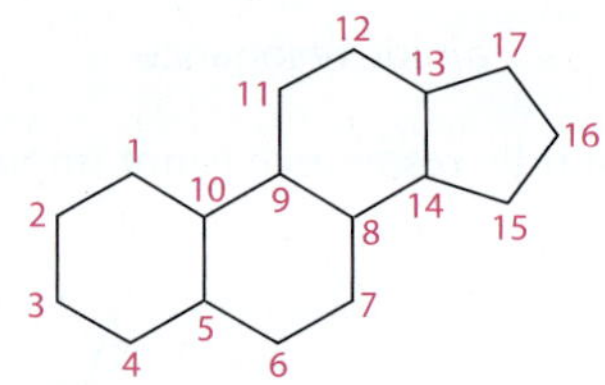

(7.39)

Steroids have a special numbering system, which is shown in the preceding structure. The various steroids differ in the functional groups that are present on this carbon skeleton.

Many important hormones and other natural products are steroids. Cholesterol occurs widely and was the first steroid to be discovered (1775). The corticosteroids and the sex hormones represent two biologically important classes of steroid hormones.

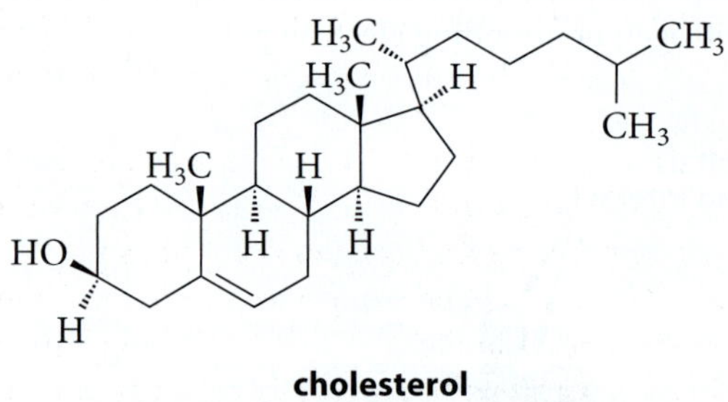

cholesterol
(important component of cell membranes: principal component of gallstones; major constituent of atherosclerotic plaques)

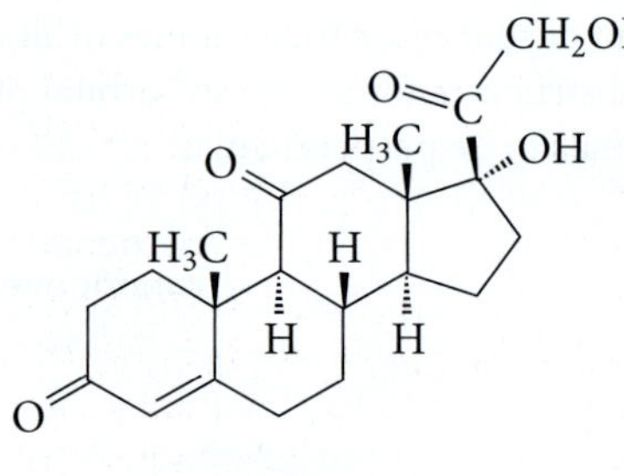

cortisone
(anti-inflammatory hormone)

(7.40a)

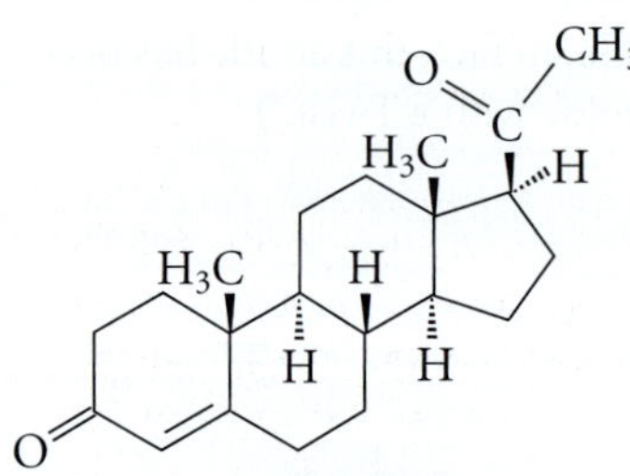

progesterone

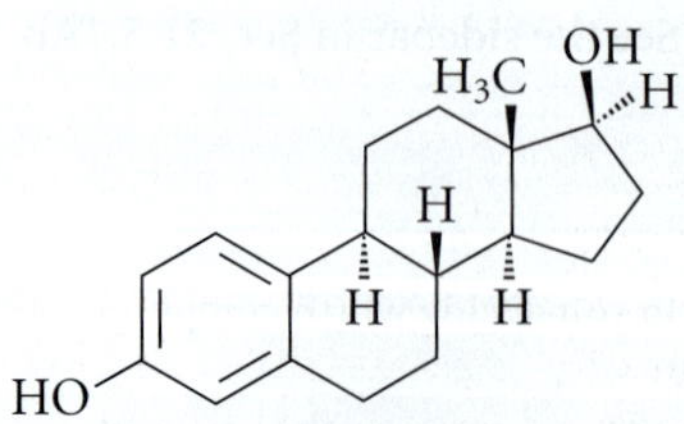

17-β-estradiol

human female sex hormones

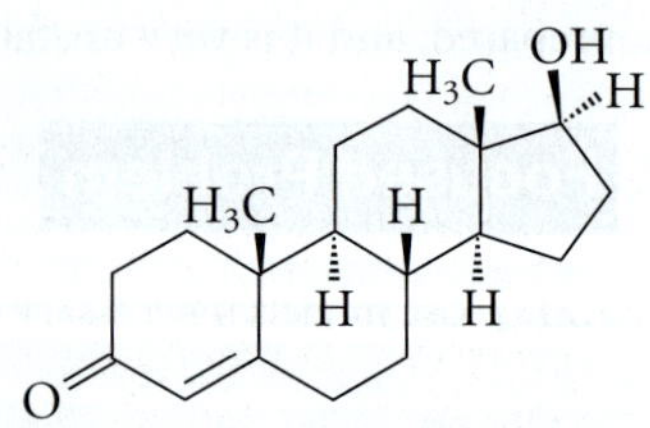

testosterone
a human male sex hormone

(7.40b)

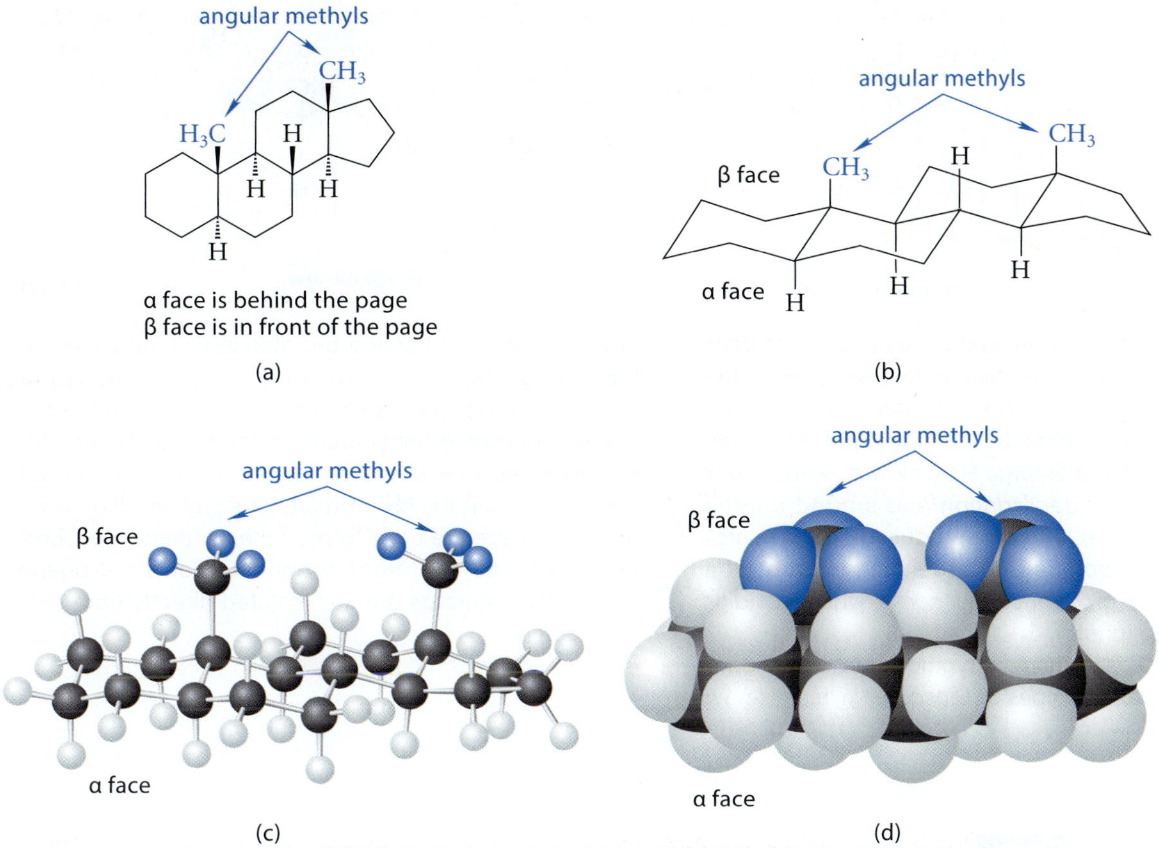

FIGURE 7.12 Four different representations of the steroid ring system. (a) A planar structure. (b) A perspective structure. (c) A ball-and-stick model. (d) A space-filling model. Notice the all-trans ring junctions and the extended, relatively flat shape. The hydrogens of the angular methyl groups are shown in blue in parts (c) and (d).

The compounds used in birth-control medications are all steroid analogs of progesterone and estradiol in which these natural steroidal female sex hormones have been synthetically modified to make them suitable for oral administration.

Two structural features are particularly common in naturally occurring steroids (**Fig. 7.12**). The first is that in many cases all ring fusions are trans (like *trans*-decalin). Because trans-fused cyclohexane rings cannot undergo the chair interconversion (see Eq. 7.32, Sec. 7.6B, and subsequent discussion), all-trans ring fusion causes a steroid to be conformationally rigid and relatively flat. This can be seen particularly with the models in Figs. 7.12c and d. Second, many steroids have methyl groups, called *angular methyls*, at carbons 10 and 13, so-called because each is located at the vertex of the angle at a bridgehead carbon. The hydrogens of these methyl groups are shown in color in Figs. 7.12c and d.

Sources of Steroids

© Nhu Nguyen

Prior to 1940, steroids were obtained only from such inconvenient sources as sows' ovaries or the urine of pregnant mares, so they were scarce and expensive. In the 1940s, however, a Pennsylvania State University chemist, Russell E. Marker (1902–1995), developed a process that could bring about the conversion of a naturally occurring compound called *diosgenin* into progesterone.

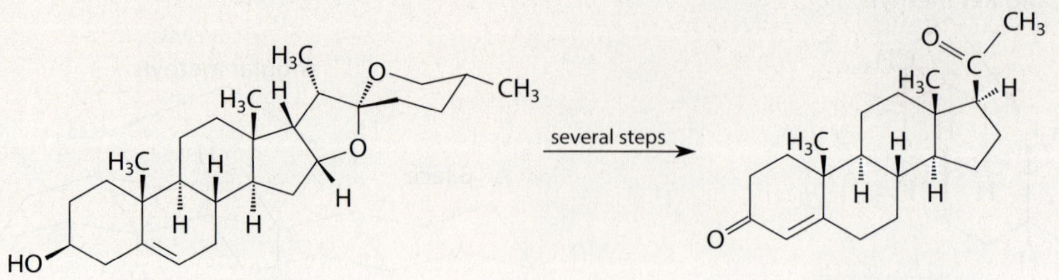

diosgenin **progesterone** (7.41)

(Various forms of this conversion, called the *Marker degradation*, are still in use.) Marker helped to found the pharmaceutical company Syntex. When he fell out with the management of Syntex, he left and took his process with him. Syntex hired George Rosenkranz, who reconstructed the Marker degradation and started a large research program that led to the production of other steroids. The natural source of diosgenin is the root of a vine, *cabeza de negro*, genus *Dioscorea*, which is indigenous to Mexico (photo).

About two-thirds of modern synthetic steroid production starts with various types of *Dioscorea*, which is now grown not only in Mexico but also in Central America, India, and China. More recently, practical industrial processes have been developed that start with steroid derivatives from other sources. In the United States, for example, a process was developed to recover steroid derivatives from the by-products of soybean-oil production, and these are used to produce synthetic glucocorticoids and other steroid hormones. Some estrogens and cardiac steroids are still isolated directly from natural sources.

7.7 REACTIONS INVOLVING STEREOISOMERS

The remainder of this chapter focuses on the importance of stereochemistry in organic reactions. To begin, we develop some general principles that apply to reactions that involve stereoisomers.

A. Reactions Involving Enantiomers

We consider two important situations in this section:

1. Reactions that involve a chiral compound as a starting material
2. Reactions that form enantiomers as products from achiral starting materials

Both situations can be illustrated by a biologically important reaction, the interconversion of malic acid and fumaric acid. (This is a reaction from the Krebs cycle, which is a sequence of reactions in biochemistry that is central to energy production in the cell.)

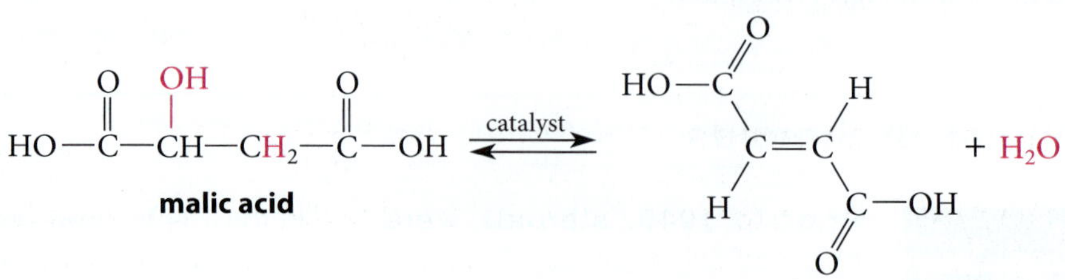

malic acid **fumaric acid** (7.42)

This is an example of alcohol dehydration (in the forward reaction) and alkene hydration (in the reverse direction). We considered this type of reaction in Secs. 4.10C and 4.9D. Because the reaction has an equilibrium constant near 1, it can be studied by starting with either fumaric acid or malic acid and allowing the two compounds to come to equilibrium. In the forward direction, a chiral compound (malic acid) reacts to give an achiral compound. In the reverse direction, an achiral compound reacts to give a chiral compound. This reaction requires an acid catalyst such as H_2SO_4 and a fairly high temperature, or it can be catalyzed by an enzyme, *fumarase*, at physiological pH (7.4) and temperature (37 °C).

Do enantiomers react at the same rate or at different rates? In other words, if we carry out the reaction in Eq. 7.42 with, first, (*R*)-malic acid, and then again with (*S*)-malic acid, do the two compounds have the same reactivity or different reactivities? A general principle applies to situations like this:

Enantiomers react at identical rates with an achiral reagent or catalyst.

This principle requires that the enantiomers of malic acid in Eq. 7.43 must react with water and an acid catalyst such as H_2SO_4—both achiral reagents—at exactly the same rates to give malic acid in exactly the same yield.

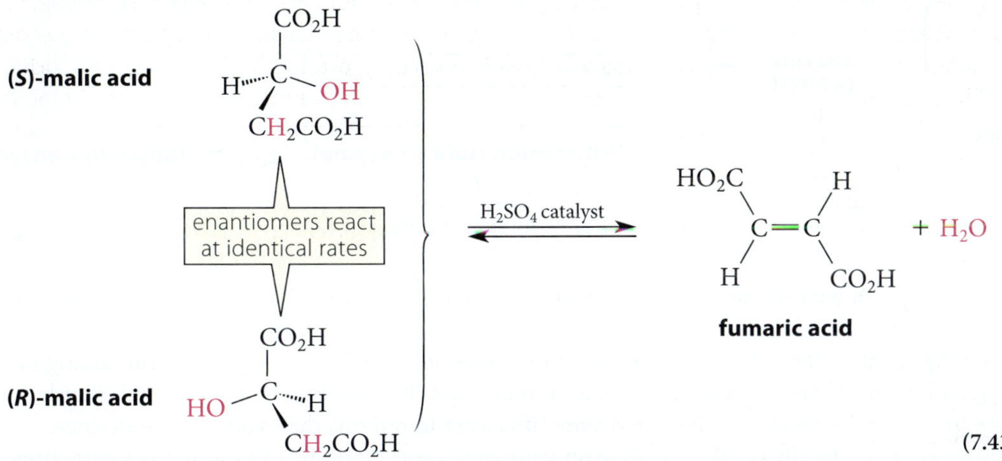

(7.43)

An analogy from common experience can help you understand why the rates are identical. Consider your feet, an enantiomeric pair of objects. Imagine placing first your right foot, then your left, in a perfectly rectangular box—an achiral object. Each foot will fit this box in exactly the same way. If the box pinches the big toe on your right foot, it will also pinch the big toe on your left foot in the same way. It should take you the same amount of time to put one foot in the box as it does the other (if we assume equal facility with both feet). Just as your feet interact in the same way with the achiral box, so enantiomeric molecules react in exactly the same way with achiral reagents. Because water and sulfuric acid are achiral reagents, the enantiomers of malic acid in Eq. 7.43 react in exactly the same way. According to the *principle of enantiomeric differentiation* (Sec. 6.10), *enantiomers behave differently only in the presence of a chiral agent.*

Enantiomers have identical free energies. That is, free energies, like boiling points and melting points, are among the properties that do not differ between enantiomers (Sec. 6.3). Both of the starting materials in Eq. 7.43 and their respective transition states are enantiomeric. The enantiomeric transition states have identical free energies, as do the enantiomeric starting materials. Because relative reactivity is determined by the difference in free energies of the transition state and starting material, and because this difference is identical for both enantiomers, enantiomers react at identical rates.

In contrast, when the reaction is catalyzed by the enzyme fumarase, the two enantiomers behave quite differently. In fact, (*S*)-malic acid reacts rapidly—the enzyme accelerates the reaction by a factor of about 10^9! However, the enzyme does *not* catalyze the reaction of (*R*)-malic acid.

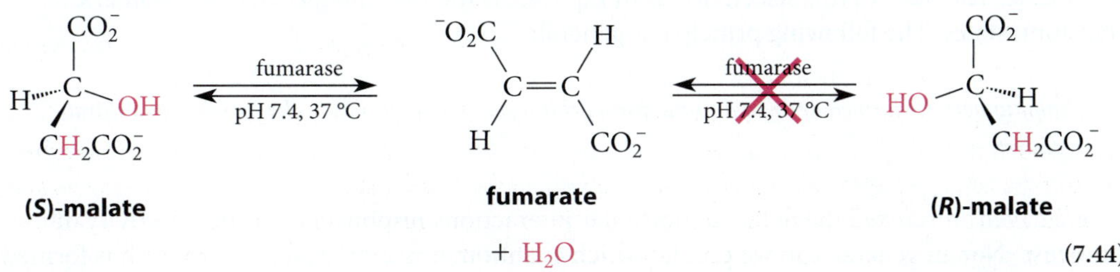

(7.44)

(The acids are shown as their ionized forms malate and fumarate because, like typical carboxylic acids, they are ionized at pH 7.4.) Why does the enzyme catalyze the reaction of one enantiomer but not the other? Fumarase and all other enzymes are *enantiomerically pure chiral molecules.* Therefore, by the principle of enantiomeric differentiation, the two enantiomers of malate react differently with the enzyme. Because the transition state for the reaction of water and malate includes the chiral enzyme, the *R* and the *S* transition states, which are enantiomers in the *absence* of the enzyme, are diastereomers in the *presence* of the enzyme, so their free energies are different. The faster reaction—the reaction of (*S*)-malate—has the transition state of lower free energy.

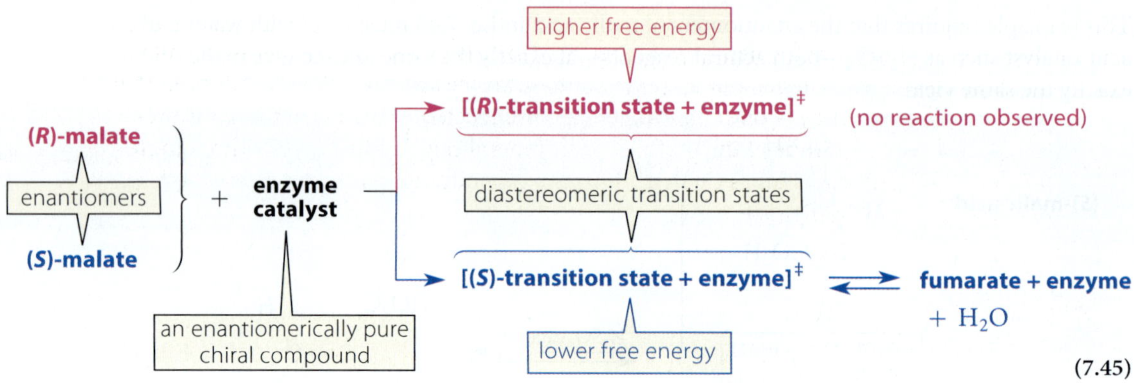

(7.45)

The principle of enantiomeric differentiation is operating here exactly as it does in enantiomeric resolution (Sec. 6.10). In fact, we can think of the enzyme fumarase in principle as a chiral resolving agent. As an analogy for this situation, imagine your left and right feet interacting in turn with your left shoe. Your feet are enantiomers, and the shoe is a "chiral reagent." Your left foot "reacts" more quickly with the left shoe (it's easier to put on) than your right foot does. Moreover, if you finally get the left shoe on your right foot, it doesn't fit as well; the interaction energy of the right foot and left shoe is higher than the interaction energy of the left foot and the left shoe. (We describe the interaction energy with the words "it doesn't fit" or "it fits.") The enzyme is analogous to the left shoe: it reacts with the two enantiomers of malic acid with different rates.

Now consider the second situation: the reaction of an achiral compound to give enantiomeric products. The principle of microscopic reversibility (Sec. 4.10C) requires that *a reaction and its reverse must have identical transition states.* Therefore, if the two enantiomers of malic acid undergo dehydration at the same rate, then the reverse reaction, hydration of fumaric acid, must give the two enantiomers of malic acid at the same rate. The following principle is general:

> *When chiral products are formed from achiral starting materials, both enantiomers of a pair are always formed at identical rates.*

In other words, the product is always the *racemate.* (This is the reason that racemates occur widely in chemistry.) A corollary of this principle is the following:

> *Optical activity never arises spontaneously in the reactions of achiral compounds.*

In the presence of an enzyme catalyst, as we have seen, only (*S*)-malate reacts to give fumarate. Therefore, in the reverse reaction, fumarate reacts in the presence of the enzyme to give only (*S*)-malate. The basis of this selectivity, as in Eq. 7.45, is the free energies of diastereomeric transition states. The following principle is general:

> *Enantiomers are formed at different rates from achiral starting materials in the presence of a chiral catalyst.*

We haven't discussed the detailed molecular interactions responsible for the selectivity of fumarase. Nor, in general, can we predict which enantiomer is more reactive, or which is formed selectively, without more information. The point here is simply to show the general principles that must be operating in cases like this.

To summarize:

1. Enantiomers react at identical rates with achiral reagents or catalysts and at different rates with chiral reagents or catalysts. Even a chiral environment, such as a chiral solvent, can in principle be enough to cause enantiomers to react at different rates. *How different* depends on the specific case and generally can't be predicted without more information. With enzymes as catalysts, the difference is in most cases so large that only one enantiomer reacts.

STUDY GUIDE LINK 7.2
Reactions of Chiral Molecules

2. When enantiomers are formed from achiral starting materials, the product is racemic *unless* the reaction is carried out under a chiral influence such as a chiral catalyst. In that case, the predominance of one enantiomer can be expected. Which enantiomer is preferred, and the magnitude of the preference, in general can't be predicted without more information. With enzymes as catalysts, however, the difference is in most cases so large that only one enantiomer is formed.

FURTHER EXPLORATION 7.3
Optical Activity

Focused Problems

7.22 Apply the principles of this section to solve each of the following problems.

(a) Assuming equal strength in both hands, would your right and left hands differ in their ability to drive a nail? To tighten a screw with a screwdriver?

(b) Imagine that a certain Mr. D. has been visited by a certain Mr. L. from elsewhere in the universe. Mr. D. and Mr. L. are alike in every way, except that they are noncongruent mirror images! You have to introduce each of them at an international press conference, but neither will agree to give his name. How would you tell them apart? (Several answers are possible.)

7.23 Tell whether the two enantiomers of alkene *A* react at the same or different rates with each of the following reagents, and explain your reasoning. Give the products of the reaction in each case.

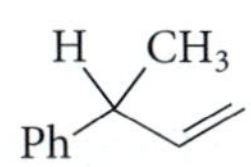

A

(a) H_2, catalyst

(b) 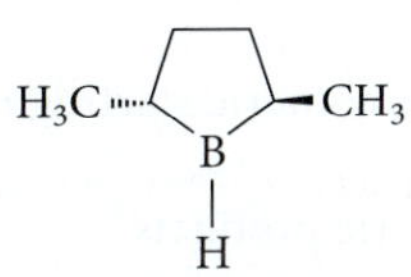

(2*R*,5*R*)-2,5-dimethylborolane
(a chiral borane)

Enantiomeric Resolution in Nature

Altin Osmanaj/Shutterstock

When a chiral compound occurs in nature, typically only one of its two enantiomers is found in a given natural source. That is, *nature is a source of optically active compounds*. For example, the sugar sucrose, produced by the sugarcane and sugar beet plants, occurs only as the dextrorotatory enantiomer, and the naturally occurring amino acid leucine is the levorotatory enantiomer.

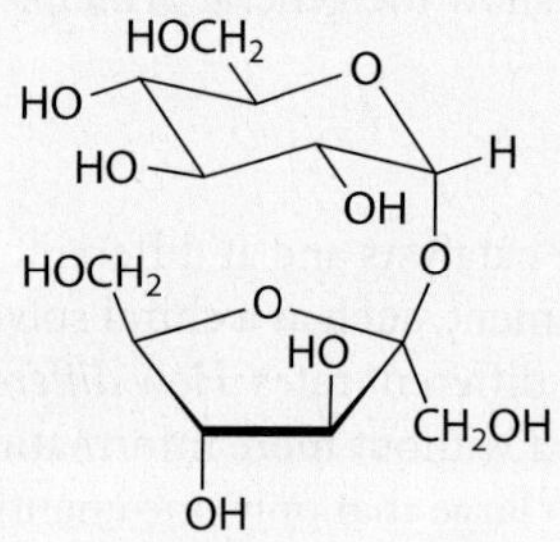

(+)-sucrose

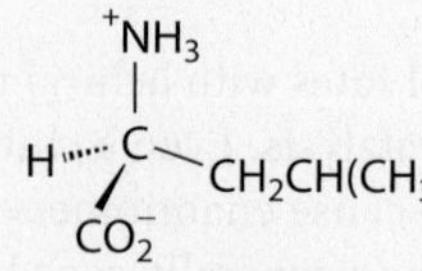

(−)-leucine

Many scientists hypothesize that, eons ago, the first chiral compounds were formed from simple, achiral starting materials, such as methane, water, and hydrogen cyanide (HCN). This hypothesis presents a problem. As shown in Sec. 7.7A, reactions that give chiral products from achiral starting materials always give the racemate; net optical activity cannot be generated in the reactions of achiral molecules. If the biological starting materials are all achiral, why is the world full of optically active compounds? Instead, it should be full of racemates! (This would mean that somewhere in the world your noncongruent mirror image is studying organic chemistry!) The only way out of this dilemma is to postulate that at some point in geologic time, one or more *enantiomeric resolutions* must have occurred. How could this have happened?

This question has generated much speculation. However, many scientists believe that the first enantiomeric resolution occurred purely by chance. Although we've said that a spontaneous enantiomeric resolution never occurs, a more accurate statement is that such an event is *highly improbable*. For example, spontaneous crystallization of one enantiomer can occur if a supersaturated solution of a racemate is seeded by a crystal of one enantiomer. Perhaps the spontaneous crystallization of a pure enantiomer took place on prebiotic Earth, seeded by a speck of dust with just the right shape. The question is an intriguing one, and no one really knows the answer.

Given that one or more enantiomeric resolutions occurred by chance at some time during the course of natural history, it is not difficult to understand how nature continues to produce enantiomerically pure compounds. You've just learned that enzymes catalyze the formation of optically active compounds from achiral starting materials, and, when the starting material of an enzyme-catalyzed reaction is chiral, an enzyme will catalyze the reaction of only one enantiomer. Such catalytic discrimination between stereoisomers guarantees a high degree of enantiomeric purity in naturally occurring compounds.

B. Reactions Involving Diastereomers

In this section, we consider two related situations:

1. The relative reactivities of diastereomeric starting materials in chemical reactions
2. Reactions that form diastereomeric products

Diastereomeric compounds in general have different reactivities toward *any* reagent, whether the reagent is chiral or achiral. The reason is that both the starting materials and the transition states are diastereomeric, and diastereomers have different free energies. Consequently, their standard free energies of activation, and therefore their reaction rates, must in principle differ. For example, the two diastereomeric alkenes *cis*- and *trans*-2-butene react at different rates with *any* reagent. We may not be able to predict which alkene is more reactive or by how much without further information. The cis stereoisomer may be more reactive with one reagent and the trans may be more reactive with another; but we can be sure that the two alkenes will not be equally reactive.

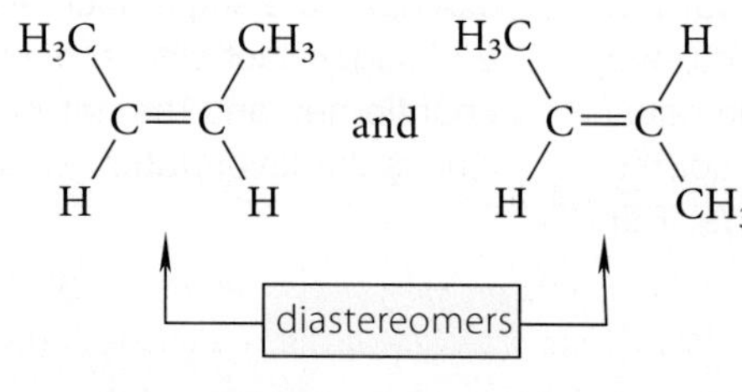

have different reactivities with any reagent

(7.46)

When reactions form diastereomeric products, the products are formed at different rates and therefore in different amounts. For example, when 1-methylcyclohexene undergoes hydroboration–oxidation, the diastereomeric 2-methylcyclohexanols—the cis and trans stereoisomers—are formed in different amounts.

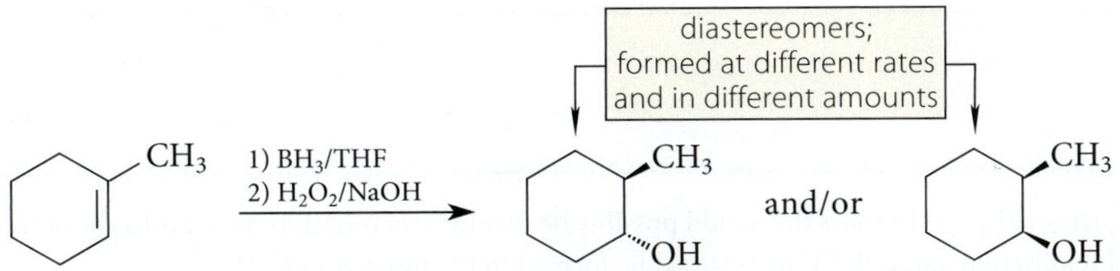

(±)-*trans*-2-methylcyclohexanol **(±)-*cis*-2-methylcyclohexanol**

Both products are racemates! (7.47)

Without knowing more about the reaction, we might not be able to predict which diastereomer is the predominant product, or by how much, but we can expect one to be formed in greater amount than the other. (In this case the trans diastereomer is the predominant one, as we discover in Sec. 7.8D.)

Diastereomers are formed in different amounts because they are formed through diastereomeric transition states. In general, one transition state has a lower standard free energy than its diastereomer. The diastereomeric reaction pathways therefore have different standard free energies of activation and therefore different rates, and their respective products are formed in different amounts.

Remember (Sec. 7.7A) that when the starting materials are achiral and the products are chiral, as in this example, each diastereomer of the product will be formed as a pair of enantiomers (the racemate). *Be aware of the following drawing convention*: For convenience we sometimes draw only one enantiomer of each product, as in this example, but in situations like this it is understood that each of these diastereomers *must* be racemic. The (±) indication in the product names reminds us that the products are racemic, but you should understand this point even if it is not specifically mentioned.

Study Problem 7.5

What stereoisomeric products could be formed in the addition of bromine to cyclohexene? Which should be formed in the same amounts? Which should be formed in different amounts?

Solution Before dealing with any issue involving the stereochemistry of a reaction, first be sure you understand the reaction itself. Bromine addition to cyclohexene gives 1,2-dibromocyclohexane:

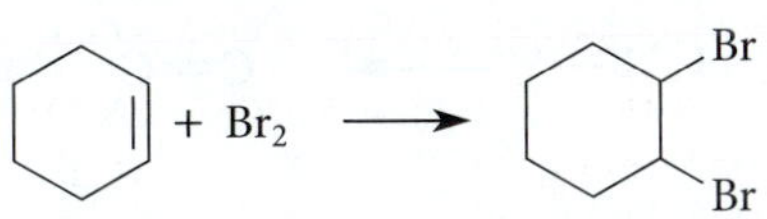

1,2-dibromocyclohexane

Next, enumerate the possible stereoisomers of the product that might be formed. The product, 1,2-dibromocyclohexane, can exist as a pair of diastereomers:

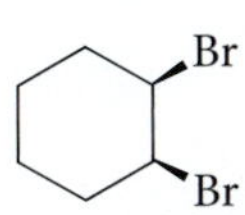

***cis*-1,2-dibromocyclohexane**

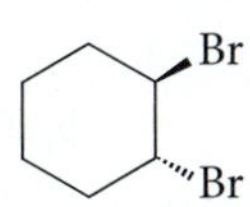

***trans*-1,2-dibromocyclohexane**

STUDY GUIDE LINK 7.3
Analysis of Reaction Stereochemistry

The trans diastereomer can exist as a pair of enantiomers, and the cis diastereomer is meso (Sec. 7.4B). Therefore, three potentially separable stereoisomers could be formed: the cis isomer and the two enantiomers of the trans isomer.

Because the cis and trans isomers are diastereomers, *they are formed in different amounts*. (You can't predict at this point which one predominates, but we'll return to that issue in Sec. 7.8C.) The two enantiomers of the trans diastereomer *must be formed in identical amounts*. Therefore, whatever the amount of the trans isomer we obtain from the reaction, it is obtained as the racemate—a 50:50 mixture of the two enantiomers.

Focused Problems

7.24 Without considering mechanisms, show all stereoisomers that could possibly be formed when *cis*-2-butene undergoes bromine addition? Which would be formed in different amounts? Which would be formed in the same amounts?

7.25 What stereoisomeric products are possible when *trans*-2-butene undergoes hydroboration–oxidation? Which are formed in different amounts? Which are formed in the same amounts?

7.26 Without considering mechanisms, draw the structures of all the possible products that might form when racemic 3-methylcyclohexene reacts with Br_2. What is the relationship of each pair? Which compounds should in principle be formed in the same amounts, and which in different amounts? Explain.

7.8 STEREOCHEMISTRY OF CHEMICAL REACTIONS

We've learned that stereochemistry adds another dimension to the study and practice of organic chemistry. No chemical structure is complete without stereochemical detail, and no chemical reaction can be planned without considering problems of stereochemistry that might arise. This section examines the possible stereochemical outcomes of two general types of reactions: addition reactions and substitution reactions. Then, some addition reactions covered in Chapter 5 will be revisited with particular attention to their stereochemistry.

A. Stereochemistry of Addition Reactions

In an *addition reaction*, a general species X—Y adds to each end of a bond. The cases we've studied so far involve addition to double bonds and triple bonds:

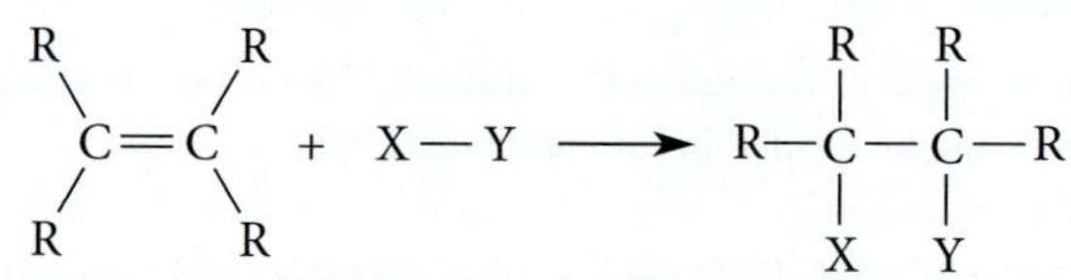

(7.48a)

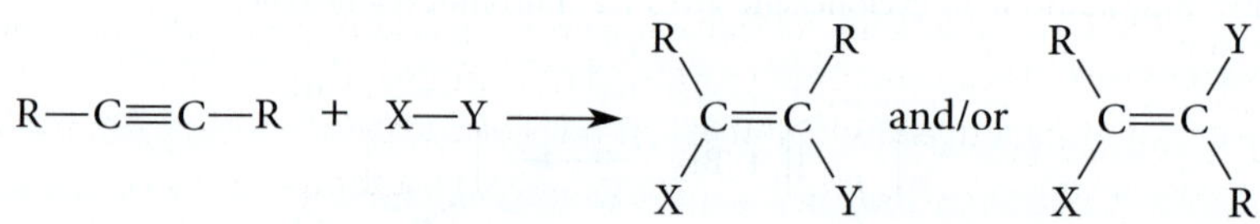

(7.48b)

Addition to a Double Bond The stereochemistry of addition to a double bond is discussed with reference to the plane that contains the double bond and its four attached groups. The sides of this plane are called **faces**. The side of the plane nearest the observer is typically called the *top face*, whereas the other side is the *bottom face*.

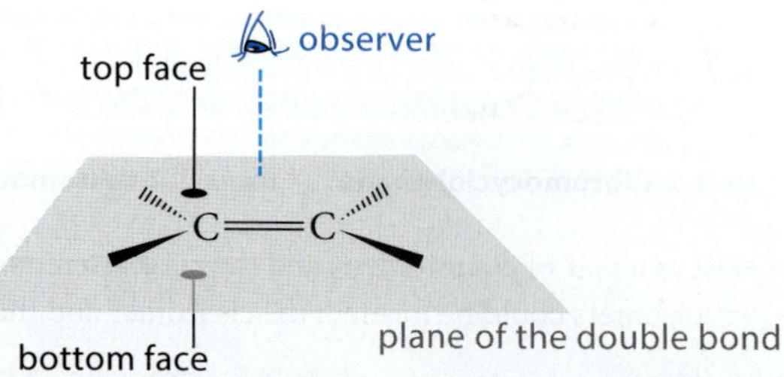

(7.49)

An addition reaction can occur in either of two stereochemically different ways, called *syn addition* and *anti addition*. These will be illustrated with cyclohexene and a general reagent X—Y. In a **syn addition**, two groups add to a double bond from the same face.

Syn addition:

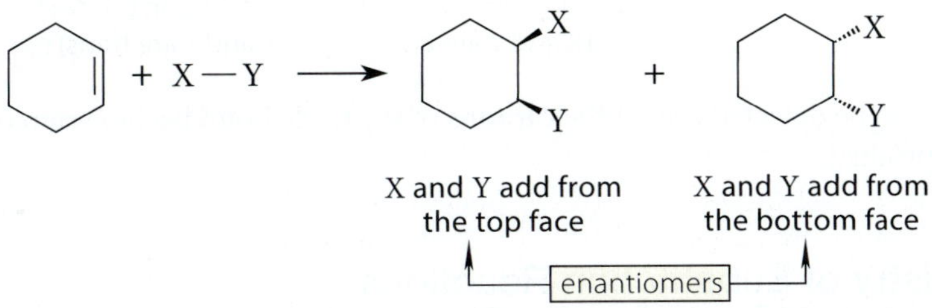

(7.50a)

When X and Y are different, the two directions of syn addition are *enantiomeric*—that is, the products from addition at the top face are enantiomers of the products formed from addition at the bottom face. (When X and Y are the same, the two syn-addition products are identical because the product is a meso compound.)

In an **anti addition**, two groups add to a double bond from opposite faces.

Anti addition:

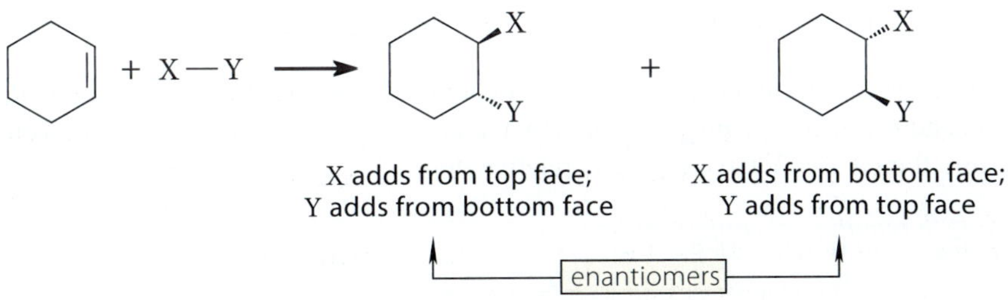

(7.50b)

The two directions of anti addition are also enantiomeric, whether X and Y are the same or different.

> *Syn addition* is sometimes called *cis addition*, and *anti addition* is sometimes called *trans addition*. The terms syn and anti are preferred because cis and trans are used for alkene stereochemistry. We want to avoid possible confusion of the stereochemistry of the addition with the stereochemistry of the alkene that is undergoing addition.

Some additions can occur as a mixture of syn and anti. In such a reaction, the products would be a mixture of all of the products in both Eqs. 7.50a and b. Examples of both syn and anti additions, as well as mixed additions, are examined later in this section.

As Eqs. 7.50a and b suggest, the syn and anti modes of addition can be distinguished *by analyzing the stereochemistry of the products*. In Eq. 7.50a, for example, the cis relationship of the groups X and Y in the product would tell us that a syn addition has occurred. *The stereochemistry of an addition can be determined only when the stereochemically different modes of addition give rise to stereochemically different products.* In contrast, when two groups X and Y add to ethylene ($H_2C{=}CH_2$), the same product (X—CH_2—CH_2—Y) results whether the reaction is a syn or an anti addition. Because this product does not exist as stereoisomers, we can't tell whether the addition is syn or anti. A more general way of stating the same point is to say that syn and anti additions give different products only when *both* carbons of the double bond become carbon stereocenters in the product. The question of syn and anti addition is a question of the relative stereochemistry at *both* carbons, so the relative stereochemistry is meaningless if both carbons aren't stereocenters.

Addition to a Triple Bond In a syn addition to a triple bond, the two groups that add end up on the same side of the double bond. In other words, the two groups that add have a cis relationship in the product. In an anti addition, the two groups that add have a trans relationship in the product.

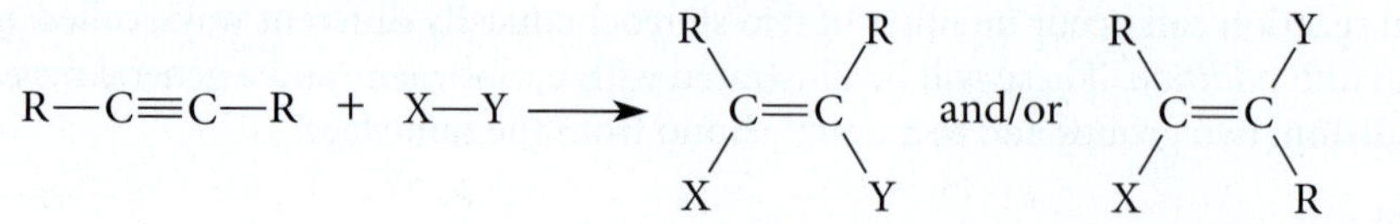

syn-addition product (X and Y are cis) | anti-addition product (X and Y are trans) (7.51)

Syn and anti addition can be detected only if the carbons of the triple bond become stereocenters in the alkene product.

B. Stereochemistry of Substitution Reactions

In a **substitution reaction**, one group is replaced by another. In the following substitution reaction, for example, the Br is replaced by OH:

$$H_3C{-}Br + {}^{-}{:}\ddot{O}H \longrightarrow H_3C{-}OH + {:}\ddot{Br}{:}^{-} \quad (7.52)$$

The oxidation step of hydroboration–oxidation is also a substitution reaction in which the boron is replaced by an OH group.

$$^{-}OH + 3HO{-}OH + (CH_3CH_2)_3B \longrightarrow 3CH_3CH_2{-}OH + \bar{B}(OH)_4 \quad (7.53)$$

A substitution reaction can occur in two stereochemically different ways, called *retention of configuration* and *inversion of configuration.* When a group Y replaces a group X with **retention of configuration**, then X and Y have the same relative stereochemical positions.

Substitution with retention of configuration:

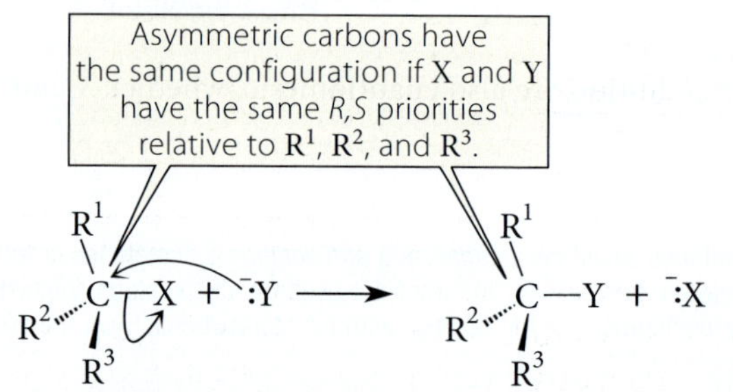

(7.54a)

Substitution with retention also implies that if X and Y have the same priorities relative to R^1, R^2, and R^3 in the *R,S* system, then the carbon that undergoes substitution will have the same configuration in the reactant and the product. So, if this carbon has, for example, the *S* configuration in the starting material, then it will have the same (that is, *S*) configuration in the product.

When substitution occurs with **inversion of configuration**, then X and Y have different relative stereochemical positions. Specifically, the incoming group Y:$^-$ must form a bond to the asymmetric carbon atom from the side *opposite* the departing group X:$^-$. To make room for Y and to maintain the tetrahedral configuration of carbon, the three R groups must move as shown by the green arrows:

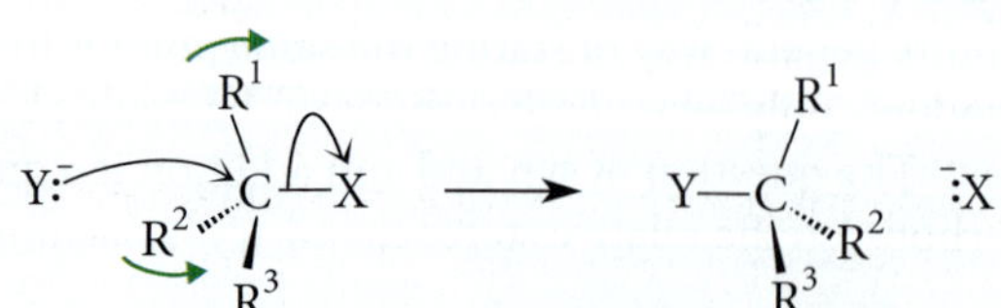

(7.54b)

This motion very much resembles what happens in amine inversion (see Fig. 6.9, Sec. 6.9B).

Substitution with inversion also implies that *if X and Y have the same priorities* relative to R^1, R^2, and R^3 in the *R,S* system, then the carbon that undergoes substitution must have opposite configurations in the reactant and the product. So, if this carbon has, for example, the *R* configuration in the starting material, then it will have the opposite (that is, *S*) configuration in the product.

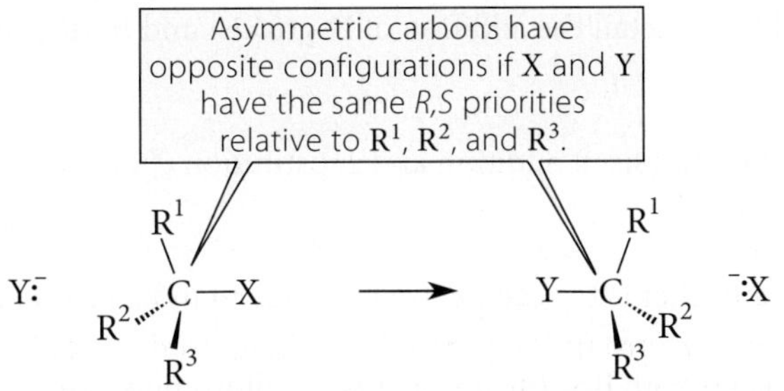

(7.54c)

As with addition, it is also possible that a reaction might occur so that both retention and inversion can occur at comparable rates in a substitution reaction. In such a case, stereoisomeric products corresponding to both pathways will be formed. Examples of substitution reactions with inversion, retention, and mixed stereochemistry are all well known.

As Eqs. 7.54a and b suggest, analysis of the stereochemistry of substitution requires that the carbon that undergoes substitution must be a stereocenter in both the reactants and the products. In the following situation, for example, the stereochemistry of substitution *cannot* be determined.

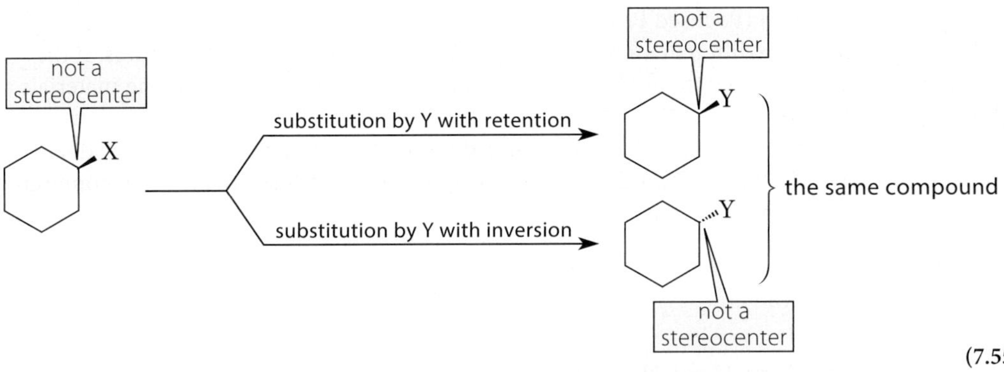

(7.55)

Because the carbon that undergoes substitution is *not* a stereocenter, the same product would be obtained from both the retention and inversion modes of substitution.

A reaction in which particular stereoisomers of the product are formed in significant excess over others is said to be a **stereoselective reaction**. Therefore, an addition that occurs only with anti stereochemistry, as shown in Eq. 7.50b, is a stereoselective addition reaction because only one diastereomer is formed to the exclusion of the other. A substitution that occurs only with inversion, as shown in Eq. 7.54b, is also a stereoselective substitution reaction because one enantiomer of the product is formed to the exclusion of the other.

To summarize, this section has established the stereochemical possibilities that might be expected in addition and substitution reactions:

1. An addition reaction can occur with syn or anti stereochemistry, or a mixture of the two.

2. A substitution reaction can occur with retention or inversion of configuration, or a mixture of the two.

3. The carbons of the π bond at which an addition reaction occurs must become stereocenters in the product in order to detect the reaction stereochemistry.

4. The carbon at which a substitution reaction occurs must be a stereocenter in both the reactant and the product in order to detect the reaction stereochemistry.

The remaining parts of this section apply these ideas in discussing the stereochemical aspects of several reactions that were first introduced in Chapter 5.

C. Stereochemistry of Bromine Addition

The addition of bromine to alkenes (Sec. 5.2A) is in many cases a highly stereoselective reaction. In this section we consider in detail the addition of Br_2 to *cis*- and *trans*-2-butene with two objectives in mind:

1. To illustrate the analysis of addition and substitution chemistry with the reactions of acyclic compounds
2. To see how stereochemistry can be used to understand a reaction mechanism

When *cis*-2-butene reacts with Br_2, the product is 2,3-dibromobutane.

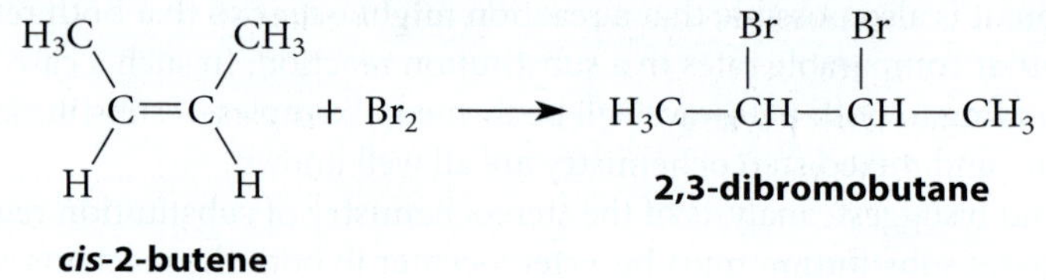

(7.56)

You should now realize that three stereoisomers of this product are possible: a pair of enantiomers and the meso compound (Focused Problem 7.24). The meso compound and the enantiomeric pair should be formed in different amounts (Sec. 7.7B) because they are diastereomers. If the enantiomers are formed, they should be formed as the racemate because the starting materials are achiral (Sec. 7.7A).

When bromine addition to *cis*-2-butene is carried out in the laboratory, the only product is the racemate. Bromine addition to *trans*-2-butene, in contrast, gives exclusively the meso compound. To summarize these results:

Experimental facts:

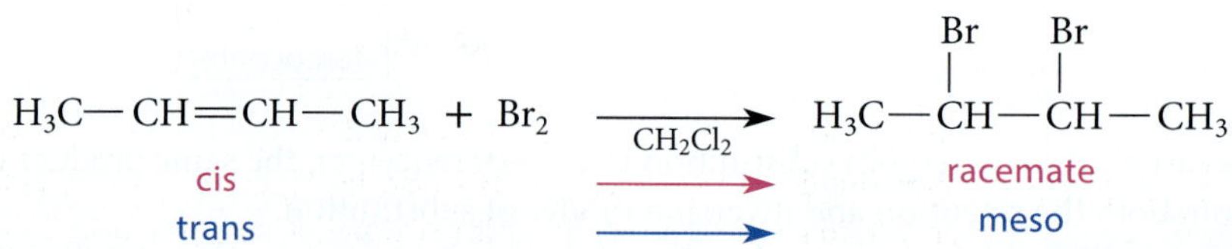

(7.57)

This information indicates that addition reactions of bromine to both *cis*- and *trans*-2-butene are highly stereoselective. Are these additions syn or anti? Because the alkene is not cyclic (unlike the alkene in Eq. 7.50a and b), the answer is not obvious. Study Problem 7.6 illustrates how to analyze the result systematically to get the answer.

Study Problem 7.6

According to the experimental results in Eq. 7.57, is the addition of bromine to *cis*-2-butene a syn or an anti addition?

Solution To answer this question, imagine *both* syn and anti additions to *cis*-2-butene and see what results would be obtained for each. Comparison of these results with the experimental facts then shows us which alternative is correct.

If bromine addition were syn, the Br_2 could add to either face of the double bond. (In the following structures, we are viewing the alkene edge-on, as in Eq. 7.49.)

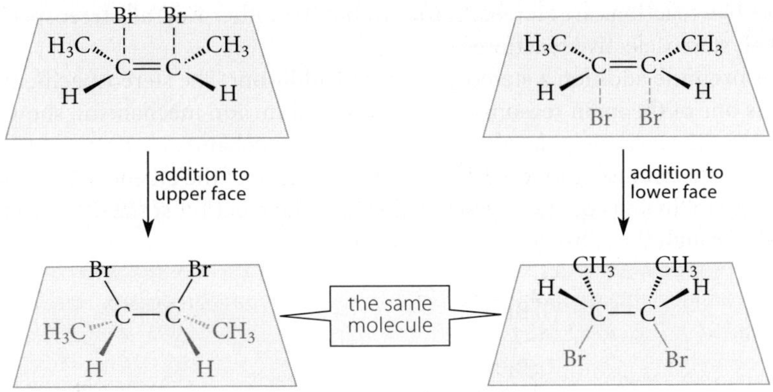

***meso*-2,3-dibromobutane** (7.58)

This analysis shows that syn addition from either direction gives the meso diastereomer. The experimental facts (Eq. 7.57) show that *cis*-2-butene does *not* give the meso isomer; so, the two bromine atoms *cannot* be adding from the same face of the molecule. Therefore, syn addition does *not* occur.

Because bromine addition is not a syn addition, presumably it is an anti addition. Let's verify this. Anti addition, too, can occur in two equally probable ways.

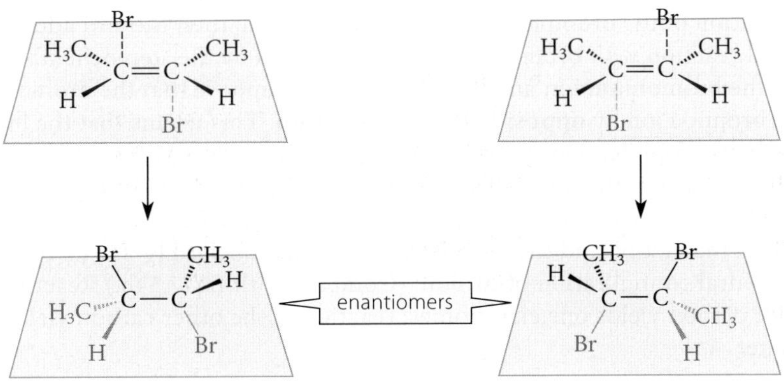

2*S*,3*S* 2*R*,3*R*

(±)-2,3-dibromobutane (7.59)

This analysis shows that each mode of addition gives the enantiomer of the other—that is, the two modes of anti addition operating at the same time should give the racemate. Because the experimental facts of Eq. 7.57 show that bromine addition to *cis*-2-butene indeed gives the racemate, this reaction is an anti addition.

You should now analyze the addition of bromine to *trans*-2-butene in a similar manner to show that this addition, too, is an anti addition.

As suggested at the end of Study Problem 7.6, you should have demonstrated to yourself that the addition of bromine to *trans*-2-butene is also a stereoselective anti addition. In fact, the bromine addition to most simple alkenes occurs exclusively with anti stereochemistry. Bromine addition is therefore a *stereoselective anti-addition reaction*.

The study of the stereochemistry of bromine addition to the 2-butenes raises an important philosophical point. To claim that bromine addition to the 2-butenes is an anti addition requires that the reaction be investigated on *both* the cis and trans stereoisomers of 2-butene. It is conceivable that, in the absence of experimental evidence, anti addition might have been observed with one stereoisomer of the 2-butenes and syn addition with the other. Had this been the result, the bromine-addition reactions would still be highly stereoselective, but we could not have made the *more general* claim that bromine addition to the 2-butenes is an anti addition.

Reactions such as bromine addition, in which different stereoisomers of a starting material give different stereoisomers of a product, are called **stereospecific reactions**. As the discussion in the previous paragraph demonstrates, all stereospecific reactions are stereoselective, but not all

STUDY GUIDE LINK 7.4
Stereoselective and Stereospecific Reactions

stereoselective reactions are stereospecific. To put it another way, all stereospecific reactions are a *subset* of all stereoselective reactions.

Why is bromine addition a stereospecific anti addition? The stereospecificity of bromine addition is one of the main reasons that the bromonium-ion mechanism, shown in Eqs. 5.10a and b in Sec. 5.2A, was postulated. Let's see how this mechanism can account for the observed results. First, the bromonium ion can form at either face of the alkene. (Reaction at one face is shown in the following equation; you should show the reaction at the other face and take your structures through the subsequent discussion.)

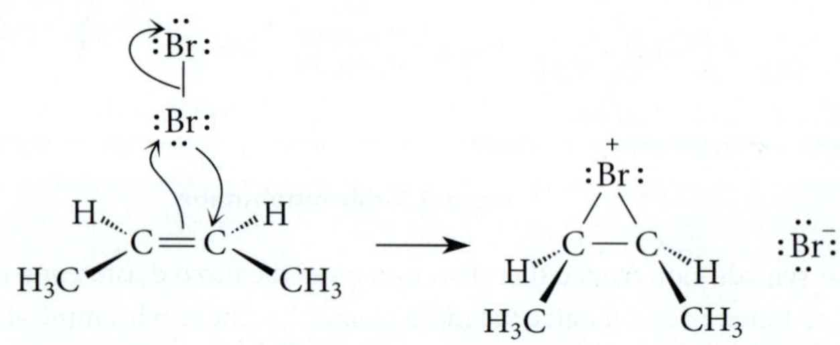

(7.60)

Bromonium-ion formation as represented here is a syn addition because both of the C—Br bonds *must be* formed at the same face of the alkene. (Formation of a bromonium ion with a 180° dihedral angle between the two carbon–bromine bonds is stereochemically impossible.)

If formation of the bromonium ion is a syn addition, then the anti addition observed in the overall reaction with bromine must be established by the stereochemistry of the reaction between the bromonium ion and the bromide ion. Suppose that the bromonium ion reacts with the bromide ion by **opposite-side substitution**. This means that the bromide ion, acting as a nucleophile, donates an electron pair to a carbon at the face opposite to the bond that breaks, which in this case is the carbon–bromine bond. *An opposite-side substitution reaction must occur with inversion of configuration* (Sec. 7.8B), because, as the substitution takes place, the methyl and the hydrogen must swing upward (*green arrows*) to maintain the tetrahedral configuration of carbon. (Compare with Eq. 7.54b.) Reaction of the bromide ion at one carbon yields one enantiomer; reaction at the other carbon yields the other enantiomer.

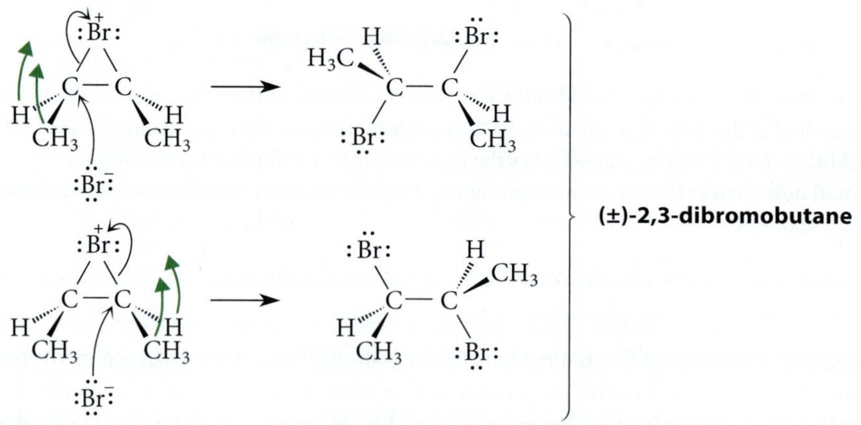

(7.61)

This analysis shows that formation of a bromonium ion followed by *opposite-side substitution* of bromide is a mechanism that accounts for the observed anti addition of Br_2 to alkenes. In general, when a nucleophile reacts at a saturated carbon atom in any substitution reaction, opposite-side substitution is observed. (Opposite-side substitution is explored further in Chapter 9.)

Could other mechanisms be consistent with the anti stereochemistry of bromine addition? Let's see what sort of prediction a carbocation mechanism makes about the stereochemistry of the reaction.

Imagine that the addition of Br_2 to *cis*-2-butene involves instead a carbocation intermediate. (Bromine addition at only the upper face is shown in Display 7.62; the equally probable addition to the lower face gives the enantiomeric carbocation.) If the carbocation lasts long enough to undergo at least one internal rotation, then both diastereomers of the products would be formed even if a bromonium ion formed subsequently:

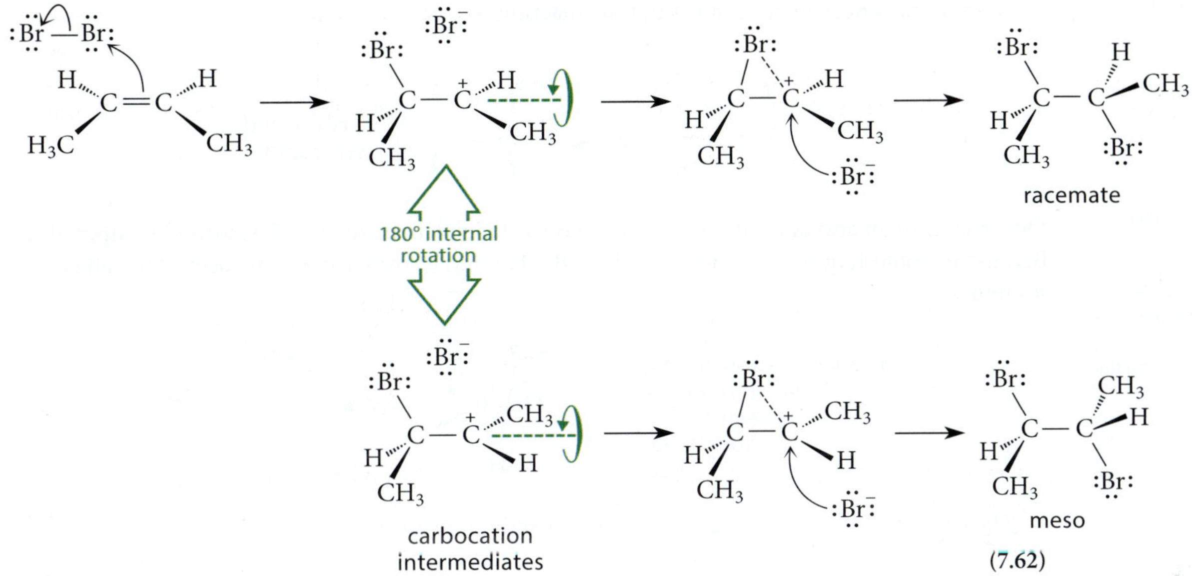

(7.62)

The reaction, then, would not be stereoselective. Because this result is not observed (Eq. 7.57), a carbocation mechanism does not agree with the data. Moreover, this mechanism is *not* in accord with the absence of rearrangements in bromine addition. The bromonium-ion mechanism, however, accounts for the results in a direct and simple way. The credibility of this mechanism has been enhanced by the direct observation of bromonium ions under special conditions. In 1985, the structure of a bromonium ion was determined by X-ray crystallography.

Does the observation of anti stereochemistry *prove* the bromonium-ion mechanism? The answer is no. *No mechanism is ever proved.* Chemists deduce a mechanism by gathering as much information as possible about a reaction, such as its stereochemistry, the presence and absence of rearrangements, and so on, thereby ruling out all mechanisms that do *not* fit the experimental facts. If someone can think of another mechanism that explains the facts, then that mechanism is just as good until someone finds a way to decide between the two by a new experiment.

Focused Problems

7.27 Assuming the operation of the bromonium-ion mechanism, give the structure of the product(s) (including all stereoisomers) expected from bromine addition to cyclohexene. (See Study Problem 7.5 in Sec. 7.7B.)

7.28 In view of the bromonium-ion mechanism for bromine addition, which of the products in your answer to Focused Problem 7.26 (Sec. 7.7B) are likely to be the major ones?

D. Stereochemistry of Hydroboration–Oxidation

Because hydroboration–oxidation involves two distinct reactions, its stereochemical outcome is a consequence of the stereochemistry of both reactions.

Hydroboration is a stereospecific syn addition.

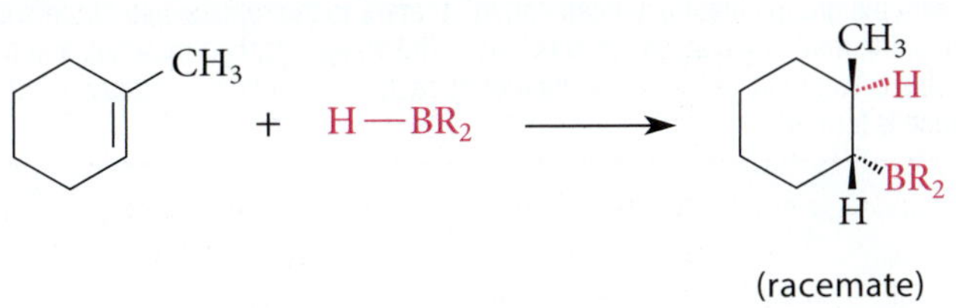

(7.63)

Notice again the structure-drawing convention used here: Even though just one enantiomer of the product is shown, the product is racemic because the starting materials are achiral (Sec. 7.7A).

The syn addition of borane, along with the absence of rearrangements, provides the major evidence for a concerted mechanism of the reaction.

The product of hydroboration is shown in an eclipsed conformation. The reaction is so favorable that it occurs *despite* the eclipsed (or nearly eclipsed) transition state for hydroboration. Internal rotation of the product to a more stable staggered conformation undoubtedly occurs instantaneously following the reaction.

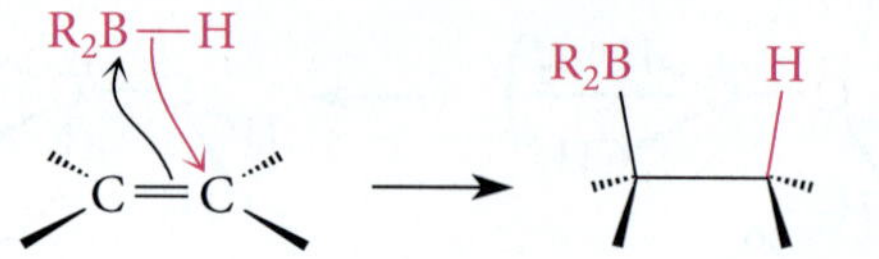

a concerted syn addition

(7.64a)

Occurrence of an anti addition by the same concerted mechanism would be virtually impossible, because it would require an abnormally long B—H bond to bridge opposite faces of the alkene π bond.

a concerted anti addition would require an unrealistic B—H bond length

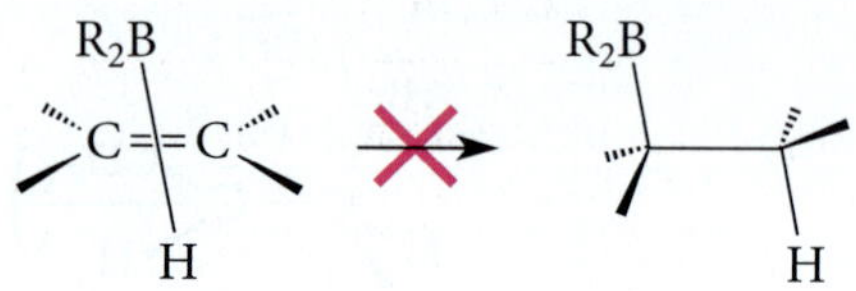

(7.64b)

The oxidation of organoboranes is a stereospecific *substitution reaction* that occurs with *retention of stereochemical configuration*.

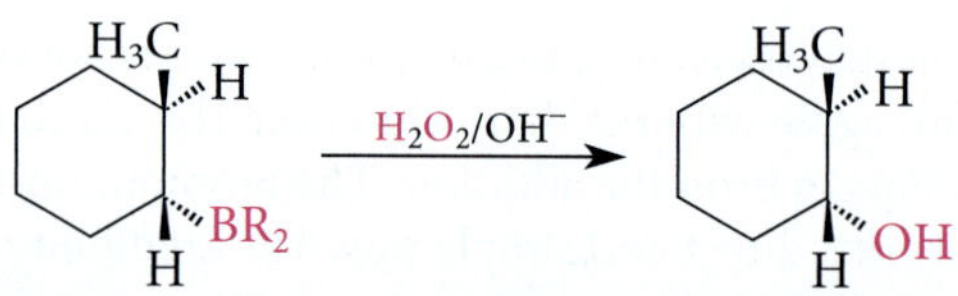

***trans*-2-methylcyclohexanol** (7.65)

FURTHER EXPLORATION 7.4
Stereochemistry of Organoborane Oxidation

We won't consider the mechanism of this substitution in detail here, but we can certainly conclude that it does *not* involve opposite-side nucleophilic substitution. (Why?) (The mechanism is shown in Further Exploration 7.4.)

The results from Eqs. 7.64a and 7.65 taken together show that hydroboration–oxidation of an alkene brings about the net syn addition of the elements of H—OH to the double bond.

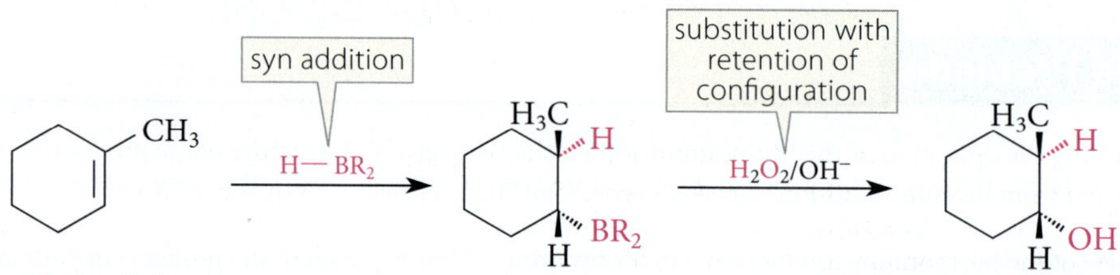

1-methylcyclohexene

(±)-*trans*-2-methylcyclohexanol

(7.66)

As far as is known, *all* hydroboration–oxidation reactions of alkenes are stereospecific syn additions.

The H and OH are added in a syn manner. The *trans* designation in the name of the product of Eq. 7.64 has nothing to do with the groups that have been added; it refers to the relationship of the substituents on the cyclohexane ring: the methyl group, which was part of the alkene starting material, and the OH group. Also, notice again the drawing convention: only one enantiomer of each chiral molecule is drawn, but it is *understood* that the racemate of each is formed.

Focused Problem

7.29 What products, including their stereochemistry, should be obtained when each of the following alkenes is subjected to hydroboration–oxidation? (D = deuterium = ^2H.)

(a) 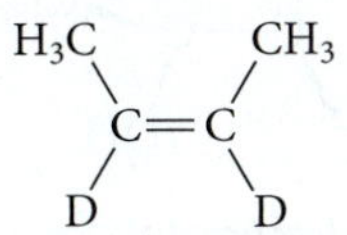

(b) 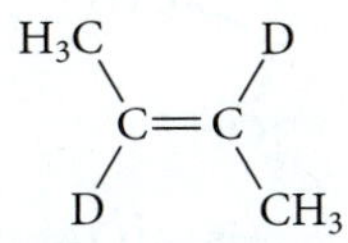

E. Stereochemistry of Other Addition Reactions

Catalytic Hydrogenation Catalytic hydrogenation of most alkenes (Sec. 4.10A) is a stereospecific syn addition. The following example is illustrative; the products are shown in eclipsed conformations for ease in seeing the stereochemical relationships.

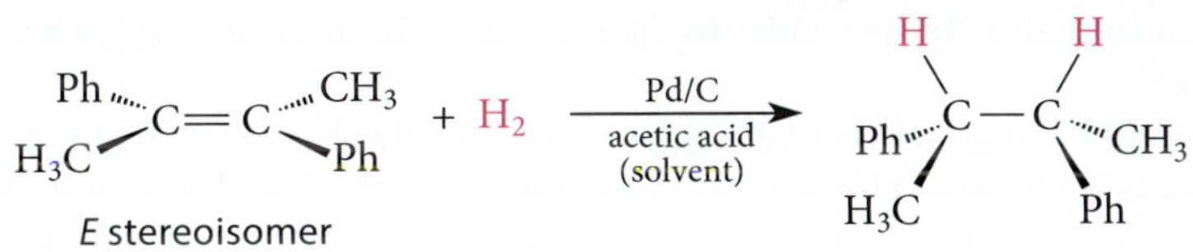

E stereoisomer

racemate (7.67a)

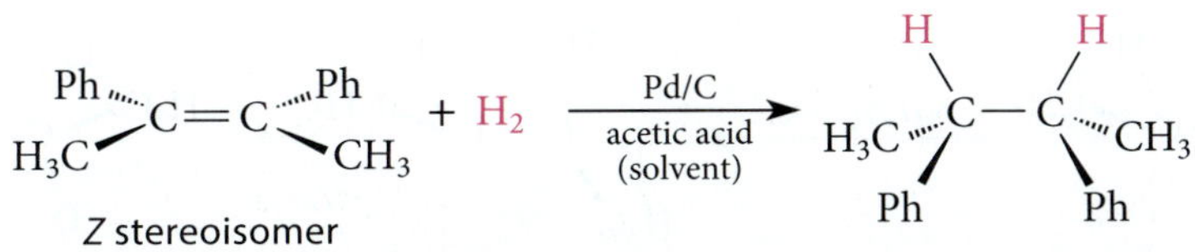

Z stereoisomer

meso stereoisomer (7.67b)

Results like these show that the two hydrogen atoms are delivered from the catalyst to the same face of the double bond. The stereospecificity of catalytic hydrogenation is one reason that the reaction is so important in organic chemistry.

Catalytic hydrogenation of alkynes is also a syn addition. As we have already seen, catalytic hydrogenation of alkynes with a poisoned catalyst gives cis alkenes (Sec. 4.10B).

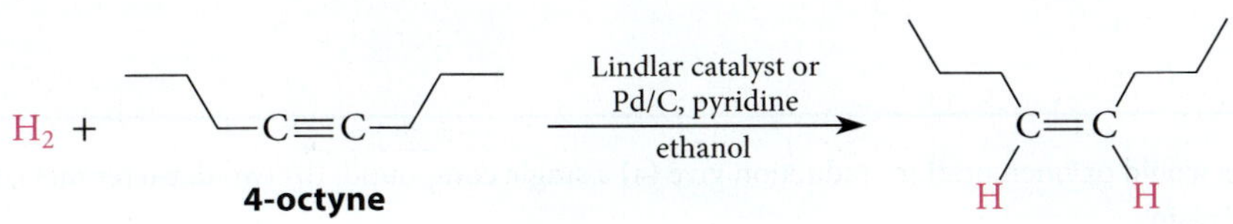

4-octyne

***cis*-4-octene** (7.68)

This reaction provides one of the best methods for preparing cis alkenes.

Oxymercuration–Reduction Oxymercuration of alkenes (Sec. 5.4A) is typically a stereospecific anti addition.

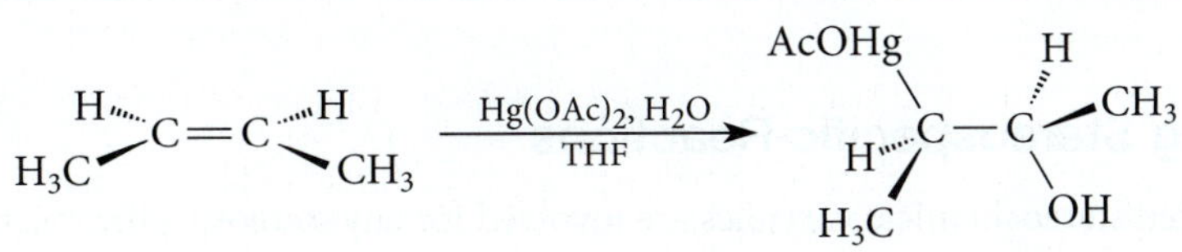

(racemic) (7.69)

(What result would you expect for the same reaction of *trans*-2-butene? See Focused Problem 7.30.) Because this reaction occurs by a cyclic-ion mechanism (Eq. 5.20a and b, Sec. 5.4A), the acetoxymercuri group blocks one face of the cyclic-ion intermediate, forcing the water to react at the opposite face.

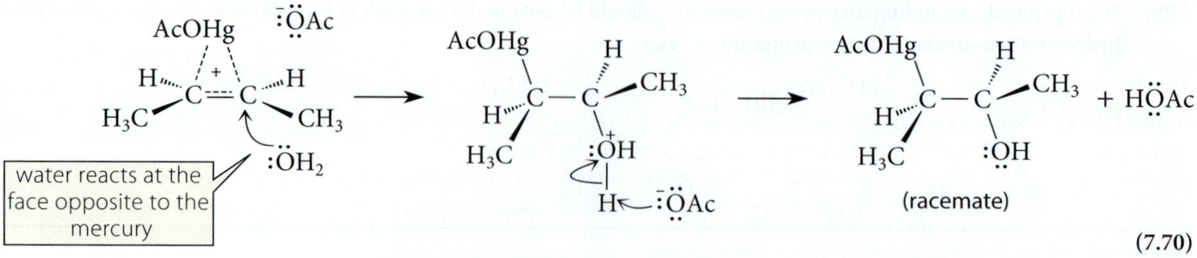

(7.70)

> **!** The reaction of water at one carbon of the mercurinium ion is shown; reaction at the other is equally probable, because the ion is a meso compound, and the two carbons are equivalent. The product of this reaction is therefore the racemate. Only one enantiomer is shown here. (You should show the reaction of water at the other carbon to give the other enantiomer.)

(In its stereochemistry, this step is similar to the reaction of bromide ion with a bromonium ion; Eq. 7.60, Sec. 7.8C).

In the reaction of the mercury-containing product with $NaBH_4$, however, the stereochemical results vary from case to case. In this example, a deuterium-substituted analog, $NaBD_4$, was used to investigate the stereochemistry, and it was found that mercury is replaced by hydrogen with *loss of stereochemical configuration*. Two diastereomers are formed.

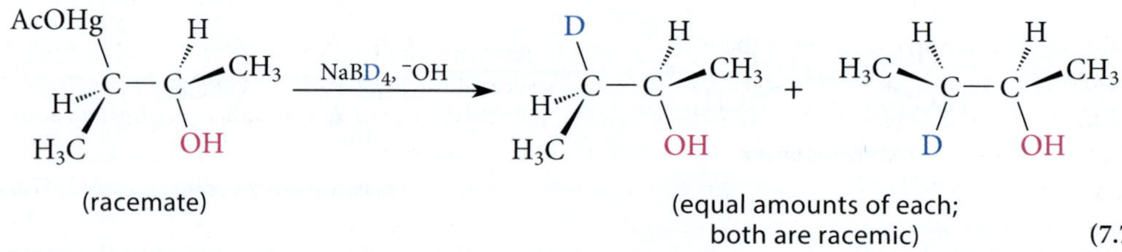

(7.71)

STUDY GUIDE LINK 7.5
When Stereoselectivity Matters

Therefore, oxymercuration–reduction is generally *not* a stereoselective reaction. Despite its lack of stereoselectivity, the reaction is highly *regioselective* and is very useful in situations in which stereoselectivity is not an issue.

Focused Problem

7.30 For which of the following alkenes would oxymercuration–reduction give (a) a single compound; (b) two diastereomers; (c) more than one constitutional isomer. Explain.

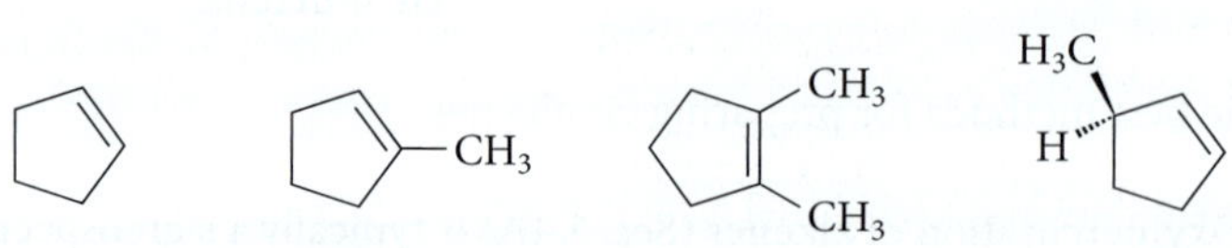

A *B* *C* *D*

F. Analyzing Stereospecific Reactions

Three interrelated stereochemical variables are involved for any stereospecific reaction:

1. The stereochemistry of the starting reactant (for example, whether an alkene is *E* or *Z*; whether an asymmetric center is *R* or *S*)

2. The stereochemistry of the reaction (syn or anti for additions, retention or inversion for substitutions)

3. The stereochemistry of the products

If you know any two of these, the third is determined. You'll find this relationship to be very useful when you start to plan syntheses of organic compounds that have stereochemical aspects. In learning about stereospecific additions, we have focused in the foregoing sections on the starting alkene and the stereochemistry of the addition. But what if you know the product of a reaction and want to determine the stereochemistry of the reactant? Study Problem 7.7 shows how to approach this problem systematically for an addition reaction.

Study Problem 7.7

What is the alkene, along with its stereochemistry, from which the following alcohol can be prepared free of diastereomeric impurities by hydroboration–oxidation?

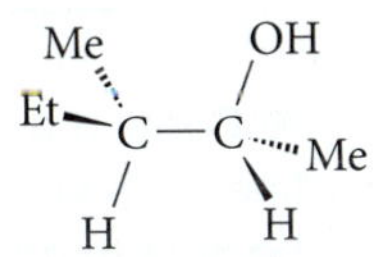

A

Solution First, any preparation of this alcohol from achiral starting materials will give the racemic mixture. We want to avoid contaminating *diastereomers.* The stereochemistry of the product is given in the problem. From Sec. 7.8D, hydroboration–oxidation is an overall syn addition. Hydroboration–oxidation installs an OH group at the alkene carbon with fewer branches, and the H is installed at the carbon with more branches. Therefore, the H and OH to be added are the ones colored blue in structure *A*. The H of the methyl group shown in red in structure *B* would have been added at a carbon with *fewer* branches, so the addition described by *B* is *not* a candidate for the hydroboration–oxidation.

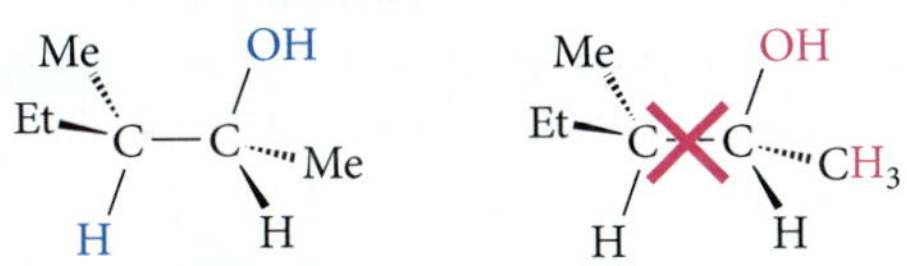

A *B*

To deduce the stereochemistry of the starting alkene, we carry out an internal rotation in the product alcohol so that the H and OH are oriented in the way that they are added to the alkene—that is, they are syn to each other. The relationship of the alkene substituents then becomes apparent when we draw the alkene in perspective.

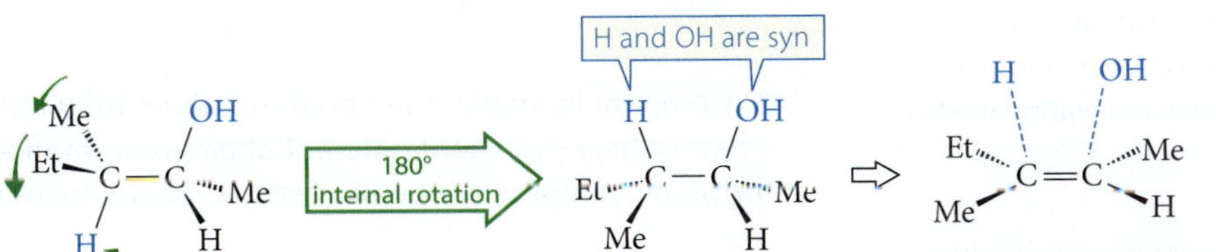

A

(*Z*)-3-methyl-2-pentene

The required alkene is then (*Z*)-3-methyl-2-pentene. (The enantiomer of the product is formed by the equally likely approach of the H and OH from the other face of the alkene.)

In summary, we carry out an internal rotation in the desired product so that the relative positions of the groups to be added correspond to the direction of addition (anti or syn) and then "read" the stereochemistry of the alkene from the remaining groups. Don't be concerned if this analysis requires an eclipsed conformation for analysis of syn addition.

In putting the molecule in an eclipsed conformation during this analysis, we recognize that it really doesn't exist that way. The stereochemistry of the addition is actually established during the hydroboration step, in which the product probably *is* in an eclipsed conformation (or something close to eclipsed) immediately after the reaction. (See Eq. 7.64a, Sec. 7.8D.) Because the oxygen replaces the boron in a subsequent step with retention of configuration, we can use the OH group in this analysis because it is stereochemically equivalent to the boron.

Focused Problem

7.31 Alkaline potassium permanganate ($KMnO_4$) can be used to bring about the addition of two OH groups to an alkene double bond. This reaction has been shown in several cases to be a stereospecific syn addition. Given the stereochemistry of the product shown in the following reaction, what stereoisomer of alkene *A* was used in the reaction? Explain.

$$\underset{A}{CH_3(CH_2)_7CH{=}CH(CH_2)_7CH_3} \xrightarrow{KMnO_4/OH^-} \underset{\text{meso stereoisomer}}{CH_3(CH_2)_7\overset{OH}{\overset{|}{C}}H{-}\overset{OH}{\overset{|}{C}}H(CH_2)_7CH_3}$$

CHAPTER SUMMARY *For a summary of the chapter, see Chapter 7 in the Study Guide and Solutions Manual.*

SKILLS OBJECTIVES WITH PROBLEMS

- **For any chair conformation of any cyclohexane derivative, draw the structure that results from the chair interconversion. 7.2B**

7.32 For each of the following compounds, draw the other chair conformation that results from the chair interconversion.

(a)

(b)

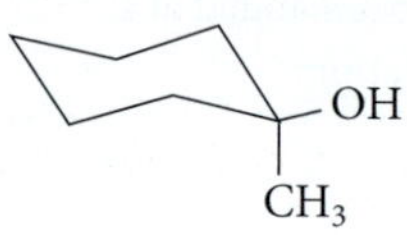

(c)

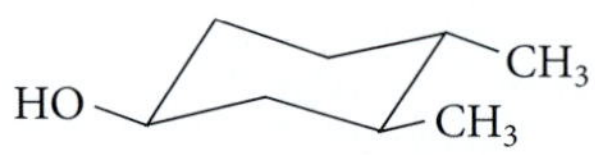

- **Determine whether the two chair conformations that result from the chair interconversion are identical, conformational enantiomers, or conformational diastereomers. 7.4A**

7.33 For each of the following compounds, draw the other chair conformation that results from the chair interconversion, and give the relationship of the two chair conformations (identical, conformational enantiomers, conformational diastereomers).

(a)

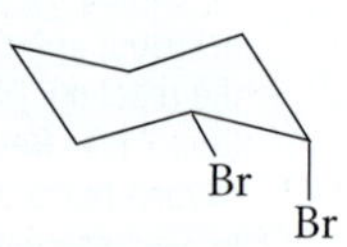

(b)

(c)

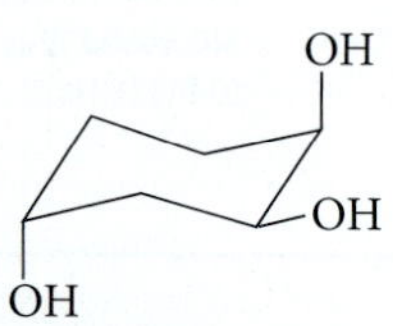

- **Convert a chair conformation into a planar-ring line-and-wedge structure. For six-membered rings, be able to convert a planar-ring structure into the two corresponding chair conformations. 7.4**

7.34 Convert each of the chair conformations in Problem 7.33 into planar-ring structures in which the stereochemistry of the substituents is represented by wedged bonds.

7.35 Convert each of the following planar-ring structures into two chair conformations that are related by the chair interconversion.

(a)

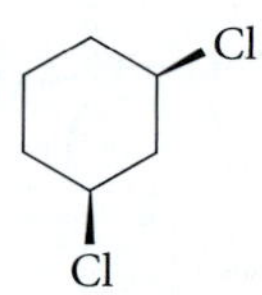

(b)

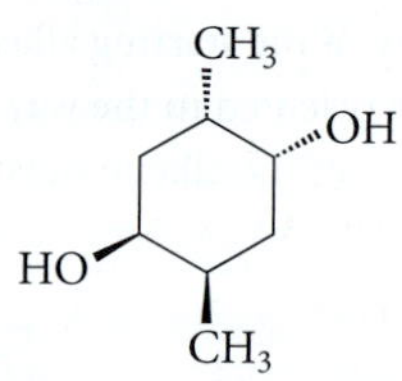

- **Compare isomeric chair conformations to determine whether they represent different chair conformations of the same compound, enantiomers, or diastereomers. 7.4**

7.36 By answering the following questions, indicate the relationship between the two structures in shown in each part. Are they chair conformations of the same molecule? If so, are they conformational diastereomers, conformational enantiomers, or identical? If they are not conformations of the same molecule, what is their stereochemical relationship? (*Hint:* Use planar-ring structures to help you.)

(a)

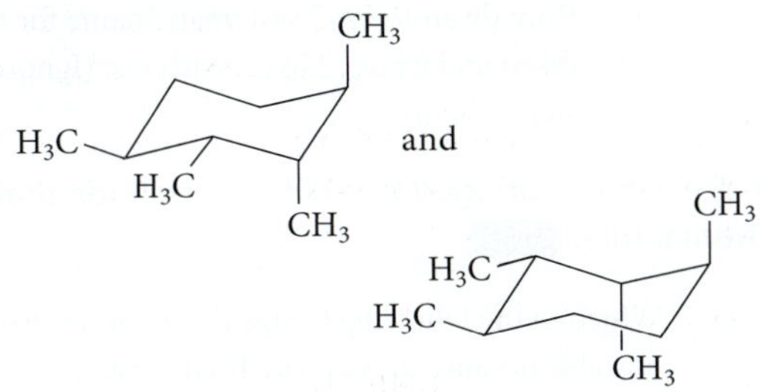

(b)

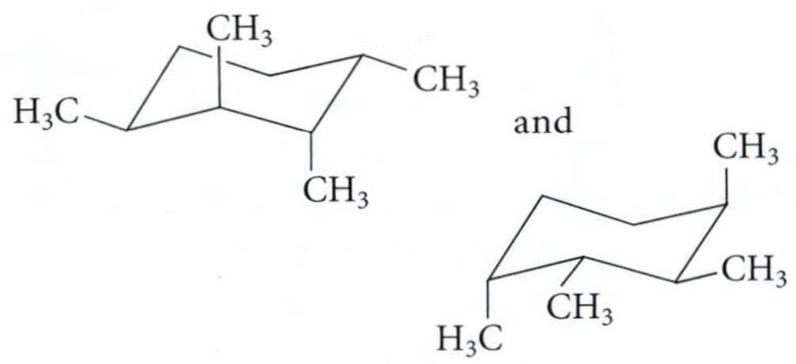

(c)

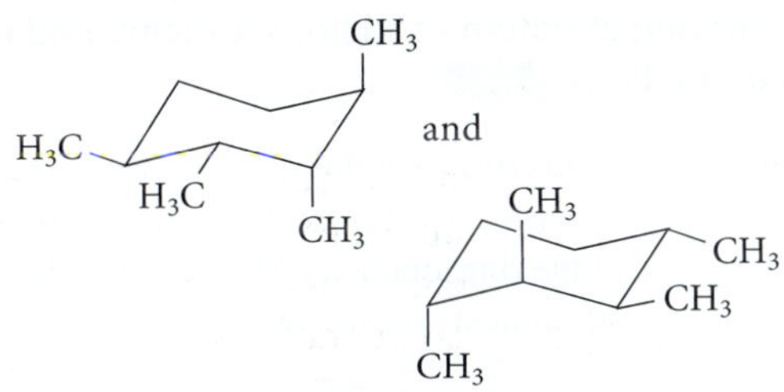

- **Determine whether a cyclic compound can be separated into enantiomers at room temperature. 7.4, 7.5**

7.37 Which of the following compounds can in principle be separated into enantiomers at room temperature? Explain how you know.

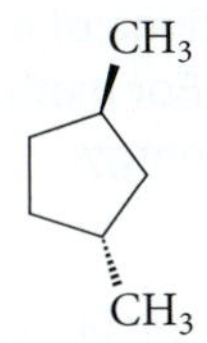

A

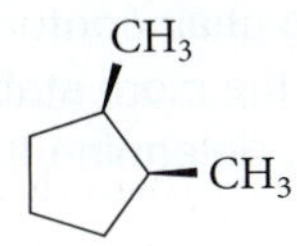

B

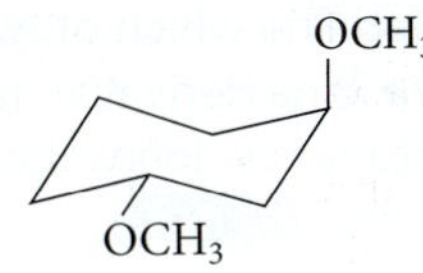

C

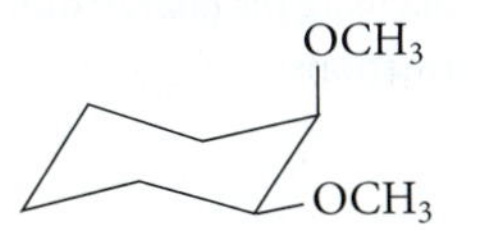

D

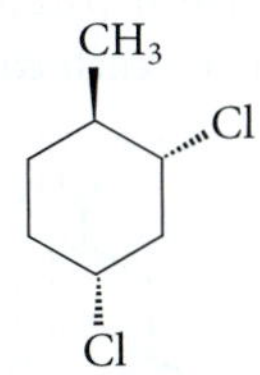

E

7.38 Determine whether each of the following compounds can be isolated in optically active form. Explain how you know.

(a) *trans*-1,2-Dimethylcyclohexane

(b) 1,1-Dimethylcyclohexane

(c) *cis*-1-Ethyl-4-methylcyclohexane

(d) *cis*-1-Ethyl-3-methylcyclohexane

- **Convert the chair conformation into a boat conformation; recognize boat and chair conformations of the same compound. 7.2C**

7.39 (a) Which structures in **Fig. P7.39** represent possible boat conformations of compound *X*?

(b) Of the correct answers to part (a), which one has the lowest energy? Explain.

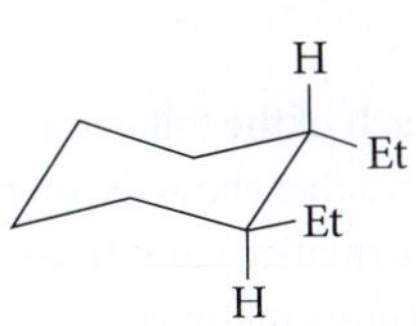

X

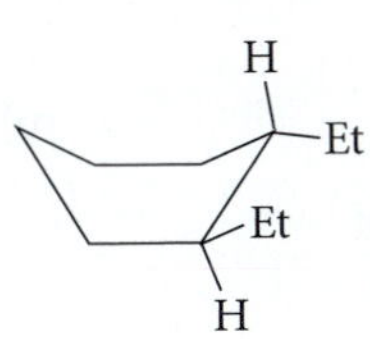

A

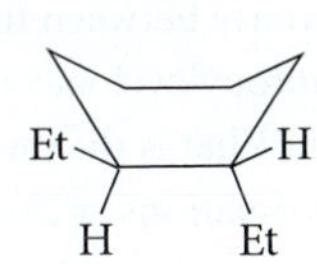

B

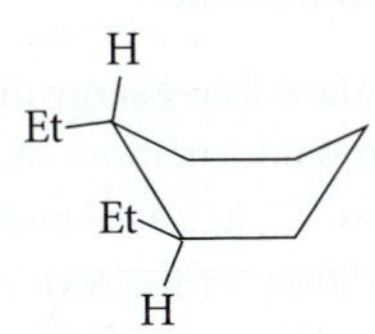

C

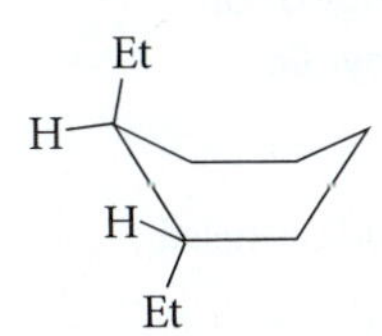

D

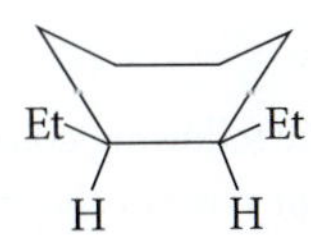

E

FIGURE P7.39

- **Determine which of two chair conformations of a cyclohexane derivative is the more stable. For methyl-substituted cyclohexanes, determine the energy difference. 7.3, 7.4**

7.40 In each part, draw the two chair conformations of the molecule shown, and determine which conformation is the more stable. In part (b), calculate the energy difference between the two chair conformations.

(a)

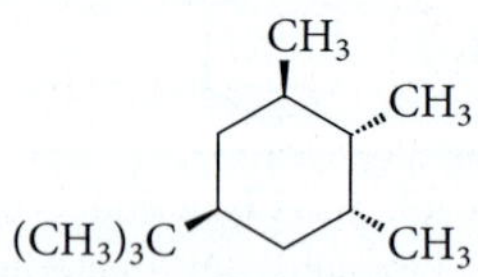

(b)

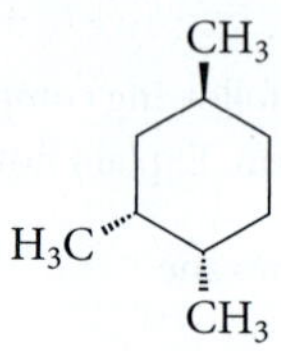

- **From a knowledge of the relative amounts of two chair conformations at equilibrium, calculate their standard free-energy difference. From a knowledge of the standard free-energy difference between two conformations, calculate the relative amounts at equilibrium. 3.5B**

7.41 (a) Chlorocyclohexane contains 2.07 times more of the equatorial form than the axial form at equilibrium at 25 °C. What is the standard free-energy difference between the two forms? Which is more stable?

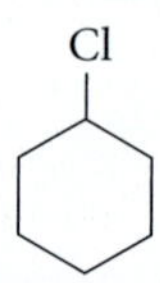

chlorocyclohexane

(b) The standard free-energy difference between the two chair conformations of isopropylcyclohexane is 9.2 kJ mol^{-1} (2.2 kcal mol^{-1}). What is the ratio of concentrations of the two conformations at 25 °C?

- **Classify bicyclic compounds as fused, bridged, or spirocyclic, and name simple bridged and bicyclic cycloalkanes. 7.6A**

7.42 (a) Classify each of the following compounds as fused, bridged, bicyclic, or none of the above.

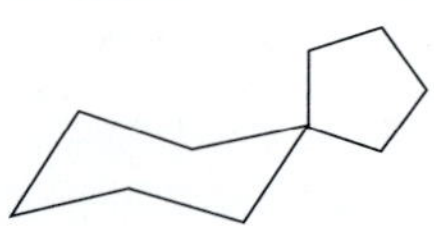

A

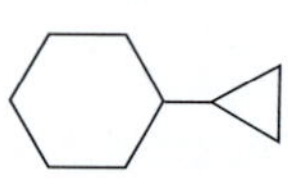

B

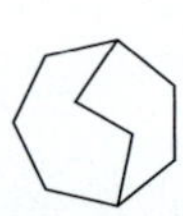

C

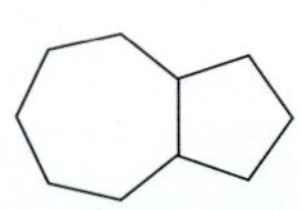

D

(b) Provide an IUPAC systematic name for the fused and bridged hydrocarbons. (Ignore stereochemistry.)

- **Recognize bridged bicyclic compounds that violate Bredt's rule. 7.6C**

7.43 Which of the following bridged compounds would be very unstable because they violate Bredt's rule?

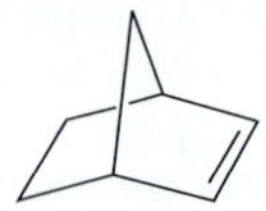

A

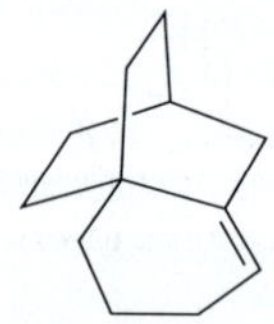

B

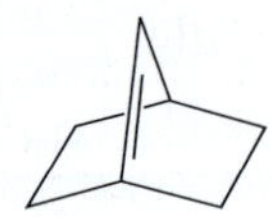

C

- **Represent cis and trans fused rings with both planar-ring structures and (for six-membered rings) chair conformations. 7.6B**

7.44 (a) Convert each of the following structures into a planar-ring structure that shows the stereochemistry of the ring junction with lines and wedges to substituents and/or hydrogens.

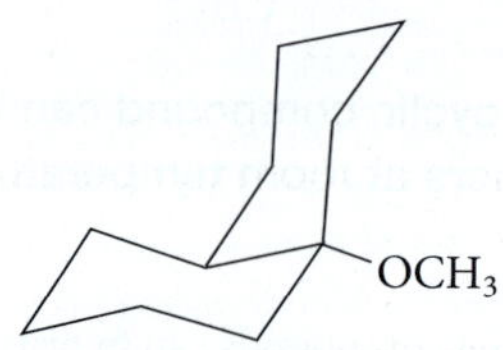

A

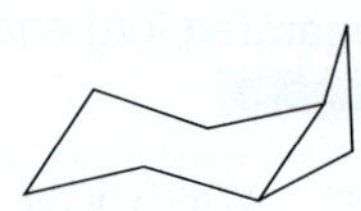

B

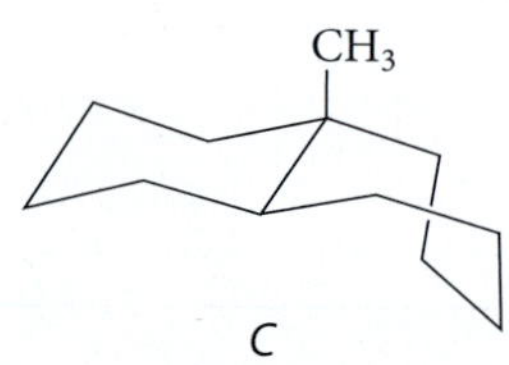

C

(b) Convert each of the following planar-ring structures to structures that show six-membered rings in chair conformations and five-membered rings in envelope conformations. Show the substituents and hydrogens at ring junctions explicitly.

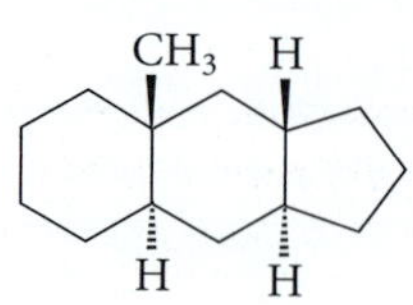

A

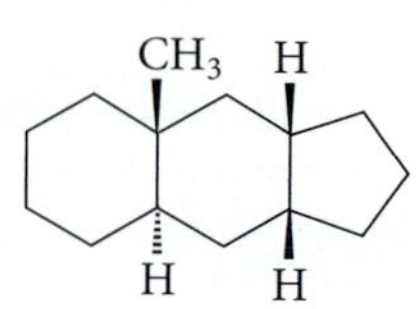

B

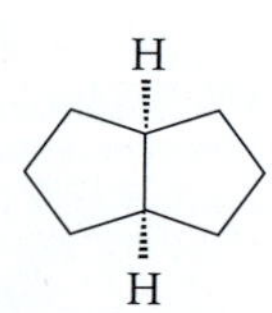

C

- **For a given reaction, state whether the products could be formed as enantiomers or diastereomers; state which stereoisomers are formed in the same amount and which in different amounts. 7.7, 7.8**

7.45 For each of the following reactions, provide the requested information. Take into account the known regiospecificity and stereospecificity (if any) of each reaction.

1. Give the structures of all products (including stereoisomers).
2. If more than one product is formed, give the stereochemical relationship (if any) of each pair of products.
3. If more than one product is formed, indicate which products are formed in identical amounts and which in different amounts.
4. If more than one product is formed, indicate which products are expected to have different physical properties (melting point or boiling point).

(a) 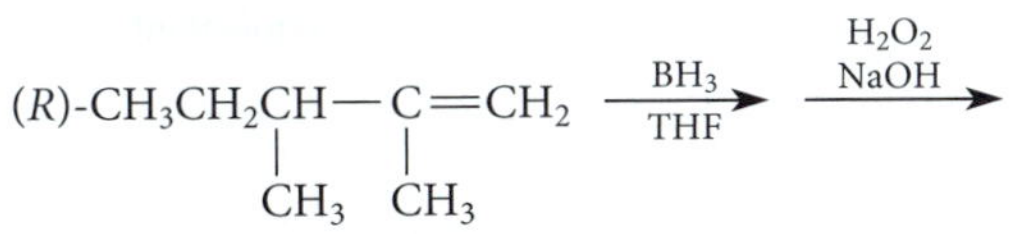

(b) 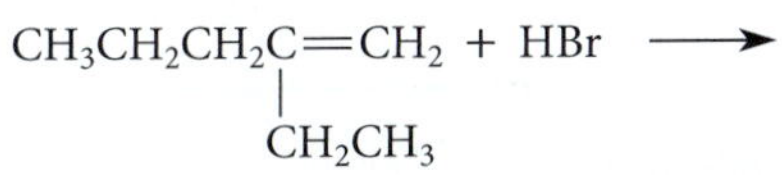

(c) $CH_3CH_2CH{=}CH_2 + Br_2 \xrightarrow[\text{(solvent)}]{CH_2Cl_2}$

(d) 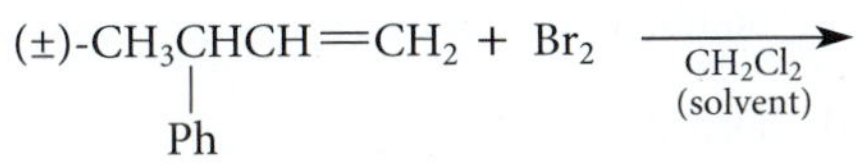

(e) 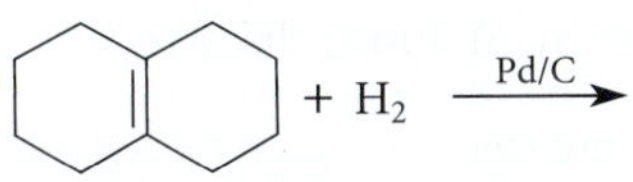

- **Given an addition reaction and the knowledge of its stereochemistry (syn or anti addition), draw a stereochemical representation of the product. 7.8F**

7.46 Bromine in water undergoes an anti addition to the following alkenes to give halohydrins. In each case, draw line-and-wedge structures for all stereoisomers of the products. Use planar-ring structures for part (b). Take into account the known regiospecificity of the reaction, if relevant.

(a) 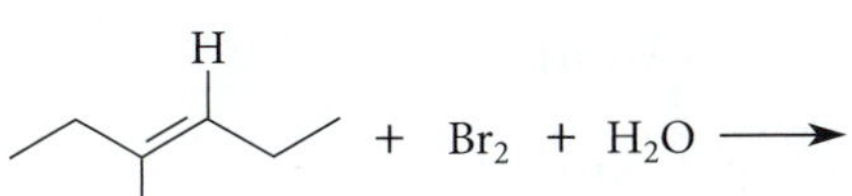

both solvent and reactant; in large excess

***trans*-2-hexene**

(b) 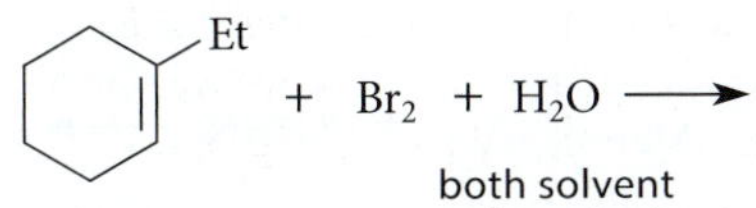

both solvent and reactant; in large excess

1-ethylcyclohexene

- **Given the reactants and the stereochemistry of the products, determine whether an addition reaction is a syn addition or an anti addition. 7.8F**

7.47 (a) When fumarate reacts with D_2O in the presence of the enzyme *fumarase* (Secs. 4.9C and 7.7A), only one stereoisomer of deuterated malate is formed, as shown in the following equation. Is this a syn or an anti addition? Explain. (D = deuterium = ^2H, an isotope of hydrogen that reacts like hydrogen.)

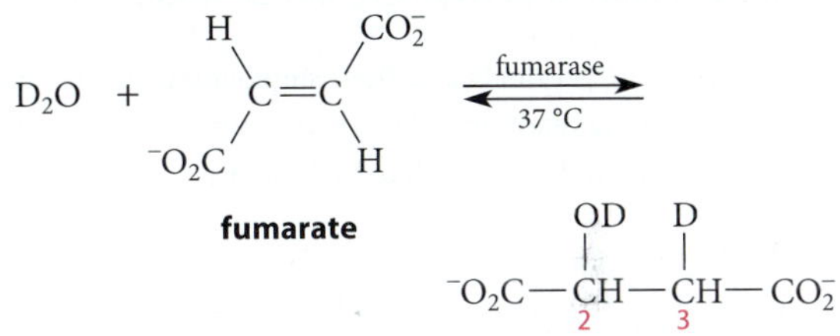

fumarate

(2*S*,3*R*)-malate-3-*d*

(b) Why is the use of D_2O instead of H_2O necessary to establish the stereochemistry of this addition?

- **Given an addition for which both the product stereochemistry and the reaction stereochemistry are known, provide missing reactants and their stereochemistry. 7.8F**

7.48 When an alkene is treated with a peroxyacid (RCO_3H) and then with aqueous acid (H_3O^+, H_2O), two OH groups are added to the double bond with anti stereochemistry. Given the following product of such an addition, give the structure of the alkene and its stereochemistry.

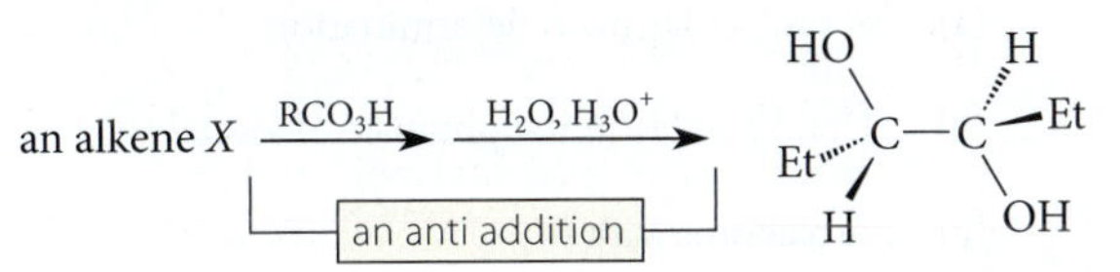

***meso*-3,4-hexanediol**

- **Given the reactants and the stereochemistry of the products, determine whether a substitution reaction has occurred with inversion or retention of stereochemistry. 7.8B**

7.49 (a) Has the following substitution reaction occurred with retention or inversion of stereochemistry? Explain.

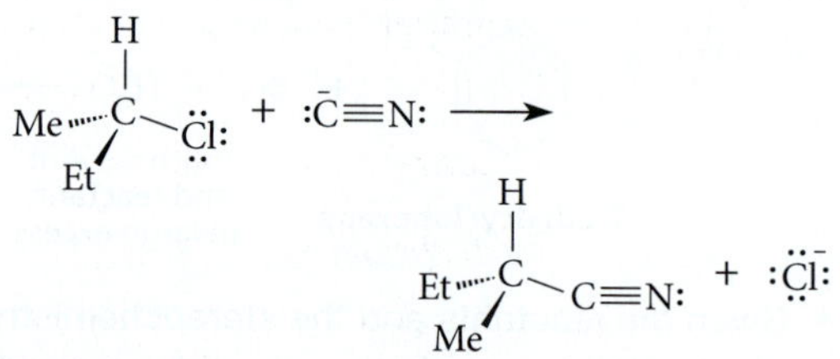

(b) Assuming that the reaction takes place in a single mechanistic step, give a curved-arrow mechanism that accounts for the stereochemistry.

• Given the reactants and their stereochemistry, complete the following addition reactions by giving missing products and their stereochemistry: bromine addition, hydroboration–oxidation, catalytic hydrogenation, and oxymercuration–reduction. 7.8A–D

7.50 Give the products and their stereochemistry for each of the following reactions. (D = deuterium = ^2H, an isotope of hydrogen that reacts like hydrogen.)

(a)

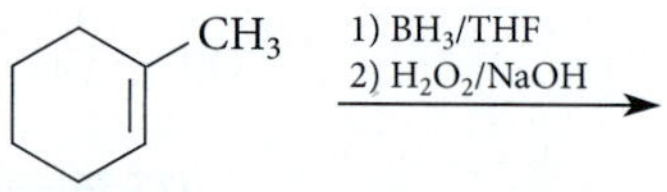

(b)

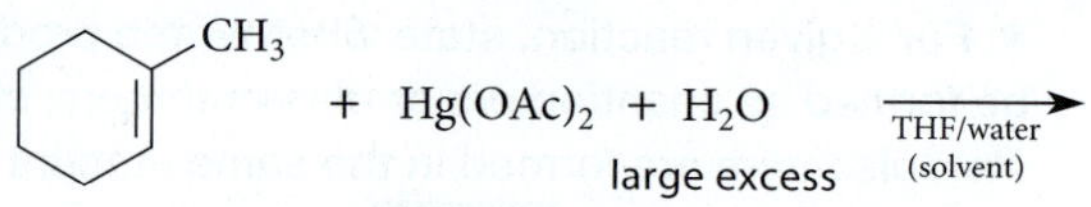

(c) Product of (b) + $NaBD_4$, NaOH

(d)

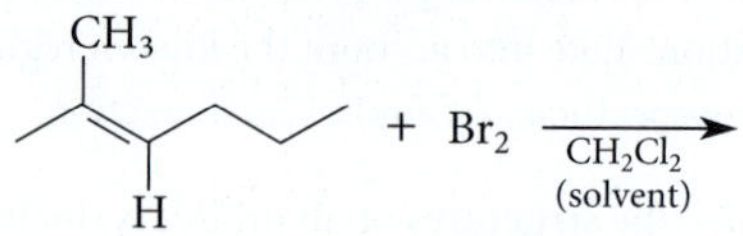

(e)

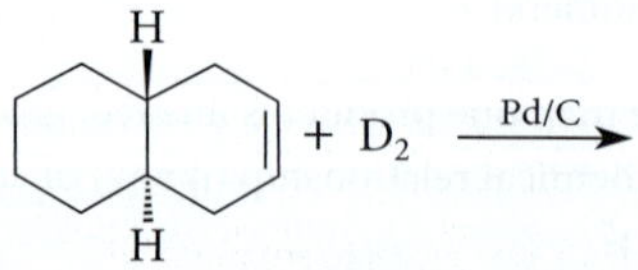

(f)

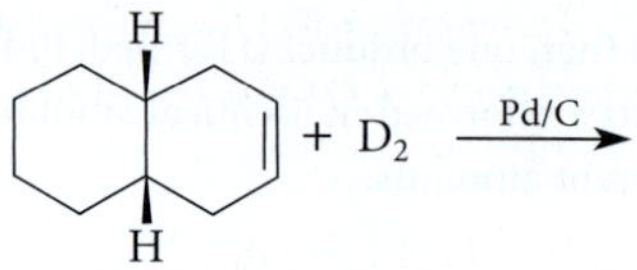

(g)

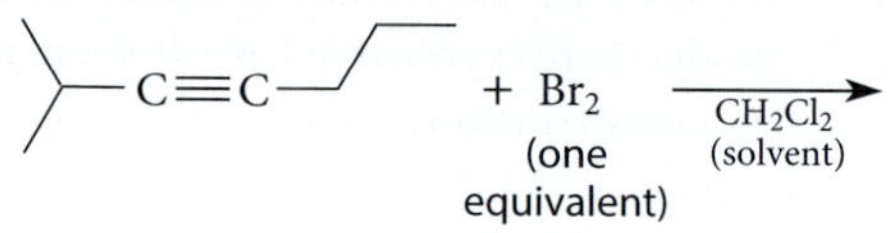

INTEGRATED PROBLEMS

7.51 Draw the structures of the following compounds, and name the compound in part (a).

(a) A bicyclic alkane with six carbon atoms

(b) (*S*)-4-Cyclobutylcyclohexene

7.52 Which of the following would distinguish (in principle) between methylcyclohexane and (*E*)-4-methyl-2-hexene? Explain your reasoning.

(a) A molecular mass determination

(b) Uptake of H_2 in the presence of a catalyst

(c) A reaction with Br_2

(d) A determination of the molecular formula

(e) A determination of the heat of formation

(f) An enantiomeric resolution

7.53 State whether you would expect each of the following properties to be identical or different for the following two stereoisomers of 1,3-cyclohexanediol, and explain your reasoning.

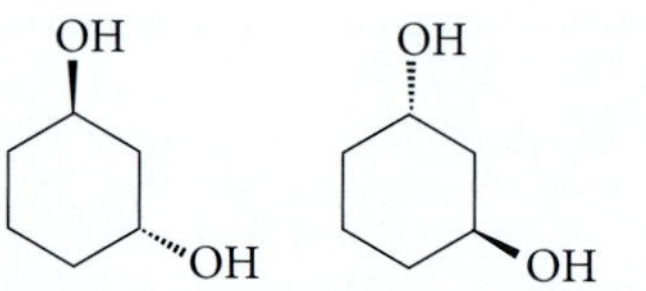

(a) Boiling point

(b) Optical rotation

(c) Solubility in hexane

(d) Density

(e) Solubility in (*S*)-3-methylhexane

(f) Dipole moment

(g) Taste (*Hint:* Your taste buds are chiral.)

7.54 Suggest a reason that the energy difference between the two chair conformations of ethylcyclohexane is about the same as that for methylcyclohexane, even though the ethyl group is larger than the methyl group.

7.55 Draw the two chair conformations of the sugar α-(+)-glucopyranose, which is one form of the sugar glucose. Which of these two conformations is the major one at equilibrium? Explain.

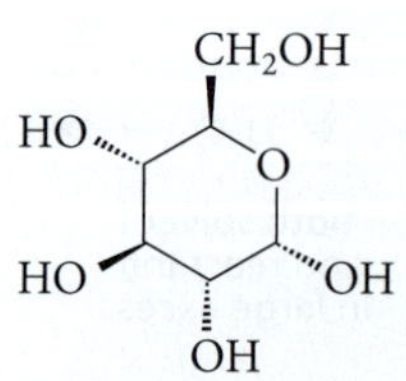

α-(+)-glucopyranose

7.56 Draw the structures of the following compounds. (Some parts may have more than one correct answer.)

(a) An achiral tetramethylcyclohexane for which the chair interconversion results in identical molecules

(b) An achiral trimethylcyclohexane with two chair forms that are conformational diastereomers

(c) A chiral trimethylcyclohexane with two chair forms that are conformational diastereomers

(d) A tetramethylcyclohexane with chair forms that are conformational enantiomers

7.57 Give the structure of every stereoisomer of 1,2,3-trimethylcyclohexane. Label the enantiomeric pairs, and show the plane of symmetry in each achiral stereoisomer.

7.58 Consider the following compound.

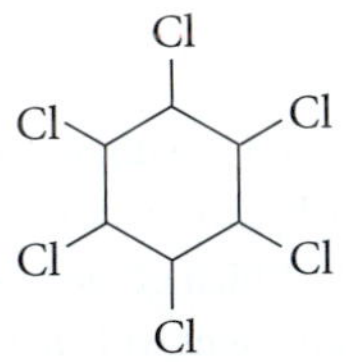

1,2,3,4,5,6-hexachlorocyclohexane

(a) Of the nine stereoisomers of this compound, only two can be isolated in optically active form under ordinary conditions. Give the structures of these enantiomers.

(b) Give the structures of the two stereoisomers for which the chair interconversion results in identical molecules.

7.59 Which of the following statements about *cis*- and *trans*-decalin (Sec. 7.6B) are true? Explain your answers.

(a) They are different conformations of the same molecule.

(b) They are constitutional isomers.

(c) They are diastereomers.

(d) Each one has two asymmetric carbons.

(e) Each one has two stereocenters.

(f) At least one chemical bond would have to be broken to convert one into the other.

(g) They are enantiomers.

(h) They interconvert rapidly.

7.60 Draw a conformational representation of the following steroid. Show the α and β faces of the steroid, and label the angular methyl groups.

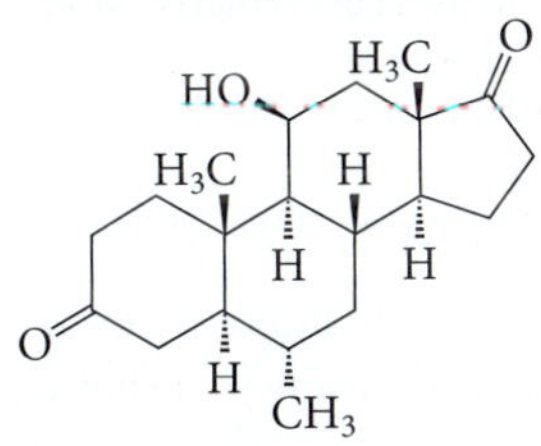

7.61 *One* of the stereoisomers given in **Fig. P7.61** exists with one of its cyclohexane rings in a twist-boat conformation. Which is it? Explain.

7.62 It has been argued that the energy difference between *cis*- and *trans*-1,3-di-*tert*-butylcyclohexane is a good approximation for the energy difference between the chair and twist-boat conformations of cyclohexane. Using models to assist you, explain why this view is reasonable.

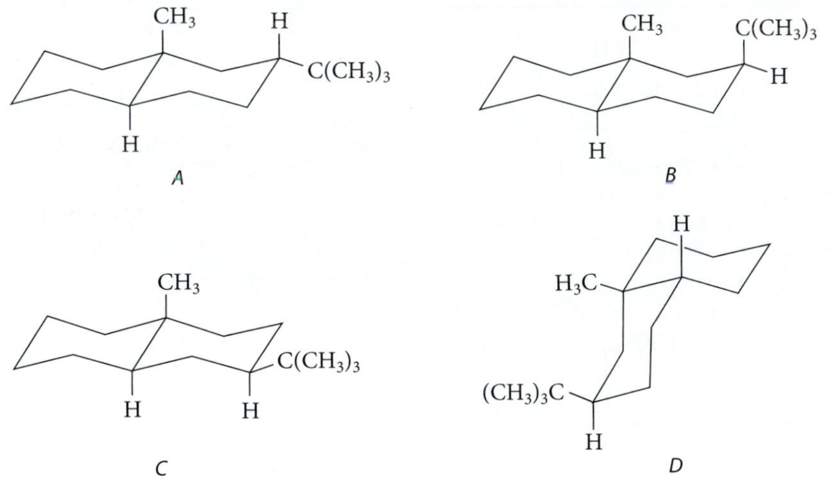

FIGURE P7.61

7.63 Olean is the sex attractant of the female fruit fly. (This is an example of a *pheromone*; Sec. 10.9.) Females secrete this compound to attract males when they are ready for mating.

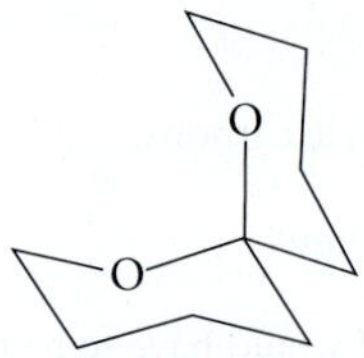

olean

Which of the following statements are true about this compound? Justify your choice(s).

(a) It is achiral and contains one or more stereocenters.

(b) It is achiral and contains no stereocenters.

(c) It is chiral and contains an asymmetric atom.

(d) It is chiral and contains one or more stereocenters.

(e) It is chiral and contains no stereocenters.

7.64 Which of the statements (a)–(e) in Problem 7.63 are true about the following compound? Justify your choice(s).

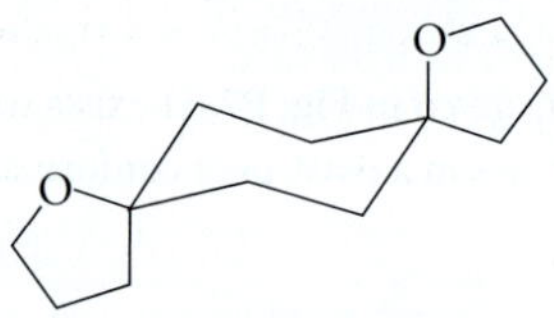

7.65 Rank the compounds given in **Fig. P7.65** according to their heats of formation (lowest first), and estimate the $\Delta H°$ difference between each pair.

7.66 From your knowledge of the mechanism of bromine addition to alkenes, give the structure and stereochemistry of the product(s) expected in each of the following reactions.

(a) Addition of Br_2 to (3*R*,5*R*)-3,5-dimethylcyclopentene

(b) Reaction of cyclopentene with Br_2 in the presence of H_2O (*Hint:* See Sec. 5.2B.)

7.67 Anti addition of bromine to the following bicyclic alkene gives two separable dibromides. Suggest structures for each. (*Hint:* Remember that *trans*-decalin derivatives can*not* undergo the chair interconversion.)

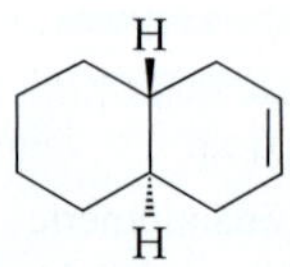

7.68 When 1,4-cyclohexadiene reacts with two equivalents of Br_2, two separable isomers with the formula $C_6H_8Br_4$ with different melting points are formed. Account for this observation.

7.69 An optically active compound *X* with molecular formula C_8H_{14} undergoes catalytic hydrogenation to give an optically inactive product. Which of the following structures for *X* is (are) consistent with all of the data?

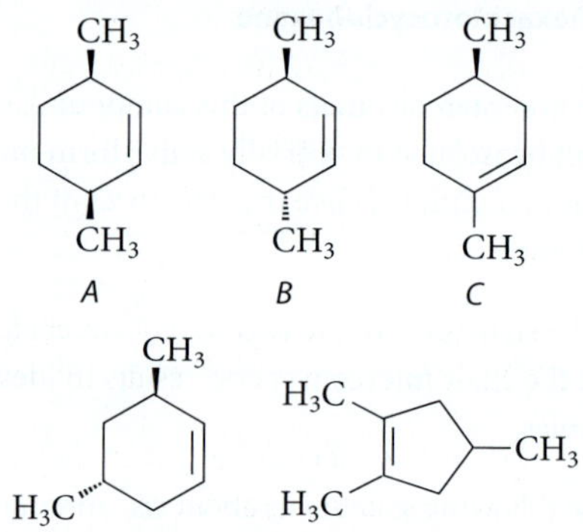

A *B* *C* *D* *E*

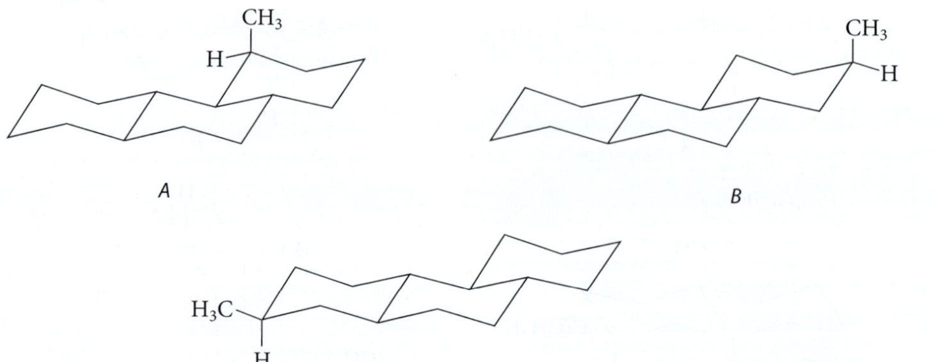

A *B* *C*

FIGURE P7.65

7.70 Which of the following alcohols can be synthesized relatively free of constitutional isomers and diastereomers by (a) hydroboration–oxidation or (b) oxymercuration–reduction? Explain. Assume that a racemate is formed if the product has asymmetric centers.

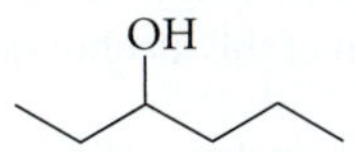
A

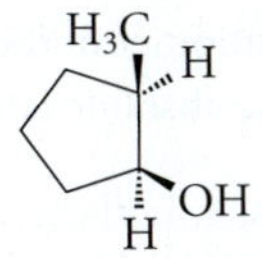
B

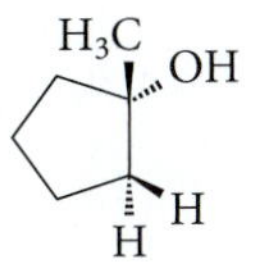
C

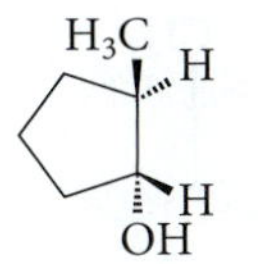
D

7.71 Identify the stereocenters (if any) in each of the following structures, and tell whether each structure is chiral.

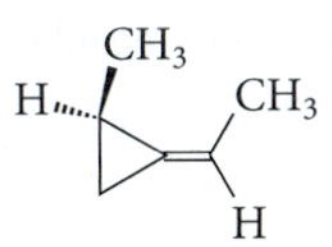
A

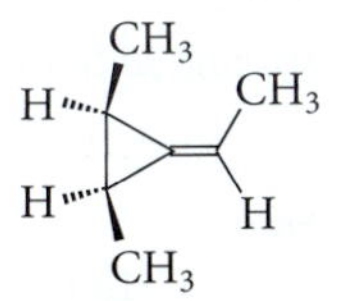
B

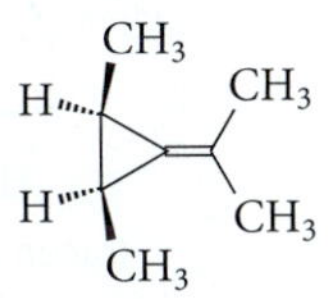
C

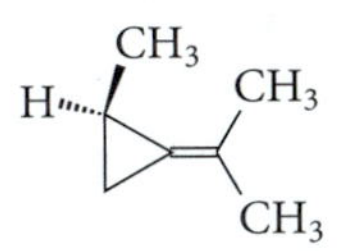
D

7.72 Draw a chair conformation for (*S*)-3-methylpiperidine showing the sp^3 orbital that contains the nitrogen nonbonding electron pair. How many chair conformations of this compound are in rapid equilibrium? (*Hint:* See Sec. 6.9B.)

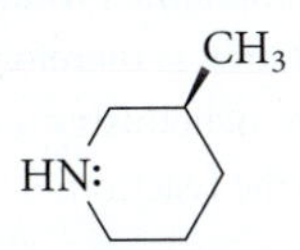
(*S*)-3-methylpiperidine

7.73 Which of the following compounds can be resolved into enantiomers at room temperature? Explain.

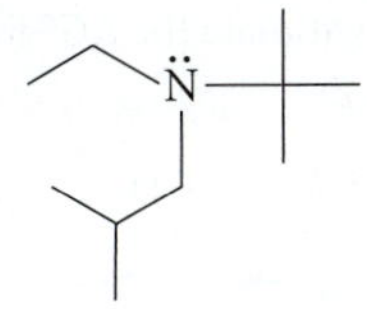
A

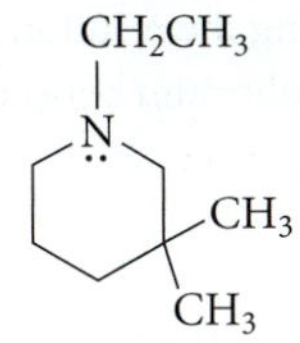
B

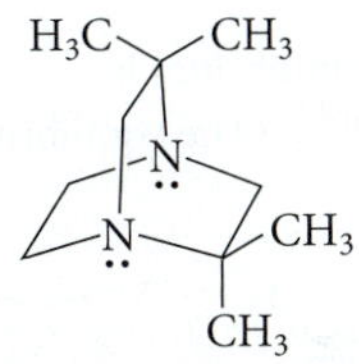
C

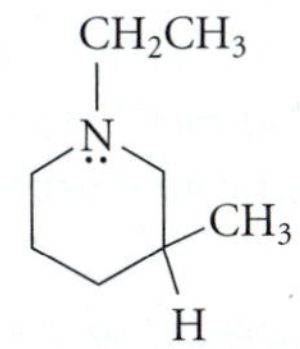
D

7.74 Explain why 1-methylaziridine undergoes amine inversion much more slowly than 1-methylpyrrolidine. (*Hint:* What are the hybridization and bond angles at nitrogen in the transition state for inversion?)

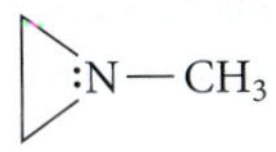
1-methylaziridine

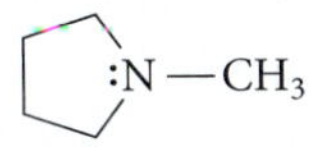
1-methylpyrrolidine

7.75 (a) Show that the dipole moment of 1,4-dioxane should be zero if it exists solely in a chair conformation.

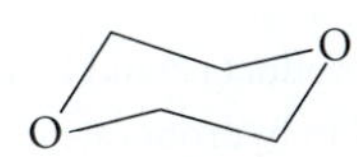
1,4-dioxane

(b) Account for the fact that the dipole moment of 1,4-dioxane, although small, is definitely not zero (0.38 D).

7.76 The $\Delta G°$ for the equilibrium between *A* and *B* shown in part (a) is 8.4 kJ mol^{-1} (2.0 kcal mol^{-1}). (Conformation *A* has lower energy.) Use this information to estimate the energy cost of a 1,3-diaxial interaction between two methyl groups:

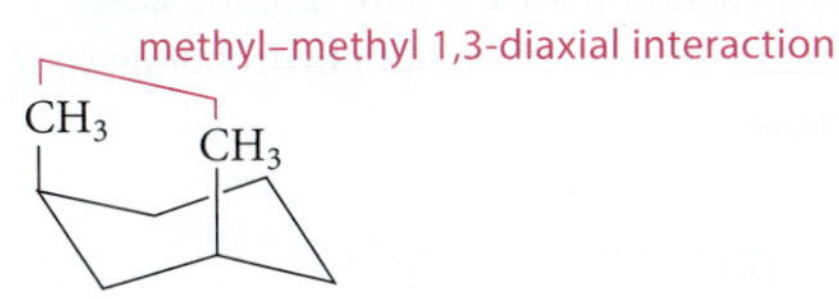

(a)

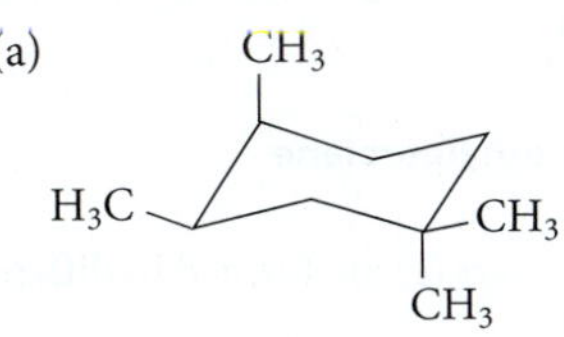
A

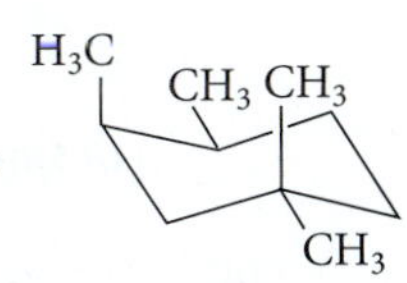
B

(b) Using the result in part (a), estimate the $\Delta G°$ for the equilibrium between *C* and *D*.

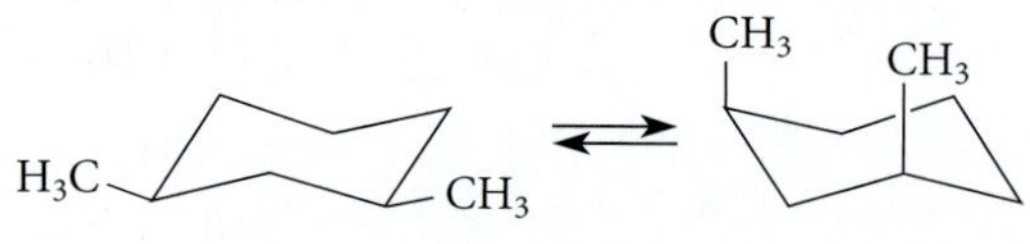

C *D*

7.77 (a) The $\Delta G°$ for the following equilibrium is 4.73 kJ mol^{-1} (1.13 kcal mol^{-1}). (The equilibrium favors conformation *A*.)

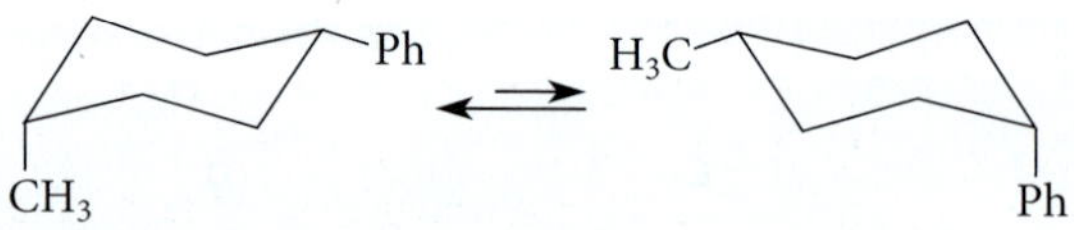

A *B*

According to these data, which group is larger, methyl or phenyl (Ph)? Why is this reasonable?

(b) Use the $\Delta G°$ given in part (a), along with any other appropriate data, to estimate the $\Delta G°$ for the following equilibrium.

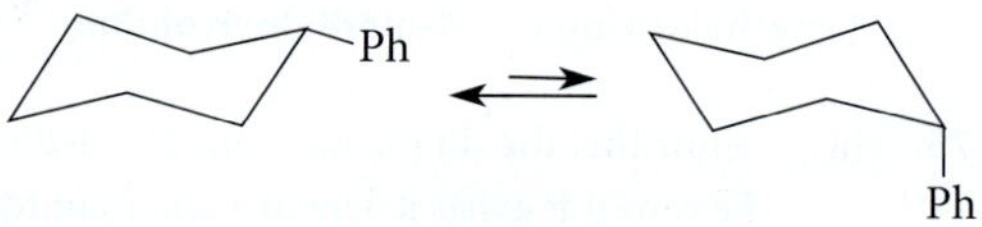

7.78 (a) What stereoisomeric products could be formed when borolane is used to hydroborate *cis*-2-butene?

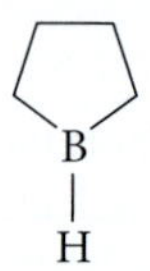

borolane

(b) Are the products in part (a) formed in the same or different amounts? Explain how you know.

(c) What stereoisomeric products could be formed when enantiomerically pure (2*R*,5*R*)-2,5-dimethylborolane is used to hydroborate *cis*-2-butene?

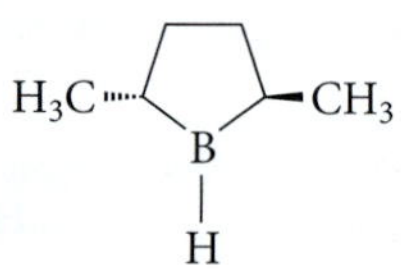

(2*R*,5*R*)-2,5-dimethylborolane

(d) The products in part (c) are formed in different amounts. Why?

(e) In fact, only one of the products in part (c) is formed in a significant amount. Which one? (*Hint:* Build models of the borane and the alkene. Let the borane model approach the alkene model from one face of the π bond, then the other. Decide which reaction is preferred by analyzing van der Waals repulsions in the transition state in each case.)

(f) When the product borane determined in part (e) is treated with alkaline H_2O_2, mostly a single enantiomer of the product alcohol is formed. What is the absolute configuration of this alcohol (*R* or *S*)?

7.79 (a) What two diastereomeric products could be formed in the hydroboration–oxidation of the following alkene?

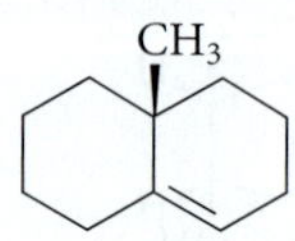

(b) Considering the effect of the methyl group on the approach of the borane–THF reagent to the double bond, suggest which of the two products you obtained in part (a) should be the major product.

7.80 Methyl hypofluorite, CH_3O—F, undergoes regiospecific and stereospecific addition to alkenes, as in the following example.

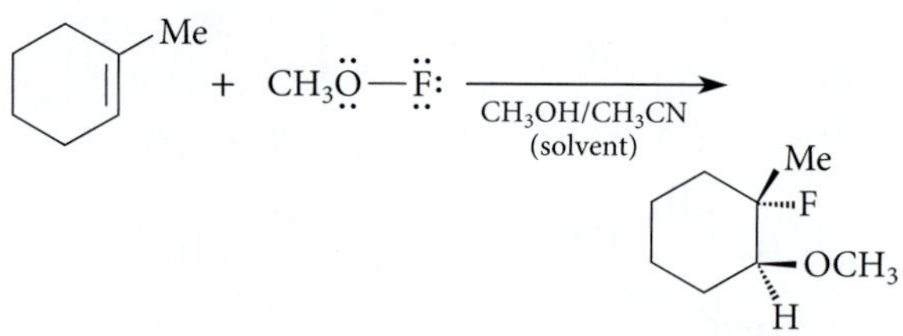

(racemate)

(a) Is this a syn or an anti addition?

(b) Provide a curved-arrow mechanism that accounts for both the regiospecificity and stereospecificity of the reaction.

7.81 When 1-methylcyclohexene undergoes acid-catalyzed alkene hydration in D_2O (Sec. 4.10C), the product is a mixture of stereoisomers; the reaction is therefore *not* stereoselective. Show that the stereochemistry is consistent with the accepted mechanism of the reaction.

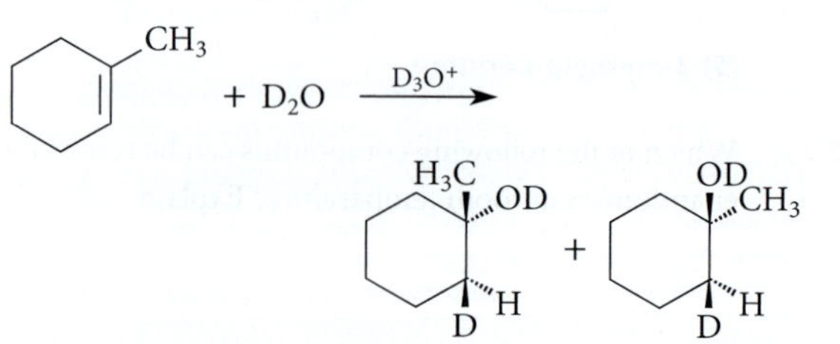

both compounds are racemates

7.82 Propose a structure for the product *X* of the following reaction, and give a curved-arrow mechanism for its formation. Pay particular attention to the stereochemistry of each step. (*Hint:* Draw the conformation of the starting material; at which face of the double bond is the bromine likely to react?)

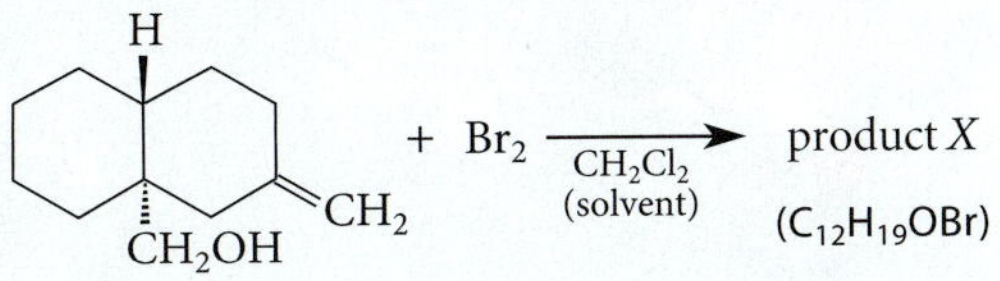

$(C_{12}H_{20}O)$

7.83 Section 3.3B (Eq. 3.18) explained that the nonbonding electron pair on the nitrogen of an amide is delocalized by resonance into the carbonyl group:

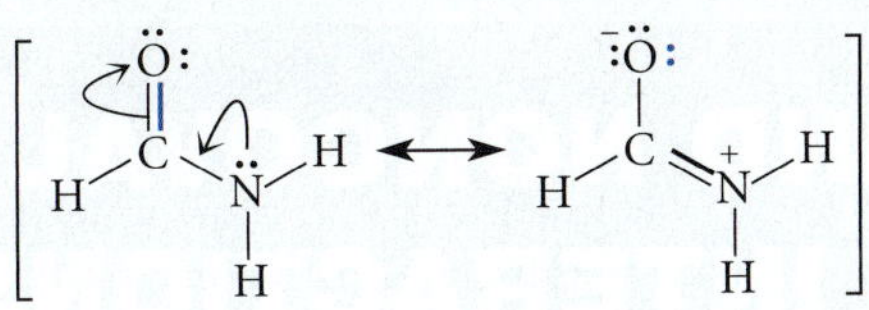

For one of the amides that follow, this resonance does not occur. Identify the amide, and explain why this type of resonance *cannot* occur.

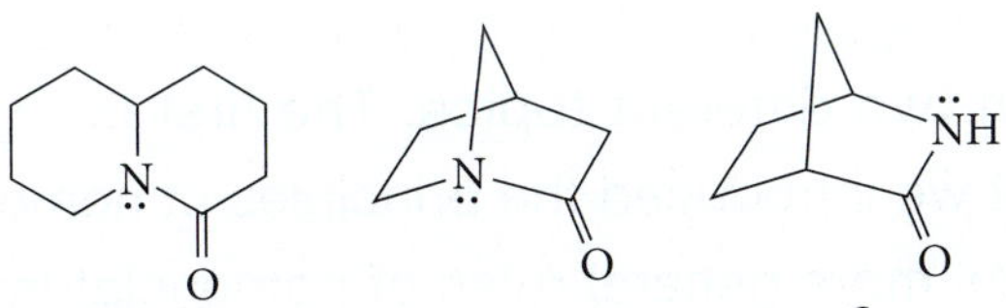

A *B* *C*

7.84 (a) The following two tricyclic compounds are examples of *propellanes* (propeller-shaped molecules). What is the relationship between these two molecules (identical, enantiomers, diastereomers)? Tell how you know.

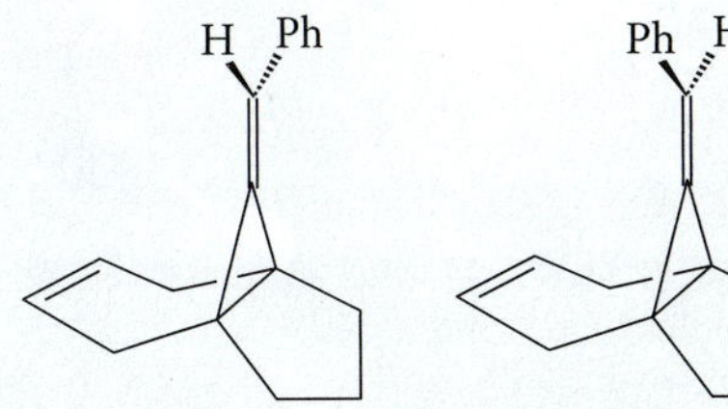

A *B*

(b) The chemist who prepared these compounds wrote that they are *E,Z* isomers. Do you agree or disagree? Explain.

7.85 (a) In how many stereochemically different ways can the two rings in a *bridged* bicyclic compound be joined?

(b) For which one of the following bridged bicyclic compounds are all such stereoisomers likely to be stable enough to isolate? Explain.

(1) Bicyclo[2.2.2]octane

(2) Bicyclo[25.25.25]heptaheptacontane (A heptaheptacontane has 77 carbons.)

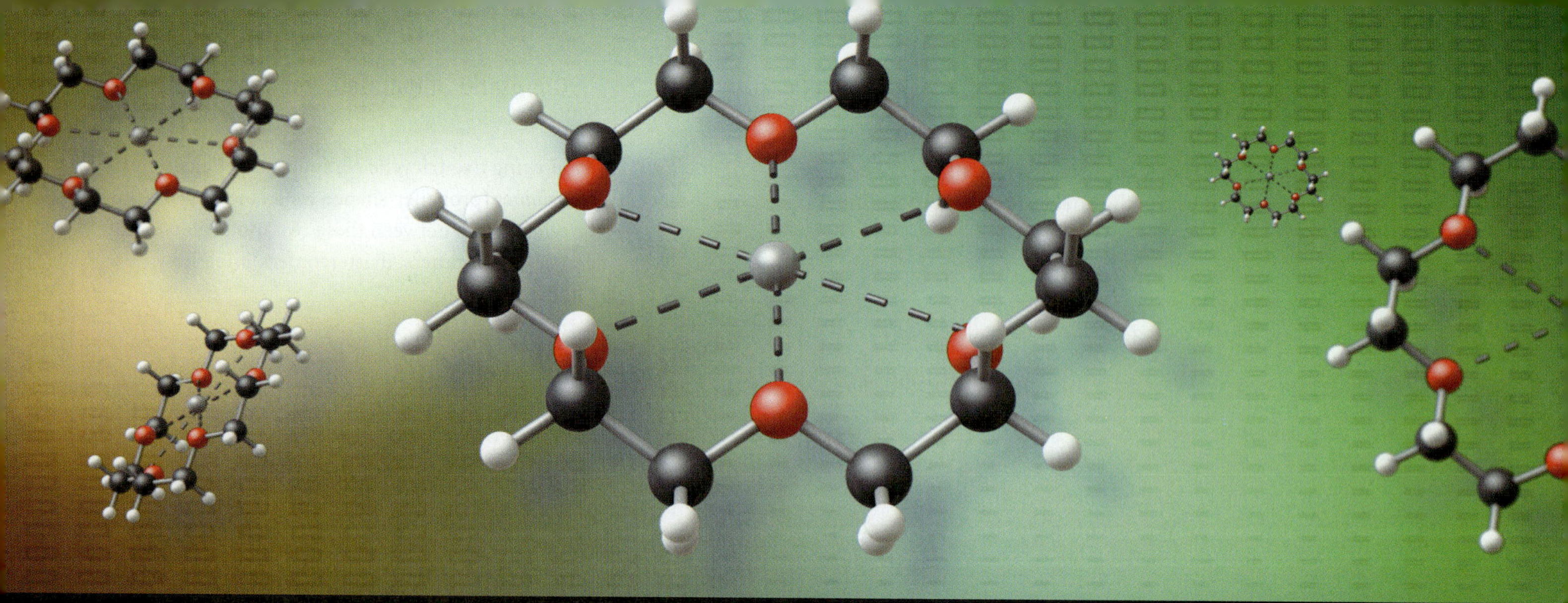

8 NOMENCLATURE AND NONCOVALENT INTERMOLECULAR INTERACTIONS

This chapter contains material on two different topics. The first is *organic nomenclature*. In Secs. 2.4 and 4.2 we introduced the principles of nomenclature. In this chapter, we present the more general rules of nomenclature that will apply throughout the remainder of this text. We also introduce several different types of organic compounds—*alcohols, thiols, alkyl halides, ethers,* and *sulfides*—which are considered together because their chemical reactions are related, as we'll see in Chapters 9–12, and because many of these compound types are used as examples in the second part of this chapter.

The second topic, and the major part this chapter, is *noncovalent intermolecular interactions*—the ways that molecules can interact without forming chemical bonds. Such interactions are fundamental to an understanding of many chemical and biological phenomena.

8.1 DEFINITIONS AND CLASSIFICATION OF ALKYL HALIDES, ALCOHOLS, THIOLS, ETHERS, AND SULFIDES

An **alkyl halide** is a compound in which a halogen (F, Br, Cl, or I) is bonded to the carbon of an alkyl group.

CH_3CH_2—Br

an alkyl bromide

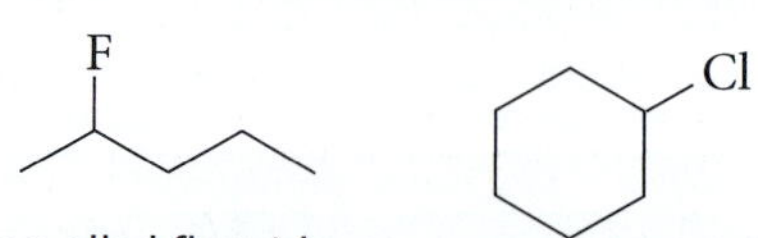

an alkyl fluoride

an alkyl chloride

alkyl halides (8.1)

An alcohol is a compound in which a **hydroxy group** (OH) is bonded to the carbon of an alkyl group, and a **thiol** is a compound in which a **mercapto group**, or **sulfhydryl group** (SH), is bonded to the carbon of an alkyl group. Thiols are sometimes called **mercaptans.**

alcohols   *thiols* (sometimes called *mercaptans*) (8.2)

Generally, the term *thio* is used to mean "sulfur in place of oxygen." For example, a thiol is an alcohol in which a sulfur has been substituted for oxygen. The term (from the Greek word *theio*, meaning "brimstone") originates from the observation that many volatile sulfur-containing organic compounds have very unpleasant odors, in some cases like the odor of burning rubber. (Many of the foul-smelling compounds emitted by an angry skunk are thiols.) The name *mercaptan* is derived from the fact that thiols form stable derivatives with mercury (and other heavy metals)—that is, a mercaptan "captures mercury."

Don't confuse an *alcohol* with a *phenol* or an *enol*. In a **phenol**, the OH group is bonded to the carbon of an *aryl group* (Sec. 2.8B).

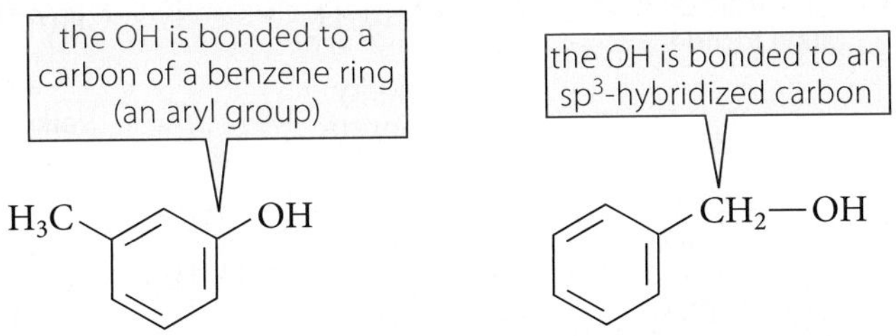

a phenol   an alcohol (8.3)

In an **enol**, the OH group is bonded to an sp^2-hybridized carbon that is part of an alkene double bond. In an alcohol, the OH group is bonded to an sp^3-hybridized carbon.

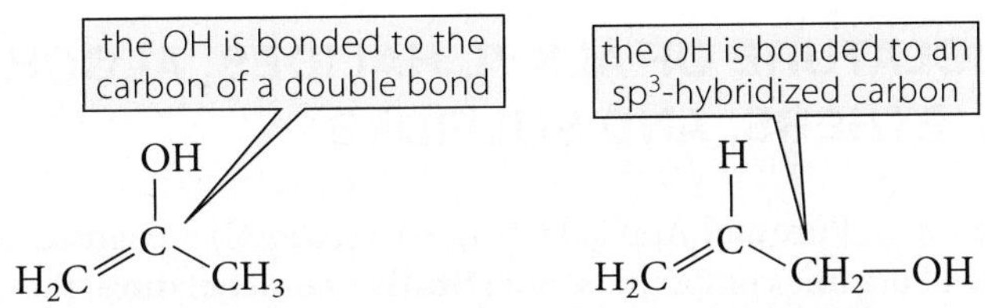

an enol   an alcohol (8.4)

Phenols, enols, and alcohols have very different properties; we discuss phenols in Chapter 18. Enols were introduced in as reaction intermediates in Secs. 5.5 and 5.6, and they are discussed more fully in Chapter 22.

Ethers are compounds in which an oxygen is bonded to two carbon groups, which can be alkyl or aryl. The carbon groups may be the same or different. **Sulfides**, which are also called **thioethers**, are the sulfur analogs of ethers.

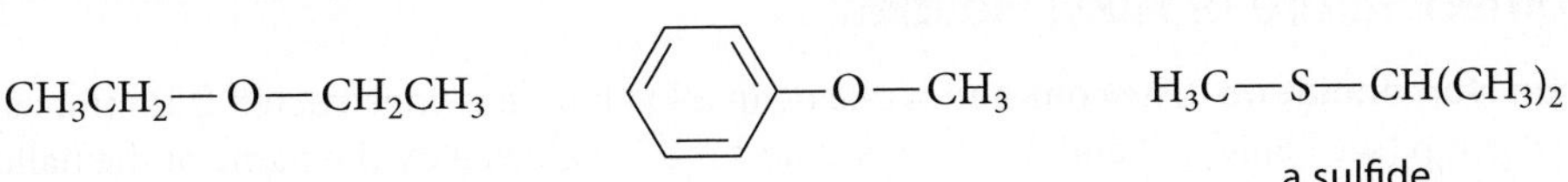

ethers   a sulfide (a thioether) (8.5)

The carbon bonded to the halogen in an alkyl halide, or to the oxygen in an alcohol or ether, is called the **alpha-carbon**, usually written with the Greek letter as **α-carbon**.

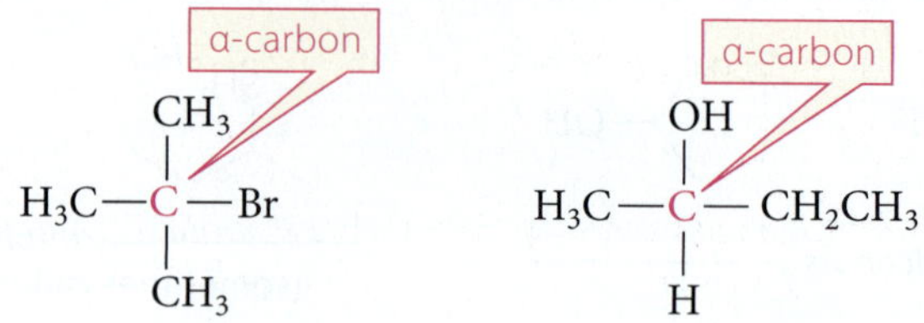

(8.6)

Alkyl halides and alcohols are classified by the *number of alkyl groups attached to the α-carbon.* A methyl halide or methyl alcohol has no alkyl groups, a **primary** halide or alcohol has one alkyl group, a **secondary** halide or alcohol has two alkyl groups, and a **tertiary** halide or alcohol has three alkyl groups. In the following examples, the alkyl substituents are shown in blue and the α-carbon in red.

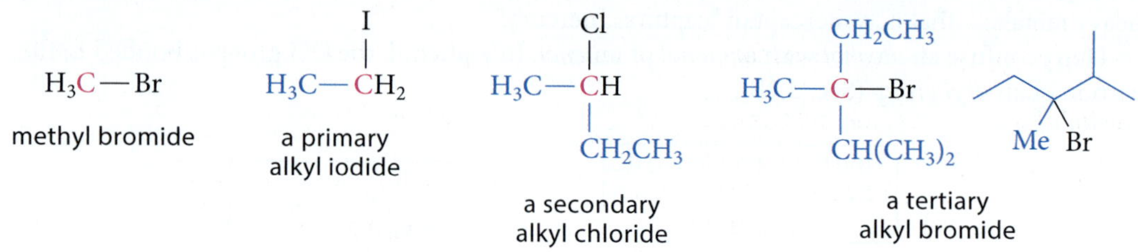

(8.7)

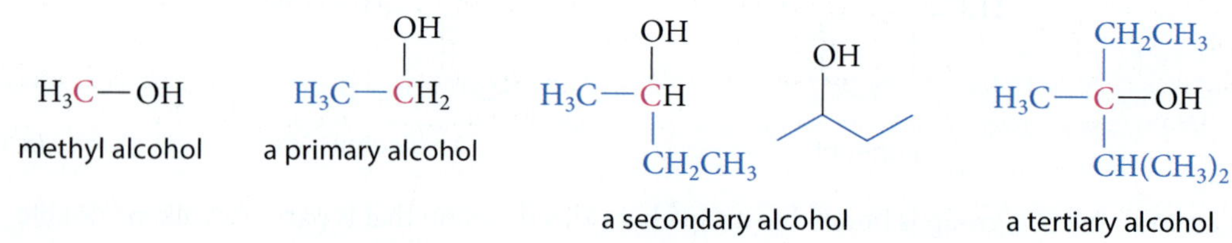

(8.8)

8.2 NOMENCLATURE OF ALKYL HALIDES, ALCOHOLS, THIOLS, ETHERS, AND SULFIDES

The International Union of Pure and Applied Chemistry (IUPAC) recognizes several systems for the nomenclature of organic compounds. **Substitutive nomenclature**, the most broadly applicable system, was introduced in the nomenclature of both alkanes (Sec. 2.4C) and alkenes (Sec. 4.2A). This system can be applied to the compound classes in this chapter as well. Another widely used system to be introduced in this chapter is called **radicofunctional nomenclature** by the IUPAC; for simplicity, this system will be called **common nomenclature**. Common nomenclature is generally used only for the simplest and most common compounds. Although the adoption of a single nomenclature system might seem desirable, historical usage and other factors have dictated the use of both common and substitutive names.

A. Nomenclature of Alkyl Halides

Common Nomenclature The common name of an alkyl halide is constructed from the name of the alkyl group (see Tables 2.2 and 2.4, Secs. 2.3 and 2.4C) followed by the name of the halide as a separate word.

$CH_3CH_2—Cl$
ethyl chloride

CH_2Cl_2
methylene chloride
($—CH_2—$ group = methylene group)

$CH_3CH_2CH_2CH_2—Br$
butyl bromide

$(CH_3)_2CH—I$
isopropyl iodide

(8.9)

The common names of the following compounds should be learned.

$H_2C{=}CH{-}CH_2{-}Cl$	$Ph{-}CH_2{-}Br$	$H_2C{=}CH{-}Cl$	CCl_4
allyl chloride	**benzyl bromide**	**vinyl chloride**	**carbon tetrachloride**

(8.10)

(Compounds with halogens attached to alkene carbons, such as vinyl chloride, are not alkyl halides, but it is convenient to discuss their nomenclature here.)

The **allyl group**, as the structure of allyl chloride implies, is the $H_2C{=}CH{-}CH_2{-}$ group. This should not be confused with the **vinyl group**, $H_2C{=}CH{-}$, which lacks the additional $-CH_2-$. Similarly, the **benzyl group**, $Ph{-}CH_2{-}$, should not be confused with the **phenyl group**.

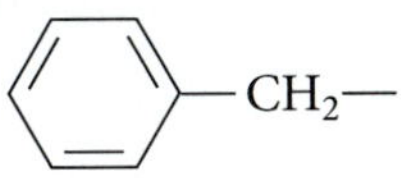 or $Ph{-}CH_2{-}$

benzyl group

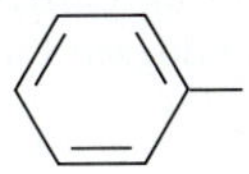 or $Ph{-}$

phenyl group (8.11)

The **haloforms** are the methyl trihalides. Chloroform is a commonly used organic solvent.

$HCCl_3$	$HCBr_3$	HCI_3
chloroform	**bromoform**	**iodoform**

(8.12)

Substitutive Nomenclature The IUPAC substitutive name of an alkyl halide is constructed by applying the rules of alkane and alkene nomenclature (Secs. 2.4C and 4.2A). Halogens are always treated as substituents; the halogen substituents are named fluoro, chloro, bromo, or iodo. Double and triple bonds have precedence in numbering just as they do in alkenes and alkynes.

$CH_3CH_2{-}Cl$

chloroethane

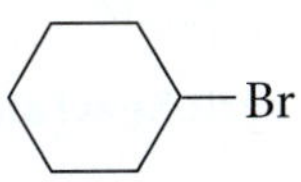

bromocyclohexane

$CH_3CH(F)CH_2CH_2CH_3$

2-fluoropentane

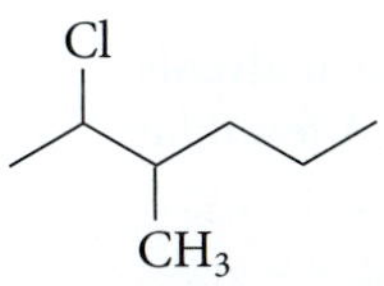

2-chloro-3-methylhexane

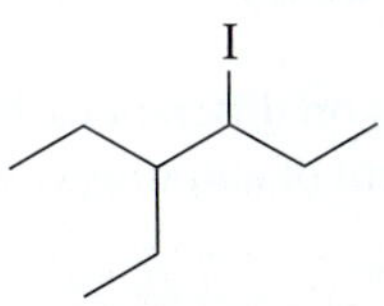

3-ethyl-4-iodohexane

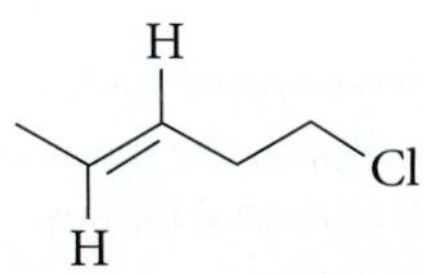

(*E*)-5-chloro-2-pentene

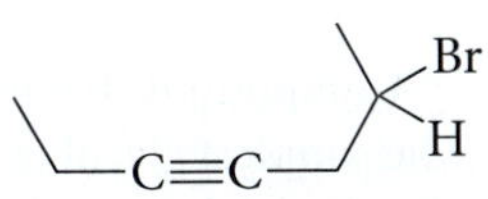

6-bromo-3-heptyne

(8.13)

Focused Problems

8.1 Give the common name for each of the following compounds, and tell whether each is a primary, secondary, or tertiary alkyl halide.

(a) 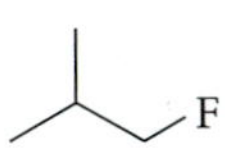

(b) 

(c) 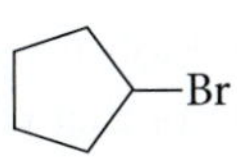

(d) $H_3C{-}C(CH_3)_2{-}CH_2Cl$

Sometimes you may see an *n* prefix used in the common nomenclature of some organic compounds, as in the following example:

$CH_3CH_2CH_2CH_2CH_3$

pentane
(sometimes called ***n*-pentane**)

$CH_3CH_2CH_2CH_2{-}Br$

1-bromobutane
(substitutive nomenclature)
butyl bromide
(common nomenclature)
(sometimes called ***n*-butyl bromide**)

$CH_3CH_2CH_2CH_2{-}OH$

1-butanol
(substitutive nomenclature)
butyl alcohol
(common nomenclature)
(sometimes called ***n*-butyl alcohol**)

The *n* prefix stands for "normal." At one time, the prefix *n* and the word *normal* were used to indicate an isomer containing a functional group at the end of an unbranched carbon chain. However, this prefix is superfluous and unnecessary. For example, the name butyl bromide itself means the structure shown in Display 8.9; an additional prefix is unnecessary. Branched-chain isomers have other names, such as isobutyl bromide, *sec*-butyl bromide, or *tert*-butyl bromide. Because common names are unambiguous without the prefix *n*-, the IUPAC recommended abandoning it. Despite this recommendation, the prefix continues to be used.

8.2 Give the structure of each of the following compounds.

(a) 2,2-Dichloro-5-methylhexane (b) Chlorocyclopropane

(c) 6-Bromo-1-chloro-3-methylcyclohexene (d) Methylene iodide

(e) 1-Bromo-1-pentyne

8.3 Give the substitutive name for each of the following compounds.

(a) 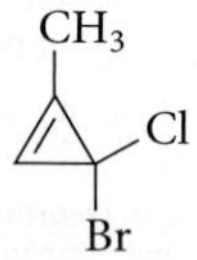

(b) 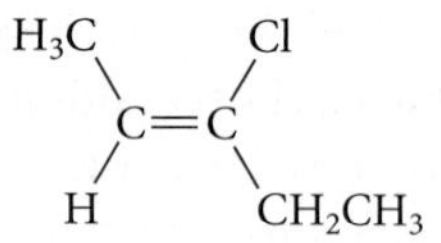

(c) 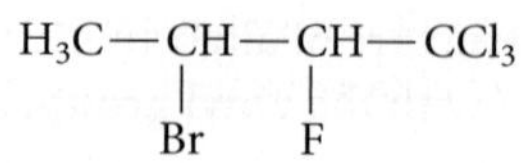

(d) Chloroform

(e) Neopentyl bromide (see Table 2.4, Sec. 2.4C)

(f) 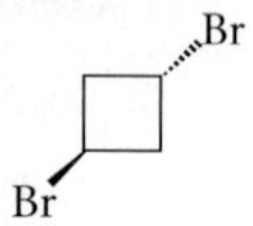

(g) 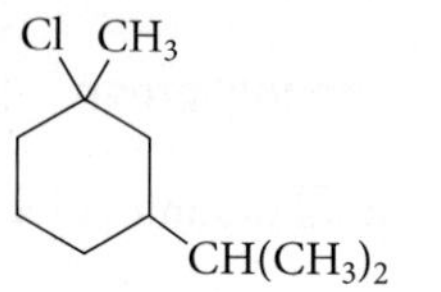

(h) 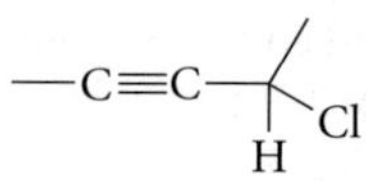

B. Nomenclature of Alcohols and Thiols; The Principal Group

Common Nomenclature The common name of an alcohol is derived by specifying the alkyl group to which the OH group is attached, followed by the separate word *alcohol.*

$H_3C—OH$ **methyl alcohol**

$(CH_3)_2CH—OH$ **isopropyl alcohol**

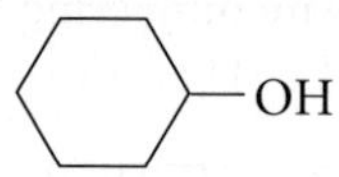

cyclohexyl alcohol

$CH_3CH_2CH_2—OH$ **propyl alcohol**

$H_2C{=}CH—CH_2—OH$ **allyl alcohol**

$Ph—CH_2—OH$ **benzyl alcohol** (8.14)

Compounds that contain two or more hydroxy groups on different carbons are called **glycols.** The simplest glycol is ethylene glycol, the main component of automotive antifreeze. A few other glycols also have widely used traditional names.

$HO—CH_2CH_2—OH$ **ethylene glycol**

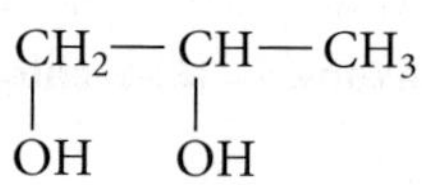

propylene glycol

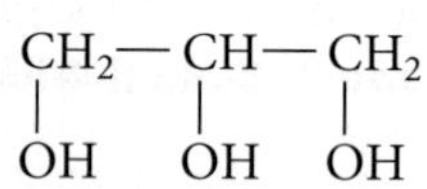

glycerol (glycerin) (8.15)

The most common glycols are those in which the OH groups are bonded to adjacent carbons. As a class, these are sometimes referred to as **vicinal glycols.** (Compare this term to the *vicinal dihalides* in Display 5.8, Sec. 5.2A.) The glycols in Display 8.15 are all vicinal glycols.

Thiols are named in the common system as *mercaptans.*

$CH_3CH_2—SH$

ethyl mercaptan (8.16)

Substitutive Nomenclature The substitutive nomenclature of alcohols and thiols involves a concept called the *principal group. This is a very important nomenclature concept* that will be used repeatedly. The **principal group** is the chemical group on which the name is based, *and it is always cited as a suffix in the name.* For example, in a simple alcohol, the OH group is the principal group, and its suffix is *ol*. The name of an alcohol is constructed by dropping the final *e* from the name of the parent alkane and adding this suffix.

$$CH_3CH_2—OH$$

ethan~~e~~ + *ol* = **ethanol** (8.17a)

The final *e* is generally dropped when the suffix begins with a vowel; otherwise, it is retained.

For simple thiols, the SH group is the principal group, and its suffix is *thiol*. The name is constructed by adding this suffix to the name of the parent alkane. Because the suffix begins with a consonant, the final *e* of the alkane name is retained.

$$CH_3CH_2—SH$$

ethane + *thiol* = **ethanethiol** (8.17b)

Only certain groups are cited as principal groups. The OH and SH groups are the only ones in the compound classes considered so far, but others are added in later chapters. If a compound does not contain a principal group, it is named as a substituted hydrocarbon in the manner illustrated for the alkyl halides in Sec. 8.2A.

The *principal group* and the *principal chain* are the key concepts defined and used in the construction of a substitutive name according to the *general rules for substitutive nomenclature of organic compounds*, which follow. The simplest way to learn these rules is to read through the rules briefly and then concentrate on the study problems and examples that follow, letting them guide you through the application of the rules in specific cases.

1. *Identify the principal group.*
   When a structure has several candidates for the principal group, the group chosen is the one given the highest priority by the IUPAC. The IUPAC specifies that the OH group receives precedence over the SH group:

   Priority as principal group: OH > SH (8.17c)

   A complete list of principal groups and their relative priorities are summarized in Appendix I. (If there is no principal group, follow rule 4b.)

2. *Identify the principal carbon chain.*
   The **principal chain** is the carbon chain on which the name is based (Sec. 2.4C). The principal chain is identified by applying the following criteria *in order* until a decision can be made:

   (a) The chain with the greatest number of principal groups

   (b) The chain with the greatest number of double and triple bonds, with double bonds having priority over triple bonds if there is an ambiguity

   (c) The chain of greatest length

   (d) The chain with the greatest number of other substituents

   These criteria cover most of the cases you'll encounter.

3. *Number the carbons of the principal chain consecutively from one end.*
   In numbering the principal chain, apply the following criteria in order until there is no ambiguity:

   (a) The lowest numbers for the principal groups

   (b) The lowest numbers for multiple bonds, with double bonds having priority over triple bonds in case of ambiguity

(c) The lowest numbers for other substituents

(d) The lowest number for the substituent cited first in the name

4. *Begin construction of the name with the name of the hydrocarbon corresponding to the principal chain.*

(a) Cite the principal group by its suffix and number; its number is the last one cited in the name. (See the examples in Study Problem 8.1.)

(b) If there is no principal group, name the compound as a substituted hydrocarbon. (See Secs. 2.4C and 4.2A.)

(c) Cite the names and numbers of the other substituents in alphabetical order at the beginning of the name.

When SH groups are cited as substituents rather than principal groups, they are called **mercapto** groups. (See Study Problem 8.1b for an example.) When OH groups are cited as substituents, they are called **hydroxy** groups. (We'll illustrate this situation when we encounter functional groups that have higher citation priority than OH.)

Study Problem 8.1

Provide an IUPAC substitutive name for each of the following compounds.

(a) $CH_3CH_2CHCH_3$ (with OH on C-2)

(b) 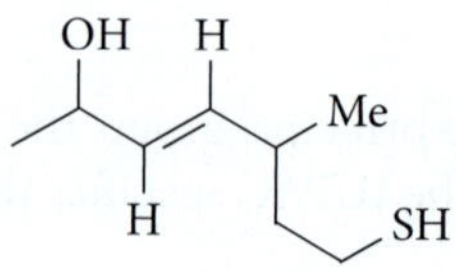

Solution

(a) From rule 1, the principal group is the OH group. Because there is only one possibility for the principal chain, rule 2 does not enter the picture. By applying rule 3a, the principal group is located at carbon-2. From rule 4a, the name is based on the four-carbon hydrocarbon, butane. After dropping the final *e* and adding the suffix *ol*, the name is obtained: 2-butanol.

4 3 2 1
$CH_3CH_2CHCH_3$ (with OH on C-2)

2-butanol

(b) From rule 1, the principal group is again the OH group, because OH has precedence over SH. From rules 2a–2c, the principal chain is the longest one containing both the OH group and the double bond, so it has seven carbons. Numbering the principal chain in accord with rule 3a gives the OH group the lowest number at carbon-2 and a double bond at carbon-3:

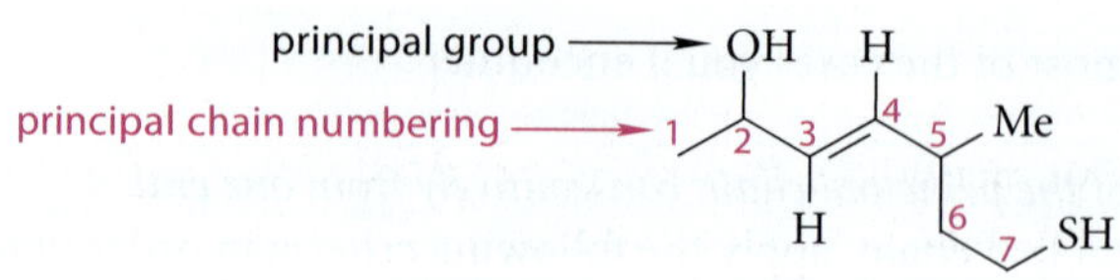

By applying rule 4a, the parent hydrocarbon is 3-heptene, from which we drop the final *e* and add the suffix *ol*, to give 3-hepten-2-ol as the final part of the name. (Because we have to cite the number of the double bond, the number for the OH principal group is located before the final suffix *ol*.) Rule 4c requires that the methyl group at carbon-5 and the SH group at carbon-7 be cited as ordinary substituents. (The substituent name of the SH group is the *mercapto* group.) The stereochemistry of the double bond is *E*. The name is

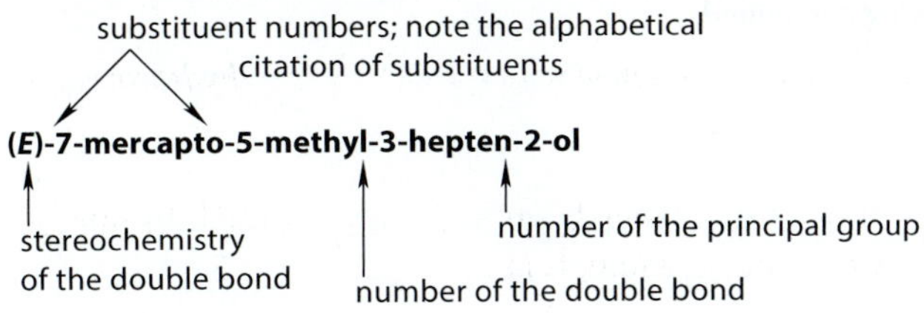

To name an alcohol containing more than one —OH group, the suffixes *diol*, *triol*, and so on are added to the name of the appropriate alkane *without* dropping the final *e*.

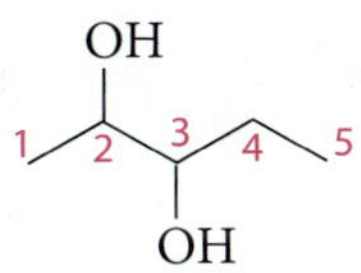

2,3-pentanediol

We introduced the 1993 IUPAC nomenclature recommendations at the end of Sec. 4.2A. Although we are continuing to use the 1979 recommendations for the reasons given there, conversion of most names to the 1993 recommendations is not difficult. The handling of double bonds, triple bonds, and principal groups is the major change introduced by the 1993 recommendations. In the 1993 system, the number of the double bond, triple bond, or principal group immediately precedes the citation of the group in the name. So, in the 1993 convention, the name of the compound in Study Problem 8.1(a) would be butan-2-ol rather than 2-butanol. The name of the compound in Study Problem 8.1(b) would be 7-mercapto-5-methylhept-3-en-2-ol. The name 2,3-butanediol would be changed to butane-2,3-diol, and the name of the compound in Study Problem 8.2 would become 6-mercaptocyclohex-4-ene-1,3-diol. As in the 1979 nomenclature, the final *e* of the hydrocarbon name is dropped when the suffix begins in a vowel.

Study Problem 8.2

Name the following compound.

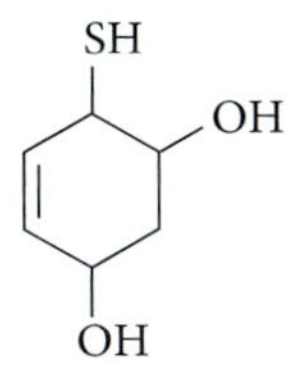

Solution From rule 1, the principal groups are the OH groups. By rule 3a, these groups are given numerical precedence, so they receive the numbers 1 and 3. Because two numbering schemes give these groups the numbers 1 and 3, we choose the scheme that gives the double bond the lower number, by rule 3b. From rule 4a, the parent hydrocarbon is cyclohexene, and because the suffix is *diol*, the final *e* is retained to give the partial name 4-cyclohexene-1,3-diol. Finally, the SH group has been eliminated from consideration as the principal group, so it is treated as an ordinary substituent group by rule 4c. The completed name is therefore

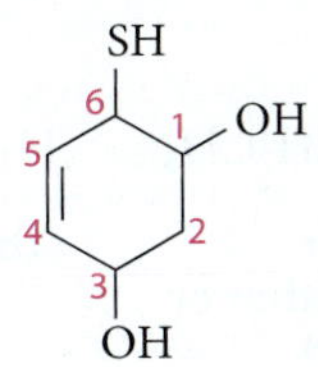

6-mercapto-4-cyclohexene-1,3-diol

Common and substitutive nomenclature should not be mixed. This rule is frequently disregarded in naming the following compounds:

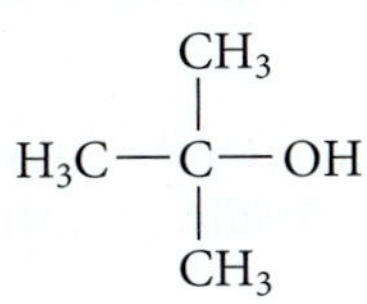

common: *tert*-butyl alcohol
substitutive: 2-methyl-2-propanol
incorrect: *t*-butanol or *tert*-butanol

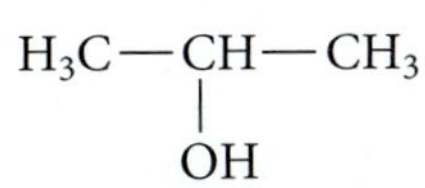

common: isopropyl alcohol
substitutive: 2-propanol
incorrect: isopropanol (8.18)

Focused Problems

8.4 Draw the structure of each of the following compounds.

(a) *sec*-Butyl alcohol

(b) 3-Ethylcyclopentanethiol

(c) 3-Methyl-2-pentanol

(d) (*E*)-6-Chloro-4-hepten-2-ol

(e) 2-Cyclohexenol

8.5 Give the substitutive name for each of the following compounds. Include stereochemistry in the name for cases in which stereochemistry is indicated in the structure.

(a) 

(b) 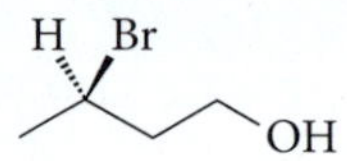

(c) 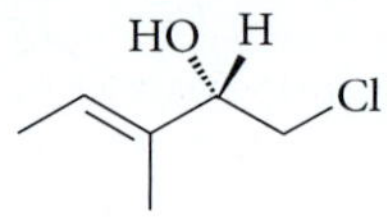

(d) 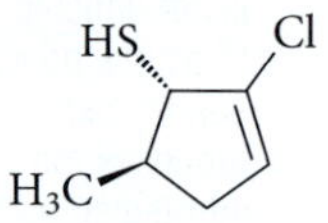

(e) 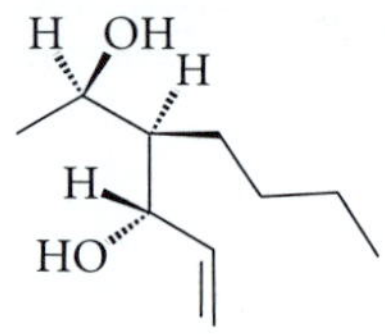

(f) 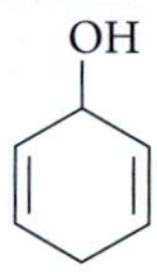

(g) 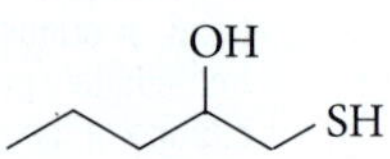

(h) 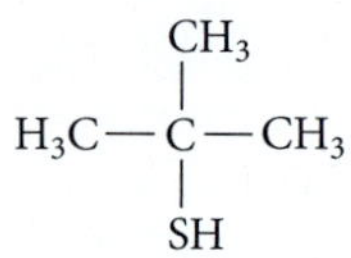

(i) 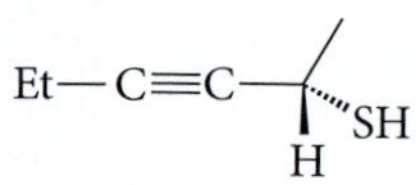

C. Nomenclature of Ethers and Sulfides

Common Nomenclature The common name of an ether is constructed by citing as separate words the two groups attached to the ether oxygen in alphabetical order, followed by the word *ether*.

$$CH_3CH_2—O—CH_2CH_3 \qquad H_3C—O—C_2H_5$$

diethyl ether (also called **ethyl ether** or simply **ether**) — **ethyl methyl ether** (8.19)

A sulfide is named in a similar manner, using the word *sulfide*. (In older literature, the word *thioether* was also used.)

$$CH_3CH_2—S—CH_3 \qquad (CH_3)_2CH—S—CH(CH_3)_2$$

ethyl methyl sulfide (also **ethyl methyl thioether**) — **diisopropyl sulfide** (8.20)

Substitutive Nomenclature In substitutive nomenclature, ethers and sulfides are never cited as principal groups. Instead, *alkoxy groups* (RO—) and *alkylthio groups* (RS—) are cited as substituents.

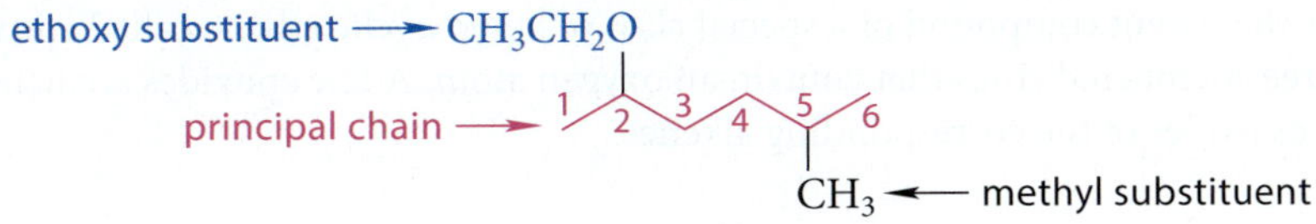

2-ethoxy-5-methylhexane (8.21)

In this example, the principal chain is a six-carbon chain. Therefore, the compound is named as a hexane, and the C_2H_5O— group and the methyl group are treated as substituents. The C_2H_5O— group is named by dropping the final *yl* from the name of the alkyl group and adding the suffix *oxy*. So, the C_2H_5O— group is the (ethy~~l~~ + oxy) = ethoxy group. The numbering follows from nomenclature rule 3d in Sec. 8.2B.

The nomenclature of sulfides is similar. An RS— group is named by adding the suffix *thio* to the name of the R group; the final *yl* is not dropped.

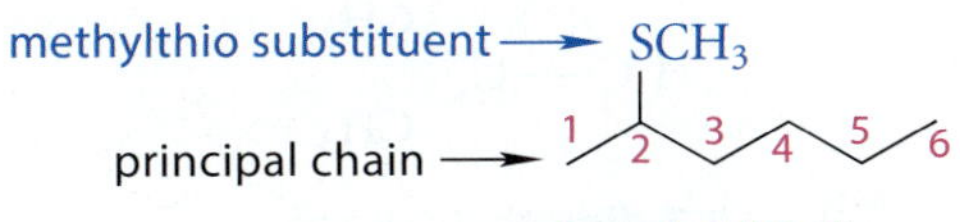

2-(methylthio)hexane (8.22)

The parentheses in the name are used to indicate that "thio" is associated with "methyl" rather than with "hexane."

Study Problem 8.3

Name the following compound.

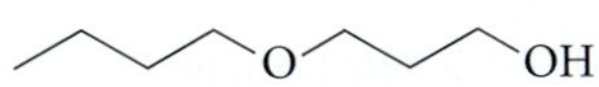

Solution The —OH group is cited as the principal group, and the principal chain is the chain containing this group. Consequently, the $CH_3CH_2CH_2CH_2O$—group is cited as a butoxy (buty~~l~~ + oxy) substituent at carbon-3 of the principal chain:

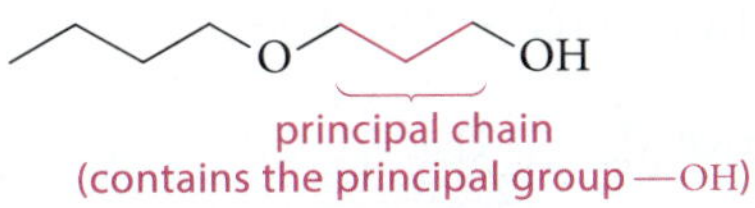

3-butoxy-1-propanol

Heterocyclic Nomenclature A number of important ethers and sulfides contain an oxygen or sulfur atom within a ring. Cyclic compounds with rings that contain at least one atom other than carbon are called **heterocyclic compounds**. The names of some common heterocyclic ethers and sulfides should be learned.

furan

tetrahydrofuran (often called **THF**)

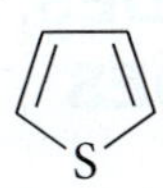

thiophene

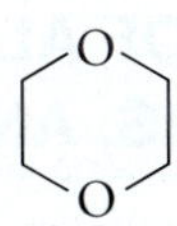

1,4-dioxane (often called simply **dioxane**)

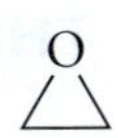

oxirane **(ethylene oxide)**

(8.23)

(The IUPAC name for tetrahydrofuran is *oxolane*, but this name is not commonly used.)

Oxirane is the parent compound of a special class of heterocyclic ethers, called **epoxides**, which are three-membered rings that contain an oxygen atom. A few epoxides are named traditionally as oxides of the corresponding alkenes:

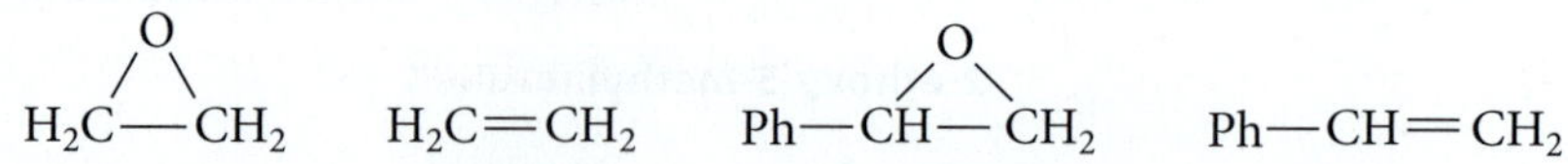

ethylene oxide **ethylene** **styrene oxide** **styrene** (8.24)

The reason for this nomenclature is that epoxides frequently are prepared from alkenes (Sec. 12.3). However, most epoxides are named substitutively as derivatives of oxirane. The atoms of the epoxide ring are numbered consecutively, with the oxygen receiving the number 1.

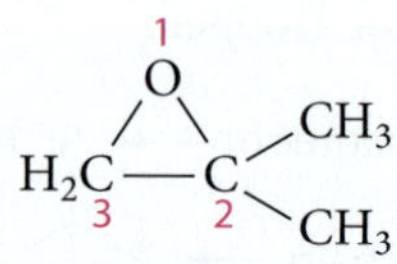

2,2-dimethyloxirane (8.25)

Focused Problems

8.6 Draw the structure of each of the following compounds.

(a) Ethyl propyl ether (b) Dicyclohexyl ether

(c) *tert*-Butyl isopropyl sulfide (d) Allyl benzyl ether

(e) Phenyl vinyl ether (f) (2*R*,3*R*)-2,3-Dimethyloxirane

(g) 5-(Ethylthio)-2-methylheptane

8.7 Give a substitutive name for each of the following compounds.

(a) $(CH_3)_3C—O—CH_3$ (b) $CH_3CH_2—O—CH_2CH_2—OH$

(c) 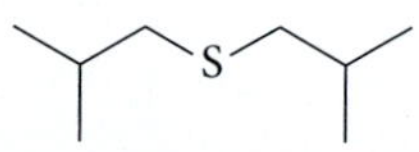 (d) 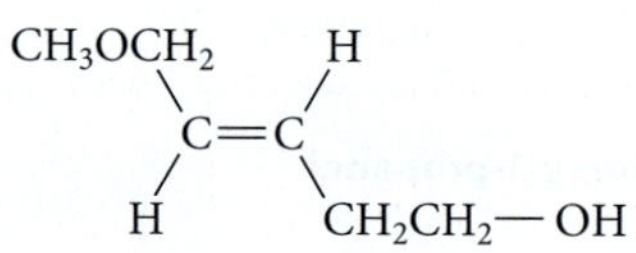

8.8 (a) A chemist used the name 3-butyl-1,4-dioxane in a paper. Although the name unambiguously describes a structure, what should the name have been? Explain.

(b) Give the structure of 2-butoxyethanol, which is an ingredient in whiteboard cleaner and kitchen cleaning sprays.

8.3 STRUCTURES OF ALKYL HALIDES, ALCOHOLS, THIOLS, ETHERS, AND SULFIDES

In all of the compounds covered in these sections, the bond angles at carbon are nearly tetrahedral, and the α-carbons are sp^3-hybridized. For example, in the simple methyl derivatives (the methyl halides, methanol, methanethiol, dimethyl ether, and dimethyl sulfide) the H—C—H bond angle in the methyl group does not deviate more than a degree or so from 109.5°.

In an alcohol, thiol, ether, or sulfide, the bond angle at oxygen or sulfur further defines the shape of the molecule. These molecules are bent at oxygen and sulfur, as you can see from the structures in **Fig. 8.1**. As we showed in Sec. 1.7E, oxygen is sp^2-hybridized; one of its nonbonding

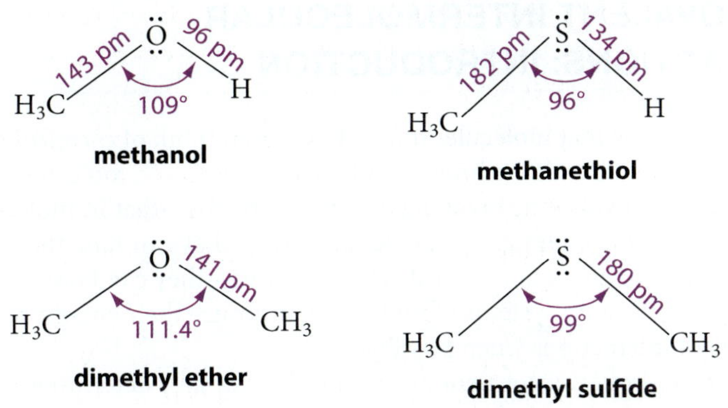

FIGURE 8.1 Bond lengths and bond angles in a simple alcohol, thiol, ether, and sulfide. Bond angles at sulfur are smaller than those at oxygen, and bonds to sulfur are longer than the corresponding bonds to oxygen.

electron pairs resides in a 2p orbital. The hybridization of the remaining nonbonding electron pair contains as much s character as possible without allowing the two bonded groups to have van der Waals repulsions. As a result, the C—O—C bond angle in dimethyl ether spreads to 111.4°, whereas the bond angle in methanol is smaller, at 109°, because hydrogen is smaller than an alkyl group.

The bond angles at sulfur are generally closer to 90° than the corresponding angles at oxygen so that the second nonbonding electron pair can have almost all of the s character; the bonds to the other groups have a high degree of p character and have C—S—C or C—S—H bond angles close to 90°. The longer bonds to sulfur allow smaller bond angles without van der Waals repulsions in the bonded groups. However, the C—S—C bond angle in dimethyl sulfide is larger than the bonds in methanethiol because a methyl group is larger than a hydrogen.

The lengths of bonds between carbon and other atoms follow the trends discussed in Sec. 1.3B. Within a column of the periodic table, bonds to atoms of higher atomic number are longer; so the C—S bond of methanethiol is longer than the C—O bond of methanol (see Fig. 8.1 and **Table 8.1**). Within a row, bond lengths decrease toward higher atomic number (that is, to the right); so the C—O bond in methanol is longer than the C—F bond in methyl fluoride (see Table 8.1); similarly, the C—S bond in methanethiol is longer than the C—Cl bond in methyl chloride. Both effects reflect decreasing atomic radii toward the right and the top of the periodic table.

Focused Problem

8.9 Using the data in Table 8.1, estimate the carbon–selenium bond length in $H_3C—Se—CH_3$.

TABLE 8.1 Bond Lengths (Picometers) in Some $CH_3—X$ Compounds

Group 4A	Group 5A	Group 6A	Group 7A
$H_3C—CH_3$	$H_3C—NH_2$	$H_3C—OH$	$H_3C—F$
153.6	147.4	142.6	139.1
		$H_3C—SH$	$H_3C—Cl$
		182	178.1
			$H_3C—Br$
			193.9
			$H_3C—I$
			212.9

8.4 NONCOVALENT INTERMOLECULAR INTERACTIONS: INTRODUCTION

When we consider the ways that molecules interact, we often think of *chemical reactions*: processes in which covalent bonds are broken and formed. However, molecules can interact in other ways that do *not* involve the breaking of covalent bonds—that is, molecules can interact *noncovalently*. These interactions can be repulsive—they can *raise* the energy of the molecules involved. These interactions can also be attractive—they can *lower* the energy of the molecules involved. Because systems of molecules minimize their energy, molecules will form attractive noncovalent interactions when possible.

Why should we care about noncovalent interactions? First, they determine a large number of molecular properties, which include boiling points, melting points, and solubility. In the rest of this chapter, you'll learn that the *structure of a molecule determines the magnitude of its noncovalent interactions* with other like molecules or with different molecules. Is a compound soluble in water? In what types of solvents will it be soluble? Will it have a low boiling point or a high boiling point? Will it be a liquid or a solid? All of these important practical questions are governed largely by noncovalent interactions among molecules. *If we can appreciate the effect of structure on noncovalent intermolecular interactions, we can then understand and even, in some cases, predict molecular properties.*

Many biological phenomena, as we'll see, owe their very existence to noncovalent intermolecular attractions: among these are the structures and conformations of biomolecules such as proteins and DNA, and biological structures such as cell membranes. The interactions of protein receptors with small molecules such as drugs are noncovalent. To catalyze a reaction, an enzyme must first associate noncovalently with the reactants (its substrates).

To understand all these phenomena, we need to understand noncovalent interactions. We're going to "start small," with interactions between identical small molecules. Then, we'll consider solutions, which involve interactions between different molecules. Finally, we'll examine a few examples from both chemistry and biology in which noncovalent interactions play a key role.

8.5 HOMOGENEOUS NONCOVALENT INTERMOLECULAR ATTRACTIONS: BOILING POINTS AND MELTING POINTS

Any condensed state of matter (a solid or a liquid) owes its existence to noncovalent intermolecular attractions. If there were no attractions between molecules in a solid or a liquid, the substance would be a gas. (Intermolecular interactions within an ideal gas don't occur, and we use the ideal-gas model as our description of gases.) The attractions between molecules in a solid or a liquid are noncovalent because no chemical reaction occurs when we convert a liquid to a gas or a solid to a liquid. That is, *chemical bonds are not broken*. Because no covalent bonds are broken, these noncovalent attractions are much weaker than the attractions that hold atoms together in covalent bonds.

We can learn a lot about noncovalent attractions by studying the conversion of a liquid—in which noncovalent intermolecular attractions exist—to a gas—in which noncovalent attractions have largely disappeared. Specifically, we can use the *boiling point* as a crude measure of noncovalent attractions. The **boiling point** is the temperature required to raise the vapor pressure of a liquid to atmospheric pressure (760 mm Hg at sea level). The use of the boiling point to measure noncovalent attractions can be justified thermodynamically (see Further Exploration 8.1), but here we justify it intuitively. The boiling point is a measure of the energy required to bring a liquid to the state in which all of the molecules want to escape from the liquid into the gas. As the boiling point increases, then, *more energy is required to break the intermolecular attractions in the liquid state*. Remember again that there are *no covalent bonds between molecules*, and intermolecular attractions have *nothing* to do with the strengths of the covalent bonds in, or the stabilities of, the molecules themselves.

FURTHER EXPLORATION 8.1
Trouton's Rule

A. Attractions between Induced Dipoles: van der Waals (Dispersion) Forces

One of the most significant observations about the boiling points of organic compounds is that *boiling points increase regularly with molecular size within a homologous series.* For example, Fig. 2.7 (Sec. 2.6A) shows that there is a fairly regular increase in the boiling points of unbranched alkanes with the number of carbon atoms. **Figure 8.2a** shows that this regular increase occurs not only for alkanes, but also for many other compound classes, and these are only a few of many examples. We consider later why the trend lines of some compound classes are displaced to higher or lower values, and why the line for alcohols (as well as a few other classes not shown) looks a little different. But generally, *the boiling points of compounds in a given class increase about 20–30 °C per carbon atom.*

Figure 8.2b shows the same data plotted against molecular mass. This plot shows that a few different compound classes have the same boiling points at a given molecular mass.

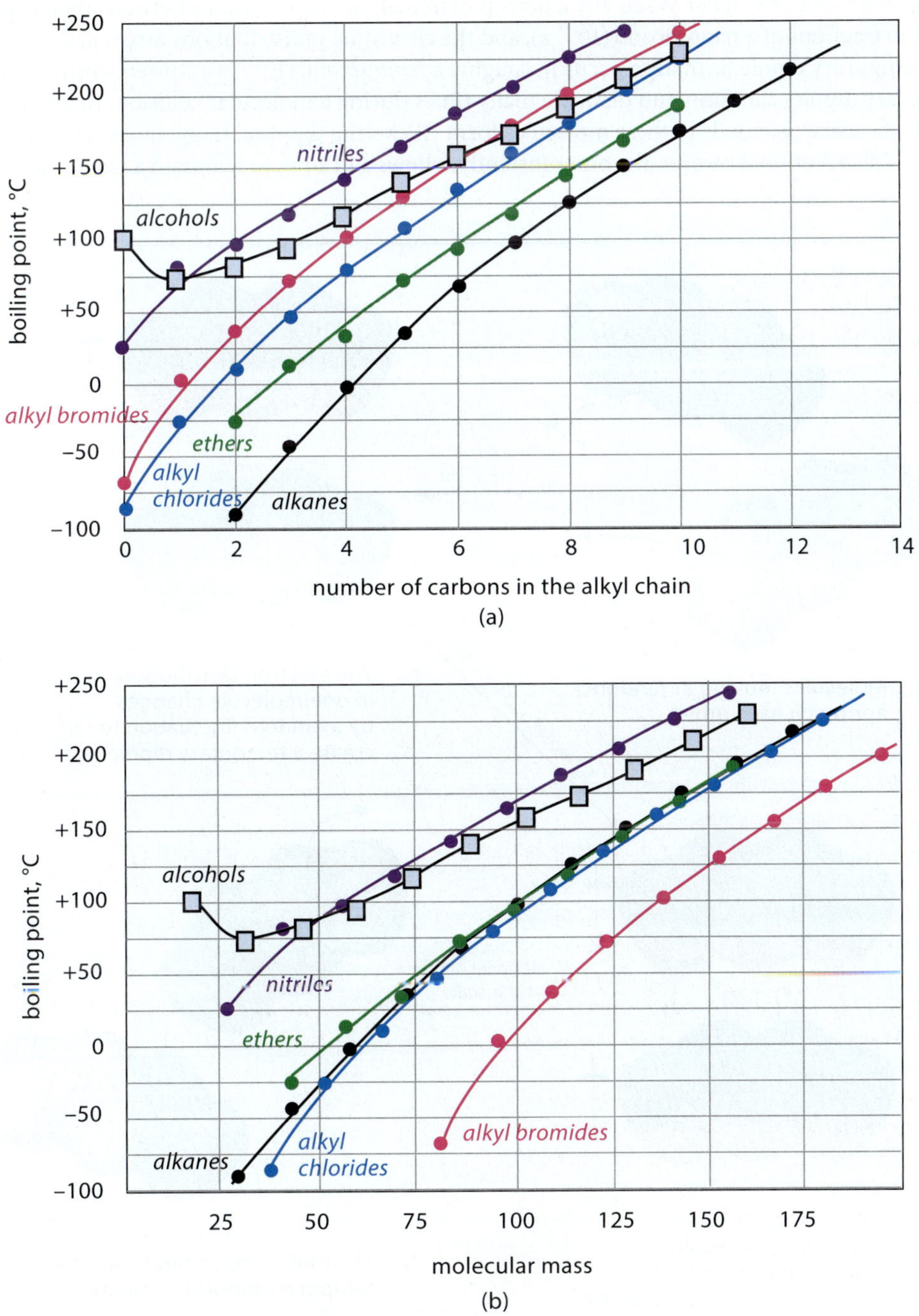

FIGURE 8.2 The boiling points of several compound classes with unbranched carbon chains plotted against (a) number of carbons and (b) molecular mass. In (a), carbon number = 0 indicates the parent compound with no carbons. For example, for alkyl halides (R—X) the parent compound is H—X, for alcohols (R—OH) the parent compound is water (H—OH), and so on. For nitriles, the carbon of the C≡N group is not included in the carbon number because it is part of the functional group.

Why should boiling points increase with increasing molecular size? In Sec. 1.8, we learned that electrons in bonds are not confined between the nuclei but rather reside in bonding molecular orbitals that surround the nuclei. We can think of the total electron distribution as an "electron cloud." The electron clouds in some molecules (such as alkanes) are rather "squishy" and can undergo distortions. Such distortions occur rapidly and *at random*, and when they occur, they result in the temporary formation of regions of local positive and negative charge—that is, these distortions cause a *temporary* dipole moment within the molecule (**Fig. 8.3**). When a second molecule is located nearby, its electron cloud distorts to form a complementary dipole, called an **induced dipole**. The positive charge in one molecule is attracted to the negative charge in the other. The attraction between temporary dipoles—called a **van der Waals attraction**, or a **dispersion interaction**—is the cohesive interaction that must be overcome to vaporize a liquid. Alkanes do *not* have significant *permanent* dipole moments. The dipoles discussed here are *temporary*, and the presence of a temporary dipole in one molecule induces a temporary dipole in another. These attractions come and go. On average, though, "nearness makes the molecules grow fonder."

The *time scale* of van der Waals attractions is extremely small. Collisions between molecules occur in fractions of a nanosecond (10^{-9} s), and the electronic redistributions associated with temporary dipole formation occur in roughly a *femtosecond* (10^{-15} s); consequently, these temporary dipoles can form and dissipate many times during a molecular collision. In other words, the electron clouds in these molecules form "flickering dipoles." If one molecule changes its electron distribution, however, the other instantly follows suit so as to maintain a net attraction.

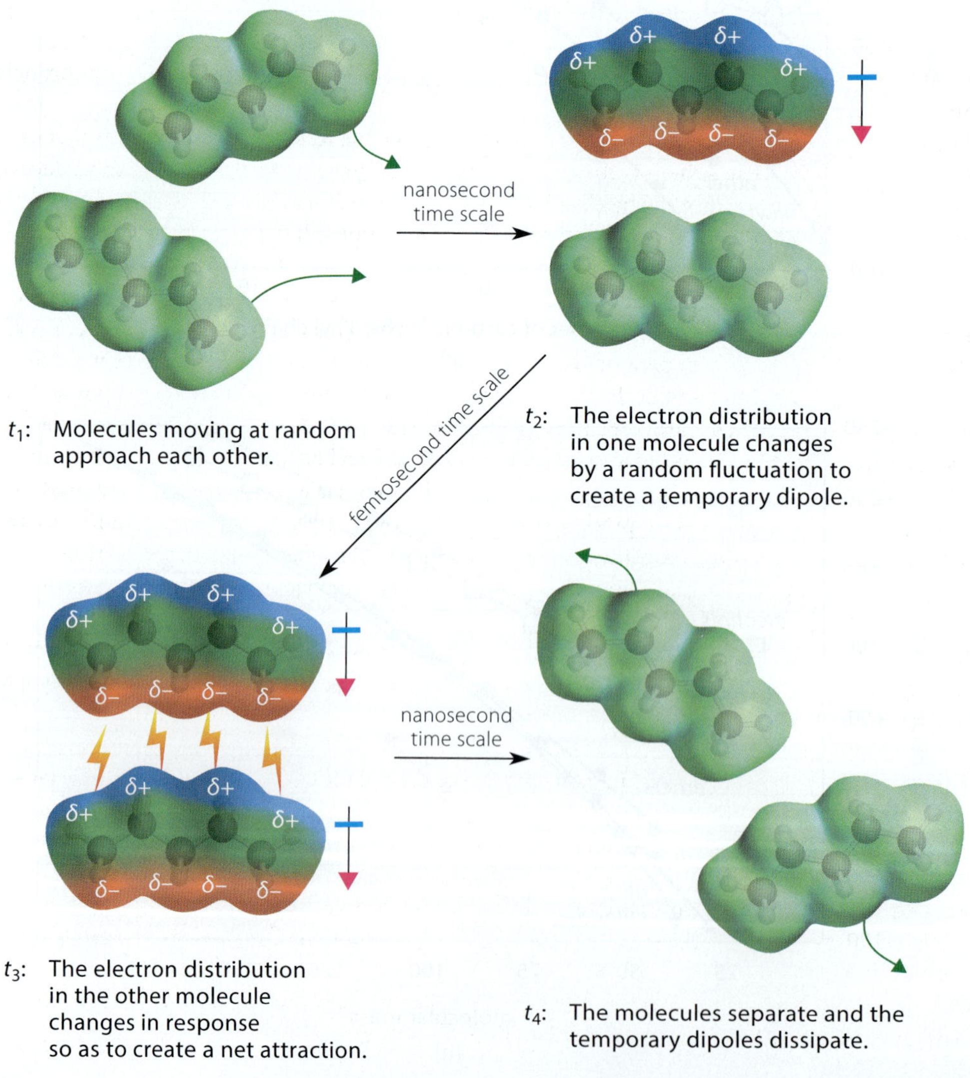

FIGURE 8.3 A cartoon showing the origin of van der Waals attractions in two pentane molecules. The frames are labeled t_1, t_2, t_3, and t_4 to show successive points in time. The colors represent electrostatic potential maps (EPMs). The green color indicates the absence of a significant permanent dipole. Although the attraction at t_3 is temporary, it recurs frequently. This attraction, averaged over time, is the source of van der Waals attractions.

Now we are ready to understand why larger molecules have higher boiling points. Van der Waals attractions increase with the *surface areas of the interacting electron clouds.* That is, the larger the interacting surfaces, the greater the magnitude of the induced dipoles. A larger molecule has a greater surface area of electron clouds and therefore greater van der Waals interactions with other molecules. It follows, then, that large molecules have higher boiling points.

To see that it is the *surface area* and not the *volume* of the molecule that controls boiling point, consider how the *shape* of a molecule affects its boiling point. For example, a comparison of the boiling points of the highly branched alkane neopentane (9.4 °C) and its unbranched isomer pentane (36.1 °C) is particularly striking. Neopentane has four methyl groups disposed in a tetrahedral arrangement about a central carbon. As the following space-filling models show, the molecule almost resembles a compact ball and could fit readily within a sphere. On the other hand, pentane is rather extended, is ellipsoidal in shape, and would *not* fit within the same sphere.

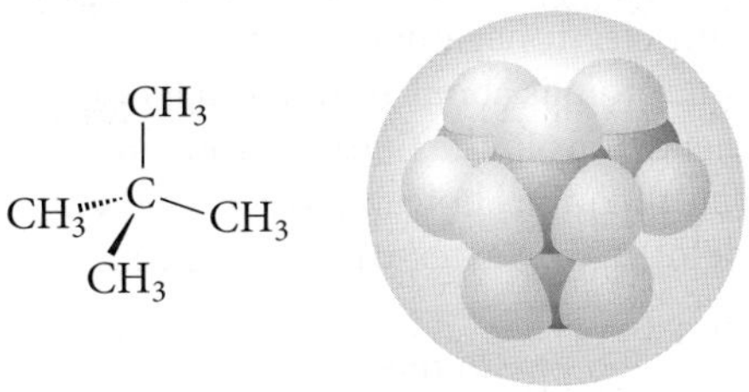

neopentane:
compact, nearly spherical
bp 9.4 °C

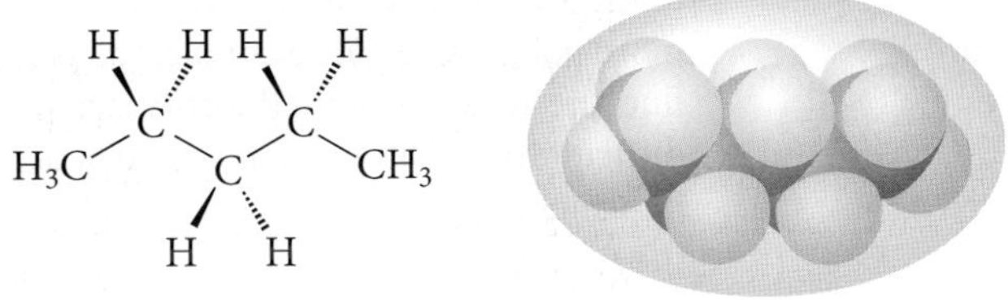

pentane:
extended, ellipsoidal
bp 36.1 °C

(8.26)

Typically, extensive branching in a compound lowers the boiling point relative to the unbranched isomer in the same class.

The more a molecule approaches spherical proportions, the less surface area it presents to other molecules, because a sphere is the three-dimensional object with the minimum surface-to-volume ratio. Because neopentane has less surface area at which van der Waals attractions with other neopentane molecules can occur, it has weaker attractions than pentane, and therefore a lower boiling point.

What sorts of energies are involved in van der Waals attractions between alkane molecules? For this information, we need the *heats of vaporization* of a few alkanes. This is the heat required to convert 1 mol of an alkane at a given temperature into 1 mol of the same alkane in the gas phase. It is a direct measure of noncovalent interactions. For a temperature of 25 °C (298 K), the heats of vaporization of hexane, heptane, and octane are 31.6, 36.6, and 41.5 kJ mol^{-1} (7.55, 8.75, and 9.92 kcal mol^{-1}), respectively. In contrast, breaking a typical alkane carbon–carbon bond requires about 380 kJ mol^{-1} (about 90 kcal mol^{-1}), and C—H bonds are even stronger. In other words, the energy required to break noncovalent attractions in a typical liquid alkane is roughly 10% of the energy required to break carbon–carbon bonds; furthermore, the heat of vaporization of an alkane only increases by about 5 kJ mol^{-1} (1.2 kcal mol^{-1}) per additional CH_2 group in the homologous series. In other words, *van der Waals attractions are weak.*

In summary, by analysis of the boiling points of alkanes, we have learned two general trends in the variation of boiling point with structure, and we have also learned the *mechanism* by which these noncovalent attractions occur:

1. Boiling points increase with increasing molecular weight within a homologous series—typically 20–30 °C per carbon atom. This increase is due to the greater van der Waals attractions between larger molecules.

2. Boiling points tend to be lower for highly branched molecules, because they have less molecular surface available for van der Waals attractions than their unbranched isomers.

Polarizability You've just learned that the deformability, or "squishiness," of electron clouds is what makes van der Waals attractions possible. The **polarizability** of a molecule is a direct measure of how easy it is energetically for an external charge (or dipole) to alter the electron distribution in a molecule or atom. In other words, more polarizable molecules have squishier electron clouds. An analogy to polarizability is the comparison of a marshmallow or a balloon to a golf ball or handball. Imagine that these objects are analogous to electron clouds. It takes very little energy to deform a marshmallow or a balloon; we can do it with our hands. These objects are squishy. Molecules or groups that have squishy, easily deformable electron clouds are polarizable and easily develop temporary dipoles when charges or other dipoles are nearby. It takes lots of energy, however, to deform a golf ball or a handball—so much that we can't do it with our hands. They're hard and not at all squishy. By analogy, molecules and groups with electron clouds that are not easily deformed are less polarizable and do *not* form temporary dipoles when other charges or dipoles are nearby.

Although polarizability can be both measured and calculated, we'll keep our discussion at a more descriptive level. Molecules (or groups within molecules) that contain very electronegative atoms are typically not very polarizable, because their electrons are held tightly and pulled closer to the nuclei. Molecules or groups that contain atoms of lower electronegativity are typically more polarizable. For example, the polarizability of the iodine atom is about 10 times that of the fluorine atom. Because fluorine is very electronegative, the electrons in the fluorine atom are difficult to pull away from the nucleus. The valence electrons in iodine, however, occupy orbitals in principal quantum level 5, and these orbitals are easily deformed by external charges because they are held less strongly.

From what we've learned, there ought to be a correspondence between molecular polarizability and boiling point. For a given shape and molecular mass, a liquid consisting of more polarizable molecules should have a higher boiling point than one consisting of less polarizable molecules because the van der Waals attractions are stronger between more polarizable molecules. We can make such a comparison using the boiling points of alkanes and perfluoroalkanes (alkanes in which all hydrogens have been replaced by fluorines). Because of the electronegativity of fluorine, perfluoroalkanes have significantly lower polarizabilities than alkanes of the same molecular mass. For example, compare the boiling points of perfluoroethane with those of the hydrocarbon 2,2,3,3-tetramethylbutane:

perfluoroethane
molecular mass = 138.0
boiling point = −78 °C

2,2,3,3-tetramethylbutane
molecular mass = 114.2
boiling point = +107 °C (8.27)

The boiling point of the fluorocarbon is *185 °C lower,* even though it has somewhat higher molecular mass than the hydrocarbon. This difference reflects the lower polarizability of the fluorocarbon, which is about 20% that of the hydrocarbon. In other words, the van der Waals attractions in the liquid fluorocarbon are *much* weaker than those in the liquid hydrocarbon; so, the van der Waals attractions between molecules in the liquid fluorocarbon are broken and its conversion into a gas occurs at a lower temperature. As another example, perfluorohexane, despite its considerably greater molecular mass, has such a low polarizability that it actually has a boiling point that is 12 °C lower than the hydrocarbon hexane:

$CF_3CF_2CF_2CF_2CF_2CF_3$
perfluorohexane
molecular mass = 338.1
boiling point = 57 °C

$CH_3CH_2CH_2CH_2CH_2CH_3$
hexane
molecular mass = 86.2
boiling point = 69 °C (8.28)

The polymer Teflon (polytetrafluoroethylene) is perhaps the ultimate fluorocarbon. It is valued precisely because it has weak noncovalent attractions with practically everything.

$$-\!\left(CF_2CF_2\right)_x\!-$$

polytetrafluoroethylene
(Teflon)

The weakness of its noncovalent attractions makes Teflon slippery because it does not adhere to other molecules, including surfaces. This property is the basis for its use in nonstick cookware. (For more about Teflon, see Sec. 10.3.)

Can a substance have such a low polarizability that it never liquefies? Helium, element 2 of the periodic table, comes close. Helium has two protons that hold its two electrons tightly in a 1s orbital. The polarizability of helium is the lowest of any element. In fact, helium is a gas down to 4.2 K (–269 °C), about 4 degrees above absolute zero. The van der Waals attractions between helium atoms in the liquid state are so weak that "heating" to 4.2 K is all that is necessary to overcome these attractions and convert the element into a gas.

Dancing on the Ceiling: The Gecko and van der Waals Attractions

Tim Vickers

The gecko (lizard), of which there are hundreds of species worldwide, is familiar to people who live in warmer climates. Geckos have the amazing ability to walk on walls or even ceilings. This capability is due to the presence of *setae* in their footpads, small hairlike structures roughly 5 micrometers (5×10^{-6} meters) in diameter. Each seta terminates with between 100 and 1000 spatulae, which are each 0.2 micrometers in diameter; the spatulae contact the surface on which the gecko moves. The protein in each spatula was shown in 2002 by a team from the University of California, Berkeley, and Stanford University to adhere to surfaces by van der Waals forces. It has been estimated that all of the spatulae on a gecko could support a weight of 60 pounds or more. There is interest in developing synthetic spatulae that could be used to allow robots (or people) to walk on walls or ceilings.

The gecko doesn't do so well on surfaces that offer only weak van der Waals attractions. When a gecko is placed on a vertical Teflon surface, it slides right off!

B. Attractions between Permanent Dipoles

Fundamentally, van der Waals attractions are due to attractions between fluctuating dipoles, as shown in Fig. 8.3. Molecules with *permanent* dipoles, then, can also show enhanced intermolecular attractions. Molecules with permanent dipoles can have higher boiling points than the alkanes of the same size and shape. For example, consider the boiling points of the following ethers and the alkanes of roughly the same shape and molecular mass:

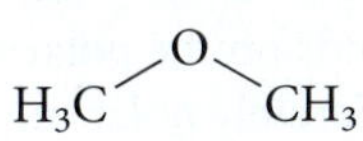

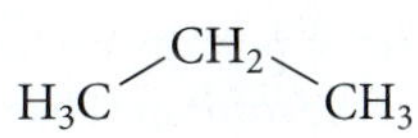

	dimethyl ether	**propane**
dipole moment	1.31 D	0.08 D
boiling point	−23.7 °C	−42.1 °C
heat of vaporization at −42.1 °C	22.7 kJ mol^{-1} (5.43 kcal mol^{-1})	19.0 kJ mol^{-1} (4.54 kcal mol^{-1})

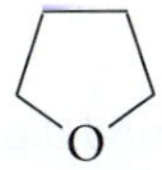

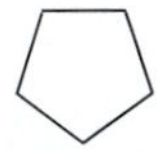

	tetrahydrofuran (THF)	**cyclopentane**
dipole moment	1.7 D	0 D
boiling point	66 °C	49.3 °C
heat of vaporization at 49.3 °C	30.7 kJ mol^{-1} (7.34 kcal mol^{-1})	27.3 kJ mol^{-1} (6.52 kcal mol^{-1})

(8.29)

A comparison of the electrostatic potential maps (EPMs) of dimethyl ether and propane shows clearly the distribution of charge that leads to the permanent dipole moment: the electronegative oxygen has a partial negative charge (red), and the methyl hydrogens have partial positive charges (blue).

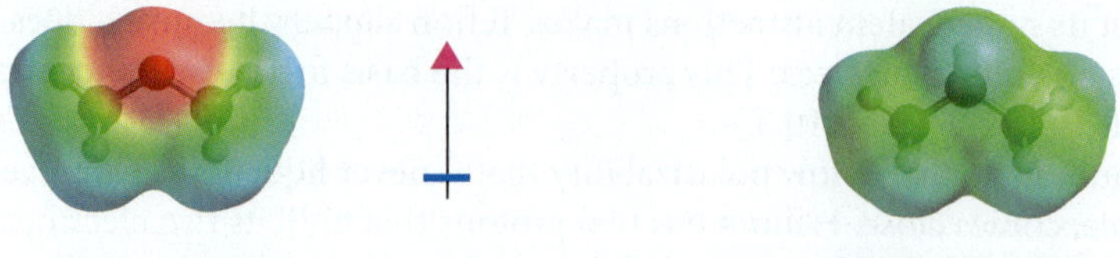

EPM of dimethyl ether (μ = 1.31 D) EPM of propane (μ ~0.1) (8.30)

The higher boiling point of the ether results from *greater attractions between molecules* in the liquid state. Molecules with permanent dipoles are attracted to each other because they can align part of the time in such a way that the negative end of one dipole is attracted to the positive end of the other. For example, two dimethyl ether molecules might align in the following way:

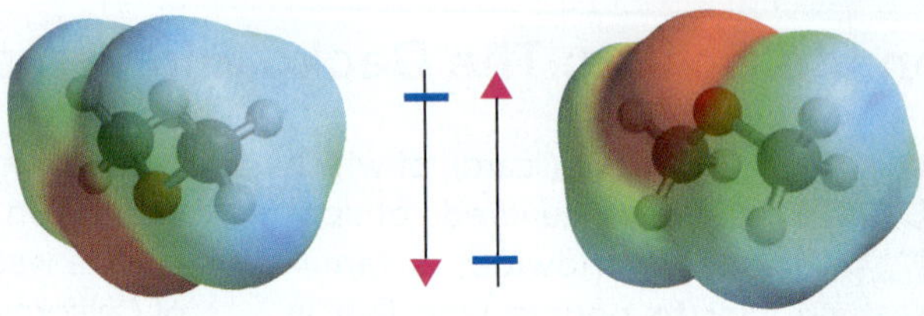

two dimethyl ether molecules
with their dipoles aligned for attraction (8.31)

Although only two molecules are shown here, attractions like this can occur among many molecules at the same time. Molecules in the liquid state are in constant motion, so their relative positions are changing constantly; on the average, however, this attraction exists and raises the boiling point of a polar compound. The electronegativity of the oxygen in an ether reduces its polarizability; but this reduction in polarizability is more than offset by the attraction between the permanent dipoles.

As we noted previously in this section, the boiling point gives a rough idea of the difference in the energy required to vaporize a molecule at its boiling point. When we compare the boiling points of two different molecules, we are in effect comparing their heats of vaporization at different temperatures. In Display 8.29, a comparison of the heats of vaporization within each pair at a common temperature gives a more precise idea of the relative strengths of noncovalent interactions under the same conditions. This comparison shows that polarity increases intermolecular attractions by 3–4 kJ mol^{-1} (about 0.75–1 kcal mol^{-1}) in these two comparisons.

Figure 8.2b (Sec. 8.5A) shows that the boiling points of ethers (green line) and alkanes (black line), except at the very small molecular masses, are not very different. This is because van der Waals attractions between the alkyl groups, and the resulting induced dipoles, become the dominant source of intermolecular attractions even in ethers of modest size. However, the effect of polarity on boiling point depends on the polarity of the molecule. Dimethyl ether and tetrahydrofuran (Display 8.29) are moderately polar molecules. Very polar molecules show significantly higher boiling points than alkanes, even at large molecular masses. For example, Fig. 8.2b shows that nitriles (purple line) have particularly high boiling points. They also have unusually large dipole moments (typically 3.7–3.8 D) that result from the parallel alignment of the bond dipoles of the two bonds to the carbon of the triple bond. (Carbons of triple bonds, like carbons of double bonds, are relatively electronegative, but more so; see Display 4.19, Sec. 4.4.)

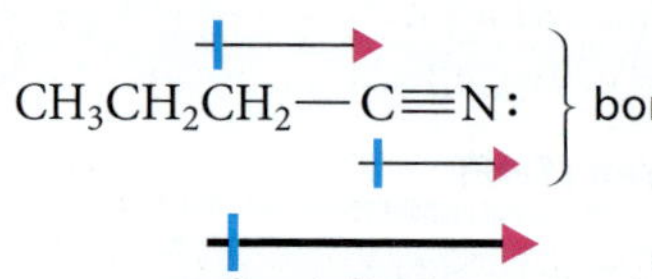

$CH_3CH_2CH_2CH_2CH_3$

butanenitrile
μ = 3.7 D
molecular mass = 69.1
boiling point = 118 °C
heat of vaporization = 39.5 kJ mol^{-1}
(9.44 kcal mol^{-1}) at 25 °C

pentane
μ = 0 D
molecular mass = 72.1
boiling point = 36 °C
heat of vaporization = 26.4 kJ mol^{-1}
(6.31 kcal mol^{-1}) at 25 °C

(8.32)

A comparison of the heats of vaporization of butanenitrile and pentane shows that the polarity of the nitrile contributes about 13 kJ mol^{-1} (3.1 kcal mol^{-1}) to its intermolecular interactions.

The tradeoff between molecular size and polarity is apparent in the boiling points of the alkyl halides. Alkyl chlorides have about the same boiling points as alkanes of the same molecular mass, and alkyl bromides and iodides have lower boiling points than the alkanes of about the same molecular mass. (The boiling points of the alkyl chlorides, alkyl bromides, and alkanes are shown in Fig. 8.2.) Notice that all of the alkyl halides have significant dipole moments. Although bromine and iodine are considerably less electronegative than chlorine, the longer carbon–halogen bond lengths increase the dipole moment, which is a product of charge separation and bond length (Eq. 1.92, Sec. 1.9A).

	$CH_3CH_2CH_2CH_2Cl$ **1-chlorobutane**	$CH_3CH_2CH_2CH_2CH_2CH_3$ **hexane**
molecular mass	92.6	86.3
dipole moment	2.0 D	0 D
boiling point	78.4 °C	68.7 °C
heat of vaporization (298 K)	33.5 kJ mol^{-1} (8.01 kcal mol^{-1})	31.4 kJ mol^{-1} (7.51 kcal mol^{-1})
density	0.886 g mL^{-1}	0.660 g mL^{-1}
	CH_3CH_2Br **bromoethane**	$CH_3CH_2CH_2CH_2CH_2CH_2CH_3$ **heptane**
molecular mass	109	100.2
dipole moment	2.0 D	0 D
boiling point	38.4 °C	98.4 °C
heat of vaporization (298 K)	28.3 kJ mol^{-1} (6.76 kcal mol^{-1})	36.6 kJ mol^{-1} (8.75 kcal mol^{-1})
density	1.46 g mL^{-1}	0.684 g mL^{-1}
	CH_3I **iodomethane**	$CH_3CH_2CH_2CH_2CH_2CH_2CH_2CH_2CH_2CH_3$ **decane**
molecular mass	142	142
dipole moment	1.7 D	0 D
boiling point	42.5 °C	174 °C
heat of vaporization (298 K)	28.0 kJ mol^{-1} (6.69 kcal mol^{-1})	51.4 kJ mol^{-1} (12.3 kcal mol^{-1})
density	2.28 g mL^{-1}	0.73 g mL^{-1}

(8.33)

The key to understanding these trends is to realize that although the molecules compared in each row have similar molecular masses, they have *very different molecular sizes and shapes*. Their relatively high densities show that alkyl halide molecules have large masses within relatively small volumes. That is, *for a given molecular mass*, alkyl halide molecules have smaller volumes than alkane molecules. Remember that the attractive forces between molecules—van der Waals forces, or dispersion forces—are greater for molecules of larger surface area. Larger intermolecular attractions translate into higher boiling points. The greater molecular volumes (and surface areas) of alkanes, then, should cause them to have *higher* boiling points than alkyl halides. The polarity of alkyl halides, in contrast, has the opposite effect on boiling points: if polarity were the only effect, alkanes would have *lower* boiling points than alkyl halides. Therefore, the effects of molecular surface area and polarity oppose each other. They nearly cancel in the case of alkyl chlorides, which have about the same boiling points as alkanes of about the same molecular mass. However, alkane molecules are so much larger than alkyl bromide and alkyl iodide molecules of the same molecular mass that the size (surface area) effect trumps polarity, and alkanes have higher boiling points. The same trends are evident with the relative heats of vaporization, which are also shown in Display 8.33. Notice in particular the significantly greater heat of vaporization of decane relative to that of iodomethane. The 23 kJ mol^{-1} (5.6 kcal mol^{-1}) difference is a reflection in energy terms of the alkane's much greater volume and covalent intermolecular attractions.

Focused Problems

8.10 The boiling points of the 1,2-dichloroethylene stereoisomers are 47.4 °C and 60.3 °C. Which stereoisomer has each boiling point? Explain. (*Hint:* Draw the structures of both stereoisomers, and consider their relative dipole moments.)

8.11 Octane and 2,2,3,3-tetramethylbutane have boiling points that differ by about 20 °C (106 °C and 126 °C). Which has the higher boiling point, and why?

8.12 (a) The dipole moment of acetaldehyde is 2.7 D and that of propene is 0.5 D. Even though they have about the same molecular mass, they differ in boiling point by about 68 °C (−47 °C and +21 °C). Which has the higher boiling point? Why?

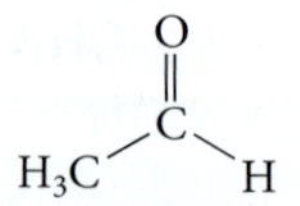

acetaldehyde

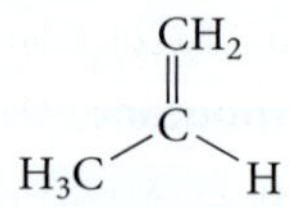

propene

(b) Given that alkenes and alkanes with the same branching pattern and number of carbons have about the same boiling points, show where you would expect the curve for aldehyde boiling points to fall in Fig. 8.2b. Explain.

(c) Where would you expect the curve for alkyl iodides to fall in Fig. 8.2b relative to alkanes with the same molecular mass? Explain.

C. Hydrogen Bonding

The boiling points of alcohols, especially alcohols of lower molecular mass, are unusually high when compared with those of other organic compounds. For example, ethanol has a much higher boiling point than other organic compounds of about the same shape and molecular mass.

	$CH_3CH_2—OH$	$CH_3CH_2CH_3$	$H_3C—O—CH_3$	$CH_3CH_2—F$	
	ethanol	**propane**	**dimethyl ether**	**ethyl fluoride**	
boiling point	78 °C	−42 °C	−24 °C	−38 °C	
dipole moment	1.7 D	0 D	1.3 D	1.8 D	(8.34)

The contrast between ethanol and both dimethyl ether and ethyl fluoride is particularly striking because all have similar dipole moments, and yet the boiling point of ethanol is much higher. The fact that something is unusual about the boiling points of alcohols is also apparent from a comparison of the boiling points of ethanol, methanol, and the simplest "alcohol," water.

	$CH_3CH_2—OH$	$H_3C—OH$	$H—OH$	
	ethanol	**methanol**	**water**	
boiling point	78 °C	65 °C	100 °C	(8.35)

Recall from Fig. 8.2 that each additional CH_2 group results in a 20–30 °C increase in the boiling points of successive members of a homologous series. Yet the difference in the boiling points of methanol and ethanol is only 13 °C; and water, although the "alcohol" of lowest molecular mass, has the highest boiling point of the three compounds. These observations are reflected in the unusual shape of the alcohol curves in Fig. 8.2 at low molecular mass. These unusual phenomena are due to a very important intermolecular attraction called *hydrogen bonding.*

Hydrogen bonding is an intermolecular attraction that results from the association of a hydrogen on one atom with a nonbonding electron pair on another. Hydrogen bonding can occur within the same molecule, or it can occur between molecules. For example, in the case of the simple alcohols, hydrogen bonding is a weak association of the O—H proton of one molecule with the oxygen of another.

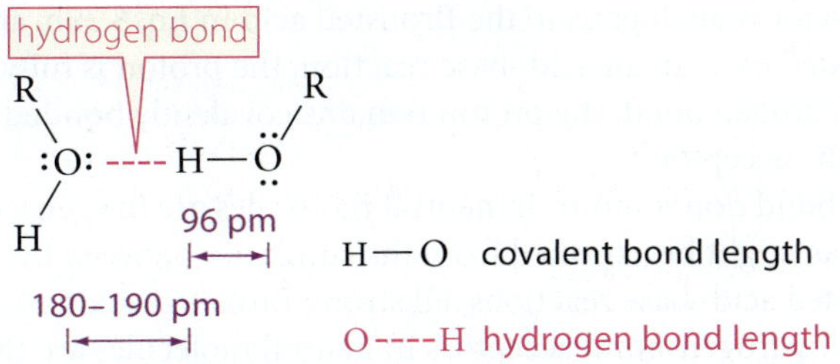

(8.36)

Formation of a hydrogen bond requires two partners: the *hydrogen-bond donor* and the *hydrogen-bond acceptor*. The **hydrogen-bond donor** is the atom to which the hydrogen is fully bonded, and the **hydrogen-bond acceptor** is the atom bearing the nonbonding electron pair to which the hydrogen is partially bonded. In this case the oxygen atom serves both roles.

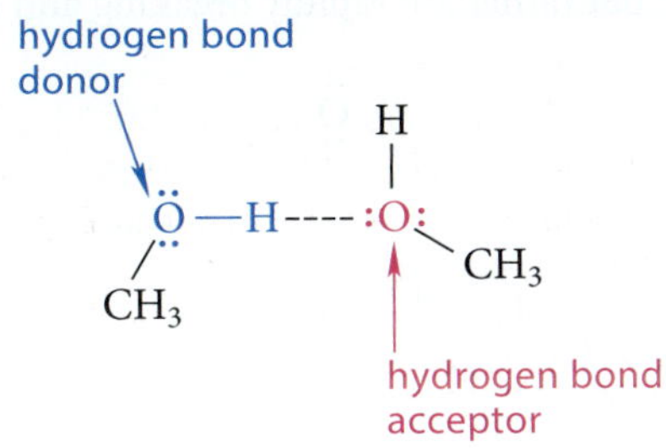

(8.37)

In a classical Lewis sense, a proton can only share two electrons, so a hydrogen bond is difficult to describe with conventional Lewis structures. Consequently, hydrogen bonds are often depicted as dashed lines. The hydrogen bond results from the combination of two factors: first, a weak covalent interaction between a hydrogen on the donor atom and nonbonding electron pairs on the acceptor atom; and second, an electrostatic attraction between oppositely charged ends of two dipoles. Opinions differ as to the relative importance of these two factors.

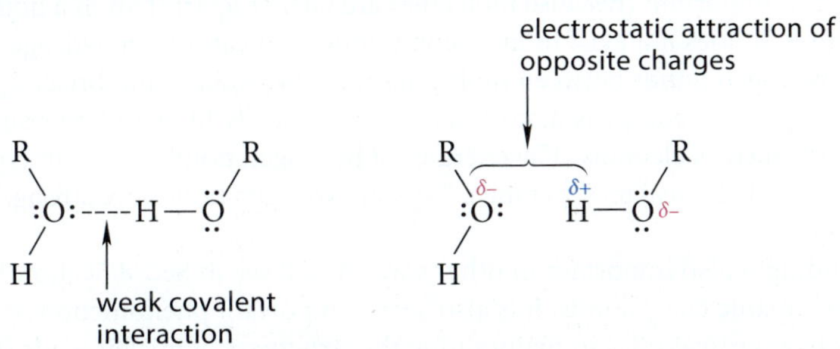

(8.38)

The hydrogen bond between two molecules resembles the same two molecules poised to undergo a Brønsted acid–base reaction:

Hydrogen bonding:

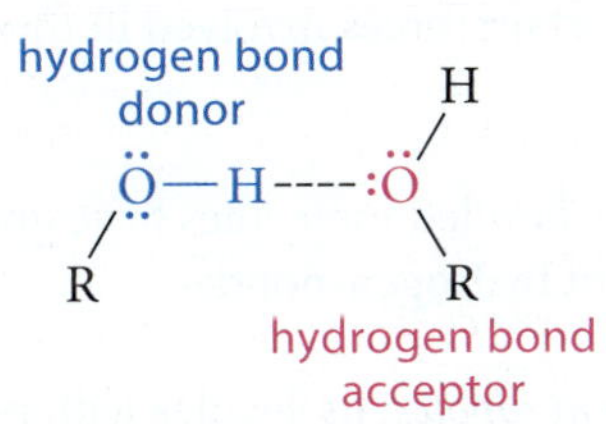

(8.39a)

Brønsted acid–base reaction:

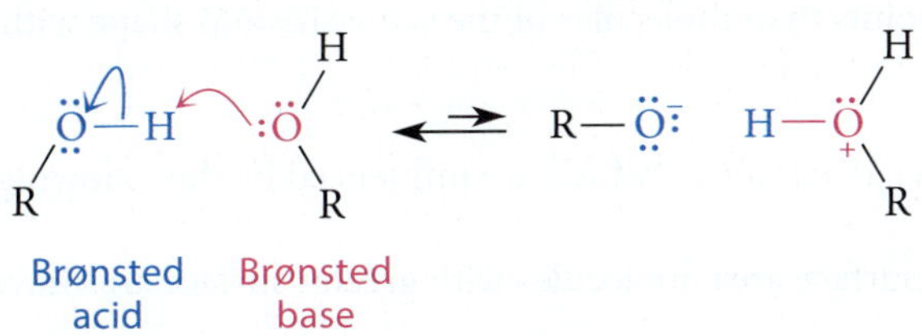

(8.39b)

The hydrogen-bond donor is analogous to the Brønsted acid in Eq. 8.39b, and the acceptor is analogous to the Brønsted base. In an acid–base reaction, the proton is fully transferred from the acid to the base; in a hydrogen bond, the proton remains covalently bonded to the donor, but it interacts weakly with the acceptor.

The best hydrogen-bond donor atoms in neutral molecules are oxygens, nitrogens, and halogens. In addition, as might be expected from the similarity between hydrogen-bond interactions and Brønsted acid–base reactions, all strong Brønsted acids are also good hydrogen-bond donors. The best hydrogen-bond acceptors in neutral molecules are the electronegative first-row atoms oxygen and nitrogen. Most anions with nonbonding electron pairs and all strong Brønsted bases are also good hydrogen-bond acceptors.

Sometimes an atom can act as both a donor and an acceptor of hydrogen bonds. For example, because the oxygen atoms in water or alcohols can act as both donors and acceptors, some of the molecules in liquid water and alcohols exist in hydrogen-bonded networks. The hydrogen bonds in these networks are not static, but rather are rapidly breaking and re-forming.

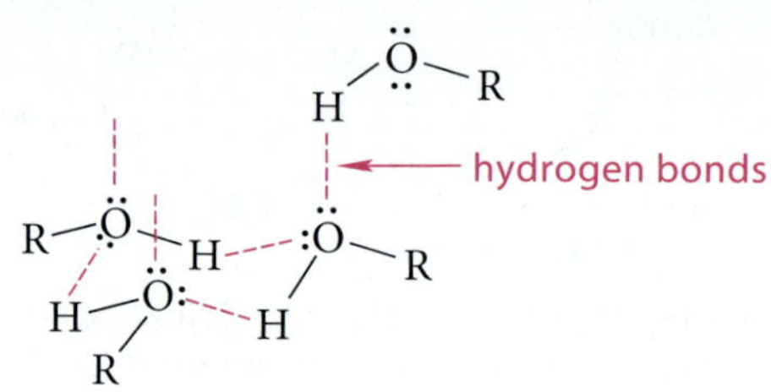

(8.40)

In contrast, the oxygen atom of an ether is a hydrogen-bond acceptor, but it is not a donor because it has no hydrogen to donate. Finally, some atoms are donors but not acceptors. The ammonium ion, $^{+}NH_4$, is a good hydrogen-bond donor; but, because the nitrogen has no nonbonding electron pair, it is not a hydrogen-bond acceptor.

Hydrogen bonding accounts for the unusually high boiling points of alcohols. In the liquid state, hydrogen bonding is an attraction that holds molecules together. In the gas phase, hydrogen bonding is much less important (because molecules are farther apart than in a liquid or solid) and, at low pressures, it does not exist in most compounds. To vaporize a hydrogen-bonded liquid, then, the hydrogen bonds between molecules must be broken, and breaking hydrogen bonds requires energy. This energy is manifested as an unusually high boiling point for hydrogen-bonded compounds such as alcohols. The energies of hydrogen bonds vary, but typically they are in the 10–40 kJ mol^{-1} (1–10 kcal mol^{-1}) range. Typical hydrogen bonds are stronger than van der Waals attractions.

Hydrogen bonding is also important in other ways. You'll see in Sec. 8.6C how it affects the water solubility of organic compounds. It is also a very important phenomenon in biology. Hydrogen bonds have critical roles in maintaining the structures of nucleic acids (such as DNA and RNA) and proteins. (These structures are covered in Secs. 26.5B and 27.9.) Without hydrogen bonds, life as we know it would not exist.

In summary, the tendency of molecules to associate noncovalently in the liquid state increases their boiling points. The most important forces involved in these intermolecular associations are the following:

1. *Hydrogen bonding*: hydrogen-bonded molecules have greater boiling points than molecules of similar polarity that are not hydrogen-bonded.

2. *Attractions between permanent dipoles*: molecules with permanent dipole moments have higher boiling points than molecules of the same size and shape with zero or small dipole moments.

3. *Attractive van der Waals forces*, which are influenced by the following:

    (a) Molecular surface area: molecules with greater surface area have greater boiling points

(b) Molecular shape: more extended, less spherical (less branched) molecules have greater boiling points

(c) Polarizability: more polarizable molecules have stronger intermolecular attractions than less polarizable ones

What is the relative importance of these effects? These effects have been listed in the approximate order of importance. For example, hydrogen bonding in small molecules usually trumps molecular polarity. In Fig. 8.2b, you can see that the curve for alcohols (the only hydrogen-bonding group in the figure) lies above all of the other curves except one. Therefore, 2-propanol (isopropyl alcohol) has a higher boiling point than acetone, which has the greater dipole moment.

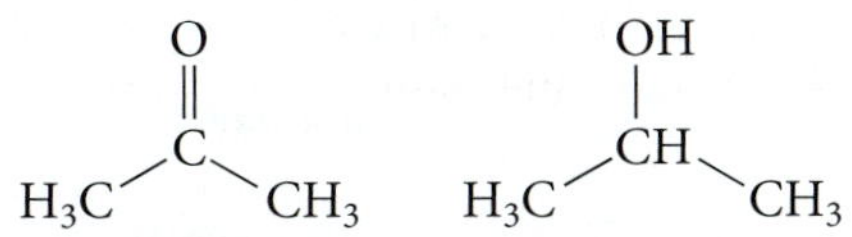

	acetone	**2-propanol**	
boiling point	56.5 °C	82.3 °C	
dipole moment	2.7 D	1.7 D	(8.41)

2-Propanol can donate and accept hydrogen bonds, but acetone lacks a donor group. (For hydrogen bonding to occur between *identical* molecules, they must contain *both* a donor and an acceptor.) In making this comparison, notice that we are holding size and shape virtually constant and comparing only polarity and hydrogen-bonding capability.

It is impossible, however, to make a general statement about the relative importance of these effects that holds true in every case. For example, nitriles, which contain no hydrogen-bond donor, are exceptionally polar molecules, with dipole moments in the 3.7–3.8 D range. (See Display 8.32, Sec. 8.5B.) They have higher boiling points than even alcohols (see Fig. 8.2b). This is an unusual, but by no means isolated, case in which polarity trumps hydrogen bonding. We've already discussed the relative boiling points of alkyl halides and alkanes (Display 8.33). Because of the density of alkyl halides, molecular surface area is more important than polarity.

Your goal should be to understand the factors that affect boiling point and the trends in boiling point rather than to split hairs. An example of the sort of reasoning to be used is shown in Study Problem 8.4.

Study Problem 8.4

Arrange the following compounds in order of increasing boiling point:

1-hexanol, 1-butanol, *tert*-butyl alcohol, pentane

Solution First, draw the structures!

$CH_3CH_2CH_2CH_2CH_2CH_2—OH$
1-hexanol

$CH_3CH_2CH_2CH_2CH_3$
pentane

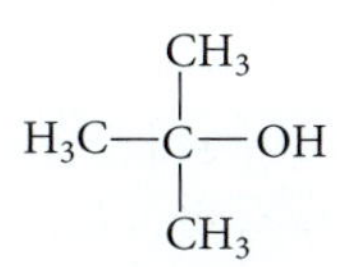
***tert*-butyl alcohol**

$CH_3CH_2CH_2CH_2—OH$
1-butanol

1-Butanol and pentane have almost the same molecular mass and about the same size and shape. 1-Butanol is a polar molecule that can both donate and accept hydrogen bonds, however; so it has a considerably higher boiling point than pentane. Because 1-hexanol, also a primary alcohol, is a larger molecule than 1-butanol, its boiling point is the highest of the three. So far, the order of increasing boiling points is pentane < 1-butanol < 1-hexanol. *Tert*-butyl alcohol has about the same molecular mass as pentane, but the alcohol has a higher boiling point because of its polarity and hydrogen bonding. However, a *tert*-butyl alcohol molecule is more branched and more nearly spherical than the isomeric 1-butanol molecule, so the boiling point of *tert*-butyl alcohol should be lower than that of 1-butanol. Therefore, the correct order of boiling points is pentane < *tert*-butyl alcohol < 1-butanol < 1-hexanol. (The respective boiling points in °C are 36, 82, 118, and 157.)

Focused Problems

8.13 Within each set, arrange the compounds in order of increasing boiling point.

(a) 4-Ethylheptane, 2-bromopropane, 4-ethyloctane

(b) 1-Butanol, 1-pentene, chloromethane

8.14 Label each of the following molecules as a hydrogen-bond acceptor, donor, or both. Indicate the hydrogen that is donated or the atom that serves as the hydrogen-bond acceptor.

(a) H—Br:

(b) H—F:

(c) 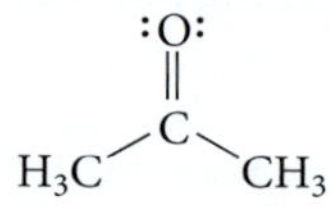

acetone

(d) 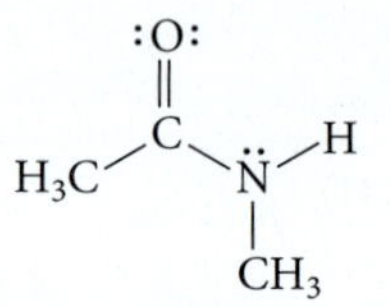

***N*-methylacetamide**

(e) 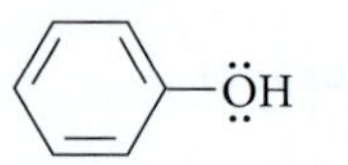

phenol

(f) $CH_3CH_2\overset{+}{N}H_3$

ethylammonium ion

D. Melting Points

We study the melting point because it is an important physical property of organic compounds, and because it has an important relationship to the solubility of solids. The **melting point** (Sec. 2.6B) is the temperature above which a solid is spontaneously transformed into a liquid. At the melting point, a solid and its liquid are in equilibrium. Melting points, like boiling points, are a reflection of noncovalent intermolecular attractions—van der Waals forces, dipole–dipole interactions, and hydrogen bonding. Boiling points provide information about a single state—the liquid state—because (to a useful approximation) there are no intermolecular interactions in the gaseous state. Melting points, however, reflect the differential effects of noncovalent interactions *in both the liquid state and the crystalline solid state*. For that reason, it isn't always possible to interpret melting points in terms of interactions within a single state of matter.

Nevertheless, there are two observations about melting points that are worth knowing about, and they have a common origin. The more useful observation is that symmetrical compounds tend to have considerably higher melting points than less symmetrical isomers. This observation has been known since the 1880s, and in some instances it can be quite significant. For example, the melting points of the following two constitutional isomers differ by nearly 72 °C.

***para*-dichlorobenzene**
(1,4-dichlorobenzene)
mp 47 °C (320.2 K)
bp 173 °C
$\Delta H_m = 18.14$ kJ mol^{-1} (4.34 kcal mol^{-1})
$\Delta S_m = 56.7$ J K^{-1} mol^{-1} (13.6 cal K^{-1} mol^{-1})

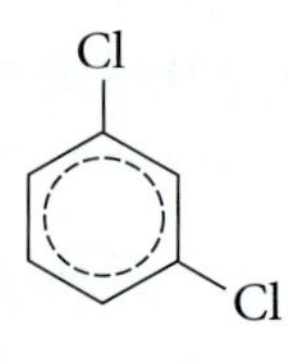

***meta*-dichlorobenzene**
(1,3-dichlorobenzene)
mp −24.7 °C (248.5 K)
bp 173 °C
$\Delta H_m = 12.60$ kJ mol^{-1} (3.01 kcal mol^{-1})
$\Delta S_m = 50.8$ J K^{-1} mol^{-1} (12.1 cal K^{-1} mol^{-1})

(8.42)

(We draw the double bonds as delocalized resonance hybrids to stress the symmetry; the benzene ring is planar.)

To say that a compound is *symmetrical* means that there are several ways that we can *rotate* the entire structure to produce an equivalent, indistinguishable structure. That is, if you turned your back and someone carried out any of these rotations with a molecular model, and then you turned around to look at the resulting structure, it would look as if nothing had been done. For example, we can rotate the structure of 1,4-dichlorobenzenebenzene three ways and obtain an indistinguishable orientation.

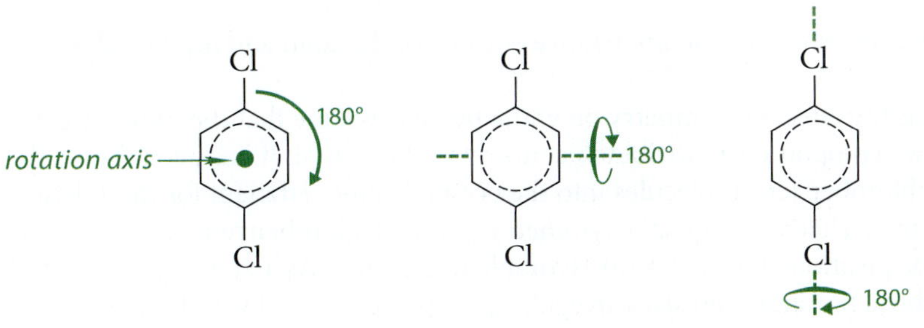

(8.43a)

If we count the equivalent structures obtained in these rotations, and include the original structure, we find that there are four equivalent structures that can be transformed into each other by these rotations. The number of equivalent, indistinguishable structures related by rotations is called the **symmetry number** (abbreviated σ). 1,4-Dichlorobenzene, then, has $\sigma = 4$.

In contrast, only one rotation can convert 1,3-dichlorobenzene into an equivalent, indistinguishable structure:

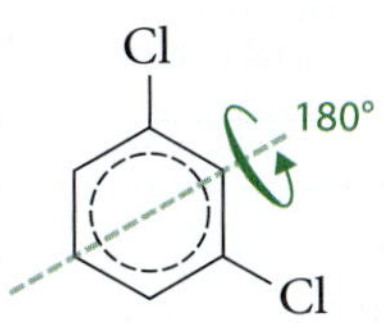

(8.43b)

1,3-Dichlorobenzene, then, has a symmetry number $\sigma = 2$. Therefore, 1,4-dichlorobenzene is more symmetrical because it has a higher symmetry number than 1,3-dichlorobenzene.

Because the boiling points of the two compounds are virtually identical (173 °C), the effect of symmetry on melting point has something to do with the crystalline state. (Identical boiling points show that the strengths of the intermolecular attractions within the liquid states of the two compounds are the same.) To understand the aspects of crystal symmetry that affect melting point, we must first recognize that the **free energy of fusion** ΔG_m of any substance at its melting temperature T_m is 0:

$$\Delta G_m = \Delta H_m - T_m \Delta S_m = 0 \quad (8.44a)$$

We use the subscript "m" (for "melting") in these quantities rather than "f" to avoid confusion with, for example, heats of formation, which are abbreviated ΔH_f°.

! Remember that when you use temperature in any equation dealing with energy, its units are *kelvins*, not °C.

This is true because, at the melting temperature, solid and melt are in equilibrium, so their free energies are equal. In this equation, ΔH_m is the **enthalpy of fusion**, which is the enthalpy difference between the solid and its liquid melt at the melting temperature. The enthalpy of fusion is a direct measure of the noncovalent interactions among molecules in the crystal relative to the interactions of molecules in the melt. ΔS_m is the **entropy of fusion**, which is the entropy difference between the solid and its liquid melt at the same temperature; we discuss the meaning of the entropy of fusion later in the chapter. Setting $\Delta G_m = 0$ as shown in Eq. 8.44a results in the following relationship between the enthalpy of fusion, the entropy of fusion, and the melting temperature:

$$T_m = \frac{\Delta H_m}{\Delta S_m} \quad (8.44b)$$

Both ΔH_m and ΔS_m are positive. ΔH_m is positive because, in every case, it takes heat to break the noncovalent interactions in the crystal to form a liquid. ΔS_m is positive because, in the liquid, the molecules are free to move around, whereas, in the crystal, their motion is constrained by the crystal structure. (As we'll see, entropy is increased by molecular motion, especially translation.) More practically, if one of the two quantities were negative, the melting point T_m would have to be negative; and a negative T_m (in kelvins) is impossible.

This equation shows that a high melting point can be caused by a large ΔH_m, a small ΔS_m, or both.

We discuss three effects of symmetry. The first two largely deal with the solid (crystalline) state:

1. The effect of symmetry on the enthalpy of fusion
2. The probabilistic effect of symmetry on the entropy of fusion

The third effect deals with both the solid and liquid states:

3. The effect of symmetry on the relative entropy of the solid and liquid states

First, what is the effect of symmetry on enthalpy? We assume that the difference in ΔH_m between the two compounds is attributable mostly to the crystal. It has been shown that the packing of dichlorobenzene molecules into the crystal is more efficient for the 1,4-isomer than for the 1,3-isomer. That is, the greater symmetry of 1,4-dichlorobenzene molecules allows the molecules to be positioned for the most favorable interaction. As an analogy, if we stack cubical blocks in a cubical box and then stack irregularly shaped chunks of wood, such as twigs (each with the same volume as a cubical block), in an identical box, the cubical blocks are on average closer together. When molecules are closer, *their noncovalent attractions are stronger*—that is, the energy of the noncovalent attractions is reduced (lower energy = greater stability). Therefore, more energy is required to melt the crystalline compound than would be required in the absence of symmetry; ΔH_m of the symmetrical solid is greater. In fact, ΔH_m for the 1,4-isomer is 1.44 times that of the 1,3-isomer.

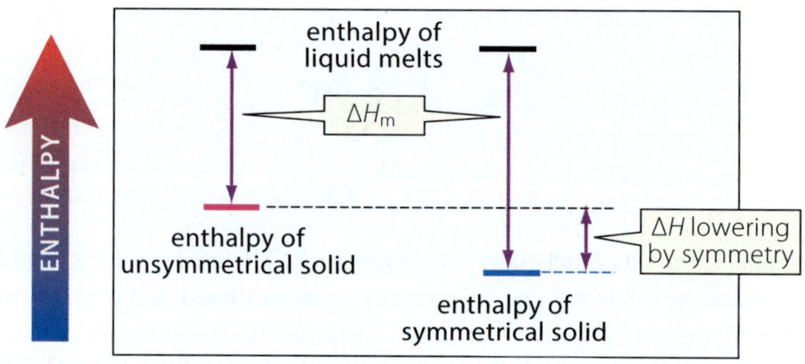

(8.45)

A comparison of ΔS_m and ΔH_m values for different compounds necessarily occurs at different temperatures because their melting points are different. To compare noncovalent interactions between the two compounds, the comparison should be at the same temperature. Because there is a temperature dependence of ΔS and ΔH for both the liquid melts and the crystals, and the data to correct for this dependence are not readily available in every case, comparing ΔS_m and ΔH_m values gives only an approximate comparison of the noncovalent interactions within the two compounds.

We pointed out the "sawtooth" melting-point behavior of unbranched alkanes (see Fig. 2.8, Sec. 2.6B), which is general for homologous series in many compound classes. The efficiency of crystal packing is known to be the cause of this phenomenon as well, and it is also a ΔH_m effect. Unbranched alkanes with odd numbers of carbons lie on a lower curve of melting point versus carbon number than those with an even number of carbons. In other words, the crystal packing of "even-carbon" alkanes is more efficient, van der Waals attractions in the solid state are somewhat greater, and melting points are higher.

What is the effect of symmetry on entropy? (We provide a more complete discussion of entropy in Sec. 8.6.) Entropy reflects *relative probability*. Symmetry makes a probability contribution to the entropy of a crystalline solid that can be calculated from its symmetry number σ:

$$S = 2.3R \log \sigma \quad \textbf{(8.46a)}$$

The crystal of the more symmetrical compound is more probable because there are more indistinguishable ways that the compound can fit into the crystal. For the two isomers in Displays 8.43a and b with $\sigma_1 = 4$ and $\sigma_2 = 2$, then, the more symmetrical isomer has the larger symmetry contribution to the entropy of its solid state by

$$\Delta S = 2.3R(\log \sigma_1 - \log \sigma_2) = 2.3R \log (\sigma_1/\sigma_2) = 2.3R \log (4/2) = 5.76 \text{ J mol}^{-1} \text{ K}^{-1} \text{ (1.38 cal mol}^{-1} \text{ K}^{-1}) \quad \textbf{(8.46b)}$$

This means that *less* entropy is gained on melting by the more symmetrical compound—that is, ΔS_m is reduced by symmetry. (Symmetry has no intrinsic effect on the entropy of the liquid state, because molecules are moving more or less independently.) This reduction of ΔS_m for the more symmetrical compound contributes to a higher melting point, by Eq. 8.44b.

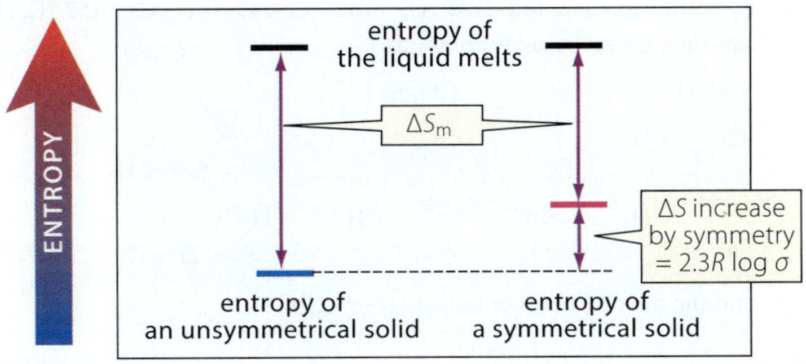

(8.47)

Because the intrinsic effect of symmetry on entropy is typically rather small, it rarely dominates the melting behavior. This is the case for 1,3-dichlorobenzene, which has a somewhat *larger* ΔS_m than the 1,4-isomer (see Display 8.42), whereas the symmetry effect alone should make it smaller. So, the greater melting point of 1,4-dichlorobenzene is caused by its greater ΔH_m. (For a comparison in which the symmetry contribution to ΔS_m plays a dominant role, see Focused Problem 8.16.)

In our third situation, the entropies of both the solid and liquid states are important. The relative melting points of pentane and neopentane show a large effect of symmetry:

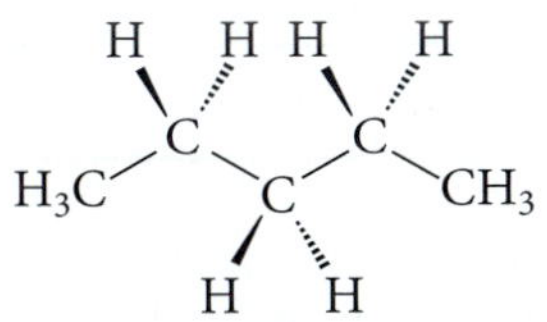

pentane
mp −129.8 °C (143 K)
ΔH_m = 8.42 kJ mol^{-1} (2.01 kcal mol^{-1})
ΔS_m = 58.8 J mol^{-1} K^{-1} (14.5 cal mol^{-1} K^{-1})
symmetry number σ = 2

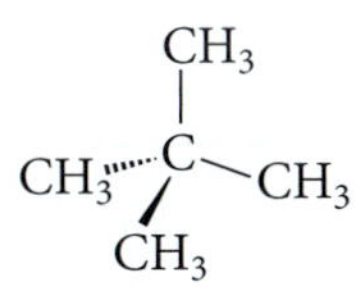

neopentane
mp −16.8 °C (256 K)
ΔH_m = 3.26 kJ mol^{-1} (0.78 kcal mol^{-1})
ΔS_m = 12.7 J mol^{-1} K^{-1} (3.04 cal mol^{-1} K^{-1})
symmetry number σ = 12

(8.48)

The data show that the relatively high melting point of neopentane is caused by its much lower ΔS_m. Although the relative symmetries of neopentane ($\sigma = 12$) and pentane ($\sigma = 2$) contribute $2.3R \log(12/2) = 2.3R \log 6 = 14.9$ J K^{-1} mol^{-1}, the large difference in the ΔS_m values of the two compounds ($58.5 - 12.7 = 45.8$ J K^{-1} mol^{-1}) shows that an effect in addition to symmetry is operating. Pentane molecules in the crystal were shown to exist in a single conformation in which there is an anti conformation about every carbon–carbon bond. When crystals of pentane melt, the liquid pentane molecules can undergo internal rotation; therefore, pentane in the liquid consists of molecules in many different conformations in which the conformation around each bond can be anti or gauche. As we show in Sec. 8.6A, *an increased number of molecular states results in an entropy increase.* In other words, the larger number of conformations available to molecules in liquid pentane causes a large increase in entropy in the transition from solid to liquid. Neopentane molecules, in contrast, have only one possible conformation in both solid and liquid because of their symmetry—the C—H bonds in each methyl group are staggered with C—CH_3 bonds on the central carbon. In other words, because of the symmetry of neopentane, it gains no entropy from a larger number of conformations when it melts. As to the relative ΔH_m values, the all-anti conformation of pentane molecules in the crystal allows them to pack very closely; and they have relatively large surface areas, which foster favorable van der Waals attractions. The van der Waals attractions between neopentane molecules in the solid are relatively weak because of its small surface area, which approaches spherical proportions. The same is true in the liquid. (Recall that the boiling point of neopentane is also unusually low.)

In summary, *the melting point of a more symmetrical isomer is typically greater than the melting point of a less symmetrical one.* This phenomenon can be due to several factors: more favorable ΔH_m because of closer packing in the crystal, an increase in the ΔS_m caused by the greater probability of forming the crystals of more symmetrical compounds, and more complex entropic factors involving both the crystal and the liquid melt.

Focused Problems

8.15 Match the structures with the following melting points. Explain. 168 – 172 °C; –74.5 °C; 143 – 147 °C; –157 °C. (*Hint:* The alkenes are liquids at room temperature, whereas the carboxylic acids are solids.)

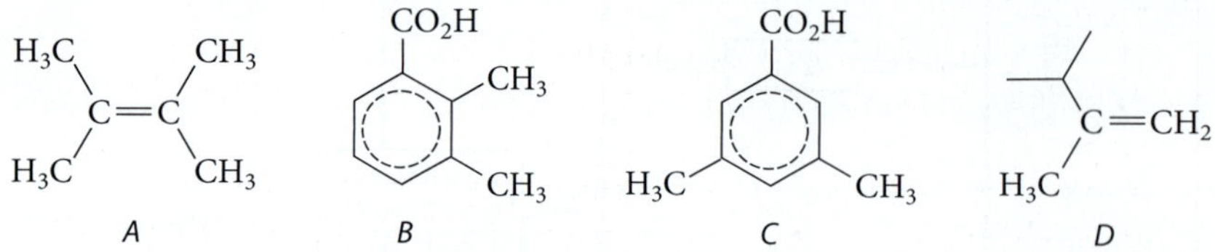

8.16 (a) The melting point of benzene is 5.5 °C, and the melting point of toluene is –95.1 °C.

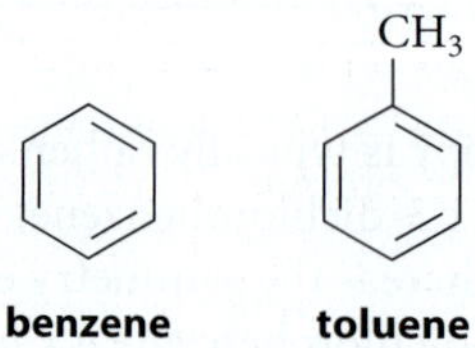

(b) $\Delta H_m = 5.73$ kJ mol^{-1} for benzene, and $\Delta H_m = 6.64$ kJ mol^{-1} for toluene. Is the effect of symmetry in the two melting points a ΔH_m effect or a ΔS_m effect? Explain. (*Hint:* Don't forget to use temperatures in K.)

(c) What is the relative symmetry contribution to the solid-state entropies of benzene and toluene? Can this account for the most of the difference in melting points? (*Hint:* Calculate symmetry numbers and apply Eq. 8.46b.)

Fats and Oils

Marc Loudon

Fats and oils used in cooking illustrate the effect of structure on melting point. Most fats, such as lard, butter, and vegetable shortening, are solids at room temperature. They contain unbranched alkyl chains that pack well into a crystal lattice. (They have a relatively high ΔH_m.) Cooking oils are essentially fats (known in commerce as "unsaturated fats") that contain one or more cis double bonds, each of which puts a "kink" in the hydrocarbon chain. This shape makes formation of the regular crystal lattice more difficult and prevents strong nonbonded attractions (van der Waals attractions) that can occur with solid fats. Oils with this structural characteristic, such as vegetable oil, canola oil, and olive oil, have much lower melting points and are therefore liquids at room temperature.

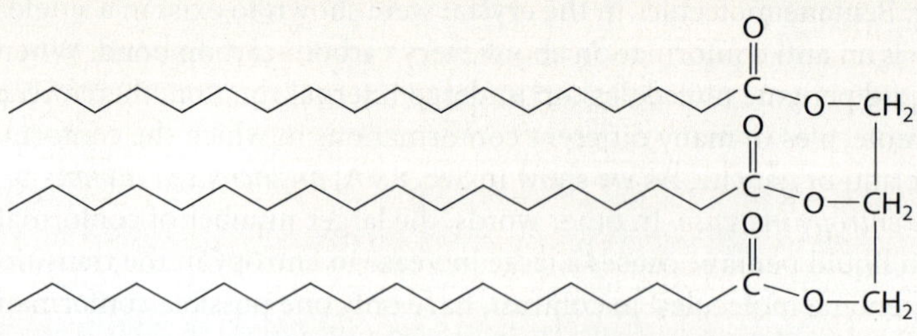

a saturated fat;
a solid at room temperature
(this structure is typical of shortening)

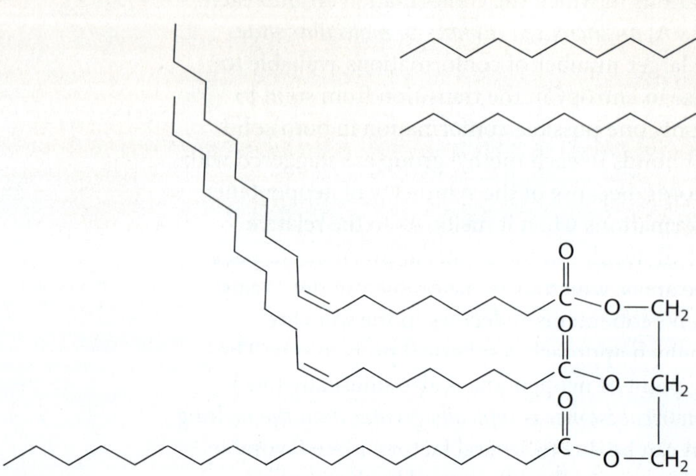

an unsaturated fat;
a liquid at room temperature
(this structure is typical of olive oil)

8.6 HETEROGENEOUS INTERMOLECULAR INTERACTIONS: SOLUTIONS AND SOLUBILITY

Now we examine the noncovalent interactions between different compounds. To do this, we consider the process of forming a solution of two liquids, in which two or more compounds coexist in the same liquid phase. First, in Sec. 8.6A, we discuss precisely what we mean by a solution, we present some terminology, and we show how we can think about the solution process in terms of energies. In Sec. 8.6B, we learn about common solvents, and how we classify them. In Sec. 8.6C, we learn some of the practical aspects and rules of thumb governing the solubility of covalent compounds. The object is to try to develop a structure-based intuition about solubilities—whether two substances will form a solution or remain in separate phases. In Sec. 8.6D, we consider specifically why hydrocarbons are insoluble in water. The principles we learn there underlie a number of important biological phenomena. All of these sections will involve the solubility behavior of liquids in other liquids. In Sec. 8.6E, we consider briefly how we have to modify our thinking when we consider the solubility behavior of solids. Finally, in Sec. 8.6F, we consider the solubility of ionic compounds. In Sec. 8.7, we then examine a number of applications in both chemistry and biology that depend on the principles that we will have covered in this section.

A. Solutions. Definitions and Energetics

In this section we define what we mean by the term *solution*, and we'll explore the energetics of solution formation. To keep things simple, we use only two components, both liquids. Imagine that a small amount of liquid *A*, the **solute**, is added to a large amount of liquid *S*, the **solvent**. We assume that the compounds do not react with each other. If the two compounds persist as separate phases, even when in contact, we say that *A* is **insoluble** in *S*. If, however, *A* and *S* form a single, clear, liquid phase, we say that *A* has formed a **solution** in *S*, and that *A* is **soluble** in *S*. We have then *dissolved* liquid *A* in solvent *S*. A diagram of the process of forming a solution at the molecular level is shown in **Fig. 8.4**.

Figure 8.4 shows that the process of forming a solution involves the interplay of *noncovalent intermolecular interactions*. In the pure compounds, on the left, *A* molecules interact with each other, and *S* molecules interact with each other. When the solution on the right is formed, some

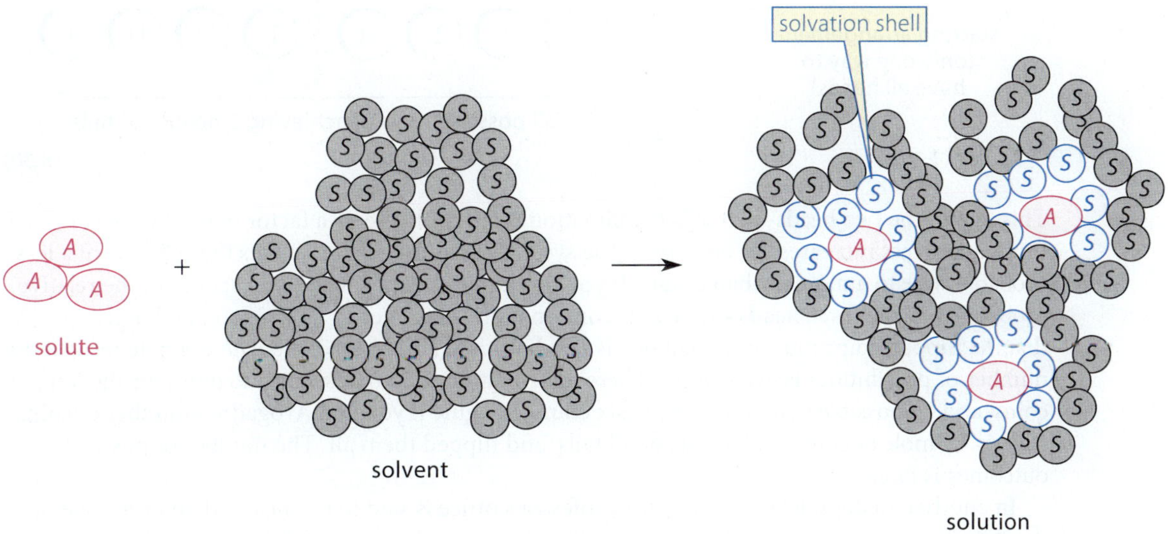

FIGURE 8.4 A schematic view of solution formation at the molecular level. Pure solute molecules *A* interact with other solute molecules, and pure solvent molecules *S* interact with other solvent molecules. When the two are mixed, solute disperses into the solvent. In dilute solution, each solute molecule is surrounded only by solvent molecules. The solvent molecules that directly interact with the solute molecules, shown in blue, constitute the solvation shell. Solvent structure is very dynamic; solvent and solvation-shell molecules exchange places rapidly.

of the interactions between S molecules and *all* of the interactions between A molecules are replaced by interactions between A and S. The molecules of solvent S that are in direct contact with the solute molecules A (shown in blue in Fig. 8.4) are called collectively the **solvation shell**, or **solvent cage**. For simplicity, we've shown the solvent shell as a single layer of solvent molecules, although the interactions between solute and solvent can in some cases extend beyond a single layer. Although the diagram in Fig. 8.4 is necessarily static, solvent structure is very dynamic, with molecules in the solvent and molecules in the solvation shell moving around and exchanging places rapidly.

When we consider the energetics of the solution process, we can think of it much as we would a chemical reaction. The free energy change is equal to the free energy of the "products" (the solution) minus the free energy of the "reactants" (the pure liquids). When a solute A is dissolved in a liter of solvent S, the free-energy change, ΔG_s, is called the **free energy of solution**.

$$\Delta G_s = G(\text{solution}) - [G(\text{pure solute}) + G(\text{pure solvent})] \quad (8.49)$$

When $\Delta G_s < 0$, the solution process is favorable; when $\Delta G_s > 0$, the solution process is unfavorable. The magnitude of ΔG_s tells us how favorable or unfavorable the process is.

The first, and very important, aspect of solution formation is that, regardless of the intermolecular interactions involved, *there is a statistical driving force for formation of the solution*, called the *entropy of mixing*, ΔS_{mixing}. The **entropy of mixing** is a quantitative description of the probability of solution formation that is completely independent of any intermolecular interactions that may be involved.

To understand the entropy of mixing, we need some more intuition about entropy. As we noted in the previous section, **entropy** is a measure of *probability.* If a system goes from a less probable to a more probable state, then the entropy of the system has increased. Let's consider a few examples of this idea.

Suppose you have four coins, all heads. Imagine a process in which you flip each coin once. Suppose you end up with two heads (H) and two tails (T). There is only one way that four coins can be all heads, but there are *six* ways that four coins could have two heads and two tails:

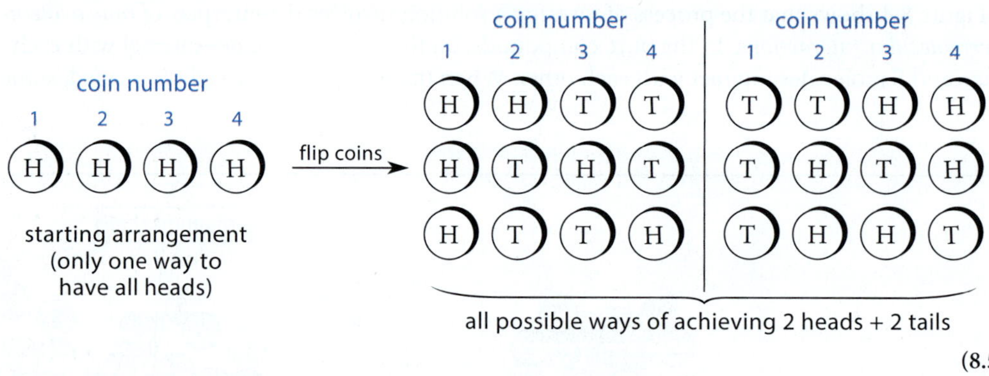

(8.50)

This makes the two heads–two tails combination more probable by a factor of 6. The coin-flipping process increases the entropy of the system of coins because it takes the coins from a less probable state to a more probable state. If you do the same experiment with *six* coins, the relative probability of the three heads–three tails combination is 20. (Verify this!) If you no longer specify that the flipped coins must have half heads and half tails, and instead allow *all* possible results, the number of possibilities is even larger. There are 16 rather than 6 possible outcomes for the four coins, and 64 possible outcomes for the six coins. Imagine if you had Avogadro's number of coins (that is, a "mole of coins"), all heads or all tails, and flipped them all. The number of possible outcomes is huge.

In another analogy, let's say that your professor's office is said to be "neat" when every one of the thousand or so items (books, papers, coffee mugs, tennis racquets) is in a specific, designated place. Over time, items are moved around, they are not put back in their proper place, and the office becomes "messy." We'll define a "messy" state to be any state not equal to the "neat" state. The "messy" states are far more probable because there are so many more of them possible than there are "neat" states. Therefore, the entropy of the office spontaneously increases with use because the "messy" states are inherently more probable (there are so many more ways that the office can be messy). The only way to overcome the effects of entropy is to do work to move objects around to reconstitute the neat state. (Entropy is sometimes linked with "disorder," but, while this is often true, it's not generally true.)

Now let's say that you have a box of 10 blue balls and another box of 1000 red balls. Imagine putting the blue balls into the box containing the red balls, covering the box, and shaking it. What would you find? Would the blue balls all stay together? Certainly not. They would disperse among the red balls, and the odds are that each blue ball would be surrounded only by red balls. A huge number of possible arrangements of balls can give this result, but only one arrangement (or only a few arrangements, if you let the blue stack move around) in which the blue balls could, by chance, stay together. The blue balls disperse into the red balls because there are so many more ways that the balls could be arranged in the dispersed state—the dispersed state is *much* more probable. The entropy change expresses this increased probability.

The foregoing analogy is *directly related* to the formation of a solution. If we squirt some water-soluble blue ink into a beaker of water, we see the ink spontaneously disperse into the water because, in the dispersed state, there are many more ways that the ink molecules can be located relative to the water molecules than the state in which the ink molecules stay together. This is a result of entropy. The dispersed state is simply more probable. If we try to envision the reverse process starting with a solution of water-soluble blue ink, wouldn't it be amazing to see all the ink aggregate spontaneously in one corner of the beaker? It doesn't happen, because this would violate the natural tendency to higher probability (higher entropy).

The ΔS_{mixing} can be calculated for mixing any number of moles of solute with any number of moles of solvent. This formula (Eq. 8.51) was derived directly from probability considerations like the ones we have been discussing.

$$\Delta S_{\text{mixing}} = -2.3R(n_1 \log x_1 + n_2 \log x_2) \tag{8.51}$$

In this equation n_1 and n_2 are the numbers of moles of the two solution components (solvent and solute) used to form the solution, x_1 and x_2 are the *mole fractions* of the two components in the solution, and R is the gas constant (8.31 J mol^{-1} K^{-1} 1.99 cal mol^{-1} K^{-1}). For example, mixing 1 mol of any solute into 1 L (55.6 mol) of water, the mole fraction of the solute is $1/(1+55.6) = 0.0177$, and the mole fraction of water is $55.6/(1+55.6) = 0.982$. From Eq. 8.51, $\Delta S_{\text{mixing}} = +41.9$ J mol^{-1} K^{-1} = 10 cal mol^{-1} K^{-1}. To convert ΔS_{mixing} to ΔG_{mixing}, we use $\Delta G_{\text{mixing}} = \Delta H_{\text{mixing}} - T\Delta S_{\text{mixing}}$. However, $\Delta H_{\text{mixing}} = 0$ because we are considering only probabilities and not energy changes associated with intermolecular interactions. Therefore,

$$\Delta G_{\text{mixing}} = -T\Delta S_{\text{mixing}} \tag{8.52}$$

where T is the temperature in kelvins. At room temperature (25 °C or 298 K),

$$\begin{aligned}\Delta G_{\text{mixing}} &= (-298\ \text{K})(41.9\ \text{J mol}^{-1}\ \text{K}^{-1}) \\ &= -12{,}500\ \text{J mol}^{-1} = -12.5\ \text{kJ mol}^{-1}\ (-2.99\ \text{kcal mol}^{-1})\end{aligned} \tag{8.53}$$

Because ΔG_{mixing} is negative, there is an intrinsic tendency to form this solution, and the magnitude of ΔG_{mixing} indicates exactly what that tendency is in free-energy terms.

If the free energy of mixing were the only free-energy change involved in solution formation, every liquid would dissolve in every other liquid! You probably know from experience that this is not the case. The reason is that noncovalent interactions also make a contribution to the free energy. Figure 8.4 shows that forming a solution involves replacing some *S–S* interactions and all of the *A–A* interactions with *S–A* interactions. Let the free-energy change associated with these interaction changes be ΔG_{inter}. The overall free energy of solution, ΔG_s, results from the balance of ΔG_{inter} against the always-favorable free-energy of mixing:

$$\Delta G_s = \Delta G_{\text{inter}} + \Delta G_{\text{mixing}} \tag{8.54a}$$

This equation is combined with Eq. 8.52 to give

$$\Delta G_s = \Delta G_{\text{inter}} - T\Delta S_{\text{mixing}} \tag{8.54b}$$

which stresses the strictly entropic origin of the mixing contribution.

We can use our previous blue ball–red ball analogy to illustrate this balance. Suppose our 10 blue balls contain embedded bar magnets so that they stick together. If we mix them with our 1000 red balls and shake, they would not disperse into the red balls if the magnets were strong enough. In this case, the attractions between the blue balls overcome the tendency toward spontaneous mixing. That is, *separating the blue balls would require energy*. Similarly, if the solute molecules, or the solvent molecules, have significant intermolecular attractions for each other that *are not replaced by compensating S–A interactions in solution*, then $\Delta G_{\text{inter}} > 0$. If ΔG_{inter} is large enough, it overcomes the entropy of mixing, and $\Delta G_s > 0$. As ΔG_s increases (becomes more positive), the amount of *A* that dissolves in *S* decreases.

Let's suppose that the solute and the solvent are so similar that the intermolecular interactions between solute and solvent in the solution are *exactly the same* as the interactions between solute molecules in the pure solute and solvent molecules in the pure solvent. For example, imagine that we dissolve 1 mol of hexane labeled at one carbon with ^{13}C (call this material hexane*) in 1 L of ordinary hexane. In this case, hexane*–hexane* interactions and some hexane–hexane interactions are replaced with hexane*–hexane interactions. The presence of a single ^{13}C in a hexane molecule has a negligible contribution to its van der Waals interactions. Therefore, in this case, $\Delta G_{\text{inter}} = 0$, and the overall ΔG_s equals the ΔG_{mixing} as calculated from Eq. 8.52. As intuition should dictate, the isotopically labeled hexane sample dissolves completely in the ordinary hexane solvent. If, however, we mix two substances that are quite different, then whether $\Delta G_s < 0$—whether a solution forms—depends on the balance between ΔG_{inter} and ΔG_{mixing}. Two substances can form a solution even if ΔG_{inter} is unfavorable (positive), provided that it is not *too* positive.

Focused Problems

8.17 In each case, which distribution has the higher entropy? Explain.

(a) Four coins in which two are heads and two are tails, or four coins in which one is heads and three are tails

(b) Six coins in which two are heads and four are tails, or six coins in which two are tails and four are heads

(c) 1 mol of a solute in 1 L of water, or 5 mol of the same solute in 1 L of water; assume that $\Delta G_{\text{inter}} = 0$

8.18 Use Eq. 8.52 to calculate the free energy of mixing of 1 mol of isotopically labeled hexane in 1 L of ordinary hexane. The molecular mass of hexane is 114 g mol^{-1}, and the density of hexane is 0.659 g mL^{-1} at 25 °C (298 K).

B. Classification of Solvents

Our goal is to understand noncovalent intermolecular interactions in terms of the structures of the interacting molecules. For this purpose, we'll find it useful to classify the most common solvents in terms of three broad categories. These categories are not mutually exclusive—that is, a solvent can be in more than one category:

1. A solvent can be *protic* or *aprotic*.
2. A solvent can be *polar* or *apolar*.
3. A solvent can be a *donor* or a *nondonor*.

A **protic solvent** consists of molecules that can act as *hydrogen-bond donors*. Water, alcohols, and carboxylic acids are protic solvents. Solvents that cannot act as hydrogen-bond donors are called **aprotic solvents**. Ether, dichloromethane, and hexane are aprotic solvents.

Unfortunately, when it comes to describing solvents, the word *polar* has a double usage in organic chemistry. One usage refers to whether the individual molecules of the solvent have a significant dipole moment (Sec. 1.9B). Recall (Sec. 8.5B) that attractions between dipoles of organic molecules can significantly affect their boiling points. As we'll learn, the attractions between the dipoles of solvent and solute molecules can also have significant effects on solubility. We'll call solvents consisting of molecules with significant dipole moments ($\mu > 1$ D) **dipolar solvents**.

The other usage of the word *polar* has to do with the solvent *dielectric constant*, for which we use the symbol ε. In this usage, a **polar solvent** is a solvent that has a high dielectric constant ($\varepsilon \geq 15$); an **apolar solvent** is one with a low dielectric constant. The **dielectric constant** is defined by the *electrostatic law*, which gives the interaction energy E between two ions with respective charges q_1 and q_2 separated by a distance r:

The dielectric constant is commonly called the *relative permittivity* in physics.

$$E = k\frac{q_1 q_2}{\varepsilon r} \qquad (8.55)$$

In this equation, k is a proportionality constant and ε is the dielectric constant of the solvent in which the two ions are imbedded. The dielectric constant is in the denominator of Eq. 8.55; so, when the dielectric constant ε is large, the magnitude of E (the energy of interaction between the ions) is small. This equation means that both attractions between ions of opposite charge and repulsions between ions of like charge are weak in a polar solvent. In other words, *a polar solvent effectively separates, or shields, ions from one another*. Therefore, *the tendency of oppositely charged ions to associate is less in a polar solvent than it is in an apolar solvent.* Water ($\varepsilon = 78$), methanol ($\varepsilon = 33$), and formic acid ($\varepsilon = 59$) are polar solvents. Hexane ($\varepsilon = 2$), ether ($\varepsilon = 4$), and acetic acid ($\varepsilon = 6$), are apolar solvents. As we'll see in Sec. 8.6F, the dielectric constant contributes significantly to the ability of a solvent to dissolve ionic compounds.

The dielectric constant is a property of many molecules of a solvent acting together, whereas the dipole moment is a property of individual molecules. Fortunately, all polar solvents consist of dipolar molecules; so, when we say "polar solvent" we can assume that the solvent molecules have a significant dipole moment. However, the converse is not true. Some *apolar* solvents (in the dielectric-constant sense) also contain dipolar molecules. The contrast between acetic acid and formic acid is a particularly striking example:

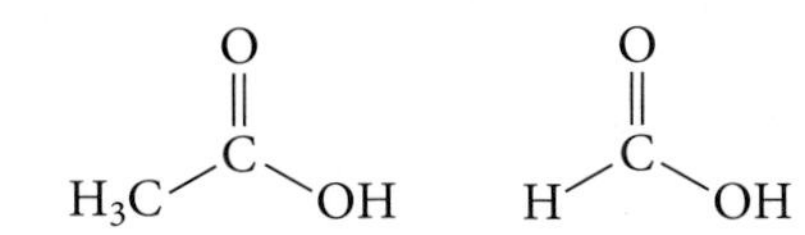

acetic acid	**formic acid**	
μ = 1.5–1.7 D	μ = 1.6–1.8 D	
ε = 6.1	ε = 59	(8.56)

These two compounds contain identical functional groups and have similar structures and dipole moments. Both are *polar molecules* and, as solvents, they can be termed *dipolar solvents*. They are also both *protic*. Yet they differ substantially in their dielectric constants and their solvent properties. Formic acid, a *polar solvent*, is much more effective in dissolving ionic compounds than acetic acid, an *apolar solvent*.

Donor solvents consist of molecules containing oxygens or nitrogens that can donate nonbonding electron pairs—that is, molecules that can act as Lewis bases. Ether, tetrahydrofuran (THF), and methanol are donor solvents. **Nondonor solvents** cannot act as Lewis bases; pentane and benzene are nondonor solvents. Although the chlorines in halogenated solvents such as dichloromethane and chloroform have nonbonding electron pairs, these solvents are poor Lewis bases and are considered to be nondonor solvents.

Table 8.2 lists some common solvents used in organic chemistry along with their abbreviations, properties, and classifications. This table shows that a solvent can have a combination of properties, as noted at the beginning of this section. For example, some polar solvents are protic (such as water and methanol), but others are aprotic (such as acetone).

TABLE 8.2 Properties of Some Common Organic Solvents
(Listed in order of increasing dielectric constant.)

Solvent	Structure	Common abbreviation	Boiling point, °C	Dielectric constant ε*	Dipole moment	Class		
						Polar	Protic	Donor
hexane	$CH_3(CH_2)_4CH_3$	—	68.7	1.9	0.08			
1,4-dioxane†	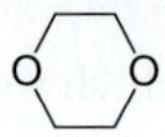	—	101.3	2.2	0.4			x
benzene†	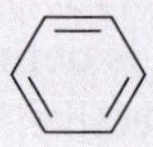	—	80.1	2.3	0.1			
diethyl ether	$(C_2H_5)_2O$	Et_2O	34.6	4.3	1.2			x
chloroform	$CHCl_3$	—	61.2	4.8	1.2			
ethyl acetate	$CH_3\overset{\overset{O}{\|}}{C}OC_2H_5$	EtOAc	77.1	6.0	1.6			x
acetic acid	$CH_3\overset{\overset{O}{\|}}{C}OH$	HOAc	117.9	6.1	1.6		x	x
tetrahydrofuran		THF	66	7.6	1.7			x
dichloromethane	CH_2Cl_2	DCM	39.8	8.9	1.1			
acetone	$CH_3\overset{\overset{O}{\|}}{C}CH_3$	Me_2CO	56.3	21	2.7	x		x
ethanol	C_2H_5OH	EtOH	78.3	25	1.7	x	x	x
N-methylpyrrolidone	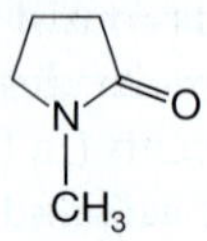	NMP	202	32	4.0	x		x
methanol	CH_3OH	MeOH	64.7	33	2.9	x	x	x
nitromethane	CH_3NO_2	$MeNO_2$	101.2	36	3.4	x		x
N, N-dimethylformamide	$H\overset{\overset{O}{\|}}{C}N(CH_3)_2$	DMF	153.0	37	3.9	x		x
acetonitrile	$CH_3C{\equiv}N$	MeCN	81.6	38	3.4	x		x
sulfolane	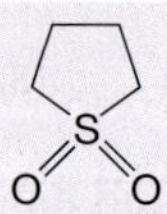	—	287 (dec)	43	4.7	x		x
dimethyl sulfoxide	$CH_3\overset{\overset{O}{\|}}{S}CH_3$	DMSO	189	47	4.0	x		x
formic acid	$H\overset{\overset{O}{\|}}{C}OH$	—	100.6	59	1.7	x	x	x
water	H_2O	—	100.0	78	1.9	x	x	x
formamide	$H\overset{\overset{O}{\|}}{C}NH_2$	—	211 (dec)	111	3.9	x	x	x

*Most values are at or near 25 °C †Known carcinogen

Focused Problem

8.19 Use their structures and dielectric constants to classify each of the following substances according to their solvent properties (as in Table 8.2).

(a) 2-Methoxyethanol ($\varepsilon = 17$) (b) 2,2,4-Trimethylpentane ($\varepsilon = 2$) (c)

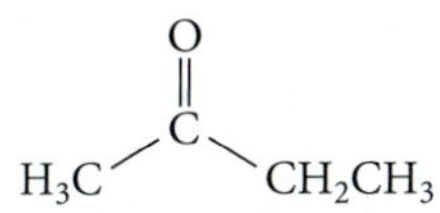

($\varepsilon = 19$)

2-butanone
(ethyl methyl ketone, MEK)

C. Solubility of Covalent Compounds

In this section, we focus on applying the principles covered in Secs. 8.6A and 8.6B with the objective of developing some intuition about the practical aspects of solubility. We start with the solubility of liquids, and then we extend what we learn to the solubility of solids.

In determining a solvent for a liquid covalent compound, a useful rule of thumb is that **like dissolves like**. That is, a good solvent usually has some of the molecular characteristics of the compound to be dissolved. For example, an apolar aprotic solvent is likely to be a good solvent for another apolar aprotic liquid. In contrast, a protic solvent in which significant hydrogen bonding occurs between molecules is likely to dissolve another liquid in which hydrogen bonding between molecules also occurs.

Let's try to understand the basis of the like-dissolves-like rule. To illustrate, imagine dissolving pentane in hexane. In terms of the energetics discussed in Sec. 8.6A, the pentane–pentane attractions and some of the hexane–hexane attractions in the pure liquids are replaced by hexane–pentane attractions in the solution. In both of the pure liquids, the major type of intermolecular attraction is van der Waals attractions, which we discussed in Sec. 8.5A. In the solution, the major type of attraction between hexane and pentane molecules should also be van der Waals attractions, *because both molecules are of the same type*. We expect (and find) that ΔG_{inter} is close to zero. From Eq. 8.54a, the ΔG_s is then determined by the free energy of mixing ΔG_{mixing}, which is always favorable (negative). In fact, pentane and hexane are **miscible** (literally, "mixable")—they form a solution when mixed in any proportions.

Here, then, is the *physical reason for the like-dissolves-like rule*: We can confidently predict that when the attractions between molecules in the pure liquids are similar to the attractions between the solvent and solute in the solution, then ΔG_{inter} will be small, ΔG_s will be dominated by ΔG_{mixing}, and a solution will be formed.

Let's examine a few more examples of "like dissolves like." The smaller alcohols (methanol, ethanol, propanol) are all miscible in water. The *major* noncovalent interaction in the pure substances, as we learned from boiling points, is hydrogen bonding. When, for example, methanol and water are mixed, water and methanol molecules can also form hydrogen bonds.

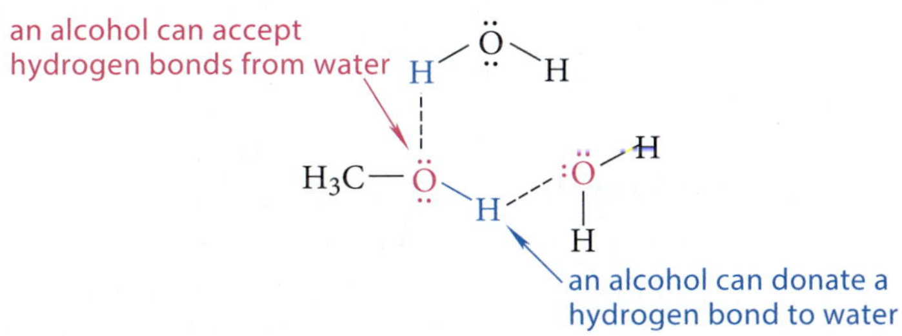

(8.57)

Because the interactions between solute and solvent are similar to the interactions between molecules in the pure liquids, we expect the free-energy cost of dissolving one in the other to be modest. As a result, the positive entropy (and resulting negative free energy) of mixing ensures the formation of a solution.

Hydrocarbons such as pentane and chlorinated solvents such as dichloromethane (CH_2Cl_2) are miscible. In pure pentane, the major source of intermolecular attractions is van der Waals forces. In dichloromethane, the major source of intermolecular attractions is both dipole–dipole attractions and van der Waals forces. When the two solvents are mixed, however, the dipoles of dichloromethane can *induce* temporary dipoles in nearby pentane molecules, and this interaction results in attractions as well. (Such interactions are sometimes called **dipole–induced dipole** attractions.)

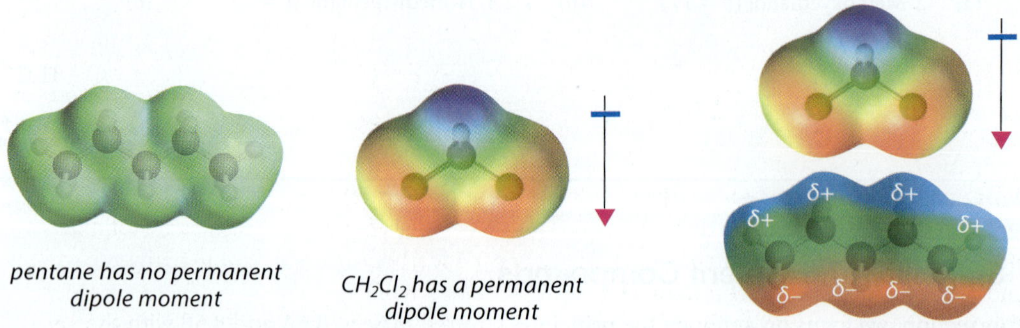

pentane has no permanent dipole moment

CH_2Cl_2 has a permanent dipole moment

separated, noninteracting molecules

the permanent dipole in CH_2Cl_2 induces a temporary dipole moment in pentane

interacting molecules

(8.58)

Such interactions are merely variations of the van der Waals forces discussed in Sec. 8.5A in which one of the molecules has a permanent dipole. In other words, the interactions between solute and solvent molecules in solution are fundamentally the same as the interactions in the pure liquids.

One of the most important practical considerations is the solubility of organic compounds in water. Whether we are dealing with environmental issues (such as the solubility of chemicals in ground water), the design of drugs (how to make a drug more water-soluble), or laboratory "green" chemistry (chemistry carried out in environmentally benign solvents), water solubility is important. So, let's look at some trends in water solubility. Consider, for example, the water solubilities of the following compounds of comparable size and molecular mass:

	$CH_3CH_2CH_2CH_3$ CH_3CH_2Cl	$CH_3CH_2—O—CH_3$	$CH_3CH_2CH_2—OH$
water solubility:	virtually insoluble	soluble	miscible

(8.59)

Of these compounds, the alcohol, 1-propanol, is most soluble; in fact, it is miscible with water. Of the compounds shown, the alcohol is also most like water because it is protic. The ability both to donate a hydrogen bond to water and to accept a hydrogen bond from water is an important factor in water solubility.

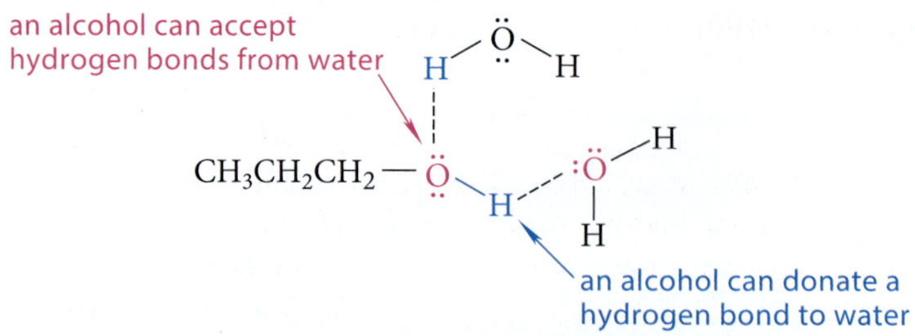

(8.60)

The ether contains an atom (oxygen) that can accept hydrogen bonds from water, although it cannot donate a hydrogen bond; that is, it has some waterlike characteristics, but it is less like water than the alcohol.

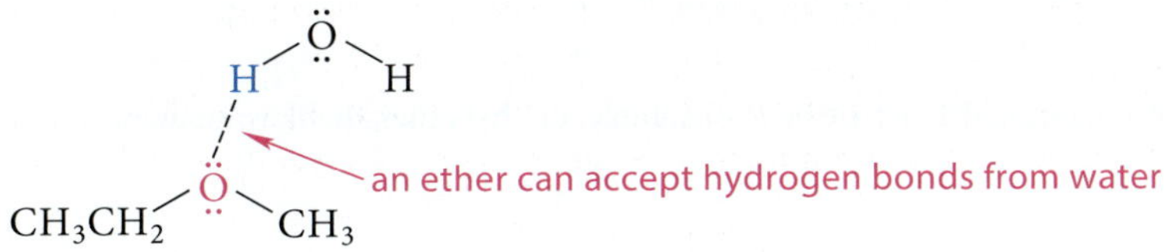

(8.61)

! The hydrogen-bond acceptor ability of ethers is relevant to their aqueous solubilities, *but not to their boiling points.* The boiling point is determined by the interactions of identical ether molecules. Ether molecules cannot form hydrogen bonds with each other because they contain no hydrogen-bond donor group.

Finally, the alkane (butane) and the alkyl halide (ethyl chloride) can neither donate nor accept hydrogen bonds and are therefore least like water; they are also the least water-soluble compounds on the list. The aqueous insolubility of hydrocarbons is important to many phenomena in biology, and we'll return to this topic in Sec. 8.6D.

The balancing of "waterlike" and "hydrocarbonlike" character is evident in the following series:

	CH_3OH	CH_3CH_2OH	$CH_3CH_2CH_2OH$	$CH_3CH_2CH_2CH_2OH$
water solubility, mol L^{-1}	miscible			0.96

	$CH_3CH_2CH_2CH_2CH_2OH$	$CH_3CH_2CH_2CH_2CH_2CH_2OH$	
water solubility, mol L^{-1}	0.23	0.0032	(8.62)

Alcohols with long hydrocarbon chains—that is, large alkyl groups—are more like alkanes than are alcohols containing small alkyl groups. Because alkanes cannot form hydrogen bonds, they are insoluble in water, but they are soluble in other apolar aprotic solvents, including other alkanes. Likewise, alcohols (as well as any other organic compounds) with long hydrocarbon chains are relatively insoluble in water and are more soluble in apolar aprotic solvents than alcohols with small alkyl chains.

One of the most important generalizations about water solubility is *the importance of hydrogen bonding*. Generally, compounds that are both donors and acceptors have better water solubility than compounds that are simply acceptors; but acceptors have greater solubility than compounds that are neither donors nor acceptors. Hydrocarbon groups reduce aqueous solubility. A useful rule of thumb is that compounds containing one OH group for every five carbons usually have significant water solubility.

In qualitative considerations of solubility, don't expect to make distinctions that are too fine. For example, diethyl ether and THF, two widely used reaction solvents, have the same number of carbons and a single oxygen.

tetrahydrofuran (THF)
miscible with water

$$CH_3CH_2—O—CH_2CH_3$$

diethyl ether
water solubility ≈ 0.8 mole L^{-1}
(forms a separate phase with water)

(8.63)

Their water solubilities, though, are significantly different. If we pour THF into water, a solution forms. If we pour diethyl ether into water, two layers form. To be sure, the water contains a significant amount of dissolved diethyl ether, and diethyl ether contains dissolved water, but qualitatively speaking, diethyl ether and water are not mutually soluble. Both ethers are apolar aprotic substances, and both dissolve a wide range of other compounds. But, because of its water solubility, THF is often used in reactions in which the presence of water is required, such as the oxymercuration of alkenes (Sec. 5.4A); it dissolves both water and alkenes.

What you should begin to see from this discussion are the *trends* to be expected in the solubility behavior of various compounds. You cannot be expected to remember absolute solubilities, but you should be able to make an intelligent guess about the relative solubilities of a given compound in different solvents or the relative solubilities of a series of compounds in a given solvent. This ability, for example, is required to solve Study Problem 8.5.

Study Problem 8.5

In which of the following solvents should 1-octene be *least* soluble: diethyl ether, dichloromethane, methanol, or 1-octanol? Explain.

Solution 1-Octene is an alkene, and the intermolecular interactions in pure 1-octene should be very similar to those in octane because both are hydrocarbons. The interaction mechanism should be van der Waals attractions. [In fact, the boiling points of 1-alkenes and alkanes, which rely on van der Waals attractions to maintain the liquid state, are nearly identical for the same carbon number—that is, in Fig. 8.2 (Sec. 8.5A), the alkene and alkane curves would virtually coincide.] We've learned that compounds with dipole moments can interact attractively with hydrocarbons by the dipole–induced dipole mechanism. Therefore, 1-octene should have significant solubility, if not miscibility, in diethyl ether and dichloromethane. However, for 1-octene to dissolve in methanol, the hydrogen bonds between methanol molecules must be disrupted, and 1-octene has no groups that can replace this interaction. The same is true in 1-octanol, except that the long alkyl chain of 1-octanol should have favorable van der Waals attractions with the alkene. Therefore, 1-octene should be least soluble in methanol.

Solubility is very important in the metabolism of xenobiotics. A **xenobiotic** (from the Greek *xenos*, meaning "alien") is any substance that is not a normal constituent of a living organism. Among the most important xenobiotics are environmental pollutants and drugs. When an organism (such as the human body) encounters a xenobiotic, it typically routes the substance into a pathway by which it can be eliminated. If the xenobiotic has very low water solubility, one of the most widely used strategies is to couple (that is, chemically attach) the xenobiotic to another group that increases its water solubility. The soluble coupling product can then be routed, for example, to the kidneys, where it can be excreted as an aqueous solution in the urine. This type of process is termed *phase II metabolism*.

One of the most common coupling strategies used in nature for this purpose is *glucuronidation*. For example, the broad-spectrum antibiotic chloramphenicol has relatively low water solubility. It is excreted largely as a glucuronide derivative. (Glucuronides are derived from glucuronic acid, a derivative of the sugar glucose.)

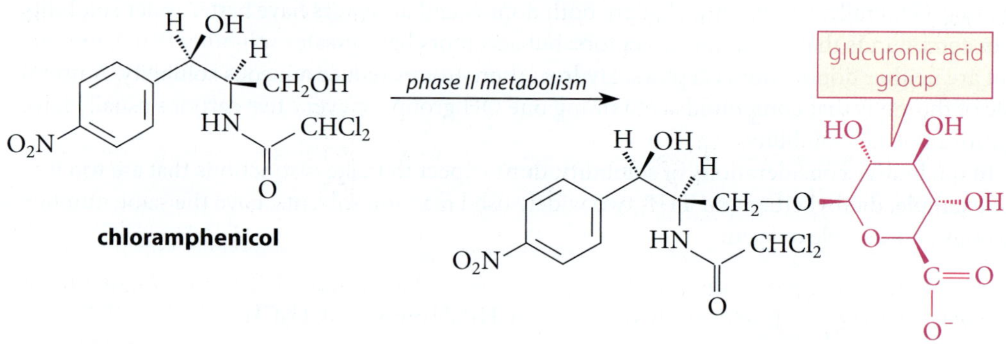

a glucuronide derivative of chloramphenicol (excreted in the urine) (8.64)

Because of its many hydroxy groups, which have the capacity to form hydrogen bonds to water, the glucuronide has greater water solubility than the uncoupled drug. In addition, as we show in Sec. 8.6E, the ionized carboxylic acid group is strongly solvated by water and also enhances aqueous solubility.

Although glucuronides are not the only derivatives used for phase II metabolism, they are among the most common. This example demonstrates that nature uses the like-dissolves-like principle just as we do in the laboratory.

Focused Problems

8.20 Into a separatory funnel is poured 200 mL of dichloromethane (density = 1.33 g mL^{-1}) and 55 mL of water. This mixture forms two layers. One mL of 1-octanol is added, and the mixture is shaken. After a time, two layers are again formed. Where is the 1-octanol—in the upper or lower layer?

8.21 A widely used undergraduate experiment is the recrystallization of acetanilide from water.

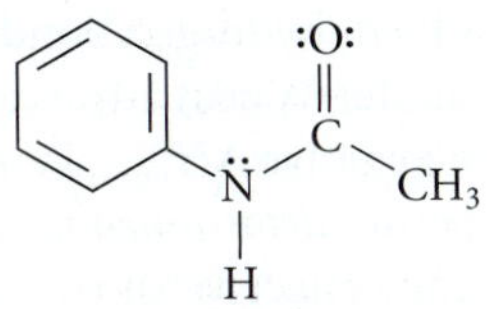

acetanilide

Acetanilide is moderately soluble in hot water, but it is much less soluble in cold water. Identify two structural features of the acetanilide molecule that would be expected to contribute positively to its solubility in water and one that would be expected to contribute negatively.

D. Solubility of Hydrocarbons in Water: Hydrophobic Bonding

Most of us know from experience that "oil and water don't mix." That is, if we pour any hydrocarbon into water, it forms a separate layer. This is one of the reasons that large-scale oil spills are such environmental disasters—water can't dissolve the oil and wash it away. At one level, we could say simply that the insolubility of hydrocarbons in water is a reverse manifestation of the like-dissolves-like rule: hydrocarbons (apolar, aprotic, nondonor) are not at all like water (polar, protic, donor); therefore, hydrocarbons don't dissolve in water. However, a deeper look at the insolubility of hydrocarbons in water will provide some insights that will help us to understand a number of biological phenomena, some of which we'll explore in Sec. 8.6E.

To say that hydrocarbons are insoluble in water is not quite accurate. Actually, hydrocarbons do have a *very low* solubility in water that can be measured. For example, imagine that we pour some pentane into 1 L of water, stir, and let the system come to equilibrium.

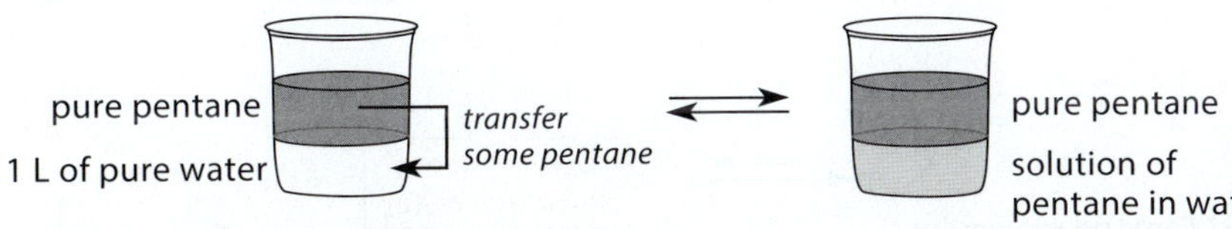

(8.65)

The concentration of pentane in the aqueous layer is small, about 5×10^{-4} M, but definitely not zero. The measured ΔG°_s for dissolving pentane in water (at 298 K) is +29.1 kJ mol^{-1}. This is a *standard* free energy: the free energy that would have to be expended to produce a standard solution in which 1 mol of pentane is dissolved in a liter of water. This large, positive ΔG°_s shows that forming such a solution is *very* unfavorable energetically. The measured standard enthalpy of solution ΔH°_s is –2.2 kJ mol^{-1}; $\Delta H^\circ_s = \Delta H^\circ_{inter}$ because, by definition, $\Delta H^\circ_{mixing} = 0$. The overall standard entropy of solution ΔS°_s is –105 J mol^{-1} K^{-1}, which contributes $-T\Delta S^\circ_s = -(298\text{ K})(-0.105\text{ J mol}^{-1}\text{ K}^{-1}) = +31.29\text{ kJ mol}^{-1}$ to the overall free energy of solution. The overall standard entropy includes contributions from both intermolecular interactions $T\Delta S^\circ_{inter}$ and the entropy of mixing $T\Delta S^\circ_{mixing}$, which we calculated in Eq. 8.53 as 12.5 kJ mol^{-1}. Separating out the entropy of mixing allows us to calculate $-T\Delta S^\circ_{inter}$, the entropy contribution to the free energy that is due solely to intermolecular interactions:

$$T\Delta S^\circ_s = T\Delta S^\circ_{inter} + T\Delta S^\circ_{mixing} \tag{8.66a}$$

or

$$T\Delta S^\circ_{\text{inter}} = -T\Delta S^\circ_{\text{s}} + T\Delta S^\circ_{\text{mixing}}$$
$$= +31.29 + 12.5 = +43.79 \text{ kJ mol}^{-1} \text{ (10.47 kcal mol}^{-1}) \quad \textbf{(8.66b)}$$

When we contrast this value with that of the interaction enthalpy $\Delta H^\circ_{\text{inter}} = -2.2$ kJ mol^{-1} (−0.53 kcal mol^{-1}), we see that *the unfavorable standard free energy for dissolving pentane in water is due to the large, negative entropy associated with intermolecular interactions.*

It might seem surprising that moving pentane into water causes an enthalpy reduction. Although pentane cannot form hydrogen bonds to water, pentane and water can interact by dipole–induced dipole (van der Waals) attractions. Although small, this interaction is stabilizing, and it is the reason for the negative $\Delta H^\circ_{\text{inter}}$. Pentane is not an isolated case: *the small solubilities of other hydrocarbons in water are determined mostly by unfavorable entropies of solution.*

To understand the entropy effect for dissolving pentane in water, consider how the solvent water changes when pentane molecules are introduced into aqueous solution (**Fig. 8.5**). When molecules of a hydrocarbon dissolve, the solvent shells of the dissolved molecules (red oxygens in Fig. 8.5) must come from the solvent water (blue oxygens in Fig. 8.5). It turns out that *water molecules in the solvation shell of a pentane molecule are different from those in ordinary water.* Specifically, the water molecules at the water–pentane interface have *reduced motional freedom* relative to solvent water. Scientists once described this water as being icelike, but more recent investigations suggest a more complex picture. A simple static picture for this "interface water" may not be possible. However, perhaps we can think of the situation in terms of the following analogy. Imagine hundreds of small children on a playground, running around at random, and a teacher comes up to some of them and says, "Form a circle around me." The children open up a "cavity" (a circle) for the teacher, and, as a

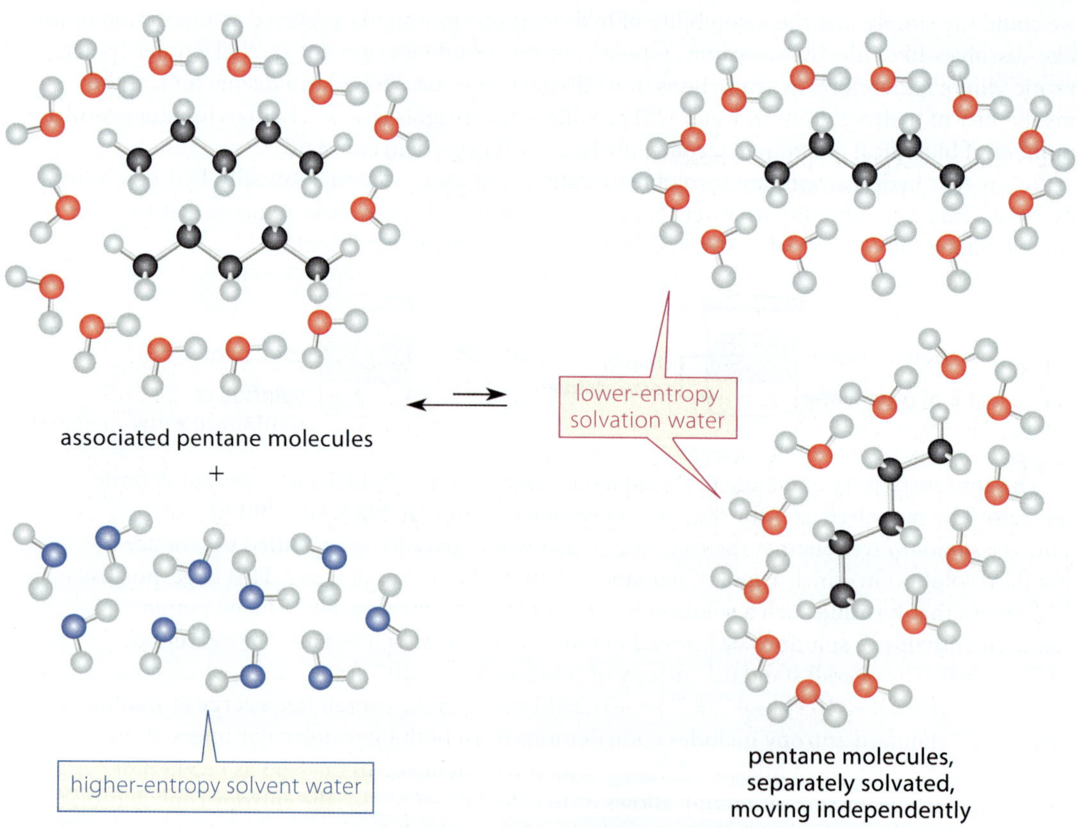

FIGURE 8.5 The role of solvation in the dissolution of hydrocarbons in water. When hydrocarbons dissolve, some solvent water must be transformed into the water of the solvent shells. Because solvation water (red oxygens) has lower entropy than solvent water (blue oxygens), the process of dissolution is accompanied by a reduction in entropy. The *reverse* of this process is essentially a model for the hydrophobic bond. That is, in the reverse direction, the association of hydrocarbon molecules releases solvation water into solvent and therefore involves an entropy increase. This diagram is necessarily two-dimensional; because solvation shells surround molecules in three dimensions, more water molecules are involved than are shown here.

result, the children on the edge of the cavity have less freedom to move around—for one thing, they have to stay out of the cavity. By analogy, the water surrounding a pentane molecule—the solvation shell—has reduced motional freedom. *Reduced motional freedom lowers entropy.* Therefore, *the water in the hydrocarbon solvent shell has lower entropy than ordinary water.* Because of the small size of a water molecule, many water molecules are involved in cavity formation (solvation) for each hydrocarbon molecule that dissolves. Therefore, the entropy *decrease* for the solvent is much greater than the positive entropy of mixing pentane and water.

Now let's look at the solubility of hydrocarbons in reverse. Envision a situation in which hydrocarbon molecules are dispersed in aqueous solution (the reverse of Fig. 8.5). As we've just learned, there is a driving force for them to associate with each other rather than with water. The association of hydrocarbon groups in aqueous solution is called **hydrophobic bonding**. This terminology, although widely used in biology, is somewhat of a misnomer. The implication in this name ("water-fearing") is that hydrocarbons do not interact favorably with water. However, as we've seen, the enthalpy (energy) of solution of hydrocarbons in water is relatively favorable; so, hydrocarbons are really not "hydrophobic" in an energetic sense. The major reason for the hydrophobic behavior of hydrocarbons is *the entropy change of the solvent water.* When hydrocarbon groups associate with each other, low-entropy solvation water is released to become ordinary, higher-entropy water. In terms of our previous analogy, when many teachers join hands on the playground, many of the "immobilized" children that formed individual cavities around each teacher are released onto the playground and can move normally. The resulting $\Delta S°$ is highly positive, and this is why $\Delta G°$ is negative for hydrocarbon association.

More sophisticated models of hydrophobic bonding have shown that there is a favorable enthalpy contribution to the interaction between water and very large "hydrophobic" surfaces in addition to the favorable entropic component. This provides a further thermodynamic drive for the formation of large "hydrophobic" clusters.

Many biological processes are driven all or in part by hydrophobic bonding—the association of hydrocarbon groups. Among these processes are membrane formation (Sec. 8.7A), the "folding" of proteins into well-defined conformations (Sec. 27.9), and the binding of many enzymes to their substrates (Sec. 27.10).

Focused Problem

8.22 How would you expect the standard entropy of solution in water to change along a series of primary alcohols (for example, from methanol to 1-octanol) as the chain length increases? Explain.

E. Solubility of Solid Covalent Compounds

When we dissolve a solid in a liquid solvent, the process is not as simple as dissolving a liquid. Let's consider the standard free-energy change in dissolving a solid. Remember that we can choose any path we wish for the calculation of free energy changes as long as the starting and ending points are the same. So, let's divide the process of dissolving a solid into two hypothetical steps. From a conceptual standpoint, two things have to happen. First, the solid has to become a liquid. Then, the resulting liquid has to dissolve in the solvent. (We assume a temperature of 25 °C, or 298 K.)

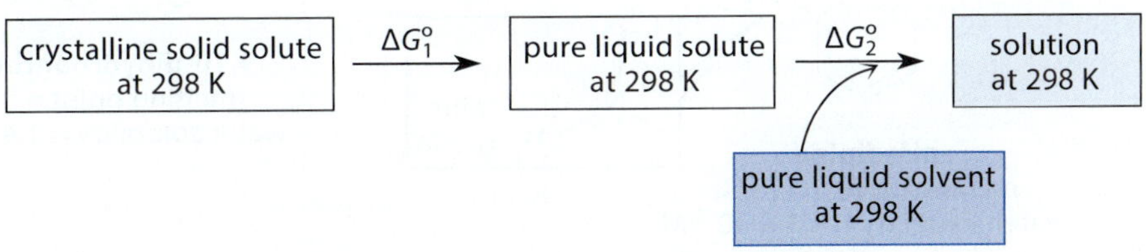

(8.67)

The standard free energy of solution for the solid, $\Delta G_s°$(solid), is the sum of the standard free energies of the individual steps, $\Delta G_1° + \Delta G_2°$. The free-energy change step 1, $\Delta G_1°$, is the free energy of fusion of the solid at 298 K, $\Delta G°_{m,298}$. (This is the free energy required to melt a mole of the compound at 298 K.) Although the free energy of fusion of a solid *at its melting point* is zero, at temperatures below the melting point the free energy of fusion *is unfavorable*—that is, $\Delta G°_{m,298} > 0$. This means that the solid is converted into a "supercooled" liquid. The greater the difference between the temperature of the solution and the melting point of the solid, the greater is $\Delta G^\circ_{m,298}$. Only after the solid is transformed into its liquid state can we then apply the solubility principles of liquids. Therefore, $\Delta G^\circ_2 = \Delta G^\circ_s(\text{liquid})$. For the overall process in Eq. 8.67, we have

$$\Delta G^\circ(\text{solid}) = \Delta G^\circ_{m,298} + \Delta G^\circ_s(\text{liquid}) \tag{8.68}$$

The point of this equation is that *melting a solid is part of the solution process* and the energy required for melting adds an additional free-energy increment to the process of dissolving a solid. When we think about the solubility of a solid, we have to take into account not only the like-dissolves-like considerations of dissolving a liquid, *but also how easy or difficult it is to melt the solid.* From this discussion we conclude that, other things being equal, solids with high melting points should have large values of $\Delta G^\circ_{m,298}$ and therefore should be less soluble in a given solvent than isomers of similar structure with a much lower melting point.

As a first example of solid solubility, recall from Sec. 8.5D that symmetrical compounds generally have higher melting points than their less symmetrical isomers. Therefore, symmetrical compounds should then have lower solubilities in *any solvent* than their less symmetrical isomers. For example, 1,4-dichlorobenzene and 1,3-dichlorobenzene are not very soluble in water; but, of the two, 1,3-dichlorobenzene has a lower melting point and is more than twice as soluble in water as its more symmetrical isomer 1,4-dichlorobenzene. 2-Methylbenzoic acid, the unsymmetrical isomer, has a lower melting point and a higher solubility in 1-octanol than its symmetrical isomer, 4-methylbenzoic acid. The solubility difference between symmetrical and unsymmetrical isomers is general.

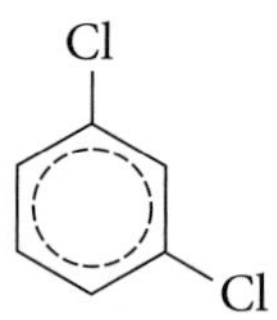

1,3-dichlorobenzene
mp −24.7 °C
solubility in water: 0.132 g L^{-1}
symmetry $\sigma = 2$

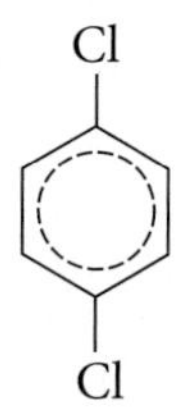

1,4-dichlorobenzene
mp 47 °C
solubility in water: 0.056 g L^{-1}
symmetry $\sigma = 4$

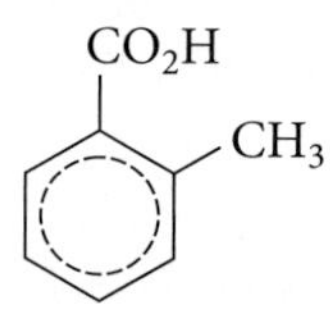

2-methylbenzoic acid
melting point = 105 °C
solubility in 1-octanol: 143 g L^{-1}
symmetry $\sigma = 1$

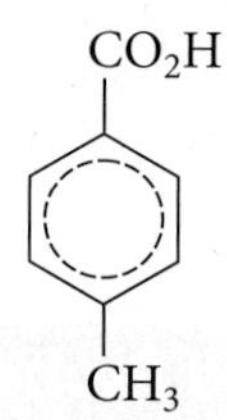

4-methylbenzoic acid
melting point = 182 °C
solubility in 1-octanol: 52.9 g L^{-1}
symmetry $\sigma = 2$

(8.69)

As another example, compare the water solubilities of nifedipine (a drug used to control angina and hypertension) and the hydrocarbon 9,10-dihydroanthracene.

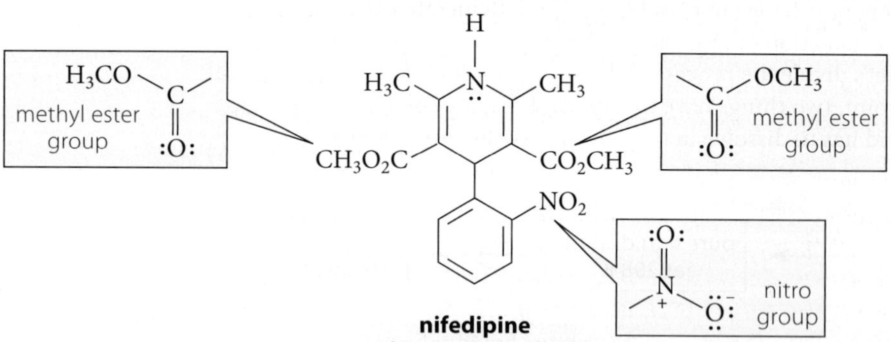

nifedipine
melting point = 173 °C
water solubility = 1.3×10^{-5} *M*

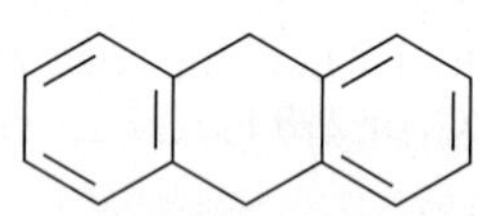

9,10-dihydroanthracene
melting point = 109 °C
water solubility = 1.4×10^{-5} *M*

(8.70)

Nifedipine has a number of hydrogen-bond acceptor sites as well as a N—H hydrogen-bond donor site. 9,10-Dihydroanthracene, a hydrocarbon, is devoid of any such sites. On the basis of "like dissolves like" alone, we would expect nifedipine to be considerably more soluble in water. In agreement with this expectation, *liquid* nifedipine is almost 30 times as soluble in water as *liquid* 9,10-dihydroanthracene. However, the water solubility of the two *solids* is about the same. The unexpectedly low solubility of *solid* nifedipine is due to its higher melting point.

The flip side of solubility is ease of crystallization. Practicing organic chemists know that it can sometimes be challenging to crystallize a solid with a low melting point. Solids with high melting points are generally easier to crystallize.

When a drug is administered in the solid state—for example, as a capsule or tablet—it has to be in solution in order to function. For this reason, the dissolution behavior of a drug in water or in aqueous buffers is crucial to its formulation, and characterization of the solubility properties of a potential new drug is a major goal of pharmaceutical research. The melting points of drug candidates are a particular concern when solubility is considered. A drug must have a high enough melting point to be crystallized, but a low enough melting point that it will dissolve. The best compromise depends on the particular drug molecule, but many potentially useful drug candidates have been rejected because of low solubility resulting from a high melting point. Even different crystal forms of the same drug can have different melting points and therefore different solubilities. A few cases are known in which one crystal form of a drug is biologically active and another is so insoluble that it is inactive even though the molecules are the same! Specific crystal forms of a drug can be patented, and patent litigations have occurred over the crystal forms of commercially important drugs.

Focused Problem

8.23 Which of these two solid compounds should have higher solubility in water? Explain.

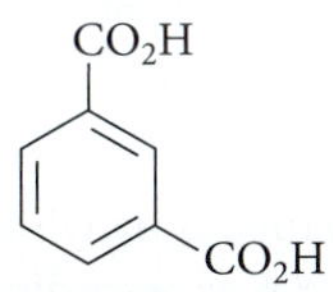

1,3-benzenedicarboxylic acid (isophthalic acid)

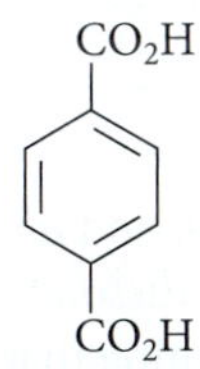

1,4-benzenedicarboxylic acid (teraphthalic acid)

F. Solubility of Ionic Compounds

Because of the importance of both ionic reagents and ionic reactive intermediates in organic chemistry, the solubility of ionic compounds is worth special attention. Ionic compounds in solution can exist in several forms, two of which, *ion pairs* and *dissociated ions*, are shown in **Fig. 8.6**. In an **ion pair**, each ion is closely associated with an ion of opposite charge. In contrast, **dissociated ions** move more or less independently in solution and are surrounded by several solvent molecules, called collectively the **solvation shell** or **solvent cage** of the ion, by analogy with the similar terms for nonionic compounds (see Fig. 8.4). **Solvation** refers to the favorable interaction of a dissolved molecule with solvent. When solvent molecules interact favorably with an ion, they are said to **solvate** the ion.

Most ionic inorganic compounds are solids. The ones with which we are most familiar have high melting points. For example, the melting point of solid sodium chloride ($Na^+ \; Cl^-$) is 801 °C. This high melting point shows that, unlike van der Waals forces or hydrogen bonds, the forces between ions in a crystal of sodium chloride are very strong. The attractions between ions are generally called **electrostatic attractions**. The energies of these attractions are governed essentially by the *electrostatic law* (Eq. 8.55), although the details of the application of this equation to the entire system of ions in a crystal is complex. But this calculation has been done, and it shows that the electrostatic attraction of ions in a sodium chloride crystal averages about –770 kJ mol^{-1} (–184 kcal mol^{-1}) per pair of ions. (The negative sign indicates an attraction.) This is a huge attraction, roughly twice the value of a covalent bond energy. For a solvent to dissolve an ionic compound, then, the solvent must provide significant, energetically favorable, solvation to replace the large attraction between ions in the crystal. The fact that ionic solids such as sodium chloride can be dissolved shows that the noncovalent forces involved in solvating ions must be considerable. Let's examine the factors involved in the dissolution of ions.

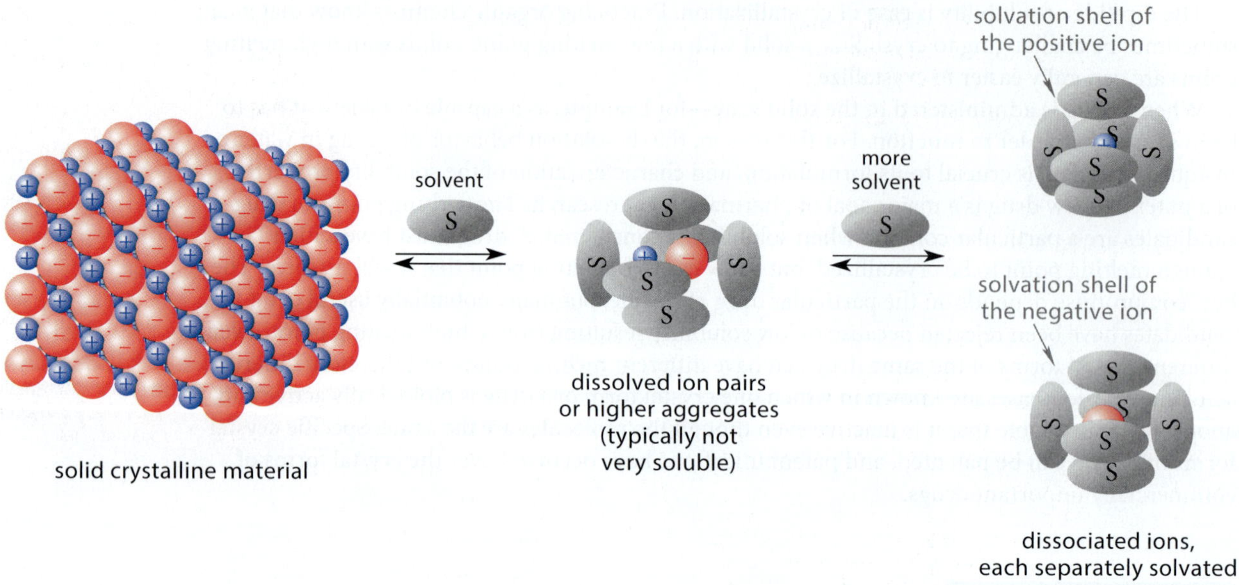

FIGURE 8.6 Ions in solution can exist as ion pairs and dissociated ions. The blue spheres are positive ions, and the red spheres are negative ions. The solubility of an ionic compound depends on the ability of the solvent to break the electrostatic attractions between ions and form separate solvation shells around the dissociated ions. Solvent molecules are represented by the gray ellipses. Solvation is very dynamic—that is, solvent molecules in the solvation shells are not fixed, but are rapidly exchanging with molecules in the bulk solvent.

Ion separation and *ion solvation* are mechanisms by which ions are stabilized in solution. If you think of the ion dissolution sequence in Fig. 8.6 as an ordinary chemical equilibrium, you will realize that anything that favors the right side of this equilibrium tends to make ions soluble. The separation and solvation of ions reduce the tendency of the ions to associate into aggregates and, ultimately, to precipitate as solids from solution. Consequently, ionic compounds are relatively soluble in solvents in which ions are *well separated and well solvated.* What solvent properties contribute to the separation and solvation of ions?

The ability of a solvent to *separate* ions is measured by its dielectric constant ε in Eq. 8.55 (Sec. 8.6B). Look carefully at this equation again. The energy of attraction of two ions of opposite charge is reduced in a solvent with a high dielectric constant. As a result, ions of opposite charge *have a reduced tendency to associate* in solvents with high dielectric constants, and consequently *have a higher solubility* in those solvents.

Solvation of dissolved ions occurs in various ways, which are illustrated in **Fig. 8.7** for the interaction of the solvent water with dissolved sodium chloride. A donor solvent can act as a Lewis base to donate its nonbonding electron pairs to a cation, which acts as a Lewis acid. This covalent interaction is called a **donor interaction**. In addition, the dipole moments of solvent molecules can interact electrostatically with the charge of the ion. This means that the water molecule is oriented so that the negative end of its dipole moment vector is pointing toward the positive ion, thus creating a favorable electrostatic interaction by Eq. 8.55. This is called a **charge–dipole interaction** or **ion–dipole interaction**. Because the orientation of the water molecule is almost the same in both charge–dipole and donor interactions, the two interactions are sometimes considered to be different aspects of the same interaction. For solvation of the anion, a favorable charge–dipole interaction can occur in which the solvent molecules are turned so that the positive ends of their dipole moment vectors are pointing toward the negative ion. Finally, if the solvent is protic and the anion can accept hydrogen bonds, the solvent can solvate a negative ion by a **hydrogen-bonding interaction**.

Solvation is *dynamic.* That is, although the solvent shells in Figs. 8.6 and 8.7 are shown as static structures, these figures represent a "snapshot" of a rapidly changing situation. The water molecules are constantly exchanging places with water molecules from bulk solvent, and the solvation mechanism within a solvent shell can change rapidly as well.

To summarize, the *high dielectric constant* of a polar solvent reduces the attraction between ions of opposite charge; as a result, these ions are easier to separate and bring into solution. Dissolved ions are stabilized (that is, kept in solution) by three general types of interaction:

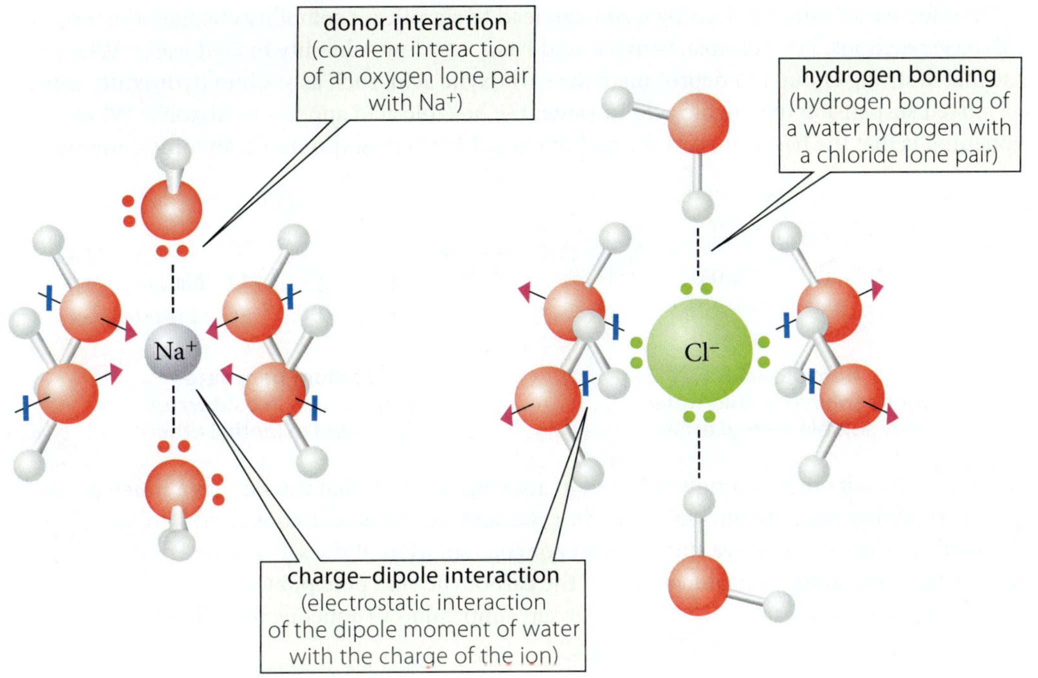

FIGURE 8.7 A "snapshot" of the interaction of solvent water molecules in the solvent shells of dissolved sodium and chloride ions. Although two donor interactions are shown for the cation and two hydrogen-bonding interactions for the anion, a greater number of such interactions can occur. Solvation shells can also contain more than six water molecules. Solvation is a dynamic process in which water molecules from the bulk solvent rapidly exchange with water molecules in the solvent shell.

1. *Charge–dipole interactions*, by which the dipole moment vectors of the molecules in a polar solvent are oriented so as to create an attractive (stabilizing) interaction with the charge of the ion
2. *Hydrogen-bonding interactions*, by which dissolved ions can be stabilized by hydrogen bonding with solvent molecules
3. *Donor interactions*, by which solvent molecules with nonbonding electron pairs can act as Lewis bases toward dissolved cations

These points show why water is the ideal solvent for ionic compounds, something you probably know from experience. First, because it is polar—it has a very large dielectric constant—water is effective in separating ions of opposite charge. Second, because it is protic—a good hydrogen-bond donor—water readily solvates anions. Third, because it is a Lewis base—an electron-pair donor—water can solvate cations by a donor interaction. Finally, its significant dipole moment enables water to provide stabilizing ion–dipole attractions to both cations and anions. In contrast, hydrocarbons such as hexane do not dissolve ordinary ionic compounds because such solvents are apolar, aprotic, and nondonor solvents. Some ionic compounds, however, have appreciable solubilities in *polar aprotic* solvents such as acetone or dimethyl sulfoxide (DMSO) (see Table 8.2, Sec. 8.6B). Although these solvents lack the protic character that solvates anions, their donor capacity solvates cations, their substantial dipole moments provide favorable charge–dipole interactions, and their high dielectric constants separate ions of opposite charge. However, it is not surprising that, because polar aprotic solvents lack the protic character that stabilizes anions, most salts are less soluble in these solvents than in water, and salts dissolved in polar aprotic solvents exist to a greater extent as ion pairs (see Fig. 8.6).

Focused Problem

8.24 DMSO (Table 8.2, Sec. 8.6B) has a very large dipole moment (4.0 D). Using structures, show the stabilizing noncovalent interactions to be expected between DMSO solvent molecules and (a) a dissolved sodium ion; (b) dissolved water; (c) a dissolved chloride ion.

The effective solvation of ions by water can lead to significant solubility changes as a result of acid–base reactions. For example, benzoic acid has a very low solubility in cold water. When a base that is strong enough to deprotonate the carboxylic acid, such as sodium hydroxide, is added to a stirred suspension of benzoic acid in water, the benzoic acid appears to dissolve. What is happening is that the base converts the carboxylic acid into its sodium salt, an ionic compound.

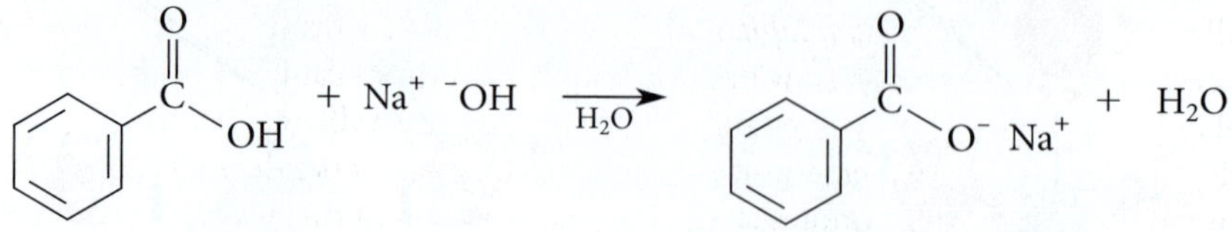

benzoic acid
slightly soluble in cold water;
very soluble diethyl ether

sodium benzoate
very soluble in cold water;
insoluble in diethyl ether (8.71)

The ions of the salt are so strongly solvated in aqueous solution that the salt has a much greater aqueous solubility than the un-ionized carboxylic acid. As the acid is converted into the salt, the salt dissolves. Conversely, if we start with an aqueous solution of the salt and add concentrated HCl, the benzoate anion is protonated, and the carboxylic acid precipitates.

A biological example in which an ionic group contributes to aqueous solubility is the ionic group of glucuronic acid (Eq. 8.64, Sec. 8.6C).

Focused Problem

8.25 (a) Triethylamine, Et_3N:, is a liquid that is insoluble in water. When HCl is added to a stirred mixture of triethylamine and water, a solution is formed. Explain.

(b) Sketch a diagram similar to Fig. 8.7 that shows the solvation of the species in the solution.

(c) What could you add to the solution formed in part (a) that would bring the triethylamine back out of solution?

8.7 APPLICATIONS OF SOLUBILITY AND SOLVATION PRINCIPLES

A. Enantiomeric Resolution: Selective Crystallization

If you return to Secs. 6.10A and 6.10B, you will see how important noncovalent intermolecular interactions, solubility, and solvation are to enantiomeric resolution. In chiral chromatography, formation of noncovalent complexes of enantiomers with the pendant groups of a chiral stationary phase (CSP) is the basis for the separation (Display 6.38, Sec. 6.10A). In the diastereomeric salt method, it is the different solubilities of diastereomeric salts that make the separation possible. Further, once one of the diastereomeric salts is crystallized and isolated, the liberation of an enantiomer from the salt depends on relative solubilities:

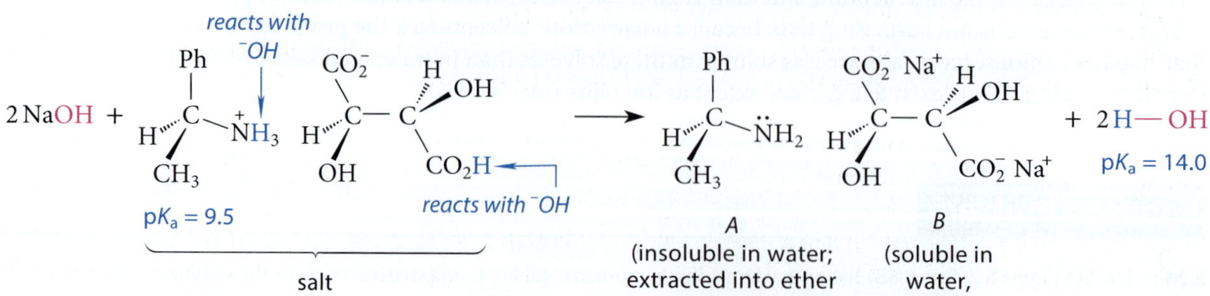

A
(insoluble in water; extracted into ether and recovered)

B
(soluble in water, insoluble in ether) (8.72)

The acid–base reaction forms amine *A*, which has a significant hydrocarbon character and is insoluble in water, and ion *B*, which, because of its ionic character (and hydroxy groups) is very soluble in water. Selective extraction of *A* into ether (in which *B* is not soluble) effects the separation.

Another method of enantiomeric resolution, and one used frequently in the pharmaceutical industry for the enantiomeric resolution of large quantities of chiral compounds that form crystalline solids, is *selective crystallization.* As you may know from your own laboratory work, crystallization is often a slow process, and it sometimes can be accelerated by adding a seed crystal of the compound to be crystallized. In **selective crystallization**, a solution of a mixture of enantiomers is cooled to supersaturation and a seed crystal of the desired enantiomer is added. In this case, the seed crystal promotes selective crystallization of the desired enantiomer.

The seed crystal contains only molecules of the pure enantiomer of the desired compound. The seed crystal can grow in two ways: it can incorporate more molecules of the same enantiomer or some molecules of the opposite enantiomer. These two possibilities generate two diastereomeric crystals. (The seed crystal in Display 8.73 is represented as a collection of molecules with the *R* configuration, symbolized by red "right hands;" and the enantiomeric *S* molecule as a "left hand," colored blue for easy recognition.)

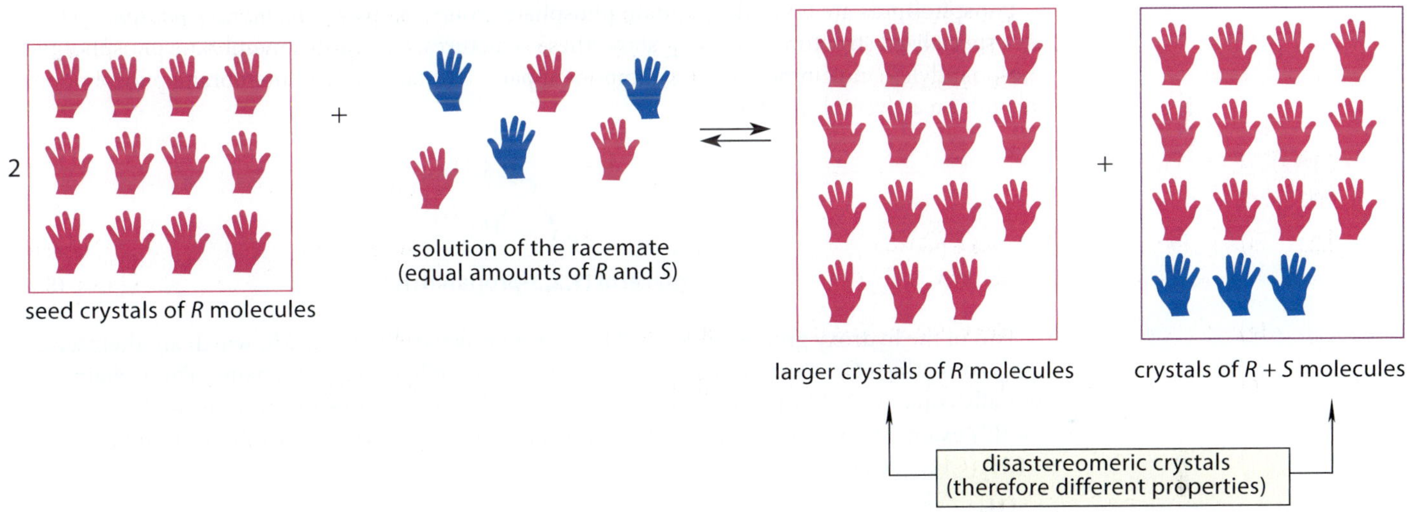

(8.73)

Because the crystals are diastereomeric, they have different properties—specifically, different solubilities. What ensures that the crystal of the pure enantiomer selectively crystallizes? As you may have learned in your laboratory work, the addition of impurities to a solid lowers its melting point. Therefore, we can expect the impure crystal to have a lower melting point than the pure crystal. We have also learned in this chapter (Sec. 8.6D) that the melting points of two isomers inversely correlate with solubility: isomers with higher melting points are less soluble. For that reason, we can expect the pure crystal to be less soluble—that is, to selectively crystallize. The growth of the enantiomeric all-*S* crystal is equally probable, but its growth is slower because there is no seed crystal provided to get the process started.

How does the principle of enantiomeric differentiation (Sec. 6.10) operate in selective crystallization? The seed crystal is the "chiral agent" required for differentiating the two enantiomers. It selectively incorporates the enantiomer already present in the crystal; if the other enantiomer is incorporated, a diastereomeric crystal of different properties (lower melting point, greater solubility) is formed.

The solvent and temperature conditions for crystallization are determined by trial and error. The seed crystal is obtained by other methods of enantiomeric resolution that can be carried out at a relatively small scale, such as chiral chromatography. Once the conditions are developed, selective crystallization can be carried out on a large scale. Selective crystallization, then, provides a mechanism for "amplification" of an enantiomeric resolution to a larger scale.

Selective crystallization is particularly useful in the pharmaceutical industry, where the preparation of large amounts of enantiomerically pure chiral compounds is important in some cases. Typically, selective crystallization is applied early in a synthesis to relatively small molecules (so as not to waste large amounts of unwanted material), and the resolved synthetic intermediate is then used for subsequent steps in the synthesis. If the preparation of a drug that can bring millions of dollars in revenue is at stake, it is worthwhile to devote significant effort and expense to determine the best conditions for selective crystallization.

Chemical Biology Topic

B. Cell Membranes and Drug Solubility

We found in Sec. 8.6E that solubility is a crucial issue in drug action. If a drug is to be administered in an aqueous solution, it must have adequate aqueous solubility. However, water solubility is not the whole story. For drugs to act, they must travel to their biological sites of action. For many drugs, this means that they must enter cells. The only way for a drug to get into a cell is for it to pass through the *cell membrane*, the "envelope" that surrounds the cell. Drugs and other substances pass through cell membranes by a variety of mechanisms; in some cases, transport requires carrier molecules imbedded in the membrane; and, in some cases, transport requires the expenditure of metabolic energy. In many cases, however, drugs simply pass unassisted through the cell membrane. (This process is called *passive diffusion.*) It turns out that the ability of a molecule to penetrate a cell membrane is essentially a solubility issue. To understand this issue, let's examine the structure of a cell membrane.

Cell membranes consist primarily of molecules called *phospholipids*. A **lipid** is a compound that shows significant solubility in apolar solvents. Because lipids are defined by a behavior rather than a precise structure, a number of different biomolecule types fit into this category. For example, steroids (Sec. 7.6D) are lipids. Even though lipids might contain polar functional groups, their solubility behavior is dominated by their significant hydrocarbon character.

Phospholipids are lipids that contain phosphate groups. However, *membrane phospholipids* have specialized structures. To understand these structures, let's build a membrane phospholipid, phosphatidylethanolamine, from its component parts. To start with, all membrane phospholipids are built on a glycerol "scaffold."

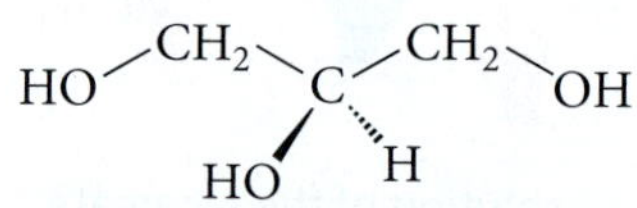

glycerol (1,2,3-propanetriol) (8.74)

Two of the hydroxy groups of glycerol form ester linkages to *fatty acids,* which are themselves lipids consisting of carboxylic acids with long, *unbranched* hydrocarbon chains. These chains typically contain 15–17 carbon atoms, and they can also contain one or more cis double bonds. For this example, we'll use a chain of 17 carbons in which $—C_{17}H_{35}$ is an abbreviation for $—CH_2(CH_2)_{15}CH_3$.

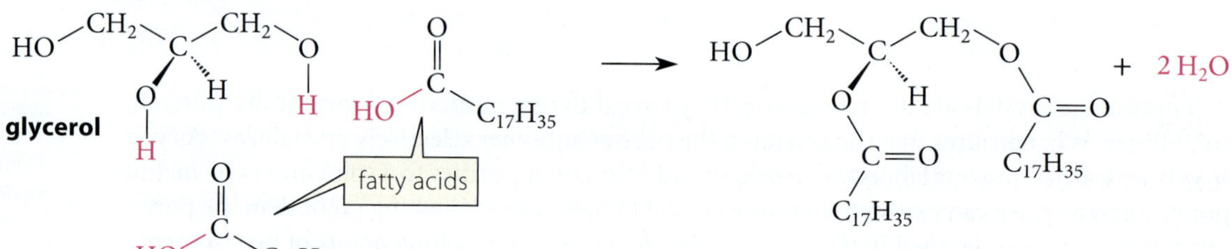

a 1,2-diacylglycerol (8.75)

The diacylglycerol is chiral even though the starting reactants are not. The production of a single stereoisomer is assured by the catalysis of this assembly process by chiral enzymes (Sec. 7.7A). The remaining hydroxy group of the glycerol backbone is connected to a phosphoric acid molecule as a phosphate ester.

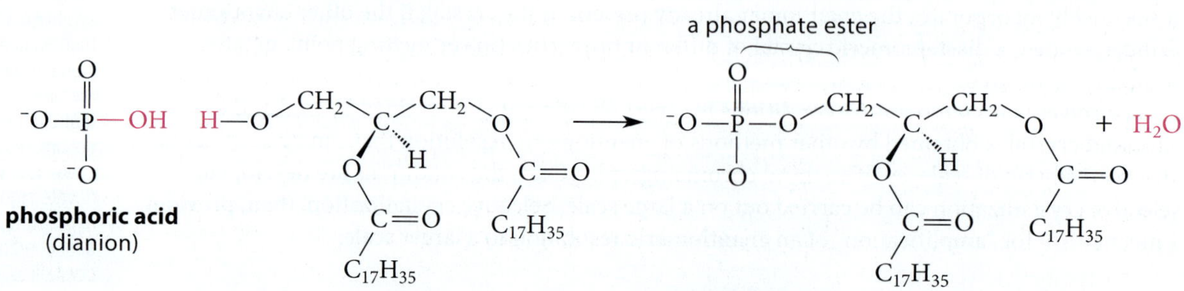

a 1,2-diacylglycerol

a 2,3-diacylglycerol-1-phosphate (8.76)

Finally, the phosphate is connected to a protonated ethanolamine molecule in another phosphate ester linkage.

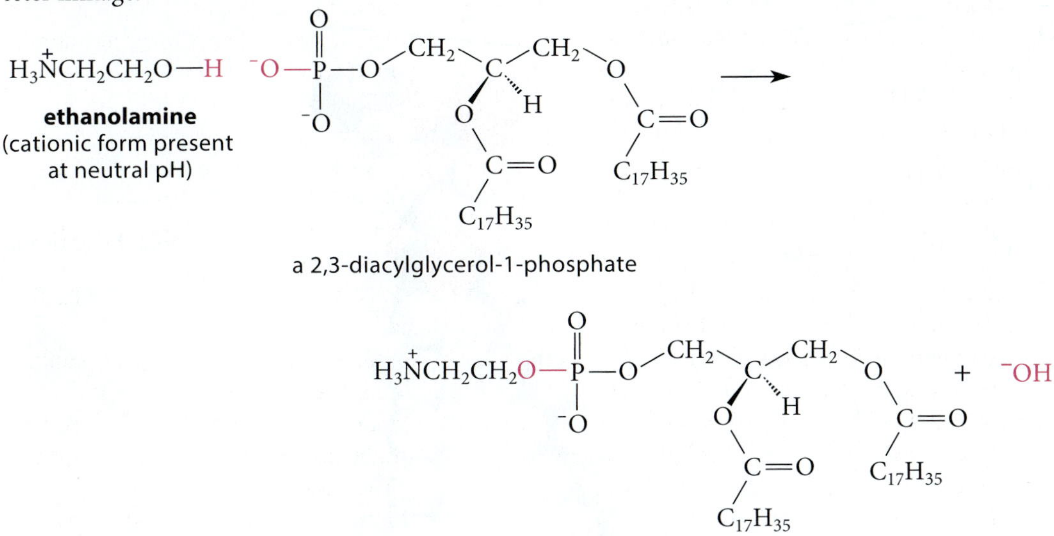

ethanolamine
(cationic form present at neutral pH)

a 2,3-diacylglycerol-1-phosphate

phosphatidylethanolamine
(a membrane phospholipid) (8.77)

A number of different compounds are utilized in this final step. Besides ethanolamine, choline and (*S*)-serine are the amino alcohols used to form the most common membrane phospholipids.

choline

(*S*)-serine (8.78)

Phosphatidylethanolamines are called *cephalins*, and phosphatidylcholines are called *lecithins*.

Focused Problem

8.26 Give the structure of (a) a phosphatidylserine; (b) a lecithin.

Two structural features of membrane phospholipids are particularly important in understanding their properties. The first is the **polar head group**, which consists of the ethanolamine, choline, or serine and the esterified phosphate. The second is the **nonpolar tails**, which are the long, unbranched hydrocarbon portions of the molecules. A space-filling model and a chemical structure of phosphatidylethanolamine are compared in **Figs. 8.8a** and **b**. The polar head group, being ionic, is well solvated by water and counterions. Groups such as the polar head group, which have stabilizing interactions with water, are sometimes called **hydrophilic groups**. The nonpolar tails are examples of **hydrophobic groups**. (More informally, hydrophobic groups might be called "greasy groups." The aqueous solvation of hydrophobic groups was discussed in Sec. 8.6D.) A membrane phospholipid, then, has a hydrophilic part and a hydrophobic part. Molecules such as phospholipids that contain discrete hydrophilic and hydrophobic regions are said to be **amphipathic**. As a reflection of this amphipathic character, membrane phospholipids are often represented in diagrams as a sphere (for the polar head group) with two "squiggly tails," as shown in **Fig. 8.8c**.

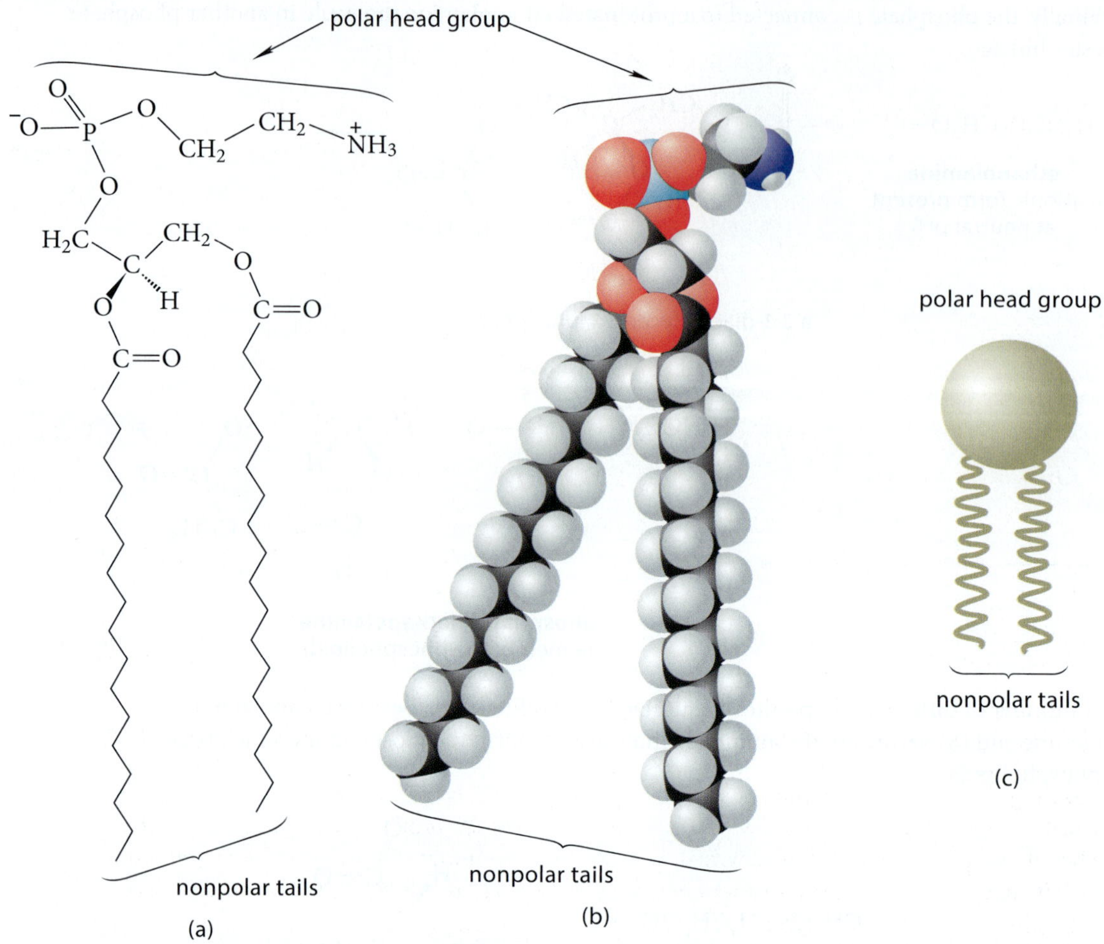

FIGURE 8.8 (a) A Lewis structure of phosphatidylethanolamine, a membrane phospholipid. (b) A space-filling model of phosphatidylethanolamine. (c) A schematic representation of a membrane phospholipid. The polar head group is represented as a sphere and the nonpolar tails by "squiggles." This representation is often used in diagrams, such as Fig. 8.10.

When membrane phospholipids are added to water, something remarkable happens: they undergo a process called *self-assembly* in which they *spontaneously* form a **phospholipid bilayer**. A phospholipid bilayer consists of many molecules in a double layer in which the nonpolar tails interact with each other on the interior of the layer, and the polar head groups interact with water on the outside of the layer (**Fig. 8.9**). The spontaneous formation of this very ordered structure would appear to be, at first sight, a violation of the tendency toward increasing entropy: a random distribution of phospholipid molecules has seemingly converted spontaneously into a seemingly improbable, ordered, bilayer structure. If we focus solely on the phospholipid molecules, their self-assembly into bilayers does indeed represent a highly negative entropy change. However, there is more to this self-assembly process than meets the eye. This is an example of *hydrophobic bonding.* The nonpolar tail of each phospholipid molecule in solution is surrounded by many waters of solvation. Remember that the water molecules in the solvation shell of a hydrocarbon have *particularly low entropy* (Sec. 8.6D, Fig. 8.5). When two (or more) hydrocarbon tails come together into the bilayer, all of the solvation water between the tails is released into the solvent. This provides a highly positive and dominating entropic contribution to the self-assembly process. This process is conceptually much like the aggregation of pentane molecules into a separate liquid phase. In other words, the self-assembly of phospholipids into bilayers is entropy-driven by the release of water.

In the phospholipid bilayer, the solvation demands of both parts of the molecule can be satisfied: hydrocarbon groups are next to hydrocarbon groups, and the polar head groups are near water. These double layers continue to grow as more phospholipid is added to form *phospholipid*

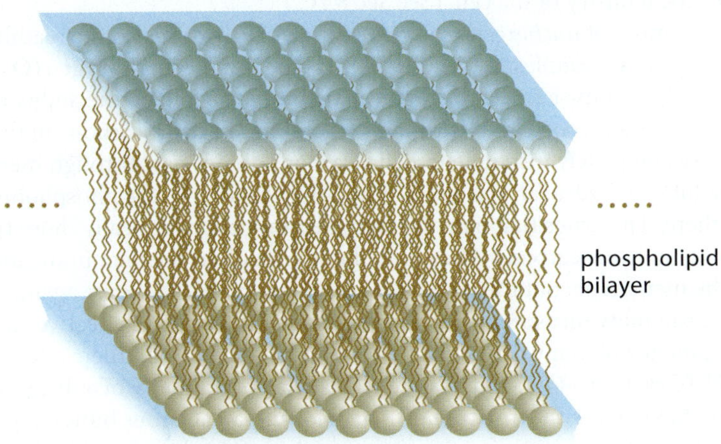

FIGURE 8.9 Part of a phospholipid bilayer. The polar head groups interact with the solvent water, whereas the nonpolar tails interact with other nonpolar tails.

vesicles—closed, more or less spherical structures in which a phospholipid bilayer encloses an inner aqueous region. The polar head groups interact with water on both the inside and outside of the vesicle. Chemically, the living cell can be thought of simplistically as a large phospholipid vesicle in which the cell membrane is a phospholipid bilayer (**Fig. 8.10**). Actual cells are more complex: they contain nuclei and other substructures (many with their own membranes), enzymes, many different biomolecules, and so on. And cell membranes contain molecules in addition to phospholipids, such as cholesterol and imbedded proteins. Nevertheless, it is the phospholipid bilayer that is primarily responsible for the unique character of the cell membrane.

The phospholipid bilayer is generally impermeable to ions. Charged molecules, as well as inorganic ions, cannot penetrate the hydrocarbonlike interior of the lipid bilayer any more than sodium chloride can dissolve in gasoline. The insolubility of ionic compounds in hydrocarbons—specifically, the phospholipid bilayer—is crucial to the cell's ability to retain proper ion balance. The transport of ions through cell membranes requires special carriers or pores, which are

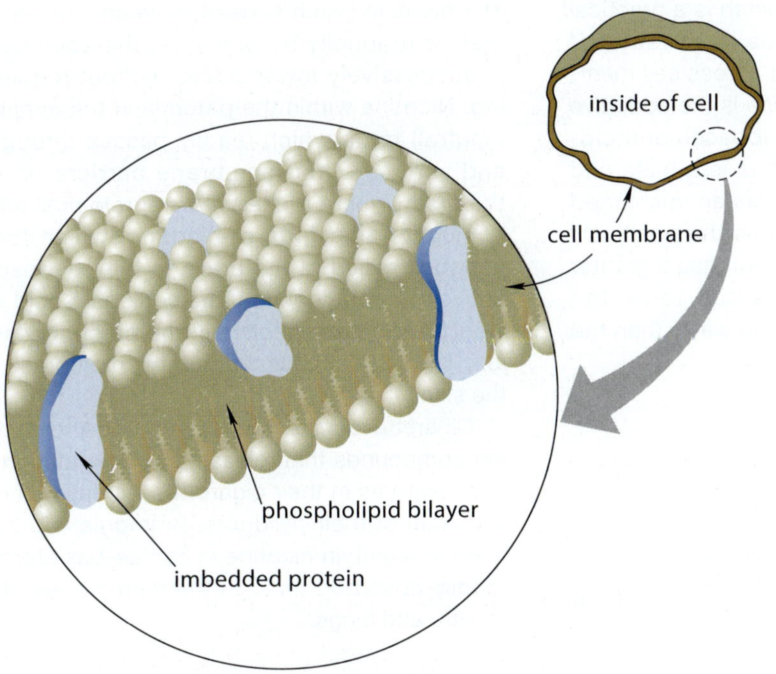

FIGURE 8.10 The cell membrane as a phospholipid vesicle. The enlargement shows a section of the phospholipid bilayer and some imbedded proteins. The phospholipid molecules are represented as shown in Fig. 8.8c, with the polar head groups represented as spheres and the hydrocarbon tails as "squiggles." The polar head groups are exposed to water, and the tails form a hydrocarbonlike region on the interior of the membrane, isolated from water.

proteins imbedded in the membrane. The operation of these ion-carrying systems is tightly regulated by the biochemistry of the cell. (See Sec 8.7C.)

Unlike ions, a number of *uncharged* molecules can passively diffuse quite readily through the cell membrane. One of the simplest molecules of this type is molecular oxygen (O_2), but many drug molecules can also diffuse freely through cell membranes. In fact, the ability of drugs to pass through the cell membranes correlates with their solubilities in hydrocarbons in the following way. Drugs that are completely insoluble in hydrocarbons do not pass through membranes. Drugs that are highly soluble in hydrocarbons don't either: they move into the phospholipid bilayer and "get stuck" there. The drugs that pass through membranes are typically those that have a moderate solubility in hydrocarbons. They are soluble enough in the membrane interior so that they can enter the membrane, but they are soluble enough in water to leave again.

In fact, simple solubility measurements have value in predicting the effectiveness of drug candidates. The potency of many drugs can be correlated, in part, with their relative solubilities in 1-octanol [$CH_3(CH_2)_6CH_2OH$] and in water. This relative solubility of a drug candidate is determined by shaking it with a mixture of 1-octanol and an aqueous buffer at pH = 7.4 (physiological pH) and then measuring the concentration of the drug in each phase. If the drug contains a group that can ionize (such as a carboxylic acid) or protonate (such as an amine), the concentration of the neutral drug molecule is calculated from its pK_a value. The ratio of concentrations of the neutral molecule in the 1-octanol and aqueous phases is called the **octanol–water partition coefficient**, often abbreviated $\boldsymbol{P_{ow}}$. Hydrophobic molecules have larger partition coefficients than hydrophilic ones, and, for a given drug class, there is an optimum value for the partition coefficient. Presumably the 1-octanol phase mimics the hydrophobic environments with which a drug must interact to exert its physiological effect. Such environments might include the phospholipid bilayer in the membrane of the target cells, the phospholipid bilayer in the intestinal epithelium (the layer of cells through which an orally administered drug must pass in order to be absorbed), and the active site of a protein target to which the drug ultimately binds, thereby exerting its effect. (See Problem 8.68 for an example.) Determining the P_{ow} is usually an obligatory part of the investigation of a molecule as a drug candidate. Remarkably, within a given drug class, P_{ow} values tend to correlate with drug activity.

The correlation of 1-octanol–water partition coefficients with drug activity was developed by Corwin Hansch (1918–2011), who was a chemistry professor at Pomona College in California.

Nicotine, the Nicotine Patch, and Cigarette Addiction

StepanPopov/Shutterstock

The nicotine patch is a practical example of the importance of drug transport across cell membranes. Nicotine is the addictive substance in tobacco and cigarettes. Nicotine is a base and can exist both as an uncharged free base and as the positively charged conjugate acid. The neutral form has a greater solubility in hydrocarbons than the salt form; the salt form, being ionic, is more soluble in water than the basic form.

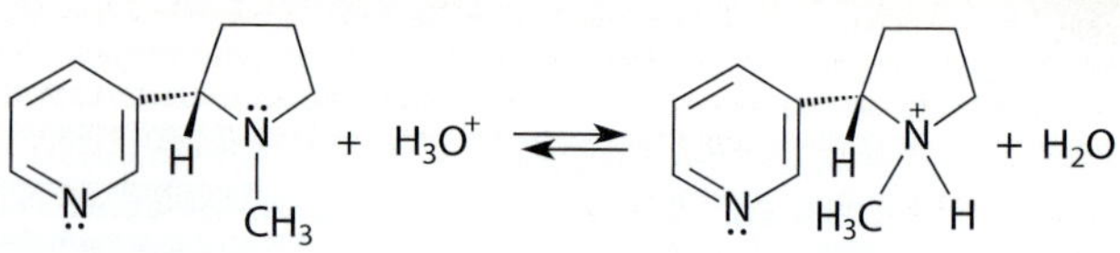

nicotine conjugate base (neutral) — **nicotine** conjugate acid (cationic) (8.79)

The nicotine patch is used to wean smokers from cigarettes gradually by providing the addictive material in successively lower doses without requiring smoking. Nicotine within the patch is in the conjugate-base (neutral) form, which readily passes through the skin and various other membrane barriers on its way to the brain, where it exerts its neurological effects. (The conjugate-acid form of nicotine, which forms in the aqueous milieu of the brain, is the pharmacologically active form of nicotine.) The conjugate acid of nicotine would not be as effective in the patch, because, as an ion, it would not pass through the membrane barriers of the skin.

Cigarette manufacturers have long known that including compounds that release ammonia at high (smoking) temperatures in their cigarettes increases the addictive potential of their products. Ammonia is a base and is used to maintain nicotine in its free-base form, which is readily absorbed through the membranes of the nose, mouth, and lungs.

C. Cation-Binding Molecules

Because we understand the mechanisms of ionic solvation, organic chemists have been able to create synthetic molecules that mimic the solvation shells of ions. These molecules are **ionophores**—molecules that form strong complexes with ions. (The word *ionophore* means "ion-bearing.") This achievement has been largely realized in the design of *crown ethers* and *cryptands,* which we examine in this section. Then, as we consider the structures of ion carriers found in nature, such as *ionophore antibiotics* and *ion channels,* we'll find that the very same mechanisms of ionic solvation are important.

Crown Ethers and Cryptands Some metal cations form stable complexes with a class of synthetic ionophores known as *crown ethers,* which were first prepared in 1967. **Crown ethers** are heterocyclic ethers containing a number of regularly spaced oxygen atoms. Some examples of crown ethers are the following:

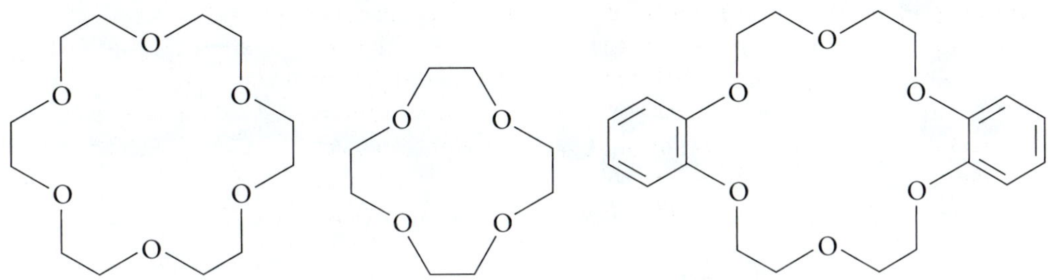

[18]-crown-6 **[12]-crown-4** **dibenzo[18]-crown-6**

(8.80)

(The number in brackets indicates the total number of atoms in the ring, and the number following the hyphen indicates the number of oxygens.) The term *crown* was suggested by the three-dimensional shape of these molecules, shown in **Fig. 8.11** for the complex of [18]-crown-6 with the potassium ion (K^+). The oxygens of the "host" crown ether wrap around the "guest" metal cation, sequestering it within the cavity of the ether using the donor and charge–dipole interactions discussed in the previous section. In fact, you can think of a crown ether molecule

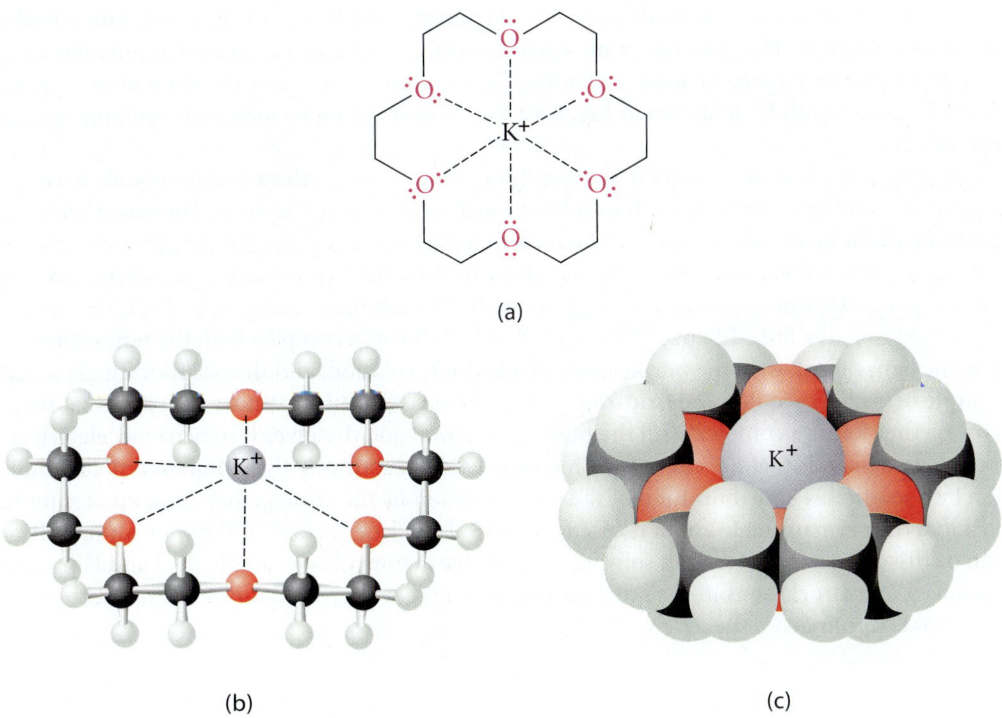

FIGURE 8.11 Structure of the [18]-crown-6 complex of the potassium ion. (a) A Lewis structure. (b) A ball-and-stick model. (c) A space-filling model. The oxygen atoms are colored red. Because the outside of the complex is essentially a hydrocarbon, crown ethers and their complexes are soluble in hydrocarbon solvents. The dashed lines indicate donor and/or charge–dipole interactions of the oxygens with the ion.

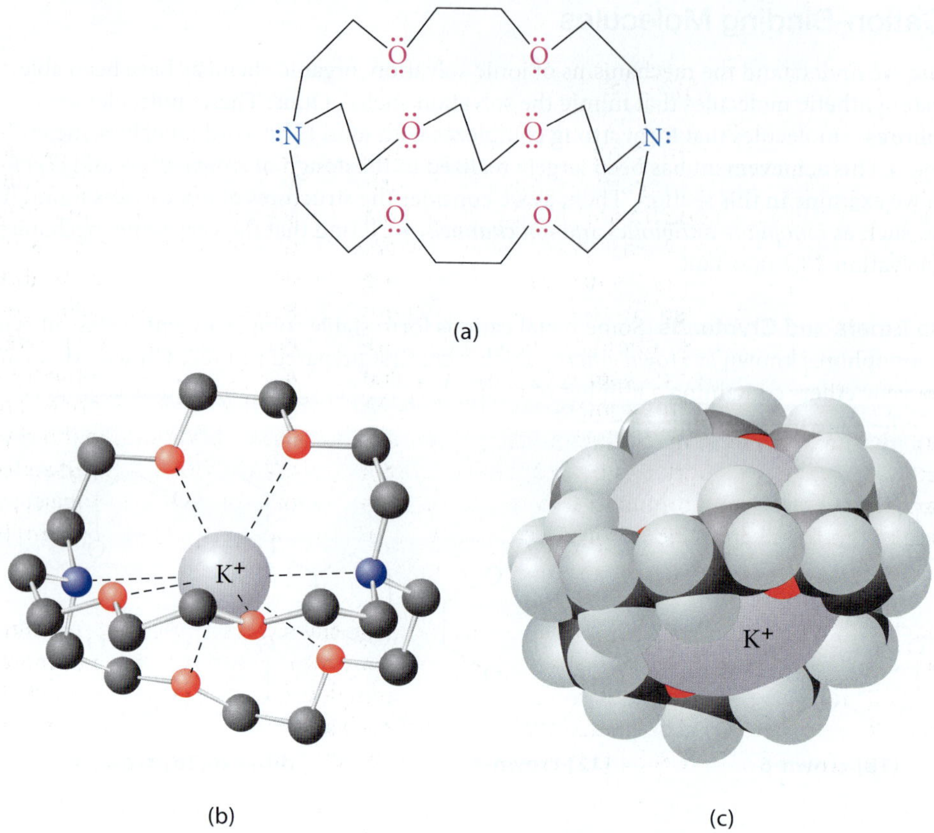

FIGURE 8.12 (a) A skeletal structure of [2.2.2]-cryptand. (The numbers refer to the number of oxygens in each of the three chains.) (b) A ball-and-stick model of a cryptate, a [2.2.2]-cryptand containing a bound potassium ion. (The hydrogen atoms have been omitted.) (c) A space-filling model of the cryptate in (b) with hydrogens shown. The dashed lines indicate donor and/or charge–dipole interactions of the oxygens (red) or nitrogens (blue) with the ion.

as a "synthetic solvation shell" for a cation. Because the metal ion must fit within the cavity, the crown ethers have some selectivity for metal ions according to size. For example, [18]-crown-6 forms the strongest complexes with K^+ and somewhat weaker complexes with Na^+, Cs^+, and Rb^+. It does not form complexes with Li^+. On the other hand, [12]-crown-4, with its smaller cavity, specifically forms complexes with Li^+.

Closely related to the crown ethers are the **cryptands**, which are nitrogen-containing analogs of the crown ethers. The presence of nitrogen allows for a bicyclic structure that provides an additional pair of oxygens to assist in binding the metal ion. The structure of a typical cryptand and its complex with K^+ is shown in **Fig. 8.12**. (Complexes of metal ions and cryptands are called *cryptates*.)

Because their structures contain hydrocarbon groups, crown ethers and cryptands have significant solubilities in hydrocarbon solvents such as hexane or benzene. The remarkable thing about the crown ethers and cryptands is that they can cause inorganic salts to dissolve in solvents in which these salts ordinarily have little or no solubility. For example, when potassium permanganate ($KMnO_4$) is added by itself to the hydrocarbon benzene, no $KMnO_4$ dissolves. Upon addition of a little dibenzo[18]-crown-6, which forms a complex with the potassium ion, the benzene takes on the purple color of a $KMnO_4$ solution, and this solution (nicknamed "purple benzene") acquires the oxidizing power typical of $KMnO_4$. What happens is that the crown ether forms a complex with the potassium cation and dissolves it in benzene; electrical neutrality demands that the colored permanganate ion accompany the complexed potassium ion into solution. The stabilization of the potassium ion by the crown ether compensates for the fact that the permanganate anion is essentially unsolvated, or "naked." Other potassium salts can be dissolved in hydrocarbon solvents in a similar manner. For example, KCl and KBr can be dissolved in hydrocarbons in the presence of crown ethers to give solutions of "naked chloride" and "naked bromide," respectively.

Host–Guest Chemistry

Crown ethers and cryptands can discriminate among various cations on the basis of *ionic size*. As a result, crown ethers and cryptands bind ions with a degree of *selectivity*. In recent years, chemists have designed other classes of molecules that can "recognize" and bind more complicated molecules on the basis of their precise structures. This type of work has been spurred, at least in part, by a desire to understand and duplicate synthetically the highly specific binding characteristic of such biological molecules as enzymes and receptors. This general field, nicknamed *host–guest chemistry*, *supramolecular chemistry,* or *molecular recognition*, was recognized with the 1987 Nobel Prize in Chemistry, which was awarded to three of its pioneers: Charles J. Pedersen (1904–1989), then a chemist with DuPont, who invented the crown ethers; Jean-Marie Lehn (b. 1939), a professor of chemistry at Université Louis Pasteur in Strasbourg, France, and the Collège de France in Paris, who devised the cryptands; and Donald J. Cram (1919–2001), a professor of chemistry at the University of California, Los Angeles, who devised a wide variety of other molecules that form host–guest complexes.

Ionophore Antibiotics The *ionophore antibiotics* are biologically important examples of ionophores. An *antibiotic* is a compound that interferes with the growth or survival of one or more microorganisms. The ionophore antibiotics form strong complexes with metal ions in much the same way as crown ethers and cryptands. An example of such a compound is nonactin, one of a group of antibiotics produced by a microorganism, *Streptomyces griseus.*

Chemical Biology Topic

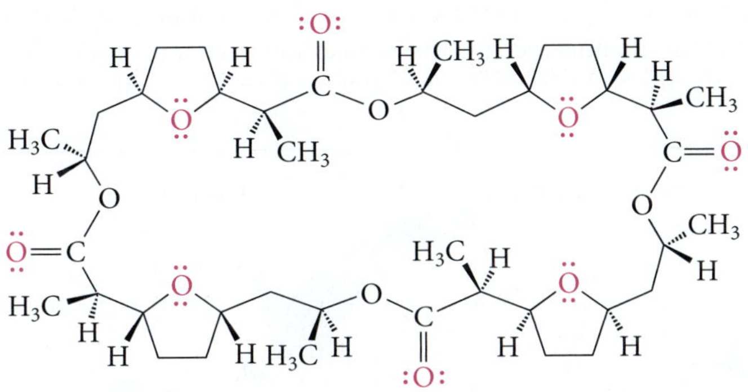

nonactin (8.81)

Nonactin has a strong affinity for the potassium ion. As shown in **Fig. 8.13**, the molecule contains a cavity in which eight of the oxygen atoms (red in Display 8.81) form a complex with the ion. In contrast, the atoms on the outside of the nonactin molecule are for the most part hydrocarbon groups. Recall from Sec. 8.7A that the interior of biological membranes consists of a phospholipid bilayer. This hydrocarbonlike region provides a natural barrier to the passage of ions that allows the cell to maintain a different concentration of ions on the inside of the cell than is present in the surrounding milieu. However, the hydrocarbon surface of nonactin allows it to enter readily into, and pass through, membranes. Because nonactin binds and transports ions, the ion concentration difference, crucial to proper cell function, is destroyed and the cell dies.

Ion Channels Ion channels, or "ion gates," provide passageways for ions into and out of cells. (Recall that ions are not soluble in membrane phospholipids.) The controlled flow of ions is essential for the transmission of nerve impulses and for other biological processes. A typical channel is a large protein molecule imbedded in a cell membrane. Through various mechanisms, ion channels can be opened or closed to regulate the concentration of ions in the interior of the cell. Ions do not diffuse passively through an open channel; rather, an open channel contains regions that bind a specific ion. Such an ion is bound specifically within the channel at one side of the membrane and is expelled from the channel on the other side.

Chemical Biology Topic

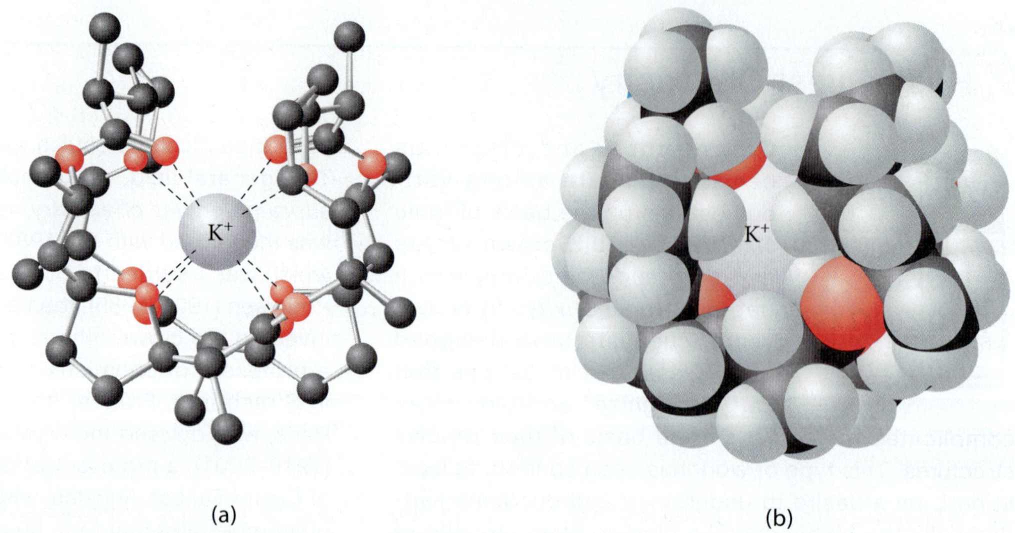

FIGURE 8.13 Models of the complex of the antibiotic nonactin with potassium ion. (a) A ball-and-stick model in which the hydrogens have been omitted. The dashed lines indicate interaction between oxygen atoms (red) and the potassium ion. (b) A space-filling model in which the hydrogens are shown. The nonactin molecule wraps around the potassium ion like a hand holding a ball. Because the outside of the complex is essentially hydrocarbon in character, the complex, like the crown ether–metal ion complexes (see Fig. 8.11), is soluble in nonpolar aprotic solvents.

Remarkably, the structures of the ion-binding regions of these channels have much in common with the structures of ionophores such as nonactin. The first X-ray crystal structure of a potassium-ion channel was determined in 1998 by a team of scientists at Rockefeller University led by Professor Roderick MacKinnon (b. 1956), who shared the 2003 Nobel Prize in Chemistry for this work. A schematic diagram of this protein and how it is situated in a membrane is shown in **Fig. 8.14a**. The exterior of the channel interacts with the hydrocarbon part of the membrane phospholipid bilayer. As we might expect from our discussion of noncovalent interactions, the groups on this part of the channel are largely hydrocarbon in character; the channel remains anchored in the membrane by hydrophobic bonding and van der Waals attractions. The entrance

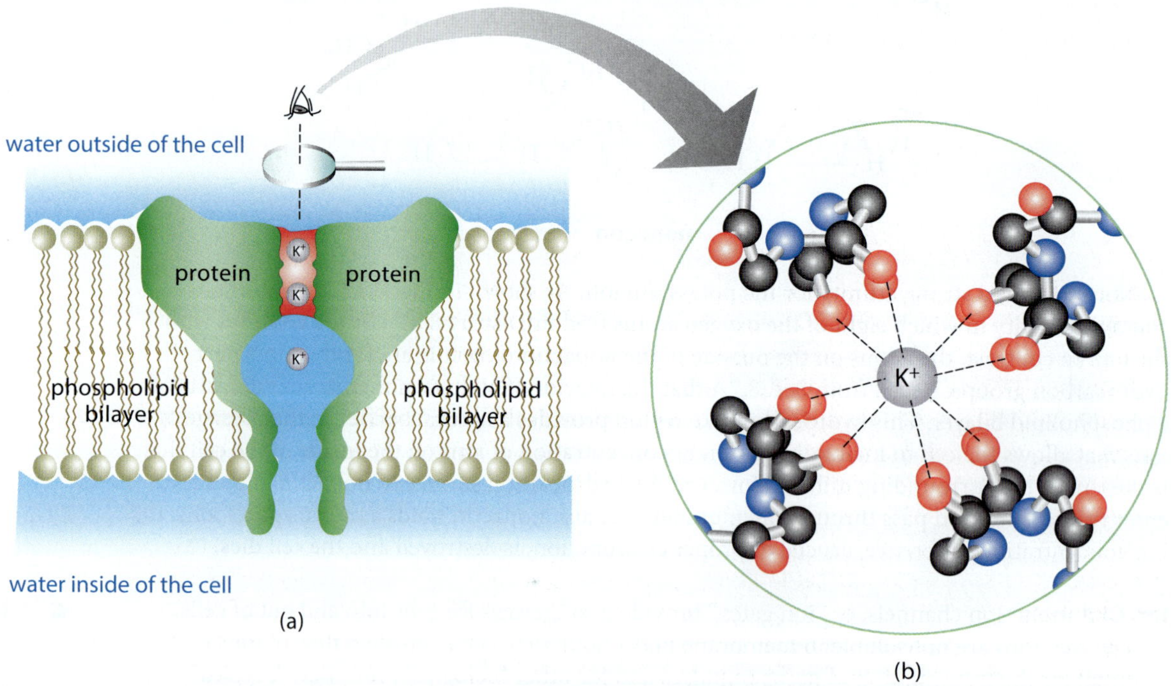

FIGURE 8.14 An open potassium ion channel. (a) A schematic cross-sectional diagram showing the location of the channel in the cell membrane. The protein is color-coded green. The protein spans the membrane, and much of the protein exterior contacts the lipid bilayer. The red region is the selectivity filter, which is the region of the protein that binds the potassium ion selectively. (b) An atomic-level view from above (eyeball) of the selectivity filter containing one potassium ion. The oxygens (*red*) are derived from C═O bonds of the protein (double bonds are not shown). Carbons (*black*) and nitrogens (*blue*) of the protein are shown without their attached hydrogens. The dashed lines indicate the ion–dipole and donor interactions of the oxygens with the ion.

to the channel on the outside of the cell contains binding sites for two potassium ions; these sites are oxygen-rich, much like the interior of nonactin. The oxygens come from C=O groups within the protein structure. They are situated so that they perfectly accommodate a potassium ion and are too far apart to interact effectively with a sodium ion. This part of the channel is called the "selectivity filter." A magnified view of the selectivity filter from above at the atomic level is shown in **Fig. 8.14b**. The oxygens in the selectivity filter provide "solvation" for the potassium ion that replaces the solvation water of the ion in solution. Release of this solvation water undoubtedly provides an entropic drive for binding of the ion to the channel. As one ion moves down the selectivity filter, a second potassium ion can enter the channel. The repulsion between the two ions then balances the ion-binding forces, and one of the ions can then leave the channel; this is postulated to be the mechanism of ionic conduction.

Focused Problem

8.27 The crown ether [18]-crown-6 (see Display 8.81) has a strong affinity for the methylammonium ion, $CH_3\overset{+}{N}H_3$. Propose a structure for the complex between [18]-crown-6 and $CH_3\overset{+}{N}H_3$.(Although the crown ether is bowl-shaped, you can draw a planar structure for the purposes of this problem.) Show the important interactions between the crown ether and the ion.

8.8 SUMMARY OF NONCOVALENT INTERMOLECULAR ATTRACTIONS

We've now seen a variety of noncovalent attractions and some of their consequences in both chemistry and biology:

1. Attractions between oppositely charged ions
2. Van der Waals attractions (dispersion forces)
3. Hydrogen bonding
4. Dipole–dipole attractions
5. Charge–dipole attractions

Although the attractions between ions can be very strong, the other attractive forces are all relatively weak. It is worth stating again that the strengths of these noncovalent attractions have *nothing* to do with the stabilities of the covalently bonded molecules involved in those interactions. The stabilities of individual molecules (as measured by heats of formation) result from the stabilities of their covalent bonds. Because melting, boiling, and dissolving compounds in solvents—processes that disrupt noncovalent intermolecular attractions—do *not* involve breaking and forming covalent bonds, noncovalent attractions must be *much weaker* than the covalent bonds.

In addition, the entropy component of noncovalent attractions can be a significant driver of processes that involve noncovalent forces, such as melting of a crystalline solid and assembly of cell membranes.

Despite the magnitudes of these "weak" noncovalent forces, they have chemically significant consequences. As you've seen, the structures of membranes, and as you will learn, the structures of proteins and DNA, are a consequence of *many weak noncovalent attractions acting together.* It's a bit like an organization in which all the members work together toward the same objectives: the efforts of a single individual may seem relatively insignificant, but all of the individuals cooperating and working together can have powerful consequences.

You will also learn in Sec. 12.8E that the biological function of enzymes and receptors hinges in part on the operation of noncovalent attractions.

To summarize, in this chapter we've shown that noncovalent attractions are a consequence of the *structures of individual molecules*, which determine how they interact with each other. These interactions determine, in turn, many of the important physical properties of chemical compounds: boiling points, melting points, and solubilities. Therefore, we have established a very important connection:

(8.82)

Focused Problem

8.28 Although we've placed noncovalent attractions into five categories, it has been claimed that all of them are fundamentally the same, because they are essentially electrostatic in nature—that is, they deal with the attractions between opposite electrical charges.

(a) For the list of five attractions given in this section, justify this view.

(b) If all of these attractions are *fundamentally* the same, how is each one different from the other?

CHAPTER SUMMARY *For a summary of the chapter, see Chapter 8 in the* Study Guide and Solutions Manual.

SKILLS OBJECTIVES WITH PROBLEMS

- **Predict and/or explain trends in bond lengths and bond angles for alcohols, ethers, alkyl halides, thiols, and sulfides. 8.3**

8.29 (a) Within the following compounds, which asterisked (*) bond is the longest? The shortest? Explain how you know.

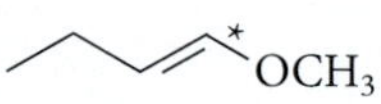

A

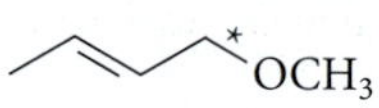

B

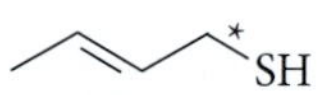

C

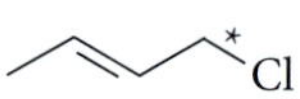

D

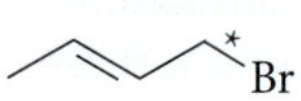

E

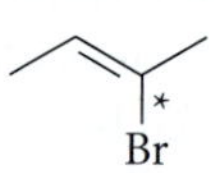

F

(b) Which one of the following has the smallest bond angle with O or S as the central atom? Explain.

methanethiol (*A*) methanol (*B*) dimethyl sulfide (*C*)

- **Classify alcohols and alkyl halides as primary, secondary, or tertiary. 8.1**

8.30 In the following group of structures, identify the one(s) (if any) that meet the following criteria.

(a) Primary alkyl halides

(b) Secondary alcohols

(c) Primary alcohols

(d) Tertiary alkyl halides

(e) Tertiary alcohols

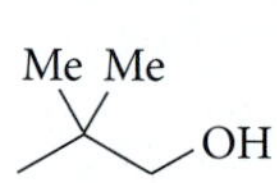

A

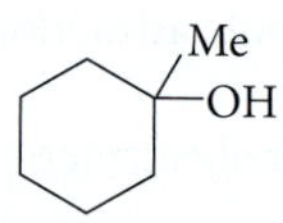

B

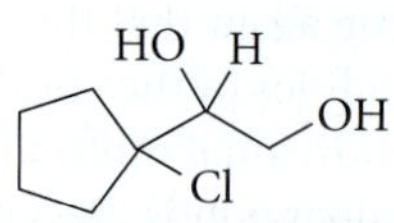

C

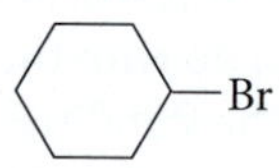

D

- **Provide common names for simple alkyl halides, alcohols, and ethers; given the common name, draw the structure. 8.2**

8.31 Give the common name for each of the following compounds.

(a) 

(b) 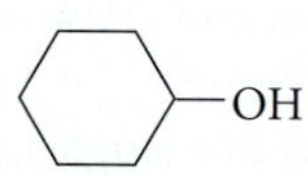

(c) 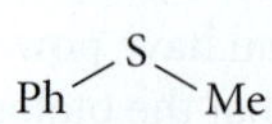

8.32 Draw a structure for the compound with each of the following names.

(a) Isopropyl methyl sulfide

(b) Methyl vinyl ether

(c) Di-*tert*-Butyl ether

(d) Allyl fluoride

• Provide IUPAC substitutive names for alcohols, alkyl halides, ethers, thiols, and sulfides; given the IUPAC name, draw the structure. 8.2

8.33 Give the IUPAC substitutive name for each of the following compounds, which have been used as general anesthetics.

(a)

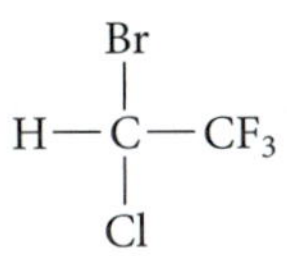

halothane

(b) $Cl_2CH—CF_2—OCH_3$

methoxyflurane

8.34 Thiols of low molecular mass are known for their extremely foul odors. In fact, the following two thiols are the active components in the scent of the skunk. Give the IUPAC substitutive names for these compounds.

(a)

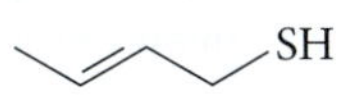

(b)

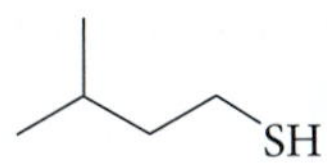

8.35 Name the following alcohol with an IUPAC substitutive name. (Ignore stereochemistry.)

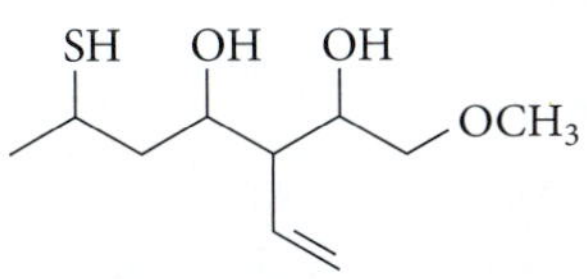

8.36 Provide a structure for the compound with each of the following names.

(a) (*S*)-2-Methyl-3-buten-1-ol

(b) 4-Mercapto-2-butyne-1-ol

• Name compounds containing the following oxygen and sulfur heterocycles: furan, thiophene, tetrahydrofuran (oxolane), 1,4-dioxane, and oxirane; given the name, draw a structure.

8.37 Name each of the following compounds.

(a)

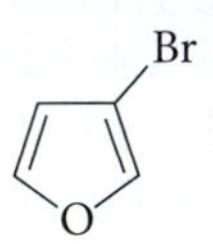

(b)

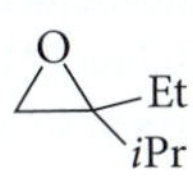

8.38 Provide a structure for a compound with each of the following names.

(a) 2-Methylthiophene

(b) *meso*-2,5-Dimethyloxolane (or *meso*-2,5-dimethyltetrahydrofuran)

(c) 2-(3-Furyl)-1-ethanol (*Hint:* The 3-furyl group is a furan ring connected at the 3-position.)

• Predict the relative order of boiling points, melting points, and solubilities of different compounds, and justify your choice with an explanation involving noncovalent intermolecular interactions. 8.5, 8.6F

8.39 Without consulting tables, arrange the compounds within each of the following sets in order of increasing boiling point, and give your reasoning.

(a) 1-Hexanol, 2-pentanol, *tert*-butyl alcohol

(b) 1-Hexanol, 1-hexene, 1-chloropentane

(c) Diethyl ether, propane, 1,2-propanediol, 1,2-propanethiol

(d) Cyclooctane, chlorocyclobutane, cyclobutane

8.40 Provide the information requested in each part, and explain your answer using noncovalent intermolecular interactions.

(a) Arrange the following compounds in order of increasing boiling point, and explain.

hexane, octane, pentane, neopentane

(b) Which compound has a melting point hundreds of degrees higher than the other, even though the two have about the same molecular mass?

chloroethane, sodium chloride

(c) Which compound has the higher boiling point?

(*Z*)-3,4-dichloro-3-hexene,
(*E*)-3,4-dichloro-3-hexene

(d) Arrange the following compounds in order of increasing solubility in water.

1,2-dichloroethane, 1-butanol, 3-octanol

• Classify a compound as a hydrogen-bond donor, a hydrogen-bond acceptor, or neither. For donors and acceptors, identify the donor and/or acceptor sites. 8.5C

8.41 Classify each of the following compounds as a hydrogen-bond donor, a hydrogen-bond acceptor, or neither.

For donors and acceptors, identify the donor and/or acceptor sites.

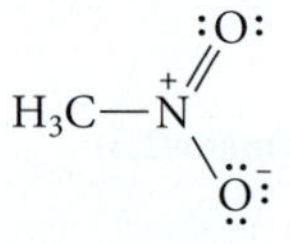

A

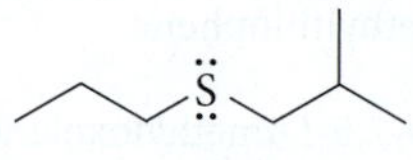

B

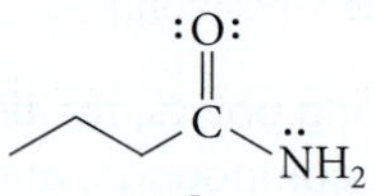

C

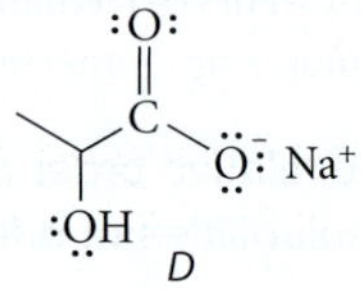

D

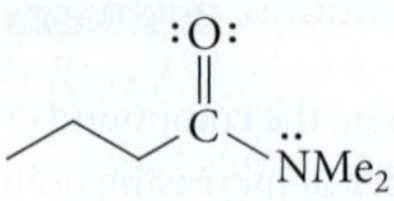

E

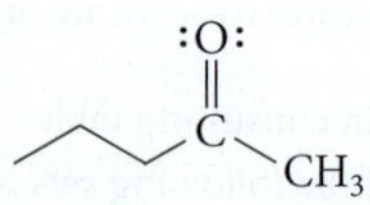

F

• Apply symmetry or other criteria to decide which of two (or more) isomers should have the greater melting point. 8.5D

8.42 In each of the following pairs, one compound has a melting point that is much higher than the other. In parts (a) and (b), the boiling points of the two compounds are similar. Choose the compound with the greater melting point, and explain your reasoning.

(a) (Reported melting points: 62–65 °C, 17 °C.)

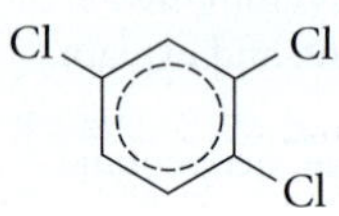

A

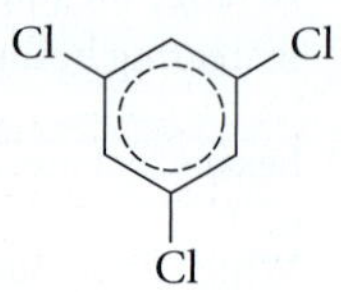

B

(b) (Reported melting points: 61–62 °C, 21 °C.)

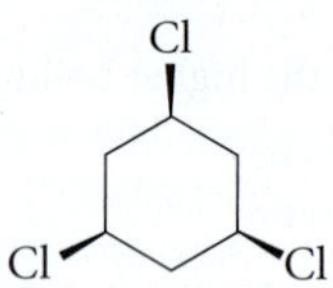

A

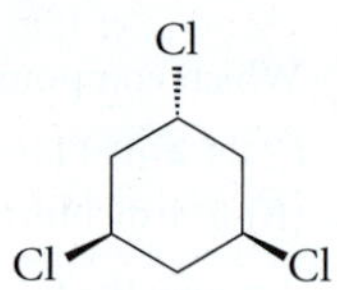

B

(c) (Reported melting points: 35.4 °C and −34.5 °C; reported boiling points: 161 °C and 185 °C. *Hint:* Compare the enthalpies and entropies of fusion with those of neopentane and pentane; see Display 8.48, Sec. 8.5D.)

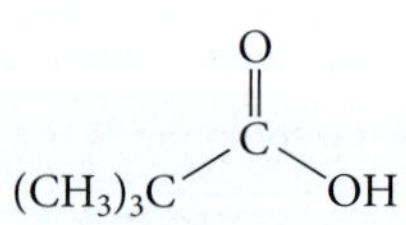

pivalic acid
$\Delta H_m = 2.1$ kJ mol^{-1} (0.50 kcal mol^{-1})
$\Delta S_m = 6.80$ J mol^{-1} K^{-1} (1.63 cal mol^{-1} K^{-1})

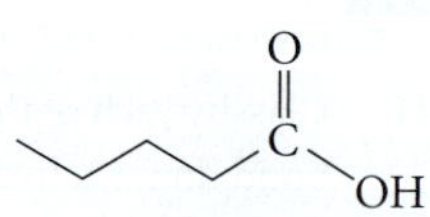

valeric acid
$\Delta H_m = 14.1$ kJ mol^{-1} (3.37 kcal mol^{-1})
$\Delta S_m = 59.1$ J mol^{-1} K^{-1} (14.1 cal mol^{-1} K^{-1})

• Given their structure and dielectric constant, classify solvents as protic/aprotic, donor/nondonor, and polar/apolar. 8.6B

8.43 Classify each of the following solvents as protic/aprotic, donor/nondonor, and polar/apolar.

(a) CF_3CH_2OH

2,2,2-trifluoroethanol
$\varepsilon = 26$

(b)

1-hexene
$\varepsilon = 2$

(c)

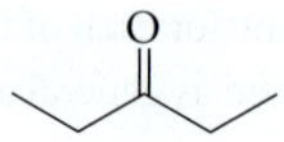

3-pentanone
$\varepsilon = 17.5$

(d)

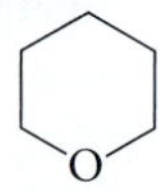

tetrahydropyran
$\varepsilon = 5.7$

(e) A 50% by volume mixture of tetrahydrofuran (THF, see Table 8.2) and water

• Apply the "like-dissolves-like" criterion to predict which solvent would be best for a given compound. 8.6C

8.44 In which of the two solvent choices given would each compound have the greater solubility? Explain your reasoning.

(a) Water or hexane

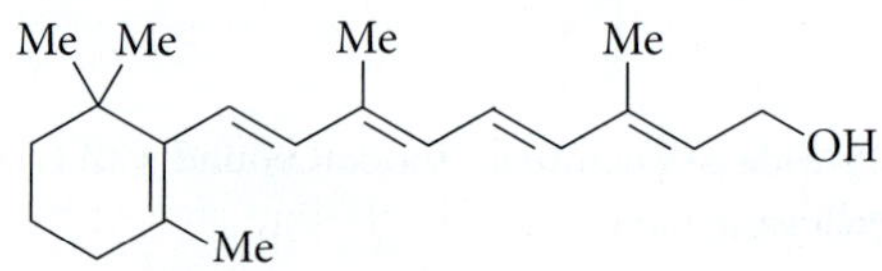

retinol

(b) Hexane or ethanol

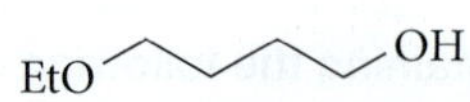

4-ethoxy-1-butanol

• Predict, among several choices, the best solvent for an ionic compound and explain. 8.6F

8.45 Which of the following solvents would be the best for dissolving solid crystalline KCl. (See Table 8.2, Sec. 8.6B, for the structures of solvent molecules.)

acetone ($\varepsilon = 21$), hexane ($\varepsilon = 2$), dimethylsulfoxide ($\varepsilon = 47$), or formic acid ($\varepsilon = 59$)

● **For cases similar to the ones discussed in this chapter, predict the relative solubilities of two solids, and give your reasoning. 8.5D, 8.6E**

8.46 In each part, which compound would have the greatest solubility in the solvent indicated?

(a) Solubility in water

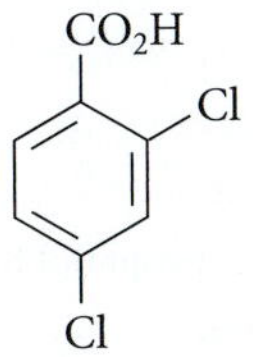

A

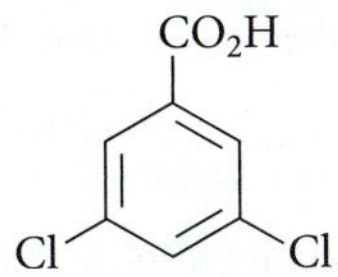

B

(b) Same compounds as in part (a), but solubility in acetonitrile

(c) Solubility in water

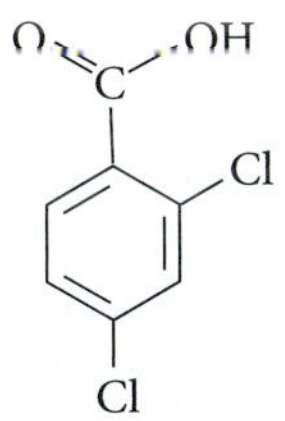

A

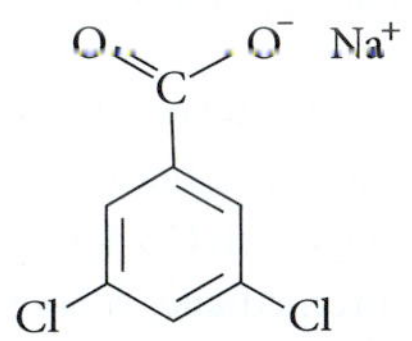

B

(d) Solubility in water at 25 °C

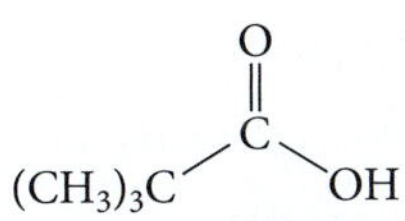

pivalic acid
mp = 35.4 °C

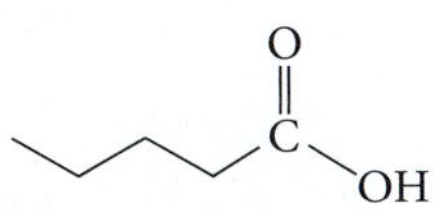

valeric acid
mp = −34.5 °C

● **Use your knowledge of noncovalent interactions involved in certain phenomena that are important in biology or medicine. 8.6, 8.7**

8.47 Offer explanations for each of the following observations.

(a) Ethanol and 1-propanol disrupt the lipid bilayer of cell membranes.

(b) Sugars such as glucose do not diffuse freely through the cell membrane.

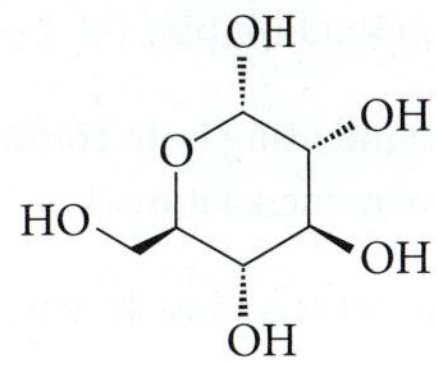

α-D-glucopyranose
(a form of glucose)

(Sugars require specific protein transporters to enter cells.)

(c) Detergents such as sodium dodecyl sulfate (SDS) are used to disrupt cell membranes.

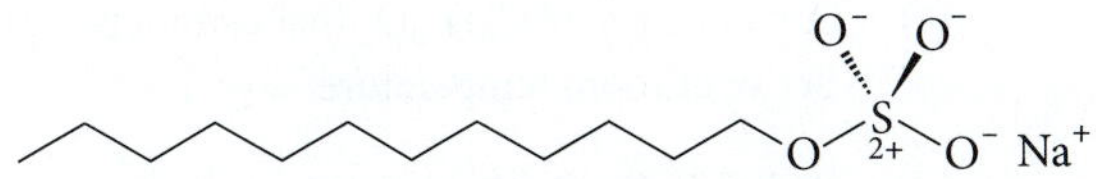

sodium dodecyl sulfate (SDS)

(d) *Propofol* is used as a general anesthetic in the first stages of surgery.

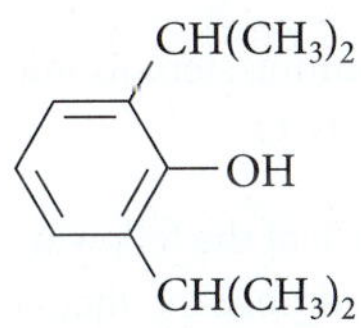

2,6-diisopropylphenol
(propofol)

Propofol is insoluble in water, but it has to be administered by intravenous injection. For this reason, propofol is formulated as a mixture with soybean oil and lecithin. (This milky formulation is sometimes nicknamed "milk of amnesia" by anesthesiologists.) (*Hint:* See the discussion of fats and oils in the sidebar at the end of Sec. 8.5D and the structure of lecithin in Sec. 8.7B.)

(e) The $\Delta S°$ for transfer of the salt $Bu_4N^+Cl^-$ from water to DMSO is strongly positive ($+130\ J\ K^{-1}\ mol^{-1}$, $+31\ cal\ K^{-1}\ mol^{-1}$). (*Hint:* Consider solvation of the cation.)

(f) Diethyl ether (Et—O—Et, bp 34.5 °C) is significantly more soluble in water (about 3.4 times) than 1-pentanol ($CH_3CH_2CH_2CH_2CH_2$—OH, bp 138 °C), the primary alcohol with about the same molecular mass and shape. (*Hint:* What do the relative boiling points imply about the interactions between molecules in the pure liquids?)

INTEGRATED PROBLEMS

8.48 (a) Give the structures of all alcohols with the molecular formula $C_5H_{11}OH$.

(b) Which of the compounds in part (a) are chiral?

(c) Name each compound using IUPAC substitutive nomenclature.

(d) Classify each as a primary, secondary, or tertiary alcohol.

(e) Identify the α-carbon on each structure.

8.49 (a) Give the structures of all ethers with the molecular formula $C_5H_{12}O$.

(b) Which of the compounds in part (a) are chiral?

(c) Name each compound using both common and IUPAC substitutive nomenclature.

(d) Identify the two α-carbons in each structure.

8.50 Give a structure for each of the following compounds. (In some cases, more than one answer is possible.)

(a) A chiral ether $C_5H_{10}O$ that has no double bonds

(b) A chiral alcohol C_4H_6O

(c) A vicinal glycol $C_6H_{10}O_2$ that cannot be optically active at room temperature

(d) A diol $C_4H_{10}O_2$ that exists in only three stereoisomeric forms

(e) A diol $C_4H_{10}O_2$ that exists in only two stereoisomeric forms

(f) The six epoxides (counting stereoisomers) with the molecular formula C_4H_8O

8.51 Give an IUPAC name for each of the following compounds, which may have been isolated from the shoes of a tennis player. Ignore stereochemistry in part (a).

(a)

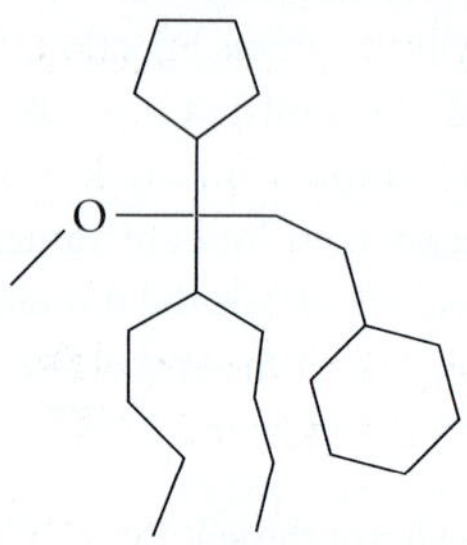

(b)

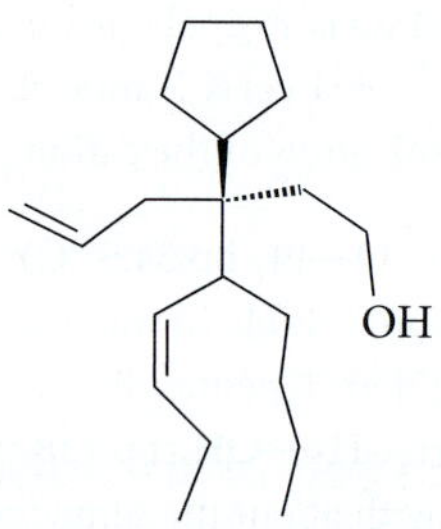

8.52 In each of the following parts, explain why the first compound has a higher boiling point than the second, despite a lower molecular mass.

(a)

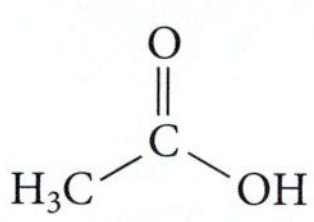

(bp 118 °C)

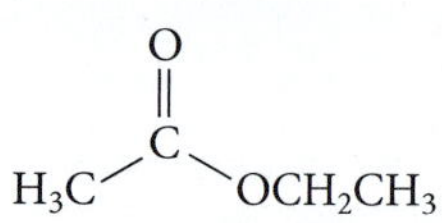

(bp 77 °C)

(b)

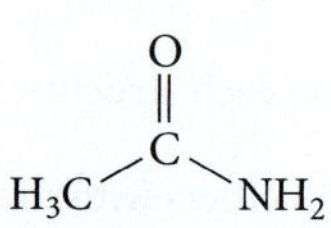

(bp 221 °C)

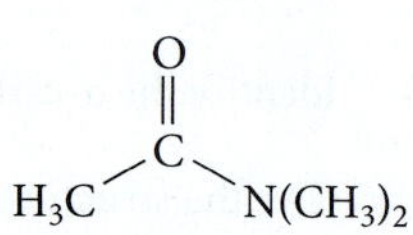

(bp 166 °C)

8.53 Suggest a structure for a constitutional isomer of the following compound that should have *greater* water solubility, and explain your reasoning. The structure should *not* be an enol (Display 8.4, Sec. 8.1), because enols are unstable. (Several correct answers are possible.)

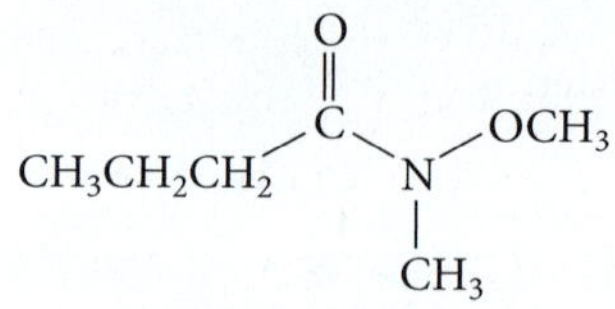

8.54 The molecules nitromethane and 2-propanol have roughly the same shape and molecular mass.

	2-propanol (isopropyl alcohol)	**nitromethane**
molecular mass	60.1	61.0
boiling point	82.4 °C	101.2 °C

Liquid 2-propanol contains hydrogen bonds, but liquid nitromethane does not. Yet nitromethane has the higher boiling point. Offer a reason for the high boiling point of nitromethane. What physical properties of the two molecules could you look up to support your answer?

8.55 (a) One of the following compounds is an unusual example of a salt that is soluble in hydrocarbon solvents. Which one is it? Explain your choice.

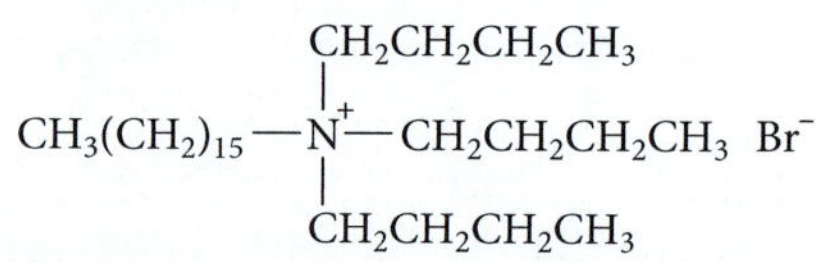

A

$\overset{+}{N}H_4$ Cl^-

B

(b) Which of the following would be present in greater amount in a hexane solution of the compound you identified in part (a): separately solvated ions, or ion pairs and higher aggregates? Explain your reasoning.

8.56 Normally, dibutyl ether is much more soluble in benzene than it is in water. Explain why this ether can be extracted from benzene into water if the aqueous solution contains moderately concentrated nitric acid. (*Hint:* The oxygen of ethers is weakly basic.)

8.57 Show with structures the types of solvation interaction that might be expected in each solution. Consider the solvation of both the cation and the anion.

(a) A solution of the salt ammonium chloride in ethanol

(b) A solution of the salt sodium chloride in *N*-methylpyrrolidone (see Table 8.2)

8.58 Which *one* of the following compounds would be expected to have a lower boiling point than isopropyl alcohol (dipole moment $\mu = 1.7$ D)? Explain your choice.

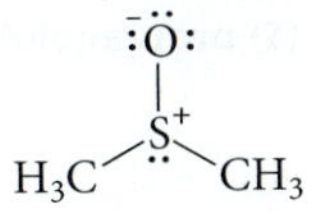

DMSO
($\mu = 3.96$)

$(CH_3)_2CH—SH$
isopropyl mercaptan
($\mu = 1.53$)

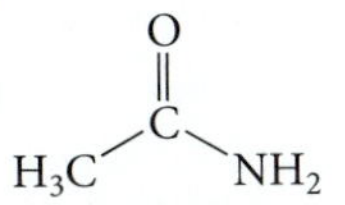

acetamide
($\mu = 3.60$)

8.59 The effectiveness of barbiturates as sedatives is directly related to their solubility in, and therefore their ability to penetrate, the lipid bilayers of membranes. Which of the following two barbiturate derivatives should be the more potent sedative? Explain.

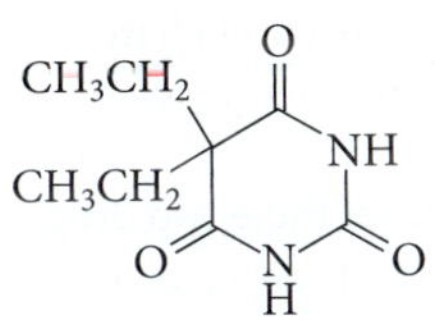

barbital

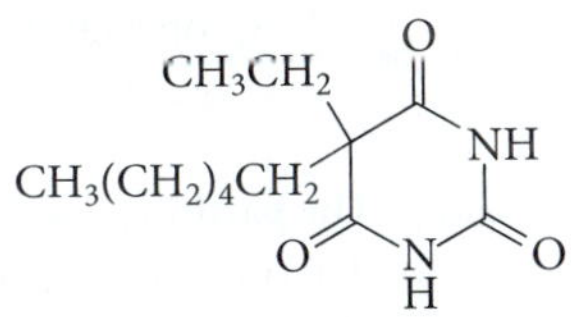

hexethal

8.60 When salad oil (see the structure of an oil in the sidebar at the end of Sec. 8.5D) is mixed with water and shaken, two layers quickly separate: the oily layer on top and the water layer on the bottom. When an egg yolk (which is rich in lecithin, a phospholipid) is added and the mixture shaken, an emulsion is formed. This means that the oil is suspended (not dissolved) in water as tiny particles that no longer form a separate layer. Explain the action of the lecithin. (Addition of romaine lettuce, garlic, croutons, and Parmesan cheese yields a Caesar salad!)

8.61 Vitamins can be classified as "fat-soluble" or "water-soluble." Fat-soluble vitamins can be stored in fatty tissues, whereas water-soluble vitamins can be excreted in the urine.

(a) The structures of some vitamins are given in **Fig. P8.61**. Using their structures to guide you, classify each as fat-soluble or water-soluble, and explain. (The structure of a fat is similar to that of an oil; see the sidebar at the end of Sec. 8.5D.)

(b) For which type of vitamin would an overdose be more dangerous? Why?

8.62 The dissociation constant K_d of the complex between a crown ether and a metal ion M^+ is given by

$$K_d = \frac{[\text{crown ether}][M^+]}{[\text{crown ether–}M^+\text{ complex}]}$$

Explain why the complex of the crown ether [18]-crown-6 with potassium ion has a much larger dissociation constant in water than it does in ether.

8.63 Nonactin (structure in Display 8.80, Sec. 8.7C) forms a strong complex with the ammonium ion, $^+NH_4$. What types of interactions are expected between the nonactin molecule and the bound ion? Contrast these interactions with those between nonactin and the bound potassium ion (see Fig. 8.13, Sec. 8.7C).

8.64 Ethyl alcohol in the solvent CCl_4 forms a hydrogen-bonded complex with an equilibrium constant $K_{eq} = 11$.

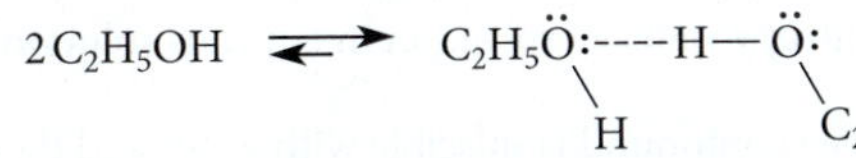

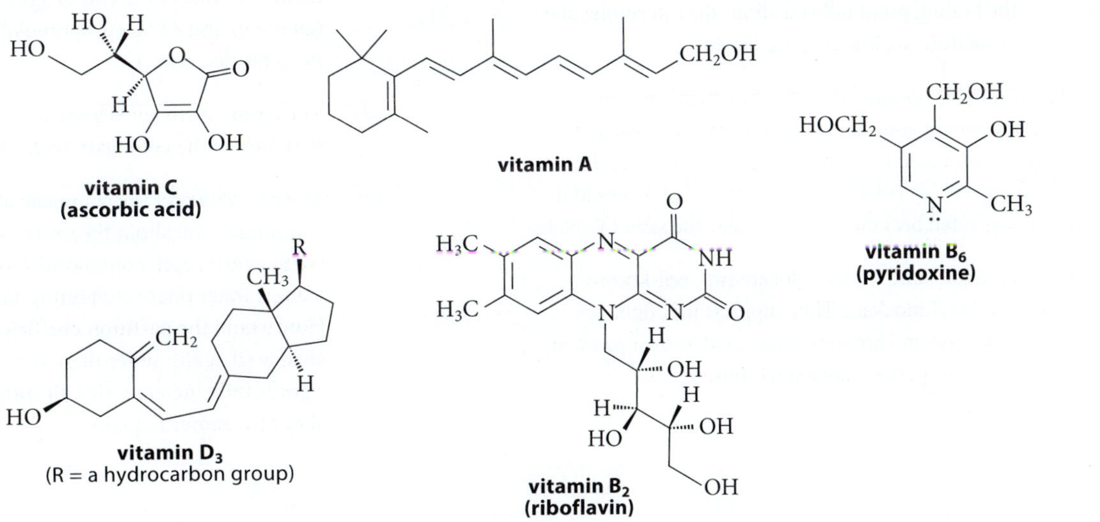

FIGURE P8.61

(a) What happens to the concentration of the complex as the concentration of ethanol is increased? Explain.

(b) What is the standard free-energy change for this reaction at 25 °C?

(c) If 1 mol of ethanol is dissolved in 1 L of CCl_4, what are the concentrations of free ethanol and of the complex?

(d) The equilibrium constant for the analogous reaction of ethanethiol is 0.004. Which forms stronger hydrogen bonds: thiols or alcohols?

(e) Which would be more soluble in water: $CH_3OCH_2CH_2SH$ or its isomer $CH_3SCH_2CH_2OH$? Explain.

8.65 If the "sawtooth" pattern of melting point behavior is explained by the efficiency of crystal packing, what would you expect to find if we were to plot the density of the *crystal* (solid) form of unbranched alkanes against carbon number?

8.66 The boiling points of *tert*-butyl alcohol (2-methyl-2-propanol) and 1-butanol differ by 36 °C (82 °C and 118 °C). Before proceeding, draw their structures.

(a) Which has the higher boiling point? Give a reason for this difference.

(b) How would the differences in their structures affect the ΔS_s° (standard entropy of solution) contribution to their water solubilities? (*Hint:* Think about the surface areas of the two alkyl groups, and decide which would require the greater amount of low-entropy water of solvation in aqueous solution.)

(c) One compound is miscible with water and the other has a limited solubility of 8 mass percent. Which compound is miscible? Explain. (*Hint:* What does the boiling point tell you about the intermolecular attractions within the pure liquids?)

8.67 If hydrophobic bonding is driven by a highly positive ΔS°, how would you expect the equilibrium constant for an association reaction driven by hydrophobic bonding to change with increasing temperature? That is, would it become greater, become smaller, or stay the same? Explain.

8.68 (*S*)-Propranolol and (*S*)-atenolol are two well-known drugs used as β-blockers. They suppress the "fight-or-flight" response and are sometimes used to treat panic in examination or performance situations.

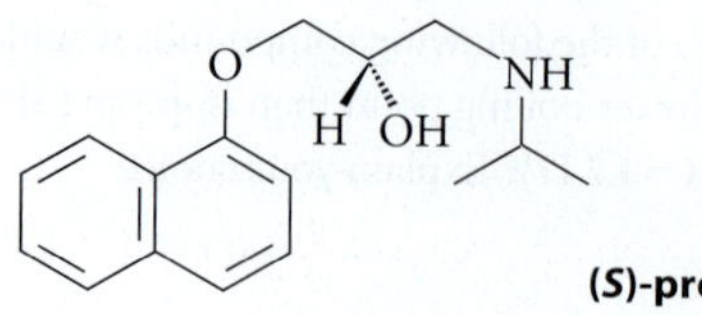

(*S*)-propranolol

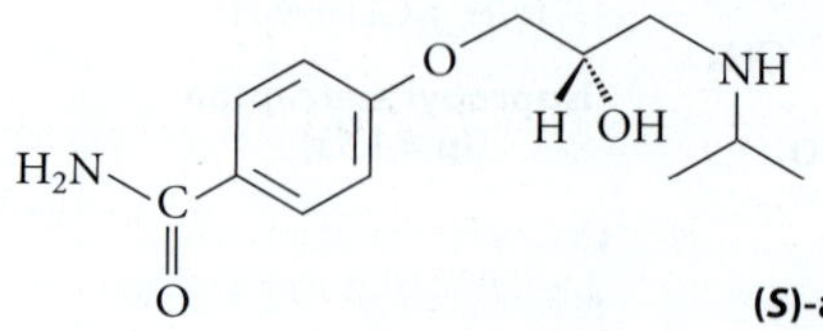

(*S*)-atenolol

Drugs such as these are often characterized by their *octanol–water partition coefficients* P_{ow} (Sec. 8.7B). A small amount of drug is dissolved in a known volume of 1-octanol and then the solution is shaken with a known amount of water. The two phases are allowed to separate, and the concentration of drug in each phase is determined. The octanol–water partition coefficient is calculated from the equilibrium concentrations in each phase:

$$P_{ow} = \frac{\text{drug concentration in 1-octanol phase}}{\text{drug concentration in water phase}}$$

(a) The partition coefficients of these two drugs were determined using water that was made basic to suppress protonation of the amine group. The partition coefficient of one drug is more than 2000 times (3.3 log units) greater than that of the other. Which drug has the greater partition coefficient? Explain.

(b) The partition coefficient of atenolol, expressed in logarithmic units, is $\log P_{ow} = 0.22$. If 1.0 g of atenolol is partitioned between equal volumes of the two phases, how much atenolol is in each phase?

(c) What does your answer to part (a) tell you, if anything, about the relative water solubilities of crystalline atenolol and propranolol? The reported melting points of the two drugs are 152–153 °C (atenolol) and 72 °C (propranolol). (*Hint:* See Eq. 8.68, Sec. 8.6E.)

(d) The amino nitrogen is basic in both drugs. Draw the structure of the conjugate acids of each.

(e) The pK_a values of the conjugate acids of both compounds are about the same (9.5). The partition coefficient of each compound was re-determined using a water phase containing a buffer at pH 7.4. How would the partition coefficient of these drugs change (if at all) under these conditions? Would it significantly increase, significantly decrease, or stay about the same? Explain.

(f) These drugs are marketed as their hydrochloride salts (their conjugate acids with a chloride counter-ion). The drugs are administered in the solid phase orally as capsules. What is one reason that the drugs are formulated in this form instead of in their basic form?

8.69 Offer an explanation for each of the following observations.

(a) Compound *A* exists mostly in a chair conformation with an equatorial OH group, but compound *B* prefers a chair conformation with an axial OH group.

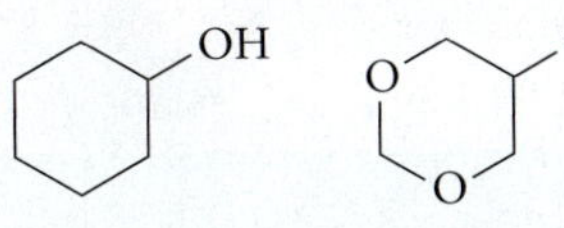

A *B*

(b) The racemate of 2,2,5,5-tetramethyl-3,4-hexanediol exists with a strong intramolecular hydrogen bond, but the meso stereoisomer has no intramolecular hydrogen bond.

8.70 (a) Use the relative bond lengths of the C—C and C—O bonds to predict which of the following two equilibria lies farther to the right. (That is, predict which of the two compounds contains more of the conformation with the axial methyl group.)

(1)

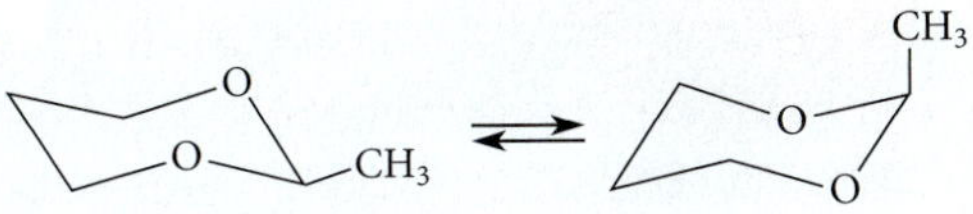

(2)

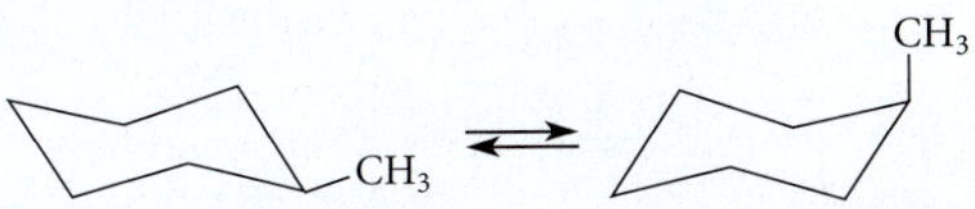

(b) Which one of the following compounds contains the greater amount of gauche conformation for internal rotation about the bond shown? Explain.

A *B*

9 THE CHEMISTRY OF ALKYL HALIDES

Substitution and Elimination Reactions

This chapter is devoted to two very important types of alkyl halide reactions: *nucleophilic substitution reactions* and *β-elimination reactions*. These are among the most common and important reactions in organic chemistry, and we cover them together because they frequently occur as competing reactions.

Compared with the total number of organic compounds, relatively few halogen-containing compounds occur naturally—around 5000. If you are a life science or premedical student, you might wonder why you should bother learning about the chemistry of compounds that you are not likely to encounter in biology. The reason is that much of our quantitative data and current understanding of reactivity comes from studies on alkyl halides. Moreover, the chemistry of alkyl halides demonstrates in a straightforward way the types of reactivity and mechanism that we'll encounter in more complex molecules, including those that occur in biological systems. In other words, *alkyl halides provide simple models from which we understand the chemistry of other compound classes*. These same considerations are equally valid for the chemistry major; furthermore, alkyl halides are important starting materials used in a wide variety of laboratory reactions.

9.1 OVERVIEW OF NUCLEOPHILIC SUBSTITUTION AND β-ELIMINATION REACTIONS

A. Nucleophilic Substitution Reactions

When a *methyl halide* or a *primary alkyl halide* reacts with a nucleophile, such as sodium ethoxide, a reaction occurs in which the nucleophilic atom of the base, in this case oxygen, displaces the halogen, which is expelled as halide ion.

$$\underset{\textbf{sodium ethoxide}}{Na^+\ CH_3CH_2\ddot{\underset{..}{O}}{:}^-} + \underset{\textbf{ethyl bromide}}{:\ddot{\underset{..}{Br}}{-}CH_2CH_3} \longrightarrow \underset{\textbf{diethyl ether}}{CH_3CH_2\ddot{\underset{..}{O}}{-}CH_2CH_3} + \underset{\textbf{sodium bromide}}{Na^+\ :\ddot{\underset{..}{Br}}{:}^-} \quad (9.1)$$

This is an example of an *electron-pair displacement reaction*, which was introduced in Sec. 3.2A. In this type of reaction, a *nucleophile* donates an electron pair to an *electrophile* to displace a *leaving group*.

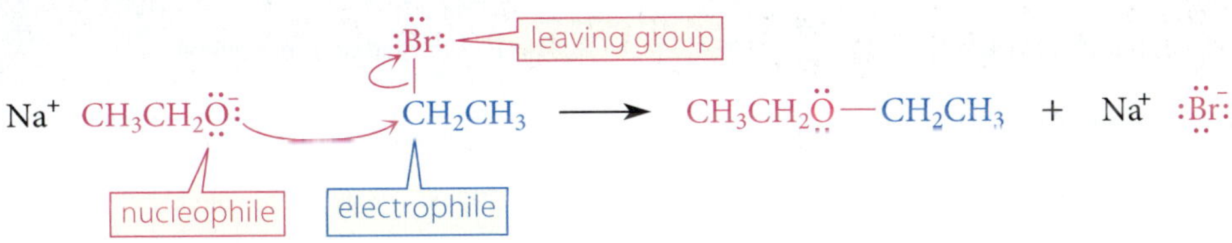

(9.2)

(You should review these terms in Sec. 3.2B if necessary.) In this chapter, the electrophilic center will typically be the α-carbon of an alkyl halide carbon and the leaving group will typically be a halide ion. Organic chemists refer to this type of reaction as a *nucleophilic substitution reaction*, or *nucleophilic displacement reaction*. A **nucleophilic substitution reaction** is an electron-pair displacement reaction by a nucleophile on an electrophilic center other than a hydrogen.

The nucleophile used in Eqs. 9.1 and 9.2, ethoxide ($CH_3CH_2O^-$, often abbreviated as EtO^-), is the conjugate base of ethanol (CH_3CH_2OH). Ethanol is a weak acid ($pK_a = 15.9$) that is somewhat less acidic than water ($pK_a = 14.0$). Therefore, ethoxide is a strong base—somewhat stronger than hydroxide ion. In general, the conjugate bases of alcohols are called **alkoxides**. You'll see many reactions in this chapter in which alkoxides are used as nucleophiles. (The acidity of alcohols and the preparation of alkoxides are covered more generally in Sec. 11.1.)

In Eqs. 9.1 and 9.2, the sodium ion is a *spectator ion*—an ion that has no overt role in the overall reaction. (Spectator ions were introduced in Sec. 3.1C; see Eq. 3.7.) Recall that a spectator ion can be removed from both sides of the chemical equation without changing the bond-making and bond-breaking chemistry. For simplicity, we often write nucleophilic substitution reactions (and elimination reactions that you'll learn about in Sec. 9.1B) as *net ionic equations*—equations in which only the reacting ionic components are shown and any spectator ions are omitted. For example, the net ionic equation corresponding to Eq. 9.1 is

$$\underset{\textbf{ethoxide ion}}{CH_3CH_2\ddot{\underset{..}{O}}{:}^-} + \underset{\textbf{ethyl bromide}}{:\ddot{\underset{..}{Br}}{-}CH_2CH_3} \longrightarrow \underset{\textbf{diethyl ether}}{CH_3CH_2\ddot{\underset{..}{O}}{-}CH_2CH_3} + \underset{\textbf{bromide ion}}{:\ddot{\underset{..}{Br}}{:}^-} \quad (9.3)$$

When you see a net ionic equation, you can assume always that a spectator ion is present—in many cases Na^+ or K^+.

Although many nucleophiles are anions, others are uncharged. Equation 9.4 contains an example of an uncharged nucleophile.

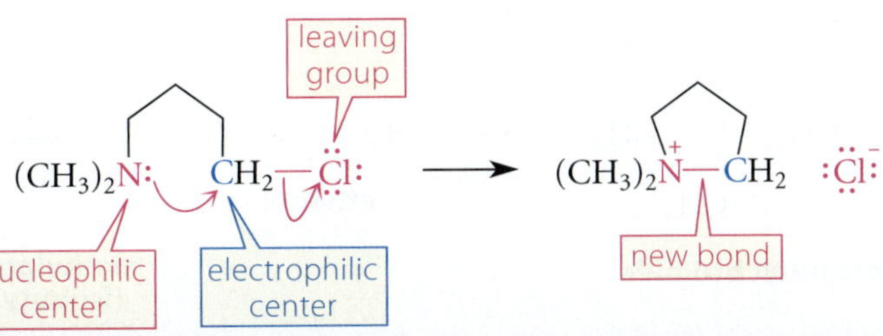

(9.4)

TABLE 9.1 Some Nucleophilic Substitution Reactions
(X = halogen or other leaving group; R = alkyl; R′ = alkyl or aryl)

R—X: + Nucleophile (name)		:X:⁻ + Product (name)
R—X: + :Y:⁻ (another halide)	⟶	:X:⁻ + R—Y: (another alkyl halide)*
+ :C≡N: ⁻ (cyanide)	⟶	+ R—C≡N: (nitrile)
+ :ÖH⁻ (hydroxide)	⟶	+ R—ÖH (alcohol)
+ :ÖR′⁻ (alkoxide)*	⟶	+ R—Ö—R′ (ether)
+ R′—C(=O:)—Ö:⁻ (carboxylate)	⟶	+ R′—C(=O:)—Ö—R (ester)
+ N_3^- (azide = ⁻:N=N⁺=N:⁻)	⟶	+ R—N_3 (alkyl azide)*
+ R′—C≡C:⁻ (alkylacetylide)*	⟶	+ R—C≡C—R′ (dialkylacetylene)*
+ :S̈R′⁻ (alkanethiolate)*	⟶	+ R—S̈—R′ (thioether or sulfide)
+ :NR'_3 (trialkylamine)*	⟶	R—$\overset{+}{N}R'_3$:X:⁻ (tetraalkylammonium salt)*
+ S̈R'_2 (dialkyl sulfide)*	⟶	R—$\overset{+}{S}R'_2$:X:⁻ (trialkylsulfonium salt)*
+ :PR'_3 (trialkylphosphine)*	⟶	R—$\overset{+}{P}R'_3$:X:⁻ (tetraalkylphosphonium salt)*

*The name is for and R, R′ = alkyl; R′ can also be an aryl group.

In addition, it illustrates an *intramolecular substitution reaction*—a reaction in which the nucleophilic center, the electrophilic center, and the leaving group are all parts of the *same* molecule. In this case, the nucleophilic substitution reaction results in the formation of a ring.

Nucleophilic substitution reactions can involve many different nucleophiles, a few of which are listed in **Table 9.1**. Notice from this table that nucleophilic substitution reactions can be used to transform alkyl halides into a wide variety of other functional groups. We discuss nucleophilic substitution reactions in more detail in Secs. 9.4 and 9.6.

Focused Problems

9.1 What is the expected nucleophilic substitution product when

(a) Methyl iodide reacts with $Na^+ \ CH_3CH_2CH_2CH_2S^-$?

(b) One equivalent of ethyl iodide reacts with ammonia?

9.2 Write the net ionic equation for the reaction in Focused Problem 9.1a.

B. β-Elimination Reactions

When a *tertiary alkyl halide* reacts with a strong Brønsted base such as sodium ethoxide, a very different type of reaction is observed.

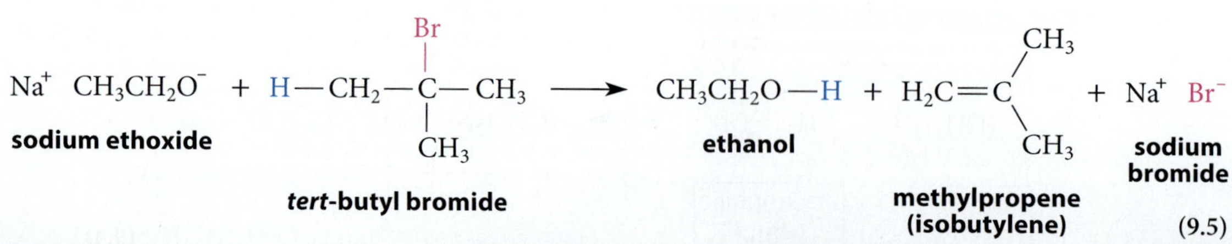

(9.5)

Equation 9.5 is an example of an **elimination reaction**, a reaction in which two or more groups (in this case H and Br) are lost from within the same molecule. We discuss the mechanism of this reaction in Sec. 9.5A.

In an alkyl halide, the carbon bearing the halogen is often referred to as the **α-carbon** (Sec. 8.1), and the adjacent carbons are referred to as the **β-carbons**. In Eq. 9.5 the halide is lost from the α-carbon and a proton is lost from a β-carbon.

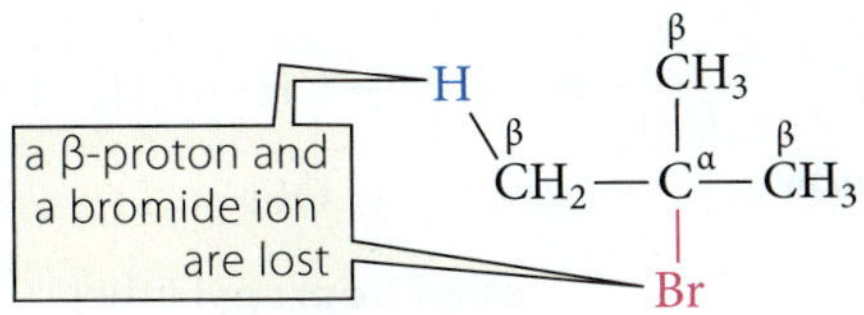

An elimination that involves loss of two groups from adjacent carbons to form a double bond is called a **β-elimination**. This is the most common type of elimination reaction in organic chemistry. Notice that a β-elimination reaction is *conceptually* the reverse of an addition to an alkene.

Strong bases promote the β-elimination reactions of alkyl halides. Among the most frequently used bases are alkoxides, which we introduced in Sec. 9.1A. The two bases you'll see most frequently in the elimination reactions discussed in this chapter are ethoxide, which is typically used as sodium ethoxide ($Na^+CH_3CH_2O^-$, abbreviated $Na^+\,^-OEt$ or simply NaOEt) and *tert*-butoxide, which is typically used as potassium *tert*-butoxide [$K^+(CH_3)_3C—O^-$, abbreviated $K^+\,^-OtBu$ or simply KO*t*Bu]. Often the conjugate-acid alcohols of these bases are used as solvents. Just as ^-OH is used as a solution in its conjugate acid water, sodium ethoxide is frequently used as a solution in ethanol and potassium *tert*-butoxide in *tert*-butyl alcohol. (The preparation of these bases is covered in Sec. 11.1B.)

If the reacting alkyl halide has more than one type of β-hydrogen atom, then more than one β-elimination reaction is possible. When these different reactions occur at comparable rates, more than one alkene product is formed, as in Eq. 9.6.

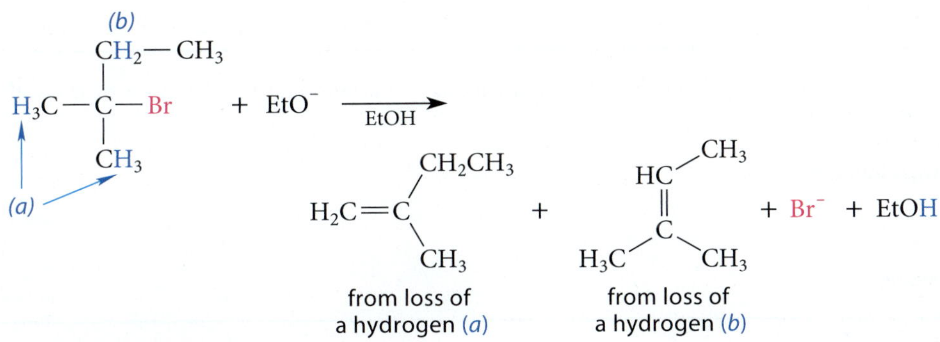

(9.6)

Note the use of a net ionic equation in this example.

Focused Problem

9.3 Draw the structures of all product(s) expected in the ethoxide-promoted β-elimination reaction of each of the following compounds.

(a) 2-Bromo-2,3-dimethylbutane

(b) 1-Chloro-1-methylcyclohexane

C. Competition between Nucleophilic Substitution and β-Elimination Reactions

In the presence of a strong base and good nucleophile such as ethoxide, primary alkyl halides typically undergo a nucleophilic substitution reaction, whereas tertiary alkyl halides undergo a β-elimination reaction. What about secondary alkyl halides? A typical secondary alkyl halide under the same conditions undergoes both reactions.

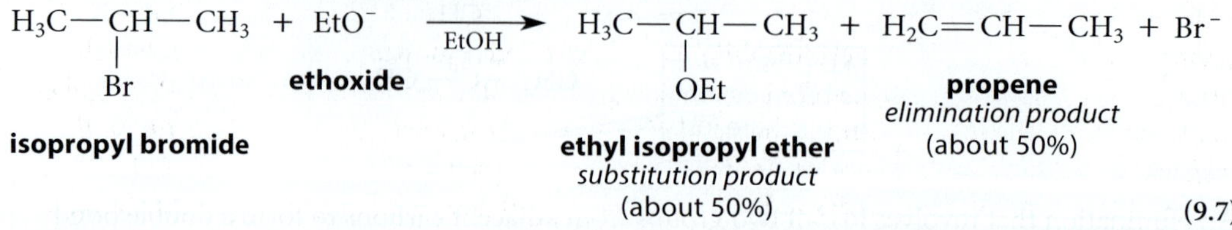

(9.7)

Because both substitution and elimination reactions are observed, the two reactions occur at comparable *rates*—in other words, the reactions are *in competition*. In fact, nucleophilic substitution and base-promoted β-elimination reactions are in competition for *all* alkyl halides with β-hydrogens, even primary and tertiary halides. In the presence of a strong Brønsted base, though, nucleophilic substitution is a faster reaction (it "wins the competition") for many primary alkyl halides, and in most cases β-elimination is a faster reaction for tertiary halides; that is why substitution predominates in the former case and elimination in the latter. Under some conditions, however, the results of the competition can be changed. For example, we can sometimes find conditions under which some primary alkyl halides give mostly elimination products. (Section 9.5G deals with this competition in more detail.)

In the following sections, we focus first on nucleophilic substitution reactions, then on β-elimination reactions. We discuss the factors that govern the reactivities of alkyl halides in each of these reaction types. Although each type of reaction is considered in isolation, keep in mind that substitutions and eliminations are always in competition.

Focused Problem

9.4 What substitution and elimination products (if any) might be obtained when each of the following alkyl halides is treated with sodium methoxide in methanol?

(a) *trans*-1-Bromo-3-methylcyclohexane

(b) Methyl iodide

(c) (Bromomethyl)cyclopentane

9.2 EQUILIBRIA IN NUCLEOPHILIC SUBSTITUTION REACTIONS

Table 9.1 (Sec. 9.1A) shows some of the many possible nucleophilic substitution reactions. How do we know whether the equilibrium for a given substitution is favorable? This problem is illustrated by the reaction of a cyanide ion with methyl iodide, which has an equilibrium constant that favors the product acetonitrile by many powers of 10.

$$:\bar{C}{\equiv}N: + H_3C{-}\ddot{\underset{..}{I}}: \longrightarrow H_3C{-}C{\equiv}N: + :\ddot{\underset{..}{I}}:^- \qquad \text{(9.8)}$$

acetonitrile

Other substitution reactions, however, are reversible or even unfavorable.

$$:\ddot{\underset{..}{I}}:^- + H_3C{-}\ddot{\underset{..}{Br}}: \rightleftharpoons H_3C{-}\ddot{\underset{..}{I}}: + :\ddot{\underset{..}{Br}}:^- \qquad \text{(9.9a)}$$

$$:\ddot{\underset{..}{I}}:^- + H_3C{-}\ddot{O}H \longleftarrow H_3C{-}\ddot{\underset{..}{I}}: + :\ddot{O}H^- \quad \text{(does not proceed to the right)} \qquad \text{(9.9b)}$$

Results such as these can be predicted by recognizing that each nucleophilic substitution reaction is conceptually similar to a Brønsted acid–base reaction. That is, if the alkyl group of the alkyl halide is replaced with a hydrogen, the substitution reaction becomes an acid–base reaction.

$$^{-}OH + H_3C—I \longrightarrow H_3C—OH + I^{-} \quad \text{(nucleophilic substitution reaction)} \quad (9.10a)$$

$$^{-}OH + H—I \longrightarrow H—OH + I^{-} \quad \text{(acid–base reaction)} \quad (9.10b)$$

In the nucleophilic substitution reaction, ^{-}OH, which acts as a nucleophile, displaces the I^{-} leaving group from the *carbon* electrophile. In the Brønsted acid–base reaction, ^{-}OH, which acts as a Brønsted base, displaces the I^{-} leaving group from the *proton* electrophile. What makes this comparison especially useful is that *it can be used to predict whether the equilibrium in the nucleophilic substitution is favorable.* To do this, we determine whether the equilibrium for the Brønsted acid–base reaction is favorable using the method described in Sec. 3.6C. Therefore, if the Brønsted acid–base reaction

$$H—X + Y:^{-} \rightleftharpoons Y—H + X:^{-} \quad (9.11a)$$

strongly favors the right side of the equation, then the analogous nucleophilic substitution reaction

$$R—X + Y:^{-} \rightleftharpoons Y—R + X:^{-} \quad (9.11b)$$

likewise favors the right side of the equation. This means that *the equilibrium in any nucleophilic substitution reaction, as in an acid–base reaction, favors release of the weaker base.* This principle, for example, explains why I^{-} will *not* displace ^{-}OH from CH_3OH in the reverse of Eq. 9.10a: I^{-} is a much weaker base than ^{-}OH (see Table 3.1, Sec. 3.5B). In fact, the reverse reaction occurs: ^{-}OH readily displaces I^{-} from CH_3I. This example illustrates how the acid–base principles discussed in Chapter 3 can prove useful in understanding nucleophilic substitution reactions.

Some equilibria that are not too unfavorable can be driven to completion by applying *Le Chatelier's principle* (Sec. 4.10C). For example, alkyl chlorides normally do not react to completion with iodide ion because iodide is a weaker base than chloride, and the equilibrium favors the formation of the weaker base (iodide). In the solvent acetone, however, potassium iodide is relatively soluble, whereas potassium chloride is relatively insoluble. So, when an alkyl chloride reacts with KI in acetone, KCl precipitates, and the equilibrium compensates for the loss of KCl by forming more of it, along with more alkyl iodide.

$$R—Cl + K^{+} I^{-} \xrightarrow[\text{acetone}]{} R—I + K^{+} Cl^{-}\downarrow \text{ (precipitates)} \quad (9.12)$$

The correspondence between equilibrium constants for nucleophilic substitution reactions and Brønsted acid–base reactions is not quantitatively exact because the electrophilic centers aren't the same (carbon versus hydrogen). In addition, the pK_a values used to predict the equilibrium constants for acid–base reactions are determined in water, whereas solvents other than water are often used for nucleophilic substitution reactions. Nevertheless, when the basicity difference between the nucleophile and the leaving group is large (as it will be in most alkyl halide reactions), we can be confident that our predictions about equilibria in nucleophilic substitution reactions will be qualitatively correct.

Focused Problem

9.5 Tell whether each of the following reactions favors reactants or products at equilibrium. (Assume that all reactants and products are soluble.) Justify your reasoning by writing the corresponding acid–base equilibrium and estimating its equilibrium constant. (*Hint:* Use Table 3.2, Sec. 3.6A.)

(a) $CH_3F + Cl^{-} \rightleftharpoons CH_3Cl + F^{-}$

(b) $CH_3Cl + N_3^{-} \rightleftharpoons CH_3N_3 + Cl^{-}$ (*Hint:* The pK_a of HN_3 is 4.72.)

(c) $CH_3Cl + {}^{-}OCH_3 \rightleftharpoons CH_3OCH_3 + Cl^{-}$

9.3 REACTION RATES

Although being able to determine whether the equilibrium for a nucleophilic substitution reaction is favorable is important, *knowing the equilibrium constant for a reaction provides no information about the rate at which the reaction takes place.* Some substitution reactions with favorable equilibria proceed rapidly, whereas others proceed slowly. For example, the reaction of methyl iodide with cyanide is a relatively fast reaction, whereas the reaction of cyanide with neopentyl iodide is so slow that it is virtually useless:

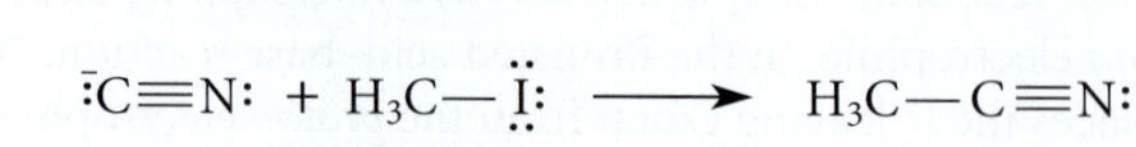

methyl iodide
(reacts rapidly) (9.13a)

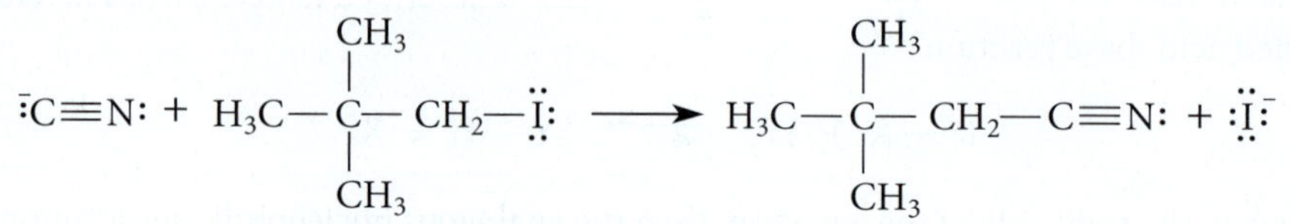

neopentyl iodide
(reacts very slowly) (9.13b)

Why do reactions that are so conceptually similar differ so drastically in their rates? In other words, what determines the *reactivity* of a given alkyl halide in a nucleophilic substitution reaction? Because this question deals with reaction rates and the concept of the transition state, you should review the introduction to reaction rates and transition-state theory in Sec. 4.8.

A. The Rate Law

The rate of a reaction is the concentration of products formed per unit time. For molecules to react with one another, they must "get together," or collide. Because molecules at higher concentrations are more likely to collide than molecules at lower concentrations, the rate of a reaction is a function of the concentrations of the reactants. The mathematical statement of how a reaction rate depends on concentration is called the **rate law**. A rate law is determined experimentally by varying the concentration of each reactant (including any catalysts) independently and measuring the resulting effect on the rate. Each reaction has its own characteristic rate law. For example, suppose that for the reaction $A + B \longrightarrow C$ the reaction rate doubles if *either* $[A]$ or $[B]$ is doubled and increases by a factor of 4 if *both* $[A]$ and $[B]$ are doubled. The rate law for this reaction is then

$$\text{Rate} = k[A][B] \tag{9.14}$$

If, in another reaction $D + E \longrightarrow F$, the rate doubles only if the concentration of D is doubled, and changing the concentration of E has no effect, the rate law is then

$$\text{Rate} = k[D] \tag{9.15}$$

The concentrations in the rate law are the concentrations of reactants at any time during the reaction, and the rate is the velocity of the reaction at that same time. The constant of proportionality, k, is called the **rate constant**. In general, the rate constant is different for every reaction, and it is a property of each reaction under particular conditions of temperature, pressure, solvent, and so on. As Eqs. 9.14 and 9.15 show, the rate constant is numerically equal to the rate of the reaction when all reactants are present under the standard conditions of 1 M concentration. *The rates of two reactions are compared by comparing their rate constants.*

The **overall kinetic order** for a reaction is the sum of the powers of all the concentrations in the rate law. For a reaction described by the rate law in Eq. 9.14, the overall kinetic order is two; the reaction described by this rate law is said to be a **second-order reaction**. The overall kinetic order of a reaction having the rate law in Eq. 9.15 is one; such a reaction is therefore a **first-order reaction**. The **kinetic order in each reactant** is the power to which its concentration is raised in the rate law. Therefore, the reaction described by the rate law in Eq. 9.14 is *first order in each reactant*. A reaction with the rate law in Eq. 9.15 is first order in D and zero order in E.

The *units* of the rate constant depend on the kinetic order of the reaction. With concentrations in moles per liter (M), and time in seconds (s), the rate of any reaction has the units of $M\,\mathrm{s}^{-1}$. For a second-order reaction, then, dimensional consistency requires that the rate constant must have the units of $M^{-1}\,\mathrm{s}^{-1}$.

$$\text{rate} = k[A][B]$$

$$M\,\mathrm{s}^{-1} = M^{-1}\,\mathrm{s}^{-1}\;M\;M \qquad (9.16)$$

Similarly, the rate constant for a first-order reaction has units of s^{-1}.

$$\text{rate} = k[D]$$

$$M\,\mathrm{s}^{-1} = \mathrm{s}^{-1}\;M \qquad (9.17)$$

FURTHER EXPLORATION 9.1
Reaction Rates

B. Relationship of the Rate Constant to the Standard Free Energy of Activation

According to transition-state theory (see Sec. 4.8), the standard free energy of activation, or energy barrier, determines the rate of a reaction under standard conditions. In Sec. 9.3A we showed that the rate constant is numerically equal to the reaction rate under standard conditions—that is, when the concentrations of all the reactants are 1 M. It follows, then, that the rate constant is related to the standard free energy of activation, $\Delta G^{\circ\ddagger}$. If $\Delta G^{\circ\ddagger}$ is large for a reaction (**Fig. 9.1a**), then the reaction is relatively slow, and the rate constant is small. If $\Delta G^{\circ\ddagger}$ is small (**Fig. 9.1b**), then the reaction is relatively fast, and the rate constant is large. The rate of a reaction, and therefore the rate constant, is related exponentially to $\Delta G^{\circ\ddagger}$ (see Eq. 4.53, Sec. 4.8B).

$$k \propto e^{-\Delta G^{\circ\ddagger}/RT} = 10^{-\Delta G^{\circ\ddagger}/2.3RT} \qquad (9.18)$$

(The symbol $\propto$ means "is proportional to.")

Table 9.2 illustrates the quantitative relationship between the rate constant and the standard free energy of activation. (For the derivation of these numbers, see Further Exploration 9.2.) Table 9.2 also translates these numbers into practical terms by giving the time required for the completion of a reaction with a given rate constant. This time is approximately $7/k$. (This is also justified in Further Exploration 9.2.)

FURTHER EXPLORATION 9.2
Absolute Rate Theory

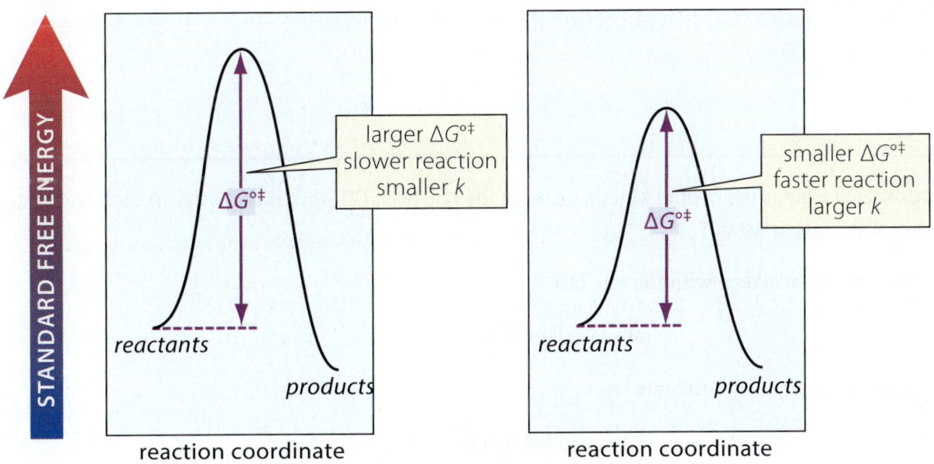

(a) (b)

FIGURE 9.1 Relationship among the standard free energy of activation ($\Delta G^{\circ\ddagger}$), the reaction rate, and the rate constant (k). (a) A reaction with a larger $\Delta G^{\circ\ddagger}$ has a smaller rate and a smaller rate constant. (b) A reaction with a smaller $\Delta G^{\circ\ddagger}$ has a larger rate and a larger rate constant.

TABLE 9.2 Relationship between Rate Constants, Standard Free Energies of Activation, and Reaction Times for First-Order Reactions

Rate constant (s^{-1}) (T = 298 K)	Time to completion*	$\Delta G^{\circ\ddagger}$	
		kJ mol^{-1}	kcal mol^{-1}
10^{-8}	22 years	119	28.4
10^{-6}	83 days	107	25.7
10^{-4}	20 hours	96.0	23.0
10^{-2}	12 minutes	84.6	20.2
1	7 seconds	73.2	17.5
10^{2}	70 milliseconds	61.7	14.8
10^{4}	700 microseconds	50.4	12.0
10^{6}	7 microseconds	38.9	9.3
6.2×10^{12}	0.01 nanosecond	0	0

*Time required for 99% completion of the reaction ≈ 7/k.

We'll most often be interested in the *relative rates* of two reactions. That is, we'll be comparing the rate of a reaction to that of a standard reaction. A **relative rate** is defined as the ratio of two rates. The relationship between the relative rate of two reactions A and B under standard conditions (that is, 1 *M* in all reactants) and their standard free energies of activation, first presented as Eq. 4.54a (Sec. 4.8B), is given in Eq. 9.19a:

$$\text{relative rate} = \frac{\text{rate}_A}{\text{rate}_B} = 10^{(\Delta G_B^{\circ\ddagger} - \Delta G_A^{\circ\ddagger})/2.3RT} \qquad \textbf{(9.19a)}$$

Because the rate constant is numerically equal to the rate under standard conditions, the relative rate is also the ratio of rate constants:

$$\text{relative rate} = \frac{k_A}{k_B} = 10^{(\Delta G_B^{\circ\ddagger} - \Delta G_A^{\circ\ddagger})/2.3RT} \qquad \textbf{(9.19b)}$$

or

$$\log\,(\text{relative rate}) = \log\left(\frac{k_A}{k_B}\right) = \frac{\Delta G_B^{\circ\ddagger} - \Delta G_A^{\circ\ddagger}}{2.3RT} \qquad \textbf{(9.19c)}$$

According to Eq. 9.19c, each increment of 2.3RT (5.7 kJ mol^{-1} or 1.4 kcal mol^{-1} at 298 K) in the $\Delta G^{\circ\ddagger}$ difference for two reactions corresponds to a one log unit (that is, a 10-fold) factor in their relative rate constants.

Focused Problems

9.6 For each of the following reactions, (1) what is the overall kinetic order of the reaction, (2) what is the order in each reactant, and (3) what are the dimensions of the rate constant?

(a) An addition reaction of bromine to an alkene with the rate law

$$\text{rate} = k[\text{alkene}][Br_2]^2$$

(b) A substitution reaction of an alkyl halide with the rate law

$$\text{rate} = k[\text{alkyl halide}]$$

9.7 (a) What is the ratio of rate constants k_A/k_B at 25 °C for two reactions A and B if the standard free energy of activation of reaction A is 14 kJ mol^{-1} (3.4 kcal mol^{-1}) less than that of reaction B?

(b) What is the difference in the standard free energies of activation at 25 °C of two reactions A and B if reaction B is 450 times faster than reaction A? Which reaction has the greater $\Delta G^{\circ\ddagger}$?

9.8 What prediction does the rate law in Eq. 9.15 make about how the rate of the reaction changes as the reactants D and E are converted into F over time? Does the rate increase, decrease, or stay the same? Explain. Use your answer to sketch a plot of the concentrations of starting materials and products against time.

9.4 THE S_N2 REACTION

A. Rate Law and Mechanism of the S_N2 Reaction

Consider now the nucleophilic substitution reaction of ethoxide ion with methyl iodide in ethanol at 25 °C.

$$\underset{\textbf{ethoxide ion}}{CH_3CH_2O^-} + H_3C\text{—}I \xrightarrow[CH_3CH_2OH]{} \underset{\substack{\textbf{ethyl methyl ether}\\ \textbf{(methoxyethane)}}}{CH_3CH_2O\text{—}CH_3} + I^- \tag{9.20}$$

The following rate law was experimentally determined for this reaction:

$$\text{rate} = k[CH_3I][CH_3CH_2O^-] \tag{9.21}$$

with $k = 6.0 \times 10^{-4}\ M^{-1}\ s^{-1}$. That is, this is *a second-order reaction that is first order in each reactant.*

The rate law of a reaction is important because it provides fundamental information about the reaction mechanism. Specifically, the concentration terms of the rate law indicate *which species are present in the transition state of the rate-limiting step*. This means that the rate-limiting transition state of reaction 9.21 consists of the elements of one methyl iodide molecule and one ethoxide ion. The rate law excludes some mechanisms from consideration. For example, any mechanism in which the rate-limiting step involves two molecules of ethoxide is *ruled out* by the rate law, because the rate law for such a mechanism would have to be second order in ethoxide.

The simplest possible mechanism consistent with the rate law is one in which the ethoxide ion *directly displaces* the iodide ion from the methyl carbon:

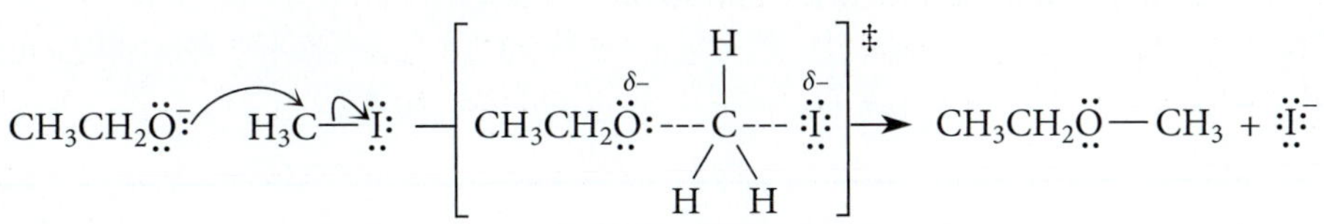

transition state (9.22)

Mechanisms like this account for many nucleophilic substitution reactions. A mechanism in which a nucleophile donates a nonbonding electron pair to an atom (usually carbon) and displaces a leaving group from the same atom in a concerted manner (that is, in one step, without reactive intermediates) is called an **S_N2 mechanism**. Reactions that occur by S_N2 mechanisms are called **S_N2 reactions**. The meaning of the "nickname" S_N2 is as follows:

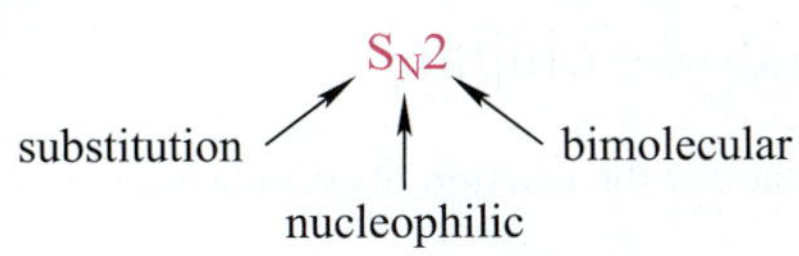

(9.23)

STUDY GUIDE LINK 9.1
Deducing Mechanisms from Rate Laws

(The word *bimolecular* means that the rate-limiting step of the reaction involves two species—in this case, one methyl iodide molecule and one ethoxide ion.) Because an S_N2 reaction is concerted, it involves no reactive intermediates.

The rate law does not reveal every detail of a reaction mechanism. Although the rate law indicates what atoms are present in the rate-limiting step, it provides no information about how they are arranged. This means that the following two mechanisms for the S_N2 reaction of ethoxide ion with methyl iodide are equally consistent with the rate law.

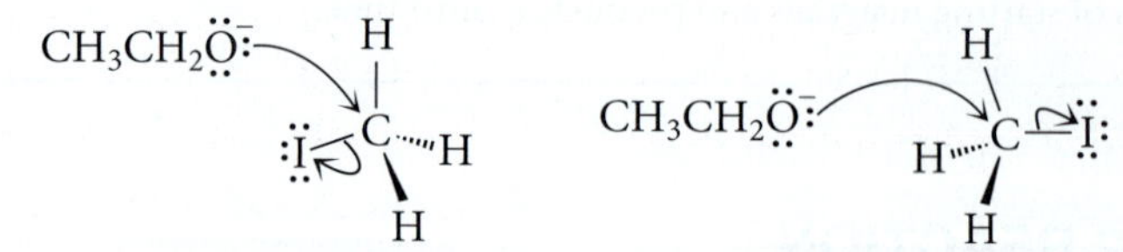

same-side substitution   opposite-side substitution   (9.24)

To decide between these two possibilities, other types of experiments are needed (Sec. 9.4C).

To summarize the relationship between the rate law and the mechanism of a reaction:

1. The concentration terms of the rate law indicate what species are involved in the rate-limiting step.

2. Mechanisms that are inconsistent with the rate law are ruled out.

3. Of the chemically reasonable mechanisms consistent with the rate law, the simplest one is provisionally adopted.

4. The mechanism of a reaction is modified or refined if required by subsequent experiments.

When scientists provide experimental data that challenge a long-held theory, their work is almost certain to be repeated and examined carefully by others. Only if the theory withstands such careful scrutiny will a new theory evolve. For an example of such a challenge, see Problem 9.75 at the end of the chapter.

Point (4) may seem disturbing because it means that a mechanism can be changed at a later time. Perhaps it seems that an "absolutely true" mechanism should exist for every reaction. A mechanism, however, can never be proved; it can only be disproved. The value of a mechanism lies not in its absolute truth but rather in its validity as a conceptual framework, or theory, that generalizes the results of many experiments and predicts the outcome of others. Mechanisms allow us to place reactions into categories and thus impose a conceptual order on chemical observations. So, when someone observes an experimental result different from that predicted by a mechanism, the mechanism must be modified to accommodate both the previously known facts and the new facts. The evolution of mechanisms is no different from the evolution of science in general. Knowledge is dynamic: theories (mechanisms) predict the results of experiments, a test of these theories may lead to new theories, and so on.

Focused Problems

9.9 The reaction of acetic acid with ammonia is rapid and follows the simple rate law shown in the following equation. Propose a curved-arrow mechanism that is consistent with this rate law.

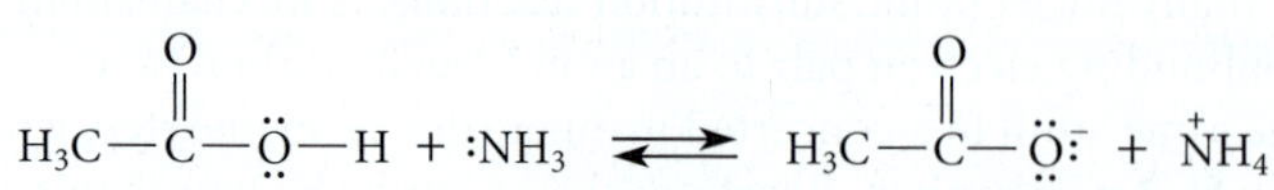

acetic acid

$$\text{rate} = k\left[\mathrm{H_3C{-}C({=}O){-}OH}\right]\left[\mathrm{NH_3}\right]$$

9.10 (a) Write the chemical equation and the expected rate law for the reaction of cyanide ion ($^-$:CN) with ethyl bromide by the S_N2 mechanism.

(b) Draw the transition state for this reaction.

B. Relative Rates of S_N2 Reactions and Brønsted Acid–Base Reactions

In Sec. 9.2, we learned about the close analogy between nucleophilic substitution reactions and acid–base reactions. The equilibrium constants for a nucleophilic substitution reaction and its acid–base analog are similar, and the curved-arrow notations for an S_N2 reaction and its acid–base analog are identical. However, their rates are different. *Most ordinary acid–base reactions occur instantaneously*—as fast as the reacting pairs can diffuse together. The rate constants for such reactions are typically in the $10^8 - 10^{10}\ M^{-1}\ s^{-1}$ range. Although many nucleophilic substitution reactions occur at convenient rates, they are *much* slower than the analogous acid–base reactions. For example, the reaction in Eq. 9.25a is completed in a little over an hour, but the corresponding acid–base reaction in Eq. 9.25b occurs within about a *billionth of a second*!

Nucleophilic substitution reaction:

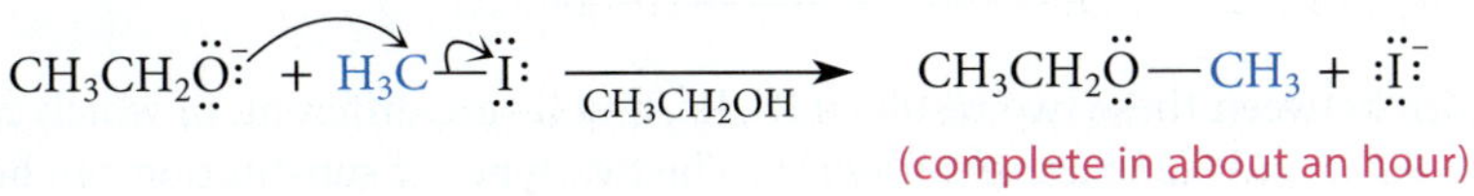

(9.25a)

Brønsted acid–base reaction:

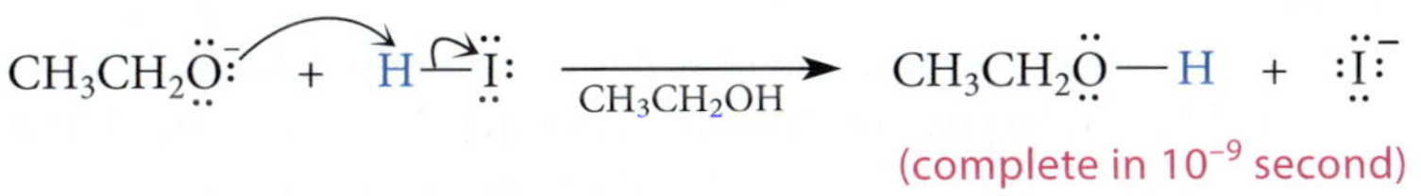

(9.25b)

This means that if an alkyl halide and a Brønsted acid are in competition for a Brønsted base, the Brønsted acid reacts much more rapidly—that is, *the Brønsted acid always wins.*

Focused Problems

9.11 Methyl iodide (0.1 *M*) and hydriodic acid (HI, 0.1 *M*) are allowed to react in ethanol solution with 0.1 *M* sodium ethoxide. What products are observed?

9.12 Ethyl bromide (0.1 *M*) and HBr (0.1 *M*) are allowed to react in aqueous THF with 1 *M* sodium cyanide ($Na^{+}\ {}^{-}CN$). What products are observed? Are any products formed more rapidly than others? Explain.

C. Stereochemistry of the S_N2 Reaction

The mechanism of the S_N2 reaction can be described in more detail by considering its *stereochemistry*. The stereochemistry of a substitution reaction can be investigated only if the carbon at which substitution occurs is a stereocenter in both reactants and products (Sec. 7.8B). A substitution reaction can occur at a stereocenter in three stereochemically different ways:

1. With *retention of configuration* at the stereocenter
2. With *inversion of configuration* at the stereocenter
3. With a combination of (1) and (2)—that is, mixed retention and inversion

If approach of the nucleophile Nuc:⁻ to an asymmetric carbon and departure of the leaving group X:⁻ occur from more or less the same direction (*same-side substitution*), then a substitution reaction would result in a product with *retention of configuration* at the asymmetric carbon.

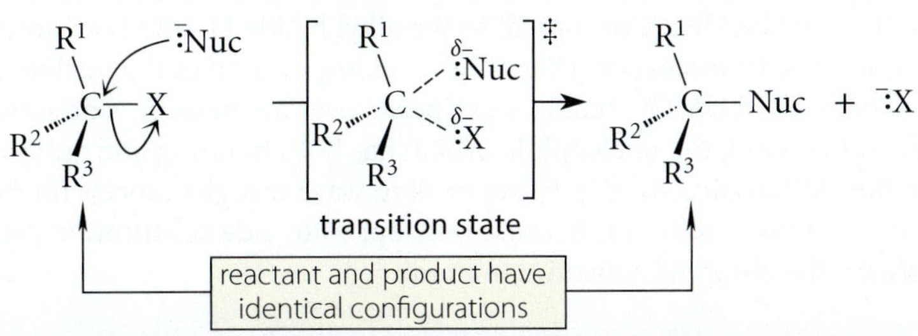

(9.26a)

In contrast, if approach of the nucleophile and loss of the leaving group on an asymmetric carbon occur from opposite directions (*opposite-side substitution*), the other three groups on carbon must invert, or "turn inside out," to maintain the tetrahedral bond angle. This mechanism would lead to a product with *inversion of configuration* at the asymmetric carbon.

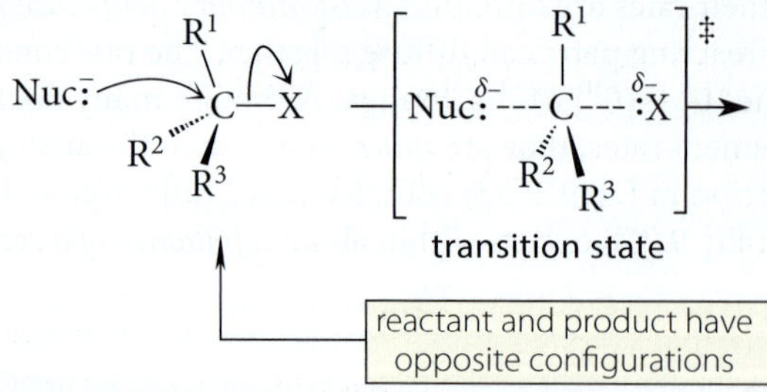

(9.26b)

We can distinguish between these two results if R^1, R^2, and R^3 are different, in which case the products of Eqs. 9.26a and 9.26b are *enantiomers*. The two types of substitution can be distinguished by subjecting one enantiomer of a chiral alkyl halide to the S_N2 reaction and determining which enantiomer of the product is formed. If both paths occur at equal rates, then the racemate will be formed.

What are the experimental results? The reaction of sodium ethanethiolate ($Na^{+}\,{}^{-}SEt$) with 2-bromooctane, a chiral alkyl halide, to give the thioether is a typical S_N2 reaction. The reaction follows a second-order rate law, first order in $Na^{+}\,{}^{-}SEt$ and first order in the alkyl halide. When (*R*)-2-bromooctane is used in the reaction, the product is the (*S*) thioether.

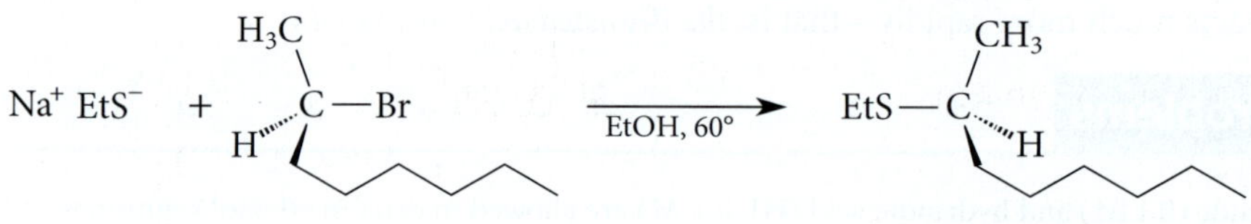

(*R*)-(−)-2-bromooctane **(*S*)-(+)-2-(ethylthio)octane** (9.27)

The stereochemistry of this S_N2 reaction shows that it proceeds with *inversion of configuration*. Therefore, the reaction occurs by *opposite-side substitution* of EtS^- on the alkyl halide.

Opposite-side substitution is also observed for the reaction of bromide ion and other nucleophiles with the bromonium ion intermediate in the addition of bromine to alkenes (Eq. 7.61, Sec. 7.8C). As you can now appreciate, that reaction is also an S_N2 reaction. In fact, *inversion of stereochemical configuration is generally observed in all S_N2 reactions at carbon stereocenters.*

The stereochemistry of the S_N2 reaction calls to mind the *inversion of amines* (see Fig. 6.9, Sec. 6.9B). In the hybrid orbital description of both processes, the central atom is turned "inside out," and it is approximately sp^2-hybridized at the transition state. In the transition state for amine inversion, the 2p orbital on the nitrogen contains a nonbonding electron pair. In the transition state for an S_N2 reaction on carbon, the nucleophile and the leaving group are partially bonded to opposite lobes of the carbon 2p orbital (**Fig. 9.2**).

Don't be concerned with the derivation of the methyl chloride LUMO shown in Fig. 9.3, which comes from a molecular orbital program. The important point is that the interaction of the nucleophile with the LUMO explains the opposite-side substitution.

Why is opposite-side substitution preferred in the S_N2 reaction? The hybrid orbital description of the reaction in Fig. 9.2 offers no answer to this question, but a molecular orbital analysis does, as shown in **Fig. 9.3** for the reaction of a nucleophile (Nuc:) with methyl chloride (CH_3Cl). When a nucleophile donates electrons to an alkyl halide, the orbital containing the donated electron pair must initially interact with an *unoccupied* molecular orbital of the alkyl halide. The MO of the nucleophile that contains the donated electron pair interacts with the unoccupied alkyl halide MO of lowest energy, called the **LUMO** (for "lowest unoccupied molecular orbital"). All of the bonding MOs of the alkyl halide are occupied, so the alkyl halide LUMO is an antibonding MO. When opposite-side substitution occurs (Fig. 9.3a), *bonding overlap* of the nucleophile orbital occurs with the alkyl halide LUMO—that is, wave peaks overlap. In same-side substitution (Fig. 9.3b), on the other hand, the nucleophile orbital has both bonding and antibonding overlap with the LUMO; the antibonding overlap (wave peak to wave trough) cancels the bonding overlap, and no net bonding can occur. Because only opposite-side substitution gives bonding overlap, this is *always* the observed substitution mode.

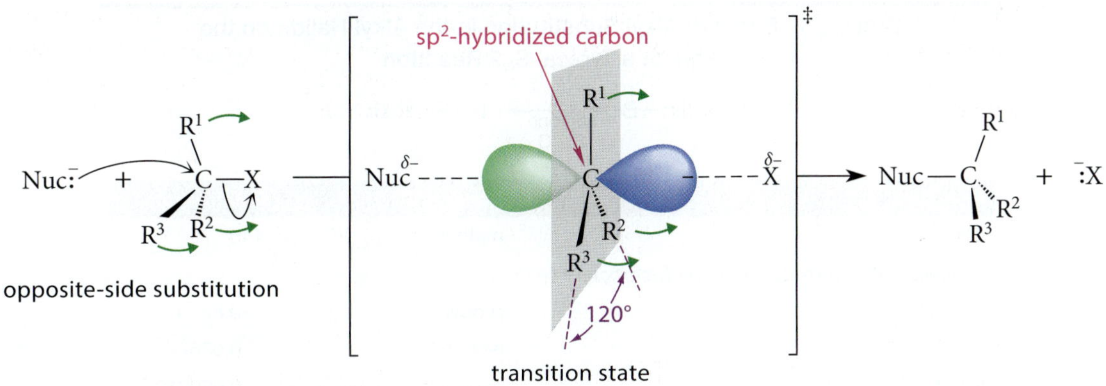

FIGURE 9.2 Stereochemistry of the S_N2 reaction. The green arrows show how the various groups change position during the reaction. (Nuc:⁻ = a general nucleophile.) The stereochemical configuration of the asymmetric carbon is inverted by the opposite-side substitution reaction.

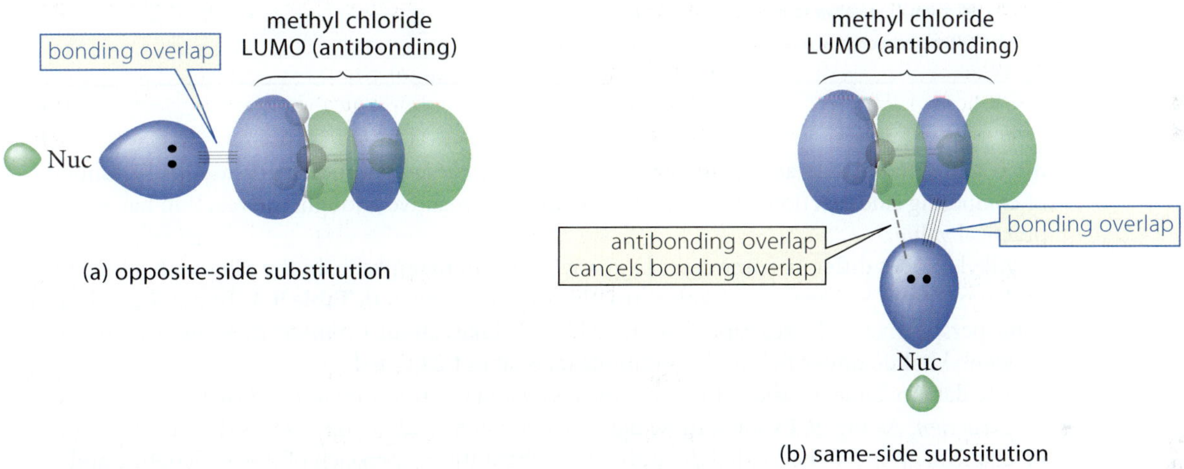

FIGURE 9.3 In the S_N2 reaction, the orbital containing the nucleophile electron pair interacts with the unoccupied molecular orbital of lowest energy (LUMO) in the alkyl halide. (a) Opposite-side substitution leads to bonding overlap. (b) Same-side substitution gives both bonding and antibonding overlaps that cancel. Therefore, opposite-side substitution is always observed.

Focused Problem

9.13 Draw a structure for the expected substitution product (including its stereochemical configuration) in the S_N2 reaction of potassium iodide in acetone solvent with the following compound? (D = ^{2}H = deuterium, an isotope of hydrogen.)

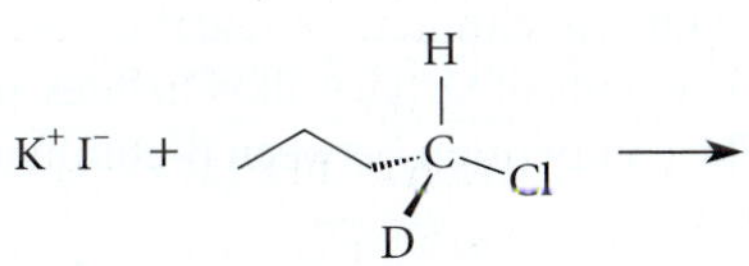

(*R*)-1-chlorobutane-1-*d*

D. Effect of Alkyl Halide Structure on the S_N2 Reaction

One of the most important aspects of the S_N2 reaction is how the reaction rate varies with the structure of the alkyl halide. (Recall Eqs. 9.13a and b in Sec. 9.3.) If an alkyl halide is very reactive, its S_N2 reactions occur rapidly under mild conditions. If an alkyl halide is relatively unreactive, then the severity of the reaction conditions (for example, the temperature) must be increased for

TABLE 9.3 Effect of Alkyl Substitution in the Alkyl Halide on the Rate of a Typical S_N2 Reaction

$$I^- + R—Br \xrightarrow[25\ ^\circ C]{acetone} I—R + Br^-$$

R—	Name of R	Relative rate*
CH_3—	methyl	145
Increased alkyl substitution at the β-carbon:		
$CH_3CH_2CH_2$—	propyl	0.82
$(CH_3)_2CHCH_2$—	isobutyl	0.036
$(CH_3)_3CCH_2$—	neopentyl	0.000012
Increased alkyl substitution at the α-carbon:		
CH_3CH_2—	ethyl	1.0
$(CH_3)_2CH$—	isopropyl	0.0078
$(CH_3)_3C$—	*tert*-butyl	~0.0005†

*All rates are relative to that of ethyl bromide.

†Estimated from the rates of closely related reactions.

the reaction to proceed at a reasonable rate. However, harsh conditions increase the likelihood of competing side reactions. So, if an alkyl halide is unreactive enough, the reaction has no practical value.

Alkyl halides differ, in some cases by many orders of magnitude, in the rates with which they undergo a given S_N2 reaction. Typical reactivity data are given in **Table 9.3**. To put these data in some perspective, if the reaction of a methyl halide takes about 1 *minute*, then the reaction of a neopentyl halide under the same conditions takes about *23 years*!

The data in Table 9.3 show, first, that *increased alkyl substitution at the β-carbon retards an* S_N2 *reaction*. As **Fig. 9.4** shows, these data are consistent with an opposite-side substitution mechanism. When a methyl halide undergoes substitution, approach of the nucleophile and departure of the leaving group are relatively unrestricted. However, when a neopentyl halide reacts with a nucleophile, both the nucleophile and the leaving group experience severe van der Waals repulsions with hydrogens of the β-methyl substituents. These van der Waals repulsions raise the energy of the transition state and therefore reduce the reaction rate. This is another example of a *steric effect*. (We first encountered an example of a steric effect in the mechanism of hydroboration; see Display 5.34b, Sec. 5.6A.) Recall that a *steric effect* is any effect on a chemical phenomenon (such as a reaction) caused by van der Waals repulsions. S_N2 reactions of branched alkyl halides, then, are retarded by a steric effect. Indeed, S_N2 reactions of neopentyl halides are so slow that they are not practically useful.

The data in Table 9.3 help explain why elimination reactions compete with the S_N2 reactions of secondary and tertiary alkyl halides with strong bases (Sec. 9.1C): these halides react so slowly in S_N2 reactions that the rates of elimination reactions are competitive with the rates of substitution. The rates of the S_N2 reactions of tertiary alkyl halides are so slow that elimination is the only reaction observed. The competition between β-elimination and S_N2 reactions is considered in more detail in Sec. 9.5G.

E. Nucleophilicity in the S_N2 Reaction

As Table 9.1 (Sec. 9.1A) illustrates, the S_N2 reaction is especially useful because of the variety of nucleophiles that can be employed. However, nucleophiles differ significantly in their reactivities. The relative reactivity of a nucleophile—how rapidly it reacts under a defined set of conditions—is called **nucleophilicity**. What factors govern nucleophilicity in the S_N2 reaction and why? Nucleophilicity is determined by three factors:

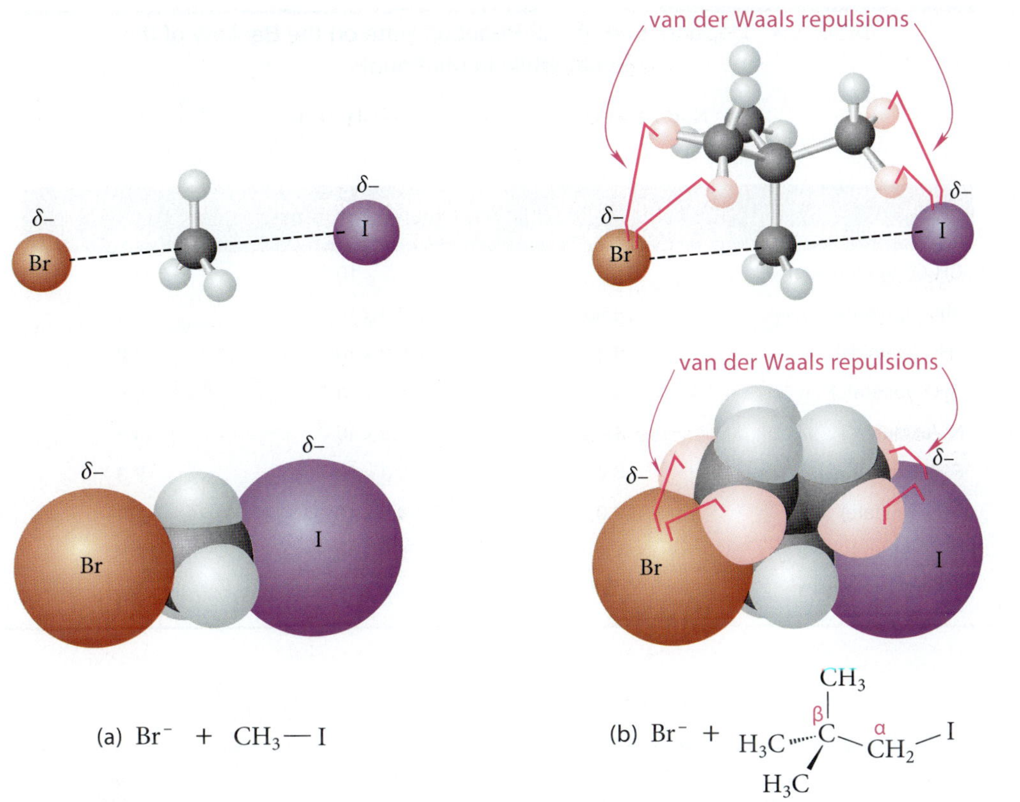

FIGURE 9.4 Transition states for S_N2 reactions. The upper panels show the transition states as ball-and-stick models, whereas the lower panels show them as space-filling models. (a) The reaction of methyl iodide with bromide ion. (b) The reaction of neopentyl iodide with bromide ion. The S_N2 reactions of neopentyl bromide are very slow because of the severe van der Waals repulsions of both the nucleophile and the leaving group with the pink hydrogens of the β-methyl substituents. These repulsions are indicated with red brackets in the models.

1. The *Brønsted* basicity of the nucleophile
2. The *solvent* in which the reaction is carried out
3. The *polarizability* of the nucleophile

Basicity and Solvent Effects on Nucleophilicity Because basicity and solvent effects on nucleophilicity are interrelated, we consider these two effects together. Remember that the Brønsted basicity of any base is measured by the pK_a of its *conjugate acid*: the higher the conjugate-acid pK_a, the greater the basicity of the nucleophile. We might expect some correlation between *nucleophilicity* and the *Brønsted basicity* of a nucleophile because both are aspects of its Lewis basicity. That is, in either role a Lewis base donates an electron pair. (Be sure to review the definitions of these terms in Sec. 3.4A.) Let's first examine some data for the S_N2 reactions of methyl iodide with anionic nucleophiles of different basicity to see whether this expectation is observed in practice. Some data for the reaction of methyl iodide with various nucleophiles in methanol solvent are given in **Table 9.4** and plotted in **Fig. 9.5**. Notice in this table that the nucleophilic atoms are all from the second period of the periodic table. Figure 9.5 shows a *very rough* trend toward faster reactions with the more basic nucleophiles, and a much clearer trend for nucleophiles in which the nucleophilic atom is an oxygen.

Let's now consider some data for the same reaction with anionic nucleophiles from different periods (rows) of the periodic table. These data are shown in **Table 9.5**. If you are expecting a similar correlation of nucleophilic reactivity and basicity, then you're in for a surprise. The sulfide nucleophile is more than three orders of magnitude *less basic* than the oxide nucleophile, and yet it is more than *four orders of magnitude more reactive*. Similarly, for the halide nucleophiles, the *least basic halide ion* (iodide) *is the best nucleophile*.

TABLE 9.4 Dependence of S_N2 Reaction Rate on the Basicity of the Nucleophile in Methanol

$$\text{Nuc:}^- + H_3C\text{—I} \xrightarrow[\text{25 °C}]{CH_3OH} \text{Nuc—}CH_3 + I^-$$

Nucleophile (name)	pK_a of conjugate acid*	*k* (second-order rate constant $M^{-1}\,s^{-1}$)	log *k*
CH_3O^- (methoxide)	15.1	2.5×10^{-4}	−3.6
PhO^- (phenoxide)	9.95	7.9×10^{-5}	−4.1
^-CN (cyanide)	9.4	6.3×10^{-4}	−3.2
AcO^- (acetate)	4.76	2.7×10^{-6}	−5.6
N_3^- (azide)	4.72	7.8×10^{-5}	−4.1
F^- (fluoride)	3.2	5.0×10^{-8}	−7.3
SO_4^{2-} (sulfate)	2.0	4.0×10^{-7}	−6.4
NO_3^- (nitrate)	−1.2	5.0×10^{-9}	−8.3

*pK_a values in water

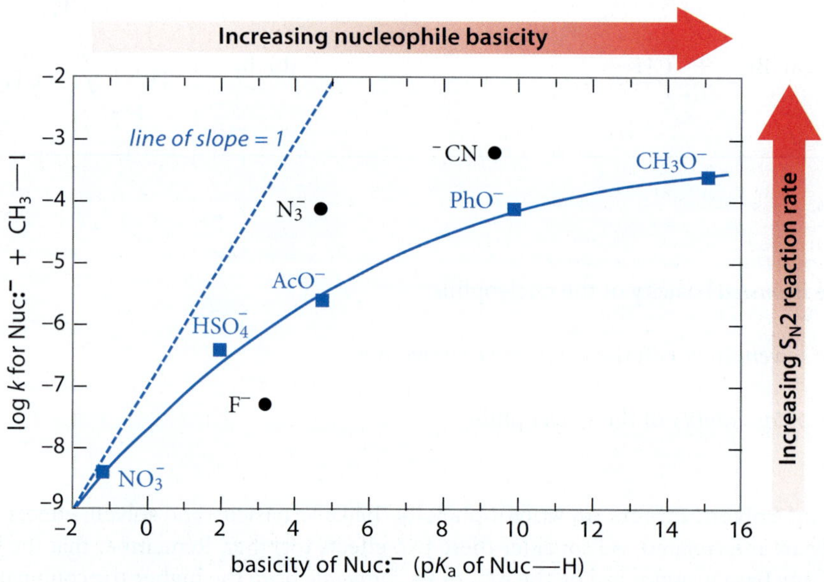

FIGURE 9.5 The dependence of nucleophile S_N2 reactivity on nucleophile basicity for a series of nucleophiles in methanol solvent. Reactivity is measured by log *k* for the reaction of the nucleophile with methyl iodide. Basicity is measured by the pK_a of the conjugate acid of the nucleophile. The blue dashed line of slope = 1 shows the trend to be expected if a change of one log unit in basicity resulted in the same change in nucleophilicity. The solid blue line shows the actual trend for a series of nucleophiles (*blue squares*) in which the nucleophilic center is —O^-. The black circles show the reactivity of other nucleophilic anions in which the reacting atoms are from period 2 of the periodic table (that is, the same period as oxygen).

Let's generalize what we've learned so far. The following apply to *nucleophilic anions* in *polar, protic solvents* (such as water–acetone and alcohols; see Sec. 8.6B):

1. In a series of nucleophiles in which the nucleophilic atoms are from the same period of the periodic table, there is a rough correlation of nucleophilicity with basicity.

2. In a series of nucleophiles in which the nucleophilic atoms are from the same group (column) but different periods of the periodic table, the less basic nucleophiles are more nucleophilic.

The interaction of the nucleophile with the solvent is the most significant factor that accounts for both of these generalizations. Let's start with generalization 2—the *inverse* relationship of basicity

TABLE 9.5 Dependence of S_N2 Reaction Rate on the Basicity of Nucleophiles from Different Periods of the Periodic Table in Methanol

$$\text{Nuc:}^- + H_3C{-}I \xrightarrow[25\,^\circ C]{CH_3OH} \text{Nuc}{-}CH_3 + I^-$$

Nucleophile	pK_a of conjugate acid*	k (second-order rate constant, $M^{-1}\,s^{-1}$)	log k
Group 6A Nucleophiles			
PhS^-	6.52	1.1	+0.03
PhO^-	9.95	7.9×10^{-5}	−4.1
Group 7A Nucleophiles			
I^-	−10	3.4×10^{-3}	−2.5
Br^-	−8	8.0×10^{-5}	−4.1
Cl^-	−6	3.0×10^{-6}	−5.5
F^-	3.2	5.0×10^{-8}	−7.3

*pK_a values in water

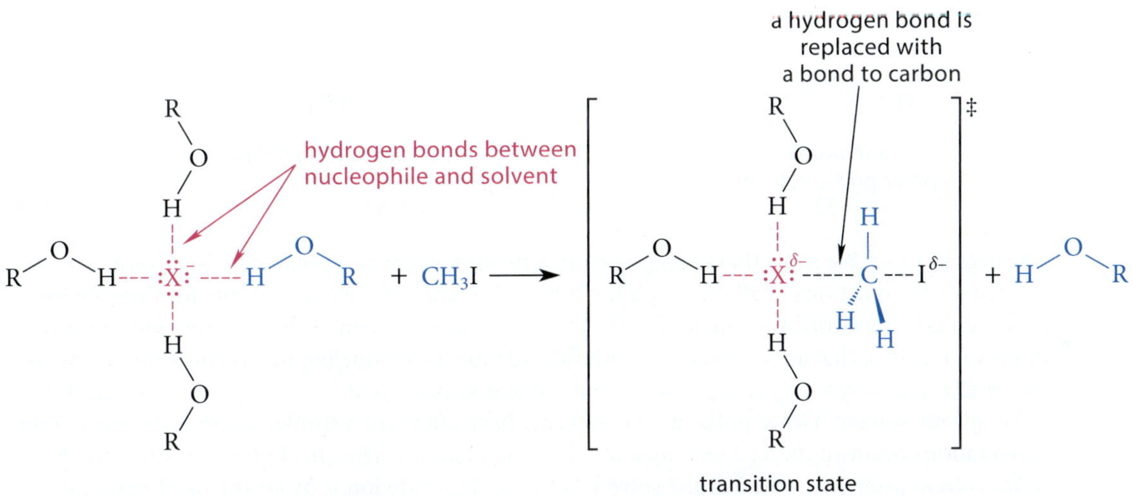

FIGURE 9.6 An S_N2 reaction of methyl iodide involving a nucleophile (:Ẍ:⁻) in a protic solvent ROH requires breaking a hydrogen bond between the solvent and the nucleophile. The energy required to break this hydrogen bond becomes part of the standard free energy of activation of the substitution reaction and thus retards the reaction.

and nucleophilicity within a group of the periodic table. The solvent in all of the cases shown in Tables 9.4 and 9.5 and Fig. 9.5 is methanol, a *protic* solvent. In a protic solvent, *hydrogen bonding* occurs between the protic solvent molecules (as hydrogen bond donors) and the nucleophilic anions (as hydrogen bond acceptors). *The strongest Brønsted bases are the best hydrogen bond acceptors.* For example, fluoride ion forms much stronger hydrogen bonds than iodide ion. When the electron pairs of a nucleophile are involved in hydrogen bonding, they are unavailable for donation to carbon in an S_N2 reaction. For the S_N2 reaction to take place, *a hydrogen bond between the solvent and the nucleophile must be broken* (**Fig. 9.6**). More energy is required to break a strong hydrogen bond to fluoride ion than is required to break a relatively weak hydrogen bond to iodide ion. This extra energy is reflected in a greater free energy of activation—the energy barrier—and, as a result, the reaction of fluoride ion is slower. To use a football analogy, the nucleophilic reaction of a strongly hydrogen-bonded anion with an alkyl halide is about as likely as a tackler bringing down a ball carrier when both of the tackler's arms are being held by opposing linemen.

The data in Fig. 9.5 and generalization 1 can be understood with a similar argument. If nucleophilicity and basicity were exactly correlated, the graph would follow the dashed blue line of Slope = 1. Focus on the blue curve, which shows the trend for nucleophiles that all have $-O^-$ as the reacting atom (*blue squares*). The downward curvature means that the nucleophiles of higher basicity do not react as rapidly with an alkyl halide as their basicity predicts, and the deviation from the line of unit slope is greatest for the most basic nucleophiles. The strongest bases form the strongest hydrogen bonds with the protic solvent methanol, and one of these hydrogen bonds has to be broken for the nucleophilic reaction to occur. As the hydrogen bond to solvent increases in strength, the rate-retarding effect on nucleophilicity also increases.

The data for nucleophiles shown with the black circles in Fig. 9.5 reflect the effects of hydrogen bonding to nucleophilic atoms that come from different groups within the same period (row) of the periodic table. For example, fluoride ion lies *below* the trend line for the oxygen nucleophiles. That is, fluoride ion is a worse nucleophile than an oxygen anion with the same basicity. The hydrogen bonds of fluoride with protic solvents are exceptionally strong, and so its nucleophilicity is correspondingly reduced. Conversely, the hydrogen bonds of azide ion and the carbon of cyanide ion with protic solvents are weaker than those of the oxygen anions, and their nucleophilicities are somewhat greater.

If hydrogen bonding by the solvent tends to reduce the reactivity of very basic nucleophiles, it follows that S_N2 reactions might be considerably accelerated if they could be carried out in solvents in which such hydrogen bonding is not possible. Let's examine this proposition with the aid of some data shown in **Table 9.6**. The two solvents, methanol ($\varepsilon = 33$) and *N,N*-dimethylformamide (DMF, $\varepsilon = 37$), were chosen for the comparison because their dielectric constants are nearly the same—that is, their polarities are very similar.

methanol
a polar protic solvent
$\varepsilon = 33$

***N,N*-dimethylformamide (DMF)**
a polar aprotic solvent
$\varepsilon = 37$ (9.28)

The data in this table show that changing from a protic solvent to a polar aprotic solvent accelerates the reactions of all nucleophiles, but the increase of the reaction rate for fluoride ion is particularly noteworthy—a factor of 10^8. In fact, the acceleration of the reaction with fluoride ion is so dramatic that an S_N2 reaction with fluoride ion as the nucleophile is converted from an essentially useless reaction in a protic solvent—one that takes years—to a rapid reaction in the polar aprotic solvent. Other polar aprotic solvents have effects of a similar magnitude, and similar accelerations occur in the S_N2 reactions of other alkyl halides. The effect on rate is due mostly to the *solvent proticity*—whether the solvent is protic. Fluoride ion is *by far* the most strongly hydrogen-bonded halide anion in Table 9.5; consequently, the change of solvent has the greatest

TABLE 9.6 Solvent Dependence of Nucleophilicity in the S_N2 Reaction

$$\text{Nuc}^{:-} + H_3C\text{—I} \xrightarrow[\text{or DMF, 25 °C}]{CH_3OH} \text{Nuc—}CH_3 + I^-$$

		In methanol		In DMF‡	
Nucleophile	pK_a*	k, $M^{-1}\,s^{-1}$	Reaction is over in—†	k, $M^{-1}\,s^{-1}$	Reaction is over in—†
I^-	−10	3.4×10^{-3}	17 min	4.0×10^{-1}	8.7 s
Br^-	−8	8.0×10^{-5}	12 h	1.3	2.7 s
Cl^-	−6	3.0×10^{-6}	13 days	2.5	1.4 s
F^-	3.2	5.0×10^{-8}	2.2 years	> 3	< 1.2 s
^-CN	9.4	6.3×10^{-4}	1.5 h	3.2×10^2	0.011 s

*pK_a values of the conjugate acid in water

† Time required for 97% completion of the reaction

‡DMF = *N, N*-dimethylformamide

effect on the rates of its S_N2 reactions. As the data demonstrate, *eliminating the possibility of hydrogen bonding to nucleophiles strongly accelerates their S_N2 reactions.*

What we've learned, then, is that S_N2 reactions of nucleophilic anions with alkyl halides are much faster in polar aprotic solvents than they are in protic solvents. If this is so, why not use polar aprotic solvents for all such S_N2 reactions? Here we must be concerned with an element of practicality. To run an S_N2 reaction in solution, we must find a solvent that dissolves an ionic compound (a salt or base) that contains the nucleophilic anion of interest. We must also remove the solvent from the products when the reaction is over. Protic solvents, precisely because they are protic, dissolve significant quantities of ionic compounds. Methanol and ethanol, two of the most commonly used protic solvents, are cheap, are easily removed because they have relatively low boiling points, and are relatively safe to use. When the S_N2 reaction is rapid enough, or if a higher temperature can be used without introducing side reactions, the use of protic solvents is often the most practical solvent for an S_N2 reaction. Except for acetone and acetonitrile (which dissolve relatively few ionic compounds), many of the commonly used polar aprotic solvents have very high boiling points and are difficult to remove from the reaction products. Furthermore, the solubility of ionic compounds in polar aprotic solvents is much more limited because they lack the protic character that solvates anions. However, for the less reactive alkyl halides, or for the S_N2 reactions of fluoride ion, polar aprotic solvents are in some cases the only practical alternative. (See the sidebar for an important medical example.)

The S_N2 Solvent Effect in Cancer Diagnosis

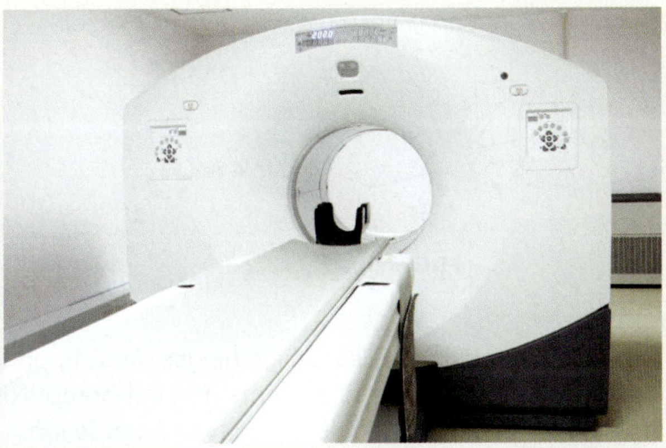

Alexander Gatsenko/iStock/Getty Images

Positron emission tomography, or "PET," is a widely used technique for cancer detection. In PET, a glucose derivative containing an isotope that emits positrons is injected into the patient. A glucose derivative is used because rapidly growing tumors have a high glucose requirement and therefore take up glucose to a greater extent than normal tissue. The emission of positrons (β^+ particles, or positive electrons) is detected when they collide with nearby electrons (β^- particles). This antimatter–matter reaction results in annihilation of the two particles and the production of two gamma rays that retreat from the site of collision in opposite directions, and these are detected ultimately as light. The light emission pinpoints the site of glucose uptake—that is, the tumor.

The glucose derivative used in PET is 2-^{18}fluoro-2-deoxy-D-glucopyranose, or FDG (known commercially as *fludeoxyglucose*), which contains the positron-emitting isotope ^{18}F ("fluorine-18"). The structure of FDG is so similar to the structure of glucose that FDG is also taken up by cancer cells.

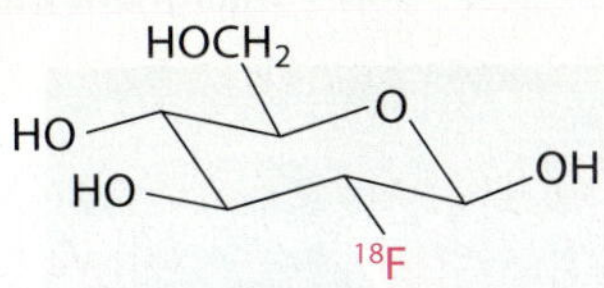

2-[^{18}F]-fluoro-2-deoxy-D-glucopyranose (FDG)

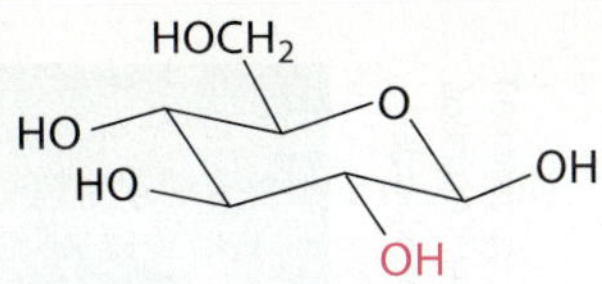

D-glucopyranose (glucose) (9.29)

The half-life of ^{18}F is only about 110 minutes. This means that half of it has decayed after 110 minutes, 75% has decayed after 220 minutes, and so on. This short half-life is good for the patient because the emitting isotope doesn't last very long in the body. But it places constraints on the chemistry used to prepare FDG. This means that ^{18}F, which is generated from $H_2{}^{18}O$ as an aqueous solution of $K^+\,{}^{18}F^-$, must be produced in or near the PET facility and used to prepare FDG quickly (in roughly 2 hours). An S_N2 reaction is used to prepare an FDG derivative using ^{18}F-fluoride as the nucleophile. Like other S_N2 reactions, this reaction occurs with inversion of configuration.

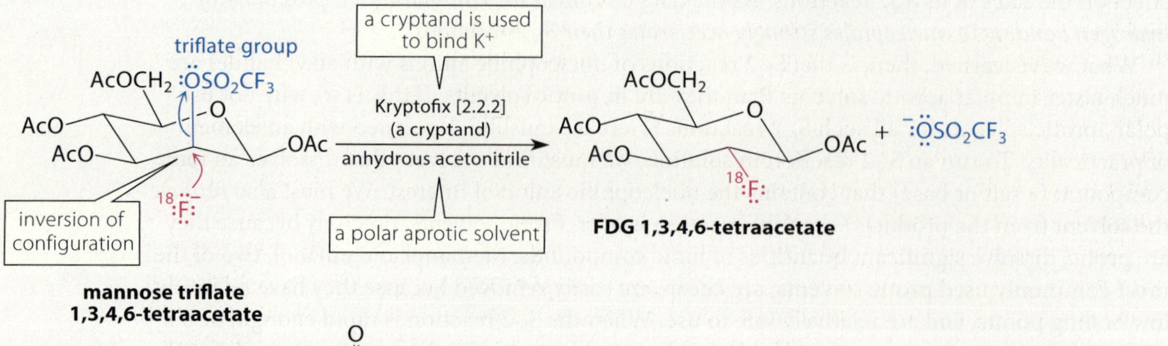

AcO— = acetate = $H_3C—\overset{O}{\overset{\|}{C}}—O—$ (9.30)

(The leaving group is a *triflate* group, which we discuss in Sec. 11.4A.) This synthesis cannot be carried out in water as a solvent because fluoride ion in protic solvents is virtually unreactive as a nucleophile. To solve this problem, water is completely removed from the aqueous fluoride solution and is replaced by acetonitrile, a polar aprotic solvent (see Table 8.2, Sec. 8.6B). Fluoride ion in anhydrous acetonitrile is a potent nucleophile, and to make it even more nucleophilic, a cryptand (see Fig. 8.12, Sec. 8.7C) is added to sequester the potassium counterion. This prevents the potassium ion from forming ion pairs with the fluoride ion. The "naked" and highly nucleophilic fluoride ion reacts rapidly with mannose triflate tetraacetate to form FDG tetraacetate, as shown in Eq. 9.30.

The acetate (—OAc) groups make the mannose derivative more soluble in acetonitrile than it would be if —OH groups were present. More important, though, O—H groups would themselves form hydrogen bonds with $^{18}F^-$, thus reducing its nucleophilicity and preventing the nucleophilic reaction from taking place. The acetate groups are rapidly removed in a subsequent ester hydrolysis reaction (Sec. 21.7A) to give FDG itself.

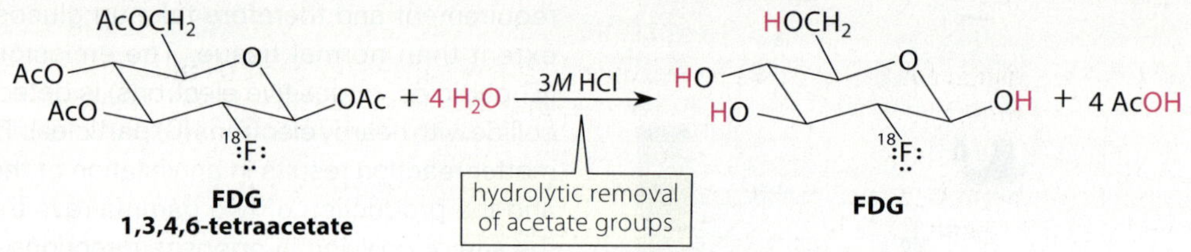

(9.31)

Figure 9.7 shows the PET image of a malignant lung tumor. PET is so sensitive that it has led to the detection of some cancers at an earlier and less invasive stage than previously possible. As we've seen, PET hinges on the rapid synthesis of FDG, which hinges, in turn, on the clever use of polar aprotic solvents and ion-complexing agents to enhance the nucleophilicity of fluoride ion.

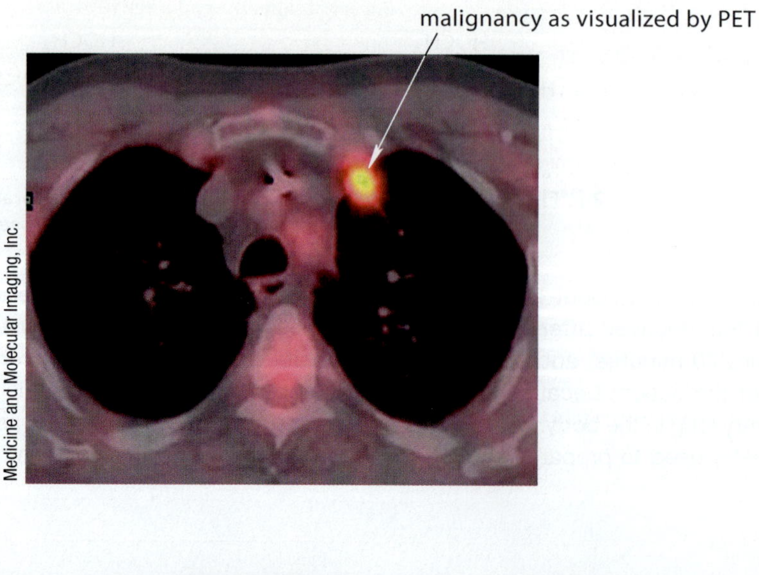

FIGURE 9.7 The PET image of a malignant lung tumor. The positron-emitting isotope ^{18}F is incorporated in the structure of FDG, a glucose derivative. FDG uptake, like glucose uptake, is enhanced in malignant tumors because they are rapidly growing and require more glucose than normal tissues.

Polarizability Effects on Nucleophilicity Examine the relative nucleophilicity of the halide ions in DMF (see Table 9.6). The most basic anions are the most nucleophilic ones, as we expect in a polar aprotic solvent, but if we compare basicity and nucleophilicity *quantitatively*, we find that the variation of nucleophilicity with basicity is not very large. Although the chloride ion is about 10,000 times as basic as iodide ion, it is only 6 times as nucleophilic. Cyanide ion is about 10^{14} times (that is, 14 pK units) as basic as bromide ion, but it is only 300 times as nucleophilic. In other words, iodide ion and bromide ion are much more nucleophilic than their basicities suggest. Why should such weak bases be good nucleophiles?

The reason is that nucleophilic atoms from higher-numbered periods of the periodic table are very *polarizable*. Recall (Sec. 8.5A) that *polarizability* is a measurement of how easily an electron cloud is distorted by an external charge—or, more intuitively, how "squishy" an electron cloud is. In nucleophiles derived from higher-numbered rows of the periodic table, the valence electron clouds are polarizable because they are screened from the nucleus by nonvalence electrons and are easily pulled away from the nucleus. For example, iodide ion is 3.1 times as polarizable as chloride ion and 6.9 times as polarizable as fluoride ion; sulfur is 3.6 times as polarizable as oxygen, and phosphorus is 3.3 times as polarizable as nitrogen. The distortion of a nucleophile's electron cloud is important as it forms a partial bond to the electrophile in a transition state; more bonding can occur at longer distances if the nucleophile is polarizable, and this leads to a transition state of lower energy.

In summary, then, *weak bases can be good nucleophiles if the nucleophilic atom is highly polarizable.* In Eq. 9.32, for example, iodide ion displaces bromide ion rapidly in the polar aprotic solvent acetone because, despite its very low basicity, the highly polarizable iodide anion is a good nucleophile.

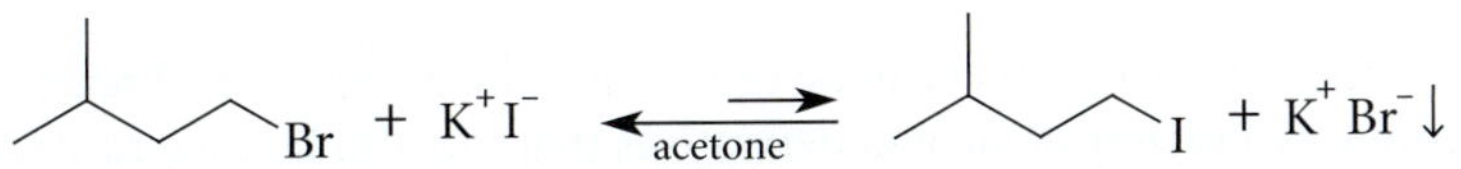

(9.32)

Although the equilibrium constant for the reaction is unfavorable, it is driven to the right by the insolubility of potassium bromide in acetone. Iodide is such a good nucleophile, in fact, that it can react in an S_N2 reaction with *tert*-butyl bromide (see Table 9.3, Sec. 9.4D).

As we've seen, *nucleophiles* and *bases* both donate electrons in their reactions. However, there are important differences between *nucleophilicity* and *basicity*. To summarize:

1. Nucleophilicity involves bond formation to atoms other than hydrogen, whereas basicity involves bond formation to hydrogen.

2. Nucleophilicities are measured with relative rates, but basicities are measured with equilibrium constants—that is, pK_a values. Polarizability has a greater effect on nucleophilicity than on basicity.

Focused Problems

9.14 When methyl bromide is dissolved in ethanol, no reaction occurs at 25 °C. When excess sodium ethoxide is added, a good yield of ethyl methyl ether is obtained. Explain.

9.15 (a) Give the structure of the S_N2 reaction product between ethyl iodide and potassium acetate.

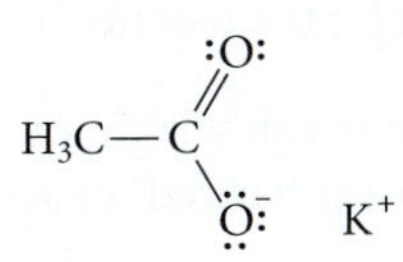

potassium acetate

(b) In which solvent would you expect the reaction to be faster: acetone or ethanol? Explain.

9.16 Which nucleophile, Et_3N: or Et_3P:, reacts most rapidly with methyl iodide in ethanol solvent? Explain, and give the product formed in each case.

F. Leaving-Group Effects in the S_N2 Reaction

In many cases, when an alkyl halide is to be used as a starting material in an S_N2 reaction, a choice of leaving group is possible. That is, an alkyl halide might be readily available as an alkyl chloride, alkyl bromide, or alkyl iodide. In such a case, the halide that reacts most rapidly is usually preferred. The reactivities of alkyl halides can be predicted from the close analogy between S_N2 reactions and Brønsted acid–base reactions. Recall that the ease of dissociating an H—X bond within the series of hydrogen halides depends mostly on the H—X bond energy (Sec. 3.6A), and, for this reason, H—I is the strongest acid among the hydrogen halides. Likewise, S_N2 reactivity depends primarily on the *carbon*–halogen bond energy, which follows the same trend: Alkyl iodides are the most reactive alkyl halides, and alkyl fluorides are the least reactive.

Relative reactivities in S_N2 reactions:

$$\text{R—I} > \text{R—Br} > \text{R—Cl} >> \text{R—F} \tag{9.33}$$

In other words, *the best leaving groups in the S_N2 reaction are those that give the weakest bases as products.* Fluoride is the strongest base of the halide ions; consequently, alkyl fluorides are the least reactive of the alkyl halides in S_N2 reactions. In fact, alkyl fluorides react so slowly that they are useless as leaving groups in most S_N2 reactions. In contrast, chloride, bromide, and iodide ions are much less basic than fluoride ion; alkyl chlorides, alkyl bromides, and alkyl iodides all have acceptable reactivities in typical S_N2 reactions, and alkyl iodides are the most reactive of these. On a laboratory scale, alkyl bromides, which are in most cases less expensive than alkyl iodides, usually represent the best compromise between expense and reactivity. For reactions carried out on a large scale, the lower cost of alkyl chlorides offsets the disadvantage of their lower reactivity.

Halides are not the only groups that can be used as leaving groups in S_N2 reactions. Section 11.4 introduces a variety of alcohol derivatives that can also be used as starting materials for S_N2 reactions.

G. Summary of the S_N2 Reaction

Most primary and some secondary alkyl halides undergo nucleophilic substitution by the S_N2 mechanism. Let's summarize five of the characteristic features of this mechanism:

1. The reaction rate is second order overall: first order in the nucleophile and first order in the alkyl halide.

2. The mechanism involves an opposite-side substitution reaction of the nucleophile with the alkyl halide and inversion of stereochemical configuration.

3. The reaction rate is decreased by alkyl substitution at both the α- and β-carbon atoms; alkyl halides with three β-branches are unreactive.

4. Nucleophilicity depends on the basicity of the nucleophile, the polarizability of the nucleophile, and the solvent.

    a. S_N2 reactions are much faster in polar aprotic solvents than in protic solvents provided that the nucleophile is soluble enough for the reaction to be practical. Protic solvents are useful if the reaction is fast enough.

    b. Nucleophiles in which the nucleophilic center is from periods ≥ 3 of the periodic table have enhanced nucleophilicity because they are highly polarizable.

    c. In polar aprotic solvents, nucleophilicity increases with the basicity of the nucleophile.

    d. In protic solvents, nucleophilicity increases with the basicity of the nucleophile if the nucleophilic atom is the same.

e. In protic solvents, nucleophilicity is significantly reduced when the nucleophilic atom is a good hydrogen-bond acceptor. For this reason, nucleophiles with period-2 nucleophilic atoms are less reactive in protic solvents than nucleophiles in which the nucleophilic atoms come from higher-numbered periods within the same group.

5. The fastest S_N2 reactions involve leaving groups that give the weakest bases as products.

9.5 THE E2 REACTION

This section discusses base-promoted β-elimination, which is a second important reaction of alkyl halides. An example of such a reaction is the elimination of the elements of HBr from *tert*-butyl bromide:

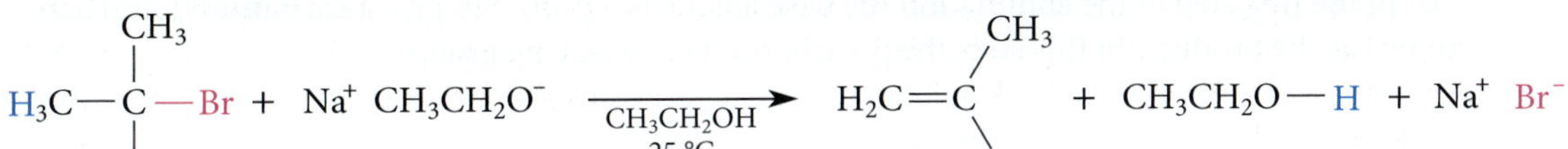

(9.34)

Recall from Sec. 9.1B that this type of elimination is a dominant reaction of tertiary alkyl halides in the presence of a strong base, and it competes with the S_N2 reaction in the case of secondary and primary alkyl halides.

A. Rate Law and Mechanism of the E2 Reaction

Base-promoted β-elimination reactions typically follow a rate law that is second order overall and first order in each reactant:

$$\text{rate} = k[(CH_3)_3C\text{—}Br][CH_3CH_2O^-] \qquad (9.35)$$

A mechanism consistent with this rate law is the following:

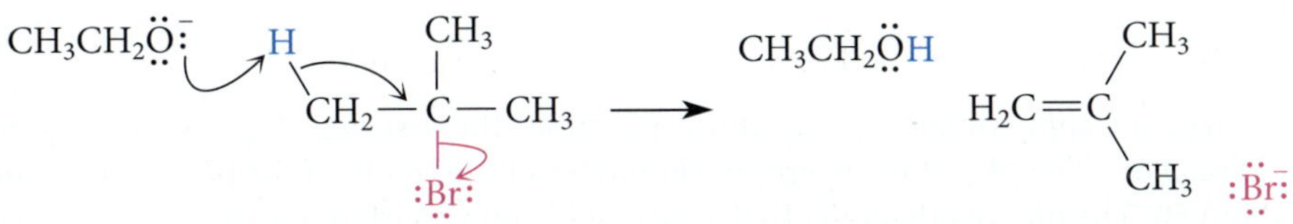

(9.36)

This type of mechanism, involving concerted removal of a β-proton by a base and loss of a halide ion, is called an **E2 mechanism**. Reactions that occur by the E2 mechanism are called **E2 reactions**. The meaning of the "nickname" E2 is as follows:

(9.37)

Bimolecular means that two molecules are involved in the rate-limiting step of the reaction. In this case, the two molecules are the base and the alkyl halide.

B. Why the E2 Reaction Is Concerted

The curved-arrow notation for the E2 mechanism in Eq. 9.36 is worth some attention. The simplest electron-pair displacement reactions we've encountered have involved the donation of an electron pair from either a Brønsted base or a nucleophile to an electrophile and simultaneous loss of a leaving group; the process is fully described by two curved arrows. However, the E2 reaction involves three curved arrows. In the E2 reaction, the base acts as a Brønsted base to remove the β-proton (first curved arrow), and the halide acts as a leaving group (last curved arrow).

How do we analyze the middle curved arrow? This curved arrow shows that the β-carbon acts *simultaneously* as a leaving group and as a nucleophile that reacts at the α-carbon. That is, the electron pair that departs from the β-hydrogen is donated to the α-carbon to expel the bromide ion.

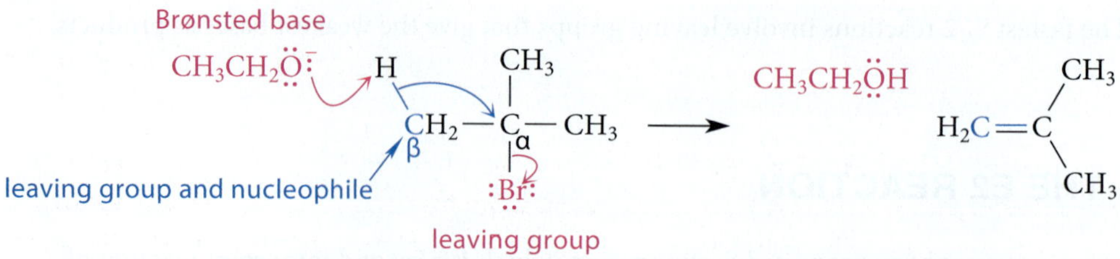

(9.38)

Let's separate this concerted mechanism into two *fictional* but more conventional steps of two curved arrows each. This will help us to understand why the reaction is concerted. Suppose that in the first step of the elimination the base abstracts a proton to give a **carbanion** (a carbon anion) as the product. In this step, the β-carbon acts as a leaving group.

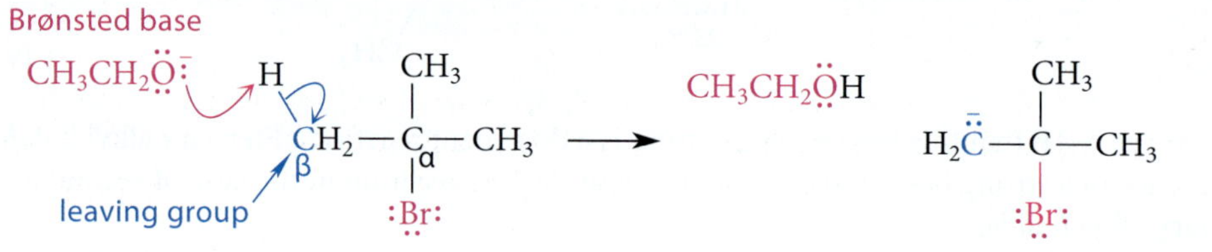

a carbanion intermediate (9.39a)

Then, in the second step, the electron pair of the carbanion acts as a nucleophile by reacting at the α-carbon, displacing bromide ion:

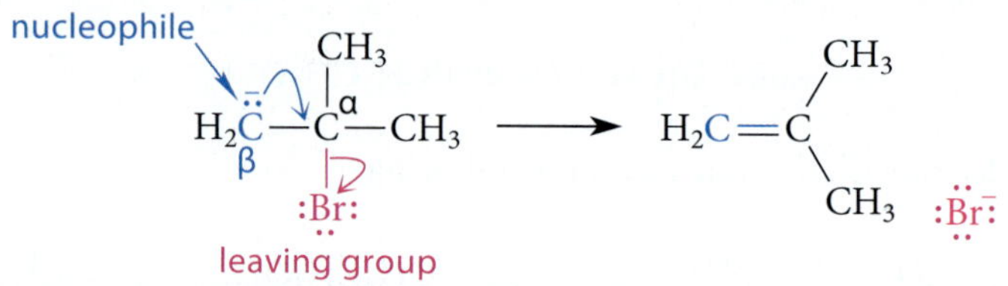

(9.39b)

Let's calculate the approximate equilibrium constant for the first step (Eq. 9.39a) using the method of Sec. 3.6C. The pK_a of the β-proton should be a little less than the pK_a of an alkane—perhaps about 50. The pK_a of ethanol is 15.9. The equilibrium constant for the first step is then $10^{(15.9-50)}$ or about 10^{-34}. The corresponding standard free-energy change is about 194 kJ mol^{-1} (46.4 kcal mol^{-1}). This means that if the reaction were to occur by this stepwise mechanism, the standard free energy of activation for the first step of this reaction must be at least 194 kJ mol^{-1}, because this is the amount of energy required to form the carbanion intermediate. The rate of such a reaction is unimaginably small: the reaction would take approximately 10^{15} years at room temperature! The elimination would not occur. In fact, typical E2 reactions occur in minutes to a few hours and have standard free energies of activation typically in the 85–95 kJ mol^{-1} (20–23 kcal mol^{-1}) range.

The concerted mechanism avoids the formation of a very unstable, strongly basic, carbanion intermediate. The concerted mechanism brings about a net transfer of electrons from the oxygen of ethoxide to bromine to form the much weaker base bromide ion. And that is why the middle curved arrow doesn't "pause" at carbon as an electron pair and "hang around" before it reacts at the α-carbon.

A carbanion mechanism can also be observed if an exceptionally strong base is used. In that case, the acid–base equilibrium shown in Eq. 9.36a is more favorable. If it is fast enough, it can then occur as a separate step.

In later sections of this text, we'll learn about β-eliminations that *do* involve carbanion intermediates. As we might expect, these reactions can take place only if the carbanion is stabilized in some way. To say that the carbanion is more stable is to say that the β-proton is much more acidic. Therefore, the stepwise β-elimination mechanism will be observed only with compounds in which the β-proton is unusually acidic (see Focused Problem 9.18).

Focused Problems

9.17 Give the possible compounds (including stereoisomers) that could be formed in the E2 reactions of each compound.

(a) 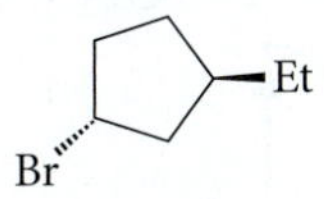

(b) 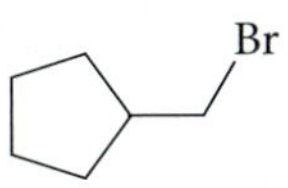

(c) 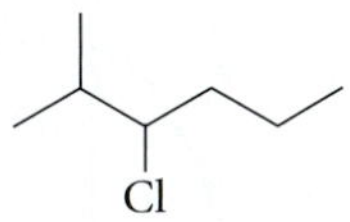

9.18 The following hydroxide-catalyzed β-elimination takes place by a carbanion stepwise mechanism. Show the carbanion intermediate, and explain its stability. Think in terms of a resonance effect (Sec. 3.7D) and a polar effect (Sec. 3.7E). Recalling also that resonance structures imply heightened stability (Sec. 1.4), draw a resonance structure for this anion as well.

$$HO{-}CH_2{-}CH_2{-}\overset{\overset{\large O}{\|}}{C}{-}CH_3 \underset{}{\overset{HO^-}{\rightleftharpoons}} H_2C{=}CH{-}\overset{\overset{\large O}{\|}}{C}{-}CH_3 + H_2O$$

C. Leaving-Group Effects on the E2 Reaction

In the mechanism of the E2 reaction, the role of the leaving halide is much the same as it is in the S_N2 reaction: its bond to carbon breaks and it takes on an additional electron pair to become a halide ion. Consequently, the rates of S_N2 and E2 reactions are affected in similar ways by changing the halide leaving group:

Relative rates of E2 reactions:

$$R{-}I > R{-}Br > R{-}Cl \qquad (9.40)$$

As in the S_N2 reaction, the reactivity difference between alkyl bromides and iodides is not great. Alkyl bromides are usually used in the laboratory for E2 reactions as the best compromise of reactivity and expense, and, when possible, the less expensive alkyl chlorides are used in large-scale reactions.

D. Deuterium Kinetic Isotope Effects in the E2 Reaction

The mechanism in Eq. 9.38 implies that a proton is removed in the transition state of the E2 reaction. This aspect of the mechanism can be tested in an interesting way. When a hydrogen is transferred in the rate-limiting step of a reaction, a compound in which that hydrogen is replaced by its isotope deuterium (^2H, usually represented as D) reacts more slowly in the same reaction. This effect of deuterium substitution on reaction rates is called a **primary deuterium kinetic isotope effect**. For example, suppose the rate constant for the following E2 reaction is k_H, and the rate constant for the reaction of the β-deuterium analog is k_D:

$$Ph{-}\overset{\overset{Me}{|}}{CH}{-}CH_2{-}Br + EtO^- \xrightarrow[EtOH]{\text{rate constant } k_H} Ph{-}\overset{\overset{Me}{|}}{C}{=}CH_2 + Br^- + EtOH \qquad (9.41a)$$

$$Ph{-}\overset{\overset{Me}{|}}{CD}{-}CH_2{-}Br + EtO^- \xrightarrow[EtOH]{\text{rate constant } k_D} Ph{-}\overset{\overset{Me}{|}}{C}{=}CH_2 + Br^- + EtOD \qquad (9.41b)$$

The primary deuterium kinetic isotope effect is the ratio of the rates for the two reactions—that is, k_H/k_D; typically such isotope effects are in the range 2.5–8. For example, k_H/k_D for the reactions in Eqs. 9.41a and 9.41b is 7.8. The observation of a kinetic isotope effect of this magnitude means that the bond to a β-hydrogen must be broken in the rate-limiting step of this reaction.

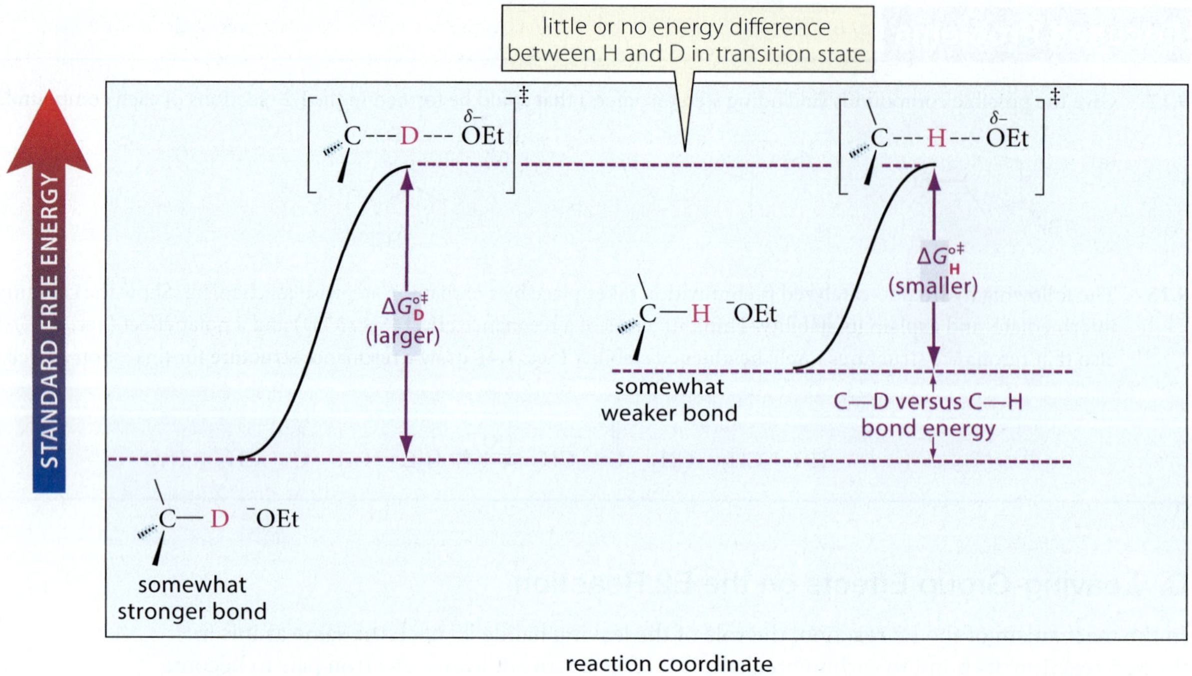

FIGURE 9.8 The source of the primary deuterium kinetic isotope effect is the stronger carbon–deuterium bond. (The difference between the bond energies of the C—H and C—D bonds [1.2% difference] is greatly exaggerated for purposes of illustration.)

The magnitude of the primary deuterium kinetic isotope effect depends on the degree of bonding to the transferred hydrogen or deuterium in the transition state, which varies from reaction to reaction. Also, small isotope effects, called *secondary isotope effects,* can occur when hydrogens or deuteriums that are not transferred are bonded to carbons that undergo hybridization changes in a reaction.

The theoretical basis for the primary kinetic isotope effect lies in the comparative strengths of C—H and C—D bonds. In the starting material, the bond to the heavier isotope D is slightly stronger (1.2% stronger). Breaking this bond therefore requires more energy than breaking the bond to the lighter isotope H. In the transition states for both reactions, however, the bond from H or D to carbon is partly broken, and the bond from H or D to the base is partly formed. To a crude approximation, the isotope undergoing transfer is not bonded to anything—it is "in flight." Because there is no bond, there is no bond-energy difference between the two isotopes in the transition state. Therefore, the compound with the C—D bond starts out at a lower energy than the compound with the C—H bond and requires more energy to achieve the transition state (**Fig. 9.8**). In other words, the energy barrier, or free energy of activation, for the compound with the C—D bond is greater; as a result, its rate of reaction is lower.

A primary kinetic deuterium isotope effect is observed only when the *hydrogen that is transferred* in the rate-determining step is substituted by deuterium. Substitution of other hydrogens with deuterium usually has little or no effect on the rate of the reaction.

Focused Problems

9.19 In each of the following series, arrange the compounds in order of increasing reactivity in the E2 reaction with Na^+ EtO^-.

(a)

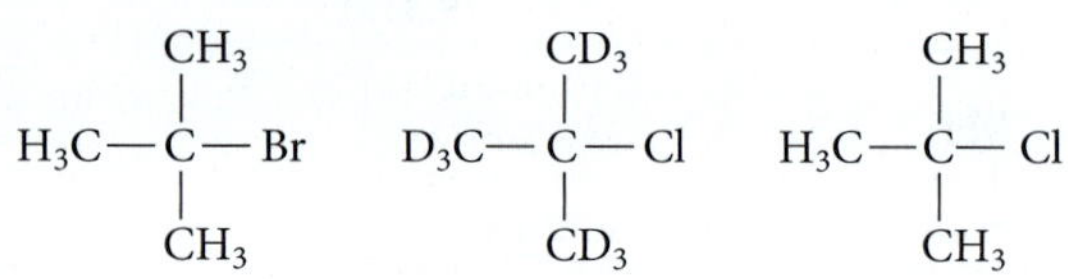

A *B* *C*

(b)

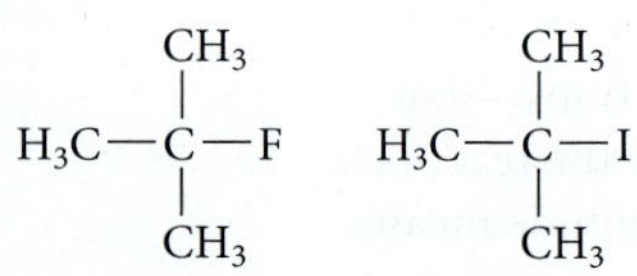

A *B*

9.20 (a) The rate-limiting step in the hydration of styrene (Ph—CH=CH_2) is the initial transfer of the proton from H_3O^+ to the alkene (Sec. 4.9B). How would you expect the rate of the reaction to change if the reaction were run in D_2O/D_3O^+ instead of H_2O/H_3O^+? Would the product be the same?

(b) How would the rate of styrene hydration in H_2O/H_3O^+ differ from that of an isotopically substituted styrene Ph—CD=CH_2? Explain.

E. Stereochemistry of the E2 Reaction

The E2 reaction can occur in two stereochemically distinct ways, illustrated as follows for the elimination of the elements of H—X from a general alkyl halide:

syn elimination:

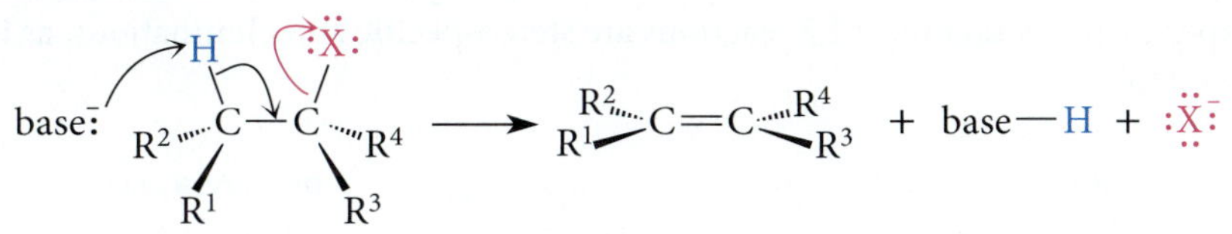

(9.42a)

anti elimination:

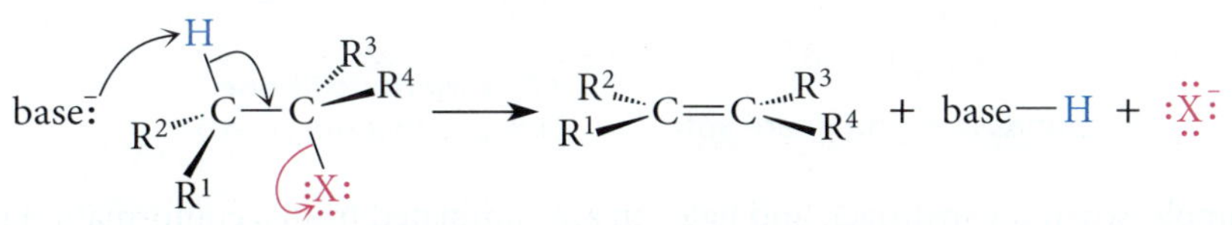

(9.42b)

As these examples illustrate, the tetrahedral α- and β-carbons become trigonal-planar when the β-proton is removed and the halide leaves. The R-groups on these two carbons move into a common plane that also contains the alkene carbons. This is the reference plane for the stereochemical designations *syn* and *anti*. This motion is shown in more detail for anti elimination in **Fig. 9.9.**

In a **syn elimination**, the dihedral angle between the C—H and C—X bonds is 0°—that is, the H and X groups leave from the *same* side of the reference plane. In an **anti elimination**, the dihedral angle between the C—H and C—X bonds is 180°, in which case the H and X groups leave from *opposite* sides of the reference plane. A Newman projection or a sawhorse projection of the transition state for anti elimination is another way to see the anti relationship of the proton that is removed and the leaving group:

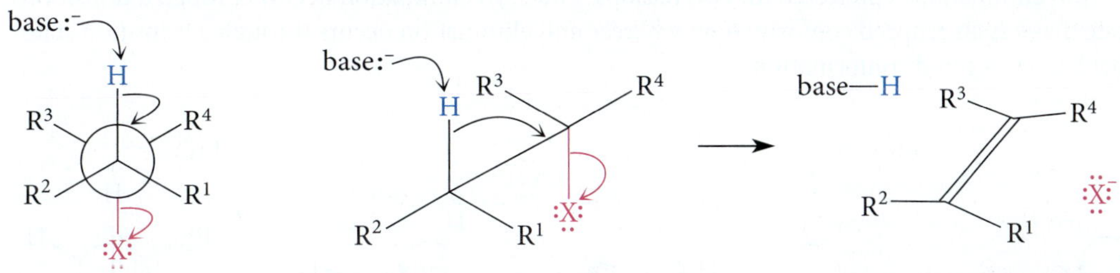

Newman projection for anti elimination

sawhorse projection for anti elimination

sawhorse projection for the alkene product of anti elimination

(9.43)

Only syn and anti eliminations are possible because only these geometries result in the planar alkene geometry that is required for π-orbital overlap. Recall from Sec. 7.8A that the terms *syn* and *anti* were used in discussing the stereochemistry of additions to double bonds. Notice that syn elimination is conceptually the reverse of a syn addition, and anti elimination is conceptually the reverse of an anti addition.

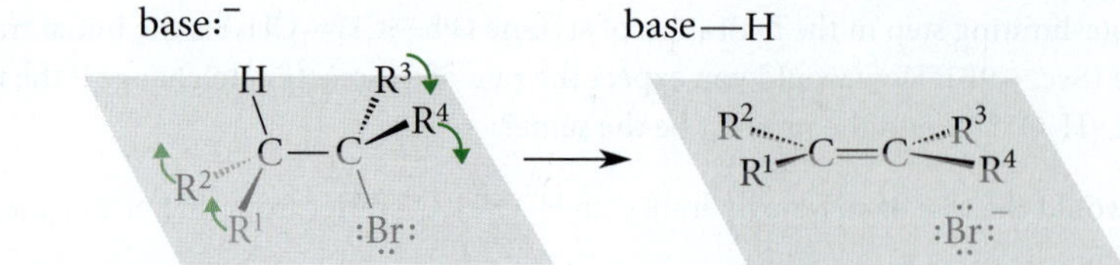

FIGURE 9.9 The stereochemical changes that occur during an anti E2 elimination. The α- and β-carbons are rehybridized from sp^3 to sp^2, and the R-groups attached to these carbons move into a common plane as shown by the green arrows. In this view, this plane is perpendicular to the page and tilted slightly downward.

Investigation of the stereochemistry of an elimination reaction requires the α- and β-carbons to be stereocenters in both the starting alkyl halide and the product alkene. In such cases, it is found experimentally that most E2 reactions are stereospecific anti eliminations, as in the following example.

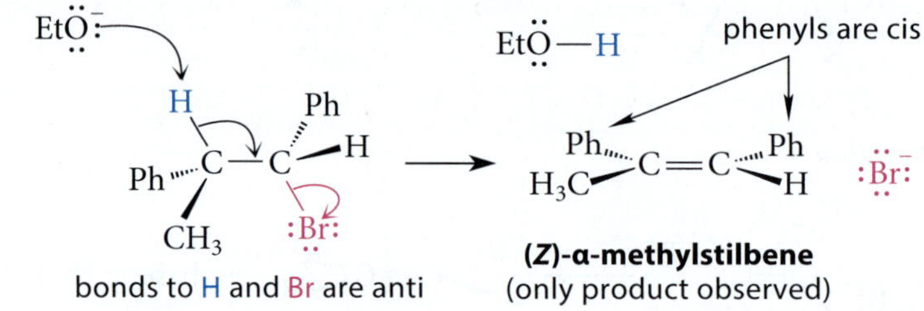

(9.44a)

In this example, when the hydrogen and halogen are eliminated from a conformation in which they are anti, the phenyl groups (Ph) are on the same side of the molecule and therefore must end up in a cis relationship in the product alkene. A syn elimination would give the other alkene stereoisomer:

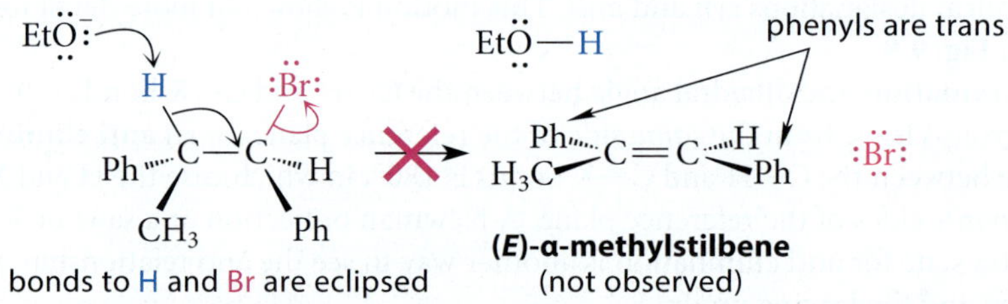

(9.44b)

Anti elimination is preferred for two reasons. First, syn elimination occurs through a transition state that has an eclipsed conformation, whereas anti elimination occurs through a transition state that has a staggered conformation.

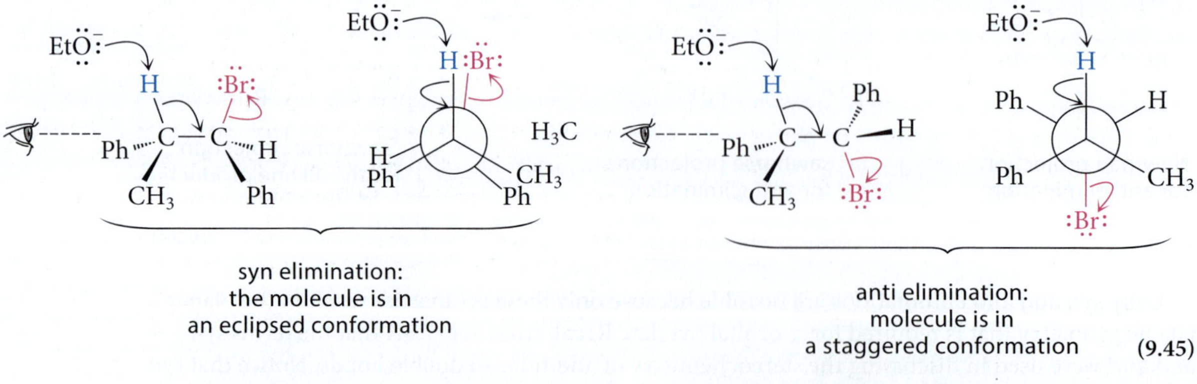

(9.45)

Because eclipsed conformations are unstable, the transition state for syn elimination is less stable than the transition state for anti elimination. As a consequence, anti elimination is faster.

The second reason is provided by molecular-orbital theory. Calculations of transition-state energies show that anti elimination is more favorable; the reasoning relates to the fact that an anti elimination involves all-opposite-side electron displacements, as in the S_N2 reaction.

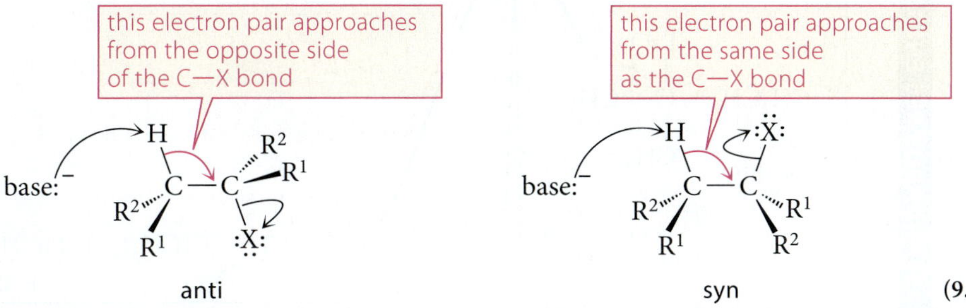

(9.46)

Focused Problems

9.21 Predict the products, including their stereochemistry, from the E2 reactions of the following diastereomers of stilbene dibromide with sodium ethoxide in ethanol. Assume that one equivalent of HBr is eliminated in each case. [*Hint for (a):* Draw either enantiomer in the correct conformation for anti elimination.]

(a) (±)-Ph—CH(Br)—CH(Br)—Ph (b) *meso*-Ph—CH(Br)—CH(Br)—Ph

9.22 Draw the structure of the starting material that would undergo anti elimination to give the *E* isomer of the alkene product in the E2 reaction of Eq. 9.44a.

F. Regioselectivity of the E2 Reaction

When an alkyl halide has more than one type of β-hydrogen, more than one alkene product can be formed (Sec. 9.1B).

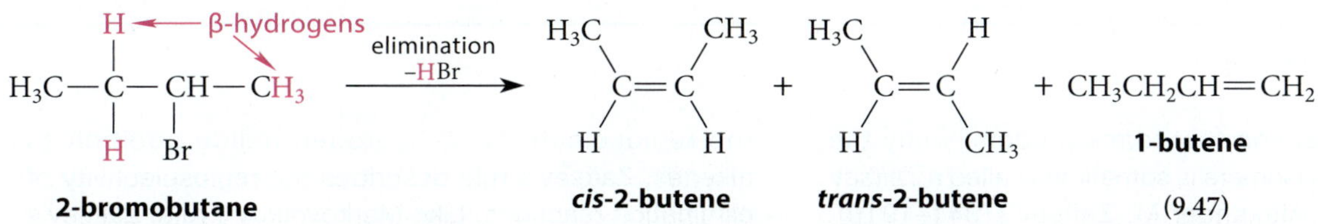

(9.47)

This section focuses on which of the possible products is preferred and why.

When simple alkoxide bases such as methoxide and ethoxide are used, *the predominant product of an E2 reaction is usually the most stable alkene isomer*. Recall (Sec. 4.5B) that the most stable alkene isomers are generally those with the most alkyl substituents at the carbons of the double bond. These isomers, then, are the ones formed in greatest amount.

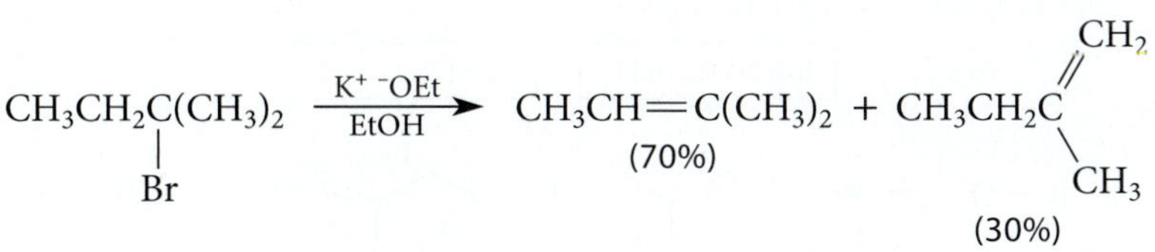

(9.48)

In this reaction, the alkene isomer formed in a smaller amount would actually be favored statistically: six equivalent hydrogens can be lost from the alkyl halide to give this alkene, but only two can be lost to give the other alkene. In the absence of a structural effect on the product distribution, three times as much of the 1-alkene would have been formed. The fact that the other alkene is the major product shows that some other factor is operating.

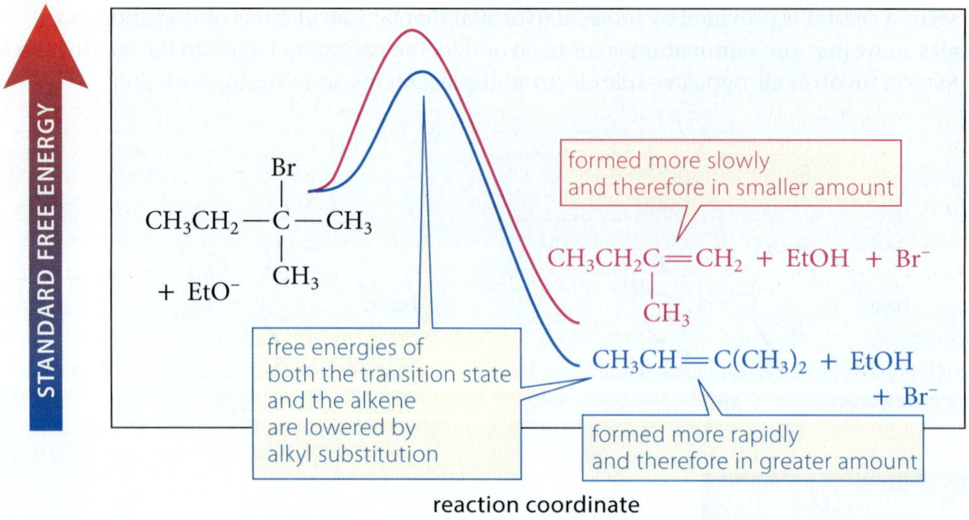

FIGURE 9.10 The alkene with more alkyl substituents on the double bond is formed more rapidly and in greater amount than the alkene with fewer substituents because the free energy of the transition state, like that of the alkene, is lowered by alkyl substitution.

The predominance of the more stable alkene isomer does *not* result from equilibration of the alkenes themselves, because *the alkene products are stable under the conditions of the reaction*. The product mixture, once formed, does not change, so the distribution of products must reflect the relative rates at which they are formed. Therefore, we look for the explanation in transition-state theory.

The transition state for the E2 reaction can be visualized as a structure that lies somewhere between alkyl halide and alkene (plus the other species present). To the extent that the transition state resembles the alkene product, it is stabilized by the same factors that stabilize alkenes—and one such factor is alkyl substitution at the double bond. A reaction that can give two alkene products is really two reactions in competition, each with its own transition state. The reaction with the transition state of lower energy—the one with more alkyl substitution at the developing double bond—is the faster reaction. Therefore, more product is formed through this transition state (**Fig. 9.10**).

Zaitsev's Rule

An elimination reaction that forms predominantly the most stable alkene isomers is sometimes called a *Zaitsev elimination*, after Alexander M. Zaitsev (1841–1910), a Russian chemist who observed this phenomenon in 1875. Just as Markovnikov's rule (Sec. 4.7A) describes the regioselectivity of hydrogen halide addition to alkenes, Zaitsev's rule describes the regioselectivity of elimination reactions. Like Markovnikov's rule, Zaitsev's rule is purely descriptive; it does not attempt to explain the reasons behind the observations.

Highly branched bases give a higher percentage of the alkene with *fewer* alkyl substituents on the double bond—that is, the *less stable* alkene.

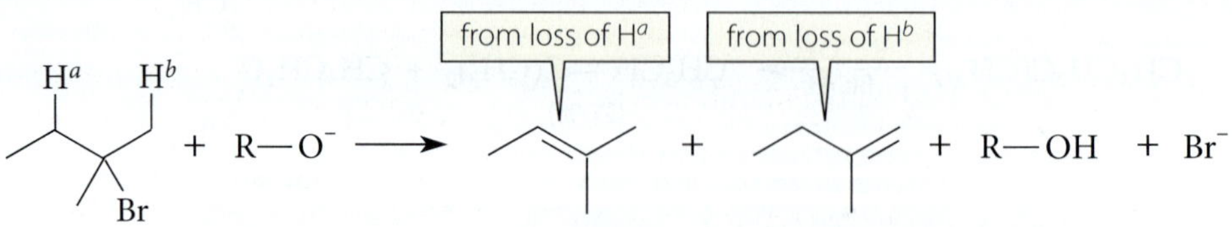

base = EtO⁻	70%	30%
base = *t*BuO⁻	28%	72%

(9.49)

This effect is probably due to the less congested environment of protons H^b at the end of the molecule. Removal of protons H^a requires somewhat greater steric repulsion of a highly branched base with the surrounding hydrogens.

We've seen that when an alkyl halide has more than one type of β-hydrogen, a mixture of alkenes is generally formed in its E2 reaction, as Eqs. 9.48 and 9.49 illustrate. The formation of a mixture means that the yield of any single alkene isomer is reduced. Furthermore, because the alkenes in such mixtures are isomers, they generally have similar boiling points, which makes them difficult to separate. Consequently, the greatest use of the E2 elimination for the preparation of alkenes occurs when the alkyl halide has only one type of β-hydrogen, and only one alkene product is possible.

Study Problem 9.1 gives you some practice in integrating the principles involved in stereochemistry and regioselectivity of eliminations with what you learned about cyclohexane conformations in Chapter 7.

Study Problem 9.1

Give the predominant product of the E2 reaction when each of the following diastereomers is allowed to react with potassium *tert*-butoxide in dimethyl sulfoxide (DMSO). Explain your reasoning.

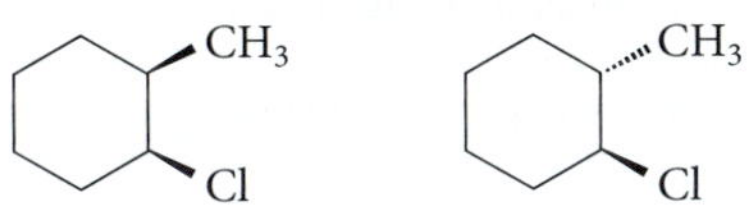

A *B*

Solution Two principles are important in solving this problem. The first is that the stereochemistry of the E2 reaction is anti. In a cyclohexane derivative, the required anti relationship of the proton that is lost and the leaving group is possible only if the two are trans. In compound *A*, only hydrogens H^x and H^y are trans to the Cl. Therefore, elimination of either proton could occur to give products *X* and *Y*, respectively. (The ring carbons are numbered for reference.)

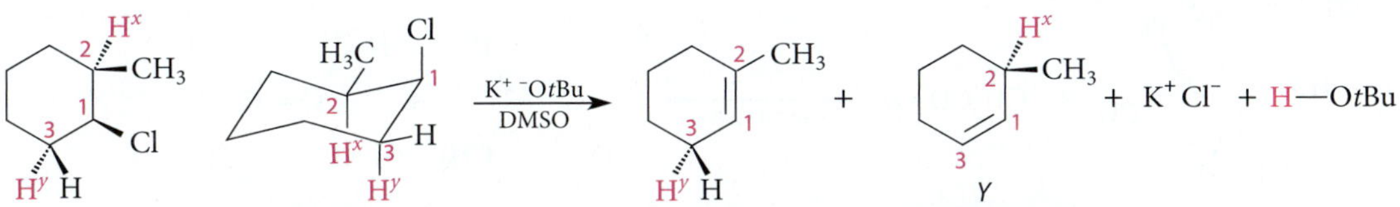

A — one chair conformation of *A* — *X* (from loss of H^x) *major product* — *Y* (from loss of H^y) *minor product*

The second principle is that the predominant product is the one with the greater number of alkyl substituents on the double bond—product *X*. Product *X*, then, is the major product, as shown in the equation.

In compound *B*, only hydrogen H^p is trans to the leaving group. Hydrogen H^q is cis to the leaving group and therefore cannot be anti. Consequently, only compound *Y* is formed. Although isomer *X* is more stable, there is no anti-elimination path by which it can be formed.

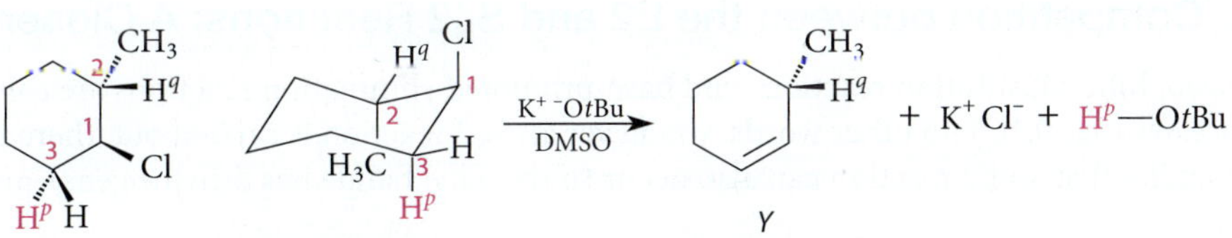

B — one chair conformation of *B* — *Y* (from loss of H^p) *only product*

In both cyclohexane derivatives *A* and *B*, notice that the Cl leaving group and the trans hydrogen are anti in *only one* of the two chair conformations of compound *A*; they are gauche in the other. Therefore, the elimination must occur from the H-anti conformation, shown here for the formation of the major product *X* from compound *A*:

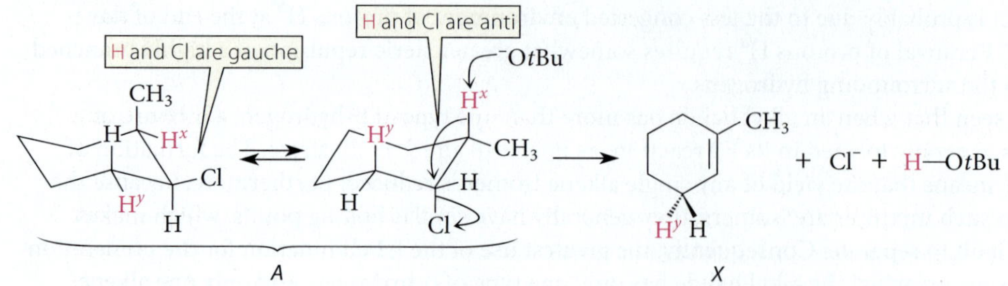

The H-anti conformation is depleted as it reacts, but it is restored rapidly by the conformational equilibrium (Le Chatelier's principle). The same is true for the formation of product *Y* from both isomers *A* and *B* (show this). (Problems 9.47, 9.71, and 9.72 at the end of the chapter provide more practice in solving this type of problem.)

Focused Problems

9.23 In the reaction shown in Eq. 9.47, which alkene is formed in *least* amount? Why?

9.24 Consider the following E2 reactions of two diastereomeric isotopically substituted derivatives of 2-bromobutane. (D = ^{2}H = deuterium.)

(1)

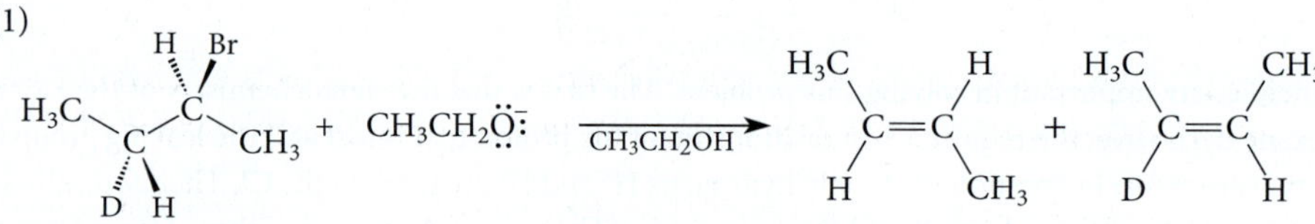

(2)

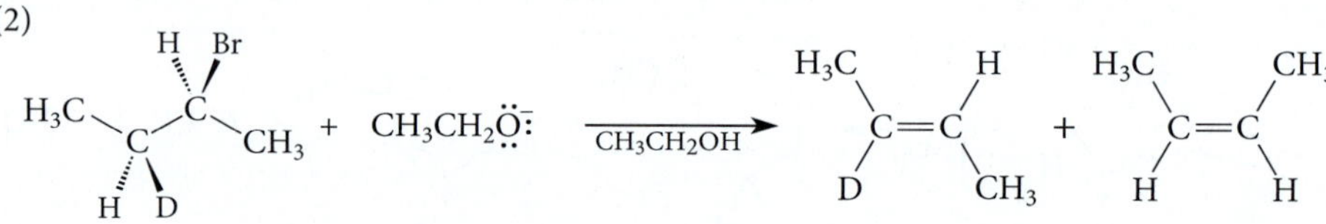

(a) Explain why, in reaction (1), only the *Z* stereoisomer of 2-butene contains deuterium, whereas, in reaction (2), only the *E* isomer contains deuterium. (*Hint:* Consider the stereochemistry of the E2 reaction.)

(b) In each reaction, a third alkene is formed in small amount. What is it? Is this alkene the same for the two reactions or different? Explain.

G. Competition between the E2 and S_N2 Reactions: A Closer Look

Nucleophilic substitution reactions and base-promoted elimination reactions are *competing* processes (Sec 9.1C). In other words, whenever an S_N2 reaction is carried out, there is the possibility that an E2 reaction can also occur (if the alkyl halide has β-hydrogens), and vice versa.

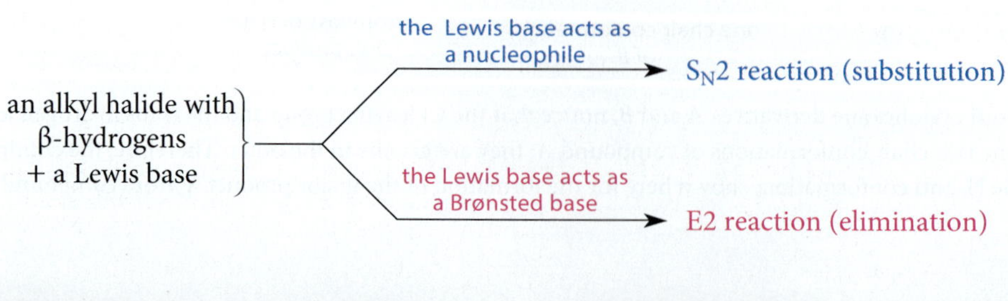

(9.50)

This competition is a matter of *relative rates*: The reaction pathway that occurs more rapidly is the one that predominates.

Two variables determine which reaction—the S_N2 reaction or the E2 reaction—will be the major process observed in a given case: (1) the structure of the alkyl halide and (2) the structure of the base.

The feature of an alkyl halide's structure that determines the amount of elimination versus substitution is the *number of alkyl substituents* at both the α- and β-carbons. For the S_N2 reaction to occur on an alkyl halide with α- or β-substituents, the nucleophile must approach through a thicket of interfering hydrogen atoms on the substituents that impede its access to the α-carbon:

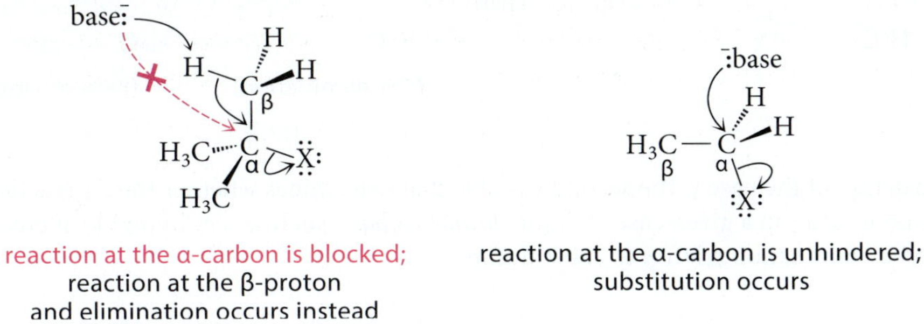

(9.51)

The resulting van der Waals repulsions create an energy barrier to the S_N2 reaction that decreases its rate. On the other hand, when a Brønsted base initiates the E2 reaction, it reacts with a β-proton that lies near the periphery of the molecule. Reaction at the β-proton is much less affected by steric repulsions than reaction at the α-carbon atom.

Another reason that alkyl substitution promotes the E2 reaction is that the standard free energy of the E2 transition state, like that of an alkene, is lowered by alkyl substitution (Sec. 9.5F). Consequently, the rate of the E2 reaction is *increased by alkyl substitution*. Two effects of alkyl substitution, then, favor the E2 reaction: the rate of the S_N2 reaction is *decreased*, and the rate of the E2 reaction is *increased*.

These same effects can be seen not only in tertiary alkyl halides, but also in secondary and even primary alkyl halides. Notice in the following examples that the alkyl halides with more β-alkyl substituents show a greater proportion of elimination product.

Secondary alkyl halides:

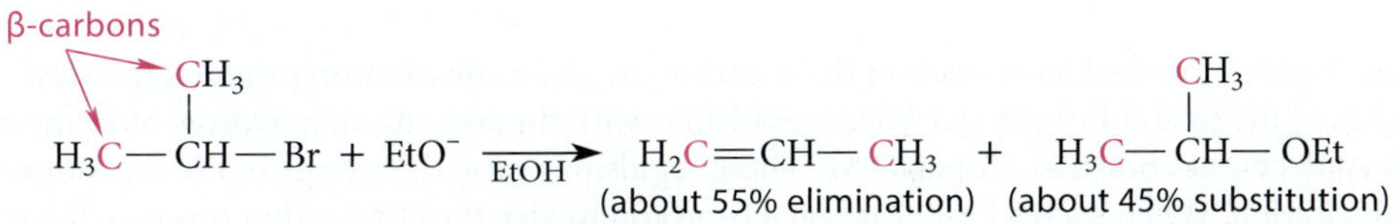

(9.52a)

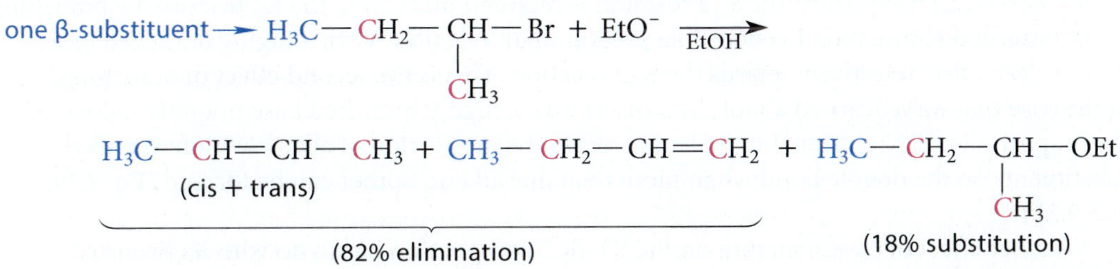

(9.52b)

Primary alkyl halides:

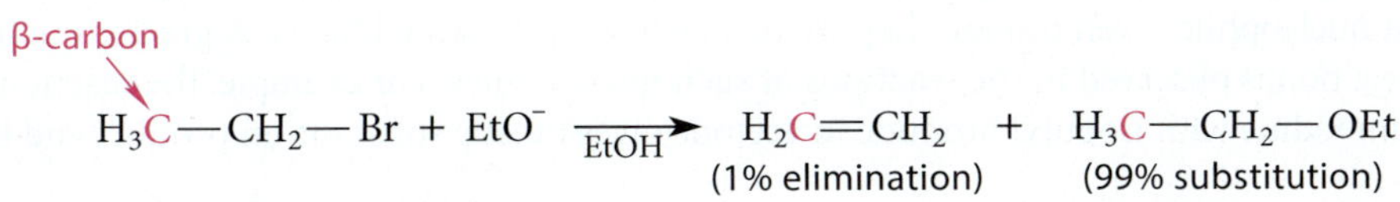

(9.53a)

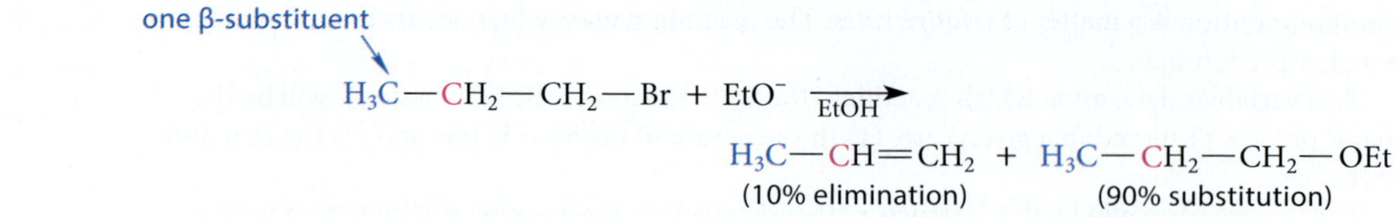

(9.53b)

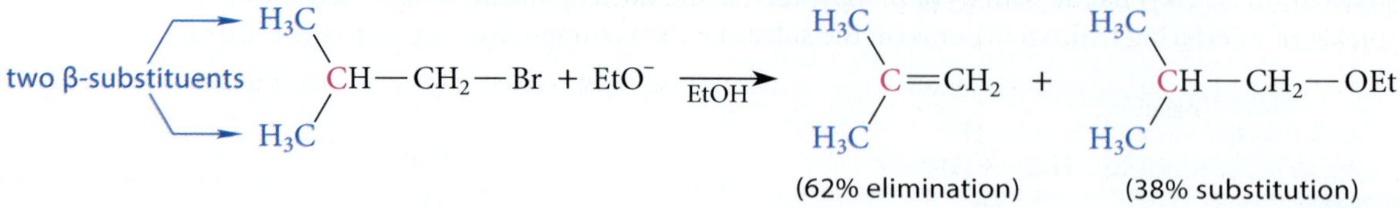

(9.53c)

The structure of the base is the second variable that determines whether the E2 reaction or the S_N2 reaction is faster in a given case. A *highly branched base*, such as *tert*-butoxide, increases the proportion of elimination relative to substitution.

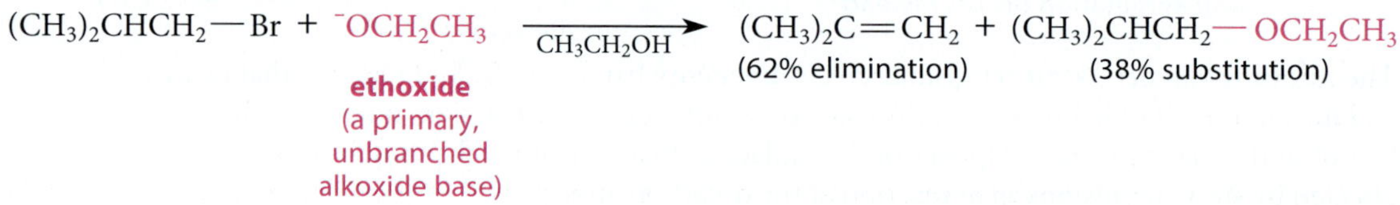

(9.54a)

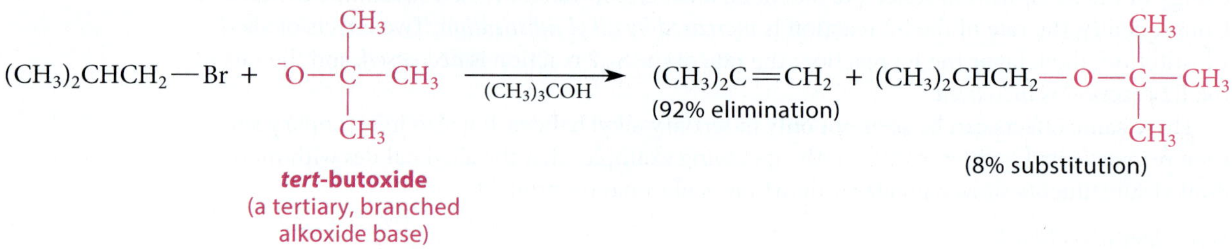

(9.54b)

When a highly branched base reacts at the α-carbon to give a substitution product, the alkyl branches of the base suffer van der Waals repulsions with the surrounding hydrogens in the alkyl halide molecule as shown in Display 9.51. These repulsions raise the energy of the transition state for substitution. When such a base reacts at a β-proton to give the elimination product, the base is further removed from the offending hydrogens in the alkyl halide, so van der Waals repulsions are less severe. Consequently, the S_N2 reaction is retarded more than the E2 reaction by branching in the base, and elimination becomes the predominant reaction. With a highly branched base, then, a *steric effect* selectively retards the S_N2 reaction. This is the second effect of branching in the base that we've learned about. In other words, a highly branched base not only increases the proportion of E2 reaction, but it also increases the proportion of alkene with fewer alkyl substituents on the double bond when more than one alkene isomer can be formed (Eq. 9.49, Sec. 9.5F).

A further effect of base structure on the E2–S_N2 competition has to do with its Brønsted basicity versus its nucleophilicity. Recall from Sec. 9.4E that the reactivity of a nucleophile—its nucleophilicity—affects the rate of its S_N2 reactions, whereas its Brønsted basicity affects the rate of its E2 reactions (because the base is reacting with a proton). Recall also that species with nucleophilic atoms from higher periods of the periodic table, such as iodide ion, are excellent nucleophiles even though they are relatively weak Brønsted bases. A greater fraction of S_N2 reaction is observed in the reactions of such nucleophiles. For example, the reaction of potassium iodide with isobutyl bromide in acetone gives mostly substitution product and little elimination, because iodide, a highly polarizable ion, is an excellent nucleophile but a weak Brønsted base:

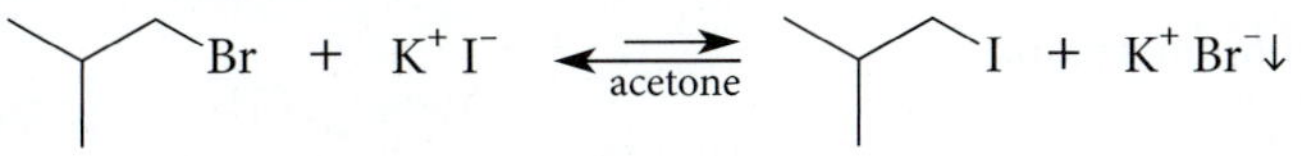

(9.55)

Contrast this reaction with that in Eq. 9.53c, in which sodium ethoxide reacts with the same alkyl halide. Ethoxide, a strong Brønsted base, gives a significant percentage of alkene and a smaller percentage of substitution product.

Let's summarize the effects that govern the competition between the S_N2 and E2 reactions.

1. *Structure of the alkyl halide:*
    a. Alkyl halides with greater numbers of alkyl substituents at the α-carbon give greater amounts of elimination. Consequently, tertiary alkyl halides give more elimination than secondary alkyl halides, which give more than primary alkyl halides.
    b. Alkyl halides with greater numbers of alkyl substituents at the β-carbon give greater amounts of elimination.
    c. Alkyl halides that have no β-hydrogens cannot undergo β-elimination.
2. *Structure of the base:*
    a. In a comparison of alkoxide bases with similar strengths, tertiary alkoxide bases such as *tert*-butoxide give a greater fraction of elimination than primary alkoxide bases. They also give higher percentages of the alkene with fewer alkyl substituents on the double bond.
    b. Weaker bases that are good nucleophiles give a greater fraction of substitution.

These ideas are applied in Study Problem 9.2.

Study Problem 9.2

Which alkyl halide and what conditions should be used to prepare the following alkene in good yield by an E2 elimination?

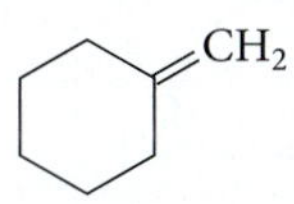

methylenecyclohexane

Solution If this alkene is to be produced in an E2 reaction from an alkyl halide, the halide must be located at one of the two carbons that eventually become carbons of the double bond. This means that there are two choices for the starting alkyl halide:

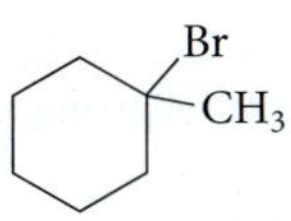

A

and

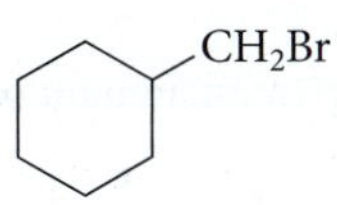

B

The advantage of alkyl halide *A* is that it is tertiary, so it poses no significant competition from the S_N2 reaction. The disadvantage of *A* is that it contains more than one type of β-hydrogen, so more than one alkene product could be formed:

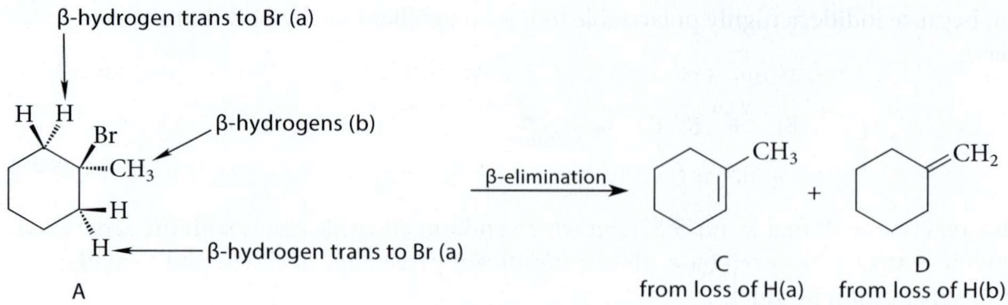

(9.56)

STUDY GUIDE LINK 9.2
Ring Carbons as Alkyl Substituents

Product *C* is the more stable alkene because its double bond has three alkyl substituents; so, if *A* is used as the starting material, a major amount of this undesired alkene will be formed. If alkyl halide *B* is the starting material, then the desired product *D* is the *only* possible product of β-elimination. Because this alkyl halide is primary, however, it is possible that some by-product derived from the S_N2 reaction will be formed. The way to minimize the S_N2 reaction is to use a tertiary alkoxide base such as *tert*-butoxide. Moreover, the extensive β-substitution in alkyl halide *B* should also minimize the substitution reaction. Therefore, a reasonable preparation of the desired alkene is the following:

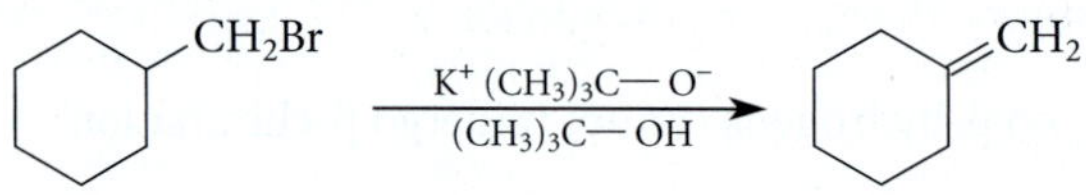

(9.57)

Focused Problems

9.25 What nucleophile or base and what type of solvent could be used for the conversion of 1-bromo-3-methylbutane into each of the following compounds?

(a) 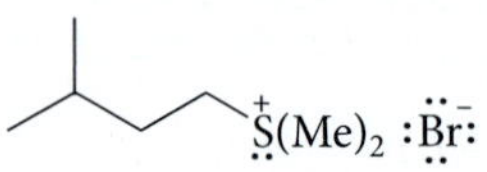

(b) 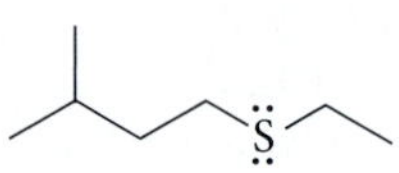

(c) 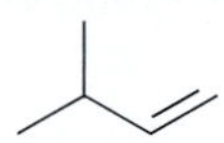

9.26 Arrange the following four alkyl halides in descending order with respect to the E2 elimination to S_N2 substitution product ratio expected in their reactions with sodium ethoxide in ethyl alcohol. Explain your answers.

CH_3I
A

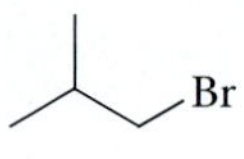

B

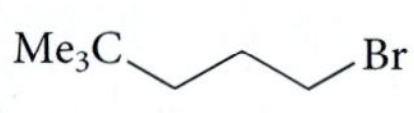

C

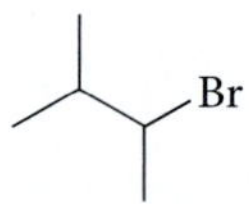

D

9.27 (a) Arrange the following four bases in descending order with respect to the E2 elimination to S_N2 substitution product ratio expected when they react with isobutyl bromide. Explain your answers.

$(CH_3)_2CH—O^-$	CH_3O^-	$(CH_3CH_2)_3C—O^-$	Cl^-
A	*B*	*C*	*D*

(b) Which base in part (a) is likely to give the greatest amount of alkene *Y* in the following reaction? Explain.

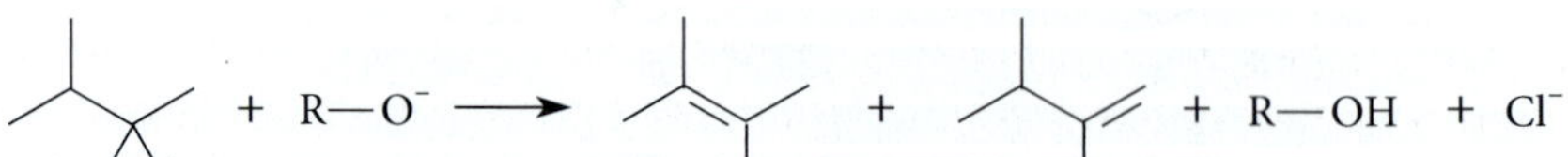

H. Summary of the E2 Reaction

The E2 reaction is a β-elimination reaction of alkyl halides that is promoted by strong bases. The following list summarizes the key points about this reaction:

1. The rates of E2 reactions are second order overall: first order in base and first order in the alkyl halide.
2. E2 reactions normally occur with anti stereochemistry.
3. The E2 reaction is faster with better leaving groups—that is, those that give the weakest bases as products.
4. The rates of E2 reactions show substantial primary deuterium kinetic isotope effects at the β-hydrogen atoms.
5. When an alkyl halide has more than one type of β-hydrogen, more than one alkene product can be formed; the most stable alkenes (the alkenes with the greatest numbers of alkyl substituents at their double bonds) are formed in greatest amount unless the base is highly branched.
6. E2 reactions compete with S_N2 reactions. Elimination is favored by alkyl substitution in the alkyl halide at the α- or β-carbon atoms, by alkyl substituents at the α-carbon of the base, and by highly branched bases.

9.6 THE S_N1 AND E1 REACTIONS

In the previous sections of this chapter, the discussion has stressed the reactions of alkyl halides with either strong bases or good nucleophiles. When a primary alkyl halide is dissolved in a *protic solv*ent such as ethanol with *no added base*, the S_N2 reaction that occurs takes 2 weeks or more (depending on the temperature and the alkyl halide), because a neutral, un-ionized alcohol is a weak base and therefore a poor nucleophile. When a tertiary alkyl halide such as *tert*-butyl bromide is subjected to the same conditions, however, both substitution and elimination reactions occur readily.

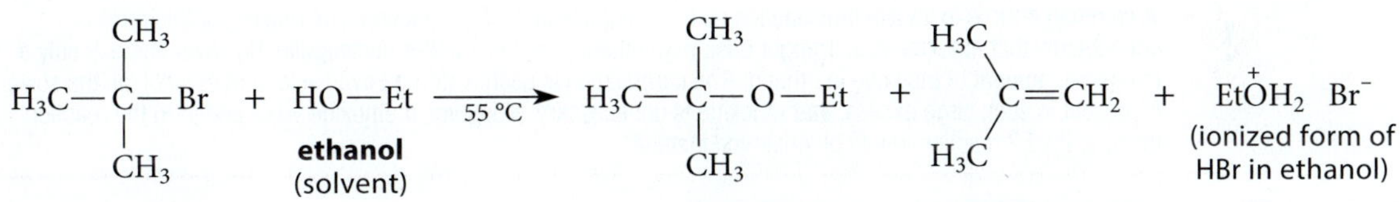

***tert*-butyl bromide** ***tert*-butyl ethyl ether** (72%) **methylpropene** (28%) (9.58)

The reaction of an alkyl halide with a solvent in which no other base or nucleophile has been added is called a **solvolysis** (literally, bond breaking by solvent). The substitution that occurs in the solvolysis of *tert*-butyl bromide cannot involve an S_N2 mechanism because chain branching at the α-carbon retards the S_N2 reaction. That is, if the solvolysis of a primary alkyl halide by an S_N2 mechanism is very slow, then the solvolysis of a tertiary alkyl halide by the same mechanism should be *even slower*. The elimination that occurs in this solvolysis cannot occur by an E2 mechanism because a strong base is not present. Because both substitution and elimination reactions occur readily, *they must then involve mechanisms that are different from the S_N2 and E2 mechanisms.* These new mechanisms are the subject of this section.

A. Rate Law and Mechanism of the S_N1 and E1 Reactions

The solvolysis of *tert*-butyl bromide follows a *first-order rate law*:

$$\text{rate} = k[(CH_3)_3CBr] \qquad (9.59)$$

Any involvement of solvent in the reaction cannot be detected in the rate law because the concentration of the solvent cannot be changed. However, the nature of the solvent does play a critical role in this reaction. The solvolysis reactions of tertiary alkyl halides are fastest in *polar, protic, donor solvents*, such as alcohols, formic acid, and mixtures of water with solvents in which the alkyl halide is soluble (for example, aqueous acetone). Recall (Sec. 8.6F) that *these solvents are the ones that are best at solvating ions.*

The occurrence of both substitution and elimination products shows that two *competing reactions* are involved. The first step in *both* reactions involves the ionization of the alkyl halide to a carbocation and a halide ion:

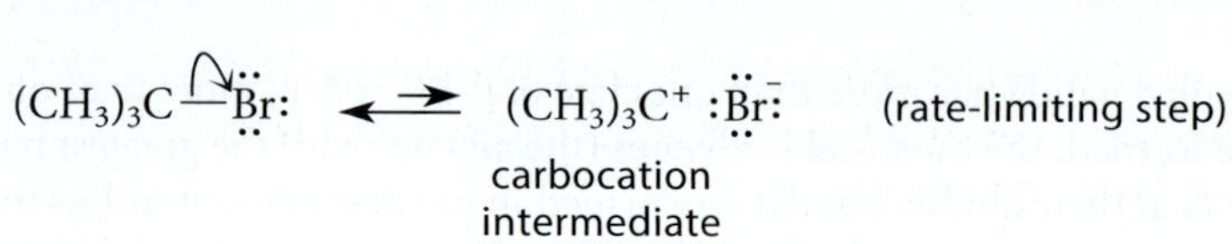

(9.60a)

This step, which is a *Lewis acid–base dissociation* (Sec. 3.1A), is the rate-limiting step of both the substitution and elimination reactions. In other words, when a tertiary alkyl halide is dissolved in a polar, protic solvent such as ethanol, it reacts by dissociating slowly into a carbocation and a halide ion. The carbocation then rapidly reacts to give both substitution and elimination products. That is, substitution and elimination products arise from *competing reactions of the same carbocation.*

The substitution product is formed by the Lewis acid–base association of a solvent molecule with the carbocation. Even though the solvent is a poor nucleophile, the reaction occurs rapidly because the solvent is present in high concentration and because the carbocation is a powerful Lewis acid.

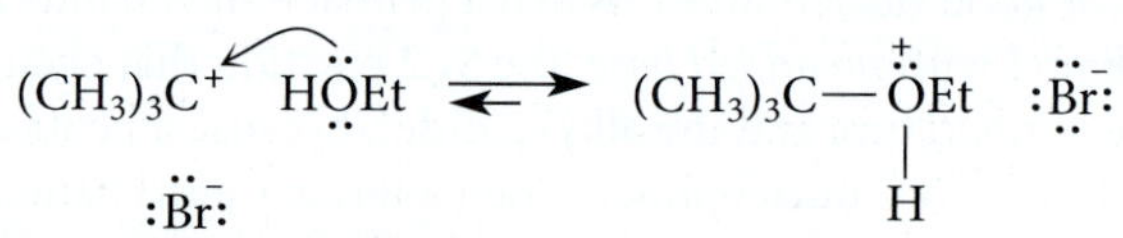

(9.60b)

> **!** A common error is to invoke ethoxide ion as a nucleophile in the S_N1 reaction. The reason for this error is undoubtedly that ethoxide is a stronger base than ethanol, so it is a better nucleophile. However, there is only a minuscule amount of ethoxide in ethanol. The carbocation is such a strong Lewis acid, and ethanol (as the solvent) is present in such large excess, that ethoxide is unnecessary. Moreover, if ethoxide were added to the reaction mixture, the E2 reaction would be observed instead.

The product of Eq. 9.60b is the conjugate acid of an ether, which is a strong acid.

The final step of the substitution reaction is a Brønsted acid–base reaction in which the protonated ether (the Brønsted acid) loses a proton to solvent (the Brønsted base) to give the ether and the conjugate acid of the solvent.

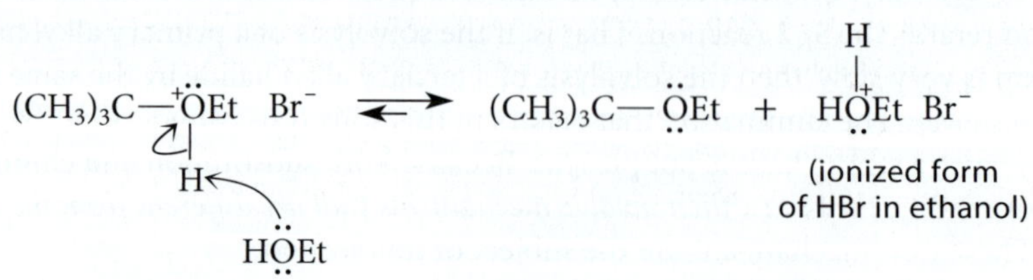

(9.60c)

The Brønsted base involved in this reaction is ethanol, not ethoxide ion. As noted in the previous step of the reaction, ethoxide ion is not present; nor is it necessary, because the protonated ether is a strong acid, and the base (the solvent) is present in large excess. The protonated solvent plus bromide ion (that is, $Et\overset{+}{O}H_2\ Br^-$) is the form of ionized HBr in ethanol solvent, just as $H_3O^+\ Br^-$ is the ionized form of HBr in water.

A substitution mechanism that involves a carbocation intermediate is called an **S_N1 mechanism**. Substitution reactions that take place by the S_N1 mechanism are called **S_N1 reactions**. The meaning of the S_N1 nickname is as follows:

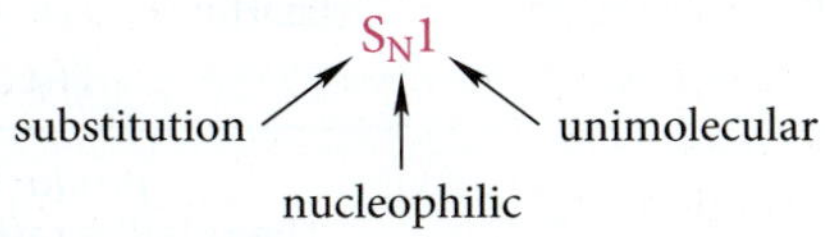

(9.61)

The word *unimolecular* means that a single molecule—the alkyl halide—is involved in the rate-limiting step.

The formation of the elimination product of Eq. 9.58 involves a different reaction of the carbocation intermediate. Loss of a β-proton (a proton from the carbon adjacent to the electron-deficient carbon) gives the alkene.

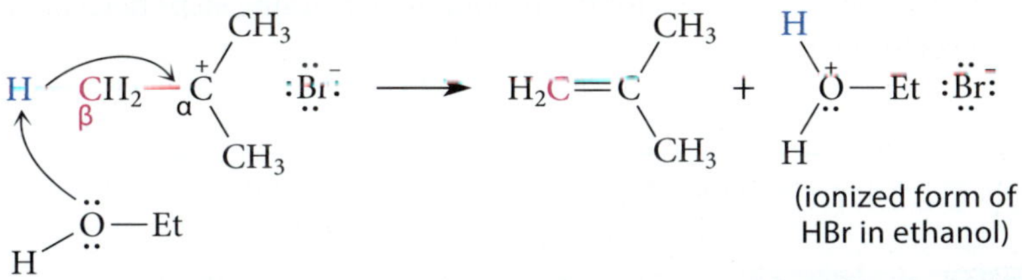

(9.62)

The base that removes a β-proton from the carbocation is typically a solvent molecule. Although ethanol is a very weak base, the reaction occurs readily because ethanol, as the solvent, is present in very high concentration and because the carbocation is a *very strong* Brønsted acid (its pK_a has been estimated to be about –8). The base is *not* ethoxide ion; no ethoxide ion is present. Notice that ionized HBr is produced in this reaction as well.

A β-elimination mechanism that involves carbocation intermediates is called an **E1 mechanism**, and reactions that occur by E1 mechanisms are called **E1 reactions**. The meaning of the E1 nickname is as follows:

(9.63)

B. Rate-Limiting and Product-Determining Steps

The S_N1 and E1 reactions have a *common rate-limiting step*. That is, the rate at which the alkyl halide disappears as it undergoes both competing reactions is determined by its *rate of ionization*—the rate at which it forms the carbocation. The relative amounts of substitution and elimination products are determined by the relative rates of the steps that *follow* the rate-limiting step: reaction of the solvent as a nucleophile with the carbocation to give a substitution product, and loss of a β-proton to solvent from the carbocation to give the elimination product. For example, more substitution than elimination product is formed in Eq. 9.58. This means that the rate of formation of the substitution product from the carbocation is greater than the rate of formation of the elimination product. Because the relative rates of these steps determine the ratio of products, they are said to be the **product-determining steps**. *The relative rates of the product-determining steps have nothing to do with the rate at which the alkyl halide dissociates into ions.*

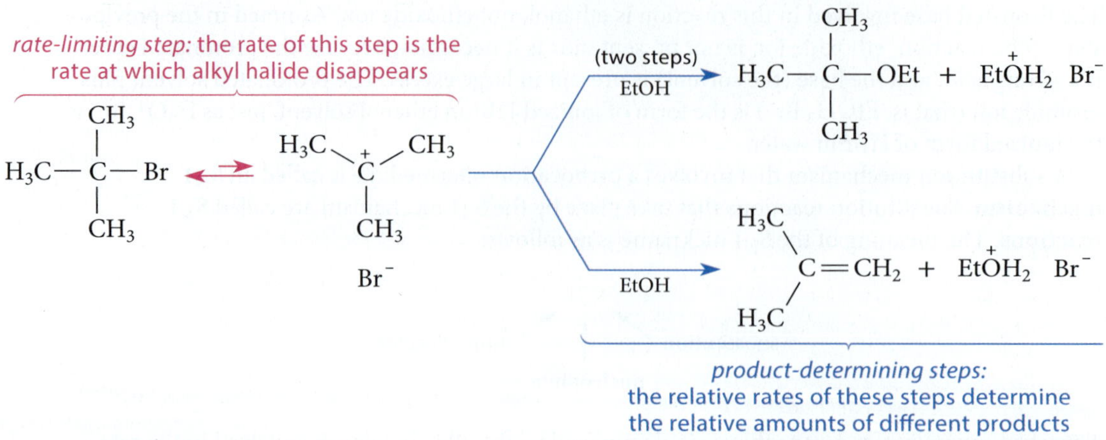

(9.64)

The reaction free-energy diagram in **Fig. 9.11** summarizes these ideas. The first step, ionization of the alkyl halide to a carbocation, is the rate-limiting step, so this step has the transition state of highest free energy. The rate of this step is the rate at which the alkyl halide reacts. The relative free-energy barriers for the product-determining steps determine the relative amounts of products formed.

An Analogy for Product-Determining Steps

Andrey Bayda/Shutterstock

Imagine that you are in the Southwest at a busy "T" intersection controlled by a traffic light. Traffic backs up during the red light, and a car perhaps has to wait for several signal changes to get through the intersection. Passage through the intersection is the *rate-limiting step* on this highway. After leaving the intersection quickly, you can turn left (to Albuquerque) or right (to Phoenix).

The total rate that cars pass through the intersection is controlled by the traffic light, not by how fast the cars make the turn after the light turns green. The relative numbers of cars that take the left (eastern) or right (western) turn determine the relative numbers of cars that end up on the highways to the two destinations. Entering a turn is analogous to the product-determining step. If more cars turn west, then the rate of cars that turn west is greater than the rate at which cars turn east. For example, if three times as many cars turn west as turn east, then the number of cars entering the western highway per unit time is three times the number entering the eastern highway. Similarly, if two products *A* and *B* are formed and if the rate of formation of *A* is three times the rate of formation of *B*, then 75% of the product is *A* and 25% is *B*. However, just as the total rate at which cars pass through the intersection is controlled by the traffic light, the total rate of a reaction is controlled by the rate of the rate-limiting step.

The competition between the S_N1 and E1 reactions is different from the competition between the S_N2 and E2 reactions. The latter two reactions share nothing in common but starting materials; they follow completely separate reaction pathways with no common intermediates.

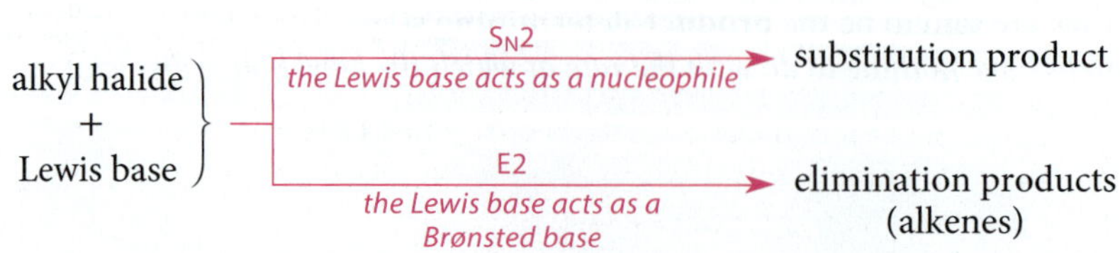

(9.65)

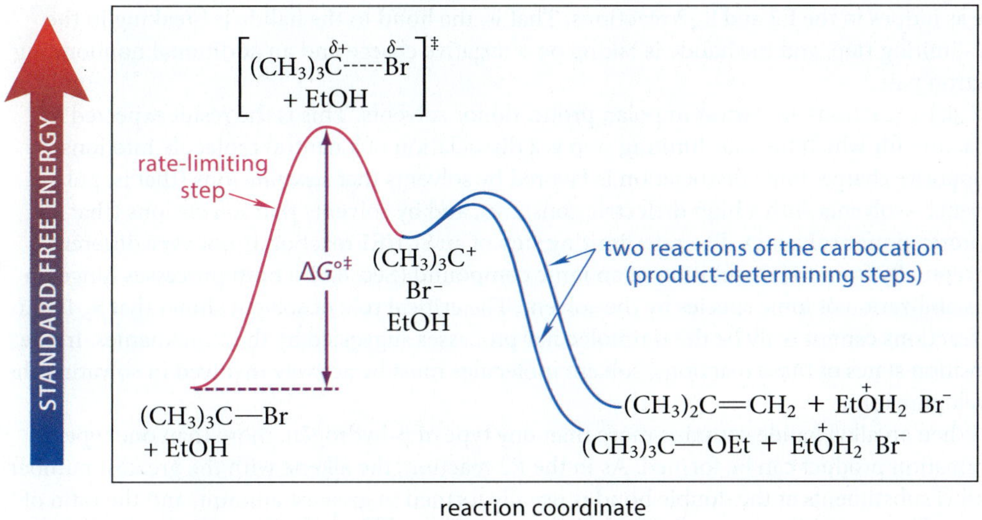

FIGURE 9.11 Reaction free-energy diagram for the S_N1/E1 solvolysis reaction of $(CH_3)_3CBr$ with ethanol. The rate-limiting step, ionization of the alkyl halide (*red curve*), has the transition state of highest standard free energy. The relative rates of the product-determining steps (*blue curves*) determine the relative amounts of substitution and elimination products. In this example, the energy barrier for the substitution reaction is lower; therefore, the substitution reaction is faster than the elimination reaction, and more substitution than elimination product is observed. (The final proton transfer required to form the substitution product is not shown explicitly.)

In contrast, the S_N1 and E1 reactions of an alkyl halide share not only common starting materials, but also a common rate-limiting step, and therefore *a common intermediate*—the carbocation.

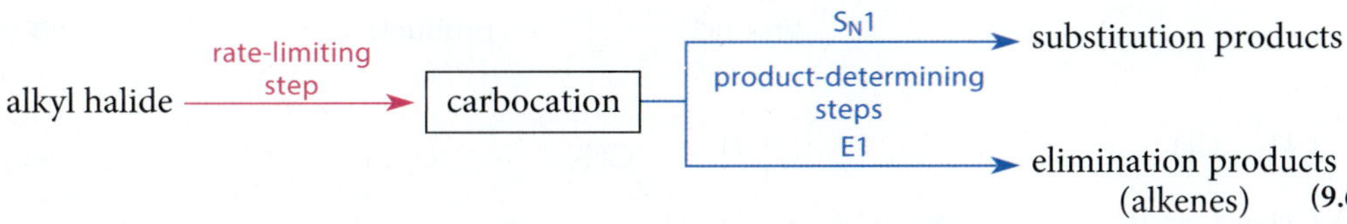

(9.66)

In the E1 reaction, the proton is not removed from the alkyl halide, as it is in the E2 reaction, but from the *carbocation*. Because the carbocation is a strong acid, a strong base is *not* required for the E1 reaction as it is for the E2 reaction.

C. Reactivity and Product Distributions in S_N1/E1 Reactions

S_N1/E1 reactions are most rapid with tertiary alkyl halides, they occur more slowly with secondary alkyl halides, and they are never observed with primary alkyl halides.

Reactivity of alkyl halides in S_N1 or E1 reactions:

$$\text{tertiary} >> \text{secondary} >> \text{primary} \qquad (9.67)$$

The exact relative reactivities vary with conditions, particularly the solvent; as an example, *tert*-butyl bromide is 1150 times more reactive than isopropyl bromide in ethanol, and about 10^5 times more reactive in water. The relative reactivity of primary alkyl halides in S_N1 or E1 reactions is not known with certainty because they do not react at all by this mechanism.

This reactivity order is expected from the relative stability of the corresponding carbocation intermediates. Hammond's postulate (Sec. 4.8D) suggests that the rate-limiting transition state of an S_N1 or E1 reaction should closely resemble a carbocation.

The reactivity order of the alkyl halides in S_N1/E1 reactions is iodides >> bromides > chlorides >> fluorides. This is the same reactivity order observed in S_N2 and E2 reactions. This relative reactivity is expected because the leaving group in the S_N1/E1 reaction has much the same role as it does in the E2 and S_N2 reactions. That is, the bond to the halide is breaking in the rate-limiting step, and the halide is taking on a negative charge and an additional nonbonding electron pair.

S_N1/E1 reactions are fastest in polar, protic, donor solvents. This is the result expected in a reaction for which the rate-limiting step is a dissociation of a neutral molecule into ions of opposite charge. Ionic dissociation is favored by solvents that *separate* ions (that is, polar solvents—solvents with a high dielectric constant), and by solvents that *solvate* ions (that is, protic, donor solvents). The rate-limiting step of an S_N1/E1 reaction is not very different conceptually from the dissolution of an ionic compound (Sec. 8.6F); both processes hinge on the stabilization of ionic species by the solvent. The critical role of solvent shows that S_N1 and E1 reactions cannot truly be the unimolecular processes suggested by their nicknames. In the transition states of these reactions, solvent molecules must be actively involved in solvating the developing ions.

When an alkyl halide contains more than one type of β-hydrogen, more than one type of elimination product can be formed. As in the E2 reaction, the alkene with the greatest number of alkyl substituents at the double bond is usually formed in greatest amount; and the ratio of alkene (E1) to substitution product (S_N1) is greater when the alkene formed contains more than two alkyl substituents at the double bond. The examples in Eqs. 9.68a and 9.68b illustrate both of these points.

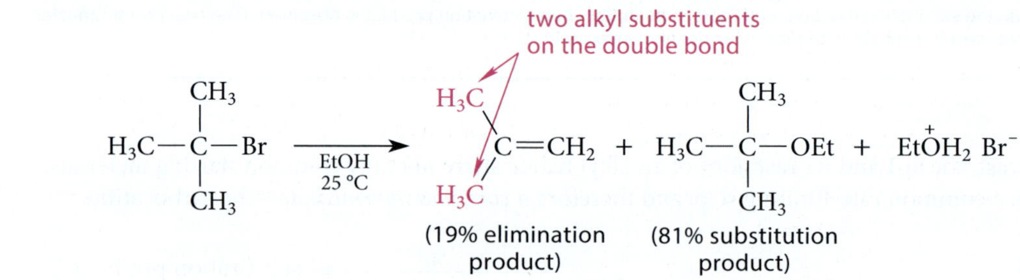

(9.68a)

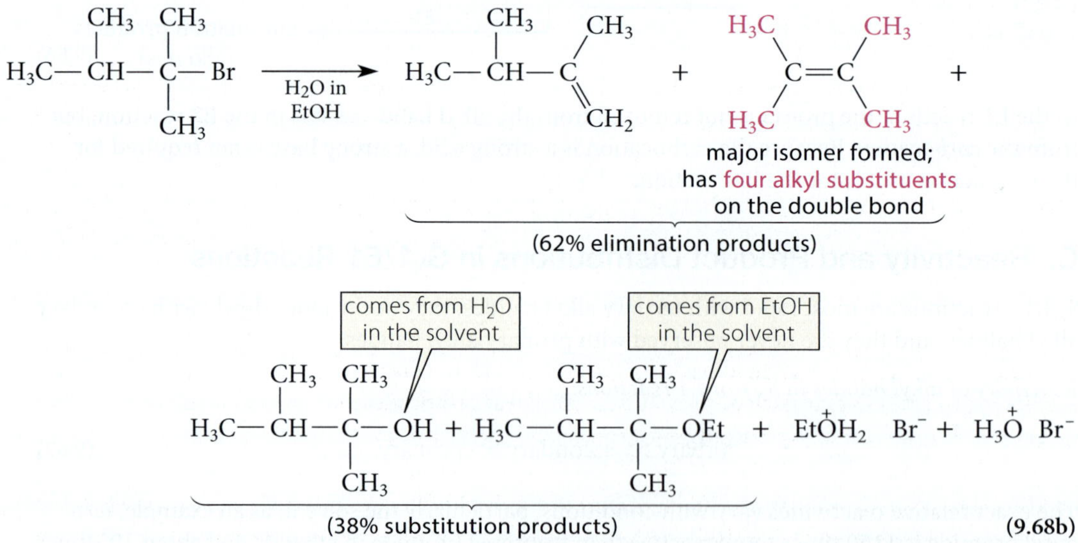

(9.68b)

In Eq. 9.68a, relatively little alkene is formed. In Eq. 9.68b, a greater proportion of alkene is formed. In Eq. 9.68b, moreover, two alkenes are formed corresponding to loss of the two types of β-hydrogens in the alkyl halide starting material; and the alkene formed in major amount is the one with the greatest number of alkyl substituents on the double bond.

Finally, rearrangements are observed in certain solvolysis reactions.

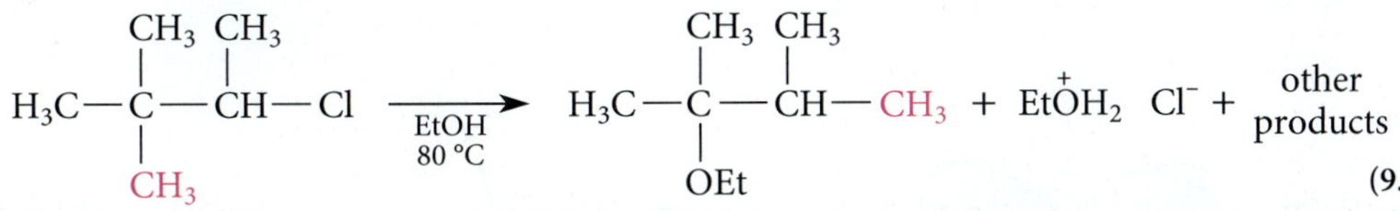

(9.69)

Recall (Sec. 4.7D) that rearrangements are a telltale sign of carbocation intermediates. For example, the secondary carbocation intermediate initially formed in Eq. 9.69 rearranges to a more stable tertiary carbocation; the nucleophilic reaction of solvent with this carbocation accounts for the product shown (see Focused Problem 9.29).

The different products that can be formed in S_N1/E1 reactions reflect three reactions of carbocation intermediates that you have now studied:

1. Reaction with a nucleophile

2. Loss of a β-proton

3. Rearrangement to a new carbocation followed by (1) or (2)

Although solvolysis reactions of alkyl halides and related compounds have been extensively studied because of their central role in the development of carbocation theory, as a practical matter S_N1/E1 reactions of alkyl halides are not very useful for preparative purposes because mixtures of products are invariably formed (unless the alkyl halide has no β-hydrogens). However, an understanding of the S_N1 and E1 mechanisms is important because these mechanisms occur in many reactions of alcohols, ethers, and amines that *are* very useful.

Focused Problems

9.28 Give all the products that might be formed when 3-chloro-2,2-dimethylbutane (the alkyl halide in Eq. 9.69) undergoes solvolysis in aqueous ethanol (that is, a solvent that contains both water and ethanol). Of the alkenes formed, which should be the major one(s)?

9.29 Write a curved-arrow mechanism for formation of the rearrangement product shown in Eq. 9.69.

D. Stereochemistry of the S_N1 Reaction

If we allow a chiral alkyl halide to undergo the S_N1 reaction, what would the reaction mechanism predict for the stereochemistry of the product? Because carbocations have trigonal planar geometry, they are achiral (assuming no asymmetric carbons elsewhere in the molecule). Therefore, the products that result from the reaction of nucleophiles with a carbocation *must* be racemic (Sec. 7.7A). The way racemic products would form mechanistically is shown in **Fig. 9.12**.

Now let's compare this prediction with the experimental facts. When (*R*)-6-chloro-2,6-dimethyloctane, a chiral tertiary alkyl halide, undergoes solvolysis in aqueous acetone, the substitution products are only *partially* racemized, and net inversion of configuration is also observed.

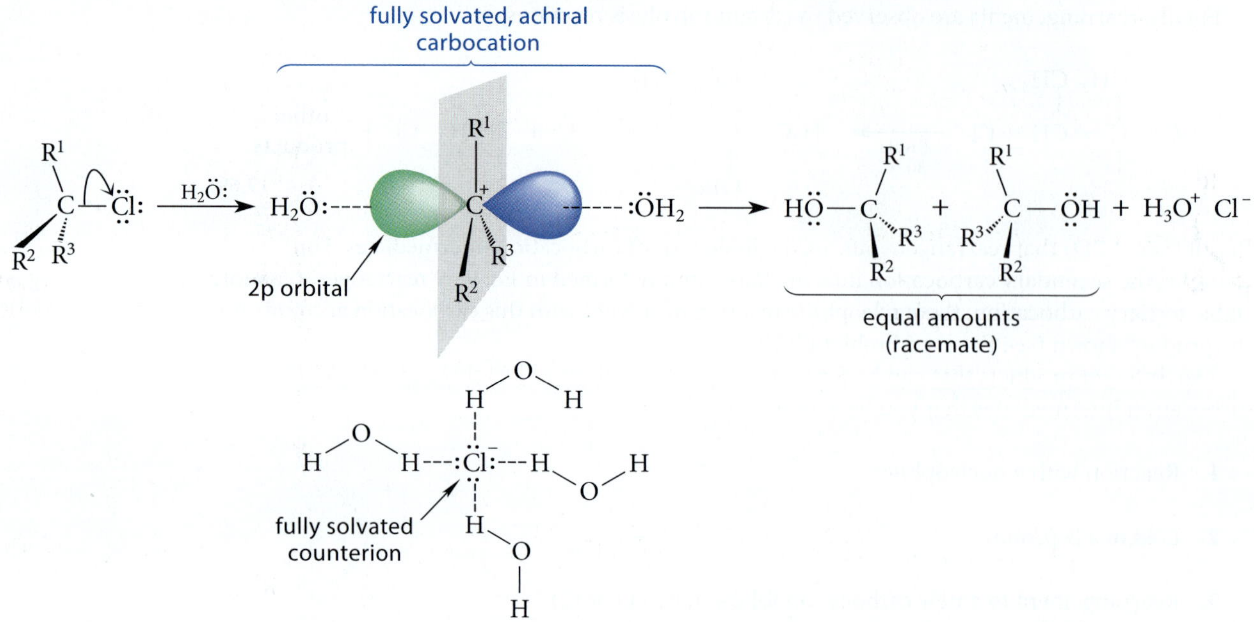

FIGURE 9.12 The stereochemical consequences of the S_N1 reaction of a chiral alkyl chloride. If the reaction proceeds through a free carbocation, the carbocation is achiral and the product of nucleophilic reaction with the carbocation must be racemic (barring asymmetric carbons in any of the three substituent groups R^1, R^2, or R^3). The racemic product results from equal probability of reaction of the nucleophile (water in this example) at each of the two lobes of the 2p orbital of the carbocation.

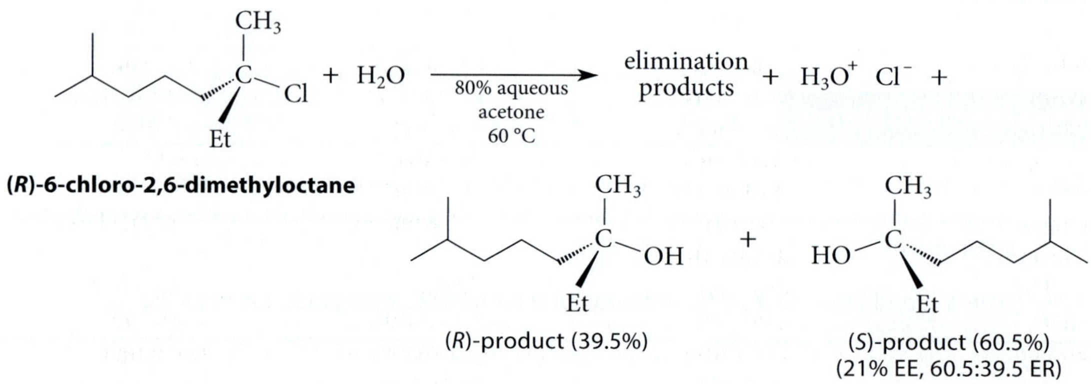

(9.70)

As Eq. 9.70 shows, the inverted product is formed in 60.5:39.5 enantiomeric ratio (ER) [21% enantiomeric excess (EE); Sec. 6.4A]. The EE corresponds to the percent inversion. The remaining percentage—79%, or (100%–EE)—corresponds to the percentage of the racemate because, by definition, it consists of equal amounts of the *R* and *S* products. How can we account for inverted product if a free carbocation is a reactive intermediate?

First, we emphasize that the stereochemical inversion *cannot* result from 21% of an S_N2 reaction because the reaction in Eq. 9.70 proceeds thousands of times faster than the S_N2 reactions of primary alkyl halides in the same solvent, and α-branching retards S_N2 reactions.

This result actually tells us something important about both the role of the solvent and the lifetime of the carbocation (**Fig. 9.13**). A mechanism that can account for this result assumes that the first reactive intermediate in the S_N1 reaction is an *ion pair* (Sec. 8.6F)—a carbocation intimately associated with its counterion, which, in this case, is the chloride ion. This ion pair (which includes the chloride ion) is still a chiral species. The chloride ion blocks the access of solvent to its side of the carbocation. Solvation of the carbocation in this ion pair occurs from the opposite side only; opposite-side substitution by the solvent molecule involved in this interaction results in inversion. However, the chloride ion might also escape from the carbocation

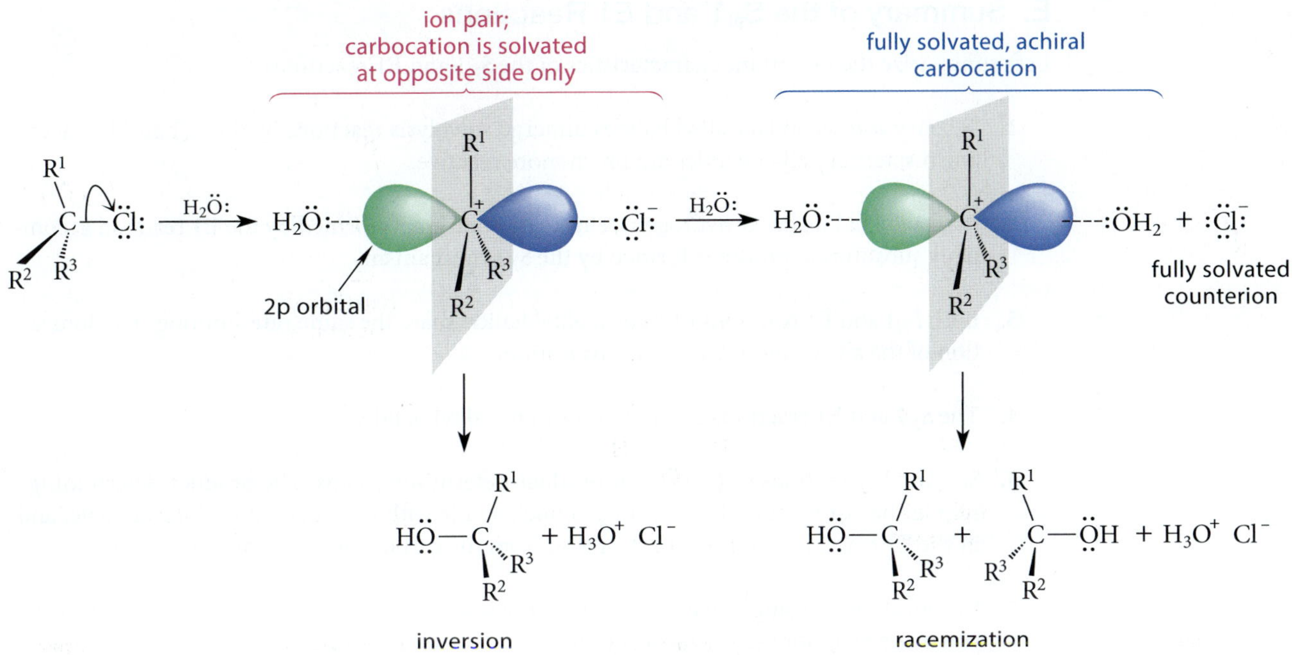

FIGURE 9.13 The ion-pair mechanism for carbocation formation in the S_N1 reaction. The reaction of the solvating solvent molecule with the carbocation in the ion pair occurs from the side opposite the departing chloride and gives inverted product. (The ion pair is chiral.) The fully solvated ion is formed when the chloride counterion diffuses away and itself becomes fully solvated. (Its solvation shell is not shown explicitly.) The fully solvated carbocation is achiral and gives racemic products.

into the surrounding solvent, leaving the carbocation solvated on both sides by solvent. This symmetrically solvated carbocation is achiral and can, with equal probability, react at either face with solvent to give racemic product. The occurrence of both racemization and inversion in Eq. 9.70 shows that both types of carbocation—ion pairs and free ions—are important in determining the products of S_N1 reactions. According to this mechanism, 21% of the product comes from the ion pair, whereas 79% (half *R*, half *S*) comes from the symmetrically solvated ion. The exact percentages of each vary from case to case.

The occurrence of some inversion also shows that the lifetime of a tertiary carbocation is very short. It takes about 10^{-8} second for a chloride counterion to diffuse away from a carbocation and be replaced by solvent. The carbocations that undergo inversion do not last long enough for this process to take place completely. The competition of opposite-side substitution (which gives inversion) with racemization shows that the lifetime of the carbocation is approximately in this range. In other words, a typical tertiary carbocation exists for about 10^{-8} second before it is consumed by its reaction with solvent. This very small lifetime provides a graphic illustration just how reactive carbocations are.

Focused Problems

9.30 The optically active alkyl halide in Eq. 9.70 reacts at 60 °C in anhydrous methanol solvent to give a methyl ether *A* plus alkenes. The substitution reaction is reported to occur with 66% racemization and 34% inversion. Give the structure of ether *A*, and state how much of each enantiomer of *A* is formed.

9.31 In light of the ion-pair hypothesis, how would you expect the stereochemical outcome of an S_N1 reaction (percent racemization and inversion) to differ from the result discussed in this section for an alkyl halide that gives a carbocation intermediate which is considerably (a) more stable or (b) less stable than the one involved in Eq. 9.70?

E. Summary of the S_N1 and E1 Reactions

Let's summarize the important characteristics of the S_N1 and E1 reactions:

1. Tertiary and secondary alkyl halides undergo solvolysis reactions by the S_N1 and E1 mechanisms; tertiary alkyl halides are much more reactive.

2. If an alkyl halide has β-hydrogens, elimination products formed by the E1 reaction accompany substitution products formed by the S_N1 mechanism.

3. Both S_N1 and E1 reactions of a given alkyl halide share the same rate-limiting step: ionization of the alkyl halide to form a carbocation.

4. The S_N1 and E1 reactions are first order in the alkyl halide.

5. S_N1 and E1 reactions differ in their product-determining steps. The product-determining step in the S_N1 reaction is reaction of a nucleophile with the carbocation intermediate, and in the E1 reaction, it is loss of a β-proton from the carbocation intermediate.

6. Carbocation rearrangements occur when the initially formed carbocation intermediate can rearrange to a more stable carbocation.

7. The best leaving groups are those that give the weakest bases as products.

8. The reactions are accelerated by polar, protic, donor solvents.

9. S_N1 reactions of chiral alkyl halides give largely racemized products, but some inversion of configuration is also observed.

9.7 SUMMARY: SUBSTITUTION AND ELIMINATION REACTIONS OF ALKYL HALIDES

This chapter has shown that substitution and elimination reactions of alkyl halides can occur by a variety of mechanisms. Although each type of reaction has been considered separately, a practical question to ask is what type of reaction is likely to occur when a given alkyl halide is subjected to a particular set of conditions.

When asked to predict how a given alkyl halide will react, you must first answer three major questions:

1. Is the alkyl halide primary, secondary, or tertiary? If primary or secondary, is there a significant amount of alkyl substitution at the β-carbon?

2. Is a Lewis base present? If so, is it a good nucleophile, a strong Brønsted base, or both? Most strong Brønsted bases, such as ethoxide, are good nucleophiles; but some excellent nucleophiles, such as iodide ion, are relatively weak Brønsted bases.

3. What is the solvent? Because of alkyl halide and ion solubility considerations, the practical choices are limited for the most part to polar protic solvents, polar aprotic solvents, or mixtures of both.

Once these questions have been answered, a satisfactory prediction in most cases can be obtained from **Table 9.7**, which is in essence a summary of this chapter. *Before using this table, you should consider each case and why the conclusions are reasonable, returning to review the material in this chapter when necessary.* Study Problem 9.3 illustrates the practical application of the table.

TABLE 9.7 Predicting Substitution and Elimination Reactions of Alkyl Halides

Entry no.	Alkyl halide structure	Good nucleophile?	Strong Brønsted base?	Type of solvent?*	Major reaction(s) expected
1	Methyl	Yes	Yes or No	PP or PA	S_N2
2	Primary, unbranched	Yes	No	PP or PA	S_N2
3		Yes	Yes, unbranched	PP or PA	S_N2
4	Primary with β-substitution	Yes	Yes, unbranched	PP or PA	E2 + S_N2
5	Any primary	Yes	Yes, branched	PP or PA	E2 + S_N2
6		No	No	PP or PA	No reaction
7	Secondary	Yes	Yes	PP or PA	E2; some S_N2 with isopropyl halides; only E2 with a branched base
8		Yes	No	PA	S_N2
9		No	No	PP	S_N1 – E1 (solvolysis)
10		No	No	PA	No reaction
11	Tertiary	Yes	Yes	PP or PA	E2
12		Yes	No	PP	S_N1 – E1 (solvolysis)
13		Yes	No	PA	no reaction, or very slow S_N2
14		No	No	PP	S_N1 – E1
15		No	No	PA	No reaction

* Solvent types are PP = polar protic; PA = polar aprotic. The S_N2, E2, S_N1, and El reactions are rarely if ever run in apolar aprotic solvents except with the most reactive alkyl halides. In these cases, the results to be expected are similar to those above with polar aprotic (PA) solvents.

Study Problem 9.3

What products are formed (if any), and by what mechanisms (S_N2, E2, S_N1, E1), in each of the following cases?

(a) Methyl iodide and sodium cyanide (NaCN) in ethanol
(b) 2-Bromo-3-methylbutane in hot ethanol
(c) 2-Bromo-3-methylbutane in anhydrous acetone at room temperature
(d) 2-Bromo-3-methylbutane in ethanol containing an excess of sodium ethoxide
(e) 2-Bromo-2-methylbutane in ethanol
(f) Neopentyl bromide in ethanol containing an excess of sodium ethoxide

Solution

(a) Methyl iodide and sodium cyanide (NaCN) in ethanol. This case corresponds to entry 1 in Table 9.7. Because a methyl halide has no β-hydrogens, it cannot undergo a β-elimination reaction. Consequently, the only possible reaction is an S_N2 reaction. Because a good nucleophile (cyanide) is present (see Table 9.4, Fig. 9.5), the product is H_3C—CN (acetonitrile), which is formed by the S_N2 mechanism. Although protic solvents are not as effective as polar aprotic ones for the S_N2 reaction, they are useful for reactive alkyl halides such as methyl iodide. The reaction would be faster, though, if it were carried out in a polar aprotic solvent (see Table 9.6).

(b) 2-Bromo-3-methylbutane in hot ethanol. This is a secondary alkyl halide. (Draw its structure if you have not done so!) The conditions involve no nucleophile or base other than the solvent, which is polar and protic. This situation is covered by entry 9 in Table 9.7. Because the solvent ethanol is a poor nucleophile and a weak base, neither S_N2 nor E2 reactions can occur. Because polar protic solvents promote the S_N1 and E1 reactions, these will be the only reactions observed:

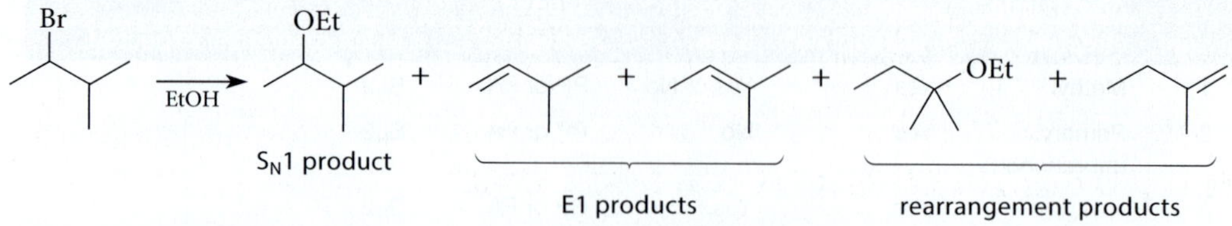

Notice the rearrangement products. (You should show how these arise from the initially formed carbocation intermediate.) Any time the S_N1 or E1 reaction is expected, the possibility of rearrangements should be considered, especially when the initially formed carbocation is secondary. Finally, "hot" ethanol is necessary because the alkyl halide is secondary and is less reactive in the S_N1/E1 reaction than a tertiary alkyl halide would be.

(c) 2-Bromo-3-methylbutane in anhydrous acetone. The alkyl halide from part (b) is subjected to conditions in which a good nucleophile is not present (no S_N2 possible), no strong base has been added (no E2 possible), and a polar aprotic solvent is used. In this type of solvent, carbocations do not form under mild conditions, so the S_N1 and E1 reactions cannot take place. Entry 10 in Table 9.7 predicts that no reaction will occur.

(d) 2-Bromo-3-methylbutane in ethanol containing an excess of sodium ethoxide. The alkyl halide from parts (b) and (c) is subjected to a strong base in a protic solvent. This situation is covered by entry 7 in Table 9.7. The S_N2 reaction is retarded by both α- and β-alkyl substitution, but the E2 reaction can take place. Although an S_N1/E1 reaction is promoted by the protic solvent, the rate of the E2 reaction is greater because of the high base concentration. The rate of the E2 reaction is first order in base (Eq. 9.35, Sec. 9.5A), whereas the rates of the S_N1 and E1 reactions are unaffected by the base concentration (Eq. 9.59, Sec. 9.6A). The products are the following two alkenes:

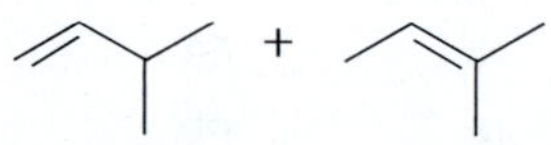

The second of these predominates because of the greater number of alkyl substituents at the double bond.

(e) 2-Bromo-2-methylbutane in ethanol. This is a tertiary alkyl halide in a polar protic solvent. Entry 14 of Table 9.7 covers this situation. The polar protic solvent promotes carbocation formation, so the S_N1 and E1 reactions are observed. The products are the following:

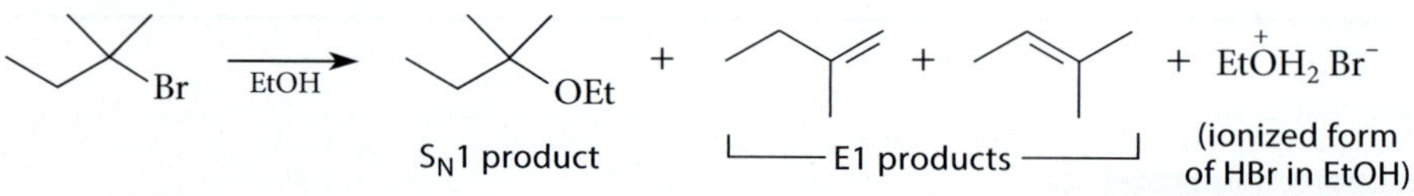

No rearrangement products are predicted, because the carbocation intermediate is tertiary.

(f) Neopentyl bromide in ethanol containing an excess of sodium ethoxide. This is a primary alkyl halide with *three* β-alkyl groups [$(Me)_3C—CH_2—Br$]. Without thinking further about the structure of this alkyl halide, you might conclude that entry 4 of Table 9.7 would cover this case. There are no β-hydrogens, however, so no elimination is possible. Neopentyl halides are essentially unreactive in S_N2 reactions (see Table 9.3, Sec. 9.4D), and, because primary alkyl halides do not form carbocations, neither an E1 nor an S_N1 reaction is possible. Therefore, this alkyl halide is essentially inert. If the reaction mixture were heated strongly, a reaction might occur after a few days, but the correct prediction is "no reaction."

STUDY GUIDE LINK 9.3
Diagnosing Reactivity Patterns in Substitution and Elimination Reactions

Focused Problem

9.32 Predict the products expected in each of the following situations, and show the mechanism of any reaction that takes place using the curved-arrow notation.

(a) 1-Bromobutane in methanol containing a large excess of sodium methoxide

(b) 2-Bromobutane in *tert*-butyl alcohol containing a large excess of potassium *tert*-butoxide

(c) 2-Bromo-1,1-dimethylcyclopentane in ethanol

(d) Bromocyclohexane in methanol, heat

CHAPTER SUMMARY *For a summary of the chapter, see Chapter 9 in the* Study Guide and Solutions Manual.

REACTION REVIEW *For a summary of reactions discussed in this chapter, see the* Reaction Review *section of Chapter 9 in the* Study Guide and Solutions Manual.

SKILLS OBJECTIVES WITH PROBLEMS

• Construct analogies between nucleophilic substitution reactions and Brønsted acid–base reactions. 9.2

9.33 Provide the acid–base analogy for each of the following nucleophilic substitution reactions.

(a)

(b)

• Use a pK_a table to predict whether an equilibrium for a nucleophilic substitution reaction is favorable. 9.2

9.34 Use the acid–base analogies you wrote in Problem 9.33 and the pK_a values in Table 3.2 (Sec. 3.6A) to determine which of the reactions in Problem 9.33 should have the equilibrium that lies furthest to the right.

• Given the rate law for a reaction, give the overall kinetic order and the order in each reactant and catalyst (if any). 9.3B

9.35 A reaction for the addition of HBr to an alkene under certain conditions has the following rate law:

$$\text{rate} = k[\text{alkene}][\text{HBr}]^2$$

(a) What is the overall kinetic order of the reaction?

(b) What is the kinetic order in alkene? In HBr?

• Evaluate possible transition states for consistency with the rate law for a reaction. 9.4A

9.36 Which of the following transition states is (are) consistent with the rate law in Problem 9.35? Explain.

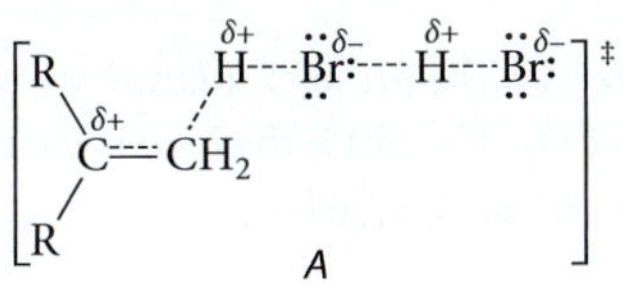

A

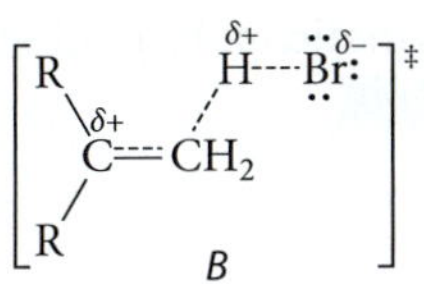

B

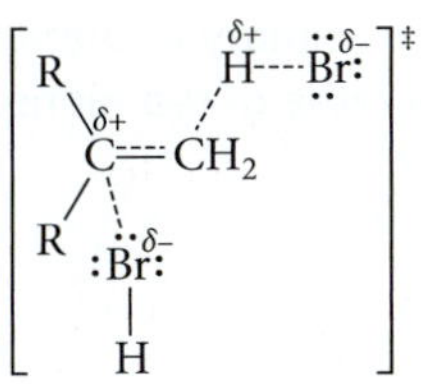

C

9.37 The rate law for the bromination of acetone is

$$\text{rate} = [\text{acetone}][H_3O^+]$$

In the following accepted mechanism for this reaction, which step is rate-limiting? Explain how you know.

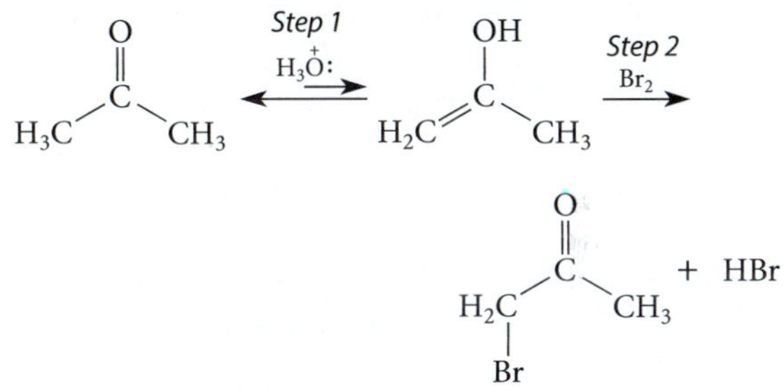

• Perform calculations that relate the relative rates of two reactions to the difference in their standard free energies of activation. 9.3C

9.38 (a) The rate of reaction (1) is 24.7 times the rate of reaction (2). Of the two reactions, which has the greater $\Delta G^{\circ\ddagger}$, and by how much?

(b) Reaction (3) has a $\Delta G^{\circ\ddagger}$ that is greater than that of reaction (2) by 24.4 kJ mol^{-1} (5.84 kcal mol^{-1}). Is this reaction faster or slower than reaction (2), and by what factor?

(1) 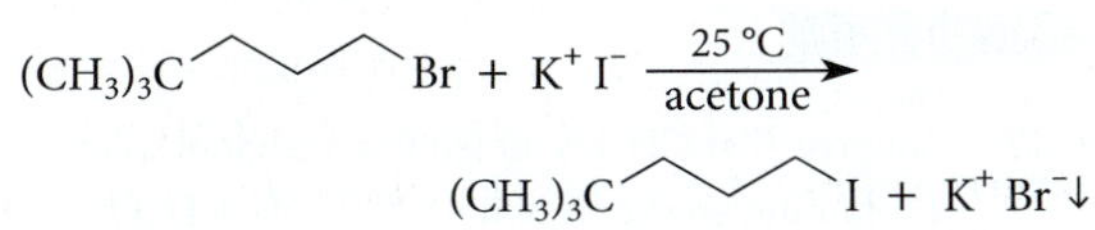

(2) 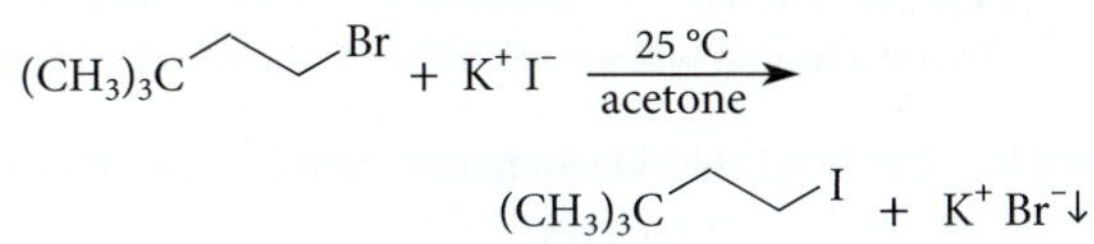

(3) 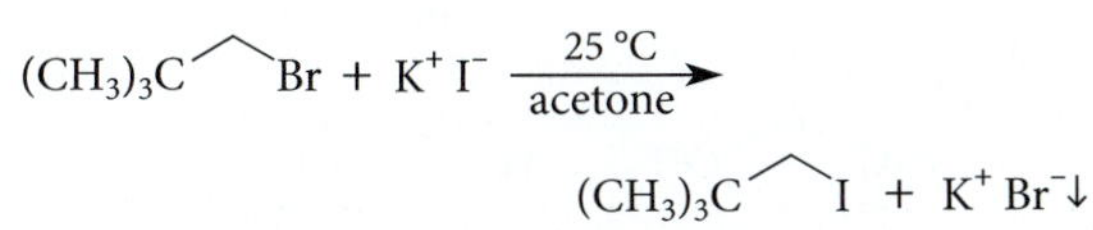

• Given an alkyl halide, complete the products of S_N2 reactions with a given nucleophile. Give the structure of the nucleophile required to convert an alkyl halide into a specified product using the S_N2 reaction. 9.1A, Table 9.1

9.39 Complete the following S_N2 substitution reactions by providing the structures of the products.

(a) $H_3C{-}I\ +\ :P(Ph)_3 \longrightarrow$

(b) 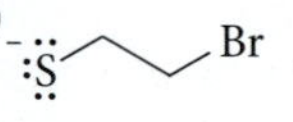 (one product contains a ring)

9.40 Give the structure of the nucleophile that could be used to convert iodoethane into each of the following compounds by an S_N2 reaction.

(a) 1-Ethoxypropane

(b) $CH_3CH_2C{\equiv}N{:}$

(c) 

(d) 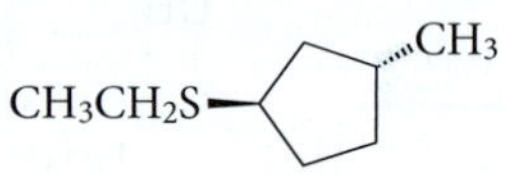

(e) $(CH_3)_3\overset{+}{N}{-}CH_2CH_3\ \ I^-$

● **Rank alkyl halides according to their reactivities in S_N2 and S_N1/E1 reactions.** **9.4D, 9.6C**

9.41 (a) Rank the following compounds in order of increasing S_N2 reaction rate with KI in acetone.

$(CH_3)_3CCl$ *A* $(CH_3)_2CHCl$ *B* $(CH_3)_2CHCH_2Cl$ *C*

$CH_3CH_2CH_2CH_2Br$ *D* $CH_3CH_2CH_2CH_2Cl$ *E*

(b) Rank the alkyl halides in part (a) in order of increasing S_N1/E1 reactivity in aqueous ethanol.

● **Determine the effect of solvent on reaction rate for selected examples of S_N2 and S_N1/E1 reactions.** **9.4E, 9.6, 9.7**

9.42 Suppose that CH_3I is added to an ethanol solution containing an excess of both $Na^+\ CH_3CH_2O^-$ and $K^+\ CH_3CH_2S^-$ in equimolar amounts. What is the major product that will be isolated from the reaction? Explain.

9.43 *Tert*-butyl chloride undergoes solvolysis in either acetic acid or formic acid solvent.

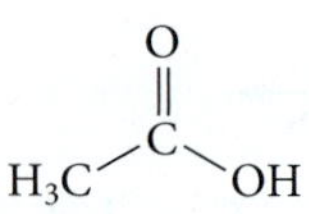

acetic acid
$\varepsilon = 6$

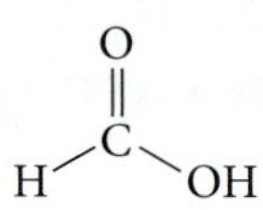

formic acid
$\varepsilon = 59$

Both solvents are protic, donor solvents, but they differ substantially in their dielectric constants ε.

(a) What is the S_N1 solvolysis product in each solvent?

(b) In one solvent, the S_N1 reaction is 5000 times faster than it is in the other. In which solvent is the reaction more rapid, and why?

● **Give the product, including its stereochemistry, of an S_N2 reaction of a chiral alkyl halide.** **9.4C**

9.44 Give all of the product(s) expected, including pertinent stereochemistry, when each of the following chiral and enantiomerically pure compounds reacts with sodium ethoxide in ethanol. (D = deuterium = ^2H, an isotope of hydrogen.)

(a) (*R*)-2-Bromopentane

(b) 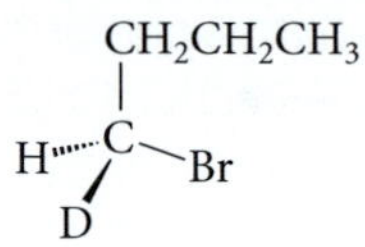

(c) 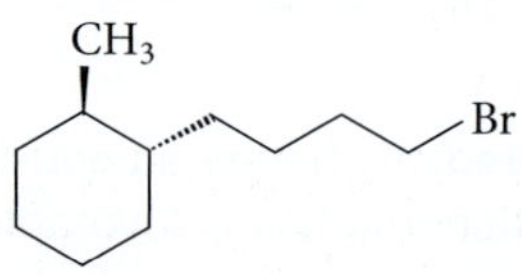

● **Give the possible products of an E2 reaction an alkyl halide; indicate which of the possible products should be formed in greatest amount and why.** **9.5D,F,H**

9.45 (a) Draw all of the alkene products formed in the E2 reactions of each of the following alkyl halides.

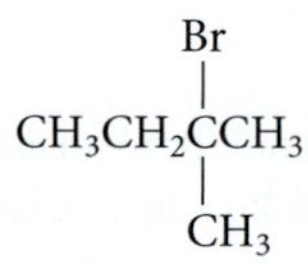

A

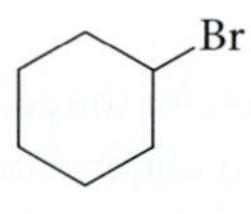

B

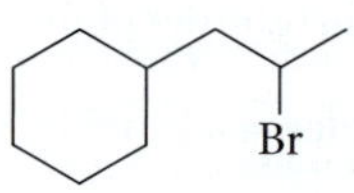

C

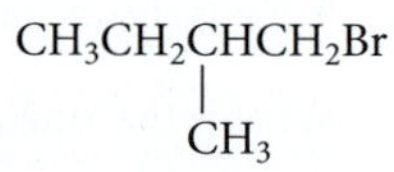

D

(b) Indicate in each case which of the products should be formed in greatest amount if the alkoxide base used is $Na^+\ {}^-OMe$ in MeOH.

(c) Indicate how the product mixture will change in each case (if at all) if the base is changed to $Et_3CO^-\ K^+$ in Et_3COH.

● **Apply the primary deuterium kinetic isotope effect to the relative rates of reactions.** **9.5D**

9.46 In each part, give the structures of the alkenes that would be formed in an E2 reaction with NaOEt in EtOH, and indicate which alkene would be formed in greatest amount.

(a) Which of the following compounds undergoes the fastest E2 elimination reaction with K^+ $^-$O—*t*Bu in *t*-BuOH? Explain.

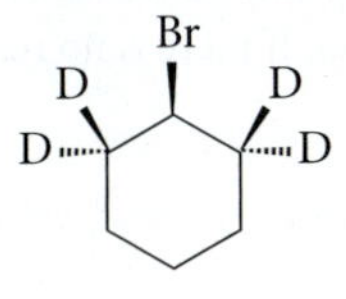

A

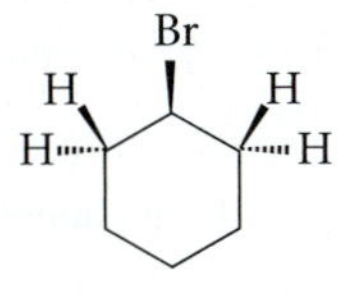

B

(b) Draw the structures of the alkenes that could be formed when the following alkene reacts with Na^+ $^-$OEt in EtOH, and indicate which alkene is formed in greater amount. Explain your choice.

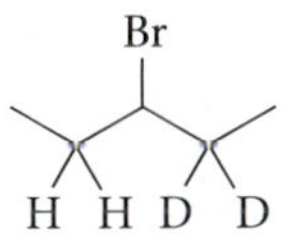

- **Given an alkyl halide with stereocenters at its α- and β-carbons, give the stereochemistry of its E2 elimination products. Given an alkene with stereocenters at its alkene carbons, give the stereochemistry of the alkyl halide required to form the alkene by an E2 reaction. 9.5E**

9.47 (a) Give the major product formed and its stereochemistry when the following alkyl halide reacts with Na^+ $^-$OMe in MeOH.

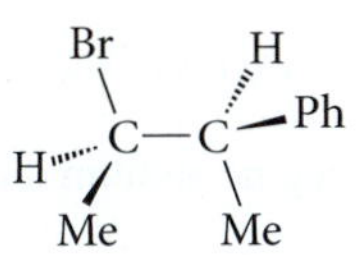

(b) What is the stereochemistry of the alkyl iodide *A* that would undergo E2 elimination with Na^+ $^-$OEt in EtOH to give the following alkene?

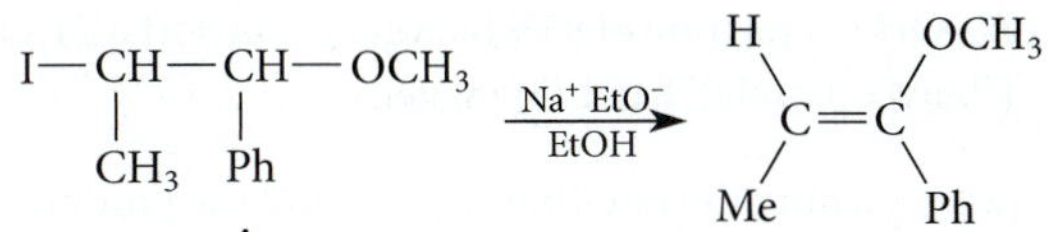

A

- **Given an alkyl halide, predict whether the E2 reaction or the S_N2 reaction will be the major process observed with a given nucleophile or base. 9.5G**

9.48 In each case, indicate whether the given alkyl halide and base will react to give mostly S_N2 product or E2 product. Give the structure of the major product(s) in each case.

(a) 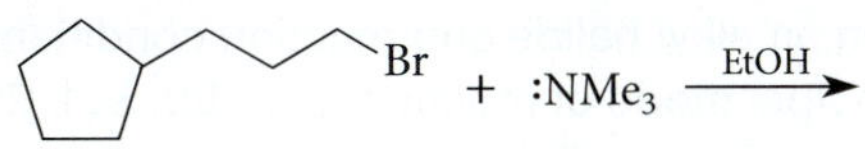

(b) 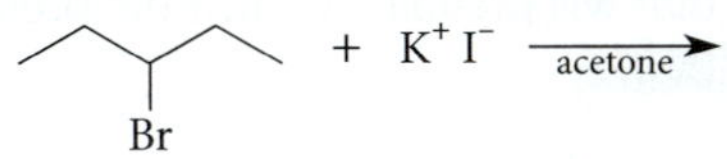

(c) 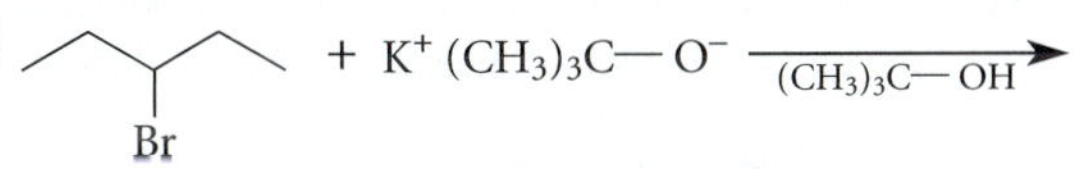

- **Given an alkyl halide and a protic solvent, predict all possible products of an S_N1/E1 solvolysis reaction, and draw the mechanisms of their formation. 9.6B,C**

9.49 Complete the following reaction by giving all substitution and elimination products, including their stereochemistry, and the curved-arrow mechanisms for their formation.

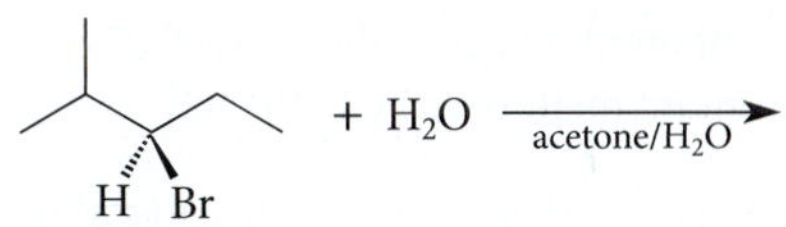

- **Given an S_N1/E1 reaction, identify the rate-limiting step and the product-determining steps. Given the relative product distributions, calculate the relative rates of their formation. 9.6A,B**

9.50 The mechanism of solvolysis of 3-bromo-3-methylpentane at 25 °C is shown in **Fig. P9.50**, in which k_1 is the rate constant for the first step. The percentage of each product in the mixture of products is also given. (The solvent

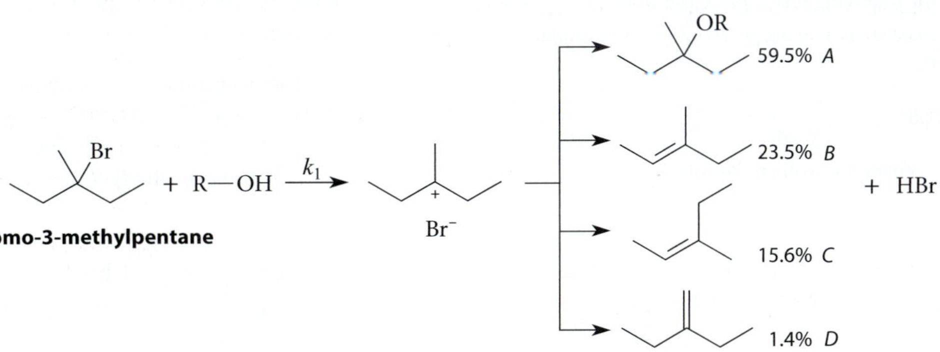

FIGURE P9.50

R—OH is a mixture of 85% by volume $BuOCH_2CH_2OH$ ["butyl cellosolve"] and 15% water.)

(a) Identify the rate-limiting step and the product-determining steps of this mechanism.

(b) The rate constant k_1 for disappearance of the reactant alkyl bromide was measured as $1.14 \times 10^{-4}\ s^{-1}$. Calculate the rate constants for the formation of each product *A* through *D* from the data provided.

(c) In this reaction, is substitution or elimination faster?

- **Given an alkyl halide and reaction conditions, identify the principal mode of reaction (S_N2, E2, S_N1, E1, or a combination) that will prevail and give the predominant products.** 9.6D, 9.7

9.51 Choose the alkyl halide(s) from the following list of $C_6H_{13}Br$ isomers that meet each criterion below.

(1) 1-Bromohexane

(2) 3-Bromo-3-methylpentane

(3) 1-Bromo-2,2-dimethylbutane

(4) 3-Bromo-2-methylpentane

(5) 2-Bromo-3-methylpentane

(a) The compound that gives the fastest S_N2 reaction with sodium methoxide in methanol

(b) The compound that is least reactive to sodium methoxide in methanol

(c) The compound(s) that give only one alkene in the E2 reaction with $K^+\ {}^-OtBu$

(d) The compound(s) that give an E2 but no S_N2 reaction with sodium methoxide in methanol

(e) The compound(s) that undergo an S_N1 reaction to give rearranged products

(f) The compound that gives the fastest S_N1 reaction

9.52 Give the products expected when isopentyl bromide (1-bromo-3-methylbutane) reacts under each of the following conditions. Identify the mechanism involved in each case with its abbreviation. If there is no reaction, give the reason.

(a) $K^+\ I^-$ in aqueous acetone

(b) $K^+\ {}^-OH$ in aqueous ethanol

(c) $K^+\ tBuO^-$ in *t*BuOH

(d) $Cs^+\ F^-$ in *N,N*-dimethylformamide (DMF; a polar aprotic solvent)

(e) $K^+\ F^-$ in aqueous acetone

(f) Sodium methoxide in methanol

9.53 Give the products expected when 2-bromo-2-methylhexane reacts with each of the following reagents. If there is no reaction, give the reason.

(a) Warm 1:1 ethanol–water

(b) Sodium ethoxide in ethanol

(c) KI in aqueous acetone

9.54 What products are expected, including their stereochemistry, when (2*S*,3*R*)-2-bromo-3-methylpentane is subjected to each of the following conditions? Explain.

(a) Methanol containing an excess of sodium methoxide

(b) Hot methanol containing no sodium methoxide

INTEGRATED PROBLEMS

9.55 In each of the following series, place the atoms, compounds, or ions in order of increasing polarizability, and explain your choices.

(a) Se, O, S

(b) Chloroform, fluoroform, iodoform

(c) I^-, Br^-, Cl^-, F^-

9.56 Rank the following compounds in order of increasing S_N2 reaction rate with KI in acetone.

methyl bromide *A* *sec*-butyl bromide *B*

3-(bromomethyl)-3-methylpentane *C*

1-bromopentane *D* 1-bromo-2-methylbutane *E*

9.57 In the *Williamson ether synthesis*, an alkoxide reacts with an alkyl halide to give an ether.

$$R-\ddot{\underset{..}{O}}:^{-} + R'-X \longrightarrow R-\ddot{\underset{..}{O}}-R' + :\ddot{\underset{..}{X}}:^{-}$$

You are in charge of a research group for a large company, Ethers Unlimited, and you have been assigned the task of synthesizing *tert*-butyl methyl ether, $(CH_3)_3C-O-CH_3$. You have decided to delegate this task to two of your staff chemists. One chemist, Ima Smart, allows $(CH_3)_3C-O^- \ K^+$ to react with H_3C-I and indeed obtains a good yield of the desired ether. The other chemist, Dimma Light, allows $CH_3O^- \ Na^+$ to react with $(CH_3)_3C-Br$. To his surprise, no ether was obtained, although the alkyl halide could not be recovered from the reaction. Explain why Dimma Light's reaction failed.

9.58 Propose a synthesis of ethyl neopentyl ether, $CH_3CH_2-O-CH_2C(CH_3)_3$, from an alkyl bromide and any other reagents.

9.59 The banned insecticide chlordane is reported to lose some of its chlorine and to be converted into other compounds when exposed to alkaline (basic) conditions. Explain.

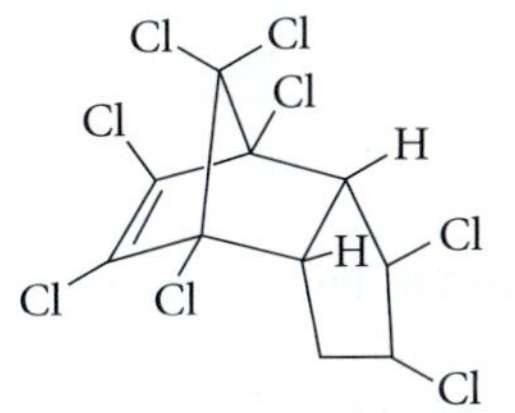

principal component of chlordane

9.60 Which of the following secondary alkyl halides reacts faster with ^-CN in the S_N2 reaction? (*Hint:* Consider the hybridization and geometry of the S_N2 transition state.)

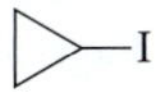 $(CH_3)_2CH-I$

A *B*

9.61 Explain why 1-chlorobicyclo[2.2.1]heptane, even though it is a tertiary alkyl halide, is virtually unreactive in the S_N1 reaction in protic solvents. (It has been estimated that it is 10^{-13} times as reactive as *tert*-butyl chloride.) (*Hint:* Consider the preferred geometry and the solvation of the reactive intermediate.)

1-chlorobicyclo[2.2.1]heptane

9.62 Explain each of the following observations.

(a) When benzyl bromide (Ph—CH_2—Br) is added to a suspension of potassium fluoride in benzene, no reaction occurs. However, when a *catalytic* amount of the crown ether [18]-crown-6 (Display 8.80, Sec. 8.7C) is added to the solution, benzyl fluoride can be isolated in high yield.

(b) If lithium fluoride is substituted for potassium fluoride, no reaction occurs even in the presence of the crown ether used in part (a).

9.63 (a) Identify the β-hydrogens in each bicyclic alkene.

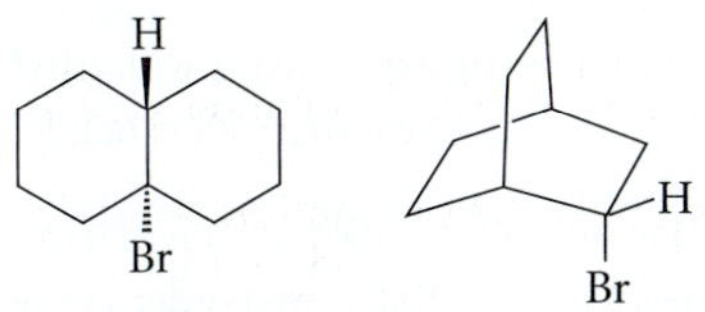

A *B*

(b) Bicyclic alkyl bromide *A* reacts with NaOEt in ethanol to give two alkenes, whereas bicyclic alkyl bromide *B* gives only one alkene under the same conditions. Give the structures of the alkene(s) formed from each compound, and explain the difference in the behavior of the two alkenes.

9.64 (a) Two isomeric S_N2 products are possible when sodium thiosulfate is allowed to react with one equivalent of methyl iodide in methanol solution. Give the structures of the two products.

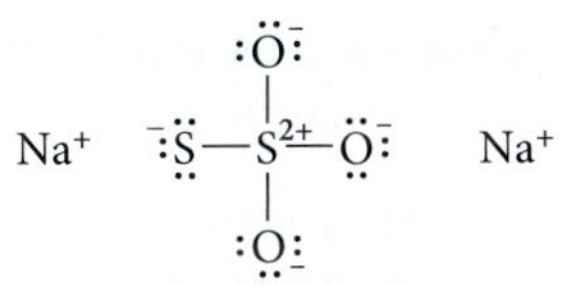

sodium thiosulfate

(Thiosulfate is an example of an *ambident*, or "two-toothed," nucleophile.)

(b) In fact, only one of the two possible products is formed. Which one is formed, and why?

9.65 Consider the following equilibrium:

$$(CH_3)_2\ddot{S}: + H_3C-\ddot{\underset{..}{Br}}: \rightleftharpoons (CH_3)_2\overset{+}{\ddot{S}}-CH_3 + :\ddot{\underset{..}{Br}}:^{-}$$

In each case (a) and (b), choose the solvent in which the equilibrium would lie farther *to the right*. Explain. (Assume that the products are soluble in all solvents considered.) (*Hint:* The reaction brings about the separation of opposite charges.)

(a) Ethanol or diethyl ether

(b) *N,N*-dimethylacetamide (a polar, aprotic solvent, ε = 38) or a mixture of water and methanol that has the same dielectric constant

9.66 When methyl iodide at 0.1 *M* concentration is allowed to react with sodium ethoxide at 0.1 *M* concentration in ethanol solution, the product ethyl methyl ether is obtained in good yield. Explain why the reaction is over much more quickly, but about the same yield of the ether is obtained, when the reaction is run with an excess (0.5 *M*) of sodium ethoxide.

9.67 When methyl bromide is dissolved in methanol and an equimolar amount of sodium iodide is added, the concentration of iodide ion quickly decreases, and then slowly returns to its original value. Explain.

9.68 Consider the following experiments with trityl chloride, $Ph_3C—Cl$, a very reactive tertiary alkyl halide:

(1) In aqueous acetone, the reaction of trityl chloride follows a rate law that is first order in the alkyl halide, and the product is trityl alcohol, $Ph_3C—OH$.

(2) In another reaction, when one equivalent of sodium azide ($Na^+\ N_3^-$; see Table 9.3, Sec. 9.4D) is added to a solution that is otherwise identical to that used in experiment (1), the reaction rate is the same as in (1); however, the product isolated in good yield is trityl azide, $Ph_3C—N_3$.

(3) In a reaction mixture in which both sodium azide and sodium hydroxide are present in equal concentrations, *both* trityl alcohol and trityl azide are formed, but the reaction rate is again unchanged.

Explain why the reaction rate is the same but the products are different in these three experiments.

9.69 The first demonstration of the stereochemistry of the S_N2 reaction was carried out in 1935 by Professor E. D. Hughes and his colleagues at the University of London. They allowed (*R*)-2-iodooctane to react with radioactive iodide ion ($*I^-$) in acetone.

2-iodooctane **2-iodooctane** (radioactive)

The rate of substitution (rate constant k_S) was determined by measuring the rate of incorporation of radioactivity into the alkyl halide. The rate of loss of optical activity from the alkyl halide (rate constant $k°$) was also determined under the same conditions.

(a) What ratio $k°/k_S$ is predicted for each of the following stereochemical scenarios: (1) retention; (2) inversion; (3) equal amounts of both retention and inversion? Explain.

(b) The experimental rate constants at 30 °C were found to be as follows:

$$k_S = (3.00 \pm 0.25) \times 10^{-5}\ M^{-1}\ s^{-1}$$
$$k° = (5.76 \pm 0.06) \times 10^{-5}\ M^{-1}\ s^{-1}$$

Which scenario in part (a) is most consistent with the data? Explain.

9.70 Which one of the following stereoisomers should undergo β-elimination most rapidly with sodium ethoxide in ethanol? Explain your reasoning. (*Hint:* Consider the optimum stereochemistry for the reaction.)

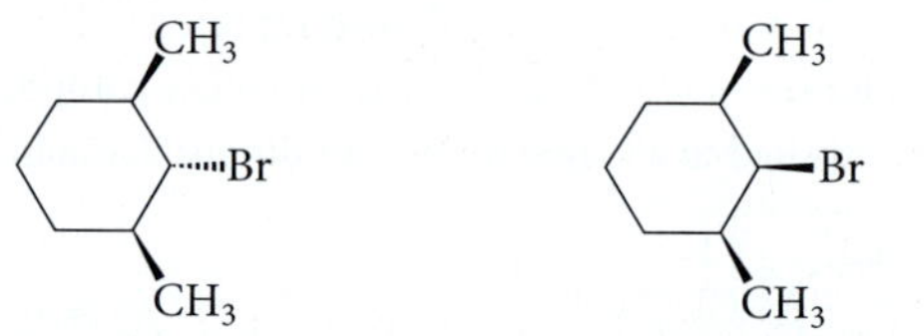

A *B*

9.71 When menthyl chloride (see **Fig. P9.71**) is treated with sodium ethoxide in ethanol, 2-menthene is the only alkene product observed. When neomenthyl chloride is subjected to the same conditions, the alkene products are mostly 3-menthene (78%) along with some 2-menthene (22%). Explain why different alkene products are formed from the different alkyl halides and why 3-menthene is the major product in the second reaction. (*Hint:* Draw the chair conformations of the starting materials, remember the stereochemistry of the E2 reaction, and don't forget about the chair interconversion of cyclohexanes.)

9.72 Explain why each alkyl halide stereoisomer gives a different alkene in the E2 reactions that follow. It will probably help to build models or draw out the conformations of the two starting materials.

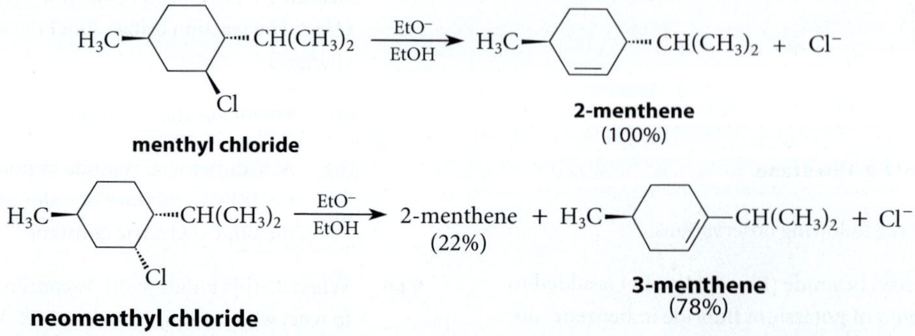

FIGURE P9.71

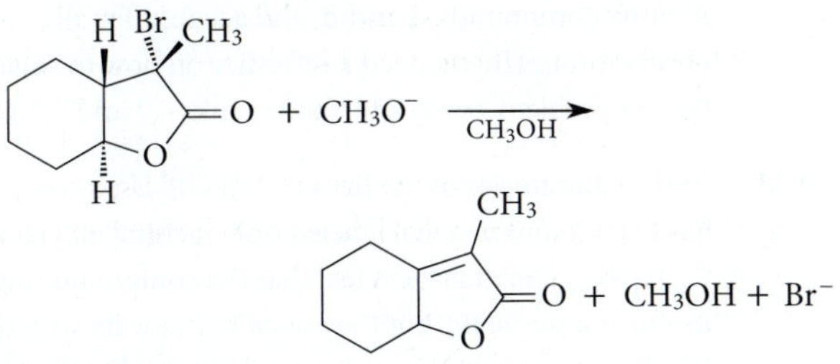

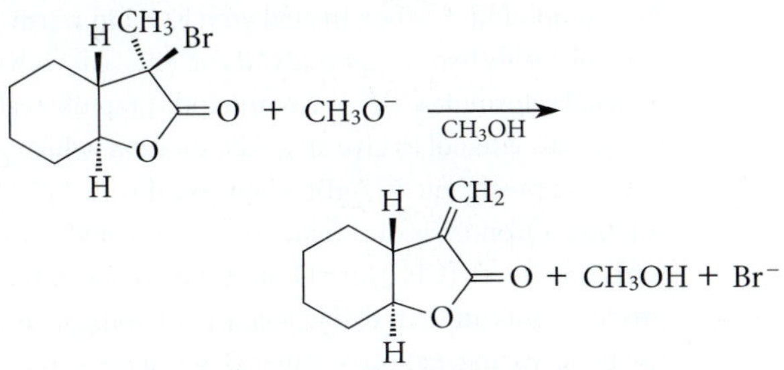

9.73 The cis and trans stereoisomers of 4-chlorocyclohexanol give different products when they react with ^-OH, as shown in the reactions given in **Fig. P9.73**.

(a) Give a curved-arrow mechanism for the formation of each product.

(b) Explain why the bicyclic compound *B* is observed in the reaction of the trans isomer or 4-chlorocyclohexanol but not in the reaction of the cis isomer.

9.74 The reaction of butylamine, $CH_3(CH_2)_3\ddot{N}H_2$, with 1-bromobutane in 60% aqueous ethanol follows the rate law

$$\text{rate} = k[\text{butylamine}][\text{1-bromobutane}]$$

The product of the reaction is $(CH_3CH_2CH_2CH_2)_2\overset{+}{N}H_2\,Br^-$. The following very similar reaction, however, has a first-order rate law:

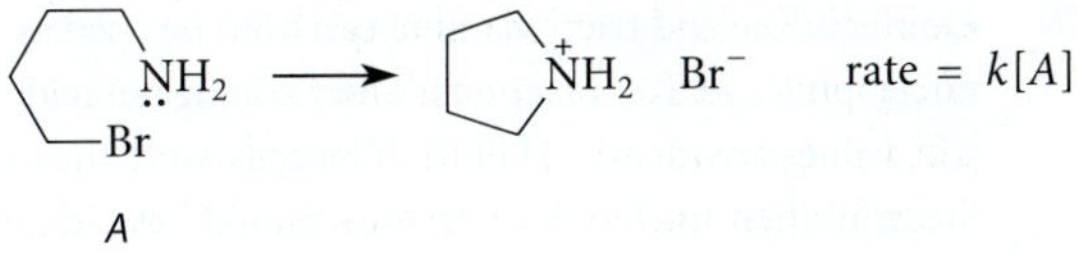

Give a mechanism for each reaction that is consistent with its rate law and with the other facts about nucleophilic substitution reactions. Use the curved-arrow notation.

9.75 In 1975, a report was published in which the reaction given in **Fig. P9.75** was observed. The —OBs (brosylate) group is a leaving group conceptually like halide. (Think of this group as you would —Br.) The reaction conditions favor an S_N2 reaction.

(a) This result created quite a stir among chemists because it seemed to question a fundamental principle of the S_N2 reaction. Explain.

(b) Because the result was potentially very significant, the work was reinvestigated soon after it was published. In this reinvestigation, it was found that after about 10 hours' reaction time, the product consisted almost completely of *trans-P*. Only on standing for a much longer time under the reaction conditions did *cis-P* form (and *trans-P* disappear) to give the product mixture shown in Fig. P9.75. Furthermore, when the trans isomer of *S* was subjected to the same conditions, mostly *cis-P* was formed after 10 hours, but, after 5 days, the same 75:25 cis:trans product mixture was formed as in Fig. P9.75. Finally, subjecting pure *cis-P* or pure *trans-P* to the reaction conditions gave, after 5 days, the same 75:25 mixture. Explain these results.

(c) Why is *cis-P* favored in the product mixture at equilibrium?

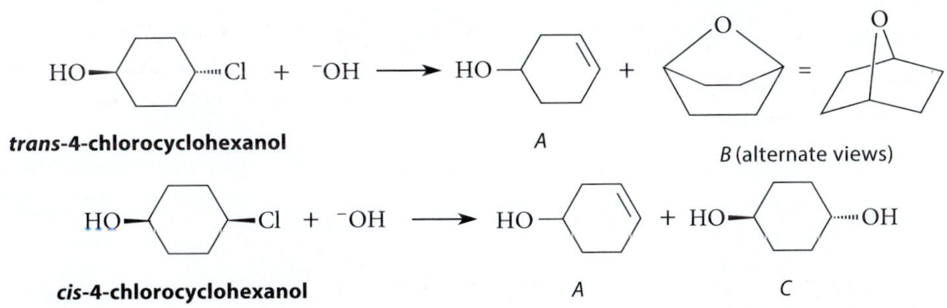

FIGURE P9.73

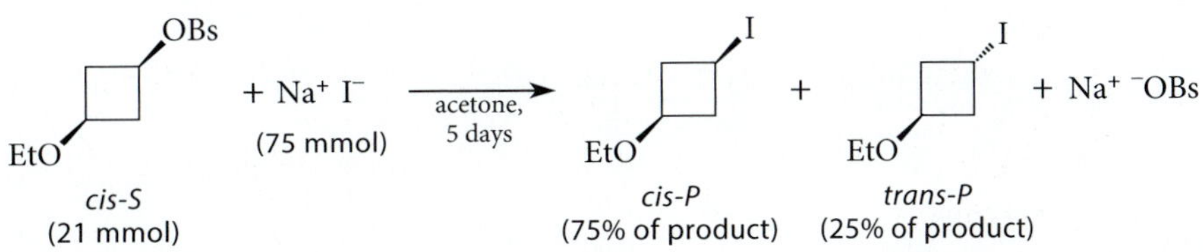

FIGURE P9.75

9.76 Quinuclidine and triethylamine can both be used as nucleophiles in S_N2 reactions. Their conjugate-acid pK_a values are identical (11.0). If basicity were the only factor in their nucleophilicity, they should have identical nucleophilicities.

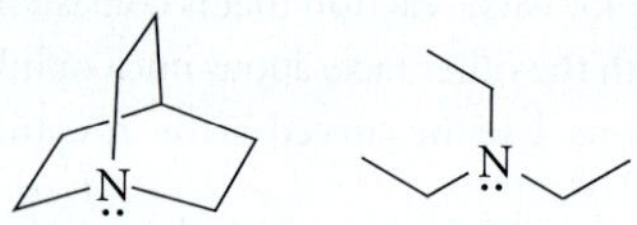

quinuclidine **triethylamine**

In their S_N2 reactions with Me—I, quinuclidine reacts 57 times faster than triethylamine. In their S_N2 reactions with isopropyl iodide, $(CH_3)_2CH—I$, quinuclidine reacts 706 times faster than triethylamine. Explain why quinuclidine reacts faster with these alkyl halides; and explain why its relative rate is much higher with isopropyl iodide than with methyl iodide.

9.77 An optically active compound *A* has the formula $C_8H_{13}Br$. Compound *A* gives no reaction with Br_2 in CH_2Cl_2 (Sec. 5.2A), but it reacts with $K^+(CH_3)_3C—O^-$ to give a single new compound *B* in good yield. Compound *B* decolorizes Br_2 in CH_2Cl_2 and takes up hydrogen over a catalyst. When compound *B* is treated with ozone followed by aqueous H_2O_2, dicarboxylic acid *C* is isolated in excellent yield; notice its cis stereochemistry.

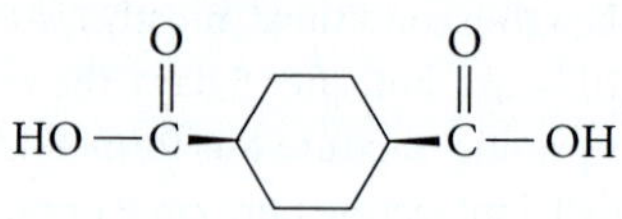

C

Identify compounds *A* and *B*, and account for all observations. (If you need a refresher on how to solve this type of problem, see Study Guide Links 4.3 and 5.2.)

9.78 In the laboratories of the firm Halides 'R' Us, a compound *A* has been found in a vial labeled only "achiral alkyl halide $C_{10}H_{17}Br$." The managers feel that the compound might be useful as a pesticide, but they need to know its structure. You have been called in as a consultant at a handsome fee. Compound *A*, when treated with KOH in warm ethanol, yields two compounds (*B* and *C*), each with the molecular formula $C_{10}H_{16}$. Compound *A* rapidly reacts in aqueous ethanol to give an acidic solution, which gives, in turn, a precipitate of AgBr when tested with $AgNO_3$ solution. Ozonolysis of *A* followed by treatment with $(CH_3)_2S$ affords $(CH_3)_2C{=}O$ (acetone) as one of the products plus unidentified halogen-containing material. Catalytic hydrogenation of either *B* or *C* gives a mixture of both *trans*- and *cis*-1-isopropyl-4-methylcyclohexane. Compound *A* reacts with one equivalent of Br_2 to give a mixture of two separable compounds, *D* and *E*, both of which can be shown to be achiral compounds. Finally, ozonolysis of compound *B* followed by treatment with aqueous H_2O_2 gives acetone and the cyclohexane-1,4-dione, *F*.

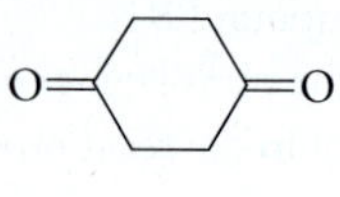

F

Propose structures for compounds *A* through *E* that best fit the data (and collect your fee).

10 FREE-RADICAL REACTIONS, MAIN-GROUP ORGANOMETALLIC COMPOUNDS, AND CARBENES

In this chapter we continue our study of alkyl halides by considering two methods for preparing alkyl chlorides and alkyl bromides. We also show how alkyl halides can be used to prepare main-group organometallic compounds—compounds with bonds between carbon and main-group metals, such as lithium, sodium, and magnesium—and we introduce a few reactions of these compounds. These compounds are important reagents throughout the rest of this text.

A major focus of this chapter is a reactive intermediate called a *free radical.* Many of the reactions of this chapter involve this intermediate. We introduce a special curved-arrow notation, called the *fishhook notation*, that is used with reactions involving free-radical intermediates. We also discuss *polymers,* the large, economically important molecules that were introduced in Sec. 5.10, and we discuss *free-radical polymerization*, a process involving free-radical intermediates used to prepare many polymers. Finally, we introduce another reactive intermediate called a *carbene*, and we show how it is formed and how it occurs as a reactive intermediate in several reactions.

10.1 FREE-RADICAL ADDITION OF HYDROGEN BROMIDE TO ALKENES

A. The Peroxide Effect

As explained in Sec. 4.7A, the addition of HBr to alkenes is a regioselective reaction in which the bromine is directed to the carbon of a double bond with the greater number of alkyl groups. For example, 1-pentene reacts with HBr to give 2-bromopentane almost exclusively:

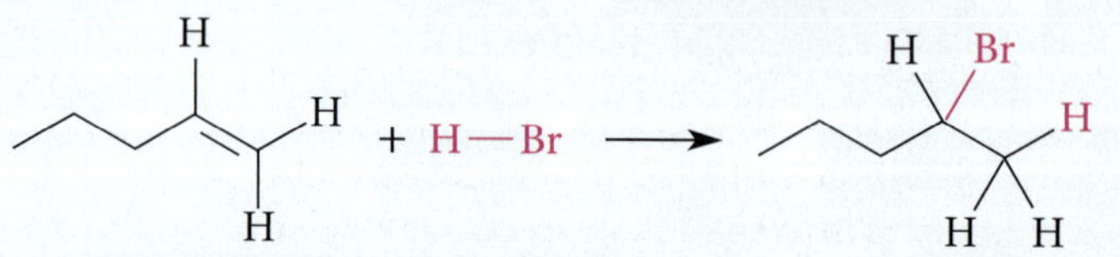

1-pentene **2-bromopentane** (79% yield) (10.1)

In the late 1920s, Morris Kharasch (1895–1957) and his students at the University of Chicago found that when traces of *peroxides* (compounds of the general structure R—O—O—R) were present in the reaction mixture, *the regioselectivity of HBr addition was reversed*. In other words, 1-pentene was found to react in the presence of peroxides so that the bromine added to the *less branched* carbon of the double bond:

Because the regioselectivity of ordinary HBr addition is described by *Markovnikov's rule* (Sec. 4.7A), the peroxide-promoted addition can be said to have *non-Markovnikov* or *anti-Markovnikov* regioselectivity. This means simply that the bromine is directed to the carbon of the alkene double bond with *fewer* alkyl substituents.

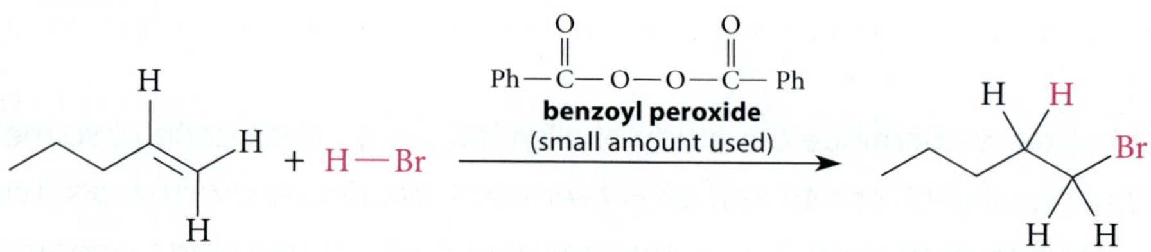

1-pentene **1-bromopentane** (96% yield) (10.2)

(Compare Eq. 10.2 to Eq. 10.1.) This reversal of regioselectivity in HBr addition is called the **peroxide effect.**

Light further promotes the peroxide effect. When Kharasch and his colleagues scrupulously excluded peroxides and light from the reaction, they found that HBr addition has the "normal" regioselectivity shown in Eq. 10.1.

The peroxide effect is observed with all alkenes in which alkyl substitution at the two carbons of the double bond is different. In other words, *in the presence of peroxides, the addition of HBr to alkenes occurs such that the bromine is bound to the carbon of the double bond bearing the smaller number of alkyl substituents, and the hydrogen is bound to the carbon of the double bond bearing the greater number of alkyl substituents.* Furthermore, the peroxide-promoted reaction is faster than HBr addition in the absence of peroxides. Very small amounts of peroxides are required to bring about this effect.

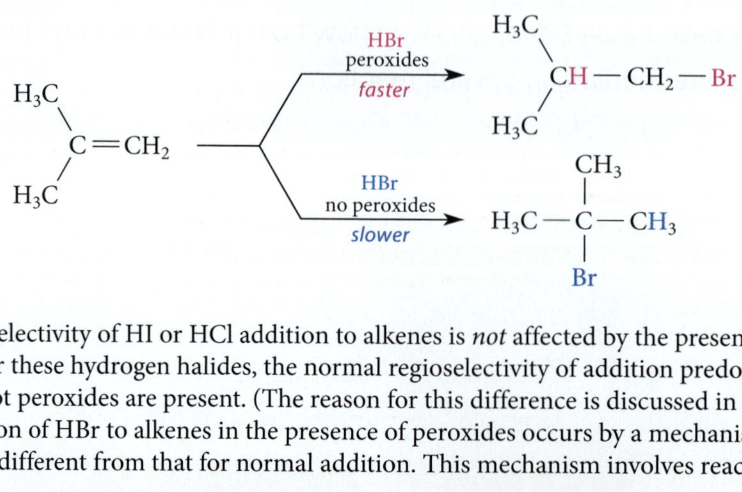

(10.3)

The regioselectivity of HI or HCl addition to alkenes is *not* affected by the presence of peroxides. For these hydrogen halides, the normal regioselectivity of addition predominates, whether or not peroxides are present. (The reason for this difference is discussed in Sec. 10.1E.)

The addition of HBr to alkenes in the presence of peroxides occurs by a mechanism that is completely different from that for normal addition. This mechanism involves reactive intermediates called *free radicals*. To appreciate the reasons for the peroxide effect, then, let's digress and learn some basic facts about free radicals.

B. Radicals and the "Fishhook" Notation

In all reactions considered previously, we used a curved-arrow notation that indicates the movement of electrons in *pairs*. The dissociation of HBr, for example, is written

heterolytic (two-electron) bond breaking (10.4)

In this reaction, bromine takes *both* electrons of the H—Br covalent bond to give a bromide ion, and the hydrogen becomes an electron-deficient species, the proton. This type of bond breaking is an example of a *heterolysis*, or *heterolytic cleavage* (*hetero* = different; *lysis* = bond breaking). In a **heterolytic process**, electrons involved in the process "move" in pairs. A **heterolysis** is a bond-breaking process that occurs with electrons "moving" in pairs.

Bond rupture, however, can occur in another way. An electron-pair bond may also break so that each bonding partner takes one electron of the chemical bond.

$$\text{H—}\ddot{\underset{..}{\text{Br}}}\text{:} \longrightarrow \text{H}\cdot + \cdot\ddot{\underset{..}{\text{Br}}}\text{:}$$

homolytic (one-electron) bond breaking (10.5)

In this process, a hydrogen *atom* and a bromine *atom* are produced. As you should verify, these atoms are uncharged. This type of bond breaking is an example of a *homolysis*, or *homolytic cleavage* (*homo* = the same; *lysis* = bond breaking). In a **homolytic process**, electrons involved in the process "move" in an unpaired way. A **homolysis** is a bond-breaking process that occurs with electrons moving in an unpaired fashion.

A different curved-arrow notation is used for homolytic processes. In this notation, called the **fishhook notation**, *electrons move individually rather than in pairs*. This type of electron flow is represented with *singly barbed arrows*, or **fishhooks**; one fishhook is used for each electron:

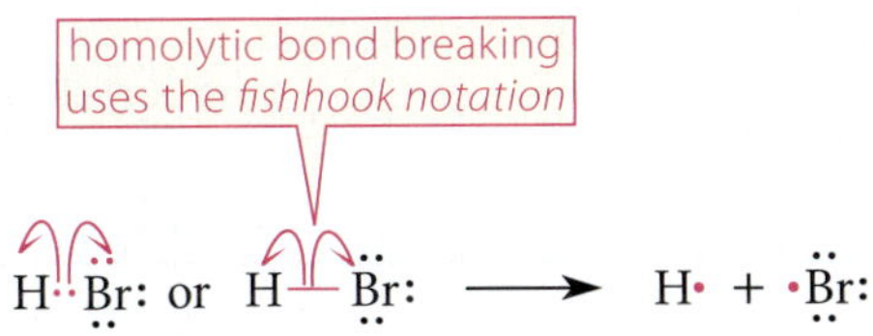

(10.6)

Homolytic bond cleavage is not restricted to diatomic molecules. Peroxides, for example, are prone to undergo homolytic cleavage because of their very weak O—O bonds:

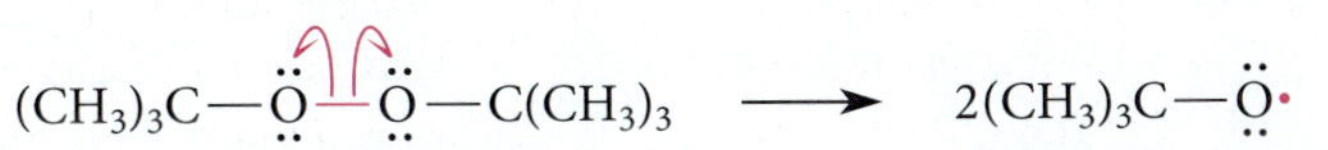

di-*tert*-butyl peroxide **_tert_-butoxy radical** (10.7)

The fragments on the right side of this equation possess unpaired electrons. Any species with at least one unpaired electron is called a **free radical**. The hydrogen atom and the bromine atom on the right side of Eq. 10.6, as well as the *tert*-butoxy radical on the right side of Eq. 10.7, are all examples of free radicals.

An important convention for representing free radicals is that we can omit the nonbonding electron pairs and show only the unpaired electron. For example, we can represent the bromine atom as Br• rather than $:\ddot{\underset{..}{\text{Br}}}\cdot$. The zero formal charge on the bromine atom tells us that it must have six (three pairs) of nonbonding valence electrons in addition to the unpaired electron. Similarly, the *tert*-butoxy radical shown in Eq. 10.7 can be represented as $(CH_3)_3C$—O•; its zero formal charge means that it must have five additional electrons: one for the covalent bond, and two nonbonding valence electron pairs.

Here are the rules for use of the fishhook notation:

Rules for unpaired electrons:

1. A fishhook tail must be located on every unpaired electron of the reactant that is involved in a transformation.

2. A fishhook barb must be located on every atom of the reactant that is to gain an unpaired electron in the product.

Rules for electron-pair bonds:

3. The tails of *two* fishhooks must be located on every bond of the reactant that is to break homolytically.

4. Two fishhook barbs must be shown together in the reactant at the location of a new electron-pair bond in the product.

Observe how these rules are followed in Study Problems 10.1 and 10.2.

Study Problem 10.1

Provide the products of the following transformation illustrated by the fishhook notation. (The bonds are labeled for reference in the solution.)

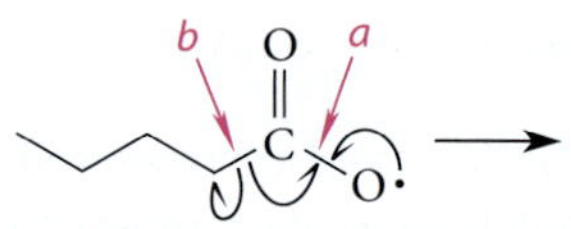

Solution Two fishhooks terminate on bond *a*. (Notice that they don't have to be drawn to the same side of the bond.) By rule 4, a new bond must be formed between the C and the O to make a carbon–oxygen double bond. One of the electrons in the new bond comes from the unpaired electron on oxygen (rule 1), and the other comes from the carbon–carbon electron-pair bond *b*. The other electron of bond *b* remains on the other carbon of bond *b* (rule 2). Because two electrons come from bond *b*, this bond breaks (rule 3). The fishhook notation therefore shows the following conversion:

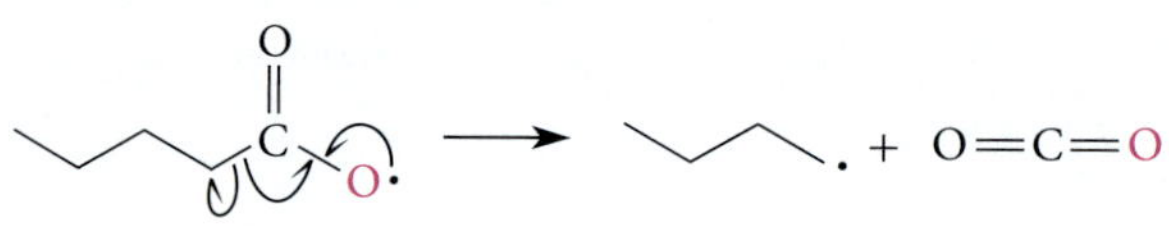

Study Problem 10.2 shows that the fishhook notation, like the curved-arrow notation, can be applied to resonance structures as well as reactions.

Study Problem 10.2

Provide the fishhook notation that converts the first resonance structure into the second. (The red letters are for reference in the solution.)

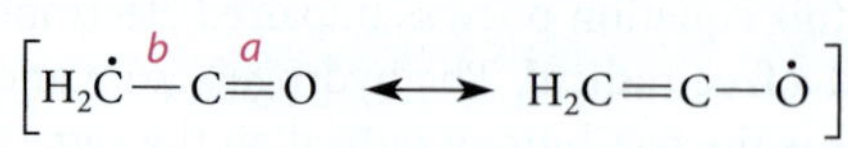

Solution A new electron-pair bond is formed at location *b*. Therefore, two barbs must point to this location (rule 4). One electron in this new bond comes from the unpaired electron on the H_2C group; therefore, the tail of one of the fishhooks on bond *b* comes from this unpaired electron (rule 1). The other electron in the new bond at *b* comes from one of the C=O bonds *a*; therefore, the tail of this fishhook starts at this C—O bond. Because this bond breaks, the tail of a second fishhook must also be located on this bond (rule 3). The barb of this fishhook points to the oxygen, which gains an unpaired electron (rule 2). In summary:

Radicals Bound and Free

The "R" used in the R-group notation (Sec. 2.8B) comes from the word *radical*. In the mid-1900s, R groups were called "radicals." For example, the CH_3 group in CH_3CH_2OH might have been referred to as the "methyl radical." Such R groups, when not bonded to anything, were then called "free radicals," so CH_3 was called the "methyl free radical." Nowadays we simply use the word *group* to refer to a group of atoms (for example, the methyl group) in a compound; we reserve the word *radical* for a species with an unpaired electron.

Focused Problems

10.1 Draw the products of each of the following transformations shown by the fishhook notation.

(a) 

(b) 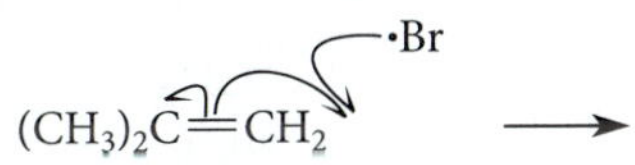

(c) 

(d) 

10.2 Indicate whether each of the following reactions is homolytic or heterolytic, and tell how you know. Write the appropriate fishhook or curved-arrow notation for each.

(a) 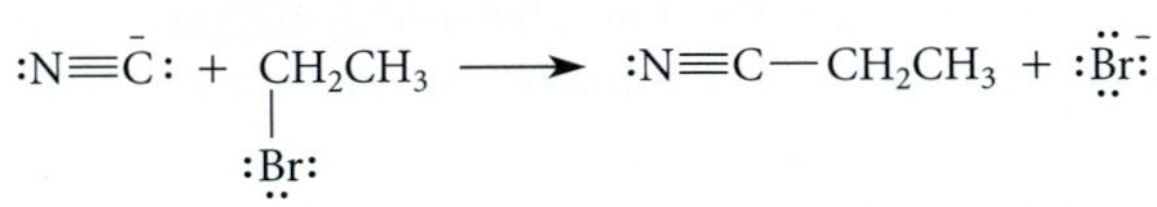

(b) $CH_3CH_2OH + CH_3O\cdot \longrightarrow CH_3\dot{C}HOH + CH_3OH$

(c) Using the fishhook notation, derive a resonance structure for the following free radical.

$$CH_3\dot{C}H—\ddot{O}H$$

C. Free-Radical Chain Reactions

Although a few stable free radicals are known, most free radicals are very reactive. When they are generated in chemical reactions, they generally behave as *reactive intermediates*—that is, they react so rapidly that they do not accumulate in significant amounts. This section shows how free radicals are involved as reactive intermediates in the peroxide-promoted addition of HBr to alkenes. This discussion provides the basis for a general understanding of free-radical reactions, as well as the peroxide effect in HBr addition, which is the subject of the next section.

Most free-radical reactions can be classified as free-radical chain reactions. A **free-radical chain reaction** involves free-radical intermediates and consists of the following three fundamental reaction steps:

1. *Initiation* steps
2. *Propagation* steps
3. *Termination* steps

> Although most peroxides can serve as free-radical initiators, a notable exception is hydrogen peroxide (H_2O_2), which is *not* commonly used as a source of initiating free radicals. The O—O bond in hydrogen peroxide is considerably stronger than the O—O bonds in most other peroxides and is therefore harder to break homolytically. The cleavage of organoboranes by hydrogen peroxide (the oxidation part of hydroboration–oxidation; Eq. 5.35, Sec. 5.6B) is *not* a free-radical reaction. (See Further Exploration 5.1.)

Let's examine each of these steps using the peroxide-promoted addition of HBr to alkenes as an example of a typical free-radical reaction. (The reason for the "chain reaction" terminology will become apparent.)

Initiation In the **initiation** steps, the free radicals that take part in subsequent steps of the reaction are formed from a **free-radical initiator**, which is a molecule that undergoes homolysis with particular ease. *The initiator is the source of free radicals.* An initiation reaction needs to occur only once to start a free-radical chain reaction. Peroxides, such as di-*tert*-butyl peroxide, are frequently used as free-radical initiators. The first initiation step in the free-radical addition of HBr to an alkene is the homolysis of the peroxide.

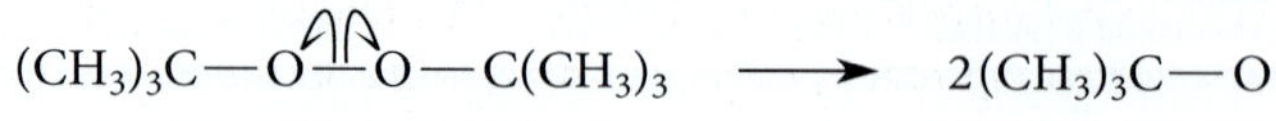

di-*tert*-butyl peroxide **_tert_-butoxy radical** (10.8)

Peroxides are not the only type of free-radical initiators. Another widely used initiator is azobis(isobutyronitrile), known by the acronym AIBN. This substance readily forms free radicals because the very stable molecule dinitrogen is liberated as a result of homolytic cleavage:

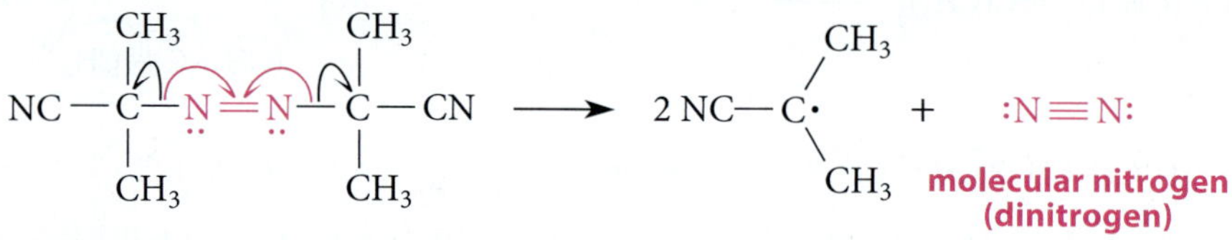

azobis(isobutyronitrile) (AIBN) (10.9)

Sometimes heat or light initiates a free-radical reaction. This usually happens because the additional energy causes homolysis of the free-radical initiator—or, in some cases, the reactants themselves—into free radicals.

The effects of initiators provide some of the best clues that a reaction occurs by a free-radical mechanism. If a reaction occurs in the presence of a known free-radical initiator but does not occur in its absence, we can be fairly certain that the reaction involves free-radical intermediates. (As noted in Sec. 10.1A, Morris Kharasch showed that the change in regiospecificity of HBr addition requires peroxides. This is the type of evidence that we now take to be strongly indicative of free-radical mechanisms.)

A second initiation step occurs in the free-radical addition of HBr to alkenes: the removal of a hydrogen atom from HBr by the *tert*-butoxy free radical that was formed in the first initiation step (Eq. 10.8).

***tert*-butoxy radical**
(from Eq. 10.8) (10.10)

This is an example of *atom abstraction*, another common type of free-radical process. In an atom abstraction reaction, a free radical removes an atom from another molecule, and a new free radical is formed (Br· in Eq. 10.10). The bromine atom is then involved in the *propagation* steps, which are the next phase of the reaction.

> **!** Don't confuse the hydrogen-abstraction reaction in Eq. 10.10 with a Brønsted acid–base reaction. A Brønsted acid–base reaction involves transfer of a *proton*. The reaction in Eq. 10.10 involves abstraction of a hydrogen *atom*.

Propagation In the **propagation steps** of a free-radical reaction, radicals react with nonradical starting materials to give other radicals; starting materials are consumed, and products are formed. *Propagation steps occur repeatedly*. When the propagation steps are considered together, there is no *net* formation or destruction of any of the radical species involved. This means that if a radical is formed, it must be consumed in a subsequent propagation step and another radical must be formed to take its place.

The first propagation step of free-radical addition of HBr to an alkene is the reaction of the bromine atom generated in Eq. 10.10 with the π bond.

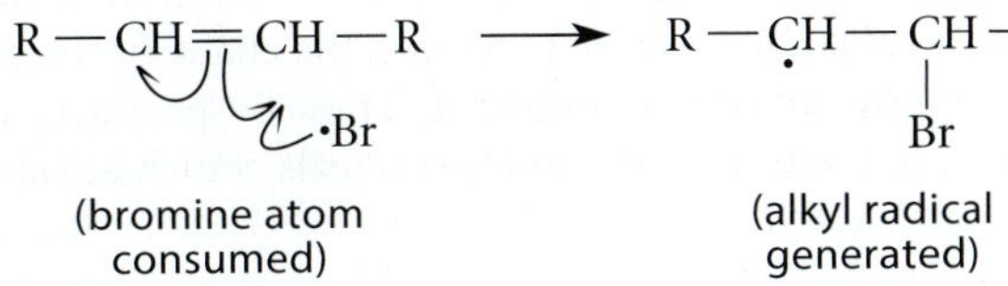

(10.11a)

Reaction of a free radical with a carbon–carbon π bond is another common process encountered in free-radical chemistry. The π bond reacts, rather than a σ bond, because carbon–carbon π bonds are weaker than carbon–carbon σ bonds.

The second propagation step is another *atom abstraction reaction*—namely, removal of a hydrogen atom from HBr by the free-radical product of Eq. 10.11a to give the addition product and a new bromine atom.

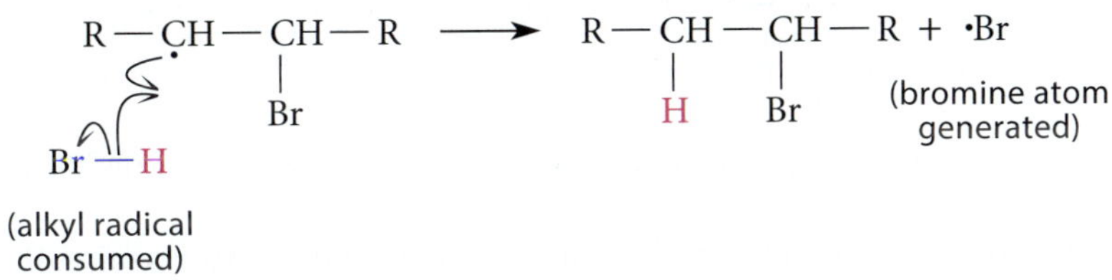

(10.11b)

The bromine atom can react, in turn, with another molecule of alkene (Eq. 10.11a), and this can be followed by the generation of another molecule of product along with another bromine atom (Eq. 10.11b). We can now see the basis of the term *chain reaction*. These two propagation steps continue in a chainlike fashion until the reactants are consumed. That is, the product free radical of one propagation step becomes the starting free radical for the next propagation step. For each "link in the chain"—each cycle of the two propagation steps—one molecule of the product is formed and one molecule of alkene starting material is consumed. For each free radical consumed in the propagation steps, one is produced. Because no net destruction of free radicals occurs, the initial concentration of free radicals provided by the initiator, so the concentration of the initiator itself, can be small. Typically, the initiator concentration is only 1–2% of the alkene concentration.

The free radicals involved in the propagation steps of a chain reaction are said to *propagate the chain*. The free radicals in Eqs. 10.11a–b are the chain-propagating radicals in the peroxide-promoted free-radical addition of HBr.

Some students think that Eq. 10.10 is part of the propagation sequence because it forms a bromine atom, which is one of the chain-propagating radicals. It is part of the initiation sequence, however, because the *tert*-butoxy radical does not reappear in the subsequent propagation steps of the reaction. Equation 10.10 has to occur only once for the subsequent propagation steps to occur repeatedly.

An Analogy for Chain Reactions

Oliver Hoffmann/ Shutterstock

An analogy for a chain reaction can be found in the world of business. A businessperson uses a little seed money, or capital, to purchase a small business. In time, this business produces profit that is used to buy another business. This second business then produces profit that can be used to buy yet another business, and so on. All this time, the businessperson is accumulating business property (instead of alkyl bromides), although the total amount of cash on hand is, by analogy to the chain-propagating free radicals in the reaction sequence described earlier, quite small compared with the total amount of the investments.

The toppling of the dominoes shown in the photograph is another example of a chain reaction from the physical world. To initiate the process, the finger needs to provide only enough energy to topple the first domino. What "propagates the chain" once it is initiated?

Termination In the **termination** steps of a free-radical chain reaction, two radicals react to give nonradical products. A common termination step involves a *radical recombination reaction,* in which two radicals come together to form a covalent bond. In other words, radical recombination is the reverse of a homolysis.

The following reactions are two examples of termination reactions that can take place in the free-radical addition of HBr to alkenes. In these reactions, the chain-propagating radicals of Eqs. 10.11a and 10.11b recombine to form by-products. These by-products are present in very small amounts because they are formed *only* from free radicals, which are also present in very small amounts.

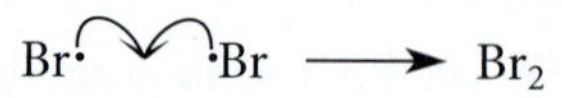

(10.12)

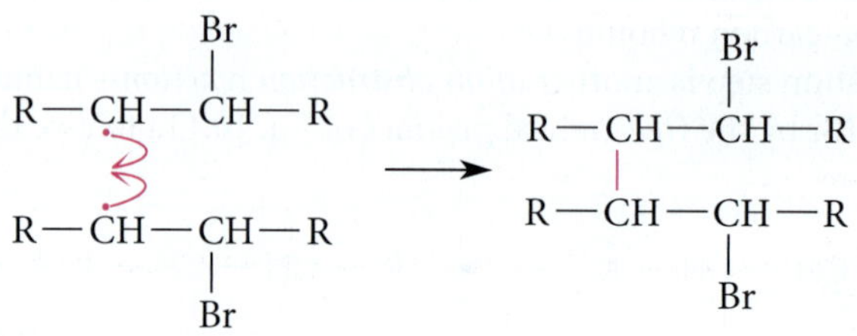

(10.13)

Because each recombination reaction takes two free radicals "out of circulation," it terminates two propagation reactions and breaks two free-radical chains.

The recombination reactions of free radicals are in general highly exothermic—that is, they have very favorable, or negative, $\Delta H°$ values. They typically occur on every encounter of two free radicals; in other words, there is no $\Delta G^{\circ\ddagger}$ for radical recombination. In view of this fact, we might ask why free radicals do not simply recombine before they propagate any chains. The answer is simply a matter of the relative concentrations of the various species involved. *Free-radical intermediates are present in very low concentration*, but the other reactants are present in much higher concentration. Consequently, it is much more probable for a bromine atom to collide with an alkene molecule in the propagation reaction of Eq. 10.11a than with another bromine atom in a recombination reaction:

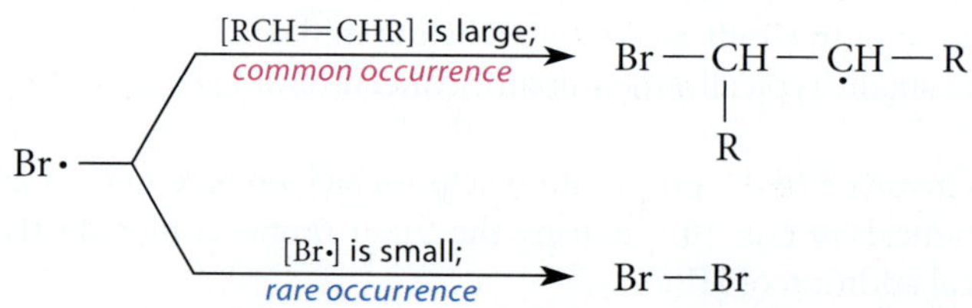

(10.14)

In a typical free-radical chain reaction, a termination reaction occurs once for every 10,000 propagation reactions. As the reactants are depleted, however, the probability becomes significantly greater that one free radical will survive long enough to wander into another free radical with which it can recombine. Small amounts of by-products resulting from termination reactions are typically observed in free-radical chain reactions.

In most cases, only exothermic propagation steps—steps with favorable, or negative, $\Delta H°$ values—occur rapidly enough to compete with the recombination reactions that terminate free-radical chain processes. Both of the propagation steps in the free-radical HBr addition to alkenes are exothermic, and they both occur readily. However, the first propagation step of free-radical HI addition, and the second propagation step of free-radical HCl addition, are quite *endothermic.* (This point is explored further in Sec. 10.1E.) For this reason, these processes occur to such a small extent that they cannot compete with the recombination processes that terminate these chain reactions. Consequently, the free-radical addition of neither HCl nor HI to alkenes is observed.

In summary: many free-radical reactions occur in three phases:

1. In the initiation steps, radicals are produced from nonradicals.

2. In the propagation steps, radicals react with nonradical starting materials to give other radicals and nonradical products. Propagation steps occur repeatedly.

3. In the termination steps, radicals react with each other to give small amounts of nonradical by-products.

In most free-radical reactions, products accumulate as the result of propagation steps because they occur much more frequently than termination steps.

This section has discussed the characteristics of free-radical chain reactions using the free-radical addition of HBr to alkenes as an example. A number of other useful laboratory reactions involve free-radical chain mechanisms. Many important industrial processes are also free-radical chain reactions (Secs. 10.3 and 10.4). Free-radical chain reactions of great environmental importance occur in the upper atmosphere when ozone is destroyed by chlorofluorocarbons (Freons), the compounds that until recently have been exclusively used as coolants in air conditioners and refrigerators. (These reactions are discussed in Sec. 10.10B.) A number of free-radical processes have also been characterized in biological systems. (See Sec. 17.5B.)

Study Problem 10.3

Alkenes undergo the addition of thiols at high temperature in the presence of peroxides or other free-radical initiators. The following reaction is an example.

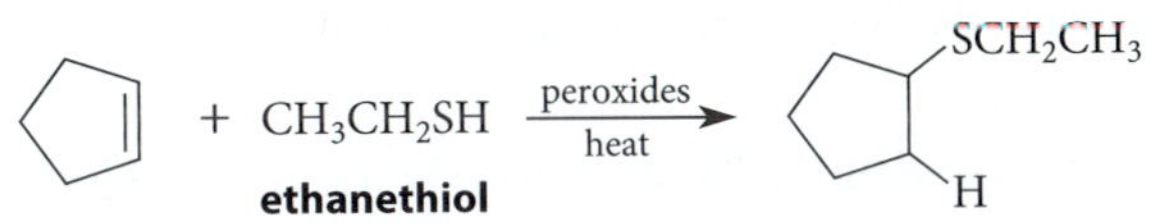

cyclopentene **ethanethiol** (a thiol) a thioether or sulfide

Propose a mechanism for this reaction.

Solution The fact that the reaction requires peroxides indicates that a free-radical mechanism is operating. The initiation step is abstraction of a hydrogen atom from the thiol by the *tert*-butoxy radical derived from homolysis of the free-radical initiator. (See Eq. 10.10.)

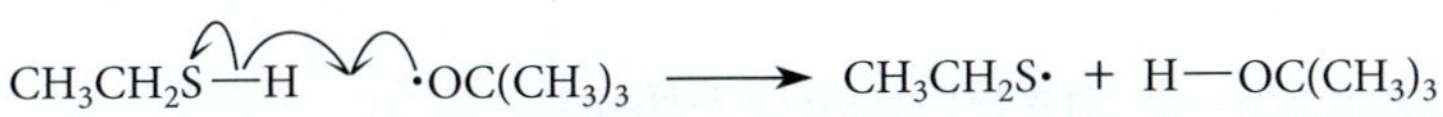

***tert*-butoxy radical**

In the first propagation step, the sulfur radical adds to the double bond of the alkene to generate a new carbon radical.

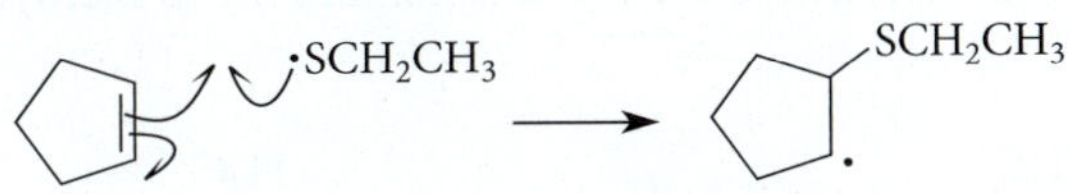

In the final propagation step, the carbon radical reacts with a thiol molecule to give the product plus a new sulfur radical, which propagates the chain by reacting with another alkene.

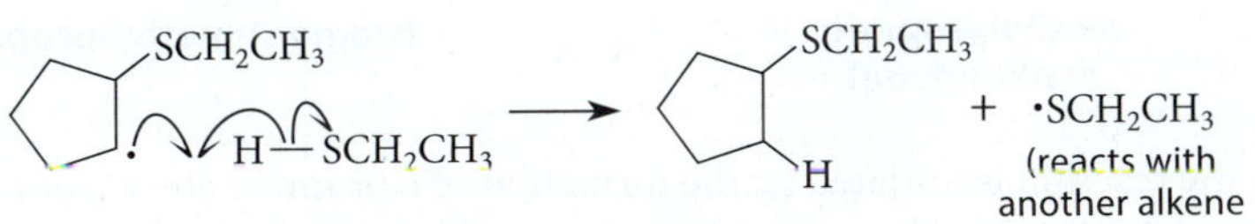

(reacts with another alkene molecule)

Some students are tempted to write a mechanism such as the following:

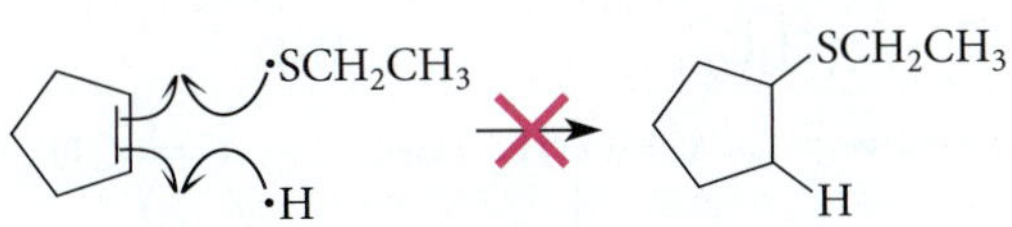

This cannot be correct for several reasons. First, this reaction destroys free radicals. If this were the mechanism, then we would need a separate initiation step for every product molecule formed, and the initiator would then have to be present at the same concentration as the reactants. Second, there is no obvious source for the hydrogen atom. Finally, and most important, free radicals are typically present in *miniscule* concentrations. A collision of three species, two of which (the radicals) are present in very low concentration, is so improbable that it doesn't occur. Therefore, *most free-radical reactions occur as chain reactions.* In a chain reaction, only one free radical reacts in a given step; it reacts with a nonradical molecule present in substantial concentration, and a new radical is produced. This last point means that initiators must supply only a small initial concentration of radicals to get things started, because a small population of free radicals is maintained until the reactants run out (at which point termination reactions can compete with propagation steps).

So, here's the message: *Write chain mechanisms as propagation reactions involving only one radical per step* when you write free-radical reactions.

Focused Problems

10.3 In the presence of light, the addition of Br_2 to ethylene can occur by a free-radical mechanism rather than a bromonium-ion mechanism. Write a free-radical chain mechanism that shows the propagation steps for the following addition:

$$Br_2 + H_2C{=}CH_2 \xrightarrow{\text{light}} BrCH_2{-}CH_2Br$$

Assume that the initiation step for the reaction is the light-promoted homolysis of Br_2:

$$Br{-}Br \xrightarrow{\text{light}} 2\,Br\cdot$$

10.4 (a) Suggest a mechanism for the free-radical addition of HBr to cyclohexene initiated by AIBN. Show the initiation and propagation steps.

(b) In the free-radical addition of HBr to cyclohexene, suggest structures for three radical recombination products that might be formed in small amounts in the termination phase of the reaction.

D. Explanation of the Peroxide Effect

The free-radical mechanism is the basis for understanding the peroxide effect on HBr addition to alkenes—that is, why the presence of peroxides reverses the normal regioselectivity of HBr addition (Eq. 10.1, Sec. 10.1A). We use the addition of HBr to methylpropene as the basis for our discussion.

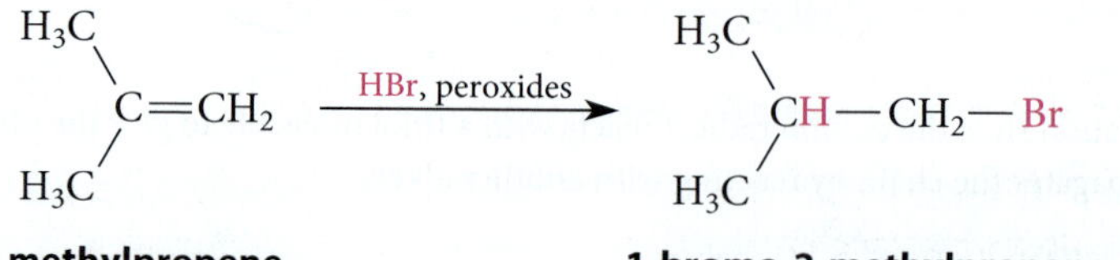

methylpropene (isobutylene) **1-bromo-2-methylpropane** (10.15)

Recall that the reaction is initiated by the formation of a bromine atom from HBr (Eq. 10.10). When the bromine atom adds to the π bond of an alkene, two reactions are in competition: the bromine atom can react at either of the two carbons of the double bond to give different free-radical intermediates.

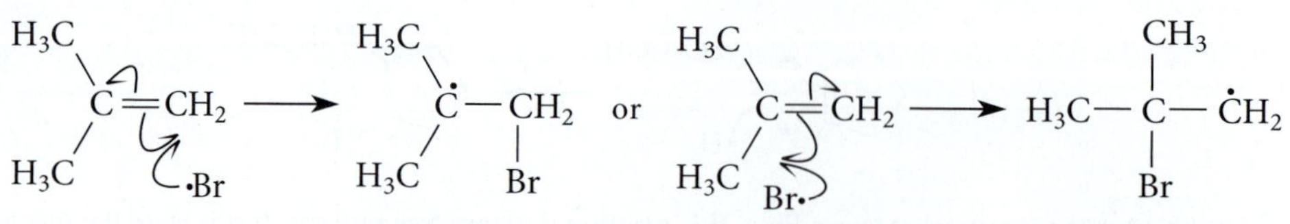

a tertiary free radical a primary free radical (10.16)

Free radicals, like carbocations, can be classified as primary, secondary, or tertiary, depending on the number of R groups bonded to the carbon with the unpaired electron.

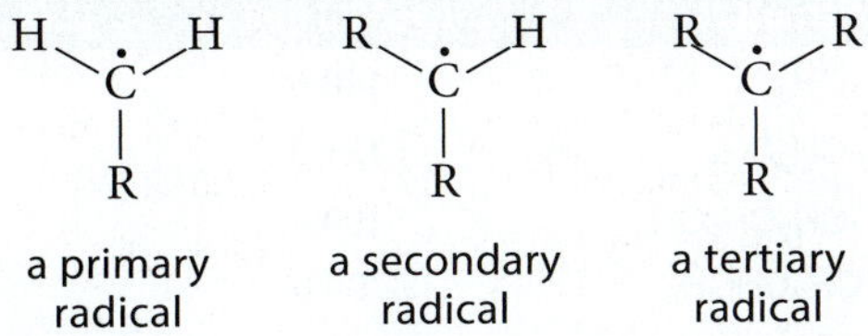

(10.17)

Equation 10.16 involves a competition between the formation of a primary free radical and the formation of a tertiary free radical. *The formation of the tertiary free radical is faster.*

The tertiary free radical is formed more rapidly for two reasons. The first reason is that when the rather large bromine atom reacts at the more-branched carbon of the double bond, it experiences van der Waals repulsions with the hydrogens in the branches (**Fig. 10.1a**). These repulsions increase the energy of this transition state. When the bromine atom reacts at the less-branched carbon of the double bond, these van der Waals repulsions are absent (**Fig. 10.1b**). Because the reaction with the transition state of lower energy is the faster reaction, reaction of the bromine atom at *the alkene carbon with fewer alkyl substituents to give the more alkyl-substituted free radical* is faster. Subsequent reaction of this free radical with HBr leads to the observed products. To summarize:

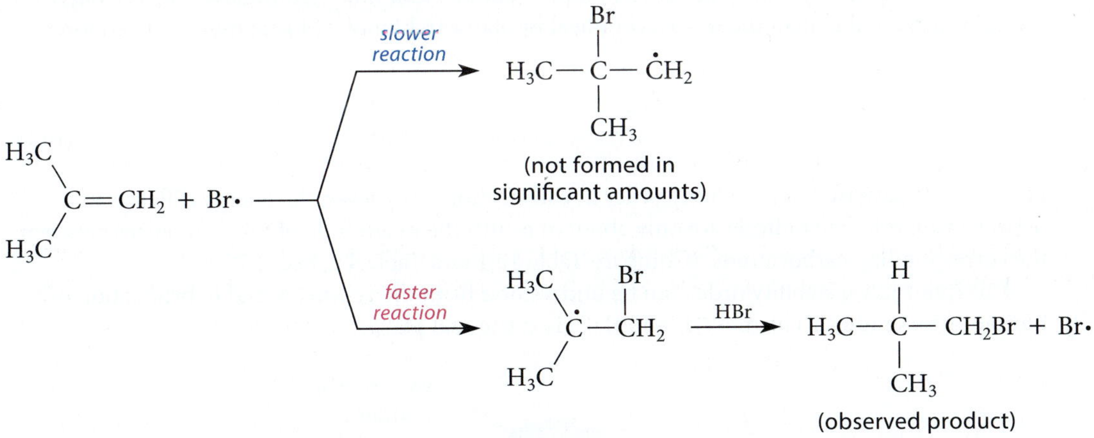

(10.18)

This is another example of a *steric effect* (Sec. 9.4D, Display 5.34b in Sec. 5.6A, and subsequent discussion).

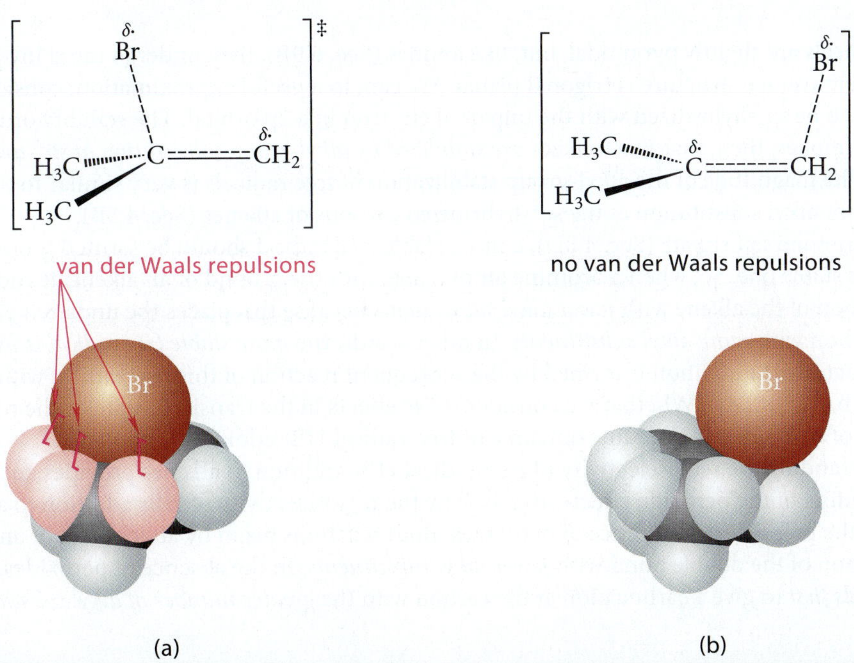

FIGURE 10.1 Space-filling models of the alternative transition states for the addition of a bromine atom to methylpropene. In part (a), the bromine is adding to the carbon of the double bond that has the two methyl substituents. This transition state contains van der Waals repulsions between the bromine and four of the six methyl hydrogens, which are shown in pink. (Three of these are shown; one is hidden from view.) In part (b), the bromine is adding to the CH_2 carbon of the double bond, and the van der Waals repulsions shown in part (a) are absent. The transition state in part (b) has lower energy, so it leads to the observed product.

TABLE 10.1 Heats of Formation of Some Free Radicals (25 °C)

Radical	Structure	ΔH_f° (kJ mol^{-1})	ΔH_f° (kcal mol^{-1})
methyl	$\cdot CH_3$	146.6	35.0
ethyl	$\cdot CH_2CH_3$	121.3	29.0
propyl	$\cdot CH_2CH_2CH_3$	100.4	24.0
isopropyl	$CH_3\dot{C}HCH_3$	90.0	21.5
butyl	$\cdot CH_2CH_2CH_2CH_3$	79.7	19.0
isobutyl	$\cdot CH_2CH(CH_3)_2$	70	17
sec-butyl	$CH_3\dot{C}HCH_2CH_3$	67.4	16.1
tert-butyl	$\cdot C(CH_3)_3$	51.5	12.3

The second reason that the tertiary radical is formed in Eq. 10.16 has to do with its relative stability. The heats of formation of several free radicals are given in **Table 10.1**. A comparison of the heats of formation for propyl and isopropyl radicals shows that the secondary radical is more stable than the primary one by about 12 kJ mol^{-1} (about 3 kcal mol^{-1}). Similarly, the *tert*-butyl radical is more stable than the *sec*-butyl radical by about 16 kJ mol^{-1} (4 kcal mol^{-1}). Therefore:

Stability of free radicals:

$$\text{tertiary} > \text{secondary} > \text{primary} \qquad \textbf{(10.19)}$$

Free radicals have the same stability order as carbocations. However, the energy differences between isomeric free radicals are only about one-fifth the magnitude of the differences between the corresponding carbocations. (Compare Table 10.1 and Table 4.2, Sec. 4.7C.)

This free-radical stability order can be understood from the geometry and hybridization of a typical carbon radical. The methyl radical $\cdot CH_3$ is trigonal planar:

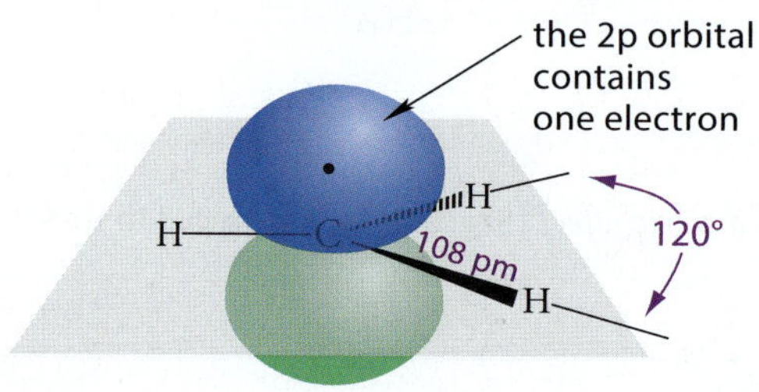

(10.20)

Other radicals are slightly pyramidal, but, like amines (Sec. 6.9B), they undergo rapid inversion, so that their average structure is trigonal planar. We can, to a useful approximation, consider them also to be sp^2-hybridized with the unpaired electron in a 2p orbital. The stability order in Eq. 10.19 implies, then, that *free radicals are stabilized by alkyl-group substitution at sp^2-hybridized carbons*. The magnitude of the alkyl-group stabilization of free radicals is very similar to that observed for alkyl substitution at the sp^2-hybridized carbons of alkenes (Sec. 4.5B).

By Hammond's postulate (Sec. 4.8D), a more stable free radical should be formed more rapidly than a less stable one. So, when a bromine atom reacts with the π bond of an alkene, it adds to the carbon of the alkene with *fewer alkyl substituents* because this places the unpaired electron on the carbon with *more alkyl substituents*. In other words, the *more stable free radical* is formed. The product of HBr addition is formed by the subsequent reaction of this free radical with HBr (Eq. 10.11b, Sec. 10.1C). Whether we consider steric effects in the transition state or the relative stabilities of free radicals, the same outcome of free-radical HBr addition is predicted.

Understanding the regioselectivity of free-radical HBr addition to alkenes provides an understanding of the peroxide effect—that is, why the regioselectivity of HBr addition to alkenes differs in the presence and absence of peroxides. Both reactions begin by attachment of an atom to the carbon of the double bond with *fewer alkyl substituents*. In the absence of peroxides, the *proton adds first* to give a carbocation at the carbon with the *greater number of alkyl substituents*.

The nucleophilic reaction of the bromide ion at this carbon completes the addition. In the presence of peroxides, the free-radical mechanism occurs; a *bromine atom adds first*, thus placing the unpaired electron on the carbon with the *greater number of alkyl substituents*. A hydrogen atom is subsequently transferred to this carbon.

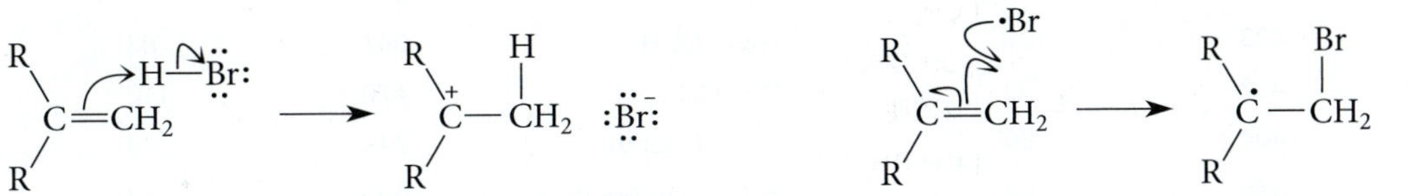

Electrophilic addition: a proton adds first to give the more stable carbocation.

Radical addition: a bromine atom adds first to give the more stable carbon radical.

(10.21)

The role of peroxides, as we've seen, is to initiate the radical reaction. Any good free-radical initiator will bring about the same effect.

The peroxide effect also operates in the addition of HBr to alkynes.

$$CH_3(CH_2)_3C{\equiv}CH + HBr \xrightarrow[0\text{–}5\ °C,\ 1\ h]{\text{peroxides}} CH_3(CH_2)_3CH{=}CHBr \quad (10.22)$$

1-hexyne → **1-bromo-1-hexene;** stereochemistry not determined; (probably *E*) (74% yield)

Focused Problems

10.5 Give the structure of the organic product(s) formed when HBr reacts with each of the following alkenes in the presence of peroxides, and explain your reasoning. If more than one product is formed, predict which one should predominate and why.

(a) 1-Pentene (b) (*E*)-4,4-Dimethyl-2-pentene

10.6 Write the propagation steps, including the fishhook notation, for the peroxide-initiated reaction of HBr with each of the following compounds.

(a) 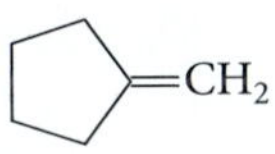

(b) 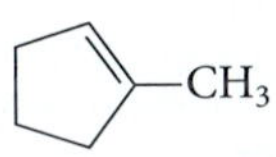

(c) 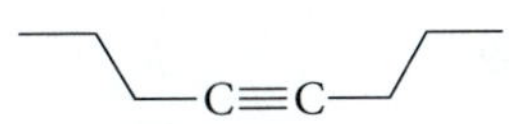

E. Bond Dissociation Energies

How easily does a chemical bond break homolytically to form free radicals? The answer depends on its *bond dissociation energy*. The **bond dissociation energy** (abbreviated **BDE**) of a bond between two atoms X—Y is defined as the standard enthalpy $\Delta H°$ of the reaction

Bond dissociation energies are often called **bond dissociation enthalpies**.

$$X{-}Y \longrightarrow X\cdot + Y\cdot \quad (10.23)$$

A bond dissociation energy always corresponds to the enthalpy required to break a bond *homolytically*. That is, the bond energy of H—Br refers to the process

$$H{-}Br \longrightarrow H\cdot + Br\cdot \quad (10.24a)$$

and *not* to the heterolytic process

$$H{-}\ddot{B}r: \longrightarrow H^+ + :\ddot{B}r:^- \quad (10.24b)$$

TABLE 10.2 Bond Dissociation Energies (Gas Phase, 25 °C)

Bond	$\Delta H°$ (kJ mol^{-1})	$\Delta H°$ (kcal mol^{-1})
C—H bonds		
H_3C—H	439	105
CH_3CH_2—H	423	101
$(CH_3)_2CH$—H	412	99
$(CH_3)_3C$—H	404	96
$PhCH_2$—H	378	90
H_2C═$CHCH_2$—H	372	89
RCH═CH—H	463	111
Ph—H	472	113
RC≡C—H	558	133
H—CN	528	126
C—Halogen bonds		
H_3C—F	481	115
$(CH_3)_2CH$—F	463	111
H_3C—Cl	350	84
CH_3CH_2—Cl	355	85
$(CH_3)CH$—Cl	356	85
$(CH_3)_3C$—Cl	355	85
H_3C—Br	302	72
CH_3CH_2—Br	303	72
$(CH_3)_2CH$—Br	309	74
$(CH_3)_3C$—Br	304	73
H_3C—I	241	58
CH_3CH_2—I	238	57
$(CH_3)_2CH$—I	238	57
$(CH_3)_3C$—I	233	56
Ph—F	531	127
Ph—Cl	406	97
Ph—Br	352	84
Ph—I	280	67
C—C bonds		
H_3C—CH_3	377	90
Ph—CH_3	433	104
$PhCH_2$—CH_3	318	78
H_3C—CN	510	122
H_2C═CH_2 (both)	728	174
H_2C═CH_2 (π bond)	243	58
HC≡CH	~966	~231
C—O bonds		
H_3C—OH	385	92
H_3C—OCH_3	347	83
Ph—OH	470	112
H_2C═O (both)	749	179
H_2C═O (π bond)	305	73
C—N bonds		
H_3C—NH_2	356	85
Ph—NH_2	435	104
H_2C═NH	~736	~176
HC≡N	~987	~236
H—X bonds		
H—OH	498	119
H—OCH_3	438	105
H—O_2CCH_3	473	113
H—OPh	377	90
H—F	569	136
H—Cl	431	103
H—Br	368	88
H—I	297	71
H—NH_2	450	108
H—SH	381	91
H—SCH_3	366	87
X—X bonds		
H—H	435	104
F—F	154	37
Cl—Cl	239	57
Br—Br	190	45
I—I	149	36
Other		
HO—OH	213	51
$(CH_3)_3CO$—$OC(CH_3)_3$	159	38
HO—Br	234	56
$(CH_3)_3CO$—Br	205	49

Although bond energies are measured and reported for the gas phase, they are generally used with the expectation that values in solution will not be very different. BDEs are generally used as comparisons; changes caused by solvent are likely to cancel in a BDE comparison.

Some bond dissociation energies are listed in **Table 10.2**. *A bond dissociation energy measures the intrinsic strength of a chemical bond in the gas phase at 25 °C (298 K)*. For example, breaking the H—H bond requires 435 kJ mol^{-1} (104 kcal mol^{-1}) of energy. It then follows that forming the hydrogen molecule from two hydrogen atoms liberates 435 kJ mol^{-1} (104 kcal mol^{-1}) of energy. Table 10.2 shows that different bonds exhibit significant differences in bond strength; even bonds of the same general type, such as the various C—H bonds, can differ considerably in bond strength.

The bond dissociation energies in Table 10.2 (or others available from compilations in the literature) can be used in a number of ways. A common use of these energies is to estimate the $\Delta H°$ of a nonradical reaction by using Hess's law to treat it as the sum of fictitious radical reactions. The sum of the $\Delta H°$ values of the radical reactions, obtained from bond dissociation energies, provides the $\Delta H°$ of the overall reaction. (See Further Exploration 10.1 and Focused Problem 10.7.) For example, if you return to the discussion of the element effect on acidity (Eqs. 3.68a–d, Sec. 3.7B), you will see that bond dissociation energies were used to show the effect of bond strength on acidity.

FURTHER EXPLORATION 10.1
(Old 5.3) Bond Dissociation Energies and Heats of Reaction

A consideration of bond dissociation energies shows why di-*tert*-butyl peroxide is an excellent free-radical initiator. The lower a bond dissociation energy, the lower the temperature required to rupture the bond in question and form free radicals at a reasonable rate. The homolysis of the O—O bond in di-*tert*-butyl peroxide requires only 159 kJ mol^{-1} (38 kcal mol^{-1}) of energy; this is one of the lowest bond dissociation energies in Table 10.2. With such a low bond dissociation energy, this peroxide readily forms small amounts of free radicals when it is heated gently or when it is subjected to ultraviolet light.

An important use of bond dissociation energies is to calculate or estimate the $\Delta H°$ of free-radical reactions. As an illustration, consider the second initiation step for the free-radical addition of HBr, in which a *tert*-butoxy radical reacts with H—Br (Eq. 10.10).

$$(CH_3)_3C\text{—}O\cdot + H\text{—}Br \longrightarrow (CH_3)_3C\text{—}O\text{—}H + \cdot Br \qquad (10.25)$$

***tert*-butoxy radical**

In this reaction, a hydrogen is abstracted from HBr by the *tert*-butoxy radical. This is not the only reaction that might occur. Instead, the *tert*-butoxy radical might abstract a bromine atom from HBr:

$$(CH_3)_3C\text{—}O\cdot + H\text{—}Br \longrightarrow (CH_3)_3C\text{—}O\text{—}Br + \cdot H \qquad (10.26)$$

***tert*-butoxy radical**

Why is hydrogen and not bromine abstracted? The reason lies in the relative enthalpies of the two reactions. These enthalpies are not known by direct measurement, but they can be calculated using bond dissociation energies. To calculate the $\Delta H°$ for a reaction, *subtract the bond dissociation energies (BDE) of the bonds formed from the bond dissociation energies of the bonds broken.*

$$\Delta H° = \text{BDE (bonds broken)} - \text{BDE (bonds formed)} \qquad (10.27)$$

This works because BDEs are the enthalpies for bond dissociation. This procedure is illustrated in Study Problem 10.4.

Study Problem 10.4

Estimate the standard enthalpies of the reactions shown in Eqs. 10.25 and 10.26.

Solution To obtain the required estimates, apply Eq. 10.27. In both equations, the bond broken is the H—Br bond. From Table 10.2, the bond dissociation energy of this bond is 368 kJ mol^{-1} (88 kcal mol^{-1}). The bond formed in Eq. 10.25 is the O—H bond in $(CH_3)_3CO$—H (*tert*-butyl alcohol). This exact compound is not found in Table 10.2, so we look for the same type of bond in as similar a compound as possible. For example, the table includes an entry for the alcohol CH_3O—H (methyl alcohol). (The O—H bond dissociation energies for methyl alcohol and *tert*-butyl alcohol differ very little.) So we use 438 kJ mol^{-1} (105 kcal mol^{-1}) for the BDE of the O—H bond. Subtracting the enthalpy of the bond formed (the O—H bond) from that of the bond broken (the H—Br bond), we obtain 368 – 438 = –70 kJ mol^{-1} (or 88 – 105 = –17 kcal mol^{-1}). This is the enthalpy of the reaction in Eq. 10.25. In other words, breaking the H—Br bond costs 368 kJ mol^{-1} (88 kcal mol^{-1}) but the formation of the O—H bond gives back 438 kJ mol^{-1} (105 kcal mol^{-1}), for a net $\Delta H°$ for hydrogen abstraction of –70 kJ mol^{-1} (–17 kcal mol^{-1}).

Following the same procedure for Eq. 10.26, use the bond dissociation energy of the O—Br bond in $(CH_3)_3CO$—Br, which is given in Table 10.2 as 205 kJ mol^{-1} (49 kcal mol^{-1}). The calculated $\Delta H°$ for Eq. 10.26 is then 368 – 205 =+163 kJ mol^{-1} (+39 kcal mol^{-1}).

These $\Delta H°$ estimates are the required solution to the problem.

> Strictly speaking, we need $\Delta G°$ values to determine whether a reaction will occur spontaneously, because $\Delta G°$ determines the equilibrium constant. However, the $\Delta S°$ values for two similar reactions, such as the two considered here, are nearly identical, so the differences between $\Delta H°$ and $\Delta G°$ tend to cancel in the comparison.

The calculation in Study Problem 10.4 shows that reaction 10.25 is highly exothermic (favorable), whereas reaction 10.26 is highly endothermic (unfavorable). The energetic advantage of hydrogen abstraction is therefore 93 kJ mol^{-1} (22 kcal mol^{-1}) over bromine abstraction. This is the reason that the abstraction of H (Eq. 10.25) occurs. Abstraction of Br (Eq. 10.26) is so unfavorable energetically that it does not occur to any appreciable extent.

Recall that a peroxide effect is not observed when HCl and HI are added to alkenes. Bond dissociation energies can explain this observation. Consider, for example, HI addition by a hypothetical free-radical mechanism, and compare the enthalpies for the addition of HBr and HI in the first propagation step. This propagation step involves the breaking of a π bond in each case (Eq. 10.28a) and the formation of a CH_2—X bond (Eq. 10.28b). Breaking the π bond requires about 243 kJ mol^{-1} (58 kcal mol^{-1}) (see Table 10.2). The energy released on formation of a CH_2—X bond is approximated by the negative bond dissociation energy of the corresponding carbon–halogen bond in CH_3CH_2—X (also from Table 10.2). Because we are making the same approximation in comparing the two halogens, any error introduced tends to cancel in the comparison. The first propagation step (Eq. 10.28c) is the sum of these two processes:

	$\Delta H°$, kJ mol^{-1} (kcal mol^{-1}) X = Br	X = I	
$H_2C{=}CH_2 \longrightarrow H_2\dot{C}{-}\dot{C}H_2$	+243 (+58)	+243 (+58)	(10.28a)
$H_2\dot{C}{-}\dot{C}H_2 + X\cdot \longrightarrow H_2\dot{C}{-}CH_2{-}X$	−303 (−72)	−238 (−57)	(10.28b)
Sum: $X\cdot + H_2C{=}CH_2 \longrightarrow H_2\dot{C}{-}CH_2{-}X$	−60 (−14) energetically favorable	+5 (+1) energetically unfavorable	(10.28c)

This calculation shows that the first propagation step is exothermic (that is, energetically favorable) for HBr, but endothermic (that is, energetically unfavorable) for HI. Remember that the propagation steps of any free-radical chain reaction are in competition with recombination steps that terminate free-radical reactions. These recombination steps are so exothermic that they occur on every encounter of two free radicals. The energy required for an endothermic propagation step, in contrast, represents an *energy barrier* that reduces the rate of this step. As a result, *only exothermic propagation steps compete successfully with recombination steps.* Therefore, HI does not add to alkenes by a free-radical mechanism because the first propagation step is endothermic, and the radical chain is terminated. In Focused Problem 10.8, you can explore for yourself from a bond-energy perspective why the addition of HCl does not occur by a free-radical mechanism either.

The use of bond dissociation energies for the calculation of $\Delta H°$ for reactions is not limited to free-radical reactions. It is only necessary that the reaction for which a $\Delta H°$ calculation is being made neither creates nor destroys ions, and that the BDEs for the appropriate bonds are known or can be closely estimated.

Another point worth noting is that bond dissociation energies apply to the gas phase. Bond dissociation energies can be used, however, to compare the enthalpies of two reactions in solution, provided that the effect of the solvent is either negligible or is the same for both of the reactions being compared (and so cancels in the comparison). This assumption is valid for many free-radical reactions, including the ones in Study Problem 10.4.

Focused Problems

10.7 Estimate the $\Delta H°$ values for each of the following gas-phase reactions using bond dissociation energies.

(a) $CH_4 + Cl_2 \longrightarrow H_3C{-}Cl + HCl$ (b) $H_2C{=}CH_2 + Cl_2 \longrightarrow Cl{-}CH_2CH_2{-}Cl$

10.8 (a) Consider the second propagation step for peroxide-promoted HBr addition to an alkene (Eq. 10.11b, Sec. 10.1C). Calculate the $\Delta H°$ for this reaction.

(b) Calculate the $\Delta H°$ for the same step using HCl instead of HBr.

(c) Use your calculation to explain why no peroxide effect is observed for the addition of HCl to an alkene.

10.9 Consider the second propagation step of peroxide-promoted HBr addition to alkenes (Eq. 10.11b). Use bond energies to explain why hydrogen, and not bromine, is abstracted from HBr by the free-radical reactant.

10.2 CONVERSION OF INTERNAL ALKYNES INTO TRANS ALKENES

The reaction of an internal alkyne (an alkyne in which the triple bond is not at the end of the carbon chain) with a solution of an alkali metal (usually sodium) in liquid ammonia gives a trans alkene.

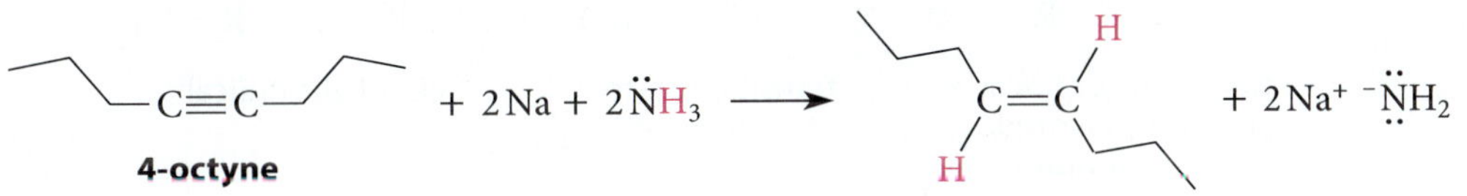

4-octyne

***trans*-4-octene**
(97% yield) (10.29)

Although ammonia is familiar to most people as a toxic gas, liquid ammonia is actually a very good protic solvent for some reactions. Gaseous ammonia is condensed to a liquid by cooling it below its boiling point of –33.3 °C. Once condensed, it is relatively easy to handle as a liquid because of its relatively high heat of vaporization.

This reaction is complementary to the catalytic hydrogenation of alkynes, which is used to prepare cis alkenes (Sec. 4.10B).

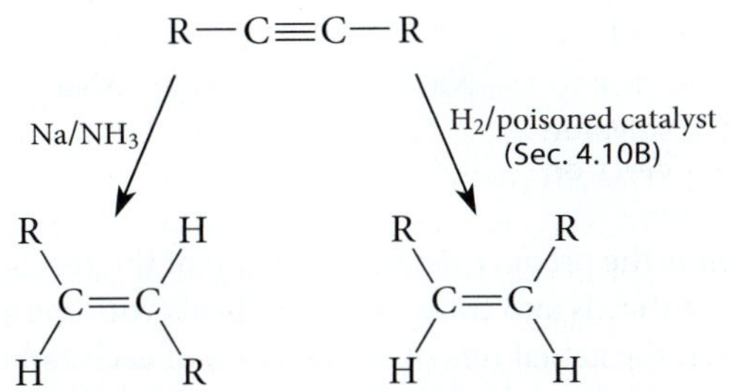

a trans alkene a cis alkene (10.30)

The stereochemistry of this reaction follows from its mechanism. If sodium or other alkali metals are dissolved in pure liquid ammonia, a deep blue solution forms that contains electrons associated noncovalently with ammonia (*solvated electrons*).

$$\text{Na}\cdot + n\,\text{NH}_3\,(\text{liq}) \longrightarrow \text{Na}^+ + \text{e}^-(\text{NH}_3)_n$$

solvated electron (10.31)

By the uncertainty principle, the solvated electron, unlike a solvated proton, can't be localized to one (or a few) solvent molecules because of its small mass. It is a negative charge "smeared out" over many solvent molecules, probably associated with the positive ends of the ammonia dipoles. But it *really is* an electron loose among the solvent molecules!

We can conceptualize the solvated electron as the simplest free radical. Recall that free radicals add to triple bonds (Eq. 10.22, Sec. 10.1D). The reaction of solvated electrons with an alkyne begins with the addition of an electron to the triple bond. The resulting species has both an unpaired electron and a negative charge. Such a species is called a **radical anion**:

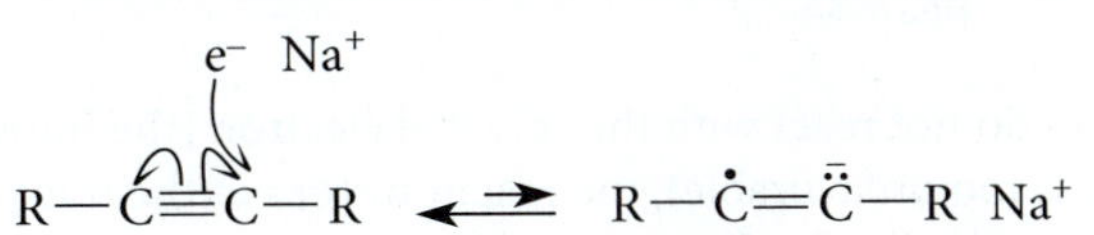

a radical anion (10.32a)

The radical anion is such a strong base that it readily removes a proton from ammonia to give a *vinylic radical*—a radical in which the unpaired electron is associated with one carbon of a double bond. This reaction is a Brønsted acid–base reaction and *not* a radical reaction.

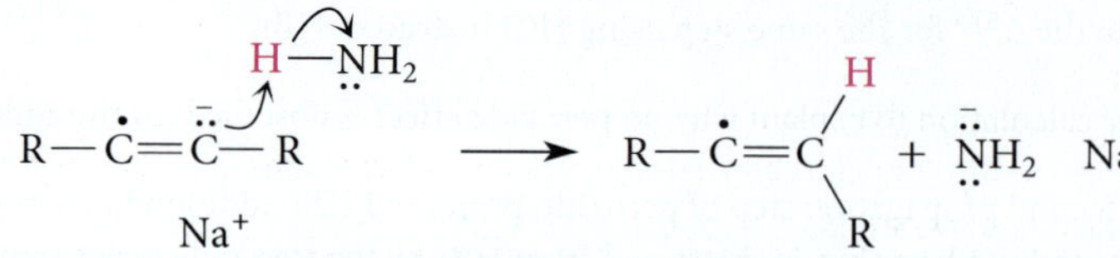

a vinylic radical **(10.32b)**

The destruction of the radical anion in this manner pulls the unfavorable equilibrium in Eq. 10.32a to the right. The vinylic radical, like the nonbonding electron pair of an amine (Sec. 6.9B), rapidly undergoes inversion, and the equilibrium between the cis and trans radicals favors the trans radical for the same reason that trans alkenes are more stable than cis alkenes: repulsions between the R groups are reduced.

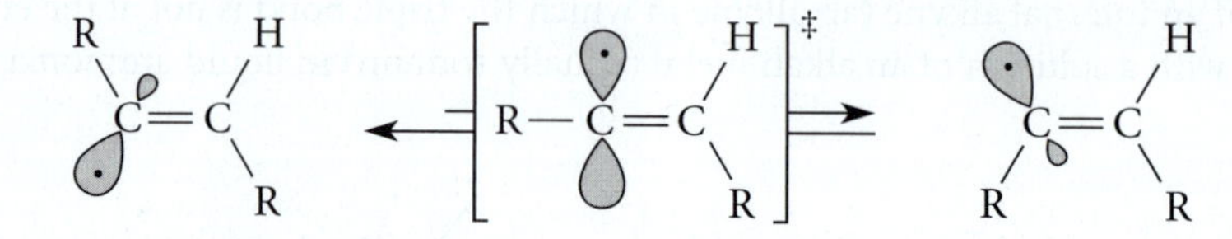

trans vinylic radical (strongly favored at equilibrium) — transition state for inversion — cis vinylic radical **(10.32c)**

Next, the vinylic radical accepts an electron to form an anion:

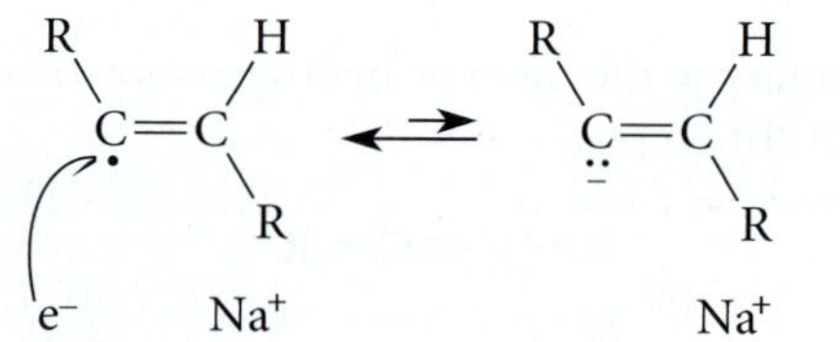

solvated electron **(10.32d)**

This step of the mechanism is the *product-determining step* of the reaction (Sec. 9.6B). The *rate constants* for the reactions of the cis and trans vinylic radicals with the solvated electron are probably the same. However, the actual *rate* of the reaction of each radical is determined by the product of the rate constant and the concentration of the radical. Because the trans vinylic radical is present in much higher concentration, the ultimate product of the reaction, the trans alkene, is derived from this radical.

The anion formed in Eq. 10.32d is also more basic than the amide anion and readily removes a proton from ammonia in another Brønsted acid–base reaction to complete the addition.

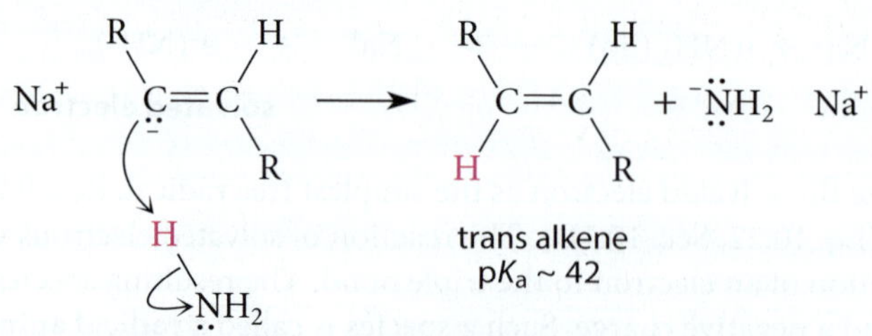

trans alkene $pK_a \sim 42$

$pK_a \sim 35$ **(10.32e)**

Because ordinary alkenes do not react with the solvated electron (the initial equilibrium analogous to Eq. 10.32a is too unfavorable), the reaction stops at the trans-alkene stage.

The Na/NH_3 reaction with alkynes does not work well on 1-alkynes unless certain modifications are made in the reaction conditions. (This is explored in Problem 10.67.) However, this is not a serious limitation for the reaction, because the reduction of 1-alkynes to 1-alkenes is easily accomplished by catalytic hydrogenation (Sec. 4.10B).

Focused Problem

10.10 What product is obtained in each case when 3-hexyne is treated in each of the following ways? (*Hint:* The products of the two reactions are stereoisomers.)

(a) With sodium in liquid ammonia and the product of that reaction with D_2 over Pd/C

(b) With H_2 over Pd/C and quinoline and the product of that reaction with D_2 over Pd/C

10.3 POLYMERS; FREE-RADICAL POLYMERIZATION OF ALKENES

A. Polymers

In the presence of free-radical initiators such as peroxides or AIBN, many alkenes react to form **polymers**, which are very large molecules composed of repeating units. Polymers are derived from small molecules in the same sense that a freight train is composed of boxcars. In a **polymerization reaction**, small molecules known as **monomers** react to form a polymer. For example, ethylene can be used as a monomer and polymerized with free-radical initiators or with special catalysts to yield an industrially important polymer called **poly(ethylene)** or, more rarely, **poly(ethene)**.

$$n\ H_2C{=}CH_2 \xrightarrow[\text{high temperature and pressure}]{\text{initiator or catalyst}} \left(CH_2{-}CH_2 \right)_n$$

ethylene (the monomer used for the production of polyethylene) — **poly(ethylene)** (the polymer of ethylene) (10.33)

Several systems are sanctioned for naming polymers, but we consider only *source-based names* here because this system is most widely used. In a **source-based name**, the polymer is named with the prefix *poly* followed by the name of the monomer in parentheses. Therefore the polymer of ethylene is called poly(ethylene), as shown in Eq. 10.33. Notice, though, that even though the polymer name contains the suffix *ene*, the polymer does *not* contain double bonds.

Equation 10.33 also shows a common convention for representing the structures of polymers. When the repeating units are joined end-to-end, as in poly(ethylene), a condensed structure is used in which the repeating polymer unit is enclosed in parentheses with bonds extending through the parentheses indicating the type of bond that connects the individual units. The subscript n written after the final parenthesis shows that we are dealing with a polymer rather than an ordinary small molecule.

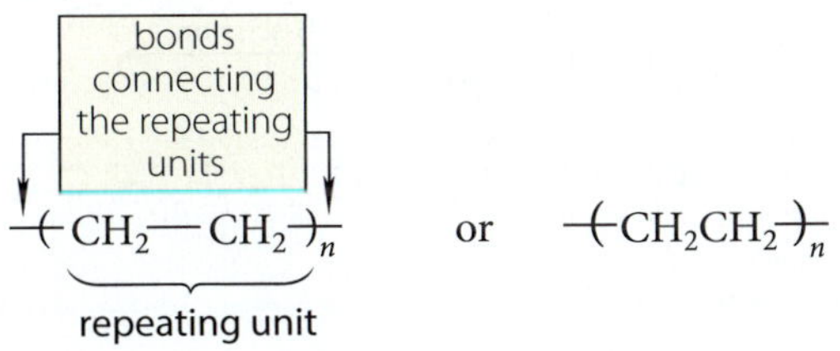

(10.34)

The subscript n means that a typical poly(ethylene) molecule contains a very large number of repeating units. Typically, n might be in the range of 3000 to 40,000, and a given sample of polyethylene contains molecules with a distribution of n values. In other words, polymers contain a distribution of molecular weights.

Polymers are large molecules, and they are members of a broader class of molecules called **macromolecules**, which are very large molecules that consist of small molecular units joined together chemically. Examples of biological macromolecules include proteins, DNA, and cellulose.

B. Free-Radical Polymerization

When the polymerization of ethylene shown in Eq. 10.33 occurs by a free-radical mechanism, it is an example of **free-radical polymerization**. The reaction is initiated when a radical R•, derived from peroxides or other initiators, adds to the double bond of ethylene to form a new radical.

$$\underset{\text{initiating radical}}{\mathrm{R\cdot}}\;\; \mathrm{H_2C{=}CH_2} \longrightarrow \mathrm{R{-}CH_2{-}\dot{C}H_2} \qquad \textbf{(10.35a)}$$

The first propagation step of the reaction involves addition of the new radical to another molecule of ethylene.

$$\mathrm{R{-}CH_2{-}\dot{C}H_2}\;\; \mathrm{H_2C{=}CH_2} \longrightarrow \mathrm{R{-}CH_2{-}CH_2{-}CH_2{-}\dot{C}H_2} \qquad \textbf{(10.35b)}$$

Ethylene (a gas) is provided under high pressure so that reaction with ethylene is the most probable fate of any radicals that are formed. Eqs. 10.35a and 10.35b are further examples of a typical free-radical reaction: reaction with a π bond. (Compare with Eq. 10.11a, Sec. 10.1C.) This process continues indefinitely as long as the ethylene monomer is present in high concentration.

$$\mathrm{R{+}CH_2CH_2{+}_{\mathit{n}}CH_2\dot{C}H_2} + \mathrm{H_2C{=}CH_2} \longrightarrow \mathrm{R{+}CH_2CH_2{+}_{\mathit{n}+1}CH_2\dot{C}H_2} \qquad \textbf{(10.35c)}$$

When the ethylene monomer is exhausted, termination reactions begin to occur. An example of a termination reaction is the radical recombination of two radical chains to form a large nonradical product:

$$\mathrm{R{+}CH_2CH_2{+}_{\mathit{n}+1}CH_2\dot{C}H_2}\quad \mathrm{\dot{C}H_2CH_2{+}CH_2CH_2{+}_{\mathit{n}+1}R} \longrightarrow$$

$$\mathrm{R{+}CH_2CH_2{+}_{\mathit{n}+1}CH_2CH_2CH_2CH_2{+}CH_2CH_2{+}_{\mathit{n}+1}R} \qquad \textbf{(10.35d)}$$

Another termination reaction is **β-scission**, in which one radical removes a β-hydrogen atom from another radical to give two nonradical products:

$$\mathrm{R{+}CH_2CH_2{+}_{\mathit{n}+1}CH(H){-}\dot{C}H_2}\quad \mathrm{\dot{C}H_2CH_2{+}CH_2CH_2{+}_{\mathit{n}+1}R} \xrightarrow{\beta\text{-scission}}$$

$$\mathrm{R{+}CH_2CH_2{+}_{\mathit{n}+1}CH{=}CH_2} + \mathrm{CH_3CH_2{+}CH_2CH_2{+}_{\mathit{n}+1}R} \qquad \textbf{(10.35e)}$$

The polymer chain is so long that the groups at its ends, in this case the initiator-derived R groups, represent an insignificant part of the total structure. Therefore, when we write the polymer structure as $\mathrm{{+}CH_2CH_2{+}_{\mathit{n}}}$, these terminal groups are ignored, just as, by analogy, we ignore the engine and the caboose when we say that a train consists of boxcars.

Polymerization in which one monomer unit adds sequentially to the growing polymer macromolecule with the production of no by-products is called **chain polymerization**.

(The word *chain* in this name refers to the chain reaction and not to the polymer chain.) Free-radical polymerization of ethylene is a typical example of chain polymerization. (This type of polymerization has also been called **addition polymerization** and **step-growth polymerization.**) In later sections of this text we consider polymerizations in which a small-molecule by-product such as H_2O is produced at each step of the reaction. These polymerizations are called **condensative chain polymerizations** (formerly called **condensation polymerizations**). Examples of this polymerization type are the production of the nylons and the polyesters (see Sec. 21.12A).

Two different monomers can be polymerized together to produce **copolymers**—molecules in which there are two different repeating units. An example of copolymers is styrene–butadiene rubber, a commercially important copolymer used in tires (see Sec. 15.5).

Focused Problem

10.11 (a) Consider the mechanism of free-radical polymerization of acrylonitrile to poly(acrylonitrile).

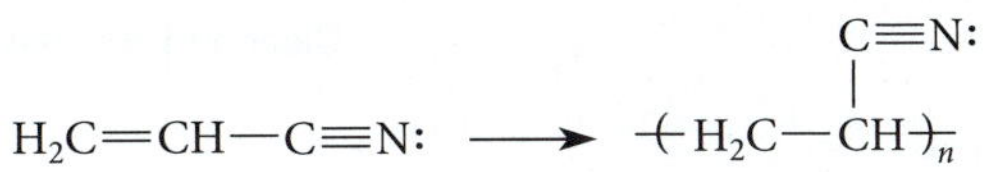

acrylonitrile **poly(acrylonitrile)**

In the first step, reaction of the initiating radical with the monomer, at which carbon of the double bond does the initiating radical react? Why? (*Hint:* Focus on resonance stabilization of the intermediate radical.)

(b) Does the regioselectivity of radical addition in part (a) matter in determining the final structure of the polymer?

C. Commercial Importance of Alkene Polymerization

Chemistry in the Real World

Alkene polymerization is very important commercially, and poly(ethylene) is the most important of the alkene polymers. About 220 billion pounds of poly(ethylene) valued at approximately $164 billion is manufactured annually worldwide, of which about 28% is made by free-radical polymerization. The free-radical process yields a transparent polymer, called *low-density poly(ethylene)* (often abbreviated *LDPE*), which is used in films and packaging. (Freezer bags and sandwich bags are usually made of LDPE.) The relatively low density of this polymer is due to the occurrence of significant branching in the polymer chains. Because branched chains do not pack so tightly as unbranched chains, the solid polymer contains lots of empty space. (If you've ever tried to stack a pile of highly branched tree limbs, you understand this point.) The mechanism of polymerization shown in Eqs. 10.35a–c does not explain how branching occurs, but this is explored in Problem 10.53 at the end of the chapter.

Other methods of poly(ethylene) manufacture produce *high-density poly(ethylene)* (*HDPE*), which consists of mostly unbranched polymer chains. HDPE is used, for example, in molded plastic containers such as milk jugs. HDPE is produced by the *Ziegler–Natta process*, which employs a titanium-based catalyst, or by the *Phillips process,* which involves a chromium oxide-based catalyst. Neither of these processes involves free-radical intermediates (Sec. 18.6D).

Many other commercially important polymers are produced from other alkene monomers by free-radical or other methods of chain polymerization. Some of these are listed in Table 10.3. Alkene polymers surround us in many everyday articles. Cell phones, computers, automobiles, sports equipment, stereo systems, food packaging, and many other items have important components fabricated from alkene polymers.

TABLE 10.3 Some Alkene Polymers Produced by Chain Polymerization

Polymer name [Trade name]	Structure of the monomer	Properties of the polymer	Uses
Poly(ethylene)	$H_2C{=}CH_2$	Flexible, semiopaque, generally inert	Containers, film
Poly(propylene)	$H_2C{=}CH{-}CH_3$	Heat resistant; fatigue resistant, lightweight	Autoclavable medical devices, carpet, packaging, housewares, automotive parts
Poly(styrene)	$H_2C{=}CH{-}Ph$	Clear, rigid; can be foamed with air	Containers, toys, packing material, insulation
Poly(vinyl chloride) [PVC]	$H_2C{=}CH{-}Cl$	Rigid, but can be plasticized with additives	Plumbing pipe, leatherette, hoses
Poly(chlorotrifluoroethylene) [Kel-F]	$F_2C{=}CF{-}Cl$	Chemically inert	Chemically inert apparatus, gaskets, and fittings
Poly(tetrafluoroethylene) [Teflon]	$F_2C{=}CF_2$	High melting point, chemically inert	Gaskets; chemically resistant apparatus and parts
Poly(methyl methacrylate) [Plexiglas, Lucite]	$H_2C{=}C(CH_3){-}C({=}O){-}OCH_3$	Clear and semiflexible	Lenses and windows; fiber optics
Poly(acrylonitrile) [PAN]	$H_2C{=}CH{-}C{\equiv}N$	Crystalline, strong, high luster	Fibers

Discovery of Teflon

Okan Metin/iStock/Getty Images

In April 1938, Roy J. Plunkett (1910–1994), who had obtained his Ph.D. only two years earlier from The Ohio State University, was working in the laboratories of the DuPont company. He decided to use some tetrafluoroethylene ($F_2C{=}CF_2$, a gas) in the preparation of a refrigerant. When he opened the valve on the cylinder of tetrafluoroethylene, no gas escaped. Because the weight of the empty cylinder was known, Plunkett was able to determine that the cylinder had the weight expected for a full cylinder of the gas. It was at this point that Plunkett's scientific curiosity paid a handsome dividend. Rather than discard the cylinder, he checked to be sure the valve was not faulty and then cut the cylinder open. Inside he found a polymeric material that felt slippery to the touch, could not be melted with extreme heat and was chemically inert to almost everything. Plunkett had accidentally discovered the polymer we know today as Teflon. At that time, no one imagined the commercial value of Teflon. Only with the advent of the atomic bomb project during World War II did it find a use: to form gaskets that were inert to the highly corrosive gas UF_6 used to purify the isotopes of uranium. In the 1960s, Teflon was introduced to consumers as a nonstick coating on cookware. Teflon is not at all polarizable, so very few things adhere to it by van der Waals attractions. (See Sec. 8.5A.)

D. Polymer Stereochemistry

When the monomer has the form $H_2C{=}CH{-}X$, asymmetric carbon stereocenters are created when polymerization occurs. Three stereochemical situations (called **tacticity**) are typically encountered. We illustrate these with poly(propylene).

1. In an **isotactic** polymer, the asymmetric carbon stereocenters have the substituents (methyl groups in this case) on the same side of the polymer chain.

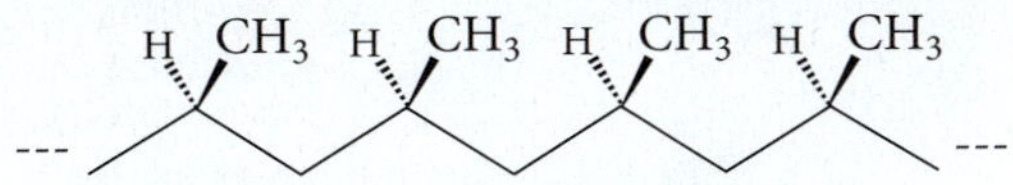

isotactic poly(propylene) (10.36)

2. In a **syndiotactic** polymer, the asymmetric carbon stereocenters have the substituents alternating regularly on opposite sides of the polymer chain.

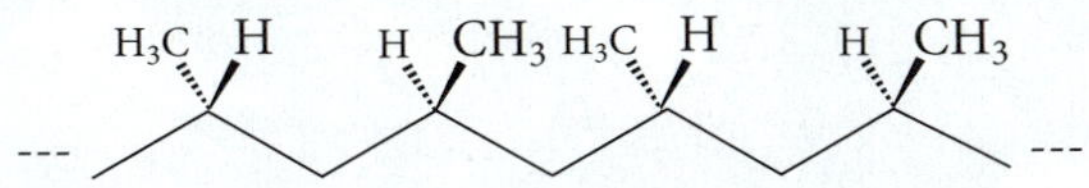

syndiotactic poly(propylene) (10.37)

Both isotactic and syndiotactic polymers are classified as **stereoregular** polymers.

3. In an **atactic** polymer, the configurations of the stereocenters are randomly distributed. These polymers are classified as **stereorandom** polymers.

The three polymer types have different properties. For example, isotactic poly(propylene) is a highly crystalline, rigid material with a higher melting point (160–184 °C, depending on the source). Syndiotactic poly(propylene) is semicrystalline with a lower melting point (130–160 °C), and it has elastomeric properties (that is, it can be stretched). Atactic poly(propylene) is amorphous (that is, it is *not* crystalline). It is a tacky, rubberlike material with no well-defined melting point.

Free-radical polymerization generally yields atactic polymers because the free-radical mechanism provides no control of stereochemistry. Transition-metal catalysts have been developed that can provide polymers with defined stereochemistry. Most commercial polypropylene is predominately isotactic. Polymers with varying degrees of tacticity can be produced for various applications.

Focused Problem

10.12 Using the monomer structures in Table 10.3, draw the structure of (a) isotactic poly(methylmethacrylate) and (b) syndiotactic poly(vinyl chloride).

E. Polymers and Global Pollution

Chemistry in the Real World

Polymers are one of the great successes of the world chemical industry. Because of the widespread use of polymers, however, disposal of polymer-containing waste has become a significant global issue. Although polymers in principle can be recycled, in practice only about 10% of polymers are actually recycled, and about 14% are burned to produce energy. The rest ends up in landfills or in the environment. As a result, large accumulations of polymer-containing waste have occurred in the world's oceans (about 200 million metric tons by one estimate). In some cases, mountains of this waste have been washed up by ocean tides onto formerly pristine beaches (**Fig. 10.2**). Ocean waste has also adversely affected wildlife; some seabirds have been found with more than 250 pieces of plastic in their stomachs. These are only a few of the problems identified with the disposal of plastics.

FIGURE 10.2 The large quantity of plastics in the global waste stream has resulted in the contamination of oceans and beaches.

Sabilin/Getty Images

Many potential solutions for this problem exist. One is to use fewer plastics. Some countries, for example, are beginning to outlaw the use of single-use plastic containers. Another solution is to incinerate the plastics and recover the energy produced. One power plant in New Jersey, for example, incinerates enough plastic (and other burnable waste) to power 30,000 homes. Incineration produces CO_2, which contributes to global warming, but the incineration process is almost as efficient as burning natural gas. For this solution to be generally viable, however, many more incineration plants in the United States are needed. (Several countries in Europe have been active in developing waste-incineration facilities because, in contrast to the situation in the United States, space for landfills is scarce.) Another solution is to recycle more plastics to provide the raw materials for end uses such as carpet fibers, outdoor furniture, and synthetic lumber.

For more on this topic, see "Chemistry May Have Solutions to Our Plastic Trash Problem," *Chemical and Engineering News,* vol. 96, No. 25 (June 18, 2018); and "Should Plastics Be a Source of Energy?" *Chemical and Engineering News,* vol. 86, No. 38 (September 24, 2018). You can probably find these issues in your university or college library and online if your institution has a subscription.

Chemists have been working on the "plastic waste" problem with promising results. One potential solution is to make biodegradable plastics. Another is to make plastics with additives that make it easier to convert polymers back into their monomers. Scientists have reported finding plastics-eating microbes. One company has developed a process for recycling some plastics into synthetic crude oil, which can then be refined.

The plastics-pollution problem presents a vast opportunity for chemists to apply innovative ideas. When these ideas are combined with the political will to solve this problem, there is reason for optimism.

10.4 FREE-RADICAL SUBSTITUTION: THE HALOGENATION OF HYDROCARBONS

A. Free-Radical Halogenation of Alkanes

Among the methods used in industry, and occasionally in the laboratory, to produce simple alkyl halides is direct halogenation of alkanes. When an alkane such as methane is treated with Cl_2 or Br_2 in the presence of heat or light, a mixture of alkyl halides is formed by successive halogenation reactions.

$$\underset{\textbf{methane}}{CH_4} + Cl_2 \xrightarrow{\text{heat or light}} \underset{\substack{\textbf{chloromethane}\\\textbf{(methyl chloride)}}}{CH_3Cl} + HCl \qquad (10.38a)$$

$$\underset{\textbf{chloromethane (methyl chloride)}}{CH_3Cl} + Cl_2 \xrightarrow{\text{heat or light}} \underset{\textbf{dichloromethane (methylene chloride)}}{CH_2Cl_2} + HCl \qquad (10.38b)$$

$$\underset{\textbf{dichloromethane (methylene chloride)}}{CH_2Cl_2} + Cl_2 \xrightarrow{\text{heat or light}} \underset{\textbf{trichloromethane (chloroform)}}{CHCl_3} + HCl \qquad (10.38c)$$

$$\underset{\textbf{trichloromethane (chloroform)}}{CHCl_3} + Cl_2 \xrightarrow{\text{heat or light}} \underset{\textbf{tetrachloromethane (carbon tetrachloride)}}{CCl_4} + HCl \qquad (10.38d)$$

The relative amounts of the various products can be controlled by varying the reaction conditions, but mixtures of them are formed, and each compound must be isolated by fractional distillation (Figs. 2.11 and 2.12, Sec. 2.9).

The products in Eqs. 10.38a–d are formed in a series of *substitution* reactions (Sec. 7.8B). For example, CH_3Cl is formed by the substitution of a hydrogen atom in methane by a chlorine atom:

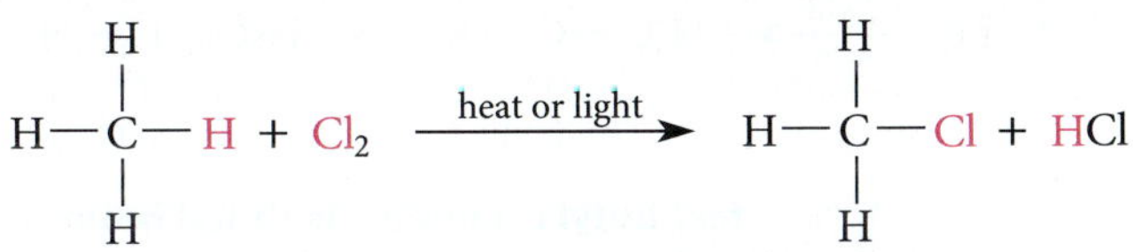

methane **chloromethane (methyl chloride)** (10.39)

The conditions of this reaction (initiation by heat or light) suggest the involvement of free-radical intermediates (Sec. 10.1C). In fact, the mechanism of this reaction follows the typical pattern of other free-radical chain reactions—namely, it has initiation, propagation, and termination steps. The reaction is initiated when a small number of halogen molecules absorb energy from heat or light and dissociate homolytically into halogen atoms:

(10.40)

The ensuing chain reaction has the following propagation steps:

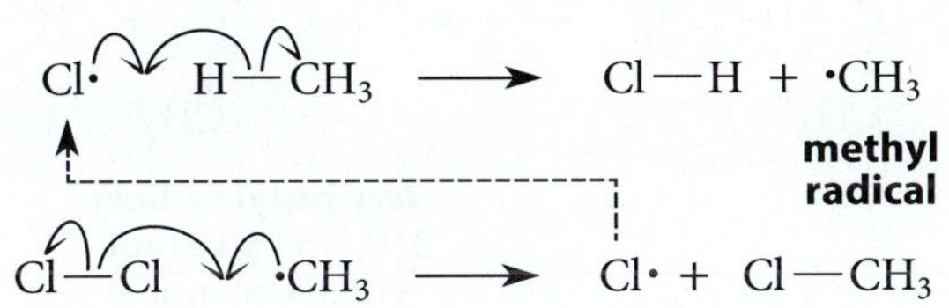

(10.41a) (10.41b)

The chlorine radical formed in Eq. 10.41b reacts with another CH_4 molecule as shown in Eq. 10.41a, and the chain reaction continues. Termination steps result from the recombination of radical species (Focused Problem 10.14) after the methane and Cl_2 concentrations are depleted.

The halogenation of alkanes by a free-radical mechanism is an example of a **free-radical substitution** reaction: a substitution reaction that occurs by a free-radical chain mechanism. (Contrast this with the *free-radical addition* mechanism for peroxide-mediated addition to alkenes in Sec. 10.1C.)

Free-radical halogenations with chlorine and bromine proceed smoothly, halogenation with fluorine is violent, and halogenation with iodine does not occur. These observations correlate with the $\Delta H°$ values for the halogenation of methane by each halogen. Fluorination is so strongly exothermic ($\Delta H° = -424$ kJ mol^{-1}, -101 kcal mol^{-1}) that the reaction is difficult to control—that is, the temperature of the reaction mixture rises more rapidly than the heat can be dissipated. Iodination, on the other hand, is endothermic ($\Delta H° = +54$ kJ mol^{-1}, $+13$ kcal mol^{-1}), so it is too unfavorable energetically to proceed to a useful extent. Chlorination

($\Delta H° = -106$ kJ mol^{-1}, -25 kcal mol^{-1}) and bromination ($\Delta H° = -30$ kJ mol^{-1}, -7 kcal mol^{-1}) are only mildly exothermic, so they proceed to completion without becoming violent.

Focused Problems

10.13 Give the free-radical chain mechanism for the formation of ethyl bromide (bromoethane) from ethane and bromine in the presence of light.

10.14 Explain why butane is formed as a minor by-product in the free-radical bromination of ethane.

B. Regioselectivity of Free-Radical Halogenation

When free-radical halogenation takes place on a hydrocarbon with more than one type of hydrogen, more than one product can be obtained. For example, bromination of isobutane could give both tertiary and primary alkyl bromides.

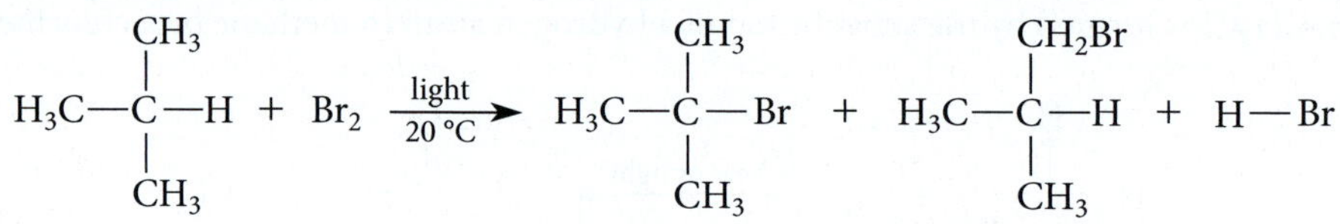

isobutane (large excess) **tert-butyl bromide** (99.5% of product) **isobutyl bromide** (0.5% of product) **(10.42)**

(A large excess of the hydrocarbon is used to avoid additional bromination.) Isobutane has nine primary hydrogens and only one tertiary hydrogen. On a purely statistical basis, then, we would expect 1/9 as much of the tertiary product as the primary product. Instead, Eq. 10.42 shows that about 200 times as much of the tertiary product is formed. On a per-hydrogen basis, then, a tertiary hydrogen in isobutane is about $(200 \div 1/9) = 1800$ times more reactive than a primary hydrogen.

This difference in reactivity is a consequence of the relative stabilities of the two possible free-radical intermediates. The rate-limiting step of the bromination reaction is the first propagation step, which is the step that forms the carbon radical in each case:

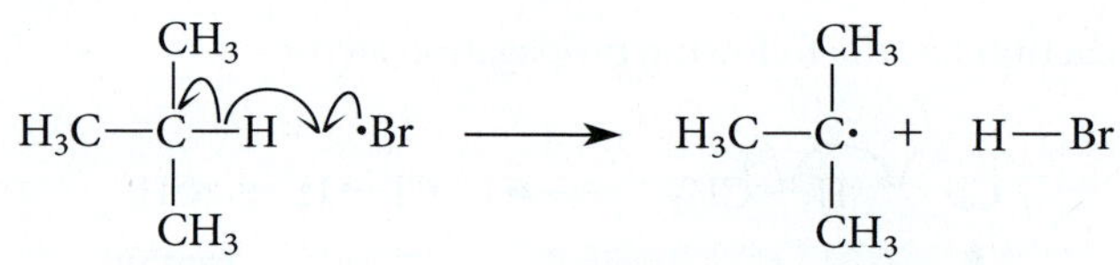

***tert*-butyl radical**
$\Delta H_f° = 51.5$ kJ mol^{-1}
(12.3 kcal mol^{-1}) **(10.43a)**

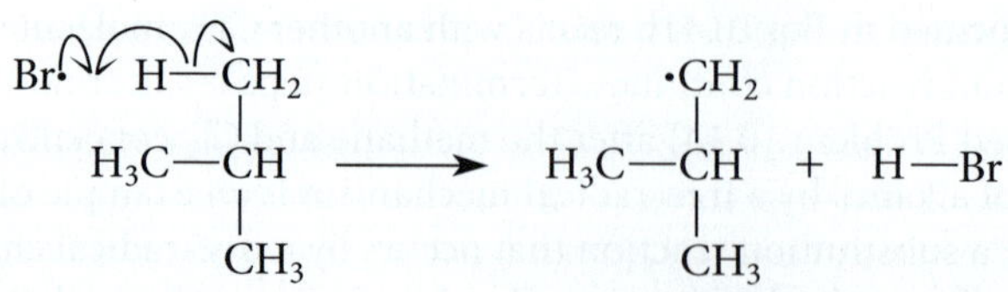

isobutyl radical
$\Delta H_f° = 70$ kJ mol^{-1}
(17 kcal mol^{-1}) **(10.43b)**

The only difference between the two steps is the free-radical intermediate. The *tert*-butyl radical is more stable than the isobutyl radical by 18.5 kJ mol^{-1} (4.4 kcal mol^{-1}) (Table 10.1, Sec. 10.1D). According to *Hammond's postulate* (Sec. 4.8D), the transition states for the two reactions resemble the reactive intermediates, which are the free radicals. Because the tertiary radical is more stable than the primary radical, the transition state leading to the tertiary radical has the lower energy, and the tertiary free radical is formed more rapidly.

Chlorination is much less selective than bromination. The light-promoted chlorination of isobutane actually gives much more of the primary alkyl chloride:

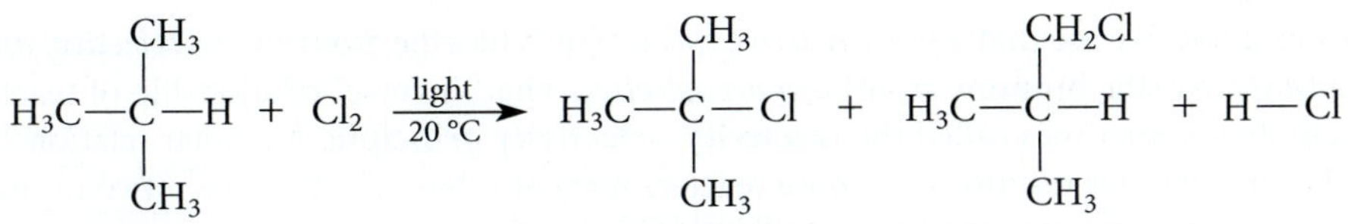

isobutane (large excess) — ***tert*-butyl chloride** (36% of product) — **isobutyl chloride** (64% of product) (10.44)

Taking into account the relative number of tertiary and primary hydrogens, we find (Focused Problem 10.15) that abstraction of the tertiary hydrogen is still preferred, but only by a factor of 5.1. Because of its greater selectivity, bromination is often the preferred method of halogenation in the laboratory. (Another selective free-radical bromination method is discussed in Sec. 17.2.)

Why is bromination so much more selective than chlorination? An exploration of this question provides a graphic illustration of Hammond's postulate. As shown in **Fig. 10.3**, the hydrogen abstraction step is *exothermic* for chlorination but is *endothermic* for bromination. (This difference between the two halogens is a direct consequence of the much greater bond dissociation energy of the H—Cl bond than the H—Br bond. Bromination goes to completion only because the second propagation step, and therefore the overall reaction, is very exothermic.) In two such closely related reactions, it is reasonable to suppose that the relative energies of the products should have some effect on the relative energy barriers that control the rate of the reaction, and this is observed experimentally: the exothermic reaction—chlorination—is much faster than the endothermic one—bromination. (Reaction 10.44 is much faster than reaction 10.42.) In other words, a chlorine atom is *much* more reactive with hydrocarbons than a bromine atom is. When we invoked Hammond's postulate to explain the selectivity of bromination, it worked because the transition state is very close in energy to the carbon radical along the reaction coordinate (red curve, Fig. 10.3). However, in chlorination, the position of the transition state along the reaction coordinate is closer to isobutane because isobutane + Cl• have a much higher energy than the carbon radical + HCl (blue curve, Fig. 10.3). In effect, isobutane and Cl• are the "unstable intermediates" in chlorination; and the transition state, by Hammond's postulate, should therefore resemble this pair more than it resembles the carbon radical.

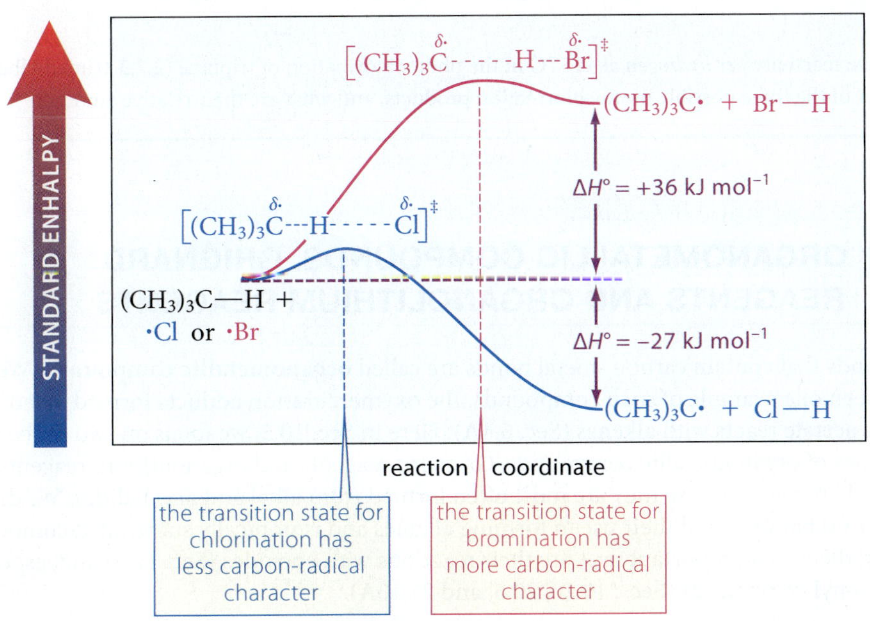

FIGURE 10.3 An enthalpy versus reaction-coordinate diagram for the first propagation step of free-radical halogenation. (The relative standard free energies follow the same pattern.) The transition state for the first propagation step of halogenation resembles the less stable species along the reaction coordinate. Because the transition state for bromination resembles the carbon radical + HBr, carbon-radical stability is important in determining the regioselectivity of bromination. Because the transition state for chlorination resembles isobutane + Cl•, it has very little carbon-radical character, and carbon-radical stability is much less important in determining regioselectivity.

Because the transition state has very little carbon-radical character, carbon-radical stability is not very important in determining the position of the chlorination reaction. In general, *when we apply Hammond's postulate to highly exothermic reactions, the transition state resembles the reactants. When we apply Hammond's postulate to endothermic reactions, the transition state resembles the products.*

In this example, we see that a *more reactive species* (the chlorine atom) *is less selective, and a less reactive species* (the bromine atom) *is more selective.* This "inverse" relationship of reactivity and selectivity is sometimes called the **reactivity–selectivity principle**. A similar relationship is observed in many other reactions—a more reactive reagent is less selective when two or more constitutionally isomeric products are possible.

An Analogy for the Reactivity–Selectivity Principle

Jason Salmon/ Shutterstock

Suppose you are very hungry—famished—and someone offers you a chocolate from a box containing different types of chocolates. You are so hungry that you cast good manners to the wind and rapidly grab the first chocolate (or chocolates) you put your hand on. Now imagine, instead, that you have finished a large and satisfying meal, and you are allowed to select a chocolate from the same box. You are now more likely to pore slowly over the different chocolates and choose a type you like the best. You are more selective when you are not so hungry. Likewise, "hungry" (reactive) reagents are in many cases less selective than "less hungry" (less reactive) reagents when they can react in different ways. In halogenation, the chlorine atom is "famished," and the bromine atom is more selective.

Focused Problems

10.15 Given the product distribution in Eq. 10.44, calculate the relative rate of abstraction (per hydrogen) of a tertiary and a primary hydrogen by a chlorine atom.

10.16 Two monobromination products *A* and *B* are obtained in a ratio of roughly 2:1 when pentane is treated with bromine and light. What are they? Explain your reasoning.

10.17 (a) What is the major monobromination product obtained when methylcyclohexane is treated with bromine and light? Explain your reasoning.

(b) The tertiary/primary relative reactivity *per hydrogen* at −15 °C in the photochlorination of triptane (2,2,3-trimethylbutane) is 4.5. What are the structures of the three possible monochlorination products, and what are their relative amounts?

10.5 ORGANOMETALLIC COMPOUNDS. GRIGNARD REAGENTS AND ORGANOLITHIUM REAGENTS

Compounds that contain carbon–metal bonds are called **organometallic compounds**. We've already seen one example of such compounds: the oxymercuration adducts formed when aqueous mercuric acetate reacts with alkenes (Sec. 5.4A). Here in Sec. 10.5, we focus on two of the most useful types of organometallic compounds, Grignard reagents and organolithium reagents. We consider them here because they are most often formed from alkyl and aryl halides. We discuss here their preparation and their use in forming alkanes and isotopically substituted compounds. However, their most important uses are their reactions with epoxides (Sec. 12.5) and, especially, with carbonyl compounds (Secs. 19.10, 20.6, and 21.10A).

A. Grignard Reagents and Organolithium Reagents

A **Grignard reagent** is a compound of the form R—Mg—X, where X═Br, Cl, or I.

Examples of Grignard reagents:

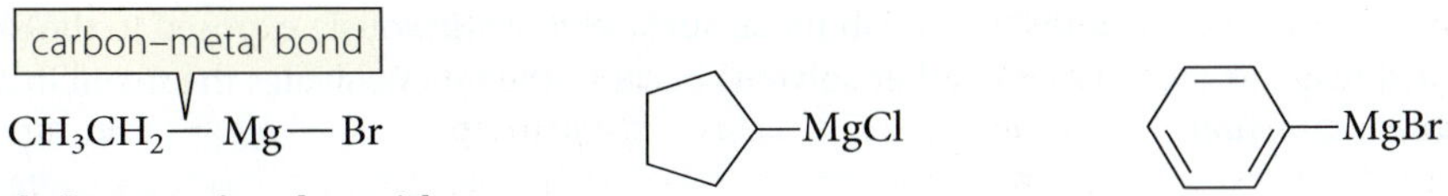

ethylmagnesium bromide **cyclopentylmagnesium chloride** **phenylmagnesium bromide** (10.45)

Development of Grignard Reagents

Editorial/Alamy

Grignard reagents are among the most versatile and important reagents in organic chemistry. The utility of these reagents was originally investigated by François Phillipe Antoine Barbier (1848–1922), a professor of general chemistry at the University of Lyon in France. However, it was Barbier's successor at Lyon, Victor Grignard (1871–1935), who developed many applications of organomagnesium halides during the early part of the twentieth century. For this work, Grignard received the Nobel Prize in Chemistry in 1912, sharing the prize with Paul Sabatier (1854–1941), a professor of chemistry at Toulouse, who received the prize for development of catalytic hydrogenation.

Organolithium reagents are compounds of the form R—Li.

Examples of organolithium reagents:

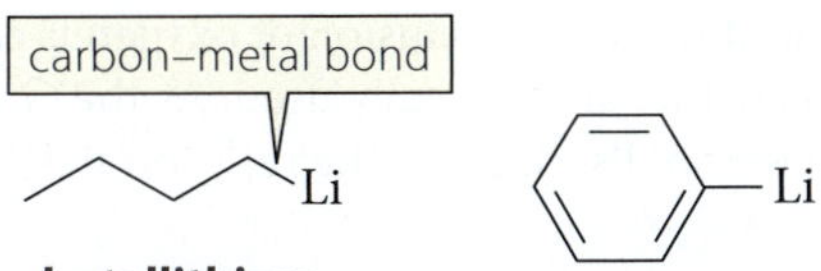

butyllithium **phenyllithium** (10.46)

Although the organolithium reagents are pictured for convenience as R—Li, many studies have shown that these reagents in solution are aggregates of several molecules [that is, $(RLi)_n$], and that the aggregation state depends on the solvent.

B. Formation of Grignard Reagents and Organolithium Reagents

Both Grignard and organolithium reagents are formed by adding the corresponding alkyl or aryl halides to rapidly stirred suspensions of small pieces of the appropriate metal. Anhydrous ether solvents must be used for the formation of Grignard reagents:

$$CH_3CH_2—Br + Mg \xrightarrow[Et_2O]{} CH_3CH_2—Mg—Br$$

bromoethane **ethylmagnesium bromide** (10.47)

chlorocyclohexane **cyclohexylmagnesium chloride** (10.48)

FURTHER EXPLORATION 10.2
Mechanism of Formation of Grignard Reagents

The formation of Grignard reagents, which occurs on the surface of the metal, involves free radicals, but not a chain reaction. (See Further Exploration 10.2 in the Study Guide and Solutions Manual.)

The solubility of Grignard reagents in ether solvents plays a crucial role in their formation. As Grignard reagents form on the metal surface, they are dissolved and washed from the metal surface by the ether solvent. As a result, a fresh metal surface is continuously exposed to the alkyl halide. Grignard reagents are soluble in ether solvents because the ether solvates the metal in a *Lewis acid–base interaction.*

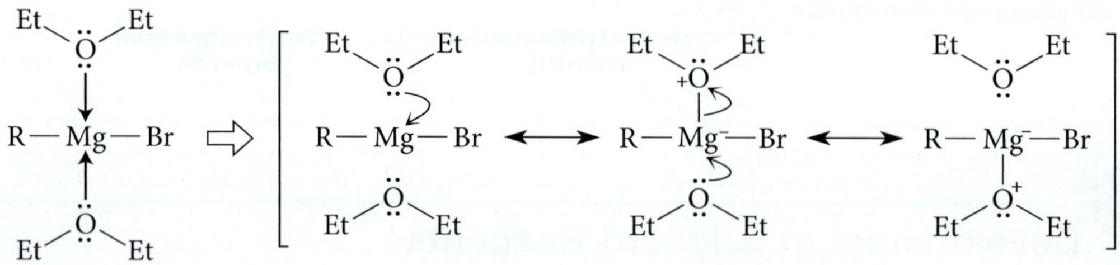

(10.49)

The magnesium of the Grignard reagent is two electron pairs short of an octet, and the oxygen of each ether molecule can donate an electron pair to the metal. (This interaction is very similar to the donor interactions that stabilize cations in solution; Sec. 8.6F.)

Organolithium reagents are typically formed in hydrocarbon solvents such as hexane:

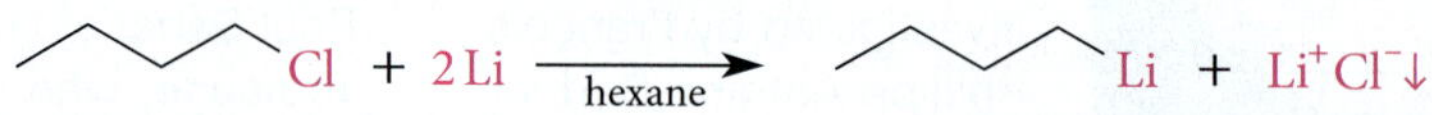

1-chlorobutane **butyllithium** (10.50)

Because organolithium reagents are soluble in hydrocarbons, ether solvents are not required for their formation.

Grignard and organolithium reagents react violently with oxygen and (as shown in Sec. 10.5C) vigorously with water, alcohols, and other compounds that have even weakly acidic hydrogens. For this reason, these reagents must be prepared under rigorously oxygen-free and moisture-free conditions. In the case of Grignard reagents, exclusion of oxygen is easily ensured by the low boiling points of the ether solvents that are normally used. As the Grignard reagent begins to form, heat is liberated and the ether boils. Because the reaction flask is filled with ether vapor, oxygen is excluded.

Focused Problems

10.18 Write an equation showing the preparation of each of the following organometallic compounds.

(a) $(CH_3)_2CH—MgBr$ (b) Ph—Li (c) 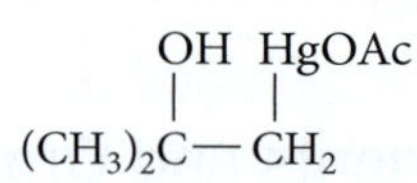

10.19 Complete each of the following equations.

(a) 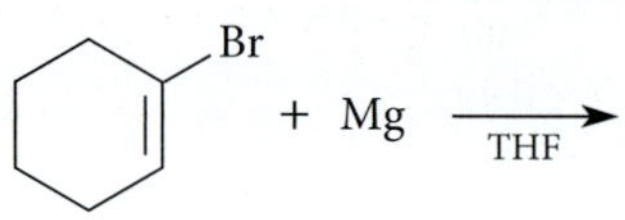

(b) $(CH_3)_3C—Cl + Li \xrightarrow{\text{hexane}}$

C. Protonolysis of Grignard Reagents and Organolithium Reagents

All reactions of Grignard and organolithium reagents can be understood in terms of the polarity of the carbon–metal bond. Because carbon is more electronegative than either magnesium or lithium, *the negative end of the carbon–metal bond is the carbon atom.* This is illustrated graphically by the electrostatic potential map (EPM) of the Grignard reagent methylmagnesium iodide, which shows the high degree of electron density on the carbon and the halogen.

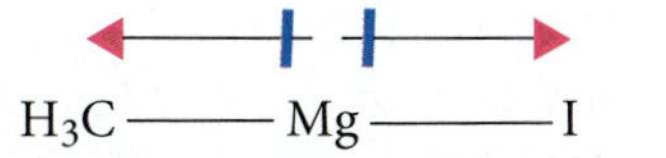

Lewis structure and bond dipoles of methylmagnesium iodide

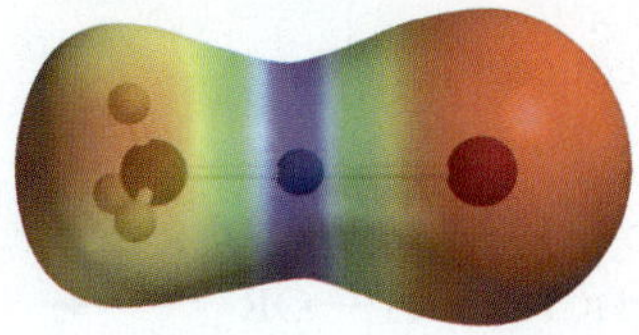

EPM of methylmagnesium iodide (10.51)

Imagine now carrying this picture of bond polarity to the extreme by breaking the carbon–metal bond of a Grignard or organolithium reagent so that the metal becomes positively charged and electron-deficient, and the pair of electrons in the bond ends up on carbon. Such a carbon, bearing three bonds, a nonbonding electron pair, and a negative formal charge, is a carbon anion, or **carbanion**. *Grignard and organolithium reagents react as if they were carbanions:*

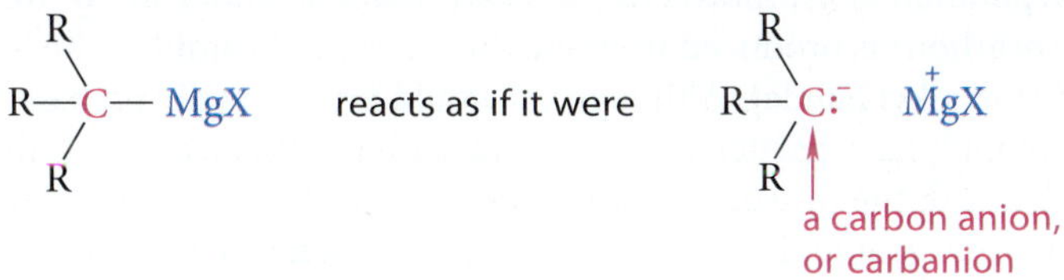

(10.52)

Grignard and organolithium reagents are not *true* carbanions because they have covalent carbon–metal bonds. However, we can predict their reactivity by treating them *conceptually* as carbanions.

For example, the view of Grignard and organolithium reagents as carbanions predicts the outcome of simple Brønsted acid–base reactions. Carbanions are powerful Brønsted bases because their conjugate acids, the corresponding alkanes, are extremely weak acids, with pK_a values estimated to be in the 55–60 range. The logic, then, is

1. R—H is a very weak acid (pK_a = 55–60); therefore,
2. $R{:}^-$ is a very strong base; therefore,
3. R—MgX and R—Li are also strong bases.

Grignard and lithium reagents are such strong bases that they react instantaneously with even weak acids such as water or alcohols. The products of such a reaction are the hydrocarbon—that is, the conjugate acid of the organometallic "carbanion"—and the conjugate base of the proton source—hydroxide ion (if the acid is water) or an alkoxide ion (if the acid is an alcohol).

$$CH_3CH_2\text{—}MgBr + H\text{—}OH \longrightarrow CH_3CH_2\text{—}H + HO^- \; {}^+MgBr \quad (10.53)$$

$$(CH_3)_3C\text{—}Li + H_2O \longrightarrow (CH_3)_3C\text{—}H + Li^+ \; {}^-OH \quad (10.54)$$

$$CH_3CH_2CH_2\text{—}MgBr + CH_3OH \longrightarrow CH_3CH_2CH_2\text{—}H + CH_3O^- \; {}^+MgBr$$

methanol (an alcohol) — **bromomagnesium methoxide** (the alkoxide conjugate base of methanol)

(10.55)

Even though Grignard reagents are not really carbanions, each of these reactions can be represented with the curved-arrow notation as if it were the reaction of a carbanion base with the proton of water or alcohol:

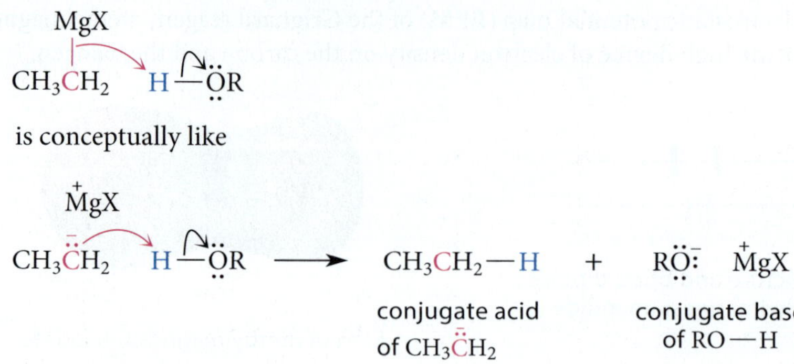

(10.56)

Reactions 10.53–10.55 are examples of *protonolysis*. A **protonolysis** is any reaction with the proton of an acid that breaks chemical bonds. For example, in the protonolysis of a Grignard reagent, the carbon–metal bond of the Grignard reagent is broken. The protonolysis reaction can be an annoyance, because it means Grignard and organolithium reagents must be prepared in the absence of moisture. The protonolysis reaction is also useful, however, because it provides *a method for the preparation of hydrocarbons from alkyl halides.* Notice in Eq. 10.55, for example, that ethane (a hydrocarbon) is produced from ethylmagnesium bromide, which, in turn, comes from ethyl bromide (an alkyl halide). Although we would not normally prepare an ordinary hydrocarbon by protonolysis, a particularly useful variation of this reaction is the preparation of hydrocarbons labeled with the hydrogen isotopes deuterium (D, or ^{2}H) or tritium (T, or ^{3}H) by reaction of a Grignard reagent with the corresponding isotopically labeled water.

$$(CH_3)_3CCH_2\text{—Br} \xrightarrow[\text{ether}]{\text{Mg}} (CH_3)_3CCH_2\text{—MgBr} \xrightarrow{D_2O} (CH_3)_3CCH_2\text{—D} \quad (10.57)$$

Focused Problems

10.20 Give the products of the following reactions. Show the curved-arrow notation for each.

(a) $H_3C\text{—Li} + CH_3OH \longrightarrow$

(b) $(CH_3)_2CHCH_2\text{—MgCl} + H_2O \longrightarrow$

10.21 (a) Give the structures of two isomeric alkylmagnesium bromides that would react with water to give propane.

(b) What compounds would be formed from the reactions of the reagents in (a) with D_2O?

D. Grignard and Organolithium Reagents from 1-Alkynes

In Sec. 3.7C we discussed the hybridization effect on acidity. In that discussion we learned that 1-alkynes have considerably greater acidity than alkenes or alkynes. The relative acidity of alkynes and the basic character of Grignard reagents play a role in the method usually used to prepare *acetylenic Grignard reagents* (reagents with the general structure R—C≡C—MgBr) and acetylenic lithium reagents (reagents with the general structure R—C≡C—Li). Section 10.5A showed that Grignard reagents are generally prepared by the reactions of alkyl halides with magnesium. The "alkyl halide" starting material for the preparation of an acetylenic Grignard reagent by this method would be a 1-bromoalkyne—that is, R—C≡C—Br. Such compounds are not generally available commercially and are difficult to prepare and store. Fortunately, acetylenic Grignard reagents are accessible by the acid–base reaction between a 1-alkyne and another Grignard reagent. Methylmagnesium bromide or ethylmagnesium bromide are often used for this purpose because the formation of a gaseous by-product (ethane in this example) ensures the irreversibility of the reaction.

$$\text{Bu—C}\equiv\text{C—H} + CH_3CH_2\text{—MgBr} \xrightarrow[\text{THF}]{} \underset{\text{an acetylenic Grignard reagent}}{\text{Bu—C}\equiv\text{C—MgBr}} + \underset{\textbf{ethane} \text{ (a gas)}}{CH_3CH_3} \quad (10.58)$$

If we again think of each Grignard reagent in terms of its carbanion character, the reaction is really just another Brønsted acid–base reaction:

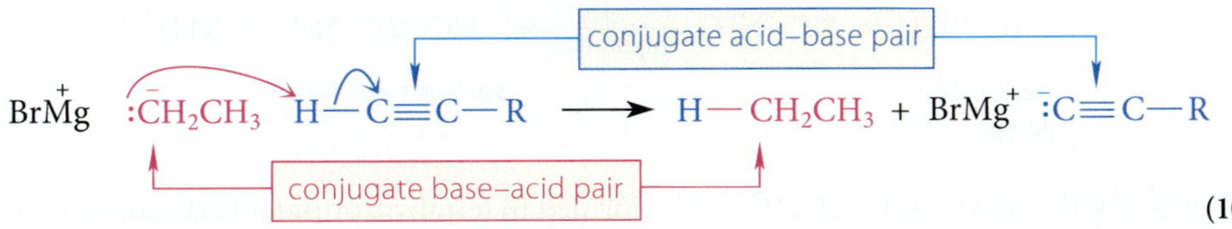

(10.59)

In this reaction, a stronger base (ethyl magnesium bromide, which resembles the ethyl anion) reacts with a stronger acid (a 1-alkyne) to give a weaker acid (ethane) and a weaker base (the alkynyl magnesium bromide, which resembles the acetylenic anion).

Other organometallic reagents, such as acetylenic organolithium reagents, can also be prepared by analogous reactions.

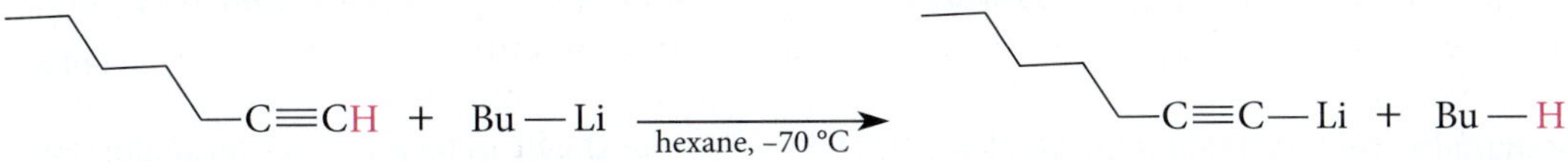

1-heptyne **butyllithium** an alkynyllithium **butane**

(10.60)

Focused Problem

10.22 Give the products identified by letter in the following sequence of reactions.

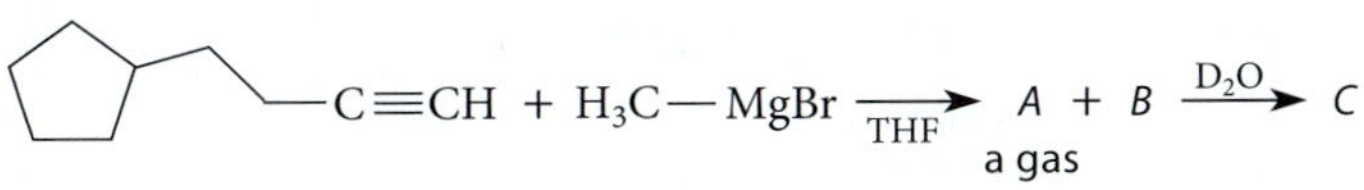

10.6 ACETYLENIC ANIONS

A. Preparation of Acetylenic Anions

With a pK_a of about 25, alkynes are acidic enough that they can be converted into their conjugate-base anions by sufficiently strong bases. One base used for this purpose is sodium amide, the conjugate base of ammonia, in liquid ammonia solvent:

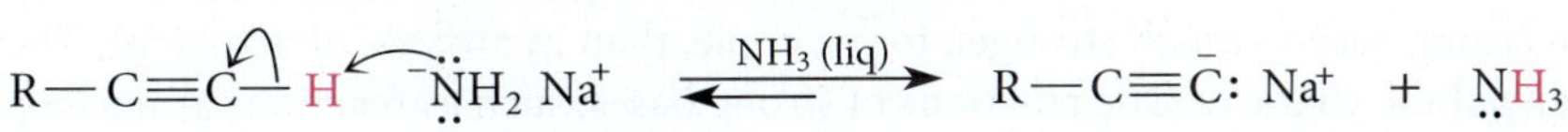

a 1-alkyne pK_a ~ 25 **sodium amide (sodamine)** a sodium acetylide **ammonia** pK_a ~ 35 (10.61)

Don't confuse the *amide ion*, which is the *conjugate base* of ammonia (pK_a ~ 35), with ammonia itself, which is a much weaker base, whose conjugate acid, *the ammonium ion* ($^+NH_4$), has a pK_a = 9.25.

Because the amide ion is a much stronger base than an acetylenic anion, the equilibrium for removal of the acetylenic proton by amide ion is very favorable. The sodium salt of an alkyne can be formed from a 1-alkyne with $NaNH_2$. Because the amide ion is a much *weaker* base than either a vinylic anion or an alkyl anion, these ions *cannot* be prepared using sodium amide.

Sodium amide is prepared by dissolving metallic sodium in liquid ammonia in the presence of a trace of a ferric ion (Fe^{3+}) source, such as iron(III) trichloride. Hydrogen gas (dihydrogen) is a by-product of this reaction.

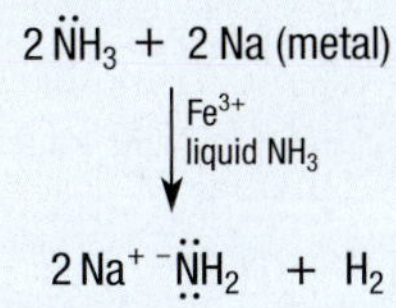

sodium amide

This reaction is conceptually like the reaction of sodium metal with water, which produces NaOH and H_2, although, in this case, ferric ion is required as a catalyst. In the absence of Fe^{3+}, *solvated electrons* (Sec. 10.2) are obtained; as we have seen, these react with alkynes in a different way.

The monosodium salt of acetylene itself can be prepared by the same reaction using a large excess of acetylene to ensure that the di-anion doesn't form:

$$\underset{\substack{\textbf{acetylene}\\\text{(large excess)}}}{HC{\equiv}CH} + Na^{+}\,{}^{-}\ddot{\underset{..}{N}}H_2 \xrightarrow[\text{liquid }NH_3]{} \underset{\textbf{sodium acetylide}}{HC{\equiv}\bar{C}\!:\ Na^{+}} + \ddot{N}H_3 \qquad (10.62)$$

Solutions of the potassium acetylide can be formed in tetrahydrofuran (THF) solution (or in solutions of other ethers) by reaction with potassium hydride, a commercially available alkali-metal hydride:

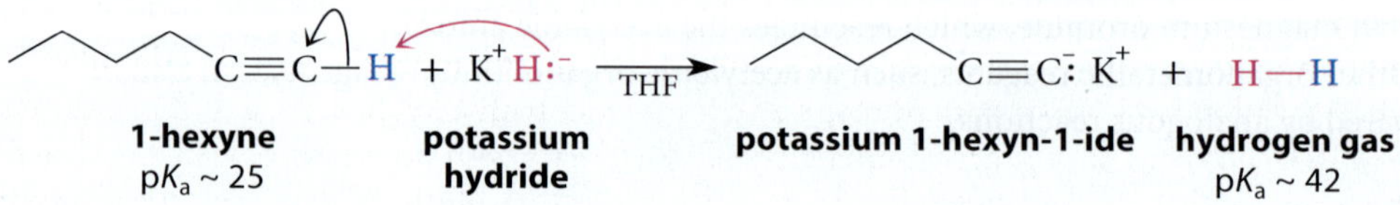

(10.63)

Sodium hydride (NaH) and lithium hydride (LiH) react too slowly to be useful in the analogous reactions.

Focused Problem

10.23 Complete each of the following reactions, and explain your reasoning.

(a) 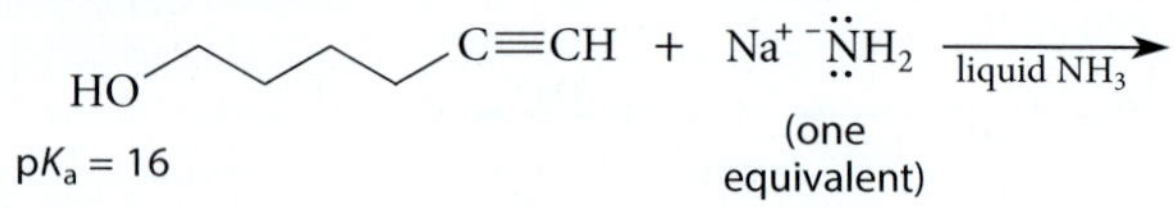

(b) 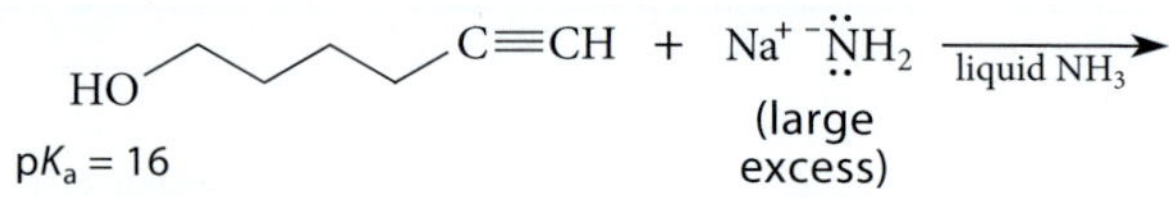

(c) 

B. Acetylenic Anions as Nucleophiles

Although acetylenic anions are the weakest bases of the simple hydrocarbon anions, they are nevertheless strong bases—much stronger, for example, than hydroxide or alkoxides. They undergo many of the characteristic reactions of strong bases, such as reactions as nucleophiles in S_N2 reactions with alkyl halides (Sec. 9.4; Table 9.1). As the following examples illustrate, acetylenic anion nucleophiles in S_N2 reactions can be used to prepare other alkynes.

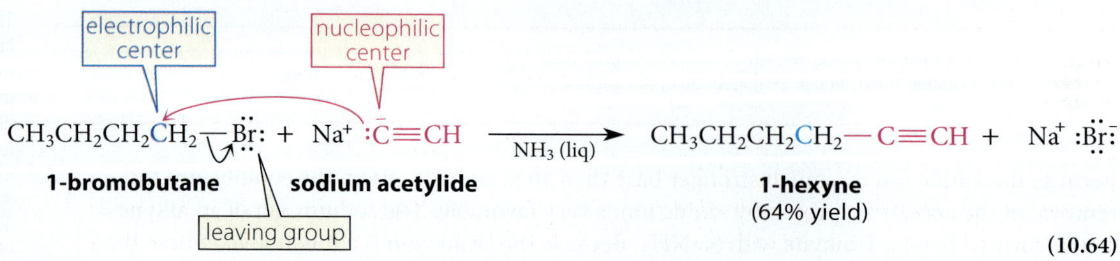

(10.64)

$$CH_3CH_2CH_2CH_2{-}C{\equiv}\ddot{C}^{-}\ Na^+ + H_3C{-}\ddot{Br}: \longrightarrow CH_3CH_2CH_2CH_2{-}C{\equiv}C{-}CH_3 + Na^+\ :\ddot{Br}:^-$$

2-heptyne

(10.65)

As in many other S_N2 reactions, the alkyl halides used in this reaction must be unhindered primary compounds. (Why? See Secs. 9.4D and 9.5G.)

This reaction is important because *it is a method of forming carbon–carbon bonds*. Reactions such as this can be used to build up carbon chains. The use of this reaction in combination with other reactions you have learned will be brought together in the discussion of organic synthesis in Sec. 10.8.

In Secs. 10.5C and 10.5D, we focused on the strongly basic character of Grignard and organolithium reagents. Could these reagents undergo S_N2 reactions with alkyl halides like acetylenic anions do? Most Grignard reagents do *not* react with alkyl halides, and the reason is the covalent character of the carbon–metal bond. They react with proton acids only because these acids are *much* more reactive than alkyl halides. Furthermore, a little reflection will convince you that, if these reagents were to react as nucleophiles with alkyl halides, they could not be prepared from alkyl halides and a metal, because the alkyl halide and Grignard reagent are present together in the reaction vessel. The result of an S_N2 reaction would be a coupling product R—R instead of a solution of the Grignard reagent:

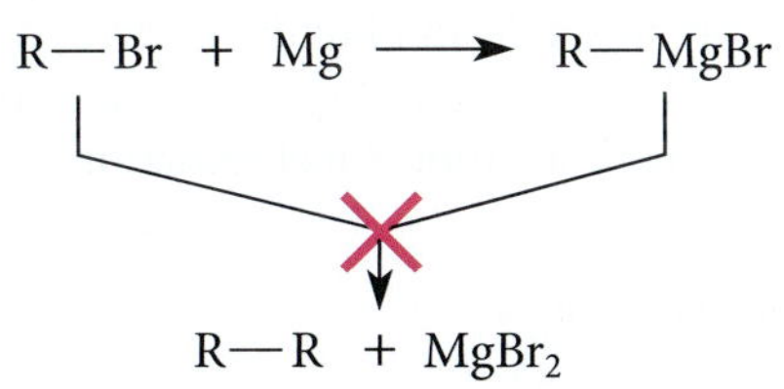

(10.66)

A few alkyl halides give Grignard reagents that have an unusually high degree of carbanion character, such as allylmagnesium bromide ($H_2C{=}CH{-}CH_2MgBr$). (Resonance stabilization of the allyl carbanion, $H_2C{=}CH{-}\ddot{C}H_2^{-}$, accounts for the greater carbanion character of the Grignard reagent.) In such cases, preparation of the Grignard reagent is difficult because the coupling reaction in Eq. 10.66 does occur. (Allylmagnesium bromide can be prepared, however, by a special experimental technique that minimizes contact of the Grignard reagent with the alkyl bromide.)

Focused Problems

10.24 Give the structures of the products in each of the following reactions.

(a) $H_3C{-}C{\equiv}\ddot{C}^{-}\ Na^+ + CH_3CH_2{-}I \longrightarrow$

(b)  $+\ HC{\equiv}\ddot{C}^{-}\ Na^+ \longrightarrow$ (excess)

10.25 Explain why a student, in attempting to synthesize 4,4-dimethyl-2-pentyne using the following reaction, obtained none of the desired product. What product did the student obtain instead? Explain.

$$(CH_3)_3C{-}Br + H_3C{-}C{\equiv}C^-\ Na^+ \not\longrightarrow H_3C{-}C{\equiv}C{-}C(CH_3)_3 + Na^+\ Br^-$$

4,4-dimethyl-2-pentyne
(none obtained)

10.26 Propose a synthesis of 4,4-dimethyl-2-pentyne (the compound in Focused Problem 10.25) from an alkyl halide and an alkyne.

10.27 Propose another pair of reactants that could be used to prepare 2-heptyne (the product in Eq. 10.65).

10.7 CARBENES AND CARBENOIDS

A. Carbenes from α-Elimination Reactions

Section 9.1B explained that β-elimination is one of the reactions that can occur when certain alkyl halides containing β-hydrogens are treated with base. When an alkyl halide contains no β-hydrogens but has an α-hydrogen, a different sort of base-promoted elimination is sometimes observed. Chloroform ($H{-}CCl_3$) is an alkyl halide that undergoes such a reaction.

Chloroform, although a weak acid (pK_a ~ 25), is a much stronger acid than an alkane because of the polar effect of the three chlorines (Sec. 3.7E). Its acidity is high enough that it can react as an acid with strong bases. Therefore, when chloroform is treated with an alkoxide base such as potassium *tert*-butoxide, a small amount of its conjugate-base anion is formed.

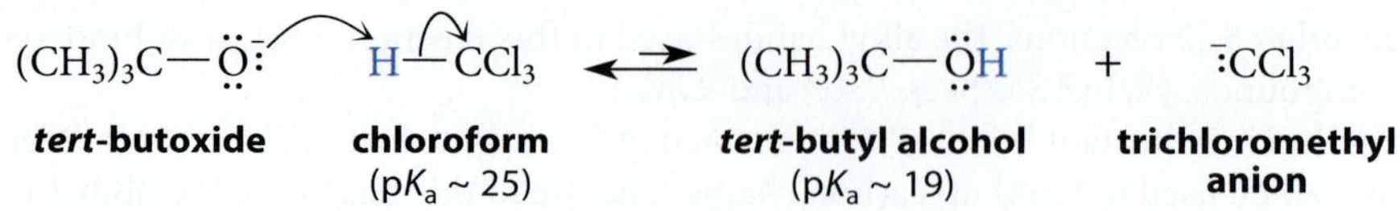

(10.67a)

This anion can lose a chloride ion to give a neutral species called *dichloromethylene*.

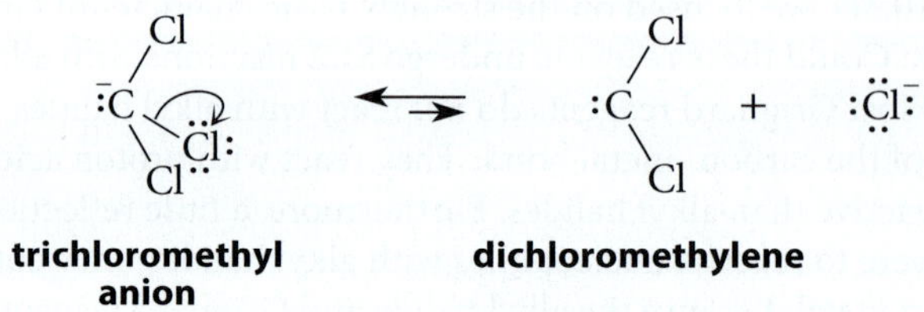

(10.67b)

Dichloromethylene is an example of a **carbene**—a species with a divalent carbon atom. Dichloromethylene has only six valence electrons on carbon, so its carbon is two electrons short of an octet. As a result, carbenes are unstable and highly reactive species.

The formation of dichloromethylene shown in Eqs. 10.67a and 10.67b involves elimination of the elements of HCl from the *same* carbon atom. An elimination of two groups from the same atom is called an **α-elimination**.

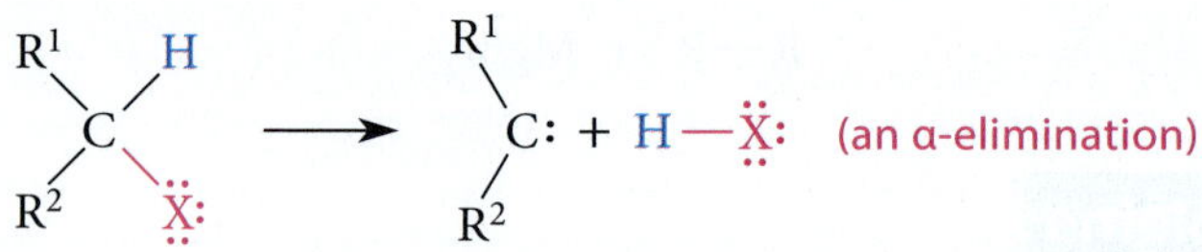

(10.68)

Chloroform cannot undergo a β-elimination because it has no β-hydrogens. When an alkyl halide has β-hydrogens, β-elimination occurs in preference to α-elimination because alkenes, the products of β-elimination, are much more stable than carbenes, the products of α-elimination. For example, CH_3CHCl_2 reacts with base to form the alkene $H_2C{=}CHCl$ rather than the carbene $CH_3—\ddot{C}—Cl$.

The reactivity of dichloromethylene follows from its electronic structure. The carbon atom of dichloromethylene bears three groups (two chlorines and the nonbonding electron pair) and therefore has approximately trigonal planar geometry. Because trigonal planar carbon atoms are sp^2-hybridized, the Cl—C—Cl bond angle is bent rather than linear, the nonbonding electron pair of electrons occupies an sp^2 orbital, and the 2p orbital is vacant:

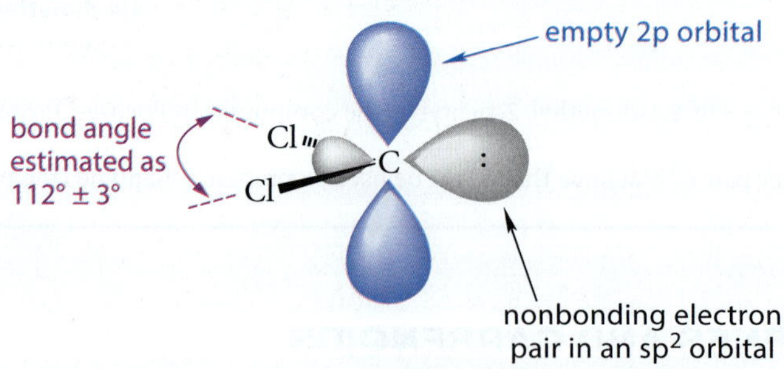

(10.69)

Because dichloromethylene lacks an electronic octet, it is an *electron-deficient compound* and can accept an electron pair; therefore, dichloromethylene is a powerful electrophile. On the other hand, an atom with a nonbonding electron pair can react as a nucleophile. The divalent carbon of dichloromethylene, with its nonbonding electron pair, fits into this category as well. Indeed, *the divalent carbon of a carbene can act as a nucleophile and an electrophile at the same time.*

An important reaction of carbenes that fits this analysis is cyclopropane formation. When dichloromethylene is generated in the presence of an alkene, a cyclopropane is formed.

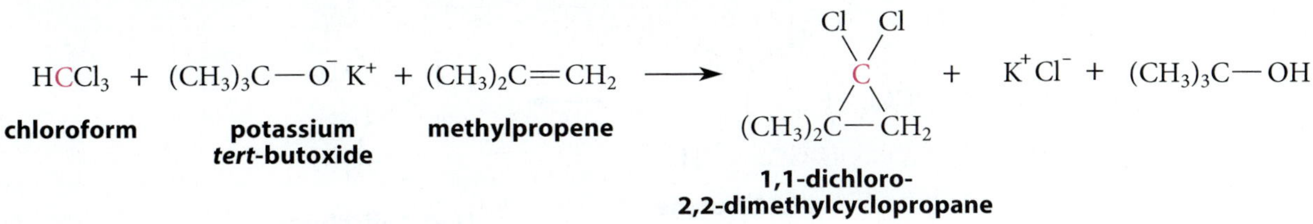

(10.70)

In general, reaction of a haloform with base in the presence of an alkene yields a 1,1-dihalocyclopropane.

Mechanistically, the reaction is a concerted syn addition. (Electron flow is shown in red, and atomic motion is shown in green.)

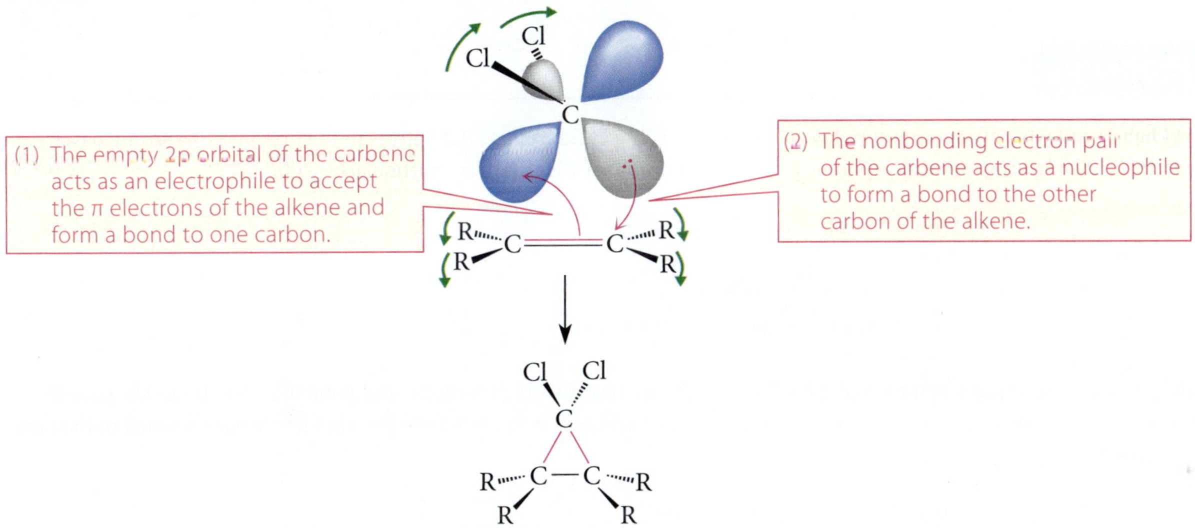

(10.71)

The empty 2p orbital of the carbene is electron-deficient and therefore acts as an *electrophile*. The π electrons of the alkene are donated to this orbital, and a bond is formed between the carbene and one carbon of the alkene. This nucleophilic reaction of one alkene carbon produces electron deficiency at the other alkene carbon. This electron deficiency is satisfied by a simultaneous reaction with the nonbonding electron pair of the carbene, which acts as a *nucleophile*, forming the other bond to the alkene.

The reaction is a stereospecific syn addition.

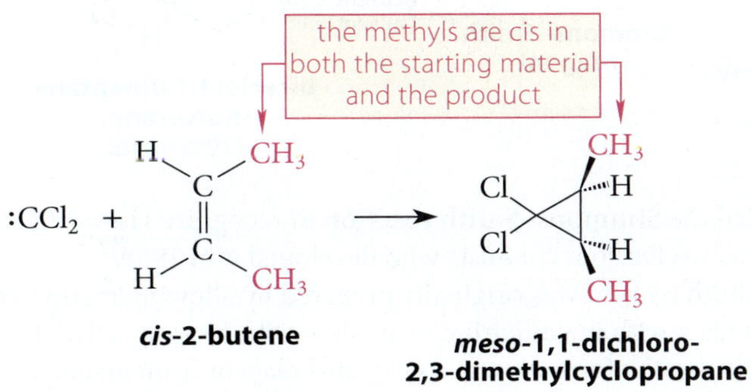

(10.72a)

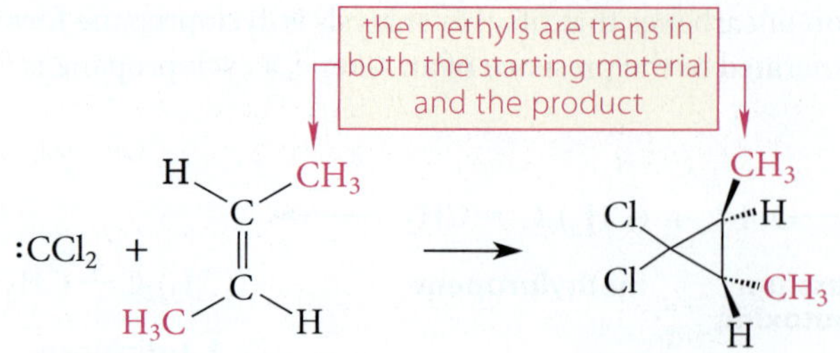

***trans*-2-butene**

(±)-1,1-dichloro-2,3-dimethylcyclopropane (10.72b)

The clean syn stereochemistry of the reaction not only is synthetically useful but also provides major evidence that the reaction is a concerted process. (The relationship between stereochemistry and mechanism was discussed in Sec. 7.8C.) A concerted anti addition is impossible because it would require the carbene carbon to add simultaneously to opposite faces of the alkene.

Focused Problems

10.28 What alkyl halide and what alkene would yield each of the following cyclopropane derivatives in the presence of a strong base? [*Hint for part (b):* The hydrogens on a carbon next to a benzene ring, or Ph group, are particularly acidic.]

(a) 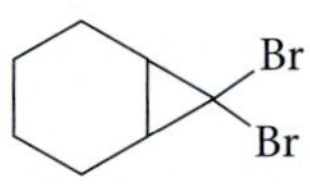

(b) 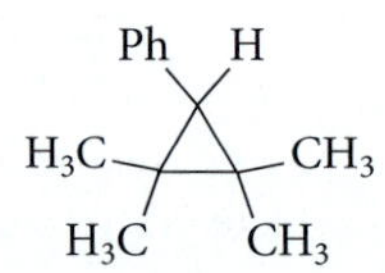

10.29 Predict the products that result when each of the following alkenes reacts with chloroform and potassium *tert*-butoxide. Give the structures of all product stereoisomers, and, if more than one stereoisomer is formed, indicate whether they are formed in the same or different amounts.

(a) Cyclopentene (b) (*R*)-3-Methylcyclohexene

B. The Simmons–Smith Reaction

Cyclopropanes without halogen atoms can be prepared by allowing alkenes to react with an organometallic reagent, iodomethylzinc iodide (I—CH_2—Zn—I), that is commonly referred to as the **Simmons–Smith reagent**.

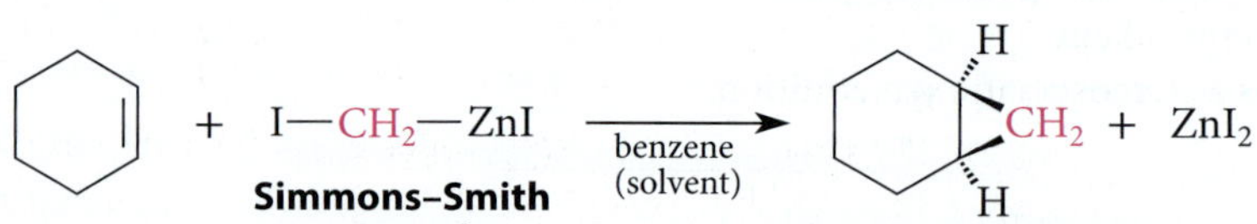

cyclohexene **Simmons–Smith reagent** **bicyclo[4.1.0]heptane (norcarane)** (79% yield) (10.73)

This reaction is called the **Simmons–Smith reaction** to recognize Howard E. Simmons and Ronald D. Smith, the two DuPont chemists who developed it in 1959.

The Simmons–Smith reagent was originally prepared by allowing methylene iodide (diiodomethane) to react with metallic zinc in an alloy with copper, called a *zinc–copper couple.* This reaction is similar to the formation of a Grignard reagent from an alkyl halide. The role of copper in the reaction is unclear.

$$CH_2I_2 + Zn \xrightarrow{\text{Zn–Cu couple}} I\text{—}CH_2\text{—}ZnI$$

diiodomethane (methylene iodide) **Simmons–Smith reagent** (10.74)

In 1967, a more convenient preparation of the Simmons–Smith reagent was discovered serendipitously by Osaka University chemist J. Furukawa. This *Furukawa modification* of the Simmons–Smith reaction utilizes the reaction of diethylzinc (Et_2Zn, another readily available organometallic reagent) with methylene iodide to prepare the Simmons–Smith reagent.

$$\underset{\textbf{diethylzinc}}{(CH_3CH_2)_2Zn} + \underset{\textbf{methylene iodide}}{CH_2I_2} \longrightarrow \underset{\textbf{Simmons-Smith reagent}}{I{-}CH_2{-}Zn{-}I} + CH_3CH_2CH_2CH_3 \qquad (10.75)$$

In the Furukawa modification of the Simmons–Smith reaction, methylene iodide is added to a solution of diethylzinc and the alkene—that is, the Simmons–Smith reagent is formed in the reaction flask without isolation.

From the discussion of the reactivity of carbenes with alkenes in the previous section, the cyclopropane product of the Simmons–Smith reaction is what would be expected if the parent carbene **methylene** ($:CH_2$) were a reactive intermediate. (We know that *free* methylene is not involved in the reaction because free methylene generated in other ways gives not only cyclopropanes, but other products as well.) However, the Simmons–Smith reagent can be conceptualized as methylene that is coordinated (loosely bound) to the Zn atom. This view is reasonable, first, because the carbon–zinc bond polarity is the same as the carbon–magnesium bond polarity in a Grignard reagent (Sec. 10.4C):

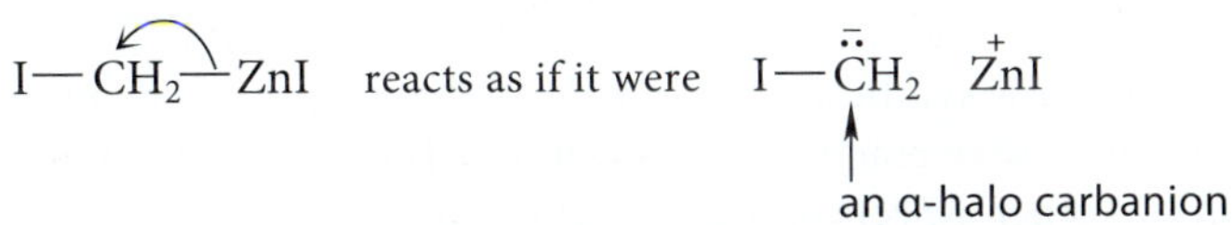

(10.76a)

and second, because an α-halo carbanion loses halide ion to give a carbene (see Eq. 10.67b):

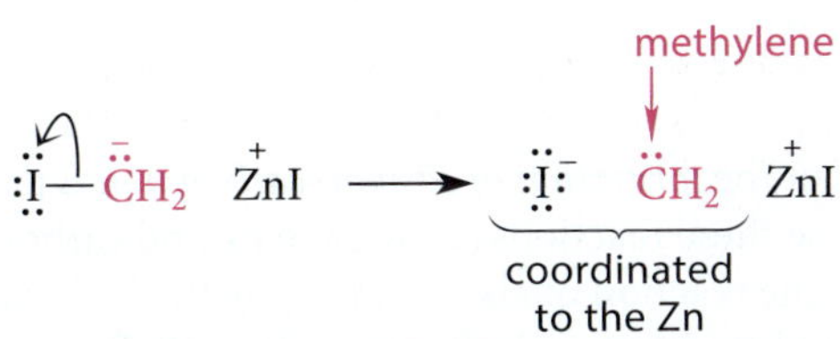

(10.76b)

Reaction of this "coordinated methylene" with the alkene double bond gives a cyclopropane. Addition reactions of methylene from Simmons–Smith reagents to alkenes, like the reactions of dichloromethylene, are stereospecific syn additions.

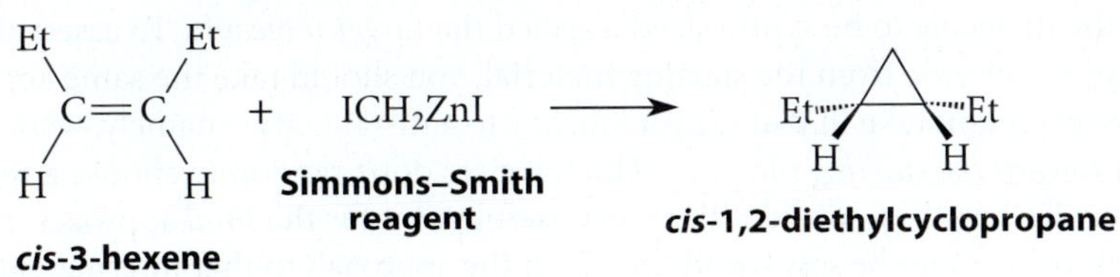

(10.77a)

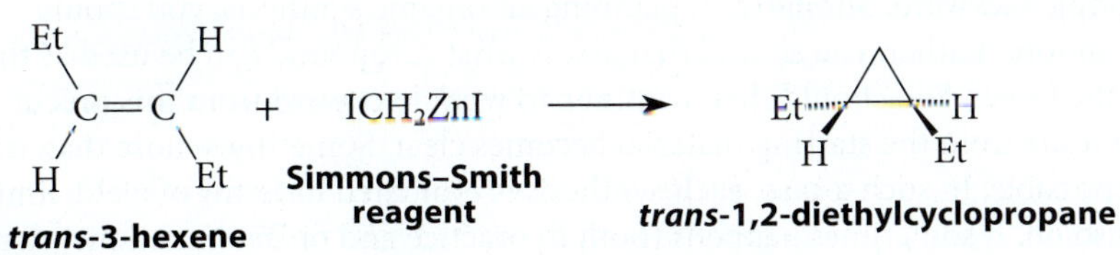

(10.77b)

Because they show carbenelike reactivity, α-halo organometallic compounds are sometimes called *carbenoids*. A **carbenoid** is a reagent that is *not* a free carbene but has carbenelike reactivity.

Addition of carbenes or carbenoids to alkenes to yield cyclopropanes is *another reaction that forms new carbon–carbon bonds*. As we noted in Sec. 10.6B, reactions that form carbon–carbon bonds are especially important in organic chemistry because they can be used to build up larger carbon skeletons from smaller ones.

Focused Problems

10.30 Give the structure of the organic product expected when the Simmons–Smith reagent reacts with each of the following alkenes:

(a) (*Z*)-4-Methyl-2-pentene

(b) 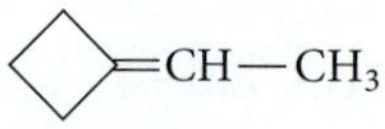

10.31 From which alkene could each of the following cyclopropane derivatives be prepared using the Simmons–Smith reaction?

(a) 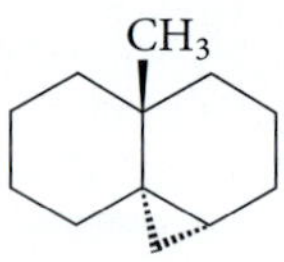

(b) 

10.8 MULTISTEP ORGANIC SYNTHESIS: RETROSYNTHETIC ANALYSIS

We began our study of organic synthesis in Sec. 5.9B. In that section we focused on *functional group transformation*—the use of reactions to convert one functional group into another. In this section, we focus on two other aspects of organic synthesis:

1. Carbon–carbon bond formation

2. Multistep synthesis

Reactions that bring about the *formation of carbon–carbon bonds* are especially important in organic chemistry, because these reactions allow us to extend carbon chains. We have now studied two such reactions: the reaction of alkyne anions with alkyl halides (Sec. 10.6B) and the use of carbenes and carbenoids to form cyclopropanes (Sec. 10.7).

Sometimes the synthesis of a compound from a given starting material can be completed with a single reaction. More often, however, the conversion of one compound into another requires more than one reaction. A synthesis involving a sequence of several reactions is called a **multistep synthesis.** Planning a multistep synthesis involves a type of reasoning that we examine here.

Recall that the molecule to be synthesized is called the *target molecule.* To assess the best route to the target molecule from the starting material, you should take the same approach that intelligent travelers might take in planning a journey to a distant city—namely, *work backward from the target toward the starting material.* The travelers don't randomly choose a route starting at the beginning. Rather, they consider the most useful point for the *final* approach to the city—perhaps a small town along the way—and then plan the approach to that intermediate objective, again by working backward. Similarly, in planning an organic synthesis, you should *not* try reactions at random. Rather, you should first assess what compound can be used as the immediate precursor of the target. You should then continue to work backward from this precursor step-by-step until the route from the starting material becomes clear. Sometimes more than one synthetic route will be possible. In such a case, each synthesis is evaluated in terms of yield, limitations, expense, and so on. It sometimes happens (both in practice and on examinations) that one synthesis is as good as another.

The process of working backward from the target is called **retrosynthetic analysis.** Study Problem 10.5 illustrates this process.

Study Problem 10.5

Using a retrosynthetic analysis, plan a synthesis of the following compound from acetylene and any other compounds containing no more than five carbons, and then outline the synthesis.

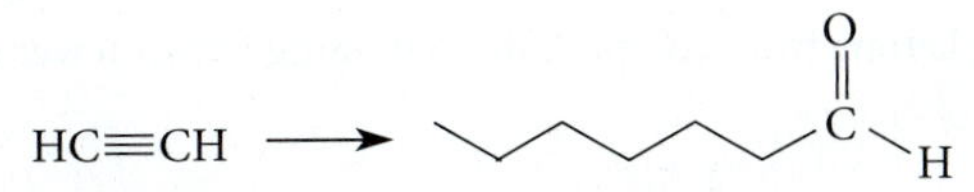

acetylene **heptanal**

(The stipulation that starting materials contain no more than five carbons is because most simple five-carbon compounds are available commercially.)

Solution To "outline a synthesis" means to provide the reagents and show the compounds produced at each step of the synthesis. (Balanced reactions aren't required.) Because the target molecule has seven carbons, and acetylene has only two, we have to use a reaction that forms carbon–carbon bonds. The only reasonable reaction that we have studied so far is the reaction of the acetylide anion (the conjugate base of acetylene) with an alkyl halide. To reveal which alkyne we need to synthesize, we start with the target molecule and work backward. What ways do we have for making aldehydes? Other than ozonolysis of alkenes (which "chops off" carbons), we have learned only one method so far: hydroboration–oxidation of 1-alkynes (Sec. 5.7). This reaction will convert 1-heptyne to the desired aldehyde:

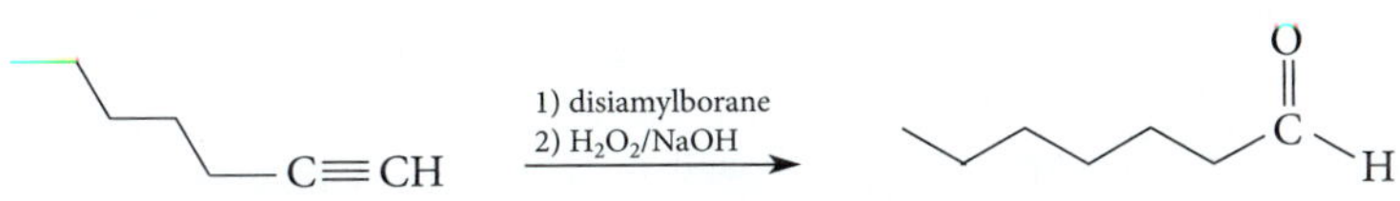

1-heptyne

The reaction of the acetylide ion and 1-bromopentane provides 1-heptyne:

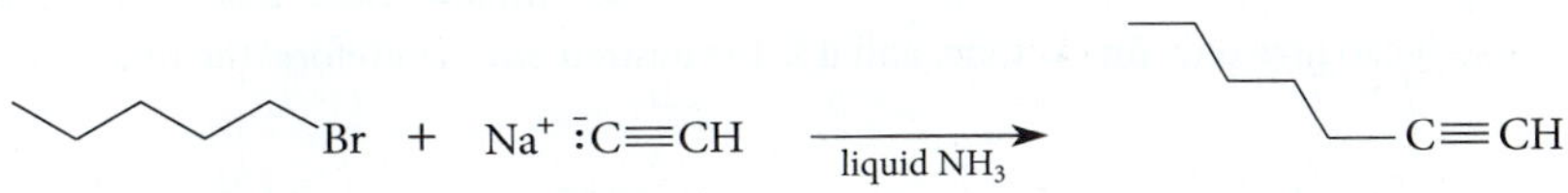

1-bromopentane **sodium acetylide** **1-heptyne**

Finally, sodium acetylide is prepared from an excess of acetylene and sodium amide in liquid ammonia:

acetylene (large excess) **sodium acetylide**

Our retrosynthetic analysis can be summarized as follows. (The symbol ⇨ means "implies as the starting material.")

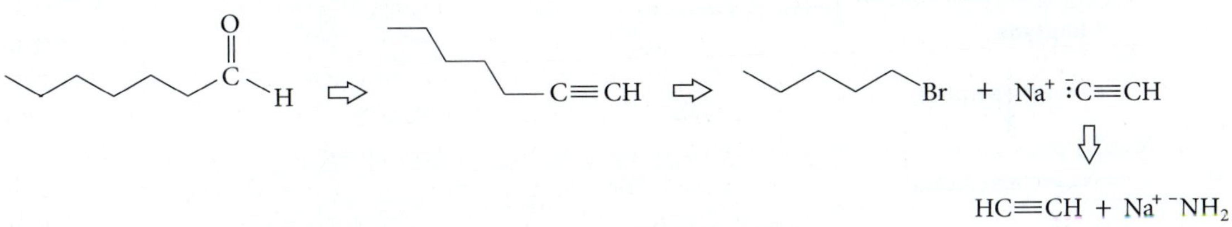

Our outlined synthesis is written as follows:

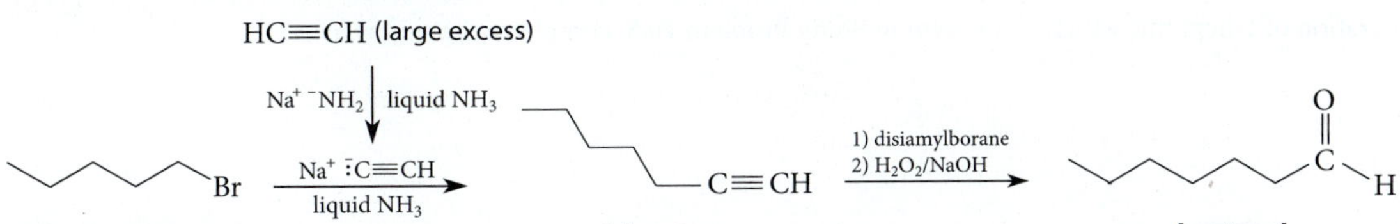

1-bromopentane **1-heptyne** **heptanal**

Study Problem 10.6 illustrates another retrosynthetic analysis.

Study Problem 10.6

Outline a synthesis of the following ketone from compounds containing five or fewer compounds and any other reagents.

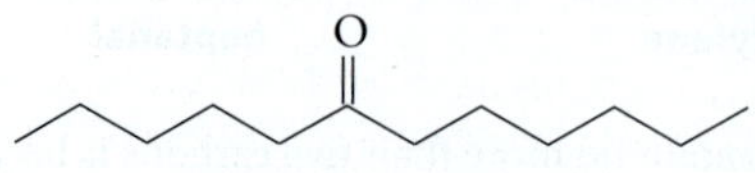

6-dodecanone

Solution Again we have to use reactions that form carbon–carbon bonds to meet the stipulation of starting materials with five or fewer carbons. We begin our retrosynthetic analysis with the target molecule. We have learned two ways of preparing ketones, both from alkynes: alkyne hydration (Sec. 5.5) and hydroboration–oxidation (Sec. 5.7). Either method is acceptable, although the requirement for water and mercuric ion for hydration could be an issue, because an alkyne with 12 carbons is not likely to have significant water solubility. However, perhaps a mixed solvent of THF/water might be satisfactory. Two possible alkynes would give the target molecule as a product of hydration (or hydroboration–oxidation):

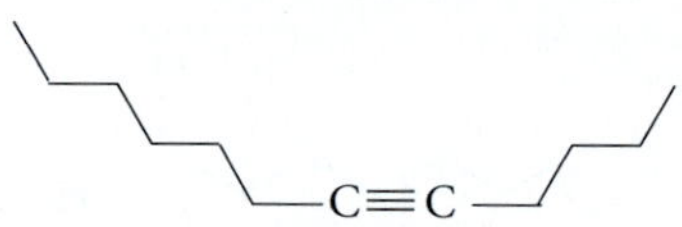

5-dodecyne

and

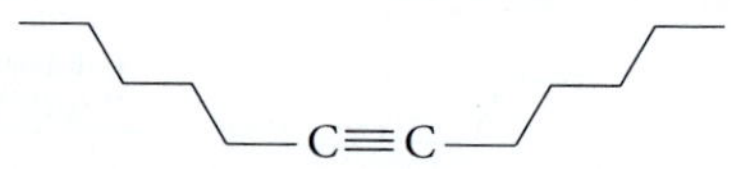

6-dodecyne

5-Dodecyne is an unsymmetrical alkyne, however, so it would give a mixture of two ketones in either reaction; and there is no reason to suppose that the desired ketone would be the predominant one. 6-Dodecyne, on the other hand, is a symmetrical alkyne, so it can give only one ketone, and it is the desired one. Therefore, the final step in our synthesis will be

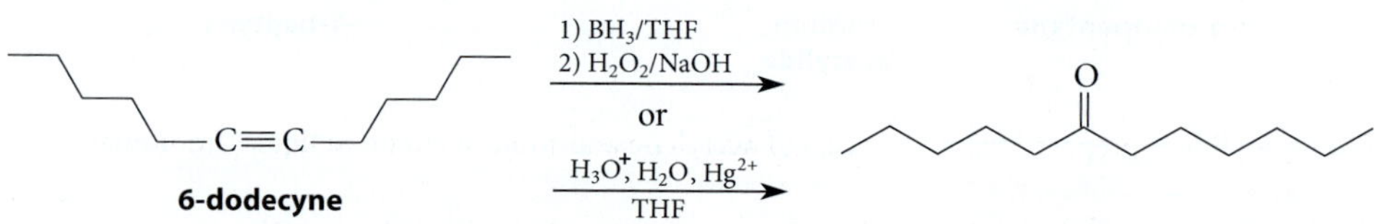

6-dodecyne

The reaction of the conjugate-base anion of 1-heptyne with 1-bromopentane will afford the desired alkyne:

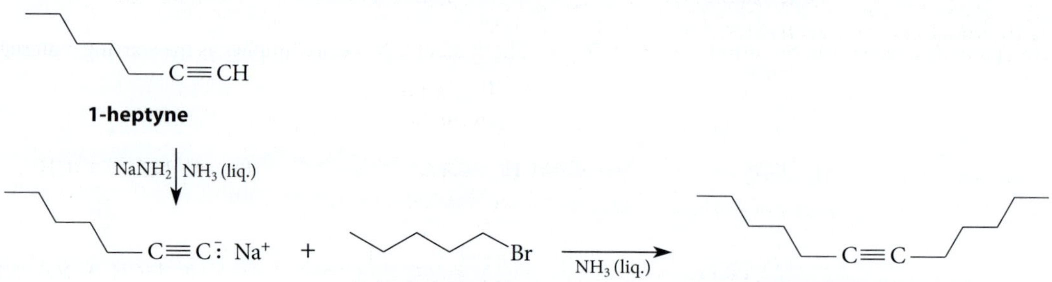

1-heptyne **1-bromopentane** **6-dodecyne**

As shown in this equation, the conjugate-base anion of 1-heptyne is prepared from the alkyne itself and $NaNH_2$. The preparation of 1-heptyne, which was shown in Study Problem 10.5, completes the synthesis.

The summary of our retrosynthetic analysis:

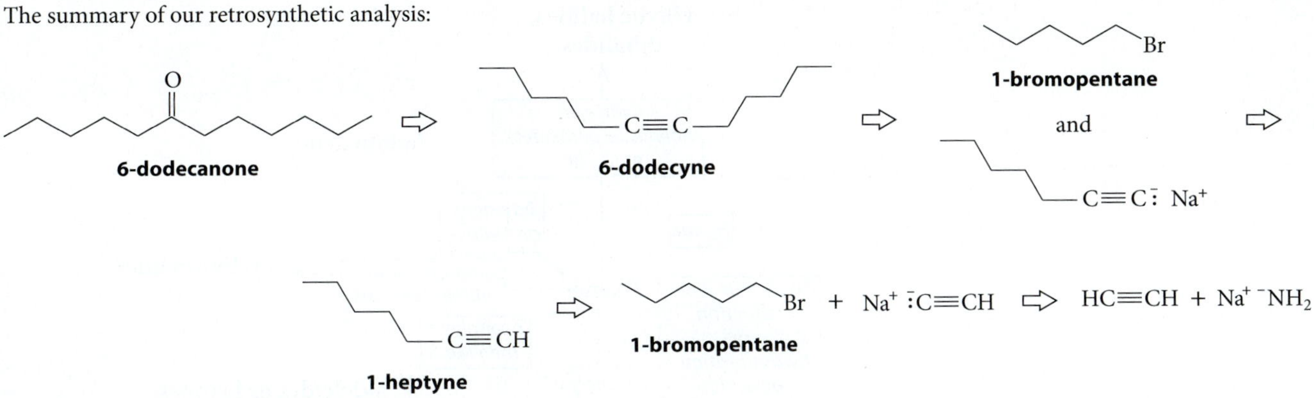

The required synthetic outline is constructed from the individual reactions:

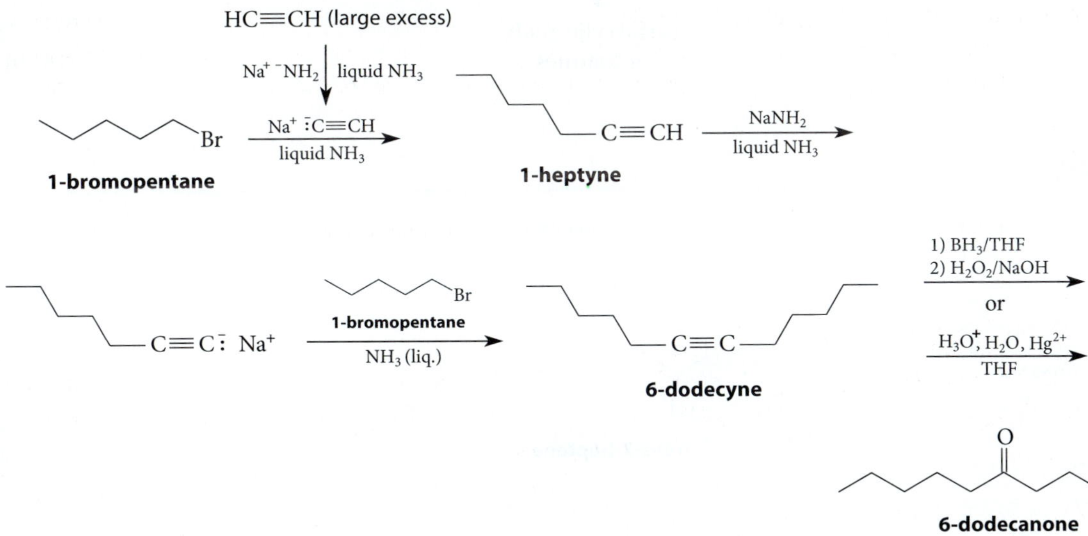

The summaries of functional group preparations in Appendix V can prove helpful as you carry out your retrosynthetic analysis. You may also want to construct diagrams of synthetic networks. Here are the ones for alkynes and alkenes from Displays 5.54 and 5.55 (Sec. 5.9B), updated to include the reactions in this chapter.

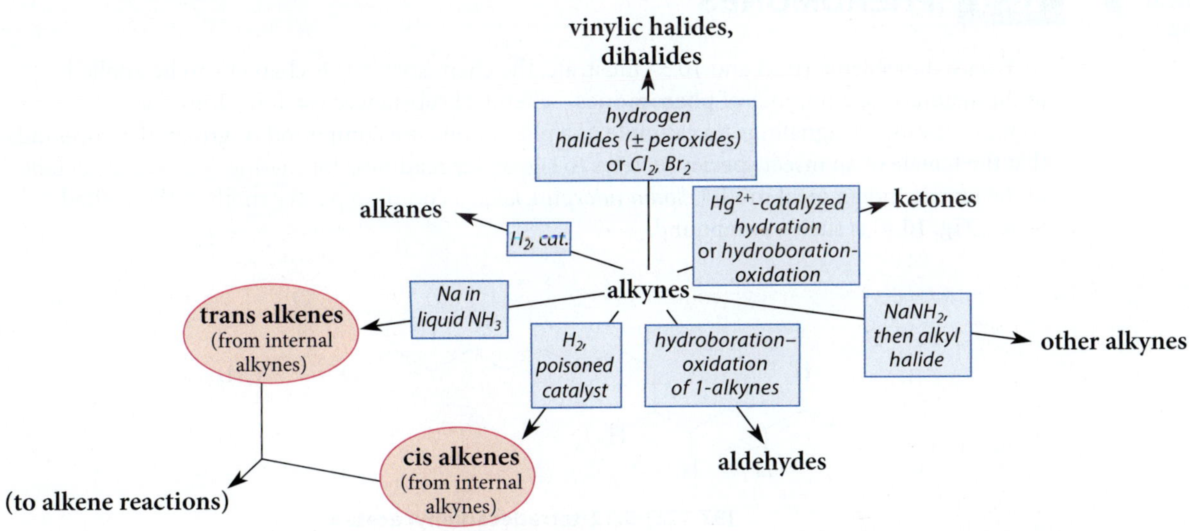

(10.78)

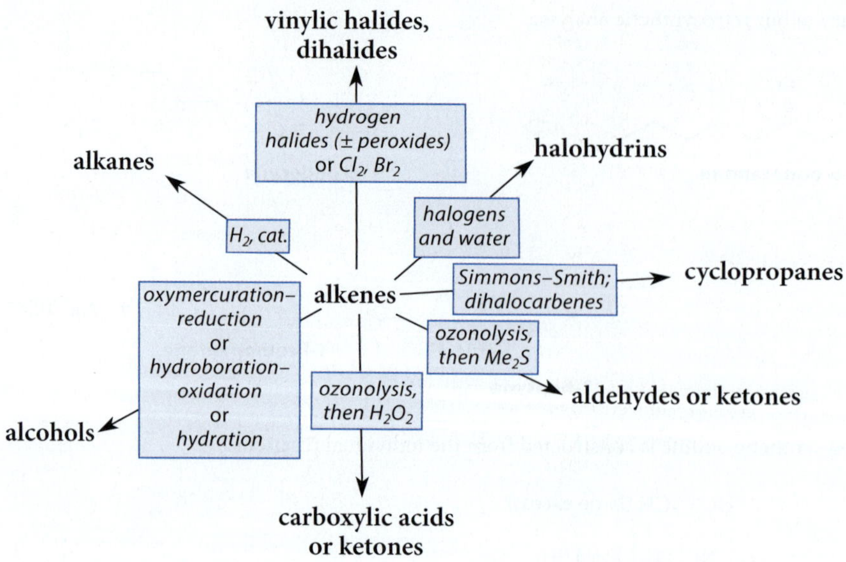

(10.79)

Focused Problem

10.32 Outline a multistep synthesis for each of the following compounds from organic compounds containing five or fewer carbons and any other reagents.

(a)

1-nonanol

(b)

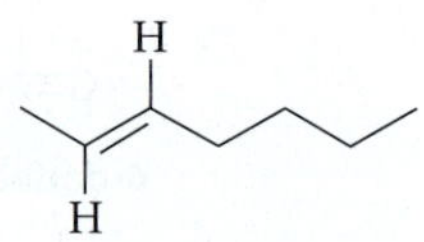

***trans*-2-heptene**

(c)

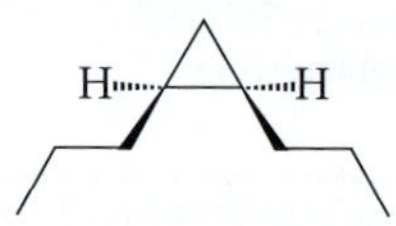

***cis*-1,2-dipropylcyclopropane**

Chemical Biology Topic

10.9 PHEROMONES

As Focused Problems 10.33 and 10.34 illustrate, the chemistry in this chapter can be applied to the synthesis of a number of **pheromones**—chemical substances used in nature for communication or signaling. An example of a pheromone is a compound or group of compounds that the female of an insect species secretes to signal her readiness for mating. The sex attractant of the female Indian meal moth (*Plodia interpunctella*, a common pantry moth in the United States; **Fig. 10.4**) is such a compound:

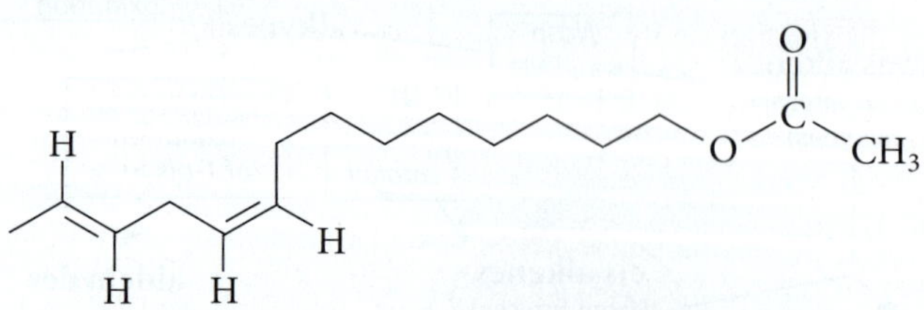

(9*Z*,12*E*)-9,12-tetradecadienyl acetate
(mating pheromone of the female Indian meal moth)

(10.80)

Marc Loudon

FIGURE 10.4 An infestation of the Indian meal moth, a common pantry moth, is controlled with a commercially available trap. The pad on the trap is impregnated with the female mating pheromone. Males attracted to the pheromone are immobilized and die on the sticky surface of the trap. The mating cycle of the moth is thus broken. Through use of these traps, fumigation of the pantry with insecticides is unnecessary.

Pheromones are also used for defense, to mark trails, and for many other purposes. It was discovered in the early 1980s that the traditional use of sows in France and Italy to discover buried truffles owes its success to the fact that truffles contain a steroid that happens to be identical to a sex attractant secreted in the saliva of boars during premating behavior!

In the early 1980s, scientists became intrigued with the idea that pheromones might be used as a species-specific form of insect control. The thinking was that a sex attractant, for example, might be used to attract and trap the male of an insect species selectively without affecting other insect populations. Alternatively, the males of a species might become confused by a blanket of sex attractant and not be able to locate a suitable female. When used successfully, this strategy would break the reproductive cycle of the insect. The harmful environmental effects and consequent banning of such pesticides as DDT stimulated interest in such highly specific biological methods.

Although experimentation has shown that these strategies are not successful for the broad control of insect populations, they are successful in specific cases. For example, local infestations of the common pantry moth can be eradicated with commercially available traps that utilize the female sex attractant (Fig. 10.4). In large-scale agriculture, traps employing mating pheromones are useful as early-warning systems for insect infestations. When this approach is used, conventional pesticides can be applied only when the target insects appear in the traps. This strategy has brought about reductions in the use of conventional pesticides by as much as 70% in many parts of the United States. Sex pheromones, aggregation pheromones (pheromones that summon insects for coordinated attack on a plant species), and kairomones (plant-derived compounds that function as interspecies signals for host plant selection) are used commercially in this manner. Attractants for more than 250 different species of insect pests are now available commercially.

Focused Problems

10.33 In the course of the synthesis of the sex attractant of the grape berry moth, both the cis and trans isomers of the following alkene were needed.

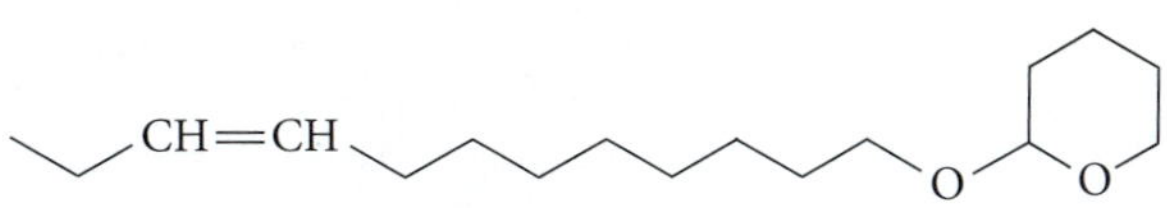

(a) Outline a synthesis of the cis isomer of this alkene from the following alkyl halide and any other organic compounds.

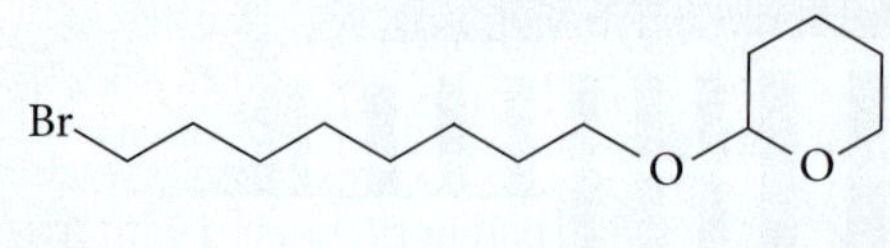

(b) Outline a synthesis of the trans isomer of the same alkene from the same alkyl halide and any other organic compounds.

10.34 The following compound is an intermediate in one synthesis of the mating pheromone of the female Indian meal moth (structure in Display 10.80). Show how this compound can be converted into the pheromone in a single reaction.

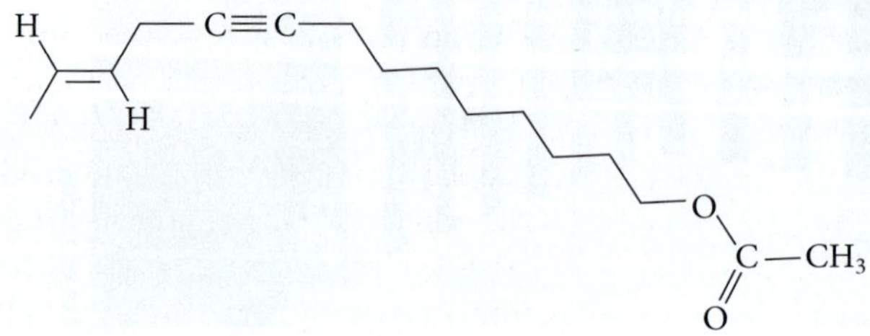

Chemistry in the Real World

10.10 USES OF HALOGEN-CONTAINING ORGANIC COMPOUNDS

A. Examples of Useful Organohalogen Compounds

Alkyl halides and other halogen-containing organic compounds have many practical uses. Methylene chloride and chloroform are important solvents (see Table 8.2, Sec. 8.6B) that are much less flammable than ethers. (Carbon tetrachloride was also important until its toxicity was recognized.) Tetrachloroethylene and trichloroethylene are used industrially as dry-cleaning solvents.

tetrachloroethylene (perchloroethylene) **trichloroethylene** (10.81)

A number of halogen-containing alkenes serve as monomers for the synthesis of useful polymers, such as PVC, Teflon, and Kel-F (see Table 10.3, Sec. 10.3C). A number of other bromine-containing organic compounds are used as commercial flame retardants. They work by producing bromine radicals under conditions of high heat. These radicals then interfere with the free-radical chain reactions associated with combustion. The environmental impact and toxicity of flame retardants has become a significant object of environmental concern.

The compound (2,4-dichlorophenoxy)acetic acid (sold as 2,4-D) mimics a plant growth hormone and causes broadleaf weeds to overgrow and eventually die. This is the dandelion killer used in commercial lawn fertilizers.

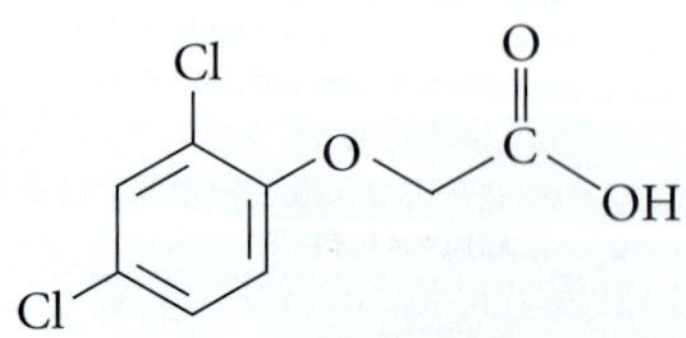

(2,4-dichlorophenoxy)acetic acid (2,4-D) (10.82)

A few alkyl halides have medical uses. Desflurane [(±)-CF_3—CHF—O—CHF_2] and sevoflurane [$(CF_3)_2$CH—O—CH_2F] are safe and inert general anesthetics that have replaced the highly flammable compounds ether and cyclopropane, which were once used widely. Certain fluorocarbons dissolve substantial amounts of oxygen, and some of these have been the subject of ongoing research as artificial blood in surgical applications.

B. Environmental Issues with Organohalogen Compounds

Because relatively few alkyl halides are found in nature, and because many are not biologically degraded, it is perhaps not surprising that some alkyl halides released into the environment have become issues of concern. The chlorofluorocarbons (CFCs, or freons), such as F_2CCl_2, and the hydrochlorofluorocarbons (HCFCs), such as $HCClF_2$ and $HCCl_2F$, are noteworthy examples. At one time these compounds were the only ones used as refrigerants in commercial cooling systems, and they were also widely used as propellants in aerosol products. Nontoxic and nonflammable, and with properties ideally suited to their applications, they seemed to be ideal industrial chemicals. During the 1970s, a number of studies implicated them in the destruction of stratospheric ozone. (The ozone layer provides an important shield against harmful ultraviolet solar radiation.) In October 1978, the United States government banned their use in virtually all aerosol products except those that are medically essential. In 1987, a number of countries, including the United States, initialed the *Montreal Protocol on Substances That Deplete the Ozone Layer*, under which industrial nations agreed to phase out the production of CFCs, carbon tetrachloride, and certain other substances by 2010. As a direct result, the production of CFCs by 1996 had dropped to about 16% of its pretreaty value. CFCs in existing refrigeration systems are recycled. HCFCs were phased out in 2015.

The problem with CFCs is that they bring about destruction of the stratospheric ozone layer by a free-radical chain reaction. The process is initiated when chlorine atoms are liberated from these compounds in upper-atmosphere *photodissociation reactions* (bond-homolysis reactions initiated by light).

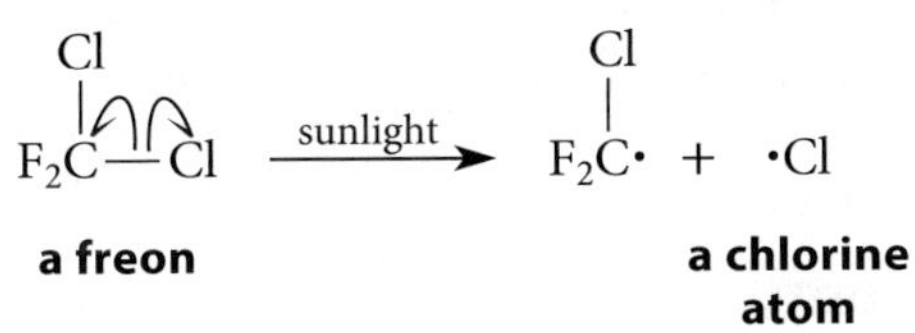

(10.83a)

It is the chlorine atom that brings about the destruction of ozone. The exact mechanism depends on the latitude and the season of the year. In one mechanism, chlorine atoms react with ozone to produce the ClO· radical and O_2:

$$Cl\cdot + O_3 \longrightarrow ClO\cdot + O_2 \qquad (10.83b)$$

Sunlight brings about the dissociation of O_2 molecules into oxygen atoms:

$$O_2 \xrightarrow{\text{sunlight}} 2\,O \qquad (10.83c)$$

These oxygen atoms react with the ClO· produced in Eq. 10.78b to give O_2 and regenerate a chlorine atom, which can then destroy another ozone molecule by Eq. 10.78b:

$$ClO\cdot + O \longrightarrow \underset{\text{(reacts with more ozone molecules)}}{Cl\cdot} + O_2 \qquad (10.83d)$$

The overall result of Eqs. 10.83b and 10.83d is the destruction of an ozone molecule.

$$O + O_3 \longrightarrow 2\,O_2 \qquad (10.83e)$$

The 1995 Nobel Prize in Chemistry was awarded for research into the chemical reactions that led to the destruction of stratospheric ozone. The recipients of the prize were Mario Molina (b. 1943), a chemist then at the Massachusetts Institute of Technology, but now at the University of California, San Diego; F. Sherwood Rowland (1927–2012), a chemist from the University of California, Irvine; and Paul Crutzen (b. 1933), a meteorologist-chemist from the Max-Planck Institute for Chemistry in Mainz, Germany.

By this mechanism, each chlorine atom (with the participation of naturally occurring oxygen atoms) *catalyzes* the conversion of an ozone molecule to an O_2 molecule. The chlorine atom can repeat this cycle many times. It has been estimated that a single chlorine atom can promote the destruction of 10^5 molecules of ozone. (Other ozone-destroying mechanisms also occur.)

One solution to this problem is to replace CFCs with related compounds that contain no chlorine. Indeed, one of the most common replacements for CFCs is the family of hydrofluorocarbons (HFCs) such as 1,1,1,2,2-pentafluoroethane (CF_3CHF_2). Although this class of compounds is less harmful to the ozone layer, HFCs nevertheless have adverse effects as greenhouse gases and ultimately exacerbate global warming. Careful recovery and recycling of HFCs from discarded refrigeration systems helps to minimize this effect of HFCs.

The most recent additions to the repertoire of refrigeration agents are the HFOs ("fluoroolefins"). Examples of these compounds are the following fluoroalkenes:

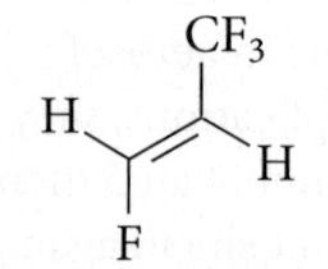

(*E*)-1,3,3,3-tetrafluoropropene
(commercial name 1234zeE)

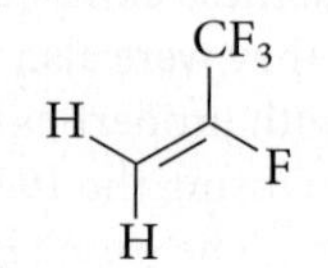

2,3,3,3-tetrafluoropropene
(commercial name 1234yf) (10.84)

These compounds possess greatly reduced global-warming potential because they decompose in a few days in the atmosphere to compounds that are believed to be environmentally benign. The issue with these compounds is that they are somewhat flammable. For this reason, there is an ongoing debate about their suitability for use in automotive air-conditioning systems.

Some potent and effective insecticides are organohalogen compounds.

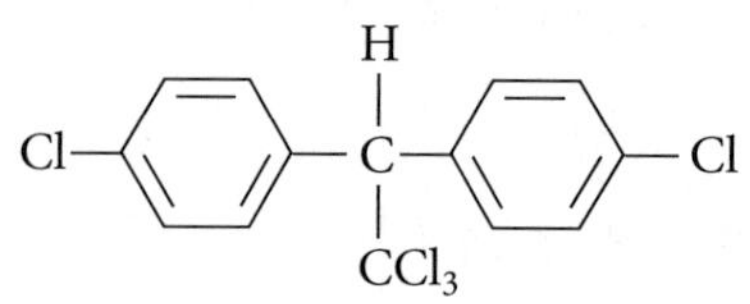

DDT

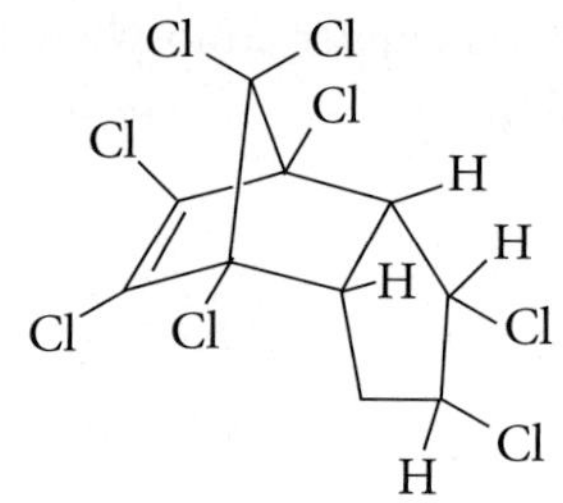

chlordane
(a mixture of isomers) (10.85)

DDT was first synthesized in 1873, but it was introduced in 1939 as a pesticide by Paul Müller (1899–1965), a Swiss chemist at the Laboratorium der Farben-Fabriken J.R. Geigy A.G., Basel. It was so effective that it was viewed for about 25 years as a savior of humankind. (For example, it virtually eliminated malaria in many areas of the world, including parts of the southern United States.) Müller received the Nobel Prize for Physiology or Medicine in 1948. Unfortunately, DDT, chlordane, and a number of other chlorinated broad-spectrum insecticides were subsequently found to accumulate in the fatty tissues of birds and fish, to be passed up the food chain, and to have harmful physiological effects. For these reasons their use has been banned or severely curtailed. The reduction in use of DDT was followed by the recovery of a number of DDT-susceptible endangered species, such as the bald eagle. However, malaria is such a large problem in certain parts of the world, and DDT is so effective against it, that it is still used in World Health Organization–endorsed spraying strategies that are designed to minimize environmental impact.

The conflict between the use of chemistry to improve humanity's living conditions and the generation of new problems caused by the release of chemicals into the environment finds real focus in the controversies surrounding the use of many organohalogen compounds. The great promise and public optimism that chemistry offered following World War II has given way to a public skepticism—or, at least, a period of public reflection and debate—as an increasing number of problems related to synthetic chemicals have surfaced. Is commercial organic chemistry in the end to be nothing but a Pandora's box of problems? Perhaps a more realistic view is that few if any human technological endeavors are without risk, and chemistry is no exception.

Each new generation of useful organic chemicals—whether they are pharmaceuticals, refrigerants, or insecticides—will likely bring some new problems with their benefits. These problems will provide a *great research opportunity* for chemists of the future who will take up the challenge of using their knowledge to improve further the benefits and to reduce or eliminate the problems.

A Nineteenth-Century Ball Ended by an Alkyl Halide Reaction

George Stuart, Artist www.galleryhistoricalfigures.com

Perhaps the first recorded instance of an environmental problem caused by alkyl halides occurred during the reign of Charles X of France. The French chemist Jean-Baptiste André Dumas (1800–1884) was asked to investigate something unusual that occurred during a ball given at the Tuileries. The candles used at the ball had sputtered and had given off noxious fumes, driving the guests from the ballroom. Dumas found that the beeswax used to make the candles had been bleached with chlorine gas. (Beeswax contains large numbers of double bonds. What reaction with chlorine took place?) The heat from the candle flame caused the chlorinated beeswax to decompose, liberating HCl gas—the noxious fumes.

CHAPTER SUMMARY *For a summary of the chapter, see Chapter 10 in the* Study Guide and Solutions Manual.

REACTION REVIEW *For a summary of reactions discussed in this chapter, see the* Reaction Review *section of Chapter 10 in the* Study Guide and Solutions Manual.

SKILLS OBJECTIVES WITH PROBLEMS

- **Provide missing reactants, reagents, or products in examples of the following reactions.**

(1) HBr addition to alkenes **4.7**

(2) Peroxide-mediated HBr addition to alkenes **10.1**

(3) Free-radical polymerization of alkenes **10.3B**

(4) Free-radical halogenation of alkanes by chlorine or bromine **10.4**

(5) Reaction of internal alkynes with Na in liquid ammonia **10.2B**; contrast with catalytic hydrogenation of the same alkynes **4.10B**

(6) Formation of Grignard and organolithium reagents **10.5B,D**

(7) Protonolysis of Grignard and organolithium reagents **10.5C**

(8) Preparation of acetylenic anions and their reactions with alkyl halides **10.6**

(9) Base-promoted formation of dihalocarbene and its reaction with alkenes **10.7A**

(10) The Simmons–Smith reaction **10.7B**

10.35 Complete each of the following reactions by providing the missing product, reactant, or reagent (indicated by "?").

(a) $CH_3CH_2MgBr + H_2O \longrightarrow ?$

(b)

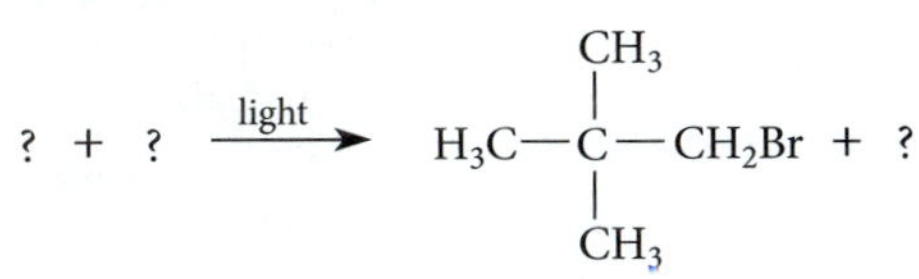

(c)

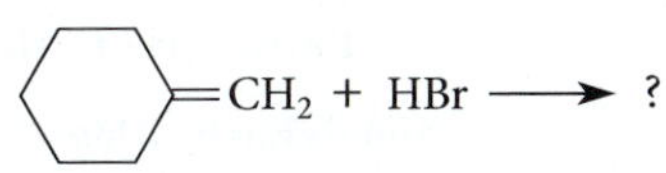

(d)

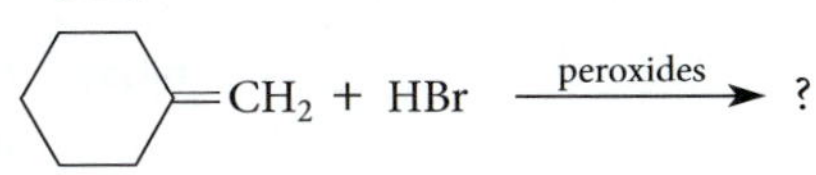

(e) $(CH_3)_2CH{-}Li + D_2O \longrightarrow ?$

(f)

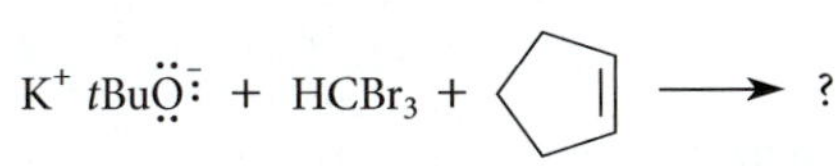

(g) 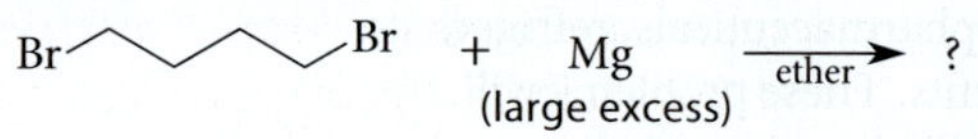

(h) 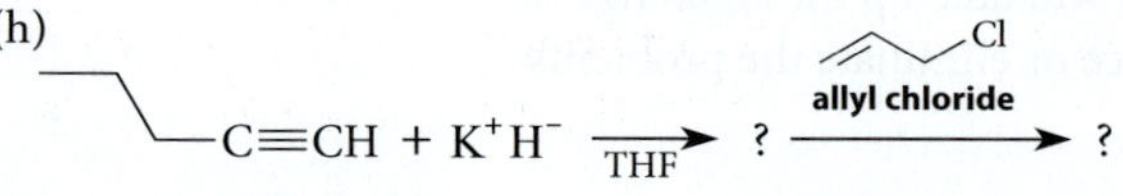

(i) 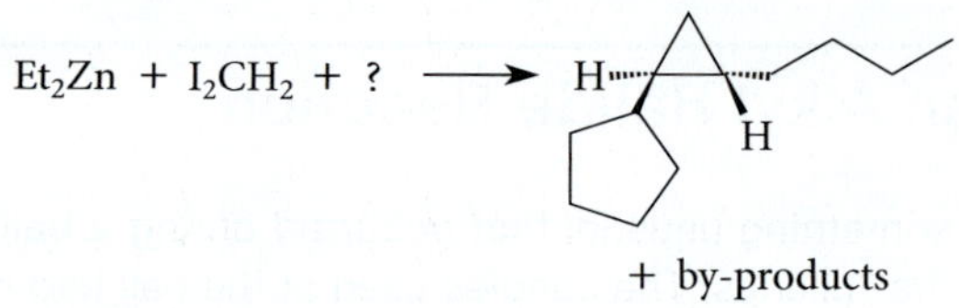

(j) Product of (g) + $D_2O \longrightarrow$?

(k) 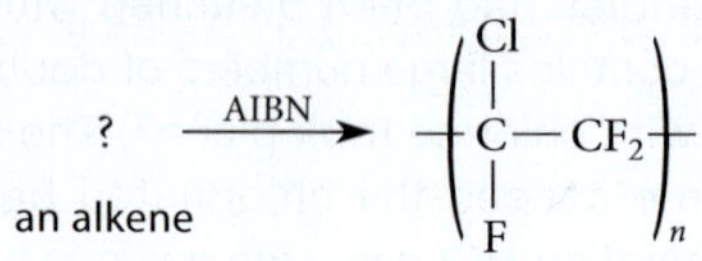

(l) $\text{MeI} \xrightarrow[\text{diethyl ether}]{\text{Mg}} ? \xrightarrow{(CH_3)_2CHCH_2CH_2C{\equiv}CH} ? + ? \text{ (a gas)} \xrightarrow{D_2O} ?$

(m) $CHCl_3 + K^+ (CH_3)_3CO^- + Ph{-}CH{=}CH_2 \longrightarrow ?$

(n) Product of (h) + Na in liquid $NH_3 \longrightarrow$

(o) Product of (h) + H_2, Lindlar catalyst $\longrightarrow$

• **Predict and explain the regiospecificity, if any, in examples of each of the following reactions.**

(1) HBr addition to alkenes **4.7**

(2) Peroxide-mediated HBr addition to alkenes **10.1**

(3) Free-radical halogenation of an alkane by chlorine or bromine **10.4B**

10.36 Predict the major constitutional isomer formed in each of the following reactions, and state the reason that it is the predominant isomer.

(a) 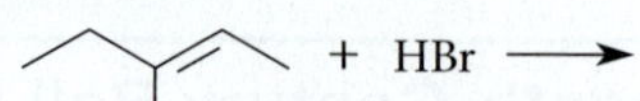

(b) 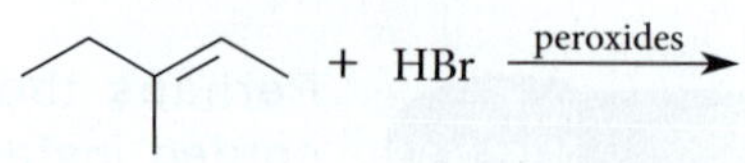

(c) 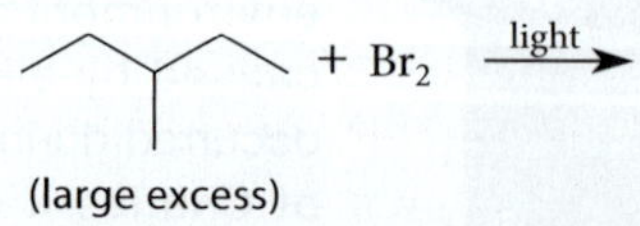

• **Provide the fishhook curved-arrow notation for any step of a free-radical reaction. Given the fishhook notation for a free-radical reaction, give the structures of the products. 10.1B**

10.37 Tributyltin hydride is a useful reagent for replacing bromine in an alkyl bromide with hydrogen. This reaction occurs by the free-radical mechanism shown in **Fig. P10.37**. Give the fishhook notation for each step.

• **List and identify the three phases of any free-radical chain reaction. 10.1C**

10.38 For the overall transformation shown in Problem 10.37, identify initiation, propagation, and termination steps of the reaction. Explain your logic for each label.

• **Using a table of bond dissociation energies, perform bond-energy calculations to determine the approximate $\Delta H°$ of a reaction. 10.1E**

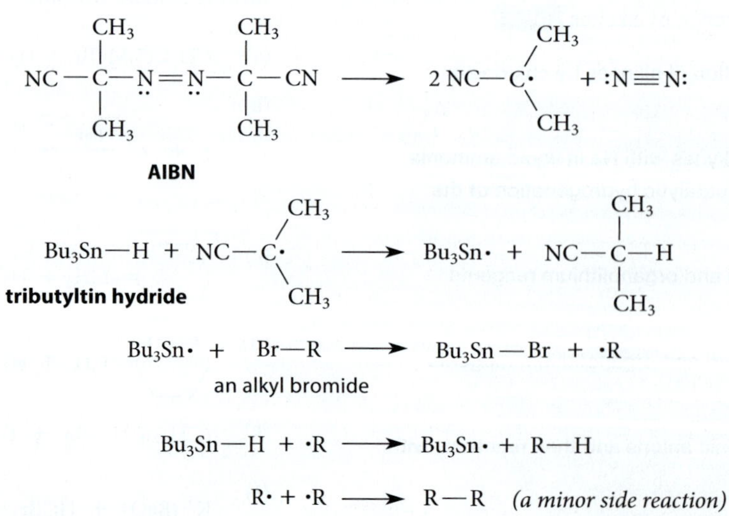

FIGURE P10.37

10.39 Estimate the $\Delta H°$ of each of the following reactions using bond dissociation energies.

(a) Bond dissociation energies (in kJ mol^{-1}): *a*, 372; *b*, 463; *c*, 320; *d*, 418; *e*, 728 (both bonds); *f*, 423; *g*, 368; *h*, 435. (For energies in kcal mol^{-1}, divide these values by 4.184.)

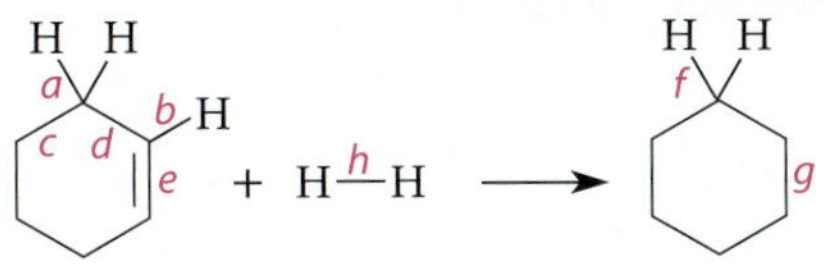

[*Hint:* Show that the $\Delta H°$ for the reaction is $8f + 5g - (4a + 2b + 2c + 2d + e + h)$].

(b) The experimental $\Delta H°_{hydrogenation}$ for cyclohexene is -120 ± 1 kJ mol^{-1} (-29 kcal mol^{-1}). Does your calculation in part (a) overestimate or underestimate the experimental value?

(c) Use the BDEs from part (a). The letters over atoms refer to C—H bond energies. (*Hint:* Treat bonds to carbons with unpaired electrons as unknowns.)

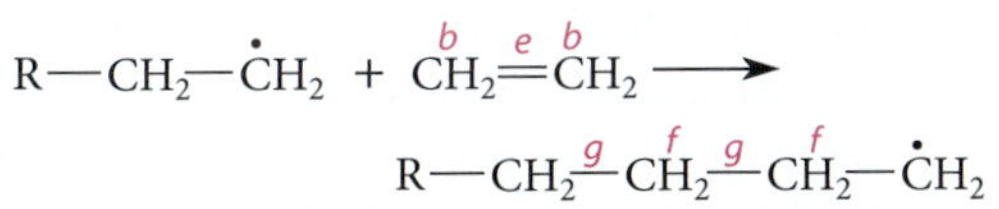

(d) The reaction in Problem 10.39(c) represents a step in the polymerization of ethylene. Given the $\Delta H°$ you calculated, should the polymerization of ethylene go to completion? If not, why does it go to completion in practice? (*Hint:* Consider the reaction conditions; see Eq. 10.33, Sec. 10.3A.)

(e) Explain why, in the free-radical addition of HBr to an alkene, the radical *A* reacts by abstracting a hydrogen from HBr to give *B* rather than adding to an alkene to give *C*.

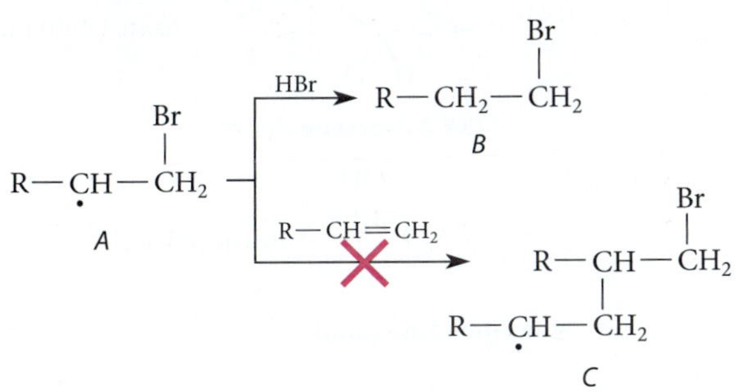

- **Given the structure of an alkene polymer, determine whether it is atactic, isotactic, or syndiotactic.**

10.40 What is the stereochemistry of a poly(styrene) with each of the following typical sections? Give your reasoning.

(a)

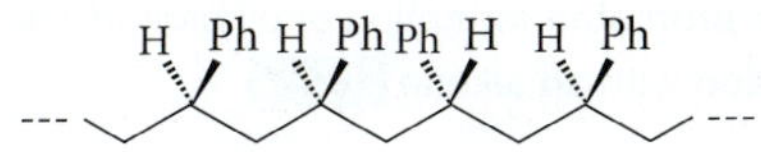

(b)

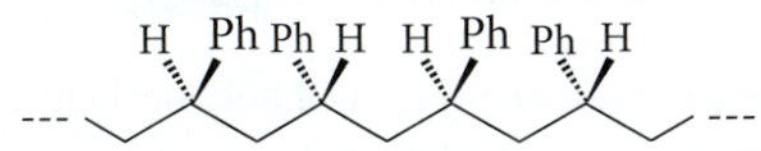

- **Given its condensed structure, draw out the structure of several contiguous units of the polymer. Given several contiguous units of a polymer, draw its condensed structure.**

10.41 (a) Polypropylene carbonate is considered to be a "green" polymer because it is made from CO_2 and is biodegradable.

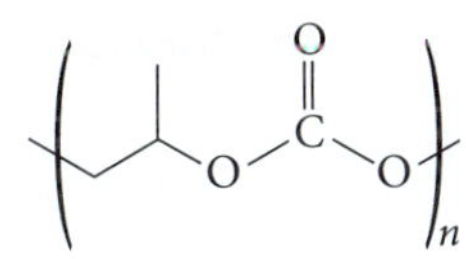

polypropylene carbonate

Draw out the structure of polypropylene carbonate for $n = 3$. (Leave the "dangling bonds" at the ends.)

(b) Draw a condensed structure for the polymer shown in **Fig. P10.41** that shows only the repeating unit.

- **Provide curved-arrow or fishhook mechanisms (as appropriate) for examples of each of the following reactions. 10.1B, 3.2, 3.3**

(1) HBr addition to an alkene **4.7**

(2) Peroxide-promoted HBr addition to an alkene **10.1C**

(3) Reaction of an internal alkyne with Na in liquid ammonia **10.2**

(4) Free-radical polymerization of an alkene **10.3B**

(5) Free-radical halogenation of an alkane **10.3A**

(6) Protonolysis of a Grignard or organolithium reagent **10.5C**

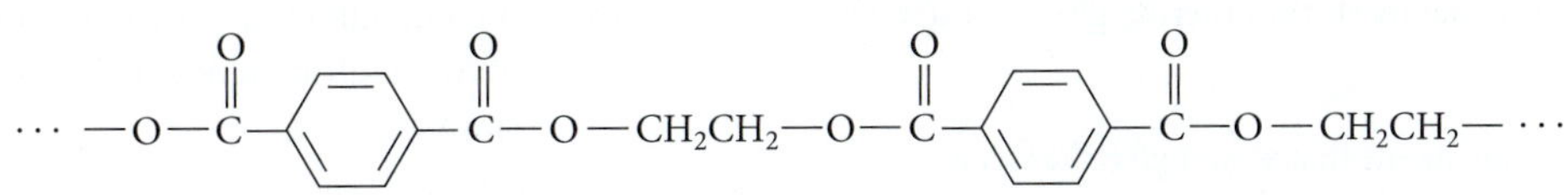

polyester

FIGURE P10.41

(7) Reaction of a 1-alkyne with a Grignard reagent **10.5D**

(8) Reaction of a 1-alkyne anion with an alkyl halide **10.6B**

(9) Base-promoted formation of dihalocarbene and its reaction with an alkene **10.7A**

(10) The Simmons–Smith reaction **10.7B**

10.42 Provide a curved-arrow or fishhook mechanism (as appropriate) for each of the following reactions.

(a) The reaction in Problem 10.35(a)

(b) The reaction in Problem 10.35(b)

(c) The reaction in Problem 10.35(c)

(d) Show fishhook mechanisms for the initiation step and for the polymerization with $n = 2$.

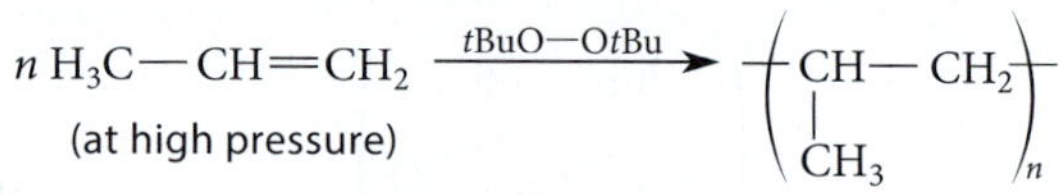

(e) The reaction in Problem 10.35(d)

(f) The reaction in Problem 10.35(f)

(g) $Me_2C{=}CMe_2 + I{-}Zn{-}CH_2I \longrightarrow$

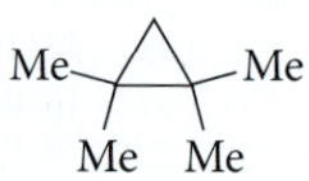

For mechanistic purposes, treat the Simmons–Smith reagent as carbene, $:CH_2$.

(h) The reactions in Problem 10.35(l)

(i)

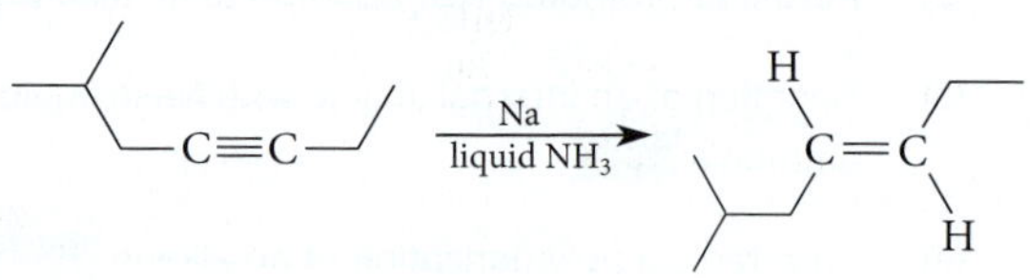

- **Determine from its structure whether an organohalogen compound is likely to undergo α-elimination, β-elimination, or neither in the presence of a strong base. 10.7A**

10.43 Tell whether each of the following alkyl bromides is likely to undergo α-elimination, β-elimination, or neither in the presence of a strong base.

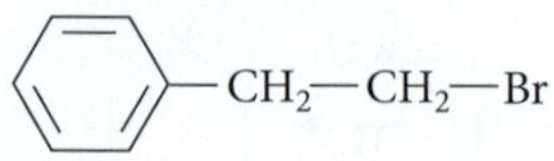

A

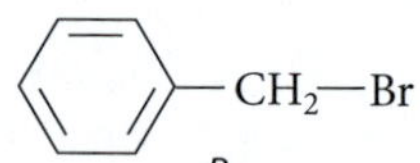

B

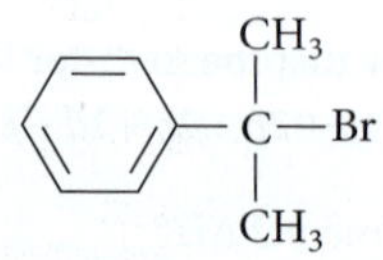

C

- **Carry out a retrosynthetic analysis of a multistep synthesis using all of the reactions that you have learned. 10.8**

10.44 Using a retrosynthetic analysis, outline a multistep synthesis for each of the following compounds from the indicated starting material, organic compounds with five or fewer carbons, and any other reagents. (Any chiral compounds will be prepared as racemates.)

(a) 1-Bromoheptane from acetylene

(b)

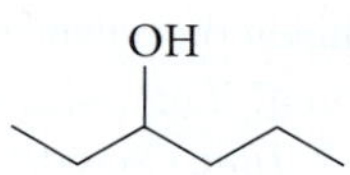

from acetylene

3-hexanol

(c)

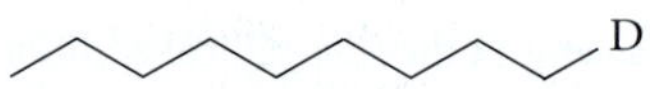

from acetylene

nonane-1-*d*

(d)

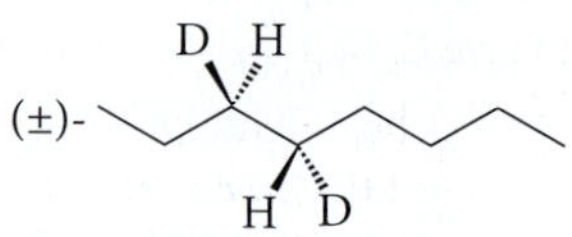

from 1-butyne

(3*R*,4*S*)-octane-3,4-d_2

(e)

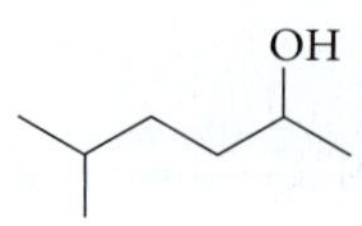

from acetylene

5-methyl-2-hexanol

INTEGRATED PROBLEMS

10.45 Draw the structure that meets the criterion given in each of the following cases.

(a) A five-carbon alkene that would give the same product of HBr addition whether or not peroxides are present.

(b) A compound C_6H_{14} that gives only two products of free-radical monochlorination, one of which is chiral.

(c) Four stereoisomeric compounds C_4H_8O, all optically active, that contain no double bonds and evolve a gas when treated with EtMgBr.

(d) A hydrocarbon C_6H_{10} that undergoes catalytic hydrogenation to hexane and liberates a gas on treatment with MeLi.

10.46 Identify the gas evolved in the following reaction. Justify your answer with a curved-arrow mechanism.

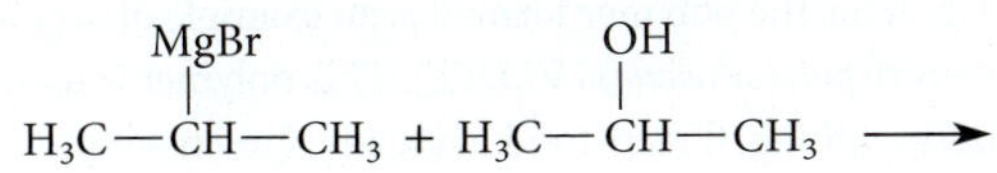

10.47 **Figure P10.47** shows the results observed when HBr is added to 3,3-dimethyl-1-butene.

(a) Explain why the different conditions give different product distributions.

(b) Write a detailed mechanism for each reaction that explains the origin of all products.

(c) Which conditions give the faster reaction? Explain.

10.48 (a) What product would be obtained from the ozonolysis of natural rubber, followed by reaction with H_2O_2? (*Hint:* Write out two units of the polymer structure.)

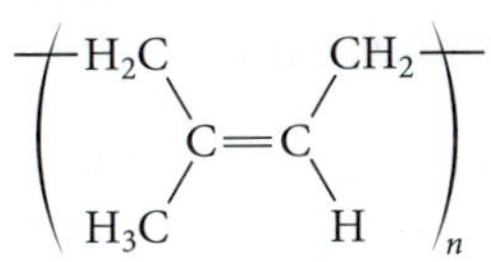

natural rubber

(b) *Gutta-percha* is a natural polymer that gives the same ozonolysis product as natural rubber. Suggest a structure for gutta-percha.

10.49 At very high temperatures, the covalent bonds in hydrocarbons will break homolytically. The bonds that break most readily (and at the lowest temperature) are those with the lowest bond dissociation energies (BDEs).

(a) Which bond in ethane should break most readily—the carbon–carbon bond or one of the carbon–hydrogen bonds? Explain.

(b) In the thermal cracking of 2,2,3,3-tetramethylbutane, which bond would be most likely to break? Explain.

(c) Which compound, 2,2,3,3-tetramethylbutane or ethane, undergoes thermal cracking more rapidly at a given temperature? Explain.

(d) Calculate the $\Delta H°$ values for the initial carbon–carbon bond breaking for the thermal cracking of both 2,2,3,3-tetramethylbutane and ethane. Use the ΔH_f° values in Table 10.1 as well as the ΔH_f° of 2,2,3,3-tetramethylbutane (-225.9 kJ mol^{-1}, -53.99 kcal mol^{-1}) and ethane (-84.7 kJ mol^{-1}, -20.24 kcal mol^{-1}). Use these calculations to justify your answer to part (c).

10.50 (a) Use bond dissociation energies to justify the statement in Sec. 10.4A that free-radical substitution of an alkane hydrogen with iodine is unfavorable and does not occur.

$$X_2 + CH_4 \longrightarrow H_3C{-}X + H{-}X$$

(does not occur with X = I)

(b) Explain why samples of CH_3I that are contaminated with traces of HI darken with the color of iodine on standing for a long time.

10.51 When 2,3-dimethylbutane is treated with Br_2 in the presence of light, the bromine-containing compounds obtained in greatest amount are compound *A* ($C_6H_{13}Br$) and compound *B* ($C_6H_{12}Br_2$). Propose structures for these compounds, and explain your reasoning.

10.52 The reagent tributyltin hydride, $Bu_3Sn{-}H$, brings about the rapid conversion of 1-bromo-1-methylcyclohexane into methylcyclohexane in the presence of AIBN.

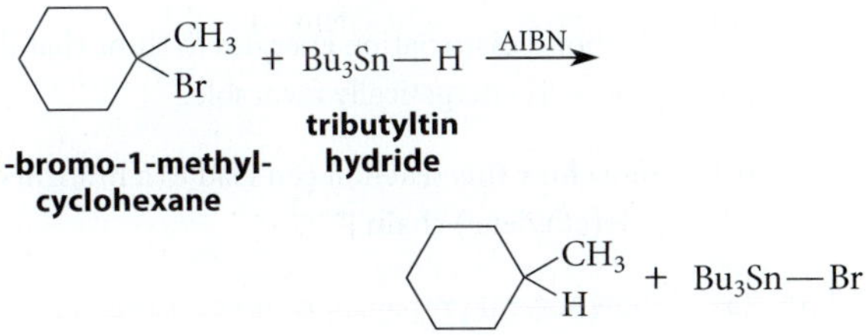

1-bromo-1-methylcyclohexane **tributyltin hydride** **methylcyclohexane**

(a) Use the general free-radical chain mechanism shown in Problem 10.37 to write out the propagation steps of this reaction.

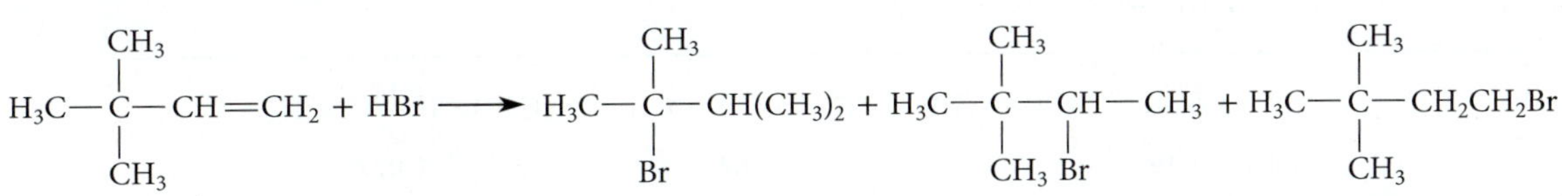

no peroxides:	71%	29%	none
with peroxides:	trace	trace	100%

FIGURE P10.47

(b) Suggest two other reaction sequences using other reagents that would bring about the same overall transformation.

(c) Using elements of the mechanism in Problem 10.37, provide a free-radical chain mechanism, complete with the fishhook notation, for the following known transformation.

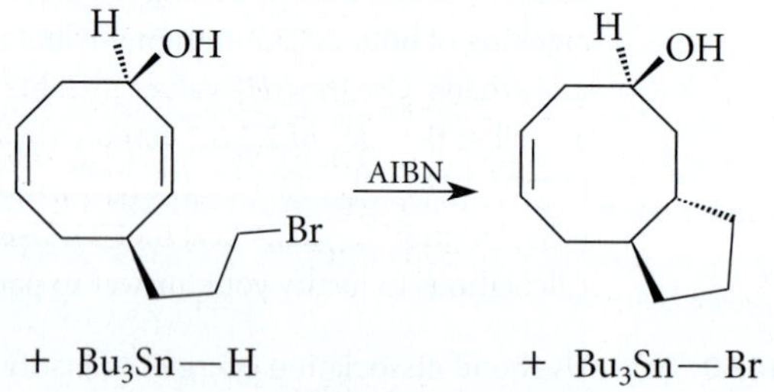

10.53 The mechanism for the free-radical polymerization of ethylene shown in Eqs. 10.35a–e (Sec. 10.3B) is somewhat simplified because it does not account for the observation that low-density poly(ethylene) (LDPE) contains a significant number of branched chains. (The branching accounts for the low density of LDPE.) It is believed that the first step that leads to branching is the following internal hydrogen abstraction reaction that occurs within the growing poly(ethylene) chain.

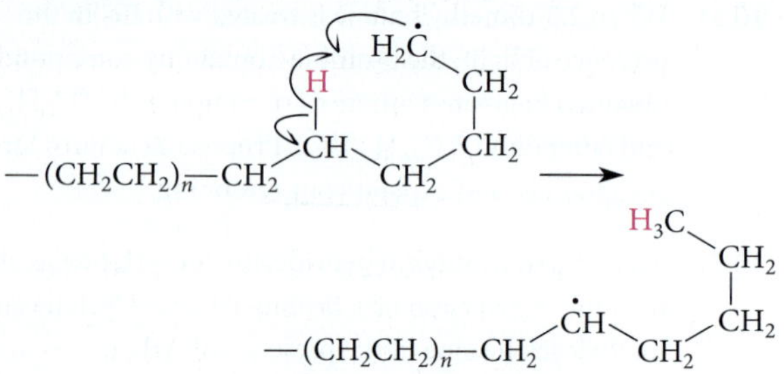

(a) Use bond dissociation energies to show that this process is energetically favorable.

(b) Show how this reaction can lead to a branched poly(ethylene) chain.

10.54 (a) Show several propagation steps for the polymerization of styrene to give poly(styrene). (See Table 10.3.)

(b) How would the structure of the polymer product differ from the one in part (a) if a few percent of 1,4-divinylbenzene were included in the reaction mixture?

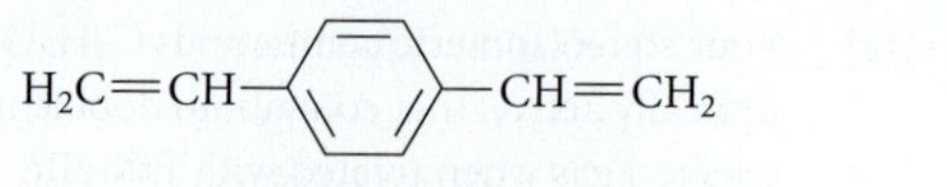

1,4-divinylbenzene

(c) How would you expect the polymer in (b) to differ in its flexibility and rigidity from ordinary polystyrene? Explain.

10.55 When ethylene is polymerized with a few percent of 1-butene, the polymer formed is an example of *very low-density poly(ethylene)* (*VLDPE*). This polymer is used for hoses, tubing, frozen-food bags, and stretch wrap.

(a) Using skeletal structures, draw out a several units of poly(ethylene) in which only one of the units is derived from 1-butene.

(b) Explain why VLDPE has lower density than poly(ethylene) itself.

10.56 Consider the reaction of a methyl radical ($\cdot CH_3$) with the π bond of an alkene:

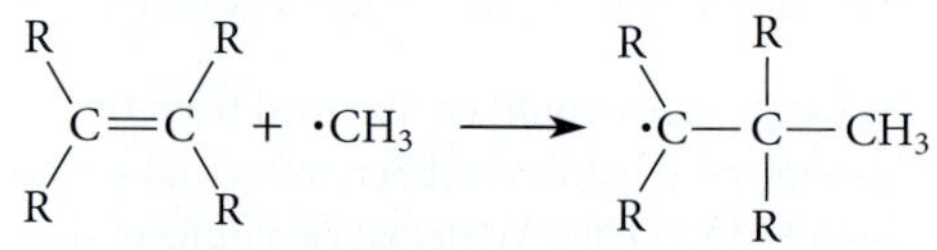

The relative rates of this reaction, shown in **Fig. P10.56**, were determined for various alkenes.

(a) Draw the free-radical product of the reaction in each case.

(b) Explain the order of the relative rates.

10.57 Free-radical addition of thiols (RSH) to alkenes is initiated by peroxides.

(a) Use data from Table 10.1 plus the following data to calculate the C—S bond dissociation energy (BDE) for ethanethiol (CH_3CH_2—SH): ΔH_f° for ethanethiol, 46.15 kJ mol^{-1}; ΔH_f° for •SH, 143.1 kJ mol^{-1}. (To convert energies in kJ mol^{-1} to kcal mol^{-1}, divide by 4.184.)

(b) Using BDEs, show that the reaction of a radical such as *t*Bu—O• (from homolysis of a peroxide) should be a good source of CH_3CH_2—S• radicals.

alkene:	$H_2C{=}CH_2$	$(CH_3)_2CH{=}CH_2$	$(CH_3)_2CH{=}CHCH_3$
relative rate:	1.0	1.4	0.077

FIGURE P10.56

(c) Using BDEs, show that each propagation step of thiol addition to an alkene such as ethylene is exothermic and therefore favorable. (For help with the mechanism, see Study Problem 10.3, Sec. 10.1C.)

(d) Using the mechanism of free-radical thiol addition to guide you, predict the product of the following reaction.

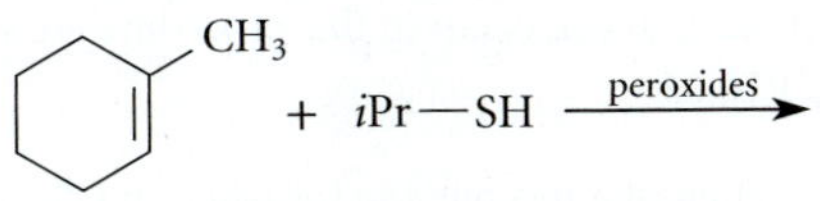

10.58 Although the addition of H—CN to an alkene could be envisioned to occur by a free-radical mechanism, this reaction does *not* occur. Support each of the following reasons using BDE calculations.

(a) The reaction of H—CN with initiating *t*Bu—O• radicals is *not* a good source of •CN radicals.

(b) The second propagation step of the free-radical addition is energetically unfavorable.

$$R\dot{C}HCH_2CN + H—CN \longrightarrow RCH_2CH_2CN + \cdot CN$$

10.59 Provide a free-radical chain mechanism, complete with fishhook notation, for each of the following known transformations.

(a) The initiation step is light-induced homolysis of the I—C bond.

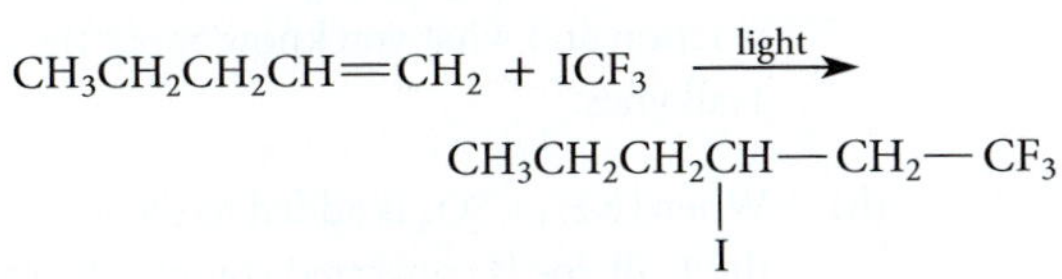

(b)

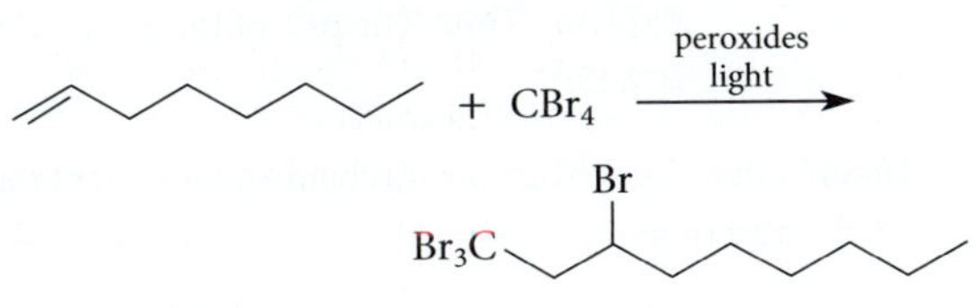

(96% yield)

(c) The initiation step is homolysis of the weakest bond in the starting material.

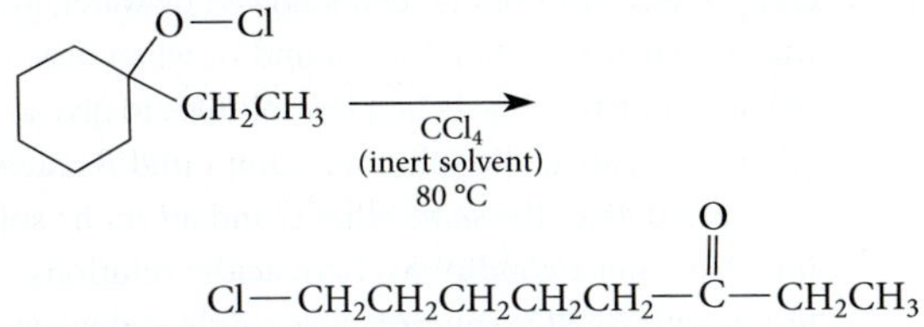

10.60 Knowing that carbocations can rearrange, chemists H. C. Brown and Glenn A. Russell at Purdue University in the early 1950s investigated whether the rearrangement of free radicals might also occur.

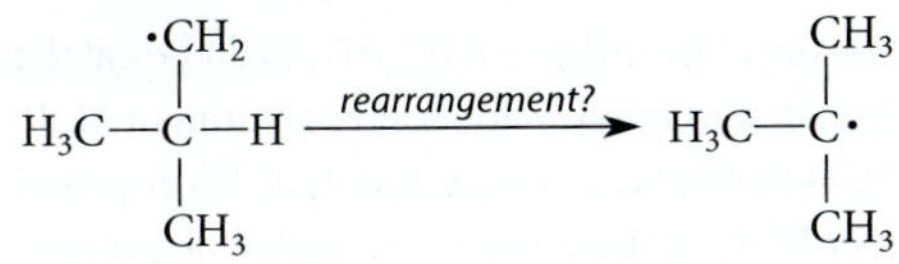

They carried out light-promoted free-radical chlorination of isobutane-2*d*, in which the tertiary hydrogen of isobutane was replaced by deuterium (D = ^{2}H), as shown in **Fig. P10.60.** The ratio of DCl to HCl produced in the reaction was found to be exactly the same as the ratio of *A* to *B*. (Any isotopically substituted *A*, if present, would not be differentiated from *A*.)

(a) This result was cited as evidence that free radicals do *not* rearrange. Explain the logic of this conclusion. (*Hint:* Compare this result with the result that would occur with rearrangement.)

(b) How would the relative percentages of *A* and *B* in this experiment compare with the percentages of *tert*-butyl chloride and isobutyl chloride that would be formed when $(CH_3)_3CH$ is subjected to the same reaction conditions? (*Hint:* See Sec. 9.5D.)

(c) Suggest a method for the preparation of the isotopically labeled starting material, isobutane-2*d*.

10.61 Three alkyl halides, each with the formula $C_7H_{15}Br$, have different boiling points. One of the compounds is optically active. Following reaction with Mg in ether, then with water, each compound gives 2,4-dimethylpentane. After the same reaction with D_2O, a different product is obtained from each compound. Suggest a structure for each of the three alkyl halides.

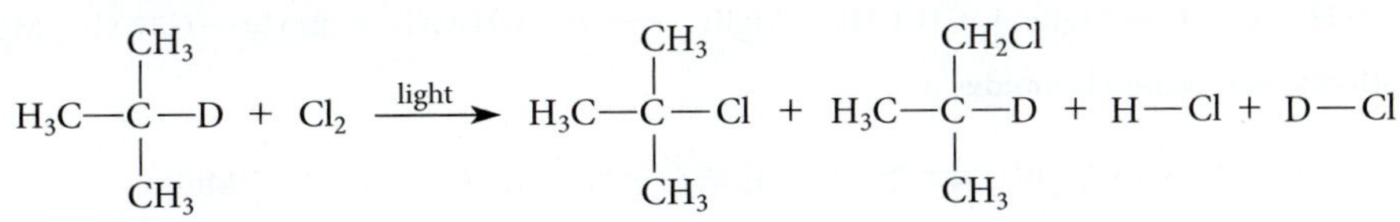

isobutane-2*d* *A* *B*

FIGURE P10.60

10.62 Samples of two liquids, *A* and *B*, are labeled only "isomeric alkyl halides $C_5H_{11}Br$." Reaction of each compound with Mg in ether, followed by water, gives the same hydrocarbon. Compound *A*, when dissolved in warm ethanol, reacts in a few minutes to give an ethyl ether *C* and an acidic solution. Compound *B* reacts more slowly and gives the *same* ether *C* and an acidic solution under the same conditions. Both acidic solutions, when tested with $AgNO_3$ solution, give a light yellow precipitate of AgBr. Reaction of compound *B* with sodium ethoxide in ethanol gives two alkenes, one of which reacts with O_3, then H_2O_2, to give acetone $(CH_3)_2C{=}O$ as one product. Give the structures of compounds *A*, *B*, and *C*, and explain your reasoning.

10.63 An optically active alkyne *A* ($C_{10}H_{14}$) can be catalytically hydrogenated to butylcyclohexane. Treatment of *A* with EtMgBr liberates no gas. Catalytic hydrogenation of *A* over Pd/C in the presence of quinoline poison and treatment of the product with O_3 and then H_2O_2 gives an *optically active* tricarboxylic acid $C_8H_{12}O_6$. (A tricarboxylic acid is a compound with three —CO_2H groups.) Give the structure of *A*, and account for all observations.

10.64 (a) Grignard reagents are normally insoluble in hydrocarbon solvents. However, they can be dissolved in these solvents if a tertiary amine (a compound of the form R_3N:) is added. Explain.

(b) Would addition of a secondary amine ($R_2\ddot{N}$—H) instead of a tertiary amine work just as well? Why or why not?

10.65 You have just been hired by Triple Bond, Inc., a company that specializes in the manufacture of alkynes containing five or fewer carbons. The company CEO, Mr. Al Kyne, needs an outlet for the company's products. You have been asked to develop a synthesis of the housefly sex pheromone, *muscalure*, with the stipulation that all of the carbon in the product must come only from the company's alkynes. The muscalure will subsequently be used in a household fly trap. You will be equipped with a laboratory containing all of the company's alkynes, requisition forms for other reagents, and one gross of fly swatters in case you are successful. Outline a preparation of muscalure that meets the company's needs.

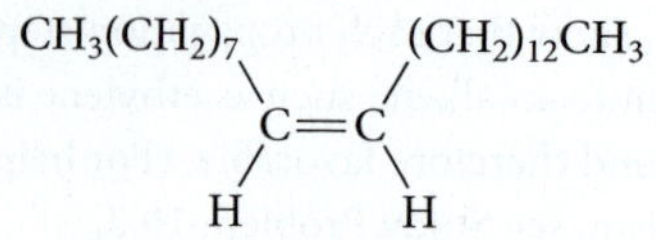

muscalure

10.66 In the preparation of ethynylmagnesium bromide by the reaction of Eq. 10.58, ethylmagnesium bromide is added to a large excess of acetylene in THF solution. Two side reactions that can occur in this procedure are shown in **Fig. P10.66.**

(a) Suggest a mechanism for reaction (1), and explain why an excess of acetylene is important for avoiding this reaction.

(b) Suggest a mechanism for reaction (2), and explain why an excess of acetylene is important for avoiding this reaction.

(c) Tetrahydrofuran (THF) is used as a solvent because the undesired by-product, BrMg—C≡C—MgBr, is relatively soluble in this solvent. Explain why it is important for this by-product to be soluble if both side reactions are to be minimized.

10.67 (a) When the conversion of alkynes into alkenes by Na in liquid ammonia is attempted with a 1-alkyne, every 3 mol of 1-alkyne give only 1 mol of alkene and 2 mol of the acetylenic anion:

$$3\,RC{\equiv}CH \xrightarrow[NH_3\text{ (liq)}]{Na} RCH{=}CH_2 + 2\,RC{\equiv}\bar{C}{:}\ Na^+$$

Explain this result using the mechanism of this reaction and what you know about the acidity of 1-alkynes.

(b) When $(NH_4)_2SO_4$ is added to the reaction mixture, the 1-alkyne is converted completely into the alkene. Explain. (*Hint:* The pK_a of the ammonium ion is 9.25.)

10.68 Give a curved-arrow mechanism for each of the following reactions.

(a) $$Ph{-}C{\equiv}C{-}H \xrightarrow[THF]{NaOD,\ D_2O\text{ (large excess)}} Ph{-}C{\equiv}C{-}D$$

(1) $$H{-}C{\equiv}C{-}MgBr + CH_3CH_2{-}MgBr \longrightarrow CH_3CH_3 + BrMg{-}C{\equiv}C{-}MgBr$$

ethynylmagnesium bromide

(2) $$2\,H{-}C{\equiv}C{-}MgBr \rightleftharpoons H{-}C{\equiv}C{-}H + BrMg{-}C{\equiv}C{-}MgBr$$

ethynylmagnesium bromide

FIGURE P10.66

(b) $HCCl_3 + Na^+ I^- \xrightarrow[35\ °C]{NaOH/H_2O} HCCl_2I + Na^+ Cl^-$

(c) 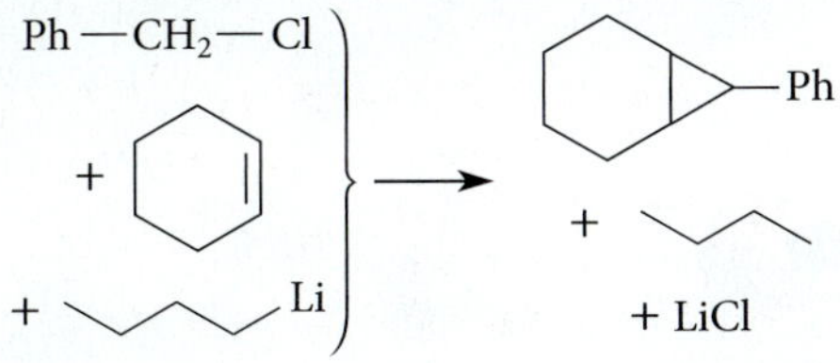

10.69 One method to form carbenes is by the photolysis (light-induced decomposition) of *diazo compounds*:

a diazo compound a carbene

The very high-energy carbene can be formed because of the input of energy from light and because of the formation of the very stable dinitrogen molecule by-product.

(a) Propose a structure for the product *X* of the following reaction, and explain your reasoning.

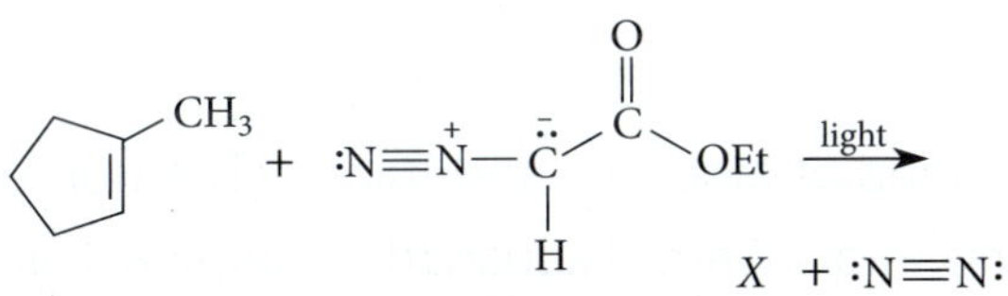

ethyl diazoacetate

(b) Organic azides decompose on photolysis to reactive intermediates called *nitrenes*, which are the nitrogen analogs of carbenes. Using the reaction in part (a) as an analogy, propose a structure of the product *Y* in the following reaction. Also give the structure of the nitrene intermediate that results from photolysis of the azide.

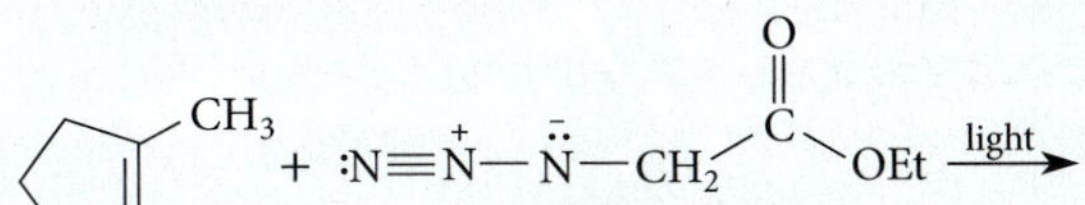

ethyl azidoacetate

Y + :N≡N:

10.70 Explain why only one stereoisomer is formed in the following Simmons–Smith reaction. (*Hint 1:* The zinc in a Simmons-Smith reagent is a Lewis acid. *Hint 2:* Draw the conformation of the starting material; use a model if necessary.)

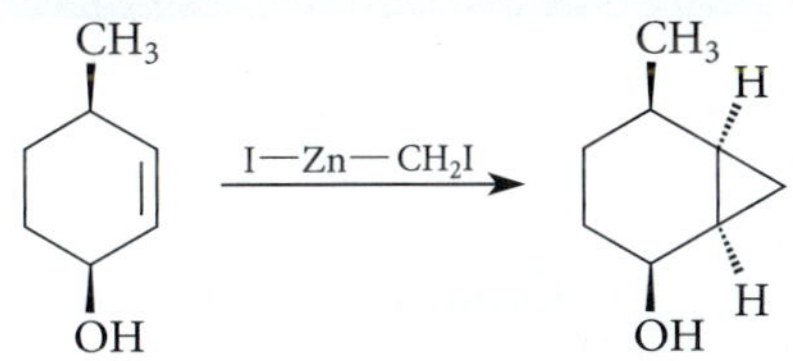

(only stereoisomer formed)

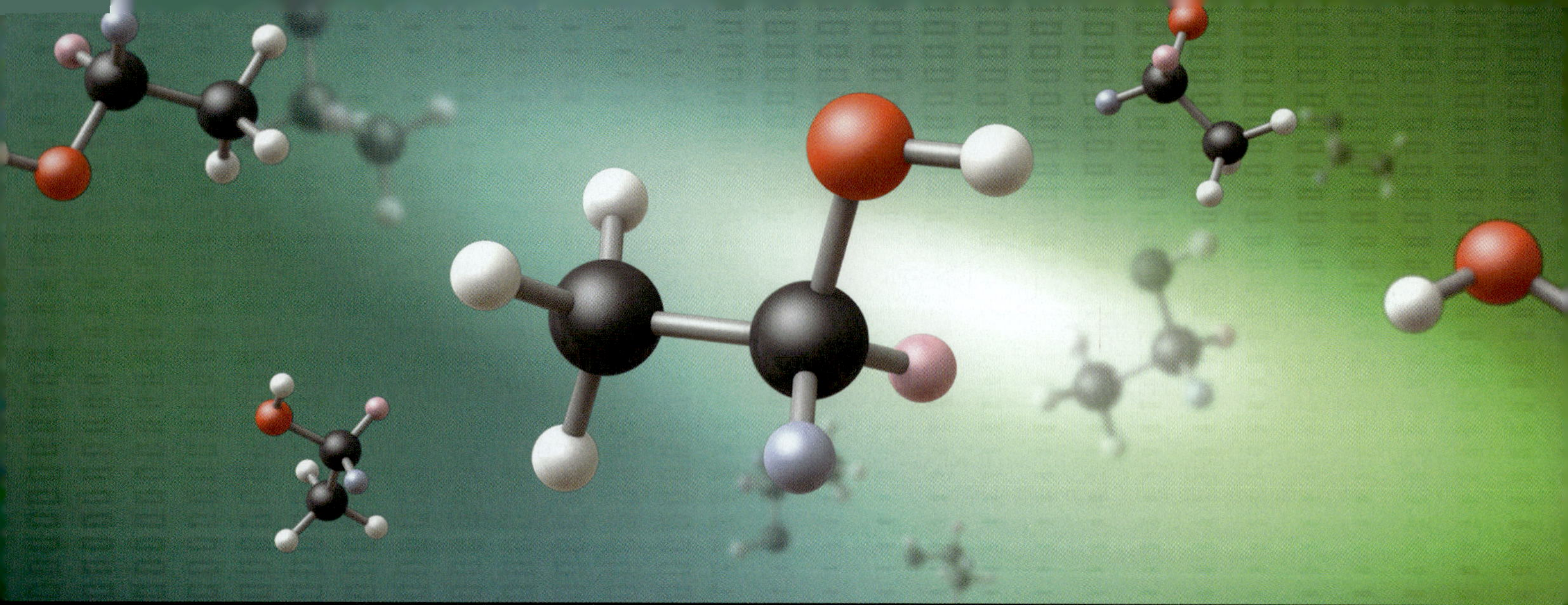

11 THE CHEMISTRY OF ALCOHOLS AND THIOLS

Chapter 11 focuses on the reactions of alcohols and thiols. (The nomenclature and classification of alcohols and thiols were discussed in Secs. 8.1 and 8.2B.) We begin with some of the simplest but most important reactions of alcohols and thiols: their Brønsted acid–base reactions. Next we consider some reactions that alcohols and thiols have in common with alkyl halides: substitution and elimination reactions. Then we study oxidation reactions of alcohols, which have no simple parallel in alkyl halide chemistry. We show how to recognize oxidations and reductions and how the oxidations of alcohols and thiols are carried out in the laboratory. A consideration of alcohol oxidation in nature leads to a discussion of our final topic in stereochemistry: the stereochemical relationships of groups within molecules. Finally, the use of alcohol reactions in organic synthesis provides more practice in retrosynthetic analysis.

11.1 ALCOHOLS AND THIOLS AS BRØNSTED ACIDS AND BASES

A. Acidity of Alcohols and Thiols

Alcohols and thiols are weak acids. Alcohols are somewhat weaker acids than water for reasons that we discuss in Sec. 11.1D.

	CH_3CH_2–O–H	H–O–H	
pK_a	15.9	14.0	(11.1)

For example, we've already encountered alkoxide bases as nucleophiles in S_N2 reactions and as bases in E2 reactions (Secs. 9.1A and 9.1B). The conjugate bases of alcohols are generally called *alkoxides*.

The common name of an alkoxide is constructed by deleting the final *yl* from the name of the alkyl group and adding the suffix *oxide*. In substitutive nomenclature, the suffix *ate* is simply added to the name of the alcohol.

$CH_3CH_2\ddot{\underset{..}{O}}{:}^-\ Na^+$ *common:* **sodium ethoxide** *substitutive:* **sodium ethanolate** (11.2)

Thiols, although weak acids, are much more acidic than alcohols, just as H_2S is a stronger acid than H_2O:

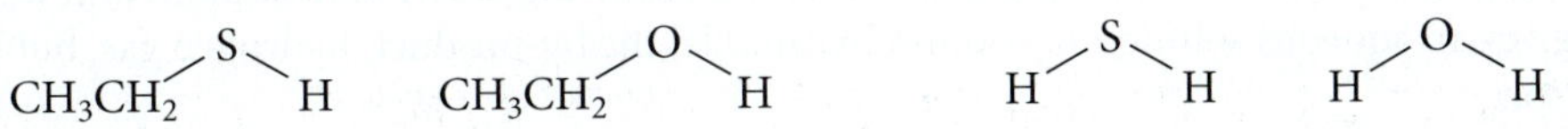

pK_a	10.5	15.9	7.0	14.0	(11.3)

The relative acidities of alcohols and thiols are a reflection of the *element effect* (Sec. 3.7B).

The conjugate bases of thiols are called *mercaptides* in common nomenclature and *thiolates* in substitutive nomenclature.

$CH_3\ddot{\underset{..}{S}}{:}^-\ Na^+$ *common:* **sodium methyl mercaptide** *substitutive:* **sodium methanethiolate** (11.4)

Focused Problems

11.1 Give the structure of each of the following compounds.

(a) Sodium isopropoxide (b) Potassium *tert*-butoxide

(c) Magnesium 2,2-dimethyl-1-butanolate (d) Lithium 1-butanethiolate

11.2 Name the following compounds.

(a) $Ca(OCH_3)_2$ (b) $CuSCH_2CH_3$

B. Formation of Alkoxides and Mercaptides

Because the pK_a of a typical alcohol is somewhat greater than that of water, the equilibrium for the reaction of hydroxide ion with an alcohol such as ethanol lies to the left.

$$\underset{\text{weaker acid}}{CH_3CH_2\ddot{O}{-}H} + {:}\ddot{O}H^- \rightleftharpoons CH_3CH_2\ddot{O}{:}^- + \underset{\text{stronger acid}}{H_2\ddot{O}} \qquad (11.5)$$

pK_a = 15.9 pK_a = 14.0

From Sec. 3.6C, the equilibrium constant for this reaction is $K_{eq} = 10^{(14-15.9)} = 10^{-1.9} = 0.013$. In a solution of sodium hydroxide in ethanol, in which the alcohol solvent is in large excess, the reaction is pushed to the right (Le Chatelier's principle), and both hydroxide and ethoxide are present.

Alkoxides can be formed irreversibly from alcohols with stronger bases. One convenient base used for this purpose is sodium hydride, NaH, which is a source of the *hydride ion*, $H{:}^-$. Hydride ion is a very strong base; the pK_a of its conjugate acid, H_2, is about 37. Therefore, its reactions

Potassium hydride and sodium hydride are supplied as dispersions in mineral oil to protect them from reaction with moisture. When these compounds are used to convert an alcohol into an alkoxide, the mineral oil is rinsed away with pentane, a solvent such as diethyl ether or tetrahydrofuran (THF) is added, and the alcohol is introduced cautiously with stirring. Hydrogen is evolved vigorously, and a solution or suspension of the pure potassium or sodium alkoxide is formed.

with alcohols go essentially to completion. In addition, when NaH reacts with an alcohol, the reaction cannot be reversed because the by-product, hydrogen gas, simply bubbles out of the solution.

2-butanol **sodium 2-butanoate** (quantitative yield) (11.6)

Solutions of alkoxides in their conjugate-acid alcohols (for example, Na^+ EtO^- in EtOH) find wide use in organic chemistry. Although these solutions can also be prepared with sodium hydride, it is more convenient to take a cue from the following reaction of sodium and water, which gives an aqueous solution of sodium hydroxide; the by-product, hydrogen gas, bubbles out of solution.

$$2\,H{-}\ddot{\underset{..}{O}}H + 2\,Na \longrightarrow 2\,Na^+ \;{:}\ddot{\underset{..}{O}}H^- + H_2 \qquad (11.7a)$$

The analogous reaction occurs with many alcohols. For example, sodium metal reacts with an alcohol to afford a solution of the corresponding sodium alkoxide:

$$2\,H{-}\ddot{\underset{..}{O}}R + 2\,Na \longrightarrow 2\,Na^+ \;{:}\ddot{\underset{..}{O}}R^- + H_2 \qquad (11.7b)$$

sodium alkoxide

The rate of this reaction depends strongly on the alcohol. The reactions of sodium with anhydrous (water-free) ethanol and methanol are vigorous, but not violent. However, the reactions of sodium with some alcohols, such as *tert*-butyl alcohol, are rather slow. The alkoxides of such alcohols can be formed more rapidly with the more reactive potassium metal. This is why you will often see potassium *t*-butoxide (K^+ $^-$O*t*Bu) used as a base.

Because thiols are much more acidic than water or alcohols, they, unlike alcohols, can be converted completely into their conjugate-base mercaptide anions by reaction with one equivalent of hydroxide or alkoxide. In fact, a common method of forming alkali-metal mercaptides is to dissolve them in ethanol containing one equivalent of sodium ethoxide:

$$CH_3CH_2\ddot{\underset{..}{S}}H + CH_3CH_2\ddot{\underset{..}{O}}{:}^- \rightleftharpoons CH_3CH_2\ddot{\underset{..}{S}}{:}^- + CH_3CH_2\ddot{\underset{..}{O}}H \qquad (11.8)$$

ethanethiol $pK_a = 10.5$ **ethoxide ion** **ethanethiolate ion** **ethanol** $pK_a = 15.9$

Because the equilibrium constant for this reaction is $> 10^5$ (Sec. 3.6C), the reaction goes essentially to completion.

Although alkali-metal mercaptides are soluble in water and alcohols, thiols form insoluble mercaptides with many heavy-metal ions, such as Hg^{2+}, Cu^{2+}, and Pb^{2+}.

$$2\,CH_3(CH_2)_9S{-}H + PbCl_2 \xrightarrow[EtOH]{} [CH_3(CH_2)_9S]_2Pb + 2\,HCl \qquad (11.9)$$

decanethiol **lead(II) decanethiolate** (87% yield)

$$2\,PhS{-}H + HgCl_2 \xrightarrow[EtOH]{} (PhS)_2Hg + 2\,HCl \qquad (11.10)$$

(98% yield)

The insolubility of heavy-metal mercaptides is analogous to the insolubility of heavy-metal sulfides [for example, lead(II) sulfide, PbS], which are among the most insoluble inorganic compounds known. One reason for the toxicity of lead salts is that the lead forms strong (stable) mercaptide complexes with the thiol groups of important biomolecules.

Curing a Disease with Mercaptides

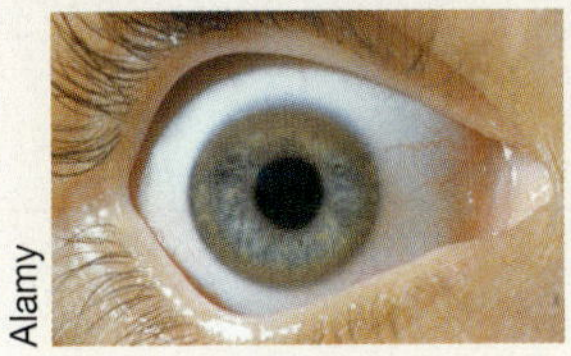

A relatively rare inherited disease of copper metabolism, Wilson's disease, can be treated by using the tendency of thiols to form complexes with copper ions. Accumulation of toxic levels of copper in the brain and liver causes the disease, which can be diagnosed by a brown ring around the iris of the eye (photo). Penicillamine is administered to form a complex with the Cu^{2+} ions:

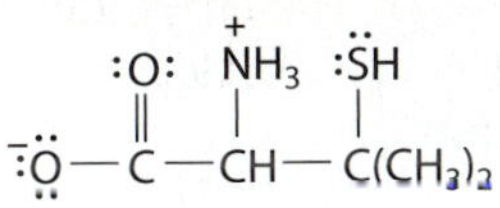

penicillamine

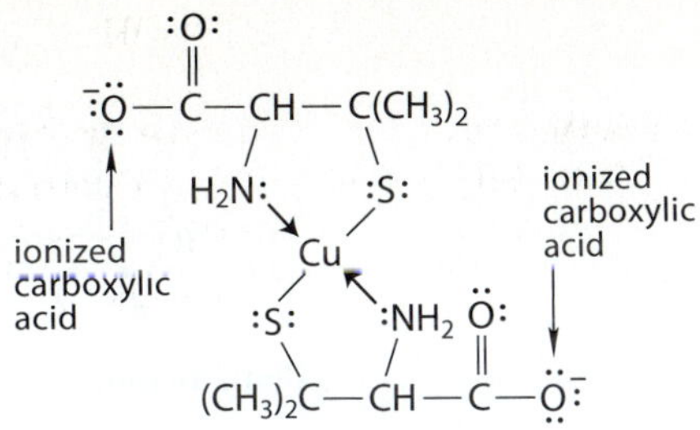

complex of two penicillamine molecules with Cu^{2+}

(The arrows ⟶ in the structure on the right symbolize Lewis acid–Lewis base interactions between the H_2N— groups and the copper.) The penicillamine–copper complex, unlike ordinary cupric thiolates, is relatively soluble in water because of the ionized carboxylic acid groups (Sec. 8.6F), and its solubility allows it to be excreted by the kidneys.

C. Polar Effects on Alcohol Acidity

Substituted alcohols and thiols show the same type of polar effect on acidity as substituted carboxylic acids do (Sec. 3.7C). For example, alcohols containing electronegative substituent groups have enhanced acidity. For example, 2,2,2-trifluoroethanol is over three pK_a units more acidic than ethanol itself.

Relative acidity:

$$\underset{15.9}{H_3C—CH_2—OH} < \underset{12.4}{F_3C—CH_2—OH} \quad pK_a \qquad (11.11)$$

As with carboxylic acids, the polar effects of electronegative groups are more important when the groups are closer to the OH group:

Relative acidity:

$$\underset{15.4}{F_3C—CH_2—CH_2—CH_2—OH} < \underset{14.6}{F_3C—CH_2—CH_2—OH} < \underset{12.4}{F_3C—CH_2—OH} \quad pK_a \qquad (11.12)$$

When the fluorines are separated from the OH group by four or more carbons, they have a negligible effect on acidity.

Focused Problem

11.3 In each of the following sets, arrange the compounds in order of increasing acidity (decreasing pK_a). Explain your choices.

(a) $ClCH_2CH_2OH$, Cl_2CHCH_2OH, $Cl(CH_2)_3OH$

(b) $ClCH_2CH_2SH$, $ClCH_2CH_2OH$, CH_3CH_2OH

(c) $CH_3CH_2CH_2CH_2OH$, $CH_3OCH_2CH_2OH$

TABLE 11.1 Acidities of Alcohols in Aqueous Solution*

Alcohol	pK_a	Alcohol	pK_a
CH_3OH	15.5	$(CH_3)_2CHOH$	17.1
CH_3CH_2OH	15.9	$(CH_3)_3COH$	19.1

*The pK_a of water is 14.0.

D. Role of the Solvent in Alcohol Acidity

Primary, secondary, and tertiary alcohols differ significantly in acidity; some relevant pK_a values are listed in **Table 11.1**. The data in this table show that the acidities of alcohols in aqueous solution are in the order methyl > primary > secondary > tertiary. The reason for this acidity order has to do with solvation of the conjugate–base anion by hydrogen bonds to solvent water.

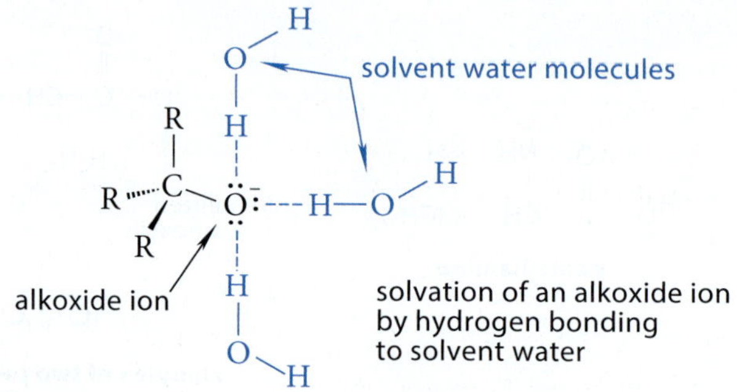

(11.13)

Stronger solvation (that is, stronger hydrogen bonding) stabilizes the conjugate-base alkoxide anion and contributes to increased acidity of the alcohol. Large branched alkyl groups interfere with this solvation, decrease the stability of the alkoxide, and therefore decrease the acidity of the alcohol.

Solvation is also the reason that water is more acidic than alcohols (Sec. 11.1A): the hydroxide ion has stronger solvation interactions with solvent water than the alcohols.

The experimental evidence that the acidity order of alcohols in solution is a solvent effect is that the relative acidities of alcohols in the gas phase (in which there is no solvent) have the opposite order—that is, tertiary and secondary alcohols are *more* acidic than primary alcohols in the gas phase.

The pK_a-increasing effect of the alkyl substituents of alcohols in solution is *not* either a simple polar effect or a steric effect caused by larger alkyl groups. (See Further Exploration 11.1 in the Study Guide and Solutions Manual for a structural description of this effect.) An important point of this discussion is that the solvent is not an idle bystander in the acid–base reaction; rather, it takes an active role in stabilizing the molecules involved, especially the charged species.

FURTHER EXPLORATION 11.1
Solvation of Tertiary Alkoxides

To summarize: *tertiary alkoxides are more basic in solution than primary alkoxides.* An equivalent statement is that *primary alcohols are more acidic in solution than tertiary alcohols.*

E. Basicity of Alcohols and Thiols

Just as water can accept a proton to form the hydronium ion, alcohols and thiols can also be protonated to form positively charged conjugate acids. Alcohols are slightly less basic than water; thiols, however, are *much* less basic.

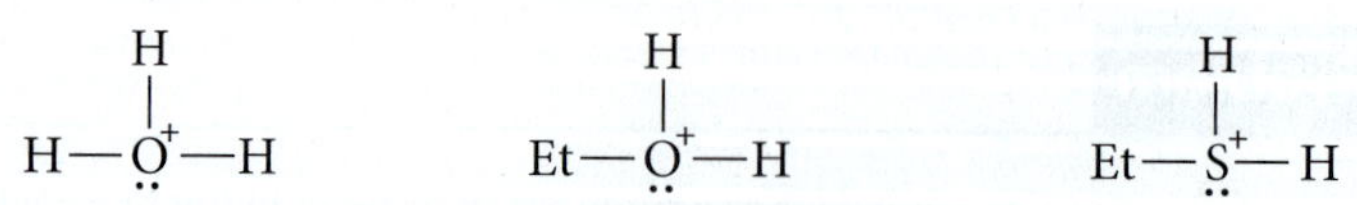

	hydronium ion	**conjugate acid of ethanol**	**conjugate acid of ethanethiol**
pK_a	0	−2 to −3	−5 to −7

(11.14)

The negative pK_a values, which reflect the *charge effect* on pK_a (Sec. 3.7A), mean that these protonated species are very strong acids, and that their neutral conjugate bases are rather weak. Nevertheless, the ability of alcohols and thiols to accept a proton plays an important role in many of their reactions, particularly those that take place in acidic solutions.

Water is a stronger base than ethanol because its conjugate acid, hydronium ion, is stabilized by donation of *three* hydrogen bonds to solvent water, whereas the conjugate acid of ethanol can only donate two.

The pK_a values in Display 11.14 refer in each case to the conjugate acid of the *neutral* base. Alcohols and thiols, like water, are *amphoteric* substances—that is, they can either gain or lose a proton. Consequently, two acid–base equilibria are associated with an alcohol:

Loss of proton:

(11.15a)

Gain of a proton:

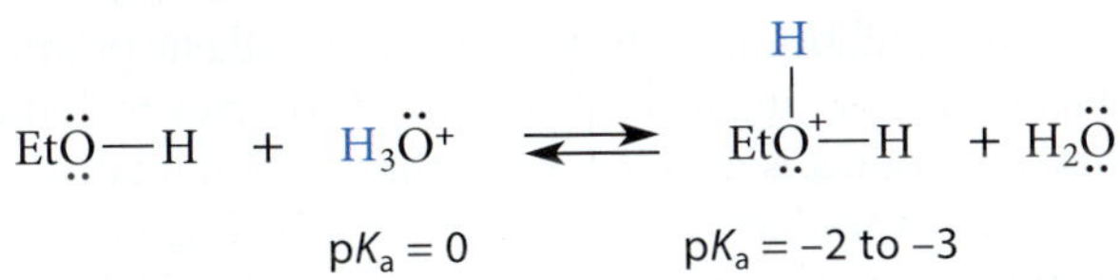

(11.15b)

The acidity of an alcohol—the loss of a proton—is exemplified by the reaction in Eq. 11.15a. Because alcohols are weak acids, this reaction occurs usually only in the presence of strong bases. The basicity of alcohols—the gain of a proton—is exemplified by the reaction in Eq. 11.15b. Because alcohols are weak bases, this reaction usually occurs significantly only in the presence of strong acids.

Thiols, too, are amphoteric: they can also act as both acids and bases. As explained in Sec. 11.1A, thiols are much more acidic than alcohols because of the *element effect* (Sec. 3.7B). The *conjugate acids* of thiols are also more acidic than the *conjugate acids* of alcohols for the same reason. In other words, *thiols are less basic than alcohols.*

Loss of a proton:

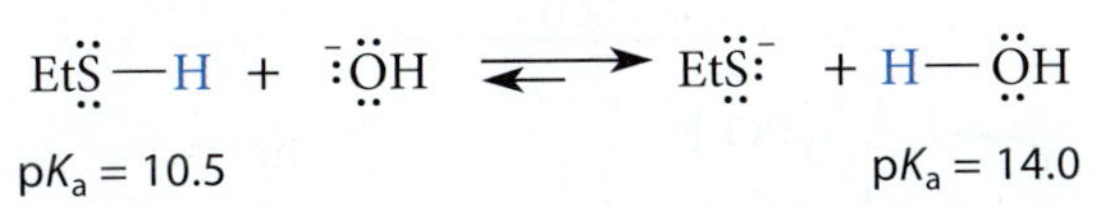

(11.16a)

Gain of a proton:

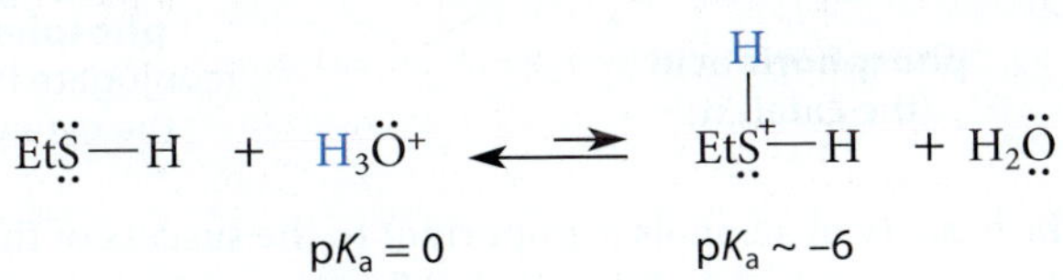

(11.16b)

Focused Problem

11.4 A small amount of HCl gas was added to a solution of 1-propanethiol in 1-propanol. After equilibrium is established, what is the major acidic species in solution? Explain.

11.2 DEHYDRATION OF ALCOHOLS

Concentrated strong acids such as H_2SO_4 and H_3PO_4 catalyze a β-elimination reaction in which water is lost from a secondary or tertiary alcohol to give an alkene. The conversion of cyclohexanol into cyclohexene is typical:

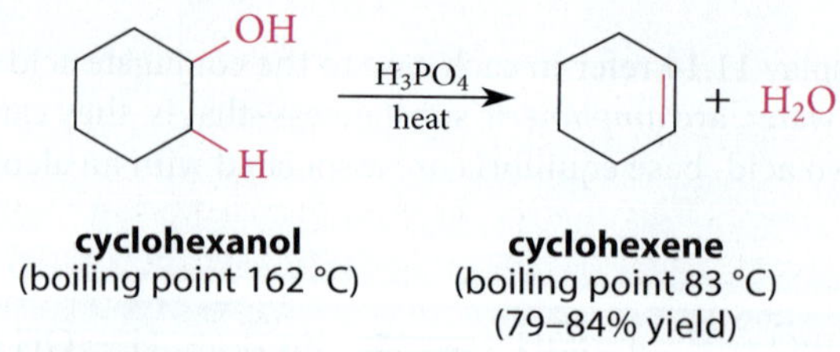

(11.17)

A reaction such as this, in which the elements of water are lost from the starting material, is called a **dehydration**. In Eq. 11.17, cyclohexanol is *dehydrated* to cyclohexene.

Most acid-catalyzed dehydrations of alcohols are reversible reactions. However, these reactions can easily be driven toward the alkene products by applying Le Chatelier's principle (Sec. 4.10C). In Eq. 11.17, for example, the equilibrium is driven toward the alkene product because the water produced as a by-product forms a strong hydrogen-bonded complex with the catalyzing acid H_3PO_4, and the cyclohexene product is distilled out of the reaction mixture. (Alkenes can be removed by distillation because they have considerably lower boiling points than alcohols with the same carbon skeleton; for example, see the boiling points in Eq. 11.17.) The dehydration of alcohols to alkenes is easily carried out in the laboratory and is an important procedure for the preparation of some alkenes.

The role of the acid catalyst in dehydration is to convert the —OH group, a poor leaving group, into the $—\overset{+}{O}H_2$ group, a good leaving group (because H_2O is a weak base). *The use of Brønsted or Lewis acids to activate the —OH group as a leaving is a central theme in alcohol chemistry.*

Alcohol dehydration occurs by a three-step mechanism that consists entirely of acid–base reactions and involves a carbocation intermediate. In the first mechanistic step, the —OH is converted into a leaving group by acting as a Brønsted base (Sec. 11.1E) and accepting a proton from the catalyzing acid:

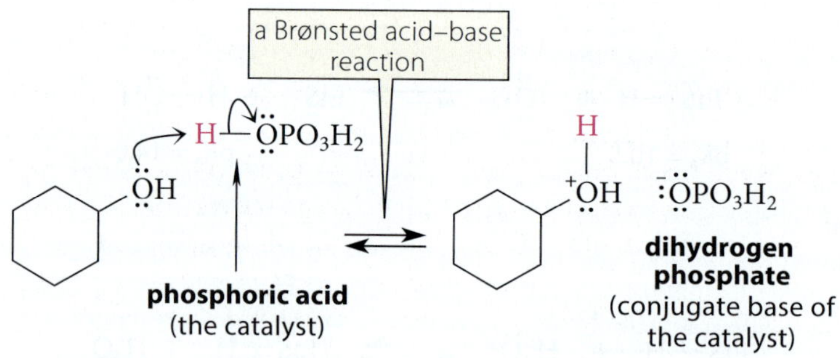

(11.18a)

This step shows why the basicity of alcohols is important to the success of the dehydration reaction. Next, the carbon–oxygen bond of the alcohol breaks in a Lewis acid–base dissociation to give water and a carbocation:

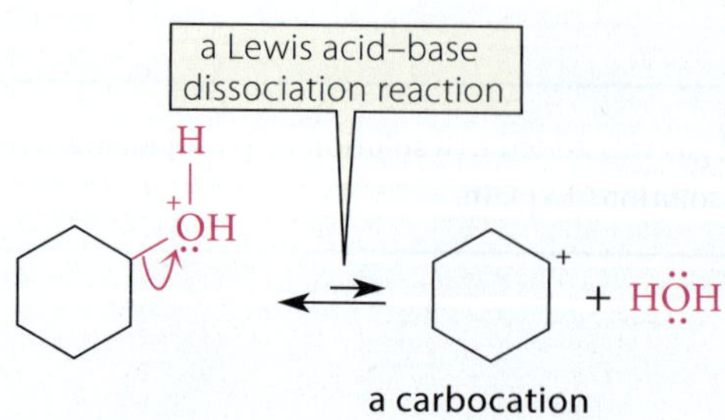

(11.18b)

Finally, the conjugate base $^-OPO_3H_2$ of the catalyzing acid removes a β-proton from the carbocation in another Brønsted acid–base reaction:

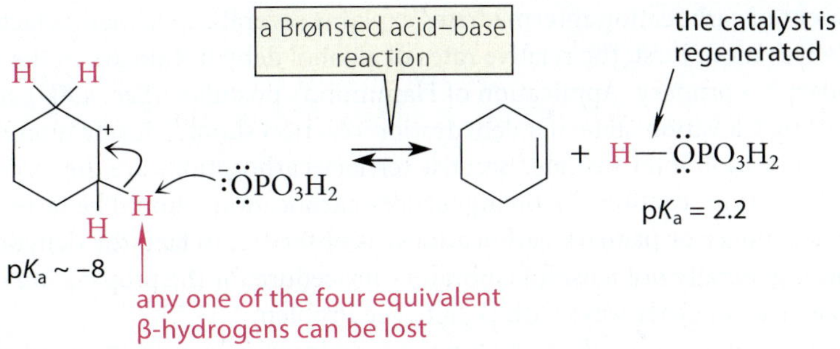

(11.18c)

This step generates the alkene product and regenerates the catalyzing acid H_3PO_4. Alternatively, the H_2O by-product generated in Eq. 11.18b or the alcohol starting material can serve as the base that removes a β-proton from the carbocation. The H_3O^+ formed in this reaction can also serve as an acid catalyst in the dehydration.

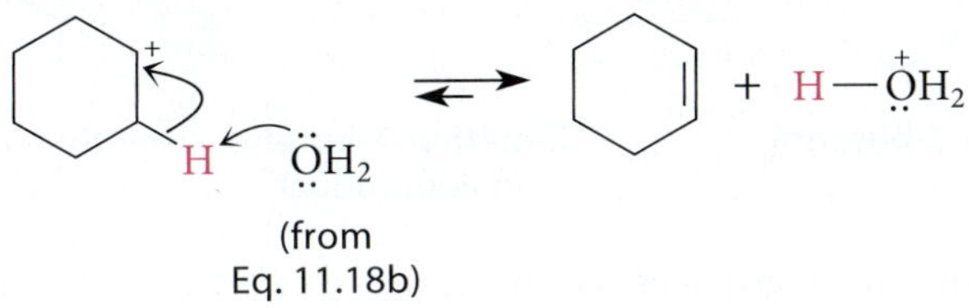

(11.18d)

It is water, not hydroxide ion, that is used as a base under acidic conditions. (See Sec. 4.10C.)

We've seen a mechanism like this twice before. First, the dehydration of alcohols is the reverse of the hydration of alkenes (Sec. 4.10C). *Hydration of alkenes and dehydration of alcohols are the forward and reverse of the same reaction.* Recall from the *principle of microscopic reversibility* (Sec. 4.10C) that the forward and reverse of the same reaction must have the same intermediates and the same rate-limiting transition states. Therefore, because protonation of the alkene is the rate-limiting step in alkene hydration, the reverse of this step—loss of the proton from the carbocation intermediate (Eq. 11.18c or 11.18d)—is rate-limiting in alcohol dehydration. This principle also requires that if a catalyst accelerates a reaction in one direction, it also accelerates the reaction in the reverse direction. Therefore, both the hydration of alkenes to alcohols and the dehydration of alcohols to alkenes are catalyzed by acids.

Second, alcohol dehydration is essentially an E1 reaction (Sec. 9.6). Once the —OH group of the alcohol is protonated, it becomes a very good leaving group (water). Like a halide leaving group in the E1 reaction, the protonated —OH departs to give a carbocation, which then loses a β-proton to give an alkene. In alcohol dehydration, however, the rate-determining and product-determining step is the same: loss of the β-proton from the carbocation.

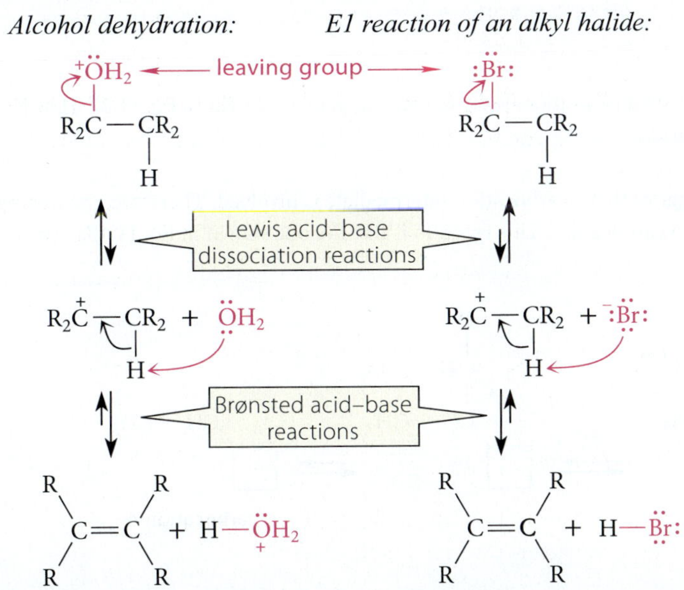

(11.19)

The involvement of carbocation intermediates explains several experimental facts about alcohol dehydration. First, the relative rates of alcohol dehydration are in the order tertiary > secondary >> primary. Application of Hammond's postulate (Sec. 4.8D) suggests that the rate-limiting transition state of a dehydration reaction should closely resemble the corresponding carbocation intermediate. Because tertiary carbocations are the most stable carbocations, dehydration reactions involving tertiary carbocations should be faster than those involving either secondary or primary carbocations, as observed. In fact, the dehydration of primary alcohols is generally not a useful laboratory procedure for the preparation of alkenes. (Primary alcohols react in other ways with H_2SO_4; see Problem 11.77.)

Second, if the alcohol has more than one type of β-hydrogen, then a mixture of alkene products can be expected. As in the E1 reaction of alkyl halides, the most stable alkene—the one with the greatest number of branches at the double bond—is the alkene formed in greatest amount:

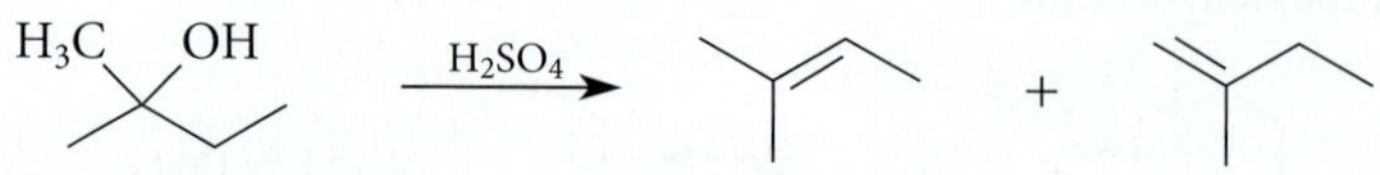

2-methyl-2-butanol **2-methyl-2-butene** major product **2-methyl-1-butene** minor product (11.20)

Finally, alcohols that react to give rearrangement-prone carbocation intermediates yield rearranged alkenes:

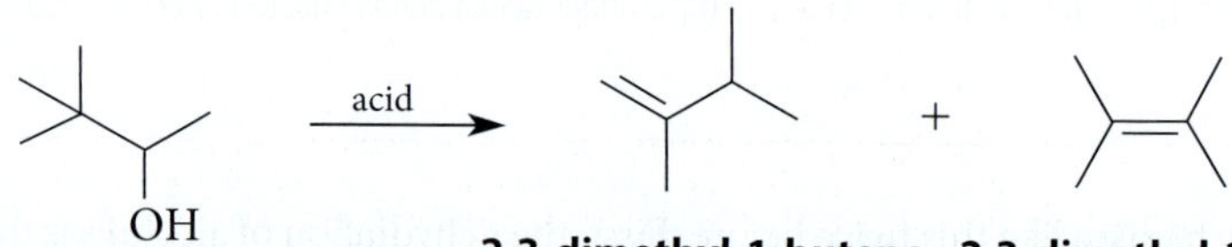

3,3-dimethyl-2-butanol **2,3-dimethyl-1-butene** (29%) **2,3-dimethyl-2-butene** (71%) (11.21)

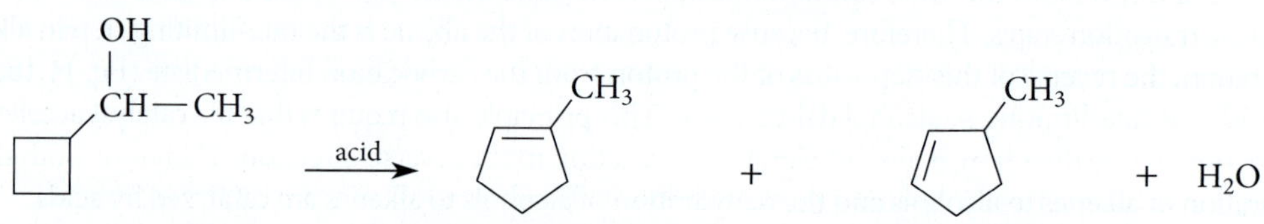

1-cyclobutylethanol **1-methylcyclopentene** **3-methylcyclopentene** (minor product) (11.22)

The mechanism of the rearrangement in Eq. 11.22 is worked through in Study Problem 11.1.

Study Problem 11.1

Provide a curved-arrow mechanism and a rationale for the rearrangement shown in Eq. 11.22. (Use H—A as a general abbreviation for the catalyzing acid.)

Solution A rearrangement suggests that a carbocation intermediate is involved. Therefore, the first part of the mechanism is the generation of a carbocation. This reaction is like the one shown in Eq. 11.18a. The resulting carbocation *A* is secondary.

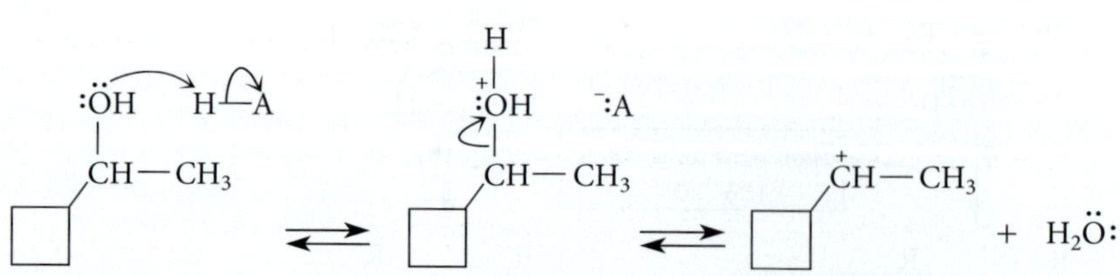

carbocation *A*

The rearrangement involves a shift of one of the *ring carbons*; this shift enlarges the ring. We have drawn out and distorted the ring structure to show this shift more clearly. (A ring carbon is just another alkyl group that happens to be "tied back.")

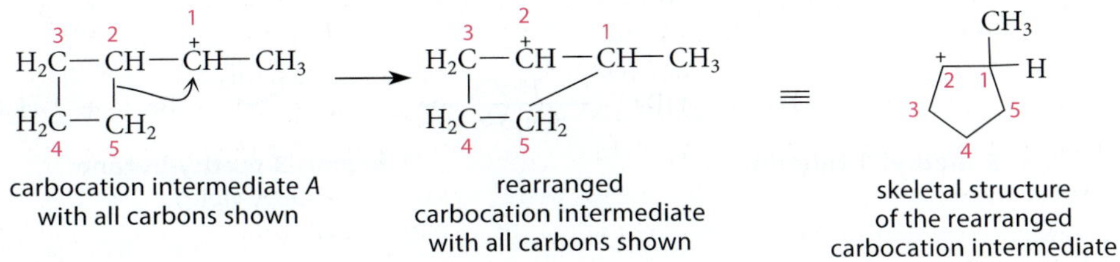

This rearrangement is analogous to the expansion of a loop in a string:

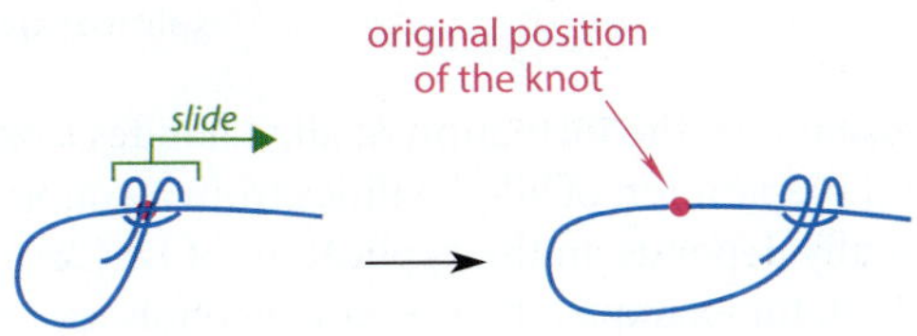

Why should a secondary carbocation rearrange to another secondary carbocation? The four-membered ring is *strained* (Sec. 7.5B), and the expansion to a five-membered ring relieves some of its ring strain. This secondary carbocation can lose a β-proton to give the major alkene product alkene and regenerate the catalyzing acid:

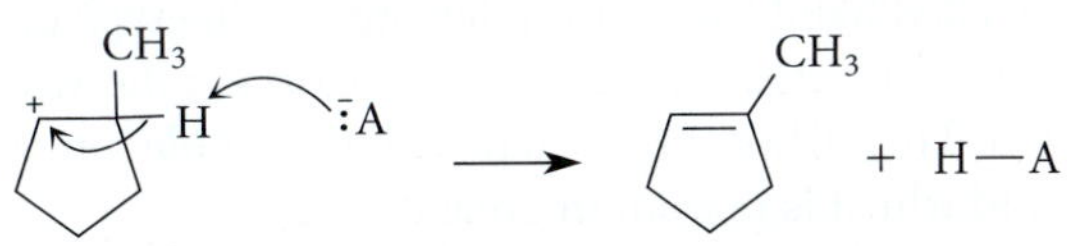

(You should show how the minor alkene product, 3-methylcyclopentene, is formed.) The secondary carbocation also can rearrange by a hydride shift to a tertiary carbocation, which can also lose a β-proton to give the major product.

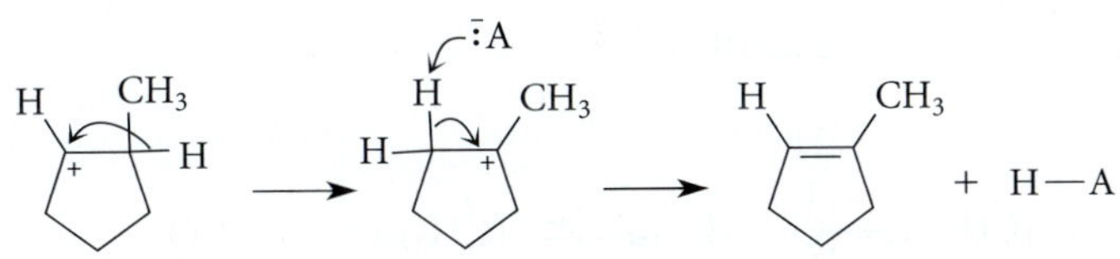

Focused Problems

11.5 What alkene(s) are formed in the acid-catalyzed dehydration of each of the following alcohols?

(a) 3-Methyl-3-heptanol

(b) PhCH(OH)CH_2Ph (OH on the central carbon of $PhCHCH_2Ph$)

11.6 Write the curved-arrow mechanism for the reaction in Focused Problem 11.5a. (Abbreviate the catalyzing acid as H—A.) In each step, identify all Brønsted acids and bases, all electrophiles and nucleophiles, and all leaving groups.

11.7 Identify the *major* alkene product(s) in part (a) of Focused Problem 11.5.

11.8 Draw the structure of the carbocation intermediate involved in the acid-catalyzed dehydration of 3-ethyl-3-pentanol.

11.9 Draw the structures of two alcohols, one secondary and one tertiary, that could give each of the following alkenes as a major acid-catalyzed dehydration product. In each case, which alcohol would dehydrate most rapidly?

(a) 1-Methylcyclohexene (b) 3-Methyl-2-pentene

11.10 Write a curved-arrow mechanism for the reaction in Eq. 11.21.

11.3 REACTIONS OF ALCOHOLS WITH HYDROGEN HALIDES

Alcohols react with hydrogen halides to give alkyl halides:

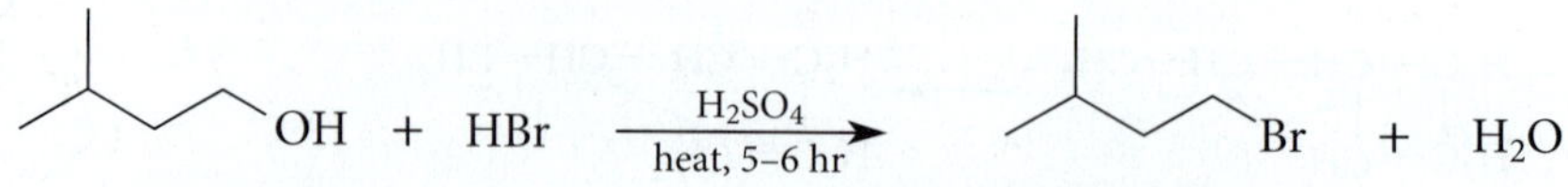

3-methyl-1-butanol **1-bromo-3-methylbutane** (93% yield) (11.23)

$$(CH_3)_3C{-}OH + HCl \xrightarrow[25\ °C,\ 20\ min]{H_2O} (CH_3)_3C{-}Cl + H_2O \qquad (11.24)$$

***tert*-butyl alcohol** ***tert*-butyl chloride** (almost quantitative)

The equilibrium constant for the formation of alkyl halides from alcohols is not large; therefore, the successful preparation of alkyl halides from alcohols, like the dehydration of alcohols to alkenes, usually depends on the application of Le Chatelier's principle (Sec. 4.10C). In both Eqs. 11.23 and 11.24, for example, the reactant alcohols are soluble in the reaction solvent, which is an aqueous acid, but the product alkyl halides are not. In each case, separation of the alkyl halide product from the reaction mixture as a water-insoluble layer drives the reaction to completion. For alcohols that are not water-soluble, a large excess of gaseous HBr, one of the reactants, can be used to drive the reaction to completion.

The mechanism of alkyl halide formation depends on the type of alcohol used as the starting material. In the reactions of tertiary alcohols, protonation of the alcohol oxygen is followed by carbocation formation. The carbocation reacts with the halide ion, which is formed by ionization of strong acid HCl, and which is present in great excess:

(excess) (11.25a)

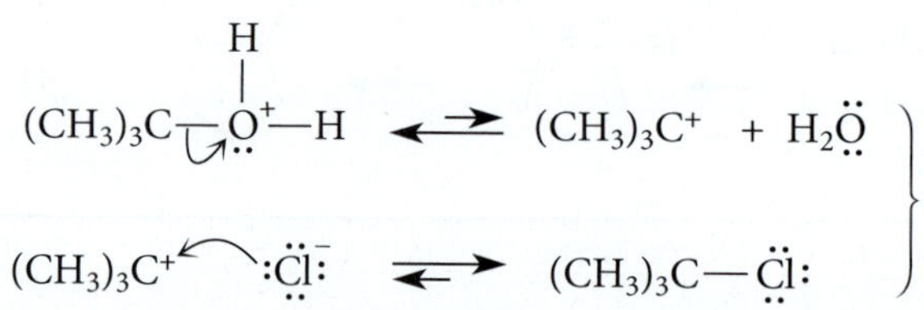

S_N1 reaction (11.25b) (11.25c)

Once the alcohol is protonated, the reaction is an S_N1 reaction with H_2O as the leaving group.

When a primary alcohol is the starting material, the reaction occurs as a concerted displacement of water from the protonated alcohol by halide ion. In other words, *it is an S_N2 reaction in which water is the leaving group.*

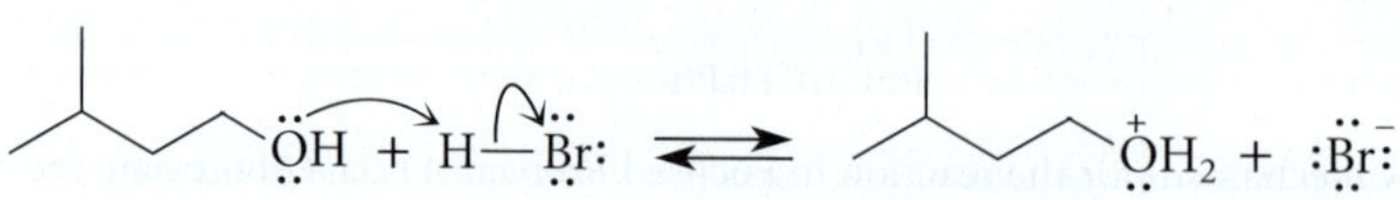

(excess) (11.26a)

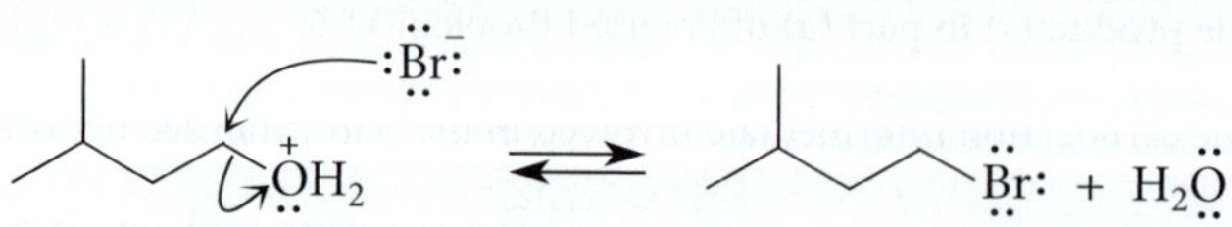

S_N2 reaction (11.26b)

Notice that the initial step in both the S_N1 and the S_N2 mechanism is protonation of the OH group.

As the different conditions of Eqs. 11.23 and 11.24 illustrate, the reactions of tertiary alcohols with hydrogen halides are much faster than the reactions of primary alcohols. Typically, tertiary alcohols react with hydrogen halides rapidly at room temperature, whereas the reactions of primary alcohols require heating for several hours. The reactions of primary alcohols with HBr and HI are satisfactory, but their reactions with HCl are very slow. Although reactions of alcohols with HCl can be accelerated with certain catalysts, other methods for preparing primary alkyl chlorides (discussed in the following section) are better.

The reactions of secondary alcohols with hydrogen halides tend to occur by the S_N1 mechanism. This means that carbocations are involved as reactive intermediates; consequently, rearrangements can occur, as in the following example:

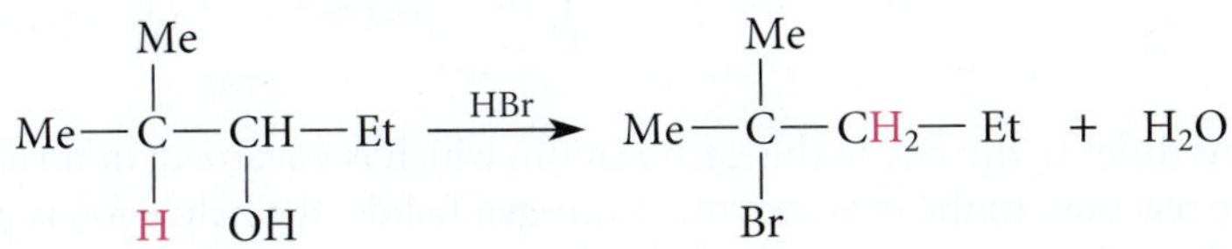

2-methyl-3-pentanol **2-bromo-2-methylpentane** (11.27)

Focused Problems

11.11 Suggest an alcohol starting material and the conditions for the preparation of each of the following alkyl halides.

(a) 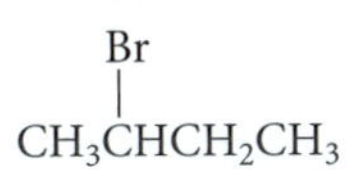

(b) 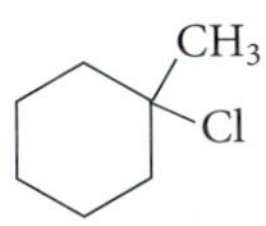

(c) 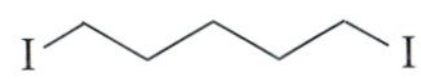

11.12 Write a curved-arrow mechanism for the rearrangement shown in Eq. 11.27.

11.13 Draw the structure of the alkyl halide product expected (if any) in each of the following reactions.

(a) 1-Propanol + HBr in the presence of H_2SO_4 catalyst

(b) $HOCH_2CH_2CH_2OH$ + excess HI $\xrightarrow{\text{heat}}$

(c) 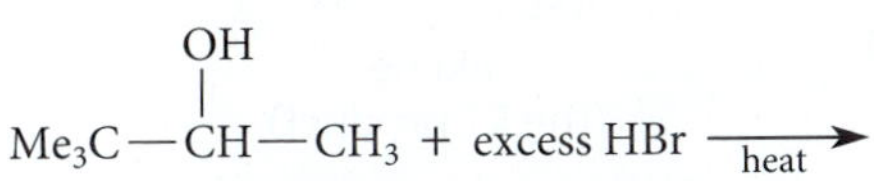

(d) $(CH_3)_3CCH_2OH$ + HCl $\xrightarrow{25\ °C}$

(*Hint:* See Fig. 9.4.)

The dehydration of alcohols to alkenes and the reactions of alcohols with hydrogen halides have some important things in common. Both take place in very acidic solution; in both reactions, the acid converts the OH group into a good leaving group. We discussed this same point for dehydrations in Sec. 11.2. For substitution reactions, if acid were not present, the halide ion would have to displace ^{-}OH to form the alkyl halide. This reaction does not take place because ^{-}OH is a much stronger base than any halide ion (Table 3.2, Sec. 3.6A), and strong bases are poor leaving groups (Sec. 9.4F).

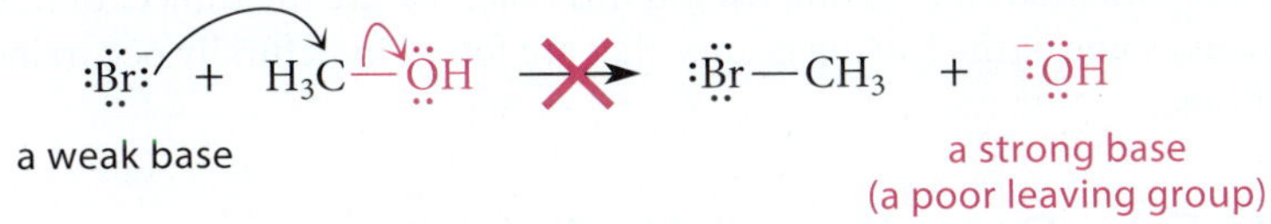

(11.28a)

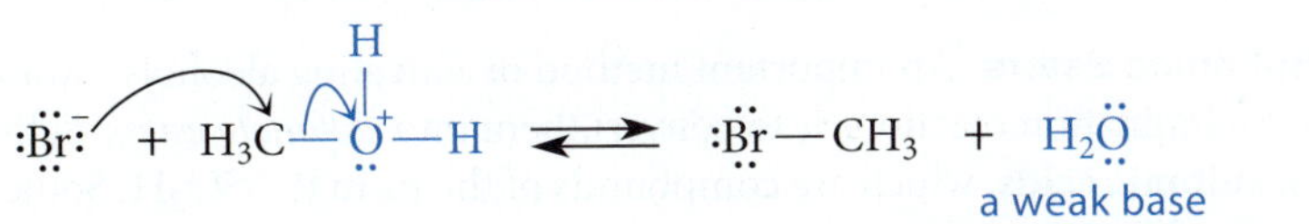

(11.28b)

Remember: *Substitution and elimination reactions of alcohols require the OH group to be converted into a better leaving group.*

The formation of secondary and tertiary alkyl halides and the dehydration of secondary and tertiary alcohols have the same initial steps: protonation of the alcohol oxygen and formation of a carbocation.

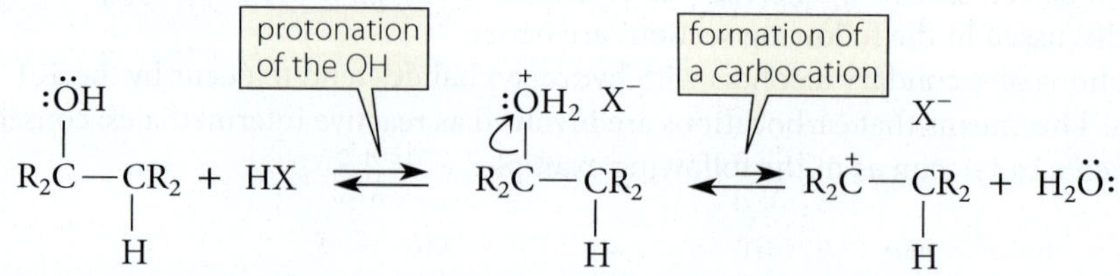

(11.29a)

The two reactions differ in the fate of this carbocation, which is governed, in turn, by the conditions of the reaction. In the presence of a hydrogen halide, the halide ion is present in excess and reacts with the carbocation to give an alkyl halide, which (if the solvent is aqueous acid) comes out of solution. In dehydration, no halide ion is present, and when the alkene forms by loss of a β-proton from the carbocation, the conditions of the dehydration reaction force the removal of the alkene product and the water by-product from the reaction mixture. It follows, then, that *alkyl halide formation and dehydration to alkenes are alternative branches of a common mechanism:*

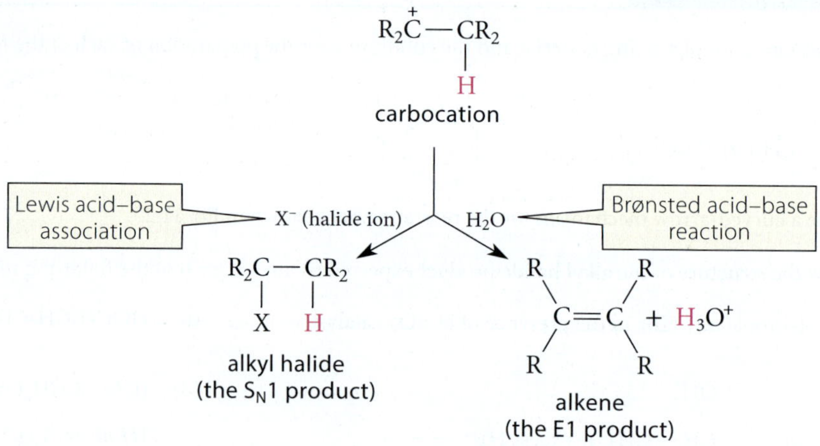

(11.29b)

These examples illustrate that the principles given in Chapter 9 for the substitutions and eliminations of alkyl halides are valid for other functional groups—in this case, alcohols.

11.4 ALCOHOL-DERIVED LEAVING GROUPS

When an alkyl halide is prepared from an alcohol and a hydrogen halide, protonation converts the OH group into a good leaving group. However, if the alcohol molecule contains a group that might be sensitive to strongly acidic conditions, or if milder or even nonacidic conditions must be used for other reasons, different ways of converting the OH group into a good leaving group are required. Laboratory methods for accomplishing this objective are the subject of this section. In addition, we discuss some of the leaving groups that are found in naturally occurring, biologically important reactions.

A. Sulfonate Ester Derivatives of Alcohols

Structures of Sulfonate Esters An important method of activating alcohols toward nucleophilic substitution and β-elimination reactions is to convert them into *sulfonate esters.* Sulfonate esters are derivatives of **sulfonic acids**, which are compounds of the form R—SO_3H. Some typical sulfonic acids are the following:

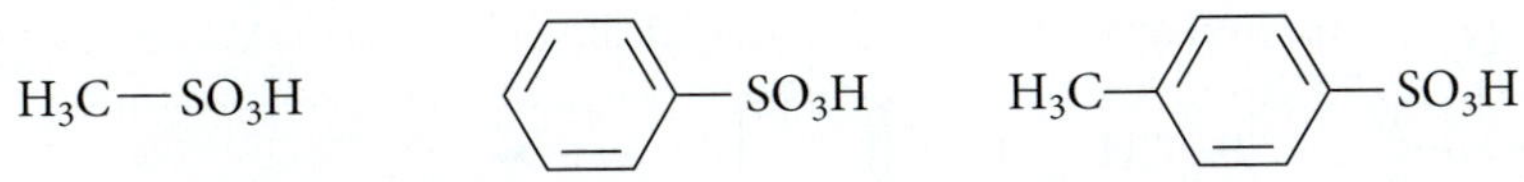

methanesulfonic acid **benzenesulfonic acid** ***p*-toluenesulfonic acid**

(11.30)

The *p* in the name of the last compound stands for *para*, which indicates the relative positions (1,4) of the two groups on the benzene ring. This type of nomenclature is discussed in Sec. 16.1.

Sulfur has an expanded electronic octet in these Lewis structures (see Secs. 1.3C and 3.3C). The octet resonance structure is much more important than the structure with S=O double bonds.

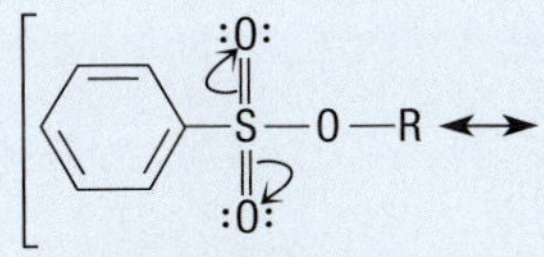

octet-expanded structure (less important)

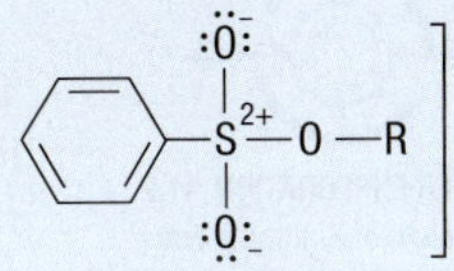

octet structure (more important)

To avoid the inconvenience of drawing many charges, however, we follow the conventional practice of using the structure with double bonds. (Bonding in sulfonic acids and their derivatives is discussed further in Sec. 11.10A.) The structure we choose to use is inconsequential to the chemistry of this section.

A **sulfonate ester** is a compound in which the acidic hydrogen of a sulfonic acid is replaced by an alkyl or aryl group. For example, in ethyl benzenesulfonate, the acidic hydrogen of benzenesulfonic acid is replaced by an ethyl group.

benzenesulfonic acid **ethyl benzenesulfonate** (a sulfonate ester)

(11.31)

Organic chemists often use abbreviated structures and names for certain sulfonate esters. Esters of methanesulfonic acid are called *mesylates* (abbreviated R—OMs), and esters of *p*-toluenesulfonic acid are called *tosylates* (abbreviated R—OTs).

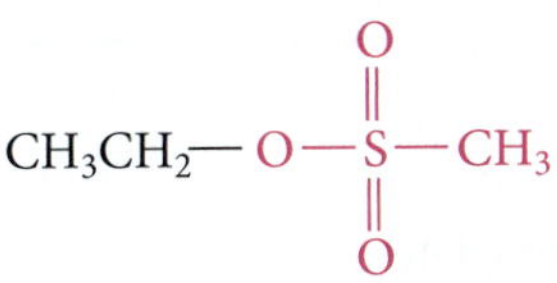 or CH_3CH_2—OSO_2CH_3 is the same as CH_3CH_2—OMs

ethyl methanesulfonate **ethyl mesylate**

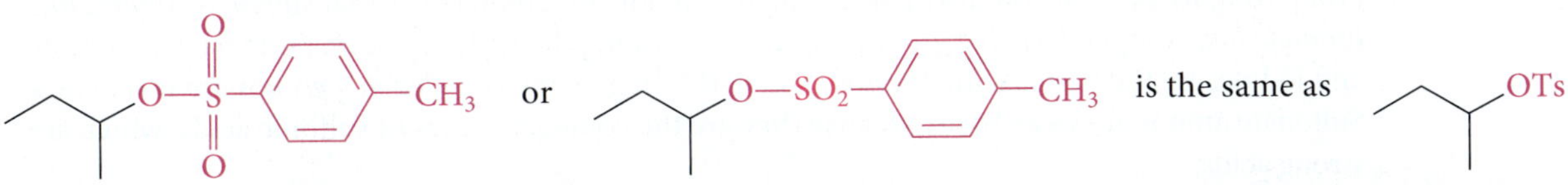

***sec*-butyl *p*-toluenesulfonate** ***sec*-butyl tosylate**

(11.32)

Focused Problem

11.14 Draw both the complete structure and the abbreviated structure, and give another name for each of the following compounds.

(a) Isopropyl methanesulfonate (b) Methyl *p*-toluenesulfonate

(c) Phenyl tosylate (d) Cyclohexyl mesylate

Preparation of Sulfonate Esters Sulfonate esters are prepared from alcohols and other sulfonic acid derivatives called sulfonyl chlorides. For example, *p*-toluenesulfonyl chloride (also known as *tosyl chloride*, TsCl) is the sulfonyl chloride used to prepare tosylate esters.

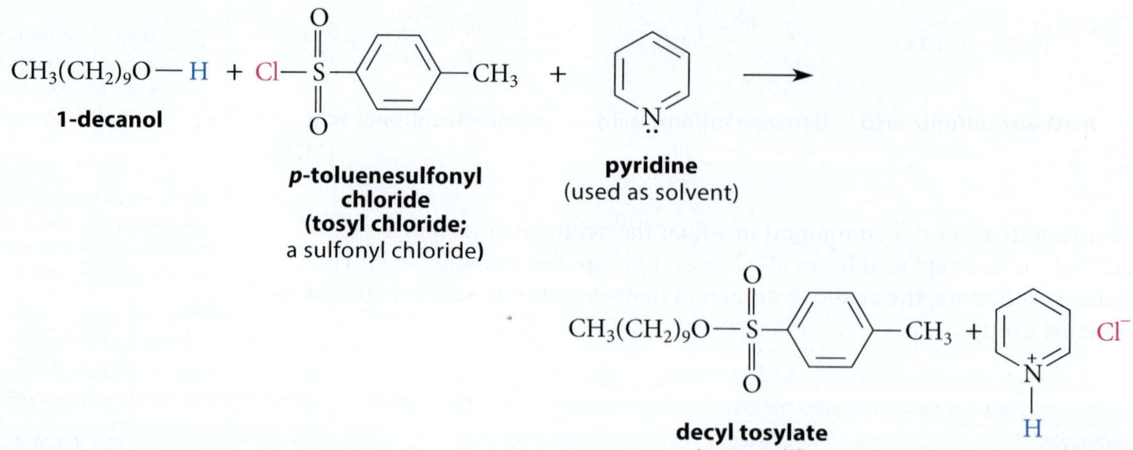

(11.33)

FURTHER EXPLORATION 11.2
Mechanism of Sulfonate Ester Formation

This is a nucleophilic substitution reaction in which the oxygen of the alcohol displaces chloride ion from the tosyl chloride. The pyridine used as the solvent is a base. Besides catalyzing the reaction, it also neutralizes the HCl that would otherwise form in the reaction (color in Eq. 11.33). (For more detail, see Further Exploration 11.2.)

Focused Problem

11.15 Suggest a preparation of cyclohexyl mesylate from the appropriate alcohol.

Reactivity of Sulfonate Esters Sulfonate esters, such as tosylates and mesylates, are useful because *they have approximately the same reactivities as the corresponding alkyl bromides in substitution and elimination reactions.* (In other words, you can think of a tosylate or mesylate ester group roughly as a "fat" bromo group.) The reason for this similarity is that *sulfonate anions, like bromide ions, are good leaving groups.* Recall that, among the halides, the weakest bases, bromide and iodide, are the best leaving groups (Sec. 9.4F). In general, *good leaving groups are weak bases.* Sulfonate anions are weak bases because they are the conjugate bases of sulfonic acids, which are strong acids.

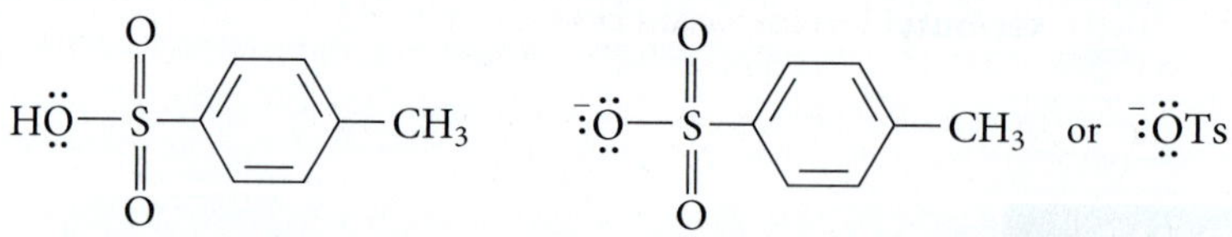

p-toluenesulfonic acid: a strong acid ($pK_a \sim -3$)

p-toluenesulfonate anion (tosylate anion): a weak base

This discussion shows why sulfonate esters prepared from primary and secondary alcohols, like primary and secondary alkyl halides, undergo S_N2 reactions in which a sulfonate ion serves as the leaving group.

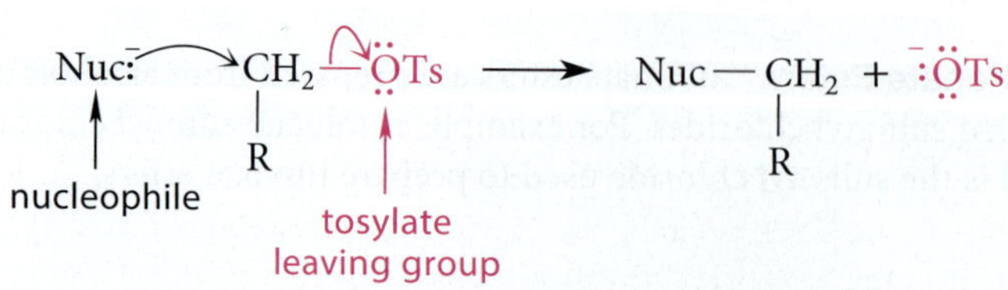

(11.34)

Secondary and tertiary sulfonate esters, like the corresponding alkyl halides, also undergo E2 reactions with strong bases, and they undergo S_N1–E1 solvolysis reactions in polar protic solvents.

Occasionally we might need a sulfonate ester that is much more reactive than a tosylate or mesylate. In such a case a *trifluoromethanesulfonate ester* is used. The trifluoromethanesulfonate group is nicknamed the **triflate** group and is abbreviated —OTf.

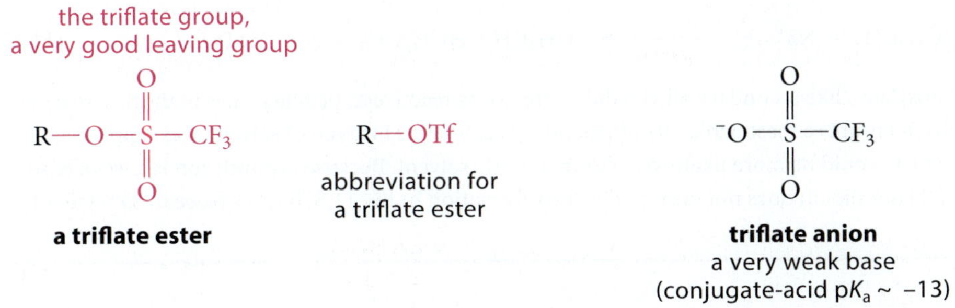

(11.35)

The triflate anion is an exceptionally weak base; the pK_a of its conjugate acid is about −13. (See Problem 11.66.) As a result, the triflate group is an exceptionally good leaving group, and triflate esters are highly reactive. For example, consider again the S_N2 reaction used to prepare FDG, an imaging agent used in positron emission tomography (PET). (See Eq. 9.30, Sec. 9.4E.) This reaction is far too slow to be useful with a tosylate leaving group. However, the triflate leaving group has considerably greater reactivity and is an ideal leaving group for this reaction, which must be carried out quickly.

Triflate esters are prepared in the same manner as tosylate esters (Eq. 11.33), except that triflic anhydride is used instead of tosyl chloride.

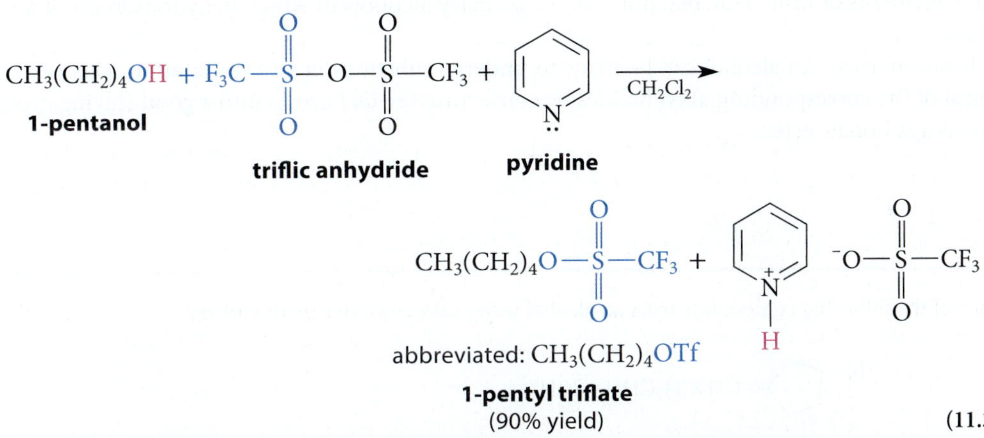

(11.36)

The use of sulfonate esters in S_N2 reactions is illustrated in Study Problem 11.2.

Study Problem 11.2

Outline a sequence of reactions for the conversion of 3-pentanol into 3-bromopentane.

Solution Before doing *anything* else, write the problem in terms of structures.

Alcohols can be converted into alkyl bromides using HBr and heat (Sec. 11.3). Secondary alcohols are prone to carbocation rearrangements and E1 elimination, however, so the HBr method is likely to give by-products. If conditions can be chosen so that the reaction will occur by the S_N2 mechanism, though, then carbocation rearrangements and elimination will not be an issue. To accomplish this objective, first convert the alcohol into a tosylate or mesylate.

$$\underset{\displaystyle CH_3CH_2\overset{|}{C}HCH_2CH_3}{OH} \xrightarrow[\text{pyridine}]{\text{TsCl}} \underset{\displaystyle CH_3CH_2\overset{|}{C}HCH_2CH_3}{OTs} \qquad (11.37a)$$

Next, displace the tosylate group with bromide ion in a polar aprotic solvent such as DMSO (Table 8.2, Sec. 8.6B).

$$\underset{\displaystyle CH_3CH_2\overset{|}{C}HCH_2CH_3}{OTs} + Na^+\ Br^- \xrightarrow[\text{DMSO}]{} \underset{\displaystyle CH_3CH_2\overset{|}{C}HCH_2CH_3}{Br} + Na^+\ {}^-OTs \qquad (11.37b)$$

Because secondary alkyl tosylates, like secondary alkyl halides, are not as reactive as primary ones in the S_N2 reaction, use of a polar aprotic solvent ensures a reasonable rate of reaction (Sec. 9.4E). This type of solvent also suppresses carbocation formation, which would be more likely to occur in a protic solvent. Because bromide ion is a weak base but a good nucleophile, E2 elimination does not occur. (The transformation in Eq. 11.37b takes place in 85% yield.)

The E2 reactions of sulfonate esters, like the analogous reactions of alkyl halides, can be used to prepare alkenes:

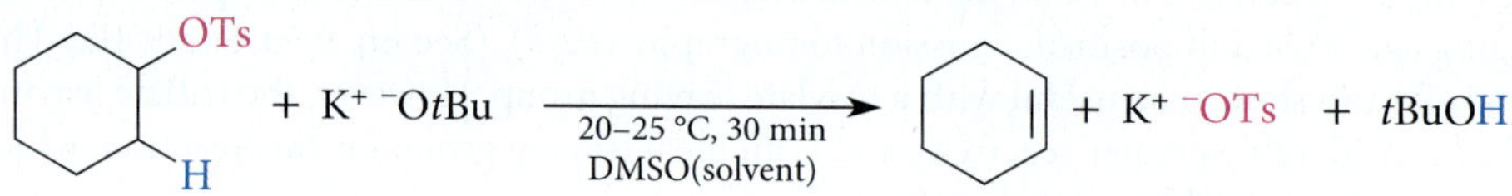

(83% yield) (11.38)

This reaction is especially useful when the acidic conditions of alcohol dehydration lead to rearrangements or other side reactions, or for primary alcohols in which dehydration is not an option.

To summarize: An alcohol can be made to undergo substitution and elimination reactions typical of the corresponding alkyl halides by converting the OH group into a good leaving group such as a sulfonate ester.

Focused Problems

11.16 Design a preparation of each of the following compounds from an alcohol using sulfonate ester methodology.

(a) 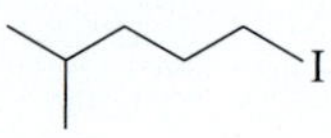

(b) 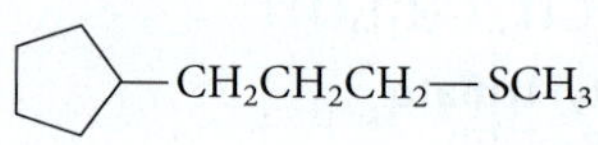

11.17 Give the product that results from each of the following sequences of reactions.

(a) $$\underset{\displaystyle CH_3\overset{|}{C}HCH_2CH_3}{OH} \xrightarrow[\text{pyridine}]{\text{TsCl}} \xrightarrow[\text{DMSO}]{\text{NaCN}}$$

(b) 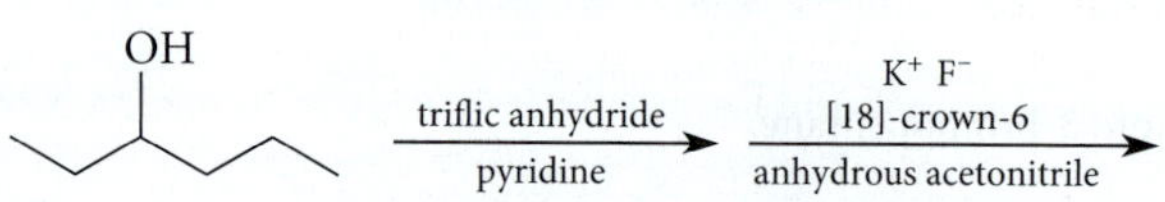

B. Alkylating Agents

Alkyl halides, alkyl tosylates, and other sulfonate esters are reactive in nucleophilic substitution reactions. In a nucleophilic substitution, an alkyl group is transferred from the leaving group to the nucleophile.

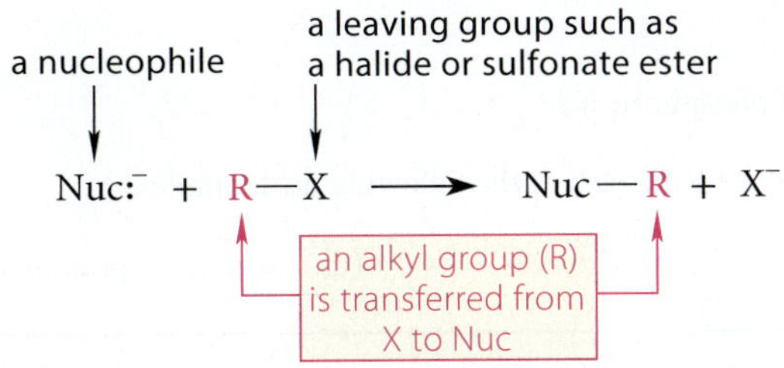

(11.39)

This type of reaction is called in general an **alkylation**. In this reaction the nucleophile is **alkylated** by the alkyl halide or the sulfonate ester in the same sense that a Brønsted base is *protonated* by a strong acid. For this reason, alkyl halides, sulfonate esters, and related compounds containing good leaving groups are sometimes referred to generically as *alkylating agents.* To say that a compound is a good **alkylating agent** means that it reacts rapidly with nucleophiles in S_N2 or S_N1 reactions to transfer an alkyl group.

C. Ester Derivatives of Strong Inorganic Acids

The esters of strong inorganic acids exemplify another type of *alkylating agent* (Sec. 11.4B). Like tosylates and mesylates, these compounds are derived conceptually by replacing the acidic hydrogen of a strong acid (in this case an inorganic acid) with an alkyl group. For example, dimethyl sulfate is an ester in which the acidic hydrogens of sulfuric acid are replaced by methyl groups.

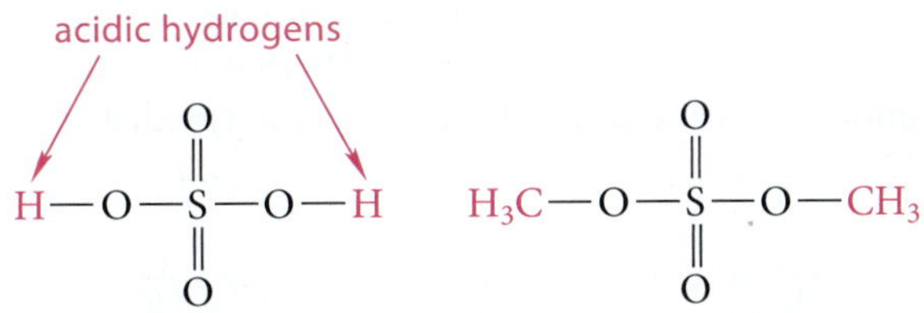

sulfuric acid **dimethyl sulfate**

Alkyl esters of strong inorganic acids are typically potent alkylating agents because they contain leaving groups that are very weak bases. For example, dimethyl sulfate is an effective methylating agent, as shown in the following example.

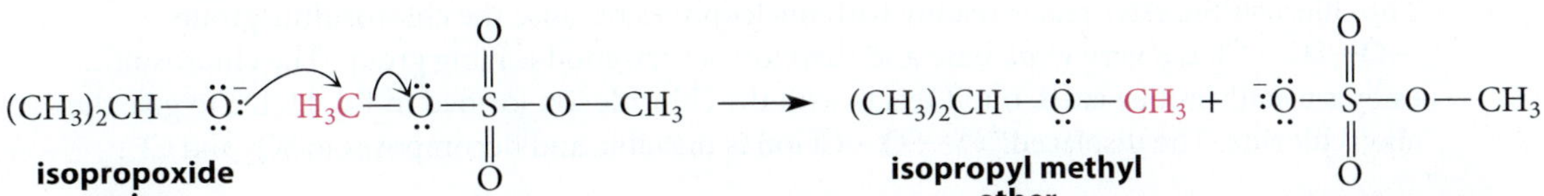

isopropoxide anion **dimethyl sulfate** **isopropyl methyl ether** a weak base; a good leaving group

(11.40)

Dimethyl sulfate and diethyl sulfate are available commercially. These reagents, like other reactive alkylating agents, are toxic because they react with nucleophilic functional groups on proteins and nucleic acids (Sec. 26.5C).

Focused Problems

11.18 Phosphoric acid (H_3PO_4) has the following structure.

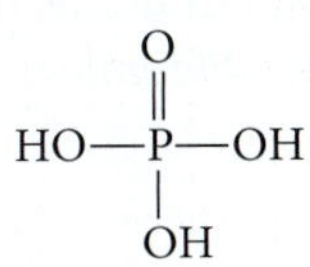

(a) Draw the structure of trimethyl phosphate.

(b) Draw the structure of the monoethyl ester of phosphoric acid.

11.19 Predict the products in the reaction of dimethyl sulfate with each of the following nucleophiles.

(a) $CH_3\ddot{N}H_2$ (b) Water (c) Sodium ethoxide (d) Sodium 1-propanethiolate

D. Reactions of Alcohols with Thionyl Chloride and Triphenylphosphine Dibromide

In most cases, the preparation of primary alkyl chlorides from alcohols with HCl is not as satisfactory as the preparation of the analogous alkyl bromides with HBr (Sec. 11.3). A better method for the preparation of primary alkyl chlorides is the reaction of alcohols with thionyl chloride:

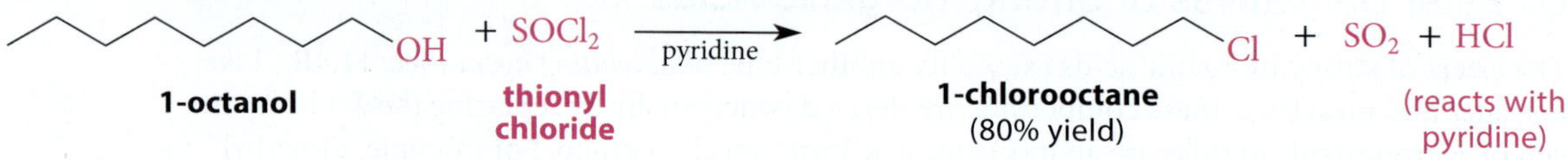

(11.41)

Thionyl chloride is a dense, fuming liquid (bp 75–76 °C). One advantage of using thionyl chloride for the preparation of alkyl chlorides is that the by-products of the reaction are HCl, which reacts with the base pyridine, and sulfur dioxide (SO_2), a gas. In many cases, then, there are no separation problems in the purification of the product alkyl chlorides.

The preparation of an alkyl chloride from an alcohol with thionyl chloride, like the use of a sulfonate ester, involves the conversion of the alcohol OH group into a good leaving group. When an alcohol reacts with thionyl chloride, a *chlorosulfite ester* intermediate is formed. (This reaction is analogous to Eq. 11.33.)

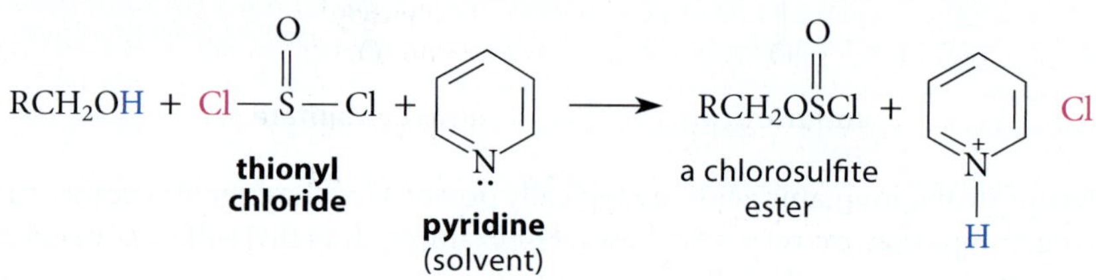

(11.42a)

The chlorosulfite ester reacts readily with nucleophiles because the chlorosulfite group, —O—SO—Cl, is a very weak base and therefore a very good leaving group. The chlorosulfite ester is usually not isolated, but it reacts with the chloride ion formed in Eq. 11.42a to give the alkyl chloride. The displaced $^{-}$O—SO—Cl ion is unstable and decomposes to SO_2 and Cl^-.

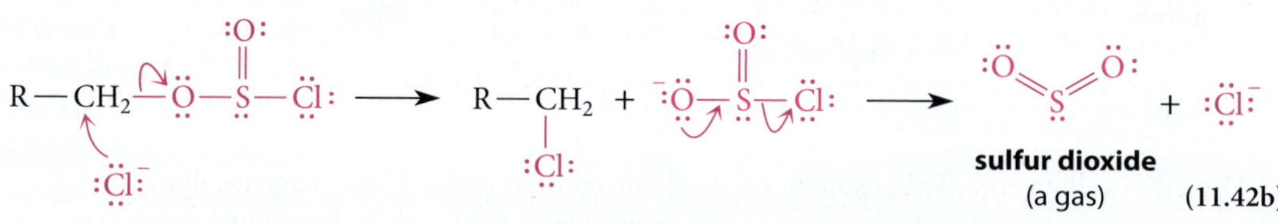

(11.42b)

The thionyl chloride method is most useful with primary alcohols, but it can also be used with secondary alcohols, although rearrangements in such cases have been known to occur. Rearrangements are best avoided in the preparation of secondary alkyl halides by using S_N2 conditions: the reaction of a halide ion with a sulfonate ester in a polar aprotic solvent (as in Study Problem 11.2).

A related method for the conversion of alcohols into alkyl bromides involves the use of Ph_3PBr_2 (dibromotriphenylphosphorane; this compound is universally known to organic chemists as *triphenylphosphine dibromide*).

cyclopentanol **dibromotriphenylphosphorane (triphenylphosphine dibromide)** **bromocyclopentane** (83% yield) **triphenylphosphine oxide**

(11.43)

Triphenylphosphine dibromide, a hypervalent compound, has a high degree of ionic character. [The ionic character of hypervalent compounds is discussed in Secs. 1.3C (Display 1.24) and 3.3B (Display 3.32).] For purposes of this mechanism, we treat the reagent as an ionic compound, as shown in Eq. 11.44a. The first mechanistic step of the reaction is an S_N2 reaction in which the oxygen of the alcohol acts as a nucleophilic center and phosphorus as the electrophilic center. The bromide ion of the reagent then acts as a base to give HBr and an intermediate:

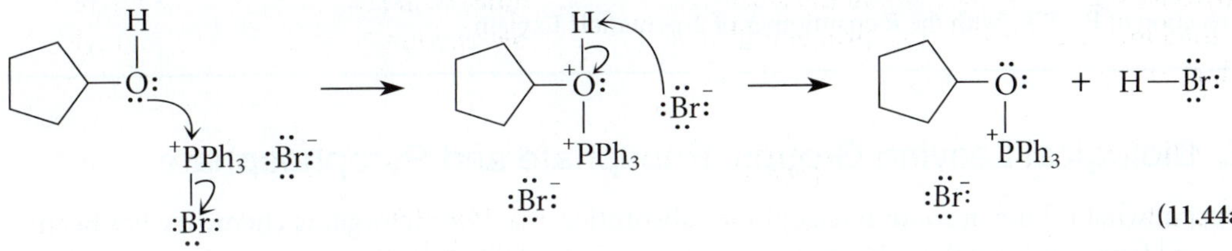

(11.44a)

The bromide ion that was displaced then acts as a nucleophile at the α-carbon of the intermediate to displace triphenylphosphine oxide as a leaving group, giving the product alkyl halide.

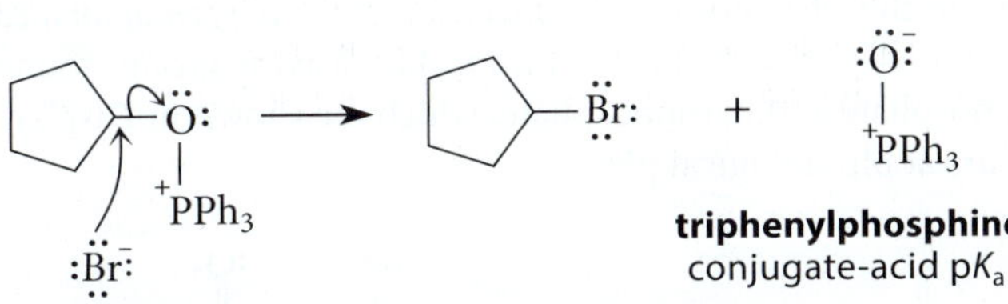

triphenylphosphine oxide
conjugate-acid pK_a ~ −2.1

(11.44b)

The triphenylphosphine oxide by-product is quite insoluble in many solvents; its precipitation helps to drive the reaction to completion, by Le Chatelier's principle.

This reaction occurs rapidly for three reasons:

1. Triphenylphosphine oxide is a *very* weak base and therefore a superb leaving group.

2. The reaction is typically carried out in polar aprotic solvents such as acetonitrile or DMF, which accelerate S_N2 reactions (Sec. 9.4E).

3. The reaction of the bromide ion nucleophile is essentially intramolecular—that is, the bromide leaving group in Eq. 11.44a reacts as a nucleophile in Eq. 11.44b before it can diffuse away. (Many intramolecular reactions are much faster than their intermolecular counterparts, as we discuss in detail in Sec. 12.8.)

The reaction of alcohols with triphenylphosphine dibromide is so fast that it can even be carried out successfully with neopentyl alcohol. Recall that neopentyl derivatives are unreactive in S_N2 reactions (Table 9.3, Fig. 9.4).

2,2-dimethyl-1-propanol (neopentyl alcohol) **1-bromo-2,2-dimethylpropane (neopentyl bromide)** (91% yield)

(11.45)

The triphenylphosphine dibromide reaction is particularly useful for the preparation of secondary bromides, as shown in Eq. 11.43. As expected for the S_N2 mechanism, this reaction occurs without rearrangement. An analogous reagent, triphenylphosphine dichloride (Ph_3PCl_2), can be used for the preparation of alkyl chlorides.

Focused Problems

11.20 Give three reactions that illustrate the preparation of 1-bromobutane from 1-butanol.

11.21 (a) According to the mechanism of the reaction shown in Eqs. 11.44a and 11.42b, what would be the absolute configuration of the alkyl chloride obtained from the reaction of thionyl chloride with (*S*)-$CH_3CH_2CH_2CHD—OH$? Explain.

(b) According to the mechanism shown in Eqs. 11.44a and b, what would be the absolute configuration of 2-bromopentane obtained from the reaction of Ph_3PBr_2 with the *R* enantiomer of 2-pentanol? Explain.

Chemical Biology Topic

E. Biological Leaving Groups: Phosphate and Pyrophosphate

Most of what is known about nucleophilic substitution reactions in organic chemistry has been learned from studying the substitution reactions of alkyl halides, sulfonate esters, and related compounds. Halogen-containing compounds are relatively rare in nature, and alkyl halides are particularly rare. Sulfonate esters don't occur at all. Nevertheless, the key concept that underlies the use of sulfonate esters in the laboratory—*the conversion of the oxygen of an alcohol into a good leaving group*—is also operative in two of the most important leaving groups found in biological substitution reactions: phosphate and pyrophosphate, which are shown in Display 11.46 in the ionization state that occurs at physiological pH.

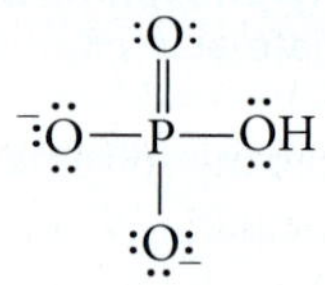

phosphate
($pK_a = 12.32$)

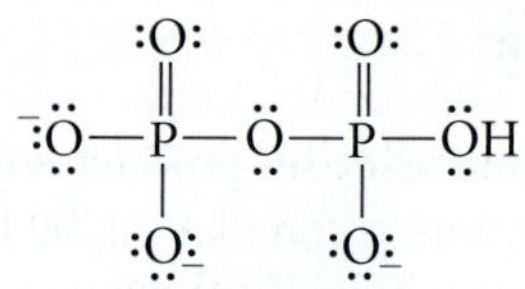

pyrophosphate
($pK_a = 9.25$)

(11.46)

Just as sulfonate esters (Sec. 11.4A) can serve as alkylating agents in laboratory reactions by loss of sulfonate groups, **pyrophosphate esters** can serve as alkylating agents by loss of the pyrophosphate leaving group.

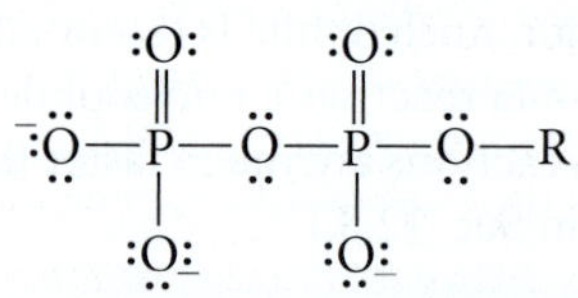

an alkyl pyrophosphate
(a pyrophosphate ester)

(11.47)

(Phosphate esters are important in other types of nucleophilic substitutions; we discuss phosphate and pyrophosphate esters more generally in Chapter 25.)

Farnesylation is an example of a biological alkylation reaction that involves a pyrophosphate leaving group. In this reaction, a 15-carbon alkyl pyrophosphate, *farnesyl pyrophosphate*, reacts in an S_N2 process with a thiolate group of a protein called *Ras*. (*Ras* is a protein that regulates cell growth. Mutations of *Ras* have been implicated in pancreatic and other cancers.) This reaction is catalyzed by a *farnesylating enzyme*, a protein distinct from *Ras* itself. The thiolate nucleophile comes from the side chain of a cysteine, one of the amino acids of *Ras*. (See Table 27.1, Sec. 27.1A.) Weak coordination of the thiolate to a Zn^{2+} ion on the farnesylating enzyme ensures that it remains ionized.

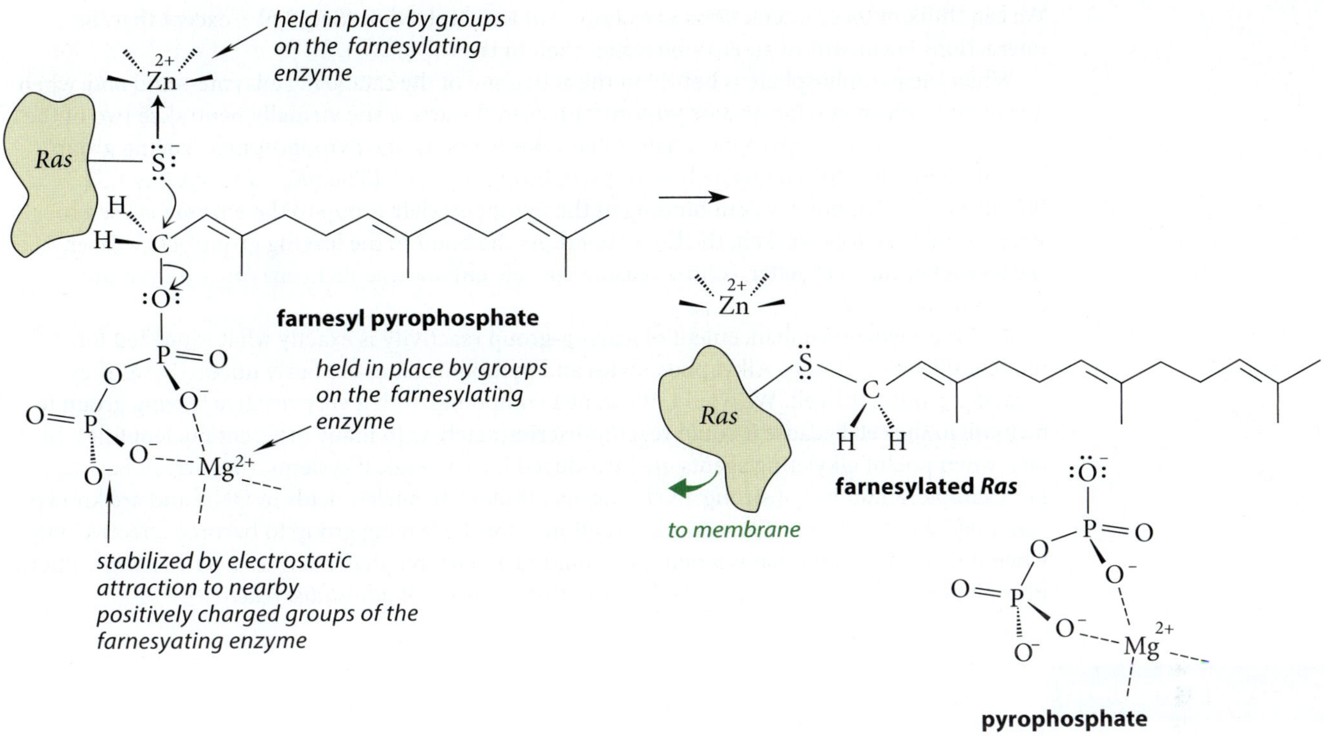

(11.48)

Through the use of deuterium substitution for one of the α-hydrogens, an isotopically chiral farnesyl pyrophosphate was synthesized. With this chiral derivative containing an asymmetric α-carbon, substitution was shown to occur with inversion of configuration, as we would expect for an S_N2 reaction. The farnesyl group, a large hydrocarbon group, moves to the cell membrane and becomes anchored in the lipid bilayer; as a result, the attached *Ras* is tethered to the membrane as well. This event is required for *Ras* to become active in promoting cell growth. (Interfering with this process has been an attractive target for anticancer drug discovery.)

In the S_N2 reaction (Sec. 9.4F), *leaving group effectiveness is inversely correlated with basicity.* That is, *the weakest bases make the best leaving groups.* Bromide (Br^-) and iodide (I^-) are excellent leaving groups for the same reason that H—Br and H—I are strong acids: the bond energies of H—Br and H—I bonds are relatively low, as are the C—Br and C—I bond energies. How basic, then, is pyrophosphate? The pK_a of the pyrophosphate tri-anion $HP_2O_7^{3-}$ is 9.25 (Display 11.46). Therefore, the pyrophosphate leaving group is *much* more basic than halide ions and tosylate ions. Therefore, the pyrophosphate group is a *very poor* leaving group. How, then, can it act as a leaving group in biological systems?

The answer is that pyrophosphate *itself* is not a leaving group. Rather, it is activated within enzyme active sites by binding to divalent metal ions—in many cases, Mg^{2+}—or by hydrogen bonding to acidic groups on the side chains of enzyme amino acids.

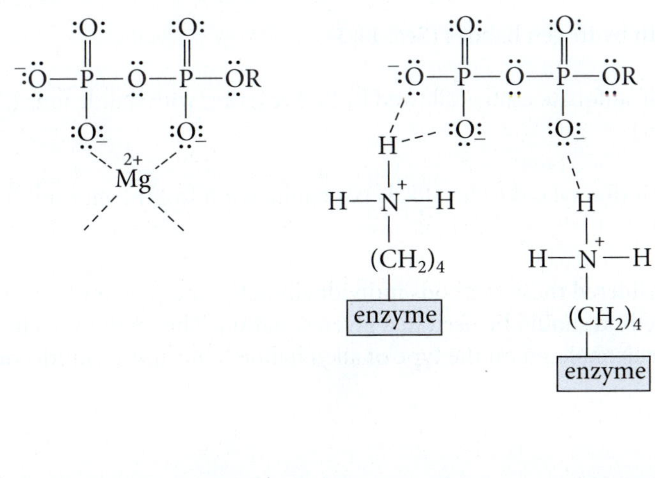

(11.49)

We can think of these interactions as examples of ionic solvation (Sec. 8.6F), except that the interactions occur within an enzyme rather than in the solvent.

When the pyrophosphate is bound to the active site of the catalyzing enzyme—and *only* when it is bound—the metal ion and/or proton sources in the active site virtually neutralize two of the negative charges on the pyrophosphate. When this happens, the pyrophosphate leaving group becomes more like the conjugate base of pyrophosphoric acid. (The pK_{a1} of $H_4P_2O_7$ is 1.52.) It is also likely that noncovalent binding of the pyrophosphate group to the enzyme is used to stretch, and therefore weaken, the C—O bond. As the bond to the leaving group gets weaker, the leaving group gets better. (Other reasons for rate enhancements in enzyme catalysis are discussed in Sec. 12.8E.)

Enzyme-mediated enhancement of leaving-group reactivity is exactly what is needed for biological leaving groups. Alkyl phosphates and pyrophosphates are fairly unreactive as they meander around the cell. We would not want a compound with a very reactive leaving group to be loose in the cell because it could react indiscriminately with many different nucleophiles. In fact, when potent alkylating agents *are* introduced into biological systems, they wreak havoc! For example, a number of strong alkylating agents alkylate nucleic acids in DNA and are known carcinogens (Sec. 26.5C). The optimal situation is for the leaving group to become effective *only* when it is needed—and that is when it is bound to an enzyme that catalyzes a *particular* reaction. Phosphates and pyrophosphates ideally meet this criterion of *adjustable reactivity.*

Focused Problem

11.22 An important class of substitution reactions covered in Chapter 21 is called *nucleophilic acyl substitution.* Substitution reactions of this type occur at the carbon of a carbonyl (C═O) group. An important laboratory example is the reaction of nucleophiles (ammonia in the example below) with acyl chlorides (also called acid chlorides).

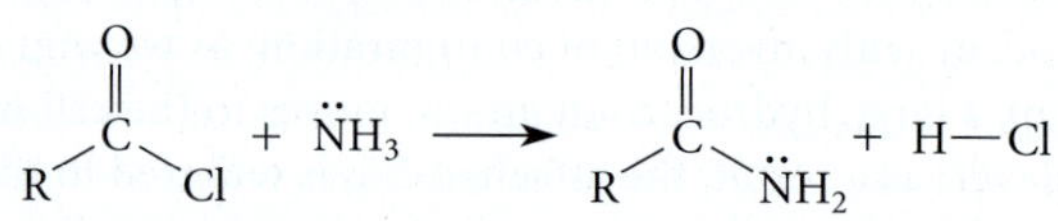

an acyl chloride

The corresponding derivatives found in biology are *acyl phosphates.* Draw the general structure of an acyl phosphate. Be sure to show its ionization state at physiological pH. Show how Mg^{2+} ions or hydrogen-bond-donating groups in an enzyme active site might enhance the reactivity of acyl phosphates.

11.5 CONVERSION OF ALCOHOLS INTO ALKYL HALIDES: SUMMARY

We have now seen a variety of reactions that can be used to convert alcohols into alkyl halides:

1. Reaction with hydrogen halides (Sec. 11.3)
2. Formation of sulfonate esters followed by S_N2 reaction with halide ions (Sec. 11.4A, Study Problem 11.1)
3. Reaction with thionyl chloride ($SOCl_2$) or triphenylphosphine dibromide (Ph_3PBr_2) (Sec. 11.4D)

Now that we've considered these methods individually, let's now view these methods holistically by asking which method should be used in a given situation. The method of choice depends on the structure of the alcohol and on the type of alkyl halide (chloride, bromide, or iodide) to be prepared.

Primary Alcohols: Alkyl bromides are prepared from primary alcohols by the reaction of the alcohol with concentrated HBr or with Ph_3PBr_2. HBr is often chosen for convenience and because the reagent is relatively inexpensive. The reaction with Ph_3PBr_2 is quite general, but it is particularly useful when the alcohol contains another functional group that would be adversely affected by the strongly acidic conditions of the HBr reaction. Primary alkyl iodides can be prepared with HI, which is usually supplied by mixing an iodide salt such as KI with a strong acid such as phosphoric acid. Thionyl chloride is the method of choice for the preparation of primary alkyl chlorides because the reactions of primary alcohols with HCl are slow. The sulfonate ester method works well with primary alcohols, but it requires two separate reactions (formation of the sulfonate ester, then reaction of the ester with halide ion). Because all of these methods have an S_N2 mechanism as their basis, alcohols with several β-alkyl substituents, such as neopentyl alcohol, do not react with any of these reagents *except* Ph_3PBr_2. This reagent does work with such alcohols because it is accelerated for the reasons given in Sec. 11.4D.

Tertiary Alcohols: Tertiary alcohols react rapidly with HCl or HBr under mild conditions to give the corresponding alkyl halides. The sulfonate ester method shown in Study Problem 11.2 is not used with tertiary alcohols because tertiary sulfonates, like tertiary alkyl halides, do not undergo S_N2 reactions.

Secondary Alcohols: If the secondary alcohol has no β-alkyl substitution, the thionyl chloride method can be used to prepare alkyl chlorides. To avoid rearrangements completely, the alcohol can be converted into a sulfonate ester, which can be treated, in turn, with the appropriate halide ion (Cl^-, Br^-, or I^-) in a polar aprotic solvent. This type of solvent provides the enhanced nucleophilicity of the halide ion necessary to overcome the relatively low S_N2 reaction rate of a secondary sulfonate ester (Sec. 9.4E). Less reactive secondary alcohols can be converted into triflates, which are much more reactive than tosylates or mesylates toward halide ions in polar aprotic solvents. The HBr method can be expected to lead to rearrangements and is therefore not satisfactory (unless rearranged products are desired). The Ph_3PBr_2 method can be used to form alkyl bromides without rearrangement from primary and secondary alcohols that have significant β-alkyl substitution.

Let's also review what we have learned mechanistically about the substitution and elimination reactions of alcohols. The —OH group itself cannot act as a leaving group because ^-OH is far too basic. To break the carbon–oxygen bond, the —OH group must first be converted into a good leaving group. Two general strategies can be used for this purpose:

1. *Protonation:* Protonated alcohols are intermediates in both the dehydration to alkenes and the reaction with hydrogen halides to give alkyl halides.

2. *Conversion into sulfonate esters, inorganic esters, or related leaving groups:* Sulfonate esters, to a useful approximation, react like alkyl halides. That is, the principles of alkyl halide reactivity covered in Chapter 9 are equally applicable to sulfonate esters. Thionyl chloride and triphenylphosphine dibromide are additional examples of this approach in which the reagent both converts the alcohol —OH into a good leaving group and provides the displacing nucleophile.

Focused Problems

11.23 Suggest conditions for carrying out each of the following conversions to yield a product that is as free of isomers as possible.

(a) 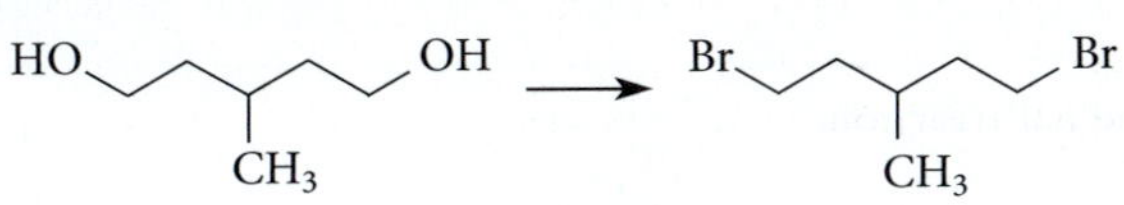

(b) 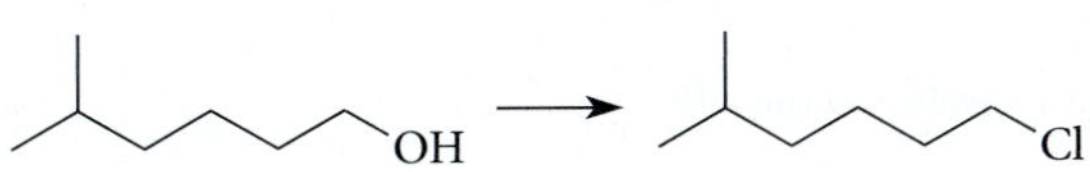

(c) 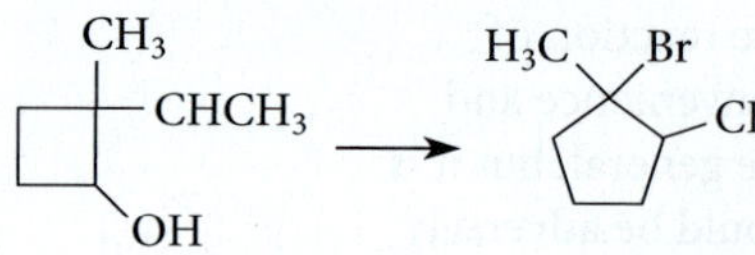

(d) 

11.24 Contrast the products expected when 2-methyl-3-pentanol is treated with (a) aqueous HBr/H_2SO_4 or (b) Ph_3PBr_2 in a polar aprotic solvent. Explain.

11.6 OXIDATION AND REDUCTION IN ORGANIC CHEMISTRY

The previous sections have discussed the substitution and elimination reactions of alcohols and their derivatives. These reactions have much in common with the analogous reactions of alkyl halides. We now turn to a different type of reaction: oxidation. Oxidation is a reaction of alcohols that has no simple analogy in alkyl halide chemistry.

A. Half-Reactions and Oxidation Numbers

An **oxidation** is a transformation in which electrons are lost; a **reduction** is a transformation in which electrons are gained. Each oxidation is accompanied by a reduction, and vice versa. The gain or loss of electrons can be illustrated with a **half-reaction**, which shows either the oxidation or the reduction but not both. An example of such a half-reaction in organic chemistry is the oxidation of ethanol to acetic acid:

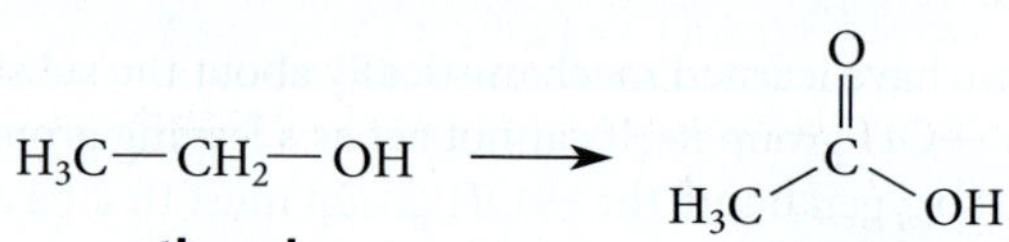

ethanol **acetic acid** (11.50)

This is a half-reaction because the reagent that brings about this oxidation (and is itself reduced) is not included. That this is an oxidation can be demonstrated by balancing the half-reaction using protons and free electrons, a technique that you may have learned in general chemistry. This process involves three steps:

Step 1. Use H_2O to balance missing oxygens.

Step 2. Use protons (that is, H^+) to balance missing hydrogens.

Step 3. Use electrons to balance charges.

This process is illustrated in Study Problem 11.3.

Study Problem 11.3

Write the transformation in Eq. 11.50 as a balanced half-reaction.

Solution

Step 1. Balance the extra oxygen on the right with a water on the left:

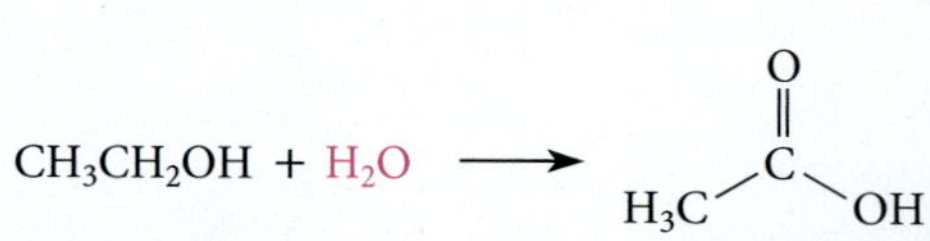

(oxygens are balanced) (11.51a)

Step 2. Balance the extra hydrogens on the left with four protons on the right:

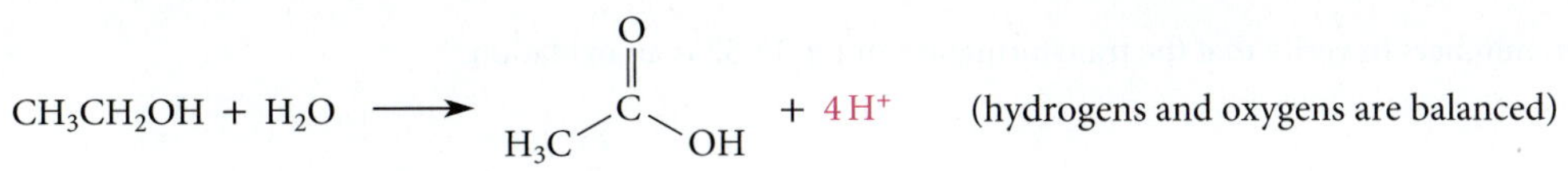

(11.51b)

Step 3. Balance the extra positive charges on the right with electrons so that the charges on both sides of the equation are equal:

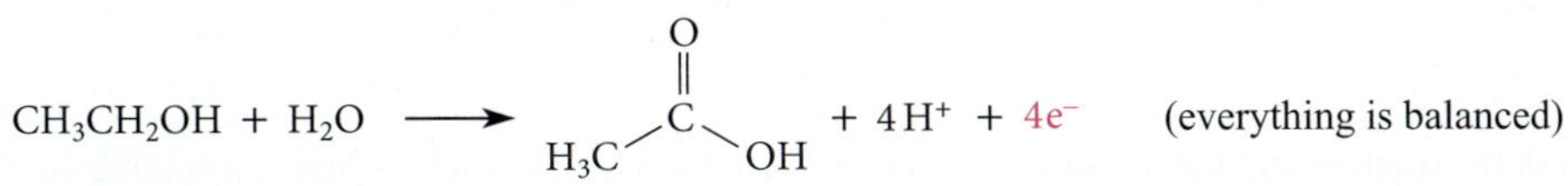

(11.51c)

The result is the balanced half-reaction.

According to this half-reaction, *four electrons are lost from the ethanol molecule when acetic acid is formed*. The loss of electrons means physically that this half-reaction can be carried out at the anode of an electrochemical cell. In most cases, though, we carry out oxidations with *reagents* (rather than anodes) that accept electrons, called *oxidizing agents,* which are discussed in Sec. 11.6B. Nevertheless, on the basis of this half-reaction, *the oxidation of ethanol to acetic acid is a four-electron oxidation*. This type of terminology is used frequently in the study of biochemistry.

A quicker way to determine whether a transformation is an oxidation or a reduction, as well as how many electrons are involved, is to determine *oxidation numbers* for the reactant and product. This is a three-step "bookkeeping" process that focuses on *individual carbon atoms* involved in the transformation. After glancing over these steps, read carefully through Study Problem 11.4, which takes you through this process.

Step 1. Assign an *oxidation level* to each carbon that undergoes a change between reactant and product by the following method:

(a) For every bond from the carbon to a less electronegative element (including hydrogen), and for every negative charge on the carbon, assign a -1.

(b) For every bond from the carbon to another carbon atom, and for every unpaired electron on the carbon, assign a 0.

(c) For every bond from the carbon to a more electronegative element, and for every positive charge on the carbon, assign a $+1$.

(d) Add the numbers assigned under parts a, b, and c to obtain the oxidation level of the carbon under consideration.

Step 2. Determine the *oxidation number* N_{ox} of both the reactant and product by adding, within each compound, the oxidation levels of all the carbons computed in step 1. Consider *only* the carbons that undergo a *change* in the reaction.

Step 3. Compute the difference N_{ox}(product) – N_{ox}(reactant) to determine whether the transformation is an oxidation, reduction, or neither.

(a) If the difference is a positive number, the transformation is an *oxidation*.

(b) If the difference is a negative number, the transformation is a *reduction*.

(c) If the difference is zero, the transformation is neither an oxidation nor a reduction.

Study Problem 11.4

Use oxidation numbers to verify that the transformation in Eq. 11.52 is an oxidation.

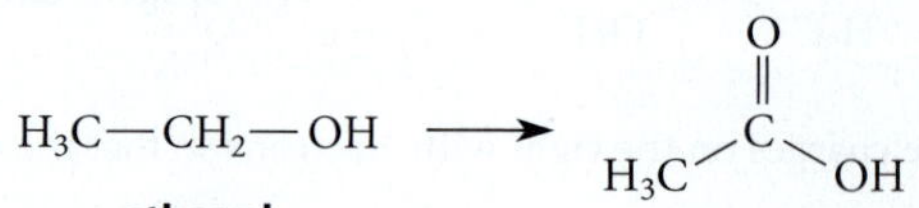

ethanol **acetic acid** (11.52)

Solution

Step 1. For both the reactant and the product, compute the oxidation level of each carbon that undergoes a change. Because the methyl group is unchanged, don't assign an oxidation level to its carbon. Only one carbon is changed. For this carbon, −1 is assigned for each bond to hydrogen; 0 is assigned to the bonds to carbon; and +1 is assigned to the bonds to oxygen. The carbon–oxygen double bond of the product is treated as *two bonds* and makes a contribution of +2.

Reactant: 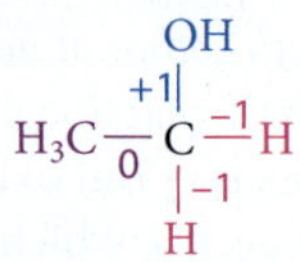 *Product:* 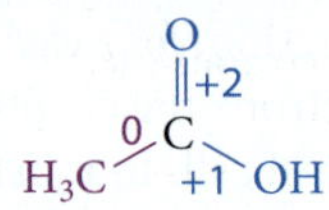

Step 2. Add the oxidation levels for each carbon that changes to determine the oxidation number. Because only one carbon changes, the oxidation level of this carbon, computed in step 1, is the oxidation number of the compound. Therefore, N_{ox}(reactant), the oxidation number of the reactant ethanol, is −1; and N_{ox}(product), the oxidation number of the product acetic acid, is +3.

Sum for the reactant: $(+1) + 0 + (-1) + (-1) = -1$

Sum for the product: $0 + (+1) + (+2) = +3$

Step 3. Compute the difference N_{ox}(product) − N_{ox}(reactant), which is $+3 - (-1) = +4$. Because this difference is positive, the transformation of ethanol to acetic acid is a four-electron oxidation.

The change in oxidation number, +4, determined in step 3 of Study Problem 11.4, is the same as the number of electrons lost, determined in Study Problem 11.3 from the balanced half-reaction. *This correspondence is general.* That is, *the change in oxidation number is always equal to the number of electrons lost or gained in the half-reaction.* If the number is positive, as in this example, the transformation is an oxidation. If it is negative, the transformation is a reduction.

In some transformations that are neither oxidations nor reductions, the oxidation level of one carbon in a molecule can be increased and the oxidation level of another carbon can be decreased by the same amount. It is the sum of *all* of the changes in carbon oxidation levels that determines whether a net oxidation or reduction has occurred. This situation is illustrated in Study Problem 11.5.

Study Problem 11.5

Verify that the acid-catalyzed hydration of methylpropene is neither an oxidation nor a reduction.

Solution First, write the structures involved in the transformation:

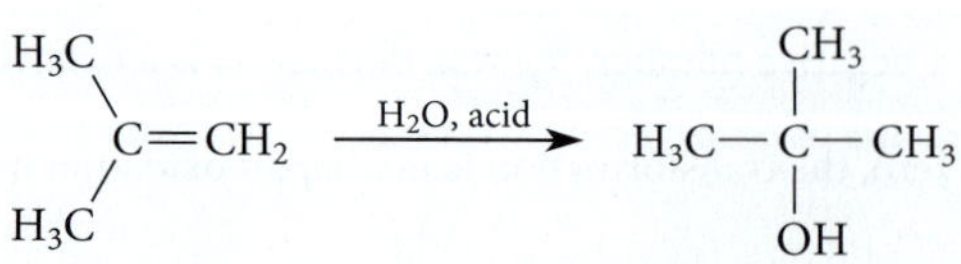

methylpropene ***tert*-butyl alcohol**

The oxidation number of the organic reactant, methylpropene, is –2.

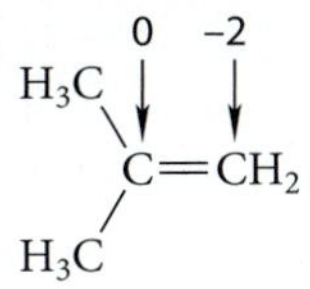

$N_{ox} = 0 + (-2) = -2$

The oxidation number of the organic product, *tert*-butyl alcohol, is also –2:

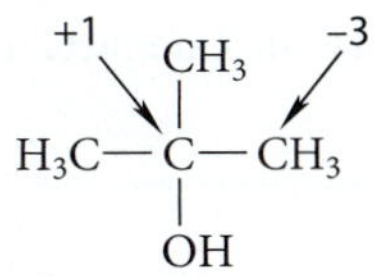

$N_{ox} = +1 + (-3) = -2$

An oxidation level is computed for only the one methyl group that was formed as a result of the transformation. (The other methyls are unchanged.) Because the oxidation numbers of the reactant and product are equal, the hydration reaction is neither an oxidation nor a reduction. The same conclusion must apply to the reverse reaction, the dehydration of the alcohol to the alkene.

The methods described here show that the addition of Br_2 to an alkene is an oxidation (the change in oxidation number is +2):

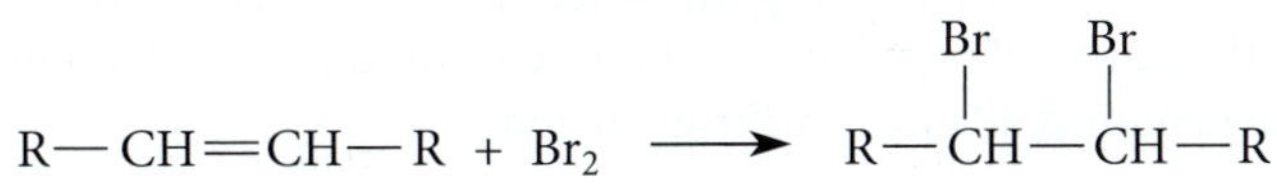

$N_{ox} = -2$ $\qquad$ $N_{ox} = 0$ (11.53)

This example demonstrates that an oxidation or a reduction does not necessarily depend on the introduction or loss of oxygen. In most oxidations of organic compounds, either a hydrogen in a C—H bond or a carbon in a C—C bond is replaced by a more electronegative element, which *may* be oxygen, but *which may also be another element such as a halogen.*

Focused Problems

11.25 Considering the organic compound, classify each of the following transformations, some of which may be unfamiliar, as an oxidation, a reduction, or neither. For those that are oxidations or reductions, tell how many electrons are gained or lost.

(a) $CH_4 \xrightarrow{Br_2,\ light} CH_3Br$

(b)

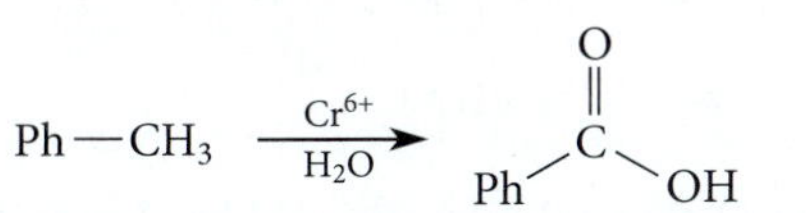

(c) $CH_3CH_2CH_2I \xrightarrow{LiAlH_4} CH_3CH_2CH_3 + I^-$

(d)

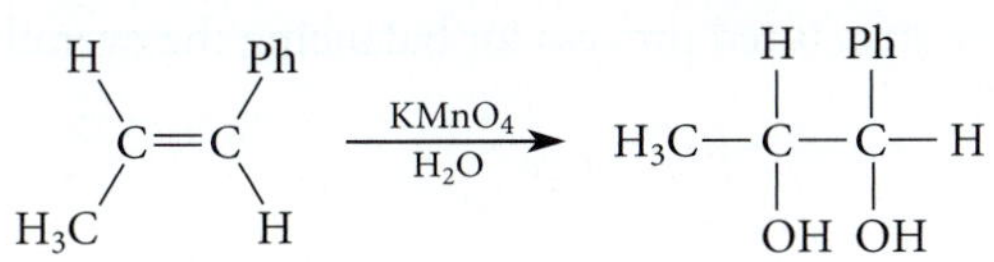

(e)

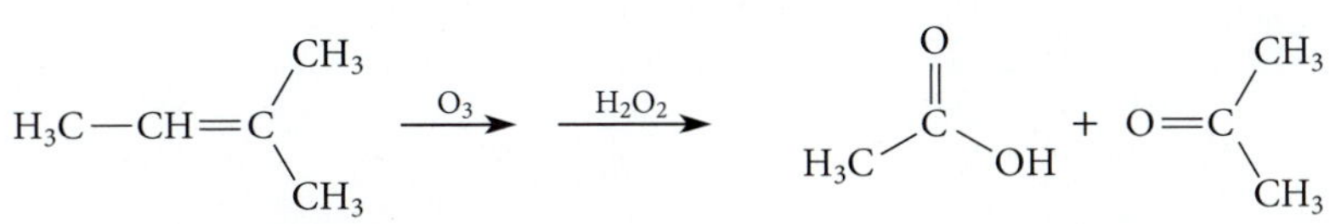

(f)

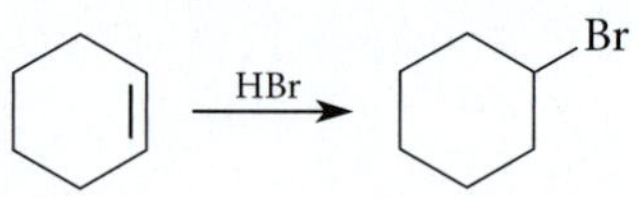

(g)

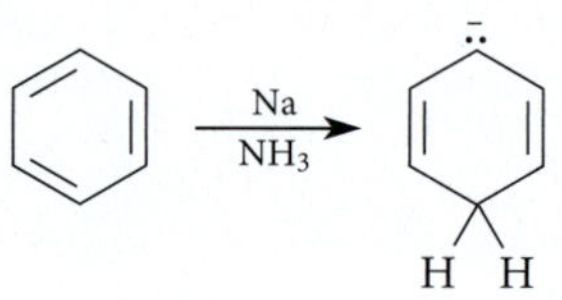

11.26 Write the transformation in Focused Problem 11.25b as a balanced half-reaction. Complete the following sentence: This reaction is a ______ (how many)-electron ______ (oxidation or reduction).

B. Oxidizing and Reducing Agents

Oxidations and reductions always occur in pairs. Therefore, *whenever something is oxidized, something else is reduced.* When an organic compound is oxidized, the reagent that brings about the transformation is called an **oxidizing agent**. Similarly, when an organic compound is reduced, the reagent that effects the transformation is called a **reducing agent**.

For example, chromate ion (CrO_4^{2-}) can be used to bring about the oxidation of ethanol to acetic acid in Eq. 11.50; in this reaction, chromate ion is reduced to Cr^{3+}. We can calculate the change in oxidation state for chromium in this reaction by calculating the oxidation states of Cr in the reactant and product and then taking the difference. To determine the oxidation state of Cr, we apply essentially the same technique we used for determining the oxidation state of carbon. Let's start with the oxidation state of Cr in chromate ion.

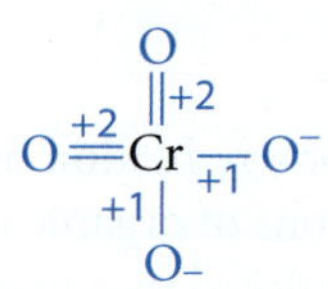

Sum = oxidation number of Cr = +6 (11.54)

Each single bond to oxygen makes a contribution of +1 and each double bond to oxygen a contribution of +2. (The negative charges do not enter into the calculation because they are on oxygen, not chromium.) The oxidation number of Cr is therefore +6. We indicate this by saying that Cr is in the +6 oxidation state and abbreviate this oxidation state as Cr(VI).

When CrO_4^{2-} is used to oxidize ethanol, Cr^{3+} is formed. To compute the oxidation state of Cr^{3+}, we count +1 for each positive charge, so Cr^{3+} has an oxidation state of +3. Therefore, chromium changes oxidation state from Cr(VI) to Cr(III), and, consequently, chromate ion undergoes a *three-electron reduction.* We can verify this by balancing the corresponding half-reaction:

$$8H^+ + 3e^- + CrO_4^{2-} \longrightarrow Cr^{3+} + 4H_2O \qquad (11.55)$$

A complete, balanced reaction for the oxidation of ethanol to acetic acid by chromate is obtained by reconciling the electrons in the half-reactions given by Eqs. 11.51c (Study Problem 11.3) and 11.55. That is, every mole of Cr(VI) ($3e^-$ gained) can oxidize three-fourths of a mole of ethanol ($0.75 \times 4e^-$ lost). A more structured process for balancing the overall reaction is illustrated in Study Problem 11.6.

Study Problem 11.6

Give a complete balanced equation for the oxidation of ethanol to acetic acid by chromate ion.

Solution The two half-reactions are

$$CH_3CH_2OH + H_2O \longrightarrow CH_3CO_2H + 4H^+ + 4e^- \qquad (11.56a)$$

$$8H^+ + 3e^- + CrO_4^{2-} \longrightarrow Cr^{3+} + 4H_2O \qquad (11.56b)$$

Multiply each equation by a factor that gives the same number of "free" electrons in both half-reactions. Thus, multiplying Eq. 11.56a by 3 and Eq. 11.56b by 4 gives 12 electrons in both reactions:

$$3CH_3CH_2OH + 3H_2O \longrightarrow 3CH_3CO_2H + 12H^+ + 12e^- \qquad (11.57a)$$

$$32H^+ + 12e^- + 4CrO_4^{2-} \longrightarrow 4Cr^{3+} + 16H_2O \qquad (11.57b)$$

Add these equations, canceling like terms on both sides. Therefore, all electrons cancel; the three water molecules on the left are canceled by three of those on the right to leave 13 water molecules on the right; and the 12 protons on the right are canceled by 12 of those on the left, leaving 20 protons on the left. The fully balanced equation is

$$20H^+ + 4CrO_4^{2-} + 3CH_3CH_2OH \longrightarrow 4Cr^{3+} + 3CH_3CO_2H + 13H_2O \qquad (11.58)$$

This equation shows that three ethanol molecules are oxidized for every four chromate ions reduced, or, as noted earlier, three-fourths of a mole of ethanol per mole of chromate ion.

STUDY GUIDE LINK 11.1
More on Half-Reactions

By considering the change in oxidation number for a transformation, you can tell whether an oxidizing or reducing agent is required to bring about the reaction. For example, the following unfamiliar transformation is neither an overall oxidation nor reduction (verify this statement):

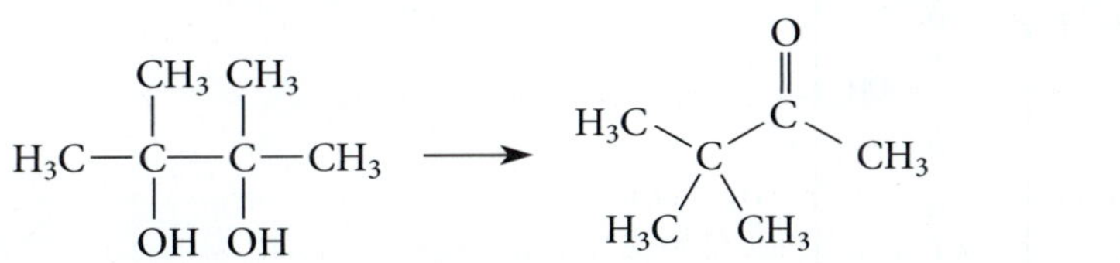

(11.59)

Although one carbon is oxidized, another is reduced. Even though you might know nothing else about the reaction, an oxidizing or reducing agent alone would not effect this transformation. (In fact, the reaction is brought about by strong acid.)

The oxidation-number concept can be used to organize organic compounds into functional groups with the same oxidation level, as shown in **Table 11.2**. Compounds *within* a given box are generally interconverted by reagents that are neither oxidizing nor reducing agents. For example, alcohols can be converted into alkyl halides with HBr, which is neither an oxidizing nor a reducing agent. On the other hand, conversion of an alcohol into a carboxylic acid involves an increase in oxidation level, and indeed this transformation requires an oxidizing agent. Notice also in Table 11.2 that carbons with larger numbers of hydrogens have a greater number of possible oxidation states. A tertiary alcohol cannot be oxidized at the α-carbon (without breaking carbon–carbon bonds) because this carbon bears no hydrogens. Methane, on the other hand, can be oxidized to CO_2. (Any hydrocarbon can be oxidized to CO_2 if carbon–carbon bonds are broken; Sec. 2.7.)

TABLE 11.2 Comparison of Oxidation States of Various Functional Groups

All molecules in the same box have the same oxidation number.
X = an electronegative group such as halogen

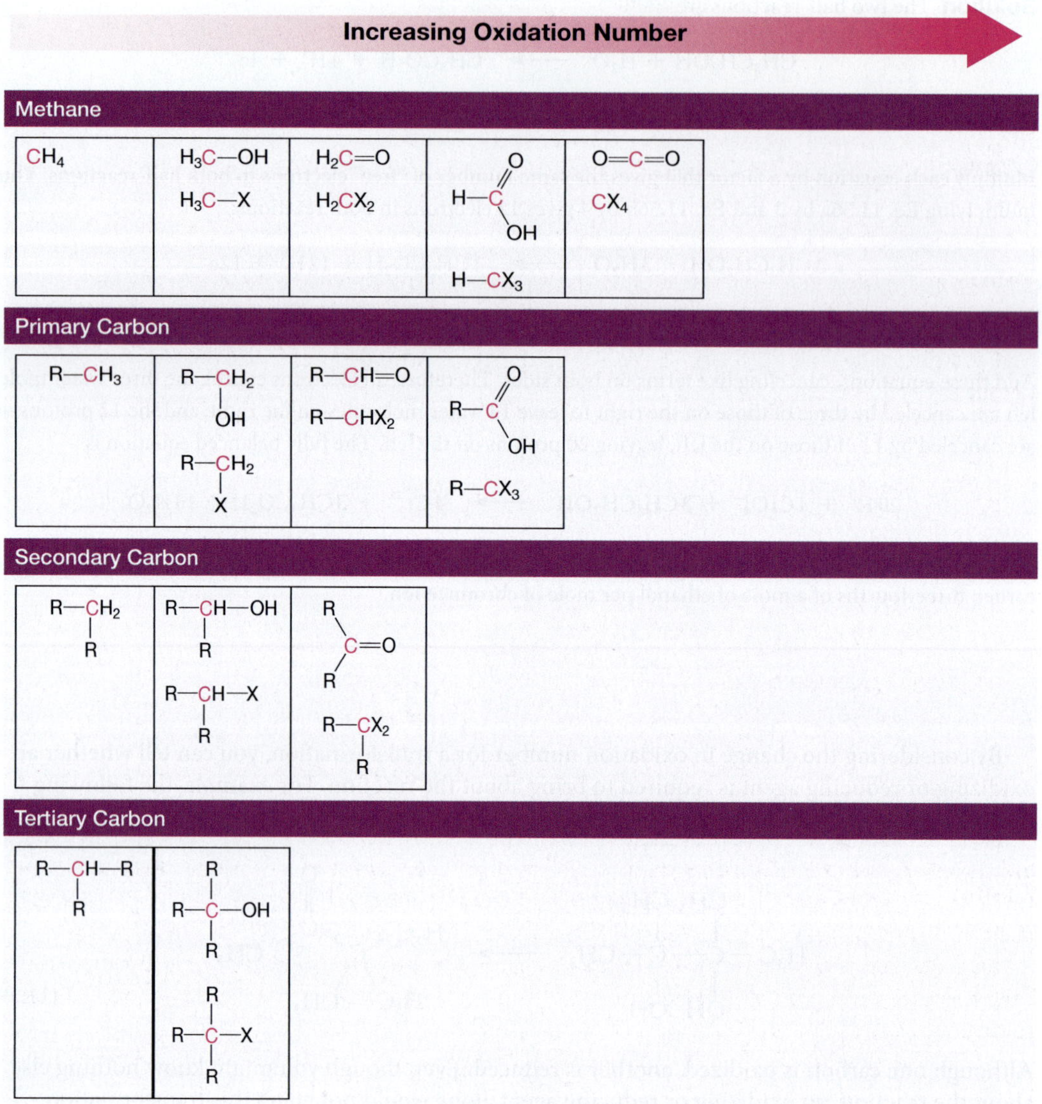

Focused Problems

11.27 For each of the following balanced oxidation–reduction reactions, indicate which compound(s) are oxidized and which are reduced. (*Hint:* Consider the change in the organic compounds in each reaction first.)

(a) $H_2C{=}CH_2 + H_2 \xrightarrow{Pd/C} H_3C{-}CH_3$

(b) $CH_3CH_2Br + Li^+\ {}^-AlH_4 \longrightarrow CH_3CH_3 + Li^+\ Br^- + AlH_3$

(c) 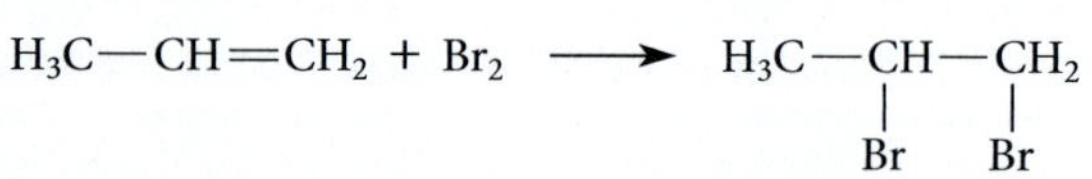

(d) 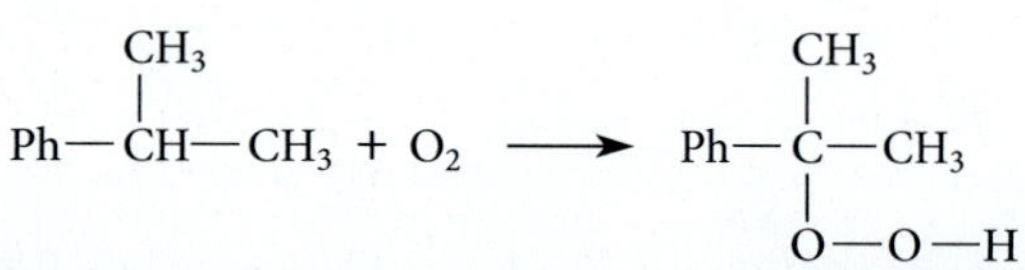

11.28 How many moles of permanganate are required to oxidize one mole of toluene to benzoic acid? (Use H_2O and protons to balance the equation.)

$$\underset{\textbf{toluene}}{Ph-CH_3} + \underset{\textbf{permanganate}}{MnO_4^-} \longrightarrow \underset{\textbf{benzoic acid}}{Ph-CO_2H} + \underset{\textbf{manganese dioxide}}{MnO_2}$$

11.7 OXIDATION OF ALCOHOLS

A. Oxidation to Aldehydes and Ketones

Primary and secondary alcohols can be oxidized by reagents containing Cr(VI)—that is, chromium in the +6 oxidation state—to give certain types of *carbonyl compounds* (compounds that contain the carbonyl group, $\rangle C{=}O$). For example, secondary alcohols are oxidized to ketones:

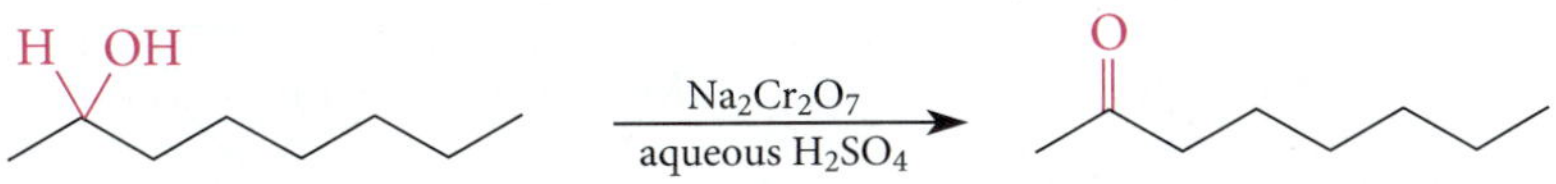

2-octanol **2-octanone** (94% yield) (11.60)

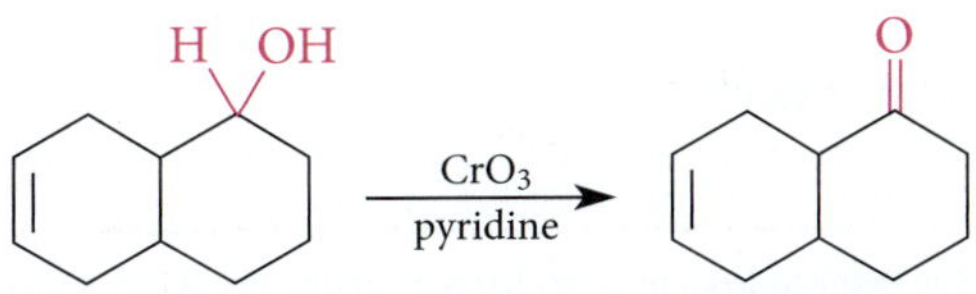

(89% yield) (11.61)

Several forms of Cr(VI) can be used to convert secondary alcohols into ketones. Three of these are chromate (CrO_4^{2-}), dichromate (CrO_7^{2-}), and chromic anhydride or chromium trioxide (CrO_3). CrO_4^{2-} and CrO_7^{2-} are customarily used under strongly acidic conditions, whereas CrO_3 is often used in pyridine. In all cases, the chromium is reduced to a form of Cr(III) such as Cr^{3+}.

Primary alcohols react with Cr(VI) reagents to give aldehydes. If water is present, however, then the reaction cannot be stopped at the aldehyde stage because aldehydes are further oxidized to carboxylic acids:

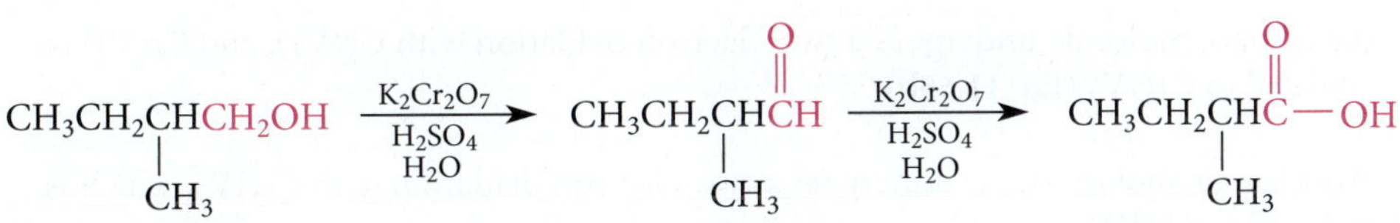

2-methyl-1-butanol **2-methylbutanal** (not isolated) **2-methylbutanoic acid**

(11.62)

For this reason, *anhydrous* preparations of Cr(VI) are generally used for the laboratory preparation of aldehydes from primary alcohols. One commonly used reagent of this type is a complex of pyridine, HCl, and chromium trioxide called *pyridinium chlorochromate*, which goes by the abbreviation "PCC." This reagent is typically used in methylene chloride solvent.

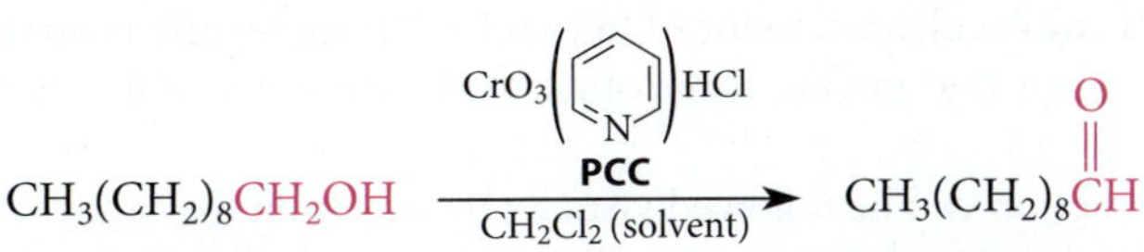

1-decanol **decanal** (92% yield) (11.63)

Water promotes the transformation of aldehydes into carboxylic acids because, in water, aldehydes are in equilibrium with hydrates formed by the addition of water across the C=O double bond. (Hydration of carbonyl groups is discussed in Sec. 19.8A.)

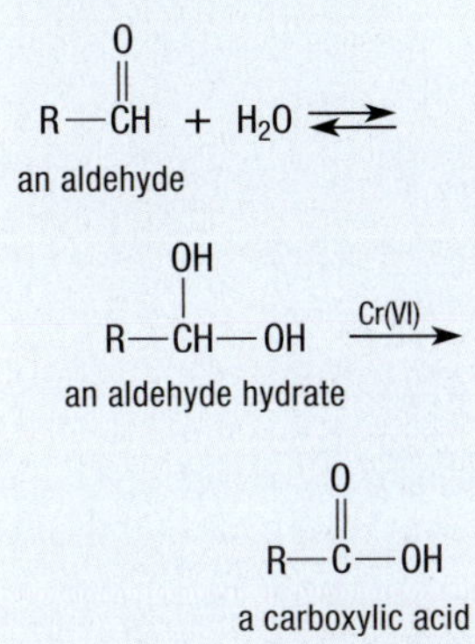

(11.64)

Aldehyde hydrates are really alcohols, so they can be oxidized just like secondary alcohols. Because of the absence of water in anhydrous reagents such as PCC, a hydrate does not form, and the reaction stops at the aldehyde.

Tertiary alcohols are not oxidized under the usual conditions. For the oxidation of alcohols at the α-carbon to occur, *the α-carbon atom must bear one or more hydrogen atoms.*

We illustrate the mechanism of alcohol oxidation with the oxidation of 2-propanol (isopropyl alcohol) to the ketone acetone. The first steps of the reaction involve an acid-catalyzed displacement of water from chromic acid by the alcohol to form a *chromate ester.* (This ester is analogous to ester derivatives of other strong acids, such as sulfate esters; Sec. 11.4C.)

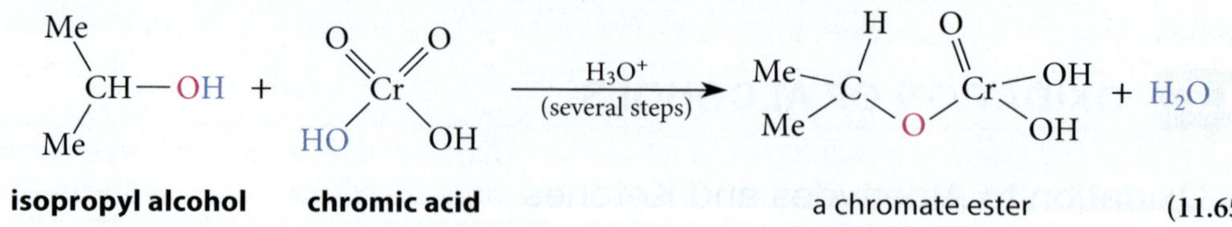

(11.65a)

The oxidation occurs within the chromate ester when *a hydrogen with its two electrons* (a hydride) is transferred to one of the oxygens (color in Eq. 11.65b). As a result, the alcohol undergoes a two-electron oxidation to a ketone, and chromium gains two electrons and is thus reduced to a Cr(IV) oxidation state.

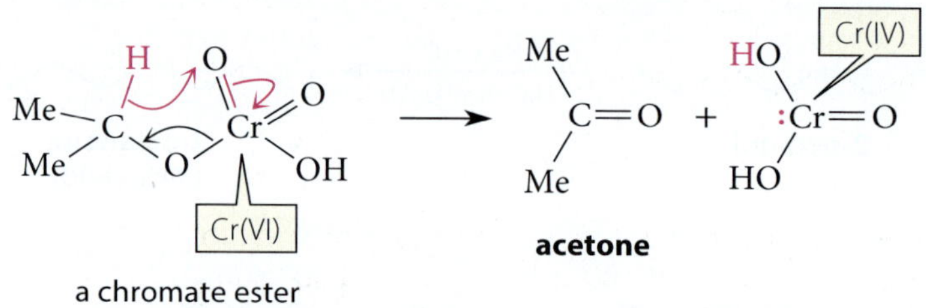

(11.65b)

> Be sure to distinguish this reaction from an acid–base reaction, which is a proton-transfer reaction; the hydrogen in this reaction is transferred not as a proton but as a *hydride:* a hydrogen with two electrons. A number of oxidation reactions involve hydride transfer.

As shown in Eq. 11.65b, the chromium in the H_2CrO_3 by-product is in a +4 oxidation state. The complete mechanism involves oxidation of alcohol molecules not only by Cr(VI), but also by Cr(IV) and Cr(V). A hydride-transfer mechanism operates in all three oxidations. The ultimate by-product of the reaction is Cr(III).

To summarize the steps in this oxidation:

1. An alcohol molecule undergoes a two-electron oxidation with Cr(VI), and Cr(VI) is reduced to Cr(IV) (Eq. 11.65b).

2. Another alcohol molecule undergoes a two-electron oxidation with Cr(IV), which is reduced to Cr(II).

3. Cr(II) and Cr(VI) react with each other in a one-electron oxidation–reduction reaction to form Cr(III) and Cr(V).

4. Another alcohol molecule undergoes a two-electron oxidation with Cr(V), which is reduced to Cr(III).

5. Overall, three alcohol molecules are oxidized to ketone (three two-electron oxidations), and two Cr(VI) molecules are reduced to two Cr(III) molecules (two three-electron reductions) by these steps that involve three other oxidation states of the metal.

The progress of the reaction can be followed visually by watching the orange color of Cr(VI) change to the blue-green color of Cr(III). (See the sidebar on the breath test for ethanol.)

The Breath Test for Ethanol

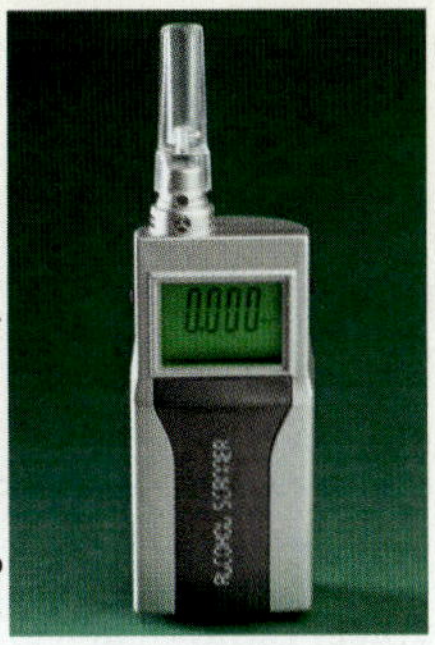
imageBROKER/Superstock

Law enforcement officers can use any of several devices to estimate a person's blood alcohol content (BAC), and two of these are based on ethanol oxidation to acetic acid. All of the devices utilize the fact that, in the lungs, a known small fraction of ethanol in the blood escapes into the air that is subsequently exhaled, and it is the alcohol content of this exhaled vapor that is analyzed. In the original measuring device, called a "breathalyzer," the exhaled air is collected and allowed to react with acidic potassium dichromate ($K_2Cr_2O_7$). The alcohol reduces the Cr(VI) to Cr(III). The resulting change in color of the chromium from the yellow-orange of the Cr(VI) oxidation state to the blue-green of Cr(III) is detected by a simple spectrometer. The amount of Cr(VI) reduced is calibrated in terms of percent blood alcohol. (See Problem 11.67.) Most modern devices use electrochemical technology. The alcohol is oxidized at a platinum electrode to produce acetic acid along with four protons and four electrons. (See Study Problem 11.3.) The resulting electron flow in the electrochemical cell is detected and calibrated in terms of blood alcohol concentration. A third type of device uses infrared spectroscopy (Chapter 13) to detect ethanol.

B. Oxidation to Carboxylic Acids

As noted in the previous section (Eq. 11.62), primary alcohols can be oxidized to carboxylic acids using *aqueous* solutions of Cr(VI), such as aqueous potassium dichromate ($K_2Cr_2O_7$) in acid. Another useful reagent for oxidizing primary alcohols to carboxylic acids is potassium permanganate ($KMnO_4$) in basic solution:

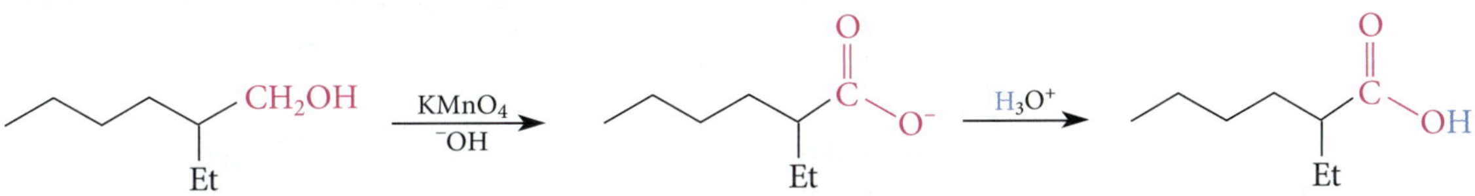

2-ethyl-1-hexanol a carboxylate ion (conjugate base of the carboxylic acid) **2-ethylhexanoic acid** (74% yield)

(11.66)

As shown in this equation, the immediate product of the permanganate oxidation is the conjugate base of the carboxylic acid because the reaction is run in alkaline solution. Isolation of the carboxylic acid itself requires addition of a strong acid such as H_2SO_4 or HCl in a second step.

The manganese in ($KMnO_4$) is in the Mn(VII) oxidation state; in the oxidation of alcohols, it is reduced to manganese dioxide (MnO_2), a common form of Mn(IV). In many cases, the reaction can be followed visually by the transformation of the deep violet color of the MnO_4^- solution to a brown precipitate of MnO_2. Because $KMnO_4$ also reacts with alkene double bonds (Sec. 12.6A), Cr(VI) is used for the oxidation of alcohols that contain double or triple bonds (see Eq. 11.61).

Potassium permanganate is *not* used for the oxidation of secondary alcohols to ketones because many ketones react further with the alkaline permanganate reagent.

Focused Problems

11.29 Give the product expected when each of the following alcohols reacts with pyridinium chlorochromate (PCC).

(a) 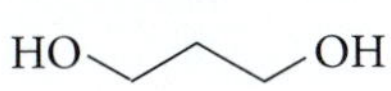

(b) 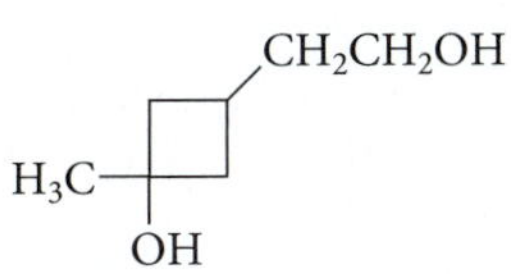

11.30 From which alcohol and by what method would each of the following compounds best be prepared by an oxidation?

(a) 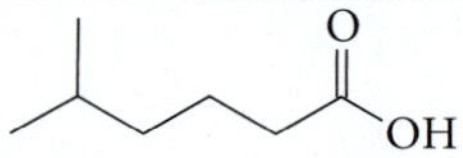

(b) $CH_3CH_2\overset{O}{\overset{\|}{C}}CH_2CH_3$

(c) 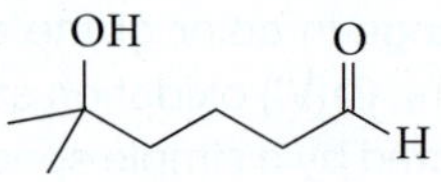

(d) 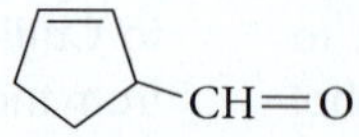

11.8 BIOLOGICAL OXIDATION OF ETHANOL

Chemical Biology Topic

Oxidation and reduction reactions are very important in living systems. A typical biological oxidation is the conversion of ethanol into acetaldehyde, the principal reaction by which ethanol is removed from the bloodstream.

$$\underset{\textbf{ethanol}}{CH_3CH_2OH} \xrightarrow[\text{(an enzyme)}]{\text{alcohol dehydrogenase}} \underset{\textbf{acetaldehyde}}{CH_3CH{=}O} \qquad (11.67)$$

The reaction is carried out in the liver and is catalyzed by an enzyme called *alcohol dehydrogenase*. (Sec. 4.10D explained that enzymes are biological catalysts.) The actual oxidizing agent is not the enzyme, but rather a large molecule tightly bound to the enzyme called *nicotinamide adenine dinucleotide*, abbreviated NAD^+; the structure of NAD^+ and a convenient abbreviated structure for it are shown in Display 11.68. In the abbreviation, the colored portion of the full structure is represented as an R group.

abbreviated structure for NAD^+ or $NADP^+$

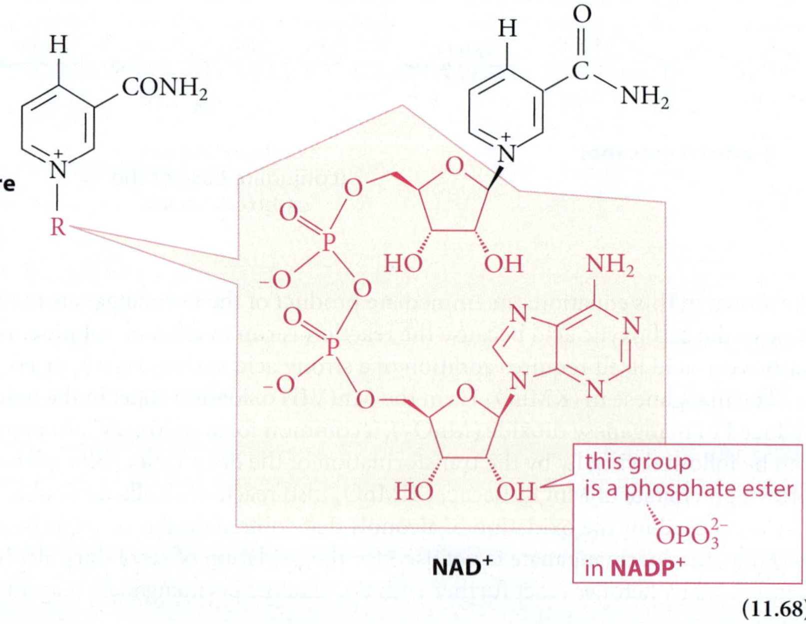

NAD^+

(11.68)

(Don't memorize the full structure of NAD^+. Rather, think of it in terms of the abbreviated structure shown in Display 11.68.)

When ethanol is oxidized, NAD^+ is reduced to a product called NADH. The hydrogen removed from carbon-1 of the ethanol ends up in the NADH; the OH hydrogen is lost as a proton.

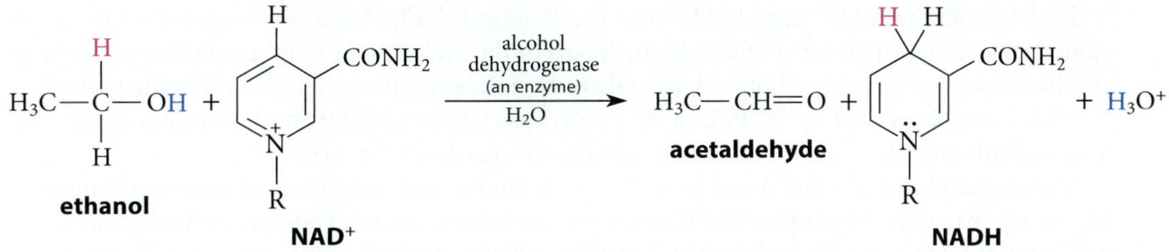

(11.69)

NAD$^+$ is one of nature's most important oxidizing agents. [From the perspective of laboratory chemistry, it might be called "nature's substitute for Cr(VI)."] NAD$^+$ is a *coenzyme*. **Coenzymes** are molecules required, along with enzymes, for certain biological reactions to occur. For example, ethanol cannot be oxidized by an enzyme unless the coenzyme NAD$^+$ is also present, because NAD$^+$ is one of the reactants. An ethanol molecule and an NAD$^+$ molecule are juxtaposed when they bind noncovalently to alcohol dehydrogenase, the enzyme that catalyzes ethanol oxidation. It is within the complex of these three molecules that ethanol is oxidized to acetaldehyde and NAD$^+$ is reduced to NADH.

How does NAD$^+$ work as an oxidizing agent? The resonance structure of NAD$^+$ shows that, because the nitrogen in the ring can accept electrons, the molecule takes on the character of a carbocation:

The coenzymes NAD$^+$ and NADH are derived from the vitamin *niacin*, a deficiency of which is associated with the disease pellagra (black tongue). Many biochemical processes employ the NAD$^+$ $\rightleftharpoons$ NADH interconversion, some of which re-oxidize the NADH formed in ethanol oxidation back to NAD$^+$.

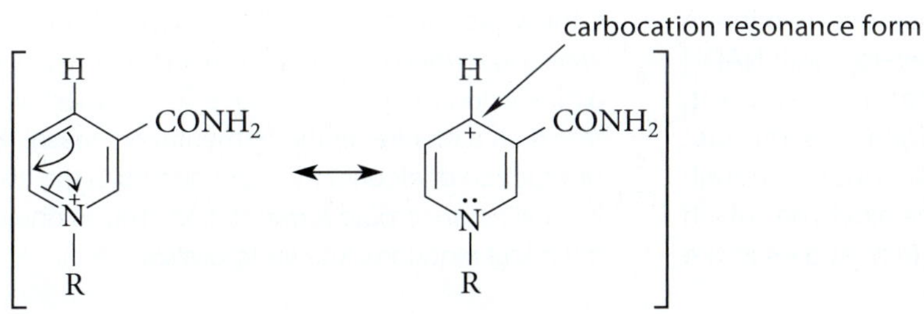

(11.70)

The electron-deficient carbon of NAD$^+$ and a α-hydrogen of ethanol (color in Eq. 11.71a) are held in proximity by the enzyme. The carbocation removes a *hydride* (a hydrogen with two electrons) from the α-carbon of ethanol:

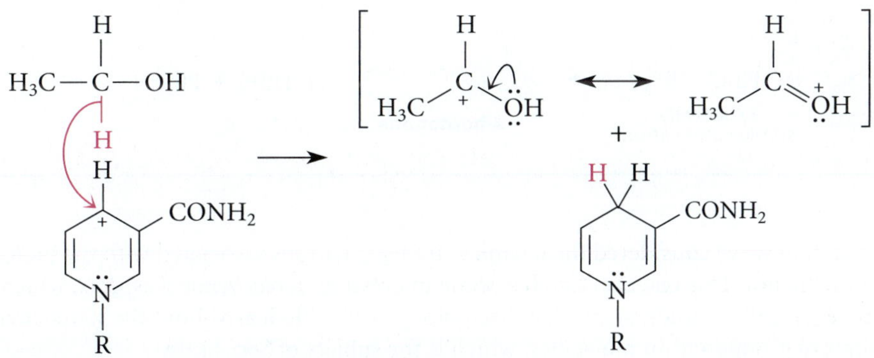

(11.71a)

This is another example of oxidation by *hydride transfer* (Sec. 11.7A). As a result, NADH and a new resonance-stabilized carbocation are formed. By loss of the proton bound to oxygen, acetaldehyde is formed.

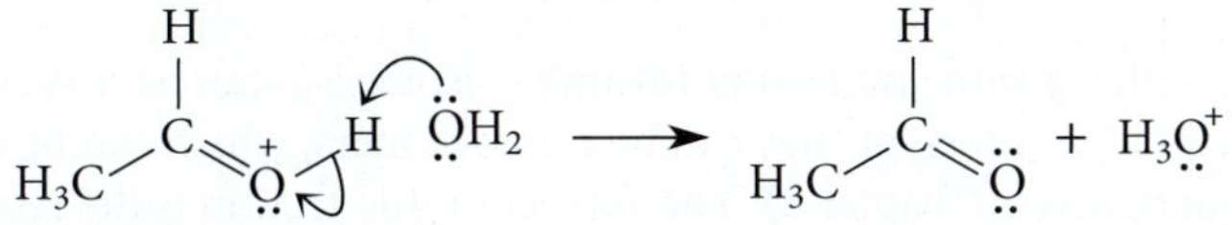

(11.71b)

The acetaldehyde and NADH dissociate from the enzyme, which is then ready for another round of catalysis. (NADH acts as a reducing agent in other reactions, by which it is converted back into NAD$^+$.)

Even though the NAD^+ molecule is large, the chemical changes that occur when it serves as an oxidizing agent take place in a relatively small part of the molecule. A number of other coenzymes also have complex structures but undergo relatively simple reactions. The groups in the part of the molecule abbreviated by "R" in Display 11.68 cause it to bind tightly to the enzyme catalyst by noncovalent attractions but remain unchanged in the oxidation reaction.

The chemical changes that occur in NAD^+-promoted oxidations have analogies in common laboratory reactions. Most other biochemical reactions have common laboratory analogies as well. Even though the molecules involved may be complex, their chemical transformations are in most cases relatively simple. Therefore, *an understanding of the fundamental types of organic reactions and their mechanisms is essential for an understanding of biochemical processes.*

Fermentation

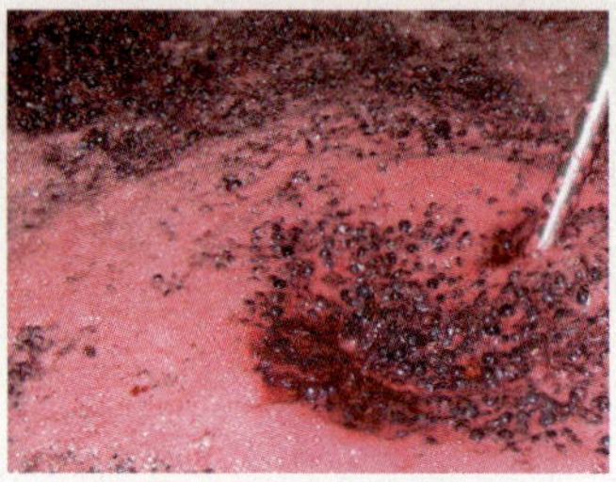

Fermenting pinot noir grapes

The human body uses the ethanol-to-acetaldehyde reaction (Eq. 11.69) to remove ethanol, but yeast cells use the reaction in reverse, with NADH as the reducing agent, as the last step in the production of ethanol. Yeast added to dough produces ethanol by the reduction of acetaldehyde, which is produced, in turn, in other reactions from sugars in the dough. Ethanol vapors, wafted away from the bread by CO_2 produced in other reactions, give rising bread its pleasant odor. Special strains of yeast ferment the sugars in grape juice or barley in the production of wine (photo), whiskey, or beer. For the fermentation reaction to take place, the reaction must occur in the absence of oxygen. Otherwise, acetic acid (CH_3CO_2H—vinegar or "spoiled wine") is formed instead by other reactions. In winemaking, air is excluded by trapping the CO_2 formed during fermentation as a blanket in the fermentation vessel. Because the production of alcohol by yeast occurs in the absence of air, it is called *anaerobic fermentation*. This is one of the oldest chemical reactions known to civilization.

Focused Problem

11.31 Write a curved-arrow mechanism for the following oxidation of 2-heptanol, which proceeds in 82% yield.

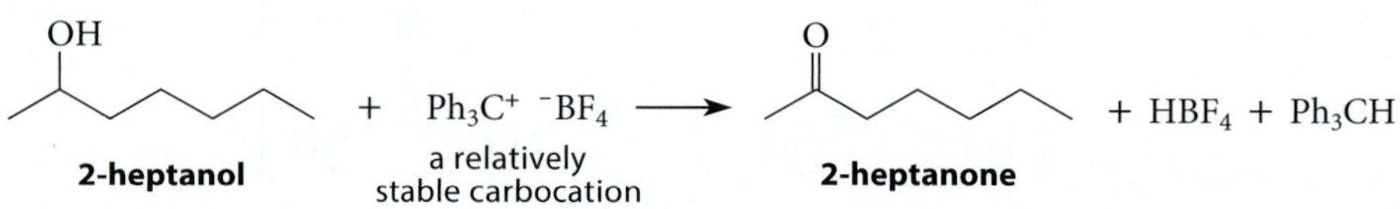

In this section we've considered the chemical transformations associated with the biological oxidation of ethanol. This reaction also has some interesting *stereochemical* aspects, which we discuss in Sec. 11.9B. To understand that discussion, we need to learn about the stereochemical relationships of groups within molecules, which is the subject of Sec. 11.9A.

11.9 CHEMICAL AND STEREOCHEMICAL GROUP RELATIONSHIPS

Different molecules with the same molecular formula—isomers—can have various relationships: they can be constitutional isomers, or they can be stereoisomers, which can be either diastereomers or enantiomers (Chapter 6). The subject of this section is the relationships that groups *within* molecules can have. This subject is particularly important in two areas. First, it is important in understanding the stereochemical aspects of enzyme catalysis. We demonstrate this by returning in Sec. 11.9B to the biological oxidation of ethanol, the reaction discussed in

Sec. 11.8. Second, the subject of group relationships is important in spectroscopy, particularly nuclear magnetic resonance spectroscopy, which is discussed in Chapter 14.

A. Chemical Equivalence and Nonequivalence

When two groups are **chemically equivalent**, they behave in exactly the same way toward a chemical reagent. Otherwise, they are **chemically nonequivalent**.

An understanding of chemical equivalence hinges, first, on the concept of *constitutional equivalence*. Groups within a molecule are **constitutionally equivalent** when they have the same connectivity relationship to all other atoms in the molecule. For example, within each of the following molecules, the hydrogens shown in color are constitutionally equivalent:

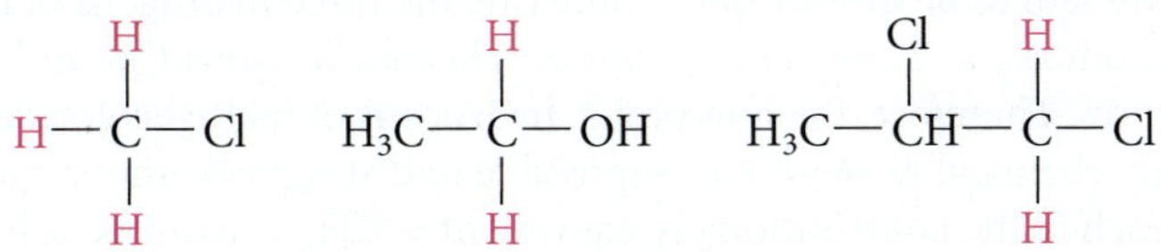

A *B* *C* (11.72)

In compound *C*, for example, each of the red hydrogens is connected to a carbon that is connected to a chlorine and to a CH_3CHCl— group. Each of these hydrogens has the same connectivity relationship to the other atoms in the molecule. On the other hand, the H_3C— hydrogens in ethanol (molecule *B*) are constitutionally nonequivalent to the —CH_2— hydrogens. The H_3C— hydrogens are connected to a carbon that is connected to a —CH_2OH, but the —CH_2— hydrogens are connected to a carbon that is connected to an —OH and a —CH_3. However, the two —CH_2— hydrogens of ethanol (*B*, red) are constitutionally equivalent to each other, as are the three H_3C— hydrogens.

In general, *constitutionally nonequivalent groups are chemically nonequivalent.* This means that constitutionally nonequivalent groups have different chemical behaviors. For example, the H_3C— and —CH_2— hydrogens of ethanol, which are constitutionally nonequivalent, have different reactivities toward chemical reagents. A reagent that reacts with one type of hydrogen will in general have a different reactivity (or perhaps none at all) with the other type. For example, the oxidation of ethanol with Cr(VI) reagents results in the loss of a —CH_2— hydrogen, but the H_3C— hydrogens are unaffected. Consequently, the two types of hydrogens have very different reactivities with the oxidizing agent.

Constitutional nonequivalence is a sufficient but not a necessary condition for chemical nonequivalence. That is, some constitutionally equivalent groups are also chemically nonequivalent. *Whether two constitutionally equivalent groups are chemically equivalent depends on their stereochemical relationship.* Therefore, to understand chemical equivalence, we need to understand the various *stereochemical relationships* that are possible between constitutionally equivalent groups.

The stereochemical relationship between constitutionally equivalent groups is revealed by a **substitution test**. In this test, we substitute each constitutionally equivalent group in turn with a fictitious circled group and compare the resulting molecules. Their stereochemical relationship determines the relationship of the circled groups. This process is best illustrated by example, starting with the molecules *A*, *B*, and *C* shown previously in Display 11.72. Substitute each hydrogen of molecule *A* with a circled hydrogen:

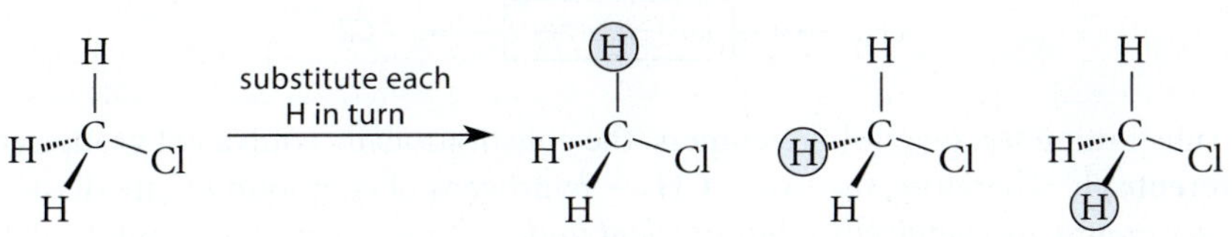

A *A1* *A2* *A3* (11.73)

Each of these "new" molecules is identical to the other. For example, the identity of *A1* and *A2* is shown in the following way:

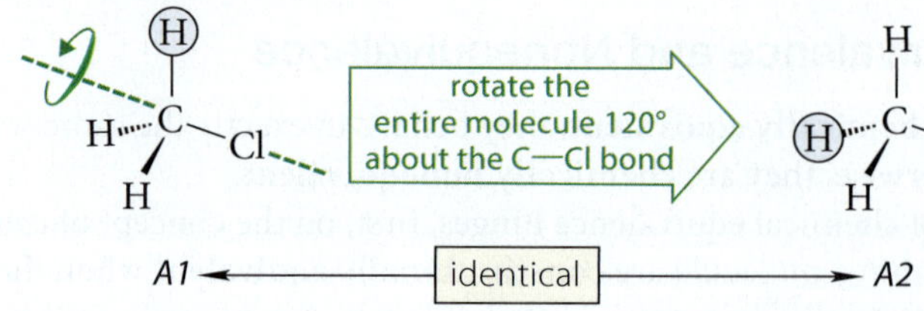

(11.74)

When the substitution test gives identical molecules, as in this example, the constitutionally equivalent groups are said to be **homotopic**. Therefore, the three hydrogens of methyl chloride (compound *A*) are homotopic. *Homotopic groups are chemically equivalent and indistinguishable under all circumstances.* Therefore, the homotopic hydrogens of methyl chloride all have the same reactivity toward any chemical reagent; it is impossible to distinguish among these hydrogens.

Substitution of each of the constitutionally equivalent —CH_2— hydrogens in molecule *B* (ethanol) gives molecules *B1* and *B2*, which are *enantiomers*:

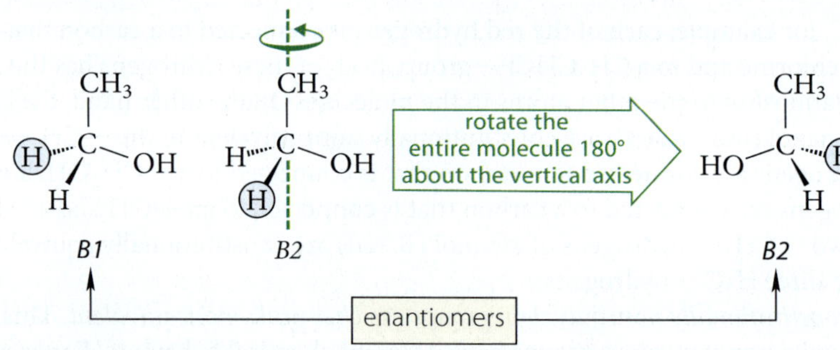

(11.75)

When the substitution test gives enantiomers, the constitutionally equivalent groups are said to be **enantiotopic**. Thus, the two —CH_2— hydrogens of compound *B* (ethanol) are enantiotopic. *Enantiotopic groups are chemically nonequivalent toward chiral reagents, but they are chemically equivalent toward achiral reagents.*

Because enzymes are chiral, they can generally distinguish between enantiotopic groups within a molecule. In the enzyme-catalyzed oxidation of ethanol, for example, one of the two enantiotopic α-hydrogens is selectively removed. (This point is further explored in Sec. 11.9B.) Achiral reagents, however, cannot distinguish between enantiotopic groups. So, in the oxidation of ethanol to acetaldehyde by chromic acid, an achiral reagent, the α-hydrogens of ethanol are removed indiscriminately.

Finally, substitution of each of the constitutionally equivalent —CH_2— hydrogens in molecule *C* gives molecules *C1* and *C2*, which are *diastereomers*.

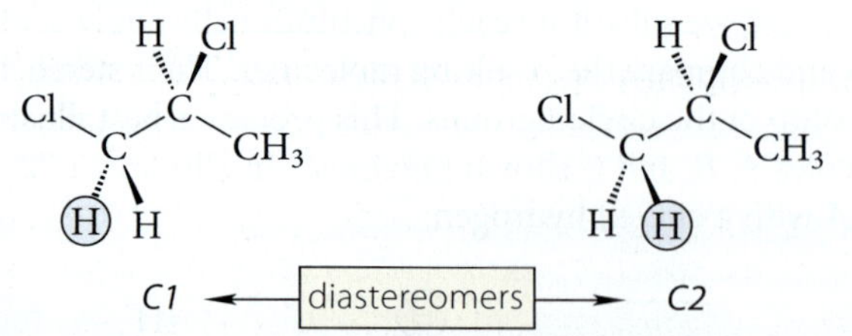

(11.76)

When the substitution test gives diastereomers, the constitutionally equivalent groups are said to be **diastereotopic**. Therefore, the two —CH_2— hydrogens of compound *C* are diastereotopic. *Diastereotopic groups are chemically nonequivalent under all conditions.* For example, the hydrogens labeled H^a and H^b in 2-bromobutane (Eq. 11.77) are also diastereotopic. In the E2 reaction of this compound, removal of the diastereotopic hydrogens gives diastereomers: anti elimination of H^b and Br gives *cis*-2-butene, whereas anti elimination of H^a and Br gives *trans*-2-butene. (Verify these statements using models, if necessary.)

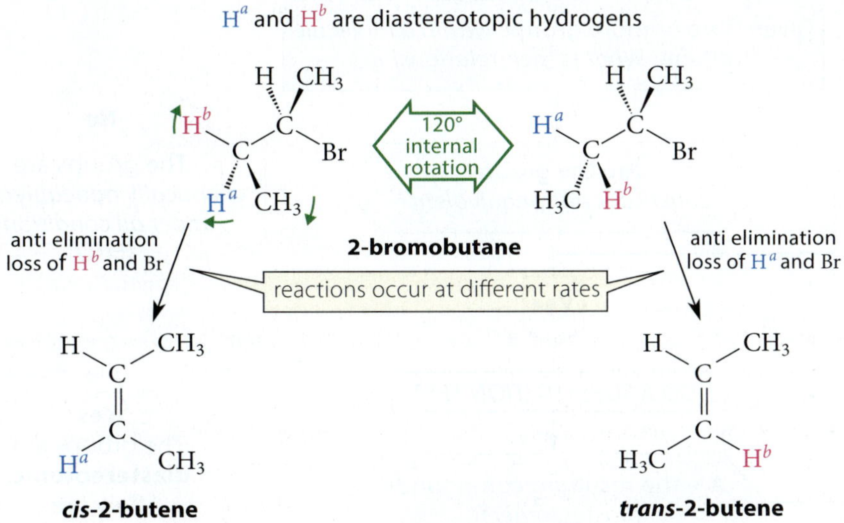

(11.77)

Different amounts of these two alkenes are formed in the elimination reaction precisely because the two diastereotopic hydrogens are removed at different rates—that is, these hydrogens are distinguished by the base that promotes the elimination.

Diastereotopic groups can be recognized at a glance in two important situations. The first occurs when two constitutionally equivalent groups are present in a molecule that contains an asymmetric carbon:

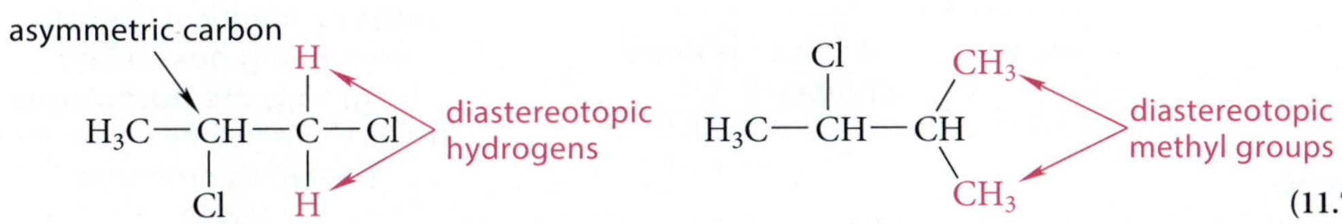

(11.78)

The carbon that bears the diastereotopic hydrogens need not be adjacent to the asymmetric carbon. In the following compound, the two hydrogens at each of the carbons marked with a diamond (◊) are diastereotopic, even though some of these carbons are not adjacent to asymmetric carbons.

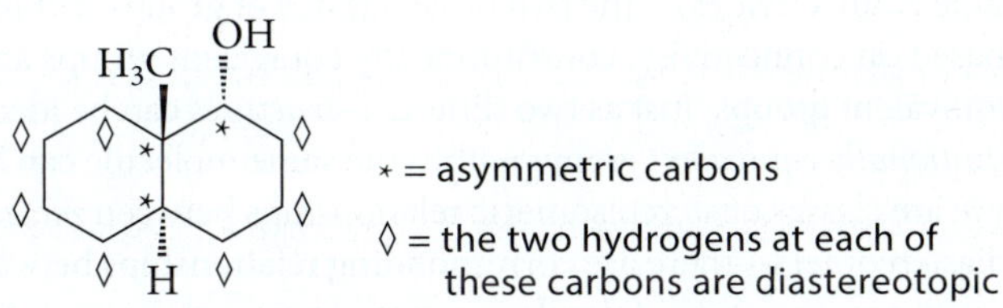

(11.79)

The second common situation occurs when two groups on one carbon of a double bond are the same and the two groups on the other carbon are different:

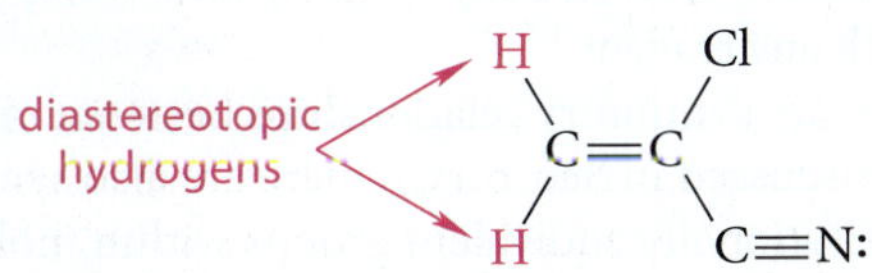

(11.80)

As you should verify, the substitution test on the red hydrogens gives *E,Z* isomers, which are diastereomers.

Just as Fig. 6.7 (Sec. 6.6B) can be used to summarize the relationships between isomeric molecules, **Fig. 11.1** can be used to summarize the relationships of groups within a molecule. Notice the close analogy between the relationships of *different molecules* and the relationships of *groups within a molecule*. Just as two broad classes of isomers are based on connectivity—constitutional isomers (isomers with different connectivities) and stereoisomers (isomers with

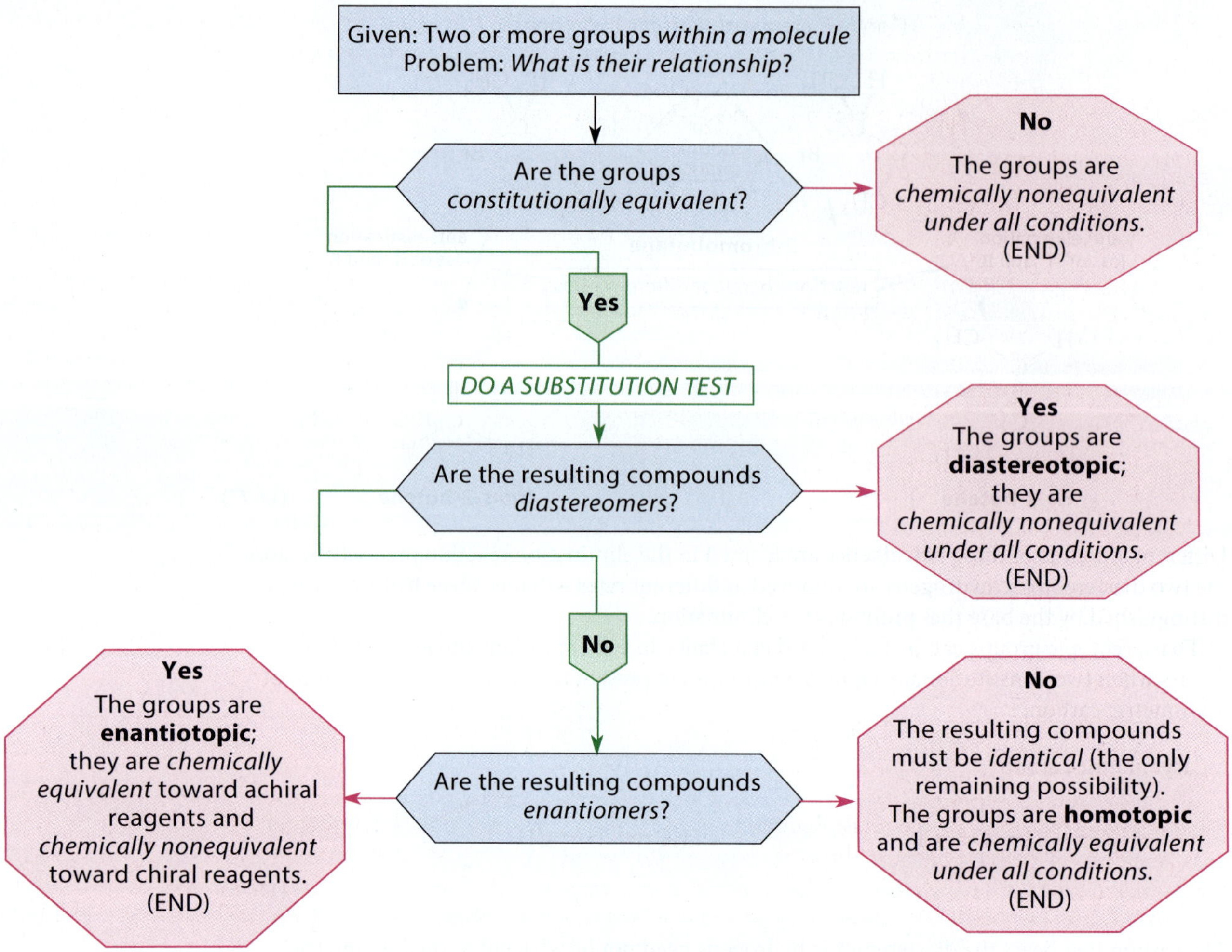

FIGURE 11.1 A flowchart for classifying groups within molecules.

Enantiotopic groups not only have different reactivities with chiral reagents, but they can be physically distinguished as well by any chiral probe. An amusing demonstration is to build a model of ethanol in which the two enantiotopic hydrogens are identical except for their color (or other distinguishing mark). Using your right or left hand as a "chiral probe," make sure you can "feel" the relationship of the colored hydrogen to the rest of the molecule, particularly the OH and the methyl group. Then have someone blindfold you and rotate the model about the C—C and C—O bonds. You should be able to pick out the colored hydrogen just by feeling its relationship to the rest of the molecule with your hand. One of the authors (ML) carried out this demonstration with his students for many years and never missed the identification.

the same connectivities)—the two broad classes of groups *within* a molecule are also based on connectivity: constitutionally equivalent groups and constitutionally nonequivalent groups. Just as two different *structures* can be identical, two *constitutionally equivalent groups* within the same molecule can be homotopic. Just as there are classes of stereoisomeric relationships between *molecules*—enantiomers and diastereomers—there are corresponding relationships between *constitutionally equivalent groups within molecules*—enantiotopic and diastereotopic relationships. Just as *enantiomers* have different reactivities only with chiral reagents, *enantiotopic groups* also have different reactivities only with chiral reagents. Just as *diastereomers* have different reactivities with any reagent, *diastereotopic groups* have different reactivities with any reagent.

Just as there are symmetry relationships between molecules that are the basis for chirality (discussed in Sec. 6.1C), there are also symmetry relationships between constitutionally equivalent groups within molecules that are associated with different group relationships. These are discussed in Further Exploration 11.3 in the *Study Guide and Solutions Manual.*

FURTHER EXPLORATION 11.3
Symmetry Relationships among Constitutionally Equivalent Groups

Let's summarize the answer to the question posed at the beginning of this section: When are two groups in a molecule chemically equivalent?

1. Constitutionally nonequivalent groups are chemically nonequivalent in all situations.

2. Homotopic groups are chemically equivalent in all situations.

3. Enantiotopic groups are chemically equivalent toward achiral reagents, but they are chemically nonequivalent toward chiral reagents (such as enzymes).

4. Diastereotopic groups are chemically nonequivalent in all situations.

Focused Problem

11.32 For each of the following molecules, state whether the groups indicated by italic letters are constitutionally equivalent or nonequivalent. If they are constitutionally equivalent, classify them as homotopic, enantiotopic, or diastereotopic. (For cases in which more than two groups are designated, consider the relationships within each pair of groups.)

(a) 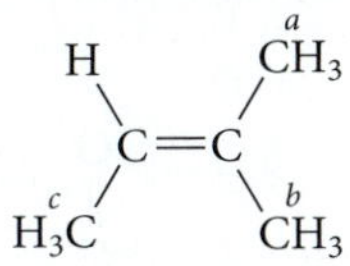
Classify each pair of methyl groups.

(b) 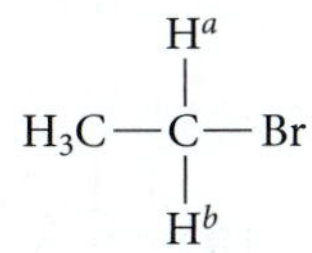

(c) 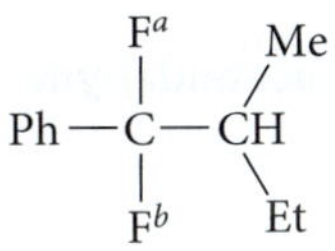

(d) 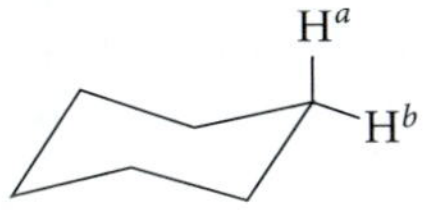

(e) 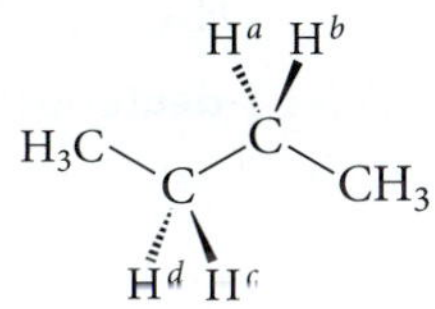

(f) 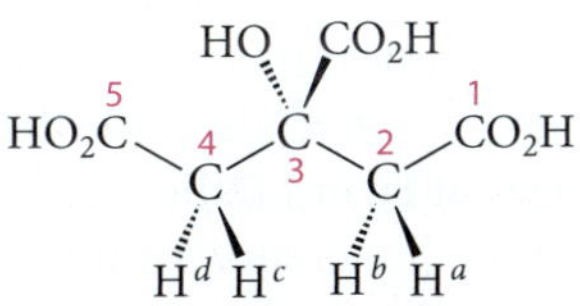
citric acid

Classify the relationship between each pair of labeled hydrogens as well as the relationship between carbon-2 and carbon-4.

B. Stereochemistry of the Alcohol Dehydrogenase Reaction

Chemical Biology Topic

In this section we use the alcohol dehydrogenase reaction from Sec. 11.8 to demonstrate not only that a chiral reagent can distinguish between enantiotopic groups, but also *how this discrimination can be detected.* Suppose each of the enantiotopic α-hydrogens of ethanol is replaced in turn with the isotope deuterium (^{2}H, or D). Replacing one hydrogen, called the pro-(*R*)-hydrogen, with deuterium gives (*R*)-1-deuterioethanol; replacing the other, called the pro-(*S*)-hydrogen, gives (*S*)-1-deuterioethanol. Although ethanol itself is *not* chiral, the deuterium-substituted analogs *are* chiral; they are a pair of enantiomers:

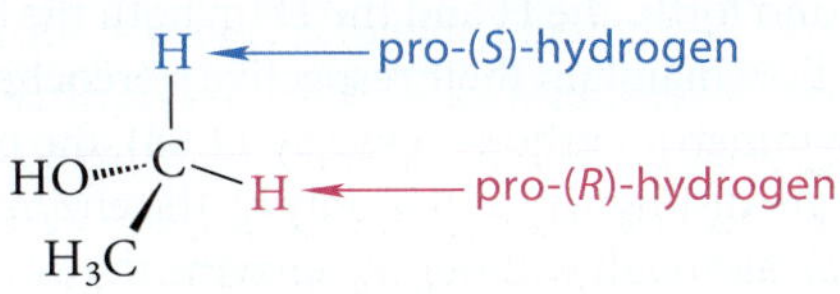

ethanol

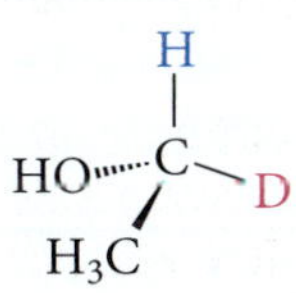
(*R*)-(+)-1-deuterioethanol

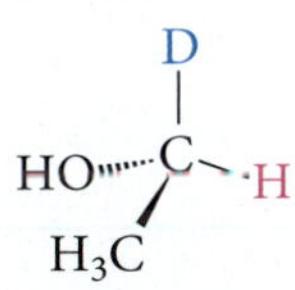
(*S*)-(−)-1-deuterioethanol

The deuterium isotope, then, provides a subtle way for the experimentalist to "label" the enantiotopic α-hydrogens of ethanol and thus to distinguish between them.

If the alcohol dehydrogenase reaction is carried out with (*R*)-1-deuterioethanol, then *only the deuterium* is transferred from the alcohol to the NAD$^+$:

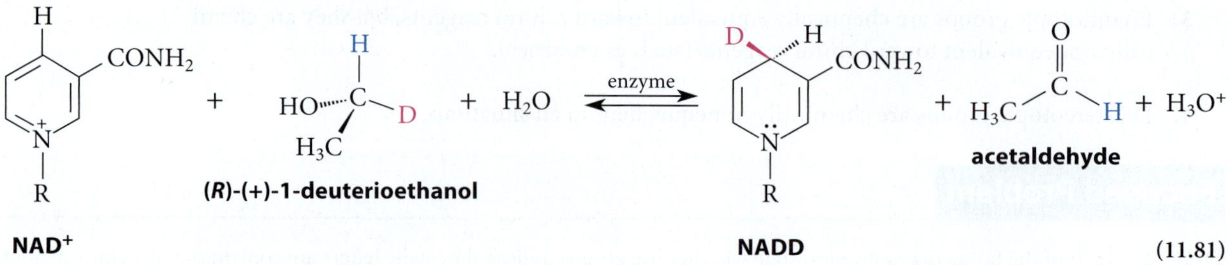

(11.81)

If the alcohol dehydrogenase reaction is carried out on (*S*)-1-deuterioethanol, however, then *only the hydrogen* is transferred.

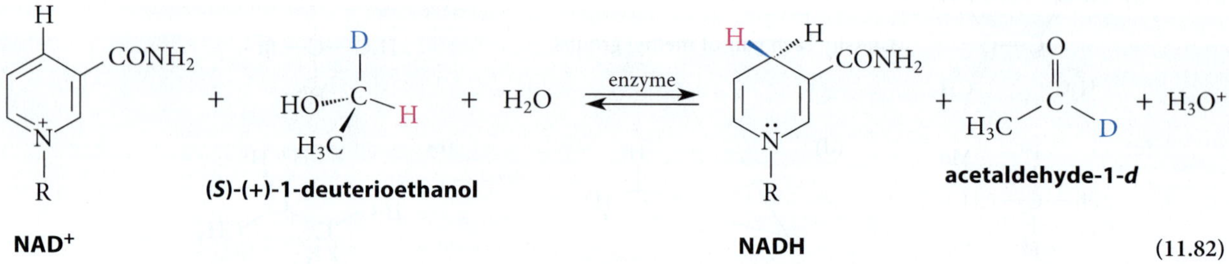

(11.82)

These two experiments show that *the enzyme distinguishes between the two α-hydrogens of ethanol.* These results cannot be attributed to a primary deuterium isotope effect (Sec. 9.5D), because an isotope effect would cause the enzyme to transfer the hydrogen in preference to the deuterium in both cases. Nor can the result be attributable to different sizes of H and D, because their size difference is minuscule. Although the isotope is used to detect the preference for transfer of one hydrogen and not the other, this experiment requires that even in the absence of the isotope the enzyme transfers the pro-(*R*)-hydrogen of ethanol.

The enzyme can distinguish between the two α-hydrogens of ethanol because the enzyme is chiral, and the two α-hydrogens are *enantiotopic*. This case, then, is an example of the principle that *enantiotopic groups react differently with chiral reagents* (Sec. 11.9A).

Equations 11.81 and 11.82 show, too, that the deuterium (or hydrogen) that is removed from the isotopically substituted ethanol molecule is transferred specifically to one particular face, or side, of the NAD^+ molecule. That is, the deuterium in the product NADD (color) occupies the position above the plane of the page. This result and the principle of microscopic reversibility (Sec. 4.10C) require that if acetaldehyde and the NADD stereoisomer shown on the right of Eq. 11.81 were used as starting materials and the reaction run in reverse, then only the deuterium should be transferred to the acetaldehyde, and (*R*)-1-deuterioethanol should be formed. Indeed, Eq. 11.81 *can* be run in reverse, and the experimental result is as predicted. No matter how many times the reaction runs back and forth, the H and the D on both the ethanol and the NADD molecules are never "scrambled"; they maintain their respective stereochemical positions. Because the R-group in NADH contains asymmetric carbons (Display 11.68), the two CH_2 hydrogens in NADH are in fact *diastereotopic*; they are distinguished not only by the enzyme, but also in the absence of the enzyme (at least in principle), although without the enzyme they might not be distinguished as thoroughly.

Focused Problem

11.33 In each of the following cases, imagine that the two reactants shown are allowed to react in the presence of alcohol dehydrogenase. Tell whether the ethanol formed is chiral. If the ethanol is chiral, draw a line-and-wedge structure of the enantiomer that is formed.

(a)

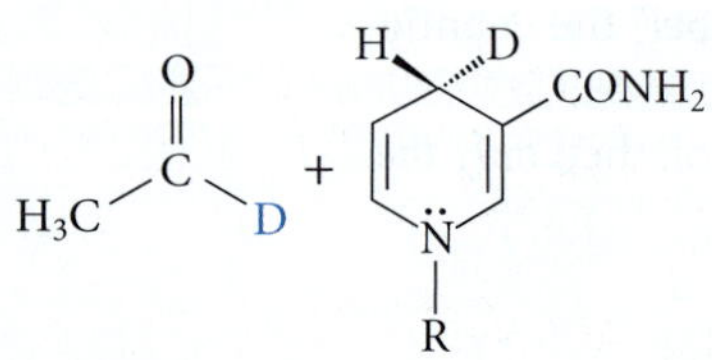

(b)

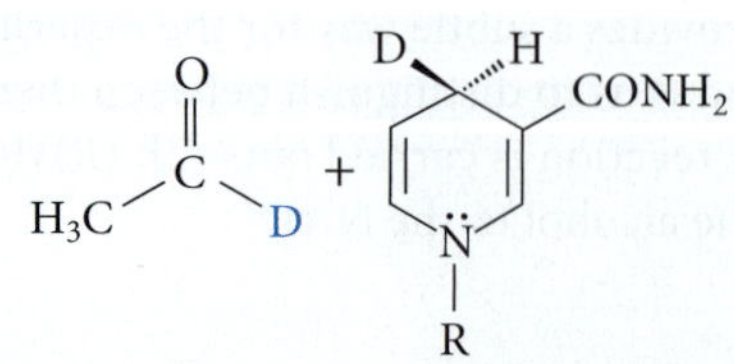

11.10 OCTET EXPANSION AND OXIDATION OF THIOLS

A. Octet Expansion and Hypervalent Compounds

Some of the chemistry of thiols is closely analogous to the chemistry of alcohols because sulfur and oxygen are in the same group of the periodic table. For oxidation reactions, however, this similarity disappears. Oxidation of an alcohol (Sec. 11.7) occurs at the *carbon* atom bearing the OH group.

$$\mathrm{RCH_2OH} \xrightarrow{\text{oxidation}} \mathrm{RCH{=}O} \xrightarrow{\text{oxidation}} \mathrm{RCO_2H} \qquad (11.83a)$$

Oxidation of a thiol, however, takes place at the *sulfur*. Although sulfur analogs of aldehydes, ketones, and carboxylic acids are known, they are *not* obtained by the simple oxidation of thiols:

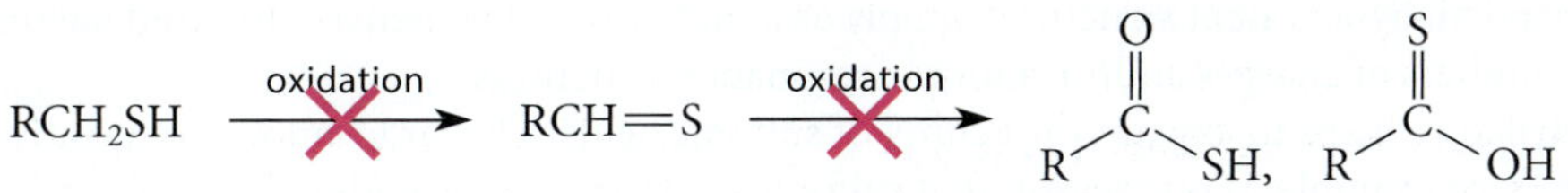

(11.83b)

Figure 11.2 shows the Lewis structures of organic compound classes containing sulfur and phosphorus in order of increasing oxidation numbers, along with the structures of the important inorganic compounds sulfuric acid and phosphoric acid. In this figure, oxidation numbers of sulfur and phosphorus are calculated by assigning a –1 to R and H, a 0 to another S, and a +1 for each bond to oxygen, and applying the "oxidation number" process in Sec. 11.6A to sulfur or phosphorus.

The Lewis structures of some of the compounds in Fig. 11.2 have more than an octet of electrons around sulfur or phosphorus. Such structures are said to have *expanded octets.* They are also referred to as *hypervalent compounds* (Secs. 1.3C and 3.3C). Each of these hypervalent structures, however, have resonance forms that conform to the octet rule but have large amounts of separated charge.

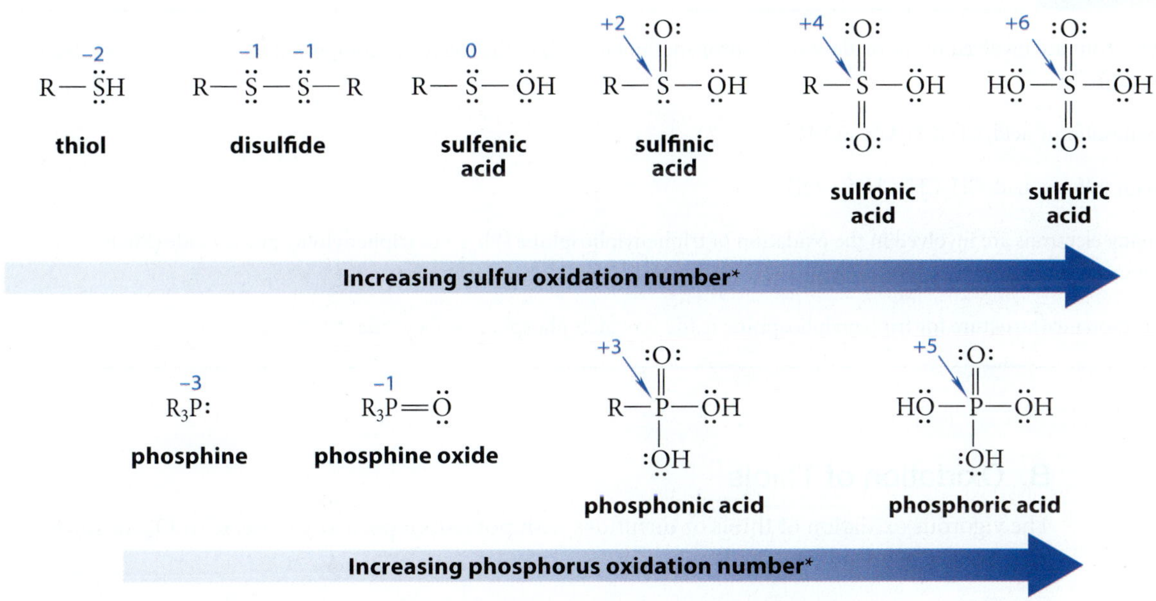

*The blue numbers are oxidation numbers (Sec. 11.6A) calculated by assigning R and H each a contribution of –1; S a contribution of 0; and each bond to O a contribution of +1.

FIGURE 11.2 Organic derivatives of sulfur and phosphorus along with the inorganic compounds sulfuric acid and phosphoric acid. These are arranged in order of increasing oxidation number of the sulfur or phosphorus. Sulfenic and sulfinic acids are not very common in organic chemistry. Sulfuric acid and phosphoric acid represent the highest oxidation states of sulfur (+6) and phosphorus (+5), respectively. One or more hydrogens of the acids can be replaced by R-groups to give organic esters; see, for example, Secs. 11.4A (sulfonate esters), 11.4C (sulfate esters), and 11.4E (phosphate esters).

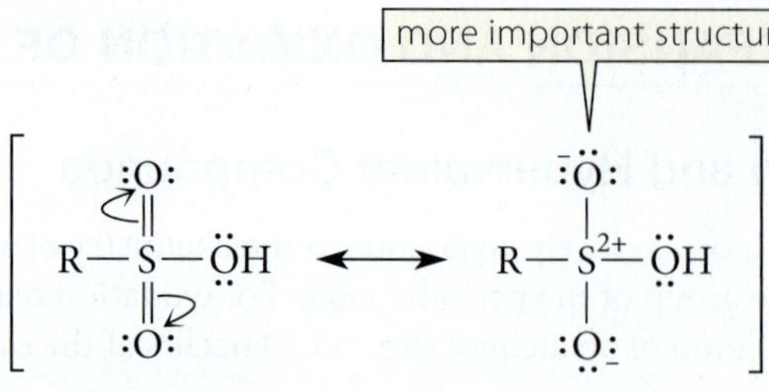

the uncharged structure has 12 electrons around sulfur

the octet structure has formal charges of opposite sign on adjacent atoms

(11.84)

It was at one time believed that the involvement of 3d orbitals on sulfur and oxygen in bonding was responsible for hypervalency. (This view was presented in the previous edition of this text.) Molecular orbital calculations have shown that the involvement of 3d orbitals in hypervalent bonding is small to negligible.

Despite the large amount of separated charge, *the octet structures are the more important resonance structures*, as discussed in Sec. 3.3C. Nevertheless, it is traditional to represent such compounds with octet-expanded, hypervalent structures simply as a matter of convenience—to avoid having to draw large numbers of charges and/or multiple resonance structures.

The octet structure helps to explain the acidity of sulfonic acids. The large amount of positive charge on sulfur, for example, is one reason that sulfonic acids are strong acids ($pK_a \sim -3$). This positive charge stabilizes the conjugate-base anion (polar effect; Sec. 3.6C).

Once we recognize that the octet structures are important, the only truly hypervalent molecules are the ones with more than four groups surrounding the central atom, such as SF_6 and PCl_5. Why can hypervalent molecules occur with atoms from periods 3 and higher of the periodic table and not with elements from period 2? The current thinking is that elements from periods ≥ 3 are larger than elements from period 2; consequently, there is simply more room around these atoms to accommodate more than four bonded groups. Recall, however, that even these hypervalent molecules have a strong contribution from ionic resonance forms that have electronic octets around the central atom (Eq. 3.24, Sec. 3.3B).

Focused Problems

11.34 How many electrons are involved in the oxidation of 1-propanethiol to each of the following compounds. (See Fig. 11.2 for detailed Lewis structures.)

(a) 1-Propanesulfonic acid, $CH_3CH_2CH_2SO_3H$

(b) 1-Propanesulfenic acid, $CH_3CH_2CH_2S{-}OH$

11.35 (a) How many electrons are involved in the oxidation of triphenylphosphine (Ph_3P) to triphenylphosphine oxide ($Ph_3P{=}O$)? Show your reasoning.

(b) Draw a resonance structure for triphenylphosphine oxide in which phosphorus obeys the octet rule.

B. Oxidation of Thiols

The vigorous oxidation of thiols or disulfides with potassium permanganate, $KMnO_4$, or nitric acid, HNO_3, gives sulfonic acids.

$$\underset{\textbf{1-butanethiol}}{CH_3CH_2CH_2CH_2{-}SH} \xrightarrow{\text{conc. } HNO_3} \underset{\substack{\textbf{1-butanesulfonic acid} \\ \text{(72–96\% yield)}}}{CH_3CH_2CH_2CH_2{-}SO_3H} \qquad (11.85)$$

As noted in Sec. 11.4A, sulfonate esters are derivatives of sulfonic acids. Other sulfonic acid chemistry is considered in Chapters 16 and 20.

Many thiols are spontaneously oxidized to disulfides merely on standing in air (O_2). Thiols can also be converted into disulfides by mild oxidants such as I_2 in base or Br_2 in an inert solvent such as CCl_4:

$$2\,CH_3(CH_2)_4SH + I_2 + 2\,NaOH \longrightarrow CH_3(CH_2)_4S{-}S(CH_2)_4CH_3 + 2\,NaI + 2\,H_2O$$

(70% yield) (11.86)

$$2\,EtSH + Br_2 \longrightarrow EtS{-}SEt + 2\,HBr$$

ethanethiol → **diethyl disulfide** (nearly quantitative yield) (11.87)

A reaction like Eq. 11.86 can be viewed as a series of S_N2 reactions in which thiolate anions, formed by thiol ionizations, react as nucleophiles, first toward a halogen, and then toward a sulfur electrophile.

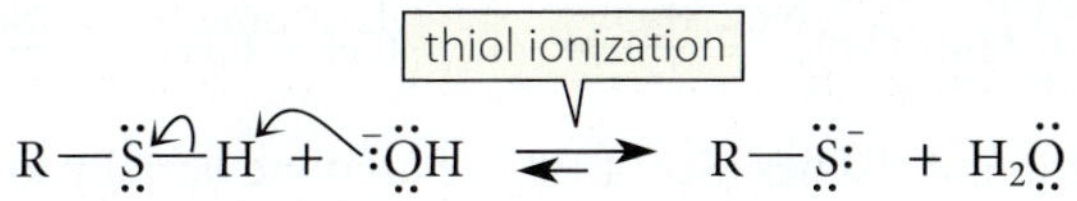

(11.88a)

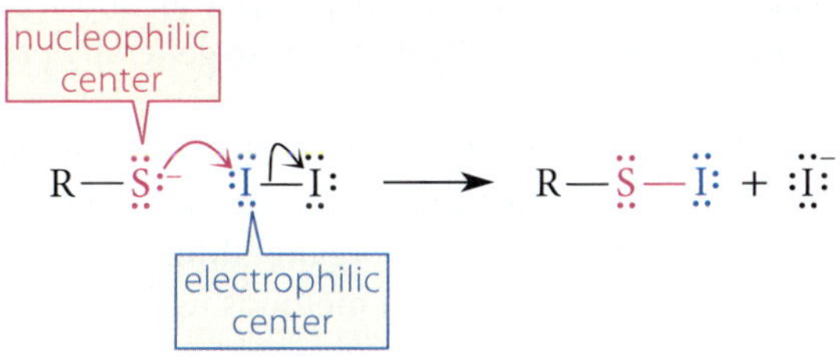

(11.88b)

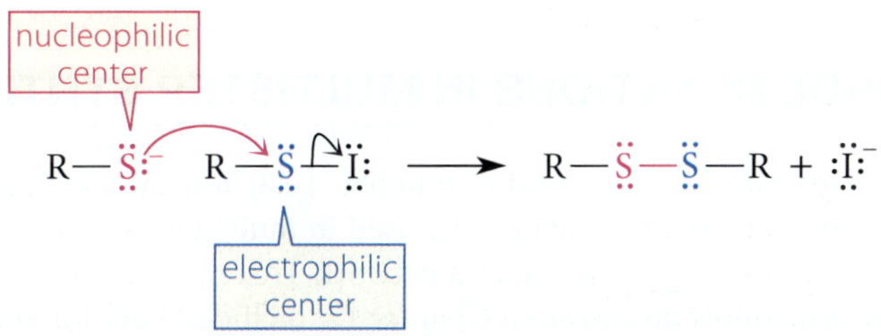

(11.88c)

When thiols and disulfides are present together in the same solution, an equilibrium among them is rapidly established. If ethanethiol and dipropyl disulfide are combined, for example, then they react to give a mixture of all possible thiols and disulfides:

$$EtSH + PrS{-}SPr \rightleftharpoons EtS{-}SPr + PrSH$$

ethanethiol + **dipropyl disulfide** ⇌ **ethyl propyl disulfide** + **propanethiol** (11.89a)

$$EtSH + EtS{-}SPr \rightleftharpoons EtS{-}SPr + PrSH$$

ethanethiol + **ethyl propyl disulfide** ⇌ **diethyl disulfide** + **propanethiol** (11.89b)

Thiols and sulfides are very important in biology. Many enzymes contain thiol groups that have catalytically essential functions, and disulfide bonds in proteins help to stabilize their three-dimensional structures (Sec. 27.8A; see the sidebar "A Practical Example of Disulfide-Bond Reduction.") The vulcanization of rubber (curing in the presence of sulfur) introduces disulfide bonds that increase the strength and rigidity of the rubber (Sec. 15.5).

Focused Problem

11.36 The rates of the reactions in Eqs. 11.89a and b are increased when the thiol is ionized by a base such as sodium ethoxide. Suggest a mechanism for Eq. 11.89a that is consistent with this observation, and explain why the presence of base makes the reaction faster.

11.11 SYNTHESIS OF ALCOHOLS: A REVIEW

We here review the methods presented in previous chapters for the synthesis of alcohols. All of these methods begin with alkene starting materials:

1. Hydroboration–oxidation of alkenes (Sec. 5.4B).

$$3\,RCH{=}CH_2 + BH_3 \xrightarrow[THF]{} (RCH_2CH_2)_3B \xrightarrow{H_2O_2/NaOH} 3\,RCH_2CH_2OH$$

2. Oxymercuration–reduction of alkenes (Sec. 5.4A).

$$R{-}CH{=}CH_2 + Hg(OAc)_2 \xrightarrow{H_2O} R{-}\underset{\large OH}{CH}{-}\underset{\large HgOAc}{CH_2} \xrightarrow{NaBH_4} R{-}\underset{\large OH}{CH}{-}CH_3$$

Acid-catalyzed hydration of alkenes (Sec. 4.10C) is used industrially to prepare certain alcohols, but this is not an important laboratory method. In principle, the S_N2 reaction of ^{-}OH with primary alkyl halides can also be used to prepare primary alcohols. This method is of little practical importance, however, because alkyl halides are generally prepared from alcohols themselves.

Some of the most important methods for the synthesis of alcohols involve the reduction of carbonyl compounds (aldehydes, ketones, or carboxylic acids and their derivatives), as well as the reactions of carbonyl compounds with Grignard or organolithium reagents. These methods are presented in Chapters 19, 20, and 21. A summary of methods used to prepare alcohols is found in Appendix V.

11.12 ALCOHOL REACTIONS IN MULTISTEP SYNTHESIS

The strategy of *multistep synthesis* was introduced in Sec. 10.8. You can now add the reactions of alcohols to your repertoire of reactions that can be used in multistep synthesis. Here is a summary of alcohol reactions from this chapter shown as a network. [Asterisked reactions (*) are methods for ether synthesis from alcohols discussed in Chapter 12, included here for your future use.]

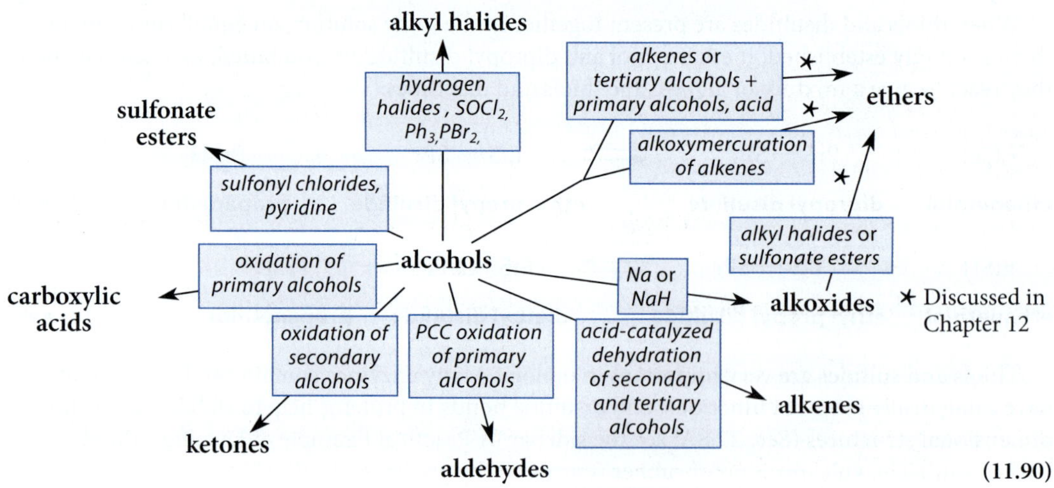

(11.90)

You will find it useful to link this network with the networks for alkenes and alkynes shown in Displays 5.54 and 5.55 (Sec. 5.9B). Study Problem 11.7 illustrates the use of such a linkage.

Study Problem 11.7

Outline a synthesis of hexanal from 1-hexene and any other reagents.

$$\underset{\textbf{1-hexene}}{\mathrm{CH_3CH_2CH_2CH_2CH{=}CH_2}} \longrightarrow \underset{\textbf{hexanal}}{\mathrm{CH_3CH_2CH_2CH_2CH_2CH{=}O}}$$

Solution Begin by working backward from hexanal, the target molecule. First, ask whether aldehydes can be directly prepared from alkenes. The answer is yes. Ozonolysis (Sec. 5.5) can be used to transform alkenes into aldehydes and ketones. However, ozonolysis breaks a carbon–carbon double bond and certainly would not work for preparing an aldehyde from an alkene *with the same number of carbon atoms*, because at least one carbon is lost when the double bond is broken. No other ways of preparing aldehydes directly from alkenes have been covered. The next step is to ask how aldehydes can be prepared from other starting materials. We have introduced two ways: hydroboration–oxidation of 1-alkynes (Sec. 5.7) and oxidation of primary alcohols (Sec. 11.7A). Because the starting material for the synthesis is an alkene, we want to use a starting material in this step that can be prepared from an alkene, if possible. This is where we use the linkage of the alkene reaction network. Primary alcohols can be prepared from 1-alkenes; therefore, oxidation of a primary alcohol could be the final step in a satisfactory synthesis:

$$\underset{\textbf{1-hexanol}}{\mathrm{CH_3CH_2CH_2CH_2CH_2CH_2{-}OH}} \xrightarrow[\mathrm{CH_2Cl_2}]{\mathrm{PCC}} \underset{\textbf{hexanal}}{\mathrm{CH_3CH_2CH_2CH_2CH_2CH{=}O}}$$

Hydroboration–oxidation will convert 1-hexene into 1-hexanol. The synthesis is now complete:

$$\underset{\textbf{1-hexene}}{\mathrm{CH_3CH_2CH_2CH_2CH{=}CH_2}} \xrightarrow[\mathrm{THF}]{\mathrm{BH_3}} \xrightarrow{\mathrm{H_2O_2},\ \mathrm{NaOH}} \underset{\textbf{1-hexanol}}{\mathrm{CH_3CH_2CH_2CH_2CH_2CH_2{-}OH}} \xrightarrow[\mathrm{CH_2Cl_2}]{\mathrm{PCC}} \underset{\textbf{hexanal}}{\mathrm{CH_3CH_2CH_2CH_2CH_2CH{=}O}}$$

We summarize the retrosynthetic process used in this solution as follows:

target molecule → hexanal ⇨ 1-hexanol ⇨ 1-hexene ← starting molecule

where the symbol ⇨ means "implies as the precursor."

Focused Problem

11.37 Outline a synthesis of each of the following compounds from the indicated starting material. Begin each synthesis with a retrosynthetic analysis.

(a) 2-Methyl-3-pentanol from 2-methyl-2-pentanol

(b) 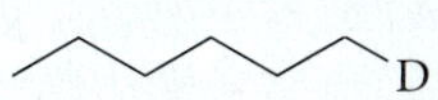 from 1-hexanol

(c) 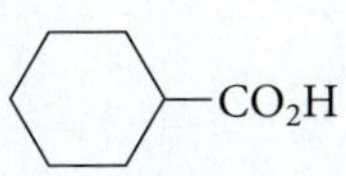 from 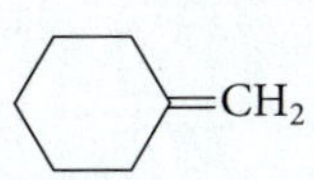

(d) $\mathrm{CH_3CH_2CH_2CH(CH_3)CH{=}O}$ from $\mathrm{CH_3CH_2CH_2CH(CH_3)CH_2Br}$

Chemistry in the Real World

11.13 PRODUCTION AND USE OF ETHANOL AND METHANOL

Ethanol A number of alcohols are important in commerce. None, however, has been affected by recent events as dramatically as ethanol. In 1980, the U.S. production of ethanol was 175 million gallons. Much of it was made industrially by the hydration of ethylene (Sec. 4.10C), which comes, in turn, from petroleum (Sec. 5.7).

$$\underset{\textbf{ethylene}}{H_2C{=}CH_2} + H{-}OH \xrightarrow[300\ ^\circ C]{H_3PO_4} \underset{\textbf{ethanol}}{CH_3CH_2OH} \qquad (11.91)$$

In recent years, the U.S. production of ethanol has climbed rapidly, reaching 16.1 billion gallons in 2018 and likely to climb significantly higher. Most of this ethanol is produced by fermentation of the sugars in grain—mostly corn (**Fig. 11.3**). In 2018, the world production of ethanol was 28.7 billion gallons, with the United States providing about 56% and Brazil about 28% of this total. (Brazil, where plants such as sugarcane that produce fermentable sugars grow rapidly, has long operated on an ethanol-based fuel economy.) Ethylene hydration now accounts for only 7% of the world's ethanol production.

The increase in ethanol production in the United States has been driven by the demand for motor fuels that are not derived from hydrocarbons, by the phase-out of methyl *tert*-butyl ether (MTBE; see the discussion that follows) and its replacement by ethanol as the main oxygen-containing component of gasoline, and by the notion that the use of ethanol and other plant-derived fuels will reduce the amount of CO_2 released into the atmosphere. (See the sidebar "Environmental Change and Biofuels," which follows.)

The relatively small amount of ethanol not used for fuel is used as a starting material for the preparation of other compounds and as a solvent for inks, fragrances, and the like. Chemically pure ethanol is subject to tight federal controls to ensure that it will not be used in beverages. In many cases the ethanol used in solvent applications is *denatured alcohol,* which is ethanol made unfit for human consumption by the addition of certain toxic additives such as methanol.

Jim Parkin/Shutterstock

FIGURE 11.3 The fermentation of the sugars in corn to ethanol is carried out in plants such as this one in Iowa. Iowa is the leading producer of corn-derived ethanol, and the United States is the leading country for production.

Beverage alcohol is produced by the fermentation of barley, grape juice, corn mash, or other sources of natural sugar. Beverage alcohol is not isolated; rather, alcoholic beverages are the mixtures of ethanol, water, and the natural colors and flavorings produced in the fermentation process and purified by sedimentation (as in wine) or distillation (as in brandy or whiskey). Industrial alcohol cannot be used legally to alter the alcoholic composition of beverages.

Ethanol is a drug, and, like many useful drugs, it is toxic when consumed in excess. Ethanol is the most abused drug in the world.

Environmental Change and Biofuels

Vaivirga/Getty Images

Global warming, caused by a rapid increase in atmospheric CO_2 levels, is discussed in Sec. 2.7 (see Fig. 2.10). The increase in CO_2 levels has resulted from the combustion of fossil fuels—oil, coal, and natural gas. The environmental impact of fossil fuels and the political instability of many of the world's oil-producing regions have conspired to make the development of alternative fuels increasingly urgent. Ideally, the goal is to produce cheap and abundant fuels that will not, on combustion, increase the *net* CO_2 content of the atmosphere. For this reason, fuels derived from plants (biomass), termed generally *biofuels,* are attractive. Plants are rich sources of glucose in various polymerized forms, such as starch (from corn) and cellulose (from stalks and grasses). Cellulose is the single most abundant organic compound on Earth. When plants produce glucose in its various forms, they remove CO_2 from the atmosphere. The energy for the plant synthesis of glucose comes from the Sun through photosynthesis. If we add the equations for the photosynthesis of glucose, the fermentation of glucose to ethanol, and the combustion of ethanol as a fuel, the result is no net change in atmospheric CO_2:

$$6\,CO_2 + 6\,H_2O \xrightarrow{\text{light}} C_6H_{12}O_6 + 6\,O_2 \quad \text{(biosynthesis of glucose, } C_6H_{12}O_6\text{)}$$

$$C_6H_{12}O_6 \longrightarrow 2\,C_2H_5OH + 2\,CO_2 \quad \text{(anaerobic fermentation)}$$

$$2\,C_2H_5OH + 6\,O_2 \longrightarrow 4\,CO_2 + 6\,H_2O \quad \text{(combustion of ethanol)}$$

$$\text{Sum: No Net Change} \qquad (11.92)$$

Photosynthesis harnessed to produce cellulose is nature's solar-energy collection and storage mechanism. However, this simple picture does not take into account the energy required to produce and transport the agricultural products, to run the fermenters, and to deliver the ethanol-containing products to the end users. Accordingly, the net energy gain from the use of various biofuels has been a matter of considerable debate.

Putting the carbon cycle of Eq. 11.92 into practice requires decisions on which fuels to use and what plants are to be their major sources. These issues are fraught with significant political and economic ramifications. For example, the diversion of grain to ethanol production has had a significant impact on the cost of food, because producers who raise poultry, pork, and beef on feed grain have seen their feed costs rise significantly, and these costs are passed on to the consumer. Therefore, finding other sources of fermentable cellulose, such as silage, waste products from wood, and grasses such as switchgrass (photo) is essential if ethanol is to become a major fuel source. Fermentation of switchgrass and related cellulose sources currently accounts for only a few million of the 16.8 billion gallons of ethanol produced in the United States.

The problem of energy is one of the greatest challenges ever to face the human race. Cheap and abundant energy—and ways to store it—would contribute significantly to solving many of the world's problems, such as poverty, the scarcity of food, and inadequate shelter. Increasingly expensive and scarce energy will lead to wars over limited resources and to environmental disaster. Biofuels will undoubtedly be only a part of the solution to the energy problem, but they will probably fill an important niche in the ultimate solution. Although it is tempting to be discouraged by the vast scale of this problem, it is encouraging to realize that immense opportunities undoubtedly await the scientifically trained citizens who can think in new, creative ways about the solutions.

Methanol Methanol is formed from a mixture of carbon monoxide and hydrogen, called synthesis gas, at high temperature over a special catalyst.

$$\underbrace{CO + H_2}_{\text{synthesis gas}} \xrightarrow[\substack{100\text{–}600\ \text{atm} \\ 250\text{–}400\ ^\circ\text{C}}]{\text{Cr-Zn catalyst}} \underset{\textbf{methanol}}{CH_3OH} \qquad (11.93)$$

Synthesis gas comes from the partial oxidation of methane, which is derived, in turn, from natural gas, the cracking of hydrocarbons (Sec. 5.7), or the gasification of coal. Methane produced by the fermentation of biological waste can also provide a link for methanol production to biofuels.

The 2020 global production of methanol is expected to be more than 20 billion gallons. Important uses of methanol include its oxidation to formaldehyde ($H_2C{=}O$) and its reaction with carbon monoxide over special catalysts to give acetic acid (CH_3CO_2H). A rapidly growing use of methanol is as a feedstock for relatively new processes for the production of alkenes, so-called methanol-to-olefin (MTO) technologies. A bright spot in methanol's future may also be its use in the production of biodiesel fuel. With an octane rating of 116, methanol itself also has a largely unrealized potential for use as a motor fuel. (It has been used as a fuel in Formula One racing engines for years, and it is used extensively in China as a motor fuel.)

In the United States in the 1990s, methanol became an important compound in the fight against urban automotive air pollution. Prior to 1990, efforts to control automotive air pollution were focused on the automobile itself—thus, the "catalytic converter." In 1990, a new strategy for reducing automotive air pollution was mandated by the Clean Air Act amendments: to add chemicals ("additives") to gasoline itself. Chief among these additives were the so-called oxygenates, and the two most important of these were ethanol and methyl *tert*-butyl ether (MTBE). Methanol is one of the two starting materials in the industrial synthesis of MTBE, an industrial example of alkene addition.

The International Union of Pure and Applied Chemistry (IUPAC) common name of MTBE is *tert*-butyl methyl ether. If the chemical industry had followed the rules for nomenclature, it would have been nicknamed TBME. However, the "incorrect" name and the abbreviation MTBE are so thoroughly entrenched that we use them in this text.

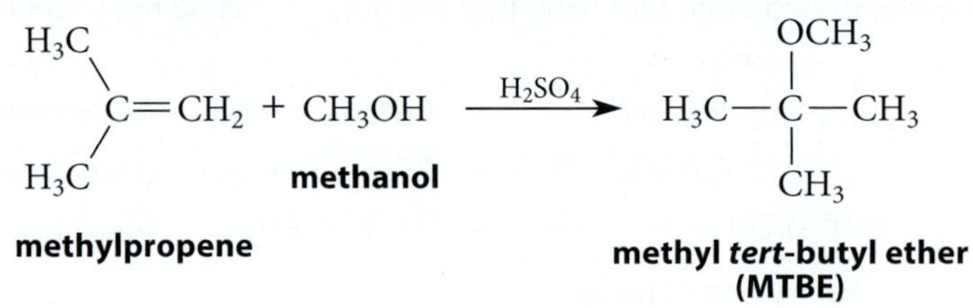

(11.94)

The use of MTBE in gasoline significantly reduced urban air pollution from automobile exhausts. It rapidly became one of the top 10 industrial organic chemicals, and it took methanol along for the ride. New methanol plants were built to feed the demand for MTBE. Then trouble started for MTBE. It was found in groundwater in California and Maine, and the source of the chemical was leakage from underground storage tanks. An advisory panel of the Environmental Protection Agency (EPA) recommended in August 1999 that Congress move to reduce substantially the use of MTBE in gasoline. A controversy developed about whether MTBE should be banned, but California mandated a phase-out by 2002. Many other states followed suit, and MTBE has largely been discontinued as a gasoline additive in the United States and Western Europe, leaving ethanol as the primary oxygenate used in gasoline. Ethanol producers are happy with this turn of events, but methanol producers were left with excess capacity. Many methanol plants were closed, and production in the United States decreased over the ensuing period.

Methanol production in the United States is now rebounding, however, because of its promise as an alternative fuel, and its potential as a starting material for hydrocarbon production via the methanol-to-olefin technologies mentioned previously.

CHAPTER SUMMARY *For a summary of the chapter, see Chapter 11 in the* Study Guide and Solutions Manual.

REACTION REVIEW *For a summary of reactions discussed in this chapter, see the* Reaction Review *section of Chapter 11 in the* Study Guide and Solutions Manual.

SKILLS OBJECTIVES WITH PROBLEMS

- **Write a reaction for the ionization of an alcohol or thiol. Answer questions involving the following aspects of alcohol and thiol acidity. 11.1**

  (1) Give the approximate pK_a of an ordinary alcohol and thiol. **11.1A**

  (2) Name the conjugate base of an alcohol or thiol. **11.1B**

  (3) Rank a series of alcohols in order of increasing acidity, and give the reasons for your choices. **11.1C,D**

11.38 (a) Write the equilibrium for the reaction of ethanol (acting as a Brønsted acid) with triethylamine, Et_3N: (acting as a Brønsted base).

(b) Estimate the equilibrium constant for this reaction. The pK_a of the conjugate acid of triethylamine is 10.65.

11.39 Name each of the following compounds.

(a) 

(b) 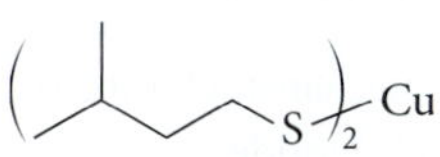

11.40 Provide structures for each of the following compounds.

(a) Magnesium methoxide

(b) Lithium cyclohexanolate

11.41 Arrange the compounds or ions within each set in order of increasing acidity (decreasing pK_a) in solution. Explain your reasoning.

(a) 2-Chloro-1-propanethiol (*A*), 2-chloroethanol (*B*), 3-chloro-1-propanethiol (*C*)

(b) Propyl alcohol (*A*), isopropyl alcohol (*B*), *tert*-butyl alcohol (*C*), 1-propanethiol (*D*)

(c) $CH_3\ddot{N}H—CH_2CH_2—\ddot{O}H$ (*A*), $CH_3\ddot{N}H—CH_2CH_2CH_2—\ddot{O}H$ (*B*), $(CH_3)_3\overset{+}{N}—CH_2CH_2—\ddot{O}H$ (*C*)

- **Write a reaction for an alcohol or thiol acting as a base. Answer questions involving the following aspects of alcohol and thiol basicity. 11.1E**

  (1) Give the approximate pK_a for the conjugate acid of an alcohol or thiol.

  (2) Rank a series of alcohol conjugate acids in order of increasing acidity, and give the reasons.

  (3) Explain the difference between an alcohol or thiol acting as a Brønsted acid and an alcohol or thiol acting as a Brønsted base.

11.42 (a) Write an equation for ethanol acting as a Brønsted base toward acetic acid (CH_3CO_2H) acting as an acid.

(b) Estimate the equilibrium constant for this reaction. The pK_a of acetic acid is 4.76.

(c) When acetic acid is dissolved in 1 L of ethanol to a concentration of 0.1 *M*, estimate the concentration of each of the species in the reaction you wrote in part (a) at equilibrium. (There are 17.04 moles of ethanol in 1 L.)

11.43 Within each set, arrange the compounds in order of increasing acidity.

(a) $CH_3\ddot{O}—CH_2CH_2—\ddot{O}H$ (*A*), $CH_3\ddot{O}—CH_2CH_2CH_2—\overset{+}{O}H_2$ (*B*), $^{-}\!:\ddot{O}—CH_2CH_2—\ddot{O}H$ (*C*), $CH_3CH_2CH_2—\ddot{O}H$ (*D*)

(b) $CH_3CH_2\ddot{O}H$ (*A*), $CH_3\overset{+}{S}(H)CH_3$ (*B*), $CH_3CH_2\overset{+}{O}(H)CH_2CH_3$ (*C*)

- **Give the names, nicknames, and structures of the most common sulfonate ester derivatives of alcohols: methanesulfonates (mesylates), *p*-toluenesulfonates (tosylates), and trifluoromethanesulfonates (triflates). 11.4A**

11.44 Draw the structure of each of the following compounds:

(a) 2-Octyl tosylate

(b) Allyl mesylate

(c) Propyl triflate

- **Complete examples of the following reactions by providing the structure of missing reactants, products, or reagents, including stereochemistry if relevant.**

  (1) Complete conversion of an alcohol or thiol to its conjugate base **11.1B**

  (2) Alcohol dehydration **11.2**

  (3) Conversion of alcohols into halides **11.3**

  (4) Conversion of alcohols into sulfonate esters **11.4A**

(5) Alkylation reactions with sulfonate and sulfate esters **11.4A,B**

(6) Oxidation of alcohols to aldehydes, ketones, or carboxylic acids **11.7A,B**

(7) Oxidation of thiols to sulfonic acids **11.10B**

(8) Oxidation of thiols to disulfides **11.10B**

11.45 Give the product expected, if any, when 1-butanol reacts with each of the following reagents.

(a) Concentrated aqueous HBr, H_2SO_4 catalyst, heat

(b) Cold aqueous H_2SO_4

(c) Pyridinium chlorochromate (PCC) in CH_2Cl_2

(d) NaH

(e) *p*-Toluenesulfonyl chloride in pyridine

(f) $SOCl_2$ in pyridine

(g) Ph_3PBr_2 in DMF

(h) Triflic anhydride and pyridine

11.46 Give the product expected when 2-methyl-2-propanol reacts with each of the following reagents.

(a) Concentrated aqueous HCl

(b) CrO_3 in pyridine

(c) H_2SO_4, heat

(d) Potassium metal

(e) Methanesulfonyl chloride in pyridine

11.47 Give the product expected, if any, when 1-propanethiol reacts with each of the following reagents.

(a) One molar equivalent of sodium methoxide

(b) Acetic acid ($pK_a = 4.76$) in methanol solution

(c) Nitric acid

(d) Br_2 with a catalytic amount of NaOH

(e) $KMnO_4$

(f) One molar equivalent of potassium hydroxide in ethanol

11.48 Complete the following reactions.

(a) The product of reaction 11.45(e) with $Na^{+}\,{}^{-}CN$ in a polar aprotic solvent

(b) The product of reaction 11.45(h) with anhydrous $K^+\ F^-$ in acetonitrile (a polar aprotic solvent)

(c) The product of reaction 11.46(d) with dimethyl sulfate

(d) The product of reaction 11.46(d) with the product of reaction 11.45(e)

- **Provide curved-arrow mechanisms for the following reactions:**

(1) Acid-catalyzed dehydration of a tertiary alcohol **11.2**

(2) Conversion of alcohols into alkyl halides with hydrogen halides; with thionyl chloride; with triphenylphosphine dibromide. **11.5**

11.49 Write the curved-arrow mechanism for the dehydration reaction of 1-methylcyclohexanol with phosphoric acid. Be sure to indicate the major product.

11.50 Alcohols react with the hypervalent compound phosphorus pentabromide (PBr_5) to give alkyl bromides. Write a curved-arrow mechanism for the reaction of 2-butanol with PBr_5. Be sure to show the by-product of the reaction (called phosphorus oxybromide), which should be revealed by your mechanism. (*Hint:* The mechanism is similar to the reaction of triphenylphosphine dibromide with alcohols.)

11.51 Write the curved-arrow mechanism for the reaction of cyclohexanol with thionyl chloride.

- **Characterize an oxidative or reductive transformation in terms of electrons gained or lost. 11.6A**

11.52 Indicate whether each of the following transformations is an oxidation, a reduction, or neither. For each oxidation or reduction process, indicate how many electrons are involved.

(a) $CH_3CH_2CH(OH)CH{=}O \longrightarrow CH_3CH_2C({=}O)CH_2OH$

(b) 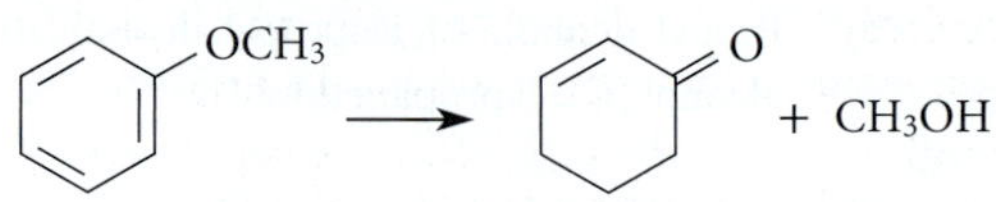

(c) $H_3C{-}C({=}O){-}NH_2 \longrightarrow CH_3NH_2 + O{=}C{=}O$

(d) $2\,Ph{-}C(CH_3)_2{-}H \longrightarrow Ph{-}C(CH_3)_2{-}C(CH_3)_2{-}Ph$

(e) $(CH_3)_3C{-}Cl \longrightarrow (CH_3)_3C^+ + Cl^-$

- **Identify the species that is oxidized and the species that is reduced in an oxidation–reduction reaction.** **11.6B**

11.53 The following is a well-known reaction of glycols (vicinal diols).

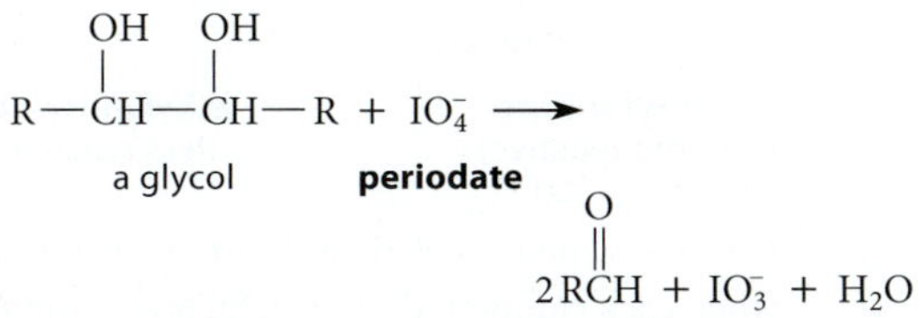

In this reaction, what species is oxidized? What species is reduced? Explain how you know.

- **Balance a complex oxidation–reduction reaction using the method of half-reactions.** **11.6A,B**

11.54 (a) In the reaction shown in Problem 11.53, how many electrons are involved in the oxidation half-reaction? In the reduction half-reaction? Explain how you arrived at your answer.

(b) How many moles of periodate are required to react completely with 0.1 mol of a glycol?

11.55 How many grams of CrO_3 are required to oxidize 10 g of 2-heptanol to the corresponding ketone? The by-product of this reaction is Cr^{3+}. (*Hint:* You don't need to balance the reaction. Consider only the electrons gained or lost in each half-reaction.)

11.56 The primary alcohol 2-methoxyethanol, $CH_3O{-}CH_2CH_2{-}OH$, can be oxidized to the corresponding carboxylic acid with aqueous nitric acid (HNO_3). The by-product of the oxidation is nitric oxide, NO. How many moles of HNO_3 are required to oxidize 0.1 mol of the alcohol?

- **Classify groups within molecules as constitutionally equivalent or nonequivalent. Classify constitutionally equivalent groups in molecules as homotopic, enantiotopic, or diastereotopic. Classify groups within molecules as chemically equivalent or nonequivalent.** **11.9**

11.57 In each compound, identify (1) the diastereotopic fluorines, (2) the enantiotopic fluorines, (3) the homotopic fluorines, and (4) the constitutionally nonequivalent fluorines.

(a) 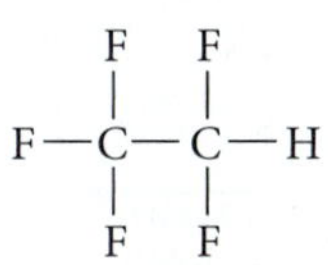

(b) 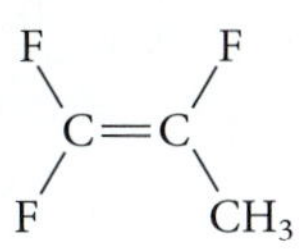

(c) 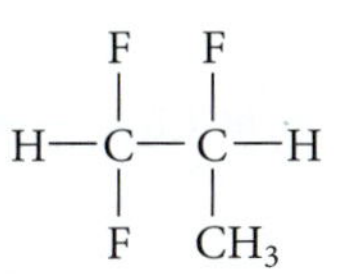

11.58 How many chemically nonequivalent sets of hydrogens are in each of the following structures?

(a) 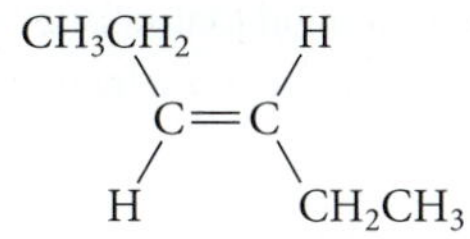

(b) 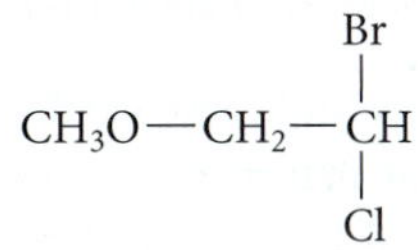

(c) 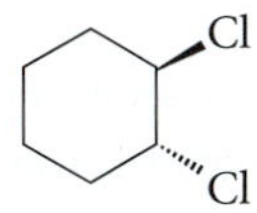

(d) 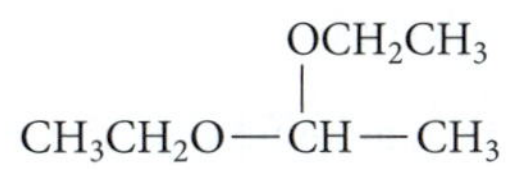

11.59 How many chemically nonequivalent sets of carbons are there in each structure of Problem 11.58?

- **Incorporate the reactions of alcohols, thiols, and sulfides into multistep syntheses.** **11.12**

11.60 Using a retrosynthetic analysis to develop your synthetic plan, outline a synthesis for each of the following compounds from the indicated starting material and any other reagents.

(a) 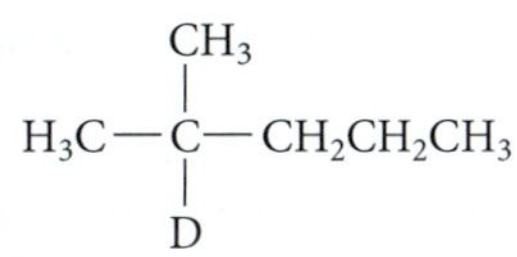 from 2-methyl-1-pentanol

(b) 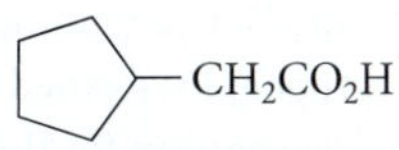 from 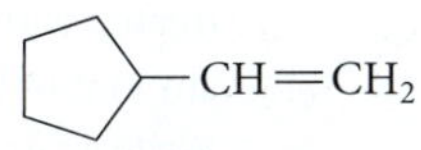

cyclopentylethylene

(c) 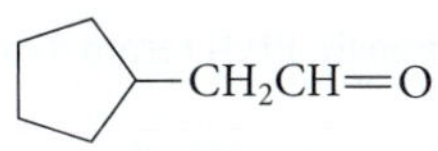 from cyclopentylethylene

(d)  from cyclopentylethylene

(e) (±)-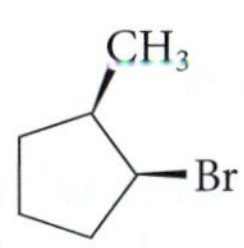 from 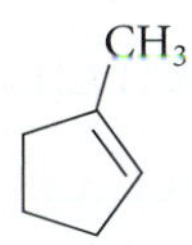

INTEGRATED PROBLEMS

11.61 Give the structure of a compound that satisfies the criterion given in each case. (There may be more than one correct answer.)

(a) A seven-carbon tertiary alcohol that yields a *single* alkene after acid-catalyzed dehydration

(b) An alcohol that, after acid-catalyzed dehydration, yields an alkene that, in turn, on ozonolysis and treatment with $(CH_3)_2S$, gives only benzaldehyde, Ph—CH═O

(c) An alcohol that gives the same product when it reacts with $KMnO_4$ as is obtained from the ozonolysis of *trans*-3,6-dimethyl-4-octene followed by treatment with H_2O_2

11.62 The following triester is a powerful explosive, but is also a medication for angina pectoris (chest pain). From what inorganic acid and what alcohol is it derived?

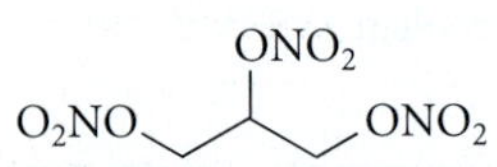

11.63 When *tert*-butyl alcohol is treated with $H_2{}^{18}O$ (water containing the heavy oxygen isotope ^{18}O) in the presence of a small amount of acid, and the *tert*-butyl alcohol is re-isolated, it is found to contain ^{18}O. Write a curved-arrow mechanism consisting of Brønsted acid–base reactions, Lewis acid–base associations, and Lewis acid–base dissociations that explains how the isotope is incorporated into the alcohol.

11.64 (a) When 1-propanol containing deuterium (D, or ^2H) rather than hydrogen at the oxygen, Pr—OD, is treated with an excess of H_2O containing a catalytic amount of NaOH, 1-propanol is formed rapidly. Write a curved-arrow mechanism for this reaction. Where does the deuterium go?

(b) How would one prepare PrOD from 1-propanol?

11.65 Outline a synthesis for the conversion of enantiomerically pure (*R*)-CH_3CH_2CHD—OH into each of the following isotopically labeled compounds. Assume that $Na^{18}OH$ or $H_2{}^{18}O$ is available as needed.

(a) (*S*)-CH_3CH_2CHD—^{18}OH

(b) (*R*)-CH_3CH_2CHD—^{18}OH (*Hint:* Two inversions of configuration correspond to a retention of configuration.)

11.66 Ethyl triflate is much more reactive than ethyl mesylate toward nucleophiles in S_N2 reactions.

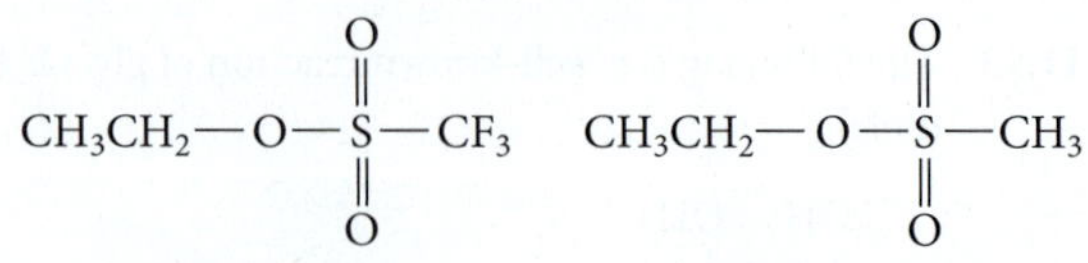

ethyl triflate (more reactive) **ethyl mesylate** (less reactive)

(a) Give the structures of all of the products formed when each compound reacts with potassium iodide in acetone (a polar aprotic solvent).

(b) Ethyl triflate is much more reactive in S_N2 reactions because the triflate group is a much better leaving group than the mesylate group. Explain in detail why the triflate group is a better leaving group.

11.67 A police officer, Lawin Order, has detained a driver, Bobbin Weaver, after observing erratic driving behavior. Administering a breathalyzer test, Officer Order collects 52.5 mL of expired air from Weaver and finds that the air reduces 0.507×10^{-6} mol of $K_2Cr_2O_7$ to Cr^{3+}. Assuming that 2100 mL of air contains the same amount of ethanol as 1 mL of blood, calculate the percent blood alcohol content (BAC), expressed as (grams of ethanol per mL of blood) $\times$ 100%. If 0.08% BAC is the lower limit of legal intoxication, should Officer Order make an arrest?

11.68 Chemist Stench Thiall, intending to prepare the disulfide *A*, has mixed 1 mol each of 1-butanethiol and 2-octanethiol with I_2 and base. Stench is surprised at the low yield of the desired compound and has come to you for an explanation. Explain why Stench should not have expected a good yield in this reaction.

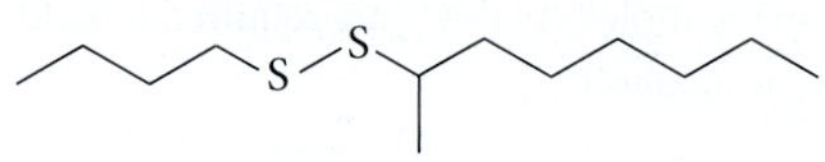

A

11.69 Complete each of the following reactions by giving the principal organic product(s) formed in each case. Explain your reasoning.

(a)
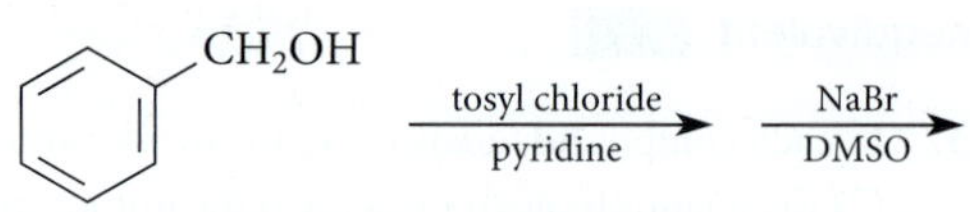

(b)
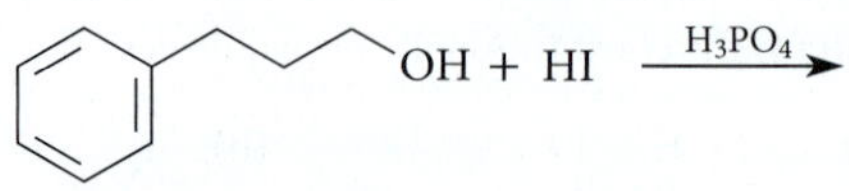

(c)
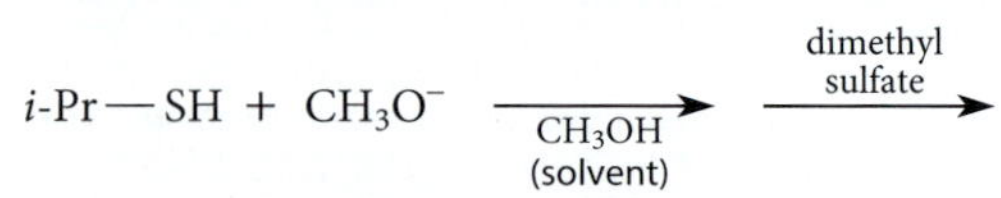

(d)
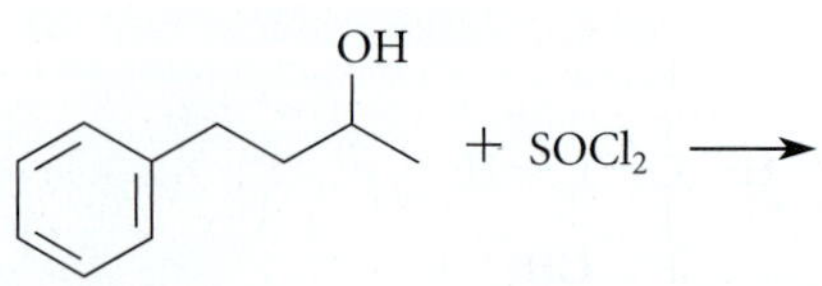

(e) 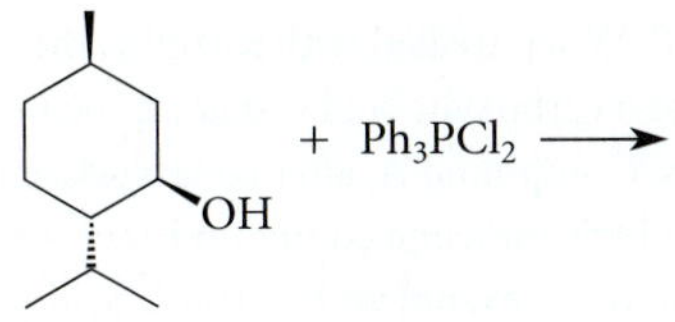

menthol

(Give the stereochemistry of the product.)

(f) 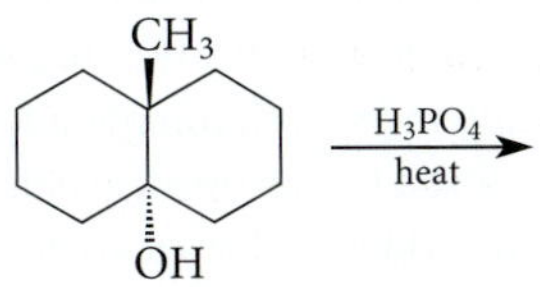

(g) 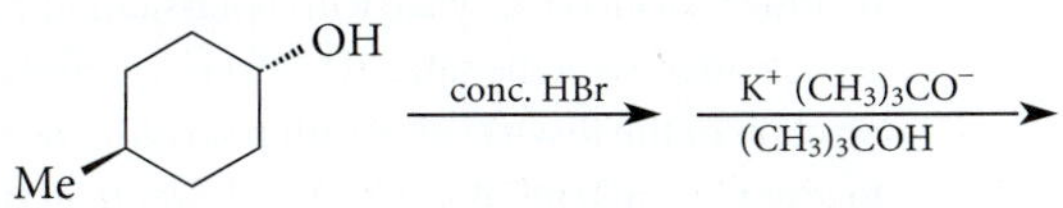

(h) 3-methyl-1-butanethiol + Et—S—S—Et $\xrightarrow[\text{(catalyst)}]{^{-}OH}$

11.70 Explain each of the following observations.

(a) When the rate of oxidation of isopropyl alcohol to acetone is compared with the rate of oxidation of a deuterated derivative, a primary isotope effect (Sec. 9.5D) is observed [see part (a) of **Fig. P11.70**]. (Assume that the mechanism is the one given in Eq. 11.65b, Sec. 11.7A.)

(b) When either (*S*)- or (*R*)-1-deuterioethanol is oxidized with PCC, the same product mixture results, and it contains significantly more of the deuterated aldehyde *A* than the undeuterated aldehyde *B* [see part (b) of Fig. P11.70].

(c) In the oxidation of (*R*)-1-deuterioethanol with alcohol dehydrogenase and NAD^+, *none* of the deuterated aldehyde is obtained.

11.71 The enzyme *aconitase* catalyzes a reaction of the Krebs cycle, an important biochemical pathway.

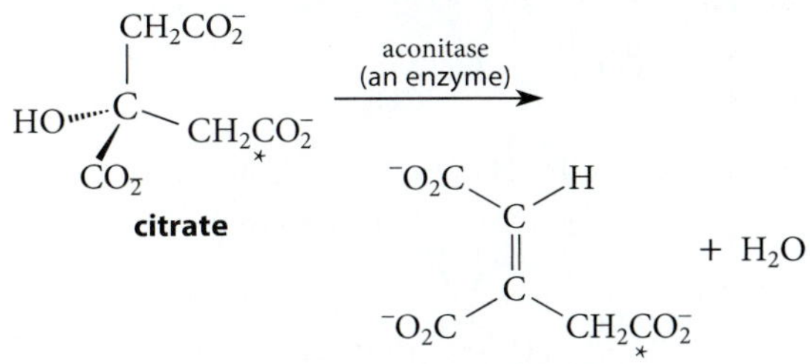

citrate **cis-aconitate**

In this equation, $^*C = {}^{14}C$, a radioactive isotope of carbon, and each $—CO_2^-$ group is the conjugate base of a carboxylic acid group. The dehydration occurs toward the branch of citrate that does *not* contain the radioactive carbon. Imagine carrying out this dehydration in the laboratory using a strong acid such as H_3PO_4 or H_2SO_4. How would the product(s) of the reaction be expected to differ, if at all, from those of the enzyme-catalyzed reaction, and why?

11.72 When the hydration of fumarate is catalyzed by the enzyme *fumarase* in D_2O, only (2*S*,3*R*)-3-deuteriomalate is formed as the product. (Each $—CO_2^-$ group is the conjugate base of a carboxylic acid group.)

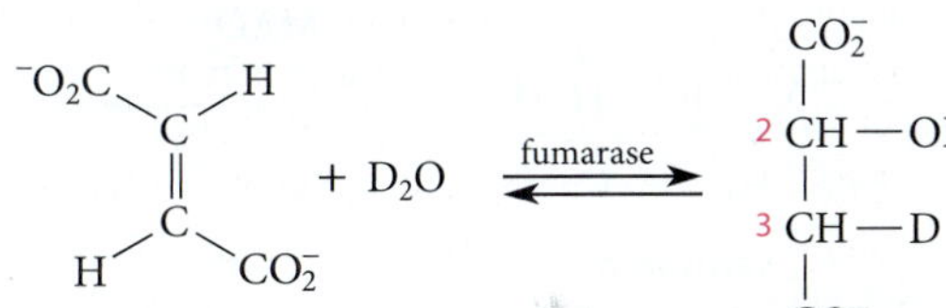

fumarate

3-deuteriomalate
(the 2*S*,3*R* stereoisomer is formed exclusively)

This reaction can also be run in reverse. By applying the principle of microscopic reversibility, predict the product

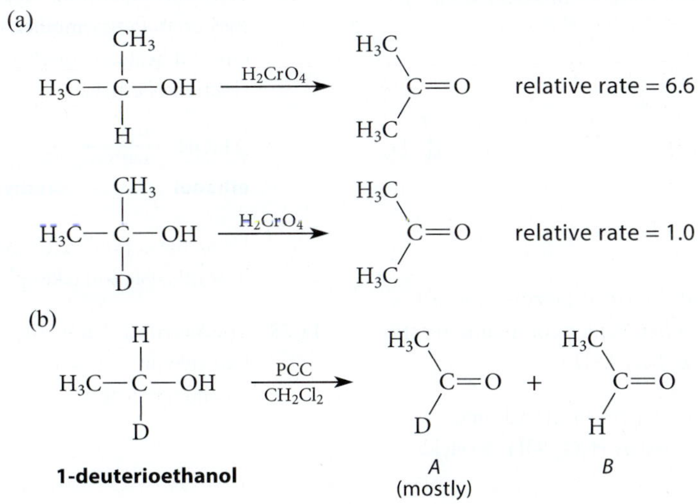

FIGURE P11.70

(if any) when each of the following compounds is treated with fumarase in H_2O:

(a)

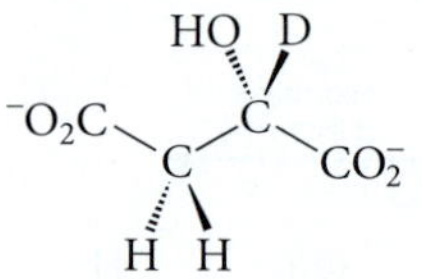

(b)

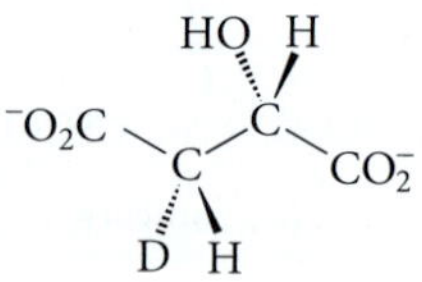

(c)

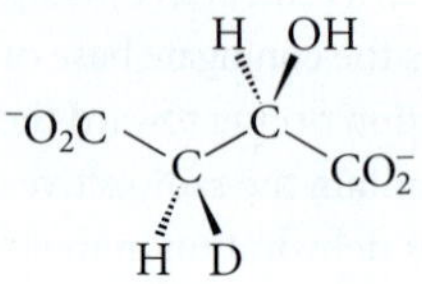

11.73 Monoamine oxidase (MAO) is an enzyme that catalyzes the oxidation of certain biologically important amines. One form of the enzyme catalyzes the following oxidation of serotonin, an important neurotransmitter. (A number of antidepressants inhibit this reaction.)

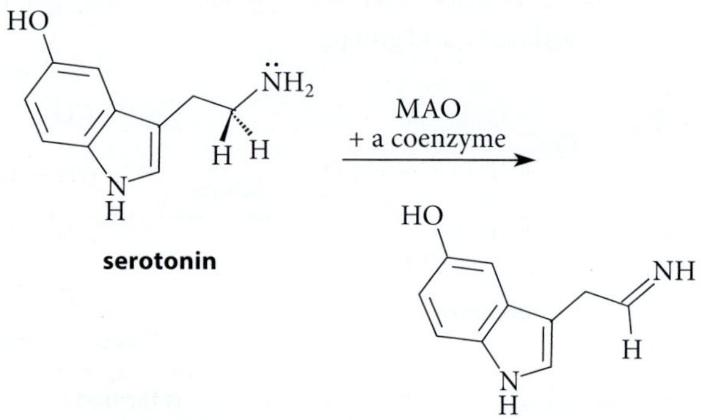

serotonin

(a) How many electrons are lost in this oxidation?

(b) The pro-*R* hydrogen is removed from carbon in this reaction. Give the product formed when each of the following compounds undergoes this oxidation.

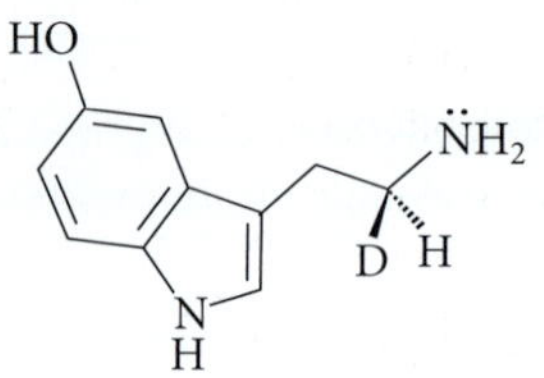

A

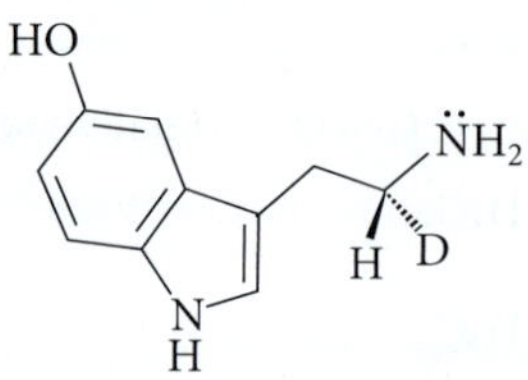

B

(c) One of the compounds in part (b) reacts about 10 times faster than the other. Which compound reacts faster? Why? (*Hint:* See Sec. 9.5D.)

11.74 Compound *A*, C_7H_{14}, decolorizes Br_2 in CH_2Cl_2 and reacts with BH_3 in THF followed by H_2O_2/OH^- to yield compound *B*. When treated with $KMnO_4$, then H_3O^+, *B* is oxidized to a carboxylic acid *C* that can be resolved into enantiomers. Compound *A*, after ozonolysis and workup with H_2O_2, yields the same compound *D* as is formed by the oxidation of 3-hexanol with chromic acid. Identify compounds *A*, *B*, *C*, and *D*.

11.75 In a laboratory are found two compounds: *A* (melting point –4.7 °C) and *B* (melting point –1 °C). Both compounds have the same molecular formula ($C_7H_{14}O$), and both can be resolved into enantiomers. Both compounds give off a gas when treated with NaH. Treatment of either *A* or *B* with tosyl chloride in pyridine yields a tosylate ester, and treatment of either tosylate with potassium *tert*-butoxide gives a mixture of the same two alkenes, *C* and *D*. However, reaction of the tosylate of *A* with potassium *tert*-butoxide to give these alkenes is noticeably slower than the corresponding reaction of the tosylate of *B*. When either *optically active A* or *optically active B* is subjected to the same treatment, *both* alkene products *C* and *D* are optically active. Treatment of either *C* or *D* with H_2 over a catalyst yields methylcyclohexane. Identify all unknown compounds, and explain your reasoning.

11.76 Organic chemistry student Buster Bluelip has observed that alcohols can be converted into alkyl bromides by treatment with concentrated HBr. He has proposed that, by analogy, alcohols should be converted into nitriles (organic cyanides, R—C≡N) by treatment with concentrated H—C≡N. Upon running the reaction, Bluelip finds that the alcohol does not react. Another student has suggested that the reason the reaction failed is the absence of a strong acid catalyst. (HCN is a weak acid; see Table 3.2, Sec. 3.6A.) Following this suggestion, Bluelip runs the reaction in the presence of H_2SO_4 and again observes no reaction of the alcohol. Explain the difference in the reaction of alcohols with HBr and HCN—that is, why the latter fails but the former succeeds.

11.77 Primary alcohols, when treated with H_2SO_4, do not dehydrate to alkenes under the usual conditions. However, they do undergo another type of "dehydration" to form ethers if heated strongly in the presence of H_2SO_4. The reaction of ethyl alcohol to give diethyl ether is typical:

$$2\,\text{EtOH} \xrightarrow[140\,°\text{C}]{H_2SO_4} \text{Et—O—Et} + H_2O$$

ethanol **diethyl ether**

Using the curved-arrow notation, show mechanistically how this reaction takes place.

11.78 The *Swern oxidation*, shown in **Fig. P11.78**, is a very mild two-step procedure for the oxidation of primary and secondary alcohols.

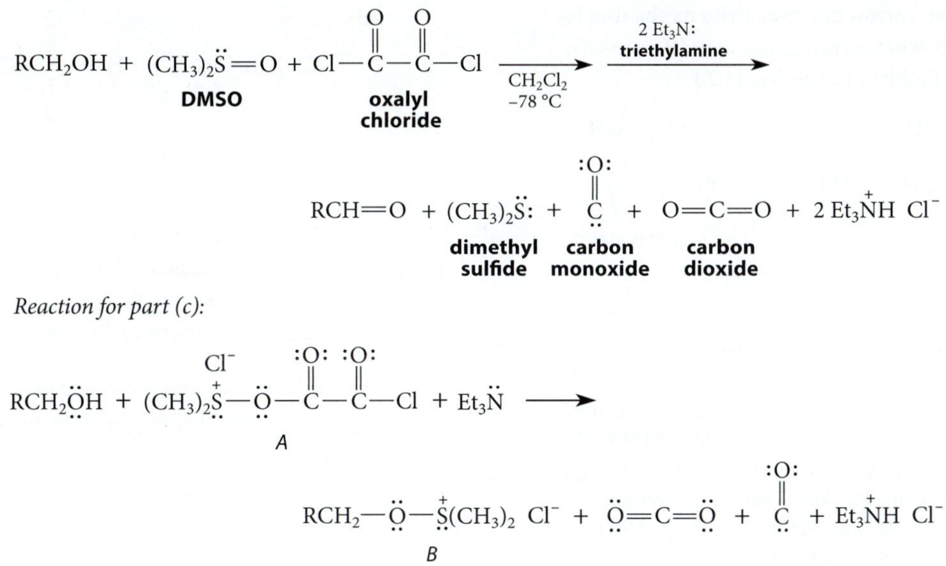

FIGURE P11.78

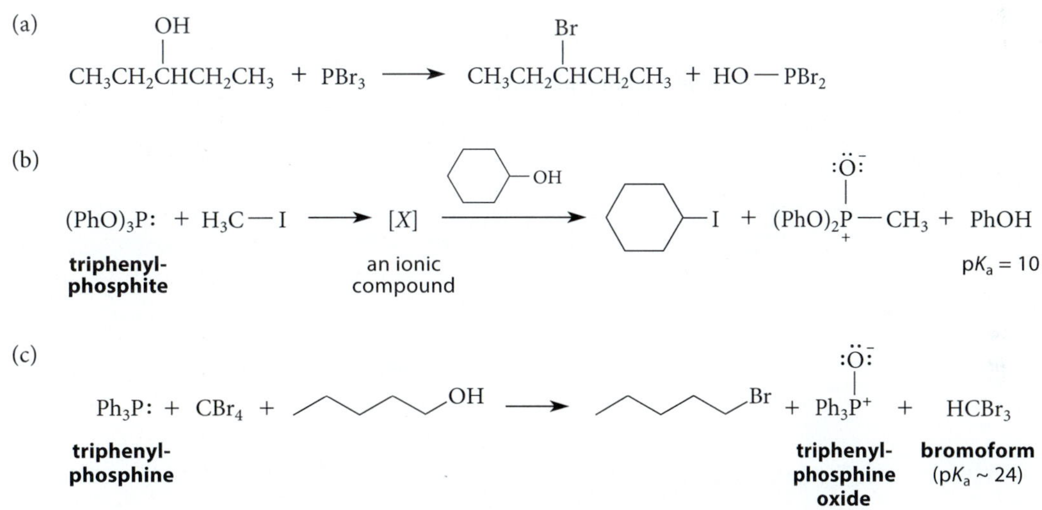

FIGURE P11.79

(a) How many electrons are involved in this oxidation? Explain.

(b) What is the oxidizing agent?

(c) DMSO reacts with oxalyl chloride to give intermediate *A*, which spontaneously reacts to form intermediate *B*. (See Fig. P11.78c.) Give a curved-arrow mechanism for the conversion of *A* into *B*.

(d) The second equivalent of triethylamine reacts with intermediate *B* to give the product aldehyde and the by-product dimethyl sulfide. When an isotopically substituted alcohol RCD_2OH is subjected to the Swern oxidation, a deuterium-substituted aldehyde R—CD=O is formed, and the by-product dimethyl sulfide contains one atom of deuterium per molecule, H_3C—S—CH_2D. Write a curved-arrow mechanism for the reaction of intermediate *B* and the base triethylamine that accounts for the isotopic distribution in the products.

11.79 The reactions shown in **Fig. P11.79** all involve the conversion of the alcohol oxygen into a good leaving group, followed by the reaction of the resulting compound with a nucleophile provided by the reagent. Write curved-arrow mechanisms for each reaction. [*Hints:* In reaction (b), the phosphorus of triphenylphosphite is nucleophilic. Identify the intermediate *X*, and show how it leads to the products. In reaction (c), let triphenylphosphine act as a nucleophile and a bromine of CBr_4 act as the electrophilic center to form an ionic intermediate, which then reacts with the alcohol. Bromoform ($HCBr_3$) has pK_a ~ 24.]

11.80 Using the curved-arrow notation, give a mechanism for each of the following known conversions (If necessary, re-read Study Problem 11.1 in Sec. 11.2.)

(a)

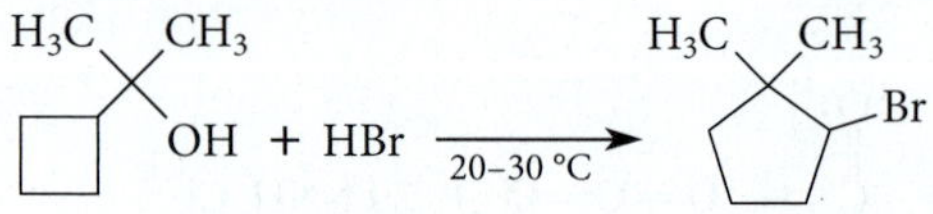

(74% yield)

(b)

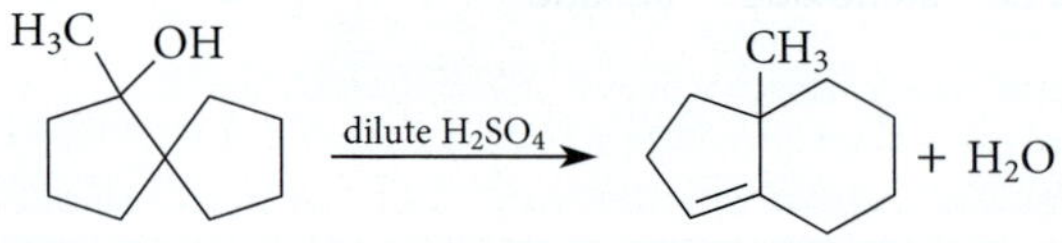

(one of several alkene products)

(c) This reaction is very important in the manufacture of high-octane gasoline. [*Hint:* If deuterated isobutane $(CH_3)_3C{-}D$ is used, the product is $(CH_3)_2CDCH_2C(CH_3)_3$.]

$$\underset{\textbf{2-methylpropene}}{(CH_3)_2C{=}CH_2} + \underset{\textbf{isobutane}}{(CH_3)_3C{-}H} \xrightarrow[\text{(catalyst)}]{\text{HF}} \underset{\textbf{2,2,4-trimethylpentane}}{(CH_3)_2CHCH_2C(CH_3)_3}$$

(d)

$$CH_3CH{=}CH_2 + HO{-}\overset{\overset{O}{\|}}{\underset{\underset{O}{\|}}{S}}{-}CF_3 \longrightarrow (CH_3)_2CH{-}O{-}\overset{\overset{O}{\|}}{\underset{\underset{O}{\|}}{S}}{-}CF_3$$

11.81 (a) Draw a resonance structure for methylenetriphenylphosphorane ($Ph_3P{=}CH_2$) that conforms to the octet rule on both carbon and phosphorus.

(b) This compound is a strong base that reacts rapidly with water. Explain by showing its reaction with D_2O.

12 THE CHEMISTRY OF ETHERS, EPOXIDES, GLYCOLS, AND SULFIDES

This chapter deals with the chemistry of ethers, epoxides, glycols, and sulfides.

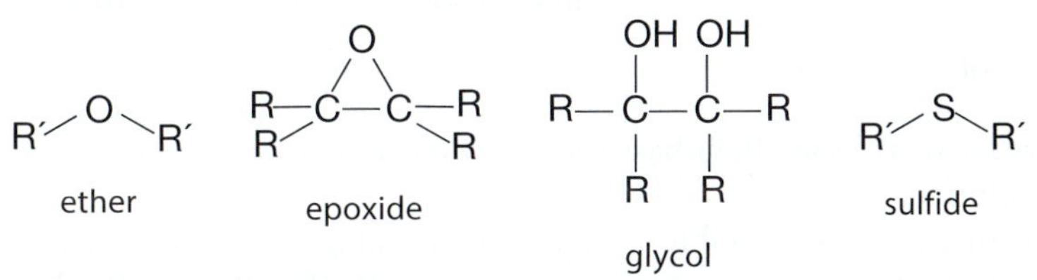

(R can be alkyl, aryl, or H; R′ can be alkyl or aryl) **(12.1)**

The nomenclature of these compound classes was introduced in Sec. 8.1.

The chemistry of ethers is closely intertwined with the chemistry of alkyl halides, alcohols, and alkenes. Ordinary ethers, however, are considerably less reactive than these other types of compounds. This chapter covers the synthesis of ethers and shows why the ether functional group is relatively unreactive.

Epoxides are heterocyclic compounds in which the ether linkage is part of a three-membered ring. Unlike ordinary ethers, epoxides are very reactive, and they are very useful in organic synthesis. This chapter presents the synthesis and reactions of epoxides.

Although glycols are diols, we consider them in this chapter because they have unique chemistry that is related to the chemistry of epoxides. For example, we'll find that epoxides are easily converted into glycols; and both epoxides and glycols can be easily prepared by the oxidation of alkenes.

Sulfides (thioethers), the sulfur analogs of ethers, are also discussed briefly in this chapter. Although sulfides share some chemistry with their ether counterparts, they differ from ethers in the way they react in oxidation reactions, just as thiols differ from alcohols (Sec. 11.10B).

In this chapter we also cover the principles governing intramolecular reactions: reactions that take place between groups in the same molecule. These principles are central to an understanding of enzyme catalysis.

Finally, the strategy of organic synthesis will be revisited with a classification of reactions according to the way they are used in synthesis and a further discussion of how to plan multistep syntheses with particular attention to stereochemistry.

12.1 BASICITY OF ETHERS AND SULFIDES

Ethers are not appreciably acidic; however, they are weakly basic and can accept a proton to form a conjugate-acid cation. Ethers are about as basic as alcohols, with conjugate-acid pK_a values of about −2 to −3.

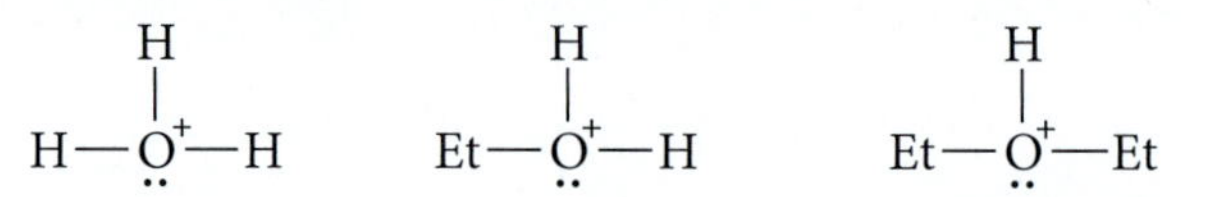

hydronium ion **conjugate acid of ethanol** **conjugate acid of diethyl ether**

pK_a 0 −2 to −3 (12.2)

Although ethers are weak bases, their basicities are important in many of their reactions that take place in acidic solution.

Sulfides, like thiols, are considerably less basic, with conjugate-acid pK_a values of about −6 to −7. This reduced basicity is a consequence of the element effect (Sec. 3.7B): the S—H bond is weaker than the O—H bond.

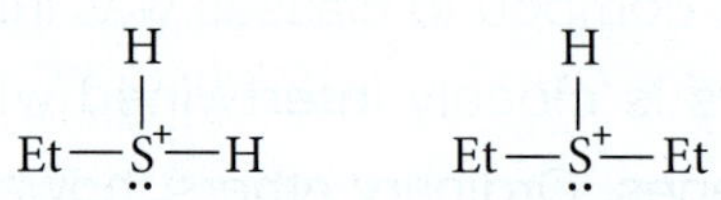

conjugate acid of ethanethiol **conjugate acid of diethyl sulfide**

pK_a ~ −6 to −7 (12.3)

Ethers are also important Lewis bases. For example, the Lewis acid–Lewis base complex of boron trifluoride and diethyl ether is stable enough to be distilled (bp 126 °C). This complex is a convenient way to handle BF_3.

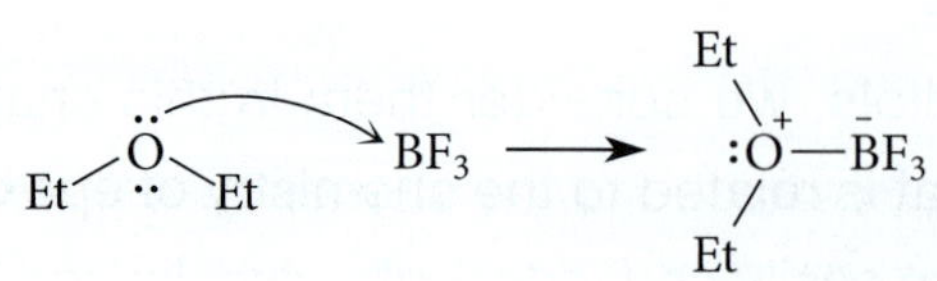

boron trifluoride etherate (12.4a)

The solvation of Grignard reagents by ether solvents (Eq. 10.49, Sec. 10.5B) is another example of the Lewis basicity of ethers.

Water and alcohols are also excellent Lewis bases, but their complexes with Lewis acids are typically unstable. The protons on water and alcohols can react further, in which case the complex is destroyed.

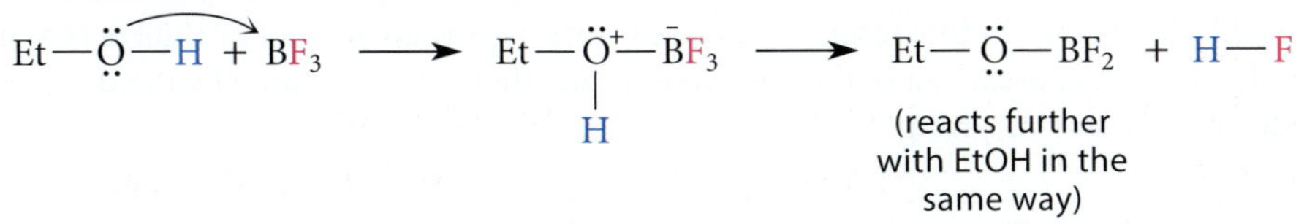

(12.4b)

Focused Problem

12.1 Arrange the following ions in order of increasing acidity, and explain your reasoning.

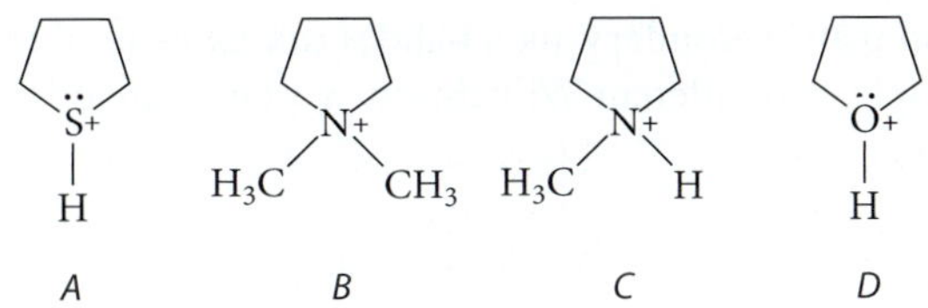

12.2 SYNTHESIS OF ETHERS AND SULFIDES

A. Williamson Ether Synthesis

Some ethers can be prepared from alcohols and alkyl halides. First, the alcohol is converted into an alkoxide (Sec. 11.1B):

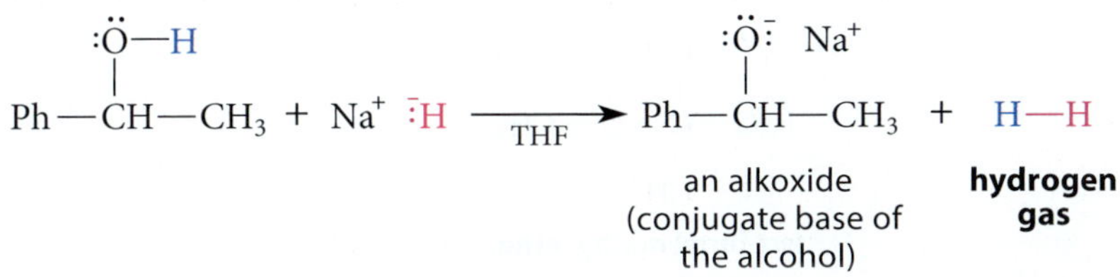

(12.5a)

(Draw the curved-arrow notation for this reaction.) Then, the alkoxide is allowed to react as a nucleophile with a methyl halide, primary alkyl halide, or the corresponding sulfonate ester to give an ether.

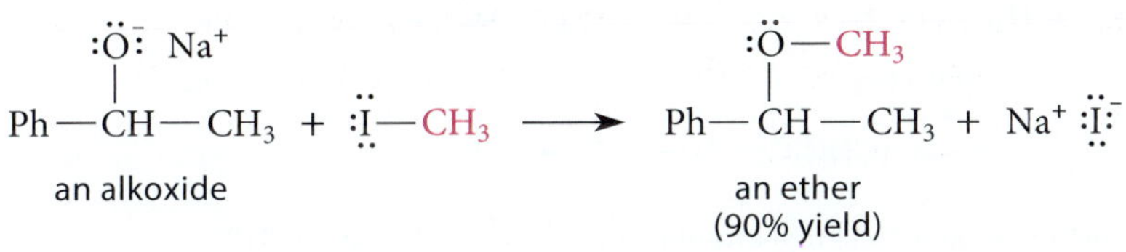

(12.5b)

Some sulfides can be prepared in a similar manner from thiolates, the conjugate bases of thiols. In the following equation, a tosylate (Sec. 11.4A) is used as the alkylating agent.

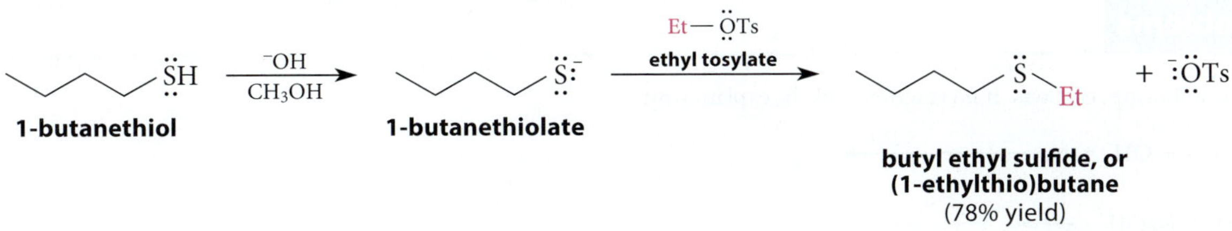

(12.6)

> The Williamson ether synthesis is named for Alexander William Williamson (1824–1904), who was Professor of Chemistry at the University of London.

(Recall that tosylates have about the same reactivity as alkyl bromides.) Both of these reactions are examples of the **Williamson ether synthesis**, which is the preparation of an ether by the alkylation of an alkoxide (and, by extension, a sulfide by the alkylation of a thiolate).

The Williamson ether synthesis is an important practical example of the S_N2 reaction (see Table 9.1, Sec. 9.1A). In this reaction, the conjugate base of an alcohol or thiol acts as a nucleophile toward the α-carbon of the alkyl halide; an ether is then formed by the displacement of a halide or other leaving group.

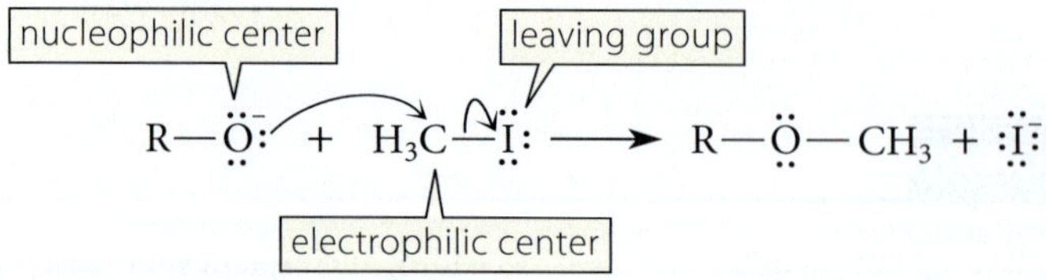

(12.7)

Tertiary and many secondary alkyl halides *cannot* be used in this reaction. (Why? See Sec. 9.5G.)

In principle, two different Williamson syntheses are possible for any ether with two different alkyl groups.

$$\left.\begin{array}{c} R^1\text{—}\ddot{\underset{\cdot\cdot}{O}}\text{:}^- + R^2\text{—X} \\ \text{or} \\ R^2\text{—}\ddot{\underset{\cdot\cdot}{O}}\text{:}^- + R^1\text{—X} \end{array}\right\} \longrightarrow R^1\text{—}\ddot{\underset{\cdot\cdot}{O}}\text{—}R^2 + X^- \qquad (12.8)$$

The preferred synthesis is usually the one that involves the alkyl halide with the greater S_N2 reactivity. This point is illustrated in Study Problem 12.1.

Study Problem 12.1

Outline a Williamson ether synthesis for *tert*-butyl methyl ether.

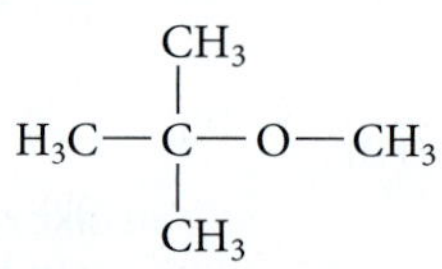

***tert*-butyl methyl ether**

Solution From Eq. 12.8, the two possibilities for preparing this compound are the reaction of methyl bromide with potassium *tert*-butoxide and the reaction of *tert*-butyl bromide with sodium methoxide. Only the former combination will work.

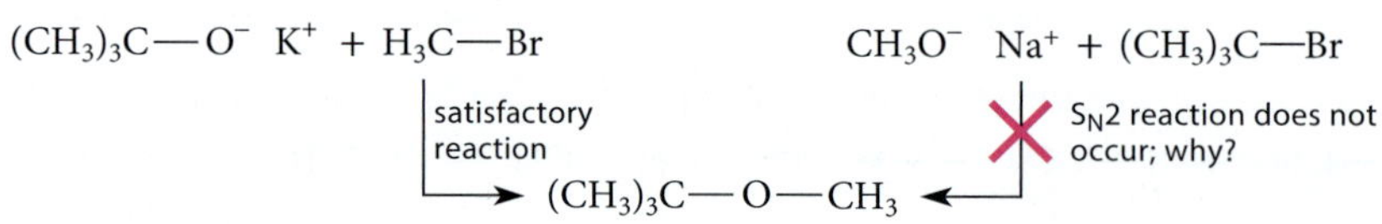

(12.9)

Do you know why sodium methoxide and *tert*-butyl bromide do not work? (See Sec. 9.5G.)

Focused Problems

12.2 Complete the following reactions. If no reaction is likely, explain why.

(a) $(CH_3)_2CH\text{—}OH + Na \longrightarrow \xrightarrow{CH_3I}$

(b) $CH_3SH + NaOH \longrightarrow$ (1 equiv.) 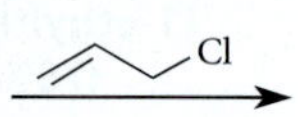

(c) $CH_3O^-\ Na^+ + (C_2H_5)_3C—Br \xrightarrow[CH_3OH]{}$

(*Hint:* What reaction competes with the S_N2 reaction?)

(d) $EtO^-\ K^+ + (CH_3)_3C—CH_2—OTs \xrightarrow[\text{25 °C, EtOH}]{}$

12.3 Suggest a Williamson ether synthesis, if one is possible, for each of the following compounds. If no Williamson ether synthesis is possible, explain why.

(a) 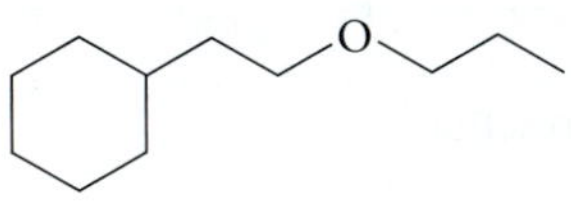

(b) $(CH_3)_2CH—S—CH_3$

(c) $(CH_3)_3C—O—C(CH_3)_3$

B. Alkoxymercuration–Reduction of Alkenes

Another method for the preparation of ethers is a variation of oxymercuration–reduction, which is used to prepare alcohols from alkenes (Sec. 5.4). If the *aqueous* solvent used in the oxymercuration step is replaced by an *alcohol* solvent, an ether instead of an alcohol is formed after the reduction step. This process is called **alkoxymercuration–reduction**:

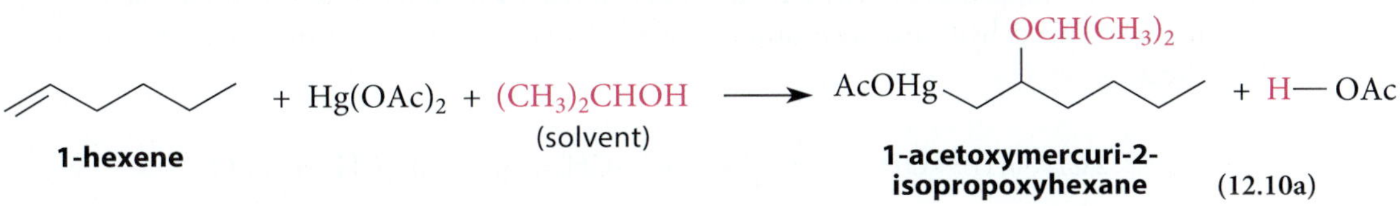

(12.10a)

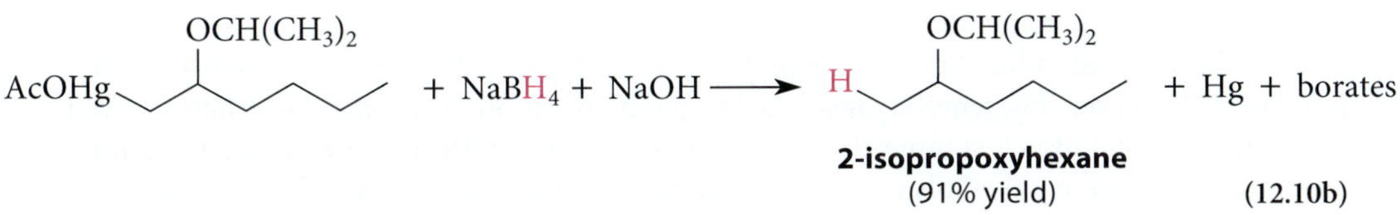

(12.10b)

Contrast:

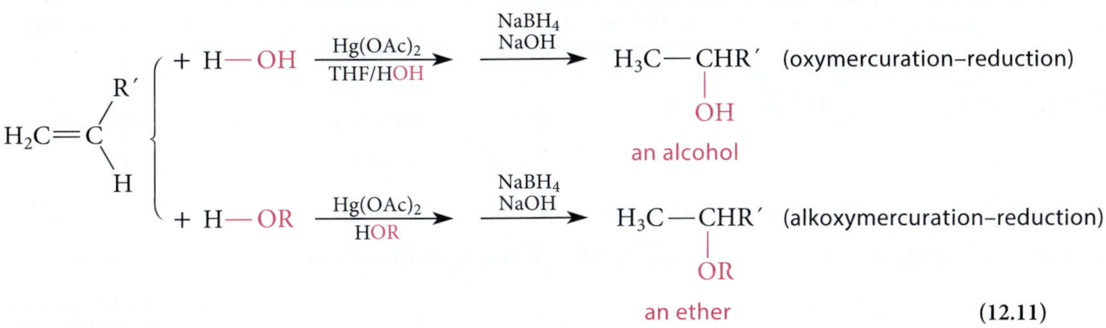

(12.11)

After reviewing the mechanism of oxymercuration in Eqs. 5.20a, 5.20b, and 5.21 (Sec. 5.4A), you should be able to write the mechanism of the reaction in Eq. 12.10a. The two mechanisms are essentially identical, except that an alcohol instead of water is the nucleophile that reacts with the mercurinium ion intermediate.

STUDY GUIDE LINK 12.1
Learning New Reactions from Earlier Reactions

Focused Problems

12.4 (a) Write the curved-arrow mechanism of Eq. 12.10a, and account for the regioselectivity of the reaction.

(b) Explain what would happen in an attempt to synthesize the ether product of Eq. 12.10b by a Williamson ether synthesis.

12.5 Complete the following reaction:

$$(CH_3)_2CH{-}CH{=}CH_2 + CH_3CH_2OH + Hg(OAc)_2 \longrightarrow \xrightarrow[NaOH]{NaBH_4}$$

12.6 Explain why a mixture of two constitutionally isomeric ethers is formed in the following reaction.

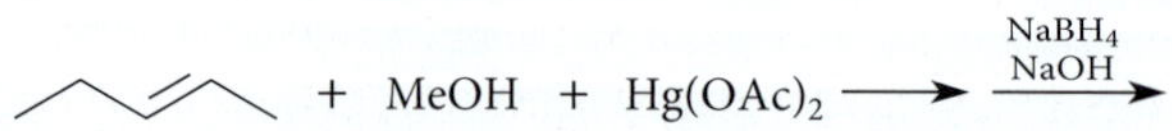

12.7 Outline a synthesis of each of the following ethers using alkoxymercuration–reduction:

(a) Dicyclohexyl ether

(b) *tert*-Butyl isobutyl ether

C. Ethers from Alcohol Dehydration and Alkene Addition

In some cases, two molecules of a *primary* alcohol can react with the loss of one molecule of water to give an ether. This dehydration reaction requires relatively harsh conditions—namely, strong acid and heat.

$$\underset{\textbf{ethanol}}{2\,CH_3CH_2{-}OH} \xrightarrow[140\,°C]{H_2SO_4} \underset{\textbf{diethyl ether}}{CH_3CH_2{-}O{-}CH_2CH_3} + H_2O \qquad (12.12)$$

This method is used industrially for the preparation of diethyl ether, and it can be used in the laboratory. However, it is generally restricted to the preparation of *symmetrical* ethers derived from *primary* alcohols. (A symmetrical ether is one in which both alkyl groups are the same.) Secondary and tertiary alcohols cannot be used because they undergo dehydration to alkenes (Sec. 11.2).

The formation of ethers from primary alcohols is an S_N2 reaction in which one alcohol displaces water from another molecule of *protonated* alcohol (see Problem 11.77).

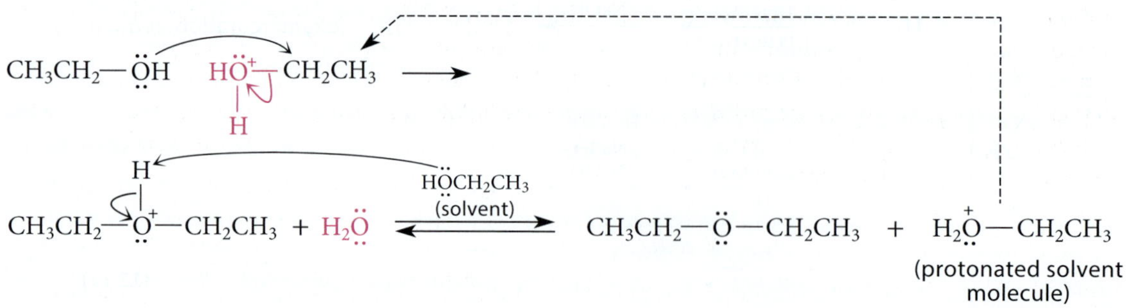

(12.13)

The protonated ethanol molecule is regenerated and is therefore a catalyst. High temperature is required because alcohols are relatively poor nucleophiles in the S_N2 reaction.

Tertiary alcohols can be converted into *unsymmetrical* ethers by treating them with dilute strong acids in an alcohol solvent. The conditions are much milder than those required for ether formation from primary alcohols. For example, ethyl *tert*-butyl ether can be prepared when *tert*-butyl alcohol is treated with ethanol (as the solvent) in the presence of an acid catalyst:

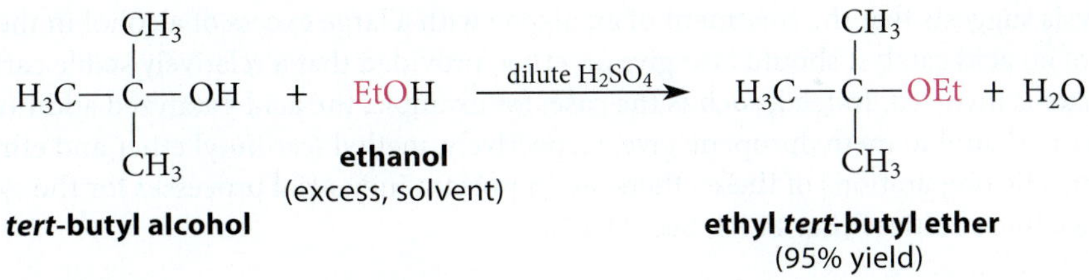

(12.14)

The key to using this type of reaction successfully is that only one of the alcohol starting materials (in this case, *tert*-butyl alcohol) can readily lose water after protonation to form a relatively stable carbocation. The alcohol that is used in excess (in this case, ethanol) must be one that either *cannot* form a carbocation by loss of water or should form a carbocation much less readily.

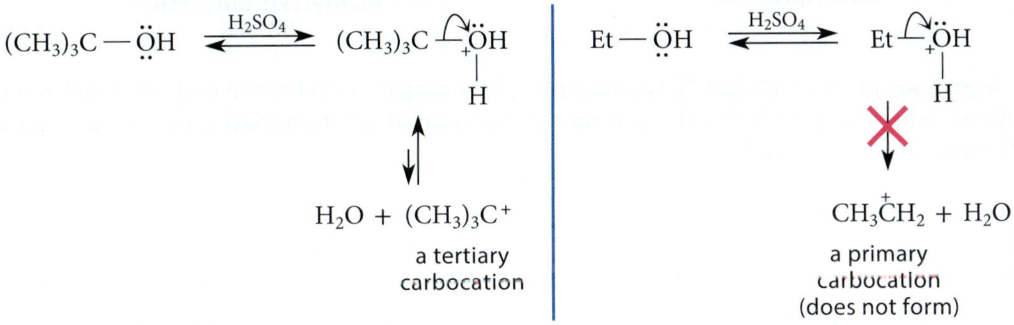

(12.15)

When the carbocation derived from the tertiary alcohol is formed, it reacts rapidly with ethanol, which is present in large excess because it is the solvent.

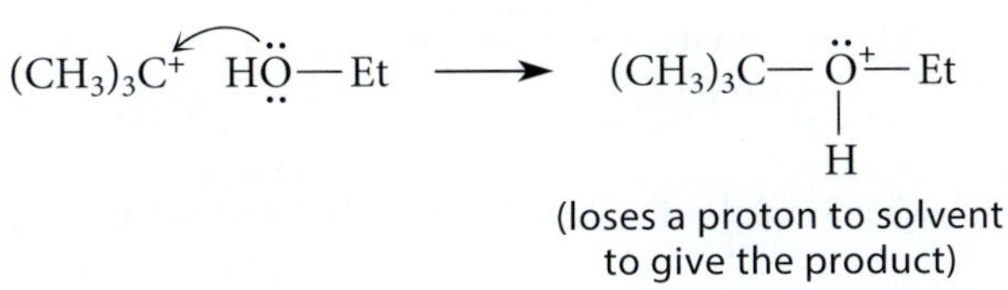

(12.16)

There is an important relationship between this reaction and alkene formation by alcohol dehydration. Alcohols, especially tertiary alcohols, undergo dehydration to alkenes in the presence of strong acids (Sec. 11.2). Ether formation from tertiary alcohols and the dehydration of tertiary alcohols are *alternate branches of a common mechanism.* Both ether formation and alkene formation involve carbocation intermediates; the reaction conditions dictate which product is obtained. The dehydration of alcohols to alkenes involves relatively high temperatures and *removal of the alkene and water products* as they are formed. Ether formation from tertiary alcohols involves milder conditions under which alkenes are *not* removed from the reaction mixture. In addition, *a large excess of the other alcohol* (ethanol in Eq. 12.16) *is used as the solvent*, so that the major reaction of the carbocation intermediate is with this alcohol. Any alkene that does form is not removed but is reprotonated to give back the same carbocation, which eventually reacts with the alcohol solvent:

STUDY GUIDE LINK 12.2
Common Intermediates from Different Starting Materials

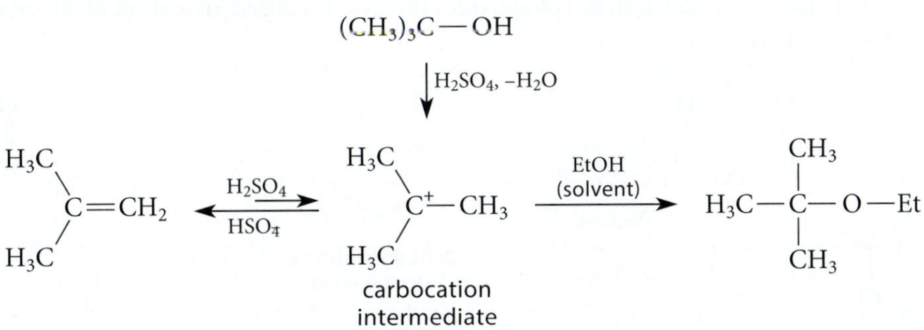

(12.17)

The notation $-H_2O$ used in Eq. 12.17 is a shorthand meaning that water is lost from the starting material and is formed as a product.

This analysis suggests that the treatment of an alkene with a large excess of alcohol in the presence of an acid catalyst should also give an ether, provided that a relatively stable carbocation intermediate is involved. Indeed, such is the case; for example, the acid-catalyzed additions of methanol or ethanol to methylpropene give, respectively, methyl *tert*-butyl ether and ethyl *tert*-butyl ether. The preparations of these ethers are important industrial processes for the synthesis of these gasoline additives (Eq. 11.94, Sec. 11.13).

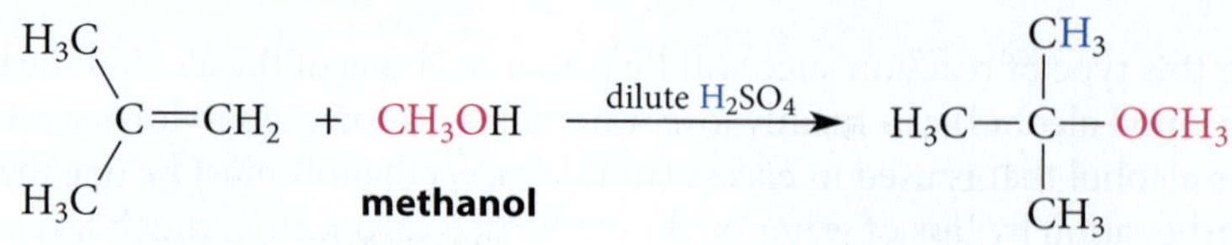

methanol

methylpropene

methyl *tert*-butyl ether (MTBE) (12.18)

The methods described in Secs. 12.2A, B, and C for the synthesis of ethers from alcohols are included as alcohol reactions in the synthetic network diagram shown in Display 11.90 (Sec. 11.12).

Equations 12.14, 12.17, and 12.18 show that for the preparation of tertiary ethers, it makes no difference in principle whether the starting material from which the tertiary group is derived is an alkene or a tertiary alcohol.

Focused Problems

12.8 Explain why the dehydration of primary alcohols can only be used for preparing *symmetrical* ethers. What would happen if a mixture of two different primary alcohols were used as the starting material in this reaction?

12.9 Complete the following reaction by giving the *major* organic product.

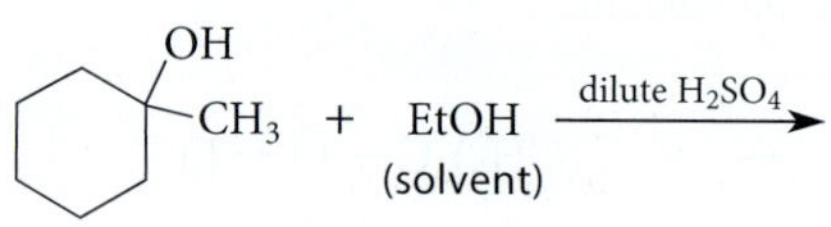

12.10 Draw the structure of two alkenes, either of which when treated with dilute H_2SO_4 and ethanol will give the same ether product as the reaction in Focused Problem 12.9.

12.11 Outline a synthesis of each of the following ethers using either alcohol dehydration or alkene addition, as appropriate.

(a) 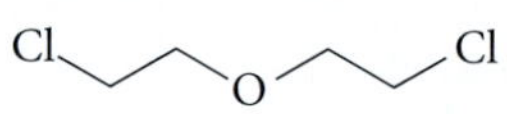

(b) 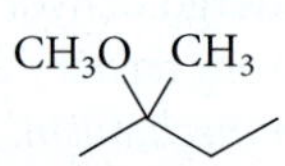

(c) Butyl *tert*-butyl ether

(d) Dibutyl ether

12.3 SYNTHESIS OF EPOXIDES

A. Oxidation of Alkenes with Peroxycarboxylic Acids

One of the best laboratory preparations of epoxides involves the direct oxidation of alkenes with peroxycarboxylic acids.

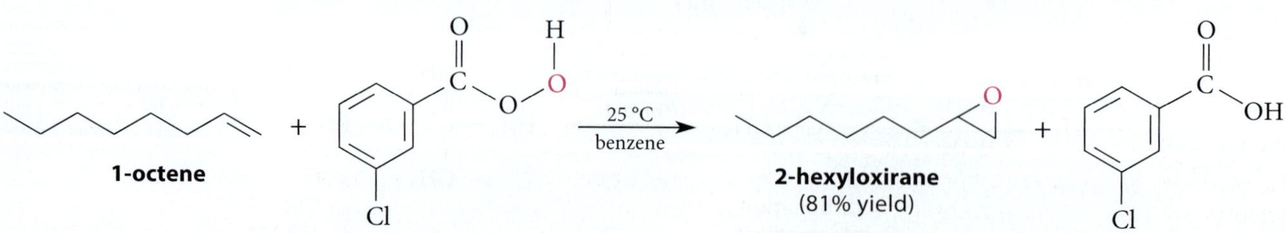

1-octene

***meta*-chloroperoxybenzoic acid (mCPBA)**
a peroxycarboxylic acid

2-hexyloxirane
(81% yield)

***meta*-chlorobenzoic acid** (12.19)

The use of alkenes as starting materials for epoxide synthesis is one reason that certain epoxides are named traditionally as oxidation products of the corresponding alkenes (Sec. 8.1C).

The oxidizing agent in Eq. 12.19, *meta*-chloroperoxybenzoic acid (abbreviated mCPBA), is an example of a **peroxycarboxylic acid**, which is a carboxylic acid that contains an —O—O—H (hydroperoxy) group instead of an —OH (hydroxy) group.

> The terms **peroxyacid** or **peracid** are sometimes used instead of *peroxycarboxylic acid*. These are actually more general terms that refer not only to peroxycarboxylic acids, but also to *any* acid containing an —O—O—H group instead of an —OH group.

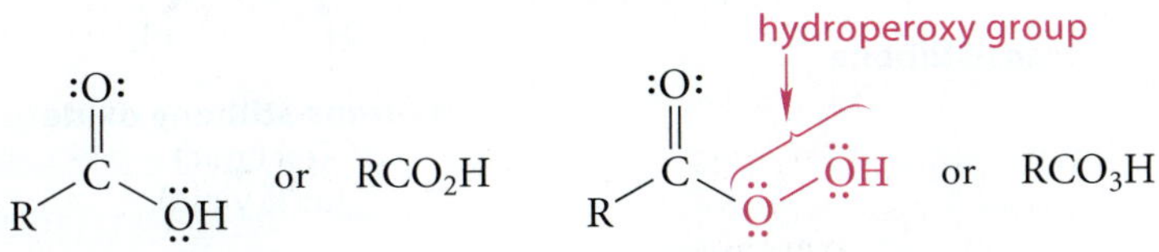

a carboxylic acid a peroxycarboxylic acid (12.20)

Many peroxycarboxylic acids are unstable, but they can be formed just prior to use by mixing a carboxylic acid and hydrogen peroxide. In principle, any one of several peroxycarboxylic acids can be used for the epoxidation of alkenes. The peroxyacid used in Eq. 12.19, mCPBA, has been popular because it is a crystalline solid that can be shipped commercially and stored in the laboratory. However, mCPBA, like most other peroxides (Sec. 12.9A), will detonate if it is not handled carefully. A less hazardous peroxycarboxylic acid that has essentially the same reactivity is the magnesium salt of monoperoxyphthalic acid, abbreviated MMPP.

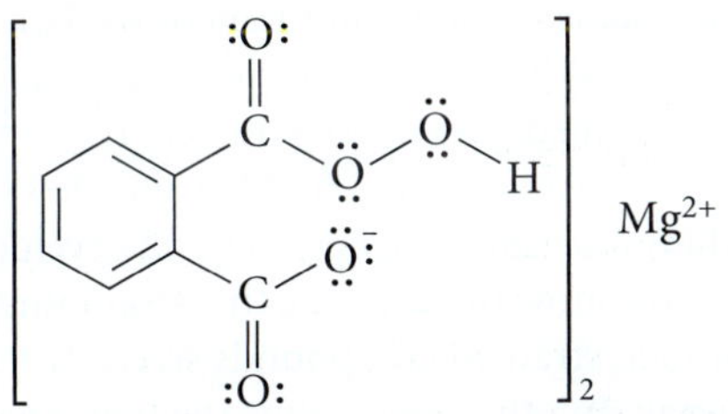

magnesium monoperoxyphthalate (MMPP) (12.21)

Epoxidation is a *concerted electrophilic addition*. The mechanism is very similar to the mechanism of bromonium-ion formation in electrophilic addition (Eq. 5.10a, Sec. 5.2A). Follow the curved arrows in the sequence of the numbered boxes:

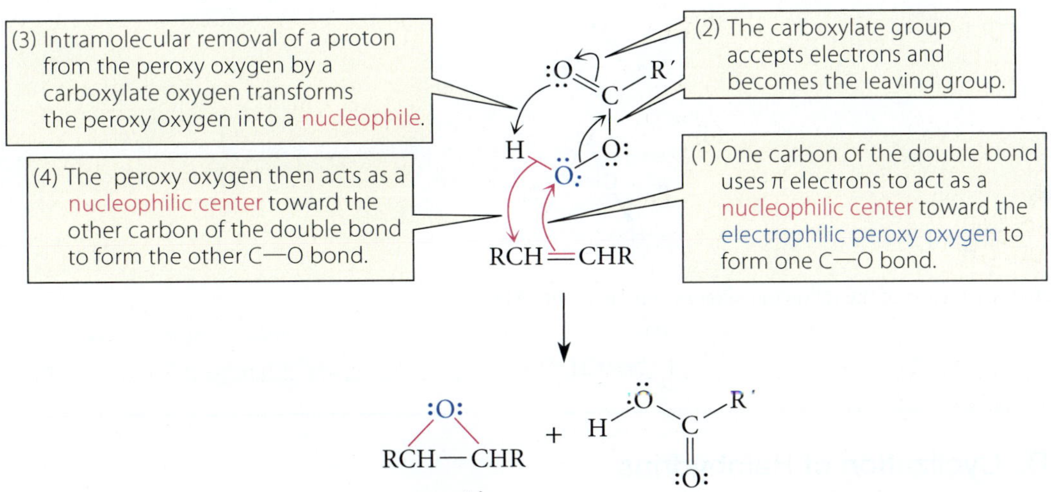

epoxide carboxylate ion (12.22)

We know this process is concerted because (1) carbocation rearrangements do not occur, and (2) the reaction is a *stereospecific* syn addition. That is, the reaction takes place with complete retention of alkene stereochemistry. (As shown in Eq. 7.62, Sec. 7.8C, a stepwise process would not be stereospecific.) This means that a cis alkene gives a cis-substituted epoxide and a trans alkene gives a trans-substituted epoxide.

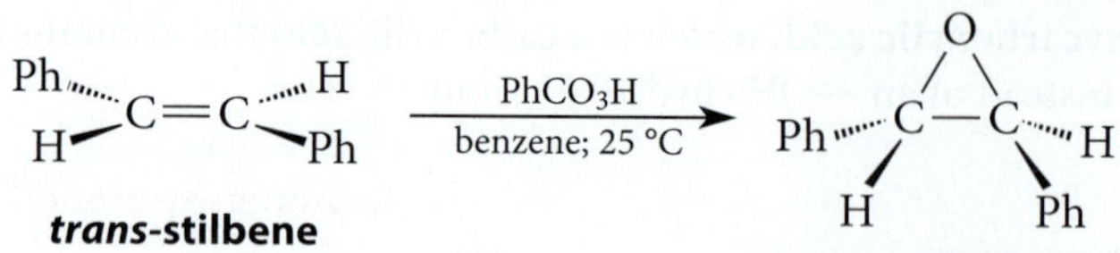

***trans*-stilbene**

(±)-*trans*-stilbene oxide
(racemic)
(55% yield)

(12.23a)

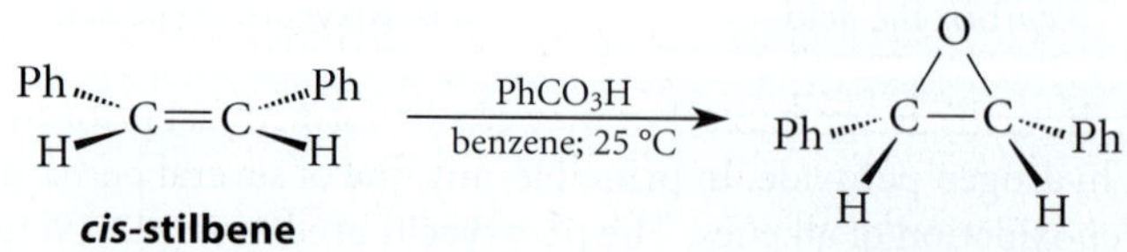

***cis*-stilbene**

***cis*-stilbene oxide**
(a meso compound)
(52% yield)

(12.23b)

The reaction is a syn addition because, in an anti addition, the epoxide oxygen would have to bridge opposite faces of the two alkene carbons simultaneously. For this to happen, the double bond in the transition state for anti addition would have to be significantly twisted and the transition state would be highly strained. As we show in Sec. 12.5, the stereospecificity of this reaction is one reason why epoxide formation is a highly valuable synthetic reaction.

Because epoxides contain three-membered rings, they, like cyclopropanes (Sec. 7.5A), have significant angle strain. As we show in Sec. 12.5, this strain imparts valuable reactivity to epoxides. It is possible to form such strained compounds so easily because the O—O bond of a peroxycarboxylic acid is *very* weak. In other words, it is the high energy of the peroxycarboxylic acid that drives epoxide formation.

Focused Problems

12.12 Give the structure of the alkene that would react with mCPBA to give each of the following epoxides.

(a) 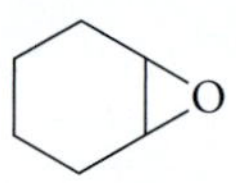

(b) 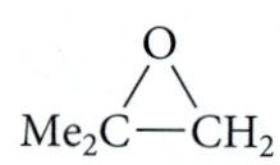

(c) 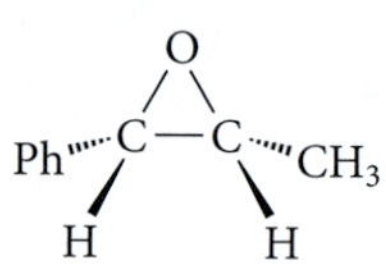

(d) 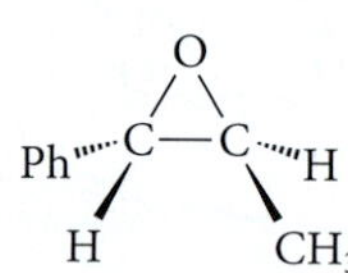

12.13 Give the product expected when each of the following alkenes is treated with MMPP.

(a) *trans*-3-Hexene

(b) 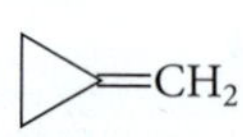

B. Cyclization of Halohydrins

Epoxides can also be synthesized by the treatment of halohydrins (Sec. 5.2B) with base:

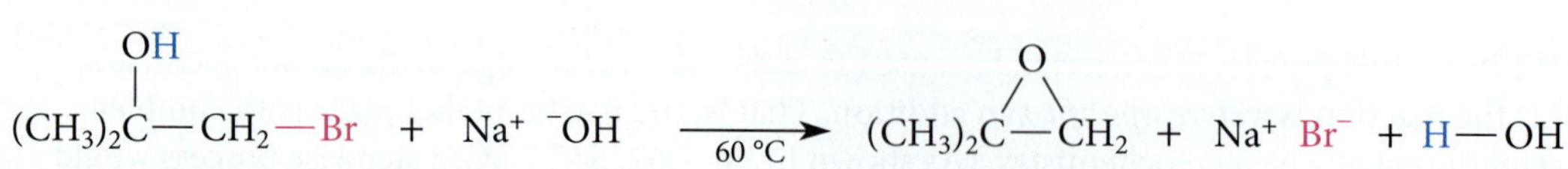

1-bromo-2-methyl-2-propanol
(a halohydrin)

2,2-dimethyloxirane
(81% yield)

(12.24)

This reaction is an intramolecular variation of the Williamson ether synthesis (Sec. 12.2A); in this case, the alcohol and the alkyl halide are part of the same molecule. The alkoxide anion, formed reversibly by reaction of the alcohol with NaOH, displaces the halide ion from the neighboring carbon:

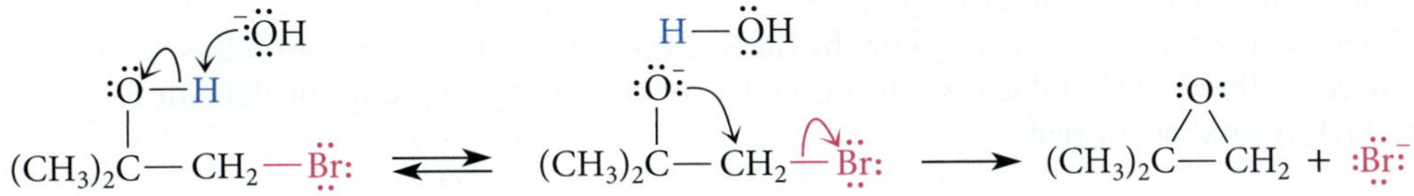

(12.25)

Like bimolecular S_N2 reactions, this reaction takes place by *opposite-side substitution* of the nucleophile—in this case, the oxygen anion—at the halide-bearing carbon (Sec. 9.4C). Such an opposite-side substitution requires that the nucleophilic oxygen and the leaving halide assume an anti relationship in the transition state of the reaction. In most noncyclic halohydrins, this relationship can generally be achieved through a simple internal rotation.

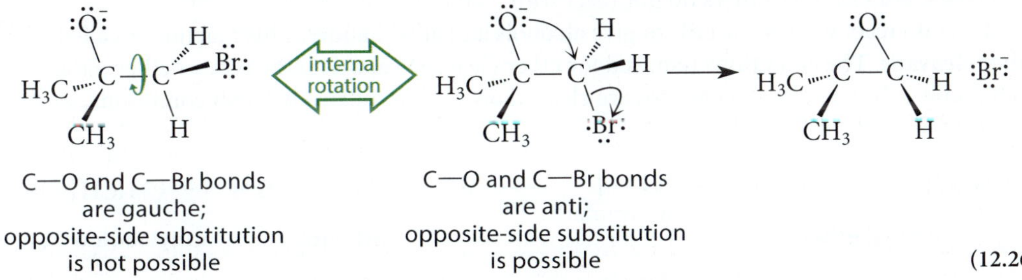

(12.26)

Halohydrins derived from cyclic compounds must be able to assume the required anti relationship through a conformational change if epoxide formation is to succeed. The following cyclohexane derivative, for example, must undergo the chair interconversion before epoxide formation can occur.

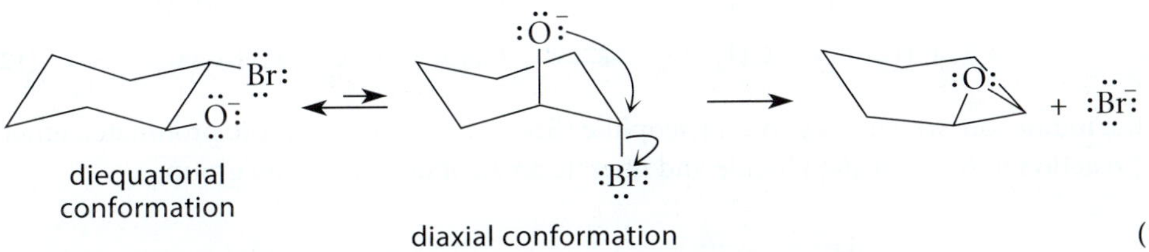

(12.27)

Even though the diaxial conformation of the halohydrin is less stable than the diequatorial conformation, the two conformations are in rapid equilibrium. As the diaxial conformation reacts to give epoxide, it is replenished by the rapidly established conformational equilibrium.

Focused Problems

12.14 From models or structures of the transition states for their reactions, predict which of the following two diastereomers of 3-bromo-2-butanol should form an epoxide at the greater rate when treated with base, and explain your reasoning. (*Hint:* Draw the conformations required for the reaction, and consider gauche interactions.)

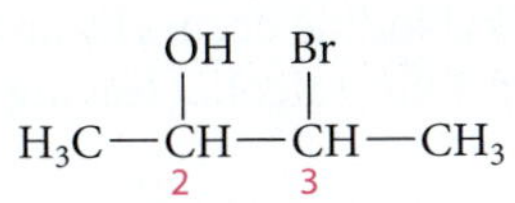

stereoisomer *A*: 2*R*,3*S*
stereoisomer *B*: 2*R*,3*R*

3-bromo-2-butanol

12.15 The chlorohydrin *trans*-2-chlorocyclohexanol reacts rapidly in base to form an epoxide. The cis stereoisomer, however, is relatively unreactive and does not give an epoxide. Explain why the two stereoisomers behave so differently.

12.4 CLEAVAGE OF ETHERS

The ether linkage is relatively unreactive under a wide variety of conditions. This is one reason ethers are widely used as solvents—that is, a great many reactions can be carried out in ether solvents without affecting the ether linkage.

Ethers do not react with nucleophiles by the S_N2 reaction for the same reason that alcohols do not react: the C—O bond is very strong, and a very basic leaving group, an alkoxide ion (Sec. 11.1), would be formed.

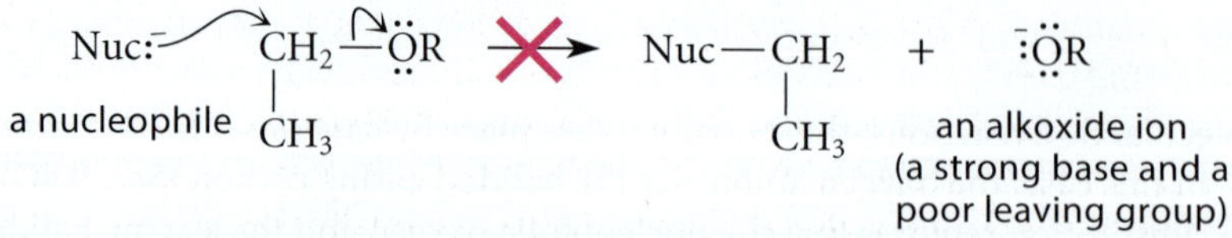

(12.28)

Remember that the S_N2 reactions of compounds with strongly basic leaving groups are very slow (Sec. 9.4F). In this case, the reaction is so slow that ethers in general are stable toward reactions with bases. For example, ethers do not react with NaOH.

Ethers do react with HI or HBr to give alcohols and alkyl halides. This reaction is called **ether cleavage**. The conditions required for ether cleavage vary with the type of ether. Ethers containing only primary alkyl groups are cleaved only under relatively harsh conditions, such as concentrated HBr or HI and heat.

$$\underset{\textbf{diethyl ether}}{CH_3CH_2—O—CH_2CH_3} + \underset{\text{(concentrated)}}{H—I} \xrightarrow{\text{heat}} \underset{\textbf{ethanol}}{CH_3CH_2—OH} + \underset{\textbf{ethyl iodide}}{I—CH_2CH_3} \qquad (12.29)$$

The alcohol formed in the cleavage of an ether (ethanol in Eq. 12.29) can go on to react with HI to give a second molecule of alkyl halide (Sec. 11.3).

The mechanism of ether cleavage involves, first, protonation of the ether oxygen:

(12.30a)

Then the iodide ion, which is a good nucleophile (Sec. 9.4E), reacts with the protonated ether in an S_N2 reaction to form an alkyl halide and liberate an alcohol as a leaving group.

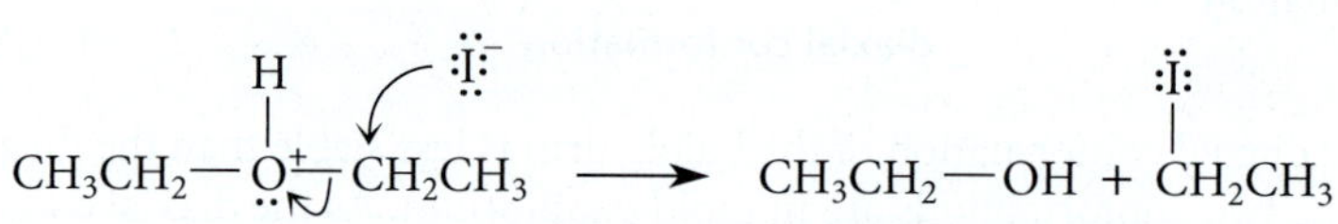

(12.30b)

If the ether is tertiary, the cleavage occurs under much milder conditions (lower temperatures, more dilute acid). The first step of the mechanism is the same—protonation of the ether linkage:

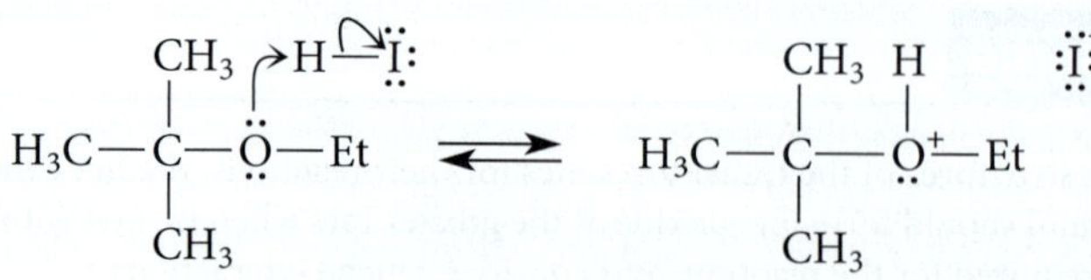

(12.31a)

In this case, the formation of the alkyl iodide occurs by an S_N1 mechanism. A *tertiary* carbocation, formed by loss of the *primary* alcohol leaving group, reacts with the iodide ion.

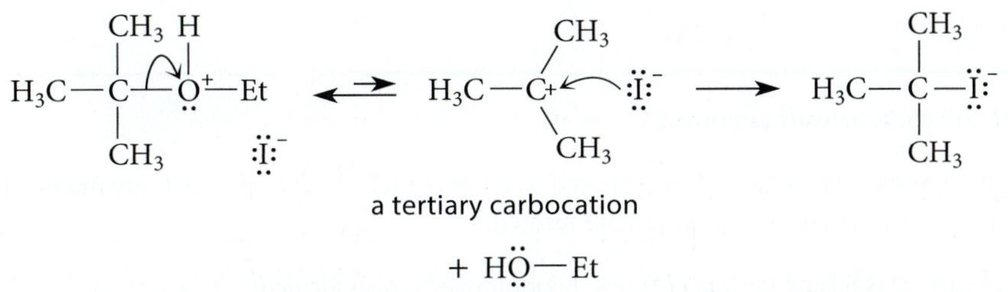

(12.31b)

Notice that the *tertiary* alkyl halide is formed along with the *primary* alcohol. Because the S_N1 reaction is faster than competing S_N2 processes, none of the primary alkyl iodide is formed.

There is great similarity in the reactions of ethers and alcohols with halogen acids (Sec. 11.3). In both cases, protonation converts a poor leaving group (—OH or —OR) into a good leaving group. In the reaction of an alcohol, *water* is the leaving group. In the reaction of an ether, an *alcohol* is the leaving group. Otherwise, the reactions are essentially the same.

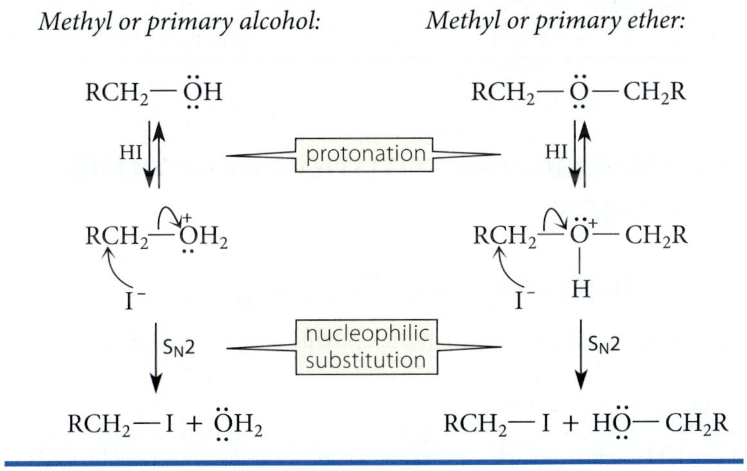

(12.32)

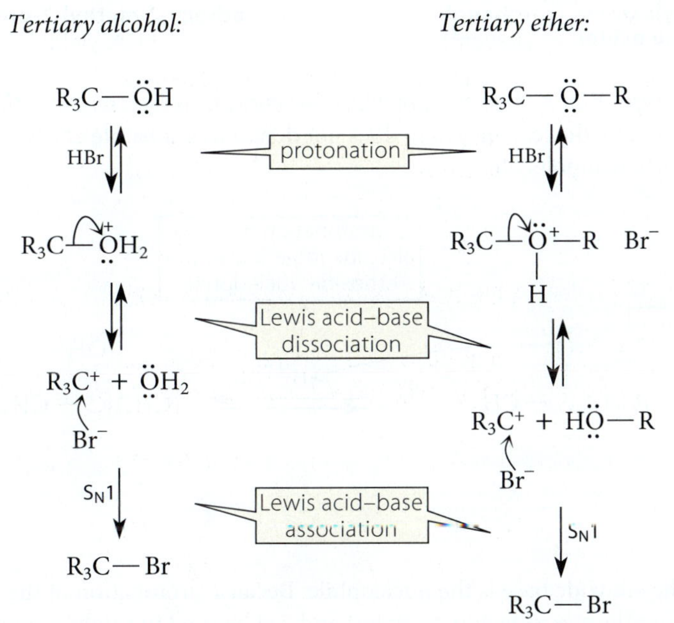

(12.33)

Secondary ethers generally follow the mechanism for tertiary ethers, but the reactions are slower because the carbocation intermediate is less stable. In addition, rearrangements can be anticipated if the secondary carbocation intermediate can rearrange to a more stable carbocation.

Although the cleavage of alkyl ethers gives alkyl halides and alcohols as products, this reaction is rarely used to prepare these compounds because ethers themselves are most often prepared from alkyl halides or alcohols, as shown in Sec. 12.2. However, it is important to appreciate these reactions because they explain the instability of ethers under acidic conditions.

Focused Problems

12.16 Explain each of the following facts with a mechanistic argument.

(a) When butyl methyl ether (1-methoxybutane) is treated with HI and heat, the initially formed products are mainly methyl iodide and 1-butanol; little or no methanol and 1-iodobutane are formed.

(b) When the reaction mixture in part (a) is heated for longer times, 1-iodobutane is also formed.

(c) When *tert*-butyl methyl ether is treated with HI, the products formed are *tert*-butyl iodide and methanol.

(d) *Tert*-butyl methyl ether cleaves much faster in HBr than its sulfur analog, *tert*-butyl methyl sulfide. (*Hint:* See Sec. 12.1.)

12.17 What products are formed when each of the following ethers reacts with concentrated aqueous HI?

(a) Diisopropyl ether

(b) 2-Ethoxy-2,3-dimethylbutane

12.5 NUCLEOPHILIC SUBSTITUTION REACTIONS OF EPOXIDES

A. Ring-Opening Reactions under Basic Conditions

Epoxides readily undergo reactions in which the epoxide ring is opened by nucleophiles.

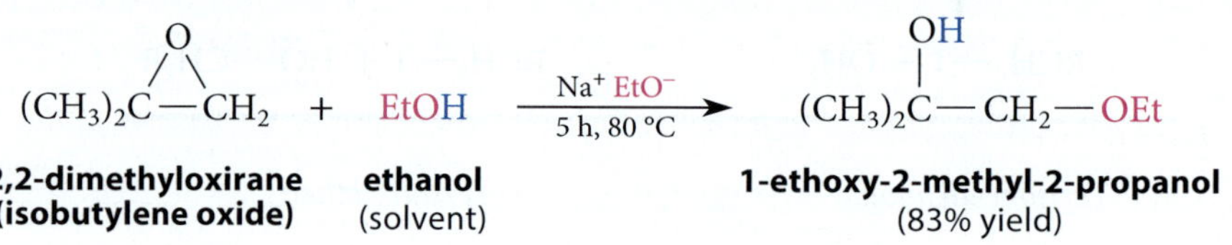

(12.34)

A reaction of this type is *an* S_N2 *reaction in which the epoxide oxygen serves as the leaving group.* In this reaction, though, the leaving group does not depart as a separate entity, but rather remains within the same product molecule.

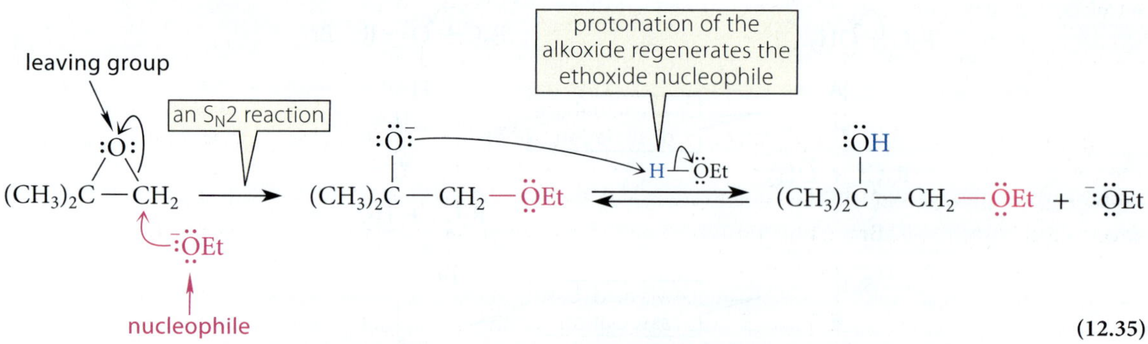

(12.35)

In this reaction the ethoxide base is the nucleophile. Because protonation of the ring-opened alkoxide regenerates the ethoxide, it is a catalyst and can be used in catalytic amounts when ethanol is the solvent.

Because an epoxide is a type of ether, the ring opening of epoxides is an ether cleavage. Recall that ordinary ethers do *not* undergo cleavage in base (Eq. 12.28, Sec. 12.4). Epoxides, however, are opened readily by basic reagents. Epoxides are reactive because they, like their carbon analogs, the cyclopropanes, possess significant *angle strain* (Sec. 7.5B). Because of this strain, the bonds of an epoxide are weaker than those of an ordinary ether, and so they are more easily broken. The opening of an epoxide relieves the strain of the three-membered ring just as the snapping of a twig relieves the strain of its bending.

In an unsymmetrical epoxide, two ring-opening products could be formed corresponding to the reaction of the nucleophile at the two different carbons of the ring. As Eq. 12.35 illustrates, *nucleophiles typically react with unsymmetrical epoxides at the carbon with fewer alkyl substituents.* This regioselectivity is expected from the effect of alkyl substitution on the rates of S_N2 reactions (Sec. 9.4D). Because alkyl substitution retards the S_N2 reaction, the reaction of a nucleophile at the unsubstituted carbon is faster and leads to the observed product.

Like other S_N2 reactions, the ring opening of epoxides by bases involves *opposite-side substitution* of the nucleophile on the epoxide carbon. When this carbon is a stereocenter, inversion of configuration occurs, as illustrated in Study Problem 12.2.

Study Problem 12.2

What is the stereochemistry of the 2,3-butanediol formed when *meso*-2,3-dimethyloxirane reacts with aqueous sodium hydroxide?

Solution First draw the structure of the epoxide. The meso stereoisomer of 2,3-dimethyloxirane has an internal plane of symmetry, and its two asymmetric carbons have opposite configurations.

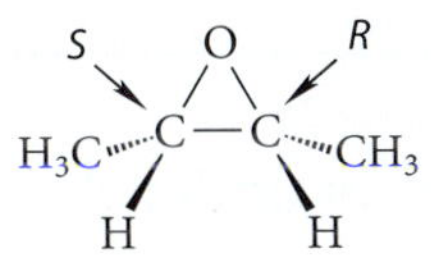

***meso*-2,3-dimethyloxirane**

Because the two different carbons of the epoxide ring are enantiotopic (Sec. 11.9A), the hydroxide ion reacts at either one at the same rate. Opposite-side substitution on each carbon should occur with inversion of configuration.

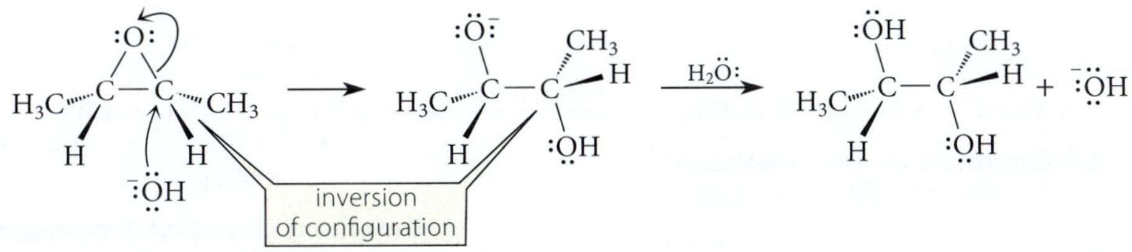

The product shown is the 2*S*,3*S* stereoisomer. Reaction of the nucleophile at the other epoxide carbon gives the 2*R*,3*R* stereoisomer. (Verify this point!) Because the starting materials are achiral, the two enantiomers of the product must be formed in equal amounts (Sec. 7.7A). Therefore, the product of the reaction is racemic 2,3-butanediol.

Although the examples in this section have involved hydroxide and alkoxides as nucleophiles, the pattern of reactivity is the same with *any* nucleophile: The nucleophile reacts at the carbon with no alkyl substituents and opens the epoxide to form an alkoxide, which then reacts in a Brønsted acid–base reaction with a proton source to give an alcohol. Letting $Nuc:^-$ be a general nucleophile, we can summarize this pattern of reactivity with Eq. 12.36:

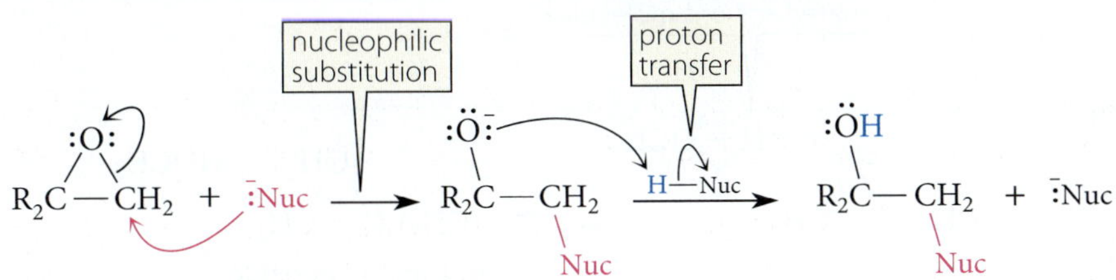

(12.36)

Focused Problems

12.18 Predict the products of the following reactions by drawing a curved-arrow mechanism for each.

(a)

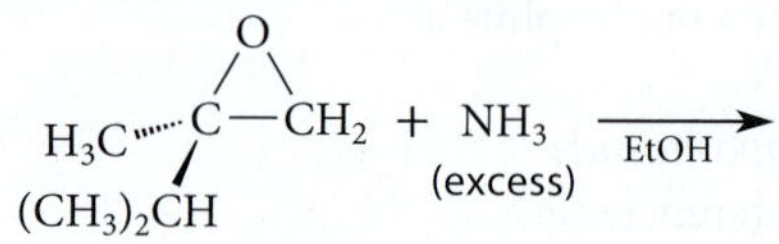

(b)

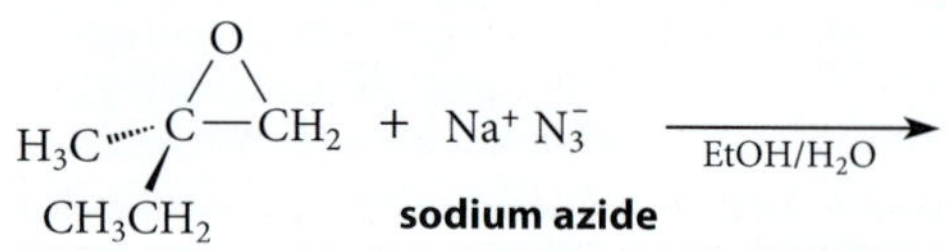

12.19 From what epoxide and what nucleophile could each of the following compounds be prepared? (Assume each is racemic.)

(a)

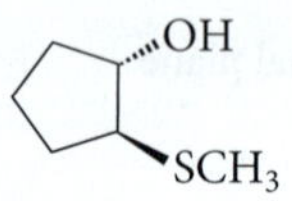

(b)

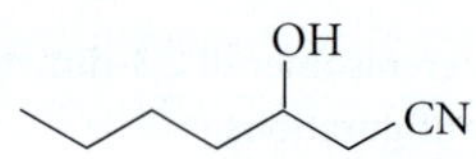

B. Ring-Opening Reactions under Acidic Conditions

Ring-opening reactions of epoxides, like those of ordinary ethers, can be catalyzed by acids. However, epoxides are *much* more reactive than ethers under acidic conditions because of their angle strain. Therefore, milder conditions can be used for the ring-opening reactions of epoxides than are required for the cleavage of ordinary ethers. For example, very low concentrations of acid catalysts are required in ring-opening reactions of epoxides.

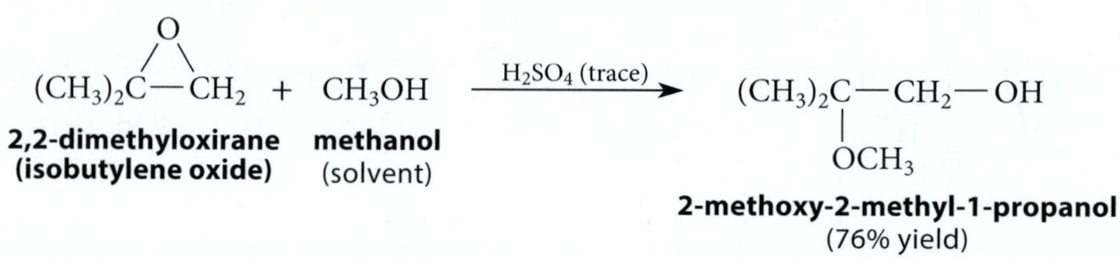

(12.37)

The regioselectivity of the ring-opening reaction is different under acidic and basic conditions. The structure of the product in Eq. 12.37 shows that the methanol nucleophile reacts at the *more substituted carbon* of the epoxide under acidic conditions. Contrast this with the result in Eq. 12.34, in which the nucleophile reacts at the *less substituted carbon* under basic conditions. In general, if one of the carbons of an unsymmetrical epoxide is tertiary, nucleophiles react at this carbon under acidic conditions.

Some insight into why different regioselectivities are observed under different conditions comes from the mechanism. The first step in the mechanism of Eq. 12.37, like the first step of ether cleavage, is protonation of the oxygen.

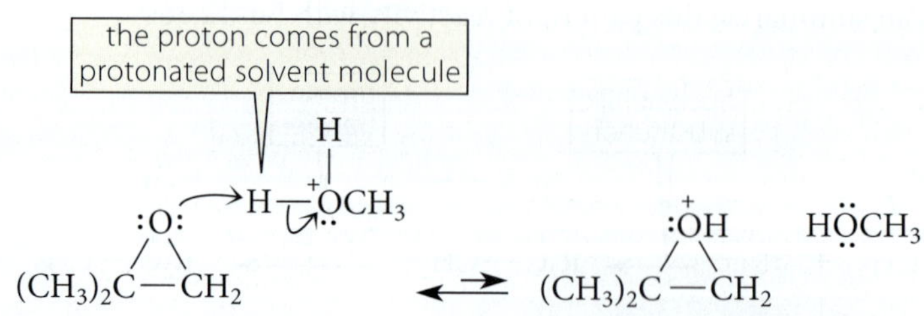

(12.38a)

The structural properties of the protonated epoxide show that it can be expected to behave like a tertiary carbocation. In this respect it is much like a bromonium ion, which has a similar ring-opening regiospecificity (Eq. 5.15, Sec. 5.2B). Both types of reactions involve the opening of strained rings containing positively charged, electronegative leaving groups.

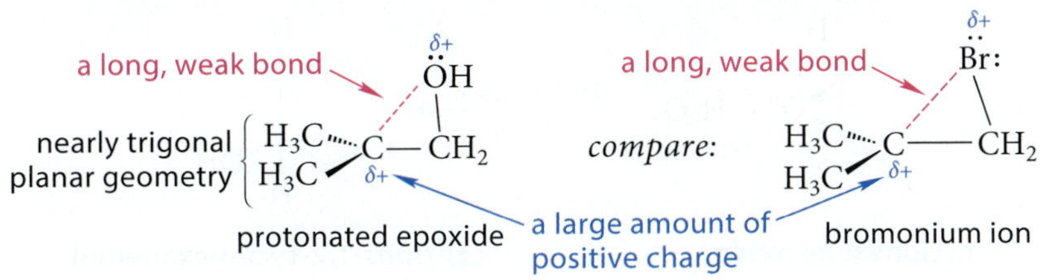

(12.38b)

First, calculations show that the tertiary carbon bears about 0.7 of a positive charge. Second, the geometry at the tertiary carbon is nearly trigonal planar. This means that the tertiary carbon and the groups around it are nearly flattened into a common plane so that little or no steric hindrance prevents the approach of a nucleophile to this carbon. Finally, the bond between the tertiary carbon and the OH group is unusually long and weak. This means that this bond is more easily broken than the other C—O bond. In fact, this cation resembles a carbocation solvated by the leaving group (see Fig. 9.13, Sec. 9.6D). The OH leaving group blocks the front side of the carbocation so that the nucleophilic reaction of a solvent molecule must occur at the side opposite to the epoxide oxygen. (This aspect of the mechanism has stereochemical consequences, as we discuss later.) Loss of a proton to solvent gives the product.

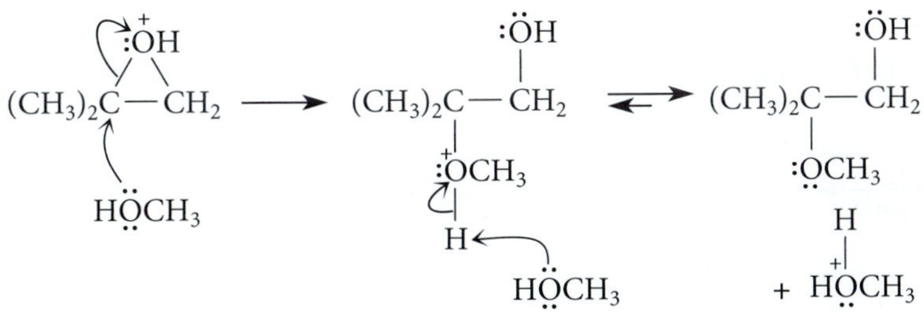

(12.38c)

It is a solvent molecule, not the alkoxide conjugate base of the solvent, which reacts with the protonated epoxide. The alkoxide conjugate base cannot exist in acidic solution; nor is it necessary, because the protonated epoxide is very reactive and because the nucleophile is also the solvent and is present in great excess.

When the carbons of an unsymmetrical epoxide are secondary or primary, there is much less carbocation character at either carbon in the protonated epoxide, and acid-catalyzed ring-opening reactions tend to give mixtures of products; the exact compositions of the mixtures vary from case to case.

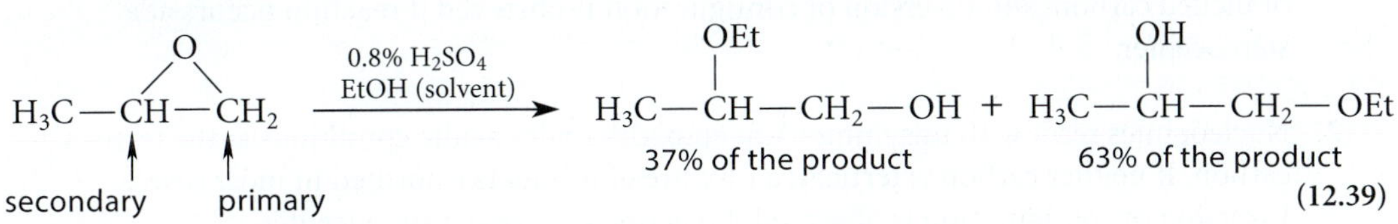

(12.39)

The mixture reflects the balance between opening of the weaker bond, which favors reaction at the carbon with more substituents, and van der Waals repulsions of the alkyl substituents with the nucleophile, which favor reaction at the carbon with fewer substituents.

When the two carbons of the epoxide are stereocenters, the reaction occurs with *inversion of stereochemical configuration,* as the example in Eq. 12.40 illustrates.

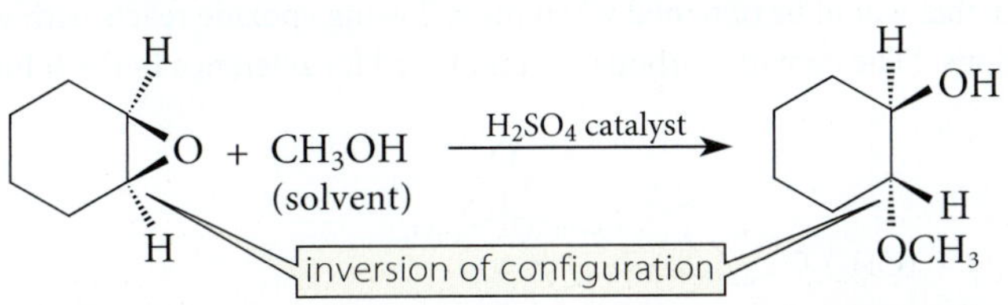

cyclohexene oxide **(±)-*trans*-2-methoxycyclohexanol** (82% yield) (12.40)

When water is used as a nucleophile in an acid-catalyzed epoxide ring opening, the product is a 1,2-diol, or *glycol*. Acid-catalyzed epoxide hydrolysis is generally a useful way to prepare glycols.

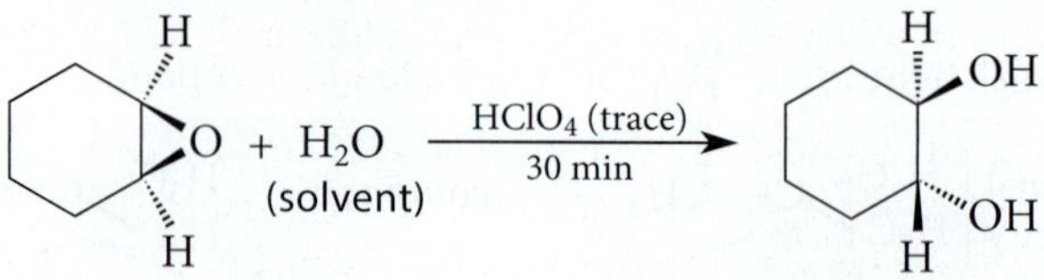

cyclohexene oxide **(±)-*trans*-1,2-cyclohexanediol** (a glycol; 80% yield) (12.41)

Notice the trans relationship of the two hydroxy groups in the product, which results from the inversion of configuration that occurs when water reacts with the protonated epoxide. It follows that *cis*-1,2-cyclohexanediol *cannot* be prepared by epoxide opening. In Sec. 12.6A, however, we show how the cis stereoisomer can be prepared by another method.

Although base-catalyzed hydrolysis of epoxides also gives glycols (see Study Problem 12.2), polymerization sometimes occurs as a side reaction under basic conditions (see end-of-chapter Problem 12.74). Consequently, acid-catalyzed hydrolysis of epoxides is generally preferred for the preparation of glycols.

Focused Problem

12.20 Predict the major product(s) of each of the following transformations.

(a)

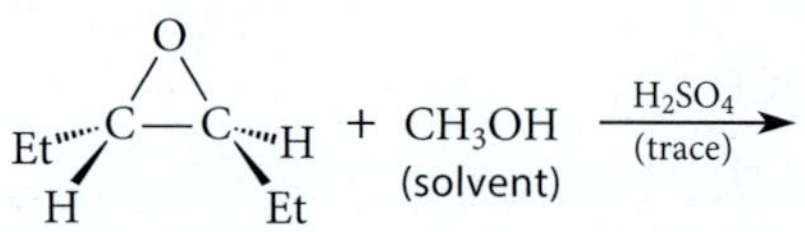

(optically active)

(b) The enantiomer of the epoxide in part (a) + CH_3OH (solvent) $\xrightarrow[\text{(trace)}]{H_2SO_4}$

Here is a summary of the facts about the regioselectivity and stereoselectivity of epoxide ring-opening reactions:

1. Nucleophiles react with unsymmetrical epoxides under basic conditions at the less branched carbon, and inversion of configuration is observed if reaction occurs at a stereocenter.

2. Nucleophiles react with unsymmetrical epoxides under acidic conditions at the tertiary carbon. If neither carbon is tertiary, a mixture of products is formed in most cases. Inversion of configuration is observed if reaction occurs at a stereocenter.

These facts are applied in Study Problem 12.3.

Study Problem 12.3

Predict the major product in each case that would be obtained when the following epoxide reacts with water under (a) basic conditions; (b) acidic conditions. (The epoxide carbons are numbered for reference in the solution.)

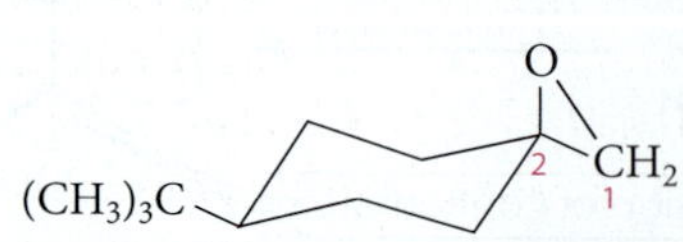

Solution As the preceding summary suggests, when attempting to predict the products of an epoxide ring-opening reaction, first decide whether the conditions of the reaction are basic or acidic. If basic, the nucleophile reacts at the *less substituted* carbon of the epoxide; if acidic, the nucleophile reacts at the *tertiary carbon* of the epoxide. Then determine whether the carbon at which the reaction occurs is a stereocenter. If so, make sure to predict the product that results from inversion of configuration.

(a) Under basic conditions, hydroxide ion is the nucleophile. It will react at the *less substituted carbon* (carbon-1) of the epoxide. (If you have difficulty seeing why this is the less substituted carbon, re-read Study Guide Link 9.2.) Because this carbon is *not* a stereocenter, the stereochemistry of the substitution does not matter. Consequently, the reaction is

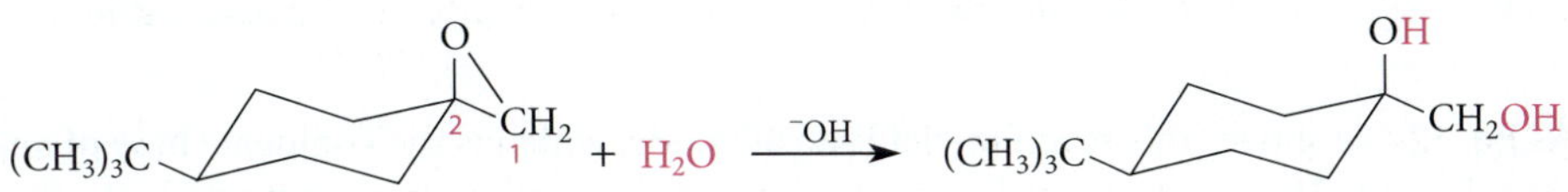

(12.42)

(b) Under acidic conditions, water is the nucleophile. It reacts with the *protonated* epoxide at the more branched carbon (carbon-2). Notice that carbon-2 is a stereocenter (even though it is *not* an asymmetric carbon); reaction of the nucleophile at carbon-2 occurs with inversion of configuration. Consequently, the product of the reaction under acidic conditions is a diastereomer of the product obtained under basic conditions.

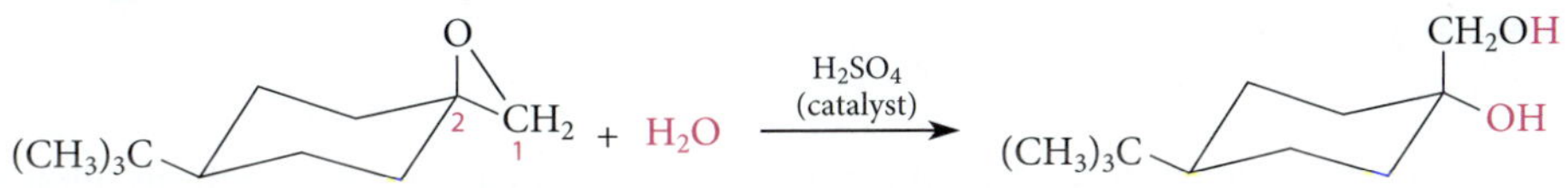

(12.43)

Focused Problem

12.21 (a) Suppose 2,2-dimethyloxirane reacts with water that has been enriched with the oxygen isotope ^{18}O. Indicate how the hydrolysis product would differ under acidic and basic conditions.

(b) Show how the stereochemistry of the products will differ (if at all) when the following enantiomerically pure epoxide reacts with water under acidic and basic conditions.

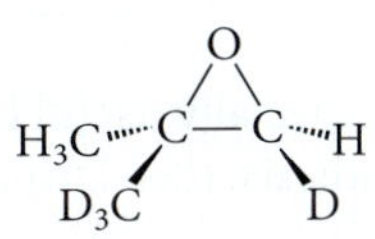

C. Reactions of Epoxides with Organometallic Reagents

Reaction of Ethylene Oxide with Grignard Reagents Grignard reagents (Sec. 10.5) react with ethylene oxide to give, after a protonation step, primary alcohols:

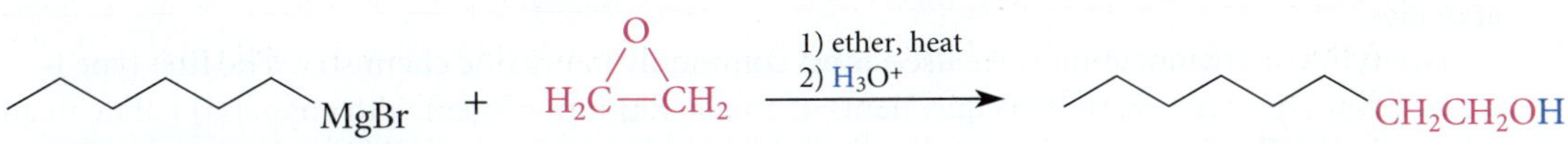

hexylmagnesium bromide (a Grignard reagent) | **ethylene oxide** | **1-octanol** (71% yield)

(12.44)

This reaction is another epoxide ring-opening reaction. To understand this reaction, recall that the carbon in the C—Mg bond of the Grignard reagent has *carbanion* character and is therefore a very *basic* carbon (Sec. 10.5C). This carbon is the nucleophilic center that reacts with the epoxide. At the same time, the magnesium of the Grignard reagent forms a Lewis acid–base complex with the epoxide oxygen. (Recall that Grignard reagents associate strongly with ether oxygens; see Eq. 10.49, Sec. 10.5B.) Just as protonation of an oxygen makes it a better leaving group, bonding of this oxygen to a Lewis acid also makes it a better leaving group. Consequently, this interaction assists the ring opening of the epoxide in much the same way that Brønsted acids catalyze ring opening (Sec. 12.5B).

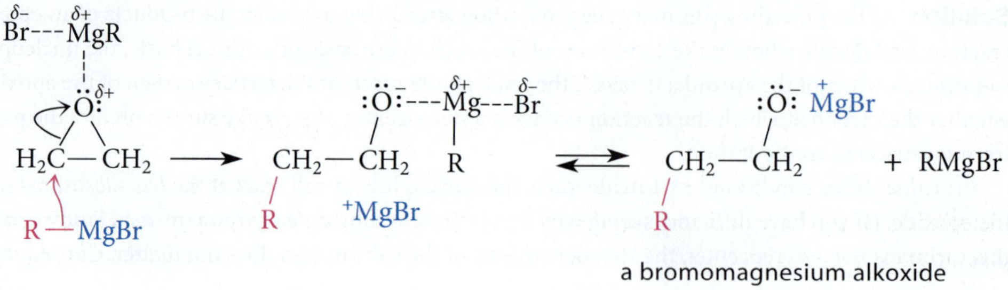

(12.45a)

As Eq. 12.45a shows, this reaction yields an alkoxide, which is the conjugate base of an alcohol (Sec. 11.1A). After the Grignard reagent has reacted, the alkoxide is converted into the alcohol product in a separate step by the addition of water or dilute acid:

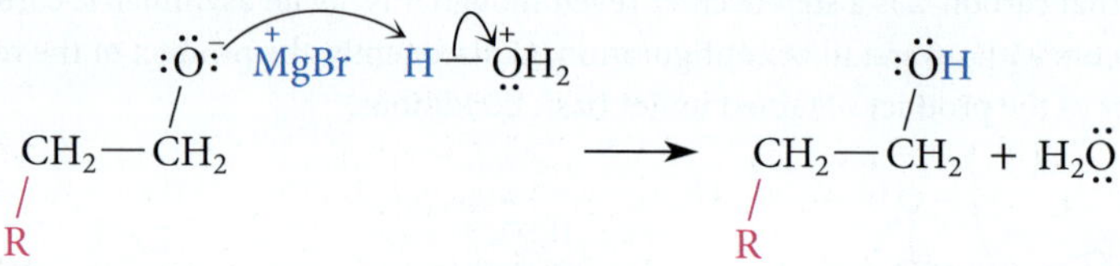

(12.45b)

This reaction accomplishes two objectives when used in an organic synthesis. First, it forms a new carbon–carbon bond, lengthening the carbon chain of the Grignard reagent by two carbons. Second, it converts the epoxide into a primary alcohol. Combining this reaction with other functional-group transformations, it can bring about the net two-carbon chain extension of an alcohol:

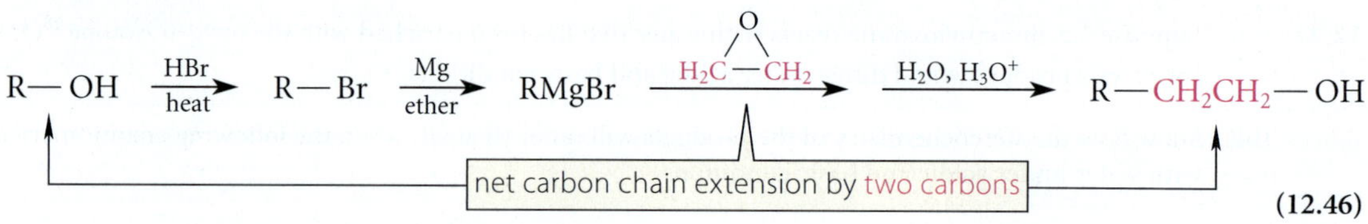

(12.46)

This transformation is especially useful because alcohols are particularly valuable starting materials in organic synthesis. (See Display 11.90, Sec. 11.12.)

Reaction of Epoxides with Lithium Organocuprate Reagents Grignard and organolithium reagents react with epoxides other than ethylene oxide, too. Many of these reactions are unsatisfactory, however, because they give not only the expected products of ring opening but also rearrangements and other side reactions as well. (Grignard reagents and organolithium reagents have some Lewis acid character that promotes such side reactions.) However, another type of organometallic reagent, the lithium organocuprate, undergoes useful ring-opening reactions with epoxides.

Two types of organocuprates are used most commonly in organic chemistry. The first type is formed from the reaction of two equivalents of an alkyllithium reagent with copper(I) halide in an ether solvent. The first equivalent of the alkyllithium reacts to form an alkylcopper reagent plus a lithium halide. The driving force for the reaction is the greater tendency of lithium, the more electropositive metal, to exist as an ion.

(12.47a)

Because the copper is a Lewis acid, the alkylcopper reagent reacts with a second equivalent of the alkyllithium to give a **lithium dialkylcuprate**.

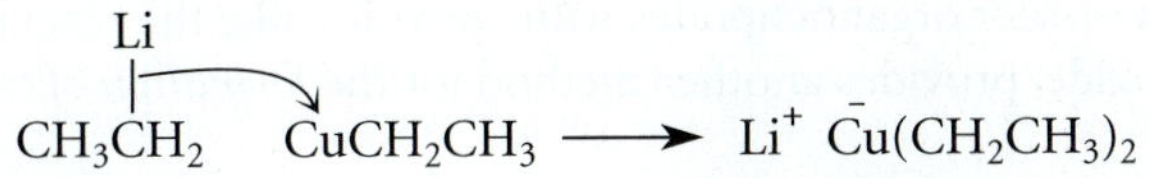

lithium diethylcuprate
(a lithium dialkylcuprate) (12.47b)

(Aryllithium reagents such as phenyllithium, Ph—Li, can also be used to prepare lithium diarylcuprates.)

Lithium dialkylcuprate or diarylcuprate reagents are sometimes called *Gilman reagents* after Henry Gilman (1893–1986), who was a professor of chemistry at Iowa State University when he discovered the reagents in 1952.

If copper(I) cyanide, CuCN, is used instead of a copper(I) halide, the cyanide group, which is much more basic than halide, remains bound to the copper, and a more complex reagent is formed:

$$2\,CH_3CH_2Li + CuCN \longrightarrow (CH_3CH_2)_2Cu(CN)Li_2 \qquad (12.48)$$

a higher-order organocuprate

Although Eq. 12.48 describes the stoichiometry of the reagent, it exists in a state (or states) of higher aggregation. Such reagents are called **higher-order organocuprates.**

Both types of organocuprate reagent are useful in organic chemistry, and both react with epoxides. However, the higher-order organocuprates are the preferred reagents for use with epoxides because they react with a wider variety of epoxides and give fewer side reactions. (Some important uses of lithium dialkylcuprates are covered in later chapters.)

An organocuprate reagent reacts as a basic nucleophile at the carbon of the epoxide with fewer alkyl substituents to give products of ring opening. Subsequent protonolysis gives the alcohol.

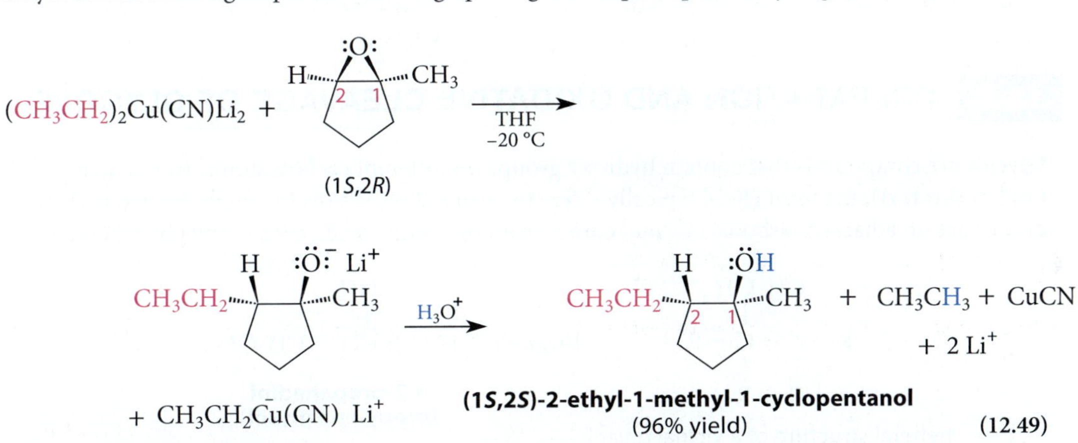

(1*S*,2*S*)-2-ethyl-1-methyl-1-cyclopentanol
(96% yield) (12.49)

Notice that, if the epoxide carbon is a stereocenter, the alkyl group from the reagent reacts at the epoxide carbon with *inversion of stereochemical configuration.*

We can think of the reaction as an S_N2 process in which a "carbanion" nucleophile is delivered from the copper to the epoxide carbon electrophile with stereochemical inversion. Epoxide opening is assisted by lithium ion, which is a "built-in" Lewis acid:

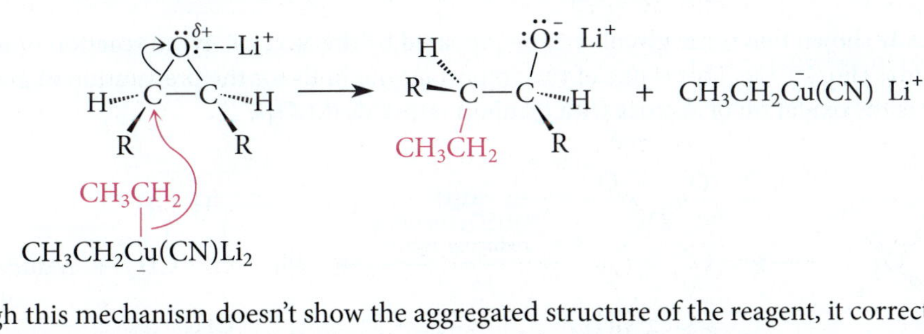

(12.50)

Although this mechanism doesn't show the aggregated structure of the reagent, it correctly predicts the chemical and stereochemical outcome of the reaction.

The reaction of higher-order lithium organocuprate reagents with epoxides is another method for the synthesis of alcohols that can be added to the list in Sec. 11.11. You should ask yourself what the limits are on the types of alcohol that can be prepared by each method.

The reaction of higher-order organocuprates with epoxides, like the reaction of Grignard reagents with ethylene oxide, provides another method for the *formation of carbon–carbon bonds.*

Focused Problems

12.22 (a) From what Grignard reagent can 3-methyl-1-pentanol be prepared by reaction with ethylene oxide, then aqueous acid?

(b) From what epoxide and what higher-order cuprate reagent can 3-ethyl-3-heptanol be prepared?

12.23 Complete the following reactions by giving the structures of the alcohol products. In part (b), show the stereochemistry of the product as well.

(a)

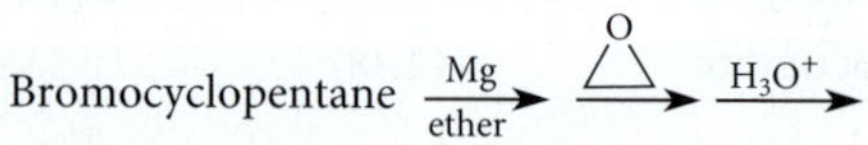

(b)

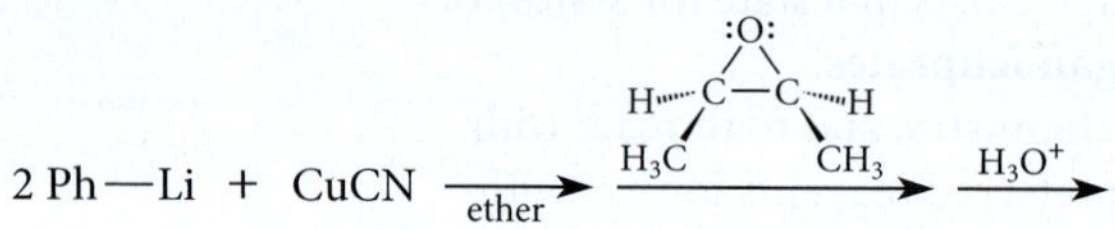

(c)

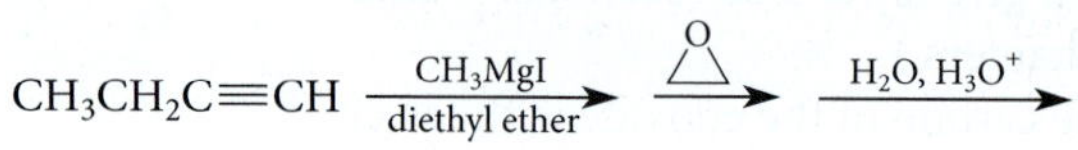

12.6 PREPARATION AND OXIDATIVE CLEAVAGE OF GLYCOLS

Glycols are compounds that contain hydroxy groups on different carbon atoms. In practice (and in this text), the term *glycol* typically refers to *vicinal* diols—diols in which the two hydroxy groups are on adjacent carbons. (*Vicinal* comes from the Latin word *vicinus* = neighborhood.)

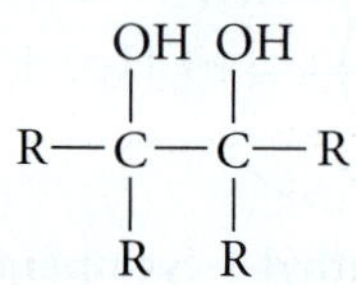

general structure of a vicinal glycol (R = alkyl, aryl, or H)

Example:

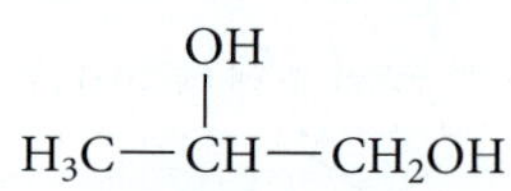

1,2-propanediol (propylene glycol)

(12.51)

Although glycols are alcohols, some glycol chemistry is quite different from the chemistry of alcohols. Some of this unique chemistry is the subject of this section.

A. Preparation of Glycols

We've already shown that some glycols can be prepared by the acid-catalyzed reaction of water with epoxides (Eq. 12.41). This is one of two important methods for the preparation of glycols. The other is the oxidation of alkenes with osmium tetroxide (OsO_4).

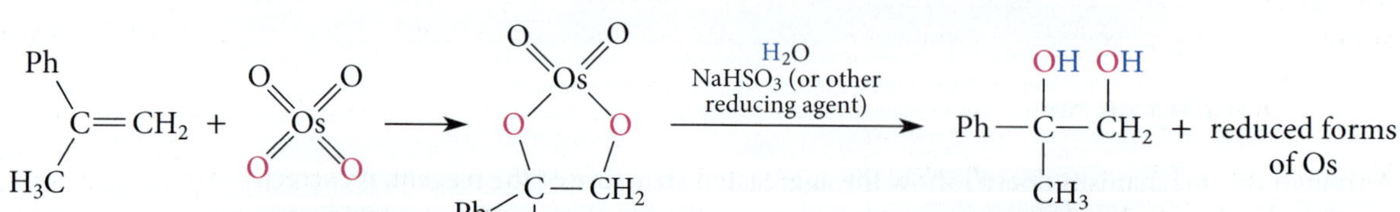

osmium(VIII) tetroxide

an osmate ester (typically not isolated)

2-phenyl-1,2-propanediol a glycol (90–95% yield)

(12.52)

The osmium in OsO_4 is in a +8 oxidation state. Metals in high oxidation states [such as Mn(VII) and Cr(VI)] are oxidizing agents because they attract electrons. This electron-attracting ability of Os(VIII) results in a concerted (that is, one-step) cycloaddition reaction between OsO_4 and an alkene to give the intermediate osmate ester:

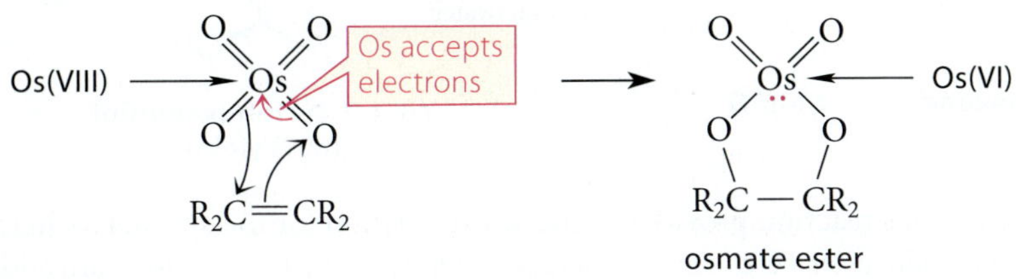

(12.53a)

OsO_4 is used as a stain in electron microscopy because it reacts with any accessible carbon–carbon double bonds in biological samples. The resulting osmate esters "fix" the osmium. The large osmium atom scatters electrons very efficiently and thus serves as an excellent contrast agent. OsO_4 is highly toxic because of its reaction with a number of biological molecules.

(The osmate ester is another example of an organic ester derivative of an inorganic acid; Sec. 11.4C.) The curved-arrow notation shows that in this reaction the osmium accepts an electron pair. As a result, its oxidation state is decreased to +6. This process closely resembles the concerted cycloaddition mechanism of ozonolysis. (See the discussion in Further Exploration 12.1.)

FURTHER EXPLORATION 12.1
Mechanism of OsO_4 Addition

A glycol is formed when the cyclic osmate ester is treated with water. Two water molecules, acting as nucleophiles, displace the glycol oxygens from the osmium. A mild reducing agent such as sodium bisulfite, $NaHSO_3$, is often added to convert the osmium-containing by-products into reduced forms of osmium that are easy to remove by filtration. (The $NaHSO_3$ is converted into sodium sulfate, Na_2SO_4.)

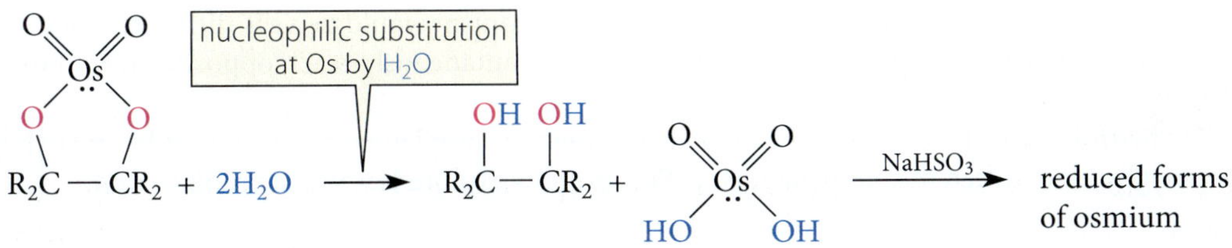

(12.53b)

Two practical drawbacks to the use of the OsO_4 oxidation are that osmium and its compounds are very toxic (see the margin box near Eq. 12.53a), and they are quite expensive. The reaction of OsO_4 with alkenes is so useful, however, that chemists have devised ways for it to be used with very small amounts of OsO_4. This is done by including in the reaction mixture an oxidant that recycles the Os(VI) by-product back into OsO_4. Among the common oxidants used for this purpose are *amine oxides*, which are compounds of the form $R_3N^+—O^-$. Two amine oxides used commonly are the following:

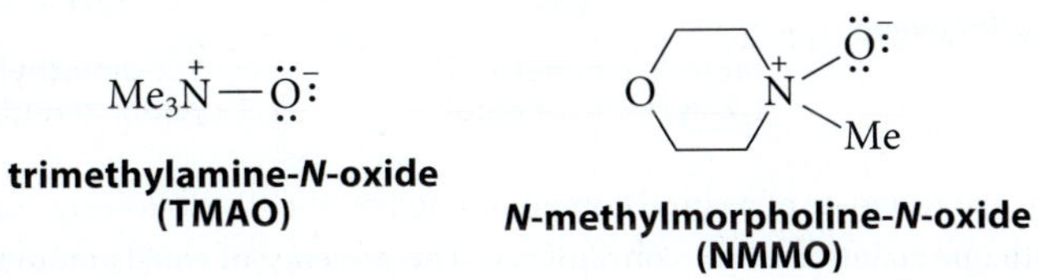

(12.54)

In other words, once a small amount of OsO_4 is used up, the Os(VI) by-product is oxidized within the reaction mixture by the amine oxide to re-form OsO_4. Consequently, a catalytic amount of OsO_4 can be used, and the amine oxide acts as the ultimate oxidant.

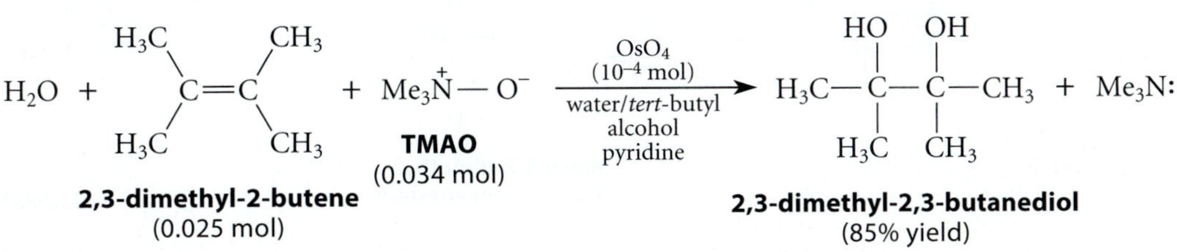

(12.55)

This reaction is sometimes called the *Upjohn dihydroxylation* because it was developed in the research laboratories of the Upjohn Company, a pharmaceutical manufacturer. When the oxidant NMMO is present, other osmium compounds such as Os(VI) salts and Os(III) salts, can be used instead of OsO_4 because they are readily oxidized to OsO_4.

The OsO_4 oxidation is particularly useful because of its stereochemistry. The formation of glycols from alkenes is a stereospecific syn addition.

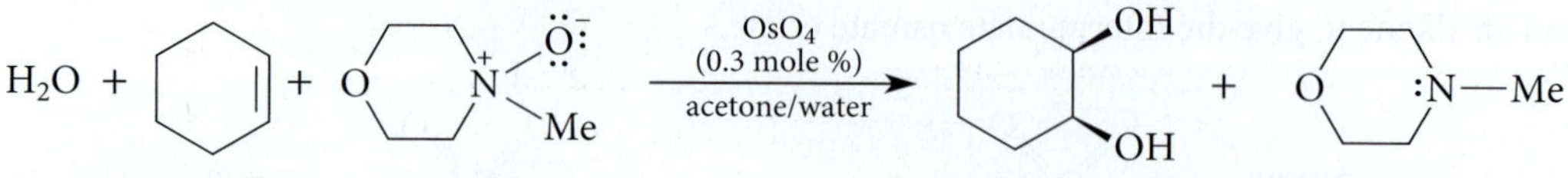

cyclohexene **NMMO** ***cis*-1,2-cyclohexanediol** (89% yield) (12.56)

The mechanism of this reaction provides a simple explanation for the syn stereochemistry. The five-membered osmate ester ring is easily formed when two oxygens of OsO_4 are added to the same face of the double bond by a concerted mechanism. Hydrolysis of the osmate ester breaks the bonds between oxygen and osmium to give the glycol.

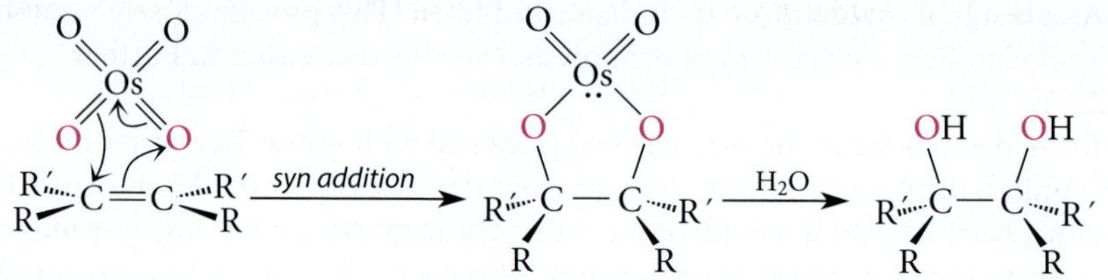

an osmate ester a 1,2-glycol (12.57)

On the other hand, an anti addition by a concerted mechanism would be very difficult, if not impossible: the two reacting oxygens of OsO_4 cannot simultaneously reach opposite faces of the π bond.

The hydrolysis of epoxides and the OsO_4 oxidation are complementary reactions because they provide glycols of different stereochemistry. This point is explored in Study Problem 12.4.

Study Problem 12.4

Outline preparations of *meso*-1,2-dimethyl-1,2-cyclohexanediol and (±)-1,2-dimethyl-1,2-cyclohexanediol from 1,2-dimethylcyclohexene.

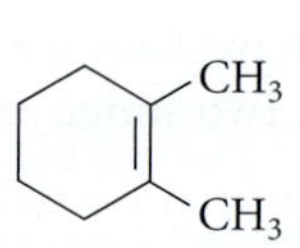

1,2-dimethylcyclohexene

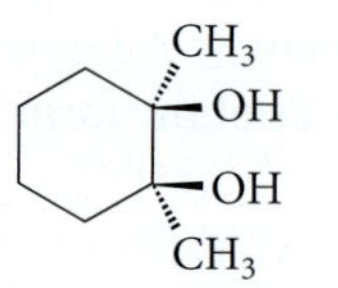

***meso*-1,2-dimethyl-1,2-cyclohexanediol**

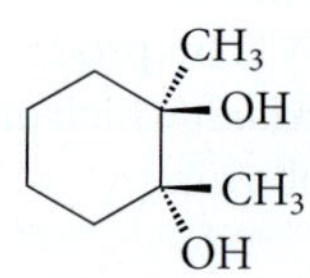

(±)-1,2-dimethyl-1,2-cyclohexanediol

Solution The (±) diastereomer is the racemate of a chiral compound. Remember the following convention: we represent racemic compounds with one enantiomer for convenience. The presence of equal amounts of the enantiomer is understood (Sec. 7.7A).

The meso stereoisomer of the product requires a syn addition of the two hydroxy groups. The osmium tetroxide method gives this addition stereochemistry.

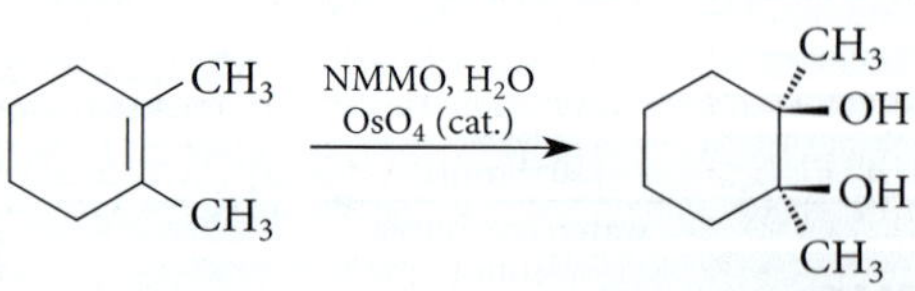

***meso*-1,2-dimethyl-1,2-cyclohexanediol** (12.58a)

Acid-catalyzed epoxide hydrolysis would provide the racemic compound because the reaction of water with the epoxide occurs with inversion of configuration.

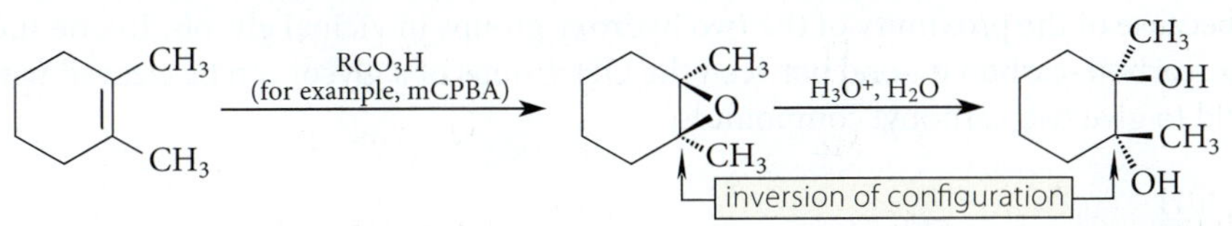

(±)-1,2-dimethyl-1,2-cyclohexanediol (12.58b)

Glycol formation from alkenes can also be carried out with potassium permanganate ($KMnO_4$), usually under aqueous alkaline conditions. This reaction is also a stereospecific syn addition, and its mechanism is probably similar to that of OsO_4 addition. (See Focused Problem 12.26.)

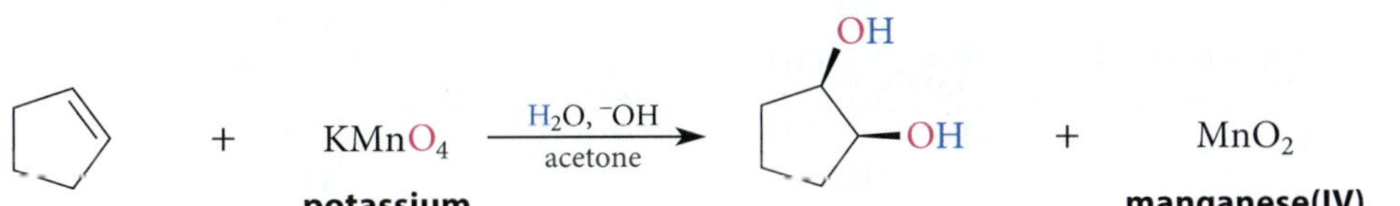

cyclopentene **potassium permanganate** (purple solution) ***cis*-1,2-cyclopentanediol** (45% yield) **manganese(IV) dioxide** (brown precipitate) (12.59)

The use of $KMnO_4$ avoids the expense and toxicity of OsO_4, but yields are low in many cases because over-oxidation occurs—that is, the glycol product is oxidized further. The reaction is often run at low temperature, and conditions have to be carefully worked out in each case to avoid this side reaction.

The manganese in MnO_4^- is in the Mn(VII) oxidation state. It is converted into Mn(IV) as a result of the reaction. Visually, when oxidation occurs, the brilliant purple color of the permanganate ion is replaced by a muddy-looking brown precipitate of manganese dioxide (MnO_2). This color change can be used as a qualitative test for double bonds and other functional groups that can be oxidized by $KMnO_4$.

Focused Problems

12.24 What organic product is formed (including its stereochemistry) when each of the following alkenes is treated with NMMO in the presence of H_2O and a catalytic amount of OsO_4?

(a) 1-Methylcyclopentene (b) *trans*-2-Butene

12.25 From what alkene could each of the following glycols be prepared by the OsO_4 or $KMnO_4$ method?

(a)

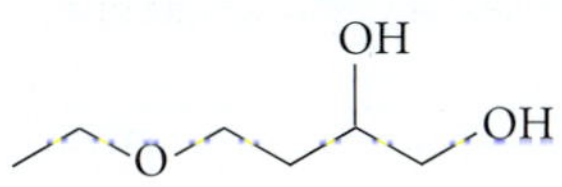

(b)

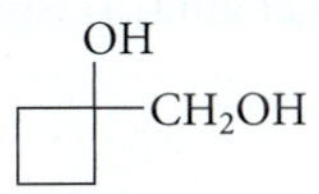

(c) *meso*-4,5-Octanediol (d) (±)-4,5-Octanediol

12.26 Show a curved-arrow mechanism for the first step, and the structure of the cyclic intermediate formed, when an alkene is treated with $KMnO_4$. A Lewis structure for the permanganate ion is as follows:

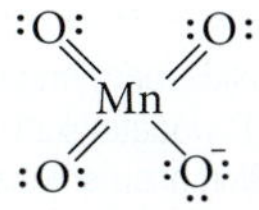

permanganate ion

B. Oxidative Cleavage of Glycols

Glycols are alcohols, so they undergo the reactions of alcohols. However, a few unique reactions can occur because of the proximity of the two hydroxy groups in vicinal glycols. In one such reaction, the carbon–carbon σ bond between the OH groups of a glycol can be cleaved with periodic acid to give two carbonyl compounds:

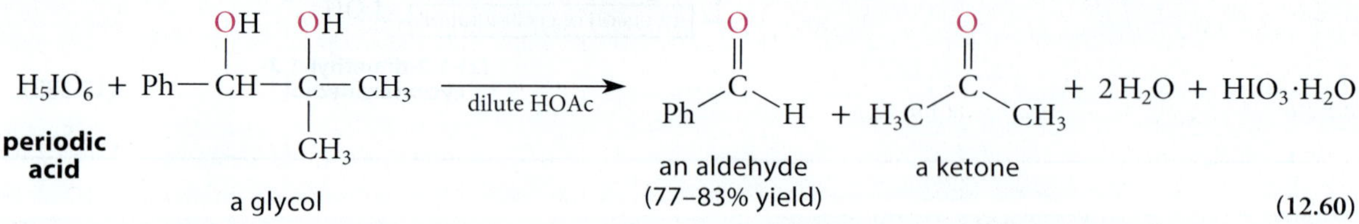

(12.60)

Periodic acid (pronounced PURR-eye-OH-dik) is the iodine analog of perchloric acid.

$HClO_4$	HIO_4
perchloric acid	**periodic acid**

Periodic acid is commercially available as the dihydrate, $HIO_4{\cdot}2H_2O$, often abbreviated, as in Eq. 12.60, as H_5IO_6 (sometimes called *para-periodic acid*). Its sodium salt, $NaIO_4$ (sodium metaperiodate), is sometimes also used. The formulas HIO_4 and H_5IO_6 are used interchangeably for periodic acid. Periodic acid is a fairly strong acid ($pK_a = -1.6$). The periodate cleavage reaction has been used as a test for glycols as well as for synthesis.

The cleavage of glycols with periodic acid takes place through a cyclic periodate ester intermediate (Sec. 11.4C) that forms when the glycol displaces two OH groups from H_5IO_6.

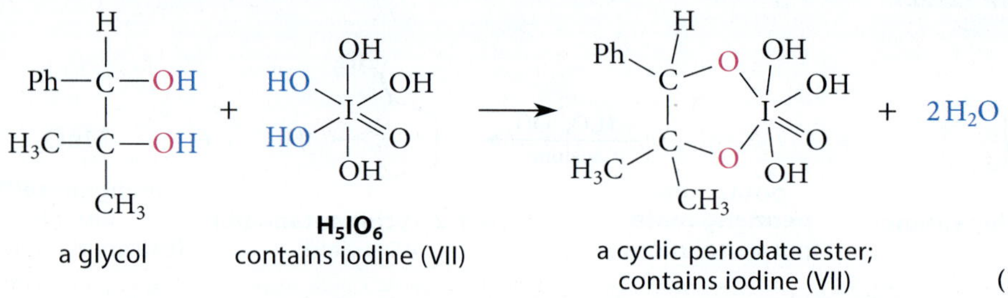

(12.61a)

The cyclic ester spontaneously breaks down by a cyclic flow of electrons in which the iodine accepts an electron pair. (The clockwise direction of electron flow is arbitrary.)

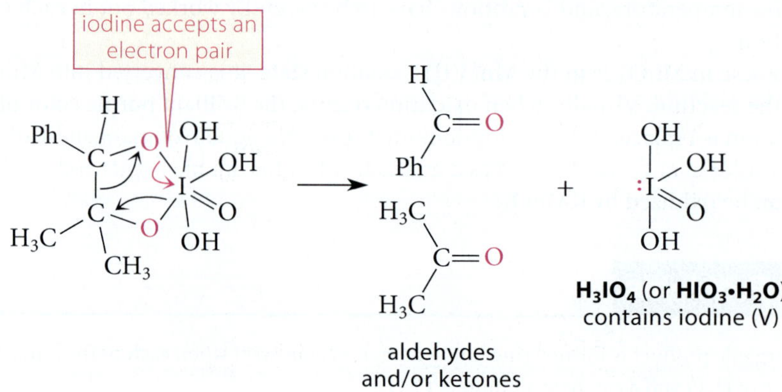

(12.61b)

A glycol that cannot form a cyclic ester intermediate is not cleaved by periodic acid. For example, the following compound is not cleaved because it is impossible for both oxygens to be part of the same cyclic periodate ester. (If you can't see why, build a model and try connecting the two oxygens with one other atom.)

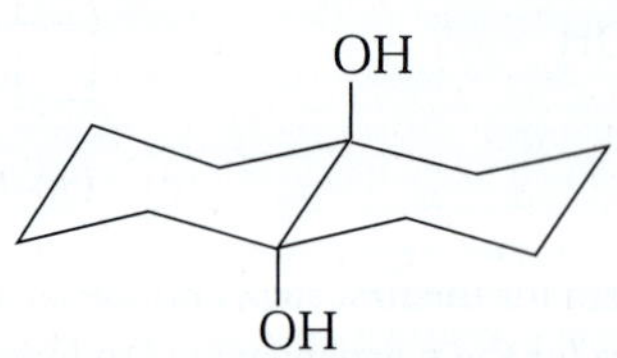

Don't confuse osmium tetroxide, permanganate, and periodate oxidations, all of which occur through cyclic ester intermediates (Sec. 12.5A). Periodate oxidizes *glycols*, but the other two reagents oxidize *alkenes* to give glycols. In all of these reactions, oxidation occurs because an atom in a highly positive oxidation state can accept an additional pair of electrons. In the periodate oxidation, the reduction of the iodine occurs during the *breakdown* of a cyclic ester; in the permanganate and osmium tetroxide oxidations, the metals are reduced during the *formation* of a cyclic ester.

Focused Problems

12.27 Give the product(s) expected when each of the following compounds is treated with periodic acid.

(a) 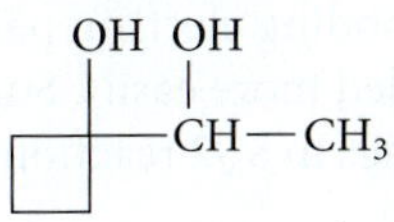

(b) 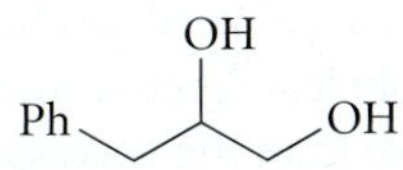

(c) 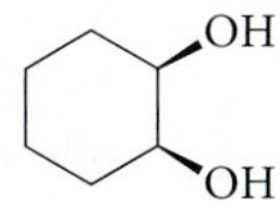

12.28 What glycol undergoes oxidation to give each of the following sets of products?

(a) 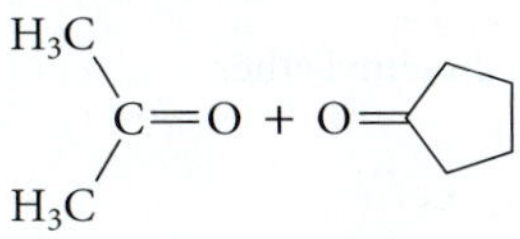

(b) 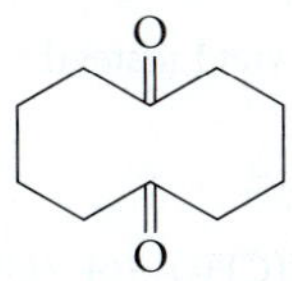

12.7 OXONIUM AND SULFONIUM SALTS

A. Reactions of Oxonium and Sulfonium Salts

If the acidic hydrogen of a protonated ether is replaced with an alkyl group, the resulting compound is called an **oxonium ion**. The sulfur analog of an oxonium salt is a **sulfonium ion**. These ions together with their counter-ions are called **oxonium salts** and **sulfonium salts**, respectively.

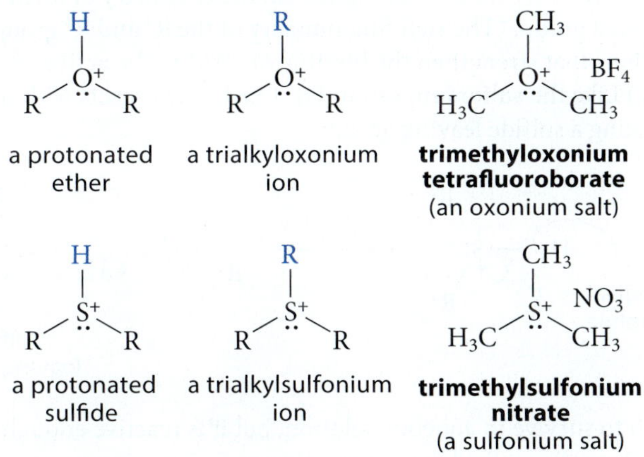

(12.62)

Oxonium and sulfonium salts react with nucleophiles in S_N2 reactions:

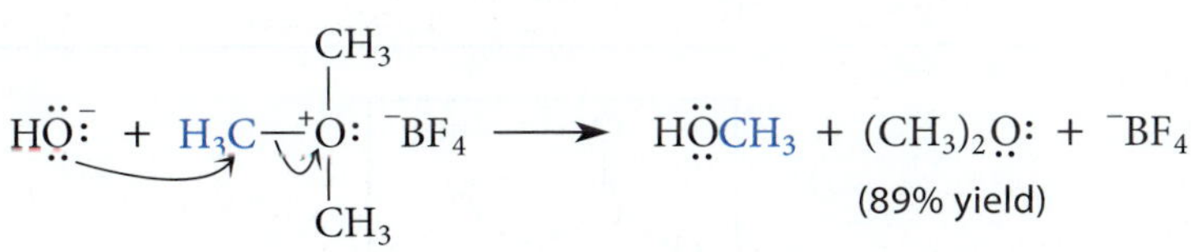

(12.63)

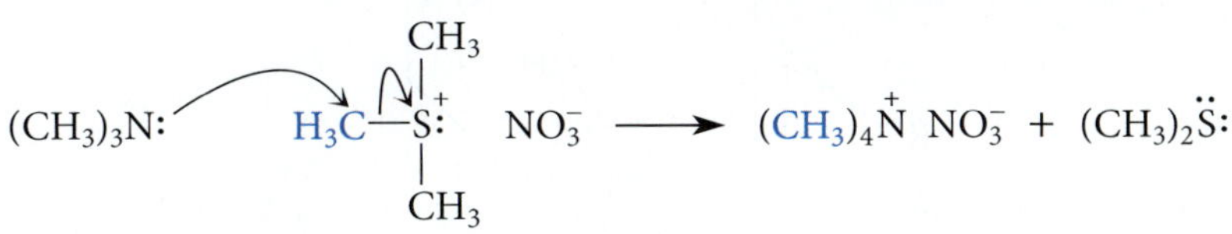

(12.64)

Oxonium salts are among the most reactive alkylating agents known, and they react rapidly with most nucleophiles. Because of their reactivity, oxonium salts must be stored in the absence of moisture. For the same reason, these salts are stable only when they contain counter-ions that are not nucleophilic, such as tetrafluoroborate ($^-BF_4$). (Tetrafluoroborate ion is not nucleophilic because, even though boron bears a negative charge, it has no nonbonding electron pairs.) *Sulfonium salts* are considerably less reactive and therefore are handled more easily. Sulfonium salts are somewhat less reactive than the corresponding alkyl chlorides in S_N2 reactions.

Focused Problems

12.29 Explain why all attempts to isolate trimethyloxonium iodide yield instead methyl iodide and dimethyl ether.

12.30 Complete the following reactions.

(a)

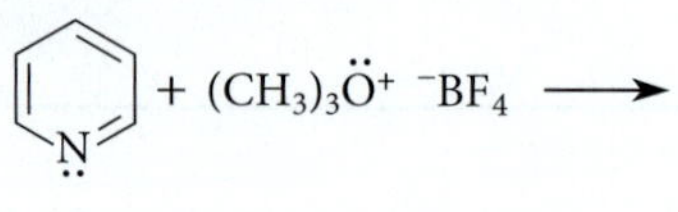

(b) $(CH_3)_2\ddot{S}: + (CH_3)_3\overset{+}{\ddot{O}}\ {}^-BF_4 \longrightarrow$

Chemical Biology Topic

B. *S*-Adenosylmethionine: Nature's Methylating Agent

A sulfonium salt, *S*-adenosylmethionine (SAM), is important in biology as a methylating agent for nucleophiles. The structure of SAM is shown in **Fig. 12.1**. SAM is an interesting example of a compound containing asymmetric sulfur. Recall that sulfonium salts, unlike amines, undergo inversion so slowly that individual stereoisomers can be isolated (Sec. 6.9B). SAM has the *S* configuration at sulfur. (The electron pair has lowest priority in the *R,S* system.)

Although the structure of SAM seems complicated, the chemistry of SAM arises solely from its sulfonium ion functional group. (The rich functionality of the R^1 and R^2 groups is utilized to form noncovalent interactions that strengthen the binding of SAM to the active sites of enzymes that catalyze its reactions.) Like the sulfonium salt in Eq. 12.64, SAM reacts with nucleophiles at the methyl carbon, liberating a sulfide leaving group:

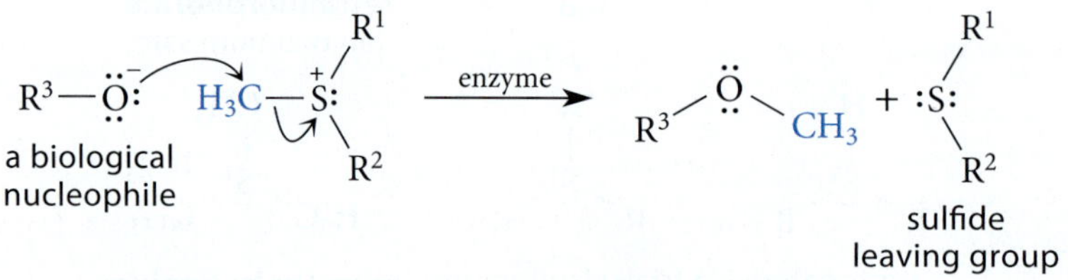

(12.65)

SAM is stable enough to survive in aqueous solution, but it is reactive enough to undergo enzyme-catalyzed S_N2 reactions. Evidence supporting an S_N2 mechanism in methylation reactions involving SAM was obtained by an elegant experiment. The methyl carbon of SAM was

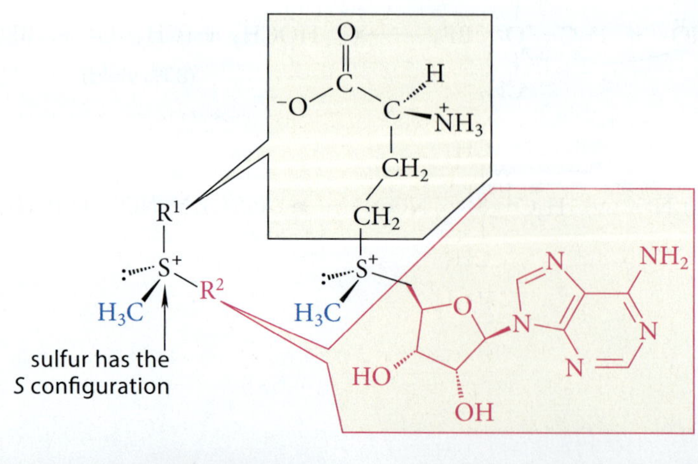

FIGURE 12.1 The structure of *S*-adenosylmethionine (SAM). The boxed parts of the structure are abbreviated R^1 and R^2 in the text. The sulfur of SAM is an asymmetric center. The solid and dashed wedges at sulfur show the orientation of the methyl group and the orbital containing the nonbonding electron pair.

made asymmetric by using the two hydrogen isotopes deuterium (D, or ^{2}H) and tritium (T, or ^{3}H). It was found that substitutions on this methyl group proceed with inversion of configuration, exactly as expected for the S_N2 mechanism.

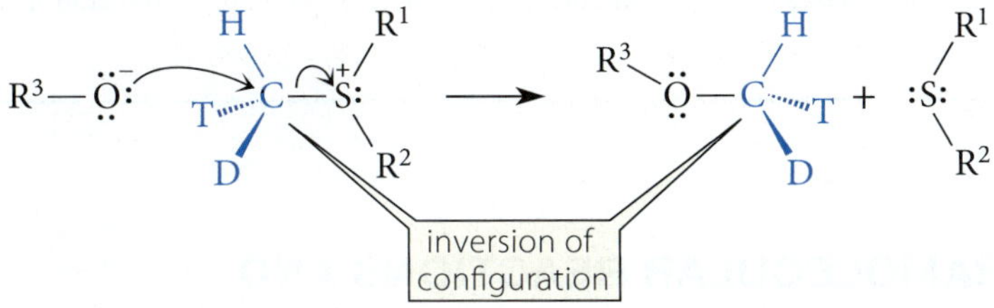

(12.66)

The compound *S*-adenosylmethionine, like NAD^+ (Sec. 11.8), is another example of a complex biological molecule that undergoes transformations that are readily understood in terms of common analogies from organic chemistry.

The Role of SAM in the Regulation of Gene Expression

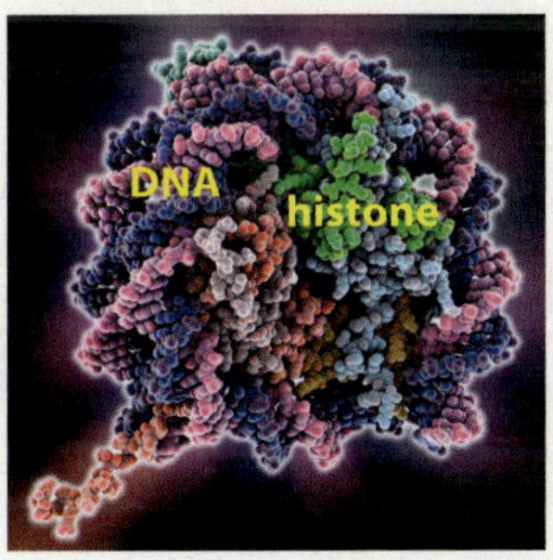

Laguna Design/Science Source

SAM has been shown to play a pivotal role in regulating the expression of genes. Within the nucleus, DNA, which encodes genetic information, is tightly packaged in *nucleosomes*. These are complexes in which the DNA is wrapped around a core of proteins called *histones* (see photo), in the same sense that thread is wrapped on a spool. In such tightly wound DNA, the DNA is "silent"—that is, it is not expressed. The histones contain large amounts of the amino acids lysine and arginine:

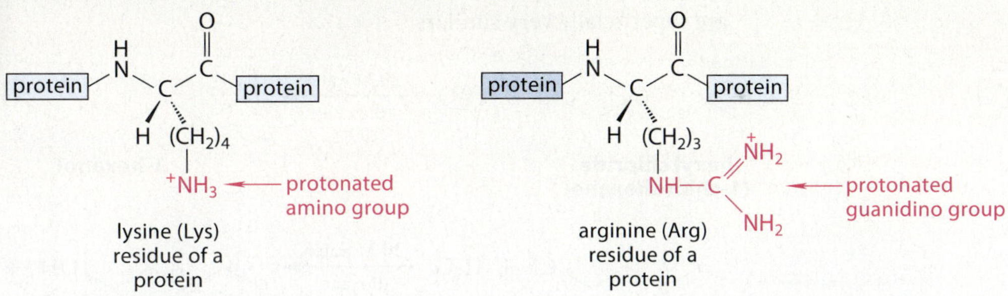

The methylation of certain lysines and arginines in histones by SAM has been shown to regulate whether a gene is actually expressed within the cell or whether it remains "silent." This methylation in some cases results in the recruitment of proteins involved in the regulation of DNA transcription. The precise mechanism of this process is yet to be unraveled. Lysine residues in histones can be methylated, once, twice, or even three times, and arginine residues can be methylated once or twice.

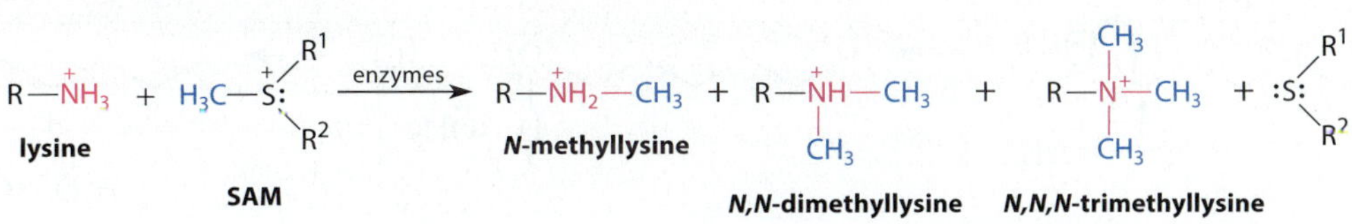

(12.67a)

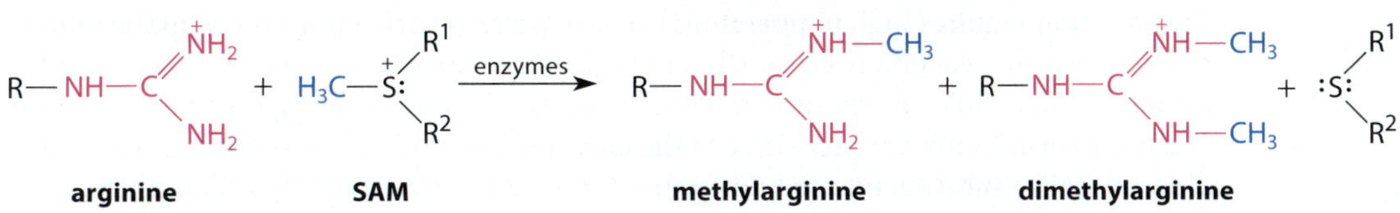

(12.67b)

The degree of methylation is controlled by the catalyzing enzymes. Because cancer is a disease characterized by aberrant gene expression, these methylations are of interest in the fight against cancer.

Focused Problem

12.31 Using the abbreviations for lysine and SAM shown in Eq. 12.67a, write a curved-arrow mechanism for the formation of *N*,*N*-dimethyllysine. Assume that acids ($^{+}$BH) and bases (:B) are available as necessary. (*Hint:* A Brønsted acid–base reaction must precede each methylation. Why?)

12.8 INTRAMOLECULAR REACTIONS AND THE PROXIMITY EFFECT

In this section we consider **intramolecular reactions**: reactions that occur between groups within the *same* molecule. Reactions between groups on *different* molecules are called **intermolecular reactions**. We've seen some examples of intramolecular reactions in previous sections, but we haven't stopped to consider how intramolecular and intermolecular reactions are different—after all, it might seem that a reaction between, for example, a nucleophile and an electrophile should be the same whether the reactions are intramolecular or intermolecular. Indeed, the *outcome* of the reaction might be the same. But what is different is that many intramolecular reactions are *faster* than their intermolecular counterparts—in some cases, *thousands of times* faster. We're going to consider some examples of such reactions and try to understand why they are so fast. The reason that we want to understand this phenomenon is that the catalytic effectiveness of enzymes—that is, biological catalysis—is largely explained by the theory of intramolecularity.

A. The Kinetic Advantage of Intramolecular Reactions

Consider the remarkable difference in the rates of the following two substitution reactions, which are superficially very similar:

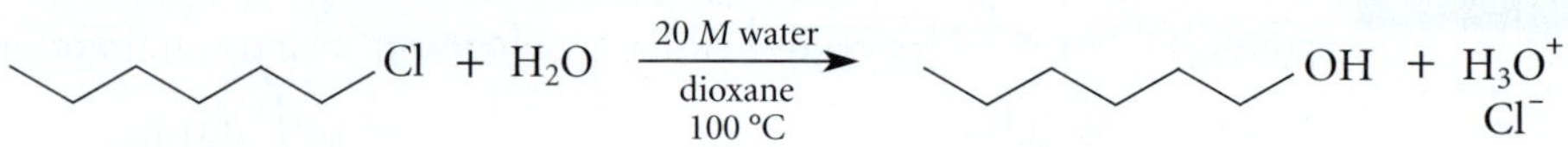

relative rate: 1

hexyl chloride (1-chlorohexane) **1-hexanol**

(12.68a)

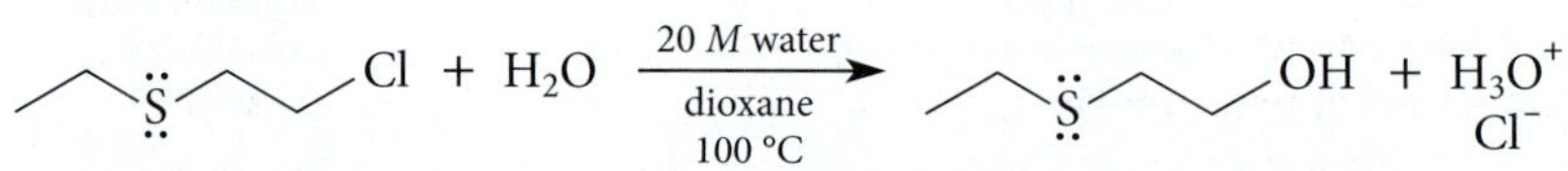

3200

β-chloroethyl ethyl sulfide **2-(ethylthio)ethanol**

(12.68b)

At first sight, both reactions appear to be simple S_N2 reactions in which chloride is displaced as a leaving group by water. In fact, this *is* the mechanism by which hexyl chloride reacts:

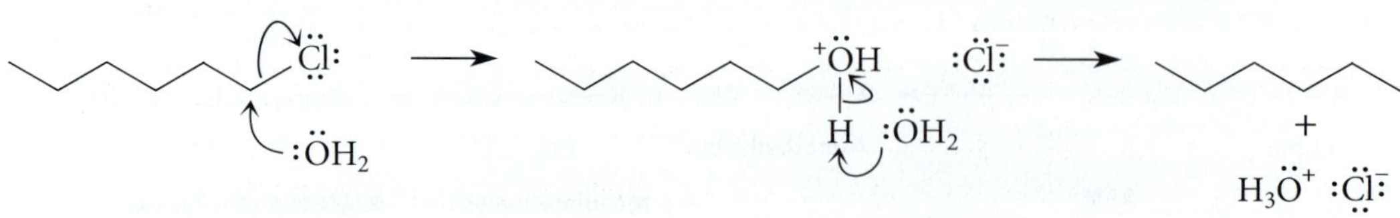

(12.69)

This reaction requires high temperature because water is such a poor nucleophile in the S_N2 reaction that the reaction is extremely slow at lower temperatures. The presence of sulfur in the alkyl halide molecule should have little effect on the rate of the S_N2 reaction, because the S_N2 mechanism is not very sensitive to the electronegativities of substituent groups. (In fact, electronegative substituents are known to *retard* S_N2 reactions slightly.) Nevertheless, the reaction in Eq. 12.68b is thousands of times faster than the reaction in Eq. 12.68a. The reaction in Eq. 12.68a takes almost two *months,* whereas the reaction in Eq. 12.68b takes about 30 *minutes.* This huge difference is due solely to the presence of sulfur in the molecule.

The rate of Eq. 12.68b is unusually large because a special mechanism facilitates the reaction, a mechanism not available to 1-chlorohexane. In the first and rate-limiting step of the mechanism, the nearby sulfur displaces the chloride *within the same molecule*:

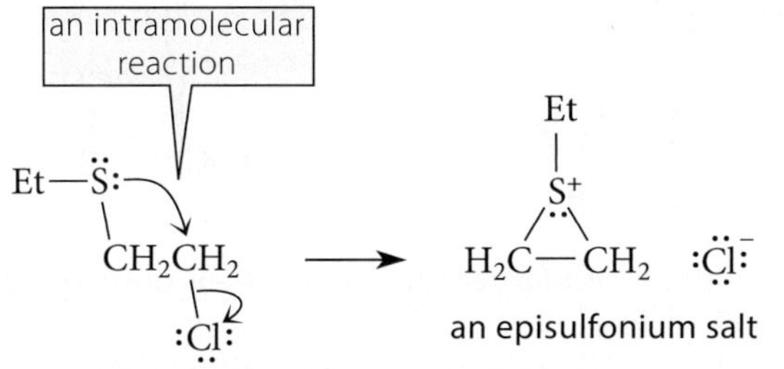

(12.70a)

The prefix *epi* is from a Greek word meaning "over," or "on top of."

The episulfonium ion that results from this internal nucleophilic substitution reaction is structurally similar to a protonated epoxide (Sec. 12.5B) or a bromonium ion (Secs. 5.2A and 7.8C). It is very reactive because it contains a strained three-membered ring and a good leaving group. Water rapidly reacts with this intermediate as it would with a protonated epoxide or bromonium ion to give the observed substitution product.

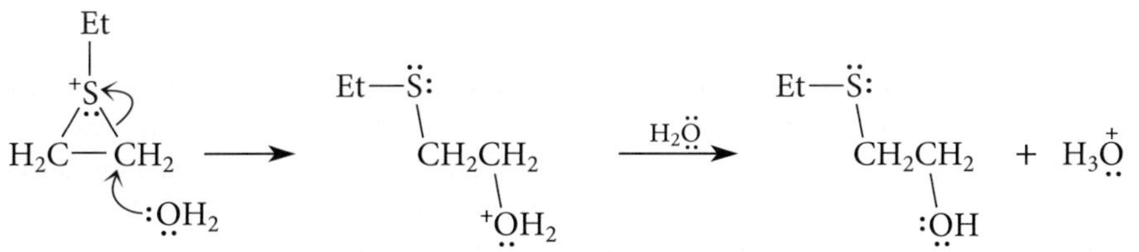

(12.70b)

This product is identical to the one that would have been formed in an ordinary S_N2 reaction in which the sulfur played no active role. In this case, the role of the sulfur is not apparent from the identity of the product. Only *the rate of the reaction* suggests that the sulfur has a special role in the mechanism.

The covalent involvement of a neighboring group—a group within the same molecule—in a chemical reaction is called **neighboring-group participation** or **anchimeric assistance** (from the Greek word *anchi*, meaning "near"). The neighboring-group mechanism of Eq. 12.70a is *in competition* with an ordinary S_N2 mechanism (Eq. 12.69) in which water reacts with the alkyl halide directly. Because the faster of two competing reactions is always the observed reaction, a reaction that involves neighboring-group participation, in order to be observed, *must* be faster than the same overall reaction that occurs by other competing mechanisms. The rate of the reaction in Eq. 12.68a provides the basis of comparison—that is, a rough idea of what rate to expect in this case for a reaction that occurs by direct substitution of water in the absence of neighboring-group participation. A large rate acceleration, such as the one in Eq. 12.68b, is typical of the experimental evidence used to diagnose the involvement of a neighboring group in a chemical reaction. Saul Winstein (1912–1969), a professor of chemistry at the University of California, Los Angeles, discovered numerous examples of neighboring-group participation and showed that all of these are associated with significant rate accelerations. (Other evidence for neighboring-group participation is discussed in Sec. 12.8C.)

A similar intramolecular substitution reaction was presented earlier in this chapter: the cyclization of halohydrins (Sec. 12.3B). The alkoxide of a bromohydrin has two competing possibilities for reaction. First, it can undergo an *intramolecular reaction* to form an epoxide:

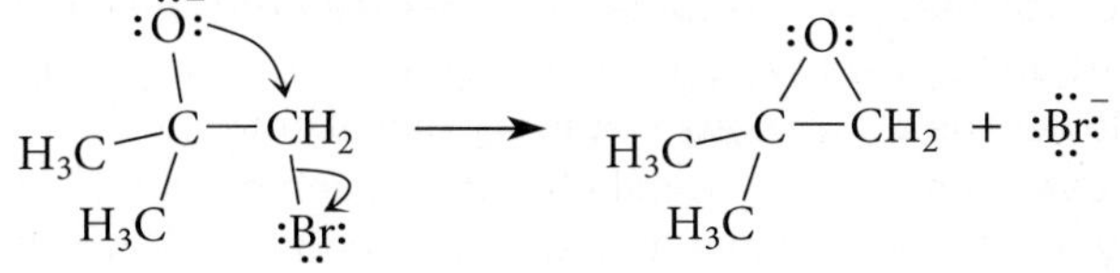

(12.71a)

This is the observed reaction. However, another possible reaction is for the alkoxide to react as a nucleophile in an *intermolecular reaction* with a *second* molecule of bromohydrin:

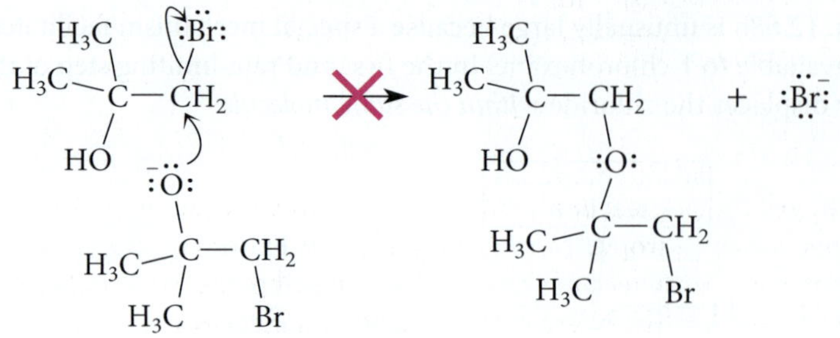

(12.71b)

Because the epoxide—the *intramolecular* reaction product—is observed, the intramolecular reaction must be significantly faster.

Why should an intramolecular reaction be faster than an intermolecular reaction? The answer has to do with the *probability* that the reaction will occur. Section 8.6A showed that the thermodynamic expression of the probability of a process is the *entropy change*, ΔS°. Because we are dealing with a question of relative rates, the answer lies in the entropic component of $\Delta G^{\circ\ddagger}$, the standard free energy of activation, which is the energy barrier for a reaction that determines the reaction rate (Sec. 4.8B).

$$\Delta G^{\circ\ddagger} = \Delta H^{\circ\ddagger} - T\Delta S^{\circ\ddagger} \qquad (12.72)$$

In this equation, $\Delta H^{\circ\ddagger}$ is the **standard enthalpy of activation**, $\Delta S^{\circ\ddagger}$ is the **standard entropy of activation**, and T is the absolute temperature in kelvins. The *standard enthalpy of activation* is determined by the strengths of the bonds broken and formed, van der Waals repulsions, and other energies that result from atomic and molecular interactions. *The standard entropy of activation reflects the intrinsic probability of forming a transition state for a reaction.* A *positive* $\Delta S^{\circ\ddagger}$, which corresponds to a *high probability of reaction*, makes $-T\Delta S^{\circ\ddagger}$ negative, *lowers* $\Delta G^{\circ\ddagger}$, and *increases the rate* of a reaction. A *negative* $\Delta S^{\circ\ddagger}$, which corresponds to a *low probability of reaction*, makes $-T\Delta S^{\circ\ddagger}$ more positive, *raises* $\Delta G^{\circ\ddagger}$, and *decreases the rate* of a reaction.

What entropic factors are important in comparing intermolecular and intramolecular reactions? The first is the *translational entropy* change. When two molecules react to become one molecule—an *intermolecular* reaction—each reactant starts with three degrees of translational freedom (that is, each molecule can move freely in the three spatial coordinates x, y, and z). Therefore, two reactants start with *six* degrees of translational freedom. As they come together to form the product—or the transition state leading to the product—the pair have to move together in a coordinated way. Three degrees of translational freedom have been lost in this process. Bringing two molecules together into one and having them *stay* together is inherently improbable. *This improbability is reflected in a high entropy cost for an intermolecular reaction.*

The next entropy factor is the *rotational entropy*. When each molecule moves randomly, the entire molecule can rotate about the three spatial axes even when it is not "translating." When two molecules come together to form one species, then, three degrees of rotational freedom are lost. The loss of rotational entropy is not as large as the loss of translational entropy because rotation does not require as much space, but it is significant.

Dancers, High-Fives, and Entropy

tarczas/Alamy

Consider two dancers alone on a crowded dance floor who decide to become a dancing couple. Becoming a couple requires a conscious decision to lower their collective entropy—to overcome the negligible probability that they might bump into each other at random. When the two dancers come together and begin moving as a couple, the six total degrees of translational freedom for the individuals become three degrees of translational freedom for the couple. In becoming a dancing couple, the two dancers have lost translational entropy.

Similarly, when each dancer is alone, he or she can spin and turn independently. When the dancers become a couple, they turn and spin as a couple (assuming they are holding each other). Becoming a couple has resulted in the loss of three degrees of rotational freedom as well. The couple has also lost rotational entropy.

Now consider the following analogy* for rates: In class, close your eyes and clap your hands. Then, with your eyes still closed, try to give one of your classmates a "high-five." Which occurs more readily?

*Dr. Lisa Bonner, a chemistry faculty member at Eckerd College in Florida, suggested the "high-five" analogy.

The third entropy factor is the *entropy of internal rotation.* Intramolecular reactions have a *disadvantage* in this type of entropy. Intramolecular reactions by definition form rings. When a ring is formed, internal rotations are lost. The larger the ring, the more internal rotations are lost. For example, when a five-membered ring is formed in a nucleophilic reaction, four internal rotations are lost.

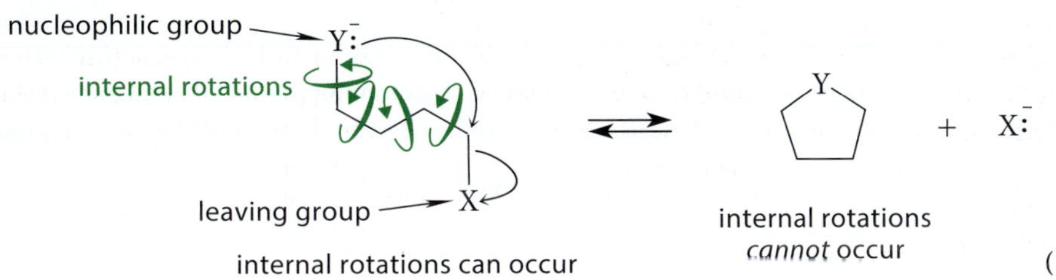

(12.73)

Internal rotations are not lost in intermolecular reactions because rings aren't formed. The loss of internal rotations lowers the entropy of a process because loss of internal rotation is a constraint on molecular freedom.

When rings have more than three members, the rings can undergo conformational changes (for example, the chair interconversion of six-membered rings). Conformational changes are coordinated partial internal rotations. The entropy of conformational changes partially offsets the loss of entropy of internal rotation, but not completely.

The $\Delta S^{\circ\ddagger}$ advantage of intramolecular reactions corresponds to the translational and rotational entropy advantage minus the internal rotation disadvantage. The translational and rotational entropy advantage is roughly 150 J K^{-1} mol^{-1}(36 cal K^{-1} mol^{-1}). (This is a rough estimate that varies with reaction type, but we can use it as a benchmark.) The corresponding effect on $\Delta G^{\circ\ddagger}$ is $-T\Delta S^{\circ\ddagger} = -(298)(150\times10^{-3})$ kJ $mol^{-1} = -45$ kJ mol^{-1} (-11 kcal mol^{-1}). Now, if this entire entropic advantage is reflected in the standard free energy of activation—the energy barrier for the reaction—the effect on rate can be calculated as a ratio of rate constants from Eq. 9.19b (Sec. 9.3B), with k_1 = the rate constant for the intramolecular reaction and k_2 = the rate constant for the intermolecular reaction.

$$\frac{k_1}{k_2} = 10^{(\Delta G_2^{\circ\ddagger}-\Delta G_1^{\circ\ddagger})/2.3RT} = 10^{45/5.71} \approx 10^8 \text{ at 298 K} \tag{12.74}$$

This means that intramolecular reactions are accelerated by factors of up to 10^8 over the corresponding intermolecular reactions. This advantage results, as we have seen, from the fact that, in an intramolecular reaction, the reactants are already tied together and do not have to find each other by random diffusion and rotation.

As we discussed previously, intramolecular reactions can have an entropic disadvantage that results from loss of internal rotations. Each internal rotation that is lost when an intramolecular reaction occurs reduces the rate of the intramolecular reaction by a factor of about 5. For example, if a five-membered ring is formed in an intramolecular reaction that results in a loss of four internal rotations (Eq. 12.72), the maximum intramolecular advantage is reduced by a factor of about $5^4 = 625$. The net intramolecular rate advantage would then be $10^8/625$ = about 160,000—a much smaller, but still considerable, intramolecular advantage. As rings get larger, however, the intramolecular advantage decreases. In practice, intramolecular reactions involving the formation of seven-membered or larger rings are not common.

As Eq. 12.72 shows, the reaction probability, or $\Delta S^{\circ\ddagger}$, is balanced against the $\Delta H^{\circ\ddagger}$ for the reaction. When a cyclic species such as the episulfonium ion in Eq. 12.70a is formed in an intramolecular reaction, its angle strain increases the $\Delta H^{\circ\ddagger}$ of the reaction. This opposes the entropic advantage of the reaction. This is why the acceleration for Eq. 12.68b is less than the theoretical maximum. Nevertheless, the entropic advantage of the intramolecular reaction is so great that *it occurs despite the strain in the three-membered ring that is formed.* In fact, the strain in the episulfonium ion in Eq. 12.70a is the reason that it reacts rapidly with water (Eq. 12.70b). If a neighboring sulfur were located four carbons away from the α-carbon of the alkyl chloride (end-of-chapter Problem 12.80), the $\Delta S^{\circ\ddagger}$ would be smaller (less positive, more negative) because of the larger number of internal rotations that are lost on forming a larger ring, but the $\Delta H^{\circ\ddagger}$ would not be increased by ring strain.

Although the kinetic advantage of intramolecular reactions was recognized intuitively by chemists for decades before, the role of proximity in understanding the rates of intramolecular reactions and the importance of intramolecularity in enzyme catalysis (Sec. 12.8E) were clarified in the period 1962–1973 by Daniel E. Koshland of the University of California, Berkeley; Thomas C. Bruice of the University of California, Santa Barbara; William P. Jencks of Brandeis University; and especially Michael I. Page of Huddersfield University in the United Kingdom.

Focused Problems

12.32 Give the structure of an *intramolecular* substitution product and an *intermolecular* substitution product that might be obtained from 4-bromo-1-butanol on treatment with one equivalent of NaOH. Which product do you think would be the major one? Why?

12.33 Two reactions, *A* and *B*, have the same $\Delta H^{\circ\ddagger}$, but the $\Delta S^{\circ\ddagger}$ of reaction *A* is – 30 J K^{-1} mol^{-1}, and the $\Delta S^{\circ\ddagger}$ of reaction *B* is – 180 J K^{-1} mol^{-1}. At 25 °C (298 K), which reaction is faster and by what factor? (*Hint:* Apply Eq. 12.74.)

12.34 Indicate which reaction in each of the following pairs should have the greater (less negative, more positive) standard entropy of activation. Explain.

(a) Formation of product *A* or formation of product *B*. (*Hint:* Remember that loss of internal rotations decreases freedom of motion and thus lowers entropy.)

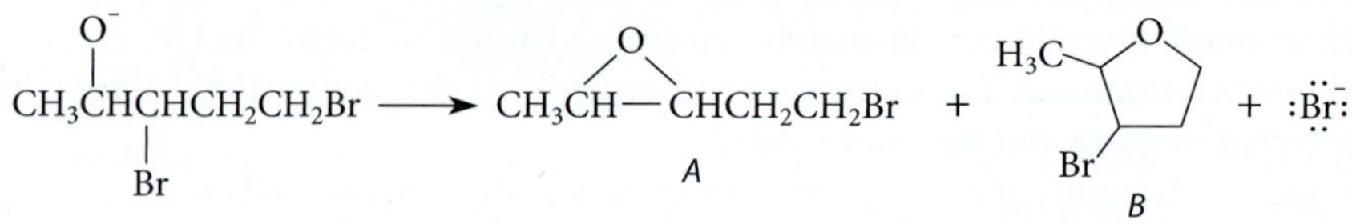

(b)

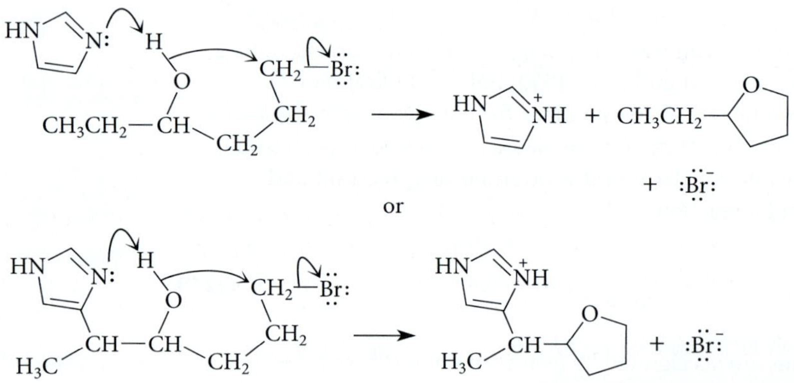

B. The Proximity Effect and Effective Molarity

The 10^8 intramolecular–intermolecular rate ratio calculated in Eq. 12.74 is the *maximum* theoretical kinetic (rate) advantage of intramolecular reactions. The actual ratio varies from reaction to reaction. The actual rate acceleration of an intramolecular reaction over its intermolecular counterpart is called the **proximity effect**. The proximity effect is expressed quantitatively as the ratio of rate constants for an intramolecular reaction (rate constant k_1) and its intermolecular counterpart (rate constant k_2).

$$\text{proximity effect} = \frac{k_1}{k_2} \qquad (12.75)$$

An intramolecular process is typically unimolecular and follows a first-order rate law. An intermolecular process is typically bimolecular and follows a second-order rate law. Because a first-order rate constant k_1 has units of s^{-1} and a second-order rate constant k_2 has units of $M^{-1}s^{-1}$ (Sec. 9.3B), the proximity effect k_1/k_2 has units of concentration (that is, *M*). For this reason, the proximity effect is sometimes called the **effective molarity**. Interpretation of the effective molarity provides us with additional insight on the advantages of intramolecular reactions. Calculation of a proximity effect and the significance of the effective molarity are explored in Study Problem 12.5.

Study Problem 12.5

Calculate the proximity effect for the intramolecular reaction in Eq. 12.68b given that its rate is 3200 times greater than its intermolecular counterpart.

Solution The intramolecular reaction (Eq. 12.68b) is 3200 times faster than the intermolecular reaction (Eq. 12.68a). The solvolysis rate of 1-chlorohexane is taken as an approximation of the intermolecular solvolysis rate of the sulfide. This approximation is necessary because the intramolecular reaction is so fast that the sulfide does not undergo the intermolecular reaction. The factor of 3200 is the ratio of rates. Taking the ratio of the rate laws for the two reactions, noting that the water concentration is 20 *M* in both reactions, and letting the alkyl halides be at the same concentration for comparison,

$$\frac{\text{rate of reaction 12.68b}}{\text{rate of reaction 12.68a}} = 3200 = \frac{k_1[\text{alkyl halide}]}{k_2[\text{alkyl halide}][\text{H}_2\text{O}]} = \frac{k_1}{k_2(20\ M)} \qquad \textbf{(12.76a)}$$

from which we calculate the proximity effect as

$$\text{proximity effect} = \frac{k_1}{k_2} = (3200)(20\ M) = 64{,}000\ M \qquad \textbf{(12.76b)}$$

The effective molarity is therefore 64,000 *M*. The effective molarity is the concentration of the nucleophile (water in this case) required for the rate of the intermolecular reaction to equal that of the intramolecular reaction. Because the concentration of water is 55.6 *M* in pure water, 64,000 *M* is an impossibly large concentration. What this large effective molarity means, then, is that it is impossible to make the water concentration high enough for the intermolecular reaction to take place; the intramolecular reaction is orders of magnitude faster than the corresponding intermolecular one regardless of the nucleophile concentration. The intermolecular reaction doesn't stand a chance in the competition with its intramolecular counterpart.

As we have seen, the rate accelerations of intramolecular reactions are described by the proximity effect. The ratio of 10^8 *M* calculated in Eq. 12.74 is the maximum value of the proximity effect. However, as Study Problem 12.5 shows, the actual proximity effect, although still considerable, can be much less.

Let's summarize what we've learned about the proximity effect:

1. Intramolecular nucleophilic substitution reactions are particularly common for cases involving the formation of three-, five-, and six-membered rings. Intramolecular nucleophilic substitution reactions that form seven-membered or larger rings are much less common.

2. The main reason that intramolecular nucleophilic substitutions are favored over their intermolecular counterparts is a more favorable (less negative or more positive) $\Delta S^{\circ\ddagger}$—the reaction probability—for intramolecular reactions.

3. The rate acceleration of intramolecular reactions over competing intermolecular reactions is measured as the ratio of rate constants for the two processes. This ratio is the *proximity effect* or the *effective molarity*.

4. The magnitude of the proximity effect varies from reaction to reaction, but the $\Delta S^{\circ\ddagger}$ contribution can theoretically be as high as 10^8 *M*.

A few intramolecular reactions are known that have effective molarities as high as 10^{13} *M*. These huge rate accelerations are caused by factors that reduce $\Delta H^{\circ\ddagger}$ in addition to high reaction probabilities.

Focused Problem

12.35 The nucleophilic substitution reaction of sodium 2-bromopropanoate with water and/or ^{-}OH can occur by both an S_N2 (intermolecular) mechanism and a mechanism that involves neighboring-group participation.

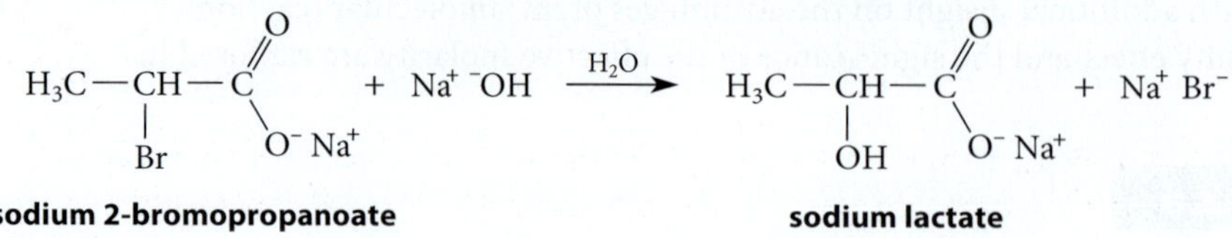

(a) Give the curved-arrow notation for the S_N2 mechanism with ^{-}OH as the nucleophile.

(b) Give the curved-arrow notation for an intramolecular mechanism. This mechanism should lead you to the structure of an unstable intermediate, which then reacts with ^{-}OH to give the product.

(c) The first-order rate constant k_1 for the intramolecular reaction is $1.2 \times 10^{-4}\ s^{-1}$. The second-order rate constant for the S_N2 reaction is $6.4 \times 10^{-4}\ M^{-1}s^{-1}$. Calculate the proximity effect for the intramolecular reaction.

(d) At what NaOH concentration does this reaction proceed by the two mechanisms at the same rate?

(e) What is the predominant mechanism in 1 *M* NaOH?

(f) Consider the structure of the intermediate you derived in part (b). The reason for the small proximity effect is that this intermediate is very unstable. Explain why this intermediate is more unstable than an ordinary epoxide. (*Hint:* Think about the preferred bond angles.)

C. Stereochemical Consequences of Neighboring-Group Participation

Neighboring-group participation can sometimes be diagnosed by considering its stereochemical outcome. Notice that the mechanism of the neighboring-group reaction given in Eqs. 12.70a and b (Sec. 12.8A) involves *two* substitution reactions. The first is the intramolecular substitution by the neighboring sulfur nucleophile to give the episulfonium ion intermediate. The second is the ring-opening reaction of water with the episulfonium ion. This double-substitution mechanism has stereochemical consequences if the reacting molecule contains appropriately situated stereocenters. This point is illustrated in Study Problem 12.6.

Study Problem 12.6

Indicate the stereochemical outcome of the following substitution reaction (a) if neighboring-group participation does *not* occur and (b) if neighboring-group participation takes place.

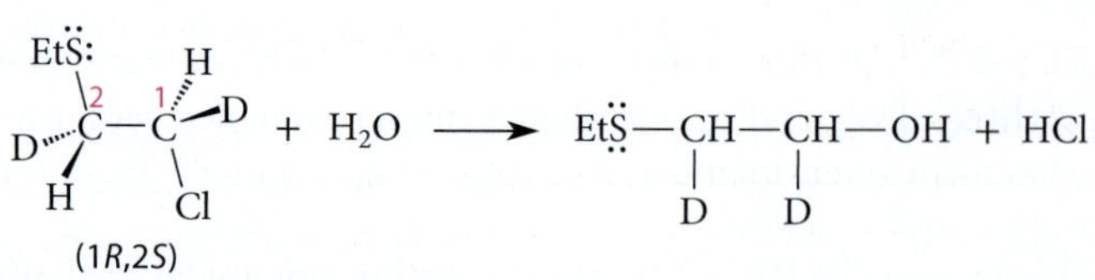

Solution

(a) If the reaction were a simple S_N2 reaction, the nucleophile would displace the chloride leaving group with inversion of configuration (Sec. 9.4C). In that case, the (1*R*,2*S*)-stereoisomer of the starting material should give the (1*S*,2*S*)-stereoisomer of the product:

The numbering of the carbons does not correspond to International Union of Pure and Applied Chemistry (IUPAC) numbering in some of the structures in Eqs. 12.77b and 12.77c. The carbon numbers are intended solely as carbon identifiers. A given carbon has the same number throughout the study problem.

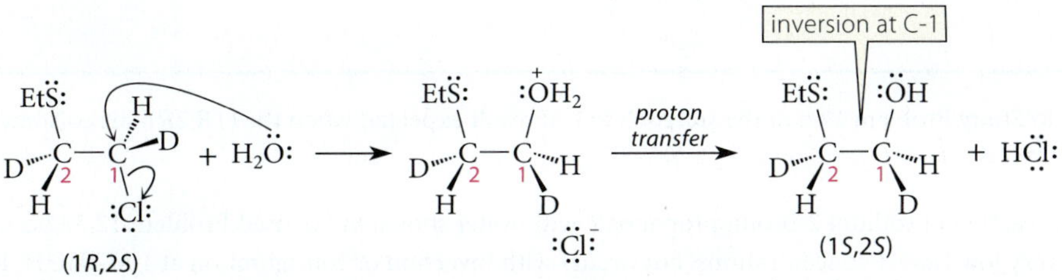

(12.77a)

(b) If the reaction involves neighboring-group participation, the intramolecular substitution occurs with inversion of configuration. In this case, the resulting episulfonium ion has the 1*S*,2*S* configuration.

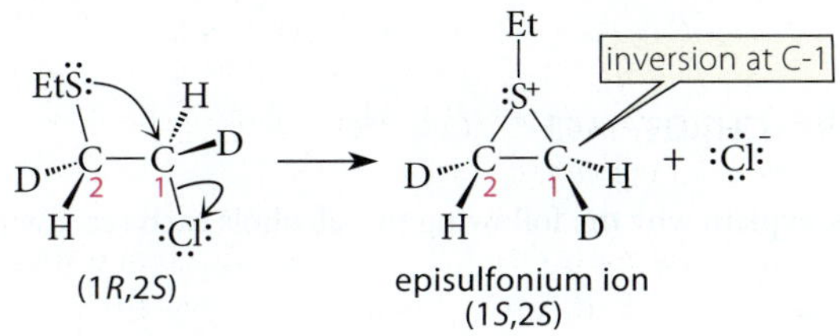

(12.77b)

The two carbons of the episulfonium ion are homotopic. Reaction of the nucleophile water at either carbon of the episulfonium ion gives the *same product*:

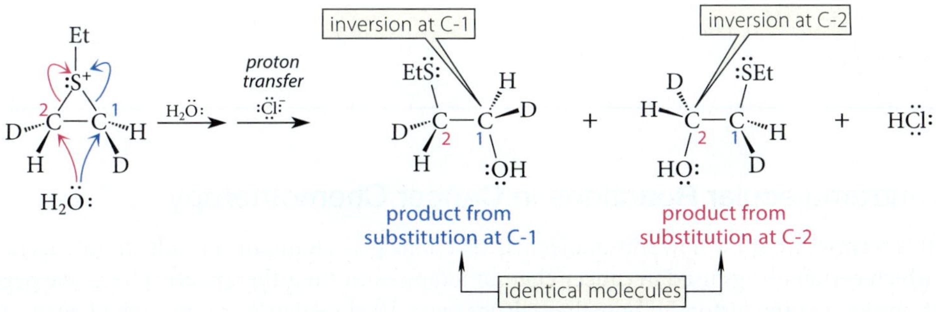

(12.77c)

If we compare the product stereochemistry with that of the starting material, we find that the neighboring-group mechanism gives net *retention of stereochemistry*.

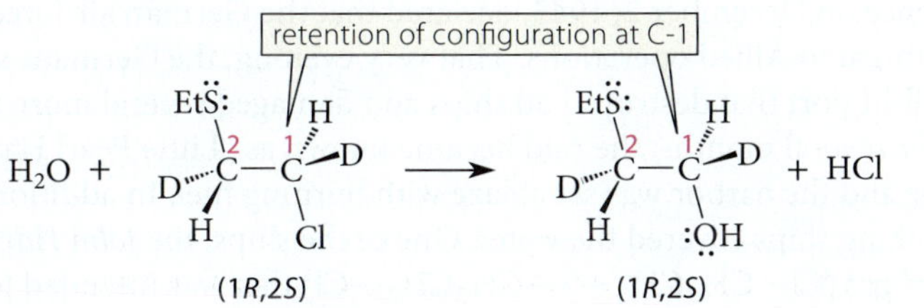

(12.77d)

Except for the deuterium labels, the reactant in this equation is the same as the starting material in Eq. 12.68b (Sec. 12.8A), for which the neighboring-group mechanism was established by relative rates. This, then, is the expected result.

Here is what we learn from this example: *When a substitution reaction occurs with overall retention of stereochemistry at an electrophilic center, we should strongly suspect that two substitutions have occurred,* because all known instances of concerted (one-step) S_N2 reactions occur with inversion. If a potential nucleophile is present in the starting material so that it can form a small ring, a cyclic intermediate is likely.

Focused Problems

12.36 Carry out an analysis similar to Study Problem 12.6 of the stereochemical result expected when the (1*R*,2*R*)-stereoisomer of the starting material is used.

12.37 The nucleophilic substitution reaction of sodium 2-bromopropanoate with water shown in Focused Problem 12.35 occurs with retention of configuration at very low NaOH concentrations, but occurs with inversion of configuration at 1 *M* NaOH. Relate this finding to your answers for Focused Problem 12.35.

12.38 In the nucleophilic substitution reaction of the following radioactively labeled compound with water, what labeling pattern should be observed in the product (a) if the neighboring-group participation does *not* occur and (b) if neighboring-group participation does occur?

$$\text{Et}\ddot{\underset{\cdot\cdot}{\text{S}}}\text{—CH}_2\overset{*}{\text{C}}\text{H}_2\text{—Cl} \qquad \overset{*}{\text{C}} = {}^{14}\text{C}$$

12.39 By drawing suitable curved-arrow mechanisms, explain why the following two alcohols each react with HCl to give the same alkyl chloride.

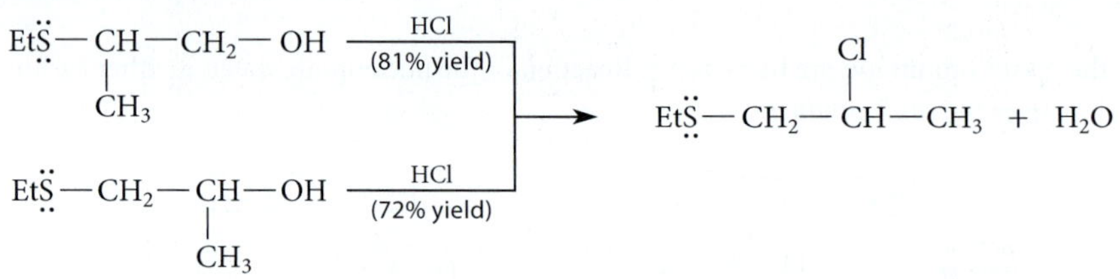

D. Intramolecular Reactions in Cancer Chemotherapy

Chemical Biology Topic

In this section, we show how intramolecular reactions play an important role in the mechanism by which certain drugs used in cancer chemotherapy exert their therapeutic effect. We begin with the fascinating history of how these drugs were developed as an outgrowth of events that occurred in World War II.

During the later part of World War II, the port of Bari in southeastern Italy was used as a logistics hub for the Allied invasion of Italy. The British commander, Sir Arthur Coningham, at a press conference on December 2, 1943, declared that the German air force had been defeated and posed no threat to Allied operations. That very evening, the Germans staged a surprise air raid on the well-lit port that destroyed 30 ships and damaged several more (**Fig. 12.2**). (The port was disabled for several months; the raid became known as "Little Pearl Harbor.") The raid also burst a fuel line and the harbor was set ablaze with burning fuel. In addition, vast amounts of oil from the sinking ships covered the water. One of the ships, the *John Harvey*, carried a secret load of mustard gas ($Cl—CH_2CH_2—S—CH_2CH_2—Cl$) that was intended for retaliatory use should the Germans resort to chemical warfare. Mustard gas was spread all over the harbor, readily dissolving in the oil that covered the water, and a cloud of gas seeped into the town, which had a population of 250,000. Mustard gas causes severe burns, and hundreds of soldiers and townspeople, especially soldiers who were covered with oil containing dissolved mustard, suffered the effects of acute mustard poisoning. Over a thousand people eventually died from mustard poisoning. Some of the survivors of this raid who had been exposed to mustard gas were subsequently found to have a greatly depleted lymphocyte (white blood cell) count and a scorched bone marrow. This observation was critical to the subsequent development of mustards for use in cancer chemotherapy.

At about the same time, the U.S. Army had awarded a research contract to two Yale University pharmacologists, Louis Goodman and Alfred Gilman, who were studying the effects of mustards and similar compounds on laboratory animals, with a view toward protecting soldiers and civilians from damage caused by the military use of mustards. They found that animals injected with the mustard compound did not show any of the burn symptoms, but the white cells in their blood and bone marrow virtually disappeared. This result was similar to the results of the mustard-gas release during the Bari bombing. Goodman and Gilman wondered whether small

Keystone/Getty Images

FIGURE 12.2 The bombing of the harbor at Bari, Italy, in December 1943. The bombing released sulfur mustard that caused a large number of both military and civilian casualties.

doses of a mustard compound might actually bring about remissions in lymphoma, a type of cancer in which abnormal white cells proliferate rapidly. A subsequent trial on a single human lymphoma patient brought about a remission.

A mustard-compound analog, mechlorethamine, in which the sulfur is replaced by H_3C—N, was subsequently developed as a drug for cancer chemotherapy. This was the first of the **nitrogen mustards**, which are used, along with other drugs, to treat lymphomas and other cancers along with other drugs. Modern nitrogen mustards include chlorambucil and cyclophosphamide. A "nitrogen mustard" functionality is common to all of these drugs.

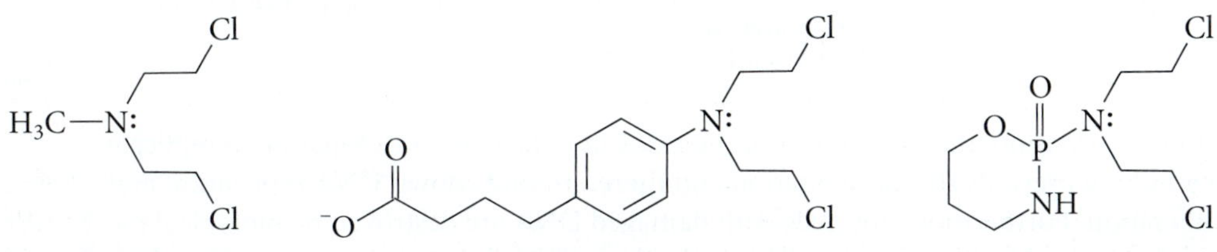

mechlorethamine (mustine) **chlorambucil** **cyclophosphamide** (12.78)

The mustards kill cells because they react with bases on DNA. The mustards, being very hydrophobic molecules, readily cross the cell and nuclear membranes, and, once in the nucleus, react with a nitrogen (*N*-7) on a guanine (G) base of DNA.

The structure of DNA is discussed in Sec. 26.5.

guanine (a DNA base)

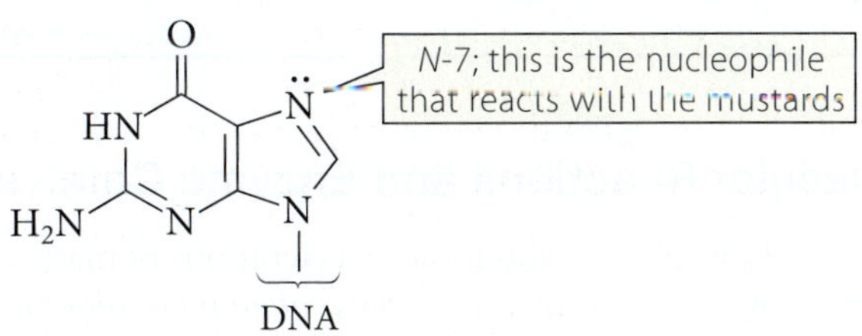

(12.79)

We've already shown why mustard gas molecules are so reactive: they rapidly form three-membered sulfonium ion rings that are then opened rapidly by nucleophiles. The nitrogen mustards react by an analogous pathway. The three-membered ring containing a positively charged nitrogen is called an *aziridinium* ring. We use mechlorethamine to illustrate this chemistry.

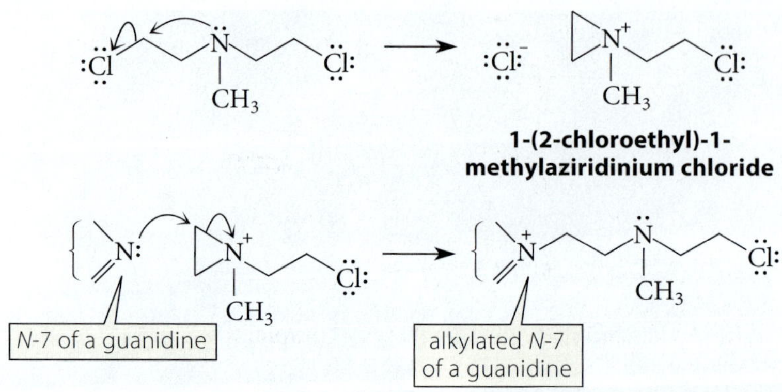

(12.80)

This alkylation reaction would be bad enough for the DNA, but the real damage hasn't yet been done. Notice that the mechlorethamine molecule contains two reactive alkyl halide groups. The two strands of the DNA double helix are held together tightly by noncovalent forces. If another guanine is nearby on the opposite strand of DNA in the double helix, it reacts at its *N*-7 with the second group on the mechlorethamine. This reaction is accelerated by a proximity effect, because the second guanine is perfectly positioned to react. As a result, the DNA strands become tethered together, or crosslinked.

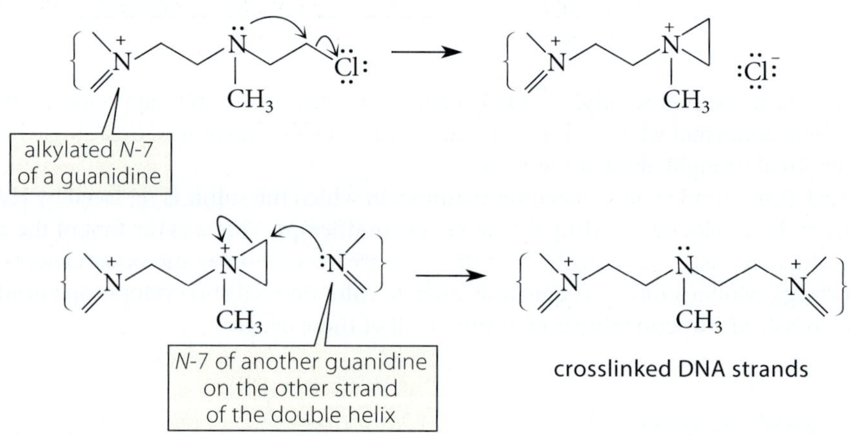

(12.81)

During cell division, the two strands of the DNA double helix must separate to replicate. Crosslinking prevents strand separation and therefore shuts down DNA replication and white-cell proliferation. Furthermore, the cells with damaged DNA are destroyed by biological mechanisms that have evolved for this purpose. This is the desired result in cancer treatment.

Focused Problem

12.40 Give the structure of the cyclophosphamide-derived DNA crosslink involving *N*-7 of a guanine base on each strand. (Use abbreviations like the ones in Eq. 12.81)

Chemical Biology Topic

E. Intramolecular Reactions and Enzyme Catalysis

Enzymes, nature's catalysts, bring about rate accelerations of many orders of magnitude in the reactions that they catalyze. Enzymes are *large protein molecules*; the smallest enzymes have molecular masses of about 8000, and most are considerably larger. The molecules on which enzyme act are called **substrates**. When an enzyme acts on its substrate, the first step is to bind the substrate tightly and *noncovalently* into a specific part of the enzyme called the *active site*. The enzyme with its bound substrate is called the **enzyme–substrate complex**. *Noncovalent binding is an obligatory step in catalysis by all enzymes*. If E is the enzyme, S is a substrate, and P is the reaction product, this scheme is as follows.

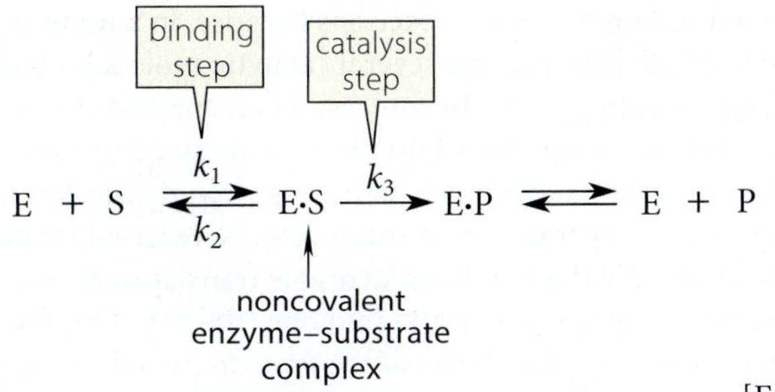

$$K_s = \text{dissociation constant of the E·S complex} = k_2/k_1 = \frac{[E][S]}{[E{\cdot}S]} \qquad (12.82)$$

A diagram of this process is shown in **Fig. 12.3**. Binding is an essential part of catalysis because the binding event positions the substrate in an optimal configuration relative to the chemical groups on the enzyme that bring about the actual reaction. The enzyme contains, as part of its structure, groups that can serve as acid catalysts, base catalysts, and nucleophiles. Metal ions, held in place by coordination with groups on the enzyme, can additionally act as Lewis acid catalysts. Here is the key point: Once the substrate is bound, we can think of the E•S complex as a single molecule. *The subsequent catalytic step* (k_3 in Eq. 12.82) *is effectively an intramolecular reaction* within the E•S complex. We've just learned about the kinetic advantage of intramolecular reactions. *The major part of the catalytic efficiency of enzymes resides in the fact that their reactions are intramolecular.* Moreover, we sometimes find that a mechanism occurring in the active site of

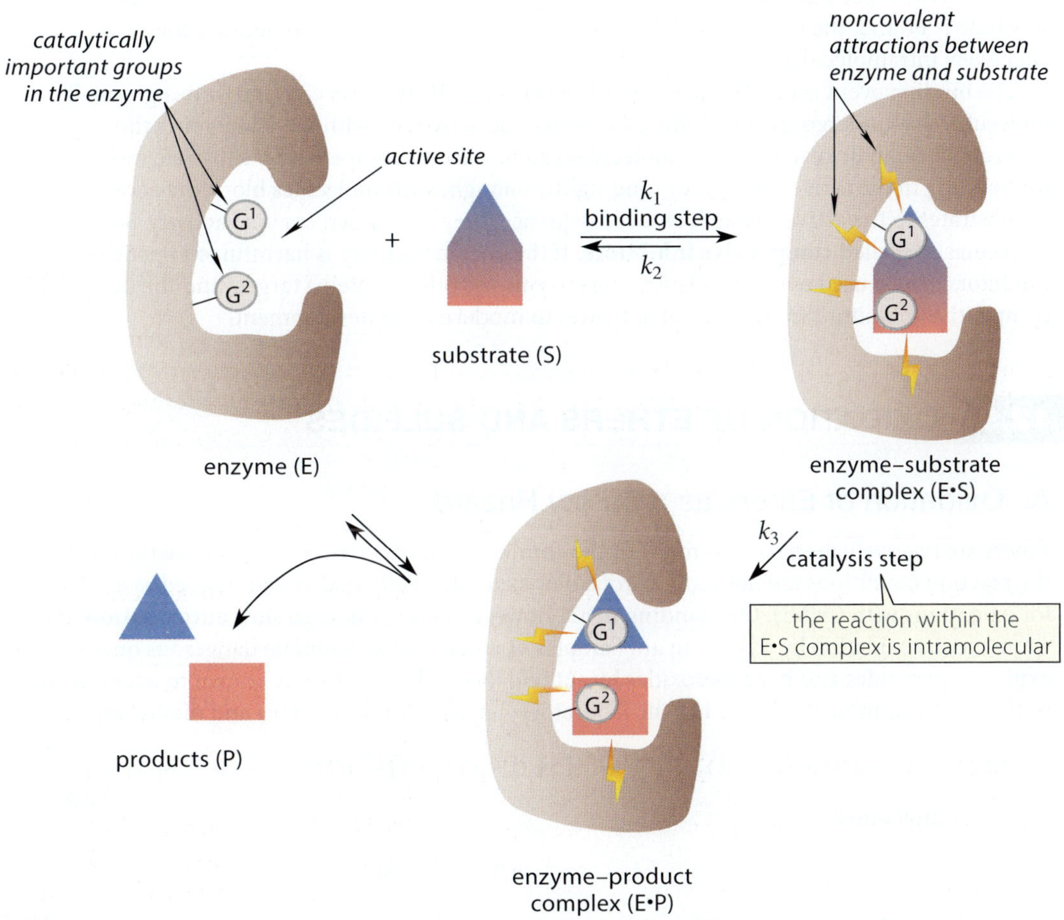

FIGURE 12.3 A diagram of enzyme catalysis that follows Eq. 12.82. The substrate S is bound to the enzyme E by noncovalent attractions to give the enzyme–substrate complex E•S. The conversion of bound substrate to bound product (E•S to E•P) is intramolecular. Release of product P from the enzyme–product complex regenerates the enzyme for further catalysis.

an enzyme is not observed in nonenzymatic reactions because, in a nonenzymatic reaction, the simultaneous positioning of the reactant and several catalytic molecules is simply too improbable (it would have a very large, negative $\Delta S^{\circ\ddagger}$). In contrast, in an enzyme-catalyzed reaction, the substrate and all of the catalytic groups built into the enzyme structure are "pre-positioned" by the binding event so that a "multi-catalyst" mechanism is strongly accelerated.

The process of bringing the substrate (or, in some cases, several substrates) together into an enzyme active site must overcome the highly unfavorable translational and rotational entropies associated with converting two species (E and S) into one (the E•S complex). How is this accomplished? Enzymes use *noncovalent forces* of the type discussed in Chapter 8—hydrogen bonding, hydrophobic bonding, and, in some cases, electrostatic and ion–dipole attractions—to pay the entropic price of binding. Electrostatic attractions—attractions between oppositely charged ionic groups on the enzyme and substrate—when they are present, are particularly strong. The noncovalent attractions involved in enzyme–substrate binding are significant enough that, even after paying the entropic cost, the binding of E and S is fairly strong. The enzyme–substrate binding is described by the dissociation constant K_s of the E•S complex (Eq. 12.82); stronger binding is characterized by smaller values of K_s. The K_s values of typical E•S complexes are in the 10^{-7} to 10^{-3} M range. Because binding is noncovalent, chemical bonds do not have to be broken and formed; so, the enzymes and substrates form E•S complexes rapidly, in many cases with rate constants k_1 in Eq. 12.82 that are at or near the rates of diffusion ($10^9 - 10^{10}$ $M^{-1}s^{-1}$).

Figure 12.3 is a "conceptual framework" for understanding enzyme catalysis. In later chapters, particularly in Sec. 27.10, we examine actual cases of enzyme binding and catalysis at the molecular level.

In summary, enzymes use noncovalent forces to overcome the entropic cost of binding their substrates. The structure of each enzyme incorporates one or more catalytic groups in proximity to the substrate. For this reason, enzymes have been referred to as "entropy traps." Once the substrate is bound, the rates of catalytic reactions are strongly enhanced because they are effectively intramolecular.

The binding event is the Achilles' heel of the enzyme. If chemists can find or design small molecules that undergo strong binding to an enzyme active site without undergoing the subsequent catalytic reaction, such molecules can be used to compete with substrate molecules for bonding in the active site. If they bind tightly enough, such molecules block the access of substrates to the active site and, as a consequence, they shut down enzyme activity. Such molecules are called **competitive inhibitors**. If the enzyme activity is harmful, competitive inhibitors can be used as drugs. Identifying enzyme (or other protein) targets and the design of competitive inhibitors are significant activities in modern drug development.

12.9 OXIDATION OF ETHERS AND SULFIDES

A. Oxidation of Ethers as a Safety Hazard

Ethers are relatively inert toward many of the common oxidants used in organic chemistry if the reaction conditions are not too vigorous. For example, diethyl ether can be used as a solvent for oxidations with Cr(VI). On standing in air, however, ethers undergo slow **autoxidation**, the spontaneous oxidation by oxygen in air. Samples of ethers can accumulate dangerous quantities of explosive peroxides and hydroperoxides by autoxidation. This reaction is known to occur in two of the most common ethers used in the laboratory: tetrahydrofuran (THF) and diethyl ether.

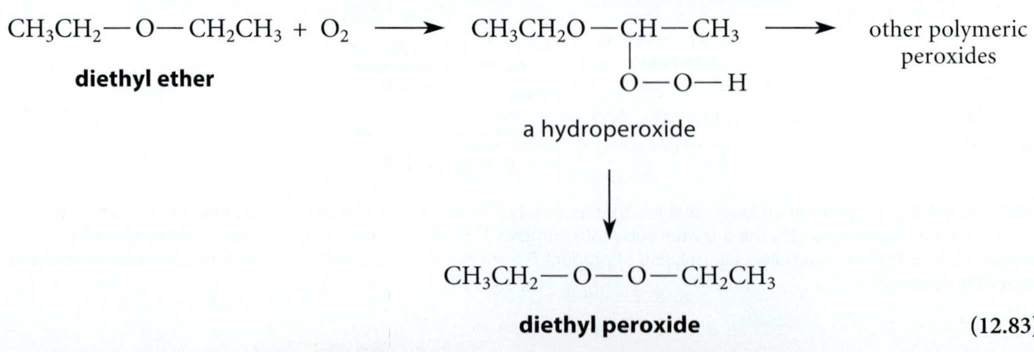

(12.83)

These peroxides can form by free-radical processes in samples of anhydrous diethyl ether, THF, and other ethers within less than two weeks. For this reason, some ethers are sold with small amounts of free-radical inhibitors, which can be removed by distilling the ether. Because peroxides are particularly explosive when heated, it is a good practice not to distill ethers to dryness. Peroxides in an ether can be detected by shaking a portion of the ether with 10% aqueous potassium iodide solution. If peroxides are present, they oxidize the iodide to iodine, which imparts a yellow tinge to the solution. Small amounts of peroxides can be removed by distillation of the ethers from lithium aluminum hydride ($LiAlH_4$), which both reduces the peroxides and removes contaminating water and alcohols.

A second oxidation reaction—combustion—is a particular hazard of diethyl ether. Its flammability is indicated by its very low flash point of –45 °C. The **flash point** of a material is the minimum temperature at which it is ignited by a small flame under certain standard conditions. (In contrast, the flash point of THF is –14 °C.) Compounding the flammability hazard of diethyl ether is the fact that its vapor is 2.6 times as dense as air. This means that vapors of diethyl ether from an open vessel will accumulate in a heavier-than-air layer along a laboratory floor or benchtop. For this reason, flames can ignite diethyl ether vapors that have spread from a remote source. Good safety practice demands that open flames or sparks not be permitted anywhere in a laboratory in which diethyl ether is in active use. Even the tiny spark from an electric switch (as on a hot plate) can ignite diethyl ether vapors. A steam bath is therefore one of the safest ways to heat this ether.

The sulfur analogs of peroxides are *disulfides* (R—S—S—R), which are the oxidation products of thiols (Sec. 11.10B). Disulfides are not explosive; in fact, they occur widely in nature within the structures of proteins (Sec. 27.8A).

B. Oxidation of Sulfides

Like thiols, sulfides oxidize at *sulfur* rather than carbon when they react with common oxidizing agents. Sulfides can be oxidized to **sulfoxides** and **sulfones**:

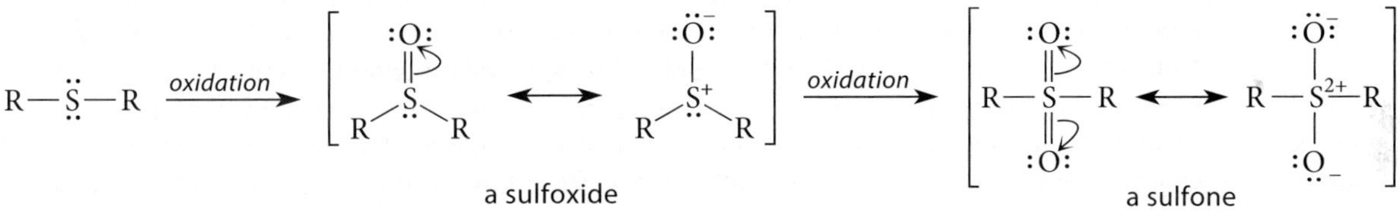

(12.84)

The nonionic Lewis structures for sulfoxides and sulfones have more than an octet of electrons on the sulfur. The bonding at sulfur in such situations was discussed in Sec. 11.10A. As with other hypervalent compounds, the ionic resonance structures are more correct, but we use the octet-expanded (nonionic) structures for convenience.

Dimethyl sulfoxide (DMSO) and sulfolane are well-known examples of a sulfoxide and a sulfone, respectively. (Both compounds are excellent polar aprotic solvents; see Table 8.2, Sec. 8.6B.)

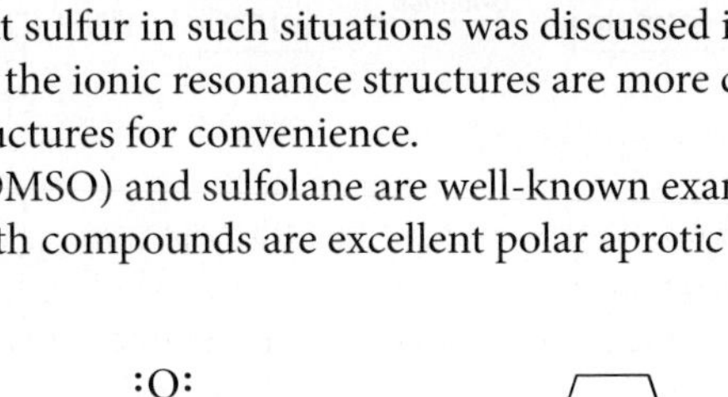

dimethyl sulfoxide (DMSO) **tetramethylene sulfone,** or **sulfolane**

Sulfoxides and sulfones can be prepared by the direct oxidation of sulfides with one and two equivalents, respectively, of hydrogen peroxide, H_2O_2:

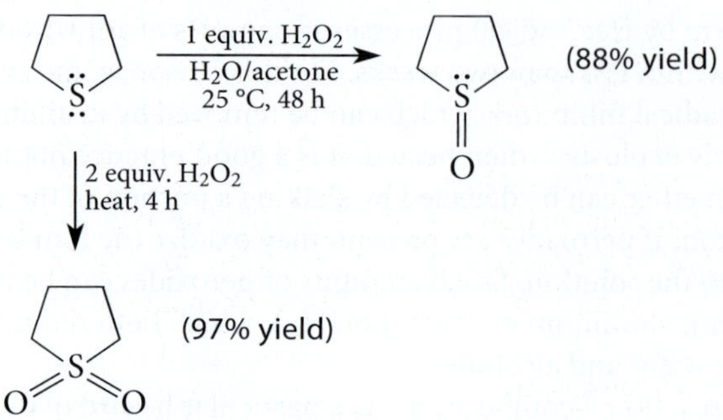

(12.85)

Other common oxidizing agents, such as $KMnO_4$, HNO_3, and peroxyacids (Sec. 11.3A), also readily oxidize sulfides.

Sulfide Oxidation and Protein-Aggregation Diseases

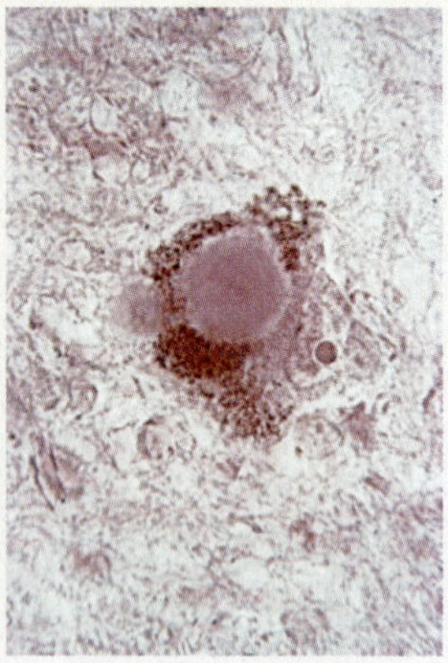

Pr. A.M. Vedrenne, CNRI/Science Source

Alzheimer's disease (AD) and Parkinson's disease (PD) are two diseases that occur mostly in the elderly. AD is characterized by increasing dementia, and PD by tremors and debilitating movement disorders. Both increase over time. As people are living longer, the incidence of both diseases is increasing. Both diseases are characterized by the aggregation of proteins into tangles or fibrils that affect neuronal function. For example, in PD, an aggregated form of a common protein α-synuclein accumulates in neuronal cells of the brain; these accumulated fibrils, when stained, appear under the microscope as "Lewy bodies" (photo). When dopamine-producing cells begin to accumulate α-synuclein fibrils, the cells begin to die. Whether the α-synuclein fibrils are causative has not been firmly established, but the process of "fibrillization" seems to be toxic in some way. An intensive research effort is under way to understand what causes formation of these fibrils.

There is some evidence that oxidation of the sulfide group of the α-amino acid methionine in certain proteins might be a primary event leading to the formation of fibrils. (Proteins are chains of α-amino acids with the general structure $H_3\overset{+}{N}—CHR—CO_2^-$ connected by amide bonds, as shown below.) The various α-amino acids differ in the structures of their R-groups. Methionine is oxidized to a sulfoxide.

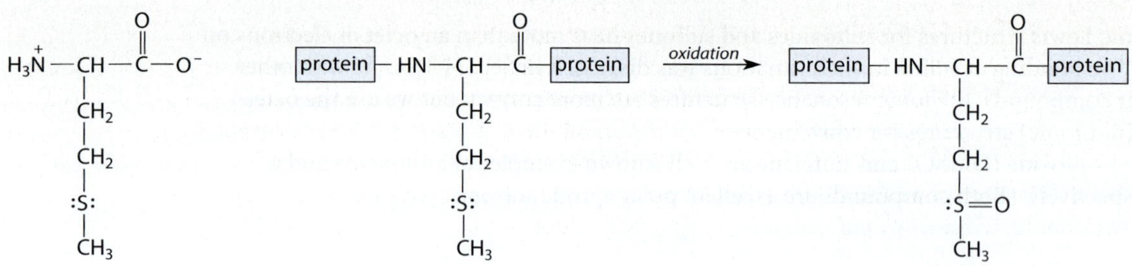

methionine (the amino acid) — a methionine residue in a peptide or protein — a methionine sulfoxide residue in a peptide or protein (12.86)

This hypothesis has led to the idea that a diet rich in antioxidants might offer some protection against AD and PD, but there is no firm evidence yet to support this idea.

Focused Problem

12.41 Draw the structure of the sulfoxide and the sulfone formed when diisopropyl sulfide is treated, respectively, with (a) one equivalent of H_2O_2 at room temperature and (b) two equivalents of H_2O_2 and heat.

12.10 THE THREE FUNDAMENTAL OPERATIONS OF ORGANIC SYNTHESIS

Sections 5.9B and 10.8 introduced organic synthesis and developed a systematic approach to solving synthesis problems. This section continues this approach by classifying the operations involved in a typical synthesis and providing some addition practice in synthetic planning.

Most reactions used in organic synthesis involve one or more of *three fundamental operations*:

1. Functional-group transformation
2. Formation of carbon–carbon bonds
3. Control of stereochemistry

Functional-group transformation—the conversion of one functional group into another—is the most common type of synthetic operation. Most of the reactions presented so far involve functional-group transformation. For example, the hydrolysis of epoxides transforms epoxides into glycols; hydroboration–oxidation converts alkenes into alcohols.

Reactions that bring about the *formation of carbon–carbon bonds* are particularly important, because these reactions must be used to add carbon atoms and "grow" larger carbon chains from smaller ones. The following reactions in this category have been presented to this point in the text:

1. Reactions of carbanions from 1-alkynes (acetylide ions) with alkyl halides and sulfonate esters (Sec. 10.6B)
2. Cyclopropane formation from carbenes or carbenoids and alkenes (Sec. 10.9)
3. Reaction of Grignard reagents with ethylene oxide (Sec. 11.5C)
4. Reaction of higher-order lithium organocuprate reagents with epoxides (Sec. 11.5C)

Control of stereochemistry is accomplished with stereoselective reactions. Whenever you have to prepare a compound that can exist as several stereoisomers, you should think in terms of these reactions. Examples of stereoselective reactions include hydroboration–oxidation, which is a syn addition, and S_N2 reactions, which occur with inversion of configuration.

Most reactions involve combinations of at least two of the three fundamental operations. For example, hydroboration–oxidation is a functional-group transformation (alkene $\longrightarrow$ alcohol) that also allows, at the same time, the control of stereochemistry. The reaction of an epoxide with a Grignard or organocuprate reagent combines all three operations: carbon–carbon bond formation, a functional-group transformation (epoxide $\longrightarrow$ alcohol), and control of stereochemistry (inversion at a carbon stereocenter).

Study Problems 12.7 and 12.8 demonstrate how to use the three fundamental operations in planning an organic synthesis.

Study Problem 12.7

Outline a synthesis of (*Z*)-3-hepten-1-ol from an organic compound with five or fewer carbons and any other reagents.

Solution As usual, first write the problem in terms of structures:

(*Z*)-3-hepten-1-ol

Next, analyze the types of operations needed. Two carbons must be added if the starting material contains five carbons, and there is an issue of stereochemistry: an alkene with *Z* (or cis) stereochemistry must be formed.

Now work backward from the product. We know of only one way to form a cis alkene, and that is by hydrogenation of an alkyne with a poisoned catalyst. Therefore, thinking retrosynthetically, we have

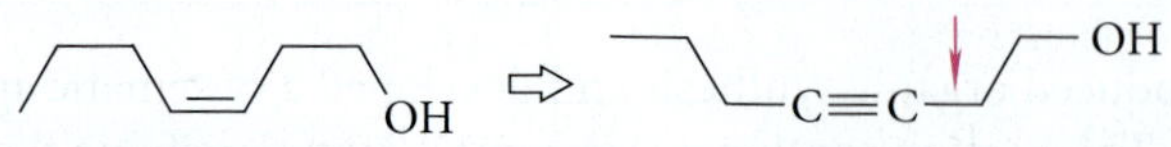

(Z)-3-hepten-1-ol **3-heptyn-1-ol**

We know two ways of preparing primary alcohols: hydroboration–oxidation of an alkene, and the reaction of a Grignard reagent with ethylene oxide. If we use the latter reaction, we can form the bond indicated by the red arrow and the alcohol at the same time. (Equation 12.46 showed that the reaction of a Grignard reagent with ethylene oxide is a good way to extend a carbon chain by two carbons.) The required acetylenic Grignard reagent is prepared from a 1-alkyne and EtMgBr or MeMgBr. This completes the retrosynthetic analysis.

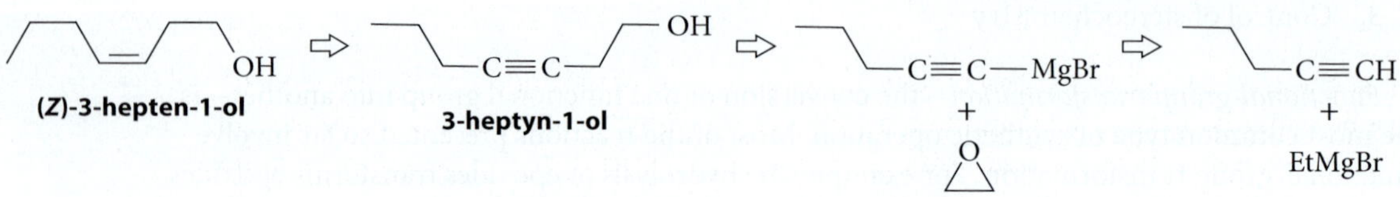

(Z)-3-hepten-1-ol **3-heptyn-1-ol**

Adding the details of each step, and writing the synthesis in the forward direction, gives the outlined synthesis.

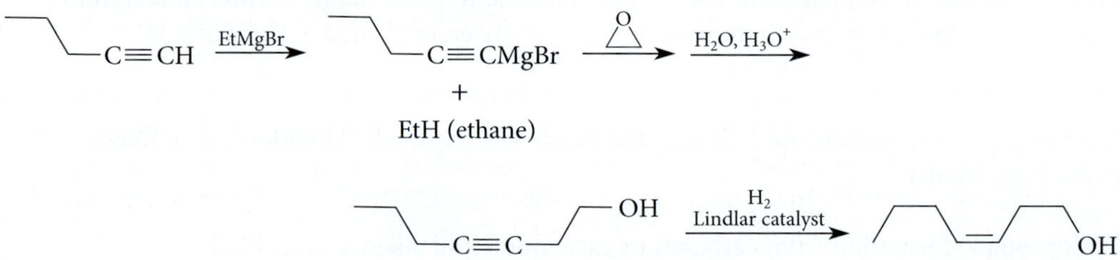

(Z)-3-hepten-1-ol

Study Problem 12.8

Outline a synthesis of (±)-*trans*-2-methoxycyclohexanol from cyclohexene.

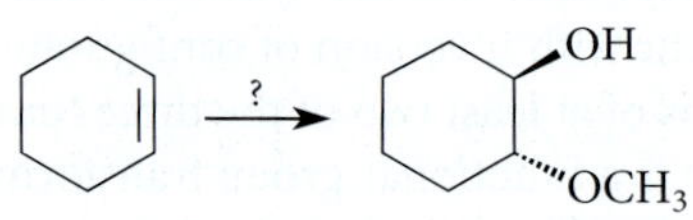

Solution No new carbon–carbon bonds are joined to the cyclohexene ring, so reactions that form carbon–carbon bonds are not likely to be useful. There is, however, a stereochemical problem: the two oxygens must be introduced in a trans arrangement. Finally, a net addition of CH_3O— and HO— to the carbon–carbon double bond is required. Such an addition cannot be completed in one step. However, in the opening of epoxides, the epoxide oxygen becomes an —OH group, and this transformation occurs with inversion of stereochemical configuration at the broken bond so that the resulting groups end up trans. Opening an epoxide with CH_3O^- in CH_3OH, or with CH_3OH and an acid catalyst, is therefore a good last step to the synthesis.

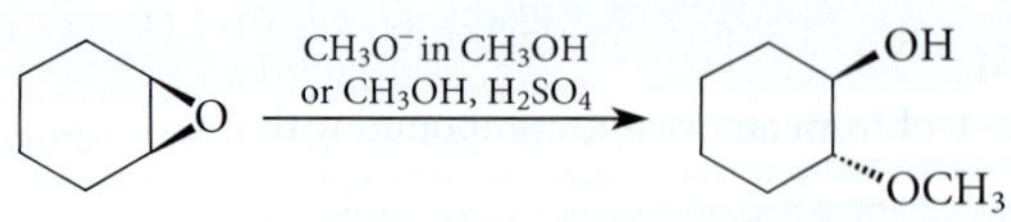

Completion of the synthesis requires only the preparation of the epoxide from cyclohexene. (How is this accomplished? See Focused Problem 12.12a.)

Focused Problem

12.42 Outline a synthesis for each of the following compounds from the indicated starting materials and any other reagents:

(a) $(CH_3)_2CHCH_2CH_2CO_2H$ from $(CH_3)_2C{=}CH_2$ (methylpropene)

(b) Dibutyl sulfone from 1-butanethiol

(c) (±)-*trans*-1-Ethoxy-2-methylcyclopentane from cyclopentene

12.11 SYNTHESIS OF ENANTIOMERICALLY PURE COMPOUNDS: ASYMMETRIC EPOXIDATION

As discussed in Sec. 12.10, control of stereochemistry is one of the important elements of an organic synthesis. Conventionally, the control of stereochemistry is limited to the synthesis of individual diastereomers. If a synthesis begins with achiral starting materials, as many syntheses do, chiral products are obtained as racemates (Sec. 7.7). In such a case, the only way to obtain pure enantiomers is to carry out an *enantiomeric resolution* (Secs. 6.10 and 8.7A) at some stage of the synthesis. When such an enantiomeric resolution is necessary, half of the material—the unwanted enantiomer—is wasted. An enantiomeric resolution can be avoided if a chiral starting material can be obtained as a single enantiomer. The enantiomerically pure chiral compounds found in nature can serve as one source of such starting materials. Occasionally, an enzyme can be found to catalyze the formation of an enantiomerically pure compound from achiral starting materials. However, such sources of pure enantiomers are relatively limited. Yet, the synthesis of pure enantiomers without enantiomeric resolution has taken on special importance as regulatory agencies have mandated that chiral pharmaceuticals be produced as pure enantiomers rather than racemates.

Chemists have learned to "custom-design" chiral catalysts that can bring about the formation of enantiomerically pure chiral compounds from achiral starting materials. This section discusses one of the most useful catalysts, which brings about the epoxidation of allylic alcohols to give enantiomerically pure epoxides. These chiral and enantiomerically pure epoxides can then be used to prepare a variety of other enantiomerically pure derivatives. An allylic alcohol contains an —OH group on a carbon *adjacent to* (not part of) a double bond. The simplest compound of this type, 2-propen-1-ol, has the common name *allyl alcohol.*

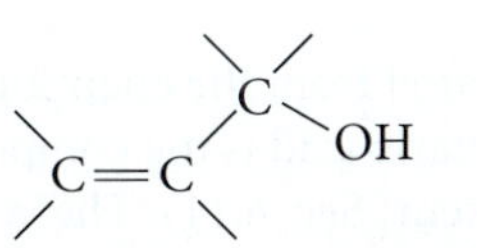

general structure of an allylic alcohol

$H_2C{=}CH—CH_2—OH$

2-propen-1-ol (allyl alcohol)

When an allylic alcohol is treated with *tert*-butyl hydroperoxide and a titanium(IV) isopropoxide catalyst, which we abbreviate as Ti(O*i* Pr)$_4$, an epoxide is formed at the double bond of the allylic alcohol functionality. Aqueous NaOH is added to destroy the catalyst and extract the by-products; the epoxide is isolated from the CH_2Cl_2 solvent. Although the reaction is a stereospecific syn addition, the product epoxide is racemic.

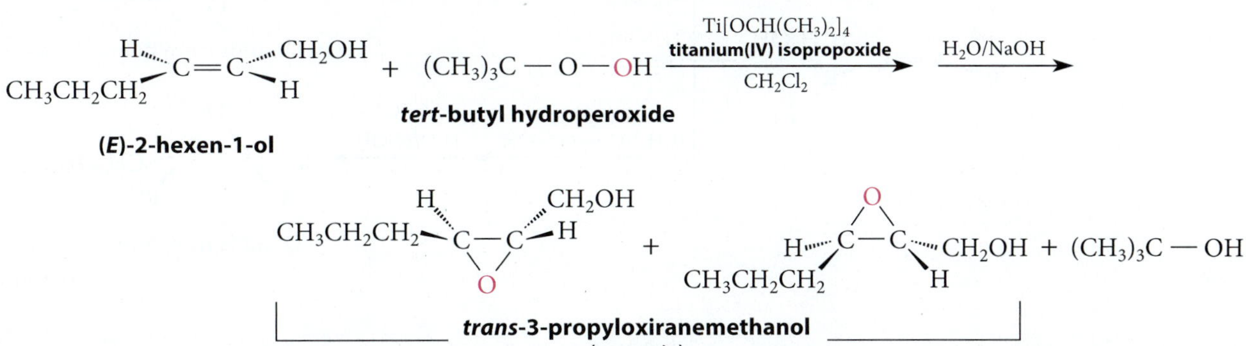

(12.87)

But when (2*R*,3*R*)-(+)-diethyl tartrate is added to the reaction mixture, the product, obtained in 97:3 enantiomeric ratio (ER), is the 2*S*,3*S* enantiomer, and the (+)-diethyl tartrate can be recovered unchanged.

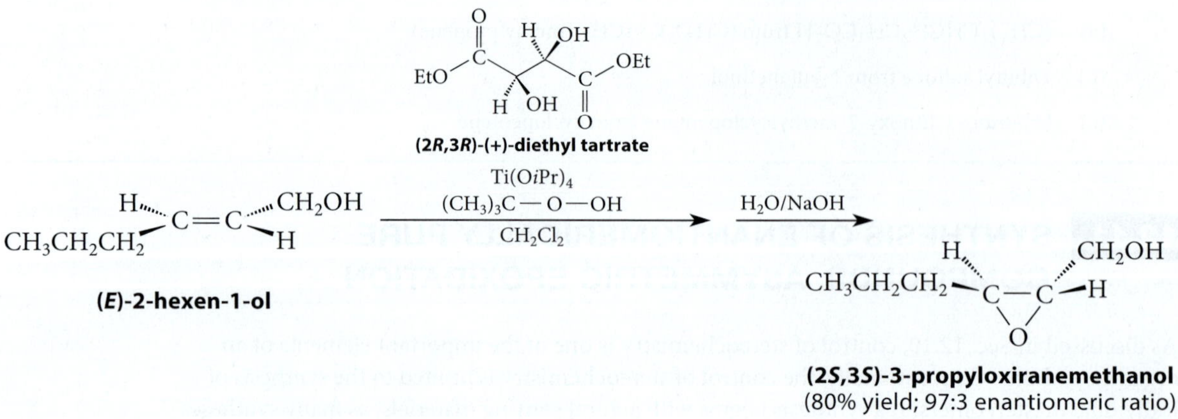

(12.88a)

When (2*S*,3*S*)- (–)-diethyl tartrate is used instead, the other enantiomer—the 2*R*,3*R* enantiomer—of the product is obtained.

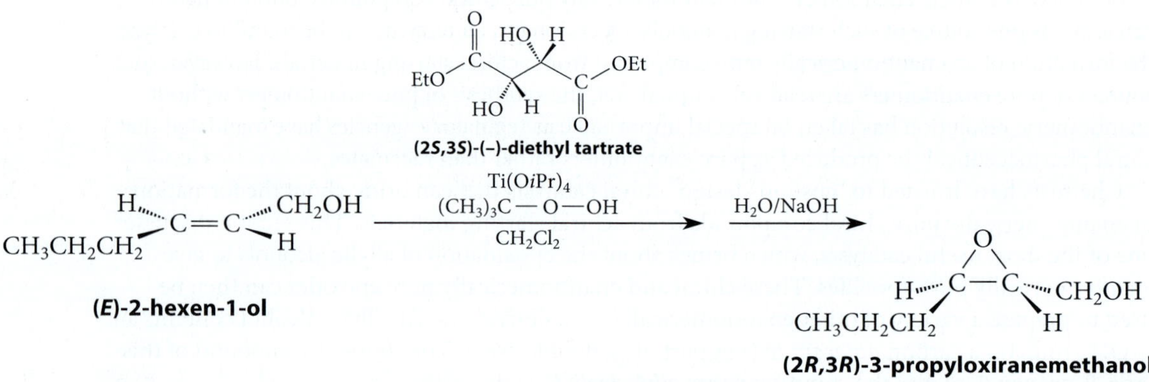

(12.88b)

The two enantiomeric tartrate esters can be easily prepared from the enantiomers of tartaric acid, a readily available, naturally occurring compound. (Tartaric acid is the compound that was the object of the first enantiomeric resolution by Louis Pasteur; Sec. 6.11). These tartrate esters are abbreviated as (+)-DET and (–)-DET, respectively.

These results are general for a large number of allylic alcohols. If we draw the allylic alcohol with the —CH_2OH group on the right and behind the page on the double bond, the results can be generalized as follows:

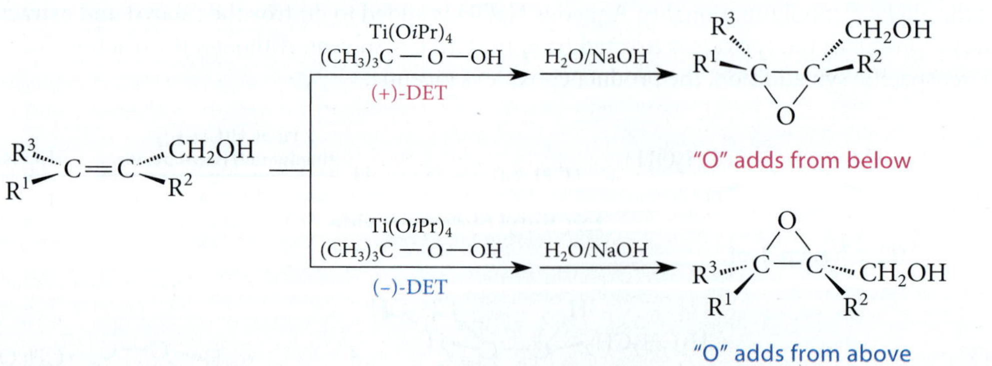

(12.89)

Only double bonds with an allylic —OH group form an epoxide; others usually don't react.

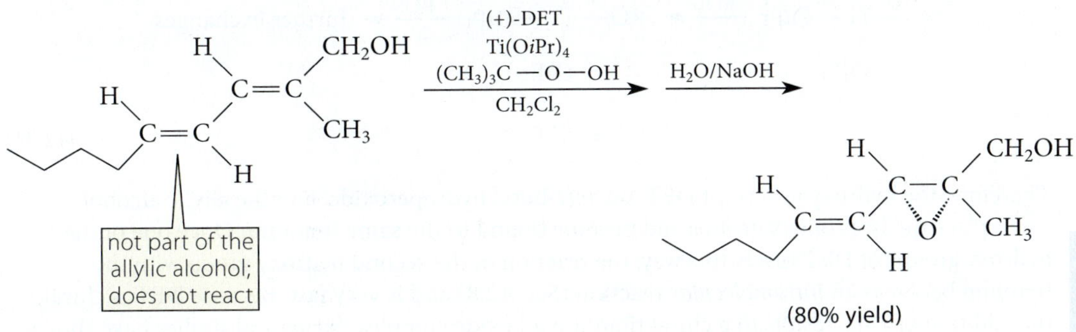

(12.90)

This reaction is called **asymmetric epoxidation** or **Sharpless epoxidation** (after its discoverer). Its importance hinges on the many stereospecific ring-opening reactions that epoxides can undergo, as illustrated in Study Problem 12.9. Following asymmetric epoxidation, stereospecific ring opening of the epoxide can be used to prepare a wide variety of enantiomerically pure organic compounds. Since its introduction, asymmetric epoxidation has been a key step in the synthesis of many important, optically pure, chiral compounds.

Asymmetric epoxidation was discovered in 1980 by K. Barry Sharpless (b. 1941), then a professor of chemistry at the Massachusetts Institute of Technology, and his student Tsutomu Katsuki. Professor Sharpless, now at The Scripps Research Institute, shared the 2001 Nobel Prize in Chemistry for this discovery.

Study Problem 12.9

Outline a synthesis of the following compound as a single enantiomer.

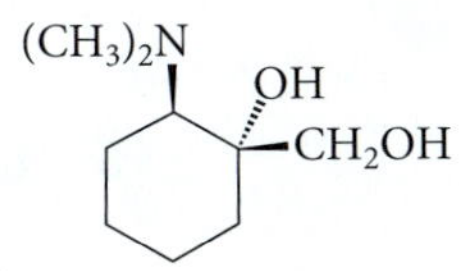

Solution When we see an —OH and another functional group, in this case the $(CH_3)_2N$— (dimethylamino) group, on adjacent carbons in a trans stereochemical relationship, we should think of an epoxide opening as a way to introduce both groups, as in Study Problem 12.8. The dimethylamine reacts at the carbon of the double bond with fewer substituents. The following reaction would work.

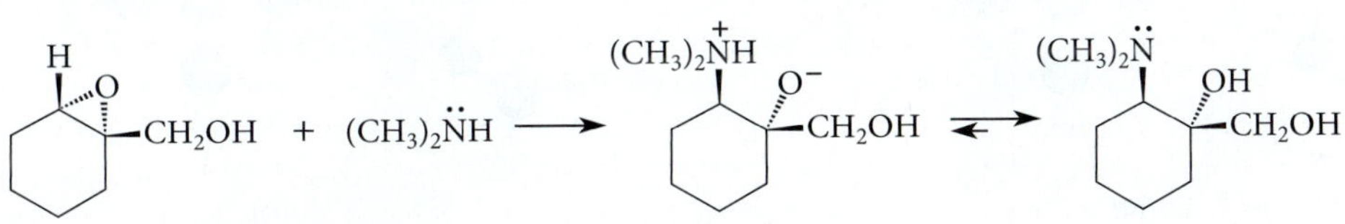

The epoxide is one that can be made by allylic epoxidation; the pattern in Eq. 12.89 shows that (+)-DET should be the chiral additive.

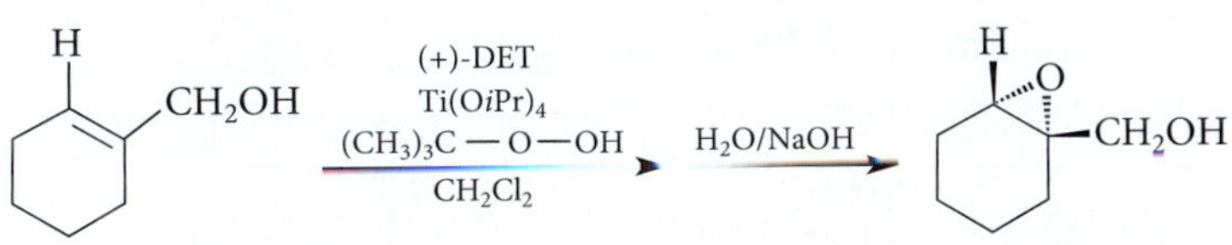

How does the chirality of a diethyl tartrate enantiomer determine the stereochemical outcome of the reaction? First of all, titanium(IV) alkoxides can rapidly exchange any or all of their alkoxide groups with other alcohols.

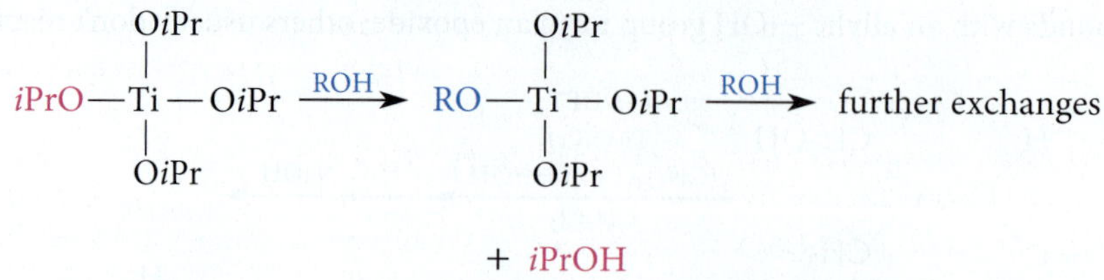

(12.91)

Therefore, the hydroxy groups of DET, the *tert*-butyl hydroperoxide, and the allylic alcohol can replace the isopropoxy groups and become bound to the same titanium. Once one of the hydroxy groups of DET reacts this way, the reaction of the second hydroxy group with the titanium becomes an *intramolecular* reaction (Sec. 12.8) and is very fast. Because DET is chiral, the addition of DET results in a chiral titanium alkoxide complex. Structural studies have shown that the complex actually involves two titanium atoms and two DET molecules with the structure shown in **Fig. 12.4a** [for (+)-DET as the additive]. In this complex, the oxygen of *tert*-butyl hydroperoxide (shown in yellow) is poised below the double bond in an optimal arrangement to receive the nucleophilic donation of π electrons, as shown in **Fig. 12.4c. Figure 12.4b** shows the same complex with the alkene oriented for reaction of the oxygen at the opposite face of the double bond—the reaction that is not observed. In this arrangement, the —CH_2— group of the allylic alcohol is forced into a more congested part of the catalyst, where it has unfavorable steric

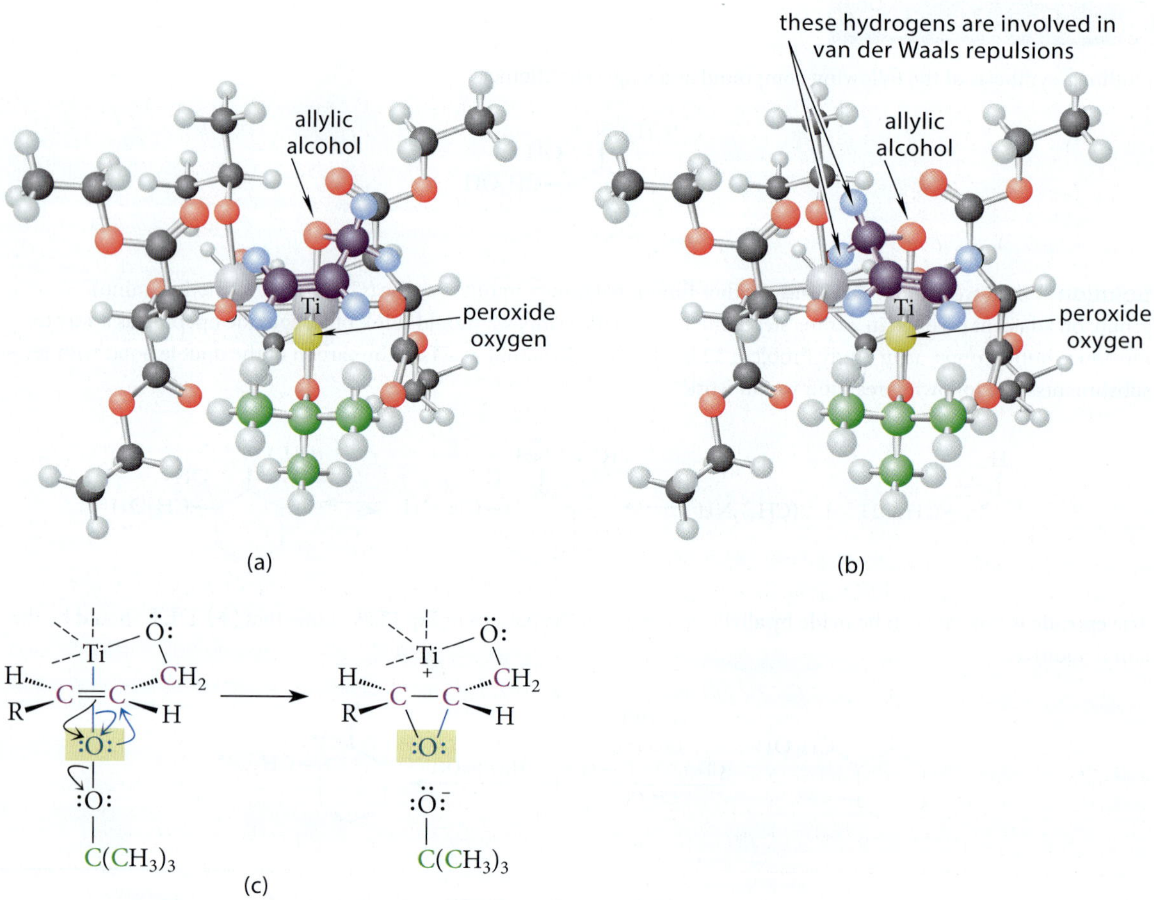

FIGURE 12.4 Molecular models of complexes for asymmetric epoxidation, shown with allyl alcohol as the reacting alcohol and (2*R*,3*R*)-(+)-diethyl tartrate as the chiral additive. The carbons and bonds of allyl alcohol are shown in purple, and the hydrogens are shown in blue. The carbons of *tert*-butyl hydroperoxide are shown in green, and the oxygen of *tert*-butyl hydroperoxide that is delivered to the alkene is shown in yellow. This oxygen is perfectly positioned to receive electrons from the π bond. (a) The complex that leads to the epoxide with the observed chirality. (b) The complex that leads to the epoxide with the chirality that is not observed. This complex is less stable because of van der Waals repulsions between the allylic —CH_2— hydrogens and one of the tartrate ester groups. (c) A curved-arrow mechanism corresponding to stereochemistry (a).

interactions (van der Waals repulsions) with one of the tartrate ester groups. This steric effect is the mechanism by which the chirality of the tartrate ester "dictates" the stereochemistry of the reaction.

The structure explains the catalyst *specificity*—that is, the catalyst is largely specific for allylic alcohols because of the relationship of the allylic —OH, the double bond, and the peroxide oxygen. Other unsaturated alcohols undoubtedly bond to the titanium, but their double bonds are not in the proper relationship to the peroxide oxygen for reaction. The reaction also doesn't work for *tertiary* allylic alcohols, because the van der Waals repulsions of the alkyl branches with the other groups bound to the catalyst prevent proper binding. (Problem 12.87 explores the epoxidation of secondary allylic alcohols.)

Focused Problems

12.43 Give the product and its stereochemistry when each of the following alcohols is subjected to asymmetric epoxidation with *tert*-butyl hydroperoxide, $Ti(Oi\,Pr)_4$, and the stereoisomer of diethyl tartrate (DET) indicated.

(a) 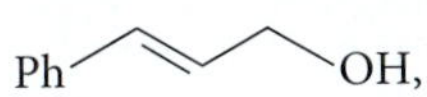 , (−)-DET

(b) 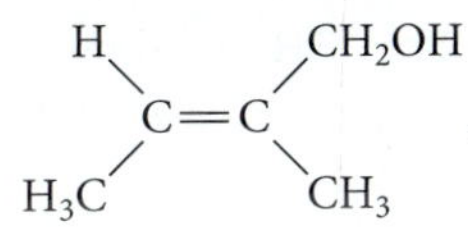 , (+)-DET

(c) 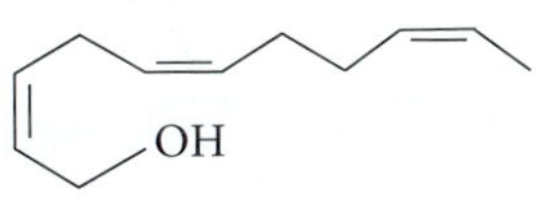 , (−)-DET

12.44 Propose a synthesis for each of the following compounds in enantiomerically pure form. Use an asymmetric epoxidation in each synthesis.

(a) 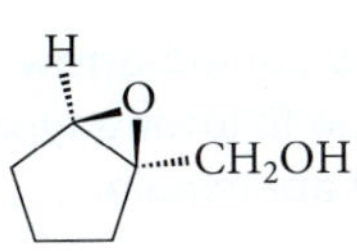

(b) 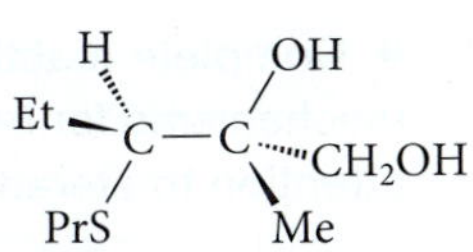

12.45 (a) Use the picture of the catalyst complex in Fig. 12.4a to explain why most *E* allylic alcohols undergo asymmetric epoxidation more rapidly than their *Z* isomers.

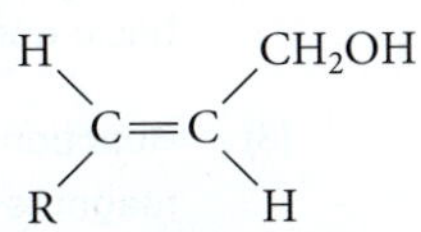

an *E* allylic alcohol

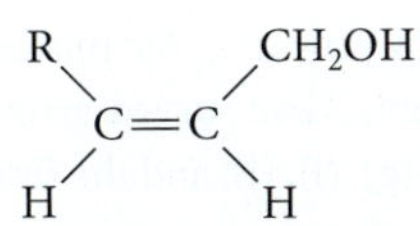

a *Z* allylic alcohol

(b) Would the same phenomenon be observed with (−)-DET, the enantiomer of the DET used in Fig. 12.4? Explain.

CHAPTER SUMMARY *For a summary of the chapter, see Chapter 12 in the* Study Guide and Solutions Manual.

REACTION REVIEW *For a summary of reactions discussed in this chapter, see the* Reaction Review *section of Chapter 12 in the* Study Guide and Solutions Manual.

SKILLS OBJECTIVES WITH PROBLEMS

- **Express the basicity of ethers and sulfides as pK_a values of their conjugate acids. Rank the basicities of ethers and sulfides along with the basicities of other compounds. 12.1**

12.46 Provide the structure of the conjugate acid of each compound below, and give the approximate pK_a values of the conjugate acids.

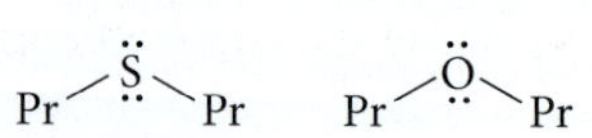

12.47 Rank the following compounds or ions in order of increasing acidity (decreasing pK_a).

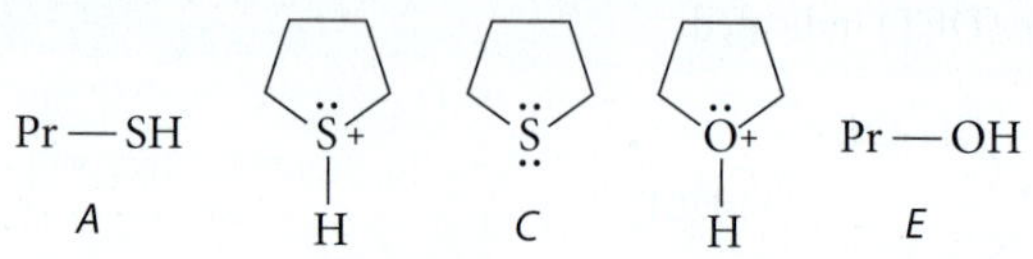

A B C D E

- **Complete reactions by providing missing starting materials or products for the following syntheses. Indicate stereochemistry (if relevant), and give curved-arrow mechanisms.**
  - (1) Synthesis of ethers and sulfides **12.2A**
  - (2) Synthesis of epoxides (including allylic epoxidation) **12.3, 12.11**
  - (3) Synthesis of glycols **12.6A**
  - (4) Synthesis of sulfoxides and sulfones **12.9B**

12.48 Complete the following reactions by providing the missing starting material, product, or reagents. Give curved-arrow mechanisms for parts (a), (b), (d)–(g), (i), (j), and the first reaction in (c).

(a) $CH_3CH_2CH_2{-}Br + Na^+\,EtO^- \xrightarrow{EtOH} ?$

(b)

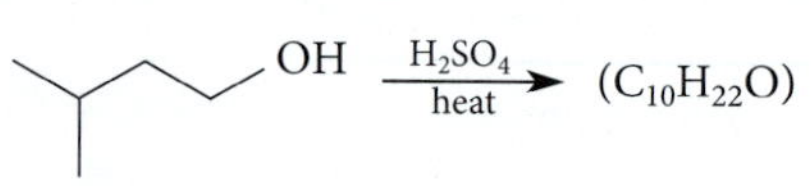

(c)

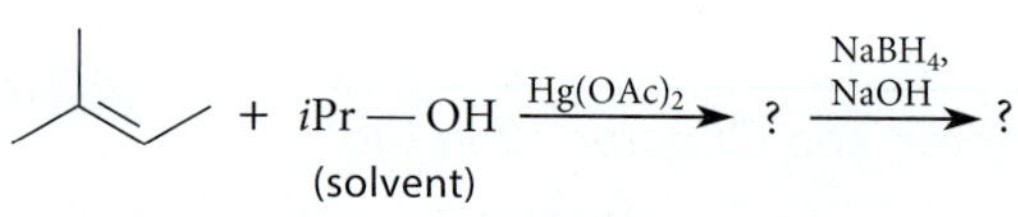

(d)

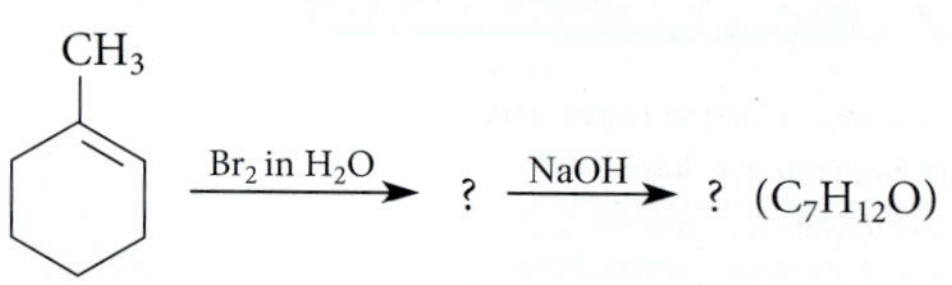

(e)

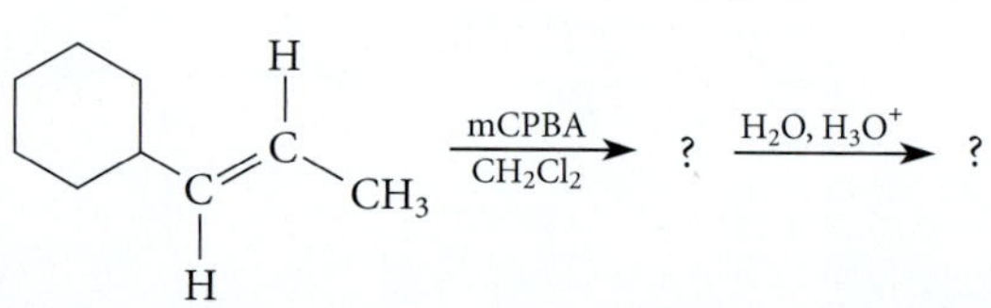

(f)

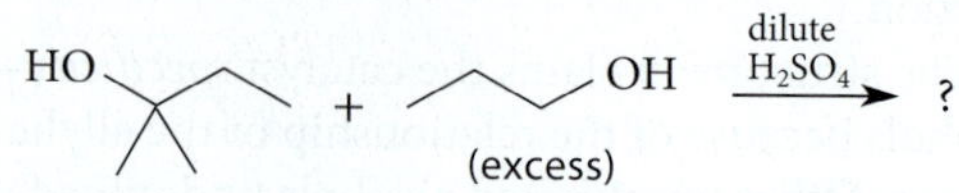

(g) $PhCH_2SH \xrightarrow[\text{MeOH (solvent)}]{NaOH} ? \xrightarrow{PhCH_2Br} ?$

(h)

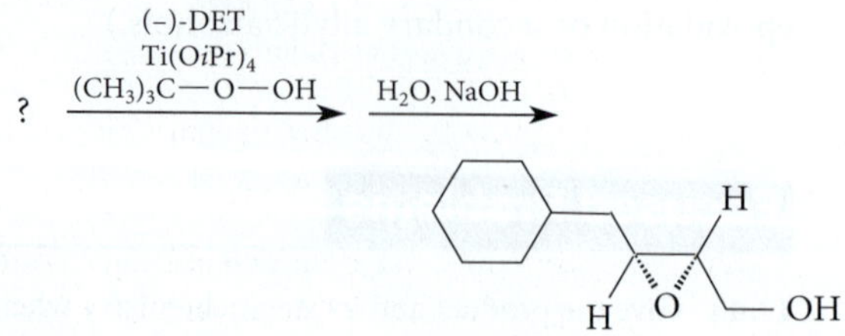

(i)

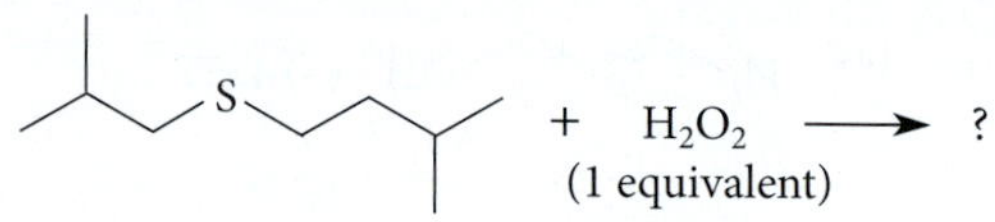

(j)

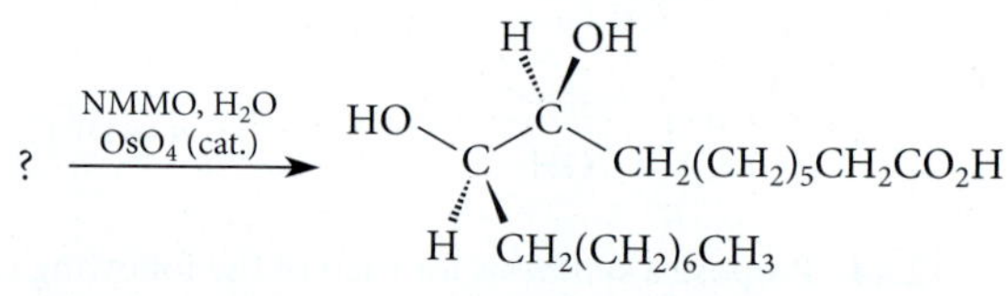

(racemate)

- **Complete reactions and give curved-arrow mechanisms for reactions of the following types with attention to stereochemistry, if applicable.**
  - (1) Ether cleavage **12.4**
  - (2) Ring opening reactions of epoxides under acidic and basic conditions **12.5A,B**
  - (3) Reactions of epoxides with organometallic reagents **12.5C**
  - (4) Periodate cleavage of glycols **12.6B**
  - (5) Reactions of oxonium and sulfonium salts with nucleophiles **12.7A**
  - (6) Reactions involving neighboring-group participation **12.8**

12.49 Give the major organic product of each of the following reactions. Include stereochemistry where relevant.

(a) Dibutyl sulfide with one equivalent of H_2O_2 at 25 °C

(b) Dibutyl sulfide with two or more equivalents of H_2O_2 and heat

(c) *cis*-3-Hexene with magnesium monoperoxyphthalate (MMPP)

(d) The product of (c) with $(CH_3)_2Cu(CN)Li_2$, then H_3O^+

(e) The product of (c) with solvent H_2O in acidic solution

(f) The product of (e) with periodic acid

(g) The product of (d) with NaH in THF followed by CH_3I

(h) (*E*)-3-Methyl-3-hexene with $Hg(OAc)_2$ in ethanol solvent followed by $NaBH_4$

(i) The product of (h) with acidic methanol

(j) (*R*)-3-Methyl-1-bromopentane with Mg in ether, then with ethylene oxide, then with CH_3I

12.50 Give the products of the reaction of 2-ethyl-2-methyloxirane (or other compound indicated) with each of the following reagents.

(a) Water, H_3O^+

(b) Water, NaOH, heat

(c) $Na^+CH_3O^-$ in CH_3OH

(d) CH_3OH and a catalytic amount of H_2SO_4

(e) Dilithium dimethylcyanocuprate, then H_3O^+

(f) The product of (c) + concentrated HBr, 25 °C

(g) The product of (d) + HBr, 25 °C

(h) The product of (c) + NaH, then CH_3I

(i) The product of (d) + NaH, then EtI

(j) The product of (a) + periodic acid

(k) The product of (f) + Mg in dry ether

(l) The product of (k) + ethylene oxide, then H_3O^+

12.51 Complete the following reactions by giving the missing products, and provide the curved-arrow mechanism of each reaction.

(a) The product of Problem 12.48(c) + concentrated HCl, 25 °C

(b) The product of Problem 12.48(b) + concentrated HBr, 25 °C

(c) 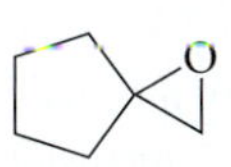 + H—C≡N $\xrightarrow{^-OH}$?

$pK_a = 9.4$

(d)  + CH_3OH $\xrightarrow{\text{strong acid (small amount)}}$?

(e) 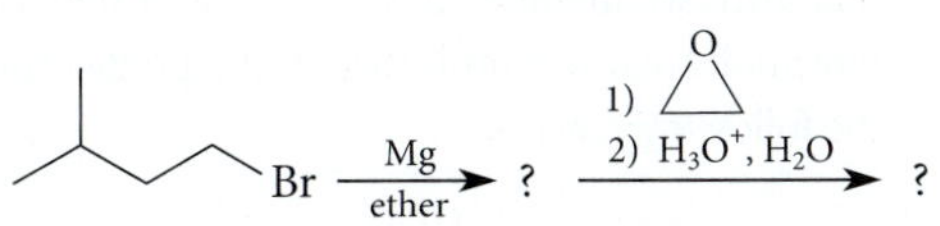

(f) 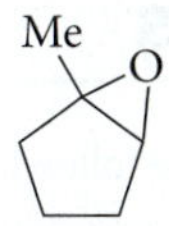 + $(CH_3)_2Cu(CN)Li_2$ ⟶ $\xrightarrow{H_2O,\ H_3O^+}$?

(g) 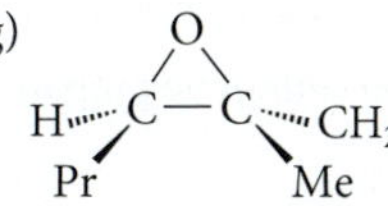 + CH_3O^- $\xrightarrow{CH_3OH}$? $\xrightarrow{H_5IO_6}$?

(h) 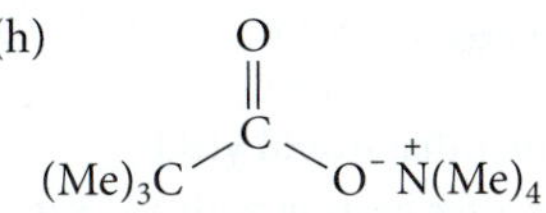 + $(CH_3)_3\overset{+}{O}\ \overset{-}{B}F_4$ $\xrightarrow{\text{ether}}$?

(i) 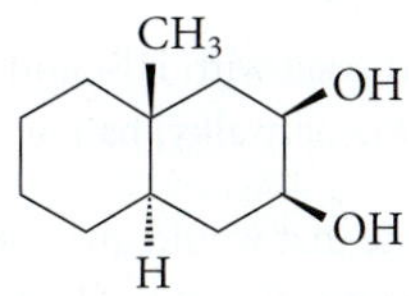 + H_5IO_6 ⟶ ?

(j) 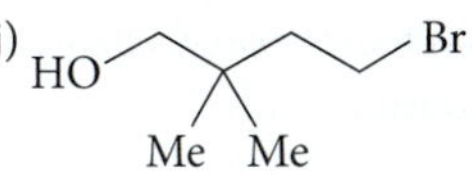 + $Na^+\ ^-OMe$ $\xrightarrow{MeOH}$? ($C_6H_{12}O$)

- **Use retrosynthetic analysis to outline multistep syntheses involving reactions in this and previous chapters, paying attention to stereochemistry where relevant. 10.8, 11.12, 12.10**

12.52 Outline short multistep syntheses of each compound from the indicated starting material and any other compounds and reagents.

(a) 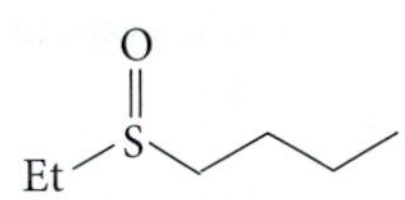 from ethanethiol

(b) 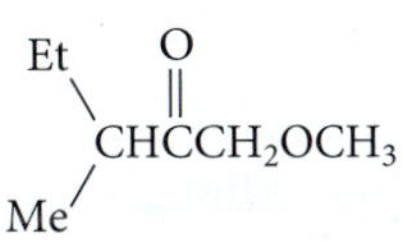 from 3-methyl-1-pentene

(c) 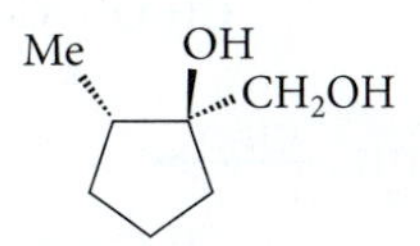 from 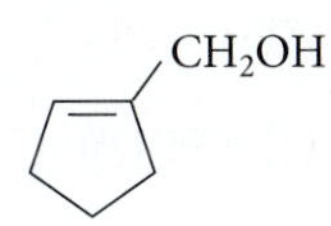

(enantiomerically pure)

(d) 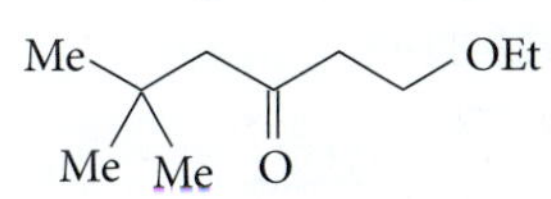 from 3,3-dimethyl-1-butyne

(*Hint:* See Eq. 5.34b, Sec. 5.6A.)

(e) 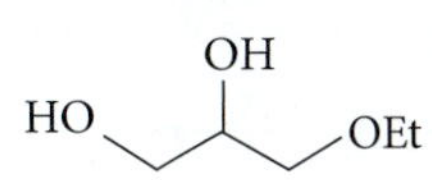 from 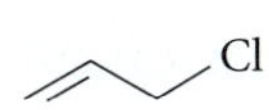

allyl chloride

(*Hint:* Allyl chloride is very reactive in S_N2 reactions.)

INTEGRATED PROBLEMS

12.53 Draw a structure that meets each of the following criteria. (Some parts may have more than one correct answer.)

(a) A nine-carbon ether that *cannot* be prepared by the Williamson synthesis

(b) A nine-carbon ether that *can* be prepared by the Williamson synthesis

(c) A four-carbon ether that would yield 1,4-diiodobutane after heating with an excess of HI

(d) An ether that would react with HBr and heat to give propyl bromide as the only alkyl halide

(e) A four-carbon alkene that would give different glycols after treatment with OsO_4/NMMO or treatment with *meta*-chloroperoxybenzoic acid followed by dilute aqueous acid

(f) A four-carbon alkene that would give the same glycol as a result of the different reaction conditions in (e)

(g) A diene (a compound with two double bonds) C_6H_8 that can form only one mono-epoxide and two di-epoxides (counting stereoisomers)

(h) An alkene C_6H_{12} that would give the same glycol either from treatment with a peroxycarboxylic acid, followed by acid catalyzed hydrolysis, or from glycol formation with OsO_4/NMMO

12.54 Which of the following ring-opening reactions should occur most readily? Explain.

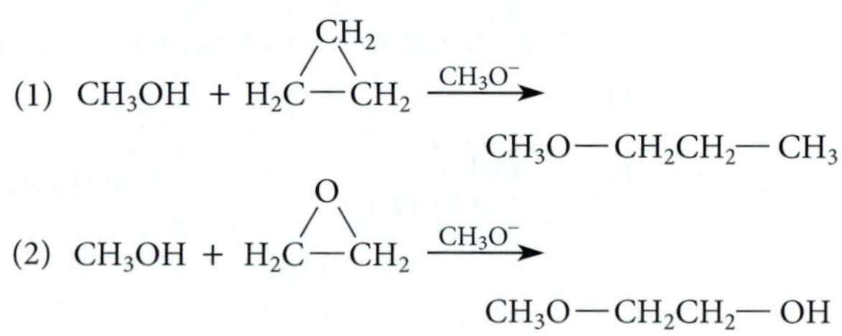

12.55 Which of the following ring-opening reactions should occur most readily? Explain.

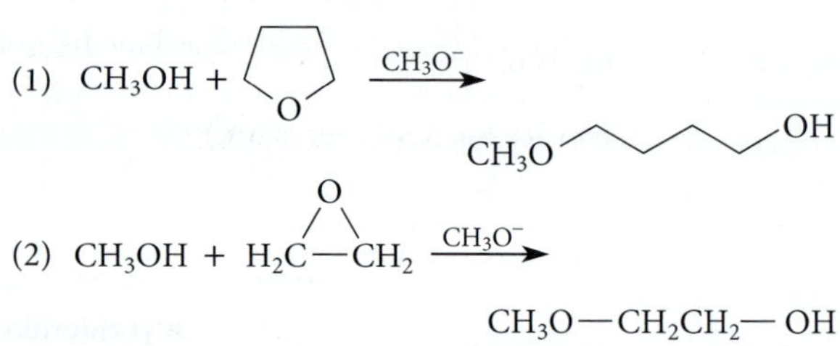

12.56 When HCl is formed as a by-product in reactions, it is usually removed from reaction mixtures by neutralization with aqueous base. At times, however, the use of base is not compatible with the products or conditions of a reaction. It has been found that propylene oxide (2-methyloxirane) can be used to remove HCl quantitatively. Explain why this procedure works.

12.57 A student has run each of the following reactions and is disappointed to find that each has given none of the desired product. Explain why each reaction failed.

(a)

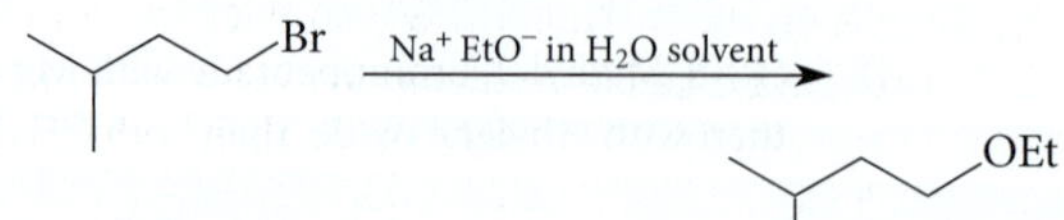

(b)

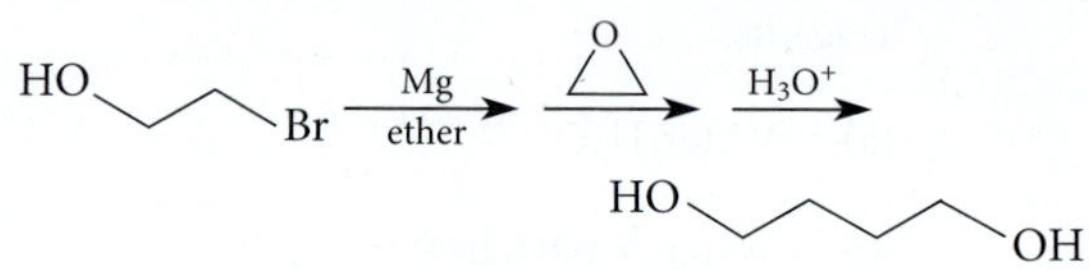

12.58 Tell whether each of the following compounds can be prepared by the reaction of a Grignard reagent with ethylene oxide. If so, show the reaction; if not, explain why and give a different synthesis that uses a different epoxide starting material.

(a) 2-Pentanol

(b) 1-Pentanol

12.59 For each of the following alkenes, state whether a reaction with OsO_4/NMMO will give a racemic mixture of products that can (in principle) be resolved into enantiomers under ordinary conditions.

(a) Ethylene

(b) *cis*-2-Butene

(c) *trans*-2-Butene

(d) *cis*-2-Pentene

12.60 The (+)-stereoisomer of 2-methyloxirane reacts with aqueous NaOH to give the (*R*)-(−)-stereoisomer of 1,2-propanediol. Use this observation to propose the absolute stereochemical configuration of (+)-2-methyloxirane.

12.61 Predict the absolute configuration of the major diol product formed by treatment of (*S*)-2-ethyl-2-methyloxirane with water in the presence of an acid catalyst.

12.62 Keeping in mind that many intramolecular reactions that form six-membered rings are faster than competing intermolecular reactions (Sec. 12.8), predict the product of the following reaction.

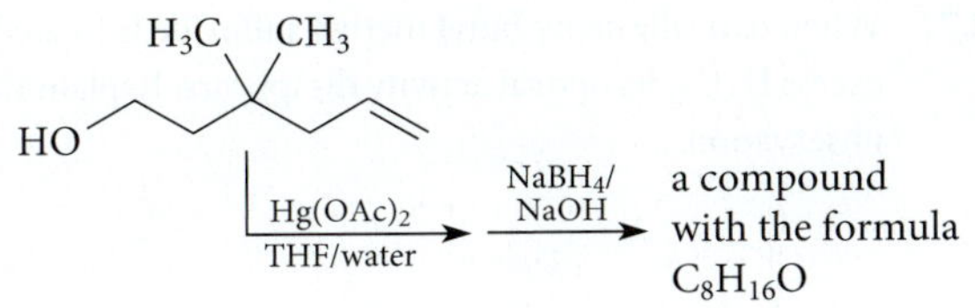

12.63 When (3*S*,4*S*)-4-methoxy-3-methyl-1-pentene is treated with mercuric acetate in methanol solvent, then with $NaBH_4$/NaOH, two isomeric compounds with the formula $C_8H_{18}O_2$ are isolated. One, compound *A*, is optically *inactive*, but the other, compound *B*, is optically active. Give the structures and absolute configurations of both compounds.

12.64 You are a manager for a company, Weighty Matters, which specializes in the manufacture of organic compounds containing ^{18}O, a heavy isotope of oxygen. You have assigned the task of preparing ether *B* to a team of two staff experts, and you have stipulated that alcohol *A* must be used as a starting material ($^*O = {}^{18}O$):

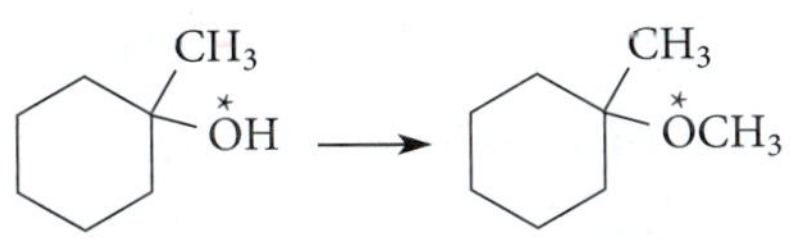

A *B*

A member of your staff, Homer Flaskclamper, has proposed the following two possible syntheses and has come to you for advice.

(1)

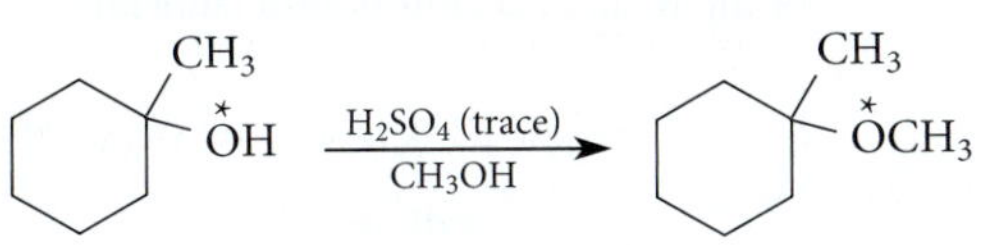

(2)

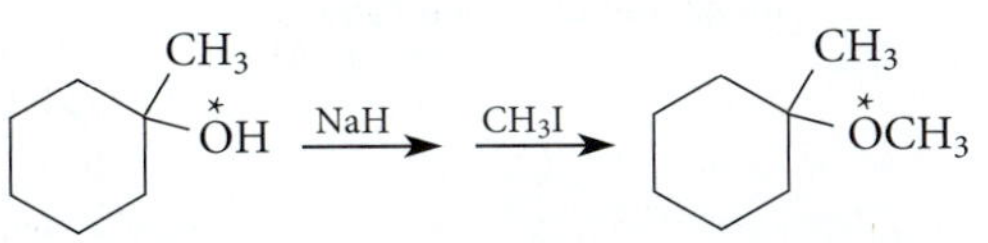

Which synthesis would you advise Flaskclamper to use and why?

12.65 Match each of the following four compounds with one of the compounds *A–D* on the basis of the following experimental facts. Compounds *A*, *B*, and *C* are optically active, but compound *D* is not. Compound *C* gives the same products as compound *D* on treatment with periodic acid, but compound *B* gives a different product. Compound *A* does not react with periodic acid.

(1) (2*S*,3*S*)-2,4-Dimethoxy-1,3-butanediol

(2) *meso*-1,4-Dimethoxy-2,3-butanediol

(3) (+)-1,4-Dimethoxy-2,3-butanediol

(4) (2*R*,3*R*)-3,4-Dimethoxy-1,2-butanediol

12.66 Complete each of the following reactions by giving the principal organic products. Indicate the stereochemistry of the products in parts (c) and (e).

(a) $H_3C{-}CH{=}CH_2 + H_2O + Br_2 \longrightarrow \xrightarrow[\text{pyridine}]{CrO_3}$

(b)

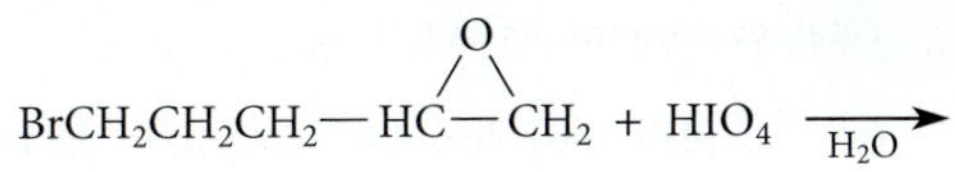

(c)

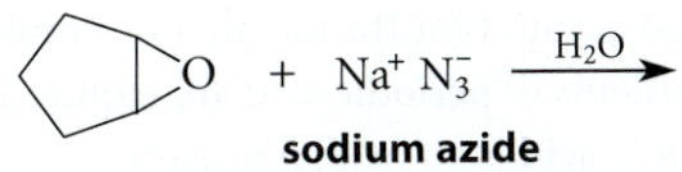

sodium azide
(see Table 9.1, Sec. 9.1A)

(d)

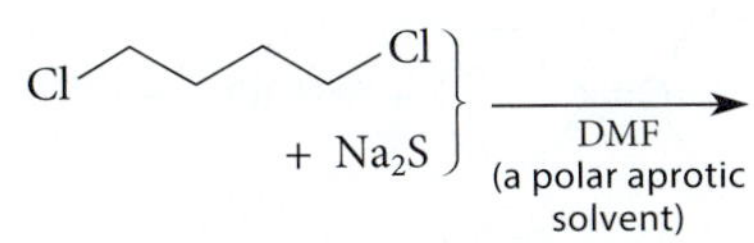

(*Hint*: Think of Na_2S as the $:\ddot{S}:^{2-}$ ion.)

(e)

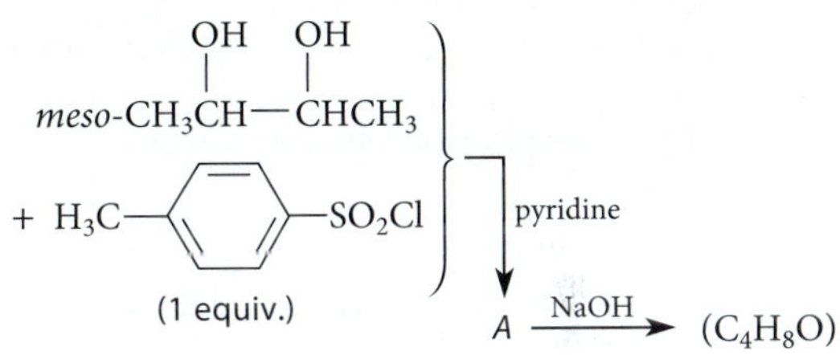

(*Hint*: Remember that the tosylate group is a good leaving group.)

12.67 Outline a synthesis for each of the following compounds in enantiomerically pure form from enantiomerically pure (2*R*,3*R*)-2,3-dimethyloxirane:

(a) (2*R*,3*S*)-3-Methoxy-2-butanol

(b)

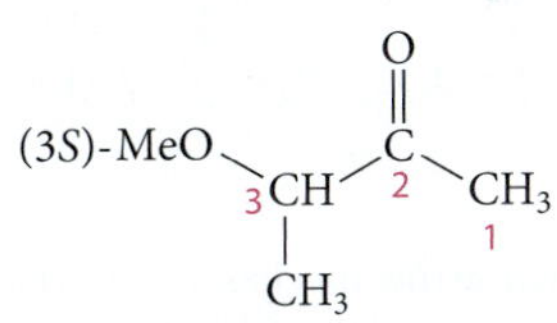

(c)

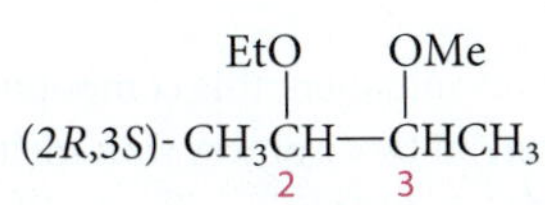

(d)

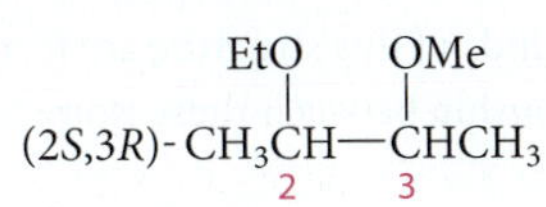

12.68 Give the structures of all epoxides that could in principle be formed when each of the following alkenes reacts with *meta*-chloroperoxybenzoic acid (mCPBA). Which epoxide should predominate in each case? Why?

(a) *cis*-4,5-Dimethylcyclohexene

(b)

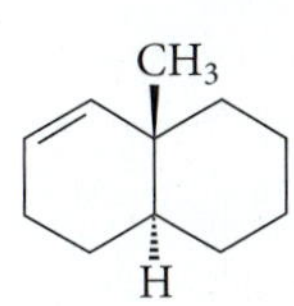

12.69 An alternative to the ozonolysis of alkenes is to treat an alkene with *two* molar equivalents of periodic acid and a catalytic amount of OsO_4.

(a) Explain the role of each reagent in the reaction shown in the following reaction. Your explanation should account for the fact that *two* molar equivalents of periodic acid are required. (*Hint:* Periodic acid is an oxidizing agent.)

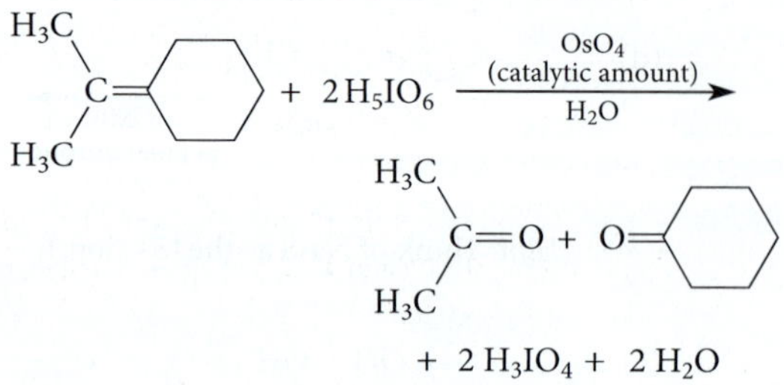

(b) Complete the following reaction.

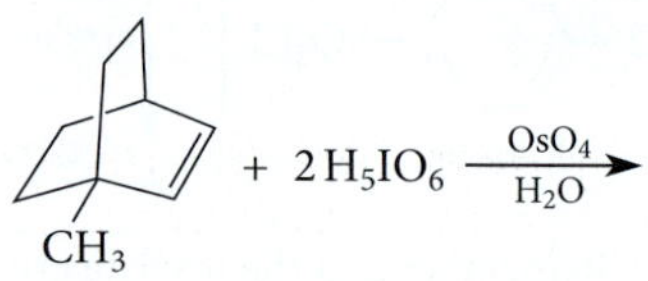

12.70 When compound *A* reacts with two equivalents of CH_3I, the double sulfonium salt *B* precipitates:

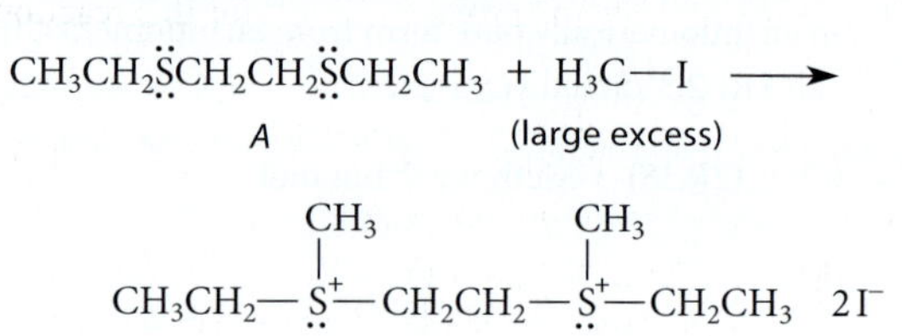

A (large excess)

B

(a) Give a curved-arrow mechanism for the formation of this salt.

(b) Upon closer examination, this compound is found to be a mixture of two isomers with melting points of 123 – 124 °C and 154 °C, respectively. Explain why *two* compounds of this structure are formed. What is the relationship between these isomers? (*Hint:* See Sec. 6.9B.)

12.71 When the naturally occurring amino acid (*S*)-methionine is converted into methionine sulfoxide, two isomers with different physical properties are formed. What are their structures, and what is their stereochemical relationship?

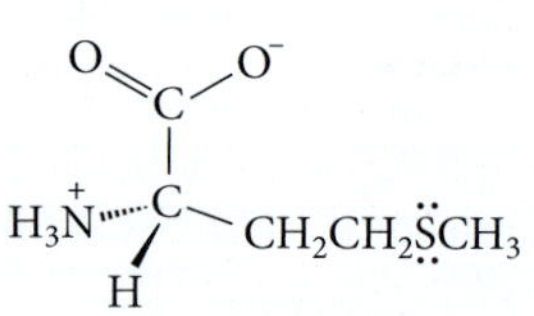

(*S*)-methionine

12.72 When *optically active* butyl methyl sulfoxide is treated with excess H_2O_2, its optical activity disappears. Explain this observation.

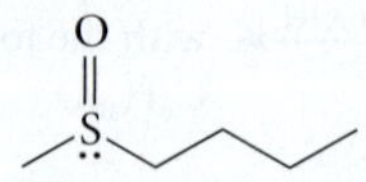

butyl methyl sulfoxide

12.73 When *S*-adenosylmethionine (Fig. 12.1, Sec. 12.7B), isolated from natural sources, is allowed to stand in aqueous solution for several weeks at room temperature, a stereoisomeric contaminant appears in solution that can be separated by ordinary methods. Suggest a structure for this contaminant and a reason that it forms. (*Hint:* No covalent bonds are broken in this process.)

12.74 One of the side reactions that occur when epoxides react with ^{-}OH is the formation of polymers. Propose a curved-arrow mechanism for the following polymerization reaction of "propylene oxide."

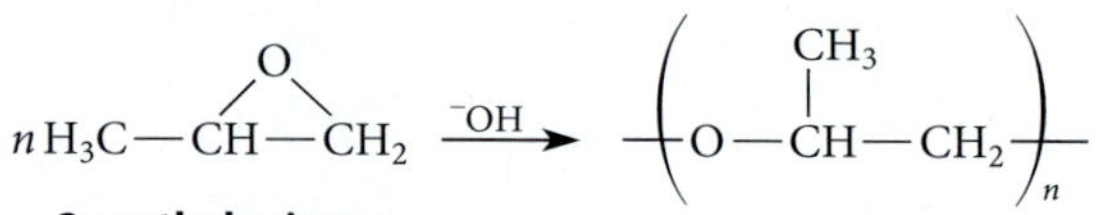

2-methyloxirane ("propylene oxide")

12.75 (a) Give a curved-arrow mechanism for the following reaction. Be sure your mechanism indicates the role of the weak acid ammonium chloride.

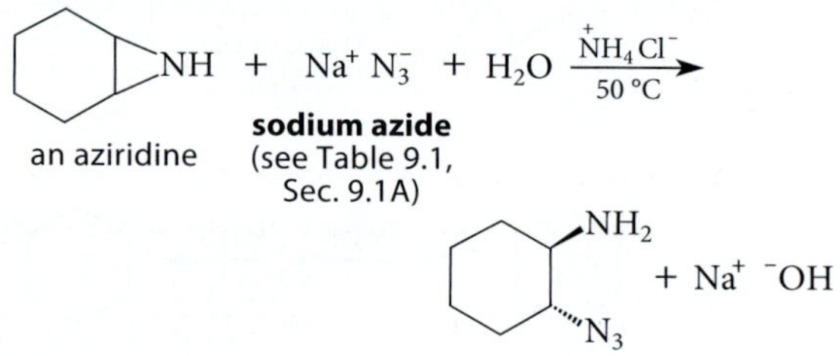

an aziridine

sodium azide (see Table 9.1, Sec. 9.1A)

(b) Why does the reaction of an aziridine require this weak acid? (*Hint:* The pK_a of an amine RNH_2 is about 32.)

(c) The pK_a of ammonium ion is 9.25; the conjugate-acid pK_a of an aziridine is about 7; and the pK_a of HN_3 is 4.2. A student has suggested that dilute (0.01 *M*) HCl should be an even better catalyst. Critique this suggestion.

12.76 In each of the following pairs, *one* of the glycols is virtually inert to periodate oxidation. Which glycol is inert? Explain why. (*Hint:* Consider the structure of the intermediate in the reaction.)

(a)

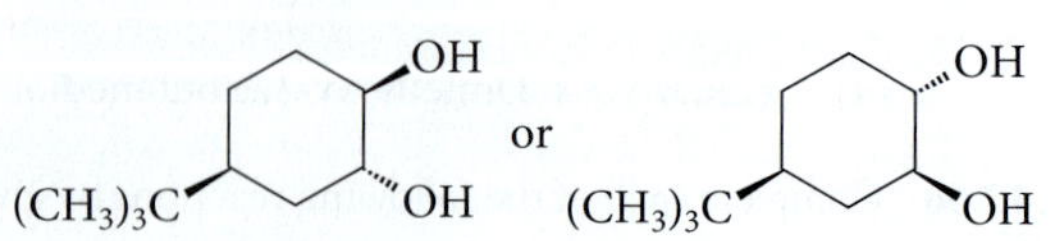

A or *B*

(b) 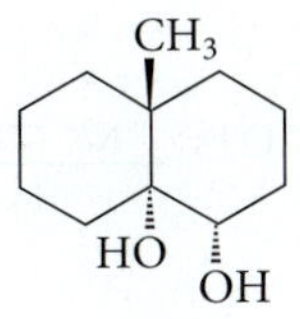 or 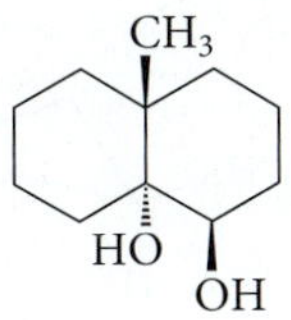

A *B*

12.77 One of the following reactions is about 2000 times faster in pure water than it is in pure ethanol. Another is about 20,000 times faster in pure ethanol than it is in pure water. The rate of the third changes very little when the solvent composition is changed from ethanol to water. Which of the reactions is faster in ethanol, which is faster in water, and which has a rate that is solvent-invariant? Explain. The solvent (ethanol, water, or a mixture of the two) in the following equations is indicated by ROH. (*Hint:* Consider the dielectric constants for ethanol and water in Table 8.2, Sec. 8.6B.)

(1) $(CH_3)_3S^+ + {}^-OH \xrightarrow{ROH} CH_3{-}OH + (CH_3)_2S$ (S_N2 reaction)

(2) $(CH_3)_3C{-}Cl + R{-}OH \longrightarrow$ (R—OH = solvent)

$(CH_3)_3C^+\ Cl^- \longrightarrow (CH_3)_3C{-}OR + HCl$ (S_N1 reaction)

(3) $(CH_3)_3C{-}\overset{+}{S}(CH_3)_2 + ROH \longrightarrow$ (ROH = solvent)

$(CH_3)_3C^+ + S(CH_3)_2 \longrightarrow (CH_3)_3C{-}OR + R\overset{+}{O}H_2$ (S_N1 reaction)

12.78 Each of three bottles, labeled respectively *A*, *B*, and *C*, contains one of the compounds shown below. On treatment with KOH in methanol, compound *A* gives no epoxide, compound *B* gives an epoxide *D*, and compound *C* gives an epoxide *E*. Epoxides *D* and *E* are stereoisomers. Under identical conditions, *C* gives *E* much more slowly than *B* gives *D*. Identify *A*, *B*, and *C*, and explain all observations.

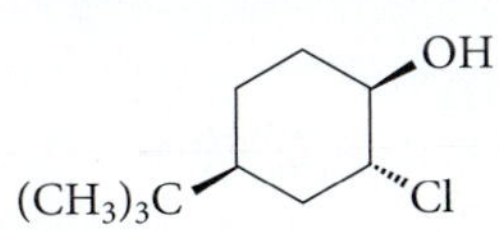

(1)

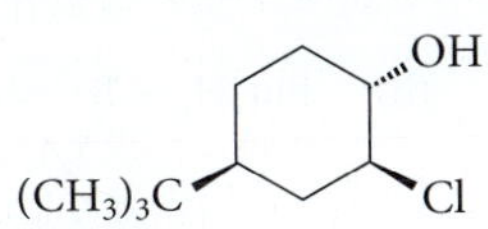

(2)

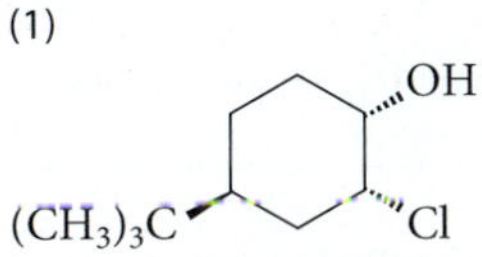

(3)

12.79 Two of the following compounds form epoxides readily when treated with ^-OH, one forms an epoxide slowly, and one does not form an epoxide at all. Identify the compound(s) in each category, and explain.

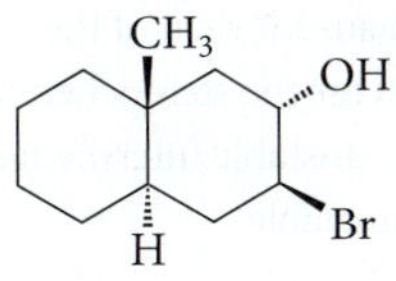

A

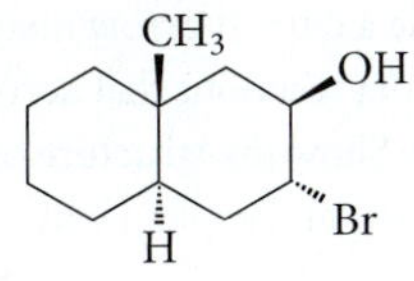

B

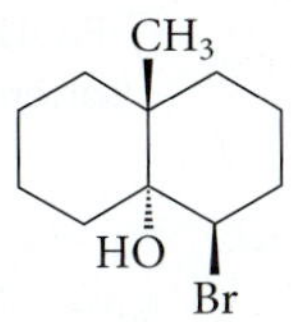

C

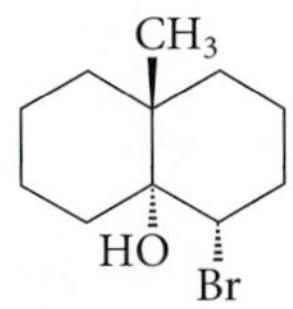

D

12.80 The reaction of δ-chlorobutyl phenyl sulfide in dioxane containing 20 *M* water at 100 °C gives a cyclic compound *X* that can be isolated.

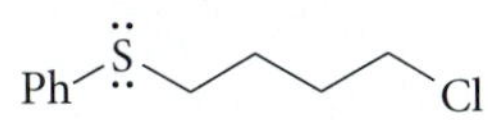

δ-chlorobutyl phenyl sulfide

$\xrightarrow[\text{dioxane, 100 °C}]{20\ M\ \text{water}}$ a cyclic compound *X*

This reaction is 21 times faster than the reaction of 1-chlorohexane under the same conditions.

(a) Deduce the structure of *X*, and give the curved-arrow mechanism for its formation.

(b) Calculate the proximity effect for this reaction.

(c) Suggest a reason that the proximity effect for this reaction is much smaller than the proximity effect calculated in Eq. 12.76b (Sec. 12.8B) for the similar reaction in Eq. 12.68b (Sec. 12.8A). (This proximity effect was calculated in Study Problem 12.5.)

12.81 Account for the following observations with a mechanism.

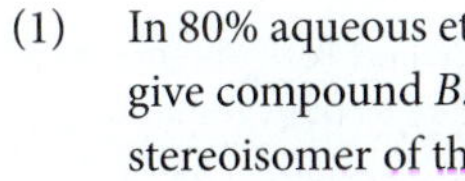

A

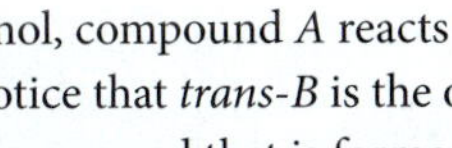

B

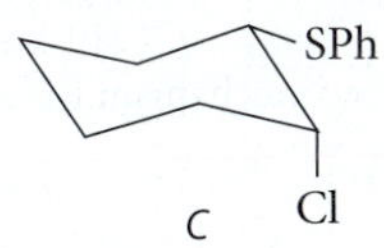

C

(1) In 80% aqueous ethanol, compound *A* reacts to give compound *B*. Notice that *trans*-*B* is the only stereoisomer of this compound that is formed.

(2) Optically active *A* gives completely racemic *B*.

(3) The reaction of *A* is about 10^5 times faster than the analogous substitution reactions of both its stereoisomer *C* and chlorocyclohexane.

12.82 Provide a curved-arrow mechanism for each of the following reactions that accounts for the stereochemical results. Show the structure of the unstable intermediate in each case, and explain why it is unstable.

(a)

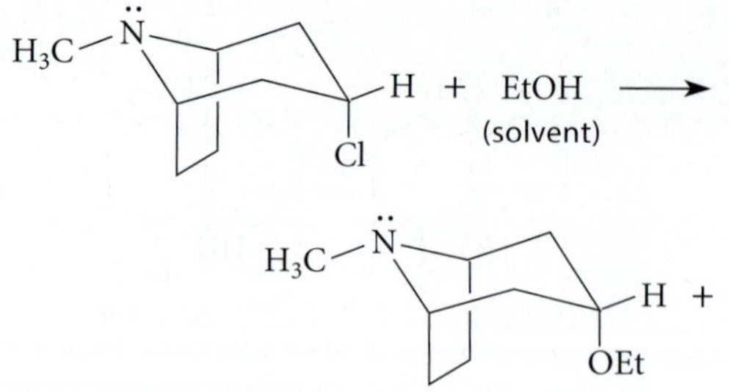

(*Hint:* Remember that twist-boat conformations of cyclohexane are in equilibrium with chair conformations.)

(b)

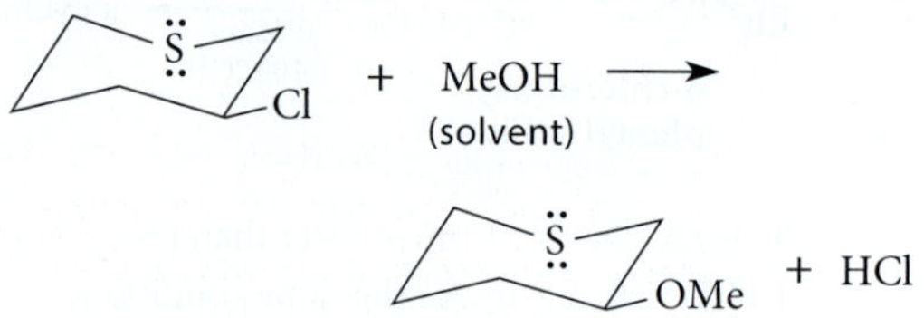

12.83 (a) Account for the stereochemical results in the following reaction with a mechanism. (*Hint:* See Study Problem 12.6, Sec. 12.8C.)

(2*S*,3*R*)-$CH_3CH(OH)CH(Br)CH_3$ + HBr ⟶ *meso*-$CH_3CH(Br)CH(Br)CH_3$ + H_2O

(2*S*,3*R*)-3-bromo-2-butanol **meso-2,3-dibromobutane**

(b) What stereochemical result would you expect if the 2*S*,3*S*-stereoisomer of 3-bromo-2-butanol undergoes the same reaction?

12.84 Draw a curved-arrow mechanism for each of the following conversions.

(a)

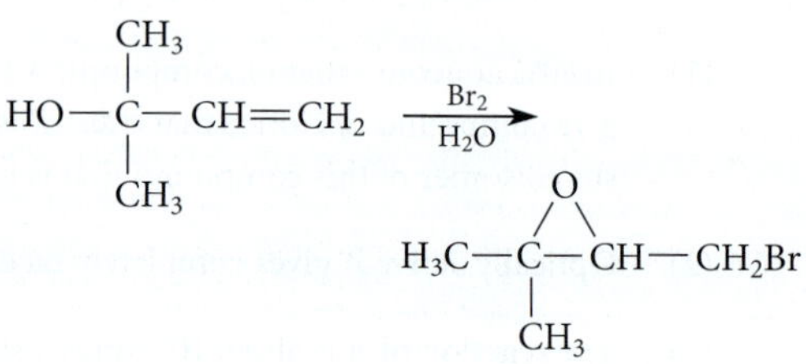

(b)

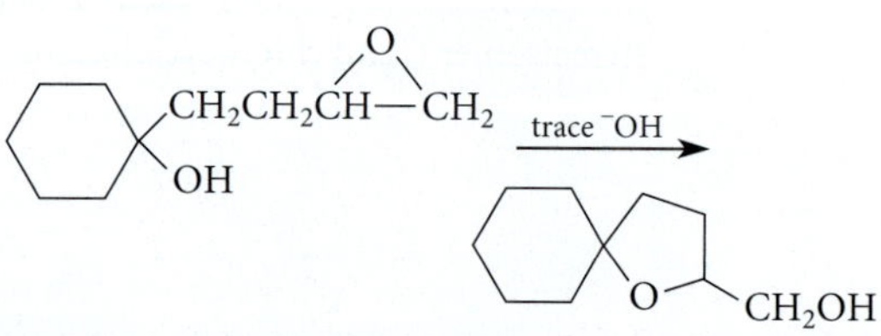

(c)

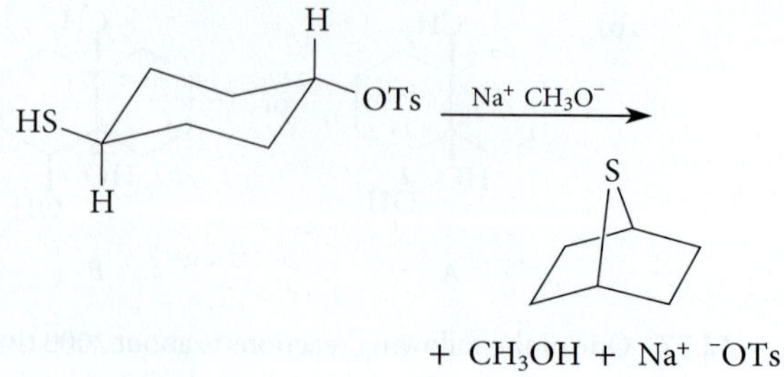

(*Hint:* The structure of the bicyclic product can also be redrawn as follows:)

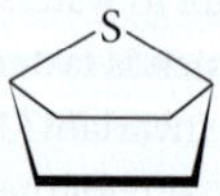

(d)

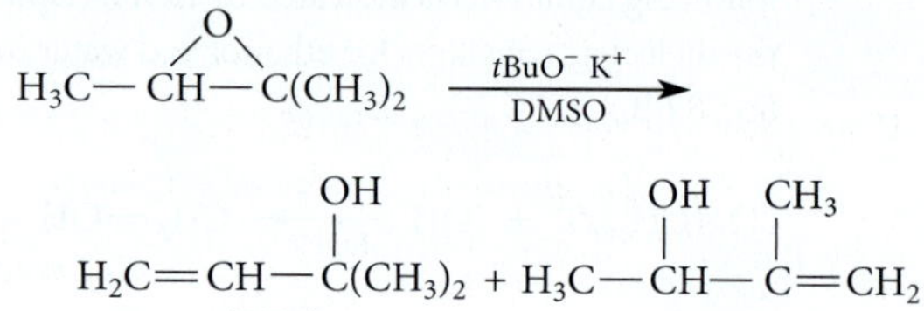

(80%) (15%)

(e)

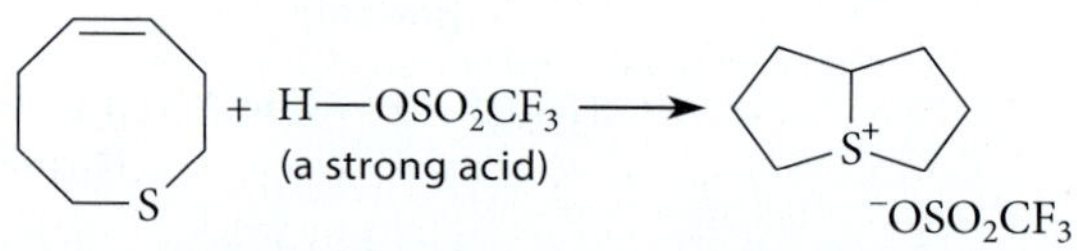

(f)

(g)

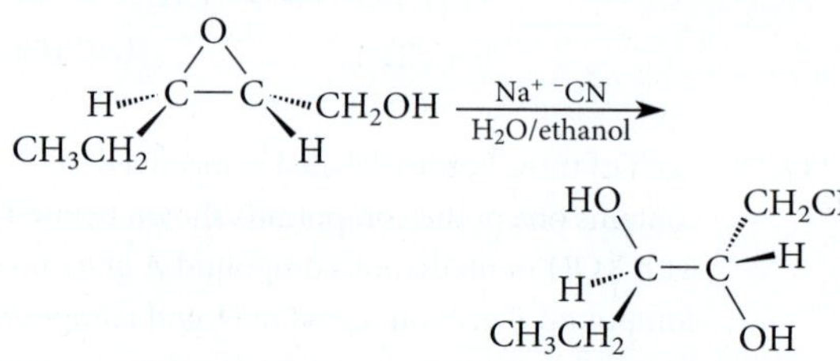

(*Hint:* Sodium cyanide in a protic solvent is fairly basic.)

(h)

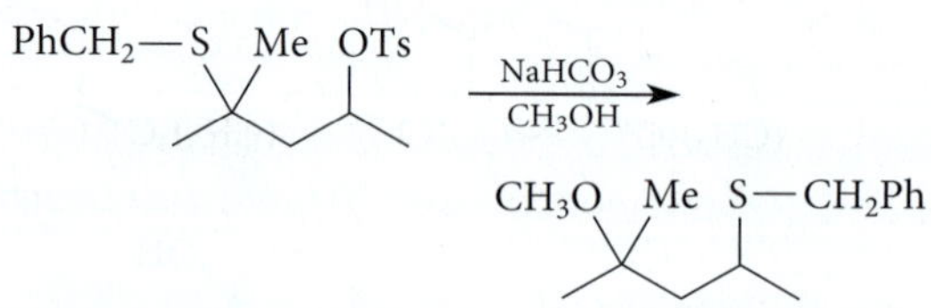

(57% yield)

(i)

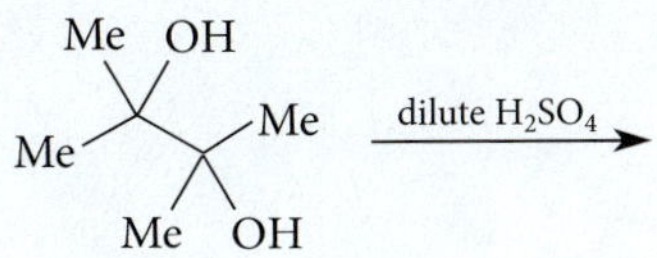

pinacol
(2,3-dimethyl-2,3-butanediol)

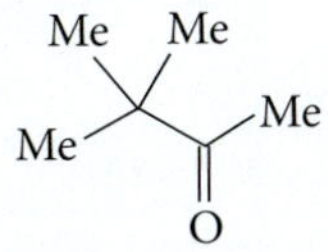

pinacolone
(3,3-dimethyl-2-butanone)

This reaction is called the *pinacol rearrangement.* [*Hints:* (1) Start the mechanism by protonating an —OH group and losing water. (2) A resonance-stabilized carbocation is involved in the mechanism.]

12.85 (a) As shown in **Fig. P12.85a**, 1,5-cyclooctadiene undergoes an electrophilic addition with SCl_2 to give compound *A*. (Notice the conformation of *A*, also shown.) Provide a curved-arrow mechanism for this transformation that accounts for the stereochemistry. (*Hint:* Start with a simple electrophilic addition of SCl_2 to one double bond.)

(b) Suggest a mechanism that accounts for the reaction of *A* shown in **Fig. P12.85b**, including the stereochemistry.

12.86 Secondary allylic alcohols react with the (+)-DET Sharpless asymmetric epoxidation reagent only if they have the (*S*) configuration at the α-carbon of the alcohol. Draw structures of the products for each of the following reactions.

(a)

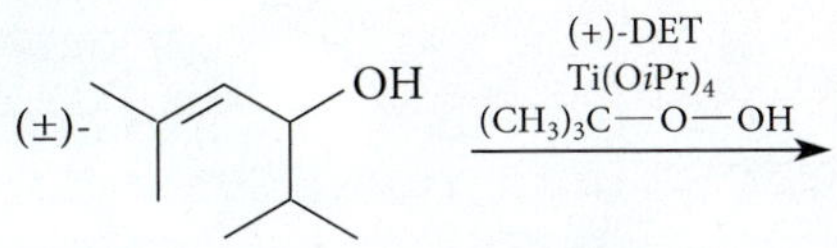

(b)

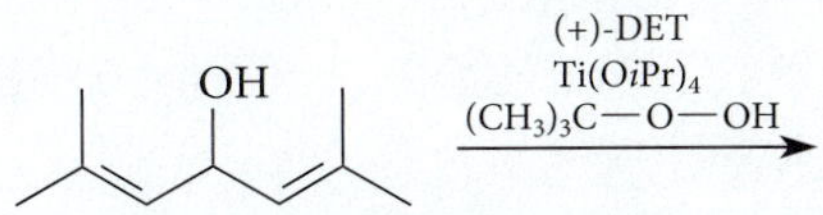

(c)

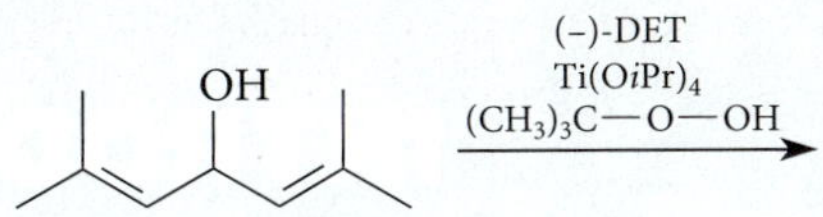

12.87 Compound *A*, C_8H_{16}, undergoes catalytic hydrogenation to give octane. When treated with *meta*-chloroperoxybenzoic acid, *A* gives an epoxide *B*, which, when treated with aqueous acid, gives a compound *C*, $C_8H_{18}O_2$, which can be resolved into enantiomers. When *A* is treated with OsO_4 followed by aqueous $NaHSO_3$, an *achiral* compound *D* (a stereoisomer of *C*) forms. Identify all compounds, including stereochemistry where appropriate.

(a)

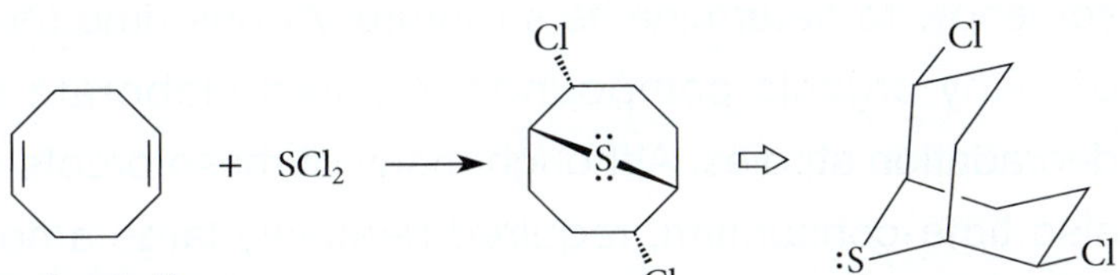

1,5-cyclooctadiene *A* conformation of *A*

(b)

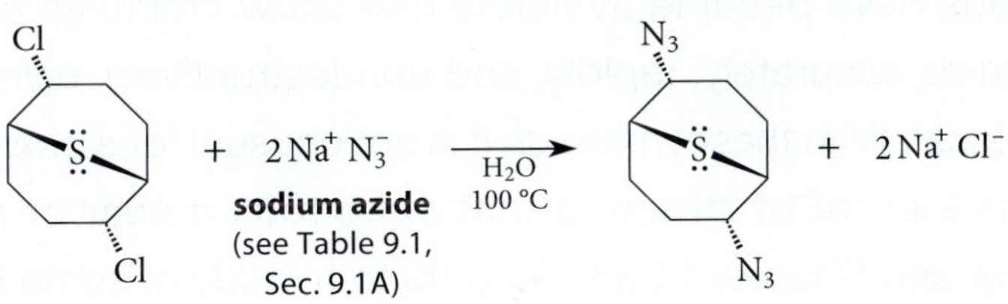

A

FIGURE P12.85

13 INTRODUCTION TO SPECTROSCOPY

Infrared Spectroscopy and Mass Spectrometry

Until this point in our study of organic chemistry, we have taken for granted that when a product of unknown structure is isolated from a reaction, it is possible somehow to determine its structure. At one time the structure determinations of many organic compounds required elaborate and laborious chemical-degradation studies. Although many of these proofs were ingenious, they were also time-consuming, required relatively large amounts of compounds, and were subject to a variety of errors. In the past 60–70 years, however, physical methods have become available that allow chemists to determine molecular structures accurately, rapidly, and nondestructively, using very small quantities of material. With these methods it is not unusual for a chemist to do in 30 minutes or less a proof of structure that once took a year or more to perform. This chapter and Chapter 14 are devoted to a study of some of these methods.

13.1 INTRODUCTION TO SPECTROSCOPY

Fundamental to modern techniques of structure determination is the field of **spectroscopy**: the study of the interaction of matter and light (or other electromagnetic radiation). Spectroscopy has been immensely important to many areas of chemistry and physics. For example, much of what is known about orbitals and bonding comes from spectroscopy. But spectroscopy is also important to the laboratory organic chemist because *it can be used to determine unknown molecular structures*. Although this presentation of spectroscopy focuses largely on its applications, some fundamentals of spectroscopy theory must be considered first.

A. Electromagnetic Radiation

Visible light is one type of **electromagnetic radiation**. Most of us are familiar with the notion that light is a *wave motion.* In other words, light can be thought of as an oscillation, like a sine or cosine wave. What is actually oscillating in a light wave? Electromagnetic radiation contains oscillating electric and magnetic field vectors that are perpendicular to each other (**Fig. 13.1**). The notion of a "field" may not be familiar. An *electric field* is capable of transferring energy to electrical charges, including charged atoms. This electrical aspect of electromagnetic radiation, as we shall see, is important in infrared (IR) spectroscopy. A *magnetic field* is capable of transferring energy to magnetic dipoles, which we can think of as tiny bar magnets. The magnetic aspect of electromagnetic radiation is important in nuclear magnetic resonance (NMR) spectroscopy, which we present in Chapter 14. Other common forms of electromagnetic radiation are X-rays, ultraviolet radiation (UV, the radiation from a tanning lamp), infrared radiation (IR, the radiation from a heat lamp), microwaves (used in radar and microwave ovens), and radiofrequency waves (rf, used to carry AM and FM radio and television signals).

Electromagnetic radiation propagates through space with a characteristic velocity, called the "speed of light," *c*, which is about 3×10^8 m s^{-1}, or 3×10^{10} cm s^{-1}. Conceptually, this means that if we were to "ride" the wave at a particular point—say, on a particular peak—we would be moving through space at the speed of light.

An important aspect of electromagnetic radiation is its **wavelength**, abbreviated with the Greek letter *lambda* (λ). This is the distance between successive peaks or successive troughs (Fig. 13.1). The various forms of electromagnetic radiation are fundamentally the same but differ in their wavelengths. The most common manifestation of wavelength to the human eye is the phenomenon of *color.* For example, blue light has a smaller wavelength than red light. The unit of distance conventionally used to express wavelength depends on the type of radiation. For example, the wavelengths of ultraviolet and visible radiation are typically given in either Ångstroms (Å) or nanometers (nm). One nm is 10^{-9} meter, and one Å is 10^{-10} meter. Red light has $\lambda = 6800$ Å (680 nm), and blue light has $\lambda = 4800$ Å (480 nm). Ultraviolet radiation has smaller wavelengths than blue light, and microwaves have much greater wavelengths than visible light.

Closely related to the wavelength of a wave is its *frequency.* The concept of frequency is necessary because electromagnetic waves are not stationary, but propagate through space with a characteristic velocity *c*. The **frequency** of a wave is the number of wavelengths that pass a point per unit time when the wave is propagated through space. The conventional symbol for frequency

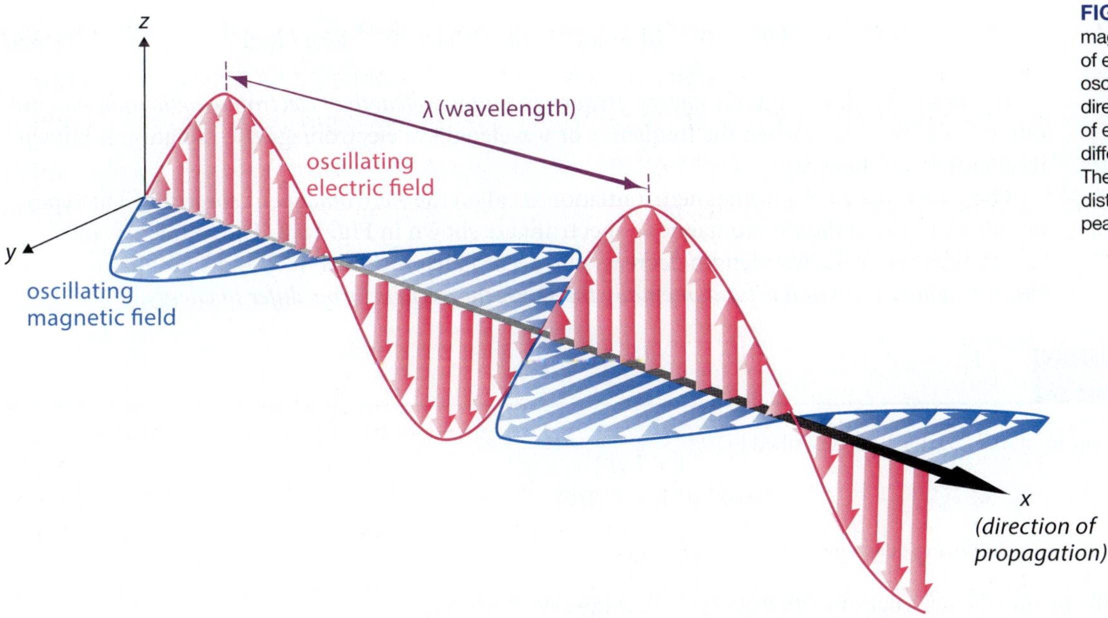

FIGURE 13.1 Electromagnetic radiation consists of electric and magnetic fields oscillating in perpendicular directions. The various forms of electromagnetic radiation differ only in their wavelengths. The wavelength (λ) is the distance between successive peaks or successive troughs.

is the Greek letter *nu* (ν). This is *not* the same as an italic *v*, even though the two letters look similar. The frequency ν of any wave with wavelength λ is

$$\nu = \frac{c}{\lambda} \tag{13.1}$$

in which c = the velocity of light. Because λ has the dimensions of length, ν has the dimensions of s^{-1}, a unit more often called *cycles per second* (cps), or *hertz* (Hz). For example, the frequency of red light is

$$\nu = \left(\frac{3 \times 10^8 \text{ m s}^{-1}}{6800 \text{ Å}}\right)\left(\frac{10^{10} \text{ Å}}{\text{m}}\right)$$

$$= 4.4 \times 10^{14} \text{ s}^{-1} = 4.4 \times 10^{14} \text{ Hz} \tag{13.2}$$

This means that 4.4×10^{14} wavelengths of red light pass a given point in 1 second.

Focused Problem

13.1 Calculate the frequency of

(a) Infrared radiation with $\lambda = 9 \times 10^{-6}$m

(b) Blue light with λ = 4800 Å

Although light can be described as a wave, it also shows particlelike behavior. The light particle is called a **photon**. The relationship between the energy of a photon and the wavelength or frequency of light is a fundamental law of physics:

$$E = h\nu = \frac{hc}{\lambda} \tag{13.3}$$

In this equation, h is *Planck's constant.* Planck's constant is a universal constant that has the value

$$h = 6.626 \times 10^{-34} \text{ J s} = 6.626 \times 10^{-37} \text{ kJ s} = 1.58 \times 10^{-37} \text{ kcal s} \tag{13.4}$$

For a mole of photons (that is, 6.022×10^{23} photons), Planck's constant has the value

$$h = 3.99 \times 10^{-13} \text{ kJ s mol}^{-1} \quad \text{or} \quad 9.54 \times 10^{-14} \text{ kcal s mol}^{-1} \tag{13.5}$$

Equation 13.3 shows that *the energy, frequency, and wavelength of electromagnetic radiation are interrelated.* Therefore, when the frequency or wavelength of electromagnetic radiation is known, its energy is also known.

The total range of electromagnetic radiation is called the *electromagnetic spectrum.* The types of radiation within the electromagnetic spectrum are shown in **Fig. 13.2.** The frequency and energy increase as the wavelength decreases, in accordance with Eqs. 13.1 and 13.3. *All electromagnetic radiation is fundamentally the same; the various forms differ in energy.*

Focused Problems

13.2 Calculate the energy in kJ mol^{-1} of the light described in

(a) Focused Problem 13.1(a)

(b) Focused Problem 13.1(b)

13.3 Use Fig. 13.2 to answer the following questions.

(a) How does the energy of X-rays compare with that of blue light (greater or smaller)?

(b) How does the energy of radar waves compare with that of red light (greater or smaller)?

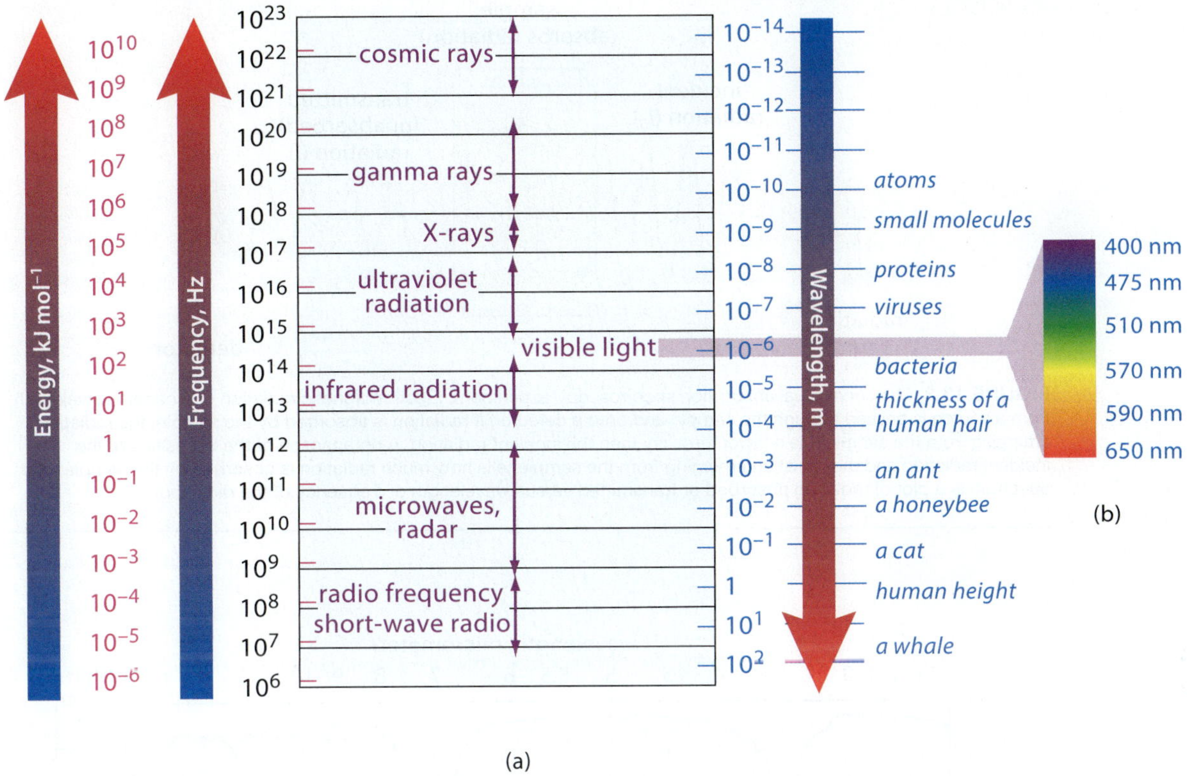

FIGURE 13.2 The electromagnetic spectrum. (a) The various forms of electromagnetic radiation as a function of energy (red scale, red ticks), frequency (black scale, black lines), and wavelength (blue scale, blue ticks). The wavelengths are compared with the lengths of various objects (blue italics). Notice the inverse relationship of wavelength with both energy and frequency—that is, longer wavelengths correspond to smaller energies and frequencies, as Eq. 13.3 indicates. (b) The range for visible light is expanded, and the numbers are the approximate wavelengths of the different colors in nanometers. (1 nanometer = 10^{-9} meter.)

B. Absorption Spectroscopy

The most common type of spectroscopy used for structure determination is *absorption spectroscopy.* The basis of absorption spectroscopy is that matter can absorb energy from electromagnetic radiation, and *the amount of absorption varies with the wavelength of the radiation used.* In an **absorption spectroscopy** experiment, the absorption of electromagnetic radiation is determined as a function of wavelength, frequency, or energy in an instrument called a **spectrophotometer** or **spectrometer**. The basic idea of an absorption spectroscopy experiment is shown schematically in **Fig. 13.3**. The experiment requires, first, a *source* of electromagnetic radiation. (If the experiment measures the absorption of visible light, the source could be a common light bulb.) The material to be examined, the *sample*, is placed in the radiation beam. A *detector* measures the intensity of the radiation that passes through the sample unabsorbed; when this intensity is subtracted from the intensity of the source, the amount of radiation absorbed by the sample is known. The wavelength of the radiation falling on the sample is then varied, and the radiation absorbed at each wavelength is recorded as a graph of either radiation transmitted or radiation absorbed versus wavelength or frequency. This graph is commonly called a **spectrum** of the sample.

The infrared spectrum of the hydrocarbon nonane, $CH_3(CH_2)_7CH_3$, is an example of such a graph. This spectrum is shown in **Fig. 13.4**. This spectrum is a plot of the radiation transmitted by a sample of nonane over a range of wavelengths in the infrared region. (We discuss the interpretation of such a spectrum in more detail in the next section.)

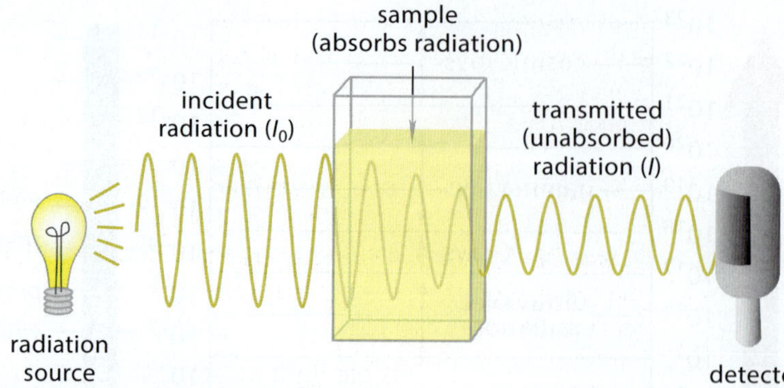

FIGURE 13.3 The conceptual absorption spectroscopy experiment. Electromagnetic radiation of a certain wavelength from a source is passed through the sample and onto a detector. If radiation is absorbed by the sample, the radiation emerging from the sample has a lower intensity than the incident radiation. A comparison of the intensities of the incident radiation and the radiation emerging from the sample tells how much radiation is absorbed by the sample. The spectrum is a plot of radiation absorbed or transmitted versus wavelength or frequency of the radiation.

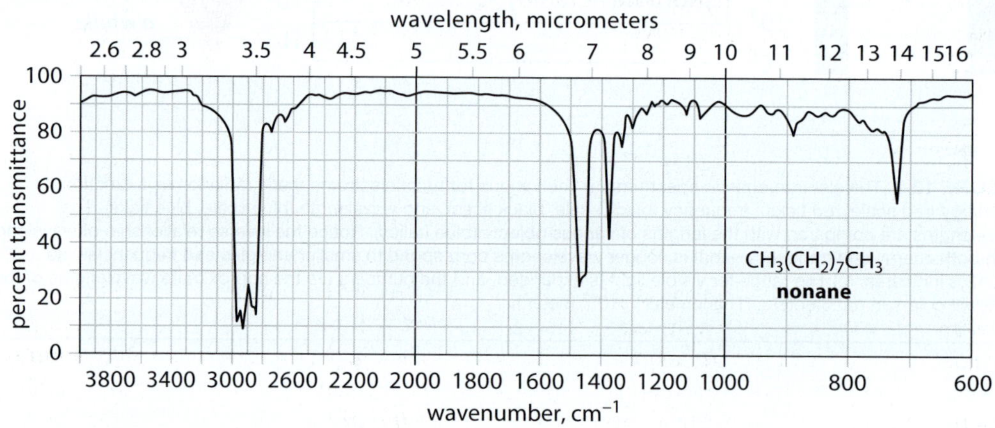

FIGURE 13.4 Infrared spectrum of nonane. The light transmitted through a sample of nonane is plotted as a function of wavelength (upper horizontal axis) or wavenumber (lower horizontal axis). (Wavenumber, which is proportional to frequency, is defined by Eq. 13.6 in Sec. 13.2A.) Absorptions are indicated by the inverted peaks. In this text, the wavenumber and wavelength scales are divided into three different regions (indicated by the vertical black lines) in which the horizontal scale is different.

An Everyday Analogy for a Spectroscopy Experiment

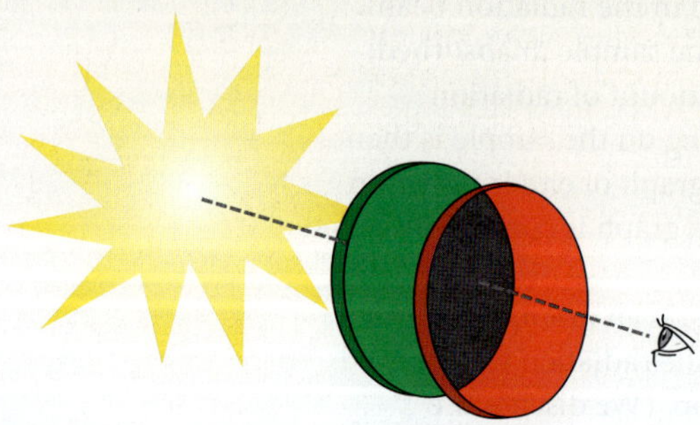

You don't need experience with a spectrophotometer to appreciate the basic idea of a spectroscopy experiment. Imagine holding a piece of green glass up to the white light of the Sun. The Sun is the source, the glass is the sample, your eyes are the detector, and your brain provides the spectrum. White light is a mixture of all wavelengths. The glass appears green because only green light is transmitted through the glass; the other colors (wavelengths) in white light are absorbed. If you view the same green glass through a red glass, no light is transmitted to your eyes—the red glass looks black—because the red glass absorbs the green light.

An important aspect of spectroscopy for the chemist is that *the spectrum of a compound is a function of its structure*. For this reason, spectroscopy can be used to determine structure. Chemists use many types of spectroscopy for this purpose. The three types of greatest use, and the general type of information each provides, are as follows:

1. *Infrared (IR) spectroscopy* provides information about what functional groups are present.

2. *Nuclear magnetic resonance (NMR) spectroscopy* provides information on the number, connectivity, and functional-group environment of carbons and hydrogens as well as certain other nuclei such as fluorine and phosphorus.

3. *Ultraviolet–visible (UV–vis) spectroscopy (often called simply UV spectroscopy)* provides information about the types of π-electron systems that are present.

These types of spectroscopy differ conceptually only in the frequency of radiation used, although their practical aspects are quite different. A fourth physical technique, *mass spectrometry*, which allows us to determine molecular masses, is also widely used for structure determination. Mass spectrometry is not a type of absorption spectroscopy, so it is fundamentally different from IR, NMR, and UV spectroscopy.

The remainder of this chapter is devoted to descriptions of IR spectroscopy and mass spectrometry. NMR spectroscopy is covered in Chapter 14 and UV spectroscopy in Sec. 15.2.

13.2 INFRARED SPECTROSCOPY

A. The Infrared Spectrum

An infrared spectrum, like any absorption spectrum, is a record of the light absorbed by a substance as a function of wavelength. In IR spectroscopy, the conventional unit of wavelength is the *micrometer.* A **micrometer**, abbreviated μm, is 10^{-6} meter. In practice, the absorption of infrared radiation with wavelengths between 2.5 μm and 20 μm is of greatest interest to organic chemists. The IR spectrum is measured in an instrument called an *infrared spectrometer*, described briefly in Sec. 13.5. Let's consider the details of an IR spectrum by returning to the spectrum of nonane in Fig. 13.4.

Reexamine this spectrum. The quantity plotted on the lower horizontal axis is the **wavenumber** $\tilde{\nu}$ of the light. (The symbol $\tilde{\nu}$ is pronounced "*nu-bar.*") The wavenumber is simply the inverse of the wavelength:

$$\text{wavenumber} = \tilde{\nu} = \frac{1}{\lambda} \tag{13.6}$$

If the unit of wavelength is the μm, then the consistent unit of wavenumber is therefore μm^{-1}, or reciprocal micrometers. Physically, the wavenumber in μm^{-1} is the number of wavelengths contained in 1 μm. For various practical reasons, the conventional unit for the wavenumber used in IR spectroscopy is *reciprocal centimeters* or *inverse centimeters* (cm^{-1}). To apply Eq. 13.6 with these units, we must include the conversion factor 10^4 μm cm^{-1}. In conventional units, then, Eq. 13.6 becomes

$$\tilde{\nu}\,(\text{cm}^{-1}) = \frac{10^4\ \mu\text{m cm}^{-1}}{\lambda\ (\mu\text{m})} \tag{13.7}$$

The micrometer and the reciprocal centimeter are the units used in Fig. 13.4. The wavenumber of the infrared radiation in cm^{-1} is plotted on the lower horizontal axis, and the wavelength in μm is plotted on the upper horizontal axis. (The vertical lines in the grid correspond to the wavenumber scale.) For example, according to Eq. 13.7, a wavelength of 10 μm corresponds to a wavenumber of 1000 cm^{-1}. You can see that in Fig. 13.4, 10 μm and 1000 cm^{-1} correspond to the same point on the horizontal axis.

Notice in Fig. 13.4 that wavenumber, across the bottom of the spectrum, increases to the left; wavelength, across the top of the spectrum, increases to the right. Finally, notice also that the *wavenumber scale is divided into three distinct regions in which the linear scale is different;* the changes in scale occur at 2200 cm^{-1} and 1000 cm^{-1}. The positions of the scale breaks can vary with different spectrometers.

The relationships among wavenumber, energy, and frequency can be derived by combining Eqs. 13.1 and 13.3 with Eq. 13.6, the definition of wavenumber.

$$\nu = \frac{c}{\lambda} = c\tilde{\nu} \tag{13.8a}$$

$$E = h\nu = \frac{hc}{\lambda} = hc\tilde{\nu} \tag{13.8b}$$

(These equations must be used with consistent units, such as λ in m, $\tilde{\nu}$ in m^{-1}, and c in m s^{-1}.) According to Eq. 13.8a, the frequency ν and the wavenumber $\tilde{\nu}$ are proportional. Because of this proportionality, you may hear the wavenumber loosely referred to as a frequency.

Now consider the vertical axis of Fig. 13.4. The quantity plotted on the vertical axis is *percent transmittance.* The *transmittance* T is defined as the ratio of the intensity I of the light emerging from the sample to the intensity I_0 of the light entering the sample (see Fig. 13.3). Transmittance by definition is a fraction that can have values between 0 and 1.

$$\text{transmittance} = T = \frac{I}{I_0} \tag{13.9a}$$

The percent transmittance is $100\% \times T$.

$$\text{percent transmittance} = \%T = 100\% \times T \tag{13.9b}$$

For example, if the sample absorbs all of the radiation, then none is transmitted, and $I = 0$; the sample has 0% transmittance. If the sample absorbs no radiation, then all of the radiation is transmitted, and $I = I_0$; the sample has 100% transmittance. Consequently, *absorptions in the IR spectrum are registered as downward deflections*—that is, "upside-down peaks." In Fig. 13.4, the absorptions in the spectrum of nonane occur at about 2850–2980, 1470, 1380, and 720 cm^{-1}. The presentation of peaks in the transmittance mode is largely for historical reasons—that is, early instruments produced data that were most conveniently plotted this way, and the practice simply hasn't changed.

In other forms of spectroscopy, absorptions are presented as "normal peaks"—that is, upward deflections. Some IR instruments produce data plotted this way as well. The quantity plotted in this case is the *absorbance A*. Absorbance is the *negative* logarithm of transmittance.

$$\text{absorbance} = A = -\log T = -\log\left(\frac{I}{I_0}\right) = \log\left(\frac{I_0}{I}\right) \tag{13.9c}$$

For example, a sample that absorbs half of the incident light has $\%T = 50$, and it has an absorbance of $-\log(0.5) = 0.3$.

Sometimes an IR spectrum is not presented in graphical form, but is summarized completely or in part using descriptions of the *positions* of the important peaks. Intensities are often expressed qualitatively using the designations vs (very strong), s (strong), m (moderate), or w (weak). Some peaks are narrow (or sharp, abbreviated sh), whereas others are wide (or broad, abbreviated br). For example, the spectrum of nonane can be summarized as follows:

$$\tilde{\nu}(\text{cm}^{-1})\text{: 2980–2850 (s); 1470 (m); 1380 (m); 720 (w)}$$

Focused Problems

13.4 (a) What is the wavenumber (in cm^{-1}) of infrared radiation with a wavelength of 6.0 μm?

(b) What is the wavelength of infrared radiation with a wavenumber of $1720\ cm^{-1}$?

13.5 In the infrared spectrum of nonane in Fig. 13.4, calculate the absorbance of the sharp peak at $1380\ cm^{-1}$.

B. The Physical Basis of IR Spectroscopy

The interpretation of an IR spectrum in terms of structure requires some understanding of why molecules absorb infrared radiation. The absorptions observed in an IR spectrum are the result of *vibrations* within a molecule. Atoms within a molecule are not stationary, but are constantly moving. For example, the C—H bonds in a typical organic compound undergo various oscillatory stretching and bending motions, called **bond vibrations**. One such vibration is the C—H stretching vibration. A useful analogy to the C—H stretching vibration is the stretching and compression of a spring (**Fig. 13.5**). This vibration takes place with a certain frequency ν—that is, it occurs a certain number of times per second. Suppose that a C—H bond has a stretching frequency of 9×10^{13} times per second; this means that it undergoes a vibration every $1/(9\times10^{13})$ second or every 1.1×10^{-14} second.

The blue line in Fig. 13.5 shows that the stretching of the C—H bond over time describes a wave motion. It turns out that a wave of electromagnetic radiation can transfer its energy to the vibrational wave motion of the C—H bond only if *there is an exact match between the frequency of the radiation and the frequency of the vibration*. For example, if a C—H vibration has a frequency of $9\times10^{13}\ s^{-1}$, then it will absorb energy from radiation with the same frequency. From the relationship $\lambda = c/\nu$ (Eq. 13.1), the radiation must then have a wavelength of

$$\lambda = (3\times10^{8}\ m\ s^{-1})/(9\ \times\ 10^{13}\ s^{-1}) = 3.33\times10^{-6}\ m = 3.33\ \mu m$$

The corresponding wavenumber of this radiation (Eq. 13.7) is $3000\ cm^{-1}$. When radiation of this wavelength interacts with a vibrating C—H bond, energy is absorbed and the intensity of the bond vibration increases. That is, after absorbing energy, the bond vibrates with the same frequency but with a larger *amplitude* (a larger stretch and tighter compression; **Fig. 13.6**). This absorption gives rise to the peak in the IR spectrum. Eventually, the bond returns to its normal, less intense, vibration, and when this happens, energy is released in the form of heat. This is why an infrared "heat lamp" makes your skin feel warm.

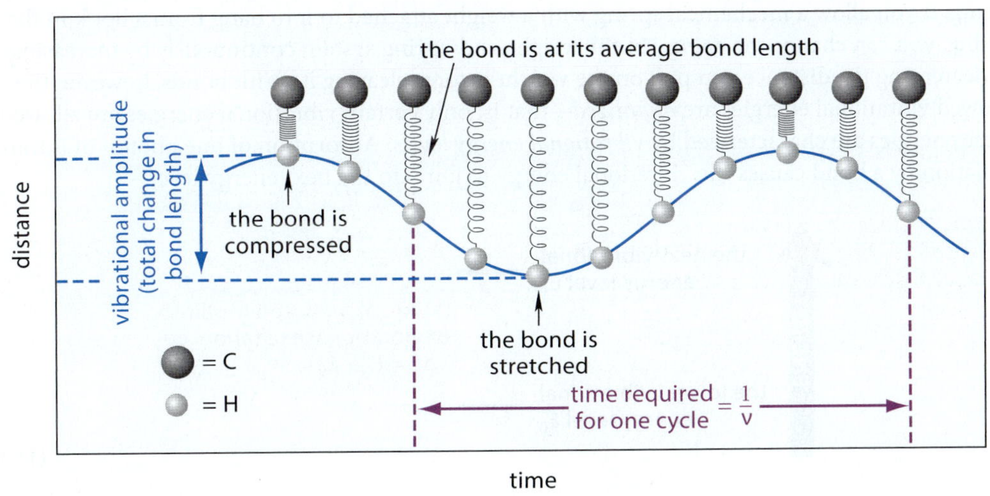

FIGURE 13.5 Chemical bonds undergo different types of vibrations. The one illustrated here is a stretching vibration. The bond, represented as a spring, is shown at various times. The bond stretches and compresses over time. The time required for one complete cycle of vibration is the reciprocal of the vibrational frequency.

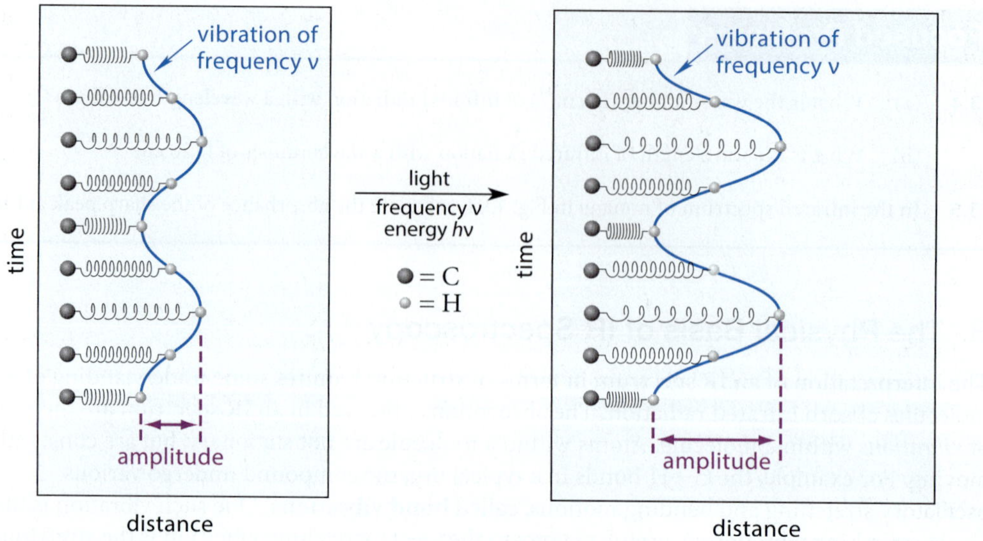

FIGURE 13.6 Light absorption by a bond vibration causes the bond (shown as a spring) to vibrate with larger amplitude. The frequency of the light must exactly equal the frequency of the bond vibration for absorption to occur. (The increase in amplitude is greatly exaggerated for emphasis.)

An Analogy to Energy Absorption by a Vibrating Bond

Imagine an unopened 12-ounce can of soda suspended from your finger with a rubber band by its pull-tab. If you move your finger up and down very slowly, the soda can scarcely moves. However, if you increase the rate of oscillation of your finger gradually, at some point the soda can will start to move up and down vigorously. The motion of the soda can will decrease as you move your finger more rapidly. The motion of your finger is analogous to electromagnetic radiation, and the rubber band–soda can combination is analogous to the vibrating bond. Energy is absorbed from the motion of your finger by the rubber band–soda can oscillator only when its natural oscillation frequency matches the oscillation frequency of your finger. Likewise, a vibrating bond absorbs energy from electromagnetic radiation only when there is a frequency match between the two oscillating systems.

A famous example of the phenomenon described in the sidebar above was the Tacoma Narrows bridge, which opened in July 1, 1940, and collapsed on the following November 7, when 40-mile-per-hour winds caused the roadway to oscillate violently at its natural vibration frequency. The continuous absorption of energy from the wind increased the amplitude of the vibrations; as a result, the bridge collapsed. (The bridge had previously received the nickname "Galloping Gertie.") Fortunately, no one was injured, and the event led to the subsequent design of long-span bridges in which such oscillations were prevented from occurring.

Although the intuitive mechanical descriptions of infrared absorption are useful, bond vibrations are subject to quantum theory. Let's contrast our "bond spring" with a mechanical spring. If you allow a mechanical spring with a weight attached to it to hang from a hook in the ceiling, you can change the energy of the weight-and-spring system continuously by increasing or decreasing the distance you pull on the weight before releasing it. With bonds, however, the allowed vibrational energies are *quantized*—that is, only certain vibrational energies are allowed; these energies are characterized by *vibrational energy levels.* Absorption of one photon of infrared radiation by a bond causes the vibrational energy to jump to the next energy level.

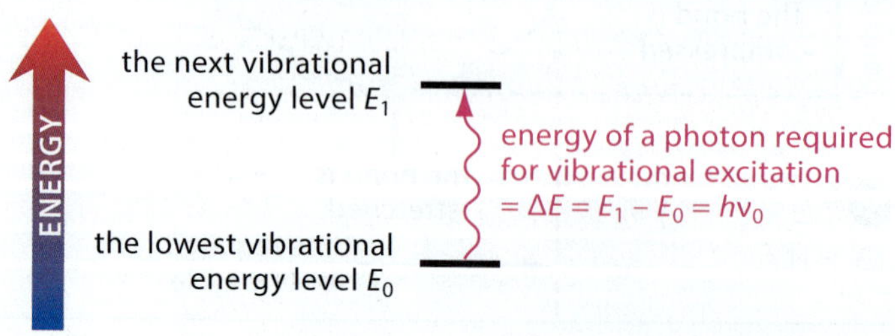

(13.10)

The energy of the photon required to bring about this transition equals the *difference* in the energy ΔE of the two vibrational levels. As the diagram in Display 13.10 shows, this difference equals $h\nu_0$, where ν_0 is the vibrational frequency of the bond. (See Further Exploration 13.1 in the *Study Guide and Solutions Manual* for the derivation of this result.) In other words, the frequency of the photon that brings about the vibrational transition is the same as the frequency of the bond vibration. If bond vibrations weren't quantized, a bond could absorb photons of any energy, and there would be no distinct peaks in the spectrum. Why different bonds have different vibrational frequencies (and therefore different IR absorption frequencies) is addressed in Sec. 13.3A.

FURTHER EXPLORATION 13.1 The Vibrating Bond in Quantum Theory

In summary:

1. Bonds vibrate with characteristic frequencies.
2. Absorption of energy from infrared radiation can occur only when the wavelength of the radiation and the wavelength of the bond vibration match.

Focused Problems

13.6 Given that a certain C—H bond vibration has a frequency of 4.41×10^{13} s^{-1}, which peak(s) in the IR spectrum of nonane (Fig. 13.4) would you assign to this vibration?

13.7 The physical basis of some carbon monoxide detectors is the infrared detection of the unique C≡O stretching vibration of carbon monoxide at 2143 cm^{-1}. How many times per second does this stretching vibration occur?

$$:\overset{-}{C}\equiv\overset{+}{O}:$$

carbon monoxide

13.3 INFRARED ABSORPTION AND CHEMICAL STRUCTURE

Each peak in the IR spectrum of a molecule corresponds to the absorption of energy by the vibration of a particular bond or group of bonds. IR spectroscopy is useful for the chemist because *in all compounds, a given type of functional group absorbs in the same general region of the IR spectrum*. The major regions of the IR spectrum are listed in **Table 13.1**.

The IR spectrum of a typical organic compound contains many more absorptions than can be readily interpreted. A major part of mastering IR spectroscopy is to learn which absorptions are important. Certain absorptions are *diagnostic*—that is, they indicate with reasonable certainty that a particular functional group is present. For example, an intense peak in the 1700–1750 cm^{-1} region indicates the presence of a carbonyl (C═O) group. Other peaks are *confirmatory*—that is, similar peaks can be found in other types of molecules, but their presence confirms a structural diagnosis made in other ways. For example, absorptions in the 1050–1200 cm^{-1} region of the IR spectrum due to a C—O bond could indicate the presence of an alcohol, an ether, an ester, or a carboxylic acid, among other things. However, if other evidence (perhaps obtained from other types of spectroscopy) suggests that the unknown molecule is, say, an ether, a peak in this region

TABLE 13.1 Regions of the Infrared Spectrum

Wavenumber range, cm^{-1}	Type of absorptions	Name of region
3400–2800	O—H, N—H, C—H stretching	
2250–2100	C≡N, C≡C stretching	Functional group
1850–1600	C═O, C═N, C═C stretching	
1600–1000	C—O, C—N, C—C stretching; various bending absorptions	Fingerprint
1000–600	C—H bending	C—H bending

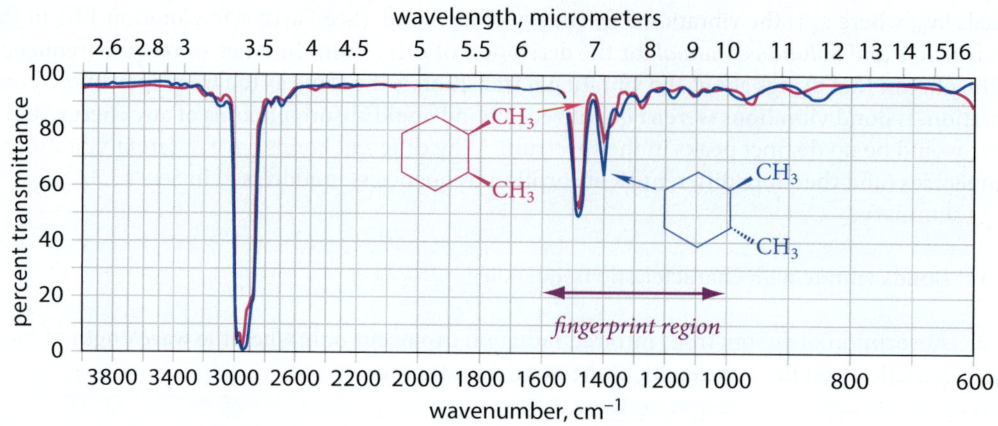

FIGURE 13.7 Comparison of the IR spectra of the cis (red) and trans (blue) stereoisomers of 1,2-dimethylcyclohexane. Although these spectra are very similar, discernible differences between them occur in the fingerprint region.

can support this diagnosis. In the sections that follow, we discuss the relatively few absorptions that are important in IR spectroscopy.

Rarely, if ever, does an IR spectrum completely define a structure; rather, it provides information that restricts the possible structures under consideration. Once the structure for an unknown compound has been deduced, a comparison of its IR spectrum with that of an authentic sample taken under the same conditions can be used as a criterion of identity. Even subtle differences in structure generally give discernible differences in the IR spectrum, particularly in the region between 1000 cm^{-1} and 1600 cm^{-1}. This point is illustrated by the superimposed IR spectra of the two diastereomers *cis*- and *trans*-1,2-dimethylcyclohexane (**Fig. 13.7**). The greatest differences between the two spectra occur in this region of the spectrum. (That these spectra are different is another illustration of the principle that diastereomers have different physical properties.) Even though absorptions in this region are generally not interpreted in detail, they serve as a valuable "molecular fingerprint." That is why this region of the spectrum is called the "fingerprint region" in Table 13.1.

A. Factors That Determine IR Absorption Position

One approach to the use of IR spectroscopy is simply to memorize the wavenumbers at which characteristic functional group absorbances appear and to look for peaks at these positions in the spectrum of unknown structures. However, you can use IR spectroscopy much more intelligently and learn the important peak positions much more easily if you understand a little more about the physical basis of IR spectroscopy. Two aspects of IR absorption peaks are particularly important. First is the *position* of the peak—the wavenumber or wavelength at which it occurs. Second is the *intensity* of the peak—how strong it is. Let's consider each of these aspects in turn.

What factors govern the position of IR absorption? Three considerations are most important:

1. Strength of the bond
2. Masses of the atoms involved in the bond
3. The type of vibration being observed

Hooke's law, which comes from the treatment of the vibrating spring by classical physics, nicely accounts for the first two of these effects. Let two atoms with masses M and m (where $M > m$) be connected by a bond, which we treat as if it were a spring. The tightness of the spring (bond) is described by a *force constant*, κ: the larger the force constant, the tighter the spring.

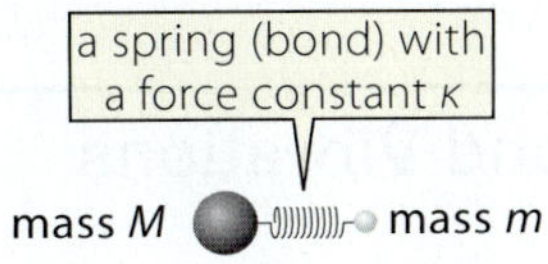

(13.11)

Equation 13.12 describes the dependence of the vibrational wavenumber on the masses and the force constant:

$$\tilde{\nu} = \frac{1}{2\pi c}\sqrt{\frac{\kappa(m+M)}{mM}} \qquad (13.12)$$

It takes more force to stretch a tighter spring than it does to stretch a looser spring. This is why *stronger bonds have larger force constants*. Because the force constant appears in the numerator of Eq. 13.12, objects connected by a tighter spring vibrate with a higher frequency or wavenumber. Likewise, atoms connected by a stronger bond also vibrate at higher frequency. A simple measure of bond strength is the energy required to break the bond, which is the *bond dissociation energy* (Table 10.2, Sec. 10.1E). It follows, then, that *the higher the bond dissociation energy, the stronger the bond*. Therefore, *the IR absorptions of stronger bonds—bonds with greater bond dissociation energies—occur at higher wavenumber*. Study Problem 13.1 illustrates this effect.

Study Problem 13.1

The typical stretching frequency for a carbon–carbon double bond is 1650 cm^{-1}. Estimate the stretching frequency of a carbon–carbon triple bond.

Solution Use Eq. 13.12. We are given $\tilde{\nu}$ for a double bond, which we'll call $\tilde{\nu}_2$, and we're asked to estimate $\tilde{\nu}_3$, the stretching frequency of the triple bond. Let the force constant for a double bond be κ_2, and that of a triple bond be κ_3. Now take the ratio of $\tilde{\nu}_3$ and $\tilde{\nu}_2$ as given by Eq. 13.12:

$$\frac{\tilde{\nu}_3}{\tilde{\nu}_2} = \sqrt{\frac{\kappa_3}{\kappa_2}} \qquad (13.13a)$$

All of the mass terms cancel because they are the same in both cases—the mass of a carbon atom. We could complete the problem if we knew the force constants, but these aren't given. Let's apply some intuition. As we just learned, force constants are proportional to bond dissociation energies. We could look these up in Table 10.2, but let's simply assume that the relative strengths of triple and double bonds are in the ratio 3:2. If this were so, then Eq. 13.13a becomes

$$\frac{\tilde{\nu}_3}{\tilde{\nu}_2} = \sqrt{\frac{3}{2}} = 1.22 \qquad \textbf{(13.13b)}$$

With $\tilde{\nu}_2 = 1650\ cm^{-1}$, we then estimate $\tilde{\nu}_3$ to be $(1.22)(1650\ cm^{-1}) = 2013\ cm^{-1}$. How close are we? See for yourself by jumping ahead to Fig. 13.12 (Sec. 13.4D, which shows the C≡C stretching absorption of an alkyne).

Now let's consider the effect of mass on the stretching frequency of a bond. Equation 13.12 also describes this effect. However, a special case of this equation is very important. Suppose the two atoms connected by a bond differ significantly in mass (for example, a carbon and a hydrogen in a C—H bond). *The vibration frequency for a bond between two atoms of different mass depends more on the mass of the lighter object than on the mass of the heavier one.* The following analogy illustrates this point.

Analogy for the Effect of Mass on Bond Vibrations

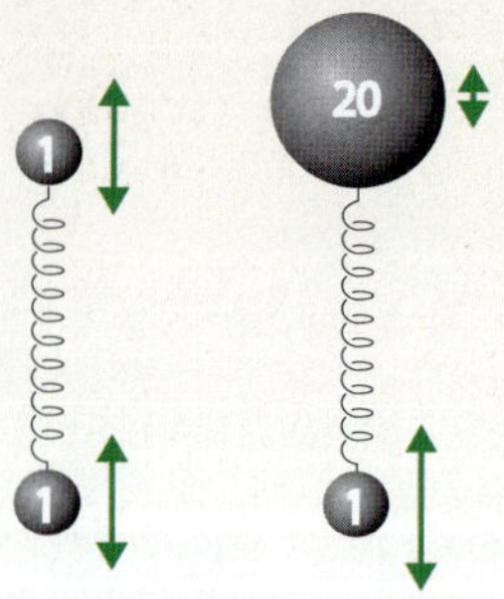

Imagine three situations: two identical light rubber balls connected by a spring, a rubber ball connected to a heavy cannonball with the same spring, and a rubber ball connected to the Empire State Building by the same spring. When the spring connecting the two rubber balls is stretched and released, both balls oscillate. That is, both masses are involved in the vibration. When the spring connecting the rubber ball and the cannonball is stretched and released, the rubber ball oscillates, and the cannonball remains almost stationary. When the spring connecting the rubber ball and the Empire State Building is stretched and released, only the rubber ball appears to oscillate; the motion of the building is imperceptible. In the last two cases, the vibrational frequencies are virtually identical, even though the larger masses differ by orders of magnitude. Therefore, changing the larger mass has no effect on the vibration frequency. If we now attach a cannonball at one end of the spring and leave the other end attached to the building, the frequency is significantly reduced, even though we haven't changed the large mass at all. Therefore, the smaller mass controls the vibration frequency.

Now let's use Eq. 13.12 to verify that the smaller mass determines the vibration frequency. Suppose in this equation that $M >> m$. In such a case, the smaller mass can be ignored in the numerator of Eq. 13.12, and the larger mass M then cancels and vanishes from the equation, leaving only the smaller mass m in the denominator. Equation 13.12 then becomes

$$\tilde{\nu} = \frac{1}{2\pi c}\sqrt{\frac{\kappa}{m}} \quad \text{for } M >> m \tag{13.14}$$

According to this equation, the vibration frequency of a bond between a heavy and a light atom depends primarily on the mass of the light atom, as our preceding intuitive argument suggested. This is why C—H, O—H, and N—H bonds all absorb in the same general region of the IR spectrum, and why C═O, C═N, and C═C bonds absorb in the same general region (see Table 13.1). In fact, the differences that do exist between the vibrational frequencies of these bonds are not primarily mass effects, but mostly bond-strength effects.

One of the best illustrations of the mass effect on vibrational frequency is a comparison of the frequencies of bond vibrations for an X—H bond with its deuterium-substituted analog X—D. (Deuterium, or ^{2}H, is the isotope of hydrogen with atomic mass = 2.) This effect is the subject of Focused Problem 13.9.

Focused Problems

13.8 The following bonds all have IR stretching absorptions in the 4000–2900 cm^{-1} region of the spectrum. Rank the following bonds in order of decreasing stretching frequencies, greatest first, and explain your reasoning. (*Hint:* Consult Table 10.2 in Sec. 10.1E.)

C—H, O—H, N—H, F—H

13.9 The ═C—H stretching absorption of 2-methyl-1-pentene is observed at 3090 cm^{-1}. If the hydrogen were replaced by deuterium, at what wavenumber would the ═C—D stretching absorption be observed? Explain. (Assume that the force constants for the C—H and C—D bonds are identical.)

The third factor that affects the absorption frequency is the *type of vibration*. The two general types of vibrations in molecules are *stretching vibrations* and *bending vibrations*. A **stretching vibration** occurs along the line of the chemical bond. A **bending vibration** is any vibration that does not occur along the line of the chemical bond. A bending vibration can be envisioned as a ball hanging on a spring and swinging side to side. In general, *bending vibrations occur at lower frequencies (higher wavelengths) than stretching vibrations of the same groups.*

An Analogy for Bending Vibrations

Imagine a ball attached to a stiff spring attached to the ceiling. A gentle tap makes it swing back and forth. It takes considerably more energy to stretch the spring.

Because the energy required to set the spring in motion is proportional to its frequency, the swinging (bending) motion has a lower frequency than the stretching motion.

The only possible type of vibration in a diatomic molecule (for example, H—F) is a stretching vibration. However, when a molecule contains more than two atoms, both stretching and bending vibrations are possible. The allowed vibrations of a molecule are called its **normal vibrational modes**. The normal vibrational modes for a $—CH_2—$ group are shown in **Fig. 13.8**. They serve as models for the kinds of vibrations that can be expected for other groups in organic molecules. The bending vibrations can be such that the hydrogens move *in the plane* of the $—CH_2—$ group or *out of the plane* of the $—CH_2—$ group. Furthermore, stretching and bending vibrations can be *symmetrical* or *unsymmetrical* with respect to a plane between the two vibrating hydrogens. The bending motions have been given names (rocking, wagging, and so on) that describe the type of motion involved. Each of these motions occurs with a particular frequency and can have an associated peak in the IR spectrum (although some peaks are weak or absent for reasons to be considered later). The $—CH_2—$ groups in a typical organic molecule undergo all of these motions simultaneously. That is, while the C—H bonds are stretching, they are also bending. The IR spectrum of nonane (Fig. 13.4, Sec. 13.1B) shows absorptions for both C—H stretching

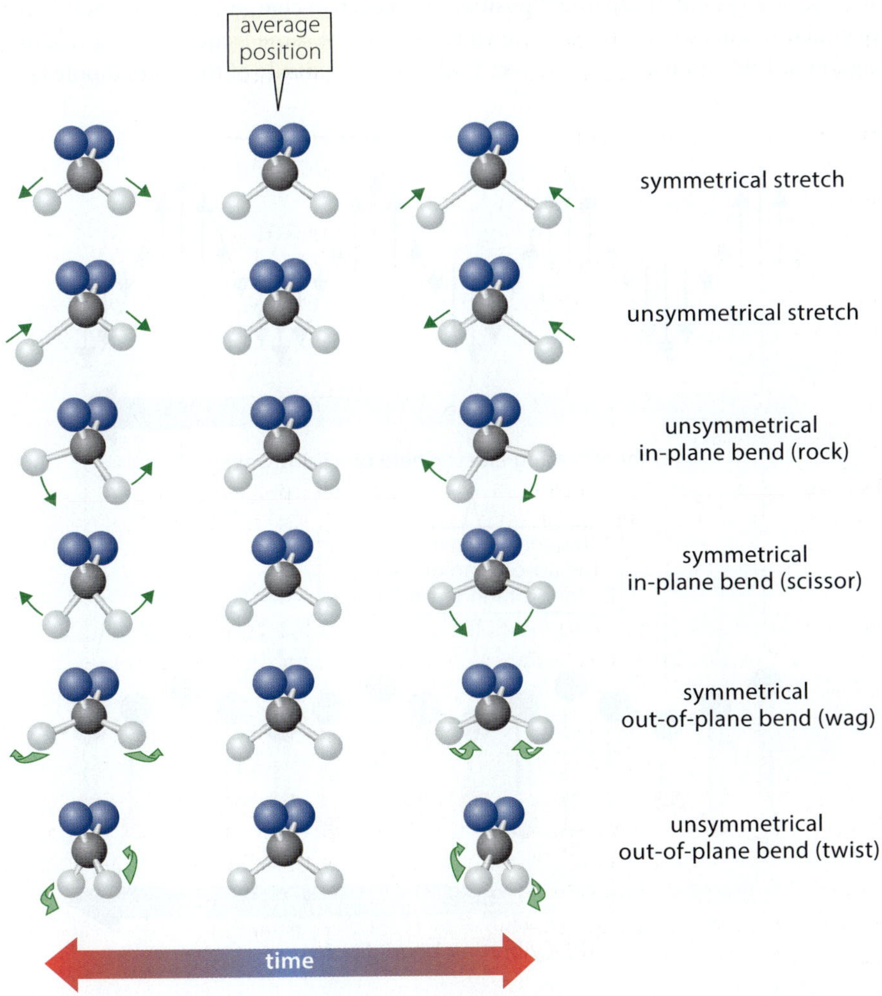

FIGURE 13.8 Normal vibrational modes of a typical CH_2 group in a structure $R—CH_2—R$. In each model, the white spheres are the hydrogens, the black sphere is the carbon of the CH_2, and the blue spheres represent the R groups. The center model in each normal mode represents the average position of the hydrogens. For each mode, start at the center and view left, center, right, center to see how the hydrogens move with time.

and C—H bending vibrations. The peak at 2920 cm^{-1} is due to the C—H stretching vibrations; the peaks at 1470 and 1380 cm^{-1} are due to various bending modes of both $—CH_2—$ and $—CH_3$ groups; and the peak at 720 cm^{-1} is due to a different bending mode, the $—CH_2—$ rocking vibration. Notice that all of the bending vibrations absorb at lower wavenumber (and therefore lower energy) than the stretching vibrations.

B. Factors That Determine IR Absorption Intensity

The different peaks in an IR spectrum typically have very different intensities. Several factors affect absorption intensity. First, a greater number of molecules in the sample and more absorbing groups within a molecule give a more intense spectrum; so, a more concentrated sample gives a stronger spectrum than a less concentrated one, other things being equal. Similarly, at a given concentration, a compound such as nonane, which is rich in C—H bonds, has a stronger absorption for its C—H stretching vibrations than a compound of similar molecular mass with relatively few C—H bonds.

The dipole moment of a molecule also affects the intensity of an IR absorption. We can see why this should be so if we think about the nature of light and how it interacts with a single vibrating bond. A light wave consists of perpendicular vibrating electric and magnetic fields, as shown in Fig. 13.1, Sec. 13.1A. Only the vibrating electric field is relevant to IR spectroscopy, so we can ignore the magnetic field for now. In **Fig. 13.9a**, the vibrating electric field of light is shown again as a vector oscillating in the plane of the page. Section 13.1A explained that *an electric field exerts a force on a charge.* This means that an electric field affects the motion of a charge. The field imposes an acceleration component on the charge in the direction of the field. In particular, if a charge is moving in the same direction as the electric field, the field increases the velocity of the charge—that is, the field makes the charge "move more."

Section 1.2D noted that a polar chemical bond has a *bond dipole.* This means that we can think of a polar bond as a system of separated positive and negative charges. If a polar bond vibrates with a particular frequency, its bond dipole vibrates with the same frequency (**Fig. 13.9b**). According to Eq. 1.92 and Display 1.93 (Sec. 1.9A), the magnitude of the bond dipole is

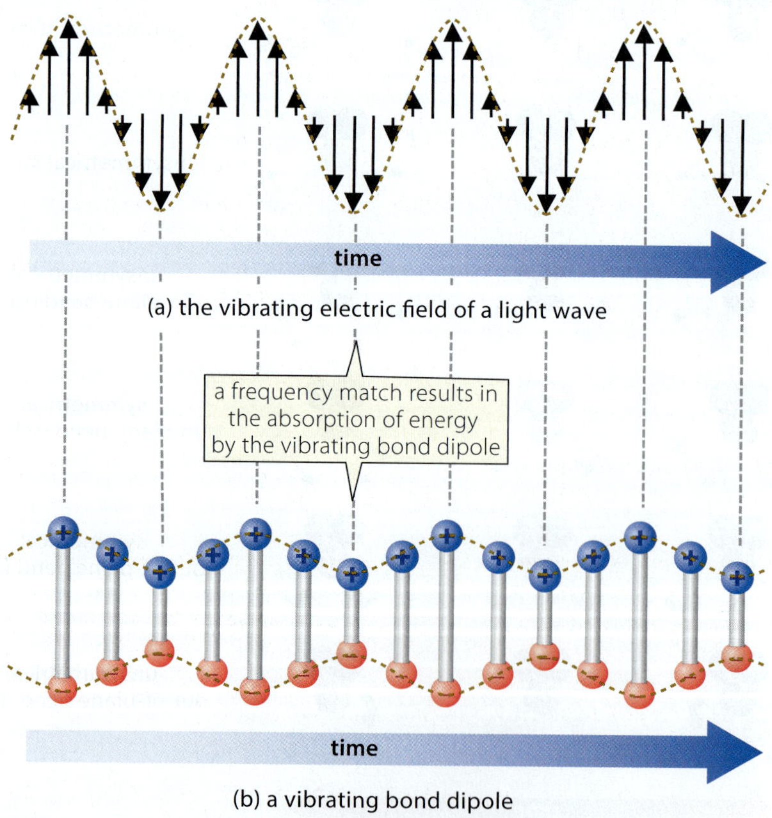

FIGURE 13.9 (a) One component of a light wave is a vibrating electric field, here represented as a vector that oscillates in a sinusoidal manner over time. (b) A bond dipole vibrating at the same frequency as the light wave in (a). When the frequencies of the light wave and the vibrating bond dipole match (as in this figure), the bond dipole absorbs energy from the light wave. (See also Fig. 13.6, Sec. 13.2B.) The alignment of peaks and troughs of the two waves is indicated with gray dashed lines.

proportional not only to the amount of charge on each bonded atom but also to the distance between the atoms—that is, the bond length. Therefore, as the bond stretches, the bond dipole increases, and as the bond compresses, the bond dipole decreases. To a light wave, a polar bond is basically a system of moving charges. Light interacts with a polar bond because its electric field exerts a force on the system of moving charges (the vibrating bond dipole). This interaction can occur only if the light wave and the charges in the bond are oscillating with the same frequency. When the electric field of the light wave exerts a force on the charges in the bond dipole, the bond dipole gains energy, as shown in Fig. 13.6, and consequently the light wave loses energy; this is the process of absorption.

If a bond has no bond dipole, then there are no moving charges with which the electric field of the light wave can interact. For example, the C═C bond in the alkene 2,3-dimethyl-2-butene has no bond dipole because of its symmetrical position in the molecule. Therefore, its stretching vibration does not interact with the electric field of light. The C═C stretching vibration *occurs*, but it *does not absorb energy* from light. This means that 2,3-dimethyl-2-butene does *not* have a C═C stretching absorption in its IR spectrum in the 1600–1700 cm^{-1} region of its IR spectrum, the region in which the double bonds of many other alkenes have such an absorption.

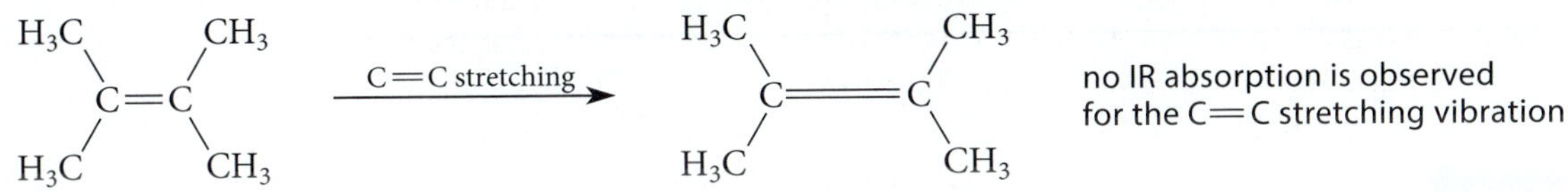

2,3-dimethyl-2-butene: zero dipole moment

"stretched" 2,3-dimethyl-2-butene: zero dipole moment

(13.15)

(This compound does have other IR absorptions.) Molecular vibrations that occur but do not give rise to IR absorptions are said to be **infrared-inactive**. (IR-inactive vibrations can be observed with *Raman spectroscopy*, another type of spectroscopy.) In contrast, any vibration that gives rise to an IR absorption is said to be **infrared-active**. We often find in practice that highly symmetrical compounds have less complex IR spectra than unsymmetrical isomers because their symmetry results in the absence of a molecular dipole moment and a relatively large number of infrared-inactive vibrations.

It is possible for a molecule that has a zero dipole moment to have a molecular vibration that creates a temporary dipole moment in the distorted molecule. Such vibrations are also infrared-active. An example of this situation is found in Study Problem 13.2. This is why some vibrations, even in symmetrical molecules, are infrared-active.

Because the intensity of an IR absorption depends on the size of the dipole moment change that accompanies the corresponding vibration, IR absorptions differ widely in intensity. Organic chemists do not try to predict intensities; rather, they rely on collective experience to know which absorptions are weaker and which are stronger. Nevertheless, for symmetrical molecules with a zero dipole moment, we must be particularly aware of the possibility of IR-inactive vibrations that would be observed in less symmetrical molecules containing the same functional groups.

Study Problem 13.2

Which one of the following molecular vibrations is infrared-inactive? (a) The C═O symmetrical stretch of CO_2; (b) The C═O unsymmetrical stretch of CO_2. (See Fig. 13.8, Sec. 13.3A.)

Solution First be sure you understand what is meant by the terms *symmetrical stretch* and *unsymmetrical stretch*. These are defined by analogy to the C—H stretching vibrations in Fig. 13.8. In the symmetrical stretch, the two C═O bonds lengthen (or shorten) at the same time so that the molecule maintains its symmetry with respect to the symmetry plane:

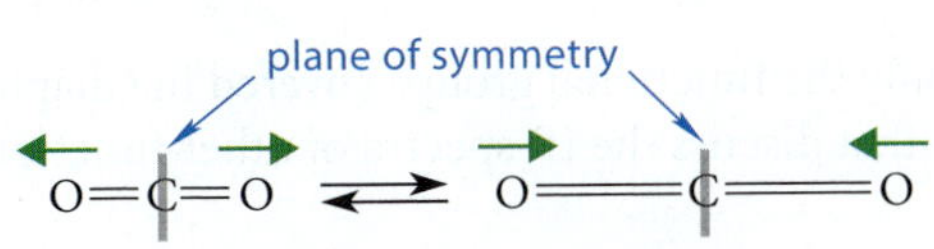

symmetrical stretch: maintains the molecular symmetry

In the unsymmetrical stretch, one C=O bond shortens when the other lengthens:

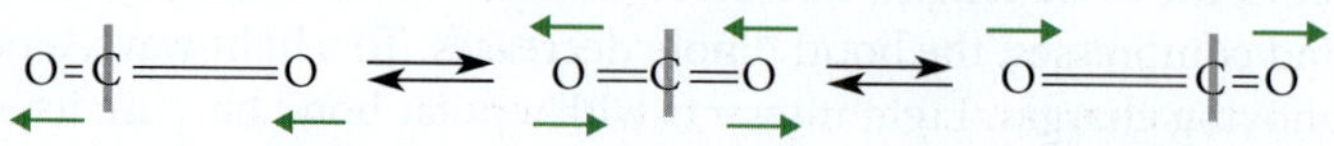

unsymmetrical stretch

Which of these vibrational modes results in a change of dipole moment? Because the CO_2 molecule is linear, the two C=O bond dipoles exactly oppose each other. Stretching a bond increases its bond dipole because the size of a bond dipole is proportional not only to the magnitudes of the partial charges at each end of the bond but also to *the distance by which the charges are separated* (Eq. 1.92, Sec. 1.9A). Consequently, after the symmetrical stretch, both bond dipoles are increased; but, because they are exactly equal and oppose each other, the net dipole moment of CO_2 remains zero. Therefore, the symmetrical stretching vibration is IR-inactive.

In an unsymmetrical stretch, one C=O bond is reduced in length while the other is increased. Because the "long" C=O bond has a greater bond dipole than the "short" C=O bond, the two bond dipoles no longer cancel. Therefore, the unsymmetrical stretch imparts a temporary dipole moment to the CO_2 molecule. Consequently, this vibration is infrared-active—it gives rise to an IR absorption.

Focused Problem

13.10 Which of the following vibrations should be infrared-active and which should be infrared-inactive (or nearly so)?

(a)  (C=C stretch)

(b)  (C=C stretch)

(c)  cyclohexane ring "breathing" (simultaneous stretch of all C—C bonds)

(d) 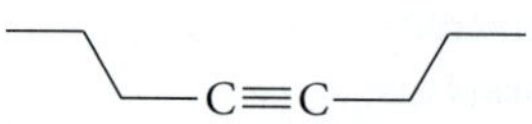 (C≡C stretch)

(e) 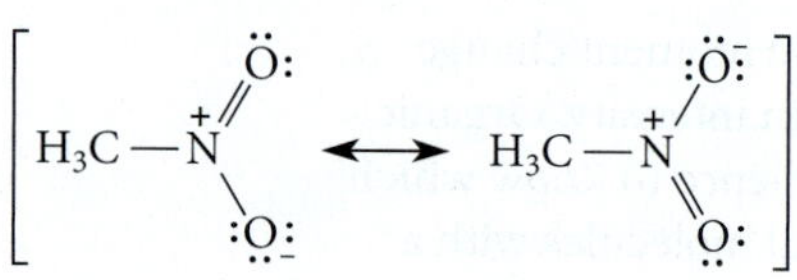 (symmetrical N—O stretch)

(f) $(CH_3)_3C—Cl$ (C—Cl stretch)

13.4 FUNCTIONAL-GROUP INFRARED ABSORPTIONS

A typical IR spectrum contains many absorptions. Chemists do not try to interpret every absorption in a spectrum. Experience has shown that some absorptions are particularly useful and important in diagnosing or confirming certain functional groups. In this section, we focus on those. We show sample spectra so that you will begin to see how key absorptions actually look.

We consider here only the functional groups covered in Chapters 1–12. Subsequent chapters contain short sections that discuss the IR spectra of other functional groups. These sections, however, can be read and understood at any time with your present knowledge of infrared spectroscopy. For example, if your laboratory course expects you to deal with the spectra of aldehydes and ketones, you can jump ahead to Sec. 19.4A to read about those spectra and to see examples. In addition, a summary of key IR absorptions is given in Appendix II.

A. IR Spectra of Alkanes

The characteristic structural features of alkanes are the carbon–carbon and carbon–hydrogen single bonds. Fortunately, the stretching of the carbon–carbon single bond is infrared-inactive (or nearly so) because this vibration is associated with little or no change of the dipole moment. (If this absorption were active, it would dominate the IR spectra of most organic compounds, and other important absorptions would be obscured.) The stretching absorptions of alkyl C—H bonds are typically observed in the 2850–2960 cm^{-1} region. The peaks near 2920 cm^{-1} in the IR spectrum of nonane (Fig. 13.4, Sec. 13.1B) are examples of such absorptions. Various bending vibrations are also observed in the fingerprint region (1380 and 1470 cm^{-1} in nonane) and in the C—H bending region (720 cm^{-1} in nonane). Absorptions in these general regions can be expected not only for alkanes but also for any compounds that contain H_3C— and —CH_2— groups. Consequently, these absorptions are not often useful, but it is important to be aware of them so that they are not mistakenly attributed to other functional groups.

B. IR Spectra of Alkyl Halides

The carbon–halogen stretching absorption of alkyl chlorides, bromides, and iodides appear in the low-wavenumber end of the spectrum, but many interfering absorptions also occur in this region. NMR spectroscopy and mass spectrometry are more useful than IR spectroscopy for determining the structures of these alkyl halides.

Alkyl fluorides, in contrast to the other alkyl halides, have useful IR absorptions. A single C—F bond typically has a very strong stretching absorption in the 1000–1100 cm^{-1} region. Multiple fluorines on the same carbon increase the frequency; for example, a CF_3 group typically has a stretching frequency in the 1300–1360 cm^{-1} region.

C. IR Spectra of Alkenes

Unlike the spectra of alkanes and alkyl halides, the infrared spectra of alkenes are very useful and can help determine not only whether a carbon–carbon double bond is present, but also the carbon substitution pattern at the double bond. Typical alkene absorptions are given in **Table 13.2**. These fall into three categories: C═C stretching absorptions, ═C—H stretching absorptions, and ═C—H bending absorptions. The stretching vibration of the carbon–carbon double bond occurs in the 1640 – 1645 cm^{-1} range; the frequency of this absorption tends to be greater, and its intensity smaller, with increased alkyl substitution at the double bond. The reason for the intensity variation is the dipole moment effect discussed in the previous section. For example, the C═C stretching absorption is clearly evident in the IR spectrum of 1-octene at 1642 cm^{-1} (see **Fig. 13.10a**), but is virtually absent in the spectrum of the symmetrical alkene *trans*-3-hexene (**Fig. 13.10b**). The C═C stretching vibration is weak or absent even in unsymmetrical alkenes that have the same number of alkyl groups on each carbon of the double bond. This absorption is weak because of the dipole moment effect discussed in Sec. 13.3B and illustrated in Eq. 13.15 (Sec. 13.3B).

NMR spectroscopy is particularly useful for observing alkene hydrogens (Sec. 14.7A). Nevertheless, a ═C—H stretching absorption can often be used for confirmation of an alkene functional group. In general, the stretching absorptions of C—H bonds involving sp^2-hybridized carbons occur at wavenumbers *greater than* 3000 cm^{-1}, and the stretching absorptions of C—H bonds involving sp^3-hybridized carbons occur at wavenumbers *less than* 3000 cm^{-1}.

TABLE 13.2 Important Infrared Absorptions of Alkenes

Functional group	Absorption*
C=C stretching absorptions	
—CH=CH$_2$ (terminal vinyl)	1640 cm^{-1} (m, sh)
C=CH$_2$ (terminal methylene)	1655 cm^{-1} (m, sh)
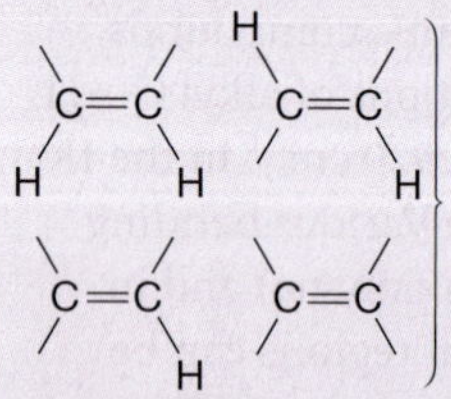	1660–1675 cm^{-1} (w) (absent in some compounds)
=C—H stretching absorptions	
=C—H, =CH$_2$	3000–3100 cm^{-1} (m)
=C—H bending absorptions	
—CH=CH$_2$ (terminal vinyl)	910, 990 cm^{-1} (s) two absorptions
C=CH$_2$ (terminal methylene)	890 cm^{-1} (s)
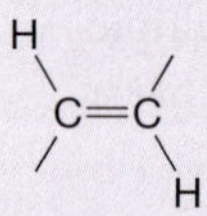 (trans alkene)	960–980 cm^{-1} (s)
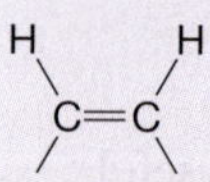 (cis alkene)	675–730 cm^{-1} (br) (ambiguous and variable for different compounds)
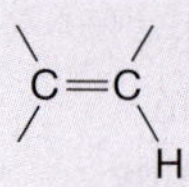 (trisubstituted)	800–840 cm^{-1} (s)

**Intensity designations:* s = strong; m = moderate; w = weak
Shape designations: sh = sharp (narrow); br = broad (wide)

For example, 1-octene has a =C—H stretching absorption at 3080 cm^{-1} (Fig. 13.10a), and *trans*-3-hexene has a similar absorption that is barely discernible at 3030 cm^{-1} (Fig. 13.10b). The higher frequency of =C—H stretching absorptions is a manifestation of the bond-strength effect: bonds to sp^2-hybridized carbons are stronger (Table 10.2, Sec. 10.1E), and stronger bonds vibrate at higher frequencies.

The alkene =C—H bending absorptions that appear in the C—H bending region of the IR spectrum at low wavenumber are in many cases very strong and can be used to determine the substitution pattern at the double bond. The first three of these absorptions in Table 13.2—the ones for terminal vinyl, terminal methylene, and trans-alkene—are the most reliable. The 910 and 990 cm^{-1} terminal vinyl absorptions are illustrated in the IR spectrum of 1-octene (Fig. 13.10a). A strong 910 cm^{-1} absorption is frequently accompanied by a weaker absorption at twice the frequency—in this case, at 1820 cm^{-1}. This is an *overtone*; it represents a vibrational transition that spans two energy levels. It shouldn't be mistaken for the absorptions of other groups. Another very useful bending absorption is the trans-alkene absorption, illustrated by the 965 cm^{-1} peak in the IR spectrum of *trans*-3-hexene (Fig. 13.10b).

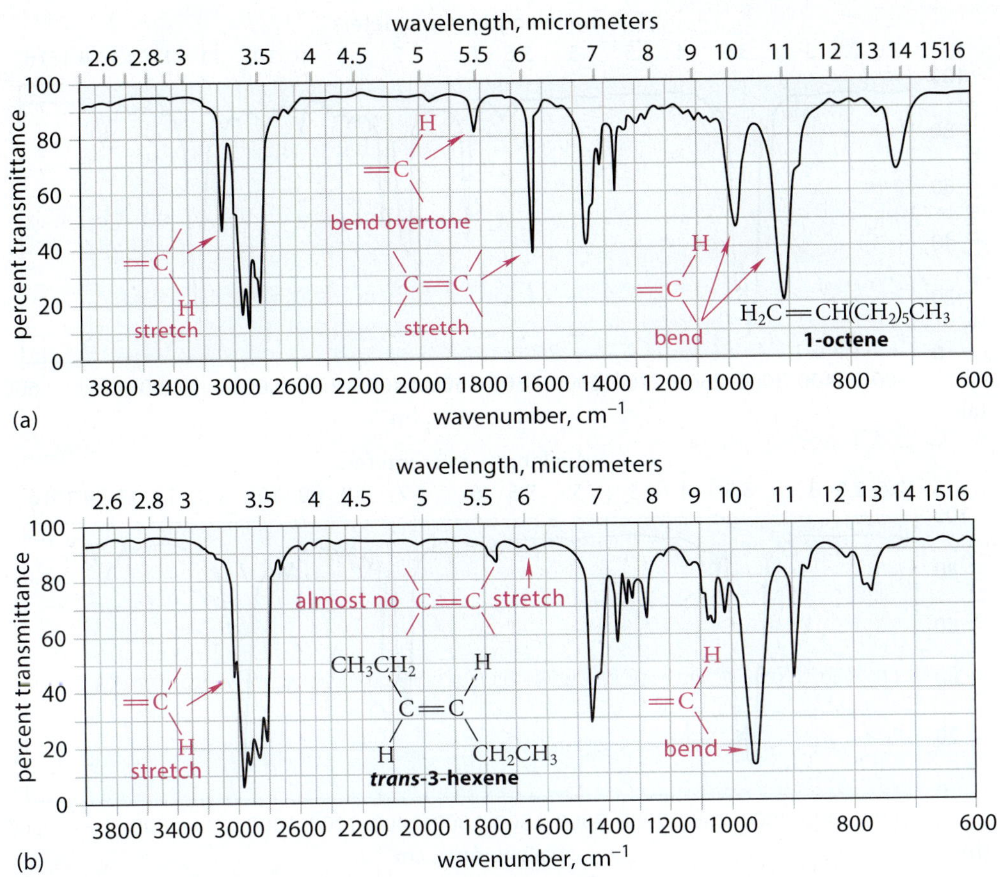

FIGURE 13.10 IR spectra of (a) 1-octene and (b) *trans*-3-hexene. Be sure to correlate the key bands indicated in these spectra with the corresponding entries in Table 13.2.

Study Problem 13.3

Each of three alkenes, *A*, *B*, and *C*, has the molecular formula C_5H_{10}, and each undergoes catalytic hydrogenation to yield pentane. Alkene *A* has IR absorptions at 1642, 990, and 911 cm^{-1}; alkene *B* has an IR absorption at 964 cm^{-1} and no absorption in the 1600–1700 cm^{-1} region; and alkene *C* has absorptions at 1658 and 695 cm^{-1}. Identify the three alkenes.

Solution In this problem, you can *write out all the possibilities* and then use the IR spectra to decide between them. The molecular formulas and the hydrogenation data show that the carbon chains of all of the alkenes are unbranched and that all are isomeric pentenes. Therefore, the *only* possibilities for compounds *A*, *B*, and *C* are the following:

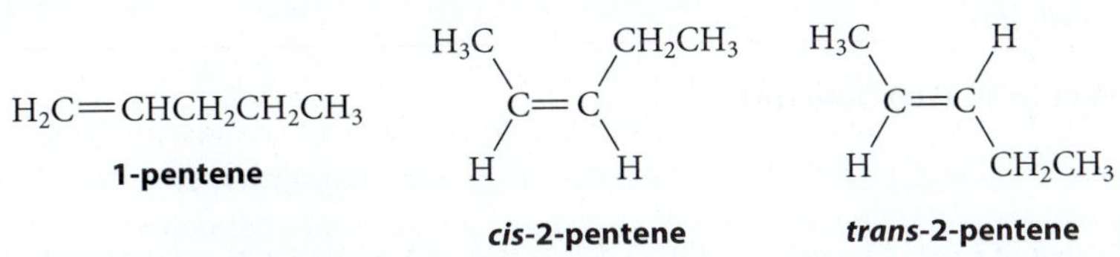

The C—H bending absorptions of *A* at 990 and 911 cm^{-1} indicate that it is a 1-alkene; therefore it must be 1-pentene. The 964 cm^{-1} C—H bending absorption of *B* shows that it is *trans*-2-pentene. (Why is the C=C stretching vibration absent?) The remaining alkene *C* must be *cis*-2-pentene; the 1658 cm^{-1} C=C stretching absorption and the 695 cm^{-1} C—H bending absorption are consistent with this assignment.

Notice that you do not need the complete IR spectrum of each compound, but only the key absorptions, to solve this problem.

FIGURE 13.11 IR spectra for Focused Problem 13.12.

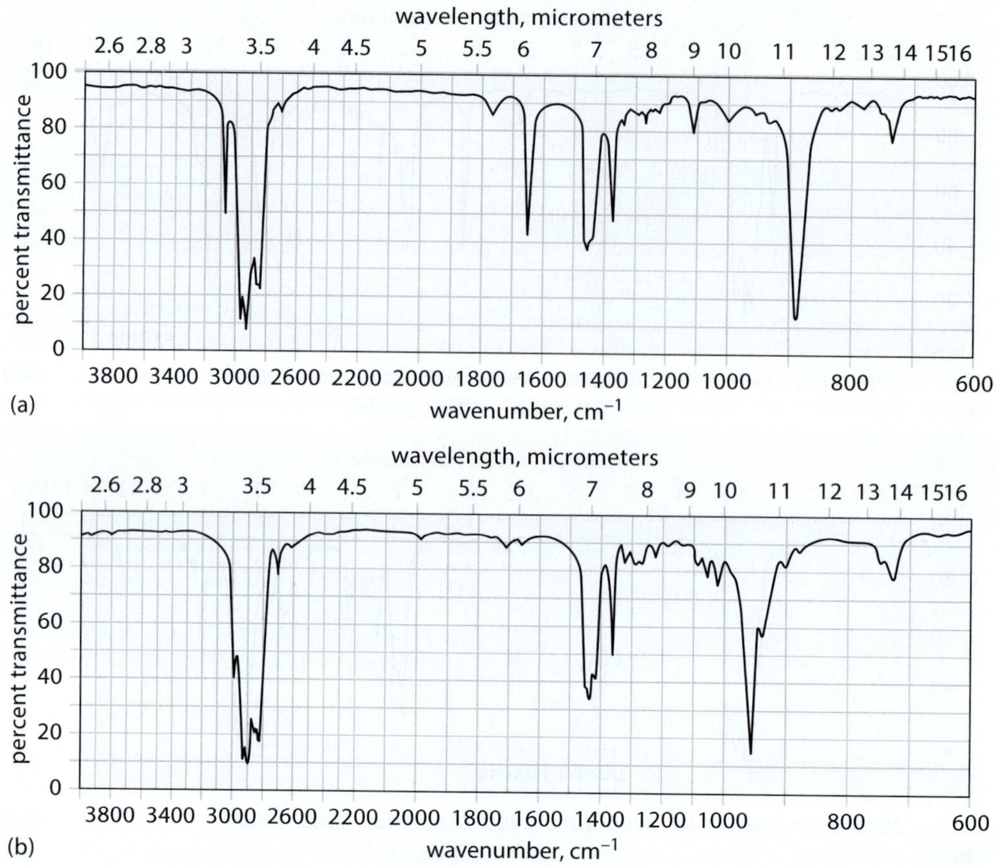

Focused Problems

13.11 Five isomeric alkenes *A–E* each undergo catalytic hydrogenation to give 2-methylpentane. The IR spectra of these five alkenes have the following key absorptions (in cm^{-1}):

Compound *A*: 912 (s), 994 (s), 1643 (s), 3077 (m)

Compound *B*: 833 (s), 1667 (w), 3050 (weak shoulder on C—H absorption)

Compound *C*: 714 (s), 1665 (w), 3010 (m)

Compound *D*: 885 (s), 1650 (m), 3086 (m)

Compound *E*: 967 (s), no absorption 1600–1700, 3040 (m)

Propose a structure for each alkene.

13.12 One of the spectra in **Fig. 13.11** is that of *trans*-2-heptene and the other is that of 2-methyl-1-hexene. Which is which? Explain.

D. IR Spectra of Alkynes

Many alkynes have a C≡C stretching absorption in the 2100–2200 cm^{-1} region of the infrared spectrum. This absorption is clearly evident, for example, at 2120 cm^{-1} in the IR spectrum of 1-octyne (**Fig. 13.12a**). However, this absorption is very weak or absent in the IR spectra of many symmetrical, or nearly symmetrical, alkynes because of the dipole moment effect (Sec. 13.3B). For example, 4-octyne has no C≡C stretching absorption at all (**Fig. 13.12b**).

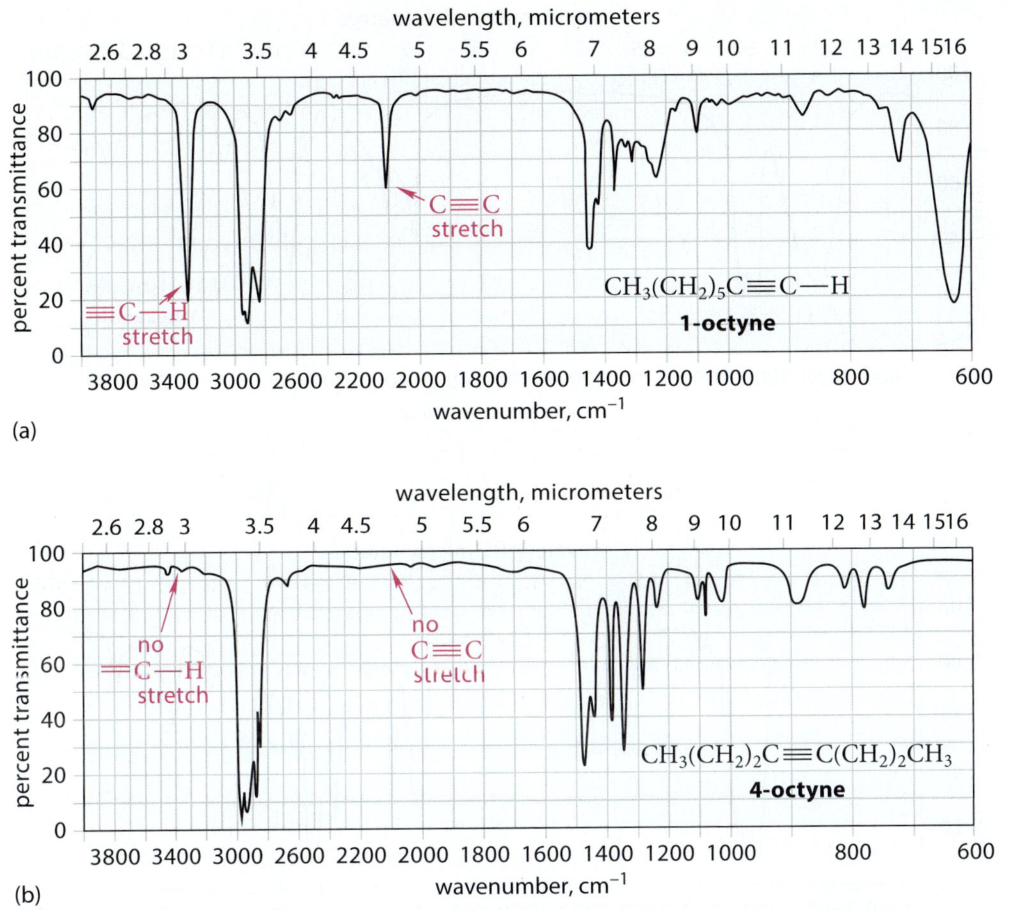

FIGURE 13.12 The IR spectrum of (a) 1-octyne and (b) 4-octyne. The two key absorptions indicated in 1-octyne are absent in the spectrum of 4-octyne.

The C≡C stretching absorption (2120 cm^{-1}) lies at considerably higher frequency than the C═C stretching frequency (1640–1675 cm^{-1}). This is a clear manifestation of the bond-strength effect on absorption frequency. (See Sec. 13.3A and Study Problem 13.1.)

A useful absorption of 1-alkynes is the ≡C—H stretching absorption, which occurs at about 3300 cm^{-1}. This strong, sharp absorption, prominent in the spectrum of l-octyne (Fig. 13.12a), is well separated from other C—H stretching absorptions. Because internal alkynes (R—C≡C—R′) lack the unique ≡C—H bond of 1-alkynes, they do not show this absorption (Fig. 13.12b).

E. IR Spectra of Alcohols and Ethers

The most characteristic absorption of alcohols is the O—H stretching absorption. The position and strength of this absorption is determined by the degree of hydrogen bonding, which depends, in turn, on the physical state of the alcohol. In the gas phase, in which O—H groups are not hydrogen-bonded to other groups, the O—H stretching absorption is a fairly sharp peak of moderate intensity that occurs near 3600 cm^{-1}. However, in most liquid and solid samples—which are more commonly used in everyday IR spectroscopy—the O—H groups are strongly hydrogen-bonded and give a broad peak of moderate to strong intensity in the 3200–3400 cm^{-1} region of the IR spectrum. Such an absorption, which is an important spectroscopic identifier for alcohols, is clearly evident in the IR spectrum of 1-hexanol (see **Fig. 13.13**). The broadness of the absorption is caused by hydrogen bonds of different strengths in a typical sample, and the strength of this absorption reflects the highly polar nature of the hydrogen bond (Display 8.38, Sec. 8.5C).

The other characteristic absorption of alcohols is a strong C—O stretching peak that occurs in the 1050–1200 cm^{-1} region of the spectrum; primary alcohols absorb near the low end of

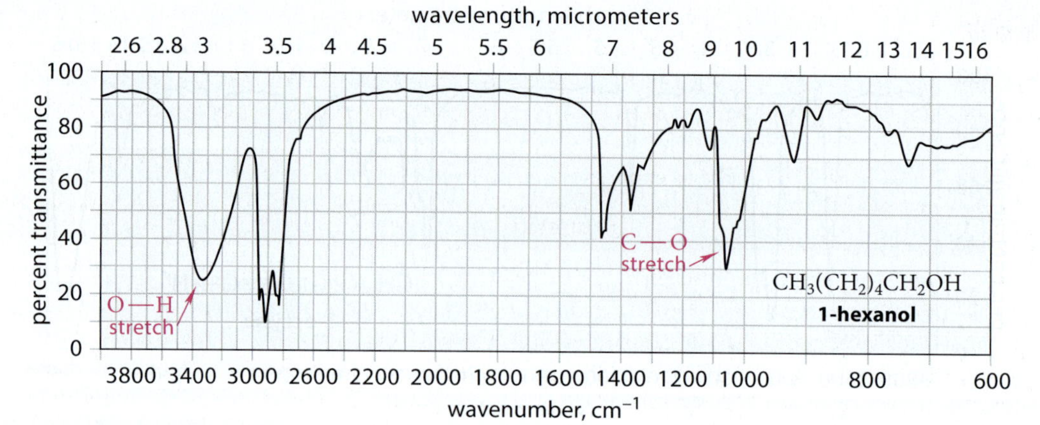

FIGURE 13.13 The IR spectrum of 1-hexanol. Notice particularly the broad O—H stretching absorption.

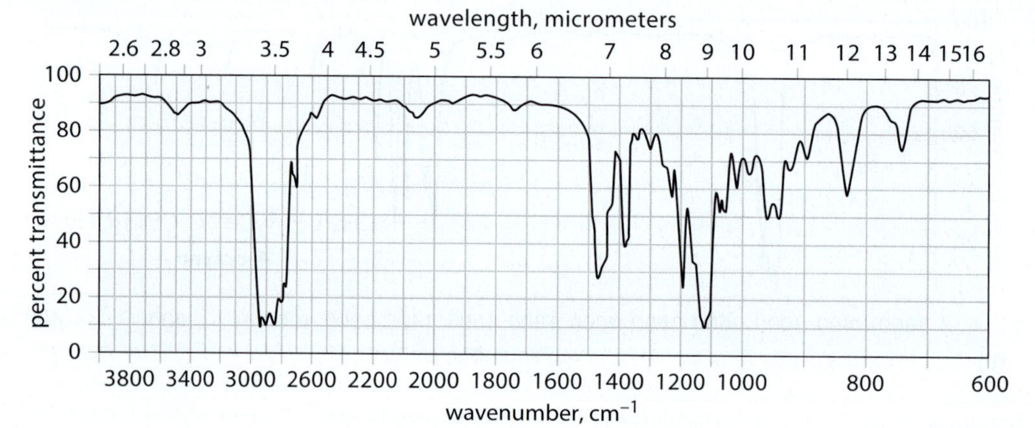

FIGURE 13.14 The IR spectrum for Focused Problem 13.13.

this range and tertiary alcohols near the high end. For example, this absorption occurs at about 1060 cm^{-1} in the spectrum of 1-hexanol. Because some other functional groups, such as ethers, esters, and carboxylic acids, also show C—O stretching absorptions in the same general region of the spectrum, the C—O stretching absorption is mainly used to support or confirm the presence of an alcohol diagnosed from the O—H absorption or from other spectroscopic evidence.

The most characteristic infrared absorption of ethers is the C—O stretching absorption, which, for the reasons just stated, is not very useful except for confirmation when an ether is already suspected from other data. For example, both dipropyl ether and an isomer 1-hexanol have strong C—O stretching absorptions near 1100 cm^{-1}.

Focused Problems

13.13 Match the IR spectrum in **Fig. 13.14** to one of the following three compounds: 2-methyl-1-octene, butyl methyl ether, or 1-pentanol.

13.14 (a) Explain why the IR spectra of some ethers have *two* C—O stretching absorptions. (*Hint:* See Fig. 13.8, Sec. 13.3A.)

(b) Give the general structure of an ether that would *not* have the second C—O stretching absorption.

13.15 Explain why the frequency of the O—H stretching absorption of an alcohol in solution changes as the alcohol solution is diluted.

13.5 OBTAINING AN INFRARED SPECTRUM

Infrared spectra are obtained with an instrument called an **infrared spectrometer**. In its simplest concept, the IR spectrometer provides a way to carry out the absorption spectroscopy experiment shown in Fig. 13.3 (Sec. 13.1B). The instrument houses a source of infrared radiation, a place to mount the sample in the infrared beam, and the optics and electronics necessary to measure the intensity of light absorbed or transmitted as a function of wavelength or wavenumber. Modern IR spectrometers, called *Fourier-transform infrared spectrometers* (FT-IR spectrometers), can provide an IR spectrum in a few seconds. (See Further Exploration 13.2.) The IR spectra in this text were obtained with an FT-IR spectrometer.

FURTHER EXPLORATION 13.2
FT-IR Spectroscopy

The sample containers ("sample cells") used in IR spectroscopy must be made of an infrared-transparent material. Because glass absorbs infrared radiation, it cannot be used. The conventional material used for sample cells is sodium chloride. The IR spectrum of an undiluted ("neat") liquid can be obtained by compressing the liquid to form a thin film between two optically polished salt plates. IR spectra can also be taken in solution cells, which consist of sodium chloride plates in appropriate holders equipped with syringe fittings for injecting the solution. If the sample is a solid, the finely ground solid can be dispersed ("mulled") in a mineral oil and the dispersion compressed between salt plates. Alternatively, a solid can be co-fused (melted) with KBr, another IR-transparent material, to form a clear pellet. Simple presses are available to prepare KBr pellets.

When mineral-oil dispersions or solvents are used, we have to be aware of the regions in which the oil or the solvents themselves absorb IR radiation, because these absorptions interfere with those of the sample. A number of solvents are commonly used; chloroform ($CHCl_3$), its isotopic analog ($CDCl_3$), and methylene chloride (CH_2Cl_2) are among them. As a few students learn the hard way every year, solvents that dissolve sodium chloride, such as water and alcohols, cannot be used.

A more recent technique for obtaining IR spectra is called *attenuated total reflectance* (ATR). In this technique (**Fig. 13.15**), a thin layer of the sample (either solid or liquid) is spread across, and pressed onto, a crystal support with a high refractive index, such as germanium, zinc selenide, or, in research environments, diamond. The IR beam is directed through the crystal from the lower surface at an angle so that it reflects from each crystal surface. The IR beam at each surface actually extends for a few micrometers beyond each surface, so that the intensity of the reflected IR beam is reduced ("attenuated") somewhat by absorption of the sample. The spectrum, as usual, is the measurement of absorption versus wavelength. The advantage of the ATR technique is that the use of expensive optically polished sodium chloride plates can be avoided, and substances that dissolve sodium chloride can be examined. Relatively inexpensive ATR instruments are available in some undergraduate laboratories.

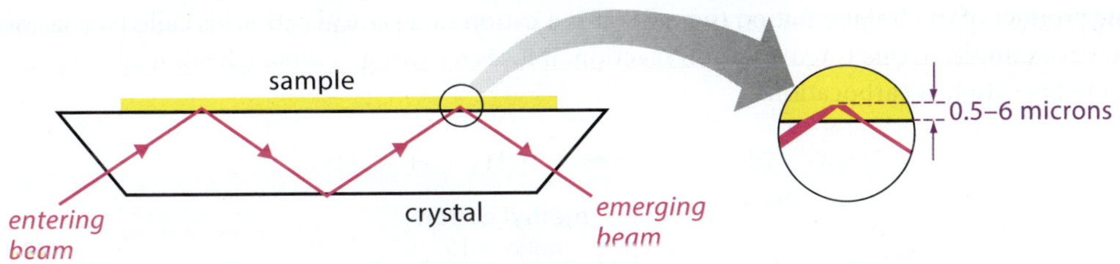

FIGURE 13.15 A diagram of the attenuated total reflectance (ATR) technique. The infrared beam is reflected from the interior surfaces of the crystal support. As the magnification shows, the IR beam at each surface actually extends for a few micrometers beyond the surface into the sample, so that the intensity of the reflected IR beam (indicated by the thickness of the red line) is reduced by sample absorption. As in other forms of absorption spectroscopy, the measurement of absorption intensity versus frequency gives the IR spectrum.

13.6 INTRODUCTION TO MASS SPECTROMETRY

In contrast to other spectroscopic techniques, mass spectrometry does not involve the absorption of electromagnetic radiation, but operates on a completely different principle. As the name implies, mass spectrometry is used to determine molecular masses, and it is the most important technique used for this purpose. It also has some use in determining molecular structure.

A. Electron-Ionization (EI) Mass Spectra

The instrument used to obtain a mass spectrum is called a **mass spectrometer**. In one type of instrument, an **electron-ionization (EI) mass spectrometer**, the organic compound to be analyzed is vaporized in a vacuum chamber and bombarded with an electron beam of high energy, typically 70 eV (electron volts)—more than 6700 kJ mol^{-1} (1600 kcal mol^{-1}). The electron cloud of a molecule repels these electrons, so these electrons don't collide with the molecule, but veer off in what might be termed a "glancing blow." As a high-energy electron passes close by, its electric field, because of its high energy, causes the ejection of an electron from the molecule. For example, if methane is treated in this manner, it loses an electron from one of the C—H bonds.

(13.16)

The product of this reaction is sometimes abbreviated as follows:

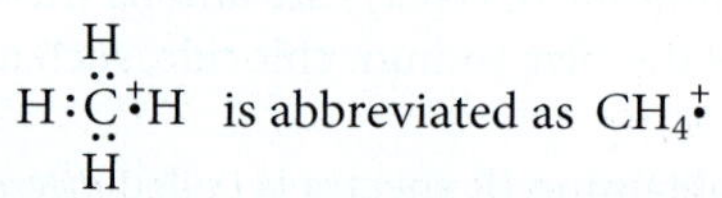

(13.17)

The symbol $\overset{+}{\cdot}$ means that the molecule is both a radical (a species with an unpaired electron) and a cation—a **radical cation**. The species $CH_4^{+\cdot}$ is called the *methane radical cation*. It is an example of an **odd-electron cation**—that is, a carbocation with an odd number of electrons. This is in contrast to our "usual" type of carbocation, such as $^+CH_3$, which is an **even-electron cation**.

an odd-electron cation   an even-electron cation   (13.18)

Although we have shown the electron being ejected from a bond, remember that the methane electrons are actually held in molecular orbitals. (See Fig. 1.14 and Display 1.88, Sec. 1.8C.) An electron is therefore ejected from an occupied molecular orbital, and the resulting positive charge is also delocalized around the molecule.

Following its formation, the methane radical cation decomposes in a series of reactions called *fragmentation reactions*. In a **fragmentation reaction**, a radical cation literally comes apart. The ionic product of the fragmentation (whether it is a cation or a radical cation) is called a **fragment ion**. For example, in one fragmentation reaction, it loses a hydrogen *atom* (the radical) to generate the methyl cation, a carbocation.

$$\underset{\text{mass = 16}}{CH_4^{+\cdot}} \longrightarrow \underset{\substack{\textbf{methyl cation}\\ \text{mass = 15}}}{^+CH_3} + H\cdot \qquad (13.19a)$$

The hydrogen atom carries the unpaired electron, and the methyl cation carries the charge; consequently, the methyl cation, now an *even-electron ion*, is the *fragment ion* in this case. The process can be represented with the free-radical (fishhook) arrow notation as follows:

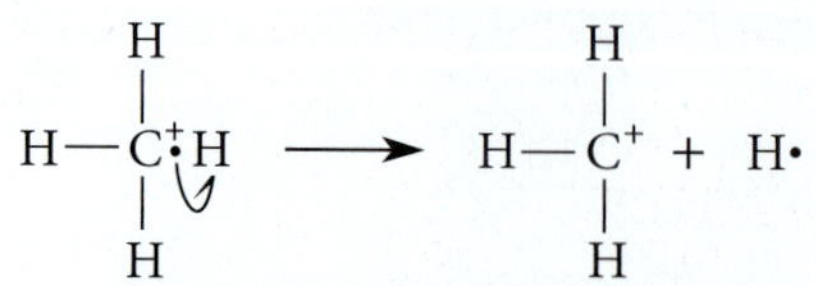

(13.19b)

Alternatively, the unpaired electron may remain associated with the carbon atom; in this case, the products of the fragmentation are a methyl radical and a proton.

$$\underset{\text{mass} = 16}{CH_4^{\cdot+}} \longrightarrow \underset{\textbf{methyl radical}}{\cdot CH_3} + \underset{\text{mass} = 1}{H^+} \quad (13.19c)$$

In this case the proton is the fragment ion. Further decomposition reactions give fragments of progressively smaller mass. (Show how these occur by using the fishhook notation.)

$$^+CH_3 \longrightarrow \underset{\text{mass} = 14}{\overset{\cdot+}{C}H_2} + H\cdot \quad (13.19d)$$

$$\overset{\cdot+}{C}H_2 \longrightarrow \underset{\text{mass} = 13}{^+CH} + H\cdot \quad (13.19e)$$

$$^+CH \longrightarrow \underset{\text{mass} = 12}{C^{\cdot+}} + H\cdot \quad (13.19f)$$

The ions formed in Eqs. 13.16 and 13.19a–f are *very* unstable species. They are not the sorts of species that would be involved as reactive intermediates in a solution reaction. Sec. 9.6, for example, explained that methyl and primary carbocations are *never* formed as intermediates in S_N1 reactions. These ions are formed in the mass spectrometer *in the gas phase* only because of the immense energy imparted to the methane molecules by the bombarding electron beam.

Thus, methane undergoes fragmentation in the mass spectrometer to give several positively charged fragment ions of differing ionic masses: $\overset{\cdot+}{C}H_4$, $^+CH_3$, $\overset{\cdot+}{C}H_2$, ^+CH, $\overset{\cdot+}{C}$, and H^+. As these examples illustrate, the **ionic mass** of a fragment ion is simply the sum of the atomic masses of its constituent atoms. In the mass spectrometer, the fragment ions are separated according to their **mass-to-charge ratio,** m/z (m = ionic mass, z = the charge of the fragment). Because most ions formed in the electron-ionization mass spectrometer have +1 charge, the m/z value can generally be taken as the ionic mass of the ion. A **mass spectrum** is a graph of the relative amount of each ion (called the **relative abundance**) as a function of the ionic mass (or m/z). When the ions are produced by electron ionization, the mass spectrum is called an **EI mass spectrum**. The EI mass spectrum of methane is shown in **Fig. 13.16.** Only the *cations* are detected by the mass spectrometer—neutral molecules and radicals do not appear as peaks in the mass spectrum. The mass spectrum of methane shows peaks at m/z = 16, 15, 14, 13, 12, and 1, corresponding to the masses of the various ionic species that are produced from methane by electron ejection and fragmentation, as shown in Eqs. 13.16 and 13.19a–f.

The mass spectrum can be determined for any molecule that can be vaporized in a high vacuum, and this includes most organic compounds. (Other techniques for vaporizing less volatile molecules have been developed and are discussed briefly in Sec. 13.6E.) Mass spectrometry is used for three purposes: (a) to determine the molecular mass of an unknown compound, (b) to determine the structure (or a partial structure) of an unknown compound by an analysis of the fragment ions in the spectrum, and (c) to confirm the structures of compounds with known or suspected structures.

The ion derived from electron ejection before any fragmentation takes place is known as the **molecular ion**, abbreviated M. *The molecular ion occurs at an m/z value equal to the molecular mass of the sample molecule.* So, in the mass spectrum of methane, the molecular ion occurs at

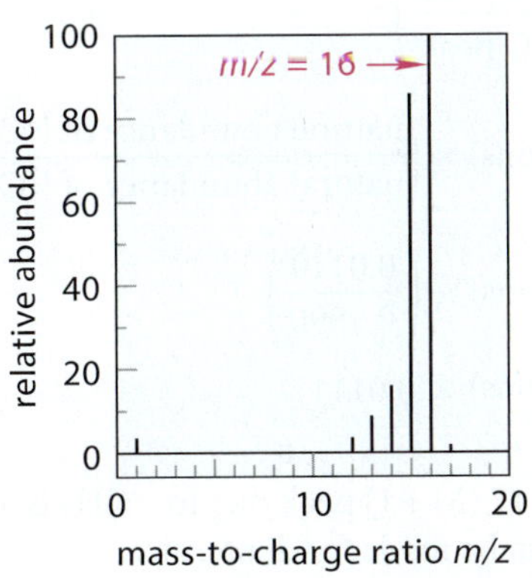

m/z	relative abundance
1	3.36
12	2.80
13	8.09
14	16.10
15	85.90
16	100.00 (base peak)
17	1.17

FIGURE 13.16 The EI mass spectrum of methane. Can you explain why there is an ion at m/z = 17? (See Sec. 13.6B for the answer.)

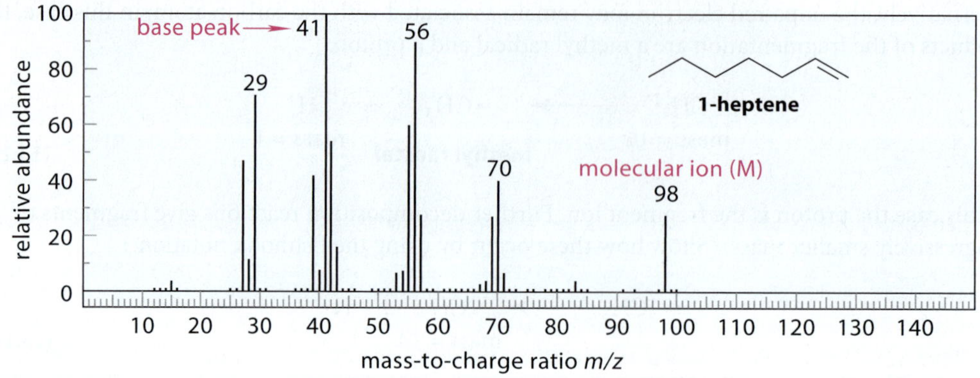

FIGURE 13.17 The EI mass spectrum of 1-heptene. In this case the molecular ion is not the base peak.

$m/z = 16$, the molecular mass of methane. In the mass spectrum of 1-heptene (see **Fig. 13.17**), the molecular ion occurs at $m/z = 98$, the molecular mass of the alkene. Except for peaks due to isotopes, discussed in the next section, the molecular ion peak is the peak of highest m/z in any ordinary mass spectrum.

The **base peak** is the ion of greatest relative abundance in the mass spectrum—that is, the ion with the largest peak. The base peak is arbitrarily assigned a relative abundance of 100%, and the other peaks in the mass spectrum are scaled relative to it. In the mass spectrum of methane, the base peak is the same as the molecular ion, but in the mass spectrum of 1-heptene (Fig. 13.17), the base peak occurs at $m/z = 41$. In the 1-heptene spectrum and in most others, the molecular ion and the base peak are different.

B. Isotopic Peaks

Look again at the mass spectrum of methane in Fig. 13.16. This mass spectrum shows a small but real peak at $m/z = 17$, a mass that is 1 unit higher than the molecular mass. This peak is called an M + 1 peak, because it occurs 1 mass unit higher than the molecular ion (M). This ion occurs because chemically pure methane is really a mixture of compounds containing the various isotopes of carbon and hydrogen.

$$\begin{array}{rccc} \text{methane} = & {}^{12}\text{CH}_4, & {}^{13}\text{CH}_4, & {}^{12}\text{CDH}_3, \text{ and others} \\ m/z = & 16 & 17 & 17 \end{array}$$

The isotopes of several elements and their natural abundances are listed in **Table 13.3**.

Possible sources of the $m/z = 17$ peak for methane are ${}^{13}CH_4$ and ${}^{12}CDH_3$. *Each isotopic compound contributes a peak with a relative abundance in proportion to its amount.* In turn, the amount of each isotopic compound is directly related to the natural abundance of the isotope involved. The abundance of ${}^{13}CH_4$ methane relative to that of ${}^{12}CH_4$ methane is then given by the following equation:

$$\text{relative abundance} = \left(\frac{\text{abundance of }{}^{13}\text{C peak}}{\text{abundance of }{}^{12}\text{C peak}}\right) \tag{13.20a}$$

$$= (\text{number of carbons}) \times \left(\frac{\text{natural abundance of }{}^{13}\text{C peak}}{\text{natural abundance of }{}^{12}\text{C peak}}\right)$$

$$= (\text{number of carbons}) \times \left(\frac{0.0110}{0.9890}\right)$$

$$= (\text{number of carbons}) \times 0.0111 \tag{13.20b}$$

Because methane has only one carbon, the $m/z = 17$ (M + 1) peak due to ${}^{13}CH_4$ is about 1.1% of the $m/z = 16$, or M, peak. A similar calculation can be made for deuterium.

TABLE 13.3 Exact Masses and Isotopic Abundances of Several Isotopes Important in Mass Spectrometry

Element	Isotope	Exact mass	Abundance,%
hydrogen	^{1}H	1.007825	99.985
	^{2}H*	2.0140	0.015
carbon	^{12}C	12.0000	98.90
	^{13}C	13.00335	1.10
nitrogen	^{14}N	14.00307	99.63
	^{15}N	15.00011	0.37
oxygen	^{16}O	15.99491	99.759
	^{17}O	16.99913	0.037
	^{18}O	17.99916	0.204
fluorine	^{19}F	18.99840	100.
silicon	^{28}Si	27.97693	92.21
	^{29}Si	28.97649	4.67
	^{30}Si	29.97377	3.10
phosphorus	^{31}P	30.97370	100.
sulfur	^{32}S	31.97207	95.0
	^{33}S	32.97146	0.75
	^{34}S	33.96787	4.22
chlorine	^{35}Cl	34.9685	75.77
	^{37}Cl	36.96590	24.23
bromine	^{79}Br	78.91834	50.69
	^{81}Br	80.91629	49.31
iodine	^{127}I	126.90447	100.

* ^{2}H is commonly known as deuterium, abbreviated D.

$$\text{relative abundance} = (\text{number of hydrogens}) \times \left(\frac{\text{natural abundance of } ^{2}\text{H}}{\text{natural abundance of } ^{1}\text{H}}\right)$$
$$= (4) \times \left(\frac{0.00015}{0.99985}\right) = 0.0006 \qquad (13.21)$$

Therefore, the CDH_3 naturally present in methane contributes 0.06% to the isotopic peak. Because the contribution of deuterium is so small, ^{13}C is the major isotopic contributor to the M + 1 peak. (We ignore contributions of ^{2}H in subsequent calculations of M + 1 peak intensities.)

In a compound containing more than one carbon, the M + 1 peak is larger than 1.1% of the M peak because there is a 1.1% probability that *each carbon* in the molecule will be present as ^{13}C. For example, cyclohexane has six carbons, and the abundance of its M + 1 ion relative to that of its molecular ion should be 6(1.1) = 6.6%. In the mass spectrum of cyclohexane, for example, the molecular ion has a relative abundance of about 70%; the relative abundance of the M + 1 ion is calculated to be (0.066)(70%) = 4.6%, which corresponds closely to the value observed. Not only the molecular ion peak, but also every other peak in the mass spectrum has isotopic peaks.

Careful measurement of the M + 1 peak (with corrections for the M + 1 contributions for isotopes of other atoms) can be used to determine the number of carbons in an unknown. This use of the M + 1 peak is less important than it once was because "carbon counting" is now possible with carbon-NMR spectroscopy (Sec. 14.9B).

Several elements of importance in organic chemistry have isotopes with significant natural abundances. Table 13.3 shows that silicon has significant M + 1 and M + 2 contributions; sulfur has an M + 2 contribution; and the halogens chlorine and bromine have very important M + 2 contributions. In fact, the naturally occurring form of the element bromine consists of about equal amounts of ^{79}Br and ^{81}Br. The mixture of isotopes leaves a characteristic trail in the mass spectrum that can be used to diagnose the presence of the element.

In the EI mass spectrum of bromomethane (**Fig. 13.18**), for example, the peaks at $m/z = 94$ and 96 result from the contributions of the two bromine isotopes to the molecular ion. They are

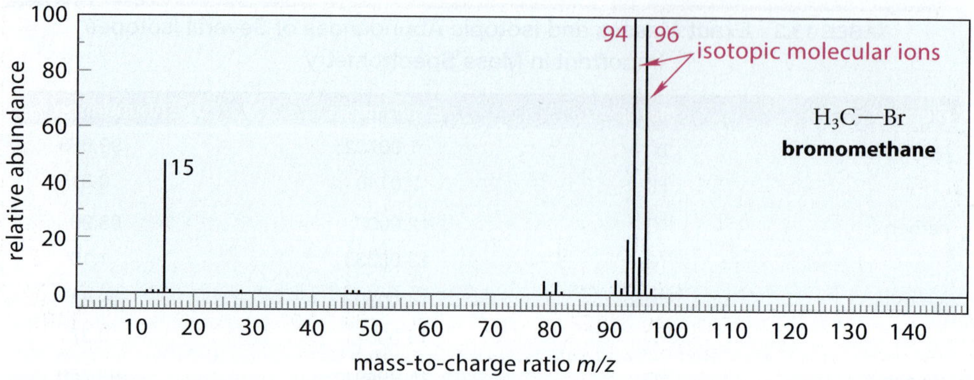

FIGURE 13.18 The EI mass spectrum of bromomethane. The two molecular ions at $m/z = 94$ and at $m/z = 96$ have nearly equal abundance and result from the presence of the two isotopes ^{79}Br and ^{81}Br.

The $m/z = 95$ peak in the mass spectrum of bromomethane is too large to contain *only* the isotopic peak. It also contains a contribution from an M − 1 peak arising the loss of a hydrogen atom in a fragmentation of bromomethane (mass = 96) containing ^{81}Br. (A similar M − 1 peak for bromomethane containing ^{79}Br is evident at $m/z = 93$.)

in the relative abundance ratio $100 : 98 = 1.02$, which is in good agreement with the ratio of the relative natural abundances of the bromine isotopes (see Table 13.3). This double molecular ion, often called a "bromine doublet," indicates a compound containing a single bromine. Notice that along with each major isotopic peak is a smaller isotopic peak 1 mass unit higher. These peaks are due to the isotope ^{13}C present naturally in bromomethane. For example, the $m/z = 95$ peak corresponds to bromomethane containing only ^{79}Br and one ^{13}C, whereas the $m/z = 97$ peak arises from bromomethane that contains only ^{81}Br and one ^{13}C.

Although isotopes such as ^{13}C and ^{18}O are normally present in small amounts in organic compounds, it is possible to synthesize compounds that are selectively enriched with these and other isotopes. Isotopes are especially useful because they provide specific labels at particular atoms without significantly changing their chemical properties. One use of such compounds is to determine the fate of specific atoms in deciding between two mechanisms. Another use is to provide nonradioactive isotopes for biological metabolic studies (studies that deal with the fates of chemical compounds when they react in biological systems). When a compound has been isotopically enriched, isotopic peaks are much larger than normal. Mass spectrometry is used to measure quantitatively the amount of such isotopes present in labeled compounds.

Focused Problems

13.16 The mass spectrum of tetramethylsilane, $(CH_3)_4Si$, has a base peak at $m/z = 73$. Calculate the relative abundances of the isotopic peaks at $m/z = 74$ and 75.

13.17 From the information in Table 13.3, predict the appearance of the molecular ion peak(s) in the mass spectrum of chloromethane. (Assume that the molecular ion is the base peak.)

C. Fragmentation

In EI mass spectrometry, the molecular ion is formed by loss of an electron. If this ion is stable, it decomposes slowly and is detected by the mass spectrometer as a peak of large relative abundance. If this ion is less stable, it decomposes, sometimes completely, into smaller pieces. Two cases of such fragmentation are most commonly observed, and in each case, two products are formed:

1. One fragmentation product can be a radical, in which case the other product is a carbocation with no unpaired electrons (an *even-electron ion*).

2. One fragmentation product can be a neutral molecule, in which case the other product must be, like the molecular ion, a radical cation (an *odd-electron ion*).

In either case, the cation is called a **fragment ion**. *Only the ion is detected in the mass spectrum*; the radical (case 1) or neutral molecule (case 2) is not detected.

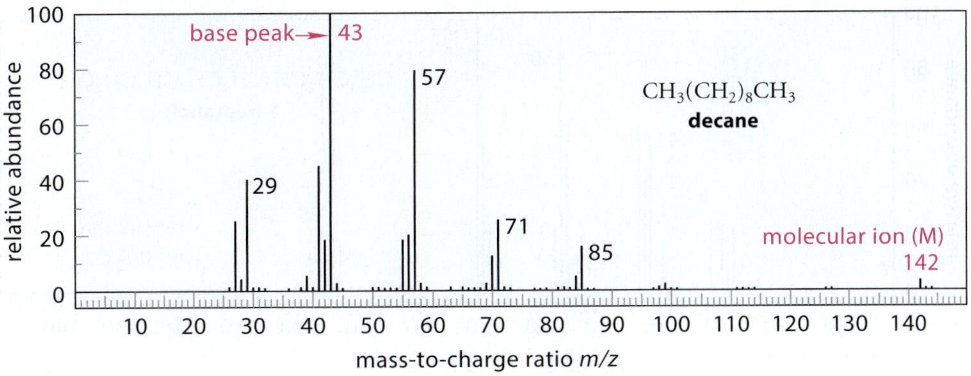

FIGURE 13.19 The EI mass spectrum of decane.

The way to discern a mechanism for a fragmentation is to consider the mass of the ion and postulate its elemental analysis on the basis of the elements present. Alternatively—or in addition—you can consider the mass of the fragment lost by subtracting the mass of the fragment observed from the mass of the molecular ion. If this corresponds to a common molecule or group, we can postulate how such a group might be lost. For example, loss of 18 mass units from a molecule containing oxygen strongly suggests that a water molecule is lost. Having postulated the source of the fragmentation, you can then write a fishhook and/or curved-arrow mechanism.

As an example of case 1, consider the $m/z = 57$ ion in the mass spectrum of decane (**Fig. 13.19**), formed in the following way. First, one of several possible molecular ions is formed by ejection of an electron from a carbon–carbon bond.

$$CH_3CH_2CH_2CH_2{-}CH_2CH_2CH_2CH_2CH_2CH_3 \xrightarrow{-e^-} CH_3CH_2CH_2CH_2\overset{+}{\cdot}CH_2CH_2CH_2CH_2CH_2CH_3$$

decane

a molecular ion of decane
(a radical cation, $m/z = 142$)

(13.22)

This ion can fragment in two ways. One mode of fragmentation gives the ion with $m/z = 57$. We show this fragmentation with the fishhook notation.

$$CH_3CH_2CH_2CH_2\overset{+}{\cdot}CH_2CH_2CH_2CH_2CH_2CH_3 \longrightarrow CH_3CH_2CH_2\overset{+}{C}H_2 + \dot{C}H_2CH_2CH_2CH_2CH_2CH_3$$

a molecular ion of decane
(a radical cation, $m/z = 142$)

a cation
$m/z = 57$
(detected by the mass spectrometer)

a radical
(not detected by the mass spectrometer)

(13.23)

Remember: only the cation is detected as a peak in the spectrum.

The radical cation can also fragment in the opposite way, with the charge on the larger fragment. The peak in Fig. 13.19 at $m/z = 85$ shows that this process occurs as well. The $m/z = 85$ fragment *does not arise from the radical in Eq. 13.23* but rather from the carbocation with $m/z = 85$ and the radical with mass = 57, as follows:

$$CH_3CH_2CH_2CH_2\overset{+}{\cdot}CH_2CH_2CH_2CH_2CH_2CH_3 \longrightarrow CH_3CH_2CH_2\dot{C}H_2 + \overset{+}{C}H_2CH_2CH_2CH_2CH_2CH_3$$

a molecular ion of decane

a radical
(not detected by the mass spectrometer)

a cation
$m/z = 85$
(detected by the mass spectrometer)

(13.24)

Other prominent peaks in the mass spectrum of decane arise from fragmentations by the same mechanism of other molecular ions formed at different carbon–carbon bonds.

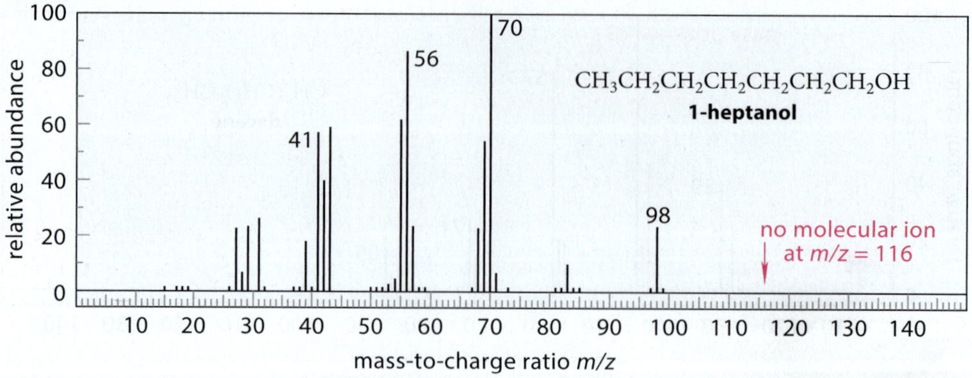

FIGURE 13.20 The EI mass spectrum of 1-heptanol.

A fragmentation of type 2 is illustrated by the mass spectra of many primary alcohols. For example, in the mass spectrum of 1-heptanol (molecular mass = 116), shown in **Fig. 13.20**, the peak of highest mass occurs at $m/z = 98$, which is 18 mass units less than the molecular mass. A fragment of this mass could arise by loss of the neutral fragment H_2O from the molecular ion. If a neutral fragment is lost from a cation radical, then, in addition to water, a new cation radical (an odd-electron ion) must be formed.

The process starts by electron ejection from one of the oxygen nonbonding electron pairs. Because they are held in orbitals of higher energy than bonding electrons, electrons from nonbonding pairs and π electrons are ejected more easily than electrons from σ bonds.

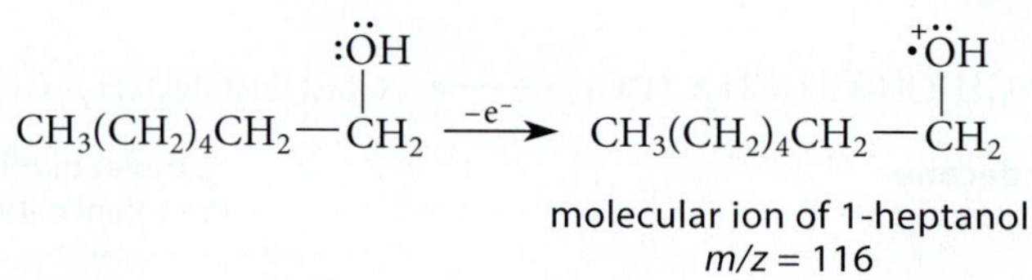

(13.25)

For H_2O to be lost as a neutral molecule, the oxygen must pick up an extra hydrogen atom from within the molecule. An intramolecular hydrogen-atom transfer through a six-membered transition state produces a radical cation of the same mass in which the oxygen is now poised to act as a leaving group. Loss of water (a stable neutral molecule, 18 mass units) leaves another odd-electron ion with $m/z = M - 18 = 98$.

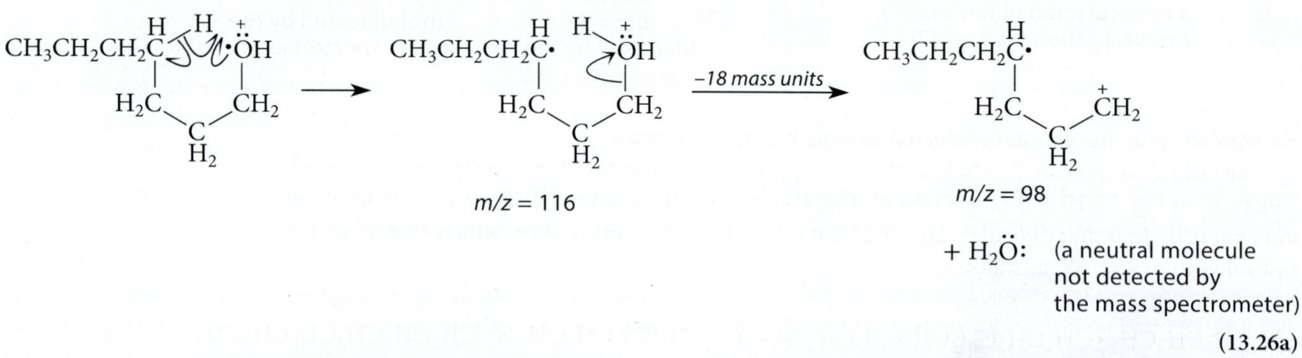

(13.26a)

The $m/z = 98$ peak is a fairly minor one in the mass spectrum because it undergoes a further type-2 fragmentation to give the neutral molecule ethylene (28 mass units) and another odd-electron radical cation, this one with five carbons.

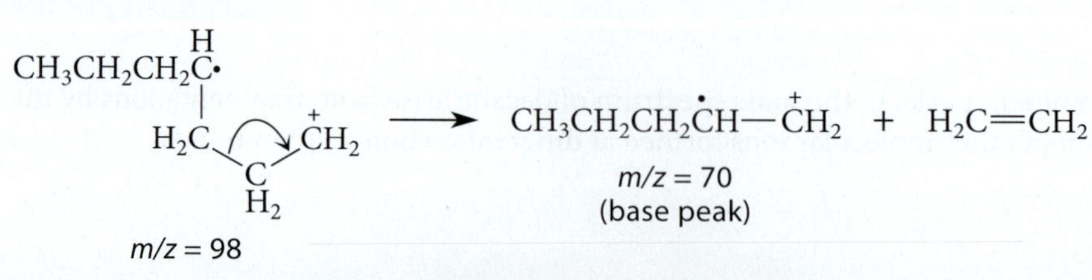

(13.26b)

The radical-cation product, in which a radical and a carbocation are located on adjacent carbons, is a type of odd-electron ion that is observed frequently in fragmentation. In this structure, both the carbon bearing the unpaired electron and the carbon bearing the positive charge contain 2p orbitals. Therefore, the unpaired electron is delocalized between them:

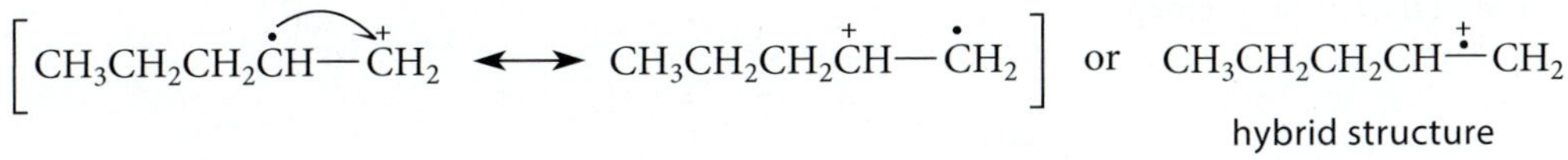

(13.26c)

If a molecule contains only C, H, O, and halogen, its even-electron fragment ions have odd mass and its molecular ion and its odd-electron fragment ions have even mass. You can verify this with the examples in Eqs. 13.23–13.26. Therefore, from the mass of the fragment ion—odd or even—you immediately know something about its structure and its origin.

This generalization is different if a molecule contains nitrogen. If a molecule or odd-electron fragment ion contains an odd number of nitrogens, it has an odd mass; but an even-electron fragment with an odd number of nitrogens has even mass. (This generalization is sometimes called the **nitrogen rule**.) For example, the amine butylisopropylmethylamine, shown in Eq. 13.27 (molecular mass = 129), has a single nitrogen and therefore its molecular ion (an odd-electron ion) has an odd mass. It undergoes fragmentation to give two prominent ions at $m/z = 114$ and $m/z = 86$. The $m/z = 114$ peak reflects a loss of 15 mass units, which is a methyl radical, and the $m/z = 86$ peak reflects a loss of 43 mass units, which is a propyl or isopropyl radical. With these losses the fragment ions must contain a nitrogen, and, as shown by the nitrogen rule, they have even masses.

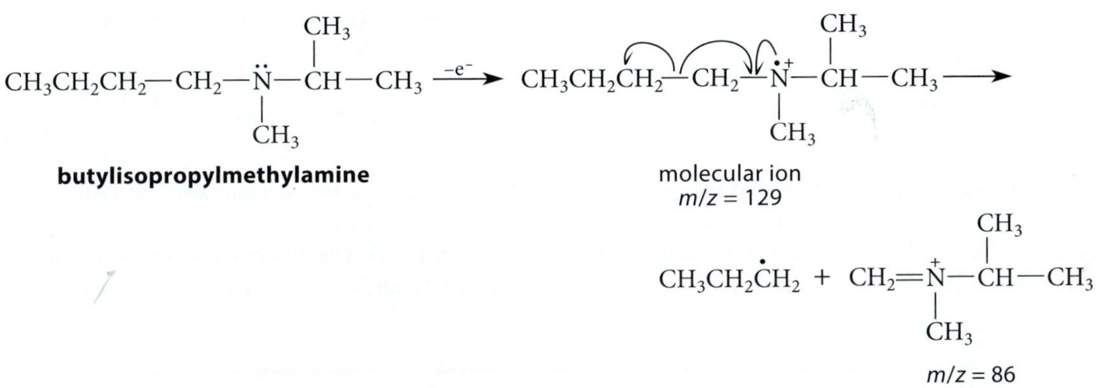

(13.27)

(We leave it to you to show the loss of the methyl radical.)

As all of the mass spectra in this section illustrate, the peaks in a mass spectrum typically do not have the same intensity. What controls the relative abundances of ions in a mass spectrum? Typically, the most stable ions appear in greatest abundance. If an ion is relatively stable, it decomposes slowly and appears as a relatively large peak. If an ion is relatively unstable, it decomposes rapidly and appears as a relatively small peak—or perhaps not at all. The principles of carbocation stability that you already know can help you to understand why certain fragment ions in a mass spectrum are prominent and others are not. This idea is illustrated in Study Problem 13.4.

Study Problem 13.4

The base peak in the mass spectrum of 2,2,5,5-tetramethylhexane (molecular mass = 142) is at $m/z = 57$, which corresponds to a composition C_4H_9. (a) Suggest a structure for the fragment that accounts for this peak. (b) Offer a reason that this fragment is so abundant. (c) Give a mechanism that shows the formation of this fragment.

Solution The first step is to draw the structure of 2,2,5,5-tetramethylhexane:

$$(CH_3)_3C\text{—}CH_2\text{—}CH_2\text{—}C(CH_3)_3$$

(a) A fragment with the composition C_4H_9 could be a *tert*-butyl cation formed by splitting the compound at the bond to either of the *tert*-butyl groups:

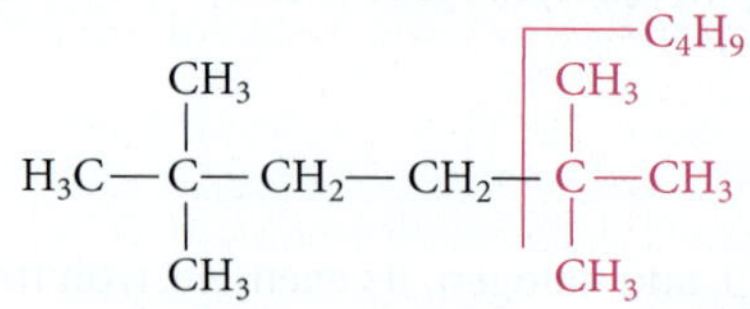

(b) The most abundant peaks in the mass spectrum result from the most stable cationic fragments. Because a *tert*-butyl cation is a relatively stable carbocation (it is tertiary), it is formed in relatively high abundance.

(c) To form this cation, one electron is ejected from the C—C bond, and the compound fragments so that the unpaired electron remains on the methylene carbon:

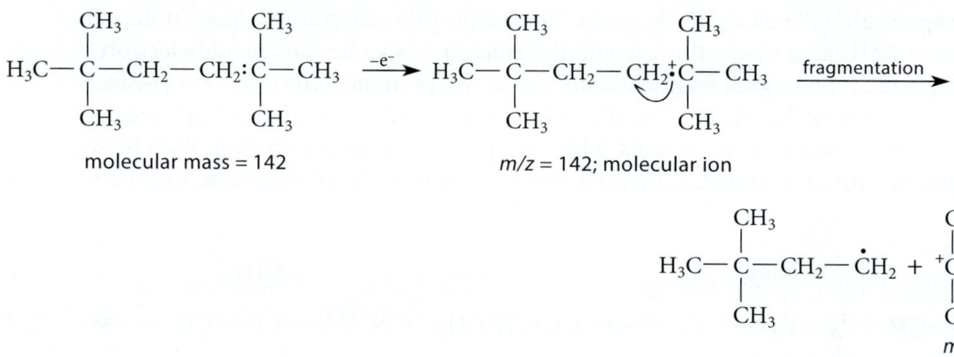

(13.28)

Fragmentation might have occurred at the same bond so that the unpaired electron remains associated with the *tert*-butyl group and a primary carbocation with $m/z = 85$ is formed. (In other words, a *more stable* free radical and a *less stable* carbocation would be formed.) There is no peak at $m/z = 85$. That this mode of fragmentation is *not* observed demonstrates that *carbocation stability is more important than free-radical stability in determining fragmentation patterns.*

Focused Problems

13.18 The peak of highest mass in the EI mass spectrum of 2,2,5,5-tetramethylhexane (the molecule discussed in Study Problem 13.4) occurs at $m/z = 71$ and has about 33% relative abundance.

(a) In a structure of the molecule, indicate the bond at which fragmentation occurs to give this ion.

(b) Give a mechanism for this fragmentation.

(c) What is the structure of the fragment ion at $m/z = 71$? (*Hint:* Apply what you know about carbocations.)

13.19 Indicate whether the following peaks in the mass spectrum of 1-heptanol are odd-electron or even-electron ions. (Don't attempt to give their structures.)

(a) $m/z = 83$ (b) $m/z = 56$ (c) $m/z = 41$

13.20 The mass spectrum of 2-chloropentane shows large and almost equally intense peaks at $m/z = 71$ and $m/z = 70$.

(a) Classify each peak as an even-electron or odd-electron ion.

(b) What stable neutral molecule can be lost to give the odd-electron ion?

(c) Write a curved-arrow or fishhook mechanism for the origin of each fragment ion.

D. The Molecular Ion; Chemical-Ionization (CI) Mass Spectra

The molecular ion is the most important peak in the mass spectrum for two reasons. First, the m/z of the molecular ion occurs at the molecular mass, and one of the most important uses of mass spectrometry is the determination of molecular mass. Second, the mass of the molecular ion is the basis for the calculation of losses due to fragmentation.

Unfortunately, a peak due to the molecular ion is weak or absent in some mass spectra. For example, the molecular mass of di-*sec*-butyl ether is 130. In its EI mass spectrum (**Fig. 13.21a**), however, a peak at this mass is absent. The three most prominent peaks in the EI mass spectrum of di-*sec*-butyl ether occur at $m/z = 101$, $m/z = 57$, and $m/z = 45$ (base peak). The $m/z = 101$ peak corresponds to a loss of 29 mass units (that is, an ethyl group) and occurs in the following way. First, the molecular ion is formed by loss of an electron from the oxygen nonbonding electron pair:

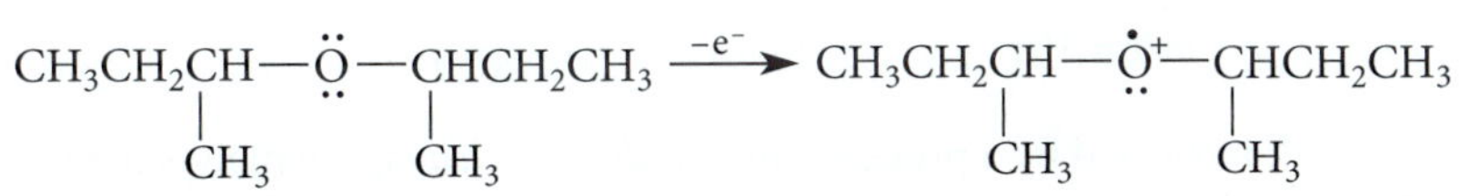

di-*sec*-butyl ether — molecular ion of di-*sec*-butyl ether ($m/z = 130$) (13.29)

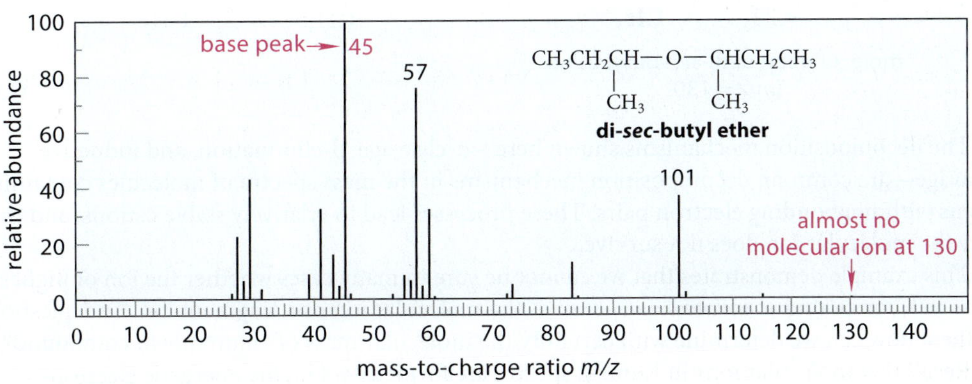

(a) EI mass spectrum

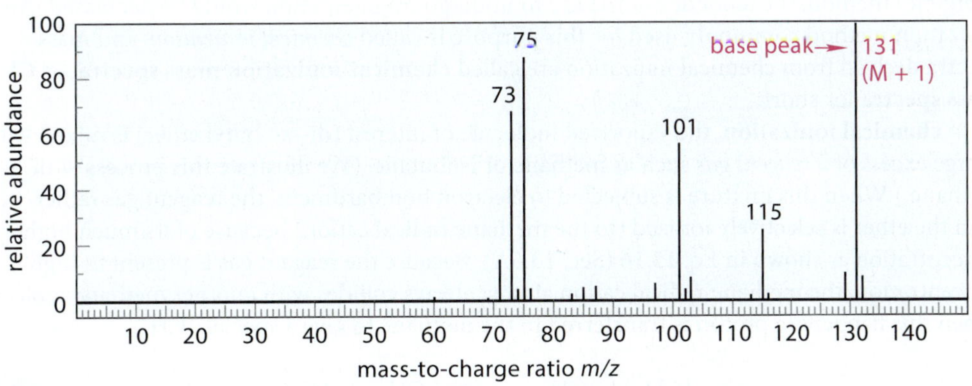

(b) CI mass spectrum

FIGURE 13.21 Mass spectra of di-*sec*-butyl ether. (a) Electron-ionization (EI) mass spectrum. (b) Chemical-ionization (CI) mass spectrum. The molecular ion at $m/z = 130$ is essentially absent in the EI spectrum, whereas the molecular ion (as the protonated ether at $m/z = 131$) is the base peak in the CI spectrum. The CI spectrum has a smaller number of fragment ions than the EI spectrum.

(The discussion of Eq. 13.25 noted that nonbonding electrons are more easily removed than bonding electrons.) Next, an ethyl radical is lost by a process that mass spectrometrists call **α-cleavage** and free-radical chemists call *β-scission*:

See Eq. 10.35e (Sec. 10.3B) for an example of β-scission in a free-radical reaction.

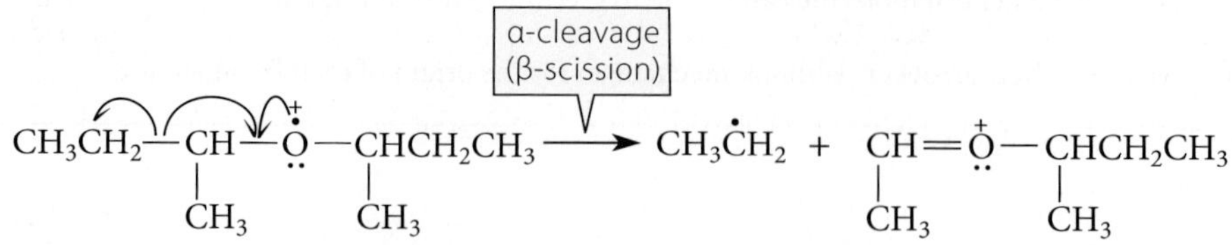

molecular ion of di-*sec*-butyl ether ($m/z = 130$) — $m/z = 101$

(13.30)

(The amine fragmentation shown in Eq. 13.27 is also an α-cleavage reaction.)

This ion reacts further by β-elimination of 2-butene (56 mass units) to give the base peak at $m/z = 45$.

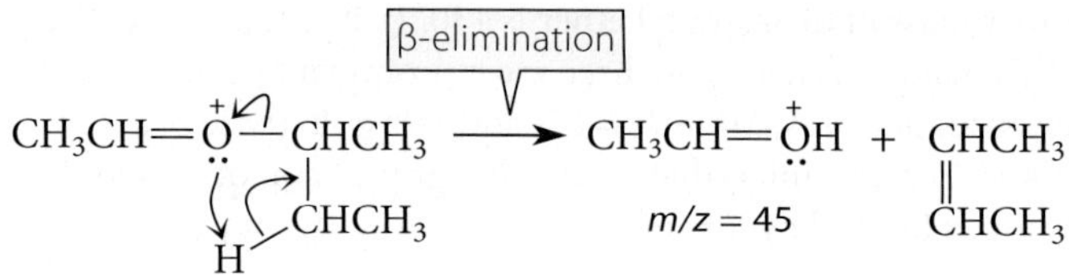

$m/z = 101$

(13.31)

The $m/z = 57$ peak is formed by a process called *inductive cleavage*, which is the radical cation version of an S_N1-like dissociation:

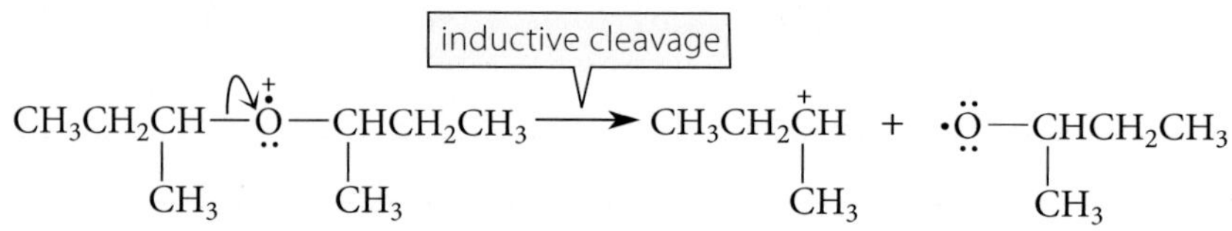

molecular ion of di-*sec*-butyl ether ($m/z = 130$)

(13.32)

The decomposition mechanisms shown here—α-cleavage, β-elimination, and inductive cleavage—are common decomposition mechanisms in the mass spectra of molecules containing atoms with nonbonding electron pairs. These processes lead to relatively stable cations, and this is why the molecular ion does not survive.

This example demonstrates that we cannot be sure in many cases whether the ion of highest mass in a compound of *unknown structure* is the molecular ion or a fragment ion. The question is, then, how can we determine with certainty the molecular mass of an unknown compound?

Recall that molecular ions in EI mass spectra are formed by a highly energetic electron-bombardment process. When a molecular ion has a very high energy, it is likely to dissipate that energy by fragmentation. However, if a molecular ion could be formed by a "softer" (less energetic) method, the tendency of the ion to undergo fragmentation would be decreased. An ionization method commonly used for this purpose is called *chemical ionization*, and mass spectra derived from chemical ionization are called **chemical-ionization mass spectra**, or **CI mass spectra** for short.

Chemical-ionization mass spectrometry was first reported in 1966 by Burnaby Munson (1933–2019) and Frank H. Field (1922–2013), chemists at the Esso Research and Engineering Company in Baytown, Texas. (Esso later became Exxon-Mobil.) Munson subsequently joined the faculty of the University of Delaware, and Field joined the faculty of The Rockefeller University.

In **chemical ionization**, the vaporized molecule of interest (di-*sec*-butyl ether) is mixed with a large excess of a *reagent gas* such as methane or isobutane. (We illustrate this process with methane.) When this mixture is subjected to electron bombardment, the reagent gas rather than the ether is selectively ionized (to the methane radical cation) because of its much higher concentration as shown in Eq. 13.16 (Sec. 13.6A). Because the reagent gas is present in high concentration, the methane radical cation almost always collides with another methane molecule. When this happens, a proton is transferred to the methane to give a species $^+CH_5$.

$$\overset{\bullet +}{}CH_4 + CH_4 \longrightarrow \dot{C}H_3 + {}^+CH_5 \qquad (13.33a)$$

In this unusual species, *five* protons share the *four* bonding electron pairs that were in the original methane. The important thing about $^{+}CH_5$ is that it is *very acidic*: in fact, *it is a source of gas-phase protons.* The $^{+}CH_5$ mostly collides with methane (giving no net reaction); but when it eventually collides with the ether, the ether is protonated at its most basic site, which is a nonbonding electron pair on the oxygen (Sec. 12.1). As a result, the ether is converted into its conjugate-base cation:

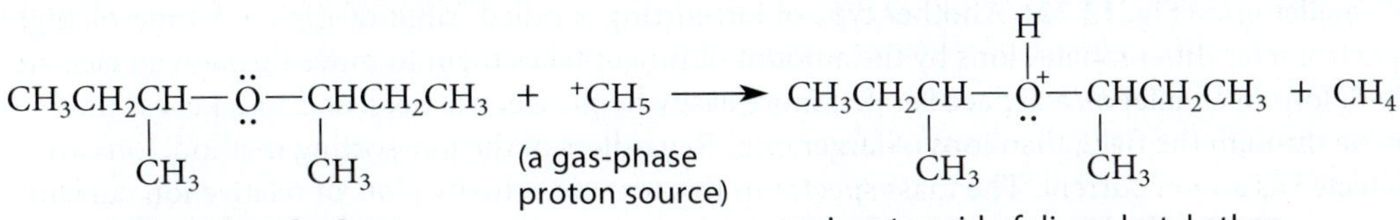

di-*sec*-butyl ether conjugate acid of di-*sec*-butyl ether $m/z = 131$ (13.33b)

This conjugate-acid cation is an even-electron ion; it is *not* a radical cation. In a CI mass spectrum, the peak for this ion occurs *1 mass unit higher* than the molecular mass of the molecule itself because of the added proton. Because this ion is formed in a relatively low-energy process, it does not fragment so readily as the molecular ion in the EI mass spectrum. The CI mass spectrum of di-*sec*-butyl ether is shown in **Fig. 13.21b**. This shows a prominent M + 1 ion at $m/z = 131$, which is also the base peak. The relatively small number of fragments come from the loss of various neutral molecules from this ion. For example, the largest fragment peak at $m/z = 75$ arises from loss of 2-butene in a β-elimination process analogous to the one in Eq. 13.30:

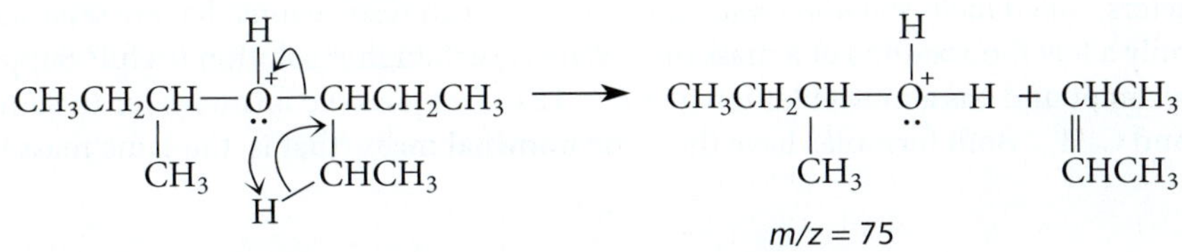

$m/z = 75$

conjugate acid of di-*sec*-butyl ether $m/z = 131$ (13.34)

Typically, a compound with unknown structure is investigated by determining both its EI and CI mass spectra. The CI mass spectrum typically gives a strong M + 1 peak that reveals the molecular mass M. The richer fragmentation pattern of the EI spectrum can then be used to deduce other aspects of the structure.

Focused Problems

13.21 (a) The EI mass spectrum of $CH_3OCH_2CH(CH_3)_2$ (methyl isobutyl ether) contains only a trace of a molecular ion and a base peak at $m/z = 45$, which arises from α-cleavage. Show the α-cleavage process, and give the structure of the ion with mass = 45.

(b) The EI mass spectrum of methyl isobutyl ether does *not* show a peak due to inductive cleavage, in contrast to the mass spectrum of di-*sec*-butyl ether (Eq. 13.31). Use what you know about carbocation stability to explain the absence of this peak.

(c) What major difference(s) would you expect to find when comparing the CI mass spectrum of methyl isobutyl ether with its EI spectrum?

13.22 Show the elimination reactions that account for each of the following fragments in the CI mass spectrum of di-*sec*-butyl ether (Fig. 13.21b). (*Hint:* β-Elimination reactions can also form C=O double bonds.)

(a) $m/z = 101$ (b) $m/z = 115$ (c) $m/z = 73$

FURTHER EXPLORATION 13.3
The Mass Spectrometer

E. The Mass Spectrometer

A mass spectrometer must produce gas-phase ions, sort them by mass, and detect the relative number of ions of each mass. We've already discussed two ways of producing ions: electron ionization and chemical ionization. Ion sorting in a conventional "magnetic-sector" mass spectrometer depends on the fact that the paths of moving ions are bent by a magnetic field. When subjected to a magnetic field, ions with large m/z traverse a path of larger radius than ions of smaller m/z (**Fig. 13.22**). Another type of ion sorting is called "time-of-flight." A time-of-flight spectrometer differentiates ions by the amount of time it takes them to move through an electric field. Ions of smaller m/z are accelerated more easily by the electric field, and so take less time to move through the field, than ions of larger m/z. Regardless of the ion-sorting method, ions are detected as an ion current. The mass spectra in this text are actually plots of relative ion currents versus m/z, with the largest ion current (that is, the base peak) assigned a relative value of 100.

A modern mass spectrometer is an extremely sensitive instrument and can readily produce a mass spectrum from amounts of material in the range of micrograms (10^{-6}g) to picograms (10^{-12}g). For this reason, the instrument is very useful for the analysis of materials available in only trace quantities. It has played a key role in such projects as the analysis of drug levels in blood serum and the elucidation of the structures of insect pheromones (Sec. 10.9) that are available only in minuscule amounts. It is also an important tool in the modern forensics laboratory ("crime lab").

One of the operating characteristics of a mass spectrometer is its *resolution*—how well it separates ions of different mass. A relatively simple mass spectrometer readily distinguishes, over a total m/z range of several hundred, ions that differ in mass by 1 unit. More complex mass spectrometers, called *high-resolution mass spectrometers*, can resolve ions that are separated in mass by only a few thousandths of a mass unit. Why is such high resolution useful? Suppose an unknown compound has a molecular ion at $m/z = 124$. Two possible formulas for this ion are $C_8H_{12}O$ and C_9H_{16}. Both formulas have the same **nominal mass** (that is, the same mass to the

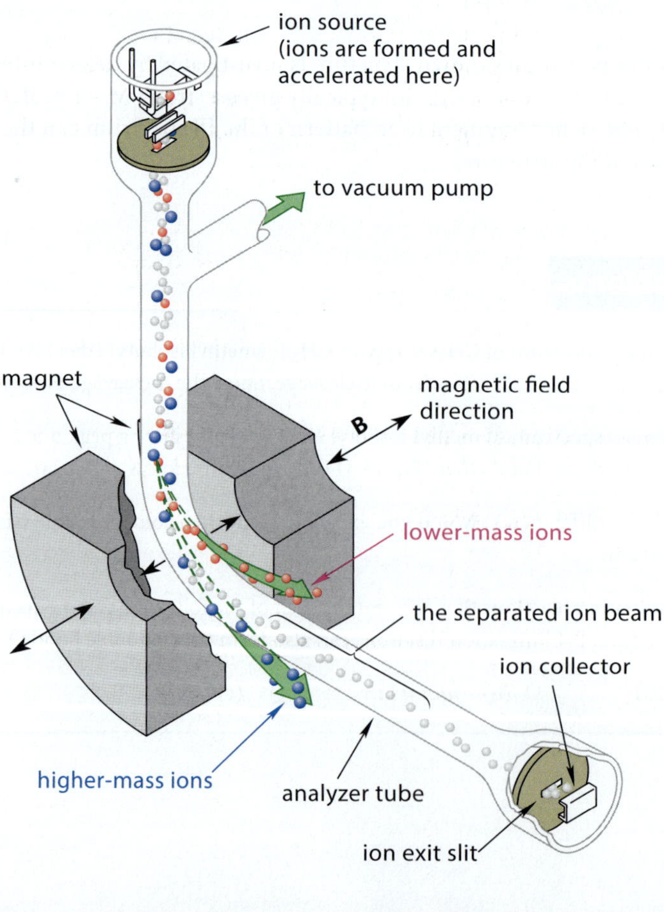

FIGURE 13.22 Diagram of a magnetic-sector mass spectrometer. After ionization of the sample by electron bombardment, the ions are accelerated by a high voltage and are passed into a magnetic field **B** along a path perpendicular to the field. The field bends the paths of the ions; the paths of lower-mass ions (red) are bent more than those of higher-mass ions (blue). (See Further Exploration 13.3.) As the field is progressively increased, ions of increasingly higher mass attain exactly the correct path to enter the ion exit slit.

nearest whole number). However, if the **exact mass** (the mass to four or more decimal places) is calculated for each formula (using the values of the most abundant isotopes in Table 13.2), then different results are obtained:

$$C_8H_{12}O\text{, exact mass: } 124.0888$$

$$C_9H_{16}\text{, exact mass: } 124.1252$$

The difference of 0.0364 mass units is easily resolved by a high-resolution mass spectrometer. Computers used with such instruments can be programmed to work backward from the exact mass and provide *an elemental analysis of the molecular ion* (and therefore the compound of interest) *as well as the elemental analysis of each fragment in the mass spectrum*! Because a modern high-resolution mass spectrometer with its associated computer and other accessories can cost several hundred thousand dollars, it is generally shared by a large number of researchers.

Before a compound can be analyzed by mass spectrometry, it must be vaporized. This presents a difficult problem for large molecules that have negligible vapor pressures. Research in mass spectrometry has focused on novel ways to produce ions in the gas phase from large nonvolatile molecules, many of which are of biological interest. In one technique, called **matrix-assisted laser desorption ionization (MALDI)**, the material to be analyzed (analyte) is co-crystallized with a material, termed a *matrix*, that can absorb radiation from a laser. In a process that is not fully understood, bombarding the matrix–analyte mixture with light from the laser ultimately produces gas-phase ions of the analyte, which are analyzed by mass spectrometry. In another technique, called **electrospray ionization (ESI)**, a solution of the analyte is atomized in highly charged droplets, much as we might atomize perfume in a sprayer. This process results in the formation of highly charged molecules in the gas phase, and these are analyzed by mass spectrometry. These techniques have made possible the analysis of materials with molecular masses in excess of 100,000, such as proteins, nucleic acids, and synthetic polymers. (Some examples of mass spectra obtained by these techniques are discussed in Sec. 27.8B.) For their discovery and development of these techniques, John P. Fenn (1917–2010), of Virginia Commonwealth University, and Koichi Tanaka (b. 1959), of the Shimadzu Corporation in Tokyo, shared part of the 2002 Nobel Prize in Chemistry.

Although the discussion in this chapter has focused on detection of cations in the mass spectrometer, anions can also be produced by electron capture during the electron-bombardment process, particularly if the energy of the electrons is reduced. These anions can be detected if the mass spectrometer is operated in negative-ion detection mode.

CHAPTER SUMMARY *For a summary of the chapter, see Chapter 13 in the* Study Guide and Solutions Manual.

SKILLS OBJECTIVES WITH PROBLEMS

- **Interconvert the wavelength, wavenumber, frequency, and energy of a wave. 13.2A**

13.23 The C═O group of acetone has an IR stretching absorption, taken in the liquid, at 1714 cm^{-1}.

(a) How many times per second does this bond vibration occur?

(b) What is the wavelength of the energy required to excite this vibration to the next vibrational energy level?

(c) What is the energy of the IR radiation required to excite this vibration to the next vibrational energy level?

- **Interconvert transmittance and absorbance data. 13.2B**

13.24 In dilute solution, the intensity of an absorption, measured as the *absorbance*, is proportional to concentration of the sample. A certain IR absorption has a % transmittance (%*T*) value of 0.1. After dilution of the sample, the %*T* of the same absorption was 0.5. By what factor was the sample diluted?

- **Know the relationship between vibrational frequency and force constant. 13.3A**

13.25 The force constant for stretching a C═C bond is 2.16 times the force constant for stretching a C—C bond. Recall that the stretching vibrations for C—C bonds are typically weak or missing in the IR spectrum. Given a C═C stretching frequency of about 1650 cm^{-1}, estimate the frequency of the C—C stretching vibration.

- **Know the relationship between vibrational frequency and masses involved in the vibrating bond. 13.3A**

13.26 If the O—H stretching frequency of an alcohol in the gas phase (where intermolecular hydrogen bonding is absent) is about 3600 cm^{-1}, estimate the stretching frequency of the O—D bond resulting from the replacement of H by D in the same compound.

- **Describe the normal vibrational modes of simple vibrating systems, and decide which should be infrared-active. 13.3A**

13.27 (a) The water molecule has three distinguishable molecular vibrations. Construct a diagram like Fig. 13.8 (Sec. 13.3A) for the three normal vibrational modes of water. (*Hint:* Each vibration at its extremes must change the molecule so that it can be distinguished from the molecule before the vibration occurs.)

(b) Classify each vibration as a stretching or bending vibration.

(c) The IR spectrum of water vapor has three absorptions: 1595, 3652, and 3756 cm^{-1}. Which are stretching vibrations, and which are bending vibrations? Explain.

13.28 Explain why a nitro compound has two N—O stretching vibrations. (These typically occur at about 1370 and 1550 cm^{-1}.)

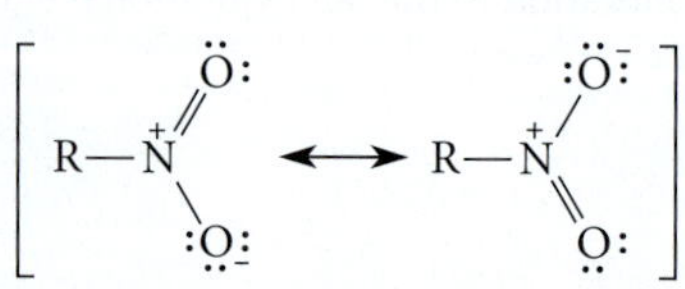

- **Indicate on the basis of symmetry whether certain vibrations should be infrared-active. 13.3B**

13.29 Classify each of the following normal vibrational modes as IR-active or IR-inactive. Explain. (The red arrows show stretching, and the blue arrows show compression.)

(a)  (acetylene C≡C stretch)

(b) 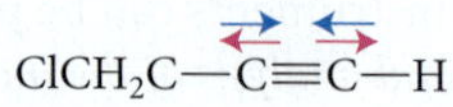 (4-chlorobutyne C≡C stretch)

(c) 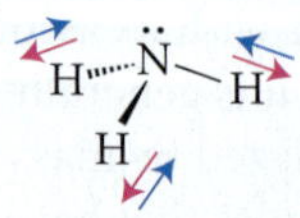 (ammonia N—H symmetrical stretch)

(d) 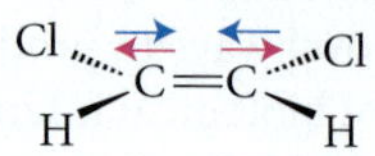 (*Z*)-1,2-dichloroethylene C═C stretch

(e) 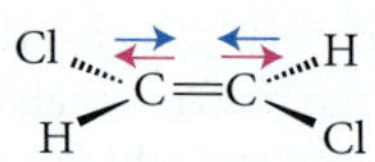 (*E*)-1,2-dichloroethylene C═C stretch

- **Know the general regions of the IR spectrum for important functional-group absorptions. 13.4**

13.30 Match each of the following approximate IR absorptions with the corresponding functional group. (Some absorptions do not have a match; the absorption is missing for one functional group.)

3400–3200 cm^{-1}	3300 cm^{-1}	3100 cm^{-1}
A	*B*	*C*

2950 cm^{-1}	1870 cm^{-1}	1750 cm^{-1}	1670 cm^{-1}
D	*E*	*F*	*G*

1235 cm^{-1}	875–990 cm^{-1}
H	*I*

(a) C≡C stretch

(b) C═C stretch

(c) O—H stretch (hydrogen-bonded)

(d) ═C—H bend

(e) C—O stretch

(f) ≡C—H stretch

(g) ═C—H stretch

- **Use IR spectra to distinguish between molecules. 13.4**

13.31 Match each of the six IR spectra in **Fig. P13.31** to one of the following compounds. (There is no spectrum for two of the compounds.)

(a) 1,5-Hexadiene

(b) 1-Methylcyclopentene

(c) 1-Hexen-3-ol

(d) Dipropyl ether

(e) *Trans*-4-octene

(f) Cyclohexane

(g) 3-Hexanol

(h) 1-Hexyne

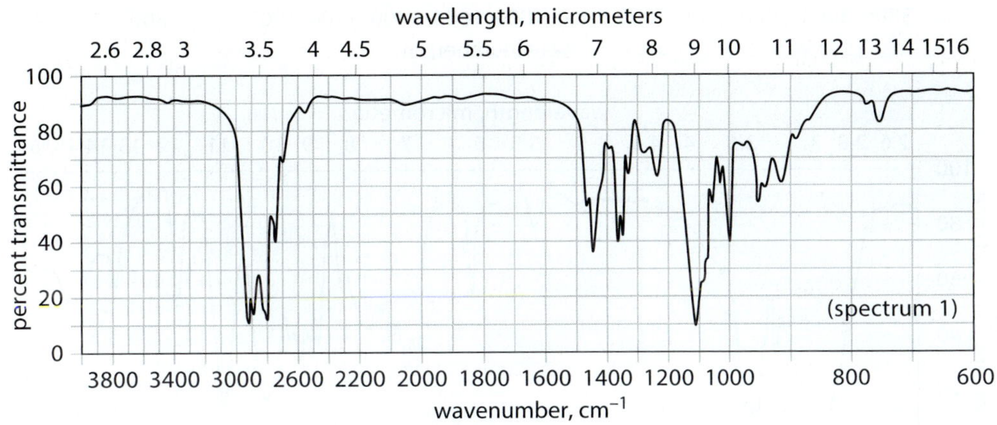

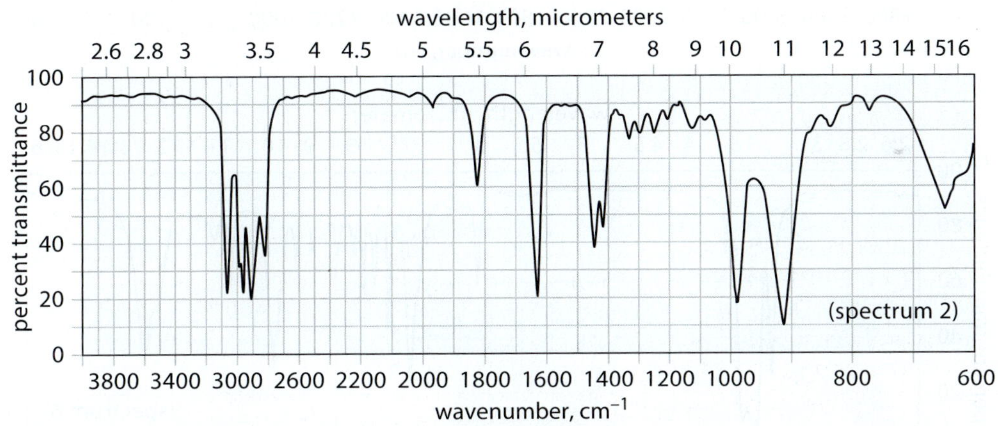

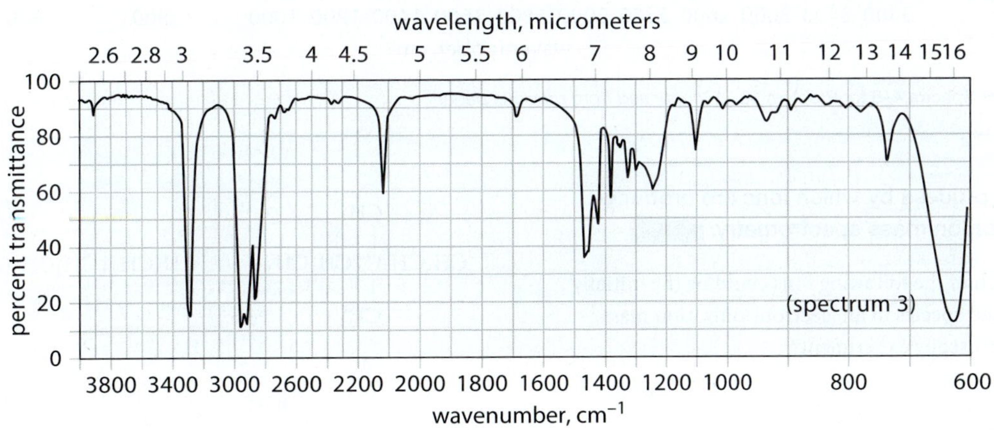

FIGURE P13.31 IR Spectra 1–3 for Problem 13.31 *(continues on next page)*

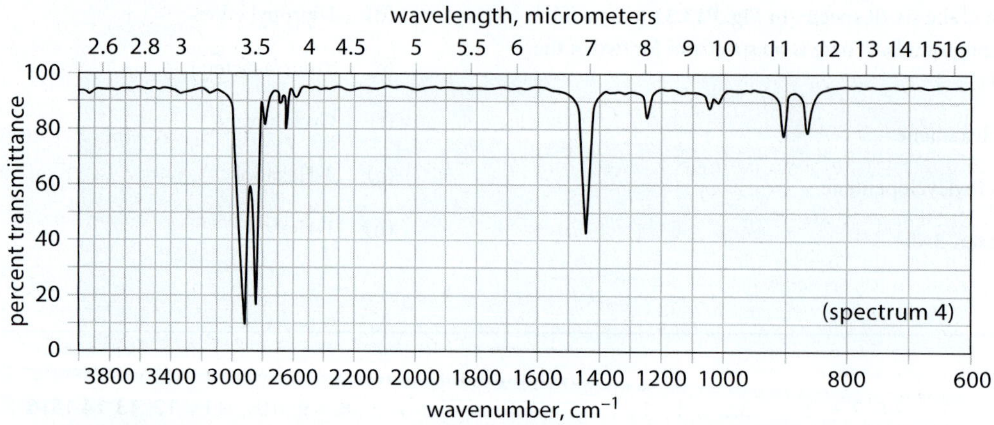

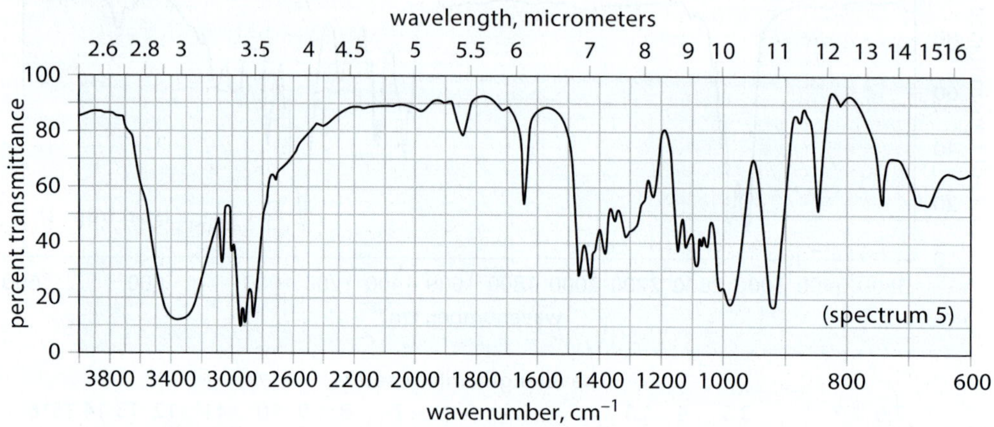

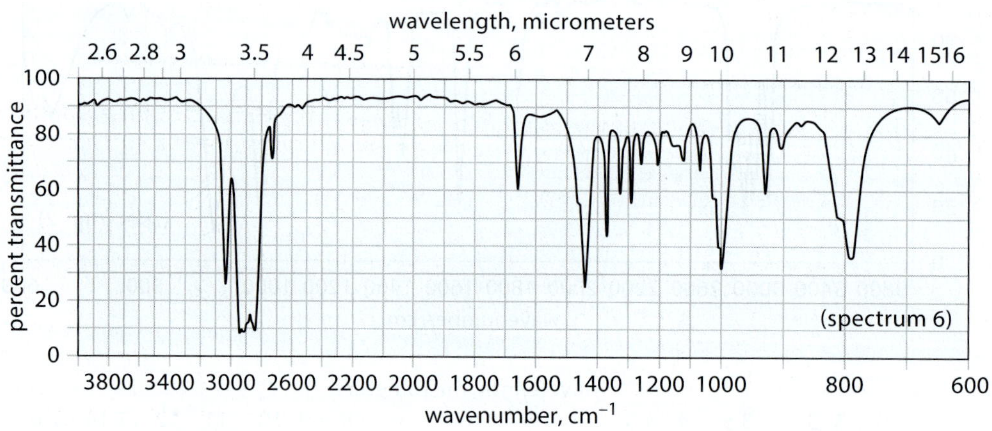

FIGURE P13.31 **IR** Spectra 4–6 for Problem 13.31 *(continued from previous page)*

- **Describe the process by which ions are produced in electron-ionization mass spectrometry.** **13.6A**

13.32 (a) Which of the following ions could be the initially formed species in an electron-ionization mass spectroscopy experiment?

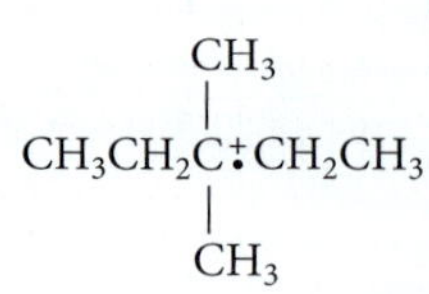

A

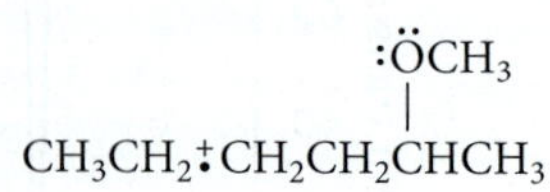

B

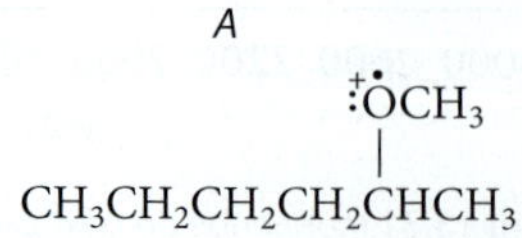

C

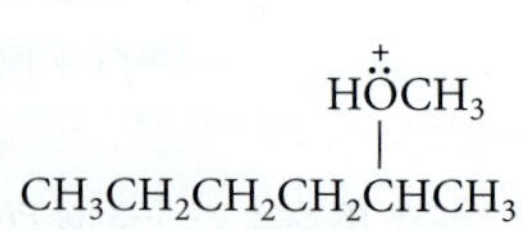

D

(b) Describe the process for the formation of any one of the ions you identified in part (a).

• Given the relative intensity of a peak in the mass spectrum and a table of relative isotopic abundances, calculate the relative intensity of an isotopic peak. 13.6B

13.33 (a) In the EI mass spectrum of 3-methyl-2,3-dihydrofuran, the molecular ion (M) (at $m/z = 84$) has a relative intensity of 55%. What is the intensity of the M + 1 ion?

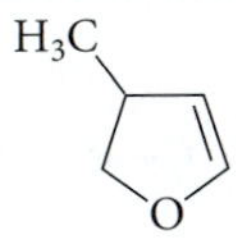

3-methyl-2,3-dihydrofuran

(b) The base peak in the mass spectrum of this compound occurs at $m/z = 69$. What is the predicted intensity of the isotopic peak at $m/z = 70$? (*Hint:* What fragment is lost to give a mass of 69?)

13.34 Predict the relative intensities of the three peaks in the mass spectrum of dichloromethane at $m/z = 84$, 86, and 88.

• Using the curved-arrow notation, show how certain fragment ions are formed in an electron-impact mass spectrum. 13.6C

13.35 Using the structure of the molecular ion in Problem 13.33a as your starting point, give a curved-arrow or fishhook notation for the formation of the fragment ion at $m/z = 69$. [*Hints:* (1) Is this an odd- or even-electron ion? (2) The molecular ion is formed by ejection of an oxygen electron.]

13.36 Rationalize the indicated fragments in the EI mass spectrum of each of the following molecules by proposing a structure of the fragment and a mechanism by which it is produced.

(a)

$$CH_3CH_2CH_2-\underset{CH_2CH_3}{\overset{CH_3}{\underset{|}{\overset{|}{C}}}}-\ddot{N}H_2, \quad m/z = 72$$

(b) 3-Methyl-3-hexanol, $m/z = 73$

(c) 1-Pentanol, $m/z = 70$

(d) Neopentane, $m/z = 57$

13.37 Rationalize each of the following observations by postulating a structure for the fragment ion(s) and the mechanisms for their formation.

(a) The EI mass spectrum of 1-methoxybutane shows fragment ions at $m/z = 56$ and $m/z = 45$ (base peak).

(b) The EI mass spectrum of 2-methoxybutane shows a base peak at $m/z = 59$.

• For a compound with known molecular formula, determine from the mass of a peak in the mass spectrum whether it corresponds to an odd-electron or even-electron ion. 13.6C

13.38 Determine whether each of the following ions is an odd- or even-electron ion.

(a) A fragment ion has a composition of C_4H_9.

(b) From a compound containing only C, H, and O, a fragment ion has $m/z = 70$.

13.39 A compound contains carbon, hydrogen, oxygen, and one nitrogen. Classify each of the following fragment ions derived from this compound as an odd-electron or an even-electron ion. Explain.

(a) The molecular ion

(b) A fragment ion of even mass containing one nitrogen

(c) A fragment ion of odd mass containing one nitrogen

• Describe the process by which molecular ions are formed in chemical-ionization mass spectra. 13.6D

13.40 Tributylamine (Bu_3N:) has a molecular mass of 185. Give the reaction by which the initially formed ion is produced in chemical-ionization mass spectrometry, and give the structure and molecular mass of the ion. Assume that CH_5^+ is the source of gas-phase protons.

13.41 Give the structure of the ion initially formed when the compound in Problem 13.33a is subjected to chemical ionization mass spectrometry. (*Hint:* The ion is resonance-stabilized.)

INTEGRATED PROBLEMS

13.42 Which of the molecules in each of the following pairs should have identical IR spectra, and which should have different IR spectra (if only slightly different)? Explain your reasoning carefully.

(a) 3-Pentanol and (±)-2-pentanol

(b) (*R*)-2-Pentanol and (*S*)-2-pentanol

(c)

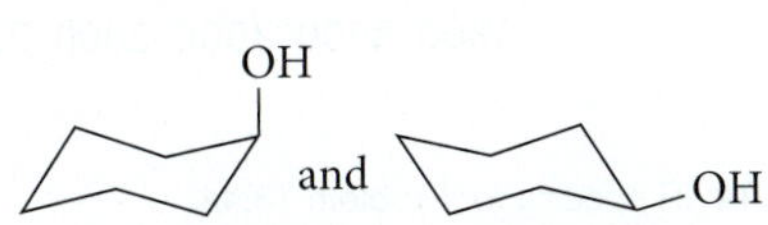

13.43 Indicate how you would carry out each of the following chemical transformations. What are some of the changes in the infrared spectrum that could be used to indicate whether the reaction has proceeded as indicated? (Your answer can include the disappearance as well as the appearance of IR absorptions.)

(a) 1-Methylcyclohexene ⟶ methylcyclohexane

(b) 1-Hexanol ⟶ 1-methoxyhexane

13.44 A former theological student, Heavn Hardley, has turned to chemistry and has carried out the following reaction:

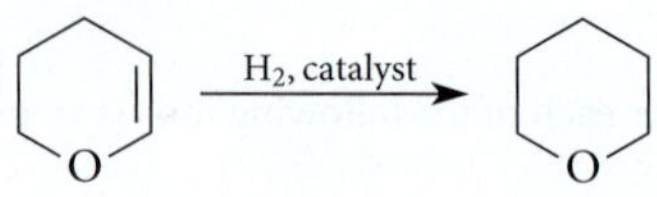

A *B*

Unfortunately, Hardley thinks he may have mislabeled his samples of *A* and *B*, but has wisely decided to take an IR spectrum of each sample. The spectra are reproduced in **Fig. P13.44.** Which sample goes with which spectrum? How do you know?

13.45 (a) Given the stretching frequencies for the C—H bonds shown in color, arrange the corresponding bonds in order of increasing strength. Explain your reasoning.

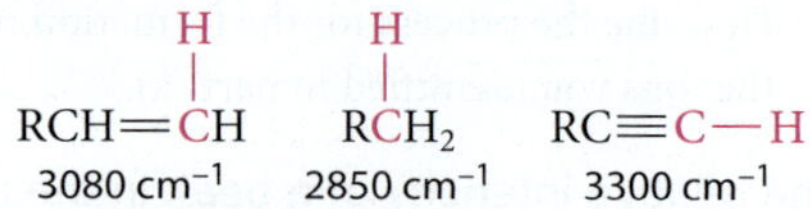

(b) If the bond dissociation energy of the ═C—H bond is 558 kJ mol^{-1} (133 kcal mol^{-1}), use the stretching frequencies in part (a) to estimate the bond dissociation energy of the C—H bond in RCH_2—H.

13.46 Arrange the following bonds in order of increasing stretching frequencies, and explain your reasoning. (*Hint:* Use Table 10.2 in Sec. 10.1E)

C═C C≡C C═O C—C

13.47 (a) Explain why the S—H stretching absorption in the IR spectrum of a thiol is less intense and occurs at much lower frequency (2550 cm^{-1}) than the O—H stretching absorption of an alcohol.

(b) Is the wavenumber difference between O—H and S—H absorptions caused primarily by the greater mass of sulfur or by the relative strengths of the two bonds? Explain how you know.

(c) Two unlabeled bottles, *A* and *B*, contain liquids. Laboratory notes suggest that one compound is

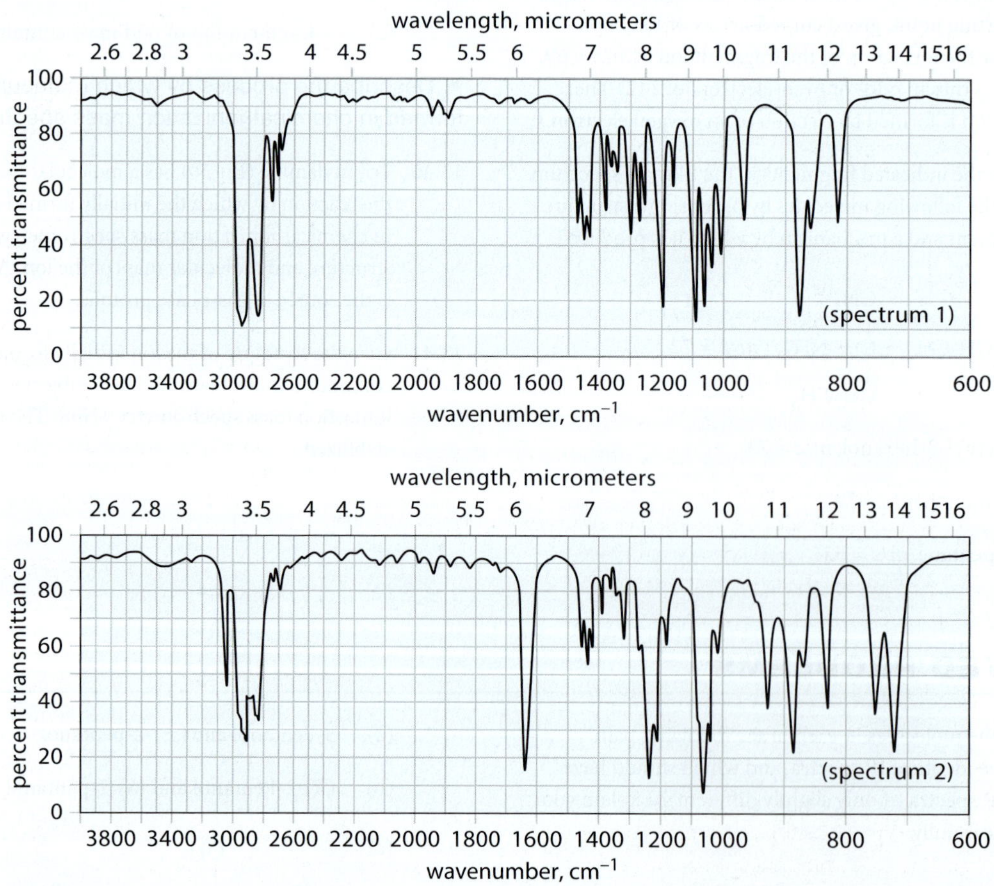

FIGURE P13.44 IR spectra for Problem 13.44.

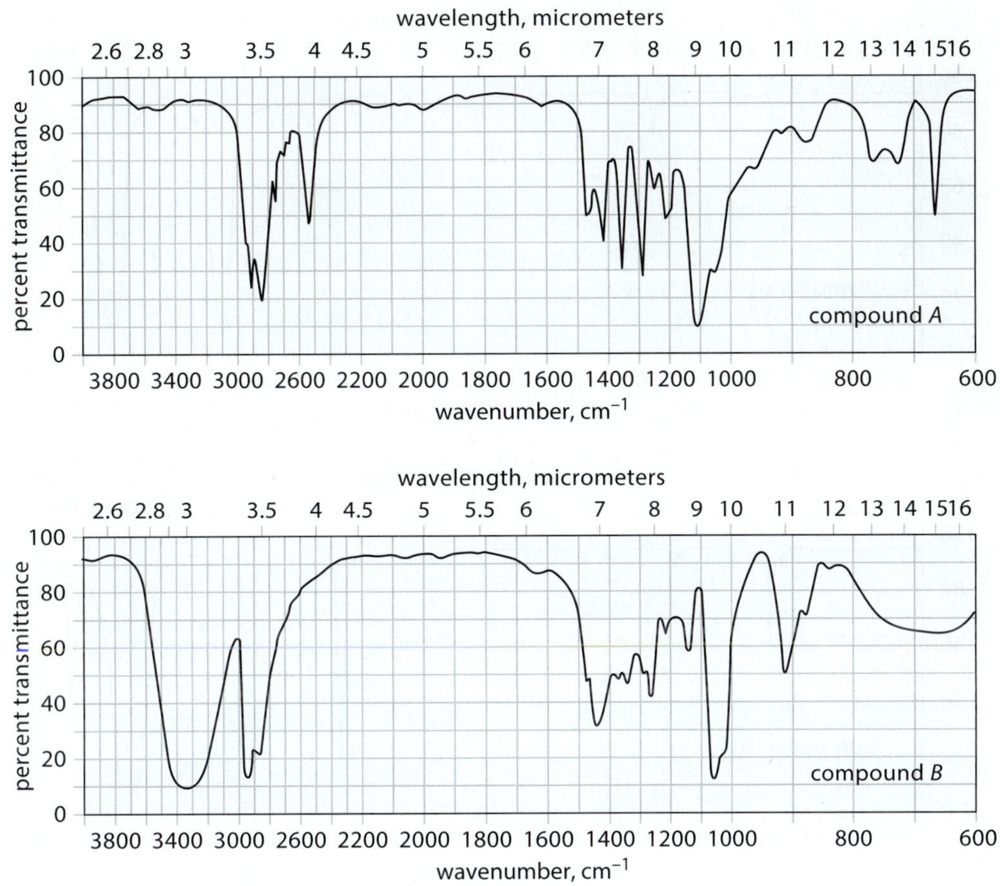

FIGURE P13.47 IR spectra for Problem 13.47.

$(HSCH_2CH_2)_2O$ and the other is $(HOCH_2CH_2)_2S$. The IR spectra of the two compounds are given in **Fig. P13.47**. Identify *A* and *B*, and explain your choice.

13.48 (a) One of the compounds with the IR spectra in **Fig. P13.48** is deuterated chloroform ($CDCl_3$) and the other is ordinary chloroform ($CHCl_3$). Which compound is which? Explain how you know.

(b) How would these compounds be distinguished by mass spectrometry?

13.49 An alcohol *A*, when treated with NaH followed by CH_3I, gives a compound *B* with a strong M + 1 peak in its CI mass spectrum at $m/z = 117$. Compound *A*, known from other evidence to be a tertiary alcohol, has prominent fragments in its EI mass spectrum at $m/z = 87$ and $m/z = 73$ (base peak). Propose structures for compounds *A* and *B*.

13.50 A chemist, Ilov Boronin, carried out a reaction of *trans*-2-pentene with BH_3 in tetrahydrofuran (THF) followed by treatment with $H_2O_2/^{-}OH$. Two product alcohols were separated and isolated. Ilov obtained their mass spectra, which are shown in **Fig. P13.50**. Suggest structures for the compounds, and indicate which mass spectrum goes with which compound. (*Hint:* The intensity of the molecular ion is negligible in both mass spectra.)

13.51 Suggest structures for the following *neutral* molecules commonly lost in mass spectral fragmentation.

(a) Mass = 28 from a compound containing only C and H

(b) Mass = 18 from a compound containing C, H, and O

(c) Mass = 36 from a compound with an M + 2 peak about one-third the size of the molecular ion

13.52 Explain why the EI mass spectrum of dibromomethane has three peaks at $m/z = 172$, 174, and 176 in the approximate relative abundances 1:2:1.

13.53 Suggest a structure for each of the ions corresponding to the following peaks in the EI mass spectrum of ethyl bromide, and give a mechanism for the formation of each ion. (The numbers in parentheses are the relative abundances.)

(a) $m/z = 110$ (98%)

(b) $m/z = 108$ (100%)

(c) $m/z = 81$ (5%)

(d) $m/z = 79$ (5%)

(e) $m/z = 29$ (61%)

(f) $m/z = 28$ (25%)

(g) $m/z = 27$ (53%)

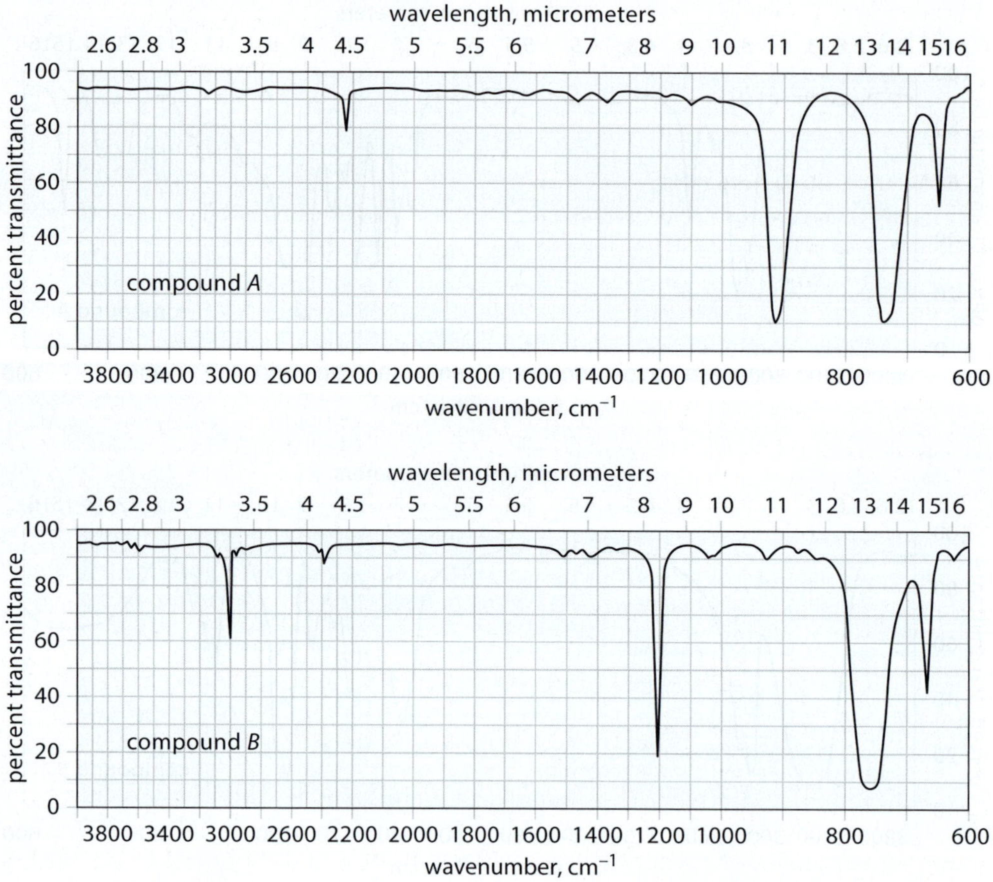

FIGURE P13.48 IR spectra for Problem 13.48.

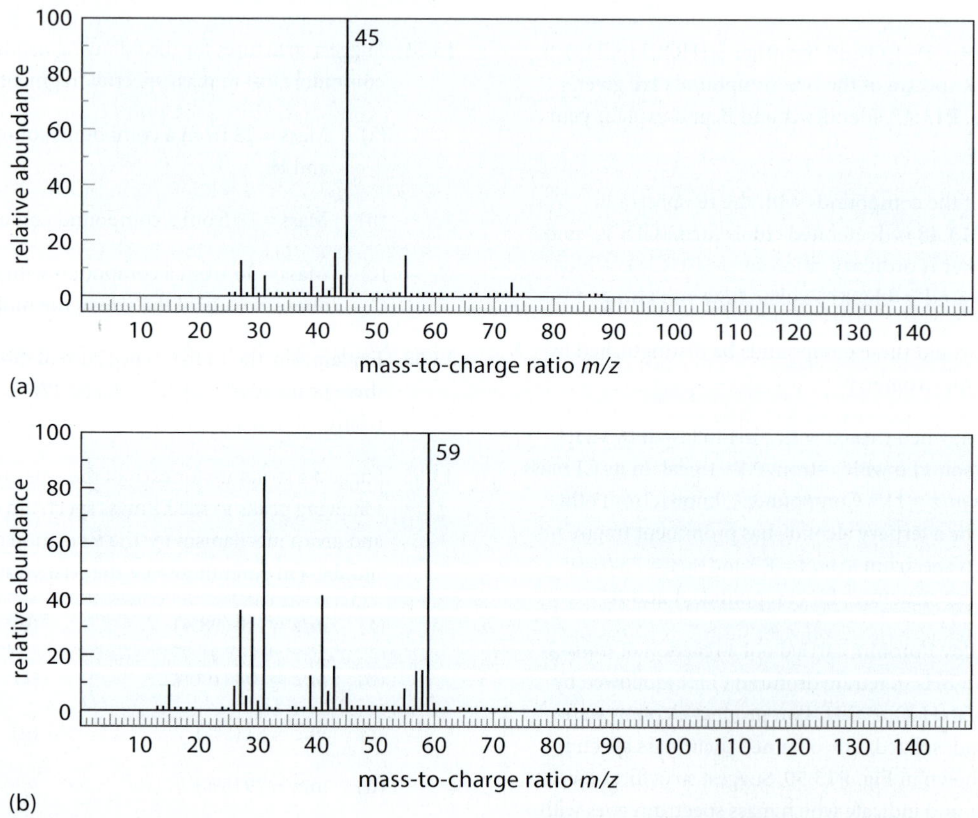

FIGURE P13.50 Mass spectra for Problem P13.50.

13.54 (a) Explain why ionization of a π electron requires less energy than ionization of a σ electron.

(b) Draw the structure of the molecular ion of 1-heptene formed by ionization of a π electron.

(c) The base peak in the EI mass spectrum of 1-heptene (Fig. 13.17 in Sec. 13.6A) occurs at $m/z = 41$, which is believed to correspond to the allyl cation, a resonance-stabilized carbocation.

$$H_2C{=}CH{-}\overset{+}{C}H_2$$

allyl cation
$m/z = 41$

Draw a curved-arrow mechanism (or fishhook mechanism) that shows the conversion of the molecular ion you drew in part (b) into the allyl cation.

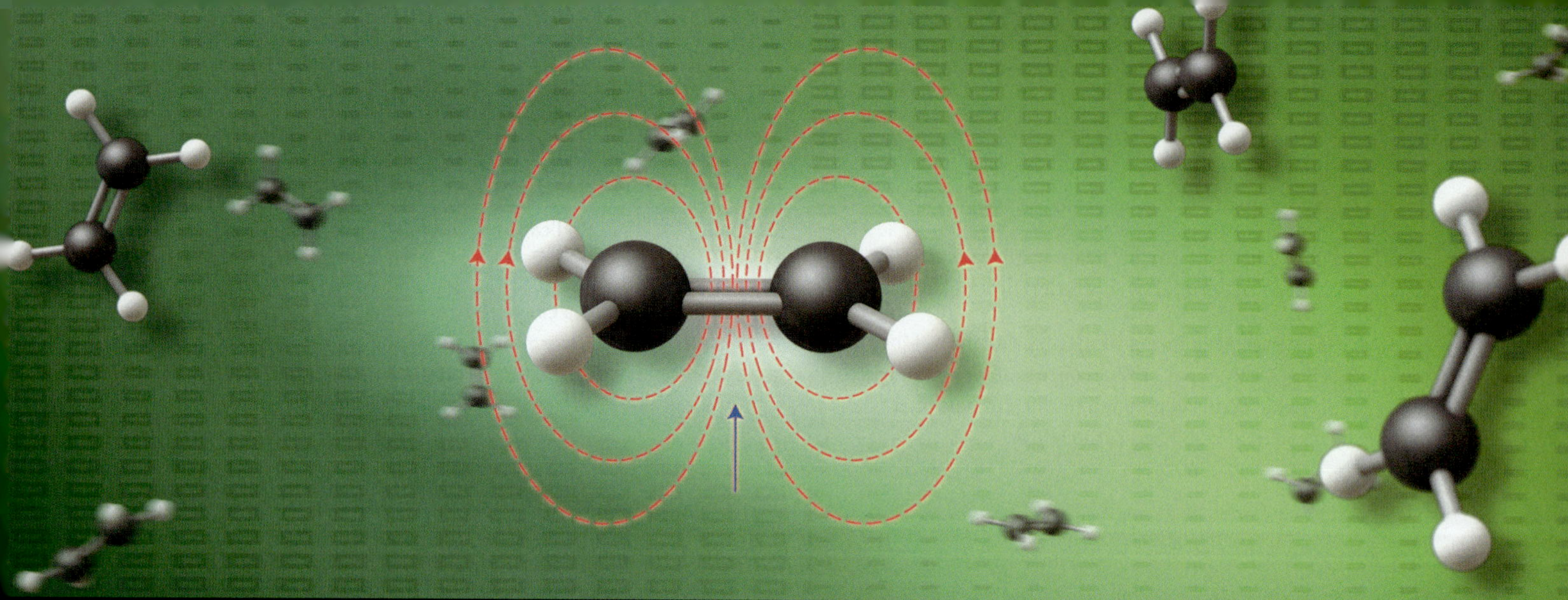

14 NUCLEAR MAGNETIC RESONANCE SPECTROSCOPY

As explained in Chapter 13, infrared spectroscopy can be used to determine the *functional groups* present in a compound, and mass spectrometry provides the *masses of a molecule and its coherent fragments*. With rare exceptions, however, neither of these techniques gives enough information to define a complete structure. Another form of spectroscopy, *nuclear magnetic resonance* (NMR) spectroscopy, enables us to probe molecular structure in much greater detail. Using NMR, sometimes in conjunction with other forms of spectroscopy but often by itself, we can in many cases determine a complete molecular structure in a very short time. Since its commercial introduction in the 1950s, NMR spectroscopy has revolutionized organic chemistry. This chapter presents the basic principles of NMR spectroscopy and shows how it is used in structure determination.

14.1 AN OVERVIEW OF PROTON NMR SPECTROSCOPY

^{13}C NMR is discussed in Sec. 14.9, and the spin properties of some other nuclei used in NMR are listed in Table 14.4, Sec. 14.9A.

NMR spectroscopy is used to detect *nuclei*, but only those nuclei that have a magnetic property called *spin*, which we discuss further in Sec. 14.2. The proton (^{1}H) and a minor isotope of carbon with atomic mass = 13 (^{13}C) have spin and can therefore be detected with NMR. The common isotope of carbon ^{12}C does *not* have spin and cannot be detected by NMR.

Historically, the first use of NMR in organic chemistry, and still an important use, is for the detection of protons—hydrogen nuclei—in organic compounds. This type of NMR is called **proton NMR,** or **^{1}H NMR**. In the first part of this chapter we deal with proton NMR.

We begin our study of NMR by looking at a simple proton NMR spectrum: the spectrum of dimethoxymethane, shown in **Fig. 14.1.**

$$CH_3O—CH_2—OCH_3$$

dimethoxymethane

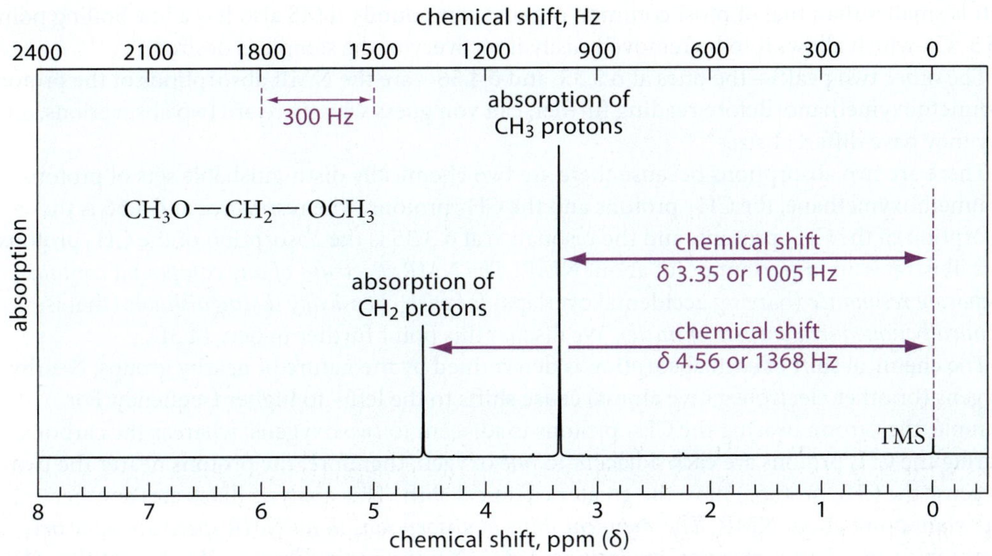

FIGURE 14.1 The proton NMR spectrum of dimethoxymethane. The lower axis is the chemical shift scale in parts per million (ppm), and the upper axis is the chemical shift scale in frequency units (Hz). The peaks represent energy absorption by the protons of each chemical type. The small peak at the far right is from the protons of tetramethylsilane (TMS), a reference standard added in a small amount.

This spectrum is a plot of energy absorption on the *y*-axis versus relative frequency of the radiation from which energy is absorbed on the *x*-axis. The absorptions detect the *protons* in the molecule. The units on the lower horizontal axis, abbreviated δ, are called *parts per million*, or *ppm*. For now, don't be concerned about how these units are derived; you should simply view them as position markers on this axis. The numbers on the upper horizontal axis are frequency units in hertz (Hz; the Hz was defined in Sec. 13.1A). Frequency decreases from left to right, as it does in an IR spectrum.

The numbers on the ppm (δ) scale and the numbers on the frequency scale are proportional. For most of the spectra in this book, the frequency numbers are exactly 300 times the δ numbers. That is, if the frequency numbers on the upper axis are symbolized by Δν, then

$$\delta = \frac{\Delta\nu}{\nu_0} \quad (\Delta\nu \text{ in Hz}, \nu_0 \text{ in MHz}) \tag{14.1}$$

where ν_0 = 300 million Hz, or 300 megahertz (MHz). The proportionality constant ν_0, in units of megahertz, is called the **operating frequency** of the NMR spectrometer. As the name implies, this is an operating characteristic of the NMR spectrometer. We discuss this relationship in Sec. 14.3B.

Peaks in an NMR spectrum are called **resonances, absorptions**, or **lines**. The position of an absorption on the horizontal axis is called its **chemical shift**. We usually express the peak positions in ppm—that is, we use the lower horizontal axis. In this case, the chemical shift of an absorption is written with a δ followed by the numerical value of the peak position. When we use the δ notation, the units of ppm are implied and are not repeated. For example, we see three peaks in Fig. 14.1: these have chemical shifts at δ 0, δ 3.35, and δ 4.56. Or we can say that the spectrum contains peaks at 0, 3.35, and 4.56 ppm.

The rightmost absorption—the one at δ 0—is not an absorption of dimethoxymethane. Rather, this is an absorption of tetramethylsilane (TMS), a compound added to each sample to provide a reference point.

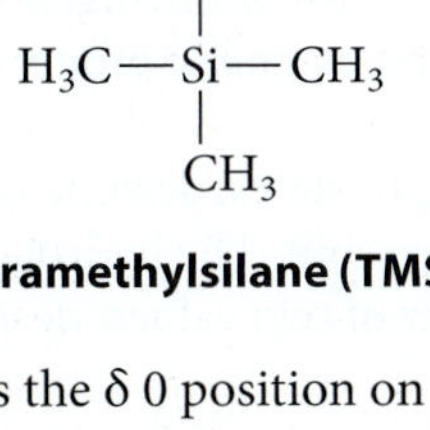

tetramethylsilane (TMS) (14.2)

The absorption position of TMS defines the δ 0 position on the *x*-axis of each spectrum. TMS is used as a standard because it has a single strong absorption, it is chemically inert, and its chemical shift is smaller than that of most common organic compounds. TMS also has a low boiling point (26.5 °C), which allows it to be removed easily if recovery of the sample is desired.

The other two peaks—the ones at δ 3.35, and δ 4.56—are the NMR absorptions of the protons of dimethoxymethane. Before reading further, can you guess why there are two absorptions, and why they have different sizes?

There are two absorptions because there are two chemically distinguishable sets of protons in dimethoxymethane, the CH_2 protons and the CH_3 protons. The resonance at δ 4.56 is the absorption of the CH_2 protons, and the resonance at δ 3.35 is the absorption of the CH_3 protons. This illustrates an important point about NMR: *The NMR spectrum of any compound contains a separate resonance* (barring accidental overlaps) *for each chemically distinguishable* (that is, *chemically nonequivalent*) *set of nuclei.* We discuss this point further in Sec. 14.3D.

The chemical shift of each absorption is determined by the nature of nearby groups. Nearby oxygens (or other electronegative atoms) cause shifts to the left—to higher frequency. For example, the carbon bearing the CH_2 protons is adjacent to *two* oxygens, whereas the carbons bearing the CH_3 protons are each adjacent to *one* oxygen; therefore, the protons nearer the two oxygens (the CH_2 protons) have the greater chemical shift. This analysis illustrates another important point about NMR: *The chemical shifts of absorptions in an NMR spectrum vary in a predictable way with the chemical environment of the corresponding protons.* We discuss the effect of structure on chemical shift in Sec. 14.3C.

The two peaks have different sizes because different numbers of protons contribute to each absorption. The resonance at δ 3.35 is larger because more protons (six) contribute to this resonance than to the resonance at δ 4.56 (two). In fact, you'll observe that the resonance at δ 3.35 is about three times as tall. This illustrates yet another important aspect of NMR spectra: *The size of a peak* (actually, the area under the peak) *is proportional to the number of protons contributing to the absorption.* This means that we can count the protons of each chemical type. We come back to this point in Sec. 14.3E.

When a compound contains hydrogens on adjacent carbons, the NMR spectrum provides additional, very powerful, information. *We can count the protons on adjacent carbons.* This aspect of NMR is not illustrated by the spectrum of dimethoxymethane, because in this molecule no two carbons are adjacent. This additional capability of NMR comes from a phenomenon called *splitting*, which we discuss in Sec. 14.4.

In summary, proton NMR provides four types of information:

1. The number of sets of chemically nonequivalent protons
2. The chemical environments of each set of protons (chemical shift)
3. The number of protons within each set
4. The number of protons in adjacent sets

With these four types of information, we can in many cases deduce completely the structures of unknown compounds. Using these ideas, we now consider the various aspects of NMR spectra in more detail. We begin by considering the NMR phenomenon itself.

14.2 THE PHYSICAL BASIS OF NMR SPECTROSCOPY

To understand NMR spectroscopy and to use it intelligently we must understand its physical basis. NMR spectroscopy is based on the magnetic properties of nuclei that result from a property called *nuclear spin.*

Just as electrons have two allowed spin states, designated by the quantum numbers $+\frac{1}{2}$ ("up") and $-\frac{1}{2}$ ("down"), some *nuclei* also have spin. The hydrogen nucleus ^1H—the proton—has a nuclear spin that also can assume either of two values, designated by quantum numbers $+\frac{1}{2}$ and $-\frac{1}{2}$.

The physical significance of nuclear spin is that the nucleus acts like a tiny magnet. You know from experience that magnets assume a preferred orientation in the presence of a magnetic field. (An example is the orientation of a compass needle in Earth's magnetic field.) The same is true of nuclei, which can be thought of as tiny magnets. Therefore, the magnetic poles of hydrogen nuclei become oriented in a magnetic field. That is, when a compound containing hydrogens is placed in a magnetic field, its hydrogen nuclei become magnetized.

Represent the hydrogen nuclei in a chemical sample with arrows indicating their magnetic ("north–south") polarity. In the absence of a magnetic field, the nuclear magnetic poles are oriented randomly. (We ignore the very small magnetic field of Earth.) After a magnetic field is applied, the magnetic poles of nuclei with spin of $+\frac{1}{2}$ are oriented parallel to the magnetic field, whereas those of nuclei with spin of $-\frac{1}{2}$ are oriented antiparallel to the field.

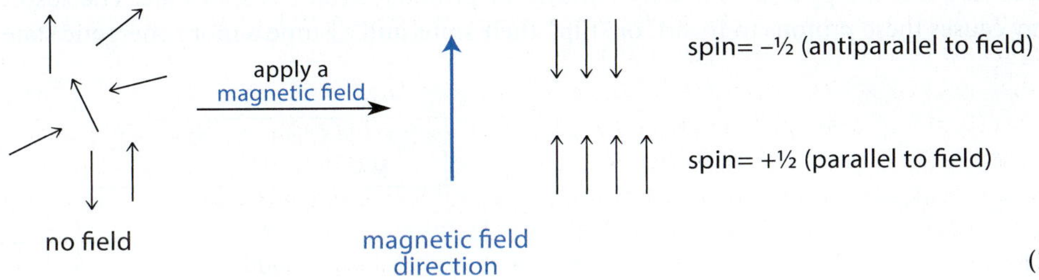

(14.3)

The most important effect of the magnetic field for NMR is how it affects the energies of the two spin states of the proton. In the absence of a field, the two spin states have the *same* energy. When a magnetic field is applied, however, the two spin states have *different* energies: the $+\frac{1}{2}$ spin state has lower energy than the $-\frac{1}{2}$ spin state. The energy difference between the two spin states of a proton p, $\Delta\varepsilon_p$, is given by the **fundamental equation of NMR**:

$$\Delta\varepsilon_p = \frac{h\gamma_{\mathrm{H}}}{2\pi}\mathbf{B}_p \tag{14.4}$$

In this equation, h is Planck's constant (Sec. 13.1A), 3.99×10^{-13} kJ s mol^{-1}; $\mathbf{B}_p$ is the magnitude of the magnetic field at the proton, in gauss (rhymes with *house*); and γ_{H} is a fundamental constant of the proton, called the **magnetogyric ratio**. The value of this constant is 26,753 radians gauss^{-1} s^{-1}. Equation 14.4 shows that when the magnetic field is zero, the spin states have identical energies, as you just learned; and as the magnetic field is increased, the energy difference between the two spin states grows, as shown in **Fig. 14.2**.

A typical field strength used in modern NMR spectrometers is 70,500 gauss. This is a *very strong* magnetic field! If you insert this value into Eq. 14.4, you can calculate that the energy

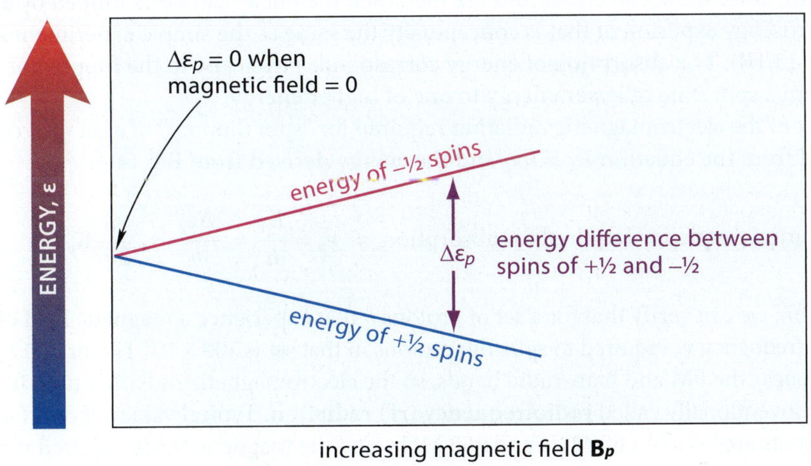

FIGURE 14.2 Effect of increasing magnetic field at a proton on the energy difference between its $+\frac{1}{2}$ and $-\frac{1}{2}$ spin states (Eq. 14.4). The two spin states have identical energies when the field is absent, and the energy difference between the two spin states grows with increasing field.

separation $\Delta\varepsilon_p$ is about 0.00012 kJ mol^{-1} (0.000029 kcal mol^{-1}). This is a *very small* energy! If we treat this energy difference as if it were $\Delta G°$ and calculate the equilibrium constant between the two spin states using Eq. 3.48b (Sec. 3.5B), we find that this energy corresponds to an equilibrium constant of 0.9999516. This equilibrium constant is so close to unity that in any sample of *1 million* protons, the difference in the populations of the two spin energy states is very small; the lower-energy spin state has roughly 20 more protons than the higher-energy spin state. Even though this difference is *minuscule*, it is physically significant and is the basis for NMR. It takes a large magnetic field to induce even this tiny energy difference between spin states.

Here is where things stand: Molecules of a sample are situated in a magnetic field; each proton is in one of two spin states that differ in energy by an amount $\Delta\varepsilon_p$; and a small excess of protons have spin $+\frac{1}{2}$. If the sample is now subjected to electromagnetic radiation with energy E_p *exactly equal* to $\Delta\varepsilon_p$, this energy is absorbed by some of the protons in the $+\frac{1}{2}$ spin state. The absorbed energy causes these protons to invert, or "flip," their spins and assume a more energetic state with spin $-\frac{1}{2}$.

Figure 13.1 (Sec. 13.1A) showed that electromagnetic radiation contains two types of oscillating fields: electric fields and magnetic fields. The electric field of infrared radiation is involved in IR spectroscopy. The magnetic field of radiofrequency radiation is involved in nuclear magnetic resonance.

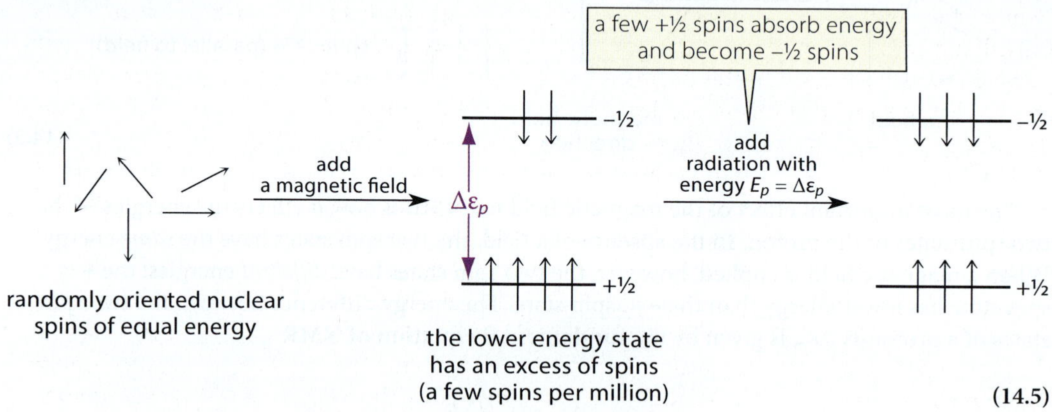

(14.5)

No radioactivity is involved in an NMR experiment. *Nuclear spin,* on which NMR depends, is a property of the nucleus, but it has nothing to do with radioactivity. The popular association of the word *nuclear* with the phenomenon of radioactivity is why magnetic resonance imaging (MRI), used extensively in medicine, was not called *nuclear* magnetic resonance imaging. MRI relies on the same NMR phenomenon used for determining molecular structures. (We discuss MRI in Sec. 14.12.)

This absorption phenomenon, called **nuclear magnetic resonance**, can be detected in a type of absorption spectrometer called a *nuclear magnetic resonance spectrometer*, or **NMR spectrometer** (Sec. 14.11). The study of this absorption is called **NMR spectroscopy**. This absorption results in the peaks we see in an NMR spectrum, such as the one in Fig. 14.1.

The resonance phenomenon in NMR has *nothing* whatsoever to do with the resonance structures discussed in Sec. 1.5.

To summarize: For nuclei to absorb energy, they must have a nuclear spin and must be situated in a magnetic field. Once these two conditions are met, then the nuclei can be examined by an absorption spectroscopy experiment that is conceptually the same as the simple experiment shown in Fig. 13.3 (Sec. 13.1B). The absorption of energy corresponds physically to the flipping of nuclear spins from a spin state of lower energy to one of higher energy.

The frequency of the electromagnetic radiation required for "spin flipping" of a set of protons p can be calculated from the equation $E_p = h\nu_p$ and the energy derived from Eq. 14.4:

The NMR phenomenon was demonstrated in 1945–1946 simultaneously in the laboratories of physicists Felix Bloch (1905–1983) at Stanford University and Edward M. Purcell (1912–1997) at Harvard University. Bloch and Purcell jointly received the 1952 Nobel Prize in Physics for their work in NMR.

$$\text{radiation frequency required for absorption} = \nu_p = \frac{E_p}{h} = \frac{\Delta\varepsilon_p}{h} = \frac{\gamma_H}{2\pi}\mathbf{B}_p \quad (14.6)$$

Using this equation, we can verify that for a set of protons p that experience a magnetic field of 70,500 gauss, the frequency ν_p required to spin-flip protons in that set is 300×10^6 Hz, or 300 MHz. This frequency is near the FM and ham-radio bands, so the electromagnetic radiation used in NMR spectroscopy is conventionally called **radiofrequency (rf) radiation**. Typical values of the rf used in NMR experiments are between 60 MHz and 950 MHz, and the magnetic fields required vary proportionately in accord with Eq. 14.6. Indeed, an ordinary radio receiver located near an NMR spectrometer and tuned to the appropriate frequency can produce audible sounds associated with an NMR experiment.

Focused Problem

14.1 (a) What radiofrequency (rf) would be required to cause spin-flipping in an NMR spectrometer in which the magnetic field at the proton is 117,400 gauss?

(b) What magnetic field at the proton would be required to cause spin-flipping in an NMR experiment in which the rf imposed on the sample is 900 MHz?

14.3 THE NMR SPECTRUM: CHEMICAL SHIFT AND INTEGRAL

A. Chemical Shift

Figure 14.1 (Sec. 14.1) shows that the two chemically distinguishable types of protons in dimethoxymethane have different *chemical shifts*. Our description of the NMR experiment in the previous section shows what this means: The two types of protons situated in a magnetic field absorb electromagnetic radiation at different frequencies. Because chemical shift tells us a great deal about structure, it is important to understand the basis of chemical shift.

The key point in understanding chemical shift is that the effective, or local, magnetic field $\mathbf{B}_p$ "sensed" by a proton is different from the external magnetic field $\mathbf{B}_0$ provided by the NMR spectrometer. In general, $\mathbf{B}_p$ is somewhat less than $\mathbf{B}_0$. The reason is that the electrons circulating in the vicinity of a proton exert their own magnetic fields that oppose the external field. The reduction of the local field by the circulation of nearby electrons is called **shielding**. Therefore, atoms of relatively low electronegativity, such as Si, near a proton increase the electron density at the proton and so increase the shielding of the proton from the external field. Conversely, more electronegative atoms such as O and Cl near a proton reduce the electron density at the proton and decrease the shielding of the proton. In summary, then, the local magnetic field at a more-shielded proton is smaller than the local field at a less-shielded proton.

An Analogy for Magnetic Shielding

Tom Wang/Shutterstock

If you use an umbrella when you go outside on a rainy day, the umbrella shields you from the rain. If you don't use an umbrella, you get soaked. If you use a very small umbrella, you get wet, but perhaps not soaked. Likewise, we can think of electrons as providing a "magnetic umbrella" that shields nuclei from the external magnetic field. If electron density is high in the vicinity of a proton, the magnetic umbrella is relatively large, and the effective field—the local field—at the proton is smaller. In this case, the chemical shift is small. If electron density is smaller at a proton, the magnetic umbrella is smaller, and the local field at the proton is higher. In this case, the chemical shift is larger.

Let's apply these ideas to the chemical shifts in the NMR spectrum of dimethoxymethane (see Fig. 14.1).

$$\overset{a}{CH_3}O—\overset{b}{CH_2}—O\overset{a}{CH_3}$$

dimethoxymethane

Let $\mathbf{B}_a$ be the local magnetic field at protons *a* and $\mathbf{B}_b$ be the local field at protons *b*. Because protons *b* are adjacent to *two* oxygens and protons *a* are adjacent to only one oxygen, protons *b* are less shielded than protons *a*. Therefore, the local field at protons *b* is greater than the local field at protons *a*:

$$\mathbf{B}_b > \mathbf{B}_a \qquad (14.7a)$$

By multiplying both sides of this inequality by $\gamma_H/2\pi$ and applying Eq. 14.4, we obtain the relative absorption frequencies:

$$\frac{\gamma_H}{2\pi}\mathbf{B}_b > \frac{\gamma_H}{2\pi}\mathbf{B}_a \quad (14.7b)$$

$$\nu_b > \nu_a \quad (14.7c)$$

This equation shows that protons *b* undergo resonance at a greater frequency than protons *a*. The chemical shift of any proton in Hz is defined as the difference between its resonance frequency and that of the protons in the reference compound TMS. Therefore, when we subtract the resonance frequency of TMS from both sides of Eq. 14.7c, we obtain the relationship between the chemical shifts of protons *a* and *b* in Hz:

$$\nu_b - \nu_{TMS} > \nu_a - \nu_{TMS} \quad (14.7d)$$

$$\Delta\nu_b > \Delta\nu_a \quad (14.7e)$$

$$\text{chemical shift of H}^b > \text{chemical shift of H}^a \text{ (in Hz)} \quad (14.7f)$$

The frequency scale across the top of the NMR spectrum in Fig. 14.1 is the scale of these $\Delta\nu$ values. As you can see in **Fig. 14.3**, $\Delta\nu_b$ is 1368 Hz and $\Delta\nu_a$ is 1005 Hz. These frequency differences are due to the different chemical environments of protons *a* and *b*. These frequency differences are the *chemical shifts* in Hz of the two proton sets. Equation 14.7f shows that the chemical shift of protons *b* is greater than that of protons *a*.

By dividing both sides of Eq. 14.7e by the operating frequency of the spectrometer, ν_0, as shown in Eq. 14.1 (Sec. 14.3B), we obtain the relationship between the two chemical shifts δ_a and δ_b in ppm:

$$\frac{\Delta\nu_b}{\nu_0} > \frac{\Delta\nu_a}{\nu_0} \quad (\Delta\nu \text{ in Hz, } \nu_0 \text{ in MHz}) \quad (14.8a)$$

$$\delta_b > \delta_a \quad (14.8b)$$

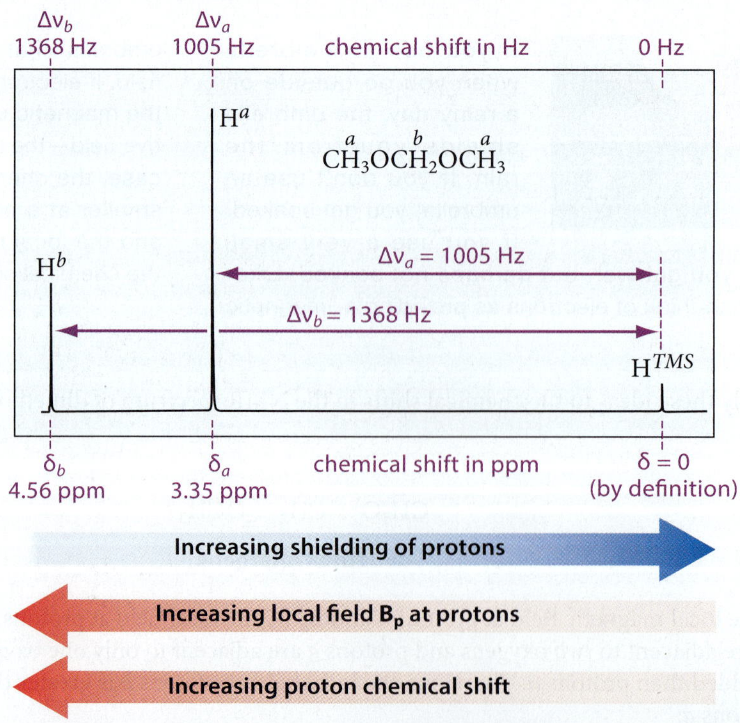

FIGURE 14.3 The spectrum of dimethoxymethane from Fig. 14.1 showing the effect of the local field on chemical shift. Different chemical shifts are a consequence of different local fields at protons in different chemical environments.

The operating frequency of the spectrometer used in most of the spectra in this text, including Fig. 14.1, is 300×10^6 Hz, or 300 MHz. Because we express $\Delta\nu$ in Hz and ν_0 in MHz, the units of δ are *parts per million*. Therefore $\delta_a = 3.35$ ppm and $\delta_b = 4.56$ ppm. We can read chemical shifts in ppm directly from the scale on the lower horizontal axis of the spectrum. Notice that chemical shifts in Hz are a *very small fraction* of the applied frequency. It follows that the relative shieldings of protons *a*, protons *b*, and the TMS protons by nearby electrons are very small fractions of the external applied field. (See Focused Problem 14.2.)

In summary, the less shielded a proton is, the greater is its chemical shift. Protons *b* are less shielded, and so have a greater chemical shift, than protons *a*. This reduction in shielding (sometimes called *deshielding*) is a consequence of decreased electron density at the *b* protons. These ideas are summarized in Fig. 14.3.

Focused Problems

14.2 What is the reduction in shielding, in magnetic field units of gauss, of (a) protons *a* and (b) protons *b* relative to the protons of TMS? The applied field in a 300-MHz NMR spectrometer is 70,500 gauss.

14.3 An NMR spectrum of a compound *X* contains four absorptions at δ 1.3, δ 4.7, δ 4.6, and δ 5.5, respectively. Which absorption comes from the most shielded protons? From the least shielded protons? Explain.

B. Chemical Shift Scales

In the previous section, we learned that a chemical shift can be expressed as a frequency difference $\Delta\nu$ in Hz or as a δ value in ppm. Typically, the δ (ppm) scale is used for citing chemical shifts for reasons that we explore here.

The operating characteristics of a given NMR spectrometer are determined by the strength of its magnet—that is, by its applied field strength $\mathbf{B}_0$. Closely related to the applied field strength is the *operating frequency* ν_0 of the NMR instrument. This is the resonance frequency of a proton if it were subjected to the full, unshielded magnetic field of the instrument. This is calculated from Eq. 14.4, with $\mathbf{B}_p = \mathbf{B}_0$:

$$\nu_0 = \frac{\gamma_H}{2\pi}\mathbf{B}_0 \qquad (14.9)$$

This equation has two important aspects. First, the operating frequency and the applied magnetic field of an instrument are proportional. Second, if you know the applied field strength, you know the operating frequency, and vice versa. For example, Eq. 14.9 shows that if a spectrometer operates at a magnetic field of 70,500 gauss, its operating frequency must be 300 million Hz, or 300 MHz. A reference to a "300-MHz proton NMR spectrum" means that the spectrum was taken on an instrument with a magnetic field of 70,500 gauss.

It turns out that chemical shifts expressed as frequency differences $\Delta\nu$ are also proportional to the field strength of the spectrometer. That is, if the applied field is doubled, the chemical shifts in Hz are also doubled. It happens that a variety of NMR instruments (and a variety of applied fields) are used in NMR spectroscopy. Therefore, if chemical shifts were cited as frequency differences, we would also have to know the applied field or operating frequency for these shifts to be meaningful. It is more convenient to tabulate chemical shifts in a way that doesn't depend on the type of spectrometer used to obtain them. This is why the δ scale is used.

$$\delta\ (\text{parts per million}) = \frac{\Delta\nu \text{ in Hz}}{\nu_0 \text{ in MHz}} \qquad (14.10)$$

This definition ensures that the field dependence is removed from δ because *both* $\Delta\nu$ and ν_0 are proportional to the applied field, and so the field dependence cancels in the ratio. Therefore, the chemical shift of a proton in ppm in a given molecule *is the same for any field strength* and

It is certainly not obvious from the NMR spectrum of dimethoxymethane why you would want to use a more powerful NMR instrument to obtain the spectrum, because the NMR spectrum of this compound is two single lines at any field strength. In Sec. 14.5B we show that the use of high-field NMR instruments gives more detailed NMR spectra for many compounds. Higher magnetic fields also result in greater instrument sensitivity (that is, the ability to detect smaller concentrations). The reason is that the sensitivity of the instrument depends on the lower excess of spins in the lower spin-energy level, which, in turn, depends exponentially on the difference in energy between the two spin energy levels (see Fig. 14.2 and the associated discussion). The problem with high-field instruments is their very high cost. "Workhorse" NMR instruments—the type used in most everyday work—represent a compromise between cost and sensitivity. The 300-MHz spectra in this text were taken on this type of instrument.

for any spectrometer (assuming the spectra are obtained under the same conditions of solvent, concentration, temperature, and so on). This means that the $\Delta\nu$ (Hz) chemical shift of the CH_2 protons of dimethoxymethane (see Fig. 14.1) depends on the field strength of the instrument, but the δ (ppm) chemical shift of these protons is 4.56 on *any* spectrometer. Therefore, *chemical shifts are conventionally cited in ppm.* The chemical shift in frequency units, if needed, can be calculated from the operating frequency ν_0 and the definition of δ in Eq. 14.10.

Focused Problems

14.4 The spectrum in Fig. 14.1 was taken on a 300-MHz NMR instrument. What is the chemical shift, in Hz, of the CH_3 protons of dimethoxymethane in a spectrum taken on different instruments with the following applied magnetic fields?

(a) 21,100 gauss (b) 141,000 gauss

14.5 What is the chemical-shift difference in ppm of two resonances separated by 45 Hz at each of the following operating frequencies?

(a) 60 MHz (b) 300 MHz

C. The Relationship of Chemical Shift to Structure

Because the chemical shift of a proton is influenced by nearby groups, the chemical shift of a proton resonance gives information about the proton's chemical environment. As noted in the previous section, one of the most important factors that affect a proton's chemical shift is the *electronegativities of nearby groups.* Some data that illustrate this idea are presented in **Table 14.1**. Examine these data using Focused Problem 14.6 as your guide.

Focused Problem

14.6 (a) Consider entries 1 through 4 of Table 14.1. How does the chemical shift of a proton vary with the electronegativity of the neighboring halogen?

(b) Compare entries 2, 5, and 6 of Table 14.1. How does chemical shift vary with the number of neighboring halogens?

(c) Compare entries 6 and 7. How is the chemical shift of a proton affected by its distance from an electronegative group?

(d) Explain why $(CH_3)_4Si$ absorbs at lower chemical shift than the other molecules in the table. Can you think of a simple molecule with protons that would have a smaller chemical shift than TMS (that is, a negative δ value)?

If you compare entries 4 and 8 (I vs. C), you will see that the chemical-shift difference is very large relative to the small electronegativity difference. Factors other than electronegativity influence chemical shift. Some of these factors are discussed in subsequent sections or chapters.

TABLE 14.1 Effect of Electronegativity on Proton Chemical Shift

Entry number	Compound	Chemical shift, δ	Electronegativity (Figure 1.15, Sec. 1.9A)
1	CH_3F	4.26	F: 3.98
2	CH_3Cl	3.05	Cl: 3.16
3	CH_3Br	2.68	Br: 2.96
4	CH_3I	2.16	I: 2.66
5	CH_2Cl_2	5.30	
6	$CHCl_3$	7.27	
7	CH_3CCl_3	2.70	
8	$(CH_3)_4C$	0.86	C: 2.55
9	$(CH_3)_4Si$	0.00*	Si: 1.90

*By definition.

You should have concluded from your examination of Table 14.1 that the following factors *increase* the proton chemical shift:

1. Increasing *electronegativity* of nearby groups
2. Increasing *number* of nearby electronegative groups
3. Decreasing *distance* between the proton and nearby electronegative groups

The effect of electropositive groups (such as Si) is opposite that of electronegative groups.

The basis of these effects, as shown in the previous section, is the different magnetic shielding of protons by surrounding electrons in different chemical environments. The spectrum of dimethoxymethane (see Fig. 14.1, Sec.14.1) also illustrates these points. The CH_2 protons, which have the larger chemical shift, are adjacent to two oxygens and are less shielded than the CH_3 protons, which are adjacent to only one oxygen.

Although methods exist for estimating chemical shifts fairly accurately (Further Exploration 14.1), it is sufficient to learn for now the general chemical-shift ranges for protons in particular environments. The chemical shifts of protons bonded to carbon in various functional environments are shown in **Fig. 14.4**. For example, Fig. 14.4 shows that the α-protons of an ether or alcohol have chemical shifts in the δ 3.2–4.2 range.

FURTHER EXPLORATION 14.1
Quantitative Estimation of Chemical Shifts

The chemical shifts in Fig. 14.4 support the idea that greater electronegativity of nearby groups corresponds to greater chemical shift. But there are a few important exceptions to this trend: the chemical shifts of vinylic protons (protons directly attached to carbons of double bonds, δ = 4.5–6.0), allylic protons (protons attached to carbons that are directly attached to double bonds, δ = 1.6–2.8), and the related protons associated with benzene rings (phenyl protons, δ = 6.5–8.0, and benzylic protons, δ = 2.3–2.9). The chemical-shift effects of these groups are much greater than would be predicted by their electronegativities. If we recognize that acetylenic carbons are *more* electronegative than vinylic carbons (Secs. 3.7E and F), the much smaller chemical shifts of acetylenic protons (δ = 1.7–2.5) relative to vinylic protons (δ = 4.5–6.0) also seem anomalous. The reason for all these exceptions is a special effect of the π electrons, which is discussed in Sec. 14.7A (see Fig. 14.15, presented later in the chapter) and Sec. 16.3B (see Fig. 16.2).

Two other general observations about chemical shift are worth remembering. The first has to do with the amount of alkyl substitution on the carbon to which a proton is bonded. The chemical shifts of **methyl protons** (that is, the protons of —CH_3 groups) are typically at the lower end of a given chemical shift range. The chemical shifts of **methylene protons** (that is, the protons of —CH_2— groups) are a few tenths of a ppm greater and are likely to be near the middle of the chemical-shift range. Finally, the chemical shifts of **methine protons** (that is, $-\overset{|}{\underset{|}{C}}-H$ protons) are typically greater still. The following chemical shifts for the α-protons of ethers illustrate this trend. (Remember in this context that α [alpha] means "next to.")

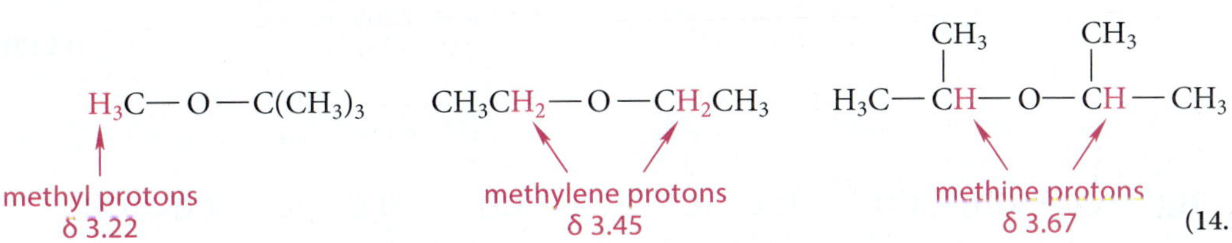

(14.11)

The second observation about chemical shift applies when a proton is near more than one functional group. In such a case, both groups affect chemical shifts. The following examples illustrate this point.

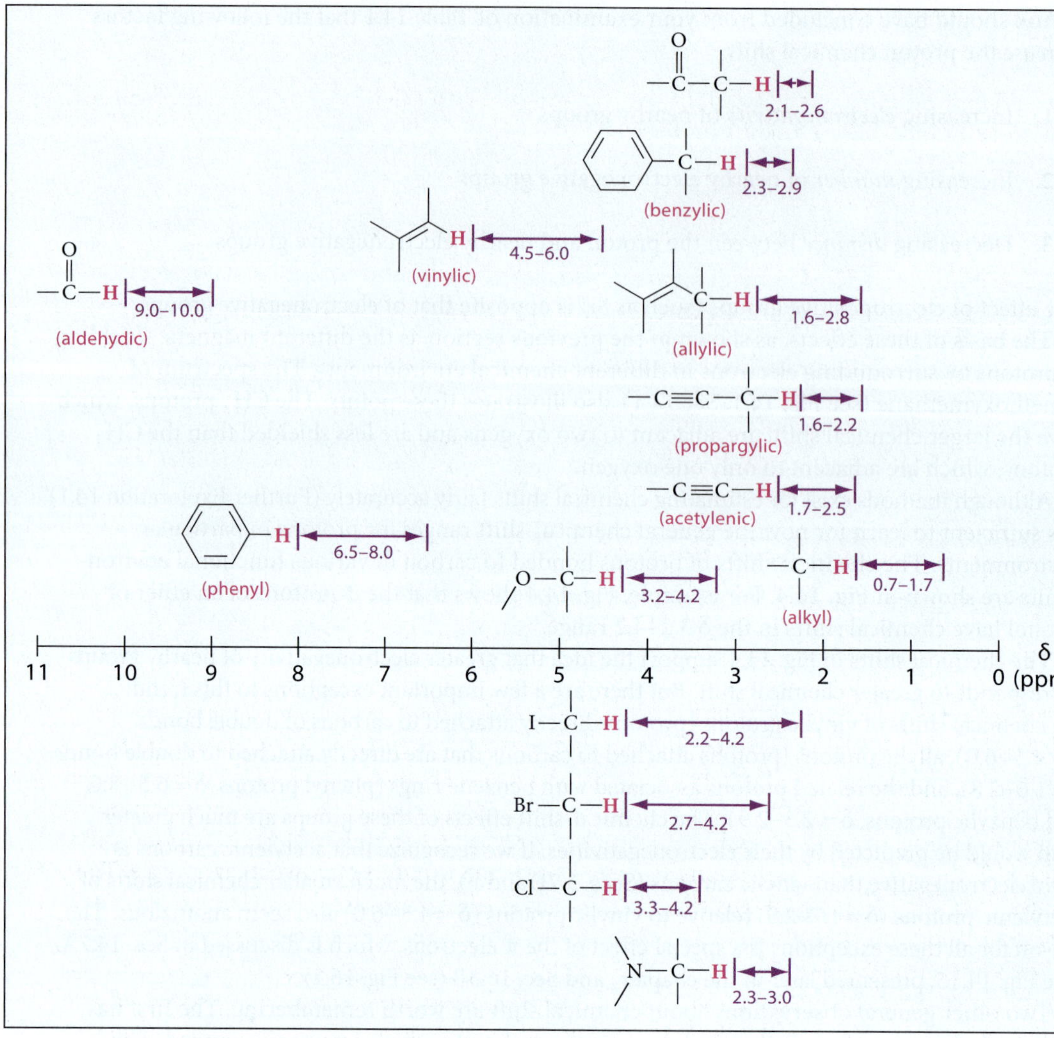

FIGURE 14.4 Approximate chemical-shift ranges for protons bonded to carbon in various chemical environments.

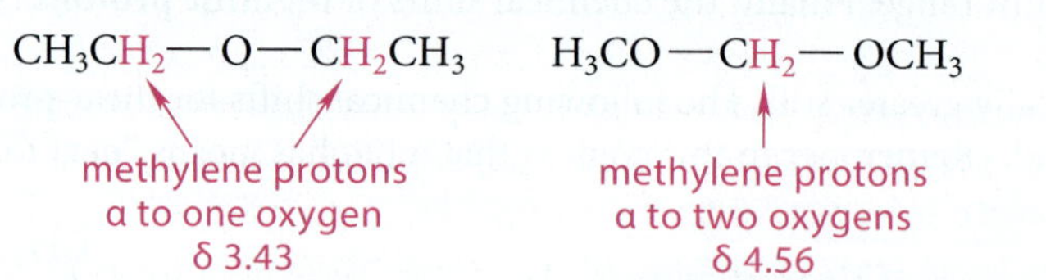

(14.12)

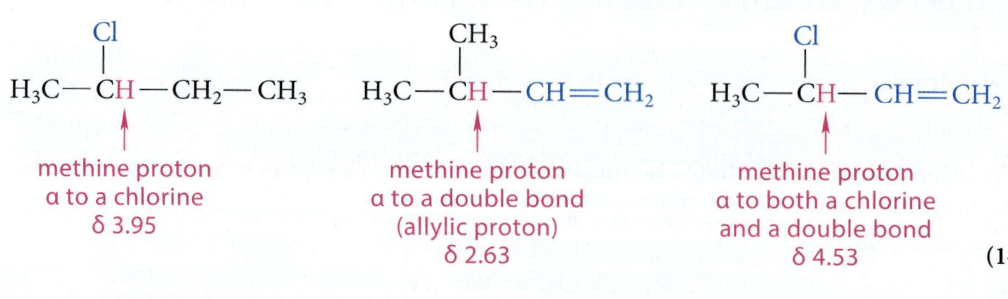

(14.13)

Entries 2, 5, and 6 in Table 14.1 illustrate this same point.

Focused Problem

14.7 In each the following sets, the NMR spectra of the compounds shown consist of a single resonance. Arrange the compounds in order of increasing chemical shift, with the smallest first.

(a) CH_2Cl_2 (*A*) CH_2I_2 (*B*) CH_3I (*C*)

(b) $Cl—CH_2CH_2—Cl$ (*A*) $Cl—C(CH_3)_2—C(CH_3)_2—Cl$ (*B*) $Cl—CH(Cl)—CH(Cl)—Cl$ (*C*) $Cl—CH_2—Cl$ (*D*)

(c) $(CH_3)_4C$ (*A*) $(CH_3)_3Al$ (*B*) $(CH_3)_4Si$ (*C*) $(CH_3)_4Pb$ (*D*)

(*Hint:* See Fig. 1.15, Sec. 1.9A. The electronegativity of Pb is 2.33.)

D. The Number of Absorptions in an NMR Spectrum

How do we know whether the different protons in a molecule will show different absorptions? This is equivalent to asking whether different protons have different chemical shifts. *Protons have different chemical shifts when they are in different chemical environments.* In many cases, deciding whether two protons are in different chemical environments is nearly intuitive. For example, in dimethoxymethane, $CH_3O—CH_2—OCH_3$, the chemical environment of the CH_2 protons is different from that of the CH_3 protons. But this distinction is not so intuitive in every case. The discussion in this section allows us to decide *rigorously* whether we can expect two protons to have different chemical shifts.

Predicting chemical-shift nonequivalence is the same as predicting chemical nonequivalence. If you have read and understood Sec. 11.9A, you already know how to do this. (If you haven't studied Sec. 11.9A, then you should do so before reading further.) *Chemically nonequivalent protons in principle have different chemical shifts.* (The qualifier "in principle" is used because it is possible for chemical-shift differences to be so small that they are undetectable.) *Chemically equivalent protons have identical chemical shifts.*

Section 11.9A noted that constitutionally nonequivalent protons are *chemically nonequivalent.* You can tell whether two protons are constitutionally equivalent by tracing their connectivity relationships to the rest of the molecule. For example, the CH_3 protons and the CH_2 protons of dimethoxymethane (see Fig. 14.1, Sec. 14.1) have a different connectivity relationship; consequently, they are constitutionally nonequivalent. This means that they are chemically nonequivalent and therefore have different chemical shifts, as Fig. 14.1 shows.

As explained in Sec. 11.9A, diastereotopic groups are constitutionally equivalent but are chemically nonequivalent. It follows, then, that diastereotopic protons in principle have different chemical shifts.

In contrast, *enantiotopic protons are chemically equivalent* as long as they are in an achiral environment. Therefore, in an achiral solvent such as CCl_4 or $HCCl_3$, enantiotopic protons have identical chemical shifts; but in an enantiomerically pure chiral solvent, or in any chiral environment such as an enzyme active site, enantiotopic protons in principle have different chemical shifts.

Finally, Sec. 11.9A showed us that *homotopic protons are chemically equivalent.* So, homotopic protons have identical chemical shifts under all circumstances.

Study Problem 14.1 shows how to apply the results of Sec. 11.9A to the determination of chemical-shift equivalence and nonequivalence in some cases involving constitutionally equivalent protons.

Study Problem 14.1

In each of the following cases, the labeled protons are constitutionally equivalent. Determine whether the labeled protons in each case are expected to have identical or different chemical shifts.

(a) 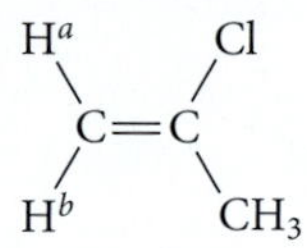

(b) 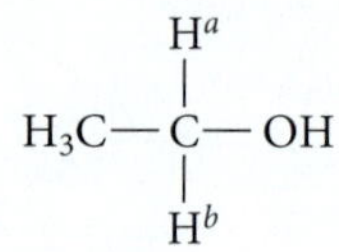

(c) 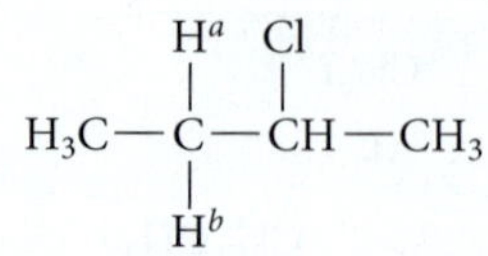

(d) 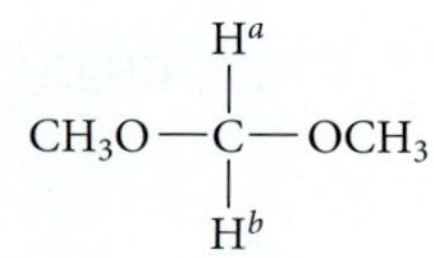

Solution Apply the principles of Sec. 11.9A to determine whether the protons in question are chemically equivalent. If the two protons are chemically equivalent, they have the same chemical shift. If not, their chemical shifts are different in principle. Because it is given that the protons in question are constitutionally equivalent, we need to determine only whether the protons are diastereotopic, enantiotopic, or homotopic.

(a) Perform the *substitution test* (Sec. 11.9A) on protons H^a and H^b—that is, replace H^a and H^b in turn with a circled proton, and examine the relationship between the resulting structures:

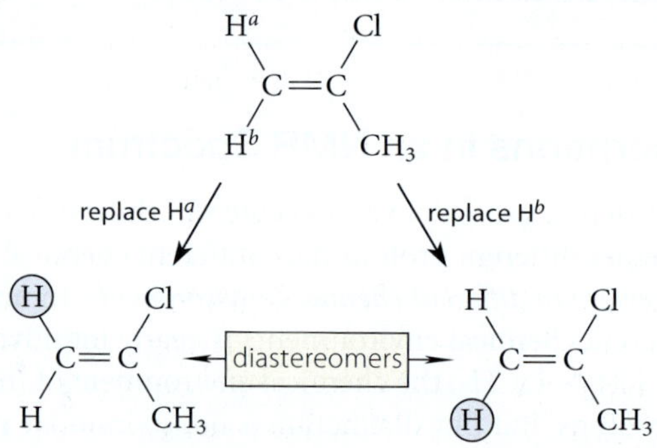

Remember to think of an “H” and a “circled H” as different atoms. The two structures that result are *E,Z* isomers and are therefore diastereomers. Consequently, H^a and H^b are diastereotopic and are therefore *chemically nonequivalent.* They are expected to have different chemical shifts.

(b) Carry out the substitution test on the two α-protons of ethanol:

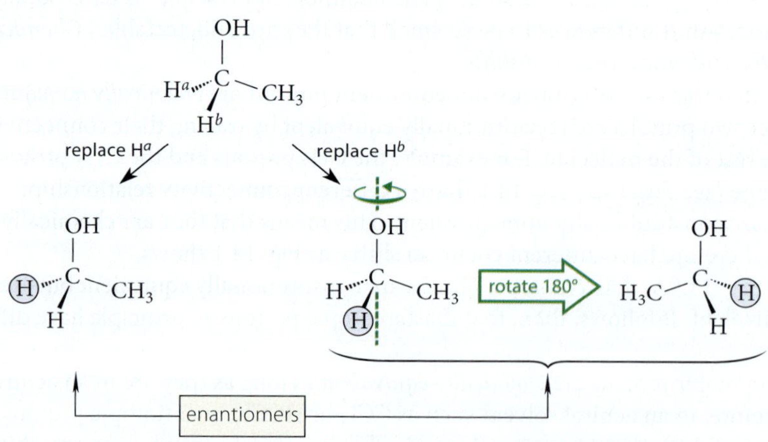

Because the resulting structures are enantiomers, the two hydrogens are enantiotopic and have the same chemical shift (in an ordinary achiral solvent).

(c) Carry out the substitution test on protons H^a and H^b of 2-chlorobutane. Notice that the molecule prior to the substitution test contains one asymmetric carbon. Choose either configuration (2*R* or 2*S*) for this carbon (2*S* is used in the following structure) and maintain whatever configuration you choose throughout the substitution test. (A circled hydrogen is assigned a higher priority than one that is not circled.)

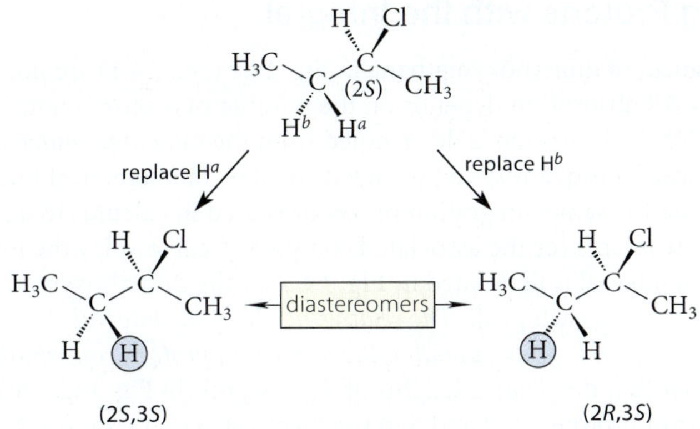

Because diastereomers are obtained in the substitution test, these protons are diastereotopic and so are chemically nonequivalent. These protons have different chemical shifts.

(d) This is dimethoxymethane (see Fig. 14.1). A substitution test shows that these two protons are homotopic. (Do this test!) Therefore, they have identical chemical shifts. That is why a single resonance is observed for these protons in the NMR spectrum. Likewise, the two CH_3 groups are homotopic, and, within each of these groups, the three protons are also homotopic. Therefore, a single resonance is observed for all six CH_3 protons.

Two of the most common situations in which diastereotopic protons are encountered are diastereotopic protons on a double bond, as in part (a) of Study Problem 14.1, and diastereotopic protons of methylene groups within a molecule containing an asymmetric carbon, as in part (c) of Study Problem 14.1. These protons are chemically nonequivalent and have different chemical shifts. Unless you are particularly alert to these situations, it is easy to regard such groups mistakenly as chemically equivalent.

It is important to understand chemical-shift equivalence and nonequivalence because the minimum number of chemically nonequivalent sets of protons in a compound of unknown structure can be determined by counting the number of different resonances in its NMR spectrum. (The "minimum" qualifier is used because of the possibility that the resonances of chemically different groups might overlap.) So, the simple act of counting resonances tells you a great deal about chemical structure.

Focused Problems

14.8 Specify whether the labeled protons in each of the following structures would be expected to have the same or different chemical shifts, and explain.

(a) 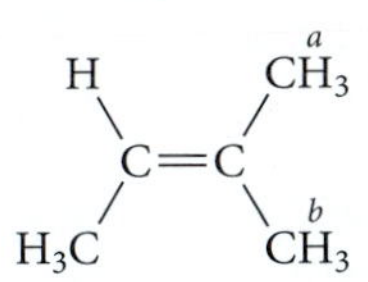

(b) 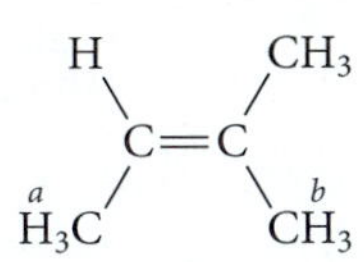

(c) 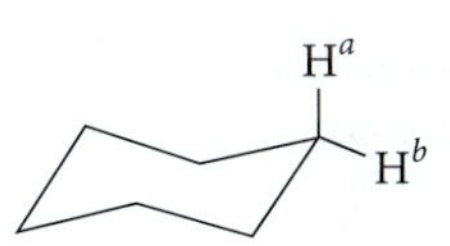

(d) 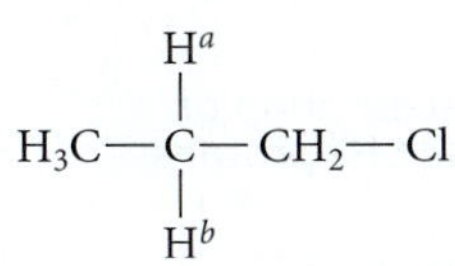

14.9 How many different absorptions are observed in the proton NMR spectrum of each of the following compounds? Explain.

(a) 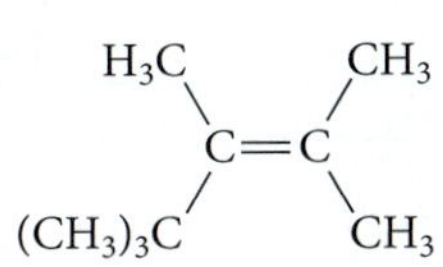

(b) 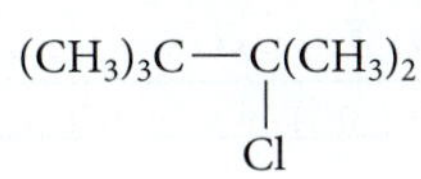

(c) 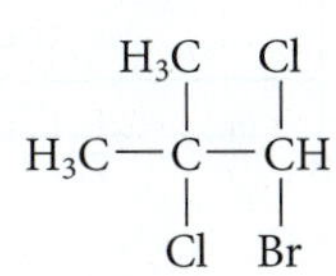

E. Counting Protons with the Integral

The two resonances of dimethoxymethane in Fig. 14.1 (Sec. 14.1) are not the same size because the size of an NMR absorption depends on the number of protons contributing to it. The intensity of an NMR absorption is determined from the *total area under the absorption peak*, called the **integral**. This quantity can be determined by mathematical integration of the peak using more or less the same integration procedures used in calculus to determine the area under a curve. NMR instruments (or the associated computers) can display the integral on the spectrum. Such a spectrum integral is illustrated in **Fig. 14.5** for the dimethoxymethane spectrum as the red curve superimposed on each peak. *The relative height of the integral (in any convenient units, such as millimeters or inches) is proportional to the number of protons contributing to the peak.* You can verify with a ruler that the relative heights of the integrals in Fig. 14.5 are in the ratio 1:3, and the relative numbers of protons are 2 and 6, respectively, also in the ratio 1:3. Errors of a few percent in the integrals are quite common.

Don't forget that the integral gives us the *ratios* of hydrogens of each type, not the absolute number. In some cases the absolute number of hydrogens is ambiguous. So, if the spectrum in Fig. 14.5 had been that of an unknown compound, the integrals in Fig. 14.5 could have corresponded to 8 hydrogens in the ratio 2H:6H, 12 hydrogens in the ratio 3H:9H, or to some other multiple of 4 hydrogens. In that situation, we would have needed a molecular formula to decide between these alternatives.

In this text, integrals of NMR spectra are presented in three ways: with the integral curve, as in Fig. 14.5; with raw numbers from the integral curve, which are *proportional* to the numbers of hydrogens (as in Study Problem 14.2); or with the actual number of hydrogens indicated in red over each resonance, as in Fig. 14.7 (Sec. 14.4A).

If a sample contains more than one compound, the spectrum intensity of each compound is proportional to its concentration. For example, the intensity of the TMS peak in most spectra is very small because it is added in very low concentration. In Fig. 14.5, however, if TMS were in the same concentration as dimethoxymethane, then its resonance would be twice the intensity of the δ 3.35 resonance because it represents twice the number of protons—that is, 12 equivalent hydrogens. In fact, another advantage of TMS is that, because of its large number of equivalent hydrogens, very little of it must be added to obtain a measurable reference line.

F. Using the Chemical Shift and Integral to Determine Unknown Structures

We now summarize the main ideas of the previous sections that will be useful in applying NMR spectroscopy to the solution of unknown structures.

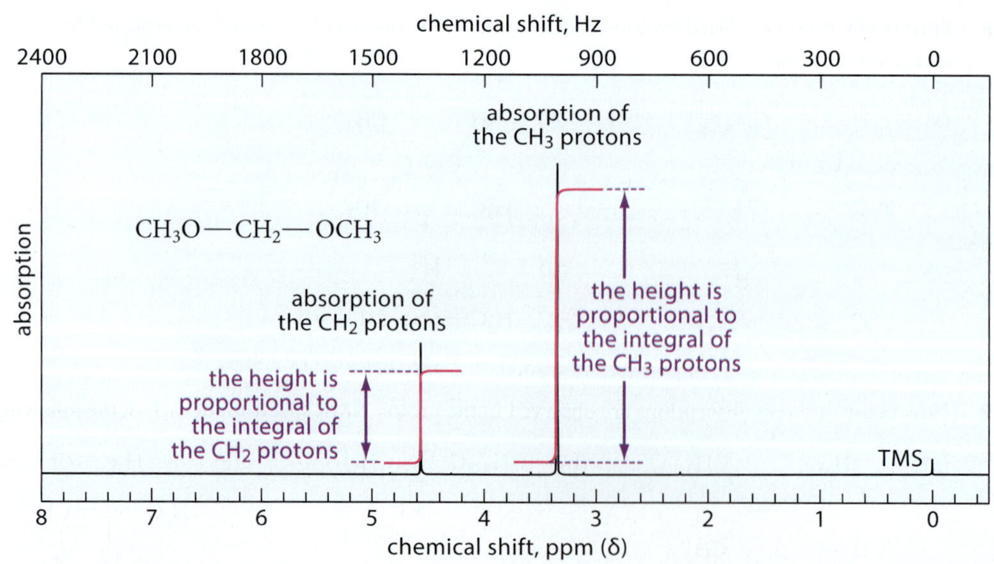

FIGURE 14.5 The NMR spectrum of Fig. 14.1 with superimposed integral. The integrals are the red curves. The heights of the integrals are indicated by the purple arrows. The integral heights have no absolute significance; their *relative* heights are proportional to the number of protons in each set.

First, each chemically nonequivalent set of protons in a molecule gives (in principle) a different resonance in the NMR spectrum. That is, the number of absorptions in principle indicates the number of chemically nonequivalent sets of protons. (If there are accidental overlaps, the number of absorptions indicates the minimum number of sets.) Next, the chemical shift of a set of protons provides information about what groups are nearby. Finally, the integral of each absorption is proportional to the number of contributing protons. Therefore, the integral indicates the relative numbers of protons in each nonequivalent set. If the total number of protons is known (from a molecular formula), then the absolute number of protons in each set can be calculated.

STUDY GUIDE LINK 14.1
Approaches to Problem Solving

These ideas are incorporated into the following systematic approach for the determination of unknown structures by NMR spectroscopy:

1. Write down everything about the molecular structure that is known from the molecular formula, including the unsaturation number and the functional groups that might be present.

2. From the number of absorptions, determine how many chemically nonequivalent sets of hydrogens are in the unknown.

3. Use the total integral of the entire spectrum and the molecular formula to determine the number of integral units per proton.

4. Use the integral of each absorption and the result from step 3 to determine the number of protons in each set.

5. From the chemical shift of each set, determine which set must be closest to each of the functional groups that are present.

6. Write down partial structures that are consistent with each piece of evidence, and then write down *all possible structures* that are consistent with *all* of the evidence.

7. Use Fig. 14.4 (Sec. 14.3C) to estimate the chemical shifts of the protons in each structure, and, if possible, choose the structure that best reconciles the predicted and observed chemical shifts. If there is only one partial structure, make sure it fits *all* of the data.

This approach is illustrated in Study Problem 14.2, where this approach is integrated into a table with three columns. In column 1, we write each observation. In column 2, we write the inference from the observation. In column 3, we write the partial structure (or final structure). This is a good process to follow as you begin to deduce structures from spectral data.

Study Problem 14.2

An unknown compound with the molecular formula $C_5H_{11}Br$ has an NMR spectrum consisting of two resonances, one at δ 1.02 (relative integral 8378 units) and the other at δ 3.15 (relative integral 1807 units). Propose a structure for this compound.

Solution Follow the seven steps.

Step 1. Make deductions from the molecular formula.

Observation	*Inference*	*Partial structure*
$C_5H_{11}Br$	Unsaturation number = 0 (Eq. 4.18, Sec. 4.3). The compound has no rings or double bonds; it is a simple alkyl halide.	(Many possibilities)

Step 2. Count the number of absorptions.

Observation	*Inference*	*Partial structure*
Two absorptions	Two nonequivalent sets of H's (barring accidental overlaps)	—

Step 3. The total integral is (8378 + 1807) = 10,185 units; because the molecular formula indicates 11 hydrogens, the integral corresponds to 10,185/11 = 926 units per hydrogen.

Step 4. The larger peak accounts for (8378/926) = 9.05 protons; the smaller peak accounts for (1807/926) = 1.95 protons. Rounding these to whole numbers, the two peaks are in the ratio 9:2. (Remember, integrals can contain errors of a few percent.) All 11 protons have been accounted for.

Observation	*Inference*	*Partial structure*
Nine equivalent H's in one set; two equivalent H's in the other	A *tert*-butyl group and a methylene group	$(CH_3)_3C$—, —CH_2—

Step 5. Because of its greater chemical shift, the two-proton set must be closer to the bromine than the nine-proton set.

Observation	*Inference*	*Partial (final) structure*
The set of two protons is closer to the Br	—CH_2—Br	$(CH_3)_3C$—CH_2—Br

Step 6. The last structure from step 5 is consistent with all of the data; so, it is the final structure: 1-bromo-2,2-dimethylpropane, or neopentyl bromide.

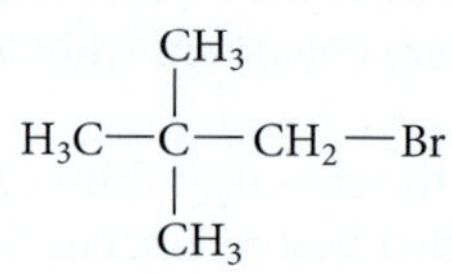

1-bromo-2,2-dimethylpropane
(neopentyl bromide)

To summarize the structure with the spectral assignments:

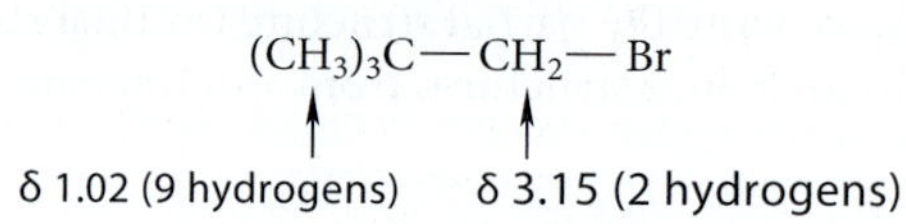

1-bromo-2,2-dimethylpropane
(neopentyl bromide)

Step 7. According to the discussion of chemical shifts, methylene protons α to a bromine should have a chemical shift in the middle of the range shown in Fig. 14.4, about δ 3.4; the observed shift is very close to this value. The shift of the methyl protons should be in the alkyl region, about δ 1.2; the observed shift is also quite close to this value.

One of the most common errors beginning students make in proposing structures from spectra is to attempt to *force-fit* an incorrect structure to the data. This is somewhat like forcing two pieces of a jigsaw puzzle together when they don't quite fit. Step 7 is particularly important: make sure that the structure you propose fits *all* the data. If it doesn't, go back and see whether there are some partial structures you haven't considered that might be better fits.

Focused Problems

14.10 In each case, give a single structure that fits the data provided. Use the process in Study Problem 14.2.

(a) A compound $C_7H_{15}Cl$ has two NMR absorptions at δ 1.08 and δ 1.59, with relative integrals of 3:2, respectively.

(b) A compound $C_5H_9Cl_3$ has three NMR absorptions at δ 1.99, δ 4.31, and δ 6.55 with relative integrals of 6:2:1, respectively.

14.11 Predict the proton NMR spectrum, including approximate chemical shifts, of the following compound. Explain your reasoning, and state any ambiguities.

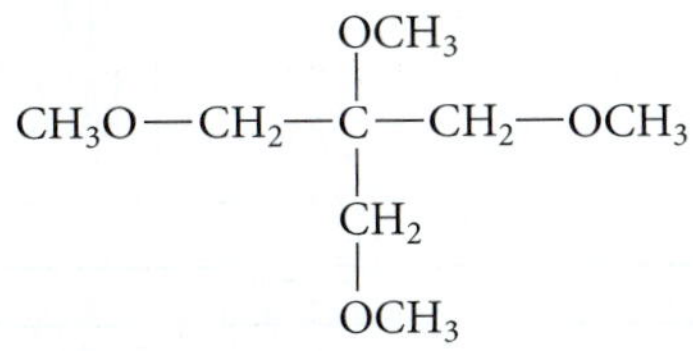

14.12 The proton NMR spectrum of a mixture of *tert*-butyl bromide and methyl iodide contains two single resonances at δ 2.2 and δ 1.8 with relative integrals of 5:1.

(a) What is the mole percent of each compound in the mixture? (*Hint:* Be sure to assign each absorption to a compound before doing the analysis.)

(b) Which would be more easily detected by proton NMR: 1 mole percent CH_3I impurity in $(CH_3)_3C—Br$, or 1 mole percent $(CH_3)_3C—Br$ impurity in CH_3I? Explain.

14.4 THE NMR SPECTRUM: SPIN–SPIN SPLITTING

Although substantial information can be obtained from the chemical shift and integral, another aspect of NMR spectra provides the most detailed information about chemical structures. Consider the compound bromoethane (ethyl bromide):

$$\overset{a}{H_3C}—\overset{b}{CH_2}—Br$$

bromoethane (ethyl bromide)

This molecule has two chemically different sets of hydrogens, labeled *a* and *b*. We expect to find NMR absorptions for these two sets of protons in the integral ratio 3:2, respectively, with the absorption of protons *a* at smaller δ. The NMR spectrum of bromoethane is shown in **Fig. 14.6**. This spectrum contains more lines than you might have expected—seven lines in all. Moreover, the lines fall into two distinct groups: a packet of three lines, or *triplet,* at smaller δ, and a packet of four lines, or *quartet,* at higher δ. These packets are expanded horizontally in boxes on the spectrum so that their details can be seen more clearly. It turns out that all three lines of the triplet are the absorption for the CH_3 protons, and all four lines of the quartet are the absorption for the CH_2 protons. The chemical shift of each packet of lines, taken at its center, is in agreement with the predictions of Fig. 14.4 (Sec. 14.3C). The quartet and the triplet have total integrals, respectively, in the ratio 2:3.

When an NMR resonance for a set of equivalent nuclei appears as more than one line, the resonance is said to be *split.* **Splitting** *arises from the effect that one set of protons has on the NMR absorption of neighboring protons.* The physical reasons for splitting are considered in Sec. 14.4B. First, we focus on the appearance of the splitting pattern and the information it provides about structure.

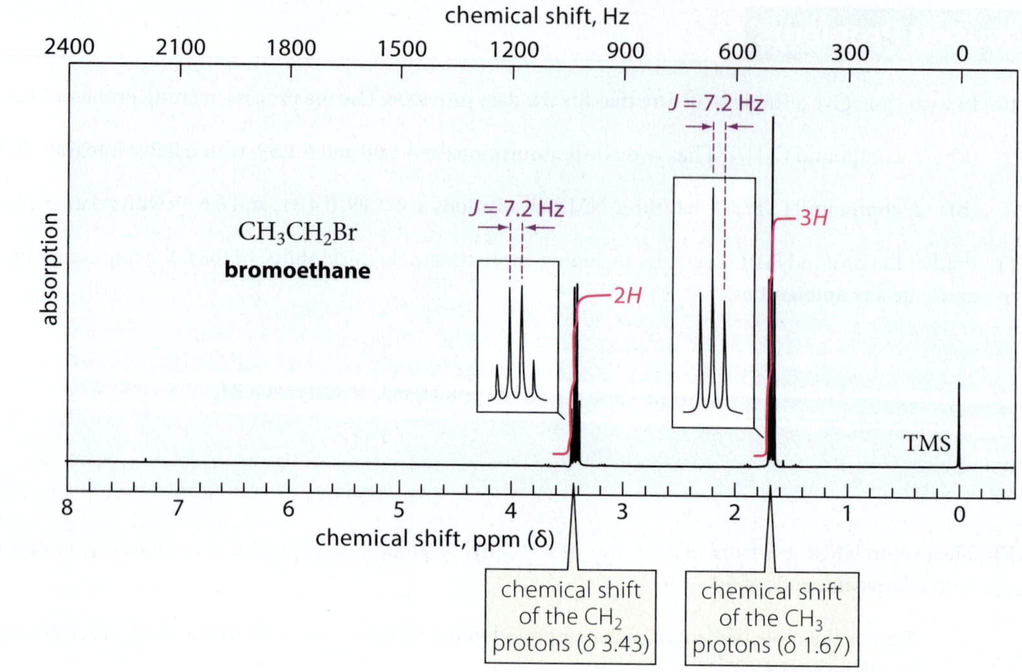

FIGURE 14.6 The NMR spectrum of bromoethane illustrates splitting. Splitting patterns contain $n + 1$ lines, where n is the number of protons on *adjacent* atoms. Therefore, the resonance for the CH_3 protons is split into a three-line packet, or triplet, because the adjacent carbon has two protons. The resonance for the CH_2 protons is split into a four-line packet, or quartet, because the adjacent carbon has three protons. The separation between lines within each packet (in this case, 7.2 Hz) is J, the coupling constant.

A. The $n + 1$ Splitting Rule

Whereas the integral gives a proton count for each resonance, the splitting pattern gives a different proton count—namely, the number of protons *adjacent* to the protons being observed. The relationship between the number of lines in the splitting pattern for an observed proton and the number of protons on adjacent atoms is given by the **$n + 1$ rule**: *n adjacent protons cause the resonance of an observed proton to be split into $n + 1$ lines.*

Let's see how the $n + 1$ rule accounts for the splitting patterns in the spectrum of bromoethane. Consider first the resonance for the methyl (CH_3) protons. Because the carbon *adjacent* to the CH_3 group has two protons, the resonance for the CH_3 protons themselves is split into a pattern of $2 + 1 = 3$ lines—that is, a *triplet.* (The fact that there are also three methyl protons is a coincidence; the number of protons determined by the integral has *nothing* to do with their splitting.)

Now consider the resonance of the methylene (CH_2) protons. Because the carbon *adjacent* to the CH_2 group has three equivalent protons, the resonance for the CH_2 protons is split into a pattern of $3 + 1 = 4$ lines—that is, a *quartet.*

Splitting is always mutual—that is, if protons *a* split protons *b*, then protons *b* split protons *a*. Therefore, in the bromoethane spectrum, the CH_3 resonance is split by the CH_2 protons, and the CH_2 resonance is split by the CH_3 protons. When two sets of protons split each other, they are said to be **coupled**. Therefore the CH_3 and the CH_2 protons of bromoethane are coupled.

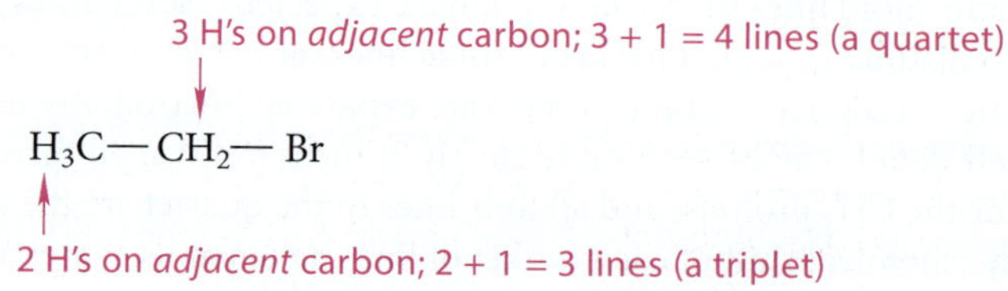

Why is no splitting observed in our previous examples of NMR spectra? First of all, *splitting is not observed between chemically equivalent hydrogens.* For example, the three hydrogens of iodomethane appear as a singlet because these hydrogens are chemically equivalent. Likewise, the two hydrogens of 1,1,2,2-tetrachloroethane also appear as a singlet because these two hydrogens, *even though they are on different carbons*, are chemically equivalent.

$H_3C—I$	$Cl_2CH—CHCl_2$
iodomethane	**1,1,2,2-tetrachloroethane**
hydrogens are chemically equivalent; no splitting is observed	hydrogens are chemically equivalent; no splitting is observed

Second, *with saturated carbon atoms, splitting is normally not observed between protons on nonadjacent carbon atoms.* Therefore, because the protons in dimethoxymethane (see Fig. 14.1, Sec. 14.1) are on *nonadjacent* carbons, their splitting is negligible; the two absorptions in the NMR spectrum of this compound are *singlets* (unsplit single lines).

Splitting provides *connectivity information*: when you observe the resonance of a proton, its splitting tells you how many protons are on *adjacent* atoms. As an analogy, suppose you were describing a puppy to a person who has never seen one. It's one thing to say that a puppy has four legs in two nonequivalent sets of two; it's much more revealing when you say that one set is attached to the body at the end near the head and the other is attached to the body at the end near the tail. It's the connectivity information that allows the more complete description of the puppy.

Now we consider some of the details of splitting. The spacing between adjacent peaks of a splitting pattern, measured in Hz, is called the **coupling constant** (symbolized by J). This spacing can be measured approximately with a ruler using the Hz scale on the upper horizontal axis of the spectrum, but the exact value is determined from analysis of the spectrum by computer. (In this text, J values are given.) For bromoethane (see Fig. 14.6), this spacing is 7.2 Hz. *Two coupled protons must have the same value of J.* Therefore, the coupling constants for both the CH_2 protons and the CH_3 protons of bromoethane are the same because these protons split each other. Letting the CH_3 protons be a and the CH_2 protons be b, then $J_{ab} = J_{ba}$. The coupling constant, unlike the chemical shift in Hz (Eq. 14.7e, Sec. 14.3A), does not vary with the operating frequency or the applied magnetic field strength. Therefore, the value of J for bromoethane is 7.2 Hz whether the spectrum is taken at 60 MHz or 300 MHz.

The chemical shift of a split resonance in most cases occurs at or near the midpoint of the splitting pattern. For example, in the bromoethane spectrum, the chemical shift of the CH_2 protons can be taken to be the midpoint of the quartet, and that of the CH_3 protons is at the middle line of the triplet.

How do you know whether a group of lines is a single split resonance rather than several individual resonances with different intensities? Sometimes there is ambiguity, but in most cases a splitting pattern can be discerned by the *relative intensities of its component lines.* These intensities have well-defined ratios, shown in **Table 14.2**. For example, the relative intensities of a triplet, such as the one in bromoethane, are in the ratio 1:2:1; the relative intensities of a quartet are 1:3:3:1. In reality, slight deviations from these ratios occur, but these deviations become significant only when the chemical shifts of the coupled protons are very similar. (We present an example of this behavior in Sec. 14.7.)

A set of protons can be split by protons on more than one adjacent carbon. An example of this situation occurs in 1,3-dichloropropane, the NMR spectrum of which is shown in **Fig. 14.7**.

$$Cl—\overset{b}{C}H_2—\overset{a}{C}H_2—\overset{b}{C}H_2—Cl$$

1,3-dichloropropane

This molecule has two chemically nonequivalent sets of protons, labeled a and b. The key to understanding this spectrum is to recognize that all four protons H^b are chemically equivalent. The absorption for H^a is therefore split into a quintet (with the intensities given in Table 14.2) because there are four protons on adjacent carbons; the fact that two of the protons are on one carbon and two are on the other makes no difference. The absorption for H^b is a triplet that appears at larger chemical shift.

If you are mathematically inclined, you might recognize that the relative intensities of a splitting pattern caused by n neighboring protons are the same as the coefficients for each term of a binomial distribution for $(x + y)^n$. For two neighboring protons, the binomial distribution for $n = 2$ is $1x^2 + 2xy + 1y^2$; the coefficients for the three lines of the splitting pattern are 1, 2, and 1. Also, from the triangular layout of splitting intensities (Pascal's triangle), you can see that a given entry is the sum of the two intensities immediately above it. The intensity 4 is the sum of 1 and 3; the intensity 10 is the sum of 6 and 4.

TABLE 14.2 Relative Intensities of Lines within Common NMR Splitting Patterns

Number of equivalent adjacent protons	Number of lines in the splitting pattern (name)	Relative line intensity within the splitting pattern
0	1 (singlet)	1
1	2 (doublet)	1 1
2	3 (triplet)	1 2 1
3	4 (quartet)	1 3 3 1
4	5 (quintet)	1 4 6 4 1
5	6 (sextet)	1 5 10 10 5 1
6	7 (septet)	1 6 15 20 15 6 1

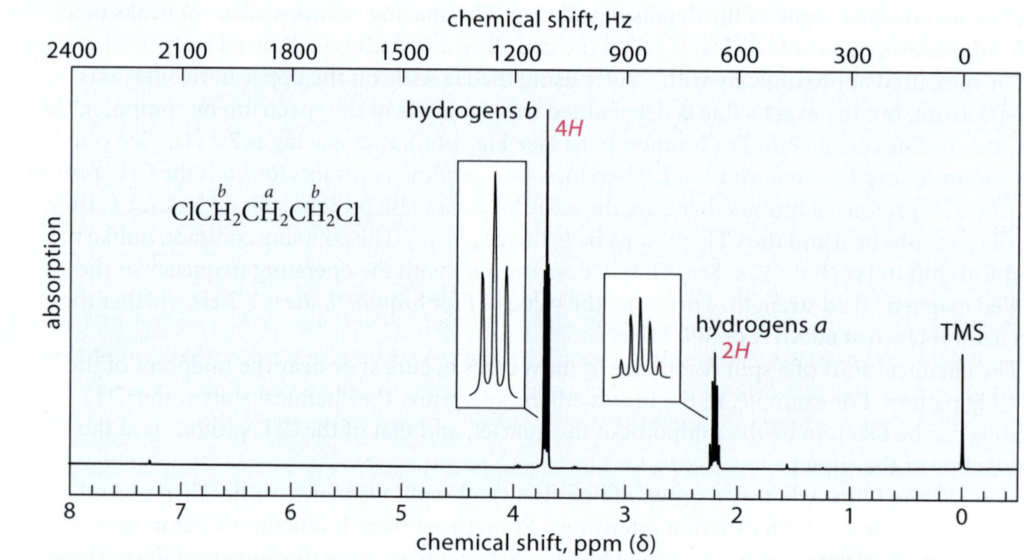

FIGURE 14.7 The NMR spectrum of 1,3-dichloropropane. (Integrals are shown in red.) Hydrogens *a* are split by all four hydrogens *b*.

Focused Problem

14.13 Predict the NMR spectrum of each of the following compounds, including splitting and approximate chemical shifts. (Assume that the coupling constants are about the same as those for bromoethane.)

(a) $H_3C—CHCl_2$

(b) $ClCH_2—CHCl_2$

(c)
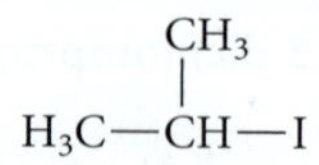

(d)
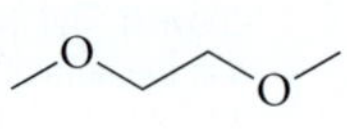

(e)
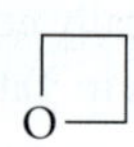

(f)
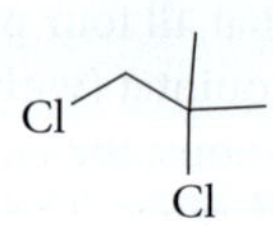

B. Why Splitting Occurs

Splitting occurs because the magnetic field caused by the spin of a neighboring proton affects the total field experienced by an observed proton. To understand, let's analyze the absorption of a set of equivalent protons *a* (such as the methyl protons of bromoethane) adjacent to *two* equivalent neighboring protons *b*. What are the spin possibilities for the neighboring protons *b*? In any one molecule, both of these protons could have spin $+\frac{1}{2}$; both could have spin $-\frac{1}{2}$; or they could have differing spins.

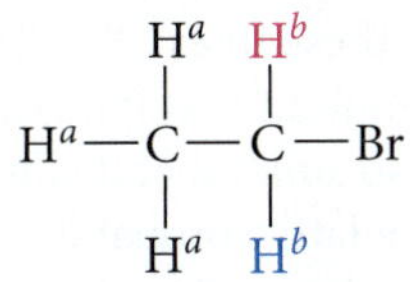

	b protons					
	1	2	1	2	1	2
spin combinations	+½	+½	+½ −½	−½ +½	−½	−½

(14.14)

If the spins of the *b* protons are different, they can differ in two ways: if proton 1 has spin $+\frac{1}{2}$, then proton 2 has spin $-\frac{1}{2}$, or vice versa.

Suppose that the *a* protons are next to two *b* protons with a spin of $+\frac{1}{2}$. Because the spins of these *b* protons are parallel to the applied field, they add to, or enhance, the applied field, and so the *a* protons are subjected to a slightly greater field than they would be in the absence of the *b* protons. Consequently, radiation of higher energy (and frequency) is required to bring the *a* protons into resonance. The result is a line in the splitting pattern at *higher* frequency (**Fig. 14.8**). Now suppose that the *a* protons are next to two *b* protons with a spin of $-\frac{1}{2}$. Because the spins of these *b* protons are antiparallel to the applied field, they subtract from the applied field, and as a result the *a* protons are subjected to a slightly smaller field than they would be in the absence of the *b* protons. Therefore, radiation of lower frequency is required to bring the *a* protons into resonance. The result is a line in the splitting pattern at *lower* frequency. Finally, suppose that

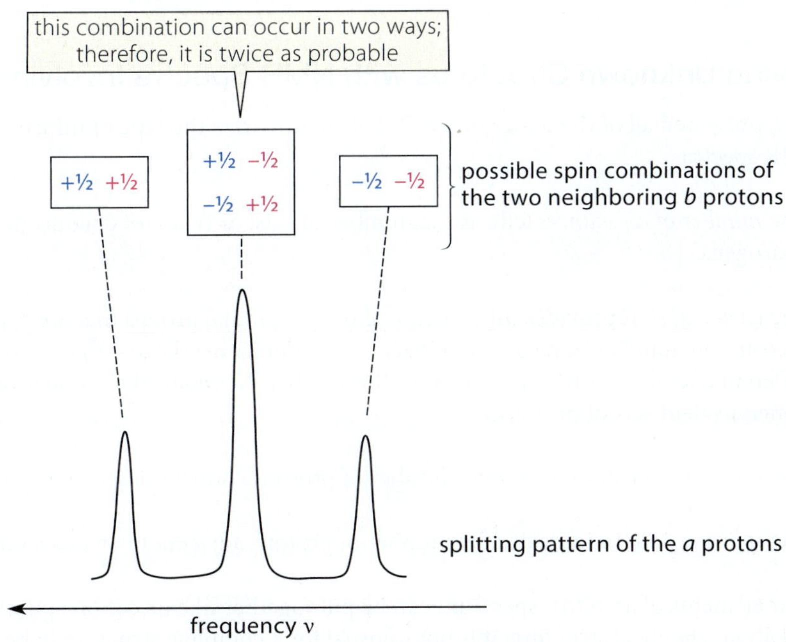

FIGURE 14.8 Analysis of the triplet splitting of a set of protons *a* by two neighboring protons *b*. When the spins of both *b* protons are aligned parallel to the applied field ($+\frac{1}{2}$ spin), they augment the local field; consequently, a greater frequency is required to bring the *a* protons into resonance. When the spins of both *b* protons are aligned against the applied field ($-\frac{1}{2}$ spin), they decrease the local field; therefore, a smaller frequency is required to bring the *a* protons into resonance. When neighboring *b* protons have opposite spin, the local field is unaffected; so, the resonance frequency of the *a* protons is unchanged. The center line of a triplet has twice the intensity of the outer lines because the corresponding spin combinations of neighboring *b* protons are twice as probable.

An analogy to the spin combinations is the combinations that can be rolled with a pair of dice. A 3 is twice as probable as a 2 because it can be rolled in two ways (2 + 1, 1 + 2), but a 2 can only be rolled in one way (1 + 1).

An analogy to the effect of distance on proton coupling can be observed with an ordinary magnet and a few paper clips. If one paper clip is held to the magnet, it may be used to hold a second paper clip, and so on. The magnetic field of the magnet dies off with distance, however, so that typically the third or fourth paper clip is not magnetized.

the two *b* protons have opposite spins; in this case, the effects of the two *b* proton spins cancel, and protons *a* are subjected to the same field that they would be in the absence of protons *b*. The result is the line in the center of the splitting pattern. These three situations correspond to the three lines in the splitting pattern.

What about the intensities of these lines? The likelihood that each proton has a spin of $+\frac{1}{2}$ is about the same as the likelihood that it has a spin of $-\frac{1}{2}$; the excess of spins in the $+\frac{1}{2}$ state (Display 14.3, Sec. 14.2) is so small that it can be ignored in the analysis of splitting. The intensities then follow from the relative probabilities of each spin combination. The probability that both *b* protons have spin $+\frac{1}{2}$ is equal to the probability that they both have spin $-\frac{1}{2}$, but the probability that these protons have opposite spins is twice as high because this situation can occur in *two ways* (Display 14.14). Therefore, the center line of the splitting pattern is twice as large as the outer lines, and a 1:2:1 triplet is observed for the absorption of protons *a*.

Try to analyze the splitting of the *b* protons of bromoethane in a similar manner (Focused Problem 14.14).

How does one proton "know" about the spin of adjacent protons? One of the most important ways that proton spins interact is *through the electrons in the intervening chemical bonds*. This interaction weakens as the protons are separated by more chemical bonds. The coupling constant between hydrogens on adjacent saturated carbons in acyclic compounds is typically 5–8 Hz, but the coupling constant between more widely separated hydrogens is normally so small that it is not observed. This is why splitting is usually not observed between protons on nonadjacent carbon atoms.

Focused Problem

14.14 Analyze the quartet splitting pattern for a set of equivalent *b* protons in the presence of *three* equivalent adjacent *a* protons. Include an analysis of the relative intensity of each line of the splitting pattern. (This is the splitting pattern for the CH_2 protons of bromoethane.)

C. Solving Unknown Structures with NMR Spectra Involving Splitting

We've now presented all of the basics of NMR. Let's summarize the type of information available from NMR spectra:

1. The *number of resonances* tells us the number of sets, or types, of chemically distinct hydrogens.
2. The *chemical shift* provides information about functional groups that are near an observed proton. The number of resonances (barring accidental overlaps)—that is, the number of different chemical shifts represented in the spectrum—equals the number of chemically nonequivalent sets of protons.
3. The *integral* indicates the relative number of protons contributing to a given resonance.
4. The *splitting pattern* indicates the number of protons adjacent to an observed proton.

These four elements of an NMR spectrum can be put together like pieces of a puzzle to deduce a great deal about chemical structure; it is not unusual for a complete structure to be determined from the NMR spectrum alone.

> Try not to confuse the *two types of proton counting* that occur in the analysis of NMR spectra. The *integral* is used to count the protons that contribute to a particular line (or group of lines in a splitting pattern) under consideration. *Splitting* is used to count the protons on atoms *adjacent to* the protons under consideration by applying the $n+1$ rule.

Because NMR spectra consume a large amount of space, it is common to see NMR spectra recorded in books and journals in an abbreviated form. In the form used in this text, the chemical shift of each resonance is followed by its integral, its splitting, and (if split) its coupling constant, if known. Abbreviations used to indicate splitting patterns are s (singlet), d (doublet), t (triplet), and q (quartet); complex patterns in which the nature of the splitting is not clear are designated m (multiplet). It is assumed that the relative intensities of each splitting pattern approximately match those in Table 14.2. For example, the spectrum of bromoethane (see Fig. 14.6) would be summarized as follows:

$$\delta\ 1.67\ (3H,\ \text{t},\ J = 7.2\ \text{Hz});\ \delta\ 3.43\ (2H,\ \text{q},\ J = 7.2\ \text{Hz})$$

- coupling constant
- type of splitting pattern (triplet; intensities in Table 14.2)
- integral
- chemical shift (ppm relative to TMS) (14.15)

You may find that you can interpret a spectrum more easily if you can see it. If so, don't hesitate to sketch the spectrum. You can do this quickly by using vertical bars for the individual lines.

We have now presented the tools needed to determine some structures using NMR spectra that contain some splitting information. The general method of problem solving given in Sec. 14.3F remains valid when splitting information is involved. Just remember to take into account splitting information when writing out partial structures (step 6). Study Problem 14.3 illustrates this approach. Once again, we use a table to record observations, inferences, and partial structures.

Study Problem 14.3

Give the structure of a compound $C_7H_{16}O_3$ with the following NMR spectrum: δ 1.30 (3*H*, s); δ 1.93 (2*H*, t, *J* = 7.3 Hz); δ 3.18 (6*H*, s); δ 3.33 (3*H*, s); δ 3.43 (2*H*, t, *J* = 7.3 Hz). Its IR spectrum shows no O—H stretching absorption.

Solution To solve the problem, apply the procedure in Sec. 14.3F. We number each line in our deduction table for reference.

Step 1. We first consider the formula and the IR spectrum.

#	*Observation*	*Inference*	*Partial structure*
1	The formula has an unsaturation number = 0	No rings or double bonds	
2	The IR spectrum shows no O—H absorption	The oxygens are in ether groups. Unsaturated groups, such as carbonyl groups, are excluded.	

Steps 2–4. The unknown contains five sets of chemically nonequivalent protons, and the numbers of protons in each resonance is given in the problem.

Steps 5–6.

#	*Observation*	*Inference*	*Partial structure*
3	There are seven carbons but only five chemically nonequivalent sets.	(a) Two of the sets must involve two equivalent carbons each; or (b) one set involves three equivalent carbons; or (c) one set involves two equivalent carbons plus a quaternary carbon (which has no hydrogens); or (d) there are two quaternary carbons.	
4	The three protons at δ 3.33 are a singlet.	Three equivalent protons on a carbon bonded to oxygen (see Fig. 14.4, Sec. 14.3C)	A methoxy group ($—OCH_3$)
5	The six protons at δ 3.18 are a singlet.	Two equivalent sets of three protons on a carbon bonded to oxygen	Two equivalent methoxy groups. This narrows the inference in #3 to (a) or (c).
6	The two-proton resonance at δ 3.43 is a triplet, and the coupling constant matches that of the δ 1.93 set.	The carbon bearing these two protons must be bonded to oxygen and must also be bonded to another carbon with two hydrogens, which can only be the δ 1.93 hydrogens.	δ 3.33 δ 3.43 δ 1.93 $CH_3O—CH_2—CH_2—$
7	The δ 1.93 hydrogens have no other splitting.	These must also be bonded to a carbon (from its chemical shift) with no hydrogens; this narrows the inference in #3 to (c).	δ 3.33 δ 3.43 δ 1.93 $CH_3O—CH_2—CH_2—C—$ (C with two further bonds, up and down)
8	δ 1.30 (3*H*, s)	A methyl is bonded to carbon with no H's.	δ 3.33 δ 3.43 δ 1.93 $CH_3O—CH_2—CH_2—C—$ (C bearing CH_3, δ 1.30, below and one further bond above)
9	Two equivalent $CH_3O—$ groups from #5 account for all remaining carbons and hydrogens.	The structure is completed.	δ 3.33 δ 3.43 δ 1.93 $CH_3O—CH_2—CH_2—C(OCH_3)_2—CH_3$ (both OCH_3 δ 3.18; CH_3 δ 1.30) **1,3,3-trimethoxybutane**

Step 7. The chemical shifts are all consistent with those given in Fig. 14.4. The shifts of the $—CH_2—$ and $—CH_3$ groups at δ 1.93 and δ 1.30 illustrate another trend in chemical shift: oxygens have a discernible chemical-shift effect roughly equal to 0.2–0.3 ppm on β-protons (protons two carbons removed from the oxygen). Typically, a $—CH_2—$ has a chemical shift of about δ 1.2, but the δ 1.94 protons are shifted another 0.7–0.8 ppm by the three β-oxygens. Similarly, a methyl group typically has a chemical shift of δ 0.9, but the δ 1.30 methyl protons are shifted by the two β-oxygens of the methoxy groups. (See Further Exploration 14.1 for a discussion of these effects.)

STUDY GUIDE LINK 14.2
More NMR Problem-Solving Hints

Focused Problems

14.15 Explain why the following two structures are ruled out by the data in Study Problem 14.3.

(a)

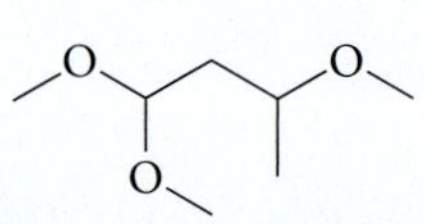

(b)

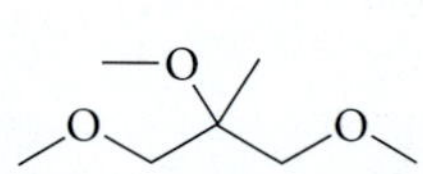

14.16 Give structures for each of the following compounds.

(a) C_3H_7Br: δ 1.03 (3*H*, t, $J = 7$ Hz); δ 1.88 (2*H*, sextet, $J = 7$ Hz); δ 3.40 (2*H*, t, $J = 7$ Hz)

(b) $C_2H_3Cl_3$: δ 3.98 (2*H*, d, $J = 7$ Hz); δ 5.87 (1*H*, t, $J = 7$ Hz)

(c) $C_5H_8Br_4$: δ 3.6 (s; only resonance in the spectrum)

14.17 A compound *X* with the molecular formula $C_{11}H_{24}O_3$ and no O—H stretching absorption in its IR spectrum has the following proton NMR spectrum: δ 1.2 (3*H*, s); δ 1.4 (6*H*, t, $J = 7$ Hz); δ 3.2 (9*H*, s); δ 3.4 (6*H*, t, $J = 7$ Hz). Systematically deduce the structure of *X*, and explain your reasoning.

An important use of spectroscopy is to confirm structures that are suspected from other information. For example, if a well-known reaction is run on a known starting material, the structure of the major product can often be predicted from a knowledge of the reaction. NMR spectroscopy can be used to confirm (or refute) the predicted structure, as shown in Focused Problem 14.18.

Focused Problem

14.18 When 3-bromopropene is allowed to react with HBr in the presence of peroxides, a compound *A* is formed that has the following NMR spectrum: δ 3.60 (4*H*, t, $J = 6$ Hz); δ 2.38 (2*H*, quintet, $J = 6$ Hz).

(a) From your knowledge of the reaction, what do you think *A* is?

(b) Use the NMR spectrum to confirm or refute your hypothesis. Identify *A*.

14.5 COMPLEX NMR SPECTRA

The NMR spectra of some compounds contain splitting patterns that do not appear to be the simple ones predicted by the $n + 1$ rule. Two common sources of such complex spectra are, first, the splitting of one set of protons by more than one other set, called *multiplicative splitting*, and, second, the breakdown of the $n + 1$ rule itself in certain cases. This section discusses each of these situations and shows how to deal with them.

A. Multiplicative Splitting

To understand multiplicative splitting, we first have to understand what controls the size of the coupling constants *J* between protons on adjacent carbons, called **vicinal coupling constants**. (Remember, the coupling constant is the separation between the individual lines of a splitting pattern.) One of the major factors that govern the size of vicinal coupling constants is the dihedral angle θ between the bonds to the coupled protons. This relationship is shown by the curve in **Fig. 14.9**, called the **Karplus curve**, after Martin Karplus (b. 1930), a professor emeritus of chemistry at Harvard University who was awarded the Nobel Prize in Chemistry in 2013 (for other work). The equation describing this curve is

$$J = J_0 \cos^2 \theta \qquad (14.16)$$

where J_0 has two different values that depend on the domain of θ. For the plot in Fig. 14.9, $J_0 = 10$ for $\theta < 90°$, and $J_0 = 14$ for $\theta > 90°$. (The value of J_0 at $\theta = 90°$ doesn't matter because $\cos^2 90° = 0$.) The exact value of J_0 depends on the nature of the substituents at the carbons bearing the coupled protons and other structural features; for example, electronegative substituents reduce this value somewhat. The important thing to notice about this curve, however, is not the exact size of J_0 but rather that (1) *J* is greatest when $\theta = 180°$—that is, when two protons have an anti relationship,

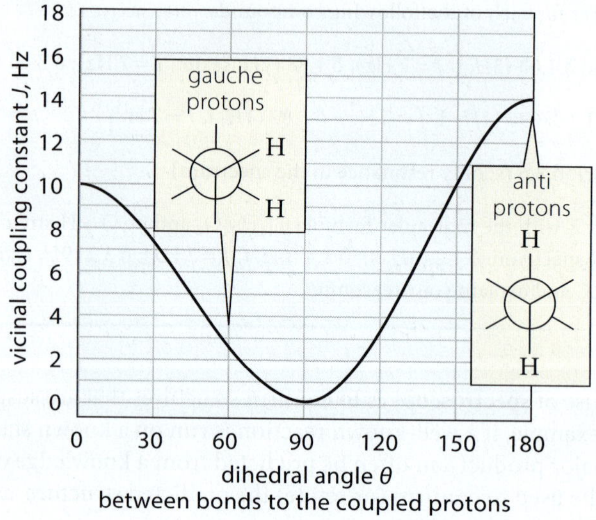

FIGURE 14.9 The Karplus curve plotted for Eq. 14.16 with $J_0 = 10$ ($\theta < 90°$) and $J_0 = 14$ ($90° < \theta < 180°$). This curve describes the coupling constant between two vicinal protons as a function of dihedral angle. The values for gauche protons ($\theta = 60°$) and anti protons ($\theta = 180°$) are marked on the graph. This curve shows that anti splitting is much greater than gauche splitting.

and (2) the value of J is much smaller when $\theta = 60°$—that is, when two protons have a gauche relationship. The following ranges for J have been observed:

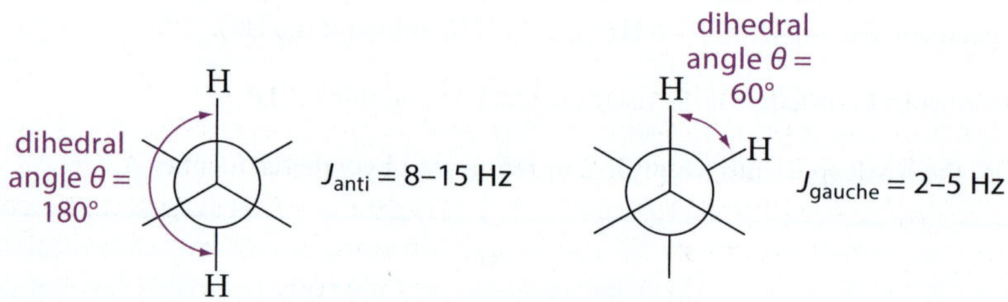

(14.17)

These ranges show that anti and gauche relationships between C—H bonds can be clearly differentiated by the coupling constants between the protons.

What is the reason for this angular dependence of splitting? Section14.4B explained that splitting between protons is mediated by the electrons in the bonds between the protons. The size of the coupling constant depends on the strength of the interactions between the bond orbitals containing these electrons. The sp^3 orbitals of C—H bonds on adjacent carbons interact most strongly when they are coplanar—that is, when they have either an anti or eclipsed relationship. Orbitals at right angles do not interact at all, and their interaction is weak when they have a gauche relationship.

The best illustrations of the Karplus relationship involve molecules that have one predominant conformation, such as the following two diastereomers. (The asterisks indicate corresponding carbons in the chair structure and the Newman projection.)

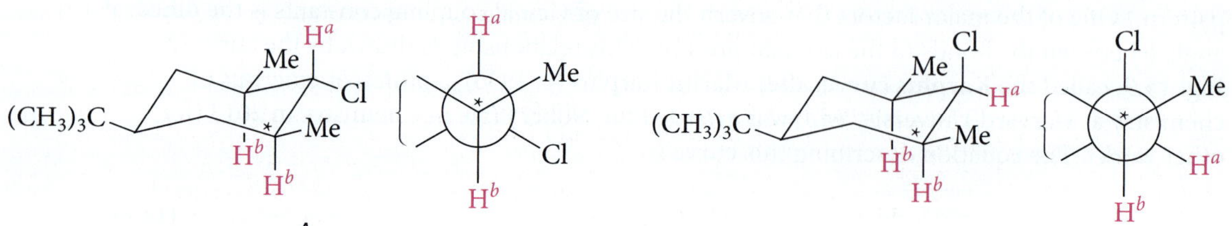

A
H^a and H^b are anti
$J_{ab} \approx 12$ Hz

B
H^a and H^b are gauche
$J_{ab} \approx 3$ Hz

(14.18)

These molecules exist almost completely in the single chair conformations shown in Display 14.18 because the large *tert*-butyl group and the methyl groups occupy equatorial positions (Eq. 7.13, Sec. 7.3). Let's focus only on the resonance of H^a in the NMR spectrum of each stereoisomer. This resonance is well separated from the other resonances in the spectrum and is easy to identify because it has the greatest chemical shift. (We ignore the remaining parts of the spectra.) Specifically, focus on the splitting of the resonance of proton H^a by the two equivalent protons H^b in each case. In the first molecule *A*, the C—H^a bond is anti to each of the equivalent C—H^b bonds. The coupling constant between these two protons is predicted by the Karplus curve to be large. (In fact, it is about 12 Hz.) Therefore, the resonance for H^a is a triplet with $J_{ab} = 12$ Hz. In the second molecule *B*, the C—H^a bond is gauche to each of the equivalent C—H^b bonds. The Karplus curve predicts a much smaller coupling constant; in fact, it is about 3 Hz. Therefore, in this compound, the resonance for H^a is a triplet with $J_{ab} = 3$ Hz. Had the relative stereochemistry of these two compounds not been known, it could have been confidently assigned by analyzing the coupling constants.

Now we are ready to consider the issue of multiplicative splitting. In the following compound (which is much like compound *A* above but without the methyl groups), the resonance of proton H^a is split by the two equivalent protons H^b and by the two equivalent protons H^c.

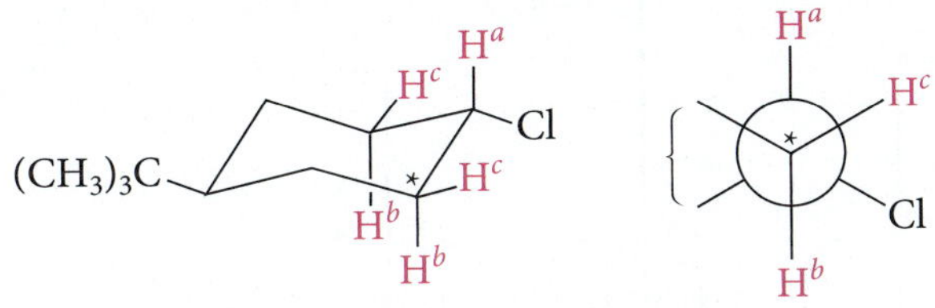

***trans*-1-*tert*-butyl-4-chlorocyclohexane**
(* indicates corresponding carbons) (14.19)

Protons H^b and H^c are *not* equivalent. Because H^a is anti to H^b but gauche to H^c, the Karplus curve indicates that J_{ab} and J_{ac} should be quite different. When a proton is coupled to two different protons with *different coupling constants*, the splitting of each is applied successively. The successive application of different splittings to the same proton is called **multiplicative splitting**. In this case, the two splittings are observed to be $J_{ab} = 11.8$ Hz and $J_{ac} = 4.3$ Hz. To see the resulting spectrum, we construct a **splitting diagram**. In this type of diagram, shown in **Fig. 14.10a**, we use single lines of the appropriate height to represent the individual resonances of a splitting pattern. Start with a single line at the chemical shift of H^a, as shown at the top of the figure. Now apply either of the two splittings—we'll start with J_{ab}. This results in a triplet (with 1:2:1 line intensity) in which the lines of the triplet are separated by 11.8 Hz, and the total intensity of the three lines equals the intensity of the starting resonance. Now we apply the second splitting $J_{ac} = 4.3$ Hz to *each* line of the splitting pattern we just created, remembering to divide the intensities appropriately. We thus obtain a *triplet of triplets*, or a nine-line pattern. The actual resonance of H^a in this compound is shown in **Fig. 14.10b**. (The order in which we apply the two splittings doesn't matter—the result is the same.)

Consider now the splitting of H^a in the cis stereoisomer of the same compound.

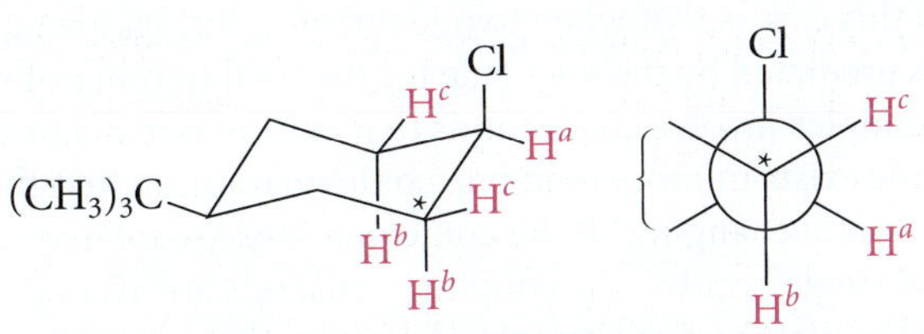

***cis*-1-*tert*-butyl-4-chlorocyclohexane**
(* indicates corresponding carbons) (14.20)

In this compound, H^a has a gauche relationship to *both* H^b and H^c. It turns out that the two splittings in this case, J_{ab} and J_{bc}, are the same (2.9 Hz). This is reasonable because the two dihedral angles (C—H^b with C—H^a and C—H^c with C—H^a) are essentially the same. A splitting diagram for this situation is shown in **Fig. 14.11a**. Successive application of two equal splittings results in the exact overlap of several lines, and the result is an apparent quintet. This is exactly the same result that we would have expected by applying the $n+1$ rule for $n =$ the total number of neighboring protons. Proton H^a has four neighboring protons, and the result is that the resonance for H^a is $4+1=5$ lines with the relative intensities 1:4:6:4:1 as shown in Table 14.2. The experimental spectrum is shown in **Fig. 14.11b**.

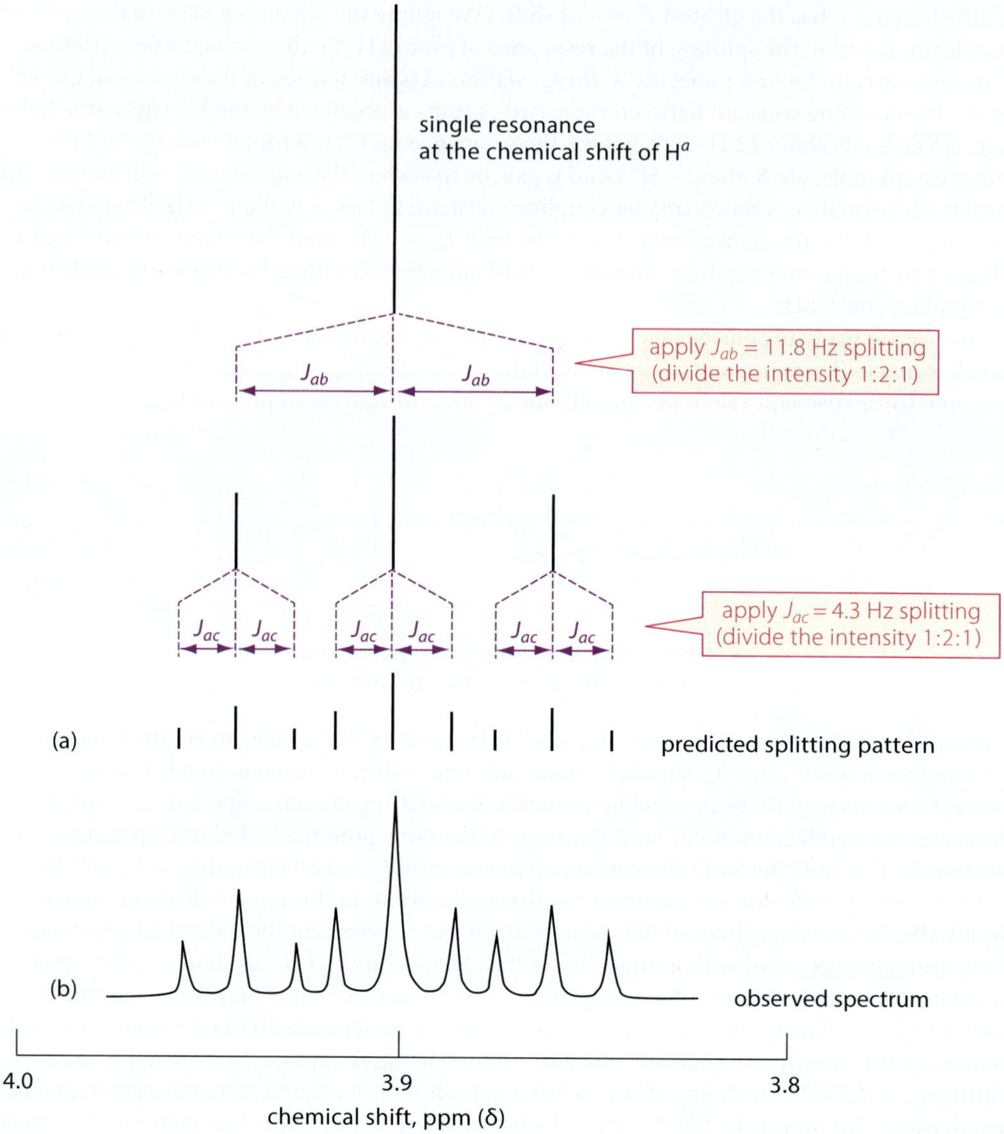

FIGURE 14.10 (a) A splitting diagram illustrating the application of two successive triplet splittings with different coupling constants to a single resonance. Resonances are depicted as vertical lines with relative intensity proportional to height. Start at the top of the figure with a single line at the chemical shift of H^a. The splitting is applied to each line of a pattern, and the intensity of that line is divided 1:2:1 in the daughter lines. Although the larger splitting has been applied first, the order that we apply the splittings makes no difference. (b) The actual spectrum corresponding to this case.

What we learn from this case is that when two identical splittings are applied multiplicatively, the result is the same as predicted by the $n + 1$ rule for the total number of neighboring protons.

The previous two examples involve compounds that exist in one major conformation. What happens when a molecule exists in more than one conformation, so that the angular relationship between protons is constantly changing? If the conformations are interconverting rapidly—as in internal rotation around single bonds—the coupling constants are averaged, and multiplicative splitting does not enter the picture. For example, in bromoethane (NMR spectrum in Fig. 14.6), if we were to "freeze" the internal rotation and analyze the splittings in one instant of time, we would obtain the following:

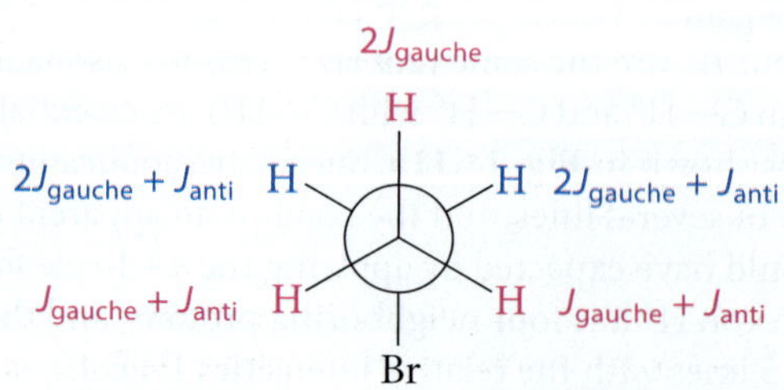

(14.21a)

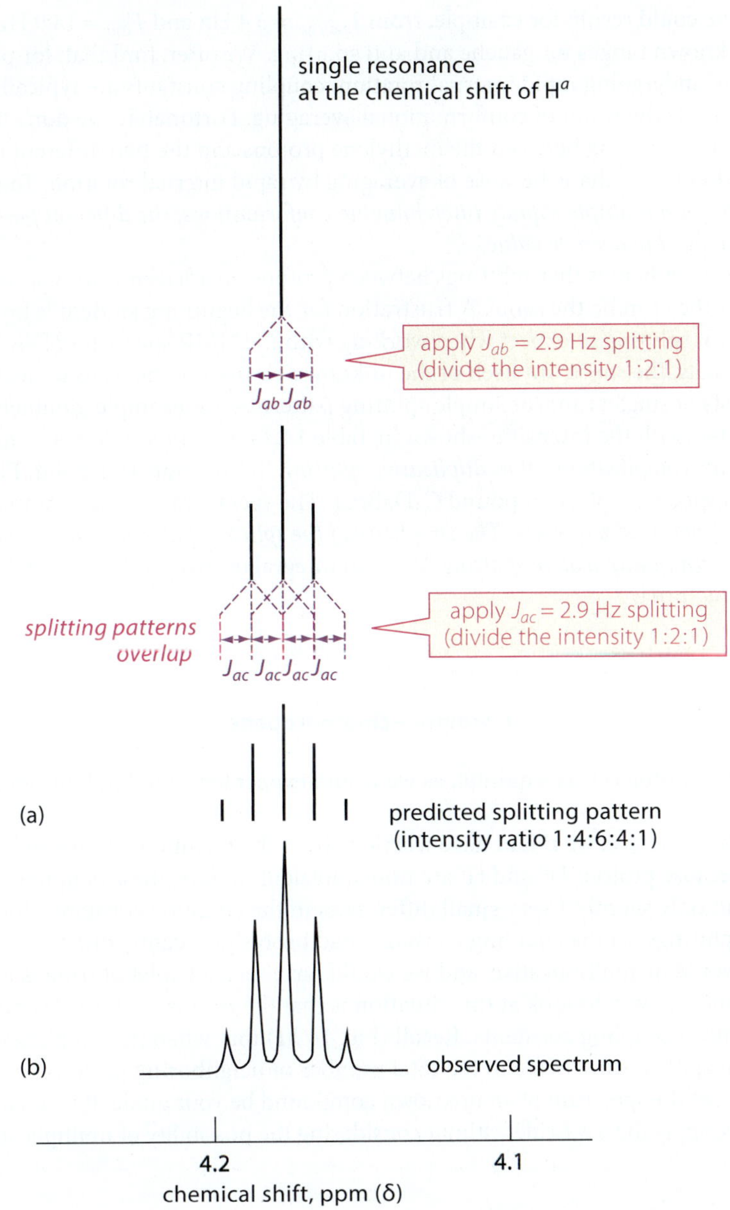

FIGURE 14.11 (a) Application of two successive equal triplet splittings to a resonance. The process used is the same as in Fig. 14.10a. In this case, overlap occurs within the inner lines; the intensities of the overlapping lines are added. As a result, the splitting pattern contains five rather than nine lines in the ratio expected from application of the $n + 1$ rule to all four of the neighboring protons. (b) The actual spectrum corresponding to this case.

The methyl protons (red) experience two different environments, one of which is twice as likely as the other. Each environment is associated with a different coupling constant with the methylene protons. Over time, however, the methyl protons rapidly exchange places by internal rotation, and, as a result, the splittings of the protons on a given carbon are averaged. For example, the six different splittings of the methyl protons are averaged to

$$J(\text{methyl, average}) = \frac{(2J_{\text{gauche}}) + (J_{\text{gauche}} + J_{\text{anti}}) + (J_{\text{gauche}} + J_{\text{anti}})}{6} = \frac{4J_{\text{gauche}} + 2J_{\text{anti}}}{6} \qquad (14.21\text{b})$$

Likewise, the six splittings of the methylene protons (blue) are averaged to

$$J(\text{methylene, average}) = \frac{(2J_{\text{gauche}} + J_{\text{anti}}) + (2J_{\text{gauche}} + J_{\text{anti}})}{6} = \frac{4J_{\text{gauche}} + 2J_{\text{anti}}}{6} \qquad (14.21\text{c})$$

The average coupling constants are the same for the two protons, as they must be. Notice in Fig. 14.6 that the observed coupling constant in bromoethane is 7.2 Hz. From Eqs. 14.21b and c, this average value could result, for example, from $J_{gauche} = 3.4$ Hz and $J_{anti} = 14.9$ Hz, values well within the known ranges for gauche and anti splitting. We often find that, for protons on adjacent carbons undergoing rapid internal rotation, coupling constants are typically in the 6–8 Hz range. This is the result of conformational averaging. Fortunately, we don't have to worry about multiplicative splitting between the methylene protons and the two different types of methyl protons in bromoethane because of averaging by rapid internal rotation. To summarize: *In molecules that contain multiple rapidly interchanging conformations, the different possible coupling constants are averaged to a single value.*

What we've shown here is that splittings between protons on adjacent carbons can be multiplicative or they can be the same. A frustration for the beginning student is how to know what type of splitting to expect when working with the NMR spectrum of an unknown compound. Actually, the spectrum itself of the unknown compound helps us to decide. If the spectrum consists of singlets and/or simple splitting patterns—for example, doublets, triplets, quartets, quintets—with the intensities shown in Table 14.2—*we can use the $n + 1$ rule without worrying about the complications of multiplicative splitting.* To illustrate this point, **Fig. 14.12** shows the NMR spectrum of a compound C_3H_6BrCl. The spectrum has three simple splitting patterns—two triplets and a quintet. *The simplicity of the splitting patterns shows that we don't need to be concerned with multiplicative splitting.* With an integral of two protons for each pattern, the only possible structure is

$$\mathrm{Br}-\overset{a}{\mathrm{CH_2}}-\overset{b}{\mathrm{CH_2}}-\overset{c}{\mathrm{CH_2}}-\mathrm{Cl}$$

1-bromo-3-chloropropane

The resonance for protons H^b is a quintet, as we would expect for a total of four neighboring protons.

Looking at the structure, we might have worried about the possibility of unequal values for J_{ab} and J_{bc} because protons H^a and H^c are nonequivalent. In fact, these coupling constants *are* unequal—but only slightly. (Very small differences in the coupling constants don't cause multiplicative splittings.) If the coupling constants had been significantly different, however, the splitting of H^b would be multiplicative, and we would have seen a triplet of triplets for H^b (up to nine lines). Another way to look at this situation is that the resonance for H^b is multiplicative with nearly identical coupling constants. Recall (Fig. 14.11) that when the coupling constants are identical, we can apply the $n + 1$ rule to the total number of neighboring protons.

In summary: Let the spectrum of an unknown compound be your guide. If it consists of common splitting patterns, apply the $n + 1$ rule without considering the possibility of multiplicative splitting.

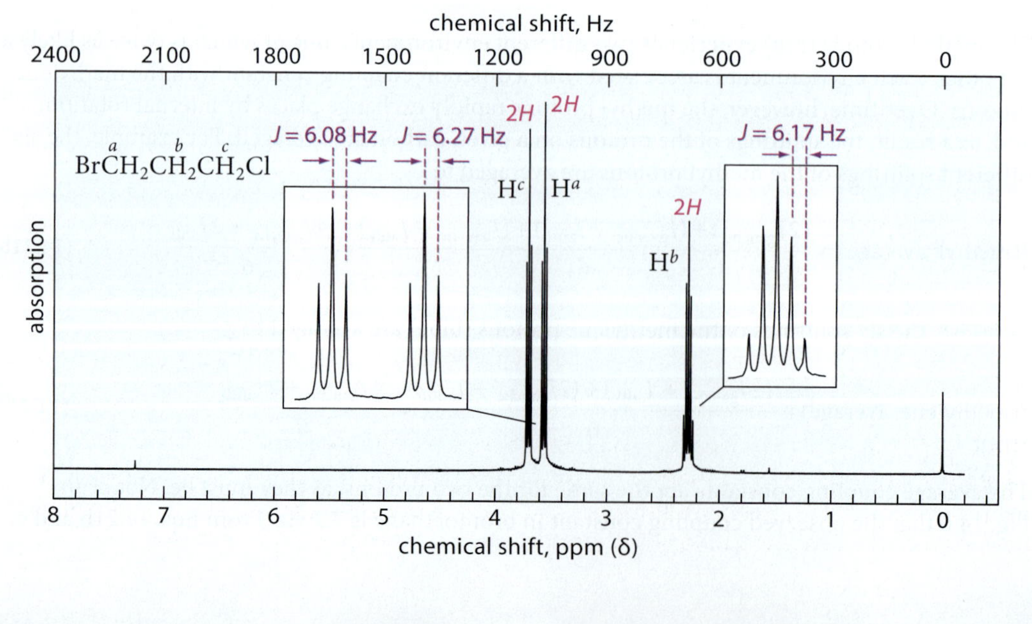

FIGURE 14.12 An NMR spectrum illustrating splitting by nonequivalent protons with nearly identical coupling constants.

What if splitting is complex? Beginning students are not expected to interpret complex splitting patterns. More experienced students in organic chemistry will begin to recognize some of the more common multiplicative patterns. Splitting diagrams can be constructed "in reverse" to analyze simpler multiplicative splittings, and there are computer-simulation programs that enable the detailed analysis of complex patterns. However, even if a spectrum is complex, chances are that it will contain some simple nonmultiplicative splitting patterns that you can readily interpret with the $n+1$ rule, and you can also use the chemical shift and integral information. Sometimes this information is sufficient to solve a structure without a detailed interpretation of all the splitting. We illustrate this approach in the next section.

Focused Problems

14.19 The following compound is unknown, but you are contemplating its synthesis and characterization. Predict its NMR spectrum under each of the following assumptions:

(a) $J_{ab} = J_{bc}$ (b) $J_{ab} \neq J_{bc}$

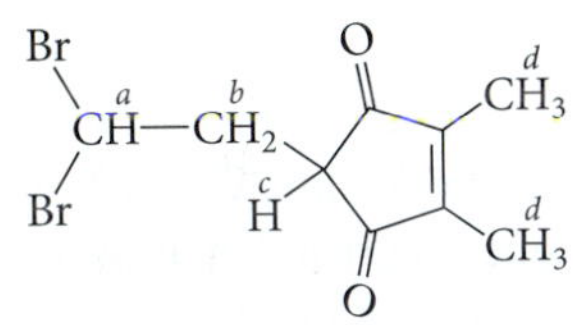

14.20 The three absorptions in the NMR spectrum of 1,1,2-trichloropropane have the following characteristics:

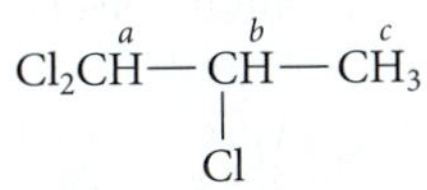

H^a: δ 5.82, $J_{ab} = 3.6$ Hz
H^b: δ 4.40, $J_{ab} = 3.6$ Hz, $J_{bc} = 6.6$ Hz
H^c: δ 1.78, $J_{bc} = 6.6$ Hz.

Using bars to represent lines in the spectrum and a splitting diagram to determine the appearance of the H^b absorption, sketch the appearance of the spectrum. Be sure to take line intensity into account when making your sketch. (Graph paper is useful in constructing splitting diagrams.)

14.21 Predict the complete NMR spectrum of 1,2-dichloropropane under each of the following assumptions. Protons H^b and H^c are diastereotopic and chemically nonequivalent.

(a) $J_{ab} = J_{ac}$ (b) $J_{ab} \neq J_{ac}$

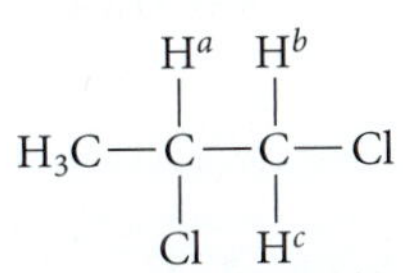

1,2-dichloropropane

B. Breakdown of the $n+1$ Rule

NMR spectra in which all resonances conform to the $n+1$ rule are called **first-order spectra.** In all of the spectra discussed to this point, even the complex multiplicative patterns discussed in the previous section, splitting patterns have been first-order. The spectra of some compounds, however, contain splitting patterns that are more complex than predicted by the $n+1$ rule.

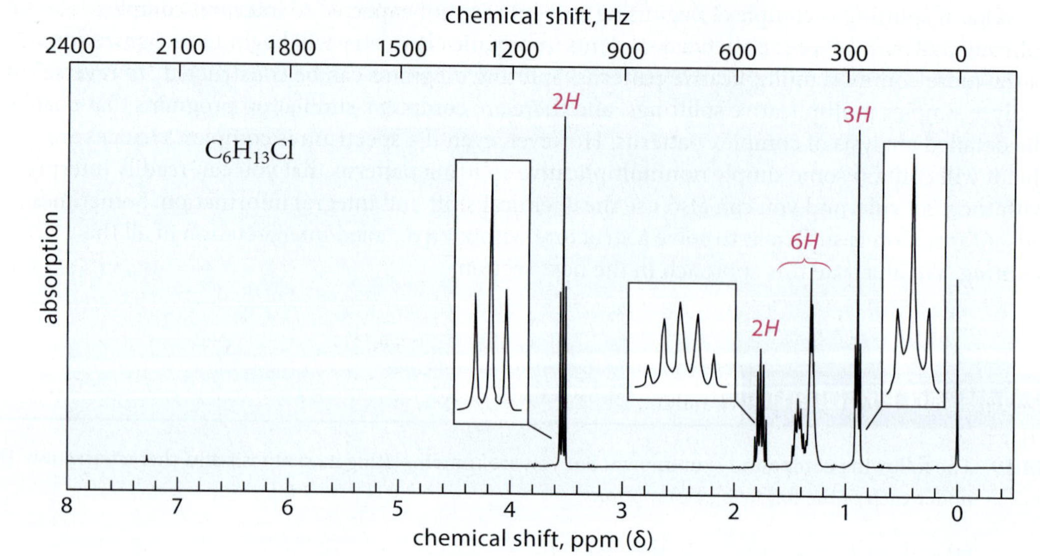

FIGURE 14.13 The NMR spectrum for Study Problem 14.4.

Although such spectra can be analyzed rigorously (in many cases) by special mathematical or instrumental techniques, a great deal of information can be obtained from them without such methods. Study Problem 14.4 illustrates a situation of this sort.

Study Problem 14.4

Determine the structure of the compound with the formula $C_6H_{13}Cl$ that has the NMR spectrum shown in **Fig. 14.13**.

Solution The unknown compound has an unsaturation number of zero and is therefore an alkyl chloride. The spectrum contains a complex splitting pattern in the δ 1.2–1.5 region that cannot be readily interpreted by the $n+1$ rule. However, three first-order features appear in the spectrum: the triplet at δ 3.52, which integrates for two protons; the triplet at δ 0.9, which integrates for three protons; and the quintet at δ 1.77, which integrates for two protons. The complex pattern in the δ 1.2–1.5 region accounts for the remaining six protons. The chemical shift of the δ 3.52 resonance indicates that this triplet must arise from protons on the carbon bonded to the chlorine. Because it accounts for two protons, we can immediately write the partial structure $—CH_2Cl$. Its splitting shows that two protons are on the *adjacent* carbon. This information gives either of the following partial structures:

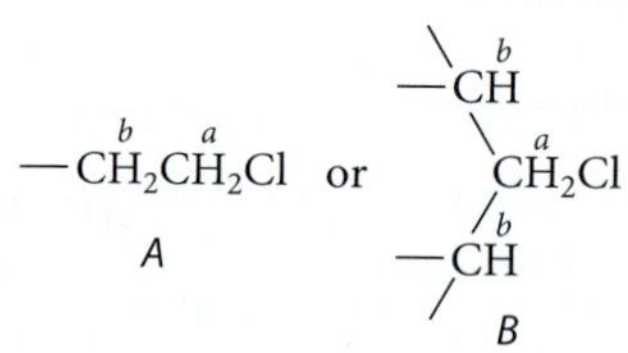

We can rule out partial structure *B* immediately, because the open bonds would have to be occupied by carbons. This would give a total of seven carbons for the molecule, which is not in agreement with the formula. From the chemical shift, the carbon with the hydrogens at δ 1.77 is the same as the carbon bonded to hydrogens *b*. The quintet splitting for H^b indicates four neighboring protons. This gives the following partial structure:

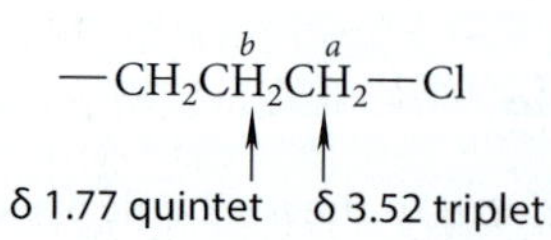

The three-proton resonance at δ 0.9 must be a methyl group. Its chemical shift shows that this group is farthest from the chlorine, and its triplet splitting indicates the partial structure CH_3CH_2—. We have accounted for five carbons. That leaves only a CH_2 unaccounted for, and only one possible structure—1-chlorohexane:

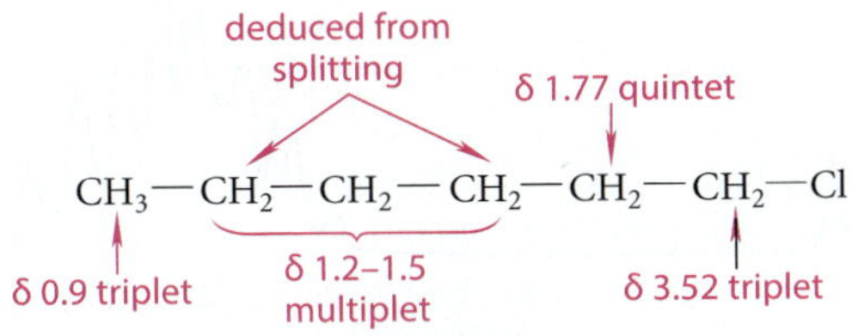

1-chlorohexane

We didn't have to interpret the 6*H* multiplet to solve the structure other than to recognize the number of protons and their chemical-shift range.

Study Problem 14.4 shows that you don't have to interpret *every* splitting pattern in a spectrum to solve a structure because *most spectra contain redundant information.*

Something very important about NMR, however, can be learned by asking why the NMR spectrum of 1-chlorohexane is so complex—why it is not first-order. It turns out that first-order NMR spectra are generally observed when *the chemical shift difference, in Hz, between coupled protons is much greater than their coupling constant.* If the difference in chemical shift of two resonances *a* and *b* is $\Delta\nu_{ab}$ (in Hz) and their coupling constant is J_{ab}, then this condition is simply expressed as follows:

$$\textit{Condition for first-order splitting: } \Delta\nu_{ab} >> J_{ab} \tag{14.22}$$

In practice, this condition means that *first-order splitting can be expected when the splitting patterns of the two coupled protons do not overlap.* For example, in the spectrum of 1-chlorohexane (see Fig. 14.13), the splitting pattern of the δ 0.9 resonance is first-order—a simple triplet—because it does not overlap with the splitting pattern of the coupled proton at δ 1.3. However, in the δ 1.2–1.5 region, the splitting patterns of three different sets of protons overlap. Consequently, the splittings of these protons are not first-order, and we can't use the $n+1$ rule to interpret them.

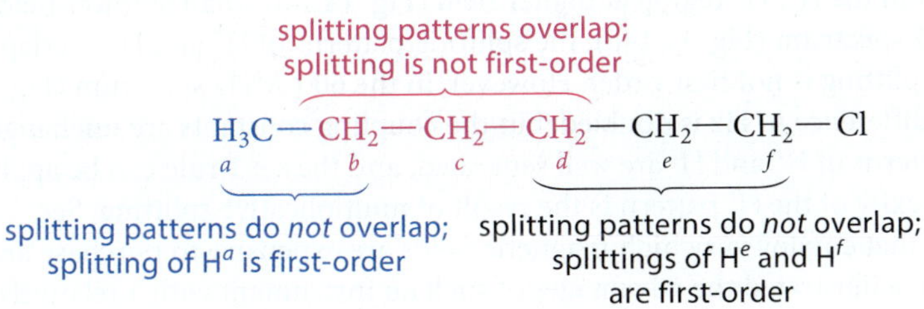

Similarly, the splitting patterns of H^e and H^f are first-order because they are well separated from each other and from the pattern for H^d.

An important aspect of NMR provides an experimental way to simplify splitting patterns that are not first-order: *Coupling constants do not vary with the magnitude of the operating frequency or the applied magnetic field.* As explained in Sec. 14.3B, NMR experiments can be run in different instruments at different operating frequencies (ν_0 in Eq. 14.10, Sec. 14.3B) and corresponding magnetic field strengths $\mathbf{B_0}$. Recall also that chemical shifts in Hz vary in proportion to the operating frequency used. Consequently, if a very large magnetic field and a correspondingly large operating frequency are used, the chemical shifts in Hz are much greater, but the coupling constants are unchanged. Consequently, the condition for first-order behavior in Eq. 14.22 is more likely to be met. This point is illustrated in **Fig. 14.14** by comparing the δ 1.2–1.8 regions of the NMR spectra of 1-chloropentane taken at two different magnetic field strengths (and

FIGURE 14.14 A contrast of the H^b, H^c, and H^d regions (δ 1.2–1.8) in the NMR spectrum of 1-chloropentane. (a) The 300-MHz spectrum. (b) The 600-MHz spectrum. In (a), the splitting patterns of H^b and H^c overlap. As a result, the splitting between these protons is *not* first-order. In (b), the splitting patterns of H^b and H^c are well separated, and their splitting patterns are first-order. (The splitting of H^c is a case of multiplicative splitting that results from the different coupling constants J_{cb} and J_{cd}; Sec. 14.5A.)

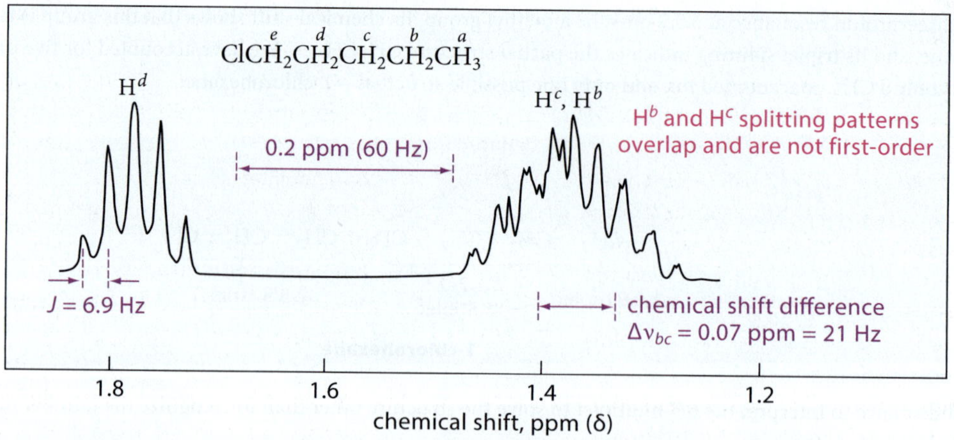

(a) Spectrum taken at ν_0 = 300 MHz (field strength $\mathbf{B_0}$ = 70,500 gauss)

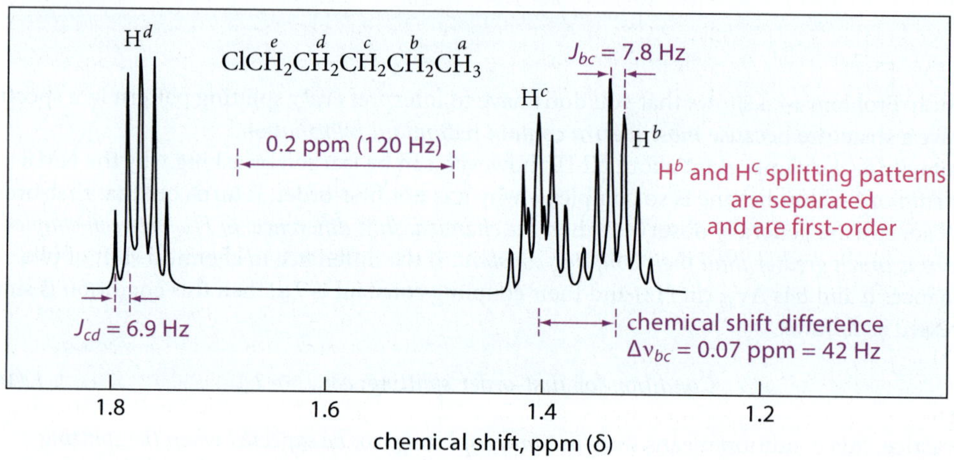

(b) Spectrum taken at ν_0 = 600 MHz (field strength $\mathbf{B_0}$ = 141,000 gauss)

therefore two different operating frequencies). Notice the greater resolution and simplification of the spectrum in the H^b/H^c region at higher field (Fig. 14.14b) than at lower field (Fig. 14.14a). In the 300-MHz spectrum (Fig. 14.14a), the splitting patterns of H^b and H^c overlap extensively. Therefore, the splitting is not first-order. However, in the 600-MHz spectrum (Fig. 14.14b), the chemical-shift difference in Hz is doubled, but the coupling constants are unchanged. As a result, the splitting patterns of H^b and H^c are well separated, and the $n+1$ rule can be applied. (The apparent complexity of the H^c pattern is the result of multiplicative splitting; Sec. 14.5A.)

Instruments that employ very high magnetic fields are expensive to purchase and maintain. Although we have illustrated the advantages of such an instrument with a relatively simple molecule, the major use of such instruments is in unraveling the structures of complex molecules whose NMR spectra would be hopelessly complicated when taken at lower field.

Focused Problem

14.22 Identify the following two isomeric alkyl halides ($C_5H_{11}Br$) from their 300-MHz NMR spectra, which are as follows:

Compound *A*: δ 0.91 (6*H*, d, J = 6 Hz); δ 1.7–1.8 (3*H*, complex); δ 3.42 (2*H*, t, J = 6 Hz)
Compound *B*: δ 1.07 (3*H*, t, J = 6.5 Hz); δ 1.75 (6*H*, s); δ 1.84 (2*H*, q, J = 6.5 Hz)

(a) Give the structure of each compound, and explain your reasoning.

(b) Predict how the spectrum of compound *A* might change if it were taken at 600 MHz.

14.6 USING DEUTERIUM SUBSTITUTION IN PROTON NMR

Deuterium (^2H, or D) finds special use in proton (^1H) NMR. Although deuterium has a nuclear spin, deuterium NMR and proton NMR require greatly different operating frequencies at a given magnetic field strength. Consequently, deuterium NMR absorptions are not detected under the conditions used for proton NMR, so deuterium is effectively "silent" in proton NMR.

One important practical application of this fact is the use of deuterated solvents in NMR experiments. (Solvents are needed for solid and viscous liquid samples because, in the usual type of proton NMR experiment, the sample must be in a free-flowing liquid state.) To ensure that the solvent does not interfere with the NMR spectrum of the sample, it must either be devoid of protons, or its protons must not have NMR absorptions that obscure the sample absorptions. Carbon tetrachloride (CCl_4) is a useful solvent because it has no protons and therefore has no ^1H NMR absorption. However, many organic compounds are not dissolved by carbon tetrachloride. In addition, this solvent has fallen out of use because of its toxicity. Many of the most useful organic solvents contain hydrogens, which have interfering absorptions. Fortunately, many such solvents are available with their hydrogens substituted by deuteriums; these "deuterated" solvents have no interfering NMR absorptions. The most widely used example of such a solvent is $CDCl_3$ (chloroform-*d*, or "deuterochloroform"), the deuterium analog of chloroform, $CHCl_3$. Most of the spectra in this text were taken in $CDCl_3$. In these spectra, you may see a tiny resonance near δ 7.3. This is due to the very small amount of $CHCl_3$ present in commercial $CDCl_3$.

The coupling constants for proton–deuterium splitting are very small. Even when H and D are on adjacent carbons, the H–D coupling is negligible. For this reason, deuterium substitution can be used to simplify NMR spectra and assign resonances. Although deuterium substitution is normally most useful in more complex molecules, let's see how it might be used to assign the resonances of bromoethane (see Fig. 14.6). If you were to synthesize CH_3CD_2Br and record its NMR spectrum, you would find that the quartet of bromoethane has disappeared from the NMR spectrum, and the remaining resonance is a singlet. This experiment would establish that the quartet is the resonance of the CH_2 group and that the triplet is the resonance of the CH_3 group.

The use of deuterium to simplify NMR spectra is particularly important for alcohols. Simply shaking the solution of an alcohol with a little D_2O results in a rapid exchange of the O—H proton for deuterium. This strategy is called **D_2O exchange** or, more colloquially, the "**D_2O shake**." This exchange eliminates the O—H resonance (and identifies it) and also eliminates any splitting between the α-protons and the O—H proton. The only splitting remaining is then the splitting with any β-protons. For example, the spectrum of dry ethanol is transformed by the D_2O shake in the following way:

The "D_2O shake" can also be used with other functional groups in which hydrogens are bonded to electronegative atoms, such as the O—H hydrogens of carboxylic acids and phenols, and the N—H hydrogens of amides and amines. In these functional groups, too, the "D_2O shake" results in the replacement of the hydrogens by deuteriums, and both the resonance for the hydrogens and splitting caused by the hydrogens are eliminated.

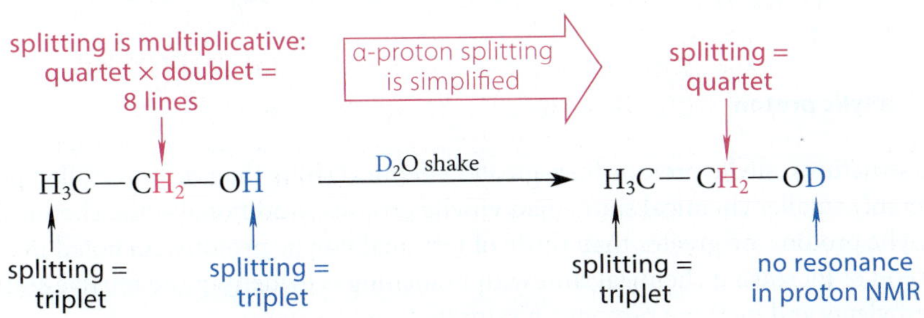

(14.23)

The spectrum of dry ethanol is shown in Fig. 14.20a (presented later in Sec. 14.7E). The spectrum of ethanol following a D_2O shake is identical to the spectrum in Fig. 14.20b, except that there is no O—H resonance.

To summarize: Substitution of a hydrogen by deuterium eliminates its resonance from the proton NMR spectrum and removes any splitting that it causes.

Focused Problems

14.23 The δ 1.2–1.5 region of the 300-MHz NMR spectrum of 1-chlorohexane, given in Fig. 14.13, is complex and not first-order. Assuming you could synthesize the needed compounds, explain how to use deuterium substitution to determine the chemical shifts of the protons that absorb in this region of the spectrum. Explain what you would see and how you would interpret the results.

14.24 Explain how the NMR spectra of each of the following compounds would change following a D_2O shake.

(a)

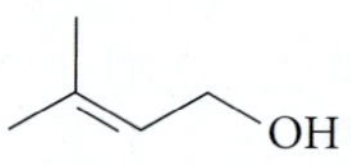

3-methyl-2-buten-1-ol

(b)

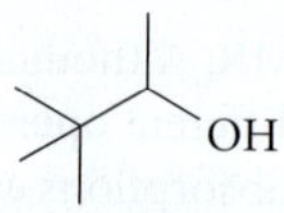

3,3-dimethyl-2-butanol

14.7 CHARACTERISTIC FUNCTIONAL-GROUP PROTON NMR ABSORPTIONS

This section surveys the important NMR absorptions of the major functional groups that were covered in previous chapters. The NMR spectra of other functional groups are considered in the chapters devoted to those groups. However, you have everything you need to read about the NMR spectra of these functional groups. For example, if you study NMR in the laboratory and need to learn about the NMR spectra of aldehydes and ketones, you can read about this topic in Sec. 19.4B. A summary table of chemical shift information is given in Appendix III.

A. Proton NMR Spectra of Alkenes

Two characteristic proton NMR absorptions for alkenes are the absorptions for the protons on the double bond, called **vinylic protons** (red in the following structures), and the protons on carbons *adjacent* to the double bond, called **allylic protons** (blue in the following structures). Don't confuse these two types of protons. Typical alkene chemical shifts are illustrated in the following structures and are summarized in Fig. 14.4 (Sec. 14.3C).

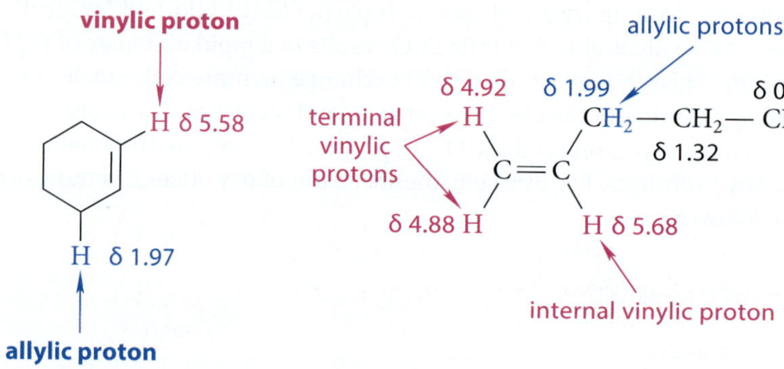

(14.24)

In these structures, allylic protons have greater chemical shifts than ordinary alkyl protons, but considerably smaller chemical shifts than vinylic protons. Additionally, the chemical shifts of internal vinylic protons are greater than those of terminal vinylic protons. As noted in Sec. 14.3C, the same trend of increasing chemical shift with branching is evident in the relative shifts of methyl, methylene, and methine protons on saturated carbon atoms.

Because molecules in solution are constantly in motion and tumbling rapidly, at any given time only a small fraction of the alkene molecules are oriented with respect to the external applied field as shown in Fig. 14.15. The chemical shift of a vinylic proton is an average over all orientations of the molecule. However, this particular orientation makes such a large contribution that it dominates the chemical shift.

The chemical shifts of vinylic protons are much greater than would be predicted from the electronegativity of the alkene functional group and can be understood in the following way. Imagine that an alkene molecule in an NMR spectrometer is oriented with respect to the external applied field $\mathbf{B}_0$ as shown in **Fig. 14.15**. The applied field induces a circulation of the π electrons in closed loops above and below the plane of the alkene. This electron circulation gives rise to an induced magnetic field $\mathbf{B}_i$ that *opposes* the applied field $\mathbf{B}_0$ at the center of the loop. This induced field can be described as contours of closed circles. Although the induced field opposes the applied field $\mathbf{B}_0$ in the region of the π bond, the curvature of the induced field causes it to lie in the same direction as $\mathbf{B}_0$ at the vinylic protons. The induced field, therefore, adds to, or enhances, the local field at the vinylic protons. As a result, the vinylic protons are subjected to a *greater* local field—that is, they are *deshielded*. This means that a greater frequency is required to bring them into resonance (Eq. 14.4, Sec. 14.2). Consequently, their NMR absorptions occur at relatively high chemical shift.

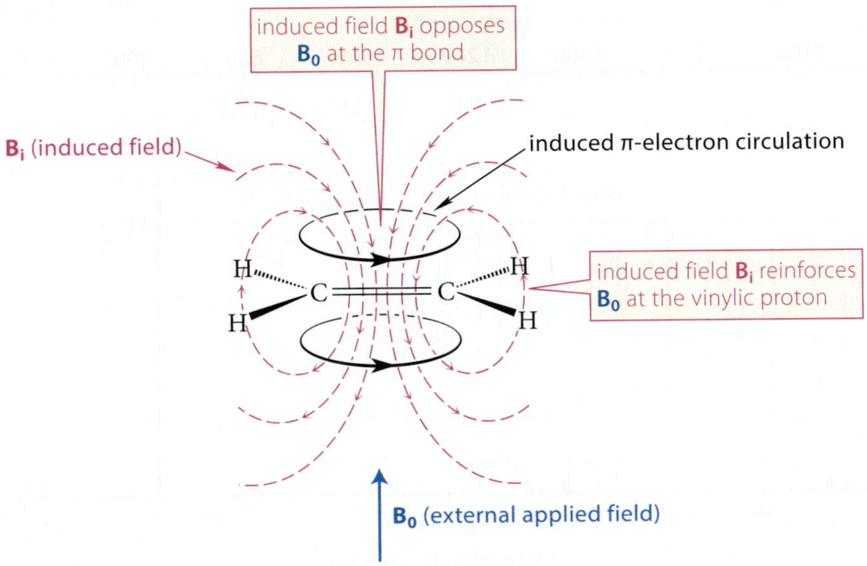

FIGURE 14.15 In an alkene, the induced field $\mathbf{B_i}$ (*red*) of the circulating π electrons augments the external applied field at the vinylic protons. As a result, vinylic protons have NMR absorptions at relatively large chemical shift (high frequency).

Splitting between vinylic protons in alkenes depends strongly on the geometrical relationship of the coupled protons. Typical coupling constants are given in **Table 14.3**. Three of the most important splitting relationships are those between trans protons, cis protons, and geminal protons. Splitting between trans protons is largest, splitting between cis protons is intermediate, and splitting between geminal protons is very small. These coupling constants, along with the characteristic =C—H bending bands from IR spectroscopy (Sec. 13.4C), provide important ways to determine alkene stereochemistry.

TABLE 14.3 Coupling Constants for Proton Splitting in Alkenes

Relationship of protons	Name of relationship	Coupling constant *J*, Hz
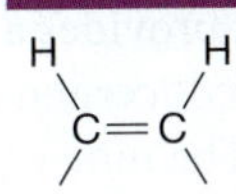	cis	6–14
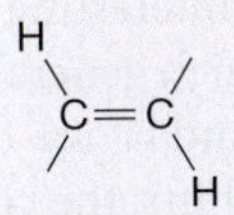	trans	11–18
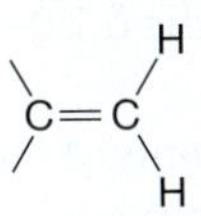	geminal	0–3.5
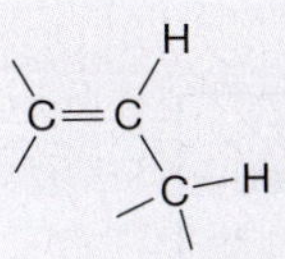	vicinal	4–10
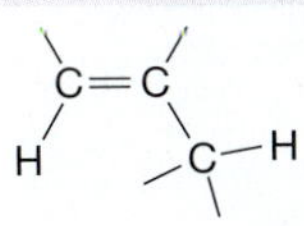 (double bond can be cis or trans)	four-bond (allylic)	0–3.0
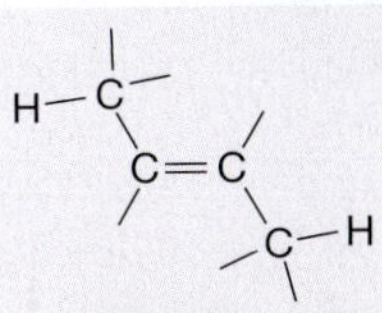 (double bond can be cis or trans)	five-bond	0–1.5

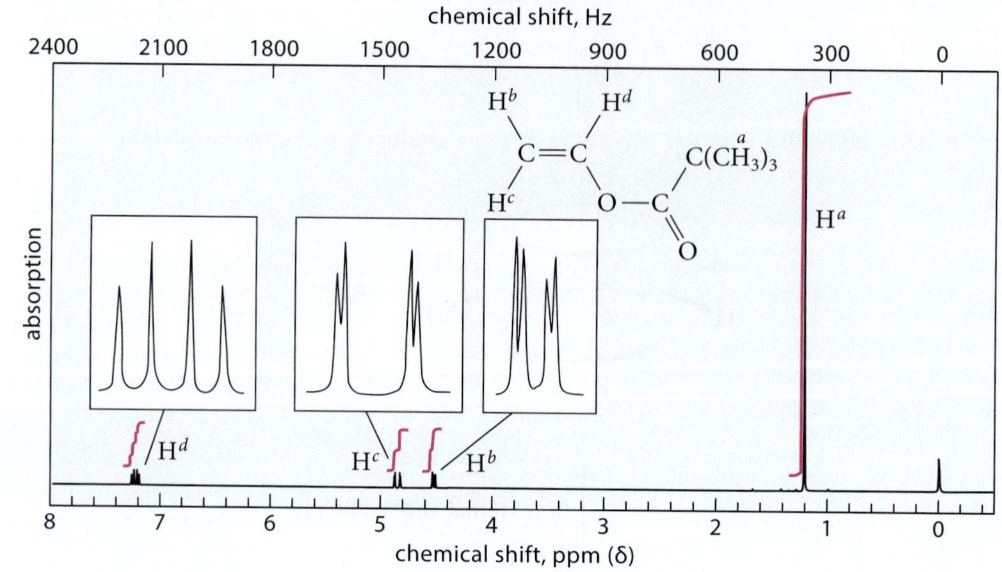

FIGURE 14.16 The proton NMR spectrum of vinyl pivalate showing expansion of the vinylic region. The resonances for these protons show multiplicative splitting, which is analyzed in Fig. 14.17.

As we noted in Sec. 14.5A, a splitting diagram can be constructed by applying successive splittings in any order.

Vinylic splitting is illustrated by **Fig. 14.16**, which is the NMR spectrum of the following compound.

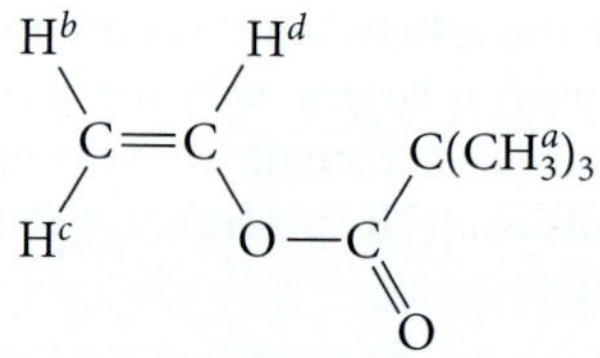

vinyl 2,2-dimethylpropanoate
(vinyl pivalate)

This spectrum illustrates three types of splitting in alkenes and, in addition, provides a useful and relatively straightforward case study for multiplicative splitting. (Do not be concerned that this molecule has an unfamiliar functional group; the principles are the same.) The nine equivalent *tert*-butyl protons of vinyl pivalate (H^a) give the large singlet at δ 1.22. The interesting part of this spectrum is the region containing the resonances of the alkene protons (which generally have chemical shifts greater than 4.5 ppm; Fig. 14.4, Sec. 14.3). The protons H^b and H^c are farthest from the electronegative oxygen and therefore have the smallest chemical shifts; the complex resonances in the δ 4.5–5.0 region are from these protons. The four lines in the δ 7.0–7.5 region are all resonances of the one proton H^d.

Each vinylic proton is split by the other two with different coupling constants. As a result, the splitting is multiplicative, and the resonance of each proton is a doublet of doublets.

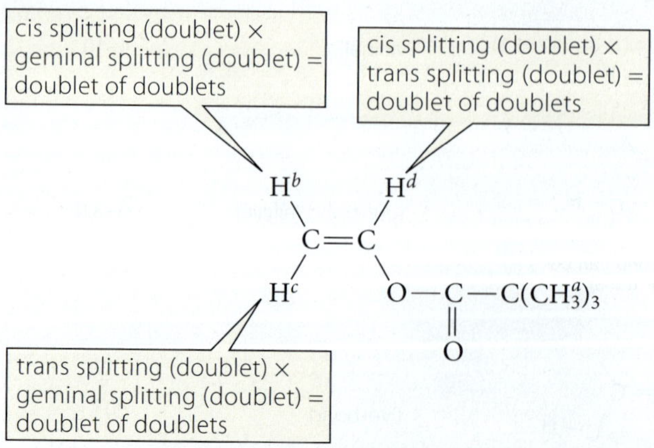

A splitting diagram for the three protons is shown in **Fig. 14.17.** The proton that corresponds to each of the resonances can be identified by its coupling constants. For example, the resonance of

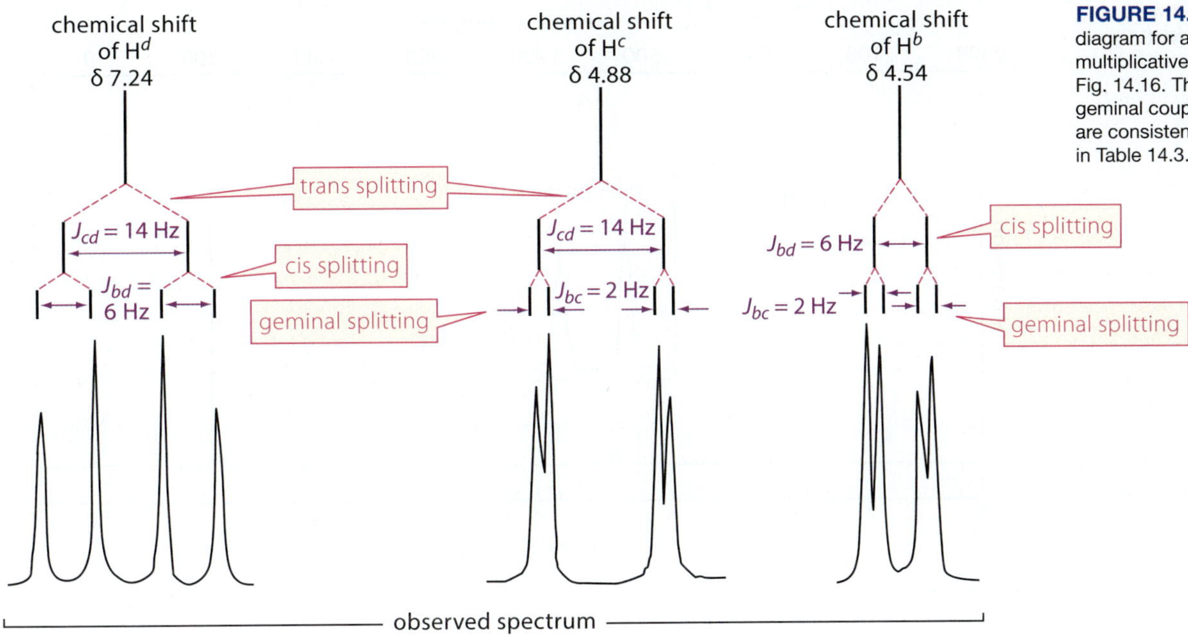

FIGURE 14.17 A splitting diagram for analysis of the multiplicative splittings in Fig. 14.16. The trans, cis, and geminal coupling constants are consistent with the values in Table 14.3.

proton H^d has the two largest (cis and trans) splittings, the resonance of proton H^b has the two smallest (geminal and cis) splittings, and the resonance of H^c has a small (geminal) and large (trans) splitting. This splitting pattern is almost always observed for a terminal vinyl group ($H_2C{=}CH{-}$).

In principle, the intensities of these lines should be equal (see Table 14.2, Sec. 14.5A). However, this is a case in which the chemical shifts of the coupled protons are very similar. In such a case, deviations from the ideal ratios in Table 14.2 are observed. The closer the chemical shifts of the coupled protons, the greater the deviations. (Such deviations are sometimes called *leaning*.) Each four-line pattern is readily distinguishable from a quartet, however, because in a quartet the spacings between the lines of the splitting pattern must be identical.

The last two entries in Table 14.3 show that small splitting in alkenes is sometimes observed between protons separated by more than three bonds. Recall that splitting over these distances is usually *not* observed in saturated compounds. These long-distance interactions between protons are transmitted by the π electrons.

In many spectra, geminal, four-bond, and five-bond splittings are not readily discernible as clearly separated lines, but instead appear as perceptibly broadened peaks. Such is the case, for example, in the NMR spectrum in **Fig. 14.18** (Focused Problem 14.25).

Focused Problem

14.25 Propose a structure for a compound with the formula C_7H_{14} with the NMR spectrum shown in Fig. 14.18. Explain in detail how you arrived at your structure.

B. Proton NMR Spectra of Alkynes

Compare the typical chemical shifts observed in the proton NMR spectra of alkynes with the analogous shifts for alkenes:

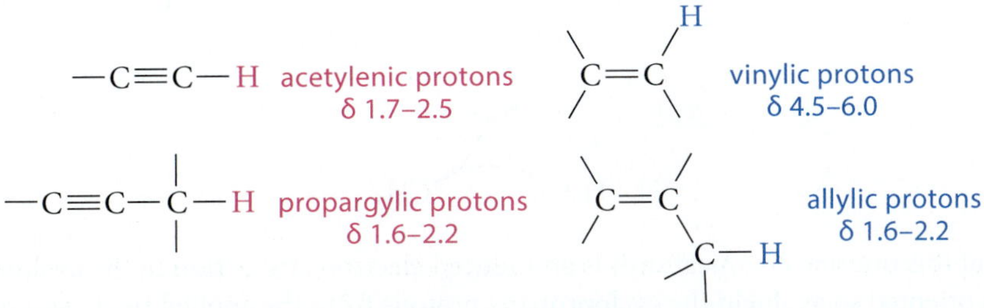

(14.25)

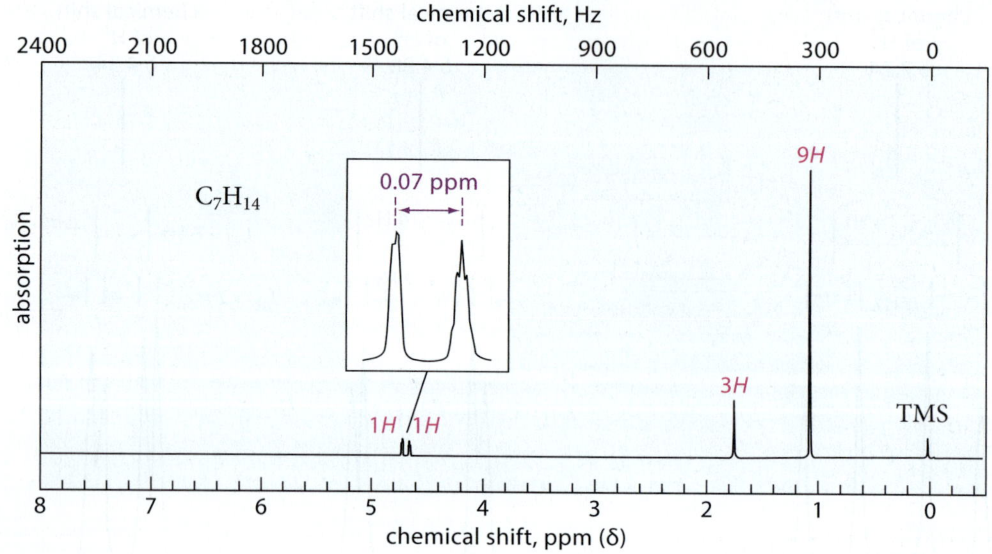

FIGURE 14.18 The 300-MHz proton NMR spectrum for Focused Problem 14.25.

Although the chemical shifts of allylic and propargylic protons are very similar (as might be expected from the fact that both double and triple bonds involve π electrons), the chemical shifts of acetylenic protons are much smaller than those of vinylic protons. Because acetylenic carbons are more electronegative than vinylic carbons (Sec. 3.7C), this is not the result expected from relative electronegativities.

The explanation for the unusual proton chemical shifts observed in alkynes is closely related to the explanation for the chemical shifts of vinylic protons (see Fig. 14.15, Sec. 14.7A), although *the effect is in the opposite direction.* Although an alkyne molecule in solution is tumbling rapidly, acetylenic proton chemical shifts are dominated by the effects resulting from one particular orientation of the alkyne molecule relative to the magnetic field, as shown in **Fig. 14.19**. When an alkyne molecule is oriented in the applied field $\mathbf{B_0}$ as shown in this figure, an induced electron circulation is set up in the cylinder of π electrons (Display 1.71, Sec. 1.7D, and Display 4.5, Sec. 4.1C) that encircles the molecule. The resulting induced field $\mathbf{B_i}$ *opposes* the applied field along the axis of this cylinder. Because the acetylenic proton lies along this axis, the local field at this proton is reduced. Consequently, by the arguments leading to Eq. 14.8b (Sec. 14.3A), acetylenic protons have NMR absorptions at smaller chemical shift than they would have in the absence of this effect.

As in alkenes, splitting in an alkyne can be observed over more than three bonds if the alkyne π bonds are between the coupled protons.

$$\overbrace{\mathrm{Cl{-}CH_2{-}C{\equiv}C{-}H}}^{J_{HH} = 2.8\text{ Hz}} \qquad (14.26)$$

C. Proton NMR Spectra of Alkanes and Cycloalkanes

Because all of the protons in a typical alkane are in very similar chemical environments, the NMR spectra of most alkanes and cycloalkanes cover a narrow range of chemical shifts, typically δ 0.7–1.7.

An interesting exception is the chemical shifts of protons on a cyclopropane ring, which are unusual for alkanes; they absorb at unusually low chemical shifts, typically δ 0–0.5. Some even have resonances at *smaller* chemical shifts than TMS (that is, negative δ values). For example, the chemical shifts of the ring protons of *cis*-1,2-dimethylcyclopropane shown in red are δ (−0.11).

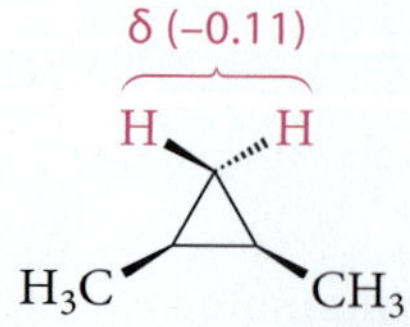

(14.27)

The cause of this unusual chemical shift is an induced electron circulation in the cyclopropane ring that is oriented so as shield the cyclopropane protons from the applied field. As a result, these

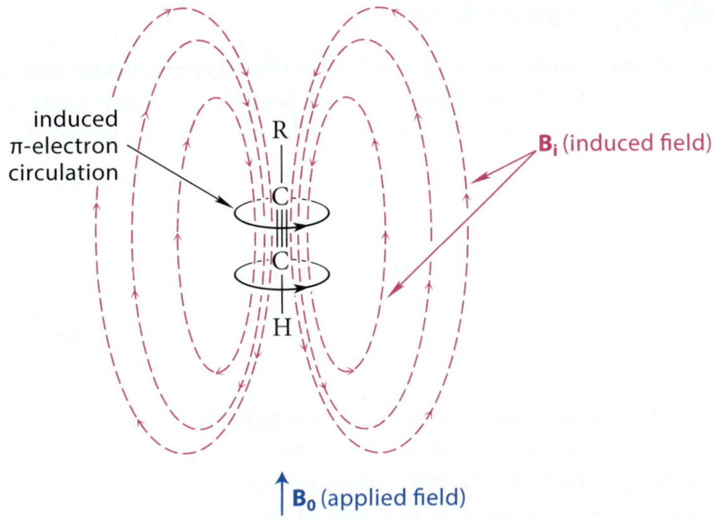

FIGURE 14.19 Explanation of the chemical shift of acetylenic protons. The induced field B_i of the circulating π electrons (*red*) opposes the applied field B_0 (*blue*) from the spectrometer in the region of space occupied by acetylenic protons. As a result, the local field at an acetylenic proton is reduced, and acetylenic protons have NMR absorptions at relatively small chemical shift.

protons are subjected to a smaller local field—that is, they are shielded—and their chemical shifts are *decreased*.

D. Proton NMR Spectra of Alkyl Halides and Ethers

Several NMR spectra of alkyl halides and ethers were presented earlier in this chapter in developing the principles of NMR. The chemical shifts caused by the halogens are usually in proportion to their electronegativities. For the most part, chloro groups and ether oxygens have about the same chemical-shift effect on neighboring protons (see Fig. 14.4, Sec. 14.3). However, epoxides, like cyclopropanes, have considerably smaller chemical shifts than their noncyclic analogs.

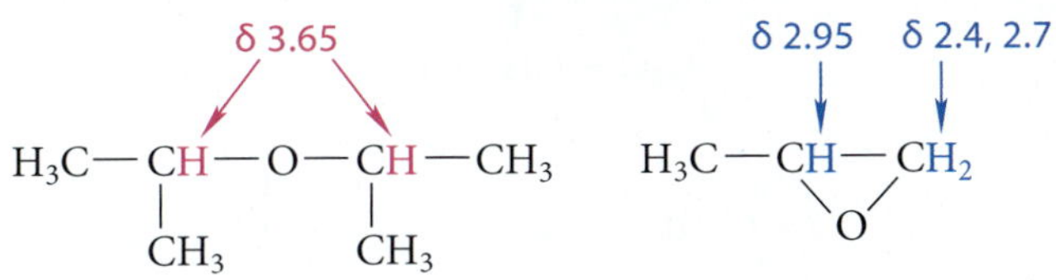

(14.28)

An interesting type of splitting is observed in the NMR spectra of compounds containing fluorine. The common isotope of fluorine (^{19}F) has a nuclear spin. Proton resonances are split by neighboring fluorine in the same general way that they are split by neighboring protons; the same $n+1$ splitting rule applies. For example, the proton in $HCCl_2F$ appears as a doublet centered at δ 7.43 with a large coupling constant J_{HF} of 54 Hz. This is *not* the NMR spectrum of the fluorine; it is the *splitting of the proton resonance caused by the fluorine*. (It is also possible to do fluorine NMR, but this requires, for the same magnetic field, a different operating frequency; the spectra of ^1H and ^{19}F do not overlap.) Values of H–F coupling constants are larger than H–H coupling constants. The J_{HF} value in $(CH_3)_3C$—F is 20 Hz; a typical J_{HH} value over the same number of bonds is 6–8 Hz. Because J_{HF} values are so large, coupling between protons and fluorines can sometimes be observed over as many as four single bonds.

Focused Problems

14.26 Suggest structures for compounds with the following proton NMR spectra.

(a) $C_4H_{10}O$: δ 1.13 (3*H*, t, $J = 7$ Hz); δ 3.38 (2*H*, q, $J = 7$ Hz)

(b) $C_3H_5F_2Cl$: δ 1.75 (3*H*, t, $J = 17.5$ Hz); δ 3.63 (2*H*, t, $J = 13$ Hz)

(c) A compound with molecular mass = 82, IR absorptions at 2100 cm^{-1} and 3300 cm^{-1}, and the following NMR spectrum: δ 1.90 (1*H*, s); δ 1.21 (9*H*, s)

14.27 How would the NMR spectrum of fluoromethane differ from that of chloroethane?

E. Proton NMR Spectra of Alcohols

Protons on the α-carbons of primary and secondary alcohols generally have chemical shifts in the same range as ethers, from δ 3.2 to δ 4.2 (see Fig. 14.4, Sec. 14.3). Because tertiary alcohols have no α-protons, the observation of an O—H stretching absorption in the IR spectrum accompanied by the *absence* of the $-\overset{|}{C}H-O$ absorption in the NMR is good evidence for a tertiary alcohol or a phenol.

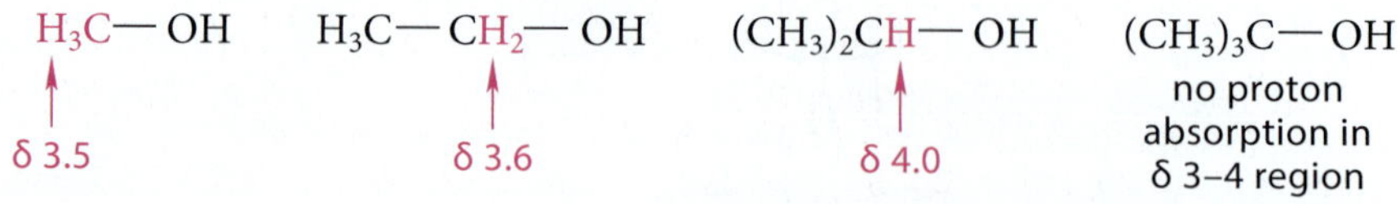

(14.29)

> The chemical shift of the O—H proton in the gas phase is not so large as might be expected for a proton bonded to an electronegative atom such as oxygen. The surprisingly small chemical shift of unassociated OH protons is probably due to the induced field caused by circulation of the nonbonding electron pairs on oxygen. This field shields the OH proton from the external applied field (Sec. 14.3A). Hydrogen-bonded protons, on the other hand, have greater chemical shifts because they bear less electron density and more positive charge (Sec. 8.5C).

The chemical shift of the OH proton in an alcohol is difficult to predict because it depends on the degree to which the alcohol is involved in hydrogen bonding under the conditions used to determine the spectrum. For example, in pure ethanol, in which the alcohol molecules are extensively hydrogen-bonded, the chemical shift of the OH proton is δ 5.3. When a small amount of ethanol is dissolved in CCl_4, the ethanol molecules are more dilute and less extensively hydrogen bonded, and the OH absorption occurs at δ 2–3. In the gas phase, there is no hydrogen bonding, and the OH resonance of ethanol occurs at δ 0.8.

The splitting between the OH proton and the α-protons of alcohols requires special attention. The NMR spectrum of dry ethanol is shown in **Fig. 14.20a**. By the $n+1$ splitting rule, the OH resonance of ethanol is a triplet, and the CH_3 resonance is also a triplet. However, the coupling constants for the two splittings are significantly different. For this reason, the resonance for the CH_2 protons, at δ 3.7 in the spectrum, shows multiplicative splitting (Sec. 14.5A) by both the adjacent CH_3 and OH protons and consists of $4 \times 2 = 8$ lines.

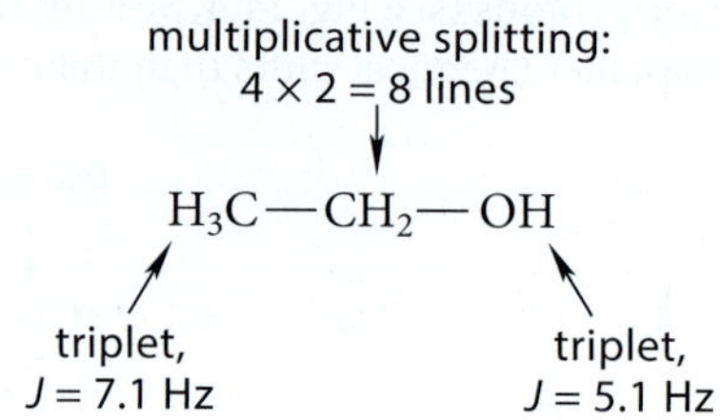

(14.30)

However, when a trace of water, acid, or base is added to the ethanol, the spectrum changes, as shown in **Fig. 14.20b**. The presence of water, acid, or base causes collapse of the O—H resonance to a single line and obliterates all splitting associated with this proton. In this case, the CH_2 proton resonance becomes a quartet, apparently split only by the CH_3 protons. This type of behavior is quite general for alcohols, amines, and other compounds with a proton bonded to an electronegative atom.

This effect of moisture on splitting is caused by a phenomenon called **chemical exchange**: an equilibrium involving chemical reactions that take place rapidly as the NMR spectrum is being determined. In this case, the chemical reaction is proton exchange between the protons of the alcohol and those of water (or other alcohol molecules). For example, acid-catalyzed proton exchange occurs as two successive acid–base reactions:

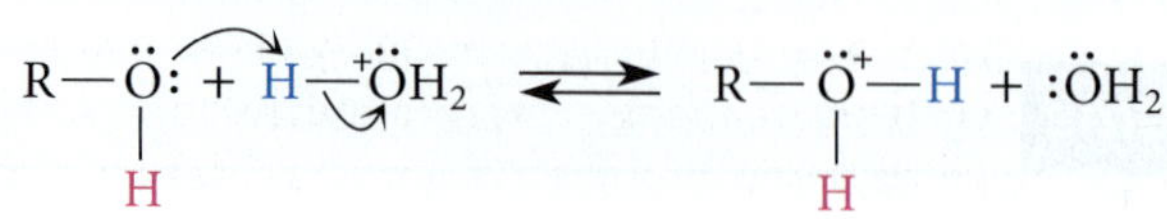

(14.31a)

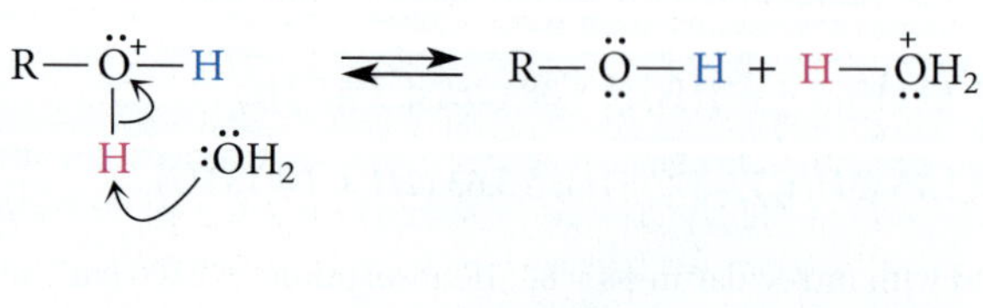

(14.31b)

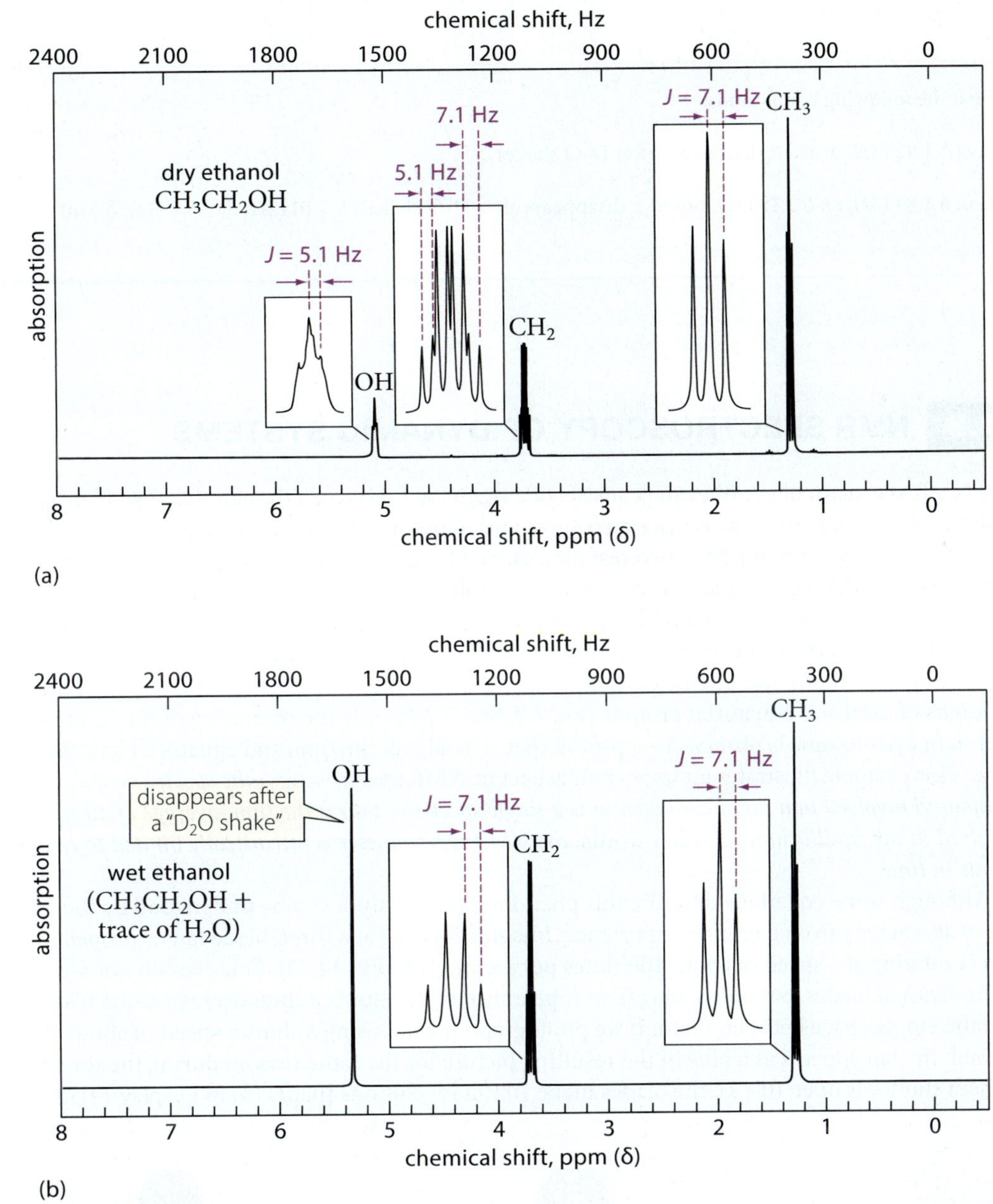

FIGURE 14.20 The NMR spectra of ethanol. (a) Absolute, or very dry, ethanol. The CH_2 resonance is split by both the CH_3 and OH protons. (b) Wet acidified ethanol. The CH_2 resonance is split only by the CH_3 protons. Notice also the shift of the OH resonance under wet and dry conditions. The more extensive hydrogen bonding under wet conditions causes a larger chemical shift. Following a "D_2O shake" (Eq. 14.23, Sec. 14.6), the spectrum in (b) is also obtained except that the O—H resonance is eliminated.

(Write the mechanism for $^-$OH-catalyzed exchange.) For reasons that are discussed in Sec. 14.8, *rapidly exchanging protons do not show spin–spin splitting with neighboring protons.* Acid and base catalyze this exchange reaction, accelerating it enough that splitting is obliterated. In the absence of acid or base, this exchange is much slower, and splitting of the OH proton and neighboring protons is observed.

As a practical matter, you have to be alert to the possibility of either fast or slow exchange when dealing with an NMR spectrum of an unknown that might be an alcohol. An intermediate situation is also common, in which the OH proton resonance is broadened but the α-protons show the splitting characteristic of fast exchange. The assignment of the OH proton can be confirmed in either of two ways. The first is by addition of a trace of acid to the NMR tube. If the α- and O—H protons are involved in splitting, the acid will obliterate this splitting and will simplify the resonances for these two protons. The second way is to use the "D_2O shake," discussed in Sec. 14.6. If a drop of D_2O is added to the NMR sample tube and the tube is shaken, the OH protons rapidly exchange with the protons of D_2O to form OD groups on the alcohol. As a result, the O—H resonance disappears when the spectrum is re-run, as shown in Fig. 14.20b. Any splitting of the α-proton caused by the O—H proton will also be obliterated because the O—H proton is no longer present.

Focused Problem

14.28 Suggest structures for each of the following compounds.

(a) $C_4H_{10}O$; δ 1.27 (9*H*, s); δ 1.92 (1*H*, broad s; disappears after D_2O shake)

(b) $C_5H_{10}O$: δ 1.78 (3*H*, s); δ 1.83 (3*H*, s); δ 2.18 (1*H*, broad s; disappears after D_2O shake); δ 4.10 (2*H*, d, $J = 7$ Hz); δ 5.40 (1*H*, t, $J = 7$ Hz)

14.8 NMR SPECTROSCOPY OF DYNAMIC SYSTEMS

The NMR spectrum of cyclohexane consists of a singlet at δ 1.4, even though the *axial* hydrogens and the *equatorial* hydrogens are diastereotopic and therefore chemically nonequivalent. Why shouldn't cyclohexane have two resonances, one for each type of hydrogen? Recall that cyclohexane undergoes a rapid conformational equilibrium, the *chair interconversion* discussed in Sec. 7.2B. The reason that the NMR spectrum of cyclohexane shows only one resonance has to do with the *rate* of the chair interconversion, which is so rapid that the NMR instrument detects only the average of the two conformations. Because the chair interconversion interchanges the positions of axial and equatorial protons (Eq. 7.9, Sec. 7.2B), only the resonance of the "average" proton in cyclohexane is observed—a proton that is axial half the time and equatorial half the time. This example illustrates an important aspect of NMR spectroscopy: *the spectrum of a compound involved in a rapid equilibrium is a single spectrum that is the time-average of all species involved in the equilibrium.* In other words, *the NMR spectrometer is intrinsically limited to resolve events in time.*

Although some equations describe this phenomenon exactly, it can be understood by the use of an analogy from common experience. Imagine looking at a three-blade fan or propeller that is rotating at a speed of about 100 times per second (Display 14.32). Our eyes do not see the individual blades, but only a blur. The appearance of the blur is a time-average of the blades and the empty space between them. If we photograph the fan using a shutter speed of about 0.1 second, the fan appears as a blur in the resulting picture for the same reason: during the time the camera shutter is open (0.1 s), the blades make 10 full revolutions [panel (a) in Display 14.32].

(a) 100 rotations per second; the image totally blurred

(b) 1 rotation per second; individual blades are visible but blurred

(c) 0.01 rotation per second; individual blades are visible and in sharp focus

(14.32)

Now imagine that we slow the fan to about one rotation per second. While the shutter is open, the fan blades make only 0.1 revolution—about 36°. The fan blades are more distinct, but still somewhat blurred [panel (b) in Display 14.32]. Finally, imagine that the fan is rotating very slowly, say, one rotation every hundred seconds. While the shutter is open, the fan blades traverse only 1/1000 of a circle—about 0.36°. In the resulting picture the individual blades are visible and in relatively sharp focus [panel (c)]. The rapid conformational equilibrium of cyclohexane is to the NMR spectrometer roughly what the rapidly rotating propeller is to the slow camera shutter.

Both axial and equatorial cyclohexane protons can be observed separately if the rate of the chair interconversion is reduced by lowering the temperature. Imagine cooling a sample of cyclohexane in which all protons but one have been replaced by deuterium. (The use of deuterium virtually eliminates splitting with neighboring protons, because splitting between H and D is very small; Sec. 14.6.)

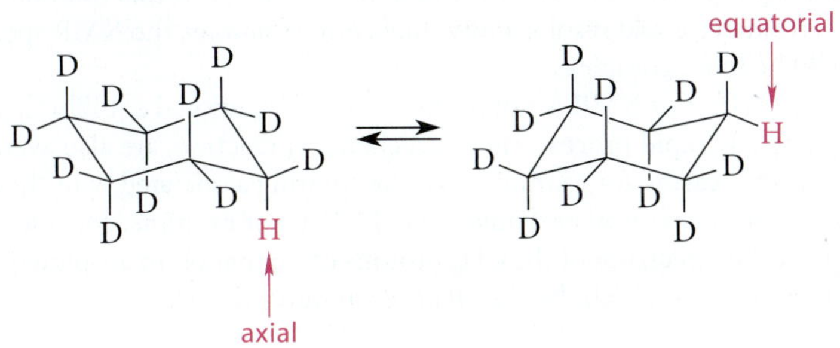

cyclohexane-d_{11} (14.33)

In the chair interconversion, the single remaining proton alternates between axial and equatorial positions.

The NMR spectrum of cyclohexane-d_{11} at various temperatures is shown in **Fig. 14.21**. At room temperature, the spectrum consists of a single line, as in cyclohexane itself. As the temperature is lowered progressively, the resonance becomes broader until, below –60 °C, it divides into two broad resonances equally spaced about the original one. When the temperature is lowered still further, the spectrum becomes two sharp single lines. Lowering the temperature has reduced the rate of the chair interconversion until, at low temperature, NMR spectrometry detects both chair forms independently. This is analogous to taking pictures of the propeller in Display 14.32 with a constant shutter speed and slowing down the propeller until the blur disappears and the individual blades become clearly separated.

The information from these spectra at different temperatures can be used to calculate the rate of the chair interconversion. The energy barrier for the chair interconversion shown in Fig. 7.5 (Sec. 7.2C) was obtained from this type of calculation.

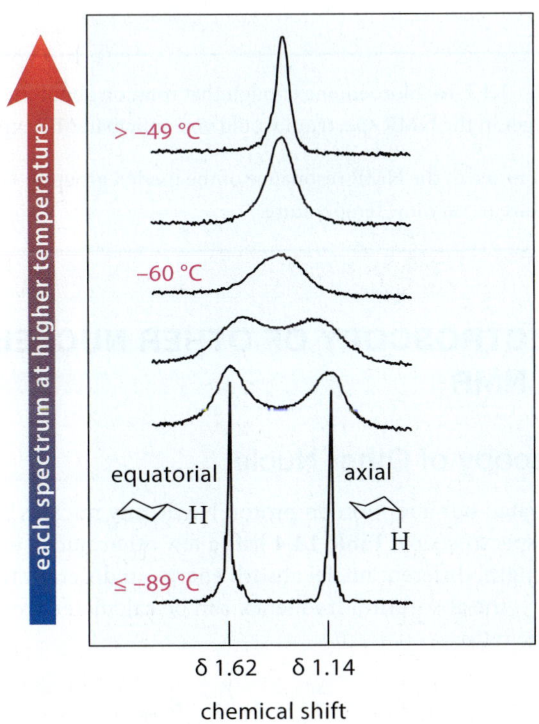

FIGURE 14.21 The 60-MHz proton NMR spectrum of cyclohexane-d_{11} (structure in Eq. 14.33) as a function of temperature. Decreasing the rate of the chair interconversion by lowering the temperature causes the axial and equatorial proton to be separately observable.

Just as NMR spectroscopy detects a time-average of the two chair conformations of cyclohexane at room temperature, it also detects an average of all conformations of any molecule undergoing rapid conformational equilibria. For example, the CH_3 protons of bromoethane (CH_3CH_2Br) give a single resonance and a single coupling constant for splitting by the CH_2 protons because the molecule undergoes rapid internal rotation about the carbon–carbon bond. (The averaging of coupling constants was discussed in Sec. 14.5A.) If this rotation were so slow that the NMR spectrometer could resolve individual conformations, the NMR spectrum of bromoethane would be more complex.

The time-averaging effect of NMR is not limited to conformational equilibria. The spectra of molecules undergoing any rapid process, such as a chemical reaction, are also averaged by NMR spectroscopy. This is the reason, for example, that the splitting associated with the OH protons of an alcohol is obliterated by chemical exchange (Sec. 14.7E). For example, consider the effects of chemical exchange on the spectrum of the CH_3 protons of methanol. In absolutely dry methanol, the resonance of these protons is split by the OH proton into a doublet.

doublet in dry methanol; singlet in wet methanol ⟶ $H_3C—OH$ ⟵ quartet in dry methanol; singlet in wet methanol

Recall from Fig. 14.8 (Sec. 14.4B) that this splitting occurs because the adjacent OH proton can have either of two spins. When acid or base added to the methanol causes the OH protons to exchange rapidly, protons of different spins jump quickly on and off the OH. As a result of this exchange, the CH_3 protons *on any one molecule* are next to an OH proton with spin $+\frac{1}{2}$ for half of the time and an OH proton with spin $-\frac{1}{2}$ for half of the time. (The minuscule difference between the numbers of protons in the two spin states can be ignored.) In other words, the CH_3 protons "see" an adjacent OH proton with a spin that averages to zero over time. Because a proton is not split by an adjacent nucleus with zero spin, rapid exchange eliminates splitting of the CH_3 protons. Similar reasoning can be applied to the spectrum of the methanol OH proton, which, in a dry sample, is a quartet, but is a singlet in a sample containing traces of moisture.

If you now return to Fig. 14.20 and examine the NMR spectra of dry and wet ethanol, you should now be able to understand why rapid chemical exchange of the OH proton eliminates the OH splitting and simplifies the splitting of the α-protons.

Focused Problems

14.29 Suppose you were able to cool a sample of 1-bromo-1,1,2-trichloroethane enough that rotation about the carbon–carbon bond becomes slow on the NMR time scale. What changes in the NMR spectrum would you anticipate? Be explicit.

14.30 Describe in detail what changes you would expect to see in the NMR resonance of the methyl group as 1-chloro-1-methylcyclohexane is cooled from room temperature to very low temperature.

14.9 NMR SPECTROSCOPY OF OTHER NUCLEI; CARBON NMR

A. NMR Spectroscopy of Other Nuclei

Although we've concentrated our attention on proton NMR, any nucleus with a nuclear spin can be studied by NMR spectroscopy. **Table 14.4** lists a few other nuclei with spin $= \pm\frac{1}{2}$. For a given magnetic field strength, different nuclei absorb energy in different frequency ranges. For a given field strength $\mathbf{B}_{nuc}$, the absorption frequency can be calculated from Eq. 14.4 using the appropriate magnetogyric ratio:

$$\nu_{nuc} = \frac{\Delta\varepsilon_{nuc}}{h} = \frac{\gamma_{nuc}}{2\pi}\mathbf{B}_{nuc} \tag{14.34}$$

TABLE 14.4 Properties of Some Nuclei with Spin ± ½

Isotope	Relative sensitivity	Natural abundance, %	Operating frequency $\nu_{0,nuc}$* MHz*	Magnetogyric ratio, γ_{nuc}‡
^{1}H	(1.00)	99.98	300	26,753
^{13}C	0.0159	1.10	75	6728
^{19}F	0.834	100	282	25,179
^{31}P	0.0665	100	122	10,840

*At magnetic field $\mathbf{B_0}$ = 70,500 gauss ‡In rad gauss^{-1} s^{-1} defined in Eq. 14.34

In this equation, $\Delta\varepsilon_{nuc}$ is the energy difference between spin energy levels for a particular nucleus ("*nuc*"), γ_{nuc} is the magnetogyric ratio of the nucleus, and $\mathbf{B}_{nuc}$ is the magnetic field at the nucleus. (The operating frequency for each nucleus can be calculated by letting $\mathbf{B}_{nuc}$ = the applied magnetic field $\mathbf{B_0}$.) This equation shows that the *absorption frequency of any nucleus at a given magnetic field strength depends on its magnetogyric ratio.* For example, the magnetogyric ratio of the proton, γ_H, is 26,753 rad gauss^{-1} s^{-1}. If the applied magnetic field is 70,500 gauss, for example, an operating frequency of 300 MHz is required for proton NMR. Because the magnetogyric ratio for ^{13}C is 6728 rad gauss^{-1} s^{-1}, or about one-quarter of that for a proton, the operating frequency required for ^{13}C NMR at the same field strength is also one-quarter of that for a proton, or about 75 MHz.

Suppose we have a molecule containing two different magnetically active nuclei, such as an alkyl fluoride that contains both ^{1}H and ^{19}F. In the *proton* NMR spectrum of such a molecule, the NMR signals of protons are observed, but not those of the fluorines, because the two nuclei require different operating frequencies. (The proton splitting *caused* by the fluorines *is* observed, however; Sec. 14.7D.) To observe *fluorine NMR,* a different frequency range is used, in which case the fluorine resonances but not the proton resonances are observed. (In this situation, the splitting of fluorine signals caused by nearby protons would be observed.)

B. Carbon-13 NMR Spectroscopy

Because all organic compounds contain carbon, the NMR spectroscopy of carbon is very important. However, the most abundant nucleus of carbon (^{12}C) does not have a nuclear spin and therefore cannot be detected by NMR. As Table 14.4 shows, the only stable isotope of carbon that has a nuclear spin is ^{13}C. The NMR spectroscopy of ^{13}C is called **^{13}C NMR spectroscopy**, which is often referred to as **carbon-13 NMR** or shortened to **carbon NMR**.

One problem with carbon NMR spectroscopy is that the resonance of a ^{13}C nucleus is *intrinsically* weaker than that of a proton because of the magnetic properties of the carbon nucleus. The intrinsic intensity of the NMR signal from each nucleus is proportional to the *cube* of its magnetogyric ratio; so the relative intensity of a proton signal versus that from the same number of ^{13}C atoms is $(\gamma_H/\gamma_C)^3 = (26{,}753/6728)^3$, or 62.9. In other words, a ^{13}C NMR resonance is about $1/62.9 = 0.0159$ times as intense as a proton resonance. Another problem is the low natural abundance of the ^{13}C isotope: organic compounds contain only about 1.1% of ^{13}C at each carbon position. The cumulative effect of intrinsic intensity and low natural abundance, therefore, is that carbon resonances are only $(0.0159)(0.011) = 0.000175$ times as intense as proton resonances.

Although the weak ^{13}C NMR resonance at one time presented a serious obstacle to detection, advances in instrumentation (Sec. 14.11) and computer power have made it possible to obtain ^{13}C NMR spectra on compounds containing the *natural abundance* of ^{13}C on a routine basis. These advances make it possible to acquire a single NMR spectrum in a fraction of a second and to store it digitally in a computer. A useful carbon-NMR spectrum of a compound is obtained by taking a few thousand individual carbon-NMR spectra of the compound and adding them together. Because electronic noise is random, it is reduced when many spectra are added together, whereas the resonances themselves are enhanced. Almost 6000 carbon NMR spectra, obtained in a few minutes, must be added together in this manner to get the same intensity that we would obtain in a single proton NMR spectrum of the same compound at the same concentration. Carbon NMR spectroscopy is an important analytical technique in organic chemistry.

Although the principles of carbon NMR and proton NMR are essentially the same, some aspects of carbon NMR are unique. First, coupling (splitting) between carbons is *not* generally observed. The reason has to do with the low natural abundance of ^{13}C. Recall that carbon NMR measures the resonance of ^{13}C, not the common isotope ^{12}C. If the probability of finding a ^{13}C at a given carbon is 0.0110, then the probability of finding ^{13}C at any two carbons in the same molecule is $(0.0110)^2$, or 0.00012. This means that *two* ^{13}C *atoms almost never occur together within the same molecule.* (The two ^{13}C atoms would have to occur in the *same* molecule for coupling to be observed.) It is possible, though, to prepare compounds that are isotopically enriched in ^{13}C, in which case the usual splitting rules apply (see Problem 14.68).

A second important aspect of carbon NMR is that the range of chemical shifts is very large compared with that in proton NMR. Typical carbon chemical shifts, shown in **Fig. 14.22**, cover a range of about 200 ppm. [Compare this range with the chemical-shift range of protons in Fig. 14.4 (Sec. 14.3).] With a few exceptions, trends in carbon chemical shifts parallel those for proton chemical shifts, but chemical shifts in carbon NMR are more sensitive to small changes

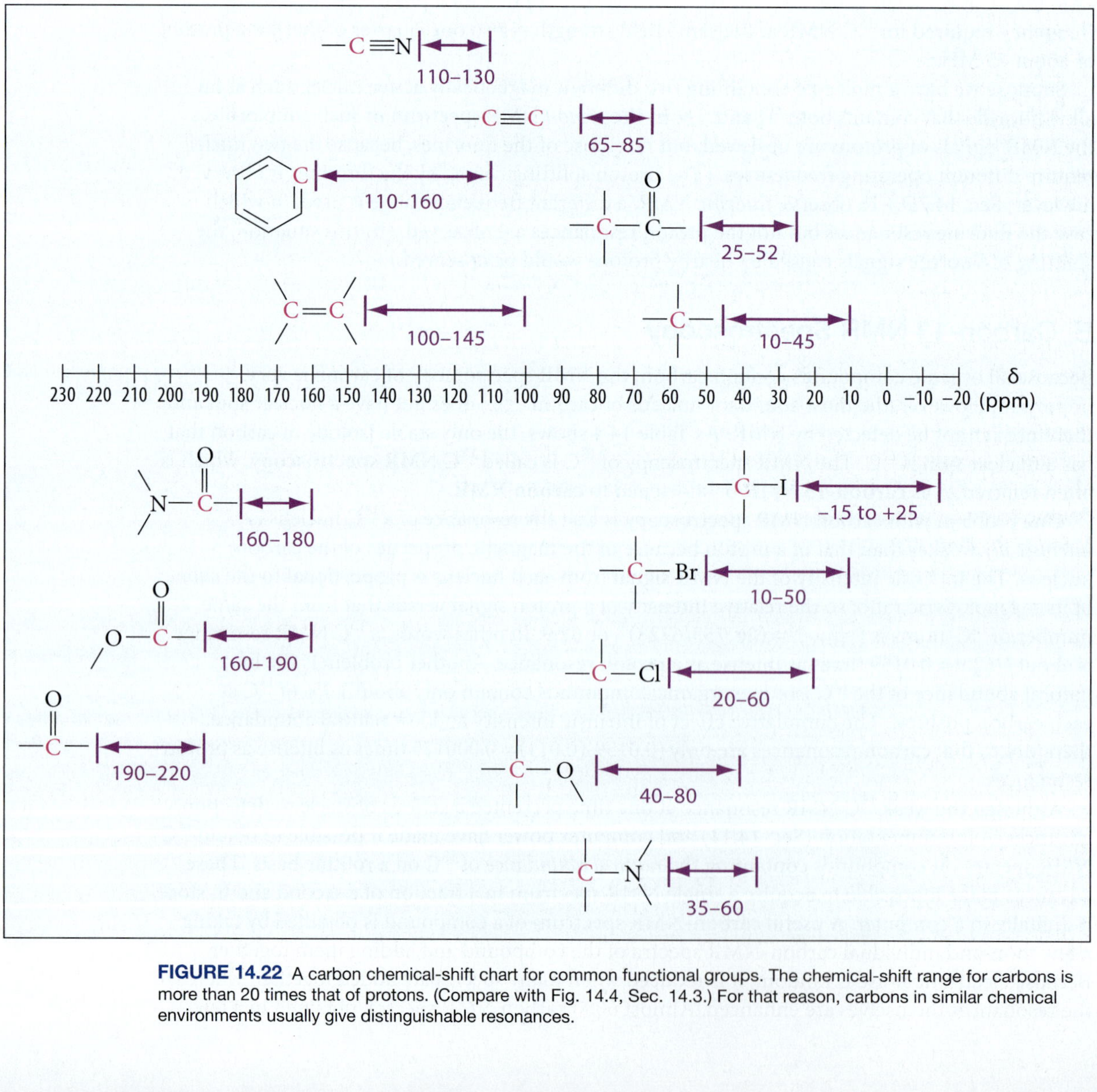

FIGURE 14.22 A carbon chemical-shift chart for common functional groups. The chemical-shift range for carbons is more than 20 times that of protons. (Compare with Fig. 14.4, Sec. 14.3.) For that reason, carbons in similar chemical environments usually give distinguishable resonances.

in chemical environment. As a result, it is often possible to observe distinct resonances for two carbons in similar chemical environments. This point is illustrated by the carbon NMR spectrum of 3-methylpentane, in which each chemically nonequivalent set of carbons gives a separate, clearly discernible resonance:

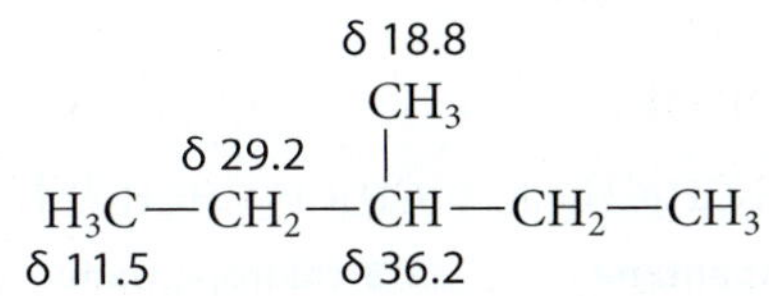

3-methylpentane

As this example illustrates, the chemical shift for a carbon depends in many cases on the number of attached carbons in the order δ (tertiary) > δ (secondary) > δ (methyl).

Carbon NMR is particularly useful for identifying carbons in double bonds—alkenes, carbonyls, and aromatic rings—because the carbons in these groups stand out at the high-shift end of the carbon-NMR spectrum. The large chemical shifts of carbons in these groups have the same explanation as the large chemical shifts of protons attached to these carbons: deshielding by the magnetic field induced by π-electron circulation (see Fig. 14.15, Sec. 14.7A, and Fig. 16.2, Sec. 16.2). Acetylenic carbons have much smaller chemical shifts, although greater than alkanes. The shielding effect of the induced π-electron circulation (see Fig. 14.19, Sec. 14.7B) accounts for the smaller chemical shifts of alkyne carbons relative to those of alkene carbons, whereas the electronegativity of alkyne carbons accounts for the greater chemical shifts of alkynes relative to alkanes.

A third unique aspect of carbon NMR is that the splitting of ^{13}C resonances by protons ($^{13}C—^{1}H$ splitting) is large; typical coupling constants are 120–200 Hz for directly attached protons. Furthermore, carbon NMR resonances are also split by more remote protons. Although such splitting can sometimes be useful, more typically it presents a serious complication in the interpretation of carbon NMR spectra, because the $^{13}C—H$ splitting patterns overlap. In most carbon NMR work, splitting is eliminated by a special instrumental technique called *proton spin decoupling*. Spectra in which proton coupling has been eliminated are called **proton-decoupled ^{13}C NMR spectra**. In such spectra a *single unsplit line* is observed for each chemically nonequivalent set of carbon atoms.

These points are illustrated by the proton-decoupled carbon NMR spectrum of 1-chlorohexane, shown in **Fig. 14.23**. The carbon NMR spectrum consists of six single lines, one for each carbon of the molecule. The assignment of the lines in Fig. 14.23 shows that carbon chemical shifts, like proton chemical shifts, decrease with distance from the electronegative chlorine.

Carbon NMR is particularly useful in differentiating closely related compounds on the basis of their molecular symmetry. The basis of this idea is that symmetrical compounds have fewer chemically nonequivalent sets of carbons than less symmetrical isomers. This point is illustrated in Study Problem 14.5.

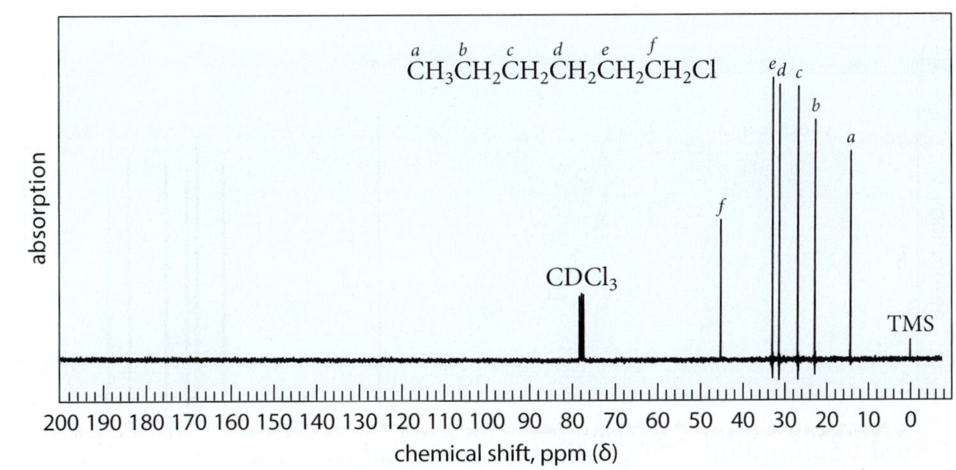

FIGURE 14.23 The proton-decoupled ^{13}C NMR spectrum of 1-chlorohexane. Notice that the resonance of each carbon is visible. (The peaks labeled "$CDCl_3$" are due to the ^{13}C resonance of the solvent. The reason that the $CDCl_3$ carbon signal is a triplet is considered in Problem 14.66b.)

Study Problem 14.5

How would you use carbon NMR spectroscopy to differentiate the two isomers 1-chloropentane and 3-chloropentane?

Solution First, draw the structures of the two compounds.

$CH_3CH_2CH_2CH_2CH_2Cl$ — **1-chloropentane**

$CH_3CH_2CH(Cl)CH_2CH_3$ — **3-chloropentane**

If we assume that a separate resonance is observed for each chemically nonequivalent set of carbons, then the proton-decoupled carbon NMR spectrum of 1-chloropentane should consist of five lines, but that of 3-chloropentane should consist of only three lines; the two CH_3 carbons are chemically equivalent, and the two CH_2 carbons are chemically equivalent. [If you need assistance with chemical equivalence, review Sec. 11.9 and Study Problem 14.1 (Sec. 14.3D)]. As this example shows, if a molecule has symmetry, it will have fewer absorptions than there are carbons.

Focused Problems

14.31 The proton-decoupled carbon NMR spectra of 3-heptanol (*A*) and 4-heptanol (*B*) are given in **Fig. 14.24**. Indicate which compound goes with each spectrum, and explain your reasoning.

14.32 Indicate two things you would look for in their carbon NMR spectra to distinguish between 1,1-dichlorocyclohexane and *cis*-1,2-dichlorocyclohexane.

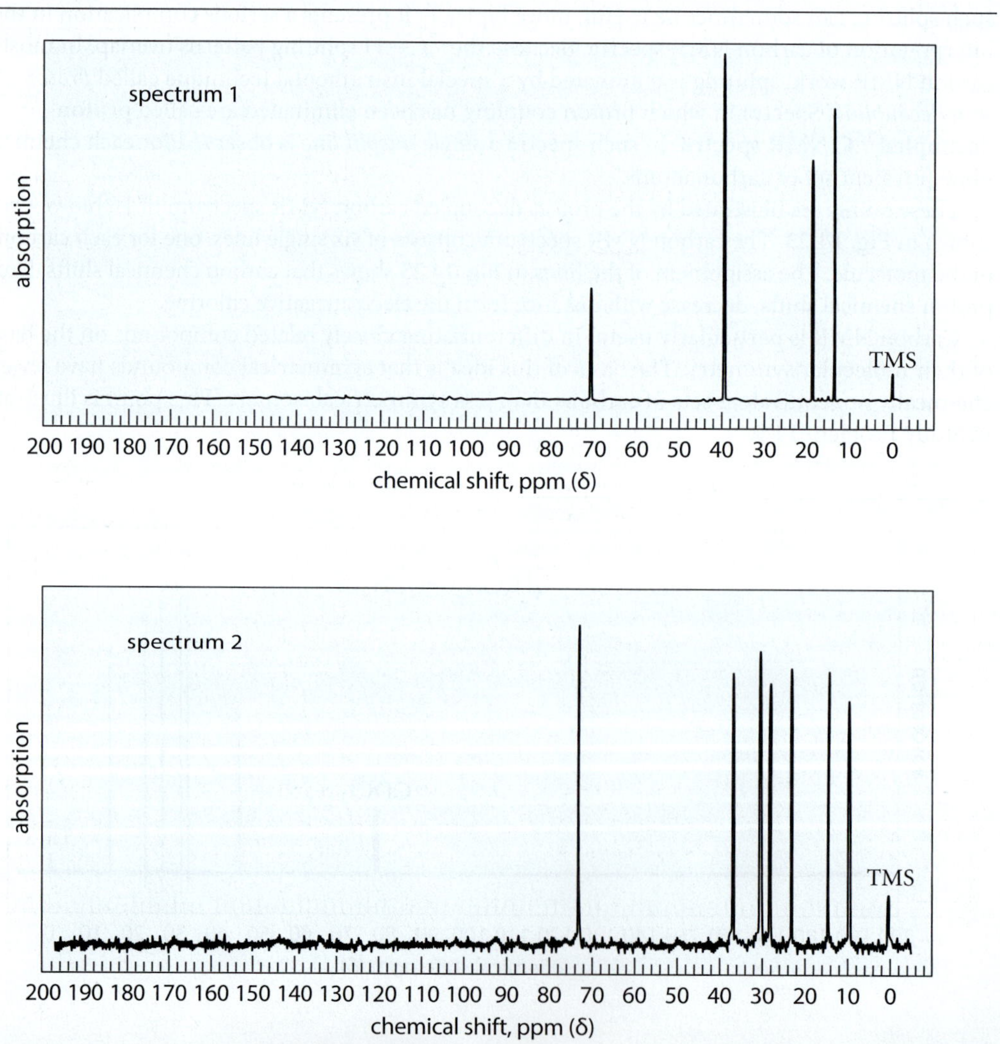

FIGURE 14.24 The proton-decoupled carbon NMR spectra for Focused Problem 14.31.

Carbon NMR spectra are generally not integrated because the instrumental technique used for taking the spectra gives relative peak integrals that are governed by factors in addition to the number of carbons. However, even this fact can be useful. For example, the proton-decoupling technique enhances the peaks of carbons that bear hydrogens; therefore, resonances for carbons that bear *no* hydrogens—such as quaternary carbons, carbons of carbonyl groups, and the α-carbons of tertiary alcohols—are usually significantly smaller than those for other carbons.

Determination of Attached Hydrogens A number of techniques add to the utility of carbon NMR by providing a count of the protons directly attached to each carbon. In other words, it is possible to determine which of the carbon signals in a carbon NMR spectrum come from methyl, methylene, methine, or quaternary carbons. One technique for making such a determination is **distortionless enhancement with polarization transfer (DEPT)**. The DEPT technique yields separate spectra for methyl, methylene, and methine carbons, and each line in these spectra corresponds to a line in the complete carbon NMR spectrum. Lines in the complete carbon NMR spectrum that do not appear in the DEPT spectra arise from carbons that have no attached hydrogens. This technique is illustrated with the DEPT spectra of camphor (**Fig. 14.25**).

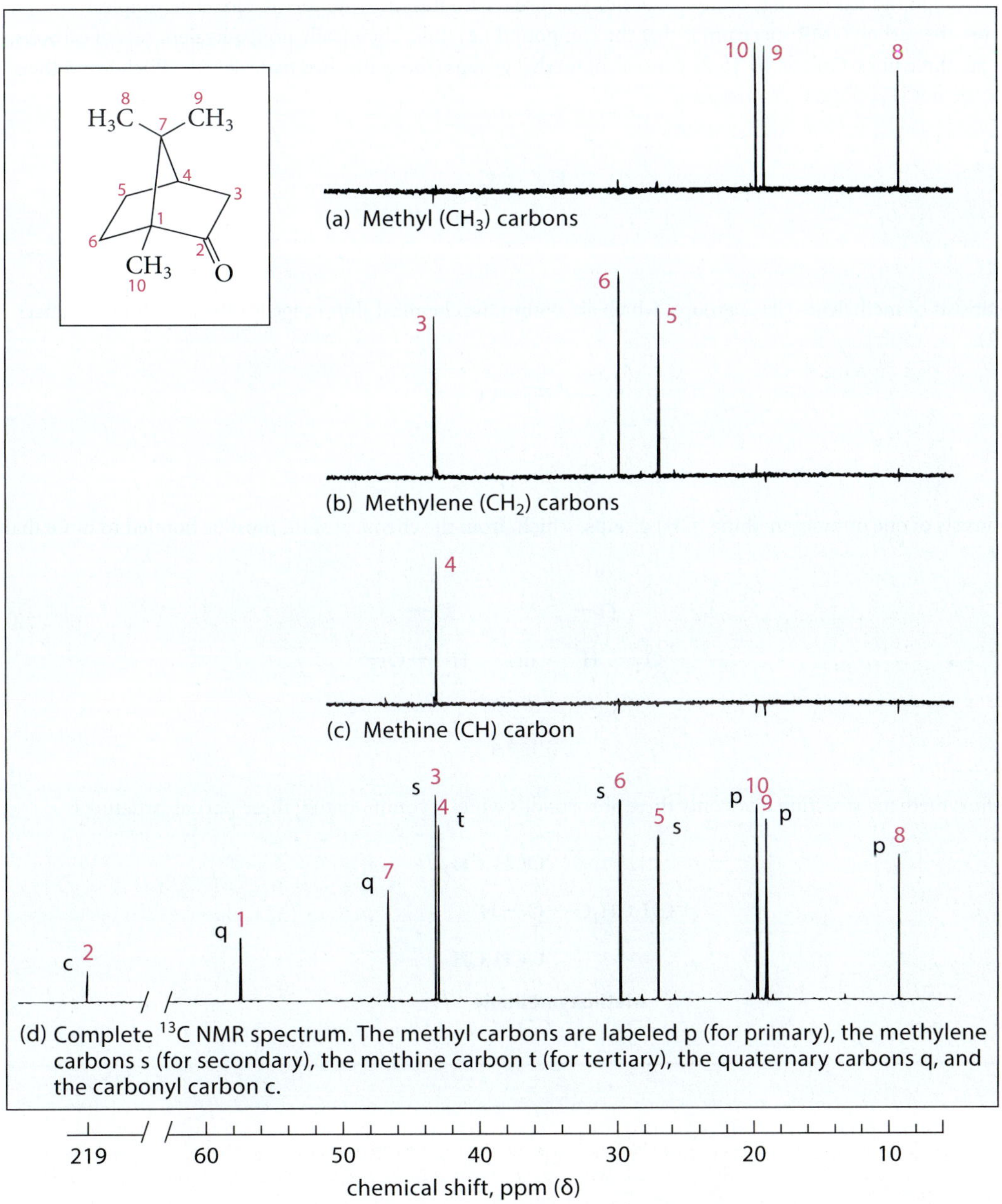

FIGURE 14.25 The ^{13}C NMR spectrum of camphor (structure in the box at the upper left) edited by the DEPT technique. The absorptions for the methyl (CH_3) carbons are given in part (a), the methylene (CH_2) carbons in part (b), and the methine (CH) carbon in part (c). Each peak in these three spectra corresponds to a peak in the full spectrum, shown in part (d). Absorptions in the full spectrum that do not appear in parts (a), (b), or (c) are due to the carbonyl carbon or the quaternary carbons. The number over each peak is the assignment using the carbon number in the camphor structure. Notice in the full spectrum that the intensities of the resonances for the carbonyl (carbon 2) and quaternary (carbons 1 and 7) carbons are lower than the intensities of carbons with attached hydrogens. (Data courtesy of John Kozlowski, Purdue University.)

Notice two other things in the full spectrum of camphor (see Fig. 14.25d). First, the intensities of the quaternary and carbonyl carbons are smaller, a point made previously. Second, the chemical shifts of the carbons remote from the carbonyl group (carbons 4–9) are in the order quaternary > tertiary > secondary > methyl. This trend was also mentioned previously in this section. (The shifts of the carbons that are part of, or near, the carbonyl group are increased by the electronegativity of oxygen and the deshielding effects of π-electron circulation, effects that are also observed in proton NMR spectroscopy; see Fig. 14.15, Sec. 14.7A.)

Study Problem 14.6

A compound with the molecular formula $C_7H_{16}O_3$ has the following ^{13}C NMR–DEPT spectrum (the numbers in parentheses indicate the number of attached hydrogens):

$$\delta\ 15.2\ (3),\ \delta\ 59.5\ (2),\ \delta\ 112.9\ (1)$$

Propose a structure for this compound.

Solution The compound has no rings or double bonds because its unsaturation number is zero. The simplest assumption from the carbon NMR spectrum is that the compound has three chemically nonequivalent sets of carbons, because there are three lines. One set (δ 15.2) consists of methyl groups (three attached hydrogens), which, from their chemical shift, are not very close to the oxygens.

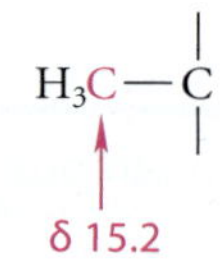

Another set consists of methylene (CH_2) groups, which are within the chemical shift range for the α-carbons of ethers (see Fig. 14.22).

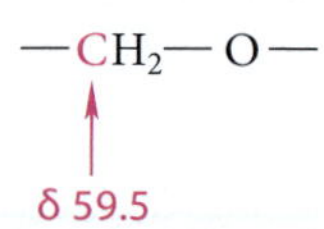

The last set consists of one or more methine (CH) groups, which, from the chemical shift, must be bonded to more than one oxygen.

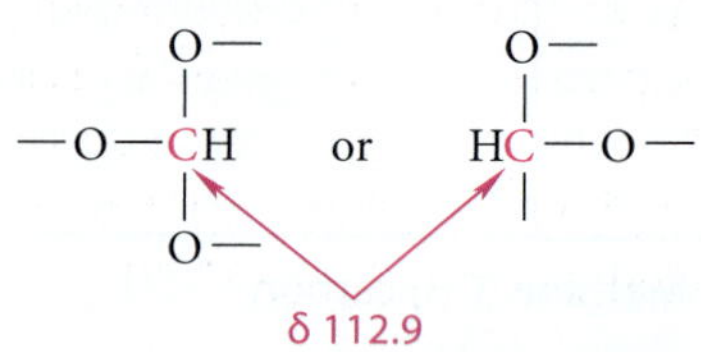

Only the triethoxymethane structure gives only three absorptions while accommodating these partial structures:

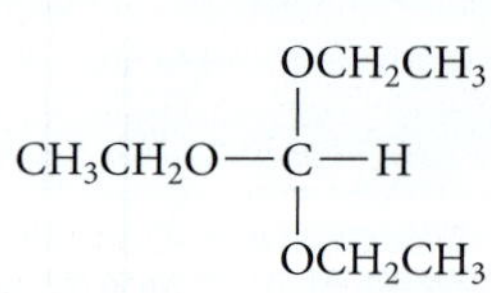

triethoxymethane

Focused Problem

14.33 Explain why each of the following structures is *not* consistent with the ^{13}C NMR data in Study Problem 14.6.

(a)

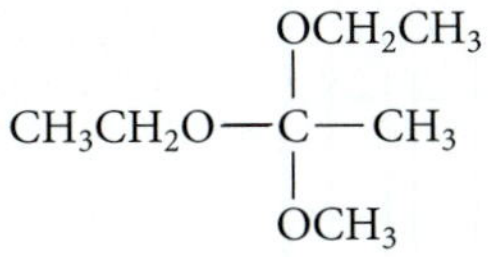

(b)

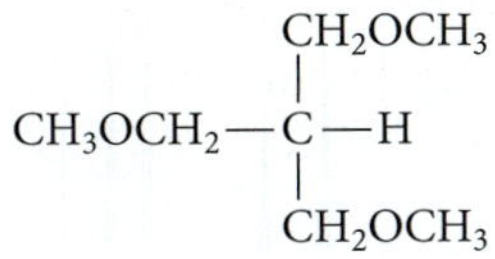

14.10 SOLVING STRUCTURE PROBLEMS WITH SPECTROSCOPY

You are now ready to use what you know about IR, NMR, and mass spectrometry to solve some problems that require more than one of these techniques. Study Problems 14.7 and 14.8 illustrate the techniques involved. Although no single method works in every case, the following suggestions should prove useful:

1. From the mass spectrum determine, if possible, the molecular mass. Identify the masses of key losses in the mass spectrum.

2. If an elemental analysis is given, calculate the molecular formula and determine the unsaturation number.

3. Look for evidence in both the IR and NMR spectra for any functional groups that are consistent with the molecular formula: OH groups, alkenes, and so on. Write down any structural fragments indicated by the spectra.

4. Use the carbon NMR spectrum and, if possible, the proton NMR spectrum to determine the number of nonequivalent sets of carbons or protons (or both). If the proton NMR spectrum is complex, this may not be possible, but you should be able to set some limits.

5. Apply the suggestions in both Sec. 14.3F and Sec. 14.4C to complete your analysis by NMR. *Be sure to write out partial structures and all possible complete structures that are consistent with your spectra.* As you write out partial structures, notice how many carbons are unaccounted for; different partial structures may have carbons in common. Decide between possible structures by asking what features of the different spectra would be expected for each, and look for those features; it is sometimes easy to overlook some feature of a spectrum that will decide between structures.

6. Finally, rationalize all spectra for consistency with the proposed structure.

Study Problem 14.7

Propose a structure for the compound with the IR, NMR, and EI mass spectra shown in **Fig. 14.26**. The compound also has a proton-decoupled carbon NMR spectrum that consists of four resonances at δ 24.6, δ 57.5, δ 115, and δ 140.

Solution The EI mass spectrum of this compound shows a pair of peaks at $m/z = 90$ and 92, with the latter peak about one-third the size of the former. This pattern indicates the presence of chlorine. Furthermore, the base peak at $m/z = 55$ corresponds to a loss of Cl (35 and 37 mass units, respectively). We adopt the hypothesis that this is a chlorine-containing compound with molecular mass of 90 (for the ^{35}Cl isotope). In the IR spectrum, the peak at 1642 cm^{-1} suggests a C═C stretch, and, in the NMR spectrum, there is a complex signal in the vinylic proton region. Moreover, the two resonances in the carbon NMR spectrum at δ 115 and δ 140 are in the region expected for alkene carbons. Evidently this compound is a chlorine-containing alkene.

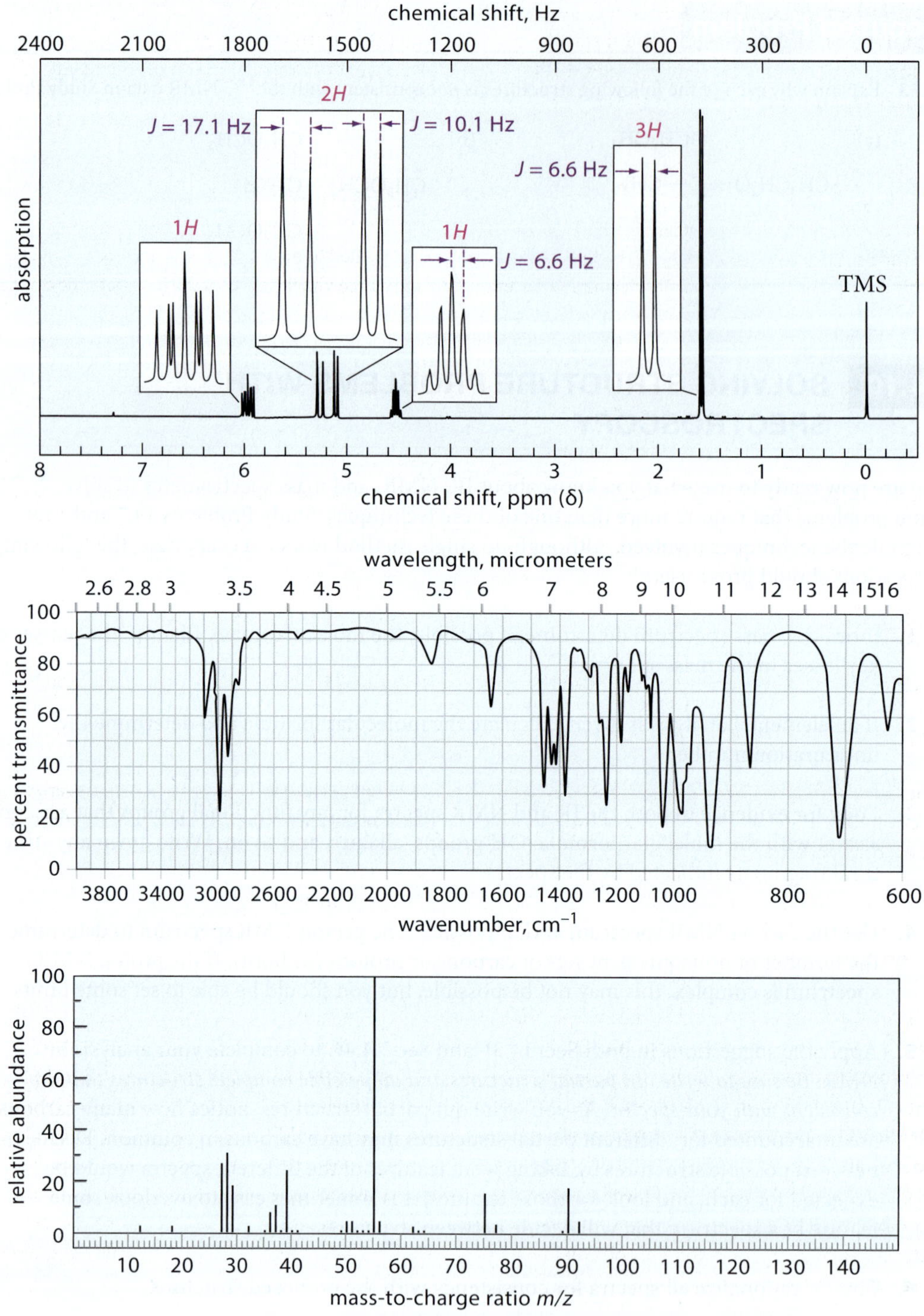

FIGURE 14.26 Proton NMR, IR, and EI mass spectra for Study Problem 14.7. The integral of each resonance in the NMR spectrum is shown in red.

In the proton NMR spectrum, the vinylic proton resonances at δ 5–6 account for three hydrogens; the quintet at δ 4.6 accounts for one hydrogen; and the doublet at δ 1.6 accounts for three hydrogens, for a total of seven hydrogens. If the compound has seven hydrogens (7 mass units) and one chlorine (35 mass units), then the remaining 48 mass units can be accounted for by four carbons, two of which are part of an alkene double bond. The presence of four carbons is also indicated by the carbon NMR spectrum. A possible molecular formula is then C_4H_7Cl. The unsaturation number for this formula is 1, so the molecule contains only one double bond. Because the NMR integral indicates three vinylic protons, the molecule must contain a —CH=CH$_2$ group. In the IR spectrum, the peaks at 930 and 990 cm^{-1} are consistent with such a group, although the former peak is at somewhat higher wavenumber than is usual for this type of alkene. The three-proton doublet at δ 1.6 suggests a methyl group adjacent to a CH group.

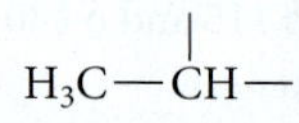

The δ 4.6 absorption accounts for one proton, and its coupling constant ($J = 6.6$ Hz) matches that of the absorption at δ 1.6. The chemical shift of the δ 4.6 absorption suggests that it is bonded to the carbon bearing the chlorine, and splitting of the δ 4.6 absorption indicates that there are four adjacent protons. The splitting and chemical shift of the δ 4.6 absorption fit the partial structure

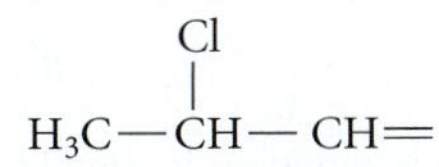

The carbon NMR resonance at δ 57.5 is consistent with a carbon bonded to the chlorine, and the resonance at δ 24.6 is consistent with a methyl carbon. With a molecular mass of 90 and three vinylic protons, the only possible complete structure is

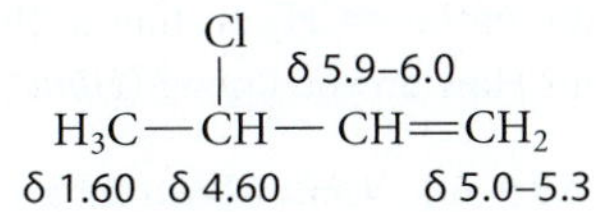

proton NMR assignments

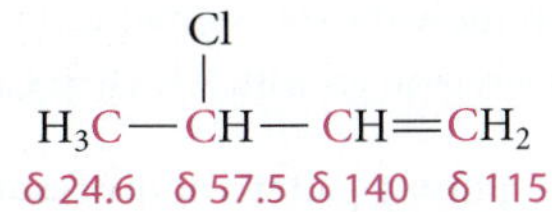

carbon NMR assignments

Study Problem 14.8

A compound $C_8H_{18}O_2$ with a strong, broad infrared absorption at 3293 cm^{-1} has the following proton NMR spectrum:

δ 1.22 (12*H*, s); δ 1.57 (4*H*, s); δ 1.96 (2*H*, s)

(The resonance at δ 1.96 disappears when the sample is shaken with D_2O.) The proton-decoupled carbon NMR spectrum of this compound consists of three lines, with the following chemical shifts and DEPT data (in parentheses) for attached protons:

δ 29.4 (3), δ 37.8 (2), δ 70.5 (0)

Identify the compound.

Solution The IR spectrum indicates the presence of an alcohol, and the disappearance of the δ 1.96 NMR absorption after the D_2O shake (Secs. 14.6 and 14.7D) provides confirmation. Furthermore, because this absorption integrates for two protons, and because the formula contains two oxygens, the compound is a diol. Because the proton NMR spectrum contains no α-hydrogen absorptions in the δ 3–4 region, both alcohols must be tertiary. The proton NMR indicates only three chemically nonequivalent sets of hydrogens, and the carbon NMR spectrum indicates only three chemically nonequivalent sets of carbons, one of which must be the two α-carbons of the tertiary alcohol groups. The DEPT data confirm that one set of carbons indeed has no attached protons, as expected for a tertiary alcohol, and the chemical shift is consistent with that expected for the α-carbon of an alcohol. The presence of only *three* nonequivalent sets of protons and *three* nonequivalent sets of carbons requires a structure of considerable symmetry. The *only* structure that fits these data is

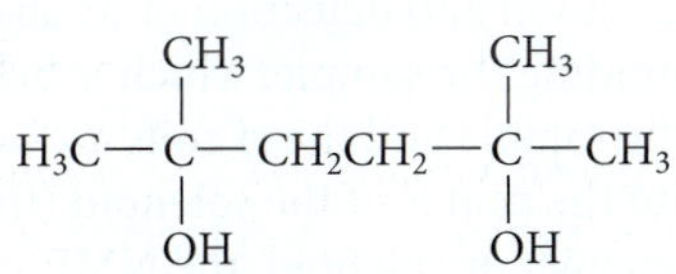

2,5-dimethyl-2,5-hexanediol

Because the two CH_2 groups are chemically equivalent, they don't show any splitting in the proton NMR spectrum.

Focused Problems

14.34 (a) Tell why each of the following structures is *not* consistent with the data in Study Problem 14.7.

trans-1-chloro-2-butene — *A*

2-chloro-1-butene — *B*

(b) Although we did not have to analyze the vinylic proton resonances in detail to determine the structure at the end of Study Problem 14.7, it is interesting to consider these resonances further. First, justify the assignments given at the end of Study Problem 14.7 for the resonances of the vinylic protons. Then notice in the NMR spectrum (see Fig. 14.26) that the $=CH_2$ proton resonances do not split each other detectably [see Table 14.3 (Sec. 14.7A), geminal protons]. Next, draw out the structure of this compound to show the stereochemical relationships of the $=CH_2$ protons to the other vinylic proton. Which resonances in the δ 5.0–5.3 region go with which $=CH_2$ proton? How do you know? (*Hint:* See Fig. 14.15, Sec. 14.7A.)

(c) Justify the number of lines in the splitting pattern of the resonance at δ 5.9–6.1. (*Hint:* Construct a splitting diagram using the three coupling constants in the spectrum.)

14.35 Tell why each of the following structures is *not* consistent with the spectroscopic data in Study Problem 14.8.

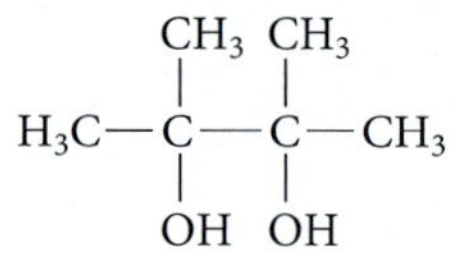

A

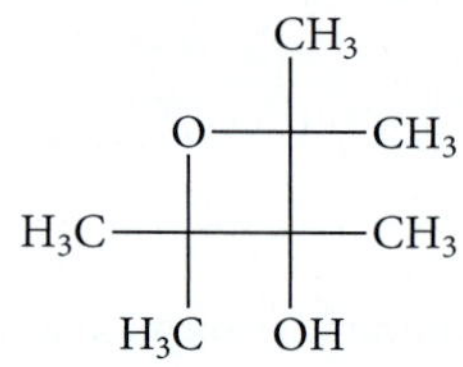

B

14.11 THE NMR SPECTROMETER

The basic components of an NMR spectrometer are shown in **Fig. 14.27**. An NMR instrument requires, first, a strong magnetic field to establish the tiny energy differences between nuclear spin states. Early NMR instruments employed electromagnets or permanent magnets that generated fields in the range of 7000–23,000 gauss. Modern instruments utilize large solenoids—essentially doughnut-shaped wire coils—fabricated of superconducting wire. Current flowing in the coil generates the magnetic field. In a superconducting wire, electric current, once established, persists indefinitely and flows without electrical resistance. Superconducting solenoids are required because the electric current required for large magnetic fields would generate far too much resistance (and therefore heat) in a conventional, nonsuperconducting solenoid. Most metals that exhibit superconductivity do so only at very low temperatures. Because liquid helium is used to maintain these low temperatures, the solenoid is housed in an elaborate cryostat (essentially, a multiwall thermos bottle). It is possible to construct superconducting solenoids that can develop magnetic fields greater than 200,000 gauss.

FURTHER EXPLORATION 14.2
Fourier-Transform NMR

The second instrumental component of the NMR experiment is the radiofrequency (rf) radiation. Application of rf radiation and detection of its absorption are managed through the use of a small wire coil surrounding the sample, which is held in a glass tube that is rapidly spun about its longitudinal axis. The sample and the rf coils are housed in a precisely constructed probe that can be inserted into the center of the solenoid (that is, into the "hole" in the wire "doughnut"). Except for the magnet or solenoid, the NMR instrument is in essence a radio transmitter and receiver.

The FT-NMR method was conceived by Richard R. Ernst (b. 1933) of the ETH (Federal Technical Institute) in Zürich, Switzerland; for this contribution, he was honored with the 1991 Nobel Prize in Chemistry.

NMR experiments in early instruments involved varying the frequency of the rf radiation slowly and detection of each resonance separately. With this technique, a spectrum was obtained in a few minutes. Although this technique was used in spectrometers through the early 1970s, modern NMR spectrometers employ a technique for taking spectra called **pulse-Fourier-transform NMR (FT-NMR)**. In FT-NMR, all of the proton spins are excited instantaneously with an rf pulse containing a broad band of frequencies, and the spectrum is obtained by analyzing the emission of rf energy as the spins return to equilibrium. (See Further Exploration 14.2.) With FT-NMR, an entire proton NMR spectrum can be obtained in less than a second. Consequently,

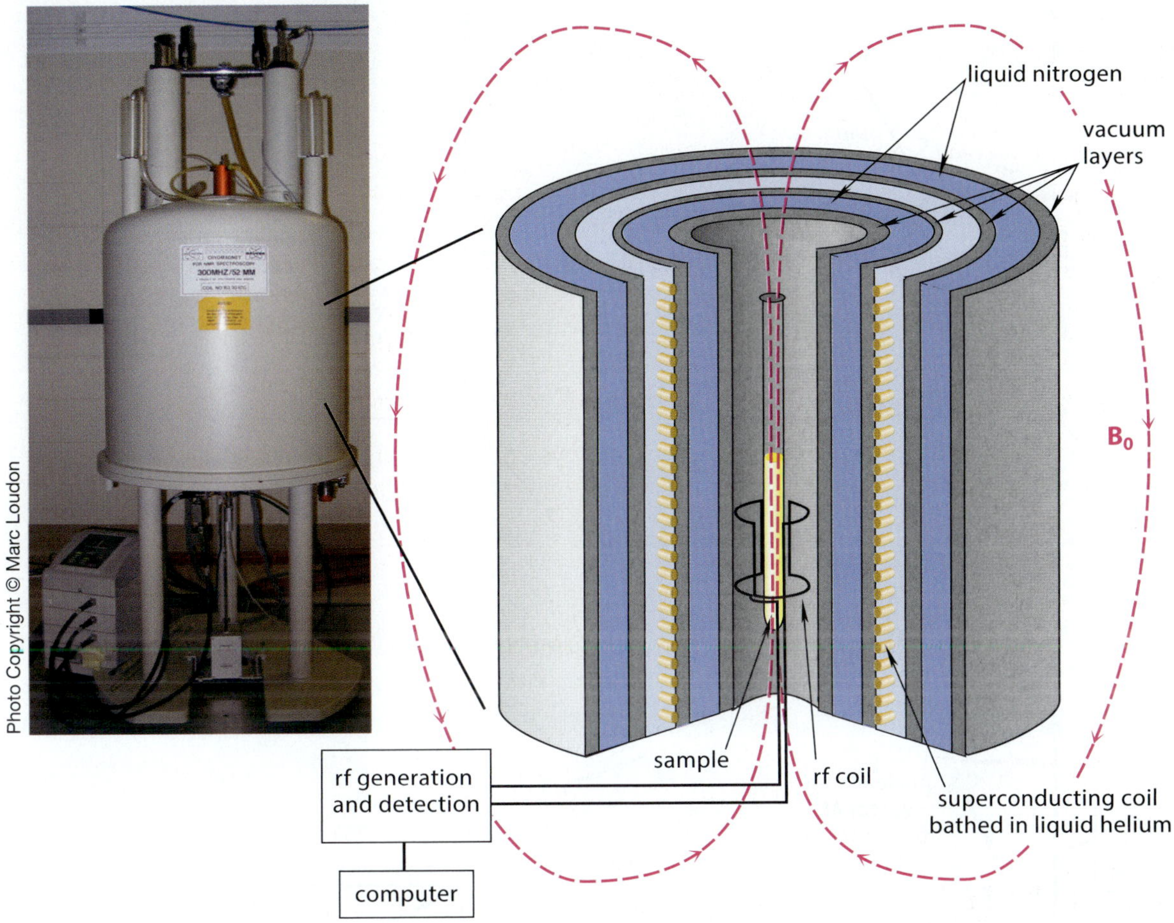

FIGURE 14.27 A 300-MHz NMR spectrometer with a cross-section of the solenoid vessel. The height of the solenoid is about 1 meter. (The size of the sample and rf coil is exaggerated.) The heart of the solenoid is a coil of superconducting wire. Because superconductivity requires extremely low temperatures, the coil is immersed in liquid helium, which has a temperature of about 4 K. This is insulated from the surroundings by an evacuated container (essentially a large thermos bottle), which itself is contained within another "thermos bottle" of liquid nitrogen. When the solenoid is energized, a large current flows around the coil with zero resistance. This current generates the magnetic field of 70,500 gauss (red dashed lines). The sample, contained in a small glass tube, is placed inside the sample probe within the gap in the center of the solenoid. Within the probe the sample is surrounded by the radiofrequency (rf) coil, which is used to transmit rf radiation and to measure rf emission. The sample tube is spun rapidly about its long axis to average out small differences in the field throughout the sample. Many of the spectra in this text were obtained with this type of instrument.

a large number of spectra of a given sample (anywhere from 50 to 20,000, depending on the sample concentration and the isotope) can be recorded in a relatively short time. A computer stores, analyzes, and mathematically sums the spectra. Because electronic noise is random, it sums to zero when averaged over many spectra, whereas the resonances of the sample reinforce to give a much stronger spectrum than could be obtained in a single experiment. The FT-NMR technique has made possible the routine use of ^{13}C NMR for structure determination. FT-NMR instruments became truly practical with the advent of relatively inexpensive dedicated small computers that are required for application of the FT-NMR technique.

14.12 MAGNETIC RESONANCE IMAGING

Chemical Biology Topic

One of the most important medical uses of NMR is *magnetic resonance imaging* (*MRI*). MRI is a noninvasive method of biomedical imaging that is particularly effective for visualizing soft tissue. Like the NMR techniques for determining structure that we've considered in this chapter, MRI involves the use of high magnetic fields and radiofrequency radiation, which are much less hazardous than X-rays. However, unlike structural NMR, MRI does not involve chemical shifts or splitting. Rather, MRI monitors the protons of a single compound—water—in various

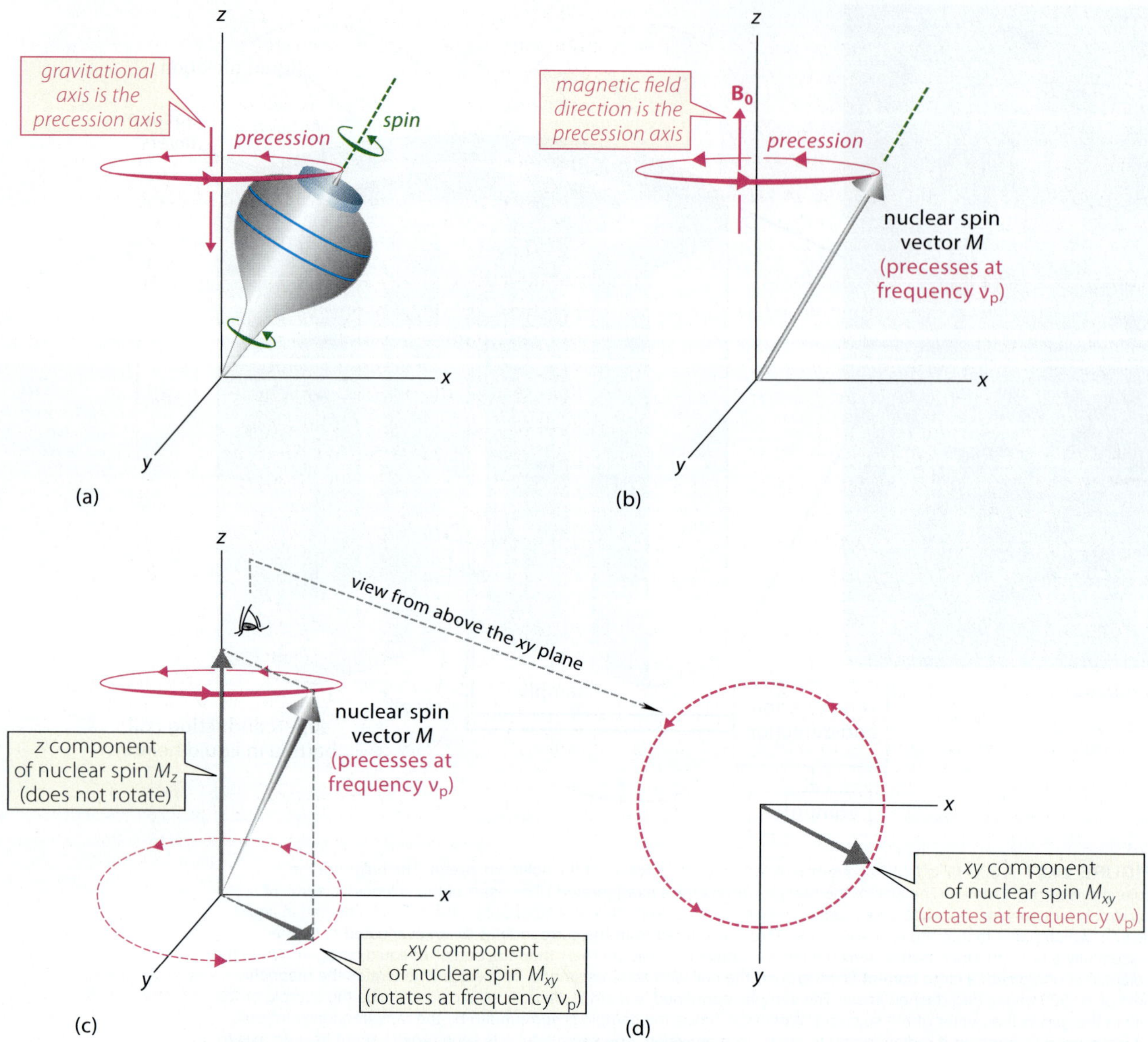

FIGURE 14.28 Spin precession. (a) A top spins about the green axis. Precession is the rotation of the spin axis itself about the *z*-axis, which is the direction of gravity. (b) The precession of nuclear spins occurs at the resonance frequency ν_p about the axis defined by the direction of the external magnetic field. (c) Resolution of the nuclear spin vector *M* into perpendicular components: a stationary component M_z along the *z*-axis, and a rotating component M_{xy} in the *xy*-plane. (d) A view of the rotating component M_{xy} from above the *xy*-plane.

physiological environments by utilizing a physical phenomenon called *nuclear relaxation.* (Because water is so abundant, we don't have to be concerned about interference from other compounds.) To understand nuclear relaxation conceptually, we have to expand our view of nuclear spins somewhat (**Fig. 14.28**). First, the term *spin* refers to the fact that nuclei behave as if they are spinning charges. Spinning charges produce magnetic fields, which have magnitude and direction. Therefore, we represent the spinning charge as a vector. This means that we can think of the spinning charge as a tiny bar magnet, with the vector running from south to north. In other words, the direction of the spin vector is the direction of the bar magnet. In our previous discussions, we have characterized nuclear spins merely as "up" ($+\frac{1}{2}$) or "down" ($-\frac{1}{2}$) relative to the magnetic field direction. However, the spin vector is not static; rather, it *precesses* about the magnetic field direction. This precessional motion is like that of a spinning gyroscope or top (Fig. 14.28a). Not only is the top spinning on its own axis, but the axis of the top is also rotating about the gravitational axis. Similarly (Fig. 14.28b), the nuclear spin vector *M* precesses about the direction of the magnetic field, which we define as the *z*-axis. The frequency of the precession—the number of times per second it precesses—is the frequency ν_p defined by Eq. 14.6 (Sec. 14.2). The spin vector M can be resolved into two components: one along the z-axis, which we'll call M_z, and one in the xy-plane, which we'll call M_{xy} (Fig. 14.28c). M_z is stationary, whereas M_{xy} is rotating, also at the frequency ν_p. Here is the connection with what we've learned previously: "up" and "down" refer to the direction of M_z. Figure 14.28c shows an "up" $(+\frac{1}{2})$ spin; a down $(-\frac{1}{2})$ spin would be described by the same figure simply "flipped" 180° relative to the field direction.

Now imagine many water molecules in a magnetic field. As we learned in Display 14.3 (Sec. 14.2), some of the protons of these water molecules have spin $= +\frac{1}{2}$, and others have spin $= -\frac{1}{2}$, and the two spin states are in equilibrium, with the lower-energy (spin $= +\frac{1}{2}$) protons in slight excess. Their M_z components of magnetization sum to a total value $\boldsymbol{M}_{z,\mathbf{eq}}$, which is in the "up" direction, because there is an excess of spins in the "up" (lower-energy) state. If this system of water molecules is pulsed briefly with a strong burst of radiofrequency (rf) radiation at the proton resonance frequency, lower-energy "up" proton spins are "flipped" and become higher-energy "down" spins. After the pulse, there is a net decrease in the value of $\boldsymbol{M}_z$, because there has been an increase in the number of "down" spins. The rf radiation is turned off and the proton spins are allowed to return to equilibrium. **Nuclear relaxation** is the term used to describe the return to equilibrium. In returning to equilibrium, the spins in the higher-energy $(-\frac{1}{2})$ state lose energy to their surroundings by various mechanisms that we don't need to consider here. At this point, however, we need to consider the relaxation of the spin components $\boldsymbol{M}_z$ and $\boldsymbol{M}_{xy}$ separately.

The relaxation of $\boldsymbol{M}_z$ occurs with a characteristic rate that has a rate constant $= 1/T_1$, where the rate constant, like a first-order reaction rate constant, has the dimension of seconds^{-1}. The reciprocal of the rate constant, T_1, then, has the dimension of seconds. The quantity T_1 is called the **longitudinal relaxation time**. (It is sometimes also called the *spin-lattice relaxation time*.) The T_1 is a measure of how long it takes the proton spins of water to return (by "flipping") back to equilibrium. The pulse–relaxation process is shown in Display 14.35:

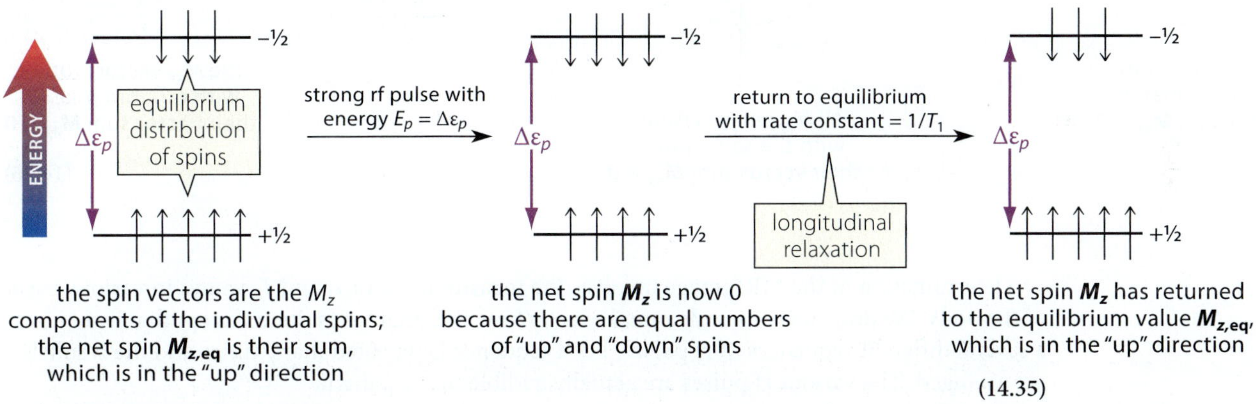

(14.35)

(The excess of spins at equilibrium is indicated in this example as 2/8, or 25%, but this is *greatly* exaggerated for illustration; the excess is only a few parts per million.)

Now consider the component of spin in the xy-plane. Before the rf pulse occurs, the xy components of the spin vectors are randomly oriented in the xy-plane. That is, the precession angles of the spins are randomly distributed. Therefore, the sum of all the xy components of many water spin vectors $\boldsymbol{M}_{xy}$ is zero. When the rf pulse occurs, the precession of the nuclear spins is brought into phase coherence. (We won't discuss the physics of this process.) That is, after the rf pulse, the M_{xy} components for the many spins all coincide at the same rotational angle relative to the x and y axes, and their sum $\boldsymbol{M}_{xy}$ is no longer zero. Over time these rotations begin to fall out of phase—that is, the rotation rates change slightly so that the angles of rotation of the different spins become randomized as they were before the rf pulse, and $\boldsymbol{M}_{xy}$ ultimately decays back to zero. This process occurs with a rate constant $= 1/T_2$. The reciprocal of this rate constant, T_2, is called the **transverse relaxation time**. Transverse relaxation is typically faster than longitudinal relaxation.

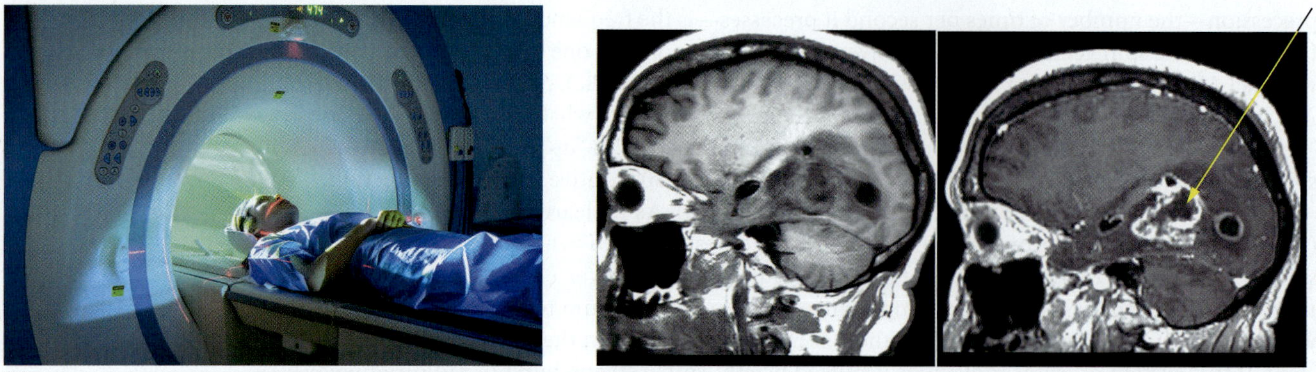

Monty Rakusen/Getty Images

(a) (b)

Living Art Enterprises, LLC/Science Source

FIGURE 14.29 (a) A modern MRI instrument. The patient, lying on the bed, is moved into the cavity of the magnet for determination of the image. (b) T_1-weighted MRI images both without *(left)* and with *(right)* Gd^{3+} contrast. The Gd^{3+} scan has revealed a glioblastoma, a type of brain cancer (yellow arrow).

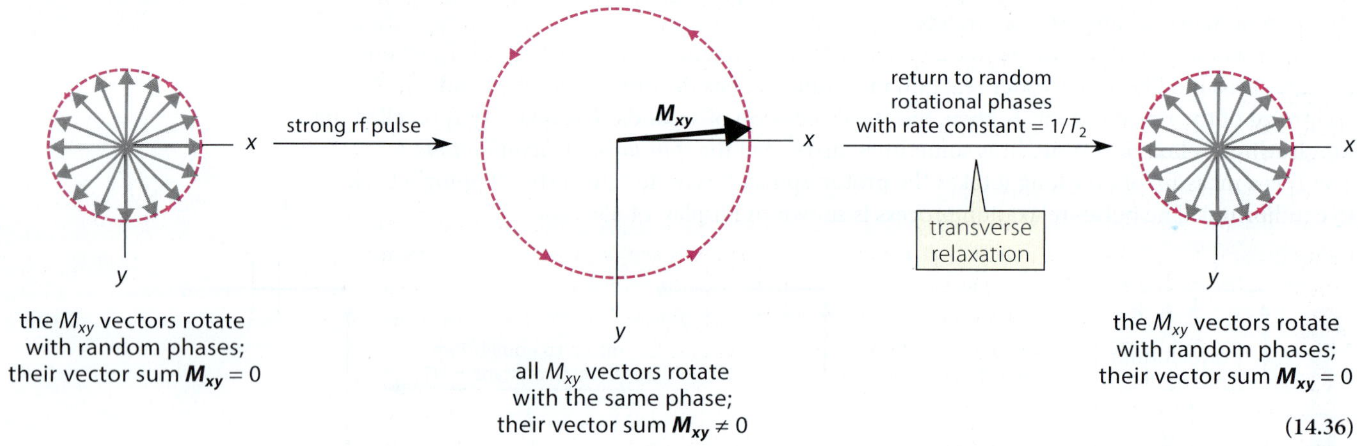

(14.36)

Determination of the MRI image involves the measurement of T_1 and T_2 values in various parts of the body. We don't need to understand exactly how this analysis is done, except that it requires a several different sequences of rf pulses. Each sequence is repeated many times and the results are averaged. The various rf pulses are actually audible to the patient.

For their research on MRI, Paul Lauterbur (1929–2007) of the University of Illinois (research done at State University of New York, Stony Brook) and Sir Peter Mansfield (1933–2017) of the University of Nottingham shared the 2003 Nobel Prize in Physiology or Medicine.

The MRI image (**Fig. 14.29**) is a map of relaxation times in various parts of the body. The central principle of MRI is that *the values of T_1 and T_2 depend on the physiological environment of water.* For example, T_1 for fluids is 1.5–2 s; for water-based tissues, it is 0.4–1.2 s; and for fats, it is 0.1–0.15 s. So, what we're viewing in an MRI image is *not* a "photograph" of, say, a brain, but rather a map of *proton relaxation rates of water in various parts the brain.* Because these rates vary with location, the MRI image is, in effect, an image of the physiological environment. When a patient undergoes MRI imaging, typically a number of images are obtained in which T_1 values are weighted in some images and T_2 values in others.

Nuclear relaxation involves loss of energy from the nuclear spins to the surroundings. This process is rather inefficient, and the T_1 process is particularly so. However, longitudinal relaxation can be accelerated by the addition of paramagnetic substances—substances with unpaired electrons. These substances serve as *contrast agents* in MRI. One of the substances most widely used for this purpose is the rare-earth ion gadolinium(III), provided in the form of various coordination compounds such as the following.

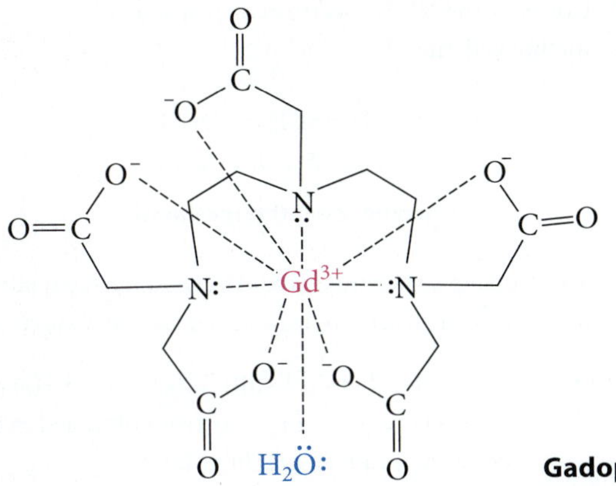

Gadopentetic acid (Optimark, Magnevist)

This ion has seven unpaired electrons (a record in the periodic table!). As a water molecule (*blue*) from solvent enters the coordination sphere of the Gd^{3+} ion, its nuclear spins can gain or lose energy by interacting with the spins of the Gd^{3+} electrons. Because water molecules interchange rapidly between solvent and the Gd^{3+}, this process in effect enhances water T_1 relaxation—that is, T_1 values are reduced. The magnitude of the effect depends on the Gd^{3+} concentration, but effects of three- to eightfold are common in practice with low concentrations of Gd^{3+}. Because the imaging agent is water-soluble, this enhancement is greatest in highly aqueous physiological regions that are penetrated well by the imaging agent. For example, the image in Fig. 14.29b shows the exposure of a glioblastoma (a type of brain tumor), which has a higher density of blood vessels than surrounding tissue.

Other Applications of NMR Other applications of NMR are worth noting. *Solid-state NMR* is being used to study the properties of important solid substances as diverse as drugs, coal, and industrial polymers. *Phosphorus NMR* (^{31}P NMR) is being used to study biological processes, in some cases using intact cells or even whole organisms. *Functional MRI* (fMRI) is a variation of MRI in which mental operations can be mapped by observing differential blood flow to various areas of the brain. This is possible because deoxygenated blood has magnetic properties that impart relatively high contrast to MRI images. The reduction in contrast that is produced with a relatively higher level of oxygenated blood can be translated into maps of brain activity associated with such processes as learning or thinking about a loved one. Functional MRI also promises to open new windows on mental illness.

Once a curiosity in a physics laboratory, nuclear magnetic resonance has revolutionized not only chemistry but also human medicine, and the end is not in sight on either frontier.

CHAPTER SUMMARY *For a summary of the chapter, see Chapter 14 in the* Study Guide and Solutions Manual.

SKILLS OBJECTIVES WITH PROBLEMS

Note: In these problems, assume that the term *NMR* refers to *proton* NMR unless otherwise indicated. Assume that all ^{13}C NMR spectra are proton-decoupled unless otherwise stated.

- **Know the four types of information obtained from a proton NMR spectrum. 14.3A,D,E; 14.4**

14.36 What four primary types of information are available from a proton NMR spectrum? How is each used?

- **Describe the physical principles on which NMR is based, specifically:**

(1) Why an external magnetic field is necessary for NMR **14.2**

(2) The fundamental equation of NMR **14.2**

(3) The condition for energy absorption in NMR **14.2**

(4) What happens physically when samples in a magnetic field absorb rf energy **14.2**

(5) The cause of chemical shift **14.3A**

(6) Why alkene and acetylenic hydrogens have unusual chemical shifts **14.7A,B**

(7) The reason for splitting **14.4A.B**

(8) The reason for the $n + 1$ splitting rule **14.4B**

(9) Why adding a trace of acid or D_2O to an alcohol sample might simplify its NMR spectrum **14.7E, 14.8**

(10) How splitting is affected by the dihedral angle between two coupled protons **14.5A**

(11) Why using a higher field simplifies complex spectra **14.5B**

(12) Why carbon NMR spectra require a different rf frequency range than proton NMR spectra **14.9A,B**

(13) How magnetic resonance imaging differs from proton NMR used for structure determination **14.12**

14.37 Answer each of the following questions.

(a) Why is subjecting a sample to an external magnetic field necessary to determine its NMR spectrum?

(b) What happens physically when protons of a molecule situated in a magnetic field absorb rf energy?

(c) Why is the local magnetic field at a proton within a molecule less than the external magnetic field?

(d) Why do vinylic protons (protons on the carbon of a double bond) have a chemical shift greater than that predicted by the electronegativity of the vinylic carbons?

(e) Why do acetylenic protons have a much smaller chemical shift than vinylic protons, even though the electronegativity of acetylenic carbons is greater?

(f) In a conformationally rigid molecule, two vicinal protons H^a and H^b have a dihedral angle of 20°. The dihedral angle between H^b and another vicinal proton H^c is 80°. Which coupling constant is greater: J_{ab} or J_{bc}? How do you know?

(g) Why does the chemical shift in ppm *not* change with the operating frequency of the NMR spectrometer?

(h) What is the relationship of the coupling constant J between two protons and the size of the magnetic field imposed by the NMR instrument?

(i) What condition must be met for the NMR splitting pattern between two protons to be first-order?

(j) What operating frequency is required for carbon NMR using a spectrometer that operates at 500 MHz for proton NMR?

(k) What do magnetic resonance imaging and ordinary proton NMR have in common? How are they different?

14.38 Consider the NMR spectrum of 1-bromo-2-methoxyethane.

$$\mathrm{Br{-}\overset{a}{C}H_2{-}\overset{b}{C}H_2{-}O\overset{c}{C}H_3}$$

δ 3.45 δ 3.60 δ 3.38

1-bromo-2-methoxyethane

The coupling constant J_{ab} is 6.0 Hz. (Assume rapid internal rotation about all carbon–carbon and carbon–oxygen bonds.)

(a) What is the chemical-shift difference (in Hz) of protons H^a and H^b in a spectrum obtained at an operating frequency of 300 MHz?

(b) What statement is true about the coupling constant J_{ba}? (1) It must be less than 6.0 Hz; (2) it depends on the dihedral angle between the coupled protons; (3) it is exactly equal to 6.0 Hz; (4) it must be greater than 6.0 Hz. Explain.

(c) By an analysis of spins similar to the one in Display 14.14 (Sec. 14.4B), derive the splitting pattern of protons H^b, including the relative intensities of the pattern lines.

(d) What is the splitting pattern of H^a?

(e) If we determine the NMR spectrum of this compound on a 300-MHz spectrometer, and then again on a 60-MHz spectrometer, which spectrometer is more likely to give first-order splitting patterns for H^a and H^b? Explain.

14.39 (a) How would the proton NMR spectrum of very dry 2-chloroethanol differ from the spectrum of the same compound to which a trace of acid has been added?

(b) How would the proton NMR spectrum of very dry 2-chloroethanol change when the sample is shaken with a drop of D_2O?

14.40 A reaction has given a mixture of the following two stereoisomers. Explain how you would use proton NMR to determine which is which. (*Hint:* Recall that trans-fused decalins do not undergo the chair interconversion; Sec. 7.6B.)

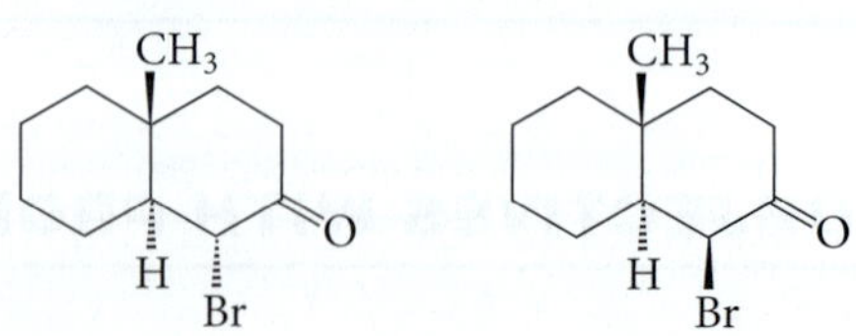

A *B*

- **Apply chemical shift and splitting information that adheres to the $n + 1$ rule from proton NMR spectra to structure determination. 14.3C, 14.4C, 14.7**

14.41 Give a partial structure that would lead to each of the following patterns in a proton NMR spectrum. The resonances refer in each case to hydrogens attached to carbon. (Assume that any splitting patterns have the same coupling constants in each case.) Make sure that every carbon atom has four bonds. (*Hint:* Use X and Y to symbolize attached groups that have no hydrogens.)

(a) Two resonances: quartet and triplet

(b) One resonance: (9*H*, s) and two resonances in carbon NMR

(c) Two resonances: septet and doublet

(d) Two resonances: triplet and quintet

(e) Two resonances: doublet and quartet

(f) Three resonances: two triplets and a quintet

(g) Three resonances: two doublets and a sextet

(h) Three resonances: two triplets and a sextet

(i) One resonance: (3*H*, s)

14.42 Give the structure that corresponds to each of the following molecular formulas and NMR spectra:

(a) C_5H_{12}; δ 0.93, s

(b) C_5H_{10}; δ 1.5, s

(c) $C_4H_{10}O_2$: δ 1.36 (3*H*, d, *J* = 5.5 Hz); δ 3.32 (6*H*, s); δ 4.63 (1*H*, q, *J* = 5.5 Hz)

(d) $C_7H_{16}O$; NMR spectrum in **Fig. P14.42d.**

(e) C_8H_{16}: NMR spectrum in **Fig. P14.42e.** This compound undergoes catalytic hydrogenation to give 2,2,4-trimethylpentane.

(f) $C_7H_{12}Cl_2$: δ 1.07 (9*H*, s); δ 2.28 (2*H*, d, *J* = 6 Hz); δ 5.77 (1*H*, t, *J* = 6 Hz)

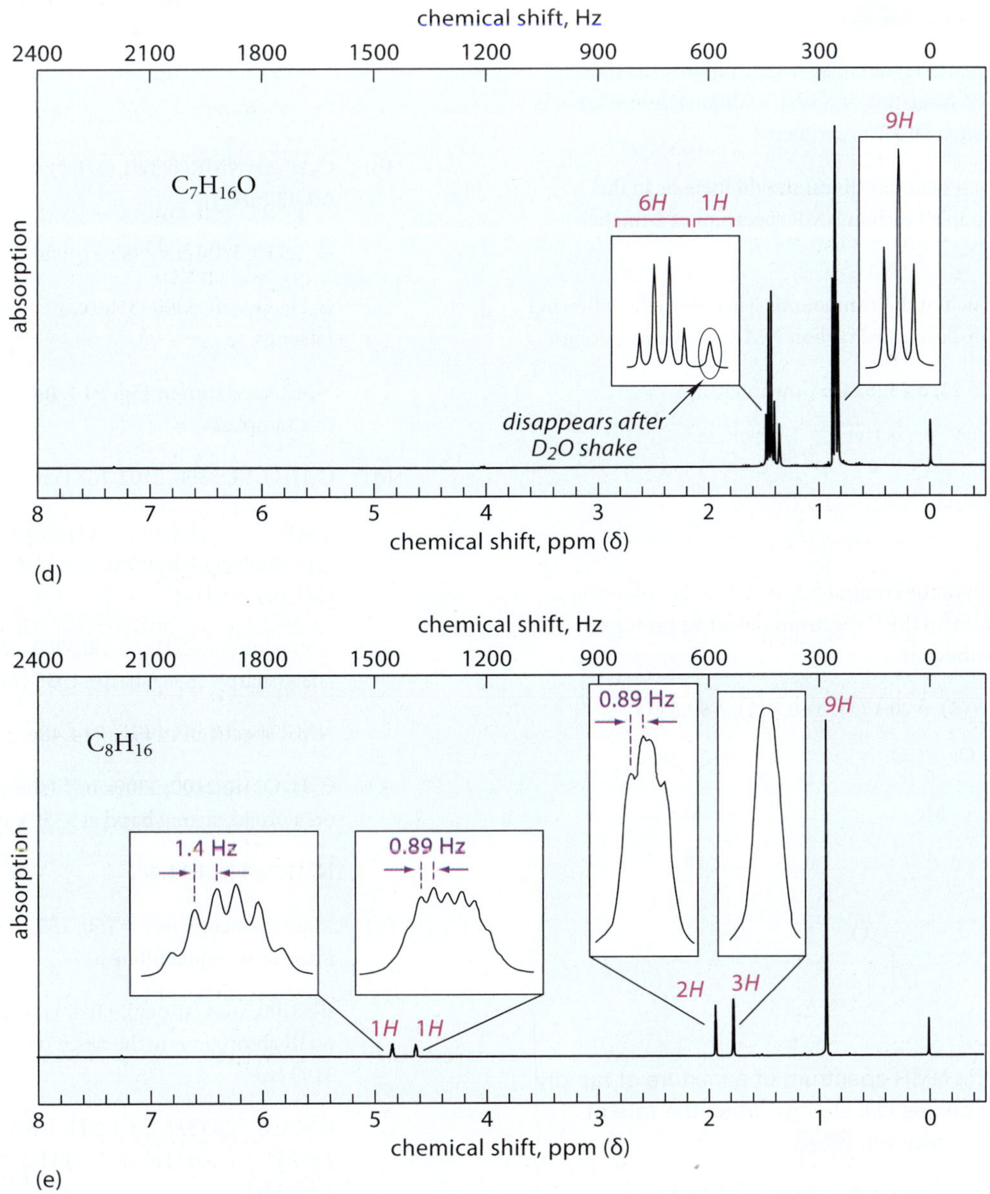

FIGURE P14.42 NMR spectra for Problems 14.42d and 14.42e. The integrals in actual numbers of hydrogens are given in red, and the coupling constants are in violet.

(g) $C_2H_2Br_2F_2$: δ 4.02 (t, $J = 16$ Hz) (*Hint:* The splitting is the proton resonance split by fluorine; the fluorine resonance is not observed in this frequency range.)

(h) $C_2H_2F_3I$: δ 3.56 (q, $J = 10$ Hz) [Same hint as part (g).]

(i) $C_7H_{16}O_4$: δ 1.93 (t, $J = 6$ Hz); δ 3.35 (s); δ 4.49 (t, $J = 6$ Hz); relative integral 1:6:1

(j) C_4H_6O: liberates a gas when treated with EtMgBr and has no O—H stretching absorption in its IR spectrum

NMR: δ 2.43 (1*H*, t, $J = 2$ Hz); δ 3.41 (3*H*, s); δ 4.10 (2*H*, d, $J = 2$ Hz)

• List the types of information that are obtained from a proton-decoupled carbon NMR spectrum, and explain what additional information is obtained from a DEPT experiment. Use a carbon NMR spectrum with DEPT to determine a structure. 14.9B

14.43 What information is obtained from a proton-decoupled carbon NMR spectrum, and what additional information is obtained from a DEPT experiment?

14.44 How many absorptions (lines) should there be in the proton-decoupled carbon NMR spectrum of 4-methyl-1-penten-3-ol?

14.45 (a) To which of the compounds listed does the following proton-decoupled carbon NMR spectrum belong?

δ 13, δ 21, δ 23, δ 80.5 (low intensity)

3-octyne	octane	1-butyne
A	*B*	*C*
trans-4-octene	4-octyne	
D	*E*	

(b) Which of the compounds *A–C* has the following ^{13}C NMR-DEPT spectrum (attached protons in parentheses)?

δ 15.5 (3), δ 20.1 (3), δ 60.7 (2), δ 99.6 (1)

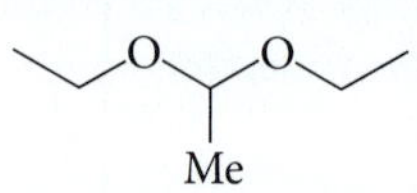

A

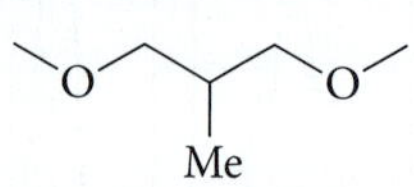

B

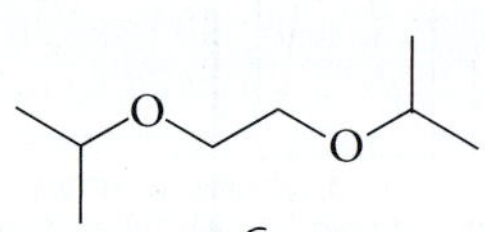

C

• Predict how the NMR spectrum of a mixture of rapidly interconverting species will change when the rate of interconversion is reduced. 14.8

14.46 The 60-MHz proton NMR spectrum of 2,2,3,3-tetrachlorobutane consists of a sharp singlet at 25 °C, but at −45 °C it consists of two singlets of different intensities separated by about 10 Hz.

(a) Explain the changes in the spectrum as a function of temperature.

(b) Explain why the two lines observed at low temperature have different intensities.

14.47 What changes would you expect in the carbon NMR spectrum of 1-bromopropane upon cooling the compound to very low temperature?

• Use information from IR, NMR (proton and carbon), and mass spectrometry together to determine unknown structures. 14.10

14.48 Determine the structure of each of the following compounds.

(a) $C_6H_{14}O$: δ 0.91 (6*H*, d, $J = 7$ Hz); δ 1.17 (6*H*, s); δ 1.48 (1*H*, s; disappears following D_2O shake); δ 1.65 (1*H*, septet, $J = 7$ Hz)

^{13}C NMR: δ 17.6, δ 26.5, δ 38.7, δ 73.2

(b) $C_6H_{10}O$: NMR: δ 3.31 (3*H*, s); δ 2.41 (1*H*, s); δ 1.43 (6*H*, s)

IR: 2110, 3300 cm^{-1} (sharp); no O—H stretch

(c) $C_6H_{14}O_2$: IR: 3200–3600 cm^{-1} (broad, strong), 1090 (strong)

NMR spectrum in **Fig. P14.48c** (*Note:* This is a very dry sample.)

(d) C_5H_6O: IR: 3300, 2102, 1634 cm^{-1}

NMR: δ 3.10 (1*H*, d, $J = 2$ Hz); δ 3.79 (3*H*, s); δ 4.52 (1*H*, doublet of doublets, $J = 6$ Hz and 2 Hz); δ 6.38 (1*H*, d, $J = 6$ Hz)

(e) Compound *B*, $C_5H_{10}O$; IR: 3200–3600 (broad, strong), 3085 (sharp), 1658 (sharp), 1055 cm, 875 cm^{-1}

NMR spectrum in **Fig. P14.48e**

(f) C_4H_6O: IR: 2100, 3300 cm^{-1} (sharp), superimposed on a broad, strong band at 3350 cm^{-1}

NMR in **Fig. P14.48f**

(g) Mass spectrum: $m/z = 150, 152$ (double molecular ions; about equal intensity)

IR: 3100, 1644 (strong), 1104, 1166, 694 cm^{-1} (strong); no IR absorptions in the range 700–1100 cm^{-1} or above 3100 cm^{-1}

NMR: δ 1.28 (3*H*, t, $J = 7$ Hz); δ 3.91 (2*H*, q, $J = 7$ Hz); δ 5.0 (1*H*, d, $J = 4$ Hz); δ 6.49 (1*H*, d, $J = 4$ Hz)

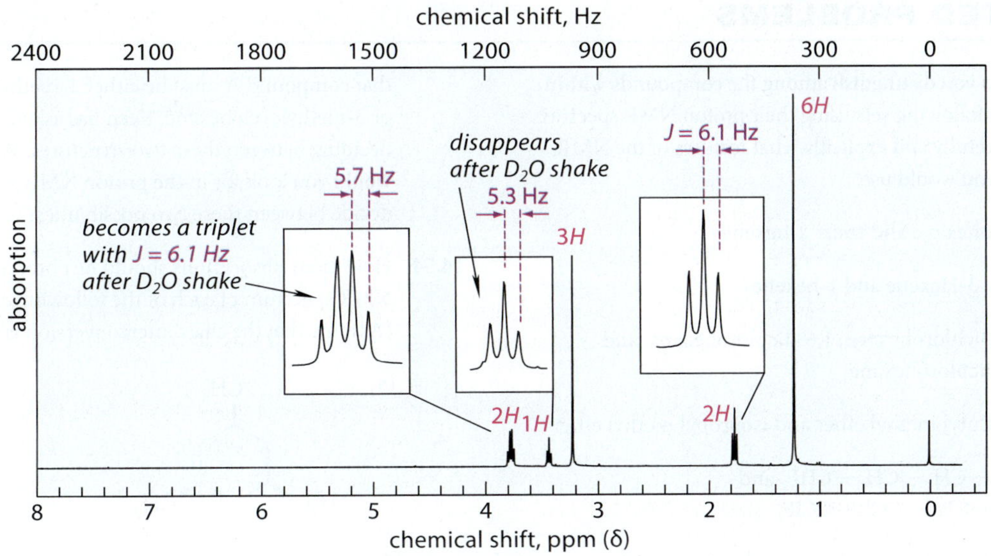

(c)

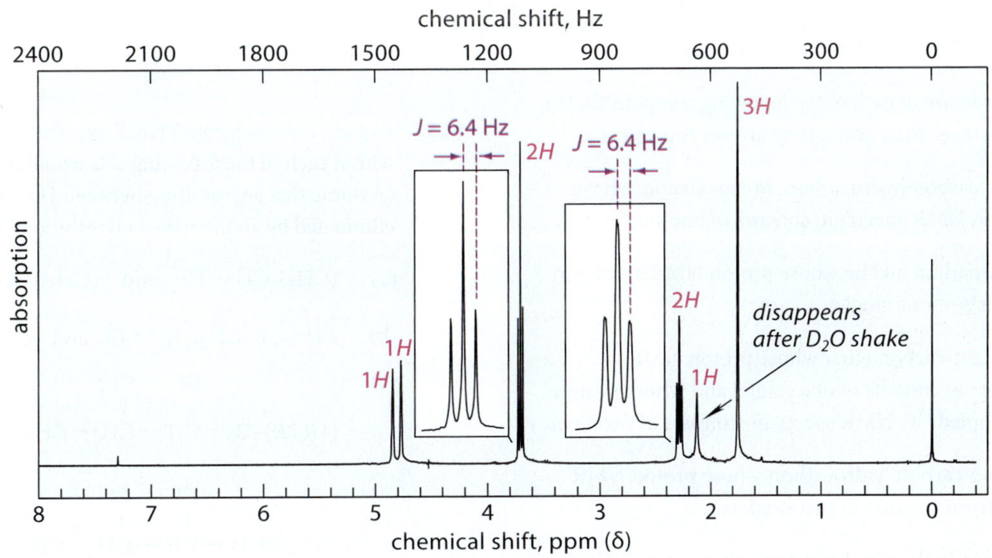

(e)

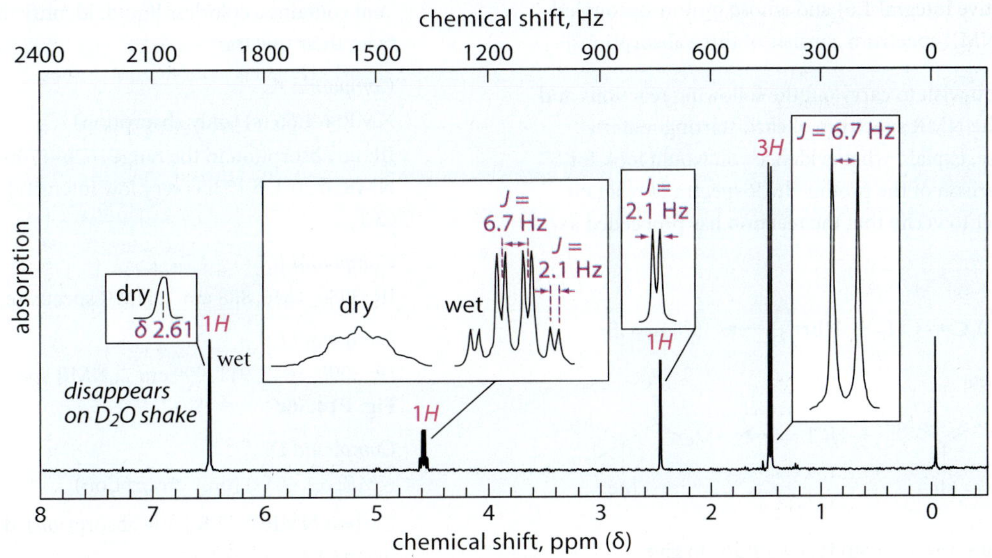

(f)

FIGURE P14.48 Proton NMR spectra for Problems 14.48c, e, and f. Integrals are shown in red, and coupling constants are shown in violet.

INTEGRATED PROBLEMS

14.49 How would you distinguish among the compounds within each of the following sets using their proton NMR spectra? Explain carefully and explicitly what features of the NMR spectrum you would use.

(a) Cyclohexane and *trans*-2-hexene

(b) *trans*-3-Hexene and 1-hexene

(c) 1,1-Dichlorohexane, 1,6-dichlorohexane, and 1,2-dichlorohexane

(d) *tert*-Butyl methyl ether and isopropyl methyl ether

(e) $Cl_3C—CH_2—CH_2—CHF_2$ and $H_3C—CH_2—CCl_2—CClF_2$

14.50 How would you distinguish among the compounds in sets (a), (b), and (c) of Problem 14.49 using proton-decoupled carbon NMR? Explain carefully and explicitly what features of the carbon NMR spectrum you would use.

14.51 Give the structure of each of the following compounds. (In some cases, more than one correct answer is possible.)

(a) A six-carbon hydrocarbon, not an alkene, whose proton NMR spectrum consists of one singlet

(b) A six-carbon alkene whose proton NMR spectrum consists of one singlet

(c) An eight-carbon ether whose proton NMR spectrum consists of one singlet and whose proton-decoupled ^{13}C NMR spectrum consists of two lines

(d) A nine-carbon hydrocarbon whose proton NMR spectrum consists of two singlets

(e) A seven-carbon hydrocarbon whose proton NMR spectrum consists of two singlets at δ 0.23 and δ 1.21 (relative integral 1:6) and whose proton-decoupled ^{13}C NMR spectrum consists of three absorptions

14.52 Suppose you wish to carry out the following reactions and you have the NMR spectrum of each starting material. In each case, explain what evidence you would look for in a comparison of the proton NMR spectra of reactant and product to verify that the reaction has proceeded as shown.

(a) $(CH_3)_2C{=}CH_2 + HBr \longrightarrow (CH_3)_3CBr$

(b)

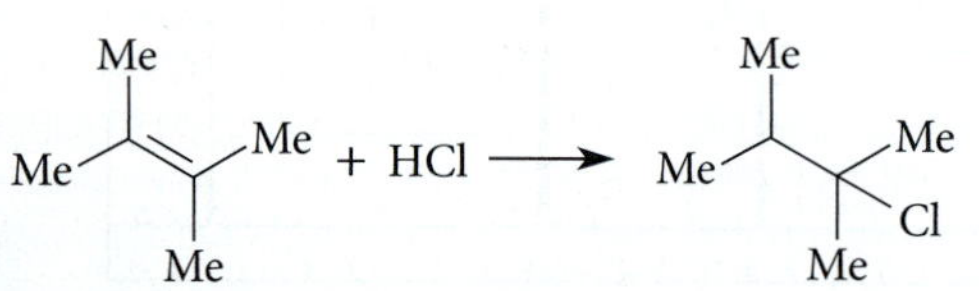

14.53 A compound *A* reacts with H_2 over Pd/C to give methylcyclohexane. A colleague, Al Keen, has deduced that compound *A* must be either 1-methylcyclohexene or 3-methylcyclohexene. Keen has asked for your help in deciding between these two structures. What evidence would you look for in the proton NMR spectrum of *A* to decide between these two possibilities?

14.54 How many absorptions should be observed in the carbon NMR spectrum of each of the following compounds? (Assume that the chair interconversion is rapid.)

(a)

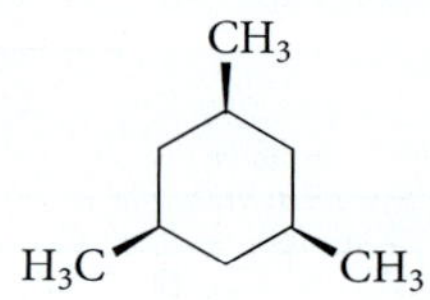

(b)

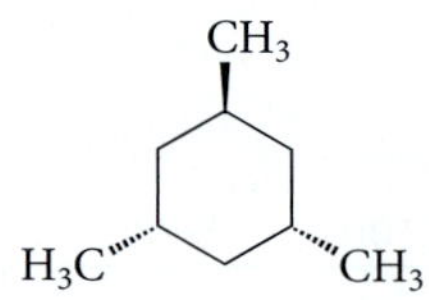

14.55 Explain how the proton NMR spectra of the compounds within each of the following sets would differ, if at all. (Assume that any splitting between H an D has been eliminated by an instrumental technique.)

(a) $(CH_3)_2CH—Cl$ and $(CH_3)_2CD—Cl$

(b) $Cl—CD_2CH_2CH_2—Cl$ and $Cl—CH_2CH_2CH_2—Cl$

(c)

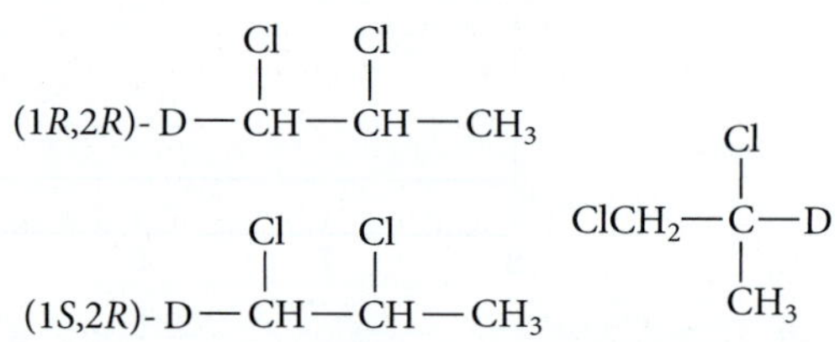

14.56 Each of four bottles, *A*, *B*, *C*, and *D*, is labeled only "C_6H_{12}" and contains a colorless liquid. Identify these compounds from their spectra:

Compound A:
NMR: δ 1.66 (s) (only absorption)
IR: no absorption in the range 1620–1700 cm^{-1} carbon NMR: δ 20.4, δ 123.5 (very low intensity); reacts with Br_2 in CCl_4

Compound B:
IR: 3080, 1646, 888 cm^{-1}; NMR spectrum in **Fig. P14.56b**

Compound C:
IR: 3090, 1642, 911, 999 cm^{-1}; NMR spectrum in **Fig. P14.56c**

Compound D:
NMR: δ 1.40 (s) (only absorption)
Carbon NMR: δ 27.8 (only absorption); does not react with Br_2 in CCl_4

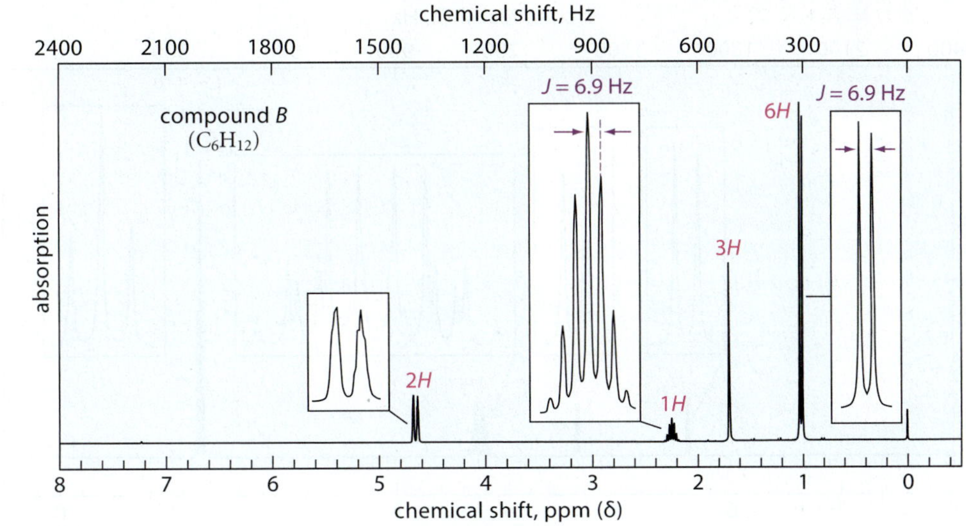

(b)

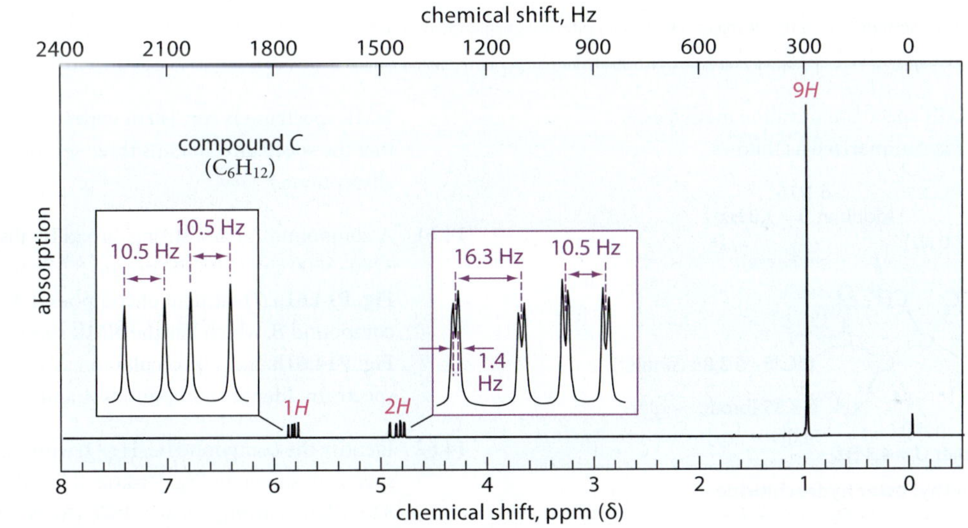

(c)

FIGURE P14.56 NMR spectra for Problem 14.56. (b) Spectrum of compound *B*. (c) Spectrum of compound *C*. The integrals are in red, and the coupling constants are in violet.

14.57 The two protons H^a and H^b in 1,2,3-trichloropropane have slightly different chemical shifts, and the splitting pattern of each is a doublet of doublets. For one proton, $J = 9.0$ Hz and 4.9 Hz; for the other, $J = 9.0$ Hz and 6.0 Hz.

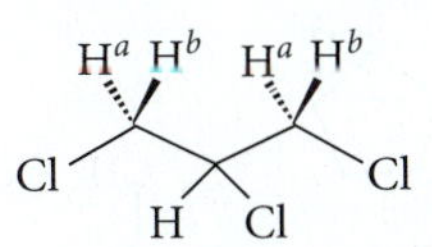

1,2,3-trichloropropane

(a) Explain why H^a and H^b have different chemical shifts. (*Hint:* What is the relationship between H^a and H^b?)

(b) Explain why the splitting pattern for each proton is a doublet of doublets.

14.58 To which of the following compounds does the NMR spectrum shown in **Fig. P14.58** belong? Explain your choice carefully. Once you have made your choice, explain why the resonance at δ 3.7 is so complex.

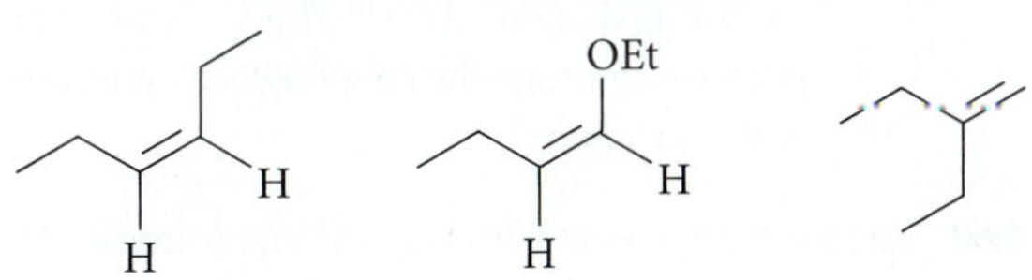

A *B* *C*

$Cl_2CH—CH(OCH_2CH_3)_2$

D

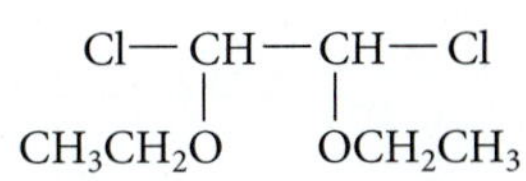

E

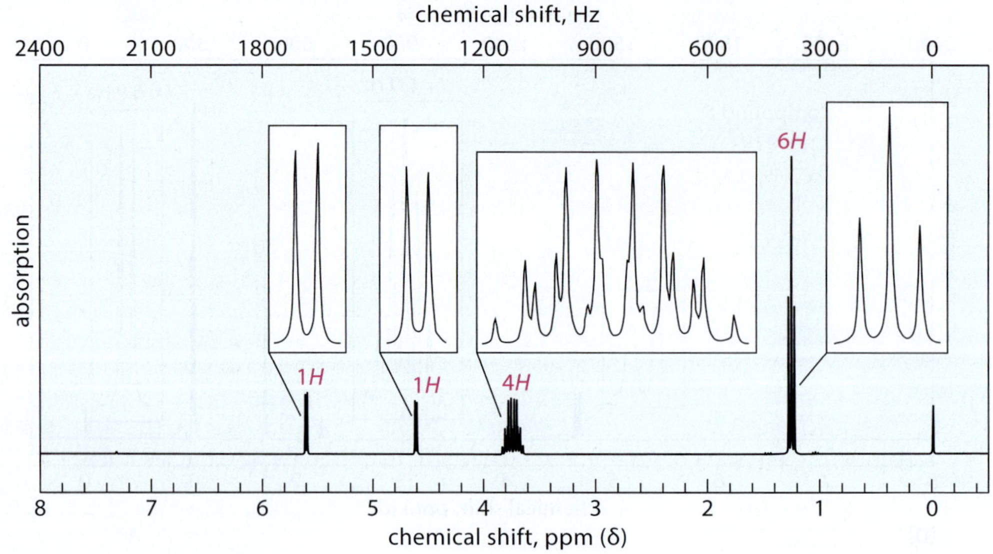

FIGURE P14.58 Proton NMR spectrum for Problem 14.58. The integrals are shown in red.

14.59 The proton NMR spectrum of valine methyl ester hydrochloride is summarized as follows:

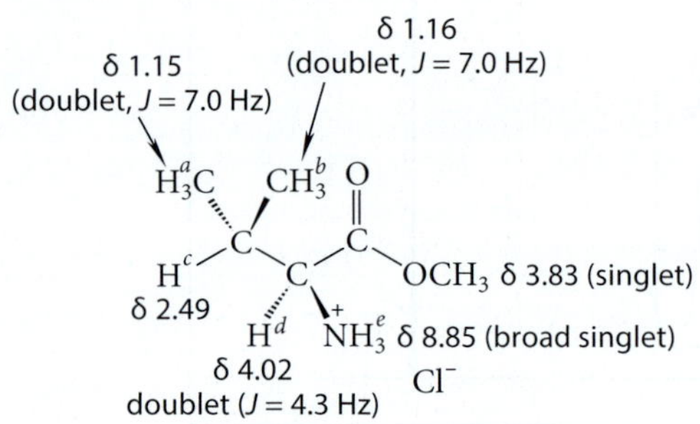

valine methyl ester hydrochloride

Protons H^e are not split by H^d (and vice versa) because protons H^e are rapidly exchanging. Also, the chemical shifts of protons H^a and H^b might be reversed; the important point is that they are different.

(a) Explain why protons H^a and H^b have different chemical shifts and why the splitting of each is a doublet.

(b) Describe the splitting of proton H^c by constructing a splitting diagram. (*Hint:* This is a case of multiplicative splitting; the coupling constants are shown in the structure.)

14.60 You work for a reputable chemical supply house. An angry customer, Fly Ofterhandle, has called, alleging that a sample of 2,5-hexanediol he purchased cannot be the correct compound. As evidence, he cites its carbon NMR spectrum:

δ 23.2, δ 23.5, δ 35.1, δ 35.8, δ 67.4, δ 67.8

After verifying the carbon NMR spectrum, you can confidently assure him that the sample he purchased contains *only* 2,5-hexanediol. Explain why the carbon NMR spectrum is consistent with this claim. (*Hint:* Notice that the spectrum contains three sets of two closely spaced absorptions.)

14.61 A compound *A* has a strong, broad IR absorption at 3200–3500 cm^{-1} and the proton NMR spectrum shown in **Fig. P14.61a**. Treatment of compound *A* with H_2SO_4 gives compound *B*, which has the NMR spectrum shown in **Fig. P14.61b** and a molecular ion at $m/z = 84$ in its EI mass spectrum. Identify compounds *A* and *B*.

14.62 Identify the compound ($C_5H_{10}O$) with the proton NMR spectrum shown in **Fig. P14.62**. It has IR absorptions at 3200–3600 (strong, broad), 1676 (weak), and 965 cm^{-1}, and it has ^{13}C NMR absorptions (attached protons in parentheses) at δ 17.5 (3), δ 23.3 (3), δ 68.8 (1), δ 125.5 (1), and δ 135.5 (1). This compound is optically inactive, but it can be resolved into enantiomers.

14.63 The proton NMR spectrum of vitamin D_3 is given in **Fig. P14.63**.

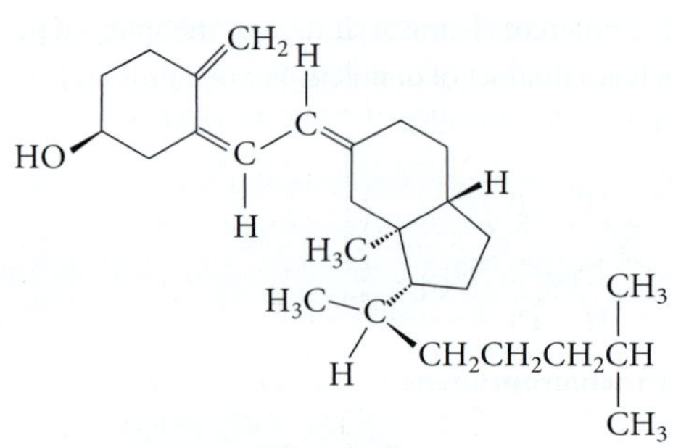

vitamin D_3

Interpret the resonances marked with asterisks (*) by indicating the part of the structure to which they correspond. (Do not try to assign the individual resonances within the groups.) Explain your choices.

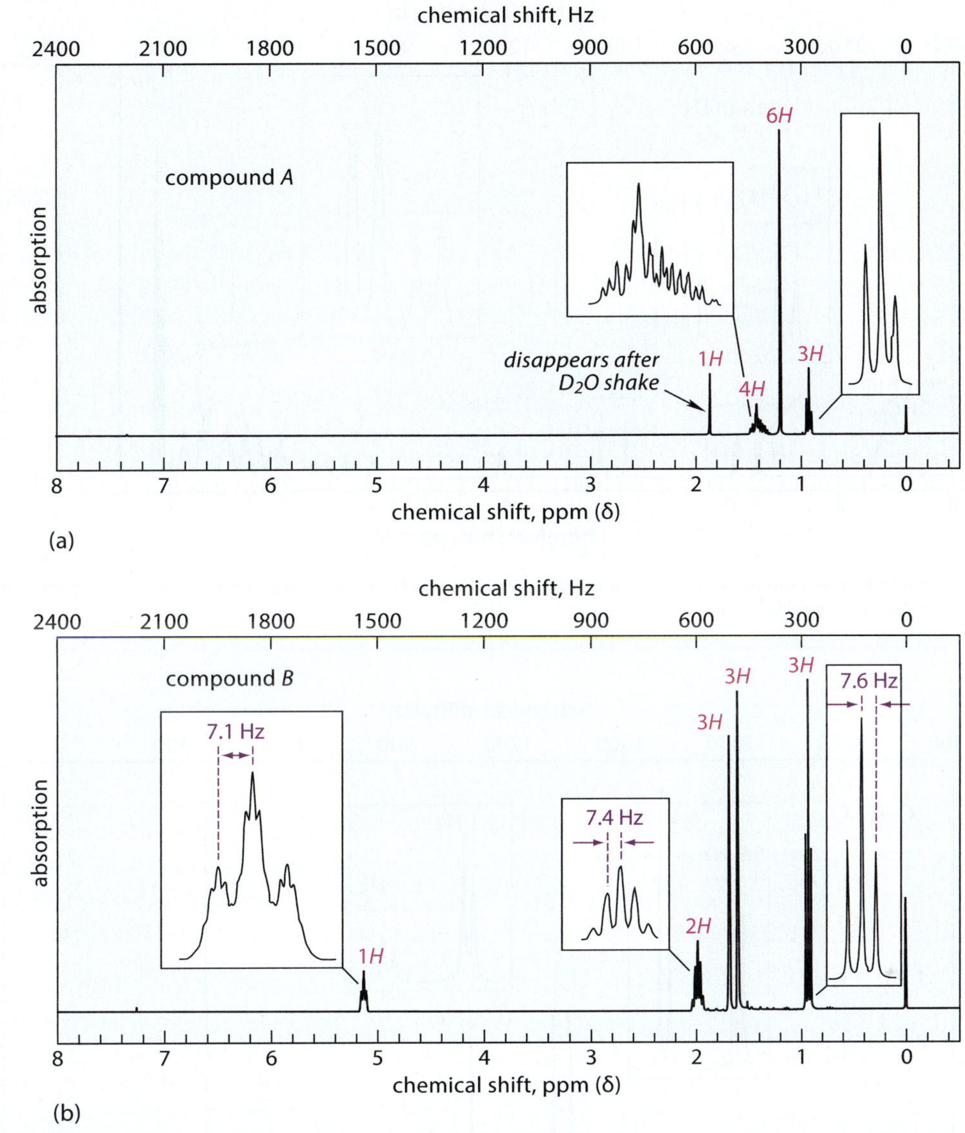

FIGURE P14.61 Proton NMR spectra for Problem 14.61. (a) Spectrum of compound *A*. (b) Spectrum of compound *B*. In the spectrum of compound *B*, the expansion of the resonance at δ 5.1 is on a larger horizontal scale than the expansions of the other resonances. The integrals are given in red, and the coupling constants in (b) are given in violet.

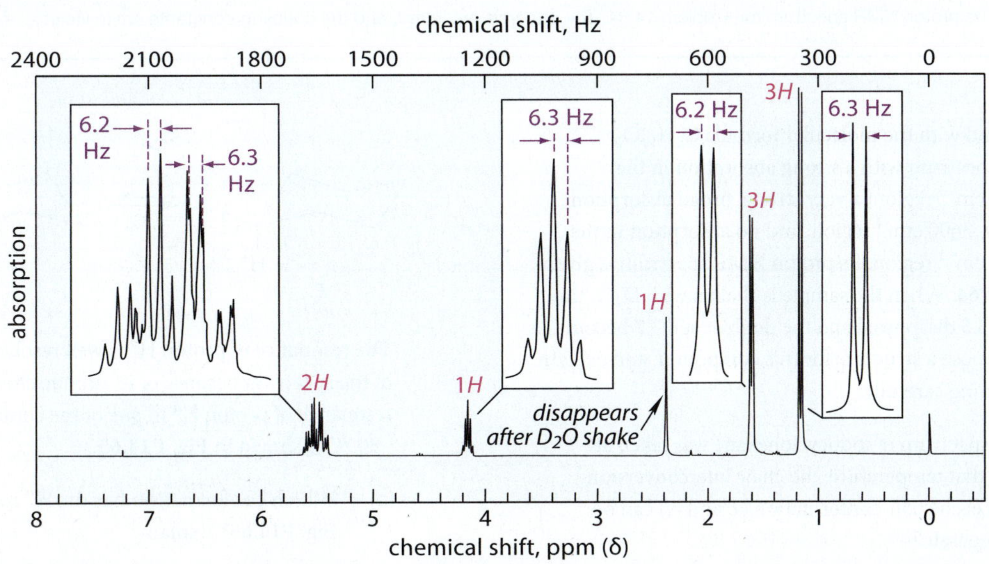

FIGURE P14.62 Proton NMR spectrum for Problem 14.62. The integrals are in red, and the coupling constants are in violet.

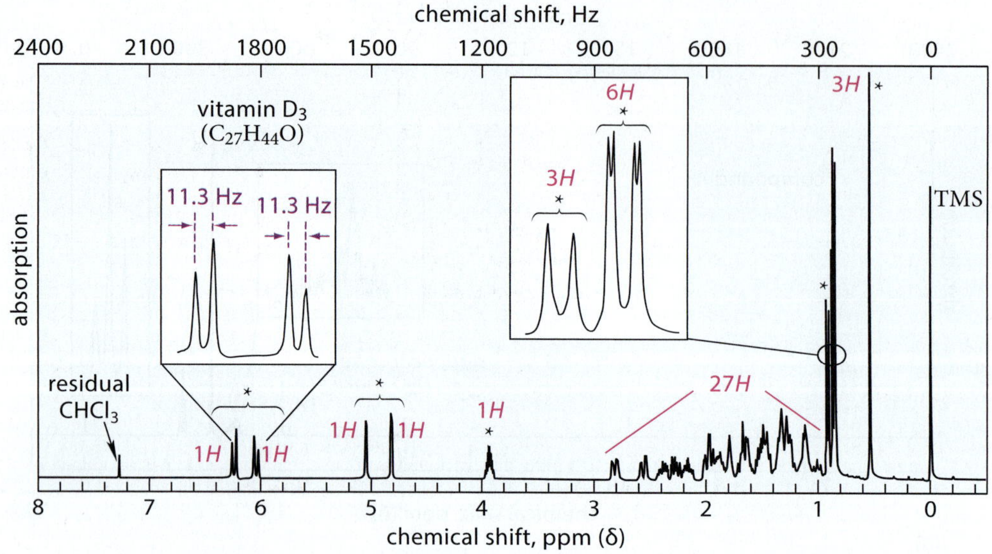

FIGURE P14.63 The proton NMR spectrum of vitamin D_3. The integrals are in red and coupling constants in violet. The expansion of the δ 6.0–6.4 region is on a different scale than the other expansion.

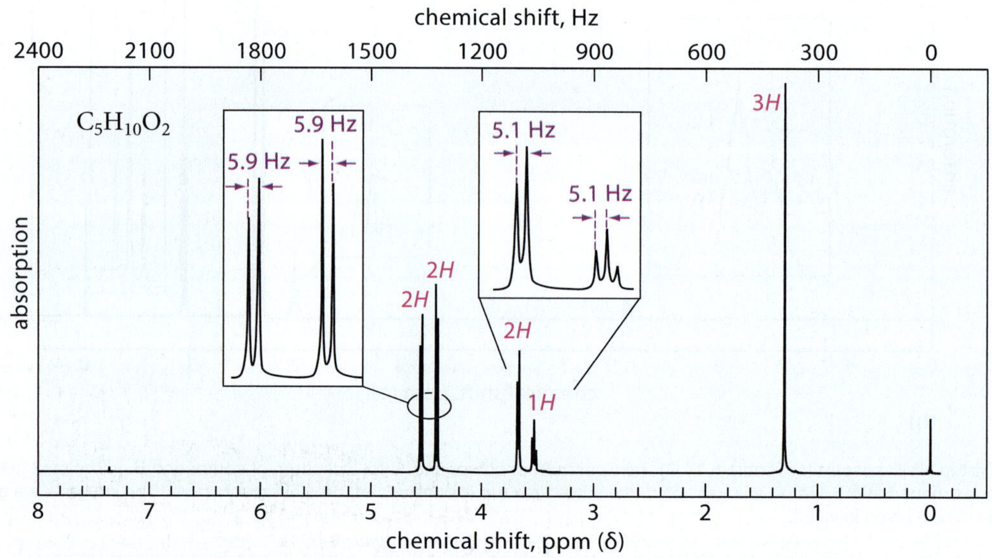

FIGURE P14.64 The proton NMR spectrum for Problem 14.64. The integrals are in red, and the coupling constants are in violet.

14.64 A compound with the molecular formula $C_5H_{10}O_2$ has an IR spectrum with a strong absorption in the 1000–1100 cm^{-1} region; a very strong, broad absorption in the 3000–3600 cm^{-1} region; and no absorption in the 1600–1700 cm^{-1} region. Its proton NMR spectrum is given in **Fig. P14.64**. When the sample is shaken with D_2O, the triplet at δ 3.5 disappears and the doublet at δ 3.7 becomes a singlet. Propose a structure for this compound, and explain your reasoning carefully.

14.65 The NMR spectrum of iodocyclohexane was taken at –80 °C. At that temperature, the chair interconversion is slow and each chair conformation (*E* and *A*) can be observed separately.

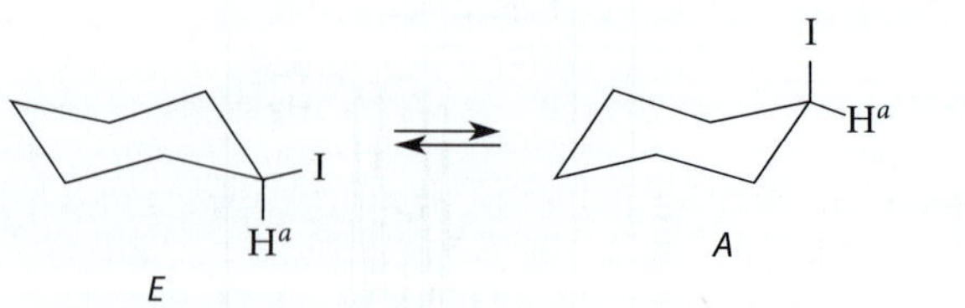

The resonance of proton H^a is well resolved from the rest of the spectrum because of its unique chemical shift. The resonance of proton H^a in *one* of the conformations at –80 °C is shown in **Fig. P14.65**.

(a) Which conformation has the H^a resonance shown in Fig. P14.65? Explain.

(b) How would you expect the resonance of H^a in the other conformation (not shown) to differ from the one shown in Fig. P14.65?

(c) The integrals of the H^a resonance in the two conformations at −80 °C are in the ratio 3.39:1. What is the $\Delta G°$ for the chair interconversion of the two conformations at this temperature? From what you know about conformational analysis, which conformation is the major one?

14.66 Although this chapter has discussed only nuclei that have spin ± ½, several common nuclei such as ^{14}N and deuterium (^2H, or D) have a spin of 1. This means that the spin has three equally probable possibilities: +1, 0, and −1.

(a) How many lines would you expect to observe in the proton NMR of $^+NH_4$? What is the theoretical relative intensity of each line?

(b) How many lines would you expect to observe in the ^{13}C NMR spectrum of $CDCl_3$? (For the answer, see Fig. 14.23, Sec. 14.9B.)

(c) Although the splitting of protons by deuteriums on *adjacent carbons* is generally negligible, the splitting of protons by deuteriums on the *same* carbon can be significant. Explain how you could tell samples of H_2CD—I, D_2CH—I, and D_3C—I apart by proton NMR. What other technique could be used for this determination?

14.67 ^{17}O is a rare isotope that has a nuclear spin of ± ½. The ^{17}O NMR of a small amount of water dissolved in CCl_4 is a triplet (intensity ratio 1:2:1). When water is dissolved in the strongly acidic HF-SbF_5 solvent, its ^{17}O NMR becomes a 1:3:3:1 quartet. Suggest a reason for these observations. (*Hint:* Think of this solvent as H^+ SbF_6^-.)

14.68 Carbon–carbon splitting is not apparent in natural-abundance ^{13}C NMR spectra because of the rarity of the ^{13}C isotope. However, it can be observed in compounds that are enriched in ^{13}C. A chemist, Buster Magnet, has just completed a synthesis of CH_3CH_2Br that contains 50% ^{13}C at each position. (What this really means is that some of the molecules contain no ^{13}C, some contain ^{13}C at one position, and some contain ^{13}C at both positions.) Buster does not know what to expect for the spectrum of this compound and has come to you for assistance. Describe the proton-decoupled ^{13}C NMR spectrum of this compound. (*Hint:* List all of the species present, and decide on their relative amounts. To determine their relative amounts, remember that the probability that two events will occur simultaneously is the product of their individual probabilities. The spectrum of a mixture shows peaks for each compound in the mixture.)

14.69 (a) Because electrons have spin, they can also undergo magnetic resonance. *Electron spin resonance spectroscopy* (ESR spectroscopy) is used to study the magnetic resonance of unpaired electrons in free radicals. (ESR spectroscopy is to unpaired electrons what NMR spectroscopy is to atomic nuclei.) Explain why the ESR spectrum of the unpaired electron in the methyl radical, $\cdot CH_3$, is a quartet of four lines in a 1:3:3:1 ratio.

(b) The magnetogyric ratio of the electron is 17.60×10^6 rad gauss^{-1} s^{-1}, 658 times greater than that of the proton. What operating frequency would be required to detect the magnetic resonance of an unpaired electron in a magnetic field of 3400 gauss (a field commonly used in ESR spectrometers)? In what region of the electromagnetic spectrum does this frequency lie? (Consult Fig. 13.2, Sec. 13.1A.)

14.70 Imagine taking the NMR spectrum of a sample of "naked" protons—that is, H^+ in the gas phase not chemically bonded to anything. In which of the following ranges of chemical shifts would you expect to find the resonance for these protons, and why?

chemical shift > δ 8 δ 8 > chemical shift > δ 0 chemical shift < δ 0

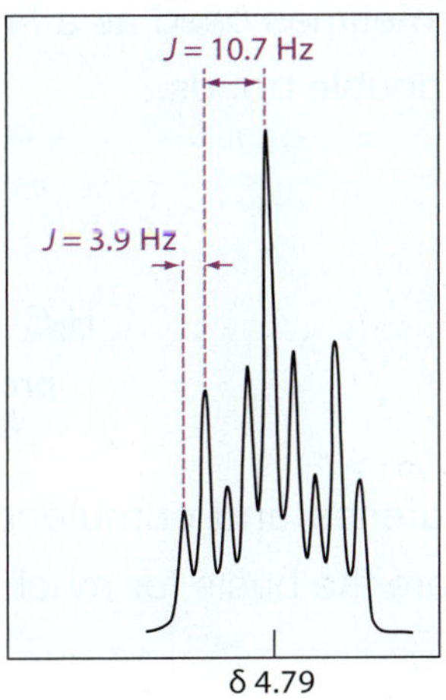

FIGURE P14.65 The proton NMR resonance of H^a in one of the two chair conformations of iodocyclohexane at −80 °C.

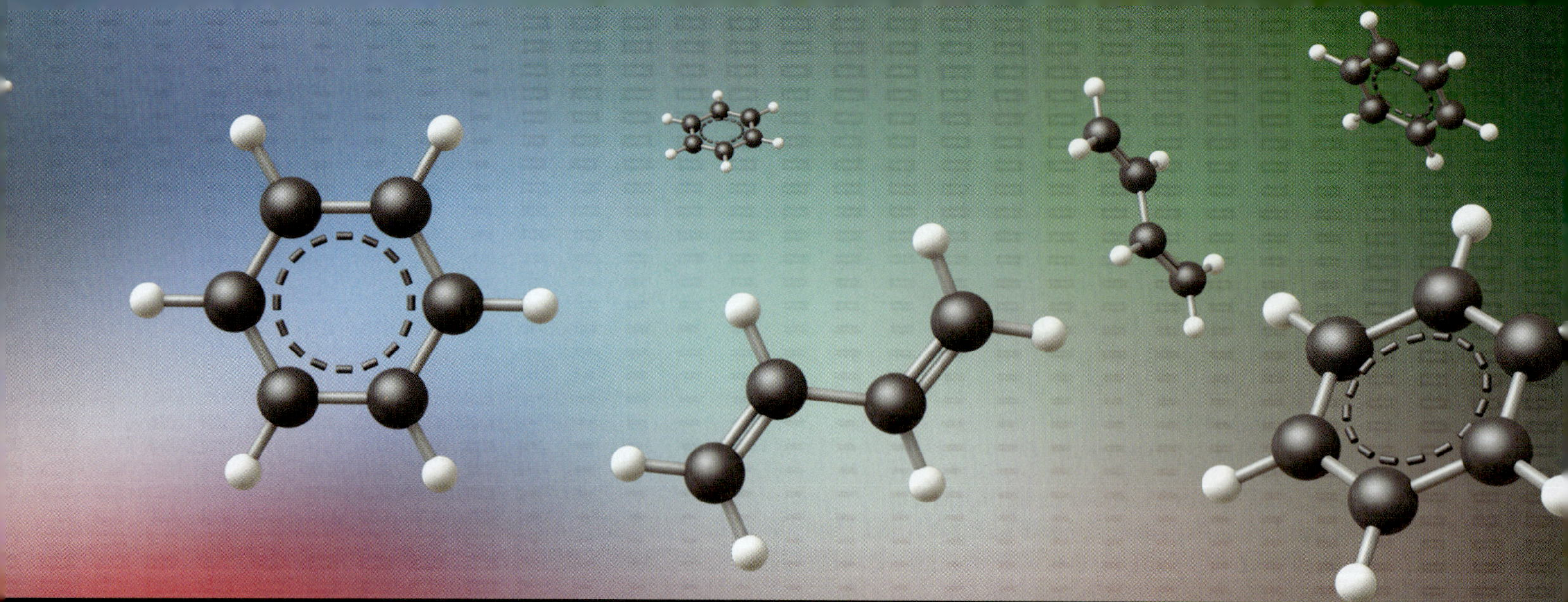

15 DIENES AND AROMATICITY

Dienes are compounds with two carbon–carbon double bonds. Their nomenclature was discussed along with the nomenclature of other alkenes in Sec. 4.2A. Dienes are classified according to the relative position of their double bonds. In **conjugated dienes**, two double bonds are separated by one single bond. These double bonds are called **conjugated double bonds**.

conjugated double bonds

$H_2C{=}CH{-}CH{=}CH_2$ 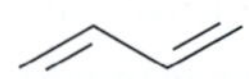

1,3-butadiene
(a conjugated diene) **(15.1)**

Cumulenes are compounds in which one carbon participates in two carbon–carbon double bonds; these double bonds are called **cumulated double bonds**. Propadiene (common name, allene) is the simplest cumulene. The term *allene* is also sometimes used as a family name for compounds containing only two cumulated double bonds.

one carbon involved in two double bonds

$H_2C{=}C{=}CH_2$

propadiene
(allene) (15.2)

Conjugated dienes and cumulenes have unique structures and chemical properties that are the basis for much of the discussion here in Chapter 15.

Dienes in which the double bonds are separated by two or more single bonds have structures and chemical properties more or less like those of simple alkenes and do not require special discussion. We refer to these dienes as "ordinary" dienes.

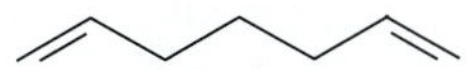

1,6-heptadiene
(an ordinary diene)

In this chapter, the interaction of two functional groups within the same molecule—in this case, two carbon–carbon double bonds—can result in special reactivity. In particular, we explain how conjugated double bonds differ in their reactivity from ordinary double bonds. This discussion leads to a consideration of benzene, a cyclic hydrocarbon in which the effects of conjugation are particularly unique. The chemistry of benzene and the effects of conjugation on chemical properties continue as central themes through Chapter 18.

15.1 STRUCTURE AND STABILITY OF DIENES

A. Stability of Conjugated Dienes. Molecular Orbitals

The heats of formation listed in **Table 15.1** provide information about the relative stabilities of dienes. The effect of conjugation on the stability of dienes can be deduced from a comparison of the heats of formation for (*E*)-1,3-hexadiene, a conjugated diene, and (*E*)-1,4-hexadiene, an unconjugated isomer. The heats of formation show that the conjugated diene is 19.7 kJ mol^{-1} (4.7 kcal mol^{-1}) more stable than its unconjugated isomer. Because the double bonds in these two compounds have the same number of branches and the same stereochemistry, this stabilization of nearly 20 kJ mol^{-1} (5 kcal mol^{-1}) is due to conjugation.

TABLE 15.1 Heats of Formation of Dienes and Alkynes

Compound	Structure	ΔH_f° (25 °C, gas phase) kJ mol^{-1}	ΔH_f° (25 °C, gas phase) kcal mol^{-1}
(*E*)-1,3-hexadiene	$H_2C{=}CH$ H C=C H CH_2CH_3	54.4	13.0
(*E*)-1,4-hexadiene	$H_2C{=}CHCH_2$ H C=C H CH_3	74.1	17.7
1-pentyne	$HC{\equiv}CCH_2CH_2CH_3$	144	34.5
2-pentyne	$CH_3C{\equiv}CCH_2CH_3$	129	30.8
(*E*)-1,3-pentadiene	$H_2C{=}CH$ H C=C H CH_3	75.8	18.1
1,4-pentadiene	$H_2C{=}CHCH_2CH{=}CH_2$	106	25.4
1,2-pentadiene	$H_2C{=}C{=}CHCH_2CH_3$	141	33.6
2,3-pentadiene	$CH_3CH{=}C{=}CHCH_3$	133	31.8

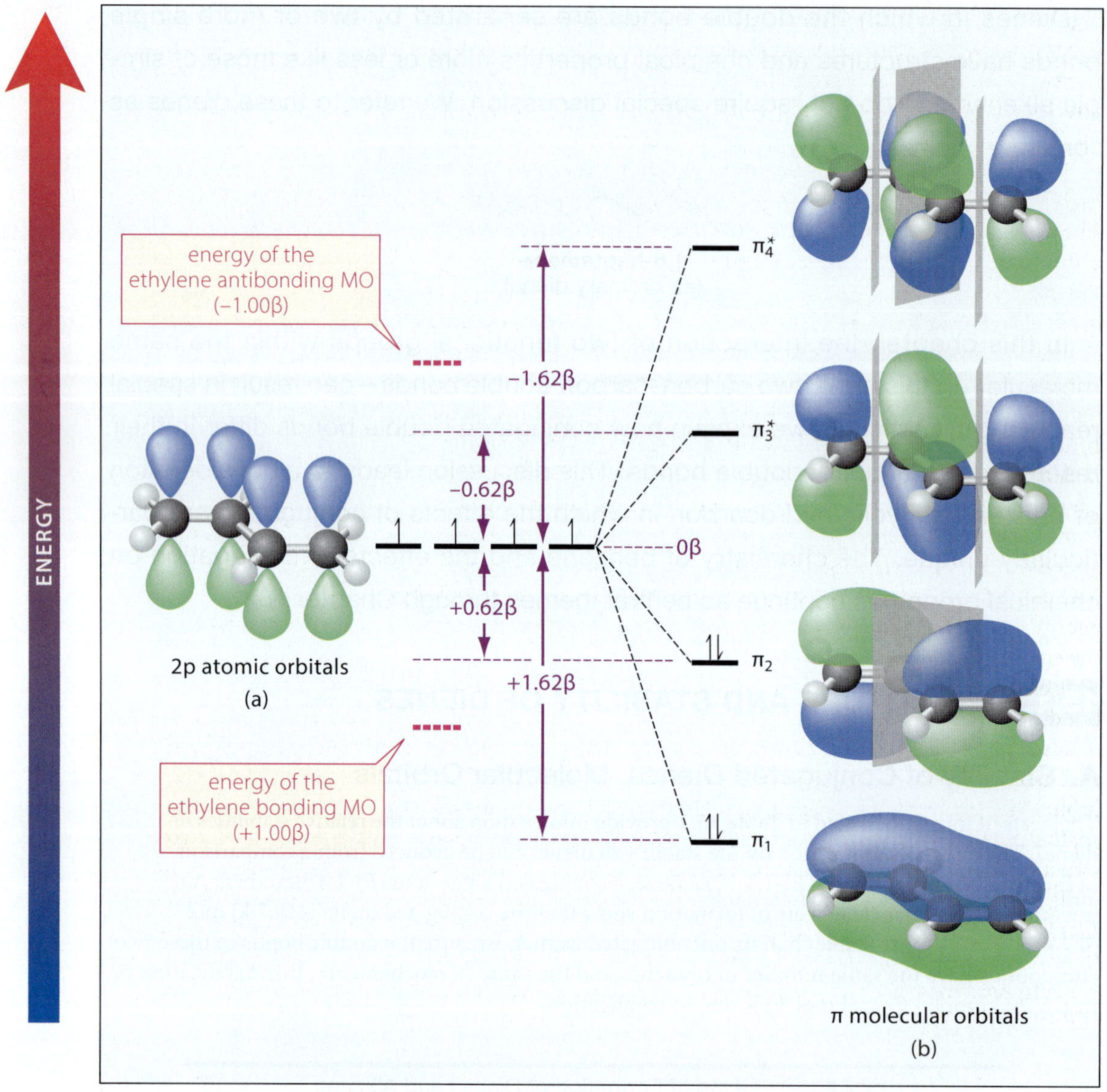

FIGURE 15.1 An orbital interaction diagram showing π molecular orbital (MO) formation in 1,3-butadiene. (a) Arrangement of 2p orbitals in 1,3-butadiene, the simplest conjugated diene. The axes of the 2p orbitals are properly aligned for overlap. (b) Interaction of the four 2p orbitals (dashed black lines) gives four π MOs. Nodal planes are shown in gray. Nodes occur between peaks and troughs in the MOs, indicated by blue and green, respectively. (The original nodal plane of the starting 2p orbitals in the plane of the molecule, which is also present in all of the MOs, is not shown.) The four 2p electrons both go into π_1 and π_2, the bonding MOs. The violet arrows and numbers show the relative energies of the MOs in β units. (β is a negative number.) The relative energies of the ethylene MOs are shown in red.

The major reason for the greater stability of conjugated dienes is the overlap of 2p orbitals across the carbon–carbon single bond connecting the two alkene units. That is, not only does π bonding occur *within* each of the alkene units, but *between* them as well. **Figure 15.1a** shows the alignment of carbon 2p orbitals in 1,3-butadiene, the simplest conjugated diene. Notice that the 2p orbitals on the central carbons are in the parallel alignment necessary for overlap.

As we showed when we considered π bonding in ethylene (see Fig. 4.3, Sec. 4.1C), the overlap of 2p orbitals results in the formation of π molecular orbitals. The overlap of *j* 2p orbitals results in the formation of *j* molecular orbitals. In the case of a conjugated diene, $j = 4$. Therefore, four molecular orbitals (MOs) are formed. For a conjugated diene, half of the MOs are bonding—they have a lower energy than an isolated 2p atomic orbital. The other half are antibonding—they have a higher energy than an isolated 2p orbital. These four MOs for 1,3-butadiene, the simplest conjugated diene, are shown in **Fig. 15.1b**. First of all, each π MO retains a node in the plane of the molecule like the 2p orbitals from which it is formed. (This node is not shown in

Fig. 15.1.) The MO of lowest energy, π_1, has no additional nodes. Each MO of successively higher energy has one additional planar node, and these nodes are symmetrically arranged within the π system. Therefore, the second bonding MO, π_2, has one additional planar node between the two interior carbons. The antibonding MOs π_3^* and π_4^* have two and three additional planar nodes, respectively. (The asterisk indicates their antibonding character.)

1,3-Butadiene has four 2p electrons; these electrons are distributed into the four MOs. Because each MO can accommodate two electrons, two electrons are placed into π_1 and two into π_2. These two bonding MOs, then, are the ones we want to examine to understand the bonding and stability of conjugated dienes. Consider first the energies of these bonding MOs. These are shown to scale relative to the energies of the ethylene MOs (see Fig. 4.3, Sec. 4.1C). The energy unit conventionally used with π MOs is called **beta (β)**, which, for conjugated alkenes, has a value of roughly -50 kJ mol^{-1} (-12 kcal mol^{-1}). By convention, β is a negative number. The π_1 MO of butadiene has a relative energy of 1.62β, and π_2 has a relative energy of 0.62β. Each π electron in butadiene contributes to the molecule the energy of its MO. Therefore, the two electrons in π_1 contribute $2 \times (1.62\beta) = 3.24\beta$, and the two electrons in π_2 contribute $2 \times (0.62\beta) = 1.24\beta$. The total π electron energy for 1,3-butadiene, then, is 4.48β.

To calculate the bonding advantage of a conjugated diene due to π-electron delocalization, we compare it to the π-electron energy of two isolated ethylene molecules—that is, two π-electron systems in which there is no overlap between the double bonds. As Fig. 15.1 shows, the bonding MO of ethylene lies at 1.00β; the two bonding π electrons of ethylene contribute a π-electron energy of 2.00β, and the π electrons of two isolated ethylenes contribute 4.00β. It follows that the energetic advantage of conjugation—orbital overlap—in 1,3-butadiene is $4.48\beta–4.0\beta = 0.48\beta$. This energetic advantage must result from π_1, which is the MO with lower energy than the bonding MO of ethylene.

Half of the total π-electron density in 1,3-butadiene is contributed by the two electrons in π_1 and half by the two electrons in π_2. Consider now the nodal structure of these two molecular orbitals. The π_2 MO has a node that divides the molecule into two isolated "ethylene halves;" so, the electrons in this MO contribute some isolated double-bond character to the π-electron structure of 1,3-butadiene. However, the π_1 MO has no node perpendicular to the plane of the molecule; consequently, the π-electron density in this MO is spread across the entire molecule—that is, the electrons in π_1 are **delocalized**. In particular, this MO contributes to bonding between the two central carbons—the carbons connected by the "single bond." The delocalization of π electrons across the central single bond is also evident from the electrostatic potential maps (EPM) of 1,3-butadiene.

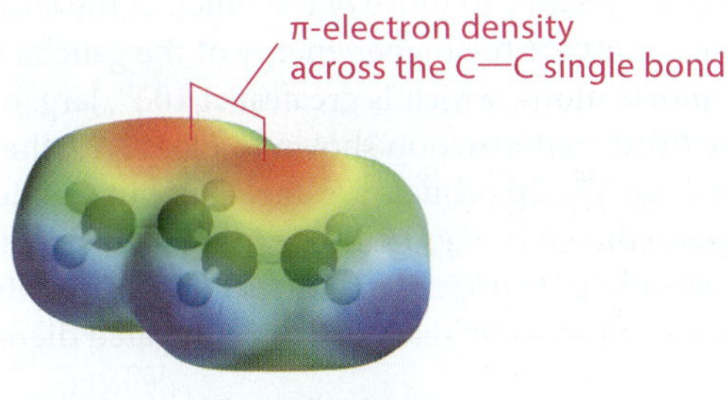

EPM of 1,3-butadiene (15.3)

This analysis shows that electron delocalization, which is not adequately conveyed by Lewis structures, is responsible for the additional stability associated with conjugation. *Conjugation results in additional bonding that makes a molecule more stable.*

The energetic advantage of conjugation is called the **delocalization energy**. This name colorfully describes its origin—the delocalization of electrons in π_1. Because β is negative, the delocalization energy describes the *reduction* in energy (that is, the increased stability) of a conjugated diene relative to two isolated, unconjugated ethylenes. In other words, the delocalization energy is energy that the molecule "doesn't have." Therefore, the delocalization energy of 0.48β for 1,3-butadiene means that this conjugated diene is more stable than two unconjugated ethylene molecules by 0.48β.

Focused Problems

15.1 The conjugated triene (*E*)-1,3,5-hexatriene has six π molecular orbitals with relative energies ±1.80β, ±1.25β, and ±0.44β. (a) Sketch these MOs. Indicate which are bonding and which are antibonding. (b) Tell how many nodes each has. (c) Show the position of the nodes in π_1, π_2, and π_6^*.

15.2 Calculate the delocalization energy for (*E*)-1,3,5-hexatriene.

15.3 Explain why there is a larger *difference* between the heats of formation of (*E*)-1,3-pentadiene and 1,4-pentadiene (29.3 kJ mol^{-1} or 7.1 kcal mol^{-1}) than between (*E*)-1,3-hexadiene and (*E*)-1,4-hexadiene (19.7 kJ mol^{-1} or 4.7 kcal mol^{-1}).

B. Structure of Conjugated Dienes

The length of the carbon–carbon single bond in 1,3-butadiene reflects the hybridization of the orbitals from which it is constructed. At 146 pm, this sp^2–sp^2 single bond is considerably shorter than both the sp^2–sp^3 carbon–carbon single bond in propene (150 pm) and the sp^3–sp^3 carbon–carbon bond in ethane (154 pm).

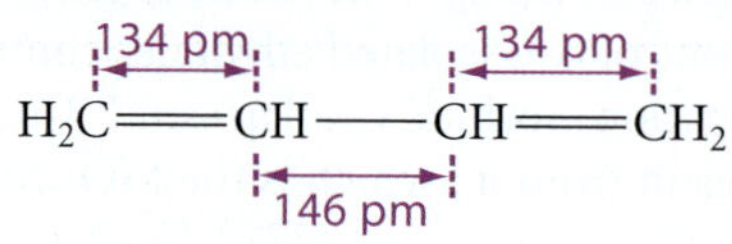

(15.4)

Section 4.1B noted that, as the fraction of s character in the component orbitals increases, the length of the bond decreases. (See Display 4.2, Sec. 4.1C.)

Conjugated dienes such as 1,3-butadiene undergo rapid internal rotation about the central single bond of the diene unit. 1,3-Butadiene has two stable conformations. The most stable conformation is the ***s*-trans conformation**. (The *s* prefix emphasizes that this refers to rotation about a *single* bond.) This conformation is sometimes called the **anti conformation**. The second conformation is the **gauche** or **skew conformation**. These conformations and their relative standard free energies are shown in **Fig. 15.2a**; Newman projections are shown in **Fig. 15.2b**. (The *s*-trans conformation is shown in Fig. 15.1 as well.) In the *s*-trans conformation, the 2p orbitals of all carbons are coplanar and can overlap. In the gauche conformation, the 2p orbitals of one double bond are twisted 38° relative to those of the other, at the cost of some orbital overlap. The partial loss of overlap accounts for the higher energy of the gauche conformation. The energy barrier between the two conformations, which is greatest at 102°, largely reflects the complete loss of overlap at this angle. The third conformation shown in Fig. 15.2a, the ***s*-cis conformation**, is unstable. In this conformation, the 2p orbitals are coplanar, but van der Waals repulsions between two of the hydrogens (shown in Fig. 15.2a) destabilize this conformation; in the gauche conformation, the offending hydrogens are further apart. Despite the instability of the *s*-cis conformation, it is important in some reactions of conjugated dienes (Sec. 15.3).

Focused Problem

15.4 Draw the *s*-cis and *s*-trans conformations of (2*E*,4*E*)-2,4-hexadiene and (2*E*,4*Z*)-2,4-hexadiene. Which diene contains the *greater* proportion of the gauche conformation? Why? (Use the *s*-cis conformation as an approximation of the gauche conformation in your drawing.)

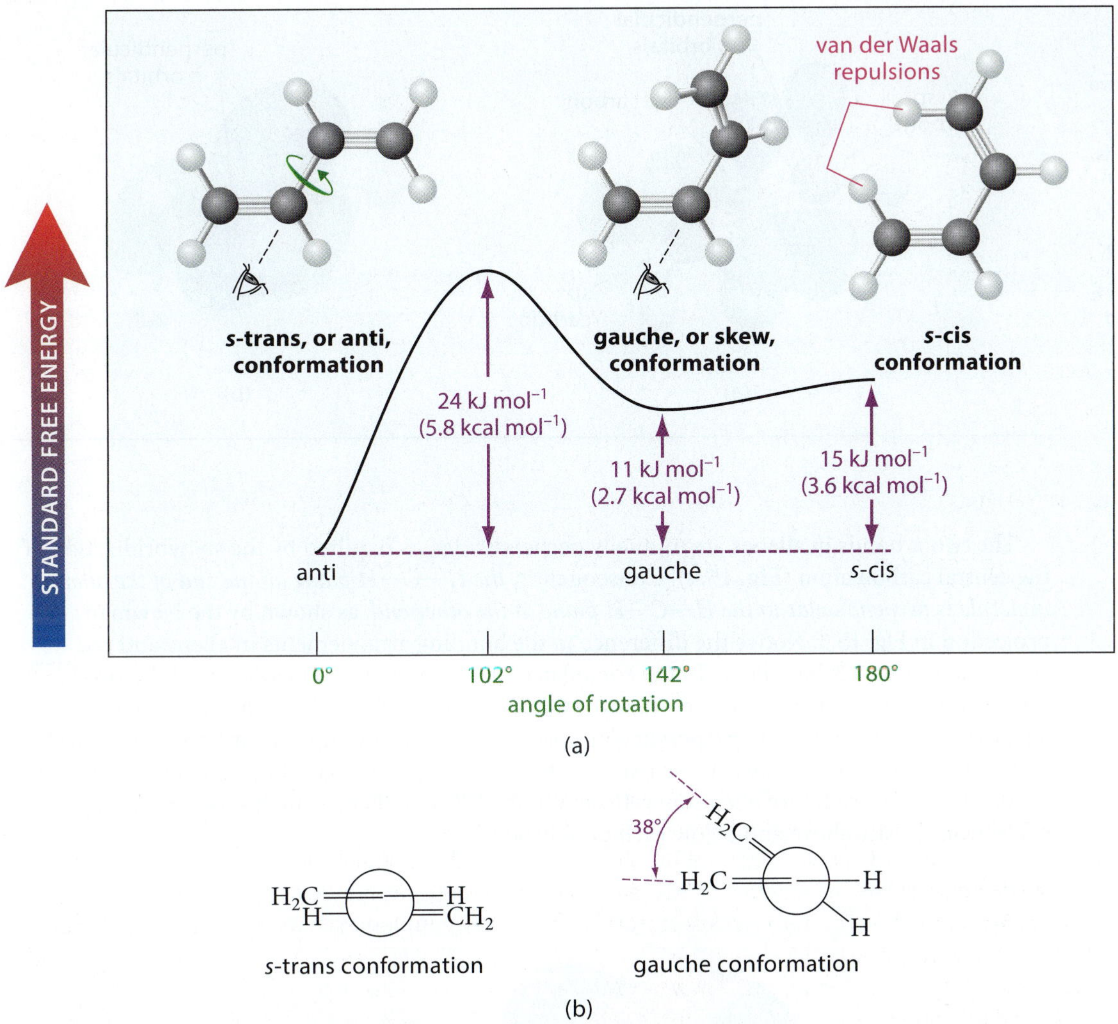

FIGURE 15.2 (a) The conformations of 1,3-butadiene and their relative standard free energies. Internal rotation occurs about the central carbon–carbon single bond (green arrow; rotation angles are shown in green along the horizontal axis). (b) Newman projections of the two stable conformations obtained by sighting along the central carbon–carbon single bond, as shown by the eyeball in (a).

C. Structure and Stability of Cumulated Dienes

The structure of allene is shown in **Fig. 15.3**. Because the central carbon of allene is bound to two groups, the carbon skeleton of this molecule is linear (Sec. 1.3B). A carbon atom with 180° bond angles is sp-hybridized (Sec. 1.7D). Therefore, the central carbon of allene, like the carbons in an alkyne triple bond, is sp-hybridized; it is much like the carbon in CO_2 (Display 1.73, Sec. 1.7D). The two remaining carbons of the cumulated diene are sp^2-hybridized and have trigonal planar geometry.

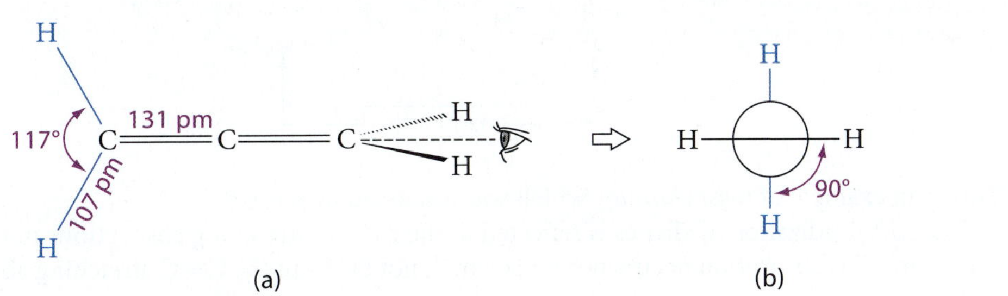

FIGURE 15.3 The structure of allene, the simplest cumulated diene. (a) Lewis structure showing the bond angles and bond lengths. (b) A Newman projection along the carbon–carbon double bonds as seen by the eyeball. The CH_2 groups at opposite ends of the molecule lie in perpendicular planes. (A model is shown in Fig. 15.4.)

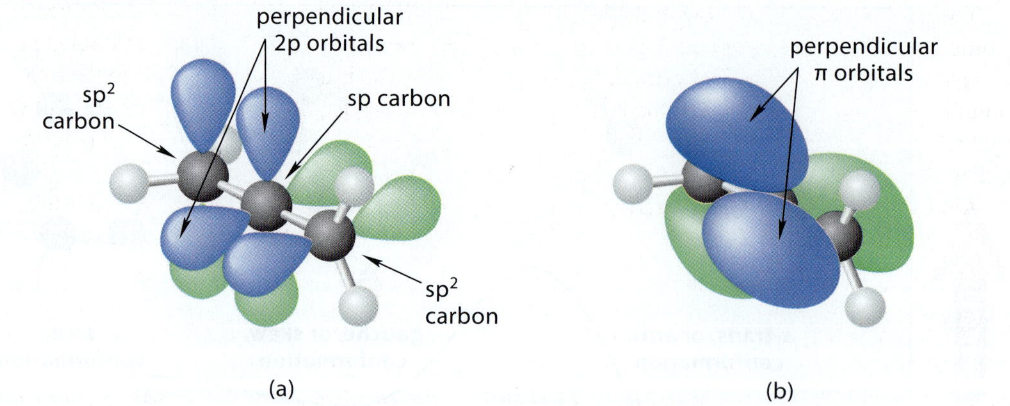

FIGURE 15.4 The π-electron structure of allene. The blue and green orbital colors represent wave peaks and wave troughs. (a) The component 2p orbitals of the double bonds. Because the central carbon is sp-hybridized, it has two mutually perpendicular 2p orbitals. (b) The π molecular orbitals that result from overlap of the 2p orbitals are mutually perpendicular and do not overlap.

The two π bonds in allenes are mutually perpendicular, as required by the sp hybridization of the central carbon atom (**Fig. 15.4**). Consequently, *the H—C—H plane at one end of the allene molecule is perpendicular to the H—C—H plane at the other end*, as shown by the Newman projection in Fig. 15.3. Notice the difference in the bonding arrangements in allene and the conjugated diene 1,3-butadiene. In the conjugated diene, the π-electron systems of the two double bonds are coplanar and can overlap; all carbon atoms are sp^2-hybridized. In contrast, allene contains two mutually perpendicular π systems, each spanning two carbons; the central carbon is part of both. Because these two π systems are perpendicular, they do *not* overlap. The perpendicular π orbitals of allene are reflected in the EPM of allene, which shows areas of π-electron density above and below each double bond.

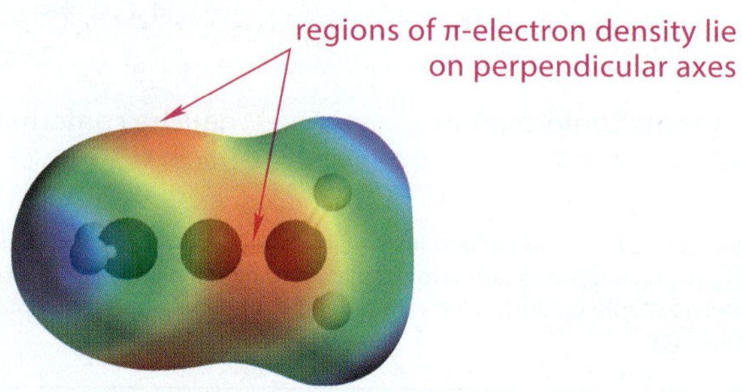

EPM of allene (15.5)

Because of their geometries, some allenes are chiral even though they do not contain an asymmetric carbon atom. 2,3-Pentadiene is a chiral allene:

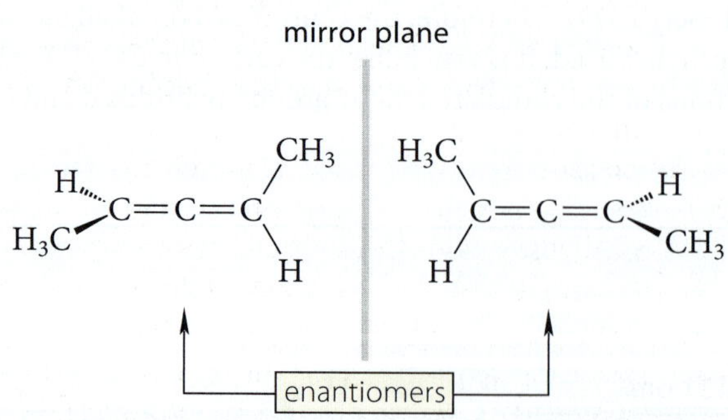

(15.6)

This is an example of *twist chirality,* which was discussed in Sec. 6.8.

The sp hybridization of allenes is reflected in their C=C stretching absorptions in the infrared spectrum. This absorption occurs near 1950 cm^{-1}, not far from the C≡C stretching absorption of alkynes.

The data in Table 15.1 show that allenes have greater heats of formation than other types of isomeric dienes. For example, 1,2-pentadiene is considerably less stable than 1,3-pentadiene or 1,4-pentadiene. Therefore, the cumulated arrangement is the least stable arrangement of two double bonds. A comparison of the heats of formation of 2-pentyne and 2,3-pentadiene shows that allenes are somewhat less stable than isomeric alkynes as well. In fact, a common reaction of allenes is isomerization to alkynes.

Although a few naturally occurring allenes are known, allenes are relatively rare in nature.

Focused Problem

15.5 Rank the following compounds in order of increasing stability (decreasing heats of formation). Give reasons for your choices.

$HC\equiv C-C\equiv C-CH_2-CH_3$ (*A*) $H_3C-C\equiv C-C\equiv C-CH_3$ (*B*) $H_2C=C=CH-CH_2-C\equiv CH$ (*C*)

15.2 ULTRAVIOLET–VISIBLE SPECTROSCOPY AND FLUORESCENCE

The IR and NMR spectra of conjugated dienes are very similar to the spectra of ordinary alkenes. However, another type of spectroscopy can be used to analyze and identify organic compounds containing conjugated double and triple bonds. In this type of spectroscopy, called **ultraviolet–visible spectroscopy**, the absorption of radiation in the ultraviolet or visible region of the spectrum is recorded as a function of wavelength. The part of the ultraviolet spectrum of greatest interest to organic chemists is the *near ultraviolet* (wavelength range 200×10^{-9} to 400×10^{-9} m). Visible light, as the name implies, is electromagnetic radiation visible to the human eye (wavelengths from 400×10^{-9} to 750×10^{-9} m). Because there is a common physical basis for the absorption of both ultraviolet and visible radiation by chemical compounds, both ultraviolet and visible spectroscopy are considered together as one type of spectroscopy, often called simply **UV–vis spectroscopy**.

Some molecules, after absorption of UV or visible radiation, can lose some of the energy gained during absorption by emission of light. We conclude this section with a description of one type of light emission, called *fluorescence*, which is a particularly important analytical technique in biology.

A. The UV–Vis Spectrum

Like any other absorption spectrum, the **UV–vis spectrum** of a substance is the graph of radiation absorption by the substance versus the wavelength of the radiation. The instrument used to measure a UV–vis spectrum is called a **UV–vis spectrophotometer**. Except for the fact that it is designed to operate in a different part of the electromagnetic spectrum, it is conceptually much like any other absorption spectrometer (see Fig. 13.3, Sec. 13.1A).

A typical UV spectrum, that of 2-methyl-1,3-butadiene (isoprene), is shown in **Fig. 15.5**. Because isoprene does not absorb visible light, only the ultraviolet region of the spectrum is shown. On the horizontal axis of the UV spectrum is plotted the wavelength λ of the radiation. In UV spectroscopy, the conventional unit of wavelength is the *nanometer* (abbreviated nm). One **nanometer** equals 10^{-9} meter. (In older literature, the term *millimicron*, abbreviated mμ, was used; a millimicron is the same as a nanometer.) The relationship between the energy of the electromagnetic radiation and its frequency or wavelength should be reviewed again (Sec. 13.1A).

The vertical axis of a UV spectrum shows the **absorbance**. (Absorbance is sometimes called **optical density**, abbreviated OD.) The absorbance is a measure of the amount of radiant energy

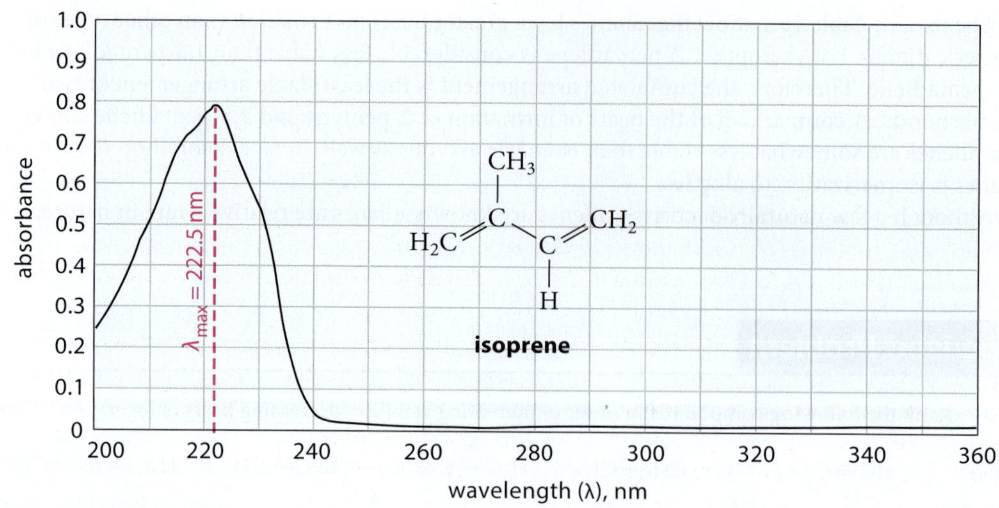

FIGURE 15.5 Ultraviolet–visible spectrum of isoprene in methanol. The λ_{max} (red) is the wavelength at which the absorption maximum occurs; for isoprene, $\lambda_{max} = 222.5$ nm.

absorbed. Suppose the radiation entering a sample has intensity I_0, and the light emerging from the sample has intensity I. The absorbance A is defined as the logarithm of the ratio I_0/I:

$$A = \log (I_0/I) \qquad (15.7)$$

According to Eq. 15.7, then, as more radiant energy is absorbed (that is, as I decreases), the ratio I_0/I increases, as does the absorbance.

Focused Problems

15.6 What is the energy of light (in kJ mol^{-1} or kcal mol^{-1}) with a wavelength of

(a) 450 nm? (b) 250 nm?

15.7 (a) What percent of the incident radiation is transmitted by a sample when its absorbance is 1.0? When its absorbance is 0?

(b) What is the absorbance of a sample that transmits one-half of the incident radiation intensity?

15.8 A thin piece of red glass held up to white light appears brighter to the eye than a piece of the same glass that is twice as thick. Which piece has the greater absorbance?

In the UV–vis spectra used in this text, absorbance increases from the bottom to the top of the spectrum. Therefore, absorption maxima occur as high points or peaks in the spectrum. Notice the difference in how UV–vis and IR spectra are presented. (Absorptions in IR spectra increase from top to bottom because IR spectra are conventionally presented as plots of *transmittance*, or percentage of light transmitted; Sec. 13.2A.) In the UV–vis spectrum shown in Fig. 15.5, the absorbance maximum occurs at a wavelength of 222.5 nm. The wavelength at the maximum of an absorption peak is called the $\boldsymbol{\lambda_{max}}$ (pronounced "lambda-max"). Some compounds have several absorption peaks and a corresponding number of λ_{max} values. Absorption peaks in the UV–vis spectra of compounds in solution are generally quite broad. That is, peak widths span a considerable range of wavelength, typically 50 nm or more. (The reason is discussed in Further Exploration 15.1 in the *Study Guide and Solutions Manual.*)

FURTHER EXPLORATION 15.1
More on UV Spectroscopy

The absorbance at a given wavelength depends on the number of molecules in the light path. If a sample is contained in a vessel with a thickness along the light path of l cm, and the absorbing compound is present at a concentration of c moles per liter, then the absorbance is proportional to the product lc.

$$A = \varepsilon lc \tag{15.8}$$

This equation is called the *Beer–Lambert law* or simply **Beer's law**. The constant of proportionality ε is called the **molar extinction coefficient** or **molar absorptivity**. The units of ε are L mol^{-1} cm^{-1}, or M^{-1} cm^{-1}; these units are sometimes omitted when values of ε are cited. Each absorption in a given spectrum has a unique extinction coefficient that depends on wavelength, solvent, and temperature. The larger is ε, the greater is the light absorption at a given concentration c and path length l. For example, the extinction coefficient of isoprene (see Fig. 15.5) at its λ_{max} of 222.5 nm is 10,750 M^{-1} cm^{-1} in methanol solvent at 25 °C; its extinction coefficient in alkane solvents is nearly twice as large.

Extinction coefficients of 10^4–10^5 M^{-1} cm^{-1} are common for molecules with conjugated π-electron systems. This means that strong absorptions can be obtained from very dilute solutions—solutions with concentrations on the order of 10^{-4} to 10^{-6} M with a typical path length of 1 cm. Because of its intrinsic sensitivity and its relatively simple instrumentation, UV–vis spectroscopy was one of the earliest forms of spectroscopy to be used routinely in the laboratory; adequate spectra could be obtained on even the most primitive spectrometers. UV–vis spectroscopy remains an important method for quantitative analysis.

Some UV–vis spectra are presented in abbreviated form by citing the λ_{max} values of their principal peaks, the solvent used, and the extinction coefficients. For example, the spectrum in Fig. 15.5 is summarized as follows:

$$\lambda_{max}(CH_3OH) = 222.5 \text{ nm } (\varepsilon = 10{,}750)$$

or

$$\lambda_{max}(CH_3OH) = 222.5 \text{ nm } (\log \varepsilon = 4.03)$$

Focused Problem

15.9 (a) From the extinction coefficient of isoprene (10,750 M^{-1} cm^{-1}) and its observed absorbance at 222.5 nm (see Fig. 15.5), calculate the concentration of isoprene in mol L^{-1} (assume a 1 cm light path).

(b) From the results of part (a) and Fig. 15.5, calculate the extinction coefficient of isoprene at 235 nm.

B. Physical Basis of UV–Vis Spectroscopy

What determines whether an organic compound will absorb UV or visible radiation? Ultraviolet and/or visible radiation is absorbed by the π electrons and, in some cases, by the nonbonding electron pairs in organic compounds. For this reason, UV–vis spectra are sometimes called *electronic spectra*. (The electrons of σ bonds absorb at much shorter wavelengths, in the far ultraviolet.) Absorptions by compounds containing only single bonds and nonbonding electron pairs are generally quite weak (that is, their extinction coefficients are small). However, intense absorption of UV and visible radiation occurs when a compound contains π electrons. The simplest hydrocarbon containing π electrons, ethylene, absorbs UV radiation at $\lambda_{max} = 165$ nm ($\varepsilon = 15{,}000$). Although this is a strong absorption, the λ_{max} of ethylene and other simple alkenes is below the usual working wavelength range of most conventional UV–vis spectrophotometers; the lower end of this range is about 200 nm. However, molecules with *conjugated* double or triple bonds (for example, isoprene; see Fig. 15.5) have λ_{max} values greater than 200 nm. Therefore, *UV–vis spectroscopy is especially useful for the diagnosis of conjugated double or triple bonds.*

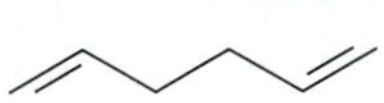

the double bonds are not conjugated; no λ_{max} above 200 nm

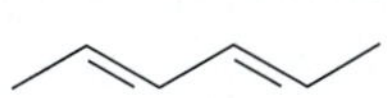

the double bonds are conjugated; this compound has a λ_{max} above 200 nm: $\lambda_{max} = 227$ nm (15.9)

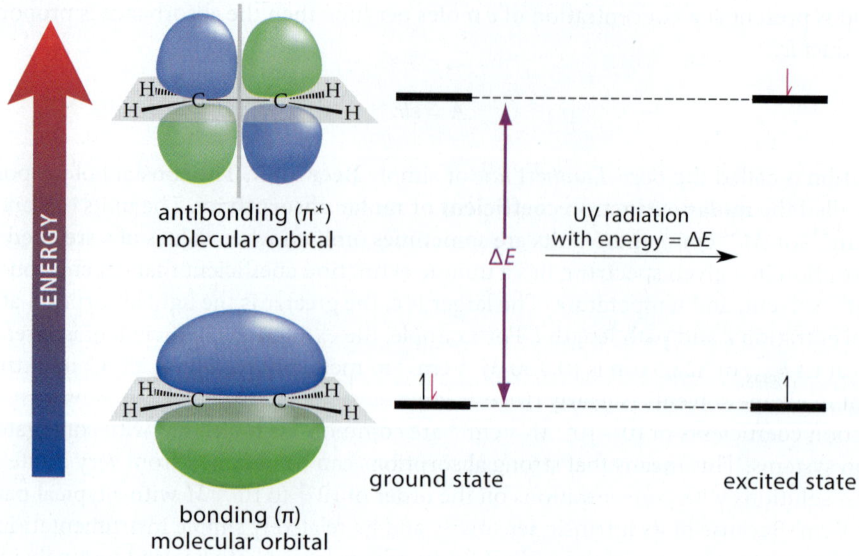

FIGURE 15.6 Absorption of UV radiation by ethylene. The molecular orbitals of ethylene are shown on the left, and the energy difference between these orbitals is shown as ΔE. In the ground state of ethylene, two electrons of opposite spin occupy the bonding (π) molecular orbital. When ethylene is subjected to UV radiation of energy = ΔE, an electron (shown in red) is promoted from the bonding molecular orbital to the antibonding (π*) molecular orbital. The product of this energy absorption is an excited state of ethylene.

The structural feature of a molecule responsible for its UV–vis absorption is called a **chromophore**, from Greek words meaning "to bear color." For example, the chromophore in isoprene (see Fig. 15.5) is the system of conjugated double bonds. Because many important compounds do not contain conjugated double bonds or other chromophores, UV–vis spectroscopy has limited utility in structure determination compared with NMR and IR spectroscopy. However, the technique is widely used for quantitative analysis in both chemistry and biology; and, when compounds do contain conjugated multiple bonds, the UV–vis spectrum can be an important element in a structure proof.

The physical phenomenon responsible for the absorption of energy in the UV–vis spectroscopy experiment can be understood from a consideration of what happens when ethylene absorbs UV–vis radiation at 165 nm. The π-electron structure of ethylene was discussed in Sec. 4.1C and is shown in **Fig. 15.6.** In the normal state of the ethylene molecule, called the *ground state*, the two π electrons occupy a *bonding* π molecular orbital. When ethylene absorbs energy from light, one π electron is elevated from this bonding molecular orbital to the *antibonding* or π* molecular orbital. This means that the electron assumes the more energetic wave motion characteristic of the π* orbital, which includes a node between the two carbon atoms. The resulting state of the ethylene molecule, in which there is one electron in each molecular orbital, is called an *excited state*. The energy required for this absorption must match ΔE, the difference in the energies of the π and π* orbitals (see Fig. 15.6). As a result, the 165 nm absorption of ethylene is called a $\pi \rightarrow \pi^*$ **transition** (read "pi to pi star"). The UV absorptions of conjugated alkenes are also due to $\pi \rightarrow \pi^*$ transitions.

C. UV–Vis Spectroscopy of Conjugated Alkenes

When UV–vis spectroscopy is used to determine chemical structure, the most important aspect of a spectrum is the λ_{max} values. The structural feature of a compound that is most important in determining the λ_{max} is the number of consecutive conjugated double (or triple) bonds. That is, *the more consecutive conjugated multiple bonds there are, the higher the wavelength of the absorption*. Molecular orbital theory provides an explanation for this effect. As shown in Fig. 15.6, the energy of the radiation required for UV–vis absorption is determined by the energy separation between the occupied (bonding) MO and the unoccupied (antibonding) MO. As shown in Sec. 13.1A, this energy is inversely proportional to the wavelength λ:

$$\Delta E = h\nu = \frac{hc}{\lambda} \qquad (15.10)$$

where h is Planck's constant, c is the speed of light, and ν is the frequency of the light.

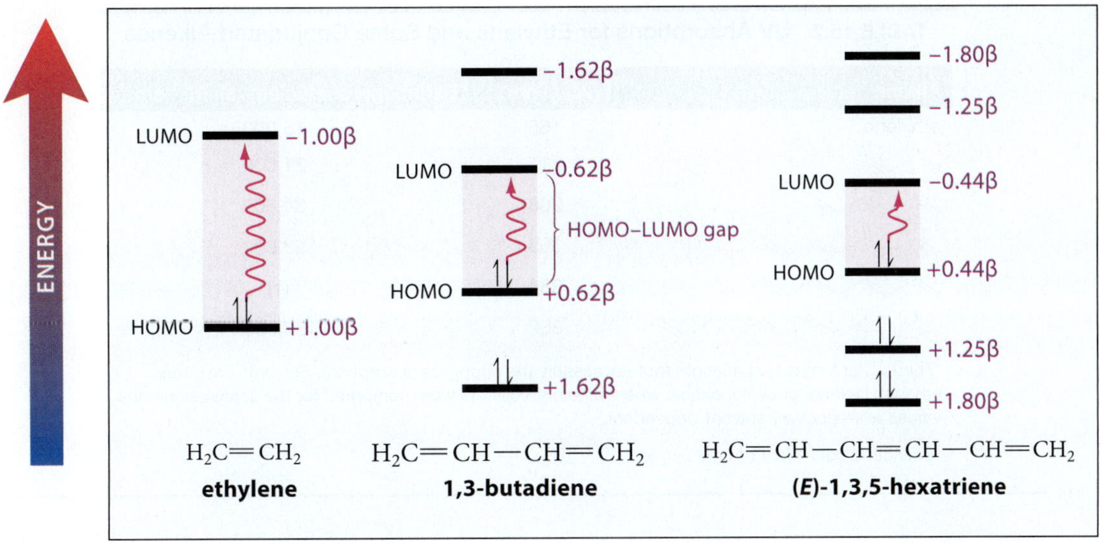

FIGURE 15.7 The relationship of the absorption energy in UV–vis spectroscopy to the number of conjugated double bonds. The energies of the π molecular orbitals for ethylene, 1,3-butadiene, and (*E*)-1,3,5-hexatriene are given in β units. The energy of the UV or visible radiation required for absorption is equal to the gap (lavender shading) between the highest occupied MO (HOMO) and the lowest unoccupied MO (LUMO). Absorption (indicated by the red "squiggly arrows") results in the promotion of an electron from the HOMO to the LUMO. As the number of double bonds increase, the size of the HOMO–LUMO gap decreases and, by Eq. 15.9, the absorption wavelength increases.

A conjugated alkene contains more than one bonding MO and more than one antibonding MO, as shown for 1,3-butadiene (see Fig. 15.1, Sec. 15.1A). In a conjugated diene, the UV–vis absorption at highest wavelength results in promotion of a π electron from the bonding MO of *highest* energy, called the **HOMO** (for "highest occupied molecular orbital") to the antibonding MO of *lowest* energy, called the **LUMO** (for "lowest unoccupied molecular orbital"). In other words, the HOMO–LUMO energy gap—the energy difference between these two MOs—determines the wavelength of the absorption. The relative energies of the π MOs for ethylene and the first two conjugated alkenes are shown in **Fig. 15.7**. This figure shows that the HOMO–LUMO gap becomes smaller as the number of conjugated double bonds increases. The energy of the radiation required for absorption, then, becomes smaller, and the wavelength greater, as the number of double bonds increases. (Other factors in addition to the HOMO–LUMO gap also contribute to the λ_{max}; see Further Exploration 15.1 in the *Study Guide and Solutions Manual*.)

Table 15.2 lists the λ_{max} values for a series of conjugated alkenes. Notice that λ_{max} (as well as the extinction coefficient) increases with an increase in the number of conjugated double bonds; each additional conjugated double bond increases λ_{max} by 30 to 50 nm. Molecules that contain many conjugated double bonds, such as the last one in Table 15.2, generally have several absorption peaks. These result not only from the HOMO–LUMO transition but also from electronic transitions involving other π MOs as well. The λ_{max} usually quoted for such compounds is the one at highest wavelength, which corresponds to the HOMO–LUMO transition.

If a compound has enough double bonds in conjugation, one or more of its λ_{max} values will be large enough to fall within the visible region of the electromagnetic spectrum, and the compound will be colored. An example of a conjugated alkene that absorbs visible light is β-carotene, which is found in carrots and is known to be a biological precursor of vitamin A:

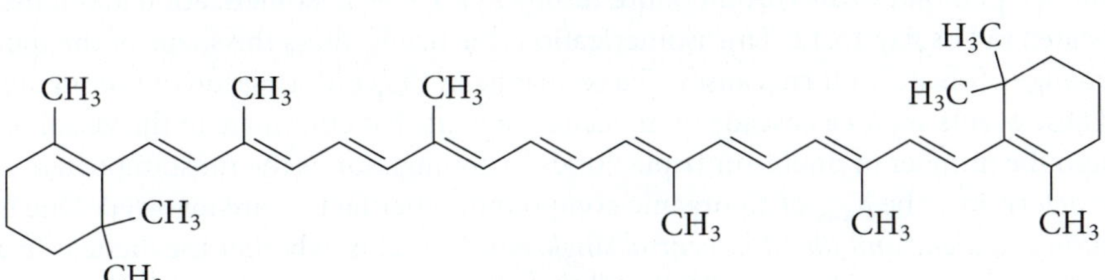

β-carotene (15.11)

TABLE 15.2 UV Absorptions for Ethylene and Some Conjugated Alkenes

Alkene	λ_{max}, nm*	Extinction coefficient (ε), M^{-1} cm^{-1}
ethylene	165	15,000
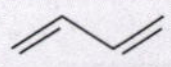	217	21,000
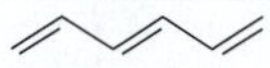	268	34,600
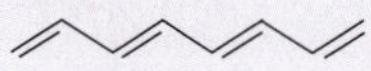	303	53,000
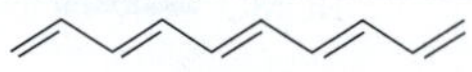	334	(†)
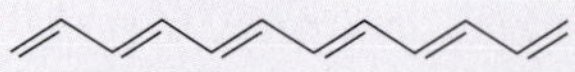	362	138,000

*The λ_{max} at longest wavelength (not necessarily the strongest absorption). Solvents are either ethanol, hexane, or cyclohexane; where different solvents were compared for the same alkene, the values were not very solvent-dependent.

†Extinction coefficient not measured.

Because of the large number of conjugated double bonds in β-carotene, it has strong absorption between 400 and 500 nm, which is in the visible (blue-green) part of the electromagnetic spectrum. When a sample of β-carotene is exposed to white light, blue-green light is absorbed, and the eye perceives the *unabsorbed* light, which is red-orange. In fact, β-carotene is responsible for the orange color of carrots. Similarly, flamingos (**Fig. 15.8**) are red-orange because of the vitamin A in their diets.

The human eye can detect visible light because the eye contains organic compounds that absorb light in the visible region of the electromagnetic spectrum. In fact, light absorption by a pigment, *rhodopsin,* in the rod cells of the eye (as well as a related pigment in the cone cells) is the event that triggers the physiological response that we know as *vision*. The chromophore in rhodopsin is its group of six conjugated double bonds (red in the following structure):

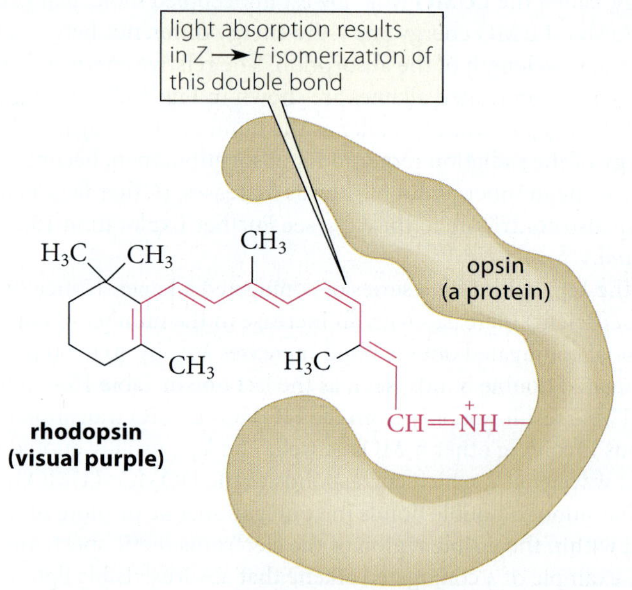

(15.12)

Absorption of a photon by this chromophore results in a $Z \longrightarrow E$ isomerization about the double bond indicated in Display 15.12. This isomerization drastically alters the shape of the molecule, and this change of shape, in turn, causes a large change in the conformation of the surrounding protein. These events set off a cascade of molecular signals that culminate in the visual response.

Although the number of double or triple bonds in conjugation is the most important thing that determines the λ_{max} of an organic compound, other factors are involved. One is *the conformation of a diene unit about its central single bond*—that is, whether the diene is in an *s*-cis or an *s*-trans conformation (Sec. 15.1B). Recall that an acyclic diene assumes the lower energy

FIGURE 15.8 The bright red-orange color of flamingos is caused by the vitamin A in their diets.

s-trans conformation. However, dienes locked into *s*-cis conformations have *higher* values of λ_{max} and *lower* extinction coefficients than comparably substituted *s*-trans compounds:

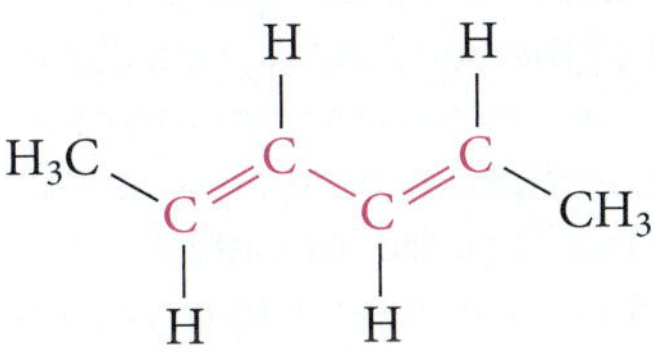

primarily *s*-trans
λ_{max} = 227 nm
(ε = 14,200)

constrained by the ring to an *s*-cis conformation
λ_{max} = 256 nm (ε = 8000)

constrained by the the ring to an *s*-cis conformation
λ_{max} = 239 nm (ε = 3400)

(15.13)

A third variable that affects λ_{max} in a less dramatic yet predictable way is the presence of substituent groups on the double bond. For example, each alkyl group (regardless of size) on a conjugated double bond adds about 5 nm to the λ_{max} of a conjugated alkene. The two methyl groups of 2,3-dimethyl-1,3-butadiene therefore add $(2 \times 5) = 10$ nm to the λ_{max} of 1,3-butadiene, which is 217 nm (see Table 15.2). The predicted λ_{max} is $(217 + 10) = 227$ nm; the observed value is 226 nm.

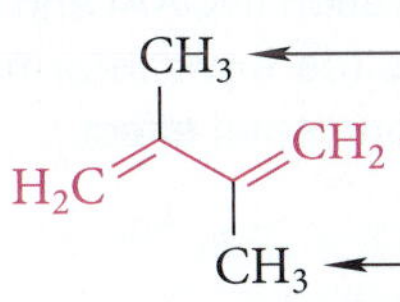

CH₃ ← one alkyl group + 5 nm
CH₂ ← basic diene unit 217 nm
CH₃ ← one alkyl group + 5 nm

227 nm = predicted λ_{max}

2,3-dimethyl-1,3-butadiene

(observed λ_{max} = 226 nm) (15.14)

Although other structural features affect the λ_{max} of a conjugated alkene, here are the two most important points to remember:

1. The λ_{max} is greater for compounds containing more conjugated double bonds.

2. The λ_{max} is affected by substituents, conformation, and other structural characteristics of the conjugated π-electron system.

Focused Problem

15.10 Predict λ_{max} for the UV absorption of each of the following compounds.

(a)

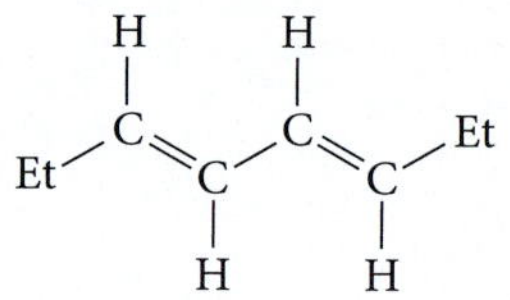

(b)

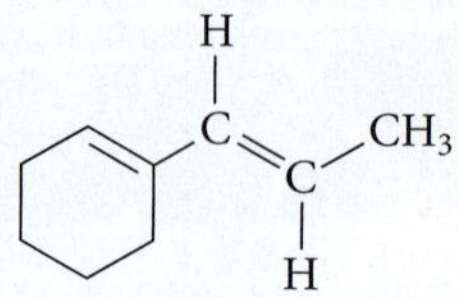

Sunscreens

Blend Images/Superstock

Sunscreens and sunblocks are used to protect the skin from harmful ultraviolet rays of the Sun. The harmful UV radiation of sunlight is often discussed in terms of the parts of the solar spectrum termed "UV-A" and "UV-B." UV-B rays, which cover roughly the 280–315 nm part of the spectrum, are mostly responsible for sunburn and have long been associated with skin cancer. UV-A rays, which cover the 315–400 nm part of the spectrum, are responsible for wrinkling and aging. Research suggests that UV-A rays can also contribute to skin cancer. Sunscreens, which have a $400 million market in the United States (part of a $2 billion annual market for all "sun-care" products), are often characterized by a sun-protection factor (SPF). The SPF typically measures protection from UV-B rays. The most effective sunscreen formulations not only have high SPF numbers but also provide protection from UV-A rays.

Sunscreens work by absorbing the UV rays of the Sun. Therefore, it should not be surprising that sunscreens have UV–visible spectra, which are direct measures of their UV-absorbing capability. The system of conjugated double bonds (red) in the structure of a typical sunscreen (shown below) is responsible for the UV absorption.

The UV spectrum of this compound (**Fig. 15.9**) shows that it is an effective absorber of UV-B rays but has only modest absorption in the UV-A region. This is why sunscreen preparations should contain UV-A blockers as well. (Most UV-A blockers also contain conjugated π-electron systems with appropriate UV absorption.)

The long hydrocarbon chain in the ester group (*blue*) reduces the water solubility of the sunscreen. This same hydrocarbon group, however, promotes absorption through the skin. The absorption of sunscreens has been a cause for concern, but their beneficial effect in reducing the incidence of skin cancer seems to outweigh the other risks of their use.

Another, more recent, concern about sunscreens such as the one in Display 15.15 is their harmful effect on coral reefs. This has led the Hawaii legislature to ban this and another ingredient, oxybenzone, in sunscreens sold in Hawaii. (Coral reefs are also threatened by climate change.) Titanium oxide and zinc oxide are mineral-based sunscreens that don't have this problem. They operate by providing a physical barrier to the Sun's rays rather than by absorbing UV radiation. [The National Oceanic and Atmospheric Administration (NOAA) and the National Geographic Society websites have information about sunscreens and their environmental effect.]

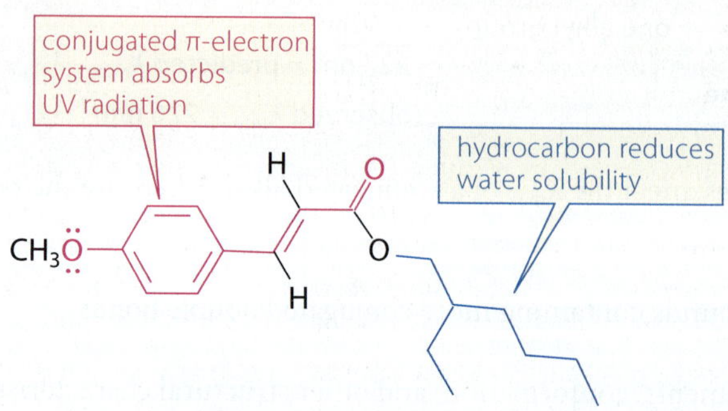

2-ethylhexyl *p*-methoxycinnamate (octinoxate)

(15.15)

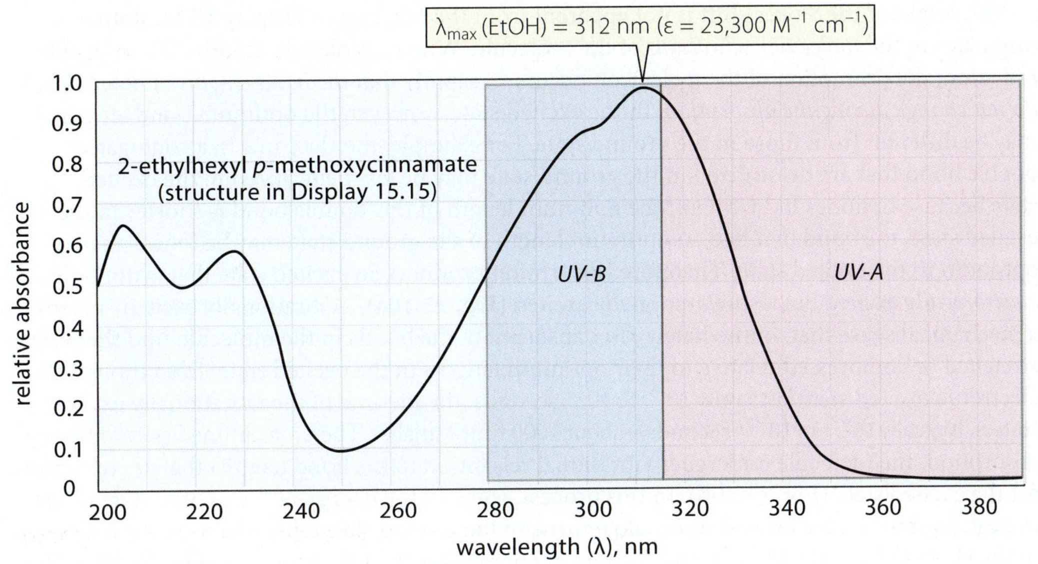

FIGURE 15.9 The UV spectrum of a typical sunscreen, 2-ethylhexyl *p*-methoxycinnamate, in ethanol with the UV-A and UV-B regions indicated.

D. Fluorescence

Chemical Biology Topic

Most of us have seen objects that "glow" with bright colors under a UV lamp. This glow is caused by the emission of light in the visible region from compounds that absorb ultraviolet light. These compounds are *fluorescent.* The purpose of this section is to explain the origin of fluorescence and why it is important, especially in biology.

To start, we consider the origin of fluorescence by looking more deeply into UV–visible absorption. Recall that when any conjugated molecule absorbs UV or visible radiation, an electron is promoted from the *highest occupied molecular orbital* (HOMO) to the *lowest unoccupied molecular orbital* (LUMO) (see Fig. 15.7). Before absorption, the molecule is said to be in its **ground state**. After absorption, the molecule is said to be in an **excited state**.

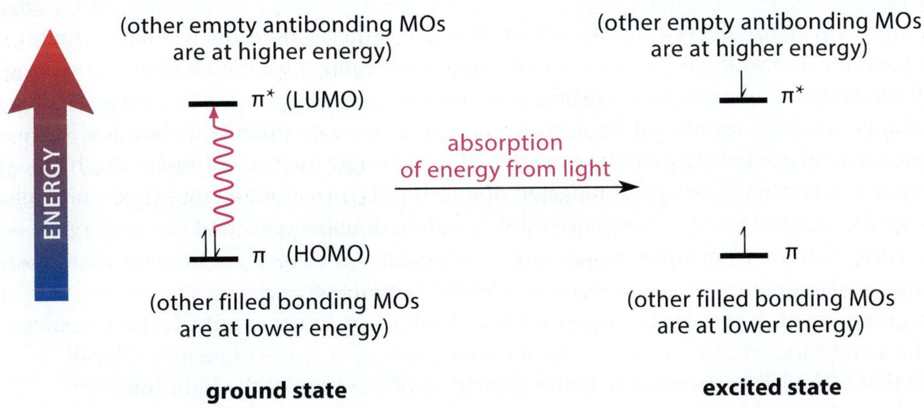

(15.16)

The molecule in its excited state now contains energy resulting from light absorption. What happens to that energy? One of the most common outcomes is that the molecule can lose its excess energy by light emission; this light emission is called **fluorescence**. In fluorescence, the electron that was promoted to the LUMO by absorption returns to the HOMO, and the energy that is lost is emitted as light. This simple picture suggests that the light emitted should have the same energy (and therefore the same wavelength) as the light absorbed. Actually, the light emitted occurs at *lower energy*, and therefore *longer wavelength*, than the light absorbed. For example, a molecule that absorbs blue or violet light might emit green fluorescence. The wavelength shift that occurs in fluorescence is called the **Stokes shift**. The Stokes shift is one of the most useful aspects of fluorescence. The Stokes shift is why certain fluorescent objects glow visibly under UV light.

The origin of the Stokes shift is not apparent from the diagram in Display 15.16. Rather, it originates in the *molecular vibrations* of the molecule. When a molecule absorbs UV or visible radiation, the promotion of the π electron occurs so rapidly that *the bond lengths of the molecule do not change during the absorption.* In the excited state, however, the optimum bond lengths may be different from those in the ground state. For example, the electrons in a conjugated double bond that are delocalized in the ground state may be less delocalized in the excited state because of nodes in LUMO. If the optimum length of this double bond is shorter in the excited state, the bond that had an optimum length in the ground state may be longer than optimum in the excited state. Therefore, absorption produces an excited state that is not only *electronically* excited but also *vibrationally* excited (**Fig. 15.10a**). "Vibrationally excited" means in a mechanical sense that, immediately after absorption, the bonds in the molecule find themselves stretched or compressed relative to their optimum lengths in the excited state. Even though the electronic excited state lasts only 10^{-9} to 10^{-8} seconds, the lifetime of the vibrationally excited state is typically 10^{-11} to 10^{-12} seconds—about 1000 times smaller. Therefore, immediately following absorption, the molecule undergoes vibrational relaxation to the bond lengths that are optimum for the excited state (**Fig. 15.10b**). In this process, some of the energy of the excited state is lost as heat. Eventually, the excited molecule returns to the ground state when the π electron returns to the HOMO from the LUMO, and the loss of energy is seen as fluorescence (**Fig. 15.10c**). Here again, however, the vibrational energy levels come into play. The fluorescence process returns the molecule to a vibrationally excited ground state, from which the remaining excess energy is lost by vibrational relaxation as heat. This description shows that *the energy lost in fluorescence is less than the energy gained in absorption.* Because $E = hc/\lambda$, *the wavelength of the emitted light is greater than that of the absorbed light.* This difference in wavelength between absorbed and emitted light is the Stokes shift.

Fluorescence intensity follows Beer's law (Eq. 15.8) just as absorption does. Molecules that fluoresce strongly must absorb UV–visible radiation strongly as well—that is, they must have large extinction coefficients for absorption. However, not all strong absorbers produce significant fluorescence because excited states can sometimes undergo reactions, internal rotations, transitions to other types of excited states, and so on—processes that we won't be concerned with here. The efficiency of fluorescence is determined by the fraction of excited states that return to ground state by fluorescence. This efficiency is called the **quantum yield** of fluorescence. Many compounds commonly used in fluorescence have quantum yields in the 0.8 to 1.0 range.

There aren't any hard-and-fast rules for determining whether a molecule will have a high quantum yield for fluorescence. To fluoresce strongly, a molecule must have a high absorbance, because fluorescence originates from absorbance. However, a high absorbance is no guarantee that a molecule will have significant fluorescence, because there are other ways besides fluorescence that an excited state can lose energy. Many strongly fluorescent molecules have chromophores consisting of several conjugated double bonds incorporated into rigid polycyclic frameworks. Fluorescent molecules typically don't contain double bonds that can undergo cis–trans isomerization. Noncarbon atoms such as nitrogen and oxygen are in many cases part of the conjugated system, as we see in the examples later in this section.

The importance of fluorescence as an analytical tool lies in its sensitivity. If fluorescence follows the same concentration dependence as absorption, why is it so sensitive? Recall (Eq. 15.8) that UV–visible absorbance is the logarithm of the *ratio* of the light intensity emerging from the sample to the intensity of the light source; $A = \log(I_0/I)$. (This ratio is usually provided physically by the optics in the spectrophotometer.) In other words, UV absorption requires comparison of the light emerging from the sample with a large background—a small difference between large numbers. In fluorescence, however, we measure the absolute intensity of light emerging from the sample, and the standard of comparison is total darkness. Typically, fluorescence detectors are set up at 90° to the light path so that the exciting light source does not interfere with detection.

An analogy to absorption sensitivity is trying to hear a few people whispering in a room full of people talking loudly. The analogy to fluorescence sensitivity is that if the same people are placed in a totally silent room, their whispering can be heard clearly.

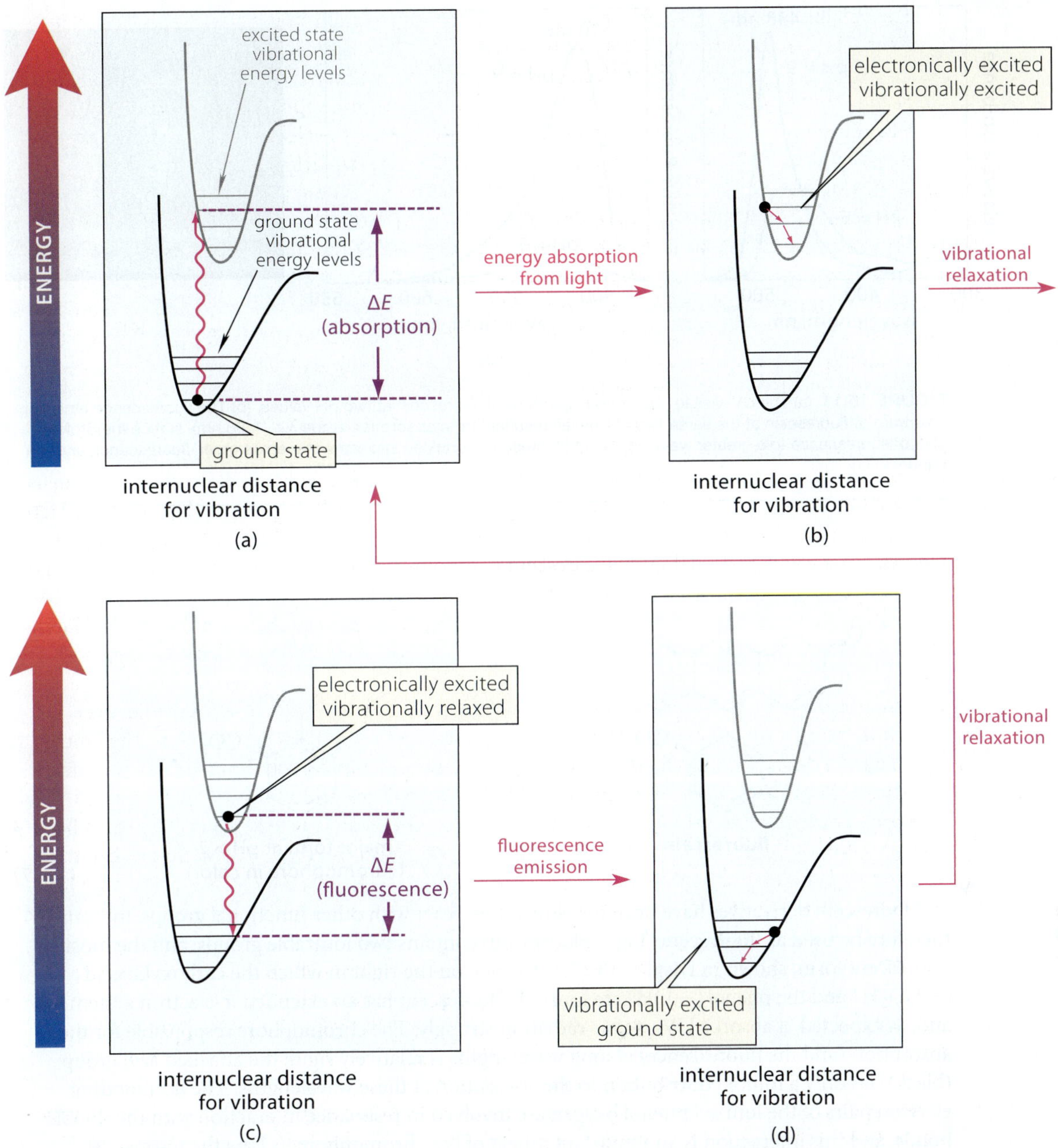

FIGURE 15.10 The absorption and fluorescence process. A molecule in its ground state (a) absorbs energy from light and is promoted to an electronically excited and vibrationally excited state (b). The electronically excited state loses vibrational energy as heat and forms an electronically excited and vibrationally relaxed state (c). This state loses energy by emitting light (fluorescence) to form a vibrationally excited ground state (d). This state then relaxes vibrationally to form the ground state shown in part (a). The energy spacing of the vibrational levels is highly exaggerated for clarity. This process is illustrated for one vibrational mode, but every vibrational mode in the molecule contributes. The energy of the light emitted, ΔE (fluorescence), is less than the energy of the light absorbed, ΔE (absorbance), and for that reason, the wavelength of the emitted light is greater than that of the light absorbed (Stokes shift).

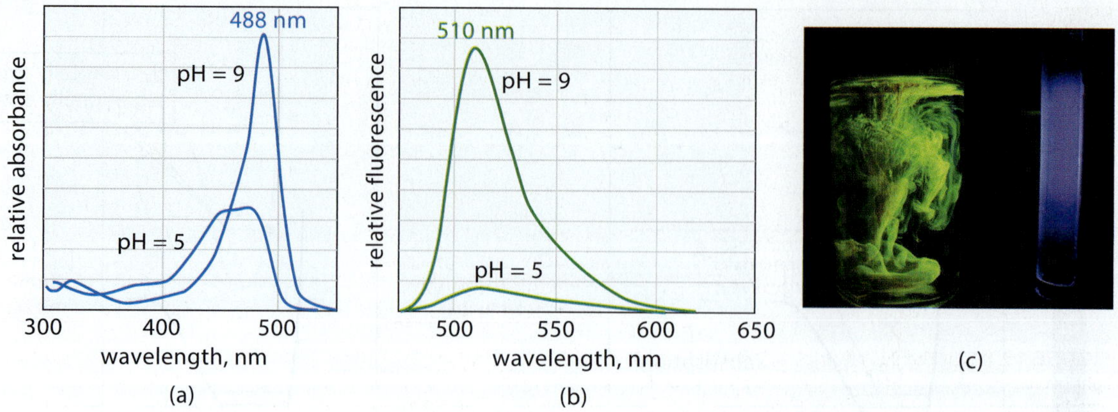

FIGURE 15.11 (a) The UV–visible absorption spectrum of fluorescein at two pH values. (b) The fluorescence emission spectrum of fluorescein at the same two pH values resulting from absorption at the λ_{max} (488 nm). Notice the Stokes shift of fluorescence to a greater wavelength. (c) Fluorescein sprinkled into tap water emits green fluorescence under a handheld UV light.

Fluorescein is a widely used, highly fluorescent compound.

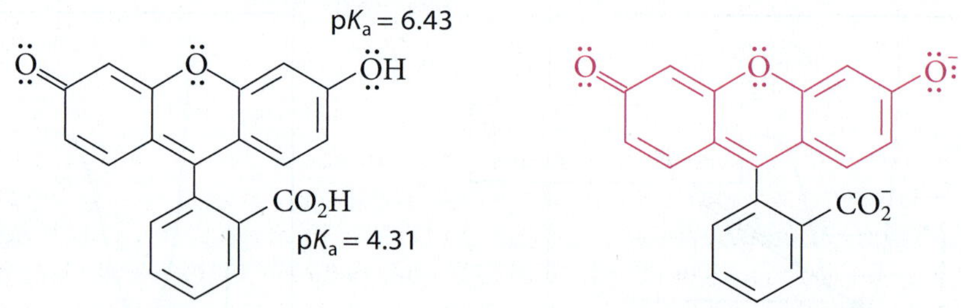

fluorescein

major form at pH > 7
(chromophore in color) (15.17)

Fluorescein derivatives have been developed that react with other functional groups and can therefore be used as fluorescent "tags." Fluorescein contains two ionizable groups, and the most fluorescent form, shown in Display 15.17, is the one on the right in which the carboxylic acid (—CO_2H) and the phenol (—OH) are ionized. Fluorescein has an extended π-electron system, and, as expected, it absorbs UV–visible radiation strongly. The chromophore responsible for the absorption (and the fluorescence) at long wavelengths is relatively rigid; the attached aryl group (black) has only a minor contribution to the absorption at these wavelengths. The nonbonding electron pairs of the ionized phenol oxygen are involved in resonance interaction with the double bonds, and this interaction is an important aspect of the chromophore. (Draw the resonance structures for the delocalization of these electrons.) The absorption and fluorescence spectra of fluorescein are shown in **Figs. 15.11a** and **15.11b**. The fluorescence spectrum is obtained by exciting the molecule at a specific wavelength (in this case, 488 nm, at the λ_{max} for absorption). The peak in the fluorescence emission occurs at 510 nm. Notice the Stokes shift in the fluorescence to higher wavelength. Notice also that the absorption intensity is smaller at lower pH, and so is the fluorescence intensity. **Figure 15.11c** shows the fluorescein fluorescence. In this photo, the electronic absorption is activated by shining a UV light on the sample. We are viewing the fluorescence at a 90° angle to the UV light path.

Fluorescence has revolutionized biology. One of the exciting applications of fluorescence resulted from the discovery of a fluorescent protein in a jellyfish (**Fig. 15.12a**), called *green fluorescent protein* (GFP). The fluorescent group in GFP is "built into" the protein structure:

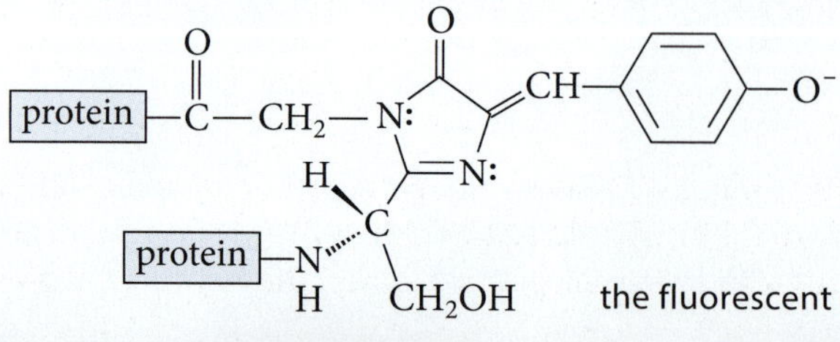

the fluorescent group of GFP (15.18)

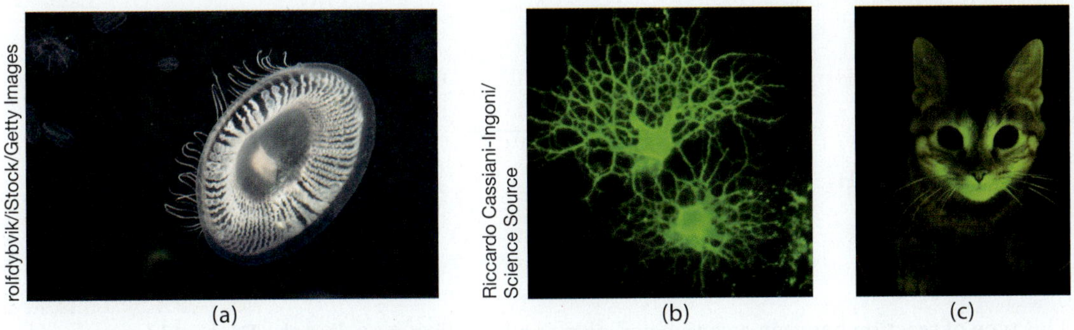

FIGURE 15.12 (a) The fluorescent jellyfish *Aequorea victoria*, source of the green fluorescent protein. (b) A fluorescent light micrograph of GFP-containing oligodendrocytes from a mouse. The gene for a protein localized in the oligodendrocyte was fused with the gene for GFP and the fusion protein was expressed within a genetically engineered mouse in its oligodendrocytes. These cells form the myelin sheath around neurons (nerve cells) in the central nervous system (CNS). (c) Scientists at the Mayo Clinic in Rochester, Minnesota, have produced transgenic cats containing a gene that imparts resistance to feline immunodeficiency virus (FIV, the cat version of HIV). The resistance gene was fused to the GFP gene so that the two genes were incorporated together. Cats containing the resistance gene are readily detected by their fluorescence under blue light.

The protein itself wraps around the fluorescent group like a barrel and excludes it from solvent water. When the protein structure is disrupted so that this group can interact with water, its fluorescence quantum yield drops to zero because the energy of the excited state is lost to solvent vibrations rather than by fluorescence.

GFP is a fairly small and relatively stable protein. The gene for GFP can in many cases be joined to the genes of other proteins of interest, and the two proteins can be co-expressed in living organisms as a single "fusion" protein. Essentially, the GFP is carried by the protein of interest as a "molecular flashlight" so that the localization of the protein of interest can be seen by viewing a cell under a fluorescence microscope (**Fig. 15.12b**). Even genetically modified whole animals carrying GFP have been produced (**Fig. 15.12c**). It is also possible to produce mutant GFPs that give fluorescence of different colors, such as red and yellow.

The discovery and development of GFP as a biological tool was recognized with the 2008 Nobel Prize in Chemistry, which was awarded to Osamu Shimomura (1928–2018) of the Marine Biological Laboratory in Woods Hole, Massachusetts; Martin Chalfie (b. 1947) of Columbia University; and Roger Y. Tsien (1952–2016) of the University of California, San Diego.

Focused Problems

15.11 Refer to Fig. 15.11. At what pH does the fluorescence of fluorescein have the greater quantum yield, pH = 9 or pH = 5? How do you know?

15.12 One of the following compounds has an intense yellow fluorescence when irradiated with UV light. Which one do you think it is, and why?

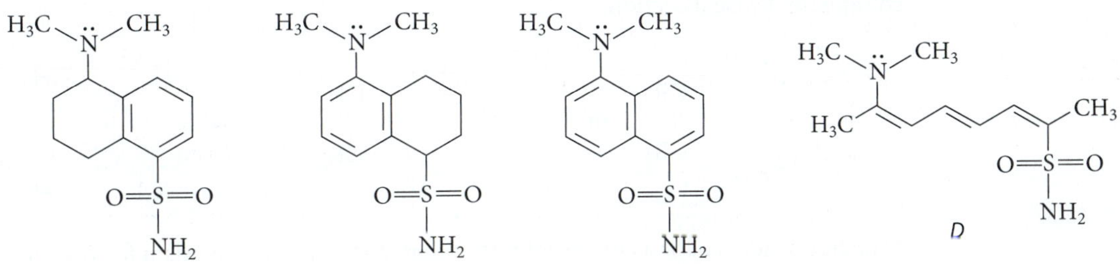

A *B* *C* *D*

15.3 THE DIELS–ALDER REACTION

A. Reaction of Conjugated Dienes with Alkenes

Conjugated dienes undergo several unique reactions. One of these was discovered in 1928, when two German chemists, Otto Diels (1876–1954) and Kurt Alder (1902–1958), showed that many conjugated dienes undergo addition reactions with certain alkenes or alkynes. The following reaction is typical:

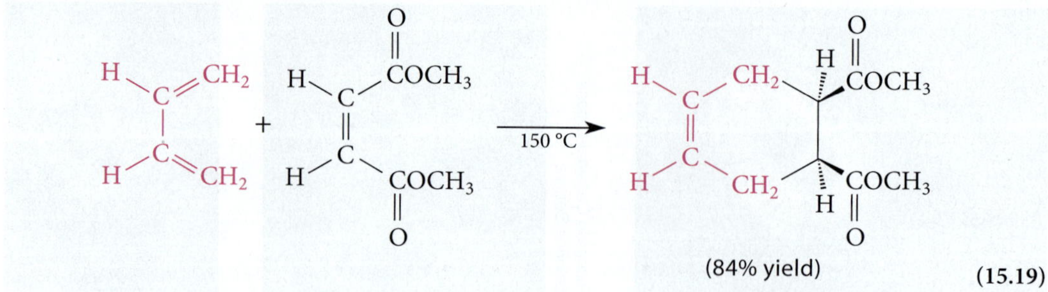

(15.19)

This type of reaction between a conjugated diene and an alkene is called the **Diels–Alder reaction**. For their extensive work on this reaction, Diels and Alder shared the 1950 Nobel Prize in Chemistry.

The Diels–Alder reaction is an example of a **cycloaddition reaction**—an addition reaction that results in the formation of a ring. Indeed, *the Diels–Alder reaction is an important method for making rings*, as the example in Eq. 15.19 illustrates.

The Diels–Alder reaction is also an example of a **1,4-addition** or **conjugate addition**. In such a reaction, addition occurs across the outer carbons (carbons 1 and 4) of the diene. *Conjugate addition is a characteristic type of reaction of conjugated dienes.* (Other conjugate additions are described in Secs. 15.4 and 22.9.) In the Diels–Alder reaction, conjugate addition also results in the formation of a double bond between carbons 2 and 3. (The numbers indicate the relative locations of the carbons involved in the addition; they have nothing to do with the numbering of the diene used in its substitutive nomenclature.)

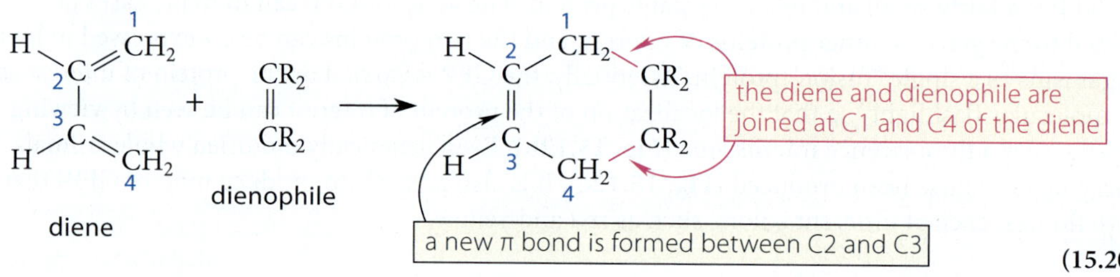

(15.20)

STUDY GUIDE LINK 15.1
A Terminology Review

When discussing the reactants in the Diels–Alder reaction, we employ the following terminology, which is illustrated in Eq. 15.20. The conjugated diene reactant is referred to simply as the *diene*, and the alkene with which it reacts is called the **dienophile** (literally, "diene-loving molecule"). 1,3-Butadiene is used as the diene for simplicity in Eq. 15.20, but as we show shortly, a wide variety of dienes can be used in this reaction.

Mechanistically, the Diels–Alder reaction occurs in a single step involving a cyclic flow of electrons. The curved arrows for this mechanism can be drawn in either a clockwise or a counterclockwise direction.

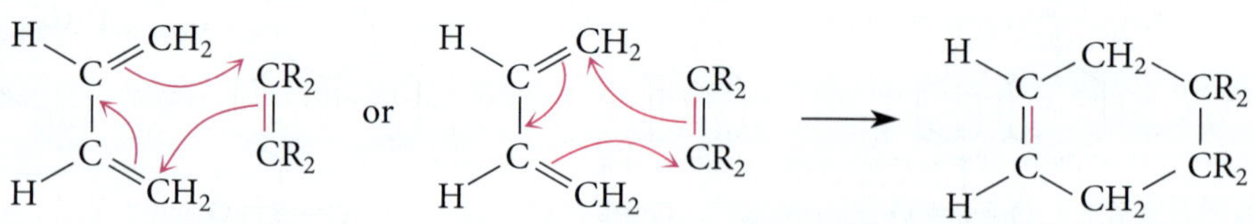

(15.21)

Pericyclic reactions as a class are discussed from the perspective of MO theory in Chapter 28.

(The best evidence for a concerted rather than a stepwise mechanism for the Diels–Alder reaction comes from the stereochemistry of the reaction, which we discuss in Secs. 15.3B and 15.3C.) A concerted reaction that involves a cyclic flow of electrons is called a **pericyclic reaction**. The Diels–Alder reaction is a pericyclic reaction, as is the hydroboration of alkenes (Sec. 5.6A). However, hydroboration is not a cycloaddition, because no ring is formed.

Some of the dienophiles that react most rapidly in the Diels–Alder reaction, as in Eq. 15.19, bear substituent groups such as esters (—CO_2R), nitriles (—CN), or certain other unsaturated, electronegative groups. However, these substituents are not strictly necessary because the reactions of many other alkenes can be promoted by heat or pressure. Some alkynes can also serve as dienophiles.

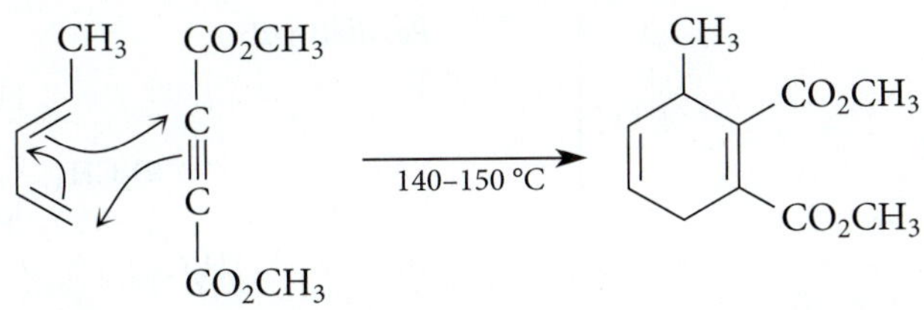

(15.22)

When a simple diene is used in the Diels–Alder reaction, a new ring is formed. When the diene is cyclic, a *second* ring is formed. In other words, the Diels–Alder reaction can be used to prepare certain *bicyclic compounds* (Sec. 7.6A).

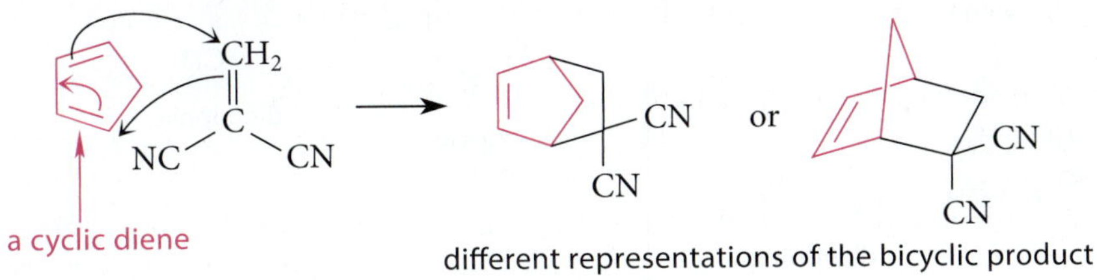

different representations of the bicyclic product (15.23)

Study Problem 15.1

Give the structure of the diene and dienophile that would react in a Diels–Alder reaction to give the following product:

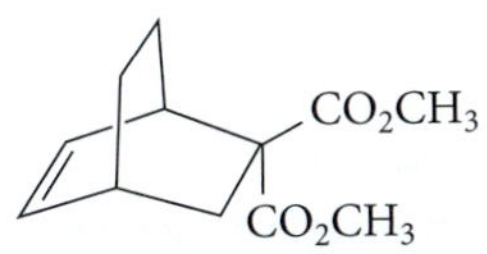

Solution In the product of a Diels–Alder reaction, the two carbons of the double bond and the two *adjacent carbons* originate from the diene. These carbons are numbered 1 through 4 in the following structure:

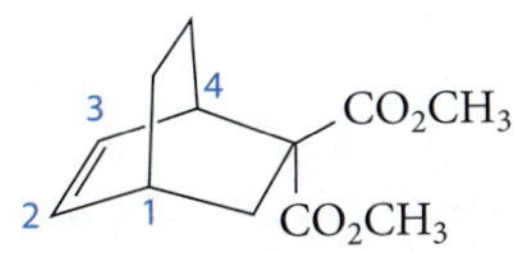

The two new single bonds formed in the reaction connect carbons 1 and 4 of the diene to the carbons of the dienophile double bond, which (because they are part of the same double bond) must be adjacent in the dienophile. This analysis reveals two possibilities, *A* and *B*, for the bonds formed in the Diels–Alder reaction:

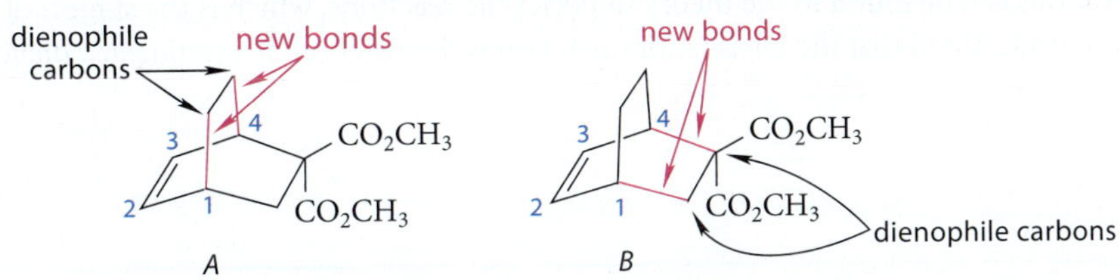

Because the product is bicyclic, the diene in either case is a *cyclic* diene. The double bonds in the diene are between carbons 1 and 2 and between carbons 3 and 4. To derive the starting materials in each case, follow these steps:

1. Disconnect the bonds between carbons 1 and 4 and their adjacent dienophile carbons.
2. Complete the diene structure by eliminating the double bond between carbons 2 and 3 and by adding the C1–C2 and C3–C4 double bonds.
3. Complete the dienophile structure by adding the double bond between its carbons.

By following these steps we find that the starting materials for possibilities *A* and *B* are as follows. (The carbon skeleton of the diene unit is first drawn exactly as it looks in the product, even though this is a distorted conformation, and then it is drawn in the more conventional way.)

Possibility A:

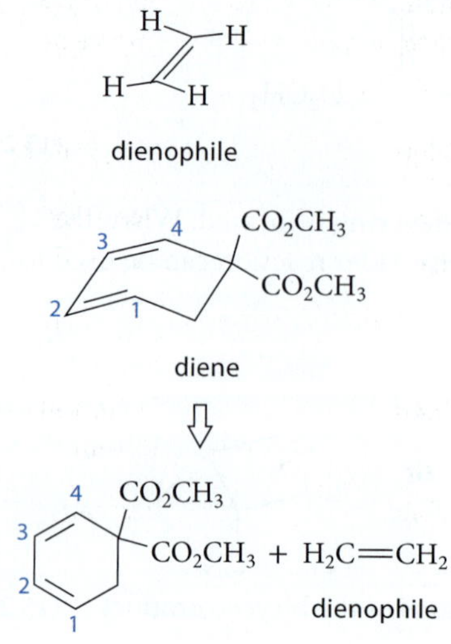

Possibility B:

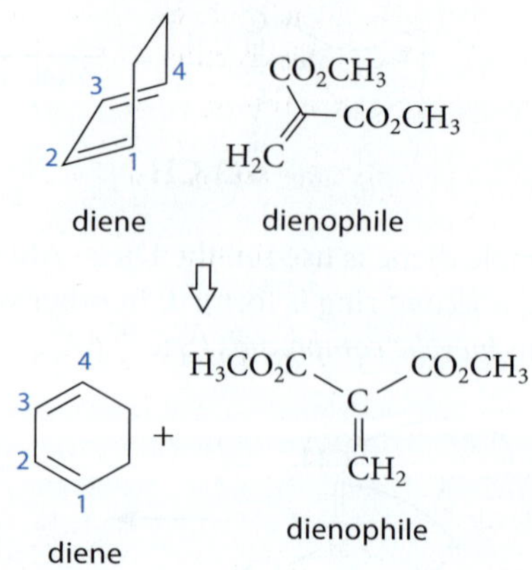

In principle, either combination *A* or *B* could serve as the starting materials in a Diels–Alder reaction. Recall, however, that dienophiles with ester groups (or other electronegative groups) react faster than those without such groups. For this reason, the reactants in *B* would be preferred.

It might seem that cycloaddition reactions between two alkenes or between two dienes could be possible:

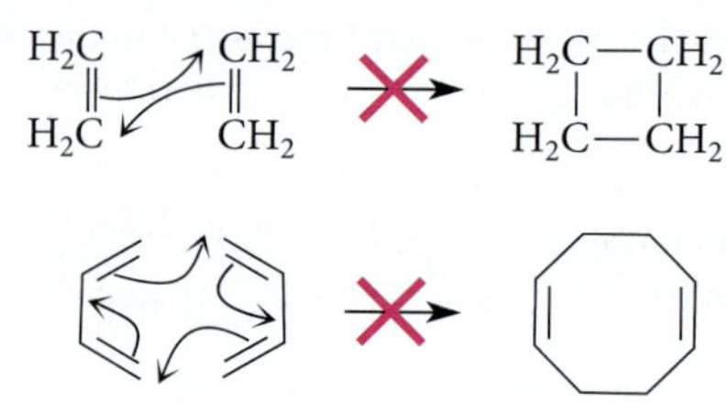

(15.24)

These reactions do *not* occur under the usual conditions. Although we might attribute the failure of the first reaction to ring strain in the product, there is more to it than that. The basis of these observations can be found in the theory of pericyclic reactions, which is the subject of Chapter 28. Be sure you understand that the Diels–Alder reaction is the reaction of a *conjugated diene* with an *alkene.*

Focused Problems

15.13 What products are formed in the Diels–Alder reactions of the following dienes and dienophiles?

(a)  and 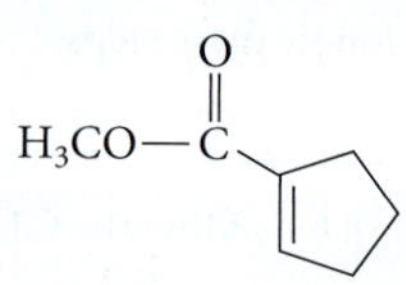

(b) 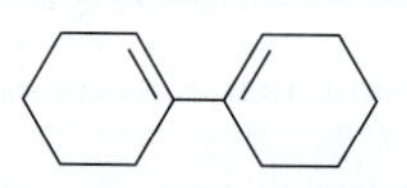 and 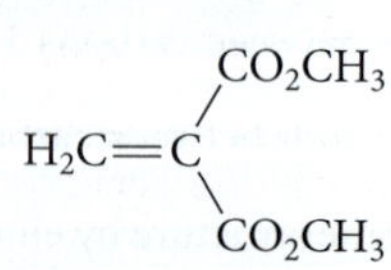

15.14 Give the diene and dienophile that would react in a Diels–Alder reaction to give each of the following products.

(a) 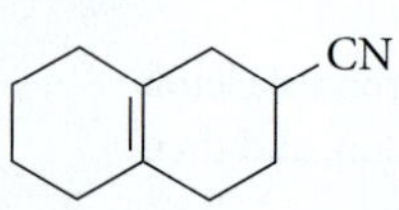

(b) 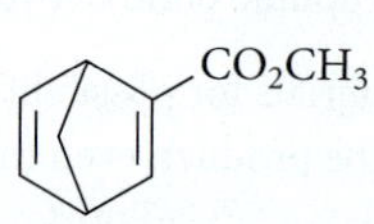

15.15 (a) What product would be expected from the Diels–Alder reaction of 1,3-butadiene as the diene and ethylene as the dienophile?

(b) This product is actually not observed under ordinary conditions because 1,3-butadiene reacts with itself faster than it does with ethylene. In this reaction, one molecule of 1,3-butadiene acts as the diene component and the other as the dienophile. Give the product of this reaction.

(c) How would you alter the reaction conditions to favor the formation of the product in part (a)?

15.16 (a) Explain why two constitutional isomers are formed in the following Diels–Alder reaction:

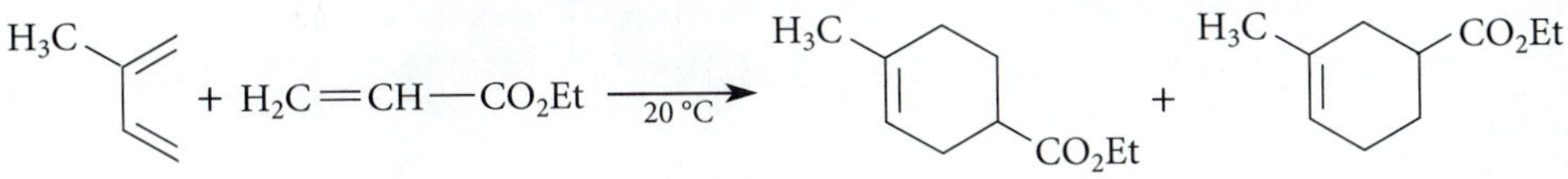

(84% of the product) (16% of the product)
(54% total yield)

(b) What two constitutional isomers could be formed in the following Diels–Alder reaction?

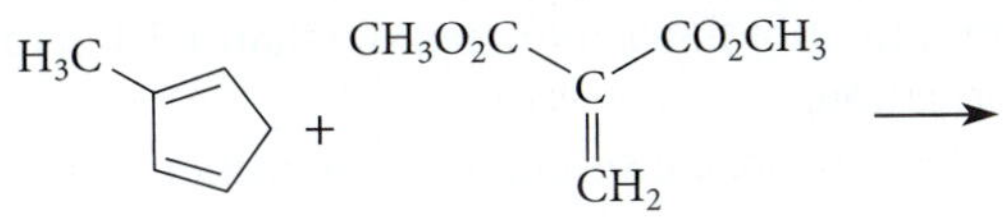

B. Effect of Diene Conformation on the Diels–Alder Reaction

Dienes that are "locked" into *s*-trans conformations are unreactive in Diels–Alder reactions:

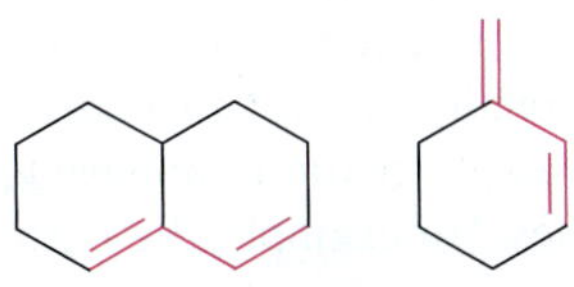

"locked" *s*-trans dienes;
unreactive in Diels–Alder reactions

The reason is that if such dienes were to form Diels–Alder products, the *s*-trans single bond of the diene would become a trans double bond in the Diels–Alder product. This means that the Diels–Alder product would contain a trans double bond in a six-membered ring. For example, consider the following reaction:

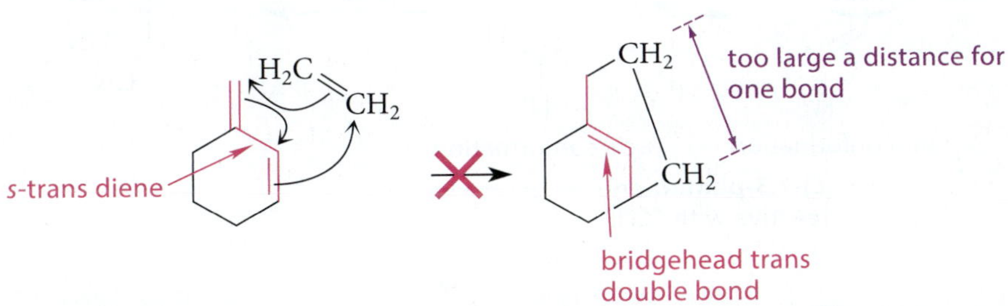

(15.25)

The product is a bicyclic compound containing a bridgehead double bond. As discussed in Sec. 7.6C, the bridgehead double bond (red in Eq. 15.25) has trans stereochemistry within one of the rings joined at the bridges, and therefore the product violates Bredt's rule and is too strained to exist. (For a graphic demonstration, try building a model of the product, but don't break your models.)

In contrast, dienes locked into *s*-cis conformations are considerably more reactive than the corresponding noncyclic dienes:

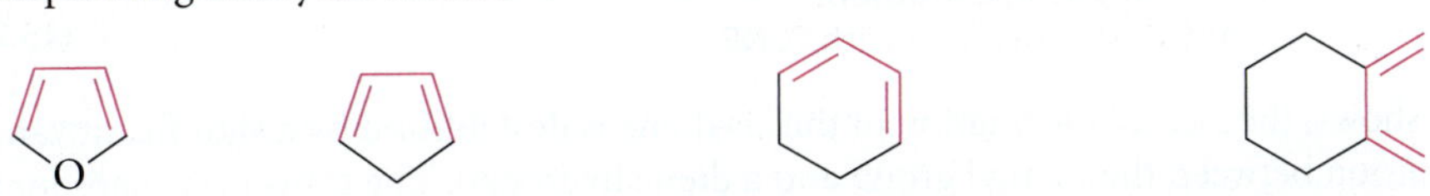

furan **1,3-cyclopentadiene** **1,3-cyclohexadiene** **1,2-dimethylenecyclohexane**

all are "locked" *s*-cis dienes;
all are reactive in the Diels–Alder reaction (15.26)

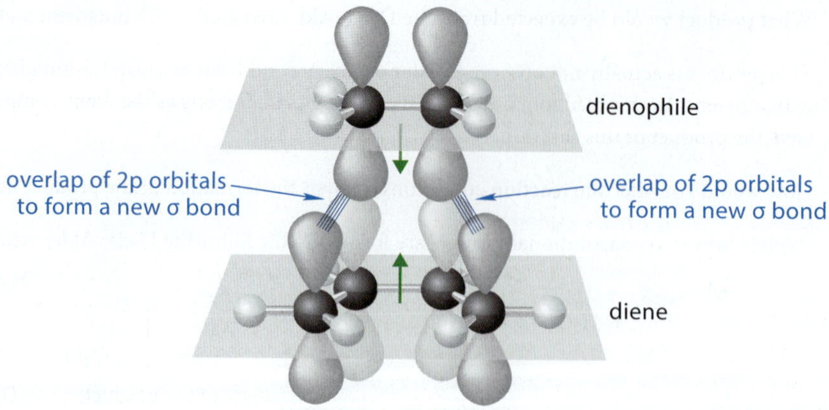

FIGURE 15.13 In the transition state for the Diels–Alder reaction, the diene (1,3-butadiene in this example) and the dienophile (ethylene in this example) approach in parallel planes (as shown by the green arrows) so that the 2p orbitals of the dienophile overlap with the 2p orbitals on carbons 1 and 4 of the diene to form the two new σ bonds. The developing overlap is indicated with blue lines. Notice that the diene is in an *s*-cis conformation.

For example, 1,3-cyclopentadiene, which is locked in an *s*-cis conformation, reacts with typical dienophiles hundreds of times more rapidly than 1,3-butadiene, which exists primarily in the *s*-trans conformation.

These observations are consistent with a transition state in which the diene component of the reaction has assumed an *s*-cis conformation. This transition state is shown in **Fig. 15.13** for the reaction of 1,3-butadiene and ethylene. In this transition state, the diene and the dienophile approach in parallel planes. The 2p orbitals on the dienophile interact with the 2p orbitals on the outer carbons of the diene to form the new σ bonds. Because 1,3-butadiene prefers the *s*-trans conformation (see Fig. 15.1, Sec. 15.1B), the energy required for it to assume the *s*-cis conformation in the transition state becomes part of the energy barrier for the reaction. In contrast, a diene that is locked by its structure into an *s*-cis conformation, such as 1,3-cyclopentadiene, does not have this additional energy barrier to climb before it can react; therefore it reacts more rapidly.

The importance of the *s*-cis diene conformation can have some fairly drastic consequences for the reactivity of some noncyclic dienes. For example, the *E* isomer of 1,3-pentadiene reacts 12,600 times more rapidly than the *Z* isomer of the same diene with tetracyanoethylene (TCNE), a very reactive dienophile:

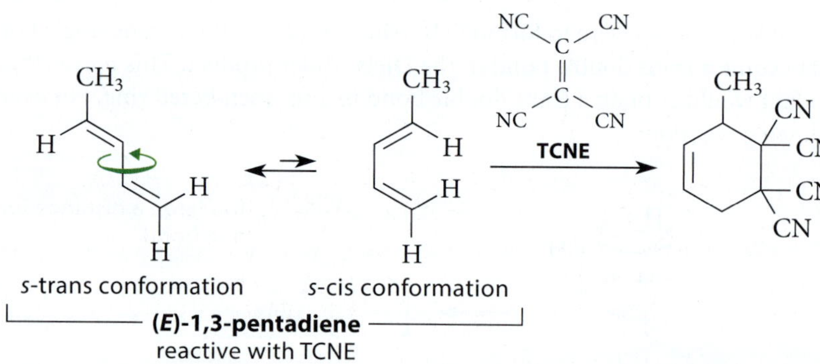

(15.27a)

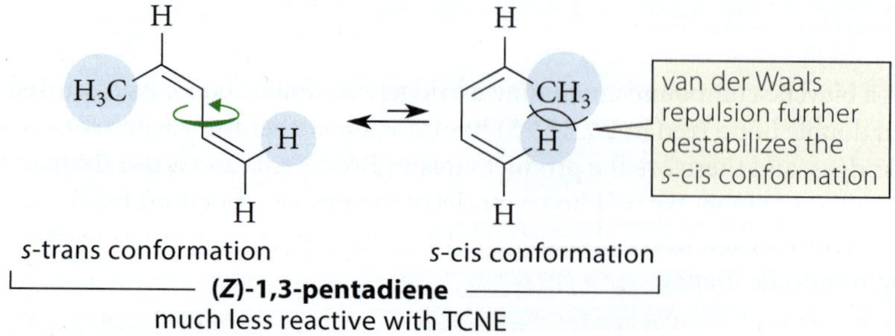

(15.27b)

As Eq. 15.27b shows, the *s*-cis conformation of the cis diene is destabilized by a significant van der Waals repulsion between the methyl group and a diene hydrogen. The transition states for the Diels–Alder reactions of this diene, which require an *s*-cis conformation, are destabilized by the same effect. Consequently, the Diels–Alder reactions of (*Z*)-1,3-pentadiene are much slower than the corresponding reactions of (*E*)-1,3-pentadiene, in which the destabilizing repulsion in its *s*-cis conformation is between hydrogens and is much less severe.

Focused Problems

15.17 A mixture of 0.1 mol of (2*E*,4*E*)-2,4-hexadiene and 0.1 mol of (2*E*,4*Z*)-2,4-hexadiene was allowed to react with 0.1 mol of TCNE. After the reaction, the unreacted diene was found to consist of only one of the starting 2,4-hexadiene isomers. Which isomer did not react? Explain.

15.18 Complete the following Diels–Alder reaction.

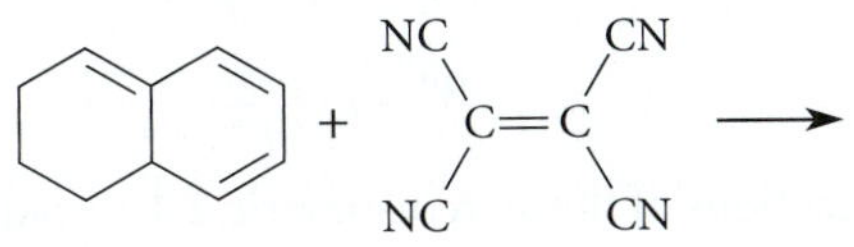

C. Stereochemistry of the Diels–Alder Reaction

If the Diels–Alder reaction takes place in a single step without reactive intermediates, and if the transition-state picture of Fig. 15.13 is correct, then the diene should undergo a syn addition to the dienophile. Likewise, the dienophile should undergo a 1,4-syn addition to the diene. Each component adds to the other at *one face* of the π system.

The stereochemistry of the Diels–Alder reaction is completely consistent with these predictions. If we use a dienophile that is a cis alkene, groups that are cis in the alkene starting material are also cis in the product.

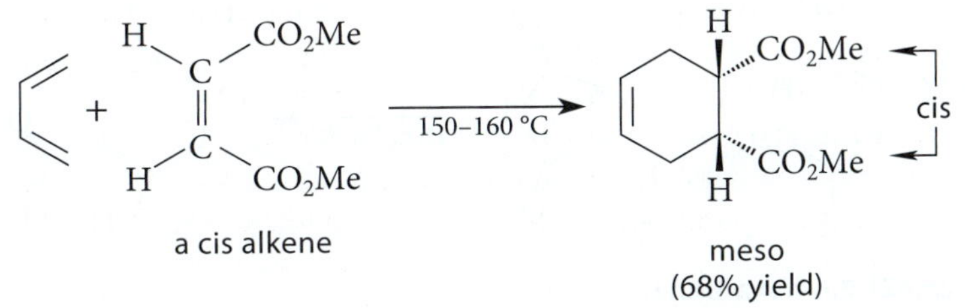

(15.28a)

Use of the trans isomer of this dienophile gives the complementary result:

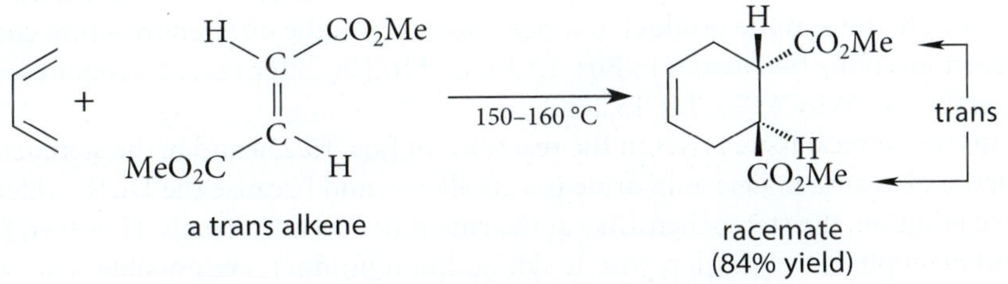

(15.28b)

Although one enantiomer of the product is shown in Eq. 15.28b, the product is the racemate, because both starting materials are achiral (Sec. 7.7A).

Syn addition to the diene is revealed if the terminal carbons of the diene unit are stereocenters. To assist in the analysis of stereochemistry, we first draw the diene in its *s*-cis conformation and then classify the groups at the terminal carbons as inner substituents (R^i) or outer substituents (R^o):

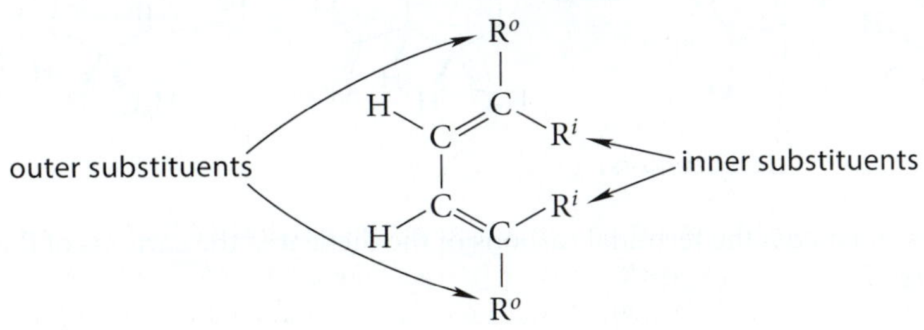

(15.29a)

A syn addition requires that in the Diels–Alder product, the two inner substituents always have a cis relationship; the two outer substituents are also cis; and an inner substituent on one carbon is always trans to an outer substituent on the other.

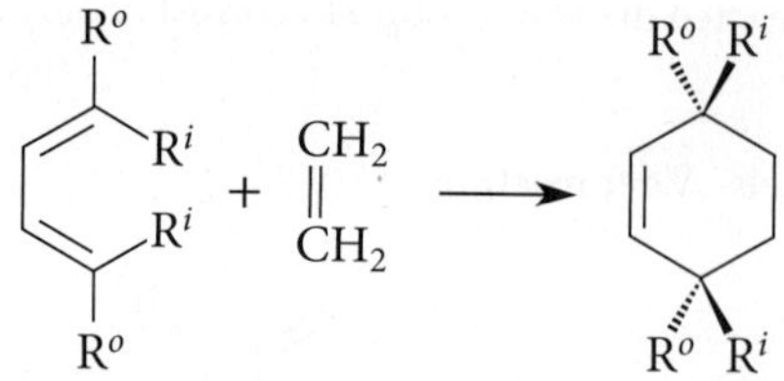

(15.29b)

The following reactions of the stereoisomeric 2,4-hexadienes with the dienophile maleic anhydride demonstrate these points.

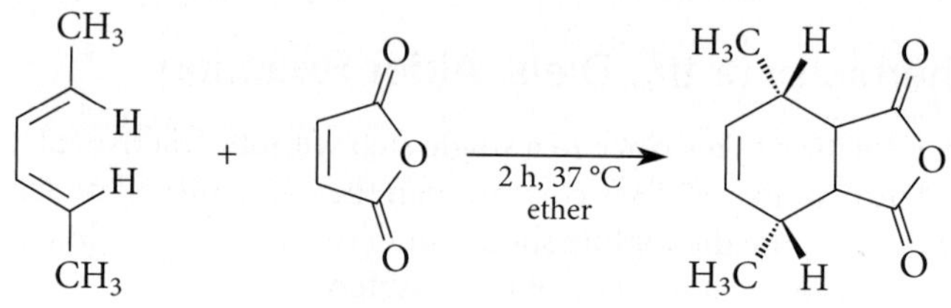

(2*E*,4*E*)-2,4-hexadiene **maleic anhydride**

(15.30a)

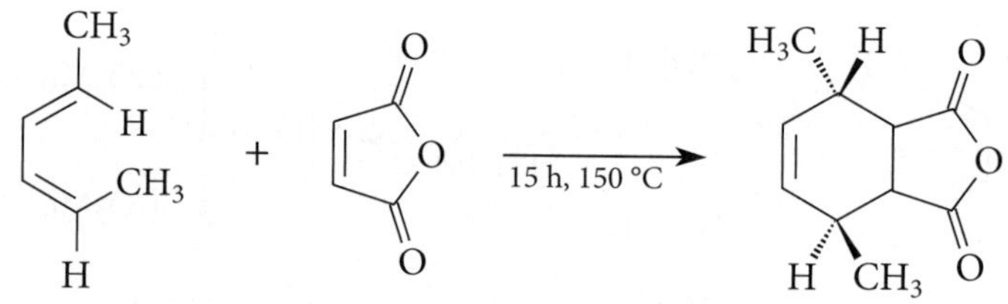

(2*E*,4*Z*)-2,4-hexadiene

(15.30b)

In Eq. 15.30a, the methyl groups in the diene are both outer substituents, and they are cis in the product. In Eq. 15.30b, one methyl group in the diene is outer and the other is inner; consequently, they are trans in the product. (Notice, incidentally, the different reaction conditions required for reactions of the two dienes in Eqs. 15.30a and b. The latter reaction requires *much* more drastic conditions. Why?) (See Eq. 15.27b.)

One other stereochemical issue arises in the reactions of Eqs. 15.27a and b: the stereochemistry at the ring junction. Because maleic anhydride is a cis alkene, and because the Diels–Alder reaction is a syn addition, the stereochemistry at the ring junction must be cis. However, for a given diene and dienophile, two diastereomeric syn-addition products are possible. The reaction of Eq. 15.31 illustrates this point:

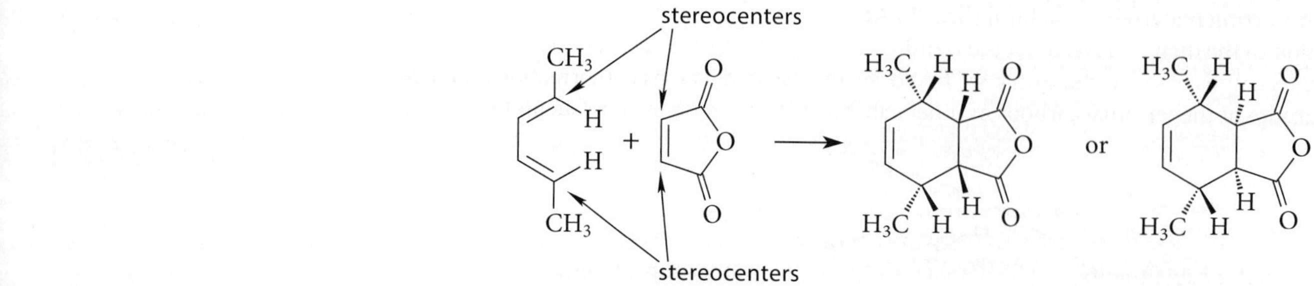

(15.31)

This issue arises when *both* the terminal carbons of the diene *and* the carbons of the dienophile are stereocenters.

We now classify these two possibilities with a more general equation in which a cis alkene reacts with a diene:

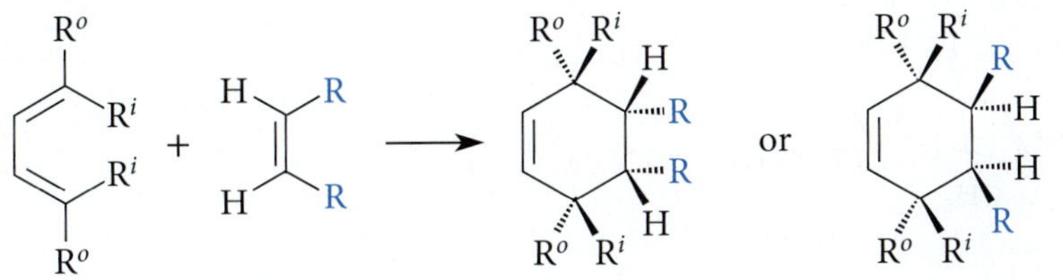

endo product exo product (15.32)

Following the diagram in Eq. 15.29a, we have drawn the diene in its *s*-cis conformation and have labeled the groups at the terminal carbons as outer or inner substituents. The product in which the alkene substituents R (shown in blue) are cis to the outer diene substituents R^o is said to have **endo** stereochemistry. The product in which the alkene substituents R are trans to the outer diene substituents R^o is said to have **exo** stereochemistry. (The terms *endo* and *exo* are from Greek roots meaning "inside" and "outside," respectively.) Because they are diastereomers, the endo and exo products are typically formed in different amounts, as in this example:

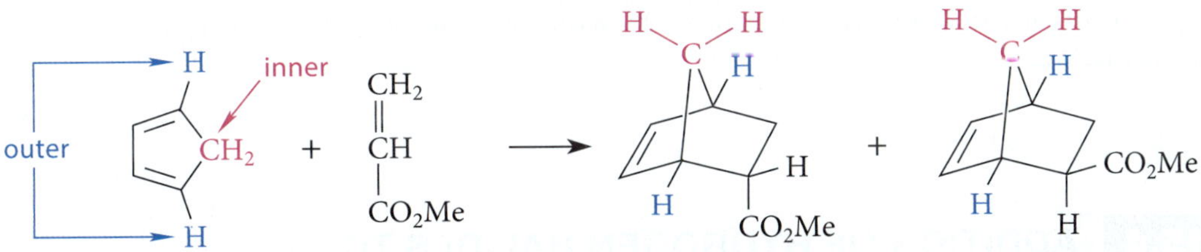

1,3-cyclopentadiene endo product (76%) exo product (24%)

(15.33)

Be sure you see the correspondence between Eq. 15.32 and Eq. 15.33. The CH_2 group of the diene (red) represents the inner groups R^i (tied together in one group as part of the ring); the hydrogens in blue are the outer groups R^o. In the predominant, or endo, product, the —CO_2Me group is cis to R^o and trans to R^i.

Focused Problems

15.19 Give the products formed when each of the following pairs reacts in a Diels–Alder reaction; show the relative stereochemistry of the substituent groups where appropriate. In part (b), show both exo and endo products, and label them.

(a)

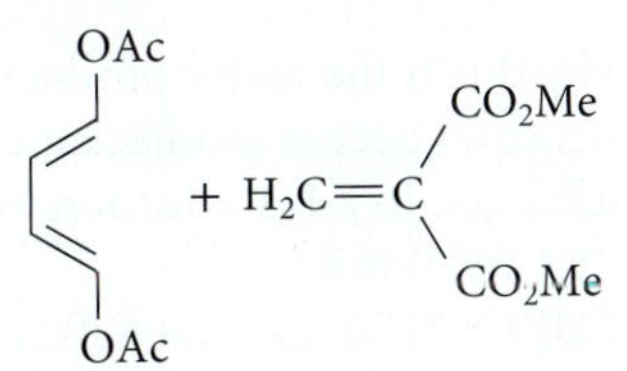

(—OAc = acetoxy = $-O-C(=O)CH_3$)

(b)

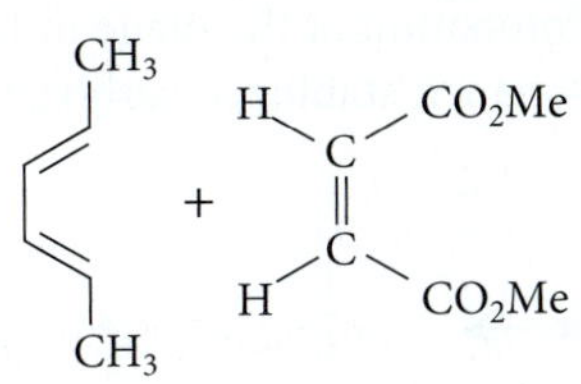

(c)

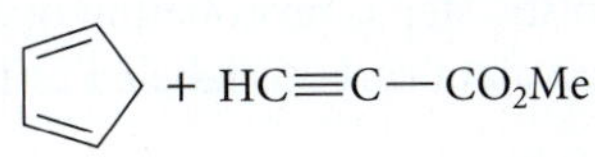

15.20 Give the structures of the starting materials that would yield each of the following compounds in Diels–Alder reactions. Pay careful attention to stereochemistry, where appropriate.

(a)

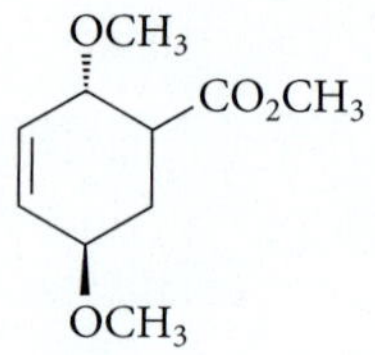

(b)

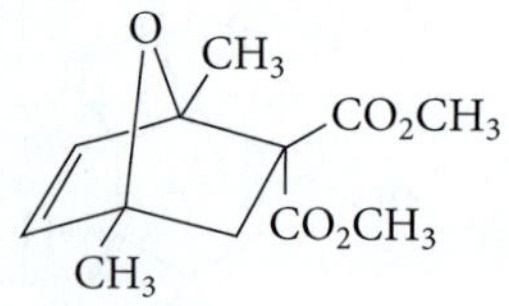

(c)

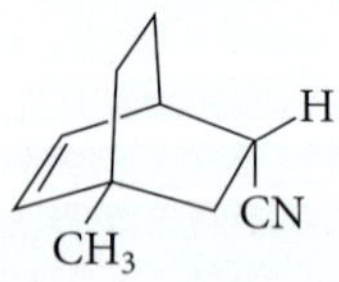

(d)

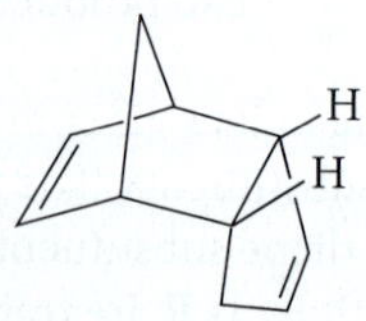

15.21 (a) In the products of Eq. 15.30a, the observed stereochemistry at the ring fusion is not specified. Show this stereochemistry, assuming that the Diels–Alder reaction gives the endo product.

(b) Sketch diagrams like the one in Fig. 15.13 (without the orbitals) that shows the approach of the diene and dienophile leading to both endo and exo products in part (a). Pay careful attention to the relative positions of substituent groups.

15.4 ADDITION OF HYDROGEN HALIDES TO CONJUGATED DIENES

A. 1,2- and 1,4-Additions

Conjugated dienes, like ordinary alkenes (Sec. 4.7), react with hydrogen halides; however, conjugated dienes give two types of addition product. (The numbers are for reference in describing the addition; they are not associated with nomenclature.)

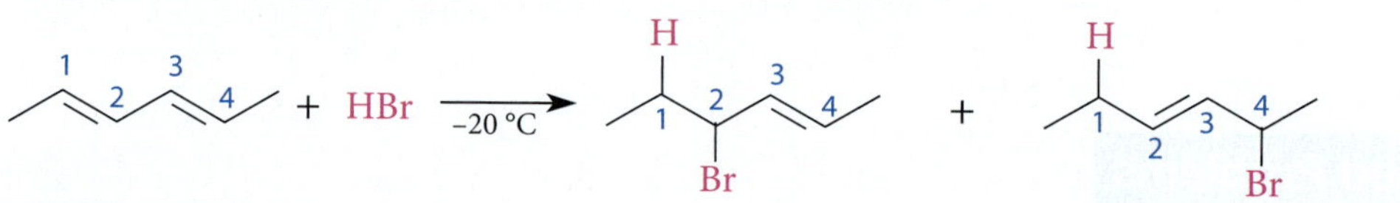

1,2-addition product (65%)

1,4-addition product (conjugate-addition product) (35%)

(15.34)

The major product is a *1,2-addition* product. (We address why this is the major product in Sec. 15.4B.) **1,2-Addition** means that addition (of HBr in this case) occurs at adjacent carbons. The minor product results from *1,4-addition*, or *conjugate addition*. In a **1,4-addition**, or **conjugate addition**, addition occurs to carbons that have a 1,4-relationship.

The 1,2-addition reaction is analogous to the reaction of HBr with an ordinary alkene. But how can we account for the conjugate-addition product? As in HBr addition to ordinary alkenes, the first mechanistic step is protonation of a double bond. Protonation of the diene in Eq. 15.34 at either of the equivalent carbons labeled as 1 and 4 gives a resonance-stabilized carbocation:

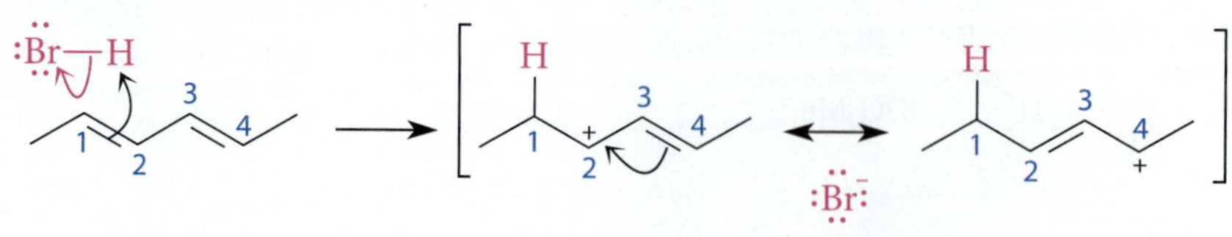

(15.35a)

The resonance structures for this carbocation show that the positive charge in this ion is not localized, but is instead *shared* by two different carbons. *Two constitutional isomers are formed in Eq. 15.34 because the bromide ion can react at either of the electron-deficient carbons:*

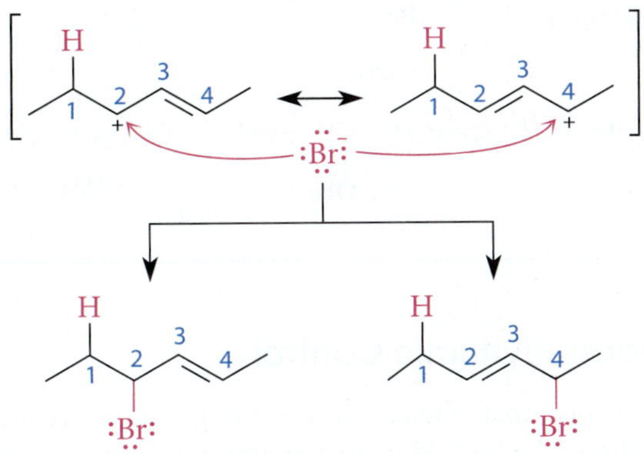

(15.35b)

Protonation at carbon-2 of the diene would give a different carbocation, which would react with bromide ion to give an alkyl halide that is different from those obtained experimentally:

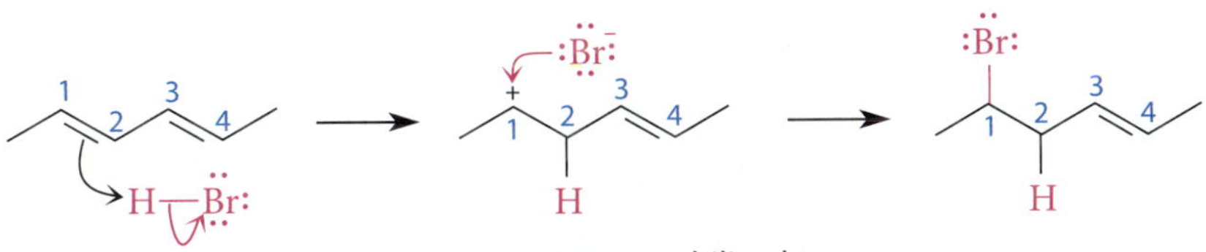

not a resonance-stabilized carbocation

(15.36)

(Compare the product of this reaction with the 1,2-addition product in Eq. 15.34 or 15.35.)

The course of addition to conjugated alkenes is suggested by Hammond's postulate (Sec. 4.8D), which predicts that the reaction pathway involving the more stable carbocation occurs more rapidly. Because the carbocation in Eq. 15.35a is more stable, it is formed more rapidly; therefore, the products derived from this carbocation are the ones observed. Although both possible carbocations are secondary, the carbocation in Eq. 15.35a is *resonance-stabilized*, but the carbocation in Eq. 15.36 is not:

more stable carbocation (Eq. 15.35a); formed more rapidly

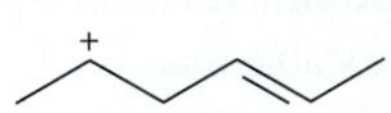

less stable carbocation (Eq. 15.36); not formed (15.37)

In other words, resonance accounts for the greater stability of the carbocation intermediate that is formed.

Focused Problems

15.22 Use the reaction mechanism, including the resonance structures of the carbocation intermediates, to predict the products of the following reactions.

(a) Addition of HCl to 1,3-butadiene

(b) S_N1 solvolysis of 3-chloro-1-methylcyclohexene in ethanol

15.23 Suggest a mechanism for each of the following reactions that accounts for both products.

(a) 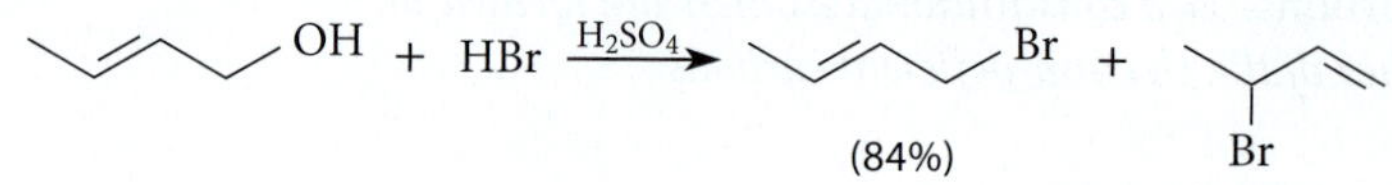

(84%) (16%)

(b) $(CH_3)_2C{=}CH{-}CH(Cl){-}CH_3 \xrightarrow{H_2O/\text{acetone}} (CH_3)_2C{=}CH{-}CH(OH){-}CH_3 + (CH_3)_2C(OH){-}CH{=}CH{-}CH_3 + HCl$

B. Kinetic and Thermodynamic Control

It would be reasonable to expect that when a reaction can give products that differ in stability, the more stable product should be formed in greater amount. However, sometimes the *less* stable product is formed in larger amount. Consider, for example, the addition of hydrogen halides to conjugated dienes. When a conjugated diene reacts with a hydrogen halide to give a mixture of 1,2- and 1,4-addition products, the 1,2-addition product predominates at low temperature:

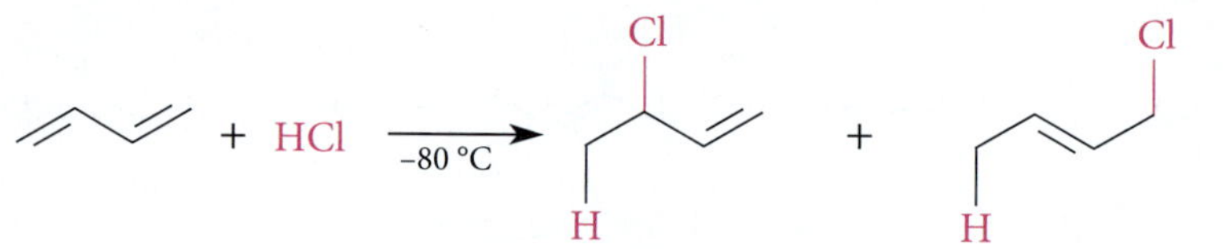

1,2-addition product (75–80%) 1,4-addition product (20–25%) **(15.38)**

In Sec. 4.5B we showed that alkenes with internal double bonds are more stable than their isomers with terminal double bonds, because the internal double bonds have more alkyl branches. Therefore, in Eq. 15.38, *the major product is the less stable one.* This can be demonstrated experimentally by bringing the two alkyl halide products to equilibrium with heat and Lewis acids:

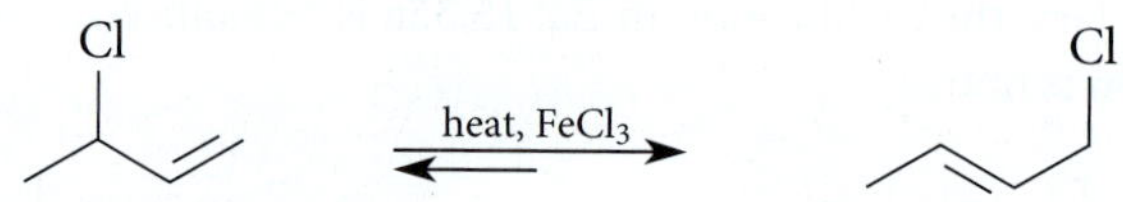

1,2-addition product the minor product at equilibrium 1,4-addition product the major product at equilibrium **(15.39)**

Because the more stable isomer always predominates in an equilibrium (Sec. 3.5), the result in Eq. 15.39 shows that the 1,4-addition product is more stable than the 1,2-addition product, as expected.

When the less stable product of a reaction is the major product, then two things must be true. First, the less stable product *must form more rapidly* than the other products. As noted in Sec. 4.8, a reaction in which two products form from the same starting material is in reality two competing reactions. Consequently, the reaction that forms the less stable product is faster. Second, the products *must not come to equilibrium* under the reaction conditions, because otherwise the more stable compound would be present in larger amount. So, in the addition of HCl to conjugated dienes, the predominance of the less stable product (Eq. 15.38) shows that 1,2-addition, which gives the less stable product, is faster than 1,4-addition:

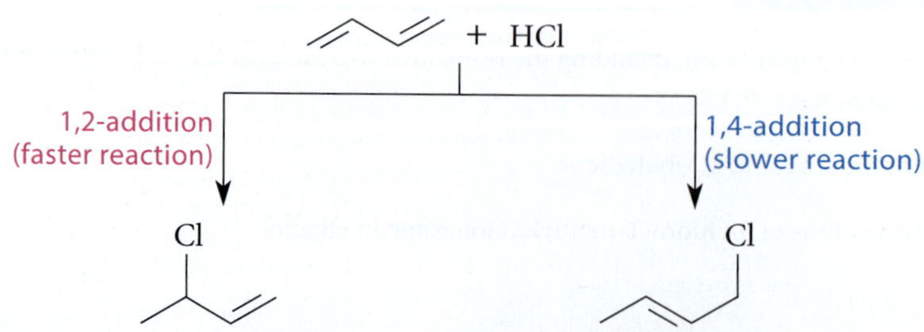

less stable product more stable product **(15.40)**

When the products of a reaction do not come to equilibrium under the reaction conditions, the reaction is said to be **kinetically controlled**. In a kinetically controlled reaction, the relative proportions of products are controlled solely by the relative rates at which they are formed. Therefore, the addition of hydrogen halides to conjugated dienes is a kinetically controlled reaction. On the other hand, if the products of a reaction come to equilibrium under the reaction conditions, the reaction is said to be **thermodynamically controlled**.

It is possible that a given kinetically controlled reaction might give about the same distribution of products as would be obtained if the products were allowed to come to equilibrium. However, it is *impossible* for a thermodynamically controlled reaction to give a product distribution other than the equilibrium distribution. So, when we obtain a product distribution that is clearly different from that obtained at equilibrium (as occurs in the addition of HCl to conjugated dienes), we know immediately that the reaction must be kinetically controlled.

An Analogy for Kinetic Control

Marius G/Shutterstock

Imagine a very disoriented steer stumbling randomly around a pasture with a shallow watering hole and a deep well with a high fence around it. Where is he likely to end up? Certainly the deep well is the state of lowest potential energy. However, because of the fence around the well, it is simply less likely that the animal will fall into the well; he is much more likely to wander into the watering hole.

Now, if you imagine a large herd of similarly disoriented steers staggering around the same (very large) pasture, you should get a reasonably good image of kinetic control. Most of the animals wander into the watering hole, even though this is not the state of lowest potential energy.

Likewise with molecules: It is possible for the formation of a more stable product to have a greater standard free energy of activation (a greater energy barrier) than the formation of a less stable product. In such a case, the less stable product forms more rapidly and in greater amount.

In hydrogen halide addition to a conjugated diene, the first and rate-limiting step in the formation of both 1,2- and 1,4-addition products is the same—protonation of the double bond. Consequently, the product distribution must be determined by the relative rates of the *product-determining steps* (Sec. 9.6B): the nucleophilic reaction of the halide ion at one or the other of the electron-deficient carbons of the allylic carbocation intermediate.

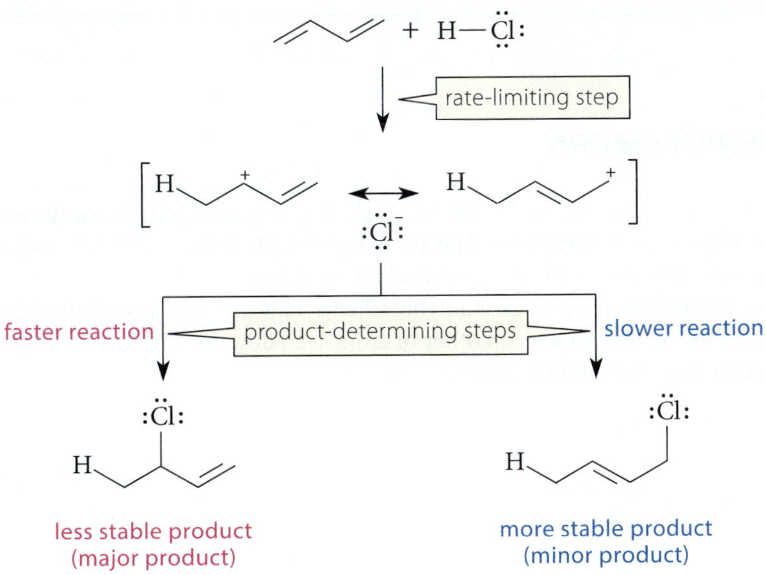

(15.41)

Why is the 1,2-addition product formed more rapidly? The reaction is typically carried out in solvents in which the HCl is not dissociated. In other words, the diene reacts with *undissociated* HCl. Consequently, the carbocation and its chloride counter-ion, when first formed, exist as an *ion pair* (see Fig. 8.6, Sec. 8.6F). That is, the chloride ion and the carbocation are closely associated. The chloride ion simply finds itself closer to the positively charged carbon adjacent to the site of protonation than to the other carbon. Addition is completed, therefore, at the nearer site of positive charge, giving the 1,2-addition product.

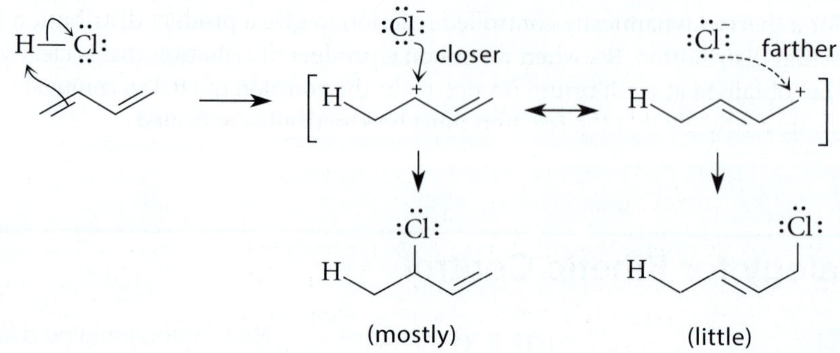

(15.42)

(The elegant experiment that suggested this explanation is described in Problem 15.74.)

The reason for kinetic control varies from reaction to reaction. Whatever the reason, the relative amounts of products in a kinetically controlled reaction are determined by the relative free energies of the *transition states* for each of the product-determining steps and *not* by the relative free energies of the products.

Focused Problem

15.24 (a) Give the structures of the 1,2- and 1,4-addition products that would be formed in the reaction of HBr with (*E*)-3-methyl-1,3-pentadiene? (*Hint:* Which double bond would be protonated more rapidly? Think about carbocation stabilities, and apply Hammond's postulate.)

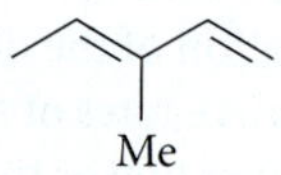

(*E*)-3-methyl-1.3-pentadiene

(b) Which of the alkyl bromide products in part (a) will predominate if they are brought to equilibrium? Why?

Chemistry in the Real World

15.5 DIENE POLYMERS

1,3-Butadiene is one of the most important raw materials of the synthetic rubber industry, which includes the production of automobile tires. The annual global production of 1,3-butadiene is about 25 billion pounds. It is also used in the production of nylon.

1,3-Butadiene can be polymerized to give polybutadiene. This polymer can result either from 1,4-addition of butadiene molecules to each other or from 1,2-addition. The 1,4-addition polymers can contain cis or trans double bonds.

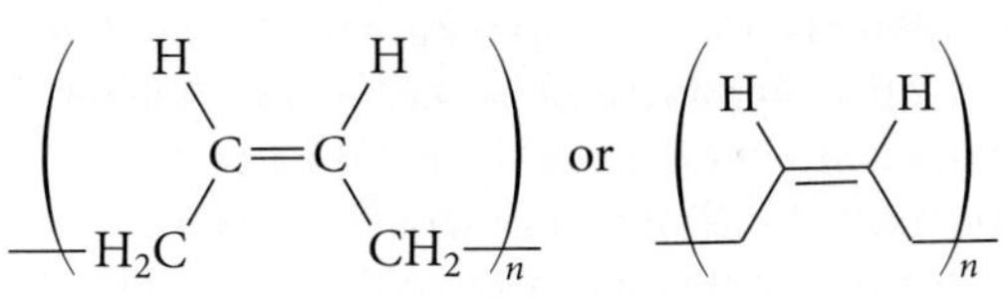

1,4-addition polymer
(cis double bond)

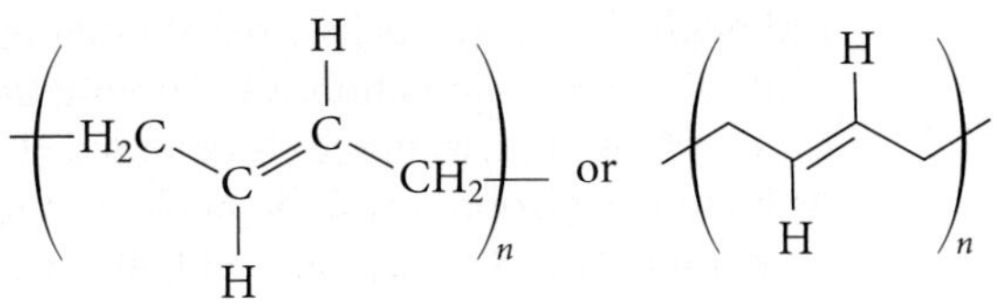

1,4-addition polymer
(trans double bond)

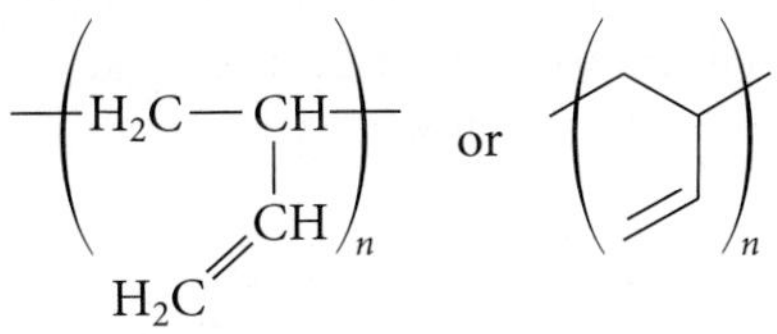

1,2-addition polymer
("vinyl" polybutadiene)

(15.43)

These different types of polymer linkages, although shown separately in the structures in Display 15.43, can be present together in various proportions in a given polymer molecule. The most useful butadiene polymers contain mostly cis double bonds. This type of polymer is produced by the action of transition-metal organometallic catalysts of a type that we present in Sec. 18.5. (The cis double bond is a consequence of the way that the π bonds of the diene interact with the catalyst.) Different types of catalysts can be used to give the different types of diene polymers.

1,3-Butadiene can also be polymerized together with styrene ($H_2C{=}CH{-}Ph$), usually in about a 3:1 ratio, to give another type of synthetic rubber called *styrene–butadiene rubber* (SBR), most of which is used for tires and tread rubber.

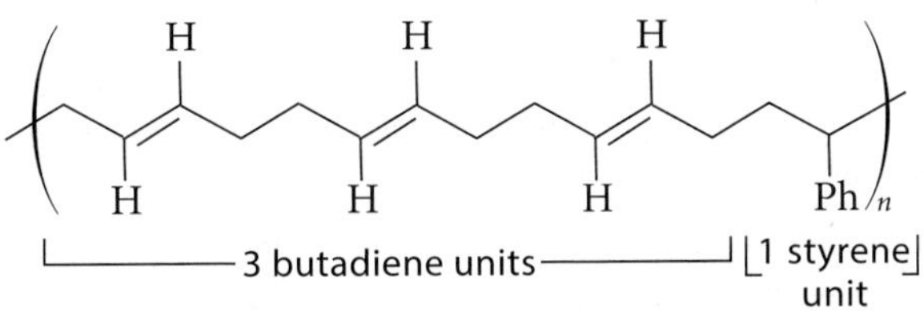

styrene–butadiene rubber (SBR)
(all-trans diene units shown)

(15.44)

As with polybutadiene, the butadiene units in SBR can be polymerized in either a 1,4- or 1,2-addition, and the 1,4-addition polymer can contain cis or trans double-bond stereochemistry. The example here is simplified because it shows an SBR unit with only trans double bonds, which is the predominant type of linkage in commercial SBR. For example, one typical formulation of SBR contains 54.5% trans double bonds in the butadiene units, 9% cis double bonds in the butadiene units, 13% "vinyl" units (resulting from 1,2-addition of butadiene), and 23.5% styrene units. Moreover, the styrene units can occur at random. SBR is produced either by organometallic catalysis or by free-radical polymerization (Sec. 10.3 and Focused Problem 15.25) in an aqueous emulsion.

SBR is an example of a **copolymer**: a polymer produced by the simultaneous polymerization of two or more monomers. The annual global demand for styrene–butadiene copolymers is more than 15 billion pounds and is growing rapidly as the demand for automobiles increases around the globe.

Natural rubber is (*Z*)-polyisoprene, another diene polymer:

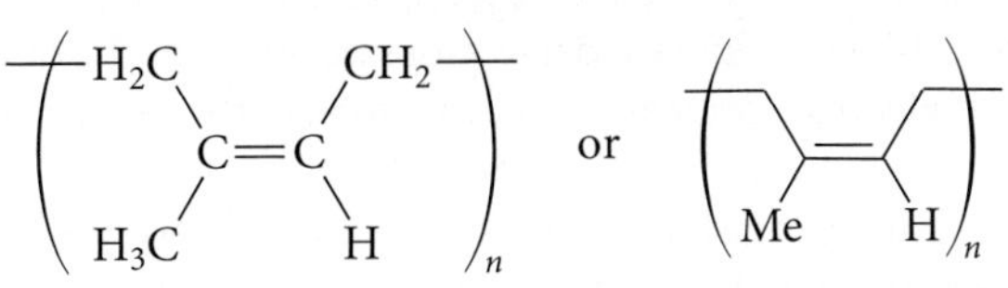

(*Z*)-polyisoprene (natural rubber)

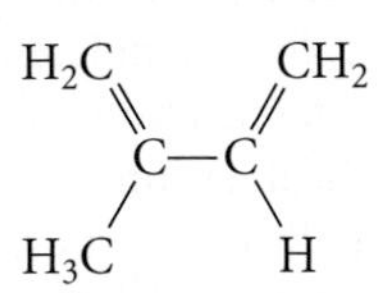

2-methyl-1,3-butadiene (isoprene)

(15.45)

Although it is conceptually a diene polymer, natural rubber is not made in nature from isoprene. (The biosynthesis of naturally occurring isoprene derivatives is discussed in Sec. 17.6B.) Rubber hydrocarbon (polyisoprene) is obtained as a 40% aqueous emulsion from the rubber tree. Although polyisoprene can be made synthetically, the natural material is generally preferred for economic reasons. Chemists and botanists are investigating the possibility of cultivating other hydrocarbon-producing plants that could become hydrocarbon sources of the future.

Crude natural rubber, SBR, and other diene polymers do not have adequate mechanical durability for their commercial applications. For this reason, they are subjected to a process called *vulcanization*. In this process, discovered in 1840 by Charles Goodyear (1800–1860), the rubber is kneaded and heated with sulfur. The sulfur forms disulfide crosslinks between the polymer chains, which can be represented schematically as follows:

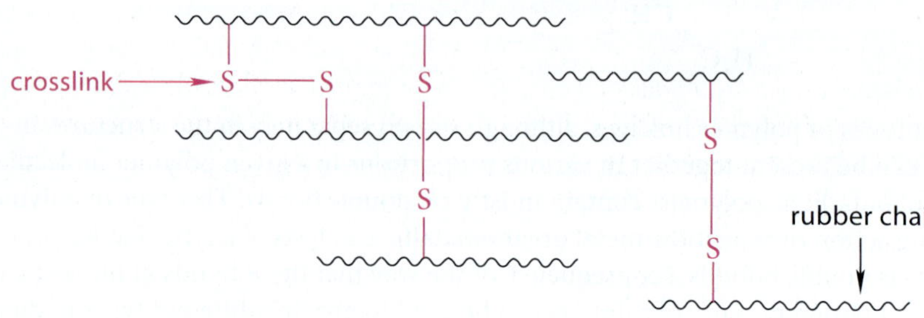

(15.46)

The crosslinks increase the rigidity and strength of the polymer at the cost of some flexibility.

Focused Problem

15.25 An initiation step (Eq. 10.8, Sec. 10.1C) in the free-radical co-polymerization of styrene and 1,3-butadiene is the free-radical addition of a peroxide-derived radical to a double bond of 1,3-butadiene:

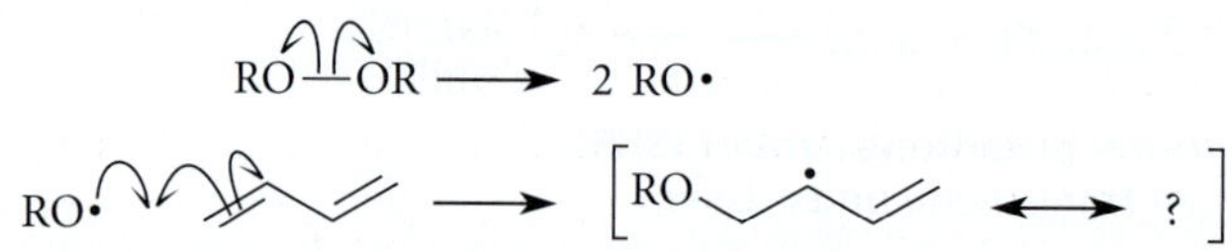

(a) Use the fishhook curved-arrow notation to derive the missing resonance structure.

(b) Use this resonance structure as part of a fishhook mechanism for free-radical co-polymerization to form SBR. (Show the incorporation of one diene unit in a 1,4-addition, one diene unit in a 1,2-addition, and one styrene unit.)

(c) Suggest a reason that the diene part of the polymer has mostly trans double-bond stereochemistry.

15.6 THE CONNECTION BETWEEN RESONANCE AND STABILITY

In Sec. 1.5, we stated that resonance is a stabilizing effect. We've shown other examples in which the resonance stabilization of a carbocation intermediate determines the course of the reaction. (See, for example, Eq. 15.35a, Sec. 15.4A.) Now we delve more deeply into the question of resonance stabilization: What is it about electron delocalization that lowers the energy of a molecule, radical, or ion?

To explore the connection between resonance and stability, we use a carbocation similar to the one involved in HBr addition to conjugated alkenes (Eq. 15.35a). This is an example of an **allylic carbocation**: a carbocation in which the positively charged, electron-deficient carbon is adjacent to a double bond.

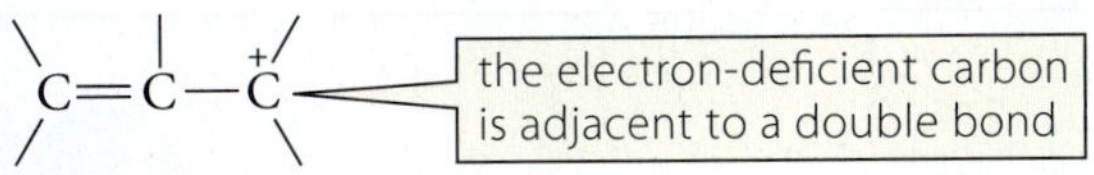

general Lewis structure of an
allylic carbocation (15.47)

The word *allylic* is a generic term applied to any functional group at a carbon adjacent to a double bond.

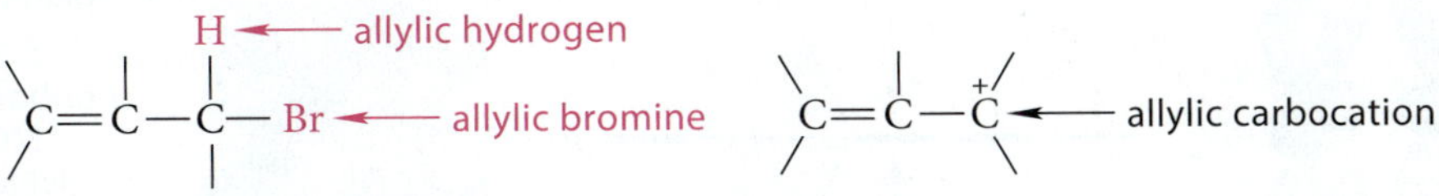

(15.48)

Allylic carbocations are more stable than comparably branched nonallylic alkyl carbocations. Roughly speaking, an allylic carbocation is about as stable as a nonallylic alkyl carbocation with one additional alkyl branch. For example, a secondary allylic carbocation is about as stable as a tertiary nonallylic one. To summarize:

Stability of carbocations:

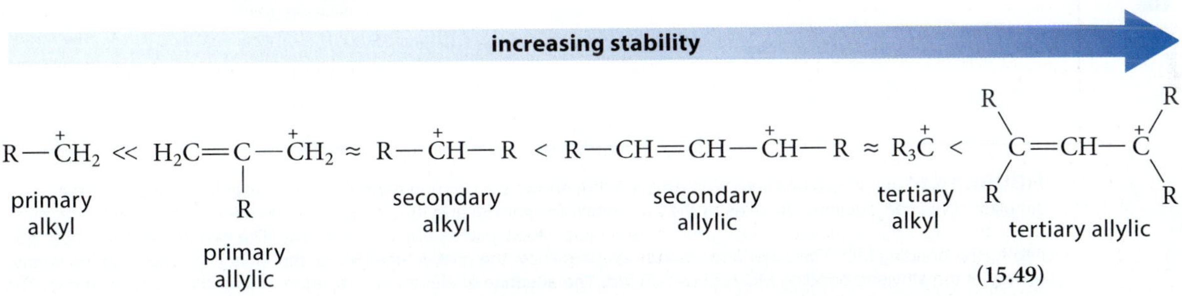

(15.49)

The stability of allylic carbocations lies in their electronic structures, which we now explore using molecular orbital (MO) theory. The π-electron structure of the allyl cation, $H_2C{=}CH{-}\overset{+}{C}H_2$ (the simplest allylic cation), is shown in **Fig. 15.14**. The electron-deficient carbon and the carbons of the double bond are all sp^2-hybridized; each carbon has a 2p orbital (Fig. 15.14a). The overlap of these three 2p orbitals results in three π molecular orbitals. The MO of lowest energy, π_1, is bonding with an energy of $+1.41\beta$, and it has no nodes. (Remember that β is a negative number.) The next MO, π_2, has the same energy as an isolated 2p orbital (0β). An MO that has the same energy as an isolated 2p orbital is called a **nonbonding MO (NBMO)**. This MO has one node. Because nodes must be placed symmetrically, this node goes through the central carbon. The position of this node determines the location of the positive charge, as we'll see shortly. The MO of highest energy, π_3^*, is antibonding. It has two nodes—one between each carbon.

The allyl cation has two electrons; both reside in π_1. Each π electron contributes $+1.41\beta$ to the energy of the molecule, for a total of $+2.82\beta$. The *delocalization energy* (Sec. 15.1A) of the allyl cation is the difference between this energy and the sum of the energies of an isolated ethylene ($+2.00\beta$) and an isolated 2p orbital (0β). The allyl cation, then, has a delocalization energy of 0.82β. Because the source of this stabilization is the "nodeless" π_1 MO, *the stabilization of the allyl cation arises from the delocalization of π electrons across the entire molecule.* This delocalization, then, is a source of *additional bonding* in the allyl cation that would not be present if the alkene π bond and the carbocation could not interact.

What is the connection between molecular stability and resonance? *Resonance structures provide a symbolic way to show electron delocalization in the low-energy bonding MOs with Lewis structures.* For example, allylic cations have two equivalent resonance structures.

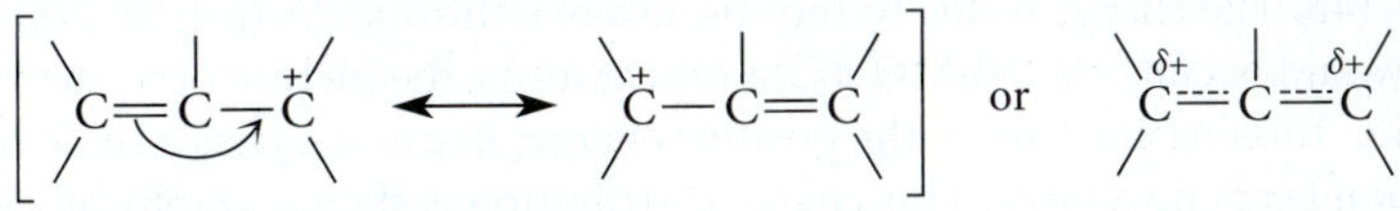

an allylic carbocation hybrid structure (15.50)

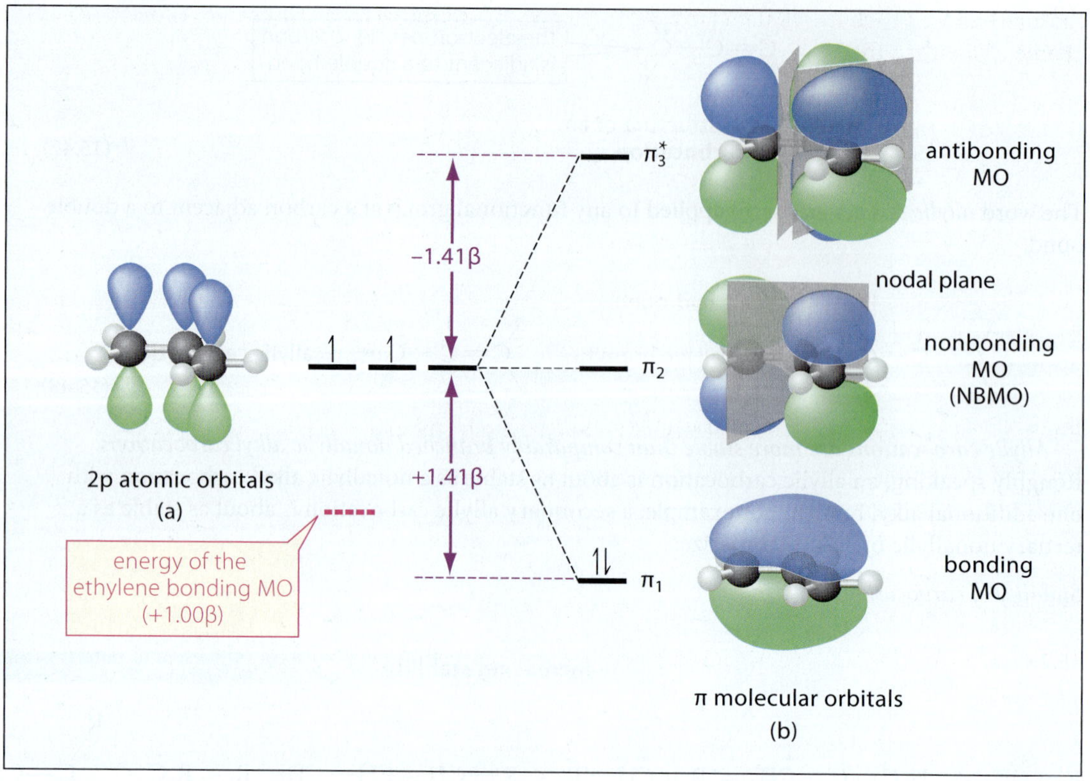

FIGURE 15.14 (a) An orbital interaction diagram that shows the arrangement of 2p orbitals in the allyl cation, the simplest allylic carbocation. The axes of the 2p orbitals are parallel and thus properly aligned for overlap. (b) Interaction of the three 2p orbitals (dashed lines) gives three π MOs. Nodal planes are shown in gray. The two 2p electrons both go into π_1, the bonding MO. The violet arrows and numbers show the relative energies of the MOs in β units, and the relative energy of the ethylene bonding MO is shown in red. The absence of electrons in π_2 accounts for the positive charge. The nodal plane in π_2 cuts through the central carbon; as a result, there is no positive charge on this carbon.

The two structures show the sharing of π-electron density and charge—in other words, electron delocalization. The hybrid structure shows the same delocalization with dashed bonds and partial charges. In summary, the logic of resonance stabilization is as follows:

1. Resonance structures symbolize electron delocalization.
2. Electron delocalization is stabilizing because it results in additional bonding associated with the formation of low-energy bonding MOs.
3. Resonance, therefore, is a stabilizing effect.

Because resonance describes electron delocalization, *delocalization energy* (Sec. 15.1B), the energetic advantage of electron delocalization, is also referred to as **resonance energy**. Section 1.4 explained that resonance-stabilized molecules are more stable than any of their fictional contributors. Resonance energy is a quantitative measure of this additional stability.

One other important aspect of resonance structures emerges from a consideration of the positive-charge distribution in the allyl cation. Notice in the resonance structures that, although the π electrons are delocalized across the entire molecule, the positive charge resides on only two of the three carbons; *there is no positive charge on the central carbon of an allylic cation.* This charge distribution is consistent with the molecular orbital description of the cation, shown in Fig. 15.14b. The charge is due to the absence of a third π electron. If a third π electron were present, it would occupy the NBMO π_2. In other words, the picture of π_2 describes the distribution of the "missing electron"—the positive charge. Because π_2 has a node on the central carbon, this carbon bears no charge. This charge distribution is shown graphically in the EPM of the allyl cation, which is calculated from MO theory. The terminal carbons have more positive charge (*blue*) than the central carbon (*red*):

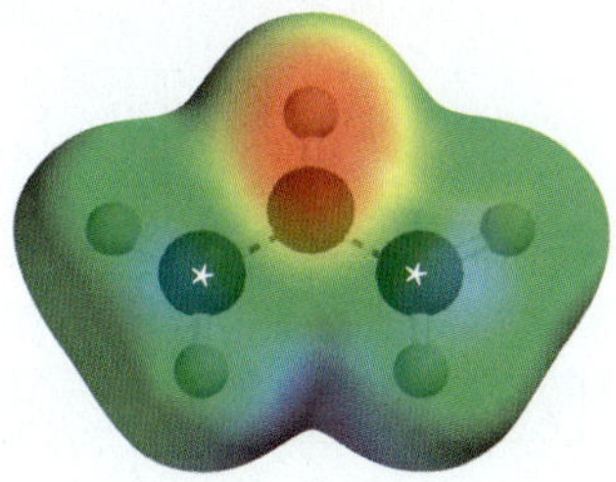

$\star$ = sites of positive charge

EPM of the allyl cation (15.51)

Resonance structures are useful because they give us the qualitative results of MO theory without any calculations!

Focused Problems

15.26 (a) The allyl anion has a nonbonding electron pair on the allylic carbon:

$$H_2C{=}CH{-}\ddot{C}H_2^{-}$$

allyl anion

This anion has two more π electrons than the allyl cation. Use the molecular orbital diagram in Fig. 15.14b to decide which molecular orbital these "extra electrons" occupy.

(b) According to the molecular orbital description, which carbons of the allyl anion bear the negative charge?

(c) Show that the same conclusion can be reached by drawing resonance structures of the allyl anion. Use the curved-arrow notation.

15.27 (a) The allyl radical has an unpaired electron on the allylic carbon:

$$H_2C{=}CH{-}\dot{C}H_2$$

allyl radical

(b) According to the molecular orbital description, which carbons of the allyl anion share the unpaired electron?

(c) Show that the same conclusion can be reached by drawing resonance structures of the allyl radical. Use the fishhook notation.

The discussion in this section shows that the phenomenon of resonance is a *symbolic depiction of orbital overlap*. Section 3.3B presented the guidelines for drawing resonance structures. To these rules we now add one more: *Do not draw resonance structures if the corresponding orbital overlap is impossible.*

As an example, consider the following carbocation in which the cationic carbon is a bridgehead carbon of a bicyclic structure:

This carbocation is formally allylic, but it is very unstable. Our first indication of this instability is that the resonance structure of this carbocation violates Bredt's rule (Sec. 7.6C) because it contains a bridgehead double bond.

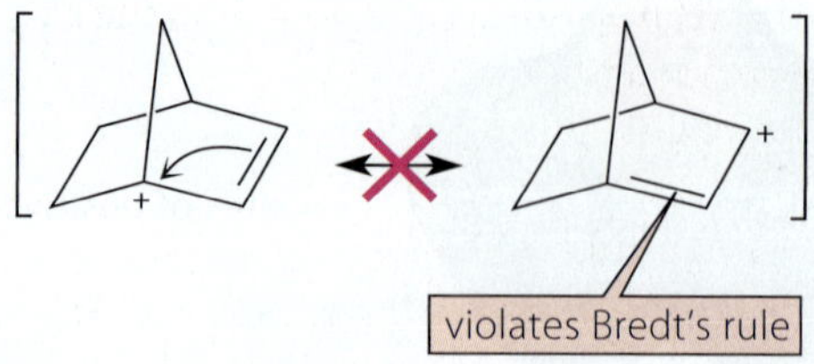

(15.52)

Another source of the instability of this carbocation is the requirement that the optimum geometry required by its sp^2 hybridization should be trigonal-planar. The cationic carbon is constrained by the ring to a somewhat pyramidal geometry, and it cannot become planar without introducing intense ring strain. However, we focus here on the absence of orbital overlap, which makes the situation that much worse.

The fundamental reason for this instability (and for Bredt's rule) is that the 2p orbitals of the double bond are perpendicular to the empty 2p orbital of the cation and therefore cannot provide the orbital overlap necessary for stabilization.

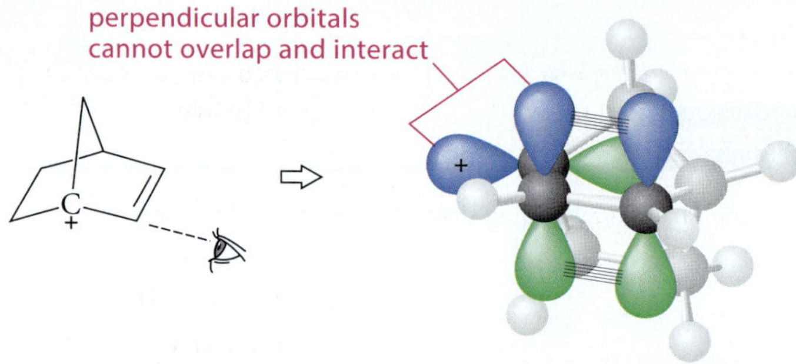

(15.53)

Moreover, the electronegativity of the sp^2-hybridized carbons of the double bond further destabilizes the positive charge.

To the rules in Sec. 3.3C for determining the importance of resonance structures, we add one more:

If the orbital overlap symbolized by a resonance structure does not occur, the resonance structure is not important.

Focused Problem

15.28 Most amides are stabilized by a resonance interaction between the nitrogen and the double bond of the carbonyl group:

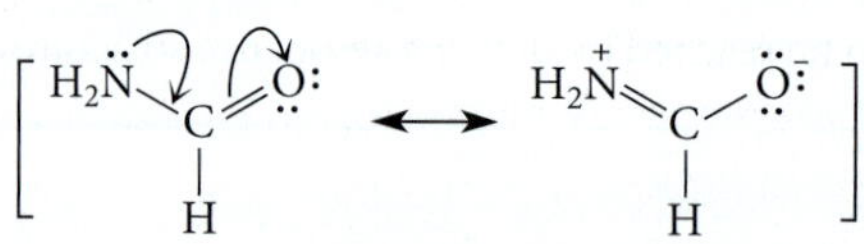

amide resonance

Is this resonance interaction possible in each of the following bicyclic amides? Why or why not?

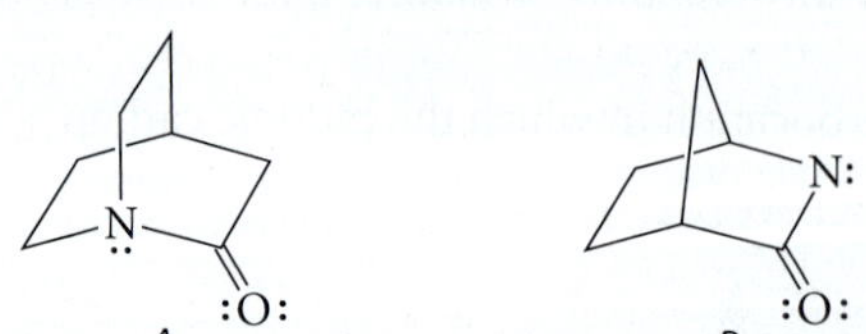

A *B*

15.7 INTRODUCTION TO AROMATIC COMPOUNDS

The term *aromatic*, as we show in this section, is a precisely defined *structural* term that applies to cyclic conjugated molecules that meet certain criteria. Benzene and its derivatives are the best-known examples of aromatic compounds.

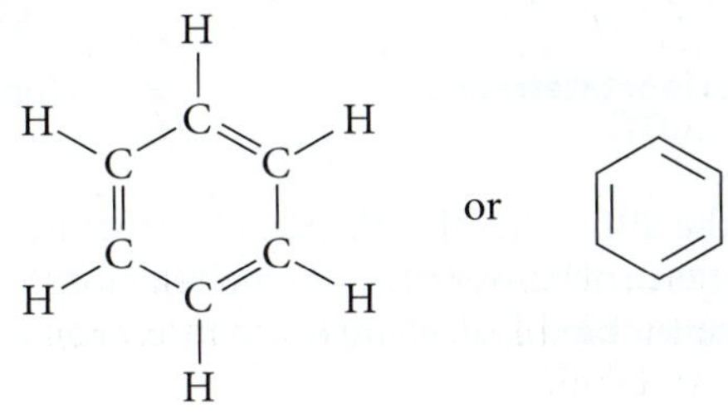

benzene (15.54)

The origin of the term *aromatic* is historical: many fragrant compounds known from earliest times, such as the following ones, proved to be derivatives of benzene.

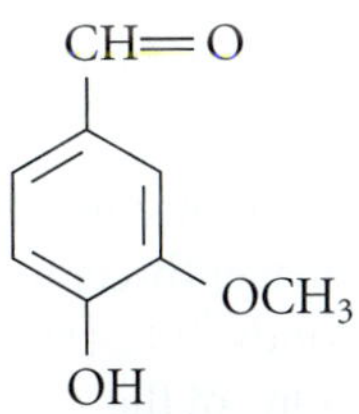

vanillin
(vanilla)

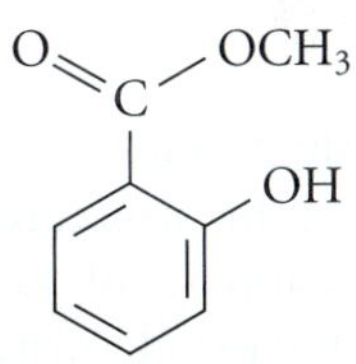

methyl salicylate
(oil of wintergreen)

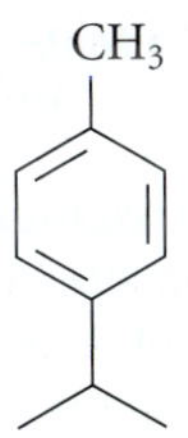

***para*-cymene**
(cumin and thyme)

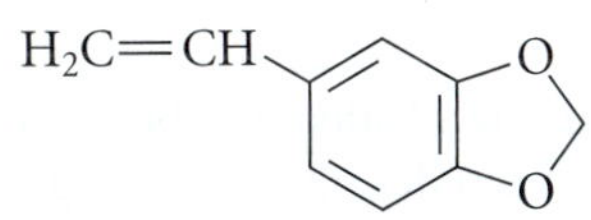

saffrole
(oil of sassafras)

Although it is known today that benzene derivatives are not distinguished by unique odors, the term *aromatic*—which has nothing to do with odor—has stuck, and it is now a class name for benzene, its derivatives, and a number of other organic compounds.

Development of the theory of aromaticity was a major theoretical advance in organic chemistry because it solved a number of intriguing problems that centered on the structure and reactivity of benzene. Before considering this theory, let's see what some of these problems were.

A. Benzene, a Puzzling "Alkene"

The structure used today for benzene was proposed in 1865 by August Kekulé, who claimed later that it came to him in a dream. The Kekulé structure portrays benzene as a cyclic conjugated triene. Yet benzene does not undergo any of the addition reactions that are associated with either conjugated dienes or ordinary alkenes. Benzene itself, as well as benzene rings in other compounds, are inert to the usual conditions of alkene addition reactions. This property of the benzene ring is illustrated by the addition of bromine to styrene, a compound that contains both a benzene ring and one additional double bond:

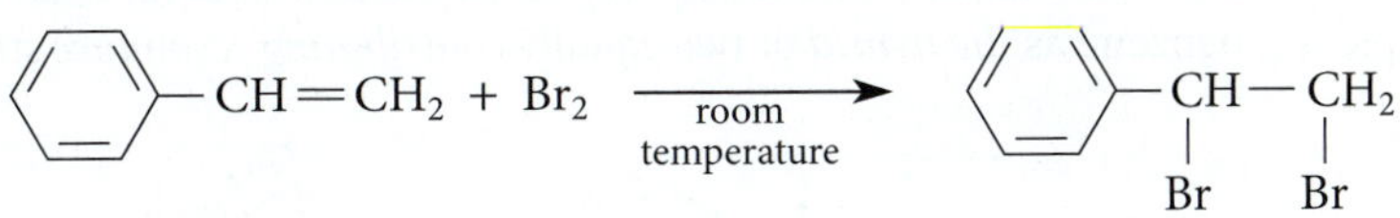

styrene (15.55)

The noncyclic double bond in styrene rapidly adds bromine, but the benzene ring remains unaffected, even if excess bromine is used. To early chemists, *this lack of alkenelike reactivity defined the uniqueness of benzene and its derivatives.*

Does benzene's lack of reactivity have something to do with its cyclic structure? Cyclohexene adds bromine readily. Perhaps, then, it is the cyclic structure and the conjugated double bonds that *together* account for the unusual behavior of benzene. However, 1,3,5,7-cyclooctatetraene (abbreviated in this text as COT) adds bromine smoothly even at low temperature.

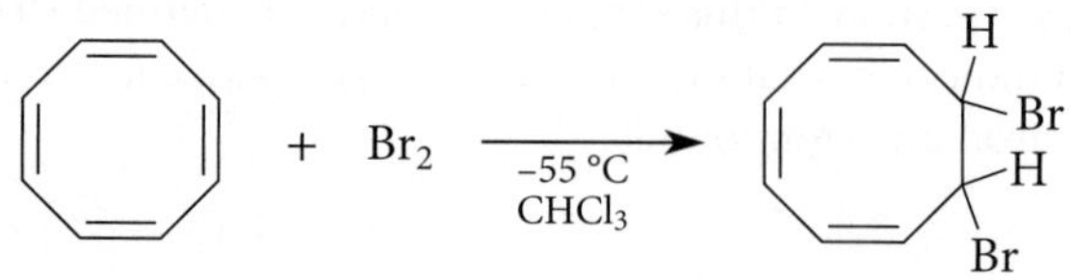

1,3,5,7-cyclooctatetraene (COT) (100% yield) (15.56)

These examples illustrate the difficulties that the Kekulé structure could not easily explain away. To account for the resistance of benzene to addition reactions, Albert Ladenburg in 1869 proposed a structure for benzene called *Ladenburg benzene* or *prismane*:

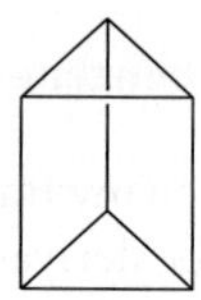

Ladenburg benzene (prismane) (15.57)

Although Ladenburg benzene is recognized today as a highly strained and unstable molecule (it has been described as a "caged tiger"), an attractive feature of this structure to nineteenth-century chemists that seemed to explain its lack of reactivity was the absence of double bonds.

Several facts, however, ultimately led to the adoption of the Kekulé structure. One of the most compelling arguments was that all efforts to prepare the alkene 1,3,5-cyclohexatriene using standard alkene syntheses led to benzene. The argument was, then, that benzene and 1,3,5-cyclohexatriene must be one and the same compound. The reactions used in these routes received additional credibility because they were also used to prepare COT, which, as Eq. 15.56 illustrates, has the reactivity of an ordinary alkene.

Although the Ladenburg benzene structure had been discarded for all practical purposes decades earlier, its final refutation came in 1973 with its synthesis by Professor Thomas J. Katz and his colleagues at Columbia University. These chemists found that Ladenburg benzene is an explosive liquid with properties that are quite different from those of benzene.

How, then, can the Kekulé "cyclic triene" structure for benzene be reconciled with the fact that benzene is inert to the usual reactions of alkenes? The answer to this question will occupy our attention in the next three parts of this section.

B. The Structure of Benzene

The structure of benzene is given in **Fig. 15.15a**. This structure shows that benzene has *one* type of carbon–carbon bond with a bond length (139.5 pm) that is the average of the lengths of sp^2–sp^2 single bonds (146 pm) and double bonds (133 pm, **Fig. 15.15c**). All atoms in the benzene molecule lie in one plane. The Kekulé structure for benzene shows *two* types of carbon–carbon bond: single bonds and double bonds. This inadequacy of the Kekulé structure can be remedied, however, by depicting benzene as *the hybrid of two equally contributing resonance structures*:

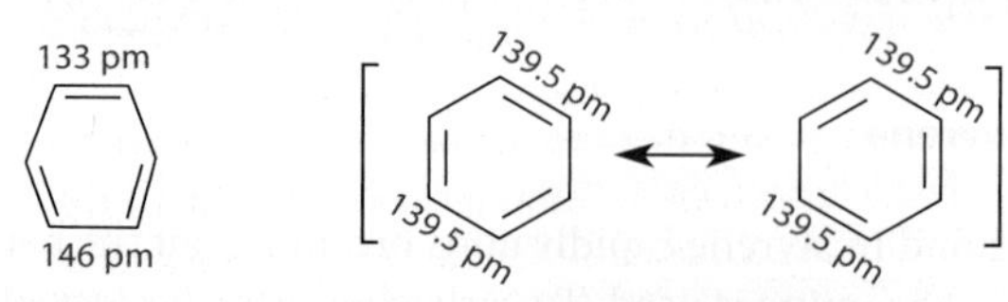

Kekulé structure resonance hybrid (15.58)

Benzene is an average of the two resonance contributors; it is *one* compound with *one* type of carbon–carbon bond with a bond order of 1.5—neither a single bond nor a double bond, but a

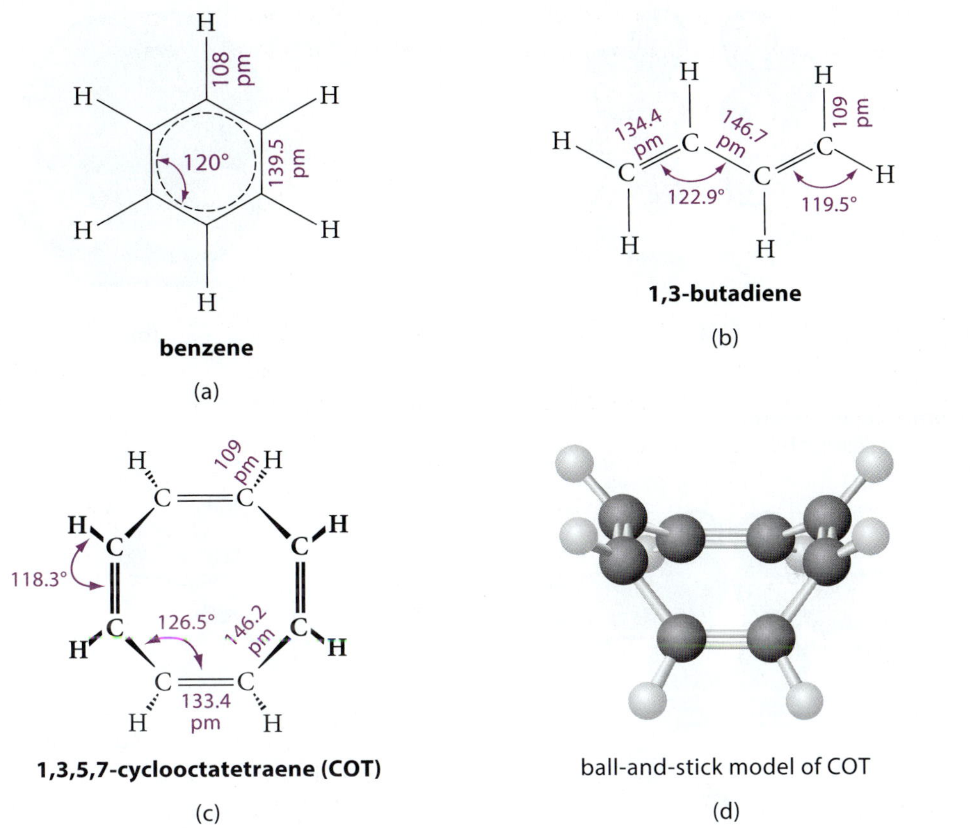

FIGURE 15.15 Comparison of the structures of benzene, 1,3-butadiene, and COT. (a) The structure of benzene. (The hybrid structure is shown.) (b) The structure of 1,3-butadiene, a conjugated diene. (c) The structure of 1,3,5,7-cyclooctatetraene (COT). (d) A ball-and-stick model of COT. The carbon skeleton of benzene is a planar hexagon, and all of the carbon–carbon bonds are equivalent with a bond length that is the average of the lengths of carbon–carbon single and double bonds in COT. In contrast, COT has distinct single and double bonds with lengths that are almost the same as those in 1,3-butadiene, and COT is tub-shaped rather than planar.

bond that is halfway in between. A benzene ring is often represented with either of the following hybrid structures, which are meant to suggest the "smearing out" of double-bond character:

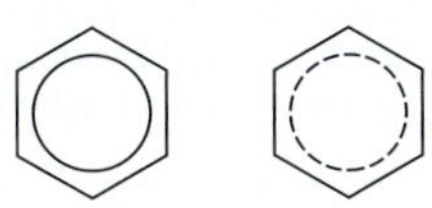

hybrid structures of benzene (15.59)

As with other resonance-stabilized molecules, we continue to represent benzene as either one of its resonance contributors because the curved-arrow notation and electronic bookkeeping devices are easier to apply to structures with fixed bonds.

It is interesting to compare the structures of benzene and 1,3,5,7-cyclooctatetraene (COT) in view of their greatly different chemical reactivities (Eqs. 15.55 and 15.56). Their structures are remarkably different (see Fig. 15.15). First, although benzene has a single type of carbon–carbon bond, COT has alternating single and double bonds, which have almost the same lengths as the single and double bonds in 1,3-butadiene. Second, COT is not planar like benzene, but instead is tub-shaped.

The π bonds of benzene and COT are also different (**Fig. 15.16**). The Kekulé structures for benzene suggest that each carbon atom should be trigonal-planar, and therefore sp^2-hybridized. This means each carbon atom has a 2p orbital (Fig. 15.16a). Because the benzene molecule is planar, and the axes of all six 2p orbitals of benzene are parallel, these 2p orbitals can overlap to form six π molecular orbitals. The bonding π molecular orbital of lowest energy is shown in Fig. 15.16b. (The other five π molecular orbitals of benzene are shown in Further Exploration 15.2 in the *Study Guide and Solutions Manual.*) This molecular orbital shows that π-electron density in benzene lies in doughnut-shaped regions both above and below the plane of the ring. *This overlap is symbolized by the resonance structures of benzene.* The π-electron overlap above and below the plane of the benzene ring is also reflected in the EPM of benzene, which shows a concentration of negative potential in these regions.

FURTHER EXPLORATION 15.2
The π Molecular Orbitals of Benzene

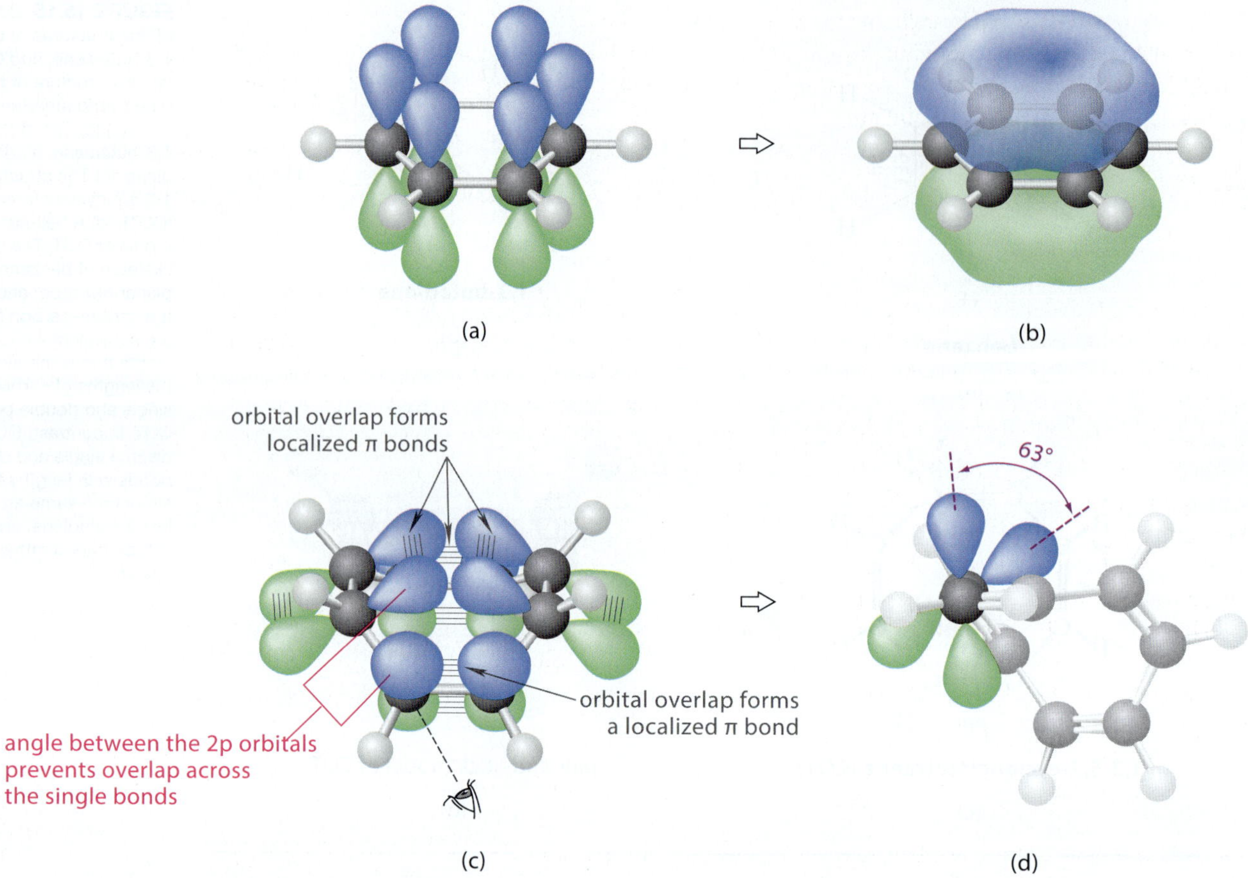

FIGURE 15.16 Comparison of the π bonds in benzene and 1,3,5,7-cyclooctatetraene (COT). (a) The carbon 2p orbitals in benzene. These orbitals are properly aligned for overlap. (b) The bonding π molecular orbital of lowest energy in benzene. This molecular orbital illustrates that π-electron density lies in a doughnut-shaped region above and below the plane of the benzene ring. (Benzene has two other occupied π molecular orbitals as well as three antibonding π molecular orbitals not shown here.) (c) The carbon 2p orbitals of COT. These orbitals can overlap in pairs to form isolated π bonds, but the tub shape prevents the overlap of 2p orbitals across the single bonds. (d) The view down a single bond indicated by the eyeball in part (c). The angle between 2p orbitals, indicated by the red bracket in part (c), is about 63°, which is too large for effective overlap.

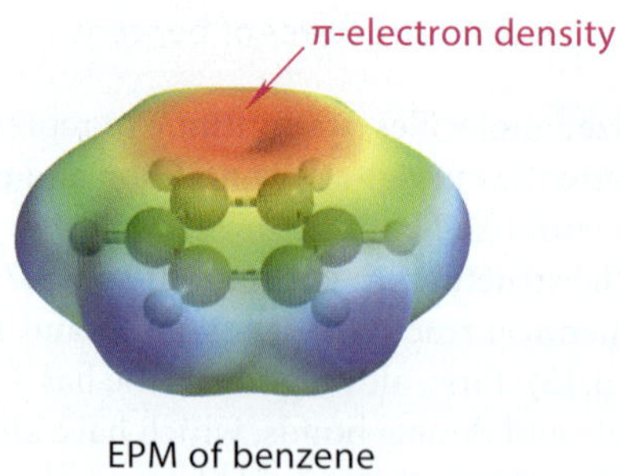

(15.60)

In contrast, the carbon atoms of COT are not all coplanar, but they are nevertheless all trigonal. This means that there is a 2p orbital on each carbon atom of COT (Fig. 15.16c). The tub shape of COT forces the 2p orbitals on the ends of each single bond to be oriented at a 63° angle, which is too far from coplanarity for effective interaction and overlap (Fig. 15.16d). Consequently, the 2p orbitals in COT cannot form a continuous π molecular orbital analogous to the one in benzene. Instead, COT contains four π-electron systems of two carbons each. As far as the π electrons are concerned, *COT looks like four isolated ethylene molecules.* Because there is no electronic overlap between the π orbitals of adjacent double bonds, *COT does not have resonance structures analogous to those of benzene.* (See the additional guideline for resonance structures in Sec. 15.6.)

(15.61)

To summarize: resonance structures can be written for benzene, because the carbon 2p orbitals of benzene can overlap to provide the additional bonding and additional stability associated with *filled bonding molecular orbitals.* Resonance structures *cannot* be written for COT because there is no overlap between 2p orbitals on adjacent double bonds.

Why doesn't COT flatten itself to allow overlap of all its 2p orbitals? We return to this point in Sec. 15.7E.

C. The Stability of Benzene

As mentioned earlier in this section, chemists of the nineteenth century considered benzene to be unusually stable because it is inert to reagents that react with ordinary alkenes. However, chemical reactivity (or the lack of it) is not the way that we measure energy content. We have shown that the more precise way to relate molecular energies is by their *standard heats of formation* ΔH_f°. Because benzene and COT have the same empirical formula (CH), we can compare their heats of formation per CH group. The ΔH_f° of benzene is 82.93 kJ mol^{-1} or $82.93/6 = 13.8$ kJ mol^{-1} per CH group. The ΔH_f° of COT is 298.0 kJ mol^{-1} or $298.0/8 = 37.3$ kJ mol^{-1} per CH group. Therefore, benzene, per CH group, is $(37.3 - 13.8) = 23.5$ kJ mol^{-1} more stable than COT. It follows that benzene is $23.5 \times 6 = 141$ kJ mol^{-1} (33.6 kcal mol^{-1}) more stable than a hypothetical six-carbon cyclic conjugated triene with the same stability as COT.

This energy difference of about 141 kJ mol^{-1} or 34 kcal mol^{-1} is called the **empirical resonance energy** of benzene. The empirical resonance energy is an experimental estimate of just how much special stability is implied by the resonance structures for benzene—thus the name "resonance energy."

The resonance energy is the energy by which benzene is *stabilized*; it is therefore an energy that benzene *doesn't have.* The empirical resonance energy of benzene has been estimated in several different ways; these estimates range from 126 to 172 kJ mol^{-1} (30 to 41 kcal mol^{-1}). (Another estimate is discussed in Sec. 16.6.) The important point, however, is not the exact value of this number, but the fact that it is *large.*

D. Aromaticity and the Hückel 4*n* + 2 Rule

We've shown that benzene is unusually stable, and that this stability is correlated with the overlap of its carbon 2p orbitals to form π molecular orbitals. In 1931, Erich Hückel (1896–1980), a German chemical physicist, elucidated with molecular orbital arguments the criteria for this sort of stability, which has come to be called *aromaticity.* Using Hückel's criteria, we can define aromaticity more precisely. (Remember again that aromaticity in this context has nothing to do with odor.) This definition has allowed chemists to recognize the aromaticity of many compounds in addition to benzene.

A compound is said to be **aromatic** when it meets *all* of the following criteria.

Criteria for aromaticity:

1. Aromatic compounds contain one or more rings that have a cyclic arrangement of p orbitals. Thus, aromaticity is a property of certain *cyclic* compounds.

2. *Every* atom of an aromatic ring has a p orbital.

3. Aromatic rings are *planar.*

4. The cyclic arrangement of *p* orbitals in an aromatic compound must contain $4n + 2$ π electrons, where *n* is any positive integer (0, 1, 2, . . .). In other words, an aromatic ring must contain 2, 6, 10, . . . π electrons.

These criteria are often called collectively the **Hückel 4*n* + 2 rule** or simply the **4*n* + 2 rule**.

The basis of the $4n + 2$ rule lies in the molecular orbital theory of *cyclic* π-electron systems. The theory holds that aromatic stability is observed only with *continuous cycles* of p orbitals—criteria 1 and 2. The theory also requires that the p orbitals must overlap to form π molecular orbitals. This overlap requires that an aromatic ring must be planar; p orbitals cannot overlap in rings significantly distorted from planarity—criterion 3. The last criterion has to do with the number of π molecular orbitals and the number of electrons they contain. Therefore, to understand criterion 4, we need to know the energies and electron occupancies of the various π molecular orbitals. Two Northwestern University physical chemists, A. A. Frost and

Boris Musulin, described in 1953 a simple graphical method for deriving the π-molecular orbital energies of cyclic π-electron systems without resorting to any of the mathematics of quantum theory. This method has come to be known as the **Frost circle**. The steps used in constructing a Frost circle follow, and they are illustrated in Study Problem 15.2.

Steps for constructing a Frost circle:

1. For a cyclic conjugated hydrocarbon or ion with *j* sides (and therefore *j* overlapping 2p orbitals), inscribe a regular polygon of *j* sides within a circle of radius 2β with one vertex of the polygon "pointing down" (in the lowest vertical position). (Remember from Sec. 15.1B that β is an energy unit used with molecular orbitals.)

2. Place one MO energy level at each vertex of the polygon. Because there are *j* vertices, there will be *j* MOs.

3. Draw a horizontal line that bisects the polygon. All MOs below the line are bonding, all MOs above the line are antibonding, and any MOs on the line are nonbonding (that is, they have the same energy as the isolated 2p orbital).

4. The lowest energy level must lie at 2β because it is at the lowest vertex, which is at the end of the vertical radius. The energies of the other levels are calculated by determining their positions along the vertical radius by trigonometry.

5. Add the π electrons to the energy levels in accordance with the Pauli principle and Hund's rules.

Study Problem 15.2 illustrates the application of the Frost circle to benzene.

Study Problem 15.2

Use the Frost circle to determine the energy levels and electron occupancies for the π MOs of benzene.

Step 1. For benzene, $j = 6$. Therefore, we inscribe a regular hexagon into a Frost circle of radius 2β with one vertex pointing down (**Fig. 15.17a**).

Step 2. Place an energy level at each vertex. This gives six energy levels.

Step 3. The horizontal dashed line in Fig. 15.17a bisects the polygon. The three MOs below the dashed line are bonding, and the three MOs above the dashed line are antibonding. There are no nonbonding MOs. (In **Fig. 15.7b**, the two MOs on the line are nonbonding.)

Step 4. Calculate the energies. The energies of π_1 and π_6^* equal the radius of the circle, so their energies must be 2β and -2β, respectively. (Remember that β is a negative number.) Then draw a perpendicular from the π_2 (or π_3) vertex to the vertical radius and calculate $E(\pi_2)$, the energy of π_2, as follows:

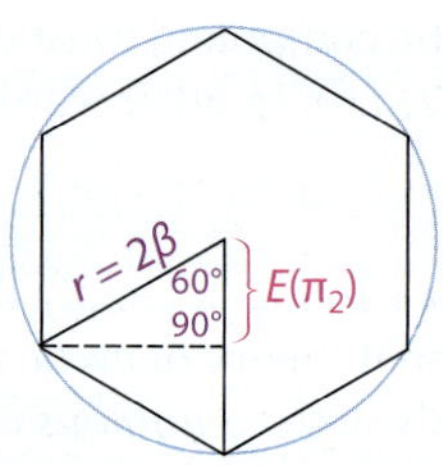

$$E(\pi_2) = r \cos 60° = (2\beta)(0.50) = 1\beta$$

From this calculation, the energies of π_2 and π_3, which are identical, equal $+1.0\beta$. By symmetry, the energies of π_4^* and π_5^* are -1.0β.

Step 5. Add the π electrons to the energy levels. Benzene has six π electrons. Because each bonding MO can accommodate two electrons, the available electrons exactly fill the bonding MOs.

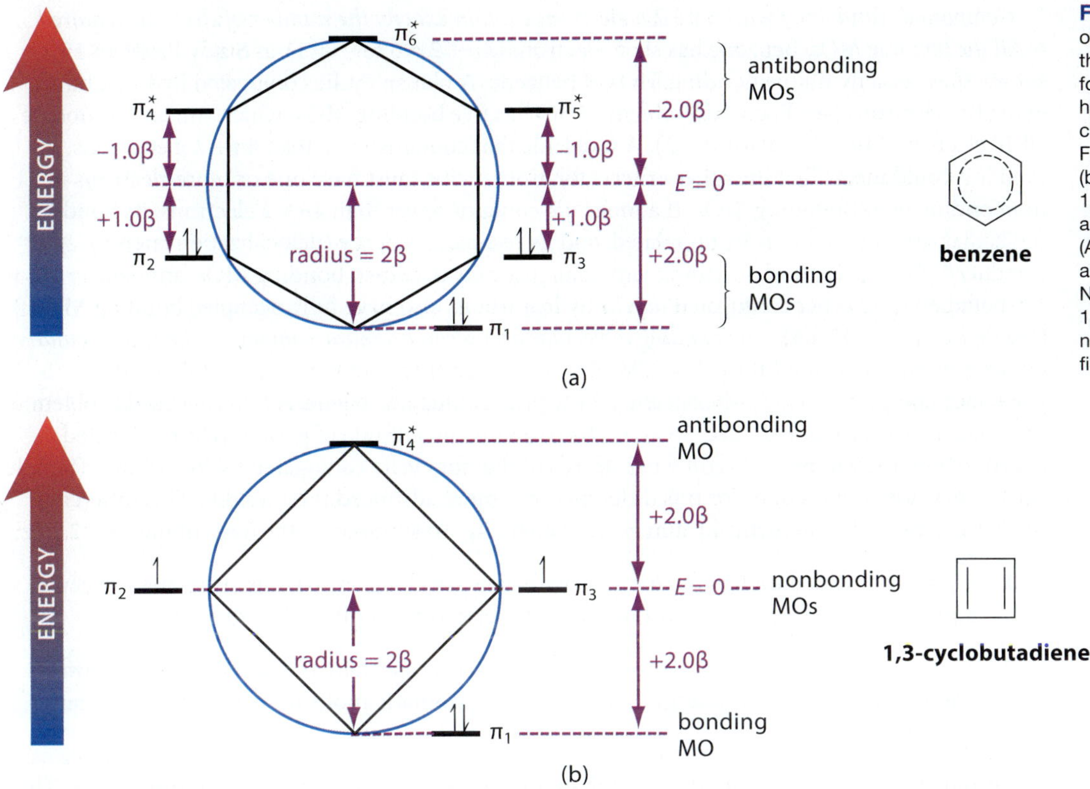

FIGURE 15.17 Application of the Frost circle to find the MOs and their energies for cyclic conjugated hydrocarbons. The Frost circle is in blue. (a) The Frost circle for benzene. (b) The Frost circle for 1,3-cyclobutadiene, an antiaromatic compound. (Antiaromatic compounds are discussed in Sec. 15.7E.) Notice that π_2 and π_3 in 1,3-cyclobutadiene are nonbonding and are only half filled.

This method shows that benzene has six MOs. (Remember, a system of *j* overlapping 2p orbitals forms *j* MOs; in this case, $j = 6$.) But it also shows that two bonding MOs, π_2 and π_3, have identical energies, and two antibonding MOs, π_4^* and π_5^*, also have identical energies. When orbitals have the same energy, they are said to be **degenerate**. Therefore π_2 and π_3 are degenerate MOs, and π_4^* and π_5^* are degenerate MOs.

The π-electron energy of benzene is 8.0β. The π-electron energy of three isolated ethylenes (with a bonding-MO energy of 1.0β) is 6.0β. The delocalization energy, or resonance energy, of benzene is then 2.0β. If we equate this to the empirical resonance energy of benzene (Sec. 15.7C), which is 141–150 kJ mol^{-1} or 34–36 kcal mol^{-1}, we find that $\beta = 70$–75 kJ mol^{-1} (17–18 kcal mol^{-1}). The resonance energy is a consequence of the very low-lying π_1 MO; the other bonding MOs, π_2 and π_3, have the same energy (1.0β) as the bonding MO of ethylene. The π_1 MO, shown in Fig. 15.16b, has no nodes, and *it most closely corresponds to the resonance-hybrid structure of benzene,* which shows the π electrons spread evenly around the entire molecule.

Focused Problems

15.29 Use a Frost circle to determine the π-electron structure of the cyclopentadienyl anion, which has a planar structure and six π electrons.

cyclopentadienyl anion

15.30 (a) How many bonding MOs are there in a planar, cyclic, conjugated hydrocarbon that contains a ring of 10 carbon atoms?

(b) How many π electrons does it have?

(c) How many of the π electrons can be accommodated in the bonding MOs?

Compounds (and ions) with $4n + 2$ π *electrons contain exactly the number of electrons required to fill the bonding MOs.* Benzene has six π electrons ($4n + 2 = 6$ for $n = 1$); as Study Problem 15.2 shows, these exactly fill the bonding MOs of benzene. A planar, cyclic conjugated hydrocarbon with 10 π electrons (see Focused Problem 15.30) has five bonding MOs, which can accommodate all 10 electrons ($4n + 2 = 10$ for $n = 2$). A molecule that contains *more* than $4n + 2$ π electrons, even if it could meet all of the other criteria for aromaticity, must have one or more electrons in nonbonding or antibonding MOs. If a molecule contains fewer than $4n + 2$ electrons, its bonding molecular orbitals are not fully populated, and its resonance energy (delocalization energy) is reduced. But there's more to aromaticity than just fully occupied bonding MOs; after all, 1,3-butadiene and other conjugated *acyclic* hydrocarbons also have fully occupied bonding MOs (see Fig. 15.1, Sec. 15.1A). *The bonding molecular orbitals in aromatic compounds have particularly low energy,* especially the MO at $E = 2.0\beta$. The resonance energy of benzene is 2.0β, but the resonance energy of (*E*)-1,3,5-hexatriene, the acyclic conjugated triene, is 1.0β (Focused Problem 15.2, Sec. 15.1A, or Fig. 15.7, Sec. 15.2C). Moreover, the magnitude of β for acyclic conjugated hydrocarbons (-50 kJ mol^{-1}) is only two-thirds of that for cyclic conjugated hydrocarbons. (Simple MO theory does not account for this difference, but more advanced theories do.) This difference further increases the energetic advantage of aromaticity. To summarize the basis of the $4n + 2$ rule:

1. Cyclic conjugated molecules and ions with $4n + 2$ π electrons (that is, aromatic molecules and ions) have exactly the right number of π electrons to fill the bonding MOs.

2. The bonding MOs in aromatic molecules and ions, especially the bonding MO of lowest energy, have a very low energy. For this reason, aromatic molecules and ions have a large resonance energy.

Recognizing aromatic compounds is a matter of applying the four criteria for aromaticity. This objective is addressed in Study Problem 15.3 and the discussion that follows it.

Study Problem 15.3

Decide whether each of the following compounds is aromatic. Explain your reasoning.

(a) 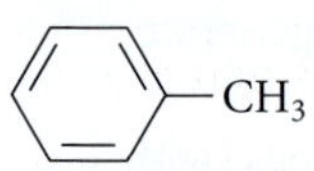

toluene

(b) 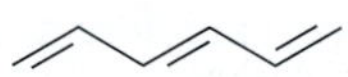
1,3,5-hexatriene

(c) 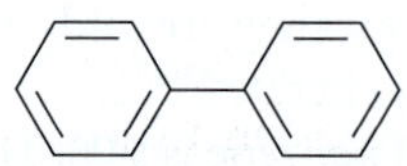
biphenyl

(d) 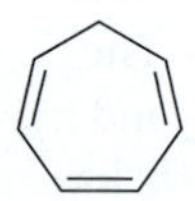
1,3,5-cycloheptatriene

(e) 
1,3-cyclobutadiene

Solution In each example, first count the π electrons by applying the following rule: *Each double bond contributes two π electrons.* Then apply *all* of the criteria for aromaticity.

(a) The ring in toluene, like the ring in benzene, is a continuous planar cycle of six π electrons, so the ring in toluene is aromatic. The methyl group is a substituent group on the ring and is not part of the ring system. Because toluene contains an aromatic ring, it is considered to be an aromatic compound. This example shows that *parts of molecules* can be aromatic or, equivalently, that aromatic rings can have nonaromatic substituents.

(b) Although 1,3,5-hexatriene contains six π electrons, it is not aromatic, because it fails criterion 1 for aromaticity: *it is not cyclic.* Aromatic species must be cyclic.

(c) Biphenyl has two rings, each of which is separately aromatic, so biphenyl is an aromatic compound.

(d) Although 1,3,5-cycloheptatriene has six π electrons, it is not aromatic, because it fails criterion 2 for aromaticity: one carbon of the ring does not have a p orbital. In other words, the π-electron system is not continuous, but is interrupted by the sp^3-hybridized carbon of the CH_2 group.

(e) 1,3-Cyclobutadiene is not aromatic. Even though it is a continuous cyclic system of 2p orbitals, it fails criterion 4 for aromaticity: it does not have $4n + 2$ π electrons. (We consider the case of 1,3-cyclobutadiene in Sec. 15.7E.)

Aromatic Heterocycles Aromaticity is not confined solely to hydrocarbons. Some *heterocyclic compounds* (Sec. 8.2C) are aromatic; for example, pyridine and pyrrole are both aromatic nitrogen-containing heterocycles.

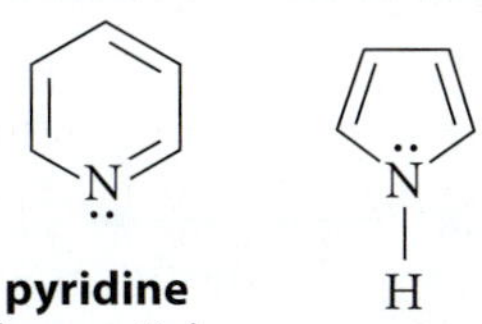

pyridine (aromatic) **pyrrole** (aromatic) (15.62)

Except for the nitrogen in the ring, the structure of pyridine closely resembles that of benzene. Each atom in the ring, including the nitrogen, is part of a double bond and therefore contributes one π electron. How does the electron pair on nitrogen figure in the π-electron count? This electron pair resides in an sp^2 orbital in the plane of the ring (see **Fig. 15.18a**). (It has the same relationship to the pyridine ring that any one of the C—H bonds has.) Because the nitrogen nonbonding pair does not overlap with the ring's π-electron system, it is not included in the π-electron count. Therefore, *vinylic electrons (electrons on doubly bonded atoms) are not counted as π electrons.*

In pyrrole, the electron pair on nitrogen is *allylic*—that is, on an atom *adjacent* to a double bond (**Fig. 15.18b**). This situation is much like that on the nitrogen in an amide. In Sec. 1.6C (Eq. 1.47) and Sec. 1.7E (Eq. 1.79) we showed that if a nitrogen can participate in resonance with adjoining π bonds, it is sp^2-hybridized. Therefore, this nitrogen in pyrrole has trigonal-planar geometry and sp^2 hybridization that allow its electron pair to occupy a 2p orbital and contribute to the π-electron count. The N—H hydrogen lies in the plane of the ring. In general, *allylic electrons are counted as π electrons when they reside in orbitals that are properly situated for overlap with the other 2p orbitals in the molecule.* Therefore, pyrrole has six π electrons—four from the double bonds and two from the nitrogen—and it is aromatic.

Notice the different ways in which we handle the electron pairs on the nitrogens of pyridine and pyrrole. The nitrogen in pyridine is part of a double bond, and the electron pair *is not* part of the π-electron system. The nitrogen in pyrrole is allylic, and its electron pair *is* part of the π-electron system.

Aromatic Ions Aromaticity is not restricted to neutral molecules; a number of ions are aromatic. One of the best characterized aromatic ions is the cyclopentadienyl anion:

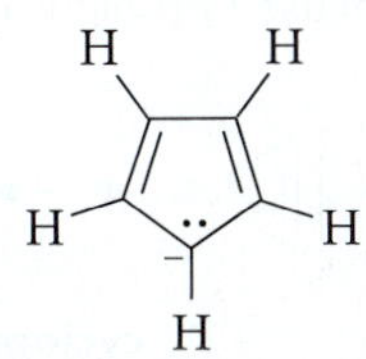

2,4-cyclopentadien-1-ide anion
(**cyclopentadienyl anion**; aromatic) (15.63)

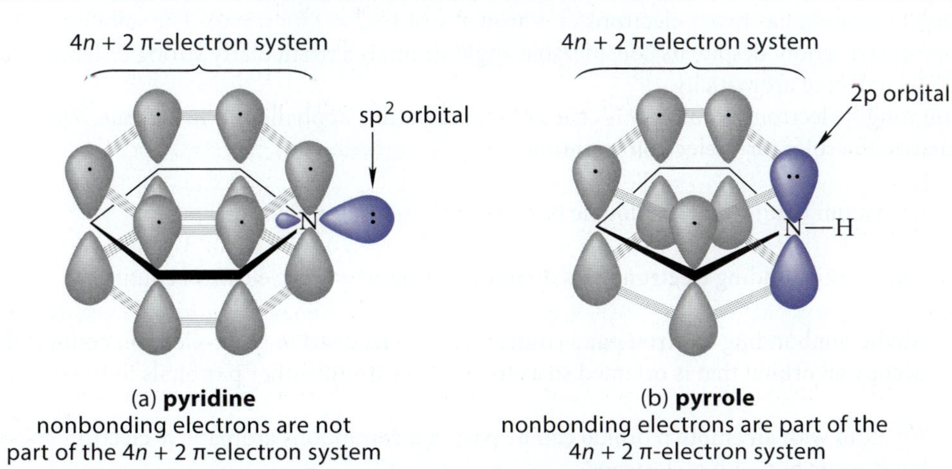

FIGURE 15.18 The 2p orbitals in pyridine and pyrrole. The gray lines represent orbital overlap. (a) The nonbonding electron pair in pyridine is vinylic and is therefore in an sp^2 orbital *(blue)* and is not part of the aromatic π-electron system. (b) The nonbonding electron pair in pyrrole is allylic and can occupy a 2p orbital *(blue)* that is part of the aromatic π-electron system.

(The Frost circle for this ion was considered in Focused Problem 15.29a.) The cyclopentadienyl anion resembles pyrrole; however, because the atom bearing the allylic electron pair is carbon rather than nitrogen, its charge is −1. One way to form this ion is by the reaction of sodium with the conjugate acid hydrocarbon, 1,3-cyclopentadiene; notice the analogy to the reaction of Na with H_2O.

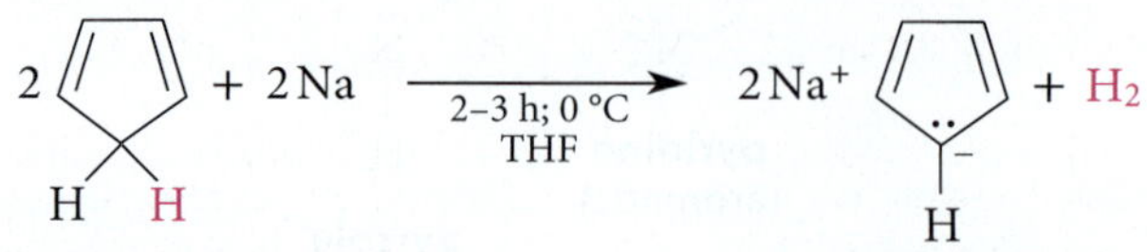

1,3-cyclopentadiene *not* aromatic — **cyclopentadienyl anion** aromatic (15.64)

The cyclopentadienyl anion has five equivalent resonance structures; the negative charge can be delocalized to each carbon atom:

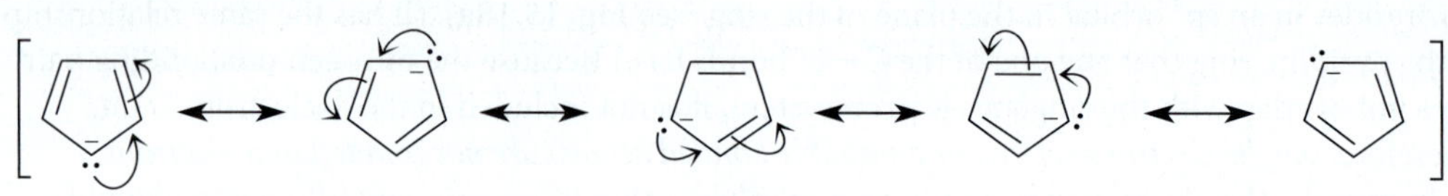

(15.65a)

These structures show that all carbon atoms of the cyclopentadienyl anion are equivalent. For this reason, the cyclopentadienyl anion is sometimes represented as a hybrid structure:

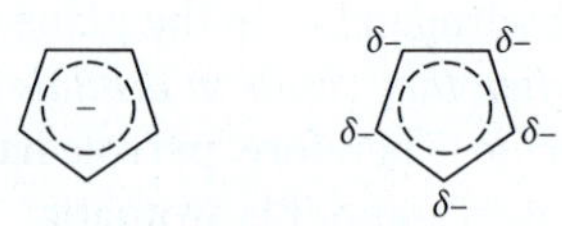

hybrid structures of the cyclopentadienyl anion (15.65b)

Because of the stability of this anion, its conjugate acid, 1,3-cyclopentadiene, is an unusually strong hydrocarbon acid. (Remember: The more stable the conjugate base, the more acidic is the conjugate acid; see Fig. 3.3, Sec. 3.7E.) With a pK_a of 15, this compound is 10^{10} times more acidic than a 1-alkyne, and almost as acidic as water!

Cations, too, may be aromatic. (See Focused Problem 15.33b.)

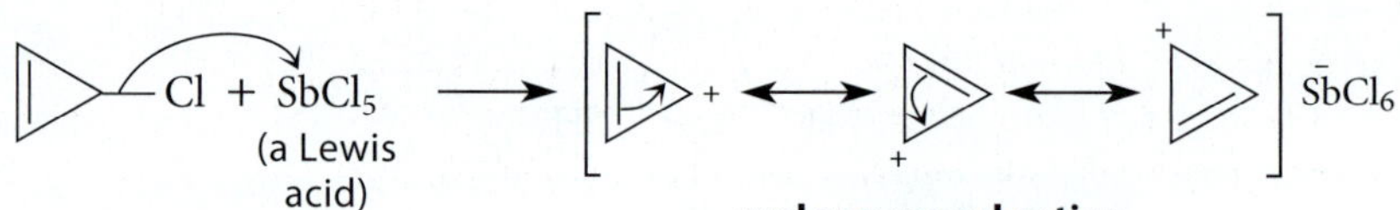

cyclopropenyl cation (aromatic) (15.66)

This example illustrates another point about counting electrons for aromaticity: *atoms with empty p orbitals are part of the π-electron system, but they contribute no electrons to the π-electron count.* Because this cation has two π electrons, it is aromatic ($4n + 2 = 2$ for $n = 0$). The stability of the cyclopropenyl cation, despite its considerable angle strain, is a particularly strong testament to the stabilizing effect of aromaticity.

Counting π electrons accurately is crucial for successfully applying the $4n + 2$ rule. We summarize the rules for π-electron counting:

1. Each atom that is part of a double bond contributes one π electron.
2. Vinylic nonbonding electron pairs do not contribute to the π-electron count.
3. Allylic nonbonding electron pairs contribute two electrons to the π-electron count if they occupy an orbital that is oriented so as to overlap with the other p orbitals in the molecule.
4. An atom with an empty p orbital can be part of a continuous aromatic π-electron system, but it contributes no π electrons.

Aromatic Polycyclic Compounds The Hückel $4n + 2$ rule applies strictly to single rings. However, a number of common fused bicyclic and polycyclic compounds, such as naphthalene, quinoline, and indole, are also aromatic:

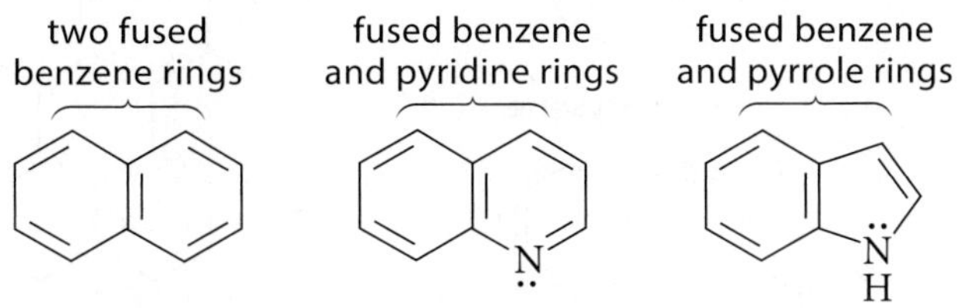

naphthalene **quinoline** **indole** (15.67)

Although rules have been devised to predict the aromaticity of fused-ring compounds, these rules are rather complex, and we need not be concerned with them. However, it shouldn't be difficult to see the resemblance of these compounds to monocyclic aromatic compounds—naphthalene to benzene, quinoline to benzene and pyridine, and indole to benzene and pyrrole.

Indole is particularly important in biology because of its presence in proteins within the structure of the amino acid *tryptophan.*

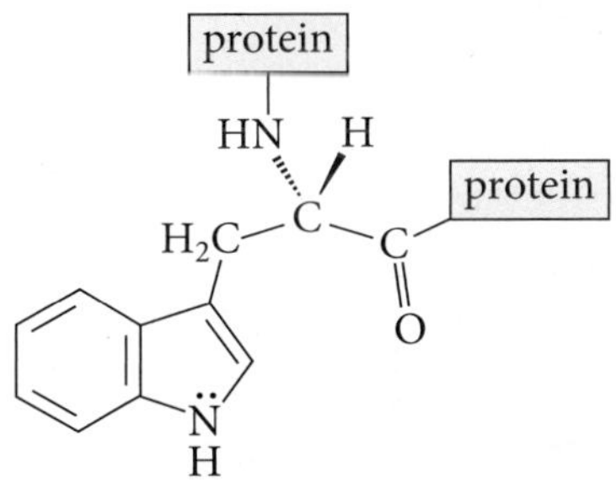

tryptophan
(as it occurs within proteins)

Fused-ring aromatic hydrocarbons with more than two rings are well known. Some of these are shown in **Fig. 15.19a**. Two of the most spectacular examples of fused-ring aromatic compounds are graphite (**Fig. 15.19b**) and buckminsterfullerene (**Fig. 15.19c**). *Graphite* is a form of elemental carbon that consists of layers of fused benzene rings. The softness of graphite and its ability to act as a lubricant can be attributed to the ease with which the layers slide past one another. Even though graphite is not an ionic compound, it is an excellent electrical conductor because its π electrons can be delocalized across its structure. Graphite is used in arc-welding electrodes and in lithium-ion and nickel–metal–hydride batteries. Single layers of graphite (called *graphene*) have been isolated and studied. (You have probably produced graphene when you have written with a lead pencil on paper. Pencil "lead" is actually graphite.)

The 2010 Nobel Prize in Physics was awarded to Andre Geim (b. 1958) and Konstantin Novoselov (b. 1974) of the University of Manchester, United Kingdom, for their experiments that characterized graphene.

Buckminsterfullerene was discovered as one component of soot and interstellar gas. The structure, which corresponds to the seams on a modern soccer ball, was proposed in 1985 by Sir Harold W. Kroto (1939–2016) of the University of Sussex, Brighton, United Kingdom, along with Richard E. Smalley (1943–2005) and Robert F. Curl (b. 1933) of Rice University. All 60 of the carbons in buckminsterfullerene are equivalent, as its single ^{13}C-NMR resonance at δ 143 shows.

The intriguing name, "buckminsterfullerene" (also sometimes nicknamed "buckyball"), came from its resemblance to the geodesic dome designed by American architect Buckminster Fuller (1895–1983). Since the discovery of buckminsterfullerene, a variety of related "buckyballs" and "buckytubes" have been discovered. Kroto, Smalley, and Curl were recognized for this and related discoveries with the 1996 Nobel Prize in Chemistry.

Aromatic Organometallic Compounds Some remarkable organometallic compounds have aromatic character. For example, the cyclopentadienyl anion, discussed previously in this section as one example of an aromatic anion, forms stable complexes with a number of transition-metal cations. One of the best known of these complexes is *ferrocene*, which is synthesized by the reaction of two equivalents of cyclopentadienyl anion with one equivalent of ferrous ion (Fe^{2+}).

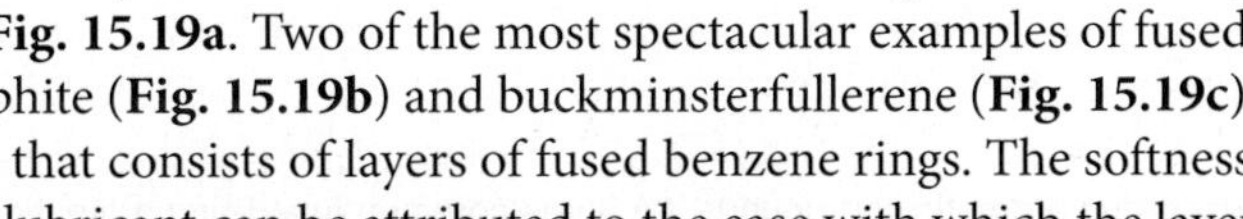

ferrocene
(90% yield) (15.68)

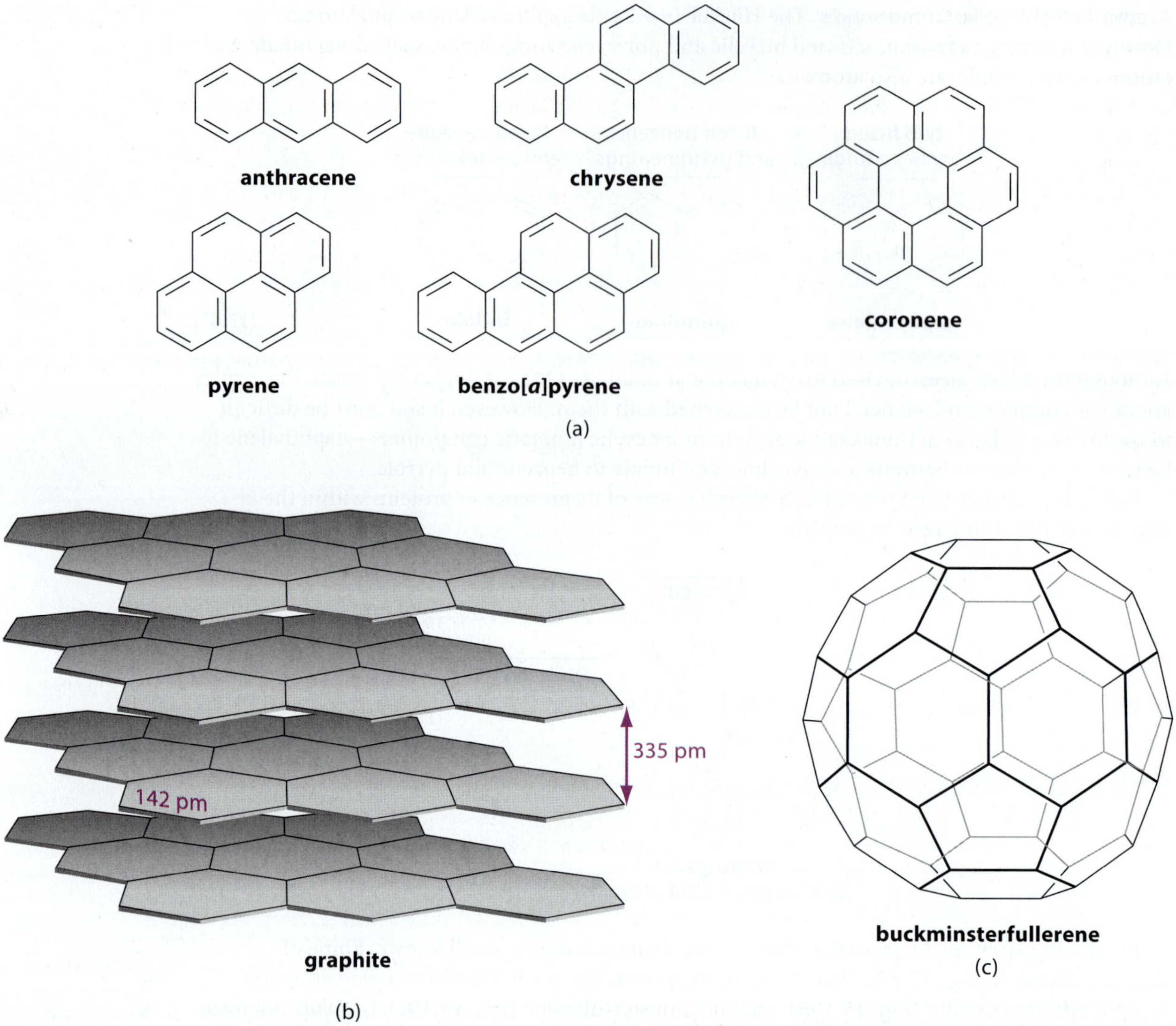

FIGURE 15.19 Some fused-ring aromatic compounds. (a) Some common fused-ring aromatic hydrocarbons containing more than two rings. (The role of benzo[*a*]pyrene in cancer is discussed in Sec. 16.7.) (b) Graphite. In this structure, the delocalized double bonds are not shown. Notice that the carbon–carbon bond length is very similar to that in benzene. (c) Buckminsterfullerene (C_{60}). In this structure the delocalized double bonds are not shown.

Although this synthesis resembles a metathesis (exchange) reaction in which two salts are formed from two other salts, ferrocene is not a salt but is a remarkable "molecular sandwich" in which a ferrous ion is imbedded between two cyclopentadienyl anions.

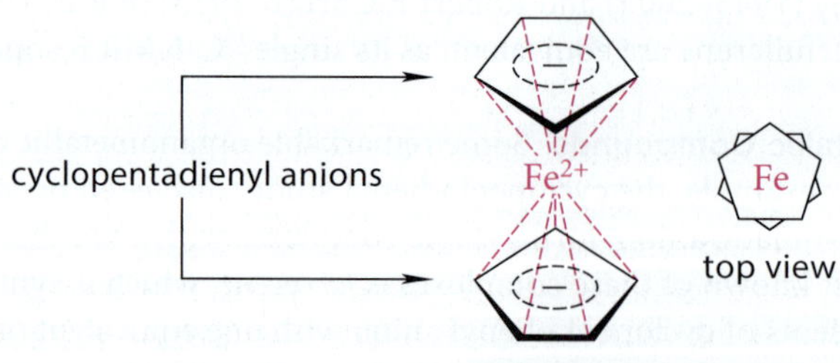

ferrocene (15.69)

The red dashed lines mean that the electrons of the cyclopentadienyl anions are shared not only by the ring carbons but also by the ferrous ion; each carbon is bonded equally to the iron.

Let's now return to the question posed near the beginning of this section: Why is benzene inert in the usual reactions of alkenes? The *aromaticity* of benzene is responsible for its unique chemical behavior. If benzene were to undergo the addition reactions typical of alkenes, its continuous cycle of $4n + 2$ π electrons would be broken; it would lose its aromatic character and much of its stability.

This is not to say, however, that benzene is unreactive under all conditions. Indeed, benzene and many other aromatic compounds undergo a number of characteristic reactions that are presented in Chapter 16. However, the conditions required for these reactions are typically much harsher than those used with alkenes, precisely because benzene is so stable. As you will also see, the reactions of benzene give very different kinds of products from the reactions of alkenes.

Focused Problems

15.31 Furan is an aromatic compound. Discuss the hybridization of its oxygen and the geometry of its two electron pairs.

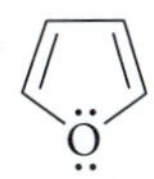

furan
(aromatic)

15.32 Do you think it would be possible to have an aromatic free radical? Why or why not?

15.33 Which of the following species should be aromatic by the Hückel $4n + 2$ rule?

(a) 

thiophene

(b) 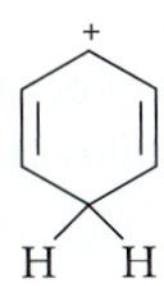

(c) 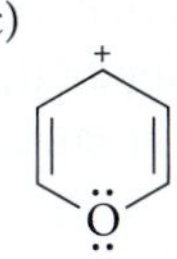

(d) 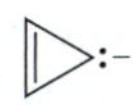

E. Antiaromatic Compounds

Compounds that contain *planar*, continuous rings of $4n$ π electrons, in stark contrast to aromatic compounds, are especially *unstable*; such compounds are said to be **antiaromatic**. 1,3-Cyclobutadiene (which we'll call simply cyclobutadiene) is such a compound; its small ring size and the sp^2 hybridization of its carbon atoms constrain it to planarity. This compound is so unstable that it can only be isolated at very low temperature, 4 K.

1,3-cyclobutadiene (15.70)

The Frost circle for 1,3-cyclobutadiene is shown in Fig. 15.17b. Two of the π MOs, π_2 and π_3, lie at $E = 0$. The π-electron energy of cyclobutadiene is 4.0β, which is *the same* as the π-electron energy of two isolated ethylenes. In other words, *cyclobutadiene has no resonance energy.* Moreover, Hund's rules require that the two degenerate MOs, π_2 and π_3, be half occupied. This means that cyclobutadiene has two unpaired electrons and is therefore a double free radical! Finally, cyclobutadiene has considerable angle strain.

The overlap of 2p orbitals in molecules with cyclic arrays of $4n$ π electrons is a *destabilizing* effect. (It could be said that antiaromatic molecules are "destabilized by resonance.") More advanced MO calculations show that cyclobutadiene, in an effort to escape this high-energy situation, distorts by lengthening its single bonds and shortening its double bonds:

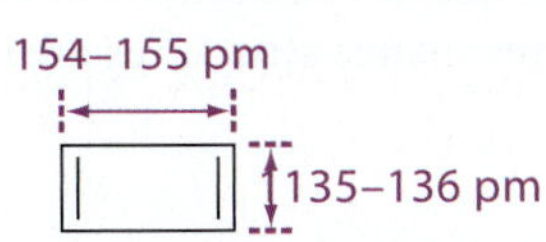

(15.71)

As a result of this distortion, the degeneracy of π_2 and π_3 is removed; so, one MO lies at slightly lower energy than the other and is doubly occupied. Cyclobutadiene, in effect, contains *localized* double bonds. This distortion, while minimizing antiaromatic overlap, introduces even more strain than the molecule would contain otherwise. The molecule can't win; it is too unstable to exist under normal circumstances.

Although cyclobutadiene is itself very unstable, it forms a very stable complex with Fe(0):

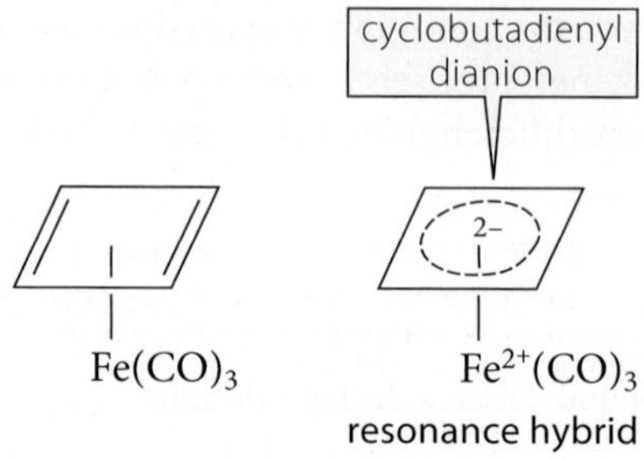

cyclobutadieneiron tricarbonyl (15.72)

(In this structure, the CO groups are neutral carbon monoxide ligands.) 1,3-Cyclobutadiene has four π electrons and is therefore two electrons short of the number (six) required for aromatic stability. These two missing electrons are provided by the iron. As the resonance hybrid structure on the right in Display 15.72 suggests, this complex in effect consists of a 1,3-cyclobutadiene with two additional electrons—a cyclobutadienyl *dianion*, a six π-electron aromatic system—combined with an iron minus two electrons—that is, Fe^{2+}. In effect, the iron stabilizes the antiaromatic diene by donating two electrons, thus making it aromatic.

This section began with a comparison of the stabilities of benzene and 1,3,5,7-cyclooctatetraene (COT). We can now recognize that COT contains a continuous cycle of $4n$ π electrons.

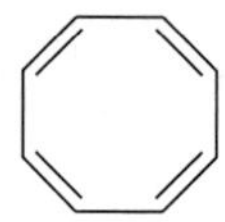

1,3,5,7-cyclooctatetraene (COT)

Is COT antiaromatic? It would be if it were planar. COT is large and flexible enough, however, that it can escape unfavorable antiaromatic overlap by folding into a tub conformation, as shown in Figs. 15.15d and 15.16c. It is believed that *planar* cyclooctatetraene, which *is* antiaromatic, is more than 58 kJ mol^{-1} (14 kcal mol^{-1}) less stable than the tub conformation.

F. Aromaticity and Resonance

This discussion of aromaticity provides our final caveats about the use of resonance structures:

1. Resonance structures are appropriate for aromatic systems. We can draw resonance structures for systems of $4n + 2$ π electrons, such as benzene and its derivatives, or other polycyclic aromatic compounds such as naphthalene.

2. Compounds with planar, cyclic arrangements of $4n$ π electrons are antiaromatic and are destabilized by resonance. Resonance structures should *not* be drawn for such compounds.

3. If compounds with $4n$ π electrons can distort from planarity, then they do so; the π orbitals of their double bonds are thus isolated from each other and do not interact. COT is an example. Because resonance structures can only be used to symbolize the stabilizing interaction of p or π orbitals, resonance structures are not appropriate for these compounds.

Focused Problems

15.34 Using the theory of aromaticity, explain the finding that *A* and *B* are different compounds, but *C* and *D* are identical. (That *A* and *B* are different molecules was established by Professor Barry Carpenter and his students at Cornell University in 1980.)

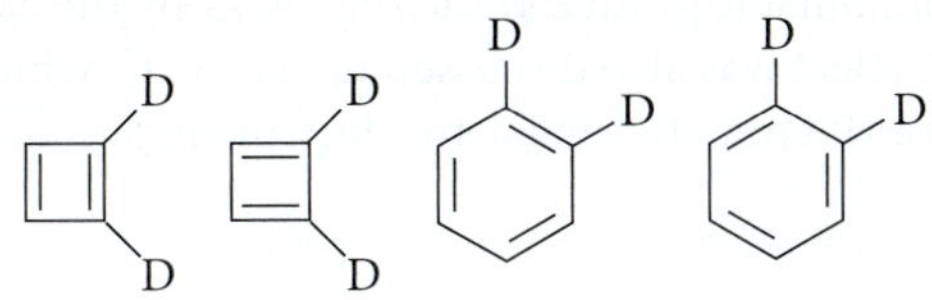

A *B* *C* *D*

15.35 Which of the compounds or ions in Focused Problem 15.33 (Sec. 15.7D), if any, are likely to be antiaromatic and therefore very unstable? Explain.

15.8 NONCOVALENT INTERACTIONS OF AROMATIC RINGS

In Secs. 8.5–8.8 we described a number of noncovalent interactions:

1. Van der Waals interactions: interactions between fluctuating dipoles and interactions between dipoles and induced dipoles
2. Electrostatic interactions: charge–charge interactions, charge–dipole interactions, and dipole–dipole interactions
3. Hydrogen bonding: the association of an O—H or N—H donor with a Lewis-base acceptor
4. "Hydrophobic bonding": the association of nonpolar groups driven by entropy changes in water

Benzene and its derivatives can interact noncovalently with other compounds. Two noncovalent interactions of aromatic rings are rather unique, and they are very important in many biomolecules and in their interactions with drug molecules. The first is the noncovalent interaction of aromatic rings with each other. The second is the noncovalent interaction of aromatic rings with cations. This section describes these noncovalent interactions and provides examples of each.

The basis of both types of interaction is the charge distribution in aromatic compounds, which we illustrate with benzene. Although benzene, because of its symmetry, has an overall zero dipole moment, it does have local regions of positive and negative charge. As we've shown in Sec. 15.7, the π electrons in benzene and other aromatic compounds are concentrated above and below the ring plane in the bonding π molecular orbitals. Associated with this π-electron density is a localized partial negative charge in these regions, illustrated by the electrostatic potential maps (EPMs) that follow. The red color is associated with regions of partial negative charge.

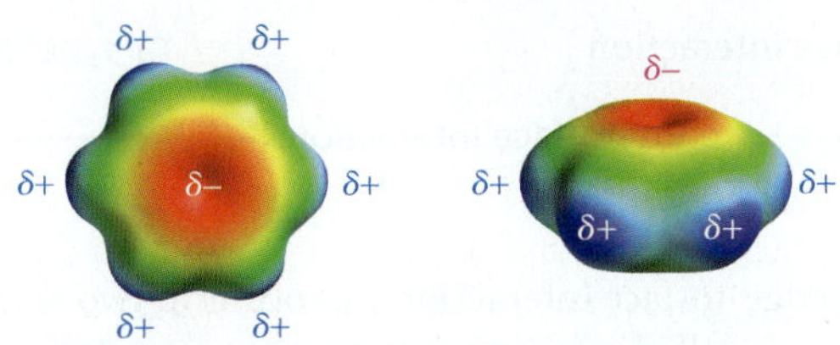

EPMs of benzene illustrating regions of partial positive and partial negative charge (15.73)

Benzene is a neutral molecule; so, corresponding to the localized partial negative charge, there must be regions of localized partial positive charge. These positively charged regions occur around the periphery of the ring, as shown by the blue areas in the EPMs. Because the π electrons are concentrated *above and below* the ring plane, the compensating partial positive charges are concentrated *in* the ring plane. This occurs because the positive charge of the carbon nucleus is screened from outside groups *in the ring plane* by the carbon sp^2 electrons but not by the π electrons. This effect was also discussed in Sec. 3.7C, which explains why sp^2-hybridized carbons (including those of aromatic rings) are electronegative and the attached hydrogens are relatively positive.

A. Noncovalent Interactions between Aromatic Rings

Aromatic rings can interact noncovalently with each other in two major ways. The first is that they can "stack" with their ring planes parallel.

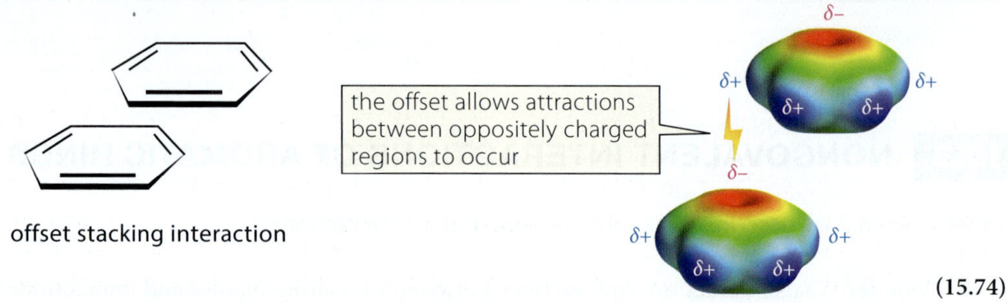

(15.74)

The centers of the rings involved in a stacking interaction are offset to avoid the juxtaposition of like charges in the two interacting rings. The offset also allows the positively charged region of one ring to interact attractively with the negatively charged region of the other. This type of interaction is called **offset stacking**.

In the second type of interaction between rings, the planes of the two interacting rings are perpendicular. This orientation also allows the positively charged region of one ring to interact attractively with the negatively charged region of the other. This type of interaction is called an **edge-to-face** ring interaction.

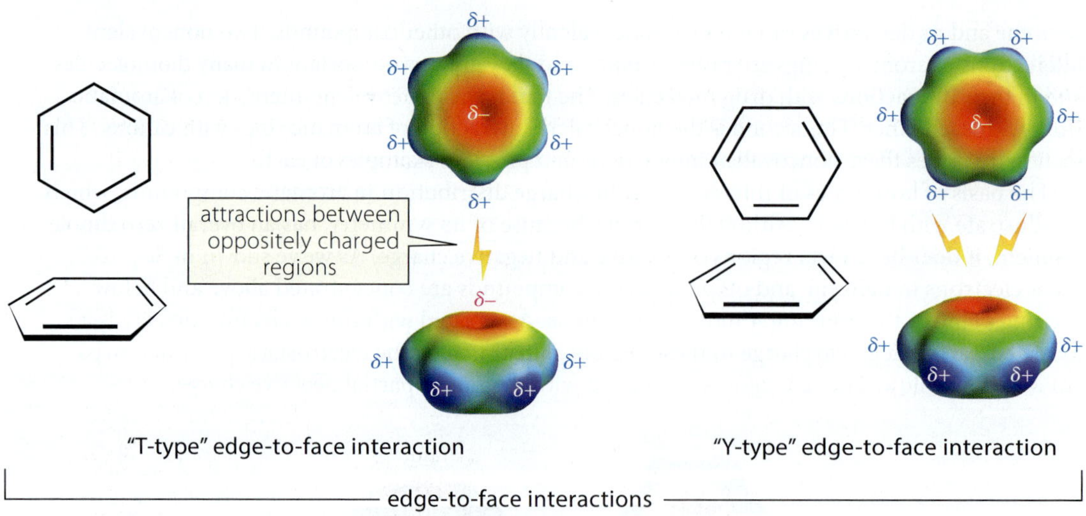

(15.75)

As the diagram shows, the edge-to-face interaction can occur in two ways; one is called a "T interaction" and the other is called a "Y interaction." Opinions differ as to the relative importance of these two orientations.

Both physical calculations and experimental evidence support the idea that the edge-to-face interaction is the more stable of the two attractions. However, there are many examples of both offset stacking and edge-to-face interactions in chemistry and biology; so, both are important.

The interactions depicted in Displays 15.74 and 15.75 are somewhat different from the "hydrophobic bonds" discussed in Sec. 8.6D, which are driven largely by a favorable entropy change associated with the solvent organization. Although an entropic component is present when two aromatic rings come together in water, a negative enthalpy change $\Delta H°$ also drives the association. The negative enthalpy is caused, at least in part, by a significant attraction between the rings, which is described in the foregoing discussion.

Aromatic rings can also interact with other polarizable but nonpolar groups (for example, alkyl groups) in conventional van der Waals attractions. Examples are known in which an alkyl group lies in the face of the ring, close to the π electrons, and other examples are known in which the hydrocarbon group lies along the edge of the ring. In either case, the localized charge in the aromatic ring induces a temporary dipole in the alkyl group, and, as a result, an attractive interaction develops between the two groups.

B. Noncovalent Interaction of Aromatic Rings with Cations

Once we understand the nature of the charge distribution in aromatic rings, we can understand that an aromatic ring can interact favorably with cations. This type of noncovalent interaction is called a **pi–cation interaction (Fig. 15.20)**. As we have shown (Display 15.73), the π-electron cloud of a benzene ring or other aromatic ring results in a region of local negative charge on the two faces of the ring. The pi-cation interaction is an electrostatic attraction of the positive cation with this region of local negative charge (Fig. 15.20a). This attraction was first discovered in a study of gas-phase interactions. In the gas phase, one or more benzene molecules can interact attractively with a potassium ion. Complexes of one to four benzene molecules with a potassium ion in the gas phase are known. Figure 15.20b shows the gas-phase pi–cation interaction of four benzene rings with a potassium ion. The formation of the complex between a potassium ion and a single benzene molecule in the gas phase has a large favorable $\Delta G°$ value of -48.5 kJ mol^{-1} (-11.6 kcal mol^{-1}), which includes the loss of entropy that typically accompanies an association reaction. The interaction energy, which has the unfavorable entropy removed, is approximated by the $\Delta H°$ value of -76.6 kJ mol^{-1} (-18.3 kcal mol^{-1}).

C. Noncovalent Interactions of Aromatic Rings in Biology

Chemical Biology Topic

Offset stacking, edge-to-face attractions, and pi–cation attractions are all very important in biology. These interactions provide mechanisms that stabilize protein structures and contribute to the formation of stable complexes between proteins (enzymes, receptors) and small molecules (substrates, inhibitors). Formation of such complexes between enzymes and substrates is an

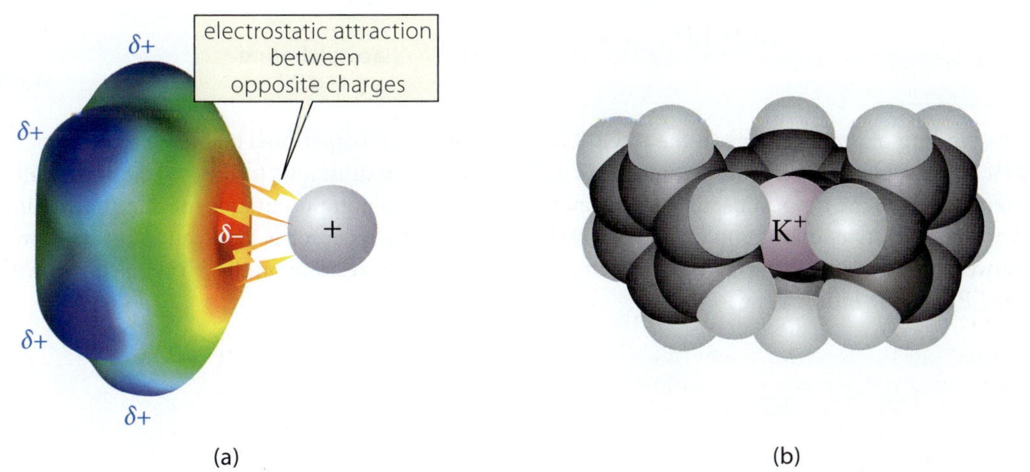

FIGURE 15.20 Pi–cation interactions. (a) The potassium ion interacts electrostatically with the region of local negative charge caused by the π electrons of a benzene ring. (b) A space-filling model showing the interaction of four benzene molecules with a potassium cation in the gas phase. In both, the cation is centered on the face of an aromatic ring.

essential prelude to catalysis, as discussed in Sec. 12.8E. This section provides a few biological examples of these interactions.

The aromatic rings in proteins are the groups in the side chains of the amino acid residues phenylalanine, tyrosine, and tryptophan.

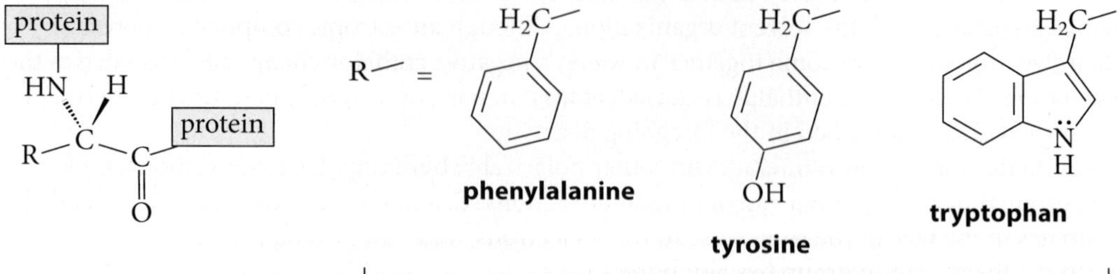

phenylalanine **tyrosine** **tryptophan**

aromatic amino acid residues

(15.76)

The amino acid histidine also has an aromatic imidazole ring in its side chain, but this ring (which is often found as its conjugate-acid cation) more often serves an acid–base role than a source of aromatic-ring interactions.

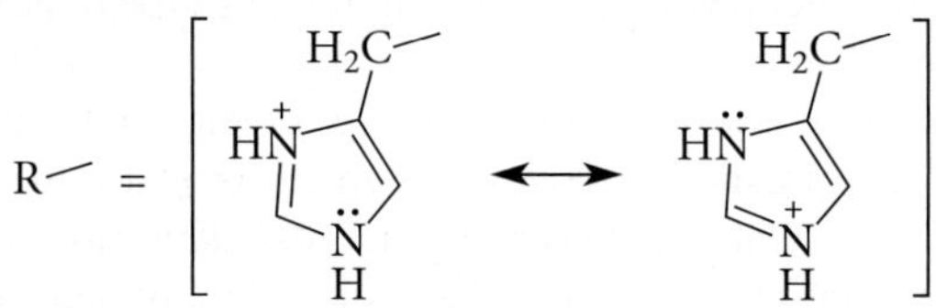

side chain of **histidine** (His, H)
conjugate-acid form (15.77)

Double-helical DNA provides a spectacular example of offset stacking. If you are familiar with the double-helical structure of DNA, you know that the bases in each strand are "stacked" perpendicular to the axis of the double helix. (The double-helical structure of DNA is shown in Fig. 26.5, Sec. 26.5B.) These bases are aromatic compounds. Their stacking is offset by the turn in the helix. The favorable interaction between the aromatic rings is an important element—possibly the *most important* element—that stabilizes the helical structure.

The structures of many proteins also contain stabilizing interactions between the rings of aromatic amino acid residues.

An important biological example of a pi–cation interaction occurs in the binding of the cationic molecule acetylcholine to the *acetylcholine receptor* (AcChR). One important form of AcChR is a large protein associated with cell membranes at the neuromuscular junction and elsewhere. Acetylcholine is a cationic molecule.

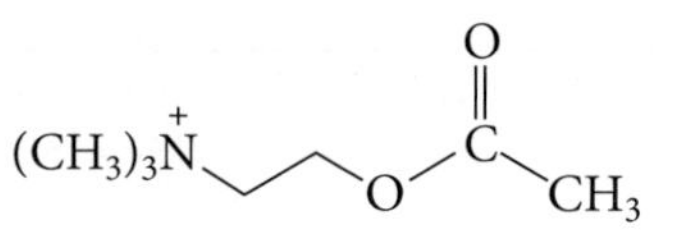

acetylcholine
(AcCh) (15.78)

The binding of acetylcholine to the AcChR causes a large conformational change in the protein that opens a pore in the protein and allows sodium and potassium ions to flow through the cell membrane. At the neuromuscular junction, this ion transmission potentiates muscle contraction. A major attractive interaction between acetylcholine and the receptor is a pi–cation interaction between the cationic group of acetylcholine and a tryptophan residue on the protein, shown with the EPM of the indole ring:

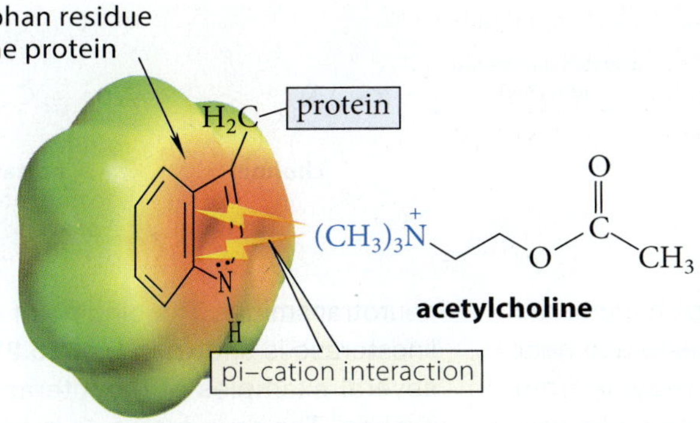

(15.79)

Binding of nicotine in its cationic, conjugate-acid form at the acetylcholine site in the AcChR of the brain is involved in the mechanism of nicotine addiction.

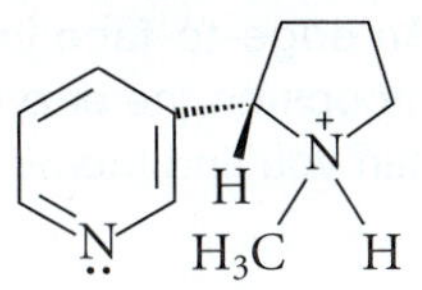

nicotine
(conjugate acid)

(15.80)

Many protein structures also contain stabilizing pi–cation interactions of the aromatic amino acid rings with the positively charged amino acid side chains of lysine and arginine.

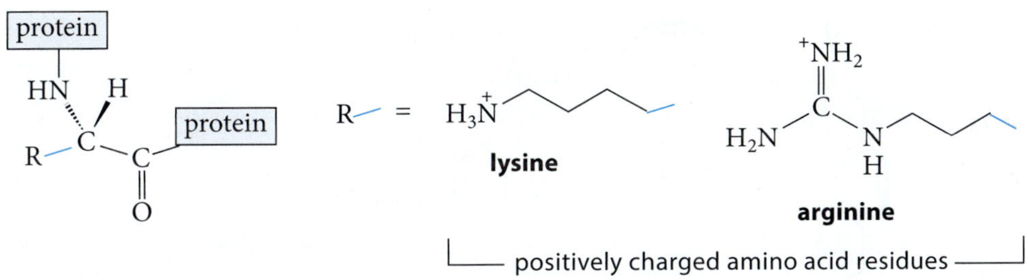

(15.81)

Aricept: Drug Therapy for Alzheimer's Disease

BSIP SA/Alamy Stock Photo

Alzheimer's disease (AD) is a condition in which a progressive and ultimately severe dementia is caused by protein misaggregation in the brain. Although largely a disease of the elderly, it can also occur in younger adults. Although much current research is focused on the prevention of AD, current drug therapy is focused largely on delaying the onset of cognitive impairment associated with the disease. One of the most widely used drugs for this purpose is *donepezil*, marketed as its hydrochloride salt under the trade name Aricept.

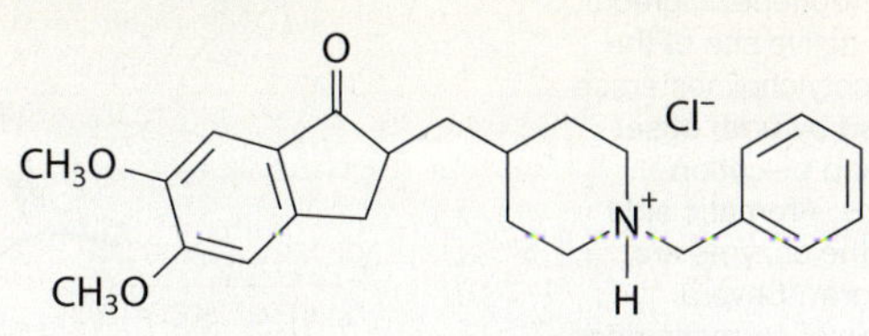

donepezil hydrochloride (Aricept)

The biological target of donepezil is the brain enzyme *acetylcholinesterase*. The normal function of this enzyme is to catalyze the hydrolysis of acetylcholine, an important neurotransmitter.

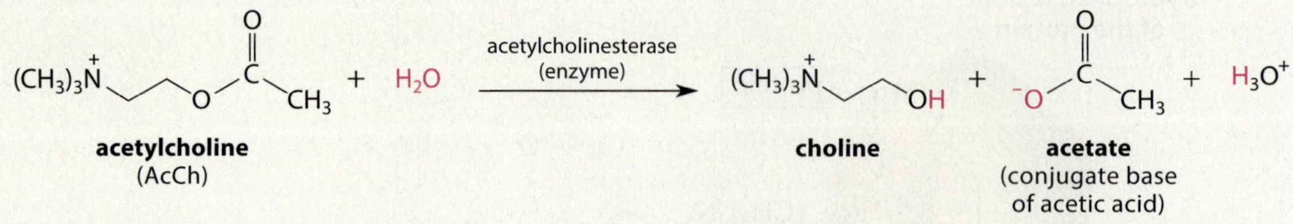

This hydrolysis terminates nerve transmission mediated by AcCh. Donepezil binds to acetylcholinesterase near its active site and thereby prevents the enzyme from binding AcCh. The inhibition of acetylcholine hydrolysis is believed to intensify neural activity in the brain.

As in all enzyme-catalyzed reactions, the hydrolysis reaction is preceded by binding of the substrate to the enzyme. Because donepezil binds close to the AcCh site in the enzyme, it blocks the binding of AcCh to the enzyme and thereby inhibits the hydrolysis of the neurotransmitter. The binding of donepezil to acetylcholinesterase is shown in **Fig. 15.21**. This binding involves several examples of the interactions discussed in this section. The drug molecule is shown in green, and various aromatic amino-acid side chains (identified by their three-letter abbreviations) are in gray. Identifiable offset stacking and pi–cation interactions are indicated. An edge-to-face interaction and offset interactions of groups on the enzyme with each other are also present; can you find them?

Focused Problem

15.36 The following molecule has a barrel shape (in which the benzene rings are the walls of the barrel). It forms a noncovalent complex with the iodide salt of acetylcholine in chloroform solvent.

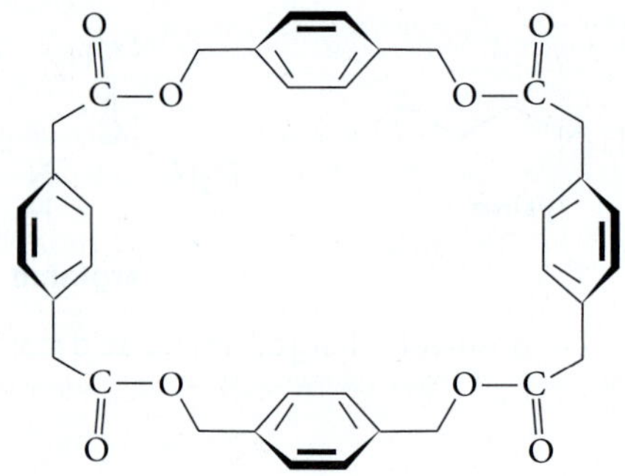

Describe the orientation of the acetylcholine molecule within the complex.

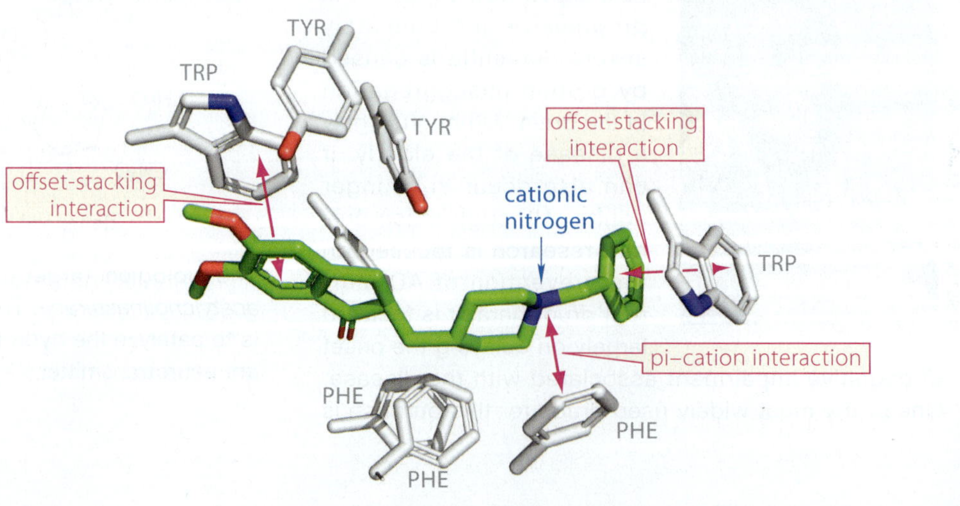

FIGURE 15.21 The binding of Aricept (donepezil, green) within the active site of the enzyme acetylcholinesterase is stabilized by both offset stacking and pi–cation interactions. Aromatic side chains of the enzyme are shown in gray. Several interactions of aromatic rings within the enzyme structure are also shown; can you identify some of them?

CHAPTER SUMMARY *For a summary of the chapter, see Chapter 15 in the* Study Guide and Solutions Manual.

REACTION REVIEW *For a summary of reactions discussed in this chapter, see the* Reaction Review *section of Chapter 15 in the* Study Guide and Solutions Manual.

SKILLS OBJECTIVES WITH PROBLEMS

- **Recognize conjugated and cumulated dienes. 15.1**

15.37 Specify which (if any) of the following compounds are conjugated alkenes and which are cumulated alkenes.

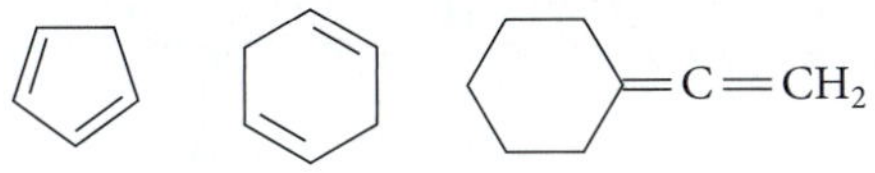

A *B* *C*

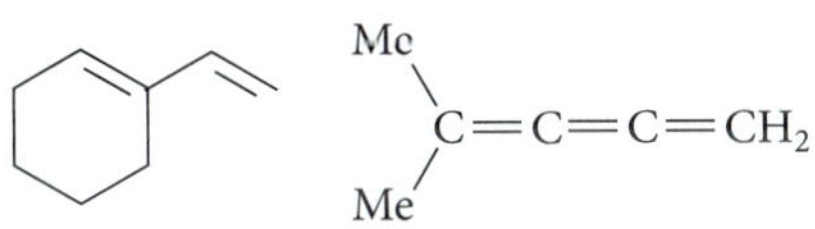

D *E*

- **Draw resonance structures for a conjugated alkene, ion, or radical, and use an MO diagram to relate the source of resonance energy to its resonance structures. 15.1A, 15.6**

15.38 (a) Derive resonance structures for the 2,4-pentadien-yl cation by using the curved-arrow notation. (*Hint:* You should have three structures, including the following one.)

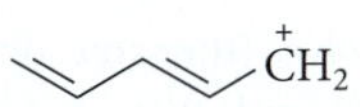

2,4-pentadien-1-yl cation

diagram for part (c)

(b) This cation has five π molecular orbitals with energies 1.73β, 1.0β, 0β, −1.0β, and −1.73β. (Remember that β is by definition a negative number.) Sketch a molecular orbital energy diagram with the energy levels arranged vertically. Add arrows into each MO to indicate electron occupancy.

(c) Represent the cation schematically as five black circles on a horizontal line (corresponding to the five carbons) next to each MO (as shown previously), and indicate the position of each node in each MO with a vertical line. (*Hint:* The nodes are placed symmetrically; the MO of lowest energy (π_1) has no nodes, and each higher-energy orbital has one additional node; the nodes for the π_2, π_3, and π_4^* MOs go through carbon atoms.)

(d) Classify each MO as bonding, antibonding, or nonbonding.

(e) Calculate the resonance energy (delocalization energy) for this cation.

(f) Explain why the MO diagram of this cation predicts the same distribution of positive charge as the resonance structures you drew in part (a).

15.39 (a) Derive resonance structures for the 2,4-pentadien-1-yl radical using the fishhook notation.

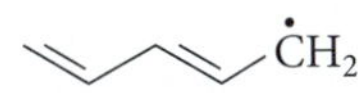

2,4-pentadien-1-yl radical

(b) The MOs of this radical are, to a useful approximation, the same as those for the cation in Problem 15.38. Populate these MOs with electrons.

(c) Calculate the resonance energy for this radical.

(d) Explain why the MO diagram of this radical predicts the same distribution of unpaired-electron character as the resonance structures you drew in part (a).

- **Given their structures, predict qualitatively the relative stability of a series of alkenes, free radicals, or ions. 15.1B,C; 15.6; 15.7D,E**

15.40 Rank the isomers within each set in order of increasing heat of formation (lowest first).

(a)

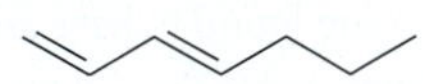

A

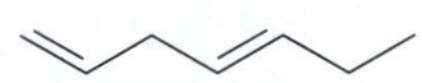

B

Et—CH=C=CH—Et

C

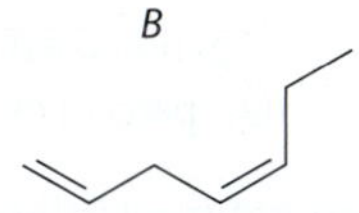

D

(b)

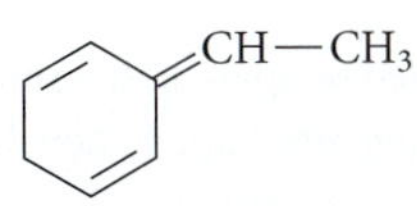

A

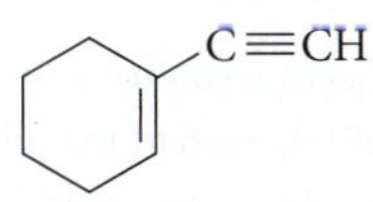

B

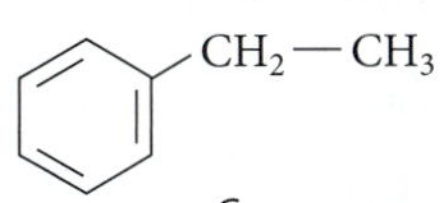

C

(c) 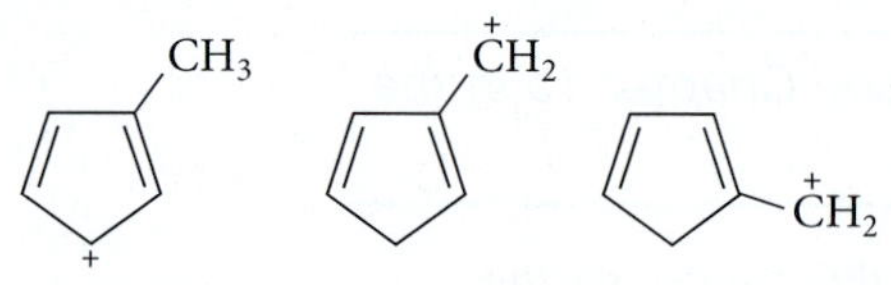

A *B* *C*

(d) 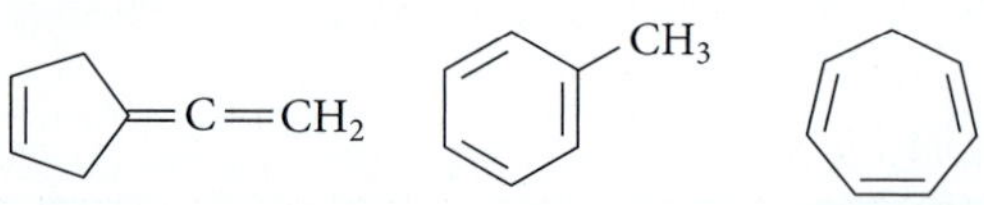

A *B* *C*

- **Explain the physical basis of UV–visible spectra. 15.2B**

15.41 (a) The molecular orbitals of 3-methylene-1,4-pentadiene occur at the following energies: 1.932β, 1β, 0.517β, −0.517β, −1β, and −1.932β.

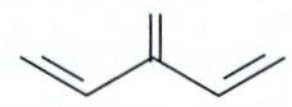

3-methylene-1,4-pentadiene

Sketch an energy diagram of these MOs, and add electrons to your diagram. Between which two MOs does the electronic transition occur when UV light is absorbed?

(b) Which has a UV absorption at higher wavelength: this compound or (*E*)-1,3,5-hexatriene? [*Hint:* See Fig. 15.7, Sec. 15.2C, for the MO energies of (*E*)-1,3,5-hexatriene.]

- **Apply the Beer–Lambert law to determine the concentration of a compound from its UV spectrum. 15.2A**

15.42 A colleague, Ima Hack, has subjected isoprene (see Fig. 15.5, Sec. 15.2A) to catalytic hydrogenation to give isopentane. Hack has inadvertently stopped the hydrogenation prematurely and wants to know how much unreacted isoprene remains in the sample. The mixture of isoprene and 2-methylbutane (75 mg total) is diluted to 1 liter with pure methanol and found to have an absorption at 222.5 nm (1 cm path length) of 0.356. Given an extinction coefficient of 10,750 at this wavelength, what mass percent of the sample is unreacted isoprene?

- **Predict qualitatively the relative λ_{max} of a series of alkenes. 15.2C**

15.43 Assume you have unlabeled samples of the compounds within each of the following sets. Explain how UV–vis spectroscopy could be used to distinguish each of the compounds from the others.

(a) 1,4-cyclohexadiene and 1,3-cyclohexadiene

(b) 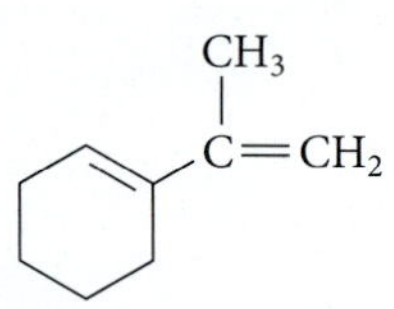 and 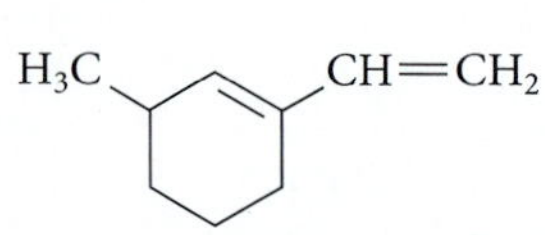

A *B*

(c) 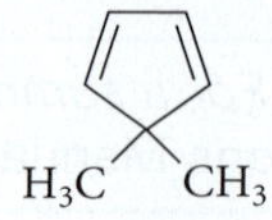 and 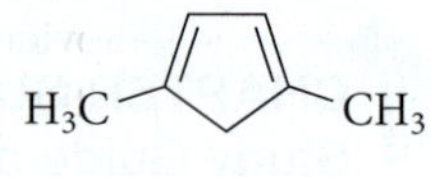

A *B*

(d) 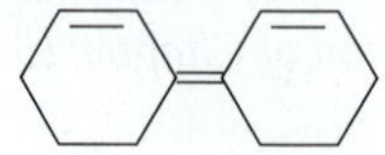 and 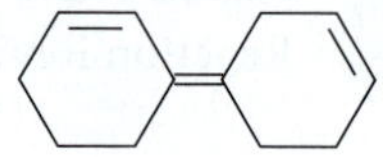

A *B*

(e) 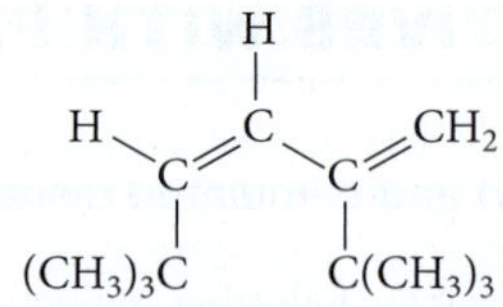 and

A

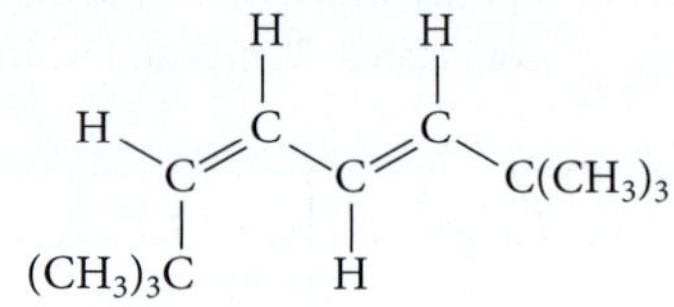

B

- **Explain the physical basis of fluorescence, including the origin of the Stokes shift. 15.2D**

15.44 Complete the following statements about fluorescence by making the proper choice between the given alternatives or by completing the sentence in the blank.

(a) The fluorescence wavelength is (greater than, the same as, less than) the absorption wavelength.

(b) Fluorescence involves (the same, different) electronic energy levels than absorption.

(c) The quantum yield of fluorescence is the ratio of ____________.

(d) Fluorescence involves (the same, different) vibrational energy levels than absorption.

(e) Fluorescence spectroscopy is a more sensitive technique than absorption spectroscopy because ____________.

- **Given the starting materials, give the products of a Diels–Alder reaction; given the products, give the structure of the dienophile and diene starting materials, including stereochemistry. 15.3**

15.45 Complete the following Diels–Alder reactions, including the stereochemistry. If endo and exo products are possible, show both and label them.

(a) 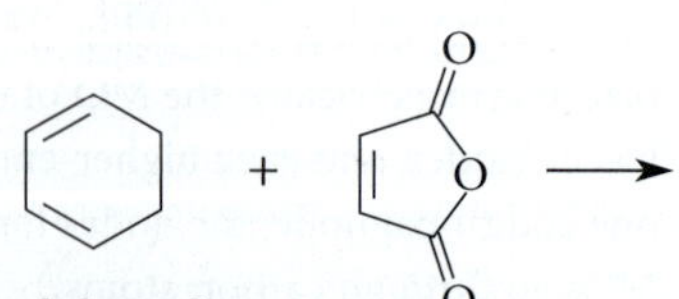

1,3-cyclohexadiene **maleic anhydride**

(b) 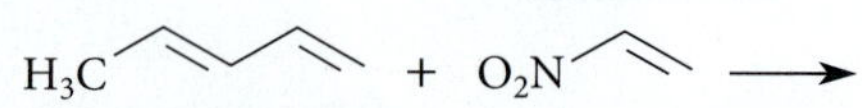

15.46 In each part, provide the structures of the starting materials that would give the product(s) shown.

(a)

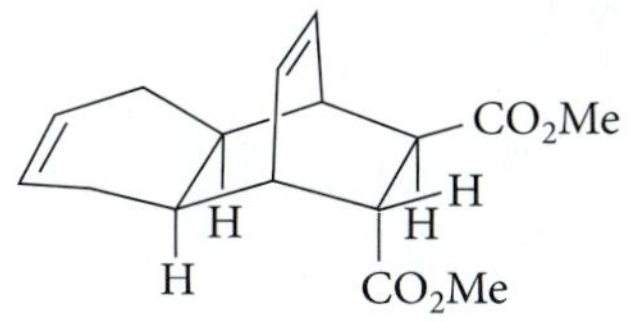

(b)

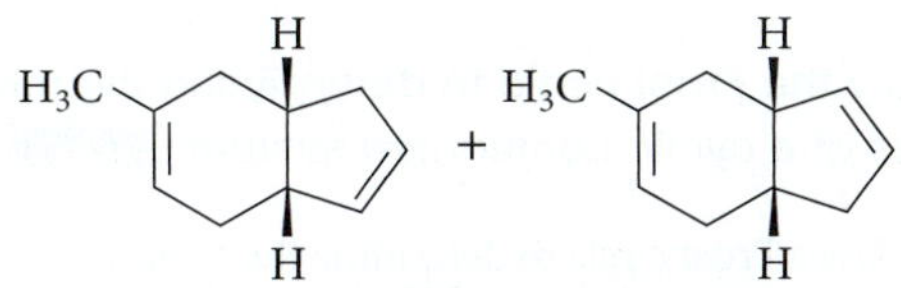

(a mixture of two products)

- **Use their reacting conformations to predict the relative reactivity of a series of conjugated dienes in the Diels–Alder reaction. 15.3A,C**

15.47 In each part, rank the dienes in order of increasing Diels–Alder reactivity with a dienophile. Explain your choices.

(a)

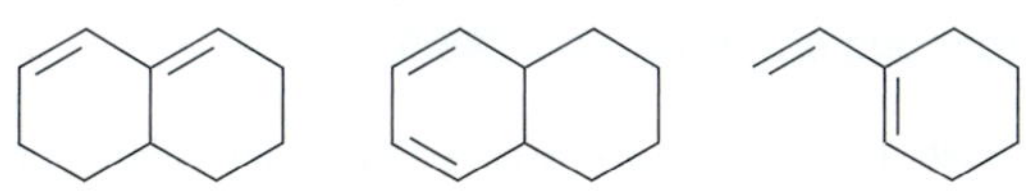

A *B* *C*

(b)

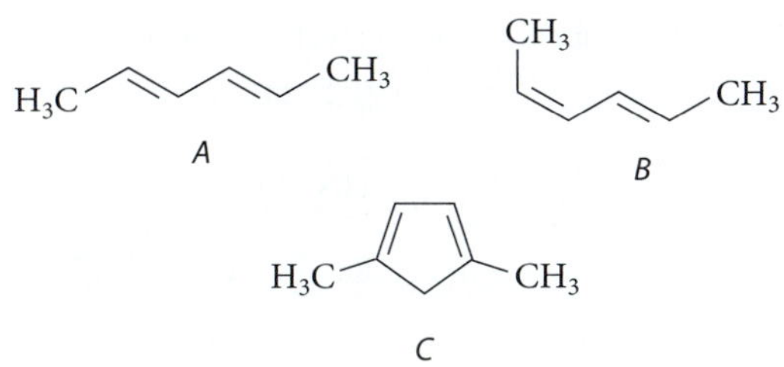

15.48 The following natural product readily gives a Diels–Alder adduct with maleic anhydride (structure in Problem 15.45a) under mild conditions. What is the most likely configuration of the two double bonds (cis or trans)? Explain.

$$CH_3(CH_2)_5CH{=}CH{-}CH{=}CH(CH_2)_7\overset{\overset{\large O}{\|}}{C}OH$$

15.49 Explain why 4-methyl-1,3-pentadiene is much less reactive as a diene in Diels–Alder reactions than (*E*)-1,3-pentadiene, but its reactivity is similar to that of (*Z*)-1,3-pentadiene.

- **Give the 1,4- and 1,2-addition products for the addition reaction of a conjugated diene. Give the curved-arrow or fishhook mechanism for each addition. 15.4A**

15.50 (a) Which carbocation is more stable: the one formed by protonation of isoprene at carbon-1 or the one formed by protonation of isoprene at carbon-4? Explain.

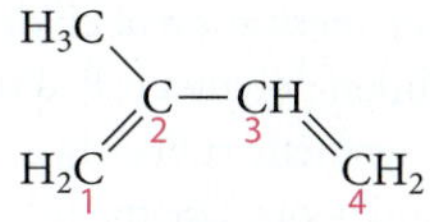

isoprene

(b) Predict the products expected from the addition of one equivalent of HBr to isoprene; explain your reasoning.

(c) Which is likely to be the major kinetically controlled product? Explain.

15.51 (a) Use mechanistic reasoning to predict the products expected from the addition of one equivalent of HBr to (*E*)-1,3,5-hexatriene; explain your reasoning.

(b) Which of the products in part (a) is likely to be the major kinetically controlled product?

15.52 The addition of HBr to 1,3-butadiene in the presence of peroxides gives mostly 1,4-addition product and a small amount of 1,2-addition product. Using the fishhook notation, show the initiation and propagation steps of this reaction that lead to both products.

- **Recognize whether the product of a reaction results from kinetic control or thermodynamic control. 15.4B**

15.53 When 1,3-cyclopentadiene and maleic anhydride (structure in Problem 15.45a) are allowed to react at room temperature, a Diels–Alder reaction takes place in which the endo product is formed as the major product. When this product is heated above its melting point of 165 °C, it is transformed into an equilibrium mixture that contains about 57% of the exo stereoisomer and 43% of the endo stereoisomer. (The equilibrium constant for interconversion of the two stereoisomers probably does not vary greatly with temperature.)

(a) Show these transformations with equations, including the structures of all compounds.

(b) According to these observations, is the Diels–Alder reaction of maleic anhydride and 1,3-cyclopentadiene at room temperature a kinetically controlled or a thermodynamically controlled reaction?

- **Given the structure of a conjugated diene, draw structures for the polymer resulting from 1,4-addition polymerization; draw free-radical mechanisms for the reaction using the fishhook notation. 15.5**

15.54 (a) When HCl and 1-buten-3-yne react, HCl adds to the triple bond to form a diene called *chloroprene.* Give the structure of chloroprene along with a curved-arrow mechanism for the reaction.

$$HC{\equiv}C{-}CH{=}CH_2 + HCl \longrightarrow \text{chloroprene}$$

(b) The free-radical polymerization of chloroprene gives (mostly) a 1,4-addition polymer called *neoprene,* which is used as a synthetic rubber in applications such as wet suits and boots. Use the fishhook mechanism for the polymerization to derive the structure of neoprene.

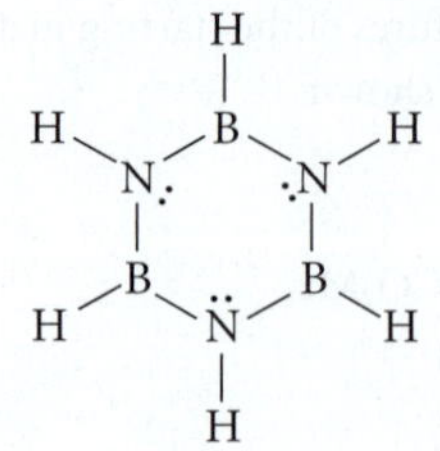

borazole

- **By applying the 4*n* + 2 rule, recognize from its structure whether a compound is likely to be aromatic, antiaromatic, or neither. 15.7D,E**

15.55 Using the Hückel $4n+2$ rule, determine whether each of the following compounds is likely to be aromatic. Explain how you arrived at the π-electron count in each case.

(a) 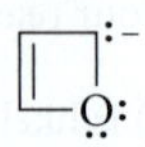 (b) 

(c)  (d) 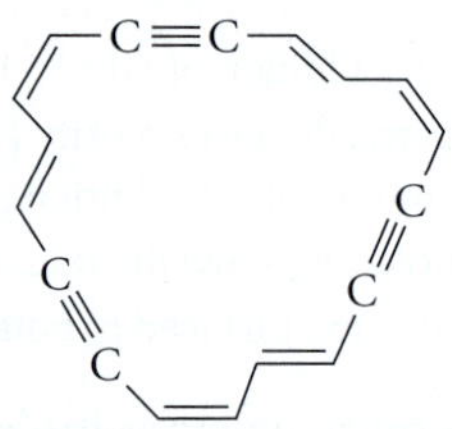

(e) 

15.56 Which of the following molecules is likely to be planar and which nonplanar? Explain.

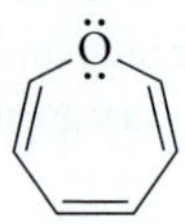

oxepin

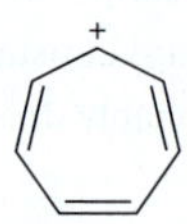

tropylium ion

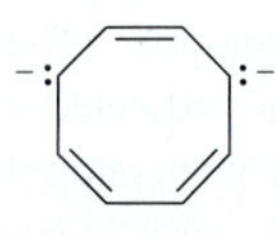

cyclooctatetraenyl dianion

15.57 Explain why borazole (sometimes called *inorganic benzene*) is a very stable compound.

- **Apply the Frost circle to determining the molecular orbitals of a cyclic conjugated system. 15.7D**

15.58 Use a Frost circle to determine the π-electron structure of the cyclopropenyl cation, which has a planar structure and two π electrons.

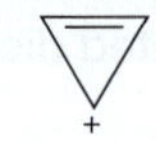

cyclopropenyl cation

- **Draw the relative orientation of an aromatic compound and cation, or two aromatic compounds, poised for noncovalent attraction. 15.8**

15.59 (a) If a lysine residue and a phenylalanine residue are located close to each other in a protein structure, describe how you would expect them to be oriented for the most energetically favorable interaction.

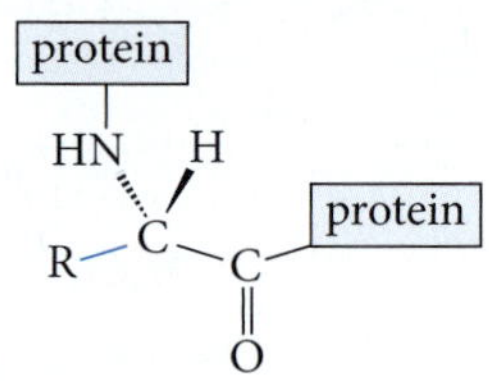

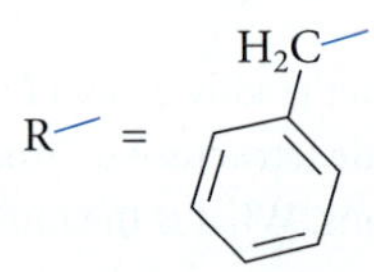

phenylalanine

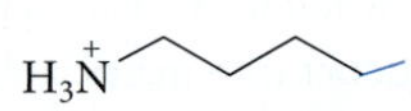

lysine

(b) If two phenylalanine residues (see the previous part) are located close to each other in a protein structure, describe how you would expect them to be oriented for the most energetically favorable interaction.

INTEGRATED PROBLEMS

15.60 Give the principal product(s) expected, if any, when *trans*-1,3-pentadiene reacts under the following conditions. Assume one equivalent of each reagent reacts unless noted otherwise.

(a) Br_2 (dark) in CH_2Cl_2

(b) HBr

(c) H_2 (two molar equivalents), Pd/C

(d) H_2O, H_3O^+

(e) Na^+ EtO^- in EtOH

(f) Maleic anhydride (structure in Problem 15.45a), heat

15.61 Explain each of the following observations.

(a) The allene 2,3-heptadiene can be resolved into enantiomers, but the cumulene 2,3,4-heptatriene cannot.

(b) The cumulene in part (a) can exist as diastereomers, but the allene in part (a) cannot.

15.62 What six-carbon conjugated diene would give the same single product from either 1,2- or 1,4-addition of HBr?

15.63 Predict the products of addition of HBr to 4-methyl-1,3-pentadiene in the presence of peroxides. Use a fishhook mechanism for the propagation steps to justify your prediction.

15.64 The following compound is not aromatic even through it has $4n + 2$ π electrons in a continuous cyclic arrangement. Explain why the compound is not aromatic. (*Hint:* Draw out the hydrogens.)

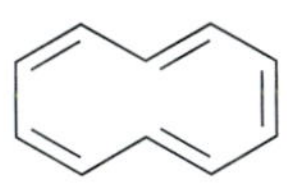

15.65 How would the color of β-carotene (structure in Display 15.11, Sec. 15.2C) be affected by treatment of the compound with a large excess of H_2 over a Pt/C catalyst? Explain.

15.66 Two of the compounds in **Fig. P15.66** are used in sunscreens, and one is not. Identify the compound that does not act as a sunscreen; explain.

15.67 Fluorescein (Display 15.17, Sec. 15.2D) was once used to color the Chicago River green on St. Patrick's Day until it was subsequently replaced with a vegetable dye. One problem with fluorescein was that the green color required bright sunshine for maximum effect—definitely a problem for Chicago in March! Explain this observation using the theory of fluorescence.

15.68 (Refer to Figs. 15.11a and b, Sec. 15.2D.) If a solution of fluorescein at pH = 9 is subjected to visible light at 488 nm, it has maximum fluorescence at 510 nm. Calculate the energy difference between the absorbed and fluorescing radiation in kJ mol^{-1}.

15.69 Draw as many resonance structures as you can for (a) the form of fluorescein present at pH = 9 (Display 15.17) and (b) the fluorescent group of the green fluorescent protein (Display 15.18). Use the curved-arrow notation to derive your structures. Be sure in both cases that a nonbonding electron pair on the anionic oxygen is involved in the resonance interaction.

15.70 A compound *A* (C_6H_{10}) is optically active and has an IR absorption at 2083 cm^{-1}. Partial hydrogenation of *A* with 0.2 equivalent of H_2 over a catalyst gives, in addition to recovered *A*, a mixture of *cis*-2-hexene and *cis*-3-hexene. Identify compound *A*, and explain your reasoning.

15.71 Explain the observation that 2,3-dimethyl-1,3-butadiene and maleic anhydride (structure in Problem 15.45a) readily react to give a Diels–Alder adduct, but 2,3-di-*tert*-butyl-1,3-butadiene and maleic anhydride do not.

15.72 Knowing that conjugated dienes react in the Diels–Alder reaction, a student, M. T. Brainpan, has come to you with an original research idea: to use conjugated alkynes as the diene component in the Diels–Alder reaction (such as the following). Would Brainpan's idea work? Explain. (*Hint:* Draw the structure of the Diels–Alder product, and determine whether the structure is reasonable.)

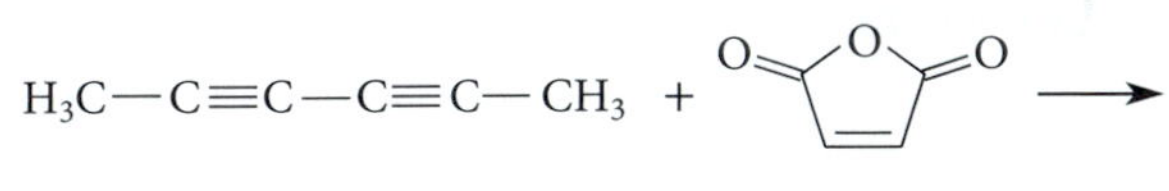

maleic anhydride

15.73 (Refer to **Fig. P15.73**.) The *N*-methylquinolinium ion forms a noncovalent complex with molecule *A* in water that has a standard free energy of dissociation $\Delta G^\circ_d = 28.9$ kJ mol^{-1} (6.9 kcal mol^{-1}). The neutral molecule 4-methylquinoline forms a noncovalent complex with molecule *A* in water with $\Delta G^\circ_d = 22.2$ kJ mol^{-1} (5.3 kcal mol^{-1}).

(a) Calculate the dissociation constant for each complex.

(b) Suggest a reason that the binding of the ion to *A* is stronger than the binding of the neutral molecule to *A*.

(c) The *N*-methylquinolinium ion forms a noncovalent complex with molecule *B* with $\Delta G^\circ_d = 35.2$ kJ mol^{-1} (8.4 kcal mol^{-1}). Suggest a reason that the complex with molecule *B* has a smaller dissociation constant than the complex with molecule *A*.

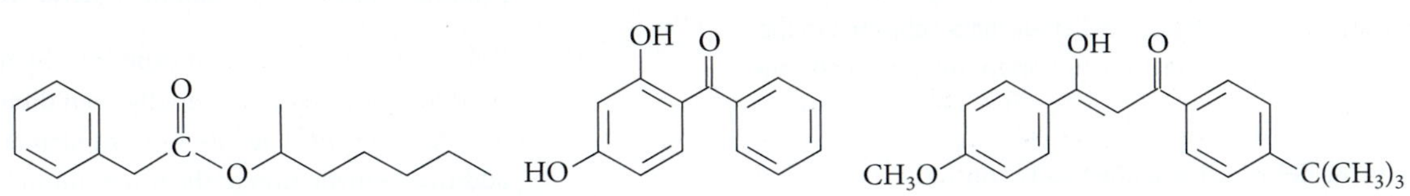

A *B* *C*

FIGURE P15.66

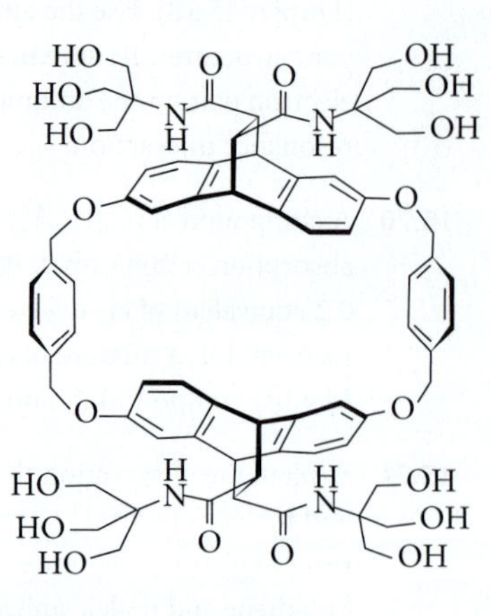

A

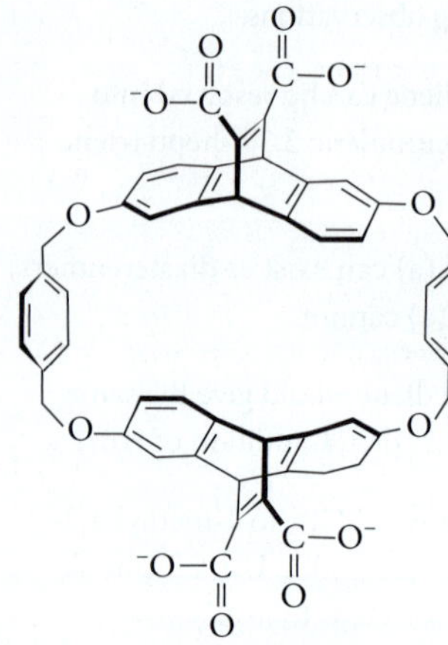

B

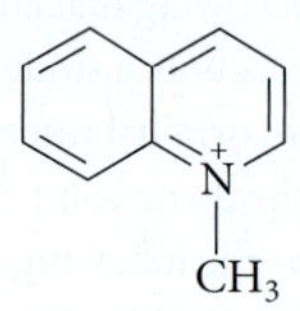

N-methylquinolinium ion

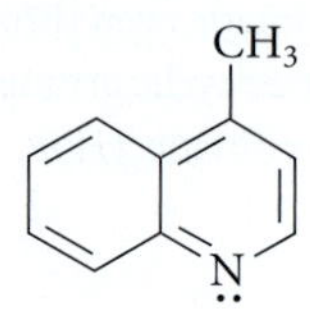

4-methylquinoline

FIGURE P15.73

15.74 This problem describes the result that established the intrinsic preference for 1,2-addition as the kinetically controlled product in the reaction of hydrogen halides with conjugated dienes.

(a) What is the relationship between the products of 1,2- and 1,4-addition in the following reaction?

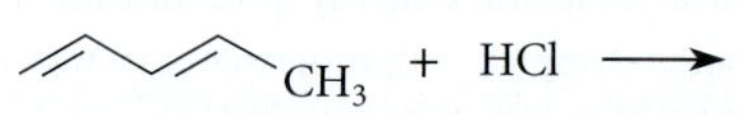

(*E*)-1,3-pentadiene

(b) How does the use of DCl change this relationship, if at all?

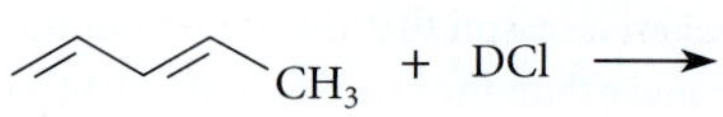

(c) The reaction with DCl gives mostly the kinetically controlled product. Give the structure of this product.

15.75 When the alcohol *A* undergoes acid-catalyzed dehydration, two isomeric alkenes are formed: *B* and *C* (see **Fig. P15.75a**). The relative percentage of each alkene formed is shown as a function of time in **Fig. P15.75b**. The composition of the alkene mixture at very long times is the equilibrium composition. Furthermore, if either alkene is subjected to the conditions of the reaction, the equilibrium mixture of alkenes is obtained.

(a) Is the dehydration a kinetically controlled or thermodynamically controlled reaction? Explain.

(b) Give a structural reason why compound *C* is favored at equilibrium.

(c) Suggest one reason why alkene *B* is formed more rapidly. (*Hint:* Think about probabilities.)

15.76 Consider the bromine addition shown in the following equation. Product *A* is the predominant product formed at low temperature. If the products are allowed to stand under the reaction conditions or are brought to equilibrium at higher temperature, product *B* is the only product formed.

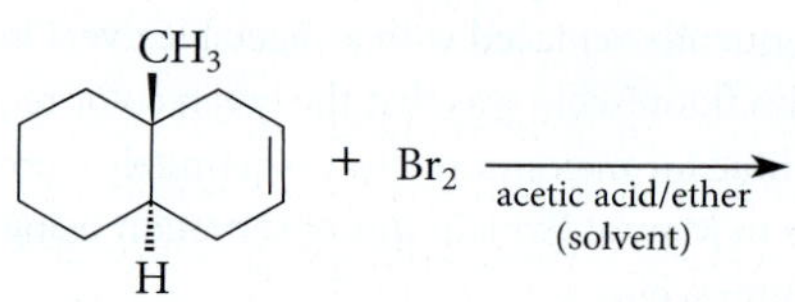

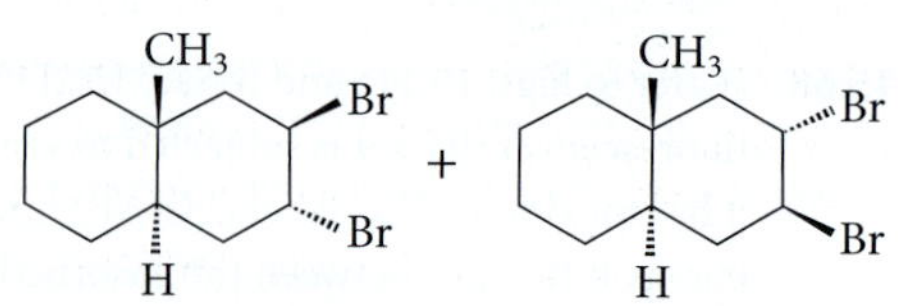

A B

(a) Which is the kinetic product, and which is the thermodynamic product?

(b) Give a structural reason that the thermodynamic product is more stable than the kinetic product.

(c) Propose a mechanism that explains why the kinetic product is formed more rapidly even though it is less stable. (*Hint*: The rate-limiting step of bromine addition is formation of the bromonium ion.)

(d) Propose a mechanism for the equilibration of the two compounds that does not involve the alkene starting material.

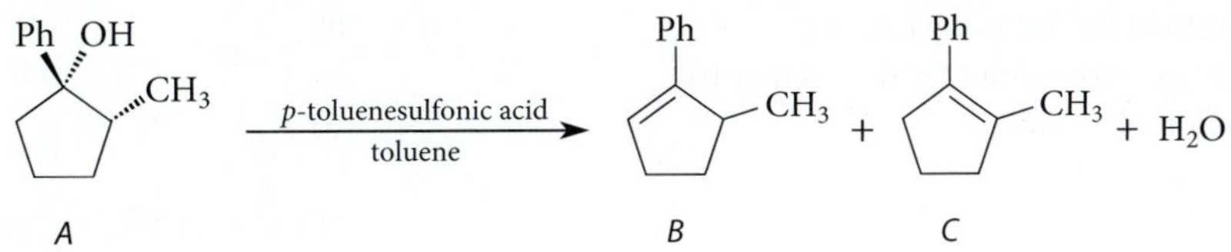

(a)

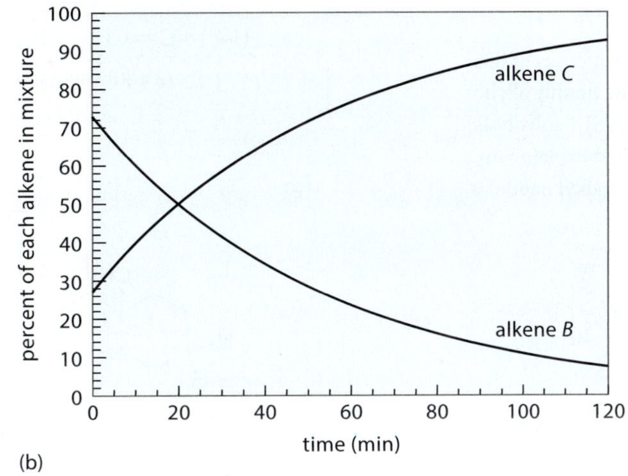

(b)

FIGURE P15.75 The relative amounts of alkene products *B* and *C* in Problem 15.75 as a function of time.

15.77 When an excess of 1,3-butadiene reacts with Cl_2 in chloroform solvent, two compounds, *A* and *B*, both with the formula $C_4H_6Cl_2$, are formed. Compound *B* reacts with more Cl_2 to form compound *C*, $C_4H_6Cl_4$, which proves to be a meso compound. Compound *A* reacts with more Cl_2 to form both *C* and a diastereomer *D*. Propose structures for *A*, *B*, *C*, and *D*, and explain your reasoning.

15.78 When 1,3-cyclopentadiene containing carbon-13 (^{13}C) only at carbon-5 (as indicated by the asterisk in the following equation) is treated with potassium hydride (KH), a species *X* is formed and a gas is evolved. When the resulting mixture is added to water, a mixture of ^{13}C-labeled 1,3-cyclopentadienes is formed as shown in the equation. Identify *X*, and explain both the origin and the percentages of the three labeled cyclopentadienes.

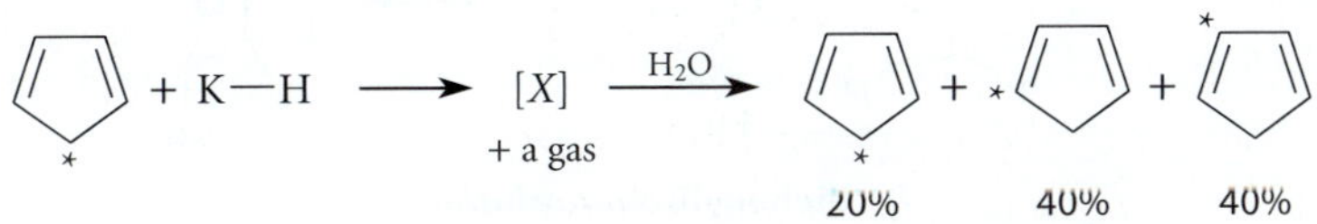

15.79 An amine R_2NH is typically more than 20 pK_a units more acidic than the hydrogens of the carbon analog, R_2CH_2 (the element effect; Sec. 3.7B).

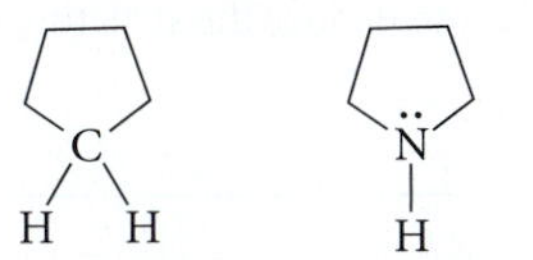

cyclopentane pK_a ~ 55 **pyrrolidine** pK_a ~ 35

However, the acidities of 1,3-cyclopentadiene and pyrrole are an exception.

1,3-cyclopentadiene pK_a = 15 **pyrrole** pK_a = 17

Use the theory of aromaticity to explain this exception. (*Hint:* Remember that the pK_a of a compound is proportional to the free energy *difference* between an acid and its conjugate base.)

15.80 (a) The α-hydrogens of aldehydes and ketones are more than 30 pK_a units more acidic than the hydrogens of alkanes.

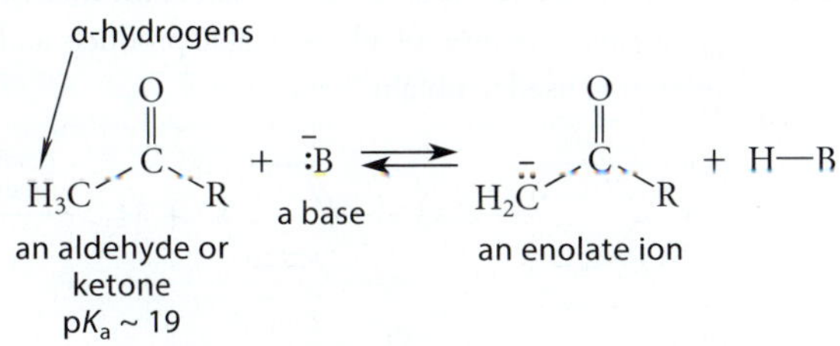

Using polar effects and resonance effects in your argument, explain the enhanced acidity of aldehydes and ketones.

(b) Which α-hydrogen of the following ketone, H^a or H^b, should be most acidic? Or should they be about equally acidic? Explain.

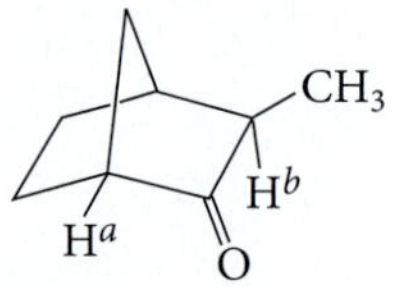

15.81 Invoking Hammond's postulate and the properties of the carbocation intermediates, explain why the doubly allylic alkyl halide *A* undergoes much more rapid S_N1 solvolysis in aqueous acetone than compound *B*. Then explain why compound C, which is also a doubly allylic alkyl halide, is solvolytically inert.

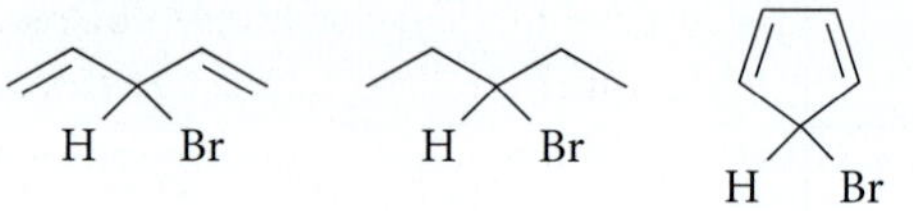

A *B* *C*

15.82 Most alkyl bromides are water-insoluble liquids. When 7-bromo-1,3,5-cycloheptatriene was first isolated, however, its high melting point (203 °C) and its water solubility led its discoverers to comment that it behaves more like a salt. Explain the saltlike behavior of this compound.

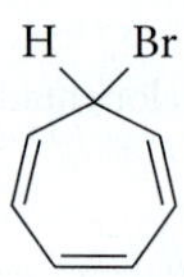

7-bromo-1,3,5-cycloheptatriene (tropylium bromide)

15.83 The hybridizations of the nitrogens in two of the following heterocyclic compounds is the same. Which one is different? What is its hybridization, and why?

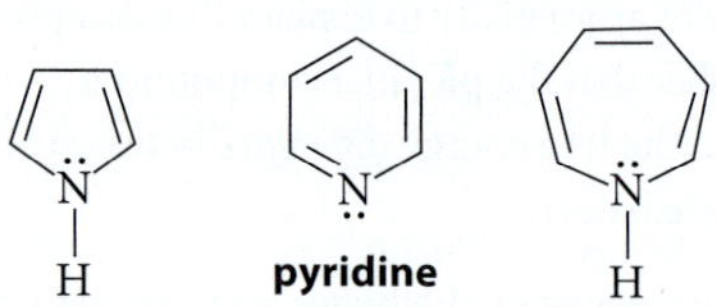

pyrrole **pyridine** **azepine**

15.84 Complete the reactions given in each of the following parts, giving the structures of all reasonable products and the reasoning used to obtain them.

(a)

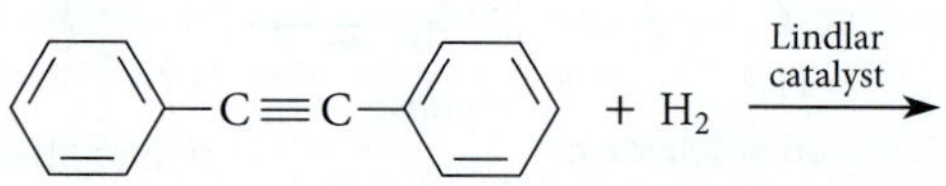

(b)

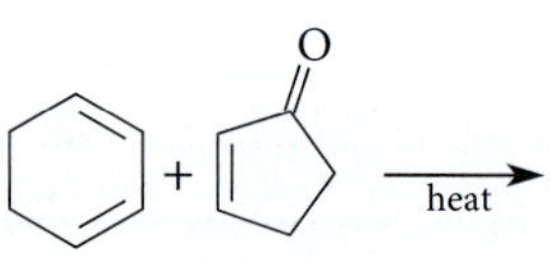

(c)

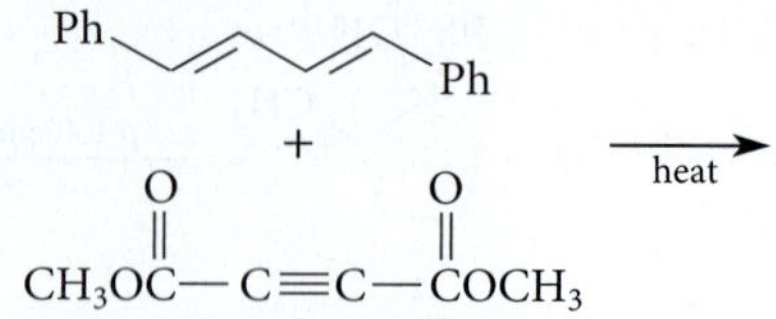

(d)

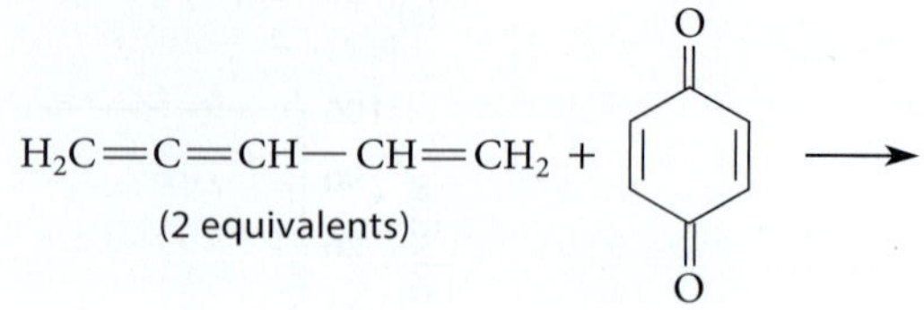

benzoquinone

(e)

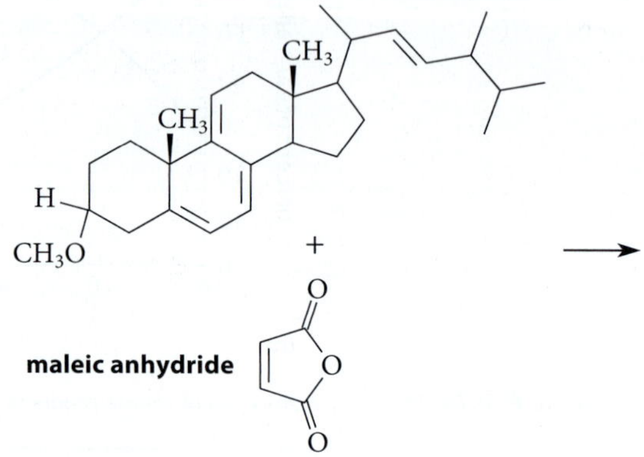

maleic anhydride

(f)

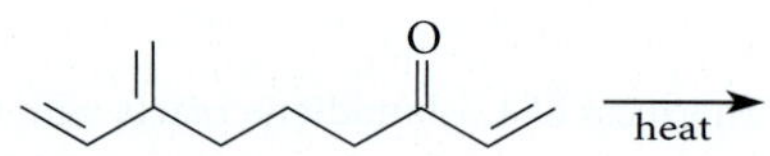

(a compound with 10 carbon atoms)

(g)

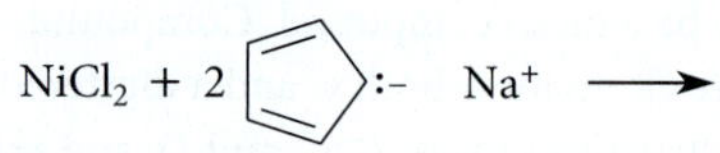

15.85 Account for the fact that the antibiotic *mycomycin* (structure in **Figure P15.85**) is optically active.

15.86 One interesting use of Diels–Alder reactions is to trap very reactive alkenes that cannot be isolated and studied directly. One compound used as a diene for this purpose is diphenylisobenzofuran, which reacts as follows:

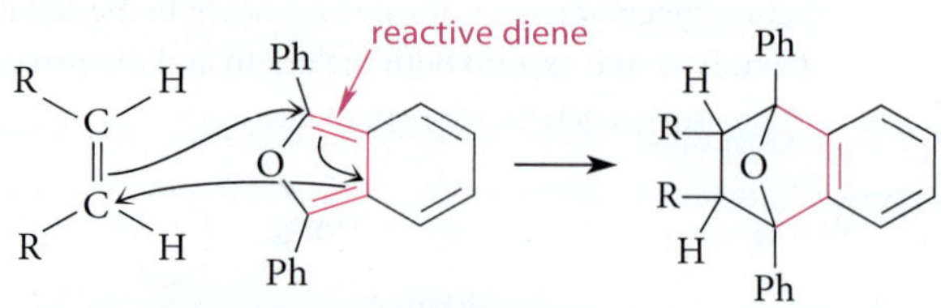

diphenylisobenzofuran

(The formation of an aromatic ring in the product helps ensure that the Diels–Alder reaction is driven to completion.) In the following reaction, use the structure of the Diels–Alder product to deduce the structure of

$$HC{\equiv}C{-}C{\equiv}C{-}CH{=}C{=}CH{-}CH{=}CH{-}CH{=}CH{-}CH_2{-}CO_2H$$

mycomycin

FIGURE P15.85

the reactive species formed in the reaction. Show by the curved-arrow notation how the reactive species is formed, and explain what makes it particularly unstable.

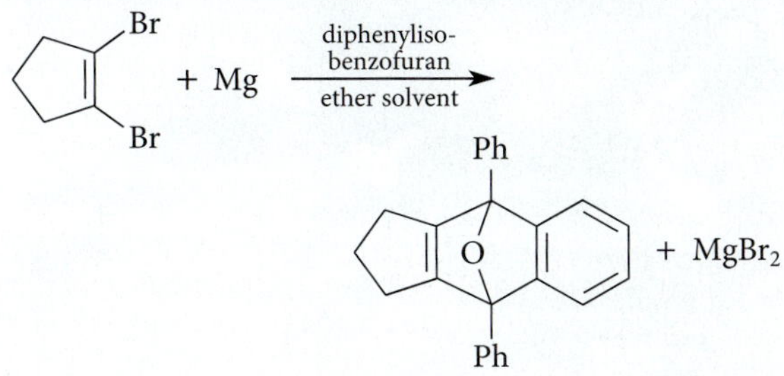

15.87 Use the structure of the Diels–Alder adduct to deduce the structure of the product *X* in the following reaction. Then give a curved-arrow mechanism for the formation of *X*.

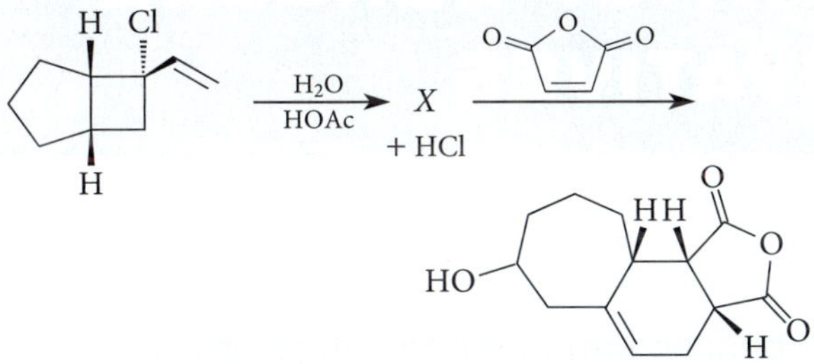

15.88 In 1991, chemists at Rice University reported that they had trapped an unstable compound called *spiropentadiene* using its Diels–Alder reaction with excess 1,3-cyclopentadiene, giving the product in the following reaction. Use the structure of this product to deduce the structure of spiropentadiene.

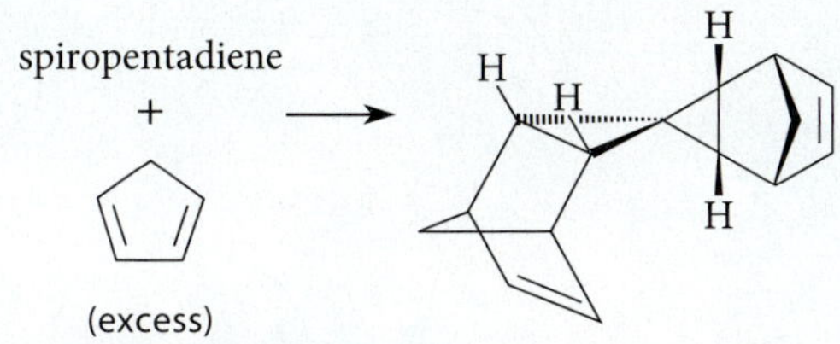

15.89 Account for each of the transformations shown in **Fig. P15.89** with a curved-arrow mechanism. (Don't try to explain any percentages.) In part (c), identify *X*, give the mechanisms for both the formation and the subsequent reaction of *X*, and explain why the equilibrium for the reaction of *X* strongly favors the products.

15.90 Account for the fact that the central "benzene ring" of [4]phenylene undergoes catalytic hydrogenation readily under conditions usually used for ordinary alkenes, but the other benzene rings do not.

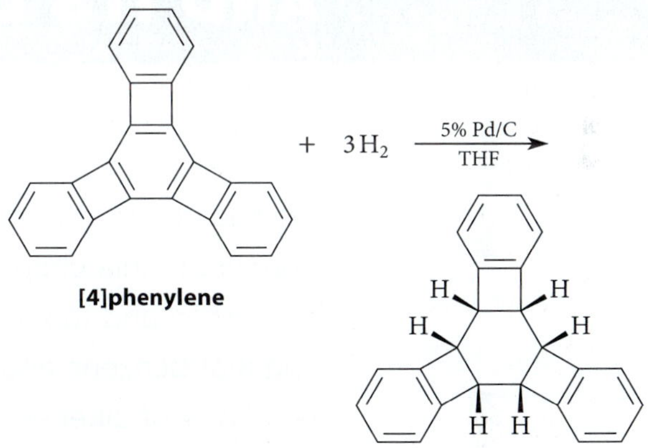

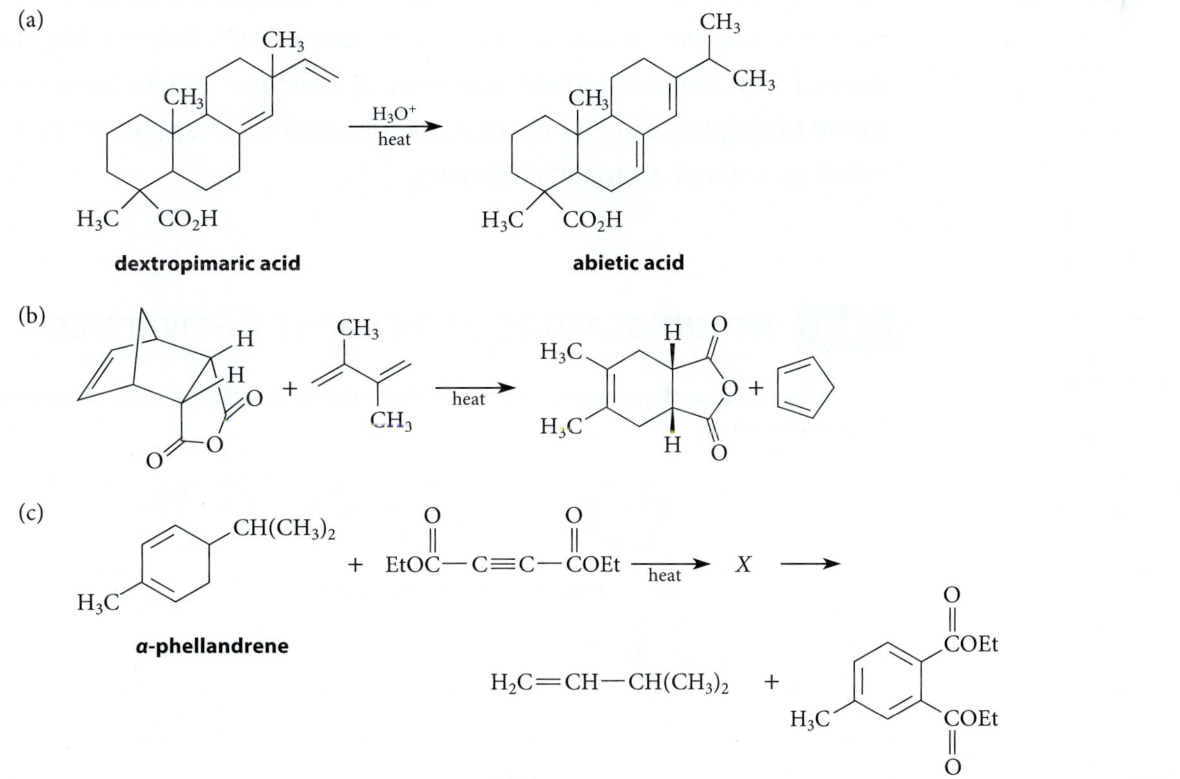

FIGURE P15.89

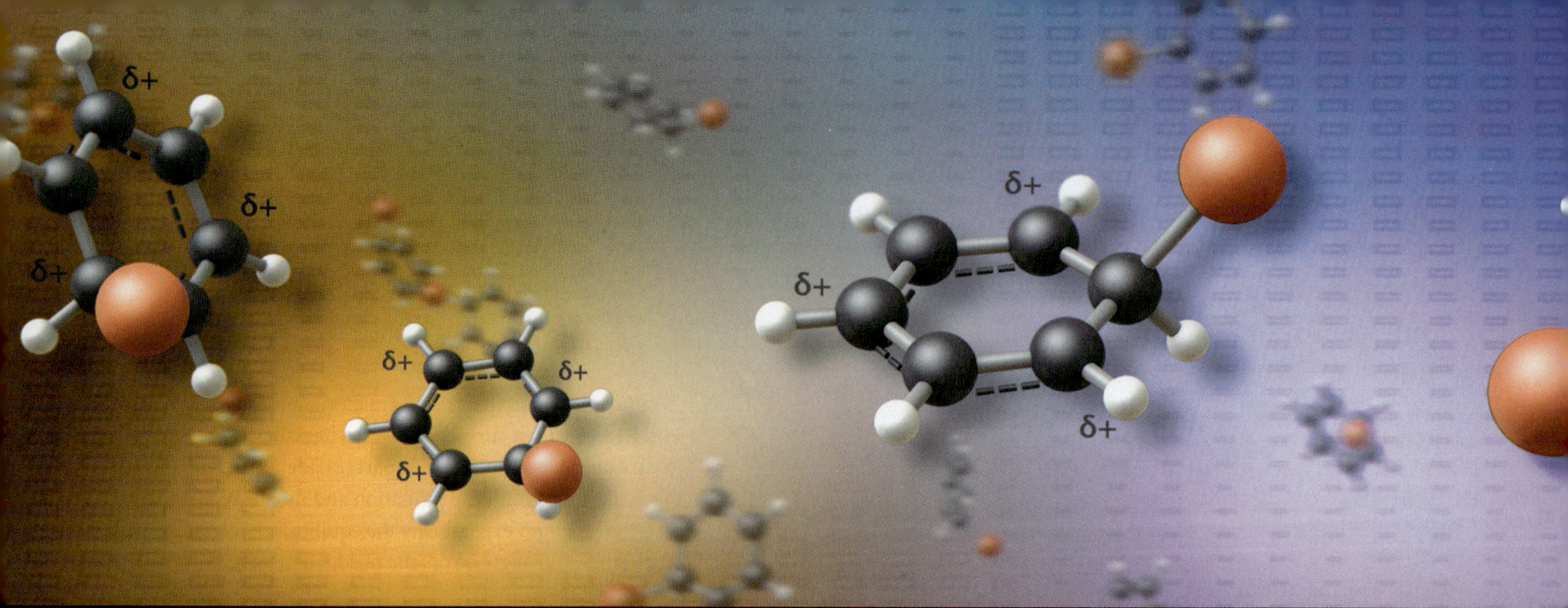

16 THE CHEMISTRY OF BENZENE AND ITS DERIVATIVES

We showed in Chapter 15 that benzene and its derivatives are *aromatic compounds*. In this chapter, we'll show how aromaticity affects the spectroscopic properties and the reactivity of benzene and its derivatives. In particular, we'll find that benzene and its derivatives do not undergo most of the usual addition reactions of alkenes. Instead, they undergo a type of reaction called *electrophilic aromatic substitution*, in which a hydrogen on an aromatic ring is substituted by another group. Such substitution reactions can be used to prepare a variety of substituted benzenes from benzene itself. Most of this chapter is concerned with the substitution reactions of benzene and its derivatives, including some biological aspects of aromatic chemistry. Chapters 17 and 18 deal with other aspects of aromatic chemistry.

16.1 NOMENCLATURE OF BENZENE DERIVATIVES

The nomenclature of benzene derivatives follows the same rules used for other substituted hydrocarbons:

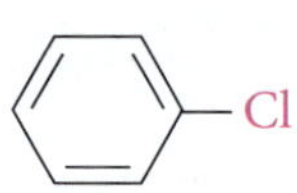

chlorobenzene

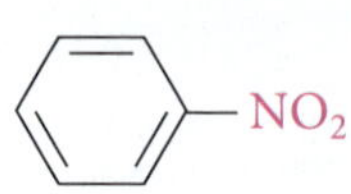

nitrobenzene

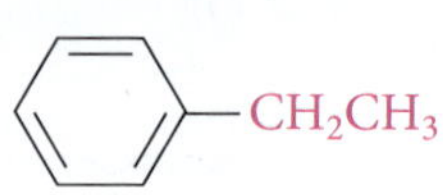

ethylbenzene

The *nitro group*, abbreviated —NO_2, which is a part of the nitrobenzene structure shown here, may be less familiar than the other substituent groups. The nitro group can be represented in more detail as a resonance hybrid of two equivalent dipolar structures:

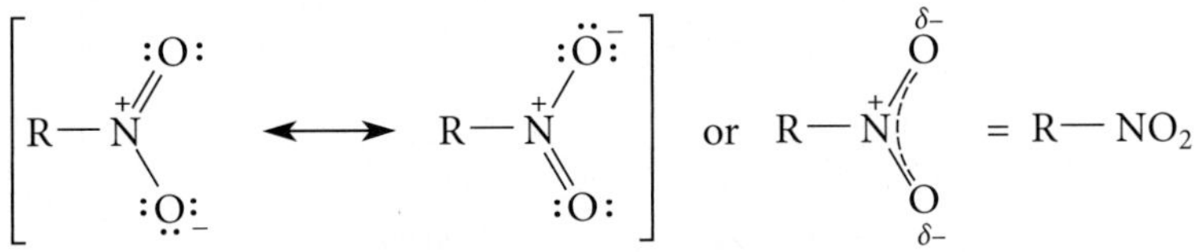

(16.1)

Some monosubstituted benzene derivatives have well-established common names that should be learned.

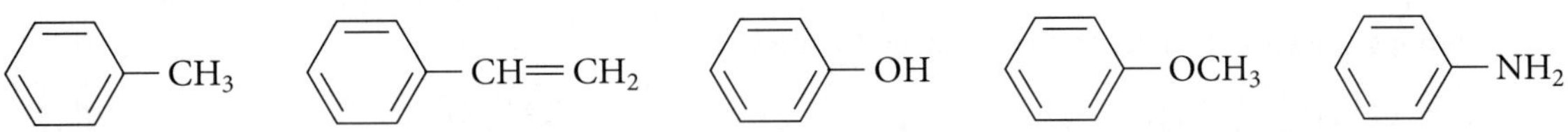

toluene **styrene** **phenol** **anisole** **aniline**

(16.2)

In fact, the IUPAC recognizes all of these compounds as **parent structures**—structures on which names are based. In other words, toluene is indexed as *toluene* rather than *methylbenzene*; styrene is indexed as *styrene* rather than *vinylbenzene*.

Other parent hydrocarbon structures recognized by the IUPAC are cumene (isopropylbenzene), cymene (the benzene substituted with both a methyl and an isopropyl group), mesitylene (1,3,5-trimethylbenzene), and the xylenes (the dimethylbenzenes).

The positions of substituent groups in disubstituted benzenes can be designated in two ways. Modern substitutive nomenclature utilizes numerical designations in the same manner as that for other compound classes. However, an older system, which is still used, employs special letter prefixes. The prefix *o* (for *ortho*) is used for substituents in a 1,2-relationship; *m* (for *meta*) for substituents in a 1,3-relationship; and *p* (for *para*) for substituents in a 1,4-relationship.

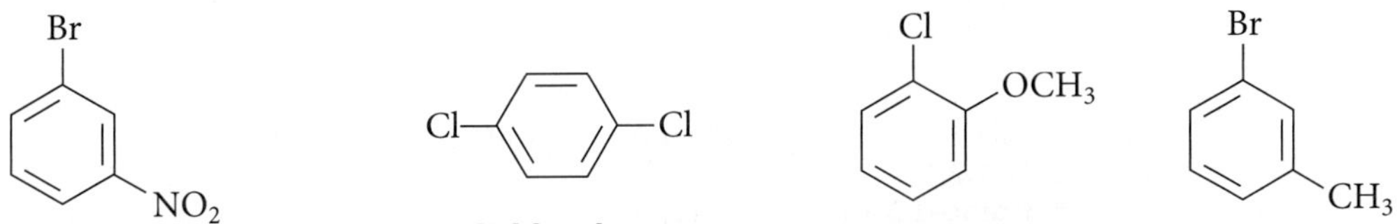

1-bromo-3-nitrobenzene (*m*-bromonitrobenzene) **1,4-dichlorobenzene (*p*-dichlorobenzene)** **2-chloroanisole (*o*-chloroanisole)** **3-bromotoluene (*m*-bromotoluene)**

(16.3)

As the first two examples illustrate, when none of the substituents qualifies as a principal group, the substituents are cited and numbered in alphabetical order. The parent structures in Display 16.2 are used in nomenclature even though they do not contain principal groups. So, in the third example of Display 16.3, anisole is recognized as the parent structure; in the last example, toluene is the parent structure used in the name. In contrast, if a substituent is eligible for citation as a *principal group* (See Appendix I), such as the OH group in the following structure, it is assumed to be at carbon-1 of the ring.

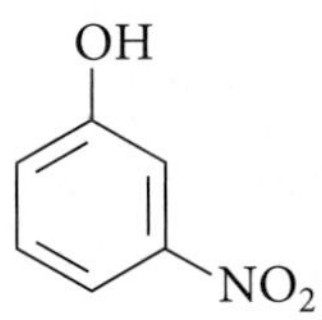

3-nitrophenol (*m*-nitrophenol)
OH group is the principal group

Some disubstituted benzene derivatives also have time-honored common names. The dimethylbenzenes are called *xylenes*, and the methylphenols are called *cresols*.

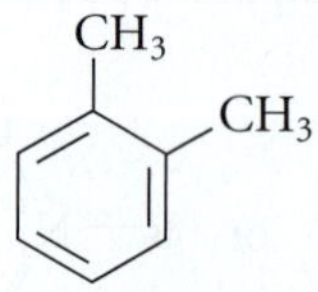

o-xylene

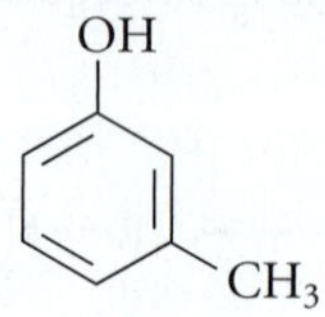

m-cresol

(16.4)

The IUPAC also recognizes the xylenes and cresols as parent hydrocarbons for indexing purposes.

The phenols with two hydroxy groups also have important common names. These are also among the parent structures recognized by the IUPAC.

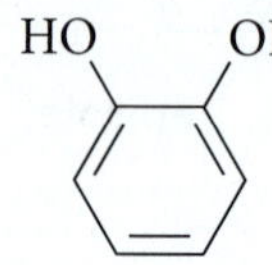

catechol

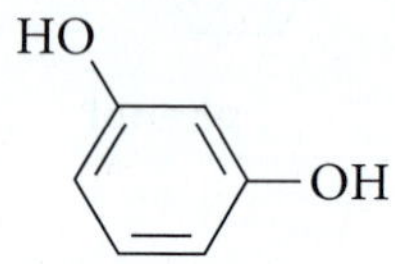

resorcinol

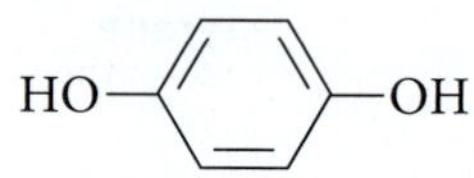

hydroquinone

(16.5)

When a benzene derivative contains more than two substituents on the ring, the *o*, *m*, and *p* designations are not appropriate; only numbers may be used to designate the positions of substituents. The usual nomenclature rules are followed (Secs. 2.4C, 4.2A, and 8.2).

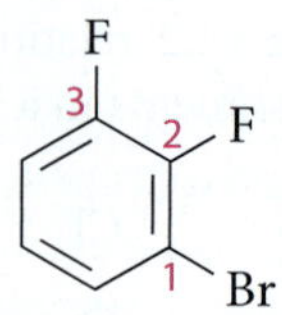

alphabetical citation: ***bromodifluoro***
numbering: **1,2,3**
name: **1-bromo-2,3-difluorobenzene**

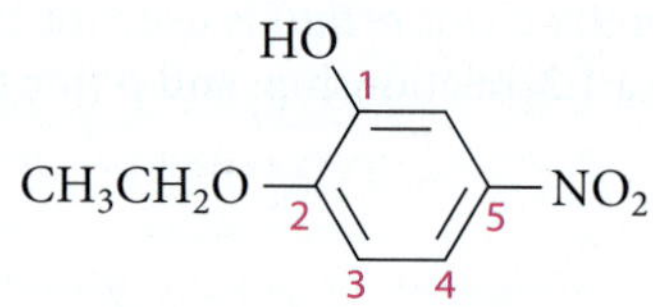

2-ethoxy-5-nitrophenol

(16.6)

Sometimes it is simpler to name a benzene ring as a substituent group. A benzene ring or substituted benzene ring cited as a substituent is referred to generally as an **aryl group**; this term is analogous to *alkyl group* in nonaromatic compounds (Sec. 2.8B). When an unsubstituted benzene ring is a substituent, it is called a **phenyl group**. This group can be abbreviated Ph—. It is also sometimes abbreviated by its group formula, C_6H_5—.

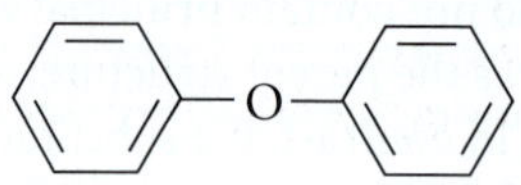

C_6H_5—O—C_6H_5 Ph—O—Ph

diphenyl ether (phenoxybenzene)
three different ways to write the structure

(16.7)

The Ph—CH_2— group is called the **benzyl group**.

PhCH_2—Cl

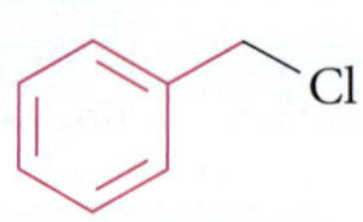

benzyl chloride or **(chloromethyl)benzene**

(16.8)

Notice the difference between the *phenyl* group, Ph—, and the *benzyl* group, Ph—CH_2—. Because the term *benzyl* suggests a benzene ring, it's easy to confuse these two.

Focused Problems

16.1 Name the following compounds.

(a) 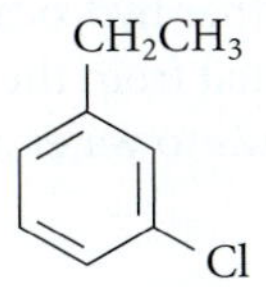
(b) 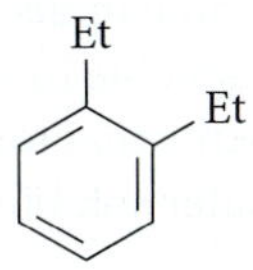
(c) 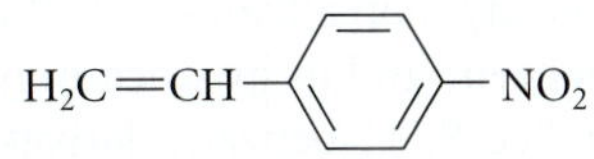

(d) 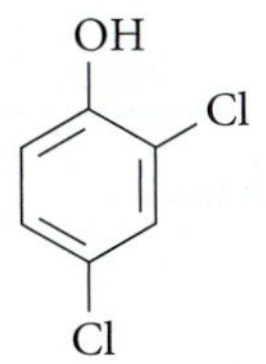
(e) 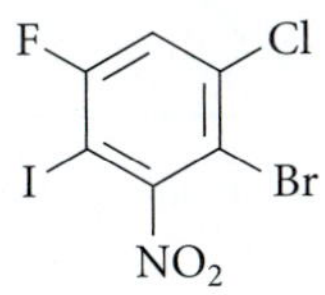
(f) 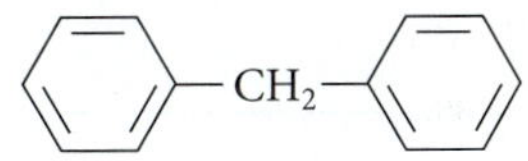

16.2 Draw the structure for each of the following compounds.

(a) *p*-Chlorophenol
(b) *m*-Nitrotoluene
(c) 3,4-Dichlorotoluene
(d) 1-Bromo-2-propylbenzene
(e) Methyl phenyl ether
(f) Benzyl methyl ether
(g) *p*-Xylene
(h) *o*-Cresol
(i) 2,4,6-Trichloroanisole (a compound present in tainted wine corks)

16.2 PHYSICAL PROPERTIES OF BENZENE DERIVATIVES

The boiling points of benzene derivatives are similar to those of other hydrocarbons with similar shapes and molecular masses.

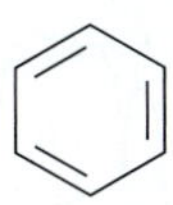  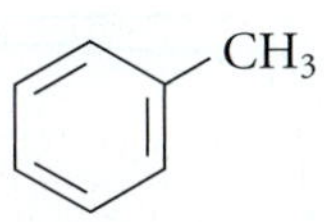

	benzene	**cyclohexane**	**toluene**
bp	80.1 °C	80.7 °C	110.6 °C
mp	5.5 °C	6.6 °C	−95 °C

The melting points of benzene and cyclohexane are unusually high because of their symmetry (Sec. 8.5D). The boiling points of toluene and benzene fit the general trend (Sec. 8.5A) that the addition of a carbon atom adds 20–30 °C to the boiling point.

The melting points of para-disubstituted benzene derivatives are typically much higher than those of the corresponding ortho or meta isomers—another effect of symmetry (Sec. 8.5D).

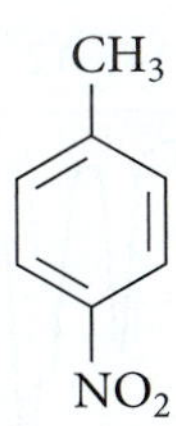 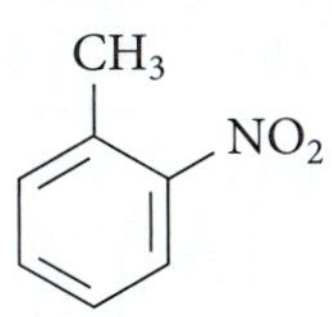 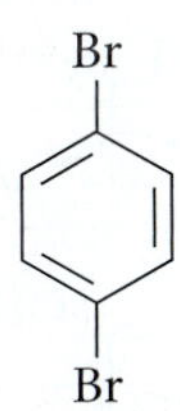 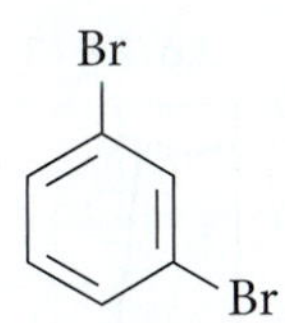

	***p*-nitrotoluene**	***o*-nitrotoluene**	***p*-dibromobenzene**	***m*-dibromobenzene**
mp	54.5 °C	−9.6 °C	87.3 °C	−7 °C

This trend can be useful in purifying the para isomer of a benzene derivative from mixtures containing other isomers. (This point is very important in the reactions of some benzene derivatives, as discussed in Sec. 16.5A.) Because the isomer with the highest melting point is usually the one that is most easily crystallized, many para-substituted compounds can be separated from their ortho and meta isomers by recrystallization.

Benzene and other aromatic hydrocarbons are not as dense as water, but they have a lower density than alkanes and alkenes of about the same molecular mass. Like other hydrocarbons, benzene and its hydrocarbon derivatives are insoluble in water. As anticipated from the discussion in Sec. 8.5C, benzene derivatives with substituents that form hydrogen bonds to water are more soluble. For example, phenol has substantial water solubility.

Focused Problem

16.3 Given the boiling point of benzene (80 °C), predict the approximate boiling points of the following compounds.

(a) Ethylbenzene (b) Propylbenzene (c) *p*-Xylene

16.3 SPECTROSCOPY OF BENZENE DERIVATIVES

A. IR Spectroscopy

The most useful absorptions in the infrared spectra of benzene derivatives are the carbon–carbon stretching absorptions of the ring, which occur at lower frequency than the C═C absorptions of alkenes. Two such absorptions are typical: one near 1600 cm^{-1} and the other near 1500 cm^{-1}. These absorptions, illustrated in the IR spectrum of toluene (**Fig. 16.1a**), occur at lower frequency than alkene C═C stretching absorptions because the carbon–carbon bonds in benzene rings have

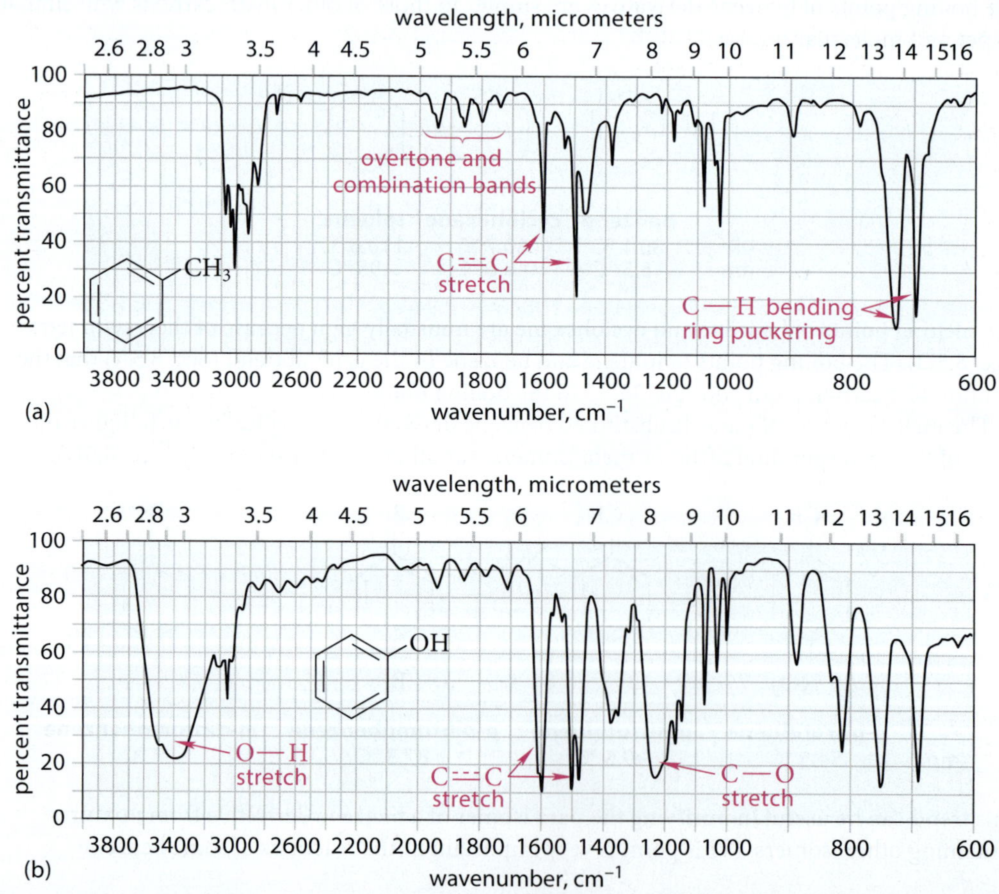

FIGURE 16.1 (a) The IR spectrum of toluene. The carbon–carbon stretching absorption is the major absorption used for diagnosing the presence of benzene rings. (b) The IR spectrum of phenol. The strong O—H and C—O stretching absorptions are much like the same absorptions of tertiary alcohols.

a bond order of 1.5—that is, they are intermediate between single bonds and double bonds. Other characteristic absorptions are also shown in Fig. 16.1a. For example, the *overtone and combination bands* in the 1660–2000 cm^{-1} region were once used to determine the substitution patterns of aromatic compounds. However, NMR spectroscopy is now a more reliable tool for this purpose. Even though the overtone bands are not used in modern structure determination, it is important to be aware of them so that they are not mistaken for other absorptions.

Phenols have not only the characteristic aromatic absorptions, but also O—H and C—O stretching absorptions, which are very much like those of tertiary alcohols. The IR spectrum of phenol is shown in **Fig. 16.1b**.

B. NMR Spectroscopy

The proton NMR spectrum of benzene consists of a singlet at a chemical shift of δ 7.4. Typical alkenes, in contrast, have chemical shifts for internal vinylic protons of δ 5.0–5.7. This example illustrates the general observation that the chemical shifts of benzene derivatives are greater than those of comparably substituted alkenes by about 1.5–2 ppm. (See also Fig. 14.4, Sec. 14.3C.) It's useful to remember that a benzene ring contributes four degrees of unsaturation. So, when dealing with an unknown for which you have deduced an unsaturation number ≥4 from the molecular formula, your eyes should move immediately to the δ 7–8 region of the NMR spectrum. Absorptions in this region immediately alert you to the likelihood of a benzene ring.

What is the reason for the unusual chemical shift of benzene? Recall that the π-electron density in benzene lies in two doughnut-shaped regions above and below the plane of the ring (see Fig. 15.16b, Sec. 15.7B). In an NMR experiment, benzene molecules in solution are moving about randomly and can assume all possible orientations relative to the applied field $\mathbf{B_0}$. However, a particular orientation dominates the chemical shift, as shown in **Fig. 16.2**. In this orientation, a circulation of π electrons around the ring, called a **ring current**, is induced by the external magnetic field. The ring current, in turn, induces a magnetic field $\mathbf{B_i}$ that forms closed loops through the ring. This induced field opposes the applied field along the axis of the ring, but it *adds to the applied field outside of the ring*, in the region occupied by the benzene protons. Therefore, the net field at these protons is higher than it would be in the absence of the ring current. As a result, a correspondingly higher frequency is required for absorption, and the chemical shifts of aromatic protons are increased (Eq. 14.4, Sec. 14.2). This explanation is similar to that for the chemical shifts of vinylic protons in alkenes (see Fig. 14.15, Sec. 14.7A), except that the effect is larger for benzene protons.

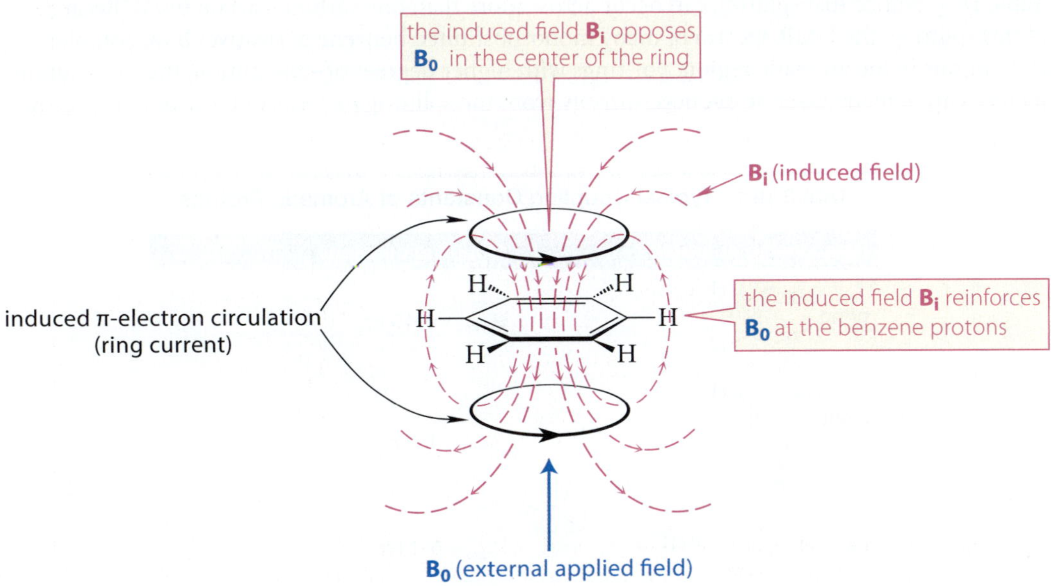

FIGURE 16.2 Origin of the large chemical shift of benzene protons. The field induced by the π-electron ring current (red dashed lines, $\mathbf{B_i}$) opposes the applied field $\mathbf{B_0}$ (blue) in the center of the ring. However, because $\mathbf{B_i}$ forms closed loops, it lies in the same direction as $\mathbf{B_0}$ in the region occupied by the benzene protons. As a result, the induced field increases the local field at the benzene protons. Consequently, these protons require a higher frequency to meet the condition for resonance. The higher resonance frequency translates into a greater chemical shift (Eq. 14.4, Sec. 14.2, and the related discussion).

The ring current and the large chemical shift are characteristic of compounds that are aromatic by the Hückel $4n+2$ rule (Sec. 15.7D). This is reasonable because the basis of both the ring current and aromaticity is the overlap of 2p orbitals in a continuous cyclic array. Many chemists believe that the existence of the ring current (detected by unusually large chemical shifts) is the best *experimental* evidence of aromatic character.

Focused Problems

16.4 Within each set, which compound (1 or 2) should show NMR absorptions with the greater chemical shifts? Explain your choices.

(a)

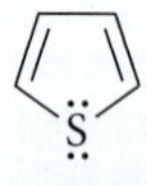

thiophene
(1)

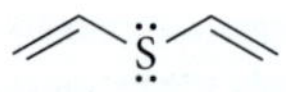

divinyl sulfide
(2)

(b)

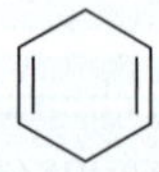

(1)

(2)

16.5 (a) Verify that the following compound meets the Hückel criteria for aromaticity.

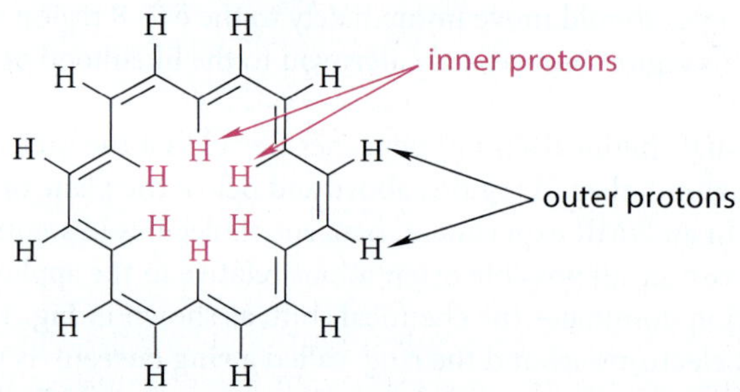

(b) The NMR spectrum of this compound consists of two sets of multiplets: one at δ 9.28 and the other at δ (−2.99); the latter resonance is at 3 ppm *lower* chemical shift than that of TMS—that is, to the right of TMS in a conventional NMR spectrum. The relative integral of the two resonances is 2:1, respectively. Assign the two sets of resonances, and explain why their chemical shifts are so different; in particular, explain why one of the chemical shifts is so small. (*Hint:* Look carefully at the direction of the induced field in Fig. 16.2.)

When the protons in a substituted benzene derivative are nonequivalent, they split each other, and their coupling constants depend on their positional relationships, as shown in **Table 16.1**. Notice that splitting can occur across more than one carbon–carbon bond. Because of this splitting, the NMR spectra of many monosubstituted benzene derivatives have complex absorptions in the aromatic region. For rings with higher degrees of substitution, the substitution pattern can in many cases be deduced directly from the splitting patterns of the aromatic protons.

TABLE 16.1 Typical Coupling Constants of Aromatic Protons

Relationship of protons	Coupling constant
ortho 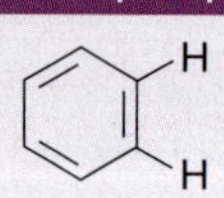	$J_{ortho} = 6–10$ Hz
meta 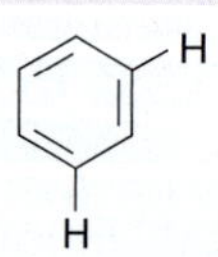	$J_{meta} = 1–3$ Hz
para 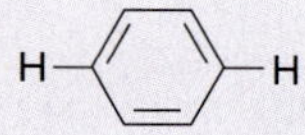	$J_{para} = 0–1$ Hz

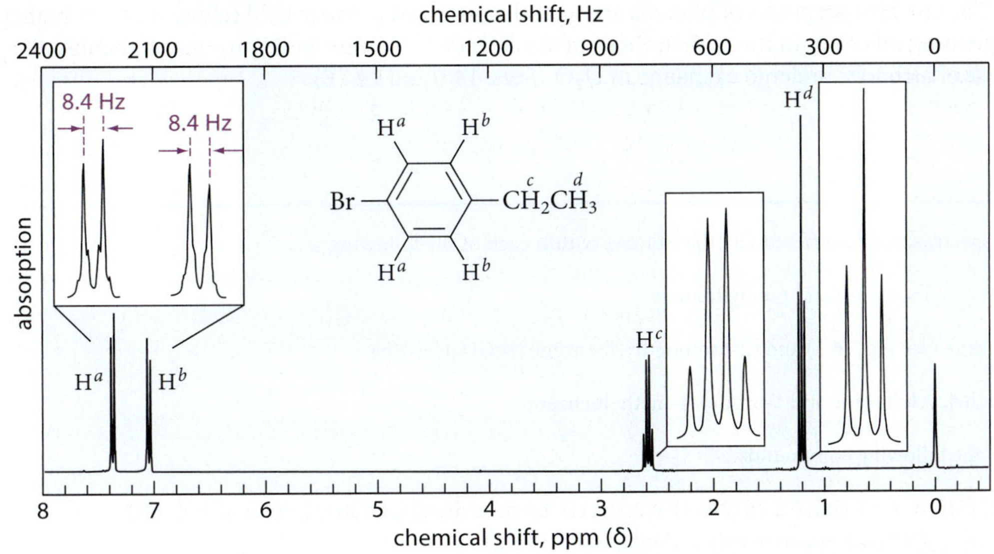

FIGURE 16.3 The proton NMR spectrum of 1-bromo-4-ethylbenzene. Notice three things about this spectrum. First, the "two-leaning-doublet" pattern near δ 7 is typical of para-disubstituted benzene derivatives in which the ring substituents are different. Second, the chemical shifts of the ring protons reflect the electronegativities of nearby groups. Therefore, the protons H^a, which are ortho to the electronegative bromine, have greater chemical shifts than protons H^b, which are ortho to the more electropositive ethyl group. Finally, the chemical shifts of the benzylic protons H^c are slightly greater than the chemical shifts of allylic protons. (See also Fig. 14.4, Sec. 14.3C.)

One particular splitting pattern in the NMR spectra of aromatic compounds occurs often enough that it is worth remembering. This pattern is illustrated by the NMR spectrum of 1-bromo-4-ethylbenzene in **Fig. 16.3**. This spectrum consists of two apparent doublets, centered near δ 7.0 and δ 7.4. Ideally, the lines in these doublets should have equal intensities (see Table 14.2, Sec. 14.4A), but because the chemical shifts of the two coupled protons are similar, the intensities differ from the 1:1 ideal. (This phenomenon is called *leaning*.) In each doublet, the major coupling constant, $J = 8.4$ Hz, reflects the large ortho coupling. A superimposed, very small, para coupling causes the additional fine structure visible in the expansions of these absorptions. *Such a "two-leaning-doublet" pattern is very typical of disubstituted benzene rings in which two different ring substituents have a para relationship.*

STUDY GUIDE LINK 16.1
NMR of Para-Substituted Benzene Derivatives

The spectrum of 1-bromo-4-ethylbenzene also shows how substituent groups can affect the chemical shifts of ring protons. The bromo group is the more electronegative group. The protons ortho to this group (H^a in Fig. 16.3) have a greater chemical shift at δ 7.4. The protons ortho to the more electropositive ethyl group (H^b in Fig. 16.3) have a smaller chemical shift at δ 7.0. (Resonance also affects ring-proton shifts; see Problem 16.57.)

The chemical shifts of *benzylic protons*—protons on carbons adjacent to benzene rings—are in the δ 2–3 region. These chemical shifts are slightly greater than those of allylic protons (see Fig. 14.4, Sec. 14.3C). The chemical shifts of benzylic protons in toluene and ethylbenzene are typical.

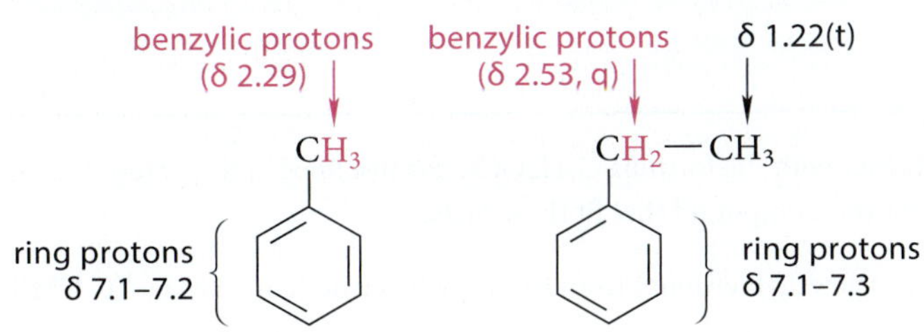

(16.9)

Also typical are the chemical shifts of the benzylic protons of 1-bromo-4-ethylbenzene in Fig. 16.3 (~δ 2.6).

The O—H absorptions of phenols are typically observed at lower field (about δ 5–6 or higher, depending on concentration) than those of alcohols (δ 2–3). The O—H protons of phenols, like those of alcohols, undergo exchange in D_2O (Secs. 14.6 and 14.7E).

Focused Problems

16.6 Explain how to use NMR spectroscopy to differentiate the isomers within each of the following sets.

(a) Mesitylene (1,3,5-trimethylbenzene) and *p*-ethyltoluene

(b) 1-Bromo-4-ethylbenzene (see Fig. 16.3) and (2-bromoethyl)benzene ($BrCH_2CH_2Ph$)

(c) *trans*-1-Bromo-4-methylcyclohexane and 1-bromo-4-methylbenzene

16.7 Give structures for each of the following compounds.

(a) $C_9H_{12}O$: NMR δ 1.27 (3*H*, d, *J* = 7 Hz); δ 2.26 (3*H*, s); δ 3.76 (1*H*, broad s, disappears after D_2O shake); δ 4.60 (1*H*, q, *J* = 7 Hz); δ 6.95, δ 7.10 (4*H*, apparent pair of doublets, *J* = 10 Hz)

(b) $C_8H_{10}O$: IR, 3150–3600 cm^{-1} (strong, broad); NMR, δ 1.17 (3*H*, t, *J* = 8 Hz); δ 2.58 (2*H*, q, *J* = 8 Hz); δ 6.0 (1*H*, broad singlet, disappears with D_2O shake); δ 6.79 (2*H*, d, *J* = 10 Hz); δ 7.13 (2*H*, d, *J* = 10 Hz)

C. ^{13}C NMR Spectroscopy

In ^{13}C NMR (carbon NMR) spectra, the chemical shifts of aromatic carbons are in the carbon–carbon double bond region (δ 110–160); the exact values depend on the ring substituents that are present. The chemical shift of benzene itself is δ 128.5. The chemical shifts of the carbons in ethylbenzene are typical:

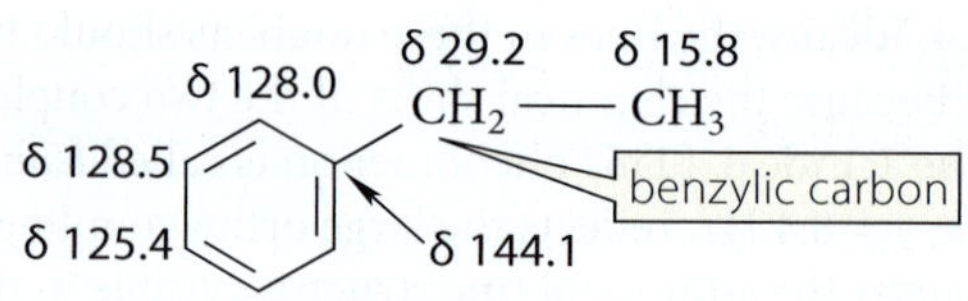

ethylbenzene (16.10)

The higher chemical shift for the quaternary ring carbon is typical. This fits the pattern of larger chemical shifts for carbons that bear no hydrogens (Sec. 14.9B). Also, because the proton-decoupling technique enhances the size of peaks of carbons that bear hydrogens, the peaks for carbons that do *not* bear hydrogens are considerably smaller. This means that the δ 144.1 resonance of ethylbenzene is the smallest peak in the spectrum.

The chemical shifts of benzylic carbons are in the δ 18–30 region—not appreciably different from the chemical shifts of ordinary alkyl carbons. The ^{13}C chemical shift for the benzylic carbon of ethylbenzene, δ 29.2, is typical.

Focused Problems

16.8 A benzene derivative known to be a methyl ether with the formula $C_7H_6OCl_2$ has five lines in its proton-decoupled ^{13}C NMR spectrum. Propose two possible structures for this compound that fit these facts.

16.9 How would you distinguish mesitylene (1,3,5-trimethylbenzene) from isopropylbenzene (cumene) by ^{13}C NMR spectroscopy?

D. UV Spectroscopy

Simple aromatic hydrocarbons have two absorption bands in their UV spectra: a relatively strong band near 210 nm and a much weaker one near 260 nm. The spectrum of ethylbenzene in methanol solvent (**Fig. 16.4**) is typical: $\lambda_{max} = 208$ nm ($\varepsilon = 7520$); 261 nm ($\varepsilon = 200$). Substituent groups on the ring alter both the λ_{max} values and the intensities of both peaks, particularly if the substituent has a nonbonding electron pair in a 2p orbital that can overlap with the π-electron system of the aromatic ring. As is also the case in alkenes, more extensive conjugation is associated with an increase in both λ_{max} and intensity. For example, 1-ethyl-4-methoxybenzene (*p*-ethylanisole) in methanol solvent has absorptions at $\lambda_{max} = 224$ nm ($\varepsilon = 10{,}100$) and 276 nm ($\varepsilon = 1930$); both absorptions occur at greater wavelengths and have greater intensities than the analogous absorptions of ethylbenzene (see Fig. 16.4). The oxygen in *p*-ethylanisole, like the oxygen of water, is sp^2-hybridized (Sec. 1.7E). In this hybridization, one of the oxygen nonbonding electron pairs occupies a 2p orbital, which has the same shape and orientation, and about the same size, as the carbon 2p orbitals of the ring. The oxygen 2p orbital can therefore overlap effectively with the carbon 2p orbitals of the ring. The UV spectrum of anisole is a direct consequence of this overlap.

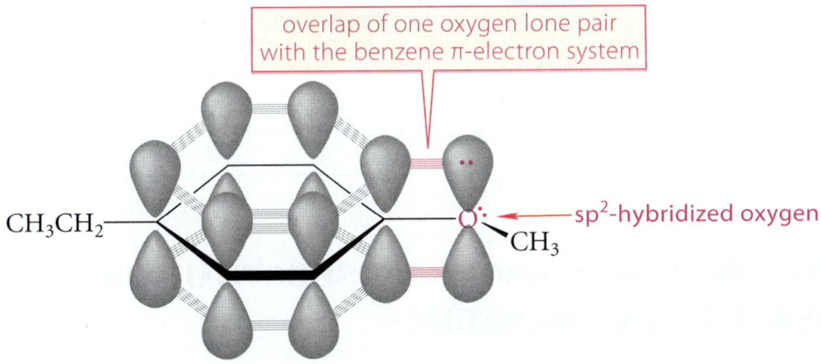

p-ethylanisole (16.11)

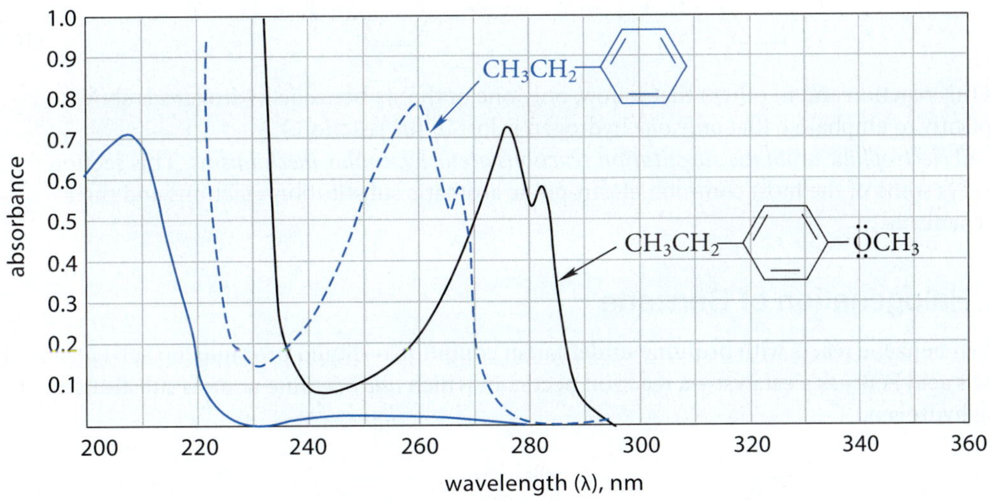

FIGURE 16.4 Comparison of the UV spectra of ethylbenzene (blue) and *p*-ethylanisole (black). The solid lines are spectra taken at the same concentrations. The dashed line is the spectrum of ethylbenzene at 50-fold higher concentration. This comparison shows that the UV spectrum of *p*-ethylanisole is generally more intense. Notice also that the λ_{max} in the *p*-ethylanisole spectrum occurs at a higher wavelength.

Focused Problems

16.10 (a) Explain why compound *A* has a UV spectrum with considerably greater λ_{max} values and intensities than are observed for ethylbenzene.

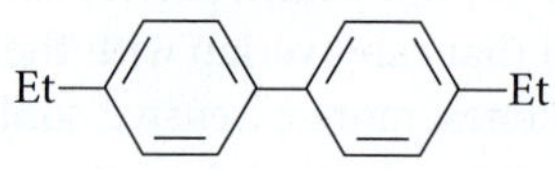

λ_{max} = 256 nm (ε = 20,000)
283 nm (ε = 5100)

A

(b) In view of your answer to part (a), explain why the UV spectra of compounds *B* and *C* are virtually identical. (*Hint:* You may want to build a model of compound *B*.)

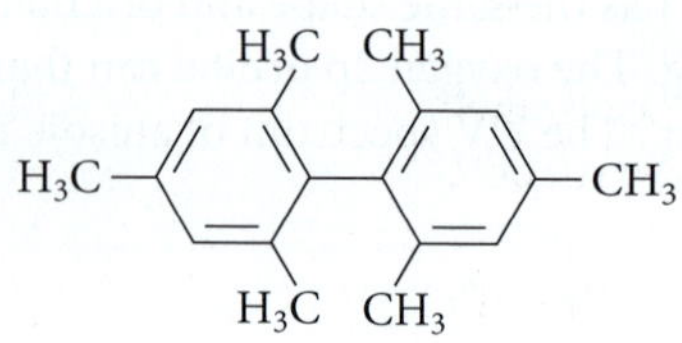

B
bimesityl
λ_{max} = 266 nm (ε = 700)

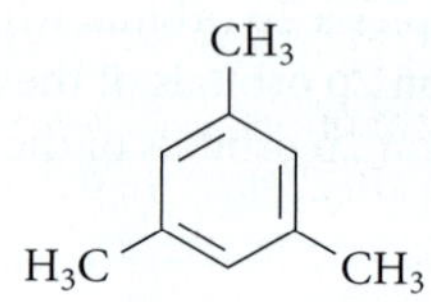

C
mesitylene
λ_{max} = 266 nm (ε = 200)

16.11 How could you distinguish styrene (Ph—CH=CH_2) from ethylbenzene by UV spectroscopy?

16.4 ELECTROPHILIC AROMATIC SUBSTITUTION REACTIONS OF BENZENE

The most characteristic reaction of benzene and many of its derivatives is *electrophilic aromatic substitution*. In an **electrophilic aromatic substitution** reaction, a hydrogen of an aromatic ring is substituted by an electrophile. The general pattern of an electrophilic aromatic substitution reaction is as follows, where E is the electrophile:

(16.12)

(In this reaction and in others that follow, only one of the six benzene hydrogens is shown explicitly to emphasize that *only one* hydrogen is lost in the reaction.)

All electrophilic aromatic substitution reactions occur by similar mechanisms. This section surveys some of the most common electrophilic aromatic substitution reactions and their mechanisms.

A. Halogenation of Benzene

When benzene reacts with bromine under harsh conditions—liquid bromine, no solvent, and the Lewis acid $FeBr_3$ as a catalyst—a reaction occurs in which *one* bromine atom is substituted for a ring hydrogen.

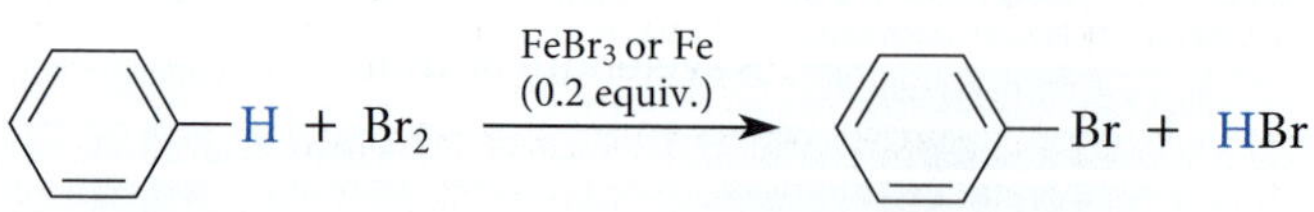

benzene **bromobenzene** (50% yield)

(16.13)

(Because iron reacts with Br_2 to give $FeBr_3$, iron filings can be used in place of $FeBr_3$.) An analogous chlorination reaction using Cl_2 and $FeCl_3$ gives chlorobenzene.

This reaction of benzene with halogens differs from the reaction of alkenes with halogens in two important ways. First is the type of product obtained. Alkenes react spontaneously with bromine and chlorine, even in dilute solution, to give *addition products*.

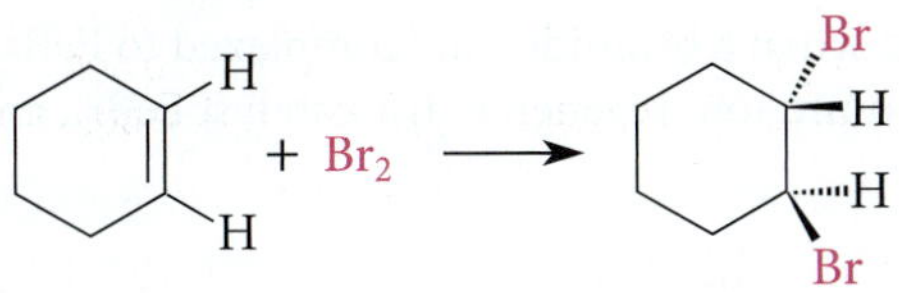

(16.14)

Halogenation of benzene, however, is a *substitution reaction*, because a ring hydrogen is *replaced* by a halogen. Second, the reaction conditions for benzene halogenation are *much* more severe than the conditions for addition of halogens to an alkene.

The first step in the mechanism of benzene bromination is the formation of a complex between Br_2 and the Lewis acid $FeBr_3$ by a Lewis acid–base association.

(16.15a)

Formation of this complex results in a formal positive charge on one of the bromines. A positively charged bromine is a better electron acceptor, and therefore a better leaving group, than a bromine in Br_2 itself. Another (and equivalent) explanation of the leaving-group effect is that $^-FeBr_4$ is a weaker base than Br^-. (Section 9.4F explained that weaker bases are better leaving groups.)

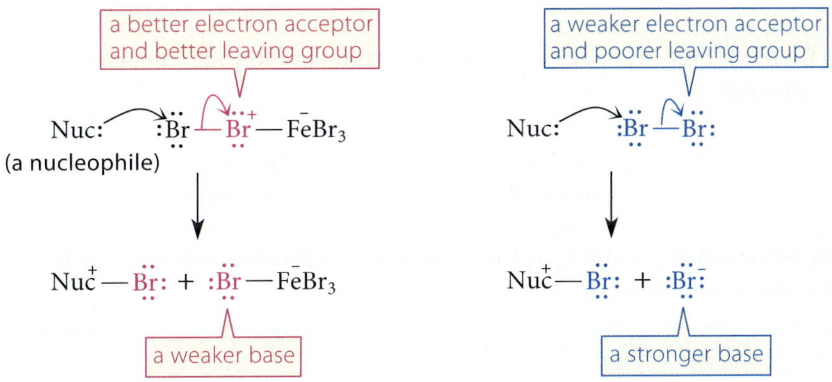

(16.15b)

As we showed in Sec. 9.4F, —Br itself is a good leaving group. The fact that a *much better* leaving group than —Br is required for electrophilic aromatic substitution illustrates how unreactive the benzene ring is.

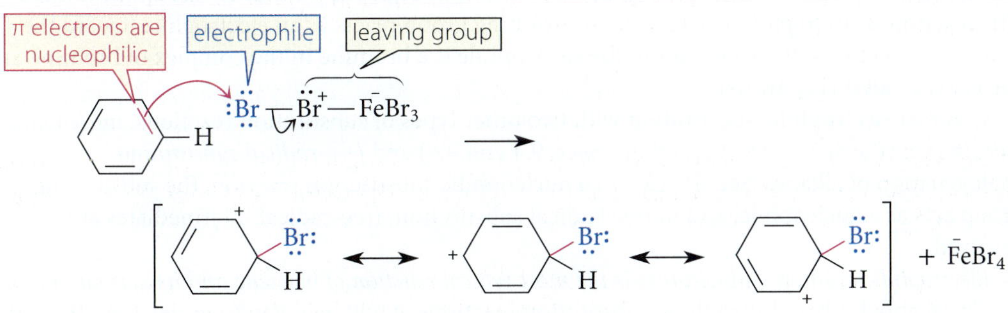

a resonance-stabilized carbocation
(an arenium ion)

(16.16a)

The carbocation intermediate formed in this mechanistic step is an example of an **arenium ion**: a carbocation that is formed by the reaction of an electrophile with a double bond of an aromatic ring. Although the arenium ion is resonance-stabilized, it is *not* aromatic. Its formation disrupts the aromatic stability of the benzene ring. For that reason, harsh conditions (high temperature and a strong Lewis acid catalyst) are required for this reaction to proceed at a useful rate. These conditions are *much* harsher than those required for bromine addition to an ordinary alkene double bond (that is, bromine dissolved in an inert solvent, no catalyst, room temperature or low temperature).

The reaction is completed when a bromide ion (complexed to $FeBr_3$) acts as a base to remove the ring proton from the arenium ion, regenerate the catalyst $FeBr_3$, and give the products bromobenzene and HBr.

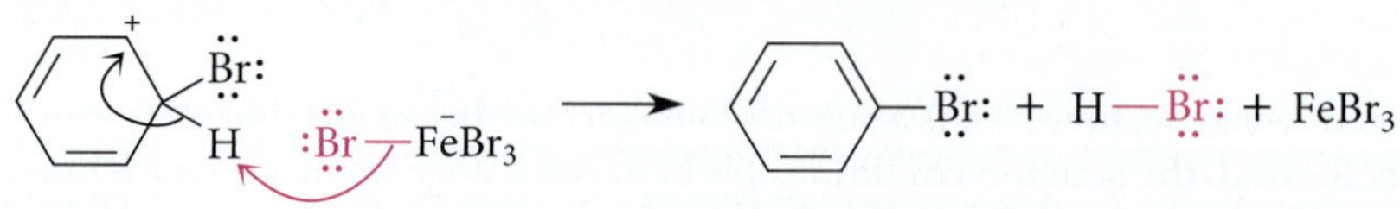

(16.16b)

Recall that loss of a β-proton is one of the characteristic reactions of carbocations (Sec. 9.6B). This step is much like the second step of an E1 reaction of an alkyl halide (Eq. 9.62, Sec. 9.6A). Another typical reaction of carbocations—reaction of bromide ion at the electron-deficient carbon itself—doesn't occur because the resulting addition product would not be aromatic:

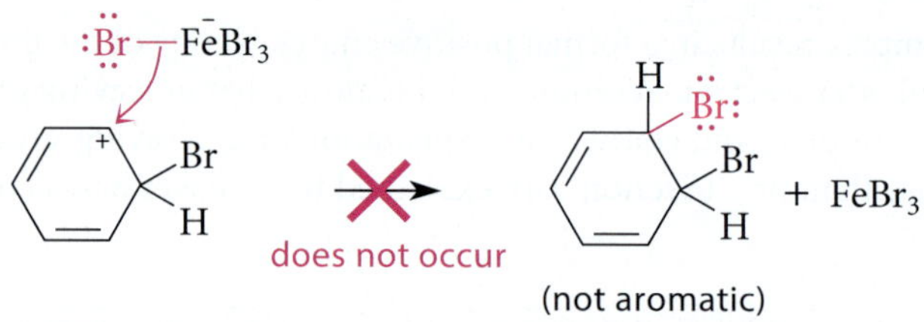

(16.16c)

By losing a β-proton instead (Eq. 16.16b), the carbocation can form bromobenzene, a stable aromatic compound.

Focused Problem

16.12 A small amount of a by-product, *p*-dibromobenzene, is also formed in the bromination of benzene shown in Eq. 16.13. Write a curved-arrow mechanism for the formation of this compound.

B. The Mechanistic Steps of Electrophilic Aromatic Substitution

Halogenation of benzene is one of many **electrophilic aromatic substitution** reactions. The bromination of benzene, for example, is an *aromatic substitution* because a hydrogen of benzene (the aromatic compound that undergoes substitution) is replaced by another group (bromine). The reaction is *electrophilic* because the substituting group reacts as an electrophile toward the benzene π electrons. In bromination, the electrophile is a bromine in the complex of bromine and the $FeBr_3$ catalyst (Eq. 16.16a).

Contrast electrophilic substitution with two other types of substitution reactions: *nucleophilic substitution* (the S_N2 and S_N1 reactions, Secs. 9.4 and 9.6) and *free-radical substitution* (halogenation of alkanes, Sec. 10.4A). In a nucleophilic substitution reaction, the substituting group acts as a nucleophile; and in free-radical substitution, free-radical intermediates are involved.

Electrophilic aromatic substitution is the most typical reaction of benzene and its derivatives. As you learn about other electrophilic substitution reactions, it will help you to understand them if you can identify in each reaction the following three mechanistic steps:

Step 1 **Generation of an electrophile.** The electrophile in bromination is the complex of bromine with $FeBr_3$, formed as shown in Eq. 16.15a.

Step 2 **Nucleophilic reaction of the π electrons of the aromatic ring with the electrophile to form a resonance-stabilized carbocation intermediate (an arenium ion).**

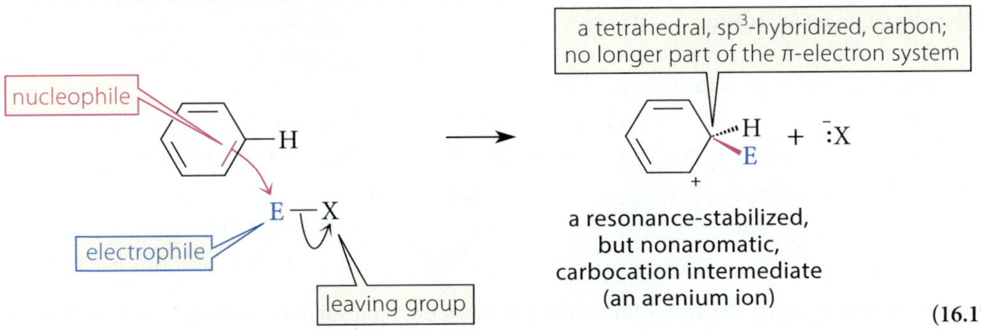

(16.17a)

The electrophile approaches the π-electron cloud of the ring above or below the plane of the molecule. In the arenium-ion intermediate, the carbon at which the electrophile reacts becomes sp^3-hybridized and tetrahedral. This step in the bromination mechanism is Eq. 16.16a.

Step 3 **Loss of a β-proton from the carbocation intermediate to form the substituted aromatic compound.** The proton is lost from the carbon at which substitution occurs. This carbon again becomes part of the aromatic π-electron system.

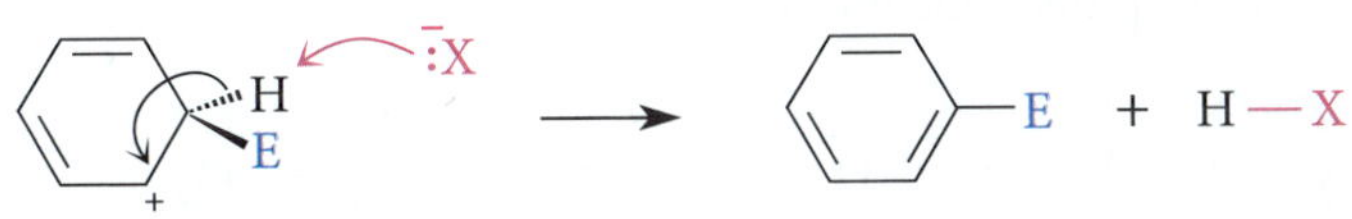

(16.17b)

This step in the bromination mechanism is Eq. 16.16b.

This sequence is classified as *electrophilic* substitution because we are focusing on the nature of the group—an electrophile—that reacts with the aromatic ring. However, the mechanism really involves nothing fundamentally new: like both *electrophilic addition* (Sec. 5.1) and *nucleophilic substitution* (Sec. 9.1), the reaction involves the reactions of nucleophiles, electrophiles, and leaving groups.

Study Problem 16.1

Give a curved-arrow mechanism for the following electrophilic substitution reaction.

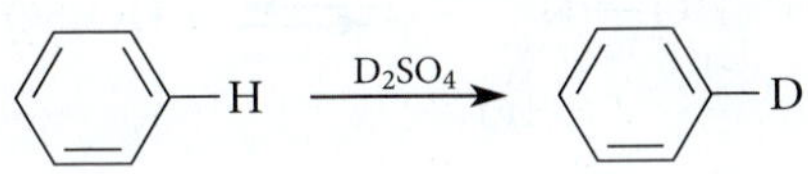

Solution Construct the mechanism in terms of the three steps given in this section.

Step 1 In this reaction, a hydrogen of the benzene ring has been replaced by an isotope D, which must come from the D_2SO_4. Because protons (in the form of Brønsted acids) are good electrophiles, the D_2SO_4 itself can serve as the electrophile.

Step 2 Reaction of the benzene π electrons with the electrophile involves protonation of the benzene ring by the isotopically substituted acid:

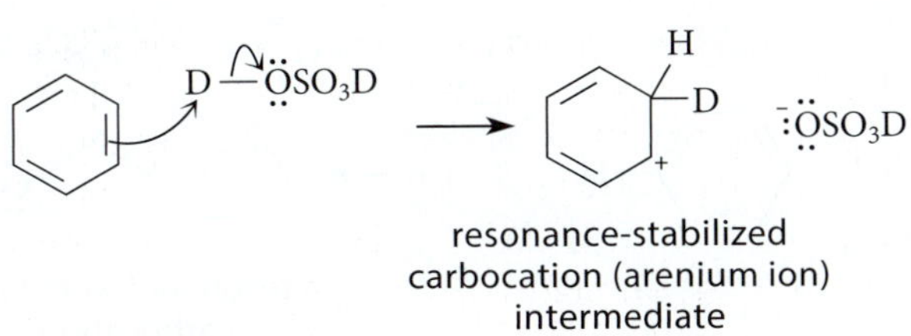

(If you're asking where that "extra" hydrogen in the carbocation came from, don't forget that each carbon of the benzene ring has a single hydrogen that is not shown explicitly in the skeletal structure. One of these is shown in the carbocation because it is involved in the next step.) You should draw the resonance structures of the carbocation intermediate.

Step 3 Removal of the β-proton gives the final product:

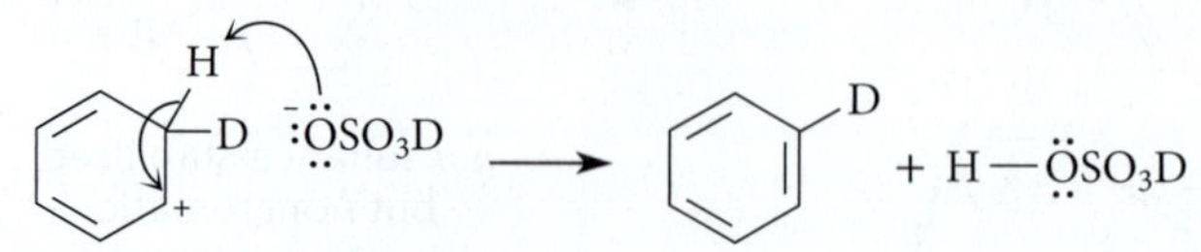

The base can remove either the deuterium or the hydrogen, but removal of the hydrogen is faster because of the primary deuterium kinetic isotope effect (Sec. 9.5D).

C. Nitration of Benzene

Benzene reacts with concentrated nitric acid, usually in the presence of a sulfuric acid catalyst, to form nitrobenzene. In this reaction, called *nitration*, the nitro group, $—NO_2$, is introduced into the benzene ring by electrophilic substitution.

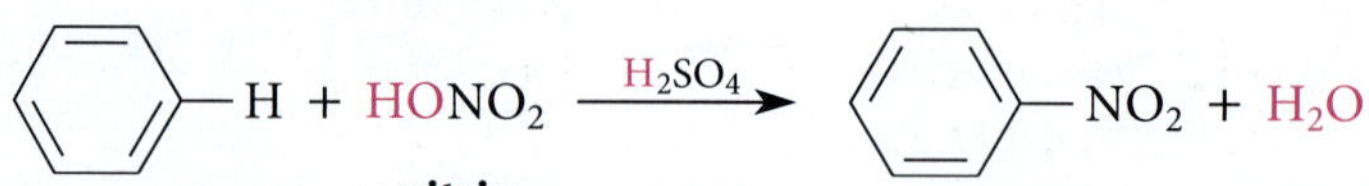

benzene **nitric acid** **nitrobenzene** (81% yield) (16.18)

This reaction fits the mechanistic pattern of the electrophilic aromatic substitution reaction outlined in the previous section:

Step 1 **Generation of the electrophile.** In nitration, the electrophile is $^+NO_2$, the *nitronium ion.* This ion is formed by the acid-catalyzed removal of the elements of water from HNO_3.

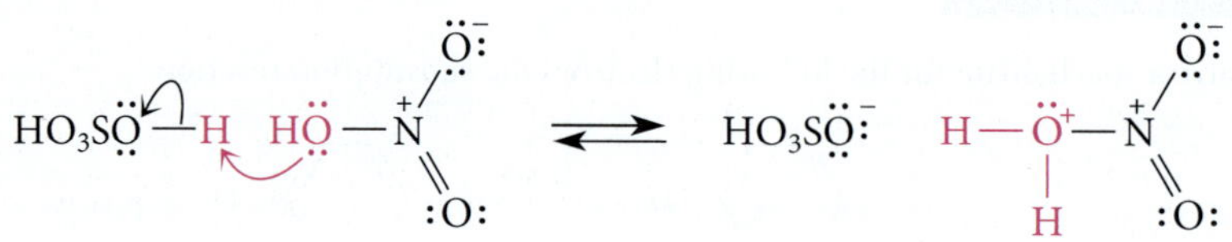

(16.19a)

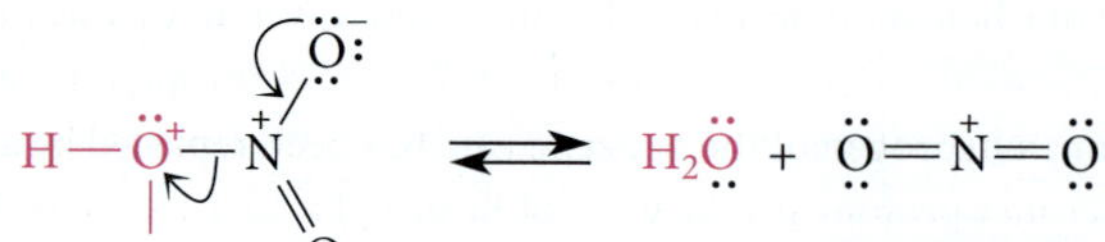

nitronium ion (16.19b)

Step 2 Reaction of the benzene π electrons with the electrophile to form a resonance-stabilized carbocation intermediate (an arenium ion).

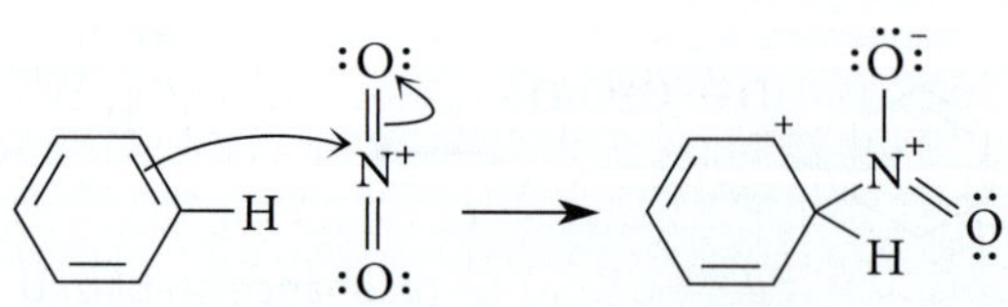

a resonance-stabilized carbocation (16.19c)

(Either of the oxygens can accept the electron pair.)

Step 3. Loss of a proton from the carbocation to give a new aromatic compound.

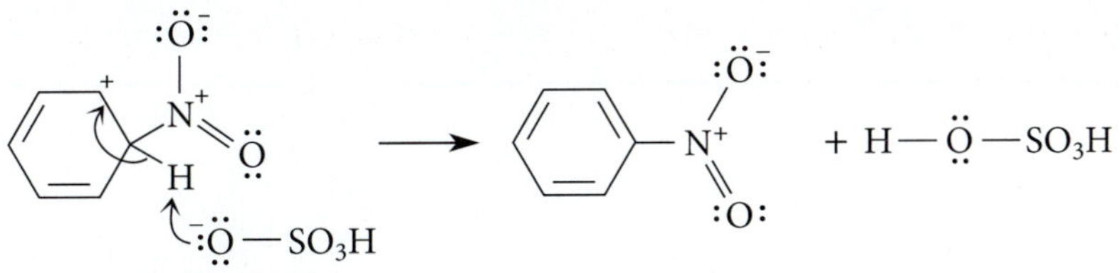

(from Eq. 16.19a) (16.19d)

Nitration is the usual way that nitro groups are introduced into aromatic rings.

D. Sulfonation of Benzene

Another electrophilic substitution reaction of benzene is its conversion into benzenesulfonic acid.

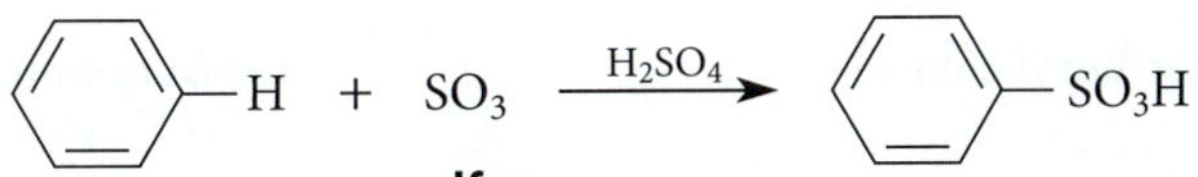

benzene **sulfur trioxide** **benzenesulfonic acid** (52% yield) (16.20)

(Sulfonic acids were introduced in Sec. 11.4A as their sulfonate ester derivatives.) This reaction, called *sulfonation*, occurs by two mechanisms that operate simultaneously. Both mechanisms involve sulfur trioxide, a fuming liquid that reacts violently with water to give H_2SO_4. The source of SO_3 for sulfonation is usually a solution of SO_3 in concentrated H_2SO_4 called *fuming sulfuric acid* or *oleum*. This material is one of the most acidic Brønsted acids available commercially.

In one sulfonation mechanism, the electrophile is neutral sulfur trioxide. When sulfur trioxide reacts with the benzene ring π electrons, an oxygen accepts the electron pair displaced from sulfur. We use the octet structure of SO_3 to stress its electrophilic character.

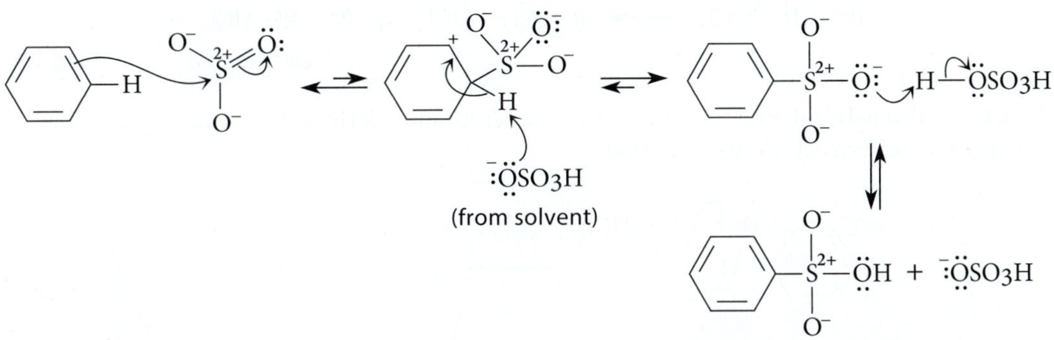

(from solvent)

benzenesulfonic acid (octet structure) (16.21)

Sulfonic acids such as benzenesulfonic acid are rather strong acids. (Notice the last equilibrium in Eq. 16.21 and the structural resemblance of benzenesulfonic acid to another strong acid, sulfuric acid; see also Fig. 11.2, Sec. 11.10A.)

Sulfonation, unlike many electrophilic aromatic substitution reactions, is reversible. The —SO_3H (sulfonic acid) group is replaced by a hydrogen when sulfonic acids are heated with steam (Problem 16.62).

Focused Problems

16.13 A second sulfonation mechanism involves the *conjugate acid* of sulfur trioxide as the electrophile. Show the protonation of SO_3, and draw a curved-arrow mechanism for the reaction of protonated SO_3 with benzene to give benzenesulfonic acid.

16.14 A compound called *p*-toluenesulfonic acid is formed when toluene is sulfonated at the para position. Draw the structure of this compound, and give the curved-arrow mechanism for its formation, including resonance structures for the carbocation intermediate.

E. Friedel–Crafts Alkylation of Benzene

The reaction of an alkyl halide with benzene in the presence of a Lewis acid catalyst gives an alkylbenzene.

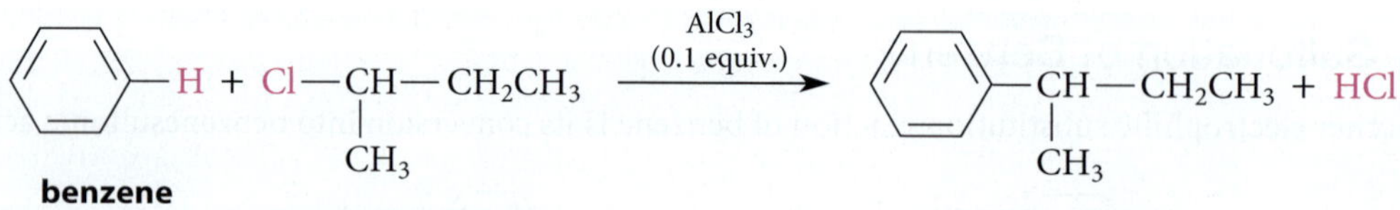

benzene (large excess) **sec-butyl chloride** **sec-butylbenzene** (71% yield) (16.22)

This reaction is an example of a *Friedel–Crafts alkylation.* Recall that an *alkylation* is a reaction that results in the transfer of an alkyl group (Sec. 11.4B). In a **Friedel–Crafts alkylation**, an alkyl group is transferred to an aromatic ring in the presence of an acid catalyst. In the preceding example, the alkyl group comes from an alkyl halide and the catalyst is the Lewis acid aluminum trichloride, $AlCl_3$.

The electrophile in a Friedel–Crafts alkylation is formed by the Lewis acid–base association reaction of the Lewis acid $AlCl_3$ with the halogen of an alkyl halide in much the same way that the electrophile in the bromination of benzene is formed by the Lewis acid–base association of $FeBr_3$ with Br_2 (Eq. 16.15a). If the alkyl halide is secondary or tertiary, this complex can further react to form carbocation intermediates.

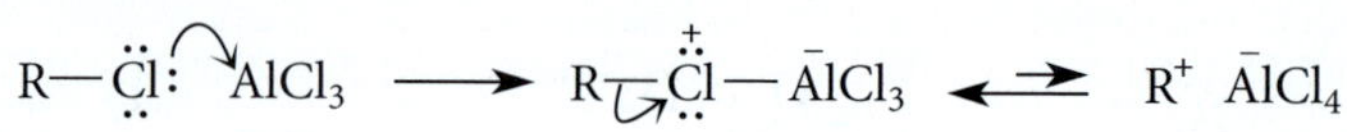

carbocation (16.23a)

Either the alkyl halide–Lewis acid complex or the carbocation derived from it can serve as the electrophile in a Friedel–Crafts alkylation.

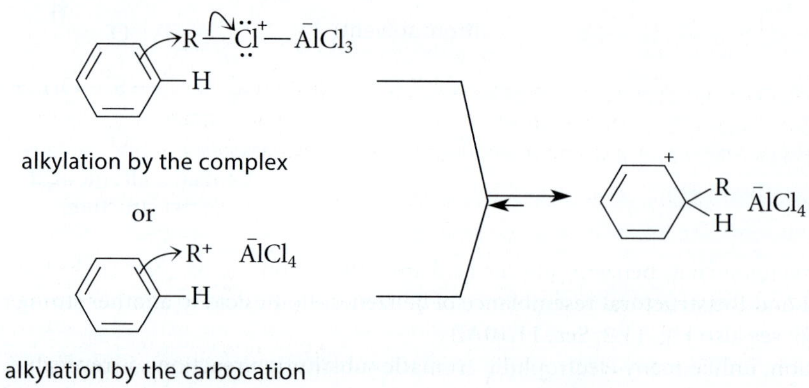

(16.23b)

Compare the role of $AlCl_3$ in enhancing the effectiveness of the chloride leaving group with the similar role of $FeBr_3$ in the bromination of benzene (Eq. 16.16a) or $FeCl_3$ in the chlorination of benzene:

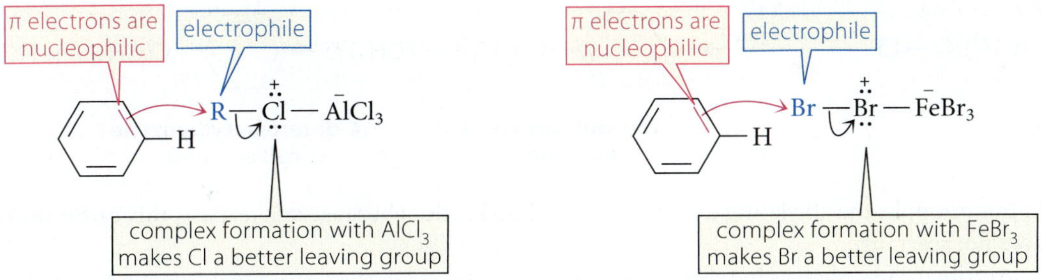

(16.23c)

We've discussed three general reactions of carbocations:

1. Reaction with nucleophiles (Secs. 4.7B and 9.6B)
2. Rearrangement to other carbocations (Sec. 4.7D)
3. Loss of a β-proton to give an alkene (Sec. 9.6B) or aromatic ring (Eq. 16.17b, Sec. 16.4A)

The reaction of a carbocation with the benzene π electrons is an example of reaction 1.
Loss of a β-proton to chloride ion completes the alkylation.

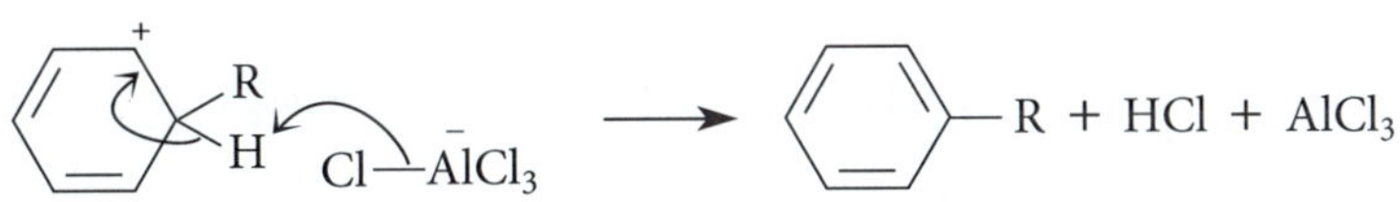

(16.23d)

Some carbocations can rearrange, so rearrangements of alkyl groups are observed in some Friedel–Crafts alkylations:

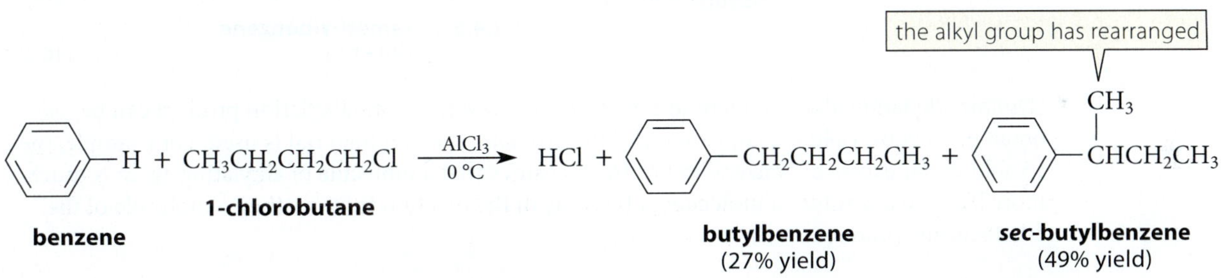

(16.24)

In this example, the alkyl group in the *sec*-butylbenzene product has rearranged. Because primary carbocations are too unstable to be involved as intermediates, it is probably the complex of the alkyl halide and $AlCl_3$ that rearranges. This complex has enough carbocation character that it behaves like a carbocation.

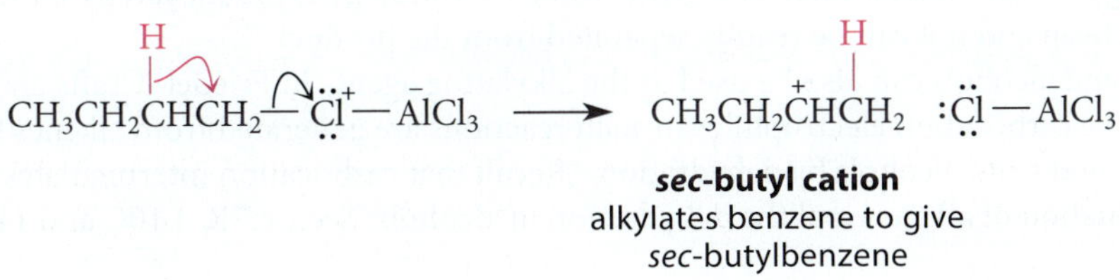

(16.25)

Rearrangement in the Friedel–Crafts alkylation is not observed if the carbocation intermediate is not prone to rearrangement.

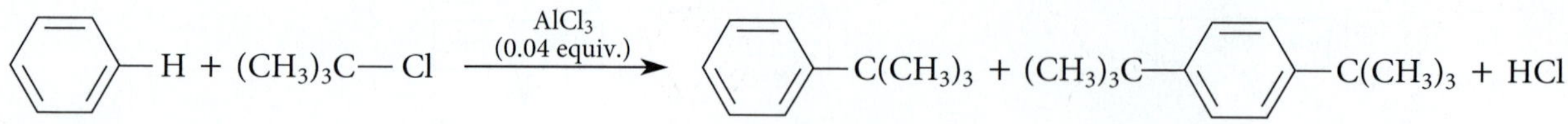

benzene (threefold excess) **tert-butylbenzene** (66% yield) **1,4-di-*tert*-butylbenzene** (small amount) (16.26)

In this example, the alkylating cation is the *tert*-butyl cation; because it is tertiary, this carbocation does not rearrange.

Alkylbenzenes, such as butylbenzene (Eq. 16.24) that are derived from rearrangement-prone alkyl halides, are generally not prepared by the Friedel–Crafts alkylation, but rather by other methods that are discussed in Secs. 18.10B and 19.12.

Another complication in Friedel–Crafts alkylation is that the alkylbenzene products are more reactive than benzene itself (for reasons discussed in Sec. 16.5B). This means that the product can undergo further alkylation, and mixtures of products alkylated to different extents are observed along with unreacted benzene.

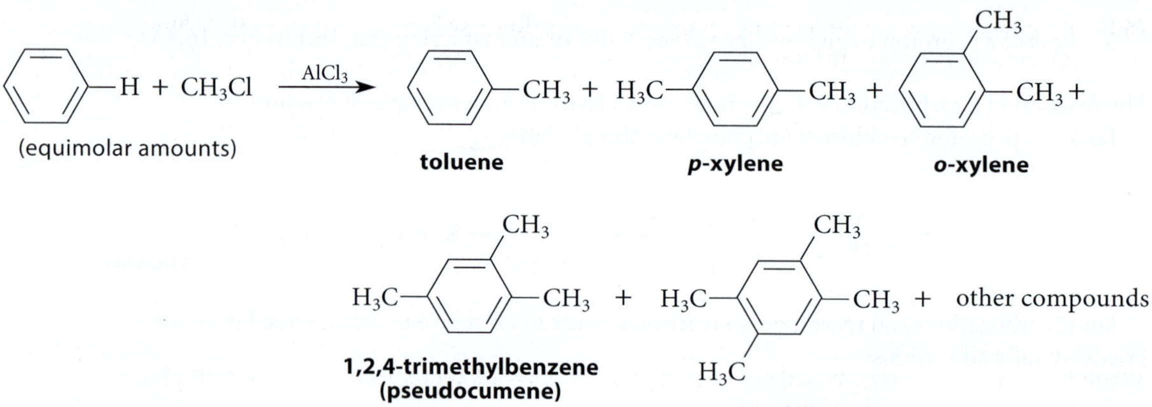

(equimolar amounts) **toluene** ***p*-xylene** ***o*-xylene** **1,2,4-trimethylbenzene (pseudocumene)** **1,2,4,5-tetramethylbenzene (durene)** (16.27)

(Double alkylation also occurs in in Eq. 16.26.) However, a monoalkylation product can be obtained in good yield if a large excess of the aromatic starting material is used. For example, the 15-fold molar excess of benzene in Eq. 16.28 ensures that a molecule of alkylating agent is much more likely to encounter a molecule of benzene in the reaction mixture than a molecule of the ethylbenzene product.

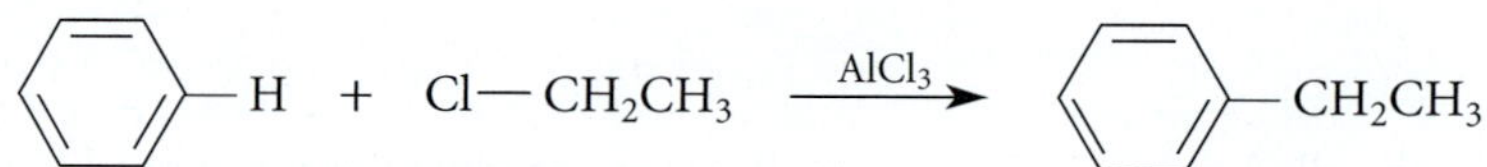

chloroethane **benzene** (15-fold excess) **ethylbenzene** (83% yield) (16.28)

STUDY GUIDE LINK 16.2
Different Sources of the Same Reactive Intermediate

(This strategy is used in Eqs. 16.22 and 16.26, too.) This strategy is practical only if the starting material is cheap and if it can be readily separated from the product.

Alkenes and alcohols can also be used as the alkylating agents in Friedel–Crafts alkylation reactions. The carbocation electrophiles in such reactions are generated from alkenes by protonation and from alcohols by dehydration. (Recall that carbocation intermediates are formed in the protonation of alkenes and the dehydration of alcohols; Secs. 4.7B, 4.10C and 11.2.)

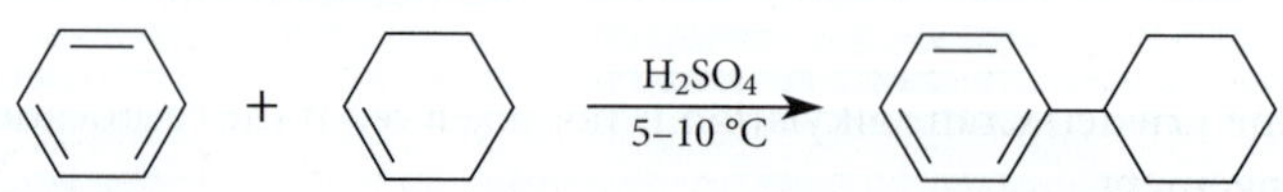

benzene (excess) **cyclohexene** **cyclohexylbenzene** (65–68% yield) (16.29a)

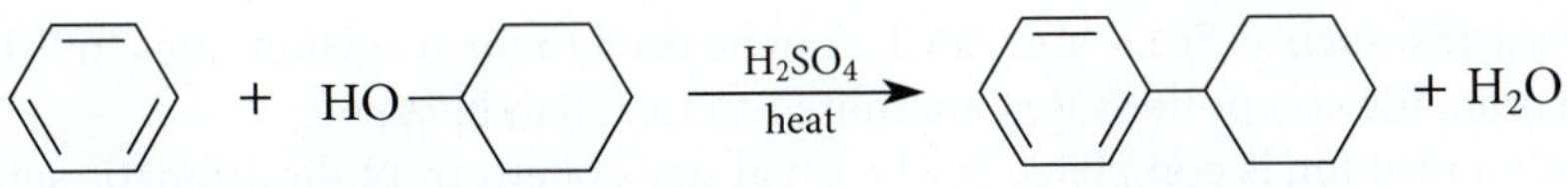

benzene (excess) **cyclohexanol** **cyclohexylbenzene** (90% yield) (16.29b)

Focused Problems

16.15 (a) Draw a curved-arrow mechanism for the reaction in Eq. 16.29a.

(b) Draw a curved-arrow mechanism for the reaction in Eq. 16.29b.

16.16 What electrophilic substitution product is formed when methylpropene is added to a large excess of benzene containing HF and the Lewis acid BF_3? By what mechanism is it formed?

16.17 Predict the product of the following reaction, and draw the curved-arrow mechanism for its formation. (*Hint:* Friedel–Crafts alkylations can be used to form rings.)

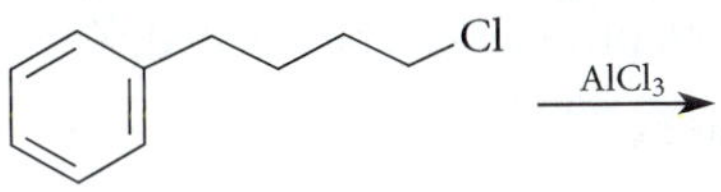

(a compound $C_{10}H_{12}$ + HCl)

F. Friedel–Crafts Acylation of Benzene

When benzene reacts with an acid chloride in the presence of a Lewis acid such as aluminum trichloride ($AlCl_3$), a ketone is formed.

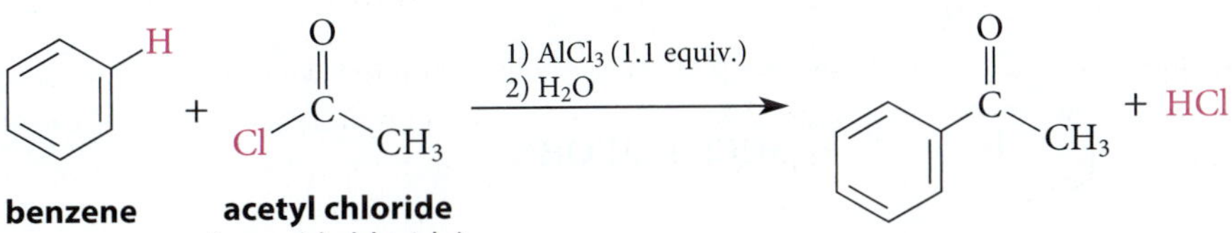

benzene **acetyl chloride** (an acid chloride) **acetophenone** (a ketone) (97% yield) (16.30)

This reaction is an example of a *Friedel–Crafts acylation* (pronounced AY-suh-LAY-shun). In an **acylation reaction**, an *acyl* (pronounced AY-sil) *group* is transferred from one group to another. In the **Friedel-Crafts acylation**, an acyl group, typically derived from an acid chloride, is introduced into an aromatic ring in the presence of a Lewis acid.

an acid chloride an acyl group

The electrophile in the Friedel–Crafts acylation reaction is a carbocation called an *acylium ion*. This ion is formed when the acid chloride reacts with the Lewis acid $AlCl_3$.

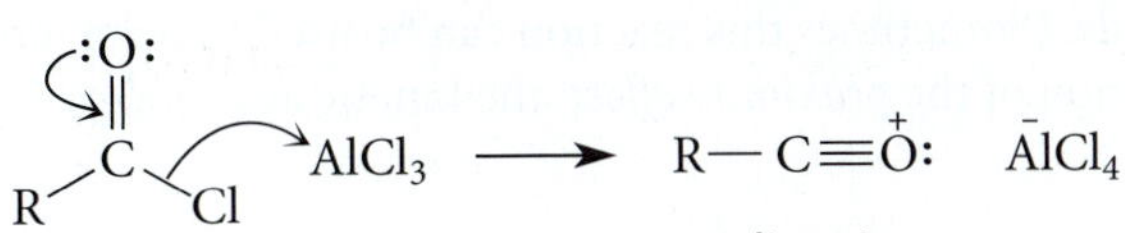

acylium ion (16.31)

Weaker Lewis acids, such as $FeCl_3$ and $ZnCl_2$, can be used to form acylium ions in Friedel–Crafts acylations of aromatic compounds that are more reactive than benzene.

The acylation reaction is completed by the usual steps of electrophilic aromatic substitution (Sec. 16.4B):

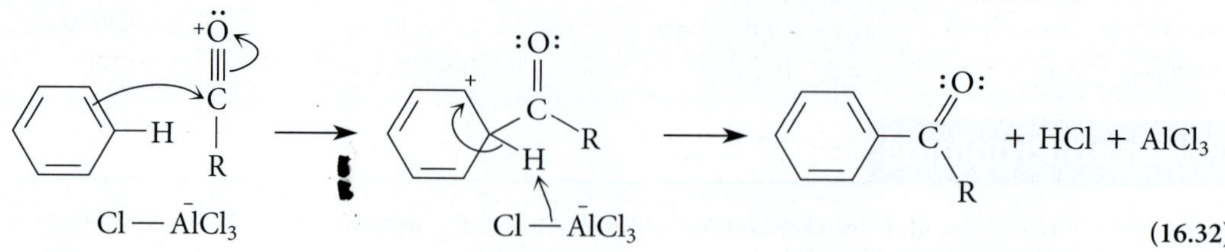

(16.32)

As we discuss in Sec. 19.7, ketones are weakly basic. Because of this basicity, the ketone product of the Friedel–Crafts acylation reacts with the Lewis acid in a Lewis acid–base association to form a complex that is catalytically inactive. This complex is the actual product of the acylation reaction. The formation of this complex has two consequences. First, at least one equivalent of the Lewis acid must be used to ensure its presence in uncomplexed form throughout the reaction. (For example, a 1.1 equivalent of $AlCl_3$ is used in Eq. 16.30.) This is in contrast to the Friedel–Crafts *alkylation*, in which $AlCl_3$ can be used in catalytic amounts. Second, the complex must be destroyed before the ketone product can be isolated. This is usually accomplished by pouring the reaction mixture into ice water.

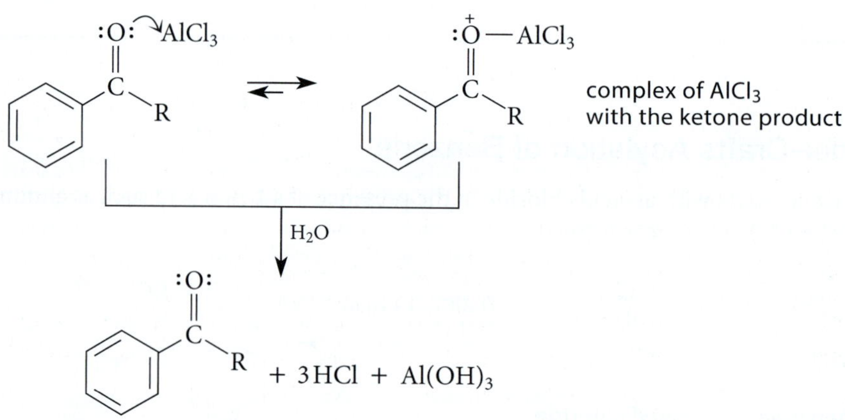

(16.33)

Both Friedel–Crafts alkylation and acylation reactions can occur *intramolecularly* when the reaction forms a five- or six-membered ring. (See also Focused Problem 16.17.)

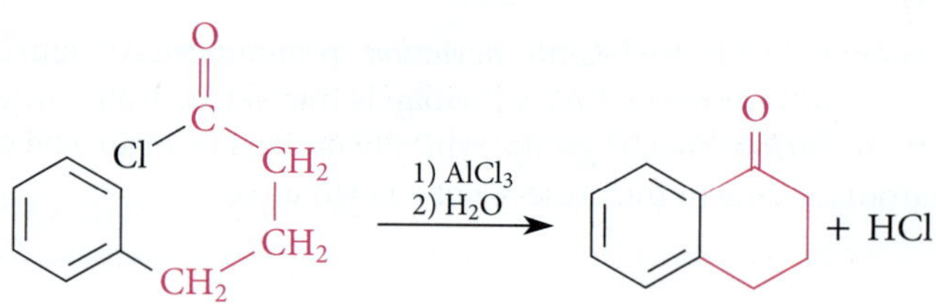

4-phenylbutanoyl chloride **α-tetralone** (74–91% yield)

(16.34)

In this reaction, the phenyl ring reacts with the acylium ion within the same molecule to form a bicyclic compound. This type of reaction can only occur at an adjacent ortho position because reaction at other positions would produce highly strained products. When five- or six-membered rings are involved, this process is much faster than reaction of the acylium ion with the phenyl ring of another molecule. (Sometimes this reaction can be used to form larger rings as well.) This is another illustration of the *proximity effect*: the kinetic advantage of intramolecular reactions (Sec. 12.8).

The formation of multiple-substitution products, as observed in Friedel–Crafts *alkylation* (Sec. 16.4E), is not a problem in Friedel–Crafts *acylation* because the ketone products of acylation are much less reactive than the benzene starting material, for reasons discussed in Sec. 16.5B.

The Friedel–Crafts alkylation and acylation reactions are important for two reasons. First, the alkylation reaction is useful for preparing certain alkylbenzenes, and the acylation reaction is an excellent method for the synthesis of aromatic ketones. Second, they provide other ways to form carbon–carbon bonds. Here is an updated list of reactions that form carbon–carbon bonds:

1. Reaction of acetylenic anions with alkyl halides or alkyl sulfonates (Sec. 10.6B)
2. Reaction of Grignard reagents with ethylene oxide and lithium organocuprate reagents with epoxides (Sec. 12.5C)
3. Addition of carbenes and carbenoids to alkenes (Sec. 10.7)
4. Diels–Alder reaction (Sec. 15.3)
5. Friedel–Crafts reactions (Secs. 16.4E and 16.4F)

Charles Friedel and James Mason Crafts

DAVID DEL RIO/Newscom/ Notimex/PARIS FRANCIA

The Friedel–Crafts acylation and alkylation (Sec. 16.4E) reactions are named for their discoverers, the French chemist Charles Friedel (1832–1899) and the American chemist James Mason Crafts (1839–1917). The two men met in Paris while in the laboratory of Charles-Adolphe Wurtz (1817–1884) at the École de Médecine (photo). Wurtz was one of the most famous chemists of that time. In 1877, Friedel and Crafts began their collaboration on the reactions that were to bear their names. Friedel subsequently became a very active figure in the development of chemistry in France, and Crafts served as a professor and chair in the Chemistry Department at Cornell University and then as president of the Massachusetts Institute of Technology.

Focused Problems

16.18 Give the structure of the product expected from the reaction of each of the following compounds with benzene in the presence of one equivalent of $AlCl_3$, followed by treatment with water.

(a)

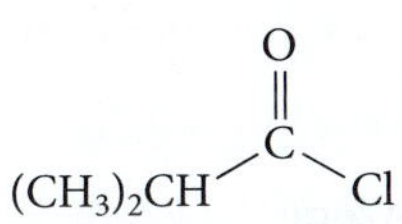

isobutyryl chloride

(b)

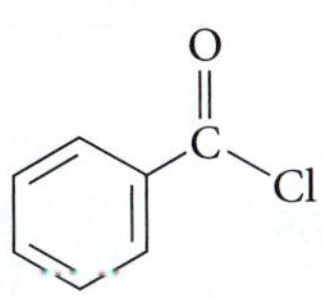

benzoyl chloride

16.19 Show two different Friedel–Crafts acylation reactions that can be used to prepare the following compound.

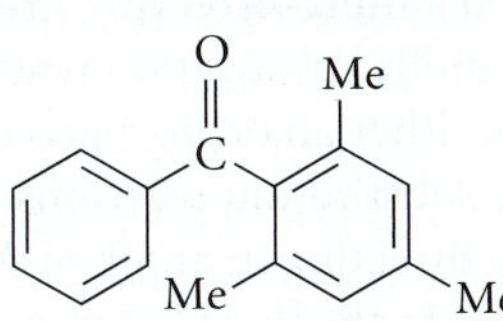

16.20 The following compound reacts with $AlCl_3$ followed by water to give a ketone *A* with the formula $C_{10}H_{10}O$. Give the structure of *A*, and draw a curved-arrow mechanism for its formation.

16.5 ELECTROPHILIC AROMATIC SUBSTITUTION REACTIONS OF SUBSTITUTED BENZENES

A. Directing Effects of Substituents

When a monosubstituted benzene undergoes an electrophilic aromatic substitution reaction, three possible disubstitution products might be obtained. For example, nitration of bromobenzene could in principle give *ortho-*, *meta-*, or *para*-bromonitrobenzene. If substitution were totally random, an ortho:meta:para product ratio of 2:2:1 would be expected. (Why?) It is found experimentally that this substitution is *not* random, but is *regioselective.*

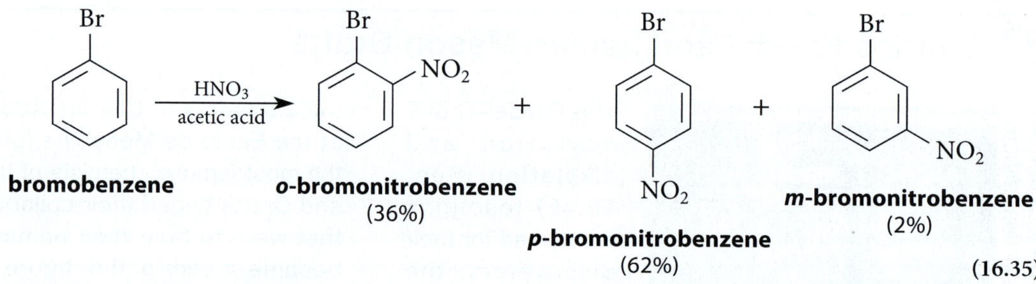

(16.35)

Other electrophilic substitution reactions of bromobenzene also give mostly ortho and para isomers. If a substituted benzene undergoes further substitution mostly at the ortho and para positions, the original substituent is called an **ortho, para-directing group**. Therefore, bromine is an ortho, para-directing group, because all electrophilic substitution reactions of bromobenzene occur predominately at the ortho and para positions.

In contrast, some substituted benzenes react in electrophilic aromatic substitution to give mostly the meta disubstitution product. For example, the bromination of nitrobenzene gives only the meta isomer.

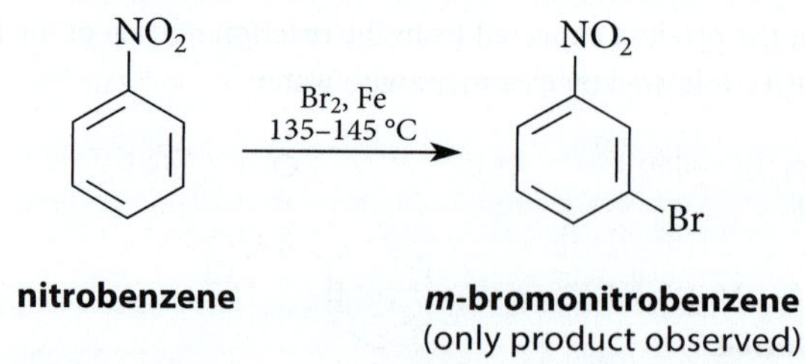

(16.36)

Other electrophilic substitution reactions of nitrobenzene also give mostly the meta isomers. If a substituted benzene undergoes further substitution mainly at the meta position, the original substituent group is called a **meta-directing group**. The nitro group, then, is a meta-directing group because all electrophilic substitution reactions of nitrobenzene occur at the meta position.

A substituent group is either an ortho, para-directing group or a meta-directing group in all electrophilic aromatic substitution reactions—that is, no substituent is ortho, para directing in one reaction and meta directing in another. A summary of the directing effects of common substituent groups is given in the third column of **Table 16.2**.

TABLE 16.2 Summary of Directing and Activating or Deactivating Effects of Some Common Functional Groups (Groups are listed in decreasing order of activation.)

Substituent group	Name of group	Directing effect	Activating or deactivating
$-\ddot{N}H_2$, $-\ddot{N}R_2$	amino, alkylamino	ortho, para directors	activating substituents
$-\ddot{O}H$	hydroxy	ortho, para directors	activating substituents
$-\ddot{O}R$	alkoxy	ortho, para directors	activating substituents
$-\ddot{N}H-C(=O)-R$	acylamino	ortho, para directors	activating substituents
$-R$	alkyl	ortho, para directors	activating substituents
$-\ddot{O}-C(=O)-R$	acyloxy	ortho, para directors	activating substituents
$-C_6H_5$	phenyl	ortho, para directors	activating substituents
$-\ddot{F}:$, $-\ddot{Cl}:$, $-\ddot{Br}:$, $-\ddot{I}:$	halogens	ortho, para directors	deactivating substituents
$-C(=O)OH$, $-C(=O)NH_2$, $-C(=O)OR$	carboxy, carboxamido, carboalkoxy	meta directors	deactivating substituents
$-C(=O)R$	acyl	meta directors	deactivating substituents
$-SO_3H$	sulfonic acid	meta directors	deactivating substituents
$-C\equiv N$	cyano	meta directors	deactivating substituents
$-NO_2$	nitro	meta directors	deactivating substituents

Focused Problem

16.21 Using the information in Table 16.2, predict the product(s) of the following.

(a) Friedel–Crafts acylation of anisole (methoxybenzene) with acetyl chloride (structure in Eq. 16.30, Sec. 16.4F) in the presence of one equivalent of $AlCl_3$ followed by H_2O

(b) Friedel–Crafts alkylation of a large excess of ethylbenzene with chloromethane in the presence of $AlCl_3$

These directing effects occur because electrophilic substitution reactions at one position of a benzene derivative are much faster than the same reactions at another position. That is, the substitution reactions at the different ring positions are in competition. For example, in Eq. 16.35, *o*- and *p*-bromonitrobenzenes are the major products because the rate of nitration is greater at the ortho and para positions of bromobenzene than it is at the meta position. Understanding these effects requires an understanding of the factors that control the rates of aromatic substitution at each position.

Ortho, Para-Directing Groups All of the ortho, para-directing substituents in Table 16.2 are either *alkyl groups* or *groups that have nonbonding electron pairs on atoms directly attached to the benzene ring.* Although other types of ortho, para-directing groups are known, the principles on which ortho, para-directing effects are based can be understood by considering electrophilic substitution reactions of benzene derivatives containing these types of substituents.

First, imagine the reaction of a general electrophile E^+ with anisole (methoxybenzene). The atom directly attached to the benzene ring (the oxygen of the methoxy group) has nonbonding electron pairs. The oxygen is sp^2-hybridized, and one of the nonbonding electron pairs resides in a 2p orbital that can overlap with the 2p orbitals of the ring carbons. This overlap is symbolized by resonance structures. Reaction of E^+ at the para position of anisole gives a carbocation (an arenium ion) intermediate with the following four important resonance structures:

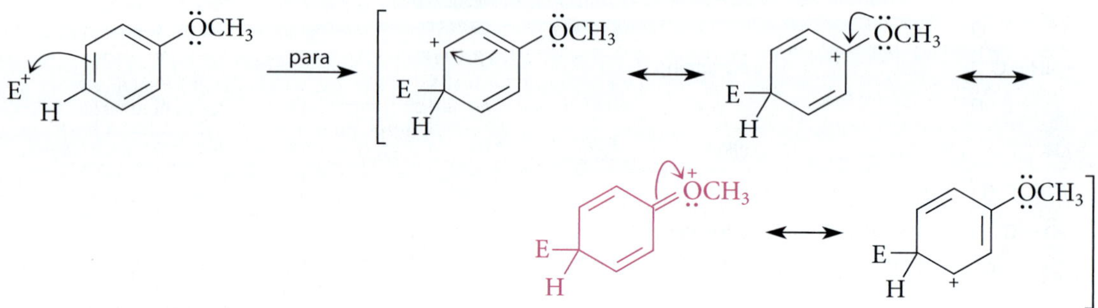

(16.37)

The colored structure shows that the nonbonding electron pair of the methoxy group can delocalize the positive charge on the carbocation. This is an especially important structure because it contains more bonds than the others, and every atom has an octet.

Focused Problem

16.22 Draw the carbocation that results from the reaction of the electrophile at the ortho position of anisole; show that this ion also has four resonance structures.

If the electrophile reacts with anisole at the meta position, the carbocation intermediate that is formed has fewer resonance structures than the ion in Eq. 16.37. *In particular, the charge cannot be delocalized onto the —OCH_3 group when reaction occurs at the meta position.* That is, no resonance structure corresponds to the colored structure in Eq. 16.37.

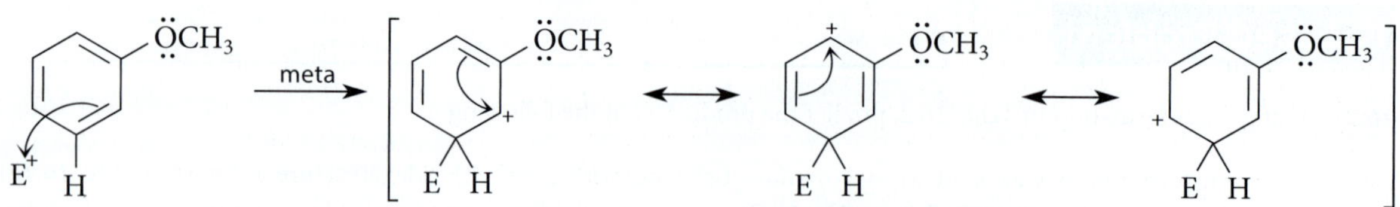

(16.38)

For the oxygen to delocalize the charge, it must be adjacent to an electron-deficient carbon, as in Eq. 16.37. The resonance structures show that the positive charge is shared on *alternate* carbons of the ring. When meta substitution occurs, the positive charge is not shared by the carbon adjacent to the oxygen.

We now use the resonance structures in Eqs. 16.37 and 16.38 (as well as those you drew in Focused Problem 16.22) to assess relative rates. The logic is to assess the relative stability of the possible carbocation intermediates and apply Hammond's postulate to determine relative rates. The reaction involving the more stable carbocation is faster. A comparison of Eq. 16.37 and the structures you drew for Focused Problem 16.22 with Eq. 16.38 shows that the reaction of an electrophile at either the ortho or para positions of anisole gives a carbocation with

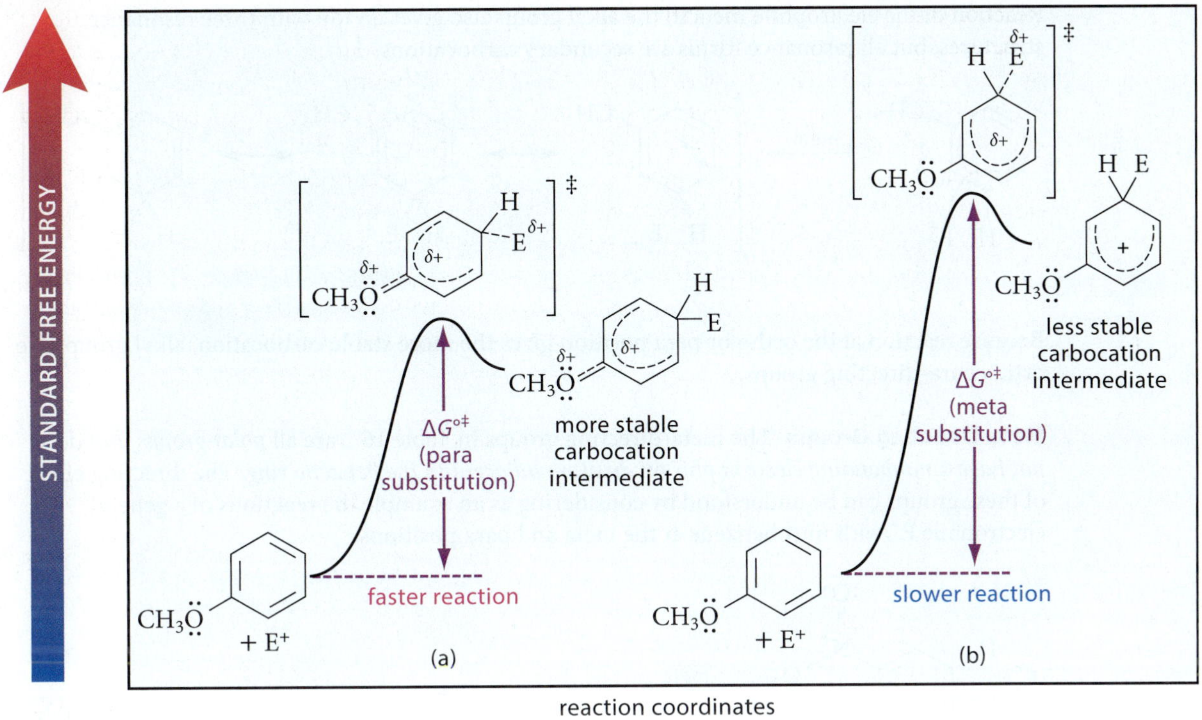

FIGURE 16.5 Basis of the directing effect of the methoxy group in the electrophilic aromatic substitution reactions of anisole. Substitution of anisole by an electrophile E^+ occurs more rapidly at (a) the para position than at (b) the meta position because a more stable carbocation intermediate is involved in para substitution. The carbocation and transition state structures are shown as hybrid structures—that is, the dashed lines within the structures symbolize the delocalization of electrons.

more resonance structures—that is, a more stable carbocation. The rate-limiting step in many electrophilic aromatic substitution reactions is *formation of the carbocation intermediate.* Hammond's postulate (Sec. 4.8D), then, indicates that the more stable carbocation should be formed more rapidly. Therefore, the products derived from the more rapidly formed carbocation—the more stable carbocation—are the ones observed. Because the reaction of the electrophile at an ortho or para position of anisole gives a more stable carbocation than the reaction at a meta position, the products of ortho, para substitution are formed more rapidly and are the products observed. This logic is diagrammed in **Fig. 16.5**. This is why the —OCH_3 group is an ortho, para-directing group.

To summarize: Substituents containing atoms with nonbonding electron pairs adjacent to the benzene ring are ortho, para directors in electrophilic aromatic substitution reactions because their electron pairs can be involved in the resonance stabilization of the carbocation intermediates.

Now imagine the reaction of an electrophile E^+ with an alkyl-substituted benzene such as toluene. Alkyl groups such as a methyl group have no nonbonding electrons, but the explanation for the directing effects of these groups involves the same logic that we used for the substitution in anisole. Reaction of E^+ at a position that is ortho or para to an alkyl group gives an ion that has one tertiary carbocation resonance structure (colored structure in the following equation).

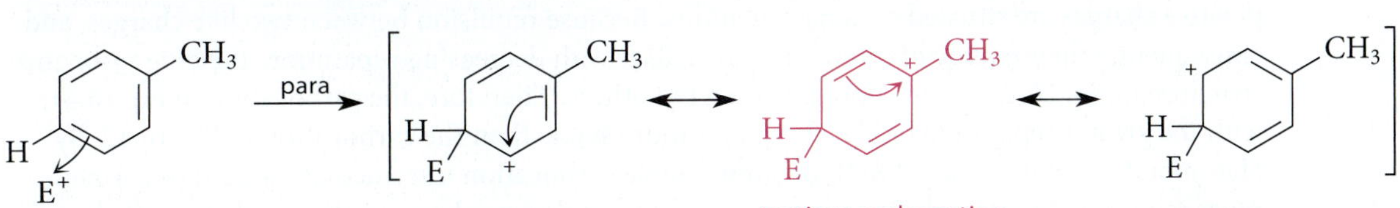

(16.39)

Reaction of the electrophile meta to the alkyl group also gives an ion with three resonance structures, but all resonance forms are secondary carbocations.

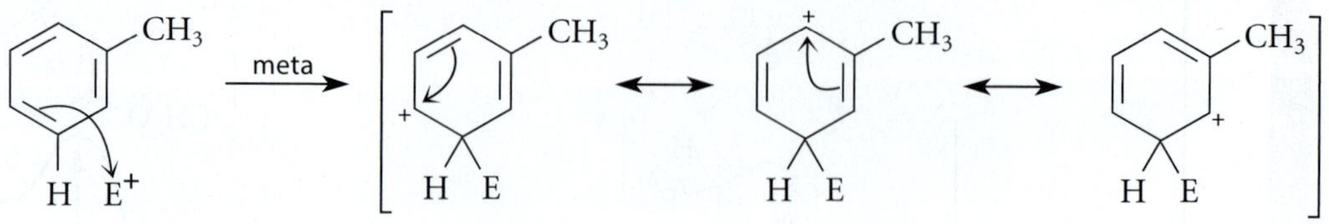

(16.40)

Because reaction at the ortho or para position gives the more stable carbocation, alkyl groups are ortho, para-directing groups.

Meta-Directing Groups The meta-directing groups in Table 16.2 are all *polar groups that do not have a nonbonding electron pair on an atom adjacent to the benzene ring.* The directing effect of these groups can be understood by considering as an example the reactions of a general electrophile E^+ with nitrobenzene at the meta and para positions.

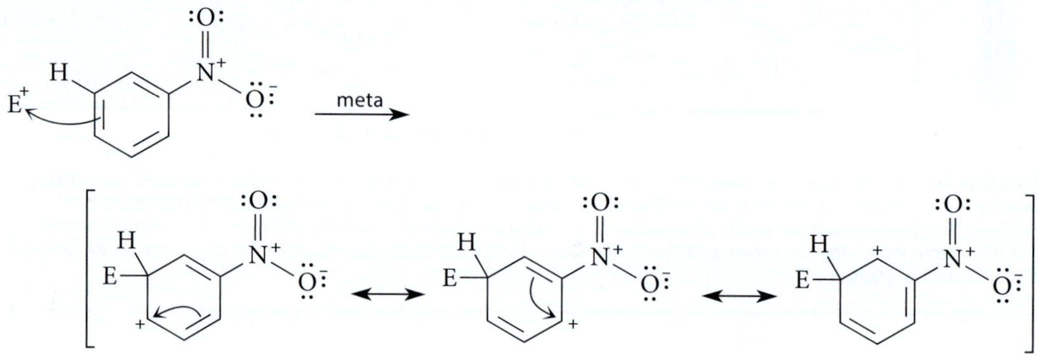

(16.41)

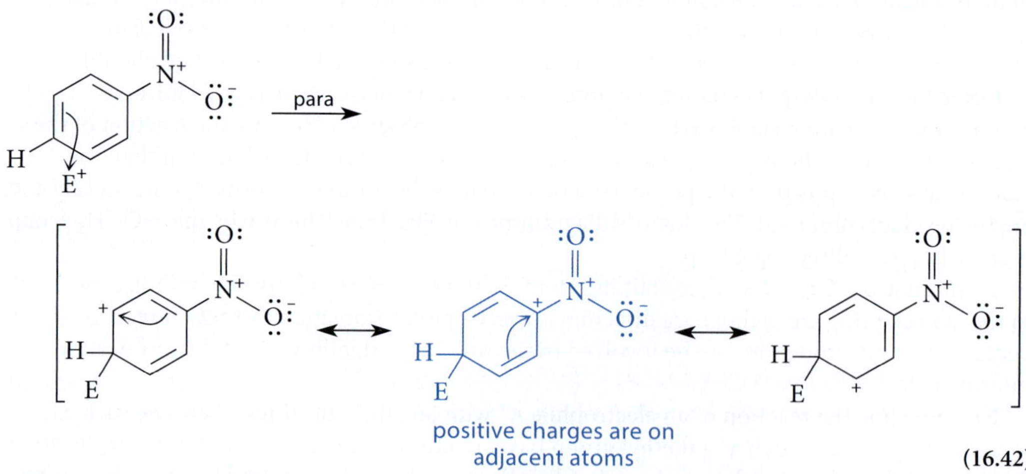

(16.42)

Both reactions give carbocations that have three resonance structures, but reaction at the para position gives an ion with one particularly unfavorable structure (*blued*). In this structure, positive charges are situated on adjacent atoms. Because repulsion between two like charges, and consequently their energy of interaction, increases with decreasing separation, the blue resonance structure in Eq. 16.42 is less important than the others. Therefore, the carbocation in Eq. 16.41, with the greater separation of like charges, is more stable than the carbocation in Eq. 16.42. By Hammond's postulate (Sec. 4.8D), the more stable carbocation intermediate should be formed more rapidly. Consequently, the nitro group is a meta director because the ion that results from meta substitution (Eq. 16.41) is more stable than the one that results from para substitution (Eq. 16.42).

In summary, substituents that have positive charges adjacent to the aromatic ring are meta directors because meta substitution gives the carbocation intermediate in which like charges are farther apart. Not all meta-directing groups have full positive charges like the nitro group, but all of them have bond dipoles that place a substantial amount of positive charge next to the benzene ring.

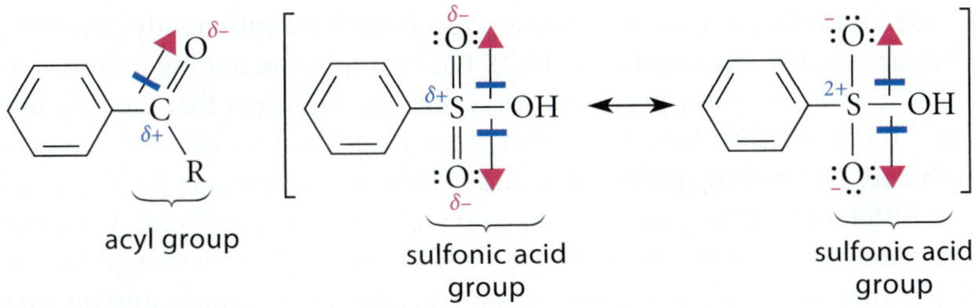

(16.43)

Focused Problems

16.23 Biphenyl (phenylbenzene) undergoes the Friedel–Crafts acylation reaction, as shown by the following example.

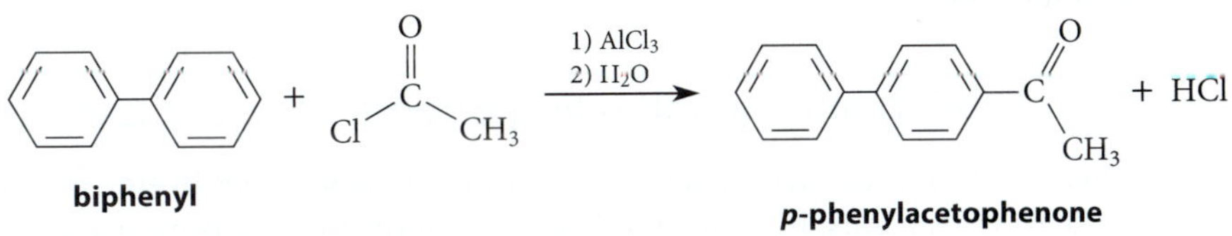

(a) On the basis of this result, what is the directing effect of the phenyl group?

(b) Using resonance arguments, explain the directing effect of the phenyl group.

16.24 Predict the predominant products that would result from bromination of each of the following compounds. Classify each substituent group as an ortho, a para director, or a meta director, and explain your reasoning.

(a) 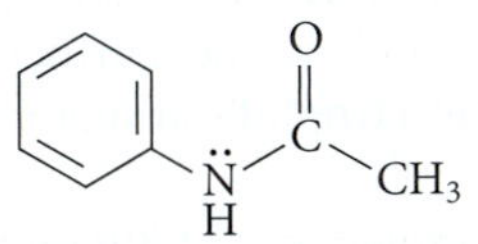

(b) 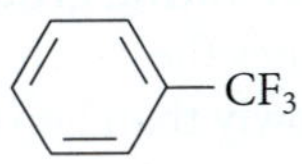

The Ortho, Para Ratio An aromatic substitution reaction of a benzene derivative bearing an ortho, para-directing group would give twice as much ortho as para product if substitution were completely random, because there are two ortho positions and only one para position available for substitution. However, this situation is rarely observed in practice; instead, it is often found that the para substitution product is the major one in the reaction mixture. In some cases, this result can be explained by the spatial demands of the electrophile. For example, Friedel–Crafts acylation of toluene gives essentially all para substitution product and almost no ortho product. The electrophile cannot react at the ortho position without developing van der Waals repulsions with the methyl group that is already on the ring. Consequently, reaction occurs at the para position, where such repulsions cannot occur.

Typically, para substitution predominates over ortho substitution, but not always. For example, nitration of toluene gives twice as much *o*-nitrotoluene as *p*-nitrotoluene. This result occurs because the nitration of toluene at either the ortho or para position is so fast that it occurs on *every encounter* of the reagents—that is, the energy barrier for the reaction is insignificant. Therefore, the product distribution corresponds simply to the relative probability of the reactions. Because the ratio of ortho and para positions is 2:1, the product distribution is 2:1. In fact, the ready availability of *o*-nitrotoluene makes it a good starting material for certain other ortho-substituted benzene derivatives.

In summary, the reasons for the ortho:para ratio vary from case to case, and in some cases these reasons are not well understood.

Whatever the reasons for the ortho:para ratio, if an electrophilic aromatic substitution reaction yields a mixture of ortho and para isomers, a problem of isomer separation must be solved if the reaction is to be useful. Usually, syntheses that give mixtures of isomers are avoided because, in many cases, isomers are difficult to separate. However, the ortho and para isomers obtained in many electrophilic aromatic substitution reactions have sufficiently different physical properties that they are readily separated (Sec. 16.2). For example, the boiling points of *o*- and *p*-nitrotoluene, 220 °C and 238 °C, respectively, are sufficiently different that these isomers can be separated by careful fractional distillation, so either isomer can be obtained relatively pure from the nitration of toluene. The melting points of *o*- and *p*-chloronitrobenzene, 34 °C and 84 °C, respectively, are so different that the para isomer can be selectively crystallized. As we noted in Sec. 16.2, the para isomer of an ortho, para pair typically has the higher melting point—in many cases, *considerably* higher. Most aromatic substitution reactions are so simple and inexpensive to run that when the separation of isomeric products is not difficult, these reactions are useful for organic synthesis despite the product mixtures obtained. You may assume in working problems involving electrophilic aromatic substitution on compounds containing ortho, para-directing groups that the para isomer can be isolated in useful amounts. For the reasons pointed out in the previous paragraph, *o*-nitrotoluene is a relatively rare example of a readily obtained ortho-substituted benzene derivative.

B. Activating and Deactivating Effects of Substituents

Different benzene derivatives have greatly different reactivities in electrophilic aromatic substitution reactions. If a substituted benzene derivative reacts more rapidly than benzene itself, then the substituent group is said to be an **activating group**. The Friedel–Crafts acylation of anisole (methoxybenzene), for example, is 300,000 times faster than the same reaction of benzene under comparable conditions. Furthermore, anisole shows a similar enhanced reactivity relative to benzene in all other electrophilic substitution reactions; so, the methoxy group is an *activating group*.

On the other hand, if a substituted benzene derivative reacts more slowly than benzene itself, then the substituent is called a **deactivating group**. For example, the rate for the bromination of nitrobenzene is less than 10^{-5} times the rate for the bromination of benzene. Furthermore, nitrobenzene reacts much more slowly than benzene in all other electrophilic aromatic substitution reactions, so the nitro group is a *deactivating group*.

A given substituent group is either activating in all electrophilic aromatic substitution reactions or deactivating in all such reactions. Whether a substituent is activating or deactivating is shown in the last column of Table 16.2, Sec. 16.5A. The most activating substituent groups are near the top of this table. Three generalizations emerge from examining Table 16.2.

1. All meta-directing groups are deactivating groups.
2. All ortho, para-directing groups except for the halogens are activating groups.
3. The halogens are deactivating groups.

Except for the halogens, there appears to be a correlation between the activating and directing effects of substituents.

This correlation suggests that the explanation of activating and deactivating effects is closely related to the explanation for directing effects. A key to understanding these effects is to understand that directing effects are concerned with the relative rates of substitution at different positions of the *same* compound, whereas activating or deactivating effects are concerned with the relative rates of substitution of *different* compounds—a substituted benzene compared with benzene itself. As in the discussion of directing effects, we consider the effect of the substituent on the stability of the intermediate carbocation, and we then apply Hammond's postulate by assuming that the stability of this carbocation is related to the stability of the transition state for its formation.

Two properties of substituents must be considered to understand activating and deactivating effects. First is the **resonance effect** of the substituent. This is the ability of the substituent to stabilize the carbocation intermediate in electrophilic substitution by delocalization of electrons from the substituent into the ring. The resonance effect is the same effect responsible for the ortho, para-directing effects of substituents with nonbonding electron pairs, such as —OCH_3 and halogen (colored structure in Eq. 16.37). We can summarize this effect with the following two of the four resonance structures for the carbocation intermediate in Eq. 16.37.

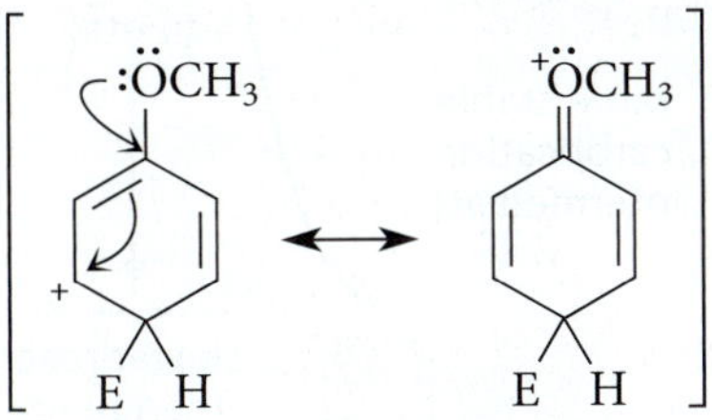

(two of the four important resonance structures)

The second property is the **polar effect** of the substituent, which is the tendency of the substituent group, because of its electronegativity, to pull electrons away from the ring. This is the same effect discussed in connection with polar effects on acidity (Sec. 3.7E). When a ring substituent is electronegative, it pulls the electrons of the ring toward itself and creates an electron deficiency, or positive charge, in the ring. In the carbocation intermediate of an electrophilic substitution reaction, the positive end of the bond dipole interacts repulsively with the positive charge in the ring, thus raising the energy of the ion:

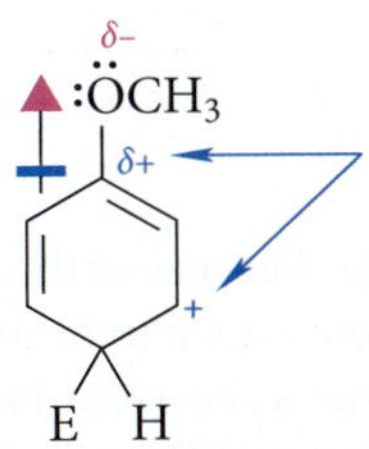

The electron-donating resonance effect of a substituent group with nonbonding electron pairs, if it were dominant, would *stabilize* positive charge and would *activate* further substitution. If such a group is electronegative, its electron-withdrawing polar effect, if dominant, would *destabilize* positive charge and would *deactivate* further substitution. These two effects operate simultaneously and in opposite directions. *Whether a substituted derivative of benzene is activated or deactivated toward further substitution depends on the balance of the resonance and polar effects of the substituent group.*

Anisole (methoxybenzene) undergoes electrophilic substitution much more rapidly than benzene because the resonance effect of the methoxy group far outweighs its polar effect. The benzene molecule, in contrast, has no substituent to help stabilize the carbocation intermediate by resonance. Therefore, the carbocation intermediate (and the transition state) derived from the electrophilic substitution of anisole is more stable relative to starting materials than the carbocation (and transition state) derived from the electrophilic substitution of benzene. In a given reaction, then, the ortho and para substitution of anisole is faster than the substitution of benzene. In other words, the methoxy group activates the benzene ring toward ortho and para substitution.

There is also an important subtlety here. Although the ortho and para positions of anisole are highly activated toward substitution, the meta position is deactivated. When substitution occurs in the meta position, the methoxy group cannot exert its resonance effect (Eq. 16.38), and only its rate-retarding polar effect is operative. So, whether a group activates or deactivates further substitution really depends on the *position* on the ring being considered. The methoxy group activates ortho, para substitution and deactivates meta substitution. But this is just another way of

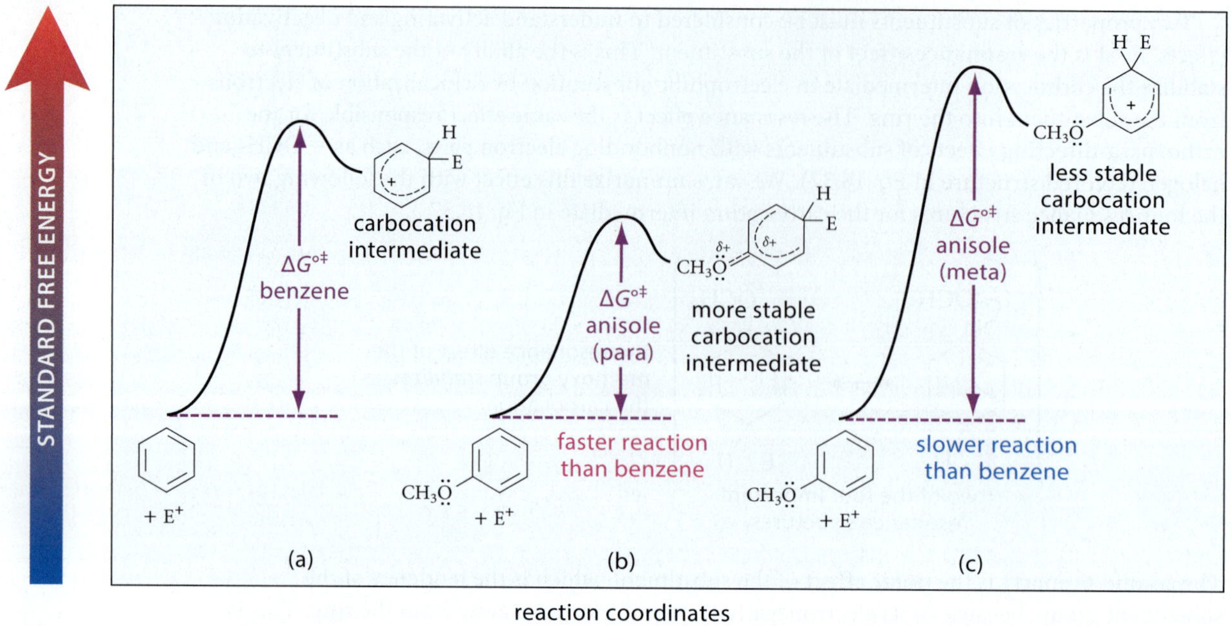

FIGURE 16.6 Basis of the activating effect of the methoxy group on electrophilic aromatic substitution in anisole. (a) The energy barrier for substitution of benzene by an electrophile E^+. (b) The energy barrier for substitution of anisole by E^+ at the para position. (c) The energy barrier for substitution of anisole by E^+ at the meta position. [The diagrams for parts (b) and (c) are the same as parts (a) and (b) of Fig. 16.5.] The substitution of anisole at the para position is faster than the substitution of benzene; the substitution of anisole at the meta position is slower than the substitution of benzene. The methoxy group is an activating group because the observed reaction of anisole—substitution at the para position—is faster than the substitution of benzene.

saying that the methoxy group is an ortho, para director. Because ortho, para substitution is the *observed* mode of substitution, the methoxy group is considered to be an activating group. These ideas are summarized in the reaction free-energy diagrams shown in **Fig. 16.6**.

The deactivating effects of halogen substituents reflect a different balance of resonance and polar effects. Consider the chloro substituent, for example. Because chlorine and oxygen have similar electronegativities, the polar effects of the chloro and methoxy groups are similar. However, the resonance interaction of chlorine electron pairs with the ring is much less effective than the interaction of oxygen electron pairs because the chlorine valence electrons reside in orbitals with higher principal quantum numbers. Because these orbitals and the carbon 2p orbitals of the benzene ring have *different sizes* and *different numbers of nodes*, they do not overlap effectively (**Fig. 16.7**). Because this overlap is the basis of the resonance effect, the resonance effect of chlorine is weak. With a weak rate-enhancing resonance effect and a strong rate-retarding polar effect, chlorine is a deactivating group. Bromine and iodine exert weaker polar effects than chlorine, but their resonance effects are also weaker. (Why?) Therefore, these groups, too, are deactivating groups. Fluorine, as a second-period element, has a stronger resonance effect than the other halogens, but, as the most electronegative element, it has a stronger polar effect as well. Fluorine is also a deactivating group. Because of its greater resonance effect, though, it is the least deactivating of the halogens.

A similar directing effect occurs in the addition of HBr to an alkyne (Sec. 4.9, Eqs. 4.63a and b). In the addition of the second mole of HBr, the polar effect of the bromine in the vinylic bromoalkene retards the reaction. This is why addition of HBr to an alkyne can be stopped after a single addition. However, the resonance effect of the bromine in the vinylic bromide controls the direction of addition, because the resonance effect only operates in the formation of the geminal dibromide.

The deactivating, rate-retarding polar effects of the halogens are similar at all ring positions but are offset somewhat by their resonance effects when substitution occurs para to the halogen. However, the resonance effect of a halogen cannot come into play at all when substitution occurs at the meta position of a halobenzene. (Why?) Therefore, meta substitution in halobenzenes is deactivated even more than para substitution is. This is another way of saying that halogens are ortho, para-directing groups.

Alkyl substituents such as the methyl group have no resonance effect (other than hyperconjugation), but the polar effect of any alkyl group toward electron-deficient carbons is

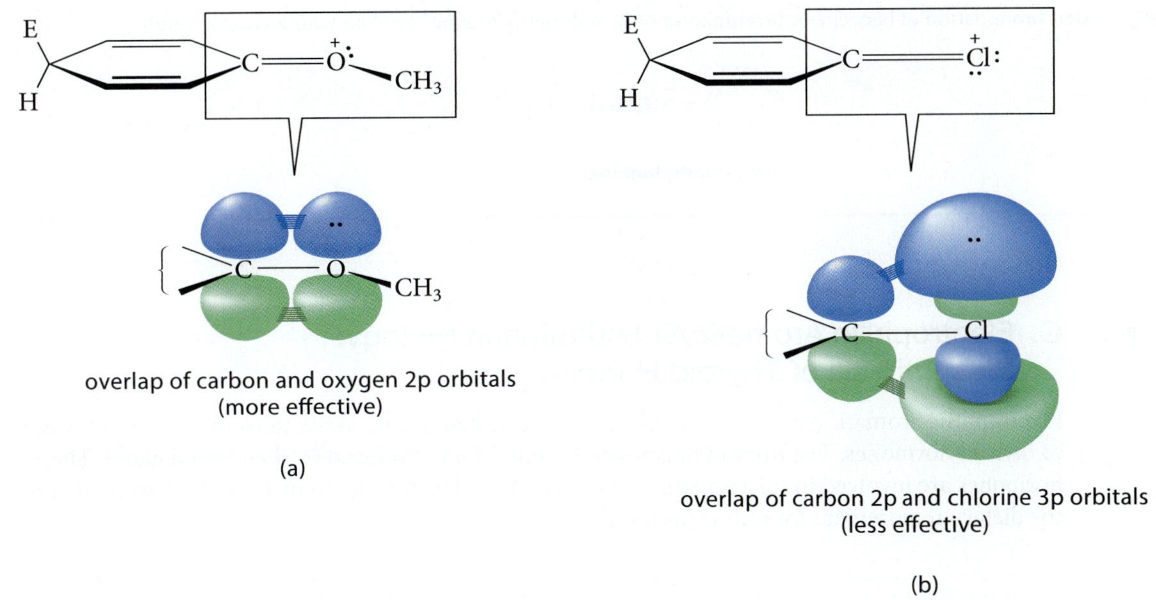

FIGURE 16.7 The carbocation (arenium ion) resulting from the reaction of an electrophile E^+ at the para position of a benzene ring substituted with (a) a methoxy group and (b) a chlorine. Below each ion is shown the orbital overlap between (a) oxygen and carbon and (b) chlorine and carbon. The other nonbonding electron pairs are not shown in the orbital diagram. This orbital overlap stabilizes the ion. The blue and green parts of the orbitals represent wave peaks and wave troughs, respectively. The overlap of carbon and oxygen 2p orbitals, shown in (a), is more effective than the overlap of carbon 2p and chlorine 3p orbitals, shown in (b). Bonding overlap occurs only when peaks overlap with peaks and troughs overlap with troughs. Because orbitals with different quantum numbers have different sizes and different numbers of nodes, part of the chlorine 3p orbital cannot overlap with the carbon 2p orbital.

an electropositive, stabilizing effect (Sec. 4.7C). It follows that alkyl substituents on a benzene ring stabilize carbocation intermediates in electrophilic substitution, and for this reason, they are activating groups. It turns out that alkyl groups activate substitution at all ring positions, but they are ortho, para directors because they activate ortho, para substitution more than they activate meta substitution (Eqs. 16.39 and 16.40).

Finally, consider the deactivating effects of meta-directing groups such as the nitro group. Because a nitro group has no electron-donating resonance effect, the polar effect of this electronegative group destabilizes the carbocation intermediate and retards electrophilic substitution at *all* positions of the ring. The nitro group is a meta-directing group because substitution is retarded more at the ortho and para positions than at the meta positions (Eqs. 16.41 and 16.42). In other words, the meta-directing effect of the nitro group is not due to selective activation of the meta positions, but rather to greater *deactivation* of the ortho and para positions. For this reason, the nitro group and the other meta-directing groups might be called *meta-allowing groups*.

Focused Problems

16.25 Draw reaction-free energy profiles analogous to that in Fig. 16.6 in which substitution on benzene by a general electrophile E^+ is compared with substitution at the para and meta positions of (a) chlorobenzene; (b) nitrobenzene.

16.26 Explain why the nitration of anisole is much faster than the nitration of thioanisole under the same conditions.

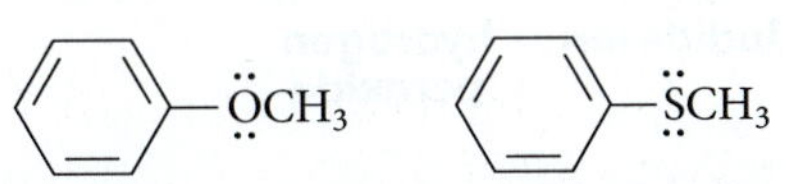

anisole **thioanisole**

16.27 Which should be faster: bromination of benzene or bromination of *N,N*-dimethylaniline? Explain your answer carefully.

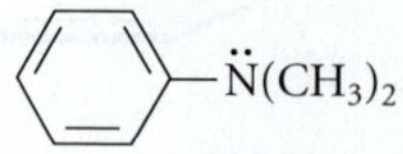

***N,N*-dimethylaniline**

Chemical Biology Topic

C. Electrophilic Aromatic Substitution in Biology: Biosynthesis of Thyroid Hormones

Electrophilic aromatic substitution in biology is illustrated by one of the steps in the biosynthesis of thyroid hormones. The thyroid hormones T3 and T4 are produced by the thyroid gland. These hormones are involved in the regulation of metabolism. The formation of T3 and T4 accounts for the dietary requirement for iodine (as iodide).

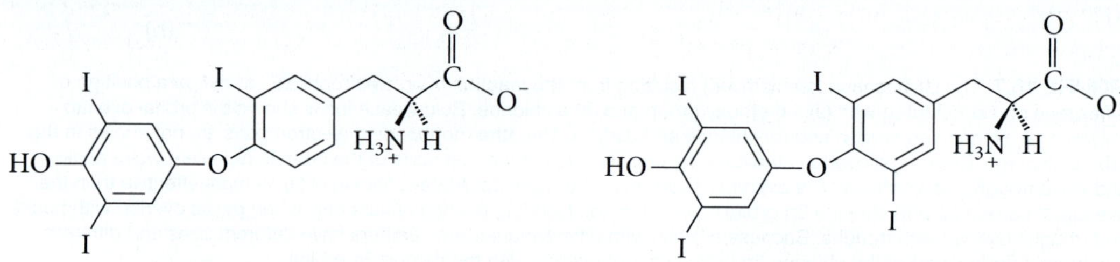

(*S*)-triiodothyronine (T3)
a thyroid hormone

(*S*)-thyroxine (T4)
a thyroid hormone (16.44)

A key step in the production of these hormones is the iodination of several tyrosine residues in a protein called *thyroglobulin.*

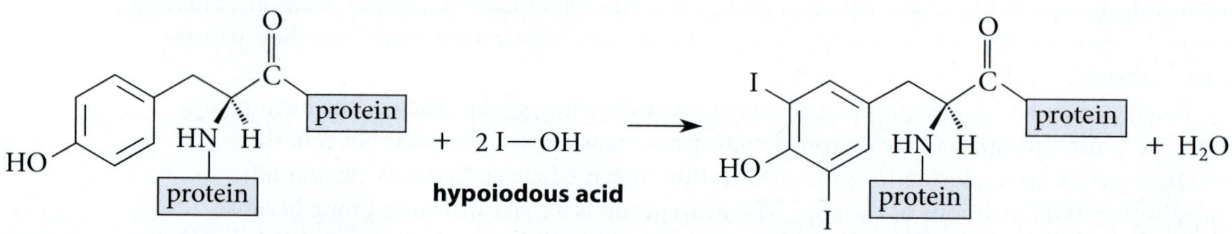

tyrosine residues in thyroglobulin

hypoiodous acid

3′,5′-diiodotyrosine residues in thyroglobulin

(16.45)

As with other electrophilic aromatic substitution reactions, the first step is formation of the electrophile. The thyroid contains an iodide transporter protein that concentrates sodium iodide within the appropriate part of the thyroid tissue. Iodide ion is not electrophilic, so it must be oxidized to become an electrophile. A thyroid enzyme called *thyroid peroxidase* catalyzes the oxidation reaction between the iodide ion and hydrogen peroxide to give the actual electrophile, which is probably hypoiodous acid (I—OH), molecular iodine (I_2), or an equivalent species. The formation of I—OH is shown here:

$$H_3O^+ + I^- + H_2O_2 \xrightarrow[\text{(enzyme)}]{\text{thyroid peroxidase}} \text{I—OH} + 2\,H_2O \qquad (16.46)$$

iodide ion **hydrogen peroxide** **hypoiodous acid**

The electrophile reacts rapidly with the aromatic ring of a tyrosine residue in an electrophilic aromatic substitution reaction.

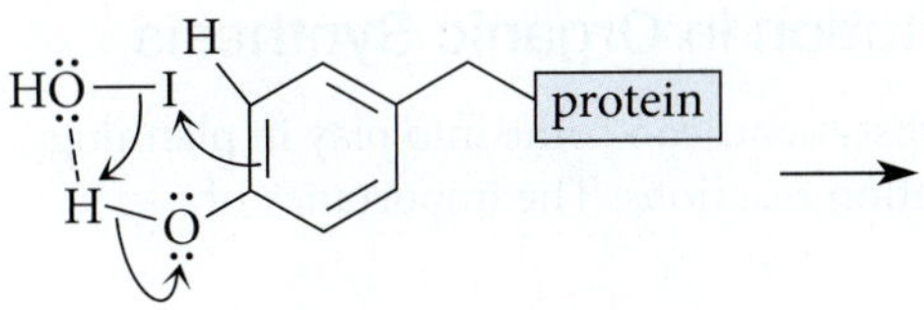

tyrosine residue in thyroglobulin

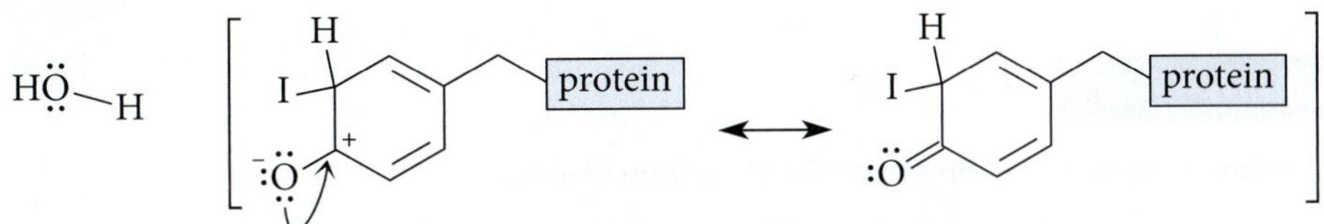

"carbocation" intermediate

(16.47a)

In this mechanism, the tyrosine OH group, which is weakly acidic, protonates the OH leaving group in hypoiodous acid. This protonation improves the leaving group (as in the protonation of alcohol OH groups in the acid-catalyzed reactions of alcohols) and results in a negative charge on the tyrosine oxygen that stabilizes the carbocation intermediate. As a result, this intermediate is actually not a carbocation but a neutral species, as the second resonance structure in Eq. 16.47a shows. The water molecule formed as a leaving group then facilitates the proton transfers to regenerate aromaticity in the substitution product.

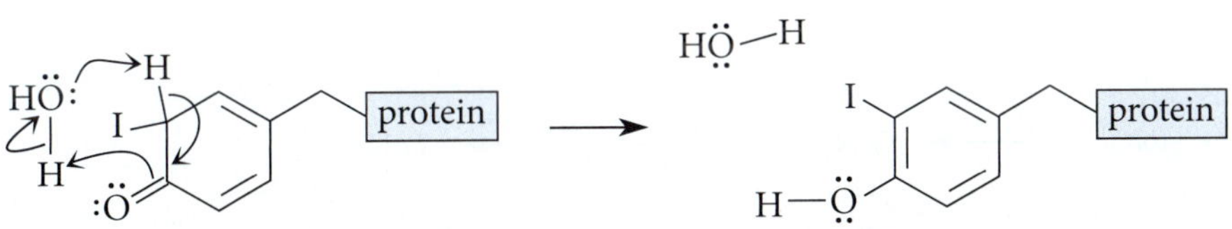

iodinated tyrosine residue (16.47b)

The second iodine atom (Eq. 16.45) is introduced into the ring by the same mechanism.

Two points about the iodination process are especially pertinent. First, the hydroxy substituent on the tyrosine ring is strongly ortho, para directing. Because the para position is already occupied in tyrosine, the two ortho positions are iodinated. Therefore, the directing effect of the hydroxy group accounts for the position of iodination. Second, the iodination of tyrosines in thyroglobulin is rapid. We expect a rapid reaction from the strong activating effect of the hydroxy substituent (see Table 16.2, Sec. 16.5A).

After the tyrosine residues in thyroglobulin are iodinated, subsequent enzyme-catalyzed oxidative reactions of thyroglobulin (which we won't consider here) form the T3 and T4 structures within the protein. In subsequent enzyme-catalyzed reactions, these are excised from the thyroglobulin protein as the free amino acids.

Focused Problems

16.28 Write the two half-reactions that correspond to the oxidation of iodide ion in Eq. 16.46. (Review Sec. 11.6A if you need help.)

16.29 The iodination reaction discussed in this section can be carried out on the amino acid tyrosine and related compounds in the laboratory with iodide ion in the presence of the enzyme thyroid oxidase. Which would undergo iodination more rapidly: tyrosine (*A*) or 3,3-difluorotyrosine (*B*)? Explain.

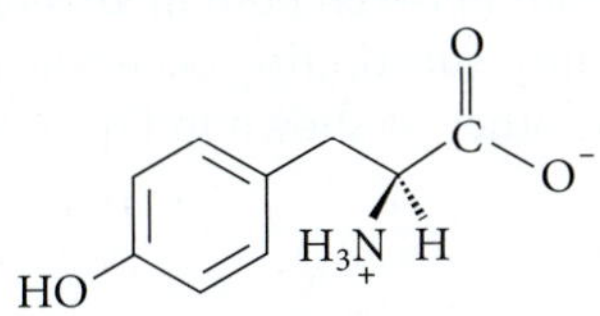

tyrosine
A

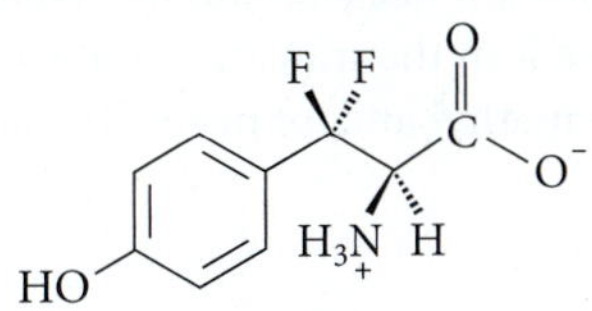

3,3-difluorotyrosine
B

D. Use of Electrophilic Aromatic Substitution in Organic Synthesis

Both activating/deactivating and directing effects of substituents can come into play in planning an organic synthesis that involves electrophilic substitution reactions. The importance of directing effects is illustrated in Study Problem 16.2.

Study Problem 16.2

Outline a synthesis of *p*-bromonitrobenzene from benzene.

Solution The key to this problem is whether the bromine or the nitro group should be the first ring substituent introduced. Introduction of the bromine first takes advantage of its directing effect in the subsequent nitration reaction:

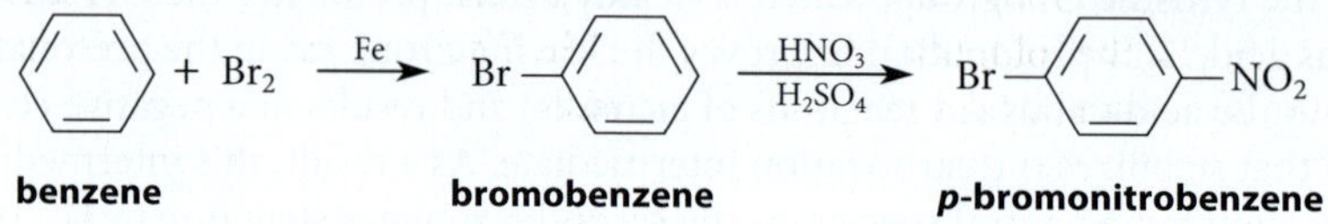

(16.48)

Introduction of the nitro group first followed by bromination would give *m*-bromonitrobenzene instead, because the nitro group is a meta-directing group.

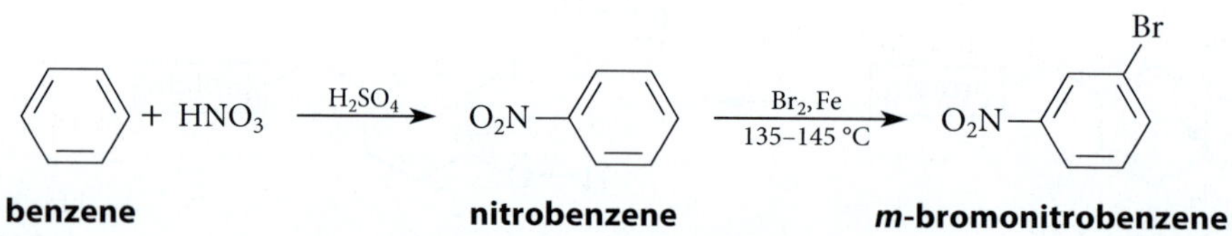

(16.49)

Therefore, to prepare the desired compound, brominate first and then nitrate the resulting bromobenzene, as shown in Eq. 16.48.

When an electrophilic substitution reaction is carried out on a benzene derivative with more than one substituent, the activating and directing effects are roughly the sum of the effects of the separate substituents. First, we consider directing effects. In the Friedel–Crafts acylation of *m*-xylene, for example, both methyl groups direct the substitution to the same positions.

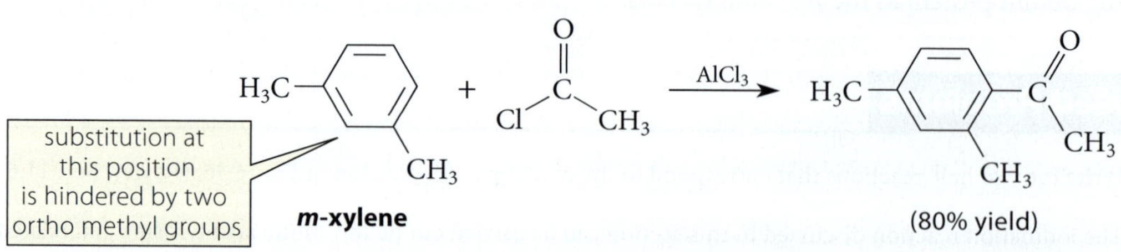

(16.50)

Methyl groups are ortho, para directors. Substitution at the position ortho to both methyl groups is difficult because van der Waals repulsions between both methyls and the electrophile would be present in the transition state. Consequently, substitution occurs at a ring position that is para to one methyl and, of necessity, ortho to the other, as shown in Eq. 16.50.

Two meta-directing groups on a ring, such as the carboxylic acid (—CO_2H) groups in the following example, direct further substitution to the remaining open meta position:

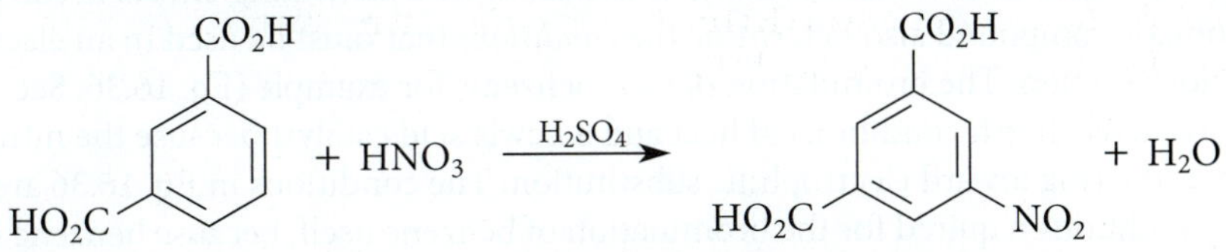

1,3-benzenedicarboxylic acid **5-nitro-1,3-benzenedicarboxylic acid** (96% of product) (16.51)

In each of the previous two examples, both substituents direct the incoming group to the same position. What happens when the directing effects of the two groups are in conflict? If one group is much more strongly activating than the other, the directing effect of the more powerful activating group generally predominates. For example, when phenol is treated with bromine in water, a small amount of the phenol ionizes to the phenoxide. The —O^- group is such a powerful activating group that phenol can be brominated three times.

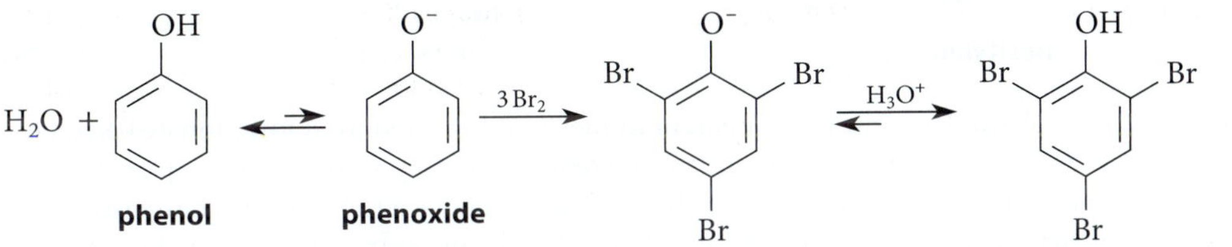

phenol **phenoxide** + H_3O^+ + 3HBr **2,4,6-tribromophenol** quantitative; virtually instantaneous (16.52)

This and other electrophilic substitution reactions of phenol are discussed in more detail in Sec. 18.9.

After the first bromination, the —O^- and —Br groups direct subsequent brominations to different positions. Each bromine that is added to the ring increases the acidity of the phenol (by a polar effect; Sec. 3.7E) and increases the formation of phenoxide. The strong activating and directing effect of the —O^- group at the ortho and para positions overrides the much weaker directing effect of the Br group. The same principle operates in the iodination of tyrosine in thyroglobulin (Eq. 16.45, Sec. 16.5C).

In other cases, mixtures of isomers are typically obtained.

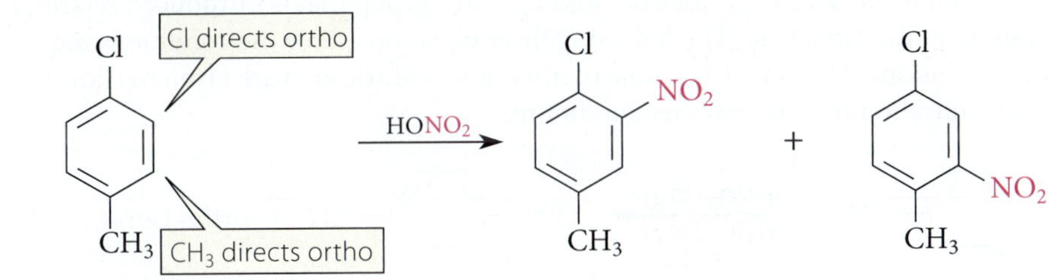

4-chlorotoluene **4-chloro-3-nitrotoluene** (42%) **4-chloro-2-nitrotoluene** (58%) (16.53)

Focused Problem

16.30 Predict the predominant product(s) from:

(a) Monosulfonation of *m*-bromotoluene

(b) Mononitration of *m*-bromoiodobenzene

STUDY GUIDE LINK 16.3
Reaction Conditions and Reaction Rate

We have just seen that the activating and directing effects of substituents must be taken into account in developing the strategy for an organic synthesis that involves a substitution reaction on an already-substituted benzene ring. The activating or deactivating effects of substituents in an aromatic compound also determine the *conditions* that must be used in an electrophilic substitution reaction. The bromination of nitrobenzene, for example (Eq. 16.36, Sec. 16.5A), requires relatively harsh conditions of heat and a Lewis acid catalyst because the nitro group deactivates the ring toward electrophilic substitution. The conditions in Eq. 16.36 are more severe than the conditions required for the bromination of benzene itself, because benzene is the more reactive compound. A dramatic example in the opposite direction is provided by the bromination of mesitylene (1,3,5-trimethylbenzene). Mesitylene can be brominated under *very* mild conditions because the ring is activated by three methyl groups; a Lewis acid catalyst is not even necessary.

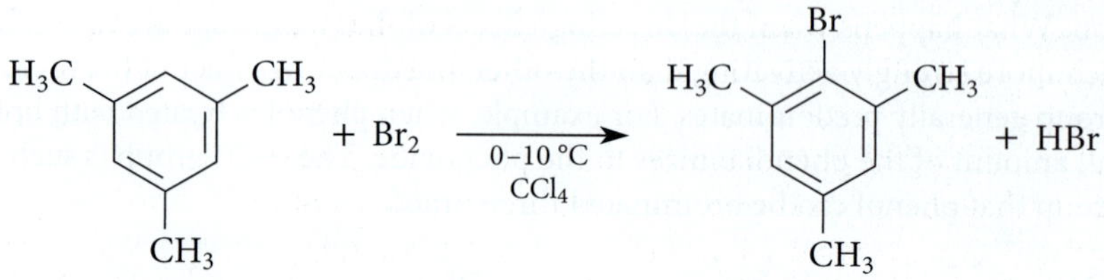

mesitylene (80% yield) (16.54)

Concentrated sulfuric acid itself without added SO_3 can be used to sulfonate more reactive aromatic compounds because small amounts of SO_3 are generated from the reaction of sulfuric acid with itself:

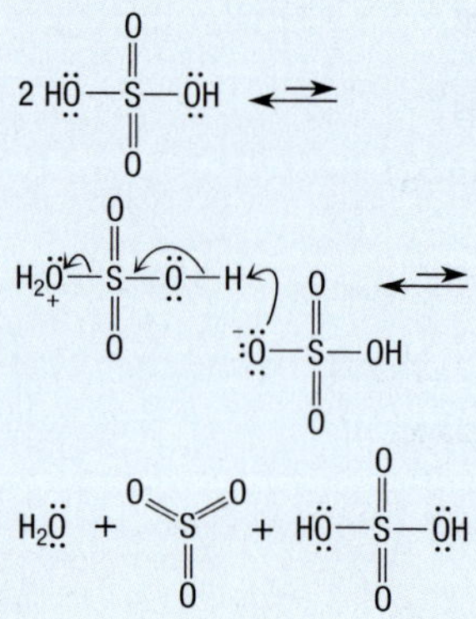

A similar contrast is apparent in the conditions required to sulfonate benzene and toluene. Sulfonation of benzene requires fuming sulfuric acid (Eq. 16.20 and the following discussion, Sec. 16.4D). However, because toluene is more reactive than benzene, toluene can be sulfonated with concentrated sulfuric acid, a milder reagent than fuming sulfuric acid.

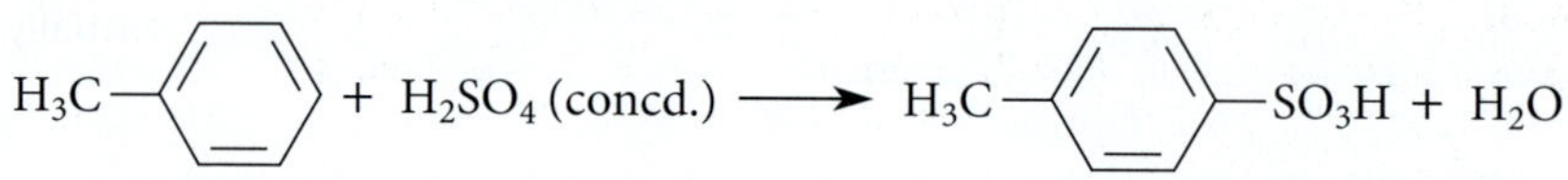

toluene ***p*-toluenesulfonic acid** (16.55)

Another very important consequence of activating and deactivating effects is that when a deactivating group—for example, a nitro group—is being introduced by an electrophilic substitution reaction, it is easy to introduce one group at a time, because the products are *less reactive* than the reactants. Therefore, toluene can be nitrated only once because the nitro group that is introduced retards a second nitration on the same ring. The following three equations show the conditions required for successive nitrations. The actual amounts of nitric and sulfuric acid are given to show that each additional nitration requires harsher conditions.

Don't memorize these conditions! They are given only as concrete examples to illustrate that it takes increasingly harsh conditions to bring about successive substitutions by groups that will become deactivating substituents. In working problems, you can indicate the relative strength of the conditions as "harsh" or "mild" without providing specifics. You can be confident that if you had to carry out the reaction in a laboratory, you could find the conditions in the chemical literature, or you could work out the conditions.

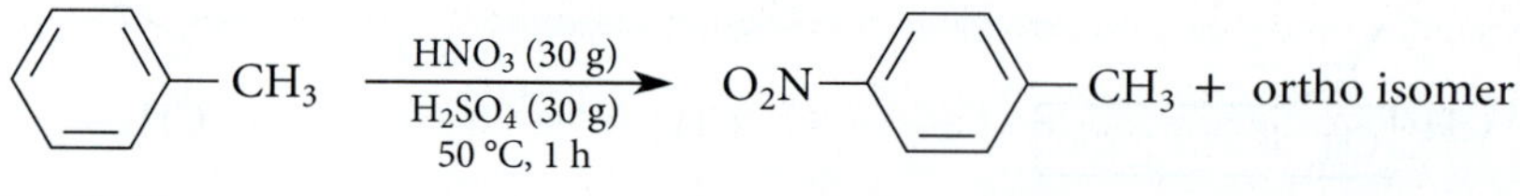

toluene 50 g **4-nitrotoluene** (16.56a)

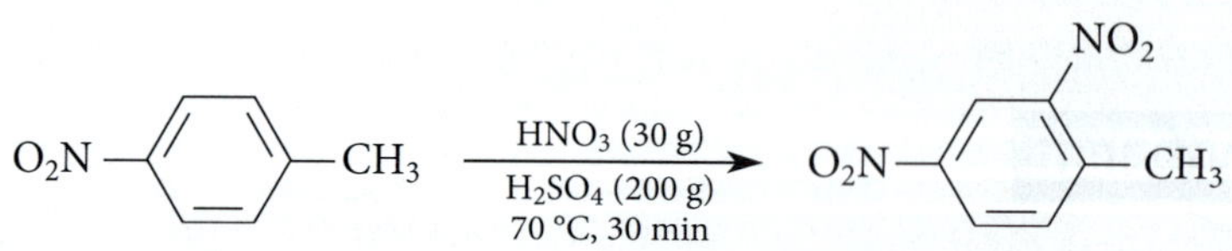

4-nitrotoluene 50 g **2,4-dinitrotoluene** (90% yield) (16.56b)

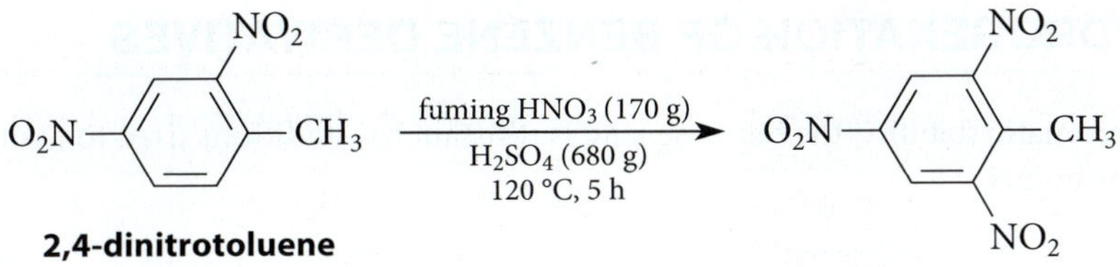

2,4-dinitrotoluene
50 g

2,4,6-trinitrotoluene ("TNT")
(90% yield) (16.56c)

Fuming nitric acid (Eq.16.56c) is an especially concentrated form of nitric acid. Ordinary nitric acid contains 68% by weight of nitric acid, whereas fuming nitric acid is 95% by weight of nitric acid. It owes its name to the layer of colored fumes usually present in the bottle of the commercial product. Fuming nitric acid is a much harsher (that is, more reactive) nitrating reagent than nitric acid itself.

In contrast, when an activating group is introduced by electrophilic substitution, the products are *more reactive* than the reactants; consequently, additional substitutions can occur easily under the conditions of the first substitution and, as a result, mixtures of products are obtained. This is the situation in Friedel–Crafts alkylation. As shown in the discussion of Eqs. 16.27 and 16.28 (Sec. 16.4E), one way to avoid multiple substitution in such cases is to use a large excess of the starting material. (Friedel–Crafts alkylation is the only electrophilic aromatic substitution reaction discussed in this chapter that introduces an activating substituent.)

Some deactivating substituents retard some reactions to the point that they are not useful. For example, Friedel–Crafts *acylation* (Sec. 16.4F) does not occur on a benzene ring substituted *solely* with one or more meta-directing groups. In fact, nitrobenzene is so unreactive in the Friedel–Crafts acylation that it can be used as the solvent in the acylation of other aromatic compounds!

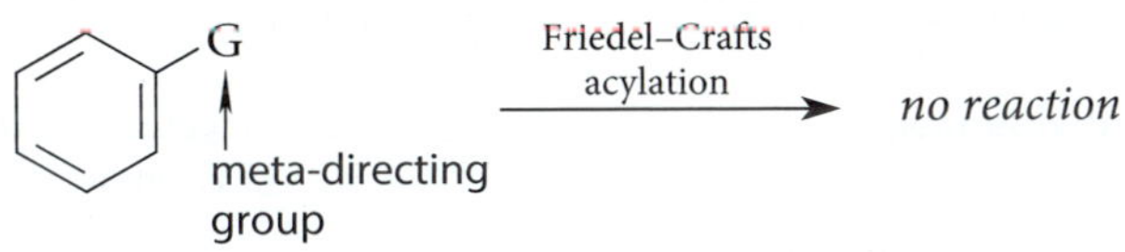

(16.57a)

Similarly, the Friedel–Crafts *alkylation* (Sec. 16.4E) is generally too slow to be useful on compounds that are more deactivated than benzene itself—even halobenzenes.

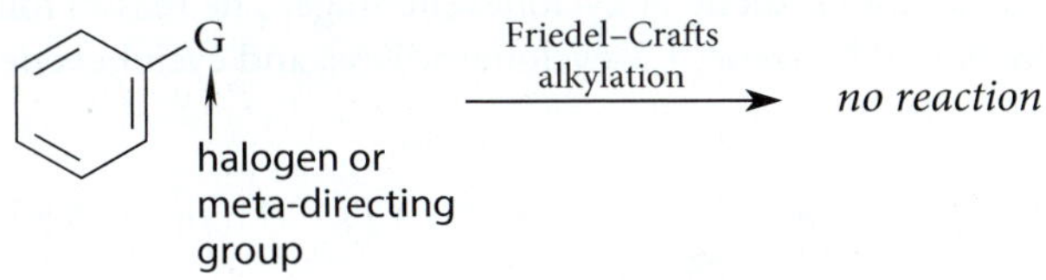

(16.57b)

Focused Problems

16.31 In each of the following sets, rank the compounds in order of increasing harshness of the reaction conditions required to accomplish the indicated reaction.

(a) Sulfonation of benzene, *m*-xylene, or *p*-dichlorobenzene

(b) Friedel–Crafts acylation of chlorobenzene, anisole, or toluene

16.32 Outline a synthesis of *m*-nitroacetophenone from benzene. Explain your reasoning.

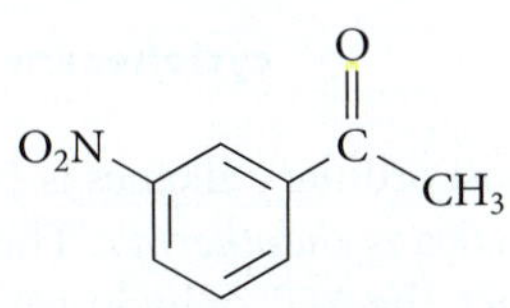

***m*-nitroacetophenone**

16.6 HYDROGENATION OF BENZENE DERIVATIVES

Because of its aromatic stability, the benzene ring is resistant to conditions used to hydrogenate ordinary double bonds.

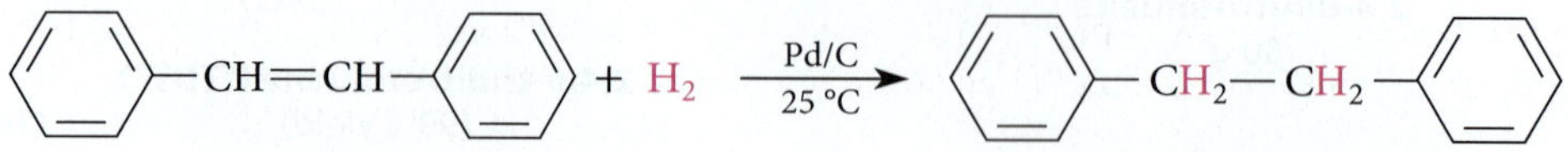

stilbene (cis or trans) → **(2-phenylethyl)benzene (bibenzyl)** (95% yield) (16.58)

Nevertheless, aromatic rings can be hydrogenated under more extreme conditions of temperature or pressure (or both), and a practical laboratory apparatus that can accommodate these conditions is readily available. Typical conditions for carrying out the hydrogenation of benzene derivatives include Rh or Pt catalysts at 5–10 atm of hydrogen pressure and 50–100 °C, or Ni or Pd catalysts at 100–200 atm and 100–200 °C. For example, compare the conditions for the following hydrogenation with those for the hydrogenation in Eq. 16.58.

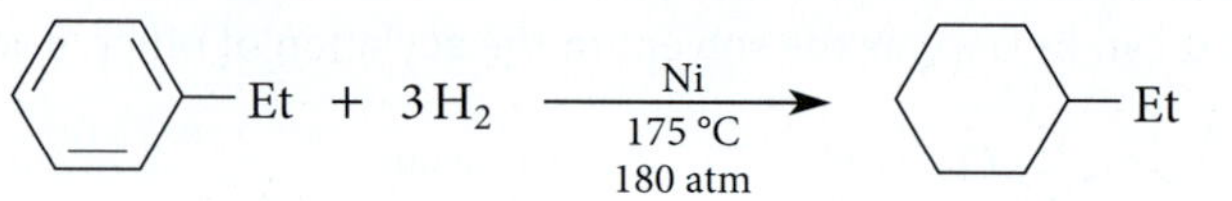

ethylbenzene → **ethylcyclohexane** (93% yield) (16.59)

As this example illustrates, a good way to prepare a substituted cyclohexane in many cases is to prepare the corresponding benzene derivative and then hydrogenate it.

Catalytic hydrogenation of benzene derivatives gives the corresponding cyclohexanes and cannot be stopped at the cyclohexadiene or cyclohexene stage. The reason follows from the enthalpies of hydrogenation of benzene, 1,3-cyclohexadiene, and cyclohexene.

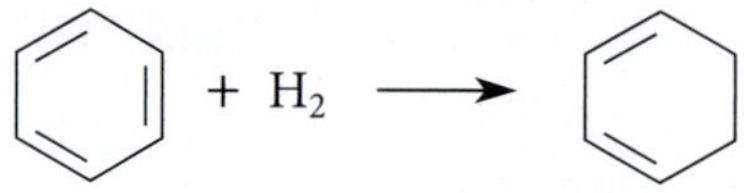 $\Delta H° = +24.3$ kJ mol^{-1} (+5.8 kcal mol^{-1})

benzene → **1,3-cyclohexadiene** (16.60a)

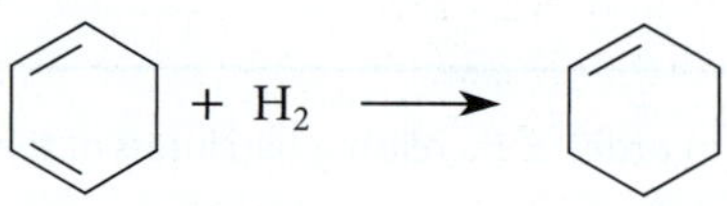 $\Delta H° = -111$ kJ mol^{-1} (–26.5 kcal mol^{-1})

1,3-cyclohexadiene → **cyclohexene** (16.60b)

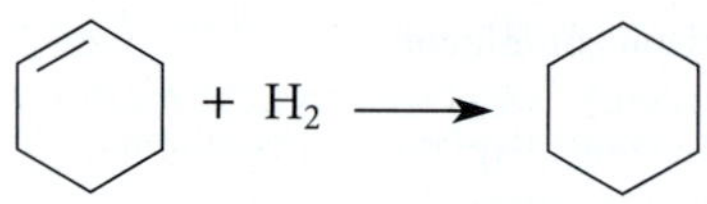 $\Delta H° = -118$ kJ mol^{-1} (−28.2 kcal mol^{-1})

cyclohexene → **cyclohexane** (16.60c)

The hydrogenation of most ordinary alkenes is *exothermic* by 113–126 kJ mol^{-1} (27–30 kcal mol^{-1}); yet the reaction in Eq. 16.60a is *endothermic*. The unusual $\Delta H°$ of this reaction reflects the aromatic stability of benzene. In fact, the $\Delta H°$ of hydrogenation of benzene can be used to provide another estimate of the *empirical resonance energy* of benzene—the aromatic stabilization energy of benzene (Sec. 15.7C). If benzene were an "ordinary" alkene, the $\Delta H°$ of hydrogenation of its three "ordinary" double bonds should be about the same as that of three cyclohexenes, which, from Eq. 16.60c, is $3 \times (-118) = -354$ kJ mol^{-1} (–85 kcal mol^{-1}). The actual $\Delta H°$ of hydrogenation of benzene is obtained by adding the $\Delta H°$ values of Eqs. 16.60a–c to obtain –205 kJ mol^{-1} (–49 kcal mol^{-1}). The difference, which is the empirical resonance energy of benzene, is 149 kJ mol^{-1} (36 kcal mol^{-1}).

This is very similar to the estimate obtained in Sec. 15.7C (141 kJ mol^{-1}, 34 kcal mol^{-1}) by comparing the heats of formation of benzene and cyclooctatetraene (COT).

Because the first hydrogenation reaction of benzene is endothermic, energy must be added for it to take place; so, the difficulty of the first hydrogenation accounts for the harsh conditions required for the hydrogenation of benzene derivatives. The hydrogenations of 1,3-cyclohexadiene and cyclohexene proceed so rapidly under these vigorous conditions that once these compounds are formed in the hydrogenation of benzene, they react instantaneously.

Focused Problem

16.33 Using benzene and any other reagents, outline a synthesis of each of the following compounds. (*Hint:* Use reactions in this chapter to form a substituted benzene first.)

(a) Cyclohexylcyclohexane (b) *tert*-Butylcyclohexane

16.7 POLYCYCLIC AROMATIC HYDROCARBONS AND CANCER

Chemical Biology Topic

Most people understand that certain chemicals are hazardous. Among the most worrisome chemical hazards is *carcinogenicity*—the proclivity of a substance to cause cancer. Cancer-causing materials are termed **carcinogens**. A few aromatic compounds are potent carcinogens. Both the historical aspects of this finding and the reasons underlying it are interesting.

After the great fire of London in 1666, Londoners began the practice of building homes with long and tortuous chimneys. The use of coal for heating resulted in deposits of black soot that had to be periodically removed from these chimneys. The passageways were so narrow that small boys, called "sweeping boys," were employed to clean the chimneys. It was common for these boys to contract a disease that we now know is cancer of the scrotum. In 1775, Percivall Pott (1714–1788), a surgeon at London's St. Bartholomew's Hospital, identified coal dust as the source of "this noisome, painful, and fatal disease," and Pott's findings subsequently led to substantial reform in the child-labor statutes in England. In 1892, Henry T. Butlin (1845–1912), also of St. Bartholomew's, pointed out that the disease did not occur in countries in which the chimney sweeps washed thoroughly after each day's work.

Why did exposure to large amounts of soot cause cancer in the sweeping boys? The source of the problem was traced to benzo[*a*]pyrene, a compound that had been isolated in 1933 from coal tar. This compound and 7,12-dimethylbenz[*a*]anthracene, both polycyclic aromatic hydrocarbons, have been found to be two of the most potent carcinogens known.

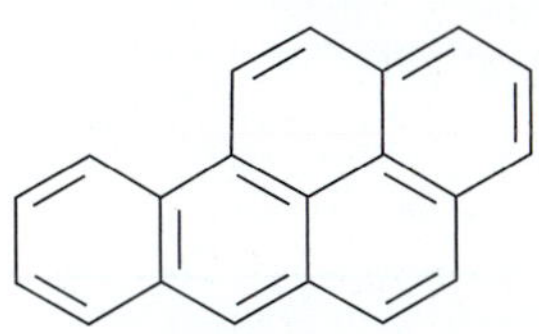

benzo[*a*]pyrene

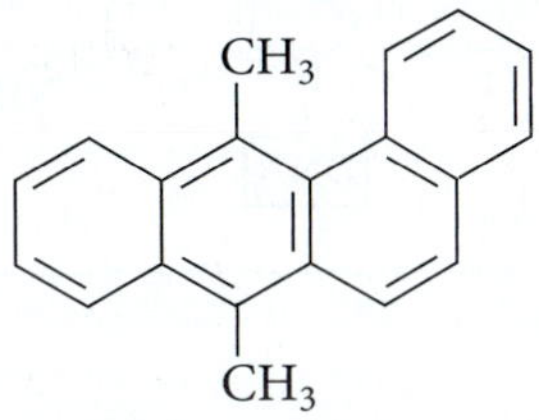

7,12-dimethylbenz[*a*]anthracene (7,12-DMBA)

These and related compounds are the carcinogens in soot. Materials such as these are also found in cigarette smoke, automobile exhaust, the smoke from wood and coal burning, and even in grilled meats. Several thousand tons of benzo[*a*]pyrene and related hydrocarbons per year—the exact amount is not known with certainty—are released into the environment. The association of lung cancer with cigarette smoking has been definitively linked to benzo[*a*]pyrene and related compounds in tobacco smoke.

A study of the carcinogenicity of benzo[*a*]pyrene and related aromatic hydrocarbons led to the finding that the hydrocarbons themselves are not the cancer-causing agents. Rather, they are metabolized in living systems to form certain epoxide derivatives, which are the *ultimate carcinogens* (the true carcinogens). The diol-epoxide shown in Eq. 16.60, formed when living cells attempt to metabolize benzo[*a*]pyrene, is the ultimate carcinogen derived from this hydrocarbon.

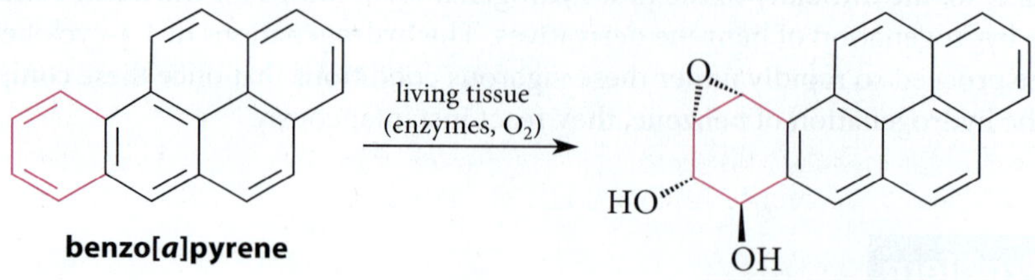

(16.61)

In phase II metabolism (Sec. 8.6C), compounds are chemically coupled to water-soluble "handles," such as glucuronides, that increase their water solubility.

This transformation is particularly remarkable in view of the usual resistance of benzene rings toward addition reactions. This metabolism is an example of *phase I metabolism*—enzyme-catalyzed reactions in which molecules are derivatized to produce more water-soluble products. The diol-epoxide in Eq. 16.60 contains groups that should enhance water solubility—particularly the two alcohol groups. The problem is the epoxide functional group. The diol-epoxide survives long enough that it can migrate into the nuclei of cells, where the epoxide group reacts with nucleophilic groups on DNA. (See Focused Problem 16.34.) When this reaction occurs in certain genes, the regulation of cell division is disrupted and cancer results. In other words, the metabolic process, in an effort to clear the polycyclic aromatic hydrocarbons from the cell, actually activates these hydrocarbons to become the ultimate carcinogens.

Benzene has also been found to be carcinogenic, and it has been supplanted for many uses by toluene, which is not carcinogenic. (Not all aromatic compounds are carcinogens.) However, benzene is much less carcinogenic than the polycyclic hydrocarbons shown here, and it continues to be used with due caution in applications for which it cannot be readily replaced—particularly in the chemical industry (see Sec. 16.8).

Focused Problem

16.34 The group in DNA that reacts with the diol-epoxide has the following general structure. Using mechanistic reasoning, show how the amino group indicated by the asterisk (*) might react with the epoxide group of the diol-epoxide in Eq. 16.60. (*Hint:* Abbreviate this structure as R—$\ddot{N}H_2$, and then fill in the completed structure after you have completed the mechanism.)

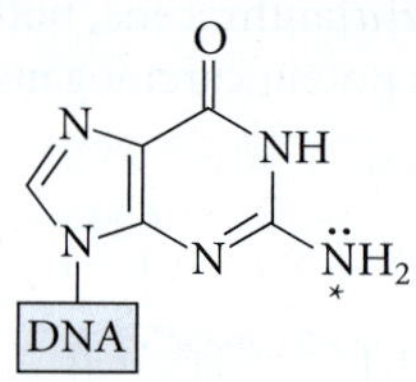

Chemistry in the Real World

16.8 SOURCE AND INDUSTRIAL USE OF AROMATIC HYDROCARBONS

The most common source of aromatic hydrocarbons is petroleum. Some petroleum sources are relatively rich in aromatic hydrocarbons, and aromatic hydrocarbons can be obtained by catalytic re-forming of the hydrocarbons from other sources. Another potentially important, but currently minor, source of aromatic hydrocarbons is *coal tar*, the tarry residue obtained when coal is heated in the absence of oxygen.

Benzene itself is obtained by separation from petroleum fractions and by demethylation of toluene. The worldwide annual production of benzene is more than 13 billion gallons. Benzene serves as a principal source of ethylbenzene, styrene, and cumene (see Eq. 16.62), and as one of the sources of cyclohexane. Because cyclohexane is an important intermediate in the production of nylon (Sec. 21.12A), benzene has substantial importance in the nylon industry.

Toluene is also obtained by separation from *reformates*, the products of hydrocarbon interconversion over certain catalysts. Although some toluene is used in the production of benzene, toluene is also used as an octane booster for gasoline and as a starting material in the polyurethane industry.

Ethylbenzene and cumene are obtained by the alkylation of benzene with ethylene and propene, respectively, in the presence of acid catalysts (Friedel–Crafts alkylation).

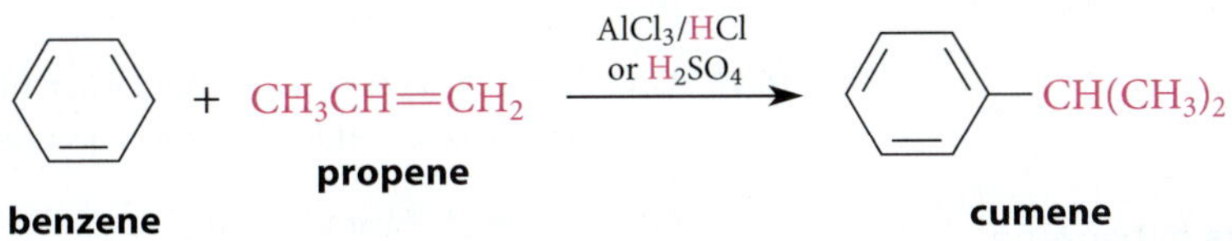

(16.62)

Cumene is an important intermediate in the manufacture of phenol and acetone (Sec. 18.11). The major use of ethylbenzene is for dehydrogenation to styrene ($PhCH{=}CH_2$), one of the most commercially important aromatic hydrocarbons. Its principal uses are in the manufacture of polystyrene (Sec. 10.3C) and styrene–butadiene rubber (Sec. 15.5).

The xylenes (dimethylbenzenes) are obtained by separation from petroleum and by re-forming C_8 petroleum fractions. Of the xylenes, *p*-xylene is the most important commercially. Virtually the entire production of *p*-xylene is used for oxidation to terephthalic acid (Eq. 16.62), an important intermediate in polyester synthesis (for example, Dacron; Sec. 21.12A). (Oxidation of alkylbenzenes is discussed in Sec. 17.5C.)

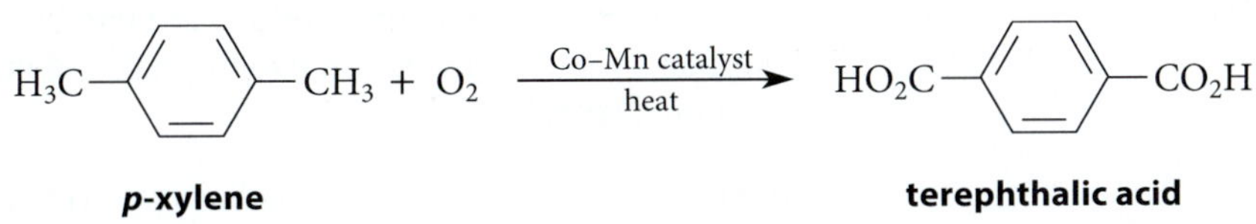

(16.63)

Focused Problem

16.35 Propose a curved-arrow mechanism for the industrial preparation of cumene (Eq. 16.62) with a sulfuric acid catalyst.

CHAPTER SUMMARY *For a summary of the chapter, see Chapter 16 in the* Study Guide and Solutions Manual.

REACTION REVIEW *For a summary of reactions discussed in this chapter, see the* Reaction Review *section of Chapter 16 in the* Study Guide and Solutions Manual.

SKILLS OBJECTIVES WITH PROBLEMS

- **Give IUPAC or common names for benzene derivatives; given the name of a benzene derivative, provide the structure. 16.1**

16.36 Give an IUPAC or common name for each of the following benzene derivatives.

(a) 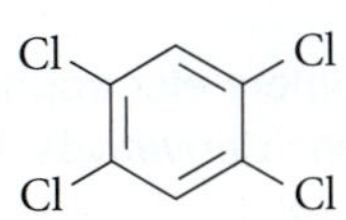

(b) 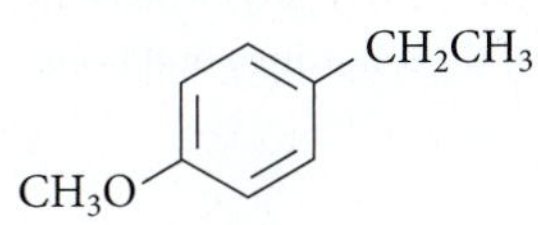

(c) 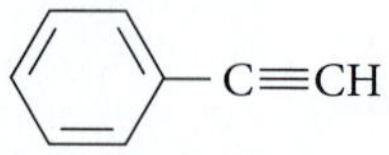

(d) 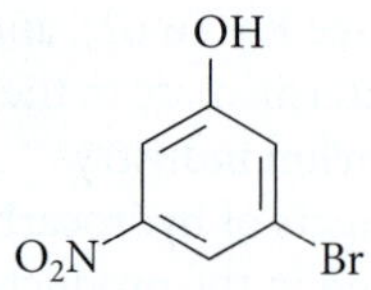

16.37 Draw a structure for each of the following compounds.

(a) 2-Chlororesorcinol

(b) *p*-Nitroaniline

(c) 1,4-Divinylbenzene

(d) 4-Methyl-1-benzenethiol

- **Interpret spectra to deduce structures of benzene derivatives. 16.3**

16.38 Explain how you would distinguish each of the following isomeric compounds from the others using proton NMR spectroscopy. Be explicit.

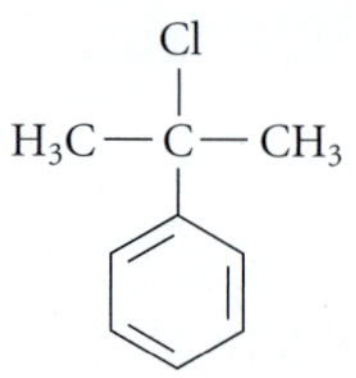

A

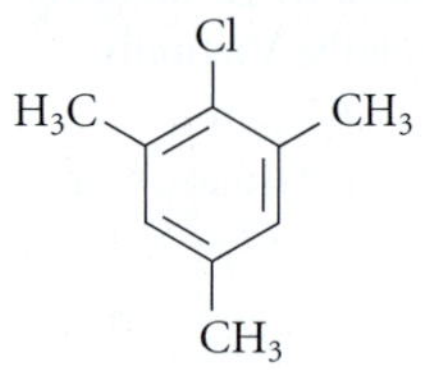

B

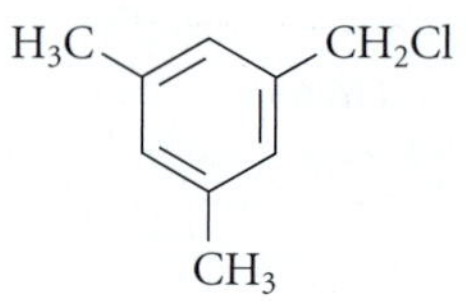

C

16.39 Explain how you would distinguish between ethylbenzene, *p*-xylene, and styrene solely by carbon NMR spectroscopy.

16.40 Identify each of the following compounds.

(a) Compound *A*: IR 1605 cm^{-1}, no O—H stretch. NMR: δ 3.72 (3*H*, s); δ 6.72 (2*H*, apparent doublet, $J = 9$ Hz); δ 7.15 (2*H*, apparent doublet, $J = 9$ Hz). Mass spectrum in **Fig. P16.40a**. (*Hint:* Notice the M + 2 peak.)

(b) Compound *B* ($C_{10}H_{12}O$): IR 965, 1175, 1247, 1608, 1640 cm^{-1} no O—H stretch. NMR in **Fig. P16.40b**. UV: λ_{max} (ethanol) 260 ($\varepsilon = 18{,}200$); this is a greater wavelength and about the same intensity as the UV spectrum of styrene.

- **Given the electrophiles, complete electrophilic substitution reactions of benzene derivatives. 16.4**

16.41 Give the products expected (if any) when ethylbenzene reacts under each of the following conditions.

(a) HNO_3, H_2SO_4

(b) Concentrated H_2SO_4

(c) 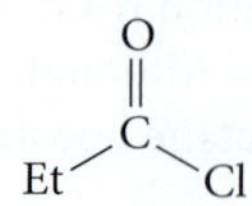 + $AlCl_3$ (1.1 equiv.), then H_2O

(d) CH_3Cl, $AlCl_3$ (large excess of ethylbenzene)

(e) Br_2, $FeBr_3$

16.42 Give the products expected (if any) when nitrobenzene reacts under each of the following conditions.

(a) Cl_2, $FeCl_3$, heat

(b) Fuming HNO_3, H_2SO_4

(c) 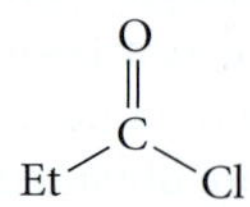 + $AlCl_3$ (1.1 equiv.), then H_2O

16.43 Complete the electrophilic substitution reactions given in **Fig. P16.43**. Explain your reasoning in each case.

- **Summarize the three mechanistic steps in electrophilic aromatic substitution, and use these to propose a mechanism for an electrophilic substitution reaction. 16.4B**

16.44 (a) Summarize the three mechanistic steps in electrophilic aromatic substitution using benzene as the aromatic compound, and E—X is the electrophile containing the electrophilic center E.

(b) Use this scheme as a guide to propose a curved-arrow mechanism for the following electrophilic aromatic substitution reaction. (*Hint:* What electrophile is generated when a strong acid reacts with a tertiary alcohol?)

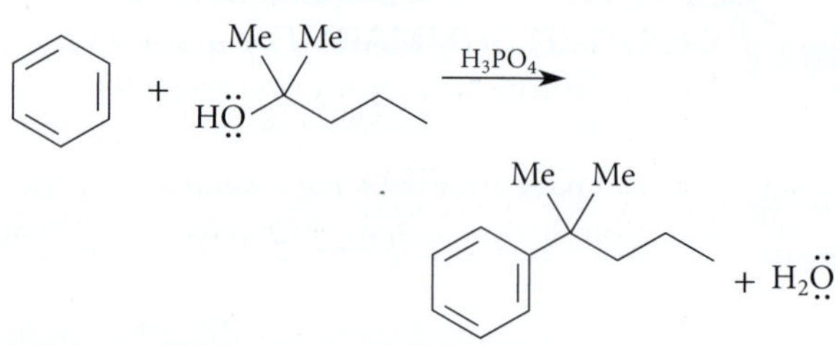

- **Draw resonance structures for the carbocation intermediates in electrophilic substitution reactions, and use them to predict relative reaction rates. 16.4A, 16.4B**

16.45 1-Methoxynaphthalene undergoes nitration primarily at carbon-4. However, smaller amounts of nitration occur in the ring not containing the methoxy group. At which carbon of the second ring should nitration nitrate more rapidly: carbon-5 or carbon-6? (*Hint:* Decide by drawing

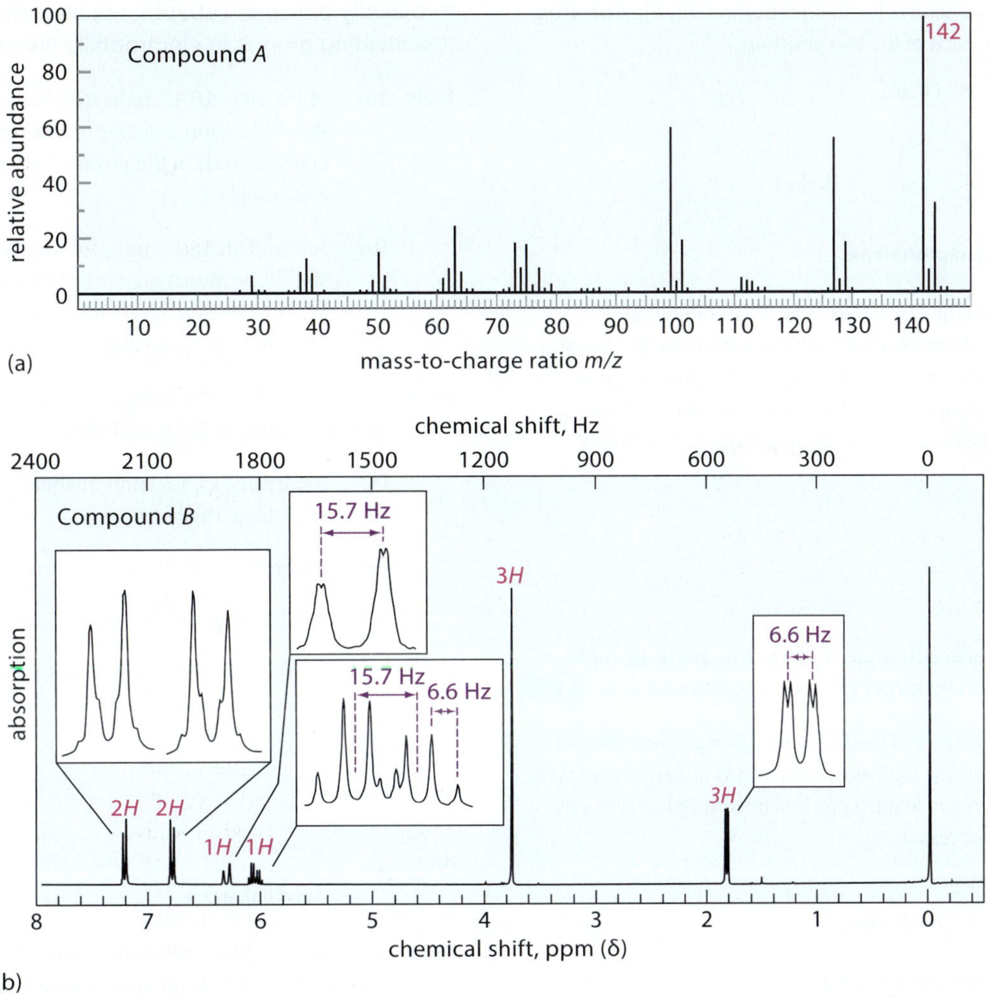

FIGURE P16.40 (a) The mass spectrum of compound *A* in Problem 16.40a. (b) The NMR spectrum of compound *B* in Problem 16.40b. The integrals are shown in red over the peaks, and the coupling constants are shown in violet.

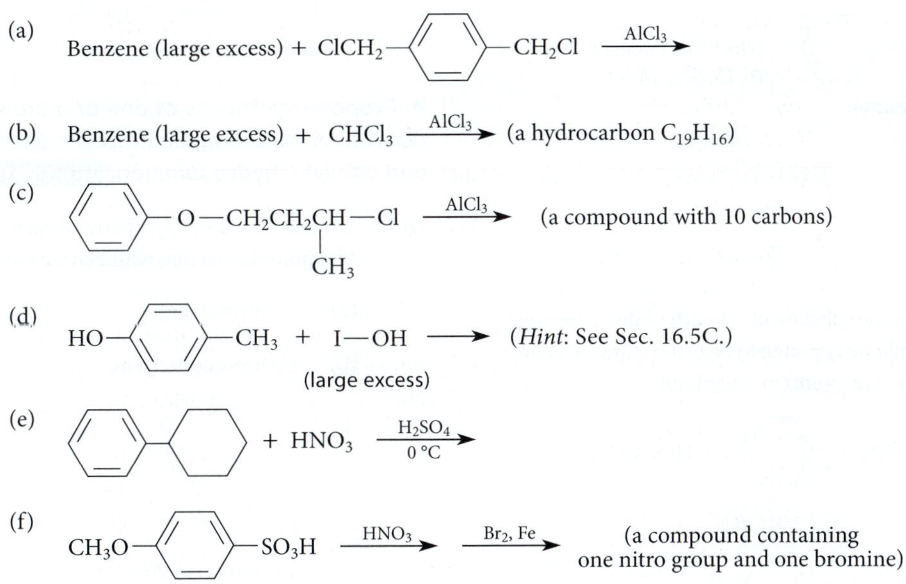

FIGURE P16.43

resonance structures for the carbocation intermediate from nitration at each of the two positions.)

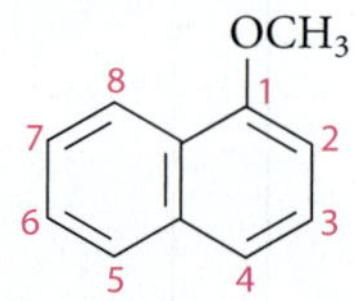

1-methoxynaphthalene

16.46 Furan is an aromatic heterocyclic compound that undergoes electrophilic aromatic substitution. By drawing resonance structures for the carbocation intermediates involved, deduce whether furan should undergo Friedel–Crafts acylation more rapidly at carbon-2 or carbon-3.

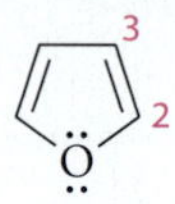

furan

• Classify benzene substituents as ortho, para- or meta-directing groups in electrophilic aromatic substitution. 16.5A

16.47 Give the structure of the major nitration product formed when each of the following compounds undergoes aromatic nitration. (A single nitro group is introduced in each case.) Explain your reasoning.

(a) 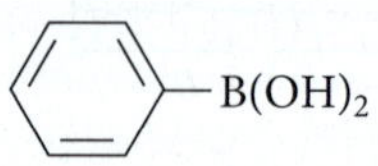

benzeneboronic acid

(b) 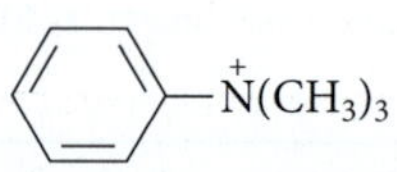

***N,N,N*-trimethylanilinium**

(c) 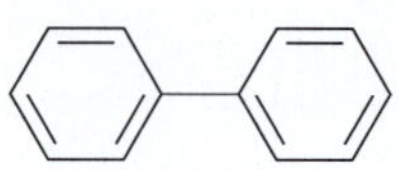

biphenyl

(*Hint*: See Focused Problem 16.23, Sec. 16.5A.)

(d) 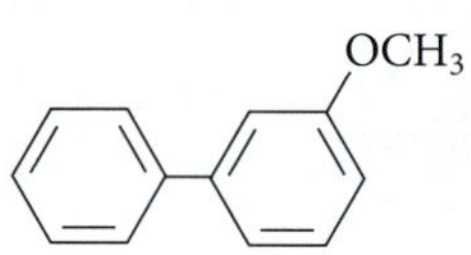

16.48 Indicate whether the substituents in each of the following compounds would be expected to be ortho, para, or meta directors. Explain any points of uncertainty.

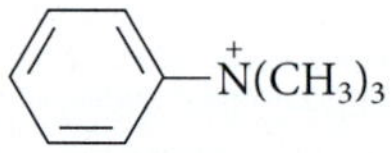

A

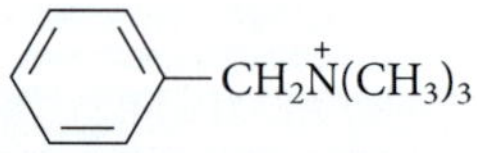

B

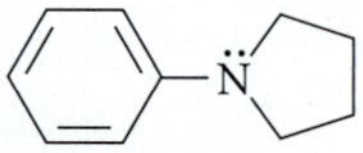

C

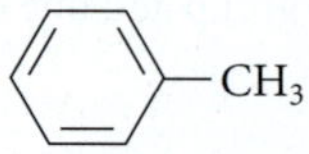

D

• Classify benzene substituents as activating or deactivating groups in electrophilic substitution. 16.5B

16.49 (a) In Problem 16.47, indicate whether each compound should be more reactive or less reactive than benzene itself in the nitration reaction. Explain your reasoning.

(b) In Problem 16.48, indicate whether each compound should be more reactive or less reactive than benzene itself in electrophilic aromatic substitution. Explain your reasoning.

16.50 Arrange the following compounds in order of increasing reactivity toward HNO_3 in H_2SO_4.

(a) Mesitylene (1,3,5-trimethylbenzene), toluene, 1,2,4-trimethylbenzene

(b) Chlorobenzene, benzene, nitrobenzene

(c) *m*-Chloroanisole, *p*-chloroanisole, anisole

(d) Acetophenone, *p*-methoxyacetophenone, *p*-bromoacetophenone

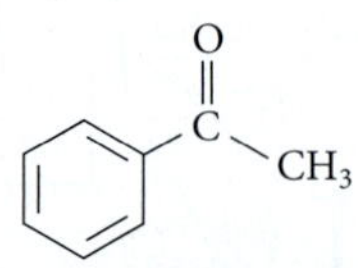

acetophenone

16.51 Give two Friedel–Crafts acylation reactions that could be used to prepare 4-methoxybenzophenone. Which reaction would be faster or occur under milder conditions? Explain.

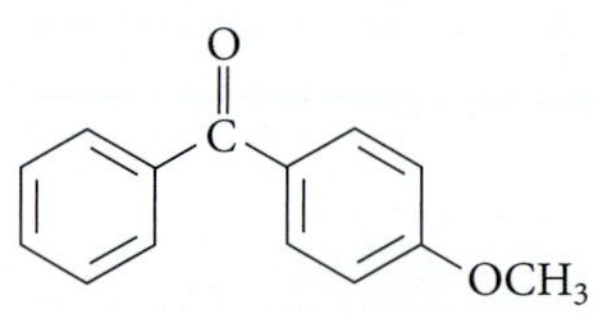

4-methoxybenzophenone

• Propose syntheses of one or more steps for benzene derivatives involving electrophilic aromatic substitution and catalytic hydrogenation. 16.5D, 16.6

16.52 Outline laboratory syntheses of each of the following compounds, starting with benzene and any other reagents.

(a) *p*-Nitrotoluene

(b) *p*-Dibromobenzene

(c) 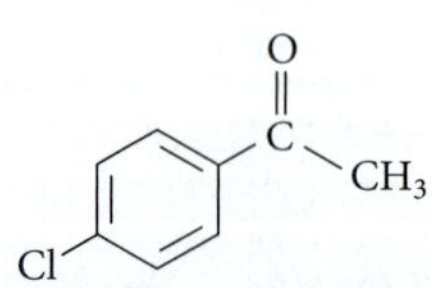

***p*-chloroacetophenone**

(d) *m*-Nitrobenzenesulfonic acid

(e) *p*-Chloronitrobenzene

(f) 1,3,5-Trinitrobenzene

(g) 2,6-Dibromo-4-nitrotoluene

(h) 2,4-Dibromo-6-nitrotoluene

(i) 4-Ethyl-3-nitroacetophenone [See Problem 16.50(d) for the structure of acetophenone.]

(j) *cis*-1,4-Di-*tert*-butylcyclohexane

(k) Cyclopentylbenzene

16.53 Give the structures of all the hydrocarbons $C_{10}H_{10}$ that would undergo catalytic hydrogenation to give *p*-diethylbenzene.

INTEGRATED PROBLEMS

16.54 Which of the following compounds *cannot* contain a benzene ring? How do you know?

$C_{10}H_{16}$	$C_8H_6Cl_2$	C_5H_4	$C_{10}H_{15}N$
A	*B*	*C*	*D*

16.55 (a) Arrange the three isomeric dichlorobenzenes in order of increasing dipole moment (smallest first).

(b) Assuming that the dipole moment is the principal factor governing their relative boiling points, arrange the compounds from part (a) in order of increasing boiling point (smallest first). Explain your reasoning.

(c) Arrange the compounds in order of increasing melting point. (*Hint:* Think about symmetry.)

16.56 Explain each of the following observations.

(a) The NMR spectrum of the sodium salt of 1,3-cyclopentadiene consists of a singlet.

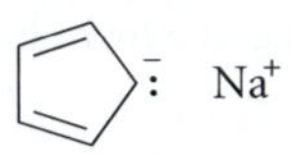

sodium salt of cyclopentadiene

(b) The methyl group in the following compound has an unusual chemical shift of δ (−1.67)—about 4 ppm lower than the chemical shift of a typical allylic methyl group. (*Hint:* See Fig. 16.2, Sec. 16.3B.)

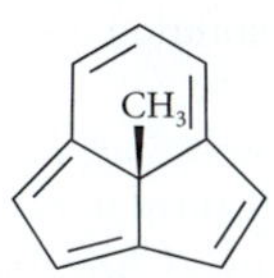

16.57 Show how resonance interaction of the electron pairs on the oxygen with the ring π electrons can account for the fact that the chemical shift of protons H^a in *p*-methoxytoluene is smaller than that of protons H^b, even though the oxygen has a greater electronegativity than the methyl carbon.

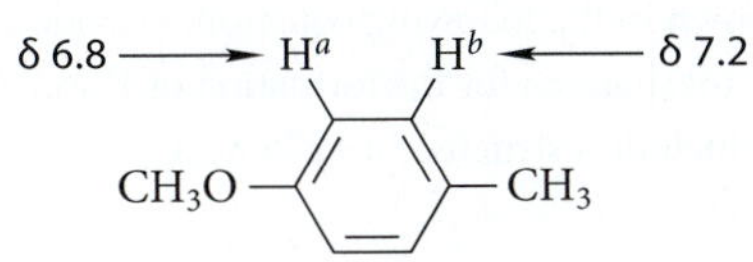

16.58 Nitration of phenyl acetate (compound *A*) results in para substitution of the nitro group. However, nitration of dimethyl phenyl phosphate (compound *B*) results in meta substitution of the nitro group. Suggest a reason that the two compounds nitrate in different positions. (*Hint:* Draw an octet structure for the phosphorus.)

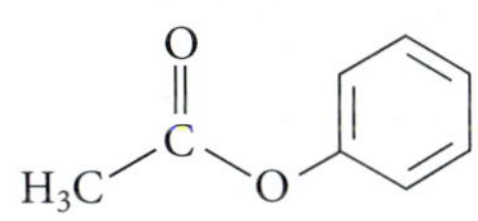

A

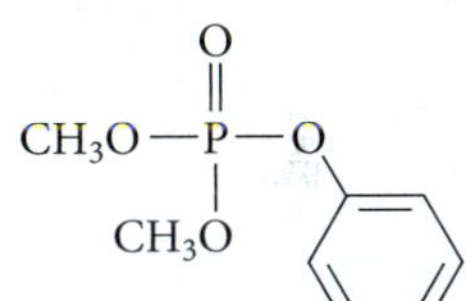

B

16.59 Two alcohols, *A* and *B*, have the same molecular formula ($C_9H_{10}O$) and react with sulfuric acid to give the same hydrocarbon *C*. Compound *A* is optically active, whereas compound *B* is not. Catalytic hydrogenation of *C* gives a hydrocarbon *D* (C_9H_{10}), which gives two and only two products when nitrated once with HNO_3 in H_2SO_4. Give the structures of *A*, *B*, *C*, and *D*.

16.60 Suggest a reason that the λ_{max} values and intensities of the UV absorptions of styrene ($PhCH{=}CH_2$) and phenylacetylene ($PhC{\equiv}CH$) are essentially identical even though phenylacetylene contains an additional π bond.

16.61 Diphenylsulfone is a by-product that is formed in the sulfonation of benzene. Give a curved-arrow mechanism for its formation from benzenesulfonic acid in the presence of *benzene*.

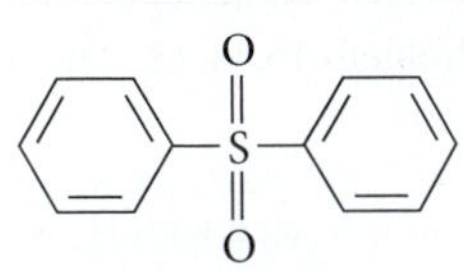

diphenylsulfone

16.62 Sulfonation, unlike most other electrophilic aromatic substitution reactions, is reversible. Benzenesulfonic acid ($Ph{-}SO_3H$) can be converted into benzene and H_2SO_4 with hot water (steam). Write a curved-arrow mechanism for this reaction.

16.63 (a) Naphthalene is sulfonated more rapidly at carbon-1 than it is at carbon-2 at 40 °C. (See **Fig. P16.63**.) By drawing resonance structures for the carbocation

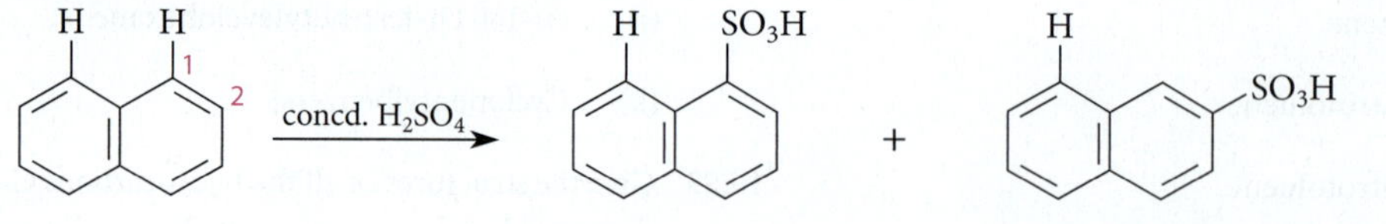

	naphthalene-1-sulfonic acid	naphthalene-2-sulfonic acid	
	84–85%	15–16%	yield at 40 °C
	7–15%	85–93%	yield at 160 °C

FIGURE P16.63

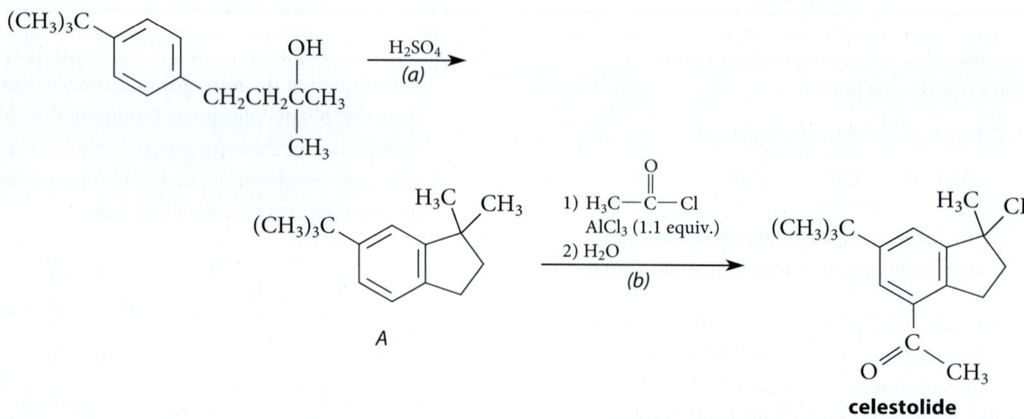

FIGURE P16.65

intermediates, explain the preference for sulfonation at carbon-1. (*Hint:* Don't forget that every intact benzene ring has two resonance structures. Furthermore, resonance structures containing intact benzene rings are more important than structures in which the cyclic π-electron system of a benzene ring is disrupted.)

(b) As shown in Fig. P16.63, if the reaction is carried out at 160 °C, or if naphthalene-1-sulfonic acid is heated under the reaction conditions to 160 °C, a mixture containing mostly naphthalene-2-sulfonic acid is obtained. Explain how this interconversion occurs, and suggest a reason that naphthalene-2-sulfonic acid is the preferred isomer at equilibrium. [*Hints:* (1) Consider Problem 16.62. (2) The sulfonic acid group is very large.]

16.64 When the following compound is treated with H_2SO_4, the product of the resulting electrophilic aromatic substitution reaction has the formula $C_{15}H_{20}$ and does not decolorize Br_2 in CH_2Cl_2.

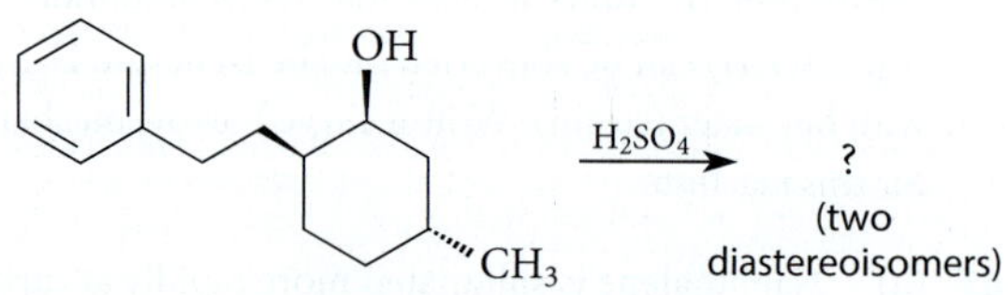

(a) Use a curved-arrow mechanism to predict the structure of the product. (*Hint:* Ignore stereochemistry while proposing the mechanism.)

(b) Explain why two diastereomeric products are possible, and suggest which should be the major one and why.

16.65 Celestolide, a perfuming agent with a musk odor, is prepared by the sequence of reactions given in **Fig. P16.65**.

(a) Give the curved-arrow mechanism for the formation of compound *A* (reaction *a*).

(b) In your mechanism, identify the three basic steps of electrophilic aromatic substitution discussed in Sec. 16.4B.

(c) What product would be obtained in reaction *b* if this reaction followed the usual directing effects of alkyl substituents? Suggest a reason that celestolide is formed instead.

16.66 When styrene is treated with a strong acid catalyst (a sulfonic acid, RSO_3H) in cyclohexane solvent, an alkene $X(C_{16}H_{16})$ is formed that is slowly transformed into isomeric compounds *Y* and *Z*, as shown in the following equation. Provide a curved-arrow mechanism for the formation of *Y* and *Z*, which should include a structure for alkene *X*.

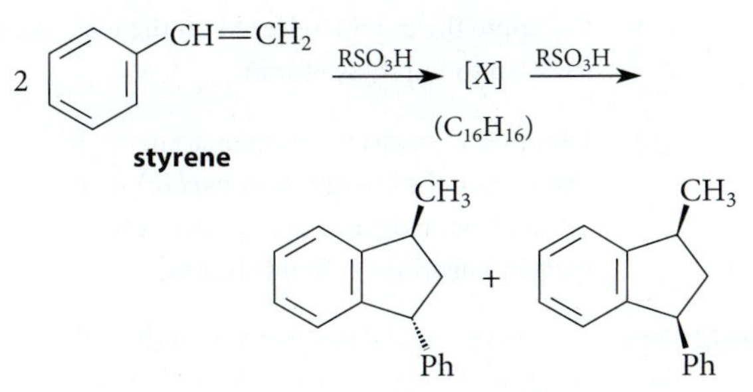

Y *Z*

16.67 An optically active compound *A* ($C_9H_{11}Br$) reacts with sodium ethoxide in ethanol to give an optically inactive hydrocarbon *B* (NMR spectrum in **Fig. P16.67**). Compound *B* undergoes hydrogenation over a Pd/C catalyst at room temperature to give compound *C*, which has the formula C_9H_{12}. Give the structures of *A*, *B*, and *C*.

16.68 A method for determining the structures of disubstituted benzene derivatives was proposed in 1874 by Wilhelm Körner of the University of Milan. Körner had in hand three dibromobenzenes, *A*, *B*, and *C*, with melting points of 89, 6.7, and –6.5 °C, respectively. He nitrated each isomer in turn and meticulously isolated *all* of the mononitro derivatives of each—even the ones formed in very small amounts. Compound *A* gave one mononitro derivative, compound *B* gave two mononitro derivatives, and compound *C* gave three. These experiments gave him enough information to assign the structures of *o*-, *m*-, and *p*-dibromobenzene.

(a) Assuming the correctness of the Kekulé structure for benzene, assign the structures of the dibromobenzene derivatives.

(b) Körner had no way of knowing whether the Kekulé or Ladenburg benzene structure (Display 15.57, Sec. 15.7A) was correct. Assuming the correctness of the Ladenburg benzene structure, assign the structures of the dibromobenzene derivatives.

(c) It is a testament to Körner's experimental skill that he could isolate all of the mononitration products. Of all the mononitration products that he isolated, which one(s) were formed in smallest amount? Explain.

(d) Jack Körner, Wilhelm's grandnephew, has decided to verify great-uncle Wilhelm's results by using ^{13}C NMR to identify the dibromobenzene isomers. What differences can he expect in the ^{13}C NMR spectra of these compounds? (Assume the Kekulé benzene structure.)

16.69 Give the structures of the principal organic product(s) expected in each of the following electrophilic aromatic substitution reactions, and explain your reasoning.

(a)

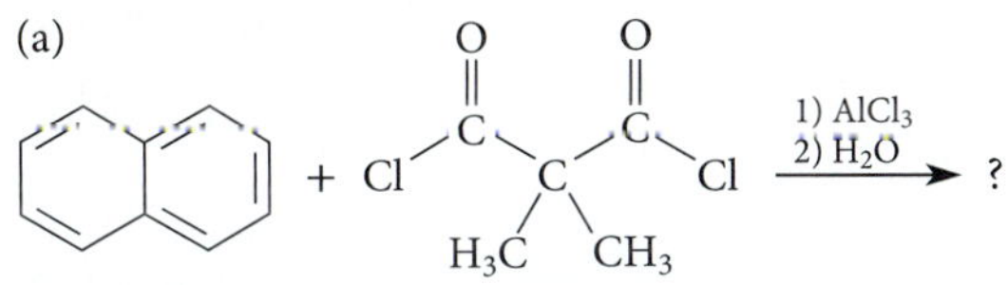

naphthalene **α,α-dimethylmalonyl dichloride**

(*Hint:* Three constitutional isomers with the formula $C_{15}H_{12}O_2$ are formed.)

(b)

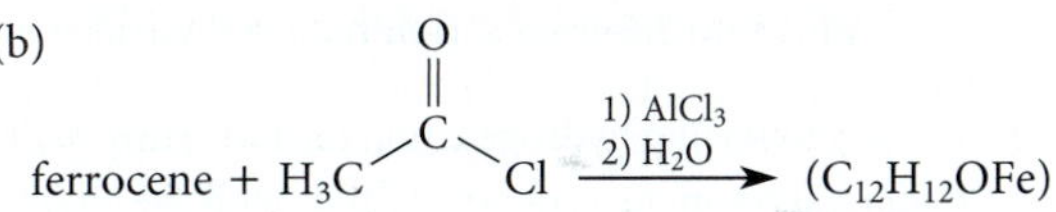

(Display 15.69, Sec. 15.7D)

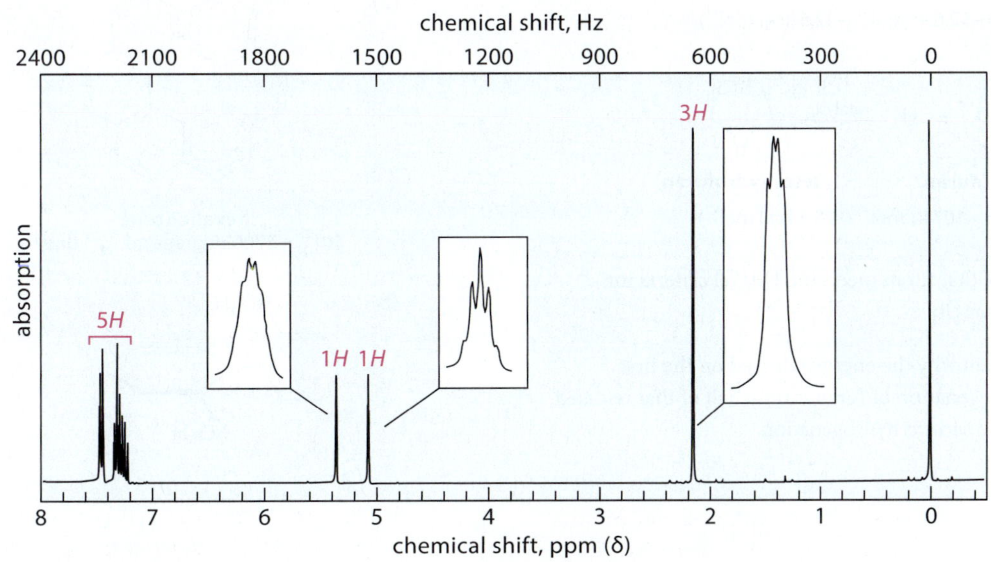

FIGURE P16.67 The NMR spectrum for Problem 16.67. The integrals are shown in red over the peaks.

16.70 Given that anisole (methoxybenzene) protonates primarily on oxygen in concentrated H_2SO_4, explain why 1,3,5-trimethoxybenzene protonates primarily on a carbon of the ring. As part of your reasoning, draw the structure of each conjugate acid.

16.71 A Diels–Alder reaction of 2,5-dimethylfuran and maleic anhydride gives a compound *A* that undergoes acid-catalyzed dehydration to give 3,6-dimethylphthalic anhydride.

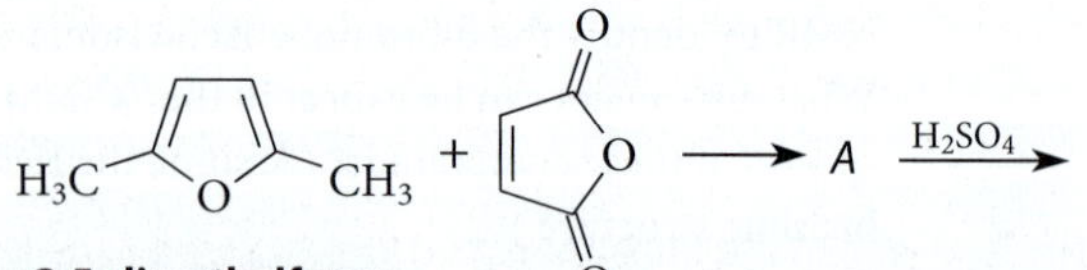

2,5-dimethylfuran

maleic anhydride

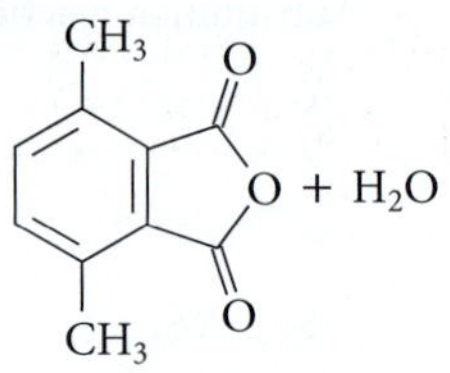

3,6-dimethylphthalic anhydride

(a) Deduce the structure of compound *A*.

(b) Give a curved-arrow mechanism for the conversion of *A* into 3,6-dimethylphthalic anhydride.

16.72 Use the following hydrogenation data to estimate the empirical resonance energy of furan with the aid of the questions that follow.

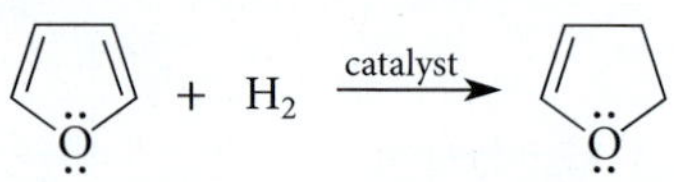

furan **2,3-dihydrofuran**

$\Delta H° = -52.6$ kJ mol^{-1} (-12.6 kcal mol^{-1})

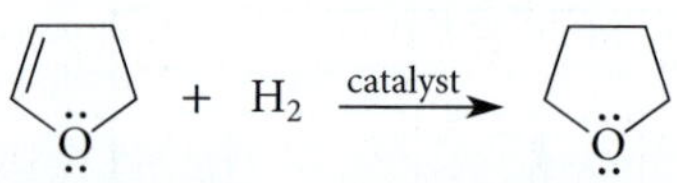

2,3-dihydrofuran **tetrahydrofuran**

$\Delta H° = -107$ kJ mol^{-1} (-25.6 kcal mol^{-1})

(a) Show that furan meets the Hückel criteria for aromaticity.

(b) Explain why the energy released on the first hydrogenation of furan is only half of that released by the second hydrogenation.

(c) Compute the energy released on the hydrogenation of furan to tetrahydrofuran.

(d) Obtain the empirical resonance energy by comparing the energy from part (c) with the energy released on hydrogenation of two "ordinary" oxygen-substituted double bonds.

16.73 Propose a curved-arrow mechanism for the following reaction. (*Hint:* Modify step 3 of the usual aromatic substitution mechanism in Sec. 16.4B.)

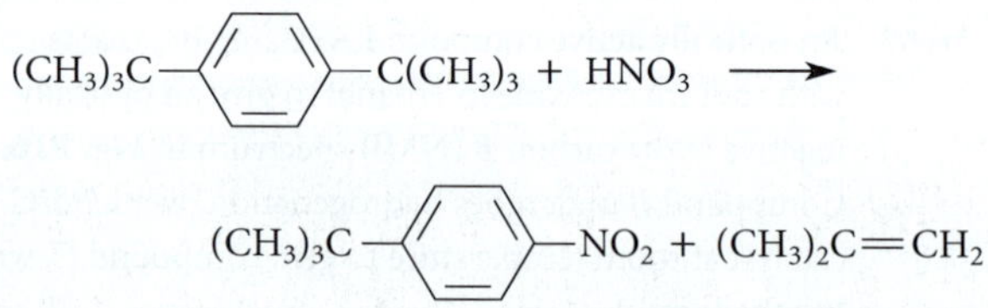

16.74 At 36 °C, the NMR resonances for the ring methyl groups of isopropylmesitylene (protons H^a and H^b in the following structure) are two singlets at δ 2.25 and δ 2.13 with a 2:1 intensity ratio, respectively. When the spectrum is taken at –60 °C, however, it shows three singlets of equal intensity for these groups at δ 2.25, δ 2.17, and δ 2.11. Explain these observations.

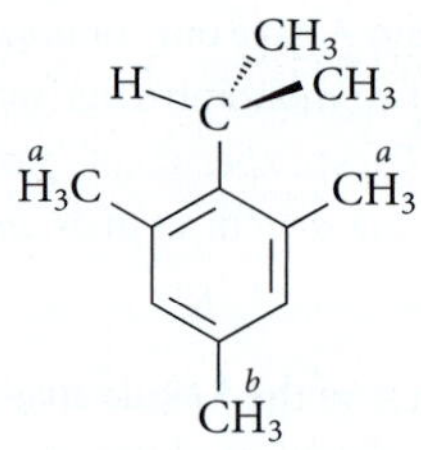

isopropylmesitylene

16.75 Each of the following aromatic compounds can be resolved into enantiomers. Explain why each is chiral, and why compound (b) racemizes when it is heated.

(a)

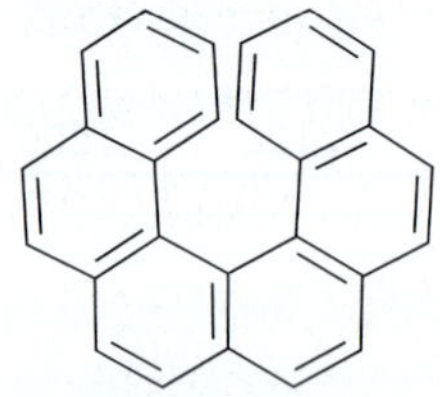

hexahelicene

$[\alpha]_D^{25} = 3700$ degrees mL g^{-1} dm^{-1}

(b)

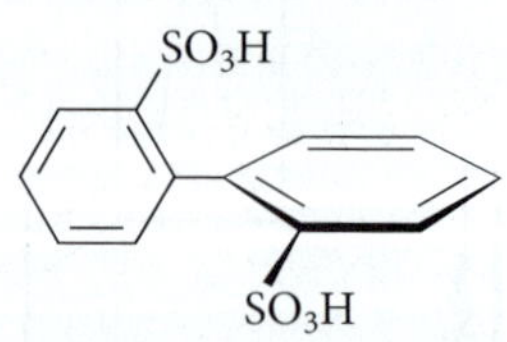

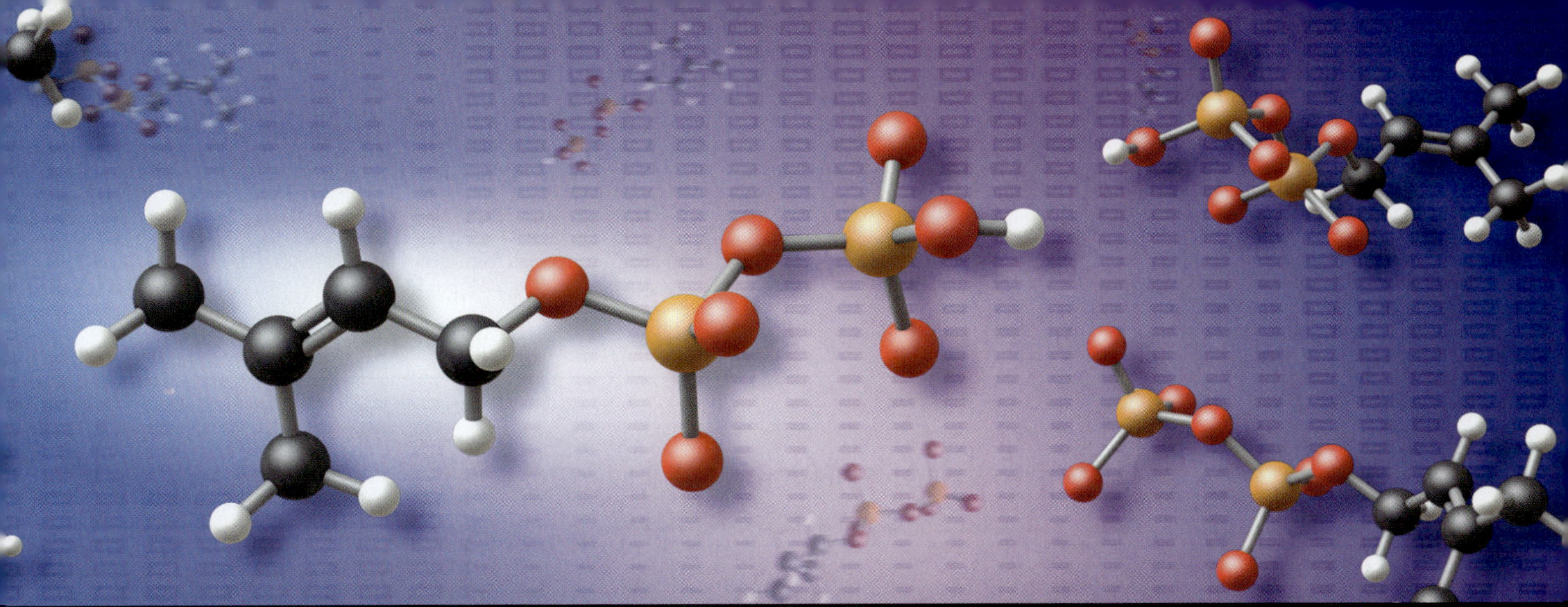

17 ALLYLIC AND BENZYLIC REACTIVITY

An **allylic group** is a group on a carbon adjacent to a double bond. A **benzylic group** is a group on a carbon adjacent to a benzene ring or substituted benzene ring.

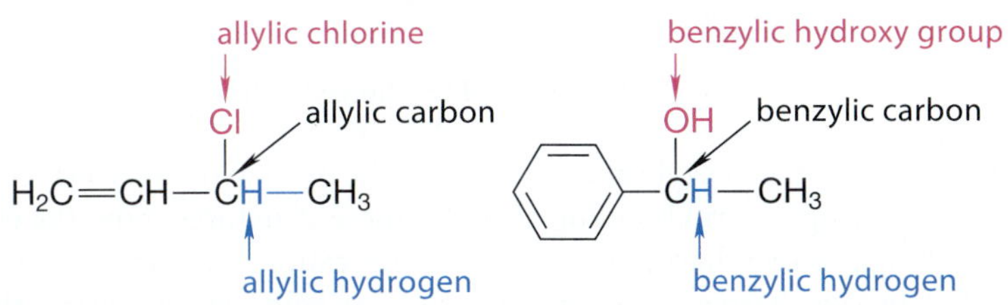

(17.1)

In many situations, *allylic and benzylic groups are unusually reactive*. This chapter examines what happens when some familiar reactions occur at allylic and benzylic positions and discusses the reasons for allylic and benzylic reactivity. This chapter also presents some reactions that occur *only* at the allylic and benzylic positions. Finally, in Secs. 17.5B and 17.6, we show that allylic reactivity is also important in some chemistry that occurs in living organisms.

Focused Problems

17.1 Identify the allylic carbons in each of the following structures.

(a)

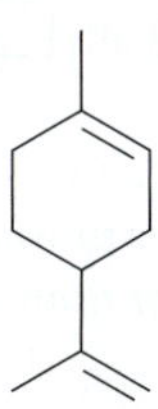

(b)

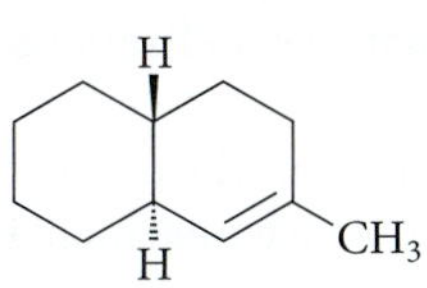

17.2 Identify the benzylic carbons in each of the following structures.

(a) 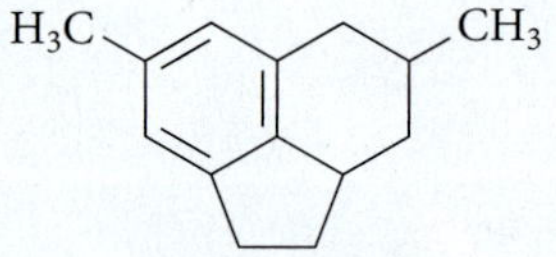

(b) 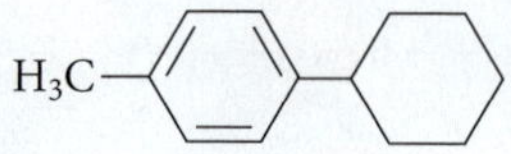

17.1 REACTIONS INVOLVING ALLYLIC AND BENZYLIC CARBOCATIONS

Recall that allylic carbocations are resonance-stabilized (Sec. 15.6). The simplest example of an allylic cation is the *allyl cation* itself:

$$\left[H_2C{=}CH{-}\overset{+}{C}H_2 \longleftrightarrow H_2\overset{+}{C}{-}CH{=}CH_2\right]$$

resonance structures of the allyl cation (17.2)

These resonance structures symbolize the delocalization of electrons and electron deficiency (along with the associated positive charge) that result from the overlap of 2p orbitals to form π molecular orbitals (MO), as shown in Fig. 15.14 (Sec. 15.6).

Benzylic carbocations are also resonance-stabilized. The *benzyl cation* is the simplest example of a benzylic cation:

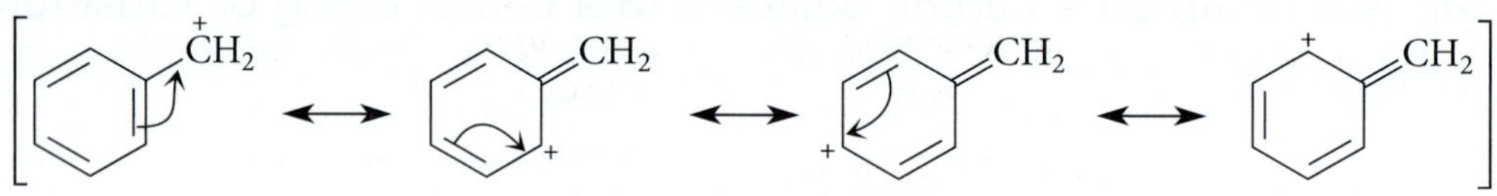

resonance structures of the benzyl cation (17.3)

The resonance structures of the benzyl cation symbolize the overlap of 2p orbitals of the benzylic carbon and the benzene ring to form bonding MOs. As these structures show, the electron deficiency and resulting positive charge on a benzylic carbocation are shared not only by the benzylic carbon, but also by alternate carbons of the ring. As with the allyl cation, the resonance structures of the benzyl cation correctly account for the distribution of positive charge that is calculated from MO theory and shown by the electrostatic potential map (EPM):

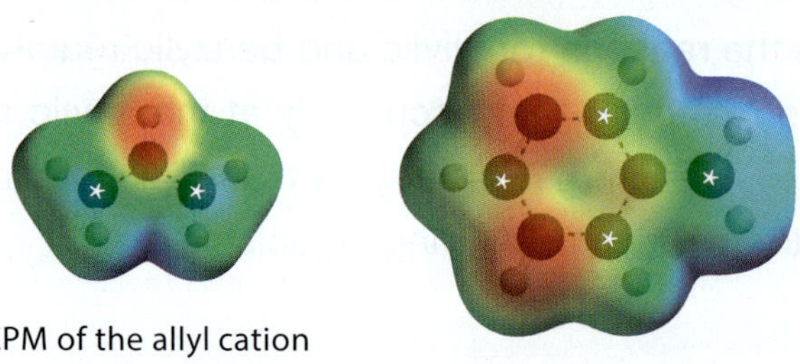

* = sites of partial positive charge

EPM of the allyl cation

EPM of benzyl cation (17.4)

The *resonance energy* (Sec. 15.6) of the benzyl cation calculated from MO theory (over and above that of the benzene ring itself) is about 0.72β, which is about 90% of the resonance energy of the allyl cation. In other words, the resonance stabilizations of the benzyl and allyl cations are about the same.

The structures and stabilities of allylic and benzylic carbocations have important consequences for reactions in which they are involved as reactive intermediates. First, *reactions in which benzylic or allylic carbocations are formed as intermediates are generally considerably faster than analogous*

TABLE 17.1 **Comparison of S_N1 Solvolysis Rates of Allylic and Nonallylic Alkyl Halides**

$$\text{R—Cl} + \text{C}_2\text{H}_5\text{OH} + \text{H}_2\text{O} \xrightarrow[\text{50\% aqueous ethanol}]{44.6\ ^\circ\text{C}} \text{R—OC}_2\text{H}_5 + \text{R—OH} + \text{HCl}$$

Alkyl chloride R—Cl	Relative rate
$H_2C{=}CH{-}C(CH_3)_2{-}Cl$	162
$(H_3C)_2C{=}CH{-}CH_2{-}Cl$	38
$CH_3CH_2{-}C(CH_3)_2{-}Cl$	1.00
$(H_3C)_2CH{-}CH_2{-}CH_2{-}Cl$	<0.00002

TABLE 17.2 **Comparison of S_N1 Solvolysis Rates of Benzylic and Nonbenzylic Alkyl Halides**

$$\text{R—Cl} + \text{H}_2\text{O} \xrightarrow[\text{90\% aqueous acetone}]{25\ ^\circ\text{C}} \text{R—OH} + \text{HCl}$$

Alkyl chloride R—Cl	Common name	Relative rate
$(CH_3)_3C{-}Cl$	*tert*-butyl chloride	1.0
$Ph{-}CH(CH_3){-}Cl$	α-phenethyl chloride	1.0
$Ph{-}C(CH_3)_2{-}Cl$	*tert*-cumyl chloride	620
$Ph_2CH{-}Cl$	benzhydryl chloride	200*
$Ph_3C{-}Cl$	trityl chloride	>600,000

*In 80% aqueous ethanol.

reactions involving comparably substituted nonallylic or nonbenzylic carbocations. This point is illustrated by the relative rates of S_N1 solvolysis reactions, shown in **Tables 17.1** and **17.2**. For example, the tertiary allylic alkyl halide in the first entry of Table 17.1 reacts more than 100 times faster than the tertiary nonallylic alkyl halide in the third entry. A comparison of the first and third entries of Table 17.2 shows the effect of benzylic substitution. *Tert*-cumyl chloride, the third entry, reacts more than 600 times faster than *tert*-butyl chloride, the first entry.

The greater reactivities of allylic and benzylic halides result from the stabilities of the carbocation intermediates that are formed when they react. For example, *tert*-cumyl chloride (the third entry of Table 17.2) ionizes to a carbocation with four important resonance structures:

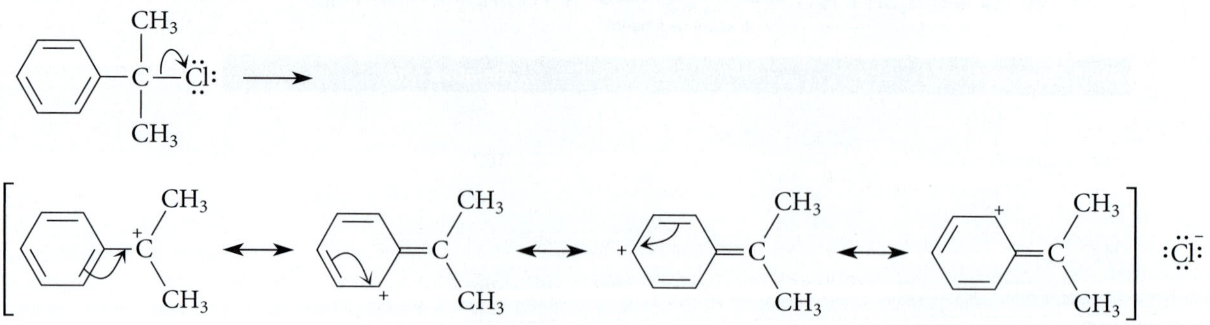

resonance-stabilized carbocation (17.5)

Ionization of *tert*-butyl chloride, on the other hand, gives the *tert*-butyl cation, a carbocation with only one important contributing structure.

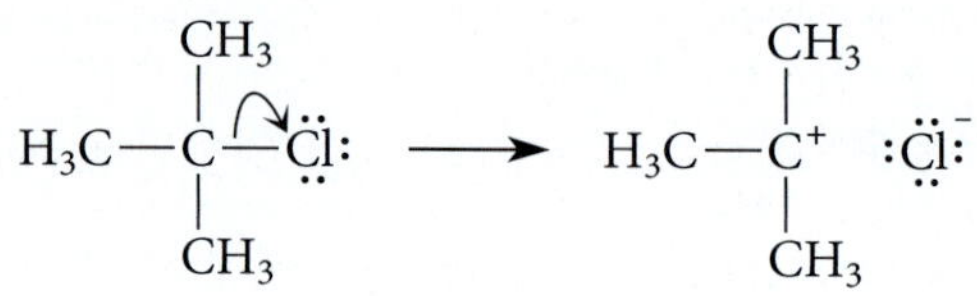

***tert*-butyl chloride** (17.6)

The benzylic cation is more stable relative to its alkyl halide starting material than is the *tert*-butyl cation, and Hammond's postulate predicts that the more stable carbocation should be formed more rapidly. A similar analysis explains the reactivity of allylic alkyl halides.

Because of the possibility of resonance, ortho and para substituent groups on the benzene ring that activate electrophilic aromatic substitution further accelerate S_N1 reactions at the benzylic position:

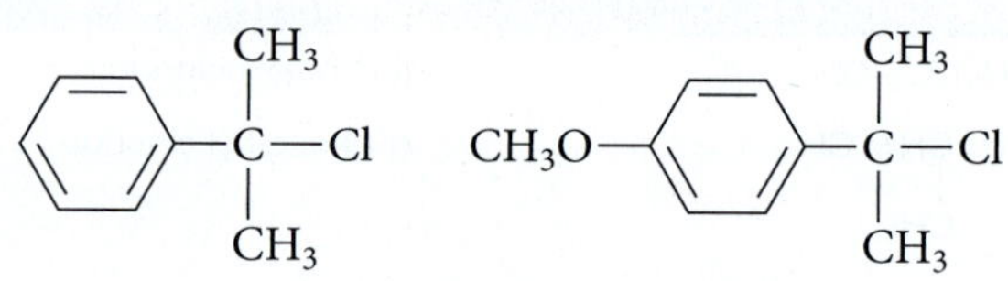

relative solvolysis rates (90% aqueous acetone, 25 °C): 1, 3400 (17.7)

The carbocation derived from the ionization of the *p*-methoxy derivative in Display 17.7 not only has the same types of resonance structures as the unsubstituted compound, shown in Eq. 17.5, but also has an additional structure (*blue*) that results from the delocalization of an oxygen lone pair (*blue arrow*) and the sharing of the positive charge by the substituent oxygen.

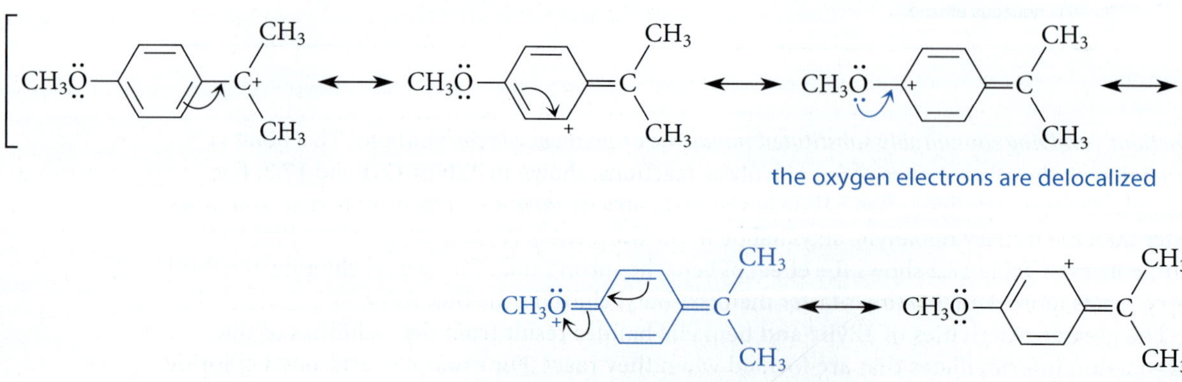

the positive charge is shared by oxygen; all atoms have octets (17.8)

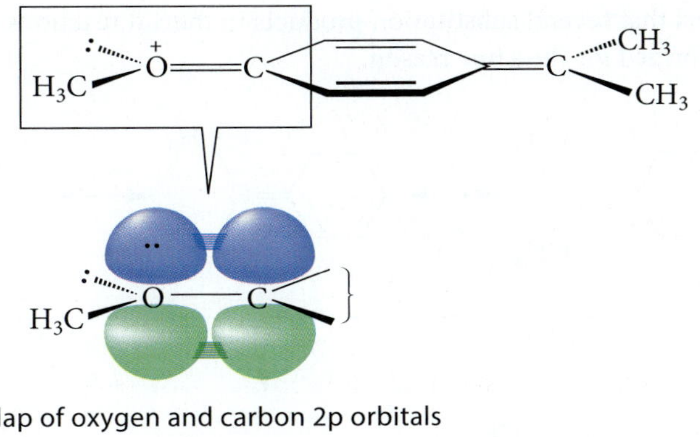

FIGURE 17.1 The delocalization of oxygen electrons in Eq. 17.8 results from the overlap of the oxygen 2p orbital containing a nonbonding electron pair with the carbon 2p orbitals of the carbocation. (Compare with Fig. 16.7a, Sec. 16.5B.)

This delocalization is the result of the overlap of a nonbonding electron pair in a 2p orbital of the sp^2-hybridized oxygen with the 2p orbitals of the ring (**Fig. 17.1**). This overlap is identical to the overlap described in Fig. 16.7a (Sec. 16.5B).

Other reactions that involve carbocation intermediates are accelerated when the carbocations are allylic or benzylic. For example, the dehydration of an alcohol (Sec. 11.2) and the reaction of an alcohol with a hydrogen halide (Sec. 11.3) are also faster when the alcohol is allylic or benzylic. Most alcohols require forcing conditions or Lewis acid catalysts to react with HCl to give alkyl chlorides, but such conditions are unnecessary when benzylic alcohols react with HCl. The addition of hydrogen halides to conjugated dienes also reflects the stability of allylic carbocations. Recall that protonation of a conjugated diene gives the allylic carbocation rather than its nonallylic isomer because the allylic carbocation is formed more rapidly (Sec. 15.4A).

A second consequence of the involvement of allylic carbocations as reactive intermediates is that in many cases *more than one product can be formed.* More than one product is possible because the positive charge (and electron deficiency) is shared between two carbons. Nucleophiles can react at either of the electron-deficient carbon atoms and, if the two carbons are not equivalent, two different products result.

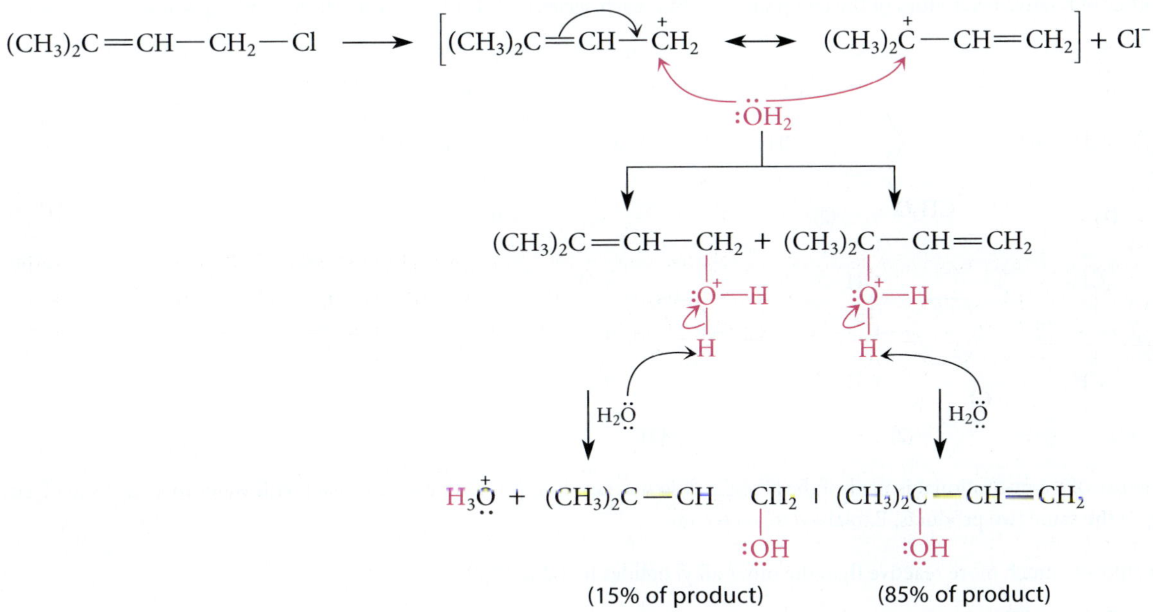

(17.9)

The two products are derived from *one* allylic carbocation that has two resonance forms. Recall that similar reasoning explains why a mixture of products (1,2- and 1,4-addition products) is obtained in the reactions of hydrogen halides with conjugated alkenes (Sec. 15.4A; Eqs. 15.35a and b).

We might expect that several substitution products in the S_N1 reactions of benzylic alkyl halides might be formed for the same reason.

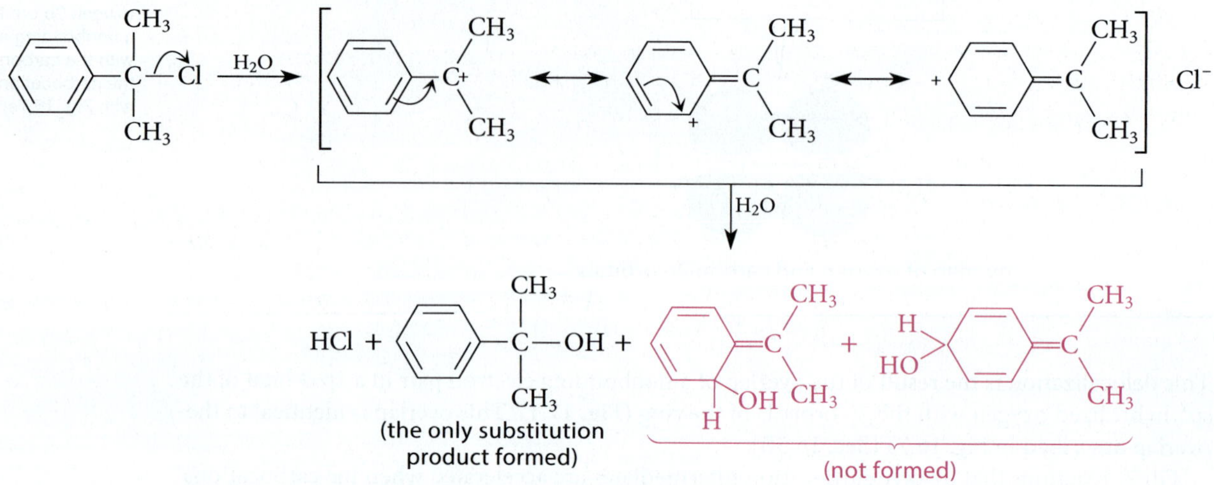

(17.10)

As Eq. 17.10 shows, the products (*red*) derived from the reactions of water at the ring carbons are not formed. The reason is that *these products are not aromatic* and so lack the stability associated with the aromatic ring. Aromaticity is such an important stabilizing factor that only the aromatic product (*black*) is formed.

Focused Problems

17.3 Predict the order of relative reactivities of the compounds within each series in S_N1 solvolysis reactions, and explain your answers carefully.

(a)

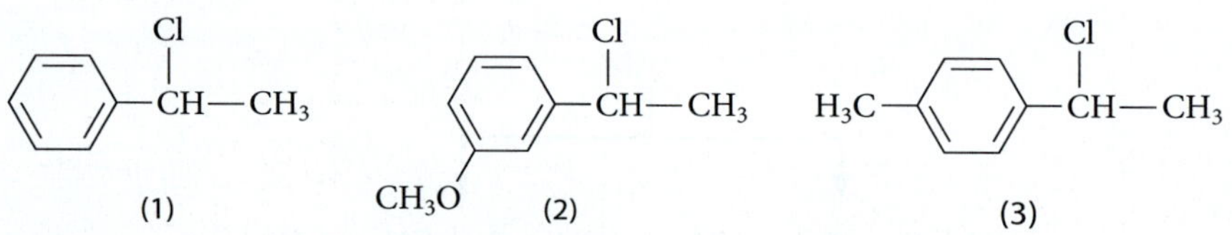

(b)

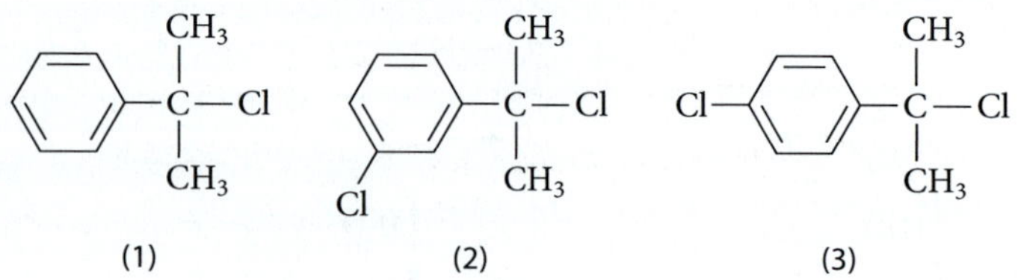

17.4 Give the structure of a constitutional isomer of the allylic halide reactant in Eq. 17.9 that would react with water in an S_N1 solvolysis reaction to give the same two products. Explain your reasoning.

17.5 Why is trityl chloride much more reactive than the other alkyl halides in Table 17.2?

17.2 REACTIONS INVOLVING ALLYLIC AND BENZYLIC RADICALS

An **allylic radical** has an unpaired electron at an allylic position. Allylic radicals are resonance-stabilized and are more stable than comparably substituted nonallylic radicals. The simplest allylic radical is the *allyl radical* itself:

resonance structures of the allyl radical (17.11)

Similarly, a **benzylic radical**, which has an unpaired electron at a benzylic position, is also resonance-stabilized. The *benzyl radical* is the prototype:

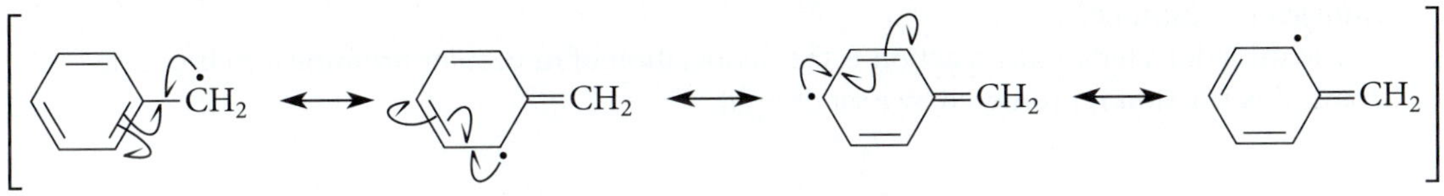

resonance structures of the benzyl radical (17.12)

These resonance structures symbolize the delocalization (sharing) of the unpaired electron that results from the overlap of carbon 2p orbitals to form molecular orbitals.

The enhanced stabilities of allylic and benzylic radicals can be experimentally demonstrated with bond dissociation energies. The bond dissociation energies of allylic and nonallylic hydrogens in 2-pentene are compared in **Fig. 17.2**. One set of methyl hydrogens is allylic and the other is not. It takes 50 kJ mol^{-1} (12 kcal mol^{-1}) less energy to remove the allylic hydrogen H^a than the nonallylic one H^b. As Fig. 17.2 shows, the difference in bond dissociation energies is a direct measure of the relative energies of the two radicals. This comparison shows that the allylic radical is stabilized by 50 kJ mol^{-1} (12 kcal mol^{-1}) relative to the nonallylic radical. The same type

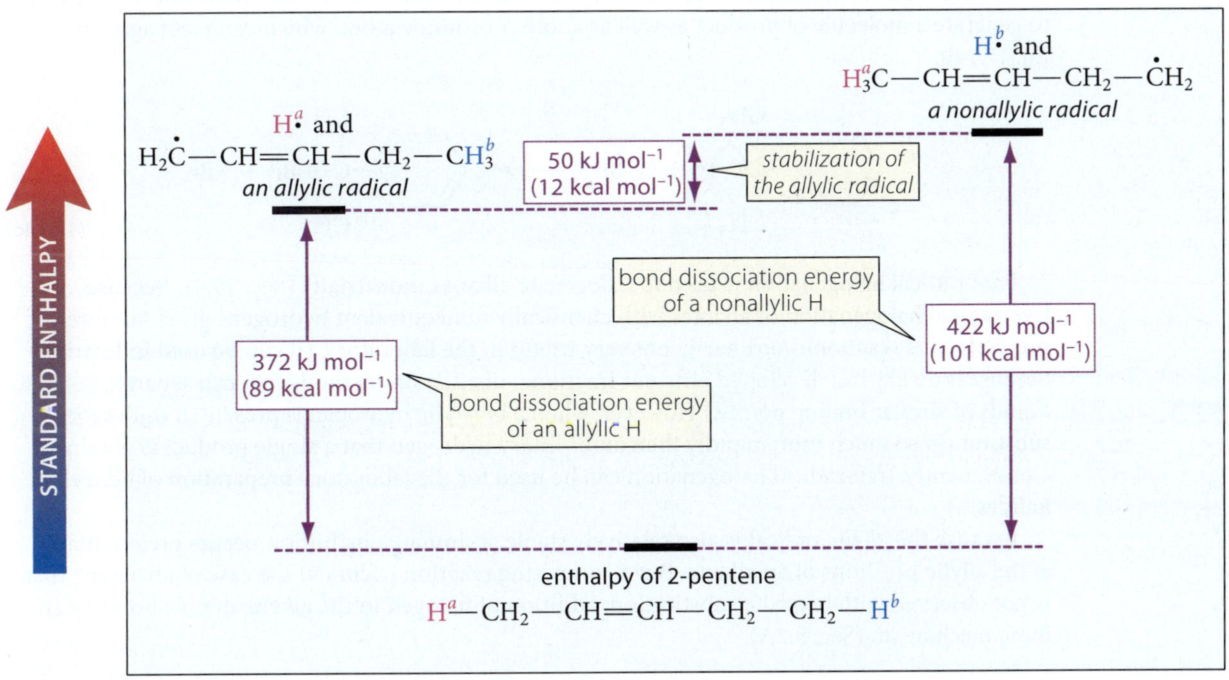

FIGURE 17.2 Use of bond dissociation energies to determine the stabilization of an allylic radical. The stabilization of the allylic radical results from the lower energy required (50 kJ mol^{-1}, 12 kcal mol^{-1}) to remove an allylic hydrogen (red) compared with a nonallylic one (blue).

of comparison shows that benzylic radicals are about 42 kJ mol^{-1} (10 kcal mol^{-1}) more stable than comparably substituted nonbenzylic radicals.

Because allylic and benzylic radicals are especially stable, they are more readily formed as reactive intermediates than ordinary alkyl radicals. Consider what happens, for example, in the bromination of cumene:

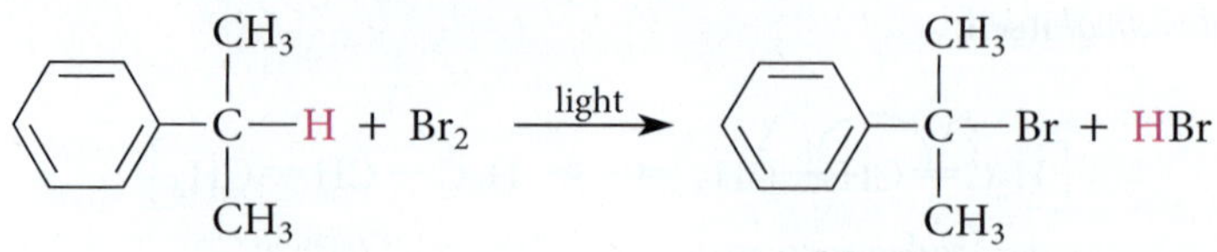

cumene **_tert_-cumyl bromide** (nearly quantitative) (17.13)

This is a free-radical chain reaction (Secs. 10.1C and 10.3B). Notice that *only the benzylic hydrogen is substituted.*

The initiation step in this reaction is the dissociation of molecular bromine into bromine atoms; this reaction is promoted by heat or light.

We remind the reader of the convention, introduced in Sec. 10.1B, that the unpaired electron is the only valence electron shown in free radicals; the other valence electrons are customarily omitted because they have no role in the chemistry.

$$Br{-}Br \longrightarrow 2\,Br\cdot \qquad (17.14a)$$

In the first propagation step, a bromine atom abstracts the one benzylic hydrogen in preference to either the six nonbenzylic methyl hydrogens or the five hydrogens of the aromatic ring. It is in this propagation step that the selectivity for substitution of the benzylic hydrogen occurs.

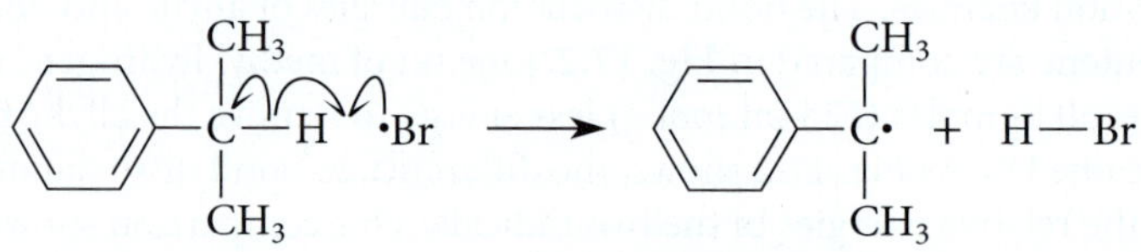

(17.14b)

The reason for this selectivity is the relative weakness of the benzylic C—H bond that is broken (see Fig. 17.1) or, equivalently, the greater stability of the benzylic radical that is formed.

In the second propagation step, the benzylic radical reacts with another molecule of bromine to generate a molecule of product as well as another bromine atom, which can react again in Eq. 17.14b.

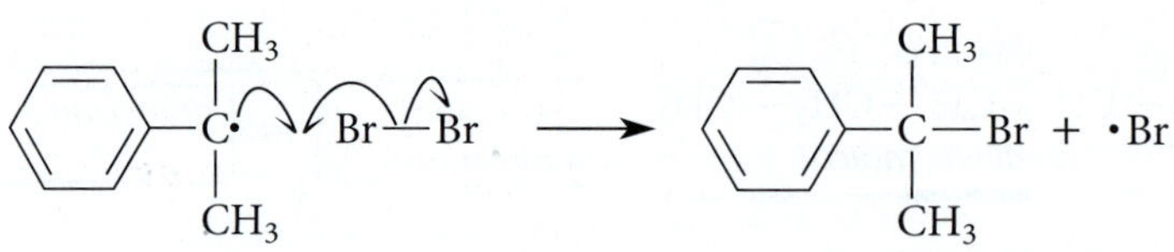

(17.14c)

Free-radical halogenation is used to halogenate alkanes industrially (Sec. 10.4). Because free-radical halogenation of alkanes with chemically nonequivalent hydrogens gives mixtures of products, this reaction is ordinarily not very useful in the laboratory. (It can be used industrially because industry has developed efficient fractional distillation methods that can separate liquids of similar boiling points.) However, when a benzylic hydrogen is present, it undergoes substitution so much more rapidly than an ordinary hydrogen that a single product is obtained. Consequently, free-radical halogenation can be used for the laboratory preparation of benzylic halides.

Because the allylic radical is also relatively stable, a similar substitution occurs preferentially at the allylic positions of an alkene. But a competing reaction occurs in the case of an alkene that is not observed with benzylic substitution: addition of halogen to the alkene double bond by an ionic mechanism (Sec. 5.2A).

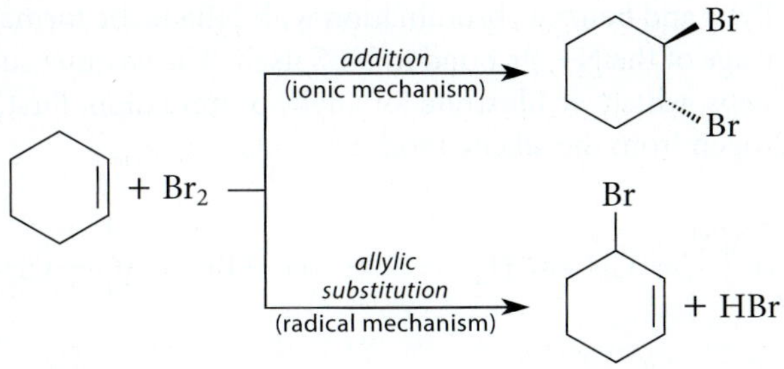

(17.15)

(Such a competing addition is not a problem in benzylic bromination because bromine doesn't add to the benzene ring in Eq. 17.13. Why?)

One reaction can be promoted over the other if the reaction conditions are chosen carefully. *Addition* of bromine is the predominant reaction if (1) free-radical substitution is suppressed by avoiding conditions that promote free-radical reactions (heat, light, or free-radical initiators) and if (2) the reaction is carried out in solvents of even slight polarity that promote the ionic mechanism for bromine addition. Therefore, addition is observed at 25 °C if the reaction is run in the dark in dichloromethane, CH_2Cl_2. On the other hand, free-radical *substitution* occurs when the reaction is promoted by heat, light, or free-radical initiators, an apolar solvent such as CCl_4 is used, and *the bromine is added slowly so that its concentration remains very low.* To summarize:

Addition:

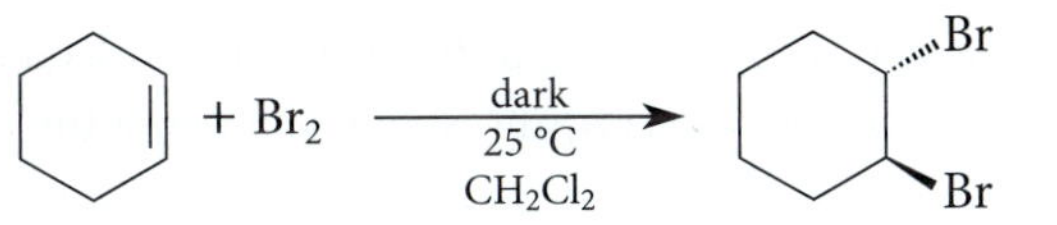

(17.16a)

Substitution:

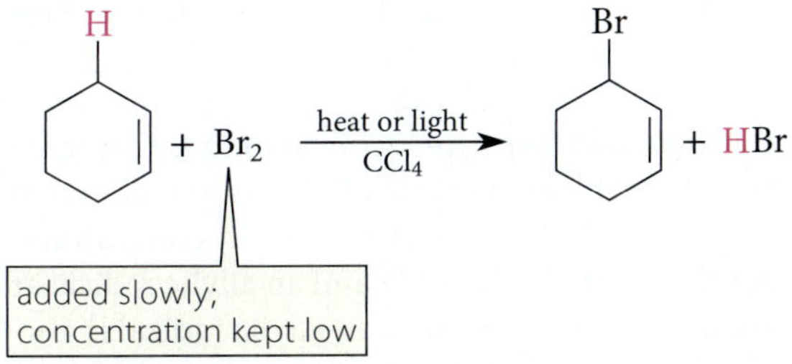

(17.16b)

The effect of bromine concentration results from the rate laws for the competing reactions. Addition has a higher kinetic order in $[Br_2]$ than substitution. Therefore, the rate of addition is decreased more than the rate of substitution by lowering the bromine concentration. This effect is discussed in Further Exploration 17.1.

FURTHER EXPLORATION 17.1
Addition versus Substitution with Bromine

Adding bromine to a reaction so slowly that it remains at very low concentration is experimentally inconvenient, but a useful reagent can be employed to accomplish the same objective: *N*-bromosuccinimide (abbreviated NBS). When a compound with allylic hydrogens is treated with NBS in CCl_4 under free-radical conditions (heat or light and peroxides), allylic bromination takes place, and addition to the double bond is not observed.

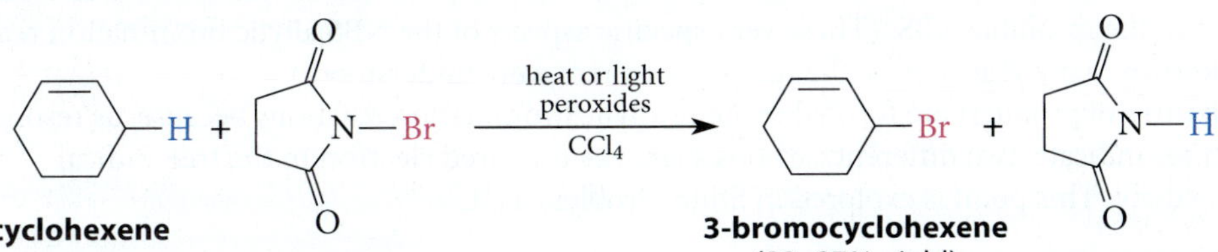

cyclohexene **N-bromosuccinimide (NBS)** **3-bromocyclohexene** (82–87% yield) **succinimide** (17.17)

NBS is also used for benzylic bromination.

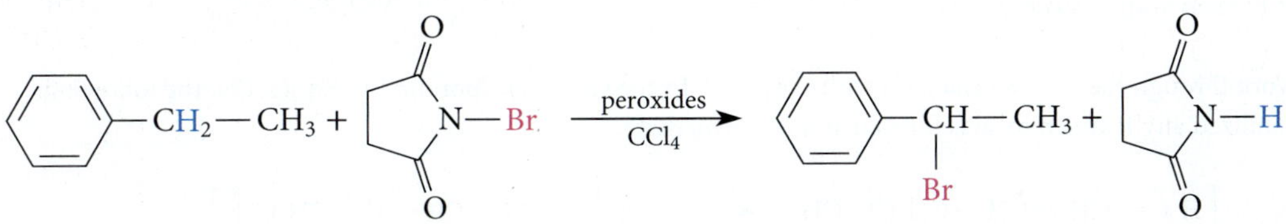

(80% yield) (17.18)

The initiation step in allylic and benzylic bromination with NBS is the formation of a bromine atom by homolytic cleavage of the N—Br bond in NBS itself. The ensuing substitution reaction has three propagation steps, which we illustrate for allylic bromination. First, the bromine atom abstracts an allylic hydrogen from the alkene molecule:

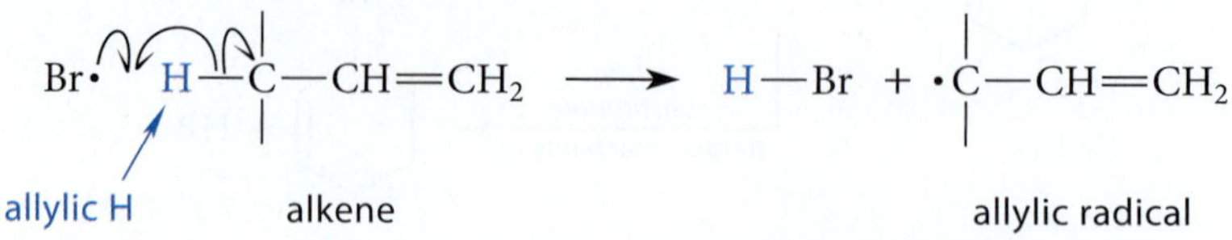

(17.19a)

The HBr formed in this reaction reacts with the NBS by an ionic mechanism to produce a Br_2 molecule.

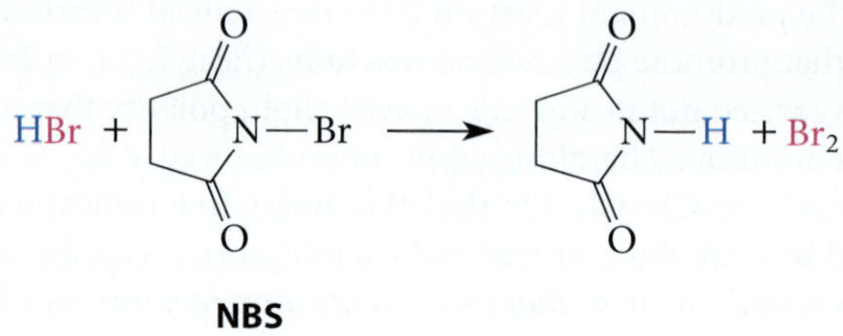

(17.19b)

The last propagation step is the reaction of this bromine molecule with the radical formed in Eq. 17.19a. A new bromine atom is produced that can begin the cycle anew.

(17.19c)

The first and last propagation steps are identical to those for free-radical substitution with Br_2 itself (Eqs. 17.14b and c). The unique role of NBS is to maintain the very low concentration of bromine by reacting with HBr in Eq. 17.19b. The Br_2 concentration remains low because it can be generated no faster than an HBr molecule and an allylic radical are generated in Eq. 17.19a. So, every time a bromine molecule is formed, an allylic radical is also formed with which the bromine can react.

The low solubility of NBS in CCl_4 (0.005 M) is crucial to the success of allylic bromination with NBS. When solvents that dissolve NBS are used, different reactions are observed. Therefore, CCl_4 *must* be used as the solvent in allylic or benzylic bromination with NBS. During the reaction, the insoluble NBS, which is denser than CCl_4, is consumed from the bottom of the flask and is replaced by the less dense by-product succinimide (Eq. 17.17), which forms a layer on the surface of the CCl_4. Equation 17.19b, and possibly other steps of the mechanism, occur at the surface of the insoluble NBS. (These very specific aspects of the NBS allylic bromination reaction were known many years before the reasons for them were understood.)

Mixtures of products are formed in some allylic bromination reactions because, as resonance structures indicate, two different carbons share the unpaired electron in the free-radical intermediate. This point is explored in Study Problem 17.1.

Study Problem 17.1

What products are expected in the reaction of $H_2C{=}CHCH_2CH_2CH_2CH_3$ (1-hexene) with NBS in CCl_4 in the presence of peroxides? Explain your answer.

Solution Work through the NBS mechanism with 1-hexene. In the step corresponding to Eq. 17.19a, the following resonance-stabilized allylic free radical is formed as an intermediate:

$$\left[\underset{A}{H_2C{=}CH{-}\dot{C}H{-}CH_2CH_2CH_3} \longleftrightarrow \underset{B}{H_2\dot{C}{-}CH{=}CH{-}CH_2CH_2CH_3}\right]$$

Because *two nonequivalent carbons* share the unpaired electron, this radical can react in the final propagation step to give *two different products.* Reaction of Br_2 at the radical site shown in structure *A* gives product (1), and reaction at the radical site shown in structure *B* gives product (2):

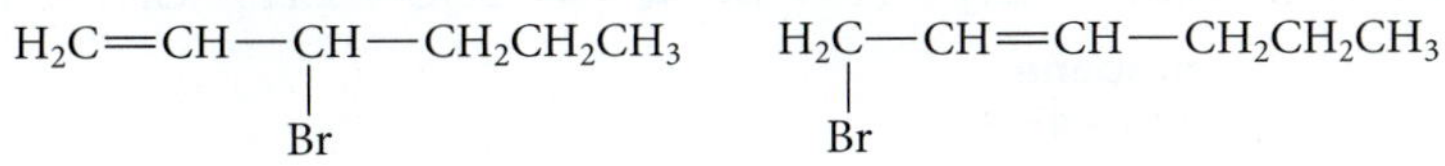

(1) (2)

Product (1) is chiral, and product (2) can exist as both cis and trans stereoisomers. Therefore, bromination of 1-hexene gives racemic (1) as well as *cis*- and *trans*-(2), although the trans isomer should predominate because of its greater stability.

Focused Problem

17.6 What product(s) are expected when each of the following compounds reacts with one equivalent of NBS in CCl_4 in the presence of light and peroxides? Explain your answers.

(a) Cyclohexene (b) 3,3-Dimethylcyclohexene

(c) *trans*-2-Pentene (d) 4-*tert*-Butyltoluene

(e) 1-Isopropyl-4-nitrobenzene

17.3 REACTIONS INVOLVING ALLYLIC AND BENZYLIC ANIONS

The prototype for allylic anions is the *allyl anion*, and the simplest benzylic anion is the *benzyl anion.*

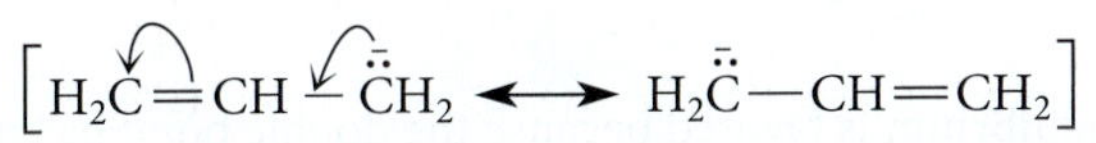

allyl anion (17.20)

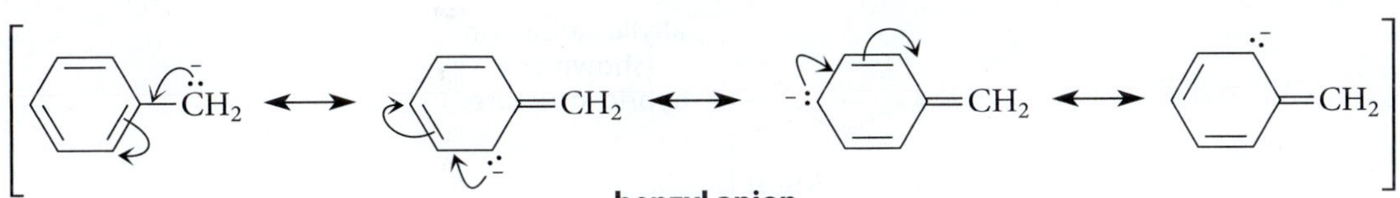

benzyl anion

(17.21)

Allylic and benzylic anions are about 59 kJ mol^{-1} (14 kcal mol^{-1}) more stable than their nonallylic and nonbenzylic counterparts. There are two reasons of roughly equal importance for the stabilities of these anions. The first is resonance stabilization, as indicated by the preceding resonance structures. The second reason is the *polar effect* (Sec. 3.7C) of the sp^2-hybridized carbons of the double bond (in the allyl anion) or the phenyl ring (in the benzyl anion). The polar effect of both groups stabilizes anions. The reason for the polar effect of both groups is the electronegativity of an sp^2-hybridized carbon, discussed in Sec. 3.7C.

The enhanced stability of allylic and benzylic anions is reflected in the pK_a values of propene and toluene ($B:^-$ = a base):

$$H_2C{=}CH{-}CH_2{-}H + B:^- \rightleftharpoons H_2C{=}CH{-}\ddot{C}H_2^- + B{-}H$$

propene
pK_a ~ 43 (17.22)

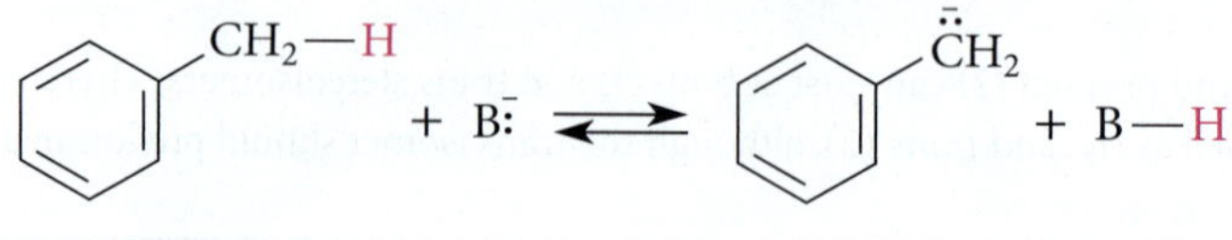

toluene
pK_a ~ 41 (17.23)

Although these compounds are very weak acids, their acidities are much greater than the acidities of alkanes that do not contain allylic or benzylic hydrogens. As noted in Sec. 3.7C, ordinary alkanes have pK_a values of about 55–60.

Free benzylic or allylic carbanions are rarely involved as reactive intermediates. However, a number of reactions involve species that have *carbanion character.* Two of these are the reactions of Grignard and related organometallic reagents, and E2 eliminations. The following sections show how these reactions are affected when carbanion character occurs at benzylic or allylic positions.

A. Allylic Grignard Reagents

The enhanced carbanion character of allylic and benzylic Grignard reagents and some of its consequences were addressed in Sec. 10.6B, Eq. 10.66. (See the margin box near the equation; see also Problem 17.24 at the end of the chapter.)

Recall that Grignard reagents have many of the properties expected of *carbanions* (Sec. 10.5C). In other words, allylic Grignard reagents resemble allylic carbanions.

$$H_2C{=}CH{-}CH_2{-}MgBr \quad \text{resembles} \quad H_2C{=}CH{-}\ddot{C}H_2^- \ \overset{+}{M}gBr \qquad (17.24)$$

Allylic Grignard reagents undergo a rapid equilibration in which the —MgBr group moves back and forth between the two partially negative carbons at a rate of about 1000 times per second.

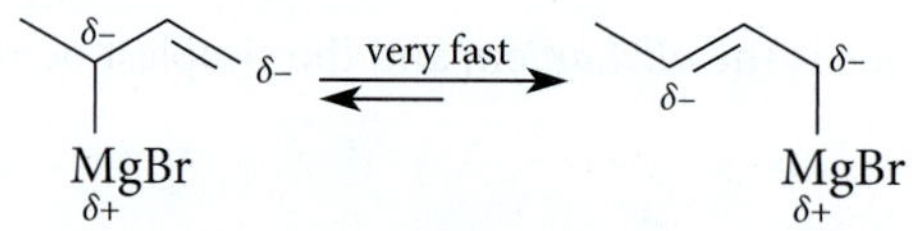

(17.25)

(The right side of the equilibrium is favored because the double bond has more alkyl substituents; Sec. 4.5B.) The transition state for this reaction can be envisioned as an ion pair consisting of an allylic carbanion and a $^+$MgBr cation.

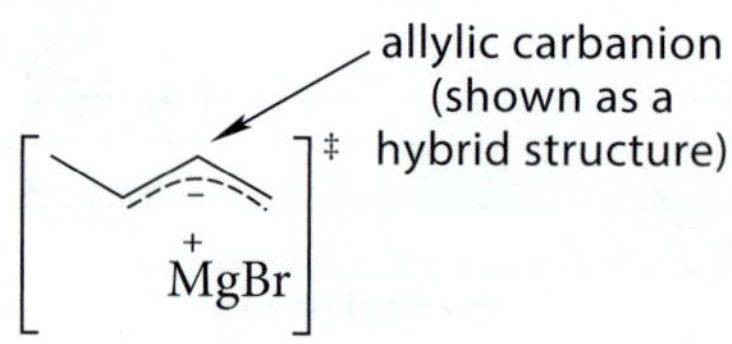

(17.26)

Because the allylic carbanion is especially stable, this transition state has relatively low energy, and, consequently, the equilibration occurs rapidly.

The equilibration in Eq. 17.25 is an example of an *allylic rearrangement.* An **allylic rearrangement** involves the simultaneous movement of a group G and a double bond so that one allylic isomer is converted into another.

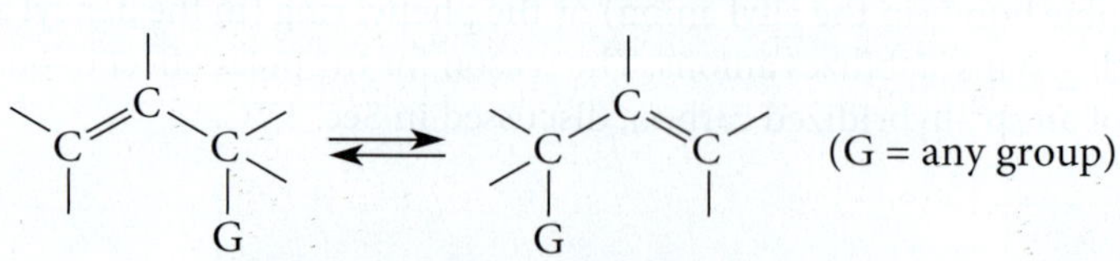

allylic rearrangement (17.27)

These two structures are *not* resonance structures; they are two *distinct isomeric* species in rapid equilibrium.

The rapid allylic rearrangement of an unsymmetrical Grignard reagent, such as the one shown in Eq. 17.25, means that the reagent is actually a mixture of two different species. This has two consequences. First, the same mixture of species is obtained from either of two allylically related alkyl halides:

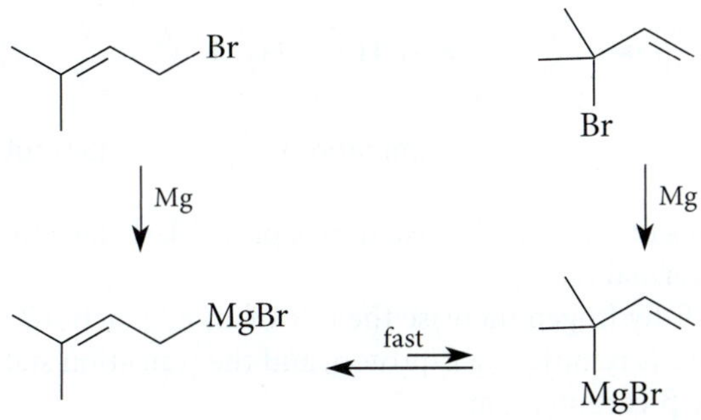

(17.28)

Second, when the Grignard reagents undergo a subsequent reaction, a mixture of products is usually obtained, and the same mixture of products is obtained regardless of the alkyl halide used to form the Grignard reagent. For example, protonolysis of the mixture of equilibrating Grignard reagents in Eq. 17.28 gives the following result:

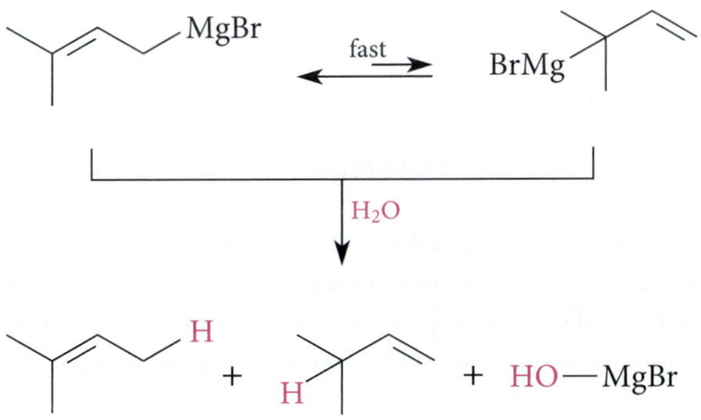

(17.29)

Focused Problem

17.7 What product(s) are formed when a Grignard reagent prepared from each of the following alkyl halides is treated with D_2O?

(a)

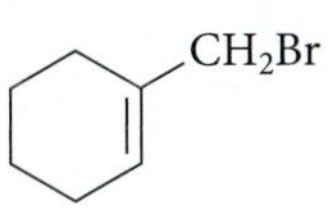

1-(bromomethyl)cyclohexene

(b)

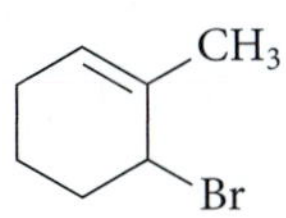

6-bromo-1-methylcyclohexene

B. E2 Eliminations Involving Allylic or Benzylic Hydrogens

Recall that the S_N2 (bimolecular substitution) and E2 (bimolecular elimination) reactions of alkyl halides are *competing reactions*, and that the structure of the alkyl halide is one of the major factors that determine which reaction is the dominant one (Sec. 9.5G). A structural effect in the alkyl halide that tends to promote a greater fraction of elimination is *enhanced acidity of the β-hydrogens*. It is found that a greater ratio of elimination to substitution is observed when the β-hydrogens of the alkyl halide have higher than normal acidity. Such a situation can occur when the β-hydrogens are allylic or benzylic. (Recall from the introduction to this section that allylic and benzylic hydrogens are more acidic than ordinary alkyl hydrogens.) For example, the E2 reaction of the alkyl bromide in Eq. 17.30 is more than 100 times faster than the E2 reaction of isopentyl bromide [$(CH_3)_2CHCH_2CH_2Br$], a comparably branched alkyl halide.

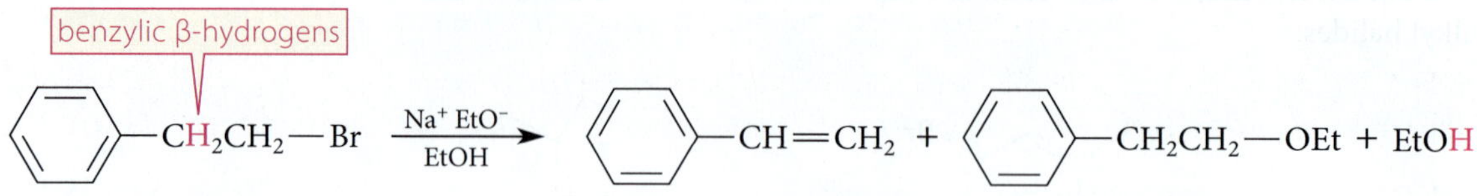

(95% elimination) (5% substitution) (17.30)

(Elimination predominates because the E2 reaction is particularly fast; the S_N2 component of the competition occurs at a normal rate.)

Why should an acidic β-hydrogen increase the rate of an E2 reaction? In the transition state of the E2 reaction, the base is removing a β-proton, and the transition state of the reaction has *carbanion character* at the β-carbon atom.

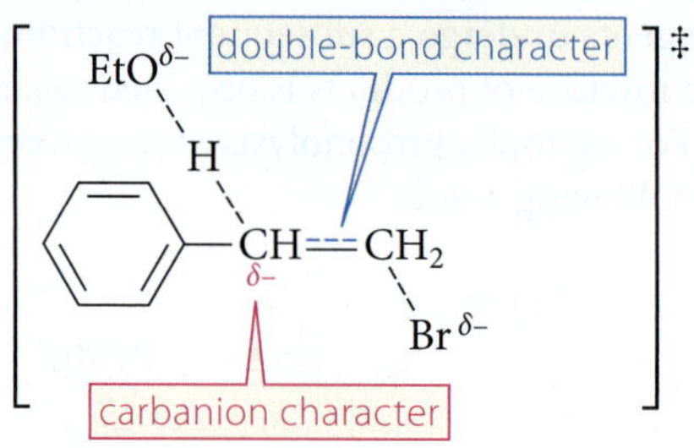

transition state for the E2 reaction (17.31)

This partially formed carbanion is stabilized in the same way that a fully formed carbanion is; a more stable transition state results in a faster reaction. Another reason that benzylic E2 reactions are faster is that the alkene double bond, which is partially formed in the transition state, is conjugated with the benzene ring; recall that conjugated double bonds are more stable than unconjugated double bonds (Sec. 15.1A).

Let's summarize the structural characteristics of alkyl halides or sulfonate esters that favor E2 reactions over S_N2 reactions:

1. Elimination reactions are favored by branching at the α-carbon (Sec. 9.5G).
2. Elimination reactions are favored by branching at the β-carbon (Sec. 9.5G).
3. Elimination reactions are favored by greater acidity of the β-hydrogens (this section).

Focused Problem

17.8 Predict the major product that is obtained when each of the following alkyl halides is treated with potassium *tert*-butoxide. Explain your reasoning.

(a)
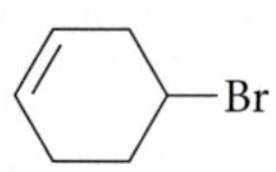

(b)
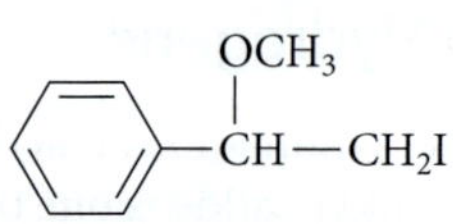

17.4 ALLYLIC AND BENZYLIC S_N2 REACTIONS

S_N2 reactions of allylic and benzylic halides are relatively fast, even though they do not involve reactive intermediates. The following data for allyl chloride are typical:

relative rate

$$H_2C{=}CH{-}CH_2{-}Cl + I^- \xrightarrow[\text{acetone}]{50\ °C} H_2C{=}CH{-}CH_2{-}I + Cl^- \qquad 73 \tag{17.32a}$$

$$H_3C{-}CH_2{-}CH_2{-}Cl + I^- \xrightarrow[\text{acetone}]{50\ °C} H_3C{-}CH_2{-}CH_2{-}I + Cl^- \qquad 1 \tag{17.32b}$$

An even greater acceleration is observed for benzylic halides.

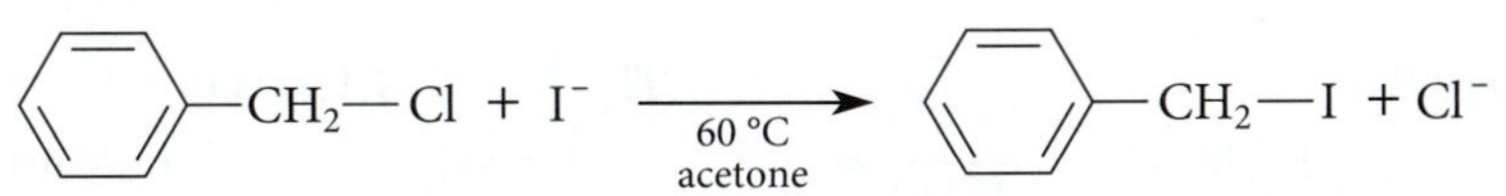

relative rate ~100,000

(17.33a)

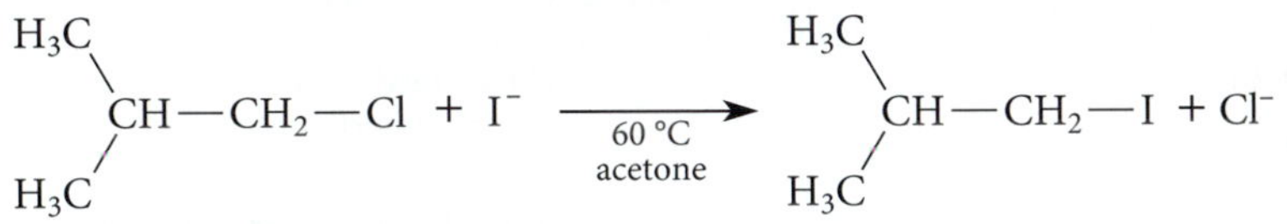

1

(17.33b)

Allylic and benzylic S_N2 reactions are accelerated because the energies of their transition states are reduced by 2p-orbital overlap, shown in **Fig. 17.3**, for an allylic S_N2 reaction. In the transition state of the S_N2 reaction, the carbon at which substitution occurs is sp^2-hybridized (see Fig. 9.2, Sec. 9.4C); the incoming nucleophile and the departing leaving group are partially bonded to a 2p orbital on this carbon. Overlap of this 2p orbital with the 2p orbitals of an adjacent double bond or phenyl ring provides additional bonding that lowers the energy of the transition state and accelerates the reaction.

Focused Problem

17.9 Explain how and why the product(s) would differ in the following reactions of *trans*-2-buten-1-ol.

(a) Reaction with concentrated aqueous HBr

(b) Conversion into the tosylate, then reaction with NaBr in acetone

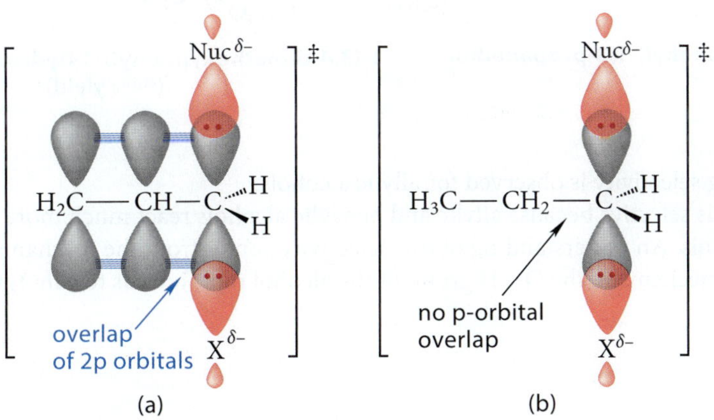

FIGURE 17.3 Transition states for S_N2 reactions at (a) an allylic carbon and (b) a nonallylic carbon. Nuc and X are the nucleophile and leaving group, respectively. The allylic substitution is faster because the transition state is stabilized by overlap (blue lines) of the 2p orbital at the site of substitution with the adjacent π bond.

17.5 ALLYLIC AND BENZYLIC OXIDATIONS

A. Oxidation of Allylic and Benzylic Alcohols with Manganese Dioxide

Allylic and benzylic alcohols are oxidized selectively by a suspension of activated manganese(IV) dioxide, MnO_2. Primary allylic alcohols are oxidized to aldehydes, and secondary allylic alcohols are oxidized to ketones.

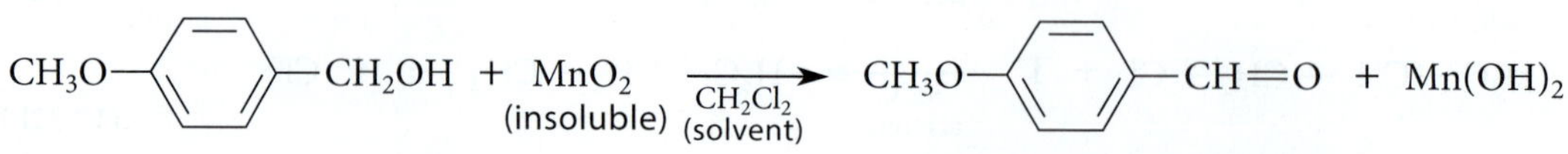

(4-methoxyphenyl)methanol (*p*-methoxybenzyl alcohol) **4-methoxybenzaldehyde** (81% yield) (17.34)

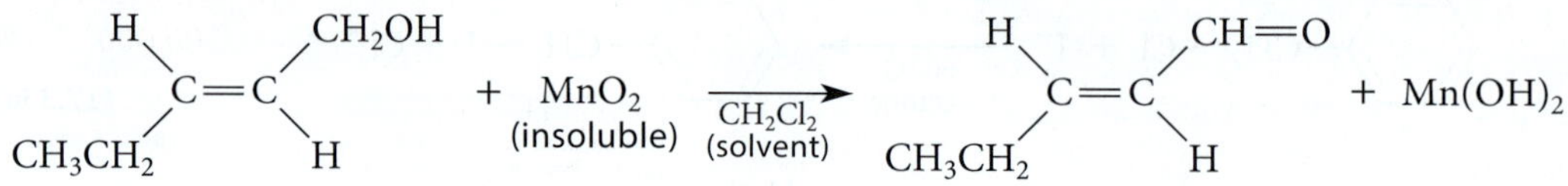

(*E*)-2-penten-1-ol **(*E*)-2-pentenal** (83% yield) (17.35)

"Activated MnO_2" is obtained by the oxidation–reduction reaction of potassium permanganate ($KMnO_4$) with an Mn^{2+} salt such as $MnSO_4$ under either alkaline or acidic conditions followed by thorough drying.

$$2\,H_2O + 3\,MnSO_4 + 2\,KMnO_4 \xrightarrow{H_2O} \underset{\textbf{manganese dioxide}}{5\,MnO_2} + 2\,H_2SO_4 + K_2SO_4 \qquad (17.36)$$

Allylic and benzylic oxidation of alcohols takes place on the surface of the MnO_2, which is insoluble in the solvents used for the reaction. Water competes with alcohol for the sites on the MnO_2, and so it must be removed by drying to produce an active oxidant.

The selectivity of MnO_2 oxidation for allylic and benzylic alcohols is illustrated by the following example, in which the ordinary (nonbenzylic) alcohol is unaffected.

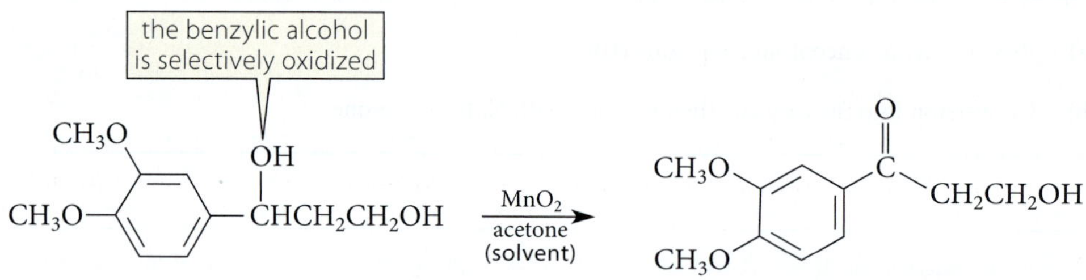

3-(3,4-dimethoxyphenyl)-1,3-propanediol **1-(3,4-dimethoxyphenyl)-3-hydroxy-1-propanone** (94% yield) (17.37)

A corresponding selectivity is observed for allylic alcohols.

The reaction is selective because allylic and benzylic alcohols react much more rapidly than "ordinary" alcohols. An understanding of this selectivity comes from the mechanism. In the first step of the mechanism, the O—H group of the alcohol rapidly adds to MnO_2 to give an ester (Sec. 11.4C).

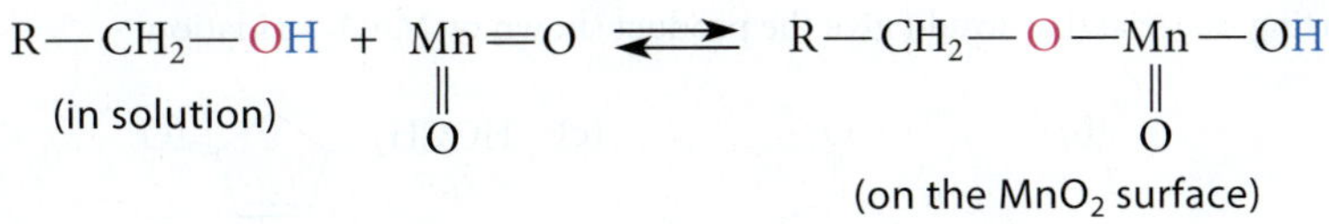

(17.38a)

(The solid-state structures of the Mn-containing species are simplified in these equations.) In the next step, which is rate-limiting, Mn(IV) accepts an electron to become Mn(III), and, at the same time, a hydrogen atom is transferred from the allylic or benzylic carbon to an oxygen of the oxidant. The product has an unpaired electron on the allylic or benzylic carbon and is therefore a resonance-stabilized radical.

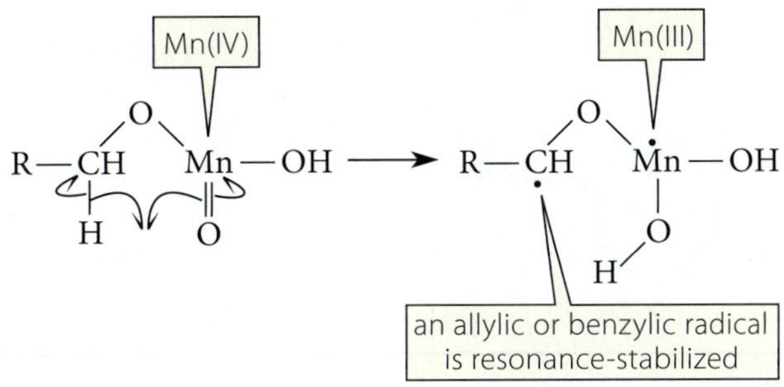

(17.38b)

The stability of the radical intermediate, by Hammond's postulate (Sec. 4.8D), increases the rate of this step. The allylic/benzylic selectivity occurs because the analogous radical intermediate in the oxidation of an alcohol that is not allylic or benzylic is less stable and is formed more slowly. In the rapid final step, Mn(III) is reduced to the more stable Mn(II), and a strong C=O double bond is formed to give the aldehyde product, which is washed away from the oxidant surface by the solvent.

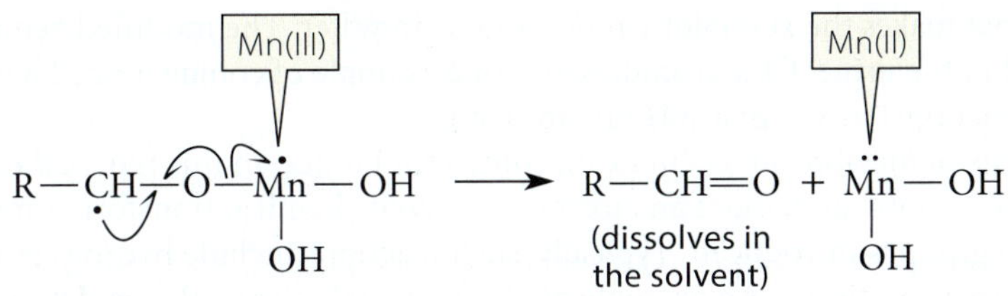

(17.38c)

Focused Problems

17.10 Give the structure of the product expected when each of the following alcohols is subjected to MnO_2 oxidation.

(a) 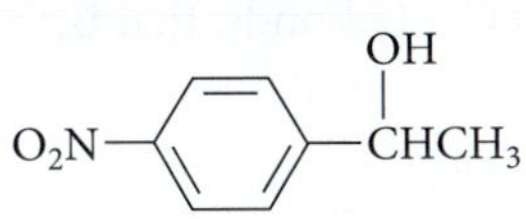

(b) 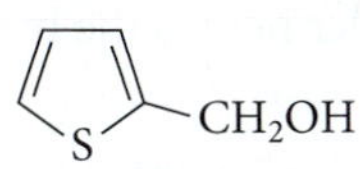

(c) 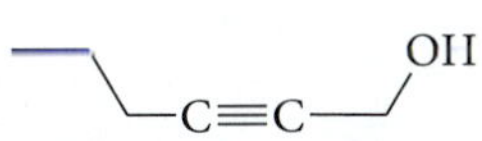

(d) 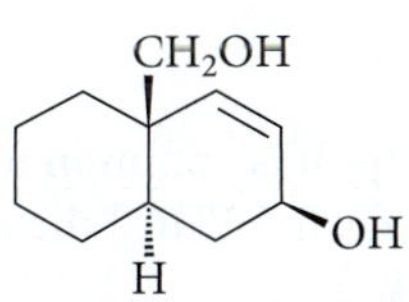

17.11 In each case, give the structure of a starting material that would give the product shown by MnO_2 oxidation.

(a) 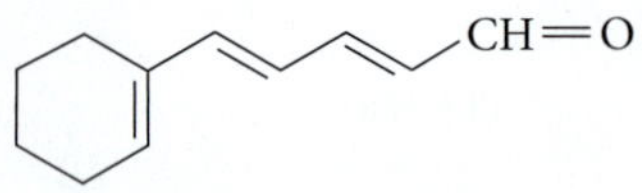

(b) 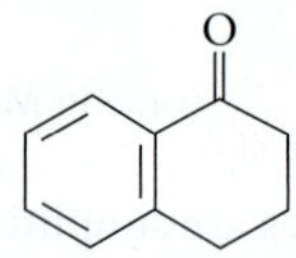

(c) 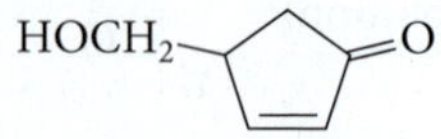

17.12 In each case, tell whether oxidation with pyridinium chlorochromate (PCC) and oxidation with MnO_2 would give the same product, different products, or no reaction. Explain.

(a) 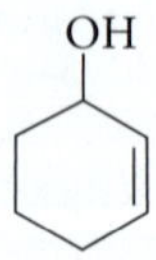

(b) 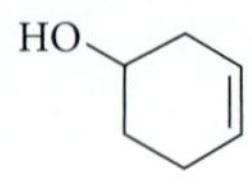

(c) 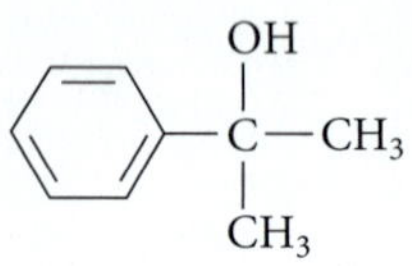

(d) 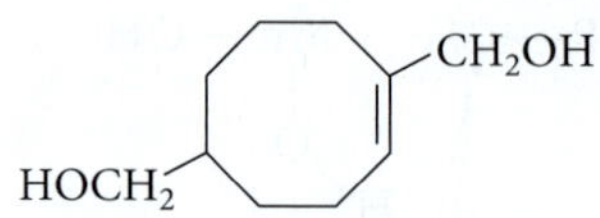

Chemical Biology Topic

B. Oxidation with Cytochrome P450

In Sec. 8.6C (Eq. 8.64), we showed that many *xenobiotics*—foreign substances introduced into biological systems—are metabolized by living organisms. There we showed that one metabolic strategy is *conjugation*, in which an existing functional group of the xenobiotic is joined to another group that makes the xenobiotic more soluble in water. The modified xenobiotic can then be excreted in the urine. Glucuronidation is one example of conjugation. The conjugation of xenobiotics is generally termed **phase II metabolism**.

CyP450 (called P450 because of its spectroscopic properties) exists in a very large number of genetic variants, and humans differ in their distribution of the different variants. One reason that drugs affect people differently is the different activities of their cytochromes. Cases of drug toxicity are known that result from a deficiency in CyP450. For example, the metabolism of the cholesterol-reducing drug atorvastatin by CyP450 is inhibited by a compound found in grapefruit; under certain circumstances, eating grapefruit while taking atorvastatin can lead to dangerously high levels of the drug that result in destruction of muscle tissue (rhabdomyolysis).

A different type of metabolism occurs with some xenobiotics. In this type, called **phase I metabolism**, a xenobiotic undergoes an enzyme-catalyzed chemical transformation in which a new functional group is introduced. Typically, such reactions include hydrolysis, oxidation, reduction, and other reactions. [An example of phase I metabolism—the oxidation of polycyclic aromatic hydrocarbons—was discussed in Sec. 16.7 (Eq. 16.61).] Although many types of phase I transformation occur in biology, oxidation plays a central role. Why should this be? Oxidation usually introduces oxygen, often in the form of a hydroxy group, which both enhances water solubility and provides a site for glucuronidation.

A number of these phase I oxidations are mediated by an enzyme called **cytochrome P450**, which we abbreviate as **CyP450**. One example is the epoxidation of double bonds. For example, CyP450 promotes the epoxidation oxidation of aromatic hydrocarbons (Sec. 16.7). Another common reaction that we consider here is the hydroxylation of C—H bonds; that is,

$$\mathrm{R{-}\underset{|}{\overset{|}{C}}{-}H} \xrightarrow{\text{CyP450}} \mathrm{R{-}\underset{|}{\overset{|}{C}}{-}\underbrace{OH}_{\text{comes from } O_2}} \qquad (17.39)$$

In CyP450-promoted oxidations, it is common to find that allylic and benzylic hydrogens are replaced selectively. For example, in "THC," the active principle in marijuana, the primary sites of oxidation are allylic hydrogens.

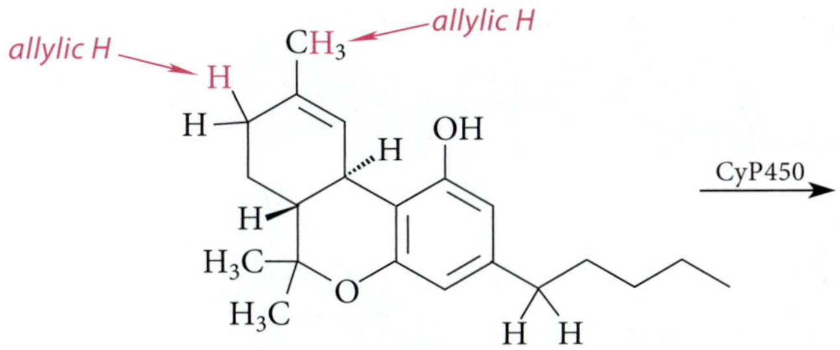

(−)-*trans*-Δ⁹-tetrahydrocannabinol ("THC")

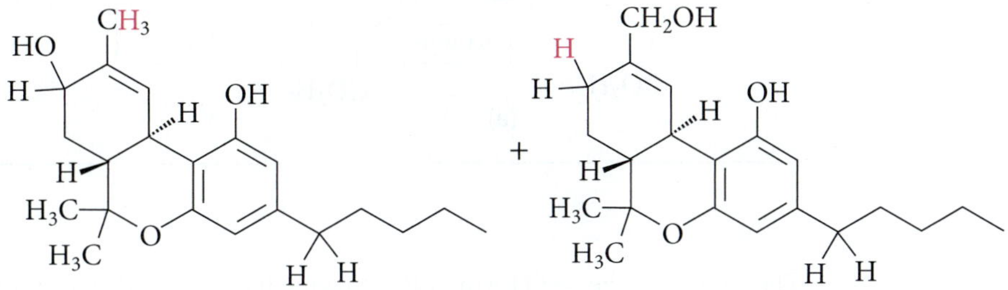

(17.40)

In this case, not all of the allylic hydrogens, and none of the benzylic hydrogens, are replaced. Presumably, this selectivity is due to the proximity of the reactive hydrogens to the active species in the CyP450 active site (discussed subsequently). Moreover, many cases are known in which "ordinary" hydrogens are replaced, as in the metabolism of the common analgesic ibuprofen:

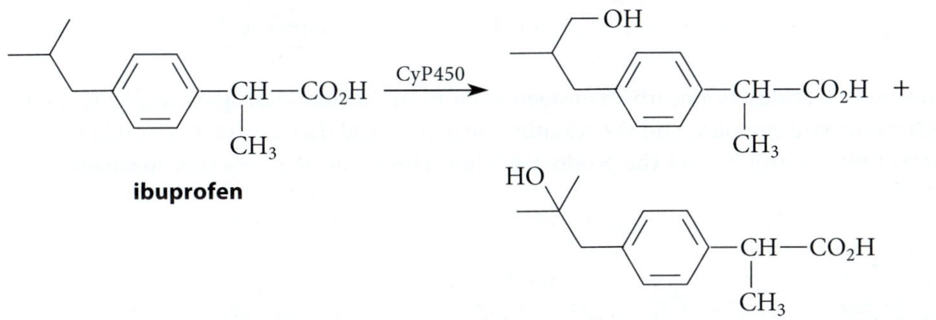

ibuprofen

major metabolites of ibuprofen (17.41)

Replacement of a hydrogen by an OH group is a remarkable process because it is not very common in the laboratory. How does this hydroxylation occur?

The active site of a cytochrome P450 contains iron in a high oxidation state that is held in place by coordination to the nitrogens of a large heterocyclic compound called *protoporphyrin IX*, which is, in turn, tightly held within the enzyme active site by both noncovalent attractions and covalent bonding of the sulfur atom of a cysteine residue to the iron. The iron is coordinated to an oxygen atom that originates from molecular oxygen. (The other oxygen of O_2 is converted into water in a reductive process preceding the formation of this species.) **Figure 17.4a** shows the detailed structure of this species, and **Fig. 17.4b** shows a convenient abbreviation used in discussing the mechanism. (Metal oxidation states in such compounds are discussed in Sec. 18.5B.) The protoporphyrin IX–Fe–oxygen group in many cases is the actual oxidizing agent in CyP450-mediated oxidations. The enzyme itself binds the substrate close to this group, and the reaction then takes place within the enzyme–substrate complex.

The mechanism of oxidation by CyP450 has been widely debated, but the currently accepted mechanism for hydroxylation involves a radical mechanism. This is *not* a radical *chain* mechanism, because the radicals involved never exit the enzyme active site; so, they are not "free" radicals. Rather, the mechanism is more like the initiation and termination steps of a radical chain reaction with no propagation steps in between.

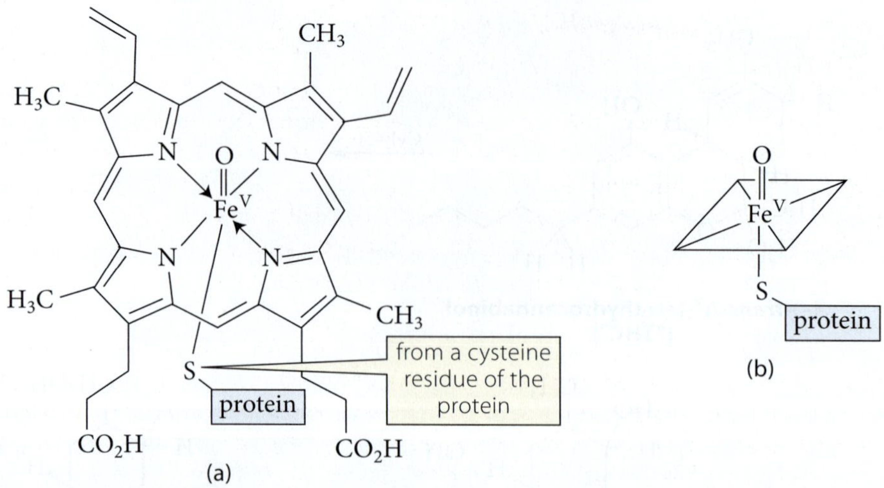

FIGURE 17.4 (a) The structure of protoporphyrin IX at the active site of CyP450, along with the coordinated oxo-iron(V) group. The cysteine bound to the iron is part of the CyP450 protein. (b) The abbreviation of CyP450 used in the text discussion.

The iron in the Fe^V=O form of the enzyme has unpaired electron spins, and the oxygen shares this unpaired-spin character. In effect, we can think of the oxygen as a bound *oxy radical*:

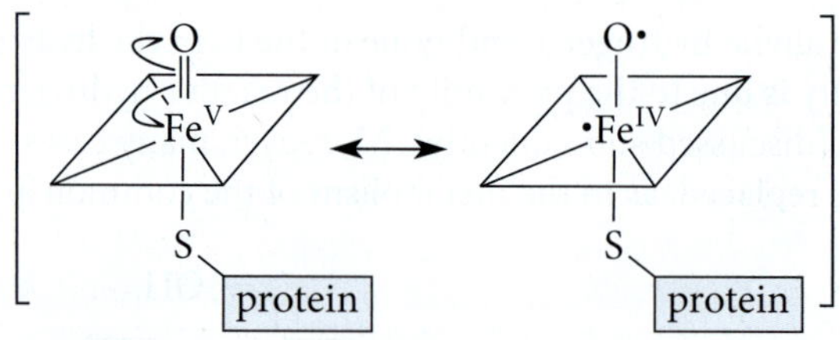

(17.42a)

The oxy radical abstracts a nearby hydrogen atom from the bound drug molecule within the CyP450–substrate complex, and the resulting carbon radical then abstracts an OH radical instantly from the iron to give the product alcohol. This is called a "rebound mechanism."

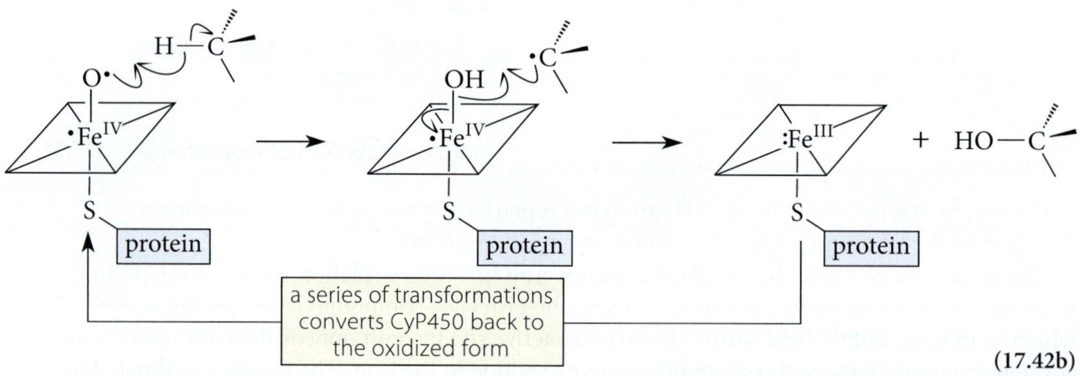

(17.42b)

As shown in the equation, the iron is reduced; therefore, it must then be re-oxidized to give the active form of the enzyme. This oxidation is a multistep, enzyme-catalyzed process that is not discussed here. Therefore, CyP450 alone is not a catalyst in Eq. 17.42b, but is actually a reactant. The catalytic cycle for hydrocarbon oxidation therefore also includes the reactions by which CyP450 is re-oxidized.

The particular hydrogen that is removed depends on which hydrogen atom is positioned closest to the reactive oxygen within the protein–substrate complex and on which genetic variant of CyP450 is involved. In other words, it is sometimes difficult to predict in a new compound which hydrogen will be abstracted. When the substrate is bound so that an allylic or benzylic hydrogen is closest to the oxygen, it is selectively removed, as in Eq. 17.40. However, if the binding of the substrate positions a nonallylic or nonbenzylic hydrogen close to the oxygen, this type of "ordinary" hydrogen can be removed as well (Eq. 17.41). Even though an ordinary

C—H bond [typical bond dissociation energy of 422 kJ mol^{-1} (101 kcal mol^{-1})] is much stronger than an allylic C—H bond [typical bond dissociation energy of 372 kJ mol^{-1} (89 kcal mol^{-1})], an O—H bond is somewhat stronger than either type of C—H bond, and a rapid rate of the abstraction reaction is ensured by the proximity of the oxygen and the C—H at the enzyme active site. This peculiar substrate-dependent selectivity demonstrates the importance of proximity (Secs. 12.8A and B) in determining relative reaction rates.

The level of a drug in the body is determined by a balance of its dose size, its absorption rate, and its metabolic rate. Strategies that reduce the rate of CyP450-promoted oxidation of drugs have the potential to reduce the size of the drug dose required for effectiveness and therefore to also reduce the potential side effects of the drug. One approach to this problem uses drugs in which deuterium has been substituted for hydrogen at the sites of CyP450-catalyzed oxidation. Such deuterium-substituted drugs should have a negligible difference in their binding to proteins or other macromolecules involved in their pharmacological activity. However, the metabolism of such a deuterium-substituted drug might be retarded if a *primary kinetic deuterium isotope effect* (Sec. 9.5D) occurs when a deuterium atom is removed at the site of metabolism.

This strategy has been realized in a number of drugs. For example, the antidepressant venlafaxine (Effexor) is metabolized by CyP450-mediated removal of the hydrogens next to O and N. (This is a common site of action for CyP450.)

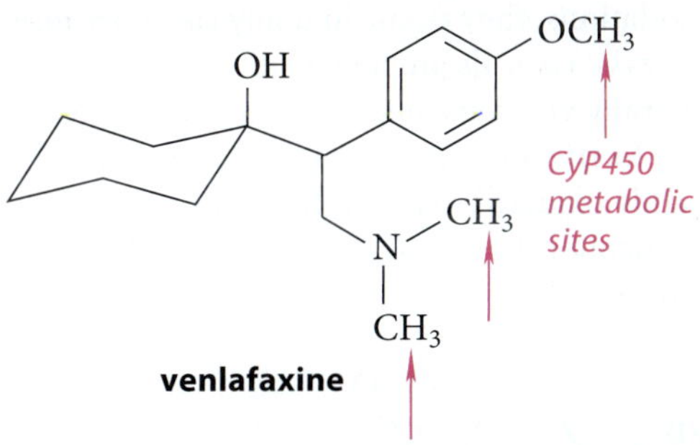

venlafaxine

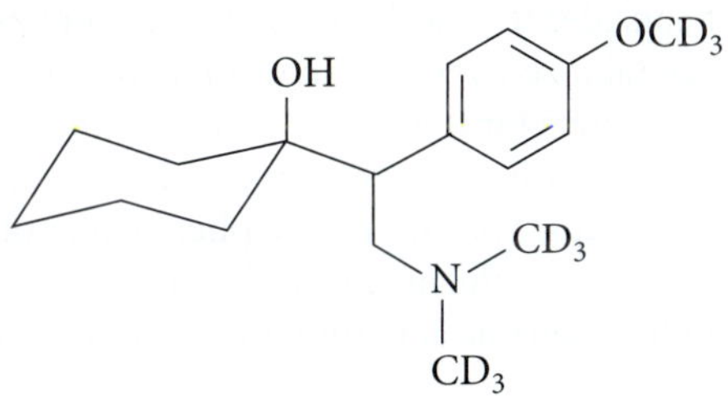

venlafaxine
(deuterium-substituted analog) (17.43)

The deuterium-substituted analog is in fact metabolized more slowly than the hydrogen analog, presumably because the removal of a hydrogen atom in the first step of the mechanism (Eq. 17.42b) is rate-limiting. Substitution of a hydrogen by deuterium results in a rate-retarding primary kinetic isotope effect.

Focused Problem

17.13 What is the major CyP450-promoted hydroxylation product of each of the following compounds? Assume that benzylic hydroxylation occurs, and neglect stereochemistry.

(a) Toluene (b)

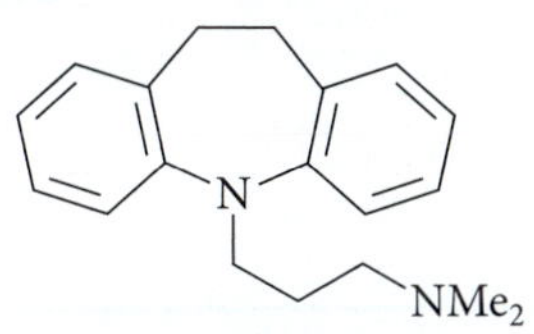

imipramine
(an antidepressant)

C. Benzylic Oxidation of Alkylbenzenes

Treatment of alkylbenzene derivatives with strong oxidizing agents under vigorous conditions converts the alkyl side chain into a carboxylic acid group. Oxidants commonly used for this purpose are Cr(VI) derivatives, such as $Na_2Cr_2O_7$ (sodium dichromate) or CrO_3; the Mn(VII) reagent $KMnO_4$ (potassium permanganate); or O_2 and special catalysts, a procedure that is used industrially (Eq. 16.63, Sec. 16.8).

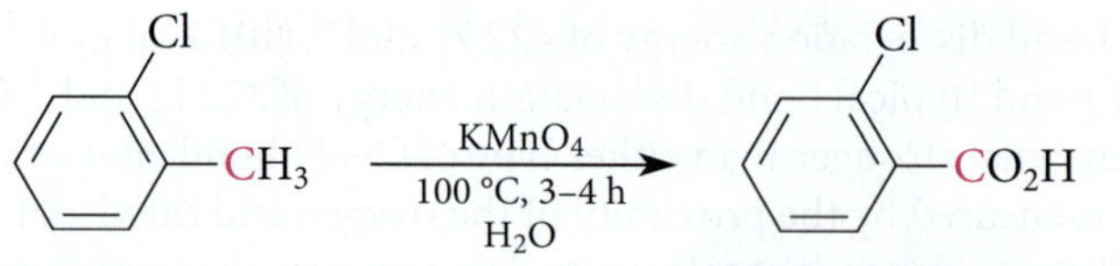

o-chlorotoluene → **2-chlorobenzoic acid** (77% yield) (17.44)

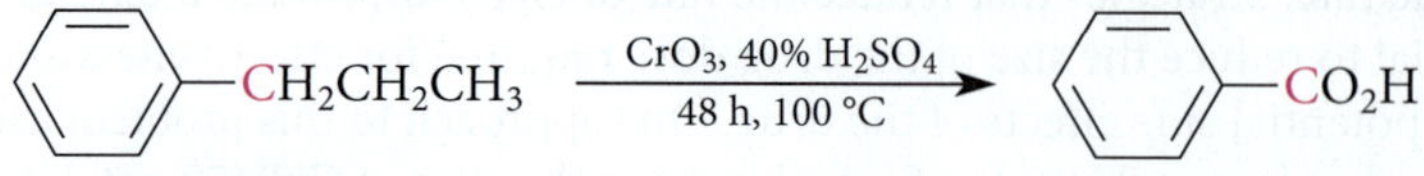

propylbenzene → **benzoic acid** (55% yield) (17.45)

STUDY GUIDE LINK 17.1
Synthetic Equivalence

In these reactions the benzene ring is left intact, and, as Eq. 17.45 demonstrates, the alkyl side chain, *regardless of length*, is converted into a carboxylic acid group. This reaction is useful for the preparation of some aryl carboxylic acids from alkylbenzenes.

Oxidation of alkyl side chains requires the presence of a benzylic hydrogen. Consequently, *tert*-butylbenzene, which has no benzylic hydrogen, is resistant to benzylic oxidation. Although we won't consider the mechanisms of these benzylic oxidations, they occur in many cases because resonance-stabilized benzylic intermediates such as benzylic radicals are involved.

You do not need to be concerned with learning the exact conditions for these reactions; rather, simply be aware that it is usually possible to find appropriate conditions for each type of oxidation. (See Study Guide Link 16.3.)

The conditions for this side-chain oxidation are generally vigorous: heat, high concentrations of oxidant, and/or long reaction times. It is also possible to effect less extensive oxidations of side-chain groups. For example, 1-phenylethanol is readily oxidized to acetophenone under milder conditions—the normal oxidation of secondary alcohols to ketones (Sec. 11.7A)—but it is converted into benzoic acid under more vigorous conditions.

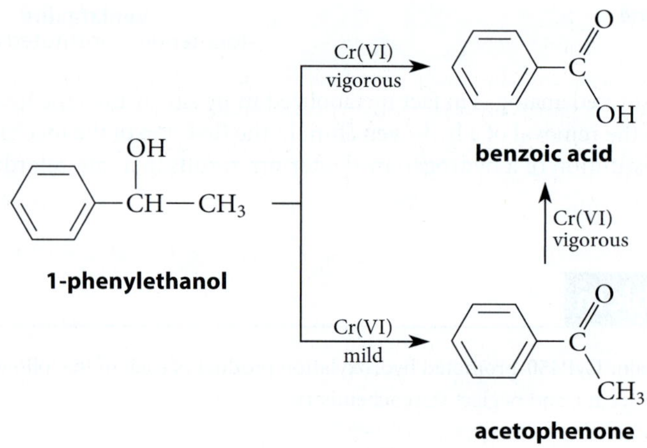

(17.46)

Focused Problems

17.14 Give the products of vigorous $KMnO_4$ oxidation of each of the following compounds.

(a)

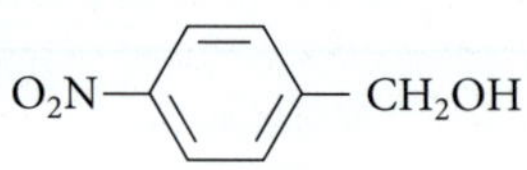

***p*-nitrobenzyl alcohol**

(b)

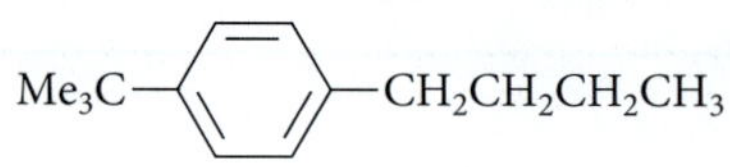

1-butyl-4-*tert*-butylbenzene

17.15 (a) A compound *A* has the formula C_8H_{10}. After vigorous oxidation, it yields phthalic acid. What is the structure of *A*?

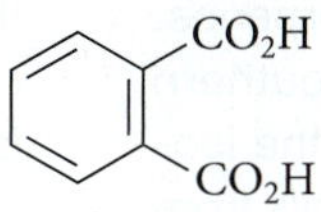

phthalic acid

(b) A compound *B* has the formula C_8H_{10}. After vigorous oxidation, it yields benzoic acid (structure in Eq. 17.46). What is the structure of *B*?

17.6 BIOSYNTHESIS OF TERPENES AND STEROIDS

Chemical Biology Topic

In this section, we show how allylic reactivity is utilized by living systems in the *biosynthesis* of a class of naturally occurring organic compounds called *terpenes*. **Biosynthesis** is the synthesis of chemical compounds by living organisms. The study of biosynthesis is an active area of research that lies at the interface of chemistry and biochemistry. This area is attracting renewed interest because biologists and chemists are collaborating to use genetic engineering technology to alter biosynthetic pathways. This technology holds the promise of using microorganisms and plants as biological factories to turn out specially engineered molecules such as drugs.

To begin with, the structure and classification of terpenes is considered in Sec. 17.6A. Then, in Sec. 17.6B, we show how allylic reactivity is brought to bear on terpene biosynthesis. Then, in Sec. 17.6C, we consider how carbocation chemistry is used in the biosynthetic conversion of terpenes into *steroids*. (The structure of steroids was introduced in Sec. 7.6D.)

A. Terpenes and the Isoprene Rule

People have long been fascinated with the pleasant-smelling substances found in plants—for example, the perfume of a rose—and have been curious to learn more about these materials, which have come to be called *essential oils*. An **essential oil** is a substance that possesses a key characteristic, such as an odor or flavor, of the natural material from which it comes. (See the sidebar.)

Essential Oils

Portrait presumed to be Paracelsus (1493–1541) (oil on panel)/Massys or Metsys, Quentin (c.1465–1530) (after)/Louvre, Paris, France/ Bridgeman Images

The history of the essential oils is an important part of the early history of both chemistry and medicine. A Swiss alchemist, Theophrastus Bombastus von Hohenheim, better known as Paracelsus (ca. 1493–1541; see photo), believed that everything had a chemical essence. For example, he could demonstrate that the odor of spearmint could be liberated from the plant as a volatile oil on heating; this was the "essence of spearmint"—its quintessence, or "reason for being." (Paracelsus tried unsuccessfully to find the essences of rocks and other refractory objects.) He believed that a person who was ill was missing part of his or her essence—and Paracelsus sought to restore the essence with chemical cures, among them mercury and sulfur, which, we now know, kill microorganisms (but with significant toxicity to the host). Paracelsus was reputed to have effected some very dramatic cures. His firm belief in the chemical essence of all things was reflected in his public tirades against physicians of the day, who (he said) "strut about with haughty gait, dressed in silk with rings upon their fingers displayed ostentatiously, or silver poniards fixed upon their loins and sleek gloves upon their hands" while chemists "sweat whole nights and days over fiery furnaces, do not waste time with empty talk, but express delight in the laboratory." Paracelsus was driven out of Basel when he angered some local clergy during a quarrel over his fees; he died in exile.

Photo copyright © Marc Loudon

Some of the most familiar essential oils turned out to be terpenes. The perfume industry in southern France developed around the isolation of fragrant essential oils from flowers such as rose, violet, and lavender, and this industry flourishes today. (The photo shows an old distillation apparatus on display at the Fragonard perfumery in Grasse, France.) The spearmint industry in the northern states of the western and midwestern United States is also based on the isolation of the "essence of spearmint," which is *R*-(–)-carvone and other terpenes. (See the sidebar, "Chiral Recognition by Scent Receptors," Sec. 6.10B.) It was the fascination with, and curiosity about, the essential oils that ultimately led chemists to their understanding of how chemical substances are synthesized in the natural world.

Essential oils, particularly oil of turpentine, were known to the ancient Egyptians. However, not until early in the nineteenth century was an effort made to determine the chemical constitution of the essential oils. In 1818, it was found that the C:H ratio in oil of turpentine was 5:8. This same ratio was subsequently found for a wide variety of natural products. These related natural products became known collectively as **terpenes**, a name coined by August Kekulé (Sec. 2.2B). The similarity in the atomic compositions of the many terpenes led to the idea that they might possess some unifying structural element.

In 1887, the German chemist Otto Wallach (1847–1931), who received the 1910 Nobel Prize in Chemistry, pointed out the common structural feature of the terpenes: they all consist of repeating units that have the same carbon skeleton as the five-carbon diene isoprene. This generalization subsequently became known as the **isoprene rule**.

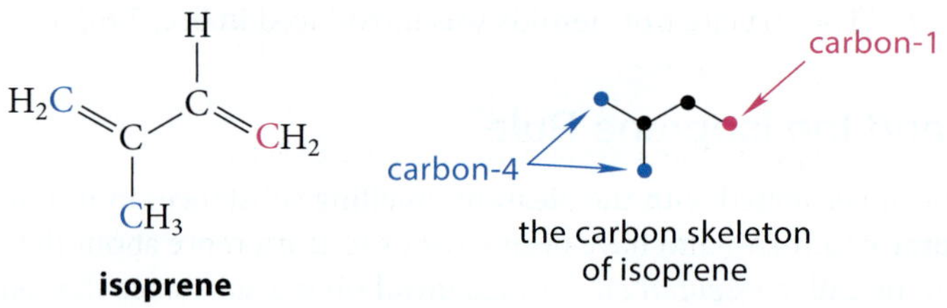

(17.47)

For example, citronellol (from oil of roses and other sources) incorporates two isoprene units:

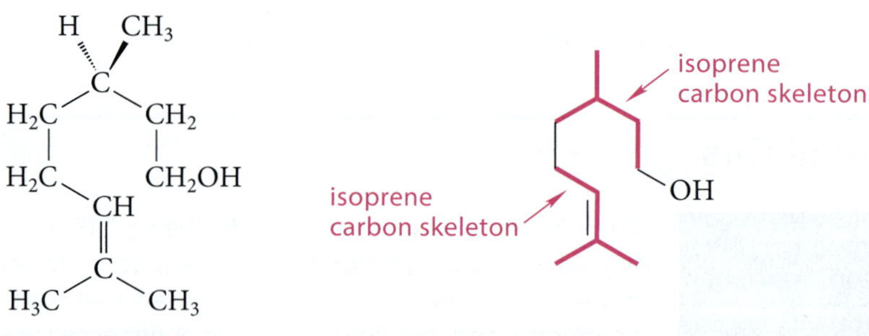

(17.48)

Because of this relationship to isoprene, terpenes are also called **isoprenoids**. The basis of the terpene or isoprenoid classification is only *the connectivity of the carbon skeleton*. The presence or the positions of double bonds and other functional groups, or the configurations of double bonds and asymmetric carbons, have nothing to do with the terpene classification.

Some notational conventions in terpene chemistry are important. As illustrated by the isoprene structure in Eq. 17.47, the carbons at the ends of the isoprene skeleton are classified as carbon-1 and carbon-4, with carbon-4 being either carbon of the dimethyl branch. (These numbers are not the same ones used in IUPAC nomenclature.) These carbons used to be called "head" and "tail," but British and American chemists confused the issue by adopting different conventions for head and tail. The Americans called carbon-1 the "head" and carbon-4 the "tail," whereas the British adopted the reverse convention. To settle the issue, Dale Poulter, a professor at the University of Utah, proposed that the carbons simply be referred to by numbers, and that is the convention adopted here.

In many terpenes, the isoprene units are connected in a 1′–4 arrangement (formerly called a "head-to-tail" or "tail-to-head" arrangement). This means that carbon-4 of one skeleton is connected to carbon-1′ of the other. The prime (′) on one number and its absence on the other mean that the connection is between *different* isoprene units. For example, this connection is readily apparent in the terpenes geraniol (from oil of geraniums) and limonene (from oil of lemons). Cyclic terpenes such as limonene have additional connections between the isoprene units (in the case of limonene, a C1-C2′ connection) that close the ring.

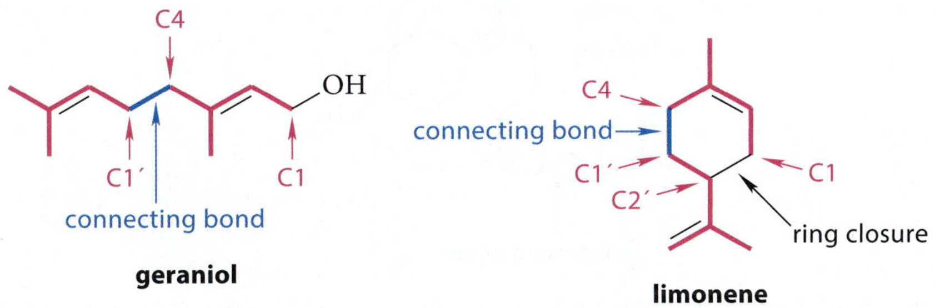

(17.49)

Citronellol (Display 17.48) has a 1′–4 connection between isoprene units as well. Because this arrangement is so common, Wallach assumed the generality of 1′–4 connectivity in his original statement of the isoprene rule. However, many examples are now known in which the isoprene units have a 1′–1 connectivity. Furthermore, some compounds are derived from the conventional terpene structures by skeletal rearrangements. Although these compounds do not have the exact terpene connectivity, they are nevertheless classified as terpenes. For our purposes, though, it will be sufficient to recognize terpenes by two criteria:

1. A multiple of five carbon atoms in the main carbon skeleton
2. The carbon connectivity of the isoprene carbon skeleton within each five-carbon unit

Because terpenes are assembled from five-carbon units, their carbon skeletons contain multiples of five carbon atoms (10, 15, 20, . . . , 5*n*). Terpenes with 10 carbon atoms in their carbon chains are classified as **monoterpenes**, those with 15 carbons as **sesquiterpenes**, those with 20 carbons as **diterpenes**, and so on. Some examples of terpenes are shown in **Fig. 17.5** in Sec. 17.6B. Many of these compounds are familiar natural flavorings or fragrances.

Study Problem 17.2

Determine whether the following compound, isolated from the frontal gland secretion of a termite soldier, is a terpene.

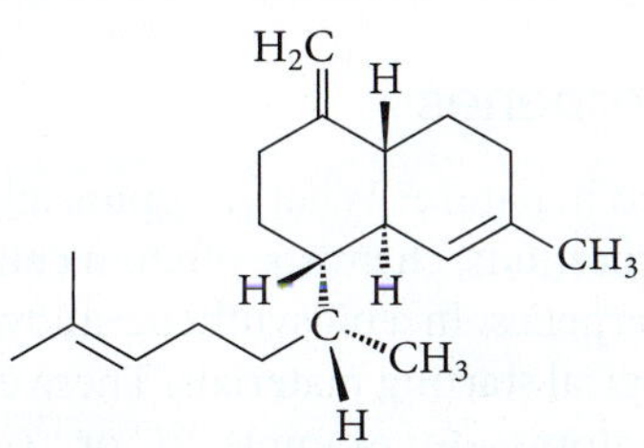

Solution Because stereochemistry is not an issue, delete all stereochemical details for simplicity. First, count the number of carbons. Because the compound has a multiple of five carbon atoms, it could be a terpene. To check for terpene connectivity, look first for a methyl branch. One is at the end of the long side chain. Identify within this group a chain of four carbons with a methyl branch at the second carbon:

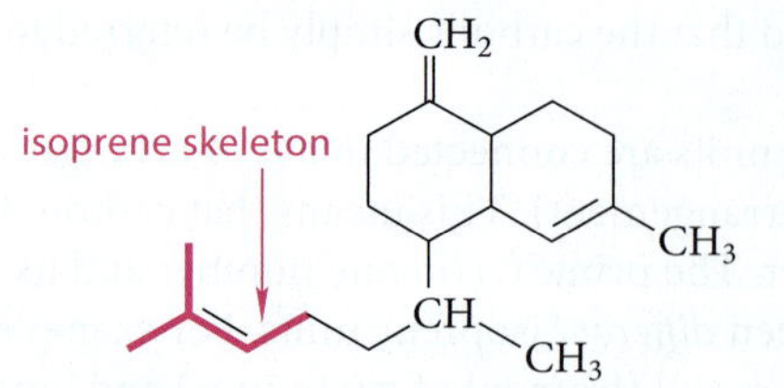

Starting at the next carbon, look for the same pattern. Remember that a bond must connect each isoprene skeleton.

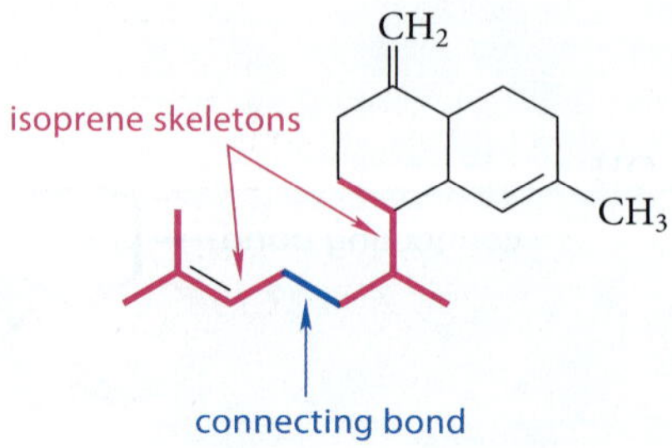

(We arbitrarily chose to proceed clockwise around the ring; you should convince yourself that in this case a counterclockwise path also works.) Continue in this fashion until either the pattern is broken or, as in this case, all carbons are included:

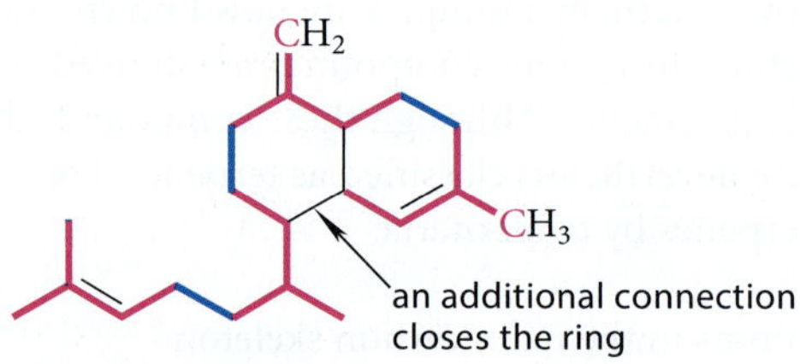

This compound incorporates four isoprene skeletons and is therefore a diterpene.

Focused Problem

17.16 Show the isoprene skeletons within the following compounds of Fig. 17.5.

(a) Vitamin A (b) Caryophyllene

B. Biosynthesis of Terpenes

How are terpenes synthesized in nature? What is responsible for the regular repetition of isoprene skeletons? To answer these questions, chemists, often in collaboration with biologists, have studied the biosynthesis of terpenes. In a biosynthetic study, chemists synthesize compounds hypothesized to be the biological starting materials. These compounds, however, are prepared with isotopically substituted atoms—for example, ^{14}C or ^{13}C substituted for natural-abundance carbon (mostly ^{12}C)—at defined positions. Investigators then feed the synthetic compounds to the organisms under study and isolate the products of biosynthesis. They determine by chemical degradations, mass spectrometry, and/or NMR whether the biosynthetic products contain the isotopic substituents and, if so, where in these products the isotopes are located. These studies can be followed in some cases by isolation and characterization of the enzymes that catalyze the transformations of interest. Eventually, plausible mechanisms are formulated for the transformations that are consistent with all of the findings. We won't be concerned here

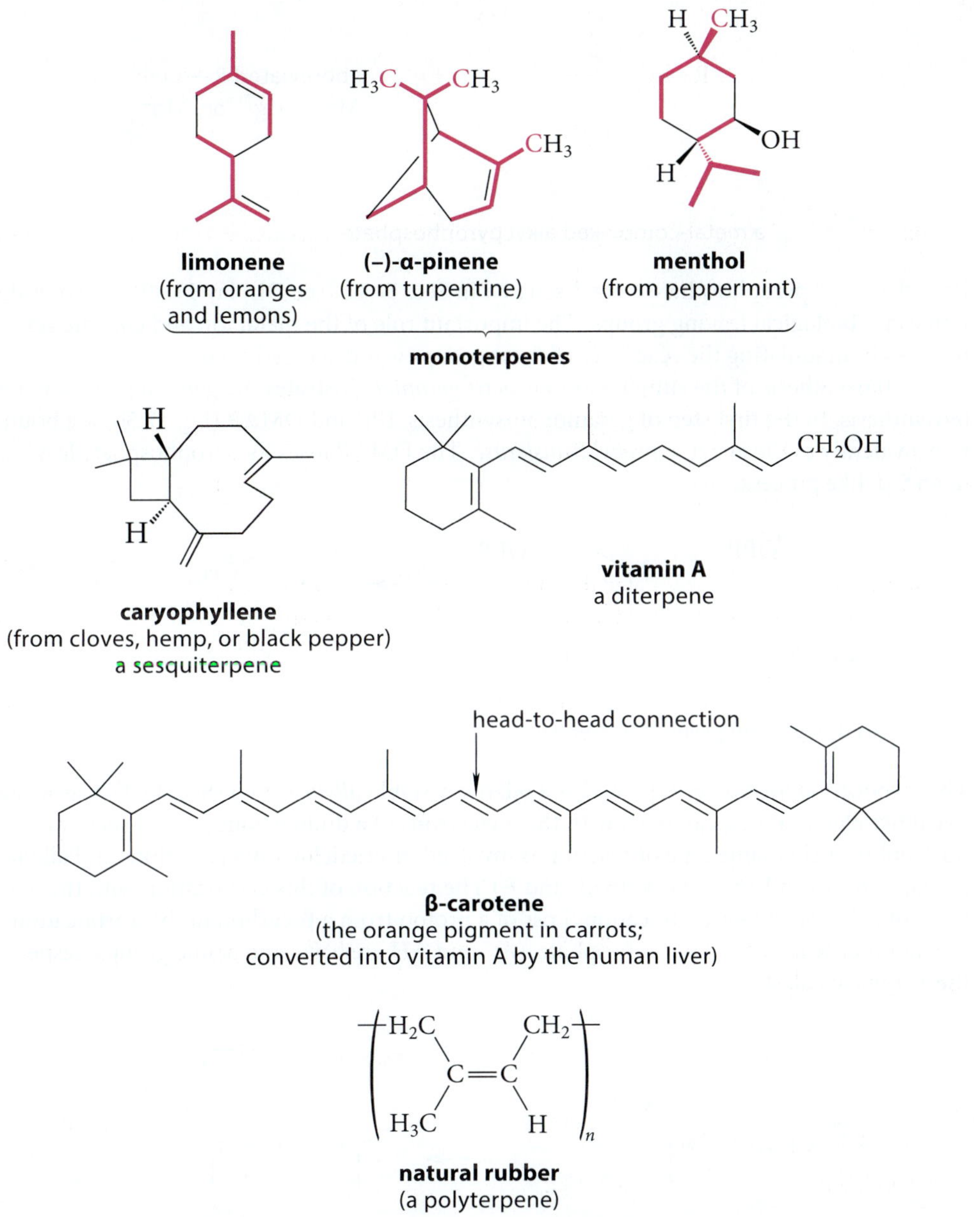

FIGURE 17.5 Examples of terpenes. In the monoterpenes, the isoprene skeletons are shown in red.

with the methodology that was used (over many years) to determine the mechanism of terpene biosynthesis; rather, we show the result and how it fits the notions of allylic reactivity that we have learned.

The repetitive isoprene skeleton in all terpenes has its origin in two simple five-carbon compounds.

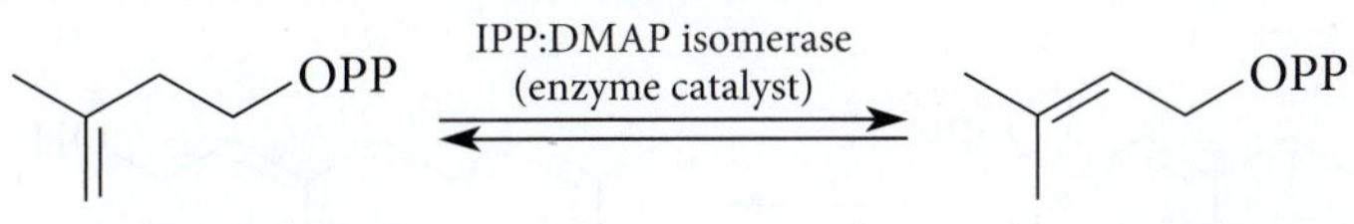

isopentenyl pyrophosphate (IPP) **γ,γ-dimethylallyl pyrophosphate (DMAP)** (17.50)

IPP and DMAP are rapidly interconverted in biological systems by an isomerization reaction (Eq. 17.50) catalyzed by an enzyme, IPP:DMAP isomerase. (For the mechanism, see Problem 17.52a.) Because of this interconversion, the presence of one compound ensures the presence of the other.

The —OPP in Eq. 17.50 is an abbreviation for the *pyrophosphate group* (*red* in the following structure), which, in nature, is usually complexed to a metal ion such as Mg^{2+} or Mn^{2+} (*blue*).

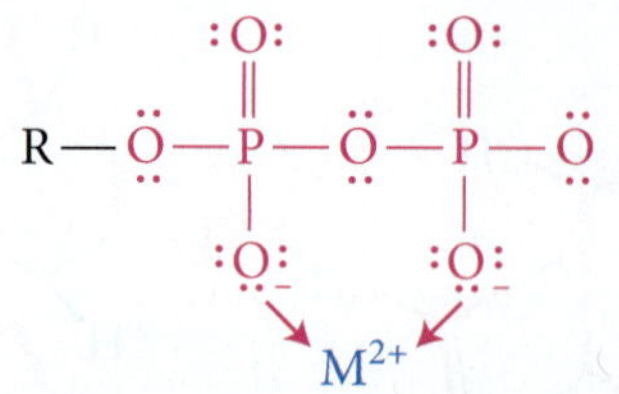

abbreviated R—OPP
$M^{2+} = Mg^{2+}$ or Mn^{2+}

a metal-complexed alkyl pyrophosphate (17.51)

Phosphate and pyrophosphate were discussed in Sec. 11.4E (see Display 11.49) as two of the most important biological leaving groups. The important role of the metal ion and enzyme active-site residues in modulating the reactivity of these groups was discussed there.

The biosynthesis of the simple monoterpene *geraniol* illustrates the general pattern of terpene biosynthesis. In the first step of geraniol biosynthesis, IPP and DMAP (Eq. 17.50) are bound noncovalently to the enzyme *prenyl transferase.* The DMAP loses its pyrophosphate leaving group in an S_N1-like process.

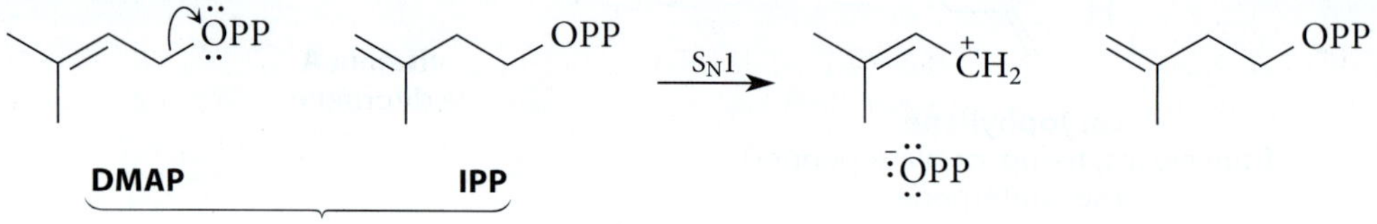

held in proximity by their noncovalent binding to the enzyme prenyl transferase (17.52)

The carbocation formed in Eq. 17.52 is a relatively stable *allylic cation* (Sec. 17.1). Carbocations, like other electrophiles, can react with the π electrons of a double bond, which serve as a nucleophile. (This same type of reaction is involved in Friedel–Crafts acylations and alkylations; see Eqs. 16.21b and 16.30, Secs. 16.4E and F.) The reaction of this carbocation with the double bond of IPP gives a new carbocation. Loss of a proton from a β-carbon of this carbocation gives the monoterpene geranyl pyrophosphate. (A:$^-$ and AH are basic and acidic groups, respectively, of the enzyme catalyst.)

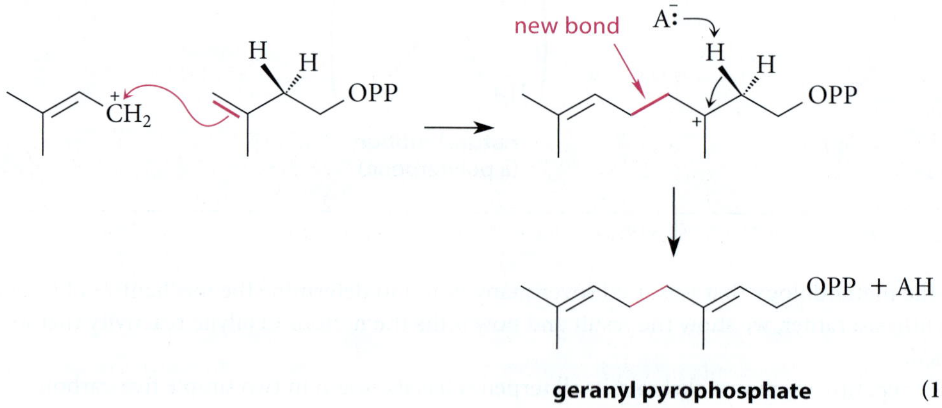

(17.53)

Geraniol is formed in an S_N2 reaction between water and geranyl pyrophosphate.

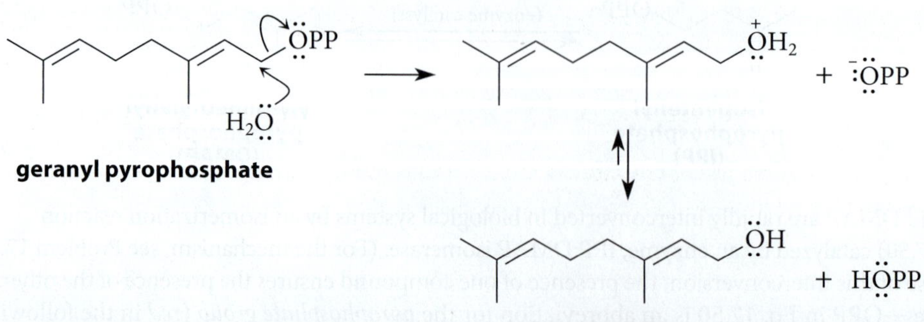

(17.54)

The S_N2 nature of this substitution reaction was suggested by the observation of stereochemical inversion at the electrophilic carbon. (See Problem 17.53a.)

All 1′–4 bonds in terpenes are formed by reactions analogous to the ones shown in Eqs. 17.53 and 17.54. As these examples illustrate, the biosynthesis of terpenes can be understood in terms of carbocation intermediates that are like those involved in laboratory chemistry. As we pointed out previously (Secs. 11.8 and 12.7B), the organic chemistry of living systems is understandable in terms of laboratory analogies.

A large body of evidence for the carbocation character of these reactions was developed in an elegant series of investigations by Professors Dale Poulter (b. 1942), Hans C. Rilling (b. 1933), and their students at the University of Utah in the period 1975–1980.

The biosynthesis of terpenes also illustrates the economy of nature: A remarkable array of substances is generated from a common starting material. This economy is evident also in other families of natural products. For example, terpenes also serve as the starting point for the biosynthesis of *steroids* (Sec. 7.6D). The isoprene rule is one of the unifying principles that underlie the chemical diversity of nature.

Focused Problem

17.17 (a) Give a biosynthetic mechanism for the formation of the cyclic terpene limonene (see Fig. 17.4) beginning with an intramolecular reaction of the following carbocation. (Assume acids and bases are present as necessary.)

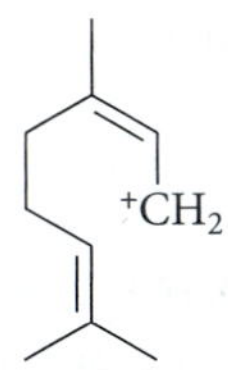

(b) Draw a curved-arrow mechanism for the biosynthesis of the carbocation intermediate in part (a) from geranyl pyrophosphate.

C. Biosynthesis of Steroids

The biosynthesis of steroids has been a very important research area for decades because atherosclerosis (deposits of cholesterol, a steroid, in arteries) is the leading cause of death in the United States. (See the sidebar, "The Statins: Blockbuster Drugs That Inhibit HMG-CoA Reductase," in Sec. 25.5C.) The starting material for the biosynthesis of steroids is the triterpene *squalene*, which comes from the head-to-head (1–1′) joining of two farnesyl pyrophosphates. (This transformation requires a reductive step that is not shown here.)

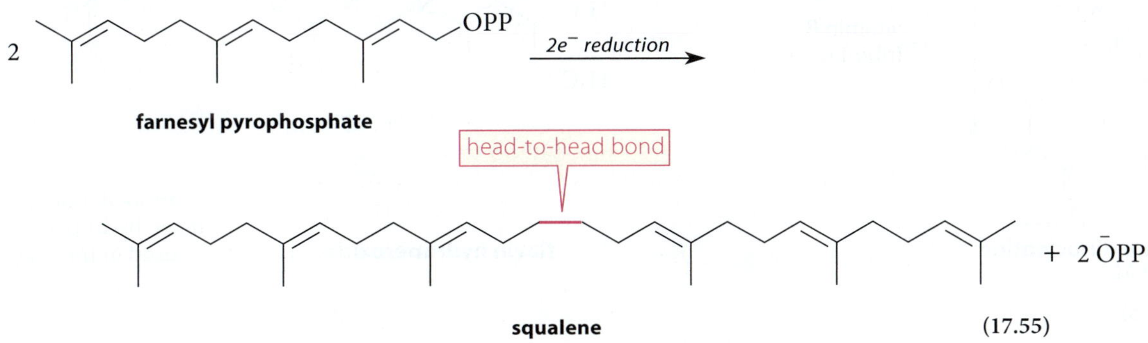

(17.55)

The biosynthesis of farnesyl pyrophosphate is an extension of the biosynthesis of geranyl pyrophosphate shown in Sec. 17.6B (see Problem 17.30a).

In mammals, squalene is converted into a steroid called *lanosterol* in just two enzymatic steps:

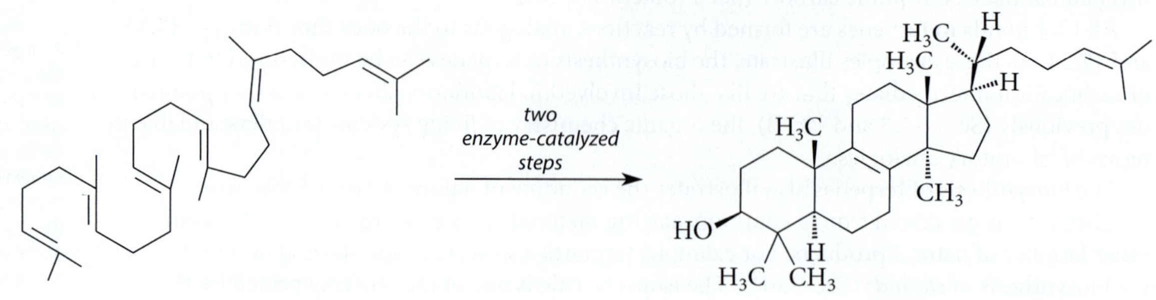

squalene
(0 rings, 0 asymmetric carbons)

lanosterol
(4 rings, 7 asymmetric carbons)

(17.56)

This conversion introduces not only four rings, but also seven asymmetric carbons, in a completely stereoselective manner.

The first of the enzymatic steps is an epoxidation catalyzed by the enzyme *squalene oxidase*. The actual oxidizing agent is *flavin hydroperoxide*, which, in turn, is derived ultimately from riboflavin, or vitamin B_2. The structure of flavin hydroperoxide is given in **Fig. 17.6**. As in many biomolecules, the reactive part of the molecule is a small part of the structure, and it will suffice for us to abbreviate flavin hydroperoxide as R—O—O—H. The source of the oxygen in flavin hydroperoxide is O_2. We can think of flavin hydroperoxide in this reaction as "nature's peroxy acid," because its role here is exactly the same as that of a peroxy acid in the laboratory epoxidation of alkenes (Sec. 12.3A). The formation of the epoxide is a concerted electrophilic addition. (Compare with Eq. 12.22, Sec. 12.3A.) This establishes the stereochemistry at one of the seven asymmetric carbons.

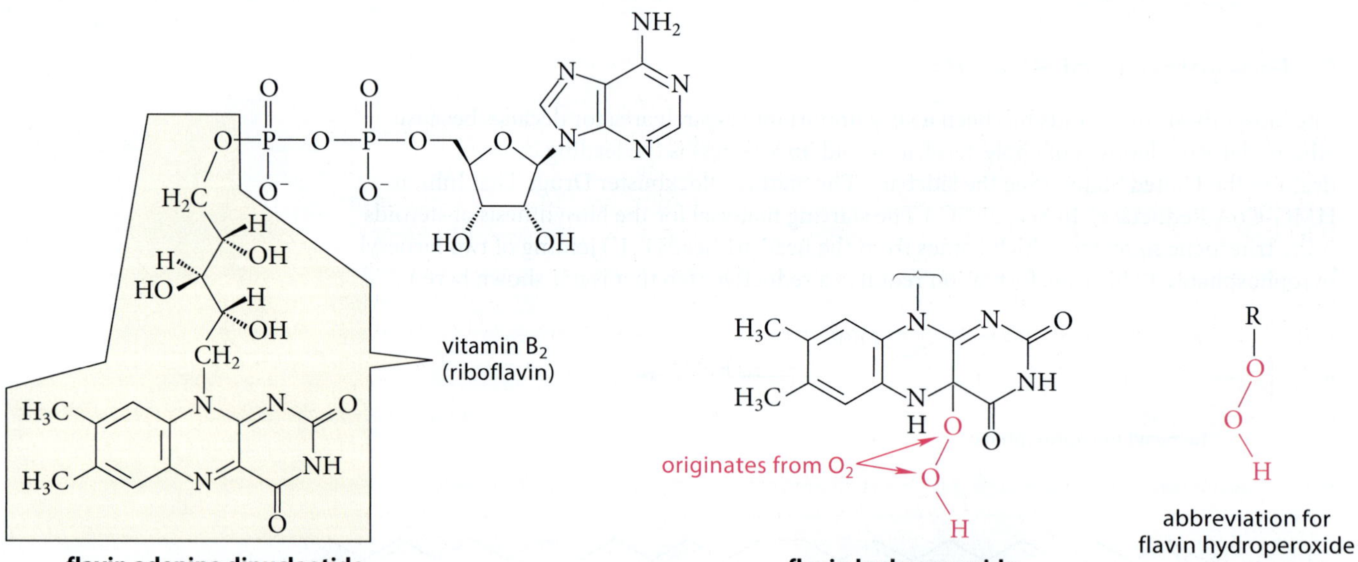

flavin adenine dinucleotide (FAD)

(a)

flavin hydroperoxide

(b)

FIGURE 17.6 (a) The structure of the coenzyme flavin adenine dinucleotide (FAD). The boxed structure (with an O—H rather than an O—P bond) is vitamin B_2 (riboflavin). FAD and a reduced derivative, $FADH_2$, are important in a number of biological oxidation–reduction reactions. (b) Flavin hydroperoxide is a modified form of FAD. (The part of the structure above the bracket is the same as in FAD.) This can be thought of in the context of steroid biosynthesis as the natural equivalent of peroxy acids. The abbreviation used in the text is shown at right.

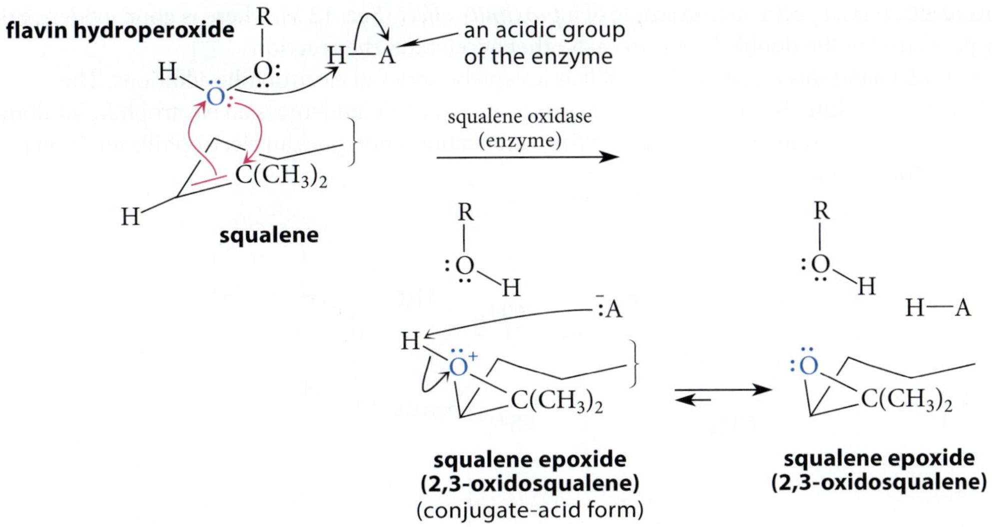

(17.57)

The second, and more remarkable, enzymatic step is the cyclization of squalene epoxide, which is catalyzed by the enzyme *2,3-oxidosqualene:lanosterol cyclase*. To understand this step, we have to consider the conformation of the epoxide 2,3-oxidosqualene as it exists in the enzyme active site. Noncovalent interactions are used to constrain 2,3-oxidosqualene to adopt the following conformation in the enzyme active site.

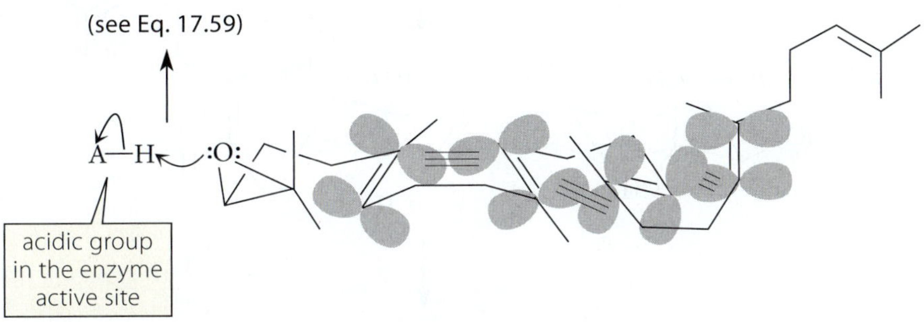

(17.58)

In this conformation, the 2p orbitals of successive double bonds are positioned so that they can react sequentially with little or no molecular realignment. The enzyme provides an acidic group that protonates the epoxide. As shown in Sec. 12.5B, protonated tertiary epoxides are essentially shielded carbocations, and they undergo reactions with nucleophiles at the more-branched carbon atom. So it is in this case. The carbocation undergoes an electrophilic addition to the neighboring double bond. This generates electron deficiency at the other carbon of the double bond, which undergoes an electrophilic addition to the next double bond, and so on. The first intermediate is a tertiary carbocation in which all the ring bonds are formed stereospecifically.

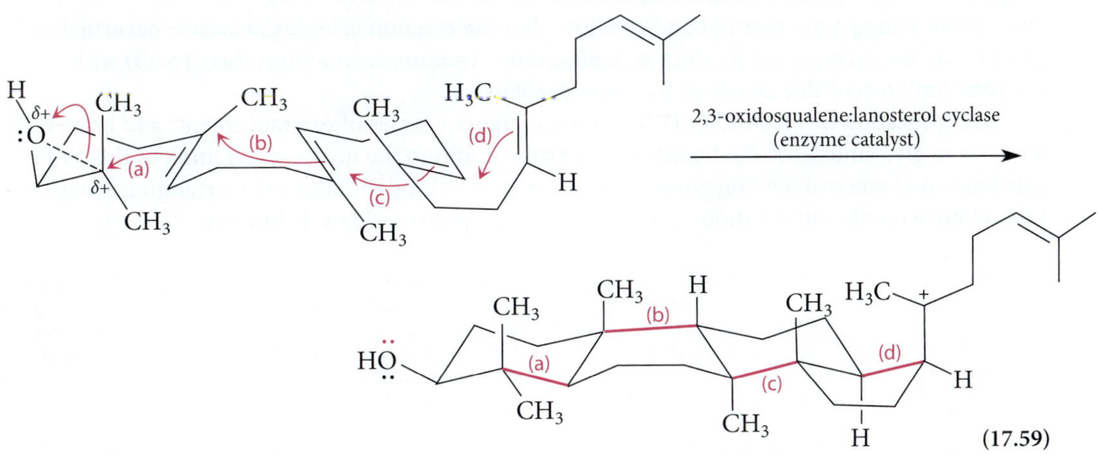

(17.59)

This reaction is a spectacular example of a *proximity effect* (Sec. 12.8). There is good evidence that the proximity of the double bonds to each other accelerates the reaction.

To understand this reaction, think of it as a stepwise series of electrophilic additions. The carbocation resulting from opening of the protonated epoxide undergoes an electrophilic addition to the neighboring double bond, the resulting carbocation undergoes an electrophilic addition to the next double bond, and so on.

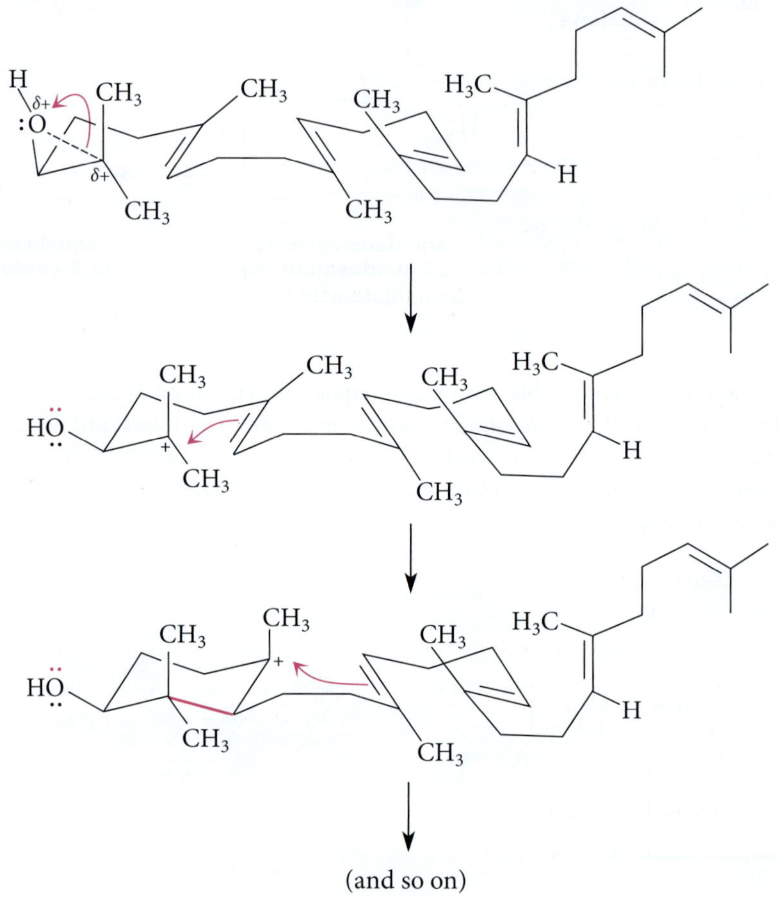

(17.60)

Although it may be helpful to view the reaction this way, the reaction appears to be a concerted process.

What is the energetic advantage of converting one tertiary carbocation into another? In this process, several π bonds are converted into σ bonds. Each π bond has a bond dissociation energy of about 243 kJ mol^{-1} (58 kcal mol^{-1}), whereas a carbon–carbon σ bond has a bond dissociation energy of about 377 kJ mol^{-1} (90 kcal mol^{-1}) (see Table 10.2, Sec. 10.1E). Therefore, the conversion of *each* π bond has a $\Delta H°$ of roughly −134 kJ mol^{-1} (−32 kcal mol^{-1}). The magnitude of this energy is reduced by the boat conformation of one ring and by the entropic cost of eliminating a number of bond rotations, but the reaction is highly favorable nevertheless. In addition, the carbocation product is stabilized by pi–cation interactions (Sec. 15.8B) with aromatic amino acid side chains of the enzyme catalyst.

The carbocation formed in Eq. 17.59 then undergoes a series of rearrangements and loss of a proton to give lanosterol. Each rearranging group is anti to the next; we can think of this as a substitution of one rearranging group with its electron pair by the adjacent rearranging group. Like all nucleophilic substitutions, each occurs by an opposite-side substitution.

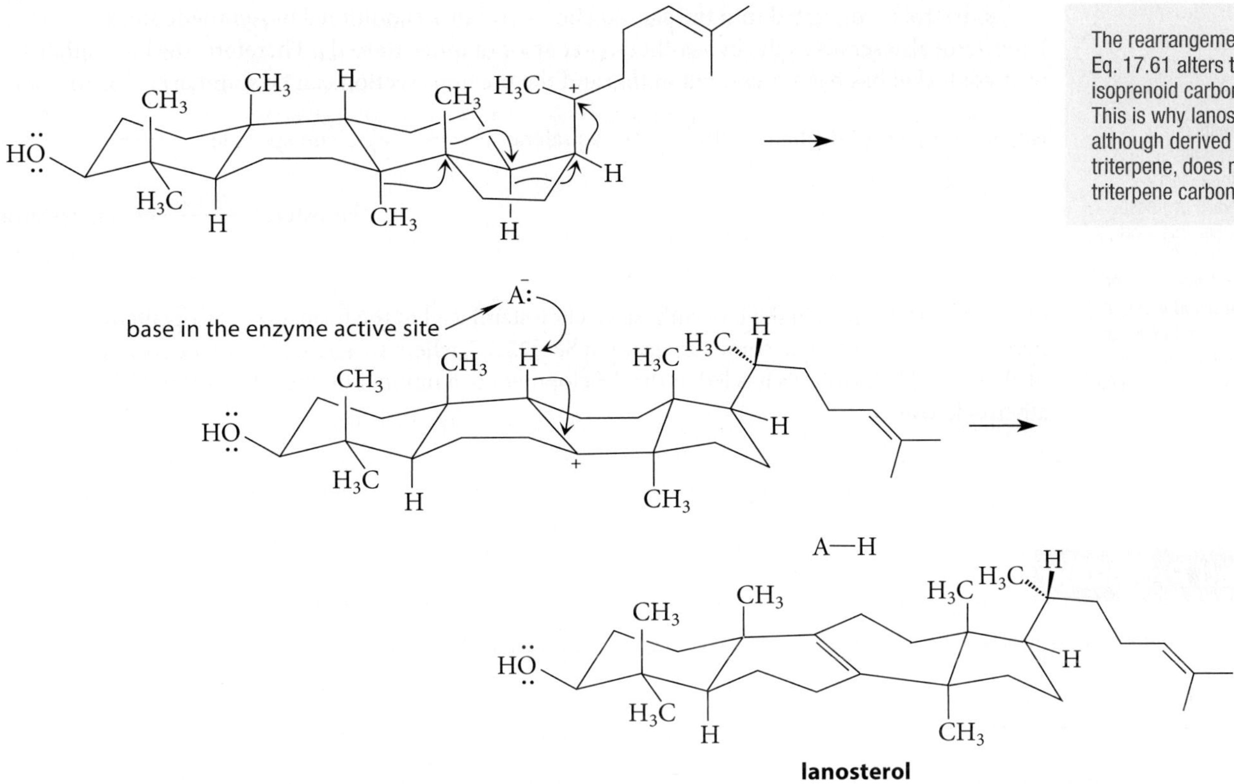

(17.61)

The rearrangement in Eq. 17.61 alters the isoprenoid carbon skeleton. This is why lanosterol, although derived from a triterpene, does not have the triterpene carbon skeleton.

As with the additions, we could rewrite the rearrangements in individual steps:

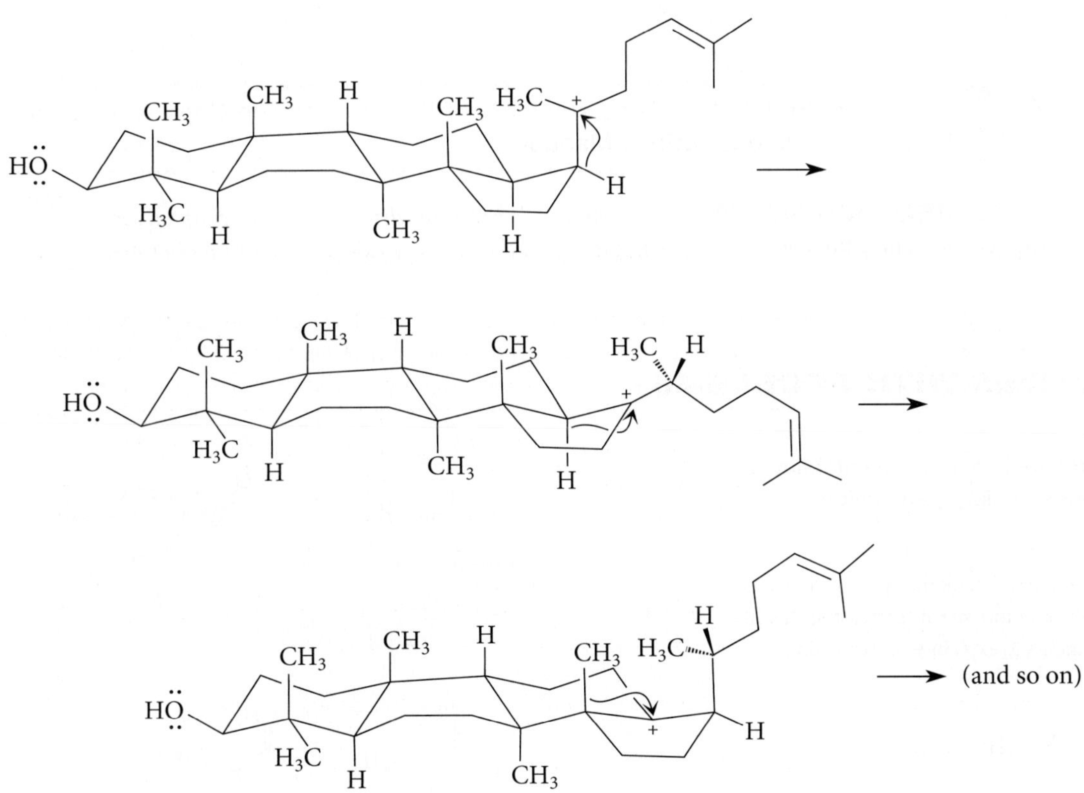

(17.62)

However, it is believed that these take place in a concerted manner.

Lanosterol is converted into the steroid cholesterol in 19 additional biosynthetic steps. Lanosterol also serves as the biosynthetic precursor of other steroids. Therefore, the biosynthesis of steroids that has been presented in this and the previous section can be summarized as follows:

$$\text{isopentenyl pyrophosphate (IPP)} \longrightarrow \text{squalene} \longrightarrow \text{squalene epoxide} \longrightarrow \text{lanosterol} \xrightarrow{\text{19 steps}} \text{cholesterol} \quad (17.63)$$

The biosynthesis of cholesterol was elucidated by Konrad Bloch (1912–2000), a Harvard professor of chemistry. For this work, Bloch shared the 1964 Nobel Prize in Physiology or Medicine with Feodor Lynen (1911–1974), a professor at the Max Planck Institute for Cellular Chemistry in Munich.

Other important steps in the biosynthesis of cholesterol lead to the formation of isopentenyl pyrophosphate. This topic comes up again in Sec. 25.5C, where we show that an understanding of cholesterol biosynthesis has led to the development of a number of important drugs to fight atherosclerosis.

Focused Problems

17.18 (a) Complete the stepwise mechanism shown in Eq. 17.60 to give the product carbocation of Eq. 17.59.

(b) Examine each of the carbocations in the stepwise process you drew in part (a). What problem do you see with the stepwise mechanism? (*Hint:* Classify the carbocation formed in each step as primary, secondary, or tertiary.)

(c) In view of your answer to part (b), can you see an advantage to a concerted, rather than a stepwise, mechanism?

17.19 Complete the stepwise process shown in Eq. 17.62 to give lanosterol, the product of Eq. 17.61.

CHAPTER SUMMARY *For a summary of the chapter, see Chapter 17 in the* Study Guide and Solutions Manual.

REACTION REVIEW *For a summary of reactions discussed in this chapter, see the* Reaction Review *section of Chapter 17 in the* Study Guide and Solutions Manual.

SKILLS OBJECTIVES WITH PROBLEMS

- **Use resonance arguments to explain the stability of allylic and benzylic cations, radicals, and anions. 17.1, 17.2, 17.3**

17.20 In each of the following parts, rank the species shown in order of decreasing heat of formation (increasing stability). Use resonance arguments to explain your reasoning.

(a)

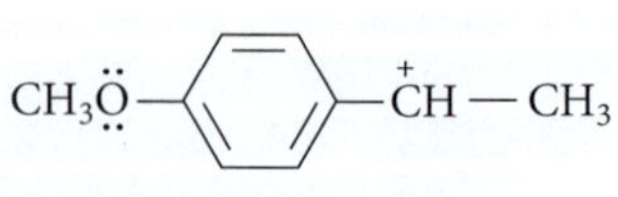

A

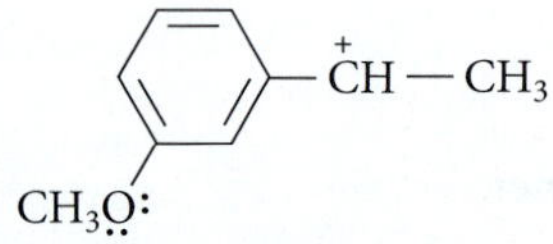

B

(b)

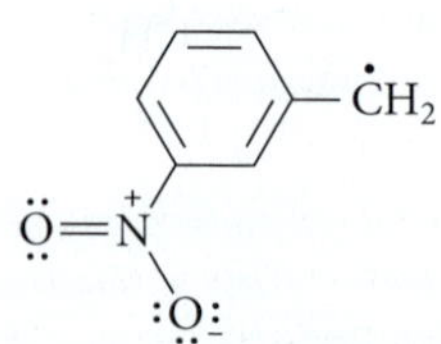

A

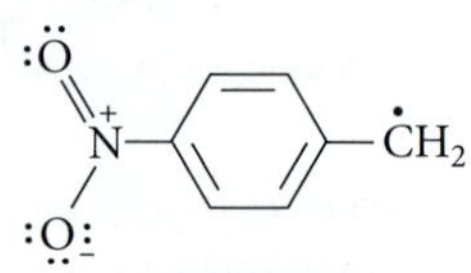

B

(c)

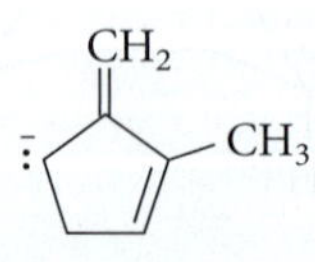

A

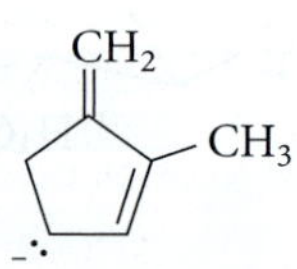

B

- **Use the stability of allylic or benzylic cations, radicals, and anions to predict or explain relative reactivity.** **17.1, 17.2, 17.3**

17.21 (a) Explain why compound *A* reacts faster than compound *B* when they undergo solvolysis by the S_N1 mechanism in aqueous acetone. Use resonance structures as part of your explanation.

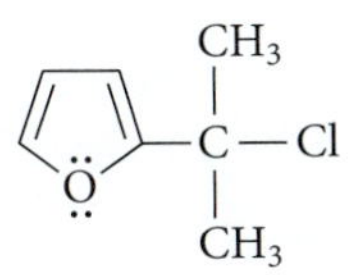

A

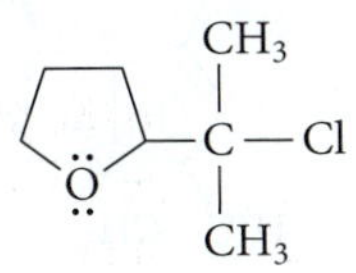

B

(b) Which of the following alkyl halides would react most rapidly in an S_N1 solvolysis reaction? Explain your reasoning.

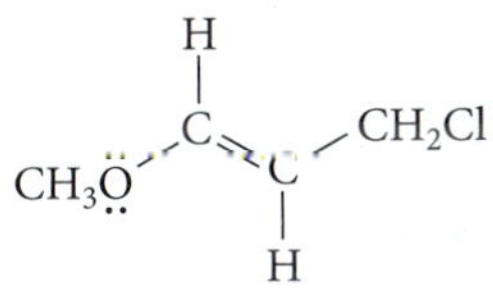

A

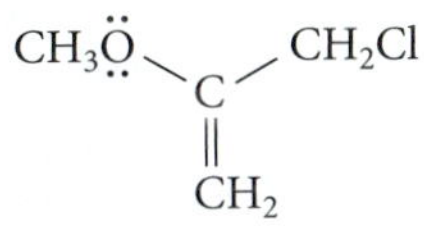

B

17.22 Predict which of the following compounds should undergo the more rapid E2 elimination reaction with $K^+ (CH_3)_3C—O^-$. Explain your reasoning, and give the product of the reaction.

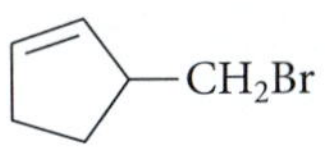

A

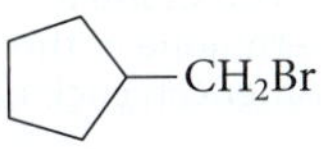

B

17.23 Which of the following compounds should undergo free-radical bromination (Br_2/light or NBS in CCl_4) most rapidly? Explain, and give the product of the faster reaction.

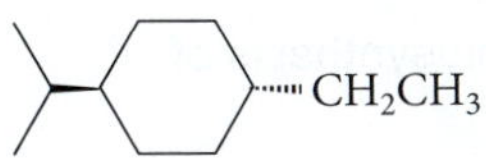

A

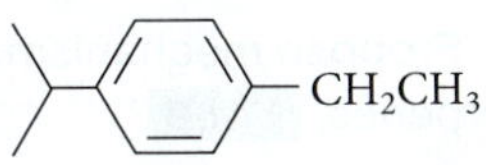

B

17.24 When a Grignard reagent is formed from benzyl chloride ($PhCH_2—Cl$), a significant amount of bibenzyl is formed in the reaction mixture. This type of reaction does not occur significantly in the formation of Grignard reagents from most other alkyl halides. Explain the origin of bibenzyl and why it forms in significant amount.

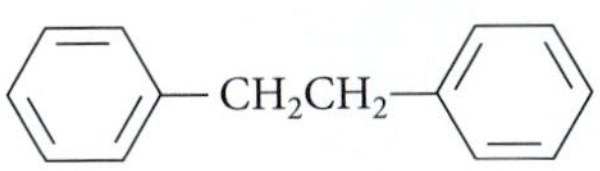

bibenzyl
[(2-phenylethyl)benzene]

- **Complete the reactions discussed in this chapter by providing the missing reactants or products.**

(1) S_N1 reactions of allylic or benzylic alkyl halides **17.1**

(2) S_N2 reactions of allylic or benzylic alkyl halides **17.4**

(3) Bromination with *N*-bromosuccinimide (NBS) **17.2**

(4) Allylic oxidation with MnO_2 **17.5A**

(5) Elimination reactions involving benzylic or allylic hydrogens **17.3B**

(6) Reactions of allylic or benzylic Grignard reagents **17.3A**

(7) Benzylic oxidations **17.5C**

17.25 Give the principal organic product(s) expected when *trans*-2-butene or another compound indicated reacts under the following conditions. Assume one equivalent of each reagent reacts in each case.

(a) *N*-Bromosuccinimide in CCl_4, light

(b) Product(s) of part (a), solvolysis in aqueous acetone

(c) Product(s) of part (a) + Mg in ether

(d) Product(s) of part (c) + D_2O

17.26 Complete each the following reactions by proposing structures for the major organic products or the missing reactant. Explain your reasoning.

(a)

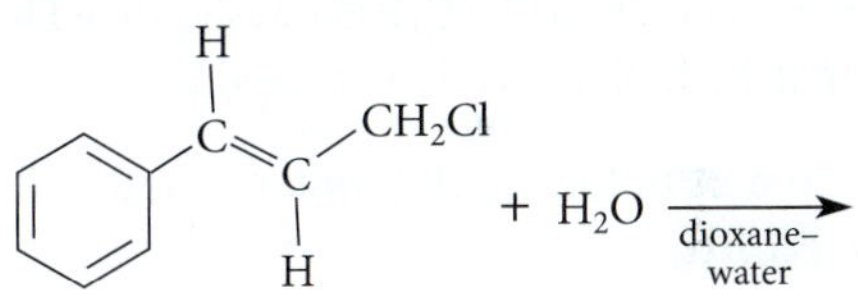

***trans*-cinnamyl chloride**

two alcohols + HCl
(constitutional isomers)

Explain how you could use UV spectroscopy to distinguish between the two products.

(b) $H_2C{=}CH—CH{=}CH—CH_2MgBr + D_2O \longrightarrow$

(c) *trans*-$BrCH_2CH{=}CHCH(CH_3)_2 + Na^+ \ CH_3CH_2S^- \xrightarrow{\text{ethanol}}$

(d) Same as part (c), but with warm ethanol only; no $Na^+ \ CH_3CH_2S^-$

(e) $H_2C{=}CH—CH{=}CH—CH_2MgBr$

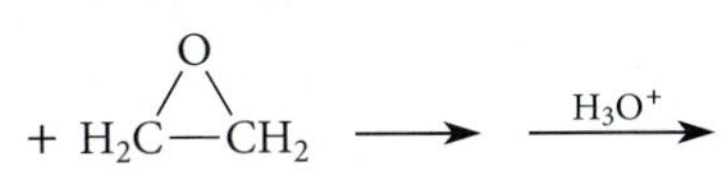

(f)

? + *N*-bromosuccinimide (1 equiv.) $\xrightarrow[CCl_4]{AIBN}$

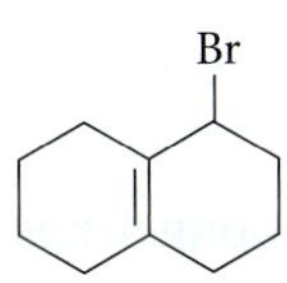

(AIBN is a free-radical initiator; see Eq. 10.9, Sec. 10.1C.)

(g) 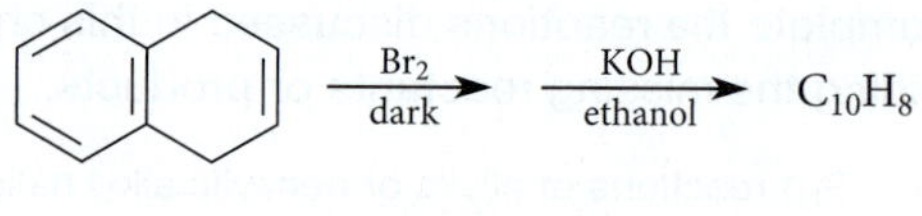

(h) 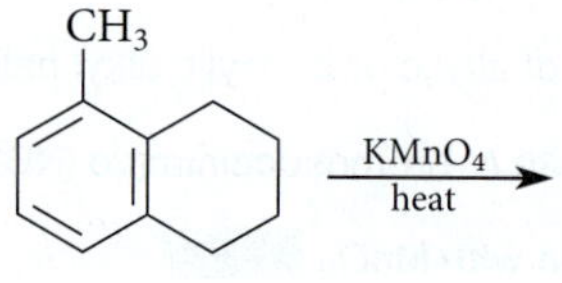

(i) 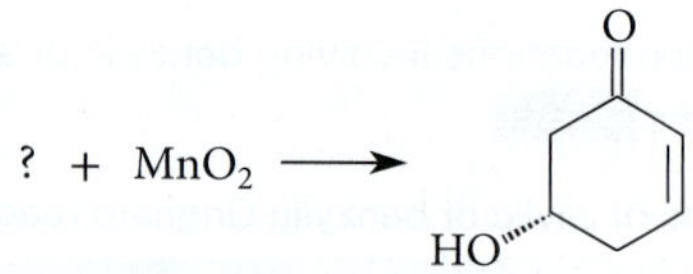

- **Use the mechanism of oxidation by cytochrome P450 to predict the products of an oxidative reaction. 17.5B**

17.27 When a deuterium-substituted cyclohexene (shown here) is subjected to CyP450-promoted oxidation, two different allylic alcohols are formed. Explain. (*Hint:* See Eq. 17.42b, and think about resonance structures for the intermediate.)

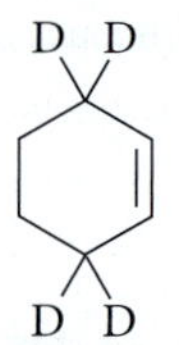

17.28 Draw a variation on the "rebound" hydroxylation mechanism that accounts for the formation of epoxides from some alkenes by CyP450, as shown in **Fig. P17.28**. (*Hint:* Radicals can add to a π bond.)

- **Analyze a structure to determine whether it is a terpene. 17.6A**

17.29 Which of the following compounds, all known in nature, can be classified as terpenes? Show the isoprene skeletons in each terpene.

(a) 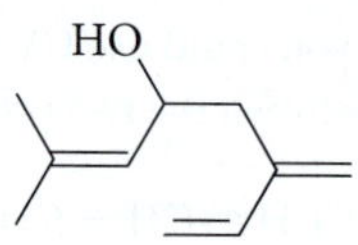

ipsdienol
(one component of the pheromone of the Norwegian spruce beetle)

(b) 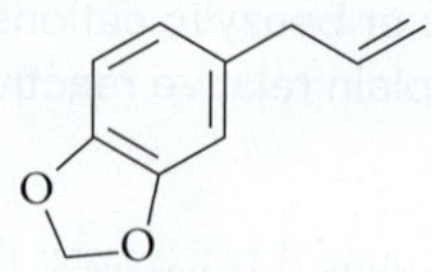

safrole
(oil of sassafras)

(c) 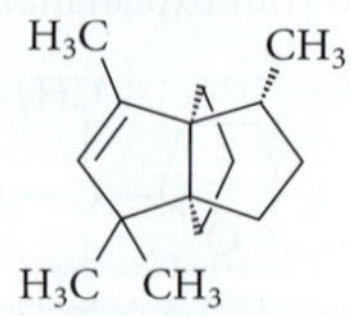

modhephene
(from rayless goldenrod)

(d) 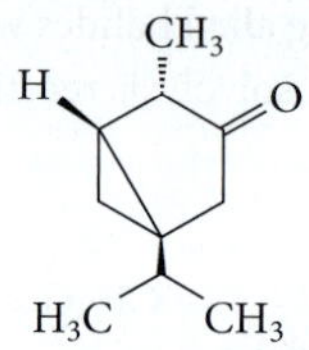

β-thujone
(from yellow cedar)

(e) 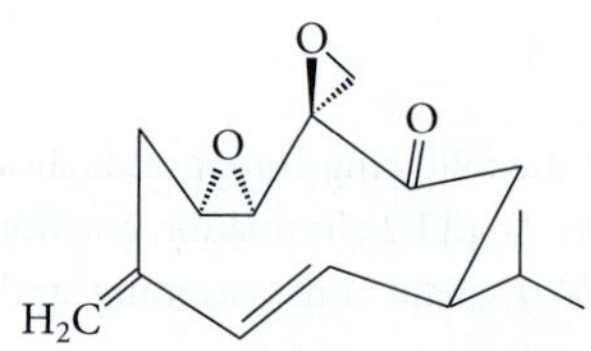

periplanone B
(pheromone of the female American cockroach)

(f) 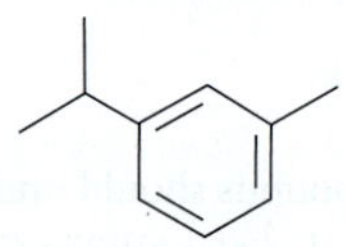

***m*-cymene**

- **Propose mechanisms for biosynthesis of terpenes. 17.6B**

17.30 Propose a biosynthetic pathway for each of the following natural products. Assume acids and bases are present as necessary.

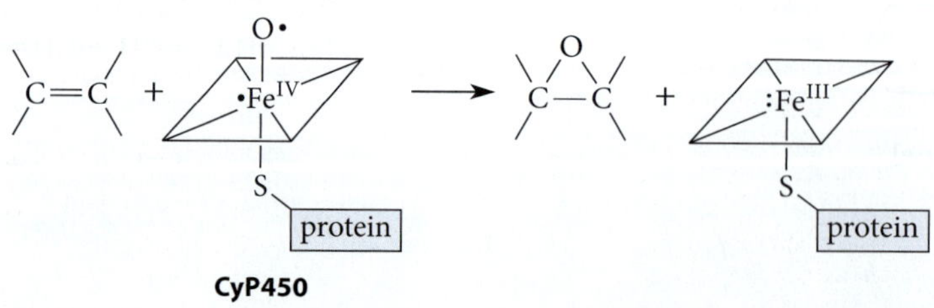

FIGURE P17.28

(a)

(b)

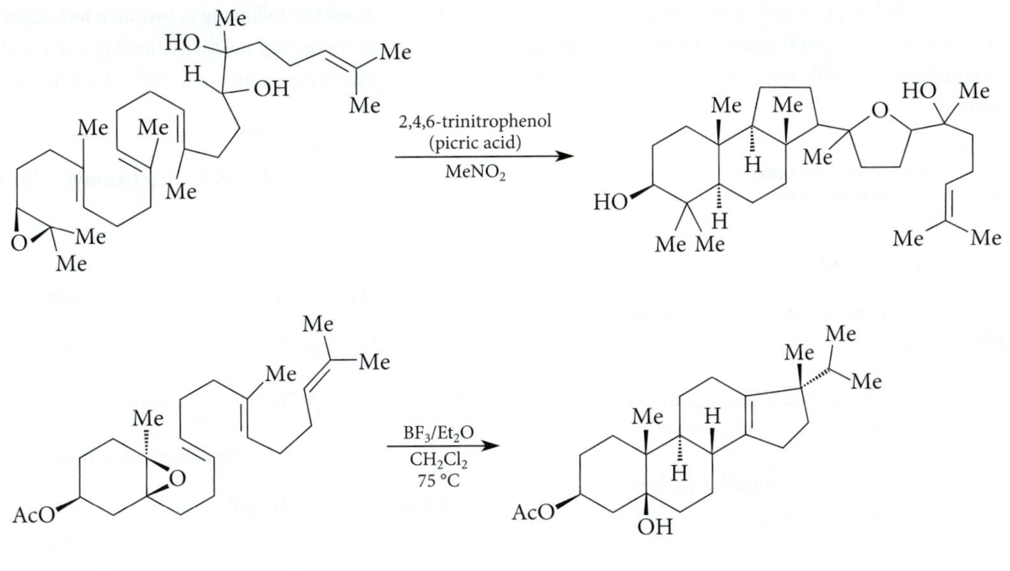

FIGURE P17.31

(a) 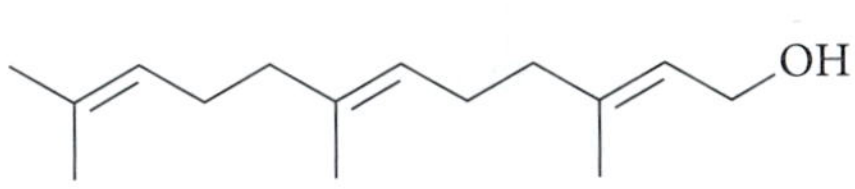

farnesol

(*Hint:* Start with geranyl pyrophosphate, Eq. 17.53, Sec. 17.6B.)

(b) 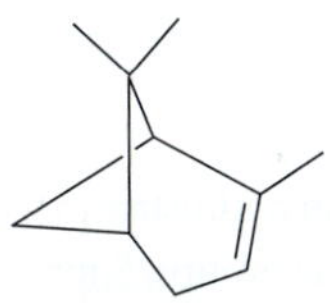

α-pinene

(*Hint:* Start with the carbocation intermediate in Focused Problem 17.17a, Sec. 17.6B.)

- **Use the mechanism of squalene oxide cyclization to propose a mechanism for cyclization of other similar epoxides. 17.6C**

17.31 In an effort to duplicate in the laboratory the natural processes that resemble the cyclization of 2,3-oxidosqualene, chemists in 1970 and 1980 carried out the conversions shown in **Fig. P17.31**. Give a curved-arrow mechanism for each. (For purposes of this problem, let the acid and base catalysts be a general acid A—H and its conjugate base A^-.) (*Hint:* See Eq. 17.59, Sec. 17.6C.)

INTEGRATED PROBLEMS

17.32 Outline a synthesis of each of the following compounds from the indicated starting materials and any other reagents.

(a) Benzyl methyl ether from toluene

(b) 3-Phenyl-1-propanol from toluene

(c) (*Z*)-1,4-Nonadiene from 1-hexyne

(d) 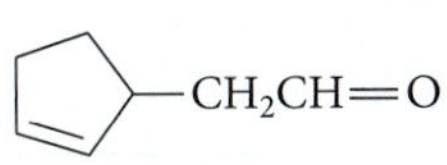 from cyclopentene

(e) 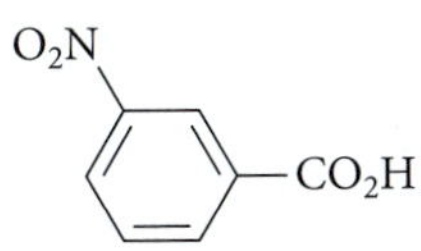 from cumene (isopropylbenzene)

(f) 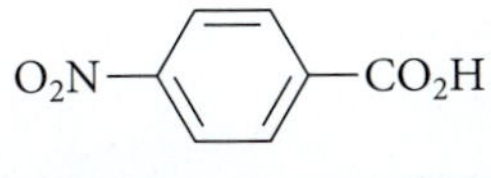 from cumene

(g) 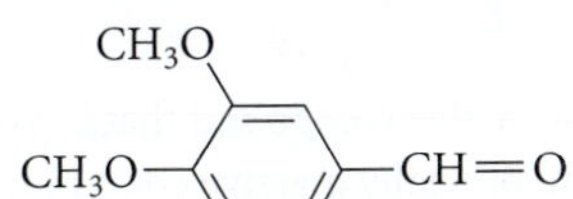

from 1,2-dimethoxy-4-methylbenzene

(h)  from 

17.33 Explain why two products are formed in the first ether synthesis, but only one in the second.

(1) 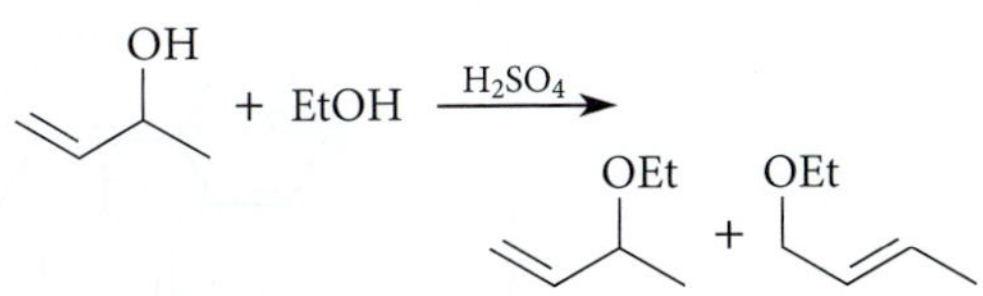

(2) 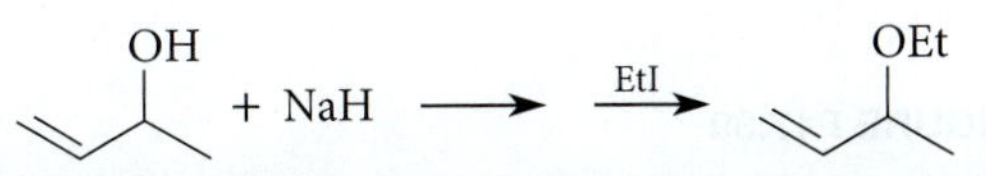

17.34 A student Al Lillich has prepared a pure sample of 3-bromo-1-butene (*A*). Several weeks later he finds that the sample is contaminated with an isomer *B* formed by allylic rearrangement.

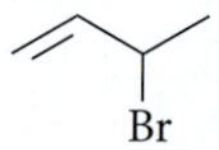

3-bromo-1-butene (compound *A*)

(a) Give the structure of *B*.

(b) Draw a curved-arrow mechanism for the formation of *B* from *A*.

(c) Which should be the major isomer at equilibrium, *A* or *B*? Explain.

17.35 (a) Determine whether zoapatanol, which is used as a fertility-regulating agent in Mexican folk medicine, is a terpene.

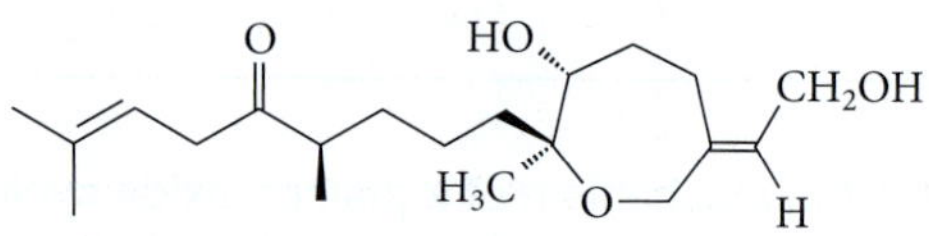

zoapatanol

(b) What product is obtained when zoapatanol is subjected to MnO_2 oxidation?

17.36 Match one or more of these structures with each of the following statements.

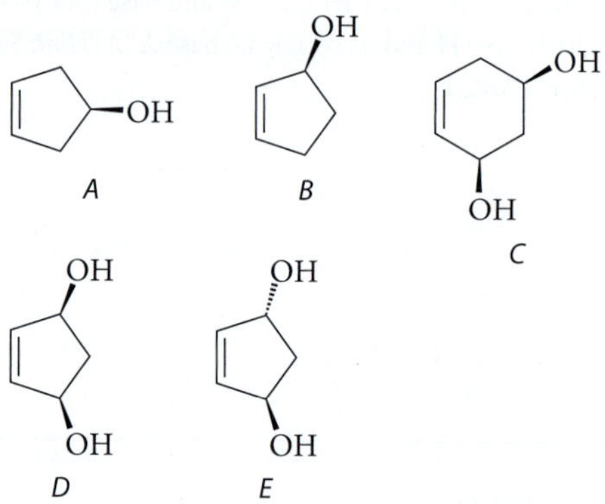

(a) An optically active compound that is oxidized by MnO_2 to an optically inactive compound

(b) An optically active compound that is oxidized by MnO_2 to an optically active compound

(c) An optically inactive compound that is oxidized by MnO_2 to an optically inactive compound

(d) A compound that is not oxidized by MnO_2

17.37 Rank the following substituted *tert*-cumyl chlorides in order of increasing reactivity (least reactive first) in an S_N1 solvolysis reaction in aqueous acetone. Explain your answers.

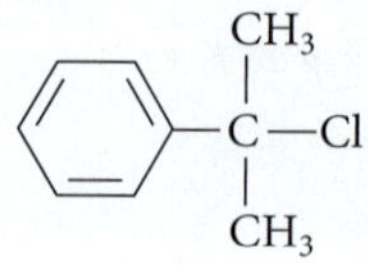

***tert*-cumyl chloride**

(1) *m*-Nitro-*tert*-cumyl chloride

(2) *p*-Methoxy-*tert*-cumyl chloride

(3) *p*-Fluoro-*tert*-cumyl chloride

(4) *p*-Nitro-*tert*-cumyl chloride

17.38 Arrange the following alcohols according to increasing rates of their acid-catalyzed dehydration to alkene (smallest rate first), and explain your reasoning.

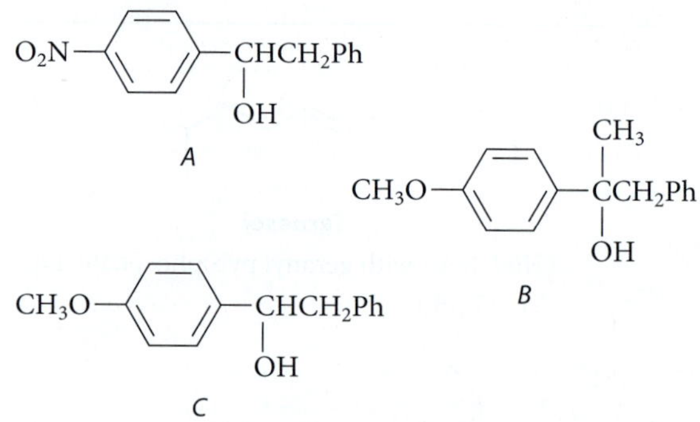

17.39 Terfenadine (**Fig. P17.39**) is an antihistaminic drug that contains two alcohol functional groups. Suppose terfenadine were to undergo acid-catalyzed alcohol dehydration (Sec. 11.2). Which alcohol would dehydrate most rapidly? Why? What would be the dehydration product?

17.40 When benzyl alcohol ($PhCH_2OH$, $\lambda_{max} = 258$ nm, $\varepsilon = 520$) is dissolved in H_2SO_4, a colored solution is obtained that has a different UV spectrum: $\lambda_{max} = 442$ nm, $\varepsilon = 53{,}000$. When this solution is added to cold NaOH, the original spectrum of benzyl alcohol is restored. Suggest a structural basis for these observations.

17.41 Account for each of the following facts with an explanation.

(a) 1,3-Cyclopentadiene is a considerably stronger carbon acid than 1,4-pentadiene, even though the acidic hydrogens in both cases are doubly allylic.

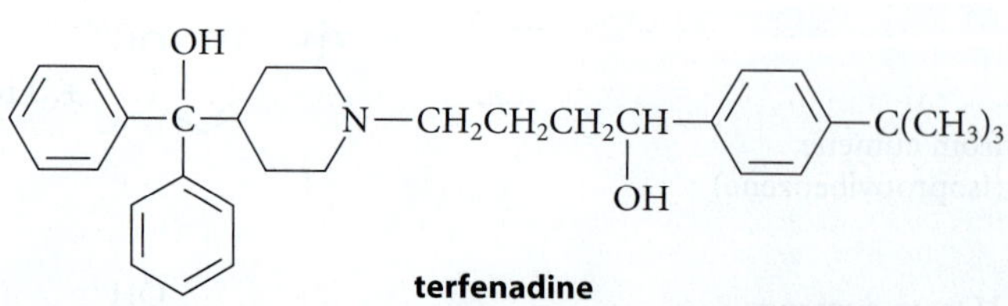

terfenadine

FIGURE P17.39

1,3-cyclopentadiene

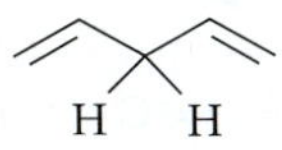

1,4-pentadiene

(b) 3-Bromo-1,4-pentadiene undergoes solvolysis rapidly in protic solvents, but 5-bromo-1,3-cyclopentadiene is virtually inert.

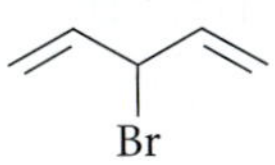

3-bromo-1,4-pentadiene

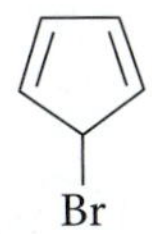

5-bromo-1,3-cyclopentadiene

17.42 Propose a curved-arrow mechanism for the following reaction, and give at least two structural reasons why the equilibrium lies to the right.

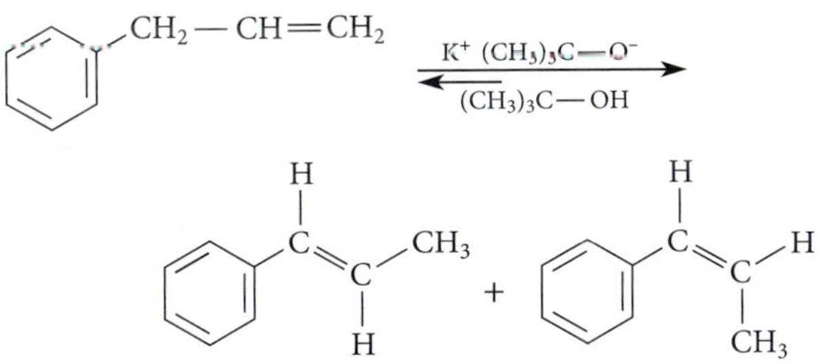

(mostly)

17.43 Propose a curved-arrow mechanism for the following reaction. Explain why the equilibrium lies to the right.

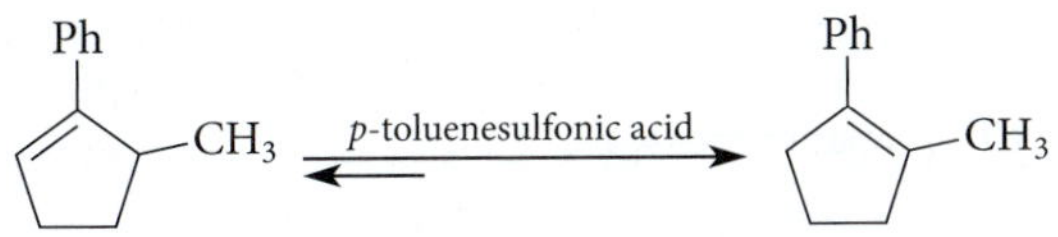

17.44 (a) When 2-hexyne is treated with certain very strong bases, it undergoes the reaction given in **Fig. P17.44**, an example of the "acetylene zipper" reaction. Give a curved-arrow mechanism for this reaction, and explain why it is irreversible.

(b) What product(s) would be expected in the same reaction of 3-methyl-4-octyne? Explain.

17.45 A hydrocarbon *A*, C_9H_{12}, is treated with *N*-bromosuccinimide in CCl_4 in the presence of peroxides to give a compound *B*, $C_9H_{11}Br$. Compound *B* undergoes rapid solvolysis in aqueous acetone to give an alcohol *C*, $C_9H_{12}O$, which cannot be oxidized with CrO_3 in pyridine. Vigorous oxidation of compound *A* with hot chromic acid gives benzoic acid, $PhCO_2H$. Identify compounds *A*–*C*.

17.46 A hydrocarbon *A*, C_9H_{10}, is treated with *N*-bromosuccinimide to give a single monobromo compound *B*. When *B* is dissolved in aqueous acetone it reacts to give two nonisomeric compounds: *C* and *D*. Catalytic hydrogenation of *D* gives back *A*, and *C* can be separated into enantiomers. When optically active *C* is oxidized with CrO_3 and pyridine, an optically inactive ketone *E* is obtained. Vigorous oxidation of *A* with $KMnO_4$ affords phthalic acid (structure in Focused Problem 17.15, Sec. 17.5C). Propose structures for compounds *A* through *E*, and explain your reasoning.

17.47 Propose a curved-arrow mechanism for each of the reactions given in **Fig. P17.47**.

17.48 Consider the relative rates of the two solvolysis reactions in acetic acid solvent shown in **Fig. P17.48**.

(a) Suggest a reason that compound *A* undergoes solvolysis *much* faster than compound *B*. (*Hint:* Consider anchimeric assistance; see Sec. 12.8A.)

(b) Account for the retention of stereochemistry observed in reaction *A* with a mechanism. (*Hint:* See Secs. 12.8A and C.)

17.49 The amount of anti addition in the chlorination of alkenes varies with the structure of the alkene, as shown in the following table. (See Sec. 7.8C.)

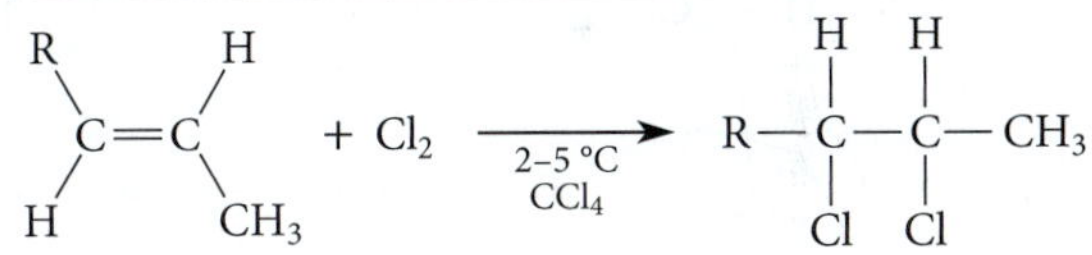

Structure of R—	Percent anti addition
H_3C—	99
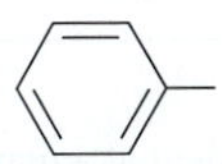	88
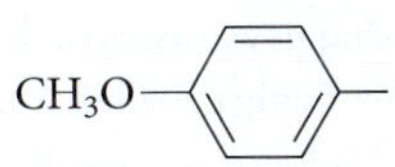	63

Suggest a reason for the variation in the stereochemistry of addition as the alkene structure is varied. (*Hint:* What types of reactive intermediate(s) could account for the stereochemical observations?)

$$H_3C—C{\equiv}C—CH_2CH_2CH_3 \; + \; B{:}^- \longrightarrow {}^-{:}C{\equiv}C—CH_2CH_2CH_2CH_3 \; + \; BH$$

$$(B{:}^- = {:}\ddot{N}H—CH_2CH_2CH_2—\ddot{N}H_2) \qquad (pK_a = 35)$$

FIGURE P17.44

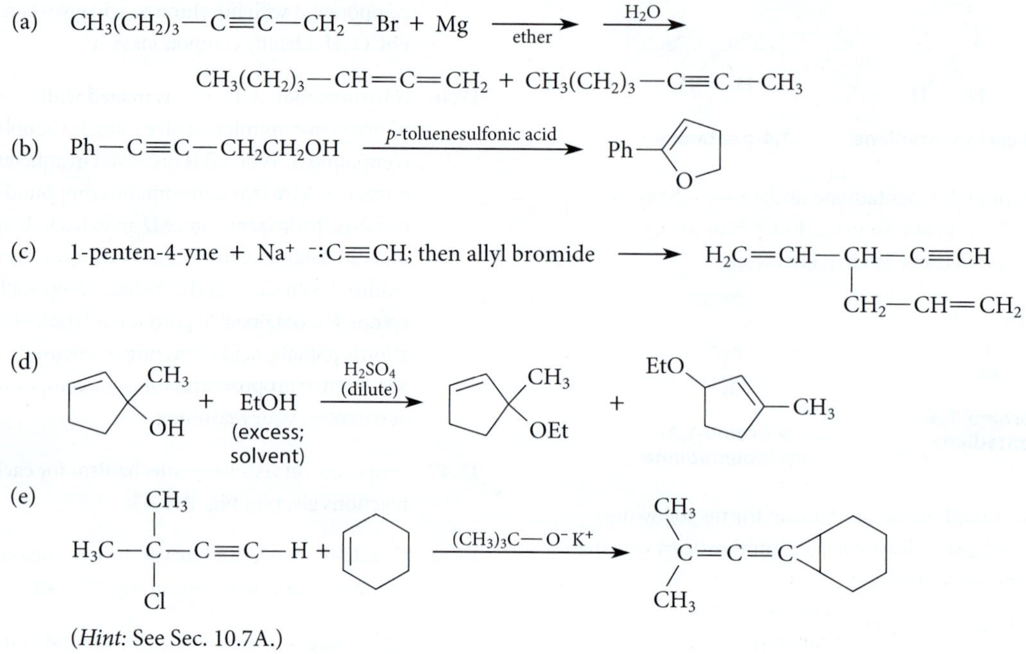

FIGURE P17.47

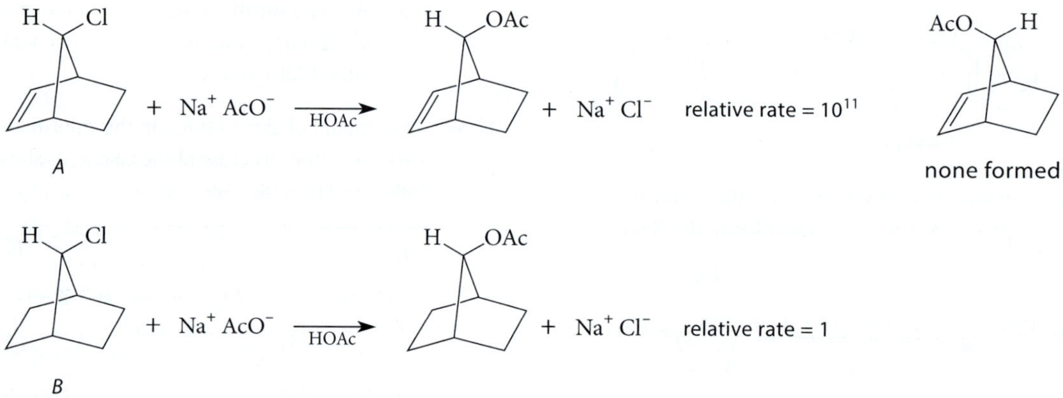

FIGURE P17.48

17.50 In the late 1970s, a graduate student at a major West Coast university began synthesizing new classes of drugs and testing them on himself. After being expelled from the university, he began making his living by illegally synthesizing and selling to heroin addicts compound *B*, a synthetic analog of meperidine (Demerol). (See **Fig. P17.50.**) After shortening his synthetic procedure and self-injecting his product, he developed severe symptoms of Parkinson's disease, as did several of his young clients; one person died. Chemists found that his compound *B* contained two by-products, alcohol *C* and another compound MPTP ($C_{12}H_{15}N$), which, when independently prepared and injected into animals, caused the same symptoms. (Ironically, this has been one of the most significant advances in Parkinsonism research.)

Given the illicit chemistry outlined in Fig. P17.50, provide the structure of compound *A*, suggest a structure for MPTP, and show how all products are formed.

17.51 When 1-buten-3-yne undergoes HCl addition, two compounds, *A* and *B*, are formed in a ratio of 2.2:1. Neither compound shows a C≡C stretching absorption in its IR spectrum. Compound *B* reacts with maleic anhydride to give compound *C*, and compound *A* undergoes allylic rearrangement to compound *B* on heating. Propose structures for compounds *A* and *B*, and explain your reasoning.

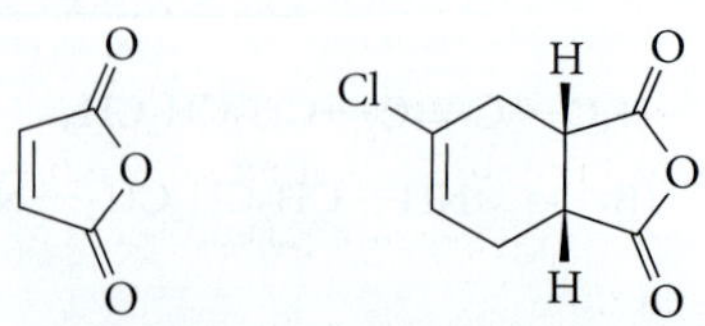

maleic anhydride *C*

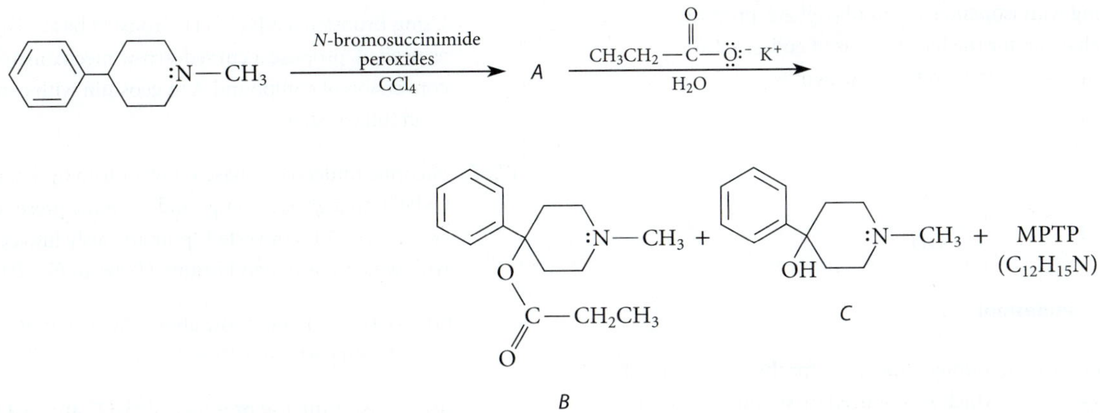

FIGURE P17.50

17.52 (a) Assuming the presence of acids (AH) and bases (A^-) as needed, give a curved-arrow mechanism for the isomerization of isopentenyl pyrophosphate into dimethylallyl pyrophosphate (Eq. 17.50, Sec. 17.6B).

(b) In some organisms, geranyl pyrophosphate (structure in Eq. 17.53, Sec. 17.6B) is converted by hydrolysis into one of two enantiomeric *tertiary* alcohols:

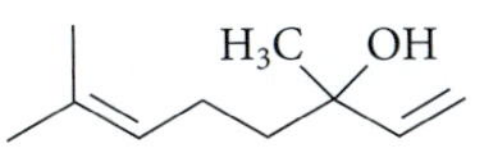

linalool

(3*R*)-(−)-linalool
(also called licareol;
has a lily-of-the-valley scent)

(3*S*)-(+)-linalool
(also called coriandrol;
has a musty herbaceous scent)

Assuming the presence of acids (AH), bases (A^-), and water as needed, give a curved-arrow mechanism for this transformation. (Ignore stereochemistry.)

(c) Each of the stereoisomeric linalools in part (b) is converted into an epoxide *A* by CyP450, and the epoxide spontaneously forms the following compound *B*. Neglecting stereochemistry, suggest a structure for *A*; and, assuming the presence of H_3O^+ and H_2O, provide a curved-arrow mechanism for the conversion of *A* into *B*.

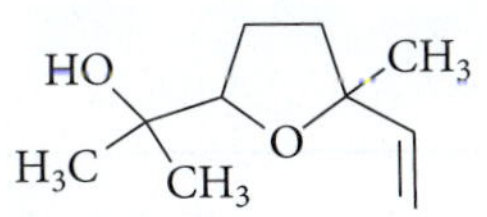

B

17.53 Consider the hydrolysis of geranyl pyrophosphate to geraniol shown in Eq. 17.54 (Sec. 17.6B).

(a) What isotopically substituted derivative of geranyl pyrophosphate could be used to establish the assertion that this hydrolysis proceeds by opposite-side substitution?

(b) Given that the isotopically substituted derivative you proposed in (a) can be prepared in enantiomerically pure form with a known absolute configuration, the absolute stereochemistry of the product geraniol must be determined to establish the stereochemistry of substitution. Explain how the alcohol dehydrogenase-catalyzed oxidation of the isotopically substituted geraniol product by NAD^+ (Sec. 11.9B) can be used to establish its stereochemistry.

17.54 (a) Geranyl pyrophosphate is methylated by *S*-adenosylmethionine (SAM; Sec. 12.7B) in the reaction shown in **Fig. P17.54**, which is catalyzed by the enzyme geranyl pyrophosphate *C*-methyl transferase. Assuming that the enzyme provides acidic and basic groups as needed, provide a curved-arrow mechanism for this transformation. (*Hint:* A carbocation intermediate is involved.)

(b) The product of the reaction in part (a) is converted into 2-methylisoborneol in an electrophilic addition reaction catalyzed by the enzyme 2-methylisoborneol synthase. Assuming the presence of acids, bases, and water as needed, suggest a curved-arrow mechanism for the reaction. (*Hint:* Before starting, relate the carbons of the product to those of the starting material to establish the connections that must be made. Then provide a plausible mechanism for those connections that involves carbocation intermediates.)

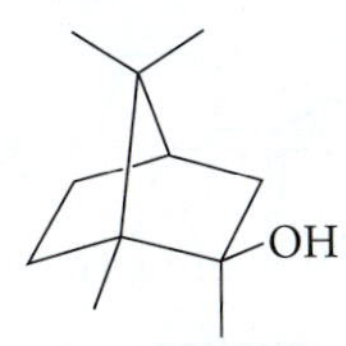

2-methylisoborneol

17.55 Starting with isopentenyl pyrophosphate, propose a mechanism for the biosynthesis of eudesmol, a sesquiterpene obtained from eucalyptus.

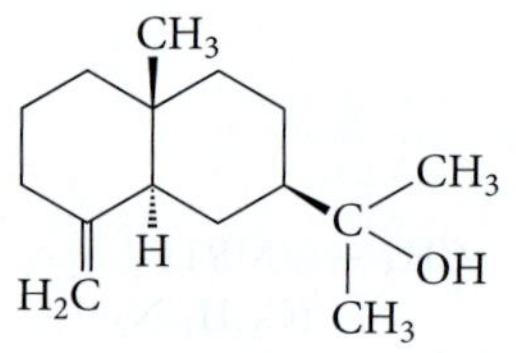

eudesmol

17.56 One of the compounds responsible for the odor of moist soil is geosmin, which is produced by streptomycetes in soil from farnesyl pyrophosphate, as shown in **Fig. P17.56.** Compounds *A* and *B* are intermediates in the biosynthesis. Using Brønsted acids (^{+}BH), Brønsted bases (B:), and water as needed, propose a curved-arrow mechanism for the conversion of compound *A* to geosmin with compound *B* as an intermediate.

17.57 Nicotine undergoes phase I metabolism by a reaction with CyP450 that gives a compound *A*. In the presence of water, compound *A* is converted spontaneously into compound *B*, with which it is in equilibrium. (Refer to **Fig. P17.57.**)

(a) Use what you know about the reactions of CyP450 to suggest a structure for compound *A*.

(b) Assuming the presence of H_3O^+ and H_2O, provide a curved-arrow mechanism for the conversion of *A* into *B*.

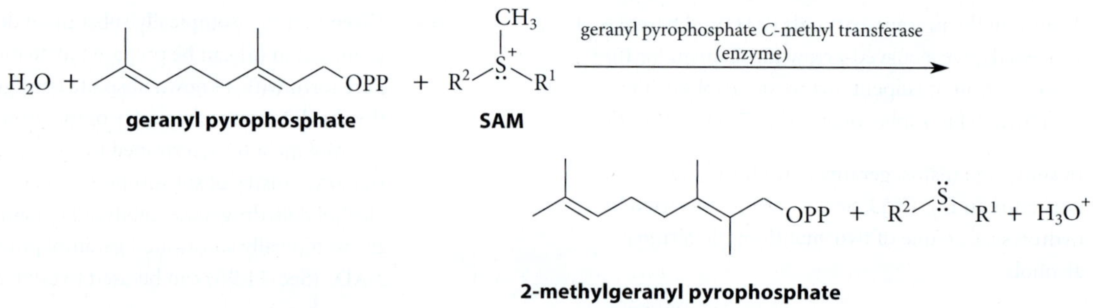

FIGURE P17.54

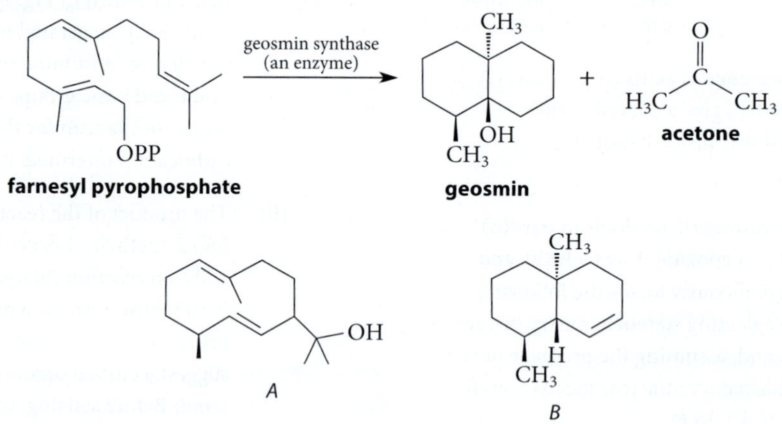

FIGURE P17.56

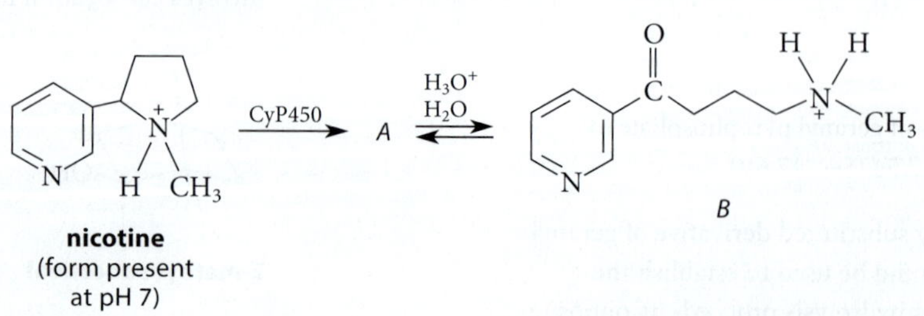

FIGURE P17.57

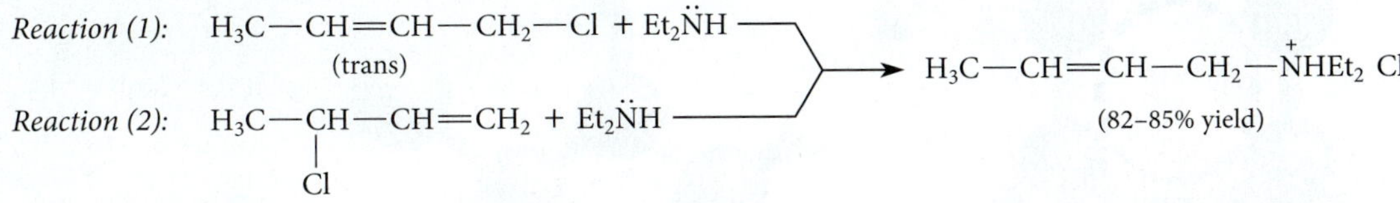

FIGURE P17.60

17.58 (a) Triphenylmethane [structure in part (b)] has a pK_a of 31.5 and, although an alkane, it is almost as acidic as a 1-alkyne. (Most alkanes have a pK_a ~ 55.) By considering the structure of its conjugate base, suggest a reason why triphenylmethane is such a strong hydrocarbon acid.

(b) Fluoradene is structurally very similar to triphenylmethane, except that the three aromatic rings are "tied together" with single bonds. Fluoradene has a pK_a = 11. Suggest a reason why fluoradene is much more acidic than triphenylmethane.

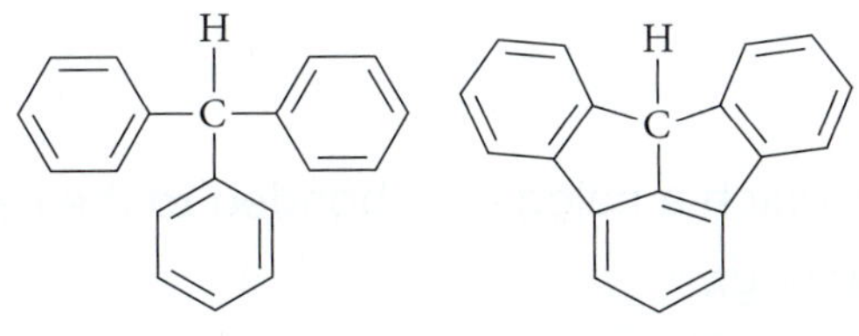

triphenylmethane p$K_a \approx 31.5$ **fluoradene** p$K_a \approx 11$

17.59 Around 1900, Moses Gomberg, a pioneer in free-radical chemistry, prepared the triphenylmethyl radical, $Ph_3C\cdot$, sometimes called the *trityl radical* (trityl = *tri*phenylme*thyl*).

(a) Explain why the trityl radical is an unusually stable radical.

(b) The trityl radical is known to exist in equilibrium with a dimer that, for many years, was assumed to be hexaphenylethane, $Ph_3C{-}CPh_3$. Show how hexaphenylethane could be formed from the trityl radical.

(c) In 1968 the structure of this dimer was investigated using modern methods and found not to be hexaphenylethane, but rather the following compound.

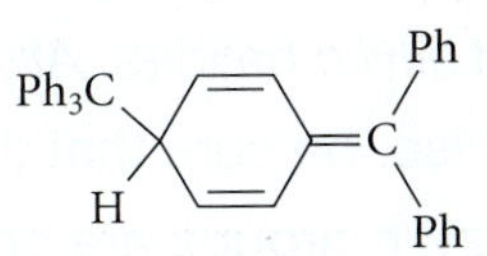

dimer of the trityl radical

Using the fishhook notation, show how this compound is formed from two trityl radicals, and explain why this compound is formed instead of hexaphenylethane. (*Hint:* Can you think of any reason why hexaphenylethane might be unstable?)

17.60 (a) For each of the two reactions shown in **Fig. P17.60**, suggest a mechanism that is consistent with all of the experimental facts given.

Experimental observations:

(1) Both reactions conform to the following rate law, although the rate constants for each reaction are different.

rate = k[alkyl halide]$Et_2\ddot{N}H$]

(2) The alkyl chloride starting materials do *not* interconvert under the reaction conditions.

(3) The following compound, prepared separately, is *not* converted into the observed product under the reaction conditions.

$$H_2C{=}CH{-}\underset{}{\overset{CH_3}{\overset{|}{C}H}}{-}\overset{+}{\ddot{N}}HEt_2 \quad Cl^-$$

In particular, explain the importance of facts (2) and (3) in understanding the mechanism.

(b) The mechanism of reaction 2 is called the S_N2' mechanism. Suggest a reason why this reaction occurs by the S_N2' mechanism and reaction 1 does not.

17.61 When the following compound is treated with a strong Brønsted acid, a stable carbocation *A* is formed.

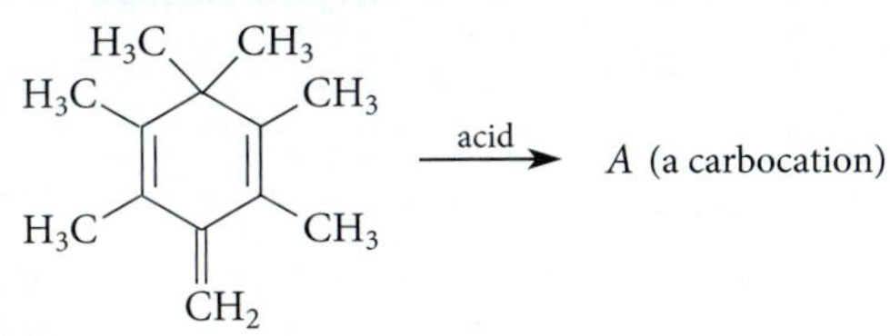

(a) Propose a structure for carbocation *A*, and draw its resonance structures.

(b) The proton NMR spectrum of carbocation *A* at –10 °C consists of four singlets at δ 1.54, δ 2.36, δ 2.63, and δ 2.82 (relative integral 2:2:2:1). Explain why the structure of *A* is consistent with this spectrum by assigning each resonance.

(c) Explain why the NMR spectrum of *A* becomes a single broad line when the temperature is raised to 113 °C. (*Hints:* [1] See Sec. 14.8; [2] this was called the "methyl walk.")

18 THE CHEMISTRY OF ARYL HALIDES, VINYLIC HALIDES, AND PHENOLS

Transition-Metal Catalysis

An **aryl halide** is a compound in which a halogen is bonded to the carbon of a benzene ring (or another aromatic ring).

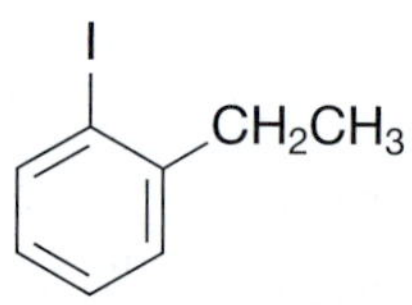

1-ethyl-2-iodobenzene
(an aryl iodide)

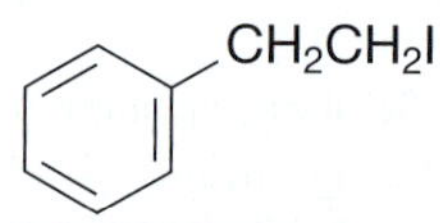

not an aryl halide;
the halogen is not attached directly
to benzene ring

(18.1)

In a **vinylic halide**, a halogen is bonded to a carbon of a double bond.

$$H_2C{=}CH{-}Cl$$

vinyl chloride
(chloroethylene)
a vinylic chloride

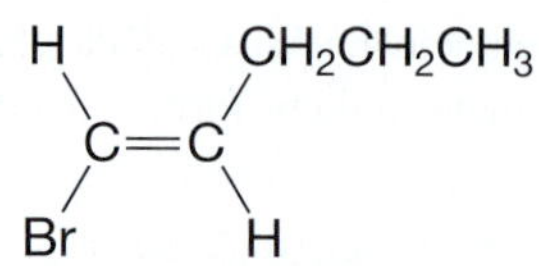

(*E*)-1-bromo-1-pentene
a vinylic bromide

(18.2)

Be sure to differentiate carefully between *vinylic* and *allylic* halides. *Allylic* groups are on a carbon *adjacent* to the double bond. Likewise, be sure that the distinction between *aryl* and *benzylic* halides is clear. *Benzylic* groups are on a carbon *adjacent* to an aromatic ring.

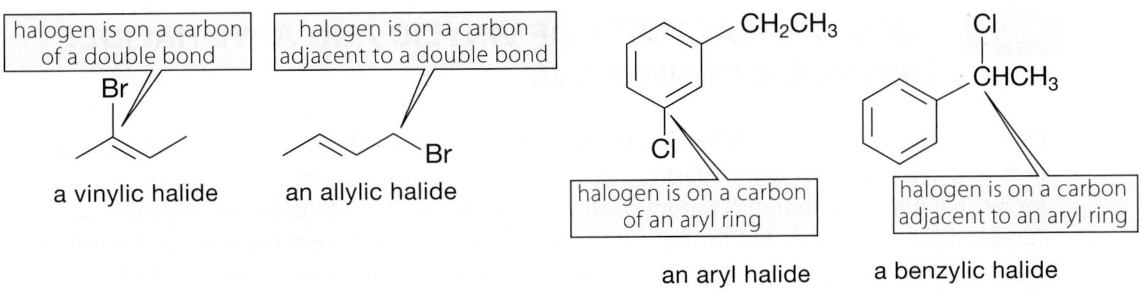

(18.3)

The reactivity of aryl and vinylic halides is quite different from that of ordinary alkyl halides. In fact, one of the major points of this chapter is that aryl halides do *not* undergo nucleophilic substitution reactions by the S_N2 or S_N1 mechanisms.

In a **phenol**, a hydroxy (OH) group is bonded to an aromatic ring. As the following structures illustrate, *phenol* is also the name given to the parent compound, and a number of phenols have traditional names that the IUPAC recognizes as *parent structures* (Sec. 16.1).

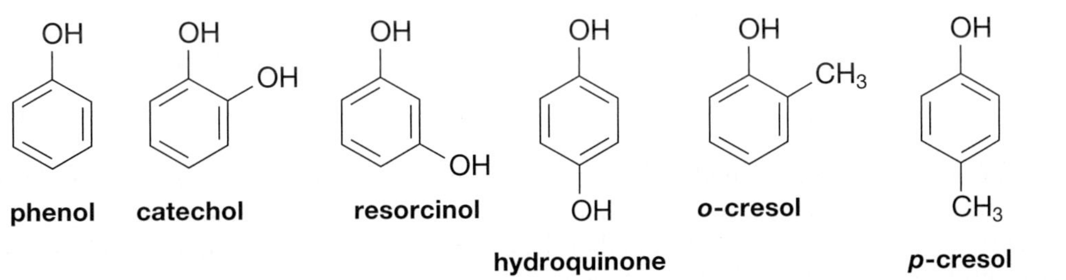

(18.4)

Although phenols and alcohols have some reactions in common, there are also important differences in the chemical behavior of these two functional groups.

A relatively recent field of organic chemistry involves the use of transition-metal catalysts in organic reactions, particularly in reactions that involve the formation of carbon–carbon bonds. Certain transition-metal catalysts dramatically increase the reactivity of aryl and vinylic halides in substitution reactions, and this heightened reactivity is the vehicle through which we present some of the basic principles involved in transition-metal catalysis.

The nomenclature and spectroscopy of aryl halides and phenols were discussed in Secs. 16.1 and 16.3, respectively. The nomenclature of vinylic halides follows the principles of alkene nomenclature (Sec. 4.2A), and the spectroscopy of vinylic halides, except for minor differences due to the halogen, is also similar to that of alkenes.

18.1 LACK OF REACTIVITY OF VINYLIC AND ARYL HALIDES UNDER S_N2 CONDITIONS

One of the most important differences between vinylic or aryl halides and alkyl halides is their reactivity in nucleophilic substitution reactions. The two most important mechanisms for nucleophilic substitution reactions of alkyl halides are the S_N2 (bimolecular opposite-side substitution) mechanism and the S_N1 (unimolecular carbocation) mechanism (Secs. 9.4 and 9.6). What happens to vinylic and aryl halides under the conditions used for S_N1 or S_N2 reactions of alkyl halides?

Consider first the S_N2 reaction. One of the most significant contrasts between vinylic or aryl halides and alkyl halides is that simple vinylic and aryl halides are inert under S_N2 conditions. For example, when ethyl bromide is allowed to react with Na^+ EtO^- in EtOH solvent at 55 °C, the following S_N2 reaction proceeds to completion in about an hour:

$$CH_3CH_2{-}Br + Na^+\ {:}\ddot{O}Et \xrightarrow[\text{EtOH}]{55\ ^\circ C} CH_3CH_2{-}\ddot{O}Et + Na^+\ Br^- \quad (18.5a)$$

Yet when vinyl bromide or bromobenzene is subjected to the same conditions, nothing happens!

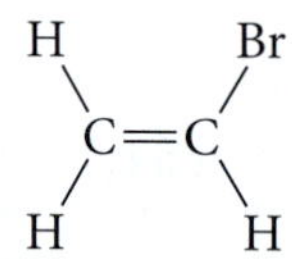 and 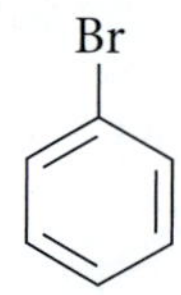 are inert to Na^+EtO^- in EtOH at 55 °C

(18.5b)

Why don't vinylic halides undergo S_N2 reactions? In Sec. 9.4C (see Fig. 9.2), we showed that in the transition state of an S_N2 reaction of an alkyl halide, the carbon undergoing substitution has a 2p orbital to which the nucleophile and the leaving group are partially bonded, and is therefore sp^2-hybridized. In other words, this carbon rehybridizes from sp^3 in the alkyl halide to sp^2 in the transition state. The carbon undergoing substitution in a vinylic halide is sp^2-hybridized in the starting material; it contains a 2p orbital involved in the double bond. If the S_N2 reaction results in the development of a second 2p orbital at this carbon, then this carbon must become sp-hybridized. Therefore, an S_N2 reaction at a vinylic carbon involves rehybridization from sp^2 in the vinylic halide to sp in the transition state.

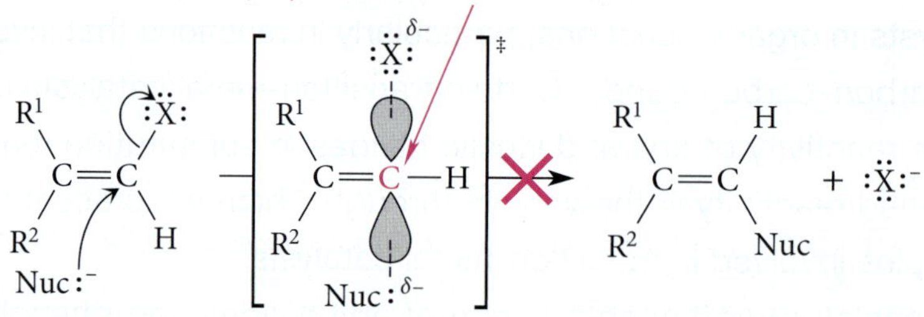

(18.6)

The sp hybridization state has such high energy (Sec. 4.5C) that conversion of an sp^2-hybridized carbon into an sp-hybridized carbon requires about 21 kJ mol^{-1} (5 kcal mol^{-1}) more energy than is required for an sp^3-to-sp^2 hybridization change. The relatively high energy of the transition state caused by sp hybridization reduces the rate of S_N2 reactions of vinylic halides by almost four orders of magnitude (by Eq. 9.19a, Sec. 9.3B). This means that, under the conditions in which the S_N2 reaction of ethyl bromide takes an hour, transition-state hybridization alone would cause the same reaction of vinyl bromide to take almost 200 days.

Rehybridization, however, is not the only reason that vinylic halides are unreactive in the S_N2 reaction. A second reason is that the C=C–halogen bond angle changes from about 120° in the vinylic halide to 90° in the sp-hybridized transition state. This brings the leaving group and

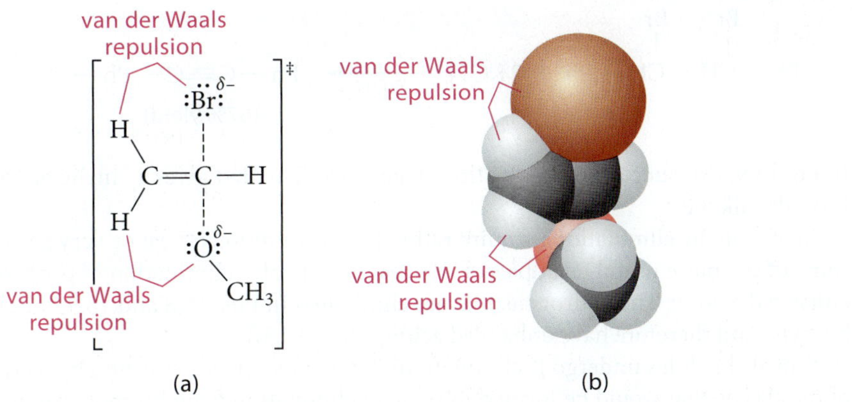

FIGURE 18.1 Van der Waals repulsions in the transition state of the S_N2 reaction of vinyl bromide with methoxide, illustrated with (a) Lewis structures and (b) space-filling models. These van der Waals repulsions raise the energy of the transition state and decrease the rate of the reaction.

the nucleophile (Nuc:⁻ in Eq. 18.6) very close to the alkene substituents on the adjacent carbon. This arrangement results in significant van der Waals repulsions (a steric effect) of both the nucleophile and the leaving group with the groups on the other vinylic carbon. This is shown for the S_N2 reaction of vinyl bromide in **Fig. 18.1**. When the groups on the other vinylic carbon are larger than hydrogen, the repulsions are even greater. These repulsions raise the energy of the transition state and decrease the reaction rate even further.

In summary, both hybridization and van der Waals repulsions (steric effects) within the transition state retard the S_N2 reactions of vinylic halides to such an extent that they do not occur.

S_N2 reactions of aryl halides have the same problems as those of vinylic halides and two other problems as well. First, an opposite-side approach to the carbon of the carbon–halogen bond would place the nucleophile on a path that goes through the plane of the benzene ring—an obvious impossibility. Furthermore, because the carbon at which substitution occurs would have to undergo stereochemical inversion, the reaction would necessarily yield a benzene derivative containing a twisted and highly strained double bond.

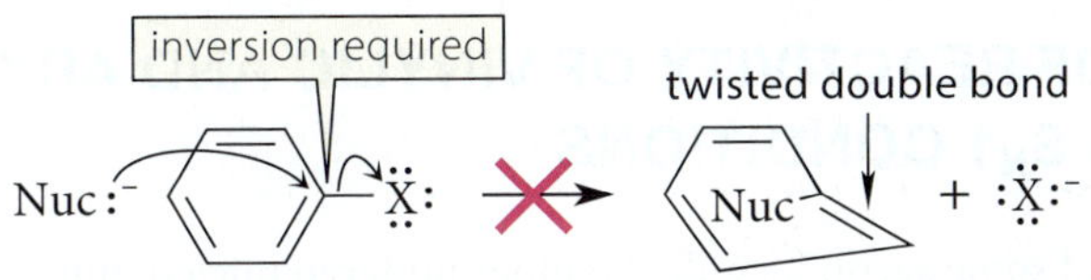

(18.7)

If the impossibility of this result is unclear, try to build a model of the product—but don't break your models!

Focused Problem

18.1 Within each set, rank the compounds in order of increasing rates of their S_N2 reactions. Explain your reasoning.

(a) Benzyl bromide, (3-bromopropyl)benzene, *p*-bromotoluene

(b) 1-Bromocyclohexene, bromocyclohexane, 1-(bromomethyl)cyclohexene

18.2 ELIMINATION REACTIONS OF VINYLIC HALIDES

Although S_N2 reactions of vinylic halides are unknown, base-promoted β-elimination reactions of vinylic halides do occur and can be useful in the synthesis of alkynes.

$$\underset{(E \text{ or } Z)}{\text{Ph—CH=CH—Br}} + \text{KOH} \xrightarrow[200\ °\text{C}]{} \text{Ph—C≡C—H} + \text{K}^+\ \text{Br}^- + \text{H}_2\text{O} \quad (18.8)$$

phenylacetylene (distills from the reaction mixture; 67% yield)

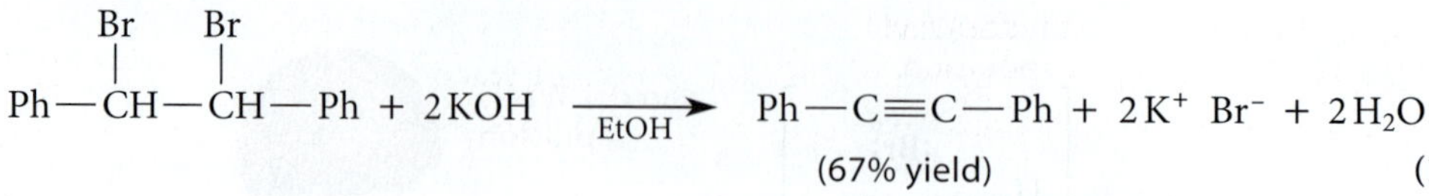

(18.9)

In Eq. 18.9, two successive eliminations take place. The first gives a vinylic halide, and the second gives the alkyne.

Many vinylic eliminations require rather harsh conditions (heat or very strong bases), and some of the more useful examples of this reaction involve elimination of β-hydrogens with enhanced acidity. The hydrogens that are eliminated in Eqs. 18.8 and 18.9, for example, are benzylic and therefore have enhanced acidity (Sec. 17.3B).

Can aryl halides undergo β-elimination? Try to answer this question by constructing a model of the alkyne that would be formed in such an elimination from bromobenzene. (This possibility is explored in Problem 18.88.)

Focused Problem

18.2 Arrange the following compounds according to increasing rate of elimination with NaOEt in EtOH. What is the product in each case? (*Hint:* Anti elimination is faster than syn elimination in vinylic eliminations as it is in ordinary E2 eliminations; Sec. 9.5E.)

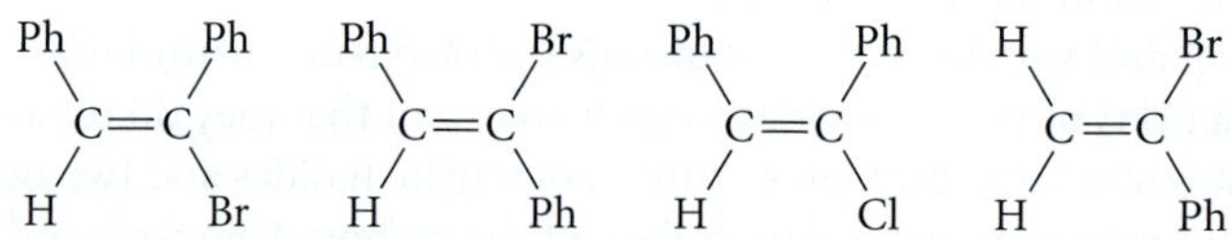

A *B* *C* *D*

18.3 LACK OF REACTIVITY OF VINYLIC AND ARYL HALIDES UNDER S_N1 CONDITIONS

Recall that tertiary and some secondary alkyl halides undergo nucleophilic substitution and elimination reactions by the S_N1 and E1 mechanisms (Sec. 9.6). For example, *tert*-butyl bromide undergoes a rapid solvolysis in ethanol to give both substitution and elimination products.

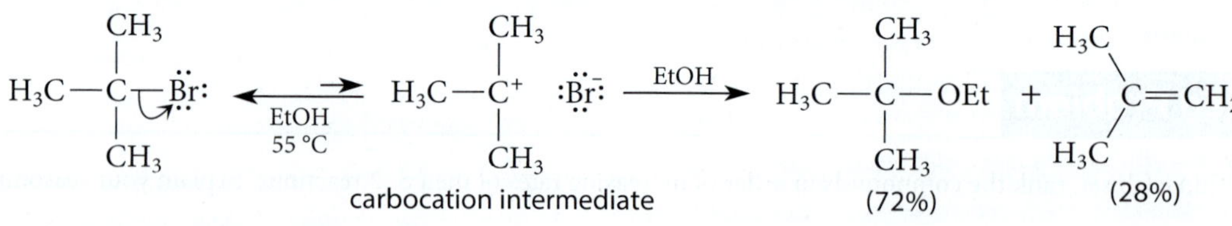

(18.10)

Vinylic and aryl halides, however, are virtually inert to the conditions that promote S_N1 or E1 reactions of alkyl halides. Certain vinylic halides can be forced to react by the S_N1–E1 mechanism under extreme conditions, but such reactions are relatively uncommon.

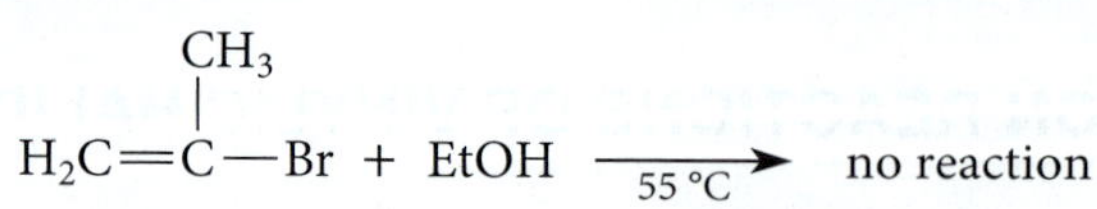

(18.11)

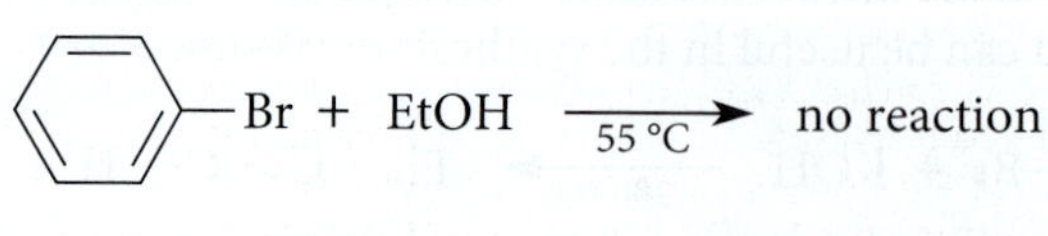

(18.12)

To understand why vinylic and aryl halides are inert under S_N1 conditions, consider what would happen if they *were* to undergo the S_N1 reaction. If a vinylic halide undergoes an S_N1 reaction, it must ionize to form a *vinylic cation.*

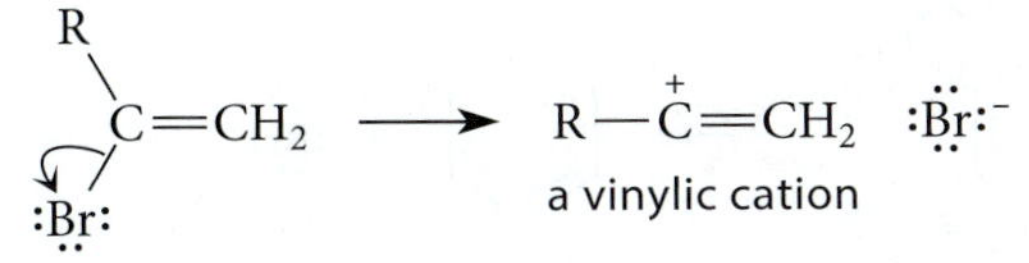

(18.13)

Vinylic cations were also encountered as reactive intermediates in the addition of HBr to alkynes (Display 4.62, Sec. 4.9) and in the hydration of alkynes (Display 5.27a, Sec. 5.5). **Figure 18.2a** shows again an orbital diagram of a vinylic cation. The geometry at the electron-deficient, positively charged carbon is *linear*, so this carbon is sp-hybridized. The vacant 2p orbital is *not* conjugated with the π-electron system of the double bond; in order to be conjugated, it would have to be coplanar with the double-bond π system. Vinylic cations are considerably less stable than alkyl carbocations because their sp hybridization has a higher energy than the sp^2 hybridization of alkyl cations (the same reason that alkynes are less stable than isomeric dienes; Sec. 4.5C) and because the electron-withdrawing polar effect of the double bond discourages formation of positive charge at a vinylic carbon. Therefore, one reason that vinylic halides do not undergo the S_N1 reaction is the *instability of the vinylic cations* that would necessarily be involved as reactive intermediates.

The second reason that vinylic halides do not undergo the S_N1 reaction is that carbon–halogen bonds are stronger in vinylic halides than they are in alkyl halides. A vinylic carbon–halogen bond involves an sp^2 carbon orbital, whereas an alkyl carbon–halogen bond involves an sp^3 carbon orbital. So, a vinylic carbon–halogen bond has more s character. Recall that bonds with more

The formation of vinylic cations in alkyne hydration is much less costly energetically than it is in S_N1 solvolysis. The reason is that the alkyne starting material is itself sp-hybridized, so the energy cost of rehybridization is not an issue. Furthermore, one of the carbons of the alkyne—the carbon that is protonated—goes from sp to sp^2 hybridization, which is energetically favorable.

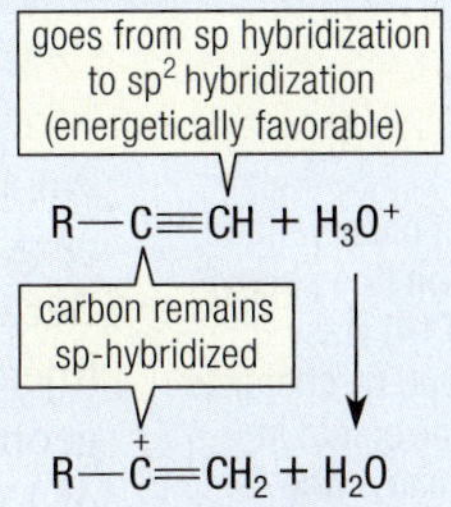

hybridization changes in the formation of a vinylic cation intermediate in alkyne hydration

Furthermore, there is no strong bond to a leaving group to be broken to form this vinylic cation as there is in Eq. 18.13. The only energetically unfavorable aspects of vinylic-cation formation in alkyne hydration, then, are the presence of an electron-deficient carbon, as in all carbocations, and the intrinsic electronegativity of the sp-hybridized carbons.

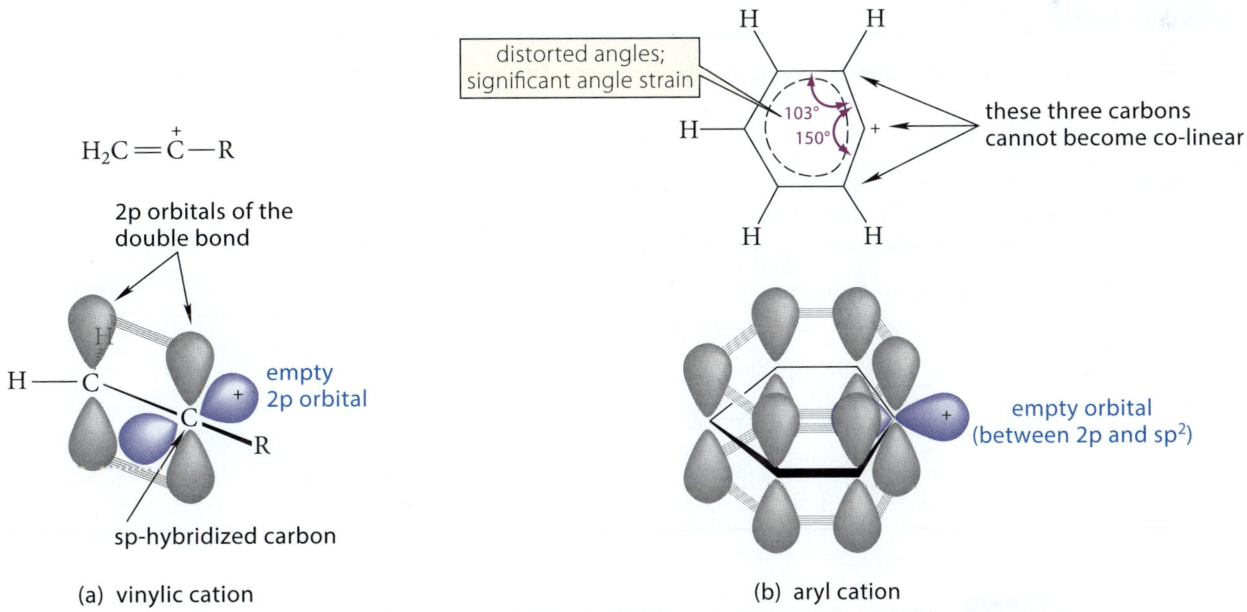

FIGURE 18.2 Lewis structures and corresponding orbital diagrams of vinylic and aryl cations. The thin gray lines indicate orbital overlap. (a) A vinylic cation. In this cation, the empty 2p orbital of the cation *(blue)* is oriented at right angles to the 2p orbitals of the alkene π bond. The carbon with the empty 2p orbital is sp-hybridized. (b) A phenyl cation, the simplest aryl cation. The structure is a compromise between the linear geometry associated with sp hybridization and the 120° bond angles that are optimal for a planar six-membered ring. Consequently, the phenyl cation has significant angle strain, and the empty orbital *(blue)* has more s character than a 2p orbital. Because hybridization and geometry are connected, the hybridization of the electron-deficient carbon is somewhere between sp and sp^2.

s character are stronger (Sec. 3.7C). Consequently, it takes more energy to break the carbon–halogen bond of a vinylic halide. This additional bond energy decreases the rate of ionization.

The S_N1 reactions of aryl halides would involve an *aryl cation* as a reactive intermediate.

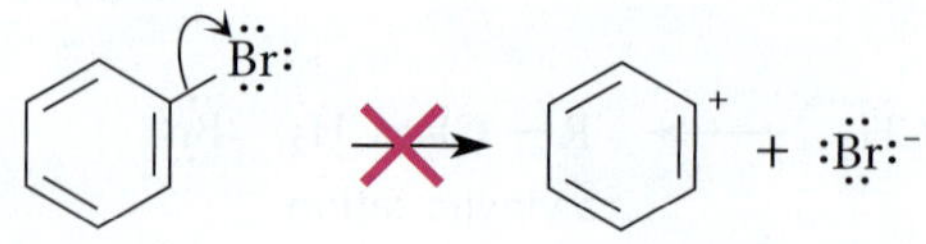

phenyl cation
(an aryl cation)

(18.14)

The first direct observation of an aryl cation (the phenyl cation, Eq. 18.14) was reported in 2000 by chemists at the Ruhr-Universität in Bochum, Germany, who trapped the cation at 4 K and observed it spectroscopically. Even though aryl cations are known species, they are *far* too unstable to form from aryl halides under typical S_N1 conditions.

An **aryl cation** is a carbocation in which the electron-deficient carbon is part of an aromatic ring. An orbital diagram of the phenyl cation, the simplest aryl cation, is shown in **Fig. 18.2b**. Because the electron-deficient carbon in an aryl cation is bonded to two groups, it should have a linear geometry. Linear geometry is associated with sp hybridization. As shown in Fig. 18.2b, the geometry at the electron-deficient carbon is distorted toward linearity with a resulting bond angle of 150°. To compensate, the neighboring angles are narrowed from the optimum 120° to 103°. The empty orbital, which would be a 2p orbital in sp hybridization, is between 2p and sp^2. Because an aryl cation is forced to assume a nonoptimal geometry and hybridization, it has a very high energy. The electron-withdrawing polar effect of the ring double bonds also destabilizes an aryl cation, just as a double bond destabilizes a vinylic cation. Therefore, S_N1 reactions of aryl halides do not occur because they would require the formation of carbocation intermediates—aryl cations—with very high energy.

! Don't confuse an aryl cation with the arenium ion formed in electrophilic aromatic substitution (Eq. 16.16a, Sec. 16.4A); the arenium-ion intermediate is stabilized by resonance. In an aryl cation, the empty orbital is *not* part of the ring π-electron system but is orthogonal (at right angles) to it. This carbocation is therefore *not* resonance-stabilized.

Focused Problem

18.3 Within each series, arrange the compounds according to increasing rates of their reactions by the S_N1–E1 mechanism. Explain your reasoning.

(a)

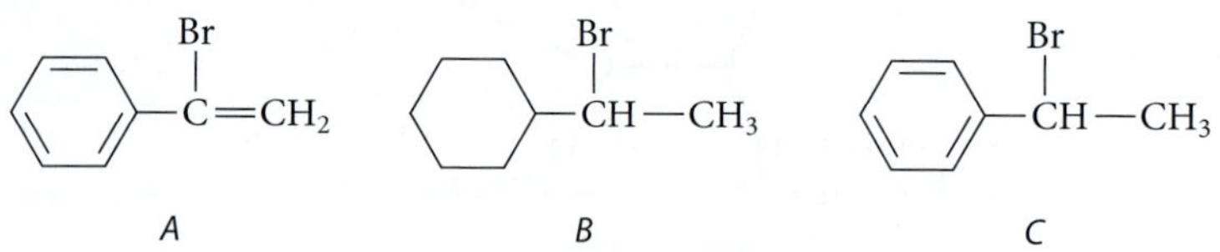

(b)

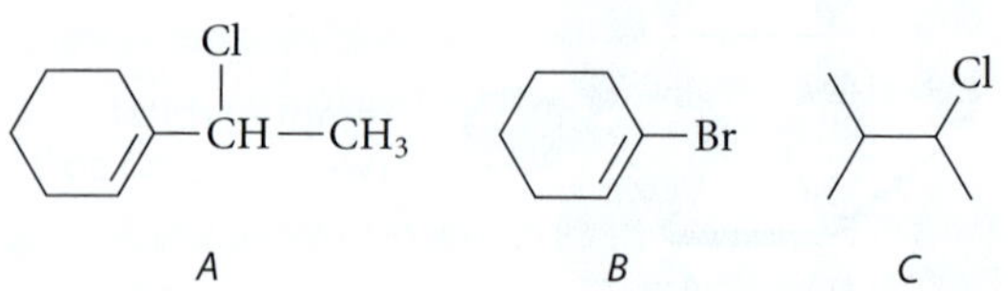

18.4 NUCLEOPHILIC AROMATIC SUBSTITUTION REACTIONS OF ARYL HALIDES

Although aryl halides do not undergo nucleophilic substitution reactions by S_N1 and S_N2 mechanisms, aryl halides that have one or more nitro groups ortho or para to the halogen undergo nucleophilic substitution reactions by a different mechanism under relatively mild conditions.

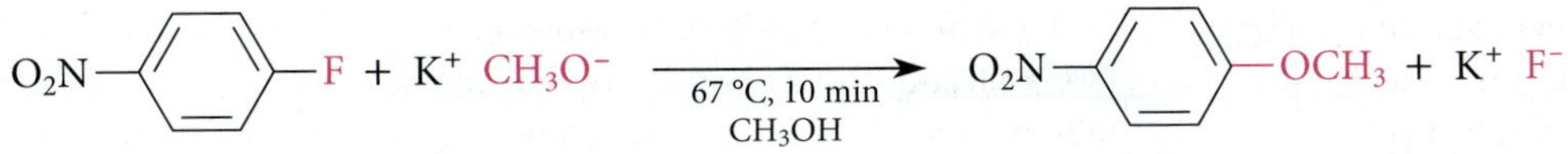

***p*-fluoronitrobenzene** ***p*-nitroanisole** (93% yield) (18.15)

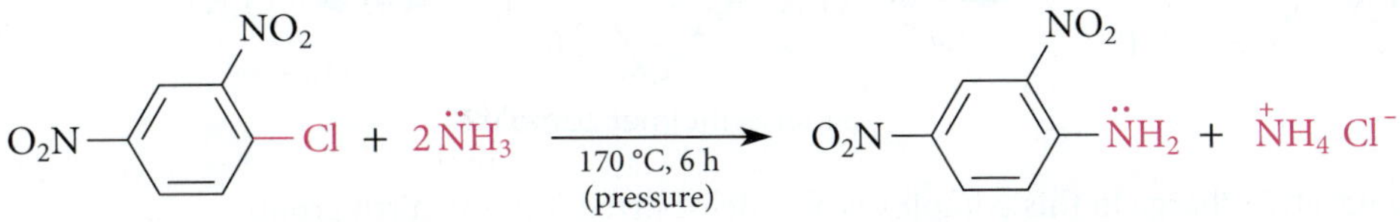

1-chloro-2,4-dinitrobenzene **2,4-dinitroaniline** (70% yield) (18.16)

These reactions are examples of **nucleophilic aromatic substitution**. Let's examine some of the characteristics of this mechanism. Like S_N2 reactions, nucleophilic aromatic substitution reactions involve nucleophiles and leaving groups, and they also obey second-order rate laws.

$$\text{rate} = k[\text{aryl halide}][\text{nucleophile}] \qquad (18.17)$$

However, nucleophilic aromatic substitution reactions do not involve a concerted opposite-side substitution for the reasons given in Sec. 18.1. Two clues about the reaction mechanism come from the reactivities of different aryl halides. First, the reaction is faster when there are more nitro groups ortho and para to the halogen leaving group:

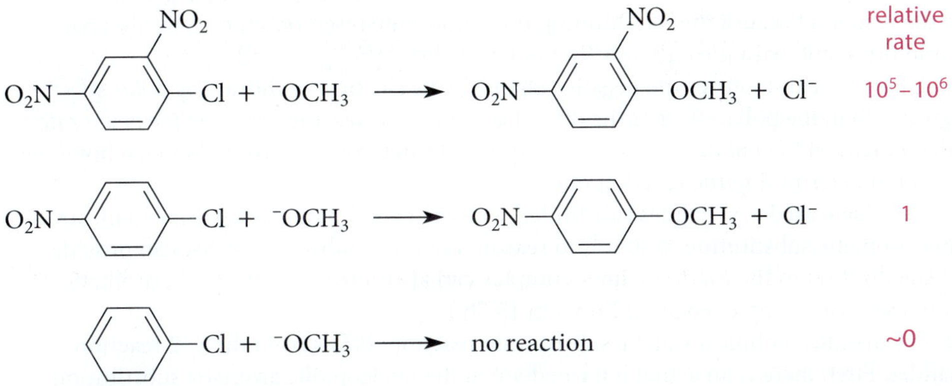

(18.18a)

(18.18b)

(18.18c)

Second, the effect of the halogen on the rate of this type of reaction is quite different from that in the S_N1 or S_N2 reaction of alkyl halides. In nucleophilic aromatic substitution reactions, aryl fluorides are most reactive.

Reactivities of aryl halides:

$$\text{Ar—F} >> \text{Ar—Cl} \approx \text{Ar—Br} \approx \text{Ar—I} \qquad (18.19)$$

In S_N2 and S_N1 reactions of *alkyl* halides, the reactivity order is exactly the reverse: alkyl fluorides are the least reactive alkyl halides (Secs. 9.4F and 9.6C).

These data are consistent with a reaction mechanism in which the nucleophile reacts at the halide-bearing carbon above or below the plane of the aromatic ring to yield a resonance-stabilized anion called a *Meisenheimer complex*. In this anion, the negative charge is delocalized throughout the π-electron system of the ring. Formation of this anion is the rate-limiting step in many nucleophilic aromatic substitution reactions.

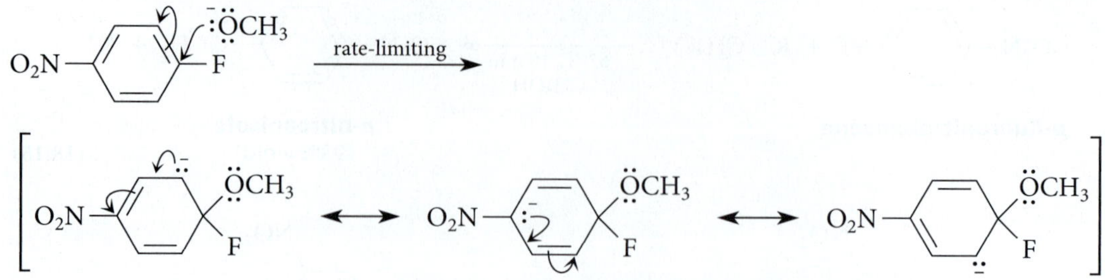

a Meisenheimer complex (18.20a)

The negative charge in this complex is also delocalized into the nitro group.

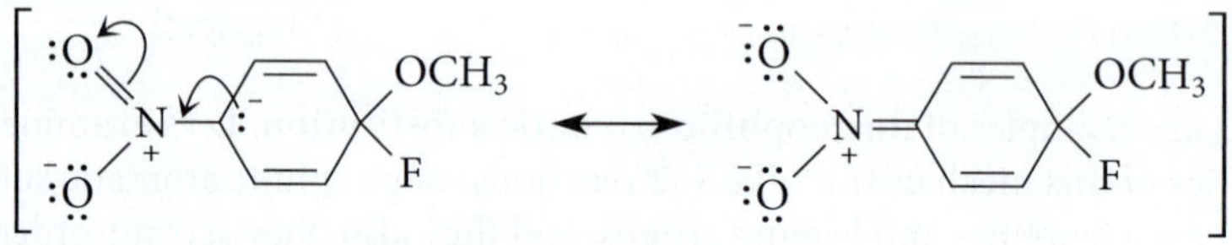

(18.20b)

The Meisenheimer complex breaks down to products by loss of the halide ion.

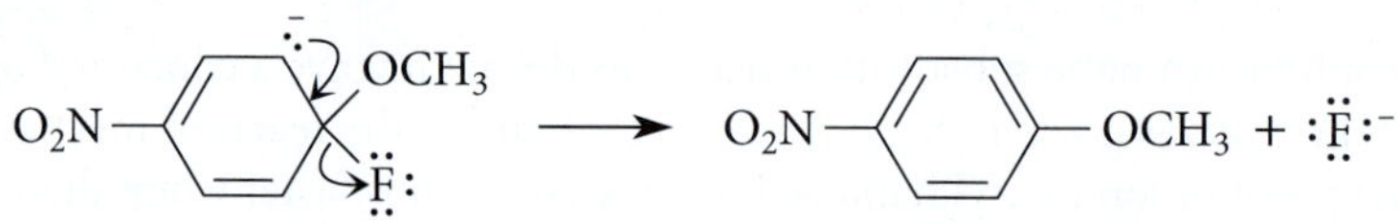

(18.20c)

Let's see how this mechanism fits the experimental facts. Ortho and para nitro groups accelerate the reaction because the rate-limiting transition state resembles the Meisenheimer complex, and ortho and para nitro groups (but not meta nitro groups) stabilize this complex by resonance. Fluorine also stabilizes the negative charge by its electron-withdrawing *polar effect*, which is greater than the polar effect of the other halogens. Because the *loss of halide is not rate-limiting*, the basicity of the halide—or, equivalently, the strength of the carbon–halogen bond—is not important in determining the reaction rate.

Although we have used aryl halides substituted with ortho and para nitro groups to illustrate nucleophilic aromatic substitution, it stands to reason that other substituents that can provide resonance stabilization to the Meisenheimer complex can also activate nucleophilic aromatic substitution. (See, for example, Focused Problem 18.5b.)

Notice how the nucleophilic aromatic substitution reaction differs from the S_N2 reaction of alkyl halides. First, there is an actual intermediate in the nucleophilic aromatic substitution reaction—the Meisenheimer complex. (In some cases, this is sufficiently stable that it can be directly observed.) There is no evidence for an intermediate in any S_N2 reaction. Second, the nucleophilic aromatic substitution reaction is a frontside substitution; it requires no inversion of configuration. The S_N2 reaction of an alkyl halide, in contrast, is an opposite-side substitution with inversion of configuration. Finally, the effect of the halogen on the reaction rate (Display 18.19) is different in the two reactions. Aryl fluorides react most rapidly in nucleophilic aromatic substitution, whereas alkyl fluorides react most slowly in S_N2 reactions.

STUDY GUIDE LINK 18.1
Contrast of Aromatic Substitution Reactions

Focused Problems

18.4 Complete the following reactions. (*No reaction* may be the correct response.)

(a)

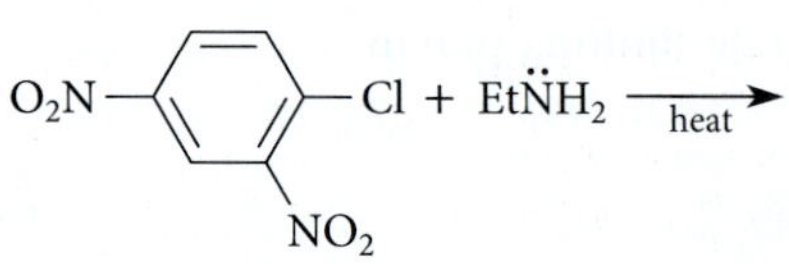

(b)

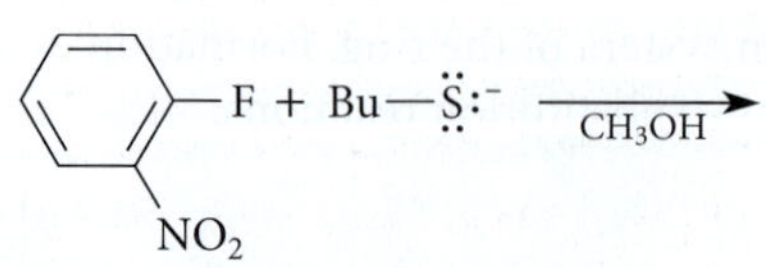

(c)

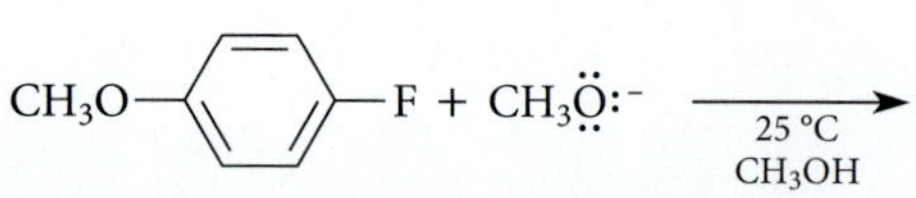

18.5 Which of the two compounds in each of the following sets should react more rapidly in a nucleophilic aromatic substitution reaction with CH_3O^- in CH_3OH? Explain your answers.

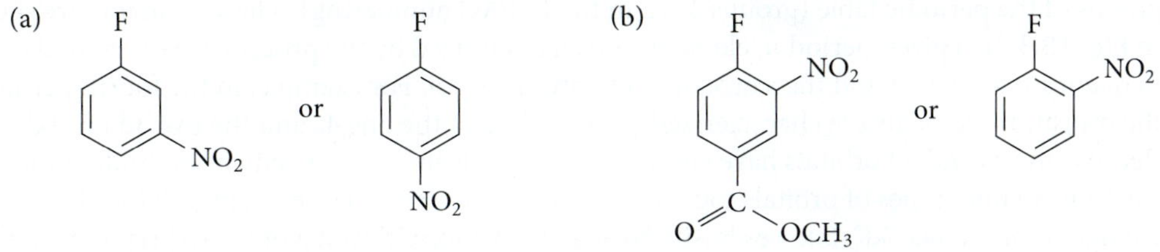

18.5 INTRODUCTION TO TRANSITION-METAL-CATALYZED REACTIONS

We just showed that aryl or vinylic halides do not undergo S_N1 or S_N2 reactions. However, reactions that *look* very much like nucleophilic substitutions *can* be carried out using certain transition-metal catalysts. Here are some examples.

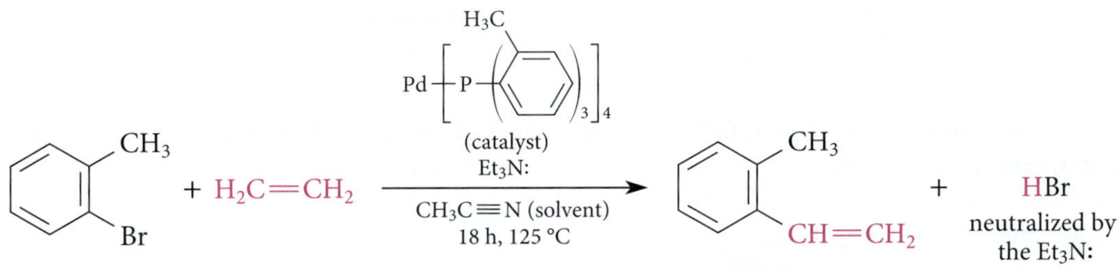

o-bromotoluene **o-methylstyrene** (86% yield) (18.21)

This reaction, called the *Heck reaction*, is very important in organic synthesis. We discuss this reaction in Sec. 18.6A. Notice the formation of the carbon–carbon bond and the release of bromide as HBr. *Superficially*, it looks as if the conjugate-base anion of ethylene displaces bromide ion from the aromatic ring. However, this reaction occurs by a very different mechanism and does not happen without the palladium catalyst. (Only about 1 mol % of the catalyst is required.)

The following reaction involves the substitution of a *vinylic* bromide by a thiolate anion.

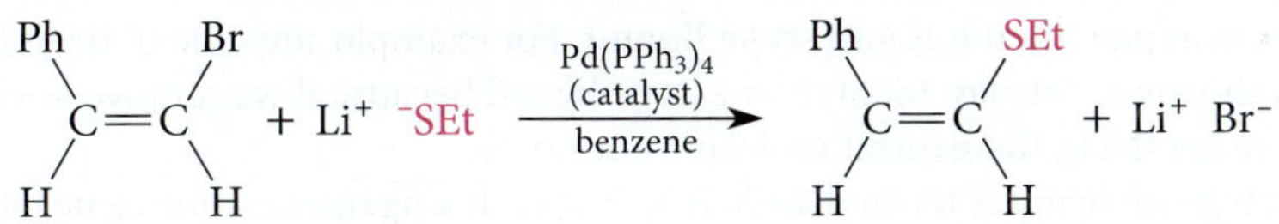

(93% yield) (18.22)

This reaction, too, looks superficially like a nucleophilic substitution reaction. However, this reaction also proceeds by a different mechanism and does not take place without the catalyst, which is present in only 1 mol %. Notice also the *retention* of alkene stereochemistry, a very different result from that expected in an S_N2 reaction.

These are only two examples of thousands known in which transition-metal catalyze seemingly impossible reactions. The field of transition-metal catalysis has exploded since the 1980s, and it has become important in both laboratory and industrial chemistry, as well as in some areas of biology. This field is part of the larger field of *organometallic chemistry*: the chemistry of carbon–metal bonds. (Grignard reagents and lithium dialkylcuprate reagents are examples of organometallic compounds that you encountered in Secs. 10.5 and 12.5C and will encounter again in subsequent chapters.) Our goal in this section is to understand some of the basic ideas of transition-metal catalysis. Then, in Sec. 18.6, we examine a few important transition-metal-catalyzed reactions in the light of these principles.

A. Transition Metals and Their Complexes

Recall from general chemistry that **transition metals** are the elements in the "d block" or "B" groups of the periodic table (groups 3–12 in the IUPAC numbering). These elements are shown in **Fig. 18.3**. In a given period *n*, elements are characterized by the progressive filling of d orbitals in quantum level $n-1$ and the s orbital in quantum level *n*. For example, in the fourth period, the transition elements are characterized by the filling of the one 4s and the five 3d orbitals. Because the 4s and 3d orbitals have similar energies, it is usually convenient to think of the electrons in both types of orbitals together as valence electrons. For example, Ni has the electronic configuration $[Ar]4s^2 3d^8$, but we classify Ni as a 10-valence-electron atom. ([Ar] represents the electronic configuration of the noble gas argon.)

Central to transition-metal chemistry are a wide variety of compounds containing transition metals surrounded by several groups, called **ligands**. Such compounds are called **coordination compounds** or **transition-metal complexes**. These can be neutral molecules, as in the first of the following examples, or *complex ions*, as in the second example.

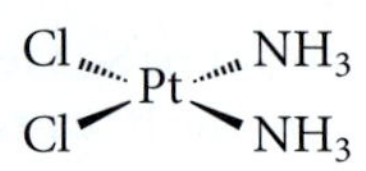

***cis*-diamminedichloroplatinum(II)**
(***cis*-platin**, an antitumor drug)
a neutral complex

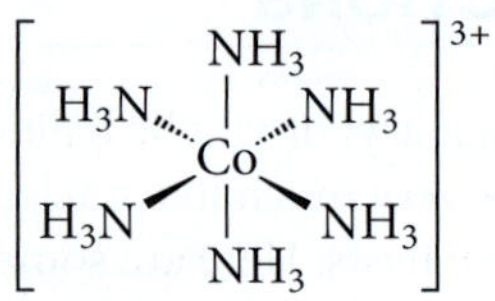

hexamminecobalt(III) ion
a complex ion

(18.23)

To deal systematically with transition-metal complexes, we must be aware of, and be able to apply, certain conventions:

1. How to classify ligands
2. How to specify formal charge on the metal
3. How to calculate the oxidation state of the metal
4. How to count electrons around the metal

Ligand Classification In transition-metal chemistry, all ligands are *Lewis bases*. That is, ligands interact with transition metals by donating electron pairs. There are two types of ligand. The first is an *L-type ligand*. If a ligand becomes a neutral molecule when it dissociates from the metal with its bonding electron pair, then it is an **L-type ligand**. For example, any one of the NH_3 ligands in the complexes shown in Display 18.23 is an L-type ligand because if we remove it with its bonding electron pair, we get $:NH_3$, the neutral molecule ammonia.

The second type of ligand is termed an *X-type ligand*. If a ligand becomes a negative ion when it dissociates from the metal with its bonding electron pair, then it is an **X-type ligand**. So, the Cl in *cis*-platin (the first example in Display 18.23) is an X-type ligand, because removing it with its bonding pair of electrons gives the chloride ion, Cl^-.

FIGURE 18.3 The transition metals. The red numbers indicate the number of valence electrons (outer shell s and d electrons) in the neutral atoms. (These are the same as the IUPAC group numbers in the periodic table; see the last page at the end of the text.)

Group number	*3B*	*4B*	*5B*	*6B*	*7B*		*8B*		*1B*	*2B*
Valence electrons in the neutral atom	**3**	**4**	**5**	**6**	**7**	**8**	**9**	**10**	**11**	**12**
Period 4 →	Sc	Ti	V	Cr	Mn	Fe	Co	Ni	Cu	Zn
Period 5 →	Y	Zr	Nb	Mo	Tc	Ru	Rh	Pd	Ag	Cd
Period 6 →	Lu	Hf	Ta	W	Re	Os	Ir	Pt	Au	Hg

The classification of ligands has implications for computing formal charge. *From a formal-charge perspective, the bonding electrons on L-type ligands are considered to "belong" completely to the ligand.* Let's see how this differs from the way we treat bonds in main-group chemistry. We know that a nitrogen with four bonds in main-group chemistry (for example, the ammonium ion, $^{+}NH_4$) has a positive formal charge. If we were to take a similar view with *cis*-platin, the nitrogens would each have a positive charge; and, because the complex is neutral, the Pt would have charge = –2. A neutral transition-metal complex ML_6 bearing six L-type ligands would have a charge of –6 on the metal and a positive charge on each ligand. It is inconvenient to draw out all of these charges; moreover, a formal charge of –6 on a metal is highly unrealistic. Instead, we adopt the *convention* that the electron pair in an L-type ligand is assigned completely to the ligand. We emphasize this point by leaving the bonding electron pair on the ligand and depicting the ligand–metal bond as an arrow from these electrons to the metal. This is called a **dative bond**.

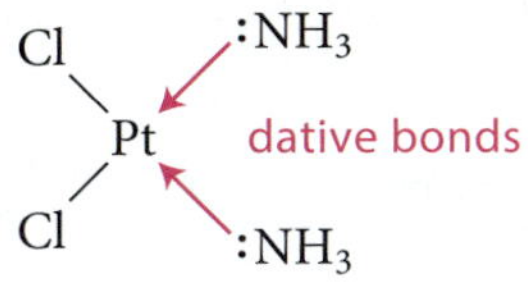

(18.24)

A dative bond is used for L-type ligands. If we choose to represent a dative bond as an ordinary bond, then we don't put a formal charge on the ligand. When the ligand dissociates, it does not gain any electrons because both electrons in the bond were assigned to the ligand in the first place.

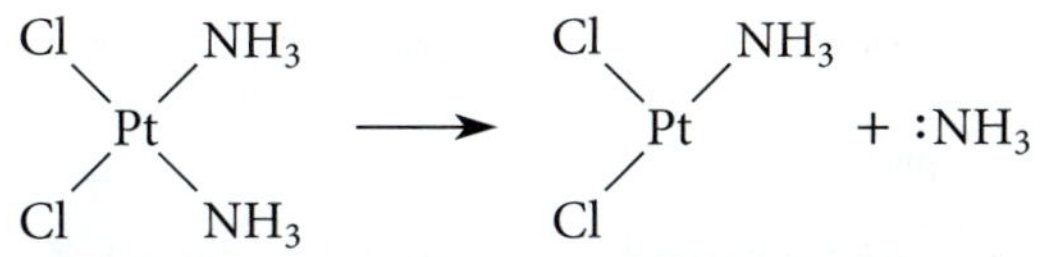
(18.25)

In contrast, electrons in the bonds to X-type ligands are assigned in the same way that we assign electrons in main-group chemistry: *one electron is assigned to the ligand and one to the metal.* This means that if we remove an X-type ligand, it takes on an additional negative charge and the metal takes on a compensating positive charge:

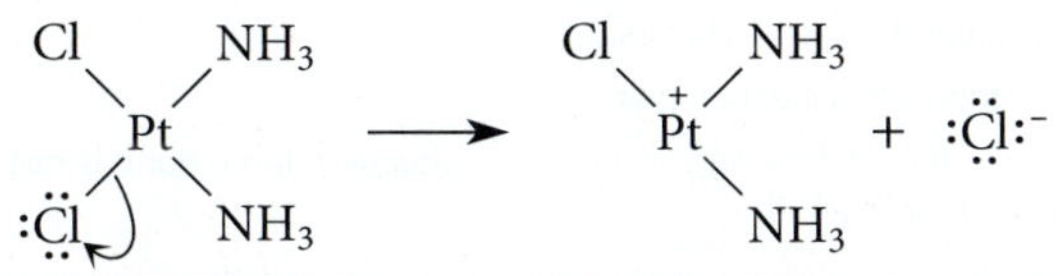
(18.26)

Differentiating between X-type and L-type bonds is a convenient bookkeeping device, but we should bear in mind that both types of bonds are covalent bonds, and the degree to which electrons (and charge) are transferred to the metal varies widely in both types of bonds, depending on the metal and the ligand.

Table 18.1 lists some of the common ligands used in transition-metal chemistry. These are classified as L-type or X-type ligands. It is worth noting two things about this table. First, alkenes or aromatic rings can act as ligands by donating their π electrons to a metal. Second, allyl and cyclopentadienyl (Cp) are classified as both L-type and X-type ligands. Let's consider the Cp case to understand this. The cyclopentadienyl anion was discussed in Sec. 15.7D (Display 15.63) as an example of an aromatic ion with six π electrons. Table 18.1 indicates that Cp is an example of an L_2X ligand. What this means is that *one* X-type bond accounts for the fact that Cp takes on *one* negative charge when removed with a bonding pair from the metal, and that the four remaining π electrons (that is, two double bonds) take part in two L-type bonds. In other words, we can think of a metal–Cp complex in the following way (M = metal):

(18.27)

The IUPAC in 2008 promulgated a set of rules for structure drawing. These rules discouraged the use of arrows for dative bonds. They proposed showing the dative bond as a normal single bond with no formal charge on the L-type ligand (as the NH_3 ligands in Eqs. 18.25 and 18.26). We disagree with this convention, and generally we show bonds to L-type ligands as dative bonds with arrows. Our convention makes it clear which ligands are L-type and which are X-type.

TABLE 18.1 Some Typical Ligands Used in Transition-Metal Chemistry

Ligand	Name	Abbreviation	Type	Electron count*
H_3N:	ammine		L	2
$H_2\ddot{O}$:	aquo		L	2†
R_3P: (R = alkyl, aryl)	trialkylphosphino, triarylphosphino		L	2
$:C{=}\ddot{O}$	carbonyl	CO	L	2‡
$H_2C{=}CH_2$§	ethylene		L	2
$CH_3C{\equiv}N$:	acetonitrile	MeCN	L	2§§
	benzene		L_3	6
F^-, Cl^-, Br^-, I^-	halo (e.g., chloro)	X	X	2†
H^-	hydrido		X	2
$H_3C{-}C({=}O){-}O^-$	acetato	AcO	X	2†
$R\bar{:}$ e.g., $H_3C\bar{:}$	alkyl (e.g., methyl)		X	2
$\bar{:}C{\equiv}N$:	cyano	CN	X	2‡
$H_2C{=}CH{-}\bar{\ddot{C}}H_2$	allyl		LX	4**
	cyclopentadienyl	Cp	L_2X	6

* The sum of all electrons in the bond(s) between the ligand and the metal.

† Only one electron pair is involved in the ligand–metal interaction.

‡ Only the electron pair on carbon is involved in the ligand–metal interaction.

§ Ethylene is listed as a prototype for many alkenes.

§§ Donation of the nitrogen nonbonding election pair.

** Allyl can also bind to metals as an X-type ligand. In such a situation, the π bond is not involved in coordination and the electron count is 2 (as with alkyl).

We know that the π electrons in Cp are completely delocalized, and they remain delocalized in metal complexes. (See the structure of ferrocene, Cp_2Fe, in Display 15.69, Sec. 15.7D, which shows this delocalization.) Consequently, a more accurate picture of such a complex would show the "L" and "X" character of the bonds parceled out equally over all five carbons, with each carbon participating in 20% of an X-type bond ($5 \times 0.20 = 1.0$ X-type bond) and 40% of an L-type bond ($5 \times 0.40 = 2.0$ L-type bonds). But this delocalization can be ignored for the bookkeeping purposes discussed in this section.

Focused Problem

18.6 Noting the LX character of the allyl ligand in Table 18.1, sketch the allyl–metal interaction, showing both L-type and X-type bonds. Use M as a general metal.

B. Oxidation State

The oxidation state of the metal is an important concept in organometallic chemistry. **Oxidation state** can be conceptualized as the charge the metal would have if all X-type ligands were dissociated with their bonding electron pairs. For example, if the oxygens of manganese dioxide, O═Mn═O, were to dissociate from the manganese with their bonding electron pairs, the oxygens would each take on a −2 charge, and the manganese would take on a +4 charge. Therefore, the manganese has an oxidation state of +4. [This process is essentially the same as the one used for assigning oxidation numbers to carbon atoms in oxidation–reduction reactions (Sec. 11.6A)]. Equation 18.28 formalizes this idea:

$$\text{Oxidation state of M} = \text{number of X-type ligands} + Q_M \qquad (18.28)$$

In this equation, Q_M is the actual charge that M has *before* the fictitious dissociation of the X-type ligands. To illustrate a situation involving Q_M, consider the hexachloroplatinate dianion, $[Pt(Cl_6)]^{2-}$. For this species, $Q_M = -2$. If the six chlorines were to dissociate with their bonding electrons, the charge on Pt would be +4, because Pt starts out with a −2 charge before the fictitious dissociation. To be sure that the oxidation state and the actual charge Q_M are not confused, the oxidation state is sometimes indicated with Roman numerals. Therefore, the name of the ion $[Pt(Cl_6)]^{2-}$ is hexachloroplatinate(IV).

L-type ligands do *not* contribute to the oxidation state. You should verify that the oxidation state of platinum in the neutral complex $Cl_2Pt(PPh_3)_2$ is +2.

Focused Problems

18.7 Calculate the oxidation state of the metal in each of the following complexes.

(a)

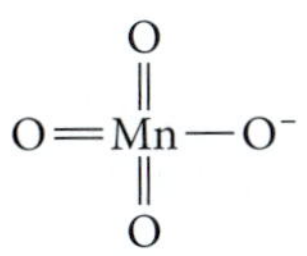

permanganate

(b) $Pd(PPh_3)_4$

tetrakis(triphenylphosphine)palladium

(c) Cp_2Fe

ferrocene

18.8 What is the oxidation state of the metal in the starting material of the following reaction? How does it change, if at all, as a result of the reaction? Is the metal oxidized, reduced, or neither?

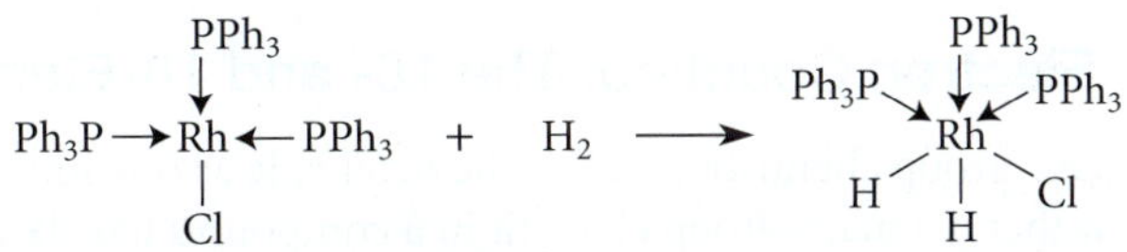

chlorotris(triphenylphosphine)rhodium

C. The d^n Notation

In understanding the reactions of main-group elements that follow the octet rule, it is important in applying acid–base concepts for us to know whether the element undergoing a transformation has nonbonding valence electrons. Often these nonbonding electrons are shown explicitly. In transition-metal chemistry, it is also important to know whether the metal has nonbonding valence electrons. In many cases, the metal has so many nonbonding valence electrons that it would be impractical or confusing to draw them all as dots. Instead, we use a convenient algorithm to calculate the number of nonbonding valence electrons. The number of nonbonding valence electrons on the metal is the number n in a notation called **d^n**. For example, if the metal in a complex has eight nonbonding valence electrons, we say that the complex is a d^8 complex.

We calculate n by determining the number of valence electrons remaining on the metal after removing all ligands with their electron pairs. We start with the number of valence electrons in the *neutral transition element* (from Fig. 18.3). We remove an electron for each positive charge, add an electron for each negative charge, and then subtract one electron for each bond to an X-type ligand. L-type ligands have no effect on d^n.

$$n = \text{valence electrons in neutral M} - Q_M - \text{number of X-type ligands} \qquad (18.29)$$

Introducing the definition of oxidation state in Eq. 18.28 (Sec. 18.5B), Eq. 18.29 becomes

$$n = \text{valence electrons in neutral M} - \text{oxidation state of M} \quad (18.30)$$

Study Problem 18.1

Calculate n in the d^n notation for ferrocene, Cp_2Fe.

Solution We'll make this calculation with both Eqs. 18.29 and 18.30.

Using Eq. 18.29: From Fig. 18.3 we see that neutral Fe has eight valence electrons. The charge of the iron is zero, and Table 18.1 shows that each Cp ligand has one X-type bond; the iron, then, has two bonds to X-type ligands. Therefore, $n = 8 - 2 = 6$, and ferrocene is a d^6 complex.

Using Eq. 18.30: We calculate that with two X-type ligands and zero charge, the Fe in ferrocene has a +2 oxidation state; therefore, Eq. 18.30 gives the value of n as $8 - 2 = 6$.

Focused Problem

18.9 What is d^n for each of the following complexes?

(a) $[W(CO)_5]^{2-}$ (b) $Pd(PPh_3)_4$ (c)

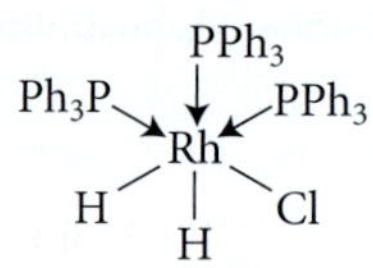

D. Electron Counting: The 16- and 18-Electron Rules

In main-group chemistry, we use the octet rule as one indicator of reactivity. For example, we know that if a main-group element in a compound has six electrons, it can accept an electron pair from a Lewis base in a Lewis acid–base association reaction. In other words, main-group elements have a tendency to complete their octets. Recall that counting for the octet involves adding an element's nonbonding valence electrons to the number of electrons in all bonds to the element. The electron count in transition-metal complexes is also important and is determined in a similar manner.

To determine the **electron count** for a transition-metal complex, we start with the n electrons in the d^n count—the nonbonding electrons—and add two electrons for every ligand (both L-type and X-type). Therefore,

$$\text{electron count} = n + 2(\text{number of all ligands}) \quad (18.31)$$

The multiplier 2 is required because there are two electrons per bond. The rationale for this formula is that the number n is the number of nonbonding valence electrons on the metal; the total electron count is the nonbonding valence electrons plus all electrons in bonds, just as in counting for the octet rule. Using Eq. 18.30, we can rewrite this formula in terms of the oxidation state of the metal:

$$\text{electron count} = \text{valence electrons in neutral M} - \text{oxidation state of M} + 2(\text{number of all ligands}) \quad (18.32)$$

By incorporating the definition of oxidation state (Eq. 18.28, Sec. 18.5B), we obtain yet another equivalent formula:

$$\text{electron count} = \text{valence electrons in neutral M} - Q_M - \text{number of X-type ligands} + 2(\text{number of all ligands}) \quad (18.33)$$

Recognizing that all ligands = X-type ligands + L-type ligands, we finally obtain the most useful formula for electron count:

$$\text{electron count} = \text{valence electrons in neutral M} - Q_M + \text{number of X-type ligands} + 2(\text{number of L-type ligands}) \quad (18.34)$$

This equation says that to obtain the electron count in a complex, we start with the electron count of the neutral metal from Fig. 18.3; we subtract the charge on the metal (taking into account its algebraic sign); we add the number of X-type ligands; and we add *twice* the number of L-type ligands. Don't let the mathematical derivation of Eq. 18.34 obscure its rationale. Remember that the goal is to count all nonbonding and bonding electrons about the metal. Because an X-type ligand *by definition* has one electron in its M—X bond assigned to X, we have to add this electron back to obtain the total number of electrons in the bond. Because both electrons in the dative bond to an L-type ligand are assigned to the ligand, we have to multiply each L-type ligand by 2 to count both of these bonding electrons.

Study Problem 18.2

Give the electron count for each of the following compounds:

(a) $Ni(CO)_4$ (b) $Cl_2Pd(PPh_3)_2$

Solution

(a) We apply Eq. 18.34. The electron count of $Ni(CO)_4$ is $10-0-0+2(4)=18$. This is an 18-electron complex. [This compound, tetracarbonylnickel(0), is a very stable complex of Ni.]

(b) The electron count of $Cl_2Pd(PPh_3)_2$ is $10-0+2+2(2)=16$. This is a 16-electron complex.

In transition-metal chemistry, the most stable complexes in many cases have electron counts of 18 electrons. This statement is called the **18-electron rule**. $Ni(CO)_4$, a very stable complex of Ni(0), is an example of the 18-electron rule. Just as the octet represents the number of valence electrons (8) in the outermost s and p orbitals of the nearest noble gas, 18 electrons is the number of total s + p + d valence electrons in the nearest noble gas. If a transition-metal complex has fewer than 18 electrons, it tends to react so as to provide the missing electrons, just as a main-group element reacts to complete the octet.

Exceptions to the 18-electron rule occur, and an important type of exception occurs frequently with transition metals in the 8–11 valence-electron group (see Fig. 18.3), which includes Ni and Pd, two metals of prime importance in the transition-metal-catalyzed reactions discussed in this section. Although a number of stable complexes of these metals have 18 electrons, others contain 16 electrons. The tendency of these metals to surround themselves with 16 electrons can be called the **16-electron rule**. The $Cl_2Pd(PPh_3)_2$ example shows the operation of the 16-electron rule.

Focused Problems

18.10 (a) What is the electron count for the Rh complex shown in Focused Problem 18.9c?

(b) *Sestamibi* (trade name Cardiolite) is a complex of $^{99}Tc(I)$ (a radioactive gamma-ray emitter) that is widely used for cardiac and parathyroid imaging.

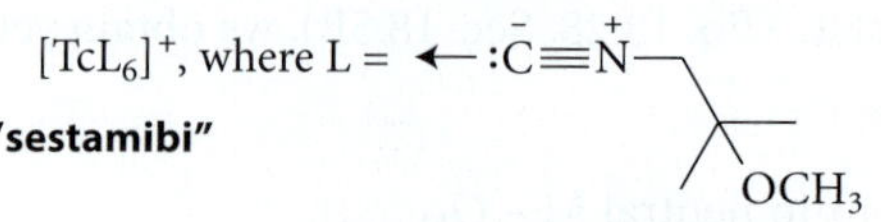

"sestamibi"

***N*-(2-methoxyisobutyl) isocyanide (MIBI)**

"MIBI" is an L-type ligand. Give the value of n for the d^n notation and the total electron count for the technetium.

18.11 How many CO ligands would be accommodated by Fe(0) if we assume that the resulting complex follows the 18-electron rule?

18.12 Using the 18-electron rule, explain why $V(CO)_6$ can be easily reduced to $[V(CO)_6]^-$.

Main-group elements have four valence orbitals (for example, four sp^3 hybrid orbitals) that can either form two-electron bonds or house nonbonding electron pairs. The maximum electron occupancy within these bonds and nonbonding electron pairs requires eight electrons. In other words, the eight-electron configuration—the octet—provides the most stable bonding arrangement. We can justify the 18-electron rule in a similar way. Consider, for example, the complex ion $[Co(CN)_6]^{3-}$. Using Eq. 18.28 (Sec. 18.5B), we find that the oxidation state of Co is +3, and, from Eq. 18.31, that this is a d^6 complex. This means that Co(III) in this complex has six nonbonding electrons. Let's imagine building this complex from a "naked" Co^{3+} ion. Start with the electronic configuration of this ion, as shown in **Fig. 18.4a**. Allow all the electrons to pair, as shown in **Fig. 18.4b**. (Although this electron pairing violates Hund's rules, the energy required is more than offset by the energy released by subsequent bond formation.). This electron pairing leaves two 3d, one 4s, and three 4p orbitals unoccupied. These are hybridized, as shown in **Fig. 18.4c**, to give *six equivalent* d^2sp^3 hybrid orbitals. Six equivalent hybrid orbitals are directed to the corners of a regular octahedron in the same sense that sp^3 carbon orbitals in methane are directed to the corners of a regular tetrahedron. Hybridization also requires energy. Each of these empty hybrid orbitals can accept an electron pair from a cyanide ion ($^-$:CN). Because these hybrid orbitals are directed in space, they can form stronger bonds to cyanide than unhybridized orbitals, and the strength of these bonds more than compensates for the energy cost of electron pairing and hybridization. The result is the octahedral $[Co(CN)_6]^{3-}$ complex shown in **Fig. 18.4d**. In other words, the 18-electron rule results from the rehybridization and maximal occupancy of all valence orbitals of the Co^{3+} ion.

The 16-electron rule is important in square planar complexes of the 10-electron elements Ni, Pd, and Pt. For example, consider the antitumor drug *cis*-platin, $Cl_2Pt(NH_3)_2$ (Display 18.24, Sec. 18.5A). This is a 16-electron d^8 complex of Pt(II) that has square planar geometry. If we start with a Pt^{2+} ion and arrange its eight electrons in pairs within four 5d orbitals, this leaves a single 5d orbital, a 6s orbital, and three 6p orbitals empty. It turns out that hybridization of four of the five empty orbitals to give four dsp^2 hybrid orbitals and one relatively high-energy 6p orbital is a particularly favorable hybridization:

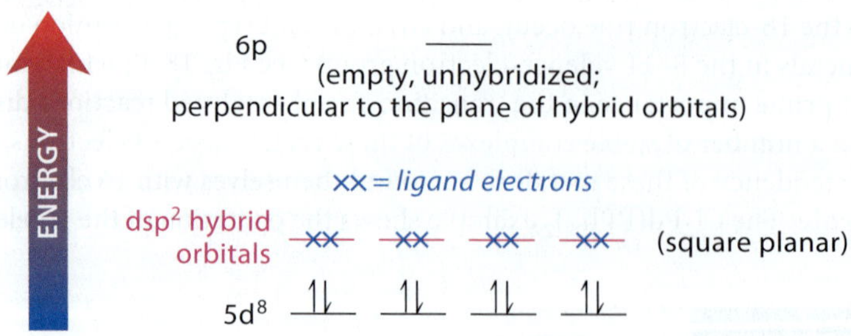

(18.35)

The four hybrid orbitals are directed to the corners of a square and accept the electron pairs from the four ligands to give a square planar complex. The element platinum can also adopt 18-electron configurations, but the point is that the 16-electron configuration is reasonably stable.

As in main-group chemistry, hybridization arguments are useful for visualizing electrons in bonds, but they are inferior to molecular orbital arguments for a detailed understanding of molecular energies. The branch of molecular orbital theory that deals with transition-metal complexes is called *ligand field theory.* We need not explore this theory here, but suffice it to say that this theory provides excellent support for the 16- and 18-electron rules.

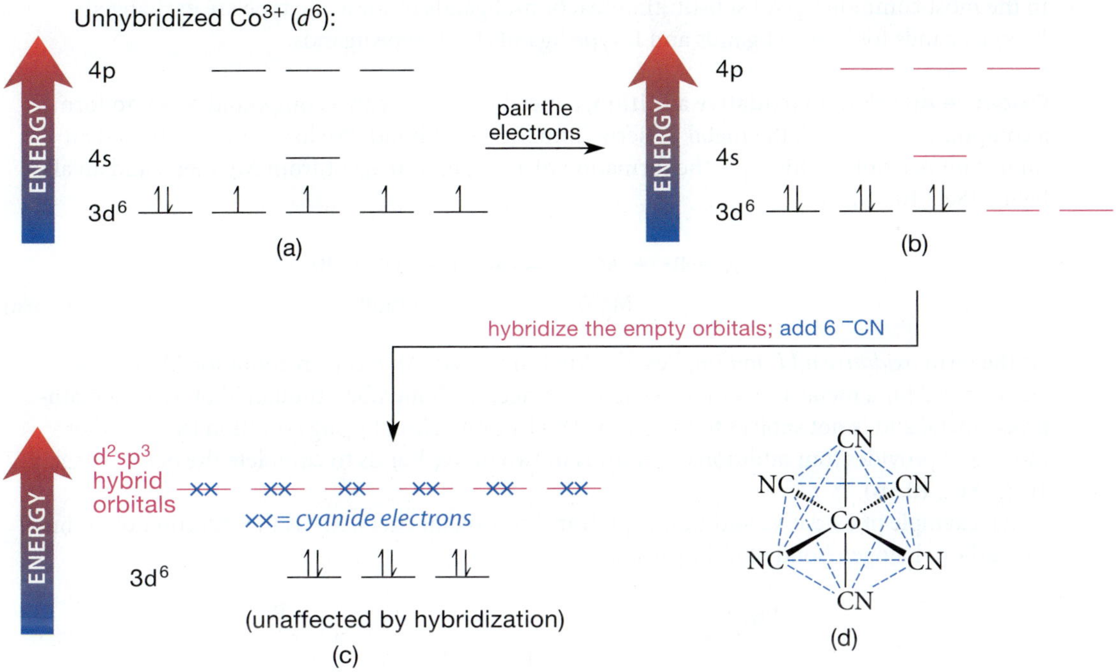

FIGURE 18.4 Development of the hybrid orbital description of $[Co(CN)_6]^{3-}$, an 18-electron complex ion. (a) The electronic configuration of Co^{3+}. (b) The Co electrons are arranged in pairs; the empty orbitals that remain *(red)* are used to form hybrid orbitals. (c) The empty orbitals are hybridized into six equivalent d^2sp^3 hybrid orbitals. Each of these orbitals can accept an electron pair (symbolized by XX) from a ^-CN ion. (d) The hybrid orbitals and consequently the six ^-CN that bind to them are oriented to the corners of a regular octahedron. (The edges of the octahedron are indicated with blue dashed lines.)

Focused Problem

18.13 Use a hybridization argument to predict the geometry of (a) the $[Zn(CN)_4]^{2-}$ ion; (b) the neutral compound $Pd(PPh_3)_4$.

E. Fundamental Reactions of Transition-Metal Complexes

We have now introduced the preliminaries needed to understand the mechanistic basis of some transition-metal-catalyzed reactions. Transition-metal complexes undergo a relatively small number of fundamental reaction types, and many reactions are readily understood simply as combinations of these fundamental processes. The goal of this section is to introduce a few of these.

Ligand Dissociation–Association; Ligand Substitution One of the most common reactions of transition-metal complexes is *ligand dissociation* and its reverse, *ligand association*. In ligand dissociation, a ligand simply departs from the metal with its pair of electrons, leaving a vacant site (orbital) on the metal.

$$\underset{\text{an 18e}^-\text{ complex}}{\underset{\text{Pd(0)}}{Pd(PPh_3)_4}} \rightleftharpoons \underset{\text{a 16e}^-\text{ complex}}{\underset{\text{Pd(0)}}{Pd(PPh_3)_3}} + :PPh_3 \qquad (18.36)$$

This process does not change the oxidation state of the metal, but it does change the electron count. A *ligand substitution* can occur by the dissociation of one ligand and the association of another, which is somewhat analogous to an S_N1 reaction in alkyl halides, or by a direct substitution, in which one ligand displaces another, somewhat analogous to the S_N2 reaction.

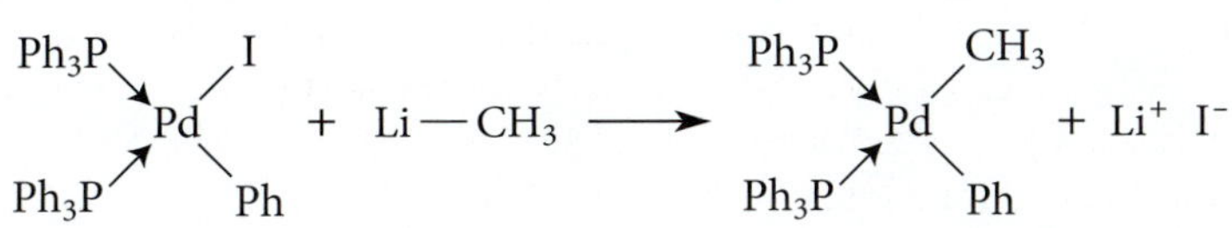

Pd(II) a 16e$^-$ complex   Pd(II) a 16e$^-$ complex   (18.37)

In the most common ligand substitution reactions, ligands of the same type are exchanged: X-type ligands for X-type ligands and L-type ligands for L-type ligands.

Oxidative Addition In **oxidative addition**, a metal M reacts with a compound X—Y to form a compound X—M—Y; the metal "inserts" into the X—Y bond. We have already studied an important reaction of this type: the formation of a Grignard reagent from Mg metal and an alkyl halide (Sec. 10.5B).

$$\underset{}{\text{R—Br}} + \underset{\text{Mg(0)}}{\ddot{\text{Mg}}} \longrightarrow \underset{\text{Mg(II)}}{\text{R—Mg—Br}} \qquad (18.38a)$$

As the term *oxidative addition* implies, the Mg is oxidized. An electron count for Mg in this reaction is 2 for a metal and 4 in the Grignard reagent. (Remember, though, that Mg is a main-group metal and is not subject to the 16- or 18-electron rules.) Strong solvation by two ether molecules provides four additional electrons in two dative bonds to complete the octet (Display 10.49, Sec. 10.5B).

An example of oxidative addition from transition-metal chemistry is the insertion of Pd into the carbon–halogen bond of iodobenzene:

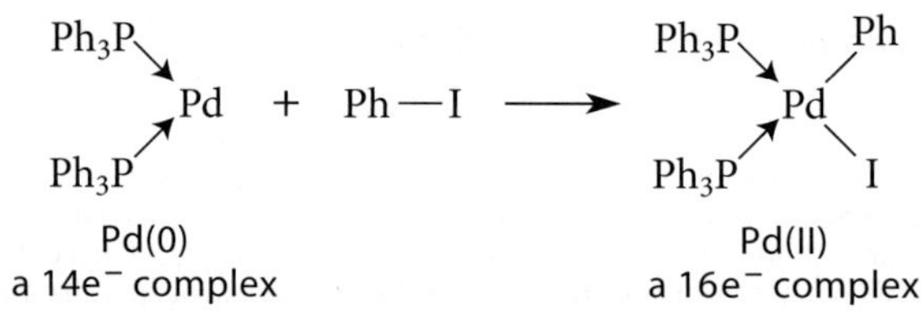

(18.38b)

Both of the new bonds are X-type bonds. As a result of this reaction, both the electron count and the oxidation number of the metal increase by 2 units.

Oxidative addition is a remarkable reaction that lies at the heart of transition-metal catalysis with aryl and vinylic halides. Why is it that a metal can break a σ bond in this way? Molecular orbital theory provides a simple way to understand this process, as shown in **Fig. 18.5**. Think of the carbon–halogen bond as a localized bond for simplicity, and imagine a molecular orbital treatment of this bond much like the molecular orbital treatment of the H—H bond in H_2 (Sec. 1.8A). The carbon–halogen bond has an associated *bonding molecular orbital*, which is occupied by the two bonding electrons, and an *antibonding molecular orbital*, which is unoccupied. The bonding molecular orbital can serve as a ligand, donating its electrons to one of the empty hybrid orbitals on the metal. At the same time, one of the filled d orbitals of the metal overlaps with the *antibonding* molecular orbital of the carbon–halogen bond. This additional overlap *strengthens* the metal–ligand interaction, but *weakens* the carbon–halogen bond, because the addition of electrons to an antibonding molecular orbital removes the energetic advantage of bonding. (See Display 1.86, Sec. 1.8A.) The carbon–halogen bond is weakened sufficiently that it actually breaks. The overall result is that electrons flow *from the aryl halide to the metal* and, at the same time, *from the metal to the aryl halide.* We can approximate the process as follows with the curved-arrow notation (L = other ligands):

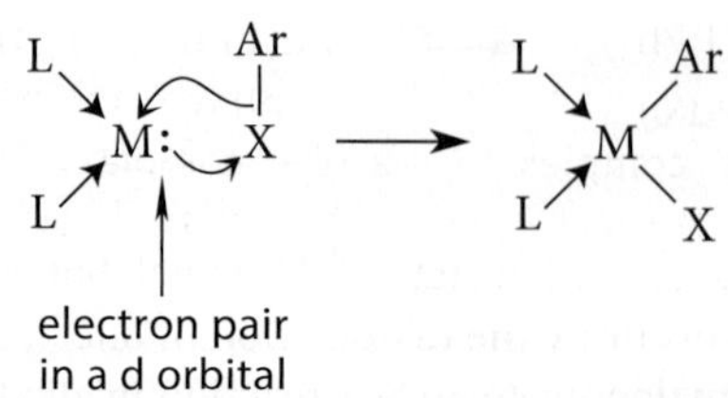

(18.39)

Oxidative addition can occur by a variety of mechanisms, but a concerted (one-step) process is fairly common.

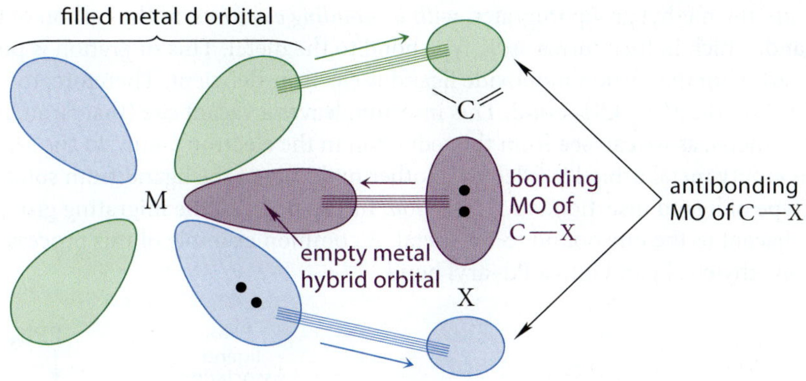

FIGURE 18.5 An orbital description of concerted oxidative addition. The bonding MO of the C—X bond donates electrons to an empty hybrid orbital on the metal. (The shape of this hybrid orbital is simplified.) These orbitals are shown in purple, the thin purple lines indicate the electronic overlap, and the purple arrow shows the direction of electron flow. At the same time, a filled d orbital on the metal donates electrons to an antibonding MO of the C—X bond. (A 3d orbital is used for simplicity.) Because the peaks and troughs of the d orbital (shown in blue and green, respectively) match the peaks and troughs of the antibonding MO of the C—X bond, this is a bonding interaction. The electronic overlap in this interaction is indicated with thin blue and green lines, and the blue and green arrows show the direction of electron flow. Because this interaction populates the antibonding MO of C—X, the C—X bond is weakened, and it breaks.

Reductive Elimination **Reductive elimination** is conceptually the reverse of oxidative addition, and the orbital interactions involved are the same, only in reverse. In reductive elimination, then, two ligands bond to each other and their bonds to the metal are broken; X—M—Y ⟶ X—Y + M. An example of this process is the formation of a carbon–carbon bond between two ligands within a Ni complex:

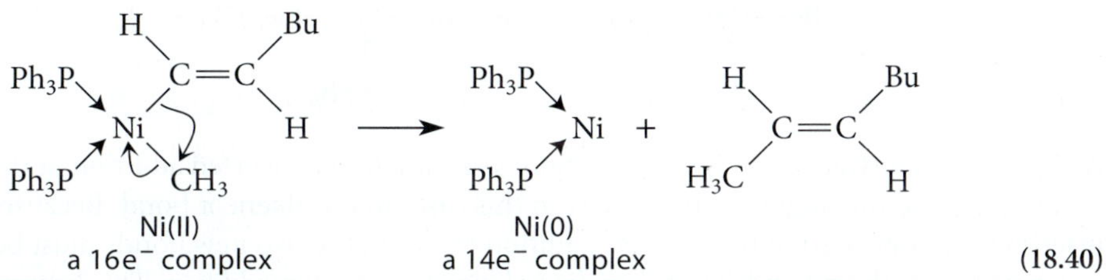

(18.40)

Because two X-type ligands are lost, the metal is reduced, and its electron count is decreased. In this particular example, the alkene stereochemistry is retained. *Reductive elimination in general occurs with retention of stereochemistry.* Because this process is the reverse of oxidative addition, it follows that *concerted oxidative addition also occurs with retention of stereochemistry.* The electron flow in Fig. 18.5 or Eq. 18.39 is consistent with this observation.

Ligand Insertion In **ligand insertion**, a ligand inserts into a metal–ligand bond—that is, L—M—R ⟶ M—L—R. Oxidative addition and ligand insertion are both insertion processes, but notice the difference between them. In an oxidative addition, the metal inserts into a chemical bond within a compound not initially associated with the metal. In a ligand insertion, a ligand L inserts into the bond between the metal and a different ligand R, and the inserting ligand gains a bond.

Two types of ligand insertion are most frequently observed in transition-metal chemistry. In a *1,1-insertion*, the new bond is formed at the same atom that was bound to the metal. Insertions of CO ligands are frequently observed examples of this type. The first reaction in Eq. 18.41 is an example.

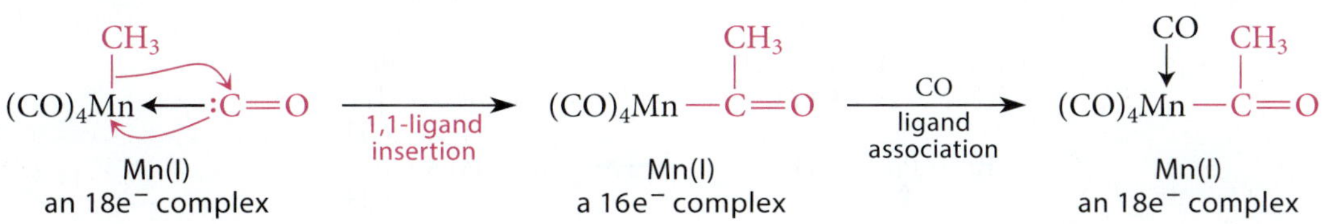

(18.41)

In this reaction, the methyl group migrates, *with its bonding electrons*, to the carbon of the carbonyl ligand, which in turn forms an X-type bond to the metal. This migration is possible because the carbon of the carbon monoxide ligand is electron-deficient. Therefore, the carbonyl carbon inserts into the Mn—CH_3 bond. This insertion leaves a vacant site (that is, an empty orbital) on the metal, as we can see from the reduction in the electron count. In the second reaction, this empty metal orbital is filled by another molecule of the ligand from solution.

Another type of ligand insertion is *1,2-insertion*. In this process, the migrating group moves to an atom adjacent to the one bound to the metal. A common example of this process is the insertion of an ethylene ligand into a Pd–aryl bond.

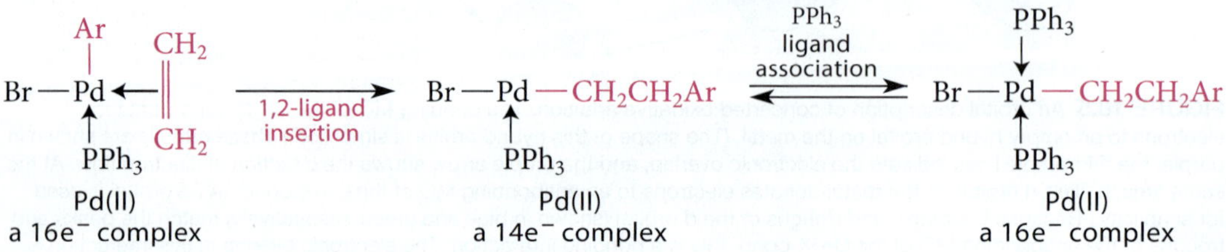

(18.42)

Again, the electron count is reduced by two—that is, the process results in an empty orbital on the metal. This orbital can then gain another electron pair by ligand association, as shown by the second step in Eq. 18.42, thus fulfilling the 16-electron rule.

We can approximate the ligand insertion process in the curved-arrow notation as follows:

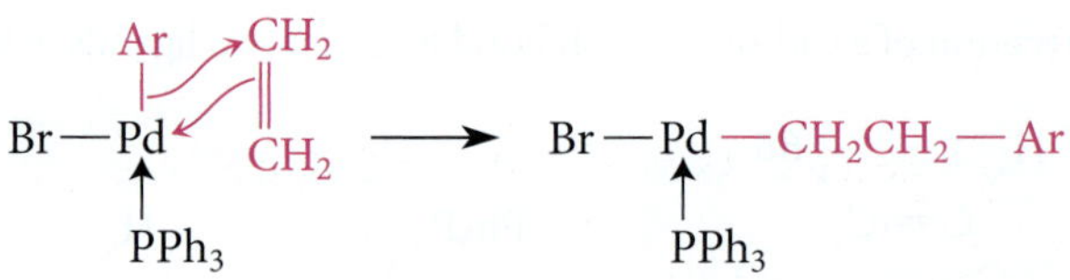

(18.43a)

As Eq. 18.43a illustrates, 1,2-ligand insertion is essentially a concerted addition of the metal (Pd in this case) and the migrating ligand (Ar in this case) to the alkene π bond. Because 1,2-ligand insertion is a concerted *intramolecular* addition reaction, the two new bonds must be formed at the same face of the π bond. Therefore, *ligand insertion is a syn addition.* This becomes evident when the carbons of the alkene double bond are stereocenters, as they are in cyclohexene. In this case, syn addition requires that the Pd and the aryl group have a cis relationship in the insertion product.

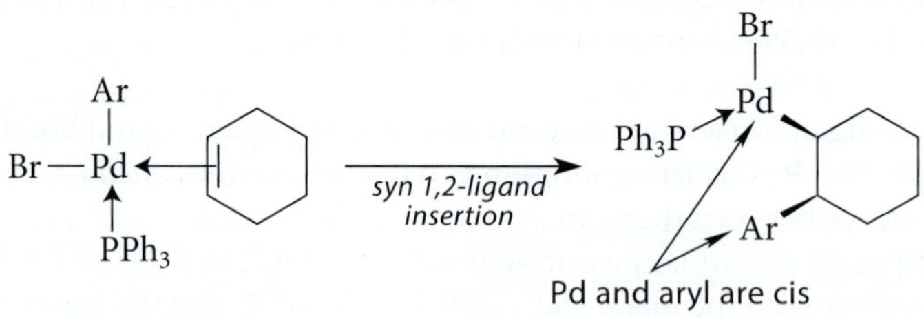

(18.43b)

β-Elimination In **β-elimination**, a group β to the metal migrates *with its bonding electron pair* to the metal. This process is conceptually the reverse of ligand insertion. (Run Eq. 18.43a backward mentally, and you will see the β-elimination of ethylene by migration of aryl.)

Many β-elimination reactions involve a hydride migration. For example, the product of Eq. 18.43a (with Ar = phenyl) can undergo β-elimination with hydride migration as follows:

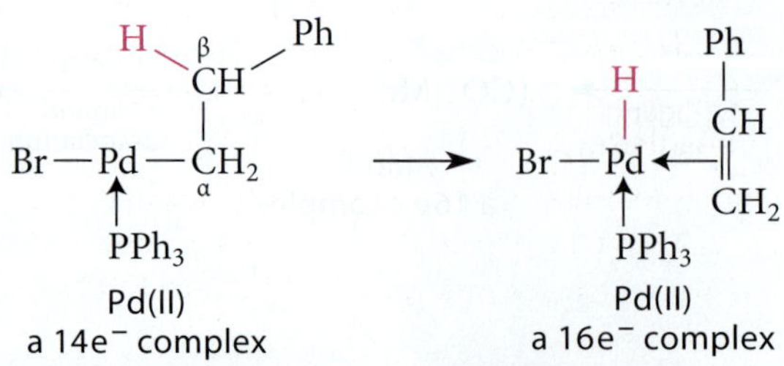

(18.44)

Notice that β-elimination requires an empty orbital on the metal, because, as a result of this process, the electron count is increased by 2 units.

We considered another type of β-elimination, the E2 reaction, in Sec. 9.5. The elimination reaction in Eq. 18.44 looks superficially similar, but *it is quite different*. In the E2 reaction, a *proton* is eliminated. In the β-elimination of Eq. 18.44, a *hydride*—a hydrogen with its bonding electrons—is eliminated by migration to the metal. We stress this point with the curved-arrow notation:

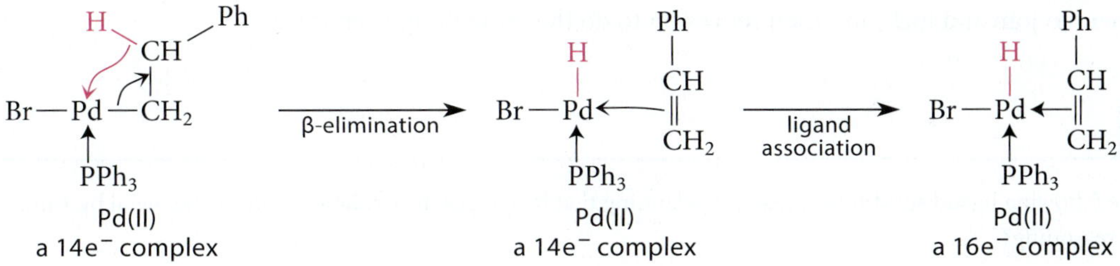

(18.45)

Because this β-elimination is *intramolecular*—within the same molecule—it must occur as a syn elimination. This makes sense because this reaction is conceptually the reverse of 1,2-ligand insertion, which is a syn addition (Eq. 18.43b). Contrast the stereochemistry of this β-elimination with that of the E2 elimination, which is a bimolecular reaction and occurs with anti stereochemistry (Sec. 9.5E).

Study Problem 18.3

Consider the following mechanism for Eq. 18.22 (Sec. 18.5). Identify the process associated with each step. Counting electrons at each stage may help you.

$$Pd(PPh_3)_4 \rightleftarrows Pd(PPh_3)_3 + PPh_3 \rightleftarrows Pd(PPh_3)_2 + PPh_3 \qquad \textbf{(18.46a)}$$

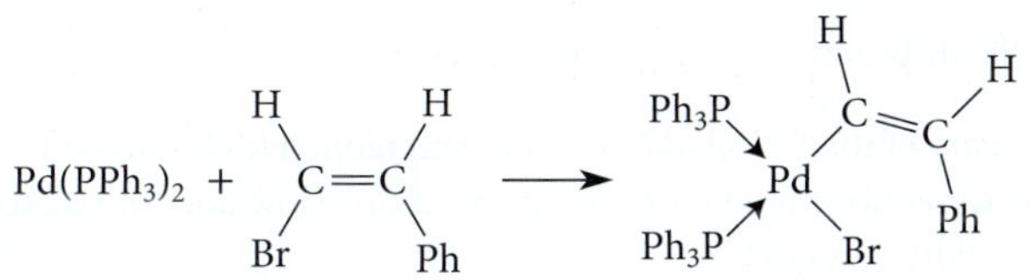

(18.46b)

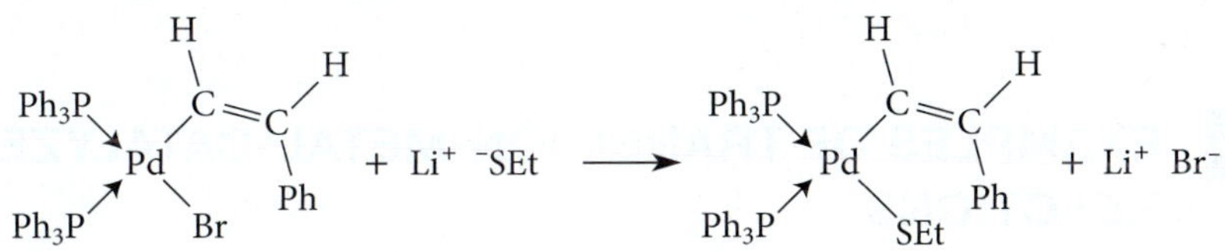

(18.46c)

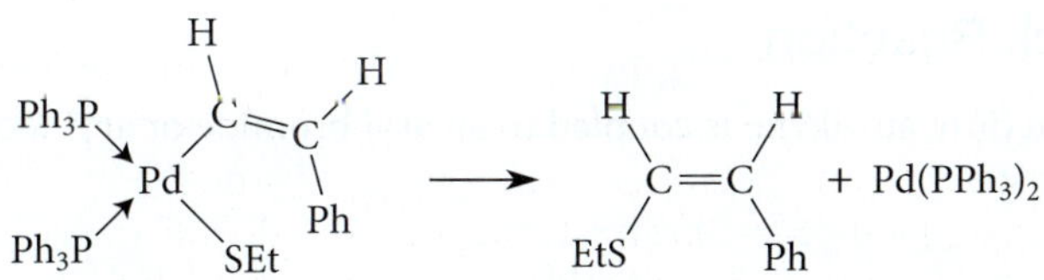

(18.46d)

Solution Step 18.46a consists of two successive *ligand dissociations* that reduce the electron count around the Pd from 18e⁻ to 14e⁻. This "makes room" for the vinylic halide, which undergoes an *oxidative addition* with retention of configuration in step 18.46b. This step takes the electron count to 16e⁻ and oxidizes the Pd(0) to Pd(II). Step 18.46c is a *ligand substitution* of a bromo ligand with an ethylthio ligand. It might occur by a prior dissociation of the Br⁻ ligand, by association of the EtS⁻ with the Pd to give an 18e⁻ complex followed by dissociation of Br⁻, or by a concerted mechanism reminiscent of the S_N2 reaction. Finally, step 18.46d is a *reductive elimination*, which forms the product with *retention of stereochemistry* and regenerates the catalytic Pd(0) species $Pd(PPh_3)_2$.

Let's use the example in Study Problem 18.3 to take stock of what the Pd is actually doing—that is, why it makes a vinylic or aryl substitution possible. Ligation to the Pd brings two groups—the vinylic group and the nucleophile—into proximity. The oxidative addition step is the key step that makes this possible, and, as we have seen (see Fig. 18.5), it is driven by the simultaneous presence of filled and empty metal orbitals that can interact with the vinylic halide so that the carbon–halogen bond is broken. The nucleophile EtS^- and the vinylic group are then connected by reductive elimination. The orbital interactions are essentially the same as in oxidative addition. The role of the metal finds analogy in the slider of a zipper: it brings two groups together, causes them to join and lock, and then moves on to do the same thing over again.

Focused Problems

18.14 A student has written the following ligand substitution reaction, claiming that it changes the oxidation state of the metal by 1 unit. What is wrong with this reasoning?

$$Cl^- + Pd(PPh_3)_4 \longrightarrow ClPd(PPh_3)_3 + :PPh_3$$

18.15 The *Wilkinson catalyst* chlorotris(triphenylphosphine)rhodium(I), $ClRh(PPh_3)_3$, brings about the catalytic hydrogenation of an alkene in homogeneous solution:

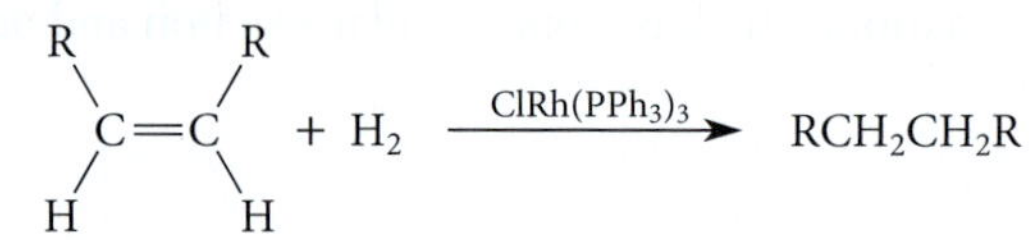

(18.47)

(a) Using the following mechanistic steps as your guide, draw structures of the transition-metal complexes involved in each step. Give the electron count and the metal oxidation state at each step.

1. Oxidative addition of H_2 to the catalyst
2. Ligand substitution of one PPh_3 by the alkene
3. 1,2-Insertion of the alkene into a Rh—H bond and re-addition of the previously expelled PPh_3 ligand
4. Reductive elimination of the alkane product to regenerate the catalyst

(b) According to the known stereochemistry of the 1,2-ligand insertion and reductive elimination steps, what would be the stereochemistry of the product if D_2 were substituted for H_2 in the reaction? How does this stereochemistry compare with that of catalytic hydrogenation over a Pd/C catalyst?

18.6 EXAMPLES OF TRANSITION-METAL-CATALYZED REACTIONS

A. The Heck Reaction

In the **Heck reaction**, an alkene is coupled to an aryl bromide or aryl iodide under the influence of a Pd(0) catalyst.

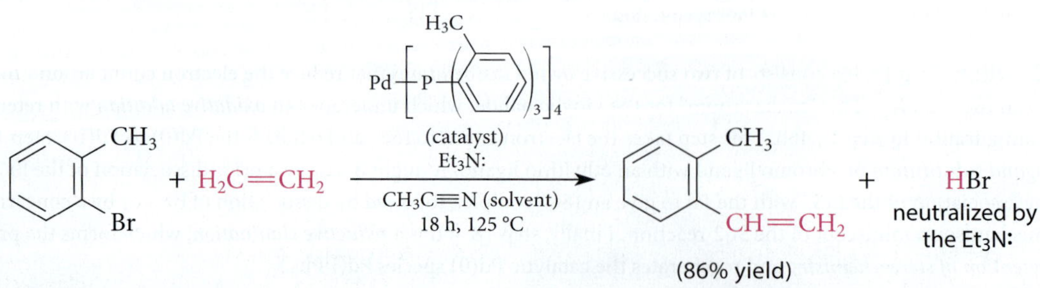

(18.48)

[The aryl substituents of the phosphine ligands used in the catalyst in this case are *o*-tolyl (that is, *o*-methylphenyl) groups rather than phenyl groups, but phenyl groups are also sometimes used.] The reaction is named for Richard F. Heck (1931–2015), who discovered the reaction in the early 1970s while a professor of chemistry at the University of Delaware. (A Japanese chemist, Tsutomu Mizoroki, simultaneously discovered the reaction, but it is generally known as the Heck reaction.) The Heck reaction has proven to be one of the most useful processes for forming carbon–carbon bonds to aromatic rings and even, occasionally, to vinylic groups. Heck shared the 2010 Nobel Prize in Chemistry with Akira Suzuki (b. 1930; see Sec. 18.6B) and Ei-ichi Negishi (b. 1935) of Purdue University (see Problem 18.84) for their discovery of transition-metal-catalyzed coupling reactions.

The mechanism of the Heck reaction is outlined in the following equations. You should identify the process or processes involved in each step (L = tri-*o*-tolylphosphine ligands; the steps in Eq. 18.49b are numbered for reference).

The actual catalytically active species is believed to be PdL_2, which is formed by two ligand dissociations:

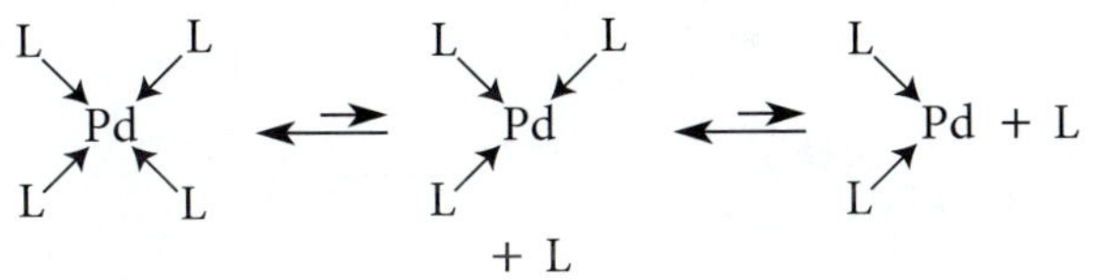

(18.49a)

The PdL_2 formed in Eq. 18.49a enters into the catalytic cycle.

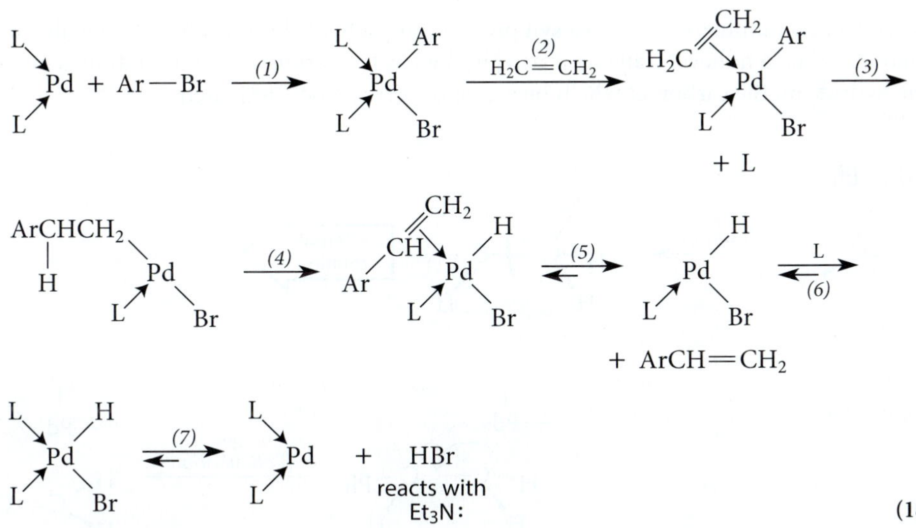

(18.49b)

Focused Problem

18.16 Characterize each step of the mechanism in Eq. 18.49b in terms of the fundamental processes discussed in Sec. 18.5E. Give the electron count and the oxidation state of the metal in each complex.

Another example of the Heck reaction illustrates two important aspects of the reaction.

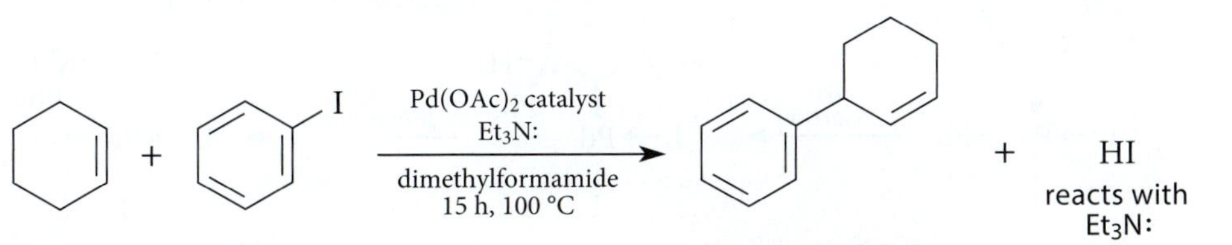

cyclohexene (excess) **iodobenzene** **(2-cyclohexenyl)benzene** (72% yield) (18.50)

First, the catalyst Pd(0) is generated from a Pd(II) compound, $Pd(OAc)_2$, which is used because it is a convenient and easily handled Pd derivative. It is believed that the Pd(II) is reduced to Pd(0), perhaps by a few molecules of alkene that are converted into vinylic acetates; Pd(0) is the actual catalyst. Addition of an oxidizable ligand such as PPh_3 can also reduce the Pd(II). Because a very small amount of Pd is used, the by-products of these reactions are also formed in very small amounts. Second, the alkene double bond in the product is *not* at the site of coupling, but rather one carbon removed. What has happened here?

This sort of product, which occurs commonly with cyclic alkenes in the Heck reaction, is a direct consequence of the stereochemistry of certain steps in the mechanism. The insertion step (step 3 in Eq. 18.49b) *must* occur in a syn manner because the reaction is intramolecular. Consequently, in the initially formed insertion complex, the Pd and the phenyl group become cis substituents on a cyclohexane ring.

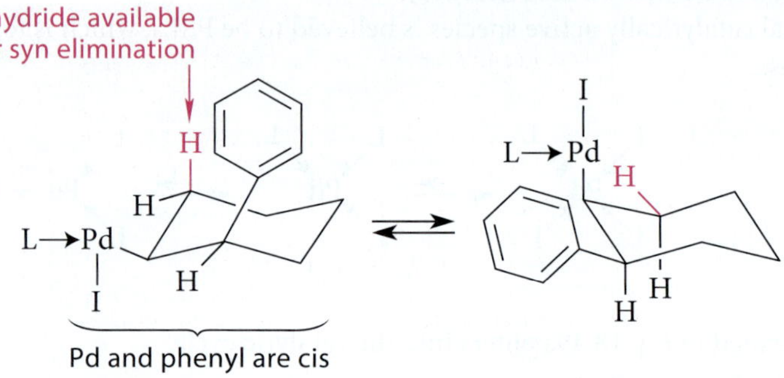

(18.51)

The subsequent β-elimination is also a syn process; only a hydride cis to the Pd is eligible for elimination. When a noncyclic alkene is used in the Heck reaction, internal rotation is possible so that the hydride on the carbon at which insertion occurs can be eliminated.

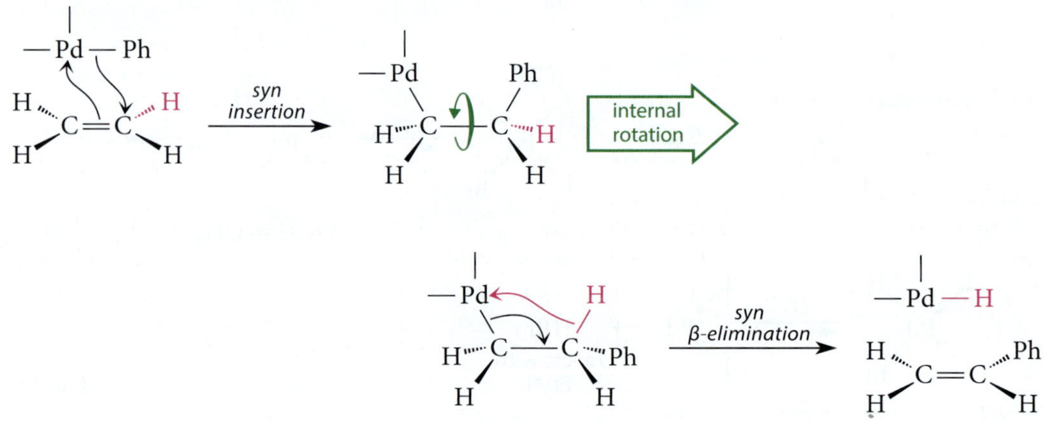

(18.52)

When the starting material is a cyclic alkene, as in Eq. 18.50, an analogous internal rotation is prevented by the ring. The only cis β-hydride available for elimination is the one (*red* in Eq. 18.51) on the *other* β-carbon. Elimination of this hydride yields an alkene in which the carbon at the insertion point—the one attached to the phenyl—is not part of the double bond, but is one carbon removed. We can summarize this in the following way, with the insertion point marked with an asterisk (*):

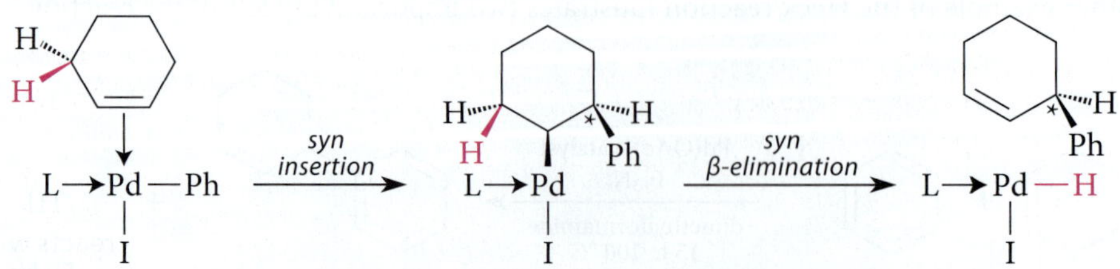

(18.53)

When the Heck reaction is applied to unsymmetrically substituted alkenes, such as an alkene of the form R—CH═CH_2, two products are in principle possible, because insertion might occur at either of the alkene carbons. It is found that when the R group is phenyl, CO_2R (ester), CN, or another relatively electronegative group, the aryl halide tends to react at the *unsubstituted carbon*—that is, the product is R—CH═CH—Ar, usually the *E* (or trans) stereoisomer. When R = alkyl, mixtures of products are often observed (Focused Problem 18.17).

Focused Problems

18.17 When iodobenzene and propene are subjected to the conditions of the Heck reaction, two constitutionally isomeric products are formed. What are they? Why are two products formed?

18.18 What *two* sets of aryl bromide and alkene starting materials would give the following compound as the product of a Heck reaction?

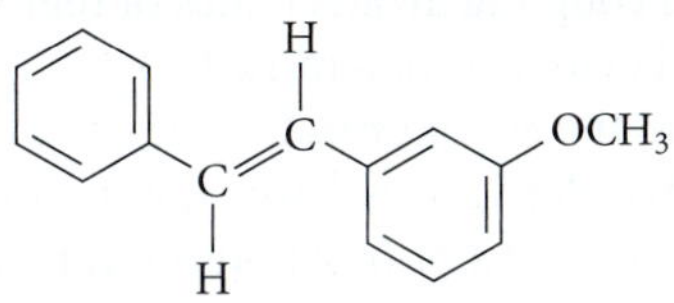

18.19 The product of a Heck reaction is, like the starting material, an alkene. Why doesn't a Heck reaction of the product compete with the reaction of the starting alkene?

18.20 What product is expected when cyclopentene reacts with iodobenzene in the presence of triethylamine and a Pd(0) catalyst?

B. The Suzuki Coupling

The Suzuki–Miyaura coupling reaction (usually called the **Suzuki reaction** or the **Suzuki coupling**) is a Pd(0)-catalyzed process in which an aryl or vinylic boronic acid (a compound of the form $RB(OH)_2$, where R = an aryl or vinylic group) is coupled to an aryl or vinylic iodide or bromide in the presence of a base, which is in many cases aqueous sodium hydroxide or sodium carbonate. The reaction can be used to prepare three compound types: **biaryls**—compounds in which two aryl rings are connected by a σ bond; aryl-substituted alkenes; and conjugated alkenes. Equation 18.54 illustrates the preparation of a biphenyl.

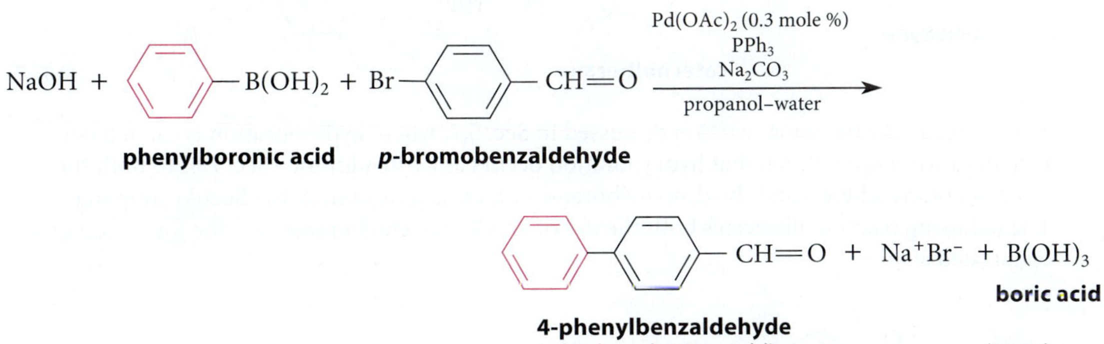

(18.54)

The Pd(0) catalyst can be $Pd(PPh_3)_4$, the same catalyst used in the Heck reaction, or the Pd(0) can be formed in the reaction flask from $Pd(OAc)_2$, a strategy that is also used in the Heck reaction, as in the preceding example.

Equation 18.55 shows the preparation of an aryl-substituted alkene.

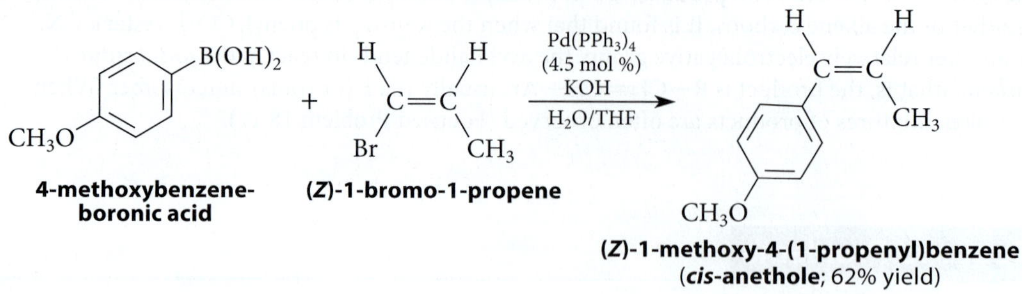

(18.55)

As this example illustrates, the coupling occurs with retention of alkene stereochemistry. You may have noticed that this is the type of compound that can be prepared by the Heck reaction. However, the Suzuki coupling avoids issues of regiochemistry that can sometimes occur with the Heck reaction. (See Focused Problem 18.17.)

The importance of the Suzuki reaction has resulted in the commercial availability of many boronic acids and their derivatives. Two ways of preparing the required boron derivatives are, first, the reaction of Grignard or organolithium reagent with trimethyl borate:

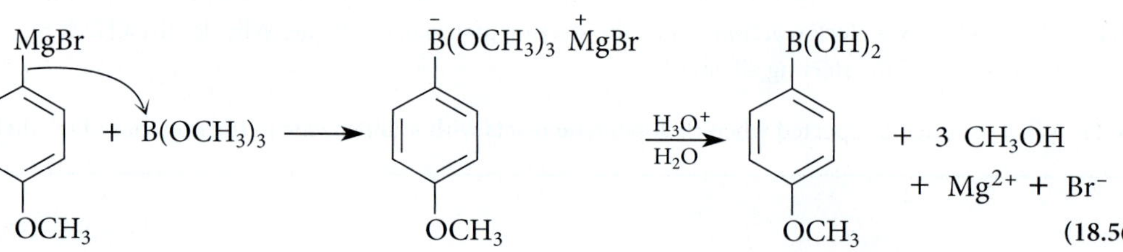

(18.56)

In this reaction, the Grignard reagent, a strong Lewis base, donates electrons to the boron in a Lewis acid–base association. A reaction with aqueous acid results in formation of the boronic acid product. (See Focused Problem 18.24.) The analogous reaction can be used to form vinylic boronic acids from vinylic Grignard reagents. A second preparation of vinylic boronic acid derivatives is the hydroboration of 1-alkynes with catecholborane:

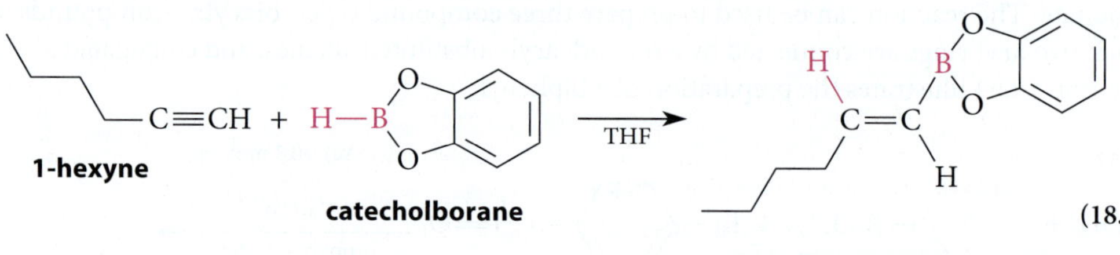

(18.57)

This is essentially the same reaction discussed in Sec. 5.7, where hydroboration is carried out with disiamylborane. Recall that hydroboration occurs as a syn addition (Sec. 7.8D). Both the catecholborane adducts and the disiamylborane adducts can be used in the Suzuki coupling. The following reaction illustrates both the use of vinylic catecholboranes and the formation of a conjugated alkene.

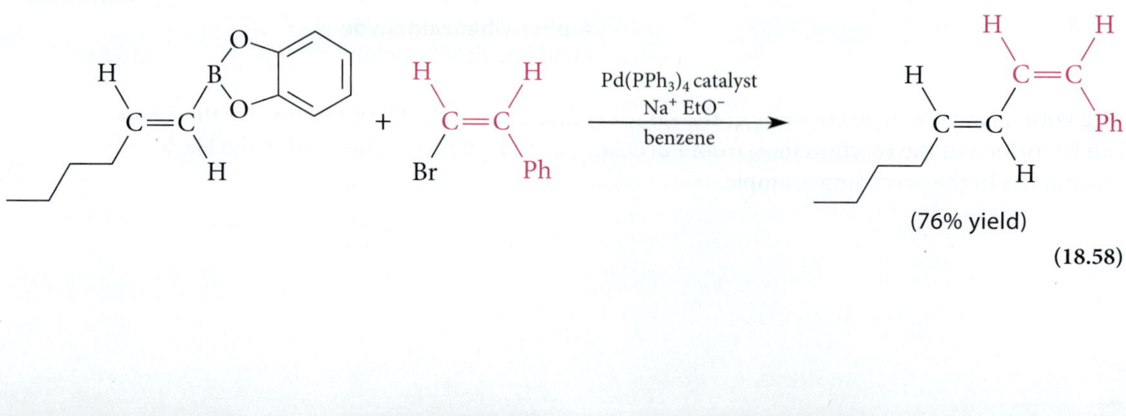

(18.58)

The mechanism of the Suzuki coupling begins exactly like that of the Heck reaction (Eqs. 18.49a and b, Sec. 18.6A) with ligand dissociation to give a $14e^-$ complex, followed by oxidative addition of the aryl halide.

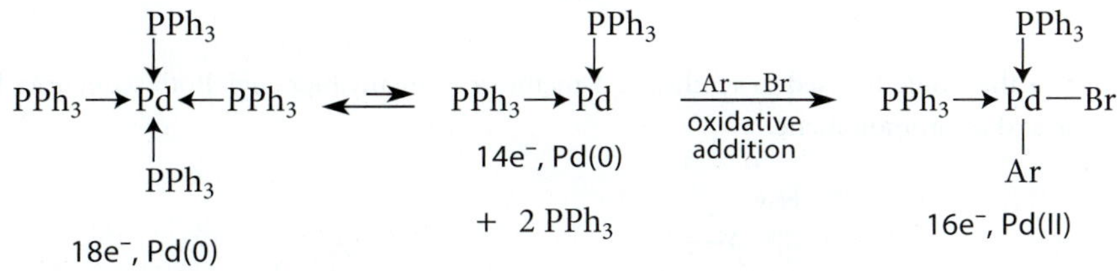

(18.59a)

From this point on, the mechanism is not definitively established, but a reasonable sequence involves another ligand substitution in which the base displaces the halide ion:

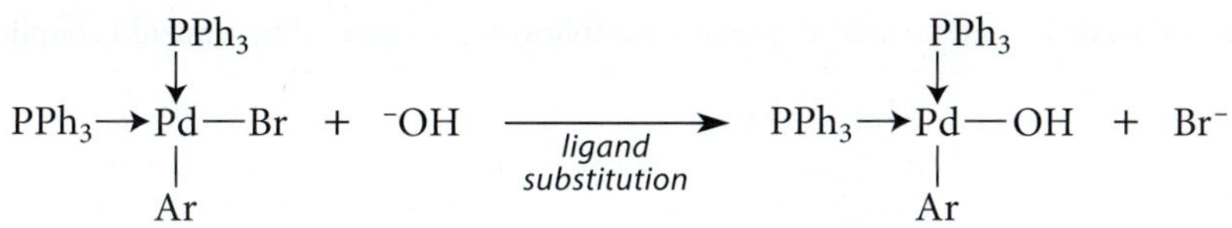

(18.59b)

A Lewis acid–base association brings the boron into the coordination sphere of the metal:

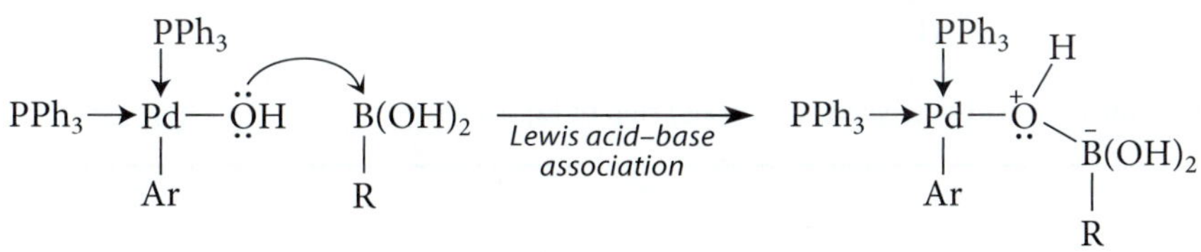

(18.59c)

This association results in a formal negative charge on boron. Remember, though, that carbon is more electronegative than boron; this means that carbon bears a significant amount of the negative charge in this complex. In other words, the Lewis acid–base association of the oxygen with boron endows the carbon in the carbon–boron bond with significant carbanion character. An intramolecular substitution of this "carbon anion," a strong base, for the weaker base HO—$B(OH)_2$ results in transfer of the R group to the metal. Reductive elimination then gives the coupling product and provides the catalyst for another cycle.

The Suzuki coupling is named for Akira Suzuki (b. 1930), who is on the faculty of the Kirashiki University of Science and the Arts in Kirashiki, Japan. Professor Suzuki had spent 2 years as a postdoctoral fellow with Herbert C. Brown (the discoverer of hydroboration), where he was immersed in the organic chemistry of boron. His intellectual synthesis of transition-metal organometallic chemistry and organoboron chemistry led to the discovery in the mid-1970s of the reaction that bears his name, in collaboration with his student Norio Miyaura, while the two were at Hokkaido University in Japan. The Suzuki coupling has become a very important reaction in both academic and industrial settings. As noted in Sec 18.6A, Suzuki shared the 2010 Nobel Prize in Chemistry.

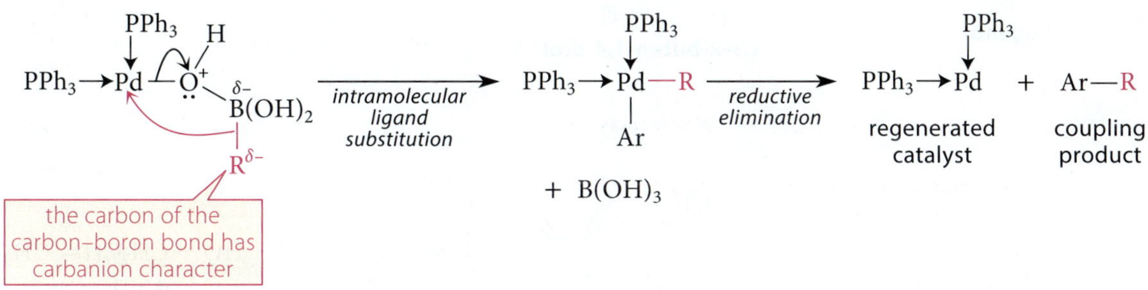

(18.59d)

Focused Problems

18.21 Complete the following Pd(0)-catalyzed Suzuki reactions by giving the coupling product. For parts (b) and (c), include the stereochemistry of the products.

(a)

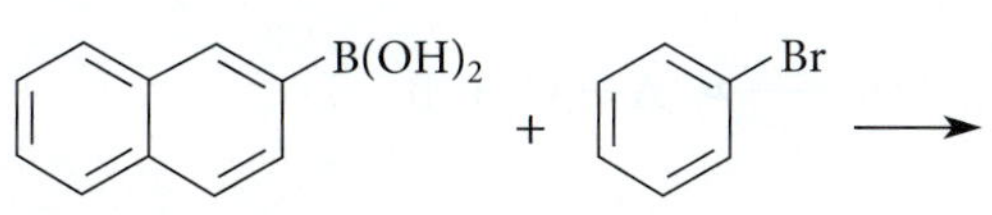

2-naphthaleneboronic acid

(b)

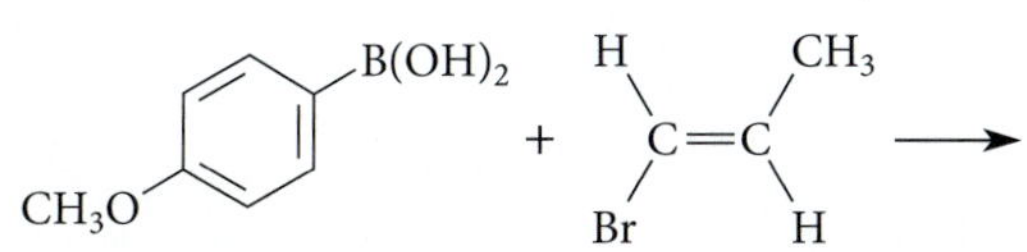

(c) 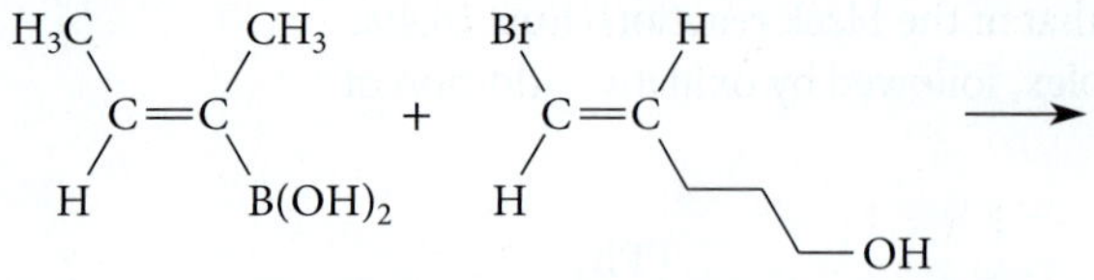

18.22 Provide two different reaction sequences that could be used to synthesize 4-methoxy-3′-methylbiphenyl. Both sequences, however, should start with both *p*-bromoanisole and *m*-bromotoluene.

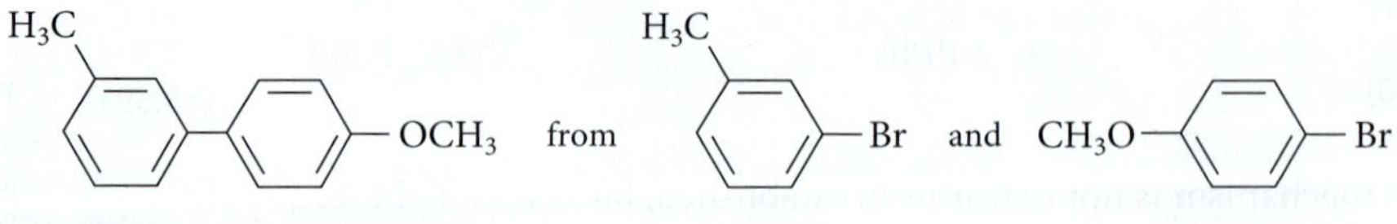

4-methoxy-3′-methylbiphenyl **m-bromotoluene** **p-bromoanisole**

18.23 Give two different pairs of starting materials that could be used to prepare the following compound by a Suzuki coupling.

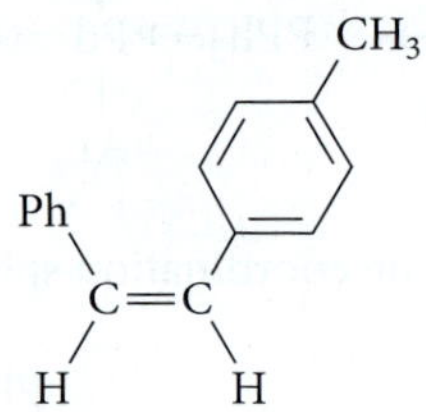

18.24 Draw a curved-arrow mechanism for the last (acid-catalyzed hydrolysis) step of Eq. 18.56.

C. Alkene Metathesis

Certain ruthenium(IV) catalysts bring about a reaction in which two alkenes are "sliced apart" at their double bonds, and the parts reassembled, to give new alkenes.

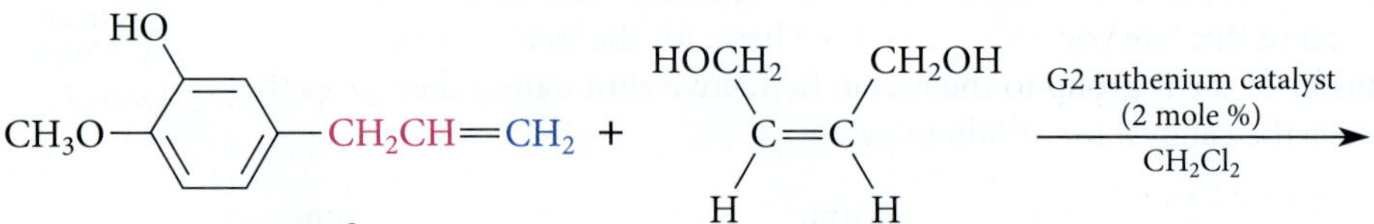

eugenol **cis-2-buten-1,4-diol**

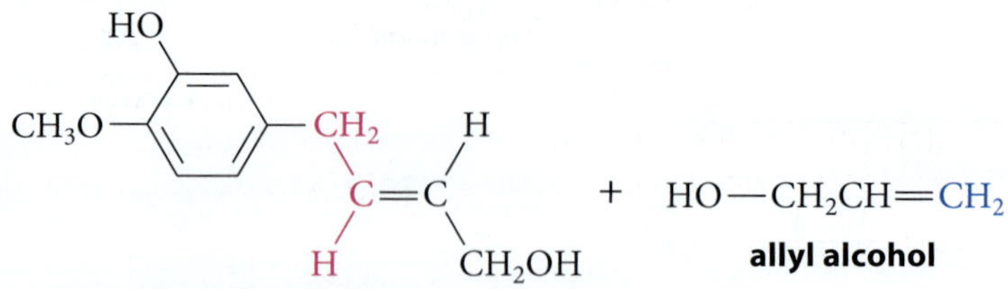

(E)-4-(3-hydroxy-4-methoxyphenyl)-2-buten-1-ol (80% yield) **allyl alcohol** (18.60)

This remarkable reaction is an example of **alkene metathesis**, also called **olefin metathesis**. A metathesis reaction (pronounced mə-tâ′-thə-sĭs, from the Greek, *meta* = change, *thesis* = place) can be represented in general as follows:

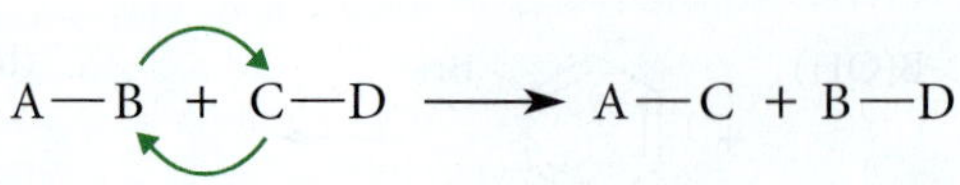

(18.61)

You are probably familiar with some inorganic examples of metathesis reactions, such as the reaction of silver nitrate with sodium chloride:

$$\overset{+}{Ag}\,\overset{-}{NO_3} + \overset{+}{Na}\,\overset{-}{Cl} \longrightarrow AgCl\downarrow + \overset{+}{Na}\,\overset{-}{NO_3} \tag{18.62}$$

silver(I) nitrate — **sodium chloride** — **silver(I) chloride** — **sodium nitrate**

In an alkene metathesis, the groups at each end of the double bonds are interchanged. For example, the reaction in Eq. 18.60 has the following form:

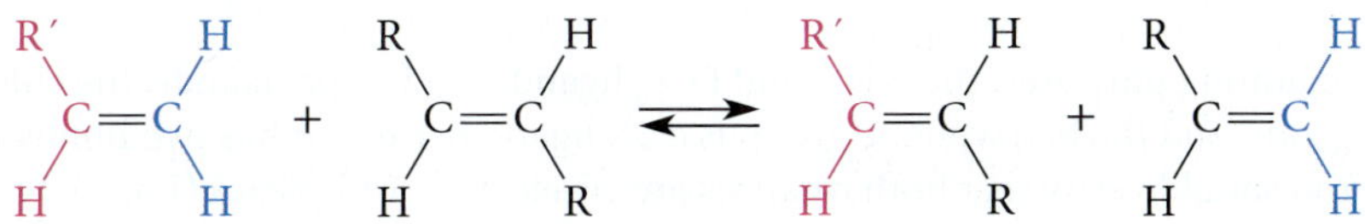

(18.63)

For a metathesis reaction of two unsymmetrical alkenes, 10 alkenes (not counting stereoisomers) can be formed. (See Focused Problem 18.25.) However, the usual applications of this reaction are typically not so complex.

A number of transition-metal catalysts have been developed for alkene metathesis. These catalysts are based on tungsten, molybdenum, and especially ruthenium. The two most widely used laboratory catalysts are the ruthenium-based catalysts G1, which stands for the *Grubbs first-generation catalyst*, and G2, the *Grubbs second-generation catalyst*. The structures of these catalysts are shown in **Fig. 18.6**. Although we need not go into detail here, the design of these catalysts was an evolutionary process that involved a consideration of steric and electronic effects of the ligands in light of the reaction mechanism (which we discuss later), as well as some outright fortuitous discoveries! Ruthenium catalysts can be easily handled in the laboratory, and, as Eq. 18.60 illustrates, they can be used in the presence of alcohols, phenols, and other functional groups. The molybdenum and tungsten catalysts are much more air-sensitive and less tolerant of other functional groups.

The ruthenium–carbon double bond plays an important role in the operation of these catalysts. The rather unusual NHC ligand in the G2 catalyst, when "dissociated" from the metal, is actually a *carbene* (a molecule containing divalent carbon; Sec. 10.7). Most carbenes are very unstable, but this carbene is stabilized by resonance.

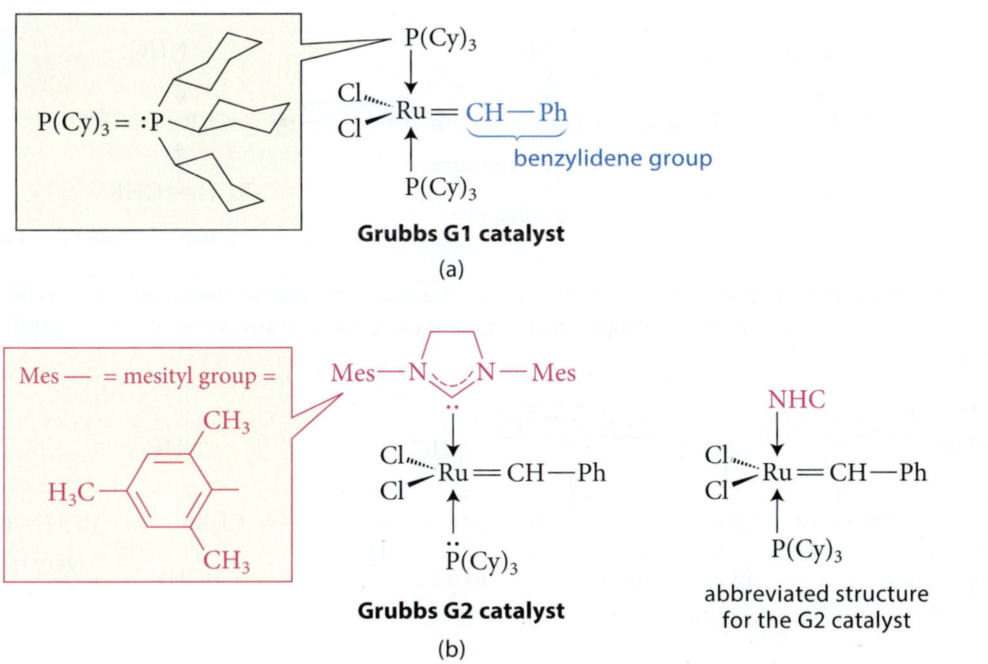

FIGURE 18.6 The structures of two ruthenium catalysts for alkene metathesis and their abbreviated structures. *Cy* is the abbreviation for the cyclohexyl group, and *Mes* is the abbreviation for the mesityl, or 2,4,6-trimethylphenyl, group. The unusual ligand in G2 is a stabilized carbene (see Eq. 18.64), and it is abbreviated *NHC* (for "*n*itrogen *h*eterocycle-stabilized *c*arbene"). Formal charges in the NHC ligand are conventionally not shown. A key aspect of these catalysts is the ruthenium–carbon double bond.

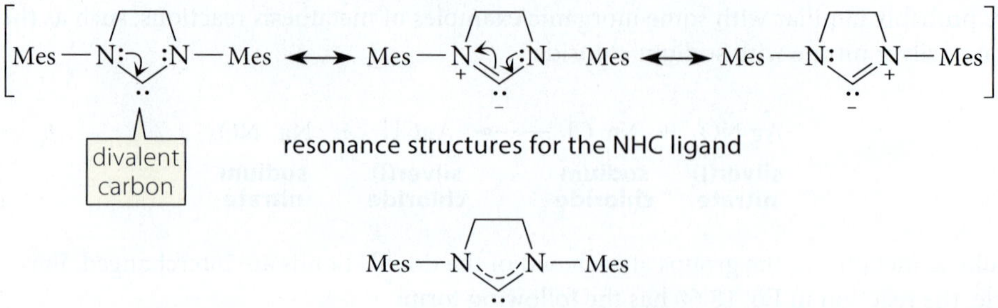

(18.64)

For electron-counting purposes, the NHC and PCy_3 ligands are L-type ligands, the chlorines are X-type ligands, and the benzylidene group is a 2X ligand because it has two bonds to the ruthenium. You should verify that both catalysts are 16-electron complexes (Eq. 18.34, Sec. 18.5D) and that the oxidation state of ruthenium (Eq. 18.28, Sec. 18.5B) is formally Ru(IV).

In many cases, the catalysts bring some or all of the possible alkenes into equilibrium. In such cases, the practical use of alkene metathesis requires the application of Le Chatelier's principle. For example, in Eq. 18.60, one of the starting materials, *cis*-2-buten-1,4-diol, is used in excess. Use of an excess of this diol is practical because it is cheap, and because both it and the by-product allyl alcohol are readily separated from the desired product.

One of the most important applications of alkene metathesis is for closing rings, and it can be used to close medium- and large-sized rings.

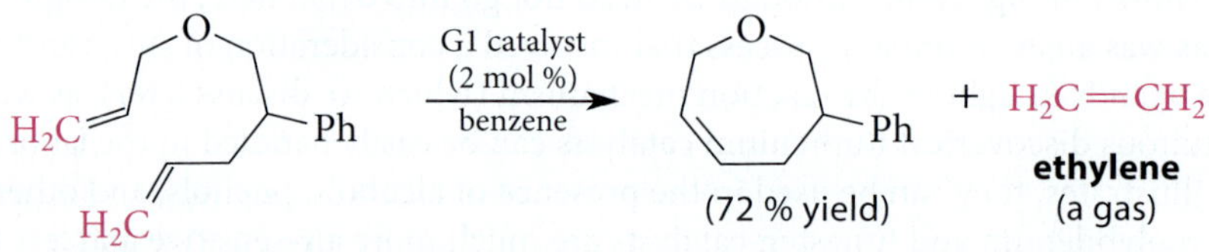

(18.65)

In this and many other ring-closing applications, the by-product is ethylene, which bubbles out of the reaction mixture, thereby driving the equilibrium toward the product—Le Chatelier's principle in operation again.

The mechanism of alkene metathesis, which we illustrate for Eq. 18.63 using the G2 catalyst, starts with loss of the PCy_3 ligand by ligand dissociation. Because ruthenium in the resulting complex has 14 electrons, it can accept 2 electrons from another ligand, in this case one of the alkenes. Let the alkene RCH=CHR be present in large excess; because of its concentration, it is more likely to interact with the catalyst.

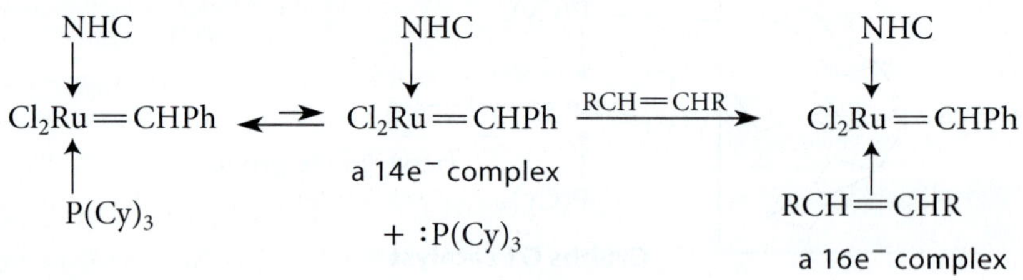

(18.66a)

Then follows a key step in alkene metathesis, a cycloaddition to form a *metallacycle* (a cyclic compound in which the metal occupies a ring position). This reaction is essentially a ligand insertion (Sec. 18.5A).

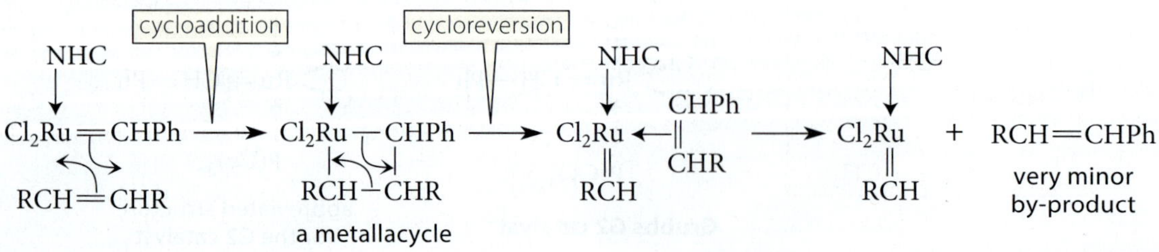

(18.66b)

As shown in this equation, the metallacycle then breaks down in the opposite sense by a *cycloreversion* (the reverse of a cycloaddition) to give a new alkene, which contains the benzylidene group. This becomes a very minor by-product, because the catalyst (and therefore the benzylidene group) is typically present in 1 or 2 mol % of the reactants. This process leaves the catalyst "primed" with the RCH═group.

The cycloaddition–cycloreversion process is now repeated. It is most probable that the process will occur with the alkene used in Eq. 18.66b, because this alkene is in excess; but the reaction with this alkene results in no change. (Be sure to demonstrate this point to yourself.) However, occasionally a molecule of the other alkene $R'CH{=}CH_2$ will bind to the catalyst, and this produces a metathesis product.

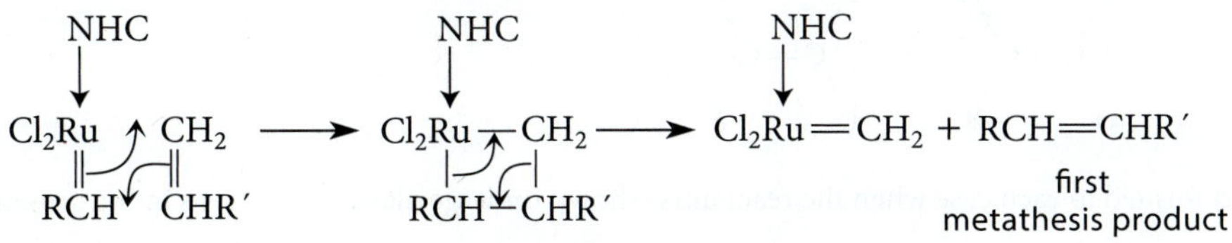

(18.66c)

This leaves the catalyst with a $={CH_2}$ group. This catalyst molecule is most likely to react with the alkene in excess, and when that happens, the second metathesis product (which is allyl alcohol in Eq. 18.60) is formed, and the catalyst is again primed with a ═CHR group.

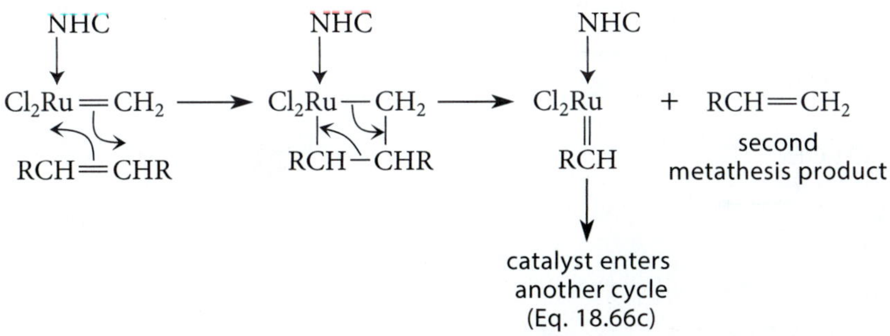

(18.66d)

The sequence in Eqs. 18.66c and d continues repeatedly until the limiting reactant is exhausted.

It is conceivable that, as the product builds up, it might undergo metathesis with itself. However, self-metathesis is largely avoided in this example because one of the starting materials is used in excess, and it intercepts the catalyst almost every time. In other words, if the product enters into the metathesis sequence and is split by the catalyst, the fragments are most likely intercepted by the alkene present in excess, and such a reaction either gives back that same alkene or the desired product.

When it is impractical to use one alkene in excess or to exploit Le Chatelier's principle in some other way, self-metathesis of the product can be a problem. However, this potential complexity of alkene metathesis is mitigated by the fact that different alkenes undergo metathesis at greatly different rates. Alkene metathesis is sensitive to the steric environment of the alkene double bonds. For example, methylpropene (isobutylene) does not undergo self-metathesis, presumably because the interaction of two $(CH_3)_2C{=}$ fragments with the catalyst results in severe van der Waals repulsions (that is, steric congestion) with the bulky catalyst ligands.

The importance of alkene metathesis was recognized with the 2005 Nobel Prize in Chemistry, which was shared by three chemists: Robert H. Grubbs (b. 1942) of the California Institute of Technology, who developed the ruthenium catalysts; Richard R. Schrock (b. 1945) of the Massachusetts Institute of Technology, who developed molybdenum-based metathesis catalysts; and Yves Chauvin (1930–2015) of the Petroleum Institute of France, who first proposed the reaction mechanism.

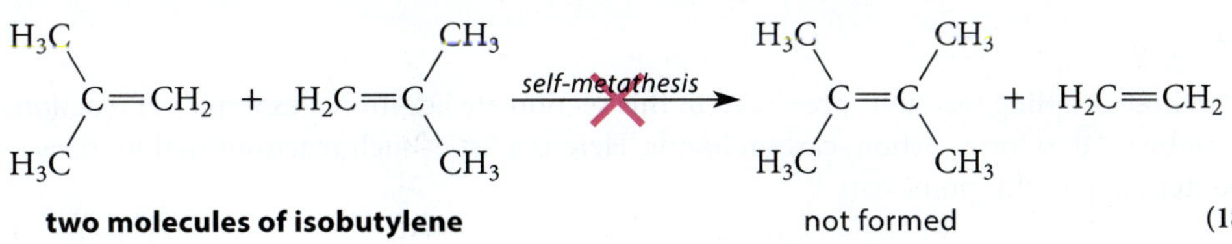

(18.67)

(Isobutylene can be used as a metathesis partner with less crowded alkenes, however.) In fact, the metathesis catalysts were designed to emphasize such differences in alkene reactivity.

Alkene metathesis plays a major role in the synthesis of complex organic molecules, and it is used industrially with increasing frequency in such diverse applications as alkene synthesis, polymer synthesis, and pheromone synthesis. Because ruthenium catalysts can be used in aqueous solution, they are providing options for "green chemistry"—chemistry that is environmentally friendly.

Focused Problems

18.25 Show that the equilibrium mixture produced by alkene metathesis of two completely different alkenes with the following general structures contains 10 different alkenes. (Assume that all alkenes have trans stereochemistry.)

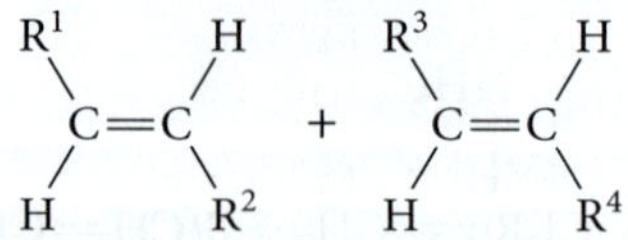

18.26 Give the structure of the major product formed in each case when the reactant(s) shown undergo alkene metathesis in the presence of an appropriate ruthenium catalyst.

(a) 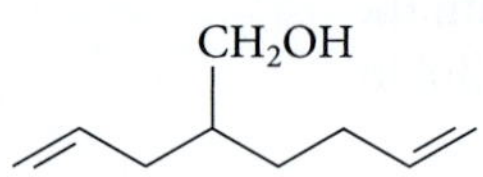 ⟶ a compound with 7 carbons

(b) 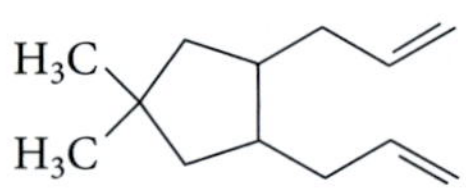 ⟶ a compound with 11 carbons

(c) 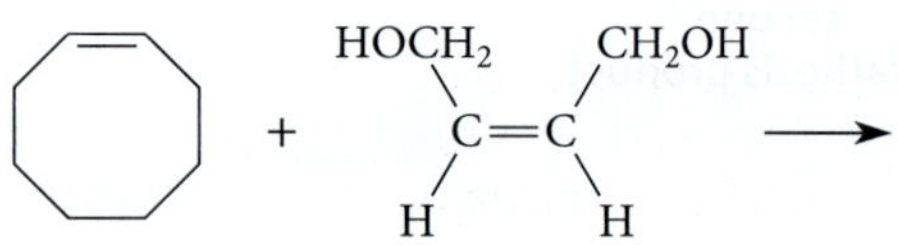

(large excess)

18.27 Suggest an alkene metathesis reaction that would yield each of the following compounds as a major product.

(a) 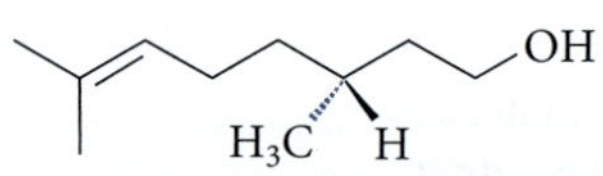

(–)-citronellol
(oil of roses)

(b) 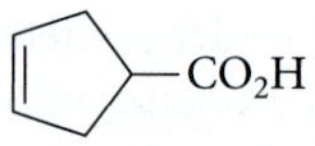

(c) 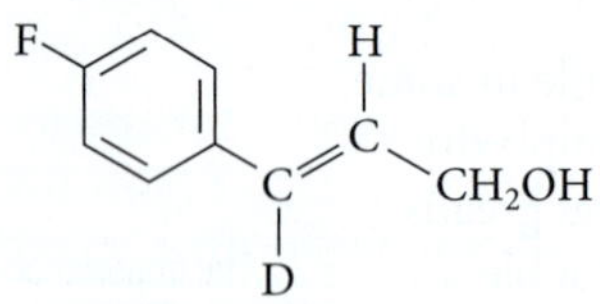

18.28 Draw structures analogous to those in Eqs. 18.66a–d for the catalytic intermediates formed in the conversion of 1,7-octadiene to cyclohexene and ethylene catalyzed by the G2 catalyst.

The three coupling reactions presented in this section are additional examples of reactions that can be used to form carbon–carbon bonds. Here is a list of such reactions that we have encountered up to this point:

1. The addition of carbenes and carbenoids to alkenes (Sec. 10.7)
2. The reaction of Grignard reagents with ethylene oxide and lithium organocuprate reagents with epoxides (Sec. 12.5C)
3. The reaction of acetylenic anions with alkyl halides or sulfonate esters (Sec. 10.6B)

4. The Diels–Alder reaction (Sec. 15.3)
5. The Friedel–Crafts alkylation and acylation reactions (Secs. 16.4E and F)
6. The Heck and Suzuki coupling reactions (Secs. 18.6A and B)
7. Alkene metathesis (this section)

D. Other Examples of Transition-Metal-Catalyzed Reactions

One of the most important transition-metal catalysts in commerce is a catalyst formed from $TiCl_3$ and $(CH_3CH_2)_2AlCl$, called the *Ziegler–Natta catalyst.* This catalyzes the polymerization of ethylene and other alkenes at 25 °C and 1 atm pressure. Although free-radical polymerization of ethylene (Sec. 10.3B) is very important, the Ziegler–Natta polymerization of ethylene accounts for most of the polyethylene produced; the resulting *high-density polyethylene* has different properties from the *low-density polyethylene* produced by free-radical processes. The discoverers of this catalyst, the German chemist Karl Ziegler (1898–1973) and the Italian chemist Giulio Natta (1903–1979), shared the 1963 Nobel Prize in Chemistry for their work. Although the mechanism of the polymerization has been enthusiastically debated, the following sequence is one possibility:

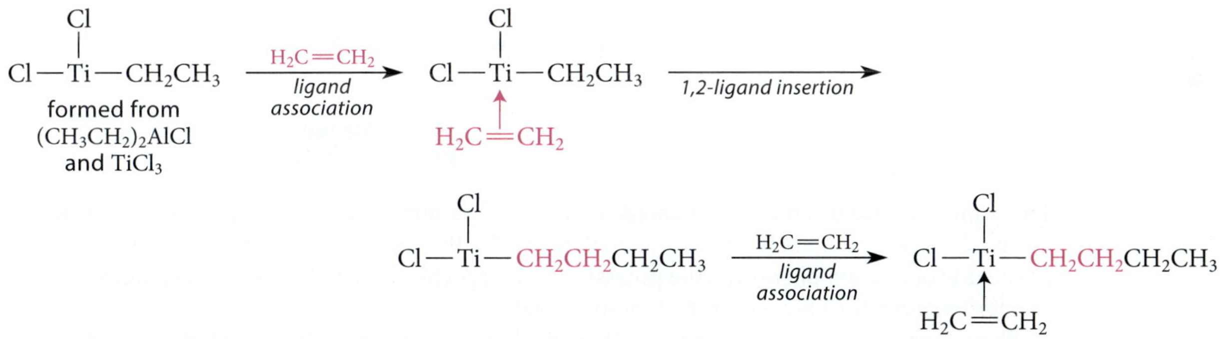

(18.68)

Continuation of the insertion–ligand-association sequence gives the polymer. It is believed that titanium brings about this reaction because a d^1 metal cannot undergo β-elimination. (β-Elimination requires some filled metal d orbitals for reasons that we haven't discussed.) The tendency toward β-elimination of other metals would terminate the reaction.

Hydroformylation is another commercially important process that involves a transition-metal catalyst, in this case a tetracarbonylhydridocobalt(I) catalyst. Propionaldehyde, for example, is produced by the hydroformylation of ethylene. (This is sometimes called the *oxo process.*)

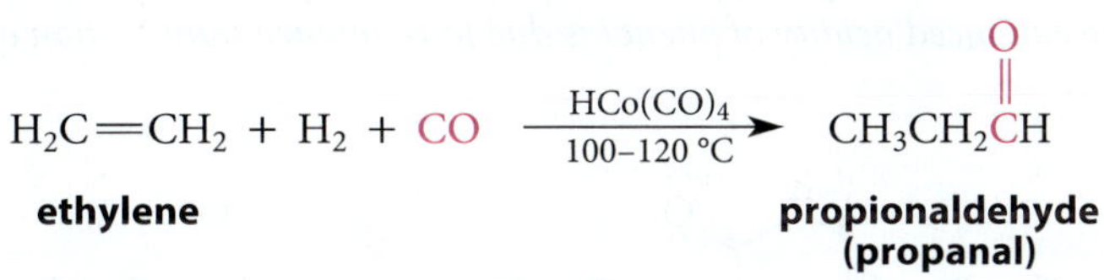

(18.69)

This process involves, among other things, a 1,2-insertion reaction of ethylene and a 1,1-insertion reaction of carbon monoxide (Focused Problem 18.29).

Yet another important transition-metal-catalyzed reaction is the *homogeneous* catalytic hydrogenation of alkenes using a soluble rhodium(I) catalyst called the *Wilkinson catalyst,* $ClRh(PPh_3)_3$. This reaction was explored in Focused Problem 18.15 (Sec. 18.5E).

And let's not forget catalytic hydrogenation (Secs. 4.9A and B), an important reaction that occurs over carbon-supported transition metals such as Ni, Pd, and Pt. The mechanism of catalytic hydrogenation is not definitively known, but it is not hard to imagine that the mechanism might involve oxidative additions and insertions on the catalyst surface much like those that take place in solution with the Wilkinson catalyst.

Many aspects of transition-metal chemistry are beyond the scope of an introduction. How does the chemist design a catalytic system and choose a catalyst? What influences the choice of ligands and solvents? These questions are sometimes addressed with a certain degree of empiricism, but the bases for the answers to these questions are becoming better understood.

Focused Problem

18.29 Suggest a mechanism for the oxo reaction (Eq. 18.69) involving intermediates that are consistent with the 16- and 18-electron rules.

18.7 ACIDITY OF PHENOLS

A. Resonance and Polar Effects on the Acidity of Phenols

Phenols, like alcohols, can ionize.

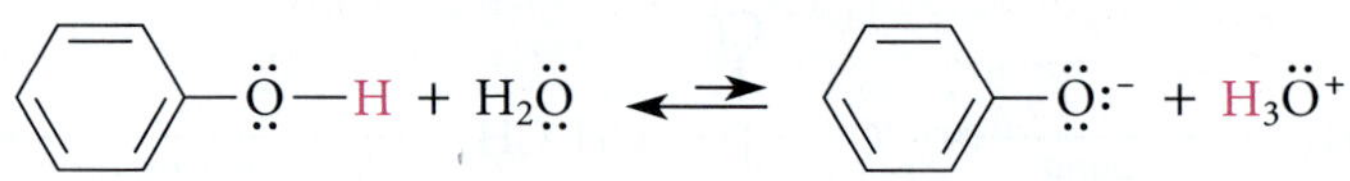

phenol **phenoxide ion (phenolate ion)** (18.70)

The conjugate base of a phenol is named, using common nomenclature, as a *phenoxide ion* or, using substitutive nomenclature, as a *phenolate ion*. So, the sodium salt of phenol is called sodium phenoxide or sodium phenolate; the potassium salt of *p*-chlorophenol is called potassium *p*-chlorophenoxide or potassium 4-chlorophenolate.

Phenols are considerably more acidic than alcohols. For example, the pK_a of phenol is 9.95, but that of cyclohexanol is about 17. In other words, phenol is approximately 10^7 times more acidic than an alcohol of similar size and shape.

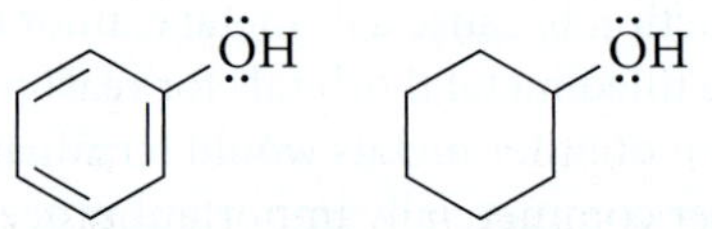

pK_a ~10 ~17 (18.71)

As explained in Sec. 3.7D, the pK_a of an acid is decreased by resonance stabilization of its conjugate base. *The enhanced acidity of phenol is due to resonance stabilization of its conjugate-base anion.*

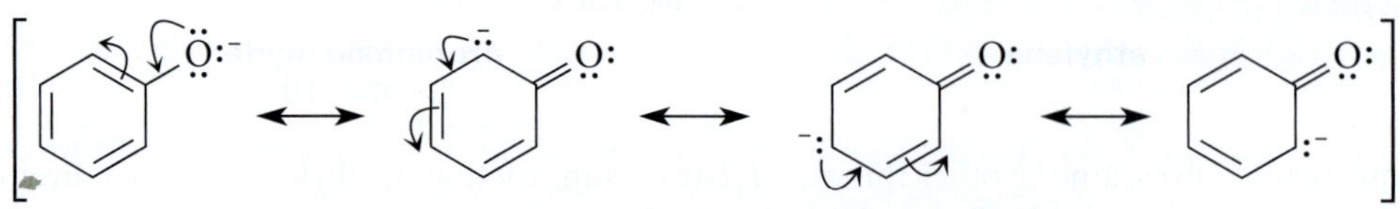

resonance structures for the phenoxide anion (18.72)

Phenol itself is also stabilized by resonance.

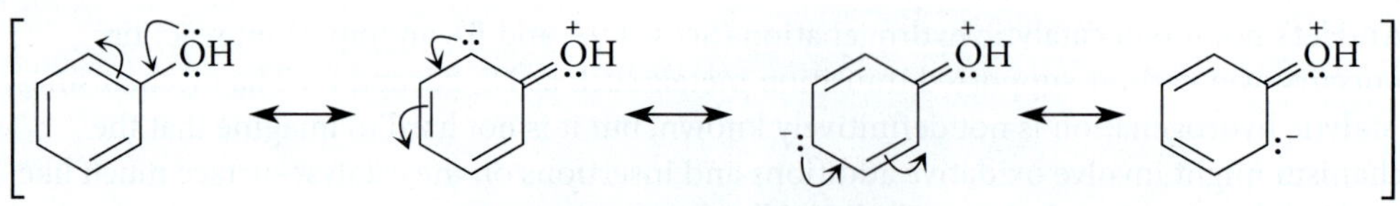

resonance structures for phenol (notice the *separation* of charge) (18.73)

However, the resonance structures for a phenolate ion *delocalize charge*, whereas those for the un-ionized phenol *separate* charge. Therefore, the resonance structures of phenolate are *more important* than the resonance structures for un-ionized phenol (Sec. 3.3C, guideline 3).

The second effect that stabilizes the conjugate base is the polar effect of the benzene ring, which also stabilizes negative charge (see Table 3.4, Sec. 3.7E). Both resonance stabilization and polar effects are the same effects that stabilize benzylic carbanions (Sec. 17.3). In fact, we can think of phenoxide as a benzylic anion in which the benzylic group is an oxygen instead of a carbon.

Alkoxides are stabilized neither by resonance nor by the polar effect of benzene rings or double bonds.

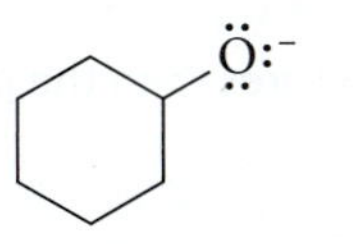

cyclohexanolate anion
no resonance structures;
no polar effect of double bonds

Because phenoxide ions are stabilized by both resonance and polar effects, less energy is required to form phenoxides from phenols than is required to form alkoxides from alcohols. *Because pK_a is directly proportional to the standard free energy of ionization* (Eq. 3.61b, Sec. 3.6D), phenols have lower pK_a values, and so are more acidic, than alcohols.

Substituent groups can also affect phenol acidity by both polar and resonance effects. For example, the relative acidities of phenol, *m*-nitrophenol, and *p*-nitrophenol reflect the operation of both effects.

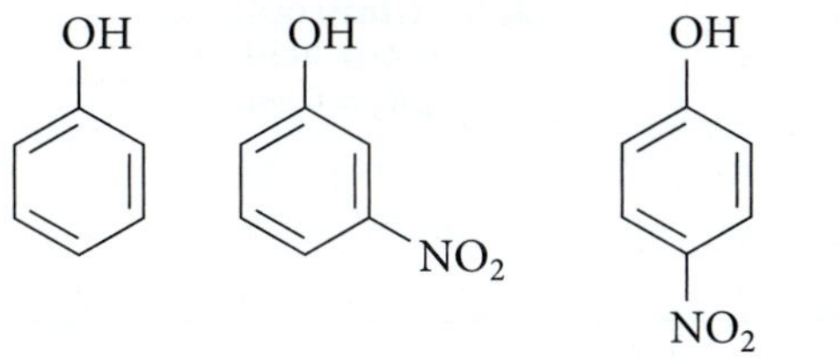

	phenol	***m*-nitrophenol**	***p*-nitrophenol**
pK_a	9.95	8.35	7.21

(18.74)

m-Nitrophenol is more acidic than phenol because the nitro group is very electronegative. The polar effect of the nitro substituent stabilizes the conjugate-base anion for the same reason that it would stabilize the conjugate base of an alcohol (Sec. 11.1C). Yet *p*-nitrophenol is more acidic than *m*-nitrophenol by more than 1 pK_a unit, even though the *p*-nitro group is farther from the phenol oxygen. This cannot be entirely the result of a polar effect, because polar effects on acidity *decrease* as the distance between the substituent and the acidic group increases. The reason for the increased acidity of *p*-nitrophenol is that the *p*-nitro group stabilizes the conjugate-base anion by resonance (red structure).

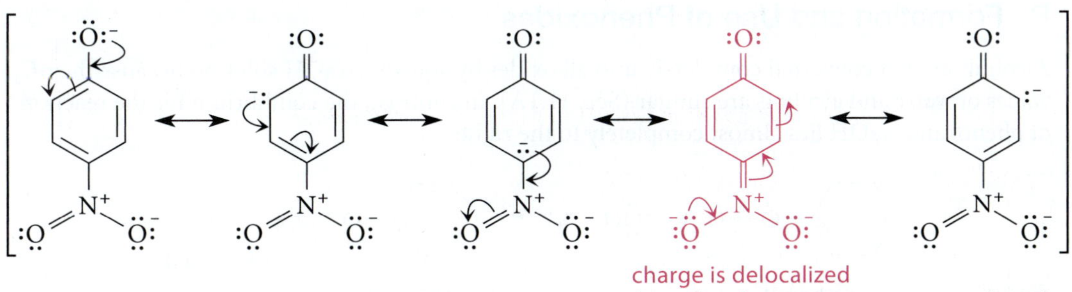

(18.75)

The structure shown in red is especially important because it places negative charge on the electronegative oxygen atom. In *m*-nitrophenol, however, it is not possible to draw a resonance structure that delocalizes the negative charge into the nitro group.

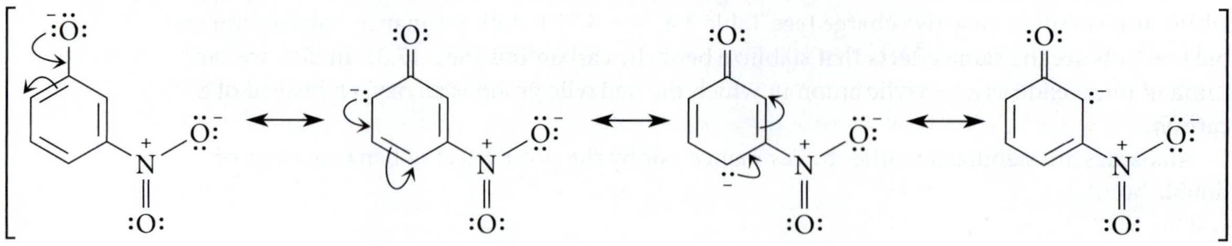

fewer important structures than the para isomer

(18.76)

Because *p*-nitrophenoxide has more important resonance structures, it is more stable relative to its corresponding phenol than is *m*-nitrophenoxide. Therefore, *p*-nitrophenol is the more acidic of the two nitrophenols. The acid-strengthening resonance effect of ortho and para nitro groups is so large that 2,4,6-trinitrophenol (picric acid) is actually a strong acid.

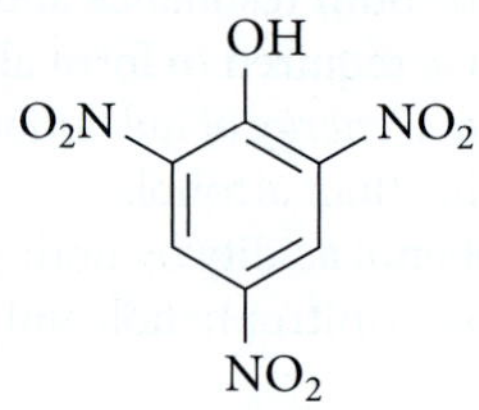

2,4,6-trinitrophenol
(picric acid)
$pK_a = 0.96$

(18.77)

Focused Problem

18.30 Which of the two phenols in each set is more acidic? Explain.

(a) 2,5-Dinitrophenol or 2,4-dinitrophenol

(b) Phenol or *m*-chlorophenol

(c)

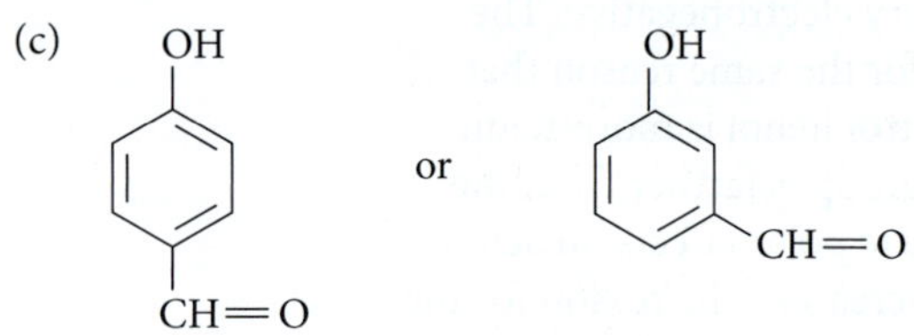

B. Formation and Use of Phenoxides

Alcohols are not converted completely into alkoxides by aqueous NaOH solution because the pK_a values of water and alcohols are similar (Sec. 11.1A). In contrast, the equilibrium for the reaction of phenol and NaOH lies almost completely to the right:

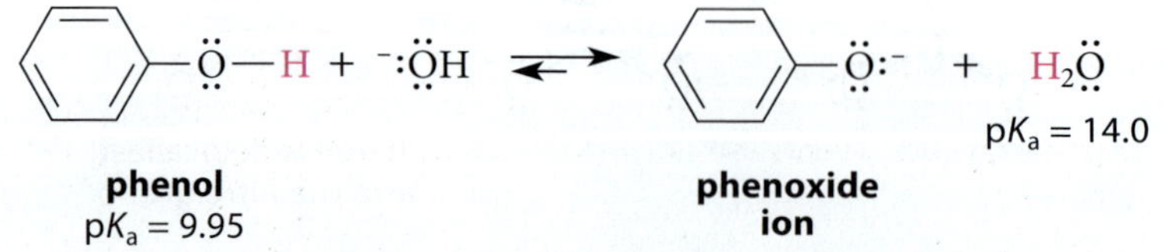

(18.78)

Because the difference in pK_a values of water and phenol is about 4, the equilibrium constant for this reaction is about 10^4 (Sec. 3.6C). This means that for all practical purposes, phenols are converted completely into their conjugate-base anions by NaOH solution. Although the stronger bases used to ionize alcohols (Sec. 10.1A) can also be used for phenols, hydroxide ion or alkoxide bases such as ethoxide ion are often adequate for the purpose. For example, when phenol is treated with one equivalent of NaOH or NaOEt in ethanol, the phenol O—H proton is ionized completely to give a solution of sodium phenoxide.

The acidities of phenols can sometimes be used to separate them from mixtures with other organic compounds. For example, suppose that we wish to separate the water-insoluble phenol, 4-chlorophenol, from the water-insoluble alcohol, 4-chlorocyclohexanol. Although the phenol itself is water-insoluble, its sodium or potassium salt, like many other alkali metal salts, has considerable solubility in water because it is an *ionic compound*. (As explained in Sec. 8.6F, water is one of the best solvents for ionic compounds.)

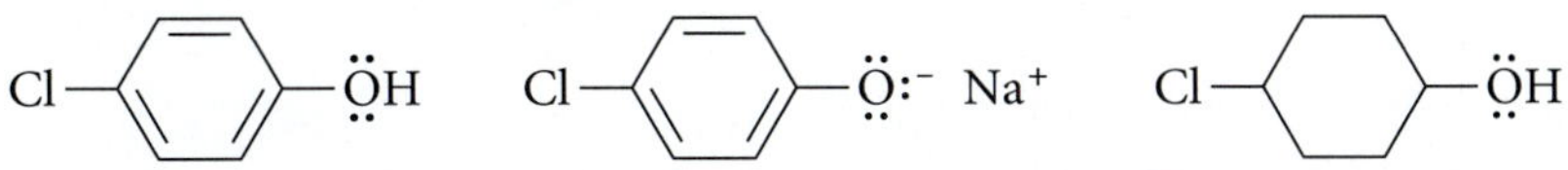

4-chlorophenol soluble in ether insoluble in water

sodium 4-chlorophenolate insoluble in ether soluble in water (an ionic compound)

4-chlorocyclohexanol soluble in ether insoluble in water

When a mixture of the phenol and the alcohol in ether solution is shaken with aqueous NaOH, the phenol is selectively extracted into the aqueous solution as its sodium salt, sodium 4-chlorophenolate, while the alcohol, which is not significantly ionized by NaOH, remains in the ether. (Although alcohols of low molecular mass are soluble in water, the chlorine and hydrocarbon parts of 4-chlorocyclohexanol dominate its solubility properties.) Acidification of the aqueous solution gives the phenol, which separates from solution because, after acidification, it is no longer ionized.

It is usually said that phenols are "soluble in sodium hydroxide solution." What is really meant by this statement is that if sodium hydroxide solution is added to a phenol, the phenol is converted into its conjugate-base phenoxide ion, which, because it is ionic, is the species that actually dissolves in the aqueous solution. Solubility in 5% NaOH solution is a qualitative test for phenols (and other compounds of equal or greater acidity).

Phenoxides, like alkoxides, can be used as nucleophiles. For example, aryl ethers can be prepared by the reaction of a phenoxide anion and an alkyl halide.

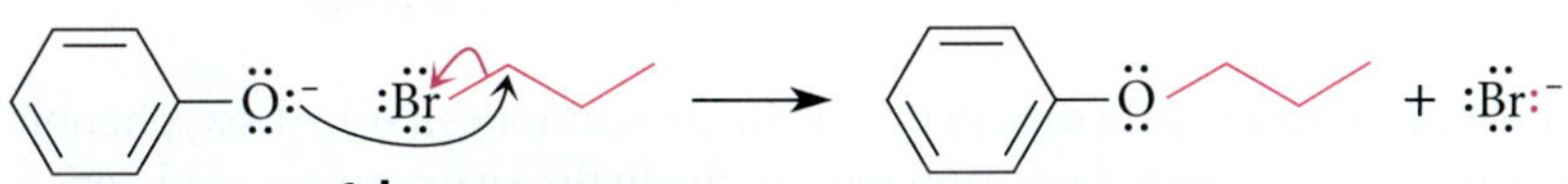

1-bromopropane

phenoxide ion

propoxybenzene (63% yield)

(18.79)

This is another example of the Williamson ether synthesis (Sec. 12.2A). However, sodium propoxide, the sodium salt of 1-propanol, and bromobenzene, could *not* be used to prepare this ether. (Why? See Sec. 18.1.)

Focused Problems

18.31 Outline a preparation of each of the following compounds from the indicated starting material and any other reagents.

(a) *p*-Nitroanisole from *p*-nitrophenol

(b) 2-Phenoxyethanol from phenol

18.32 The following compound, unlike most phenols, is *soluble* in neutral aqueous solution, but *insoluble* in aqueous base. Explain this unusual behavior.

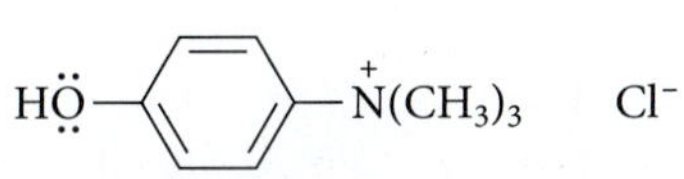

18.8 QUINONES AND SEMIQUINONES

A. Oxidation of Phenols to Quinones

Even though phenols do not have hydrogen at their α-carbon atoms, they do undergo oxidation. The most common oxidation products of phenols are quinones.

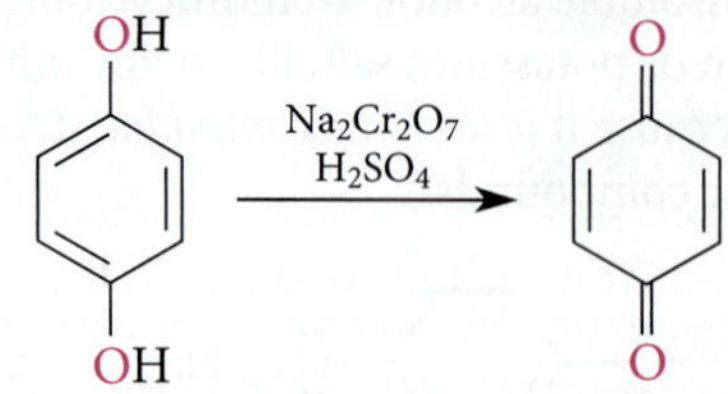

hydroquinone ***p*-benzoquinone** (86–92% yield) (18.80)

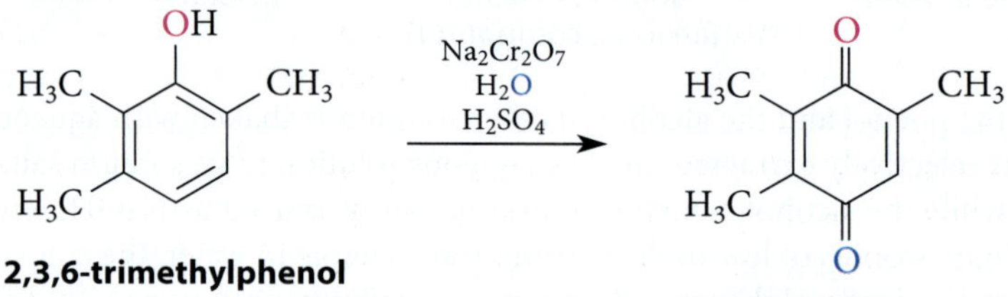

2,3,6-trimethylphenol **2,3,5-trimethyl-1,4-benzoquinone** (50% yield) (18.81)

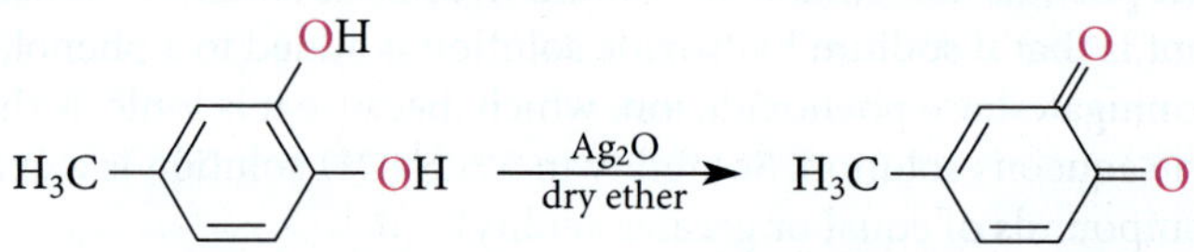

4-methylcatechol **4-methyl-1,2-benzoquinone** (unstable red crystals) (68% yield) (18.82)

As Eqs. 18.80–18.82 illustrate, *p*-hydroxyphenols (hydroquinones), *o*-hydroxyphenols (catechols), and phenols with an unsubstituted position para to the hydroxy group are oxidized to quinones. A **quinone** is any compound containing either of the following structural units.

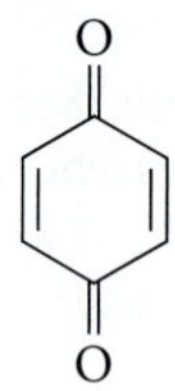

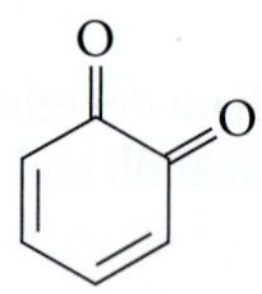

an *ortho*-quinone

a *para*-quinone (18.83)

If the quinone oxygens have a 1,4 (para) relationship, the quinone is called a ***para*-quinone**; if the oxygens are in a 1,2 (ortho) arrangement, the quinone is called an ***ortho*-quinone**. The following compounds are typical quinones.

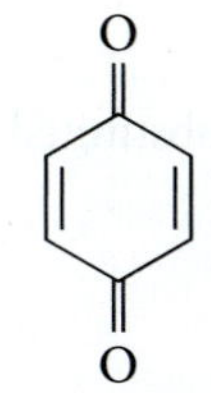

***p*-benzoquinone (1,4-benzoquinone)**

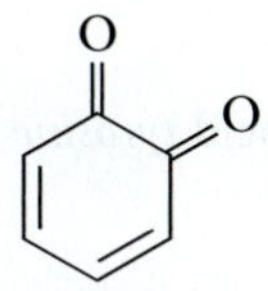

***o*-benzoquinone (1,2-benzoquinone)**

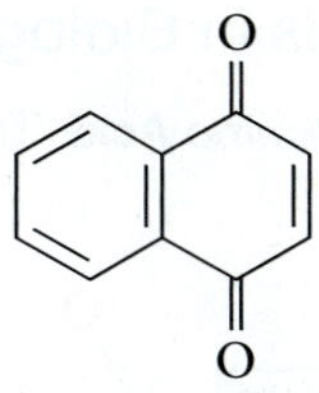

1,4-naphthoquinone

derived from

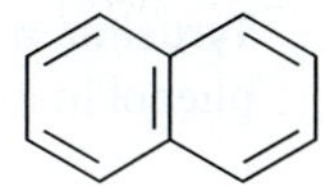

naphthalene

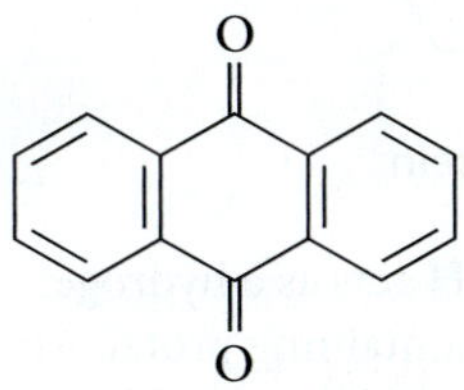

9,10-anthraquinone

derived from

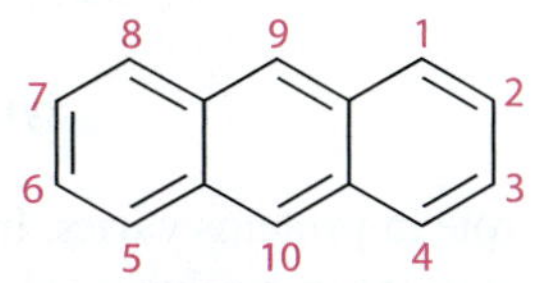

anthracene

(18.84)

As illustrated by the preceding structures, the names of quinones are derived from the names of the corresponding aromatic hydrocarbons: *benzo*quinone is derived from benzene, *naphtho*quinone from naphthalene, and so on.

Ortho-quinones, particularly *ortho*-benzoquinones, are typically considerably less stable than their *para*-quinone isomers. One reason for this difference is that in *ortho*-quinones, the ends of the C═O bond dipoles with like charges are close together and therefore have a repulsive, destabilizing interaction. In *para*-quinones these dipoles are pointed in opposite directions and are farther apart.

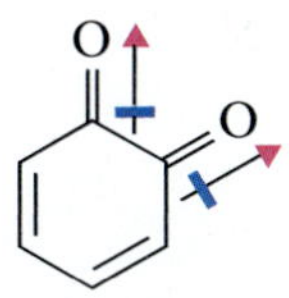

***o*-benzoquinone**
like charges in the bond dipoles are close together

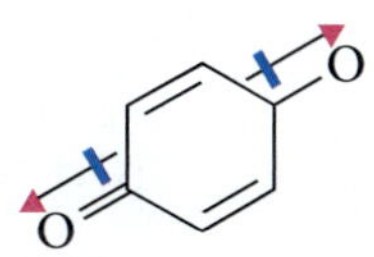

***p*-benzoquinone**
bond dipoles have opposite directions and are farther apart

(18.85)

Focused Problems

18.33 The structure of phenanthrene is as follows:

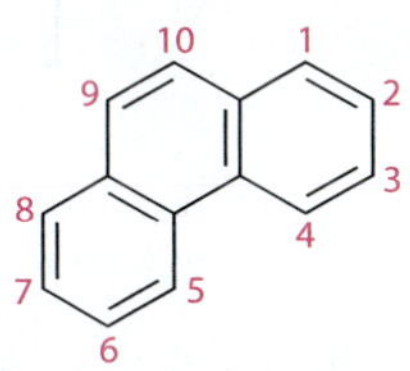

phenanthrene

Draw a structure for each of the following:

(a) 9,10-Phenanthroquinone (b) 1,4-Phenanthroquinone

18.34 Complete the following reactions:

(a)

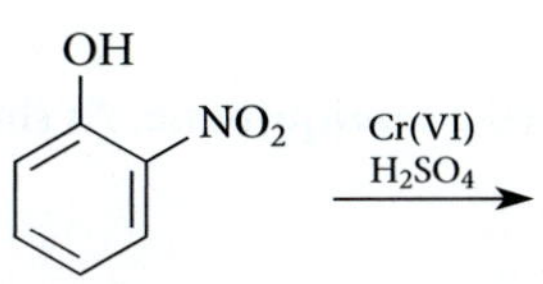

(b)

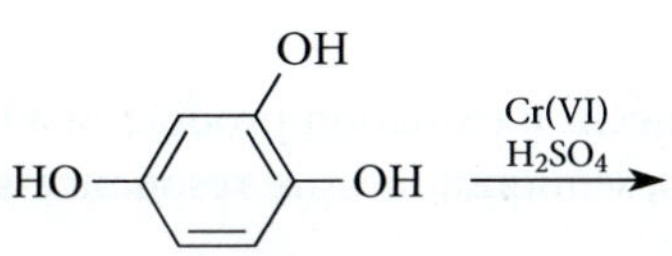

(*Hint:* The product is a hydroxy quinone; but which one?)

Chemical Biology Topic

B. Quinones and Phenols in Biology

Tyrosine, a Phenol-Containing Amino Acid The amino acid *tyrosine* has a para-substituted phenol in its side chain.

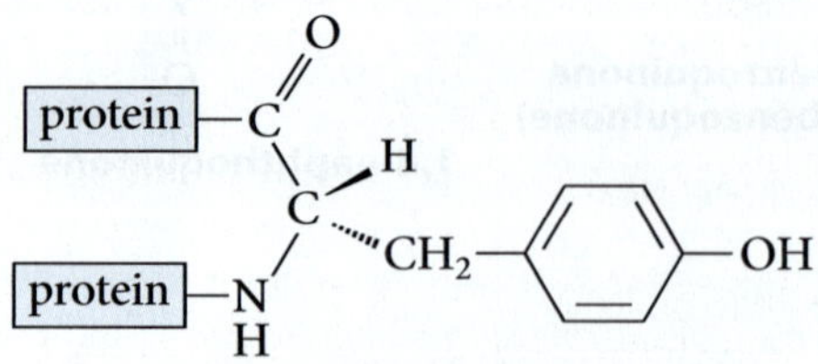

tyrosine residue in a protein (18.86)

The role of a tyrosine residue in enzyme catalysis is illustrated in Eq. 22.65c, Sec. 22.5.

Its functional role in proteins varies. In some cases, the O—H acts as a hydrogen-bond donor that can be important in catalyzing chemical reactions or maintaining protein structure; it can be involved in the noncovalent interactions between rings (offset-stacking or edge-to-face interactions; Sec. 15.8A) or as the "pi" component in pi–cation interactions (Sec. 15.8B). In other cases, the conjugate base of the O—H group is involved as a basic group in enzyme catalysis.

Oxidation and Reductions of Phenols in Biology The oxidation–reduction reactions and free-radical reactions of phenols are very important in biology. To understand this role, think of hydroquinone oxidation as a half-reaction (Sec. 11.6A).

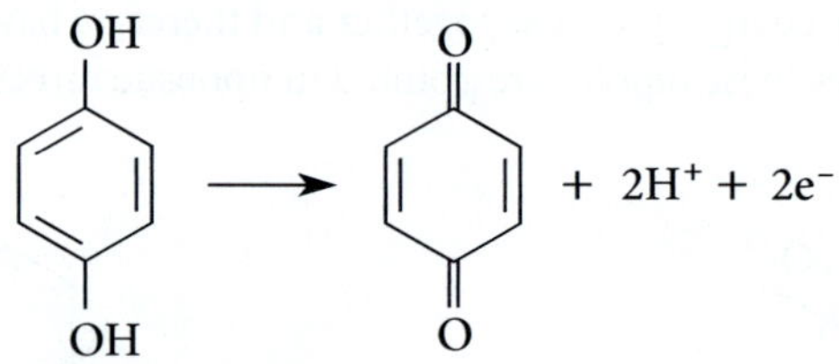

(18.87a)

(In fact, this half-reaction can be carried out reversibly in an electrochemical cell.) We can think of this oxidation mechanistically as two individual one-electron steps:

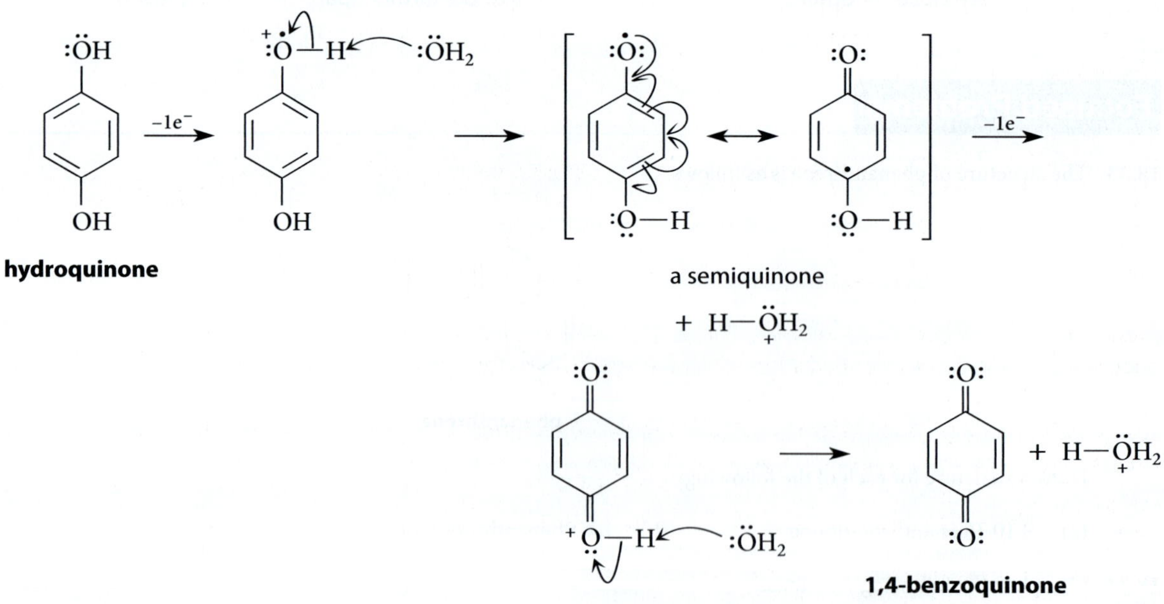

(18.87b)

The one-electron oxidation product of a hydroquinone is called a **semiquinone**. As shown in Eq. 18.87b, semiquinones are resonance-stabilized radicals.

The view of hydroquinone oxidation (or quinone reduction) as a half-reaction is biologically significant in the process known as *oxidative phosphorylation*. In oxidative phosphorylation, O_2 is reduced to water:

$$4e^- + 4H^+ + O_2 \longrightarrow 2H_2O \qquad (18.88)$$

This half-reaction has a standard free energy (at pH 7) of −157.6 kJ (−37.7 kcal) per mole of O_2 consumed. This very large free energy is captured and ultimately used to drive the synthesis of adenosine triphosphate (ATP) (Secs. 25.2B and 25.8B), the source of chemical energy in the cell. (You will study oxidative phosphorylation if you take a biochemistry course.) The free energy liberated by the reduction of O_2 is not captured in one step; rather, it is captured incrementally in a series of steps in which electrons are transferred from cellular reducing agents to oxygen. The reactions in which these electron transfers take place are called collectively the **electron-transport chain**. A biological quinone, **ubiquinone**, sometimes called **coenzyme Q**, or simply **Q**, is involved in electron transport.

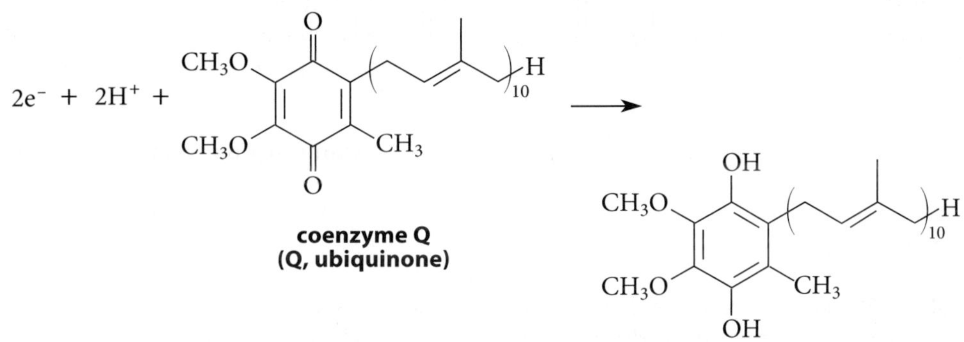

coenzyme Q
(Q, ubiquinone)

reduced coenzyme Q
(QH_2) (18.89)

Q is reduced to QH_2 in a two-electron process, as shown in Eq. 18.89. (The reducing agents are NADH and succinate, but we won't detail this process here.) The QH_2 then delivers an electron to Fe^{3+}, which is part of a heterocycle–protein complex called *cytochrome c*. In this reaction, the ferric iron (Fe^{3+}) undergoes a one-electron reduction to ferrous iron (Fe^{2+}), and QH_2 is converted into its semiquinone. Reduction of a second cytochrome-bound Fe^{3+} by the semiquinone gives Q and another Fe^{2+}.

$$\underset{}{QH_2} + \underset{\text{(in cytochrome c)}}{Fe^{3+}} \longrightarrow \underset{\substack{\text{semiquinone}\\ \text{form of Q}}}{\cdot QH} + Fe^{2+} \qquad (18.90a)$$

$$\cdot QH + Fe^{3+} \longrightarrow Q + Fe^{2+} \qquad (18.90b)$$

Notice that *two* Fe^{3+} can be reduced by each QH_2, each in a one-electron step. For this reason, coenzyme Q is sometimes called a two-electron/one-electron bridge in electron transport.

Another important quinone is **vitamin K**, which is important in a group of biochemical reactions involved in blood clotting. (Different forms of vitamin K, which are all active, differ in the detailed structure of the isoprenoid side chain.)

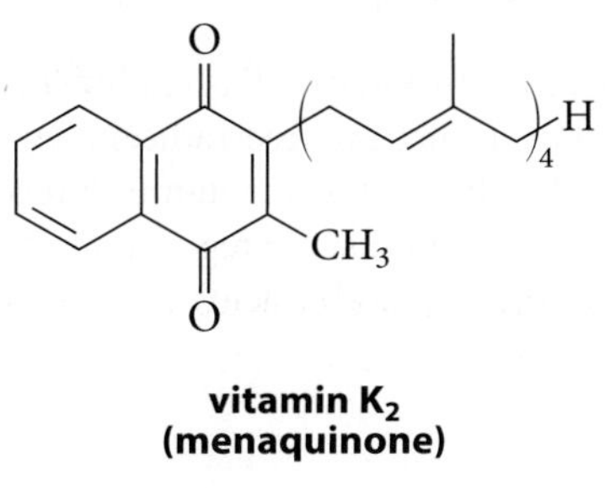

vitamin K_2
(menaquinone) (18.91)

Its reduction product, often called KH_2, undergoes oxidation by O_2 to vitamin K-2,3-epoxide as part of the blood-clotting process. The conversion of the epoxide back into KH_2 is essential for proper clotting activity. The widely used anticoagulant warfarin (Coumadin) inhibits this process.

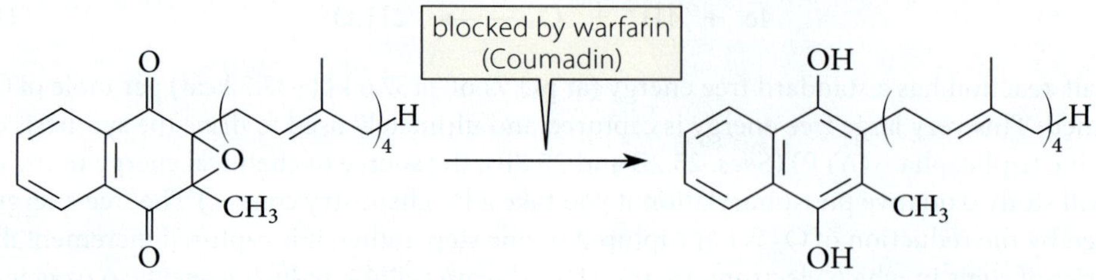

vitamin K_2-2,3-epoxide **reduced vitamin K_2 (KH_2)**

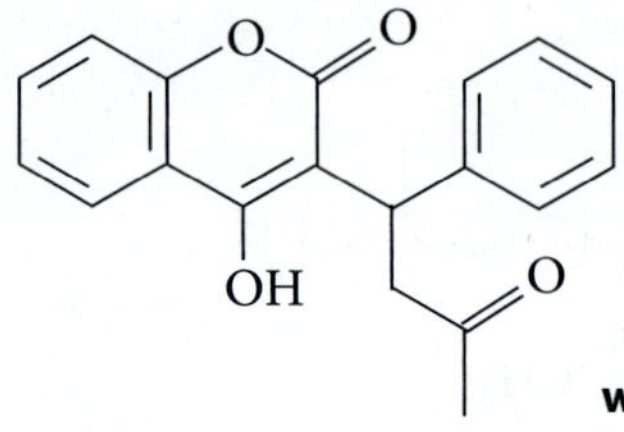

warfarin (Coumadin) (18.92)

Warfarin is the active ingredient in rat poison, but is also used in human medicine as an anticoagulant.

Phenols as Radical Scavengers Many phenols and quinones are inhibitors of free-radical reactions. The basis of this effect with phenols is that a free radical (R• in Eq. 18.93) can abstract a hydrogen atom from a phenol to give a resonance-stabilized free radical. [A semiquinone (Eq. 18.87b) is an example of such a radical.]

a resonance-stabilized free radical (18.93)

Vitamin E, a phenol, serves as a free-radical inhibitor in biological membranes and fats.

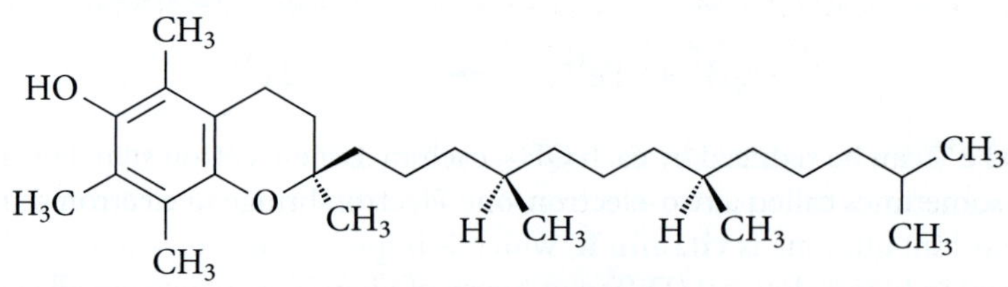

α-tocopherol
a major form of vitamin E (18.94)

Because of its long hydrocarbon side chain, vitamin E is readily incorporated into the phospholipid bilayer of cell membranes, which it protects from free-radical damage by terminating radical chains as shown in Eq. 18.93. The radical thus formed from vitamin E reacts with vitamin C, a water-soluble vitamin, at the membrane–water interface to regenerate vitamin E and form a resonance-stabilized radical derived from vitamin C (ascorbic acid).

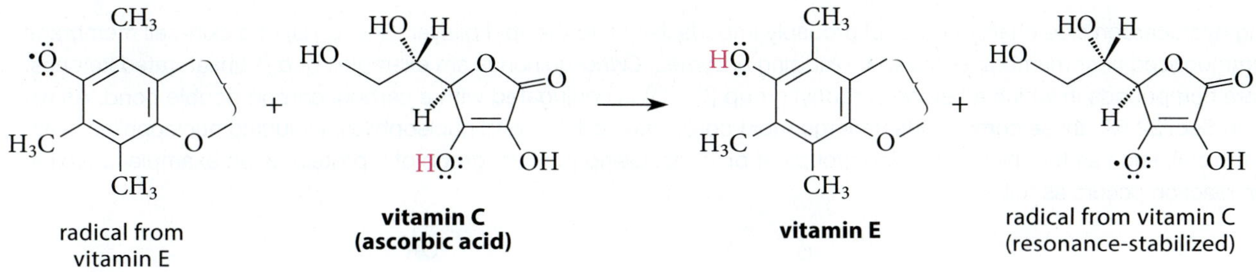

(18.95)

The vitamin C-derived radical is ultimately reduced back to vitamin C by other biological reducing agents. As this example illustrates, vitamin C is also an important reducing agent in biology.

The effectiveness of several widely used food preservatives is based on reactions such as these. Examples of such preservatives are "butylated hydroxytoluene" (BHT) and "butylated hydroxyanisole" (BHA).

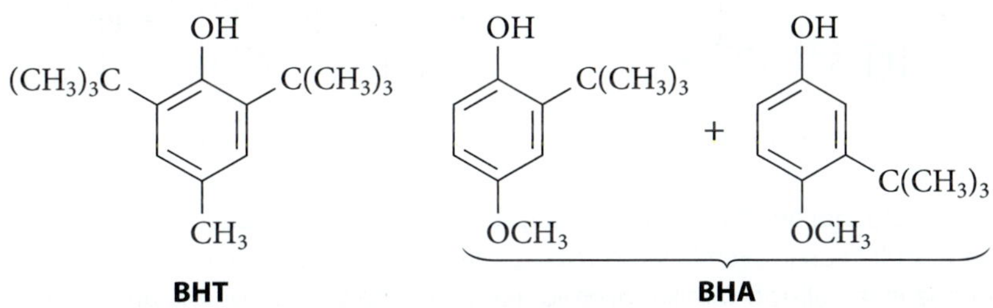

(18.96)

Oxidation involving free-radical processes is one way that foods discolor and spoil. A preservative such as BHT inhibits these processes by donating its OH hydrogen atom to free radicals in the food (as in Eq. 18.93). The BHT is thus transformed into a phenoxy radical, which is too stable and unreactive to propagate radical chain reactions. Although the use of BHT and BHA as food additives has generated some controversy because of their potential side effects, without such additives, foods could not be stored for any appreciable length of time or transported over long distances.

Poison Ivy and Itchy Quinones

Chris Hill/Shutterstock

Over half of the people in the United States suffer from an allergy to poison ivy, poison oak, and poison sumac. The active principle in these plants is a family of catechol derivatives known collectively as urushiol.

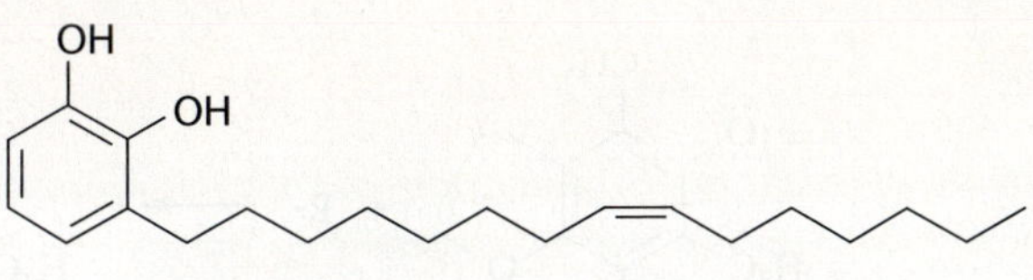

urushiol (one of several components) **(18.97a)**

(The various urushiol components differ in the number, positions, and possibly the stereochemistry of the side-chain double bonds.) The allergic reaction is not caused by the catechol itself, but rather by its oxidation product, an *ortho*-quinone:

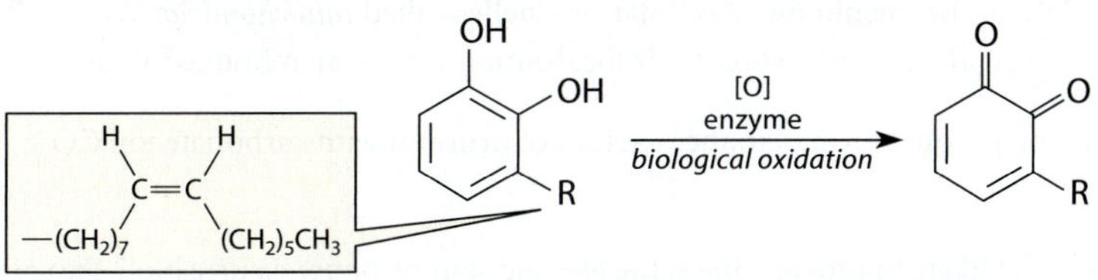

(18.97b)

The long hydrocarbon side chain of urushiol probably imbeds itself into the lipid bilayer (Sec. 8.7B) of a skin-cell membrane and is immobilized near membrane-localized oxidizing enzymes. *Ortho*-quinones are examples of α,β-unsaturated ketones, which are compounds in which a ketone carbonyl group (C═O) is conjugated with a carbon–carbon double bond. As we explore in Sec. 22.9A, these compounds undergo rapid conjugate addition with nucleophiles, including nucleophiles available in the cell, such as the thiol and amino groups of proteins. Using the thiol group of a protein as an example, a typical addition reaction occurs as follows:

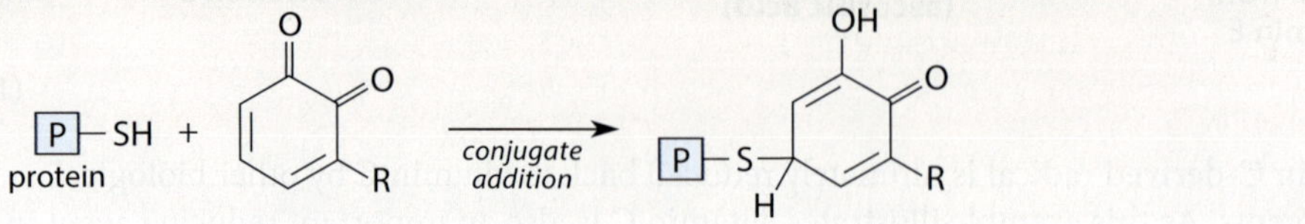

(18.97c)

A subsequent rapid reaction with Brønsted acids and bases forms a substituted catechol.

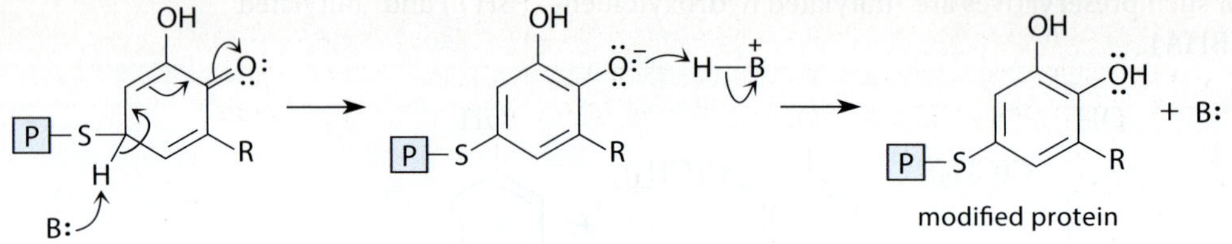

(18.97d)

Ortho-quinones are not aromatic, but the catechol addition products are; the aromatic stabilization of the product makes the reaction irreversible. These reactions result in an irreversibly modified protein, which the immune system now senses as "foreign." The resulting biological response is the all-too-familiar allergic reaction—the skin eruptions and the intense itch.

Focused Problems

18.35 Draw the important resonance structures of the radicals formed when each of the following reacts with R•, a general free radical.

(a) Vitamin E (b) BHT

18.36 Consider the detailed structure of the semiquinone •QH shown in Eq. 18.90a.

(a) There are two possible structures for this semiquinone; draw them both.

(b) Show the resonance structures for either of the structures you gave in part (a).

18.37 (a) Using the fishhook notation, derive the important resonance structures of the vitamin C-derived radical in Eq. 18.95.

(b) In the laboratory, the radical derived from vitamin E can react with a second free radical R• to give the following oxidation product (among others).

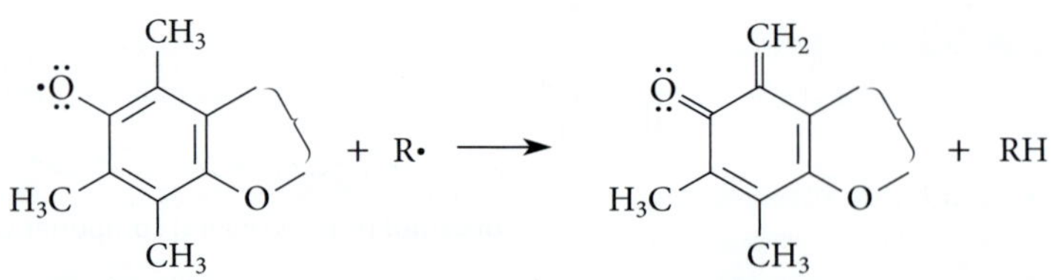

vitamin E radical

Using the fishhook notation, give a mechanism for this reaction.

18.38 Electron transport takes place in the membrane of cellular organelles called *mitochondria*. What is it about the structure of ubiquinone and its reduction products that ensures their localization within membranes? Explain.

18.39 (a) Give the structure of the product formed in the reaction of urushiol with carbonate ion CO_3^{2-} (a base) and a large excess of methyl iodide.

(b) Would this compound be likely to provoke the same allergic skin response as urushiol? Explain.

18.9 ELECTROPHILIC AROMATIC SUBSTITUTION REACTIONS OF PHENOLS

Phenols are aromatic compounds, and they undergo electrophilic aromatic substitution reactions such as those described in Sec. 16.4. In some of these reactions, the OH group has special effects that are not common to other substituent groups.

Because the OH group is a strongly activating substituent, phenol can be halogenated once under mild conditions that are completely ineffective for benzene itself.

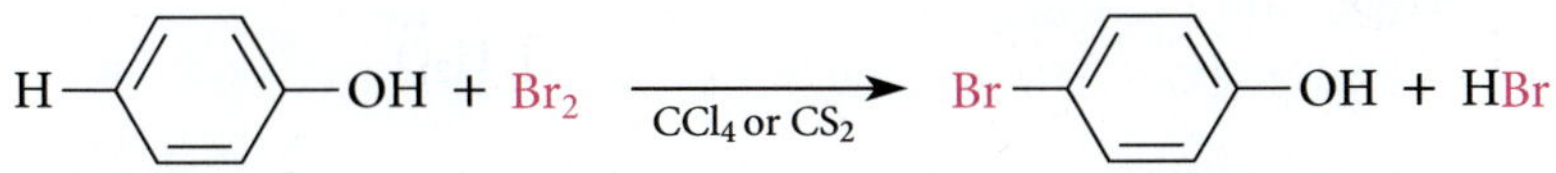

phenol **p-bromophenol** (82% yield) (18.98)

Notice the mild conditions of this reaction. A Lewis acid such as $FeBr_3$ is not required. (A solution of Br_2 in CCl_4 is the reagent usually used for adding bromine to alkenes.) But when phenol reacts with Br_2 in H_2O (bromine water), more extensive bromination occurs and 2,4,6-tribromophenol is obtained.

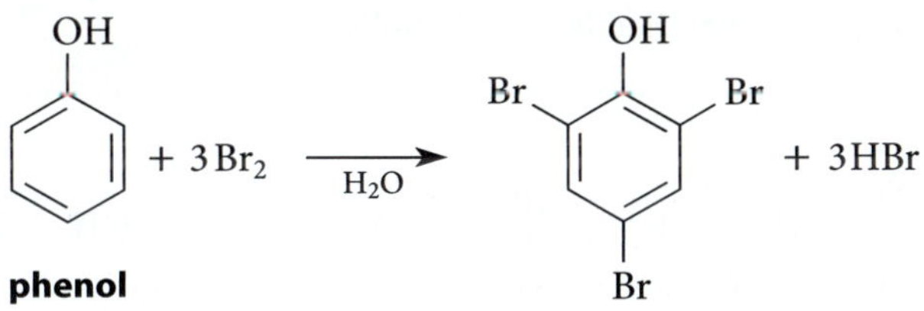

phenol

2,4,6-tribromophenol (~100% yield; (precipitates) (18.99a)

This more extensive bromination occurs for two reasons. First, bromine reacts with water to give protonated hypobromous acid, a more potent electrophile than bromine itself.

$$Br_2 + H_2\ddot{O}: \rightleftharpoons Br^- + Br{-}\overset{+}{\ddot{O}}H_2 \rightleftharpoons HBr + Br{-}\ddot{\underset{..}{O}}H$$

protonated hypobromous acid; hypobromous acid (18.99b)

Second, in aqueous solutions near neutrality, phenol partially ionizes to its conjugate-base phenoxide anion. Although only a small amount of this anion is present, it is very reactive and brominates instantly, thereby pulling the phenol–phenolate equilibrium to the right.

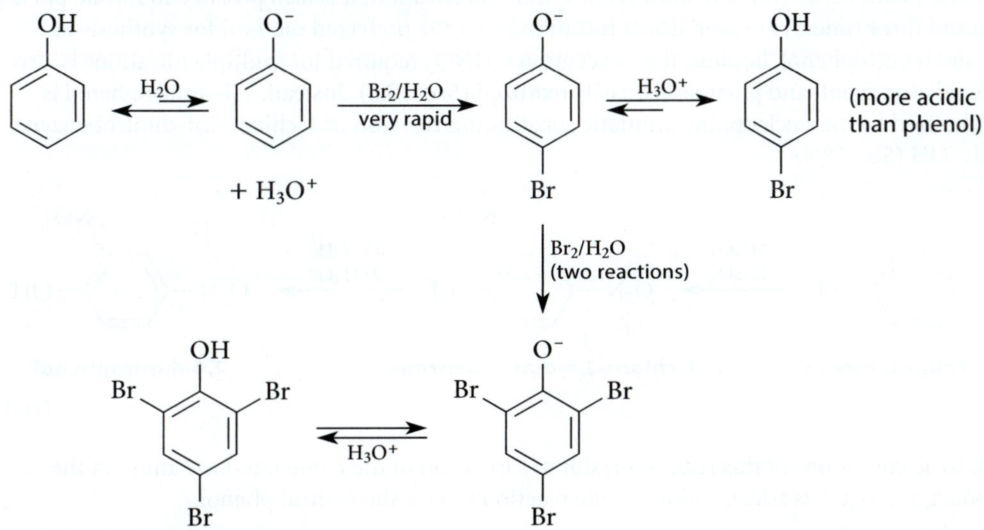

(very insoluble; precipitates) (18.99c)

Phenoxide ion is much more reactive than phenol because the reactive intermediate is not a carbocation, but is instead a more stable neutral molecule (red structure).

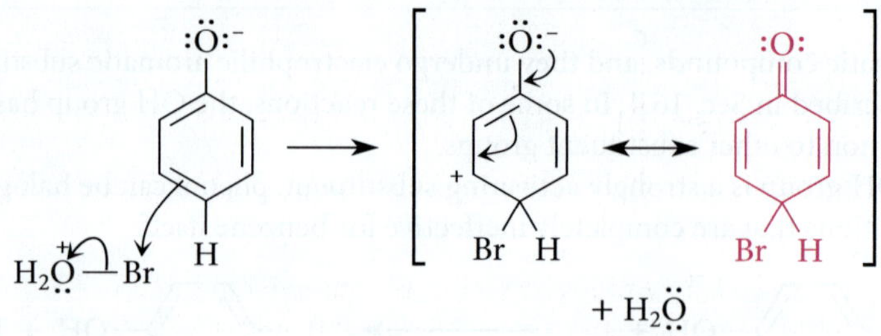

(18.99d)

p-Bromophenol is also in equilibrium with its conjugate base *p*-bromophenoxide anion, which brominates again until all ortho and para positions have been substituted. Notice in Eq. 18.99c that in the second and third substitutions the powerful ortho, para-directing and activating effects of the —O⁻ group override the weaker deactivating and directing effects of the bromine substituents. [The same activating and directing effects of the phenolic OH group apply to the iodination of tyrosine (Sec. 16.5C).]

In strongly acidic solution, in which formation of the phenolate anion is suppressed, bromination can be stopped at the 2,4-dibromophenol stage.

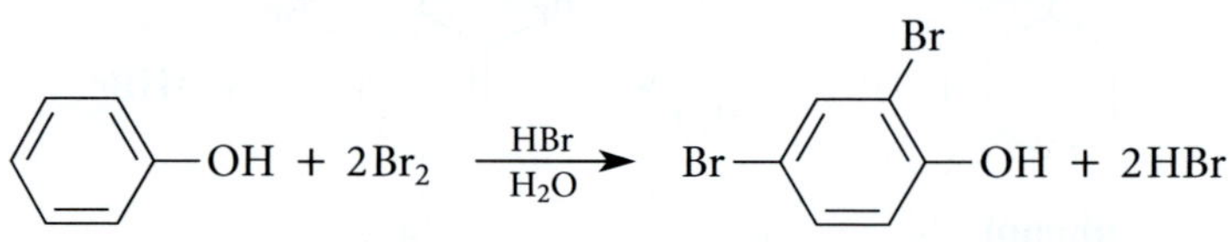

phenol **2,4-dibromophenol** (87% yield) (18.100)

Phenol is also very reactive in other electrophilic substitution reactions, such as nitration. Phenol can be nitrated once under mild conditions. (Notice in the following reaction that H_2SO_4 is *not* present, whereas it is in the nitration of benzene; Eq. 16.18, Sec. 16.4C.)

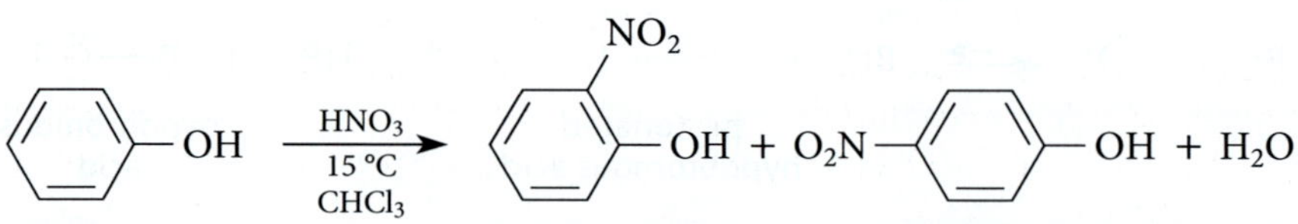

phenol ***o*-nitrophenol** (26% yield) ***p*-nitrophenol** (61% yield) (18.101)

Because phenol is activated toward electrophilic substitution, it is also possible to nitrate phenol two and three times. However, direct nitration is *not* the preferred method for synthesis of di- and trinitrophenol, because the concentrated HNO_3 required for multiple nitrations is also an oxidizing agent, and phenols are easily oxidized (Sec. 18.8). Instead, 2,4-dinitrophenol is synthesized by the nucleophilic aromatic substitution reaction of 1-chloro-2,4-dinitrobenzene with ^{-}OH (Sec. 18.4A).

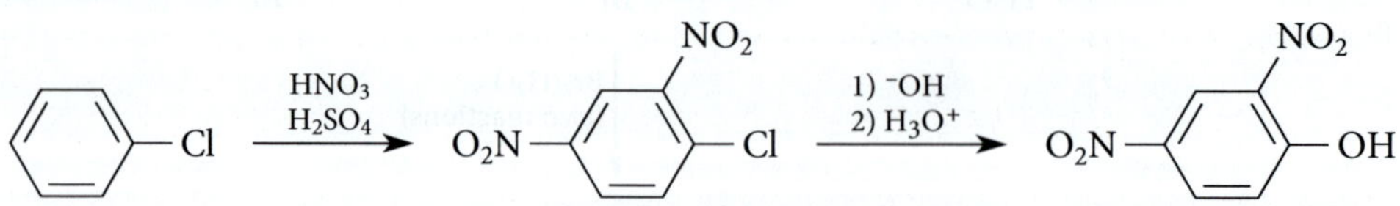

chlorobenzene **1-chloro-2,4-dinitrobenzene** **2,4-dinitrophenol**

(18.102)

The basic conditions of this reaction result in formation of the conjugate-base anion of the product; the H_3O^+ is added following the reaction to give the neutral phenol.

The great reactivity of phenol in electrophilic aromatic substitution does not extend to the Friedel–Crafts acylation reaction, because phenol reacts rapidly with the $AlCl_3$ catalyst.

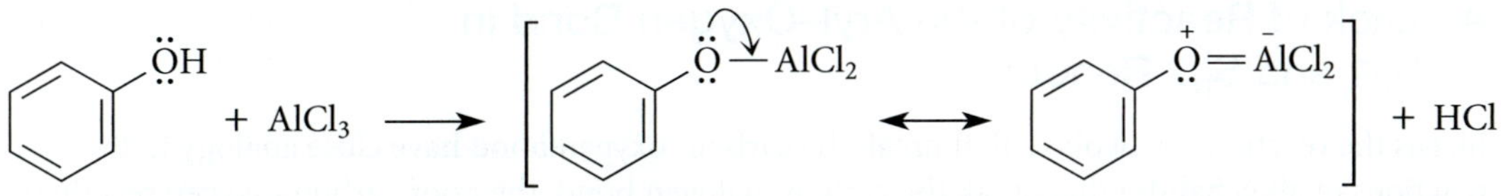

(18.103)

The adduct of phenol and $AlCl_3$ is much less reactive than phenol itself in electrophilic aromatic substitution reactions because, as shown in Eq. 18.103, the oxygen electrons are delocalized onto the electron-deficient aluminum. Because of their delocalization away from the benzene ring, these electrons are less available for resonance stabilization of the carbocation intermediate formed within the ring during Friedel–Crafts acylation. Therefore, Friedel–Crafts acylation of phenol occurs slowly, but can be carried out successfully at elevated temperatures. Because it is not highly activated, the ring is acylated only once.

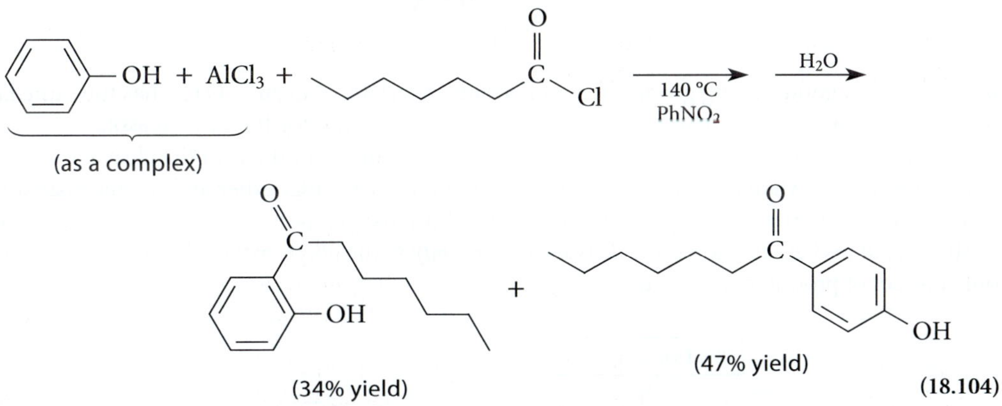

(18.104)

FURTHER EXPLORATION 18.1
The Fries Rearrangement

Friedel–Crafts alkylation of phenol is also possible. This is similar to the acid-catalyzed Friedel-Crafts reactions of benzene with alcohols and alkenes (Eqs. 16.29a and b, Sec. 16.4E). In these cases, the catalyst is a proton acid, which brings about the generation of a carbocation from the alcohol. This catalyst avoids the deactivating effect caused by $AlCl_3$ complexation with the OH group of the phenol. (The OH group of the phenol is not removed by the acid; why?)

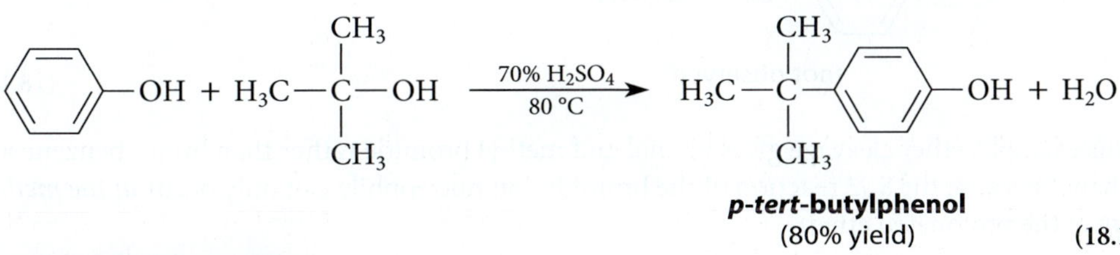

(18.105)

Focused Problems

18.40 Give the principal organic product(s) formed in each of the following reactions.

(a) *o*-Cresol + Br_2 in CCl_4 ⟶

(b) *m*-Chlorophenol + HNO_3, low temperature ⟶

(c)

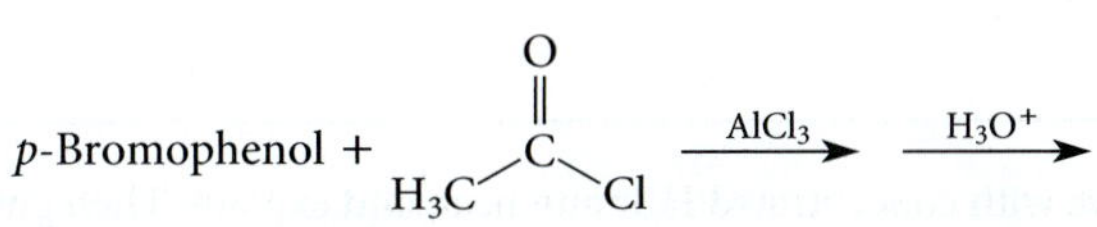

18.41 Give a curved-arrow mechanism for the reaction in Eq. 18.105. Be sure to identify the electrophilic species in the reaction and to show how it is formed.

18.10 REACTIVITY OF THE ARYL–OXYGEN BOND

A. Lack of Reactivity of the Aryl–Oxygen Bond in S_N1 and S_N2 Reactions

Just as the reactions of alcohols that break the carbon–oxygen bond have close analogy to the reactions of alkyl halides that break the carbon–halogen bond, the poor carbon–oxygen reactivity of phenols is analogous to the poor carbon–halogen reactivity of *aryl* halides. Recall that aryl halides do not undergo S_N1 or E1 reactions (Sec. 18.3); for the same reasons, phenols also do not react under conditions used for the S_N1 or E1 reactions of alcohols. Therefore, phenols do *not* form aryl bromides with concentrated HBr, and they do *not* dehydrate with concentrated H_2SO_4. (Instead, they undergo sulfonation; see Sec. 16.4D.) The reasons for these observations are exactly the same as those that explain the lack of reactivity of aryl halides (see Secs. 18.1 and 18.3).

More generally, *any* derivative of the form

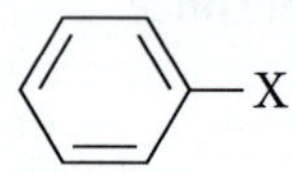

in which X is a good leaving group, such as tosylate, mesylate, or even $-\overset{+}{O}H_2$, has the same lack of reactivity toward S_N1 and S_N2 conditions as aryl halides—and for the same reasons.

The lack of reactivity of the aryl–oxygen bond can be put to good use in the cleavage of aryl ethers. Recall that when ethers cleave—depending on the particular ether and the mechanism—products resulting from cleavage at either of the carbon–oxygen bonds are possible (Sec. 12.4). In the case of aryl ethers, however, cleavage occurs *only* at the alkyl–oxygen bond; consequently, only one set of products is formed:

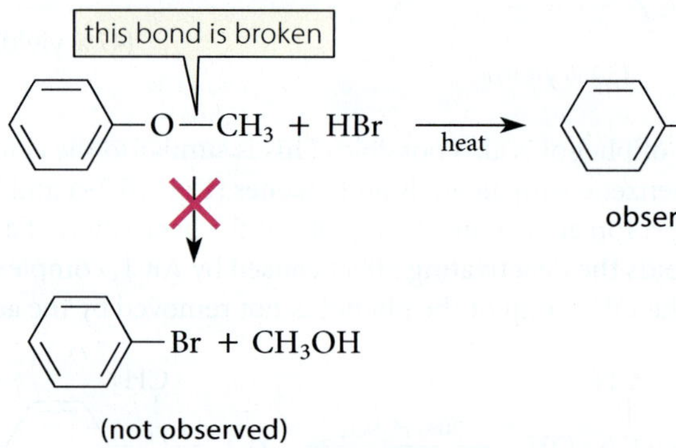

(18.106a)

In this example, ether cleavage gives phenol and methyl bromide rather than bromobenzene and methanol because the S_N2 reaction of the bromide-ion nucleophile can only occur *at the methyl group* of the protonated ether:

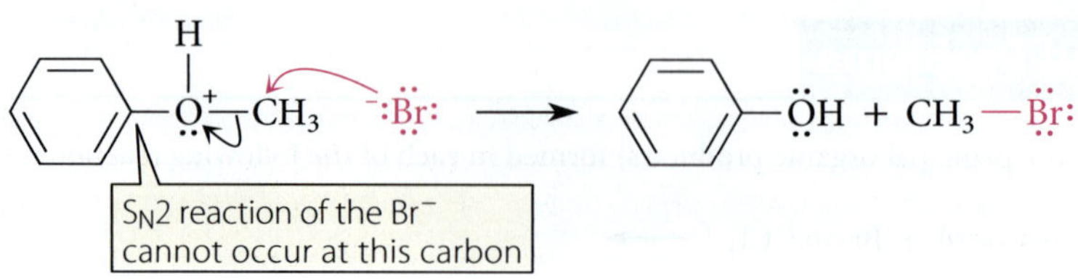

(18.106b)

Focused Problem

18.42 Within each set, identify the ether that would *not* readily cleave with concentrated HBr and heat, and explain. Then give the products of ether cleavage and the mechanisms of their formation for the other ether(s) in the set.

(a)

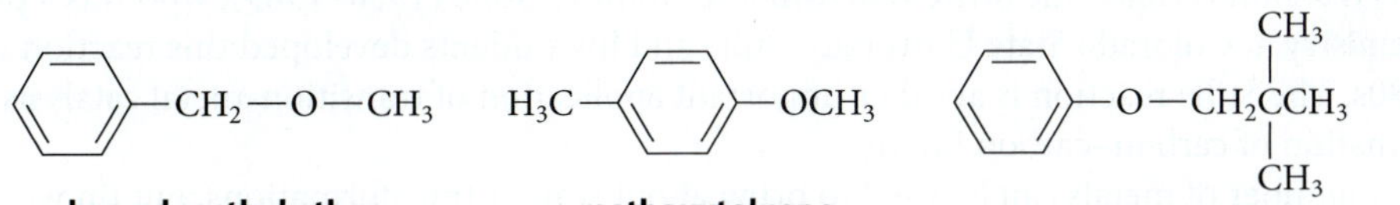

benzyl methyl ether
A

***p*-methoxytoluene**
B

(2,2-dimethylpropoxy)benzene
(neopentyl phenyl ether)
C

(b)

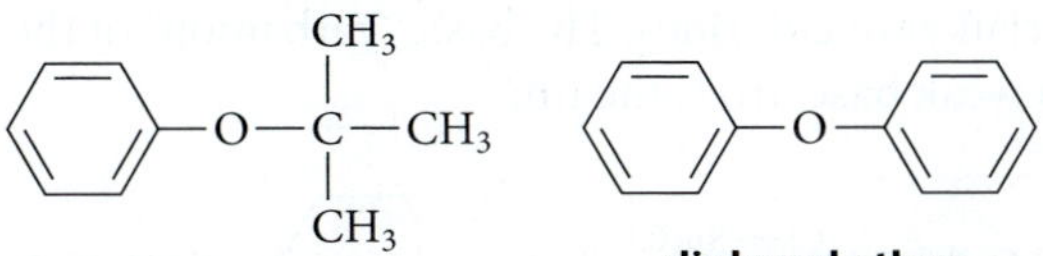

***tert*-butyl phenyl ether**
(*tert*-butoxybenzene)
A

diphenyl ether
(phenoxybenzene)
B

B. Substitution at the Aryl–Oxygen Bond: The Stille Reaction

We showed in Secs. 18.5 and 18.6 that certain transition-metal catalysts, particularly Pd(0), can catalyze the substitution of aryl halides at aryl carbons. Pd(0) catalysts can also catalyze substitution at the aryl–oxygen bond. The first requirement for this to occur is conversion of the phenolic —OH group into a triflate, a very reactive leaving group. [Triflates were introduced in Sec. 11.4A (Display 11.35).]

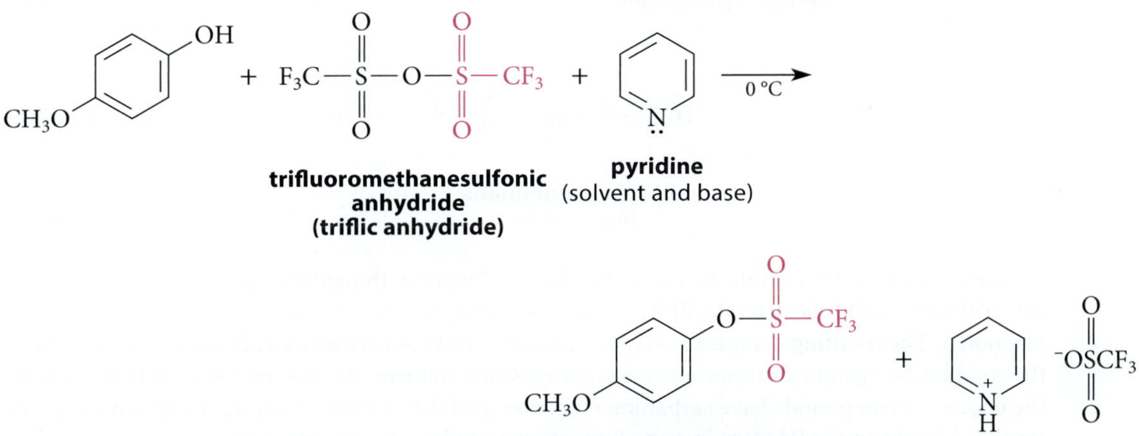

trifluoromethanesulfonic anhydride
(triflic anhydride)

pyridine
(solvent and base)

4-methoxyphenyl trifluoromethanesulfonate
(*p*-methoxyphenyl triflate)
(93% yield)

(18.107)

Aryl triflates react readily with organotin derivatives in the presence of Pd(0) catalysts to give coupling products.

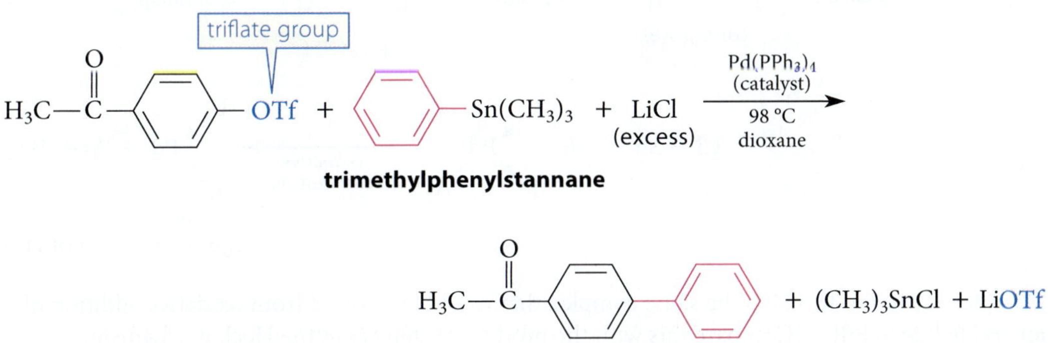

trimethylphenylstannane

(85% yield)

(18.108)

This reaction is called the **Stille reaction** after John K. Stille (1930–1989), who was a professor of chemistry at Colorado State University. Stille and his students developed this reaction in the early 1980s. The Stille reaction is another important application of transition-metal catalysis for the formation of carbon–carbon bonds.

A number of metals can be used to bring about similar transformations, but tin was chosen because the organotin compounds used in the Stille reaction are relatively insensitive to moisture, tolerant of other functional groups, and easy to handle. (Some can be distilled or crystallized.) Many of them are either commercially available, or they are readily prepared from Grignard reagents and commercially available trialkyltin chlorides. The basic "carbanion" of the Grignard reagent displaces chloride, a weak base, from the tin:

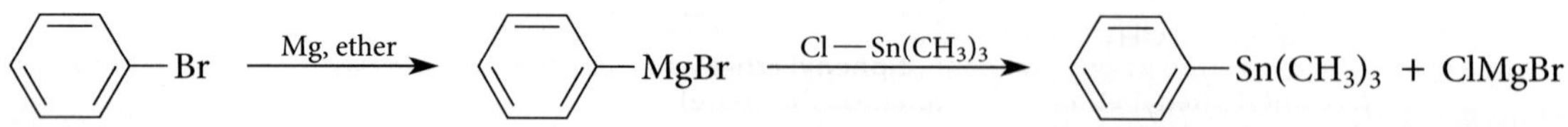

(18.109)

In Eq. 18.108, notice that the phenyl group is transferred in preference to the methyl groups in the Stille reaction. In general, vinylic groups, aryl groups, and other unsaturated groups are transferred preferentially. However, if a tetraalkylstannane is used, alkyl groups can also be transferred. This provides an excellent way to prepare alkylbenzenes. In contrast to the Friedel–Crafts alkylation reaction (Sec. 16.4E), the Stille reaction is *not* plagued by rearrangements.

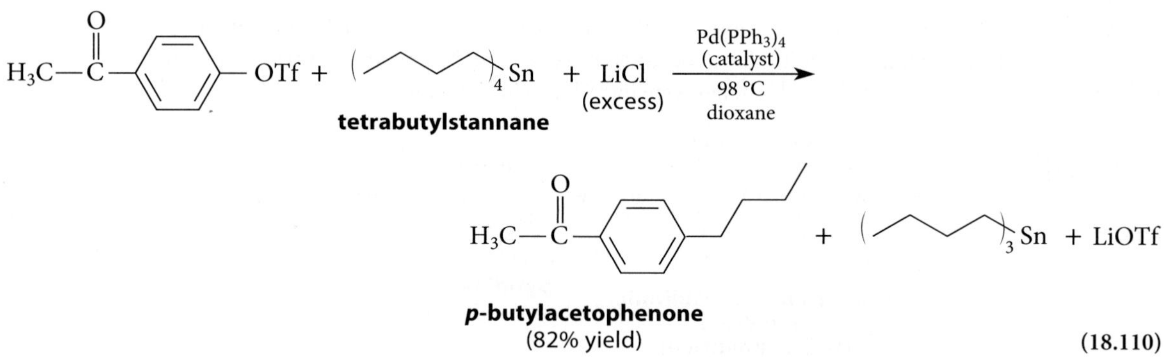

(18.110)

The mechanism of the Stille reaction (Eq. 18.111) begins with oxidative addition of the aryl triflate to the 14-electron $Pd(PPh_3)_2$ (the same catalytic species as in the Heck and Suzuki reactions). The resulting complex *(1)* is very unstable, and the excess chloride ion (as LiCl) rescues the complex by ligand substitution from decomposition to form *(2)*. The aryl or alkyl R-groups on the organotin compounds have carbanion character, and they are nucleophilic enough to substitute for the chloride on the Pd. A reductive elimination completes the mechanism.

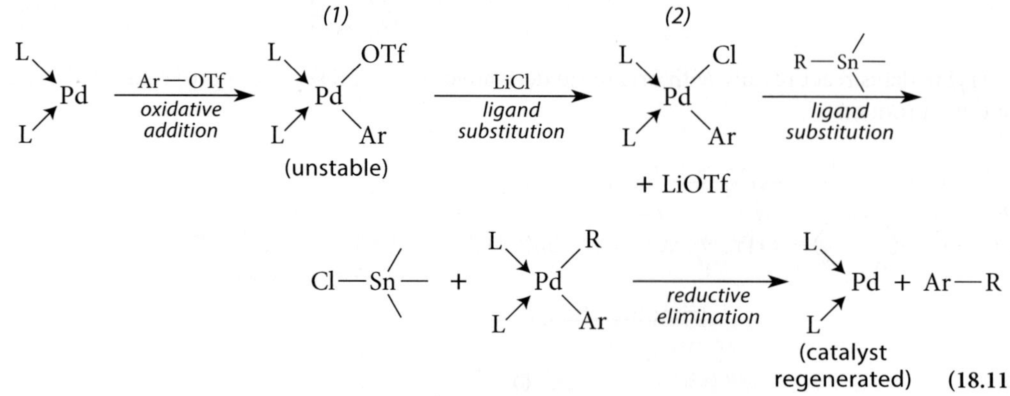

(18.111)

Complex *(2)* in Eq. 18.111 is the same complex that would be formed from oxidative addition of an aryl halide to PdL_2. [Compare this with the product of step *(1)* in the Heck mechanism, Eq. 18.49b (Sec. 18.6A).] Therefore, the Stille reaction can be carried out with aryl halides instead of aryl triflates; and the Heck reaction can be carried out with aryl triflates instead of aryl halides.

The Stille reaction adds another method to the group of reactions that can be used to form carbon–carbon bonds:

1. The addition of carbenes and carbenoids to alkenes (Sec. 10.7)
2. The reaction of Grignard reagents with ethylene oxide and lithium organocuprate reagents with epoxides (Sec. 12.5C)
3. The reaction of acetylenic anions with alkyl halides or sulfonates (Sec. 10.6B)
4. The Diels–Alder reaction (Sec. 15.3)
5. The Friedel–Crafts alkylation and acylation reactions (Secs. 16.4E and F)
6. The Heck and Suzuki coupling reactions (Secs. 18.6A and B)
7. Alkene metathesis (Sec. 18.6C)
8. The Stille reaction (this section)

Focused Problems

18.43 Predict the product of the Stille reaction between ethynyltrimethylstannane, $HC{\equiv}C{-}Sn(CH_3)_3$, and phenyl triflate, PhOTf, in the presence of $Pd(PPh_3)_4$ and excess LiCl.

18.44 What reactants would be required to form the following compound by the Stille reaction?

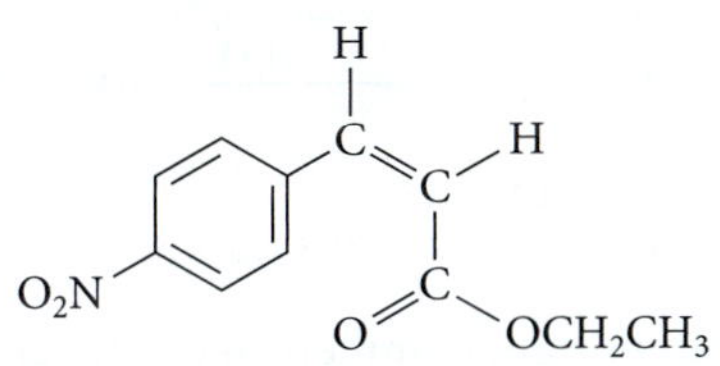

18.11 INDUSTRIAL PREPARATION AND USE OF PHENOL

Chemistry in the Real World

Historically, phenol has been made in a variety of ways, but the principal method used today is an elegant example of a process that gives two industrially important compounds, phenol and acetone, $(CH_3)_2C{=}O$, from a single starting material. The starting material for the manufacture of phenol is cumene (isopropylbenzene), which comes from benzene and propene, two compounds obtained from petroleum (Sec. 16.8). The production of phenol and acetone is a two-stage process. In the first stage, cumene undergoes an *autoxidation* to form cumene hydroperoxide. (An **autoxidation** is an oxidation reaction involving molecular oxygen as the oxidizing agent.)

Autoxidation:

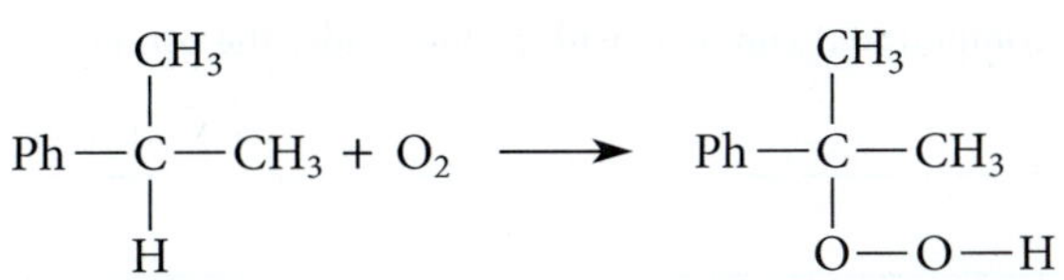

cumene **cumene hydroperoxide** (18.112)

Autoxidation is a free-radical chain reaction. Dioxygen is actually a *diradical*—a double free radical. (For the reason, see Problem 1.70.) The initiation step involves abstraction of the benzylic hydrogen of cumene:

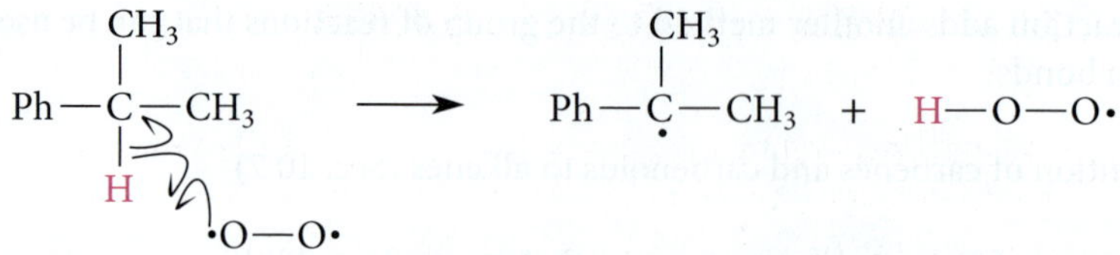
(18.113a)

In the propagation steps, the resulting benzylic radical then reacts with another molecule of dioxygen, and the resulting radical abstracts a benzylic hydrogen from another molecule of cumene to form cumene hydroperoxide.

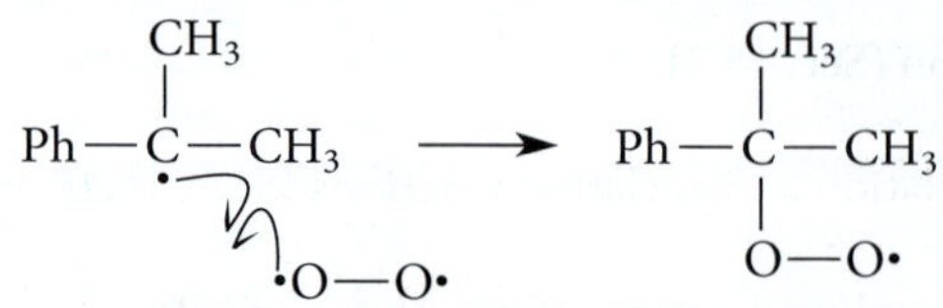
(18.113b)

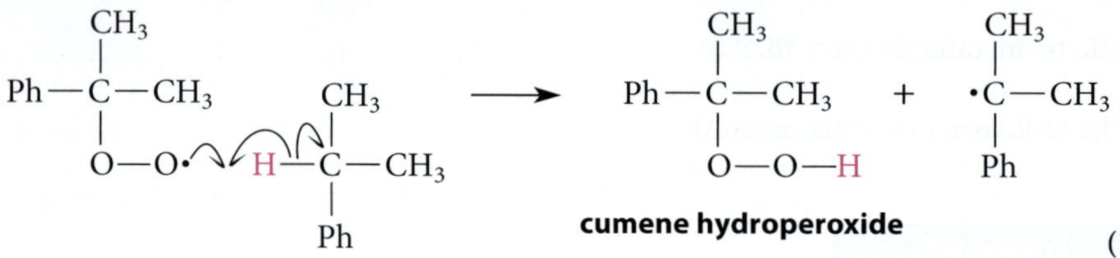
(18.113c)

In the second stage of the phenol–acetone synthesis, cumene hydroperoxide is subjected to an acid-catalyzed rearrangement that yields both acetone and phenol.

Rearrangement:

STUDY GUIDE LINK 18.2
The Cumene Hydroperoxide Rearrangement

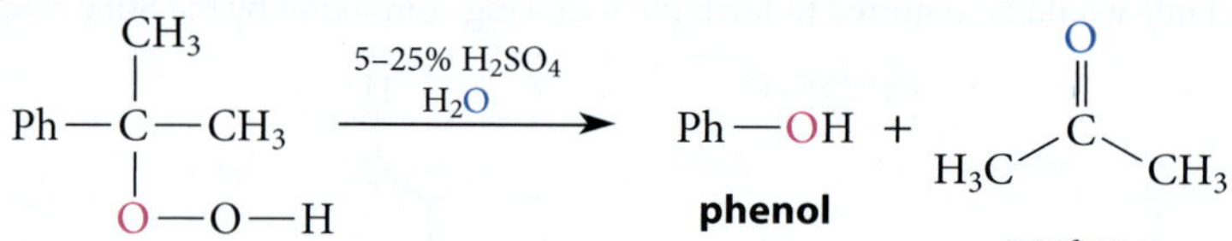
(18.114)

Phenol is a very important commercial chemical. It is a starting material for the production of phenol–formaldehyde resins (Sec. 19.17), which are polymers that have a variety of uses, including plywood adhesives, glass fiber (fiberglass) insulation, molded phenolic plastics used in automobiles and appliances, and many others.

Focused Problem

18.45 Compound *A* is a by-product of the autoxidation of cumene, and compound *B* is a by-product of the acid-catalyzed conversion of cumene hydroperoxide to phenol and acetone.

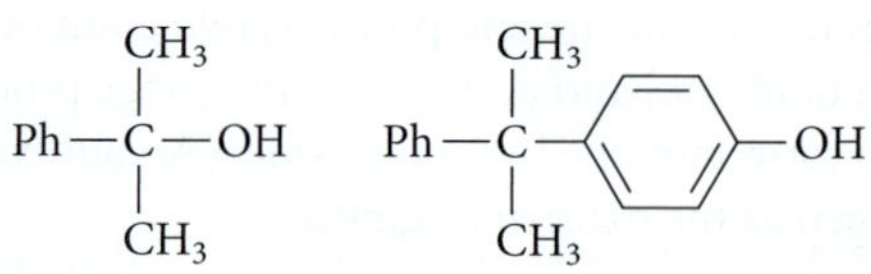
A *B*

Draw a curved-arrow mechanism that shows how compound *A* can react with phenol under the conditions of Eq. 18.114 to give compound *B*.

CHAPTER SUMMARY *For a summary of the chapter, see Chapter 18 in the* Study Guide and Solutions Manual.

REACTION REVIEW *For a summary of reactions discussed in this chapter, see the* Reaction Review *section of Chapter 18 in the* Study Guide and Solutions Manual.

SKILLS OBJECTIVES WITH PROBLEMS

- **Explain why vinylic halides, aryl halides, and other aryl—X derivatives (where X is a good leaving group) are very unreactive in S_N2 reactions.** **18.1**

18.46 In each part, rank the compounds in order of increasing rate of their S_N2 reactions with potassium iodide in acetone. Give the reasons for your ranking.

(a)

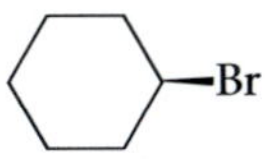

A

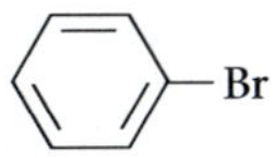

B

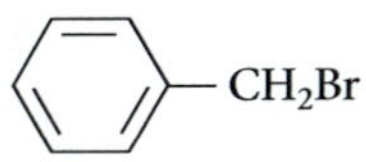

C

(b)

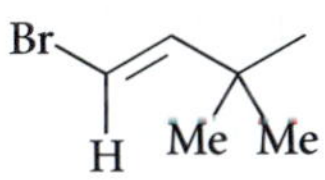

A

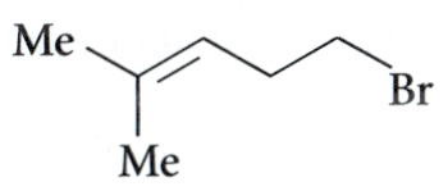

B

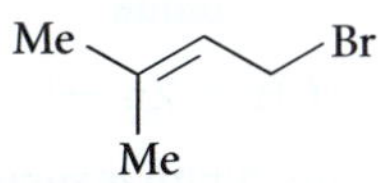

C

- **Explain why vinylic halides, aryl halides, and other aryl—X derivatives (where X is a good leaving group) are very unreactive in S_N1 reactions.** **18.3**

18.47 In each part of Problem 18.46, rank the compounds in order of increasing rate of their S_N1 reactions. Give the reasons for your ranking.

- **Explain why aryl halides, especially aryl fluorides, bearing nitro groups in the ortho and/or para positions are reactive in nucleophilic substitution.** **18.4**

18.48 Rank the following compounds in order of increasing rate of reaction with Na^+ EtO^- in EtOH. Explain your reasoning.

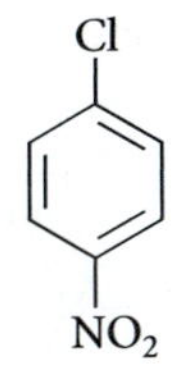

A

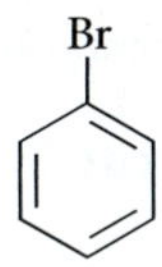

B

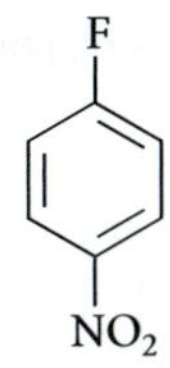

C

- **Calculate the following numbers for transition-metal complexes:**
  - **(1) Oxidation state of the metal** **18.5B**
  - **(2) The d^n configuration of the metal** **18.5C**
  - **(3) The electron count** **18.5D**

18.49 (a) Heme (**Fig. P18.49a**) is the iron-containing compound that is part of hemoglobin, the oxygen-carrying protein in blood. Calculate each of the following for the iron:

(1) The oxidation state

(2) The d^n configuration

(3) The electron count for the complex

(b) When an oxygen molecule binds to hemoglobin, it becomes an L-type ligand on the iron, as shown in **Fig. P18.49b**. From the perspective of the 18-electron rule, explain why this binding should be favorable. (Other interactions of the oxygen molecule with the protein, not shown, also stabilize the complex.)

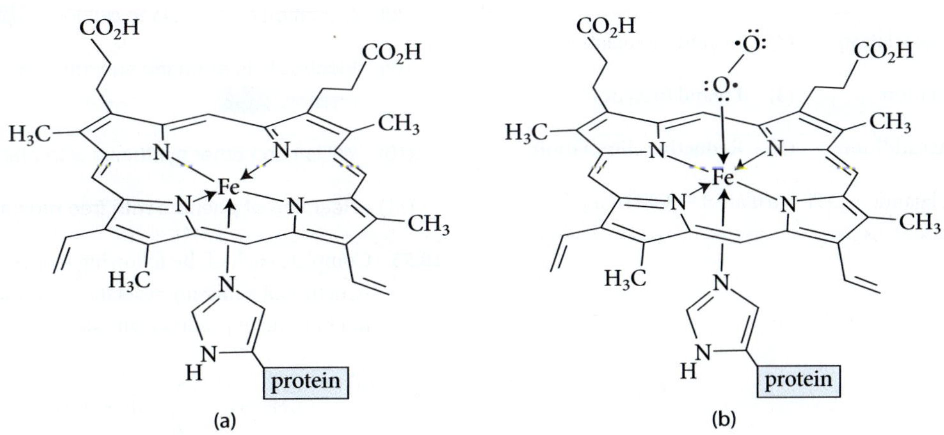

FIGURE P18.49 The heme molecule (a) without and (b) with a coordinated oxygen molecule. The ligand labeled "protein" is a group from the globin protein in which the heme is imbedded to form hemoglobin.

- **Predict the relative acidities of phenols and alcohols using the effects on acidity covered in Chapter 3. 18.7A**

18.50 Arrange the compounds within each set in order of increasing acidity, and explain your reasoning. If you use resonance arguments, draw resonance structures.

(a) Cyclohexanol, phenol, benzyl alcohol

(b)

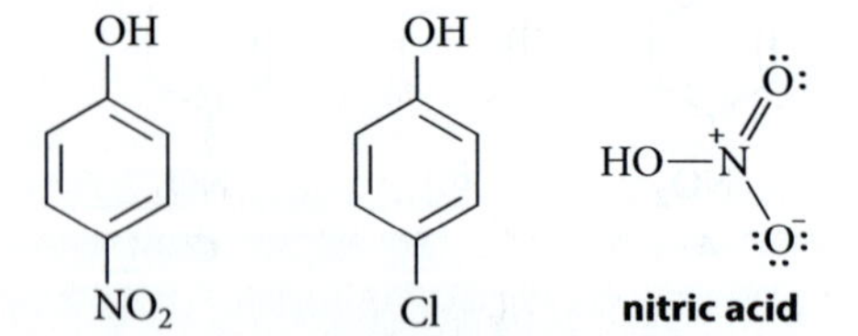

p-nitrophenol **p-chlorophenol** **nitric acid**

(c) Cyclohexyl mercaptan, cyclohexanol, benzenethiol

(d) 4-Nitrobenzenethiol, 4-nitrophenol, phenol

(e)

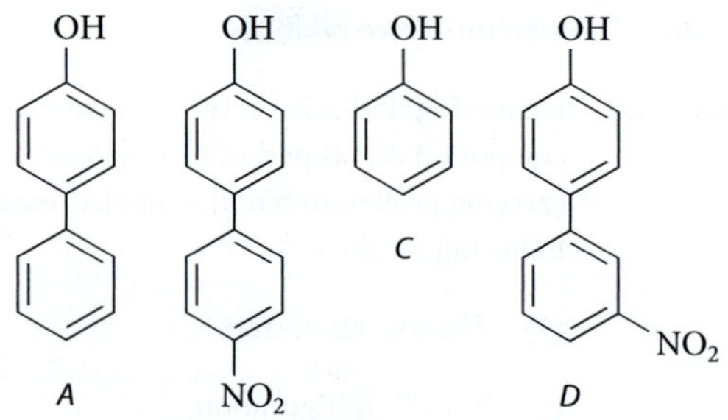

A *B* *C* *D*

- **Identify the following elementary reactions in the mechanisms of transition-metal catalyzed reactions: 18.5E**
  - (1) Ligand association and dissociation
  - (2) β-Elimination and ligand insertion
  - (3) Oxidative addition and reductive elimination

18.51 In the following mechanism, identify each step as one of the following:

(1) Ligand association (2) Ligand dissociation

(3) β-Elimination (4) Ligand insertion

(5) Oxidative addition (6) Reductive elimination

L = an L-type ligand; —OTf = triflate = $—OSO_3CF_3$

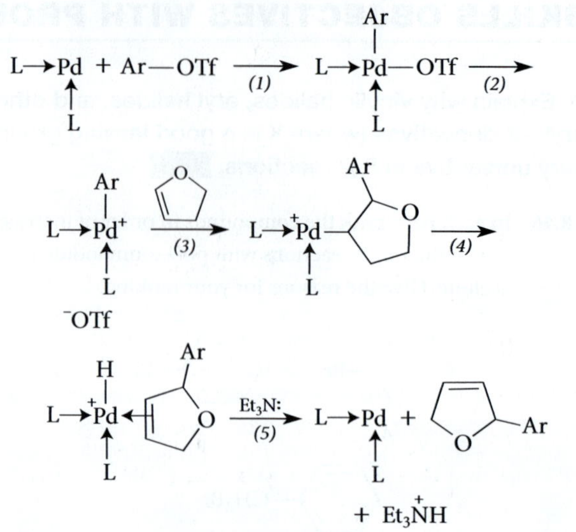

18.52 Outline a mechanism for formation of the Simmons–Smith reagent (Sec. 10.7B) by the following equation. Use some of the steps listed in Problem 18.51.

$$(CH_3CH_2)_2Zn + CH_2I_2 \longrightarrow$$

diethylzinc **methylene iodide**

$$I—CH_2—Zn—I + CH_3CH_2CH_2CH_3$$

Simmons–Smith reagent

- **Provide either a missing starting material or a missing product for examples of the following reactions.**
  - (1) Heck reaction 18.6A
  - (2) Suzuki reaction 18.6B
  - (3) Alkene metathesis 18.6C
  - (4) Stille reaction 18.10B
  - (5) β-Elimination reactions of vinylic halides 18.2
  - (6) Nucleophilic aromatic substitution 18.4
  - (7) Cleavage of phenol ethers 18.10A
  - (8) Oxidation of phenols to quinones 18.8A
  - (9) Electrophilic aromatic substitution reactions of phenols 18.9
  - (10) Williamson ether synthesis with phenolate ions 18.7B
  - (11) Reaction of phenols with free radicals 18.8B

18.53 Complete each of the following reactions by providing the structure of a missing reactant or product. ("No reaction" might be an appropriate answer.)

(a)

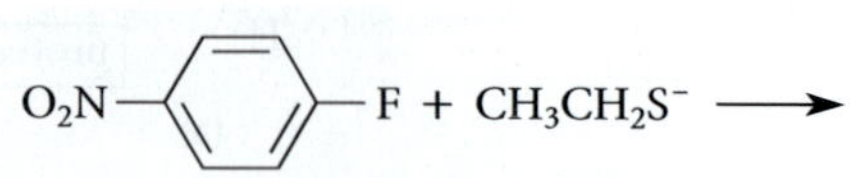

(b)

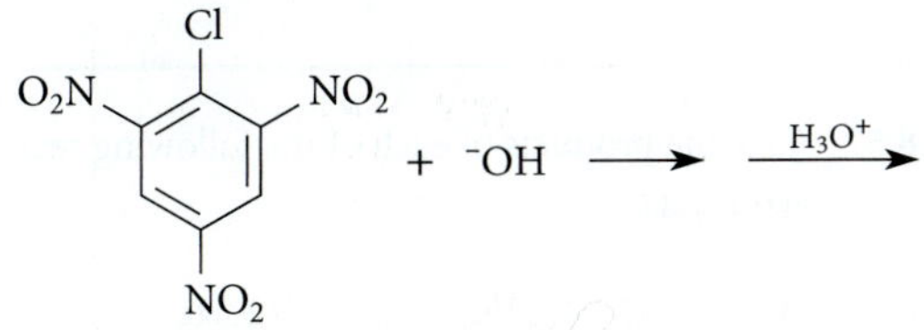

(c)

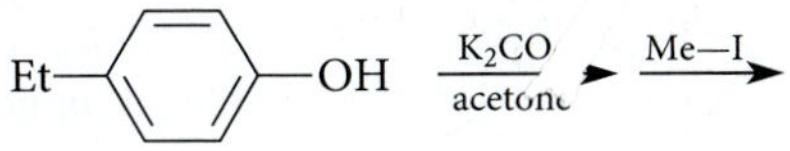

(d)

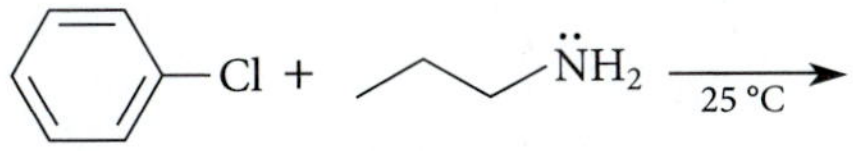

(e)

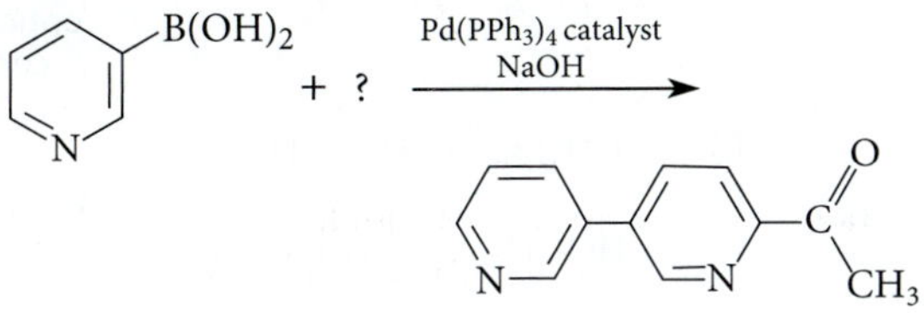

(f)

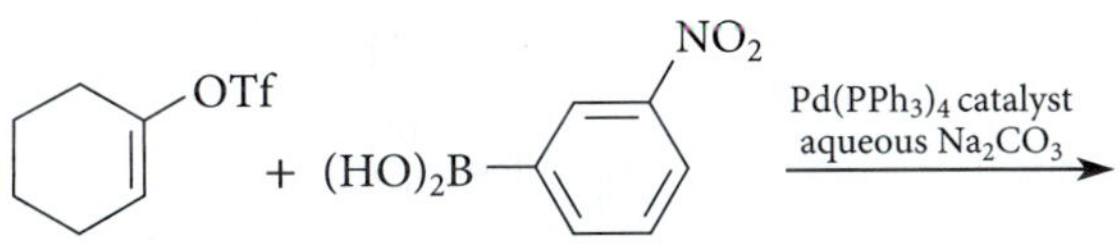

(g)

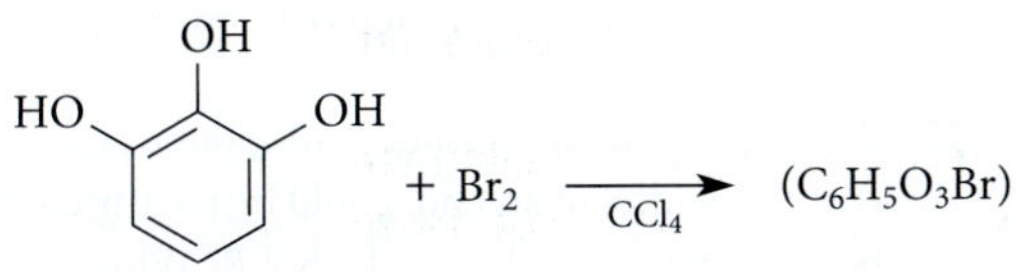

pyrogallol

(h)

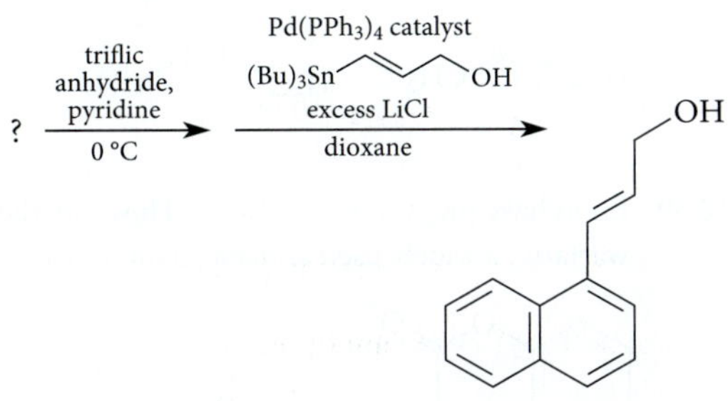

(i)

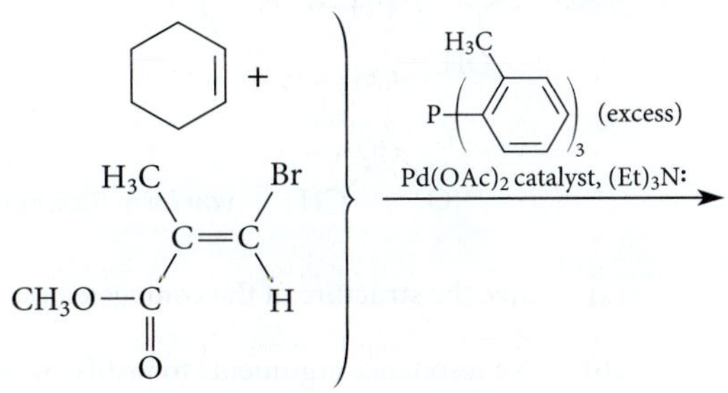

(j)

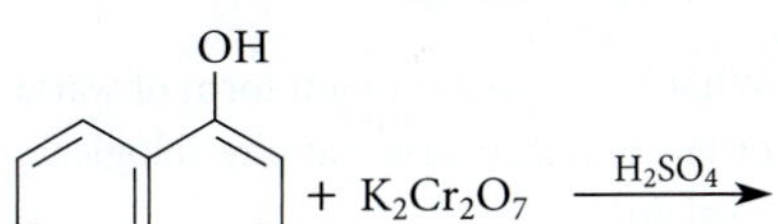

(k)

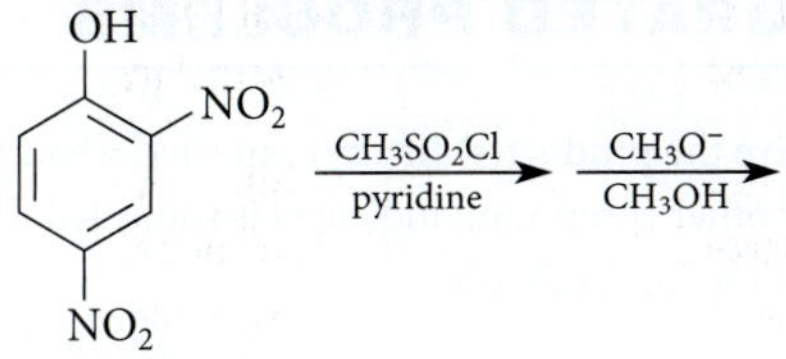

2,4-dinitrophenol

(l)

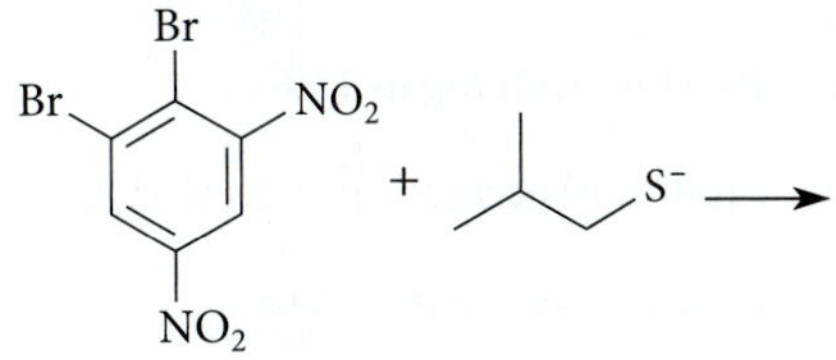

(m)

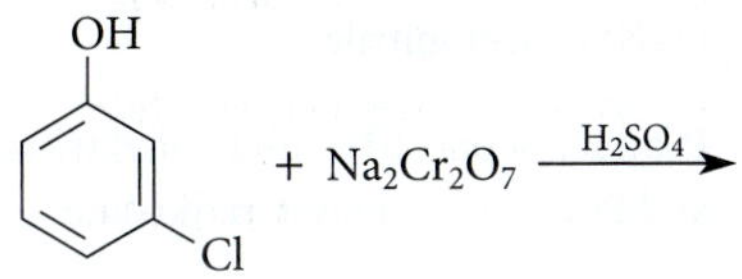

***m*-chlorophenol**

(n)

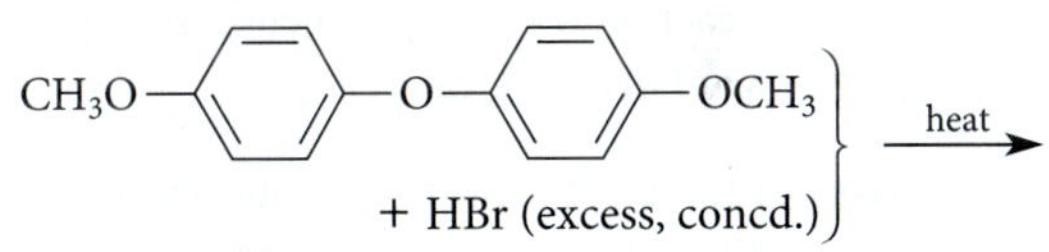

(o)

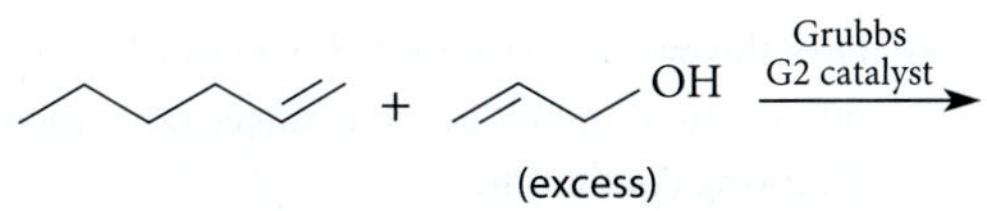

(p)

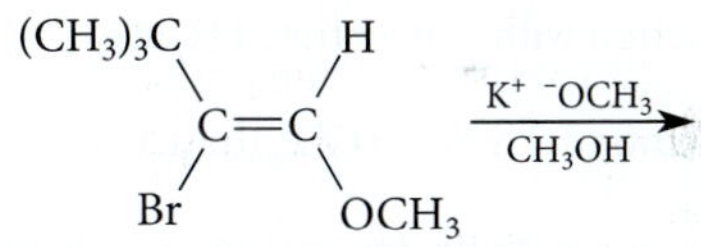

(q)

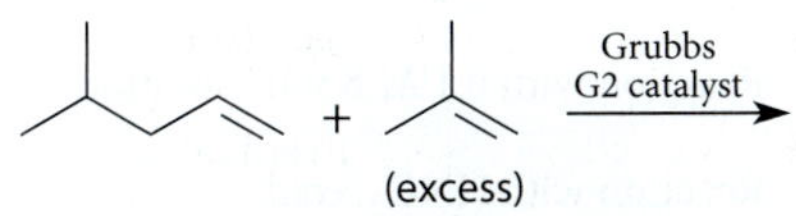

18.54 Ferulic acid is a potent antioxidant found in tomatoes and other vegetables that protects against oxidative stress in neuronal cells, and it is of some interest in research on Alzheimer's disease. (See the sidebar, "Sulfide Oxidation and Protein-Aggregation Diseases," Sec. 12.9A.)

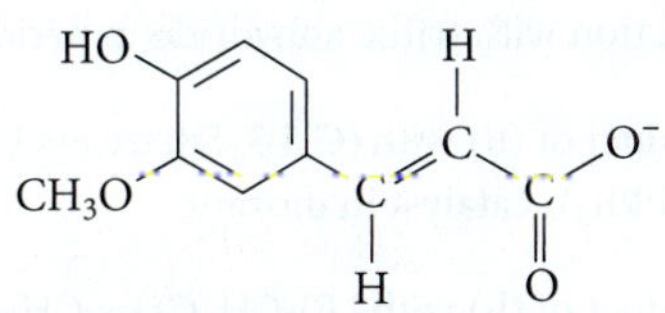

4-hydroxy-3-methoxycinnamic acid (ferulic acid)
(conjugate base)

(a) Show how a free radical R· would react with ferulic acid, and include resonance structures of the product radical.

(b) Would the isomer of ferulic acid, 3-hydroxy-4-methoxycinnamic acid, be equally effective as a free-radical inhibitor? Explain.

INTEGRATED PROBLEMS

18.55 Give the product(s) (if any) expected when *p*-iodotoluene or other compound indicated is subjected to each of the following conditions.

(a) Reaction with CH_3OH, 25 °C

(b) Reaction with CH_3O^- in CH_3OH, 25 °C

(c) Reaction with Mg in THF

(d) Product of part (c) with $ClSn(CH_3)_3$

(e) Reaction with Li in hexane

(f) Reaction with $H_2C{=}CH_2$, $Pd(PPh_3)_4$ catalyst, and Et_3N: in acetonitrile

(g) Product of part (d) with phenyl triflate, excess LiCl, and $Pd(PPh_3)_4$ catalyst in dioxane

(h) Product of part (c) with $B(OMe)_3$, then H_3O^+/H_2O

(i) Reaction with the product of (h), aqueous Na_2CO_3, and $Pd(PPh_3)_4$ catalyst

(j) Product of (h) with (*E*)-1-bromopropene, aqueous Na_2CO_3, and $Pd(PPh_3)_4$ catalyst

18.56 Give the product(s) expected (if any) when *m*-cresol or other compound indicated is subjected to each of the following conditions.

(a) Reaction with concentrated H_2SO_4

(b) Reaction with Br_2 in CCl_4 (dark)

(c) Reaction with Br_2 (excess) in CCl_4, light

(d) Reaction with 0.1 *M* NaOH solution

(e) Reaction with HNO_3, cold

(f) Reaction with 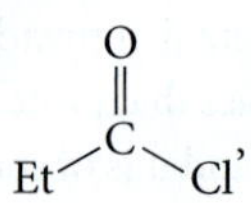

, $AlCl_3$, heat; then H_2O

(g) Reaction with $Na_2Cr_2O_7$ in H_2SO_4

(h) Reaction with triflic anhydride in pyridine, 0 °C

(i) Product of (h) with $(CH_3)_4Sn$, excess LiCl, and $Pd(PPh_3)_4$ catalyst in dioxane

(j) Product of (h) with (*E*)-$CH_3CH{=}CH{-}B(OH)_2$, aqueous NaOH, and $Pd(PPh_3)_4$ catalyst

18.57 Give the products of each of the following reaction sequences.

(a)

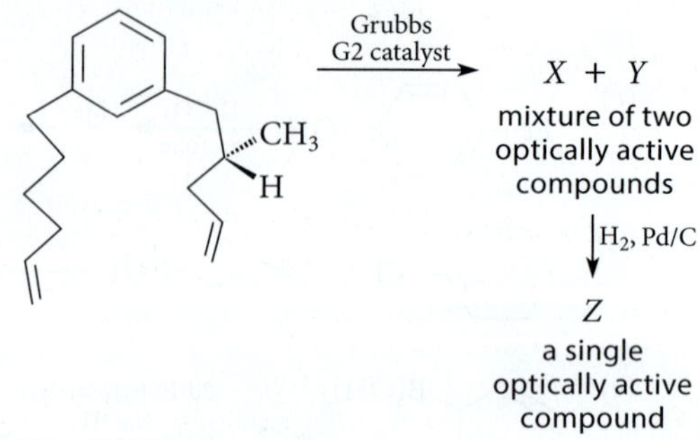

(b)

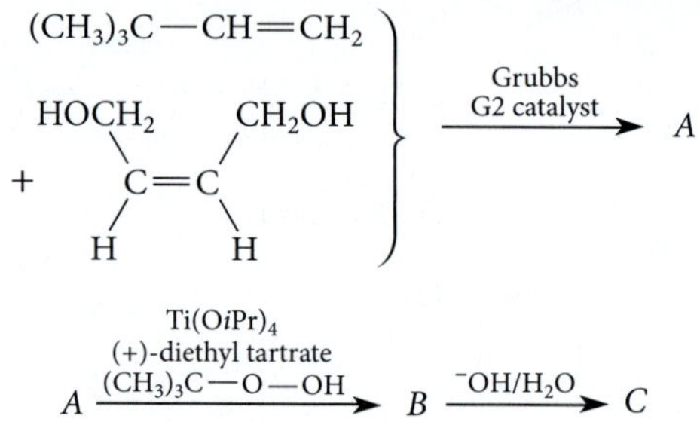

(Give the stereochemistry of *C*)

18.58 Although enols are unstable compounds (Sec. 5.5), suppose that the acidity of an enol could be measured. Which would be more acidic: enol *A* or alcohol *B*? Why?

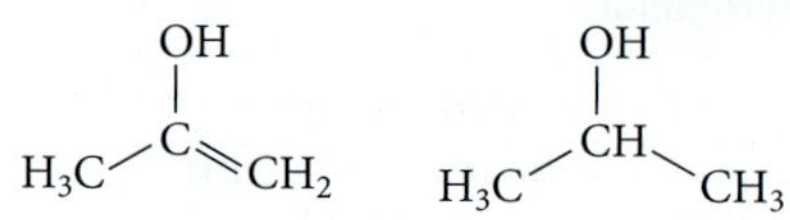

A *B*

18.59 Enols have pK_a values of ~10–11. However, the pK_a of warfarin, a widely used anticoagulant, is 5.0.

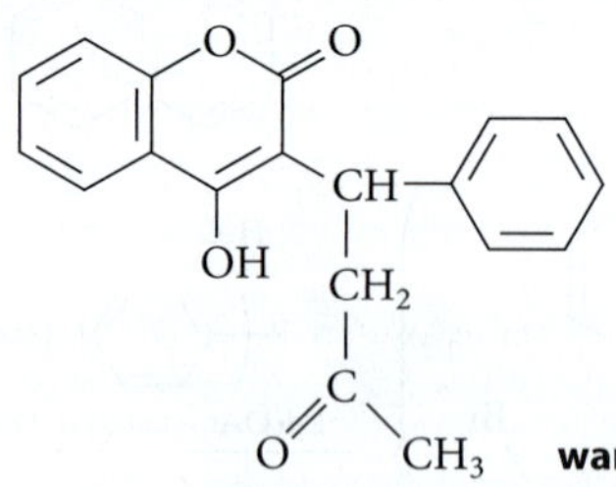

warfarin (Coumadin)

(a) Give the structure of the conjugate base of warfarin.

(b) Use resonance arguments to justify the unusually low pK_a of warfarin.

(c) What is the predominant form of warfarin—the enol or its conjugate base—at physiological pH (7.4)? Explain.

18.60 Identify compounds *A*, *B*, and *C* from the following information.

(a) Compound *A*, $C_8H_{10}O$, is insoluble in water but soluble in aqueous NaOH solution, and it yields 3,5-dimethylcyclohexanol when hydrogenated over a nickel catalyst at high pressure.

(b) Aromatic compound *B*, $C_8H_{10}O$, is insoluble in both water and aqueous NaOH solution. When treated successively with concentrated HBr, then Mg in THF, then water, it gives *p*-xylene.

(c) Compound *C*, $C_9H_{12}O$, is insoluble in water and in NaOH solution, but reacts with concentrated HBr and heat to give *m*-cresol and a volatile alkyl bromide.

18.61 Contrast the reactivities of cyclohexanol and phenol with each of the following reagents, and explain.

(a) Aqueous NaOH solution

(b) NaH in THF

(c) Triflic anhydride in pyridine, 0 °C

(d) Concentrated aqueous HBr, H_2SO_4 catalyst

(e) Br_2 in CCl_4, dark

(f) $Na_2Cr_2O_7$ in H_2SO_4

(g) H_2SO_4, heat

18.62 Choose the one compound within each set that meets the indicated criterion, and explain your choice.

(a) The compound that reacts with alcoholic KOH to liberate fluoride ion.

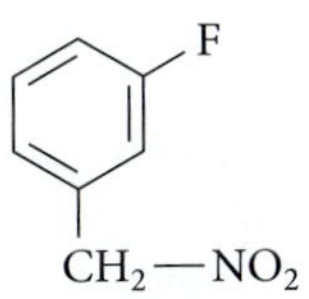

A

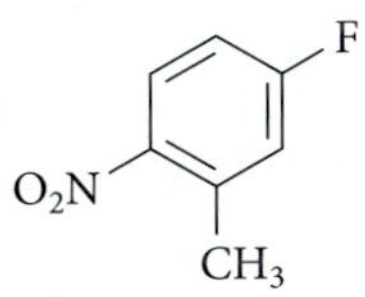

B

(b) The compound that cannot be prepared by a Williamson ether synthesis.

A

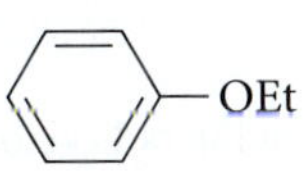

B

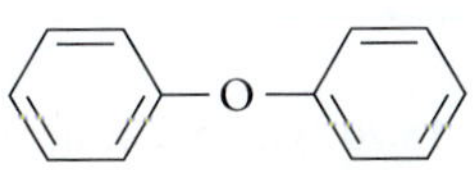

C

(c) The compound that gives an acidic solution when allowed to stand in aqueous ethanol.

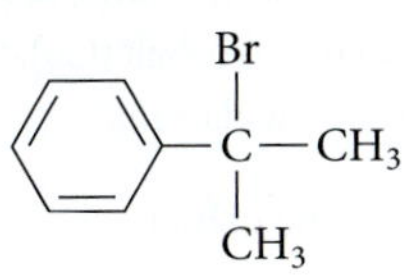

A

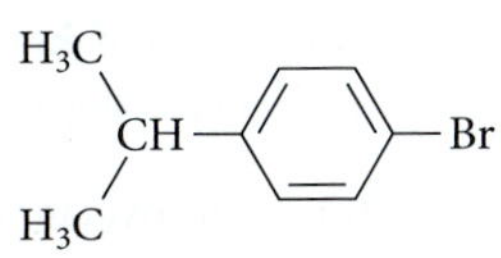

B

(d) The ether that cleaves more rapidly in HI.

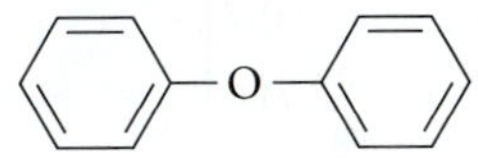

A

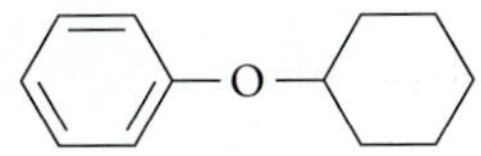

B

18.63 Give the products (if any) when each of the following isomers reacts with HBr and heat.

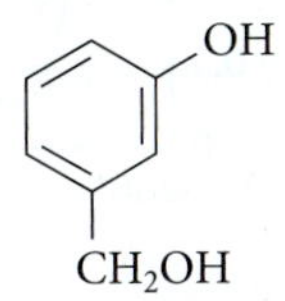

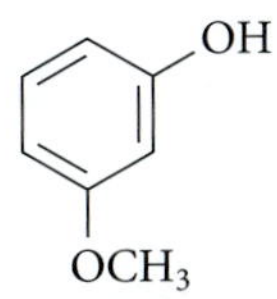

3-(hydroxymethyl)phenol **3-methoxyphenol**

18.64 What products (if any) are formed when 3,5-dimethylbenzenethiol is treated first with one equivalent of Na^+ EtO^- in ethanol and then with each of the following?

(a) Allyl bromide

(b) Bromobenzene

18.65 Phenols, like alcohols, are weak Brønsted bases.

(a) Write the reaction in which the oxygen of phenol reacts as a base with the acid H_2SO_4. (For convenience you can abbreviate H_2SO_4 as H_2A.)

(b) On the basis of resonance and polar effects, decide whether phenol or cyclohexanol should be the stronger base.

18.66 The UV spectrum of *p*-nitrophenol in aqueous solution is shown in **Fig. P18.66** (spectrum *A*). When a few drops of concentrated NaOH are added, the solution turns yellow and the spectrum changes (spectrum *B*). On addition of a few drops of concentrated acid, the color disappears and spectrum *A* is restored. Explain these observations.

18.67 Vanillin is the active component of natural vanilla flavoring.

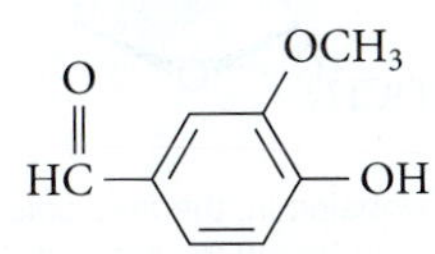

vanillin

When a few drops of vanilla extract (an ethanol solution of vanillin) are added to an aqueous NaOH solution, the characteristic vanilla odor is not present. Upon acidification of the solution, a strong vanilla odor develops. Explain.

18.68 A mixture of *p*-cresol (4-methylphenol), $pK_a = 10.2$, and 2,4-dinitrophenol, $pK_a = 4.11$, is dissolved in ether. The ether solution is then vigorously shaken with one of the following aqueous solutions. Which solution effects the best separation of the two phenols by dissolving one in the water layer and leaving the other in the ether solution? Explain. (*Hint:* Apply Eqs. 3.67a and b, Sec. 3.6E.)

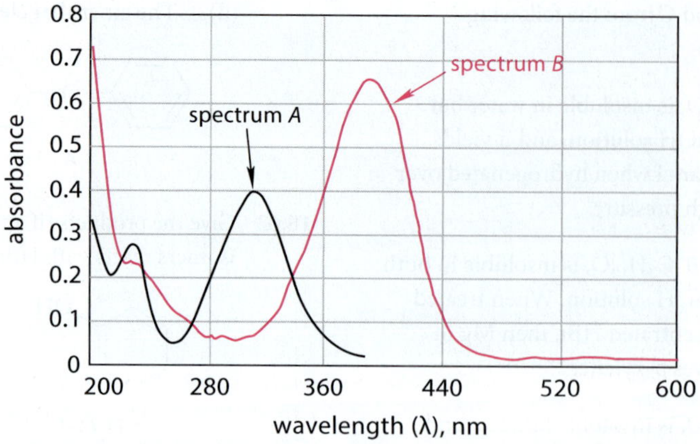

FIGURE P18.66 UV spectra for Problem 18.66. Spectrum *A* was taken in acidic solution, whereas spectrum *B* was taken in basic solution.

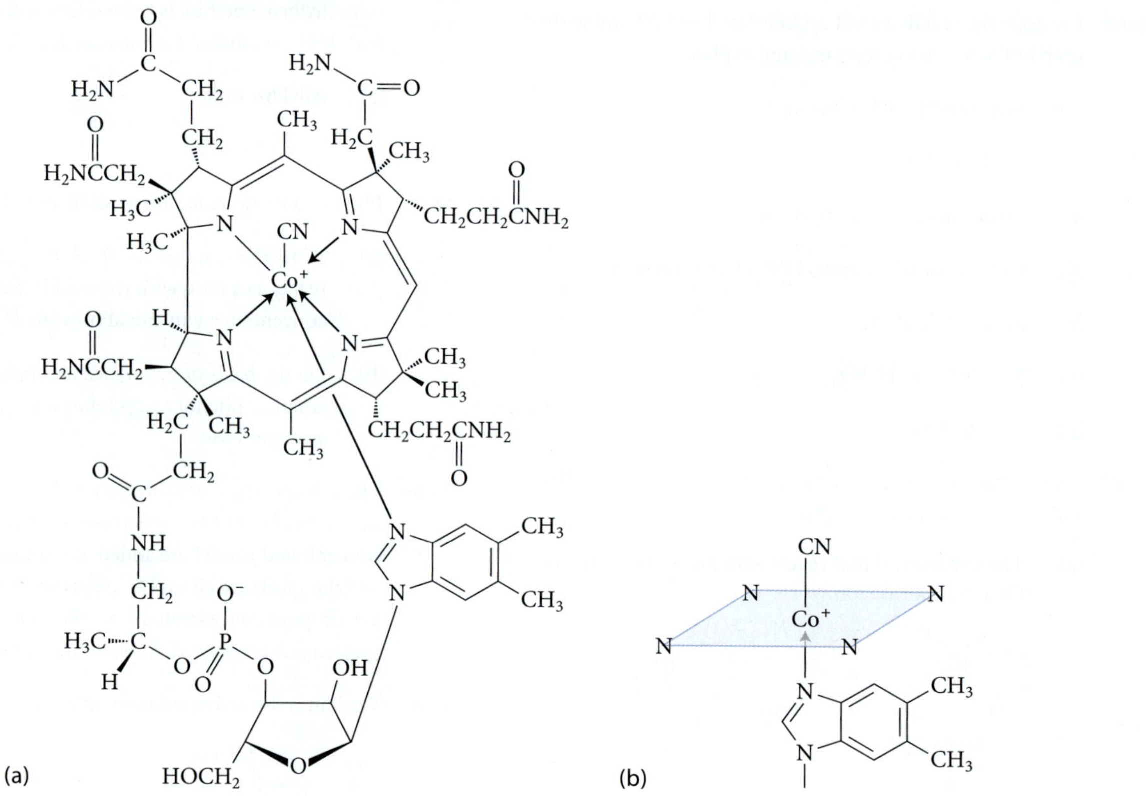

FIGURE P18.71 The structure of cyanocobalamin, the first isolated form of vitamin B_{12}. (a) A Lewis structure. (b) A perspective drawing that shows the position of the two ligands that are above and below the plane defined by the four ring nitrogens.

(1) A 0.1 M aqueous HCl solution

(2) A solution that contains a large excess of pH = 4 buffer

(3) A solution that contains a large excess of pH = 7 buffer

(4) A 0.1 M NaOH solution

18.69 1-Haloalkynes are known compounds.

(a) Would 1-bromo-2-phenylacetylene (Br—C≡C—Ph) be likely to undergo an S_N2 reaction? Explain.

(b) Would the same compound be likely to undergo an S_N1 reaction? Explain.

18.70 Give the structure of the radical formed, and its resonance structures, when warfarin (the structure in Problem 18.59) reacts with R•, a general free radical.

18.71 The structure of cyanocobalamin, one form of vitamin B_{12}, is given in **Fig. P18.71**. As the structure shows, cyanocobalamin is a complex of the transition metal cobalt (Co). Calculate each of the following for cyanocobalamin:

(a) The oxidation state of the cobalt

(b) The d^n count (that is, the value of n)

(c) The total electron count around the metal

18.72 Provide the following information, and show how you obtained it, for the Fe(V) in the protoporphyrin IX group of cytochrome P450 (see Fig. 17.4a, Sec. 17.5B). In addition, verify the oxidation state of the iron.

(a) The d^n count (that is, the value of n)

(b) The total electron count around the metal

18.73 When a suspension of 2,4,6-tribromophenol is treated with an excess of bromine water, the white precipitate of 2,4,6-tribromophenol disappears and is replaced by a precipitate of a yellow compound that has the following structure. Give a curved-arrow mechanism for the formation of this compound.

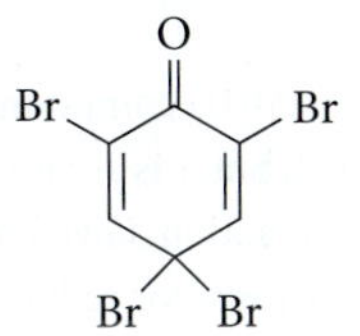

18.74 It has been suggested that the solvolysis of 2-(bromomethyl)-4-nitrophenol at alkaline pH values involves the intermediate shown in brackets. Give a curved-arrow mechanism for the formation of this intermediate and for its reaction with aqueous hydroxide ion to give the final product.

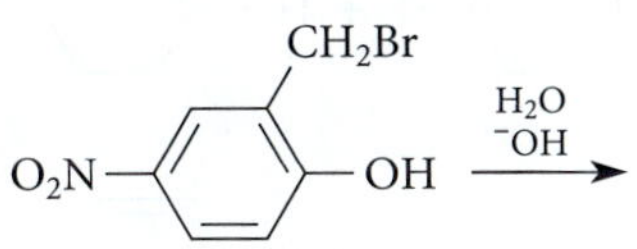

2-(bromomethyl)-4-nitrophenol

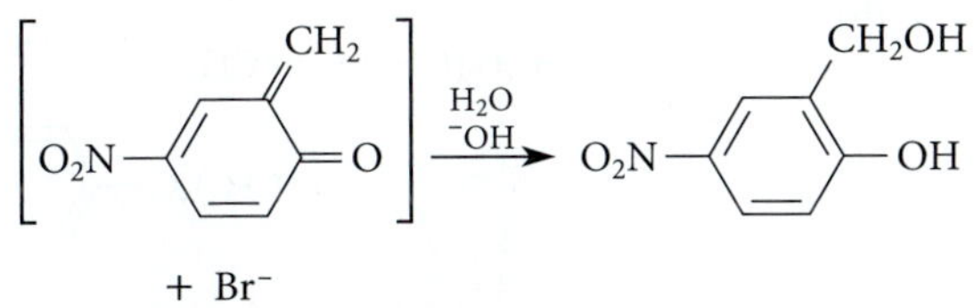

18.75 Outline a synthesis for each of the following compounds from the indicated starting material and any other reagents.

(a) 1-Chloro-2,4-dinitrobenzene from benzene

(b) 1-Chloro-3,5-dinitrobenzene from benzene

(c) Ph—C≡C—Ph from 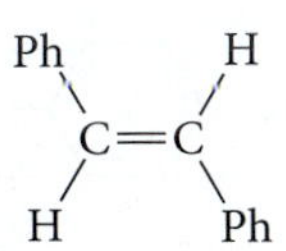

(d) 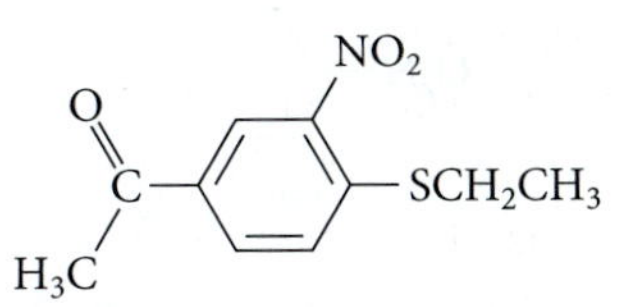 from chlorobenzene

(e) 2-Chloro-4,6-dinitrophenol from chlorobenzene

(f) "Butylated hydroxytoluene" (BHT; Display 18.96, Sec. 18.8B) from *p*-cresol (4-methylphenol)

(g) 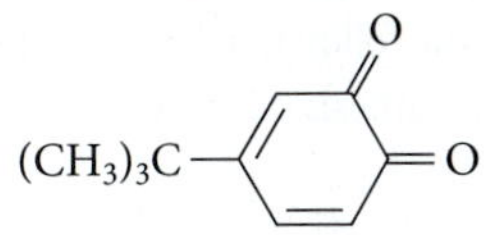 from catechol (benzene-1,2-diol)

(h) 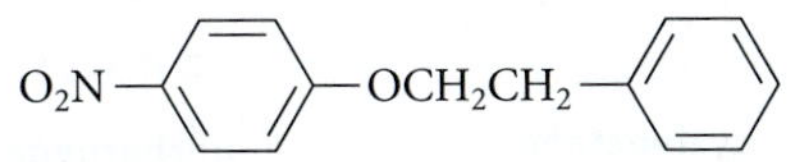

from $PhCH_2CH_2OH$ and fluorobenzene

(i) 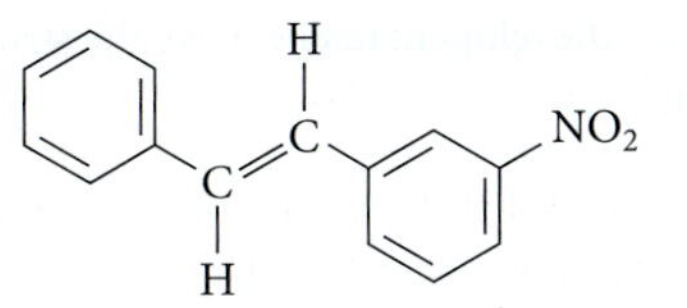

from benzene and ethylene as the only sources of carbon

(j) 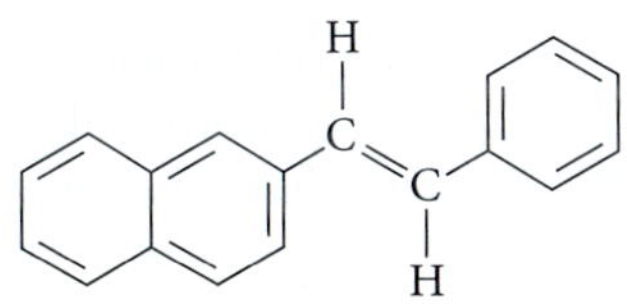

from 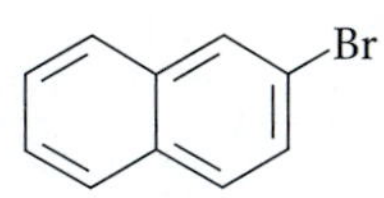

(k) 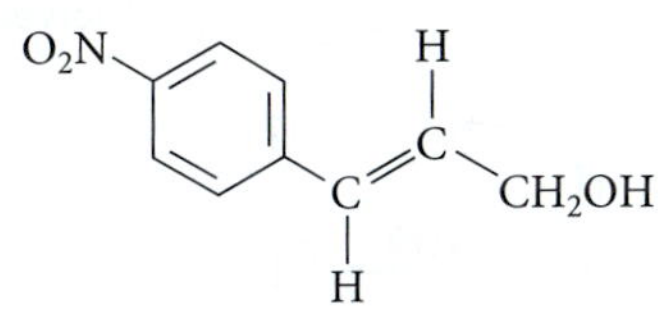

from bromobenzene, ethylene, and any other reagents

(l) 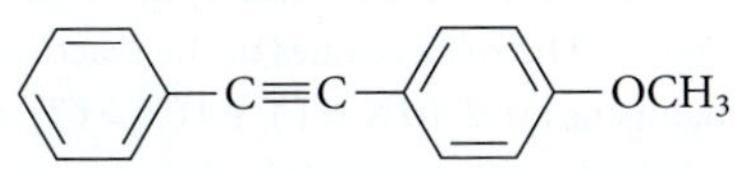

from PhC≡CH and 4-methoxyphenol (*Hint*: See. Eq. 10.58, Sec. 10.5D, and Eq. 18.109, Sec. 18.10B.)

(m) 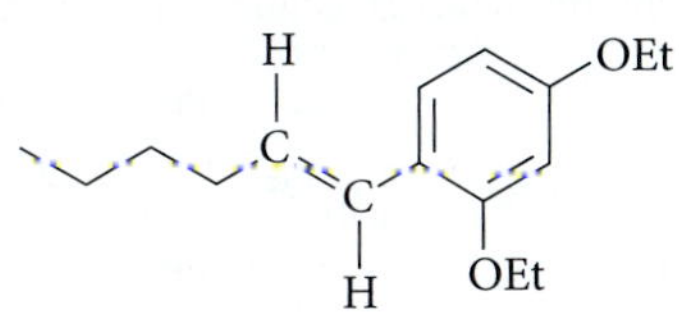

from 1-hexyne and resorcinol

(n) 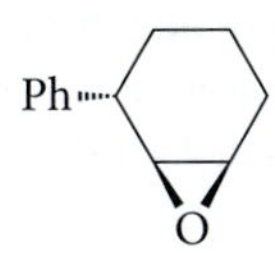 from iodobenzene

(racemate)

18.76 One use of alkene metathesis is to form polymers from cyclic alkenes using ring-opening metathesis polymerization (ROMP). Give the structure of the polymer formed when each of the following alkenes is polymerized with an appropriate metathesis catalyst.

(a) 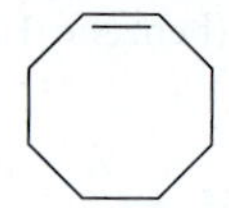

cyclooctene

(b) 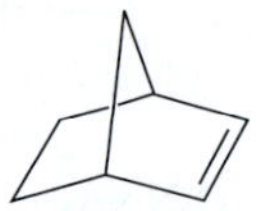

norbornene

18.77 (a) 1,3-Cyclopentadiene undergoes a Diels–Alder reaction with itself to give a diene known commonly as *endo*-dicyclopentadiene. Give the structure of this diene.

(b) When *endo*-dicyclopentadiene is subjected to ROMP (see Problem 18.76), a polymer is formed that is so strong that a 1-inch-thick block will stop a 9 mm bullet. Give a structure for this polymer, and suggest one reason why it is so strong.

18.78 Explain why, in the following two reactions, different stereoisomers of the starting material give different products. (*Hint:* Think about the stereochemistry of elimination.)

(1) 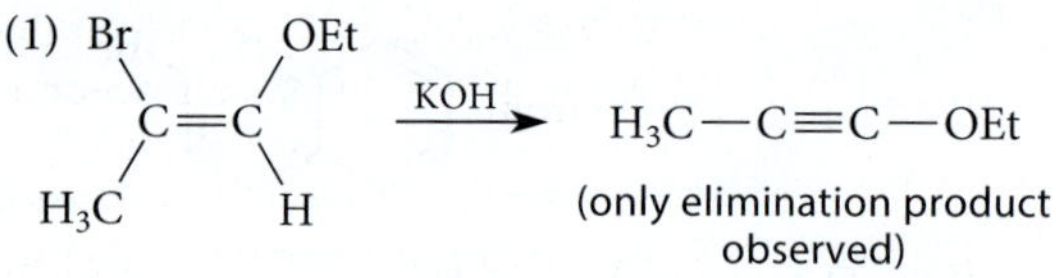

(2) 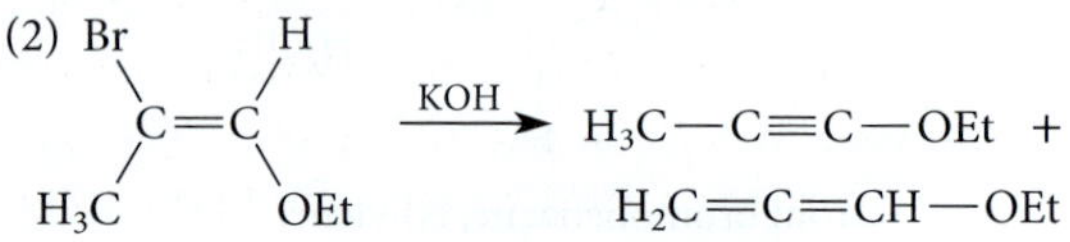

18.79 The following reaction occurs readily at 95 °C (X = halogen). The relative rates of the reaction for the various halogens are 290 (X = F), 1.4 (X = Cl), and 1.0 (X = Br). When a nitro group is in the para position of each benzene ring, the reaction is substantially accelerated. Give a detailed mechanism for this reaction, and explain how it is consistent with the experimental facts. Your mechanism should involve resonance structures for an intermediate.

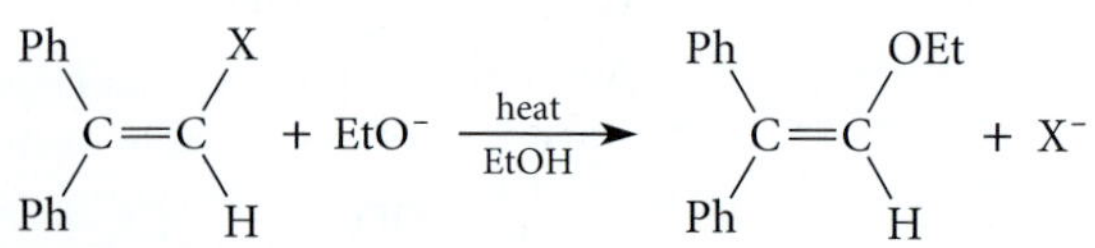

18.80 When 1,3,5-trinitrobenzene [NMR: δ 9.1 (s)] is treated with $Na^+ \, CH_3O^-$, an ionic compound is formed that has the following NMR spectrum: δ 3.3 (3*H*, s); δ 6.3 (1*H*, t, $J = 1$ Hz); δ 8.7 (2*H*, d, $J = 1$ Hz). Suggest a structure for this compound.

18.81 Suggest structures for *X* and *Y* in the following reaction sequence, and suggest a mechanism for the formation of *Y*. Notice that aluminum (Al) is just below boron (B) in the periodic table; to predict *X*, imagine that the Al is a B and ask how that compound would react with an alkene. (*Hint:* See Sec. 5.7.)

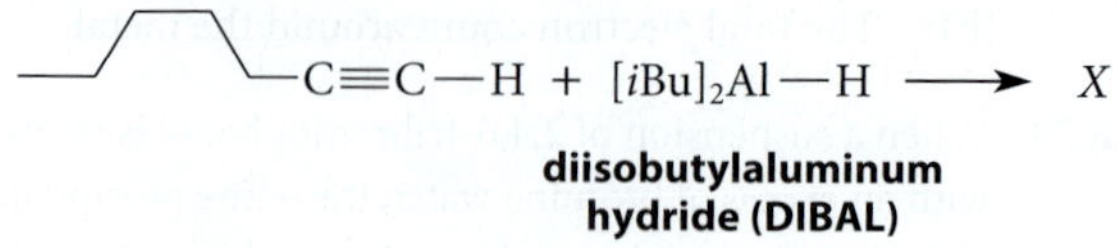

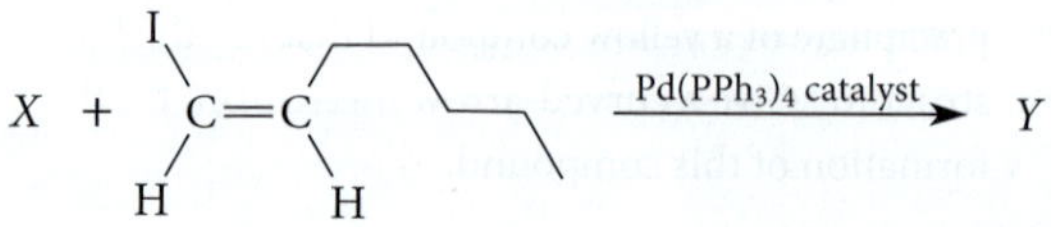

18.82 In some Pd(0)-catalyzed reactions, a Pd(II) compound such as $PdCl_2$ can be used instead of Pd(0), but it is assumed that the Pd(II) is reduced to Pd(0) in the reaction. Give both the product and the curved-arrow mechanism for reduction of $PdCl_2$ to Pd(0) by $:PPh_3$. (*Hint:* If Pd is reduced, something has to be oxidized.)

18.83 Draw curved-arrow mechanisms for the following reactions.

(a) 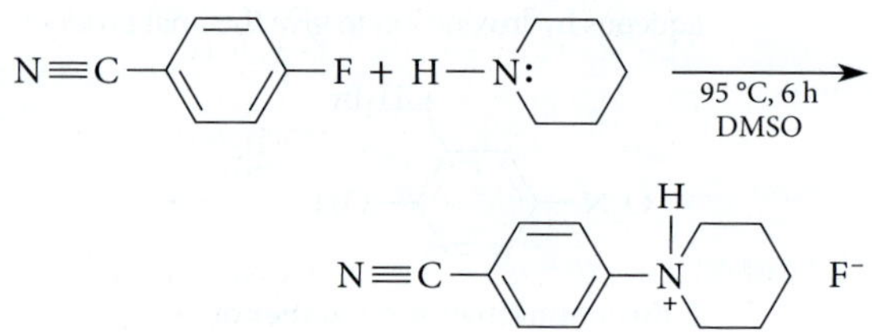

(b) 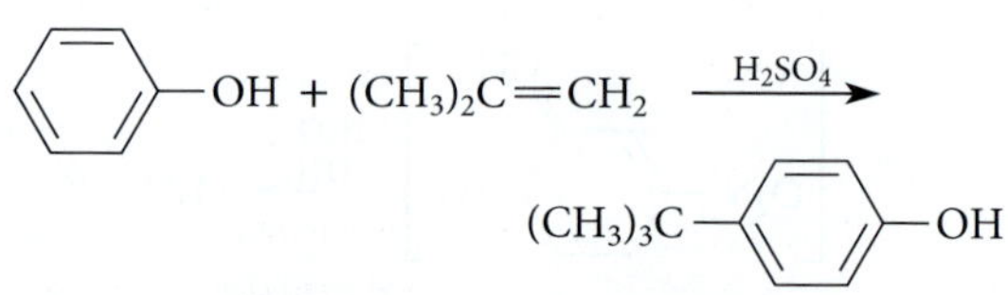

18.84 The *Negishi reaction* is a Pd(0)-catalyzed cross-coupling reaction of organozinc compounds and aryl or vinylic iodides, as in the example that follows. [Ei-ichi Negishi (b. 1935), a professor of chemistry at Purdue University, shared the 2010 Nobel Prize with Suzuki and Heck for his research into cross-coupling reactions.]

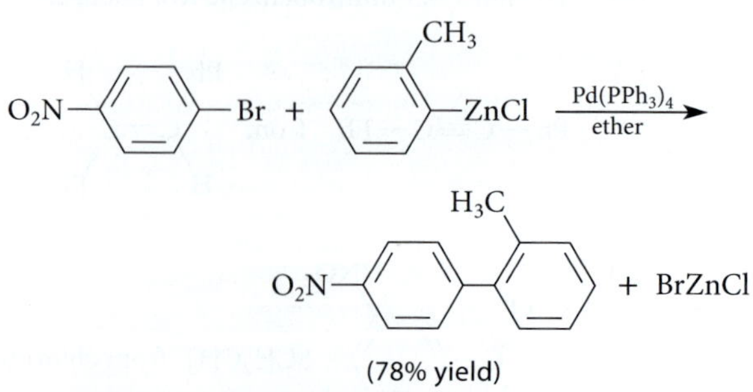

(a) Outline a mechanism of this coupling reaction by showing the important catalytic intermediates.

(b) Show how the organozinc compound could be prepared from *o*-iodotoluene. (*Hint:* See Eq. 18.109, Sec. 18.10B.)

18.85 In 1991, chemists at the University of Iowa were able to prepare trifluoromethylcopper(I), $F_3C—Cu$. This compound could be used to prepare trifluoromethyl-substituted aromatic compounds by reactions like the following:

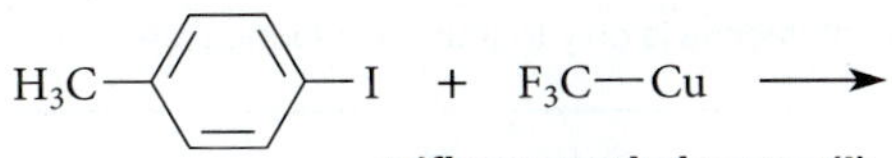

trifluoromethylcopper(I)

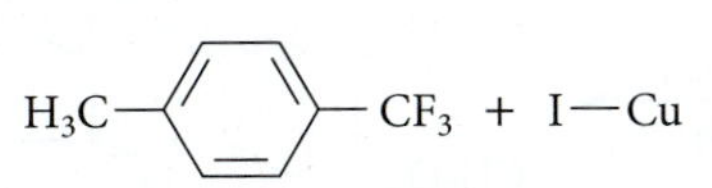

(a) Suggest a stepwise mechanism for this transformation.

In 2013, chemists at Imperial College (London) showed that they could prepare $F_3C—Cu$ containing the isotope ^{18}F. They could use this compound to prepare a number of important drugs that contain trifluoromethyl groups. Each synthesis took less than an hour.

(b) Why might this advance be medically important? (*Hint:* See the sidebar, "The S_N2 Solvent Effect in Cancer Diagnosis," Sec. 9.4E.)

(c) Why is it important that these drugs be prepared and used quickly?

18.86 Complete the reaction sequence in **Fig. P18.86** by giving the structures of *A* and *B*. Give the stereochemistry of the product *B*, and explain your reasoning. (*Note:* 9-BBN is a sterically hindered hydroboration reagent.)

18.87 The reaction given in **Fig. P18.87**, used to prepare the drug mephenesin (a skeletal muscle relaxant), appears to be a simple Williamson ether synthesis. During this reaction, however, a precipitate of NaCl forms after only about 10 minutes, but a considerably longer reaction time is required to obtain a good yield of mephenesin. Taking these facts into account, suggest a mechanism for this reaction. (*Hint:* See Sec. 12.8.)

18.88 In 1960, a group of chemists led by Professor John D. Roberts at Caltech reported that when chlorobenzene containing a ^{14}C isotopic label at carbon-1 is treated with the very strong base potassium amide, a substitution product, aniline, is formed in which the isotopic label is equally distributed between carbon-1 and carbon-2.

* = carbon-14

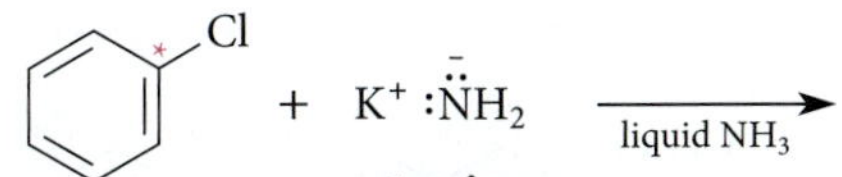

chlorobenzene **potassium amide**

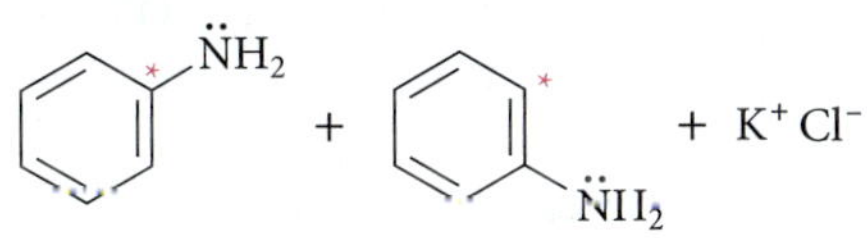

aniline
(about 50% of each)

This result was interpreted as evidence for a very interesting unstable intermediate called *benzyne.* Propose a mechanism for this substitution that accounts for the isotopic labeling result. (*Hint:* Think about a β-elimination reaction.)

18.89 In the conversion shown in **Fig. P18.89**, the Diels–Alder reaction is used to trap a very interesting intermediate by its reaction with anthracene. From the structure of the product, deduce the structure of the intermediate. Then write a curved-arrow mechanism that shows how the intermediate is formed from the starting material. (*Note:* The anthracene has nothing to do with generation of the intermediate; its role is only to trap the intermediate by reacting with it.)

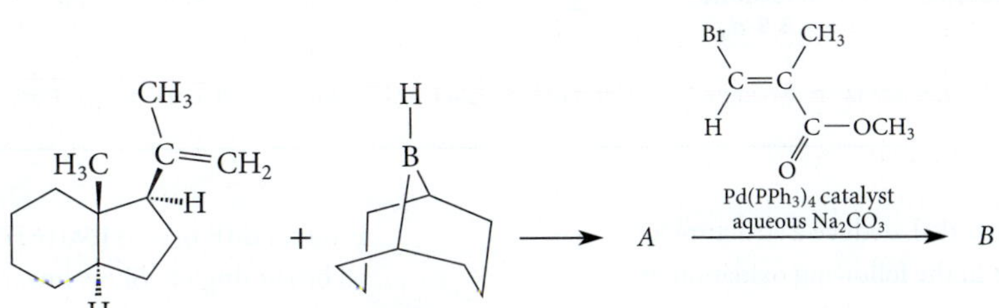

9-borabicyclononane (9-BBN)

FIGURE P18.86

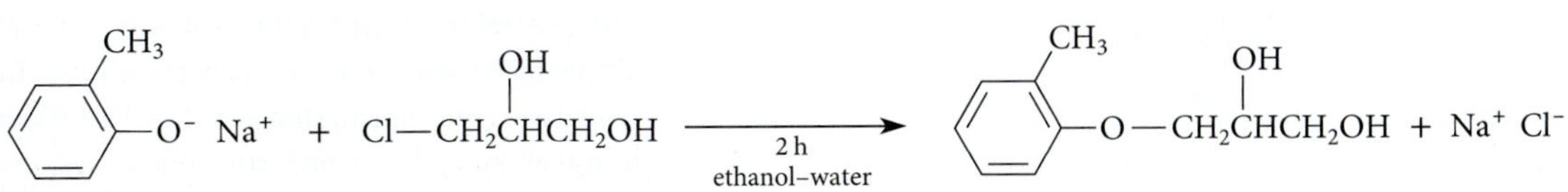

mephenesin

FIGURE P18.87

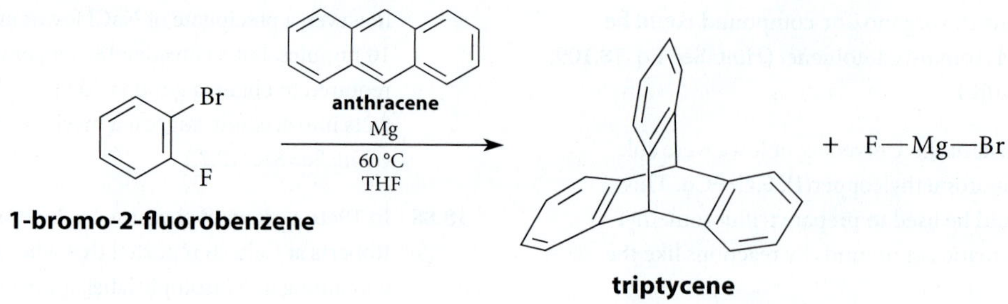

FIGURE P18.89 The intermediate is generated following the reaction with Mg. The role of the anthracene is only to trap the intermediate.

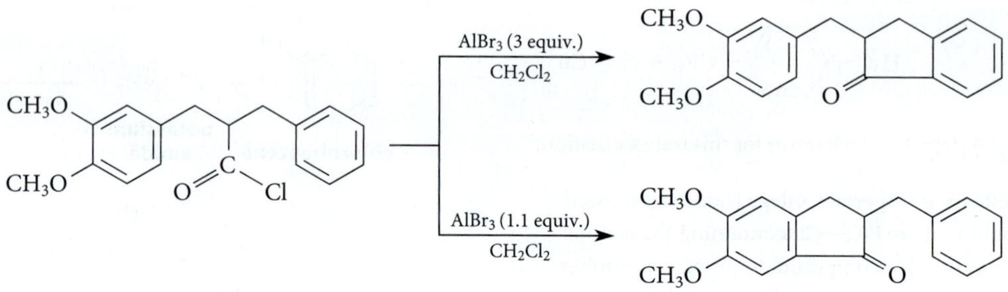

FIGURE P18.91

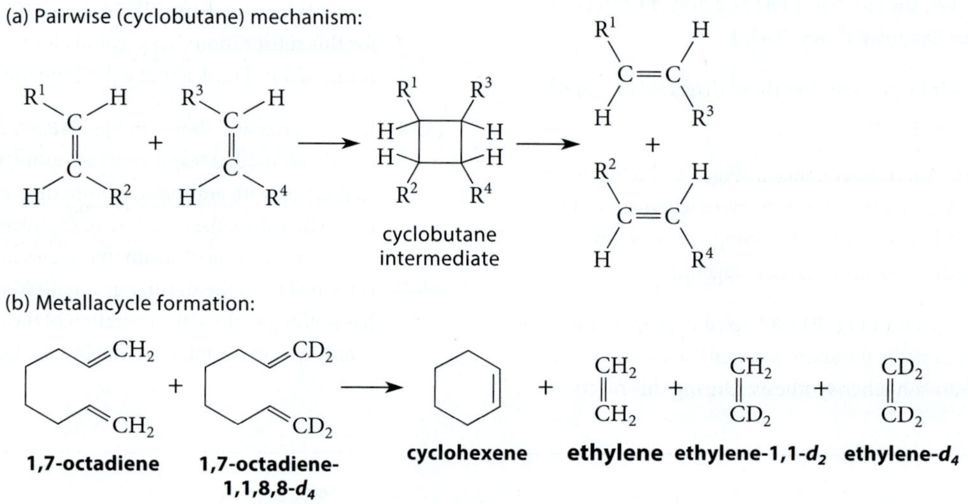

FIGURE P18.92 The pairwise mechanism is shown in part (a). The metallacycle mechanism is given in the text in Eqs. 18.66b–d in Sec. 18.6C.

18.90 From the NMR data provided, deduce a structure for the product A obtained in the following oxidation of 2,4,6-trimethylphenol. (Compound A is an example of a rather unstable type of compound called generally a *quinone methide*.)

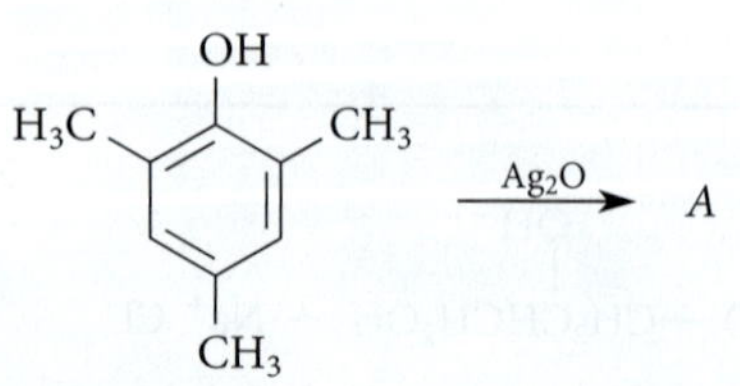

Proton NMR of A: δ 1.90 (6H), δ 5.49 (2H), δ 6.76 (2H), all broad singlets. Show how the NMR spectrum is consistent with your structure.

18.91 For the reactions given in **Fig. P18.91**, explain why different products are obtained when different amounts of $AlBr_3$ catalyst are used. (*Hint:* See Eq. 18.104, Sec. 18.9.)

18.92 Two general mechanisms (or various versions of them) for alkene metathesis were originally considered. In the first (pairwise) mechanism, shown in **Fig. P18.92a**, the catalyst brings about cyclobutane formation between two alkenes. The second mechanism involves metallacycle formation, as shown in Eqs. 18.66b–d in Sec. 18.6C.

(a) The metathesis reaction of cyclopentene and (*E*)-2-pentene gives three products (not counting stereoisomers): 2,7-nonadiene, 2,7-decadiene, and 3,8-undecadiene, in a ratio of 1:2:1. Show how this result can be used to decide between the two mechanisms. Ignore self-metathesis products of cyclopentene and (*E*)-2 pentene.

(b) What is predicted by the two mechanisms for the ratio of the three ethylene products in the reaction shown in **Fig. P18.92b**? Assume that once an ethylene molecule is formed, it undergoes no further reaction.

18.93 When vinylic boronic acids are treated with Br_2 and then with NaOH, vinylic bromides are formed with inversion of configuration, as shown in **Fig. P18.93a**. But when vinylic boronic acids are treated with I_2 in the presence of NaOH, vinylic iodides are formed with *retention* of configuration, as shown in **Fig. P18.93b**. Suggest mechanisms that account for these results. [*Hint:* Iodine does not add to double bonds, but it does form iodohydrins (Sec. 5.2B). Remember the stereochemistry of halogen addition (Sec. 7.8C).]

18.94 *Bombykol* is the mating pheromone of the female silkworm moth.

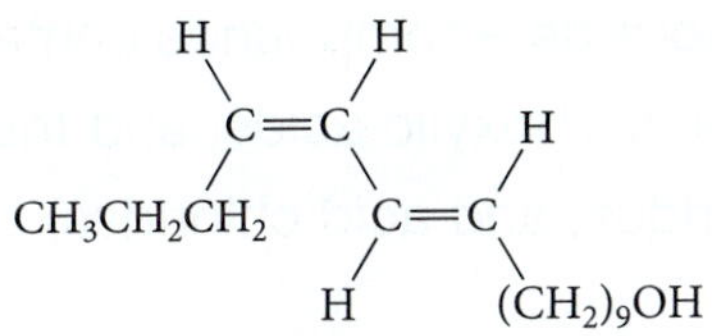

bombykol

Using the result in Problem 18.93, propose a synthesis of bombykol, using both H—C≡C$(CH_2)_9$OH (10-undecyne-1-ol) and 1-pentyne as starting materials.

18.95 Potassium tri(isopropoxy)borohydride sometimes finds use as a source of nucleophilic $H:^-$ (hydride ion).

$$K^+\ H—\bar{B}(O\mathit{i}Pr)_3$$

potassium tri(isopropoxy)borohydride

Suggest a mechanism for the following substitution reaction that accounts for the stereochemistry of the reaction. (*Hint:* See Sec. 12.8.)

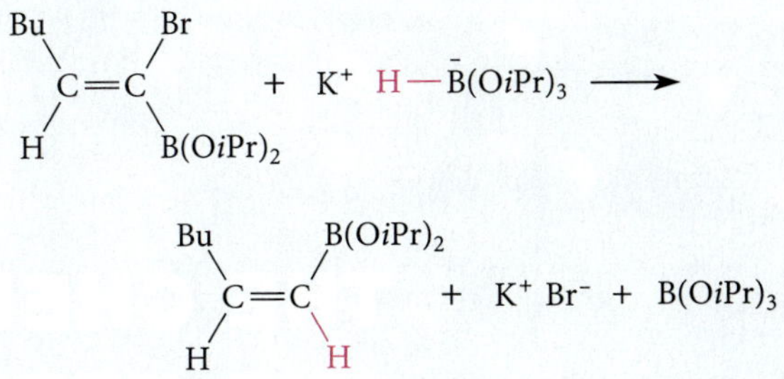

18.96 Citalopram is used as an antidepressant.

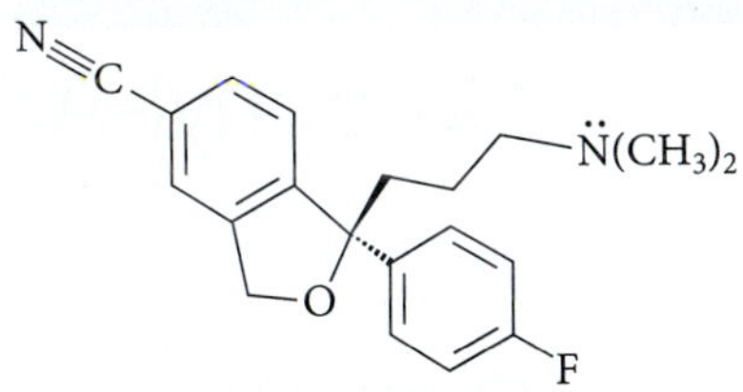

citalopram

A biologist, Heywood U. Clonum, has argued that this compound could be hazardous because it could release toxic cyanide ($^-$C≡N) and fluoride ions by nucleophilic substitution reactions with biological nucleophiles such as water. Is this argument chemically reasonable? Explain.

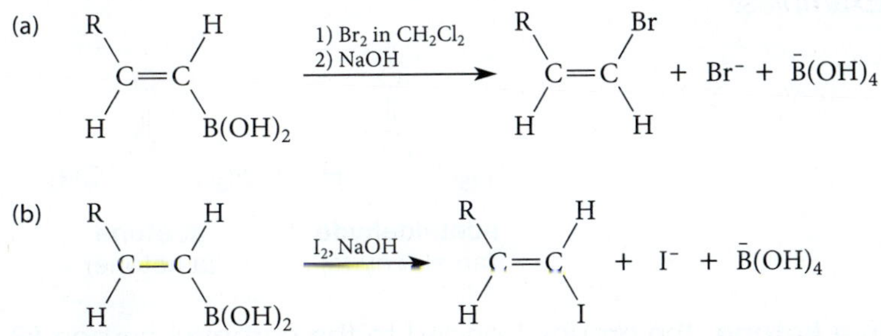

FIGURE P18.93

19 THE CHEMISTRY OF ALDEHYDES AND KETONES

Carbonyl-Addition Reactions

Chapter 19 begins the study of **carbonyl compounds**—compounds containing the **carbonyl group**, C=O. Aldehydes, ketones, carboxylic acids, and the carboxylic acid derivatives (esters, amides, anhydrides, and acid chlorides) are all carbonyl compounds.

Aldehydes and ketones have the following general structures:

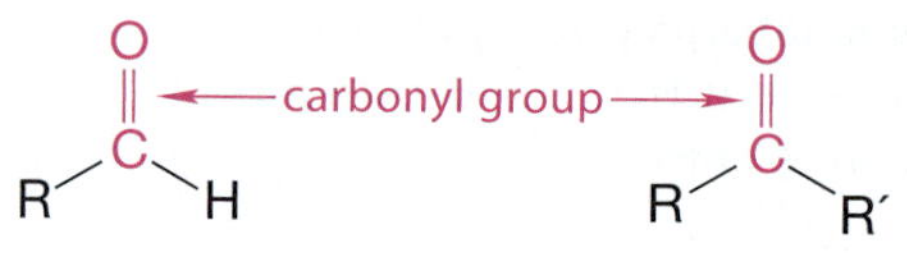

aldehyde **ketone** **(19.1)**

Examples:

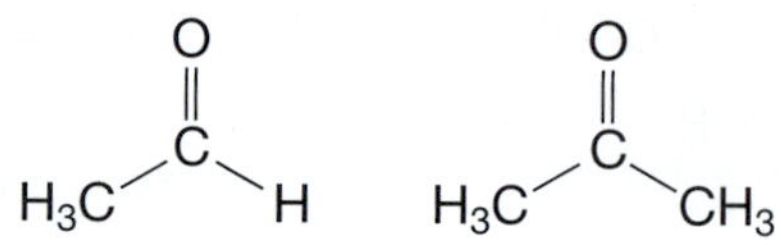

acetaldehyde (an aldehyde) **acetone** (a ketone) **(19.2)**

In a **ketone**, the groups bonded to the carbonyl carbon (R and R′ in Display 19.1) are alkyl or aryl groups. In an **aldehyde**, at least one of the groups at the carbonyl carbon atom is hydrogen, and the other may be alkyl, aryl, or a second hydrogen.

This chapter focuses on the nomenclature, properties, and characteristic carbonyl-group reactions of aldehydes and ketones. Chapters 20 and 21 cover carboxylic acids and carboxylic acid derivatives, respectively. Chapter 22 deals with ionization, enolization, and condensation reactions, which are common to the chemistry of all classes of carbonyl compounds.

Many biologically important molecules contain carbonyl groups, and carbonyl-group reactivity plays an important role in many biological reactions. We consider a number of these reactions in Chapters 19–22 and how they relate to the laboratory reactions that you will learn.

19.1 STRUCTURE AND BONDING IN ALDEHYDES AND KETONES

The carbonyl carbon of a typical aldehyde or ketone is sp^2-hybridized with bond angles approximating 120°. The carbon–oxygen double bond consists of a σ bond and a π bond (**Fig. 19.1a**), much like the double bond of an alkene (Display 4.1, Sec. 4.1B). The carbonyl oxygen, like all terminal atoms, is sp-hybridized (Display 1.74, Sec. 1.7D). One of the two nonbonding electron pairs on oxygen resides in 2p orbital, and the other resides in an sp orbital (**Fig. 19.1b**).

The structures of some simple aldehydes and ketones are given in **Fig. 19.2**. Just as C—O single bonds are shorter than C—C single bonds, C═O bonds are shorter than C═C bonds (Sec. 1.6B).

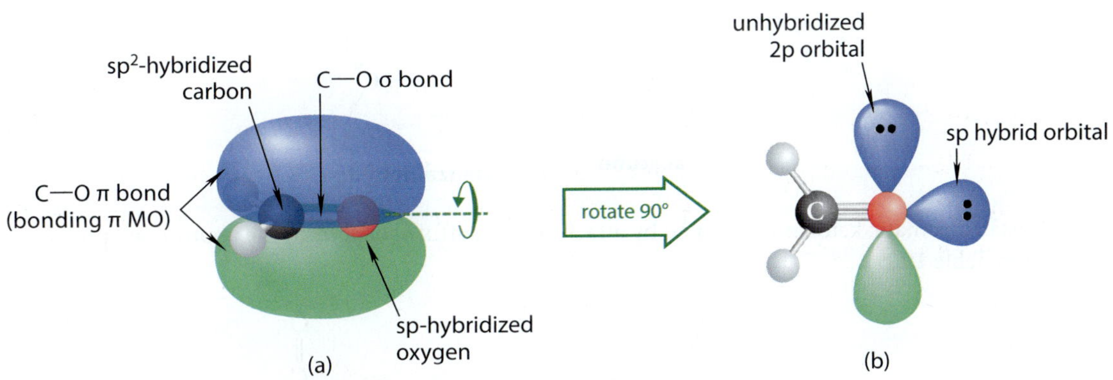

FIGURE 19.1 Bonding in formaldehyde, the simplest aldehyde, is typical of bonding in aldehydes and ketones. (a) The carbonyl group and the two hydrogens lie in the same plane, and both a σ bond and a π bond connect the carbonyl carbon and oxygen. The π bond is a bonding π molecular orbital (MO) occupied by two electrons. (b) The oxygen is sp-hybridized; the nonbonding electron pairs occupy a 2p orbital and an sp orbital.

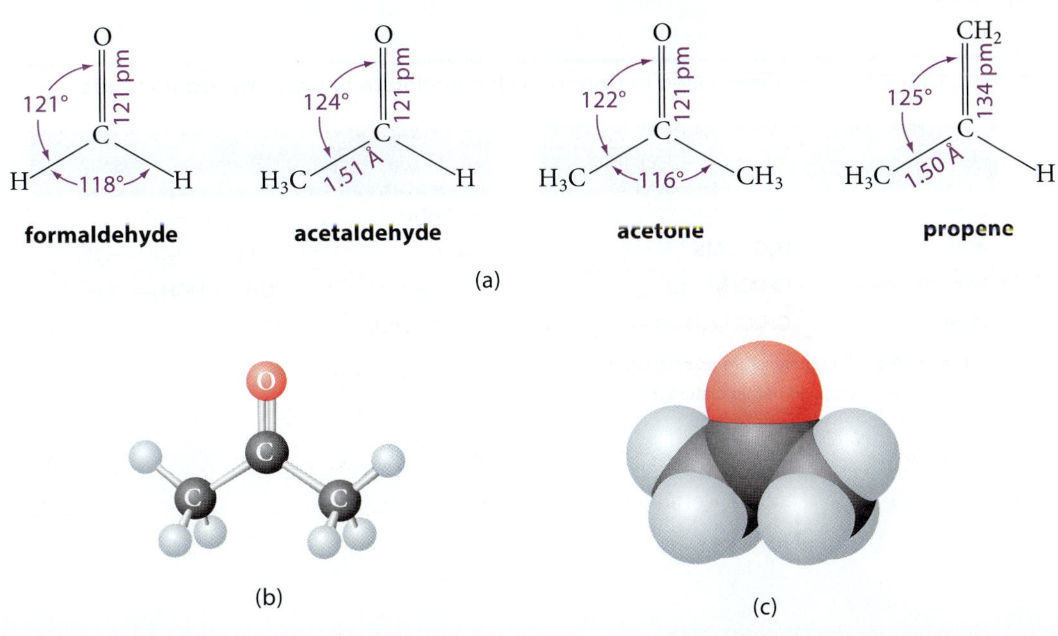

FIGURE 19.2 Structures of aldehydes and ketones. (a) The structures of formaldehyde, acetaldehyde, and acetone are compared with the structure of propene. The C═O bonds are shorter than the C═C bonds, and the carbonyl carbon is trigonal planar with bond angles very close to 120°. (b) A ball-and-stick model of acetone. (c) A space-filling model of acetone.

The reduced length of the C=O bond is due both to the smaller atomic size of oxygen relative to carbon and to the sp hybridization of oxygen, which contributes greater s character to the σ bond.

19.2 NOMENCLATURE OF ALDEHYDES AND KETONES

A. Common Nomenclature

Common names are almost always used for the simplest aldehydes. In common nomenclature, the suffix *aldehyde* is added to a prefix that indicates the chain length. Appropriate prefixes are listed in **Table 19.1**.

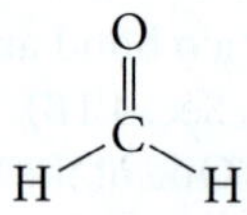

formaldehyde

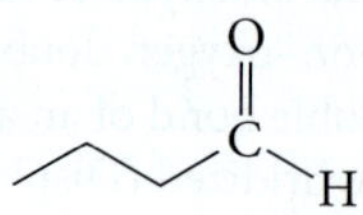

butyr + *aldehyde* = **butyraldehyde**

Acetone is the common name for the simplest ketone, and benzaldehyde is the simplest aromatic aldehyde.

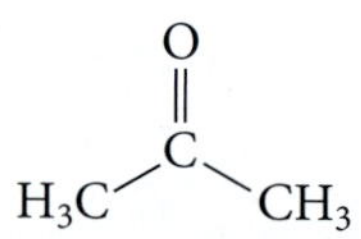

acetone

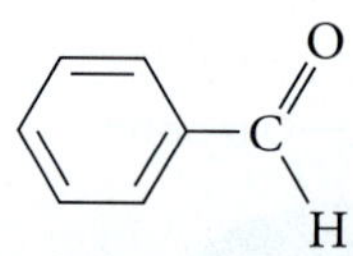

benzaldehyde

Certain aromatic ketones are named by attaching the suffix *ophenone* to the appropriate prefix from Table 19.1.

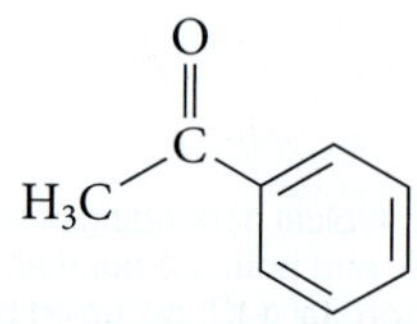

acet + *ophenone* = **acetophenone**

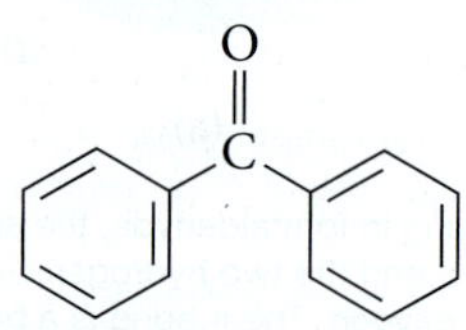

benzophenone (19.3)

TABLE 19.1 Prefixes Used in Common Nomenclature of Carbonyl Compounds

Prefix	R— In R—CH=O or R—CO_2H	Prefix	R— In R—CH=O or R—CO_2H
form	H—	isobutyr	$(CH_3)_2CH$—, *i*Pr
acet	H_3C—, Me	valer	$CH_3CH_2CH_2CH_2$—, Bu
propion, propi*	CH_3CH_2—, Et	isovaler	$(CH_3)_2CHCH_2$—, *i*Bu
butyr	$CH_3CH_2CH_2$—, Pr	benz, benzo†	Ph—

*Used in phenone nomenclature as discussed in the text.

†Used in carboxylic acid nomenclature (Sec. 20.1A).

The common names of some ketones are constructed by citing the two groups on the carbonyl carbon in alphabetical order followed by the word *ketone*.

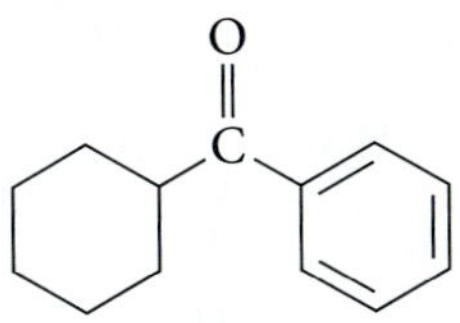

cyclohexyl phenyl ketone

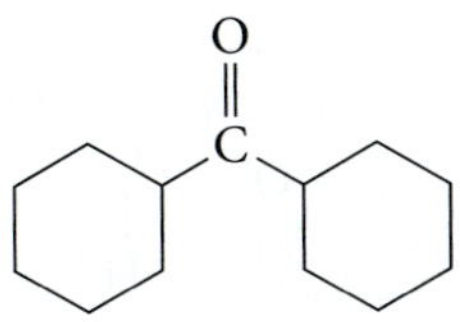

dicyclohexyl ketone

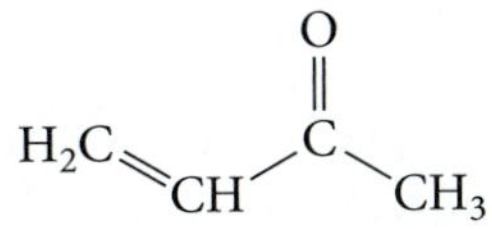

methyl vinyl ketone

Simple substituted aldehydes and ketones can be named in the common system by designating the positions of substituents with Greek letters, beginning at the position *adjacent* to the carbonyl group.

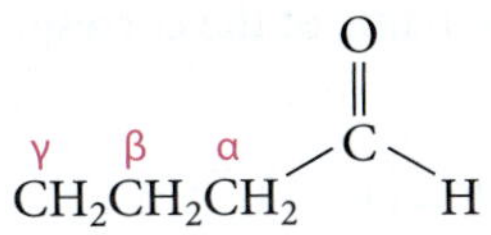

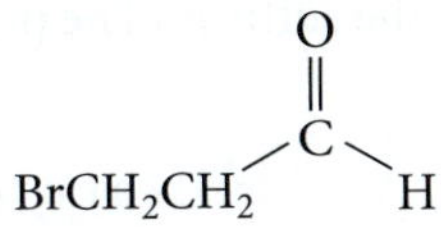

β-bromopropionaldehyde (19.4)

As suggested by this nomenclature, a carbon *adjacent* to the carbonyl group is called the **α-carbon**, and the hydrogens on the α-carbon are called **α-hydrogens**.

Many common carbonyl-containing substituent groups are named by a simple extension of the terminology in Table 19.1: the suffix *yl* is added to the appropriate prefix. The following names are examples:

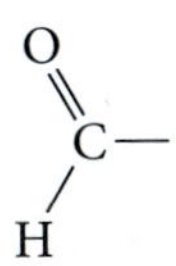

formyl group

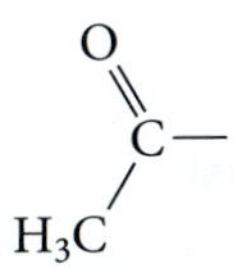

acetyl group

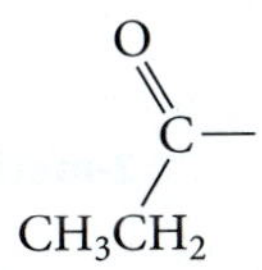

propionyl group

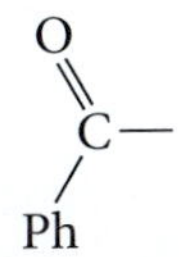

benzoyl group (19.5)

These are generally referred to as **acyl groups**. (This is the source of the term *acylation*, used in Sec. 16.4F.) To be named as an acyl group, a substituent group must be connected to the remainder of the molecule *at its carbonyl carbon*.

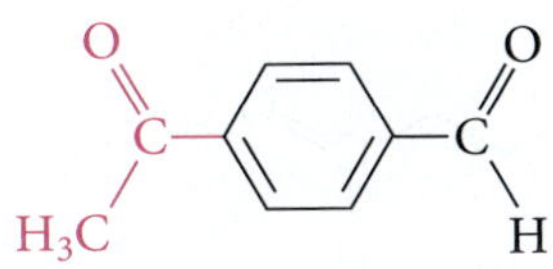

***p*-acetylbenzaldehyde**

! Do not confuse the *benzoyl* group (an acyl group) with the *benzyl* group (an alkyl group). The benzoyl group has an "o" in both the name and the structure.

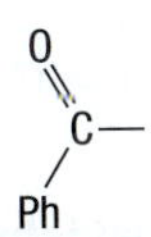

benzoyl group

$Ph—CH_2—$

benzyl group

(19.6)

A great many aldehydes and ketones were well known long before any system of nomenclature existed. These are known by the traditional names illustrated by the following examples:

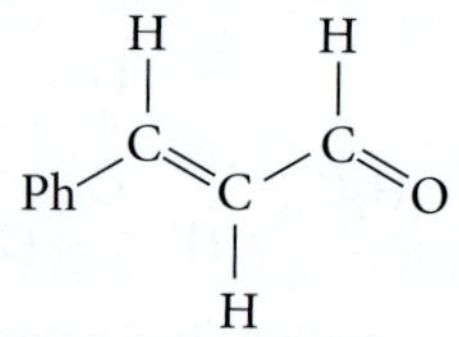

***trans*-cinnamaldehyde**

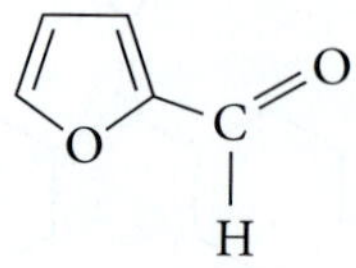

furfural

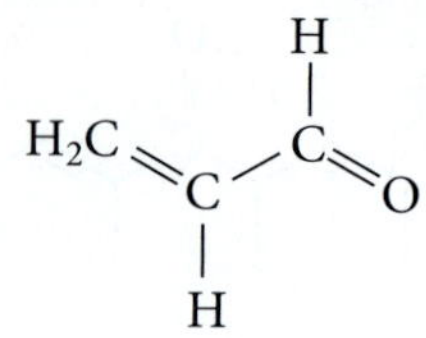

acrolein (19.7)

B. Substitutive Nomenclature

The substitutive name of an aldehyde is constructed from a prefix indicating the length of the carbon chain followed by the suffix *al*. The prefix is the name of the corresponding hydrocarbon without the final *e*.

$$CH_3CH_2CH_2CH{=}O$$

butan~~e~~ + *al* = **butanal** (19.8)

In numbering the carbon chain of an aldehyde, the carbonyl carbon receives the number 1. The carbon number of the aldehyde group is not written in the name because the aldehyde must be at the end of a carbon chain.

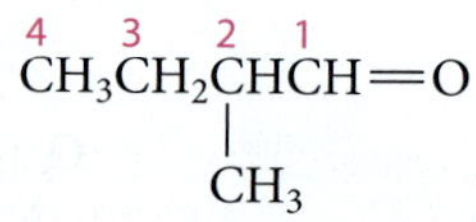

2-methylbutanal

Notice the difference in chain numbering of aldehydes in common and substitutive nomenclature. In common nomenclature, numbering begins at the carbon *adjacent* to the carbonyl (the α-carbon); in substitutive nomenclature, numbering begins at the carbonyl carbon itself.

As with diols, the final *e* is *not* dropped when the carbon chain has more than one aldehyde group.

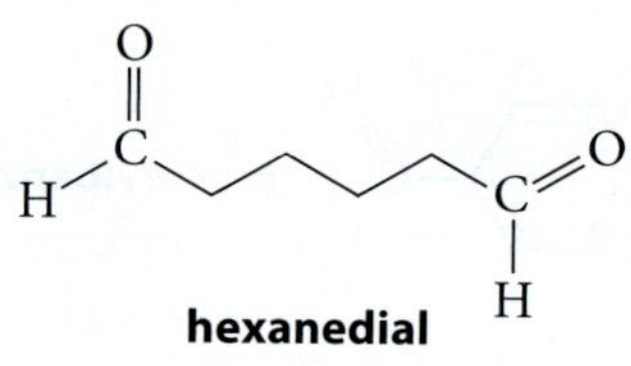

hexanedial

When an aldehyde group is attached to a ring, the suffix *carbaldehyde* is appended to the name of the ring. (In older literature, the suffix *carboxaldehyde* was used.)

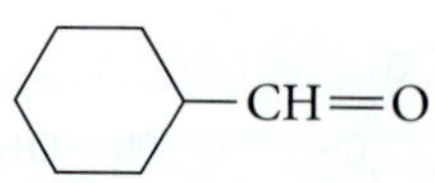

cyclohexanecarbaldehyde

In aldehydes of this type, carbon-1 is not the carbonyl carbon, but rather the ring carbon attached to the carbonyl group.

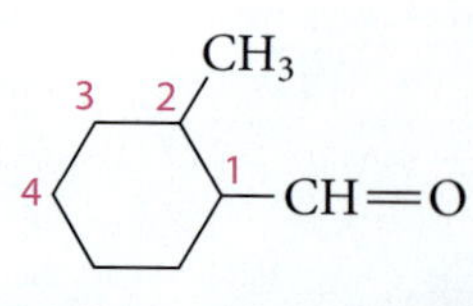

2-methylcyclohexanecarbaldehyde

The name *benzaldehyde* (Sec. 19.2A) is used in both common and substitutive nomenclature.

A ketone is named by giving the hydrocarbon name of the longest carbon chain containing the carbonyl group, dropping the final *e*, and adding the suffix *one*. The position of the carbonyl group is given the lowest possible number.

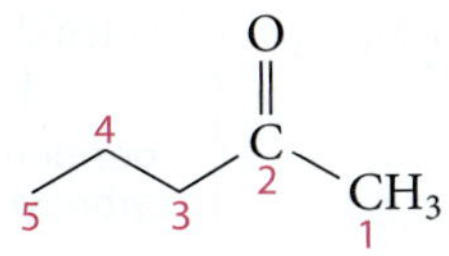

five-carbon chain: pentane + *one* = **pentanone**
position of carbonyl: **2-pentanone**

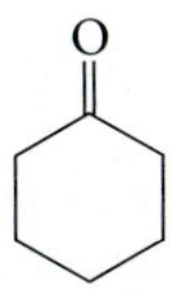

cyclohexane + *one* = **cyclohexanone**

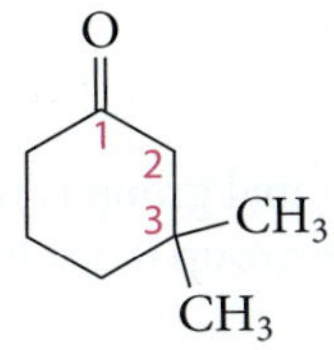

3,3-dimethylcyclohexanone

As with diols and dialdehydes, the final *e* of the hydrocarbon name is *not* dropped in the nomenclature of diones, triones, and so on.

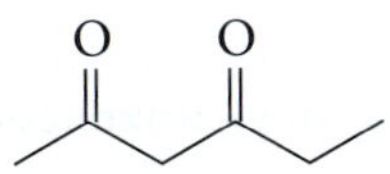

six-carbon chain: hexane + *dione* = **hexanedione**
positions of carbonyls: **2,4-hexanedione**

Aldehyde and ketone carbonyl groups receive higher priority than OH or SH groups for citation as *principal groups* (Sec. 8.2B).

Priority for citation as a principal group:

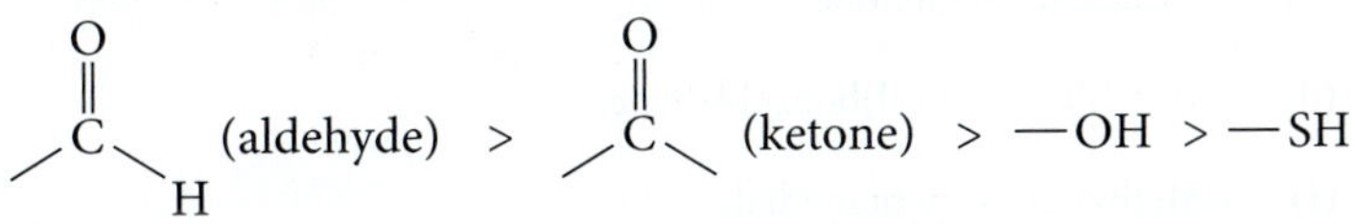

(19.9)

Study Problem 19.1

Provide a substitutive name for the following compound. (The numbers are used in the solution that follows.)

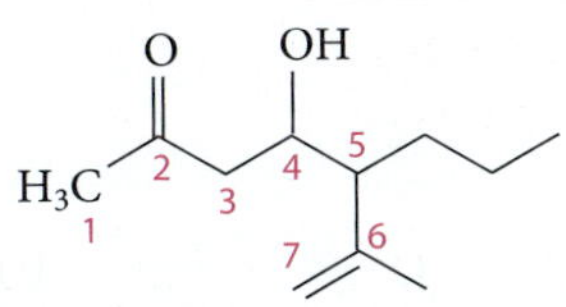

Solution To name this compound, use the nomenclature rules in Sec. 8.2B. First, *identify the principal group*. Possible candidates are the carbonyl group at carbon-2 and the hydroxy group at carbon-4. Because ketones have a higher citation priority than hydroxy groups (Eq. 19.9), the compound is named as a ketone with the suffix *one*. Next, *identify the principal chain*. This is the longest carbon chain containing the principal group and the greatest number of double and triple bonds. This chain (numbered in the preceding structure) contains seven carbons. The longer carbon chain within the molecule is not the principal chain because the presence of double bonds takes precedence over length. So, the compound is named as a heptenone with hydroxy, methyl, and propyl substituents cited in alphabetical order. Finally, *number the principal chain*. Number from the end of the chain so that the principal group—the carbonyl group—receives the lowest possible number. Therefore, the carbonyl carbon is carbon-2, the hydroxy carbon is carbon-4, the carbon bearing the propyl group is carbon-5, and the first alkene carbon is carbon-6 (see numbering in the preceding structure). Carbon-6 is also the carbon bearing the methyl group. Cite the substituent groups in alphabetical order. The name is therefore:

In the 1993 IUPAC revised system (Sec. 4.2A, "1993 IUPAC Recommendations"), the name is 4-hydroxy-6-methyl-5-propylhept-6-en-2-one.

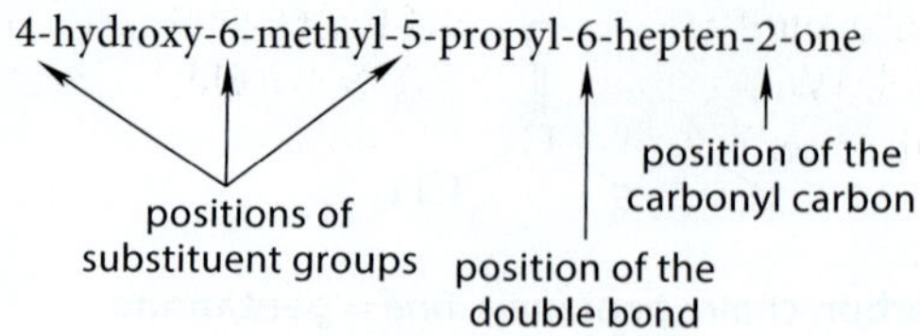

When a ketone carbonyl group is treated as a substituent, its position is designated by the term *oxo*. When an aldehyde group is a substituent, it is called the *formyl* group (Display 19.5).

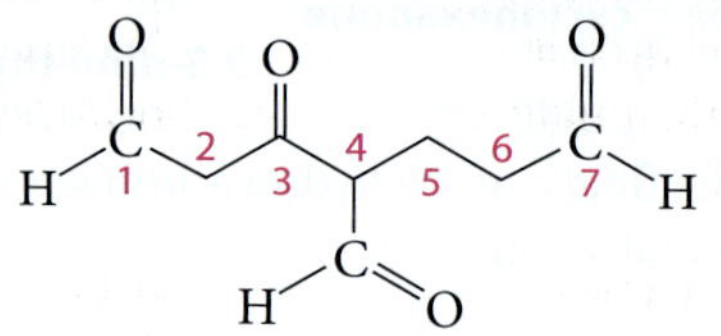

principal group: aldehyde carbonyl
name: **4-formyl-3-oxoheptanedial**

Focused Problems

19.1 Draw a structure for each of the following compounds.

(a) Isobutyraldehyde (b) *o*-Bromoacetophenone

(c) γ-Chlorobutyraldehyde (d) 4-(2-Chlorobutyryl)benzaldehyde

(e) 3-Cyclohexenone (f) 3-Methyl-2-oxopentanedial

19.2 Give the substitutive name for each of the following compounds.

(a) Diisopropyl ketone

(b) 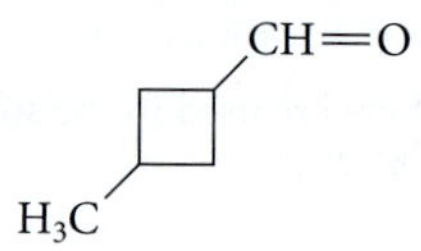

(c) 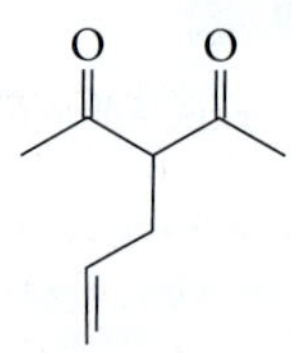

(d) 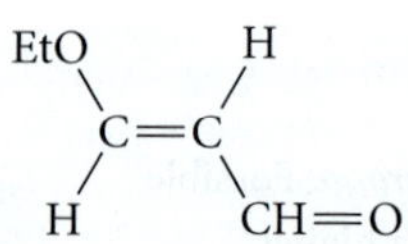

(e) 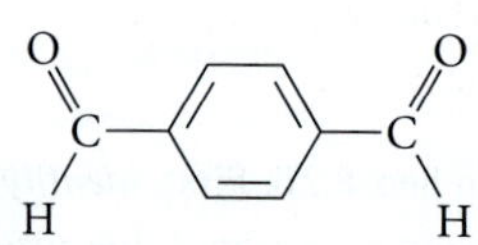

(f) 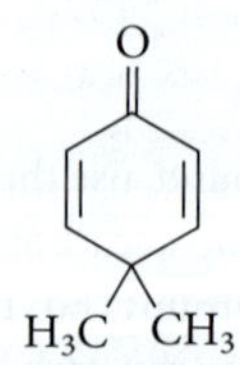

19.3 PHYSICAL PROPERTIES OF ALDEHYDES AND KETONES

Most simple aldehydes and ketones are liquids. However, formaldehyde is a gas, and acetaldehyde has a boiling point (20.8 °C) very near room temperature, although it is usually sold as a liquid.

Aldehydes and ketones are polar molecules because of their C=O bond dipoles. The charge separation that results from this dipole is evident in the EPMs of aldehydes and ketones, illustrated in Display 19.10 for acetone.

bond dipole of the C=O bond — EPM of acetone (19.10)

Because of their polarities, aldehydes and ketones have higher boiling points than alkenes or alkanes with similar molecular masses and shapes. But because aldehydes and ketones are not hydrogen-bond donors, their boiling points are considerably lower than those of the corresponding alcohols.

	$CH_3CH{=}CH_2$	$CH_3CH{=}O$	CH_3CH_2OH
boiling point	−47.4 °C	20.8 °C	78.3 °C
dipole moment	0.4 D	2.7 D	1.7 D

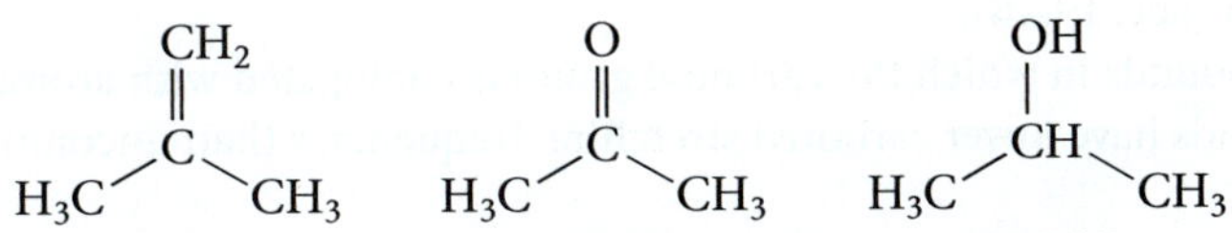

boiling point	−6.9 °C	56.5 °C	82.3 °C
dipole moment	0.5 D	2.7 D	1.7 D

(19.11)

Aldehydes and ketones with four or fewer carbons have considerable solubilities in water because they can accept hydrogen bonds from water at the carbonyl oxygen.

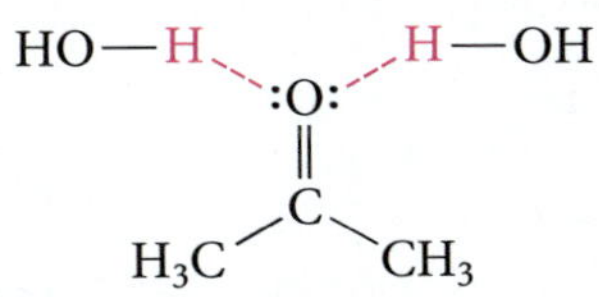

(19.12)

Because they are hydrogen-bond acceptors, acetaldehyde and acetone are miscible with water (that is, soluble in all proportions). The water solubility of aldehydes and ketones along a series diminishes rapidly as the number of carbons increases.

Acetone and 2-butanone are especially valued as solvents because they dissolve not only water but also a wide variety of organic compounds. These solvents have sufficiently low boiling points that they can be easily separated from other less volatile compounds. Acetone, with a dielectric constant of 21, is a polar aprotic solvent and is often used as a solvent or co-solvent for nucleophilic substitution reactions.

19.4 SPECTROSCOPY OF ALDEHYDES AND KETONES

A. IR Spectroscopy

The principal infrared absorption of aldehydes and ketones is the C=O stretching absorption, a strong absorption that occurs in the vicinity of 1700 cm^{-1}. In fact, this is one of the most

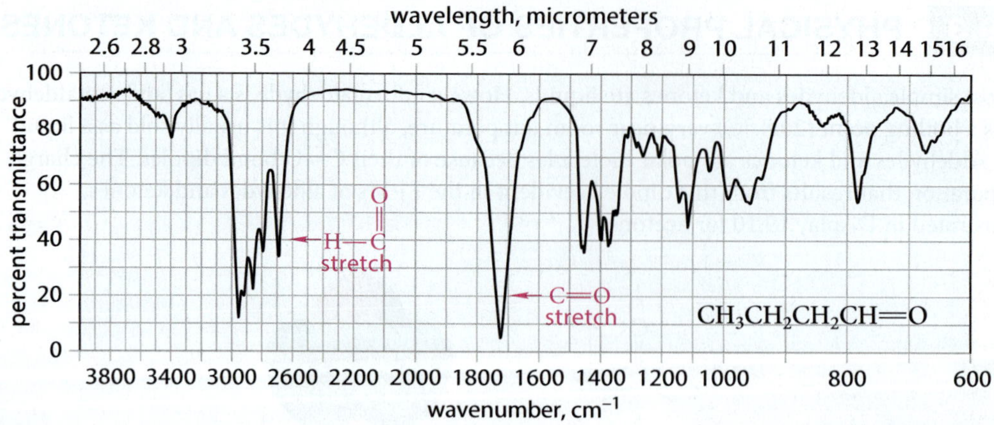

FIGURE 19.3 The infrared spectrum of butyraldehyde with key absorptions indicated in red.

important of all infrared absorptions. Because the C=O bond is stronger than the C=C bond, the stretching frequency of the C=O bond is greater (Eq. 13.12, Sec. 13.3A).

The position of the C=O stretching absorption varies predictably for different types of carbonyl compounds. It generally occurs at 1710–1715 cm^{-1} for simple ketones and at 1720–1725 cm^{-1} for simple aldehydes. The carbonyl absorption is clearly evident, for example, in the IR spectrum of butyraldehyde (**Fig. 19.3**). The stretching absorption of the carbonyl–hydrogen bond of aldehydes near 2710 cm^{-1} is another characteristic absorption. NMR spectroscopy, however, provides a more reliable way to diagnose the presence of this type of hydrogen (Sec. 19.4B).

Compounds in which the carbonyl group is conjugated with aromatic rings, double bonds, or triple bonds have lower carbonyl stretching frequencies than unconjugated carbonyl compounds.

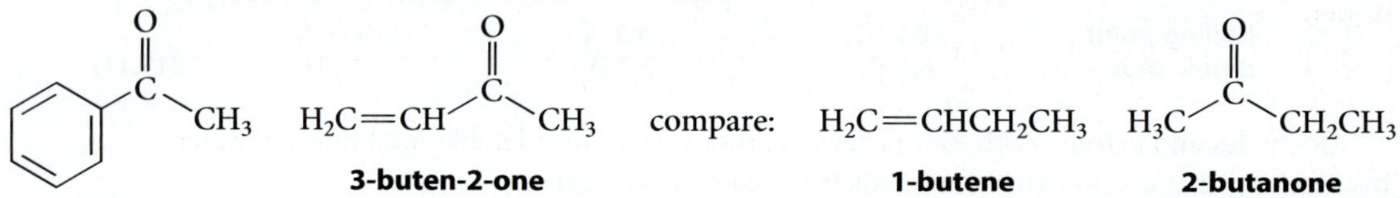

	acetophenone	**3-buten-2-one**	compare: **1-butene**	**2-butanone**
C=O	1685 cm^{-1}	1670 cm^{-1}	—	1715 cm^{-1}
C=C	1600 cm^{-1}	1613 cm^{-1}	1642 cm^{-1}	—

(19.13)

The carbon–carbon double-bond stretching frequencies are also lower in the conjugated molecules. These effects can be explained by the resonance structures for these compounds. The C=O and C=C bonds have some single-bond character, as the following resonance structures indicate, so they are somewhat weaker than ordinary double bonds, in which case they absorb in the IR at lower frequency.

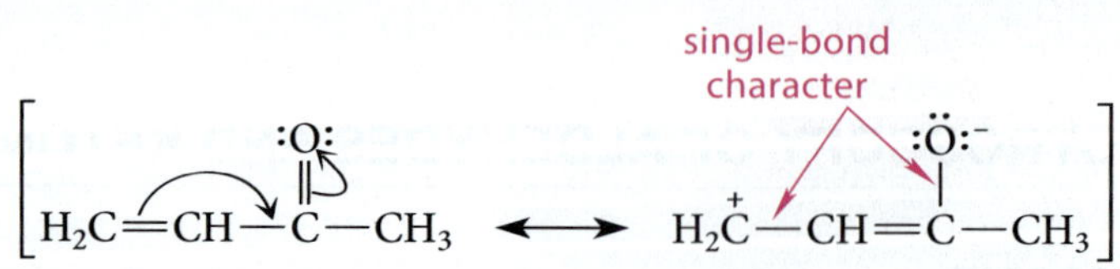

(19.14)

In cyclic ketones with rings containing fewer than six carbons, the carbonyl absorption frequency increases significantly as the ring size decreases. (See Further Exploration 19.1.)

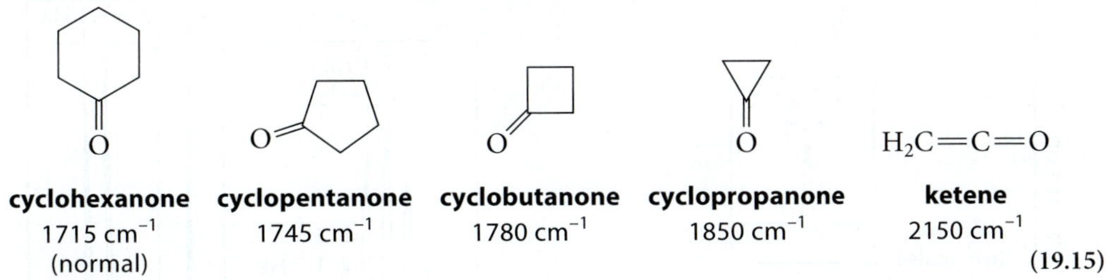

(19.15)

FURTHER EXPLORATION 19.1
IR Absorptions of Cyclic Ketones

Focused Problem

19.3 Explain how IR spectra could be used to differentiate the isomers within each of the following pairs.

(a) Cyclohexanone and 4-hexenal

(b) 3-Cyclohexenone and 2-cyclohexenone

(c) 2-Butanone and 3-buten-2-ol

B. Proton NMR Spectroscopy

The characteristic proton NMR absorption common to both aldehydes and ketones is that of the protons on the carbons *adjacent* to the carbonyl group—the α-protons. This absorption is in the δ 2.0–2.5 region of the spectrum (see also Fig. 14.4, Sec. 14.3C). This absorption occurs at somewhat greater chemical shift than the absorptions of allylic protons because the C=O group is more electronegative than the C=C group. In addition, the absorption of the aldehydic proton is quite distinctive, occurring in the δ 9–10 region of the NMR spectrum, at a greater chemical shift than most other absorptions.

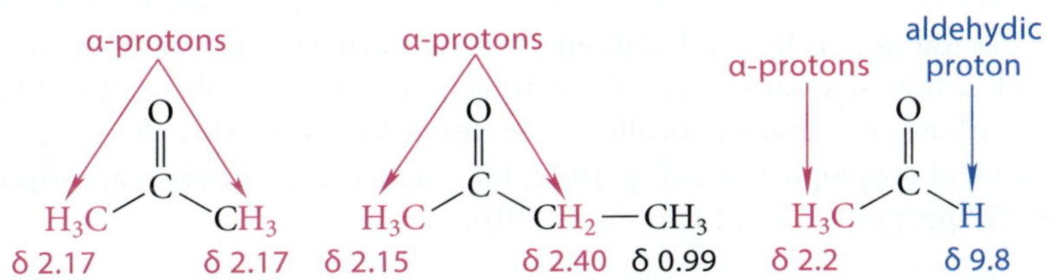

(19.16)

In general, aldehydic protons have very large chemical shifts. The explanation is the same as that for the large chemical shifts of protons on a carbon–carbon double bond (Sec. 14.7A). However, the carbonyl group has a greater effect on chemical shift than a carbon–carbon double bond because of the added effect of the electronegativity of the carbonyl oxygen.

Focused Problem

19.4 Deduce the structures of the following compounds.

(a) C_4H_8O: IR 1720, 2710 cm^{-1}
Proton NMR in **Fig. 19.4**

(b) C_4H_8O: IR 1717 cm^{-1}
Proton NMR δ 0.95 (3*H*, t, *J* = 8 Hz); δ 2.03 (3*H*, s); δ 2.38 (2*H*, q, *J* = 8 Hz)

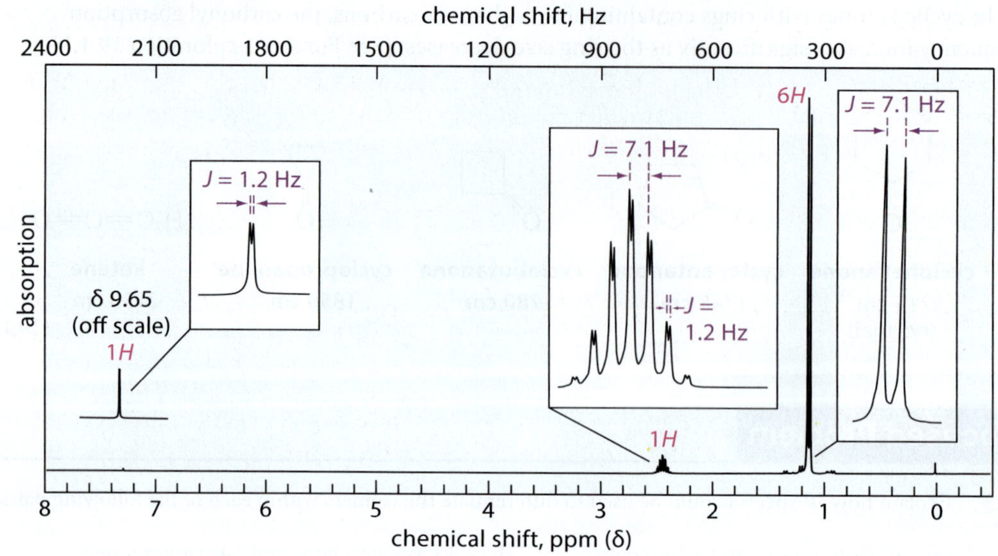

FIGURE 19.4 The NMR spectrum for Focused Problem 19.4(a). The relative integrals are indicated in red over their respective resonances.

(c) A compound with molecular mass = 70.1, IR absorption at 1780 cm^{-1}, and the following proton NMR spectrum: δ 2.01 (quintuplet, J = 7 Hz); δ 3.09 (t, J = 7 Hz). The integral of the δ 3.09 resonance is twice as large as that of the δ 2.01 resonance.

(d) $C_{10}H_{12}O_2$: IR 1690 cm^{-1}, 1612 cm^{-1}
Proton NMR δ 1.4) (3H, t, J = 8 Hz); δ 2.5 (3H, s); δ 4.1 (2H, q, J = 8 Hz); δ 6.9 (2H, d, J = 9 Hz); δ 7.9 (2H, d, J = 9 Hz)

C. Carbon NMR Spectroscopy

The most characteristic absorption of aldehydes and ketones in ^{13}C NMR spectroscopy is that of the carbonyl carbon, which occurs typically in the δ 190–220 range (see Fig. 14.22, Sec. 14.9B). This large downfield shift is due to the induced π-electron circulation, as in alkenes (see Fig. 14.15, Sec. 14.7A), and to the additional chemical-shift effect of the electronegative carbonyl oxygen. Because the carbonyl carbon of a ketone bears no hydrogens, its ^{13}C NMR absorption, like that of other quaternary carbons, is characteristically rather weak (Sec. 14.9). This effect is evident in the ^{13}C NMR spectrum of propiophenone (**Fig. 19.5**). Like quaternary carbons, carbonyl carbons are absent from DEPT spectra (see Fig. 14.25, Sec. 14.9B).

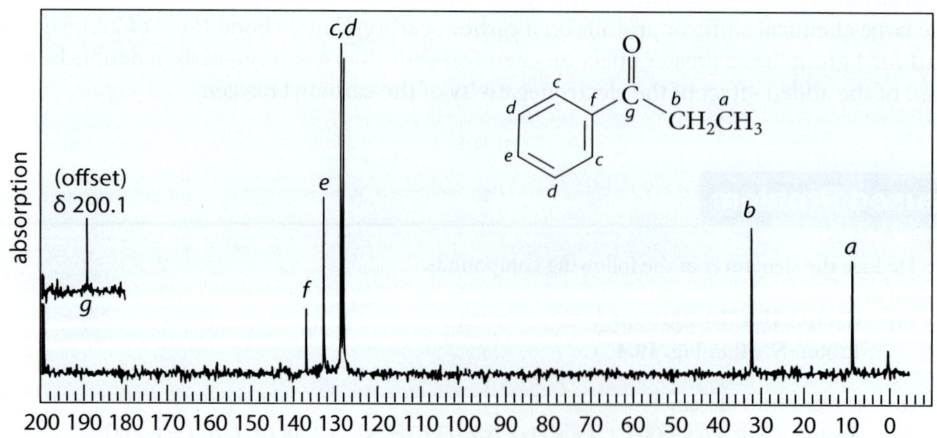

FIGURE 19.5 The ^{13}C NMR spectrum of propiophenone. Two particularly important features of the spectrum are the large chemical shift of the carbonyl carbon g and the small resonances for the two carbons (f and g) that bear no protons. As explained in Sec. 14.9, absorption intensities in ^{13}C NMR spectra generally do not accurately correspond to numbers of carbons, and quaternary carbons generally have considerably weaker absorptions than proton-bearing carbons.

The α-carbon absorptions of aldehydes and ketones show modest chemical shifts, typically in the δ 30–50 range, with, as usual, greater shifts for more branched carbons. The α-carbon shift of propiophenone, 31.7 ppm (see Fig. 19.5, carbon *b*), is typical. Because shifts in this range are also observed for other functional groups, these absorptions are less useful than the carbonyl carbon resonances for identifying aldehydes and ketones.

Focused Problems

19.5 Propose a structure for a compound $C_6H_{12}O$ that has IR absorption at 1705 cm^{-1}, no proton NMR absorption at a chemical shift greater than δ 3, and the following ^{13}C NMR spectrum: δ 24.4, δ 26.5, δ 44.2, and δ 212.6. The resonances at δ 44.2 and δ 212.6 have very low intensity.

19.6 The ^{13}C NMR spectrum of 2-ethylbutanal consists of the following absorptions: δ 11.5, δ 21.7, δ 55.2, and δ 204.7. Draw the structure of this aldehyde, label each chemically nonequivalent set of carbons, and assign each absorption to the appropriate carbon(s).

D. UV Spectroscopy

The $\pi \rightarrow \pi^*$ absorptions (Sec. 15.2B) of unconjugated aldehydes and ketones occur at about 150 nm, a wavelength well below the operating range of common UV spectrometers. Simple aldehydes and ketones also have another, much weaker, absorption at higher wavelength, in the 260–290 nm region. This absorption is caused by excitation of the nonbonding electrons on oxygen (sometimes called the *n* electrons). This high-wavelength absorption is usually referred to as an $\boldsymbol{n \rightarrow \pi^*}$ **absorption.**

$$(CH_3)_2C{=}\ddot{\underset{..}{O}} \qquad n \rightarrow \pi^* \qquad 271\ \text{nm}\ (\varepsilon = 16)\ (\text{in ethanol}) \tag{19.17}$$

n electrons (arrow to O)

This absorption is easily distinguished from a $\pi \rightarrow \pi^*$ absorption because it is only 10^{-2} to 10^{-3} times as strong. However, it is strong enough that aldehydes and ketones cannot be used as solvents for UV spectroscopy.

Like conjugated dienes, the π electrons of compounds in which carbonyl groups are conjugated with double or triple bonds have strong absorption in the UV spectrum. The spectrum of 1-acetylcyclohexene (**Fig. 19.6**) is typical. The 232-nm peak is due to UV absorption by the

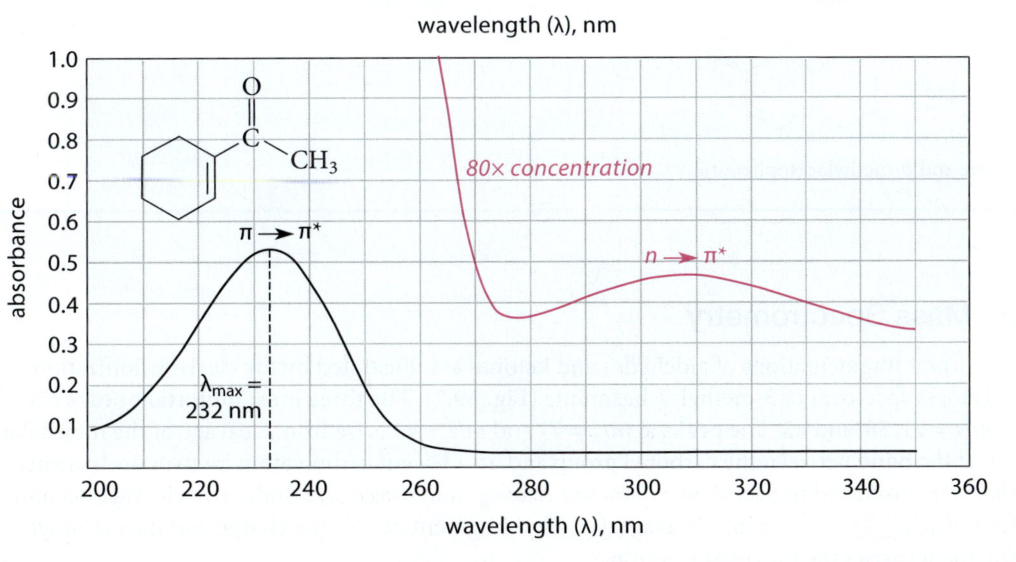

FIGURE 19.6 The ultraviolet spectrum of 1-acetylcyclohexene [1-(cyclohexen-1-yl)ethanone] in methanol. The spectrum of a more concentrated solution *(red)* reveals the "forbidden" $n \rightarrow \pi^*$ absorption, which is so weak that it is not apparent in the spectrum taken on a more dilute solution *(black)*.

conjugated π-electron system and is thus a $\pi \longrightarrow \pi^*$ absorption. It has a very large extinction coefficient, much like that of a conjugated diene. The weak 308-nm absorption is an $n \longrightarrow \pi^*$ absorption.

conjugated π-electron system

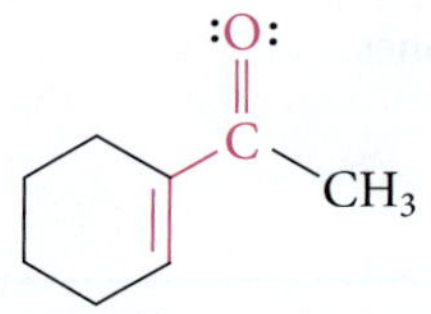

λ_{max} = 232 nm (ε = 13,200)
λ_{max} = 308 nm (ε= 150)
(in methanol)

1-acetylcyclohexene
[1-(cyclohexen-1-yl)ethanone] (19.18)

The λ_{max} of a conjugated aldehyde or ketone is governed by the same variables that affect the λ_{max} values of conjugated dienes: the number of conjugated double bonds, substitution on the double bond, and so on. When an aromatic ring is conjugated with a carbonyl group, the typical aromatic absorptions are more intense and shifted to higher wavelengths than those of benzene.

λ_{max} = 204 nm (ε = 7900)
254 nm (ε = 212)

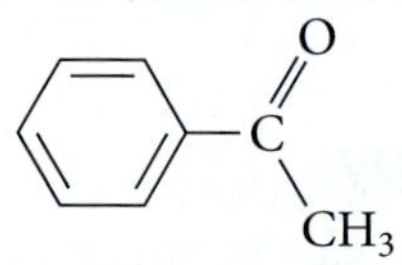

λ_{max} = 240 nm (ε = 13,000)
278 nm (ε = 1100)
319 nm (ε = 50) ($n \longrightarrow \pi^*$) (19.19)

The $\pi \longrightarrow \pi^*$ absorptions of conjugated carbonyl compounds, like those of conjugated alkenes, arise from the promotion of a π electron from a bonding to an antibonding (π*) molecular orbital (MO) (Sec. 15.2B). An $n \longrightarrow \pi^*$ absorption arises from promotion of one of the *n* (nonbonding) electrons on a carbonyl oxygen to a π* molecular orbital. There are theoretical reasons that this transition should be forbidden, but in practice it is present but very weak. The 261-nm $\pi \longrightarrow \pi^*$ absorption of ethylbenzene (see Fig. 16.4, Sec. 16.3D), which has ε = 200, and a similar absorption of benzene itself, at 254 nm (ε = 212), are other examples of low-intensity "forbidden" absorptions.

Focused Problem

19.7 Explain how the compounds within each set can be distinguished using only UV spectroscopy.

(a) 2-Cyclohexenone and 3-cyclohexenone

(b) 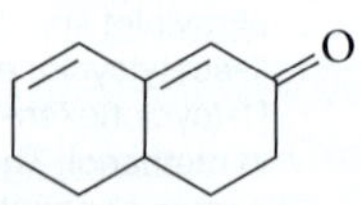 and 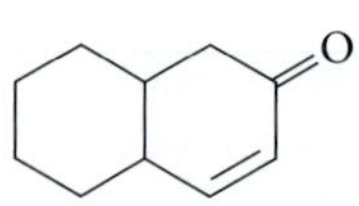

(c) 1-Phenyl-2-propanone and *p*-methylacetophenone

E. Mass Spectrometry

Important fragmentations of aldehydes and ketones are illustrated by the electron-ionization (EI) mass spectrum of 5-methyl-2-hexanone (**Fig. 19.7**). The three most important peaks occur at $m/z = 71$, 58, and 43. The peaks at $m/z = 71$ and $m/z = 43$ arise from cleavage of the molecular ion at the bond between the carbonyl group and an adjacent carbon atom by two mechanisms that were discussed in Sec. 13.6C: *inductive cleavage* and *α-cleavage*. Inductive cleavage accounts for the $m/z = 71$ peak. In this cleavage, the alkyl fragment carries the charge and the carbonyl fragment carries the unpaired electron.

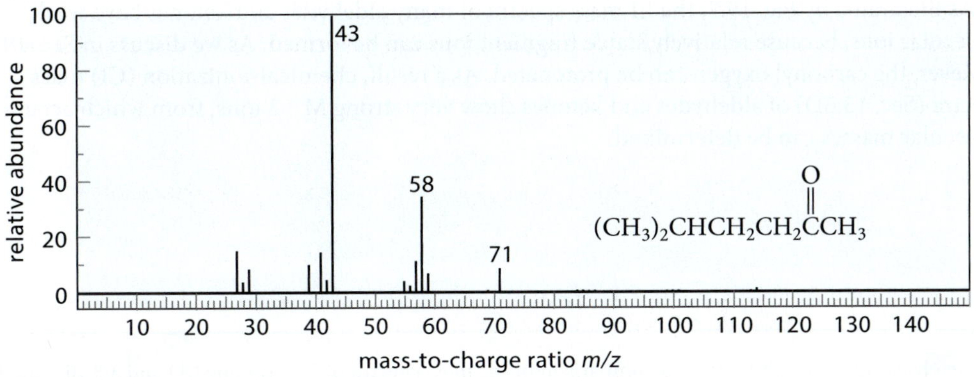

FIGURE 19.7 The EI mass spectrum of 5-methyl-2-hexanone. The odd-electron ion at $m/z = 58$ results from a McLafferty rearrangement of the molecular ion.

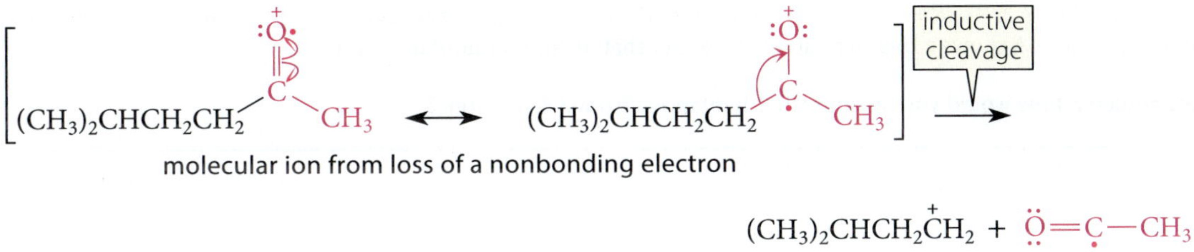

(19.20)

α-Cleavage accounts for the $m/z = 43$ peak. In this case the same molecular ion undergoes fragmentation in such a way that the carbonyl fragment carries the charge and the alkyl fragment carries the unpaired electron:

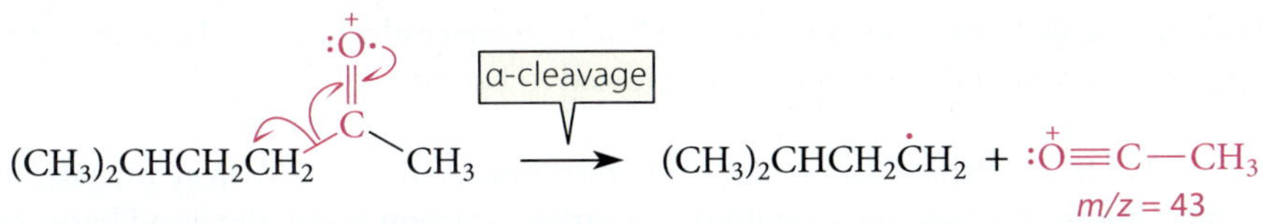

(19.21)

An analogous cleavage at the carbon–hydrogen bond accounts for the fact that many aldehydes show a strong M − 1 peak.

What accounts for the $m/z = 58$ peak? A common mechanism for the formation of odd-electron ions is hydrogen transfer followed by loss of a stable neutral molecule. In this case, the oxygen radical in the molecular ion abstracts a hydrogen atom from a carbon *five atoms away*, and the resulting radical then undergoes α-cleavage.

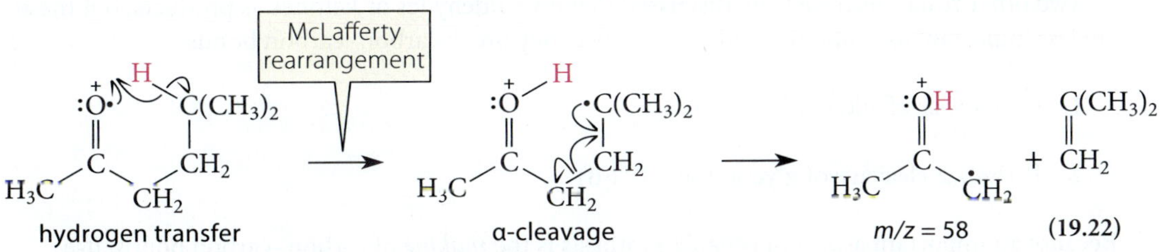

(19.22)

If we count the hydrogen that is transferred, the first step occurs through a transient six-membered ring. This process is called a **McLafferty rearrangement**, after Professor Fred McLafferty (b. 1923) of Cornell University, who investigated this type of fragmentation extensively. The McLafferty rearrangement and subsequent α-cleavage constitute a common mechanism for the production of odd-electron fragment ions in the mass spectrometry of carbonyl compounds.

As illustrated by Fig. 19.7, the EI mass spectra of many aldehydes and ketones have weak molecular ions, because relatively stable fragment ions can be formed. As we discuss in Sec. 19.7, however, the carbonyl oxygen can be protonated. As a result, chemical-ionization (CI) mass spectra (Sec. 13.6D) of aldehydes and ketones show very strong M + 1 ions, from which accurate molecular masses can be determined.

Focused Problems

19.8 Explain each of the following observations resulting from a comparison of the mass spectra of 2-hexanone (*A*) and 3,3-dimethyl-2-butanone (*B*).

(a) The $m/z = 57$ fragment peak is much more intense in the spectrum of *B* than it is in the spectrum of *A*.

(b) The spectrum of compound *A* shows a fragment at $m/z = 58$, but that of compound *B* does not.

19.9 Using only mass spectrometry, how would you distinguish 2-heptanone from 3-heptanone?

19.5 SYNTHESIS OF ALDEHYDES AND KETONES

Several reactions presented in previous chapters can be used for the preparation of aldehydes and ketones. The following are the four most important of these:

1. Oxidation of alcohols (Secs. 11.7A and 17.5A). Primary alcohols can be oxidized to aldehydes, and secondary alcohols can be oxidized to ketones.

2. Friedel–Crafts acylation (Sec. 16.4F). This reaction provides a way to synthesize aryl ketones. It also involves the formation of a carbon–carbon bond, the bond between the aryl ring and the carbonyl group.

3. Hydration of alkynes (Sec. 5.5A)

4. Hydroboration–oxidation of alkynes (Sec. 5.7). This reaction, when applied to 1-alkynes, gives aldehydes and, when applied to internal alkynes, gives ketones.

Two other reactions have been discussed that give aldehydes or ketones as products, but these are less important as synthetic methods because they break carbon–carbon bonds:

1. Ozonolysis of alkenes (Sec. 5.8)

2. Periodate cleavage of glycols (Sec. 12.6B)

Because an important aspect of organic synthesis is the *making* of carbon–carbon bonds, use of these reactions in effect wastes some of the effort that goes into making the alkene or glycol starting materials. Nevertheless, these reactions can be used synthetically in certain cases.

Other important methods of preparing aldehydes and ketones start with carboxylic acid derivatives; these methods are discussed in Chapter 21. Appendix V summarizes all of the synthetic methods for aldehydes and ketones, arranged in the order in which they appear in the text.

19.6 INTRODUCTION TO ALDEHYDE AND KETONE REACTIONS

The reactions of aldehydes and ketones can be grouped into two categories: (1) reactions of the carbonyl group, which are considered in this chapter, and (2) reactions involving the α-carbon, which are presented in Chapter 22.

The great preponderance of carbonyl-group reactions of aldehydes and ketones fall into three categories:

1. *Reactions with acids.* The carbonyl oxygen is weakly basic and reacts with Lewis and Brønsted acids. With E^+ as a general electrophile, this reaction can be represented as follows:

(19.23)

Carbonyl basicity is important because it plays a role in several other carbonyl-group reactions.

2. *Addition reactions.* The most important carbonyl-group reaction is addition to the C=O double bond. With E—Y symbolizing a general reagent, addition can be represented in the following way:

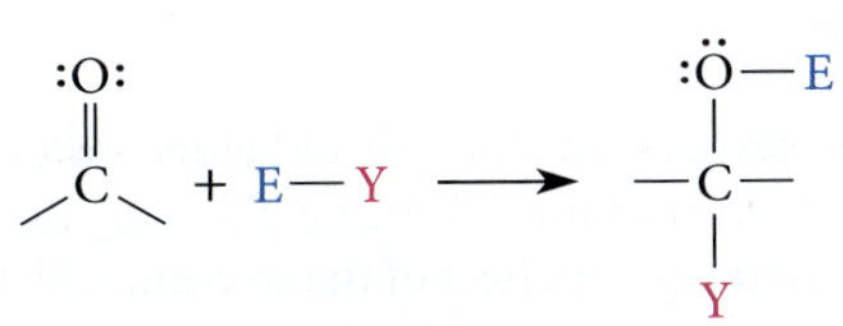

(19.24)

Superficially, carbonyl addition is analogous to alkene addition (Sec. 4.6).

Many reactions of aldehydes and ketones are simple additions that conform exactly to the model in Eq. 19.24. Others are multistep processes in which an initial addition reaction is followed by other reactions.

3. *Oxidation of aldehydes.* Aldehydes can be oxidized to carboxylic acids:

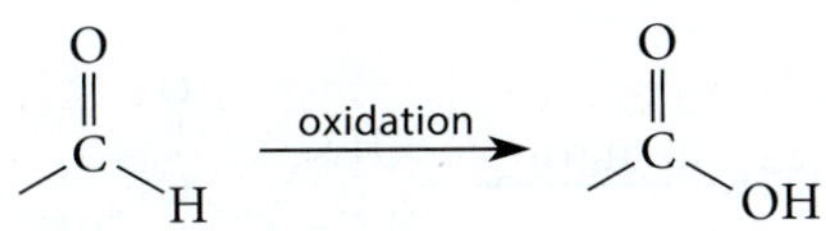

(19.25)

19.7 BASICITY OF ALDEHYDES AND KETONES

Aldehydes and ketones are weakly basic and react at the carbonyl oxygen with protons or Lewis acids.

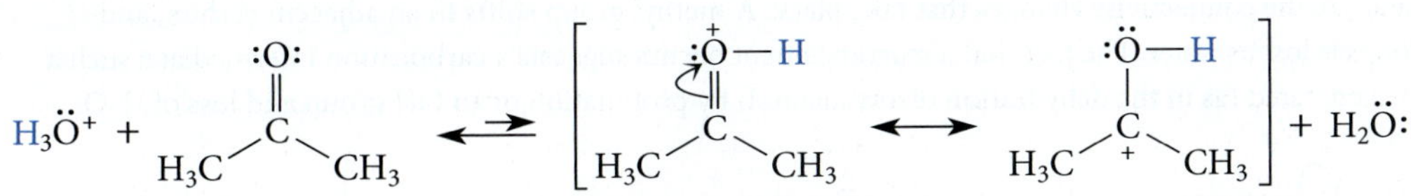

acetone all atoms have an octet carbocation structure

conjugate acid of acetone
$pK_a \sim -7$

(19.26)

As Eq. 19.26 shows, the conjugate acid of an aldehyde or ketone is resonance-stabilized. The resonance structure on the left is more important because all atoms have octets. However, the resonance structure on the right shows that the protonated carbonyl compound has carbocation character. In fact, in some cases, the conjugate acids of aldehydes and ketones undergo typical carbocation reactions.

The conjugate acids of aldehydes and ketones can be viewed as *α-hydroxy carbocations*. If we conceptually replace the acidic proton in a protonated aldehyde or ketone with an alkyl group, we get an *α-alkoxy carbocation.*

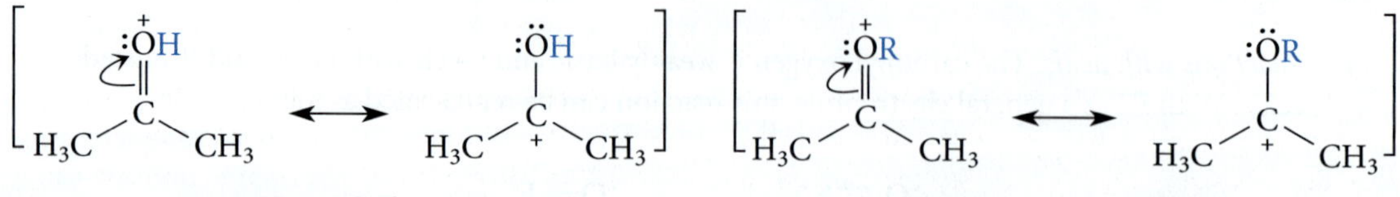

conjugate acid of acetone (an α-hydroxy carbocation) (an α-alkoxy carbocation) (19.27)

α-Hydroxy carbocations and α-alkoxy carbocations are considerably more stable than ordinary carbocations. For example, a comparably substituted α-alkoxy carbocation is about 100 kJ mol^{-1} (24 kcal mol^{-1}) more stable than an ordinary tertiary carbocation in the gas phase.

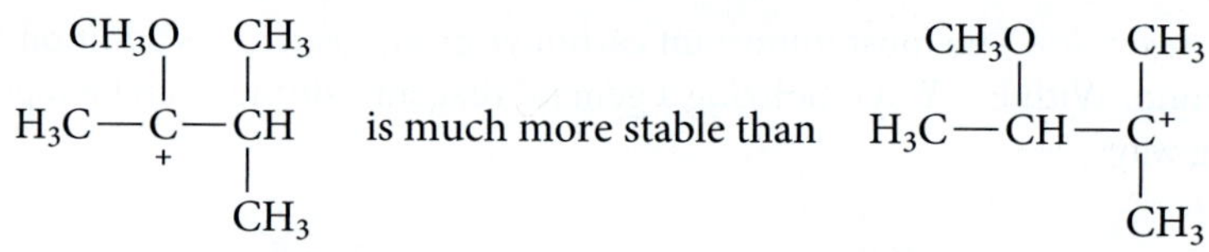

(19.28)

An α-alkoxy carbocation, like a protonated aldehyde or ketone, owes its stability to the resonance interaction of the electron-deficient carbon with the neighboring oxygen. This resonance effect far outweighs the electron-attracting polar effect of the oxygen, which, by itself, would destabilize the carbocation.

Study Problem 19.2

Many 1,2-diols, under the acidic conditions used for dehydration of alcohols, undergo a reaction called the *pinacol rearrangement*:

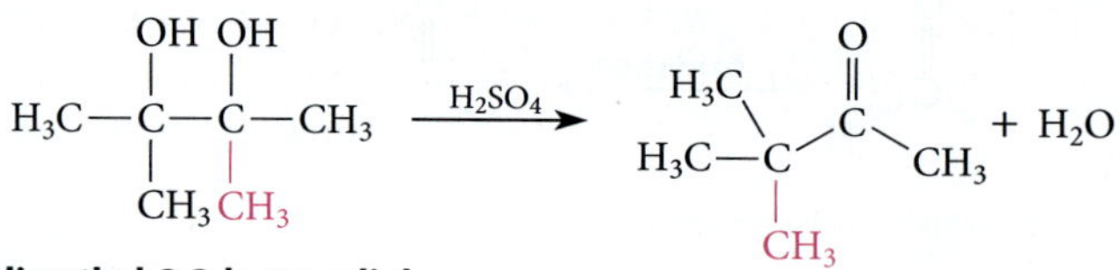

2,3-dimethyl-2,3-butanediol (pinacol) **3,3-dimethyl-2-butanone (pinacolone)** (65–72% yield) (19.29)

Propose a curved-arrow mechanism for this reaction, and explain why the rearrangement step is energetically favorable.

Solution First, analyze the connectivity changes that take place. A methyl group shifts to an adjacent carbon, and one of the OH groups is lost as water. The fact that a rearrangement occurs suggests a carbocation intermediate; such a carbocation can be generated (as in the dehydration of any alcohol) by protonation of an OH group and loss of H_2O:

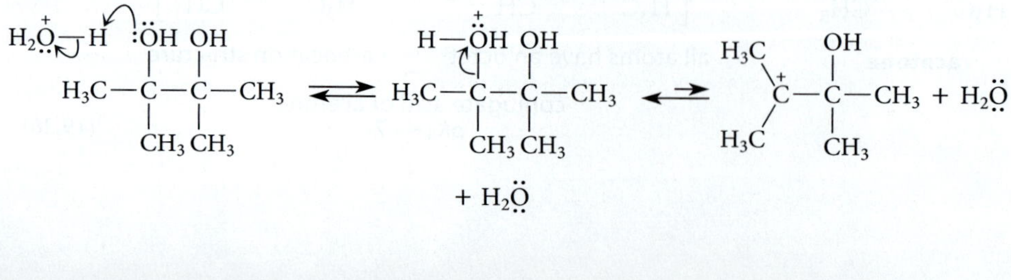

(19.30a)

The rearrangement can now take place. The product of the rearrangement is an α-hydroxy carbocation, which, as we have seen, is the same thing as the conjugate acid of a ketone.

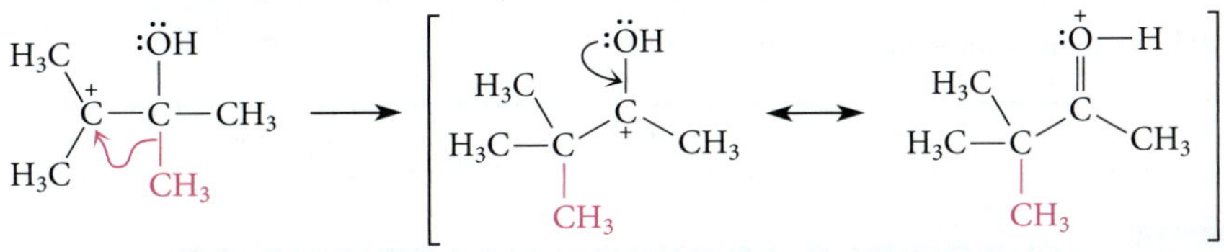

both an α-hydroxy carbocation and a ketone conjugate acid

(19.30b)

We just showed that such carbocations are especially stable. So, the rearrangement step is favorable because the α-hydroxy carbocation is more stable than the tertiary carbocation. The second resonance structure in Eq. 19.30b shows that the α-hydroxy carbocation is also the conjugate acid of the ketone. Removal of a proton from the carbonyl oxygen gives the product.

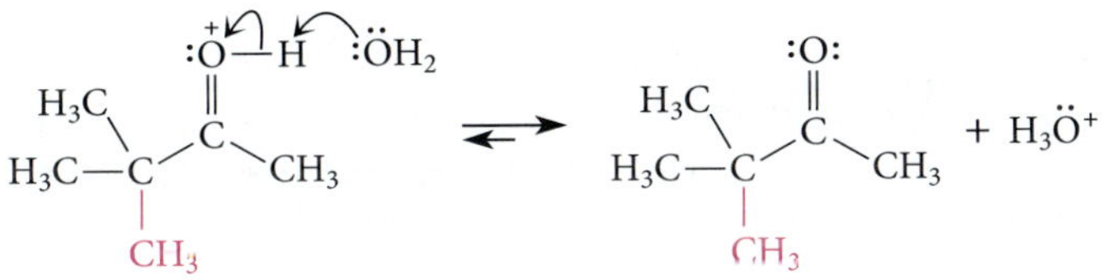

(19.30c)

Aldehydes and ketones in solution are considerably less basic than alcohols (Sec. 11.1E). In other words, their conjugate acids are more acidic than those of alcohols.

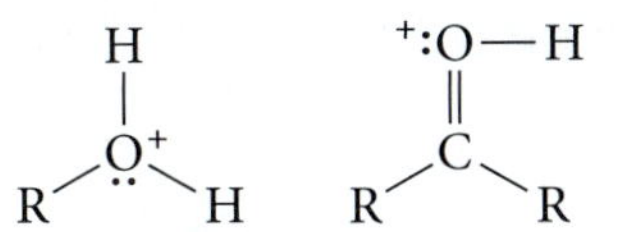

$pK_a \sim -2.5$ $pK_a \sim -7$ (19.31)

Because protonated aldehydes and ketones are resonance-stabilized and protonated alcohols are not, we might have expected protonated carbonyl compounds to be *more stable* relative to their conjugate bases and therefore *less acidic*. The relative acidity of protonated alcohols and carbonyl compounds is an example of a solvent effect. In the *gas phase*, aldehydes and ketones *are* indeed more basic than alcohols. In solution, the solvation of a protonated alcohol by hydrogen bonding is evidently so effective that it outweighs the resonance stabilization of a protonated aldehyde or ketone.

Focused Problems

19.10 (a) Write an S_N1 mechanism for the solvolysis of CH_3OCH_2Cl [(chloromethoxy)methane] in ethanol; draw appropriate resonance structures for the carbocation intermediate.

(b) Explain why the alkyl halide in part (a) undergoes solvolysis much more rapidly than 1-chlorobutane. (In fact, it reacts in ethanol more than 100 times more rapidly.)

19.11 Predict the product when each of the following diols undergoes the pinacol rearrangement.

(a) 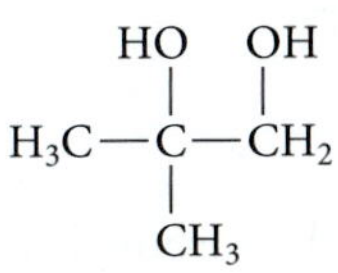

(b) 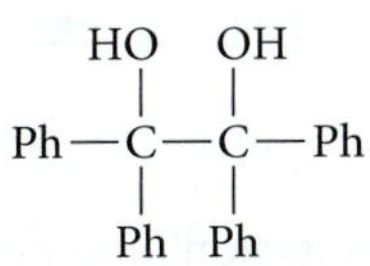

(c) 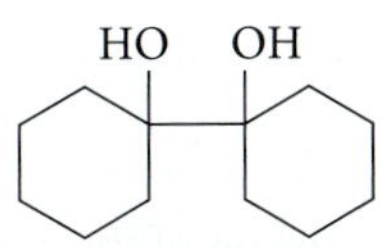

19.12 Use resonance arguments to explain each of the following observations:

(a) *p*-Methoxybenzaldehyde is more basic than *p*-nitrobenzaldehyde.

(b) 3-Buten-2-one is more basic than 2-butanone.

19.8 REVERSIBLE ADDITION REACTIONS OF ALDEHYDES AND KETONES

One of the most typical reactions of aldehydes and ketones is *addition* to the carbon–oxygen double bond. To begin with, we focus on two simple addition reactions: addition of hydrogen cyanide (HCN) and hydration (addition of water).

Addition of HCN:

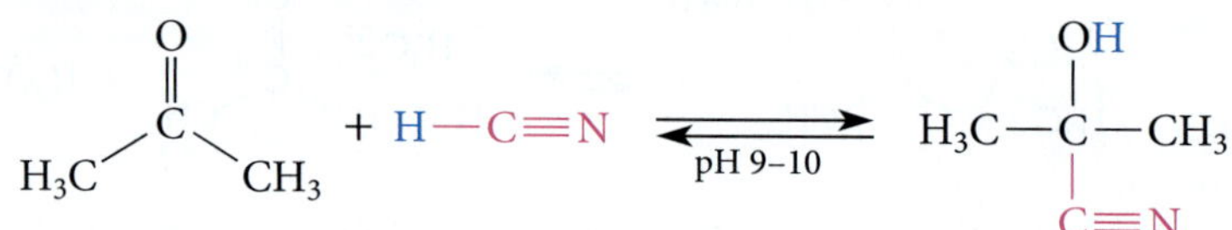

acetone **acetone cyanohydrin** (77–78% yield) (19.32)

The product of HCN addition to an aldehyde or ketone is called a **cyanohydrin**. Cyanohydrins are a special class of *nitriles* (compounds of the form R—C≡N). (The chemistry of nitriles is discussed in Chapter 21.) The preparation of cyanohydrins is another method of forming carbon–carbon bonds.

Addition of water (hydration):

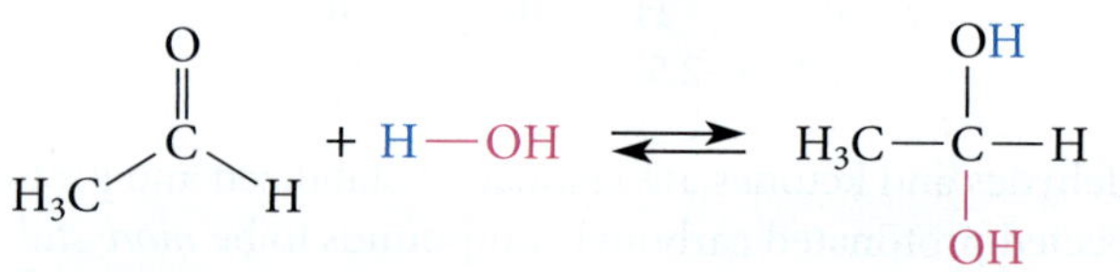

acetaldehyde **acetaldehyde hydrate** (19.33)

The product of water addition to an aldehyde or ketone is called a **hydrate**, or gem-diol. As we noted in Eq. 4.63a, Sec, 4.9, the prefix *gem* (for *geminal*) is used in chemistry when two identical groups are present on the same carbon.

In all carbonyl-addition reactions, the more electropositive species (for example, the hydrogen of H—CN or H—OH) adds to the carbonyl oxygen, and the more electronegative species (for example, the CN or the OH) adds to the carbonyl carbon.

A. Mechanisms of Carbonyl-Addition Reactions

Carbonyl-addition reactions occur by two general types of mechanism. The first mechanism, called **nucleophilic carbonyl addition**, involves the reaction of a nucleophile at the carbonyl carbon. In cyanohydrin formation (Eq. 19.32), the nucleophile is cyanide ion, which is formed by the ionization of HCN:

$$\underset{pK_a = 9.4}{\text{H—CN}} + {}^{-}\text{OH} \rightleftharpoons \underset{\textbf{cyanide ion}}{{}^{-}\text{:CN}} + H_2O \qquad (19.34a)$$

The carbon of the cyanide ion donates its nonbonding electron pair to the carbonyl carbon of the aldehyde or ketone, and the carbonyl oxygen accepts the displaced electron pair and assumes a negative charge. We can think of this process in the same way that we think of nucleophilic reactions at saturated carbon atoms (Sec. 3.2A). The electrophilic center is the carbonyl carbon, and the "leaving group" is the π bond of the carbonyl group. Because the carbon–oxygen σ bond remains intact, the "leaving group" doesn't actually leave.

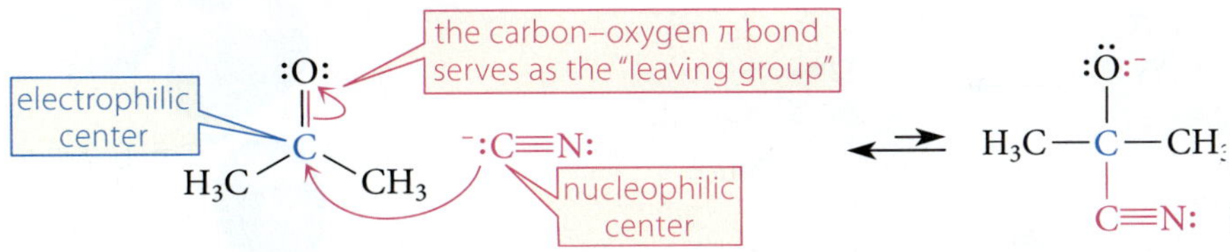

(19.34b)

To complete the nucleophilic addition, the negatively charged oxygen—an alkoxide ion, and a relatively strong base—is protonated by either water or HCN.

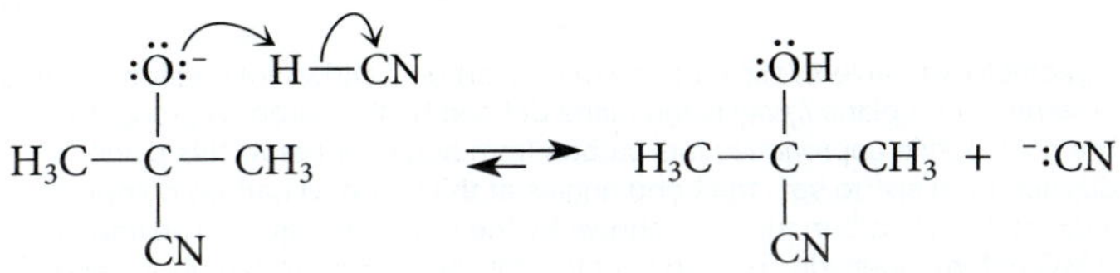
(19.34c)

The nucleophilic carbonyl-addition mechanism finds no analogy in additions to ordinary carbon–carbon double bonds. Yet, nucleophilic carbonyl addition occurs even though the carbon–oxygen π bond is 62 kJ mol^{-1} (15 kcal mol^{-1}) *stronger* than the carbon–carbon π bond of an alkene. The stronger bond is more reactive because the nonbonding electron pair (and negative charge) formed in the carbonyl-addition mechanism is transferred to a very electronegative atom, oxygen. The same reaction of an alkene would place a nonbonding electron pair and negative charge on a carbon, a much less electronegative atom.

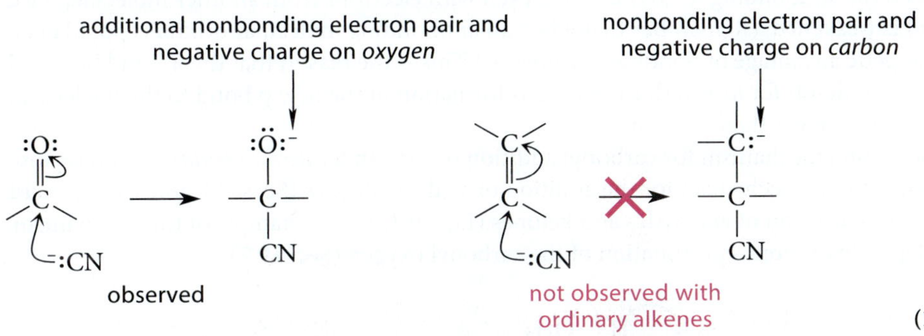

(19.35)

The reaction of a nucleophile with the carbonyl group, then, is driven by the ability of oxygen to accept the nonbonding electron pair. For this reason, a nucleophile cannot add to the carbonyl oxygen. *Nucleophiles always react with carbonyl groups at the carbonyl carbon.*

The geometry of nucleophilic addition is shown in **Fig. 19.8**. The carbonyl group and the two atoms bonded to the carbonyl carbon define a reference plane. The nucleophile approaches the carbonyl carbon from above or below this plane, as shown in Fig. 19.8a. As a result of this reaction, the carbonyl carbon changes hybridization from sp^2 to sp^3, the oxygen accepts an electron pair, and the geometry at the carbonyl carbon changes from trigonal planar to tetrahedral. In other words, the angle between the bonds to the carbonyl carbon, initially about 120°, compresses to about 109°, the tetrahedral angle. As a result of the reaction, then, the groups bonded to the carbonyl carbon become closer together.

The reason for the addition geometry is similar to the reason for opposite-side substitution in the S_N2 reactions of alkyl halides (see Fig. 9.3, Sec. 9.4C). The curved-arrow notation might convey the impression that the nucleophile reacts at the π bond. However, the bonding π molecular orbital of the carbonyl group (see Fig. 19.1, Sec. 19.2) is fully occupied with two electrons and cannot accommodate any more electrons. The electron pair of the nucleophile interacts instead with the *unoccupied* MO of lowest energy (LUMO), which, in the case of the carbonyl group, is the antibonding π* molecular orbital. This MO is shown in Fig. 19.8b.

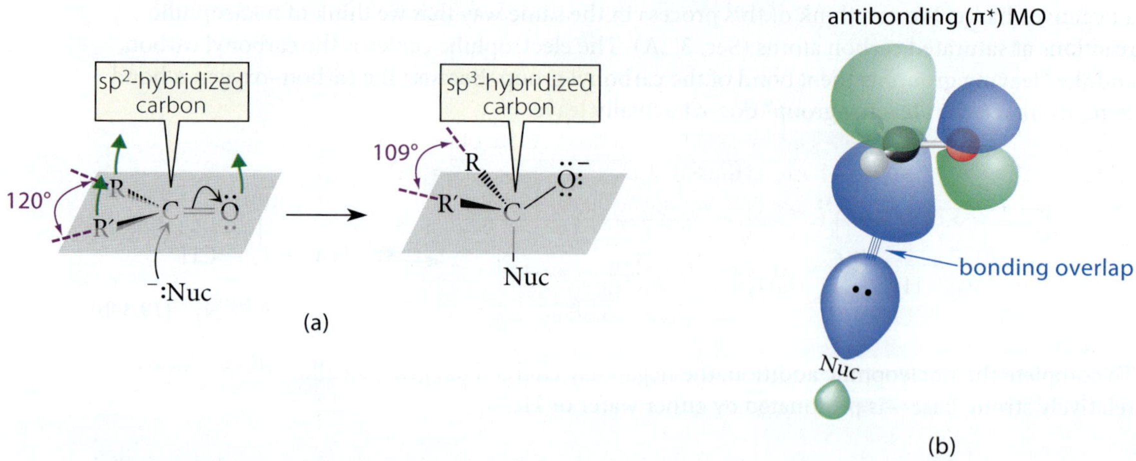

FIGURE 19.8 (a) The geometry of nucleophilic reaction to the carbonyl carbon of a ketone, with the nucleophile represented by Nuc:⁻. The reference plane *(gray)* is the plane defined by the carbonyl group and the atoms attached to the carbonyl carbon. The nucleophile approaches this carbon from above or below this plane. As a result, the carbonyl carbon changes hybridization from sp^2 to sp^3, the bond angles at the carbonyl carbon compress from 120° to 109°, and the groups attached to the carbonyl carbon move as shown by the green arrows. (b) The interaction of a nucleophile with the π^* antibonding MO of formaldehyde. The lobes of this MO are concentrated above and below the plane of the molecule. The interaction of the nucleophile with this MO defines the direction of nucleophilic approach to the carbonyl carbon.

This MO has lobes above and below the reference plane. The nucleophile, then, must begin its bonding interaction with the carbonyl carbon from the direction along which the LUMO is concentrated, as shown in Fig. 19.8a.

When the antibonding π^* MO is filled, even with electrons from another molecule, the C═O π bond is weakened. (Remember that when an antibonding molecular orbital is populated, the energetic advantage of bonding disappears.) This is the reason that the π bond breaks. The energetic trade-off for loss of this bonding is formation of the new σ bond to the nucleophile; a σ bond is stronger than a π bond.

The second mechanism for carbonyl addition occurs under *acidic conditions* and is closely analogous to the mechanism for the addition of acids to alkenes (Secs. 4.7 and 4.10C). **Acid-catalyzed hydration** of aldehydes and ketones (Eq. 19.33) is an example of this mechanism. The first step in hydration is protonation of the carbonyl oxygen (Sec. 19.7).

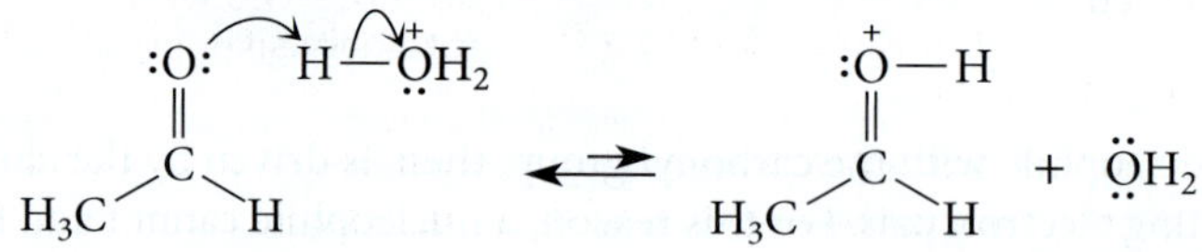

(19.36a)

A positively charged oxygen attracts electrons more strongly than the oxygen of an unprotonated carbonyl group. In other words, the protonated carbonyl compound is a much stronger *Lewis acid* (electron acceptor) than an unprotonated carbonyl compound. As a result, even the relatively weak base H_2O can react at the carbonyl carbon. Loss of a proton to solvent completes the reaction.

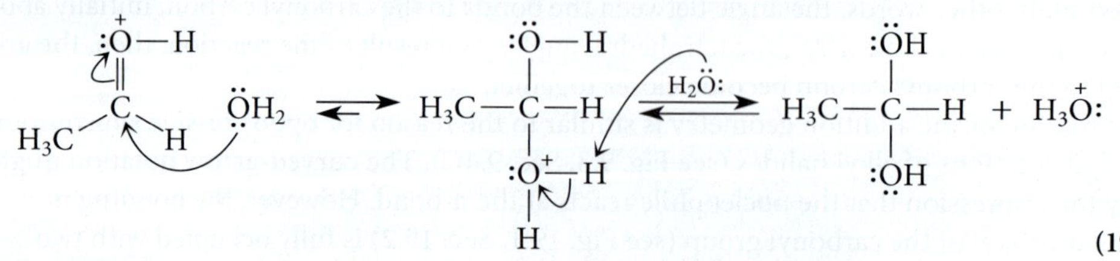

(19.36b)

Hydration of aldehydes and ketones also occurs in neutral and basic solution (Focused Problem 19.13).

The direction of approach of the nucleophile to the protonated carbonyl group is the same as it is for approach to the unprotonated carbonyl group (see Fig. 19.8)—from above or below the reference plane. This is explained by the shape of the LUMO of the protonated carbonyl group, which is very similar to the shape of the LUMO of the carbonyl group itself.

Carbonyl Resonance and Carbonyl Reactivity It is tempting to use the following resonance structures to rationalize the reactivity of a carbonyl group:

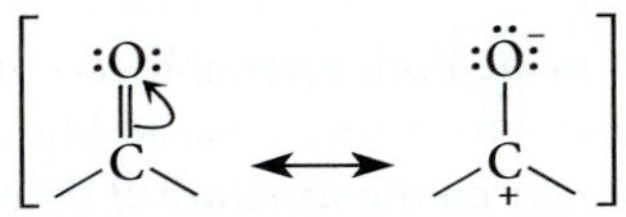

(19.37)

It is sometimes erroneously said that carbonyl compounds react with nucleophiles at the carbonyl carbon because there is positive charge at this carbon and this positive charge attracts the electron pairs of nucleophiles.

This argument has two fallacies. First, when two species collide, they do so randomly, not from any preferred direction. For example, a nucleophile in solution can collide randomly with any of the atoms in a carbonyl compound. In other words, a nucleophile is not selectively directed to a carbonyl carbon. Rather, reaction occurs the way it does because when the nucleophile happens to collide with the carbonyl carbon from the proper direction, *electrons (and charge) can be shifted onto the electronegative carbonyl oxygen.*

The second and greater fallacy in this argument is its implication that resonance increases reactivity. In fact, *resonance stabilization has the opposite effect.* Carbonyl compounds are actually *less reactive* than they would be if such resonance stabilization did not exist. Many studies have shown that when molecules are prepared in which resonance stabilization of double bonds cannot occur (for example, molecules in which π-electron systems are twisted out of coplanarity), these bonds become more reactive than double bonds in similar molecules in which resonance interaction can occur. The carbon–oxygen π bond is actually about 62 kJ mol^{-1} (15 kcal mol^{-1}) *stronger* than a carbon–carbon π bond; this fact, taken alone, means that it should be *less reactive* than a carbon–carbon double bond, exactly as the resonance structures suggest.

If the carbonyl π bond is so stable, then, why does it react? Remember from Sec. 3.6A that when a Brønsted base reacts with a Brønsted acid, it is not only the strength of the bond to the hydrogen, but also *how well the other atom in the bond accepts electrons* that governs how easily the reaction takes place. A nucleophilic reaction at a carbonyl carbon is no different, except that the electrophile is a carbon rather than a hydrogen, and a π bond rather than a σ bond is broken. *Carbonyl compounds react with nucleophiles at the carbonyl carbon because the electronegative oxygen readily accepts negative charge.*

It nevertheless is true that resonance structures *do* suggest ways that molecules can react. In the case of a carbonyl group, the charge-separated resonance structure in Display 19.37 suggests that the carbon, with its partial positive charge, is the electrophilic site. The reason this works is that the atom (oxygen) that accepts electrons in the resonance structure is the same one that accepts electrons in the transition state for nucleophilic addition. *Such correspondences between reactivity patterns and resonance structures invariably occur throughout organic chemistry.* For this reason, organic chemists find themselves using resonance structures to predict sites of reactivity in molecules. Resonance structures are undeniably useful for this purpose; but when we use them this way, we must remember that resonance is not the *reason for reactivity.*

Focused Problems

19.13 (a) Write a curved-arrow mechanism for the hydroxide-catalyzed hydration of acetaldehyde.

(b) Write a curved-arrow mechanism for the decomposition of acetone cyanohydrin (Eq. 19.15) in aqueous hydroxide. Explain why the ketone–cyanohydrin equilibrium favors the ketone at high pH.

19.14 Write a curved-arrow mechanism for each of the following reactions.

(a) The acid-catalyzed addition of methanol to benzaldehyde

(b) The methoxide-catalyzed addition of methanol to benzaldehyde

B. Equilibria in Carbonyl-Addition Reactions

Hydration and cyanohydrin formation are both reversible reactions. (Not all carbonyl additions are reversible.) Whether the equilibrium for a reversible addition favors the addition product or the carbonyl compound *depends strongly on the structure of the carbonyl compound.* For example, cyanohydrin formation favors the cyanohydrin addition product in the case of aldehydes and methyl ketones, but the equilibrium favors the carbonyl compound when aryl ketones are used.

The effect of aldehyde or ketone structure on the addition equilibrium for hydration is shown in **Table 19.2**. The data in the table illustrate the following trends:

1. Addition is more favorable for aldehydes than for ketones.
2. Electronegative groups near the carbonyl carbon make carbonyl addition more favorable.
3. Addition is less favorable when groups are present that donate electrons by resonance to the carbonyl carbon.

The trends in this table and the reasons behind them are important for two reasons. First, the equilibria for *all* addition reactions show similar effects of structure. Second, and more important, the *rates* of carbonyl-addition reactions—that is, the *reactivities* of carbonyl compounds—follow similar trends (Sec. 19.8C).

What is the reason for the effect of structure on carbonyl addition? The relative stabilities of the carbonyl compound and the addition product govern the $\Delta G°$ for addition. As shown in **Fig. 19.9**, the primary effect on the hydration equilibrium is the *difference in the stabilities of the carbonyl compounds.* Added stability in the carbonyl compound increases the energy change $\Delta G°$, *and therefore decreases the equilibrium constant*, for formation of an addition product.

TABLE 19.2 Equilibrium Constants for the Hydration of Aldehydes and Ketones

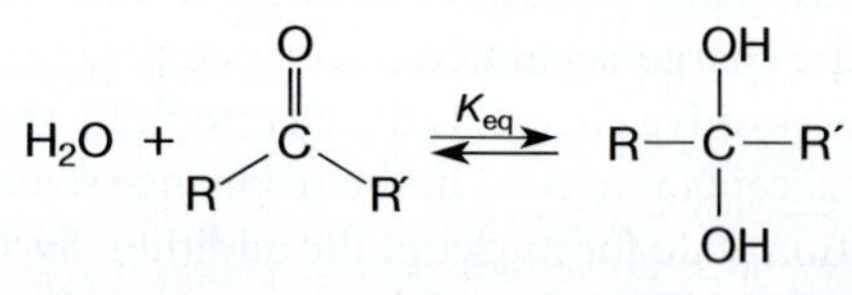

Aldehydes	K_{eq}	Ketones	K_{eq}
$H_2C{=}O$	2.2×10^{3}	$(CH_3)_2C{=}O$	1.4×10^{-3}
$CH_3CH{=}O$	1.0		
$(CH_3)_2CHCH{=}O$	0.5–1.0	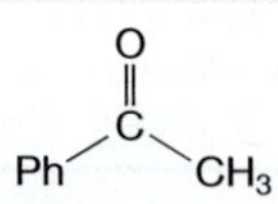	6.6×10^{-6}
$PhCH{=}O$	8.3×10^{-3}	$Ph_2C{=}O$	1.2×10^{-7}
$ClCH_2CH{=}O$	37	$(ClCH_2)_2C{=}O$	10
$Cl_3CCH{=}O$	2.8×10^{4}	$(F_3C)_2C{=}O$	too large to measure

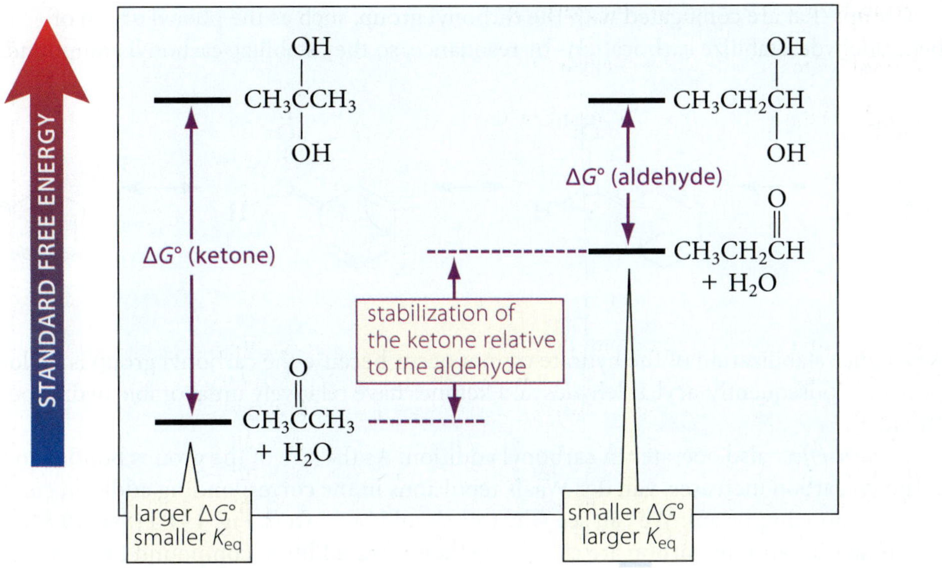

FIGURE 19.9 Greater stability of a ketone relative to an aldehyde causes the ketone to have greater standard free energy of hydration and therefore a smaller equilibrium constant for hydration. (The two hydrates have been placed at the same energy level for comparison purposes.)

The major factors that stabilize carbonyl compounds can be understood by considering the resonance structures of the carbonyl group:

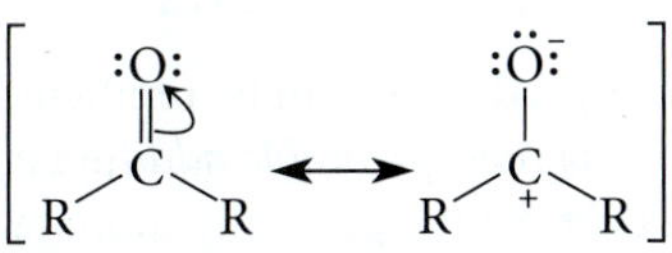
(19.38a)

The structure on the right, although not as important a contributor as the one on the left, reflects the polarity of the carbonyl group and has the characteristics of a carbocation. Therefore, anything that stabilizes carbocations also tends to stabilize carbonyl compounds. Because alkyl groups stabilize carbocations, ketones (R = alkyl) are more stable than aldehydes (R = H). This stability is reflected in the relative heats of formation of aldehydes and ketones. For example, acetone is more stable than its isomer propionaldehyde:

acetone
$\Delta H_f^\circ = -218$ kJ mol^{-1}
(-52.1 kcal mol^{-1})

propionaldehyde
$\Delta H_f^\circ = -189$ kJ mol^{-1}
(-45.2 kcal mol^{-1})

(19.38b)

Because alkyl groups stabilize carbonyl compounds, the equilibria for additions to ketones are less favorable than those for additions to aldehydes (trend 1). Formaldehyde, with *two* hydrogens and no alkyl groups bonded to the carbonyl carbon, has a very large equilibrium constant for hydration.

Electronegative groups such as halogens destabilize carbocations by their polar effect and for the same reason destabilize carbonyl compounds. Consequently, halogens make the equilibria for addition more favorable (trend 2). In fact, chloral hydrate (known in medicine as a hypnotic) is a stable crystalline compound.

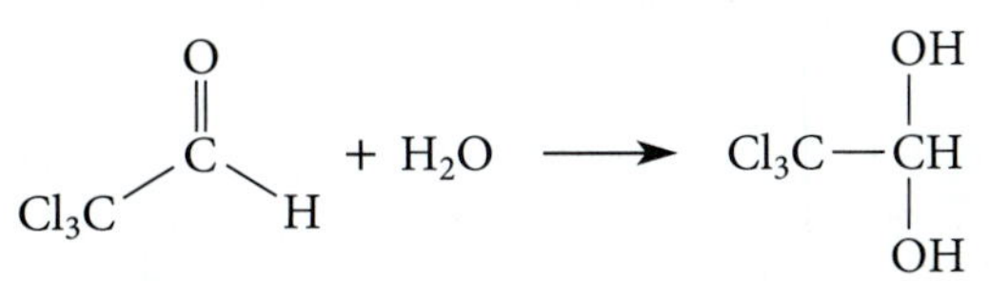

chloral
(2,2,2-trichloroethanal)

chloral hydrate

(19.39)

Groups that are conjugated with the carbonyl group, such as the phenyl group of benzaldehyde, stabilize carbocations by resonance, so they stabilize carbonyl compounds.

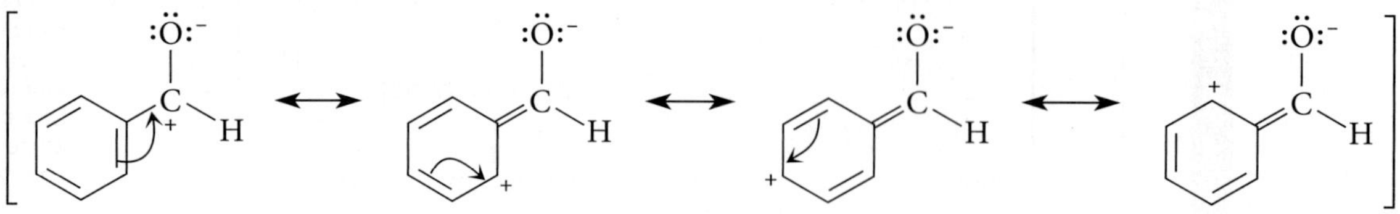

(19.40)

Resonance stabilization of the hydrate cannot occur because the carbonyl group is no longer present. Consequently, aryl aldehydes and ketones have relatively unfavorable hydration equilibria (trend 3).

A *steric effect* also operates in carbonyl addition. As the size of the groups bonded to the carbonyl carbon increases, van der Waals repulsions in the corresponding addition compounds become more important. We can see why this should be so from Fig. 19.8a (Sec. 19.8A). The groups at the carbonyl carbon are closer together in the addition compound than they are in the carbonyl compound; so, van der Waals repulsions are more pronounced in the addition compound. These van der Waals repulsions *raise* the energy of the addition compound relative to the carbonyl compound and *increase* the $\Delta G°$ for addition.

C. Rates of Carbonyl-Addition Reactions

The trends in relative rates of carbonyl addition can be predicted from the trends in equilibrium constants. That is, *compounds with the most favorable addition equilibria tend to react most rapidly in addition reactions.* Therefore, aldehydes are generally more reactive than ketones in addition reactions; formaldehyde is more reactive than many other simple aldehydes.

The reason for the parallel trends in rates and equilibria is that the transition states for addition reactions resemble addition products. In other words, it is a convenient approximation to think of the addition compounds in Fig. 19.9 as transition states, and the standard free energies $\Delta G°$ as standard free energies of activation $\Delta G^{\circ\ddagger}$. Just as the destabilization of aldehydes or ketones decreases the $\Delta G°$ for their addition reactions, the same destabilization decreases the free energies of activation $\Delta G^{\circ\ddagger}$ for addition and thus increases the rate of addition.

Notice that *we are analyzing carbonyl reactivity differently than we have analyzed other forms of reactivity.* In many cases, we gauge reactivity by applying Hammond's postulate to the relative stability of *reactive intermediates* such as carbocations. The more stable the intermediate, the more reactive a compound is. However, remember that reactivity involves the *difference* in two standard free energies: the standard free energy of the transition state and the standard free energy of the reactant (see Fig. 4.4, Sec. 4.8A). In carbonyl chemistry, it is often the relative free energies of the *reactants* that provide the major effects on reactivity. When one reactant is less stable than another relative to its transition state, the compound is more reactive because it is pushed further up the "free-energy hill" toward its transition state (**Fig. 19.10**). Because we'll continue to return to this point in Chapters 21 and 22, it is important to understand the relative reactivities of carbonyl compounds in these terms.

This section has covered two examples of addition to the carbonyl group. Subsequent sections deal with other addition reactions as well as more complex reactions that have mechanisms in which the initial steps are addition reactions. These addition reactions all have mechanisms similar to the ones discussed in this section, and the trends in reactivity are the same. *Addition to the carbonyl group is a common thread that runs throughout most of aldehyde and ketone chemistry.*

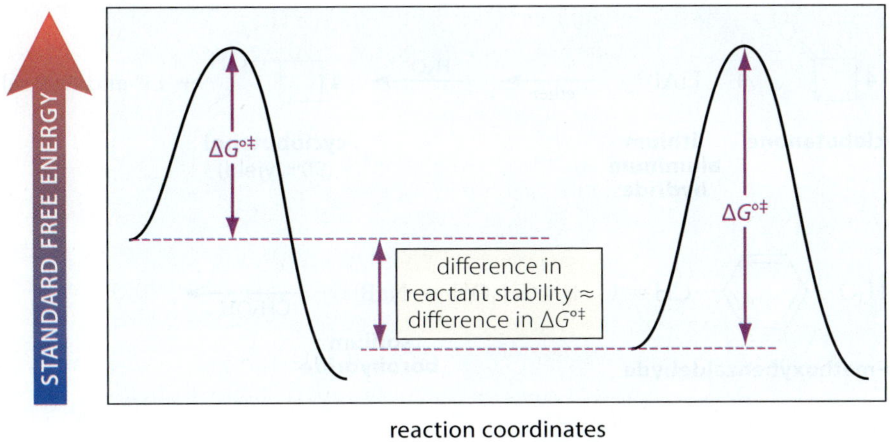

FIGURE 19.10 The effect of reactant stability on reactivity when two reactions are compared at the same transition-state free energy. In many carbonyl reactions, the relative stability of the reactants determines relative reactivity. In these cases, the less stable compound reacts more rapidly. In many carbonyl reactions, the effect of reactant stability on rate parallels the effect of reactant stability on equilibrium constant (see Fig. 19.9). Therefore, we can use equilibrium data, such as that in Table 19.2, to estimate relative reactivity.

Focused Problems

19.15 Which carbonyl compound should form the greater proportion of cyanohydrin at equilibrium? Draw the structure of the cyanohydrin, and explain your reasoning.

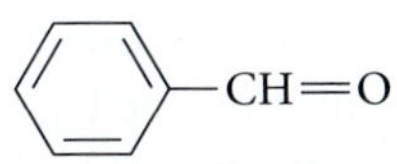

benzaldehyde

or $CH_3CH_2CH{=}O$

propanal

19.16 Within each set, which compound should be more reactive in carbonyl-addition reactions? Explain your choices.

(a)

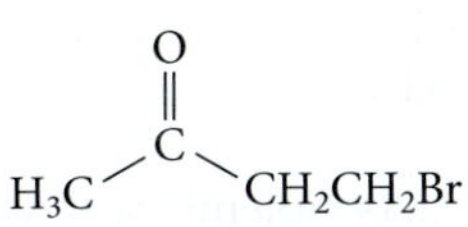

A

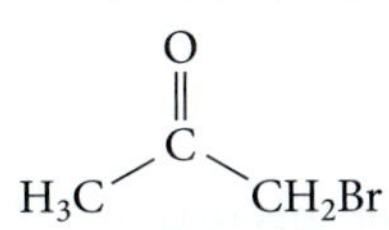

B

(b)

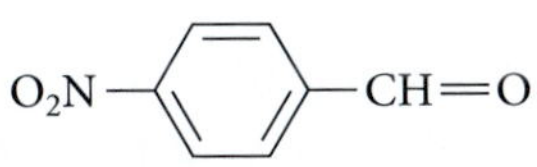

A

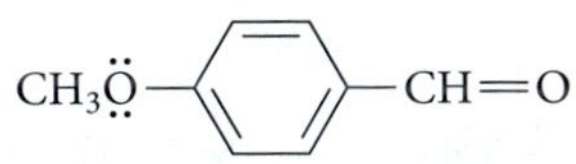

B

(c)

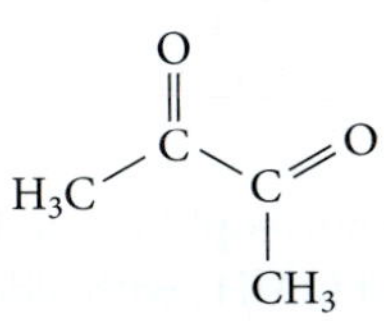

A

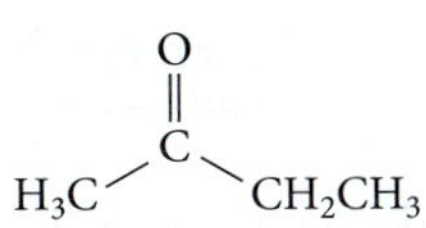

B

19.9 REDUCTION OF ALDEHYDES AND KETONES TO ALCOHOLS

A. Reduction with Lithium Aluminum Hydride and Sodium Borohydride

Aldehydes and ketones are reduced to alcohols with either lithium aluminum hydride ($LiAlH_4$) or sodium borohydride ($NaBH_4$). These reactions result in the net *addition* of the elements of H_2 across the C=O bond.

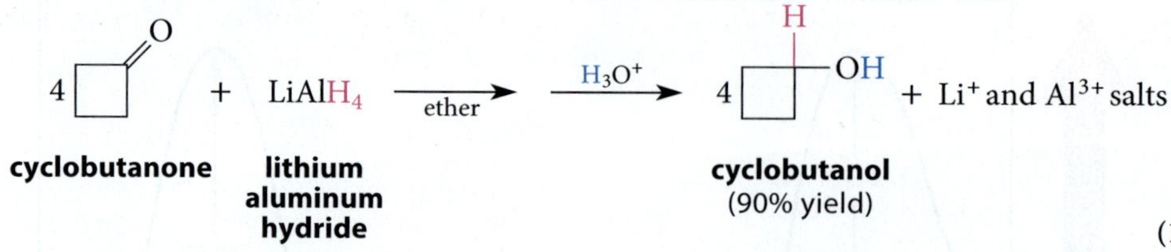

(19.41)

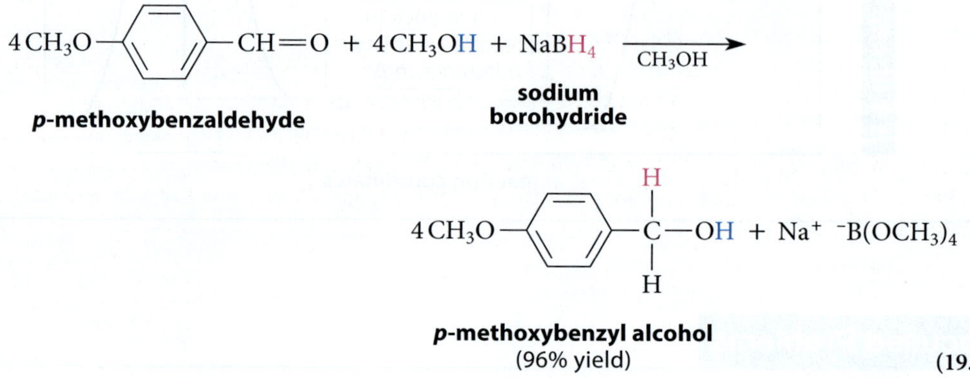

(19.42)

As these examples illustrate, reduction of an aldehyde gives a primary alcohol, and reduction of a ketone gives a secondary alcohol.

Lithium aluminum hydride serves generally as a source of H:$^-$, the *hydride ion*. Because hydrogen is more electronegative than aluminum (see Fig. 1.15, Sec. 1.9A), the Al—H bonds of the $^-AlH_4$ ion carry a substantial fraction of the negative charge. In other words,

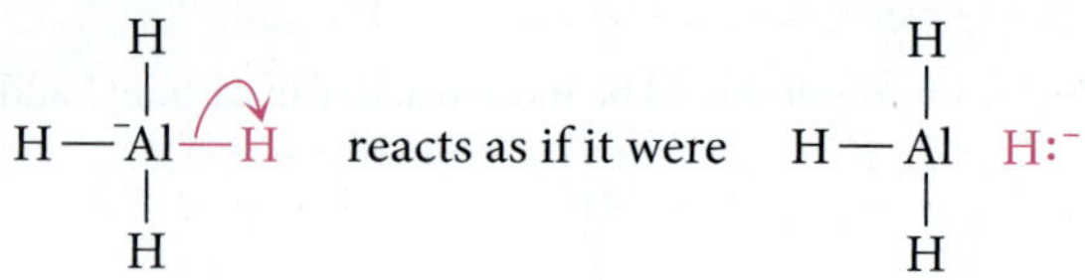

(19.43)

The hydride ion in $LiAlH_4$ is very basic. For this reason, $LiAlH_4$ reacts violently with water and therefore must be used in dry solvents such as anhydrous ether and THF.

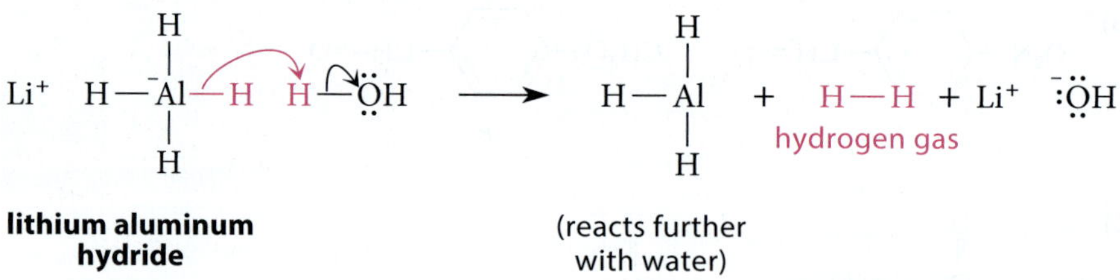

(19.44)

Like many other strong bases, the hydride ion in $LiAlH_4$ is a good nucleophile, and $LiAlH_4$ contains its own "built-in" Lewis acid, the lithium ion. The reaction of $LiAlH_4$ with aldehydes and ketones involves the nucleophilic reaction of hydride (delivered from $^-AlH_4$) at the carbonyl carbon. The lithium ion acts as a Lewis acid catalyst by coordinating to the carbonyl oxygen.

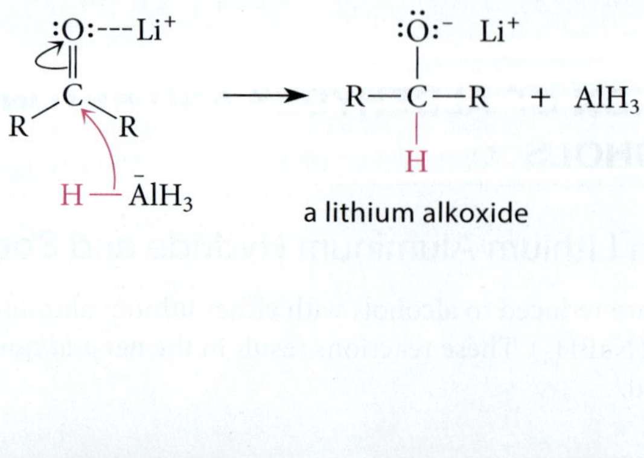

(19.45a)

The addition product, an alkoxide salt, can react with AlH_3, and the resulting product can also serve as a source of hydride.

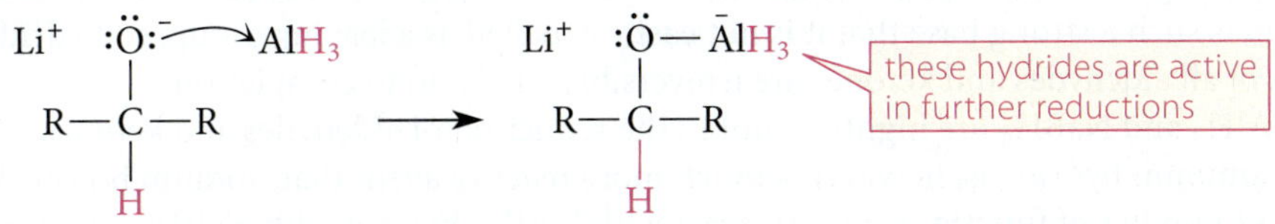

(19.45b)

Similar processes occur at each stage of the reduction until all of the hydrides are consumed. Therefore, as shown in the stoichiometry of Eq. 19.41, all four hydrides of $LiAlH_4$ are active in the reduction. In other words, it takes one-fourth of a mole of $LiAlH_4$ to reduce a mole of aldehyde or ketone.

After the reduction is complete, the alcohol product exists as an alkoxide addition compound with the aluminum. This is converted by protonation in a separate step into the alcohol product. The proton source can be an aqueous HCl solution or even an aqueous solution of a weak acid such as ammonium chloride.

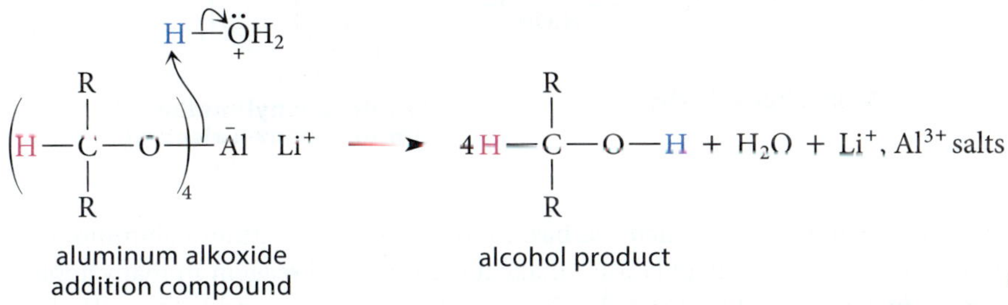

(19.45c)

The reaction of $NaBH_4$ with aldehydes and ketones is conceptually similar to that of $LiAlH_4$. The sodium ion is a much weaker Lewis acid than the lithium ion. For this reason, $NaBH_4$ reductions are carried out in protic solvents such as alcohols. Hydrogen bonding between the alcohol solvent and the carbonyl group serves as a weak acid catalysis that activates the carbonyl group. Unlike $LiAlH_4$, $NaBH_4$ reacts only slowly with alcohols and can even be used in water if the solution is not acidic.

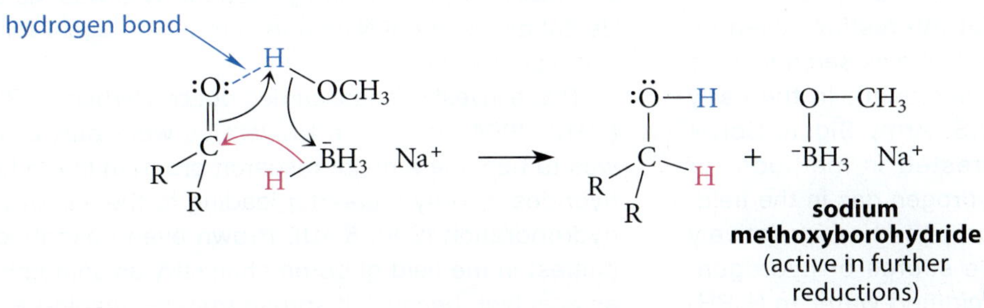

(19.46)

As Eq. 19.42 shows, all four hydride equivalents of $NaBH_4$ are active in the reduction.

Because $LiAlH_4$ and $NaBH_4$ are hydride donors, reductions by these and related reagents are generally referred to as **hydride reductions**. The important mechanistic point about these reactions is that they are further examples of *nucleophilic addition*. Hydride ion from $LiAlH_4$ or $NaBH_4$ is the nucleophile, and the proton is delivered from acid added in a separate step (in the case of $LiAlH_4$ reductions) or solvent (in the case of $NaBH_4$ reductions).

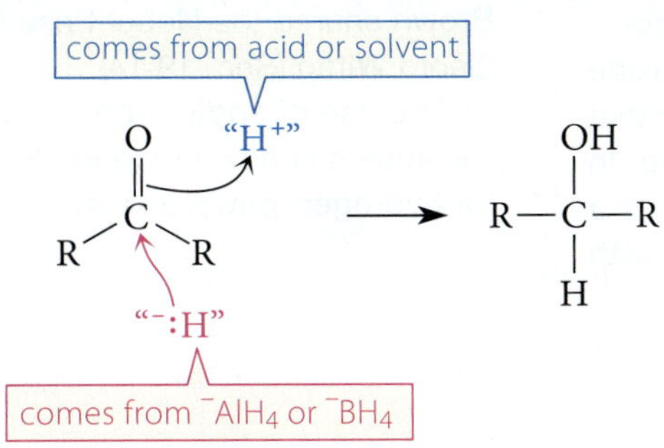

(19.47)

Unlike the additions discussed in Sec. 19.8, hydride reductions are *not* reversible. Reversal of carbonyl addition would require the original nucleophilic group, in this case, $H:^-$, to be expelled as a leaving group. As in S_N1 or S_N2 reactions, the best leaving groups are the weakest bases. Hydride ion is such a strong base that it is not easily expelled as a leaving group, so hydride reductions of all aldehydes and ketones are irreversible—they go to completion.

Both $LiAlH_4$ and $NaBH_4$ are highly useful in the reduction of aldehydes and ketones. Lithium aluminum hydride is, however, a much more *reactive* agent than sodium borohydride. A significant number of functional groups react with $LiAlH_4$ but not with $NaBH_4$; among them are alkyl halides, esters, alkyl tosylates, and nitro groups. Sodium borohydride can be used as a reducing agent in the presence of these groups.

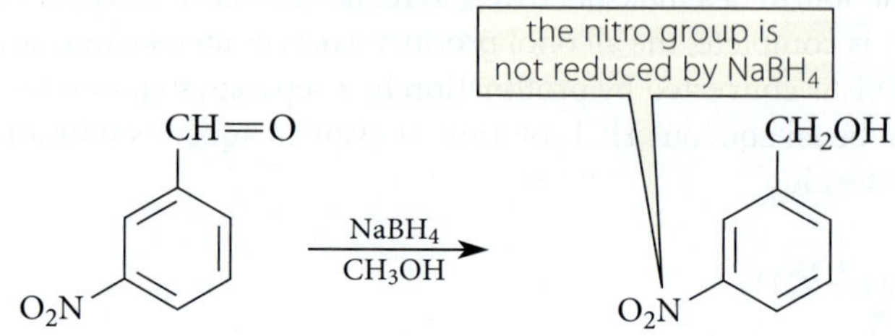

3-nitrobenzaldehyde

(3-nitrophenyl)methanol
(*m*-nitrobenzyl alcohol)
(82% yield) **(19.48)**

Sodium borohydride is also a much less hazardous reagent than lithium aluminum hydride. The greater selectivity and safety of $NaBH_4$ make it the preferred reagent in many applications, but either reagent can be used for the reduction of simple aldehydes and ketones. Both are very important in organic chemistry.

Discovery of $NaBH_4$ Reductions

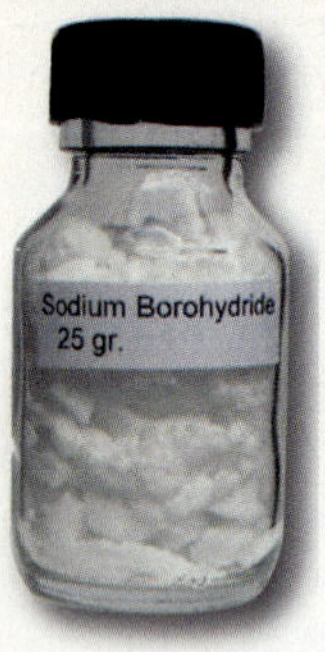

Wildo Oelen

The discovery of $NaBH_4$ reductions illustrates that interesting research findings are sometimes serendipitous (obtained by accident). In the early 1940s, the U.S. Army Signal Corps became interested in methods for generating hydrogen gas in the field. $NaBH_4$ was proposed as a relatively safe, portable source of hydrogen: addition of acidified water to $NaBH_4$ results in the evolution of hydrogen gas at a safe, moderate rate. To supply the required quantities of $NaBH_4$, a large-scale synthesis was necessary. The following reaction appeared to be suitable for this purpose.

$$4\,NaH + B(OCH_3)_3 \longrightarrow NaBH_4 + 3\,NaOCH_3 \quad \textbf{(19.49)}$$

The problem with this process was that the sodium borohydride had to be separated from the sodium methoxide by-product. Several solvents were tried in the hope that a significant difference in solubilities could be found. In the course of this investigation, acetone was tried as a recrystallization solvent, and it was found to react with the $NaBH_4$ to yield isopropyl alcohol. This was the accidental discovery of $NaBH_4$ as a reducing agent for carbonyl compounds.

These investigations, carried out by Herbert C. Brown (1912–2004) at Purdue University, were part of what was to become a major research program in the boron hydrides, shortly thereafter leading to the discovery of hydroboration (Sec. 5.4B). Brown even described his interest in the field of boron chemistry as something of an accident, because it sprung from his reading a book about boron and silicon hydrides that his girlfriend (who later became his wife) gave him as a graduation present. Mrs. Brown observed that the choice of this particular book was dictated by the fact that it was among the least expensive chemical titles in the bookstore; in the depression era, students had to be careful how they spent their money! For his work in organic chemistry, Brown shared the Nobel Prize in Chemistry in 1979 with Georg Wittig (Sec. 19.14).

The use of sodium borohydride has come full circle, because it is now being explored as a hydrogen source for hydrogen-powered fuel cells.

Focused Problems

19.17 From what aldehyde or ketone could each of the following be synthesized by reduction with either $LiAlH_4$ or $NaBH_4$?

(a) 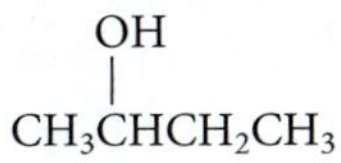

(b) 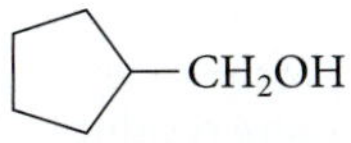

(c) 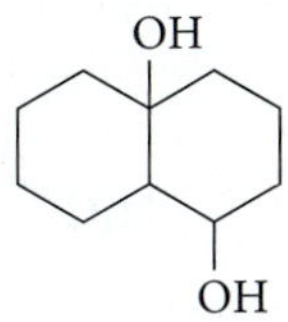

19.18 Which of the following alcohols could *not* be synthesized by a hydride reduction of an aldehyde or ketone? Explain.

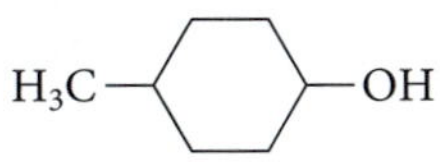

A

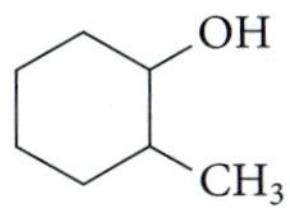

B

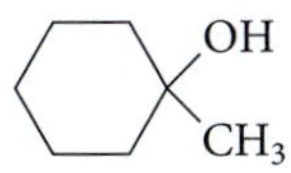

C

B. Hydride Reduction in Biology

Chemical Biology Topic

In Sec. 11.8 we introduced NAD^+ as an important biological oxidizing agent. (The structure of NAD^+ is shown in Display 11.68, Sec. 11.8.) When NAD^+ is used for an oxidation, its reduced form, NADH, is formed. NADH is "nature's hydride reagent." It acts as a biological reducing agent in the same sense that $LiAlH_4$ and $NaBH_4$ act as laboratory reducing agents, as we illustrate in this section.

One of the most important biological functions of NADH is *carbonyl reduction*. An example of this reaction is the conversion of pyruvate into (*S*)-lactate under anaerobic conditions in muscle tissue. The pro-(*R*) hydride of NADH (*red*) is transferred stereospecifically.

Both NAD^+ and NADH exist in phosphorylated forms, $NADP^+$ and NADPH; the position of phosphorylation is also shown in Display 11.68, Sec. 11.8. These forms differ in their cellular location and the general role that they play in cellular biochemistry. You will learn these details if you study biochemistry. However, from a chemical standpoint, they function in the same way.

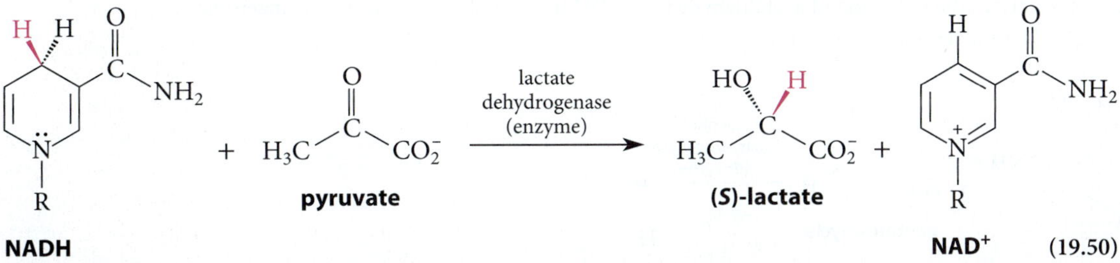

(19.50)

(The lactate produced in this reaction is the source of the burning in muscle that occurs after very intense exercise; the conditions of oxygen deprivation promote this reaction as a way of generating NAD^+ required in the metabolism of glucose.) This reaction is catalyzed by the enzyme *lactate dehydrogenase*. Although the enzyme is commonly called a "dehydrogenase," metabolically the reaction normally proceeds in the reductive direction as shown in Eq. 19.50. [The enzyme is "officially" named (*S*)-lactate:pyruvate oxidoreductase; it catalyzes the reaction in both directions, as it must by the principle of microscopic reversibility.]

In the reduction of pyruvate to lactate, the ketone carbonyl group is activated by hydrogen bonding from a protonated histidine residue of the enzyme. In this respect, the reaction resembles sodium borohydride reduction, in which the carbonyl group is activated by hydrogen-bond donation from a protic solvent (Eq. 19.46, Sec. 19.9A).

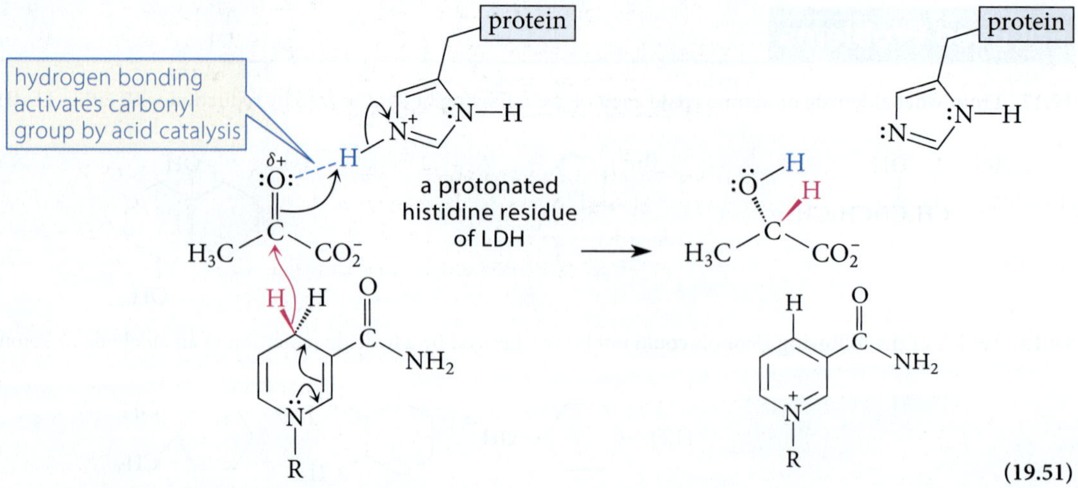

(19.51)

This basic theme—delivery of a hydride from NADH (or NADPH) to a carbonyl-activated substrate in an enzyme active site—is played out in all of the carbonyl reductions in which NADH or NADPH is involved.

It is worth noting that, in other reductions involving NADH, the pro-(*S*) hydride of NADH is delivered instead of the pro-(*R*) hydride. Although these hydrides are diastereotopic, their intrinsic reactivity is virtually the same. The stereochemical differences among the reactions catalyzed by the various enzymes has to do with the way the NADH or NADPH binds in the enzyme active site relative to the substrate.

Focused Problem

19.19 Yeast alcohol dehydrogenase catalyzes the reduction of acetaldehyde by NADH to ethanol in the last step of anaerobic fermentation. (See the sidebar, "Fermentation," in Sec. 11.8.)

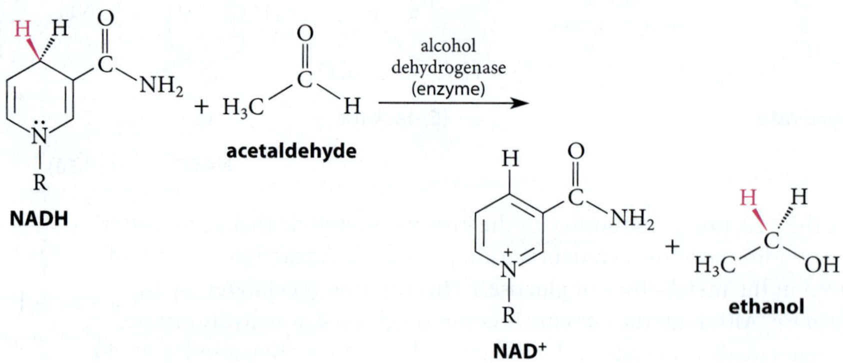

(a) The structure of this enzyme contains a Zn^{2+} ion (held in place by coordination with several amino acid side chains of the enzyme) that is required for catalysis near the carbonyl of acetaldehyde. Propose a role for this metal ion in catalysis. (*Hint:* See Eq. 19.45a, Sec. 19.9A.)

(b) Using your answer to part (a), draw a curved-arrow mechanism for this reduction in which the role of the Zn^{2+} ion is explicitly shown.

C. Reduction by Catalytic Hydrogenation

Aldehydes and ketones can also be reduced to alcohols by catalytic hydrogenation. This reaction is analogous to the catalytic hydrogenation of alkenes (Sec. 4.10A).

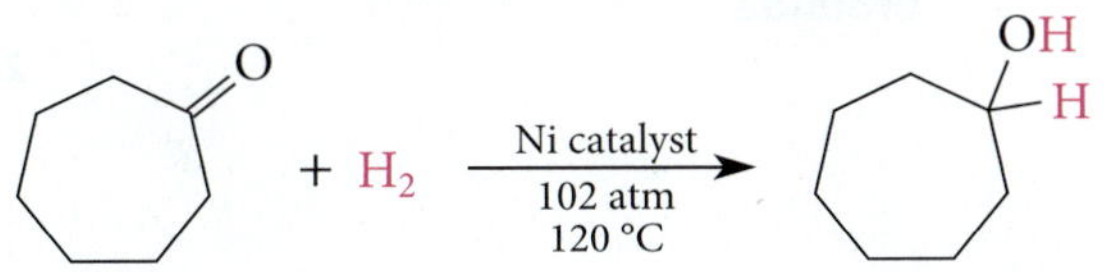

cycloheptanone **cycloheptanol** (92% yield) (19.52)

Catalytic hydrogenation of carbonyl groups requires typically higher pressure and temperature than hydrogenation of carbon–carbon π bonds because of the greater bond strength of the carbon–oxygen π bond. For this reason, it is usually possible to use catalytic hydrogenation for the selective reduction of an alkene double bond in the presence of a carbonyl group. Palladium catalysts are particularly effective for this purpose.

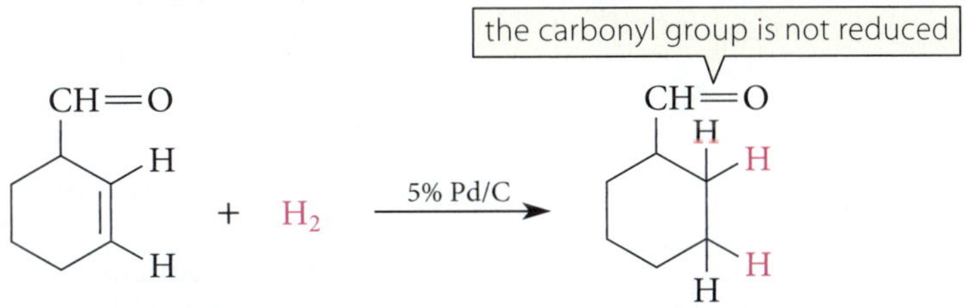

2-cyclohexenecarbaldehyde **cyclohexanecarbaldehyde** (81% yield) (19.53)

Focused Problem

19.20 Give three different starting materials ($C_9H_{14}O$) that might be used to obtain the following product by selective catalytic hydrogenation. (*Hints:* The alkene must bind to the catalyst surface; remember the stereospecificity of the reaction.)

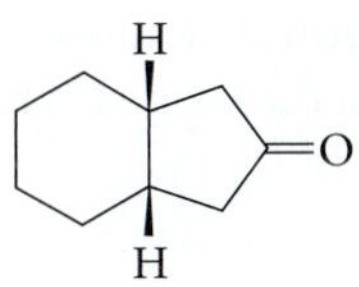

19.10 REACTIONS OF ALDEHYDES AND KETONES WITH GRIGNARD AND RELATED REAGENTS

The formation and electronic characteristics of Grignard and organolithium reagents were discussed in Sec. 10.5. The reaction of these reagents with carbonyl groups is their most important application in organic chemistry. Addition of Grignard reagents to aldehydes and ketones in an ether solvent, followed by protonolysis, gives alcohols.

$$\underset{\textbf{2-methylpropanal}}{(CH_3)_2CH\overset{O}{\overset{\|}{C}}H} + \underset{\textbf{ethylmagnesium bromide}}{BrMg-CH_2CH_3} \xrightarrow{\text{ether}} \xrightarrow{H_3O^+} \underset{\textbf{2-methyl-3-pentanol (68\% yield)}}{(CH_3)_2CH\overset{OH}{\overset{|}{C}}HCH_2CH_3} \quad (19.54)$$

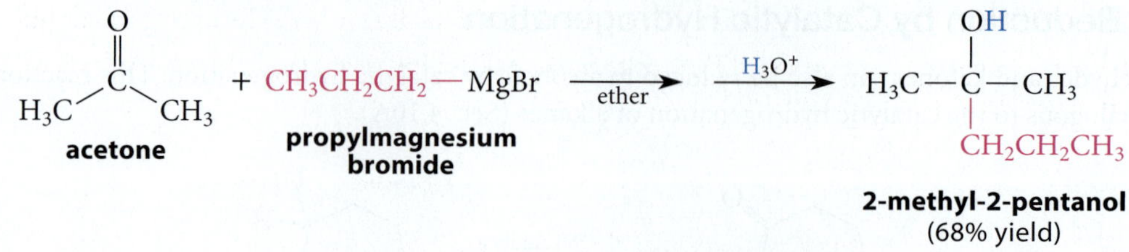

(19.55)

STUDY GUIDE LINK 19.1
Lewis Acid Catalysis

The reaction of Grignard reagents with aldehydes and ketones is another example of *carbonyl addition.* In this reaction, the magnesium of the Grignard reagent, a Lewis acid, bonds to the carbonyl oxygen. This bonding, much like protonation in acid-catalyzed hydration, makes the carbonyl carbon more electrophilic (that is, makes it more reactive toward nucleophiles) by making the carbonyl oxygen a better acceptor of electrons. The carbon group of the Grignard reagent reacts as a nucleophile at the carbonyl carbon. Recall that this group is a strong base that behaves much like a *carbanion* (Display 10.52, Sec. 10.5C).

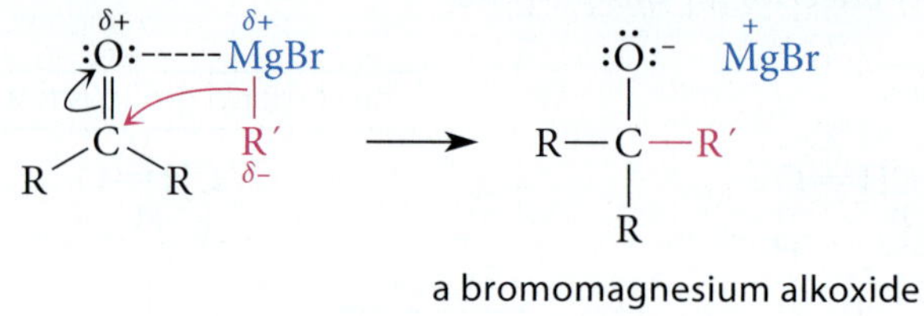

(19.56a)

The product of this addition, a bromomagnesium alkoxide, is essentially the magnesium salt of an alcohol. Addition of dilute acid to the reaction mixture gives an alcohol.

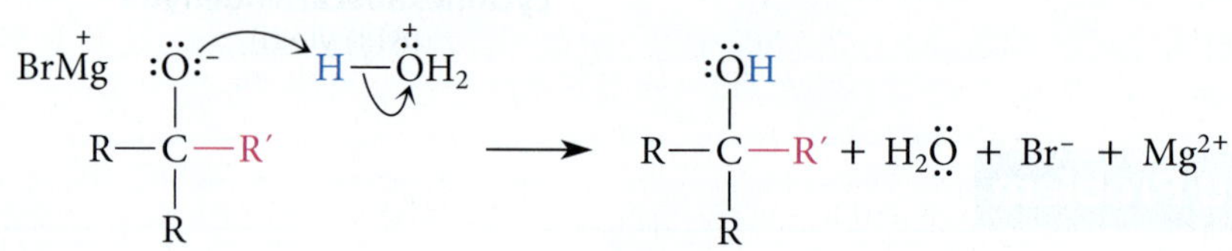

(19.56b)

This addition, like hydride reductions, is irreversible, and it works with just about any aldehyde or ketone.

The reactions of organolithium and sodium acetylide reagents with aldehydes and ketones are fundamentally similar to the Grignard reaction.

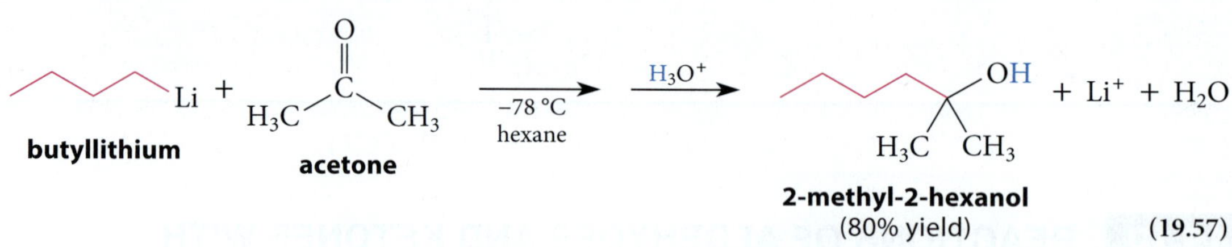

(19.57)

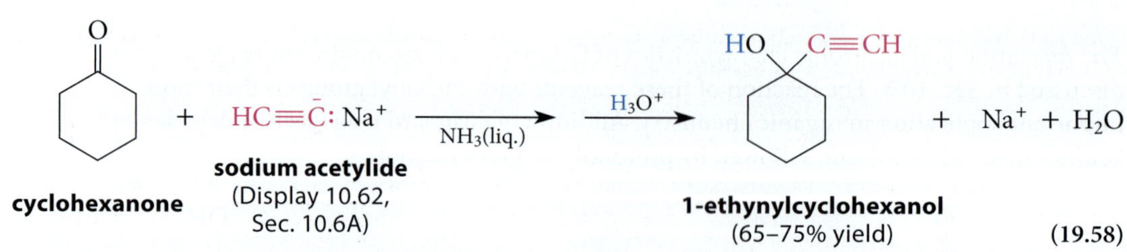

(19.58)

The reaction of Grignard and related reagents with aldehydes and ketones is important not only because it can be used to convert aldehydes or ketones into alcohols, but also because it is an excellent method of *carbon–carbon bond formation.*

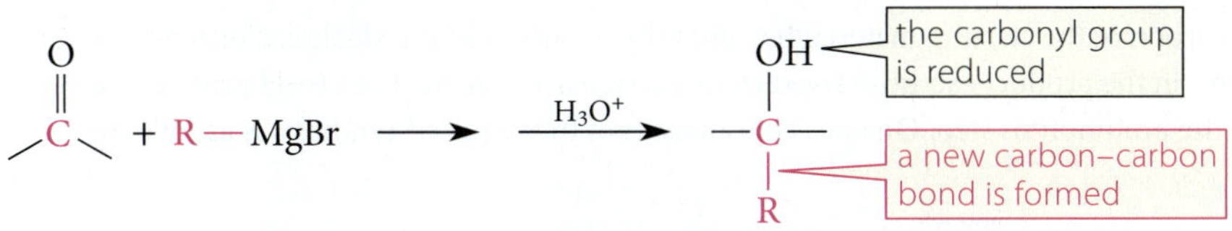

(19.59)

A complete list of reactions that form carbon–carbon bonds is given in Appendix VI.

The possibilities for alcohol synthesis with the Grignard reaction are almost endless. Primary alcohols are synthesized by the addition of a Grignard reagent to formaldehyde.

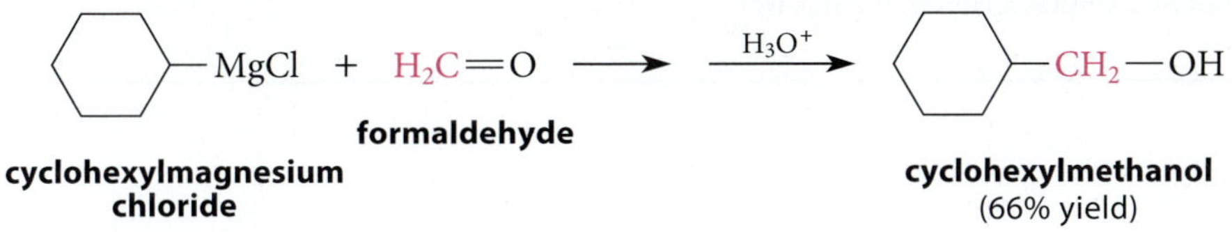

(19.60)

Because Grignard reagents are made from alkyl halides, which in many cases can be synthesized from alcohols, this reaction can be incorporated as a key element in a one-carbon chain extension of an alcohol:

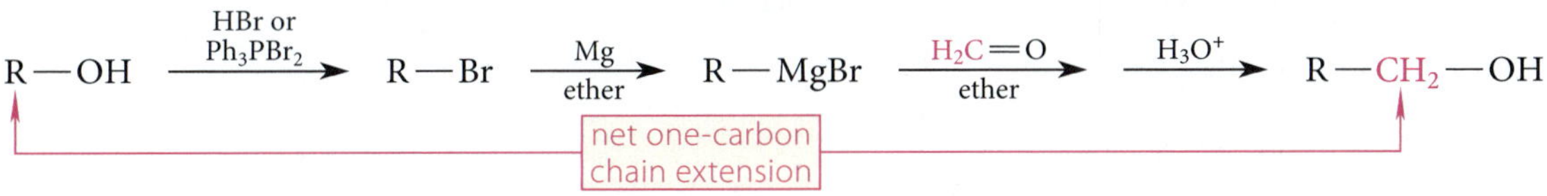

(19.61)

Addition of a Grignard reagent to an aldehyde other than formaldehyde gives a secondary alcohol (Eq. 19.54), and addition to a ketone gives a tertiary alcohol (Eq. 19.55). The Grignard synthesis of a tertiary alcohol and, in some cases, a secondary alcohol can also be extended to an alkene synthesis by dehydration of the alcohol with strong acid during the protonolysis step (Sec. 11.2).

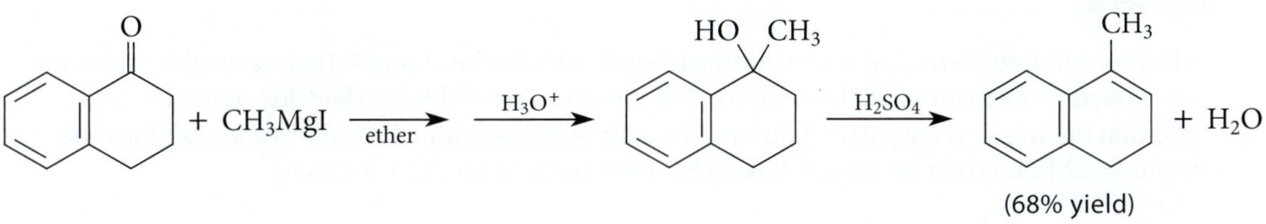

(19.62)

When you are asked to prepare an alcohol, you can determine whether it can be synthesized by the reaction of a Grignard reagent with an aldehyde or ketone if you understand that the *net effect* of the Grignard reaction, followed by protonolysis, is addition of R—H (R = an alkyl or aryl group) across the C=O double bond:

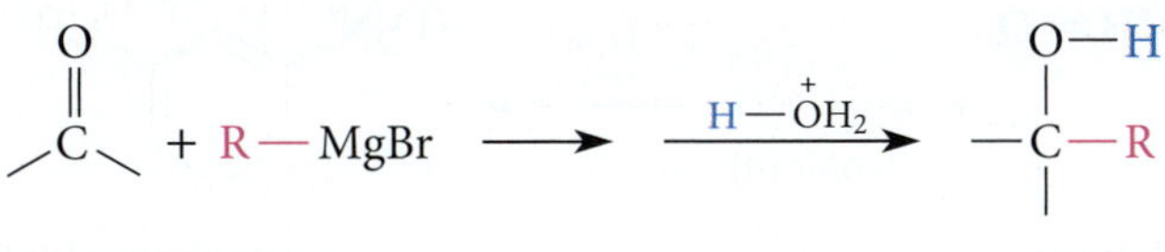

(19.63)

Once you grasp this relationship, you can determine the starting materials for a particular synthesis by mentally subtracting R and H from the target alcohol. This approach is illustrated in Study Problem 19.3.

Study Problem 19.3

Propose a synthesis of 2-butanol by the reaction of a Grignard reagent with an aldehyde or ketone.

Solution The carbonyl carbon of the starting material becomes the α-carbon of the alcohol. Consequently, any alkyl group bound to this carbon in the product can be derived from a Grignard reagent. The O—H proton is derived from the water or acid used in the protonolysis step. One possible analysis of the required synthesis is as follows:

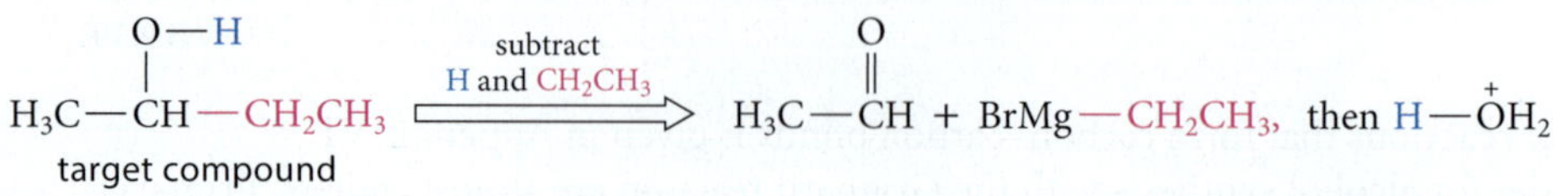

(19.64)

Remember that the open arrow ⇨ means, "Implies as starting materials." Another possibility for a Grignard synthesis of 2-butanol can be found by a similar analysis. What is it?

Focused Problems

19.21 Show how bromoethane can be used as a starting material in the preparation of each of the following compounds. (*Hint:* How are Grignard reagents prepared? See Sec. 10.5B.)

(a) 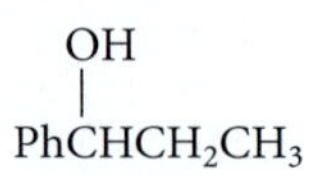

(b) 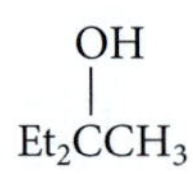

(c) 1-Butanol

(d) 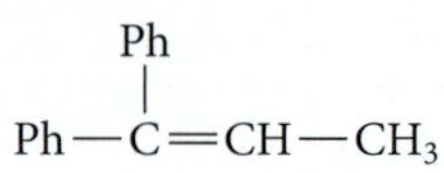

(e) 

19.22 Outline three different Grignard syntheses for 3-methyl-3-hexanol.

19.11 ACETALS AND THEIR USE AS PROTECTING GROUPS

The preceding sections deal with simple carbonyl-addition reactions—first, reversible additions (cyanohydrin formation and hydration); then, irreversible additions (hydride reduction and addition of Grignard reagents). This and the subsequent sections consider some reactions that begin as additions that are then followed by other types of mechanistic steps.

A. Preparation and Hydrolysis of Acetals

Acetals derived from ketones were once called *ketals*, but this name is no longer used.

When an aldehyde or ketone reacts with a large excess of an alcohol in the presence of a trace of strong acid, an *acetal* is formed.

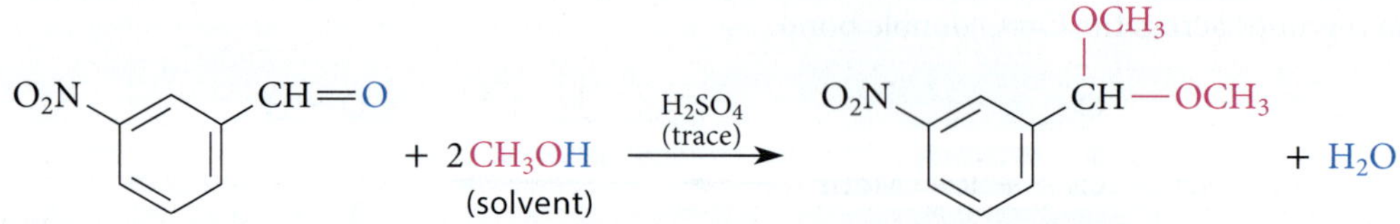

m-nitrobenzaldehyde

m-nitrobenzaldehyde dimethyl acetal (76–85% yield)

(19.65)

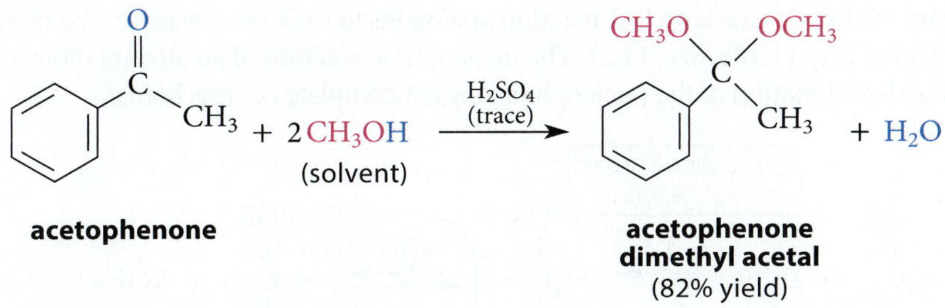

(19.66)

An **acetal** is a compound in which two ether oxygens are bound to the same carbon. In other words, acetals are the ethers of carbonyl hydrates, or geminal diols (Sec. 19.8A).

Two equivalents of alcohol are consumed in each of the preceding reactions. Because 1,2- and 1,3-diols contain two OH groups within the same molecule, one equivalent of a 1,2- or 1,3-diol can react to form a *cyclic acetal*, in which the acetal group is part of a five- or six-membered ring, respectively.

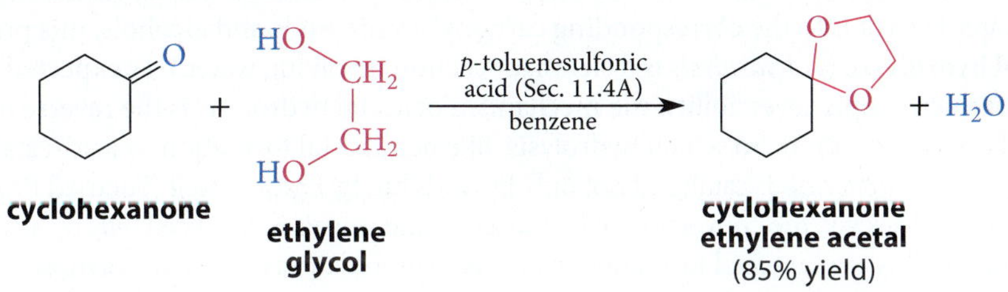

(19.67)

The formation of acetals is readily reversible. The reaction is driven to the right by the use of excess alcohol as the solvent, by removal of the water by-product, or both. This strategy is another application of Le Chatelier's principle. In Eq. 19.67, for example, the water can be removed by distillation as an *azeotrope* with benzene. (The benzene–water azeotrope is a mixture of benzene and water that has a lower boiling point than either benzene or water alone.)

The first step in the mechanism of acetal formation is acid-catalyzed *addition* of the alcohol to the carbonyl group to give a **hemiacetal**—a compound with OR and OH groups on the same carbon (*hemi* = half; *hemiacetal* = half acetal).

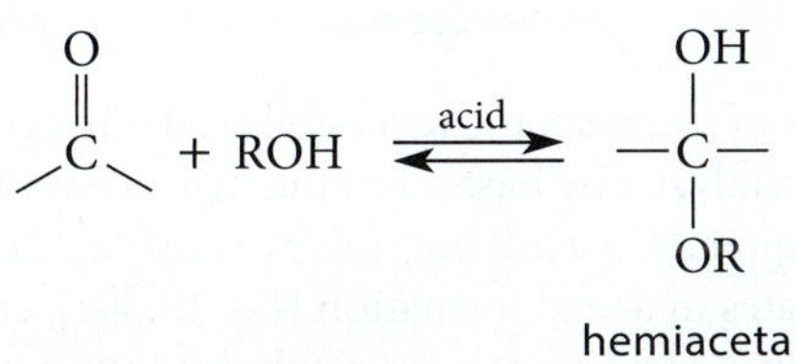

(19.68a)

The mechanism of hemiacetal formation is analogous to that of acid-catalyzed hydration. (Write the stepwise mechanism of this reaction; see Focused Problem 19.14a, Sec. 19.8A.)

The hemiacetal reacts further when the OH group is protonated and water is lost to give a relatively stable carbocation, an α-alkoxy carbocation (Sec. 19.7).

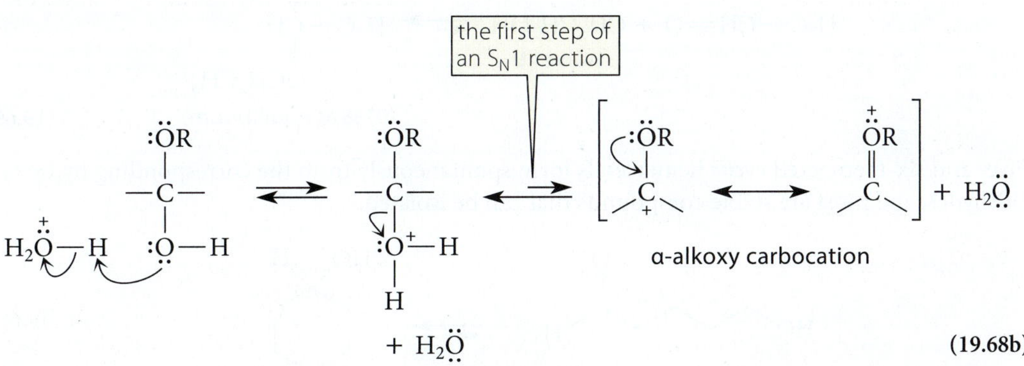

(19.68b)

STUDY GUIDE LINK 19.2
Hemiacetal Protonation

Loss of water from the hemiacetal is an S_N1 reaction analogous to the loss of water in the dehydration of an ordinary alcohol (Eq. 11.18b, Sec. 11.2). The nucleophilic reaction of an alcohol molecule with the cation and the deprotonation of the nucleophilic oxygen complete the mechanism.

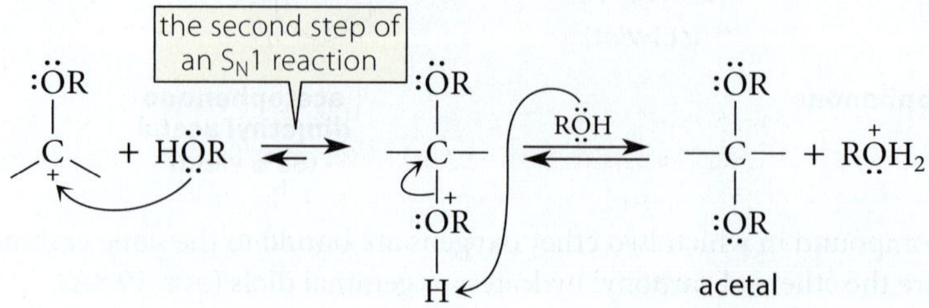

(19.68c)

As we have just shown, the mechanism for acetal formation is really a combination of other familiar mechanisms. It involves an *acid-catalyzed carbonyl addition* followed by a *substitution* that occurs by the S_N1 solvolysis mechanism.

Because the formation of acetals is reversible, acetals in the presence of acid and excess water are transformed rapidly back into the corresponding carbonyl compounds and alcohols; this process is called **acetal hydrolysis**. (A *hydrolysis* is a cleavage reaction involving water.) As expected from the principle of microscopic reversibility, the mechanism of acetal hydrolysis is the reverse of the mechanism of acetal formation. So, acetal hydrolysis, like hemiacetal formation, is acid-catalyzed.

The formation of *hemiacetals* is catalyzed not only by acids but by bases as well (Focused Problem 19.14b, Sec. 19.8A). However, the conversion of hemiacetals into acetals is catalyzed *only* by acids (Eqs. 19.68b and c). This is why acetal formation, which is a combination of the two reactions, is catalyzed by acids but not by bases.

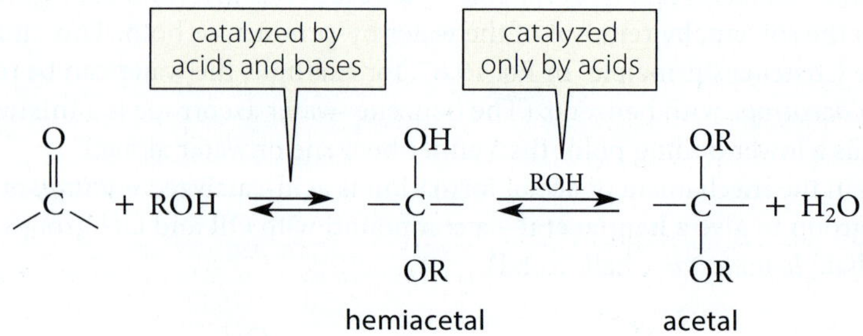

(19.68d)

As expected from the principle of microscopic reversibility, the hydrolysis of hemiacetals to aldehydes and ketones is also catalyzed by bases, but the hydrolysis of acetals to hemiacetals is catalyzed *only* by acids. Consequently, *acetals are stable in basic and neutral solution.*

Hemiacetals, the intermediates in acetal formation (Eq. 19.68a), cannot be isolated in most cases because they react further to yield acetals (in alcohol solution under acidic conditions) or decompose to aldehydes or ketones and an alcohol. Simple aldehydes, however, form appreciable amounts of hemiacetals in alcohol solution, just as they form appreciable amounts of hydrates in water (see Table 19.2, Sec. 19.8B).

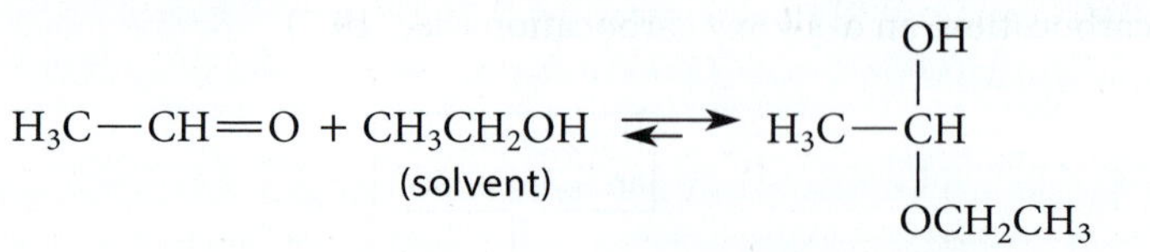

(19.69)

Five- and six-membered *cyclic* hemiacetals form spontaneously from the corresponding hydroxy aldehydes, and most are stable compounds that can be isolated.

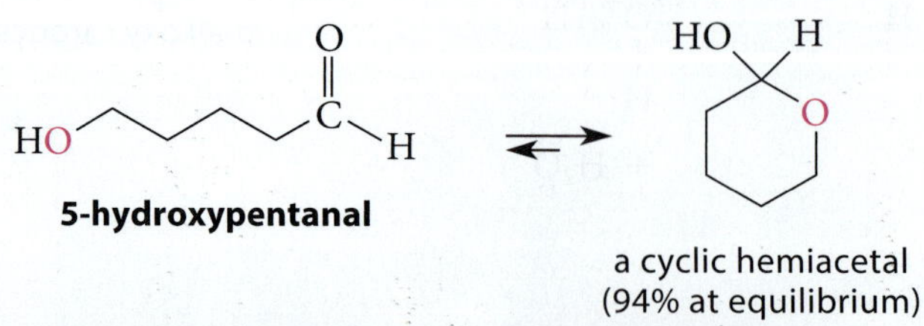

(19.70)

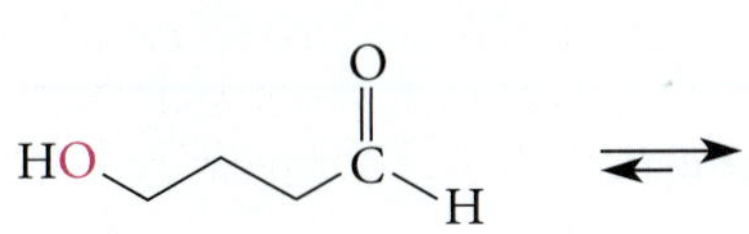

4-hydroxybutanal (89% at equilibrium) (19.71)

We showed in Sec. 12.8 that intramolecular reactions that give six-membered or five-membered rings are faster than the corresponding intermolecular reactions. Such intramolecular reactions are also more favored thermodynamically—that is, they have larger equilibrium constants, because an intramolecular OH group simply has a greater probability of reaction (that is, a greater $\Delta S°$) than an OH group in a different molecule.

The five- and six-carbon sugars are important biological examples of cyclic hemiacetals.

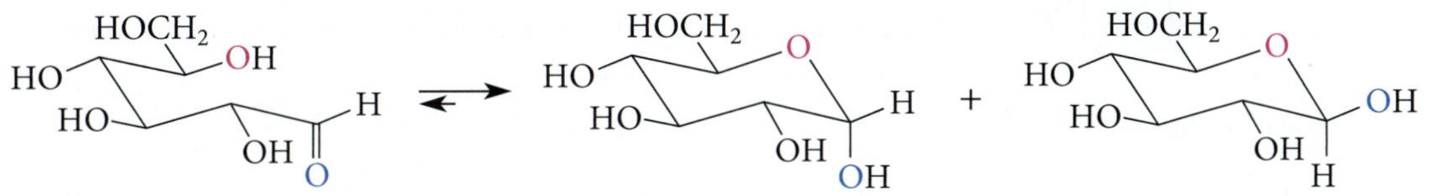

(+)-glucose **α-(+)-glucopyranose** **β-(+)-glucopyranose**

cyclic forms of glucose

(19.72)

(This reaction and its stereochemistry are discussed in Sec. 24.3B.)

Storage of Aldehydes as Acetals

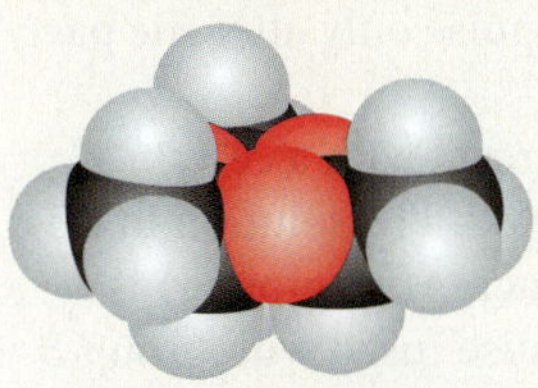

Some aldehydes are stored as acetals. Acetaldehyde, when treated with a trace of acid, readily forms a cyclic acetal called *paraldehyde*. Each molecule of paraldehyde is formed from three molecules of acetaldehyde. (An alcohol is not involved in formation of paraldehyde.) Paraldehyde, with a boiling point of 125 °C, is a particularly convenient way to store acetaldehyde, which itself boils near room temperature. Upon heating with a trace of acid, acetaldehyde can be distilled from a sample of paraldehyde. (See Problem 19.72e.)

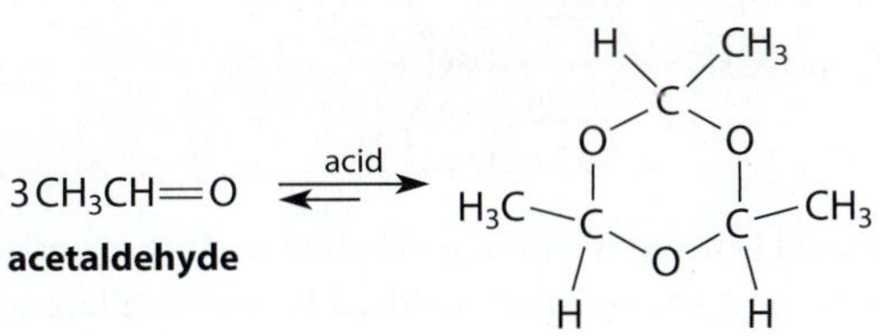

paraldehyde (19.73)

Formaldehyde can be stored as the acetal polymer *paraformaldehyde*, which precipitates from concentrated formaldehyde solutions.

$$HO\!\left(CH_2-O\right)_n\!H$$

paraformaldehyde

(An alcohol is not involved in paraformaldehyde formation either.) Because it is a solid, paraformaldehyde is a useful form in which to store formaldehyde, itself a gas. Formaldehyde is liberated from paraformaldehyde by heating.

Focused Problems

19.23 Write the structure of the product formed in each of the following reactions.

(a)

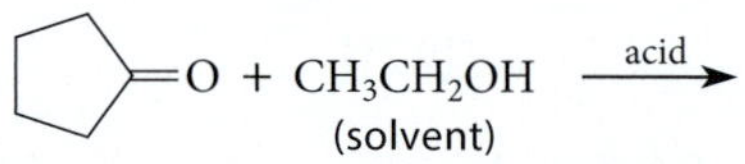

(b)

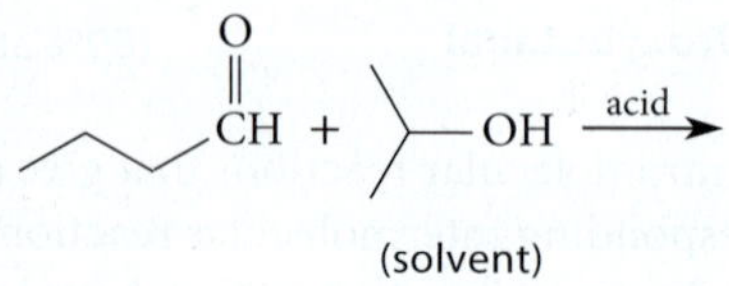

19.24 Propose syntheses of each of the following acetals from carbonyl compounds and alcohols.

(a)

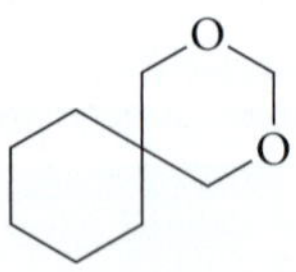

(b)

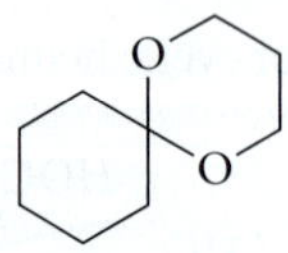

19.25 Suggest a structure for the acetal product of each reaction.

(a)

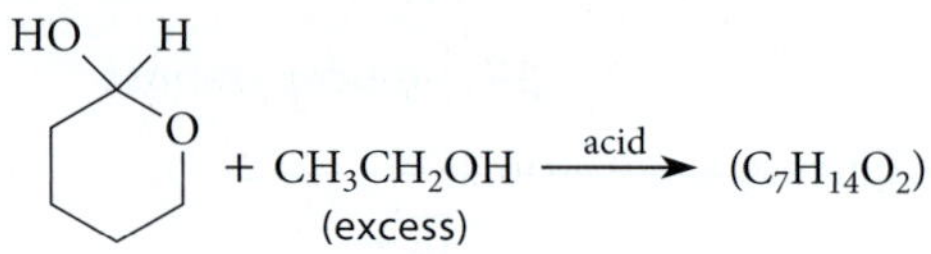

(b)

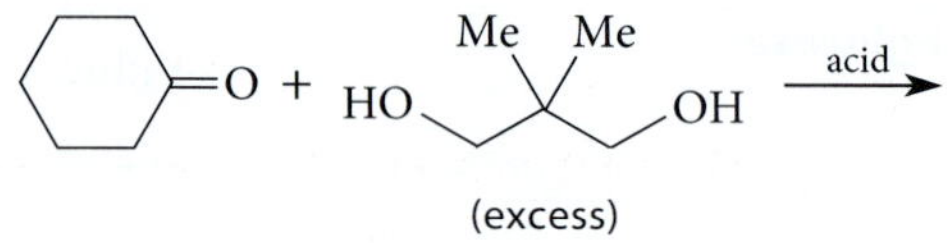

B. Protecting Groups

The following analogy illustrates the use of *protecting groups* in organic synthesis. Suppose you and a friend haven't been invited to a party but are determined to attend it anyway. To avoid recognition and confrontation, you wear a disguise, which might be a wig, a false mustache, or even more drastic disguises. Your friend doesn't bother with such deception. The host recognizes your friend and insists that he leave the party, but, because you are not recognized, you avoid such a confrontation and can remain to enjoy the evening, removing your disguise only after the party is over.

Now, suppose two groups in a molecule, *A* and *B*, are both known to react with a certain reagent, but we want to let only group *A* react and leave group *B* unaffected. The solution to this problem is to *disguise*, or *protect*, group *B* in such a way that it cannot react. After group *A* is allowed to react, the disguise of group *B* is removed. The "chemical disguise" used with group *B* is called a **protecting group** or **protective group**. Acetals are among the most commonly used protecting groups for aldehydes and ketones. Study Problem 19.4 illustrates the use of an acetal as a protecting group.

Study Problem 19.4

Propose a sequence of reactions for carrying out the following conversion.

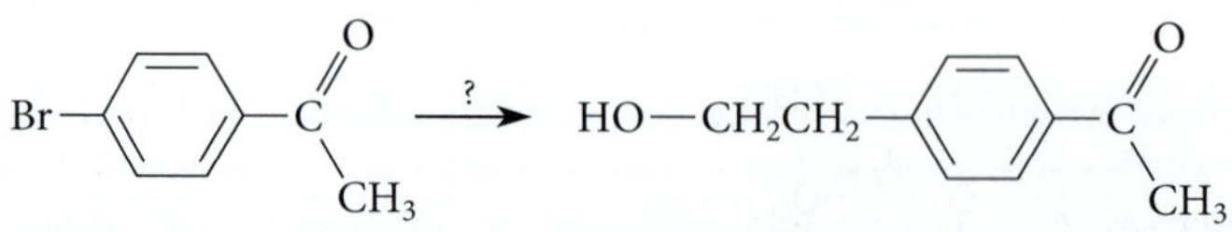

Solution It might seem that the way to effect this conversion would be to convert the aryl halide starting material into the corresponding Grignard reagent, and then allow this reagent to react with ethylene oxide, followed by dilute aqueous acid (Sec. 12.5C). However, Grignard reagents also react with ketones (Sec. 19.10); so, the Grignard reagent derived from one molecule of the starting material would react with the carbonyl group of another molecule. The ketone group, then, would not survive this reaction. However, the ketone can be *protected* as an acetal, which does *not* react with Grignard reagents. (An acetal is a type of ether, and ethers are unaffected by Grignard reagents.) The following synthesis incorporates this strategy.

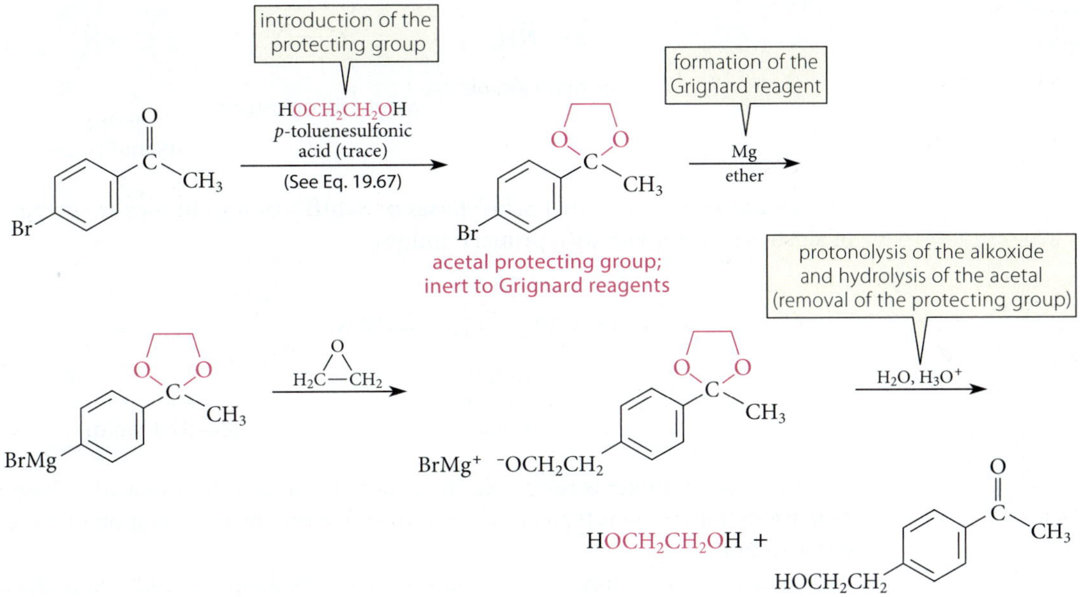

Notice in this synthesis that all steps following acetal formation involve basic or neutral conditions. Acid can be used only when removal of the acetal protecting group is desired.

Although any acetal group can in principle be used, the five-membered cyclic acetal is frequently employed as a protecting group because it forms rapidly (proximity effect; Sec. 12.8) and it introduces relatively little steric congestion into the protected molecule.

A number of reagents that react with carbonyl groups also react with other functional groups. Acetals are commonly used to protect the carbonyl groups of aldehydes and ketones from basic, nucleophilic reagents. Once the protection is no longer needed, the acetal protecting group is easily removed and the carbonyl group is re-exposed by treatment with dilute aqueous acid. Because acetals are unstable in acid, they do *not* protect carbonyl groups under acidic conditions.

Focused Problem

19.26 Outline a synthesis of the following compound from *p*-bromoacetophenone and any other reagents.

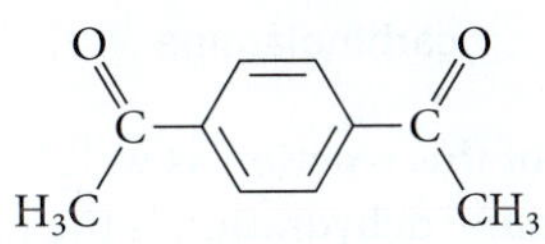

19.12 REACTIONS OF ALDEHYDES AND KETONES WITH AMINES

A. Reaction with Primary Amines and Other Monosubstituted Derivatives of Ammonia

A **primary amine** is an organic derivative of ammonia in which one ammonia hydrogen is replaced by an alkyl or aryl group. An **imine** is a nitrogen analog of an aldehyde or ketone in which the C=O group is replaced by a C=NR group, where R = alkyl, aryl, or H.

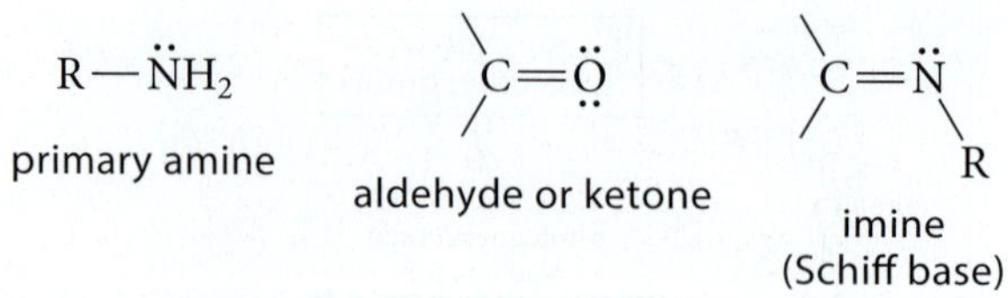

(19.74)

(Imines are sometimes called **Schiff bases** or **Schiff's bases.**) Imines are prepared by the reaction of aldehydes or ketones with primary amines.

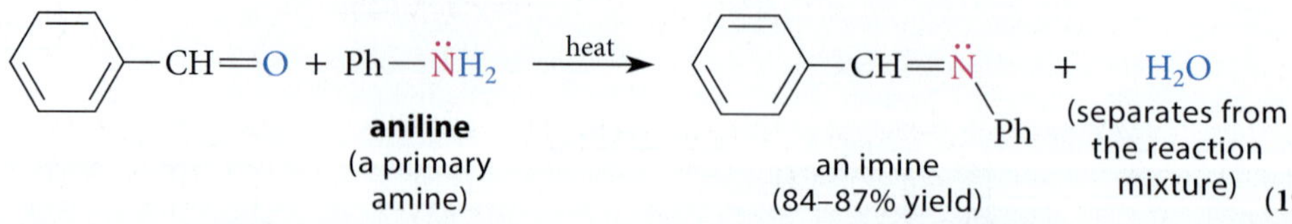

(19.75)

Formation of imines is reversible and generally takes place with acid or base catalysis or with heat. Imine formation is typically driven to completion by precipitation of the imine, removal of water, or both.

The mechanism of imine formation begins as a nucleophilic addition to the carbonyl group. In this case, the nucleophile is the amine, which reacts with the aldehyde or ketone to give an unstable addition product called a *carbinolamine*. A **carbinolamine** is a compound with an amine group (—NH_2,—NHR, or —NR_2) and a hydroxy group on the same carbon.

STUDY GUIDE LINK 19.3
Mechanism of Carbinolamine Formation

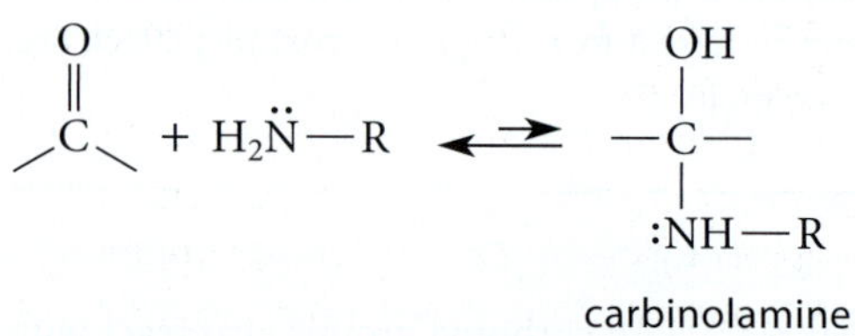

(19.76a)

(You should write the detailed curved-arrow mechanism, which is analogous to the mechanism of other reversible additions.) Carbinolamines are not isolated but undergo acid-catalyzed dehydration to form imines. This reaction is essentially an alcohol dehydration (Sec. 11.2), except that it is typically much faster than dehydration of an ordinary alcohol.

STUDY GUIDE LINK 19.4
Dehydration of Carbinolamines

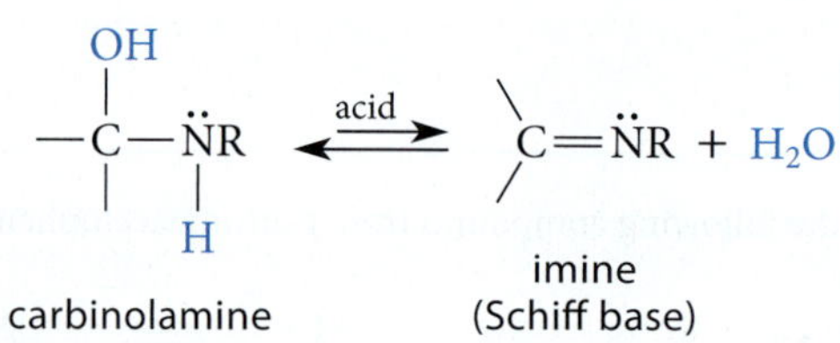

(19.76b)

(Write the mechanism of this reaction as well.)

Typically, carbinolamine dehydration is the rate-limiting step of imine formation. This is why acids catalyze imine formation. Yet the acid concentration cannot be too high because amines are basic compounds and because protonated amines cannot act as nucleophiles.

$$R\ddot{N}H_2 + H_3\ddot{O}^+ \rightleftharpoons R\overset{+}{N}H_3 + H_2\ddot{O}: \qquad (19.76c)$$

Protonation of the amine pulls the equilibrium in Eq. 19.76a to the left; consequently, if the acid concentration is high enough, carbinolamine formation cannot occur. For this reason, many imine syntheses are carried out in very dilute acid. (Stronger acid can be used if the amine is not very basic.)

To summarize: Imine formation is a sequence of two reactions that have close analogies to familiar reactions—namely, *carbonyl addition* followed by *β-elimination*.

How are imines used? One important use of imines is in the preparation of amines. This process, called *reductive amination*, is discussed in Sec. 23.7B. Another use was more important

TABLE 19.3 Some *N*-Substituted Imine Derivatives of Aldehydes and Ketones

Amine	Name	Carbonyl Derivative	Name
$H_2\ddot{N}-\ddot{O}H$	hydroxylamine	$R_2C=\ddot{N}-\ddot{O}H$	oxime
$H_2\ddot{N}-\ddot{N}H_2$	hydrazine	$R_2C=\ddot{N}-\ddot{N}H_2$	hydrazone
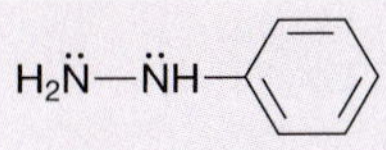	phenylhydrazine	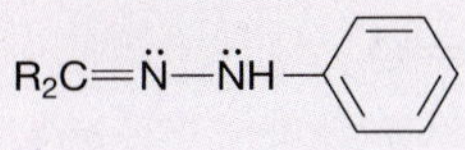	phenylhydrazone
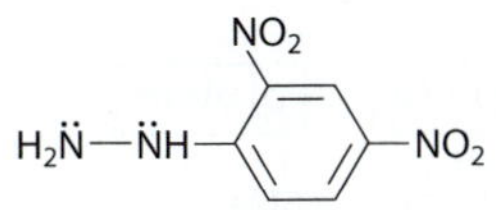	2,4-dinitrophenylhydrazine (2,4-DNP)	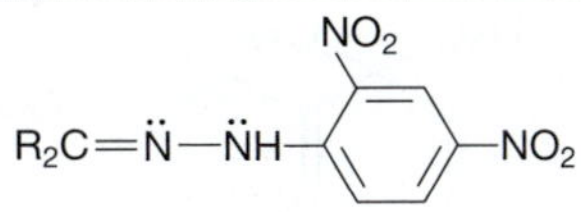	2,4-dinitrophenylhydrazone (2,4-DNP derivative)
$H_2\ddot{N}-\ddot{N}H-C(=\ddot{O})-\ddot{N}H_2$	semicarbazide	$R_2C=\ddot{N}-\ddot{N}H-C(=\ddot{O})-\ddot{N}H_2$	semicarbazone

before the advent of spectroscopy. Certain *N*-substituted imines were used to characterize or identify the aldehydes or ketones from which they were derived. (This use of imines is discussed in Further Exploration 19.2.) Some common *N*-substituted imines that were used for this purpose appear often in older chemical literature. These imines, and the *N*-substituted amines used to prepare them from carbonyl compounds, are listed in **Table 19.3**.

FURTHER EXPLORATION 19.2
Use of Imines for Characterization of Aldehydes and Ketones

Some *N*-substituted imines are important in their own right. For example, we'll encounter hydrazones as intermediates in Sec. 19.13. Cyclohexanone oxime is an intermediate in the industrial synthesis of one type of nylon. All *N*-substituted imines, like other imines, are prepared from the parent carbonyl compound and the appropriate *N*-substituted amine. For example, cyclohexanone oxime is prepared from the reaction of cyclohexanone and hydroxylamine.

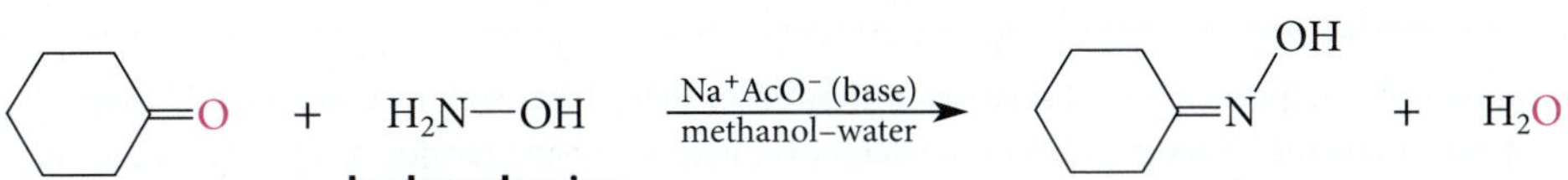

cyclohexanone **hydroxylamine** **cyclohexanone oxime** (92% yield) (19.77)

Focused Problems

19.27 Draw the structure of each of the following:

(a) The oxime of acetone

(b) The imine formed in the reaction between 2-methylhexanal and ethylamine ($EtNH_2$)

19.28 Write a curved-arrow mechanism for the reaction in Eq. 19.77. (This is a base-catalyzed process; the base is sodium acetate, $Na^+ AcO^-$.)

19.29 Write a curved-arrow mechanism for the acid-catalyzed hydrolysis of the imine derived from benzaldehyde and ethylamine ($EtNH_2$). Use the principle of microscopic reversibility (Sec. 4.10C) to guide you.

B. Imines in Biology

Chemical Biology Topic

Imines are well known in biology. Typically, imines are employed as a tether when an aldehyde group on a small molecule is connected to an enzyme or other protein. An amino group in the side chain of a lysine residue of the protein reacts with the aldehyde to give the imine linkage. An example is the connection between retinal, the visual pigment, and its protein, opsin, to give rhodopsin (see Display 15.12, Sec. 15.2C).

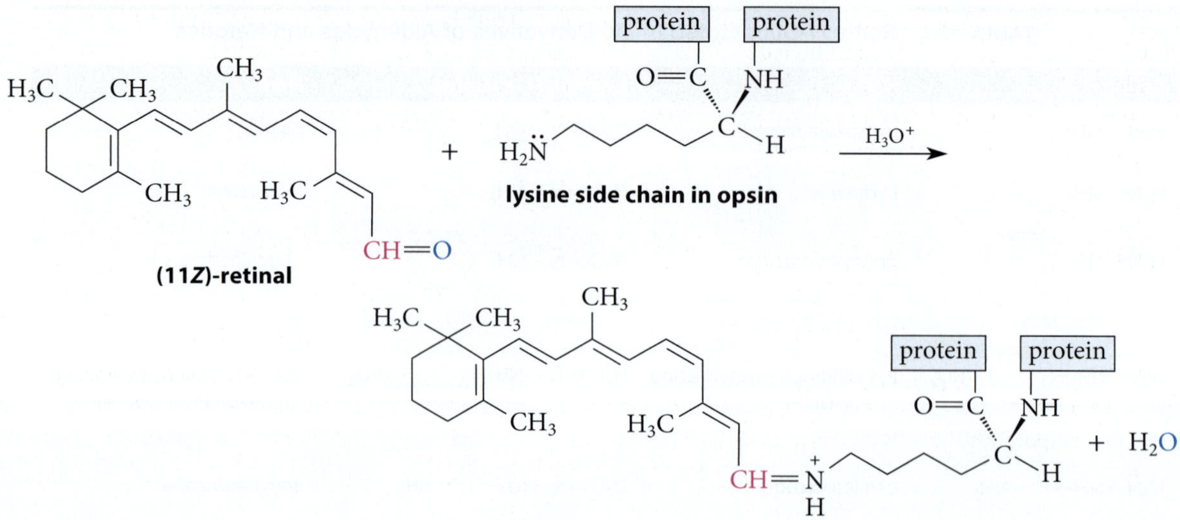

rhodopsin (an imine; shown in its conjugate-acid form)
(the light receptor in vision) (19.78)

Because the bond between the small molecule and the protein is covalent, it is particularly strong. However, because imine formation is reversible, this linkage can be hydrolyzed back to the aldehyde and the free lysine amino group when required. For example, after the imine of (11*Z*)-retinal is converted into the imine of (11*E*)-retinal, the retinal must be removed from the protein by imine hydrolysis. The direction in which the reaction runs depends on the conformation of the protein and, undoubtedly, the local availability of water in the vicinity of the imine linkage.

Imines are particularly important in the biological reactions of pyridoxal phosphate, a form of vitamin B_6; we discuss these in Sec. 26.4E. (See Focused Problem 19.30.)

Focused Problem

19.30 A form of vitamin B_6 (pyridoxal phosphate) forms imine derivatives with most proteins that catalyze its reactions. Given the structure of pyridoxal phosphate, draw the structure of its imine derivative with the lysine residue of a protein. (Use an abbreviation for the lysine residue like the one shown in Eq. 19.78.)

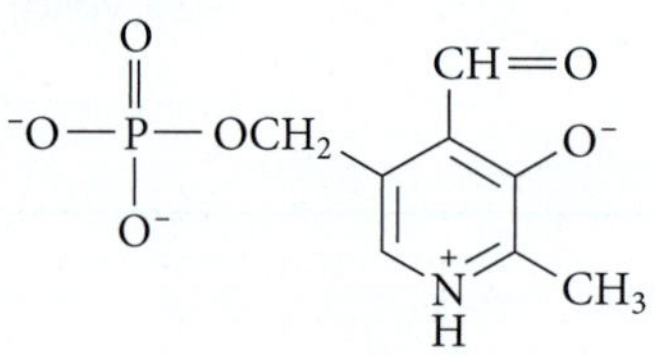

pyridoxal phosphate
(a form of vitamin B_6)

C. Reaction with Secondary Amines

A **secondary amine** has the general structure $R_2\ddot{N}H$, in which alkyl or aryl groups replace two ammonia hydrogens. An **enamine** (pronounced ēn′-ə-mēn′) has the following general structure:

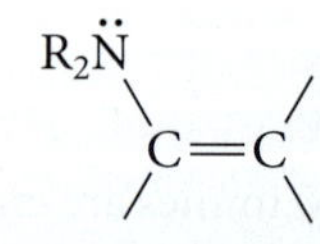

general enamine structure (19.79)

The name *enamine* is a contraction of the word *amine* (a compound of the form R_3N:) and the suffix *ene*, which is used for naming alkenes. The name recognizes that an *amine* nitrogen is bonded to a carbon that is part of a double bond (that is, an alk*ene*).

Formation of an enamine occurs when a secondary amine reacts with an aldehyde or ketone, provided that the carbonyl compound has an α-hydrogen.

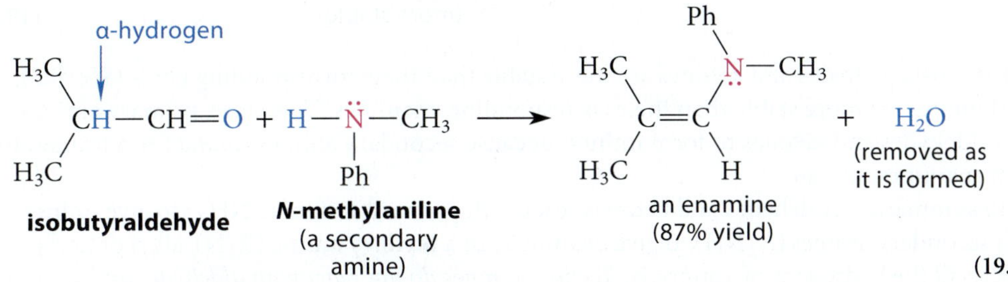

(19.80)

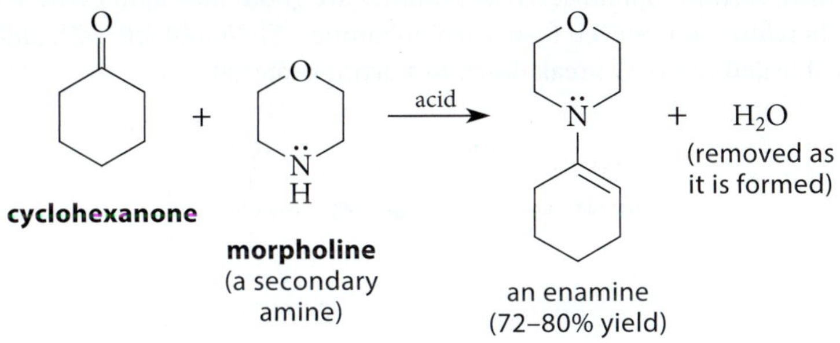

(19.81)

As Eq. 19.81 illustrates, the two alkyl groups of a secondary amine may be two carbons of a ring.

Like imine formation, enamine formation is reversible and must be driven to completion by the removal of one of the reaction products (usually water, as in Eq. 19.81). Enamines, like imines, revert to the corresponding carbonyl compounds and amines in aqueous acid.

The mechanism of enamine formation begins, like the mechanism of imine formation, as a nucleophilic addition to give a carbinolamine intermediate. (Write the mechanism of this reaction.)

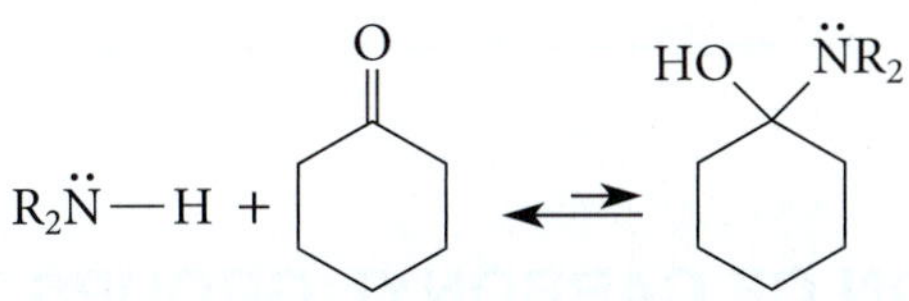

(19.82a)

Because no hydrogen remains on the nitrogen of this carbinolamine, imine formation cannot occur. Instead, dehydration of the carbinolamine involves loss of a hydrogen from an adjacent *carbon*.

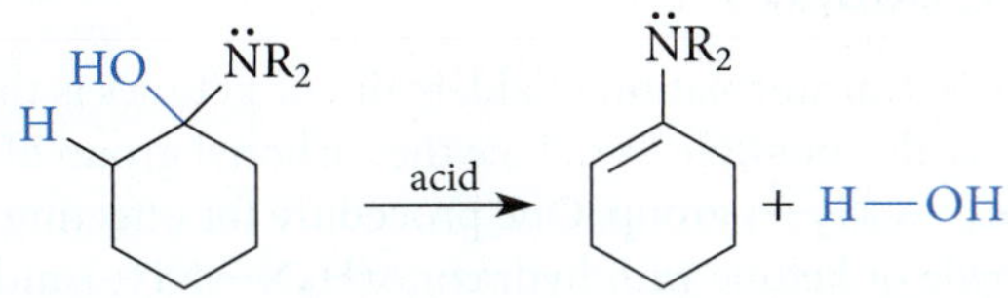

(19.82b)

Why don't primary amines react with aldehydes or ketones to form enamines rather than imines? The answer is that the enamines bear the same relationship to imines that *enols* bear to ketones.

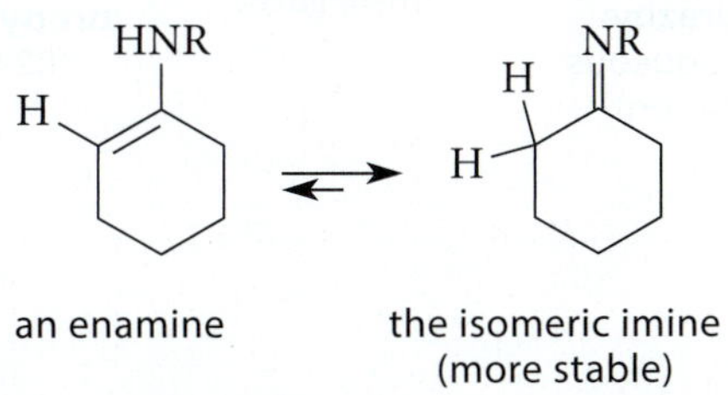

(19.83a)

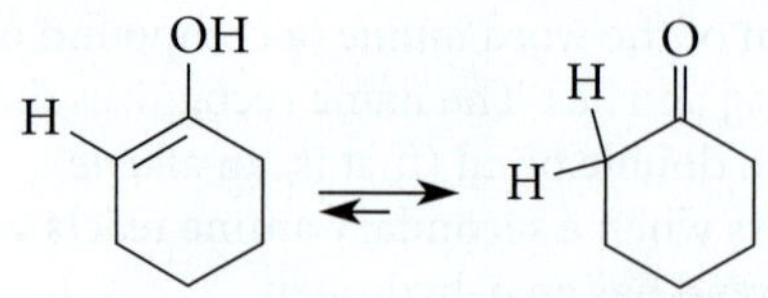

an enol the isomeric ketone (more stable) (19.83b)

Just as most aldehydes and ketones are more stable than their corresponding enols (Sec. 5.5), most imines are more stable than their corresponding enamines. Therefore, primary amines react with aldehydes and ketones to form imines. Because secondary amines *cannot* form imines, they form enamines instead.

To summarize: Aldehydes and ketones react with primary amines (RNH_2) to give imines and with secondary amines (R_2NH) to give enamines. In a **tertiary amine** (R_3N), alkyl or aryl groups replace all the hydrogens of ammonia. *Tertiary amines do not react with aldehydes and ketones to form stable derivatives.* Although most tertiary amines are good nucleophiles, they have no NH hydrogens and therefore cannot even form carbinolamines. Their adducts with aldehydes and ketones are unstable and can only break down to starting materials.

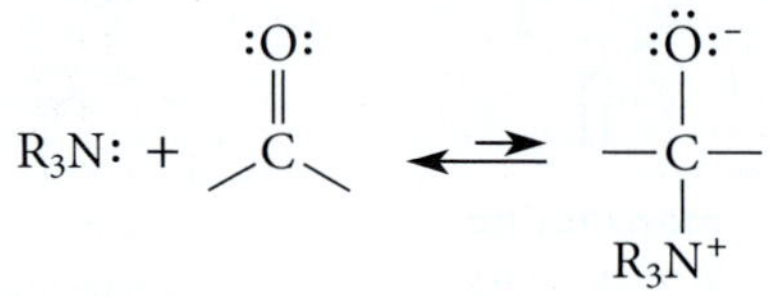

(19.84)

Focused Problem

19.31 Give the enamine product formed when each of the following pairs reacts.

(a) Acetone and 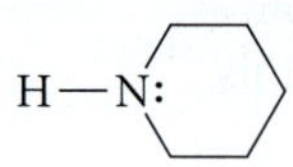

(b) $PhCH_2CH{=}O$ and $(CH_3)_2\ddot{N}H$

19.13 REDUCTION OF CARBONYL GROUPS TO METHYLENE GROUPS

A. Wolff–Kishner Reduction

The most common reductive transformation of aldehydes or ketones is their conversion into alcohols (Sec. 19.9). But it is also possible to reduce the carbonyl group of an aldehyde or ketone completely to a methylene (—CH_2—) group. One procedure for effecting this transformation involves heating the aldehyde or ketone with hydrazine ($H_2N{-}NH_2$) and strong base.

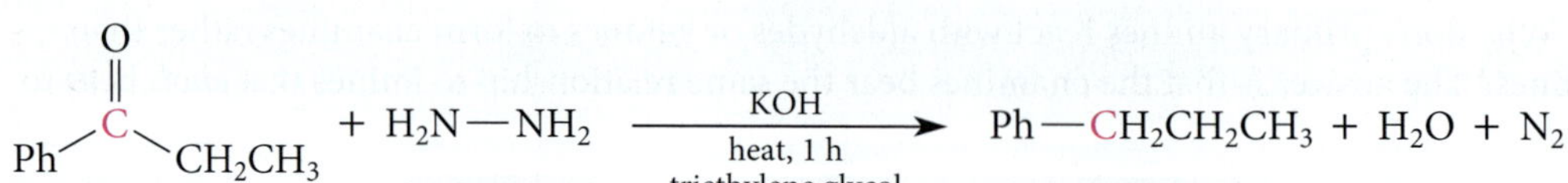

propiophenone **hydrazine** (85% aqueous solution) **propylbenzene** (82% yield)

(19.85)

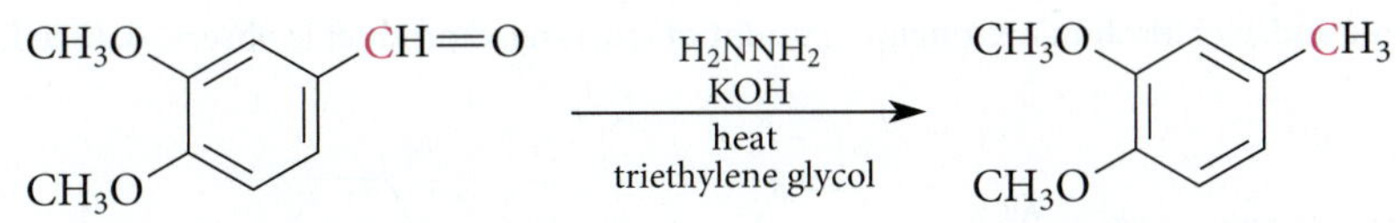

3,4-dimethoxybenzaldehyde **3,4-dimethoxytoluene** (81% yield) (19.86)

This reaction, called the **Wolff–Kishner reduction**, typically uses ethylene glycol or similar high-boiling compounds as co-solvents. (Triethylene glycol, which has the structure $HOCH_2CH_2OCH_2CH_2OCH_2CH_2OH$ and a boiling point of 278 °C, is used in Eqs. 19.85 and 19.86.) The high boiling points of these solvents allow the reaction mixtures to reach the high temperatures required for the reduction to take place at a reasonable rate.

The Wolff–Kishner reduction is an extension of imine formation (Sec. 19.12A) because a *hydrazone* (see Table 19.3) is an intermediate in the reaction. A series of Brønsted acid–base reactions lead ultimately to expulsion of dinitrogen gas and formation of the product. (Try writing the curved-arrow mechanism for this reaction; check your work in Study Guide Link 19.5.)

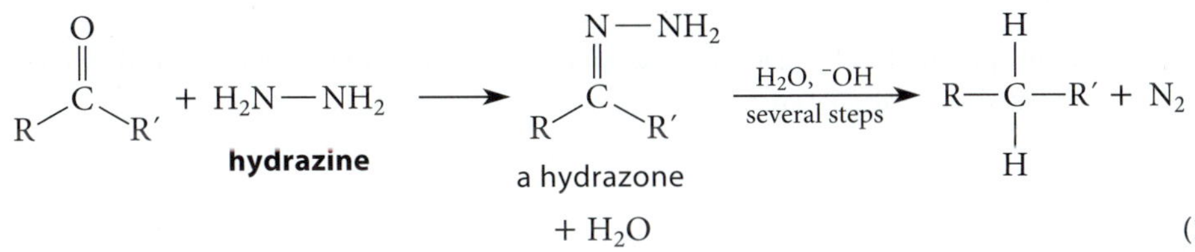

(19.87)

STUDY GUIDE LINK 19.5
Mechanism of the Wolff–Kishner Reaction

B. The Clemmensen Reduction

The Wolff–Kishner reduction takes place under strongly basic conditions. The same overall transformation can be achieved under acidic conditions by a reaction called the **Clemmensen reduction**. In this reaction, an aldehyde or ketone is reduced with zinc amalgam (a solution of zinc metal in mercury) in the presence of HCl. The reduction takes place on the surface of the Zn metal.

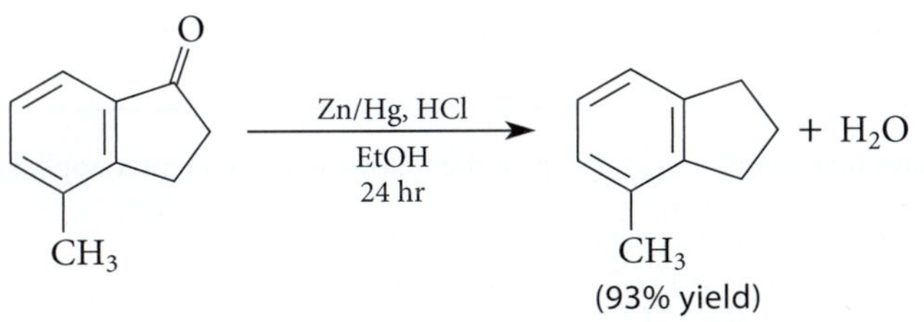

(19.88)

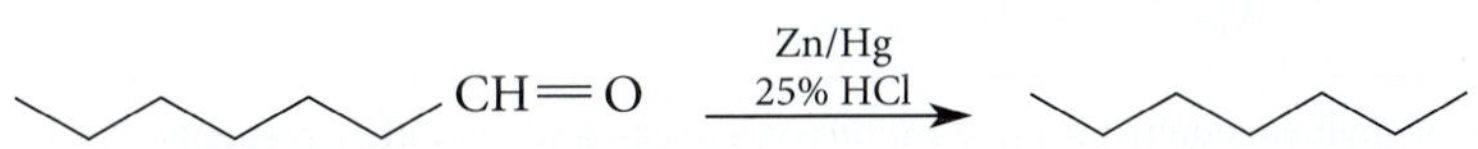

heptanal **heptane** (87% yield) (19.89)

The mechanism of the Clemmensen reduction is uncertain.

One of the most useful applications of the Wolff–Kishner and Clemmensen reductions is the introduction of alkyl substituents into benzene rings. This is illustrated in Study Problem 19.5.

Study Problem 19.5

Outline a synthesis of butylbenzene from benzene and any other reagents.

Solution When you are asked to prepare an alkylbenzene from benzene, Friedel–Crafts alkylation (Sec. 16.4E) should come to mind. Indeed, the Friedel–Crafts alkylation reaction is useful for introducing groups that do not rearrange, such as methyl groups, ethyl groups, and *tert*-butyl groups, into benzene rings. But when this reaction is used to prepare butylbenzene from benzene and 1-chlorobutane, a major amount of rearranged product is observed. (See Eq. 16.24, Sec. 16.4E.)

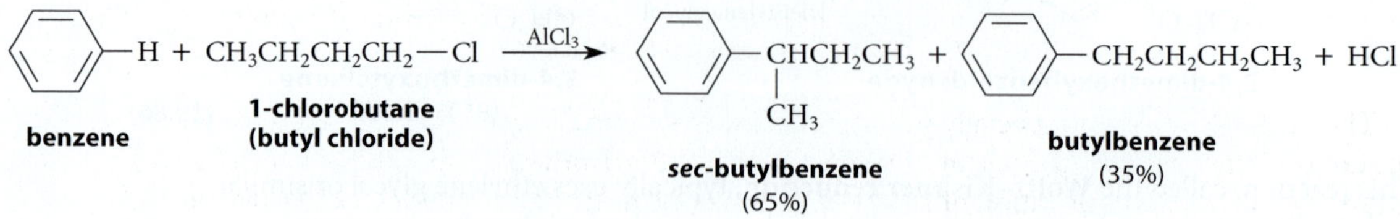

Butylbenzene can be easily prepared free of isomers, however, by either the Wolff–Kishner reduction or the Clemmensen reduction of butyrophenone. (We show the Wolff-Kishner reduction here.)

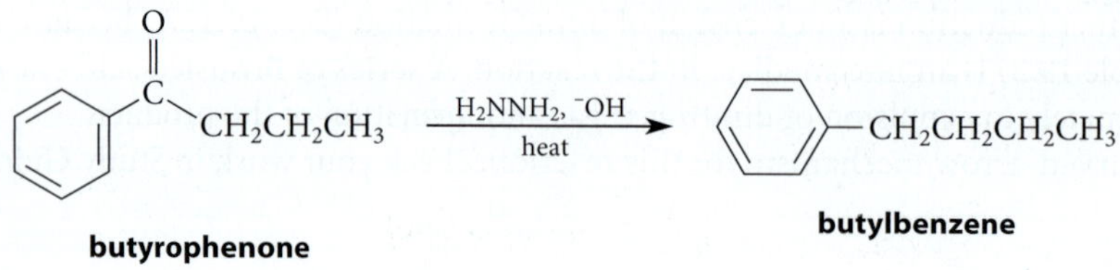

In turn, butyrophenone is readily prepared by Friedel–Crafts *acylation* (Sec. 16.4F), which is not plagued by the rearrangement problems associated with the *alkylation*.

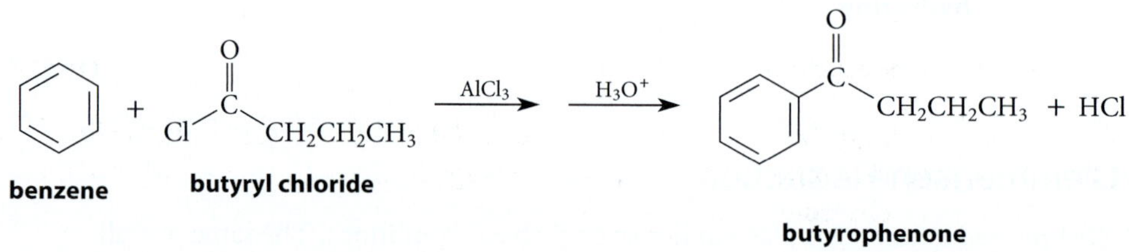

(Butylbenzene can also be prepared by the Stille reaction; Sec. 18.10B.)

Focused Problems

19.32 Draw the structures of all aldehydes or ketones that could in principle give the following product after application of either the Wolff–Kishner or Clemmensen reduction.

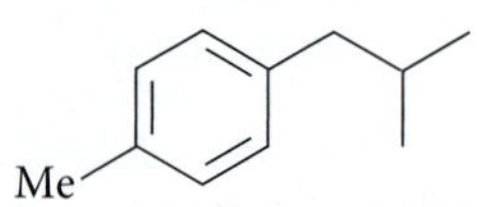

19.33 Outline a synthesis of 1,4-dimethoxy-2-propylbenzene from 1,4-dimethoxybenzene and any other reagents.

19.14 THE WITTIG ALKENE SYNTHESIS

The **Wittig alkene synthesis** is an important addition–elimination reaction sequence used for preparing alkenes from aldehydes and ketones. An example is the preparation of methylenecyclohexane from cyclohexanone.

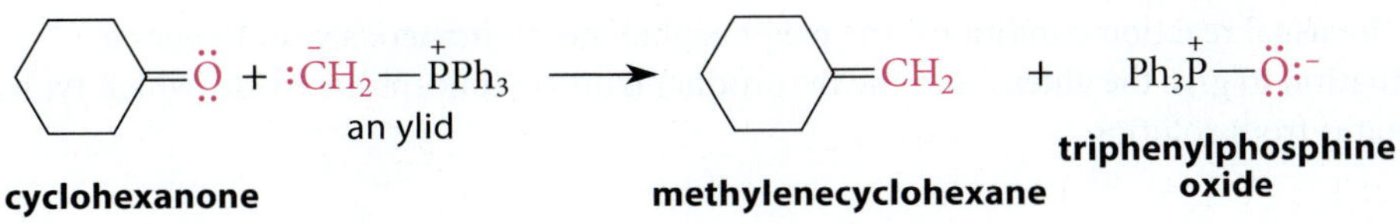

(19.90)

The Wittig synthesis is especially important because it gives alkenes in which the *position* of the double bond is unambiguous; in other words, the Wittig synthesis is completely *regioselective*. It can be used for the preparation of alkenes that would be difficult to prepare by other reactions. For example, methylenecyclohexane, which is readily prepared by the Wittig synthesis (Eq. 19.90), cannot be prepared by dehydration of 1-methylcyclohexanol; 1-methylcyclohexene is obtained instead, because alcohol dehydration gives the alkene isomer(s) in which the double bond has the greatest number of alkyl substituents (Sec. 11.2).

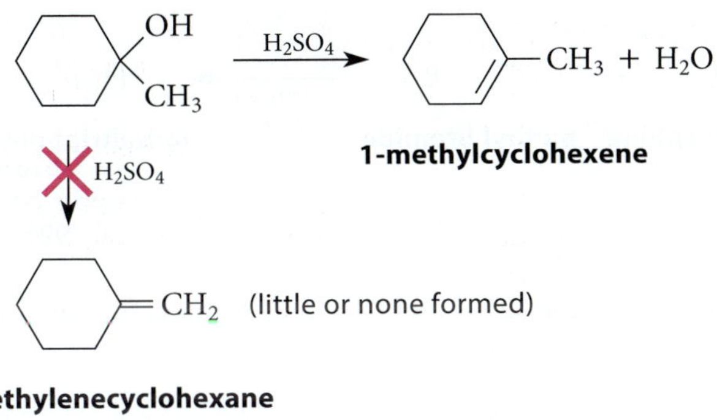

(19.91)

The nucleophile in the Wittig alkene synthesis is a type of *ylid* (pronounced ĭ′ ləd). An **ylid** (sometimes spelled *ylide*) is any compound with opposite charges on adjacent, covalently bonded atoms, each of which has an electronic octet.

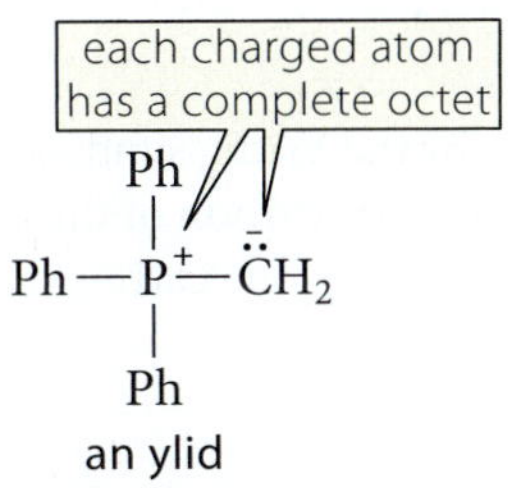

(19.92)

Although we can write a structure for the ylid with a hypervalent phosphorus, the charge-separated octet structure is more important (Sec. 11.10A).

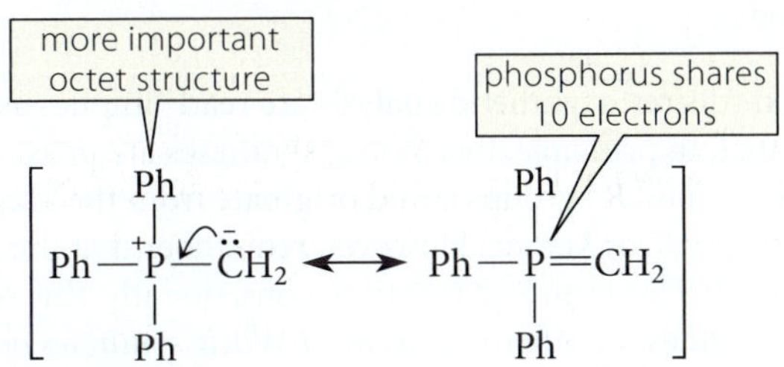

(19.93)

The mechanism of the Wittig reaction starts as a nucleophilic addition in which the anionic carbon of the ylid reacts at the carbonyl carbon.

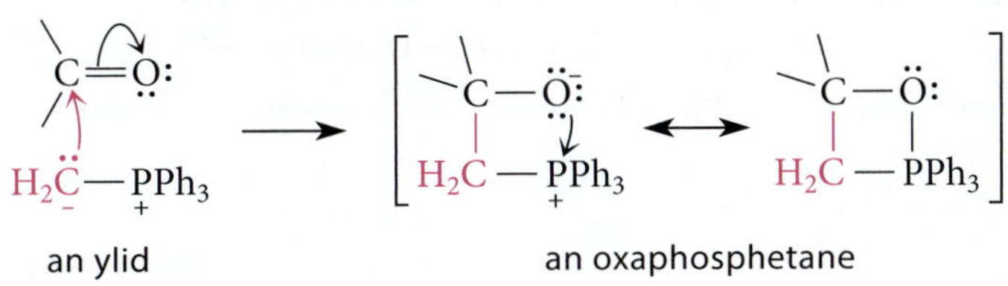

(19.94a)

Under the usual reaction conditions, the oxaphosphetane spontaneously undergoes a β-elimination to give the alkene and the by-product triphenylphosphine oxide, which typically precipitates from solution.

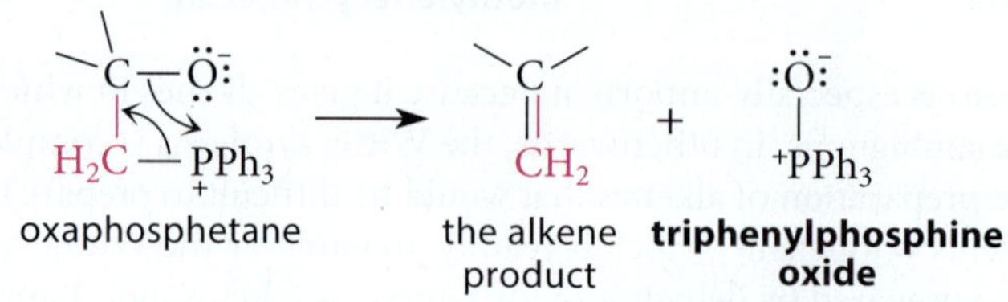

(19.94b)

The ylid starting material in the Wittig synthesis is prepared by the reaction of an alkyl halide with triphenylphosphine (Ph_3P:) in an S_N2 reaction to give a *phosphonium salt*.

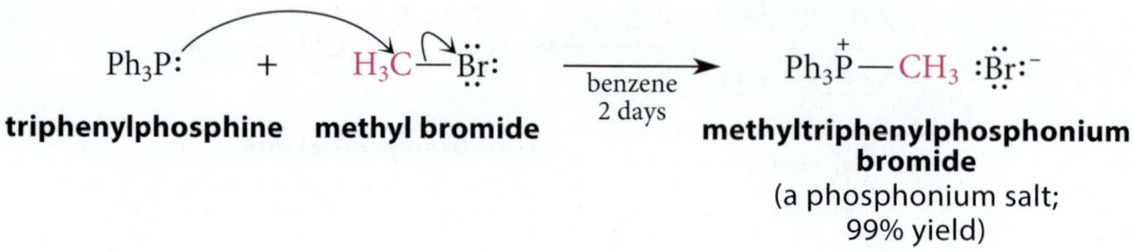

(19.95a)

The phosphonium salt can be converted into its conjugate base, the ylid, by reaction with a strong base such as an organolithium reagent.

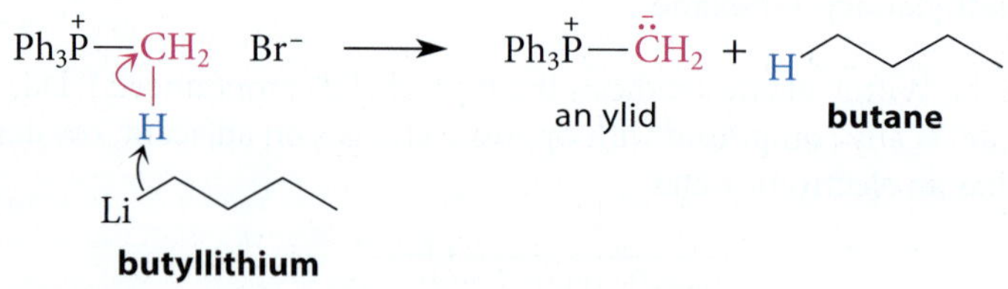

(19.95b)

To plan the preparation of an alkene by the Wittig synthesis, consider the origin of each part of the product, and then reason deductively. One carbon of the alkene double bond originates from the alkyl halide used to prepare the ylid; the other is the carbonyl carbon of the aldehyde or ketone:

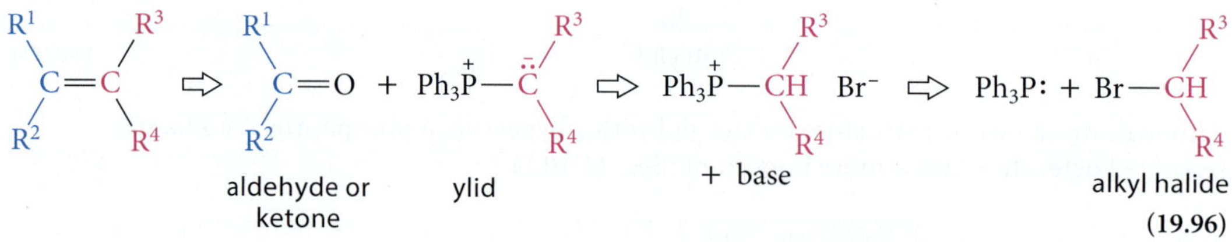

(19.96)

(Again, the arrows used in this retrosynthetic analysis are read "implies as starting material.") This analysis also shows that, in principle, two Wittig syntheses are possible for any given alkene; in the other possibility, the R^1 and R^2 groups could originate from the alkyl halide and the R^3 and R^4 groups from the aldehyde or ketone. However, remember that the reaction used to form the phosphonium salt (Eq. 19.95a) is an S_N2 reaction; consequently, this reaction is fastest with methyl and primary alkyl halides. In other words, *most Wittig syntheses are planned so that the most reactive alkyl halide can be used as one of the starting materials.*

One complication with the Wittig alkene synthesis is that it gives mixtures of *E* and *Z* alkene isomers.

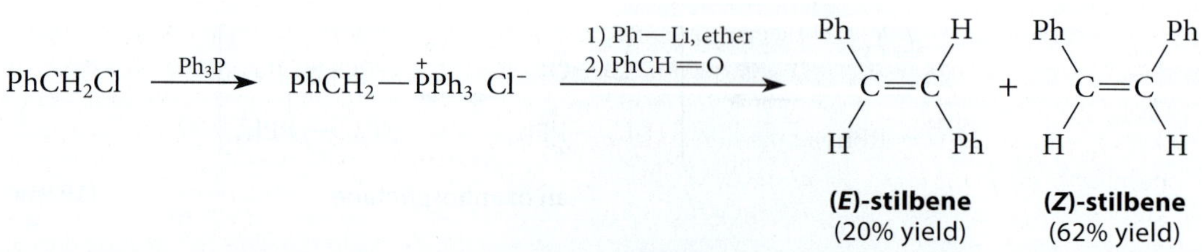

(19.97)

In many cases, the *Z* isomer predominates, as in this example. However, modifications of the Wittig reaction have been discovered (the use of different bases, lower temperatures, and other modifications) that can give either nearly pure *Z* isomer or nearly pure *E* isomer, depending on the modification used. However, the details of these modifications are outside of the scope of our discussion.

Study Problem 19.6

Outline two Wittig alkene syntheses of 2-methyl-1-hexene. Is one synthesis preferred over the other? Why?

Solution The analysis in Eq. 19.96 suggests that the "right-hand" part of the alkene can be derived from the ketone 2-hexanone:

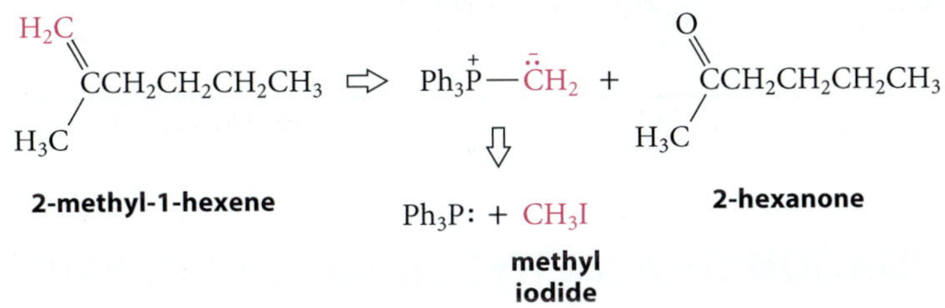

(19.98a)

Another possibility, however, is that the "left-hand" part of the alkene is derived from formaldehyde:

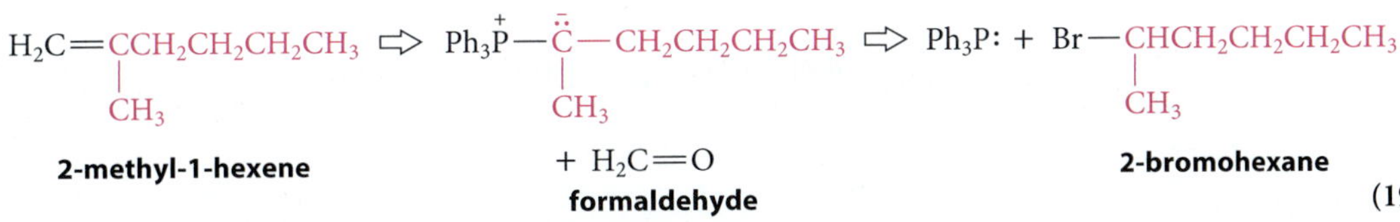

(19.98b)

Although both syntheses seem reasonable, the one in Eq. 19.98a requires an S_N2 reaction of triphenylphosphine with a methyl halide, whereas the one in Eq. 19.98b requires an S_N2 reaction of triphenylphosphine with a secondary alkyl halide. The first reaction is preferred because methyl halides are much more reactive than secondary alkyl halides (Sec. 9.4C).

The outlined synthesis is as follows:

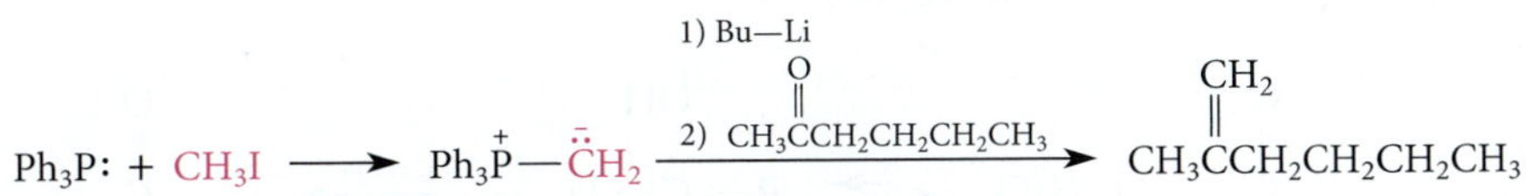

Discovery of the Wittig Alkene Synthesis

Editorial/Alamy

The Wittig alkene synthesis is named for Georg Wittig (1897–1987), who was a professor of chemistry at the University of Heidelberg. Wittig and his co-workers discovered the alkene synthesis in the course of other work in phosphorus chemistry; they had not set out to develop this reaction explicitly. Once the significance of the reaction was recognized, it was widely exploited. Wittig shared the 1979 Nobel Prize in Chemistry with Herbert C. Brown (Sec. 19.9A).

The Wittig reaction is not only important as a laboratory reaction; it has also been industrially useful. For example, it is an important reaction in the industrial synthesis of vitamin A derivatives.

Focused Problems

19.34 Give the structure of the alkene(s) formed in each of the following reactions.

(a)

$$\mathrm{CH_3CH_2I} \xrightarrow{\mathrm{Ph_3P}} \xrightarrow{\text{butyllithium}} \xrightarrow[\textbf{acetone}]{\mathrm{(CH_3)_2C{=}O}}$$

(b)

$$\mathrm{CH_3Br} \xrightarrow{\mathrm{Ph_3P}} \xrightarrow{\text{butyllithium}} \xrightarrow[\textbf{benzaldehyde}]{\mathrm{Ph{-}CH{=}O}}$$

19.35 Outline a Wittig synthesis for each of the following alkenes; give two Wittig syntheses of the compound in part (a).

(a) (b) (c)

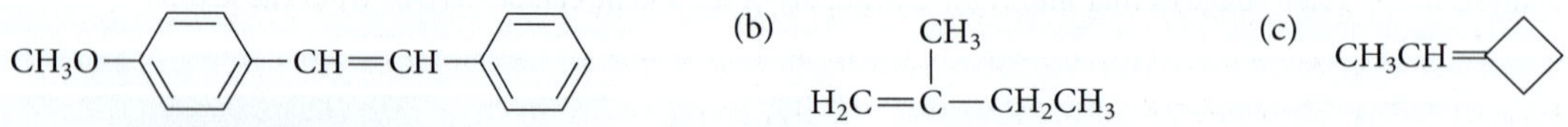

(mixture of cis and trans)

19.15 OXIDATION OF ALDEHYDES TO CARBOXYLIC ACIDS

Aldehydes can be oxidized to carboxylic acids.

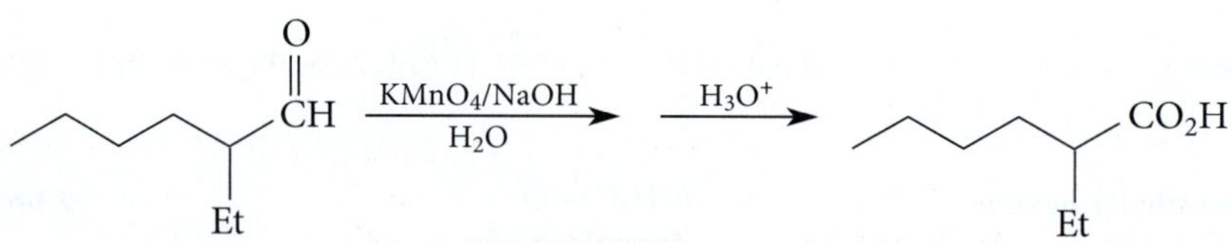

2-ethylhexanal **2-ethylhexanoic acid** (78% yield) **(19.99)**

Other common oxidants, such as aqueous Cr(VI) reagents, also work in this reaction. These oxidizing agents are the same ones used for oxidizing alcohols (Sec. 11.7A). However, in the oxidation of aldehydes by Cr(VI) reagents, the *hydrate*, not the aldehyde, is actually the species oxidized. (See Eq. 11.64, Sec. 11.7A.)

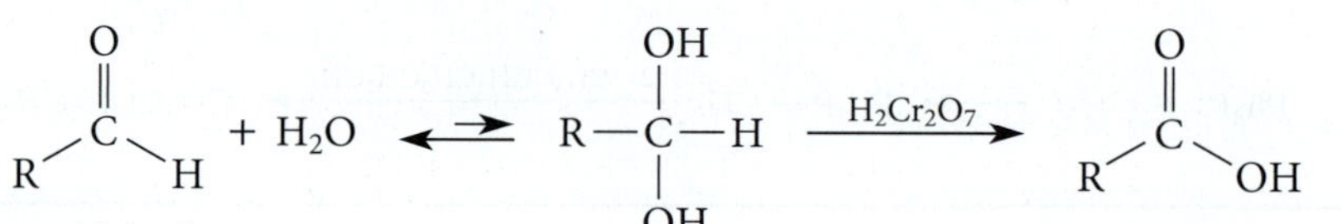

an aldehyde — the aldehyde hydrate — a carboxylic acid **(19.100)**

That is, the "aldehyde" oxidation is really an "alcohol" oxidation, the "alcohol" being the hydrate formed by addition of water to the aldehyde carbonyl group. For this reason, some water should be present in solution so that aldehyde oxidations with Cr(VI) occur at a reasonable rate.

In the laboratory, aldehydes can be conveniently and selectively oxidized to carboxylic acids with Ag(I) reagents.

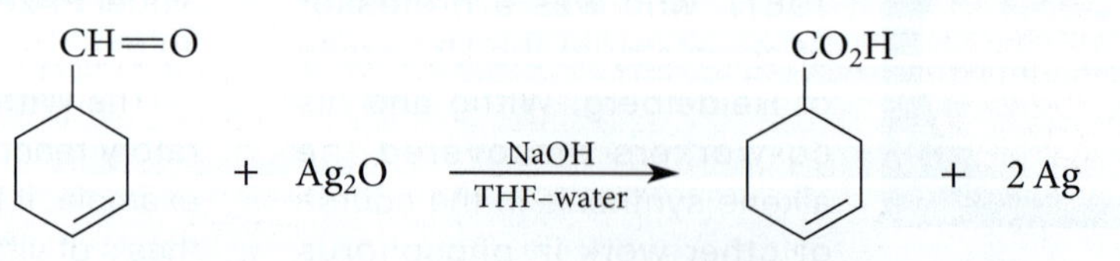

3-cyclohexenecarbaldehyde **3-cyclohexenecarboxylic acid** (75% yield) **(19.101)**

The expense of silver limits its use to small-scale reactions, as a rule. However, the Ag_2O oxidation is especially handy when the aldehyde to be oxidized contains double bonds or alcohol OH groups—that is, functional groups that react with other oxidizing reagents but do not react with Ag_2O.

Sometimes, as in Eq. 19.101, the Ag(I) is used as a slurry of brown Ag_2O, which changes to a black precipitate of silver metal as the reaction proceeds. If the silver ion is solubilized as its ammonia complex, $^{+}Ag(NH_3)_2$, oxidation of the aldehyde is accompanied by the deposition of a metallic silver mirror on the walls of the reaction vessel. This observation can be used as a convenient test for aldehydes, known as the **Tollens test**.

Many aldehydes are slowly oxidized by the oxygen in air upon standing for a long time. This process, another example of *autoxidation* (Secs. 12.9A and 18.11), is responsible for the contamination of some aldehyde samples with appreciable amounts of carboxylic acids.

Ketones cannot be oxidized without breaking carbon–carbon bonds (see Table 11.2, Sec. 11.6B). Ketones are resistant to mild oxidation with Cr(VI) reagents, and acetone can even be used as a solvent for oxidations with these reagents. Potassium permanganate, however, oxidizes ketones by breaking carbon–carbon bonds, and it is therefore not useful as an oxidizing reagent in the presence of ketones.

Focused Problems

19.36 Give the structure of an aldehyde $C_8H_8O_2$ that would be oxidized to terephthalic acid by $KMnO_4$.

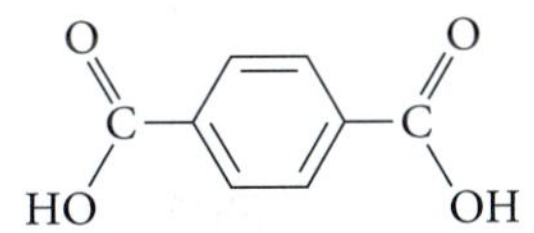

terephthalic acid

19.37 What product is formed when the following compound is treated with Ag_2O?

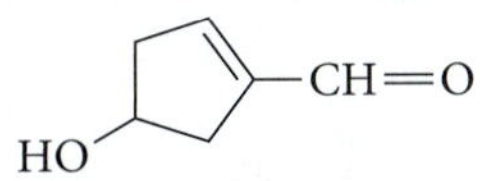

19.16 ALDEHYDES AND KETONES IN ORGANIC SYNTHESIS

Aldehydes and ketones are very important starting materials in organic synthesis because many functional types can be prepared from them. We summarize aldehyde reactions from this chapter with the following diagram:

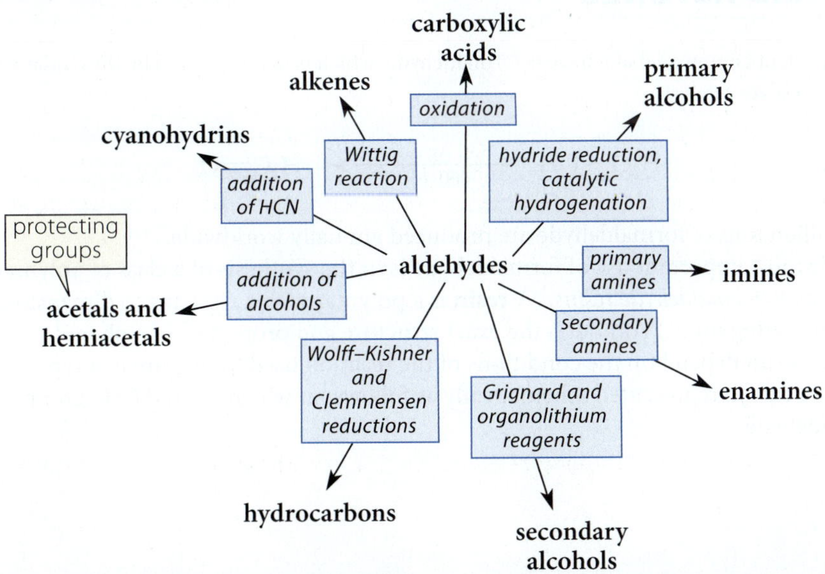

(19.102)

A similar diagram summarizes ketone reactions:

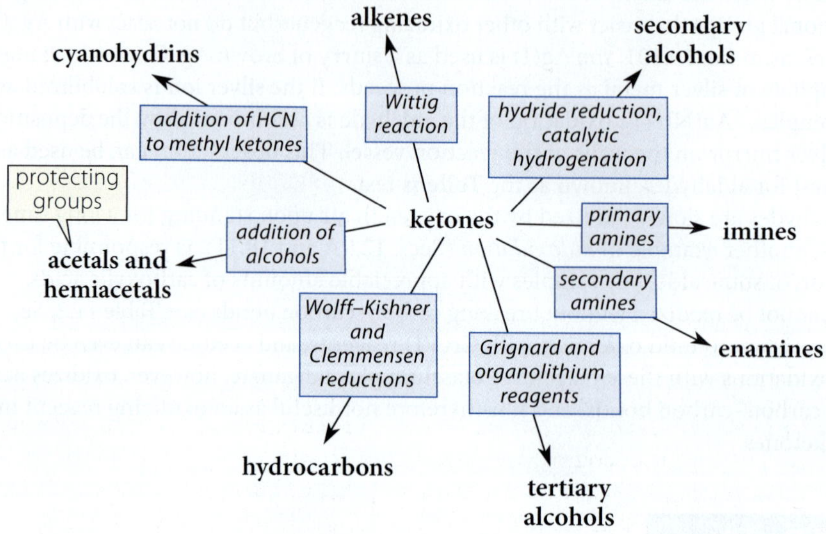

(19.103)

Combining these diagrams with the ones for other functional groups, such as alkenes (Display 5.54, Sec. 5.9B), alkynes (Display 5.55, Sec. 5.9B), and alcohols (Display 11.90, Sec. 11.12) unveils the large number of possibilities for multistep synthesis that are available from these functional-group interconversions.

Focused Problem

19.38 Outline a multistep synthesis for each of the following compounds from cyclopentanol.

(a) 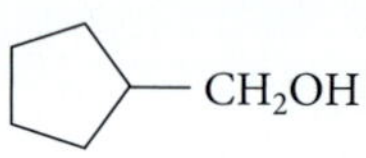

(b) 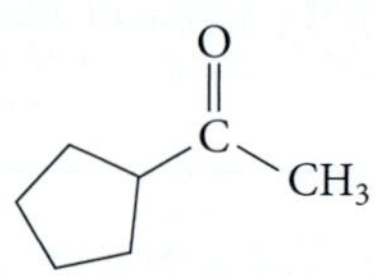

Chemistry in the Real World

19.17 MANUFACTURE AND USE OF ALDEHYDES AND KETONES

The most important commercial aldehyde is formaldehyde, which is manufactured by the oxidation of methanol over a silver catalyst.

$$H_3C{-}OH \xrightarrow[600\text{–}650\ °C]{O_2,\ Ag\ catalyst} H_2C{=}O \tag{19.104}$$

About 23 million tons of formaldehyde are produced annually worldwide.

The single most important use of formaldehyde is in the synthesis of a class of polymer known as *phenol–formaldehyde resins.* (A **resin** is a polymer with a rigid three-dimensional network of repeating units.) Although the exact structure and properties of a phenol–formaldehyde resin depend on the conditions of the reaction used to prepare it, a typical segment of such a resin can be represented schematically as follows, in which the red CH_2 groups come from formaldehyde.

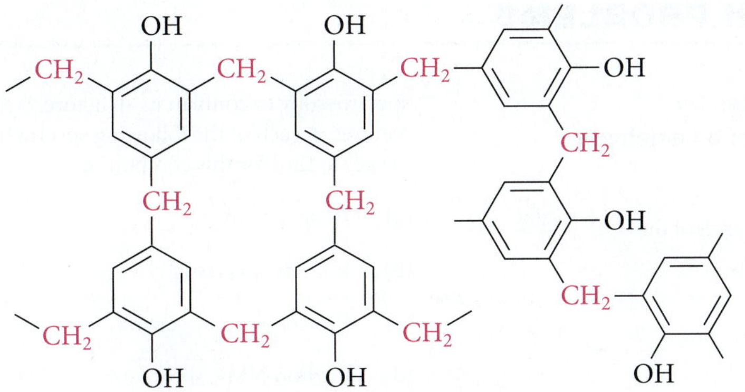

(19.105)

Phenol–formaldehyde resins are produced by a variation of the Friedel–Crafts alkylation in which phenol and formaldehyde are heated with acidic or basic catalysts. The formaldehyde in some cases is supplied in the form of its addition product with ammonia. Various formulations of these resins are used for telephones, adhesives in exterior-grade plywood, and heat-stable bondings for brake linings. A phenol–formaldehyde resin called *Bakelite*, patented in 1909 by the Belgian immigrant Leo H. Baekeland, was the first useful synthetic polymer.

Because formaldehyde, a known carcinogen, is used extensively in the manufacture of some construction materials, some concern has developed about formaldehyde release in confined environments in which such materials have been used. This concern received particular focus when it was discovered in 2006 that the trailers provided as temporary housing to victims of Hurricane Katrina by the Federal Emergency Management Agency (FEMA) contained formaldehyde levels that were 4–7 times the federally mandated limit. Formaldehyde-based construction materials were used extensively in these trailers.

Acetone, the simplest ketone, is co-produced with phenol by the autoxidation–rearrangement of cumene (Sec. 18.11). The worldwide production of acetone is estimated at roughly 7–8 million tons. Acetone itself finds use as an important solvent, and acetone cyanohydrin, which is produced from acetone (Eq. 19.32, Sec. 19.8), is a starting material for the production of poly(methyl methacrylate), an important polymer (Table 10.3, Sec. 10.3C). Acetone and phenol are used together as starting materials for the production of bisphenol A, which, in turn, is used as a monomer in the production of polycarbonate plastics.

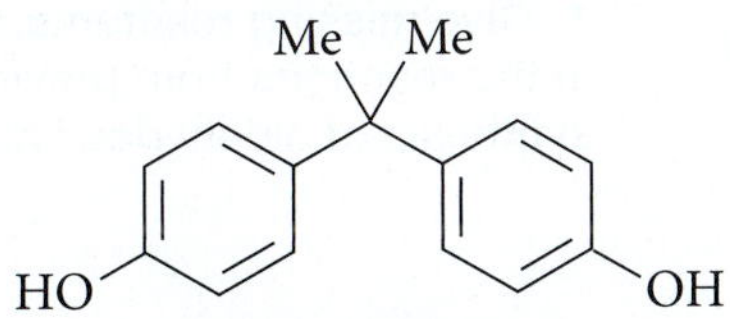

bisphenol A (BPA)

(19.106)

(The reaction of acetone and phenol to give bisphenol A is explored in Problem 19.75.)

CHAPTER SUMMARY *For a summary of the chapter, see Chapter 19 in the* Study Guide and Solutions Manual.

REACTION REVIEW *For a summary of reactions discussed in this chapter, see the* Reaction Review *section of Chapter 19 in the* Study Guide and Solutions Manual.

SKILLS OBJECTIVES WITH PROBLEMS

• Provide common and substitutive names for aldehydes and ketones; given the name of an aldehyde or ketone, draw its structure. 19.2A,B

19.39 Provide an IUPAC substitutive name for each of the following compounds.

(a)

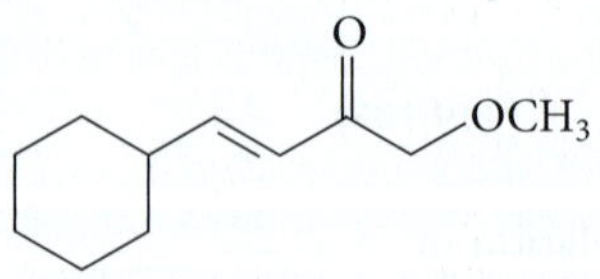

(b)

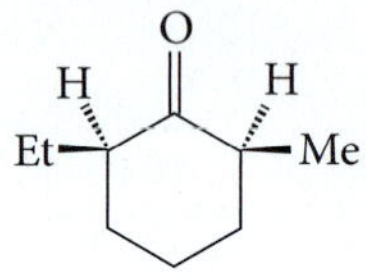

(c) $(CH_3)_2CH-C\equiv C-CH_2-CH=O$

(d)

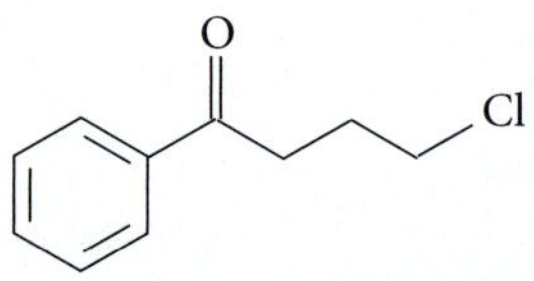

19.40 Give a structure for each of the following compounds.

(a) Valerophenone

(b) 3-Hydroxy-2-butanone

(c) 2-Oxocyclopentanecarbaldehyde

• State the unique features of aldehyde and ketone IR, proton NMR, carbon NMR, UV–vis, and mass spectra. 19.4

19.41 You believe that you have prepared a pure sample of 4-methylpentanal, and you want to use different forms of spectroscopy to confirm its structure. State one or more features in each of the following spectra that you would expect to find for this compound.

(a) IR spectrum

(b) UV–vis spectrum

(c) Proton NMR spectrum

(d) Carbon NMR spectrum

(e) EI mass spectrum

(f) CI mass spectrum

• Use IR, proton NMR, carbon NMR, UV–vis, and mass spectra to identify aldehydes and ketones. 19.4

19.42 Identify each of the following compounds.

(a) $C_{10}H_{10}O_2$ ^{1}H NMR: δ 2.82 (6*H*, s), δ 8.13 (4*H*, s)

IR: 1681 cm^{-1}, no O—H stretch

(b) $C_5H_{10}O$ ^{1}H NMR: δ 9.8 (1*H*, s), δ 1.1 (9*H*, s)

(c) $C_6H_{10}O$ ^{1}H NMR in **Fig. P19.42**

IR: 1701 cm^{-1}, 970 cm^{-1}

UV: λ_{max} = 215 (ε = 17,400), 329 (ε = 26)

19.43 Identify the compound with the mass spectrum and proton NMR spectrum shown in **Fig. P19.43**. This compound has IR absorptions at 1678 cm^{-1} and 1600 cm^{-1}, and UV absorptions at λ_{max} (hexane) = 259 nm (ε = 18,100) and 321 nm (ε = 64).

• Give missing reactants, starting materials, or reagents in the reactions from previous chapters used for the synthesis of aldehydes and ketones. 19.5

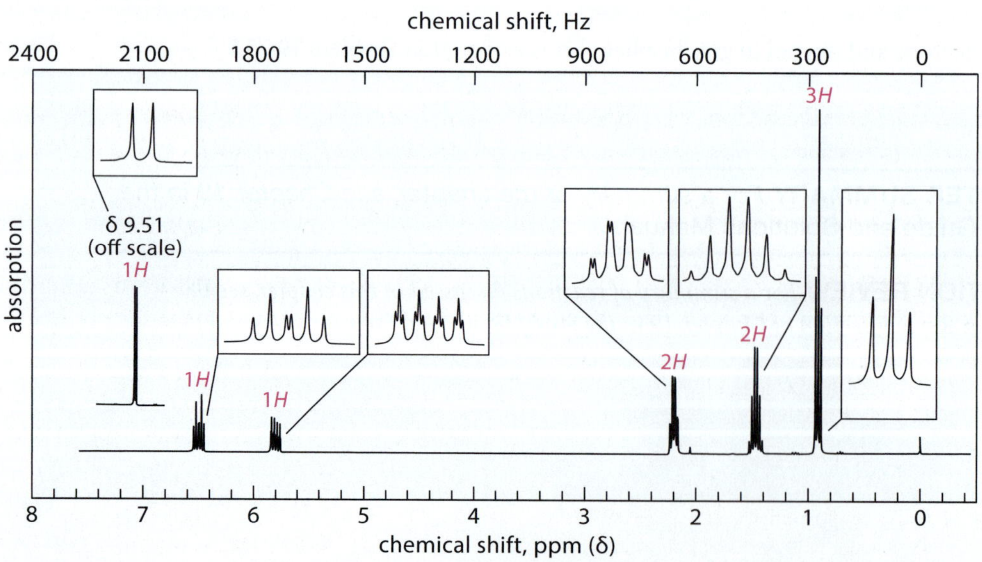

FIGURE P19.42 Proton NMR spectrum for Problem 19.42(c). The relative integrals are indicated in red over their respective resonances. The horizontal scales of the insets are identical.

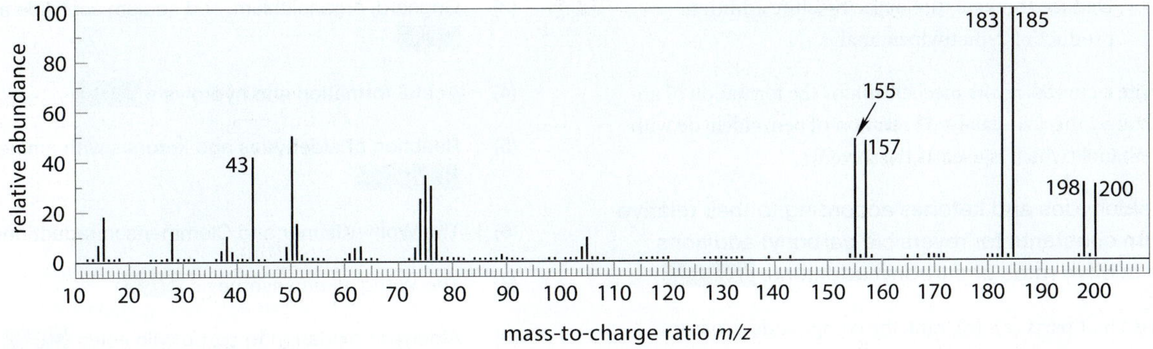

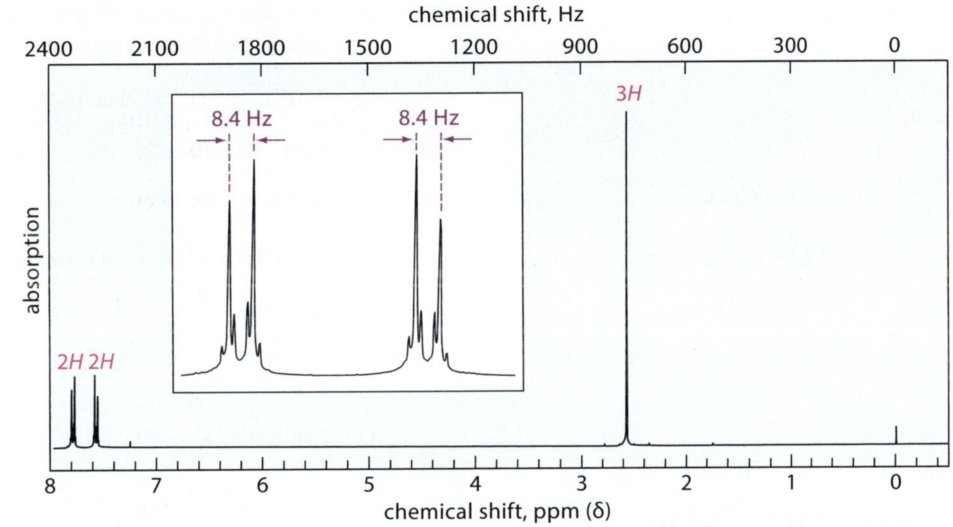

FIGURE P19.43 The mass spectrum and proton NMR spectrum for Problem 19.43. The relative integrals in the NMR spectrum are indicated in red over their respective resonances.

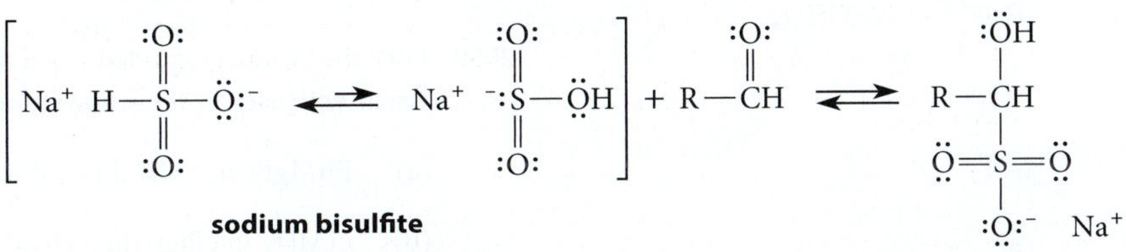

FIGURE P19.45

19.44 Give the missing starting materials, reagents, or products in each of the following reactions.

(a)

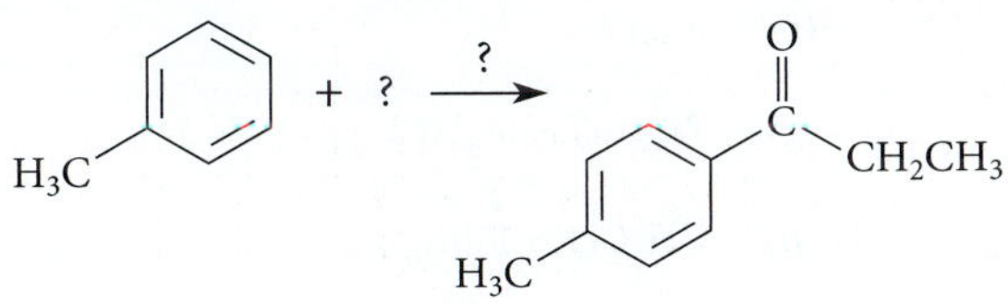

(b)

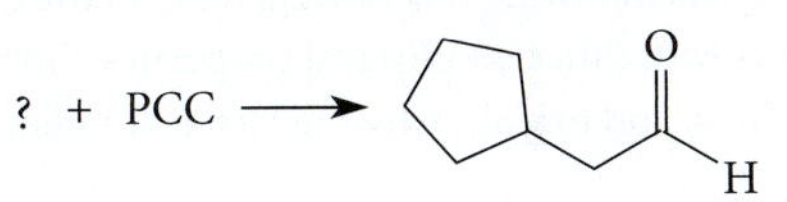

(c) $Me_2CH{-}C{\equiv}CH + H_2O \xrightarrow{H_3O^+,\ Hg^{2+}}$?

(d) $Me_2CH{-}C{\equiv}CH \xrightarrow[\text{2) } H_2O_2,\ NaOH]{\text{1) disiamylborane}}$?

- **Write curved-arrow mechanisms for reactions that involve carbonyl–addition steps. 19.8A**

19.45 Sodium bisulfite adds reversibly to aldehydes and a few ketones to give *bisulfite addition products* (**Fig. P19.45**).

(a) Write a curved-arrow mechanism for this addition reaction; assume water is the solvent.

(b) The reaction can be reversed by adding either H_3O^+ or ^-OH. Explain this observation using Le Chatelier's principle and the following equilibria.

$$\underset{\substack{\textbf{sulfurous}\\ \textbf{acid}\\ pK_a = 1.76}}{H_2SO_3} \underset{H_3O^+}{\overset{H_2O}{\rightleftharpoons}} \underset{\substack{\textbf{bisulfite}\\ pK_a = 7.0}}{HSO_3^-} \underset{H_3O^+}{\overset{H_2O}{\rightleftharpoons}} \underset{\textbf{sulfite}}{SO_3^{2-}}$$

(c) Deduce the structure of the bisulfite addition product of 2-methylpentanal.

19.46 Write a curved-arrow mechanism for the formation of an acetal by the acid-catalyzed reaction of benzaldehyde with methanol (which is used as the solvent).

- **Rank aldehydes and ketones according to their relative equilibrium constants for reversible carbonyl additions and their relative rates of carbonyl additions. 19.8B,C**

19.47 In each of parts (a)–(c), rank the compounds in order of their increasing rate of reaction in carbonyl-addition reactions, and explain your reasoning.

(a)

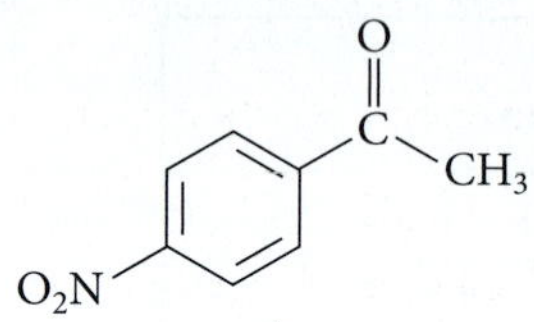

A

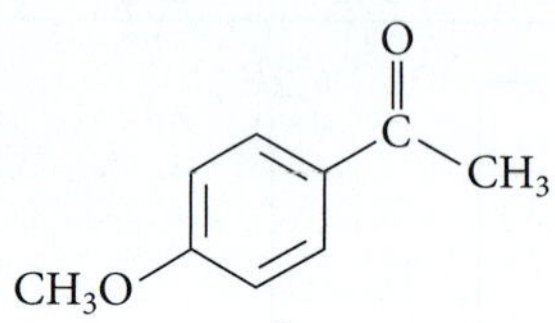

B

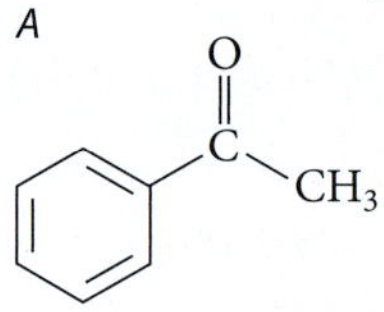

C

(b)

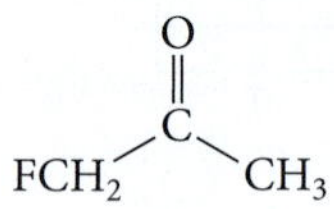

A

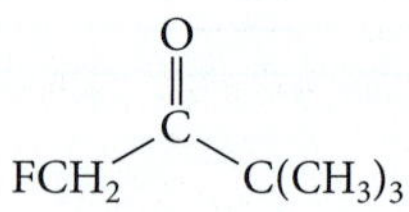

C

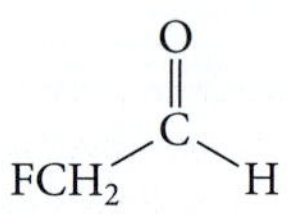

B

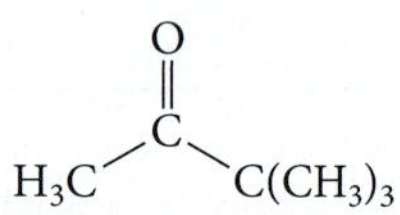

D

(c)

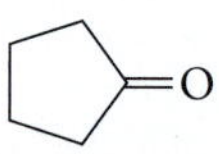

A

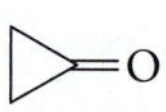

B

(*Hint:* Notice the change of bond angles in Fig. 19.8a.)

19.48 The compound *ninhydrin* exists almost entirely as a hydrate. Explain, and draw the structure of the hydrate.

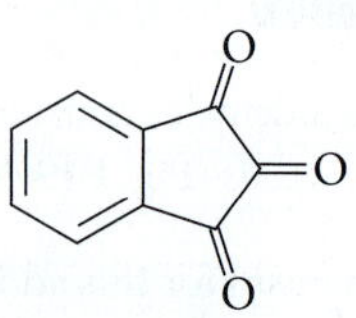

ninhydrin

- **Complete aldehyde and ketone reactions discussed in this chapter by providing the missing starting materials, products, or reagents.**

(1) Hydride reduction **19.9A**

(2) Catalytic hydrogenation **19.9C**

(3) Grignard, organolithium, and sodium acetylide additions **19.10**

(4) Acetal formation and hydrolysis **19.11**

(5) Reaction of aldehydes and ketones with amines **19.12A,C**

(6) The Wolff–Kishner and Clemmensen reductions **19.13**

(7) The Wittig alkene synthesis **19.14**

(8) Aldehyde oxidation to carboxylic acids **19.15**

19.49 Give the products expected (if any) when acetone reacts with each of the following reagents.

(a) $NaBH_4$ in CH_3OH, then H_2O

(b) CrO_3, pyridine

(c) NaCN, pH 10, H_2O

(d) CH_3OH (solvent), H_2SO_4 (trace)

(e) , trace of acid

(f) Semicarbazide, dilute acid

(g) CH_3MgI in ether, then H_3O^+

(h) H_2, Ni catalyst

(i) $H_2\ddot{\bar{C}}-\overset{+}{P}Ph_3$

(j) Zn amalgam, HCl

19.50 Give the product expected when butyraldehyde (butanal) reacts with each of the following reagents.

(a) PhMgBr in THF, then dilute H_3O^+

(b) $LiAlH_4$ in ether, then H_3O^+

(c) Alkaline $KMnO_4$, then H_3O^+

(d) Aqueous $H_2Cr_2O_7$

(e) NH_2OH, pH = 5

(f) Ag_2O

(g) Zn amalgam, HCl

(h) $CH_3\ddot{\bar{C}}H-\overset{+}{P}Ph_3$

19.51 Each of the following reactions gives a mixture of two isomers with different physical properties. Give their structures, and explain why they have different properties.

(a)

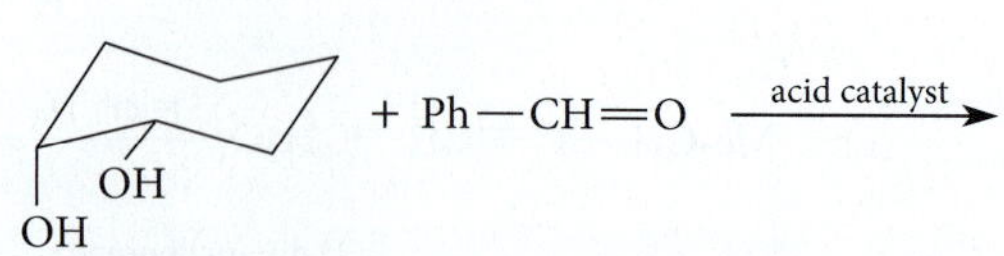

(b)

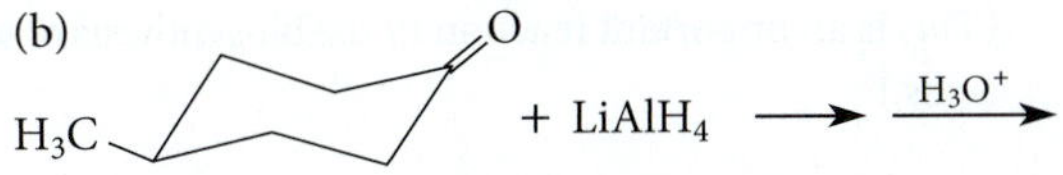

(c) $Ph—CH=O + CH_3\ddot{C}H—\overset{+}{P}Ph_3 \longrightarrow$

19.52 Complete each of the following reactions by giving the principal organic product(s).

(a)

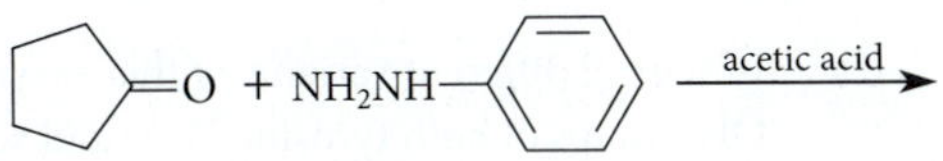

phenylhydrazine
(see Table 19.3)

(b)

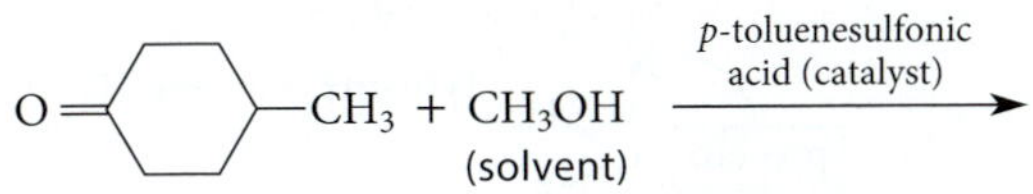

(c)

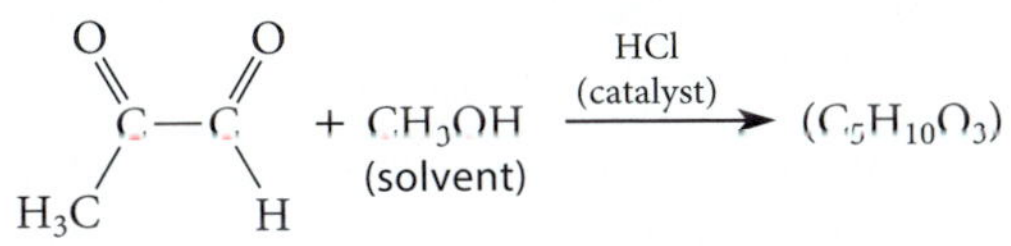

(d)

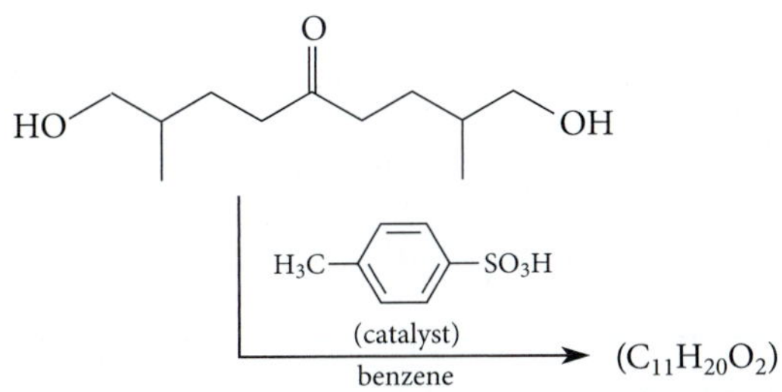

(e)

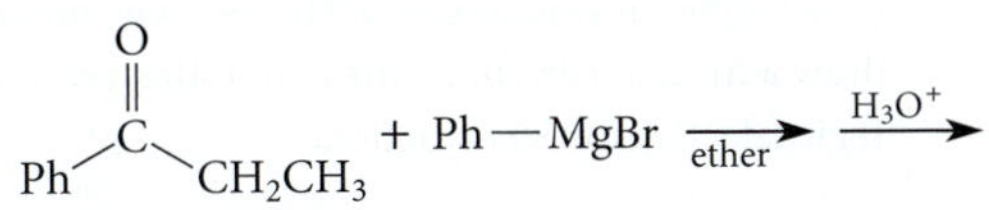

propiophenone

(f) $CH_3I + Ph_3P \longrightarrow \xrightarrow[\text{benzene}]{\text{Bu—Li}} \xrightarrow{Ph_2C=O}$

(g)

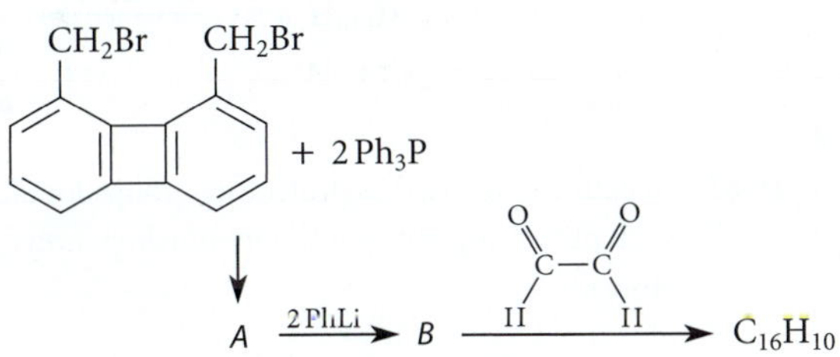

19.53 What are the starting materials for the synthesis of each of the following imines?

(a)

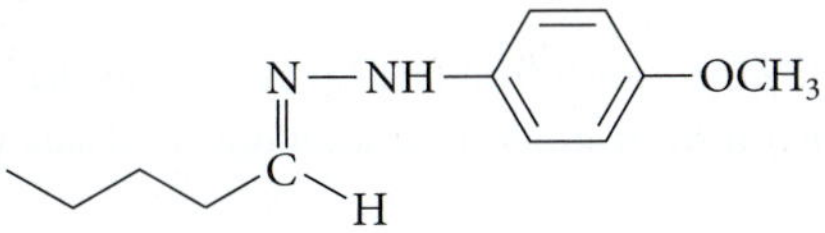

(b)

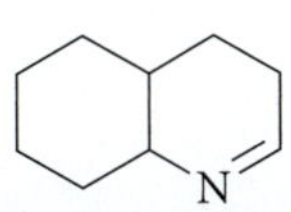

- **Use an acetal as a protecting group in a synthetic scheme. 19.11B**

19.54 Propose a multistep synthesis for the following conversion. (*Hint:* The source of deuterium is D_2O.)

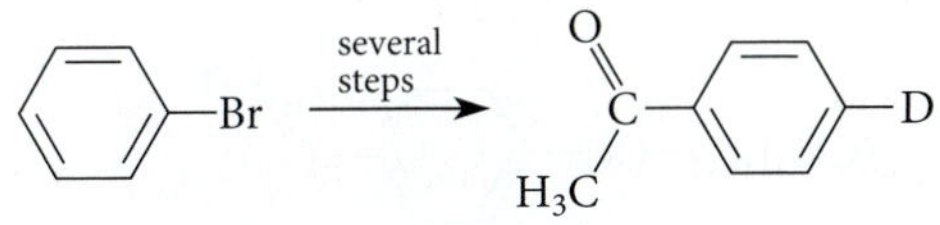

- **Use the reactions of aldehydes and ketones in multistep syntheses. 19.15**

19.55 Outline a synthesis for each of the following compounds from the indicated starting materials and any other reagents.

(a) 1-Phenyl-1-butanone (butyrophenone) from butyraldehyde

(b) 2-Cyclohexyl-2-propanol from cyclohexanone

(c) Cyclohexyl methyl ether (methoxycyclohexane) from cyclohexanone

(d) $PhCH_2OCH_2Ph$ (dibenzyl ether) from benzaldehyde as the only carbon source

(e) 2,3-Dimethyl-2-hexene from 3-methyl-2-hexanone

(f) 2,3-Dimethyl-1-hexene from 3-methyl-2-hexanone

(g) 1,6-Hexanediol from cyclohexene

(h)

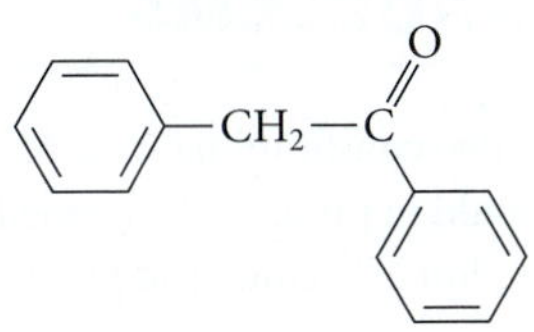

from benzaldehyde as the only source of carbon

(i) 1-Butyl-4-methoxybenzene from anisole (methoxybenzene)

(j)

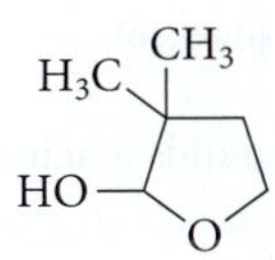

from

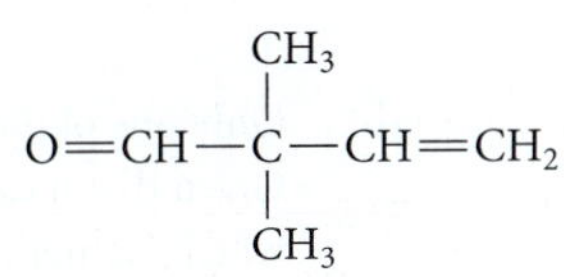

(*Hints:* [1] BH_3 in THF reduces aldehydes and ketones to alcohols; you need a protecting group. [2] Can you find an aldehyde lurking somewhere in the target molecule?)

19.56 Outline a synthesis for each of the following compounds from the indicated starting materials and any other reagents. Each sequence involves a transition-metal-catalyzed reaction.

(a)

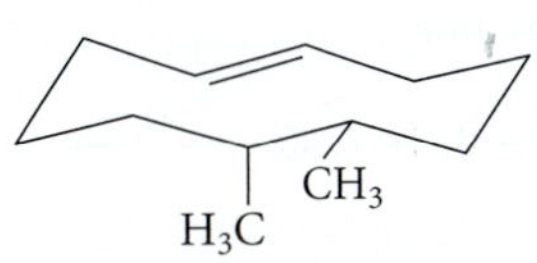

from (*Z*)-5,6-dimethyl-5-decen-1,10-diol

(b) 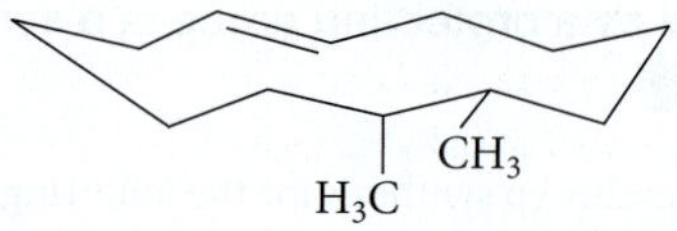

from the starting material for part (a)

(c) 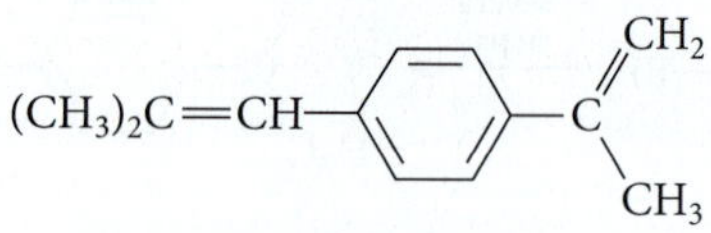

from bromobenzene using a Heck reaction

(d) 1-Butyl-4-methoxybenzene from *p*-methoxyphenol (hydroquinone monomethyl ether)

• **Give the general mechanism for a biological reduction using NADH or NADPH. 19.9B**

19.57 The enzyme 3-ketobutanoyl thioester reductase catalyzes the reduction of the ketone carbonyl group of the following compound with NADPH to give the *R* stereoisomer of the product.

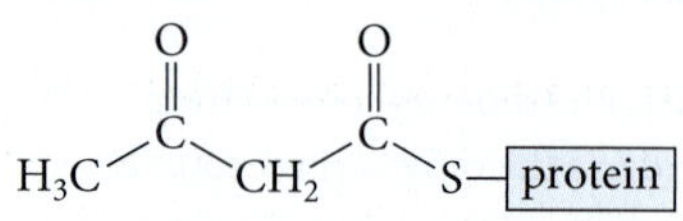

3-ketobutanoyl thioester

(This is an important reaction in the biosynthesis of fatty acids.)

(a) Give the structure of the product, including stereochemistry.

(b) The pro-(*R*) hydrogen of NADPH is transferred in the reaction. Show the relative positions of the NADPH and the 3-ketobutanoyl thioester molecules in the transition state of the reaction.

(c) In several different variants of this enzyme, the OH groups of both tyrosine (Tyr) and serine (Ser) residues are positioned near the ketone carbonyl group.

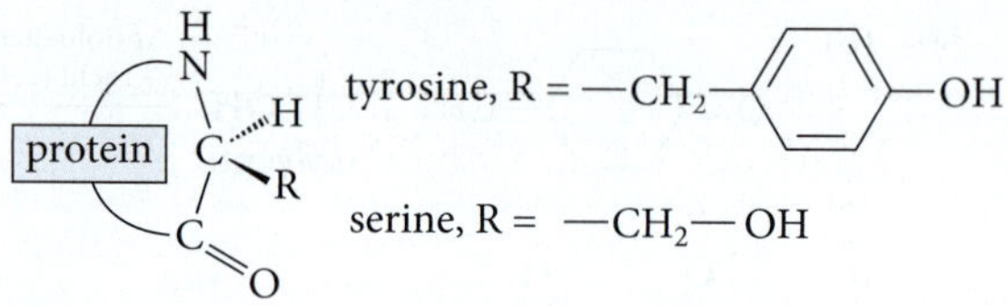

What is the likely role of these groups? Add these amino acid side chains to the diagram you drew in part (b).

INTEGRATED PROBLEMS

19.58 (a) What are the two constitutionally isomeric cyclic acetals that could in principle be formed in the acid-catalyzed reaction of acetone and glycerol?

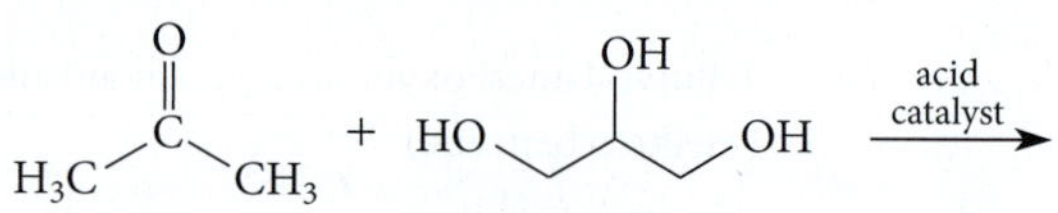

glycerol

(b) Only one of the two compounds is actually formed. Given that it can be resolved into enantiomers, which isomer in part (a) is the one that is produced?

19.59 (a) The following compound is unstable and spontaneously decomposes to acetophenone and HBr. Give a mechanism for this transformation.

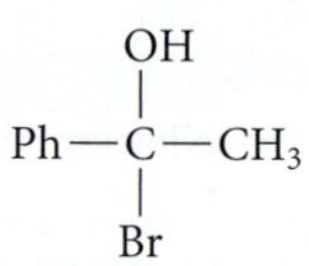

(b) Use the information in part (a) to complete the following reaction:

$$H_2C{=}CH{-}Br + OsO_4 \xrightarrow{H_2O}$$

19.60 The product *A* of the following reaction hydrolyzes in dilute aqueous acid to give acetophenone. Identify *A*, and draw a mechanism for its formation that accounts for the regioselectivity of the reaction.

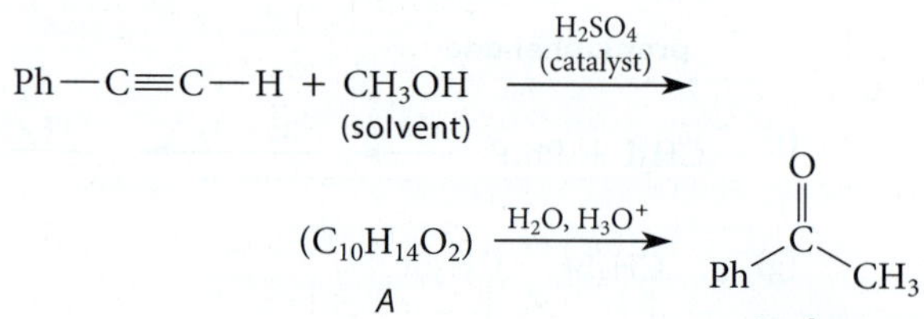

acetophenone

19.61 Acetals can be used as protecting groups for alcohols. One such protecting group is the *tetrahydropyranyl ether* (THP ether).

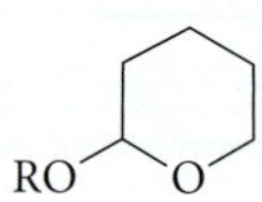

a THP ether

THP ethers are introduced by treating an alcohol with dihydropyran and *p*-toluenesulfonic acid catalyst.

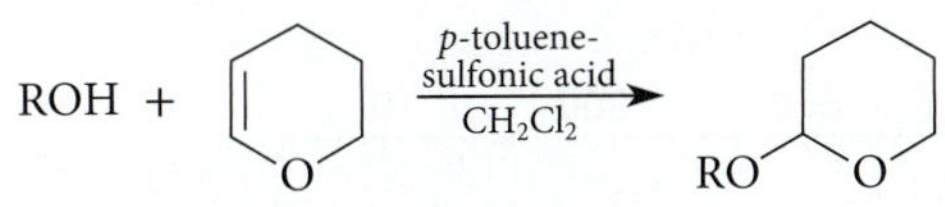

dihydropyran

THP ethers are stable to base but are rapidly removed by dilute aqueous acid.

(a) Give the structure of the product formed (in addition to the alcohol ROH) when a THP ether is treated with aqueous acid.

(b) Using the THP protecting group as part of your strategy, outline a synthesis of 2-methyl-2,6-hexanediol from 4-bromo-1-butanol.

19.62 From your knowledge of the reactivity of $LiAlH_4$, as well as the reactivity of epoxides with nucleophiles, predict the product (including stereochemistry, if appropriate) in each of the following reactions:

(a)

(b)

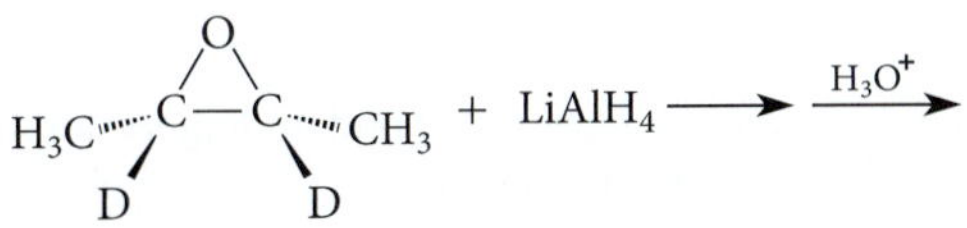

19.63 Thumbs Throckmorton, a graduate student in his 12th year of study, has designed the synthetic procedures shown in **Fig. P19.63**. Indicate the problems (if any) that each synthesis is likely to encounter.

19.64 (a) You are the chief organic chemist for Bugs and Slugs, Inc., a firm that specializes in environmentally friendly pest control. You have been asked to design a synthesis of 4-methyl-3-heptanol, the aggregation pheromone of the European elm beetle (the carrier of Dutch elm disease). Outline a synthesis of this compound from starting materials containing five or fewer carbons.

(b) After successfully completing the synthesis in part (a) and delivering your compound, you are advised that it appears to be a mixture of isomers. Assuming that you have prepared the correct compound, provide an explanation.

19.65 Starting with any organic compound you wish, outline synthetic procedures for preparing each of the following isotopically labeled materials using the indicated source of the isotope.

(a) Ph—CH(^{18}OH)—CH_2—Ph using $H_2{}^{18}O$

(b) Ph—CD(OH)—CH_2—Ph using $LiAlD_4$

19.66 Trichloroacetaldehyde, Cl_3C—CH=O, forms a cyclic trimer analogous to paraldehyde (Display 19.73, Sec. 19.11A).

(a) Account for the fact that two forms of this trimer are known (α, bp 223 °C and mp 116 °C; β, bp 250 °C and mp 152 °C).

(b) Which of your structures is likely to be the one with the higher melting point? Explain.

(c) Assuming you have in hand a sample of both forms, show how NMR spectroscopy could be used to verify your hypothesis in part (b).

19.67 Offer a rational explanation for each of the following observations.

(a) Although biacetyl (2,3-butanedione) and 1,2-cyclopentanedione have the same type of functional group, their dipole moments differ substantially.

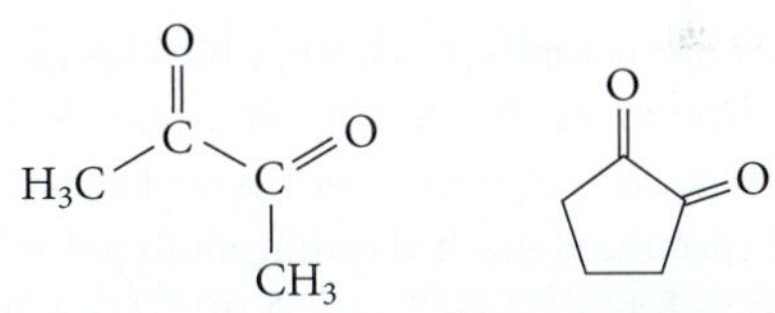

biacetyl μ = 1.04 D

1,2-cyclopentanedione μ = 2.21 D

(b) When acetaldehyde is mixed with a 10-fold excess of ethanethiol, its $n \rightarrow \pi^*$ absorption at 280 nm is nearly eliminated.

(a)

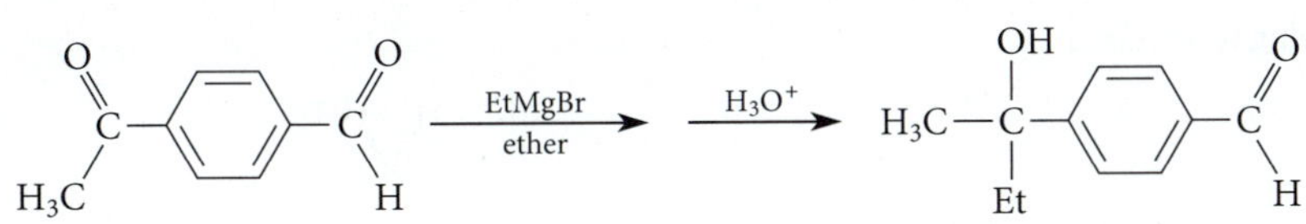

(b) $Ph_3P + (CH_3)_3CCH_2Br \longrightarrow \xrightarrow{CH_3(CH_2)_3Li} \xrightarrow{Ph—CH=O} PhCH=CHC(CH_3)_3$

FIGURE P19.63

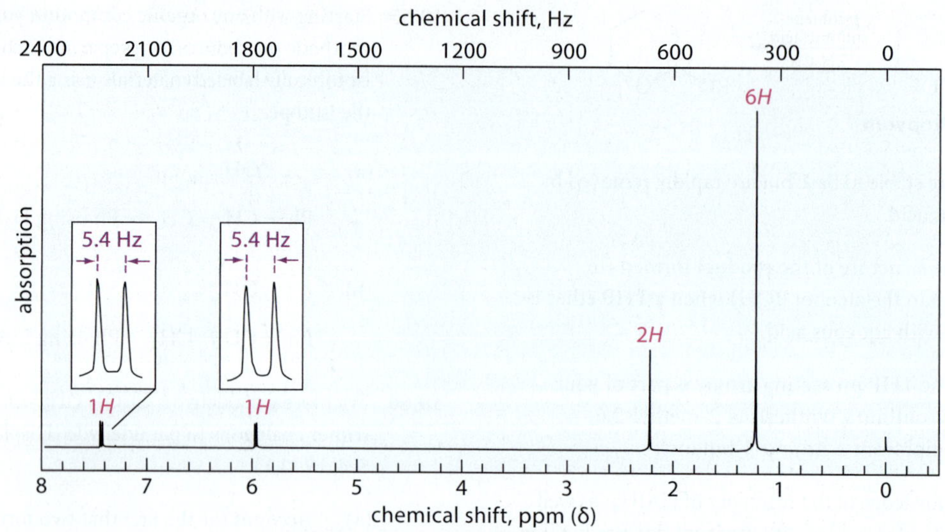

FIGURE P19.68 The proton NMR spectrum for Problem 19.68. The relative integrals are indicated in red.

(c) Compound *A* has a *much* weaker IR carbonyl absorption than compound *B*.

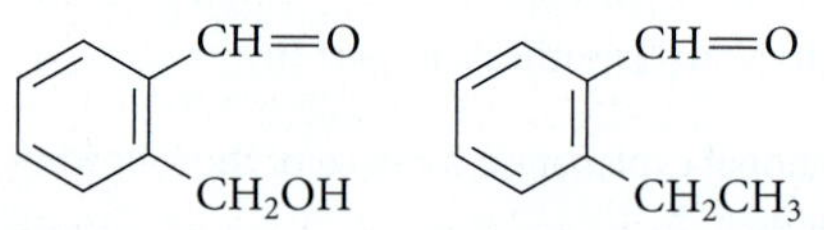

A *B*

19.68 Identify the compound $C_7H_{10}O$ that has an IR absorption at 1703 cm^{-1} and the proton NMR spectrum shown in **Fig. P19.68.**

19.69 Identify compound *A*, $C_6H_{12}O_3$, which has an IR absorption at 1710 cm^{-1} (no absorption in the 3200–3400 cm^{-1} region), as well the following ^{13}C NMR-DEPT spectrum (attached hydrogens in parentheses): δ 30.6 (3), δ 47.2 (2), δ 53.5 (3), δ 101.7 (1), δ 204.9 (0). One of the proton NMR absorptions of compound *A* is a singlet at δ 2.1.

19.70 In neutral alcohol solution, the UV spectra of *p*-hydroxyacetophenone and *p*-methoxyacetophenone are virtually identical. When NaOH is added to the solution, the λ_{max} of *p*-hydroxyacetophenone increases by about 50 nm, but that of *p*-methoxyacetophenone is unaffected. Explain these observations.

19.71 Using known reactions and mechanisms discussed in the text, complete the following reactions.

(a)

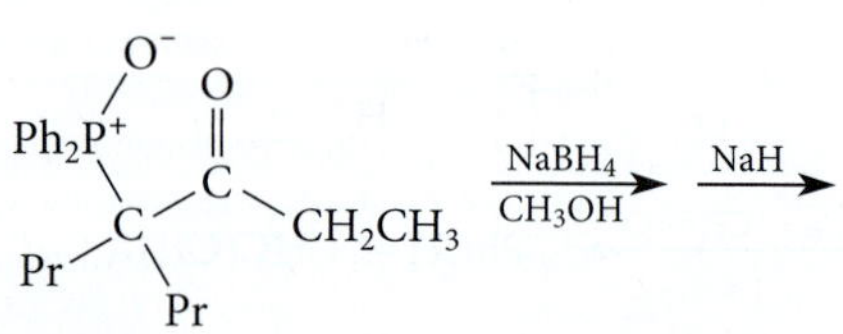

(*Hint:* See Eq. 19.94b, Sec. 19.14.)

(b)

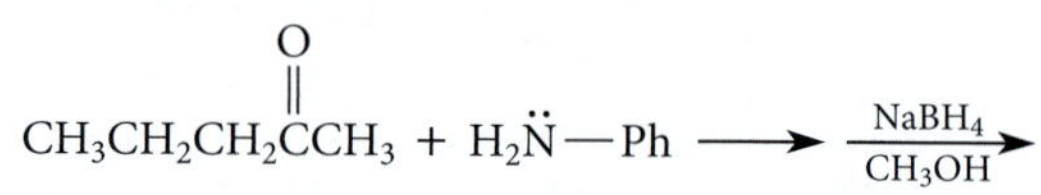

(*Hint:* The C=N bond undergoes addition much like the C=O bond.)

19.72 Give curved-arrow mechanisms for the following reactions.

(a)

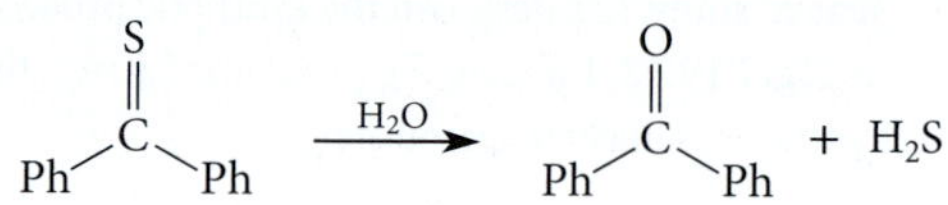

(b)

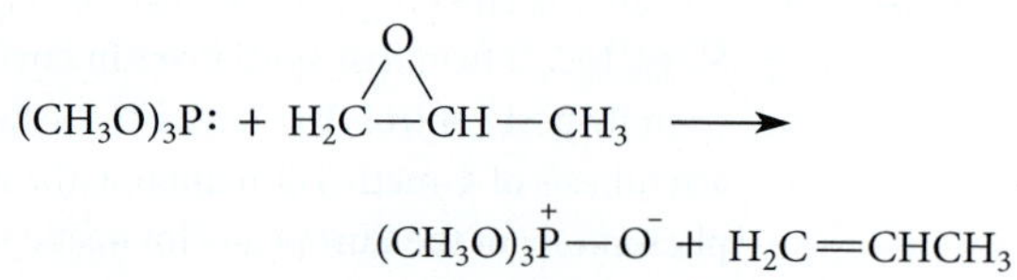

(c)

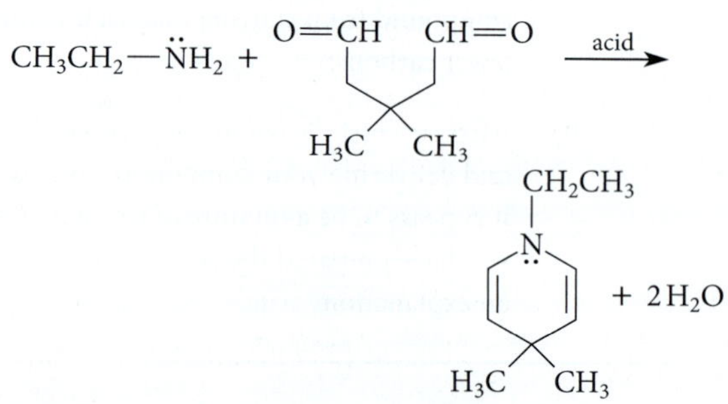

(d)

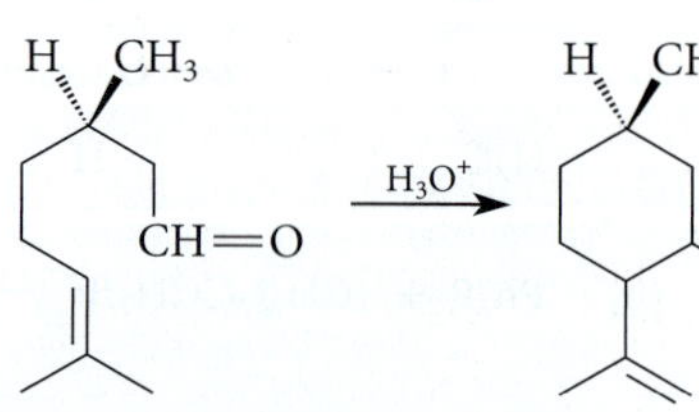

(e)

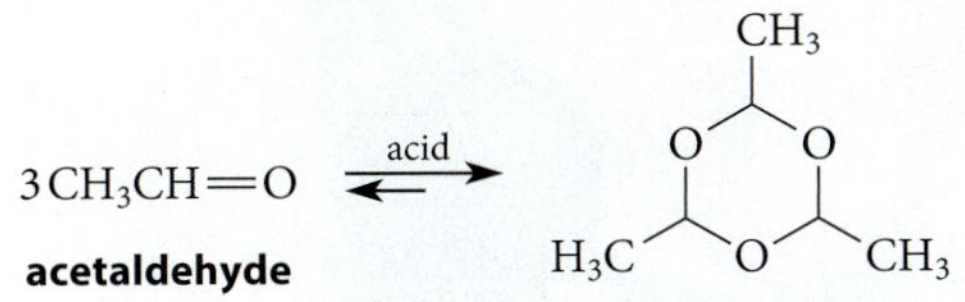

acetaldehyde **paraldehyde**

19.73 A compound A, C_8H_8O, when treated with Zn amalgam and HCl, gives a xylene (dimethylbenzene) isomer that in turn gives only one ring monobromination product with Br_2 and Fe. Propose a structure for A.

19.74 Compound A, $C_{11}H_{12}O$, which gave a negative Tollens test, was treated with $LiAlH_4$ followed by dilute acid to give compound B, which could be resolved into enantiomers. When optically active B was treated with CrO_3 in pyridine, an optically inactive sample of A was obtained. Heating A with hydrazine in base gave hydrocarbon C, which, when heated with alkaline $KMnO_4$, gave carboxylic acid D. Identify all of the compounds, and explain your reasoning.

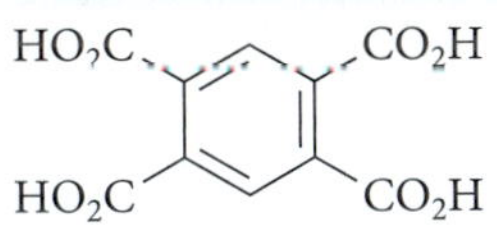

D

19.75 Compound A, $C_6H_{12}O_2$, was found to be optically active, and it was slowly oxidized to an optically active carboxylic acid B, $C_6H_{12}O_3$, by $^{+}Ag(NH_3)_2$. Oxidation of A by anhydrous CrO_3 gave an optically inactive compound that reacted with Zn amalgam/HCl to give 3-methylpentane. With aqueous H_2CrO_4, compound A was oxidized to an optically inactive dicarboxylic acid C, $C_6H_{10}O_4$. Give structures for compounds A, B, and C.

19.76 (a) The insecticide DDT can be prepared by the reaction shown in **Fig. P19.76a**. Remembering that a protonated aldehyde or ketone is a type of carbocation, and that carbocations are electrophiles, draw a curved-arrow mechanism for this electrophilic aromatic substitution reaction. (See Sec. 16.4E.)

(b) Bisphenol A (Sec. 19.17) is used in the manufacture of polycarbonate plastics. (Its carcinogenicity has become a cause for concern.) Bisphenol A is prepared by the acid-catalyzed reaction of acetone with two equivalents of phenol (**Fig. P19.76b**). Draw a curved-arrow mechanism for this reaction.

19.77 Salsolinol is formed in the brain when acetaldehyde reacts with dopamine.

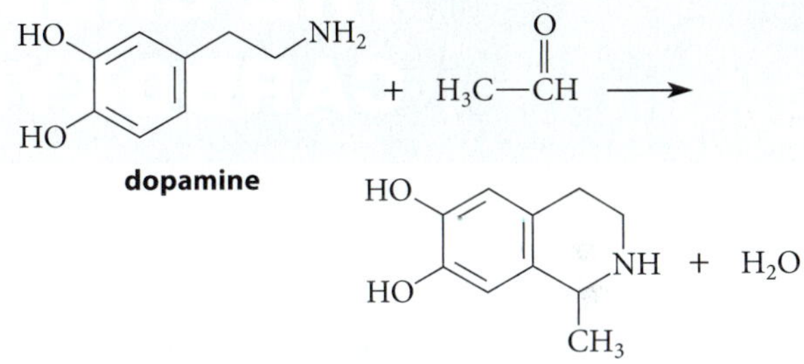

dopamine **salsolinol**

Because acetaldehyde is a biological oxidation product of ethanol (Sec. 10.8), it has been suggested that salsolinol might be used as a biological marker for alcohol consumption. Draw a curved-arrow mechanism for the formation of salsolinol from acetaldehyde and dopamine. Assume acids and bases are present as needed. (This is an example of the *Pictet–Spengler reaction.*)

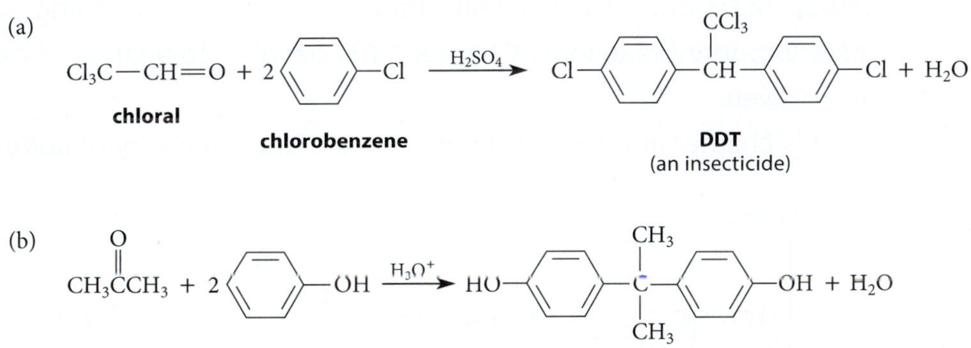

FIGURE P19.76

20 THE CHEMISTRY OF CARBOXYLIC ACIDS

The characteristic functional group in a **carboxylic acid** is the **carboxy group**.

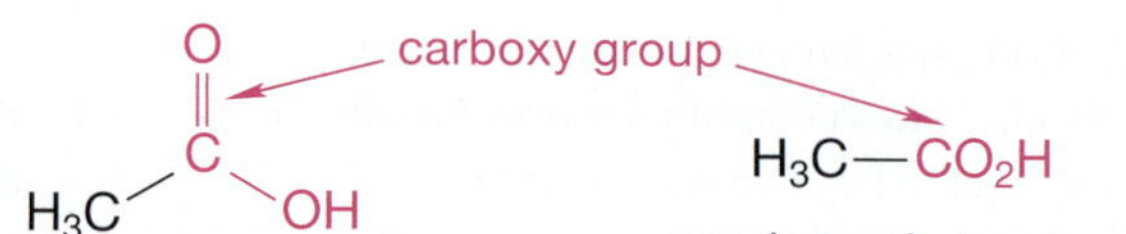

acetic acid
(a simple carboxylic acid)

condensed structure

(20.1)

Carboxylic acids and their derivatives rank with aldehydes and ketones among the most important organic compounds because they occur widely in living organisms and because they serve important roles in organic synthesis. This chapter is concerned with the structures, properties, acidities, and carbonyl-group reactions of carboxylic acids themselves, including some biologically important reactions. Chapter 21 is devoted to a study of carboxylic acid derivatives.

This chapter also surveys briefly some of the chemistry of **sulfonic acids**.

As we have pointed out elsewhere (Sec. 11.10), the octet structure is very important, but the hypervalent structure is often drawn for convenience.

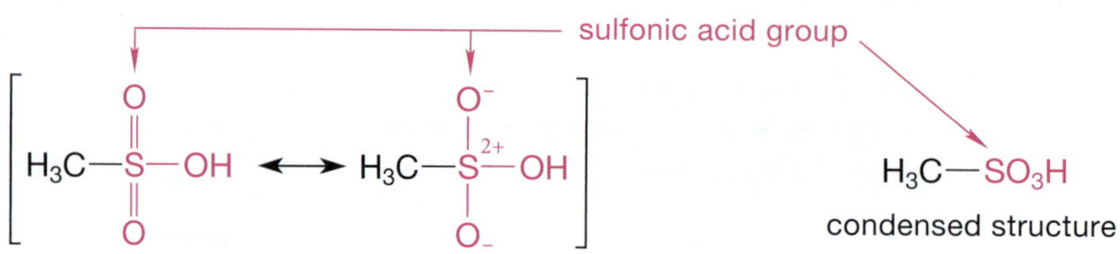

methanesulfonic acid
(a simple sulfonic acid)

condensed structure

(20.2)

20.1 NOMENCLATURE OF CARBOXYLIC ACIDS

A. Common Nomenclature

Common nomenclature is widely used for the simpler carboxylic acids. A carboxylic acid is named by adding the suffix *ic* and the word *acid* to the prefix for the appropriate group given in Table 19.1 (Sec. 19.2A).

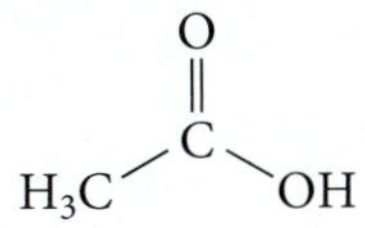

prefix (Table 19.1): acet + *ic acid* = **acetic acid**

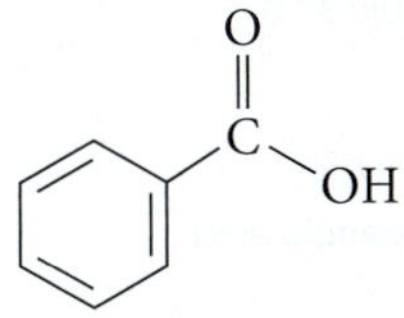

benzo + *ic acid* = **benzoic acid**

Some of these names owe their origin to the natural source of the acid. For example, formic acid occurs in the venom of the red ant (from the Latin *formica*, meaning "ant"), acetic acid is the acidic component of vinegar (from the Latin *acetus*, meaning "vinegar"), and butyric acid is the foul-smelling component of rancid butter (from the Latin *butyrum*, meaning "butter"). The common names of carboxylic acids, listed in **Table 20.1**, are used as much or more than the substitutive names.

As with aldehydes and ketones, substitution in the common system is denoted with Greek letters rather than numbers. The position *adjacent* to the carboxy group is designated as α.

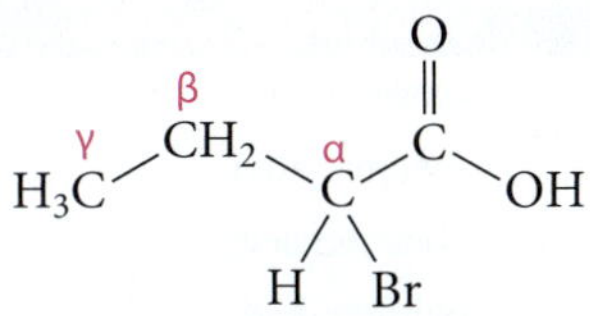

α-bromobutyric acid

In common nomenclature, the position of the substituent is omitted if it is unambiguous. For example, $ClCH_2CO_2H$ is named chloroacetic acid rather than α-chloroacetic acid.

Carboxylic acids with two carboxy groups are called **dicarboxylic acids**. The unbranched dicarboxylic acids are particularly important and are invariably known by their common names. Some important dicarboxylic acids are also listed in Table 20.1.

$HO_2C—CH_2CH_2—CO_2H$

succinic acid

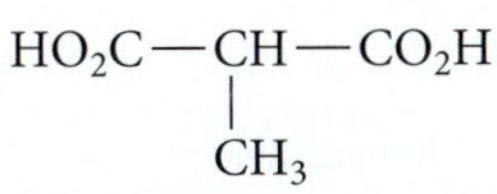

methylmalonic acid

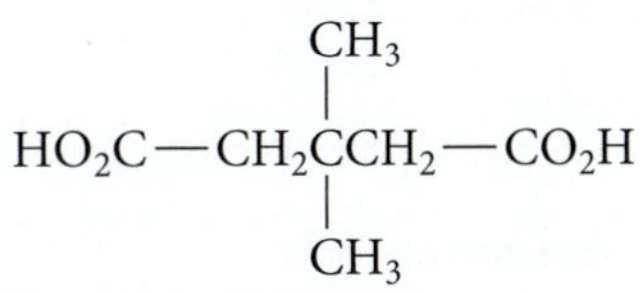

β,β-dimethylglutaric acid

Phthalic acid and terephthalic acid are important aromatic dicarboxylic acids.

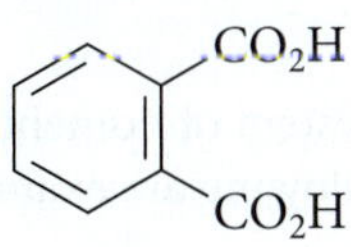

phthalic acid

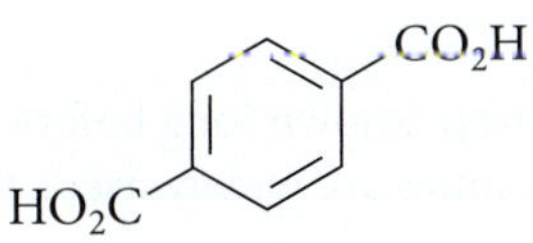

terephthalic acid

A mnemonic device used by generations of organic chemistry students for remembering the names of the dicarboxylic acids is the phrase "*Oh*, *My*, *Such* *Good* *Apple* *Pie*," in which the first letter of each word corresponds to the name of successive dicarboxylic acids: oxalic, malonic, succinic, glutaric, adipic, and pimelic acids.

TABLE 20.1 Names and Structures of Some Carboxylic Acids

Systematic name	Common name	Structure
methanoic* acid	formic acid	HCO_2H
ethanoic* acid	acetic acid	CH_3CO_2H
propanoic acid	propionic acid	$CH_3CH_2CO_2H$
butanoic acid	butyric acid	$CH_3CH_2CH_2CO_2H$
2-methylpropanoic acid	isobutyric acid	$(CH_3)_2CHCO_2H$
pentanoic acid	valeric acid	$CH_3(CH_2)_3CO_2H$
3-methylbutanoic acid	isovaleric acid	$(CH_3)_2CHCH_2CO_2H$
2,2-dimethylpropanoic acid	pivalic acid	$(CH_3)_3CCO_2H$
hexanoic acid	caproic acid	$CH_3(CH_2)_4CO_2H$
octanoic acid	caprylic acid	$CH_3(CH_2)_6CO_2H$
decanoic acid	capric acid	$CH_3(CH_2)_8CO_2H$
dodecanoic acid	lauric acid	$CH_3(CH_2)_{10}CO_2H$
tetradecanoic acid	myristic acid	$CH_3(CH_2)_{12}CO_2H$
hexadecanoic acid	palmitic acid	$CH_3(CH_2)_{14}CO_2H$
octadecanoic acid	stearic acid	$CH_3(CH_2)_{16}CO_2H$
2-propenoic* acid	acrylic acid	$H_2C{=}CHCO_2H$
2-butenoic* acid	crotonic acid	$CH_3CH{=}CHCO_2H$
benzoic acid	benzoic acid	$PhCO_2H$
	Dicarboxylic acids	
ethanedioic* acid	oxalic acid	$HO_2C{-}CO_2H$
propanedioic* acid	malonic acid	$HO_2CCH_2CO_2H$
butanedioic* acid	succinic acid	$HO_2C(CH_2)_2CO_2H$
pentanedioic* acid	glutaric acid	$HO_2C(CH_2)_3CO_2H$
hexanedioic* acid	adipic acid	$HO_2C(CH_2)_4CO_2H$
heptanedioic* acid	pimelic acid	$HO_2C(CH_2)_5CO_2H$
1,2-benzenedicarboxylic* acid	phthalic acid	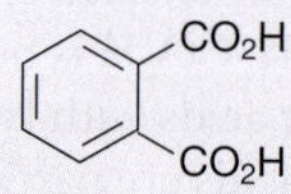
(*Z*)-2-butenedioic* acid	maleic acid	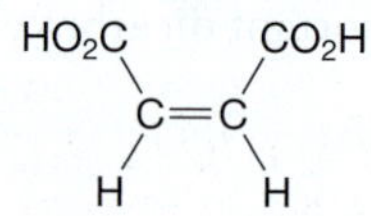
(*E*)-2-butenedioic* acid	fumaric acid	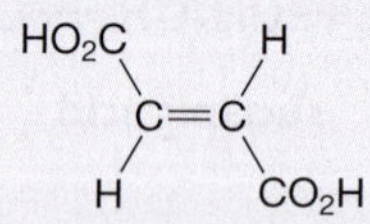

*The common name is almost always used instead.

Many carboxylic acids were known long before any system of nomenclature existed, and their time-honored traditional names are widely used. The following are examples of these.

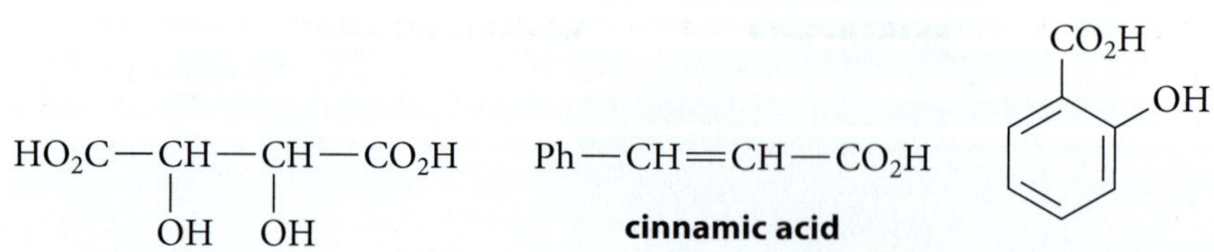

tartaric acid **cinnamic acid** **salicylic acid** (20.3)

B. Substitutive Nomenclature

A carboxylic acid is named systematically by dropping the final *e* from the name of the hydrocarbon with the same number of carbon atoms and adding the suffix *oic* and the word *acid*.

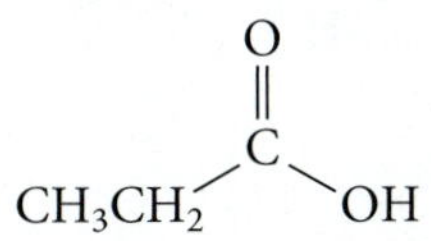

propane + *oic acid* = **propanoic acid**

The final *e* is not dropped in the name of dicarboxylic acids.

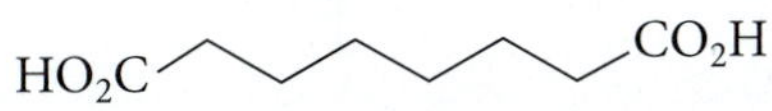

octanedioic acid

When a carboxylic acid is derived from a cyclic hydrocarbon, the suffix *carboxylic* and the word *acid* are added to the name of the hydrocarbon. (This nomenclature is similar to that for the corresponding aldehydes; Sec. 19.2B.)

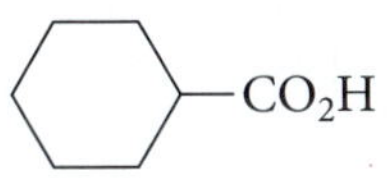

cyclohexanecarboxylic acid

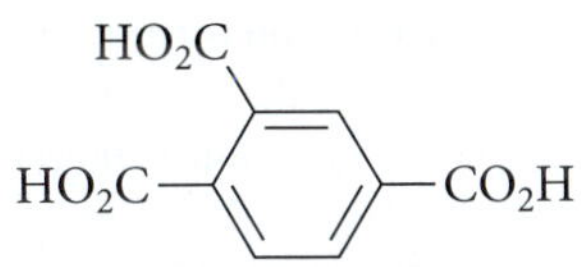

1,2,4-benzenetricarboxylic acid

One exception to this nomenclature is benzoic acid (Ph—CO_2H), for which the IUPAC recognizes the common name.

The principal chain in substituted carboxylic acids is numbered, as in aldehydes, by assigning the number 1 to the carbonyl carbon.

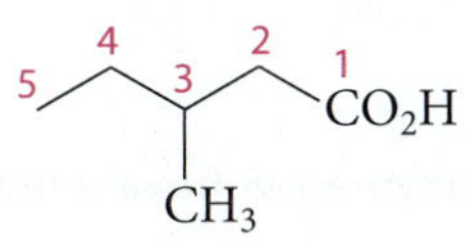

3-methylpentanoic acid

This numbering scheme should be contrasted with that used in the common system, in which numbering begins with the Greek letter α at carbon-2 (Sec. 20.1A).

In carboxylic acids derived from cyclic hydrocarbons, numbering begins at the ring carbon bearing the carboxy group.

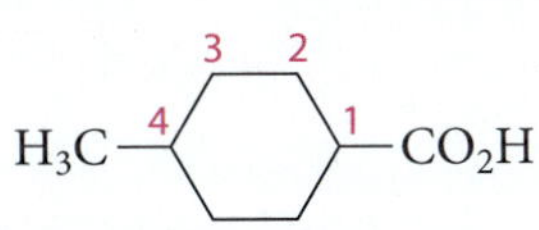

4-methylcyclohexanecarboxylic acid

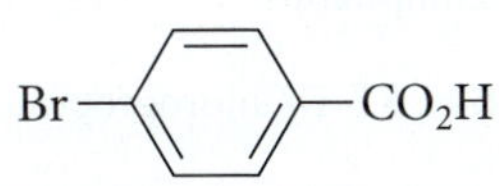

4-bromobenzoic acid
or ***p*-bromobenzoic acid**

When carboxylic acids contain other functional groups, the carboxy groups receive priority over aldehyde and ketone carbonyl groups, hydroxy groups, and mercapto groups for citation as the principal group.

Priority for citation as the principal group:

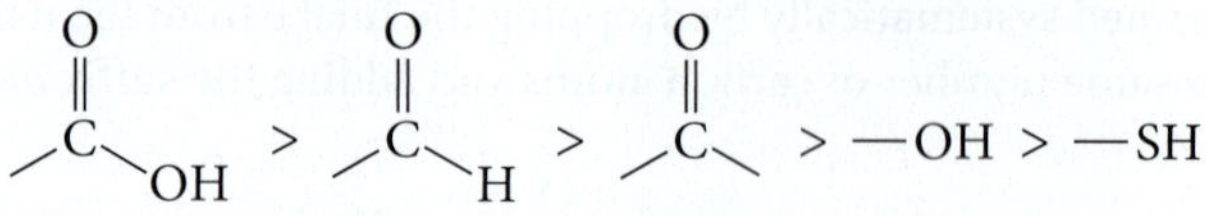

(20.4)

Study Problem 20.1

Provide a substitutive name for the following compound, including stereochemistry.

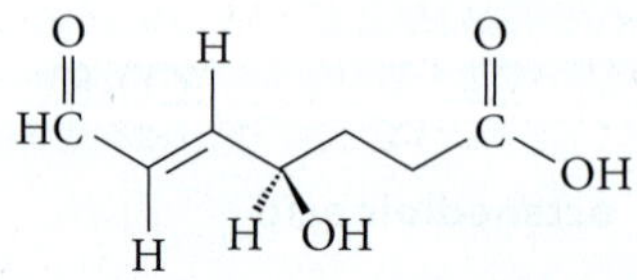

Solution First, decide on the principal group. From the order in Display 20.4, the carboxy group has highest priority. The aldehyde oxygen and the hydroxy group are treated as substituents. The structure has seven carbons and one double bond, so it is a heptenoic acid. The carboxy group is given the number 1; therefore, the double bond is at carbon-5, and the molecule is a 5-heptenoic acid. The OH group is named as a 4-hydroxy substituent and the aldehyde oxygen as a 7-oxo substituent. Application of the priority rules (Sec. 4.2B) shows that the double-bond stereochemistry is *E* and the stereochemistry at carbon-4 (Sec. 6.2) is *R*. The name is therefore (4*R*,5*E*)-4-hydroxy-7-oxo-5-heptenoic acid.

The carboxy group is sometimes named as a substituent:

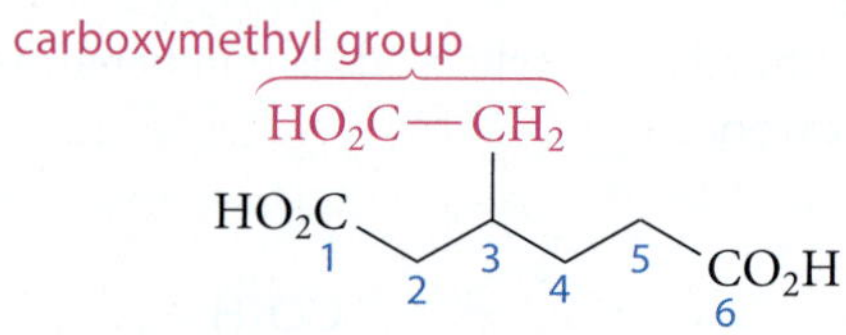

3-(carboxymethyl)hexanedioic acid (20.5)

A complete list of nomenclature priorities for all of the functional groups covered in this text is given in Appendix I.

Focused Problems

20.1 Draw a structure for each of the following compounds.

(a) γ-Hydroxybutyric acid (b) β,β-Dichloropropionic acid

(c) (*Z*)-3-Hexenoic acid (d) 4-Methylhexanoic acid

20.2 Name each of the following compounds. Use a common name for at least one compound.

(a)

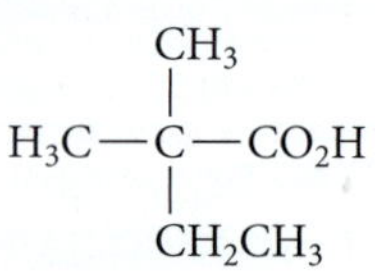

(b)

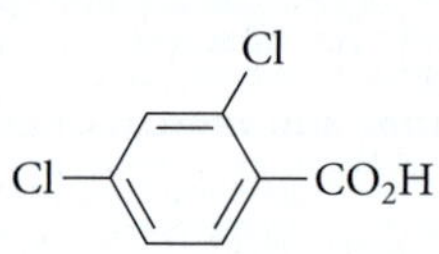

(c)

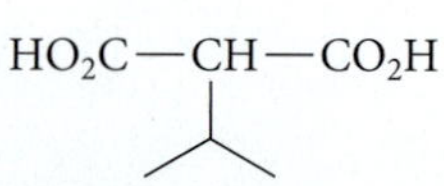

(d)

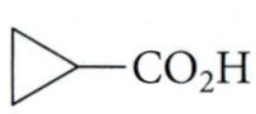

20.2 STRUCTURE AND PHYSICAL PROPERTIES OF CARBOXYLIC ACIDS

The structure of a simple carboxylic acid, acetic acid, is compared with the structures of other oxygen-containing compounds in **Fig. 20.1**. Carboxylic acids, like aldehydes and ketones, have trigonal geometry at their carbonyl carbons. The two oxygens of a carboxylic acid are quite different. One, the **carbonyl oxygen**, is the oxygen involved in the C═O double bond. This oxygen, being a terminal atom, is sp-hybridized (Sec. 1.7D). Figure 20.1 shows that the C═O bonds of aldehydes, ketones, and carboxylic acids have the same length. The other oxygen, called the **carboxylate oxygen**, is the oxygen involved in the C—O single bond. Like the oxygen in water, this oxygen is sp^2-hybridized. Figure 20.1 shows that the C—O bond in a carboxylic acid is considerably shorter than the C—O bond in an alcohol or ether (about 136 pm vs about 142 pm). The reason for this difference is that the C—O bond in an acid is an sp^2–sp^2 single bond, whereas the C—O bond in an alcohol or ether is an sp^3–sp^2 single bond.

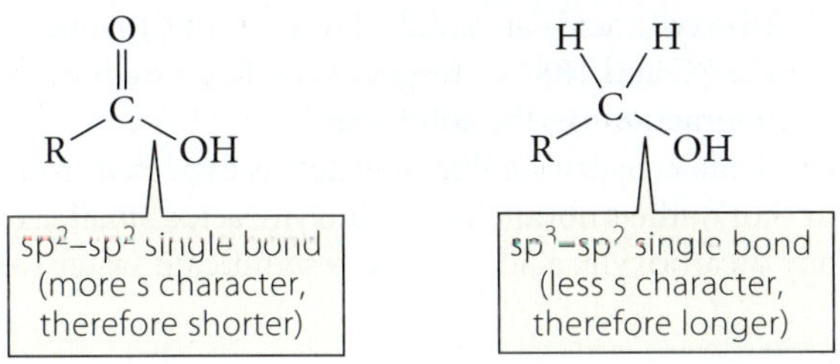

(20.6)

The carboxylic acids of lower molecular mass are high-boiling liquids with acrid, piercing odors. They have considerably higher boiling points than many other organic compounds of about the same molecular mass and shape:

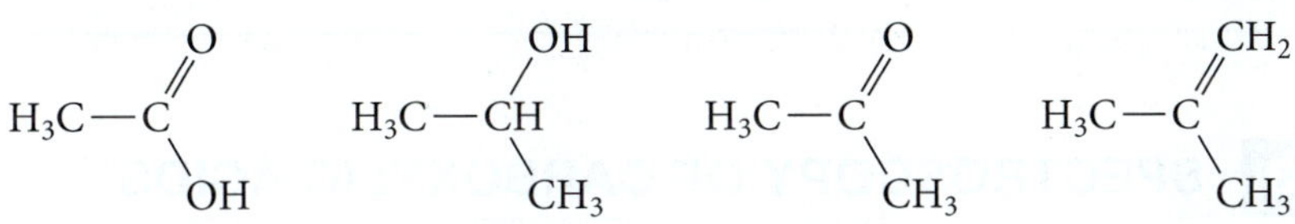

	acetic acid	**isopropyl alcohol**	**acetone**	**isobutylene**
boiling point	117.9 °C	82.3 °C	56.5 °C	−6.9 °C

(20.7)

The high boiling points of carboxylic acids can be attributed not only to their polarity, but also to the fact that they form very strong hydrogen bonds (Sec. 8.5C). In the solid state, and under some

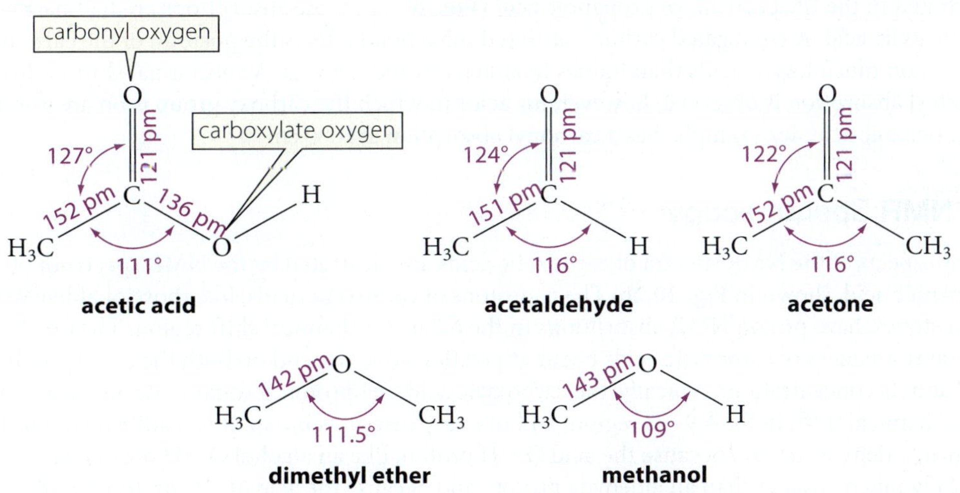

FIGURE 20.1 Comparison of the structures of acetic acid and other oxygen-containing compounds. The carbonyl compounds have identical C═O bond lengths, and the C—O single bond in a carboxylic acid is shorter than that in an ether or an alcohol.

conditions in both the gas phase and solution, carboxylic acids exist as hydrogen-bonded dimers. (A **dimer** is any structure derived from two identical smaller units.)

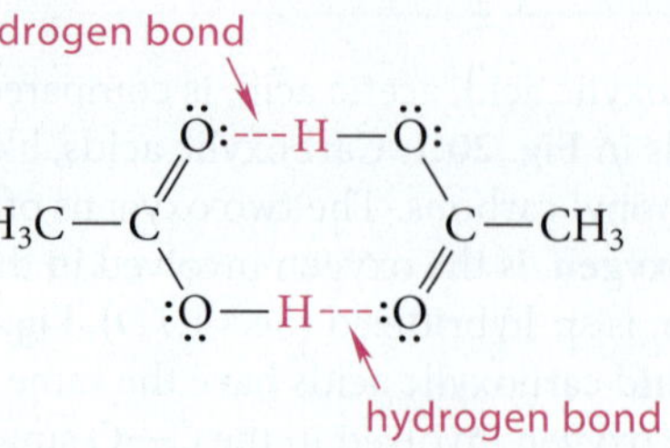

acetic acid dimer (20.8)

The equilibrium constants for the formation of such dimers in solution are very large—on the order of 10^6 to 10^7 M^{-1}. (The equilibrium constant for hydrogen-bond dimerization of ethanol, in contrast, is 11 M^{-1}.)

Many aromatic and dicarboxylic acids are solids. For example, the melting points of benzoic acid and succinic acid are 122 °C and 188 °C, respectively. Their high melting points reflect strong hydrogen-bonding interactions in the solid state.

The simpler carboxylic acids are very soluble in water, as expected from their hydrogen-bonding capabilities (Sec 8.6D); the unbranched carboxylic acids smaller than pentanoic acid are miscible with water. Many dicarboxylic acids also have significant water solubilities.

Focused Problem

20.3 At a given concentration of acetic acid, in which solvent would you expect the amount of acetic acid dimer to be greater: CCl_4 or water? Explain.

20.3 SPECTROSCOPY OF CARBOXYLIC ACIDS

A. IR Spectroscopy

The IR spectra of carboxylic acids are nearly always run under conditions such that they are in the dimer form. The carbonyl absorptions of carboxylic acid monomers occur near 1760 cm^{-1} but are rarely observed.

Two important absorptions are found in the infrared spectrum of a typical carboxylic acid. One is the C═O stretching absorption, which occurs near 1710 cm^{-1} for carboxylic acid dimers. The other is the O—H stretching absorption. This absorption is much broader than the O—H stretching absorption of an alcohol or phenol and covers a very wide region of the spectrum—typically 2400–3600 cm^{-1}. (In many cases this absorption obliterates the C—H stretching absorption of the acid.) The carbonyl absorption and this broad O—H stretching absorption are illustrated in the IR spectrum of propanoic acid (**Fig. 20.2a**); these absorptions are hallmarks of a carboxylic acid. A conjugated carbon–carbon double bond affects the position of the carbonyl absorption much less in acids than it does in aldehydes and ketones. A substantial shift in the carbonyl absorption is observed, however, for acids in which the carboxy group is on an aromatic ring. Benzoic acid, for example, has a carbonyl absorption at 1680 cm^{-1}.

B. NMR Spectroscopy

Many aspects of the NMR spectra of carboxylic acids are illustrated by the NMR spectrum of propanoic acid, shown in **Fig. 20.2b**. The α-protons of carboxylic acids, like those of aldehydes and ketones, have proton NMR absorptions in the δ 2.0–2.5 chemical shift region. The O—H proton resonances of carboxylic acids occur at positions that depend on both the acidity of the acid and its concentration. Typically, the carboxylic acid OH proton resonance occurs at a very large chemical shift, in the δ 9–13 region, and in many cases it is broad. It is readily distinguished from an aldehydic proton because the acid O—H proton, like an alcohol O—H proton, is typically much broader than an aldehyde proton, and because the acid O—H proton rapidly exchanges with D_2O (Sec. 14.6).

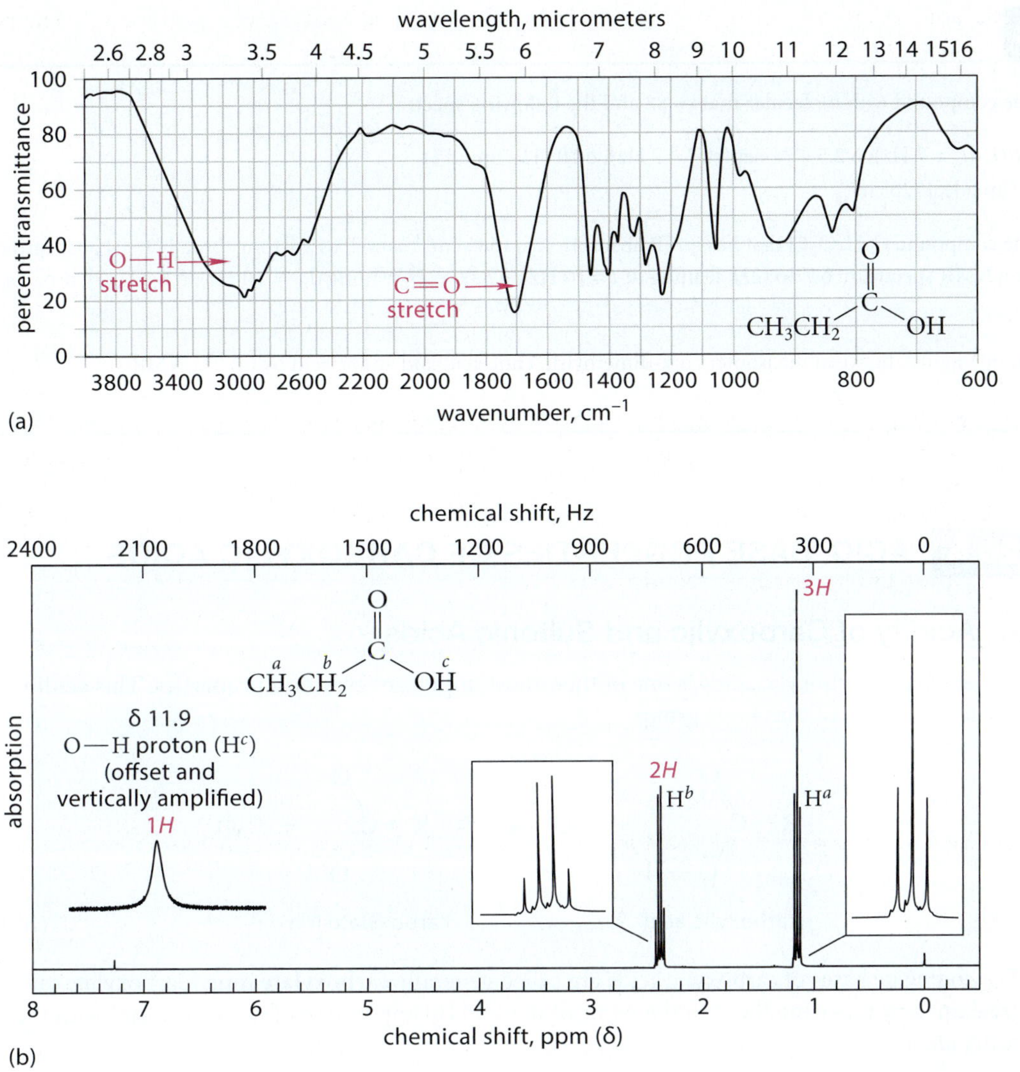

FIGURE 20.2 The spectra of propanoic acid illustrate typical characteristics of carboxylic acid spectra. (a) IR spectrum of propanoic acid. Notice particularly the very broad O—H stretching absorption. (b) Proton NMR spectrum of propanoic acid. The O—H absorption occurs at a very large chemical shift, and the chemical shifts of the other hydrogens are in about the same positions as shifts of the corresponding protons in aldehydes and ketones.

The ^{13}C NMR absorptions of carboxylic acids are similar to those of aldehydes and ketones, although the carbonyl carbon of an acid has a somewhat *smaller* chemical shift than that of an aldehyde or ketone.

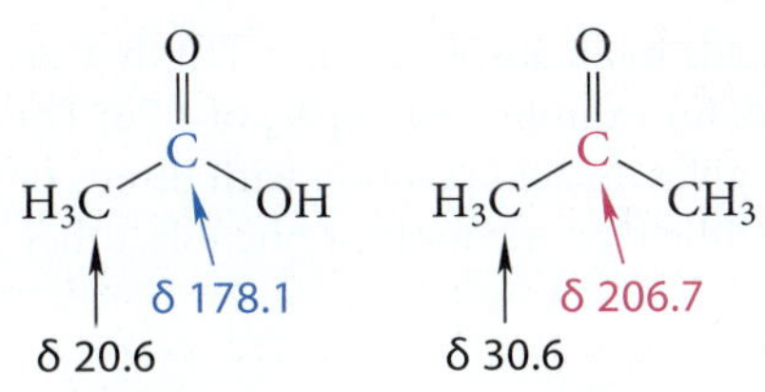

acetic acid **acetone** (20.9)

This carbonyl shift is contrary to what is expected from the relative electronegativities of oxygen and carbon; electronegative atoms generally cause *greater* chemical shifts. This unusual chemical shift is caused by shielding effects of the nonbonding electron pairs on the carboxylate oxygen.

FURTHER EXPLORATION 20.1 Chemical Shifts of Carbonyl Carbons

Focused Problems

20.4 Give the structure of the compound with molecular mass = 88 and the following spectra.

Proton NMR: δ 1.2 (6*H*, d, $J = 7$ Hz); δ 2.5 (1*H*, septet, $J = 7$ Hz); δ 10 (1*H*, broad s)
IR: 2600–3400 cm^{-1} (broad), 1720 cm^{-1}

20.5 Give the structure of the compound $C_7H_5O_2Cl$ that has an IR absorption at 1685 cm^{-1} as well as a strong, broad O—H absorption, and the following proton NMR spectrum: δ 7.56 (2*H*, leaning d, $J = 10$ Hz); δ 8.00 (2*H*, leaning d, $J = 10$ Hz); δ 8.27 (1*H*, broad s, exchanges with D_2O).

20.6 Explain how you would distinguish between the isomers α,α-dimethylsuccinic acid and adipic acid by (a) ^{13}C NMR; (b) proton NMR.

20.4 ACID–BASE PROPERTIES OF CARBOXYLIC ACIDS

A. Acidity of Carboxylic and Sulfonic Acids

The acidity of carboxylic acids is one of their most important chemical properties. This acidity is due to ionization of the O—H group.

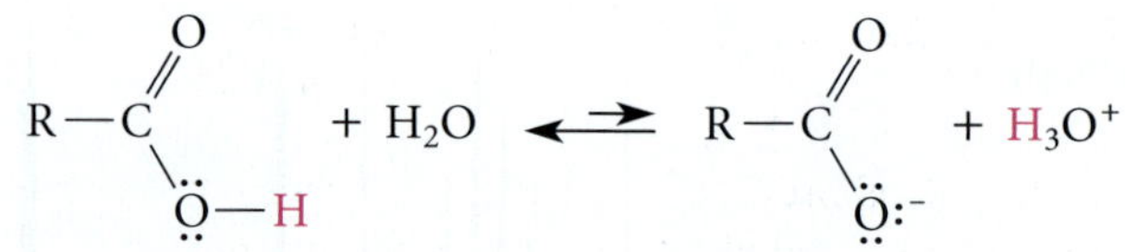

carboxylic acid carboxylate ion (20.10)

The conjugate bases of carboxylic acids are called generally **carboxylate ions**. Carboxylate salts are named by replacing the *ic* in the name of the acid (in any system of nomenclature) with the suffix *ate*.

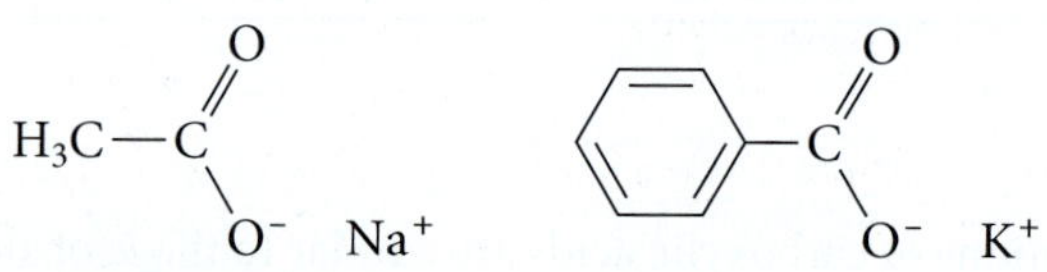

sodium acetate (acet~~ic~~ + *ate* = acetate) **potassium benzoate**

STUDY GUIDE LINK 20.1 Reactions of Bases with Carboxylic Acids

Carboxylic acids are weak acids in an absolute sense, but they are among the most acidic organic compounds; acetic acid, for example, has a pK_a of 4.76. This pK_a is low enough that an aqueous solution of acetic acid gives an acid reaction with litmus or pH paper.

Carboxylic acids are more acidic than alcohols or phenols, other compounds with O—H bonds.

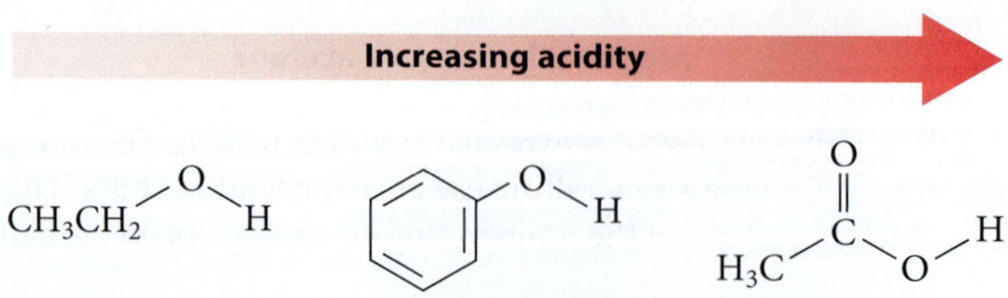

pK_a 15.9 9.95 4.76 (20.11)

The acidity of carboxylic acids is due to two factors. The first is the *polar effect* of the carbonyl group itself. The carbonyl group, because of its sp^2-hybridized carbon and sp-hybridized oxygen atom, as well as the partial positive charge on the carbonyl carbon, is a very electronegative group—much more electronegative than the phenyl ring of a phenol or the alkyl group of an alcohol. The polar effect of the carbonyl group stabilizes charge in the carboxylate ion. Remember that *stabilization of a conjugate base enhances acidity* (see Fig. 3.3, Sec. 3.7E).

The second factor that accounts for the acidity of carboxylic acids is the resonance stabilization of their conjugate-base carboxylate ions.

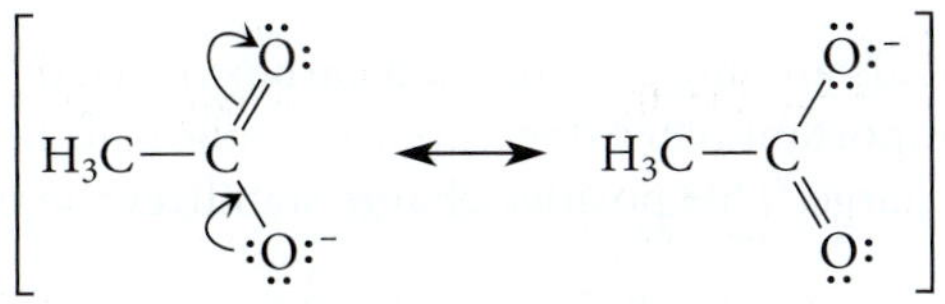

resonance structures of the acetate ion (20.12)

STUDY GUIDE LINK 20.2
Resonance Effect on Carboxylic Acid Acidity

This was used as an example in the discussion of the resonance effect on acidity (Display 3.75, Sec. 3.7D).

Although typical carboxylic acids have pK_a values in the 4–5 range, the acidities of carboxylic acids vary with structure. Recall, for example (Display 3.80, Sec. 3.7E), that halogen substitution within the alkyl group of a carboxylic acid enhances acidity by a polar effect.

	$H_3C—CO_2H$	$FCH_2—CO_2H$	$F_2CH—CO_2H$	$F_3C—CO_2H$
	acetic acid	**fluoroacetic acid**	**difluoroacetic acid**	**trifluoroacetic acid**
pK_a	4.76	2.66	1.24	0.23

(20.13)

Trifluoroacetic acid, commonly abbreviated TFA, is such a strong acid that it is often used in place of HCl and H_2SO_4 when an acid of moderate strength is required.

The pK_a values of some carboxylic acids are given in **Table 20.2**, and the pK_a values of some simple dicarboxylic acids are given in **Table 20.3**. The data in these tables provide some idea of the range over which the acidities of carboxylic acids vary.

TABLE 20.2 pK_a **Values of Some Carboxylic Acids**

Acid*	pK_a
formic	3.75
acetic	4.76
propionic	4.87
2,2-dimethylpropanoic (pivalic)	5.05
acrylic	4.26
chloroacetic	2.85
phenylacetic	4.31
benzoic	4.18
p-methylbenzoic (*p*-toluic)	4.37
p-nitrobenzoic	3.43
p-chlorobenzoic	3.98
p-methoxybenzoic (*p*-anisic)	4.47
2,4,6-trinitrobenzoic	0.65

*See Table 20.1 for structures.

TABLE 20.3 pK_a **Values of Some Dicarboxylic Acids**

Acid*	First pK_a	Second pK_a
carbonic	3.77‡	10.33
oxalic	1.27	4.27
malonic	2.86	5.70
succinic	4.21	5.64
glutaric	4.34	5.27
adipic	4.41	5.28
phthalic	2.95	5.41

*See Table 20.1 for structures.

‡This value, which corrects for the amount of H_2CO_3 in aqueous CO_2, is the actual pK_a of carbonic acid. An often-cited value of 6.4 treats all dissolved CO_2 as H_2CO_3.

Sulfonic acids are much stronger than comparably substituted carboxylic acids.

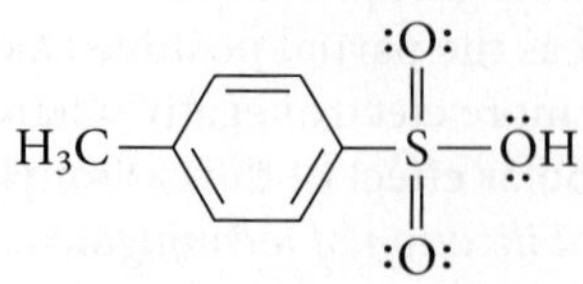

***p*-toluenesulfonic acid**
(**TsOH,** or **tosic acid**)
a strong acid; $pK_a \sim -3$ (20.14)

One reason that sulfonic acids are more acidic than carboxylic acids is the high oxidation state of sulfur. The more important octet structure for a sulfonate anion indicates that sulfur has considerable positive charge. This positive charge stabilizes the negative charge on the oxygens.

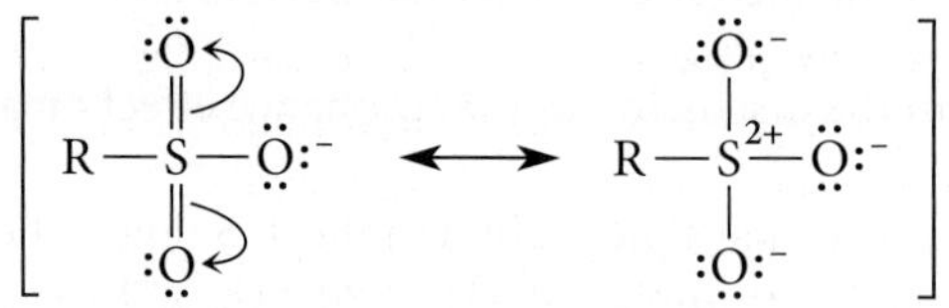

expanded octet at sulfur — octet at sulfur (more important structure)
a sulfonate anion (20.15)

Sulfonic acids are useful as acid catalysts in organic solvents because they are more soluble than most inorganic acids. For example, *p*-toluenesulfonic acid is moderately soluble in benzene and toluene and can be used as a strong acid catalyst in those solvents. (Sulfuric acid, in contrast, is completely insoluble in benzene and toluene.)

Many carboxylic acids of moderate molecular mass are insoluble in water. Their alkali metal salts, however, are ionic compounds, and in many cases they are much more soluble in water (Sec. 8.7F; Eq. 8.71). Therefore, many water-insoluble carboxylic acids dissolve in solutions of alkali metal hydroxides (NaOH, KOH) because the insoluble acids are converted completely into their soluble salts.

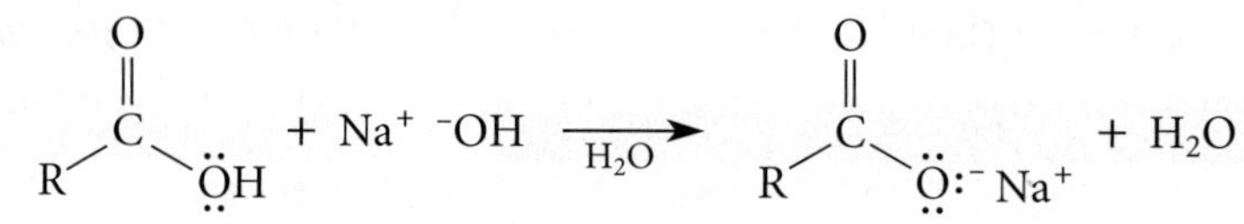

more soluble than the carboxylic acid in water (20.16)

Even a 5% sodium bicarbonate ($Na^+ \ HCO_3^-$) solution is basic enough (pH ~ 8.5) to dissolve a carboxylic acid. This statement follows from Eq. 3.67a. Remember that the fraction of an acid that is ionized, f_A, is determined by the difference $\Delta = \text{pH} - pK_a$. With pH = 8.5 and $pK_a < 4.5$, then $\Delta = 4$, and the fraction ionized is essentially 1:

$$f_A = \frac{1}{1 + 10^{-\Delta}} = \frac{1}{1 + 10^{-4}} = 0.9999 \approx 1 \tag{20.17}$$

Provided that the bicarbonate is in excess (so that it is not completely consumed by the carboxylic acid), the carboxylic acid, then, is completely converted into its conjugate-base anion, which is soluble in water.

A typical carboxylic acid can be separated from mixtures with other water-insoluble, nonacidic substances by extraction with NaOH, Na_2CO_3, or $NaHCO_3$ solution. The acid dissolves in the basic aqueous solution, but nonacidic compounds do not. After separating the basic aqueous solution, it can be acidified with a strong acid to yield the carboxylic acid, which may be isolated by filtration or by extraction with organic solvents. (A similar idea was used in the separation of phenols; Sec. 18.7B.) Carboxylic acids can also be separated from phenols by extraction with 5% of $NaHCO_3$ if the phenol is not unusually acidic. Because the pK_a of a typical phenol is about 10, it remains largely un-ionized and therefore insoluble in an aqueous solution with a pH of 8.5. (This conclusion follows from an equation for phenol ionization analogous to Eq. 20.17.)

Focused Problems

20.7 (a) Write the equations for the first and second ionizations of succinic acid. Label each with the appropriate pK_a values from Table 20.3.

(b) Why is the first pK_a value of succinic acid lower than the second pK_a value?

20.8 Imagine that you have just carried out a conversion of *p*-bromotoluene into *p*-bromobenzoic acid and wish to separate the product from the unreacted starting material. Design a separation of these two substances that would enable you to isolate the purified acid starting with a solution of both compounds in methylene chloride.

B. Basicity of Carboxylic Acids

Although we think of carboxylic acids primarily as acids, the carbonyl oxygens of acids, like those of aldehydes or ketones, are weakly basic.

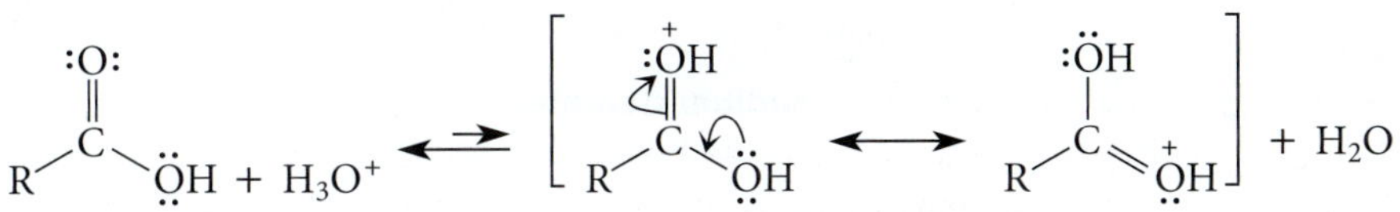

protonated carboxylic acid
pK_a ~ −6

(20.18)

As we show in subsequent sections, the basicity of carboxylic acids plays an important role in many of their reactions.

Protonation of an acid on the *carbonyl oxygen* occurs because, as Eq. 20.18 shows, a resonance-stabilized cation is formed. Protonation on the *carboxylate oxygen* is much less favorable because it does not give a resonance-stabilized cation and because the positive charge on oxygen is destabilized by the polar effect of the carbonyl group.

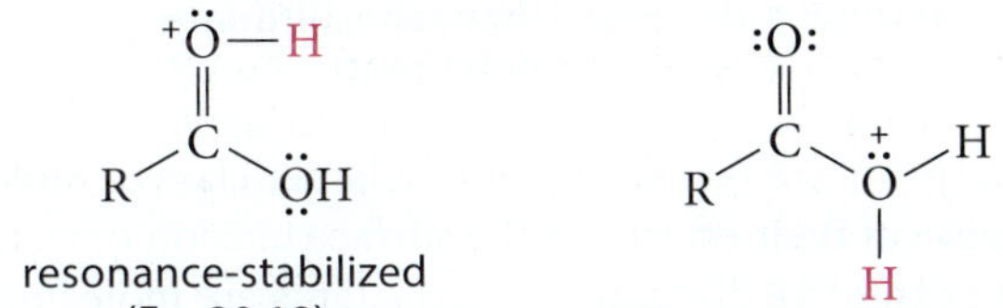

resonance-stabilized (Eq. 20.18)

not resonance-stabilized; does not form

(20.19)

20.5 FATTY ACIDS, SOAPS, AND DETERGENTS

Carboxylic acids with long, unbranched carbon chains are called **fatty acids** because many of them are liberated from fats and oils by a hydrolytic process called *saponification* (Sec. 21.7A). Some fatty acids contain carbon–carbon double bonds. Fatty acids with cis double bonds occur widely in nature, but those with trans double bonds are rare. The following compounds are examples of common fatty acids:

$CH_3(CH_2)_{14}CO_2H$ or $CH_3CH_2CH_2CH_2CH_2CH_2CH_2CH_2CH_2CH_2CH_2CH_2CH_2CH_2CH_2CO_2H$

or 

palmitic acid
(from palm oil) (20.20a)

$CH_3(CH_2)_{16}CO_2H$ or 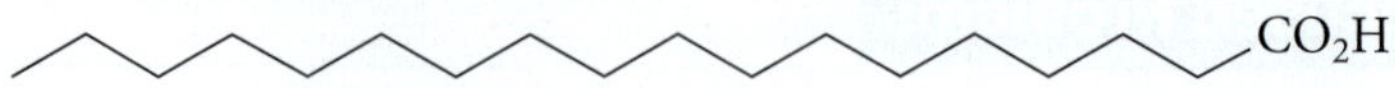

stearic acid
(Greek *stear*, meaning "tallow," or "beef fat")

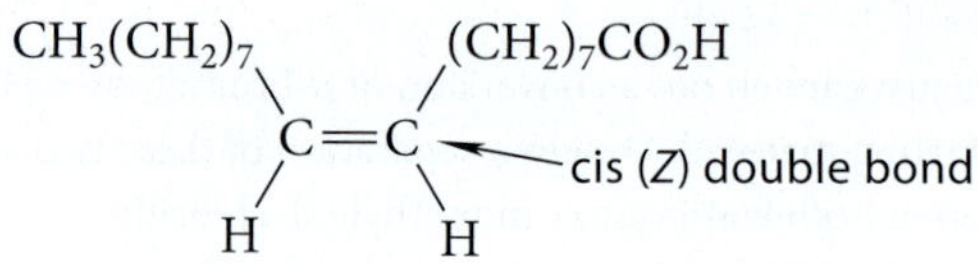

oleic acid (20.20b)

The sodium and potassium salts of fatty acids, called **soaps**, are the major ingredients of commercial soap.

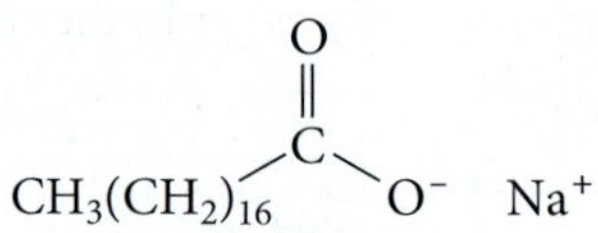

sodium stearate
(a soap) (20.21)

Closely related to soaps are **detergents**. A detergent, like a soap, has a long hydrocarbon tail and an ionic head group, but its polar head group is something other than a carboxylate group. For example, the following compound, the sodium salt of a sulfonic acid, is used in household laundry detergent formulations.

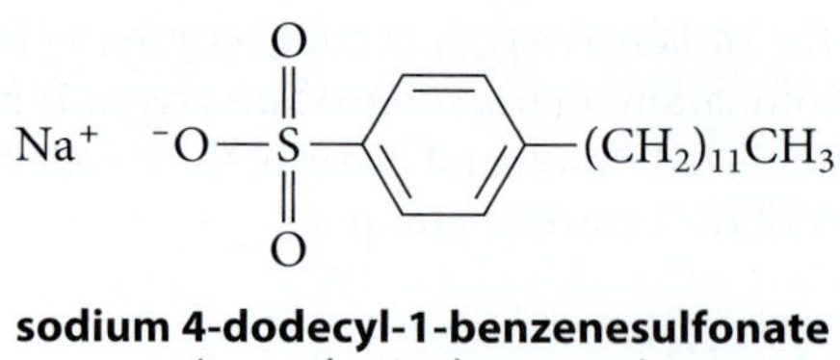

sodium 4-dodecyl-1-benzenesulfonate
(a synthetic detergent) (20.22)

FURTHER EXPLORATION 20.2
More on Surfactants

Soaps and synthetic detergents are two examples of a larger class of molecules known as **surfactants**, so-called because of their effects on the surface tension of water. (See Further Exploration 20.2 for a more extensive discussion.) Surfactants are molecules with two structural parts that interact with water in different ways: *a polar head group*, which is readily solvated by water, and a *hydrocarbon tail*, which, like a long alkane, is not readily solvated by water. In Sec. 8.7B we discussed phospholipids, the major components of cell membranes, which also have polar head groups and hydrocarbon tails. Phospholipids are also surfactants. In a soap, the polar head group is the carboxylate anion, and the hydrocarbon tail is the carbon chain. The soap and the detergent shown Displays 20.21 and 20.22 are examples of *anionic* *surfactants*—that is, surfactants with an anionic polar head group. *Cationic surfactants* are also known:

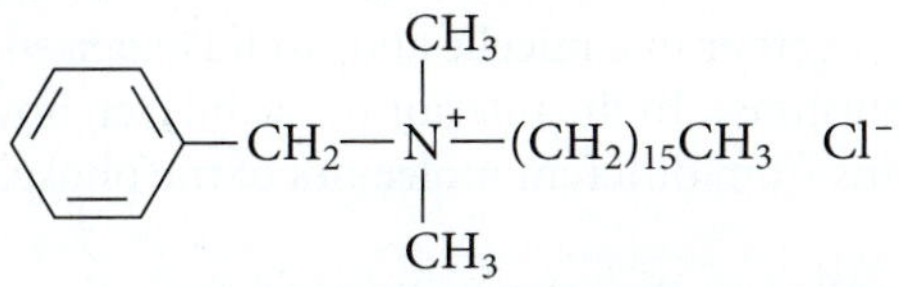

benzylcetyldimethylammonium chloride
(benzalkonium chloride)
a cationic detergent and germicide (20.23)

Although small amounts of soap and detergent molecules dissolve in water, when their concentrations are raised above a certain value, called the **critical micelle concentration (CMC)**, the soap or detergent molecules spontaneously form **micelles**, which are approximately spherical aggregates of 50–150 molecules (**Fig. 20.3**). Think of a micelle as a large ball in which the polar head groups, along with their counterions, are exposed on the outside of the ball and the nonpolar tails are buried on the inside of the ball. The micellar structure satisfies the solvation requirements of both the polar head groups, which are close to water, and the "greasy groups"—the nonpolar tails—which associate with each other on the inside of the micelle.

The spontaneous assembly of micelles calls to mind the spontaneous formation of phospholipid bilayers in aqueous media (Sec. 8.7B). Both phenomena have similar causes: the entropy-driven release of solvation water as the hydrocarbon chains come together and the van der Waals attractions between the hydrocarbon chains (Sec. 8.6D). Why don't soaps and detergents form the same sort of bilayers that phospholipids do? Or, to turn the question around, why don't phospholipids form micelles instead of bilayers? Recall that phospholipids contain two long hydrocarbon chains per molecule, whereas soaps and detergents have only one. If a bilayer were to form from a soap, the alkyl chains would be relatively far apart, and their favorable van der Waals interactions would be relatively weak. (Van der Waals forces increase strongly with decreasing distance.) In a micelle, the hydrocarbon chains of a soap are packed more closely,

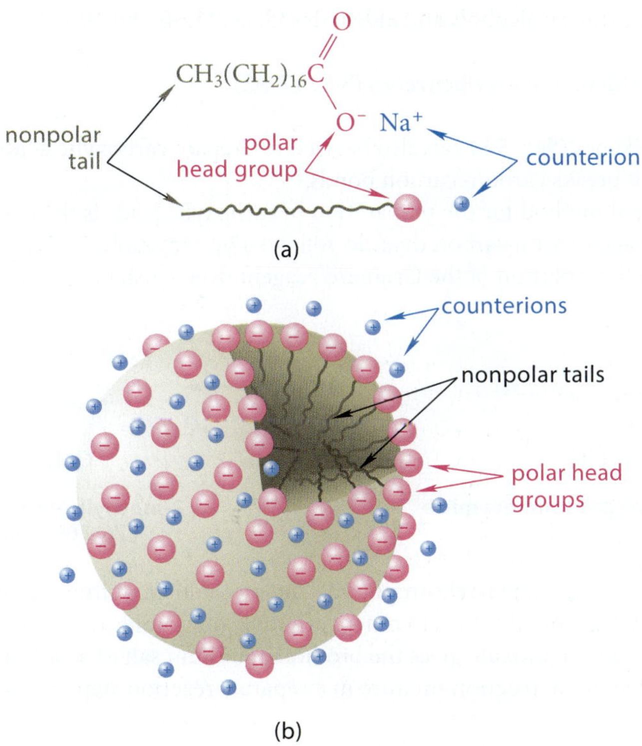

FIGURE 20.3 (a) Schematic diagram of a soap. The polar head group (the carboxylate group) is represented by a red sphere, the nonpolar tail by a wiggly line, and the counterion as a blue sphere. (b) A cutaway diagram of a micelle structure. Each micelle contains 50–150 molecules and is approximately spherical. The polar head groups within the micelle are directed outward toward the solvent water, whereas the nonpolar tails interact with one another and are isolated from water within the interior of the micelle.

so their attractive van der Waals interactions are stronger. On the other hand, in the interior of a micelle derived from a phospholipid, there isn't sufficient space to accommodate the two hydrocarbon chains per molecule. In other words, the hydrocarbon chains of phospholipid molecules would be *too close* together in a micelle and would experience unfavorable van der Waals interactions (steric repulsions). In the interior of the bilayer, however, there is enough room for the two hydrocarbon chains from different molecules of the phospholipid to interact favorably without steric repulsion.

The use of soaps and detergents in cleaning applications is understandable once the rationale for the formation of micelles is clear. When a fabric with greasy dirt is exposed to an aqueous solution containing micelles of a soap or detergent, the dirt associates with the "greasy" hydrocarbon chains on the interior of the micelle and is incorporated into the micellar aggregate. The dirt is thus lifted away from the surface of the fabric and carried into solution. Many soaps and detergents also have germicidal properties that owe their success to a similar phenomenon. When a bacterial cell is exposed to a solution containing a detergent, phospholipids of the cell membrane tend to associate with the detergent. In some cases, this disrupts the membrane enough that the cell can no longer function, and it dies.

So-called *hard water*, which contains Ca^{2+} and Mg^{2+}, interferes with the cleaning action of soaps. Hard-water scum ("bathtub ring") is a precipitate of the calcium or magnesium salts of fatty acids, which (unlike the sodium and potassium salts) do not form micelles because they are too insoluble in water. These offending ions can be solubilized and removed by complexation with phosphates. Phosphates, however, cause excessive growth of algae in rivers and streams, and their use has been banned. (This is the reason for "phosphate-free detergents.")

Surfactants are used not only in "soap-and-water"–type cleaning operations. They are also extremely important as components of fuels and lubricating oils. For example, detergents in engine oils assist in keeping deposits suspended in the oil and prevent them from building up on engine surfaces.

20.6 SYNTHESIS OF CARBOXYLIC ACIDS

Two reactions covered in previous chapters are especially important for the preparation of carboxylic acids:

1. Oxidation of primary alcohols and aldehydes (Secs. 11.6B and 19.15)
2. Side-chain oxidation of alkylbenzenes (Sec. 17.5C)

The ozonolysis of alkenes (Sec. 5.8) can also be used to prepare carboxylic acids, although it is less important because it breaks carbon–carbon bonds.

Another important method for the preparation of carboxylic acids is the reaction of Grignard or organolithium reagents with carbon dioxide, followed by protonolysis. Typically, the reaction is run by pouring an ether solution of the Grignard reagent over crushed dry ice.

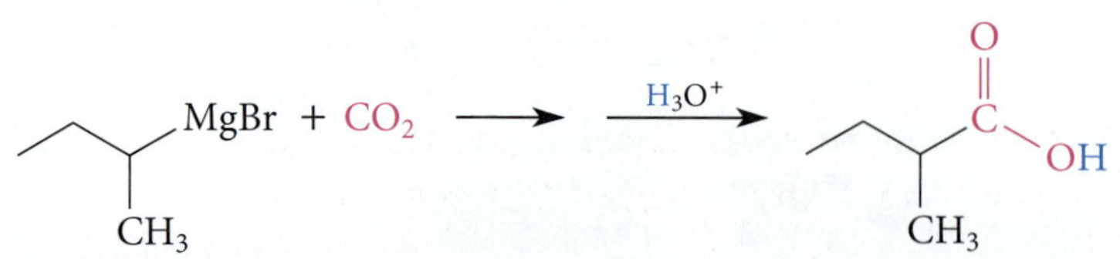

***sec*-butylmagnesium bromide** **2-methylbutanoic acid** (76–86% yield) (20.24)

Carbon dioxide is itself a carbonyl compound. The mechanism of this reaction is much like the mechanism for Grignard additions to other carbonyl compounds (Sec. 19.10). Addition of the Grignard reagent to carbon dioxide gives the bromomagnesium salt of a carboxylic acid. When aqueous acid is added to the reaction mixture in a separate reaction step, the neutral carboxylic acid is formed.

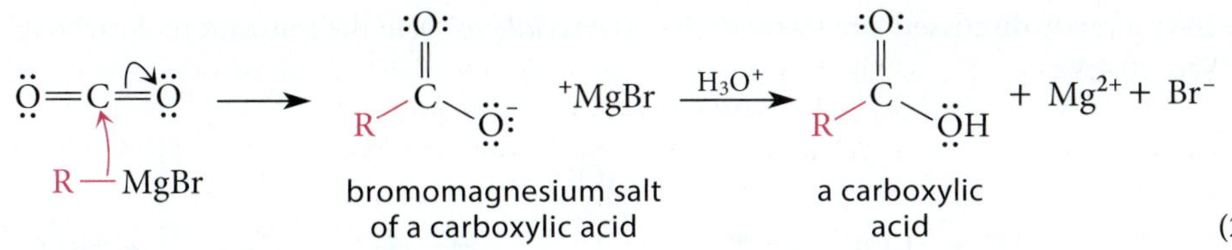

(20.25)

The reaction of Grignard reagents with CO_2, unlike the other reactions listed at the beginning of this section, is another method for forming carbon–carbon bonds. (Be sure to review the others; Appendix VI.)

All of the methods used for preparing carboxylic acids are summarized in Appendix V. A number of these are discussed in Chapters 21 and 22.

Focused Problem

20.9 Outline a synthetic scheme for each of the following transformations.

(a) Cyclopentanecarboxylic acid from cyclopentanol

(b) Octanoic acid from 1-heptene

20.7 INTRODUCTION TO CARBOXYLIC ACID REACTIONS

The reactions of carboxylic acids can be categorized into four types.

1. Reactions at the carbonyl group
2. Reactions at the carboxylate oxygen
3. Loss of the carboxy group as CO_2 (decarboxylation)
4. Reactions involving the α-carbon

The most typical reaction at the carbonyl group is *substitution at the carbonyl carbon*. Let E—Y be a general reagent in which E is an electrophilic group (for example, a hydrogen) and Y is a nucleophilic group. Typically, the OH of the carboxy group is substituted by the group Y.

(20.26)

The carbonyl oxygen of carboxylic acids and their derivatives also react with electrophiles (Lewis acids or Brønsted acids)—that is, the carbonyl oxygen reacts as a *base*:

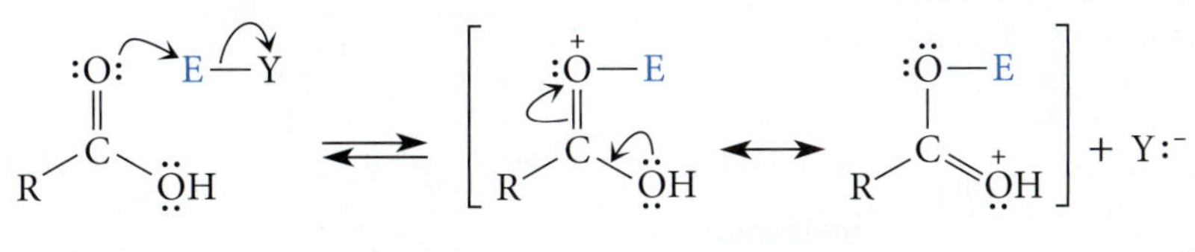

(20.27)

We discussed one such reaction, protonation of the carbonyl oxygen, in Sec. 20.4B. Many substitution reactions at the carbonyl carbon are acid-catalyzed—that is, the reactions of nucleophiles at the carbonyl *carbon* are catalyzed by the reactions of acids at the carbonyl *oxygen*.

We have already discussed a *reaction at the carboxylate oxygen*: the ionization of carboxylic acids (Sec. 20.4A):

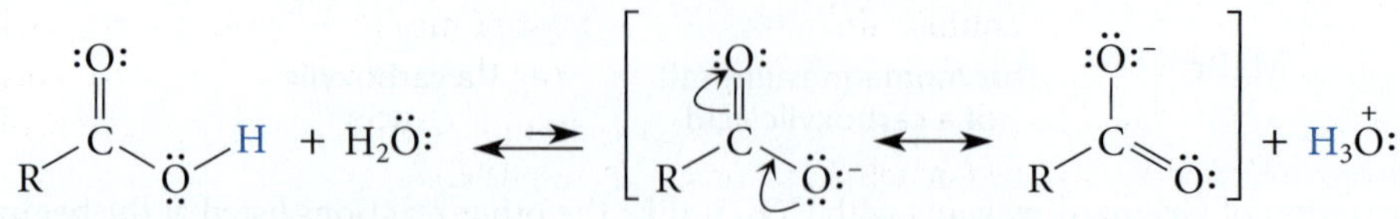

(20.28)

Another general reaction involves reaction of the carboxylate oxygen as a nucleophile ($Y:^-$ = halide, sulfonate ester, or other leaving group).

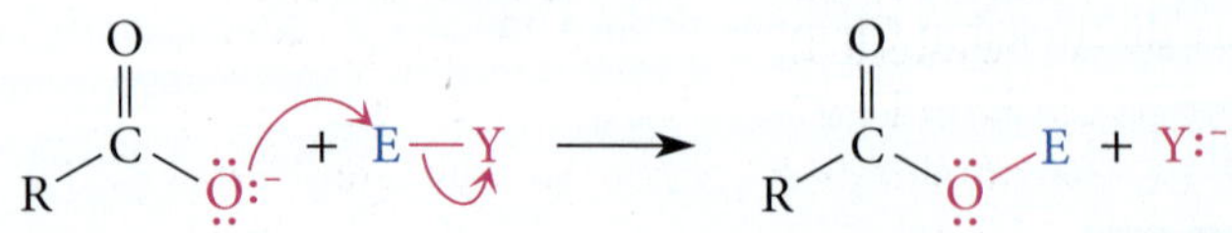

(20.29)

Decarboxylation is loss of the carboxy group as CO_2.

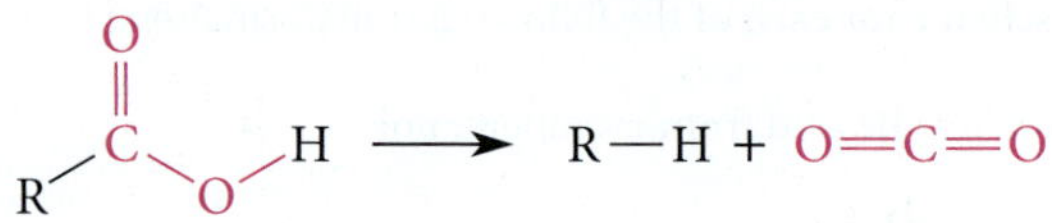

(20.30)

This reaction is more important for some types of carboxylic acids than for others (Sec. 20.11), and some decarboxylation reactions are biologically important.

This chapter concentrates on the first three types of reaction: reactions at the carbonyl group, reactions at the carboxylate oxygen, and decarboxylation. For the most part, Chapter 21 considers reactions at the carbonyl group of carboxylic acid *derivatives*. Chapter 22 takes up the fourth type of reaction—reactions involving the α-carbon—for both carboxylic acids and their derivatives, as well as for other types of carbonyl compounds.

20.8 CONVERSION OF CARBOXYLIC ACIDS INTO ESTERS

A. Acid-Catalyzed Esterification

Carboxylate esters are carboxylic acid derivatives with the following general structure:

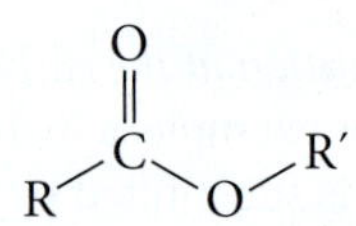

(R = H, alkyl, or aryl; R′ = alkyl or aryl)

general structure of a carboxylate ester

(20.31)

As shown in Secs 11.4A and 11.4C, ester derivatives of other types of acids are also well known. In this section, however, we'll use the simpler and more general term *ester* to mean a carboxylate ester.

When a carboxylic acid is treated with a large excess of an alcohol in the presence of a strong acid catalyst, an ester is formed.

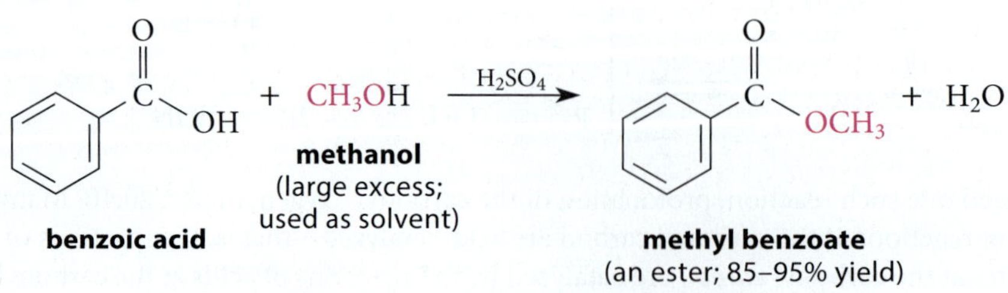

(20.32)

This reaction is called **acid-catalyzed esterification**, or sometimes **Fischer esterification**, after the renowned German chemist Emil Fischer (1852–1919).

The equilibrium constants for esterifications with most primary alcohols, although favorable, are not large; for example, the equilibrium constant for the esterification of acetic acid with ethyl alcohol is 3.38. The reaction is driven to completion by using the reactant alcohol as the solvent. Because the alcohol is present in large excess, the equilibrium is driven toward the ester product. This is another application of Le Chatelier's principle (Sec. 4.10C).

Acid-catalyzed esterification *cannot* be applied to the synthesis of esters from phenols or tertiary alcohols. Tertiary alcohols undergo dehydration (Sec. 11.2) and other reactions under the acidic conditions of the reaction, and the equilibrium constants for the esterification of phenols are much less favorable than those for the esterification of alcohols by a factor of about 10^4. Although it is possible in principle to drive the esterification of phenols to completion, there are simpler ways for preparing esters of both phenols and tertiary alcohols that are discussed in Sec. 21.8A.

In acid-catalyzed esterification, does the oxygen of the water liberated in the reaction come from the carboxylic acid or the alcohol?

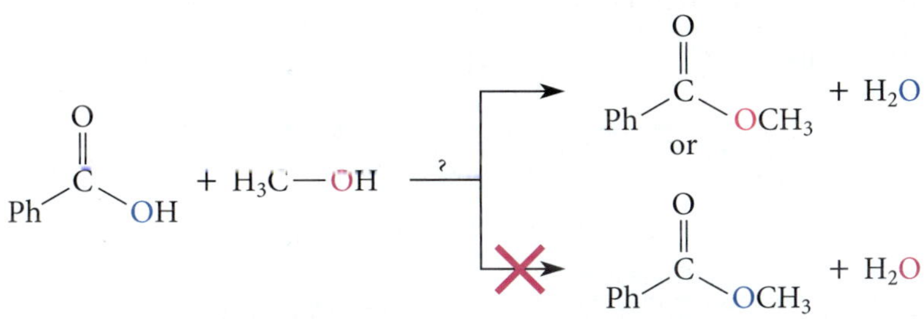

(20.33a)

(20.33b)

This question was answered in 1938, when it was found, using the ^{18}O isotope to label the alcohol oxygen, that the OH of the water produced comes exclusively from the carboxylic acid. Acid-catalyzed esterification is therefore a *substitution of OH at the carbonyl group of the acid by the oxygen of the alcohol* (Eq. 20.33a). Therefore, acid-catalyzed esterification is an example of *substitution at a carbonyl carbon.*

The mechanism of acid-catalyzed esterification is important because it serves as a model for the mechanisms of other acid-catalyzed reactions of carboxylic acids and their derivatives. The mechanism that follows shows the formation of a methyl ester in the solvent methanol. The first step of the mechanism is protonation of the carbonyl oxygen (Sec. 20.4B):

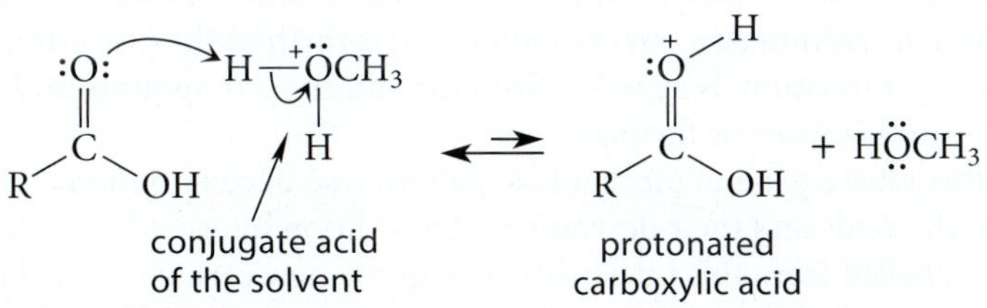

(20.34a)

The catalyzing acid is the conjugate acid of the solvent; this is the actual acid present when a strong acid such as H_2SO_4 is dissolved in methanol.

As explained in Sec. 19.8A, *protonation of a carbonyl oxygen makes the carbonyl carbon more electrophilic* because the carbonyl oxygen becomes a better electron acceptor. The carbonyl carbon of a protonated carbonyl group is electrophilic enough to react with the weakly basic methanol molecule. A nucleophilic reaction of methanol at the carbonyl carbon, followed by loss of a proton, gives a *tetrahedral addition intermediate.*

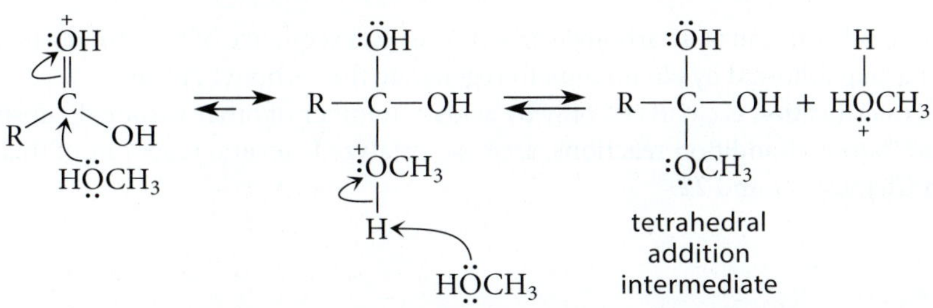

(20.34b)

A **tetrahedral addition intermediate** is simply the product of carbonyl addition. In aldehyde and ketone reactions, the product of carbonyl addition is in many cases a stable compound that can be isolated. (See, for example, Eqs. 19.32 and 19.33, Sec. 19.8.) In the case of carboxylic acid derivatives, it is called an *intermediate* because it reacts further, as we shall see. In esterification, formation of the tetrahedral addition intermediate is essentially the same reaction as the acid-catalyzed reaction of an alcohol with a protonated aldehyde or ketone to form a hemiacetal (Sec. 19.11A).

The tetrahedral addition intermediate, after protonation, loses water to give the conjugate acid of the ester:

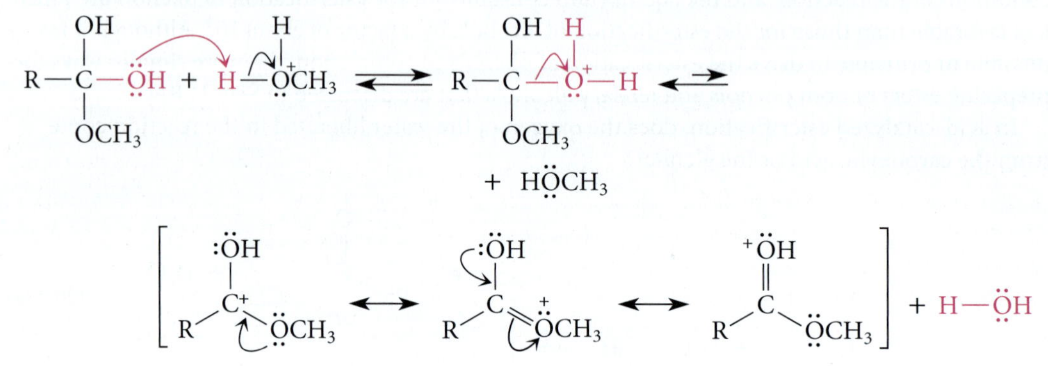

conjugate acid of the ester (20.34c)

Loss of a proton gives the ester product and regenerates the acid catalyst.

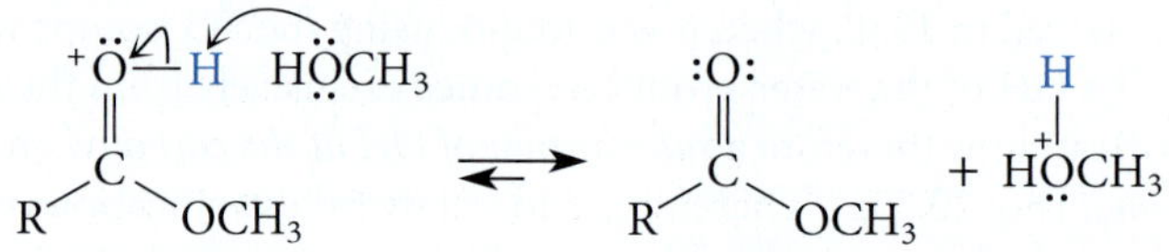

(20.34d)

The *mechanism of esterification is an extension of the mechanism of carbonyl addition.* In esterification, a nucleophile approaches the carbonyl carbon from above or below the plane of the carbonyl group and interacts with the π^* (antibonding) molecular orbital, just as in the addition reactions of aldehydes and ketones (see Fig. 19.8, Sec. 19.8A). Reaction of the nucleophile at the carbonyl carbon gives the addition compound. In esterification, however, the addition compound—the *tetrahedral addition intermediate*—reacts further; the OH group from the carboxylic acid, after protonation, is expelled from the addition compound, and a carboxylic acid derivative—an ester in this case—is formed.

Esterification illustrates a general mechanistic pattern that occurs in many substitution reactions of carboxylic acids and their derivatives. An addition intermediate is formed, and a leaving group X is expelled from this intermediate to give a new carboxylic acid derivative.

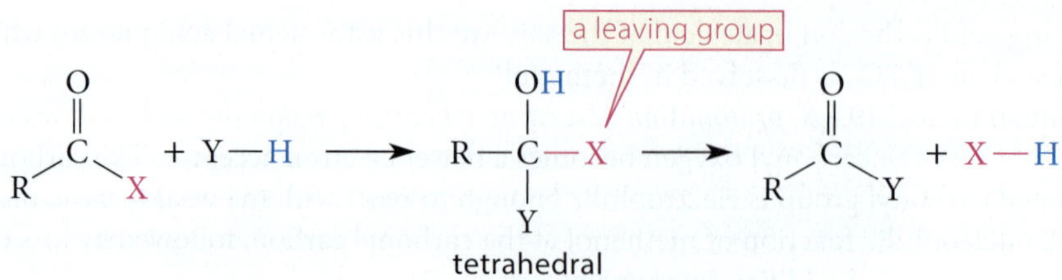

tetrahedral addition intermediate (20.35)

In other words, substitution at a carbonyl carbon is really a sequence of two processes: *addition* to the carbonyl group followed by *elimination* to regenerate the carbonyl group.

Although esterification is catalyzed only by acids, a number of other carbonyl-substitution reactions, like carbonyl-addition reactions, are base-catalyzed. Several reactions of this type are discussed in Chapters 21 and 22.

Why don't aldehydes and ketones undergo substitution at their carbonyl carbons? After a nucleophile reacts at the carbonyl carbon of an aldehyde or ketone, *neither of the groups attached to the carbonyl carbon can be a leaving group.*

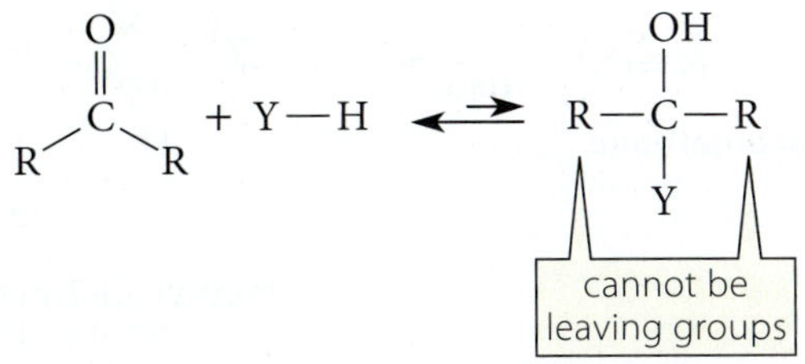

(20.36)

Tetrahedral addition compounds derived from carboxylic acid derivatives are known, but they are not formed in carbonyl-addition reactions. (See Further Exploration 20.3)

The reason is that the H— or the R— group of an aldehyde or ketone, to be a leaving group, would be expelled as H:⁻ or R:⁻, respectively, either of which is a *very strong* base. In carbonyl-substitution reactions, as in S_N2 and S_N1 reactions, the best leaving groups are generally the *weakest* bases (Secs. 9.4F and 9.6C). In contrast, one of the groups attached to the carbonyl carbon of a carboxylic acid derivative is either a weak base or (as in the case of esterification, Eq. 20.34c) is converted by protonation into a weak base; in either case, such a group can act as a good leaving group, and substitution results.

FURTHER EXPLORATION 20.3
Orthoesters

Focused Problems

20.10 Give the structure of the product formed when the following occurs:

(a) 3-Methylhexanoic acid is heated with a large excess of ethanol (as solvent) with a sulfuric acid catalyst.

(b) Adipic acid (Table 20.1) is heated in a large excess of 1-propanol (as solvent) with a sulfuric acid catalyst.

20.11 (a) Using the principle of microscopic reversibility (Sec. 4.10C), give a detailed mechanism for the acid-catalyzed hydrolysis of methyl benzoate (structure in Eq. 20.32) to benzoic acid and methanol.

(b) Given that ester formation is reversible, what reaction conditions would you use to bring about the acid-catalyzed hydrolysis of methyl benzoate to benzoic acid and methanol?

20.12 (a) The discussion in Sec. 19.8 shows that carbonyl-addition reactions can occur under basic conditions. The hydrolysis of methyl benzoate is also promoted by $^-$OH. Write a mechanism for the hydrolysis of methyl benzoate in NaOH solution.

(b) A student has suggested the following transformation, arguing that it can be driven to completion with a large excess of methanol and sodium methoxide.

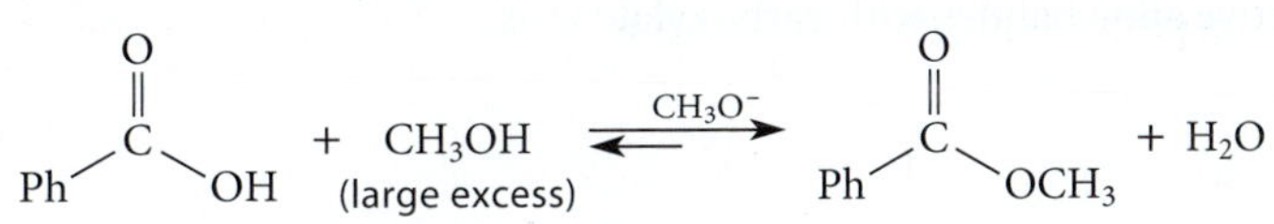

In fact, this reaction does *not* occur because, under the basic conditions, the carboxylic acid undergoes a different reaction. What is that reaction?

B. Esterification by Alkylation

The esterification discussed in the previous section involves the reaction of a nucleophile at the carbonyl carbon. This section considers a different method of forming esters that illustrates another mode of carboxylic acid reactivity: nucleophilic reactivity of the *carboxylate oxygen* (Eq. 20.29).

When a carboxylic acid is treated with diazomethane in ether solution, it is rapidly converted into its methyl ester.

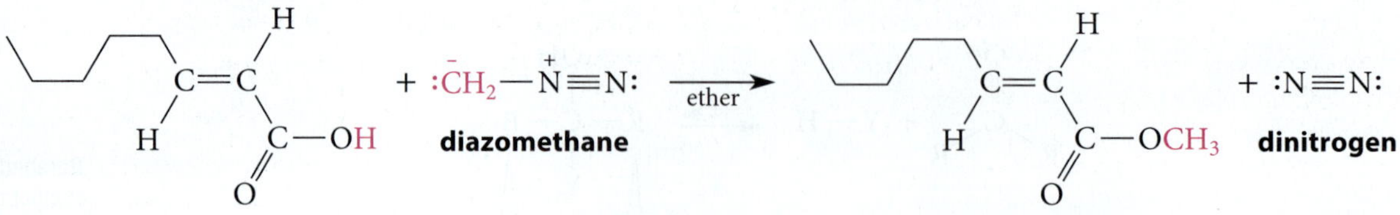

(E)-2-octenoic acid **diazomethane** **methyl (E)-2-octenoate** (91% yield) **dinitrogen** (20.37)

Diazomethane, a toxic yellow gas (bp = −23 °C), is usually generated chemically as it is needed from a commercially available precursor and is co-distilled with ether into a flask containing the carboxylic acid to be esterified. Diazomethane is both explosive and allergenic and is therefore only used in small quantities under conditions that have been carefully established to maintain safety. Nevertheless, esterification with diazomethane is so mild and free of side reactions that in many cases it is the method of choice for the synthesis of methyl esters in small-scale reactions.

The acidity of the carboxylic acid is important in the mechanism of this reaction. Protonation of diazomethane by the carboxylic acid gives the methyldiazonium ion.

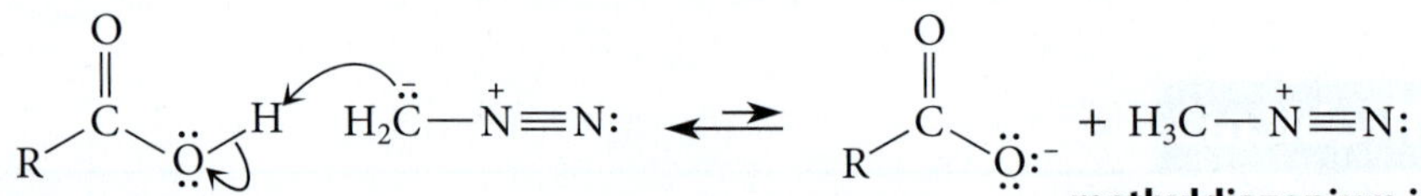

methyldiazonium ion (20.38a)

This ion contains dinitrogen, one of the best leaving groups. An S_N2 reaction of the methyldiazonium ion with the carboxylate oxygen results in the displacement of N_2 and formation of the ester.

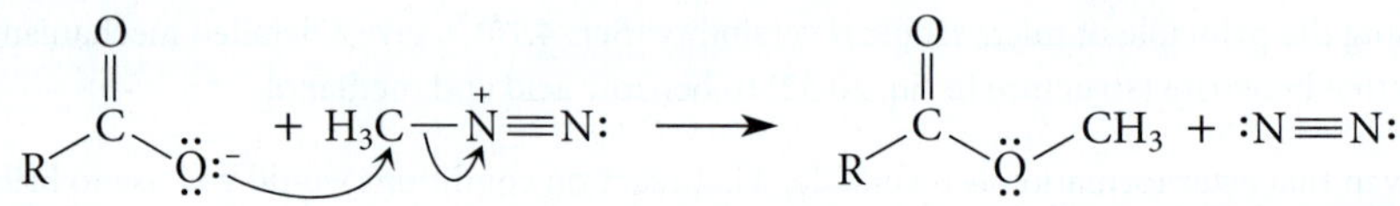

(20.38b)

Carboxylate ions are less basic, and therefore less nucleophilic, than alkoxides or phenoxides, but they do react with especially reactive alkylating agents. The methyldiazonium ion formed by the protonation of diazomethane is one of the most reactive alkylating agents known. Notice, though, that the carboxylic acid, not the carboxylate salt, is required for the reaction with diazomethane because protonation of diazomethane by the acid is the first step of the reaction.

The nucleophilic reactivity of carboxylates is also illustrated by the reaction of certain very reactive alkyl halides with carboxylate ions.

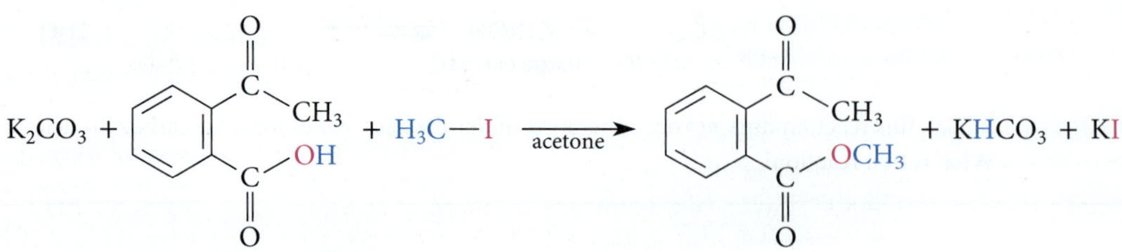

2-acetylbenzoic acid **methyl 2-acetylbenzoate** (65% yield) (20.39)

This is an S_N2 reaction in which the carboxylate ion, formed by the acid–base reaction of the acid and K_2CO_3, reacts as a nucleophile with the alkyl halide. Because carboxylate ions are weak nucleophiles, this reaction works best on alkyl halides that are especially reactive in S_N2 reactions, such as methyl iodide and benzylic or allylic halides (Sec. 17.4). This reaction is typically carried out in polar aprotic solvents that accelerate S_N2 reactions, such as acetone, as Eq. 20.39 illustrates.

Let's contrast the esterification reactions in Eqs. 20.37 and 20.39 with the acid-catalyzed esterification in Sec. 20.8A. In all of the reactions discussed in this section, the carboxylate oxygen of the acid acts as a nucleophile. This oxygen is alkylated by an alkyl halide or diazomethane. In acid-catalyzed esterification, the carbonyl carbon, after protonation of the carbonyl oxygen, acts as an electrophile (a Lewis acid). The nucleophile in acid-catalyzed esterification is the oxygen atom of the solvent alcohol molecule.

Focused Problems

20.13 Give the structure of the ester formed in each of the following reactions.

(a) The reaction of isobutyric acid with diazomethane in ether

(b) The reaction of succinic acid with a large excess of diazomethane in ether

(c) The reaction of isobutyric acid with benzyl bromide and K_2CO_3 in acetone

(d) The reaction of benzoic acid with allyl bromide and K_2CO_3 in acetone

20.14 *Tert*-butyl esters can be prepared by the acid-catalyzed reaction of methylpropene (isobutylene) with carboxylic acids.

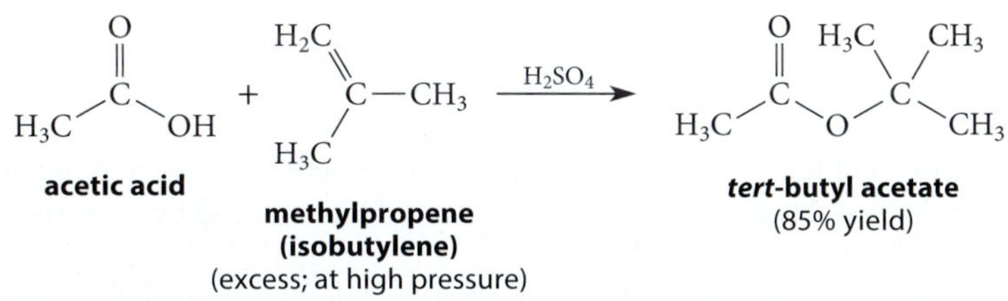

Suggest a mechanism for this reaction that accounts for the role of the acid catalyst.

20.9 CONVERSION OF CARBOXYLIC ACIDS INTO ACID CHLORIDES AND ANHYDRIDES

Acid chlorides and acid anhydrides are very reactive carboxylic acid derivatives that play an important role in the synthesis of other carboxylic acid derivatives, such as esters and amides. This section shows how these derivatives are prepared from carboxylic acids. Their use in the synthesis of other carboxylic acid derivatives is discussed in Chapter 21, starting in Sec. 21.8.

A. Synthesis of Acid Chlorides

Acid chlorides are carboxylic acid derivatives with the following general structure:

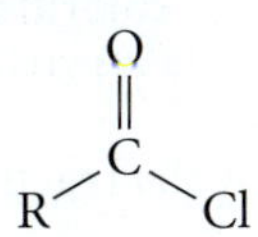

(R = H, alkyl, or aryl)

general structure of an acid chloride (20.40)

Acid chlorides are almost always prepared from carboxylic acids. Two reagents used for this purpose are thionyl chloride, $SOCl_2$, and phosphorus pentachloride, PCl_5.

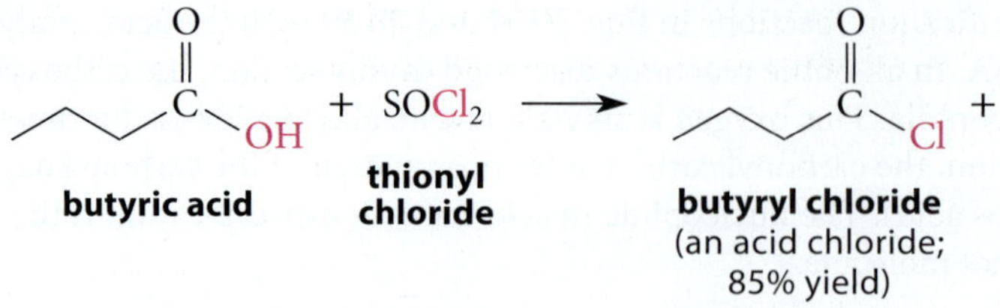

(20.41)

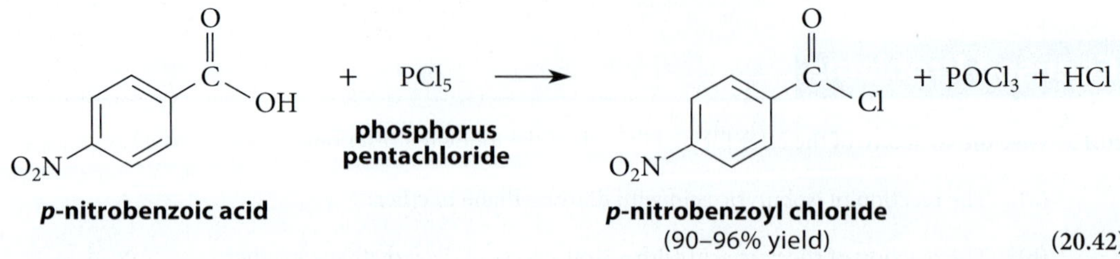

(20.42)

STUDY GUIDE LINK 20.3
Mechanism of Acid Chloride Formation

Acid chloride synthesis fits the general pattern of substitution at a carbonyl group; in this case, OH is substituted by Cl. (For the mechanistic details, see Study Guide Link 20.3.)

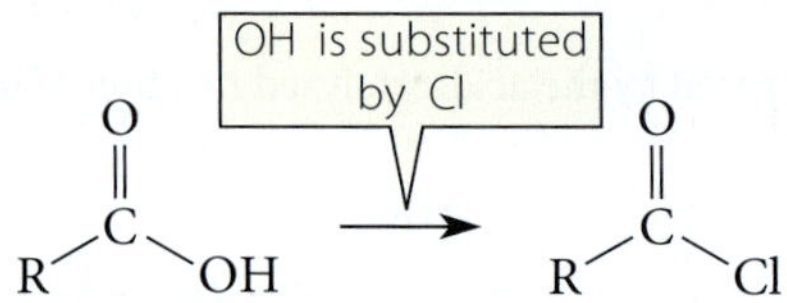

(20.43)

Thionyl chloride (Eq. 20.41) is the same reagent used for making alkyl chlorides from alcohols (Sec. 11.4D), a reaction in which OH is replaced by Cl at the carbon of an *alkyl* group. (Study Guide Link 20.3 provides the mechanistic details.) Phosphorus pentachloride, like triphenylphosphine dibromide (Ph_3PBr_2), is a hypervalent phosphorus compound with significant ionic character (Sec. 11.4D). The mechanism of its reaction with carboxylic acids no doubt resembles the mechanism of the reaction of alcohols with Ph_3PBr_2 (Eqs. 11.44a and b). (After reading Study Guide Link 20.3, try to write the mechanism of the reaction with PCl_5.)

STUDY GUIDE LINK 20.4
More on Synthetic Equivalents

Recall that acid chlorides are one of the starting materials in the Friedel–Crafts acylation reaction (Sec. 16.4F), which is used for the preparation of aromatic ketones. We show in Chapter 21 that acid chlorides are very reactive; for this reason, they are also very useful for the synthesis of other carbonyl compounds. As Study Guide Link 20.4 discusses, they are another example of synthetic equivalents.

Sulfonyl chlorides, the acid chlorides of sulfonic acids, are prepared by the treatment of sulfonic acids or their sodium salts with PCl_5.

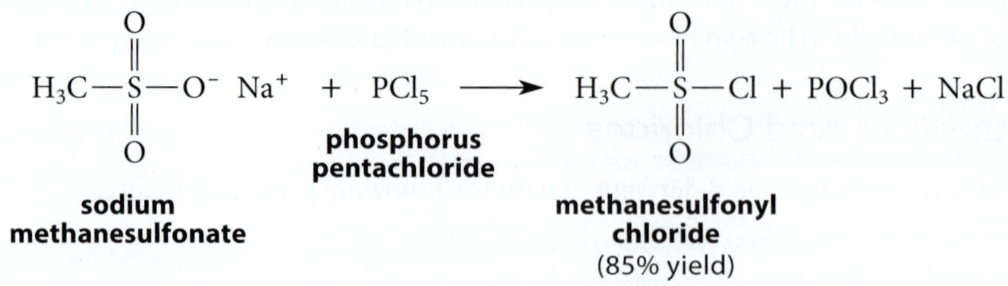

(20.44)

Aromatic sulfonyl chlorides can be prepared directly by the reaction of aromatic compounds with chlorosulfonic acid.

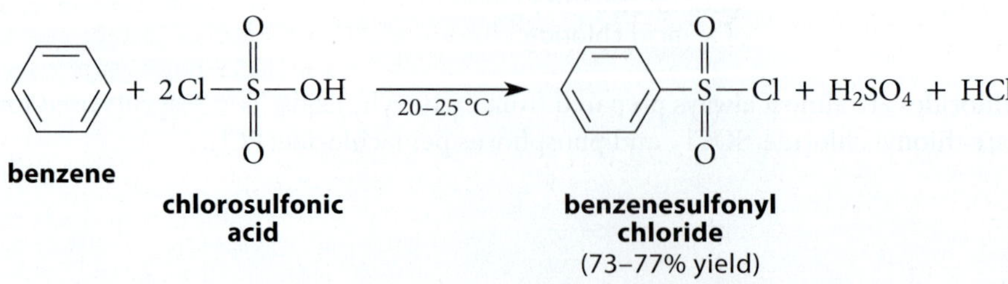

(20.45)

This reaction is a variation of aromatic sulfonation, an electrophilic aromatic substitution reaction (Sec. 16.4D). In this reaction, chlorosulfonic acid, the acid chloride of sulfuric acid, acts as an electrophile in this reaction just as SO_3 does in sulfonation.

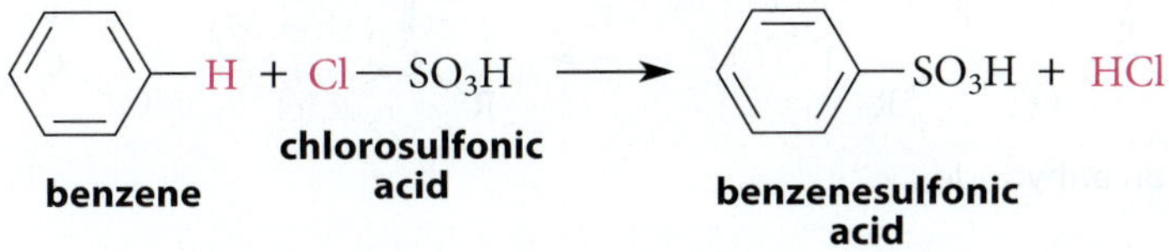

(20.46a)

The sulfonic acid produced in the reaction is converted into the sulfonyl chloride by reaction with another equivalent of chlorosulfonic acid.

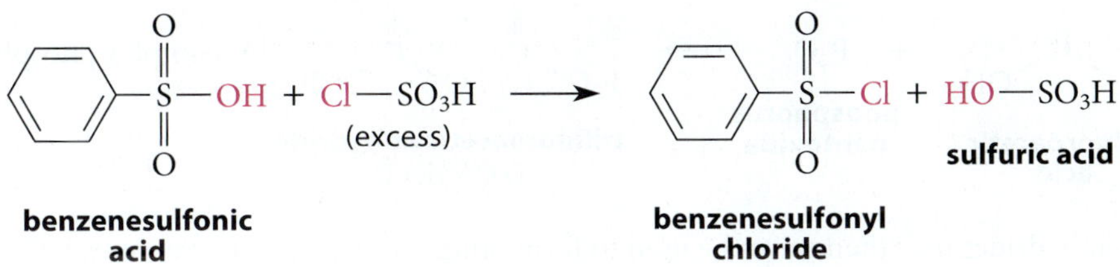

(20.46b)

This part of the reaction is analogous to the reaction of a carboxylic acid with thionyl chloride (Eq. 20.41).

Focused Problems

20.15 Draw the structures of the acid chlorides derived from (a) 2-methylbutanoic acid; (b) *p*-methoxybenzoic acid; (c) 1-propanesulfonic acid.

20.16 Give the structure of the acid chloride formed in each of the following transformations.

(a) Sodium ethanesulfonate + PCl_5 →

(b) Benzoic acid + $SOCl_2$ →

(c) *p*-Toluenesulfonic acid + excess chlorosulfonic acid →

20.17 Outline a synthesis of *p*-methoxybenzophenone from benzoic acid and any other reagents.

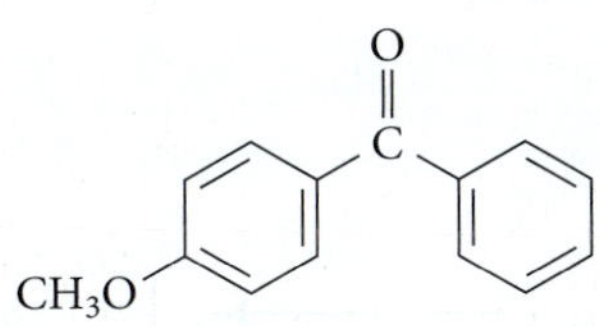

p-methoxybenzophenone

(*Hint:* See Study Guide Link 20.4.)

B. Synthesis of Anhydrides

Carboxylic acid *anhydrides* have the following general structure:

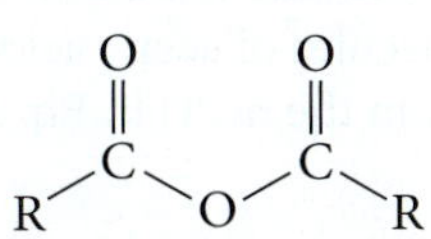

(R = H, alkyl, or aryl)

general structure of an anhydride

(20.47)

The name *anhydride*, which means "without water," comes from the fact that an anhydride reacts with water to give two equivalents of a carboxylic acid.

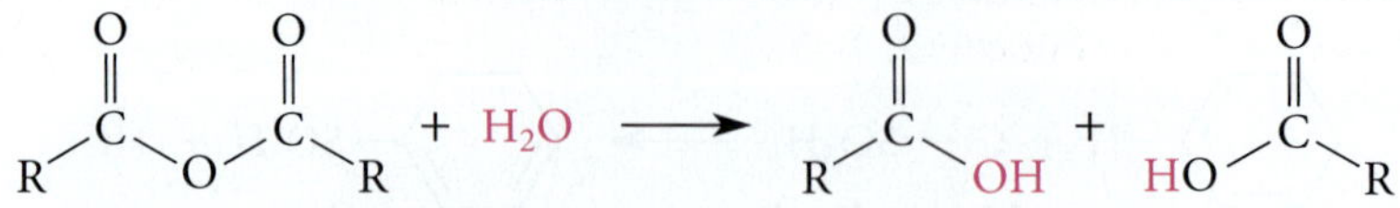

an anhydride (20.48)

The name *anhydride* also graphically describes one of the ways that anhydrides are prepared: the treatment of carboxylic acids with strong dehydrating agents.

Phosphorus pentoxide (actual formula P_4O_{10}) is a white powder that rapidly absorbs, and reacts violently with, water. It is also used as a potent desiccant. This compound is a complex anhydride of phosphoric acid, because it gives phosphoric acid when it reacts with an excess of water.

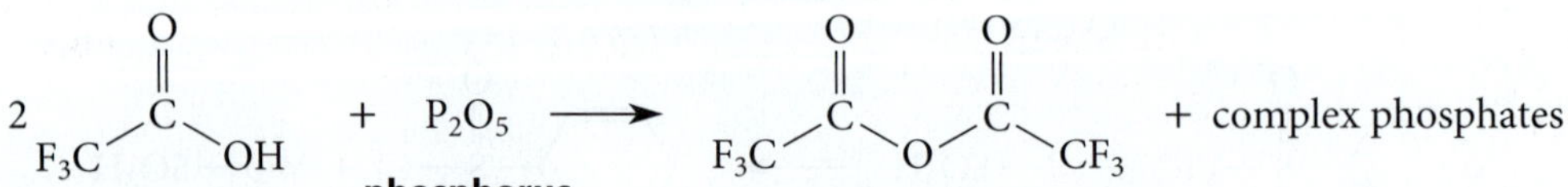

trifluoroacetic acid **phosphorus pentoxide** **trifluoroacetic anhydride** (74% yield) (20.49)

Most anhydrides may themselves be used to form other anhydrides. For example, P_2O_5 is an inorganic anhydride that is used to form anhydrides of carboxylic acids (Eq. 20.49). In the following example, a dicarboxylic acid reacts with acetic anhydride to form a *cyclic anhydride*—a compound in which the anhydride group is part of a ring:

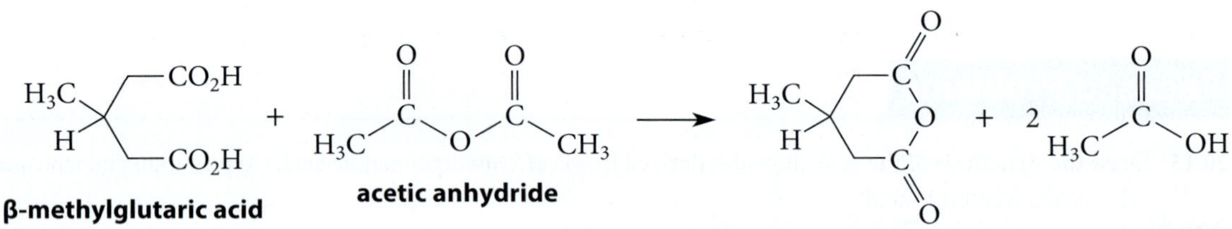

β-methylglutaric acid **acetic anhydride** **β-methylglutaric anhydride** (>90% yield; a cyclic anhydride) (20.50)

Phosphorus oxychloride ($POCl_3$) and P_2O_5 (Eq. 20.49) can also be used for the formation of cyclic anhydrides. Cyclic anhydrides containing five- and six-membered anhydride rings are readily prepared from their corresponding dicarboxylic acids. Compounds containing either larger or smaller anhydride rings generally cannot be prepared in this way. Formation of cyclic anhydrides with five- and six-membered rings is so facile that in some cases it occurs on heating the dicarboxylic acid.

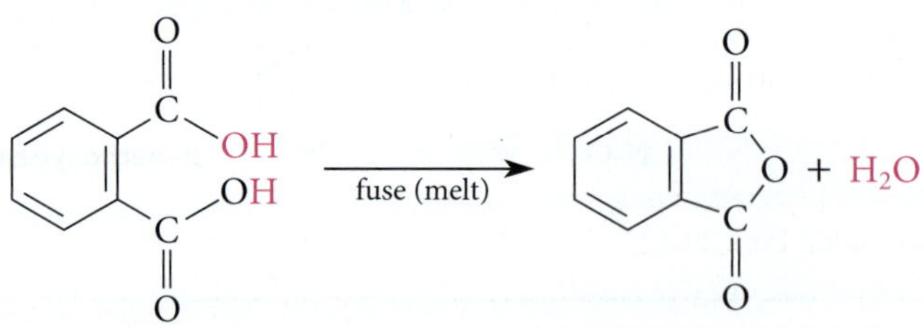

phthalic acid **phthalic anhydride** (20.51)

Cyclic anhydride formation is another example of a *proximity effect* (Secs. 12.8A and B). These reactions are accelerated because they are intramolecular. Furthermore, they have a particularly favorable $\Delta G°$ for the same reason. The equilibrium benefits in addition from a positive $\Delta S°$ component of both reactions: the release of two molecules of acetic acid in Eq. 20.50 and the expulsion of a water molecule (removed as steam from the melt) in Eq. 20.51.

The formation of anhydrides from carboxylic acids, like many other carboxylic acid reactions that have been discussed, fits the pattern of substitution at the carbonyl carbon: the —OH of one carboxylic acid molecule is substituted by the *acyloxy group* (*red*) of another. (The mechanistic details are provided in Study Guide Link 20.5.)

STUDY GUIDE LINK 20.5
Mechanism of Anhydride Formation

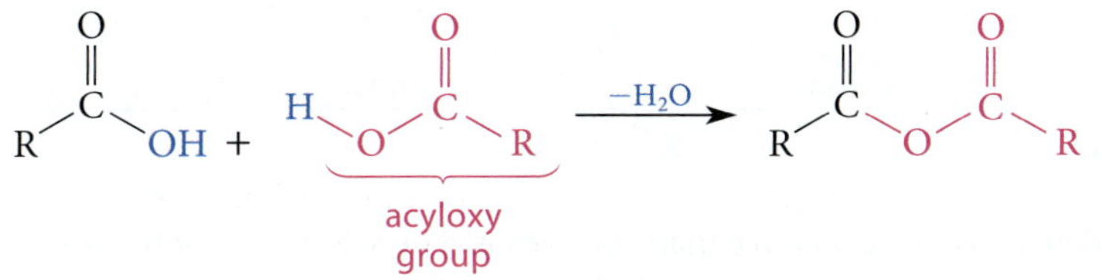

(20.52)

As we show in Sec. 21.8B, anhydrides, like acid chlorides, are used in the synthesis of other carboxylic acid derivatives.

Focused Problems

20.18 Give the structures of the products formed when (a) chloroacetic acid and (b) *p*-chlorobenzoic acid react with P_2O_5.

20.19 (a) Fumaric and maleic acids (Table 20.1, Sec. 20.1A) are *E* and *Z* isomers. One forms a cyclic anhydride on heating, and one does not. Which one forms the cyclic anhydride? Explain.

(b) Which one of the following compounds forms a cyclic anhydride on heating: methylmalonic acid or 2,3-dimethylbutanedioic acid? Explain.

20.10 HYDRIDE REDUCTION OF CARBOXYLIC ACIDS TO PRIMARY ALCOHOLS

When a carboxylic acid is treated with lithium aluminum hydride, $LiAlH_4$, then dilute acid, a primary alcohol is formed.

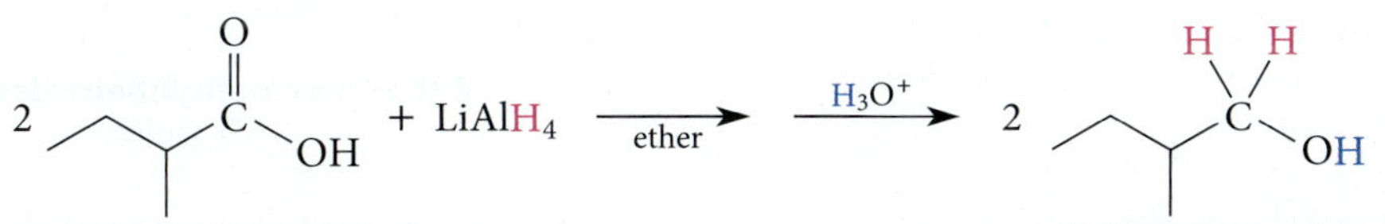

2-methylbutanoic acid **2-methyl-1-butanol** (83% yield) (20.53)

This is an important method for the preparation of primary alcohols.

Before the reduction itself takes place, $LiAlH_4$, a source of the strongly basic hydride ion (H:⁻), reacts with the acidic hydrogen of the carboxylic acid to give the lithium salt of the carboxylic acid and one equivalent of hydrogen gas (that is, dihydrogen).

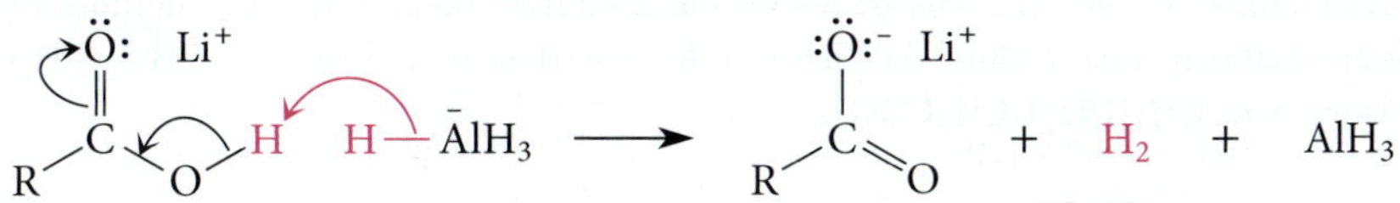

dihydrogen (20.54a)

The lithium salt of the carboxylic acid is the species that is actually reduced.

FURTHER EXPLORATION 20.4
Mechanism of the $LiAlH_4$ Reduction of Carboxylic Acids

The reduction occurs in two stages. In the first stage, the AlH_3 formed in Eq. 20.54a reduces the carboxylate ion to an aldehyde. (The detailed mechanism is discussed in Further Exploration 20.4.) The aldehyde is rapidly reduced further to give, after protonolysis, the primary alcohol (Sec. 19.9A).

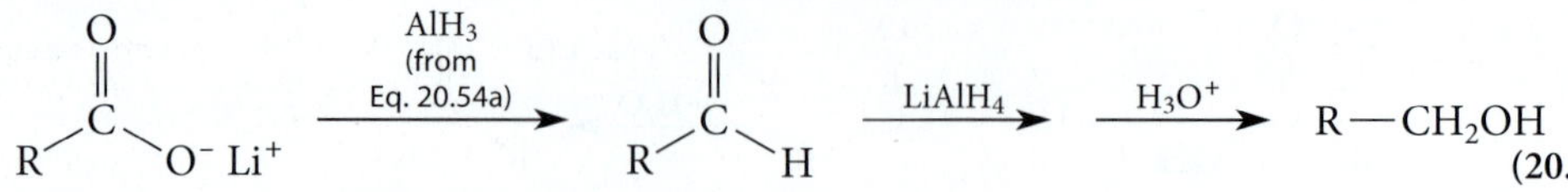

(20.54b)

The reactivity differences among the different carbonyl compounds, and the reasons for them, are considered in Sec. 21.7E.

Because the aldehyde is more reactive than the carboxylate salt, it *cannot* be isolated.

The $LiAlH_4$ reduction of a carboxylic acid incorporates two different types of carbonyl reaction. The first is a net *substitution* at the carbonyl carbon to give the aldehyde intermediate. The second is an *addition* to the aldehyde.

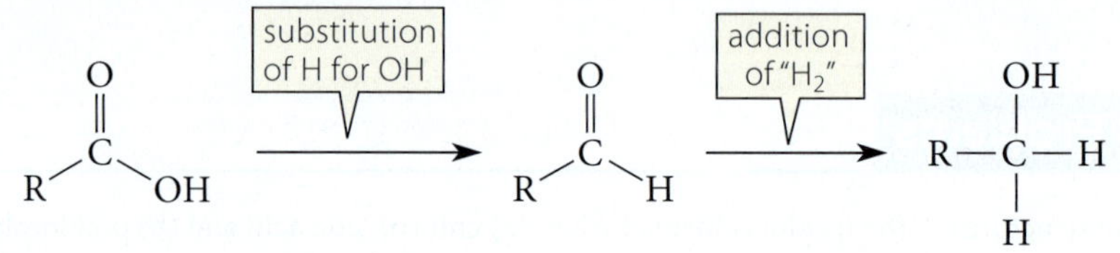

(20.55)

Many of the reactions of carboxylic acid derivatives discussed in Chapter 21 also fit this same pattern of substitution followed by addition.

Sodium borohydride, $NaBH_4$, another important hydride reducing agent (Sec. 19.9A), does *not* reduce carboxylic acids, although it does react with the acidic hydrogens of acids in a manner analogous to Eq. 20.54a. This selectivity of $NaBH_4$ allows the reduction of aldehydes and ketones in the presence of carboxylic acids.

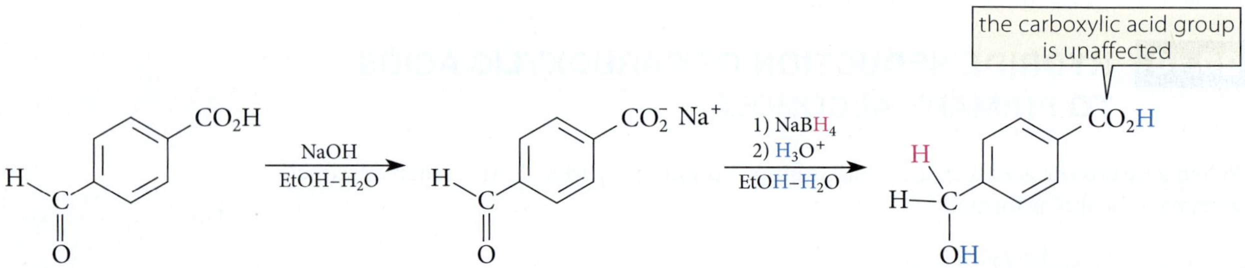

4-formylbenzoic acid **4-(hydroxymethyl)benzoic acid**
(75% yield) (20.56)

The $LiAlH_4$ reduction of carboxylic acids can be combined with the Grignard synthesis of carboxylic acids (Sec. 20.6) to provide a one-carbon chain extension of carboxylic acids, as illustrated by Study Problem 20.2.

Study Problem 20.2

Fatty acids containing an even number of carbon atoms are readily obtained from natural sources, but those containing an odd number of carbons are relatively rare. Outline a synthesis of the rare tridecanoic acid, $CH_3(CH_2)_{10}CH_2CO_2H$, from the readily available lauric acid, $CH_3(CH_2)_9CH_2CO_2H$.

Solution The problem requires the synthesis of a carboxylic acid with the addition of one carbon atom to a carbon chain. The Grignard synthesis of carboxylic acids (Sec. 20.6) will accomplish this objective:

$$\underset{\textbf{1-bromododecane}}{CH_3(CH_2)_{10}CH_2Br} \xrightarrow[\text{2) } CO_2\text{, then } H_3O^+]{\text{1) Mg, ether}} \underset{\textbf{tridecanoic acid}}{CH_3(CH_2)_{10}CH_2CO_2H}$$

The required alkyl bromide, 1-bromododecane, comes from treatment of the corresponding alcohol with concentrated HBr; and the alcohol comes, in turn, from the $LiAlH_4$ reduction of lauric acid:

$$\underset{\textbf{lauric acid}}{CH_3(CH_2)_{10}CO_2H} \xrightarrow[2)\ H_3O^+]{1)\ LiAlH_4 \text{ in ether}} \underset{\textbf{1-dodecanol}}{CH_3(CH_2)_{10}CH_2OH} \xrightarrow{\text{conc. HBr}} \underset{\textbf{1-bromododecane}}{CH_3(CH_2)_{10}CH_2Br}$$

Focused Problems

20.20 In each case, give the structure of a compound with the indicated formula that would give the following diol in a $LiAlH_4$ reduction followed by protonolysis.

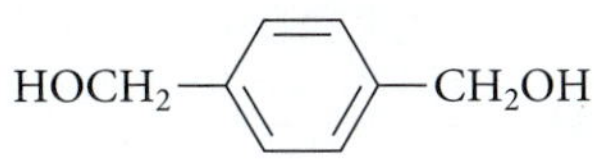

(a) $C_8H_6O_3$ (b) $C_8H_6O_4$

20.21 Propose reaction sequences for each of the following conversions:

(a) Benzoic acid into phenylacetic acid, $PhCH_2CO_2H$

(b) Benzoic acid into 3-phenylpropanoic acid

20.11 DECARBOXYLATION OF CARBOXYLIC ACIDS

The loss of carbon dioxide from a carboxylic acid is called **decarboxylation**.

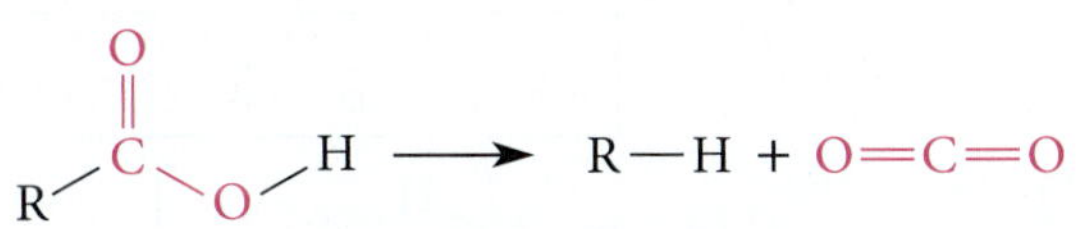

(20.57)

Although decarboxylation is not an important reaction for most ordinary carboxylic acids, certain types of carboxylic acids are readily decarboxylated. Among these are the following:

1. β-Keto acids
2. Malonic acid derivatives
3. Carbonic acid derivatives

In Sec. 20.11A we examine the decarboxylation reactions of each type of carboxylic acid. Then we use the mechanistic ideas developed there as we consider biological decarboxylation reactions in Sec. 20.11B.

A. Decarboxylation of β-Keto Acids, Malonic Acid Derivatives, and Carbonic Acid Derivatives

Decarboxylation of β-Keto Acids *β-Keto acids*—carboxylic acids with a keto group in the β-position—readily decarboxylate at room temperature in *acidic* solution.

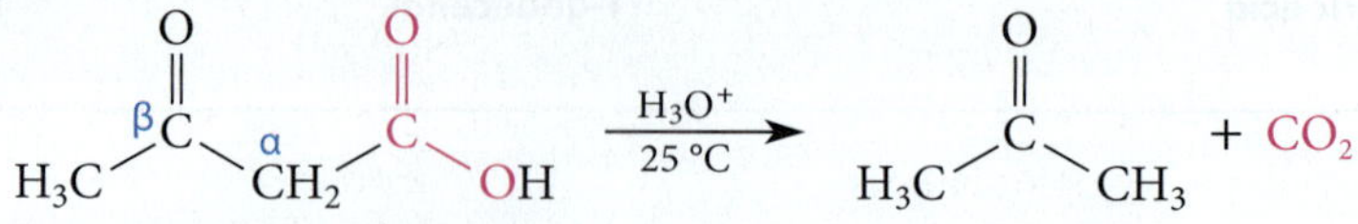

acetoacetic acid (a β-ketoacid) **acetone** (20.58a)

Decarboxylation of a β-keto acid involves an *enol intermediate* that is formed by an internal proton transfer from the carboxylic acid group to the carbonyl oxygen atom of the ketone. The enol is transformed spontaneously into the corresponding ketone (Sec. 5.5).

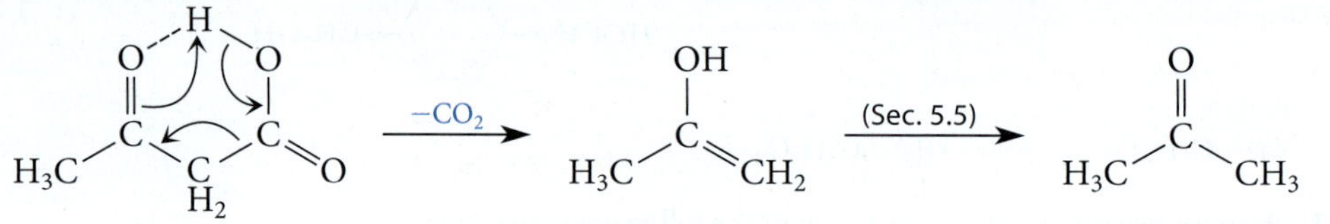

acetone enol **acetone** (20.58b)

The *acid* form of the β-keto acid decarboxylates more readily than the conjugate-base carboxylate form because the latter has no acidic proton that can be donated to the β-carbonyl oxygen. In effect, the carboxy group promotes its own removal.

Why should this internal proton transfer promote decarboxylation? If we represent the internally hydrogen-bonded structure of the β-keto acid in Eq. 20.58b with resonance structures (Eq. 20.59), then we can see that the keto oxygen is partially protonated. To the extent that it is protonated, it becomes a very electronegative, positively charged atom, which attracts electrons ultimately from the carboxylate anion. In other words, the protonated keto carbonyl group is the destination, or "**electron sink**," for the electrons released by loss of the carboxylate group.

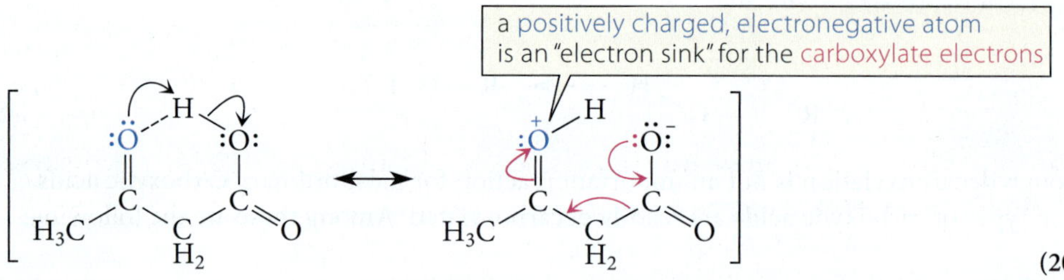

(20.59)

The presence of an "electron sink" is an important aspect of biological decarboxylations as well (Sec. 20.11B).

Decarboxylation of Malonic Acid Derivatives *Malonic acid* and its derivatives readily decarboxylate upon heating in acidic solution.

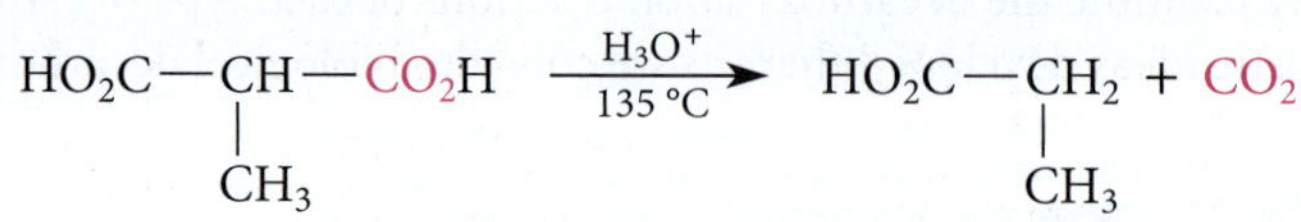

methylmalonic acid **propionic acid** (20.60)

This reaction, which also does not occur in base, bears a close resemblance to the decarboxylation of β-keto acids because both malonic acids and β-keto acids have a carbonyl group β to the carboxy group.

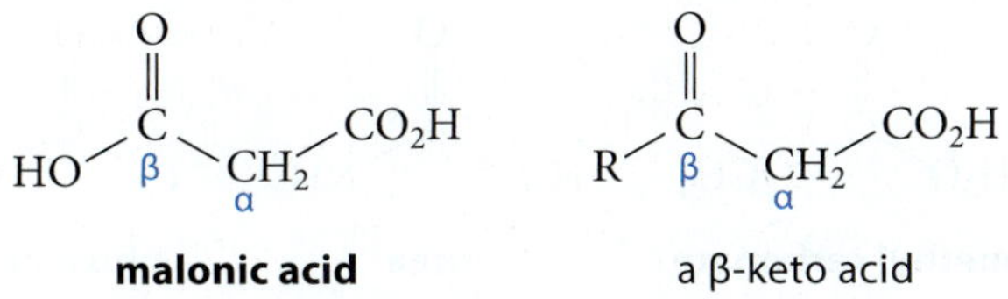

(20.61)

Because decarboxylation of malonic acid and its derivatives requires heating, the acids themselves can be isolated at room temperature.

Decarboxylation of Carbonic Acid Derivatives *Carbonic acid* is unstable and decarboxylates spontaneously in acidic solution to carbon dioxide and water. (Carbonic acid is formed reversibly when CO_2 is bubbled into water; carbonic acid gives carbonated beverages their acidity, and CO_2 gives them their "fizz.")

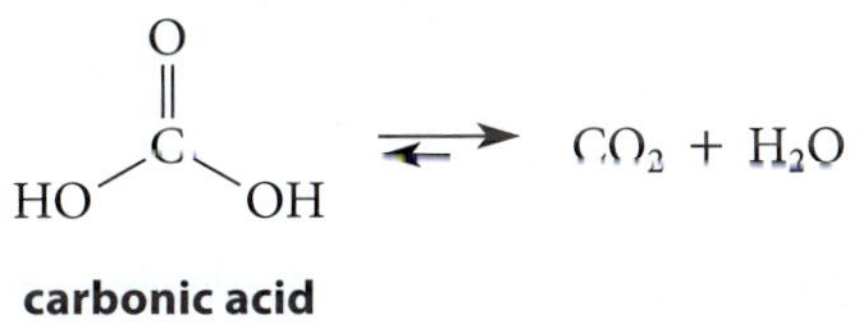

(20.62a)

In this decarboxylation, a water molecule of the solvent serves simultaneously as both an acid and a base. Loss of a water molecule as a leaving group provides the "electron sink" for the carboxy-group electrons.

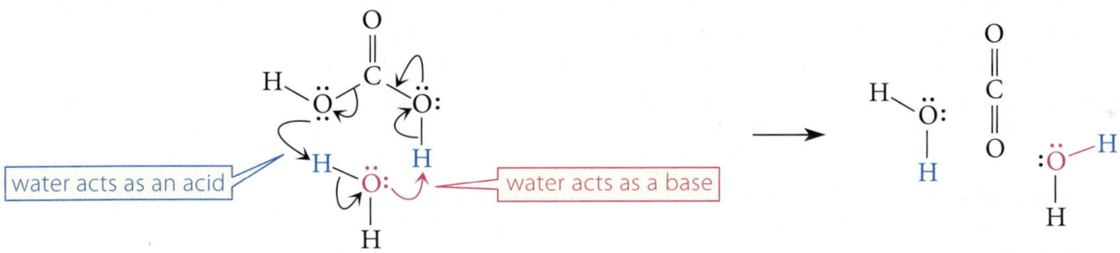

(20.62b)

Similarly, any carbonic acid derivative with a free carboxylic acid group will also decarboxylate under acidic conditions.

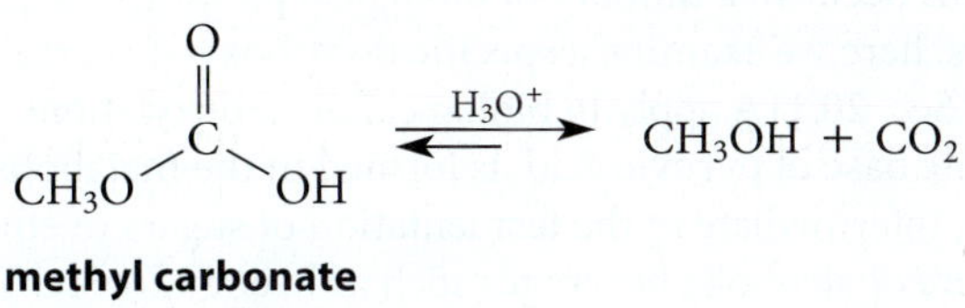

(20.63a)

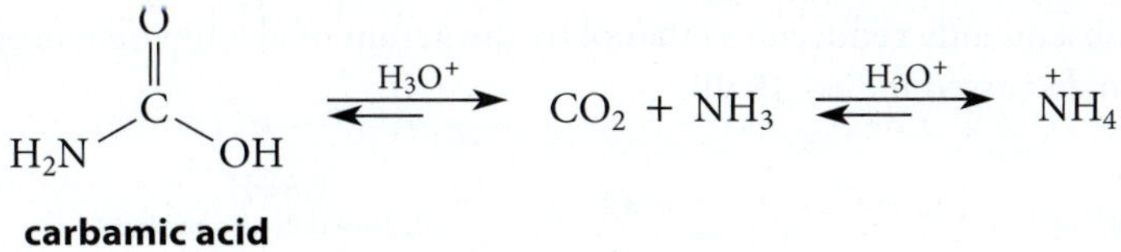

(20.63b)

Under basic conditions, carbonic acid and its derivatives exist as carboxylate salts and do not decarboxylate. For example, the sodium salts of carbonic acid, such as sodium bicarbonate ($NaHCO_3$) and sodium carbonate (Na_2CO_3), are familiar stable compounds.

Carbonic acid diesters and diamides are stable. Dimethyl carbonate (a diester of carbonic acid) and urea (the diamide of carbonic acid) are examples of such stable compounds. Likewise, the toxic acid chloride phosgene is also stable.

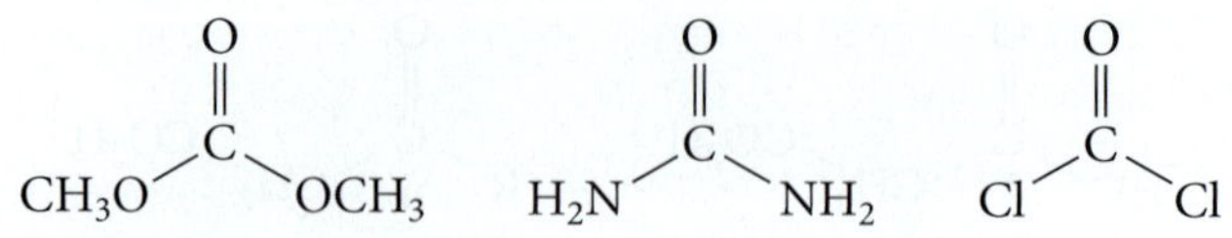

dimethyl carbonate **urea** **phosgene** (20.64)

Focused Problems

20.22 Give the product expected when each of the following compounds is treated with acid.

(a) 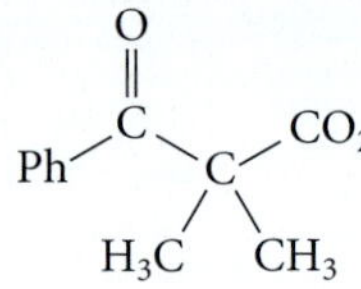

(b) 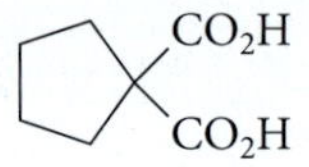
(+ heat)

(c) $CH_3CH_2NHCO_2^- \ Na^+$

20.23 Give the structures of all of the β-keto acids that will decarboxylate to yield 2-methylcyclohexanone.

20.24 One piece of evidence supporting the enol mechanism in Eq. 20.58b is that β-keto acids that cannot form enols are stable to decarboxylation. For example, the following β-keto acid can be distilled at 310 °C without decomposition. Attempt to construct a model of the enol that would be formed when this compound decarboxylates. Use your model to explain why this β-keto acid resists decarboxylation. (*Hint:* See Sec. 7.6C.)

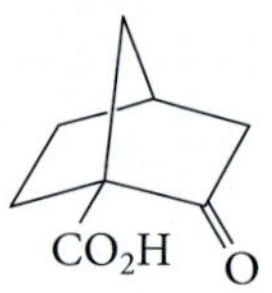

Chemical Biology Topic

B. Decarboxylations in Biology; Thiamin Pyrophosphate

Decarboxylation reactions occur in a number of biological pathways. Although they occur in many different situations, here we examine a specific decarboxylation reaction to illustrate how the mechanistic ideas of Sec. 20.11A apply in biological decarboxylations.

Pyruvate, the conjugate base of pyruvic acid, is formed in the metabolism of glucose. In yeast, pyruvate is an important intermediate in the fermentation of sugars to ethanol, a process that is central to the manufacture of alcoholic beverages such as beer and wine. The decarboxylation of pyruvate to acetaldehyde in yeast is catalyzed by the yeast enzyme *pyruvate decarboxylase.* The acetaldehyde is subsequently reduced to ethanol by the action of alcohol dehydrogenase and NADH, a reaction discussed in Sec. 19.9B.

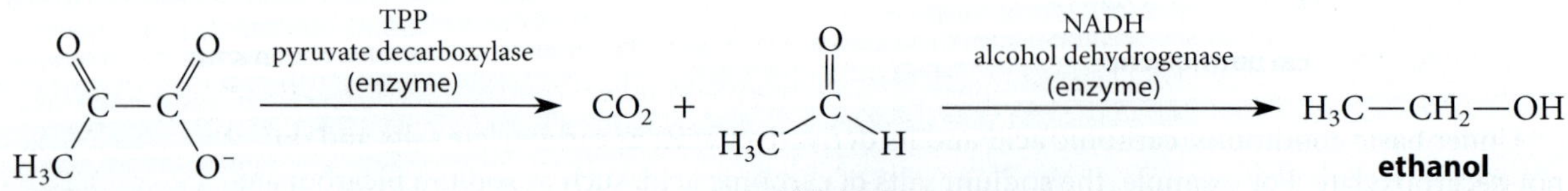

pyruvate **acetaldehyde** **ethanol** (20.65)

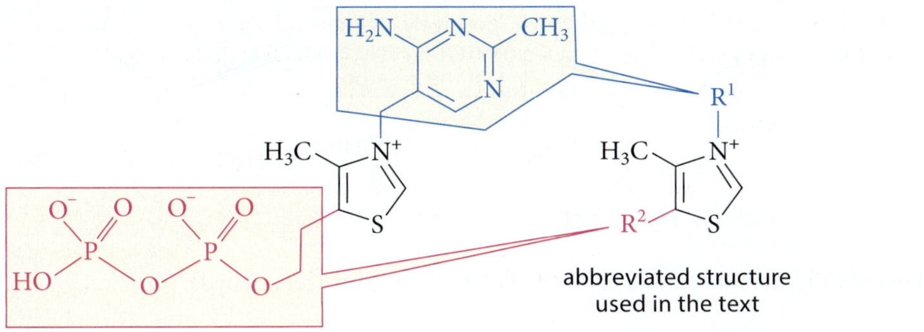

FIGURE 20.4 The structure of thiamin pyrophosphate (TPP) and the abbreviation for TPP used in the text. TPP is a form of vitamin B_1; vitamin B_1 itself is the structure without the pyrophosphate group—that is, $—R^2 = —CH_2CH_2OH$.

We showed in the previous section that decarboxylation requires an "electron sink" for the electrons provided by decarboxylation. From this perspective, the decarboxylation of pyruvate presents a mechanistic problem, because the keto group of pyruvate is *not* a β-keto group, but rather an α-keto group. The ketone is therefore in the wrong position to serve as the "electron sink." This problem is solved by the involvement of a coenzyme, **thiamin pyrophosphate** (abbreviated **TPP**) in the decarboxylation reaction. (This is sometimes called *thiamin diphosphate*. You will also sometimes see "thiamin" spelled "thiamine.") The structure of TPP is shown in **Fig. 20.4**. TPP is a form of vitamin B_1.

TPP is weakly acidic ($pK_a \sim 18$) and can ionize to form an *ylid* in the active site of the alcohol dehydrogenase enzyme. As noted in Sec. 19.14, an **ylid** is a species in which adjacent atoms have opposite charges and complete octets. Because both charged atoms have octets, the charges can't neutralize each other by resonance (Focused Problem 20.25). The ylid is stabilized by the positive charge on the adjacent nitrogen.

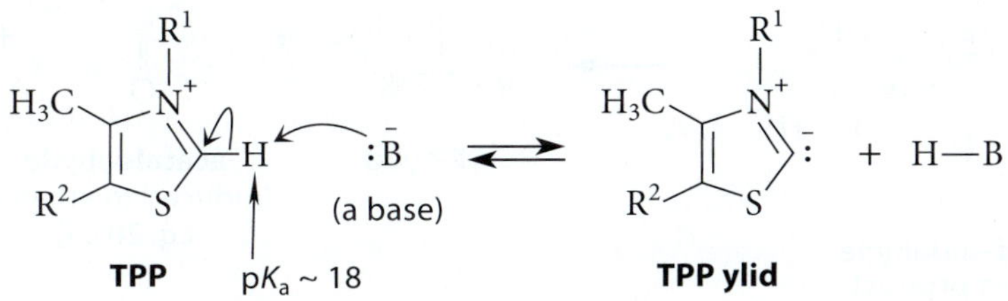

(20.66a)

The TPP ylid reacts instantly by adding to the ketone carbonyl group of pyruvate, which is bound nearby on the enzyme active site. (In Sec. 21.7E we show that ketone carbonyl groups are *much* more reactive than carboxylate carbonyl groups.) This addition is assisted by an acidic group of the enzyme present within the active site.

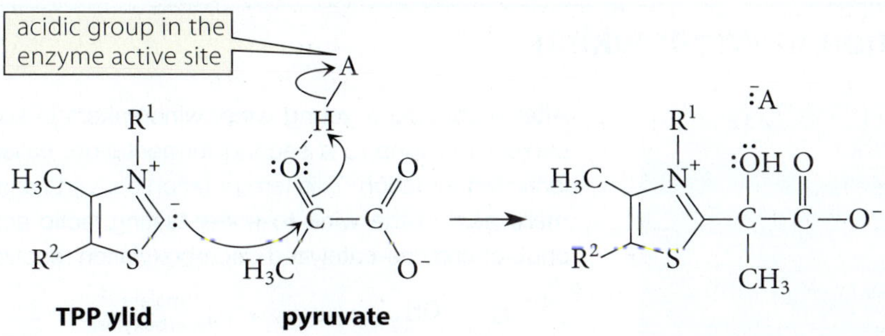

(20.66b)

This addition provides the "electron sink" that can receive the carboxylate electrons produced in the decarboxylation process. Like the internally hydrogen-bonded β-carbonyl of a β-keto acid, the addition product contains a double bond to a positively charged, electronegative atom at the β-carbon.

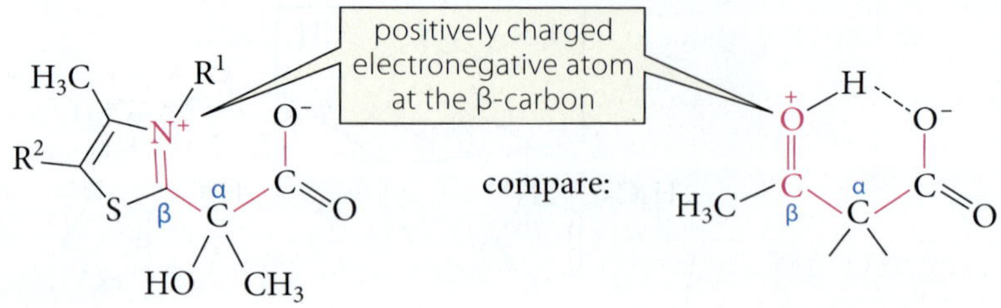

(20.66c)

Loss of CO_2 gives the addition product of TPP ylid and acetaldehyde.

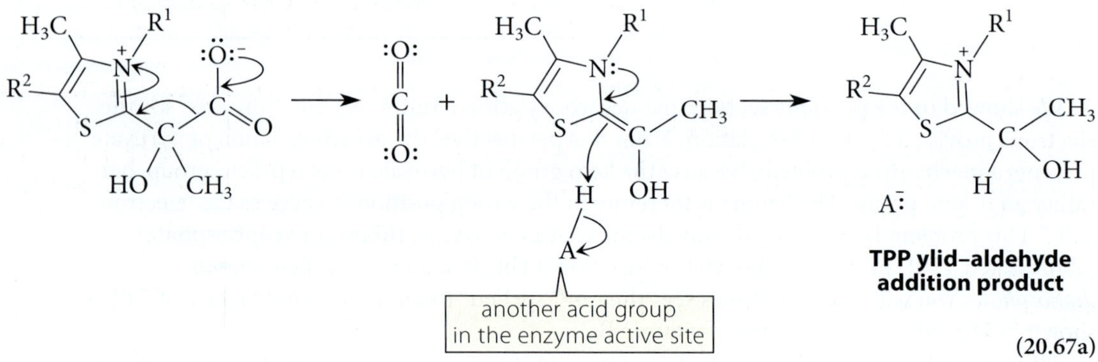

(20.67a)

This addition product breaks down to acetaldehyde and the TPP ylid.

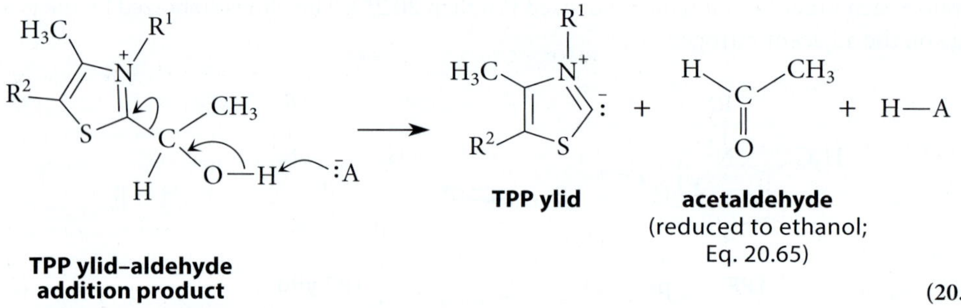

(20.67b)

The acetaldehyde is reduced to ethanol by NADH (Eq. 20.65), and the TPP ylid is ready for another cycle of catalysis by reacting with pyruvate (Eq. 20.66b).

Malolactic Fermentation in Winemaking

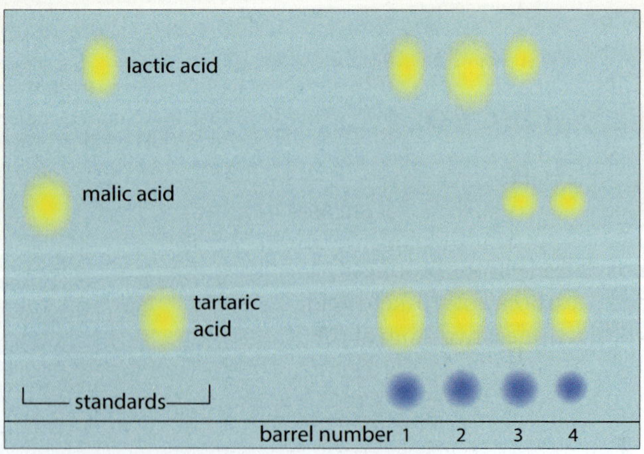

After producing a young wine, winemakers in some cases subject the wine to a second fermentation, called a malolactic fermentation. This fermentation converts sharp-tasting malic acid in the wine to softer-tasting lactic acid. This is another enzyme-catalyzed decarboxylation reaction.

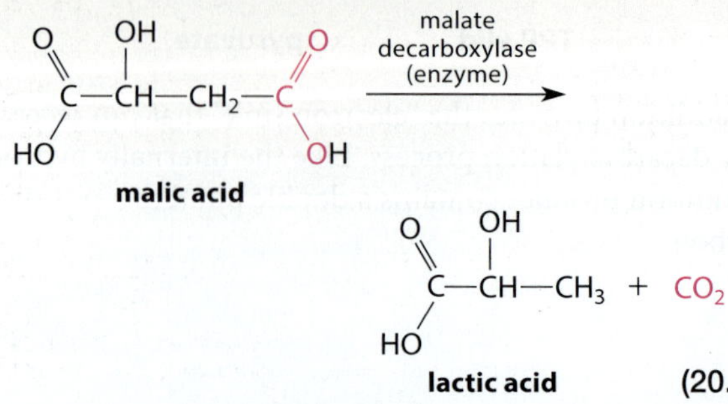

(20.68)

The progress of this reaction is typically followed by paper chromatography (illustration), in which a drop of each wine is spotted at the bottom of absorbent paper, and a solvent, moving up the paper by capillary action, separates the malic acid from lactic acid and tartaric acid (another dicarboxylic acid present in grape juice). The solvent contains bromocresol green, an acid–base indicator that allows visualization of the carboxylic acids as yellow spots against a blue-green background. The illustration shows that the wines in barrels 1 and 2 have undergone complete malolactic fermentation, whereas the wine in barrel 3 has undergone partial malolactic fermentation, and the wine in barrel 4 has not undergone malolactic fermentation. Lactic acid has a higher pK_a than malic acid (why?), so the pH of the wine is somewhat greater after the malolactic fermentation. (Besides reducing wine acidity, the fermentation also brings about other subtle changes in the wine that affect its flavor.) As Eq. 20.68 shows, malic acid lacks the β-keto group necessary for decarboxylation. Why, then, does the decarboxylation reaction occur? Focused Problem 20.26 addresses this issue.

Focused Problems

20.25 Suppose that the following curved-arrow notation and resonance structures for TPP ylid have been proposed in an attempt to demonstrate that the charges in the ylid can be neutralized by resonance. Explain what is wrong with each resonance structure.

(a)

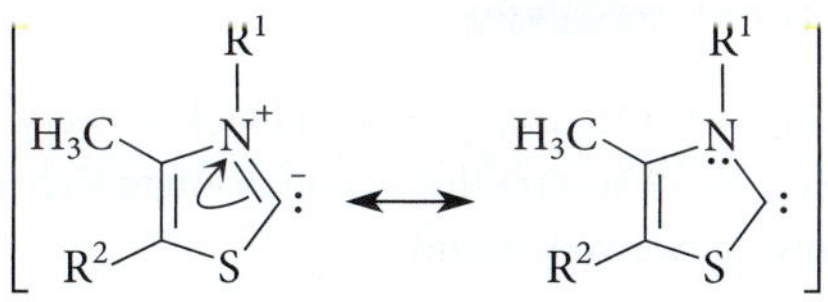

(b)

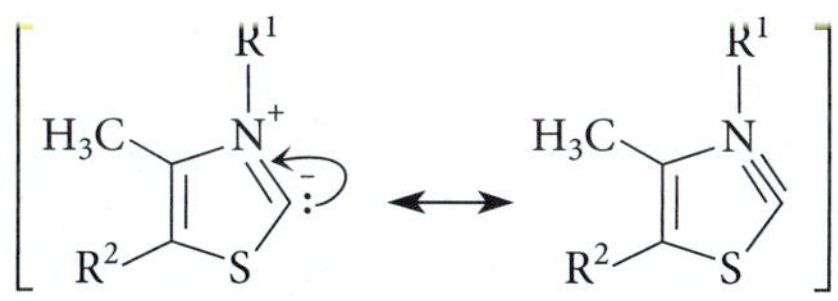

20.26 The enzyme malate decarboxylase, which catalyzes the reaction shown in Eq. 20.68, contains a molecule of tightly bound NAD^+ (Display 11.68 and Sec. 19.9B), which undergoes no *net* change as a result of the reaction. Without the NAD^+, the enzyme is completely inactive as a catalyst. Explain the role of the NAD^+ in facilitating the decarboxylation of malic acid. (*Hint:* Remember that NAD^+ is an oxidizing agent.)

CHAPTER SUMMARY *For a summary of the chapter, see Chapter 20 in the* Study Guide and Solutions Manual.

REACTION REVIEW *For a summary of reactions discussed in this chapter, see the* Reaction Review *section of Chapter 20 in the* Study Guide and Solutions Manual.

SKILLS OBJECTIVES WITH PROBLEMS

- **Given the name of a carboxylic acid, draw its structure. Given the structure of a carboxylic acid, provide its name using IUPAC substitutive nomenclature or, when appropriate, common nomenclature. 20.1A,B**

20.27 Draw a structure for each of the following compounds.

(a) 1,4-Cyclohexanedicarboxylic acid

(b) α,α-Dichloroadipic acid

(c) Oxalic acid

20.28 Provide an IUPAC substitutive name for each of the following carboxylic acids.

(a)

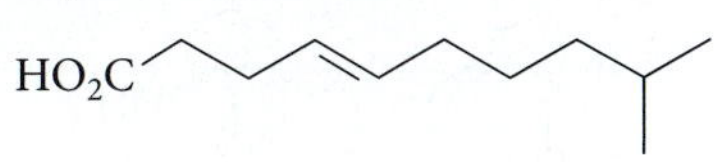

(b)

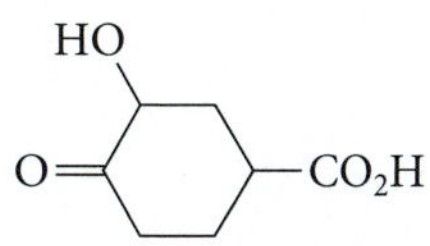

(c)

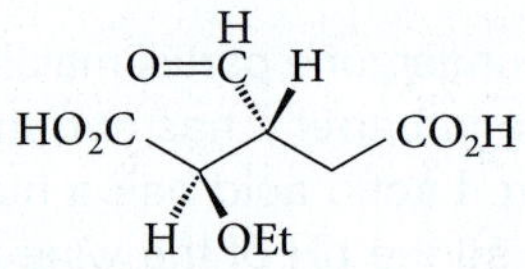

(*Hint:* See Display 19.5 in Sec. 19.2A.)

- **Discuss the hybridization of atoms in a carboxylic acid as well as the trends in bond length and bond angles. 20.2**

20.29 Give the hybridization of atoms *a–e* in the following structure of (*E*)-2-butenoic acid. Also give the approximate bond angles *abc* and *abd*.

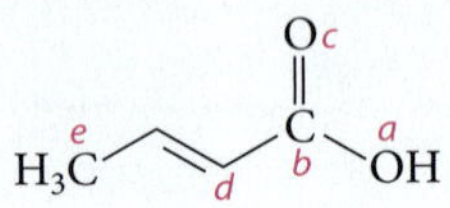

20.30 Arrange the bonds marked *a–g* in the following structure in order of increasing bond length.

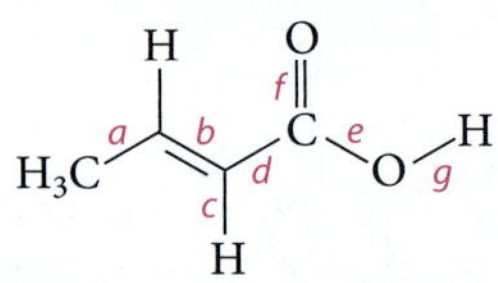

- **Relate the physical properties (particularly melting point and boiling point) of carboxylic acids to those of other functional classes. 20.2**

20.31 In each of the following parts, arrange the three compounds in order of increasing boiling point. Explain your choices.

(a)

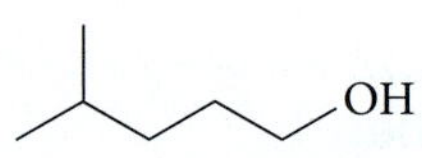

A

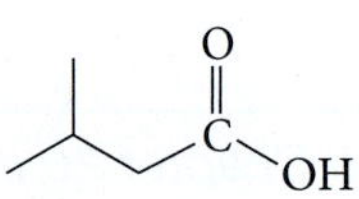

B

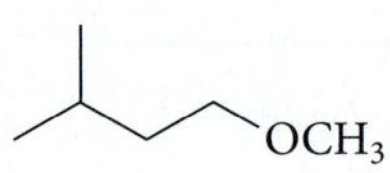

C

(b)

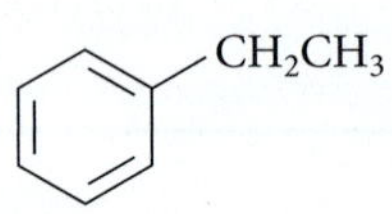

A

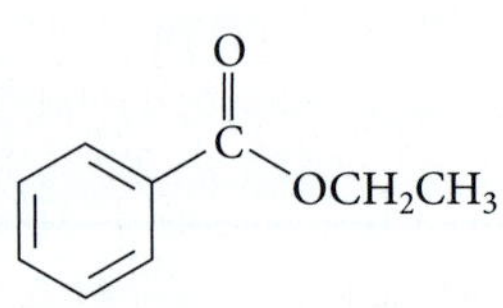

B

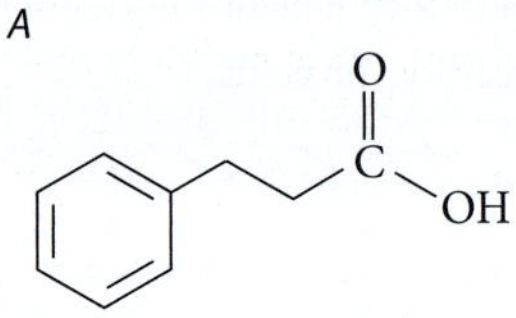

C

20.32 Arrange the following compounds in order of increasing melting point. Explain your choices.

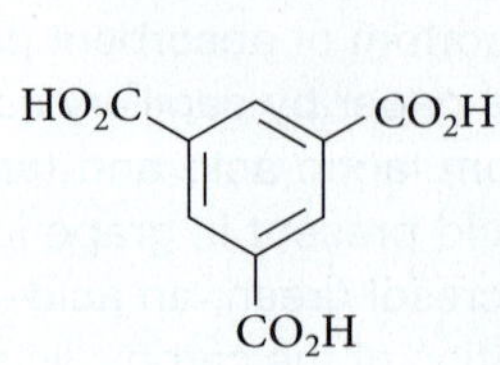

A

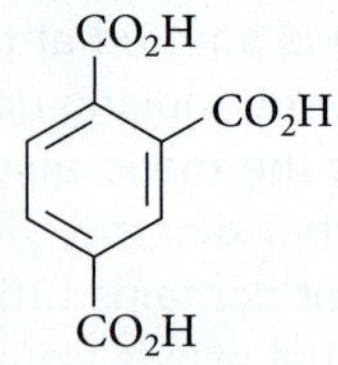

B

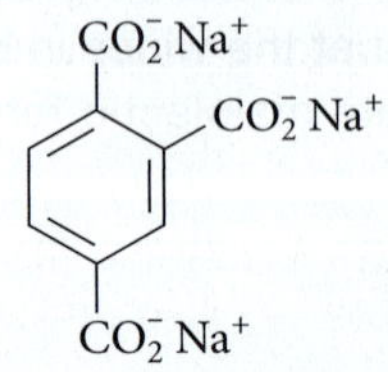

C

- **Know the unique absorptions characteristic of carboxylic acids in their IR and NMR spectra. Be able to use IR and NMR spectra to determine the structure of a carboxylic acid. 20.3A,B**

20.33 Explain what features you would look for in their IR and proton NMR spectra that would distinguish between the following two compounds.

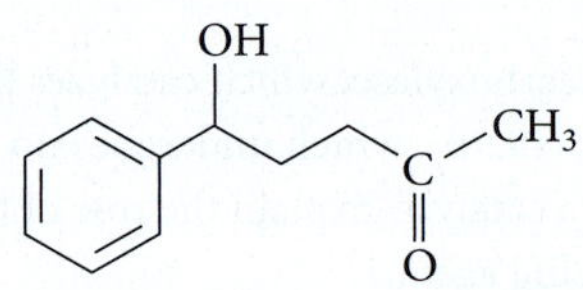

A

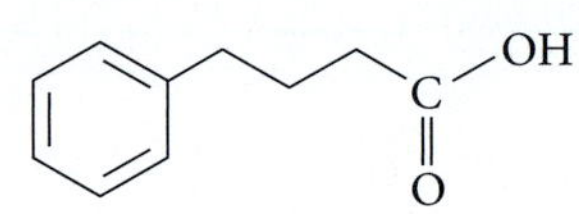

B

20.34 Propose structures for each of the following compounds.

(a) A compound *A* with the formula $C_9H_{10}O_3$ that has the following spectra:

IR 2400–3200, 1700, 1630 cm^{-1} NMR: δ 1.53 (3*H*, t, $J = 8$ Hz); δ 4.32 (2*H*, q, $J = 8$ Hz); δ 7.08, δ 8.13 (4*H*, pair of leaning doublets, $J = 10$ Hz); δ 10 (1*H*, broad, disappears with D_2O shake)

(b) An optically inactive compound *B* ($C_6H_{10}O_4$) that can be resolved into enantiomers and has the following NMR spectra:

^{13}C NMR: δ 13.5, δ 41.2, δ 177.9
proton NMR: δ 1.13 (6*H*, d, $J = 7$ Hz); δ 2.65 (2*H*, quintet, $J = 7$ Hz); δ 9.9 (2*H*, broad s, disappears after a D_2O shake)

- **Know the structural differences between fatty acids, soaps, phospholipids, and detergents. 20.5**

20.35 Match each of the following statements with one or more of the following: phospholipid (*A*), soap (*B*), detergent (*C*), fatty acid (*D*)

(a) Its structure contains a long unbranched hydrocarbon component.

(b) When naturally occurring, its double bonds almost always have the *Z* configuration.

(c) It readily forms micelles.

(d) It has a long unbranched hydrocarbon chain attached to a sulfonate group.

(e) Its structure has the general form R_4N^+ in which one of the R groups is a long unbranched alkyl group.

(f) Its structure contains two fatty acyl groups.

(g) Its structure contains a diester of phosphoric acid.

- **Know the approximate pK_a value of a carboxylic acid, and be able to rank various acids (including carboxylic acids) in order of acidity. 20.4A**

20.36 In what pK_a range do typical carboxylic acids lie?

(a) 1–3 (b) 4–5 (c) 6–8 (d) 9–11

20.37 Rank the following compounds in order of increasing acidity (decreasing pK_a). Explain your choices.

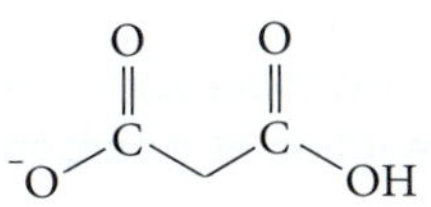

A

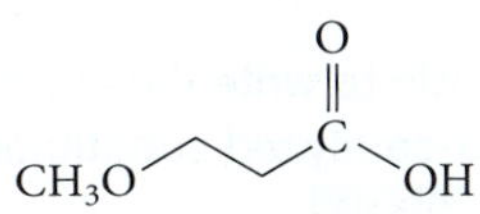

B

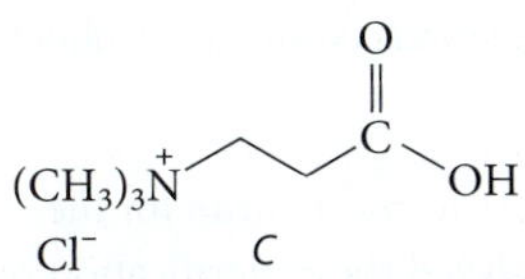

C

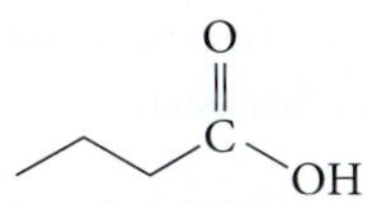

D

20.38 Draw the structure of the major species present in solution when 0.01 mol of the following acid in aqueous solution is treated with 0.01 mol of NaOH. Explain.

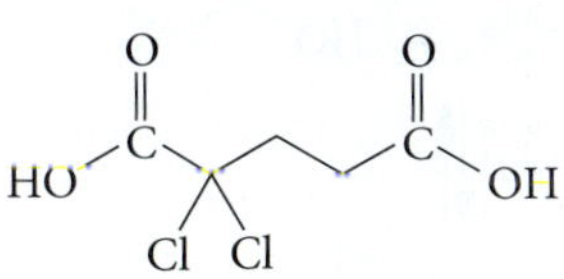

20.39 Explain why the *differences* between the first and second pK_a values of the dicarboxylic acids become smaller as the lengths of their carbon chains increase (Table 20.3).

- **Show which of the two oxygens of a carboxylic acid is more basic, and explain. 20.4B**

20.40 (a) Complete the following acid–base reaction, and explain how you deduced the structure of the conjugate acid AH.

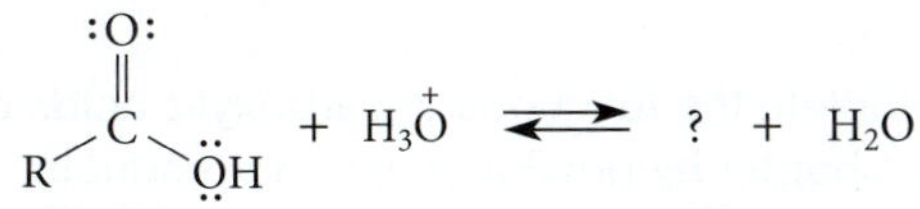

A *AH*

(b) Use your reasoning from part (a) to give the structure of the conjugate acid of an amide (*BH*).

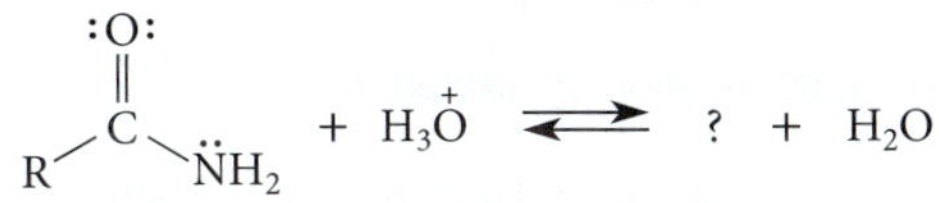

B (an amide) *BH*

- **Identify the carboxylic acids that are likely to undergo decarboxylation at room temperature or with gentle heating. 20.11**

20.41 Arrange the following carboxylic acids in order of their rates of decarboxylation, smallest first.

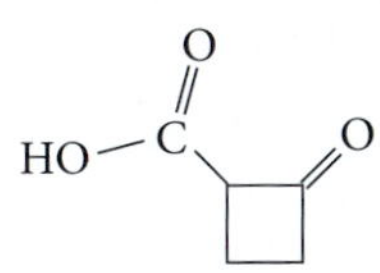

A

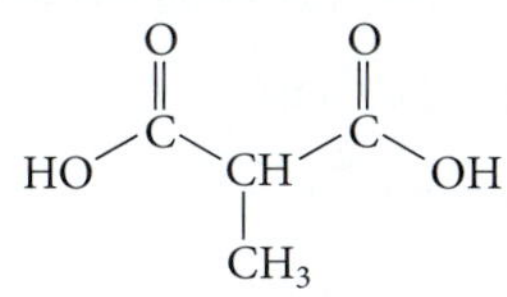

B

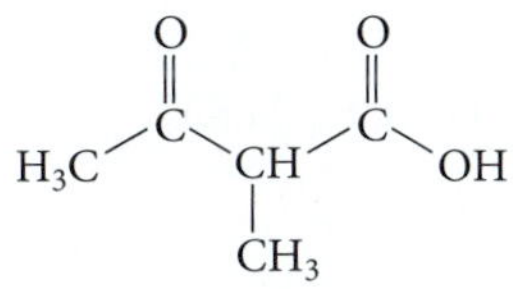

C

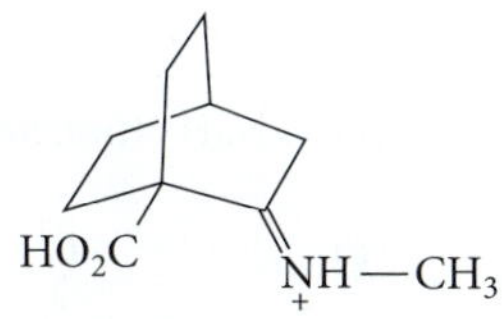

D

20.42 β-Imino carboxylic acids such as the one in **Fig. P20.42** decarboxylate rapidly provided that the pH is neither too low nor too high.

(a) Give a curved-arrow mechanism for the decarboxylation reaction shown in Fig. P20.42.

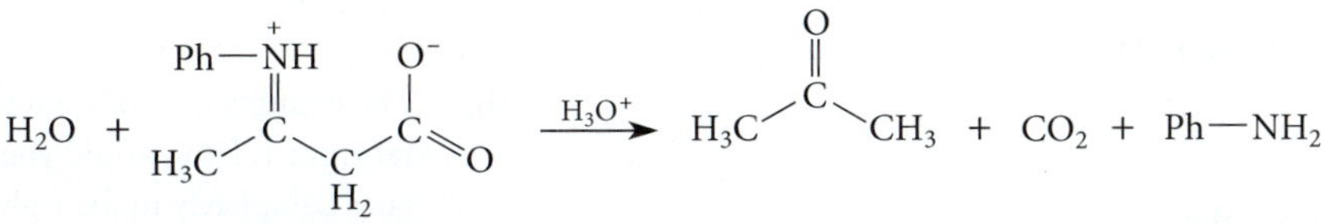

3-(*N*-phenylimino)butanoic acid
(a β-imino carboxylic acid)
zwitterion form

FIGURE P20.42

(b) Explain why very low or very high pH values reduce the rate of this decarboxylation.

(c) Primary amines (amines of the form R—NH_2) catalyze the decarboxylation of acetoacetic acid. Explain. (*Hint:* See Sec. 19.12A.)

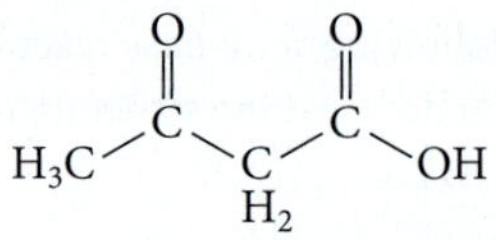

acetoacetic acid

- **Complete the reactions of carboxylic acids discussed in this chapter by providing missing products, starting materials, or reagents. 20.4, 20.8A,B, 20.9, 20.10, 20.11**

(1) Acid–base reactions

(2) Acid-catalyzed esterification

(3) Esterification by alkylation

(4) Acid chloride and anhydride formation

(5) Hydride reduction to alcohols

(6) Decarboxylation

20.43 Give the product(s) formed and the curved-arrow notation for the reaction of 0.01 mol of each of the following reagents with 0.01 mol of acetic acid.

(a) Na^+ CH_3O^- in methanol

(b) Cs^+ ^-OH

(c) EtMgBr

(d) H_3C—Li

(e) :NH_3

(f) NaH (*Hint:* See Sec. 11.1B, Eq. 11.6.)

(g)

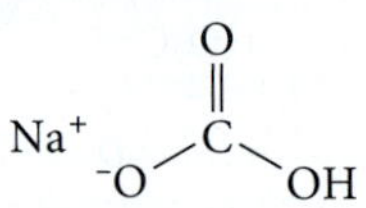

20.44 Give the product expected when butyric acid (or other compound indicated) reacts with each of the following reagents. [Parts (g) and (h) involve reactions from previous chapters.]

(a) Ethanol (solvent), H_2SO_4 catalyst

(b) Aqueous NaOH solution

(c) $LiAlH_4$ (excess), then H_3O^+

(d) $SOCl_2$

(e) Diazomethane in ether

(f) Product of part (c) (excess) + $CH_3CH{=}O$, HCl (catalyst) (Sec. 19.11)

(g) Product of part (d), $AlCl_3$, benzene, then H_2O (Sec. 16.4F)

(h) Product of part (g), H_2NNH_2, KOH, ethylene glycol (solvent), heat (Sec. 19.13A)

20.45 Give the product expected when benzoic acid reacts with each of the following reagents.

(a) $H_2C{=}CH{-}CH_2{-}Br$, K_2CO_3, acetone solvent

(b) PCl_5

(c) Concentrated HNO_3, H_2SO_4 (Sec. 16.5)

(d) P_2O_5, heat

20.46 Which of the following compounds give off water (or steam) when heated, which give off CO_2, and which give off neither? Explain your choices.

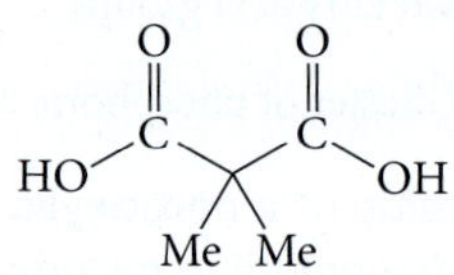

A

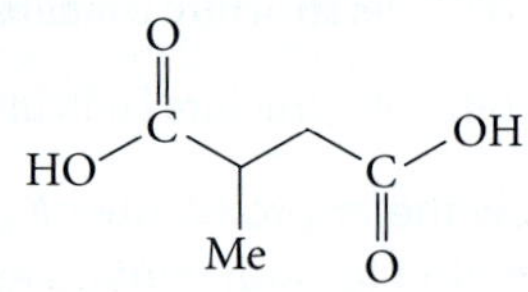

B

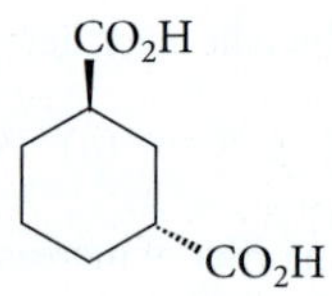

C

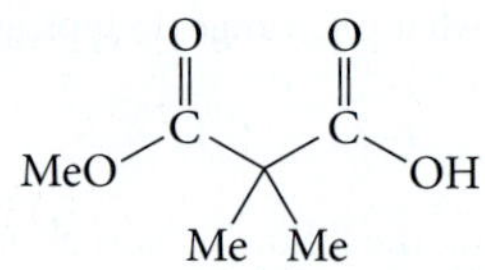

D

- **Be able to write the curved-arrow mechanism of an acid-catalyzed substitution reaction at a carbonyl group. 20.8A**

20.47 In *transesterification* an ester reacts with an alcohol to give another ester.

(a) Provide a curved-arrow mechanism for the following acid-catalyzed transesterification reaction. (Use H—A as an abbreviation for your acid.)

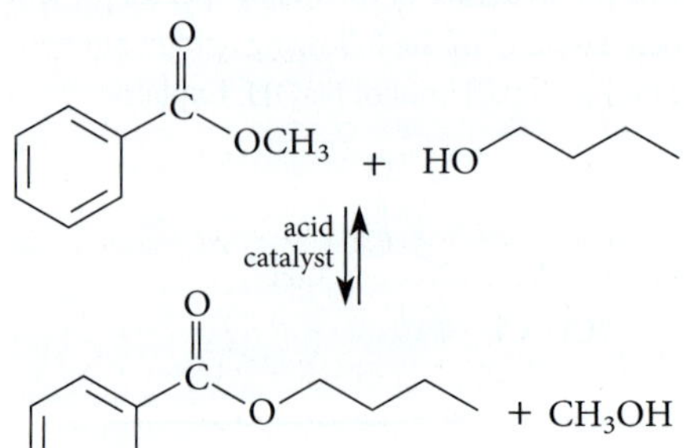

(b) The equilibrium constant for this reaction is not far from 1. How would you ensure that the reaction goes completely to the right?

- **Incorporate both the synthesis of carboxylic acids and their reactions into multistep synthetic sequences.** **20.6, 20.8–20.11**

20.48 Outline a synthesis of each of the following compounds from isobutyric acid (2-methylpropanoic acid) and any other necessary reagents.

(a) 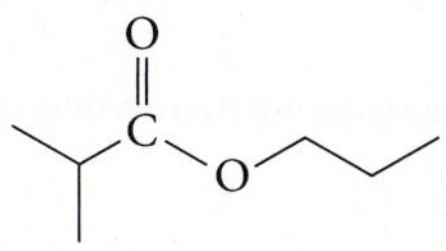

(b) 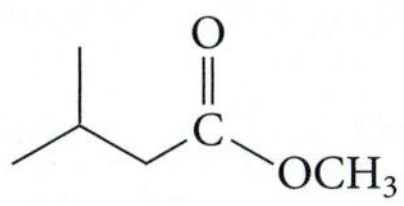

(c) 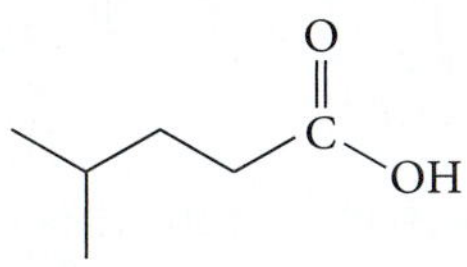

(d) 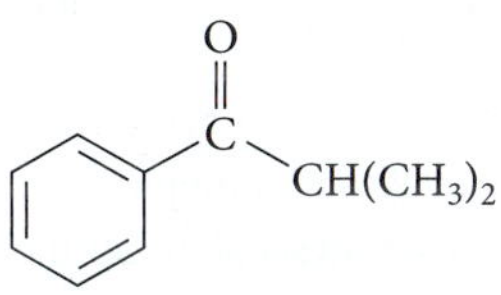

isobutyrophenone

(e) 3-Methyl-2-butanone

(f) $(CH_3)_2CHCH{=}CH_2$

20.49 You have been employed by a biochemist, Fungus P. Gildersleeve, who has given you a very expensive sample of benzoic acid labeled equally in both oxygens with ^{18}O. He asks you to prepare methyl benzoate (structure in Eq. 20.32), preserving as much ^{18}O label in the ester as possible. Propose a synthesis of ^{18}O-labeled methyl benzoate that meets this objective.

20.50 The sodium salt of *valproic acid* is a drug that has been used in the treatment of epilepsy. (Valproic acid is a name used in medicine.)

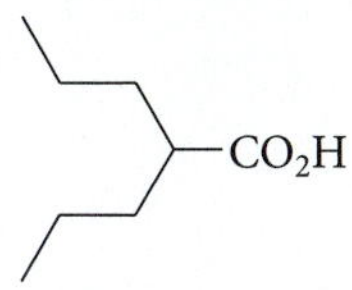

valproic acid

(a) Give the IUPAC substitutive name of valproic acid.

(b) Give the common name of valproic acid.

(c) Outline a synthesis of valproic acid from carbon sources containing fewer than five carbons and any other necessary reagents.

INTEGRATED PROBLEMS

20.51 (a) What is the molecular mass of a carboxylic acid containing a single carboxylic acid group if 8.61 mL of 0.1 *M* NaOH solution is required to neutralize 100 mg of the acid?

(b) How many mL of aqueous 0.1 *M* NaOH are required to form the disodium salt from 100 mg of succinic acid?

20.52 Ordinary litmus paper turns red at pH values below about 3. Show that a 0.1 *M* solution of acetic acid ($pK_a = 4.76$) will turn litmus red, and that a 0.1 *M* solution of phenol ($pK_a = 9.95$) will not.

20.53 (a) Explain why the most acidic species that can exist in significant concentration in any solvent is the conjugate acid of the solvent.

(b) Show why HBr is a stronger acid in acetic acid solvent than it is in water.

20.54 Penicillin-G is one of the penicillin family of drugs. In which fluid would you expect penicillin-G to be more soluble: stomach acid (pH = 2) or the bloodstream (pH = 7.4)? Explain.

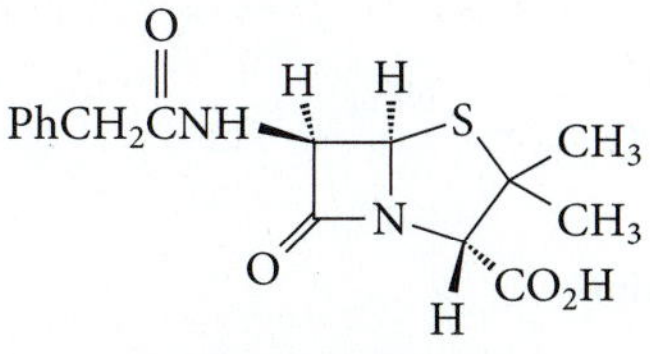

penicillin-G

20.55 You are a chemist for Chlorganics, Inc., a company specializing in chlorinated organic compounds. A process engineer, Turner Switchback, has accidentally mixed the contents of four vats containing, respectively, *p*-chlorophenol, 4-chlorocyclohexanol, *p*-chlorobenzoic acid, and chlorocyclohexane. The president of the company, Haley Ojinn (green with anger), has ordered you to design an expeditious separation of these four compounds. Success guarantees you a promotion; accommodate her.

20.56 (a) Squaric acid has pK_a values of about 1 and 3.5. Draw the reactions corresponding to the two successive ionizations of squaric acid, and label each with the appropriate pK_a value.

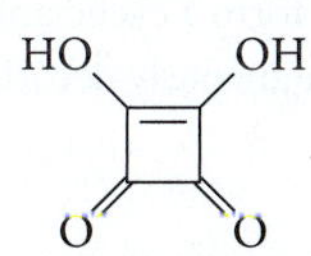

squaric acid

(b) Most enols have pK_a values in the 10–12 range. Using polar and resonance effects, explain why squaric acid is much more acidic.

(c) Given that squaric acid behaves like a dicarboxylic acid, draw structures for the products formed when it reacts with (i) excess $SOCl_2$ and (ii) ethanol solvent in the presence of an acid catalyst.

20.57 (a) The relatively stable carbocation *crystal violet* has a deep blue-violet color in aqueous solution. When NaOH is added to the solution, the blue color fades because the carbocation reacts with sodium hydroxide in about 1–2 min to give a colorless product. Show the reaction of crystal violet with NaOH, and explain why the color changes.

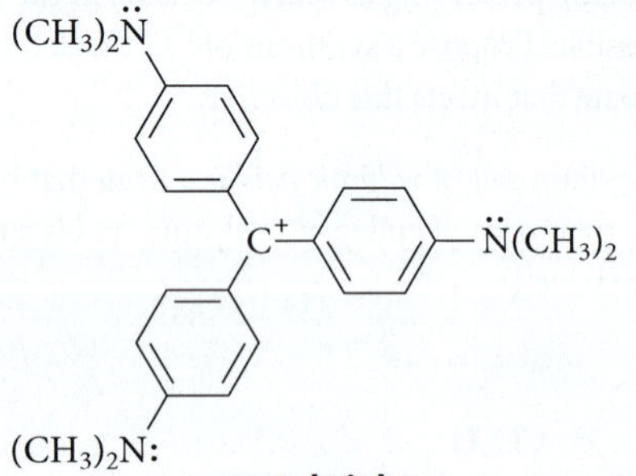

crystal violet

(b) When the detergent sodium dodecyl sulfate (SDS) is present in solution above its critical micelle concentration, the bleaching of crystal violet with NaOH takes several days. Account for the effect of SDS on the rate of this bleaching reaction.

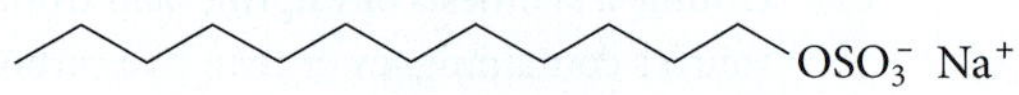

sodium dodecyl sulfate (SDS)

20.58 In each case, draw the structure of the cyclic anhydride that forms when the dicarboxylic acid is heated.

(a)

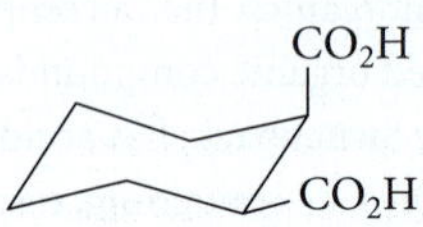

(b) *meso*-α,β-Dimethylsuccinic acid

(c)

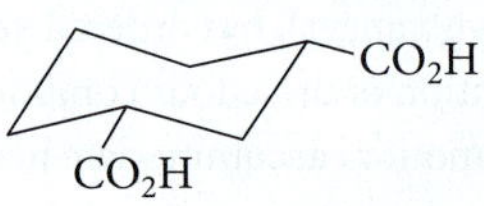

(*Hint:* Don't forget about the chair interconversion.)

20.59 Draw the structures and give the names of all dicarboxylic acids with the molecular formula $C_6H_{10}O_4$. Indicate which are chiral, which would readily form a cyclic anhydride on heating, and which would spontaneously decarboxylate on heating.

20.60 (a) Decarboxylation of compound *A* gives *two* separable products; draw their structures and explain.

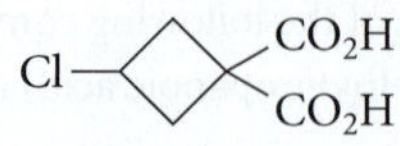

A

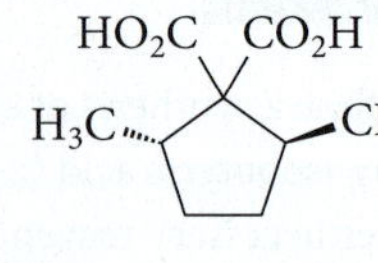

B

(b) How many products are formed when compound *B* is decarboxylated?

20.61 Pyridoxal phosphate (PLP) is a form of vitamin B_6.

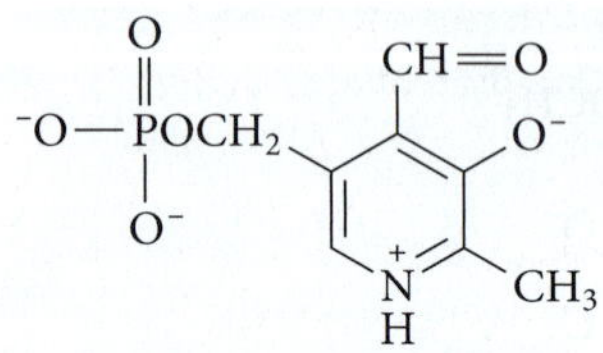

pyridoxal phosphate (PLP) (zwitterion form)

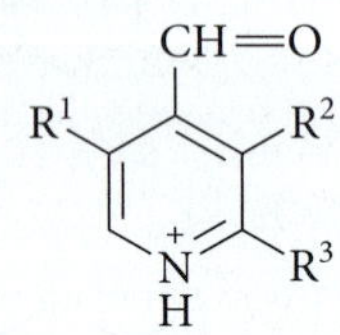

abbreviated structure of PLP

PLP serves as a coenzyme in the enzyme-catalyzed decarboxylation of *meso*-2,6-diaminopimelic acid, which is the last step in the biosynthesis of the amino acid *lysine*. The sequence of reactions involved in this process is shown in **Fig. P20.61**.

(a) In the first step of this reaction, a molecule of PLP forms an imine (Schiff base) adduct *A* with either of the amino groups of *meso*-2,6-diaminopimelate. Give the structure of adduct *A*. [*Hint:* The amino group that forms a Schiff base must be in its neutral (unprotonated) form to react with PLP.]

(b) Carbon dioxide is lost from *A* to form a new imine intermediate *X*. Draw a curved-arrow mechanism for this decarboxylation reaction. This mechanism should lead you to the structure of *X*. Show that the formation of *A* provides the "electron sink" that is crucial to this decarboxylation reaction.

(c) Imine *X* is transformed by a series of acid–base reactions to a different imine derivative *Y*, which is identical to the imine that can be formed separately from PLP and one of the amino groups of lysine. Imine *Y* undergoes imine hydrolysis to PLP and lysine. Draw a structure for *Y* and a curved-arrow

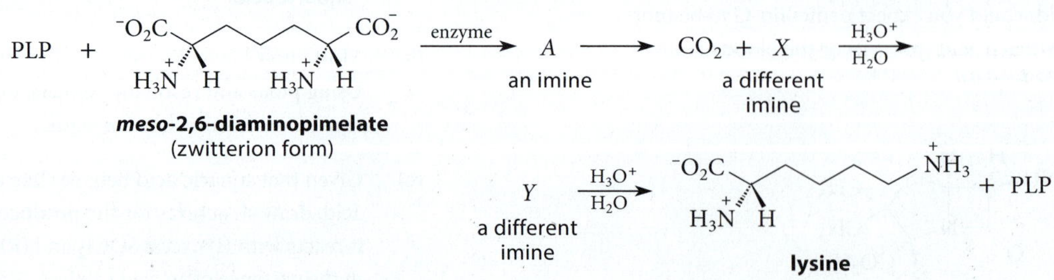

FIGURE P20.61

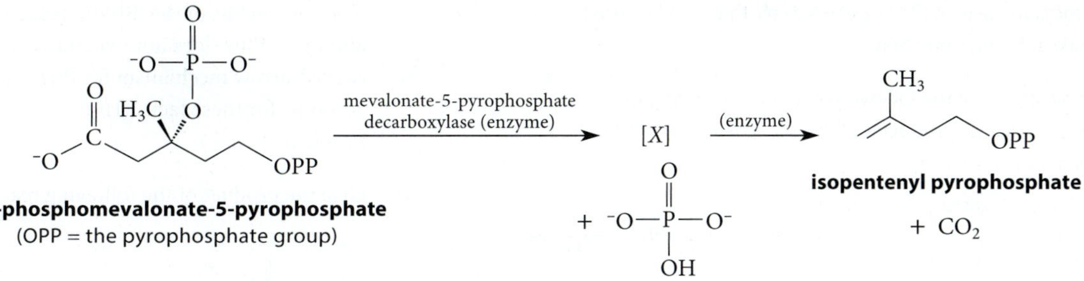

(*R*)-phosphomevalonate-5-pyrophosphate
(OPP = the pyrophosphate group)

FIGURE P20.62

mechanism for its formation from *X*. (Assume acids and bases are present as necessary; these are provided by the enzyme.)

20.62 Isopentenyl pyrophosphate, the starting material for isoprenoid and steroid biosynthesis (Sec. 17.6B), is formed biosynthetically by the decarboxylation reaction of (*R*)-phosphomevalonate-5-pyrophosphate catalyzed by the enzyme mevalonate-5-pyrophosphate decarboxylase, shown in **Fig. P20.62**. When a different starting material is used in which the CH_3— group is replaced with a fluoromethyl (FCH_2—) group, the reaction takes place 2500 times more slowly.

(a) Draw a two-step curved-arrow mechanism that includes the structure of the intermediate *X*. Your mechanism should account for the large effect of fluorine substitution on the rate of the reaction. (Assume acids and bases are available as needed.)

(b) What is the "electron sink" for the decarboxylation reaction?

20.63 Propose a synthesis of each of the following compounds from the indicated starting material(s) and any other necessary reagents.

(a) 2-Pentanol from propanoic acid

(b) 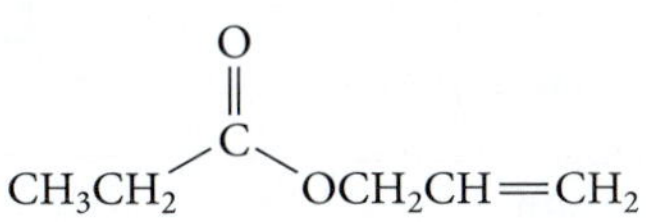

from allyl alcohol ($H_2C{=}CH{-}CH_2OH$) as the only carbon source

(c) 2-Methylheptane from pentanoic acid

(d) *m*-Nitrobenzoic acid from toluene

(e) $PhCH_2CH_2CH_2Ph$ from benzoic acid

(f) 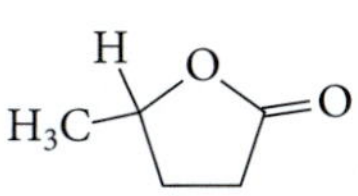 from 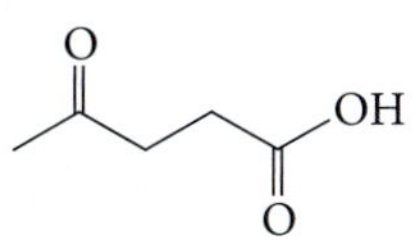

γ-butyrolactone

(*Hint:* Cyclic esters in a 5-membered ring, called γ-lactones, form spontaneously from γ-hydroxy acids.)

(g) 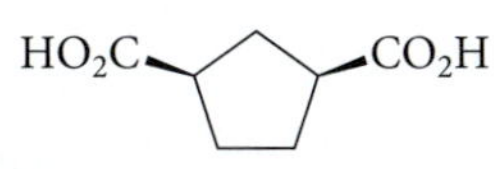 from 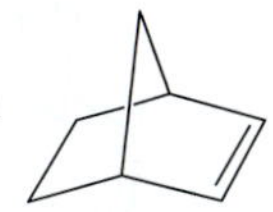

norbornene

(*Hint:* Relate the carbons of the product to the carbons of the alkene starting material.)

(h) 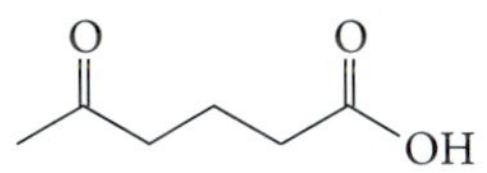 from 5-bromo-2-pentanone

5-oxohexanoic acid

(*Hint:* Use a protecting group.)

(i) 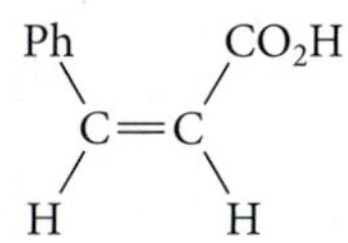 from Ph—C≡C—H

***cis*-cinnamic acid**

20.64 A graduate student, Al Kane, has been given a precious sample of (−)-3-methylhexane, along with optically active samples of both enantiomers of 4-methylhexanoic acid, each of known absolute configuration. Kane has been instructed to determine the absolute configuration of (−)-3-methylhexane. Show what Kane should do to deduce the configuration of the optically active hydrocarbon from the acids of known configuration. Be specific. (*Hint:* See Sec. 6.5.)

20.65 Because the radioactive isotope carbon-14 is used at very low ("tracer") levels, its presence cannot be detected by spectroscopy. It is generally detected by counting its radioactive decay in a device called a scintillation counter. The location of carbon-14 in a chemical compound must be determined by carrying out chemical degradations, isolating the resulting fragments that represent different carbons in the molecule, and counting them.

A well-known biologist, Fizzi O. Logicle, has purchased a sample of phenylacetic acid ($PhCH_2CO_2H$) advertised to be labeled with the radioactive isotope carbon-14 *only* at the carbonyl carbon. Before using this compound in experiments designed to test a promising theory of biosynthesis, she has wisely decided to be sure that the radiolabel is located only at the carbonyl carbon as claimed. Knowing your expertise in organic chemistry, she asks that you devise a way to determine what fraction of the ^{14}C is at the carbonyl carbon and what fraction is elsewhere in the molecule. Outline a reaction scheme starting with Dr. Logicle's sample that could be used to make this determination.

20.66 Complete each of the following reactions by giving the principal organic product(s). Give the reasons for your answers.

(a)
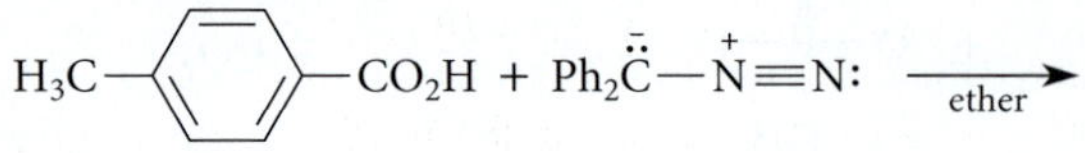

(b)

(c)
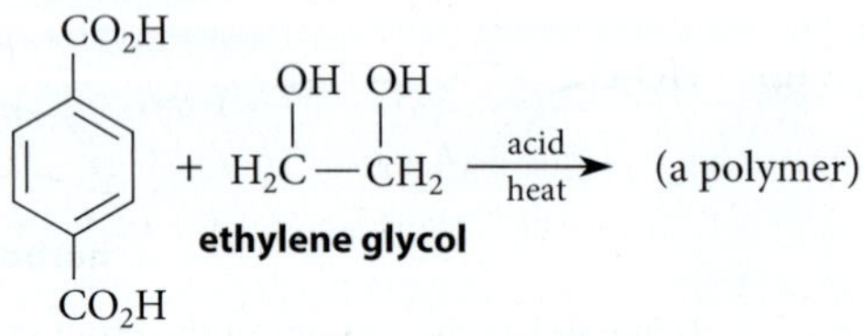

(d)
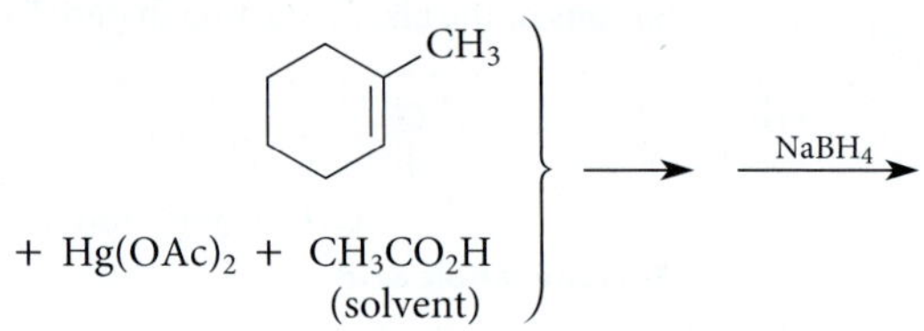

(e)
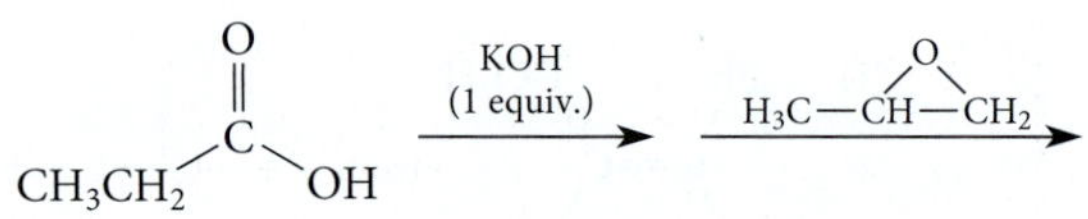

(f)
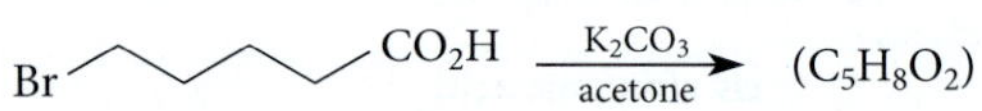

(g)
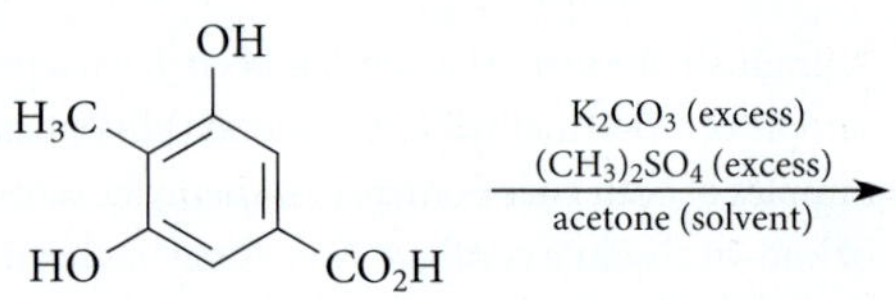

(h)

20.67 Explain why all efforts to synthesize a carboxylic acid containing the isotope ^{18}O at *only* the carbonyl oxygen fail and yield instead a carboxylic acid in which the labeled oxygen is distributed equally between both oxygens of the carboxy group ($*O = {}^{18}O$).

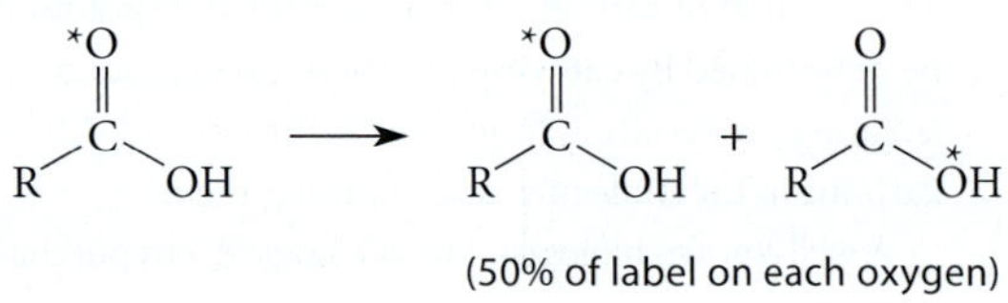

20.68 (a) Organolithium reagents such as methyllithium (CH_3Li) react with carboxylic acids to give ketones.

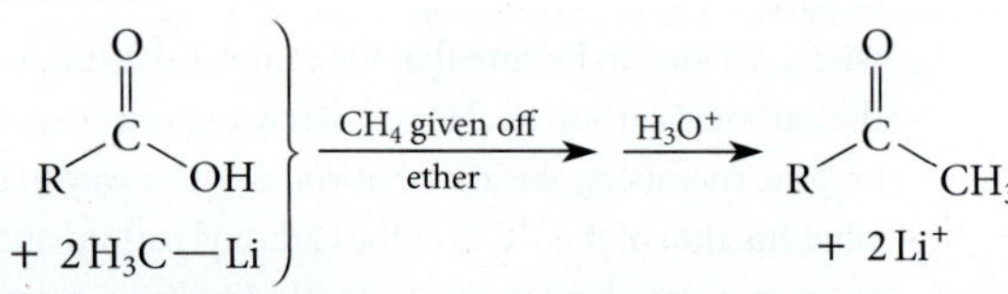

Two equivalents of the lithium reagent are required, and the ketone does not react further. Provide a curved-arrow mechanism for this reaction that accounts for these facts. (*Hint:* Start by looking at Problem 20.43d.)

(b) Give the product of the following reaction.

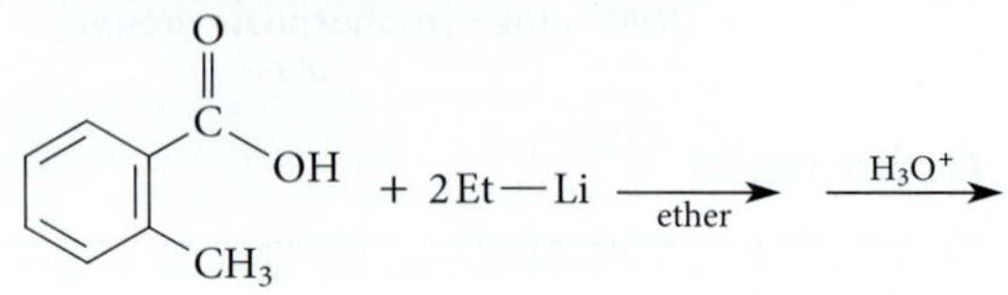

20.69 Give a curved-arrow mechanism for each of the following reactions.

(a)
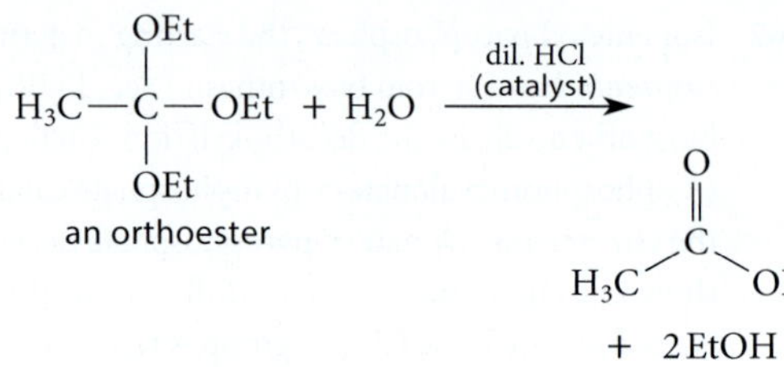

(b)
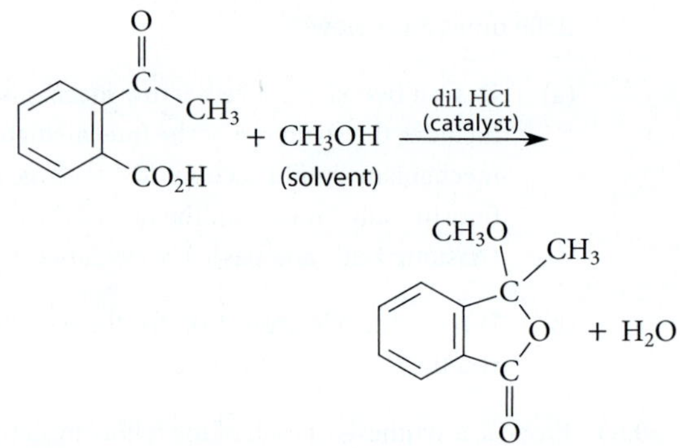

(c)
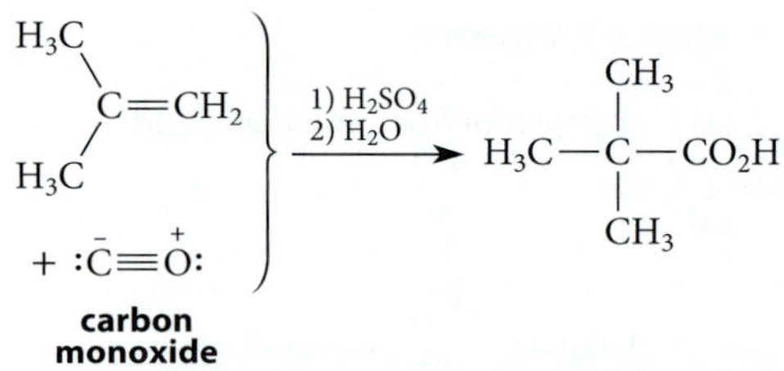

(d)
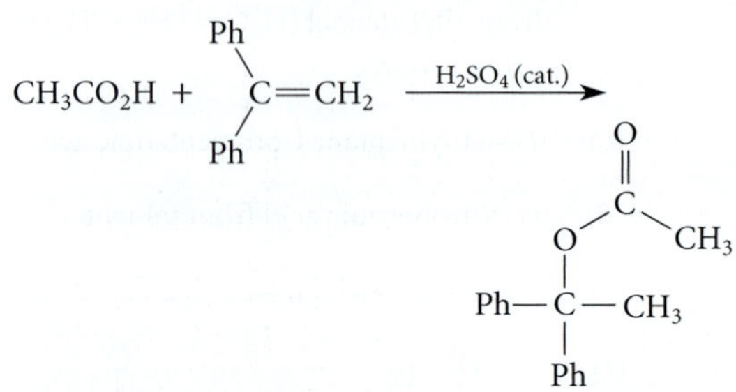

(e)
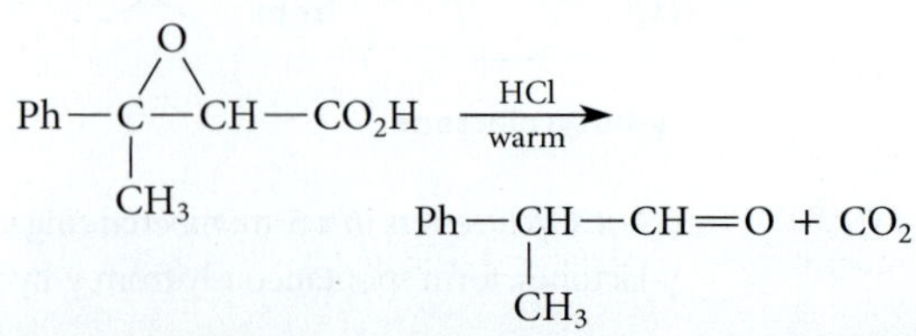

(f) 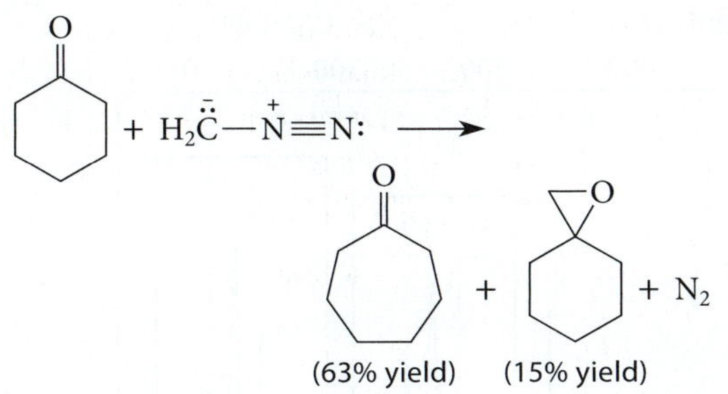

(g) 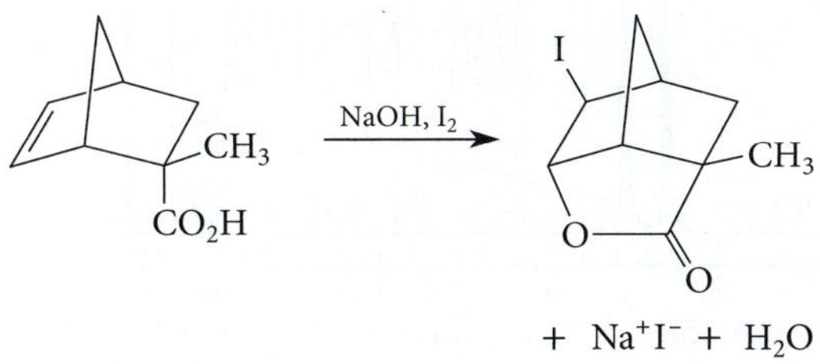

20.70 Propose reasonable fragmentation mechanisms that explain each of the following observations.

(a) The EI mass spectrum of 2-methylpentanoic acid has a strong peak at $m/z = 74$.

(b) The EI mass spectrum of benzoic acid shows major peaks at $m/z = 105$ and $m/z = 77$.

20.71 Propose structures for the compounds that have the following spectral data.

(a) $C_9H_{10}O_3$: IR 2300–3200 (broad, strong), 1710, 1600 cm^{-1}; proton NMR spectrum in **Fig. P20.71a**

(b) IR 3000–3580 (broad), 1698, 981 cm^{-1} UV: λ_{max} = 212 nm (ε = 10,800); CI mass spectrum: M + 1 peak at $m/z = 115$; proton NMR spectrum in **Fig. P20.71b**

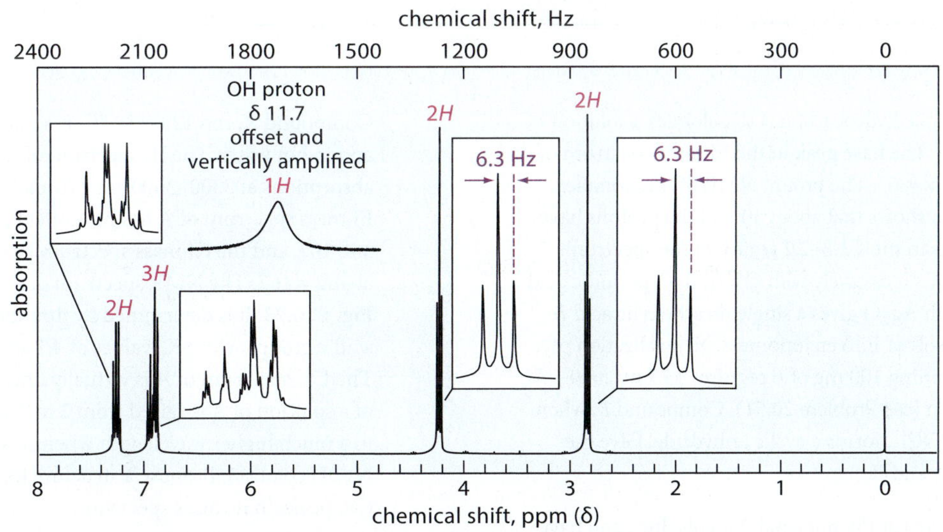

(a)

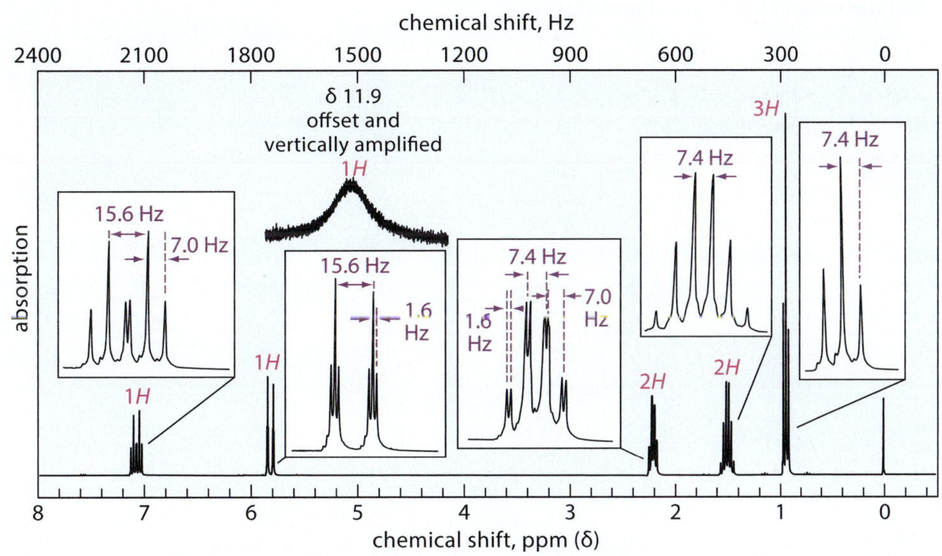

(b)

FIGURE P20.71 Proton NMR spectra for Problem 20.71a and b. The relative values of the integrals are given in red over their respective resonances.

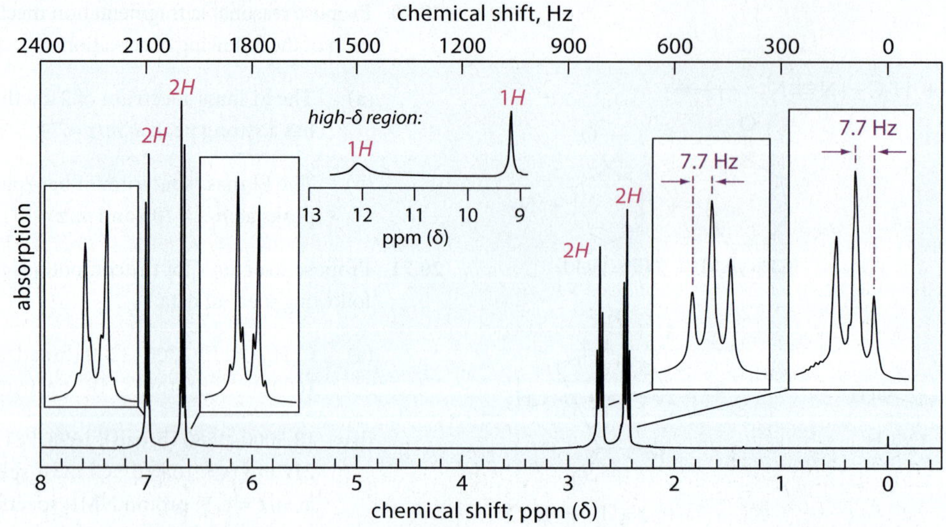

FIGURE P20.73 The proton NMR spectrum for Problem 20.73. The part of the spectrum beyond δ 8 is traced separately. The relative values of the integrals are given in red over their respective resonances.

20.72 A water-insoluble hydrocarbon *A* decolorizes a solution of Br_2 in CH_2Cl_2. The base peak in the EI mass spectrum of *A* occurs at $m/z = 67$. The proton NMR of *A* is complex, but integration shows that about 30% of the protons have chemical shifts in the δ 1.8–2.2 region of the spectrum. Treatment of *A* successively with OsO_4, then periodic acid, and finally with Ag_2O gives a single dicarboxylic acid *B* that can be resolved into enantiomers. Neutralization of a solution containing 100 mg of *B* requires 13.7 mL of 0.1 *M* NaOH solution (see Problem 20.51). Compound *B*, when treated with $POCl_3$, forms a cyclic anhydride. Give the structures of *A* and *B*.

20.73 You are employed at Phenomenal Phenols, Inc., and have been asked by your supervisor, O. H. Gruppa, to identify a compound *A* that was isolated from natural sources. Compound *A*, mp 129–130 °C, is soluble in NaOH solution and in hot water. The IR spectrum of *A* shows prominent absorptions at 3300–3600 cm^{-1} (broad) and 1680 cm^{-1}; the EI mass spectrum of *A* has prominent peaks at $m/z = 166$ and 107, and the CI mass spectrum has an M + 1 peak at $m/z = 167$. The NMR spectrum of *A* is given in **Fig. P20.73**. It is determined by titration that *A* has two acidic groups with pK_a values of 4.7 and 10.4, respectively. The UV spectrum of *A* is virtually unchanged as the pH of a solution of *A* is raised from 2 to 7, but the λ_{max} shifts to a much higher wavelength when *A* is dissolved in 0.1 *M* NaOH solution. Propose a structure for *A*. Rationalize the two peaks in its mass spectrum.

21 THE CHEMISTRY OF CARBOXYLIC ACID DERIVATIVES

A **carboxylic acid derivative** is a compound that can be hydrolyzed under acidic or basic conditions to give the parent carboxylic acid. All carboxylic acid derivatives can be conceptually derived by replacing a portion of the carboxylic acid structure with other groups, as shown in **Table 21.1**.

Carboxylic acids and their derivatives have not only structural similarities, but also close relationships in their chemistry. With the exception of nitriles, all carboxylic acid derivatives contain a *carbonyl group*. Many important reactions of these compounds occur at the carbonyl group. Furthermore, the —CN (cyano) group of nitriles has reactivity that resembles that of a carbonyl group. So, the chemistry of carboxylic acid derivatives, like that of aldehydes, ketones, and carboxylic acids, involves the chemistry of the carbonyl group.

21.1 NOMENCLATURE AND CLASSIFICATION OF CARBOXYLIC ACID DERIVATIVES

A. Esters and Lactones

Carboxylic esters are named as derivatives of their parent carboxylic acids by applying a variation of the system used in naming carboxylate salts (Sec. 20.4A). The group attached to the carboxylate oxygen is named first as a simple alkyl or aryl group. This name is followed by the name of the parent carboxylate, which, as we have shown, is constructed by dropping the final *ic* from the name of the acid and adding the suffix *ate*. This procedure is used in both common and substitutive nomenclature.

TABLE 21.1 Structures of Carboxylic Acid Derivatives

General structure, name of derivative	Condensed structure	Derivation* Replace—	Derivation* With—	Example
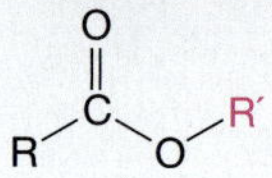 ester	R—CO_2R′	—H	—R′	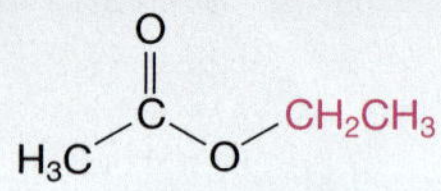 **ethyl acetate**
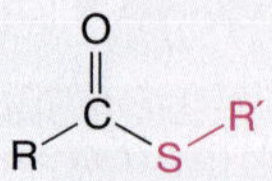 thioester	R—CO—SR′	—OH	—SR	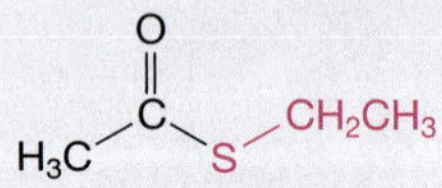 **ethyl thioacetate (S-ethylthioacetate)**
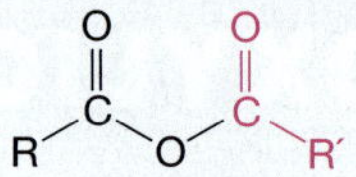 anhydride	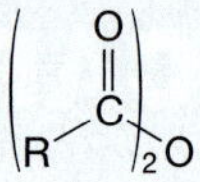	—H	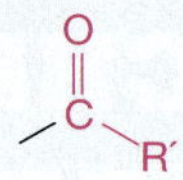 (acyl group)	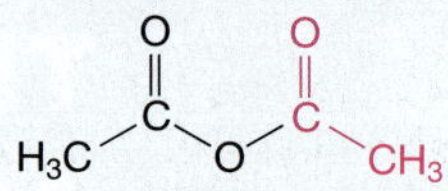 **acetic anhydride**
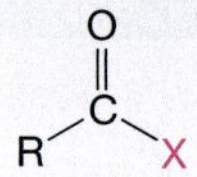 acid halide	R—CO—X	—OH	—X (halogen)	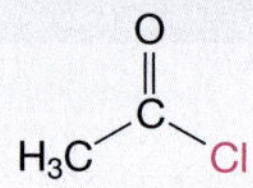 **acetyl chloride**
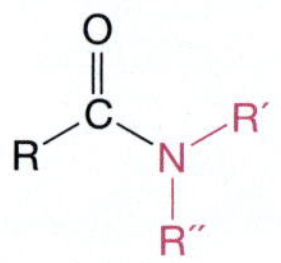 amide	R—CO—NR′R″	—OH	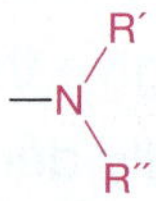	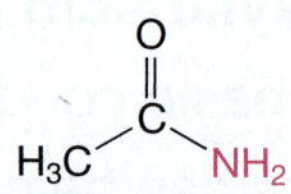 **acetamide**
R—C≡N nitrile	R—CN	—CO_2H	—CN (cyano group)	H_3C—C≡N **acetonitrile**

*Within the carboxylic acid structure 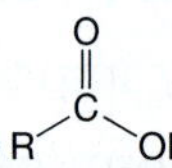 , replace the group in column 3 with the group in column 4 to obtain the derivative. (Note that this shows the relationship of structures, but not necessarily how they are interconverted chemically.)

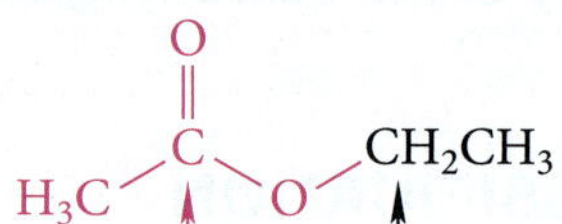

ethyl acetate
(common)

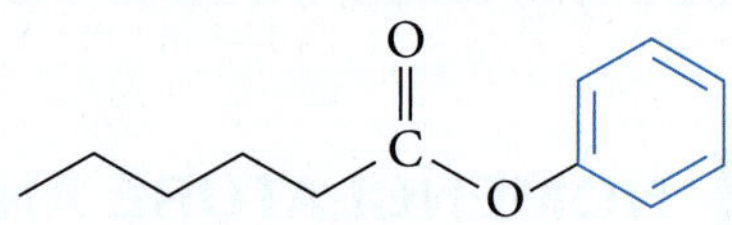

phenyl hexanoate
(substitutive)

(21.1)

Esters of acetic acid (acetate esters) are so common that the structure of the acetate group is often abbreviated as —OAc.

acetate = —OAc = 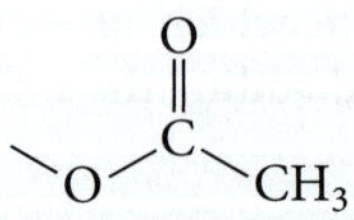 (21.2)

Ethyl acetate could then be abbreviated CH_3CH_2OAc or EtOAc. (This group is called *acetoxy* when it is a substituent; see Table 21.2, Sec. 21.1F.)

In ester nomenclature, substituents can occur in the acyl part of the ester or in the carboxylate part. In substitutive nomenclature, the position of a substituent in the acyl part of an ester is indicated by number as it is in carboxylic acids, in which the carbonyl carbon receives the number 1. In common nomenclature, the position of a substituent is indicated by a Greek letter, and numbering starts at the carbon *adjacent* to the carbonyl carbon.

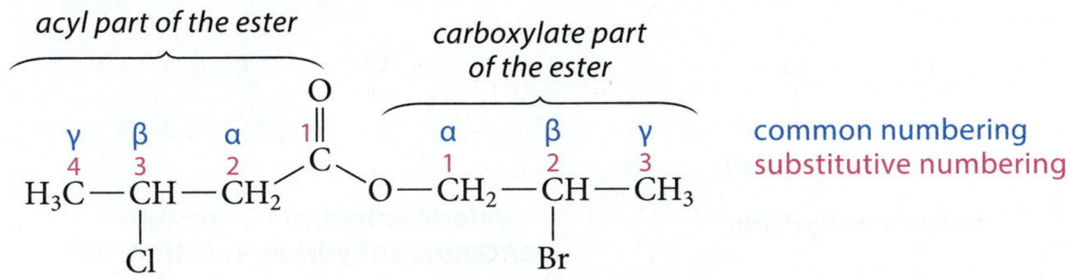

common name: **β-bromopropyl β-chlorobutyrate**
substitutive name: **2-bromopropyl 3-chlorobutanoate** (21.3)

As this example shows, , substitution in the carboxylate part of the ester is also indicated by number (in substitutive nomenclature) or by Greek letter (in common nomenclature). In this part of the ester, numbering starts at the carbon attached to the ester oxygen.

Esters of other acids are named by analogous extensions of acid nomenclature.

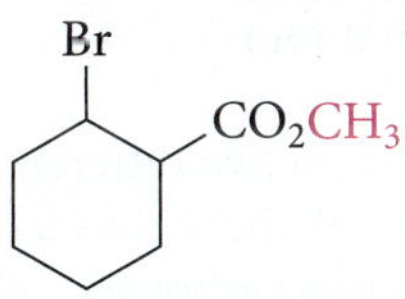

methyl 2-bromocyclohexanecarboxylate

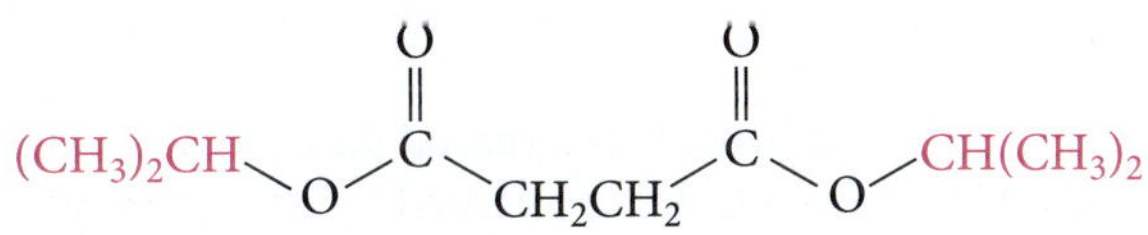

diisopropyl succinate

Cyclic esters are called **lactones**.

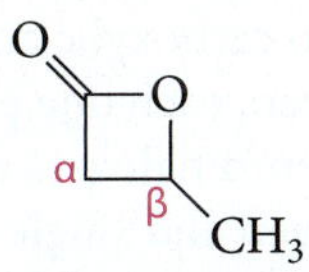

β-butyrolactone
(a β-lactone)

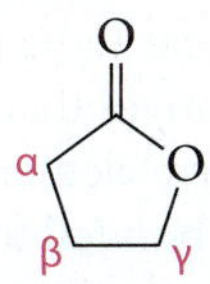

γ-butyrolactone
(a γ-lactone) (21.4)

In common nomenclature, illustrated in these examples, the *name* of a lactone is derived from the acid with the same number of carbons in its principal chain; the *ring size* is denoted by a Greek letter corresponding to the point of attachment of the lactone ring oxygen to the carbon chain. For example, in a β-lactone, the ring oxygen is attached at the β-carbon to form a four-membered ring.

The substitutive nomenclature of lactones is a specialized extension of heterocyclic nomenclature that we will not consider.

B. Acid Halides

Acid halides are named in any system of nomenclature by replacing the *ic* ending of the acid with the suffix *yl*, followed by the name of the halide.

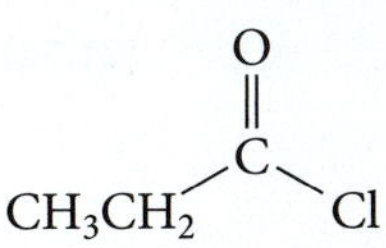

propion~~ic~~ + yl =
propionyl chloride
(common)
propanoyl chloride
(substitutive)

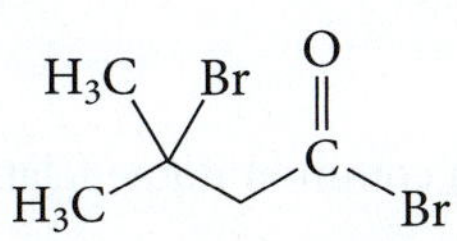

β-bromo-β-methylbutyryl bromide (common)
3-bromo-3-methylbutanoyl bromide (substitutive)

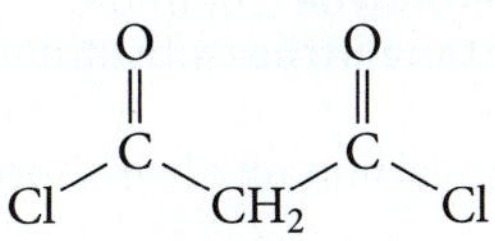

malonyl dichloride

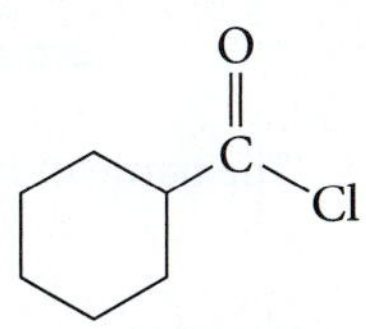

cyclohexanecarbonyl chloride

(21.5)

Notice the special nomenclature required when the acid halide group is attached to a ring: the compound is named as a cycloalkanecarbonyl halide.

C. Anhydrides

To name an anhydride, the name of the parent acid is followed by the word *anhydride*.

benzoic anhydride

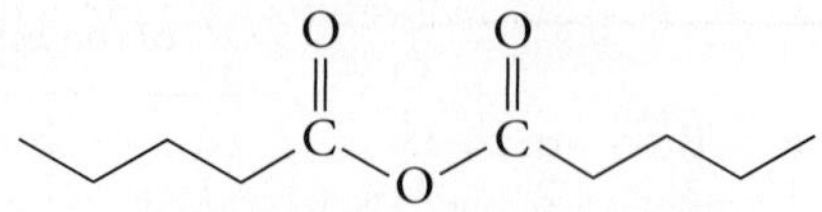

valeric anhydride (common)
pentanoic anhydride (substitutive)

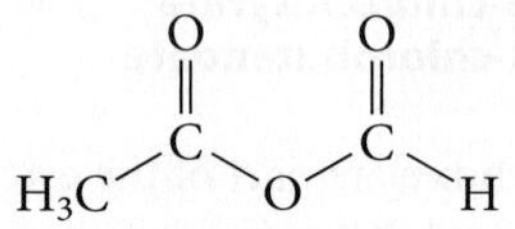

acetic formic anhydride
(a mixed anhydride)

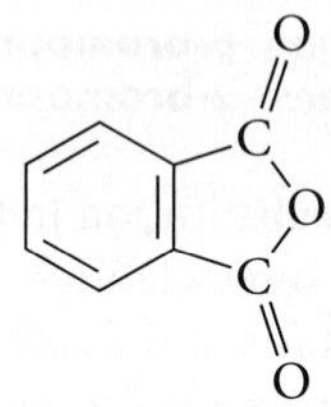

phthalic anhydride
(a cyclic anhydride) (21.6)

Acetic formic anhydride is a **mixed anhydride**, because it is derived from two different carboxylic acids. Mixed anhydrides are named by citing the two parent acids in alphabetical order. Phthalic anhydride is a **cyclic anhydride**, which is an anhydride derived from two carboxylic acid groups within the same molecule [in this case, phthalic acid (Table 20.1)].

D. Nitriles

At first glance, nitriles may not appear to be related to carboxylic acids. However, in carboxylic acids and the carboxylic acid derivatives that we've seen, each one possesses a carbon that has three bonds to a more electronegative element (oxygen, a halogen, and so on). Carboxylic acids, for example, have a carbon double-bonded to an oxygen and single-bonded to another oxygen. Nitriles, then, fit the definition because the carbon has three bonds to nitrogen. In other words, the nitrile carbon has the same oxidation state as the carbonyl carbon of a carboxylic acid (Sec. 11.6A).

Nitriles are named in the common system by dropping the *ic* or *oic* from the name of the acid *with the same number of carbon atoms* (counting the nitrile carbon) and adding the suffix *onitrile*. In substitutive nomenclature, the suffix *nitrile* is added to the name of the hydrocarbon with the same number of carbon atoms.

$$Ph-C\equiv N:$$

benzonitrile (benz~~oic~~ + onitrile)

$$H_3C-C\equiv N:$$

acetonitrile (acet~~ic~~ + onitrile)

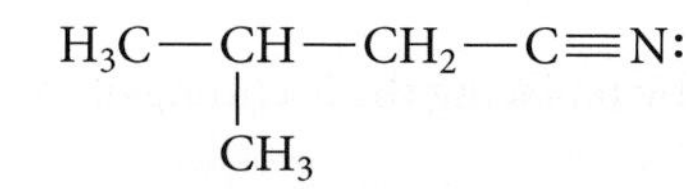

isovaleronitrile (common)
3-methylbutanenitrile (substitutive)

$$:N\equiv C-CH_2-CH_2-C\equiv N:$$

succinonitrile (common)
butanedinitrile (substitutive) (21.7a)

The name of the three-carbon nitrile is shortened in common nomenclature:

$$CH_3CH_2-C\equiv N$$

propionitrile (*not* propiononitrile) (21.7b)

When the nitrile group is attached to a ring, a special *carbonitrile* nomenclature is used.

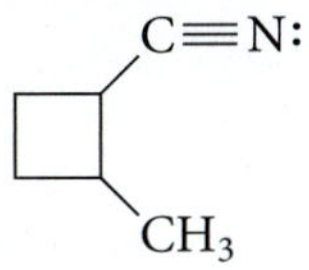

2-methylcyclobutanecarbonitrile (21.7c)

E. Amides, Lactams, and Imides

Simple amides are named in any system by replacing the *ic* or *oic* suffix of the acid name with the suffix *amide*.

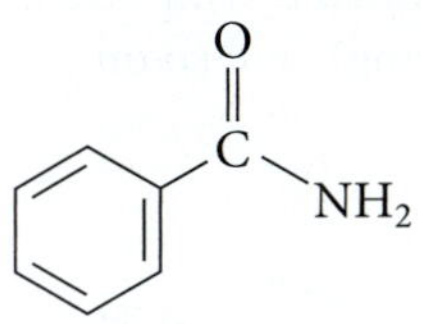

benzamide (benzoic + amide)

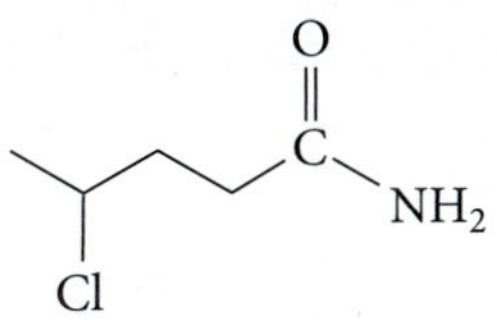

γ-chlorovaleramide (common)
4-chloropentanamide (substitutive) (21.8a)

When the amide functional group is attached to a ring, the suffix *carboxamide* is used.

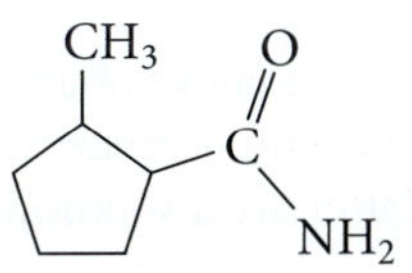

2-methylcyclopentanecarboxamide (21.8b)

Like amines (Sec. 19.12A; Chapter 23 Introduction), amides are classified as *primary*, *secondary*, or *tertiary* according to the number of hydrogens on the amide nitrogen.

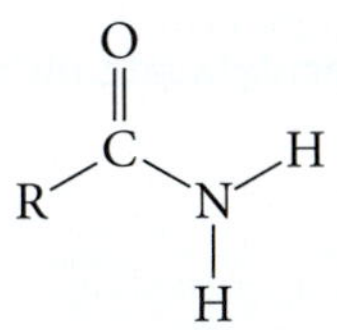

primary amide

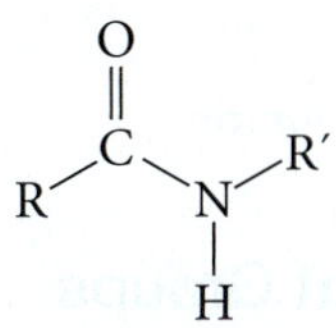

secondary amide

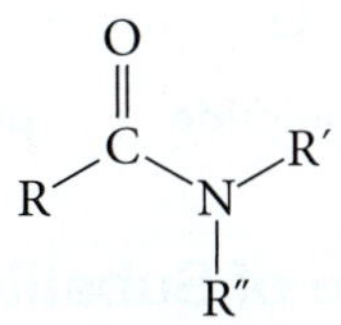

tertiary amide (21.9a)

This classification, *unlike that of alkyl halides and alcohols*, refers to substitution *at nitrogen* rather than substitution at carbon. For example, the following compound is a *secondary amide*, even though a tertiary alkyl group is bound to nitrogen.

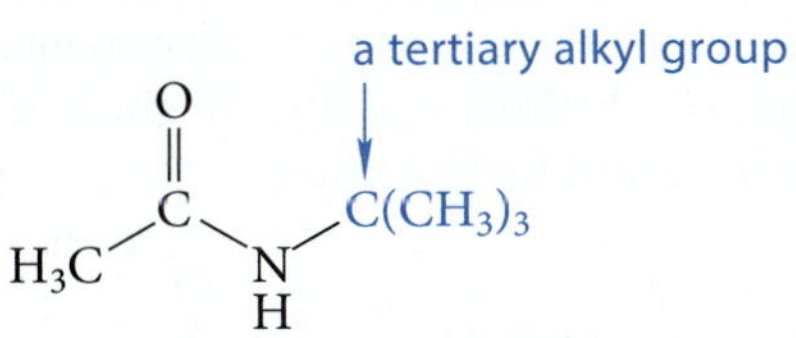

a secondary amide (21.9b)

Substitution on nitrogen in secondary and tertiary amides is designated with the letter *N* (italicized or underlined).

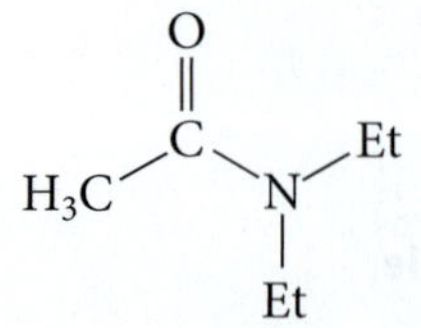

***N,N*-diethylacetamide**
(the double *N* designation means that both ethyl groups are on nitrogen)

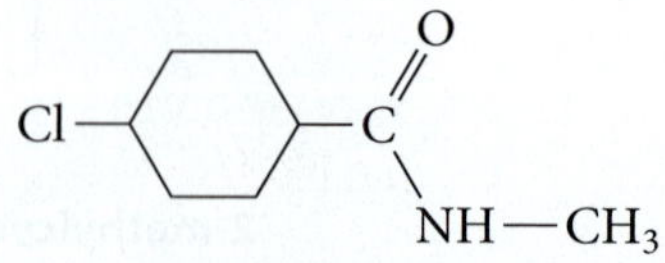

4-chloro-*N*-methylcyclohexanecarboxamide

(21.10)

Cyclic amides are called **lactams**, and the common nomenclature of the simple lactams is analogous to that of lactones. Lactams, like lactones, are classified by ring size as γ-lactams (five-membered lactam ring), β-lactams (four-membered lactam ring), and so on.

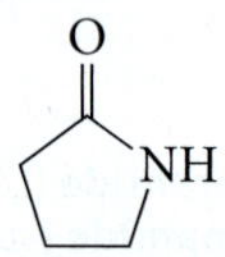

γ-butyrolactam
(a γ-lactam)

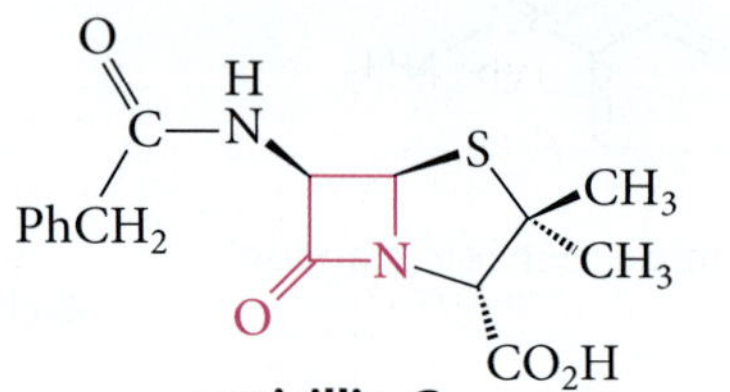

penicillin-G
(a β-lactam)

(21.11)

Imides can be thought of as the nitrogen analogs of anhydrides. Cyclic imides, such as succinimide and phthalimide, are of greater importance than noncyclic imides, such as the last example in Display 21.12, although the latter are also known compounds.

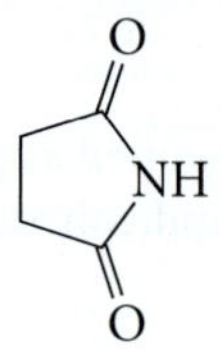

succinimide

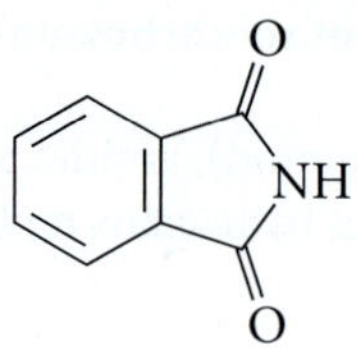

phthalimide

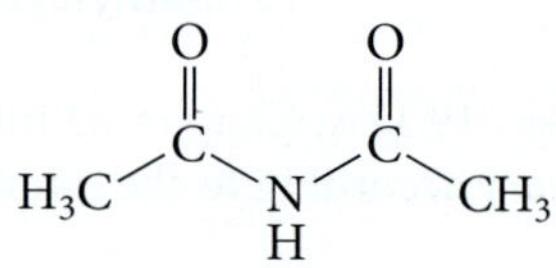

***N*-acetylacetamide**

(21.12)

F. Nomenclature of Substituent Groups

The priorities for citing principal groups in a carboxylic acid derivative are as follows:

$$\text{acid} > \text{anhydride} > \text{ester} > \text{acid halide} > \text{amide} > \text{nitrile} \quad (21.13)$$

All of these groups have citation priority over aldehydes and ketones, as well as the other functional groups considered in previous chapters. (A complete list of group priorities is given in Appendix I.) The names used for citing these groups as substituents are given in **Table 21.2**. (The names of other carbonyl-containing substituent groups are given in Display 19.5, Sec. 19.2A.) The following compounds illustrate the use of these names:

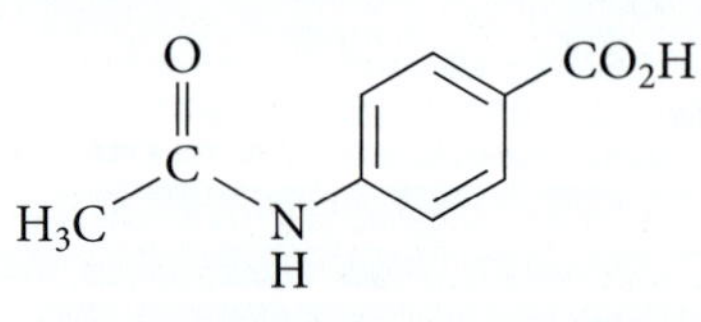

4-(acetylamino)benzoic acid
[*p*-acetamidobenzoic acid]

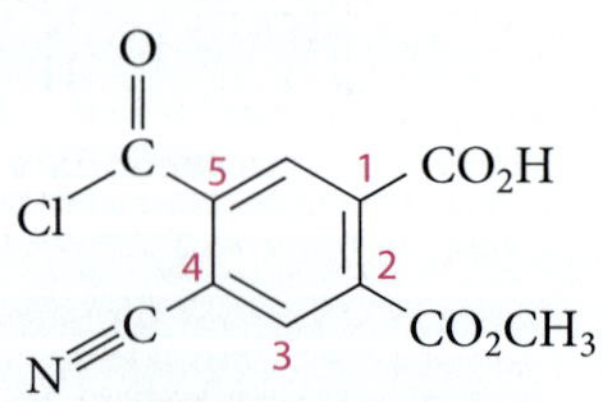

5-chloroformyl-4-cyano-2-methoxycarbonylbenzoic acid

(21.14)

TABLE 21.2 Names of Carboxylic Acid Derivatives When Used as Substituent Groups

Group	Name	Group	Name
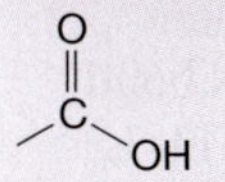	carboxy	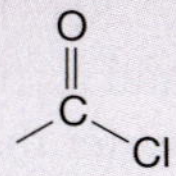	chloroformyl
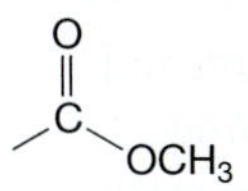	methoxycarbonyl	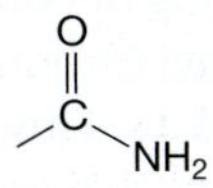	carbamoyl
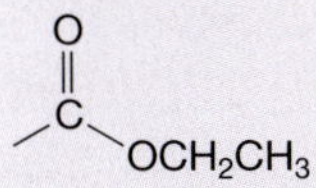	ethoxycarbonyl	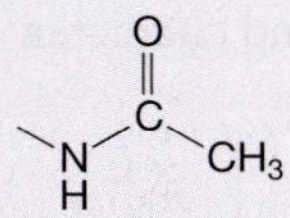	acetamido or acetylamino*
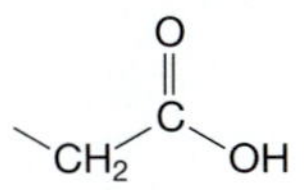	carboxymethyl	—C≡N	cyano
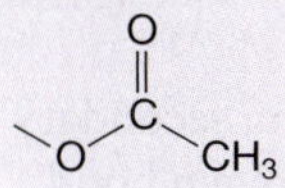	acetoxy or acetyloxy*		

*Used by Chemical Abstracts.

G. Carbonic Acid Derivatives

Esters of carbonic acid (Sec. 20.11A) are named like any other ester, but other important carbonic acid derivatives have special names that should be learned.

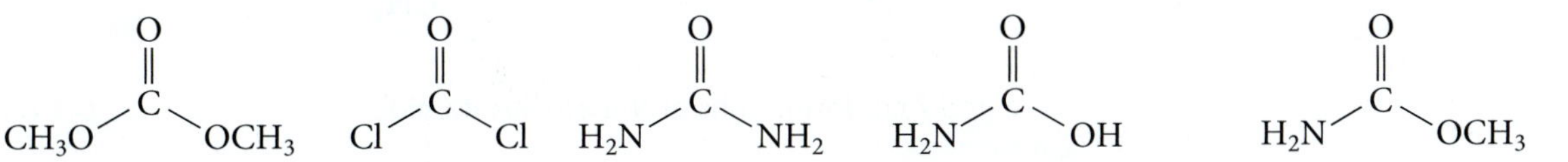

dimethyl carbonate **phosgene** **urea** **carbamic acid** (unstable, but has many stable derivatives) **methyl carbamate** (a stable carbamic acid derivative)

(21.15)

Focused Problems

21.1 Give a structure for each of the following compounds. (Refer to Table 20.1, Sec. 20.1A, for the common names of carboxylic acids.)

(a) 5-Cyanopentanoic acid (b) Isopropyl valerate (c) Ethyl methyl malonate

(d) Cyclohexyl acetate (e) *N,N*-Dimethylformamide (f) γ-Valerolactone

(g) Glutarimide (h) α-Chloroisobutyrylchloride (i) 3-Ethoxycarbonylhexanedioic acid

21.2 Name the following compounds.

(a) $CH_3CH_2CH_2CN$ (b) 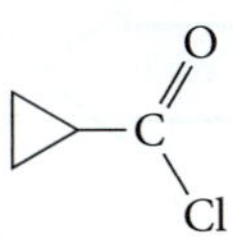 (c) 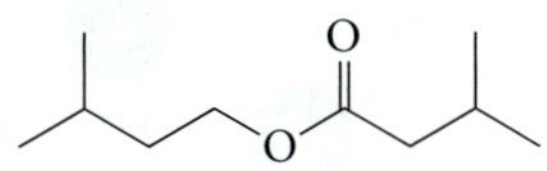 (d) 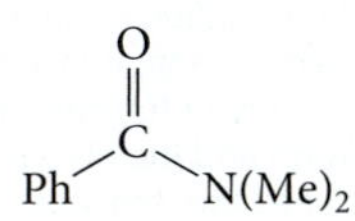

(e) 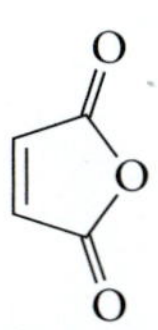 (f) 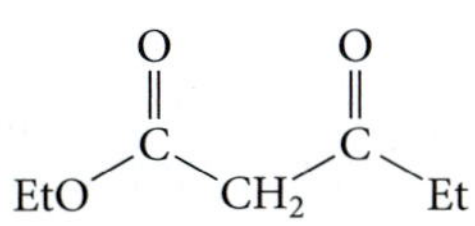 (g) 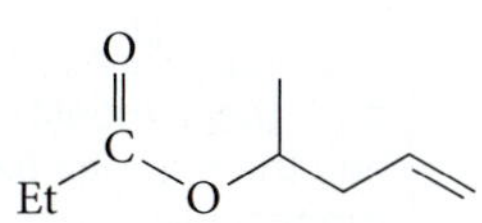

21.2 STRUCTURES OF CARBOXYLIC ACID DERIVATIVES

The structures of many carboxylic acid derivatives are very similar to the structures of other carbonyl compounds (see Fig. 19.2, Sec. 19.1). For example, the C═O bond length is about 121 pm, and the carbonyl group and its two attached atoms are planar. The nitrile C≡N bond length, 116 pm, is significantly shorter than the acetylene C≡C bond length, 120 pm. This is another example of the shortening of bonds to smaller atoms (Sec. 1.6B).

In an amide, both the carbonyl carbon and the amide nitrogen have sp^2 hybridization and trigonal planar bonding (**Fig. 21.1**). This geometry can be understood from the following resonance structures, which show that the bond between the nitrogen and the carbonyl carbon has considerable double-bond character.

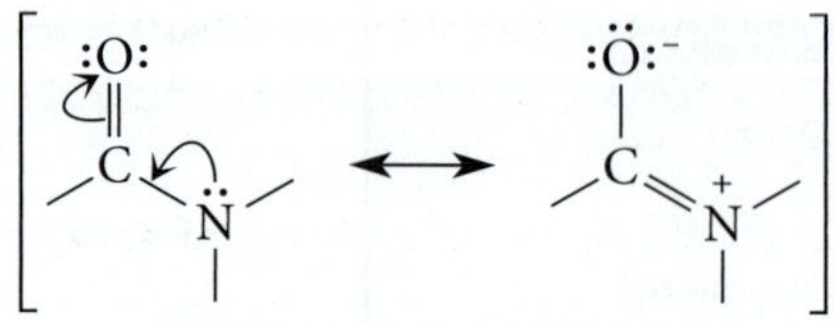

amide resonance structures (21.16)

(See also Display 1.47, Sec. 1.6C.) Because of the trigonal planar geometry at nitrogen, secondary and tertiary amides can exist in both *E* and *Z* conformations about the carbonyl–nitrogen bond; the *Z* conformation predominates in most secondary amides because, in this form, van der Waals repulsions between the largest groups—in this case, the methyl groups—are avoided.

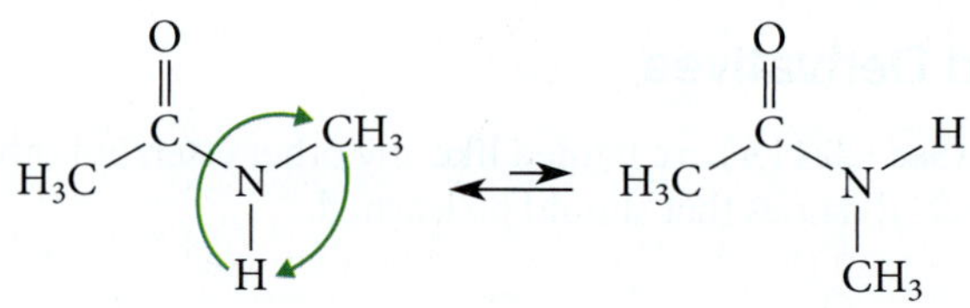

E and *Z* conformations of *N*-methylacetamide (21.17)

FURTHER EXPLORATION 21.1
NMR Evidence for Internal Rotation in Amides

The interconversion of the *E* and *Z* forms of amides is too rapid at room temperature to permit their separate isolation, but it is very slow compared with rotation about ordinary carbon–carbon single bonds. A typical energy barrier for rotation about the carbonyl–nitrogen bond of an amide is 71 kJ mol^{-1} (17 kcal mol^{-1}), which results in an internal rotation rate of about 10 times per second. (In contrast, internal rotation in butane occurs about 10^{11} times per second.) The relatively low rate of internal rotation is caused by the significant double-bond character in the carbon–nitrogen bond. We showed in Sec. 4.1D (Display 4.7) that rotation about carbon–carbon double bonds does not occur because rotation would require breaking the π bond. Rotation about a "partial double bond" is slow for the same reason.

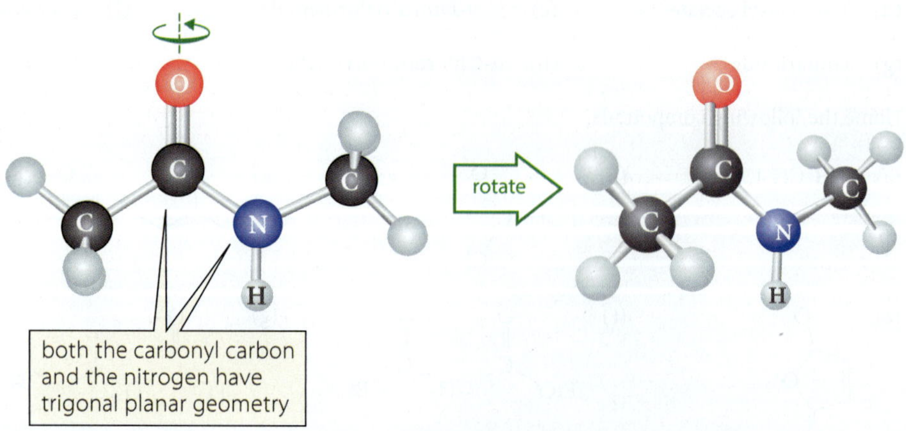

FIGURE 21.1 A ball-and-stick model of *N*-methylacetamide, shown in two perspectives. The perspective on the right is obtained from the one on the left by rotation about the axis shown. The trigonal planar geometry of the carbonyl carbon and the nitrogen requires that the labeled atoms lie in the same plane.

Focused Problems

21.3 Shown below is one conformation of the amino acid derivative *N*-acetylproline about the amide bond.

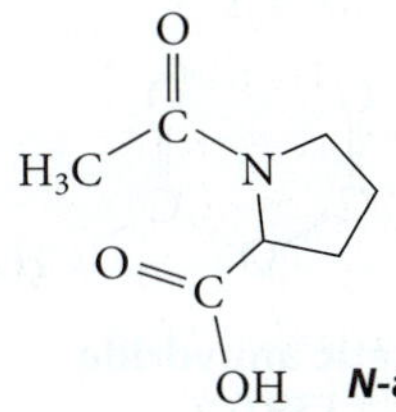

***N*-acetylproline**

(a) Is this the *E* or the *Z* conformation about the amide bond?

(b) Draw the other conformation about the amide bond.

21.4 Draw the structure of an amide that must exist in an *E* conformation about the carbonyl–nitrogen bond. (*Hint:* Think about a ring.)

21.3 PHYSICAL PROPERTIES OF CARBOXYLIC ACID DERIVATIVES

A. Esters

Esters are polar molecules, but they lack the capability to donate hydrogen bonds that carboxylic acids and amides have. The smaller esters are typically volatile, fragrant liquids that have lower densities than water. Most esters are insoluble in water. The low boiling point of a typical ester (*red*) is illustrated by the following comparison:

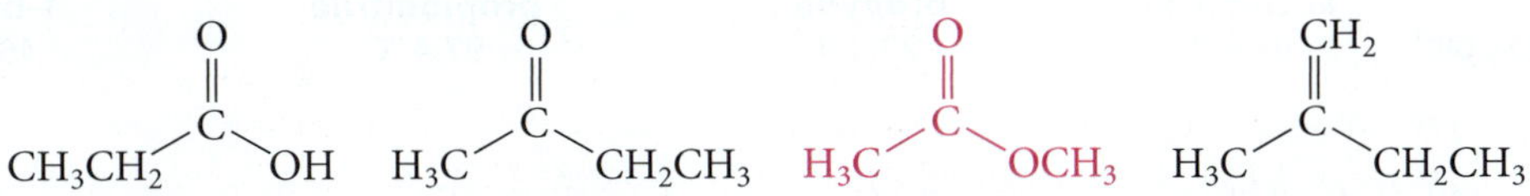

	propionic acid	**2-butanone**	**methyl acetate**	**2-methyl-1-butene**
boiling point	141 °C	80 °C	57 °C	31.2 °C

(21.18)

Focused Problems

21.5 Pentanoic acid and methyl butyrate are constitutional isomers. Which has the higher boiling point and why?

21.6 (a) Assuming that the difference in the relative boiling points of methyl acetate and 2-butanone (see Display 21.18) is caused by the difference in their dipole moments, predict which compound has the greater dipole moment.

(b) Use a vector analysis of bond dipoles to show why your answer to part (a) is reasonable.

B. Anhydrides and Acid Chlorides

Most of the lower anhydrides and acid chlorides are dense, water-insoluble liquids with acrid, piercing odors. Their boiling points are not very different from those of other polar molecules of about the same molecular mass and shape.

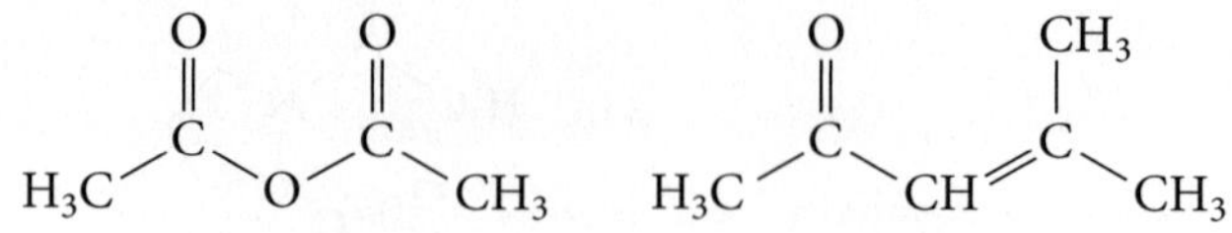

	acetic anhydride	4-methyl-3-penten-2-one
boiling point	139.6 °C	129.8 °C
density	1.082 g mL^{-1}	0.86 g mL^{-1}

(21.19a)

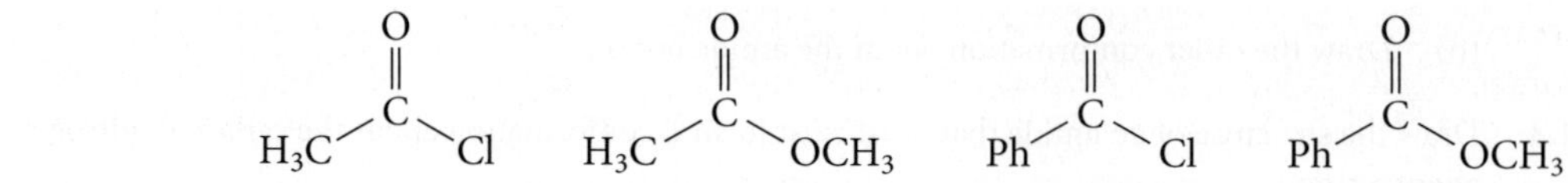

	acetyl chloride	methyl acetate	benzoyl chloride	methyl benzoate
boiling point	50.9 °C	57 °C	197.2 °C	213 °C
density	1.051 g mL^{-1}	0.93 g mL^{-1}	1.212 g mL^{-1}	1.09 g mL^{-1}

(21.19b)

The simplest anhydride, formic anhydride, and the simplest acid chloride, formyl chloride, are unstable and cannot be isolated under ordinary conditions.

C. Nitriles

Nitriles are among the most polar organic compounds. Acetonitrile, for example, has a dipole moment of 3.4 D. The polarity of nitriles is reflected in their boiling points, which are rather high despite the absence of hydrogen bonding. (See Fig. 8.2, Sec. 8.5A.)

	$H_3C{-}C{\equiv}N{:}$	$H_3C{-}C{\equiv}C{-}H$	$CH_3CH_2{-}C{\equiv}N{:}$	$CH_3CH_2{-}C{\equiv}C{-}H$
	acetonitrile	**propyne**	**propionitrile**	**1-butyne**
boiling point	81.6 °C	−23.3 °C	97.4 °C	46.5 °C

(21.20)

Although nitriles are very poor hydrogen-bond acceptors (because they are very weak bases; see Sec. 21.6), acetonitrile is miscible with water and propionitrile has a moderate solubility in water. Nitriles with more carbons are insoluble in water. Acetonitrile serves in some cases as a useful polar aprotic solvent because of its moderate boiling point and its relatively high dielectric constant of 38 (see Table 8.2, Sec. 8.6B).

D. Amides

The amides of lower molecular mass are water-soluble, polar molecules with high boiling points. Primary and secondary amides, like carboxylic acids (Sec. 20.2), tend to associate into hydrogen-bonded dimers or higher aggregates in the solid state, in the pure liquid state, or in solvents that do not form hydrogen bonds. This association has a noticeable effect on the properties of amides and is of substantial biological importance in the structures of proteins (Sec. 27.9A). For example, simple amides have very high boiling points; many are solids.

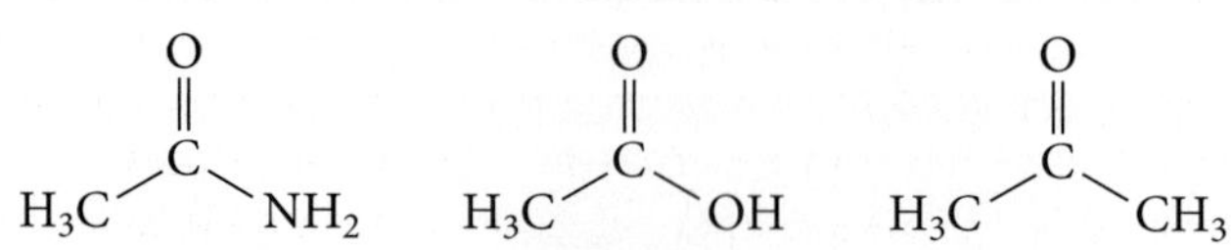

	acetamide	acetic acid	acetone
boiling point	221.2 °C	117.9 °C	56.5 °C
melting point	82.3 °C	16.7 °C	−94 °C

(21.21a)

Primary amides have two hydrogens on the amide nitrogen that can form hydrogen bonds. Along a series in which these hydrogens are replaced by methyl groups, the capacity for hydrogen bonding is reduced, and boiling points decrease in spite of the increase in molecular mass.

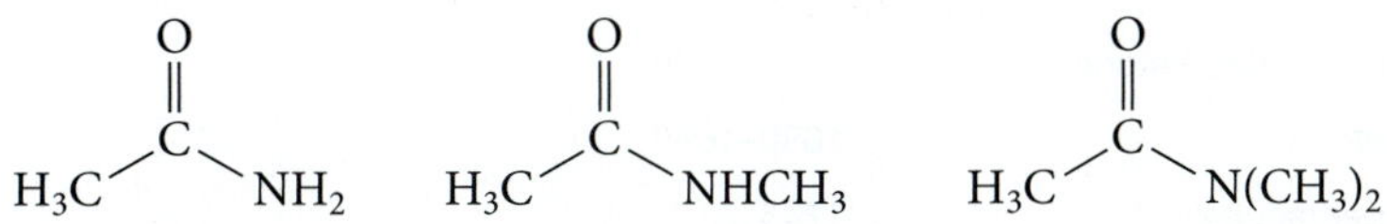

	acetamide	*N*-methylacetamide	*N,N*-dimethylacetamide
boiling point	221.2 °C	204–206 °C	166.1 °C
melting point	82.3 °C	28 °C	−20 °C

(21.21b)

A number of amides have high dielectric constants (see Table 8.2, Sec. 8.6B). *N,N*-Dimethylformamide (DMF), which has a dielectric constant of 37, for example, dissolves a number of inorganic salts and is widely used as a polar aprotic solvent, despite its high boiling point.

21.4 SPECTROSCOPY OF CARBOXYLIC ACID DERIVATIVES

A. IR Spectroscopy

The most important feature in the IR spectra of most carboxylic acid derivatives is the C═O stretching absorption. For nitriles, the most important feature in the IR spectrum is the C≡N stretching absorption. These absorptions are summarized in **Table 21.3**, along with the absorptions of other carbonyl compounds. Some of the noteworthy trends in this table are the following:

1. Esters are readily differentiated from carboxylic acids, aldehydes, or ketones by the unique ester carbonyl absorption at 1735–1745 cm^{-1}.

2. Lactones, lactams, and cyclic anhydrides, like cyclic ketones, have carbonyl absorption frequencies that increase significantly as the ring size decreases. (See Further Exploration 19.1 in the *Study Guide and Solutions Manual* for the explanation.)

3. Anhydrides have two carbonyl absorptions, which are due to the symmetrical and unsymmetrical stretching vibrations of the carbonyl groups (see Fig. 13.8, Sec. 13.3A).

4. Acid chlorides have a carbonyl absorption around 1800 cm^{-1}.

5. The carbonyl absorptions of amides occur at much lower frequencies than those of other carbonyl compounds.

6. The C≡N stretching absorptions of nitriles generally occur in the triple-bond region of the spectrum. These absorptions are stronger than the C≡C absorptions of alkynes because of the large bond dipole of the carbon–nitrogen triple bond, and they occur at higher frequencies.

Carboxylic acids in the *gas phase* have about the same carbonyl absorptions as esters. However, routine IR spectra of carboxylic acids are typically determined in solution or in the solid state, in which extensive hydrogen bonding lowers the carbonyl stretching frequency.

The IR spectra of some carboxylic acid derivatives are shown in **Figs. 21.2a–c.**

Other useful absorptions in the IR spectra of carboxylic acid derivatives are also summarized in Table 21.3. For example, primary and secondary amides show an N—H stretching absorption in the 3200–3400 cm^{-1} region of the spectrum. Many primary amides show two N—H absorptions (**Fig. 21.2d**), and secondary amides show a single strong N—H absorption. In addition, a strong N—H bending absorption occurs in the vicinity of 1640 cm^{-1}, typically appearing as a shoulder on the low-frequency side of the amide carbonyl absorption. Tertiary amides lack both of these N—H vibrations (**Fig. 21.2e**).

TABLE 21.3 Important Infrared Absorptions of Carbonyl Compounds and Nitriles

Compound	Carbonyl absorption, cm^{-1}	Other absorptions, cm^{-1}
ketone	1710–1715	
α,β-unsaturated ketone	1670–1680	
aryl ketone	1680–1690	
cyclopentanone	1745	
cyclobutanone	1780	
aldehyde	1720–1745	aldehydic C—H stretch at 2720
α,β-unsaturated aldehyde	1680–1690	
aryl aldehyde	1700	
carboxylic acid (dimer)	1710	OH stretch at 2400–3000 (strong, broad); C—O stretch at 1200–1300
aryl carboxylic acid	1680–1690	
ester or six-membered lactone (δ-lactone)	1735–1745	C—O stretch at 1000–1300
α,β-unsaturated ester	1720–1725	
5-membered lactone (γ-lactone)	1770	
4-membered lactone (β-lactone)	1840	
acid chloride	1800	a second weaker band is sometimes observed at 1700–1750
anhydride	1760, 1820 (two absorptions)	C—O stretch as in an ester
6-membered cyclic anhydride	1750, 1800	
5-membered cyclic anhydride	1785, 1865	
amide	1650–1655	N—H bend at 1640 N—H stretch at 3200–3400; double absorption for a primary amide
6-membered lactam (δ-lactam)	1670	
5-membered lactam (γ-lactam)	1700	
4-membered lactam (β-lactam)	1745	
nitrile		C≡N stretch at 2200–2250

B. NMR Spectroscopy

Proton NMR Spectroscopy The α-proton resonances of all carboxylic acid derivatives are observed in the δ 1.9–3 region of the proton NMR spectrum (see Fig. 14.4, Sec. 14.3C). In esters, the chemical shifts of protons on the alkyl carbon adjacent to the carboxylate oxygen are about 0.6 ppm greater than the chemical shifts of the analogous protons in alcohols and ethers. This shift is attributable to the electronegative character of the carbonyl group.

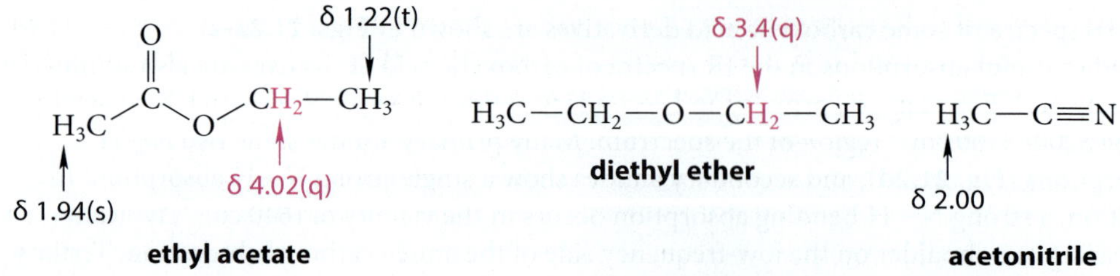

(21.22)

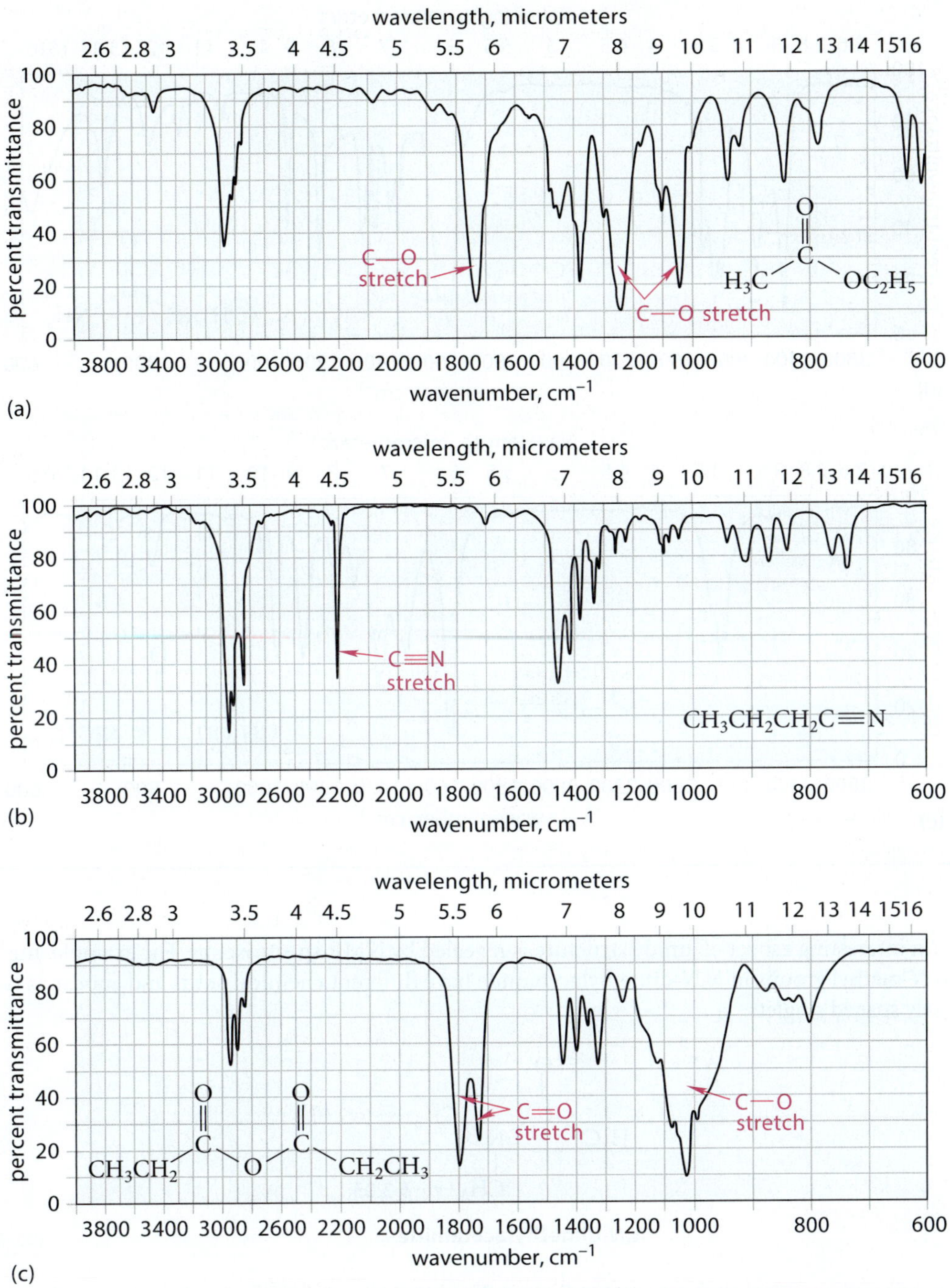

FIGURE 21.2 Infrared spectra of some carboxylic acid derivatives. (a) Ethyl acetate. The distinguishing feature in the IR spectra of esters is the position of the carbonyl stretching absorption. (b) Butyronitrile. The distinguishing feature in the IR spectra of nitriles is the C≡N stretching absorption. (c) Propionic anhydride. The distinguishing feature in the IR spectra of anhydrides is the double carbonyl stretching absorption. *(Continued next page)*

The *N*-alkyl protons of amides have chemical shifts in the δ 2.6–3 chemical shift region, and the NH proton resonances of primary and secondary amides are observed in the δ 7.5–8.5 region. The resonances for these protons, like those of carboxylic acid OH protons, are sometimes broad. This broadening is caused by a slow chemical exchange with the protons of other protic substances (such as traces of moisture) and by unresolved splitting with ^{14}N, which has a nuclear spin. Amide NH resonances, like the OH signals of acids and alcohols, can be eliminated by exchange with D_2O ("D_2O shake"; Sec. 14.6).

STUDY GUIDE LINK 21.1
Solving Structure Problems Involving Nitrogen-Containing Compounds

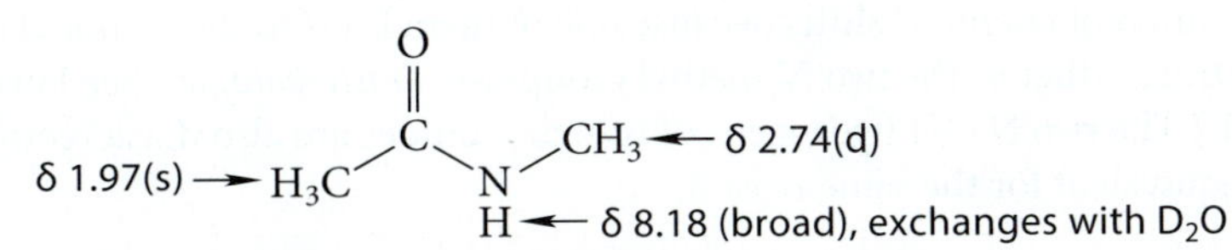

***N*-methylacetamide** (21.23)

FIGURE 21.2 *(continued)* Infrared spectra of amides. (d) 2,2-Dimethylpropanamide (pivalamide); primary amides typically have two N—H stretching absorptions. (e) *N,N*-Dimethylpropanamide. The N—H stretching and bending absorptions seen in primary amides are absent in tertiary amides.

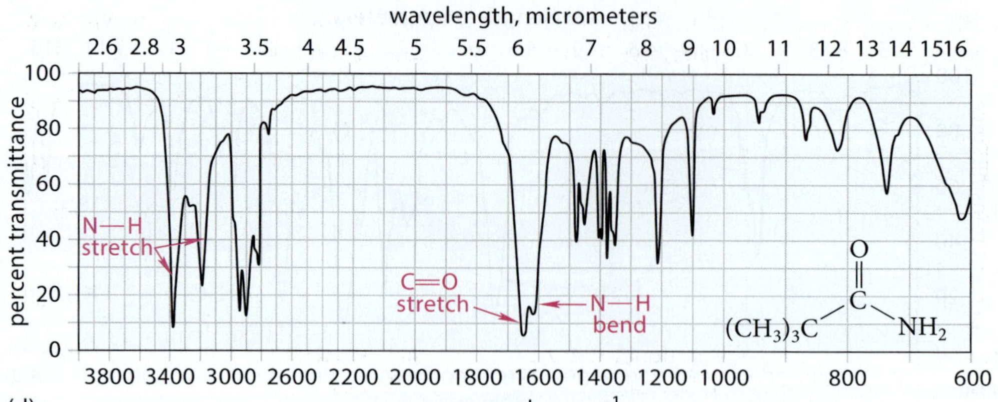

(d)

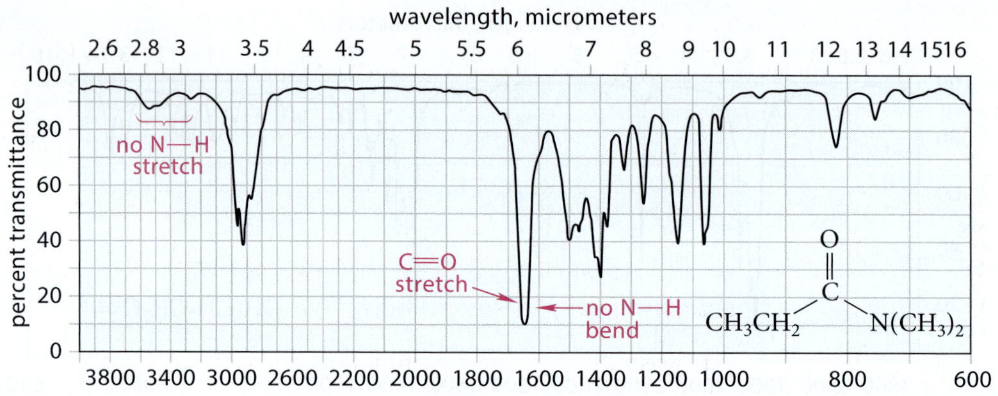

(e)

An interesting aspect of amide structure is revealed by NMR spectroscopy. For example, the two *N*-methyl groups in *N,N*-dimethylacetamide have different chemical shifts and appear as two closely spaced singlets:

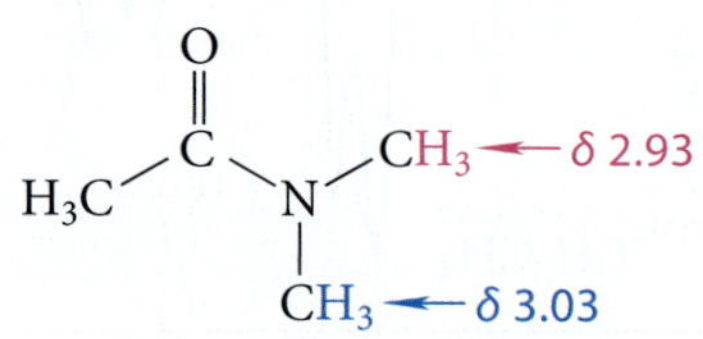

***N,N*-dimethylacetamide** (21.24)

The different chemical shifts show that the two *N*-methyl groups are *chemically nonequivalent.* Why should this be so?

As discussed in Sec. 21.2 (Display 21.16), there is a significant amount of double-bond character in the bond between the nitrogen and the carbonyl group, and this causes a considerably reduced rate of internal rotation about this bond. Although the internal rotation occurs about 10 times per second, a rate that is large on the human time scale, this rate is very small in the context of an NMR experiment. That is, the time scale of the NMR measurement is so small that the internal rotation about the carbonyl–nitrogen bond appears to be frozen. (See Sec. 14.8 for a discussion of the effect of slow internal rotations on NMR spectra.) Therefore, the *N*-methyl groups behave in the NMR experiment like substituents on a double bond. The *N*-methyls have different chemical shifts because one of them is cis to the carbonyl oxygen and the other is trans—that is, the two *N*-methyl groups are *diastereotopic.* (See Further Exploration 21.1.) The two N—H hydrogens of primary amides are also diastereotopic and chemically nonequivalent for the same reason.

^{13}C NMR Spectroscopy In ^{13}C NMR spectra, the carbonyl chemical shifts of carboxylic acid derivatives are in the δ 165–180 range, very much like those of carboxylic acids.

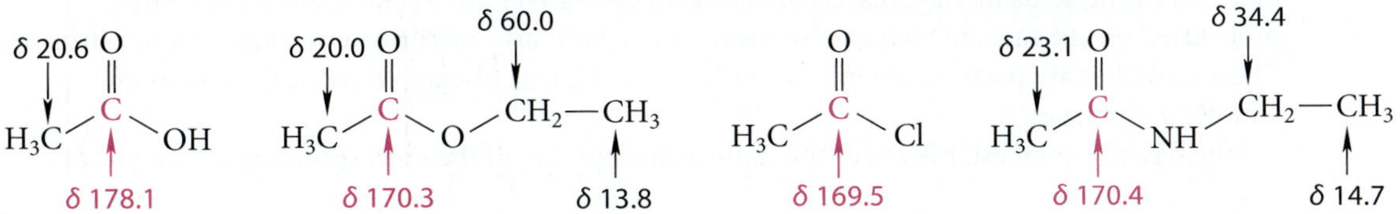

(21.25)

The chemical shifts of nitrile carbons are considerably smaller, occurring in the δ 115–120 range. These shifts are much greater, however, than those of acetylenic carbons.

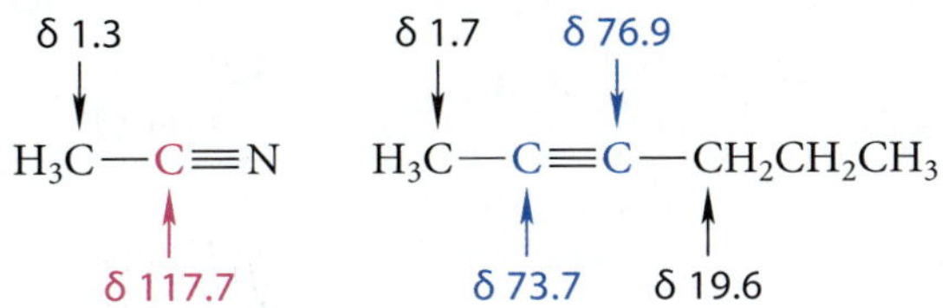

(21.26)

Focused Problems

21.7 How would you differentiate between the compounds in each of the following pairs?

(a) *p*-Ethylbenzoic acid and ethyl benzoate by IR spectroscopy

(b) 2,4-Dimethylbenzonitrile and *N*-methylbenzamide by proton NMR spectroscopy

(c) Methyl propionate and ethyl acetate by proton NMR spectroscopy

(d) *N*-Methylpropanamide and *N*-ethylacetamide by proton NMR spectroscopy

(e) Ethyl butyrate and ethyl isobutyrate by carbon NMR spectroscopy

21.8 Identify the compound C_4H_9NO with the proton NMR spectrum given in **Fig. 21.3**. This compound has IR absorptions at 3300 and 1650 cm^{-1}.

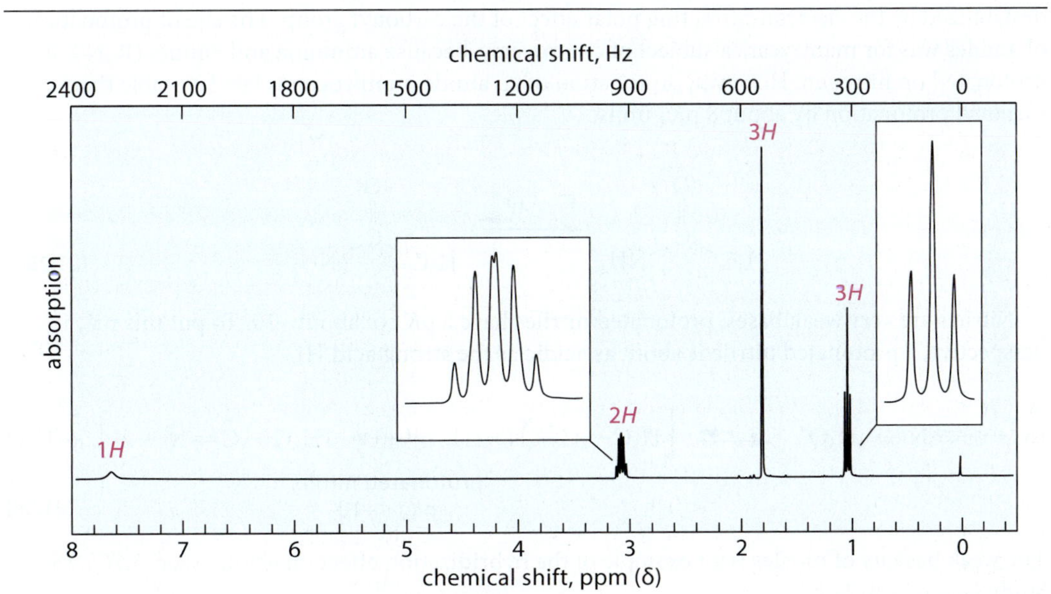

FIGURE 21.3 The NMR spectrum for Focused Problem 21.8. The integrals are shown in red over their respective absorptions. The broad resonance at δ 7.6 disappears after a D_2O shake, and the multiplet at δ 3.05 simplifies to a quartet.

21.5 BASICITY OF CARBOXYLIC ACID DERIVATIVES

Like carboxylic acids themselves, carboxylic acid derivatives are weakly basic and can be protonated on the carbonyl oxygen by strong acids. Similarly, nitriles are weakly basic at nitrogen. These basicities are particularly important in some of the acid-catalyzed reactions of esters, amides, and nitriles.

The basicity of an ester is about the same as the basicity of the corresponding carboxylic acid.

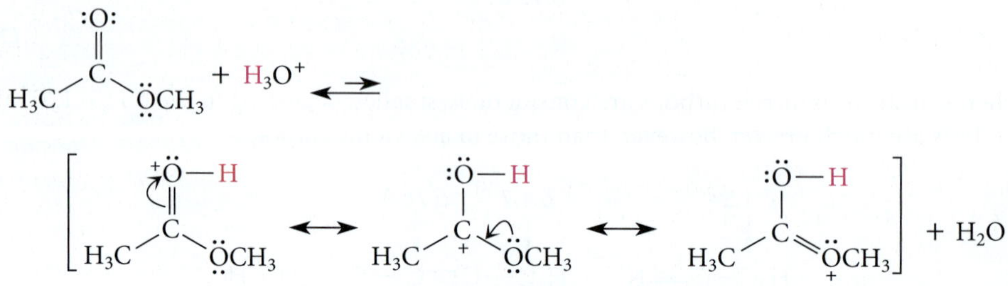

protonated ester; $pK_a \sim -6$

(21.27)

Amides are considerably more basic than other carboxylic acid derivatives. This basicity, relative to esters, is a reflection of the reduced electronegativity of nitrogen relative to oxygen. That is, the resonance structures in which positive charge is shared on nitrogen are particularly important for a protonated amide.

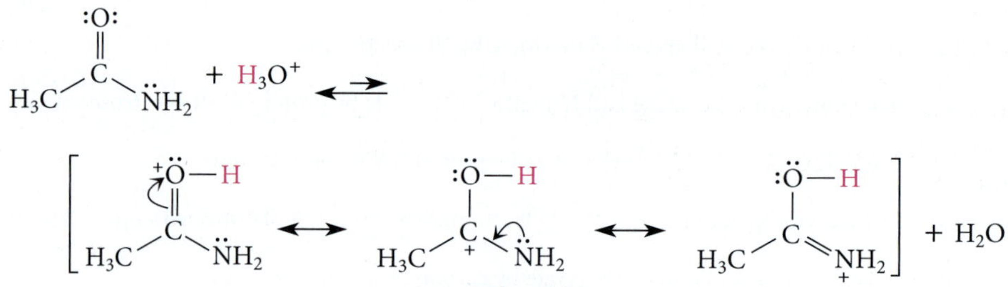

protonated amide; $pK_a \sim -0.5$ to -1

(21.28a)

Both esters and amides, like carboxylic acids (Sec. 20.4B), are protonated on the *carbonyl oxygen*. Protonation of esters on the carboxylate oxygen, or protonation of amides on the nitrogen, would give a cation that is *not* resonance-stabilized and, additionally, one that is destabilized by the electron-attracting polar effect of the carbonyl group. The site of protonation of amides was for many years a subject of controversy, because ammonia and amines (R_3N:) are protonated on nitrogen. However, protonation of an amide on nitrogen is less favorable than carbonyl protonation by about 8 pK_a units.

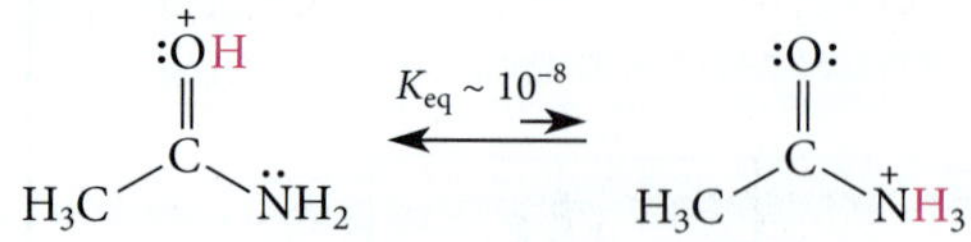

(21.28b)

Nitriles are *very* weak bases; protonated nitriles have a pK_a of about −10. To put this pK_a in perspective, a protonated nitrile is about as acidic as the strong acid HI.

protonated nitrile; $pK_a \sim -10$

(21.29)

STUDY GUIDE LINK 21.2
Basicity of Nitriles

The weak basicity of nitriles is an example of the hybridization effect on acidity (Sec. 3.7C). (See Study Guide Link 21.2.)

An Amide with a Twist

A graphic demonstration of the importance of resonance in amides and their conjugate acids was provided when, in 2006, chemists at Cal Tech led by Professor Brian Stoltz synthesized the conjugate acid of the cyclic amide 2-quinuclidone.

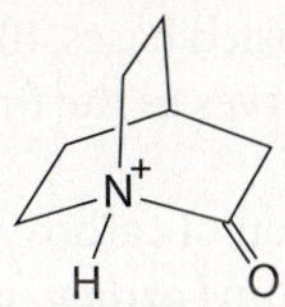

2-quinuclidone
(conjugate acid)

They showed that this amide, unlike ordinary amides, protonates on nitrogen rather than on the carbonyl oxygen. The reason for this difference is that the ion resulting from protonation on the carbonyl oxygen is not resonance-stabilized, because the resonance structure of the oxygen-protonated amide violates Bredt's rule (Sec. 7.6C).

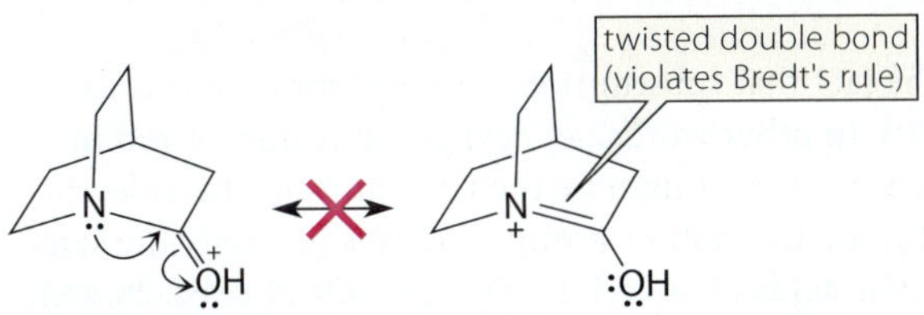

oxygen-protonated
2-quinuclidone

In terms of the orbitals involved, the orbitals containing the nitrogen nonbonding pair and the π orbital of the carbonyl group are perpendicular and cannot overlap. In other words, the amide bond is twisted because of the bicyclic ring structure, and because a rotation about the nitrogen—carbon bond that would allow orbital overlap to occur would introduce a large degree of strain. (See Sec. 7.6C, Displays 7.37a, b, and c.)

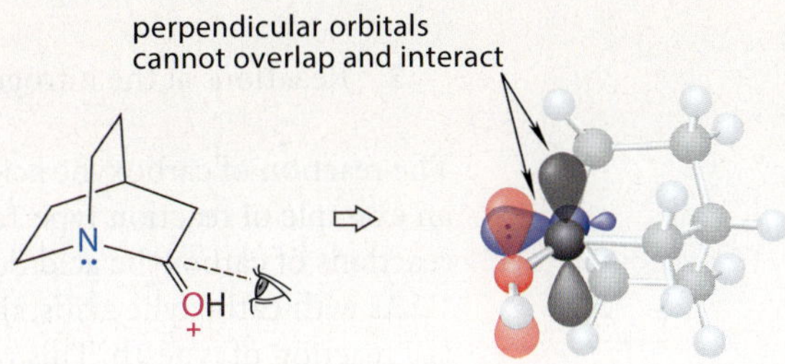

As a result, the nitrogen behaves more like an amine nitrogen than an amide nitrogen. (Amines are basic and are readily protonated by acids.) An attempt to form unprotonated 2-quinuclidone by treatment of its conjugate acid with base in aqueous solution resulted in its rapid hydrolysis to form the cyclic amino acid. As we show in Sec. 21.7, ordinary amides hydrolyze rather slowly because they are stabilized by the resonance interaction within the amide bond. Lacking this stabilization, 2-quinuclidone is unusually reactive.

Focused Problem

21.9 Which of the two isomers in each of the following sets should have the greater basicity at the carbonyl oxygen? Explain.

(a)

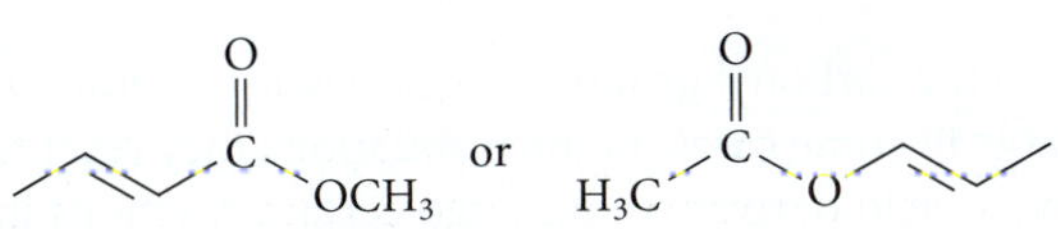

(b)

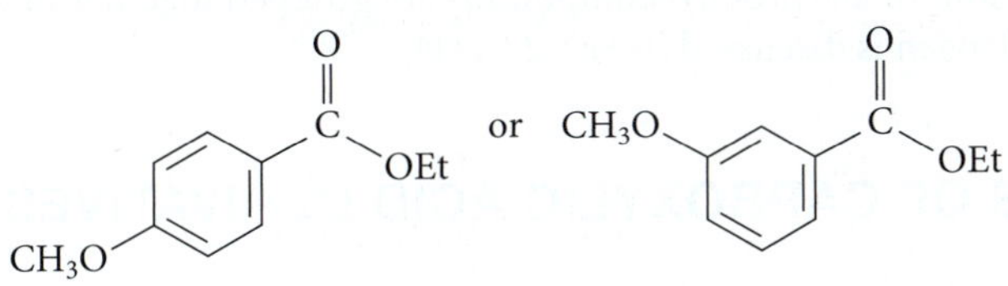

(c)

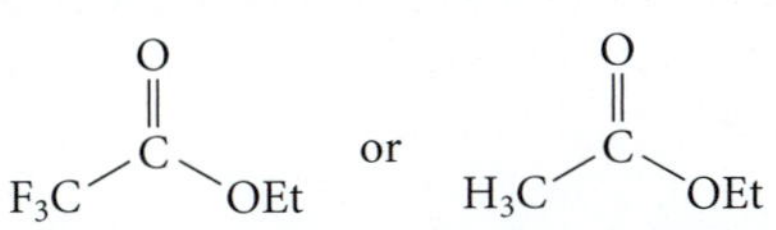

21.6 INTRODUCTION TO THE REACTIONS OF CARBOXYLIC ACID DERIVATIVES

The reactions of carboxylic acid derivatives can be categorized as follows:

1. Reactions at the carbonyl group, or the cyano group of a nitrile
   a. Reactions at the carbonyl oxygen or cyano nitrogen
   b. Reactions at the carbonyl carbon or cyano carbon
2. Reactions involving the α-carbon
3. Reactions at the nitrogen of amides

The reaction of carboxylic acids and their derivatives as Brønsted bases, illustrated in Sec. 21.5, is an example of reaction type 1a. This type of reaction often serves as the first step in acid-catalyzed reactions of carboxylic acid derivatives.

As with carboxylic acids, the major carbonyl-group reaction of carboxylic acid derivatives is a reaction of type 1b. This reaction, *substitution at the carbonyl carbon*, is also called **acyl substitution**. Acyl substitution can be represented generally as follows, with E = an electrophilic group and Y = a nucleophilic group:

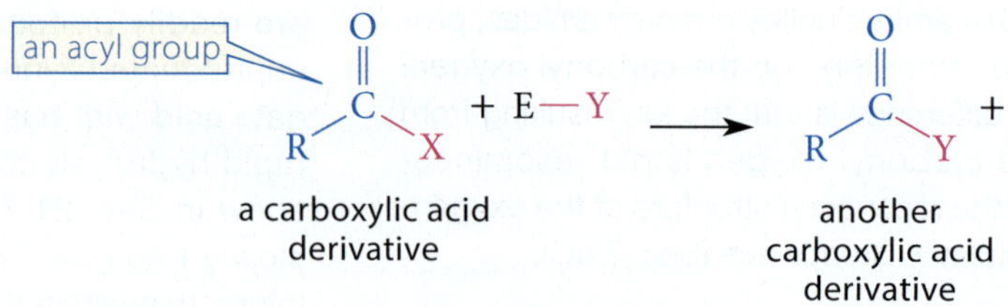

(21.30)

The term *acyl substitution* comes from the fact that substitution occurs at the carbonyl carbon of an *acyl group* (the blue group in Eq. 21.30). In other words, an acyl group is transferred in Eq. 21.30 between an X and a Y group. The group X might be the Cl of an acid chloride, the OR of an ester, and so on; this group is substituted by another group Y. This is precisely the same type of reaction as the esterification of carboxylic acids (X = OH, E—Y = H—OCH_3; Sec. 20.8A). Acyl substitution reactions of carboxylic acid derivatives are the major focus of this chapter.

Although nitriles are not carbonyl compounds, the C≡N bond behaves chemically much like a carbonyl group. For example, a typical reaction of nitriles is *addition.*

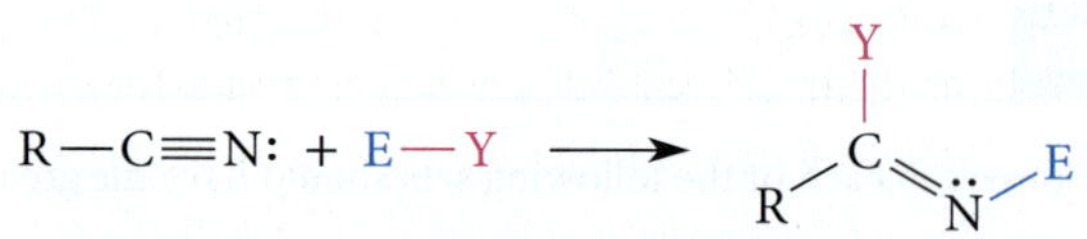

(21.31)

(Compare this reaction with addition to the carbonyl group of an aldehyde or ketone.) Although the resulting addition products are stable in some cases, in most situations they react further.

Like aldehydes and ketones, carboxylic acid derivatives undergo certain reactions involving the α-carbon. The α-carbon reactions of all carbonyl compounds are grouped together in Chapter 22. The reactivity of amides at nitrogen is discussed in Sec. 23.11D.

21.7 HYDROLYSIS OF CARBOXYLIC ACID DERIVATIVES

All carboxylic acid derivatives have in common the fact that they undergo *hydrolysis* (a cleavage reaction with water) to yield carboxylic acids.

A. Hydrolysis of Esters and Lactones

Base-Promoted Hydrolysis (Saponification) of Esters One of the most important reactions of esters is the cleavage reaction with hydroxide ion to yield a carboxylate salt and an alcohol. Strong acid is required in a second step to form the carboxylic acid.

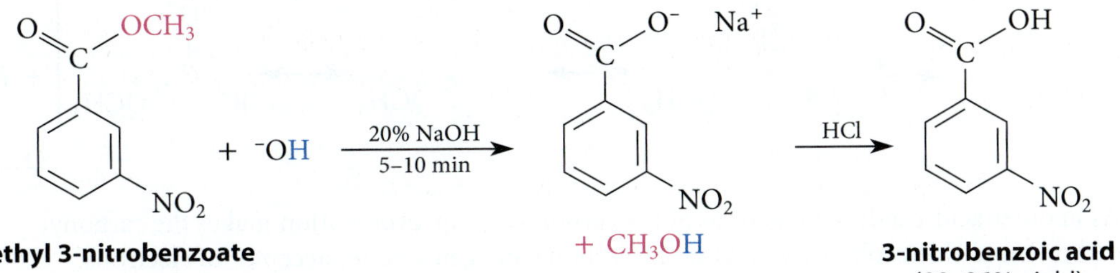

methyl 3-nitrobenzoate

3-nitrobenzoic acid
(90–96% yield)

(21.32)

Ester hydrolysis in aqueous hydroxide is called **saponification** because it is used in the production of soaps from fats (Sec. 21.12C). Despite its association with fatty-acid esters, the term *saponification* can be used to refer to the hydrolysis in base of any carboxylic acid derivative.

The mechanism of ester saponification involves the reaction of the nucleophilic hydroxide ion at the carbonyl carbon to give a *tetrahedral addition intermediate* from which an alkoxide ion is expelled.

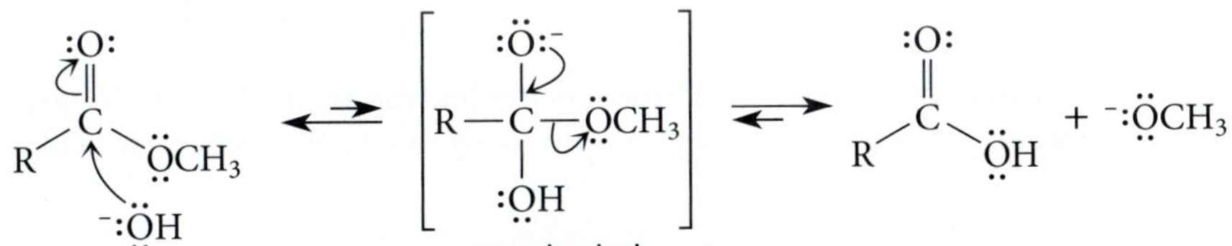

tetrahedral addition intermediate

(21.33a)

The alkoxide ion, after expulsion as a leaving group (methoxide in Eq. 21.33a), reacts with the acid to give the carboxylate salt and the alcohol.

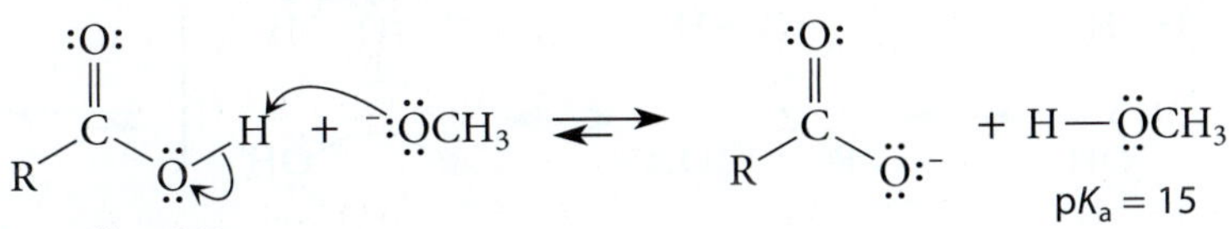

$pK_a = 4.5$ $pK_a = 15$

(21.33b)

The equilibrium in this reaction lies far to the right because the carboxylic acid is a much stronger acid than methanol. Le Chatelier's principle operates: The reaction in Eq. 21.33b removes the carboxylic acid from the equilibrium in Eq. 21.33a as its salt and thus drives the hydrolysis to completion. Therefore, *saponification is effectively irreversible.* Although an excess of hydroxide ion is often used as a matter of convenience, many esters can be saponified with just one equivalent of ^{-}OH. Saponification can also be carried out with ^{-}OH in an alcohol solvent, even though an alcohol is one of the products of the reaction. If saponification were reversible, an alcohol could not be used as the solvent because the equilibrium would be driven toward starting materials.

Acid-Catalyzed Ester Hydrolysis Because esterification of an acid with an alcohol is a reversible reaction (Sec. 20.8A), esters can be hydrolyzed to carboxylic acids in aqueous solutions of strong acids. In most cases, this reaction is slow and must be carried out with an excess of water, in which most esters are insoluble. Saponification, followed by acidification, is a much more convenient method for hydrolysis of most esters because it is faster, it is irreversible, and it can be carried out not only in water but also in a variety of solvents—even alcohols.

As expected from the principle of microscopic reversibility (Sec. 4.10C), the mechanism of acid-catalyzed hydrolysis is the exact reverse of the mechanism of acid-catalyzed esterification (Sec. 20.8A, Eqs. 20.34a–d). The ester is first protonated by the acid catalyst:

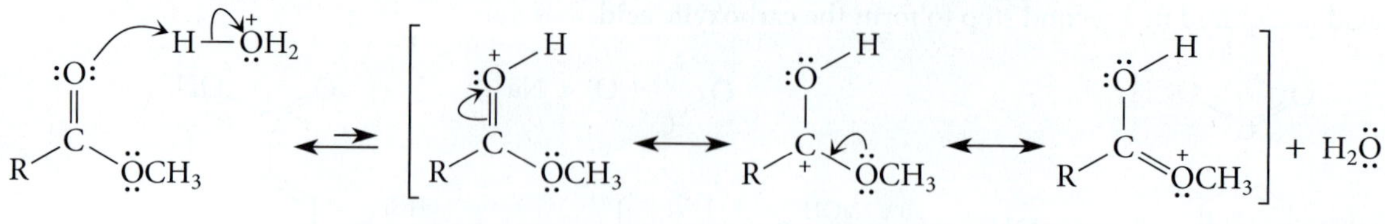

(21.34a)

As in other acid-catalyzed reactions at the carbonyl group, protonation makes the carbonyl carbon more electrophilic by making the carbonyl oxygen a better acceptor of electrons. Water, acting as a nucleophile, reacts at the carbonyl carbon and then loses a proton to give the tetrahedral addition intermediate:

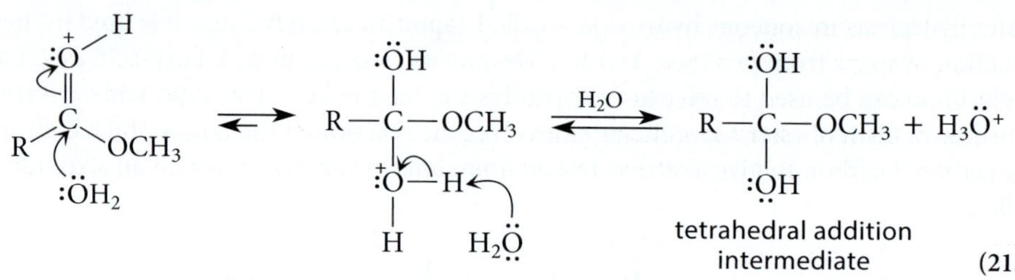

(21.34b)

Protonation of the leaving oxygen converts it into a better leaving group. Loss of this group gives a protonated carboxylic acid, from which a proton is removed to give the carboxylic acid itself.

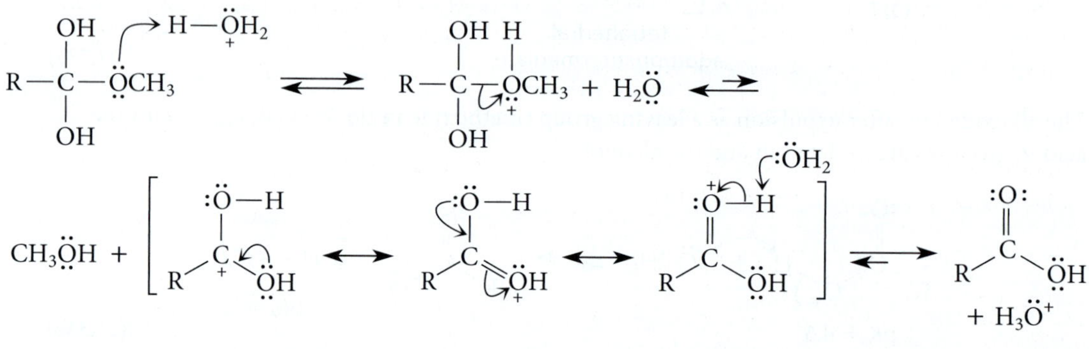

(21.34c)

STUDY GUIDE LINK 21.3
Mechanism of Ester Hydrolysis

Let's summarize the important differences between acid-catalyzed ester hydrolysis and ester saponification:

1. In acid-catalyzed hydrolysis, the carbonyl carbon can react with the relatively weak nucleophile water because the carbonyl oxygen is protonated. In base, the carbonyl oxygen is not protonated; consequently, a much stronger base than water—namely, hydroxide ion—is required to react at the carbonyl carbon.

2. Acid *catalyzes* ester hydrolysis, but base *is not a catalyst* because it is consumed by the reaction in Eq. 21.33b.

3. Acid-catalyzed ester hydrolysis is reversible, but saponification is irreversible, again because of the ionization in Eq. 21.33b.

FURTHER EXPLORATION 21.2
Cleavage of Tertiary Esters and Carbonless Carbon Paper

Ester hydrolysis and saponification are both examples of *acyl substitution* (Sec. 21.6). Specifically, the mechanisms of these reactions are classified as **nucleophilic acyl substitution** mechanisms. In a nucleophilic acyl substitution reaction, the substituting group reacts as a nucleophile at the carbonyl carbon. As in the reactions of aldehydes and ketones

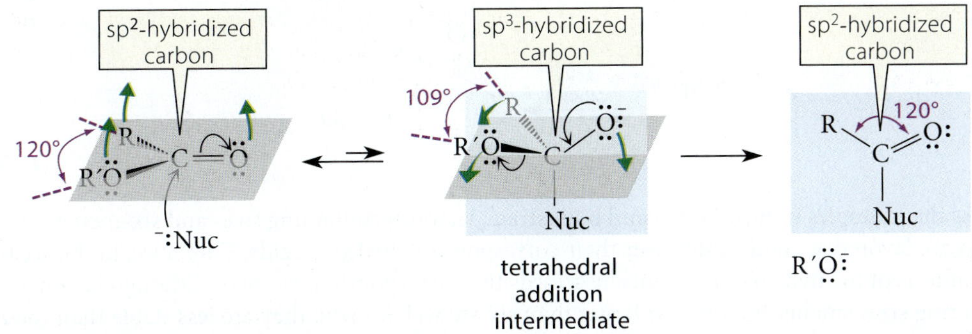

FIGURE 21.4 The geometry of nucleophilic acyl substitution. The nucleophile (Nuc:$^-$) approaches the carbonyl carbon above or below the carbonyl plane (gray) to form a tetrahedral addition intermediate. The leaving group (R′O$^-$) departs from above or below the plane of the newly formed carbonyl group (blue). The green arrows show the movement of the various groups.

(see Fig. 19.8, Sec. 19.8A), nucleophiles approach the carbonyl carbon from above or below the plane of the carbonyl group (**Fig. 21.4**), first interacting with the π^* (antibonding) molecular orbital of the carbonyl group (or the protonated carbonyl group in acid-catalyzed reactions). As the result of this reaction, a *tetrahedral addition intermediate* is formed. The leaving group is expelled from this intermediate, departing from above or below the plane of the new carbonyl group. In saponification, the nucleophile is $^-$OH, and in acid-catalyzed hydrolysis, the nucleophile is water. In saponification, the OR group of the ester is displaced as an alkoxide $^-$OR. In acid-catalyzed hydrolysis, the OR group of the ester is protonated and expelled as the alcohol HOR. With the exception of the reactions of nitriles, most of the reactions of carboxylic acid derivatives in the remainder of this chapter are nucleophilic acyl substitution reactions. They follow the same patterns as saponification or acid-catalyzed hydrolysis, the only substantive difference being the identity of the nucleophiles and the leaving groups.

Hydrolysis and Formation of Lactones Because lactones are cyclic esters, they undergo the reactions of esters, including saponification. Saponification converts a lactone completely into the carboxylate salt of the corresponding hydroxy acid.

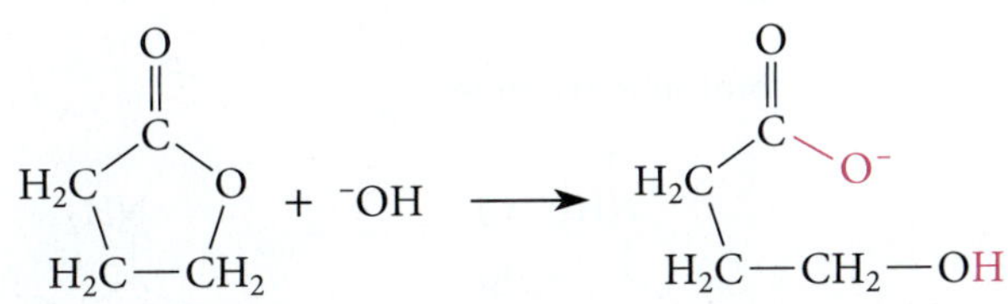

γ-butyrolactone **γ-hydroxybutyrate** (21.35)

Upon acidification, the hydroxy acid forms. However, *if a hydroxy acid is allowed to stand in acidic solution, it comes to equilibrium with the corresponding lactone.* The formation of a lactone from a hydroxy acid is an *intramolecular* esterification (an esterification within the same molecule) and, like esterification, the lactonization equilibrium is acid-catalyzed. The formation of a 5- or 6-membered lactone ring is considerably faster than ordinary esterification because of the proximity effect (Sec. 12.8).

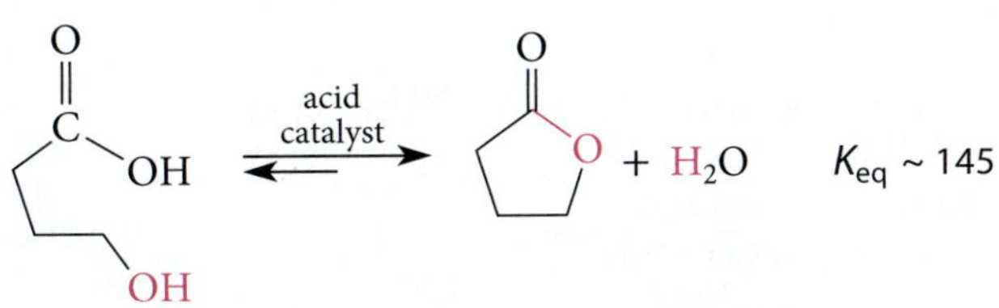

(21.36a)

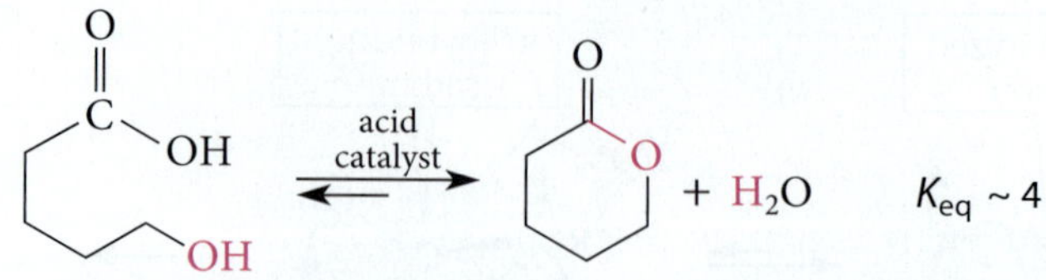

(21.36b)

As the examples in Eqs. 21.36a and b illustrate, lactones containing five- and six-membered rings are favored at equilibrium over their corresponding hydroxy acids. Therefore, in these cases a synthesis of the hydroxy acid typically affords the corresponding lactone. Although lactones with ring sizes smaller than five or larger than six are well known, they are less stable than their corresponding hydroxy acids. Consequently, the lactonization equilibria for these compounds favor the hydroxy acids instead.

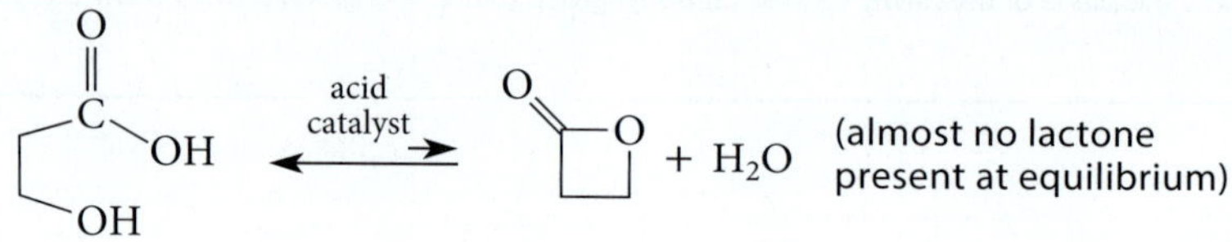

(21.37)

B. Hydrolysis of Amides

Amides can be hydrolyzed to carboxylic acids and ammonia or amines by heating them in acidic or basic solution.

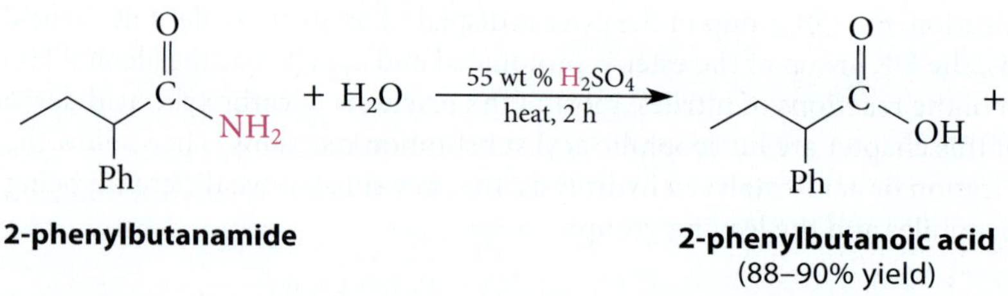

(21.38)

In acid, protonation of the ammonia or amine product drives the hydrolysis equilibrium to completion. The amine can be isolated, if desired, by addition of base to the reaction mixture following hydrolysis, as in Eq. 21.39.

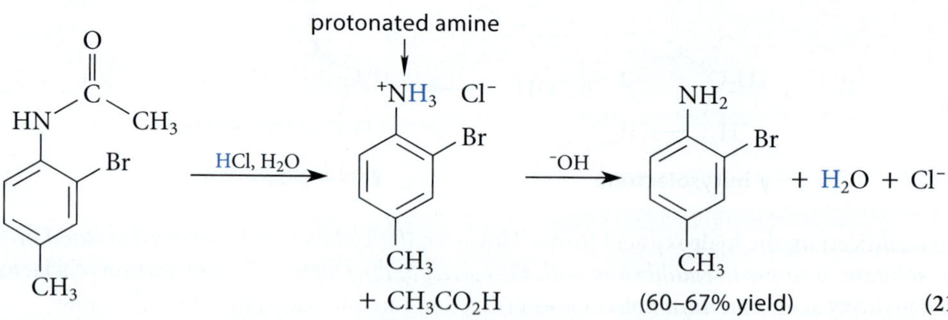

(21.39)

The hydrolysis of amides in base is analogous to the saponification of esters. In base, the reaction is driven to completion by formation of the carboxylic acid salt.

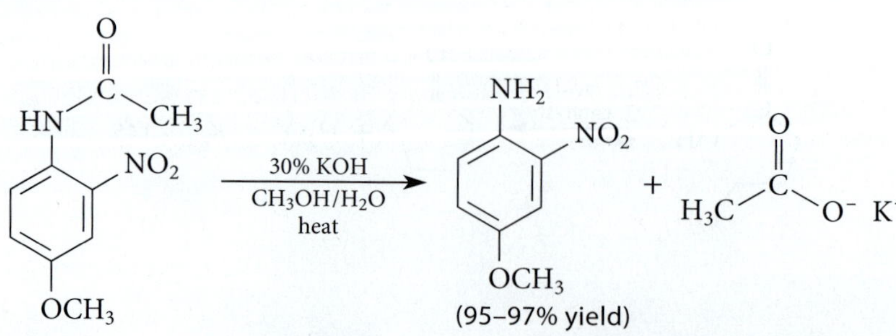

(21.40)

The conditions for both acid- and base-promoted amide hydrolysis are considerably more severe than the corresponding reactions of esters. That is, amides are considerably *less reactive* than esters. The relative reactivities of carboxylic acid derivatives are discussed in Sec. 21.7E.

The mechanisms of amide hydrolysis are typical nucleophilic acyl substitution mechanisms; you are asked to explore this point in Focused Problem 21.10.

Focused Problems

21.10 Show in detail the curved-arrow mechanism for the hydrolysis of *N*-methylbenzamide (a) in acidic solution; (b) in aqueous NaOH. Assume that each mechanism involves a tetrahedral addition intermediate.

21.11 Give the structures of the hydrolysis products that result from each of the following reactions. Be sure to show the product stereochemistry in part (b).

(a)

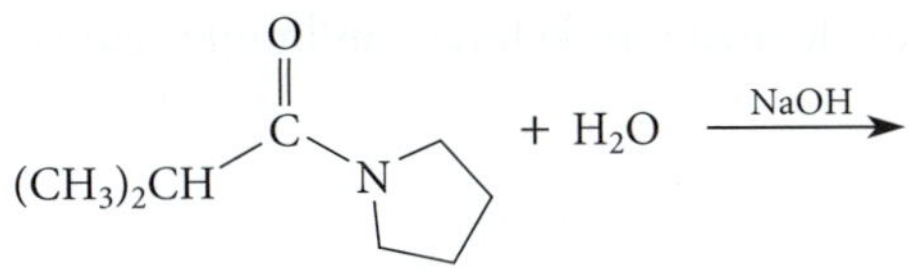

(b)

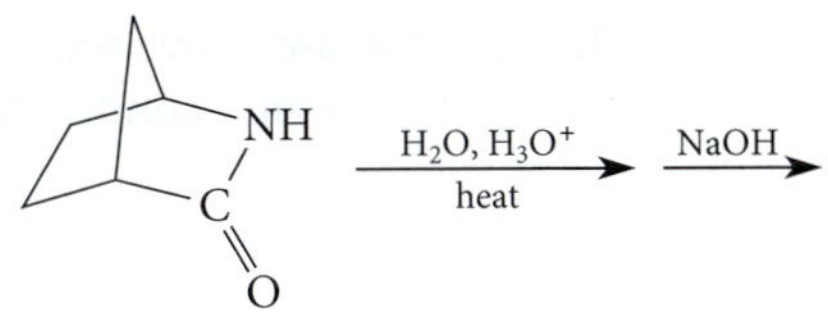

C. Hydrolysis of Nitriles

Nitriles are hydrolyzed to carboxylic acids and ammonia by heating them in strongly acidic or strongly basic solution.

$$\underset{\textbf{phenylacetonitrile}}{PhCH_2—C\equiv N} + 2H_2O + \underset{(57\text{ wt }\%)}{H_2SO_4} \xrightarrow[\text{3 h}]{\text{heat}} \underset{\substack{\textbf{phenylacetic acid}\\(78\%\text{ yield})}}{PhCH_2—CO_2H} + \overset{+}{N}H_4\ HSO_4^- \qquad (21.41)$$

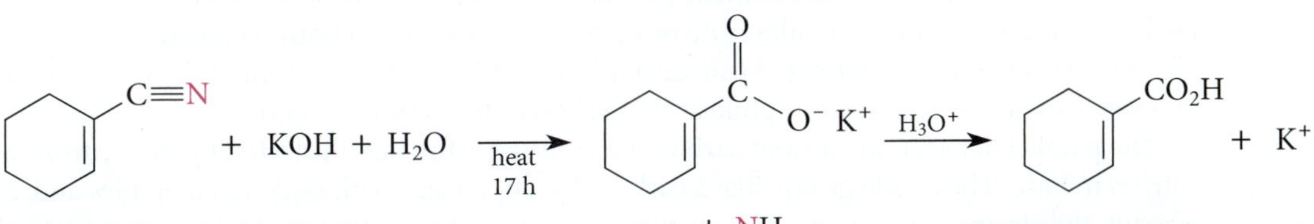

1-cyclohexenecarbonitrile **1-cyclohexenecarboxylic acid** (79% yield) (21.42)

Nitriles hydrolyze more slowly than esters and amides. Consequently, the conditions required for the hydrolysis of nitriles are more severe.

The mechanism of nitrile hydrolysis in acidic solution involves, first, protonation of the nitrogen (Sec. 21.5):

$$R—C\equiv N\!:\ H—\overset{+}{\ddot{O}}H_2 \rightleftarrows R—C\equiv\overset{+}{N}—H + :\ddot{O}H_2 \qquad (21.43a)$$

This protonation makes the nitrile carbon much more electrophilic, just as protonation of a carbonyl oxygen makes a carbonyl carbon more electrophilic. A nucleophilic reaction of water at the nitrile carbon and loss of a proton gives an intermediate called an *imidic acid.*

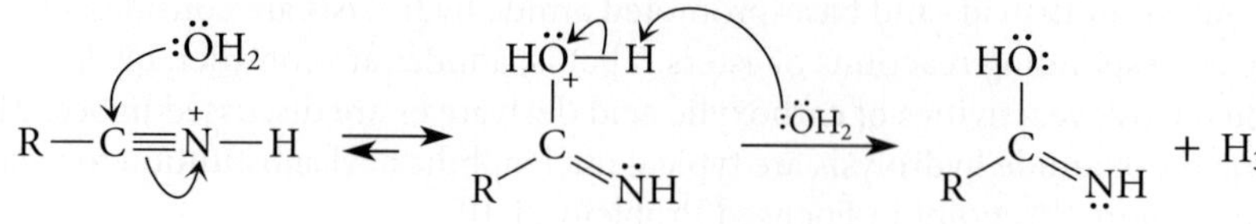

(21.43b)

An imidic acid is the nitrogen analog of an enol (Sec. 5.5). That is, an imidic acid is to an amide as an enol is to a ketone.

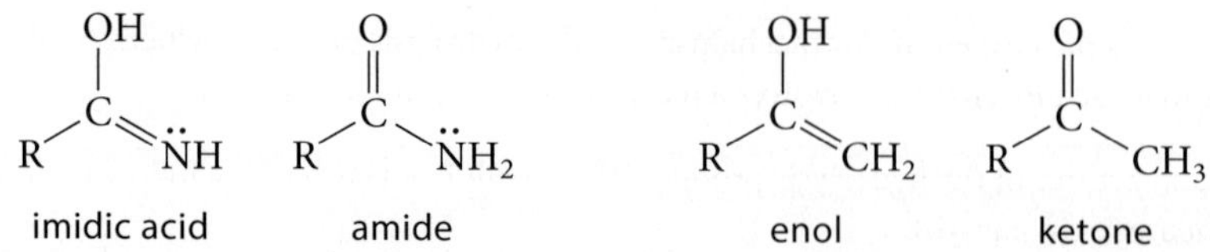

(21.43c)

Just as enols are converted spontaneously into aldehydes or ketones, an imidic acid is converted under the reaction conditions into an amide:

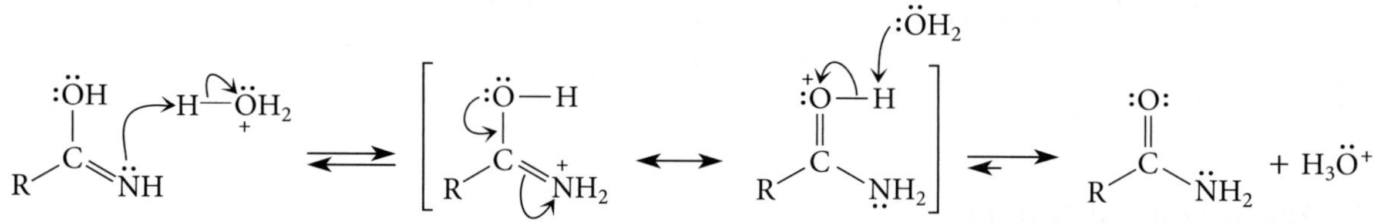

(21.43d)

Because amide hydrolysis is faster than nitrile hydrolysis, the amide formed in Eq. 21.43d does not survive under the vigorous conditions of nitrile hydrolysis and is therefore hydrolyzed to a carboxylic acid and ammonium ion, as discussed in Sec. 21.7B. Consequently, the ultimate product of nitrile hydrolysis in acid is a carboxylic acid.

Nitriles behave mechanistically much like carbonyl compounds. Compare, for example, the mechanism of acid-promoted nitrile hydrolysis in Eqs. 21.43a and b with that for the acid-catalyzed hydration of an aldehyde or ketone (Sec. 19.8A). In both mechanisms, an electronegative atom is protonated (nitrogen of the C≡N bond, or oxygen of the C═O bond), and water then reacts as a nucleophile at the carbon of the resulting cation.

The parallel between nitrile and carbonyl chemistry is further illustrated by the hydrolysis of nitriles in base. The nitrile group, like a carbonyl group, reacts with basic nucleophiles and, as a result, the electronegative nitrogen assumes a negative charge. Proton transfer gives an imidic acid (which, like a carboxylic acid, ionizes in base).

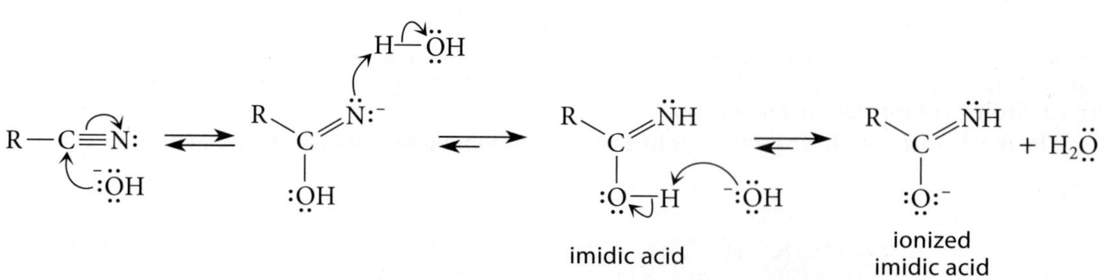

(21.44a)

As in acid-promoted hydrolysis, the imidic acid reacts further to give the corresponding amide, which, in turn, hydrolyzes under the reaction conditions to the carboxylate salt of the corresponding carboxylic acid (Sec. 21.7B).

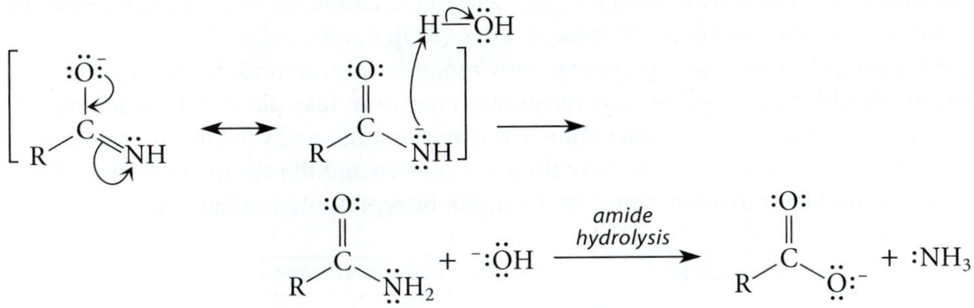

(21.44b)

D. Hydrolysis of Acid Chlorides and Anhydrides

Acid chlorides and anhydrides react *rapidly* with water, even in the absence of acids or bases.

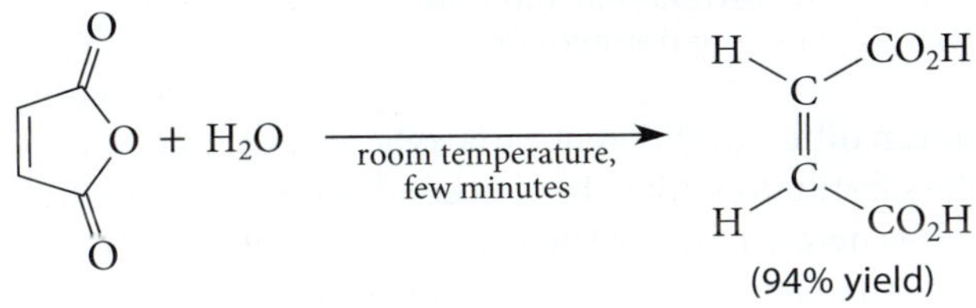

(21.45)

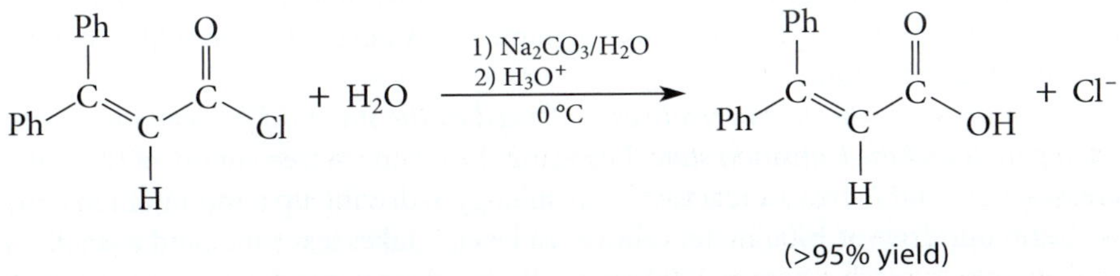

(21.46)

The hydrolysis reactions of acid chlorides and anhydrides are almost never used for the preparation of carboxylic acids because these derivatives are themselves usually prepared from acids (Sec. 20.9). Rather, these reactions serve as reminders that if samples of acid chlorides and anhydrides are allowed to come into contact with moisture, they will rapidly become contaminated with the corresponding carboxylic acids.

E. Mechanisms and Reactivity in Nucleophilic Acyl Substitution Reactions

All carboxylic acid derivatives can be hydrolyzed to carboxylic acids; however, the *conditions* under which the different derivatives are hydrolyzed differ considerably. Hydrolysis reactions of amides and nitriles require heat as well as acid or base; hydrolysis reactions of esters require acid or base, but require heating only briefly, if at all; and hydrolysis reactions of acid chlorides and anhydrides occur rapidly at room temperature even in the absence of acid and base. These trends in reactivity, which are observed not only in hydrolysis but in *all* nucleophilic acyl substitution reactions, can be summarized as follows:

Reactivities of carboxylic acid derivatives in nucleophilic acyl substitution reactions:

nitriles < amides < esters, thioesters, acids << anhydrides < acid chlorides

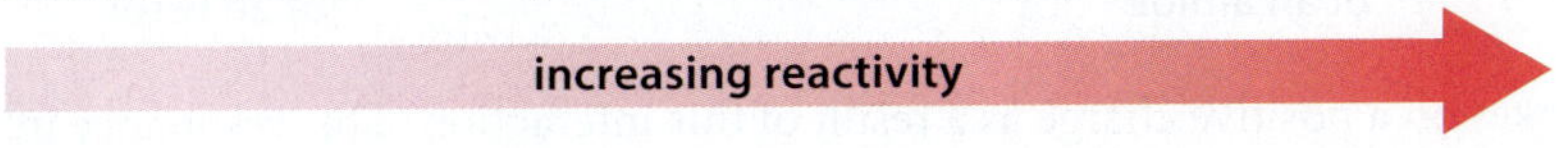

(21.47)

(The reactions of nitriles are additions, not substitutions, but they are included for comparison.)

The practical significance of this reactivity order is that selective reactions are possible. In other words, an ester can be hydrolyzed under conditions that will leave an amide in the same molecule unaffected; likewise, nucleophilic substitution reactions on an acid chloride can be carried out under conditions that will leave an ester group unaffected.

Understanding the trends in relative reactivity requires, first, an understanding of the mechanisms by which nucleophilic acyl substitution reactions take place. (The reactivity of nitriles is considered later.) As we have shown, the nucleophilic acyl substitution reaction mechanism typically consists of two steps: the addition step and the elimination step. These steps, and the corresponding transition states for them, can be represented as follows:

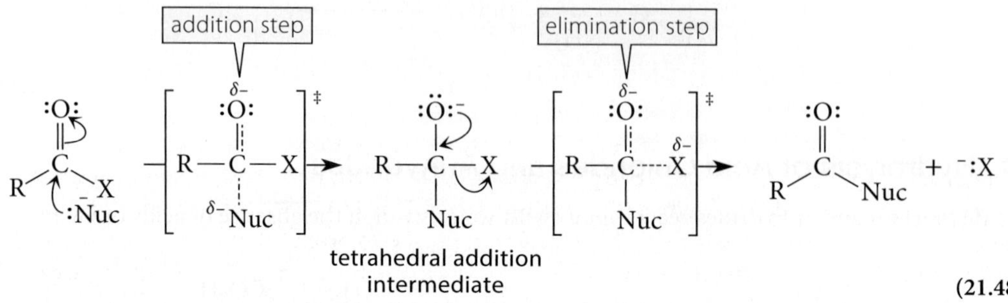

(21.48)

The rate-limiting steps can differ for different carboxylic acid derivatives; and, in the more reactive derivatives, the substitution might be concerted, with no tetrahedral intermediate at all. However, regardless of the details of the mechanism, we'll adopt the view that *the structure of the tetrahedral intermediate is an approximation of the transition-state structure,* and, therefore, *the standard free energy of the tetrahedral intermediate is an approximation of the standard free energy of the transition state.* Essentially, we're invoking Hammond's postulate (Sec. 4.8D) to describe the transition state of the reaction.

The reactivity of carboxylic acid derivatives is affected by the standard free energies of both the carbonyl compound and the transition state. Lowering the standard free energy of the transition state decreases $\Delta G^{\circ\ddagger}$ and increases reactivity. An analogy is driving up a mountain to a mountain pass. If we begin our drive at 1000 meters above sea level, it takes less time (and a smaller increase in potential energy) to reach a pass at 2500 meters than it does to reach a pass at 3000 meters. Lowering the standard free energy of the starting material (the carbonyl compound) increases the standard free energy of activation $\Delta G^{\circ\ddagger}$ and therefore decreases reactivity. By analogy, if we are driving to a particular mountain pass at 3000 meters, it takes more time (and a larger gain in potential energy) if we start at 1000 meters than it does if we start at 1500 meters.

First consider how the stability of the carbonyl compound varies among the different carboxylic acid derivatives. The major factor in the stability of the carbonyl compound is *the resonance interaction of the potential leaving group X with the carbonyl group.* The basis of this resonance interaction is the overlap of the nonbonding electrons of the group X with the carbonyl π molecular orbital.

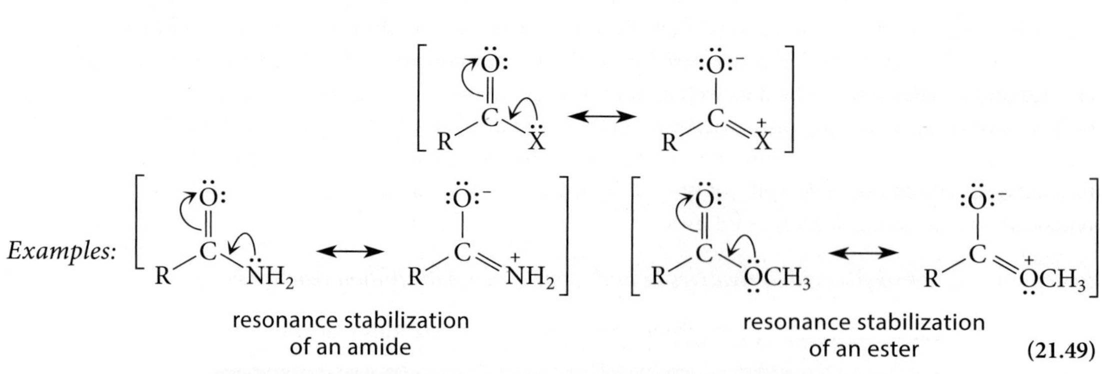

(21.49)

The X atom takes on a positive charge as a result of this interaction. This resonance interaction is important in both esters and amides; however, *it is less important in an ester than it is in an amide* because the greater electronegativity of oxygen compared with nitrogen opposes electron donation by resonance.

Comparing the resonance interaction in an anhydride with that in an ester, we find that the resonance interaction in an anhydride creates repulsion between the positively charged oxygen and the bond dipole of the adjacent carbonyl group:

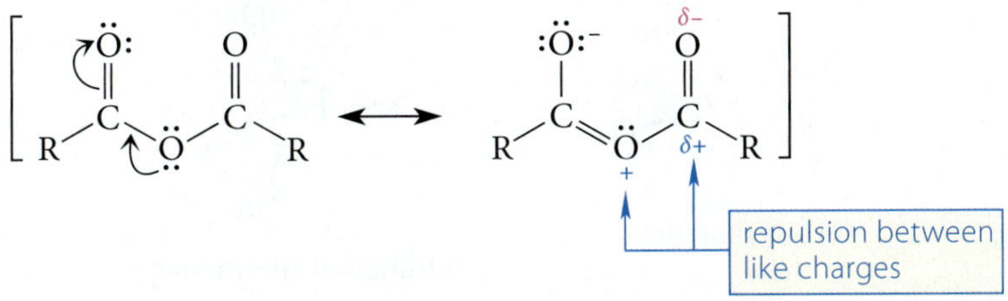

(21.50)

Therefore, *an anhydride is stabilized less by resonance than an ester is.*

In an acid chloride, the resonance interaction between the chlorine and the carbonyl group is opposed by the electronegativity of the leaving group, as in an ester. However, an even more important factor is the weaker electronic overlap between a 3p orbital of the chlorine with the 2p orbital of the carbonyl carbon (see Fig. 16.7, Sec. 16.5B), which is the basis of the resonance effect. *Consequently, resonance stabilization in an acid chloride is less important than it is in an anhydride.*

Considering the stabilization of the carbonyl compound, the reactivity order of the various derivatives should be the following:

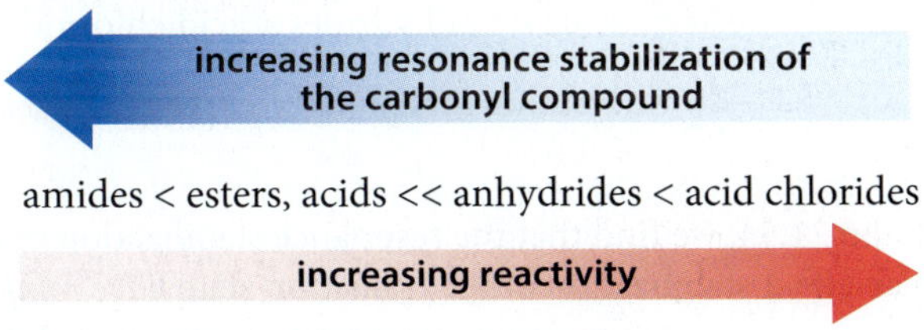

(21.51)

This is exactly the order observed.

Resonance stabilization of the carbonyl compound, however, is only half of the story. Now consider the *polar effect* of the leaving group X on the stability of the transition state. (Remember, we are using the tetrahedral addition intermediate as an approximation for the transition state.) A greater electronegativity of the X group results in a greater C—X bond dipole moment. A greater C—X bond dipole results in more partial positive charge on the carbon end of the dipole and therefore greater electrostatic stabilization of the transition state.

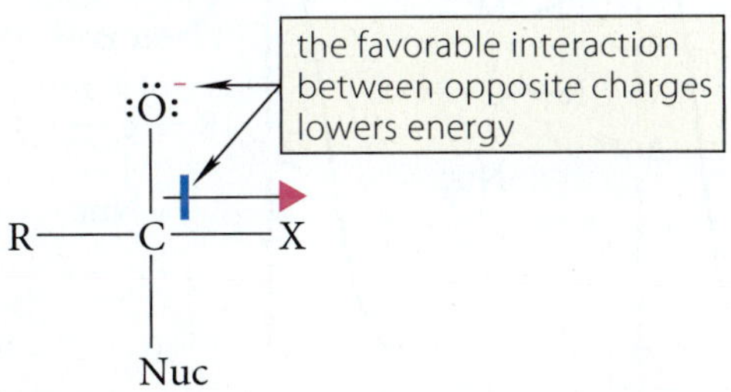

(21.52)

This polar effect is exactly the same as the effect we discussed for the pK_a values of carboxylic acids (Sec. 3.7E). Recall that an electronegative substituent lowers the pK_a of a carboxylic acid because its bond dipole interacts favorably with the negative charge of the carboxylate anion (Eq. 3.84, Sec. 3.7E).

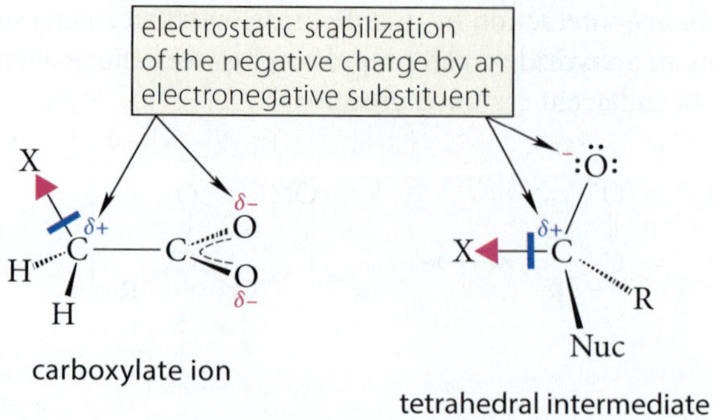

(21.53)

The polar effect on the stability of the tetrahedral intermediate should be even greater than the polar effect on the stability of a carboxylate anion because the X group is closer to the negatively charged oxygen in the tetrahedral intermediate than it is in a substituted carboxylate ion. (Recall that polar effects increase with decreasing distance.)

Therefore, the polar effect should increase reactivity in the following order:

> Although the analysis in this section applies to negatively charged nucleophiles, the same reactivity order is observed for any nucleophile. The same order is also observed for acid-catalyzed acyl substitutions. You can try an analysis for an acid-catalyzed hydrolysis in Focused Problem 21.12.

increasing polar stabilization of the transition state →

amides < esters, acids << anhydrides < acid chlorides

increasing reactivity →

(21.54)

Comparing Eqs. 21.51 and 21.54, we find that the resonance stabilization of the carbonyl compound and the polar-effect stabilization of the transition state *have reinforcing effects on reactivity*. Both effects contribute to the reactivity order that is observed (**Fig. 21.5**).

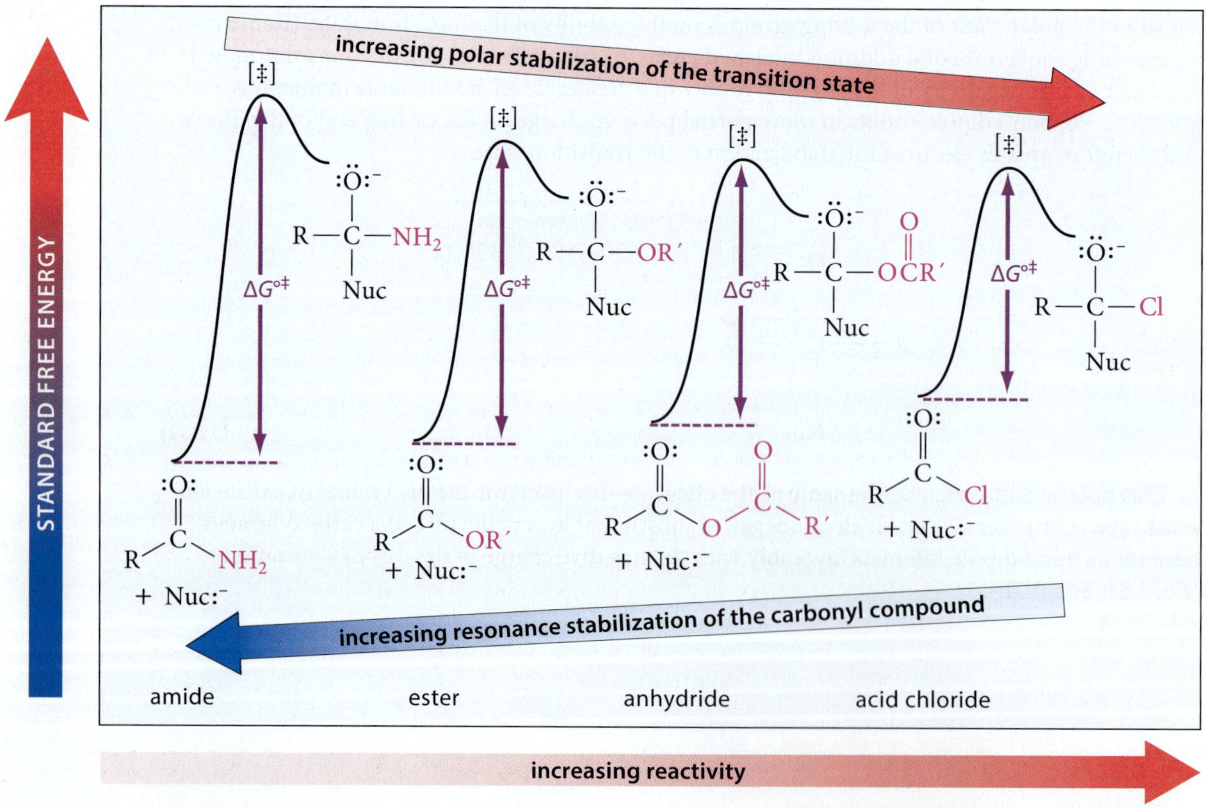

FIGURE 21.5 The effect of structure on the reactivity of carboxylic acid derivatives in nucleophilic acyl substitution. Within the diagram, the blue arrow maps increasing importance of an effect that decreases reactivity, and the red arrow maps increasing importance of an effect that increases reactivity. The overall reactivity increases to the right, as shown on the label for the x-axis.

What about nitriles? Reactions of nitriles in base are slower than those of other acid derivatives because nitrogen is less electronegative than oxygen and accepts additional electrons less readily. Reactions of nitriles in acid are slower because of their extremely low basicities. It is the protonated form of a nitrile that reacts with nucleophiles in acid solution, but so little of this form is present (Sec. 21.5) that the rate of the reaction is very small.

Focused Problems

21.12 Use an analysis of resonance effects and leaving-group basicities to explain why acid-catalyzed hydrolysis of esters is faster than acid-catalyzed hydrolysis of amides.

21.13 The hydrolysis of acetyl chloride is 7800 times faster than the hydrolysis of acetyl fluoride in 25% water –75% acetone solution.

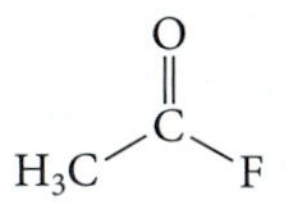
acetyl fluoride

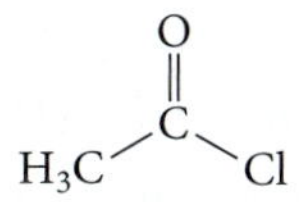
acetyl chloride

Which factor is more important in determining the relative hydrolysis rate: resonance stabilization of the carbonyl compound or polar stabilization of the transition state? Explain how you know.

21.14 Complete the following reactions.

(a)
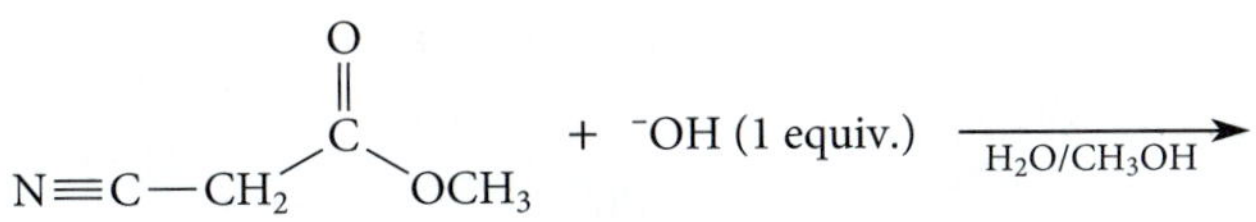

(b)
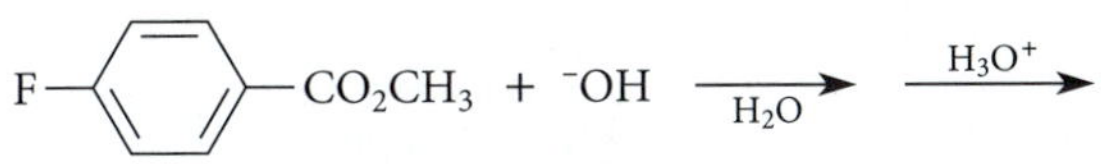

(c)
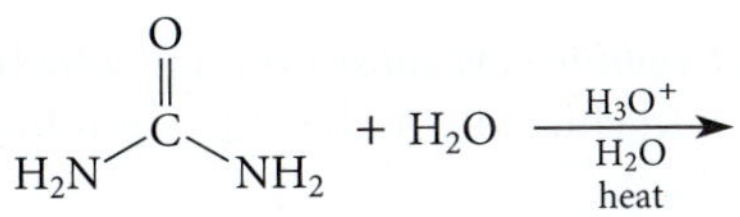
(*Hint:* See Sec. 20.11A.)

21.8 REACTIONS OF CARBOXYLIC ACID DERIVATIVES WITH NUCLEOPHILES

Section 21.7 showed that all carboxylic acid derivatives hydrolyze to carboxylic acids. Water and hydroxide ion, the nucleophiles involved in hydrolysis, are only two of the nucleophiles that react with carboxylic acid derivatives. This section shows how the reactions of other nucleophiles with carboxylic acid derivatives can be used to prepare other carboxylic acid derivatives. As you proceed through this section, notice how all of the reactions fit the pattern of nucleophilic acyl substitution.

A. Reactions of Acid Chlorides with Nucleophiles

Among the most useful ways of preparing carboxylic acid derivatives are the reactions of acid chlorides with various nucleophiles. Because of the great reactivity of acid chlorides, such reactions are typically rapid and can be carried out under mild conditions. Recall that acid chlorides are readily prepared from the corresponding carboxylic acids (Sec. 20.9A).

Reactions of Acid Chlorides with Ammonia and Amines Acid chlorides react rapidly and irreversibly with ammonia or amines to give amides. Reaction of an acid chloride with *ammonia* yields a primary amide:

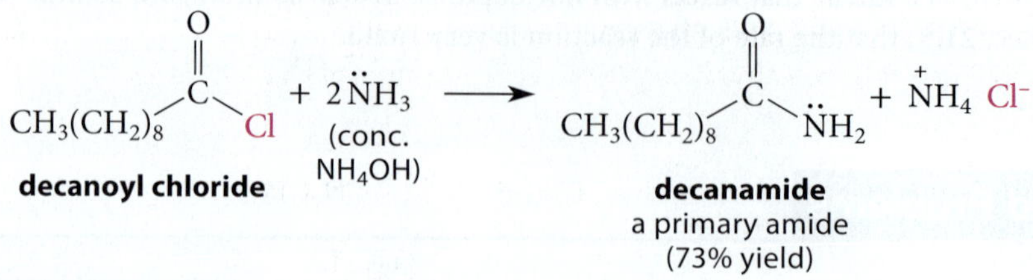

(21.55)

The reaction of an acid chloride with a *primary amine* (an amine of the form RNH_2) gives a secondary amide:

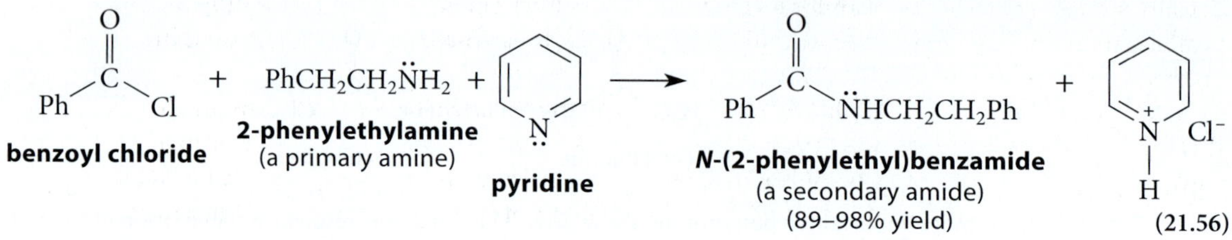

(21.56)

The reaction of an acid chloride with a *secondary amine* (an amine of the form R_2NH) gives a tertiary amide:

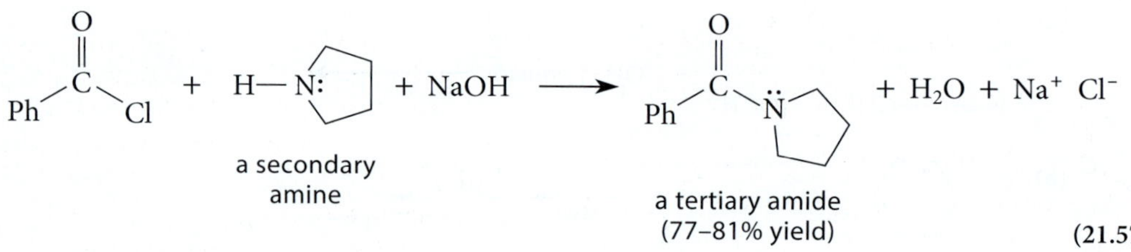

(21.57)

These reactions are all additional examples of *nucleophilic acyl substitution* in which an amine reacts as a nucleophile at the carbonyl carbon of the acid chloride. A chloride ion is lost from the tetrahedral intermediate.

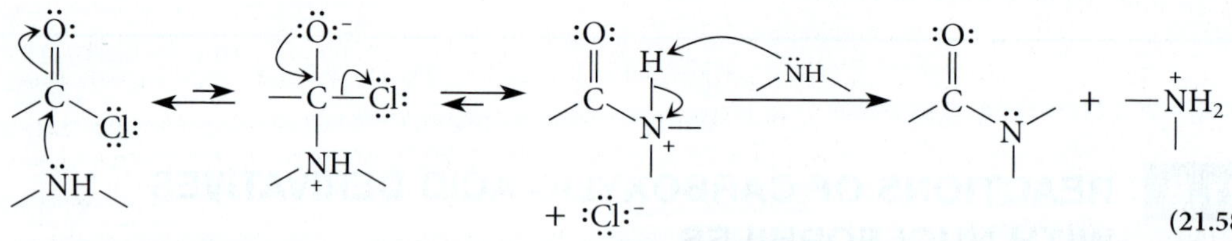

(21.58)

A proton is removed from the amide nitrogen in the last step of the mechanism. Unless another base is added to the reaction mixture, *the starting amine acts as the base in this step*. Therefore, for each equivalent of amide that is formed, an equivalent of amine is protonated. When the amine is protonated, its electron pair is taken "out of action," and the amine is no longer nucleophilic.

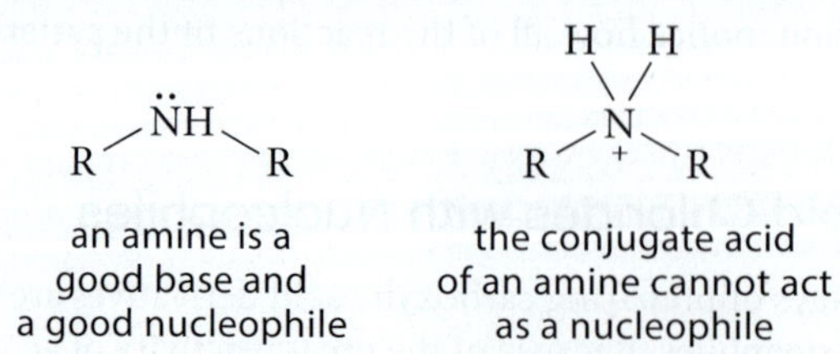

(21.59)

Therefore, if the only base present is the amine nucleophile (for example, as in Eq. 21.55), then at least *two* equivalents must be used—one equivalent as the nucleophile and one as the base in the final proton-transfer step.

The use of excess amine is practical when the amine is cheap and readily available. Another alternative is to use a *tertiary amine* (an amine of the form R_3N:), such as triethylamine or pyridine, as the base. The reaction in Eq. 21.56 is an example of this strategy.

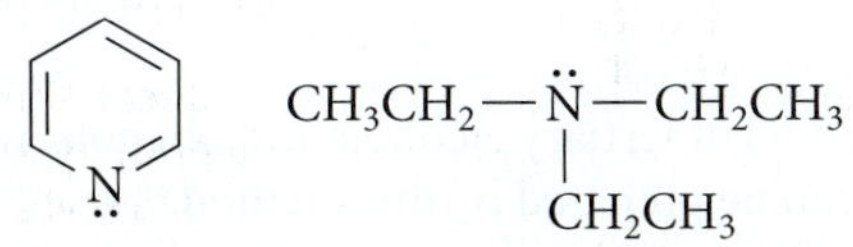

pyridine **triethylamine**

The presence of a tertiary amine does not interfere with amide formation by another amine because a tertiary amine itself cannot form an amide. (Why?) The use of a tertiary amine is particularly practical if the amine used to form the amide is expensive and cannot be used in excess.

FURTHER EXPLORATION 21.3
Reaction of Tertiary Amines with Acid Chlorides

Yet another alternative for amide formation is to use the *Schotten–Baumann* technique. In this method, the reaction is run with an acid chloride in a separate layer (either alone or in a solvent) over an aqueous solution of NaOH (Eq. 21.57). Hydrolysis of the acid chloride by NaOH is avoided because acid chlorides are typically insoluble in water and therefore are not in direct contact with the water-soluble hydroxide ion. The amine, which is soluble in the acid chloride solution, reacts to yield an amide. The aqueous NaOH extracts and neutralizes the protonated amine that is formed.

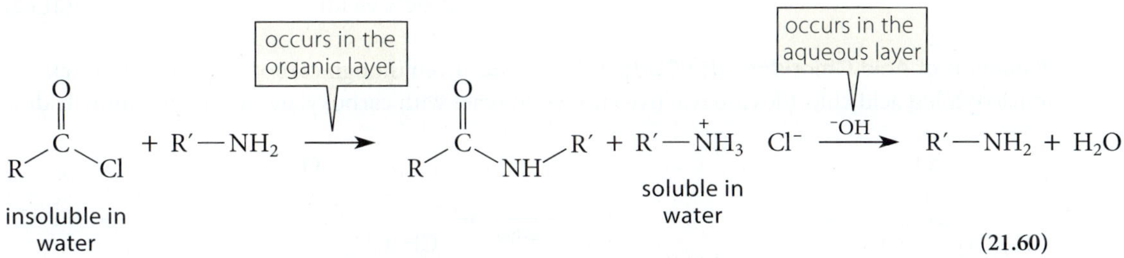

(21.60)

The important point about all of the methods for preparing amides is that either two equivalents of amine must be used, or an equivalent of base must be added to bring about the final neutralization.

Reaction of Acid Chlorides with Alcohols and Phenols Esters are formed rapidly when acid chlorides react with alcohols or phenols. In principle, the HCl liberated in the reaction need not be neutralized because alcohols and phenols are not basic enough to be extensively protonated by the acid. However, some esters (such as *tert*-butyl esters; see Further Exploration 21.2) and alcohols (such as tertiary alcohols; Secs. 11.2 and 11.3) are sensitive to acid. In practice, a tertiary amine like pyridine is added to the reaction mixture or is even used as the solvent to neutralize the HCl.

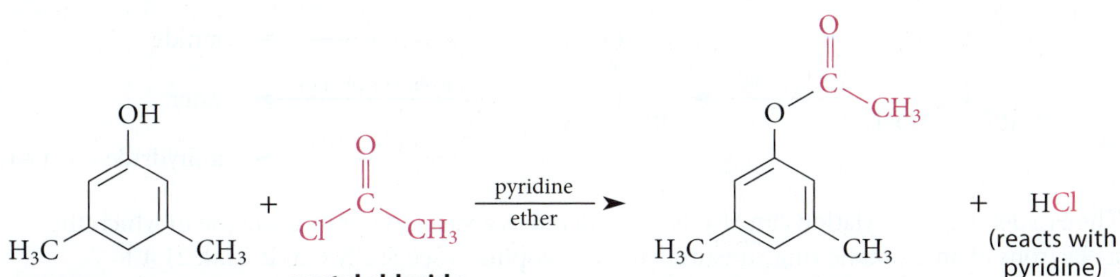

3,5-dimethylphenol **acetyl chloride** **3,5-dimethylphenyl acetate**

(21.61a)

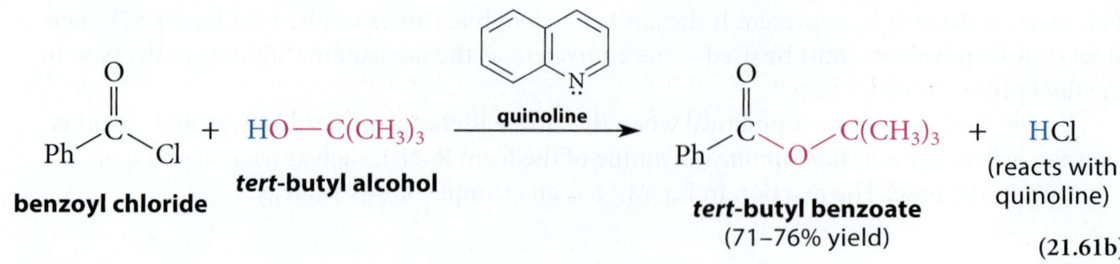

(21.61b)

As these examples illustrate, esters of tertiary alcohols and phenols, which cannot be prepared by acid-catalyzed esterification, can be prepared by this method.

Sulfonate esters (esters of sulfonic acids) are prepared by the analogous reactions of sulfonyl chlorides (the acid chlorides of sulfonic acids) with alcohols. This reaction was introduced in Eq. 11.33, Sec. 11.4A.

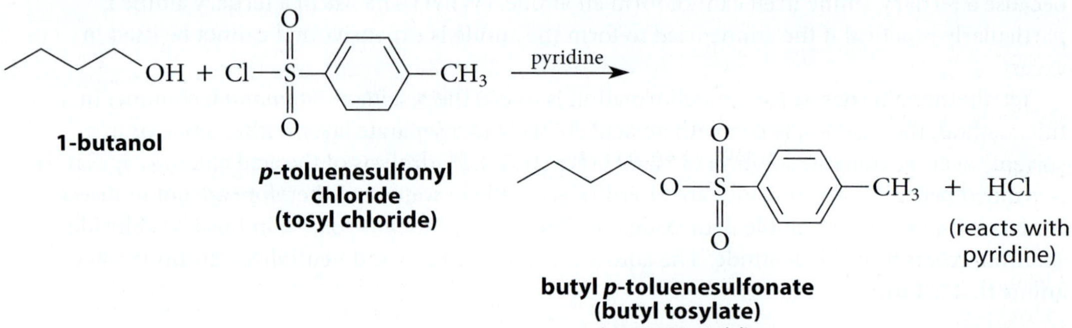

(21.62)

Reaction of Acid Chlorides with Carboxylate Salts Even though carboxylate salts are weak nucleophiles, acid chlorides are reactive enough to react with carboxylate salts to give anhydrides.

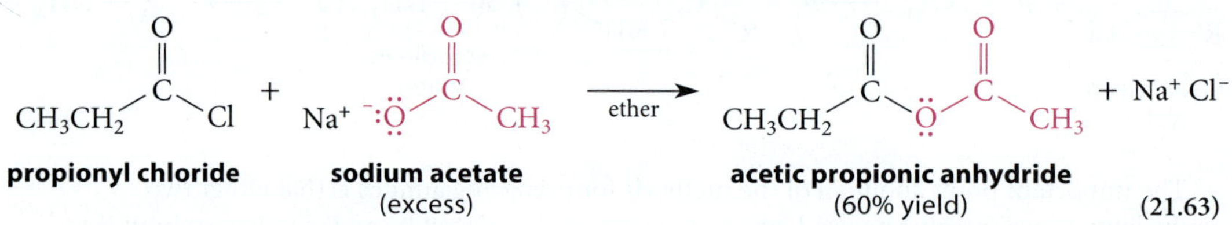

(21.63)

This is a second general method for the synthesis of anhydrides. The anhydride synthesis discussed in Sec. 20.9B (dehydration of carboxylic acids) can only be used for the synthesis of symmetrical anhydrides. However, the reactions of acid chlorides with carboxylate salts can be used to prepare mixed anhydrides, as the example in Eq. 21.63 illustrates.

Summary: Use of Acid Chlorides in Organic Synthesis One of the most important general methods for converting a carboxylic acid into an ester, amide, or anhydride is first to convert the carboxylic acid into its acid chloride (Sec. 20.9A) and then use one of the acid chloride reactions discussed in this section to form the desired carboxylic acid derivative. To summarize:

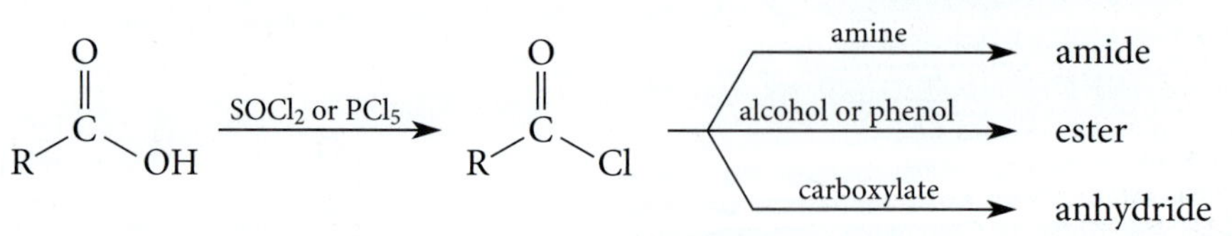

(21.64)

STUDY GUIDE LINK 21.4
Another Look at the Friedel-Crafts Reaction

The Friedel–Crafts acylation can also be considered as a variation of this scheme in which the π electrons of an aromatic ring constitute the nucleophile. (See Study Guide Link 21.4.)

B. Reactions of Anhydrides with Nucleophiles

Anhydrides react with nucleophiles in much the same way as acid chlorides—that is, the reaction with an amine yields an amide, the reaction with an alcohol yields an ester, and so on.

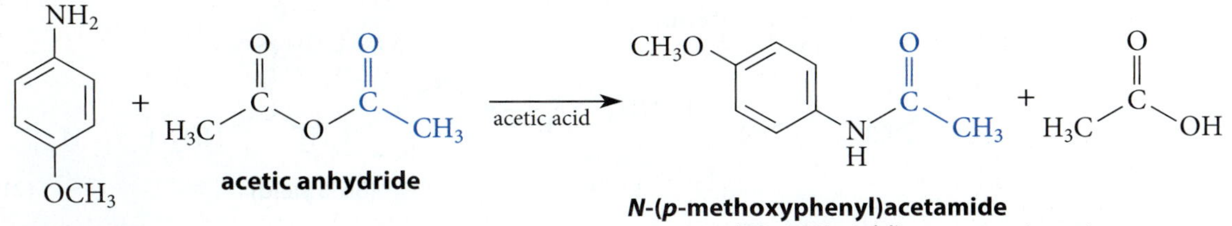

p-methoxyaniline + **acetic anhydride** → ***N*-(*p*-methoxyphenyl)acetamide** (75–79% yield)

(21.65)

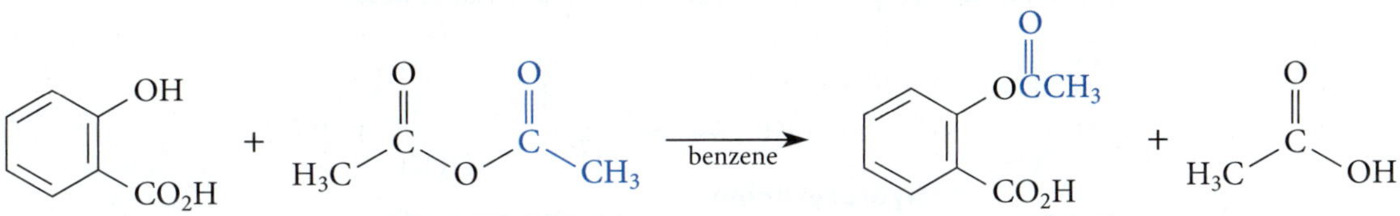

2-hydroxybenzoic acid (salicylic acid) + **acetic anhydride** → **2-acetoxybenzoic acid (aspirin)** (80–90% yield)

(21.66)

Because most anhydrides are prepared from the corresponding carboxylic acids, the use of an anhydride to prepare an ester or amide wastes one equivalent of the parent acid as a leaving group. (For example, acetic acid is a by-product in Eqs. 21.65 and 21.66.) Therefore, this reaction in practice is used only with inexpensive and readily available anhydrides, such as acetic anhydride. However, one exception is the formation of half-esters and half-amides from cyclic anhydrides:

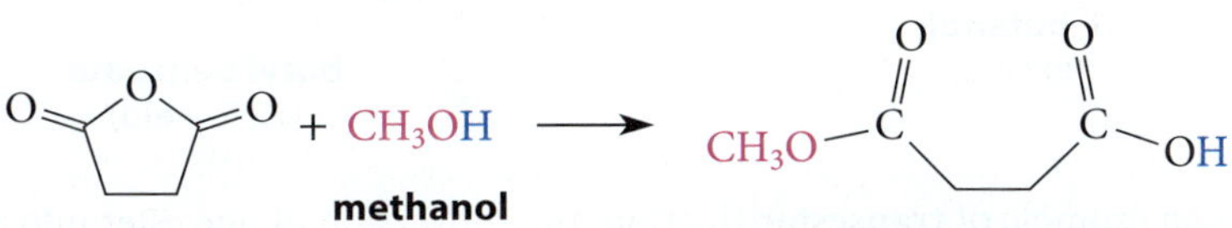

succinic anhydride + **methanol** → **methyl hydrogen succinate** (95–96% yield)

(21.67)

Half-amides of dicarboxylic acids are produced in analogous reactions of amines and cyclic anhydrides. These compounds can be cyclized to imides by treatment with dehydrating agents such as anhydrides, or, in some cases, just by heating, when five- or six-membered rings are formed. This reaction is the nitrogen analog of cyclic anhydride formation (Sec. 20.9B), and its facility is yet another example of the proximity effect (Sec. 12.8).

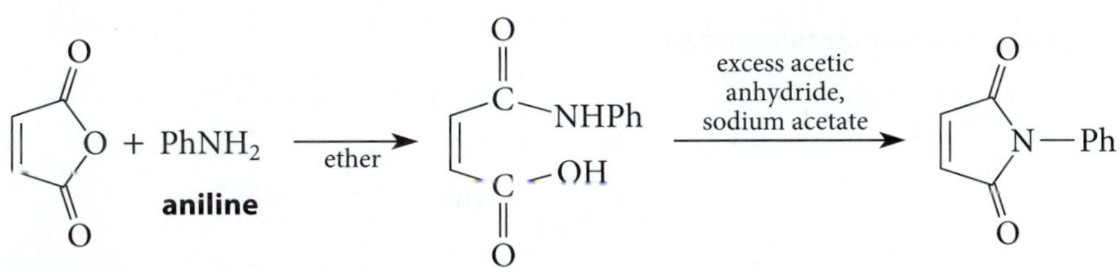

maleic anhydride + **aniline** → (97% yield) → ***N*-phenylmaleimide** (75–80% yield)

(21.68)

C. Reactions of Esters with Nucleophiles

Just as esters are much less reactive than acid chlorides toward hydrolysis, they are also much less reactive toward amines and alcohols. Nevertheless, reactions of esters with these nucleophiles are sometimes useful. The reaction of esters with ammonia or amines yields amides.

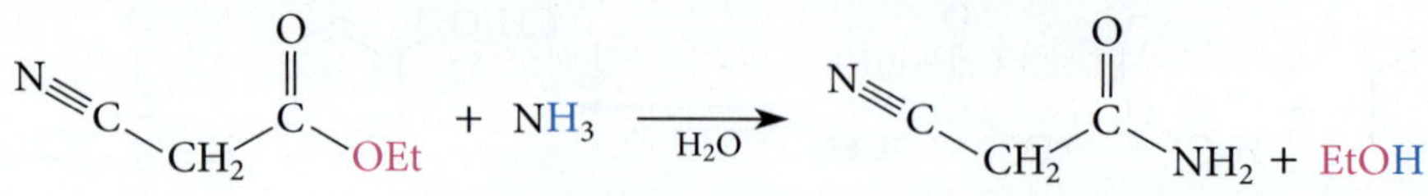

(86% yield) (21.69)

The reaction of esters with *hydroxylamine* (NH_2OH, Table 19.3, Sec. 19.12A) gives *N*-hydroxyamides; these compounds are known as **hydroxamic acids**.

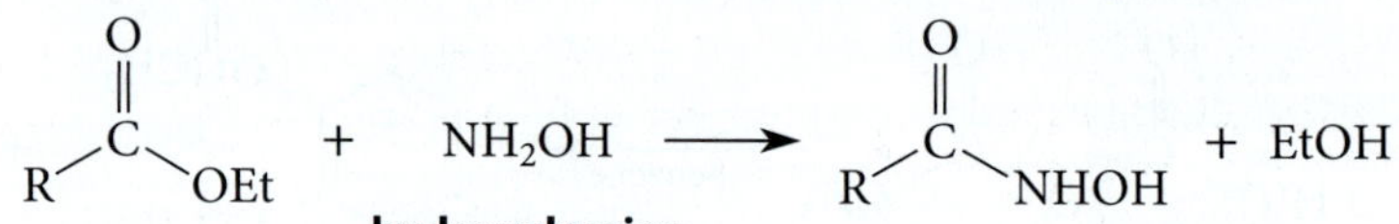

an ester **hydroxylamine** a hydroxamic acid (21.70)

(Acid chlorides and anhydrides also react with hydroxylamine to form hydroxamic acids.) This chemistry is the basis of the *hydroxamate test*, used mostly for esters. The hydroxamic acid products are easily recognized because they form highly colored complexes with ferric ion.

When an ester reacts with an alcohol under acidic conditions, or with an alkoxide under basic conditions, a new ester is formed.

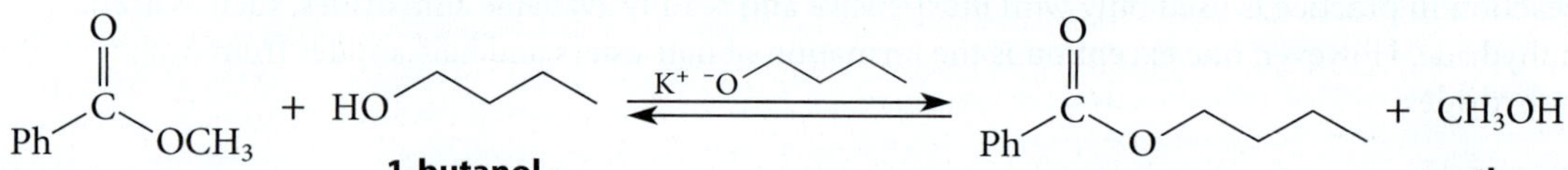

methyl benzoate **1-butanol** (excess) **butyl benzoate** (72% yield) **methanol** (21.71)

This reaction is an example of **transesterification**: the conversion of one ester into another by reaction with an alcohol. In a typical transesterification, neither ester is strongly favored at equilibrium. The reaction is driven to completion by the use of an excess of the alcohol nucleophile or by removal of the alcohol by-product by distillation—Le Chatelier's principle in action once again.

Focused Problems

21.15 Complete the following reactions by giving the major organic products.

(a) $CH_3CH_2CO_2H \xrightarrow{\text{SOCl}_2\ \text{(excess)}} \xrightarrow{(CH_3)_2NH\ \text{(excess)}}$

(b)

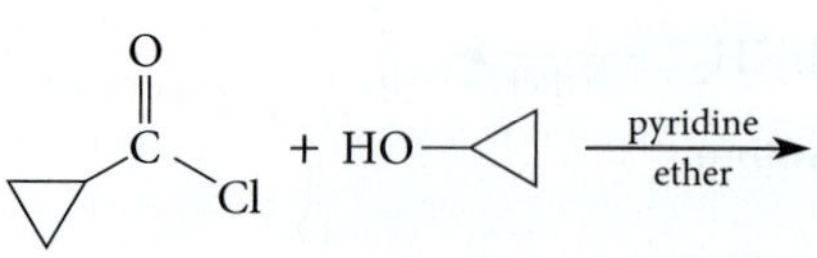

(c)

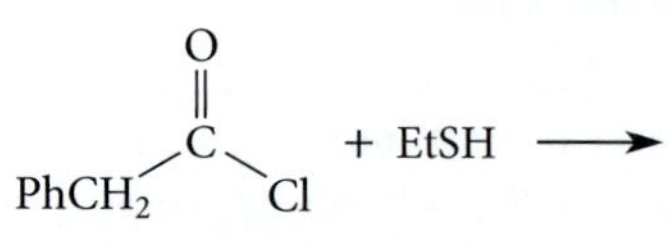

(d)

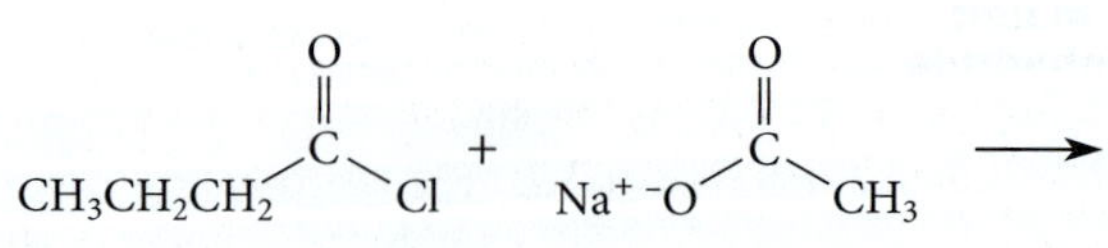

(e) 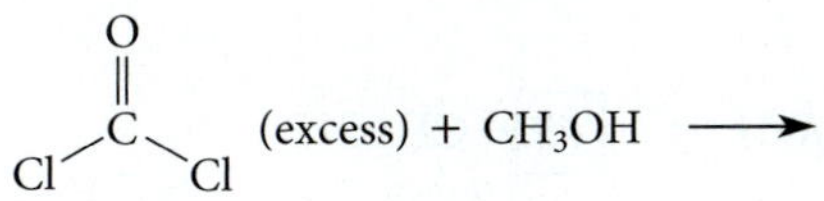

(f) 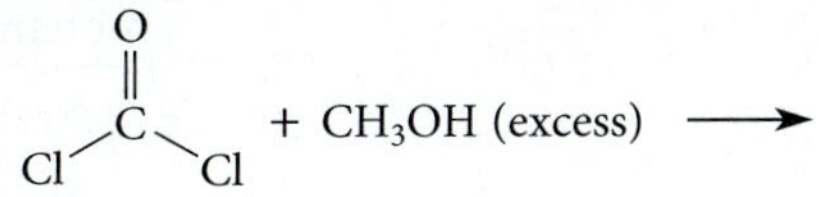

(g) 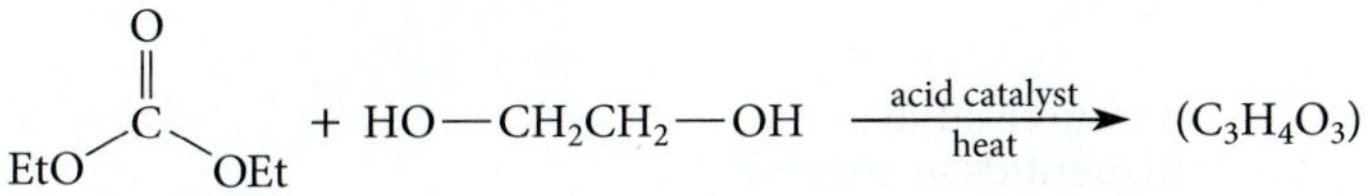

(h) 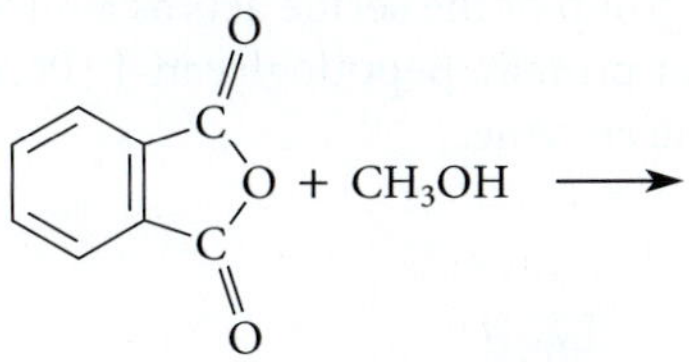

21.16 How would you synthesize each of the following compounds from an acid chloride?

(a) 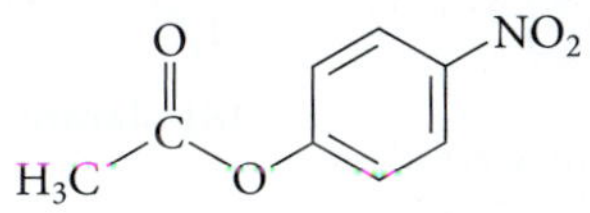

(b) 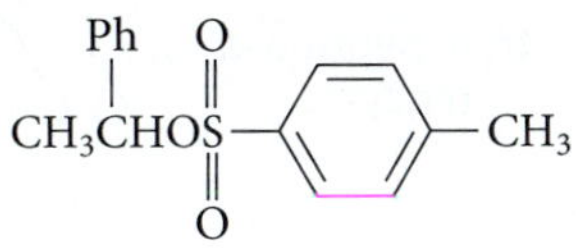

(c) 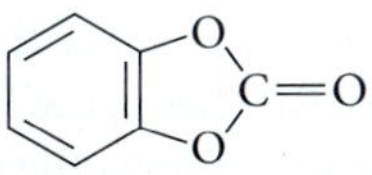

(d) 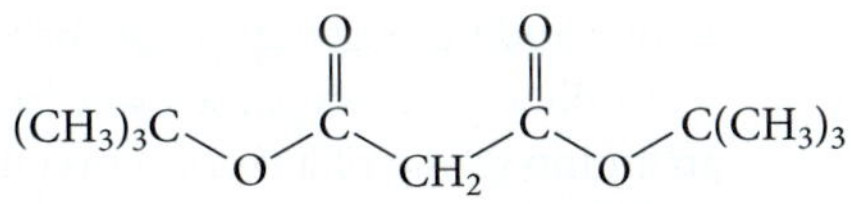

D. Reaction of Amides with Nucleophiles: Penicillin

Chemical Biology Topic

Amides, like other carboxylic acid derivatives, can react with nucleophiles. Amides, however, are very unreactive, so weak nucleophiles such as water and alcohols typically do not react with amides at a useful rate at room temperature in the laboratory. However, the reactions of some amides with nucleophiles are catalyzed by enzymes. The most ubiquitous example is the hydrolysis of the amide bonds that connect the amino acids in proteins. This hydrolysis is catalyzed by enzymes called *proteases.* In this case, the nucleophile is water.

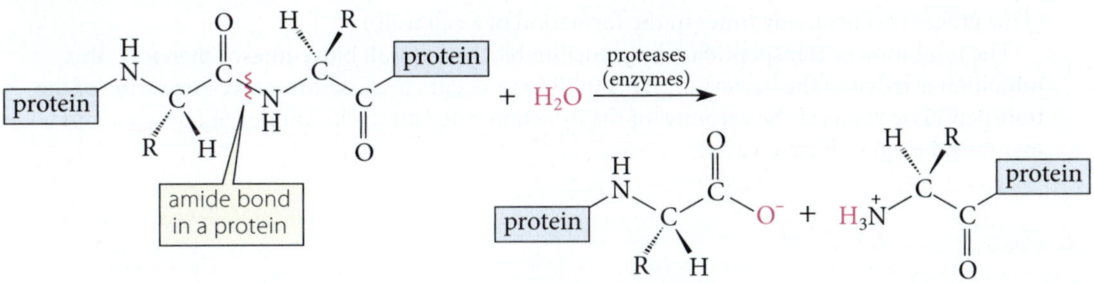

(21.72)

(We discuss proteases and their catalytic mechanism in more detail in Sec. 27.10A.)

A biological example of *enzyme-catalyzed* amide reactivity with an alcohol is found in the mode of action of *penicillin*, the first general-use antibiotic, discovered in 1928 by Sir Alexander Fleming. Penicillin inhibits *transpeptidase* (TP), an important bacterial enzyme involved in the construction of cell walls, which protect the microorganism from bursting due to the high osmotic pressure within the cell. Transpeptidase catalyzes the formation of amide bonds in peptidoglycan, a component of the cell walls. This enzyme contains a particular amino acid, serine, in its active site that operates as a nucleophile in one of the reactions of peptidoglycan synthesis.

Sir Alexander Fleming (1881–1955), who spent much of his research career at St. Mary's Medical School in London, shared the 1945 Nobel Prize in Physiology or Medicine with Lord Howard Walker Florey and Sir Ernest Boris Chain for their discovery and research on penicillin and its effect in curing various diseases.

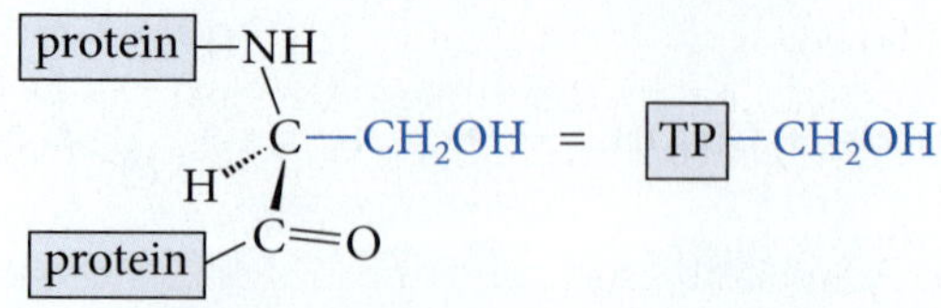

a serine residue in the transpeptidase enzyme

In the first step of peptidoglycan synthesis, the OH group of the serine acts as a nucleophile toward the terminal amide bond in a cell-wall precursor protein, peptidoglycan-1 (PG1), to form an ester in which the peptidoglycan-1 is esterified to the enzyme:

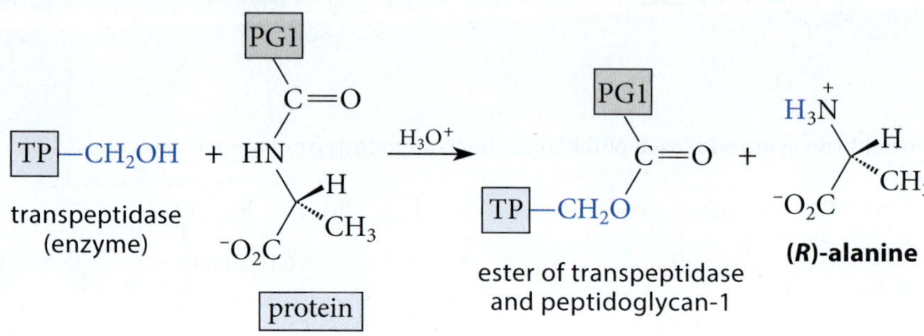

(21.73a)

Acidic and basic groups (not shown) in the active site of the enzyme help to catalyze this reaction.

In the second step, the ester from the first step undergoes an aminolysis (Sec. 21.8C) with an amino group of a second precursor, peptidoglycan-2 (PG2). This reaction links the two peptidoglycan chains in an amide bond to generate the completed peptidoglycan linkage of the cell wall and regenerates the active-site serine.

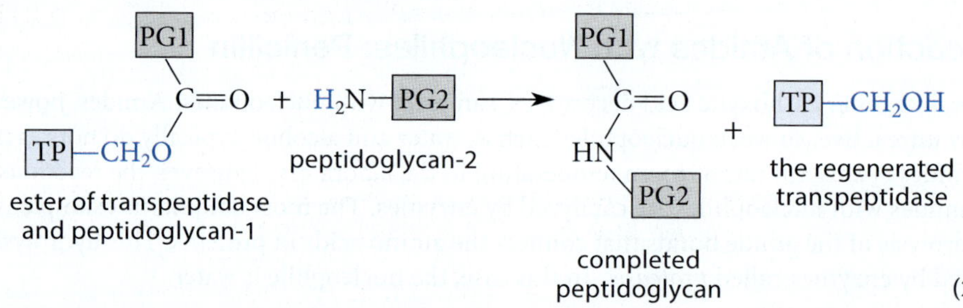

(21.73b)

(This process occurs many times in the formation of a cell wall.)

The inhibition of transpeptidase by penicillin blocks cell wall biosynthesis; therefore, this inhibition is lethal to the bacterium. This inhibition occurs because the active-site serine of the transpeptidase reacts at the carbonyl of the β-lactam ring (the cyclic amide containing a four-membered ring) to form an ester.

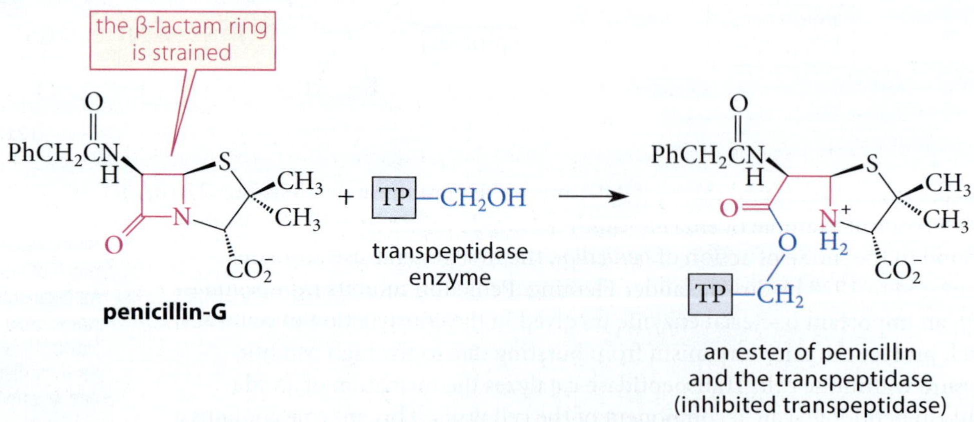

(21.74)

This reaction is unusually favorable thermodynamically, and it also occurs rapidly, because the β-lactam ring is *strained*; opening the ring releases free energy by relieving ring strain. The penicillin ester formed this way blocks the active site of the transpeptidase, and cell-wall biosynthesis cannot occur.

Bacteria have developed penicillin resistance by evolving another enzyme, *β-lactamase* (β-Lac), which intercepts the penicillin and opens the β-lactam ring to form its own serine ester. However, β-lactamase also catalyzes the hydrolysis of its own serine ester to give an inactive form of penicillin and regenerate the enzyme.

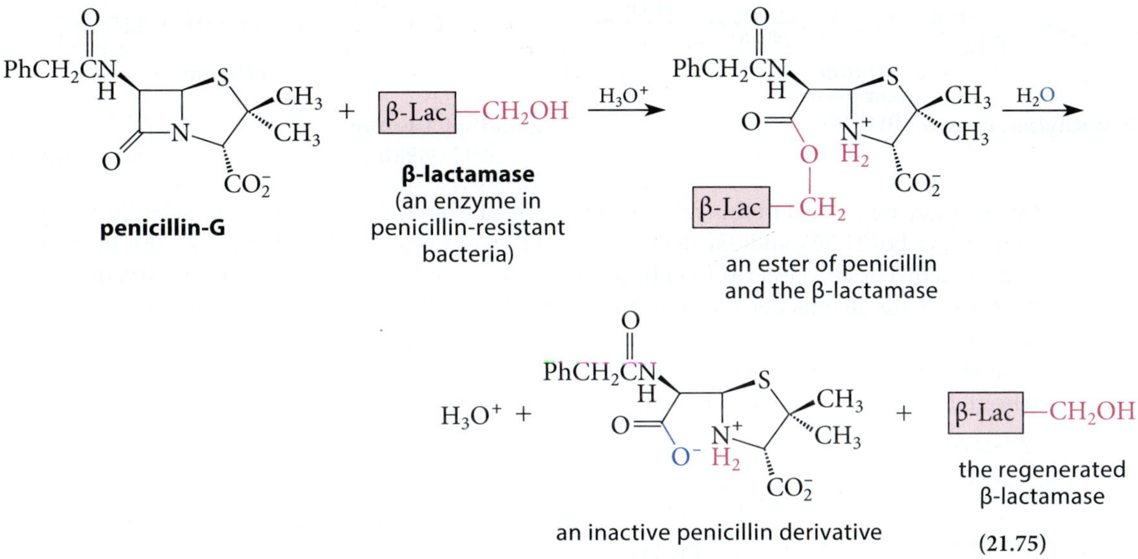

(21.75)

The β-lactamase, then, protects the process of cell-wall biosynthesis by inactivating penicillin.

Focused Problem

21.17 (a) Draw the structure of the product that is formed when the following compound is heated with *one equivalent* of sodium methoxide in methanol. Explain your reasoning.

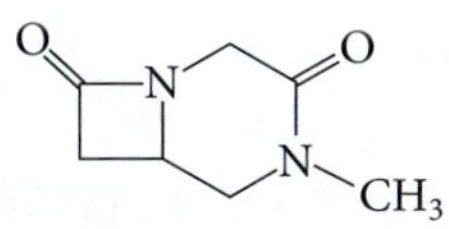

(b) What is the product if the same compound is heated with a large excess of aqueous H_2SO_4?

Antibiotic Resistance

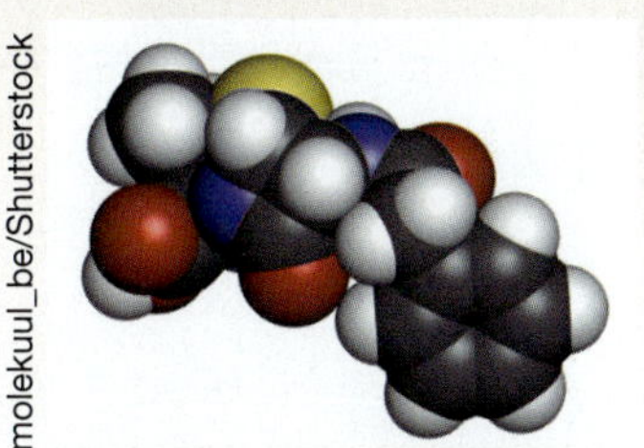

molekuul_be/Shutterstock

Space-filling model of penicillin-G

Chemists have developed inhibitors of the β-lactamase that can be administered along with penicillins. (See Problem 22.102.) However, new bacterial strains have evolved that are resistant to this inhibitor as well. As antibiotics are used more and more, the pressure for the evolution of antibiotic-resistant bacterial strains increases. (The widespread use of antibiotics in feed grain has been a particular area of concern.) For this reason, it is important to limit antibiotic use to cases in which they are truly needed. For example, the common cold is caused by a virus and cannot be cured with antibiotics; so, taking antibiotics for a cold has no effect. The continuous cycle of antibiotic discovery and the evolution of resistance can be expected to recur indefinitely. This cycle means that the development of antibiotics will always be a new opportunity in the pharmaceutical industry.

21.9 REDUCTION OF CARBOXYLIC ACID DERIVATIVES

A. Reduction of Esters to Primary Alcohols

Lithium aluminum hydride, $LiAlH_4$, reduces all carboxylic acid derivatives. Reduction of esters with this reagent, like the reduction of carboxylic acids, gives primary alcohols.

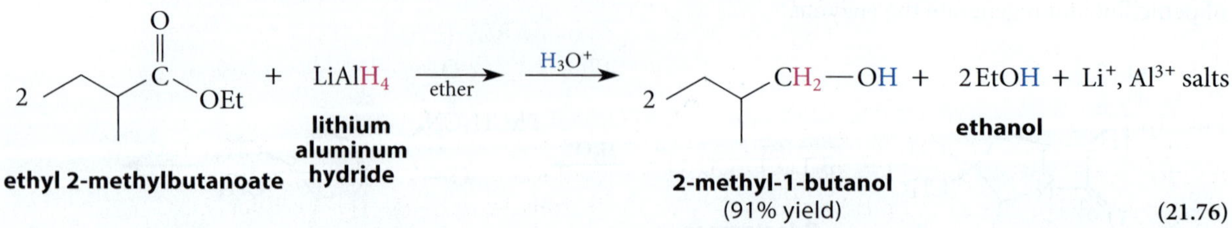

(21.76)

Two alcohols are formed in this reaction, one derived from the *acyl group* of the ester (2-methyl-1-butanol in Eq. 21.76), and one derived from the alkoxy group (ethanol in Eq. 21.76). In most cases, a methyl or ethyl ester is used in this reaction, and the by-product (methanol or ethanol) is discarded; the alcohol derived from the acyl portion of the ester is typically the product of interest.

As shown in Sec. 19.9A, the active nucleophile in $LiAlH_4$ reductions is the *hydride ion* ($H{:}^-$) delivered from $^-AlH_4$, and this reduction is no exception.

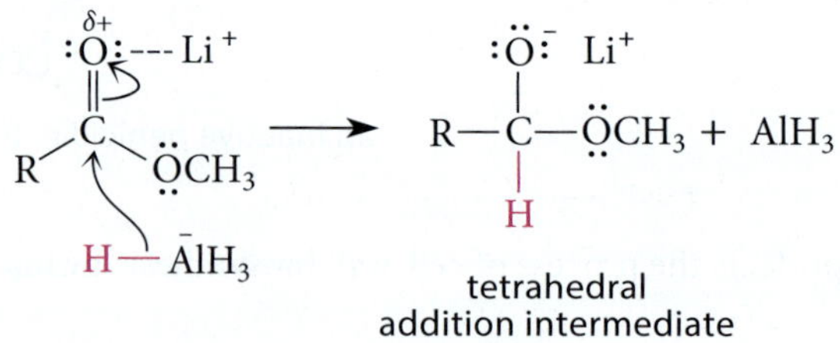

(21.77a)

The tetrahedral addition intermediate eliminates methoxide ion to form an intermediate aldehyde, which forms an addition complex with the Lewis acid AlH_3.

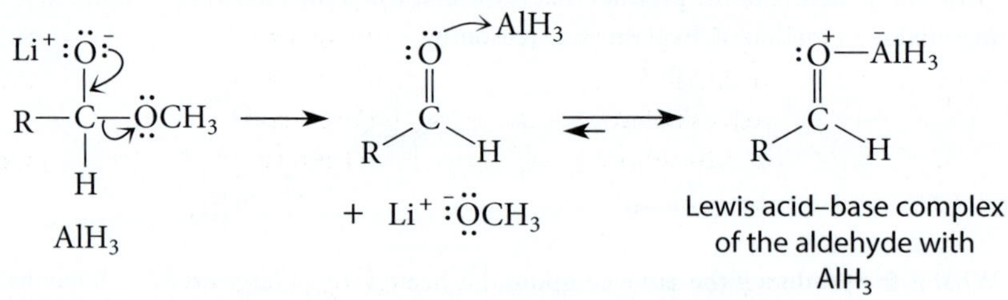

(21.77b)

Formation of the aldehyde completes the *nucleophilic acyl substitution* part of the mechanism. The aldehyde complex, with its positively charged oxygen, is very reactive and undergoes a rapid intramolecular *carbonyl-addition* reaction with another hydride equivalent. The remaining two hydride equivalents react by a similar mechanism with another ester molecule to give a Lewis acid–base addition complex of the two alkoxides with Al^{3+}. Addition of acid in a separate step gives the two alcohol products.

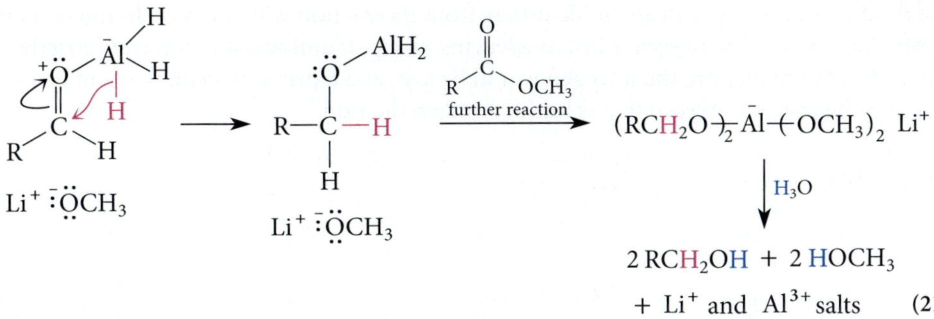

(21.77c)

To summarize: The reduction of esters to alcohols involves a *nucleophilic acyl substitution* reaction followed by a *carbonyl-addition* reaction. The substitution product, an aldehyde (most likely as its AlH_3 addition complex), is far too reactive to be isolated and undergoes addition to give the final product.

Sodium borohydride, $NaBH_4$, another useful hydride reducing agent, is much less reactive than lithium aluminum hydride. It reduces aldehydes and ketones, but it reacts sluggishly with most esters. In fact, $NaBH_4$ can be used to reduce aldehydes and ketones selectively in the presence of esters.

Acid chlorides and anhydrides also react with $LiAlH_4$ to give primary alcohols. However, because acid chlorides and anhydrides are usually prepared from carboxylic acids, and because carboxylic acids themselves can be reduced to alcohols with $LiAlH_4$ (Sec. 20.10), the reduction of acid chlorides and anhydrides is seldom used.

B. Reduction of Amides to Amines

Amines are formed when amides are reduced with $LiAlH_4$.

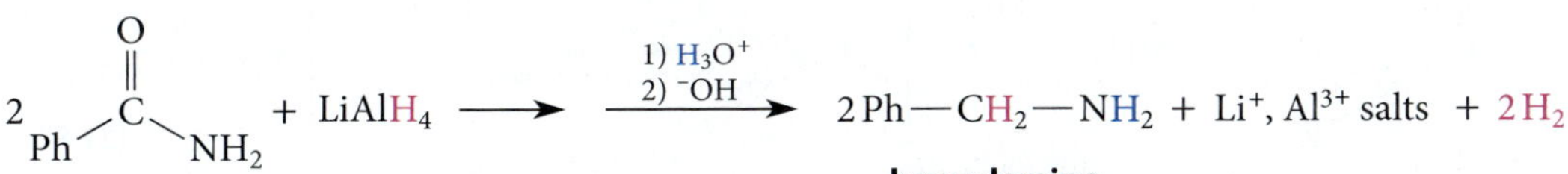

benzamide **benzylamine** (80% yield) (21.78)

In the workup conditions, H_3O^+ is followed by ^-OH. An aqueous acidic solution is often used to carry out the protonolysis step that follows the $LiAlH_4$ reduction (as shown in the mechanism discussed subsequently). The excess of acid that is typically used converts the amine, which is a base, into its conjugate-acid ammonium ion. Hydroxide is then required to neutralize this ammonium salt and give the neutral amine.

$$^-OH + RCH_2\overset{+}{N}H_3 \rightleftharpoons RCH_2\ddot{N}H_2 + H_2O$$

conjugate-acid ammonium ion (typical pK_a = 8–11); amine; (pK_a = 14.0) (21.79)

Water itself, rather than acid, can be used in the protonolysis step, but the acidic workup is more convenient for experimental reasons. Consequently, the extra neutralization step is required.

Amide reduction can be used not only to prepare primary amines from primary amides, but also to prepare secondary and tertiary amines from secondary and tertiary amides, respectively.

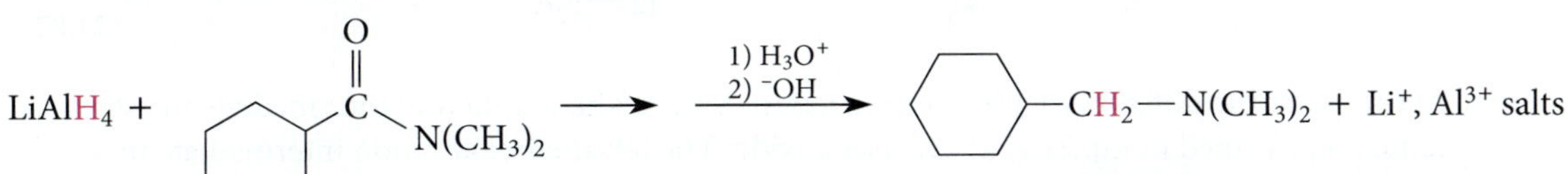

***N,N*-dimethylcyclohexanecarboxamide** a tertiary amide; a tertiary amine (88% yield) (21.80)

The reaction of $LiAlH_4$ with an amide differs from its reaction with an ester. In the reduction of an ester, the *carboxylate oxygen* is lost as a leaving group. If amide reduction were strictly analogous to ester reduction, the nitrogen would be lost, and a primary alcohol would be formed. Instead, it is the *carbonyl oxygen* that is lost in amide reduction.

Ester reduction:

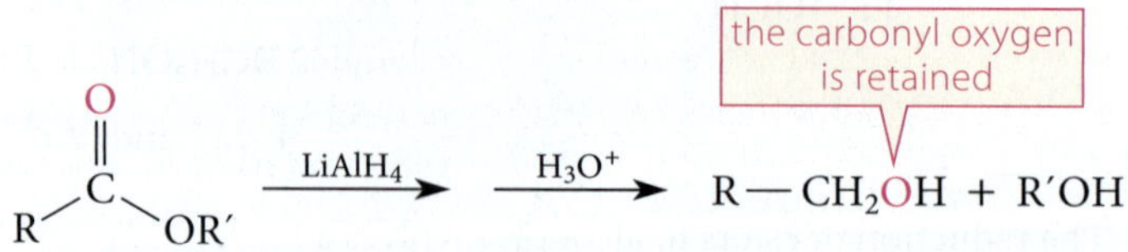

(21.81a)

Amide reduction:

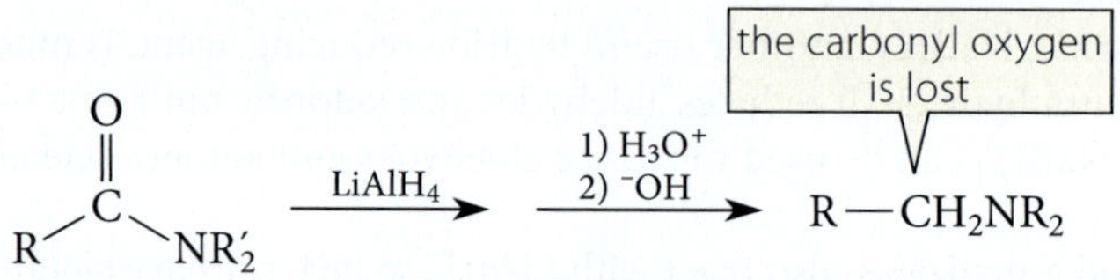

(21.81b)

Let's consider the reason for this difference, using as a case study the reduction of a secondary amide. (The mechanisms of the reduction of primary and tertiary amides are somewhat different, but they have the same result.)

In the first step of the mechanism, the weakly acidic amide proton reacts with an equivalent of hydride, a strong base, to give hydrogen gas, AlH_3, and the lithium salt of the amide.

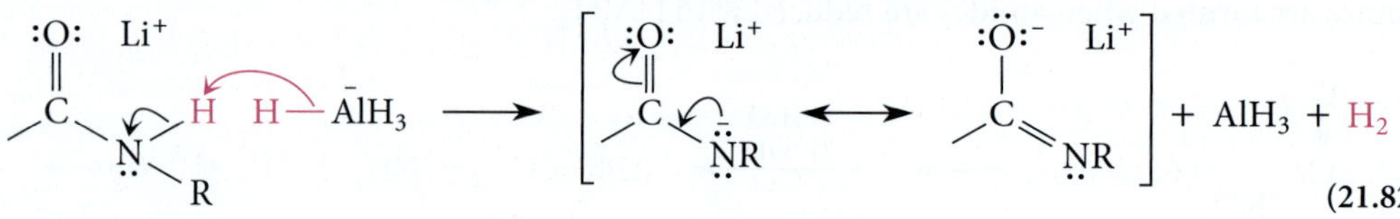

(21.82a)

The lithium salt of the amide, a Lewis base, reacts with the Lewis acid AlH_3 in a Lewis acid–base association.

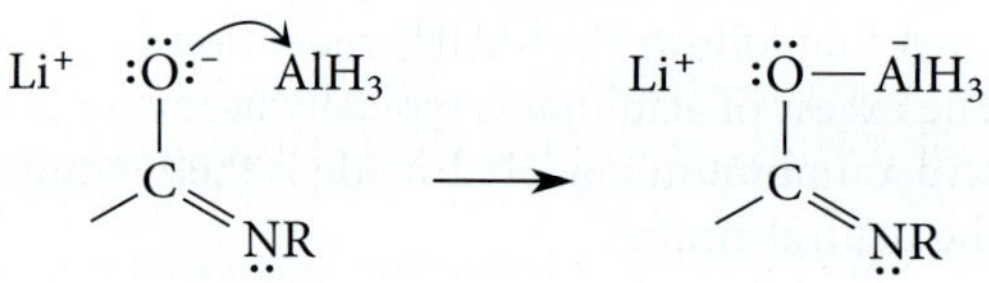

(21.82b)

The resulting species is an active hydride reagent, conceptually much like $LiAlH_4$, and it can deliver hydride to the C═N double bond.

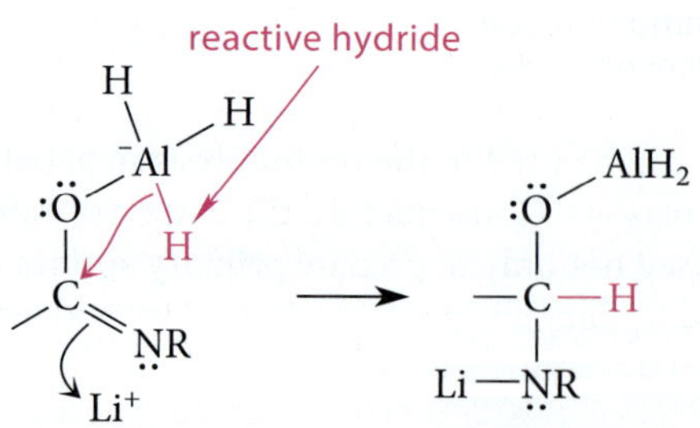

(21.82c)

At this point the difference with ester reduction occurs. The tetrahedral intermediate in ester reduction (formed in Eq. 21.77a) expels alkoxide. The tetrahedral addition intermediate in Eq. 21.82c can expel either of two leaving groups: the $^{-}OAlH_2$ group or the $RN^-\ Li^+$ group. The $^{-}OAlH_2$ group is *much* less basic than the nitrogen leaving group by 20 pK_a units or more. Therefore, the $^{-}OAlH_2$ group is expelled to give an *imine* (Sec. 19.12A).

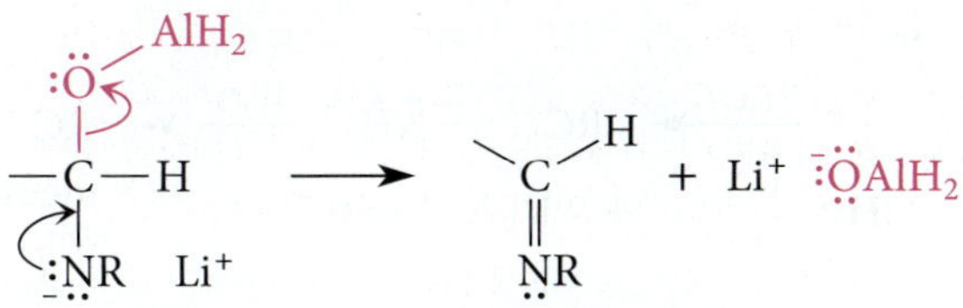

an imine (21.82d)

The C═N of the imine, like the C═O of an aldehyde, undergoes nucleophilic addition with "H:⁻" from $^{-}AlH_4$ or from one of the other hydride-containing species in the reaction mixture. Addition of acid to the reaction mixture converts the addition intermediate into an amine by protonolysis and then into its conjugate-acid ammonium ion.

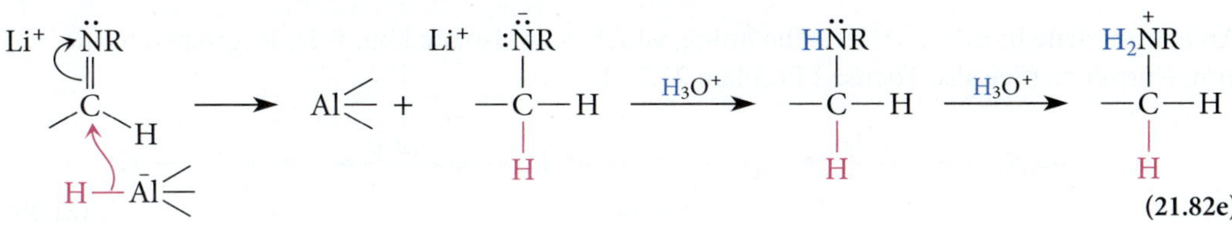

(21.82e)

The ammonium ion is neutralized to the free amine when ^{-}OH is added in a subsequent step (Eq. 21.81b).

C. Reduction of Nitriles to Primary Amines

Nitriles are reduced to primary amines by reaction with $LiAlH_4$, followed by the usual protonolysis step.

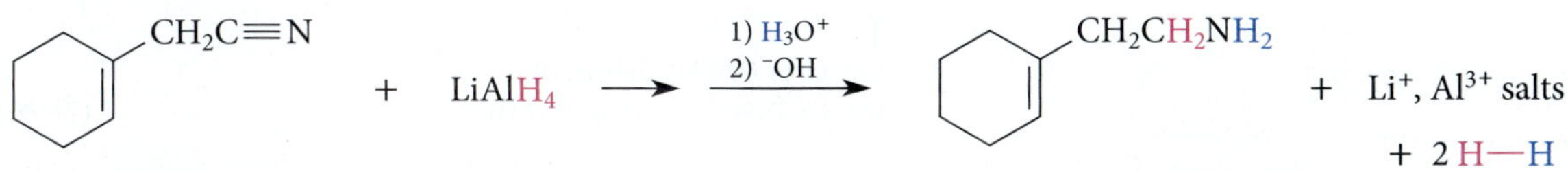

2-(1-cyclohexenyl)eth-anenitrile **2-(1-cyclohexenyl)ethanamine** (74% yield) (21.83)

As in amide reduction, isolation of the neutral amine requires addition of ^{-}OH at the conclusion of the reaction.

The mechanism of this reaction illustrates again how the C≡N and C═O bonds react in similar ways. This reaction probably occurs as two successive *nucleophilic additions.*

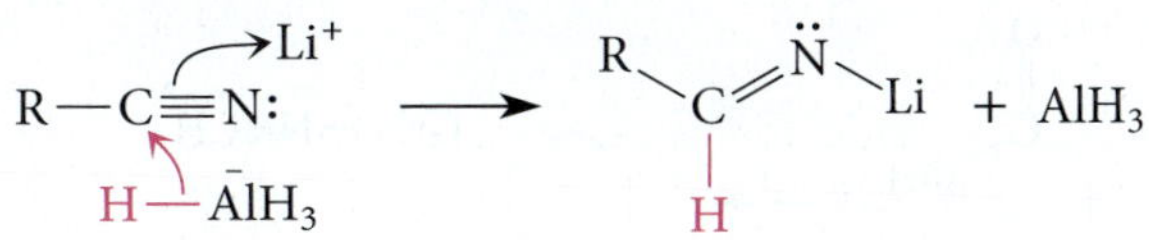

imine salt (21.84a)

In the second addition, the imine salt reacts in a similar manner with AlH_3 (or another equivalent of $^{-}AlH_4$).

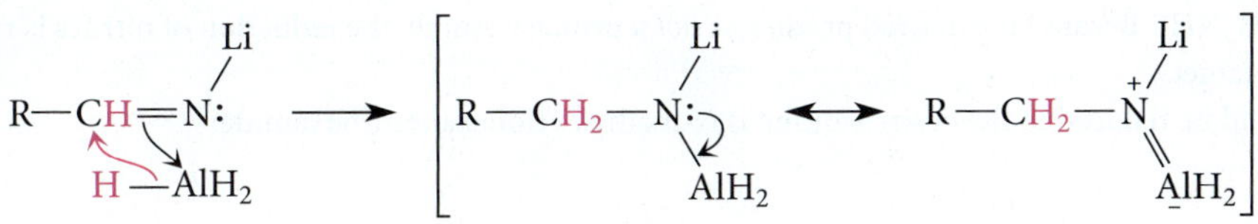

(21.84b)

In the resulting derivative, both the N—Li and the N—Al bonds are very polar, and the nitrogen has a great deal of anionic character. Both bonds are susceptible to protonolysis, as are the remaining Al—H bonds. Therefore, an amine, and then an ammonium ion, is formed when aqueous acid is added to the reaction mixture.

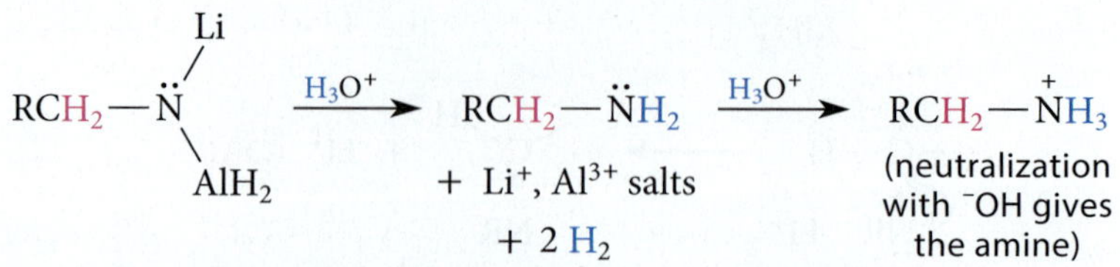

(21.84c)

Nitriles are also reduced to primary amines by catalytic hydrogenation using Raney nickel, a type of nickel–aluminum alloy.

$$CH_3(CH_2)_4—C{\equiv}N + 2H_2 \xrightarrow[\text{2000 psi, 120–130 °C}]{\text{Raney Ni}} CH_3(CH_2)_4—CH_2NH_2 \quad (21.85a)$$

hexanenitrile → **1-hexanamine**

An intermediate in the reaction is the imine, which is not isolated but is hydrogenated to the amine product. (See also Focused Problem 21.21.)

$$R—C{\equiv}N \xrightarrow{H_2,\text{ catalyst}} [R—CH{=}NH] \xrightarrow{H_2,\text{ catalyst}} R—CH_2—NH_2 \quad (21.85b)$$

imine

The reductions discussed in this and the previous section describe the formation of the *amine* functional group from amides and nitriles, the nitrogen-containing carboxylic acid derivatives. Therefore, any synthesis of a carboxylic acid can be used as part of an amine synthesis, but it is important to remember that the amine prepared by these methods must have the following form:

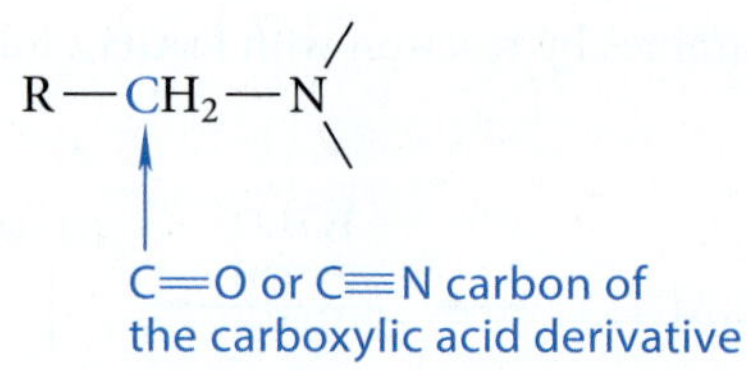

(21.86)

As this diagram shows, the carbon of the carbonyl group or cyano group in the carboxylic acid derivative ends up as a —CH_2— group adjacent to the amine nitrogen.

Study Problem 21.1

Outline a synthesis of (cyclohexylmethyl)methylamine from cyclohexanecarboxylic acid.

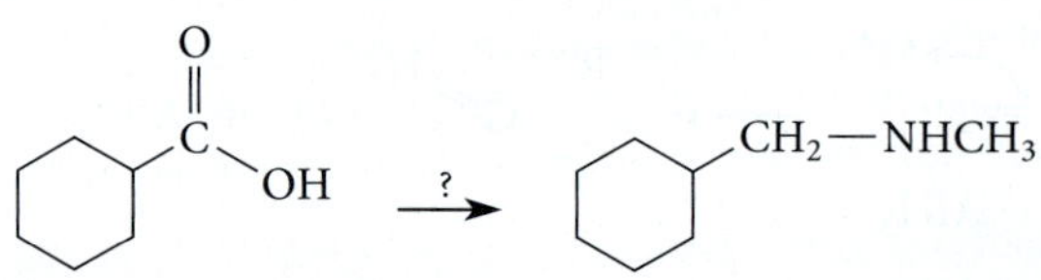

cyclohexanecarboxylic acid **(cyclohexylmethyl)methylamine**

Solution Any carboxylic acid derivative used to prepare the amine must contain nitrogen; the two such derivatives are amides and nitriles. However, the only type of amine that can be prepared directly by nitrile reduction is a primary amine of the form RCH_2NH_2. Because the desired product is *not* a primary amine, the reduction of nitriles is ruled out as an approach to this target.

The amide that could be reduced to the desired amine is *N*-methylcyclohexanecarboxamide:

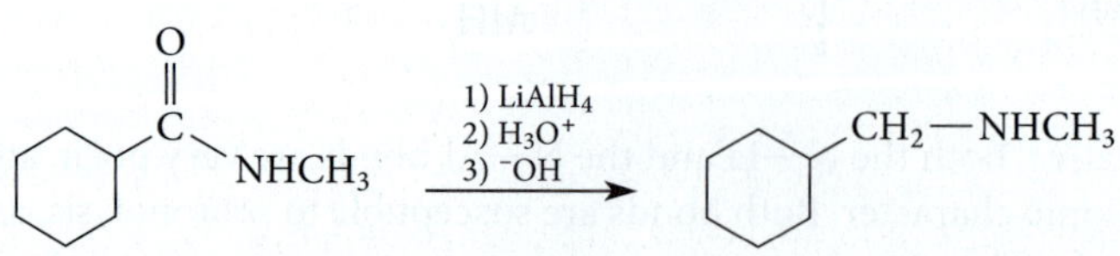

***N*-methylcyclohexanecarboxamide**

This amide can be prepared, in turn, by reaction of methylamine with an acid chloride:

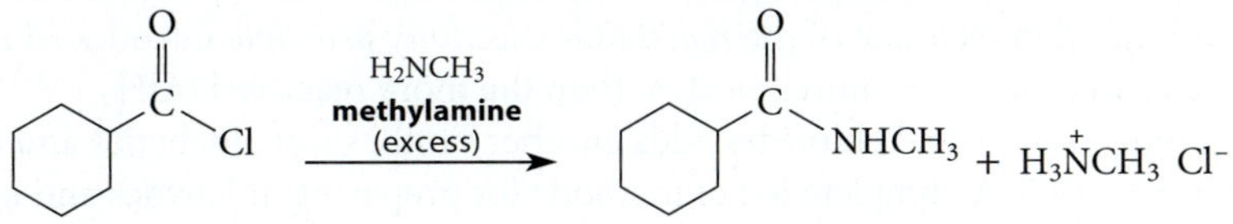

cyclohexanecarbonyl chloride

Finally, the acid chloride is prepared from the carboxylic acid (Sec. 20.9A).

D. Reduction of Acid Chlorides to Aldehydes

Acid chlorides can be reduced to aldehydes by either of two procedures. In the first, the acid chloride is hydrogenated over a catalyst that has been deactivated, or *poisoned*, with an amine, such as quinoline, that has been heated with sulfur. (Amines and sulfides are catalyst poisons.) This reaction is called the **Rosenmund reduction**.

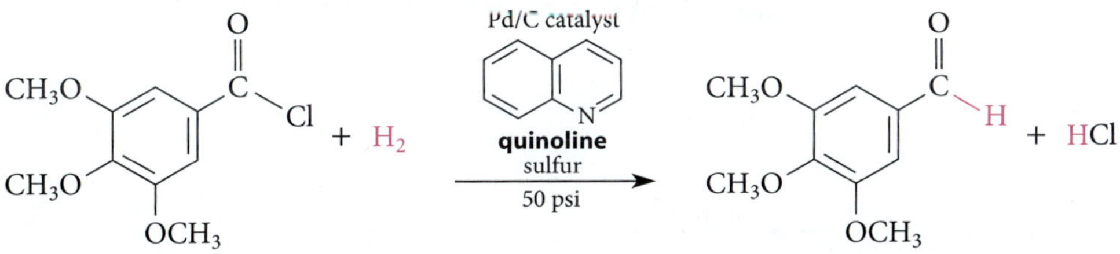

3,4,5-trimethoxybenzoyl chloride **3,4,5-trimethoxybenzaldehyde** (54–83% yield) (21.87)

The poisoning of the catalyst prevents further reduction of the aldehyde product.

A second method of converting acid chlorides into aldehydes is the reaction of an acid chloride at low temperature with lithium tri(*tert*-butoxy)aluminum hydride, a "cousin" of $LiAlH_4$.

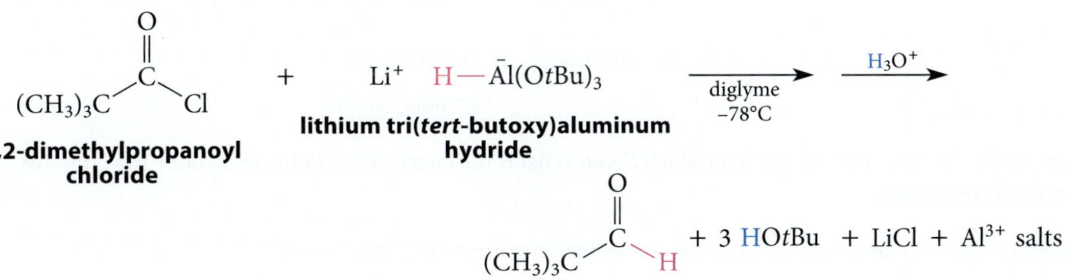

2,2-dimethylpropanal (21.88)

The hydride reagent used in this reduction is derived by the replacement of three hydrogens of lithium aluminum hydride by *tert*-butoxy groups. As the hydrides of $LiAlH_4$ are replaced successively with alkoxy groups, less reactive reagents are obtained. In fact, the preparation of $LiAlH(OtBu)_3$ owes its success to the poor reactivity of its hydride: the reaction of $LiAlH_4$ with *tert*-butyl alcohol stops after 3 moles of alcohol have been consumed.

$$Li^+ \; {}^-AlH_4 + 3tBu—O—H \longrightarrow Li^+ \; H—\bar{A}l(O—tBu)_3 + 3H—H \qquad (21.89)$$

The one remaining hydride reduces only the most reactive functional groups. Because *acid chlorides are more reactive than aldehydes toward nucleophiles*, the reagent reacts preferentially with the acid chloride reactant rather than with the product aldehyde. In contrast, lithium aluminum hydride is so reactive that it fails to discriminate to a useful degree between the aldehyde and acid chloride groups, and it reduces acid chlorides to primary alcohols.

This is another example of the *reactivity–selectivity principle* introduced in Sec. 10.4B. The less reactive $LiAlH(OtBu)_3$ is more selective than the more reactive $LiAlH_4$.

The reduction of acid chlorides adds another synthesis of aldehydes and ketones to those given in Sec. 19.5. A complete list of methods for preparing aldehydes and ketones is given in Appendix V.

Focused Problems

21.18 Show how benzoyl chloride can be converted into each of the following compounds.

(a) Benzaldehyde (b) 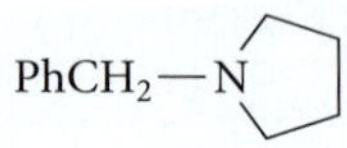

21.19 Complete the following reactions by giving the principal organic product(s).

(a) $PhCH_2C{\equiv}N + H_2 \xrightarrow[\text{heat}]{\text{Raney Ni (catalyst)}}$

(b) 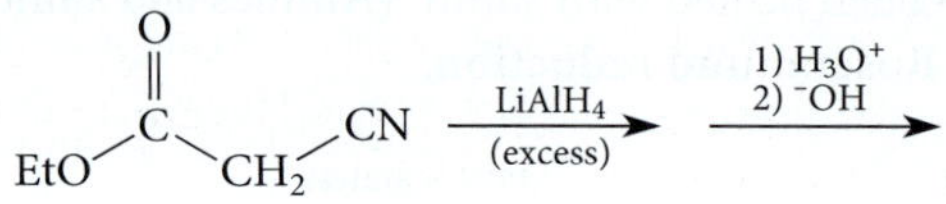

(c) 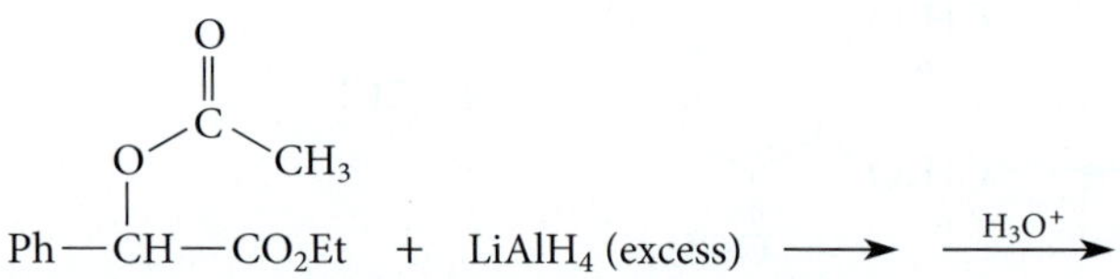

21.20 Give the structures of two compounds that would give the following amine after $LiAlH_4$ reduction.

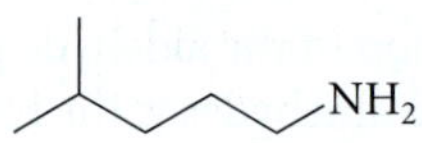

21.21 (a) In the catalytic hydrogenation of some nitriles to primary amines, secondary amines are obtained as by-products:

$$R{-}C{\equiv}N \xrightarrow{H_2/\text{catalyst}} RCH_2NH_2 + \underset{\text{a secondary amine}}{(RCH_2)_2NH}$$

Suggest a mechanism for the formation of this by-product. (*Hint:* What is the intermediate in the reduction? How can this intermediate react with an amine?)

(b) Explain why ammonia added to the reaction mixture prevents the formation of this by-product.

E. Relative Reactivities of Carbonyl Compounds

Recall that the reaction of $LiAlH_4$ with a carboxylic acid (Sec. 20.10) or ester (Sec. 21.9A) involves an aldehyde intermediate. But the product of such a reaction is a primary alcohol, not an aldehyde, because *the aldehyde intermediate is more reactive than the acid or ester*. The instant a small amount of aldehyde is formed, it is in competition with the remaining acid or ester for the $LiAlH_4$ reagent. Because it is more reactive, the aldehyde reacts faster than the remaining ester reacts. Therefore, the aldehyde cannot be isolated under such circumstances. On the other hand, the $LiAlH(OtBu)_3$ reduction of acid chlorides (Sec. 21.9D) can be stopped at the aldehyde because acid chlorides are more reactive than aldehydes. When the aldehyde is formed as a product, it is in competition with the remaining acid chloride for the hydride reagent. Because the acid chloride is more reactive, it is consumed before the aldehyde has a chance to react.

These examples show that the outcomes of many reactions of carboxylic acid derivatives are determined by the *relative reactivities of carbonyl compounds* toward nucleophilic reagents, which can be summarized as follows. (Nitriles are included as "honorary carbonyl compounds.")

Relative reactivities of carbonyl compounds:

nitriles < amides < esters, acids << ketones < aldehydes < acid chlorides

increasing reactivity → (21.90)

The explanation of this reactivity order is the same one used in Sec. 21.7E—namely, the stability of each type of carbonyl compound relative to its transition state for addition or substitution determines relative reactivity. *The more a compound is stabilized, the less reactive it is; the more a transition state for nucleophilic addition or substitution is stabilized, the more reactive the compound is* (see Fig. 21.5, Sec. 21.7E). For example, esters are stabilized by resonance (Eq. 21.49, Sec. 21.7E) in a way that aldehydes and ketones are not. Therefore, esters are less reactive than aldehydes. In contrast, resonance stabilization of acid chlorides is much less important, and acid chlorides are destabilized by the electron-attracting polar effect of the chlorine. Moreover, the transition-state energies for nucleophilic substitution reactions of acid chlorides are lowered by the polar effect of chlorine. For these reasons, acid chlorides are more reactive than aldehydes, in which these effects of the chlorine are absent.

21.10 REACTIONS OF CARBOXYLIC ACID DERIVATIVES WITH ORGANOMETALLIC REAGENTS

A. Reaction of Esters with Grignard and Organolithium Reagents

Most carboxylic acid derivatives react with Grignard or organolithium reagents. One of the most important reactions of this type is the reaction of esters with Grignard reagents. In this reaction, a tertiary alcohol is formed after protonolysis. (Secondary alcohols are formed from esters of formic acid; see Focused Problem 21.23a.)

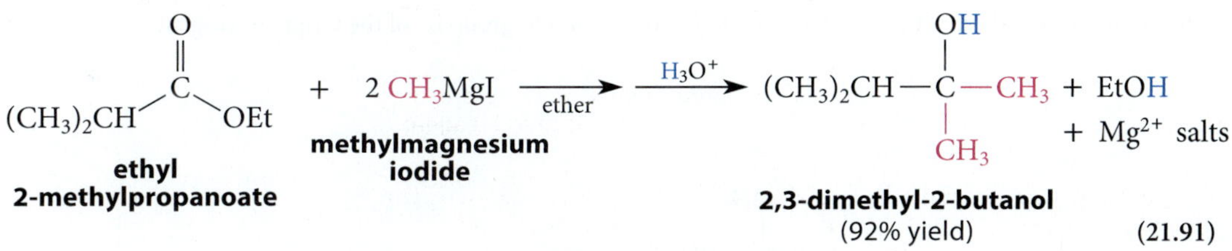

(21.91)

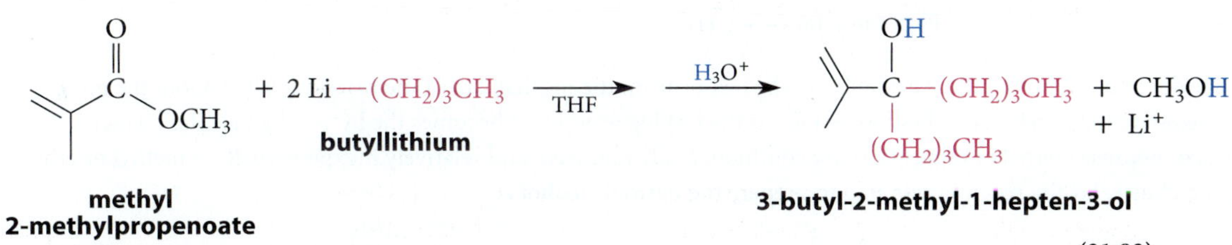

(21.92)

Two moles of organometallic reagent react per mole of ester, and a second alcohol is produced in the reaction (ethanol and methanol in Eqs. 21.91 and 21.92, respectively). Recall from Sec. 21.9A that a similar situation occurs in the $LiAlH_4$ reduction of esters. This alcohol is typically not the one of interest and is discarded as a by-product.

Like the $LiAlH_4$ reduction of esters, this reaction is a nucleophilic acyl substitution followed by an addition. A ketone is formed in the substitution step. (Fill in the details of the mechanism.)

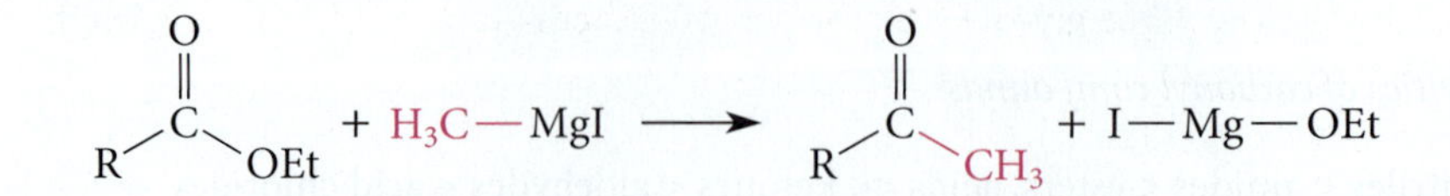

(21.93a)

The ketone intermediate is not isolated because *ketones are more reactive than esters toward nucleophilic reagents* (Eq. 21.90, Sec. 21.9E). Consequently, the ketone reacts with a second equivalent of the Grignard reagent to form a halomagnesium alkoxide, which, after protonolysis, gives the alcohol (Sec. 19.10).

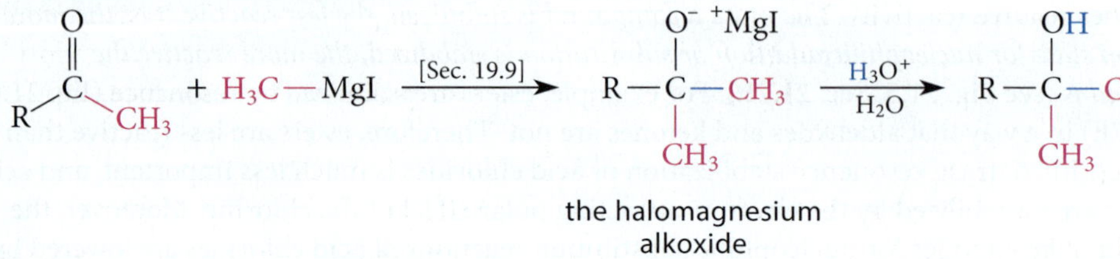

(21.93b)

Study Problem 21.2

From what ester and Grignard reagent could 3-methyl-3-pentanol be prepared by a single reaction, followed by protonolysis?

Solution Rephrase the problem in terms of structures:

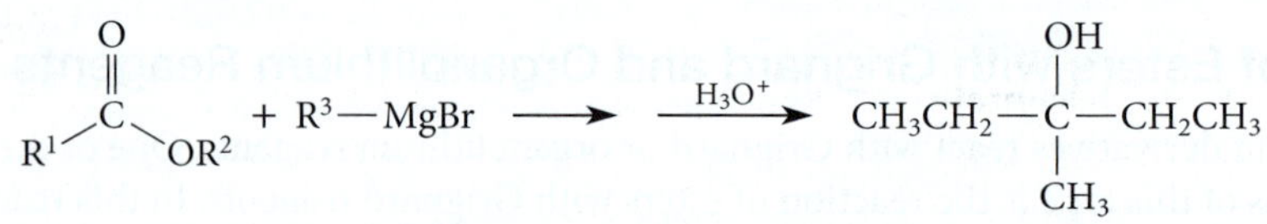

What choices should be made for R^1, R^2, and R^3? There are two keys to solving this problem. First, the carbonyl carbon of the ester starting material becomes the α-carbon of the target alcohol; therefore, R^1 must be attached to this carbon. Second, the two *identical* groups on the α-carbon of the target alcohol *must* correspond to group R^3 of the Grignard reagent.

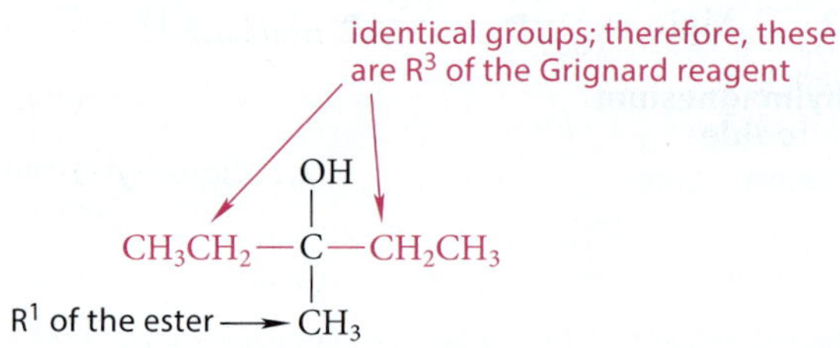

These deductions follow from the examples in the text or from the mechanism of the reaction. What about the group R^2 in the ester? It doesn't matter, because —OR^2 is the leaving group that becomes the by-product alcohol, which is discarded. Because methyl or ethyl esters are common, easily removed, and relatively inexpensive, R^2 = methyl or ethyl is a good choice. So, the reaction required to prepare the desired alcohol is

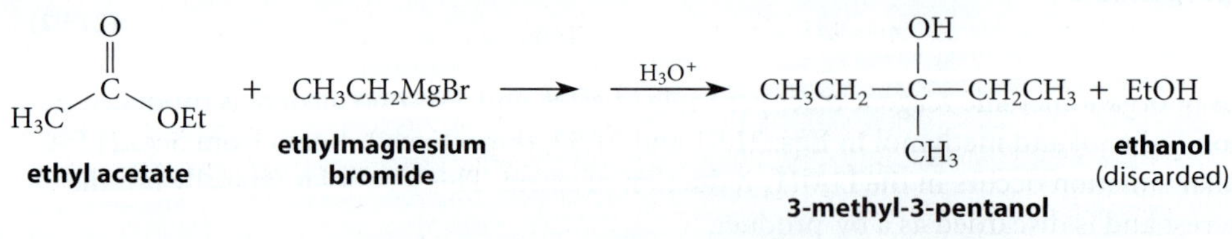

As Study Problem 21.2 demonstrates, the reaction of a Grignard reagent with an ester is an important way to prepare alcohols in which at least two of the groups on the α-carbon are identical. (A complete list of methods for preparing alcohols is found in Appendix V.)

B. Reaction of Acid Chlorides with Lithium Dialkylcuprates

Because acid chlorides are more reactive than ketones, the reaction of an acid chloride with a Grignard reagent can in principle give a ketone without further reaction of the ketone itself.

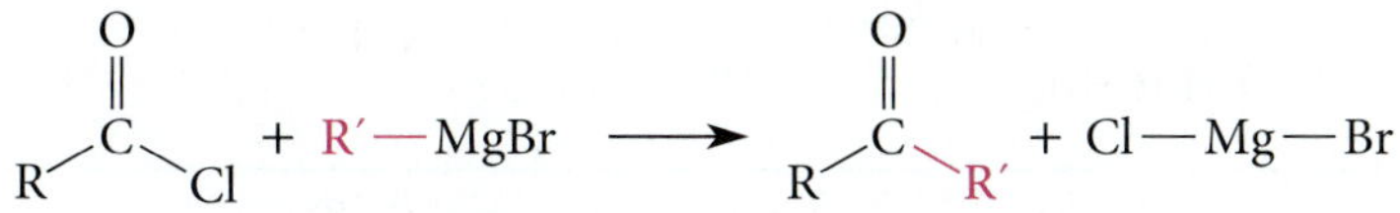

(21.94)

However, Grignard reagents are so reactive that this transformation is difficult to achieve in practice without careful control of the reaction conditions—that is, it is hard to prevent the further reaction of the product ketone with the Grignard reagent to give an alcohol. However, we can exploit the *reactivity-selectivity principle* (Sec. 10.4B) by using a less reactive organometallic reagent—a lithium dialkylcuprate ($R_2Cu^- \, Li^+$)—to bring about the conversion of acid chlorides into ketones. We showed in Sec. 12.5C that these reagents are prepared from organolithium reagents and copper(I) chloride:

$$2R{-}Li + CuCl \longrightarrow \underset{\text{lithium dialkylcuprate}}{R{-}\bar{C}u{-}R \;\; Li^+} + Li^+ \, Cl^- \qquad (21.95)$$

Lithium dialkylcuprates typically react readily with acid chlorides, aldehydes, and epoxides, very slowly with ketones, and not at all with esters. The reaction of lithium dialkylcuprates with acid chlorides gives ketones in excellent yield.

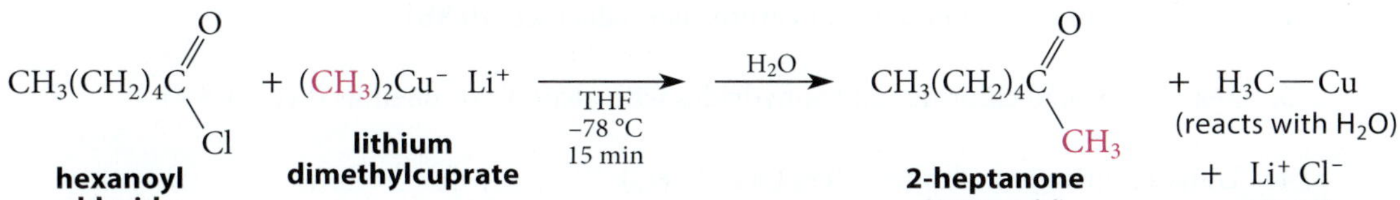

(21.96)

Because ketones are much less reactive than acid chlorides toward lithium dialkylcuprates, they do not react further.

The reaction of acid chlorides with lithium dialkylcuprates is one of several good methods for the preparation of ketones. Be sure to review the others in Appendix V. Both this reaction and the reaction of Grignard reagents with esters also provide additional methods for the formation of carbon–carbon bonds. We urge you to review the other reactions used for carbon–carbon bond formation, found in Appendix VI.

Focused Problems

21.22 Suggest a sequence of reactions for carrying out each of the following conversions.

(a) Benzoic acid to Ph_3C—OH (triphenylmethanol)

(b) Butyric acid to 3-methyl-3-hexanol

(c) Isobutyronitrile to 2,3-dimethyl-2-butanol (two ways)

(d) Propionic acid to 3-pentanone

21.23 (a) What is the general structure of the alcohols obtained by the reaction of Grignard reagents of the form RMgBr with ethyl formate, followed by protonolysis?

(b) Outline a synthesis of 3-pentanol from ethyl formate and a Grignard reagent.

21.24 Predict the product when each of the following compounds reacts with one equivalent of lithium dimethylcuprate, followed by protonolysis. Explain.

(a) 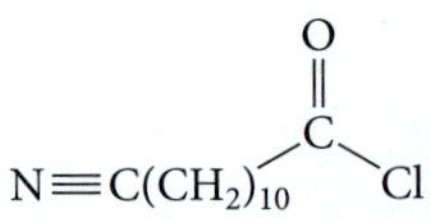 (b) 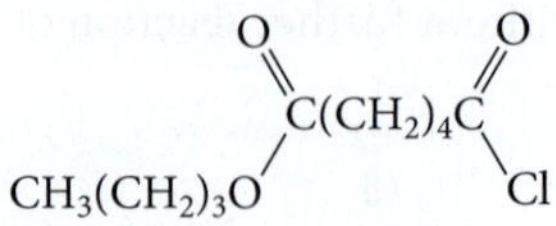

21.11 CARBOXYLIC ACID DERIVATIVES IN ORGANIC SYNTHESIS

A. Synthesis of Carboxylic Acid Derivatives

Reactions in this and the previous chapter demonstrate that many syntheses of carboxylic acid derivatives begin with other carboxylic acid derivatives. Let's review the methods that have been covered:

Synthesis of esters:

1. Acid-catalyzed esterification of carboxylic acids (Sec. 20.8A)
2. Alkylation of carboxylic acids or carboxylate salts (Sec. 20.8B)
3. Reaction of acid chlorides and anhydrides with alcohols or phenols (Sec. 21.8A)
4. Transesterification of other esters (Sec. 21.8C)

Synthesis of acid chlorides:

Reaction of carboxylic acids with $SOCl_2$ or PCl_5 (Sec. 20.9A)

Synthesis of anhydrides:

1. Reaction of carboxylic acids with dehydrating agents (Sec. 20.9B)
2. Reaction of acid chlorides with carboxylate salts (Sec. 21.8A)

Synthesis of amides:

Reaction of acid chlorides, anhydrides, or esters with amines (Secs. 21.8A and C)

Synthesis of nitriles:

The synthesis of nitriles is an important exception to the generalization that carboxylic acid derivatives are usually prepared from other carboxylic acid derivatives. Two syntheses of nitriles are the following:

1. Cyanohydrin formation (Sec. 19.8A)
2. S_N2 reaction of cyanide ion with alkyl halides or sulfonate esters (Sec. 9.4)

The S_N2 reaction was discussed thoroughly in Sec. 9.4, and the reaction of alkyl halides with cyanide ion is used as an example in Table 9.1 (Sec. 9.1A). Let's now focus on that reaction as part of a useful organic synthesis. An S_N2 reaction of cyanide ion, like all S_N2 reactions, requires a primary or unbranched secondary alkyl halide or sulfonate ester, as in Eqs. 21.97 and 21.98.

$$\underset{\textbf{benzyl chloride}}{PhCH_2Cl} + Na^{+}\ {}^{-}\!:C\equiv N \xrightarrow[EtOH/H_2O]{} \underset{\substack{\textbf{phenylacetonitrile}\\ \text{(80–90\% yield)}}}{PhCH_2C\equiv N} + Na^{+}\ Cl^{-} \qquad (21.97)$$

$$\underset{\textbf{1,3-dibromopropane}}{Br(CH_2)_3Br} + 2\ Na^{+}\ {}^{-}\!:C\equiv N \xrightarrow[EtOH/H_2O]{} \underset{\substack{\textbf{glutaronitrile}\\ \text{(77–86\% yield)}}}{N\equiv C(CH_2)_3C\equiv N} + 2\ Na^{+}\ Br^{-} \qquad (21.98)$$

Both the S_N2 reaction of cyanide and the synthesis of cyanohydrins are noteworthy because they provide additional ways to form carbon–carbon bonds. See how many reactions you can list that form carbon–carbon bonds; check yourself against the summary in Appendix VI.

Because nitriles are prepared from compounds other than carboxylic acid derivatives, the preparation of a nitrile can be particularly useful as an intermediate step in the preparation of a carboxylic acid. Study Problem 21.3 illustrates this approach.

Study Problem 21.3

Outline a synthesis of pentanoic acid (valeric acid) from 1-butanol.

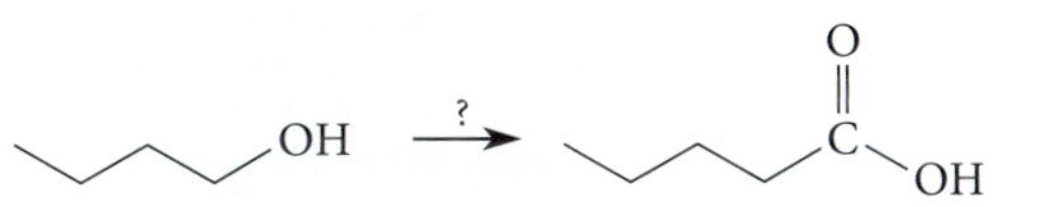

1-butanol **pentanoic acid**

Solution A new carbon–carbon bond must be formed at some point in this synthesis. Because nitriles can be hydrolyzed to carboxylic acids, the immediate precursor of the carboxylic acid can be the corresponding nitrile.

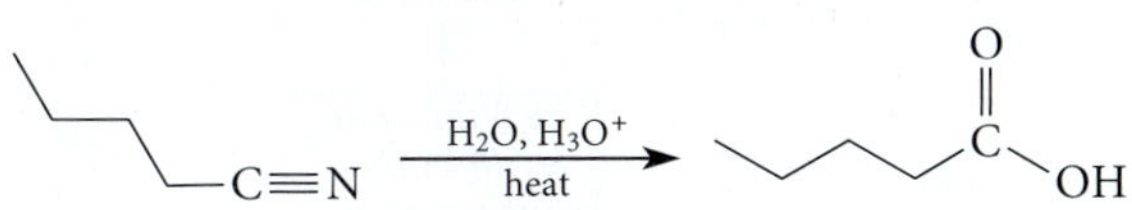

The nitrile, in turn, can be prepared from the corresponding alkyl halide:

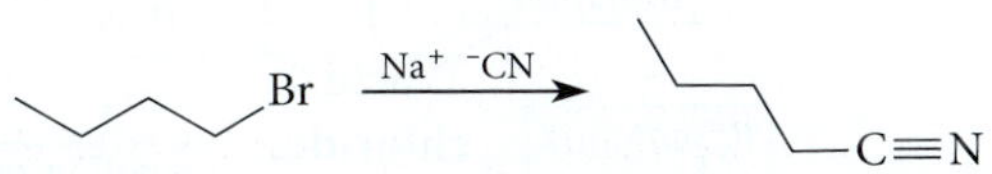

The new carbon–carbon bond is formed at this point. The alkyl halide is formed from the alcohol by any of the methods summarized in Sec. 11.5, such as by reaction with concentrated HBr:

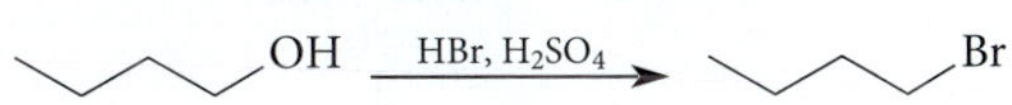

This alkyl halide could also be converted into the target carboxylic acid by converting it into a Grignard reagent and then treating the Grignard reagent with CO_2 (Sec. 20.6). That method is particularly valuable when we must use an alkyl halide that is relatively unreactive in an S_N2 reaction, a vinylic halide, or an aryl halide.

Focused Problems

21.25 Outline two methods for the preparation of 5-methylhexanoic acid from 1-bromo-4-methylpentane. (See Sec. 20.6.)

21.26 From what nitrile can 2-hydroxypropanoic acid (lactic acid) be prepared? How can this nitrile be prepared from acetaldehyde? (*Hint:* See Sec. 19.8A.)

B. Use of Carboxylic Acid Derivatives in Organic Synthesis

Carboxylic acid derivatives have a very important role in organic synthesis because many of them are cheap and readily available. Also, they can be used to prepare other types of organic compounds, which themselves are very useful as synthetic intermediates. Summary diagrams for the uses of esters, acid chlorides, amides, and nitriles as synthetic starting materials are given in Displays 21.99–29.101. Esters and amides are shown as starting materials for carboxylic acids because a number of esters and amides are available from natural sources. Nitriles are shown as a source of carboxylic acids because they are typically prepared from other types of compounds. However, most amides and esters, and virtually all acid chlorides and anhydrides, are prepared from carboxylic acids.

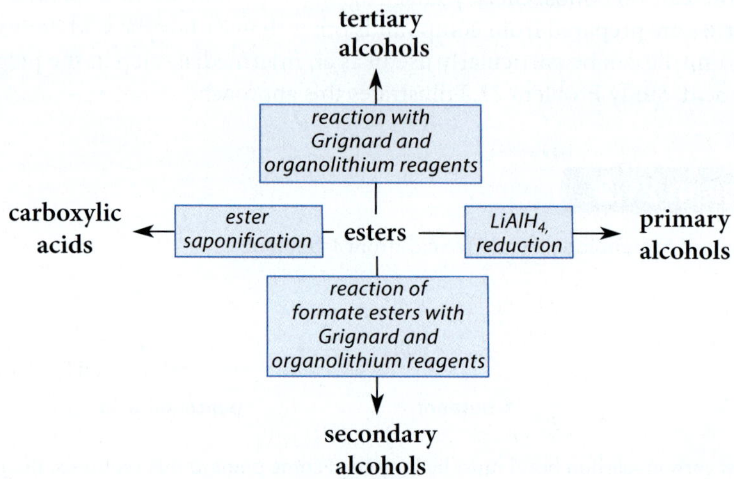

(21.99)

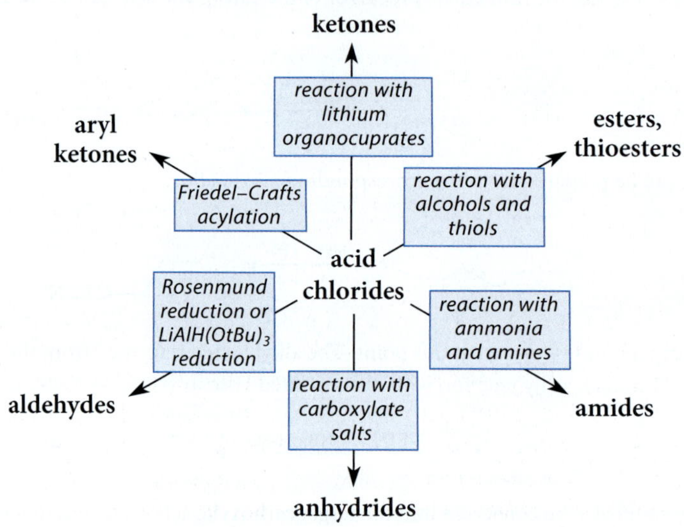

(21.100)

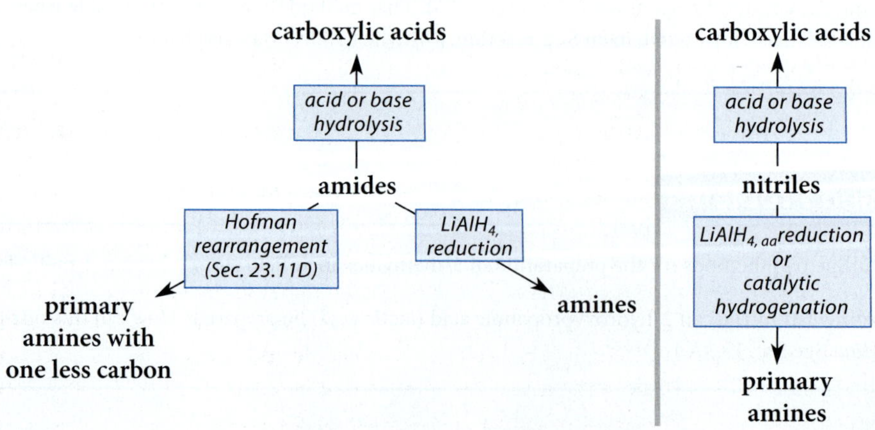

(21.101)

21.12 USE AND OCCURRENCE OF CARBOXYLIC ACIDS AND THEIR DERIVATIVES

A. Nylon and Polyesters

Two of the most important polymers produced on an industrial scale are *nylon* and *polyesters*. The chemistry of carboxylic acids and their derivatives plays an important role in the synthesis of these polymers.

Nylon is the general name given to a group of polymeric amides, or *polyamides*. The two most widely used are nylon-6,6 and nylon-6.

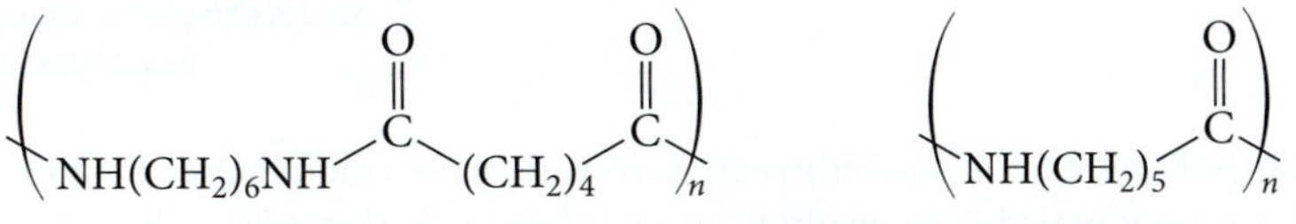

nylon-6,6 **nylon-6** (21.102)

Nylon-6,6 was invented by a DuPont chemist, Wallace H. Carothers (1896–1937), in the early 1930s. About 20 billion pounds of nylon are produced annually worldwide. Nylon is used in tire cord, carpet, and apparel.

The starting material for the industrial synthesis of nylon-6,6 is adipic acid. In one process, adipic acid is converted into its dinitrile and then into 1,6-hexanediamine (hexamethylenediamine).

$$\underset{\textbf{adipic acid}}{HO_2C(CH_2)_4CO_2H} \xrightarrow{\text{several steps}} \underset{\textbf{adiponitrile}}{N{\equiv}C(CH_2)_4C{\equiv}N} \xrightarrow{H_2/\text{catalyst}} \underset{\textbf{hexamethylenediamine}}{H_2N(CH_2)_6NH_2} \qquad (21.103)$$

The hexamethylenediamine product is mixed with more adipic acid to form a salt. Heating the salt forms the polymeric amide.

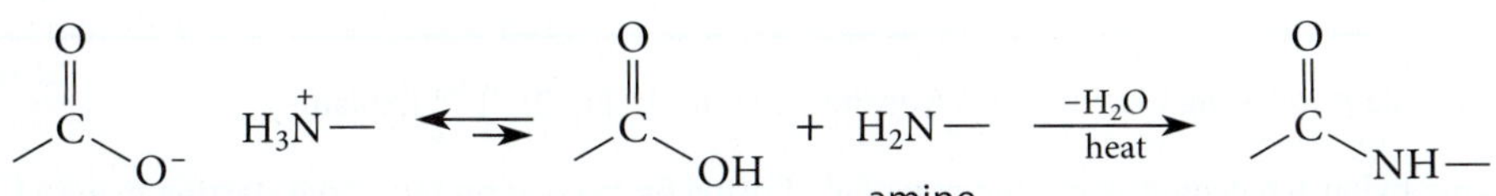

carboxylate salt acid amine nylon (an amide) (21.104)

The reaction of an amine with a carboxylic acid to form an amide is analogous to the reaction of an amine with an ester (Sec. 21.8C). However, much more vigorous conditions are required because the amine is basic, and so the equilibrium on the left of Eq. 21.104 strongly favors the salt. In the salt, the amine is protonated and therefore not nucleophilic, and the carboxylate ion is very unreactive toward nucleophiles. (Why?) The small amount of amine and carboxylic acid in equilibrium with the salt react when the salt is heated, pulling the equilibrium to the right.

The starting material for nylon-6 is ε-caprolactam. (For the structure of ε-caprolactam and its polymerization to nylon-6, see Focused Problem 21.29.) Both adipic acid and ε-caprolactam are prepared from cyclohexanone (see Problem 21.61), which, in turn, is prepared by the oxidation of cyclohexane. Cyclohexane comes from petroleum. So, an important segment of the chemical economy depends on petroleum feedstocks.

Nylon-6,6 is a **condensative-addition polymer**—a polymer formed in a reaction that liberates a small molecule. The polymerization process is called **condensative chain polymerization**, where the term *condensative* refers to the liberation of a small molecule at each step of the polymerization. In the synthesis of nylon-6,6 in Eq. 21.104, for example, formation of each amide bond is accompanied by the loss of the small molecule H_2O. Contrast this type of polymerization with *chain polymerization* (Sec. 10.3B), in which one molecule adds to another without liberation of a by-product small molecule. (Ethylene polymerization is an example of chain polymerization. The polymerization of ε-caprolactam is another example—see Focused Problem 21.29.)

Polyesters are condensative-addition polymers derived from the reaction of diols and dicarboxylic acids. One widely used polyester, poly(ethylene terephthalate), can be produced by the esterification of ethylene glycol and terephthalic acid.

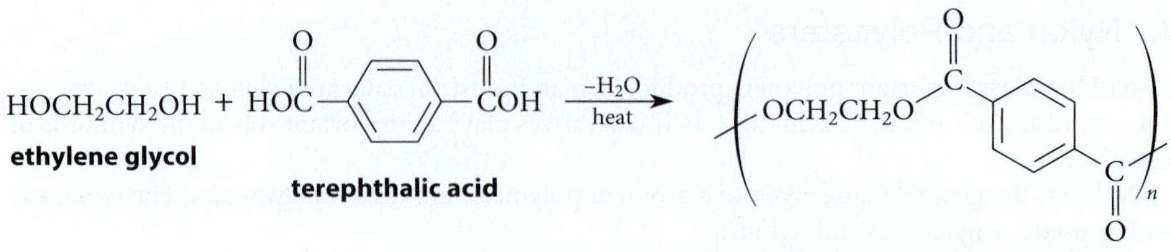

poly(ethylene terephthalate)
(a polyester) **(21.105)**

Some familiar polyester fibers and films are sold under the trade names Dacron and Mylar, respectively. The annual worldwide production of polyester is about 80 million tons. As the synthetic scheme in Eq. 21.106 shows, polyester production also depends on raw materials derived from petroleum.

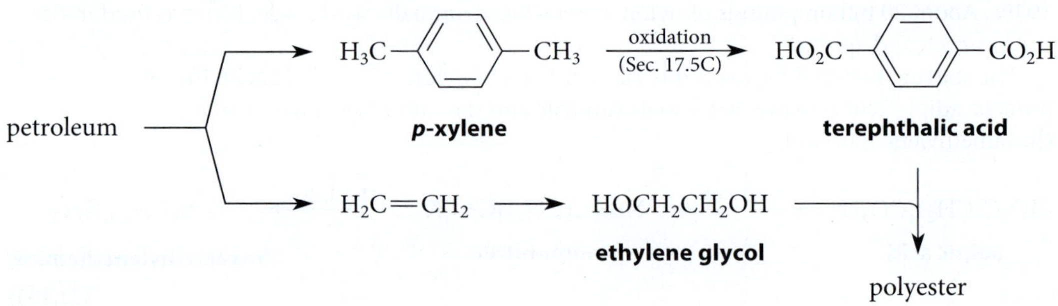

(21.106)

Focused Problems

21.27 Which polymer should be more resistant to strong base: nylon-6,6 or the polyester in Eq. 21.105? Explain.

21.28 One interesting process for making nylon-6,6 demonstrates the potential of using biomass as an industrial starting material. The raw material for this process, outlined in the following reaction, is the aldehyde furfural, obtained from sugars found in oat hulls. Suggest conditions for carrying out each of the steps in this process indicated by the italicized letters *a–f*.

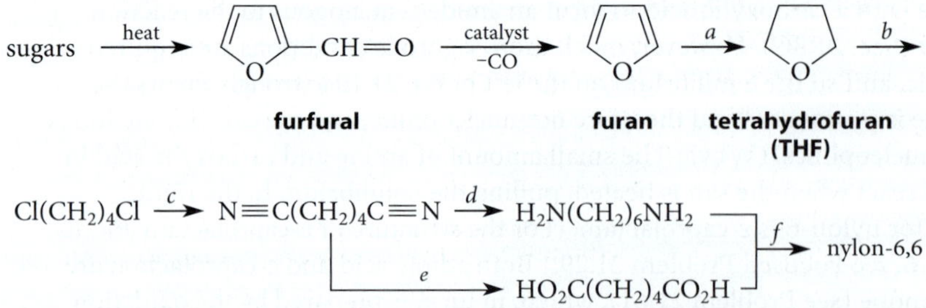

21.29 ε-Caprolactam is polymerized to nylon-6 when it is heated with a *catalytic amount* of water.

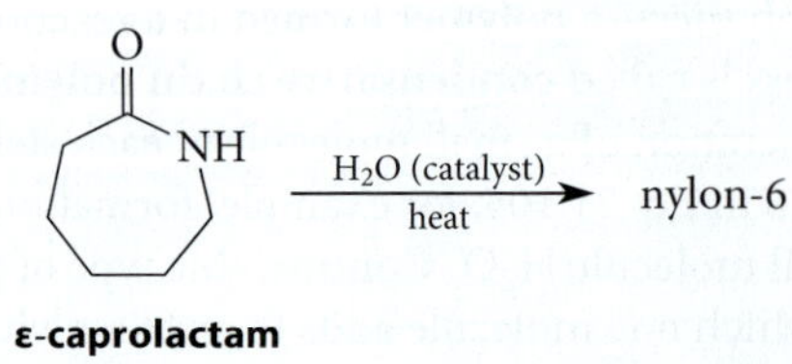

Write a mechanism for the polymerization that clearly shows the role of the water catalyst.

B. Proteins

Proteins are naturally occurring condensative-addition polymers of **alpha-amino acids (α-amino acids)**.

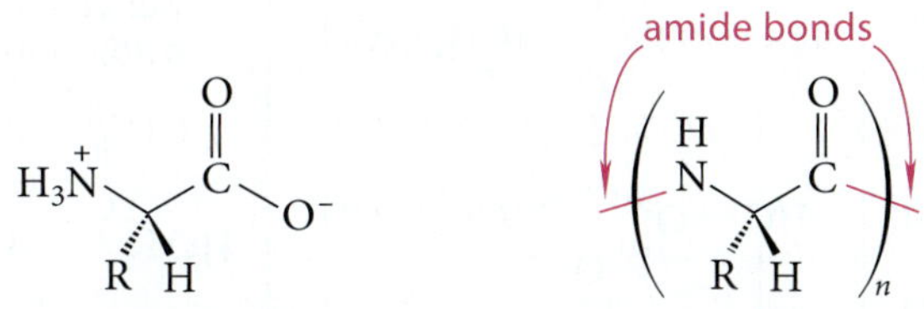

an α-amino acid (the form at pH 7.4)

a polymeric α-amino acid (a protein)

(21.107)

As in nylon, the bonds joining the polymeric units are amide bonds. In fact, the impetus for developing nylon was to produce a "synthetic silk"; silk is a protein. Although a few proteins are polymers of a single α-amino acid, most proteins are **heteropolymers**. This means that each unit of the polymer contains a constant part and a variable part. The variable part of each unit is the side chain of the amino acid. There are 20 different naturally occurring α-amino acids. (See Table 27.1, Sec. 27.1A.)

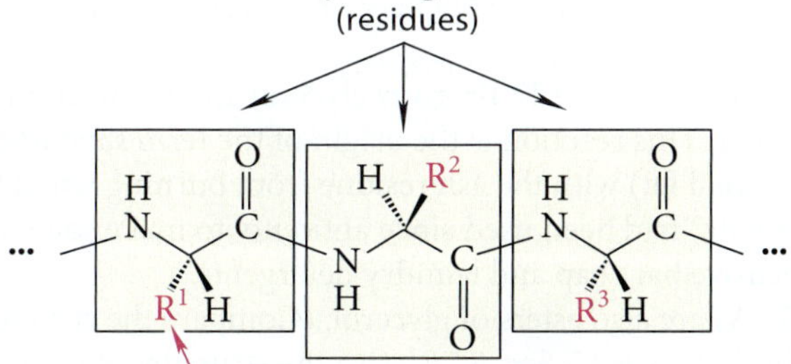

(21.108)

Proteins serve multiple roles in biology. In this and previous chapters, we have focused largely on their role as enzyme catalysts. Chapter 27 is devoted to the structure and stereochemistry of amino acids and proteins, including protein conformation. You have the background to study Chapter 27 at any time.

C. Waxes, Fats, and Phospholipids

Waxes, fats, and phospholipids are all important naturally occurring ester derivatives of fatty acids. A **wax** is an organic compound that consists of large hydrocarbon groups, sometimes modified with functional groups. For example, paraffin wax, sold commercially as "paraffin," is a long unbranched alkane of 30–36 carbons. Many waxes are esters of a fatty acid and a "fatty alcohol," a primary alcohol with a long unbranched carbon chain. For example, carnauba wax, obtained from the leaves of the Brazilian carnauba palm and valued for its hard, brittle characteristics, consists of about 80% of esters derived from C_{24}, C_{26}, and C_{28} fatty acids and C_{30}, C_{32}, and C_{34} alcohols. The following compound is a typical constituent of carnauba wax.

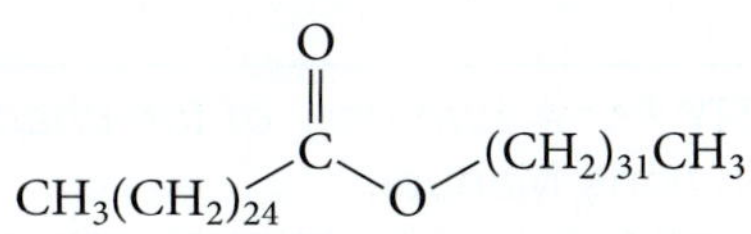

a constituent of carnauba wax

A **fat** is an ester derived from a molecule of glycerol and three molecules of fatty acid.

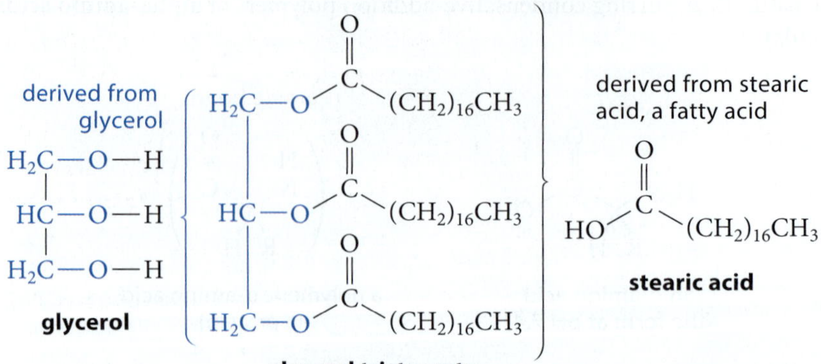

glyceryl tristearate (a typical fat) (21.109)

The three acyl groups in a fat (shown in black Display 21.109) may be the same, as in a glyceryl tristearate, or different, and they may contain unsaturation, which is typically in the form of one or more cis double bonds. Fats with no double bonds, called **saturated fats**, are typically solids; lard is a saturated fat. Fats containing cis double bonds, called **unsaturated fats**, are in many cases oily liquids; olive oil is an unsaturated fat. Fats, which are stored in highly concentrated form in the body, serve as the biological storehouse of energy reserves. (See the sidebar, "Fats and Oils," in Sec. 8.5D.)

The treatment of fats with NaOH or KOH gives glycerol and the sodium or potassium salts of fatty acids (soaps; Sec. 20.5). This reaction is the origin of the term *saponification* (Sec. 21.7A). The treatment of lard (animal fat) with the ash residue from burning wood (potash, an impure form of potassium carbonate) had been used since antiquity to make soap until commercial soap making provided inexpensive bar soap and laundry detergents.

Phospholipids (Sec. 8.7A) are also esters of glycerol. (Compare the structure of phosphatidylethanolamine in Display 8.77, Sec. 8.7B, with the structure of the fat glyceryl tristearate.) The only structural difference between a fat and a phospholipid is that, in a phospholipid, one of the glyceryl primary hydroxy groups is esterified to a polar phosphoric acid derivative rather than to a fatty acid.

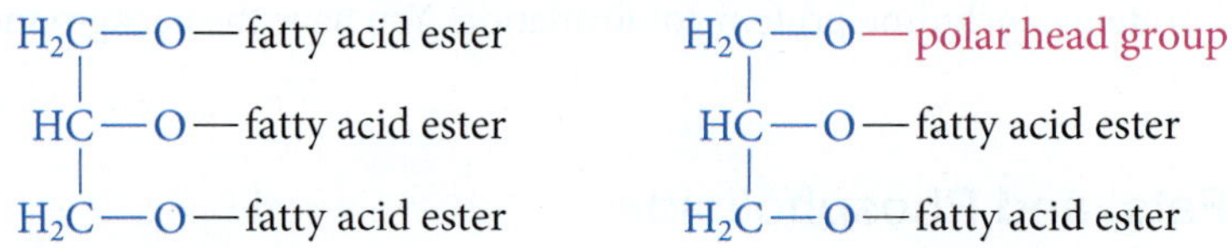

a fat a phospholipid (21.110)

This difference is responsible for the amphipathic behavior of phospholipids. Because fats lack polar head groups, they do not form vesicles when added to water; instead, a fat forms a separate, insoluble layer like any other water-insoluble compound. (If you have ever mixed cooking oil and water, you have observed this behavior.)

CHAPTER SUMMARY *For a summary of the chapter, see Chapter 21 in the* Study Guide and Solutions Manual.

REACTION REVIEW *For a summary of reactions discussed in this chapter, see the* Reaction Review *section of Chapter 21 in the* Study Guide and Solutions Manual.

SKILLS OBJECTIVES WITH PROBLEMS

• Given the structure of a carboxylic acid derivative, provide a name. Given the name, draw the structure. Know the priority of carboxylic acid derivatives as principal groups and how they are named as substituents. 21.1

21.30 Draw a structure for each of the following compounds.

(a) Ethyl *p*-acetylbenzoate

(b) Isopropyl methyl carbonate

(c) β-Ethyl-β-propionolactone

(d) 3-Bromocyclopentanecarbonitrile

(e) *N*-Ethyl-*N*-methylcyclohexanecarboxamide

21.31 Provide a name for each of the following compounds. Use a common name for at least one compound.

(a)

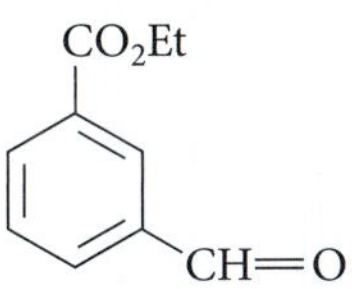

(b)

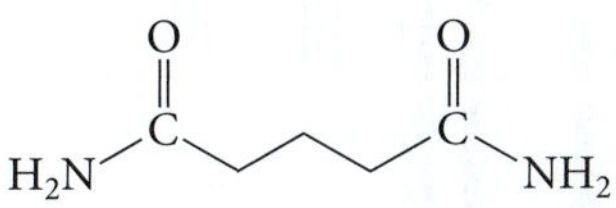

(c)

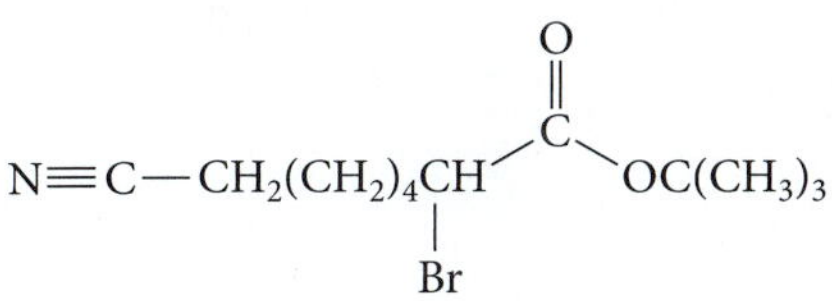

(d)

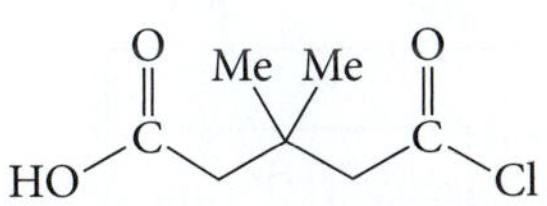

• Give the general structures of carboxylic acid derivatives; explain why rotation around the carbonyl–nitrogen bond of an amide is relatively slow. 21.2

21.32 (a) Draw the structures of the two conformations of each of the following compounds that are interconverted by the internal rotations shown.

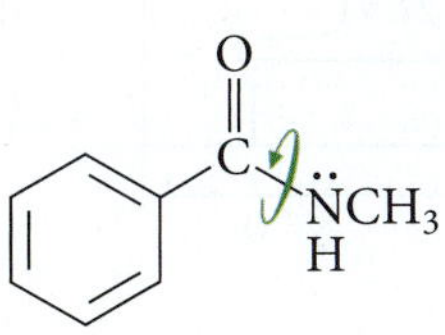

A

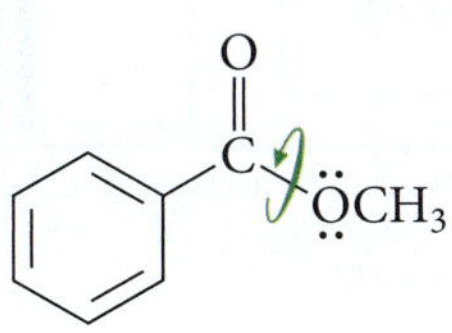

B

(b) Use resonance structures to show why internal rotation about the bond shown in the amide *A* is slower than typical rotations about single bonds.

(c) Use resonance structures to explain why internal rotation of the ester *B* is much faster than internal rotation in the amide *A*.

• In a general way, show how the physical properties of carboxylic acid derivatives compare with those of other functional classes. 21.3

21.33 Rank the following compounds in order of increasing boiling point; explain your reasoning.

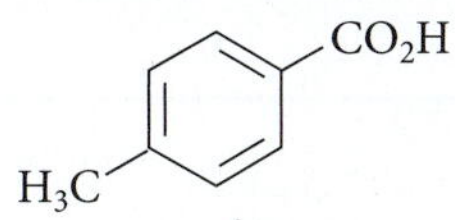

A

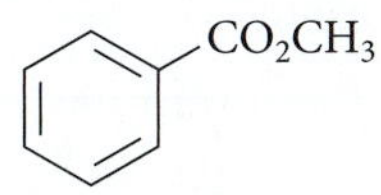

B

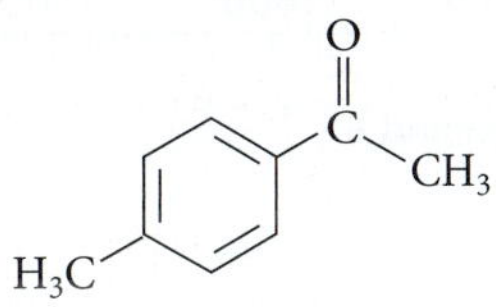

C

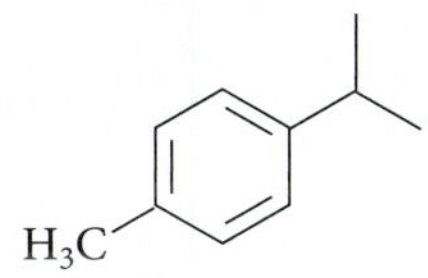

D

• Recognize the unique characteristics of the spectra of carboxylic acid derivatives. Use spectra to determine the structures of unknown carboxylic acid derivatives. 21.4

21.34 Show how you would use IR spectroscopy to distinguish between the compounds in each of the following sets.

(a)

$CH_3CH_2—C{\equiv}C—CH_2CH_3$ (*A*) $(CH_3)_2CHCH_2—C{\equiv}N$ (*B*)

(b)

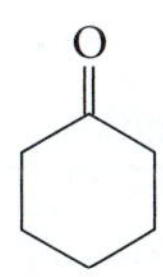

A

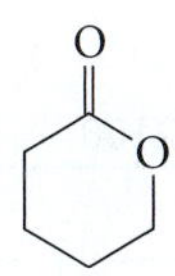

B

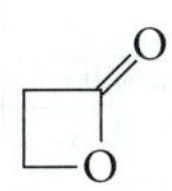

C

(c)

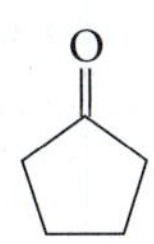

A

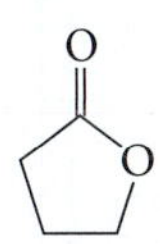

B

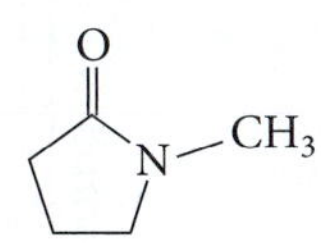

C

21.35 Identify each of the following compounds from their spectra.

(a) Compound *A*: molecular mass 113; gives a positive hydroxamate test; IR 2237, 1733, 1200 cm^{-1}; proton NMR: δ 1.33 (3*H*, t, $J = 7$ Hz), δ 3.45 (2*H*, s), δ 4.27 (2*H*, q, $J = 7$ Hz).

(b) Compound *B*: $C_6H_{12}O_2$; IR: 1743 cm^{-1}; proton NMR spectrum shown in **Fig. P21.35a**.

(c) Compound *C*: molecular mass 71; IR: 3200 (strong, broad), 2250 cm^{-1}; no absorptions in the 1500–2250 cm^{-1} range; proton NMR: δ 2.62 (2*H*, t, $J = 6$ Hz), δ 3.42 (1*H*, broad s; eliminated by D_2O shake), δ 3.85 (2*H*, t, $J = 6$ Hz).

(d) Compound *D*: EI mass spectrum: two molecular ions of about equal intensity at $m/z = 180$ and 182; IR: 1740 cm^{-1}; proton NMR: δ 1.30 (3*H*, t, $J = 7$ Hz); δ 1.80 (3*H*, d, $J = 7$ Hz); δ 4.23 (2*H*, q, $J = 7$ Hz); δ 4.37 (1*H*, q, $J = 7$ Hz).

(e) Compound *E*: UV spectrum: $\lambda_{max} = 272$ nm ($\varepsilon = 39{,}500$); EI mass spectrum: $m/z = 129$ (molecular ion and base peak); IR: 2200, 970 cm^{-1}; proton NMR: δ 5.85 (1*H*, d, $J = 17$ Hz); δ 7.35 (1*H*, d, $J = 17$ Hz); δ 7.4 (5*H*, apparent s).

(f) Compound *F*: $C_{10}H_{13}NO_2$; IR: 3285, 1659, 1246 cm^{-1}; proton NMR spectrum shown in **Fig. P21.35b**.

(g) Compound *G*: molecular mass 101; IR: 3397, 3200, 1655, 1622 cm^{-1}; ^{13}C NMR: δ 27.5, δ 38.0 (weak), δ 180.5 (weak).

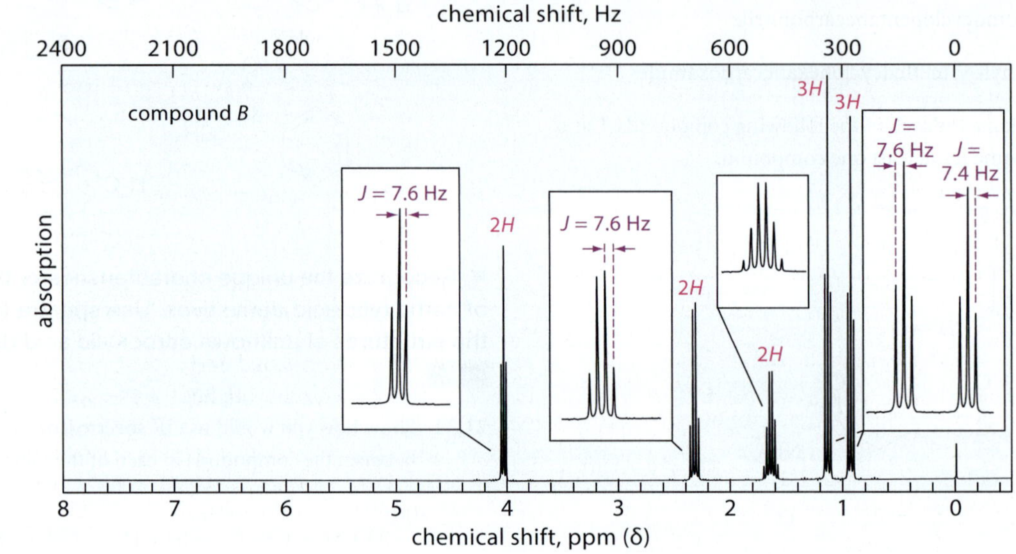

(a)

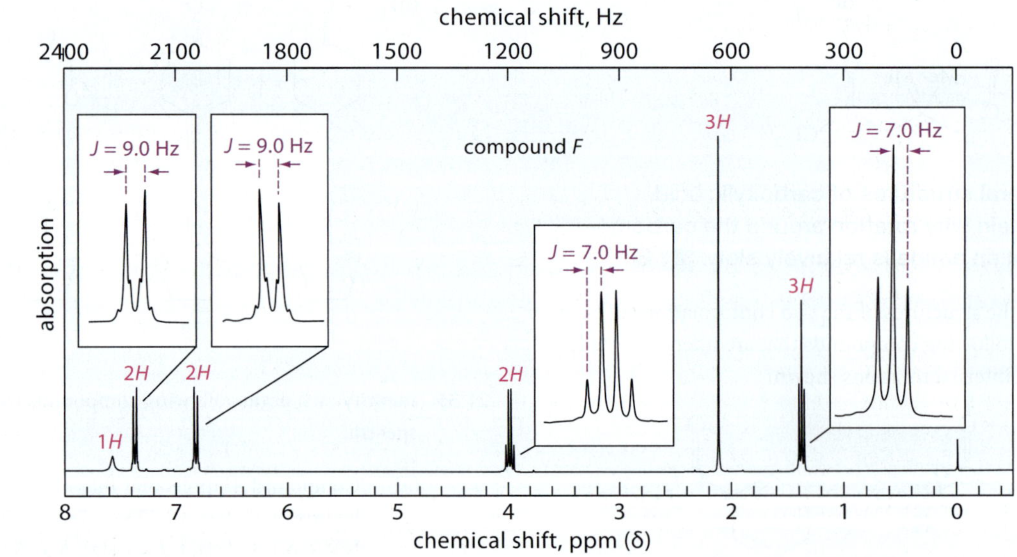

(b)

FIGURE P21.35 (a) The NMR spectrum of compound *B* for Problem 21.35b. The integrals are shown in red over their respective resonances. The resonance at about δ 1.6 is coupled to both of the resonances at δ 1.0 and δ 4.0; the coupling constants of 7.4 Hz and 7.6 Hz are sufficiently close that the $n+1$ rule is followed. (b) The NMR spectrum of compound *F* in Problem 21.35f. The integrals are shown in red over their respective resonances.

- **Know the approximate pK_a values for the conjugate acids of esters, amides, and nitriles, and know the position of protonation of esters and amides. 21.5**

21.36 Rank each of the following ions according to their pK_a, with the least acidic ion first and the most acidic last, and explain.

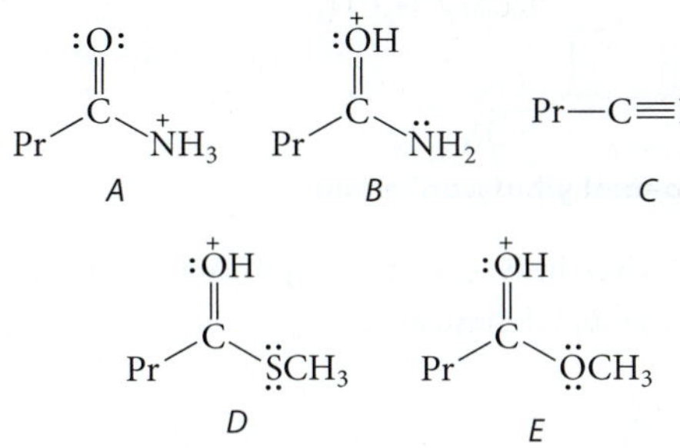

- **Draw the curved-arrow mechanism for an acyl substitution reaction under either basic or acidic conditions. 21.7A**

21.37 (a) Draw a curved-arrow mechanism for the following reaction in methanol solvent at a pH high enough that dimethylamine is not converted into its conjugate acid.

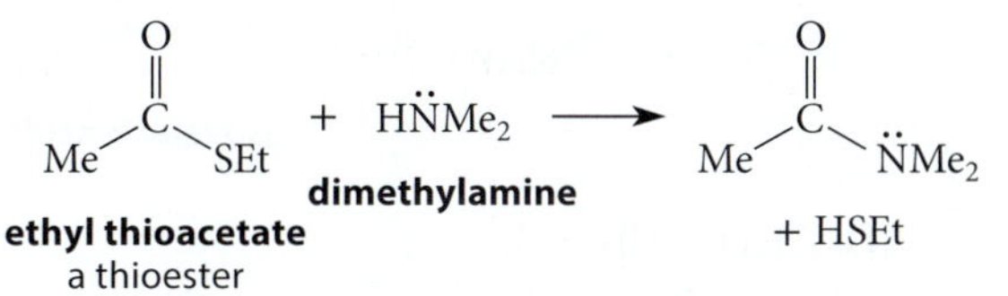

(b) Draw a curved-arrow mechanism for the acid-catalyzed hydrolysis of the thioester in part (a). (Represent a strong acid as H—A.)

- **Provide missing starting materials or products in the reactions of carboxylic acid derivatives.**
  - (1) Hydrolysis **21.7**
  - (2) Reactions with nucleophiles **21.8**
  - (3) Hydride reductions of esters, amides, acid chlorides, and nitriles; catalytic hydrogenation of nitriles **21.9**

21.38 Give the principal organic product(s) expected when propionyl chloride reacts with each of the following reagents.

(a) H_2O

(b) Ethanethiol, pyridine, 0 °C

(c) $(CH_3)_3COH$, pyridine

(d) $(CH_3)_2CuLi$, −78 °C, then H_2O

(e) H_2, Pd catalyst (quinoline/sulfur poison)

(f) $AlCl_3$, toluene, then H_2O

(g) $(CH_3)_2CHNH_2$ (2 equiv.)

(h) Sodium benzoate

(i) *p*-Cresol (4-methylphenol), pyridine

21.39 Give the principal organic product(s) expected (if any) when ethyl benzoate or the other compound indicated reacts with each of the following reagents.

(a) H_2O, heat, acid catalyst

(b) NaOH, H_2O

(c) Aqueous NH_3, heat

(d) $LiAlH_4$, then H_2O

(e) Excess $CH_3CH_2CH_2MgBr$, then H_2O

(f) Product of part (e) + acetyl chloride, pyridine, 0 °C

(g) Product of part (e) + benzenesulfonyl chloride

(h) Na^+ $CH_3CH_2O^-$

21.40 Give the structure of a compound that satisfies each of the following criteria.

(a) A compound C_3H_5N that liberates ammonia on treatment with hot aqueous KOH

(b) A compound C_3H_7ON that liberates ammonia on treatment with hot aqueous KOH

(c) A compound that gives 1-butanol and 2-methyl-2-propanol on treatment with excess CH_3MgI followed by protonolysis

(d) A compound that gives equal amounts of 1-hexanol and 2-hexanol on treatment with $LiAlH_4$ followed by protonolysis

(e) A compound $C_3H_2N_2$ that gives CO_2, $^+NH_4$, and acetic acid after boiling in concentrated aqueous HCl

21.41 A major component of olive oil is glyceryl trioleate.

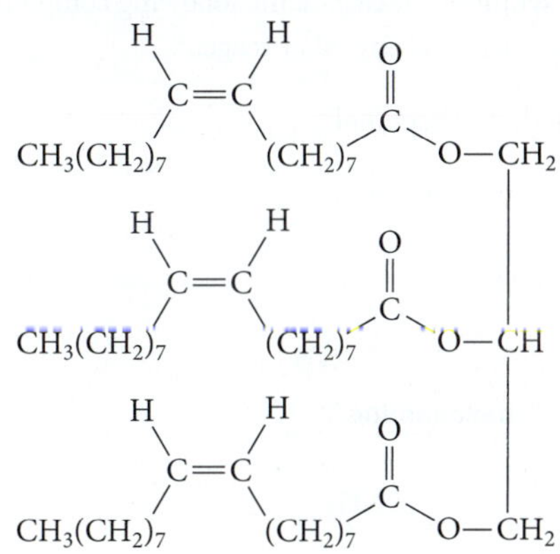

glyceryl trioleate

(a) Give the structures of the products expected from the saponification of glyceryl trioleate with excess aqueous NaOH.

(b) Give the structure of the product formed when glyceryl trioleate is subjected to catalytic hydrogenation. How would the physical properties of this product differ from those of the starting material? (*Hint:* See the sidebar, "Fats and Oils," Sec. 8.5D.)

• Rank carboxylic acid derivatives according to their reactivities in nucleophilic acyl substitution reactions, and give the reasons for the ranking. 21.7E, 21.9D

21.42 Rank the following compounds in order of increasing rate of their nucleophilic acyl substitution reactions with excess ammonia to give the amide.

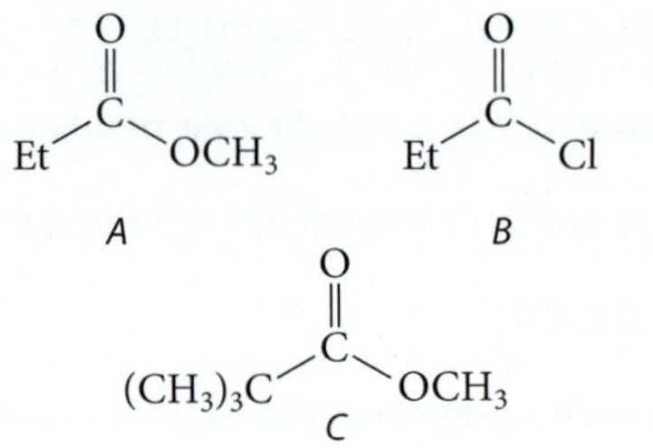

A *B* *C*

21.43 Analyze esters *A* and *B* in terms of the following two factors discussed in Sec. 21.7E.

(a) Resonance stabilization of the carbonyl compounds

(b) Polar stabilization of the transition state

Explain why their saponification rates are almost the same.

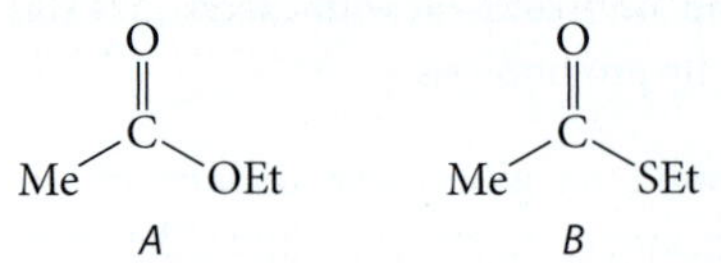

A *B*

(The hydrolysis of thioesters is discussed in Sec. 25.5A.)

• Use the reactions of carboxylic acids and their derivatives in multistep organic syntheses. 21.11B

21.44 Propose a synthesis for each of the following compounds from butyric acid and any other reagents.

(a) 4-Methyl-4-heptanol

(b) 2-Methyl-2-pentanol

(c) 4-Heptanol

(d) 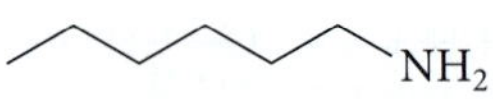

1-hexanamine

(e) 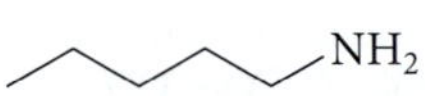

1-pentanamine

(f) 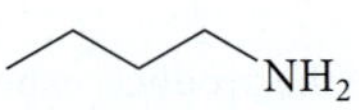

1-butanamine

21.45 Propose a multistep synthesis of each of the following compounds from the given starting material.

(a) *N*-Methylhexanamide from hexanoic acid

(b) 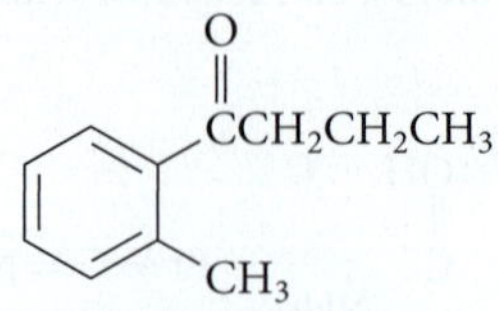 from *o*-bromotoluene

o-methylbutyrophenone

(c) 1-Cyclohexyl-2-methyl-2-propanol from bromocyclohexane

(d) 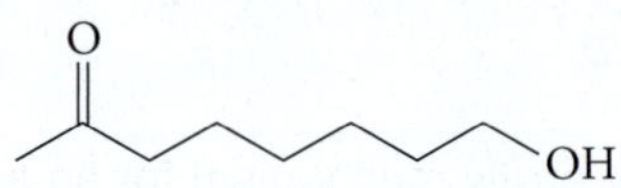

from

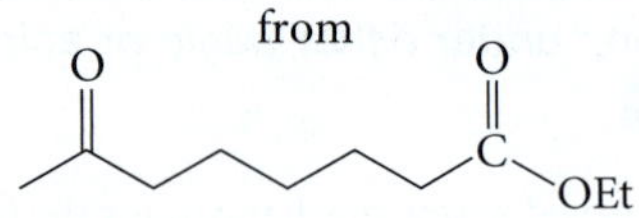 (*Hint:* Use a protecting group.)

(e) 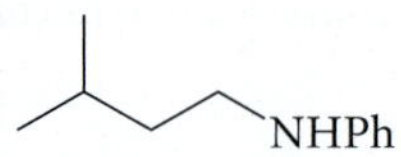 from 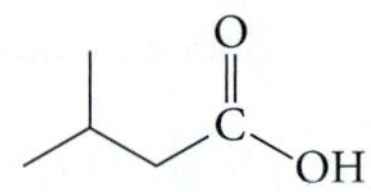

isovaleric acid

(f) 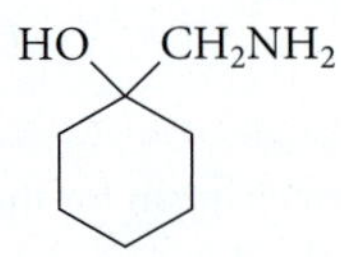 from cyclohexanone

(g) 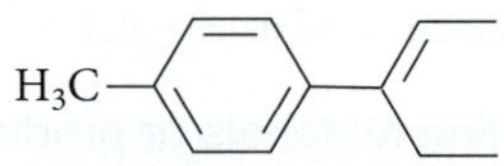 from toluene

(h) 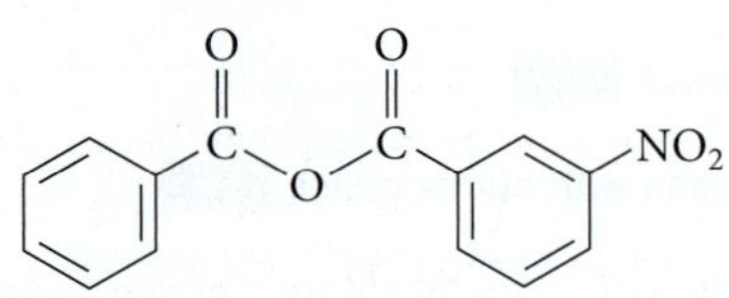

from benzoic acid as the only carbon source

• Analyze a condensative addition polymer to determine the monomeric starting materials. Given the starting materials, give the structure of a polymer in polymer notation. 21.12A

21.46 You are employed by Fibers Unlimited, a company specializing in the manufacture of specialty polymers. The vice president for research, Strong Fishlein, has asked you to design laboratory preparations of the following polymers. You are to use as starting materials the company's extensive stock of dicarboxylic acids containing six or fewer carbon atoms. Accommodate him.

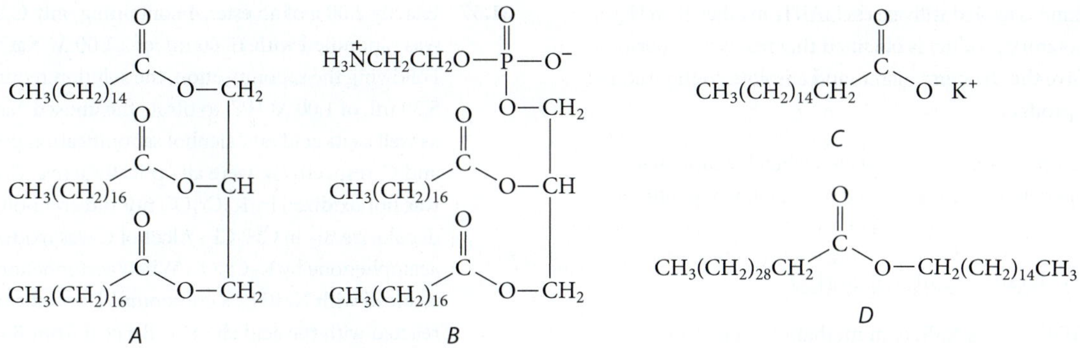

FIGURE P21.48

(a)

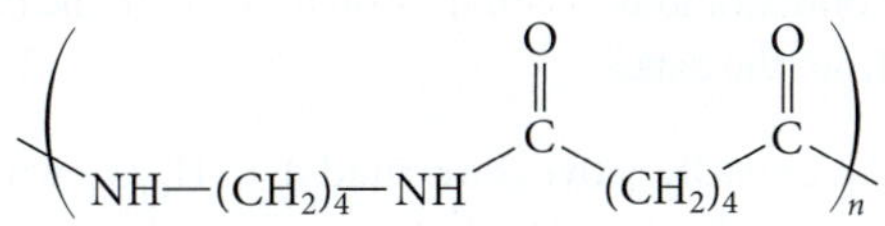

nylon-4,6

(b)

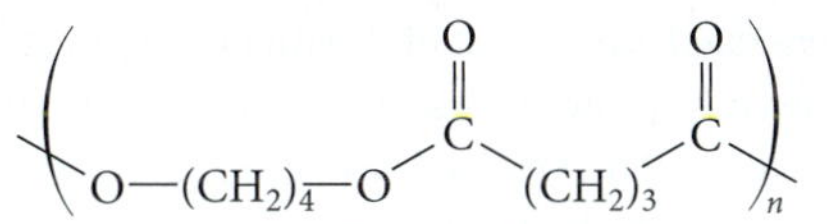

21.47 (a) Kevlar is a polymer made from the reaction of the following two compounds:

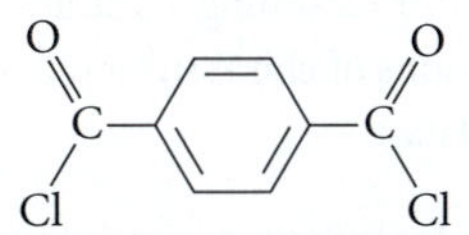

1,4-benzenedicarbonyl chloride

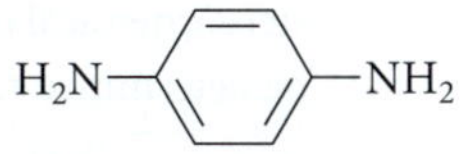

1,4-benzenediamine

Give the structure of Kevlar in polymer notation.

(b) Kevlar has such high strength that it can be used for bulletproof vests and other applications in which penetration resistance is important. This strength is due to inter-chain hydrogen bonds. Draw a few repeating units of two Kevlar chains, and show the hydrogen bonds that connect the two chains. Indicate the hydrogen-bond donors and acceptors.

- **Classify a molecule from its structure as a wax, a fat, or a phospholipid. 21.12C**

21.48 (a) Classify each of the molecules shown in **Fig. P21.48** as a wax, a phospholipid, or a fat. One of the molecules has a different classification. Which one is it, and what is its classification?

(b) Which of the molecules are amphipathic?

INTEGRATED PROBLEMS

21.49 (a) Draw the structures of all of the products that would be obtained when (±)-α-phenylglutaric anhydride is treated with (±)-1-phenylethanol. Which products should be separable without enantiomeric resolution? Which products should be obtained in equal amounts? In different amounts?

(b) Answer the same question for the reaction of the *S* enantiomer of the same alcohol with (±)-α-phenylglutaric anhydride.

21.50 Treatment of acetic propionic anhydride with ethanol gives a mixture of two esters consisting of 36% of the higher boiling one, *A*, and 64% of the lower boiling one, *B*. Give the structures of *A* and *B*.

21.51 Complete the diagram shown in **Fig. P21.51** by filling in all missing reagents or intermediates.

21.52 When (*R*)-(−)-mandelic acid (α-hydroxy-α-phenylacetic acid) is treated with CH_3OH and H_2SO_4, and the resulting

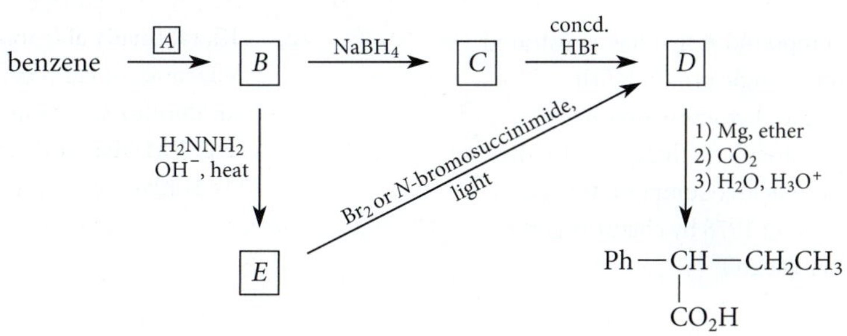

FIGURE P21.51

compound is treated with excess $LiAlH_4$ in ether, then H_2O, a levorotatory product is obtained that reacts with periodic acid. Give the structure, name, and absolute configuration of this product.

21.53 Contrast the results to be expected when levulinic acid (4-oxopentanoic acid) is treated in the following different ways.

(a) With excess $LiAlH_4$, then H_3O^+

(b) With excess $NaBH_4$ in methanol, then H_3O^+

21.54 You are a chemist working for the Imahot Pepper Company and have been asked to provide some information about *capsaicin*, the active ingredient of hot peppers.

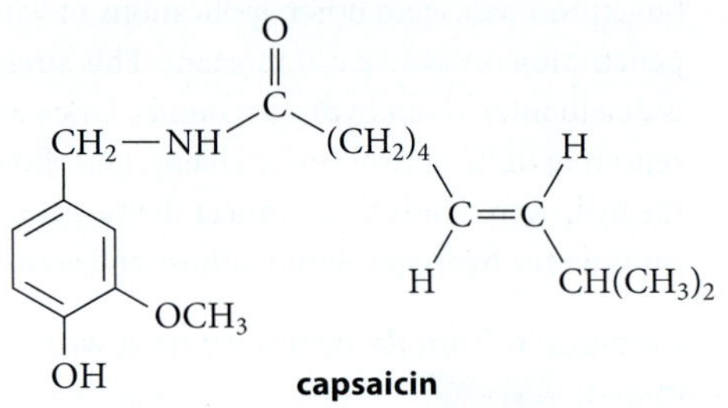

capsaicin

How should capsaicin or the other compounds indicated react under each of the following conditions?

(a) Br_2/CH_2Cl_2

(b) Dilute aqueous NaOH

(c) Dilute aqueous HCl

(d) H_2, catalyst

(e) Product of part (d) + 6 *M* HCl, heat

(f) Product of part (b) + CH_3I

(g) Product of part (d) + concentrated aqueous HBr, heat

21.55 A compound *A* has prominent infrared absorptions at 1050, 1786, and 1852 cm^{-1} and shows a single absorption in the proton NMR at δ 3.00. When heated gently with methanol, compound *B*, $C_5H_8O_4$, is obtained. Compound *B* has IR absorption at 2500–3000 (broad), 1730, and 1701 cm^{-1}, and its proton NMR spectrum in D_2O consists of resonances at δ 2.7 (complex splitting) and δ 3.7 (a singlet) in the intensity ratio 4:3. Identify *A* and *B*, and explain your reasoning.

21.56 Propose a structure for a compound *A* that has an infrared absorption at 1820 cm^{-1} and a single proton NMR absorption at δ 1.5. Compound *A* reacts with water to give dimethylmalonic acid and with methanol to give the monomethyl ester of the same acid. (Compound *A* was unknown until its preparation in 1978 by chemists at the University of California, San Diego.)

21.57 Exactly 2.00 g of an ester *A* containing only C, H, and O was saponified with 15.00 mL of a 1.00 *M* NaOH solution. Following the saponification, the solution required 5.30 mL of 1.00 *M* HCl to titrate the unused NaOH. Ester *A*, as well as its acid and alcohol saponification products *B* and *C*, respectively, were all optically active. Compound *A* was not oxidized by $K_2Cr_2O_7$, nor did compound *A* decolorize Br_2 in CH_2Cl_2. Alcohol *C* was oxidized to acetophenone by $K_2Cr_2O_7$. When acetophenone was reduced with $NaBH_4$, a compound *D* was formed that reacted with the acid chloride derived from *B* to give two optically active compounds: *A* (identical to the starting ester) and *E*. Propose a structure for each compound that is consistent with the data. (The absolute stereochemical configurations of chiral substances cannot be determined from the data.)

21.58 An optically active compound *A*, $C_6H_{10}O_2$, when dissolved in NaOH solution, consumed one equivalent of base. On acidification, compound *A* was slowly regenerated. Treatment of *A* with $LiAlH_4$ in ether followed by protonolysis gave an optically inactive compound *B* that reacted with acetic anhydride to give an acetate diester derivative *C*. Compound *B* was oxidized by aqueous chromic acid to β-methylglutaric acid (3-methylpentanedioic acid). Identify compounds *A*, *B*, and *C*, and explain your reasoning. (The absolute stereochemical configurations of chiral substances cannot be determined from the data.)

21.59 Klutz McFingers, a graduate student in his ninth year of study, has suggested the following synthetic procedures and has come to you in the hope that you can explain why none of them works very well (or not at all).

(a) Noting the fact that primary alcohols + HBr give alkyl halides, Klutz has proposed by analogy the nitrile synthesis shown in **Fig. P21.59a**.

(b) Klutz has proposed the synthesis for the half-ester of adipic acid shown in **Fig. P21.59b**.

(c) Klutz has proposed the synthesis of acetic benzoic anhydride shown in **Fig. P21.59c**.

(d) Noting correctly that methyl benzoate is completely saponified by one molar equivalent of NaOH, Klutz has suggested that methyl salicylate (structure in **Fig. P21.59d**) should also undergo saponification with one equivalent of NaOH.

(e) Klutz, finally able to secure a position with a pharmaceutical company working on β-lactam antibiotics, has proposed the reaction given in **Fig. P21.59e** for deamidation of a cephalosporin derivative.

(a) $Ph{-}CH_2{-}OH + HCN \xrightarrow[H_2O]{} Ph{-}CH_2{-}CN + H_2O$

(*Hint:* HCN is a weak acid; its pK_a is 9.4.)

(b) $\underset{(1.0\ mol)}{HO_2C(CH_2)_4CO_2H} + \underset{(1.0\ mol)}{CH_3OH} \xrightarrow{H_2SO_4} HO_2C(CH_2)_4CO_2CH_3 + H_2O$

(c)

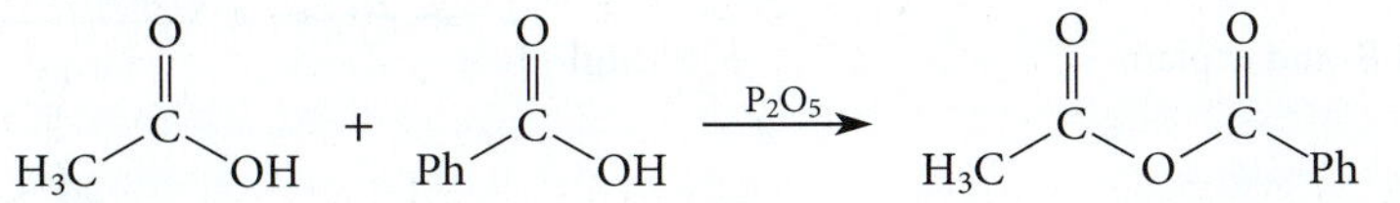

(d)

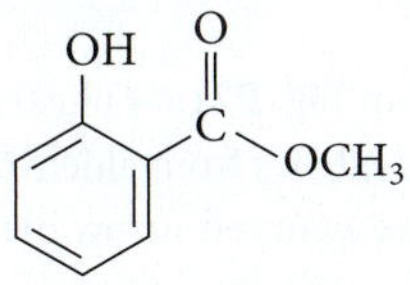

methyl salicylate

(e)

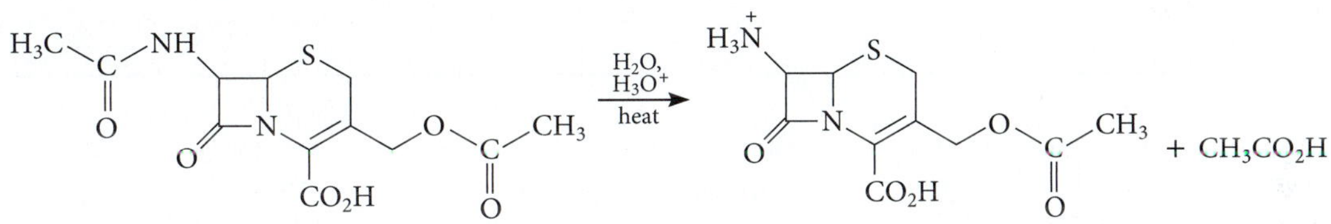

FIGURE P21.59

21.60 Complete each of the following reactions by giving the structures of the principal organic products. Explain how you arrived at your answers.

(a)

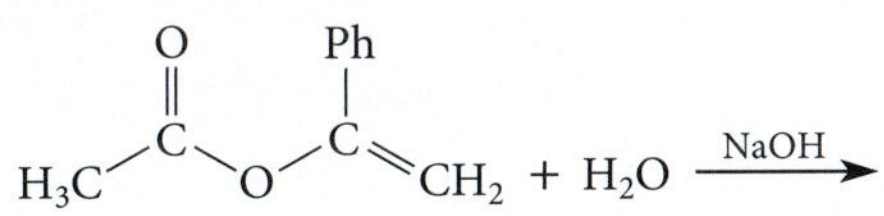

(b)

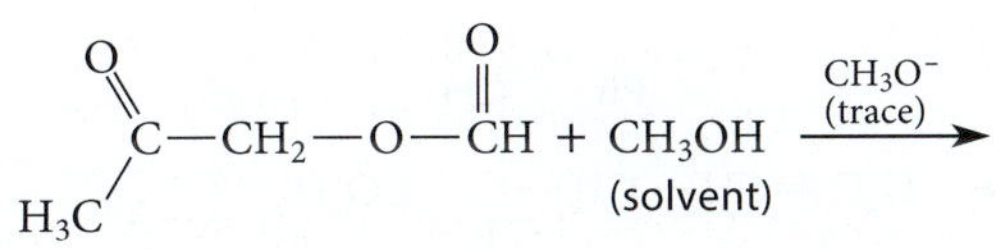

(c)

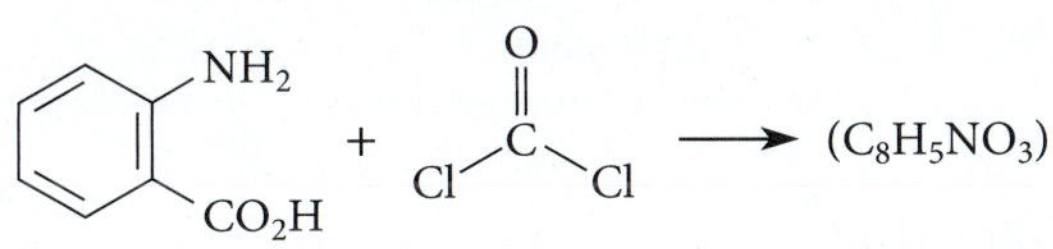

(d)

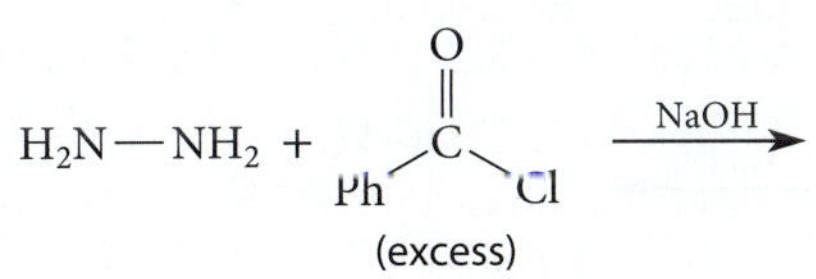

(e)

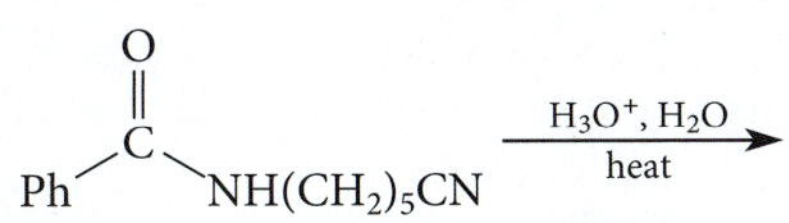

(f)

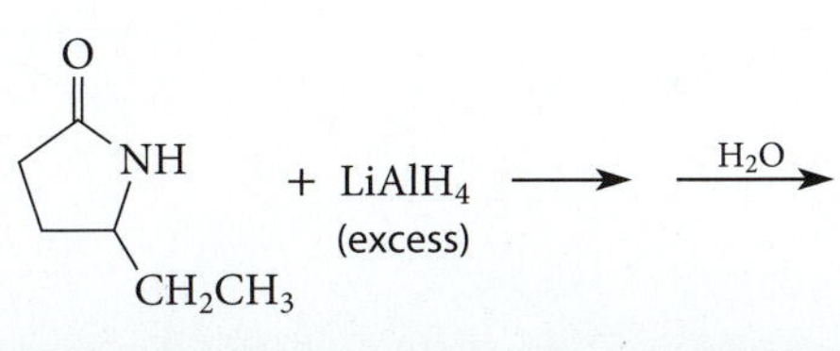

(g)

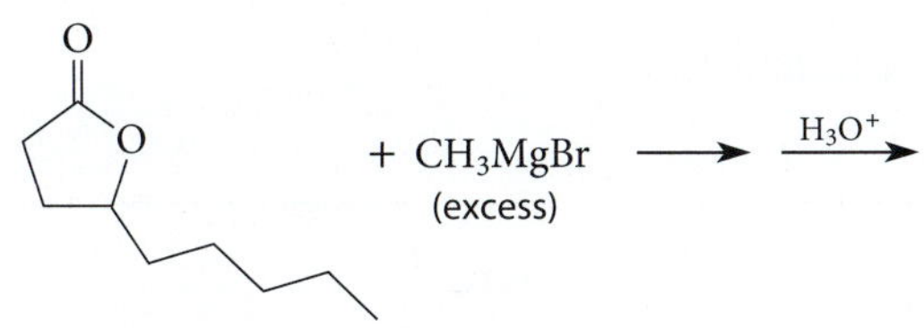

(h)

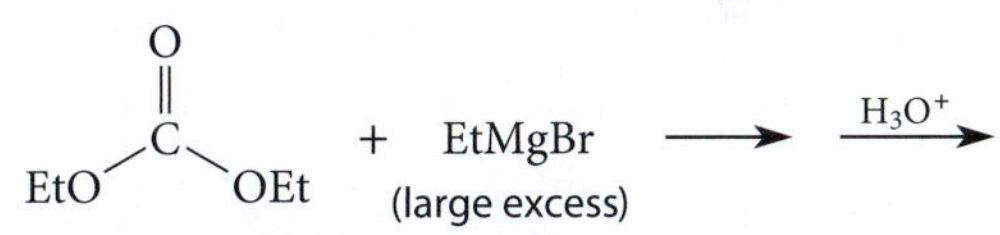

(i)

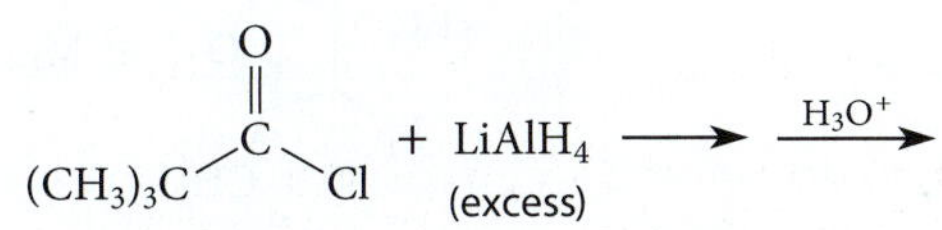

(j)

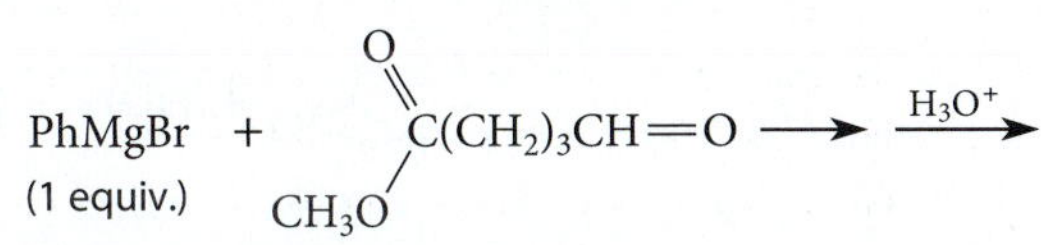

(k) glyceryl tristearate (structure in Problem 21.41) $+ CH_3OH \xrightarrow{CH_3O^-}$

(l) $(CH_3)_2C(NH_2)CH_2CH_2CO_2CH_3 \xrightarrow{\text{stand in } CH_3OH} (C_6H_{11}NO)$

21.61 In clinical studies of patients with atherosclerosis (hardening of the arteries) it was found that one of the metabolites of the hyperlipidemia drug (*Z*)-3-methyl-4-phenyl-3-butenamide (*A*, **Fig. P21.61**) is a compound *B*, which has the formula $C_{11}H_{15}NO_2$. When compound *B* is heated in aqueous acid, lactone *C* is formed along with the ammonium ion ($^+NH_4$).

(a) Propose a structure for compound *B*, and explain your reasoning.

(b) Give a curved-arrow mechanism for the conversion of *B* into *C*.

21.62 Rationalize each of the reactions in **Fig. P21.62** with a mechanism, using the curved-arrow notation where possible. In part (d), identify compound *A*, and show the mechanism for its formation. (Do not give the mechanism of the $NaBH_4$ step.)

21.63 The reaction of Grignard reagents with nitriles is another method of preparing ketones. The following sequence is an example of this synthesis. Identify compound *A*, and give the curved-arrow mechanisms for its formation and conversion into the ketone product.

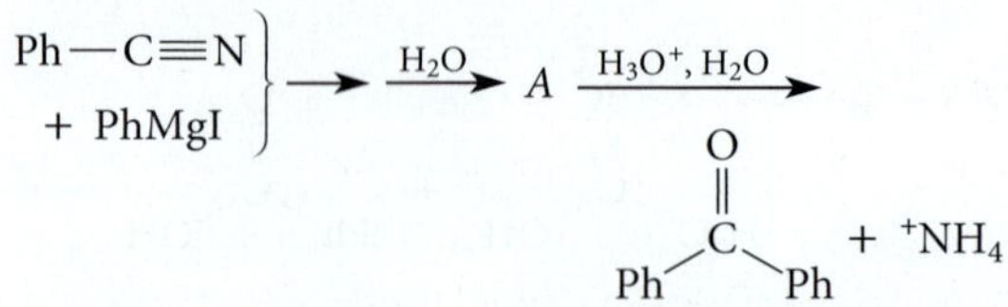

21.64 The sequence shown in **Fig. P21.64** illustrates a method for the preparation of nitriles from aldehydes. Identify compound *A*, and give a curved-arrow mechanism for the conversion of *A* to the products.

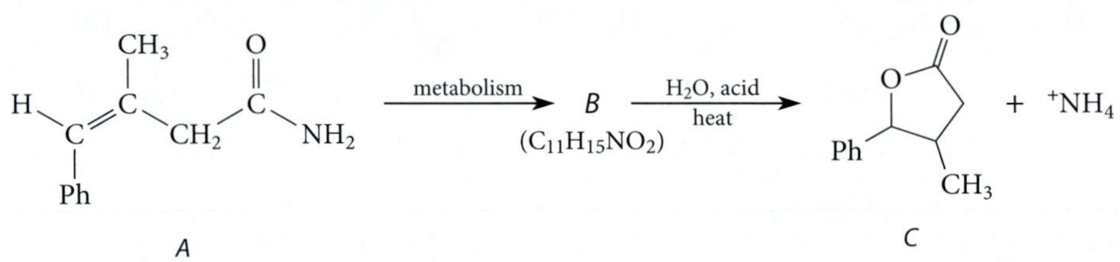

FIGURE P21.61

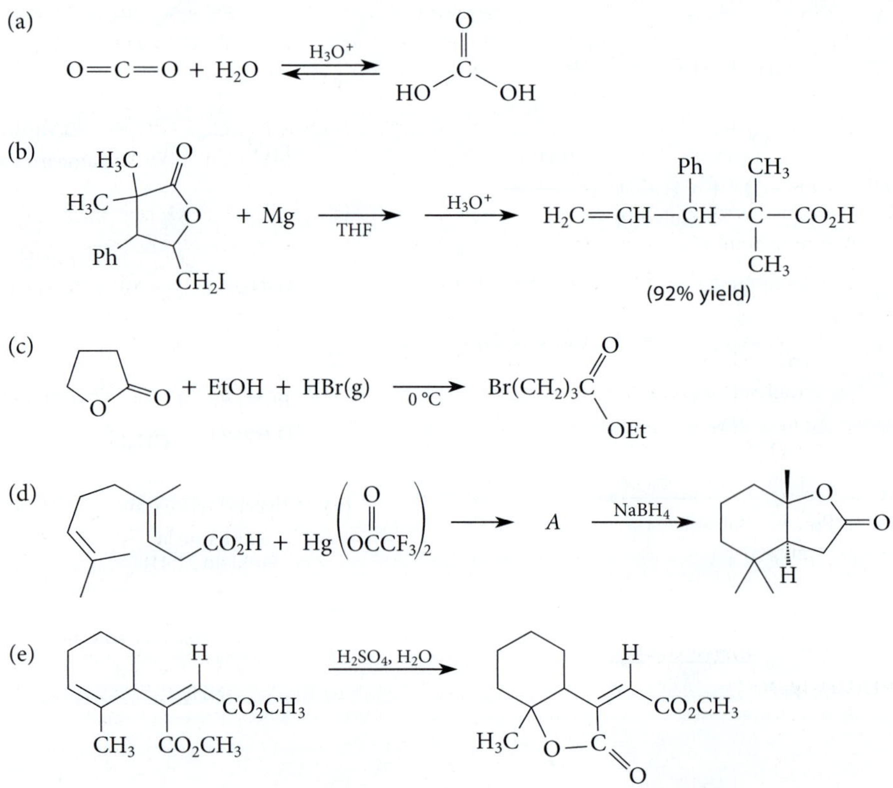

FIGURE P21.62

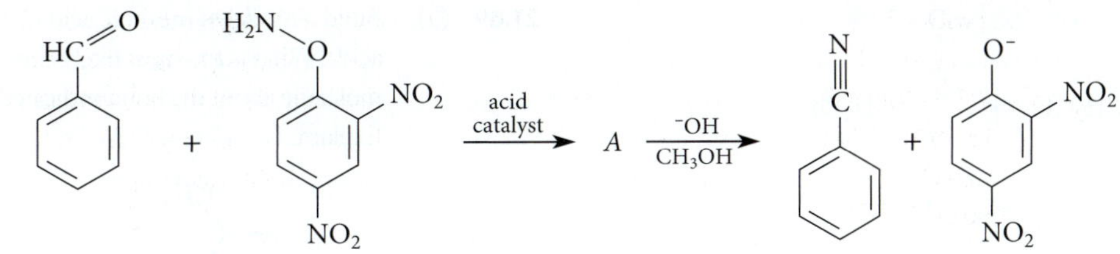

FIGURE P21.64

21.65 The *Weinreb amide* method is a good way to prepare ketones in high yields from acid chlorides.

(a) In the first step of this method, an acid chloride is converted into an amide with *N,O*-dimethylhydroxylamine. The product is known generically as a *Weinreb amide,* after its inventor Steven Weinreb (b. 1941), a Pennsylvania State University chemist. Give the structure of the Weinreb amide by completing the following reaction.

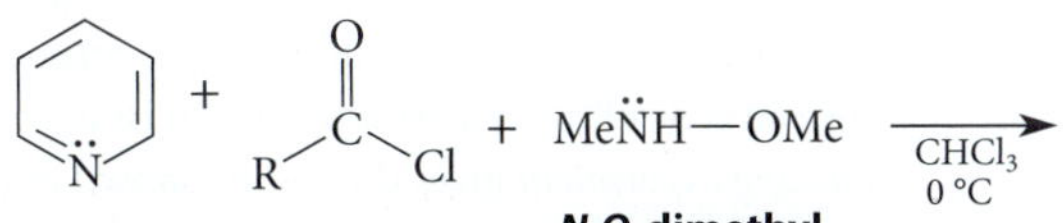

pyridine **N,O-dimethylhydroxylamine**

(b) The Weinreb amide is allowed to react with a Grignard or organolithium reagent to form a *stable* tetrahedral addition intermediate. Using a generic structure for the organolithium reagent R′—Li, draw a structure for this intermediate. This intermediate is stable to decomposition for two reasons. What are they? (*Hint:* The intermediate contains the metal ion from the Grignard or lithium reagent.)

(c) Only after aqueous acid is added does the tetrahedral intermediate break down to the ketone product and other by-products. Write a curved-arrow mechanism for this process and show all the products.

(d) Give the products formed when cyclohexanecarbonyl chloride (*A*) is converted into a Weinreb amide (*B*) and this amide is allowed to react with propylmagnesium bromide (*C*) followed by protonolysis.

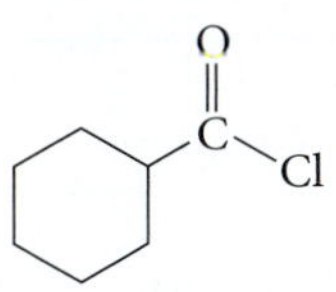

A
cyclohexanecarbonyl chloride

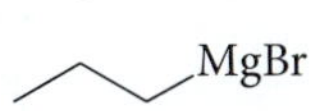

C
propylmagnesium bromide

(e) Why not simply allow the acid chloride to react with the Grignard or organolithium reagent? Why is the Weinreb amide used instead?

21.66 ε-Caprolactam is the starting material for nylon-6 preparation (Sec. 21.12A, Focused Problem 21.29). ε-Caprolactam can be prepared from cyclohexanone in a reaction sequence that involves the *Beckmann rearrangement,* the second of the two reactions shown in the following sequence.

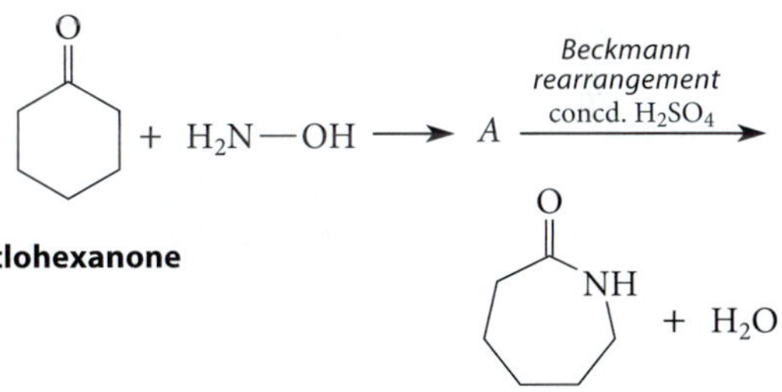

cyclohexanone **ε-caprolactam**

Identify compound *A*, and suggest a curved-arrow mechanism for its conversion to ε-caprolactam in strong acid.

21.67 Outline a synthesis of each of the following compounds from the indicated starting material.

(a) The following compound, the active ingredient in some insect repellents, from 3-methylbenzaldehyde

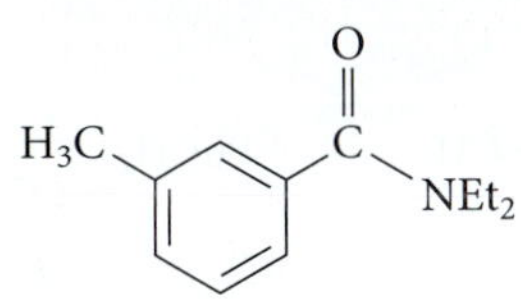

(b) 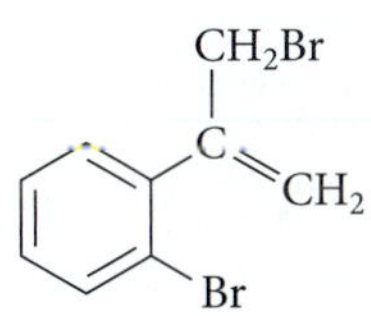 from 2-bromobenzoic acid

(c) 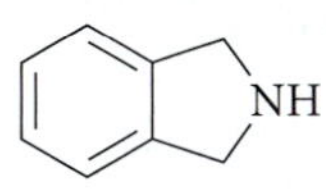 from 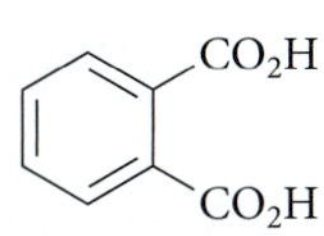

phthalic acid

(d)

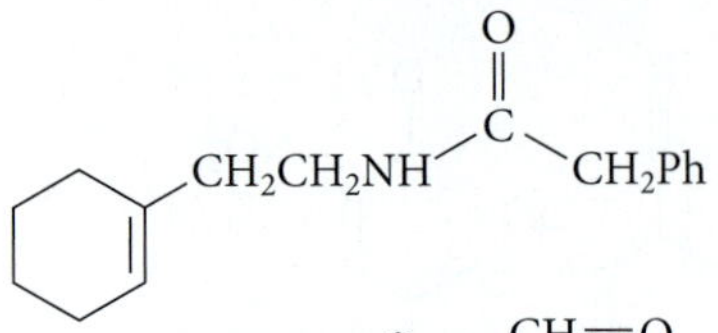

from

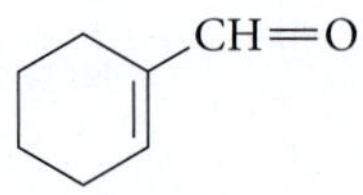

(e)

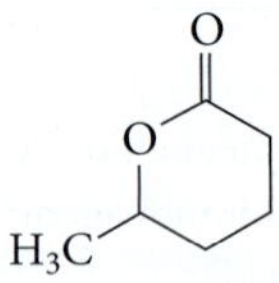

from

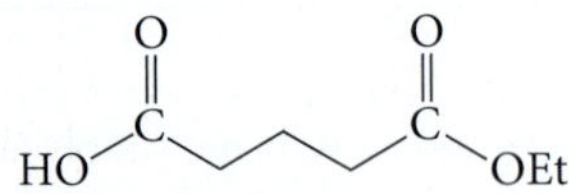

adipic acid monoethyl ester

(*Hint:* See Eq. 21.36b, Sec. 21.7A.)

(f)

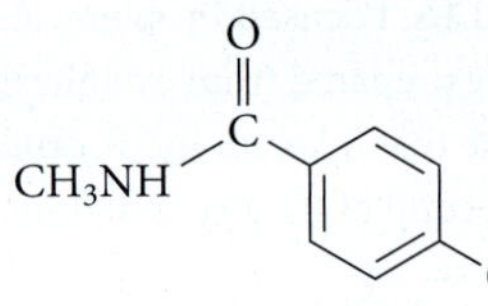

from *p*-methylanisole (1-methoxy-4-methylbenzene)

(g)

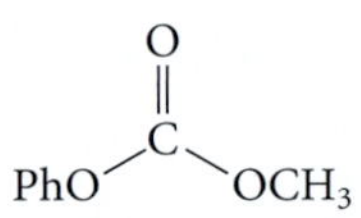

from phosgene (Sec. 21.1G)

(h)

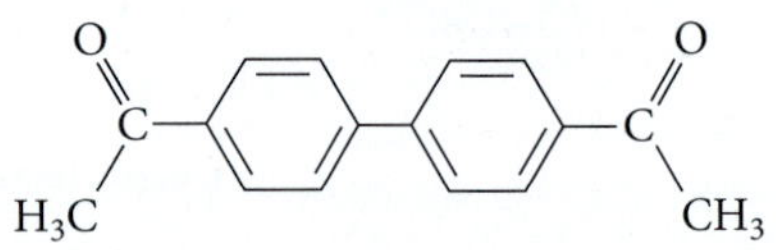

from *p*-bromotoluene (*Hint:* See Sec. 18.6B.)

21.68 Give the structure of the product in the reaction of each of the following esters with isotopically labeled sodium hydroxide, Na^+ $^{18}OH^-$, and explain your reasoning. (*Hint:* See Sec. 11.4.)

STUDY GUIDE LINK 21.5
Esters and Nucleophiles

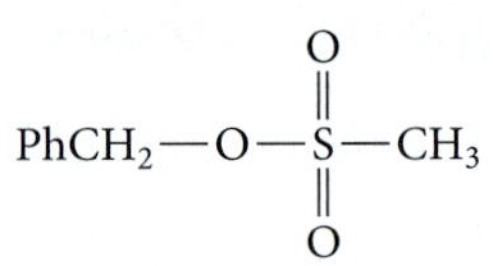

A

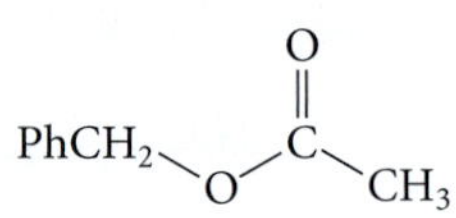

B

21.69 (a) Build a model of mesitoic acid (2,4,6-trimethylbenzoic acid). What is the most likely conformation of the molecule about the bond indicated by the arrow? Explain.

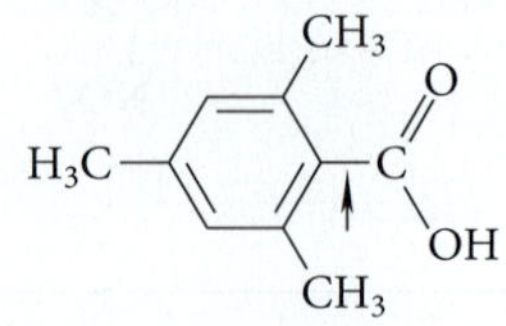

mesitoic acid

(b) Explain why the acid-catalyzed hydrolysis of the methyl ester of mesitoic acid does not occur at a measurable rate.

(c) Which *one* of the following methods should be used to make the methyl ester of mesitoic acid: acid-catalyzed esterification in methanol or esterification with diazomethane? Explain your choice.

21.70 In aqueous solution at pH 3, the hydrolysis of phthalamic acid to phthalic acid (see the reaction in **Fig. P21.70**) is 10^5 times faster than the hydrolysis of benzamide under the same conditions. Furthermore, an isotope *double-labeling experiment* gives the results shown in Fig. P21.70 (* = ^{18}O, # = ^{13}C). Phthalic anhydride (Display 21.6, Sec. 21.1C) was postulated as an intermediate in this reaction.

(a) Using the curved-arrow notation, show how phthalic anhydride is formed from the starting materials.

(b) Show how the intermediacy of phthalic anhydride can explain the double-labeling experiment.

(c) Explain why this reaction is much faster than the hydrolysis of a typical amide. (*Hint:* See Sec. 12.8.)

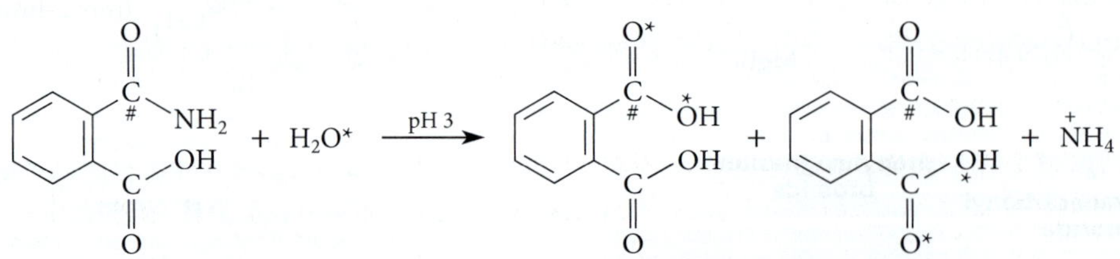

phthalamic acid

equal amounts of each
phthalic acid (isotopically labeled)

FIGURE P21.70

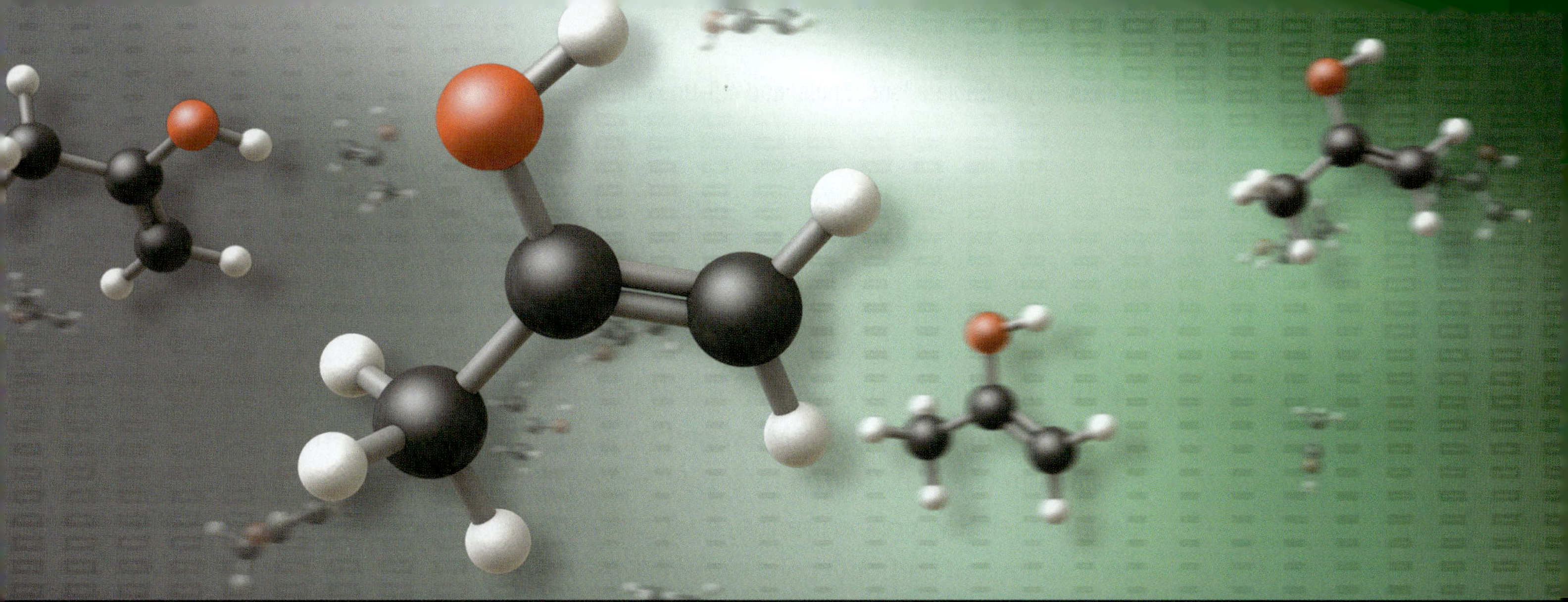

22 THE CHEMISTRY OF ENOLATE IONS, ENOLS, AND α,β-UNSATURATED CARBONYL COMPOUNDS

Chapters 19–21 examined the chemistry of carbonyl compounds, concentrating largely on reactions at the carbonyl group. This chapter completes the survey of carbonyl compounds by considering reactions involving the α-carbon. Hydrogens at the α-carbons of carbonyl compounds are acidic enough to be removed by strong bases. When an α-proton is removed, a conjugate-base anion is formed at the α-carbon. The conjugate-base anion of a carbonyl compound formed by removal of an α-hydrogen is called an **enolate ion** or sometimes simply an **enolate**.

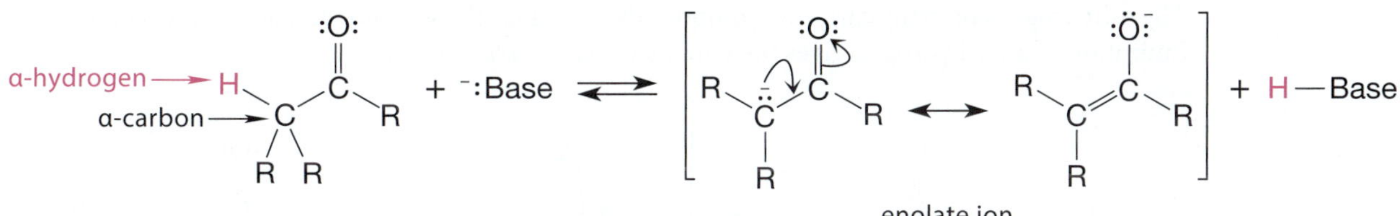

(22.1a)

Enolate ions are bases, and, like most bases, they can act as nucleophiles. Consequently, the α-carbon of a carbonyl compound, as the site of a conjugate-base enolate ion, is a site of *nucleophilic reactivity*. Much of this chapter deals with aspects of this reactivity.

The α-carbon and α-hydrogens of carbonyl compounds are also involved in the formation of *enols*, which were first introduced in Sec. 5.5. An **enol** is any compound in which a hydroxy group is on a carbon of a carbon–carbon double bond—that is, an enol is a vinylic alcohol.

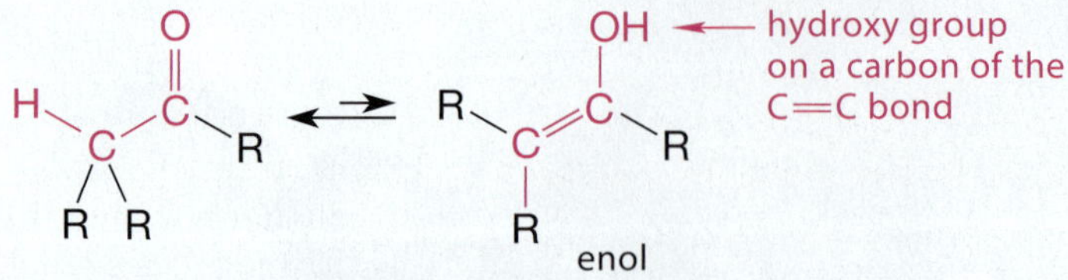

(22.1b)

As this equation shows, an enol and the corresponding carbonyl compound are *constitutional isomers.* Most carbonyl compounds with α-hydrogens are in equilibrium with small (in many cases, *very* small) amounts of their enol isomers. Despite their low concentration, enols are intermediates in a number of important reactions of carbonyl compounds. The chemistry of enols is another topic of this chapter.

This chapter also covers some unique chemistry of **α,β-unsaturated carbonyl compounds**—compounds in which a carbonyl group is conjugated with a carbon–carbon double bond.

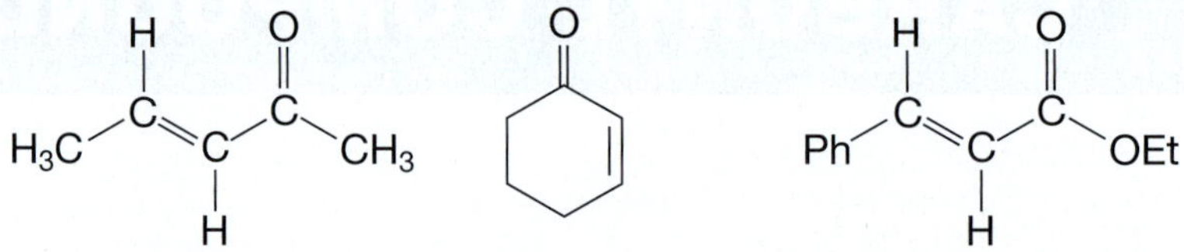

α,β-unsaturated ketones (sometimes called enones) an α,β-unsaturated ester (22.2)

In this chapter we also present some examples of the chemistry of enolates, enols, and α,β-unsaturated carbonyl compounds in biology.

22.1 ACIDITY OF CARBONYL COMPOUNDS

A. Formation of Enolate Anions

The α-hydrogens of many carbonyl compounds, as well as those of nitriles, are weakly acidic. Ionization of an α-hydrogen gives the conjugate-base *enolate anion.*

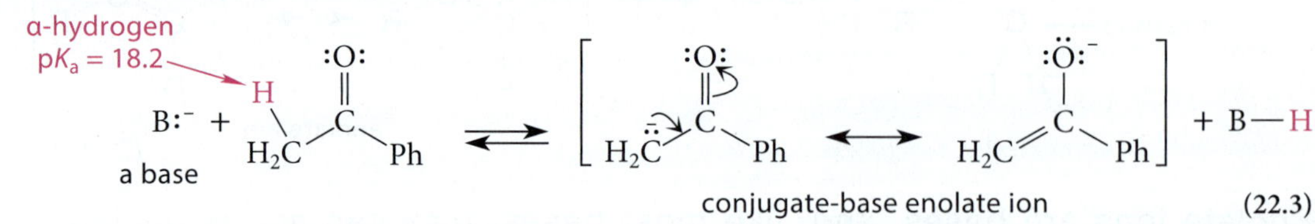

(22.3)

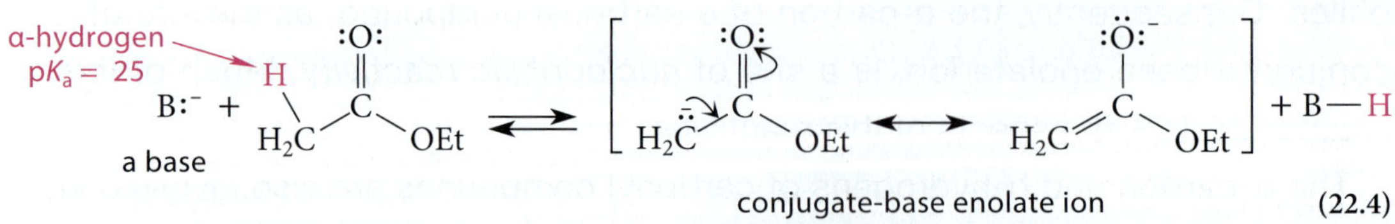

(22.4)

STUDY GUIDE LINK 22.1
Ionization versus Nucleophilic Reaction at the Carbonyl Carbon

The pK_a values of simple aldehydes or ketones are in the range 16–20, and the pK_a values of esters are about 25. The α-hydrogens of nitriles and tertiary amides also have acidities similar to those of esters.

Although carbonyl compounds are classified as weak acids, their α-hydrogens are much more acidic than other types of hydrogens bonded to carbon. For example, the dissociation constants of carbonyl compounds are greater than those of alkanes by about 30 powers of 10! To understand the greater acidity of carbonyl compounds, first recall that the stabilization of a base lowers the pK_a of its conjugate acid (Display 3.74, Sec. 3.7D). Enolate ions are resonance-stabilized, as shown in Eqs. 22.3 and 22.4. Therefore, carbonyl compounds have lower pK_a values—greater acidities—than carbon acids whose conjugate bases lack this stabilization. As discussed in Sec. 15.6, resonance is a symbolic way of depicting orbital overlap. In an enolate ion, the anionic α-carbon is sp^2-hybridized. This hybridization allows the nonbonding pair of electrons to occupy a 2p orbital, which is aligned for overlap with the 2p orbitals of the carbonyl group. **Figure 22.1a** shows this 2p-orbital alignment for the conjugate-base enolate ion of acetaldehyde. This overlap results in the formation of three π molecular orbitals (MOs), which are shown in **Fig. 22.1b**. [Compare Fig. 22.1 with Fig. 15.15 (Sec. 15.6), for the allyl cation.] Because the enolate ion has four π electrons, two of its MOs, π_1 and π_2, are fully occupied. In the occupied MO of lowest energy (π_1), the π electrons extend across all three constituent atoms. *This additional overlap provides additional bonding and, therefore, additional stabilization.* The occupied MO of higher energy (π_2), however, has a node that is more or less at the carbonyl carbon. The electrons in this molecular orbital are the ones involved in the chemical reactions of enolate ions. In this molecular orbital, *the α-carbon and the carbonyl oxygen are the major sites of electron density.* This is exactly the same conclusion that we reach from the resonance structures of the enolate ions in Display 22.1. These sites of negative charge are also confirmed by the red regions in the electrostatic potential map (EPM) of this enolate ion:

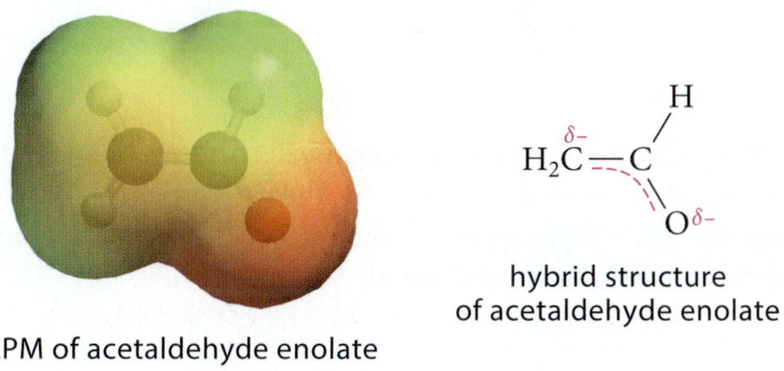

(22.5)

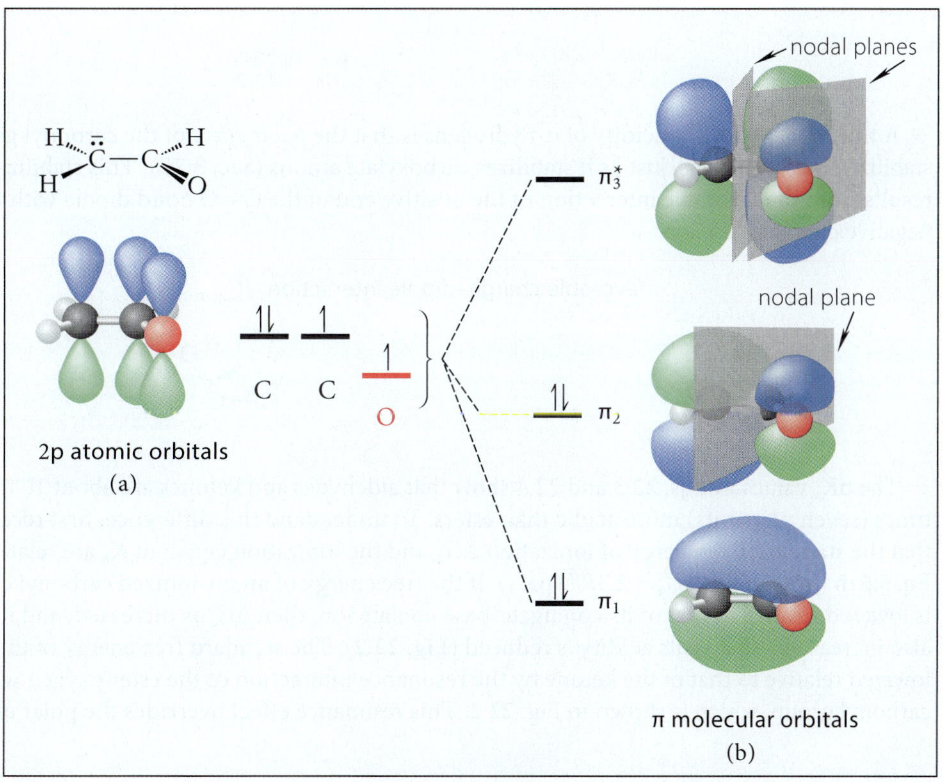

FIGURE 22.1 An orbital interaction diagram for the conjugate-base enolate ion of acetaldehyde, which is shown in the upper left. (a) The 2p orbitals are shown aligned for overlap. The energy of the 2p orbital of oxygen *(red)* is lower than the energy of the carbon 2p orbitals. (b) The three π MOs of the ion. The presence of oxygen lowers the energy of π_2. [Compare with the orbital interaction diagram in Fig. 15.15 (Sec. 15.6) for the allyl cation.] In the π_1 MO, electrons are delocalized across the entire molecule, and this MO has the lowest energy. This MO is the major source of stabilization of the ion.

If the structure of an enolate ion constrains the geometry of the component 2p orbitals so that they *cannot* overlap, the enolate ion is no longer stabilized. For example, an enolate ion *cannot* form at the bridgehead carbon of the following bicyclic ketone because the resonance structure of the enolate ion violates Bredt's rule (Sec. 7.6C).

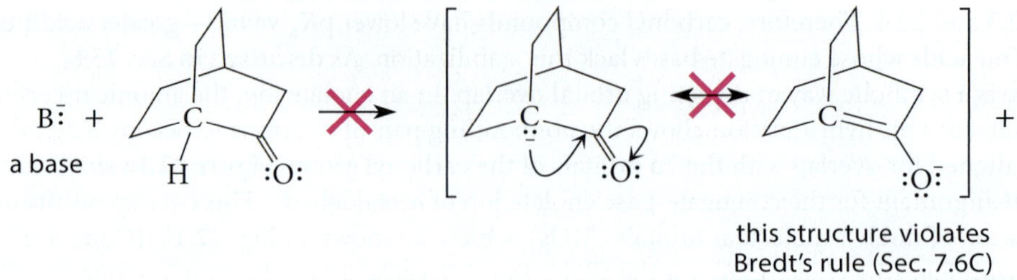

(22.6a)

In other words, the bicyclic structure forces the sp^2 orbital containing the nonbonding pair of the enolate to be oriented so that it cannot overlap with the π-electron system of the carbonyl group.

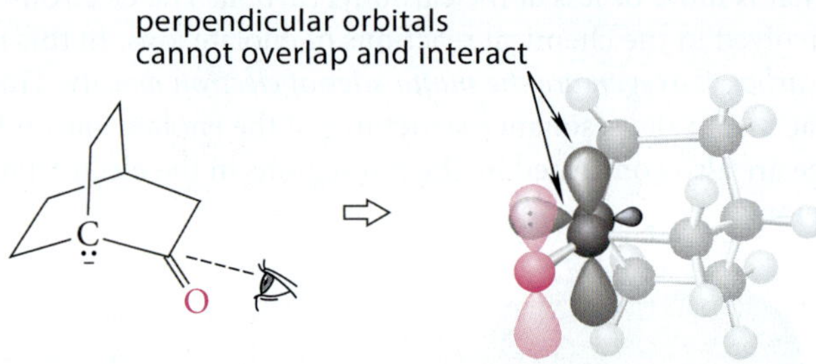

(22.6b)

The α-hydrogens are acidic, too, because the negative charge in an enolate ion is delocalized onto oxygen, an electronegative atom. Consequently, the α-hydrogens of carbonyl compounds are much more acidic than the allylic hydrogens of alkenes, even though the conjugate-base anions of both types of compounds are resonance-stabilized.

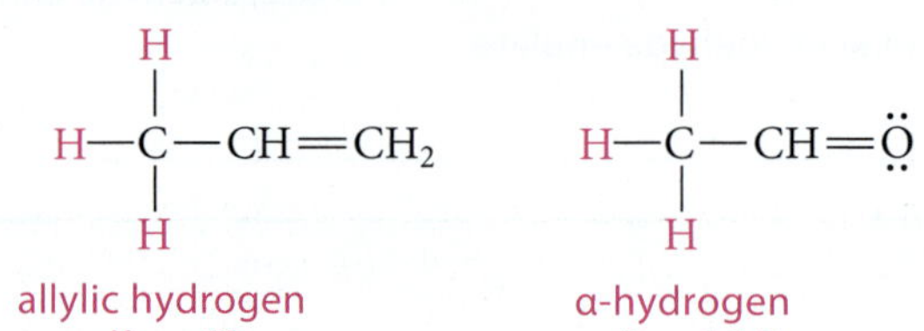

(22.7)

A third reason for the acidity of α-hydrogens is that the *polar effect* of the carbonyl group stabilizes enolate anions, just as it stabilizes carboxylate anions (Sec. 3.7E). This stabilization results from the favorable interaction of the positive end of the C=O bond dipole with the negative charge of the ion:

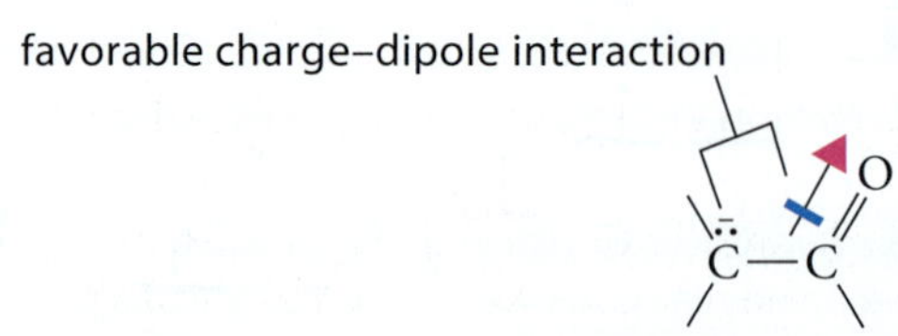

(22.8)

The pK_a values in Eqs. 22.3 and 22.4 show that aldehydes and ketones are about 10 million times (seven pK_a units) more acidic than esters. To understand this difference, first recall that the standard free energy of ionization ΔG_a° and the ionization constant K_a are related by Eq. 3.61b (Sec. 3.6D): $\Delta G_a^\circ = 2.3RT(pK_a)$. If the free energy of an un-ionized carbonyl compound is lowered relative to that of its conjugate-base enolate ion, then ΔG_a° is increased, and its pK_a is also increased—that is, its acidity is reduced (**Fig. 22.2**). The standard free energy of the ester is lowered relative to that of the ketone by the resonance interaction of the ester oxygen with the carbonyl group, which is shown in Fig. 22.2. This resonance effect overrides the polar effect of the

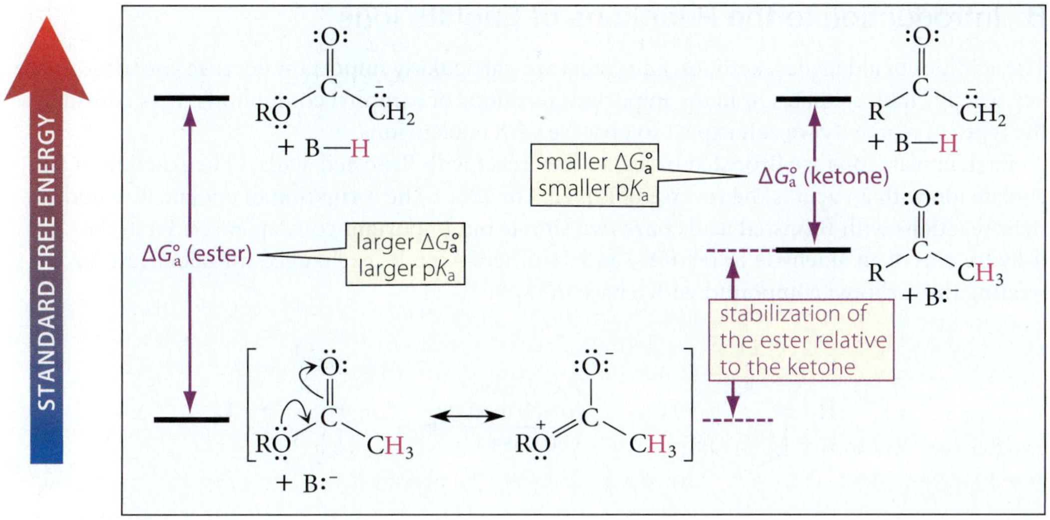

FIGURE 22.2 Resonance stabilization of an ester increases its standard free energy of ionization relative to that of a ketone and raises its pK_a. The free energies of the conjugate-base enolate ions have been placed at the same level for comparison purposes, and the resonance structures of the enolate ions (Eqs. 22.3 and 22.4) are not shown.

ester oxygen, which, in the absence of resonance, would increase the acidity of esters relative to ketones. In the enolate ion, the analogous resonance structure is *much less important* because of the repulsion between negative charges.

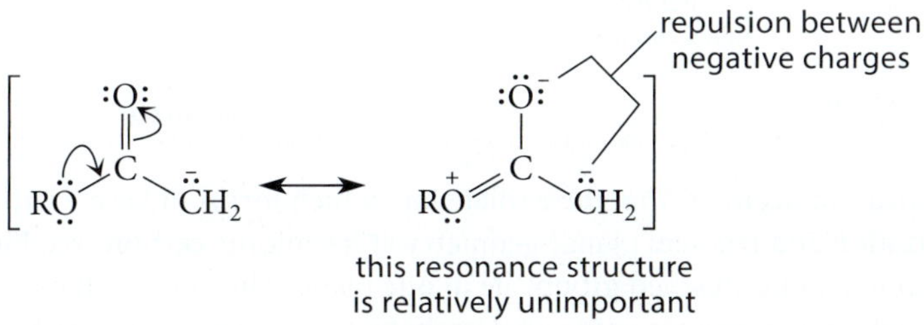

(22.9)

Therefore, the partial loss of "ester resonance" on ionization increases ΔG°_a for the ionization of the ester.

Amide N—H hydrogens are also α-hydrogens—that is, they are attached to an atom that is adjacent to a carbonyl group. The N—H hydrogens are the most acidic hydrogens in primary and secondary amides.

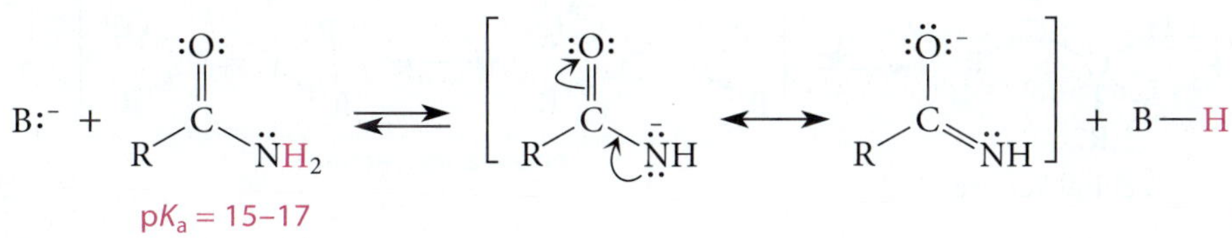

(22.10)

Similarly, carboxylic acid OH hydrogens (pK_a ~ 4–5) are also α-hydrogens. We can think of amide conjugate-base anions as nitrogen analogs of enolate ions and carboxylate anions as oxygen analogs of enolate anions. The acidity order carboxylic acids > amides > (aldehydes, ketones) corresponds to the relative electronegativities of the atoms to which the acidic hydrogens are bonded—oxygen, nitrogen, and carbon, respectively (element effect; Sec. 3.7B).

Focused Problems

22.1 Give the structures of (a) diethyl malonate (pK_a = 12.9) and (b) ethyl acetoacetate (ethyl 3-oxobutanoate, pK_a = 10.7), identify the acidic hydrogen in each, and explain why these compounds are much more acidic than ordinary esters.

22.2 Which is more acidic: the diamide of succinic acid or the imide succinimide (Display 21.12, Sec. 21.1E)? Why?

B. Introduction to the Reactions of Enolate Ions

The acidities of aldehydes, ketones, and esters are particularly important because enolate ions are key reactive intermediates in many important reactions of carbonyl compounds. Let's consider the types of reactivity we can expect to observe with enolate ions.

First, enolate ions are Brønsted bases, so they react with Brønsted acids. (The reaction of an enolate ion with an acid is the reverse of Eq. 22.3 or 22.4.) The formation of enolate ions and their reactions with Brønsted acids have two simple but important consequences. First, the α-hydrogens of an aldehyde or ketone—and no others—can be exchanged for deuterium by treating the carbonyl compound with a base in D_2O.

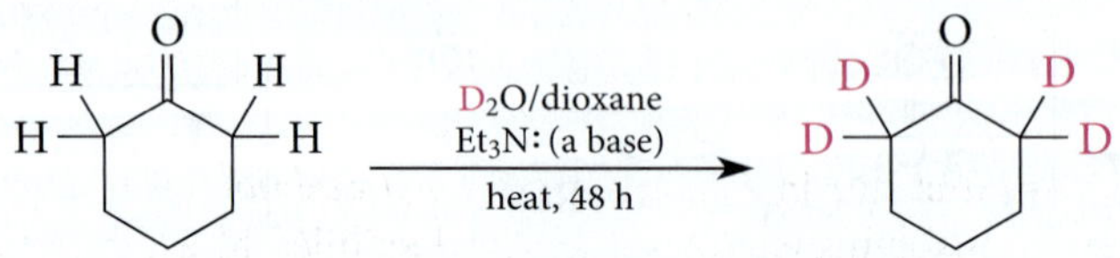

(22.11)

The second consequence of enolate-ion formation and protonation is that if an optically active aldehyde or ketone owes its chirality solely to an asymmetric α-carbon, and if this carbon bears a hydrogen, the compound will be racemized by base.

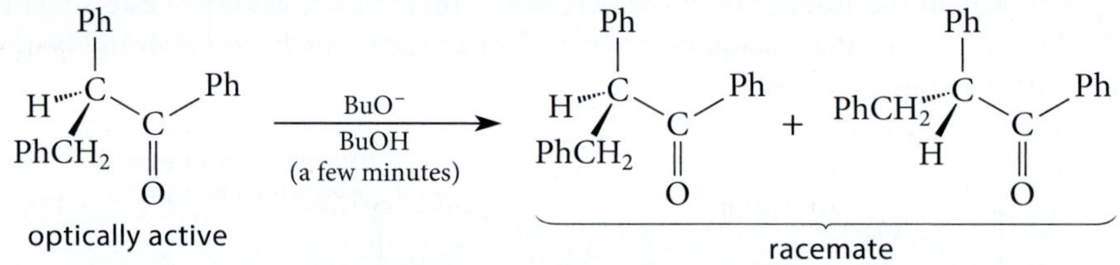

(22.12)

The reason racemization occurs is that the enolate ion, which forms in base, is *achiral* because of the sp^2 hybridization and trigonal planar geometry at its anionic carbon (see Fig. 22.1). That is, the ionic α-carbon and its attached groups lie in one plane. The anion can be reprotonated at either face to give either enantiomer with equal probability. Although not very much enolate ion is present at any one time, the reactions involved in the ionization equilibrium are relatively fast, and racemization occurs relatively quickly if the carbonyl compound is left in contact with base.

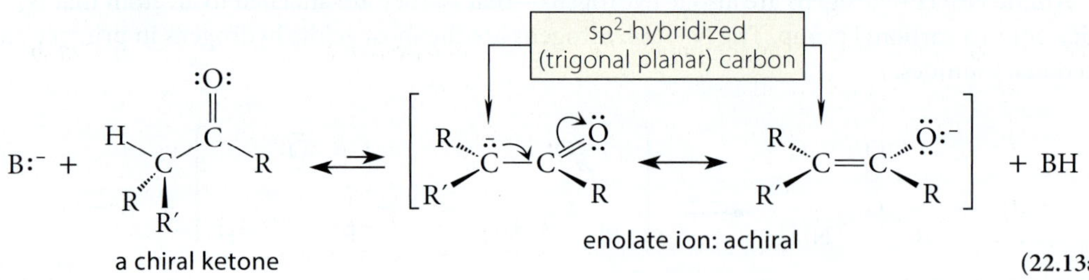

(22.13a)

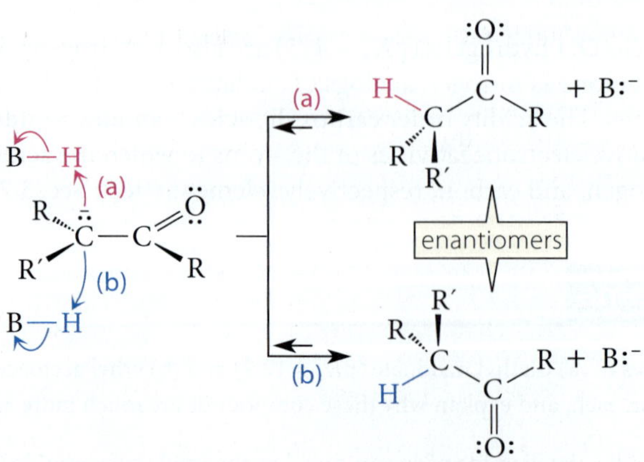

(22.13b)

The α-hydrogens of aldehydes and ketones are more acidic than the α-hydrogens of esters, so aldehydes and ketones form enolate ions more rapidly and under milder conditions. As a result, α-hydrogen exchange and racemization reactions of aldehydes and ketones occur much more readily than those of esters.

Enolate ions are not only Brønsted bases but Lewis bases as well. Consequently, enolate ions react as *nucleophiles*. Like other nucleophiles, enolate ions react at the carbons of carbonyl groups:

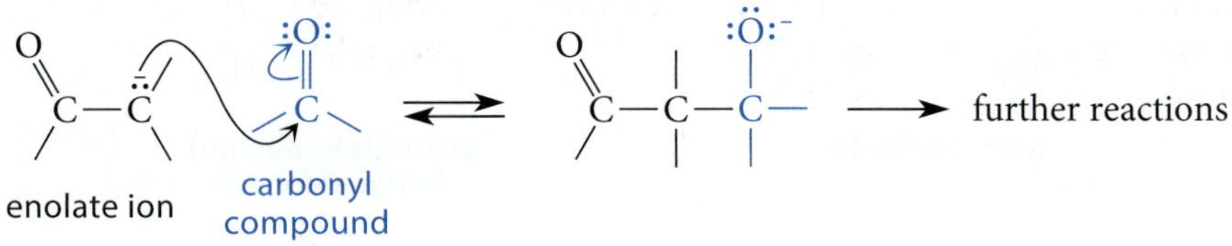

(22.14)

This type of process is the first step in a variety of *carbonyl-addition* reactions and *nucleophilic acyl substitution* reactions involving enolate ions as nucleophiles. Much of this chapter is devoted to a study of such reactions.

Enolate ions, like other nucleophiles, are alkylated by nucleophilic substitution reactions with alkyl halides and sulfonate esters:

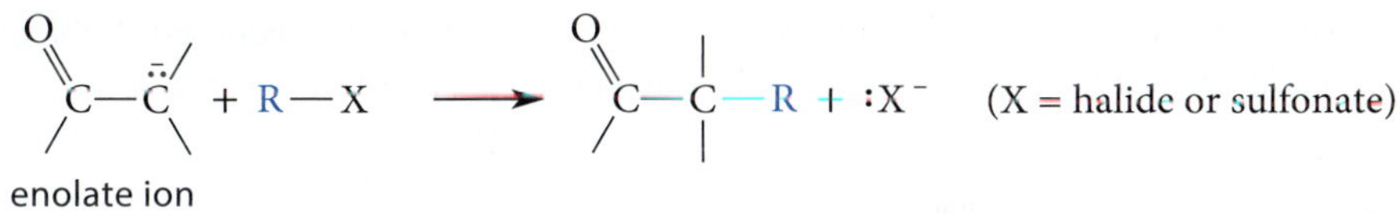

(22.15)

This type of reaction, too, is an important part of the chemistry discussed in this chapter.

Focused Problems

22.3 Write a mechanism involving an enolate ion intermediate for the reaction shown in Eq. 22.11. Explain why *only* the α-hydrogens are replaced by deuterium.

22.4 Describe or sketch the proton NMR spectrum of 2-butanone, and explain how this spectrum would change if the compound were treated with D_2O and a base.

22.5 Explain why the following compound does *not* undergo base-catalyzed exchange in D_2O even though it has an α-hydrogen. (*Hint:* See Eq. 22.6a.)

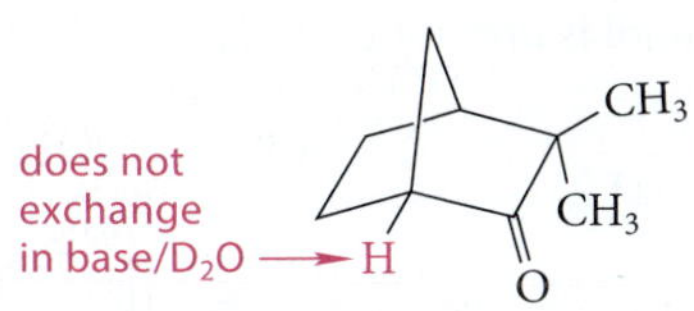

22.6 Indicate which hydrogen(s) in each of the following molecules (if any) would be exchanged for deuterium following base treatment in D_2O.

(a)
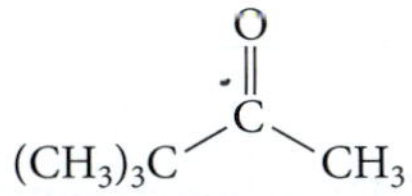

(b)
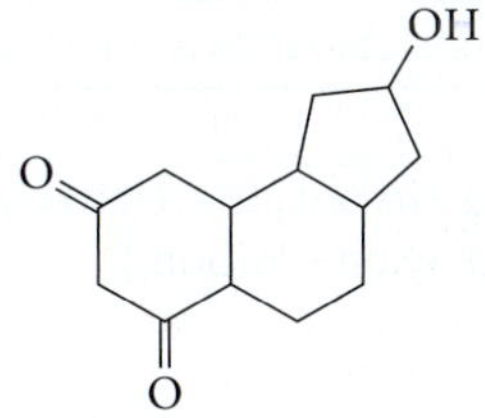

(c)

22.2 ENOLIZATION OF CARBONYL COMPOUNDS

Carbonyl compounds with α-hydrogens are in equilibrium with small amounts of their enol isomers. The equilibrium constants shown in the following equations are typical.

The term **tautomers**, which means "constitutional isomers that undergo such rapid interconversion that they cannot be independently isolated," is sometimes used to describe the relationship between enols and their corresponding carbonyl compounds. Indeed, under most common circumstances, carbonyl compounds and their corresponding enols are in rapid equilibrium. However, the interconversion of enols and their corresponding carbonyl compounds is catalyzed by acids and bases (see following discussion). This reaction can be very slow in dilute solution in the *absence* of acid or base catalysts, and indeed, enols have actually been isolated under very carefully controlled conditions. Therefore, the term *tautomers* is not very accurate and is of such limited utility that it is falling into disuse.

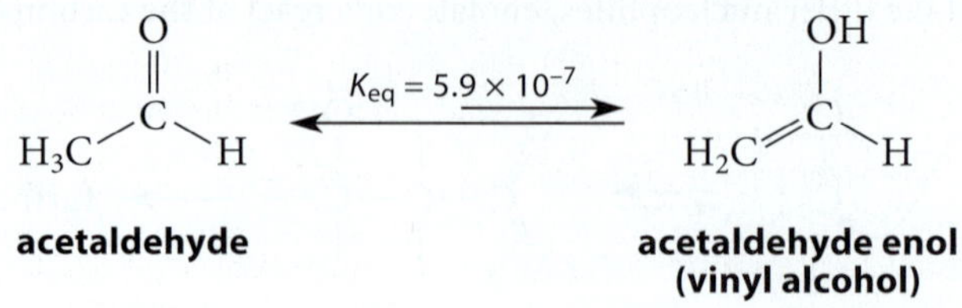

(22.16)

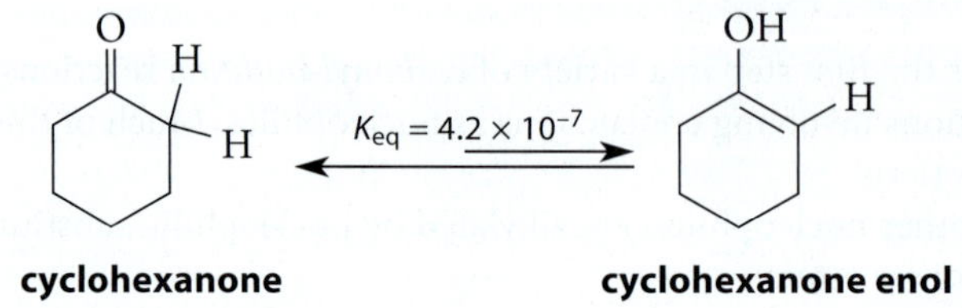

(22.17)

Unsymmetrical ketones are in equilibrium with more than one enol. (See Focused Problem 22.8.)
Esters contain even smaller amounts of enol isomers than aldehydes or ketones.

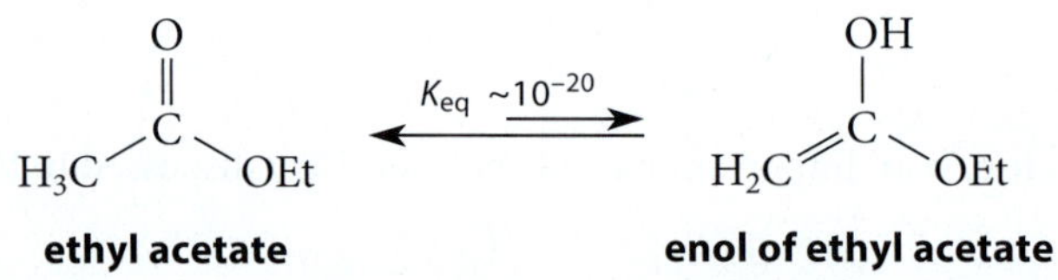

(22.18)

As the equilibrium constants in Eqs. 22.16–22.18 suggest, most carbonyl compounds are considerably more stable than their corresponding enols. Furthermore, these equations illustrate the fact that enolizations of esters and carboxylic acids are even less favorable than enolizations of most aldehydes and ketones. The major reason for the instability of enols is that the C═O double bond of a carbonyl group is a stronger bond than the C═C double bond of an enol. With esters and acids, the additional instability of enols results from loss of the stabilizing resonance interaction between the carboxylate oxygen and the carbonyl π electrons that is present in the carbonyl forms. (See Eq. 21.49, Sec. 21.7E.)

A few enols are more stable than their corresponding carbonyl compounds. Phenol, for example, is conceptually an enol—a "vinylic alcohol." However, it is more stable than its keto isomers because phenol is *aromatic*.

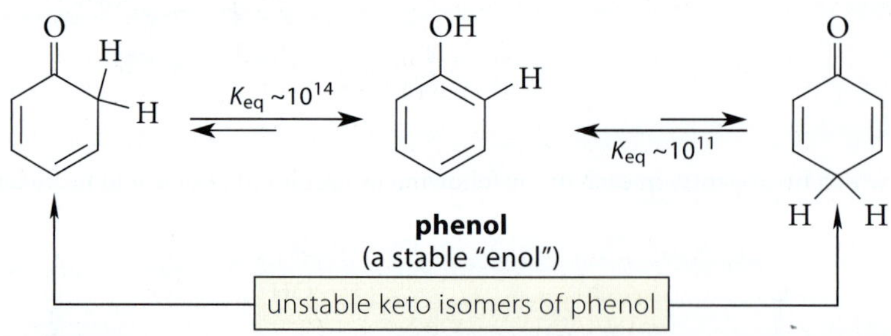

(22.19)

The enols of *β-dicarbonyl compounds* are also relatively stable. (**β-Dicarbonyl compounds** have two carbonyl groups separated by one carbon.)

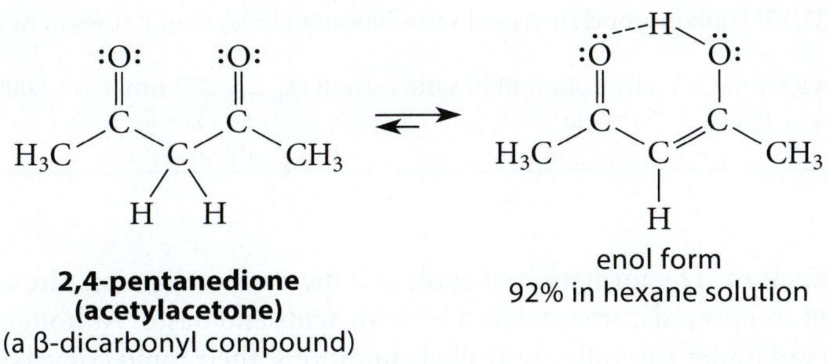

(22.20a)

These enols are stable for two reasons. First, they are conjugated, whereas their parent carbonyl compounds are not. The resonance stabilization (π-electron overlap) associated with conjugation provides additional bonding that stabilizes the enol.

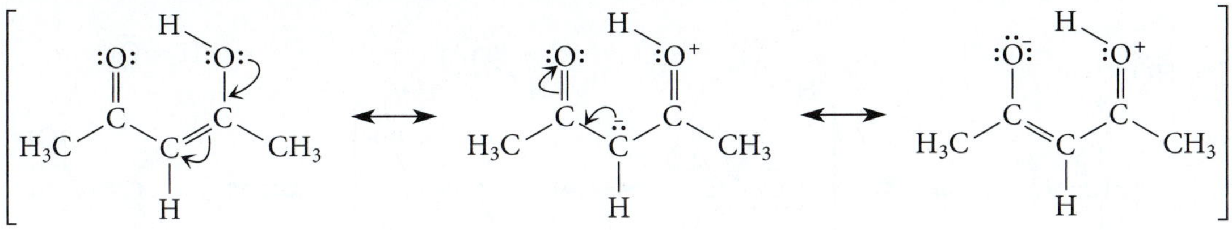

(22.20b)

The second stabilizing effect is the intramolecular hydrogen bond present in each of these enols. This provides another source of increased bonding and, as a result, increased stabilization.

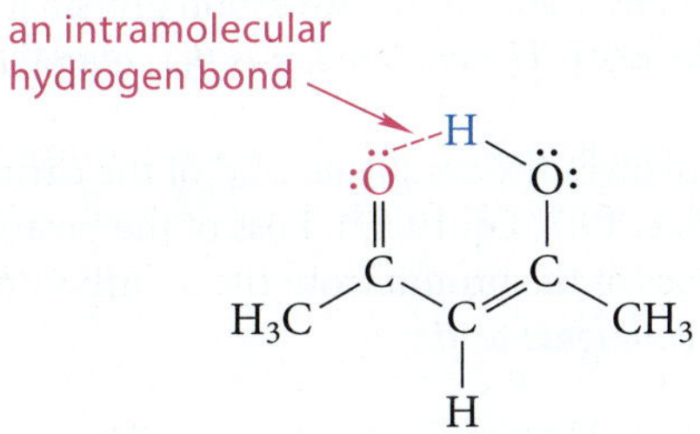

(22.20c)

Focused Problems

22.7 Draw all enol isomers of the following compounds. If there are none, explain why.

(a) 2-Methylcyclohexanone (b) 2-Methylpentanoic acid

(c) Benzaldehyde (d) *N,N*-Dimethylacetamide

(e) *N,N*-Dimethylformamide

22.8 Draw all of the enol forms of 2-butanone. Which is the least stable? Explain why. (*Hint:* Apply what you know about alkene stability.)

22.9 Draw the "enol" isomers of the following compounds. (The "enol" of a nitro compound is called an *aci*-nitro compound, and the "enol" of an amide is called an *imidic acid.*)

(a) 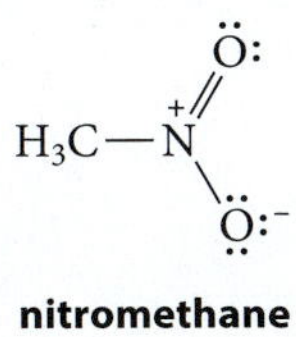

nitromethane

(b) 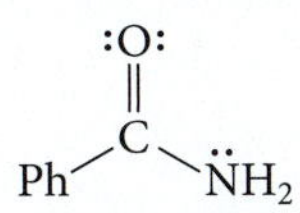

benzamide

22.10 (a) Explain why 2,4-pentanedione (Eq. 22.15) contains much less enol form in water (15%) than it does in hexane (92%).

(b) Explain why the same compound has a strong UV absorption in hexane solvent (λ_{max} = 272 nm, ε = 12,000), but a weaker absorption in water (λ_{max} = 274 nm, ε = 2050).

STUDY GUIDE LINK 22.2
Kinetic versus Thermodynamic Stability of Enols

Catalysis of Enolization The formation of enols and the reverse reaction, the conversion of enols into carbonyl compounds, are catalyzed by both acids and bases. Although enols have been isolated and observed under carefully controlled conditions, their rapid conversion into carbonyl compounds under most ordinary circumstances accounts for the fact that enols are difficult to isolate as pure compounds. (See Study Guide Link 22.2.)

The conversion of a carbonyl compound into its enol is called **enolization**. *Base-catalyzed enolization* involves the intermediacy of an *enolate ion*.

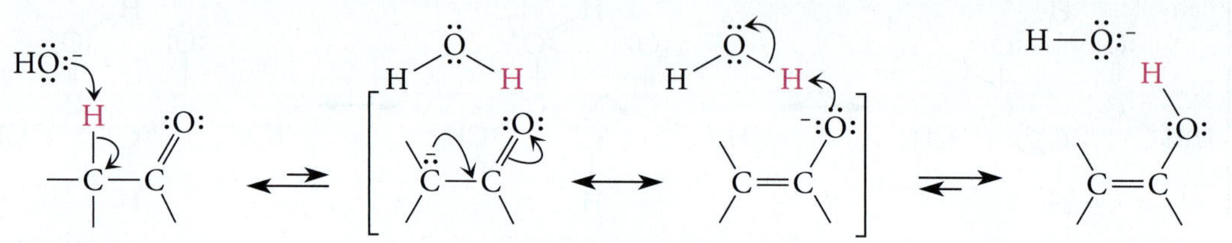

aldehyde or ketone — enolate ion (conjugate base of both the carbonyl compound and the enol) — enol **(22.21a)**

Protonation of the enolate anion by water on the α-carbon gives back the carbonyl compound; protonation on oxygen gives the enol. *The enolate ion is the conjugate base of both the carbonyl compound and the enol.*

Acid-catalyzed enolization involves the conjugate acid of the carbonyl compound. Recall that this ion is also a carbocation (Sec. 19.7, Eq. 19.27). Loss of the proton from oxygen gives back the starting carbonyl compound; loss of the proton from the α-carbon gives the enol. *An enol and its carbonyl isomer have the same conjugate acid.*

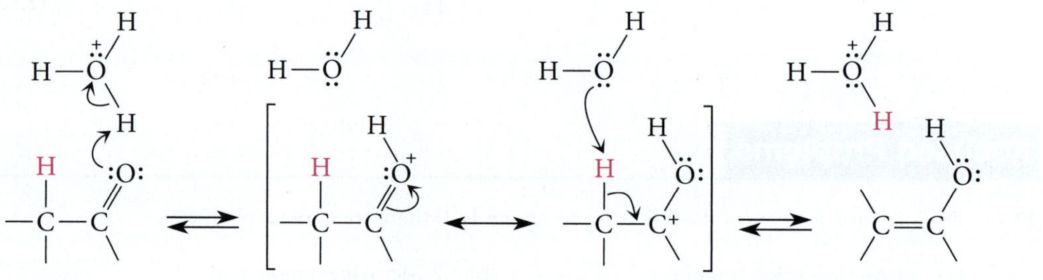

aldehyde or ketone — (conjugate acid of both the carbonyl compound and the enol) — enol **(22.21b)**

Exchange of α-hydrogens for deuterium, as well as racemization at the α-carbon, are catalyzed not only by bases (Sec. 22.1B) but also by acids.

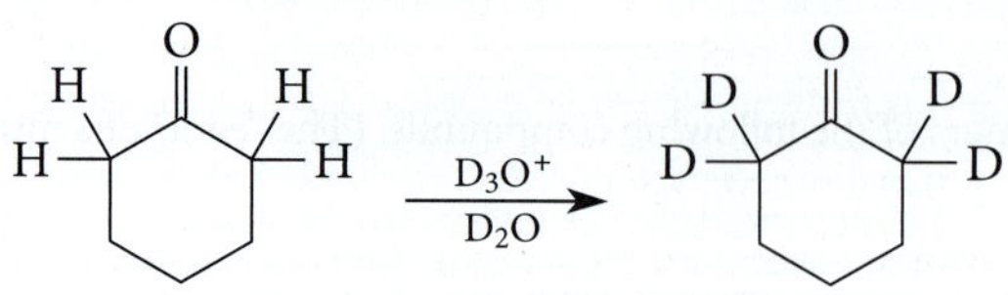

(22.22)

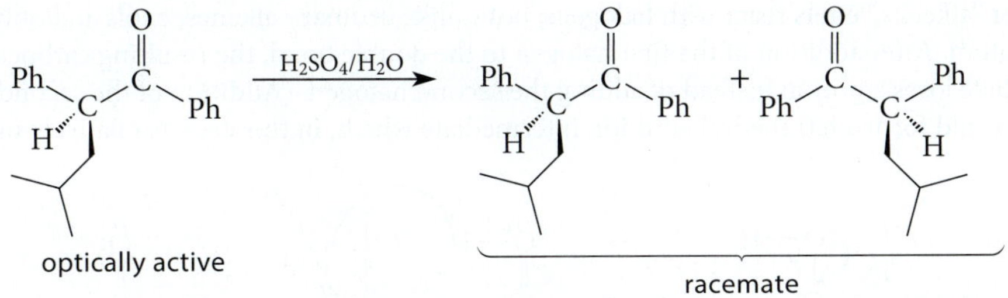

(22.23)

Both acid-catalyzed processes can be explained by the intermediacy of enols. As you can see by following Eq. 22.21b in the reverse direction, formation of a carbonyl compound from an enol introduces hydrogen from solvent at the α-carbon; this fact accounts for the observed isotope exchange. This carbon of an enol, like that of an enolate ion, has planar geometry and is therefore not an asymmetric carbon. The absence of chirality in the enol accounts for the racemization of aldehydes and ketones observed in acid.

Focused Problem

22.11 Give a curved-arrow mechanism for the racemization shown in Eq. 22.23.

22.3 α-HALOGENATION OF CARBONYL COMPOUNDS

A. Acid-Catalyzed α-Halogenation

This section begins a survey of reactions that involve enols and enolate ions as reactive intermediates. Halogenation of an aldehyde or ketone in *acidic* solution usually results in the replacement of *one* α-hydrogen by halogen.

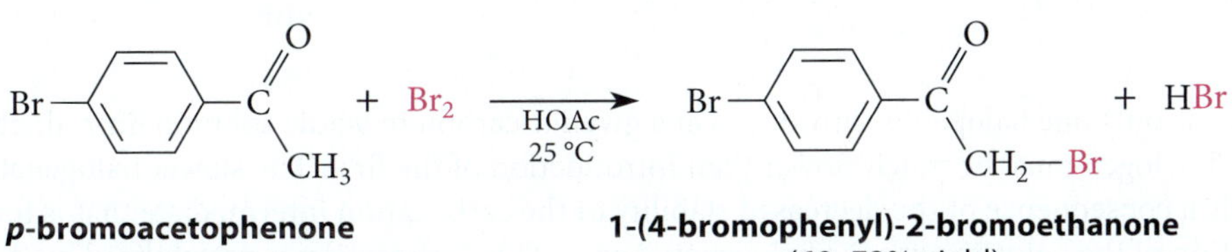

(22.24)

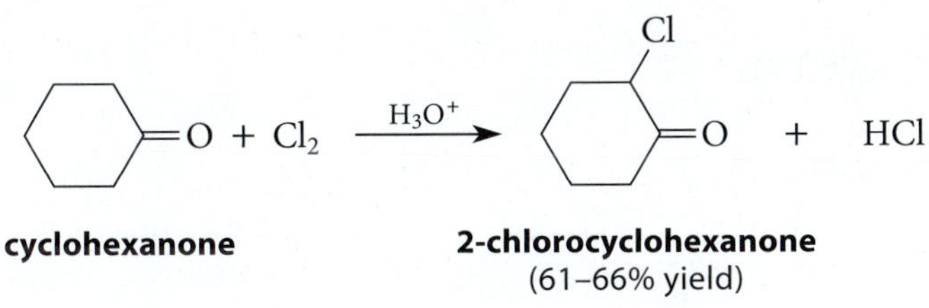

(22.25)

Enols are reactive intermediates in these reactions.

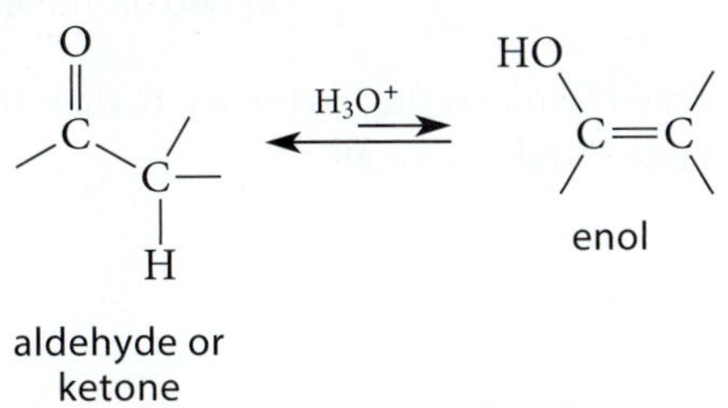

(22.26a)

Like other "alkenes," enols react with halogens; but unlike ordinary alkenes, enols add only one halogen atom. After addition of the first halogen to the double bond, the resulting carbocation intermediate loses a proton instead of adding the second halogen. (Addition of the second halogen would form a tetrahedral addition intermediate which, in this case, is relatively unstable.)

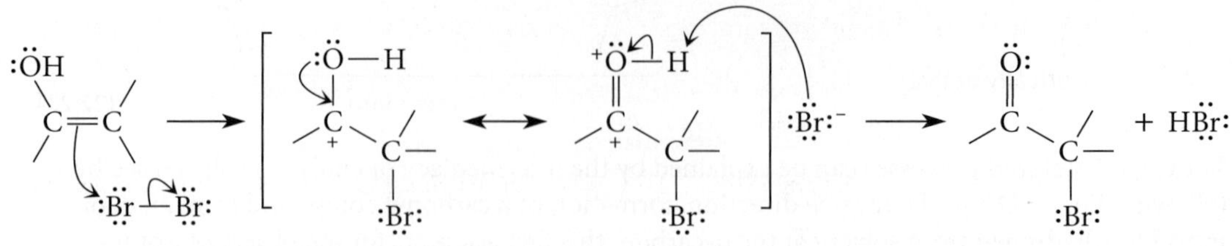

(22.26b)

Acid-catalyzed halogenation provides a particularly instructive case study that shows the importance of the rate law in determining the mechanism of a reaction. Under the usual reaction conditions, the rate law for acid-catalyzed halogenation is

$$\text{rate} = k[\text{ketone}][\text{H}_3\text{O}^+] \quad (22.27)$$

where k is the rate constant. This rate law implies that, even though the reaction is a halogenation, *the rate is independent of the halogen concentration.* Consequently, halogens *cannot* be involved in the transition state for the rate-limiting step of the reaction (Sec. 9.3B). From this observation and others, it was deduced that *enol formation* (Eq. 22.26a) *is the rate-limiting process in the acid-catalyzed halogenation of aldehydes and ketones.* Because the halogen is not involved in enol formation, it does not appear in the rate law at the concentrations of halogen ordinarily used.

Enol formation is described in this equation as the rate-limiting *process.* This process consists of two elementary steps, as shown in Eq. 22.21b (Sec. 22.2B). The rate-limiting *step* of acid-catalyzed enolization is the second step, removal of the α-proton. The same step, therefore, is also the rate-limiting *step* of α-halogenation.

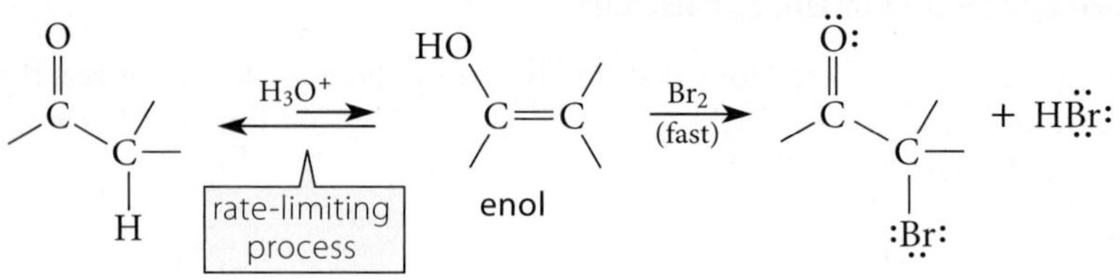

(22.28)

Because only one halogen is introduced at a given α-carbon in acidic solution, introduction of a second halogen must be much slower than introduction of the first. The slower halogenation is probably a consequence of the decreased stability of the carbocation intermediate that is formed by reaction of the halogen with the halogenated enol. This carbocation is destabilized by the electron-withdrawing polar effect of *two* halogens:

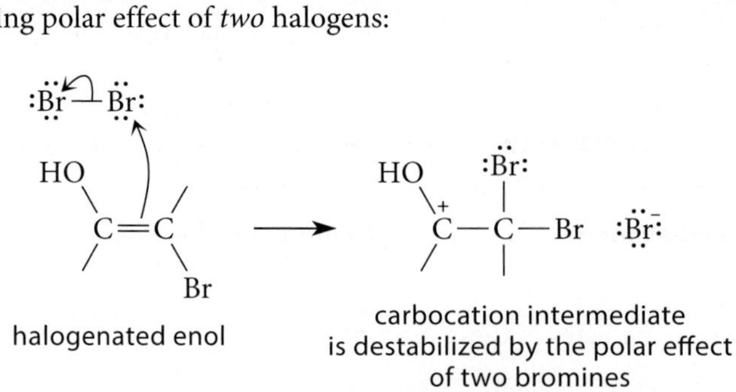

(22.29)

If the rate-limiting transition state resembles this carbocation, then the transition state should have very high energy and the rate should be small.

Focused Problems

22.12 (a) Sketch a reaction free-energy diagram for acid-catalyzed enol formation using the mechanism in Eq. 22.21b as your guide. Assume that the second step, proton removal from the α-carbon, is rate-limiting.

(b) Incorporating the results of part (a), sketch a reaction free-energy diagram for the acid-catalyzed halogenation of an aldehyde or ketone.

22.13 Explain each of the following observations.

(a) The rate of iodination of optically active 1-phenyl-2-methyl-1-butanone in acetic acid/HNO_3 is identical to its rate of racemization under the same conditions.

(b) The rates of bromination and iodination of acetophenone are identical at a given acid concentration.

B. Halogenation of Aldehydes and Ketones in Base: The Haloform Reaction

Halogenation of aldehydes and ketones with α-hydrogens also occurs in base. In this reaction, *all* α-hydrogens are substituted by halogen.

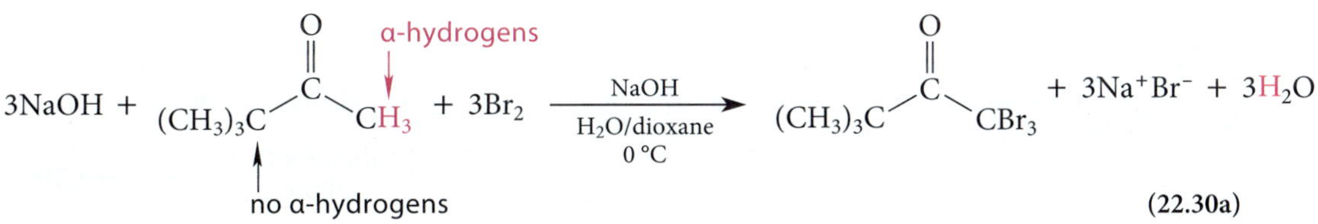

(22.30a)

When the aldehyde or ketone starting material is either acetaldehyde or a methyl ketone (as in Eq. 22.30a), the product of halogenation is a trihalo carbonyl compound, which is unstable under the reaction conditions. This compound reacts further to give, after acidification of the reaction mixture, a carboxylic acid and a haloform. (As explained in Sec. 8.2A, a *haloform* is a trihalomethane—that is, a compound of the form HCX_3, where X = halogen.)

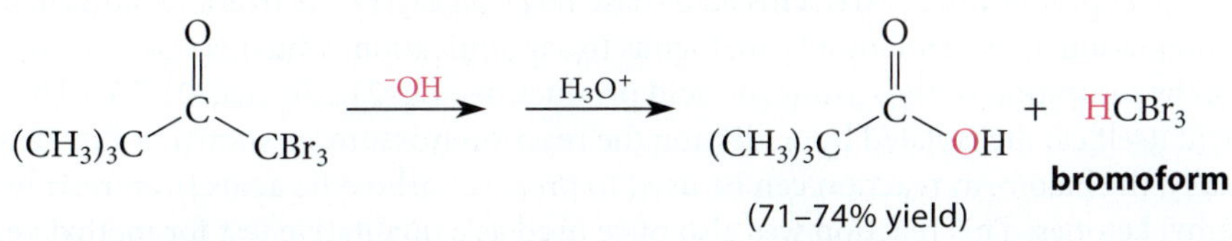

(22.30b)

The conversion of acetaldehyde or a methyl ketone into a carboxylic acid and a haloform by halogen in base, followed by acidification, as illustrated by Eqs. 22.30a and b, is called the **haloform reaction**. Notice that a carbon–carbon bond is broken in a haloform reaction.

The mechanism of the haloform reaction involves the formation of an *enolate ion* as a reactive intermediate.

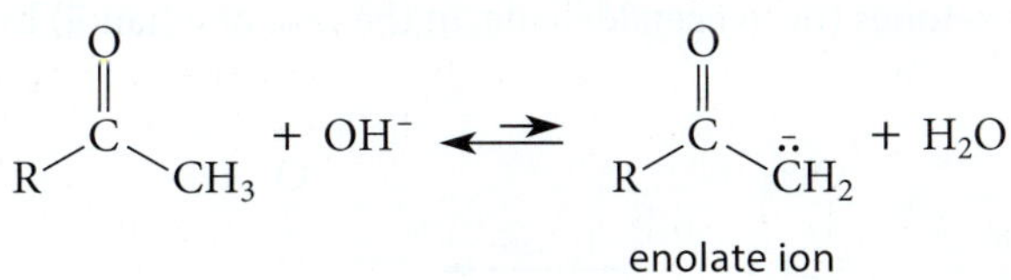

(22.31a)

The enolate ion reacts as a nucleophile with halogen to give an α-halo carbonyl compound.

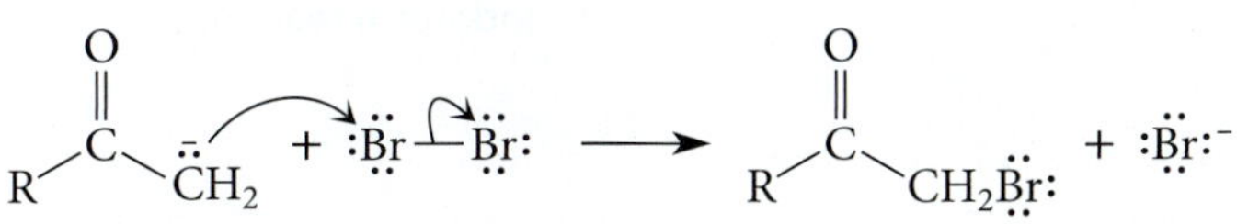

(22.31b)

Halogenation does not stop here, however, because the enolate ion of the α-halo ketone is formed even more rapidly than the enolate ion of the starting ketone. The polar effect of the halogen stabilizes the enolate ion and, by Hammond's postulate, the transition state for enolate-ion formation. Consequently, a second bromination occurs.

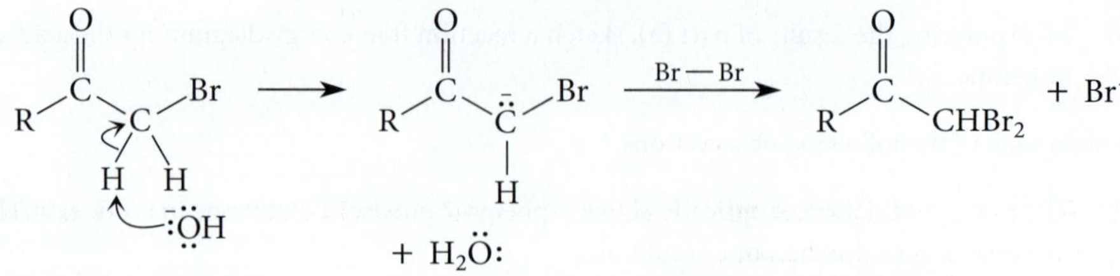

(22.31c)

The dihalo carbonyl compound brominates again, even more rapidly. (Why?)

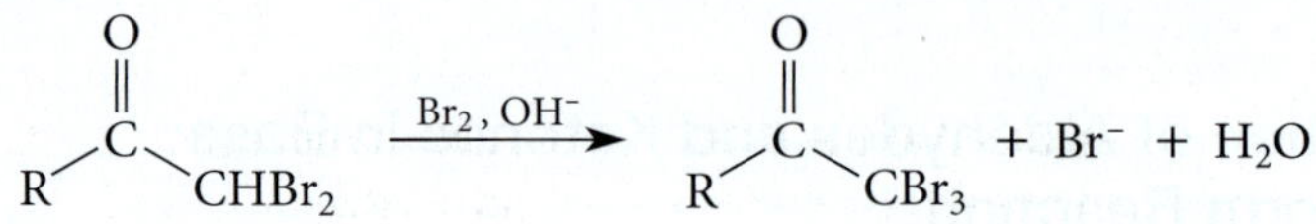

(22.31d)

A carbon–carbon bond is broken when the trihalo carbonyl compound undergoes a *nucleophilic acyl substitution reaction.*

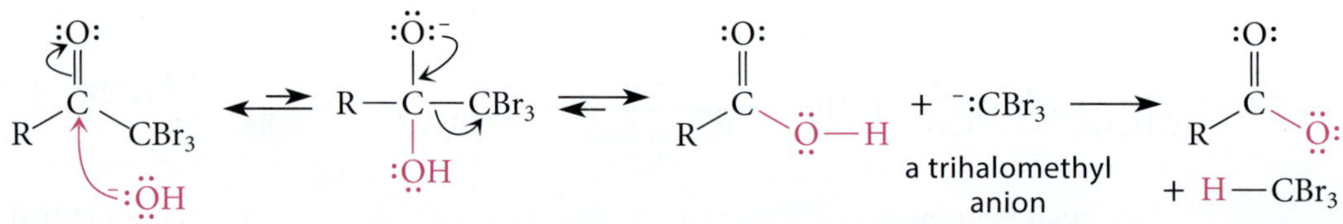

(22.31e)

The leaving group in this reaction is a trihalomethyl anion. Usually, carbanions are too basic to serve as leaving groups; but trihalomethyl anions are much less basic than ordinary carbanions. (Why?) However, the basicity of trihalomethyl anions, although low enough for them to act as leaving groups, is high enough for them to react irreversibly with the carboxylic acid by-product, as shown in the last part of Eq. 22.31e. This acid–base reaction drives the overall haloform reaction to completion. (This situation is analogous to saponification, which is also driven to completion by ionization of the carboxylic acid product; see Eq. 21.33b, Sec. 21.7A.) The carboxylic acid itself can be isolated by acidifying the reaction mixture, as shown in Eq. 22.30b.

There are a few unusual situations in which ketones other than methyl ketones undergo the haloform reaction, but these can be understood from mechanistic considerations. (See, for example, Problem 22.82.)

Occasionally, the haloform reaction can be used to prepare carboxylic acids from readily available methyl ketones. This reaction was also once used as a qualitative test for methyl ketones, called the **iodoform test**. In the iodoform test, a compound of unknown structure is mixed with alkaline I_2. A yellow precipitate of iodoform (HCI_3) is taken as evidence for a methyl ketone (or acetaldehyde, the "methyl aldehyde"). The iodoform test is specific for methyl ketones because only by replacement of *three* hydrogens with halogen does the carbon become a good enough leaving group for the nucleophilic acyl substitution reaction shown in Eq. 22.31e to occur.

Alcohols of the form shown in Eq. 22.32 also undergo the iodoform reaction because they are oxidized to methyl ketones (or to acetaldehyde, in the case of ethanol) by the basic iodine solution.

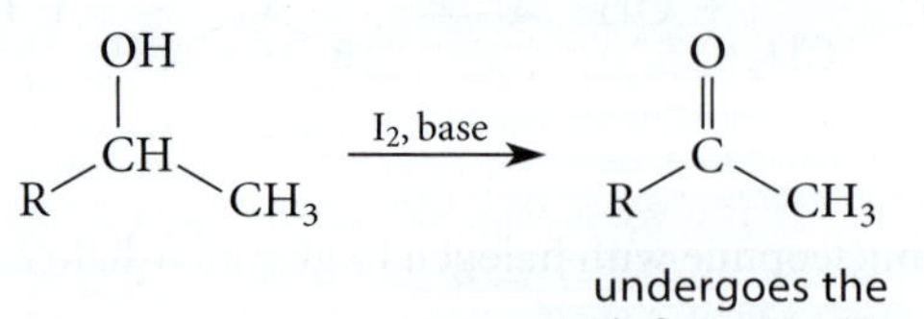

(22.32)

Focused Problems

22.14 Give the products expected (if any) when each of the following compounds reacts with Br_2 in NaOH.

(a) 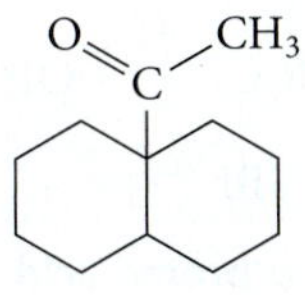 (b) 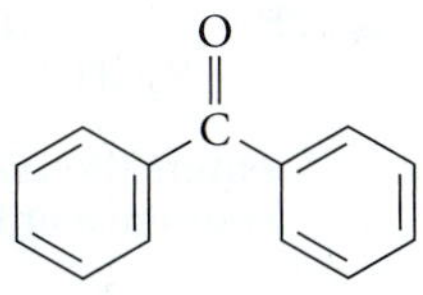 (c) Ph—CH(OH)—CH_3

22.15 Give the structure of a compound $C_6H_{10}O_2$ that gives succinic acid and iodoform on treatment with a solution of I_2 in aqueous NaOH, followed by acidification.

C. α-Bromination of Carboxylic Acids

Carboxylic acids can be brominated at their α-carbons. A bromine is substituted for an α-hydrogen when a carboxylic acid is treated with Br_2 and a catalytic amount of red phosphorus or PBr_3. (The actual catalyst is PBr_3; phosphorus can be used because it reacts with Br_2 to give PBr_3.)

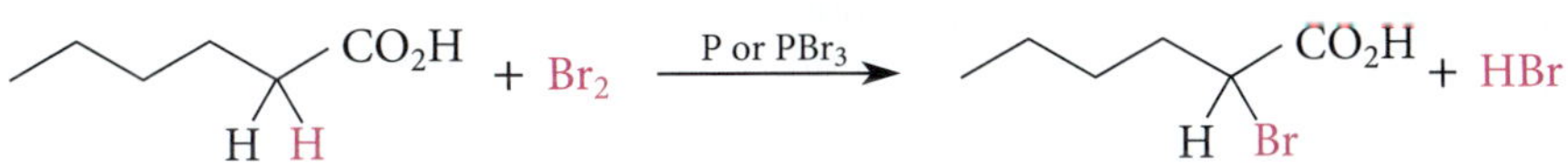

hexanoic acid **2-bromohexanoic acid** (83–89% yield) (22.33)

This reaction, called the **Hell–Volhard–Zelinsky reaction**, is sometimes nicknamed the **HVZ reaction.**

The HVZ reaction was discovered by German chemists Carl Magnus von Hell (1849–1926) and Jacob Volhard (1834–1910), and Russian chemist Nicolai Zelinsky (1861–1953), and named after them.

The first stage in the mechanism of the HVZ reaction is the conversion of a small amount of the carboxylic acid into the acid bromide by the catalyst PBr_3 (Sec. 20.9A).

carboxylic acid with α-hydrogens (R = H, alkyl, or aryl) an acid bromide (22.34a)

From this point, the mechanism closely resembles that for the acid-catalyzed bromination of ketones (Eqs. 22.26a and b). The *enol* of the acid bromide is the species that actually brominates.

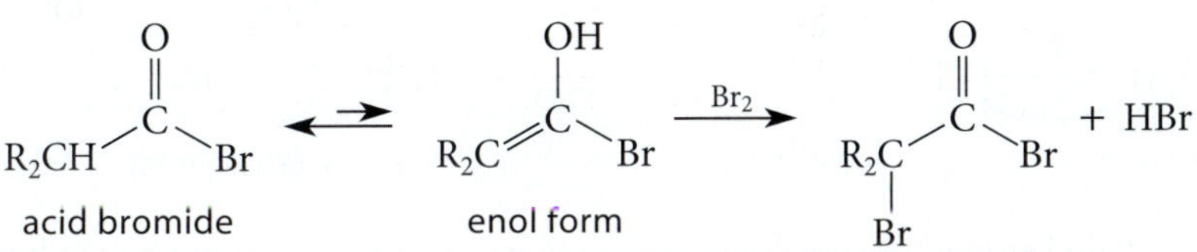

acid bromide enol form α-bromo acid bromide (22.34b)

When *a small amount* of PBr_3 catalyst is used, the α-bromo acid bromide reacts with the carboxylic acid to form more acid bromide, which is then brominated as shown in Eq. 22.34b.

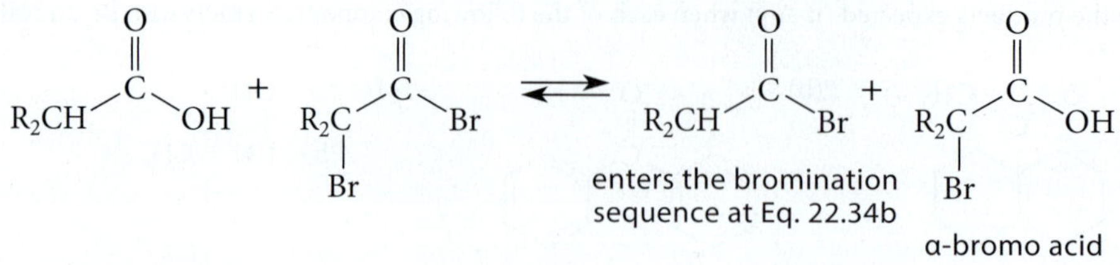

(22.35a)

The mechanism of this reaction probably involves the conjugate acid of a mixed anhydride, as shown in the following equation. (The oxygens are color-coded to assist in following the mechanism.)

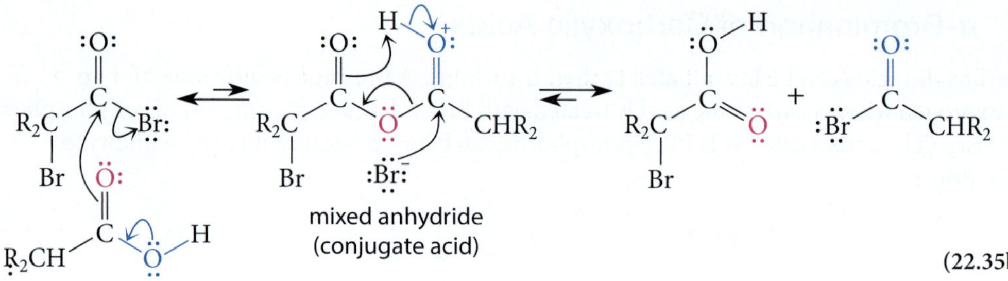

(22.35b)

As Eq. 22.35a shows, when a catalytic amount of PBr_3 is used, the reaction product is the α-bromo acid.

If one full equivalent of PBr_3 is used, the α-bromo acid bromide is the reaction product; this can be used in many of the reactions of acid halides discussed in Sec. 21.8A. For example, the reaction mixture can be treated with an alcohol to give an α-bromo ester:

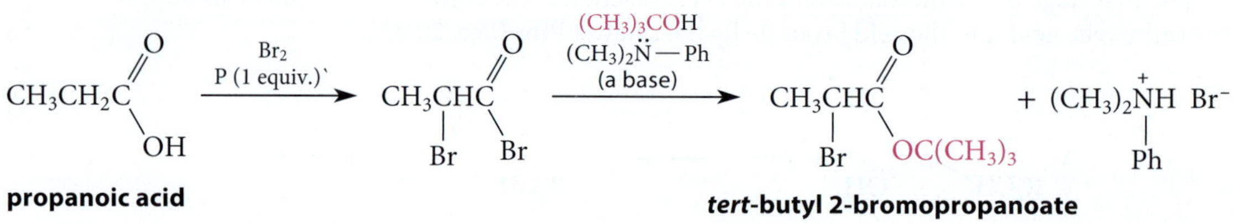

(22.36)

D. Reactions of α-Halo Carbonyl Compounds

Most primary α-halo carbonyl compounds are very reactive in S_N2 reactions and can be used to prepare other α-substituted carbonyl compounds.

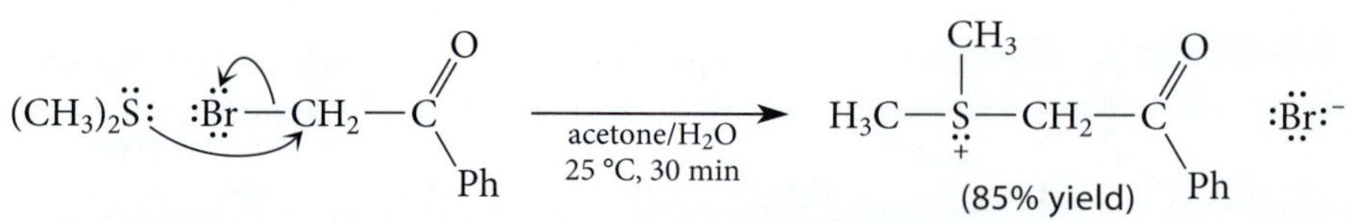

(22.37)

In the case of α-halo ketones, nucleophiles used in these reactions must not be too basic. For example, dimethyl sulfide, used in Eq. 22.37, is a very weak base but a fairly good nucleophile. (Stronger bases promote enolate-ion formation, and the enolate ions of α-halo ketones undergo other reactions.) More basic nucleophiles can be used with α-halo acids because, under basic conditions, α-halo acids are ionized to form their carboxylate conjugate-base anions; a second ionization to give an enolate ion, which would introduce a second negative charge into the molecule, does not occur.

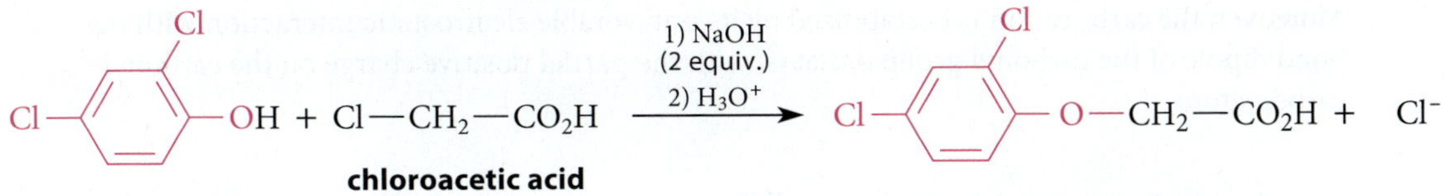

2,4-dichlorophenol (ionized by NaOH) **chloroacetic acid** **2,4-dichlorophenoxyacetic acid** (2,4-D, a selective herbicide; 87% yield)

(22.38)

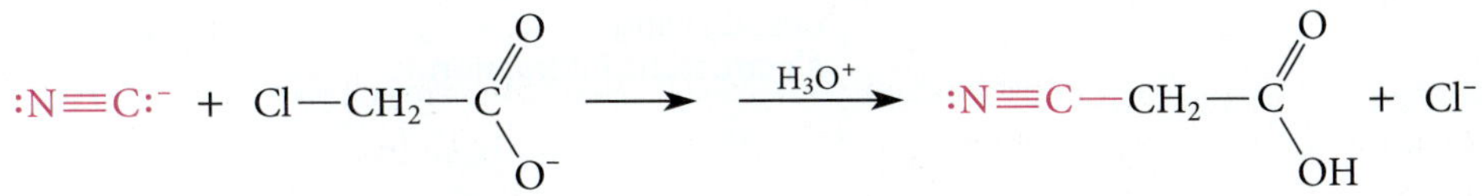

chloroacetate anion **cyanoacetic acid** (77–80% yield)

(22.39)

The following comparison gives a quantitative measure of the S_N2 reactivity of α-halo carbonyl compounds:

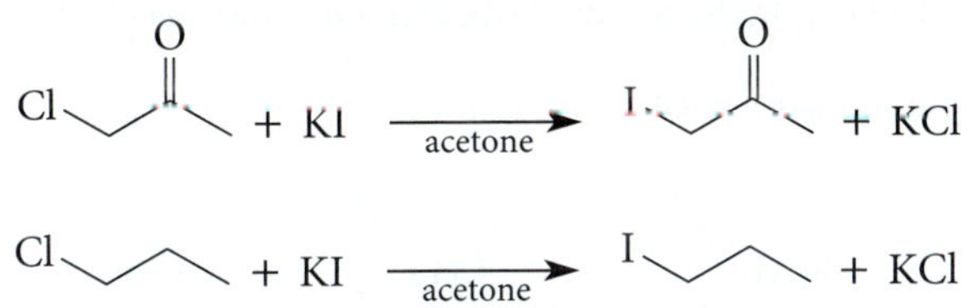

relative rate:

35,000 (22.40a)

1 (22.40b)

The explanation for the enhanced reactivity is probably similar to that for the increased reactivity of allylic alkyl halides in S_N2 displacements (see Fig. 17.3, Sec. 17.4).

In contrast, α-halo carbonyl compounds react so slowly by the S_N1 mechanism that this reaction is not useful.

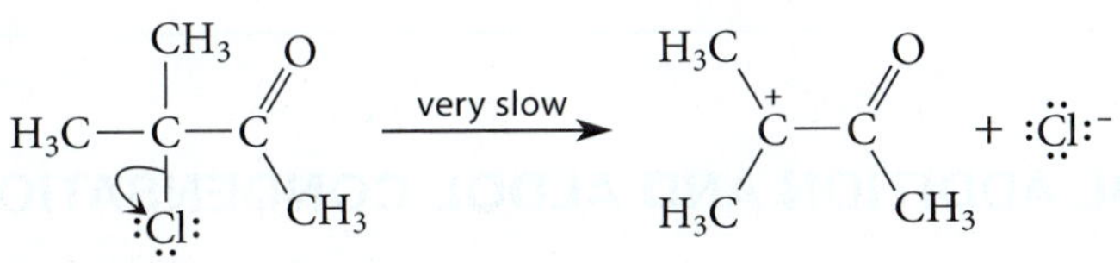

(22.41)

In fact, *reactions that require the formation of carbocations alpha to carbonyl groups generally do not occur.* Although it might seem that an α-carbonyl carbocation should be resonance-stabilized like an allylic cation, its resonance structure is not important because it involves an electron-deficient oxygen.

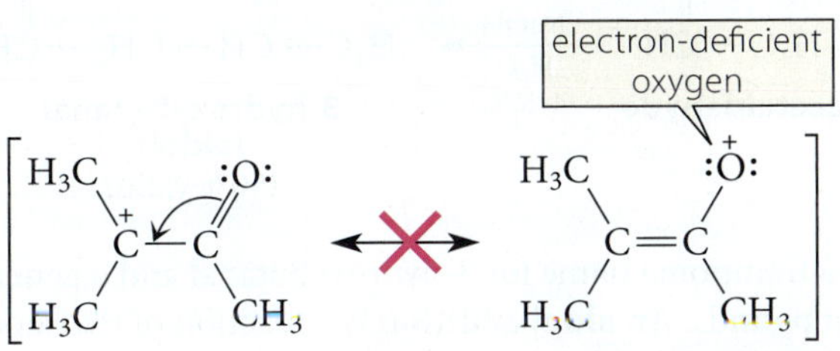

not an important resonance structure

(22.42)

Moreover, the carbocation is destabilized by its unfavorable electrostatic interaction with the bond dipole of the carbonyl group—that is, with the partial positive charge on the carbonyl carbon atom.

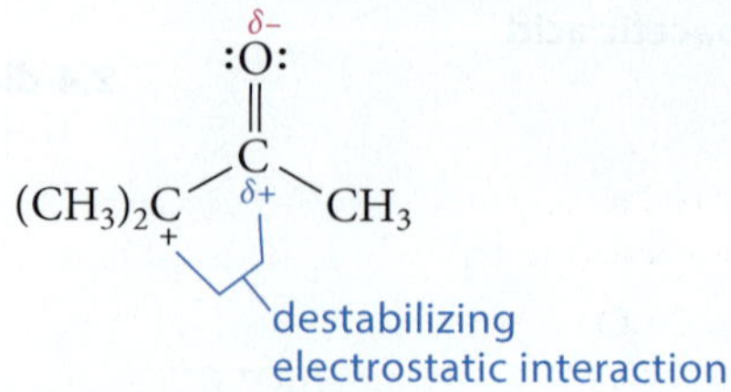

(22.43)

Focused Problems

22.16 What product is formed in each of the following situations?

(a) Phenylacetic acid is treated first with Br_2 and one equivalent of PBr_3, then with a large excess of ethanol.

(b) Propionic acid is treated first with Br_2 and one equivalent of PBr_3, then with a large excess of ammonia.

22.17 Give the structure of the product expected in each of the following reactions.

(a)

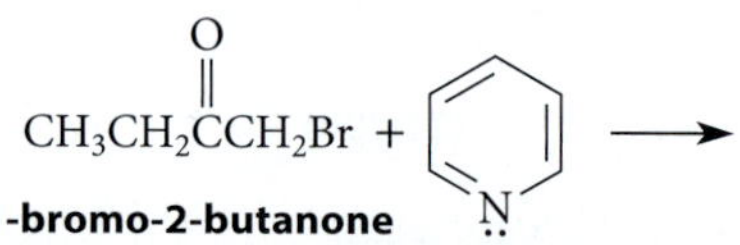

1-bromo-2-butanone **pyridine**

(b)

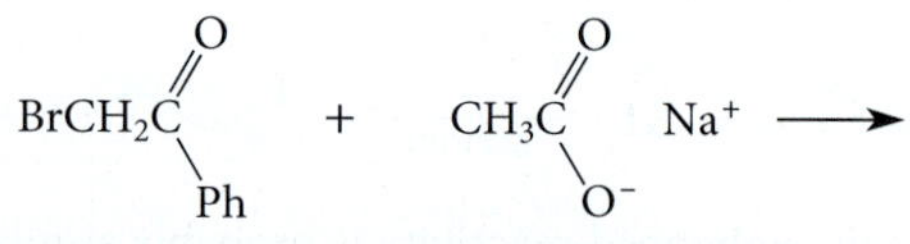

α-bromoacetophenone **sodium acetate**

22.18 Give a curved-arrow mechanism for the reaction in Eq. 22.38. Your mechanism should show why *two equivalents* of NaOH must be used.

22.4 ALDOL ADDITION AND ALDOL CONDENSATION

A. Base-Catalyzed Aldol Reactions

In aqueous base, acetaldehyde undergoes a reaction called the *aldol addition.*

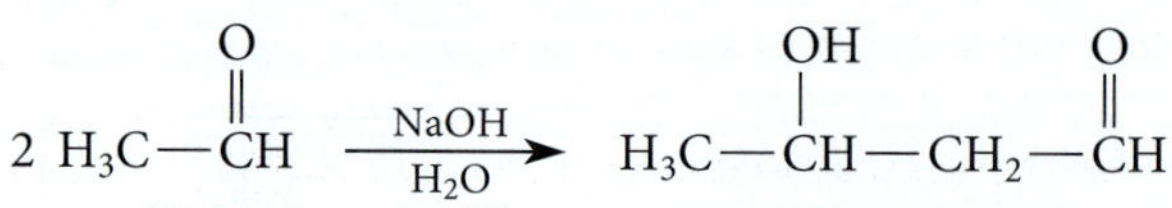

acetaldehyde **3-hydroxybutanal (aldol)** (50% yield)

(22.44)

The term **aldol** is both a traditional name for 3-hydroxybutanal and a generic name for β-hydroxy carbonyl compounds. An **aldol addition** is a reaction of two aldehyde molecules to form a β-hydroxy aldehyde. The aldol addition is a very important and general reaction of aldehydes and ketones that have α-hydrogens. This reaction provides another method of forming carbon–carbon bonds.

The base-catalyzed aldol addition involves an *enolate ion* as an intermediate. In this reaction, an enolate ion, formed by the reaction of acetaldehyde with aqueous NaOH, adds to a second molecule of acetaldehyde.

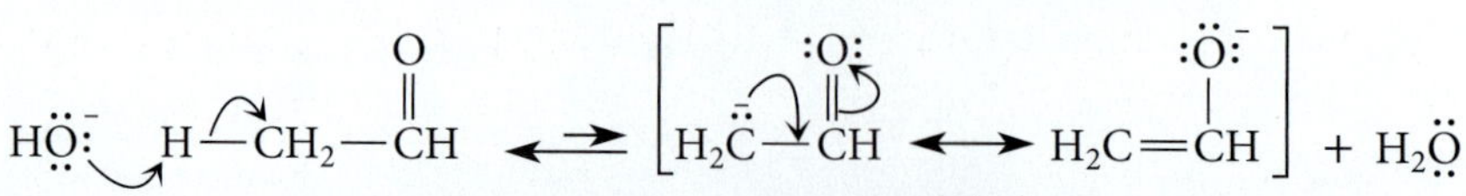

enolate ion

(22.45a)

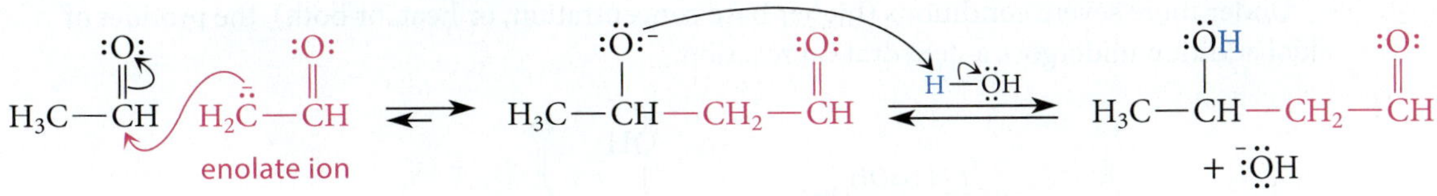

(22.45b)

The aldol addition is another nucleophilic addition to a carbonyl group. In this reaction, the nucleophile is an enolate ion. The reaction may *look* more complicated than some additions because of the number of carbon atoms in the product. However, it is not conceptually different from other nucleophilic additions, such as cyanohydrin formation.

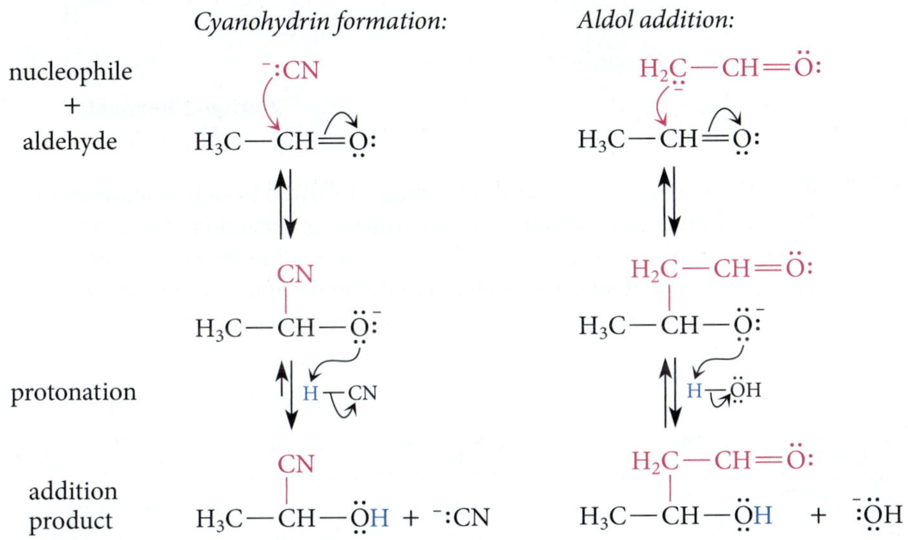

(22.46)

Focused Problem

22.19 Use the reaction mechanism to deduce the product of the aldol addition reaction of (a) $PhCH_2CH{=}O$ (phenylacetaldehyde); (b) propionaldehyde.

The aldol addition is reversible. Like many other carbonyl-addition reactions (Sec. 19.8B), the equilibrium for the aldol addition is more favorable for aldehydes than for ketones.

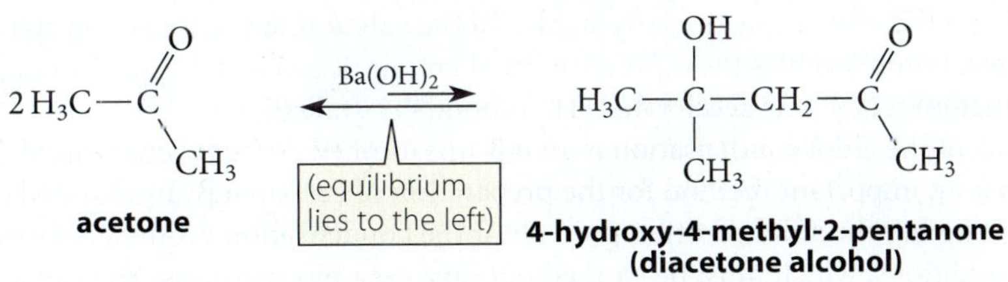

(22.47)

In this aldol addition reaction of acetone, the equilibrium favors the ketone reactant rather than the addition product, diacetone alcohol. This product can be isolated in good yield only if an apparatus is used that allows the product to be removed from the base catalyst as it is formed. This strategy drives the reaction toward formation of product by Le Chatelier's principle.

Under more severe conditions (higher base concentration, or heat, or both), the product of aldol addition undergoes a dehydration reaction.

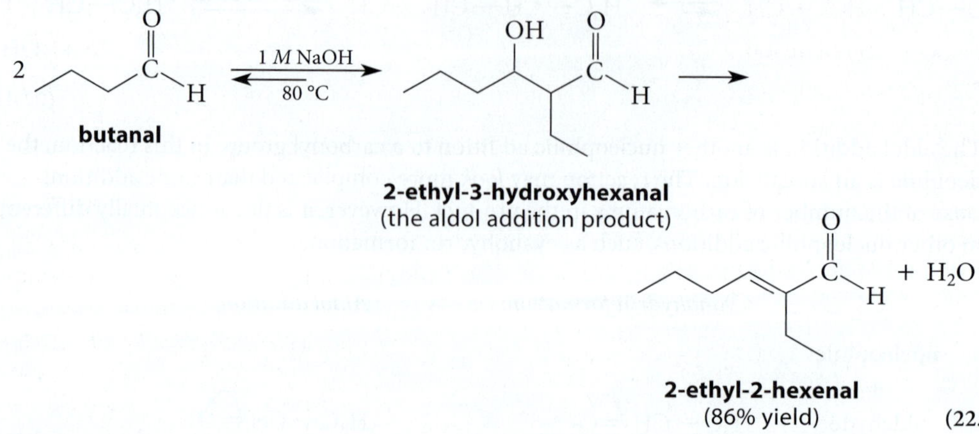

(22.48)

The term *aldol condensation* has been used historically to refer to the aldol addition reaction as well as to the addition and dehydration reactions together. To eliminate ambiguity, *aldol condensation* is used in this text only for the addition–dehydration sequence. The term *aldol reaction* is used to refer generically to both addition and condensation reactions.

The sequence of reactions consisting of the aldol addition followed by dehydration, as in Eq. 22.48, is called the **aldol condensation.** (A **condensation** is a reaction in which two molecules combine to form a larger molecule with the elimination of a small molecule, in many cases water.)

The dehydration part of the aldol condensation is a β-elimination reaction catalyzed by base, and it occurs in two distinct steps through an *enolate-ion intermediate.*

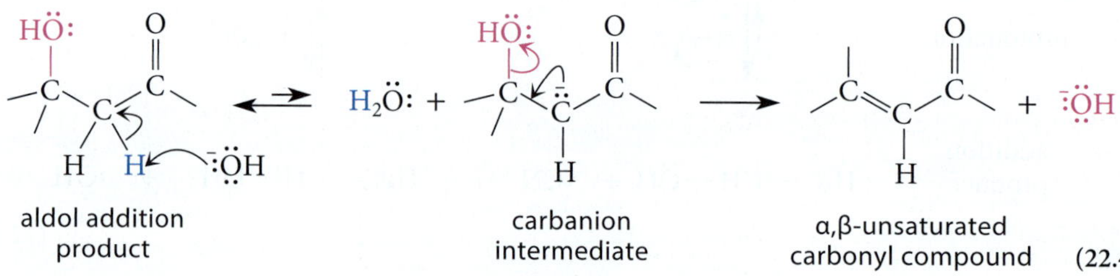

(22.49)

This is *not* a concerted β-elimination. In this respect, it differs from the E2 reaction. (See Focused Problem 9.18, Sec. 9.5B.) A β-elimination mechanism that involves the unimolecular decomposition of the conjugate base is sometimes called an **E1cB mechanism.**

elimination ⟶ E1cB ⟵ conjugate base
↑
unimolecular

(22.50)

Ordinary alcohols do *not* dehydrate in base because hydroxide ion is such a poor leaving group. β-Hydroxy aldehydes and β-hydroxy ketones, however, do dehydrate for two reasons. First, their α-hydrogens are relatively acidic. Recall that base-promoted β-eliminations are particularly rapid when acidic hydrogens are involved (Secs. 17.3B and 9.5B). Second, the product is conjugated and therefore is particularly stable. To the extent that the transition state of the dehydration reaction resembles the α,β-unsaturated ketone, it too is stabilized by conjugation, and the elimination reaction is accelerated (Hammond's postulate).

The product of the aldol condensation is an α,β-unsaturated carbonyl compound. The aldol condensation is an important method for the preparation of certain α,β-unsaturated carbonyl compounds. Whether the aldol addition product or the condensation product is formed depends on reaction conditions, which must be worked out on a case-by-case basis. You can assume for purposes of problem-solving, unless stated otherwise, that either the addition product or the condensation product can be prepared.

Musical History of the Aldol Condensation

Pictorial Press Ltd/Alamy

Discovery of the aldol condensation in 1872 is usually attributed solely to Charles-Adolphe Wurtz (1817–1884), a French chemist who trained Friedel and Crafts. However, the reaction was first investigated during the period 1864–1873 by Aleksandr Borodin (1833–1887), a Russian chemist who was also a self-taught and proficient musician and composer. (Borodin was one of "The Five," a circle of self-trained St. Petersburg composers who sought to reshape Russian music along nationalistic lines.) Borodin found it difficult to compete with Wurtz's large, modern, well-funded laboratory. Borodin also lamented that his professional duties so burdened him with "examinations and commissions" that he could only compose when he was at home ill. Knowing this, his musical friends used to greet him, "Aleksandr, I hope you are ill today!"

B. Acid-Catalyzed Aldol Condensation

Aldol condensations are also catalyzed by acid.

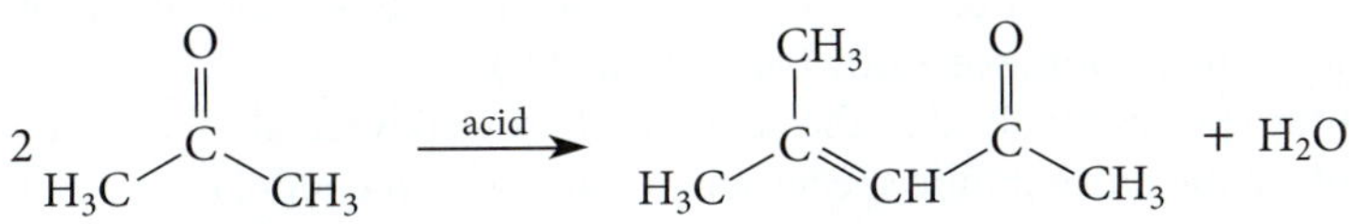

acetone **mesityl oxide (4-methyl-3-penten-2-one)** (79% yield) (22.51)

Acid-catalyzed aldol condensations, as in this example, generally give α,β-unsaturated carbonyl compounds as products; addition products cannot be isolated.

In acid-catalyzed aldol condensations, the conjugate acid of the aldehyde or ketone is a key reactive intermediate.

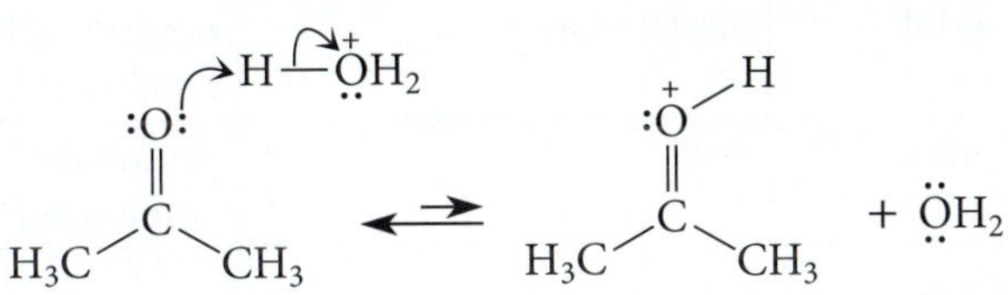

(22.52a)

This protonated ketone plays two roles. First, it serves as a source of the *enol*, as shown in Eq. 22.21b in Sec. 22.2. Second, the protonated ketone is the electrophilic species in the reaction. It reacts as an electrophile with the π electrons of the enol to give the conjugate acid of the addition product:

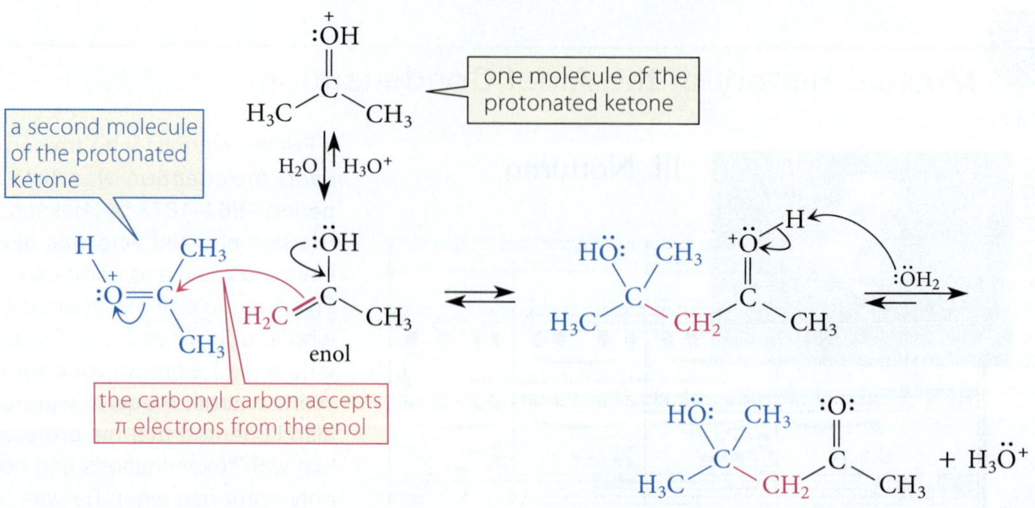

(22.52b)

As the second part of Eq. 22.52b shows, the loss of a proton gives the β-hydroxy ketone product. Under the acidic conditions, this material spontaneously undergoes acid-catalyzed dehydration to give an α,β-unsaturated carbonyl compound:

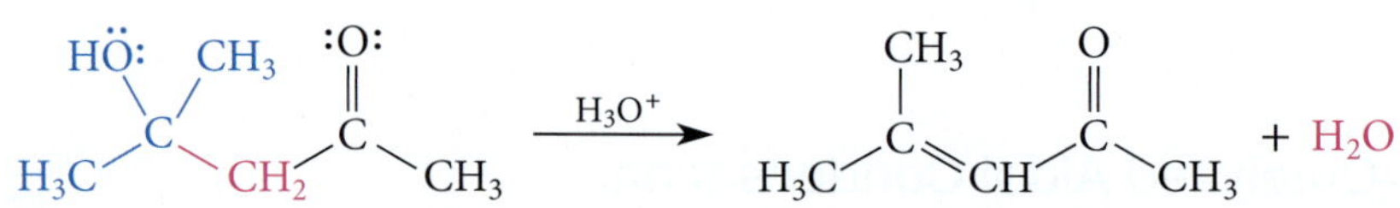

(22.52c)

STUDY GUIDE LINK 22.3
Dehydration of β-Hydroxy Carbonyl Compounds

This dehydration drives the aldol condensation to completion. (Recall that without this dehydration, the aldol condensation of ketones is unfavorable; Eqs. 22.47 and 22.48.) For the mechanism of this dehydration, see Study Guide Link 22.3.

Let's contrast the species involved in the acid- and base-catalyzed aldol reactions. An *enol*, not an enolate ion, is the nucleophilic species in an acid-catalyzed aldol condensation. *Enolate ions are too basic to exist in acidic solution.* Although an enol is much less nucleophilic than an enolate ion, it reacts at a useful rate because the protonated carbonyl compound (an α-hydroxy carbocation) with which it reacts is a potent electrophile. In a base-catalyzed aldol reaction, an *enolate ion* is the nucleophile. A protonated carbonyl compound is *not* an intermediate because it is too acidic to exist in basic solution. The electrophile that reacts with the enolate ion is a *neutral* carbonyl compound. To summarize:

Reaction	*Nucleophile*	*Electrophile*
Base-catalyzed aldol reaction	enolate ion	neutral carbonyl compound
Acid-catalyzed aldol condensation	enol	protonated carbonyl compound

C. Special Types of Aldol Reactions

Crossed Aldol Reactions The preceding discussion considered only aldol reactions between two molecules of the same aldehyde or ketone. When two *different* carbonyl compounds are used, the reaction is called a **crossed aldol reaction**. In many cases, the result of a crossed aldol reaction is a difficult-to-separate mixture, as Study Problem 22.1 illustrates.

Study Problem 22.1

Give the structures of the aldol addition products expected from the base-catalyzed reaction of acetaldehyde and propionaldehyde.

Solution Such a reaction involves four different species: acetaldehyde (*A*) and its enolate ion (*A′*), as well as propionaldehyde (*P*) and its enolate ion (*P′*):

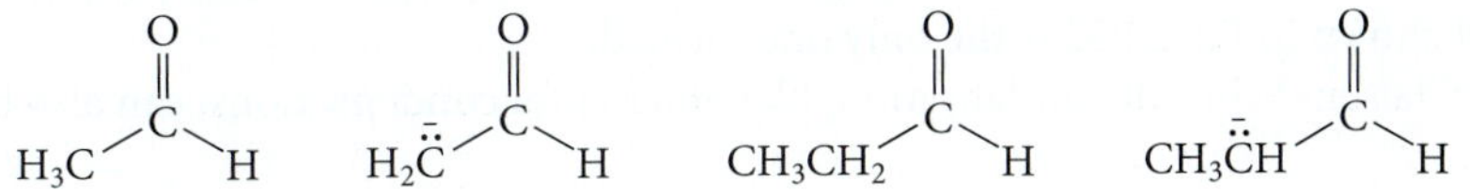

A *A′* *P* *P′*

Four possible addition products can arise from the reaction of each enolate ion with each aldehyde:

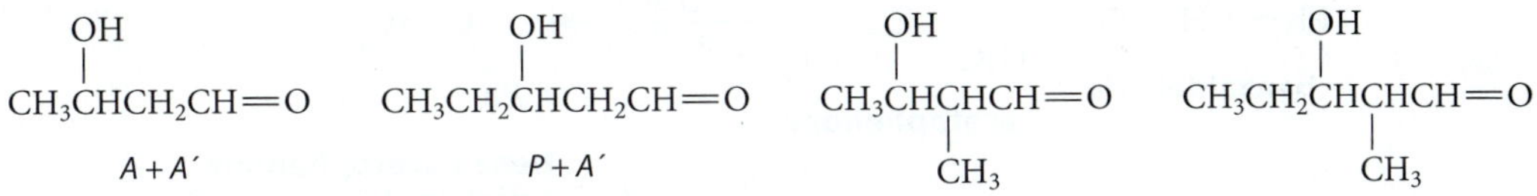

A + *A′* *P* + *A′* *A* + *P′* *P* + *P′*

(Be sure you see how each product is formed; write a mechanism for each, if necessary.) To complicate the situation even further, diastereomers are possible for the last two products because each has two asymmetric carbons.

Crossed aldol reactions that provide complex mixtures, such as the one in Study Problem 22.1, are not very useful because the product of interest is not formed in very high yield, and because isolation of one product from a complex mixture is in most cases extremely tedious. Conditions that favor one product or another in crossed aldol reactions have been worked out in specific cases. As a practical matter, however, under the usual conditions (aqueous or alcoholic acid or base), useful crossed aldol reactions are limited to situations in which *a ketone with α-hydrogens is condensed with an aldehyde that has no α-hydrogens*. An important example of this type is the **Claisen–Schmidt condensation**. In a Claisen–Schmidt condensation, a ketone with only one type of α-hydrogen—acetone in the following example—is condensed with an aromatic aldehyde that has no α-hydrogens—benzaldehyde in this case.

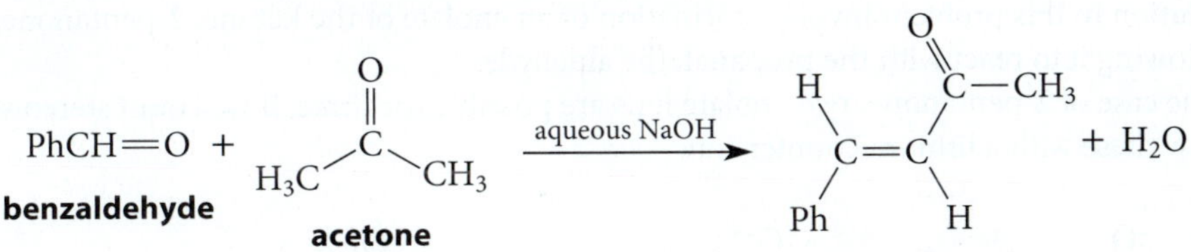

benzaldehyde **acetone** (excess) **benzalacetone** **[(*E*)-4-phenyl-3-buten-2-one]** (65–78% yield) (22.53)

The addition product is not isolated in this reaction; only the condensation product is isolated because it is highly conjugated and therefore very stable. Additionally, only the trans (or *E*) stereoisomer is isolated because it is much more stable than the cis stereoisomer.

In view of the complex mixture obtained in the example used in Study Problem 22.1, it is reasonable to ask why only one product is obtained from the crossed aldol condensation in Eq. 22.53. The analysis of this case highlights several important principles of carbonyl-compound reactivity. First, *because the aldehyde in the Claisen–Schmidt reaction has no α-hydrogens, it cannot act as the enolate component of the aldol condensation*; consequently, two of the four possible crossed aldol products cannot form. The other possible side reaction is the aldol addition reaction of the ketone with itself, as in Eq. 22.51; why doesn't this reaction occur? The enolate ion from acetone can react either with another molecule of acetone or with benzaldehyde, but *addition to a ketone occurs more slowly than addition to an aldehyde* (Sec. 19.8C). Furthermore, even if addition to acetone does occur, *the aldol addition reaction of two ketones is reversible* (Eq. 22.47), whereas *addition to an aldehyde has a more favorable equilibrium constant* (Sec. 19.8B). So, in Eq. 22.53, both the rate and equilibrium for addition of the acetone enolate to benzaldehyde are more favorable than they are for addition to a second molecule of acetone. Consequently, the product shown in Eq. 22.53 is the only one formed.

The Claisen–Schmidt condensation, like other aldol condensations, can also be catalyzed by acid.

STUDY GUIDE LINK 22.4
Understanding Condensation Reactions

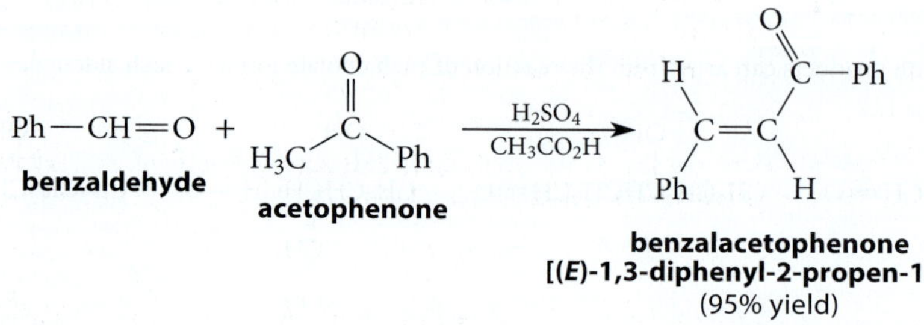

(22.54)

Directed Aldol Reactions There have been many approaches to the crossed-aldol problem. A well-established approach is to *pre-form* an enolate ion, or a compound that behaves like an enolate ion, from one carbonyl compound. This enolate component is then allowed to react with the carbonyl group of a second carbonyl compound. If we can dictate successfully, by a suitable choice of reagents and conditions, which component is to be the enolate component and which is to be the carbonyl component, we have solved the crossed-aldol problem. An aldol reaction that meets these criteria is usually called a **directed aldol reaction.**

To illustrate a directed aldol reaction, we use the reaction of an enolate ion derived from a ketone with the carbonyl group of an aldehyde. Consider, for example, the reaction of 2-pentanone with propanal.

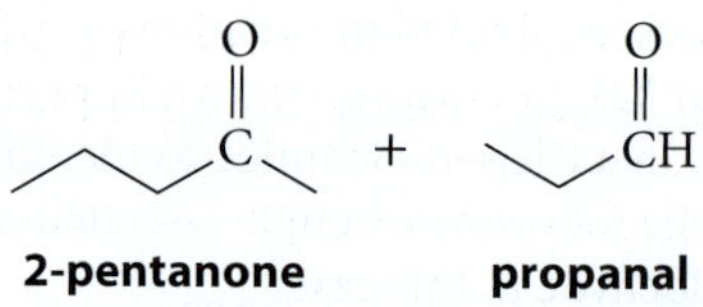

The solution to this problem involves formation of an enolate of the ketone, 2-pentanone, and then allowing it to react with the propanal, the aldehyde.

In the case of 2-pentanone, two enolate ions are possible (or three, if we count stereoisomers). We show these with a lithium counter-ion:

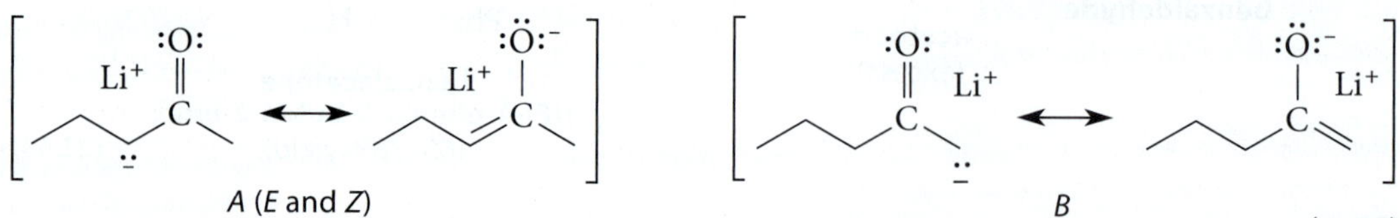

(22.55)

Enolate ions *A* are more stable than enolate ion *B* because the double bond in *A* has more alkyl substituents. The question is whether any of these enolates can be formed selectively.

It was discovered in the early 1970s that a family of very strong, highly branched nitrogen bases, such as the following two examples, can be used to form stable enolate ions rapidly and irreversibly at –78 °C from ketones (and esters, as we shall see in Sec. 22.8B).

A temperature of –78 °C is used often with chemical reactions for which very cold temperatures are necessary. This temperature is conveniently achieved with a slurry of dry ice (that is, solid CO_2) in acetone or isopropyl alcohol in a Dewar vessel (an insulated container, which is the laboratory version of a thermos). Slurries of other solvents with dry ice can be used to achieve other temperatures.

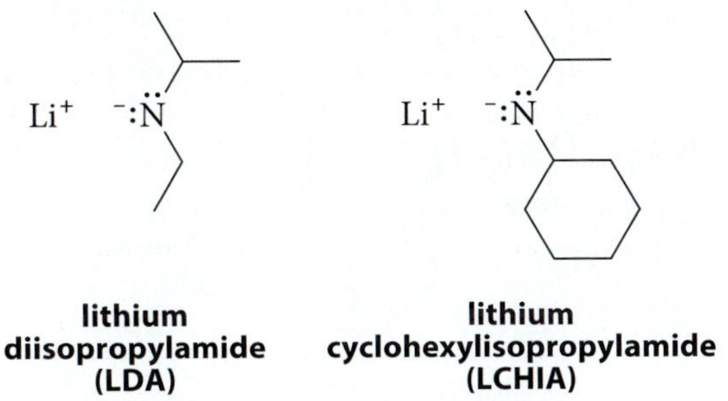

(22.56)

> **!** Don't confuse the term *amide* in the names of these bases with the carboxylic acid derivative. This term has a double usage. As used here, an *amide* is the conjugate-base anion of an amine.

These amide bases themselves are generated from the corresponding amines and butyllithium (a commercially available organolithium reagent) at –78 °C in tetrahydrofuran (THF) solvent.

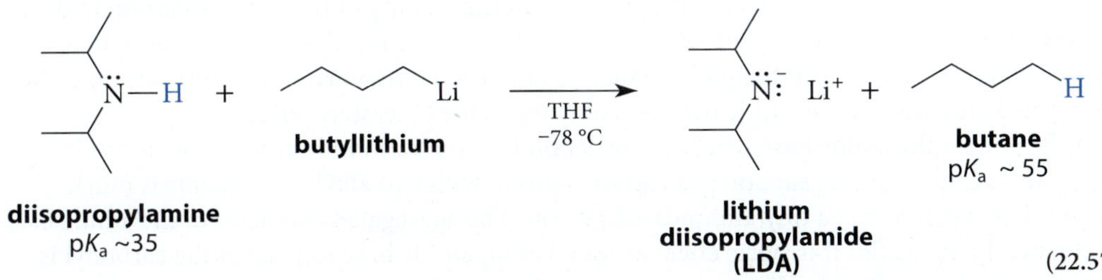

(22.57)

Because ketones have pK_a values near 20, these amide bases are strong enough to convert ketones completely into their conjugate-base enolate ions. When LDA reacts with 2-pentanone, the α-proton of the methyl group is removed *very rapidly* to give the conjugate-base lithium enolate *B*:

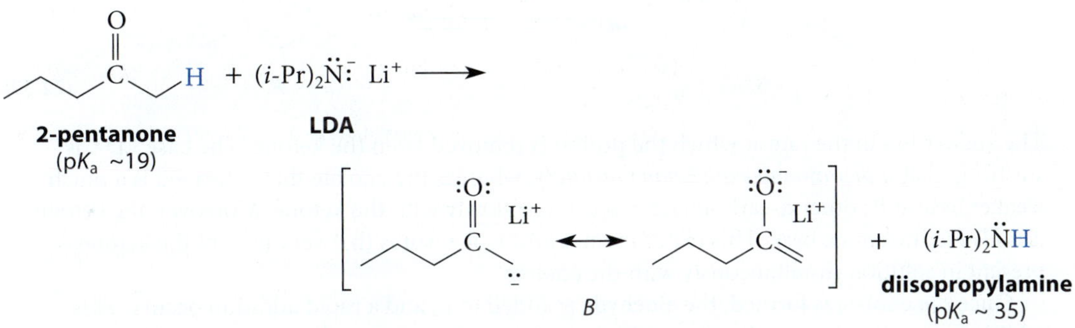

(22.58)

This is the *less stable* enolate ion of 2-pentanone, because the double bond is on the α-carbon with no alkyl branches.

For simplicity, we almost always draw both the enolate ion and LDA as simple ionic compounds, but they are actually more complex. They can have several different structures that depend on concentration and solvent, but they typically are multimolecular aggregates that include several solvent molecules. (For the aggregated structure of the amide base, see Further Exploration 23.2 in Chapter 23 of the *Study Guide and Solutions Manual.*) The enolate structure shown in **Fig. 22.3** is typical. Notice that the oxygens of the enolate are strongly coordinated to

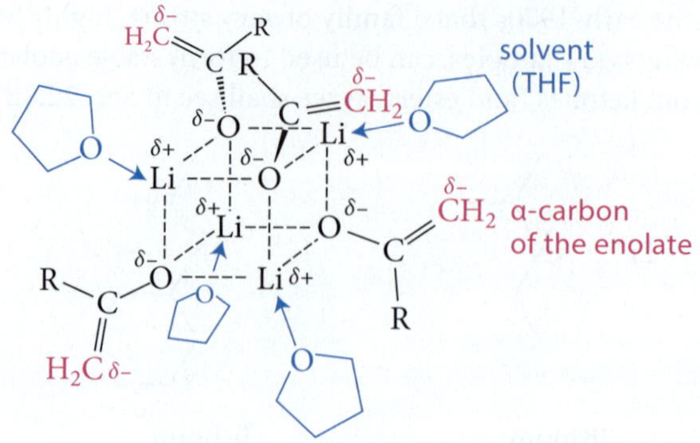

FIGURE 22.3 Typical aggregated structure of a lithium enolate in THF solvent. Solvent molecules are shown in blue. Each enolate oxygen is coordinated to three lithium ions. (This bonding, indicated by dashed lines, is similar to a strong hydrogen bond with lithium instead of hydrogen.) The α-carbons of the enolate *(red)* are accessible along the periphery of the aggregate.

the lithium atoms with partial covalent bonds, and the α-carbons are exposed along the periphery. The aggregated character of these reagents is important to their reactivity, as we explain in the subsequent discussion.

The *less stable* enolate ion is formed selectively in Eq. 22.58 because it is *formed more rapidly*. This is another example of *kinetic versus thermodynamic control* (Sec. 15.4B). The isopropyl groups of the amide base play a key role in this selectivity. The large, highly branched base reacts with a methyl proton—that is, a proton at the carbon with fewer alkyl substituents—more rapidly for steric reasons. Van der Waals repulsions occur in the vicinity of the other α-carbon to decrease the reaction rate of LDA with the proton at that carbon. Steric repulsion, then, is used to *decrease* the rate of proton removal at the more branched (and therefore more congested) carbons. The aggregated character of the amide base probably intensifies this steric effect.

Why doesn't the amide base react as a nucleophile at the carbonyl carbon? Again, steric repulsion is important. A reaction at a carbonyl group with two alkyl substituents is much slower than a reaction at a tiny, unhindered proton. The aggregated character of the amide base undoubtedly intensifies this steric effect as well. For an amide base to react at the carbonyl is analogous to trying to put a dinner plate in a vending machine slot intended for a credit card.

Why couldn't the enolate *B*, once formed, react with un-ionized ketone to form a mixture of *A* and *B*?

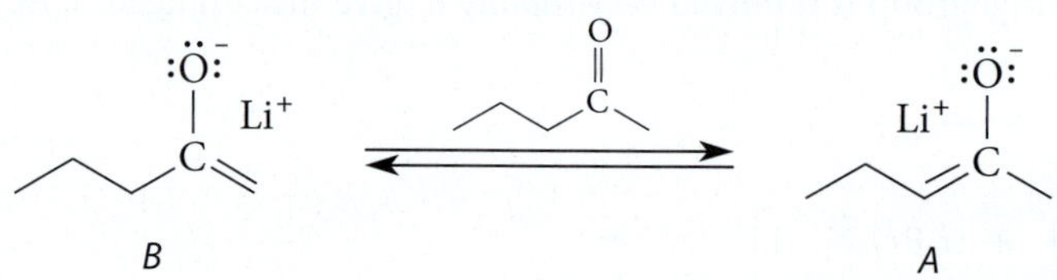

(22.59)

The answer lies in the rate at which the proton is removed from the ketone. The base LDA is so strong that *it deprotonates the ketone instantly*, whereas the enolate that is formed is a much weaker base at its other α-carbon and reacts more slowly with the ketone. Moreover, the ketone is *added to* the amide base. This experimental protocol ensures that very little of the ketone is present in solution simultaneously with the enolate.

Once the enolate is formed, the aldehyde is added to it, and a rapid addition occurs. This addition is assisted by coordination of both oxygens—the oxygen of the enolate and the oxygen of the aldehyde—to the lithium. (The aldehyde oxygen probably replaces a solvent molecule in the aggregate structure of Fig. 22.3.) This coordination brings the exposed enolate α-carbon (*red*) into proximity with the aldehyde carbonyl carbon (*blue*) through a six-membered cyclic transition state:

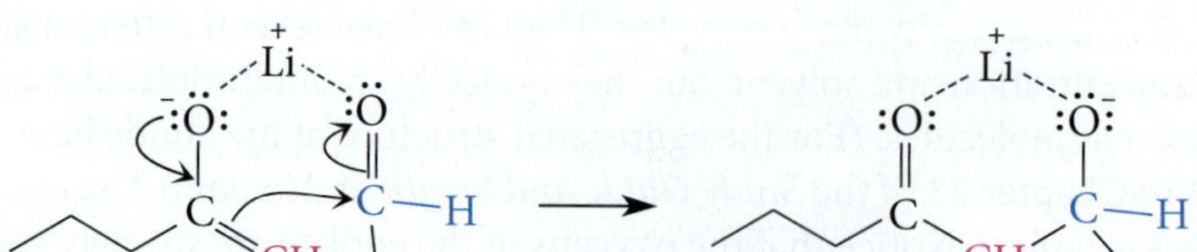

(22.60a)

This equation shows why the α-carbon rather than the oxygen is the nucleophilic site of the enolate. Protonation of the lithium alkoxide affords the aldol addition product when dilute aqueous acid is added.

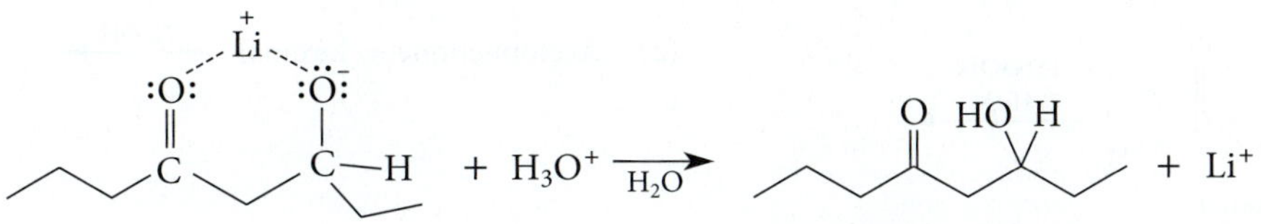

6-hydroxy-4-octanone
(62% yield) (22.60b)

(As with other aldol additions, stronger acid can bring about dehydration.)

We could go on to ask why the addition product of Eq. 22.60a does not react with another equivalent of ketone enolate. Remember that ketone carbonyl groups are much less reactive than aldehyde carbonyl groups; therefore, the reaction of the ketone enolate with the aldehyde is the faster addition reaction.

This discussion shows that all of the potential side reactions are largely avoided by the character of the reagents and the experimental protocol—that is, addition of aldehyde to preformed enolate at low temperature.

In summary, we can expect the directed aldol addition to be useful when the enolate component is derived from a methyl ketone or another ketone in which a single enolate constitutional isomer can be formed, and the carbonyl component is derived from an aldehyde or another unusually reactive carbonyl compound.

Focused Problem

22.20 (a) Give the simplified (nonaggregated) ionic structure of the enolate formed by treatment of 2,2-dimethylcyclohexanone with LDA; then, give the product of its directed aldol reaction with acetaldehyde following acidic workup.

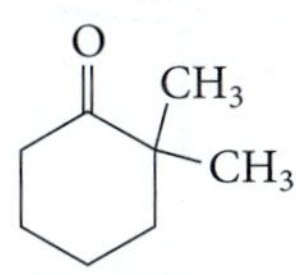

2,2-dimethylcyclohexanone

(b) Two diastereomers of the product in part (a) are formed. Explain. Are they formed in equal amounts or different amounts?

Intramolecular Aldol Condensations When a molecule contains more than one aldehyde or ketone group, an *intramolecular* reaction (a reaction within the same molecule) is possible. In such a case, the aldol condensation results in formation of a ring. Intramolecular aldol condensations are particularly favorable when five- and six-membered rings can be formed because of the *proximity effect* (Sec. 12.8).

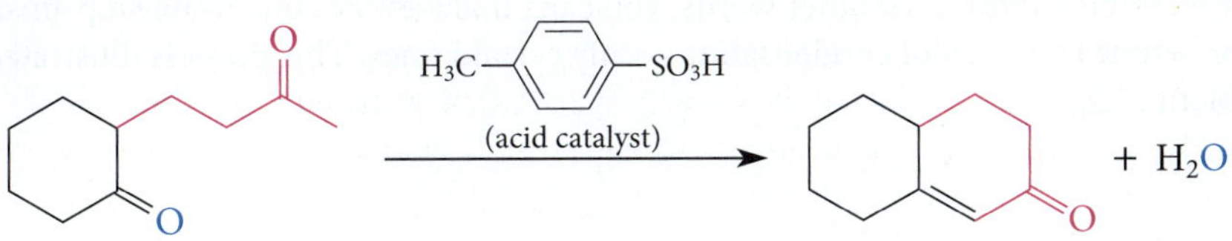

(22.61)

Focused Problem

22.21 Predict the product(s) in each of the following aldol condensations.

(a)

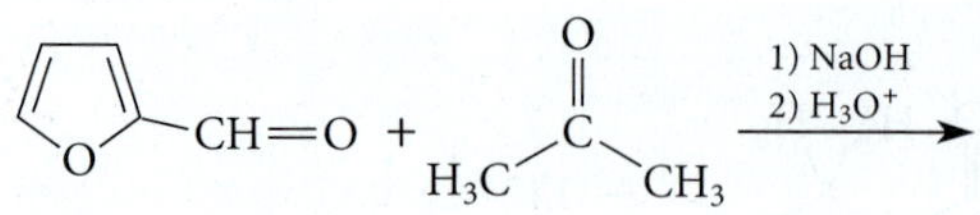

(equal molar amounts)

(b) Acetophenone + hexanal $\xrightarrow{\text{NaOH}}$

(c)

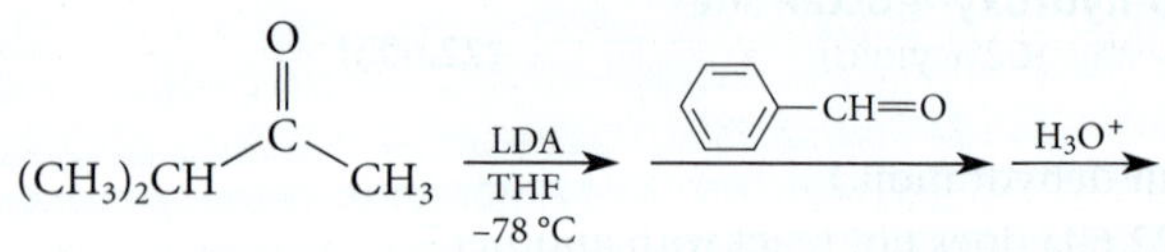

(d)

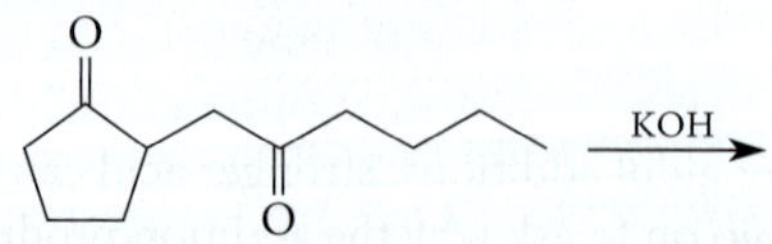

D. Synthesis with the Aldol Condensation

The aldol condensation can be applied to the synthesis of a wide variety of α,β-unsaturated aldehydes and ketones, and it is also another method for the formation of carbon–carbon bonds. (See the complete list in Appendix VI.) If you want to prepare a particular α,β-unsaturated aldehyde or ketone by the aldol condensation, you must be able to answer two questions: (1) What starting materials are required in the aldol condensation? (2) With these starting materials, is the aldol condensation of these compounds a feasible one?

The starting materials for an aldol condensation can be determined by mentally "splitting" the α,β-unsaturated carbonyl compound at the double bond:

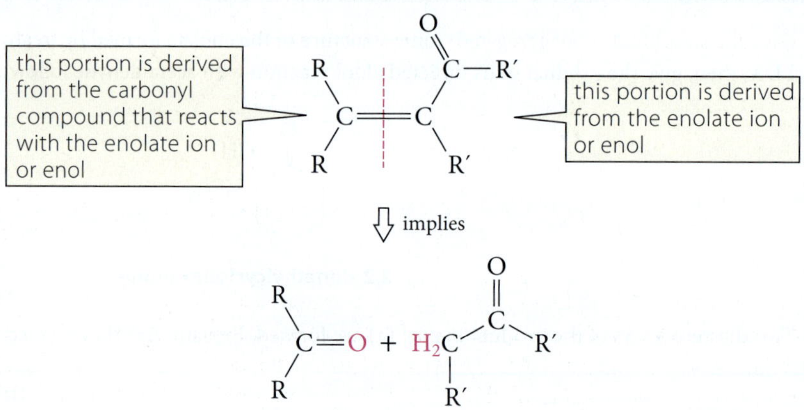

(22.62)

That is, work backward from the desired synthetic objective by replacing the double bond on the carbonyl side by two hydrogens and on the other side by a carbonyl oxygen (=O) to obtain the structures of the starting materials in the aldol condensation.

Knowing the potential starting materials for an aldol condensation is not enough; you must also know whether the condensation is one that works, or whether instead it is one that is likely to give troublesome mixtures. In other words, you can't make every conceivable α,β-unsaturated aldehyde or ketone by the aldol condensation—only certain ones. This point is illustrated in Study Problem 22.2.

Study Problem 22.2

Determine whether the following α,β-unsaturated ketone can be prepared by an aldol reaction.

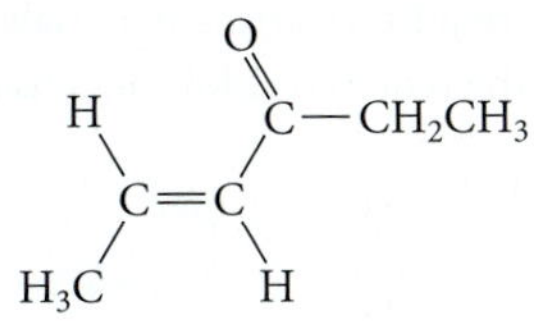

Solution Following the procedure in Display 22.62, analyze the desired product as follows:

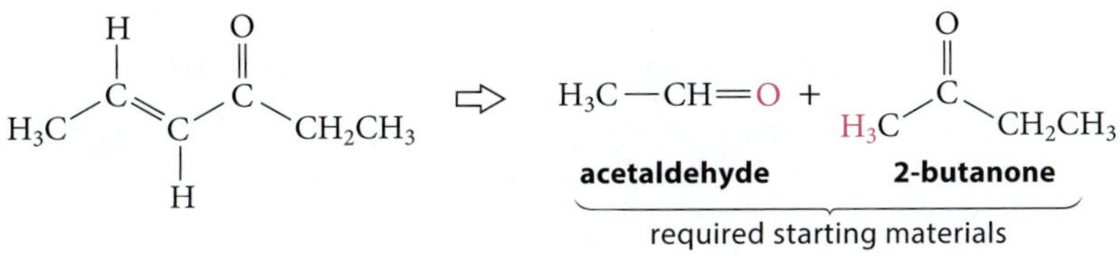

The desired product requires a crossed aldol condensation between an aldehyde and a ketone: acetaldehyde and 2-butanone. The question, then, is whether the desired product is the only one that could form, or whether other competing aldol reactions would occur.

First, either acetaldehyde or 2-butanone could serve as the enolate component of an aldol addition. Although the aldehyde should be more reactive toward addition of the enolate than the ketone, its reaction with the enolate ions from both 2-butanone and another acetaldehyde molecule would give a mixture. To complicate matters even more, 2-butanone has two nonequivalent α-carbons at which enolate ions (or enols) could form. This opens up yet other possibilities for aldol reactions and for the formation of complex product mixtures. To summarize all of these possibilities:

desired:

but also possible:

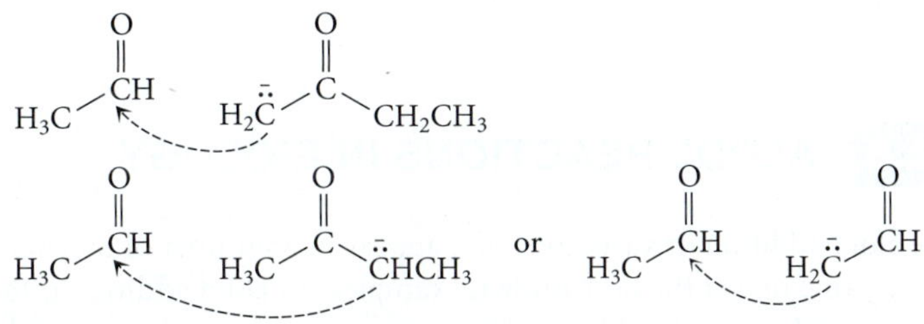

Therefore, the base-catalyzed or acid-catalyzed aldol reaction of acetaldehyde and 2-butanone would *not* be useful for preparing the desired ketone because a large number of other products would be expected.

However, because 2-butanone is a methyl ketone, and because the carbonyl component is an aldehyde, a directed aldol addition followed by dehydration should work. Therefore, we could form the less stable (kinetic) lithium enolate of 2-butanone with LDA in THF at –78 °C and allow it to react with acetaldehyde. Acid- or base-catalyzed dehydration of the addition product should give the desired product in the more stable *E* configuration.

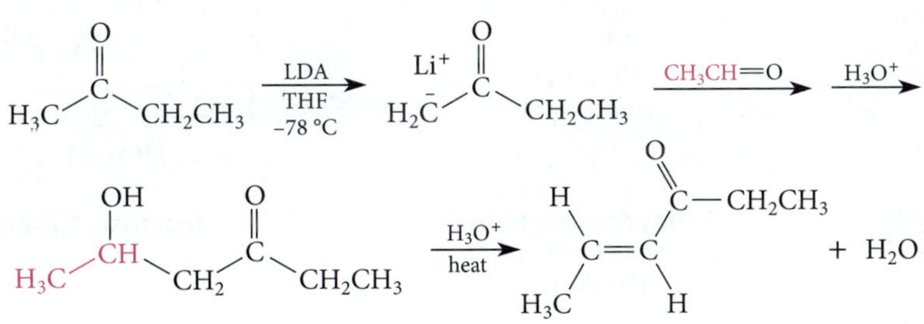

Focused Problems

22.22 Some of the following molecules can be synthesized in good yield using an aldol condensation (or an aldol addition followed by a dehydration). Identify these, and give the structures of the required starting materials. Others cannot be synthesized in good yield by an aldol condensation. Identify these, and explain why the required aldol condensation would not be likely to succeed.

(a)

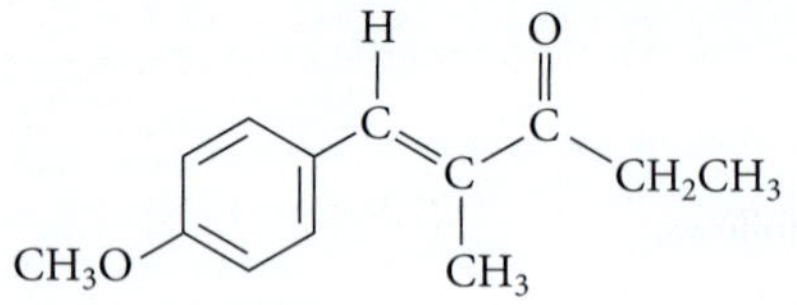

(b)

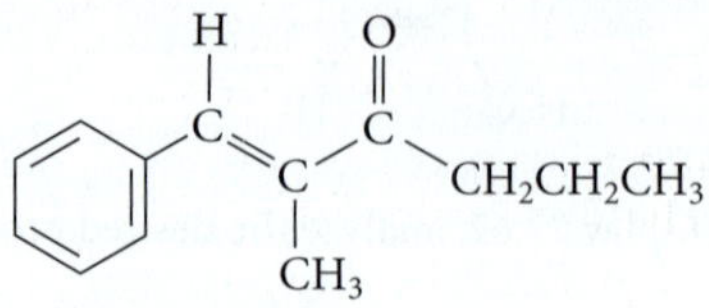

(c)

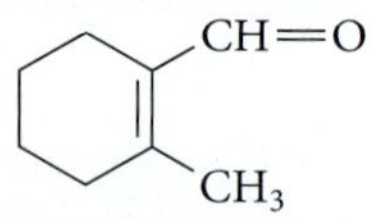

(d)

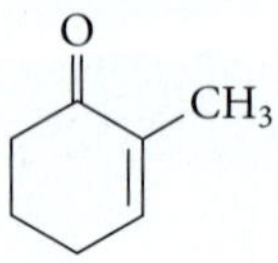

(e)

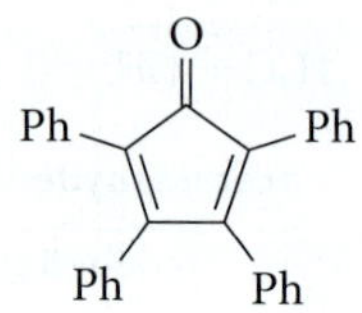

(f)

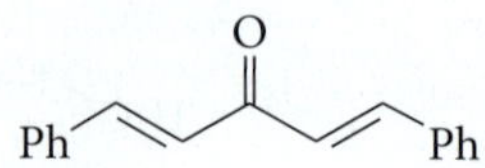

(g)

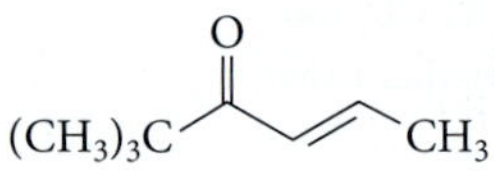

(h)

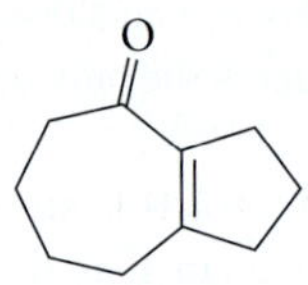

22.23 Analyze the aldol condensation in Eq. 22.61 (Sec. 22.4C) using the method given in Display 22.62. Show that four possible aldol condensation products might in principle result from the starting material. Explain why the observed product is the most reasonable one.

Chemical Biology Topic

22.5 ALDOL REACTIONS IN BIOLOGY

The aldol addition or variations of it appear in a number of biochemical pathways. In this section we examine one of the best-known examples, an aldol addition catalyzed by an enzyme called, appropriately enough, *aldolase.* There are two major variations of this enzyme, and we discuss the mechanism of the reaction catalyzed by the enzyme (Class I aldolase) found in animals (including humans) and plants. (A different aldolase, Class II aldolase, is found in fungi and bacteria.)

Aldolase catalyzes the formation of fructose-1,6-bisphosphate from a ketone, dihydroxyacetone phosphate (DHAP), and an aldehyde, glyceraldehyde-3-phosphate (G3P).

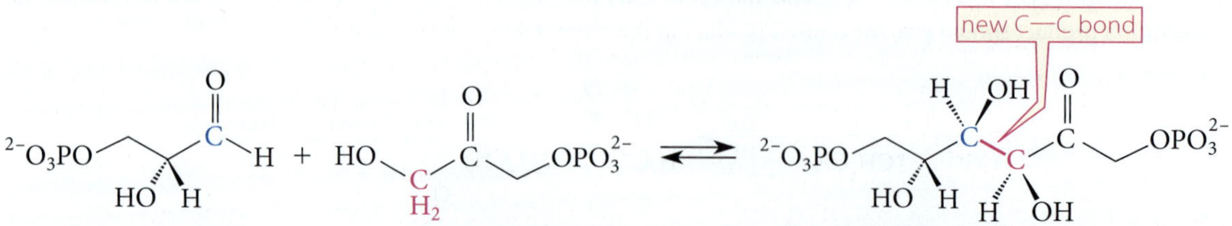

(*R*)-glyceraldehyde-3-phosphate (G3P) **1,3-dihydroxyacetone phosphate (DHAP)** **fructose-1,6-bisphosphate** (22.63)

Don't be distracted by the stereochemical details in this reaction; notice primarily that *a new carbon–carbon bond is formed at the α-carbon of the ketone.* We focus on how this bond is formed.

The aldolase active site contains a lysine residue, which has a primary amine in its side chain:

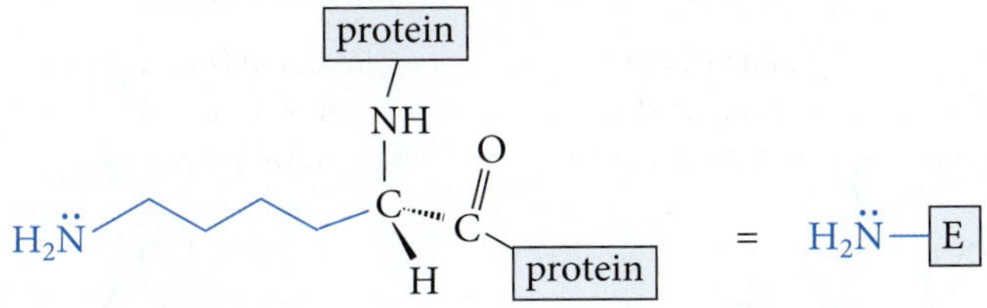

structure of a lysine residue in a protein (conjugate-base form)

(22.64)

This lysine forms an *imine*, or *Schiff base*, derivative with DHAP (Sec. 19.12A). This imine can isomerize to an *enamine* (Sec. 19.12C). Recall that enamines have the same relationship to imines as enols have to aldehydes and ketones.

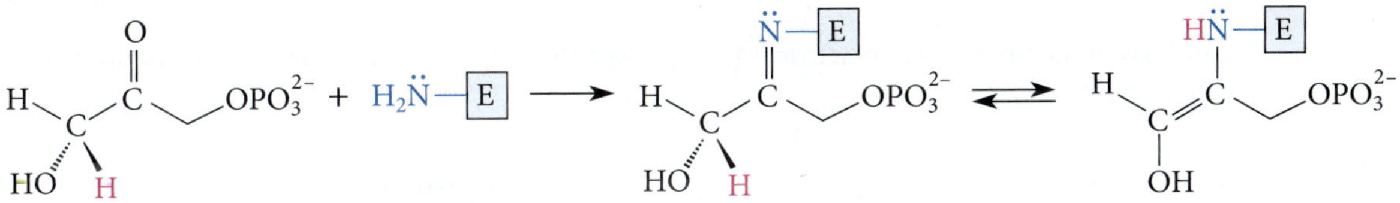

DHAP

imine (Schiff base)

+ H_2O

enamine

(22.65a)

In the enamine, the nonbonding electron pair on the nitrogen is delocalized by resonance onto the α-carbon.

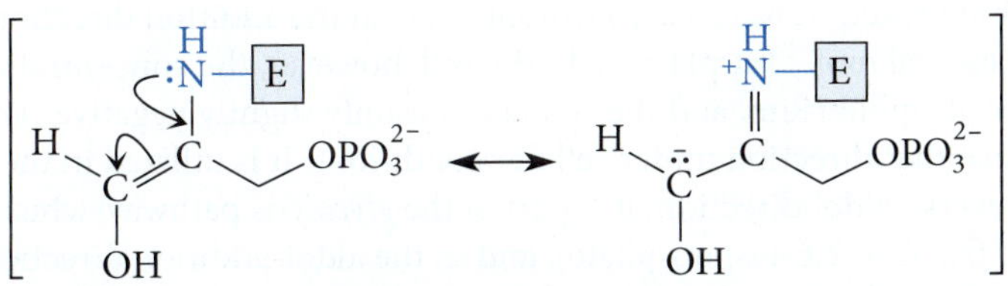

(22.65b)

In other words, this enamine has some *carbanion character* and can be thought of as a *disguised carbanion.* The negative charge at the α-carbon is further stabilized by the polar effect of the OH group. When the enamine is brought together in proximity to the carbonyl group of glyceraldehyde-3-phosphate in the enzyme active site, the aldol addition occurs. In this addition, the carbanion reacts as a nucleophile at the carbonyl carbon of glyceraldehyde. Hydrogen bonding of the carbonyl oxygen to a proton from a tyrosine OH group in the enzyme active site increases the electrophilic character of the carbonyl group:

This mechanism and the other enzyme catalysis mechanisms discussed in the text require a precise placement of the substrate functional groups near the appropriate groups in the enzyme active site that are responsible for the catalysis. Enzymes have evolved to exhibit exactly this sort of specificity. Our knowledge of the details of catalysis is typically based on one or more X-ray crystal structures of the enzyme, in many cases with a substrate or other closely related structure bound tightly, but noncovalently, in the active site.

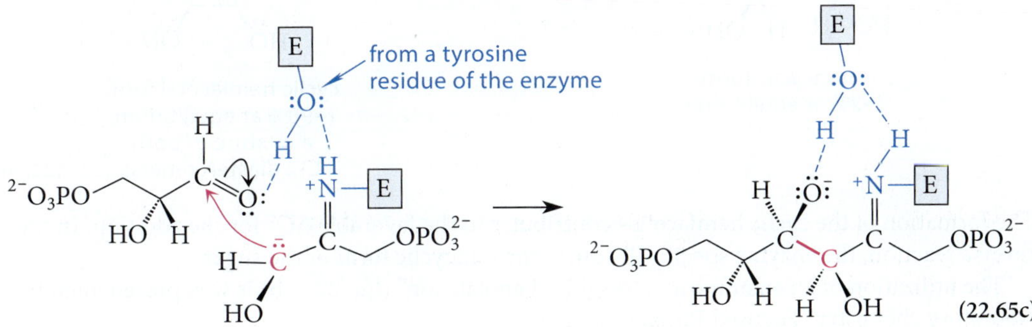

(22.65c)

The oxygen from the active-site tyrosine residue facilitates the proton transfers that complete the addition. An imine derivative of the product is formed in the active site.

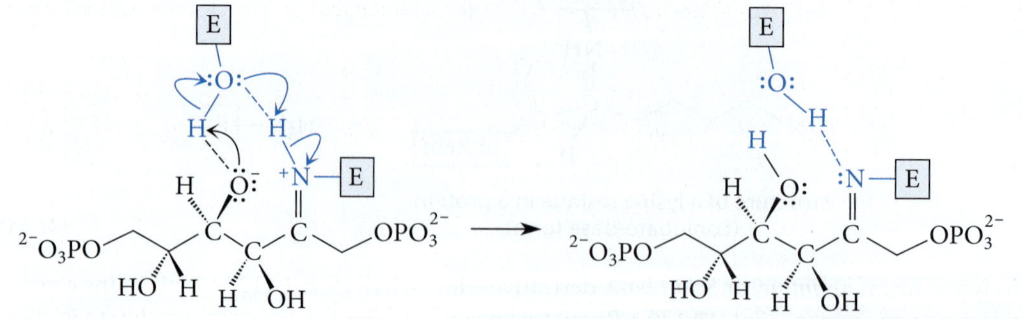

imine derivative of fructose-1,6-bisphosphate

(22.65d)

Hydrolysis of the imine gives fructose-1,6-bisphosphate and regenerates the enzyme for another catalytic cycle.

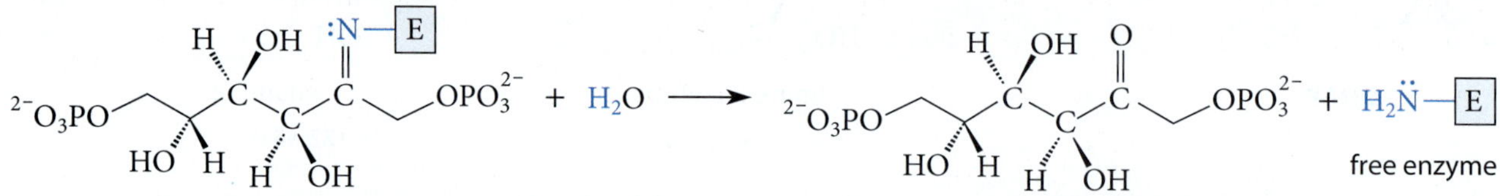

imine derivative of fructose-1,6-bisphosphate **fructose-1,6-bisphosphate** free enzyme (22.65e)

This aldol addition reaction has a very favorable $\Delta G°$ in the addition direction [−23.8 kJ mol^{-1} (5.69 kcal mol^{-1}) at pH = 7]. In the cell, however, the concentrations of various metabolites are not at equilibrium, and the actual ΔG is only slightly negative. Therefore, the reaction can run in either direction under cellular conditions. It is utilized in the cell in two pathways. In the reverse-aldol direction, it is part of the *glycolysis* pathway, which breaks down glucose (by way of fructose-1,6-bisphosphate); and in the aldol-addition direction shown in the foregoing equations, it is part of the *gluconeogenesis* pathway, which operates in periods of glucose deprivation to form glucose (also by way of fructose-1,6-bisphosphate) from three-carbon compounds.

Fructose-1,6-bisphosphate spontaneously forms cyclic hemiacetals (Sec. 19.11A), which are the major forms in solution.

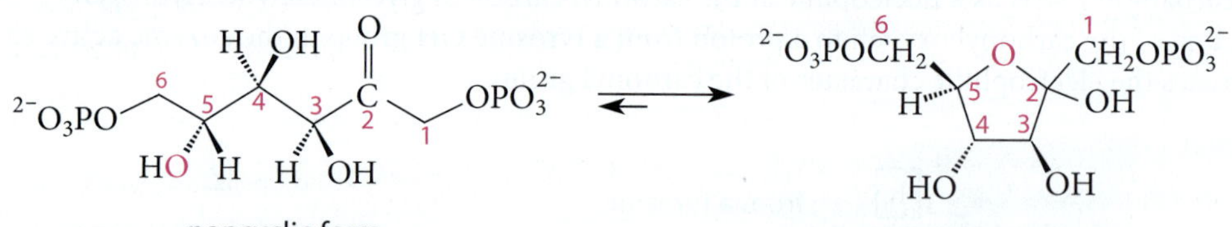

noncyclic form (~2% at equilibrium)

cyclic hemiacetal form (~98% at equilibrium; a mixture of both C-2 diastereomers) (22.66)

The formation of the cyclic hemiacetals contributes to the favorable $\Delta G°$ for the addition. In the reverse reaction, the enzyme specifically binds the noncyclic form of the sugar.

The utilization of an enamine as a "disguised enolate ion" (Eq. 22.65b) is well precedented in laboratory chemistry (Focused Problem 22.25).

Focused Problems

22.24 (a) The enzyme "KDPG aldolase" catalyzes the aldol addition reaction between pyruvate and glyceraldehyde-3-phosphate.

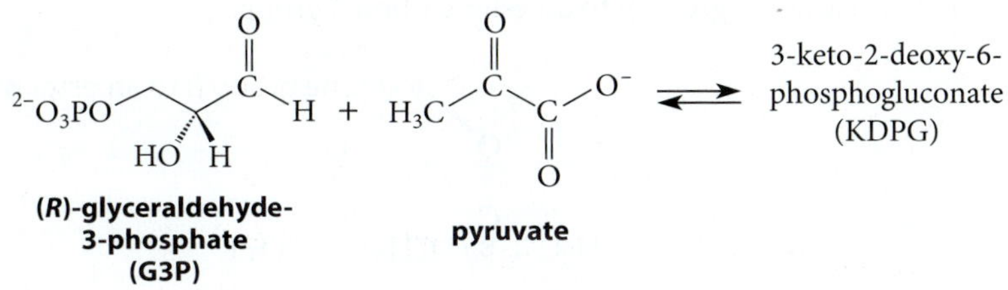

The reaction is known to involve the formation of an imine (Schiff base) between a lysine residue of the enzyme and pyruvate. Give the structure of the reaction product (KDPG), and outline a curved-arrow mechanism for its formation. Assume that acids and bases are present as needed. (The stereochemistry of the asymmetric carbons is the same as in Eq. 22.63.)

(b) The product of this reaction exists as two diastereomeric cyclic acetals. Give their structures.

22.25 The following reaction is known to involve enamine formation between the amine catalyst and the carbonyl marked with an asterisk (*).

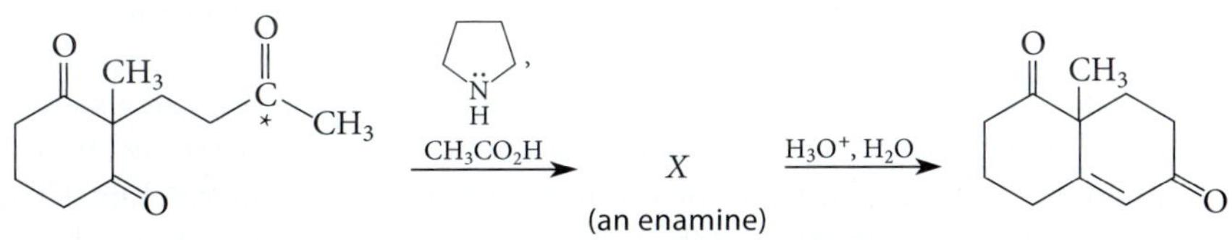

Draw a curved-arrow mechanism for this reaction, using the following steps as your guide.

(a) Begin your mechanism by showing the formation of enamine *X*.

(b) Complete the mechanism by showing the acid-catalyzed aldol condensation of the enamine that forms the ring.

(c) Draw the hydrolysis mechanism for the reaction that gives the ketone and regenerates the catalyst.

22.6 CONDENSATION REACTIONS INVOLVING ESTER ENOLATE IONS

A. Claisen Condensation

The base-catalyzed aldol reactions discussed in Sec. 22.4A involve enolate ions derived from *aldehydes* and *ketones*. This section discusses condensation reactions that involve the enolate ions of *esters*.

Ethyl acetate undergoes a condensation reaction in the presence of one equivalent of sodium ethoxide in ethanol to give ethyl 3-oxobutanoate, which is known commonly as ethyl acetoacetate.

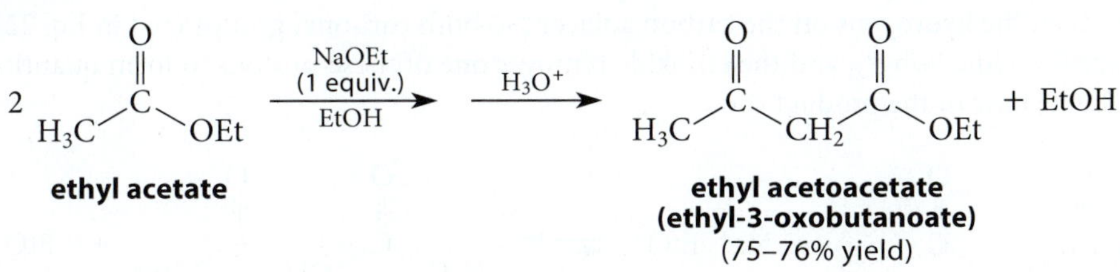

(22.67)

This is the best-known example of the *Claisen condensation*, which is named for German chemist Rainer Ludwig Claisen (1851–1930). (See also Sec. 28.4B. Don't confuse this reaction with the Claisen–Schmidt condensation in the previous section—same Claisen, different reaction.) The product of this reaction, ethyl acetoacetate, is an example of a **β-keto ester**: a compound with a ketone carbonyl group β to an ester carbonyl group.

We discuss reactions in which the carbonyl group at the β-position is derived from an aldehyde in Sec. 22.6C.

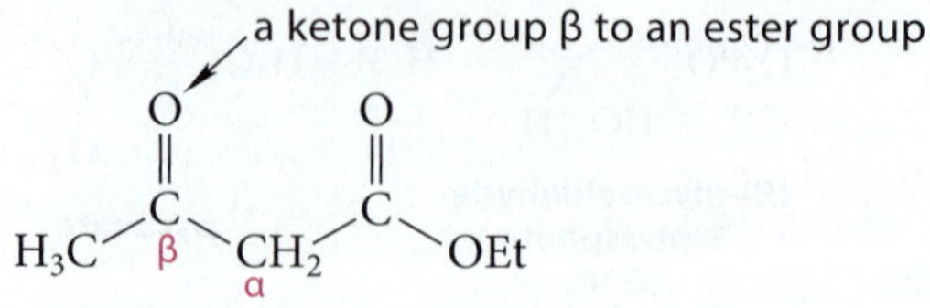

(22.68)

A **Claisen condensation** is the base-promoted condensation of two ester molecules to give a β-keto ester.

The first step in the mechanism of the Claisen condensation is formation of an *enolate ion* by the reaction of the ester with the ethoxide base.

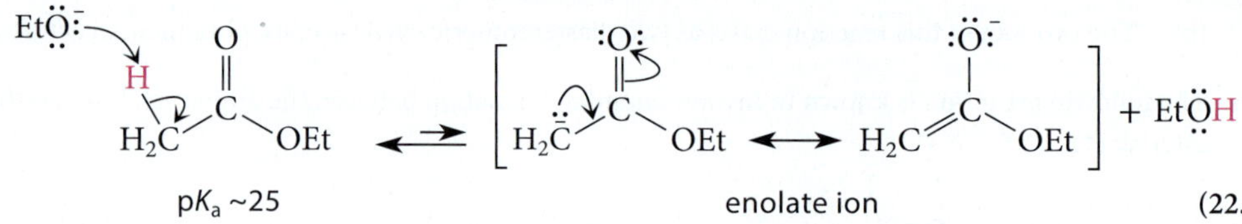

(22.69a)

Because ethoxide ion is a nucleophile, we might ask whether it can also react at the carbonyl group of the ester to give the usual nucleophilic acyl substitution reaction. This reaction undoubtedly takes place, but the products are the same as the reactants! This is why ethoxide ion is used as a base with ethyl esters in the Claisen condensation (see Study Guide Link 22.1 and Focused Problem 22.27).

Although the ester enolate ion is formed in very low concentration, it is a strong base and good nucleophile, and it undergoes a *nucleophilic acyl substitution reaction* with a second molecule of ester (Eq. 22.69b). The usual two-step substitution mechanism is observed—that is, formation of a tetrahedral addition intermediate followed by loss of a leaving group:

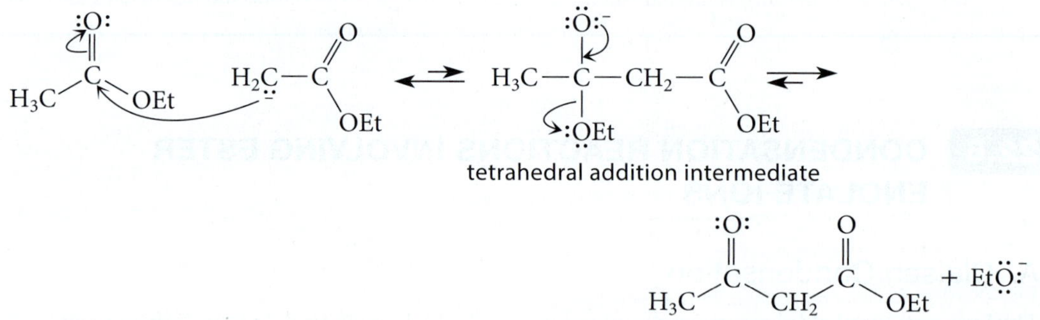

(22.69b)

The overall equilibrium as written in Eqs. 22.69a and b lies far on the side of the reactants—that is, *all β-keto esters are less stable than the esters from which they are derived.* For this reason, the Claisen condensation must be driven to completion by applying Le Chatelier's principle. The most common technique is to use one full equivalent of ethoxide in the reaction. In the β-keto ester product, the hydrogens on the carbon adjacent to both carbonyl groups (red in Eq. 22.69c) are especially acidic (why?), and the ethoxide removes one of these protons to form quantitatively the conjugate base of the product.

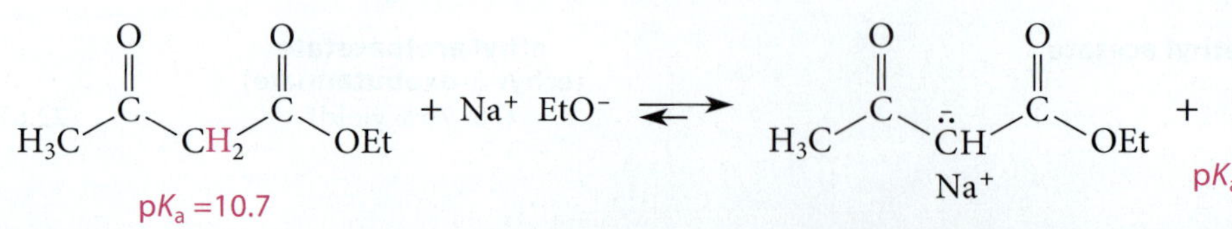

(22.69c)

The un-ionized β-keto ester product in Eq. 22.67 is formed when acid is added subsequently to the reaction mixture.

Notice that ethoxide ion is a *catalyst* for the reactions in Eqs. 22.69a and b, but it is consumed in Eq. 22.69c. Therefore, ethoxide is a reactant rather than a catalyst in the overall reaction, and this is the reason that *one full equivalent* of ethoxide must be used in the Claisen condensation.

The removal of a product by ionization is the same strategy employed to drive ester saponification to completion (Sec. 21.8A). The importance of this strategy in the success of the Claisen condensation is evident if the condensation is attempted with an ester that has only one α-hydrogen: *no condensation product is formed.* In this case, the desired condensation product has a quaternary α-carbon, so it has no α-hydrogens acidic enough to react completely with ethoxide.

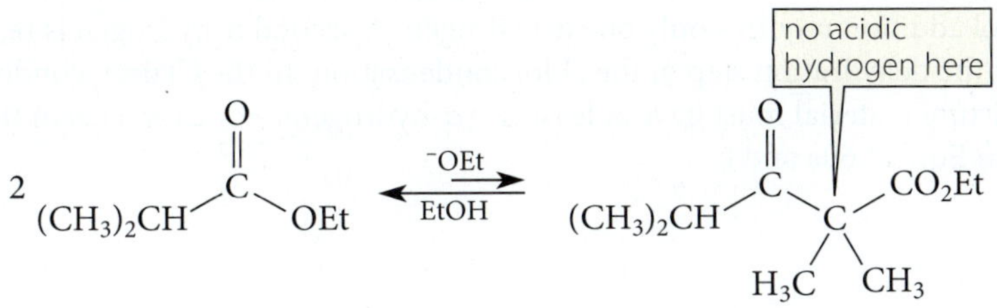

(no product observed) (22.70)

Furthermore, if the product of Eq. 22.70 [prepared by another method (Eq. 22.93, Sec. 22.8B)] is subjected to the conditions of the Claisen condensation, it readily decomposes back to starting materials because of the reversibility of the Claisen condensation.

The Claisen condensation is another example of *nucleophilic acyl substitution.* In this reaction, the nucleophile is an enolate ion derived from an ester. Although the reaction may seem complex because of the number of carbon atoms in the product, it is not conceptually different from other nucleophilic acyl substitutions, such as ester saponification:

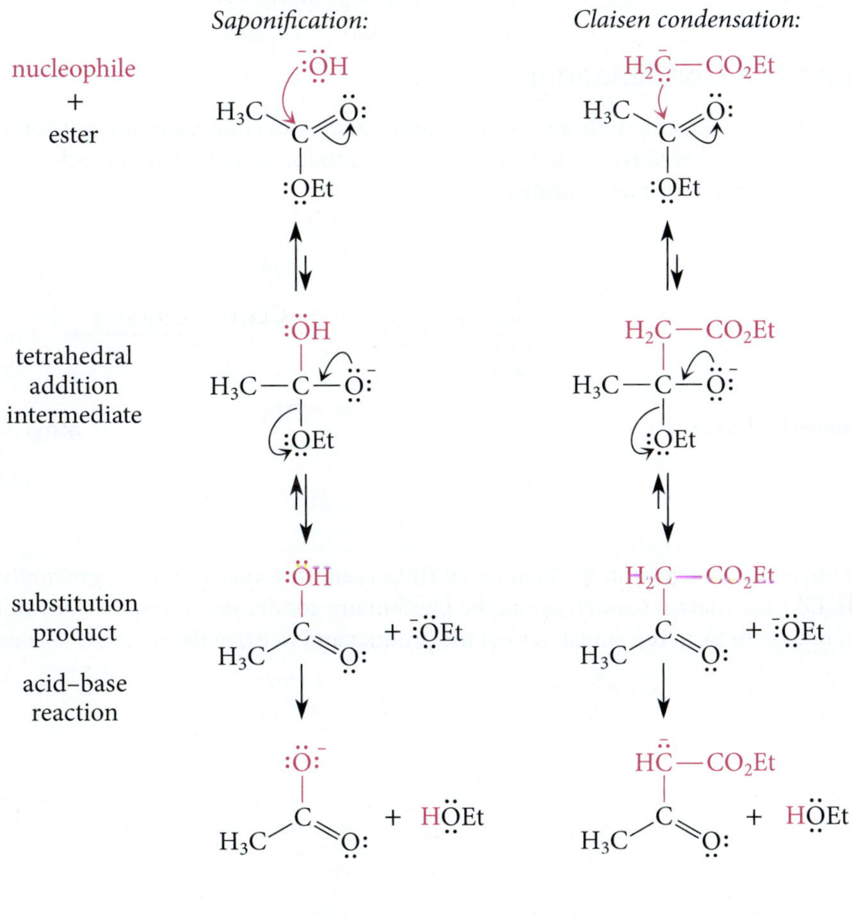

(22.71)

We have now discussed two types of condensation reactions: the aldol condensation and the Claisen condensation. These condensations are quite different and should not be confused. To compare:

1. The aldol condensation is an addition reaction of an enolate ion or an enol with an aldehyde or ketone followed by an alcohol dehydration. The Claisen condensation is a nucleophilic acyl substitution reaction of an enolate ion with an ester group.

2. The aldol condensation is catalyzed by both base and acid. The Claisen condensation requires a full equivalent of base and is *not* catalyzed by acid.

3. The aldol addition requires only one α-hydrogen. A second α-hydrogen is required, however, for the dehydration step of the aldol condensation. In the Claisen condensation, the ester starting material must have at least *two* α-hydrogens—one for each of the ionizations shown in Eqs. 22.69a and c.

Focused Problems

22.26 Give the Claisen condensation product formed in the reaction of each of the following esters with one equivalent of NaOEt, followed by neutralization with acid.

(a) Ethyl phenylacetate

(b) Ethyl butyrate

22.27 Hydroxide ion is about as basic as ethoxide ion. Would NaOH be a suitable base for the Claisen condensation of ethyl acetate? Explain by writing suitable equations. (*Hint:* See Study Guide Link 22.1.)

B. Dieckmann Condensation

Intramolecular Claisen condensations, like intramolecular aldol condensations, take place readily when five- or six-membered rings can be formed. The intramolecular Claisen condensation reaction is called the **Dieckmann condensation**.

The numbers in Eq. 22.72 are for purposes of carbon identification; they do not correspond to IUPAC numbering except in the reactant. In the product, carbon-5 is carbon-1 in the IUPAC numbering, carbon 1 is carbon-2, and so on.

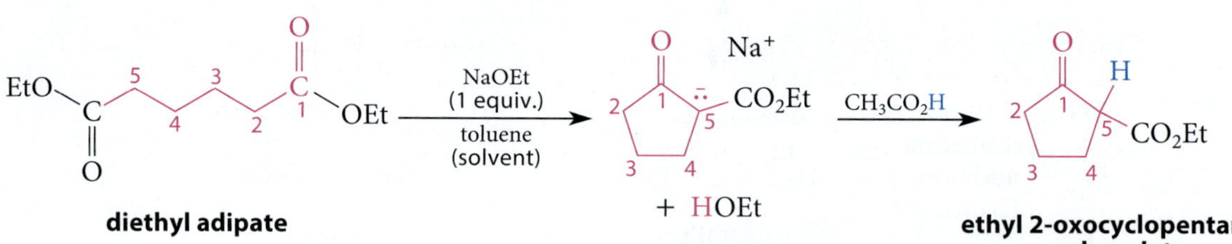

diethyl adipate

ethyl 2-oxocyclopentane-carboxylate
(74–81% yield)

(22.72)

The facile formation of five- and six-membered rings is another example of the *proximity effect* (Sec. 12.8). Like the Claisen condensation, the Dieckmann condensation requires one full equivalent of base to form the enolate ion of the product and to drive the reaction to completion.

Focused Problem

22.28 (a) Explain why compound *A*, when treated with one equivalent of NaOEt, followed by acidification, is completely converted into compound *B*.

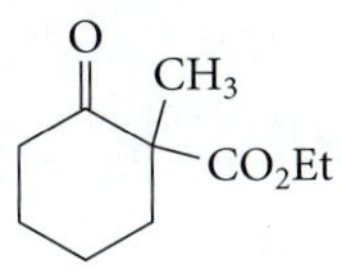

A

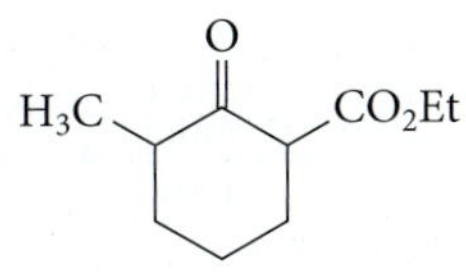

B

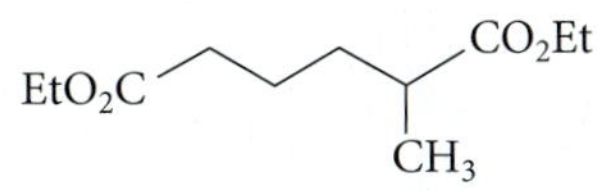

C

(b) Write a curved-arrow mechanism for this conversion.

(c) Give the structure of the only product formed when diethyl α-methyladipate (compound *C*) reacts in the Dieckmann condensation. Explain your reasoning.

C. Crossed Claisen Condensation

The Claisen condensation of two *different* esters is called a **crossed Claisen condensation.** The crossed Claisen condensation of two esters that both have α-hydrogens gives a mixture of four compounds that are typically difficult to separate. Such reactions in most cases are not synthetically useful.

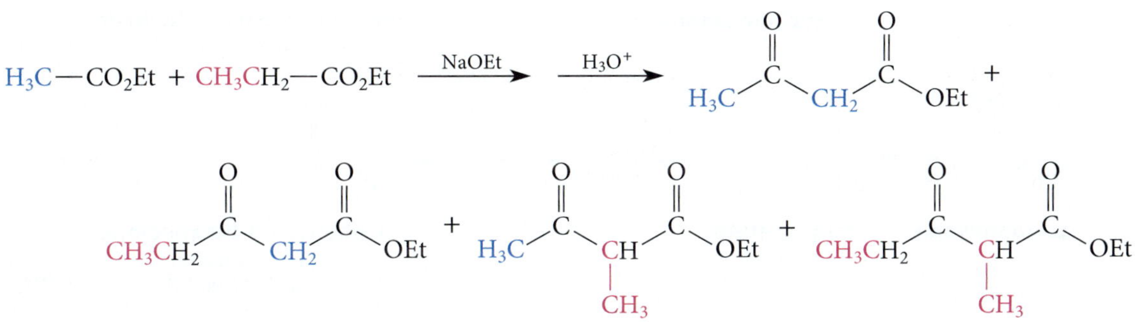

(22.73)

This problem is conceptually similar to the problem with crossed aldol reactions, discussed in Study Problem 22.1 (Sec. 22.4C).

Crossed Claisen condensations are useful, however, if one ester is especially reactive or has no α-hydrogens. For example, formyl groups (—CH=O) are readily introduced with esters of formic acid such as ethyl formate:

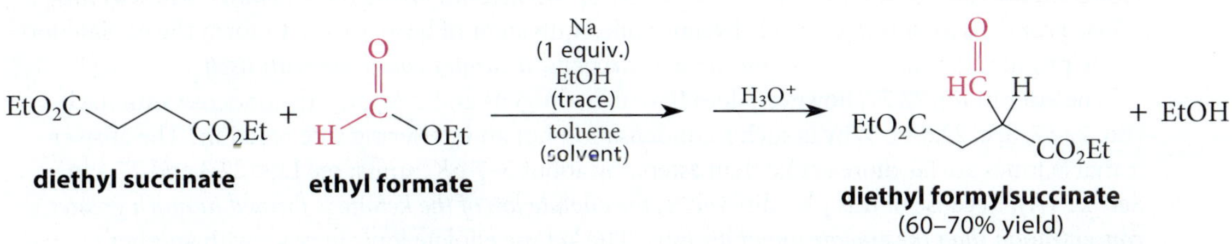

(22.74)

This product is an example of an ester in which the carbonyl group at the β-position is derived from an aldehyde rather than a ketone. Formate esters fulfill both of the criteria for a crossed Claisen condensation. First, they have no α-hydrogens; second, their carbonyl reactivity is considerably greater than that of other esters. The reason for their higher reactivity is that the carbonyl group in a formate ester is "part aldehyde," and aldehydes are particularly reactive toward nucleophiles (Eq. 21.90, Sec. 21.9E).

A less reactive ester without α-hydrogens can be used if it is present in excess. For example, an ethoxycarbonyl group can be introduced with diethyl carbonate.

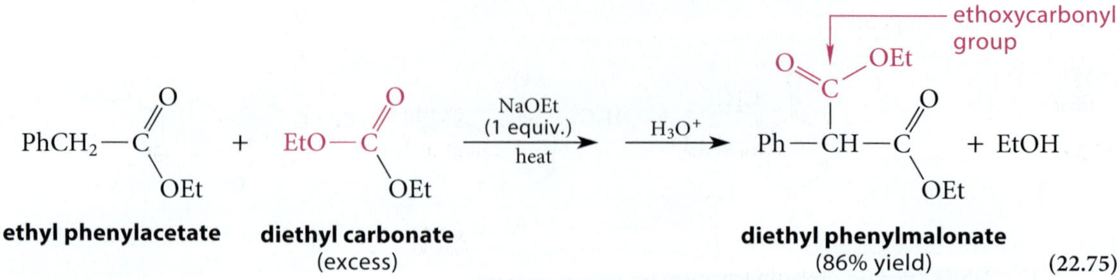

(22.75)

In this example, the enolate ion of ethyl phenylacetate condenses preferentially with diethyl carbonate rather than with another molecule of itself because of the much higher concentration of diethyl carbonate. The excess diethyl carbonate must then be separated from the product.

Another type of crossed Claisen condensation is the reaction of ketones with esters. In this type of reaction, the enolate ion of a ketone reacts at the carbonyl group of an ester.

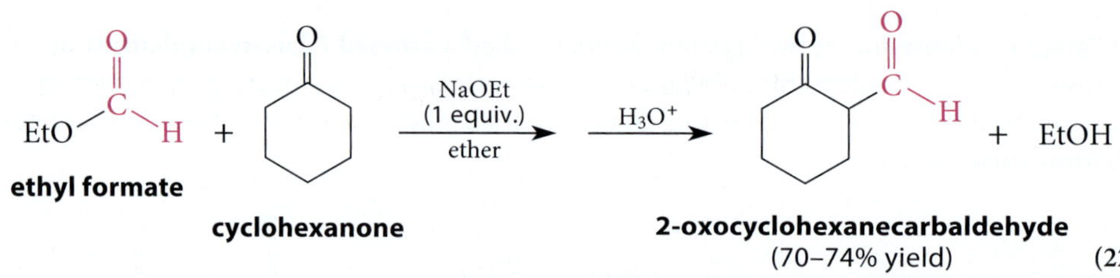

(22.76)

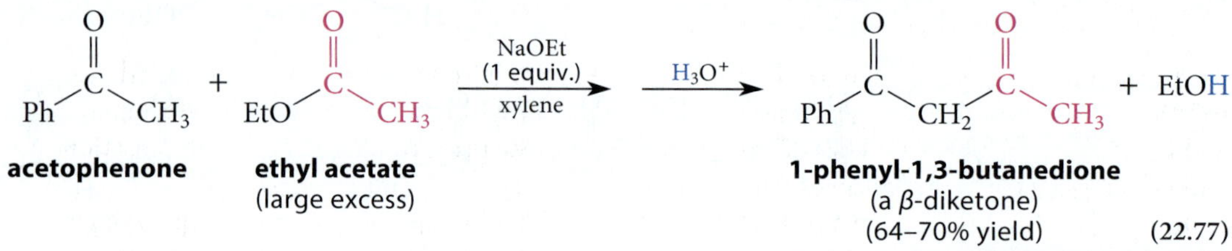

(22.77)

In Eq. 22.76, the enolate ion derived from the ketone cyclohexanone is acylated by the ester ethyl formate. This is a good way to introduce an aldehyde carbonyl group β to a ketone carbonyl group. In Eq. 22.77, the enolate ion of the ketone acetophenone is acylated by the ester ethyl acetate. In these reactions, several side reactions are possible in principle but in fact do not interfere. The analysis of these cases again highlights important principles of carbonyl-compound reactivity.

In Eq. 22.76, a possible side reaction is the aldol addition of cyclohexanone with itself. However, *the equilibrium for the aldol addition of two ketones favors the reactants*, whereas *the Claisen condensation is irreversible* because one equivalent of base is used to form the enolate ion of the product. *Because the ester has no α-hydrogens, it cannot condense with itself.*

The ester in Eq. 22.77, however, does have α-hydrogens and is known to condense with itself (Eq. 22.67, Sec. 22.6A). Why is such a condensation not an interfering side reaction? The answer is that ketones are far more acidic than esters (by about 5–7 pK_a units; see Eqs. 22.3 and 22.4 in Sec. 22.1A). Because of this pK_a difference, *the enolate ion of the ketone is formed in much greater concentration than the enolate ion of the ester.* The ketone enolate ion can react with another molecule of ketone—an unfavorable equilibrium—or it can be intercepted by the excess of ethyl acetate to give the observed product, which is a β-diketone. Even though esters are less reactive than ketones, a β-diketone is especially acidic (like a β-keto ester) and is ionized completely by the one equivalent of NaOEt. (Be sure to identify the most acidic hydrogens of the product in Eq. 22.77.) Therefore, *β-diketone formation is observed because ionization makes this an irreversible reaction.*

These examples illustrate that the crossed Claisen condensation can be used for the synthesis of a wide variety of β-dicarbonyl compounds.

Focused Problem

22.29 Complete the following reactions. Assume that one equivalent of NaOEt is present in each case.

(a)

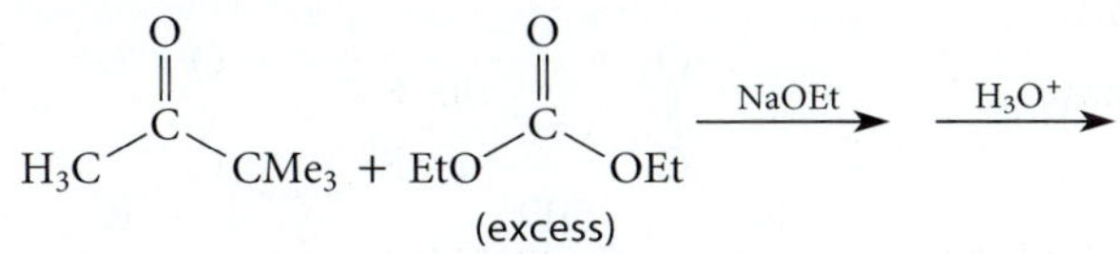

(b)

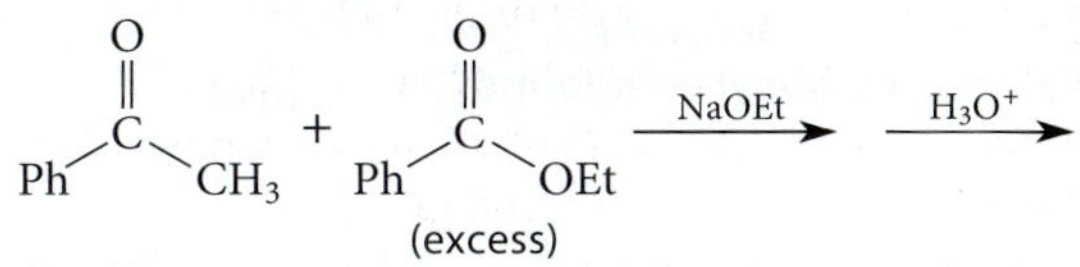

(c)

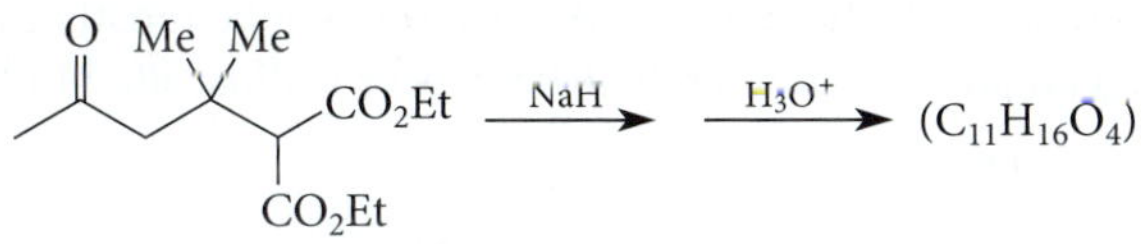

D. Synthesis with the Claisen Condensation

As the examples in the previous sections have shown, the Claisen condensation and related reactions can be used for the synthesis of β-dicarbonyl compounds: β-keto esters, β-diketones, and the like. Compare these types of compounds with those prepared by the aldol condensation, and notice carefully the differences.

In planning the synthesis of a β-dicarbonyl compound, we adopt the usual two-step strategy: first examine the target molecule and work backward to reasonable starting materials. Then analyze the reaction of these starting materials to see whether the desired reaction is reasonable or whether other reactions will occur instead.

To determine the starting materials for a Claisen condensation, mentally reverse the condensation by adding the elements of ethanol (or another alcohol) across either of the carbon–carbon bonds *between* the carbonyl groups. Because there are two such bonds, we will generally find two possible disconnections [labeled (a) and (b) in the following display] and two corresponding sets of starting materials by this procedure.

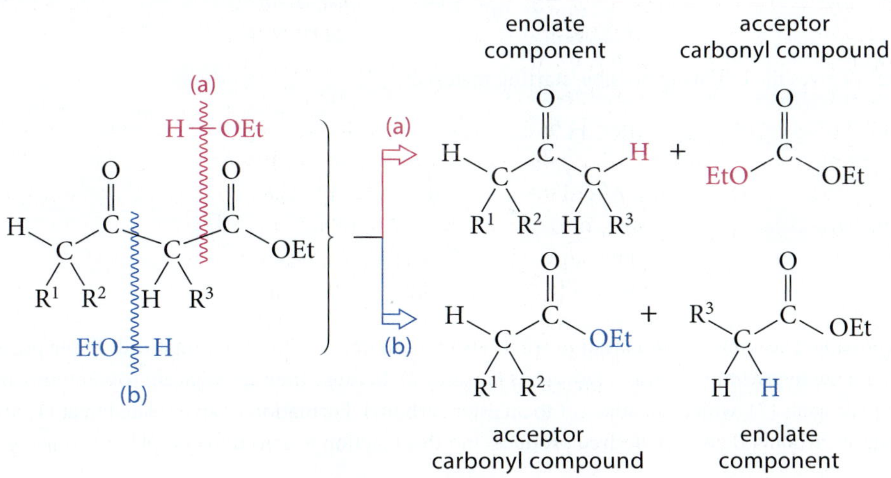

(22.78)

A β-diketone can be similarly analyzed in two different ways:

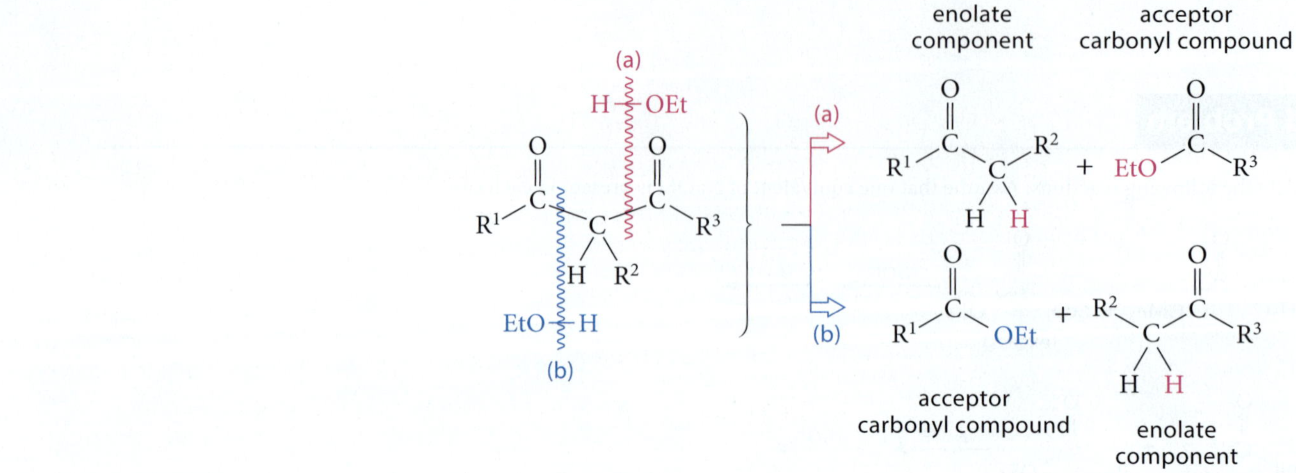

(22.79)

Having determined the possible starting materials required in a Claisen condensation, we then ask whether the Claisen condensation of the required materials will give mostly the desired product or a complex mixture. Such an analysis of a target β-keto ester is illustrated in Study Problem 22.3.

Study Problem 22.3

Determine whether the following compound can be prepared by a Claisen condensation or one of its variations; if so, give the possible starting materials.

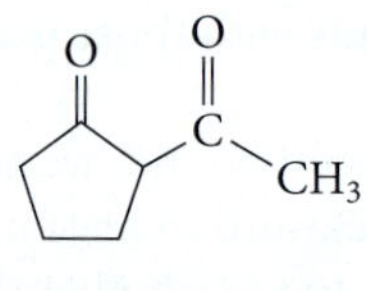

Solution This is a β-diketone, a type of compound for which a Claisen or Dieckmann condensation might be appropriate. To determine the possible starting materials, follow the procedure in Display 22.79: add EtOH in turn across each of the bonds indicated:

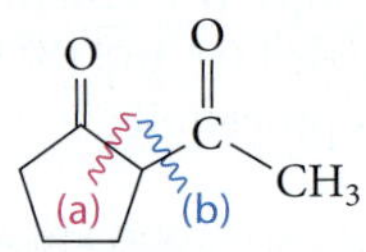

Addition across bond (a) gives the following possible starting material:

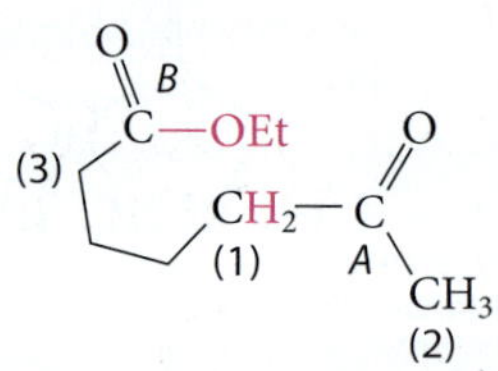

Now think about *all* possible Dieckmann condensation reactions that can occur with this compound. Three possible sets of α-hydrogens could ionize to give enolate ions. Hydrogens (1) and (2), because they are adjacent to a ketone carbonyl, are more acidic than hydrogens (3), which are adjacent to an ester carbonyl. Formation of an enolate ion at (1) and reaction of this enolate at carbonyl *B* give the desired product, and this reaction is driven to completion by using one equivalent of NaOEt. Formation of an enolate ion at (2) and reaction of this enolate at carbonyl *B* would give a β-diketone product containing a seven-membered ring:

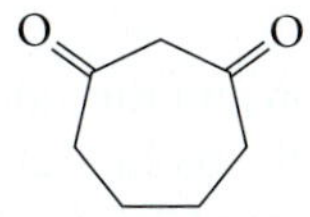

Because five-membered rings typically form much more rapidly than seven-membered rings (Sec. 12.8A), the desired product should be the major one, although formation of the seven-membered ring is a potential complication.

Breaking bond (b) in the target gives the following starting materials:

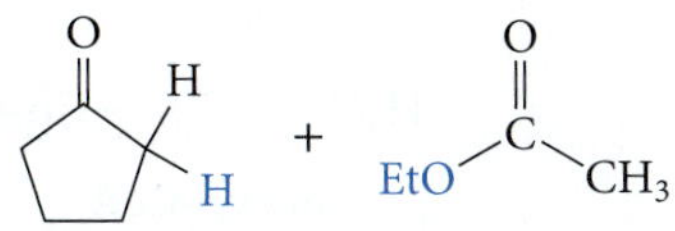

cyclopentanone **ethyl acetate**

In this case, the ketone, cyclopentanone, is more acidic than the ester, ethyl acetate. Because of its symmetry, cyclopentanone can give only one enolate ion. Aldol addition of this enolate ion to another molecule of cyclopentanone is an unfavorable equilibrium; recall that the equilibria for aldol additions of ketones are unfavorable. If an excess of ethyl acetate is used, this potential side reaction can be further suppressed, if it occurs at all. The desired Claisen condensation can be made irreversible by use of one equivalent of NaOEt to ionize the products. Consequently, this set of starting materials—cyclopentanone and ethyl acetate—should give the desired reaction.

Evidently, *both* sets of potential starting materials would work and in fact are acceptable answers. Which would be best in practice? Cyclopentanone and ethyl acetate are inexpensive and readily available. The other starting material would probably have to be prepared in a multistep synthesis. Consequently, cyclopentanone and ethyl acetate are the starting materials of choice. (This synthesis is conceptually the same as the one in Eq. 22.77.)

STUDY GUIDE LINK 22.5
Variants of the Aldol and Claisen Condensations

Focused Problems

22.30 Analyze each of the following compounds, and determine what starting materials would be required for its synthesis by a Claisen condensation. Then decide which if any of the possible Claisen condensations would be a reasonable route to the desired product.

(a) 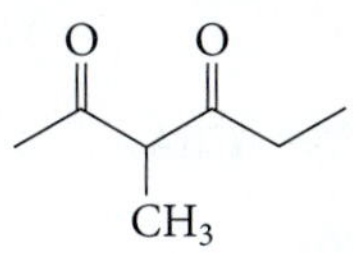

(b) 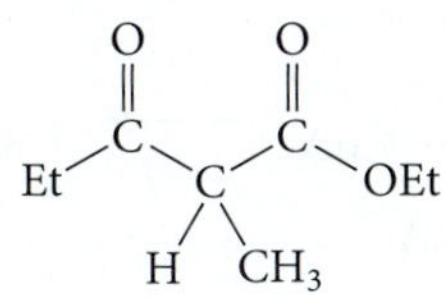

(c) 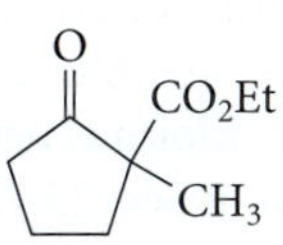

(d) 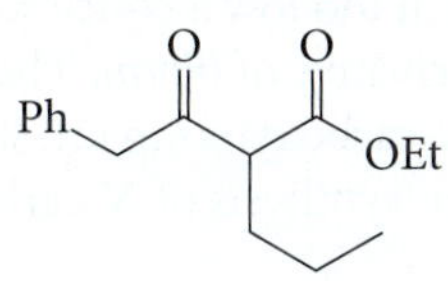

22.31 Give the starting material required for the synthesis of each of the following compounds by a Dieckmann condensation.

(a) 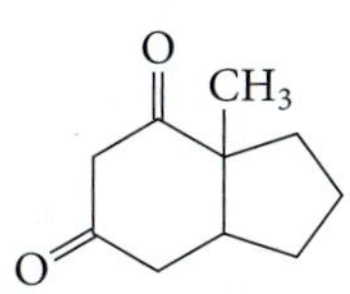

(b) 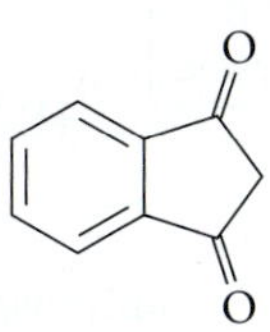

Chemical Biology Topic

22.7 THE CLAISEN CONDENSATION IN BIOLOGY: BIOSYNTHESIS OF FATTY ACIDS

The utility of the Claisen condensation and the aldol reactions is not confined to the laboratory; these reactions are also important in the biological world. The biosynthesis of *fatty acids* (Sec. 20.5) illustrates how nature uses the Claisen condensation to build long carbon chains.

The carbons of fatty acids are all derived from **acetyl-coenzyme A**, abbreviated *acetyl-CoA*, a thiol ester of acetic acid.

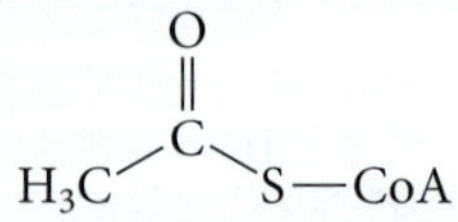

acetyl-CoA

The full structure of acetyl-CoA is given in Fig. 25.1 (Sec. 25.1). (There is more about the role of acetyl-CoA and other thioesters in biology in Chapter 25.) The complexity of the CoA part of the molecule is important for its binding interactions with proteins, but can be ignored in the chemical transformations we're going to learn about here.

The overall chemical transformation of acetyl-CoA to the 16-carbon fatty acid, palmitic acid, is as follows:

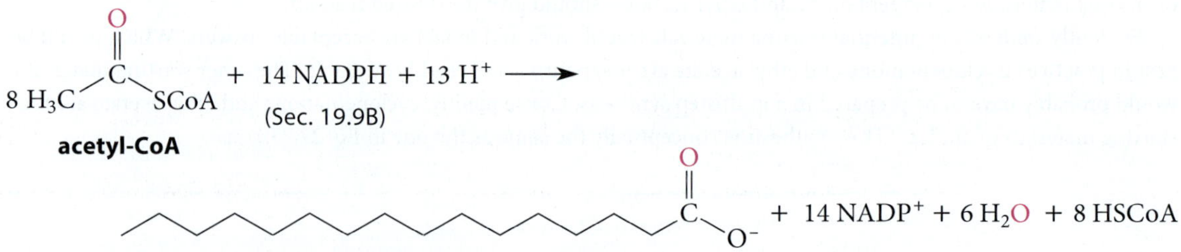

palmitate (the conjugate base of palmitic acid)
(all carbons are derived from acetyl-CoA) (22.80)

Prior to the actual condensation that forms the new carbon–carbon bonds of fatty acids, a molecule of acetyl-CoA is activated by carboxylation to form malonyl-CoA.

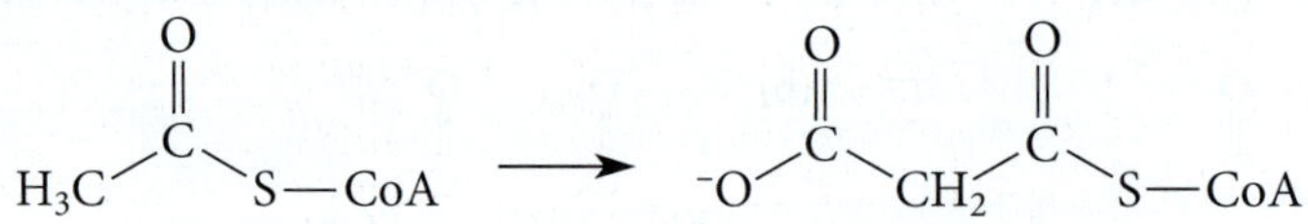

acetyl-CoA **malonyl-CoA** (22.81)

Carbon dioxide itself is present in too low a concentration to be useful in this conversion; rather, it is present as an *N*-carboxy derivative of *biotin*. The structures of biotin (sometimes called vitamin H or vitamin B_7) and *N*-carboxybiotin are shown in Display 22.82. Biotin is nature's "carbon dioxide carrier." (The biosynthesis of *N*-carboxybiotin is discussed in Sec. 25.7B.)

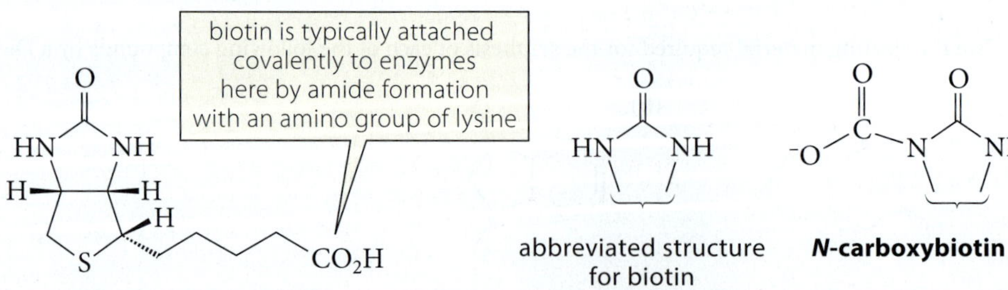

biotin abbreviated structure for biotin used in the text **N-carboxybiotin** (22.82)

The enzyme that catalyzes the carboxylation of acetyl-CoA is *acetyl-CoA carboxylase*. In the active site of this enzyme, *N*-carboxybiotin decomposes to CO_2 and the biotin anion; this transformation is assisted by hydrogen-bond donation (*blue*) from the enzyme to the carbonyl oxygen, which reduces the basicity of the biotin anion and makes it a better leaving group. The biotin anion abstracts an α-proton from the methyl group of acetyl-CoA to give the conjugate-base enolate ion, which is stabilized by hydrogen-bond donation from the enzyme. Before the just-liberated CO_2 can escape from the active site of the enzyme, it reacts with the enolate ion in a carbonyl-addition reaction to give malonyl-CoA.

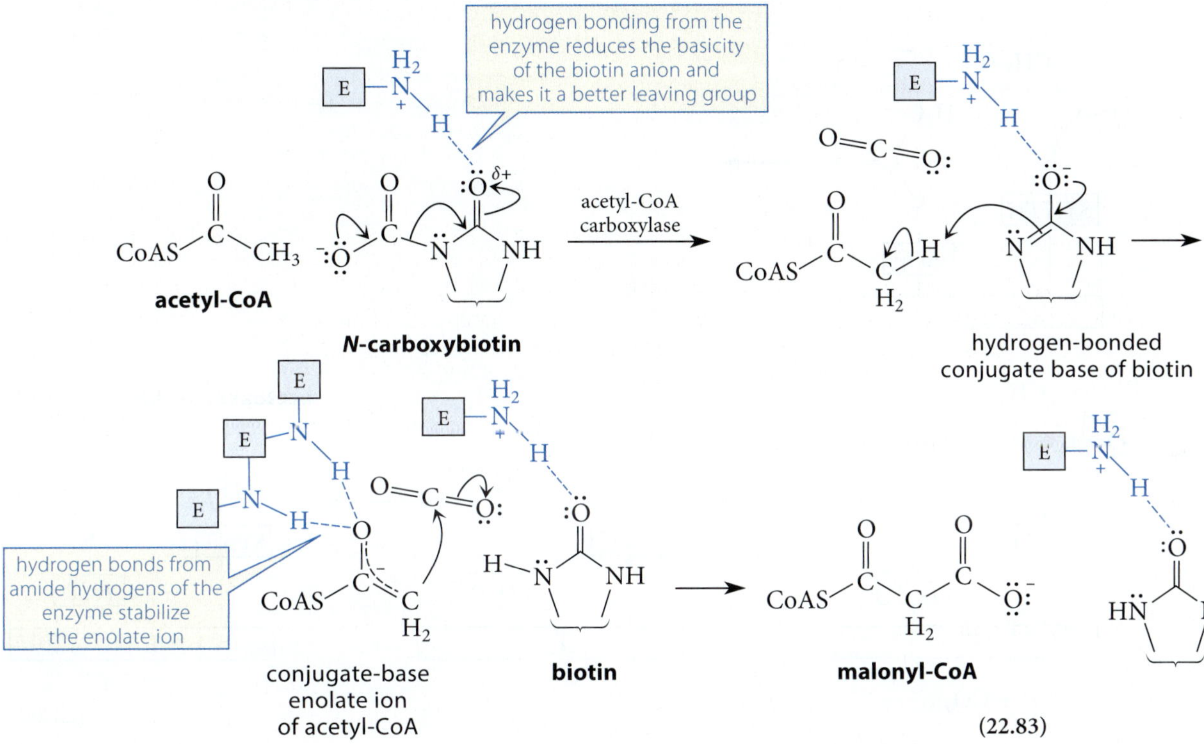

(22.83)

Malonyl-CoA and acetyl-CoA are then transferred to two different thiol groups on the next enzyme by *transthioesterification* reactions. This enzyme, fatty acid synthase, is a large protein that, in animals, consists of several tightly associated component proteins, each of which catalyzes one of the subsequent reactions. In one transthioesterification reaction, the malonyl-CoA is transferred to a thiol-containing protein subunit of the enzyme called *acyl carrier protein* (HS—ACP) to give malonyl-ACP. The acetyl-CoA is transferred in another transthioesterification reaction to a different thiol of the protein that we abbreviate simply as "SYNTH—SH." Each transfer liberates a molecule of CoA—SH as a by-product. (E = the overall enzyme.)

Transesterification, the carbonyl-substitution reaction of an ester with an alcohol to give a different ester, was discussed in Sec. 21.8C. *Transthioesterification* is the sulfur analog of this reaction—the reaction of a thioester with a thiol to give a different thioester.

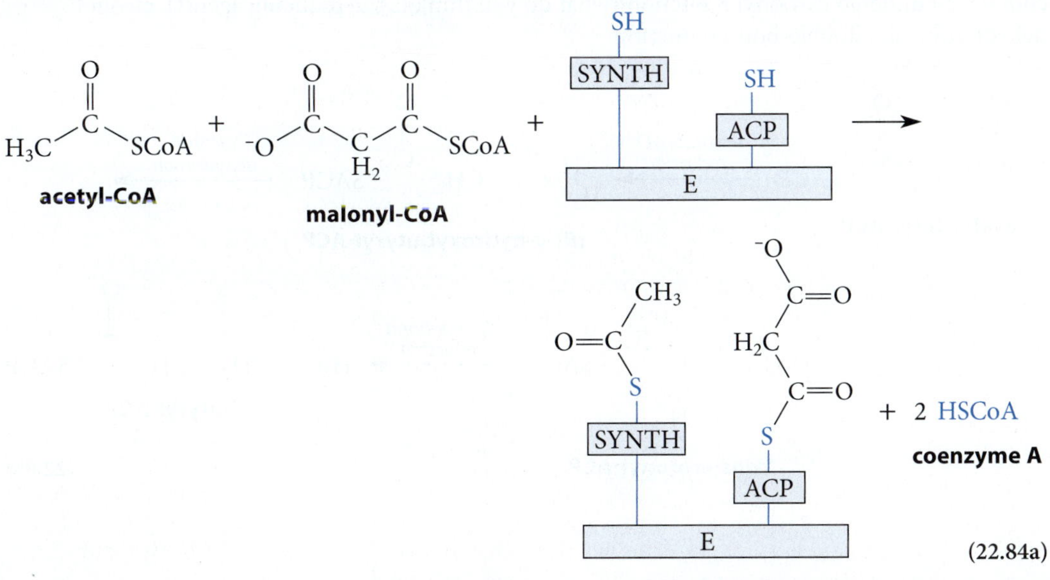

(22.84a)

The purpose of these transthioesterifications is to position the α-carbon of the malonyl thioester and the carbonyl group of the acetyl thioester in proximity for the next reaction, which is the Claisen condensation.

Loss of CO_2 from the malonyl group gives a transient enolate ion, which immediately reacts at the carbonyl of the nearby acetyl thioester to displace the thiolate anion and form the condensation product, acetoacetyl-ACP.

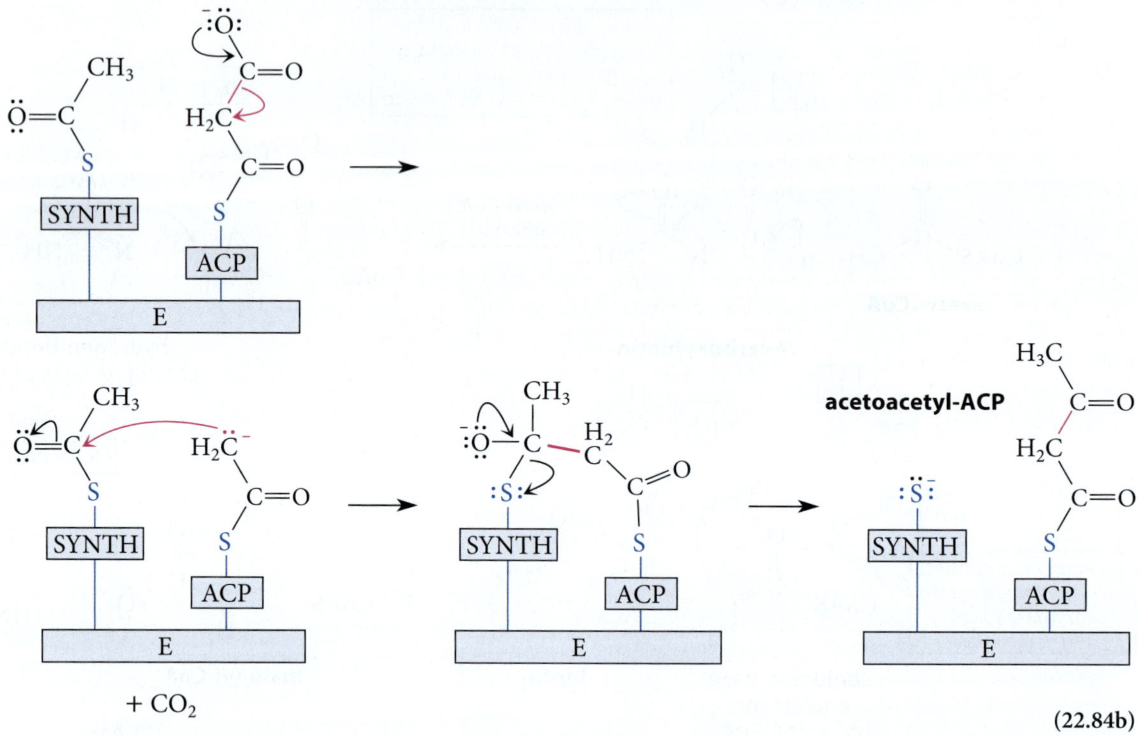

(22.84b)

It may seem odd that nature would install a CO_2 group on acetyl-CoA (Eq. 22.83) only to lose it again in the Claisen condensation. Recall, however, that the Claisen condensation in the laboratory is thermodynamically unfavorable, and it is driven to completion by ionization of an α-hydrogen of the acetoacetate ester product (Eq. 22.69c, Sec. 22.6A). In the enzyme-catalyzed Claisen condensation, no group in the active site of the enzyme or in solution is basic enough to ionize this hydrogen completely. Instead, the loss of CO_2, which ultimately escapes as a gas, drives the reaction to completion.

In subsequent steps of fatty-acid biosynthesis, the ACP serves as a remarkable "swinging arm" that rotates the acetoacetyl group successively into the other active sites on the fatty acid synthase complex to undergo carbonyl reduction (what do you think is the reducing agent?), alcohol dehydration, and double-bond reduction.

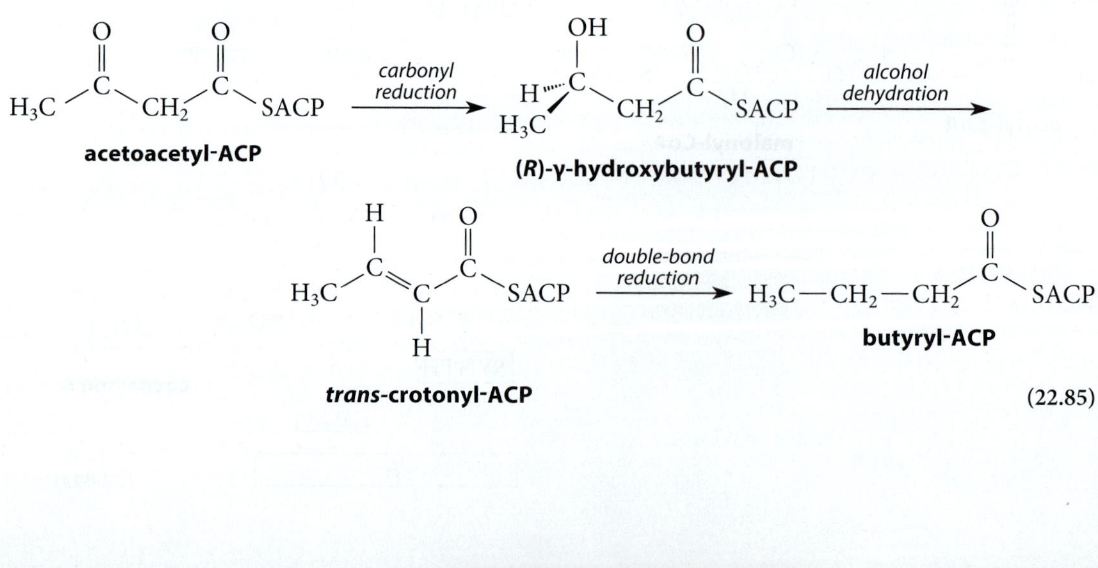

(22.85)

The overall result is the conversion of a two-carbon thioester (acetyl-CoA) into a four-carbon thioester (butyryl-ACP).

The butyryl-ACP is finally transthioesterified onto the thiol of the SYNTH—SH protein, the site previously occupied by the acetyl group. The ACP—SH is "recharged" with another malonyl group from malonyl-CoA, and an analogous series of reactions takes place again that converts the four-carbon thioester into a six-carbon thioester.

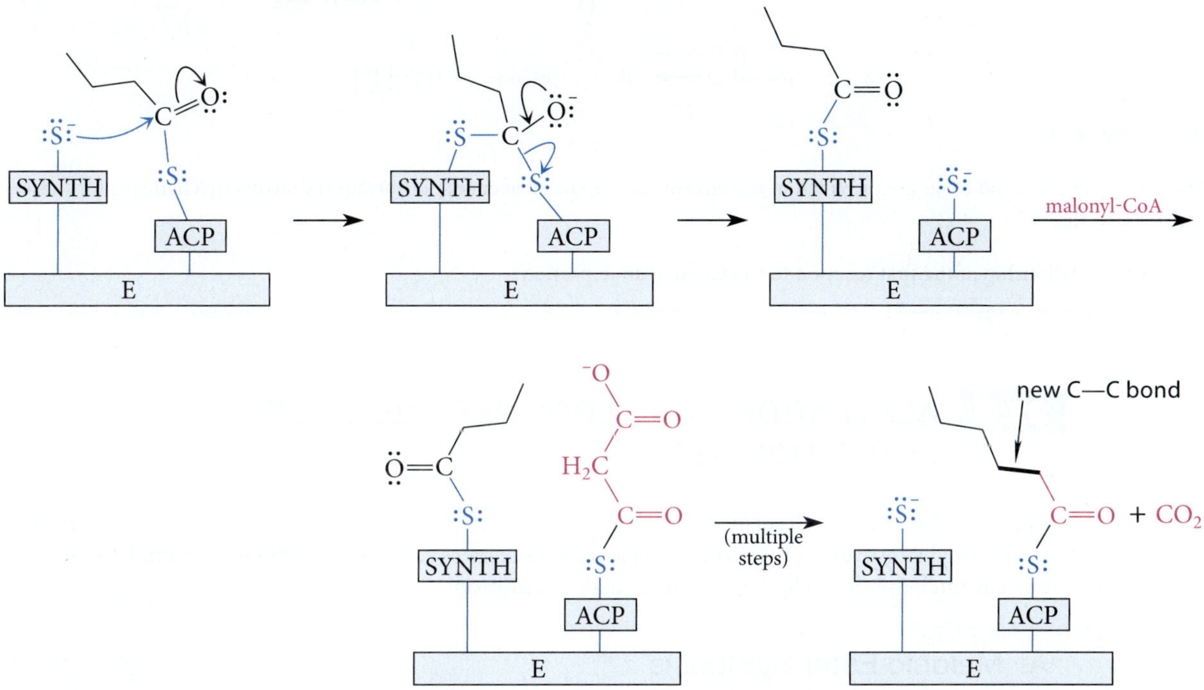

(22.86)

This process continues until the growing chain contains 16 carbons. The addition of carbons to the growing chain two carbons at a time accounts for the fact that *ordinary fatty acids have even numbers of carbons.* Hydrolysis of the 16-carbon thioester gives palmitic acid, the 16-carbon fatty acid.

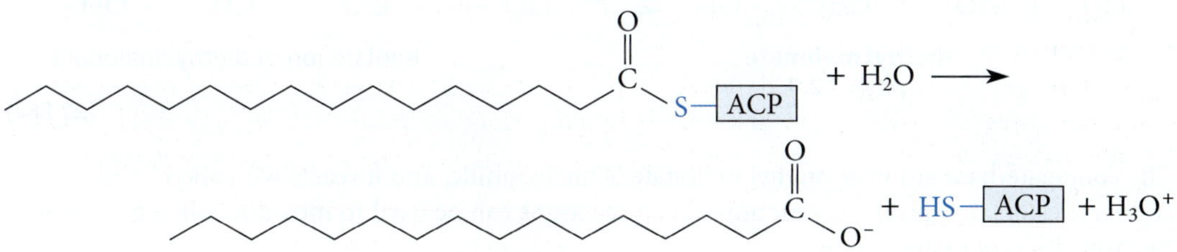

palmitate
(conjugate-base anion of **palmitic acid**)

(22.87)

Longer fatty acids and unsaturated fatty acids are produced by other reactions.

Focused Problems

22.32 Suppose a sample of acetyl-CoA labeled at the carbonyl carbon with the radioisotope ^{14}C is introduced into the fatty-acid synthase system, and palmitic acid (the 16-carbon fatty acid; Eq. 22.87) is isolated. Which carbons of palmitic acid should be radiolabeled?

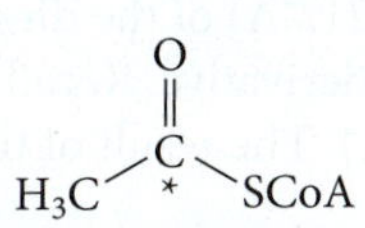

acetyl-CoA (carbonyl ^{14}C-labeled)

22.33 Fatty acids are degraded to acetyl-CoA in fatty-acid metabolism. The enzyme that catalyzes this conversion, *acyl-CoA acetyl transferase*, contains a nucleophilic thiol group in its active site and catalyzes the following reactions (enzyme = E):

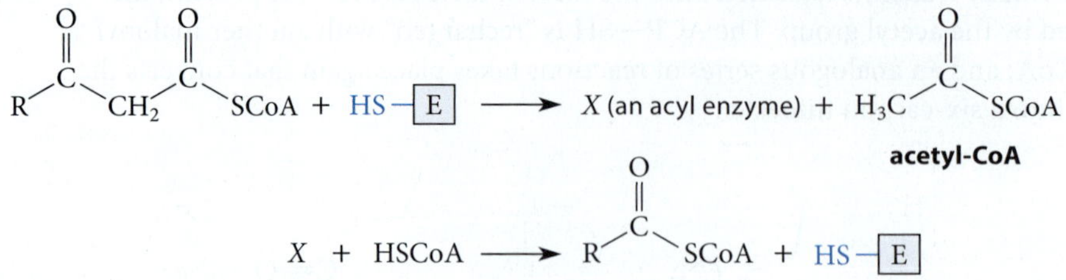

(a) What is species *X*?

(b) Assuming that acids and bases are provided in the enzyme active site as needed, outline curved-arrow mechanisms for these transformations.

(c) What is the relationship of the first reaction to the Claisen condensation?

22.8 ALKYLATION AND ALDOL REACTIONS OF ESTER ENOLATE IONS

Sections 22.6 and 22.7 described reactions in which the enolate ions of esters react as nucleophiles in carbonyl substitution reactions. This section considers reactions in which ester enolates are used as nucleophiles in S_N2 reactions and aldol additions.

A. Malonic Ester Synthesis

Diethyl malonate (malonic ester), like many other β-dicarbonyl compounds, has unusually acidic α-hydrogens. (Why?) Consequently, its conjugate-base enolate ion can be formed nearly completely with alkoxide bases such as sodium ethoxide.

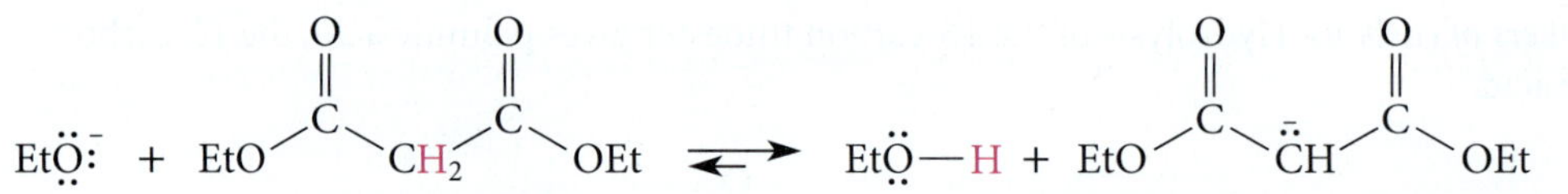

diethyl malonate
$pK_a = 12.9$

enolate ion of diethyl malonate

(22.88a)

The conjugate-base anion of diethyl malonate is nucleophilic, and it reacts with alkyl halides and sulfonate esters in typical S_N2 reactions. Such reactions can be used to introduce alkyl groups at the α-position of malonic ester.

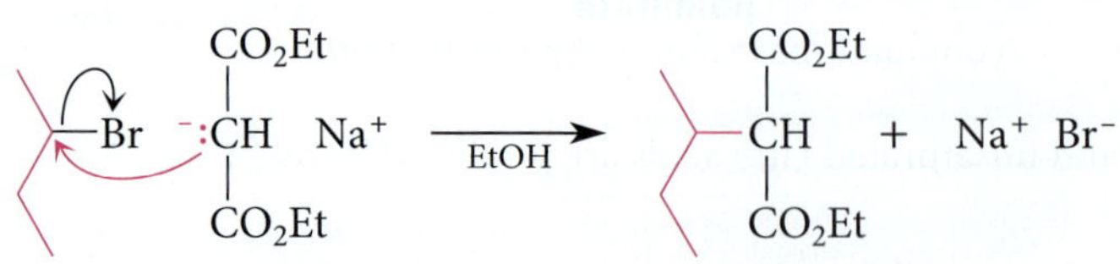

(83% yield) (22.88b)

FURTHER EXPLORATION 22.1
Malonic Ester Alkylation

As this example shows, even secondary halides can be used in this reaction. (See Further Exploration 22.1.)

The importance of this reaction is that it can be extended to the preparation of carboxylic acids. Saponification (Sec. 21.7A) of the diester and acidification of the resulting solution give a substituted malonic acid derivative. Recall that heating any malonic acid derivative causes it to *decarboxylate* (Sec. 20.11). The result of the alkylation, saponification, and decarboxylation sequence is a carboxylic acid that conceptually is a substituted acetic acid—an acetic acid molecule with an alkyl group on its α-carbon.

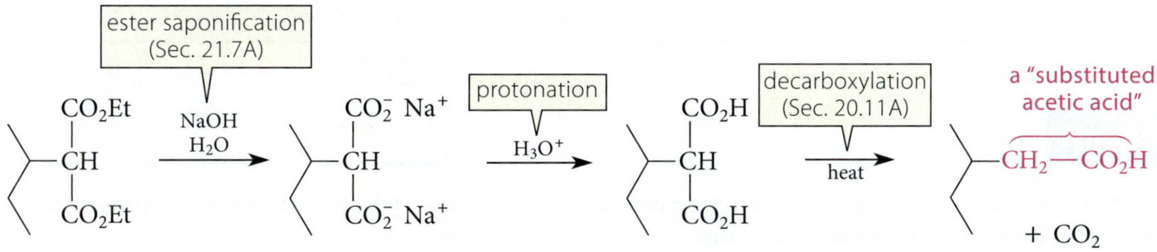

(22.88c)

The overall sequence of ionization, alkylation, saponification, and decarboxylation starting from diethyl malonate (Eqs. 22.88a–c) is called the **malonic ester synthesis**. Notice that the alkylation step of the malonic ester synthesis (Eq. 22.88b) results in the formation of a new carbon–carbon bond.

The anion of malonic ester can be alkylated twice in two successive reactions with different alkyl halides (if desired) to give, after hydrolysis and decarboxylation, a *disubstituted* acetic acid. This possibility allows us to think of any disubstituted acetic acid in terms of diethyl malonate and two alkyl halides, as follows (X = halogen):

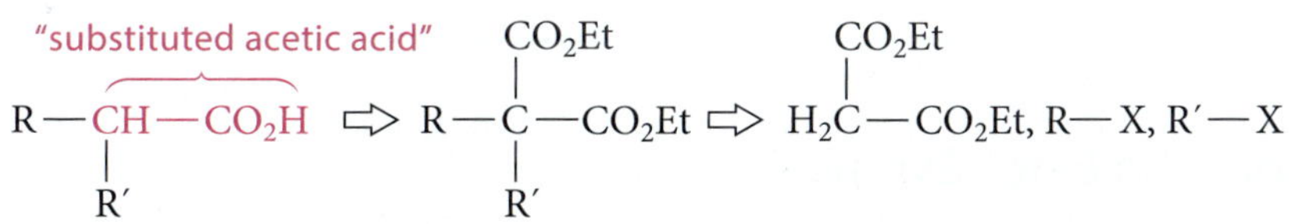

(22.89)

If the alkyl halides R—X and R′—X are among those that will undergo the S_N2 reaction, then the target carboxylic acid can in principle be prepared by the malonic ester synthesis. This analysis is illustrated in Study Problem 22.4.

Study Problem 22.4

Outline a malonic ester synthesis of the following carboxylic acid:

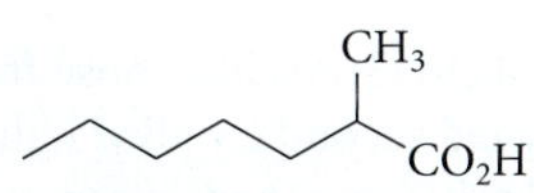

2-methylheptanoic acid

Solution Using the analysis in the text, identify the "acetic acid" unit in the carboxylic acid. The two alkyl groups—in this case, a methyl group and a pentyl group—are derived from alkyl halides.

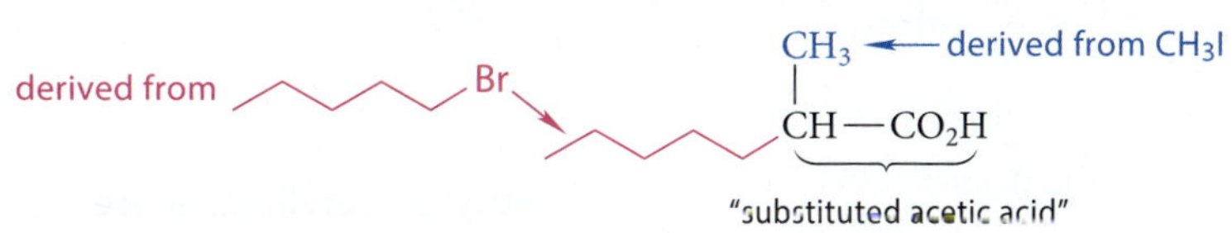

This analysis leads to the following synthesis:

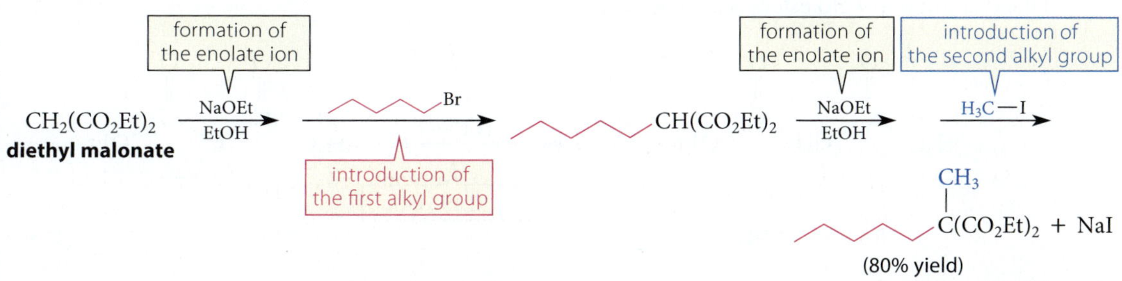

(22.90)

Ester saponification, acidification, and decarboxylation, as in Eq. 22.88c, give the desired product.

The two enolate-forming and alkylation reactions must be performed as *separate steps*. Adding two different alkyl halides and two equivalents of NaOEt to malonic ester at the same time would give a mixture of products. (Why?)

Focused Problems

22.34 Indicate whether each of the following compounds could be prepared by a malonic ester synthesis. If so, outline a preparation from diethyl malonate and any other reagents. If not, explain why.

(a) 3-Phenylpropanoic acid

(b) 2-Ethylbutanoic acid

(c) 3,3-Dimethylbutanoic acid

22.35 (a) Give the product of the following reaction sequence, and explain your answer.

$$CH_2(CO_2Et)_2 + BrCH_2CH_2CH_2Cl \xrightarrow[\text{EtOH}]{\text{2 NaOEt}} \xrightarrow{\text{NaOH}} \xrightarrow[\text{heat}]{\text{HCl}} (C_5H_8O_2)$$

(b) Why is $BrCH_2CH_2CH_2Cl$ rather than $BrCH_2CH_2CH_2Br$ used as an alkylating agent in this reaction?

B. Acetoacetic Ester Synthesis

Recall that β-keto esters, like malonic esters, are substantially more acidic than ordinary esters (Eq. 22.69c, Sec. 22.6A) and are completely ionized by alkoxide bases.

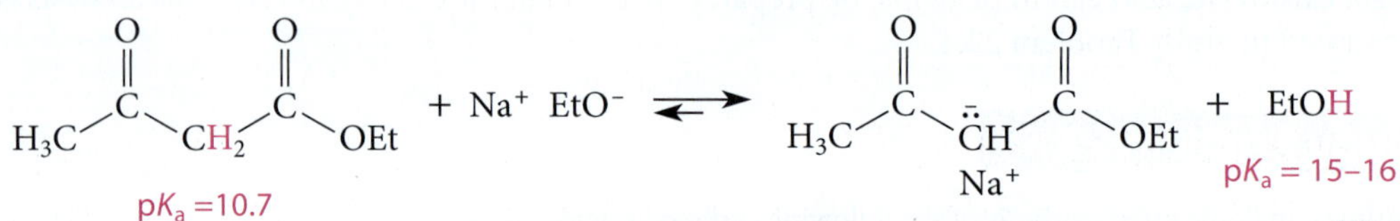

(22.91)

The enolate ions derived from β-keto esters, like those from malonate ester derivatives, can be alkylated by primary or unbranched secondary alkyl halides or sulfonate esters. In this case, the product of Eq. 22.91 is alkylated following the Claisen condensation without isolation of the β-keto ester.

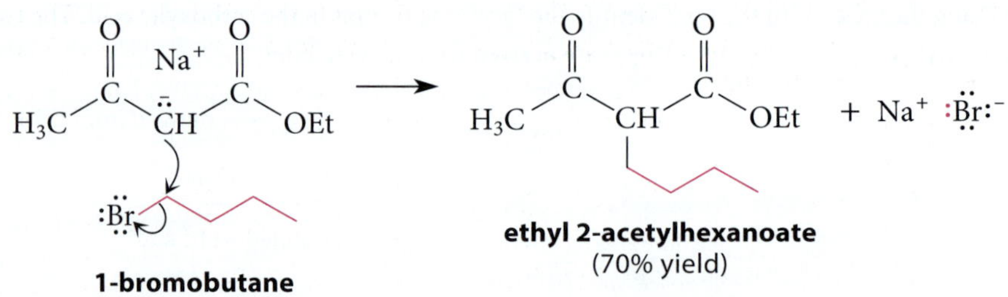

(22.92)

Dialkylation of β-keto esters is also possible.

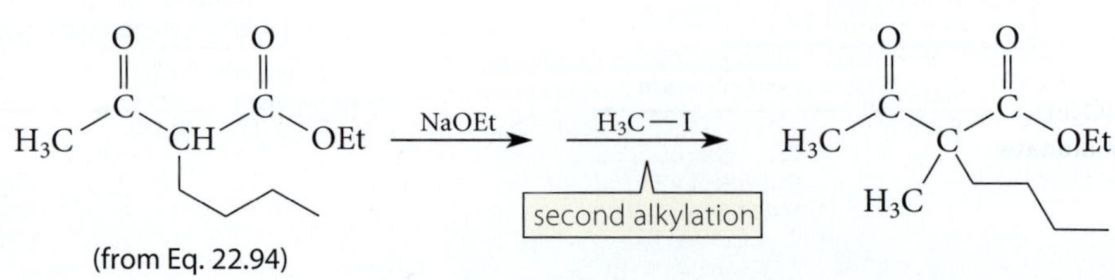

(22.93)

Alkylation of a Dieckmann condensation product is the same type of reaction:

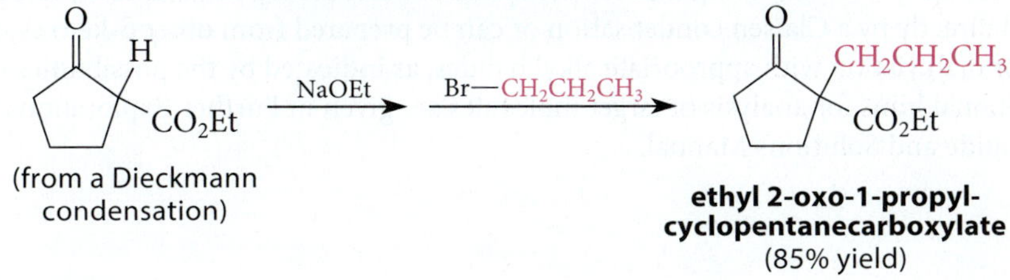

(22.94)

Like esters of substituted malonic acids, the alkylated derivatives of ethyl acetoacetate can be hydrolyzed and decarboxylated to give ketones. Ester saponification and protonation give a substituted β-keto acid; and β-keto acids spontaneously decarboxylate at room temperature (Sec. 20.11). This series of reactions is illustrated as carried out on the alkylation product in Eq. 22.92:

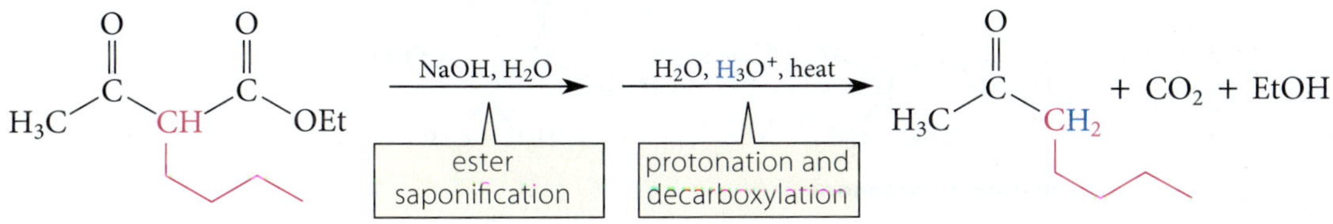

(22.95a)

If the β-keto ester alkylation product doesn't have an especially acidic α-hydrogen at carbon-2, as in the product of Eq. 22.93, the basic conditions of saponification will bring about a reverse Claisen condensation (see Eq. 22.70, Sec. 22.6A, and the accompanying discussion). In cases like this, *acid-catalyzed* ester hydrolysis should be used. The resulting β-keto acid decarboxylates spontaneously after it forms under the acidic conditions.

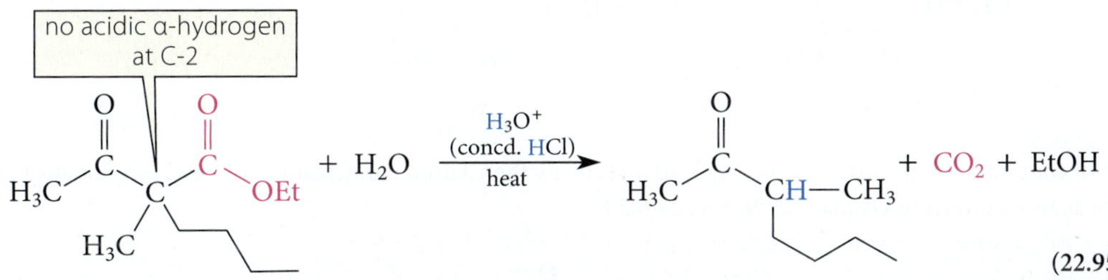

(22.95b)

The alkylation of ethyl acetoacetate followed by saponification, protonation, and decarboxylation to give a ketone is called the **acetoacetic ester synthesis**. The alkylation part of this sequence, like the alkylation of diethyl malonate, involves the construction of new carbon–carbon bonds.

Whether a target ketone can be prepared by the acetoacetic ester synthesis can be determined by mentally reversing the synthesis.

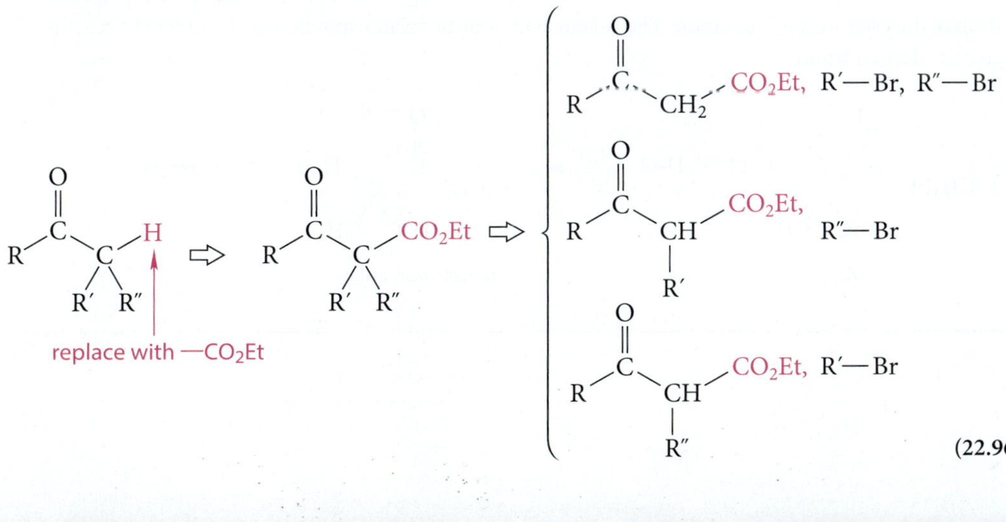

(22.96)

STUDY GUIDE LINK 22.6
More Target Analysis for the Claisen Condensation

This analysis involves replacing an α-hydrogen of the target ketone with a —CO_2Et group. This process unveils the β-keto ester required for the synthesis. The β-keto ester, in turn, can either be prepared directly by a Claisen condensation or can be prepared from other β-keto esters by alkylation or dialkylation with appropriate alkyl halides, as indicated by the possibilities in Display 22.96. Additional hints for analysis of target molecules are given in Further Explorations 22.6 in the Study Guide and Solutions Manual.

Study Problem 22.5

Outline a preparation of 2-methyl-3-pentanone by a reaction sequence that involves at least one Claisen condensation.

Solution The discussion in the text leads to the following analysis:

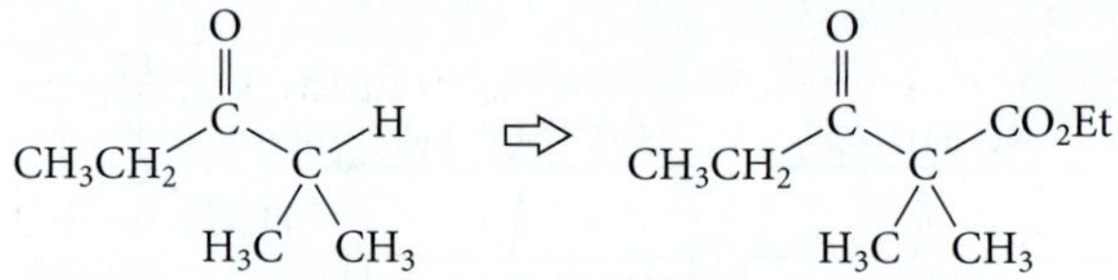

2-methyl-3-pentanone *A*

The symbol ⇨, as usual, means "implies as a starting material." The β-keto ester *A* cannot be prepared directly by a Claisen condensation because it would require a crossed Claisen condensation (see Display 22.79, Sec. 22.6D), and because the reaction could not be made irreversible by deprotonation. A second option is to provide one of the methyl groups by alkylation of the enolate ion derived from β-keto ester *B*:

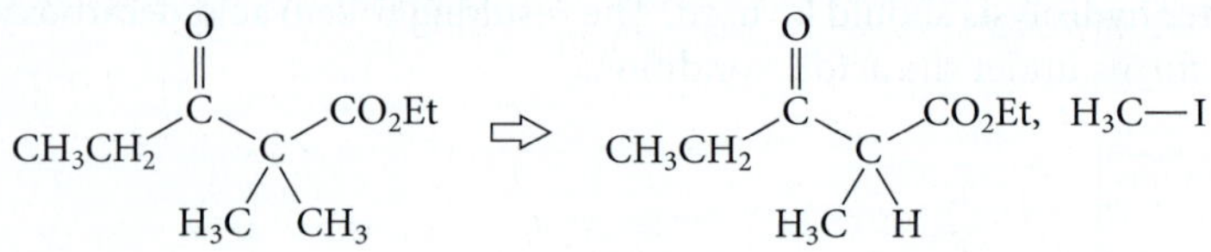

A *B*

The enolate ion of compound *B*, in turn, can be prepared directly by the Claisen condensation of ethyl propionate. (This follows from the analysis shown in Display 22.79, Sec. 22.6D.)

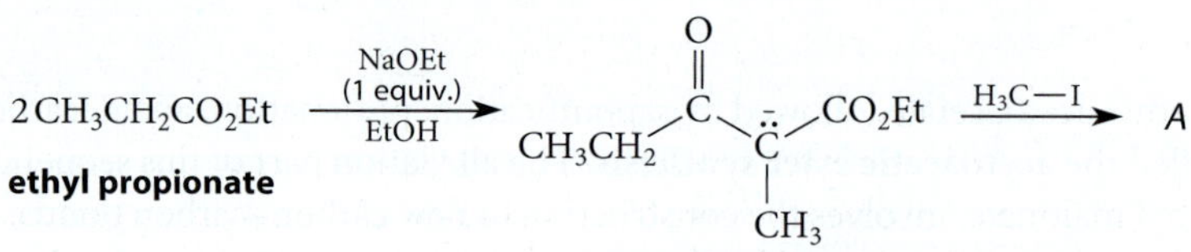

ethyl propionate enolate ion of *B*

Because *A* doesn't have an acidic α-hydrogen, treatment with base will bring about a reverse Claisen condensation. So, instead, we hydrolyze the ester in acidic solution. The β-keto acid decarboxylates spontaneously under the reaction conditions to give the desired ketone.

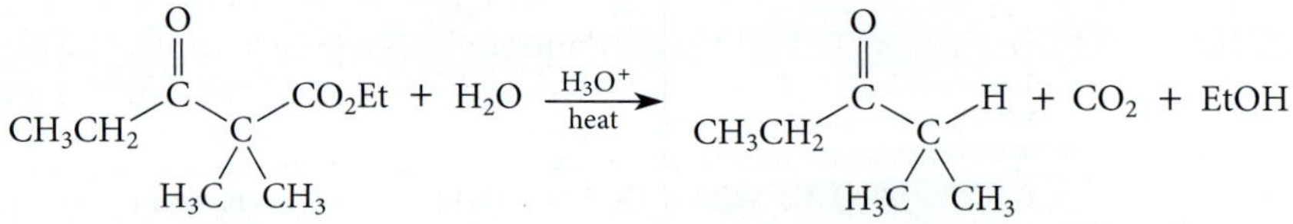

A target molecule

This section has discussed the reactions of ester enolates with alkylating agents such as alkyl halides. Conceptually, we might ask whether the enolate of an aldehyde or a ketone might also be alkylated. They can be alkylated, but such alkylations are less useful because they occur at both the α-carbon and the oxygen of the enolate.

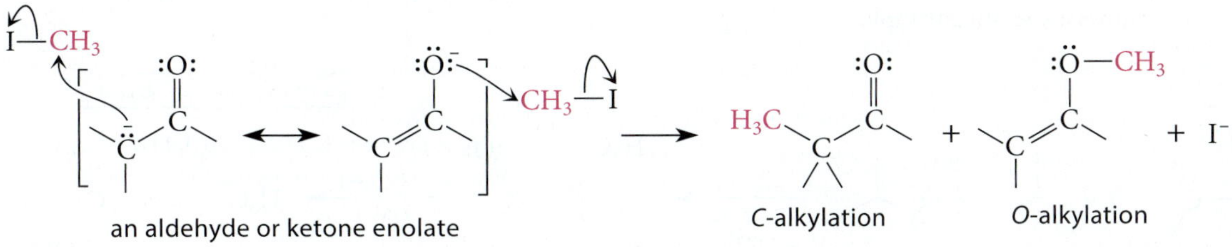

(22.97)

O-alkylation is not a problem with ester enolates. (Further Exploration 22.2 explains this aspect of enolate chemistry in greater detail.)

Therefore, alkylation at the α-carbon of a ketone requires further activation of the α-carbon by an ester group, which is, after alkylation, removed by saponification and decarboxylation.

FURTHER EXPLORATION 22.2
Alkylation of Enolate Ions Derived from Ketones

Focused Problems

22.36 Outline a synthesis of each of the following compounds from ethyl acetoacetate and any other reagents.

(a) 5-Methyl-2-hexanone (b) 4-Phenyl-2-butanone

22.37 Outline a synthesis of each of the following compounds from a β-keto ester; then show how the β-keto ester itself can be prepared.

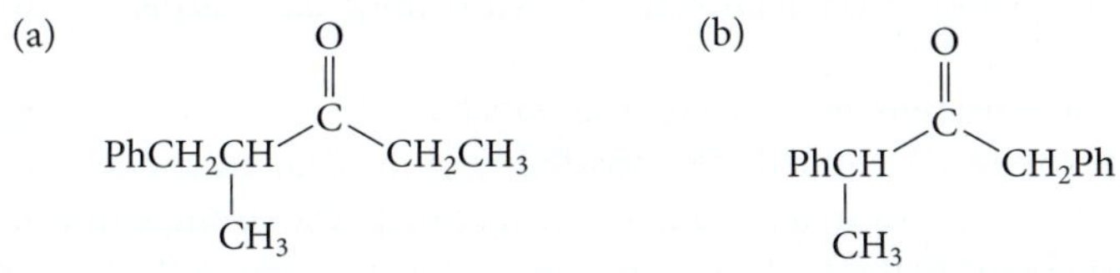

22.38 Predict the outcome of the following reaction by identifying *A*, then *B*, then the final product. (*Hint:* How do nucleophiles react with epoxides under basic conditions? See Sec. 12.5A.)

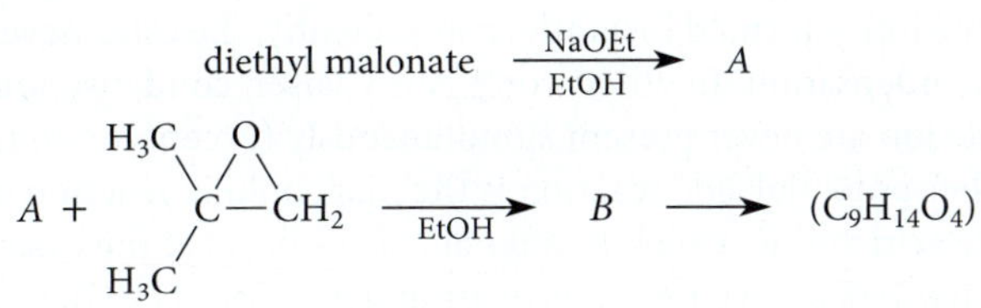

C. Direct Alkylation of Enolate Ions Derived from Monoesters

In the synthesis of carboxylic acids by malonic ester alkylation, a $—CO_2Et$ group is "wasted" because it is later removed. Why not avoid this altogether and alkylate directly the enolate ion of an acetic acid ester?

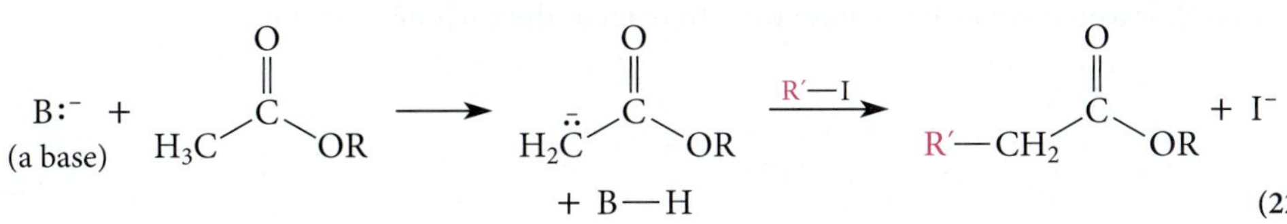

(22.98)

At one time this idea could not be used in practice because enolate ions derived from esters, once formed, undergo another, faster reaction: Claisen condensation with the parent ester (Sec. 22.6A). However, the development of the strong amide bases introduced in the discussion of the directed aldol addition (Sec. 22.4C) made it possible to form the lithium enolates of esters. These enolates can be alkylated with alkyl halides as shown conceptually in Eq. 22.98. The following equation shows a specific example.

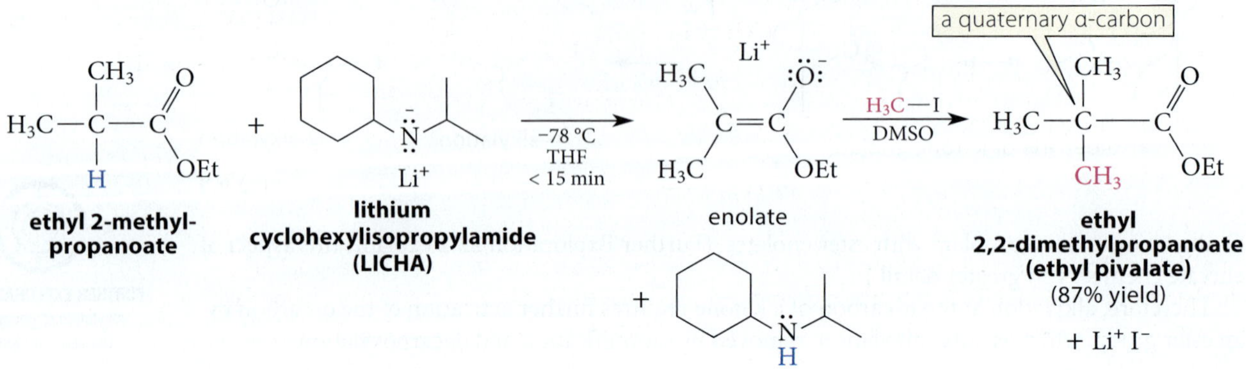

(22.99)

Notice in this example that a third alkyl substituent can be introduced at the ester α-carbon. This is not possible with the malonic ester synthesis. (Why?)

This method of ester alkylation is considerably more expensive than the malonic ester synthesis. It also requires special inert-atmosphere techniques because the strong bases that are used react vigorously with both oxygen and water. For these reasons, the malonic ester synthesis remains very useful, particularly for large-scale syntheses. However, for the preparation of laboratory samples, or for the preparation of compounds that are unavailable from the malonic ester synthesis, the preparation and alkylation of enolate ions with amide bases is particularly valuable.

As with the directed aldol condensation, we consider the possible side reactions that might occur with this method and why they are avoided. The possibility of the Claisen condensation as a side reaction was noted in the discussion of Eq. 22.98. The use of a very strong amide base avoids the Claisen condensation because the reaction is run by *adding the ester to the base.* When a molecule of ester enters the solution, it can react either with the strong base to form an enolate ion or with a molecule of already formed enolate ion in the Claisen condensation. The reaction of esters with strong amide bases is so much faster at −78 °C than the Claisen condensation that the enolate ion is formed instantly; consequently, the ester never has a chance to undergo the Claisen condensation. In other words, the Claisen condensation is avoided because the ester and its enolate ion are never present simultaneously (except for an instant) in the reaction flask.

Another potential side reaction is the nucleophilic reaction of the amide base (or even its conjugate acid amine, which is, after all, also a base) at the ester carbonyl group. Because amines react with esters to give products of aminolysis (Sec. 21.8C), it might be reasonable to expect the *conjugate bases* of amines—very strong bases, indeed—to react even more rapidly as nucleophiles with esters. That this reaction does not happen is once again the result of a competition. When an amide base reacts with the ester, it can either remove a proton or react at the carbonyl carbon. The reaction rate at the carbonyl carbon is reduced by van der Waals repulsions between groups on the carbonyl compound and the large branched groups on the bases, as in the directed aldol reaction. If the amide base could be in contact with the ester long enough, it would eventually react at the carbonyl carbon; but the base instead reacts more rapidly in a different way: it abstracts an α-proton. Reaction with a tiny hydrogen does not involve the van der Waals repulsions that would occur if the base were to react at the carbonyl carbon.

Focused Problem

22.39 Outline a synthesis of each of the following compounds from either diethyl malonate or ethyl acetate. Because the branched amide bases are relatively expensive, you may use them in only one reaction.

(a) 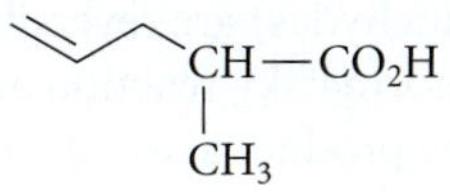

(b) 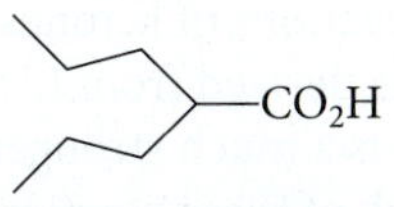

valproic acid
(used in treatment of epilepsy)

(c) 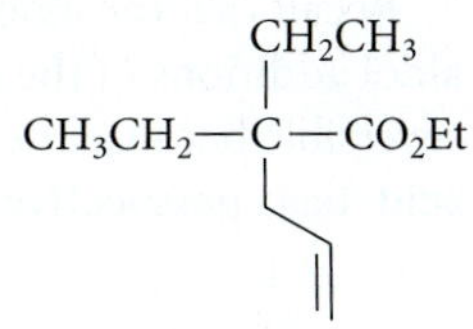

D. Aldol Reactions of Ester Enolates

Ester enolates react as nucleophiles with aldehyde and ketone carbonyl groups and give useful aldol addition or aldol condensation reactions. For example, the lithium enolate formed from an acetate ester gives an aldol addition with aldehydes and ketones.

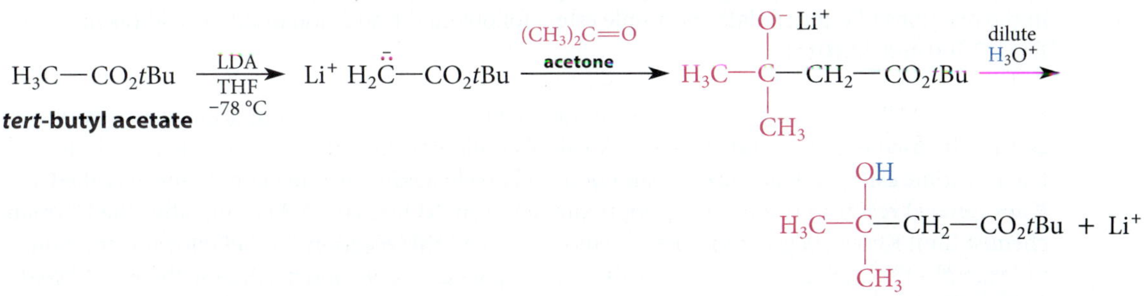

tert-butyl acetate

tert-butyl 3-hydroxy-3-methylbutanoate
(>90% yield)

(22.100)

Reformatsky Reaction One of the oldest but nevertheless widely used examples of an aldol addition is called the **Reformatsky reaction** after its discoverer, Sergei Nikolaevich Reformatsky (1860–1934), a Russian chemist who worked at the University of Kiev in Ukraine.

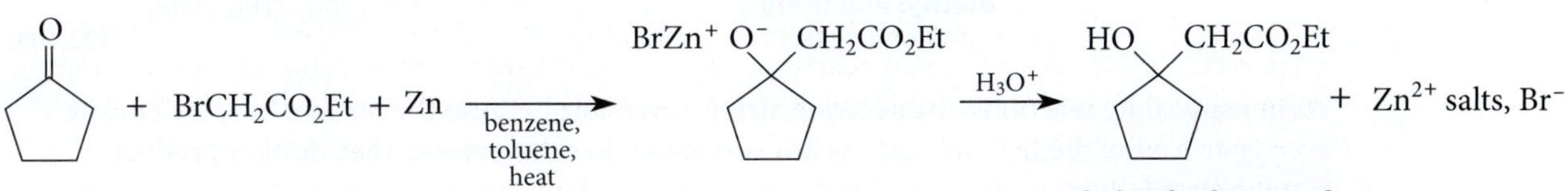

ethyl (1-hydroxycyclopentyl)acetate
(72% yield)

(22.101a)

In this reaction, an enolate is formed by the reaction of powdered zinc with an α-bromo ester. (See Sec. 22.3C for the preparation of α-bromo esters.) The zinc metal undergoes an insertion reaction with an α-bromo ester to form the zinc analog of a Grignard reagent (Eq. 10.47, Sec. 10.5B). Unlike Grignard reagents, which have to be formed in a separate step before they are used, Reformatsky reagents are formed in the presence of the aldehyde or ketone. Although the actual structure of the reagent is more complex, we can think of it as a zinc enolate of the ester.

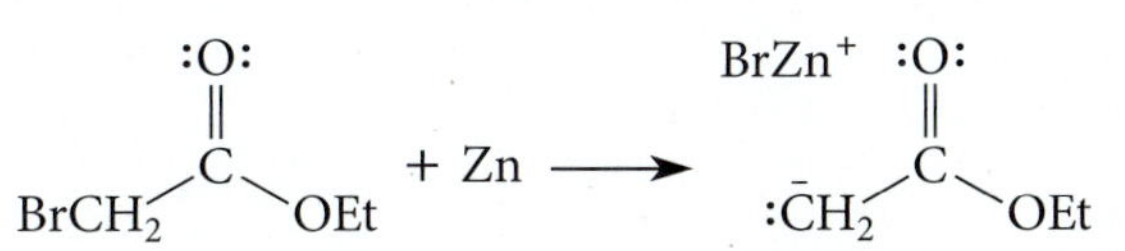

conceptual enolate structure
of the Reformatsky reagent

(22.101b)

Organozinc compounds are much less reactive than Grignard reagents; consequently, the Reformatsky reagent reacts with aldehydes and ketones, but not with esters. Therefore, the ester group of the reagent is unaffected. In Eq. 22.101a, the zinc enolate undergoes an addition reaction with the carbonyl group of the ketone to give a zinc alkoxide. Addition of aqueous acid gives the aldol addition product.

Recall that the aldol addition reactions of ketones (and some aldehydes) are reversible. The aldol additions of the ester enolates derived from LDA and the Reformatsky reaction are *not* reversible because the nucleophile is a much stronger base than the product alkoxide. From an acid–base perspective, we can think of these reactions in the following way (M^+ = metal):

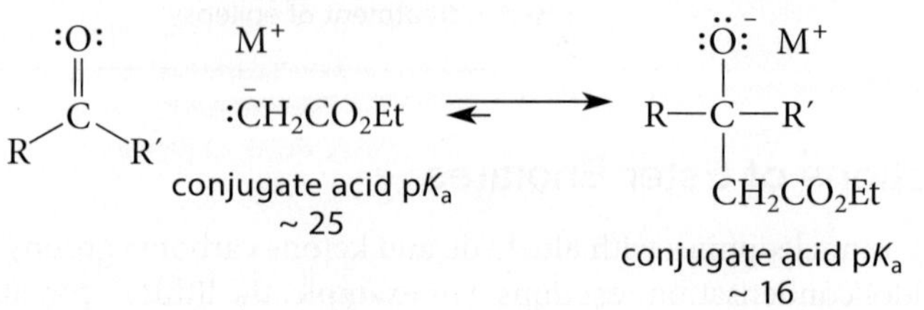

(22.102)

The practical significance of this irreversibility is that aldol addition products can be isolated in the reactions of ester enolates of simple esters following the addition of dilute acid (as in Eqs. 22.100 and 22.101a).

Knoevenagel Condensation The enolates derived from malonic esters or acetoacetic esters can also be used in aldol condensations. An aldol condensation reaction of enolate ions derived from malonic ester, acetoacetic ester, and other relatively acidic carbonyl compounds is called a **Knoevenagel reaction** (pronounced approximately kuh-NOER-vuh-NAH-gul), after the German chemist Emil Knoevenagel (1865–1921), who developed this reaction. In the following reaction, for example, the enolate ion of malonic ester undergoes an aldol condensation with benzaldehyde.

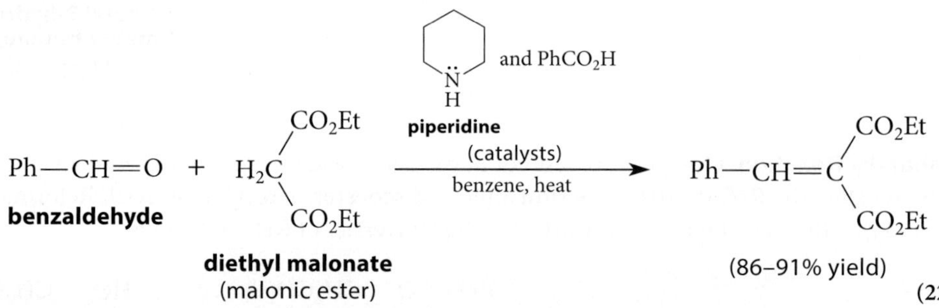

(22.103)

As in many aldol reactions, the addition step is reversible because the nucleophile, the enolate conjugate base of diethyl malonate, is not very basic. For that reason, the addition product is not isolated; the reaction is driven to completion by dehydration of the addition product (by an E1cB mechanism; Display 22.50, Sec. 22.4A) to give an aldol condensation product. (B: = the amine catalyst.)

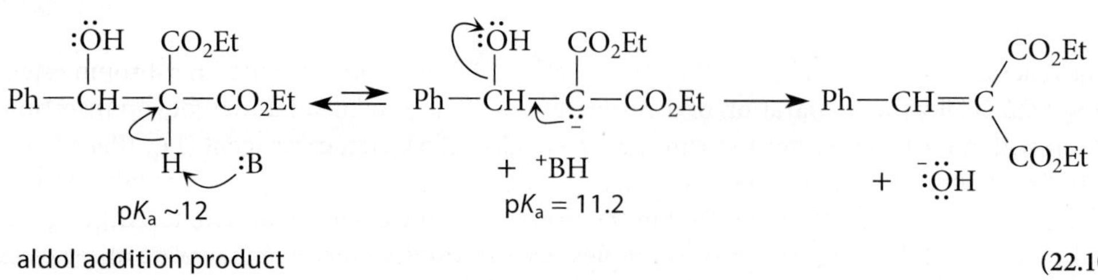

(22.104a)

Reaction of the hydroxide by-product with the conjugate acid of the catalyst is a favorable reaction that regenerates the catalyst.

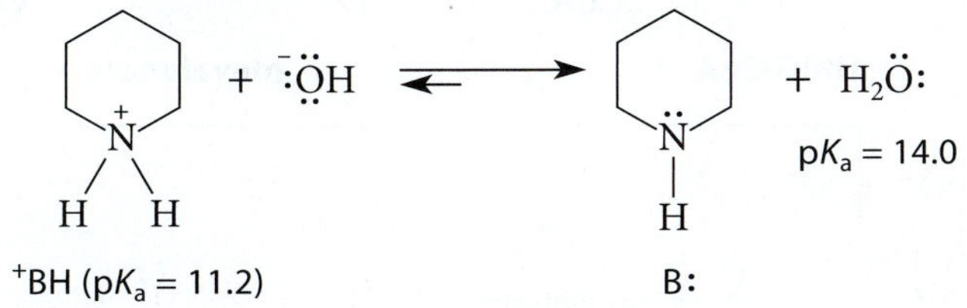

(22.104b)

Ester hydrolysis of the diester product of a Knoevenagel condensation gives a malonic acid derivative, and decarboxylation of this derivative (Sec. 20.11A) gives an α,β-unsaturated carboxylic acid.

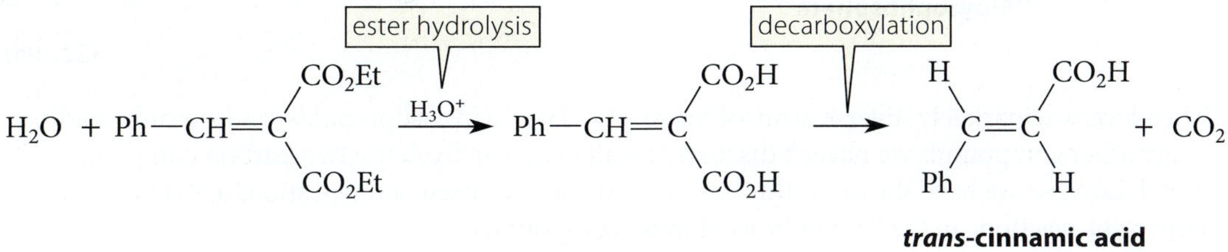

(22.104c)

Aldol Addition of Ester Enolates in Biology: HMG-CoA Biosynthesis The aldol addition reaction of ester enolates serves as a model for similar reactions in biology. An important example is the reaction of acetoacetyl-CoA with acetyl-CoA to give (*S*)-3-hydroxy-3-methylglutaryl-CoA, known in biology as *HMG-CoA*.

Chemical Biology Topic

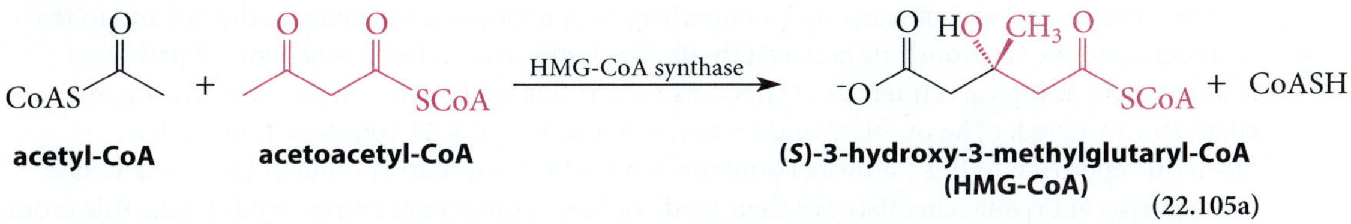

(22.105a)

We can conceptualize this reaction mechanistically as the aldol addition reaction of an enolate derived from acetyl-CoA with the ketone carbonyl group of acetoacetyl-CoA.

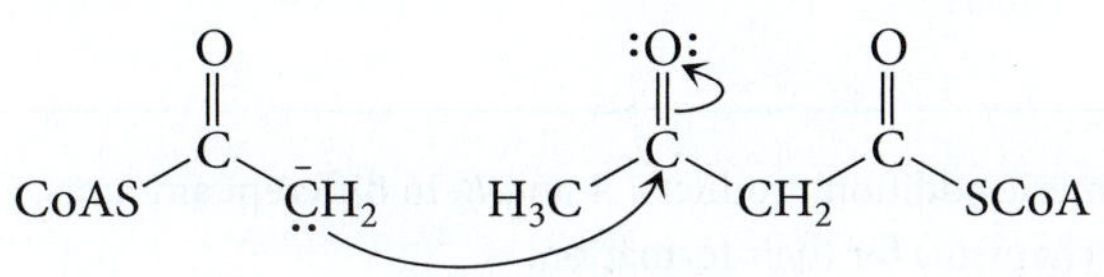

(22.105b)

(The mechanism of this reaction is explored in Focused Problem 22.42.) The reduction of HMG-CoA is the rate-limiting step in isoprenoid and cholesterol biosynthesis, as we show in Sec 25.5C. In this reduction, HMG-CoA is reduced to mevalonate, which is converted, in turn, into isopentenyl pyrophosphate (Problem 20.62), the key starting material in isoprenoid and steroid biosynthesis (Secs. 17.6B and C). Acetoacetyl-CoA is itself generated from fatty-acid metabolism (Focused Problem 22.33), and fatty acids ultimately originate from acetyl-CoA (Sec. 22.7).

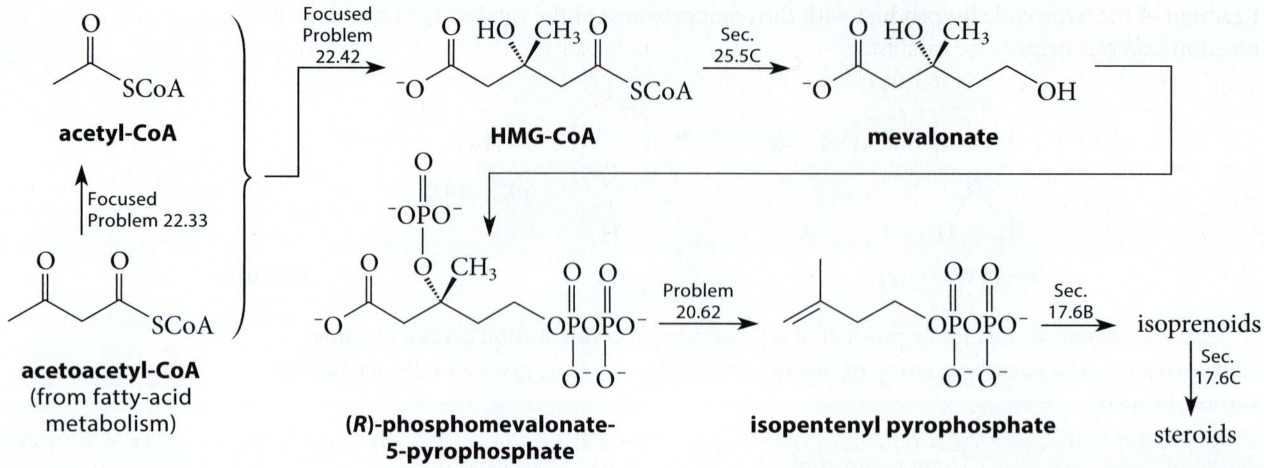

(22.106)

Therefore, a remarkably diverse array of materials—fatty acids, isoprenoids, and steroids (and many other compounds we haven't discussed)—all originate from the two-carbon compound acetyl-CoA. As we have shown in this chapter, aldol and Claisen condensations are very important reactions in the biosynthesis of these compounds.

In this text we've presented a number of examples of how chemistry is carried out in living systems, and we have shown that *all of these processes have close analogies in laboratory chemistry*. With the benefit of hindsight, it might seem obvious that natural chemistry and laboratory chemistry should be closely related. However, this point was far from obvious to early chemists. The serendipitous synthesis of urea by Friedrich Wöhler in 1828 (Eq. 1.1, Sec. 1.1C) signaled the beginning of an age in which the chemistry of living systems and laboratory chemistry are regarded as branches of the same basic science.

The "traditional" way of learning biochemistry is to memorize the many pathways and to try to understand the relationships between them. The better way to learn biochemical pathways is to see them as logical sequences of transformations that make sense in terms of the organic chemistry involved. (The problem and section references in Eq. 22.106 show how we have taken this approach to this pathway.) Students who bring an understanding of the fundamental mechanisms of organic chemistry to their study of biochemistry are empowered to take this more logical, and certainly less tedious, approach.

Focused Problems

22.40 The following aldol addition gives two diastereomeric addition products, *A* and *B*, in different amounts. Compounds *A* and *B* are both racemates. Give their structures and the mechanisms for their formation.

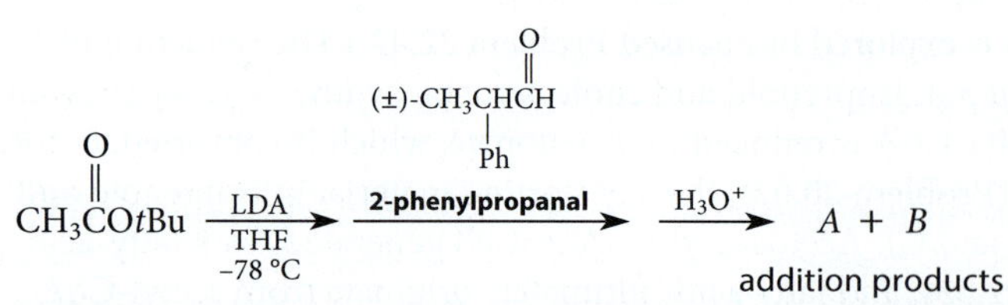

22.41 Give the addition or condensation products of each of the following reactions indicated by letter, and explain your reasoning.

(a)

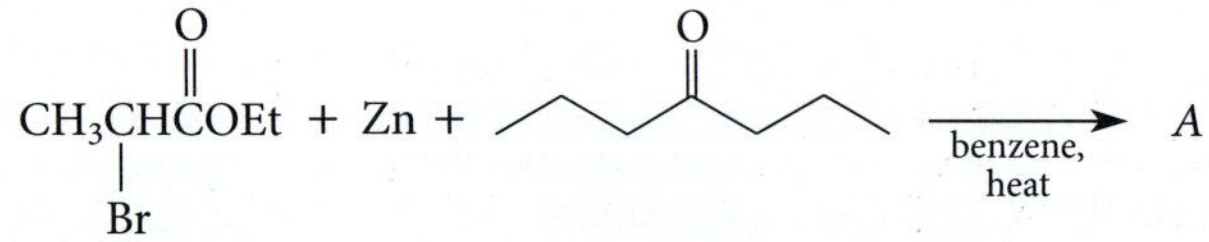

(b)

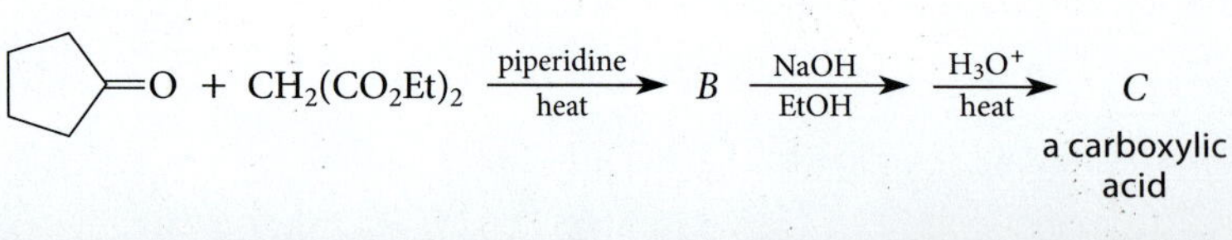

22.42 In HMG-CoA biosynthesis (Eq. 22.105a), the first step is transfer of the acetyl group from acetyl-CoA to a thiol group of HMG-CoA synthase, the catalyzing enzyme (= E in the equations that follow). The resulting thioester undergoes aldol addition to acetoacetyl-CoA followed by hydrolysis of the enzyme thioester:

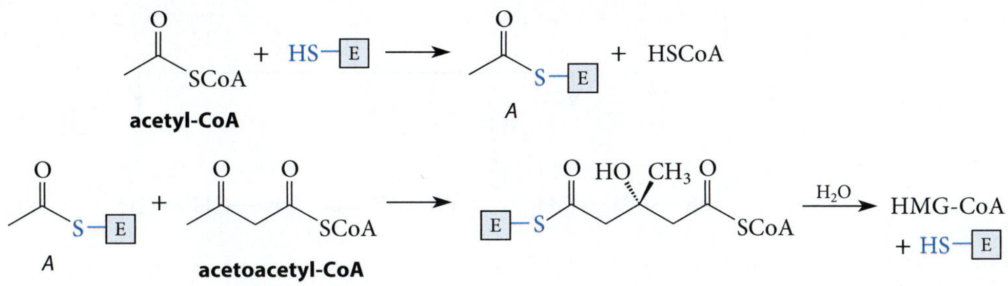

Assuming that acids and bases are provided by groups on the enzyme as needed, give a curved-arrow mechanism for each of these reactions.

22.9 CONJUGATE-ADDITION REACTIONS

A. Conjugate Addition to α,β-Unsaturated Carbonyl Compounds

The conjugated arrangement of C=C and C=O bonds endows α,β-unsaturated carbonyl compounds with unique reactivity, which is illustrated by the reaction of an α,β-unsaturated ketone with HCN.

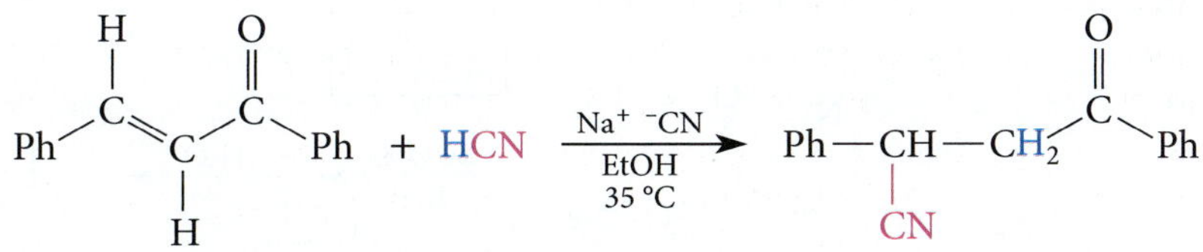

(trans isomer) (93–96% yield) (22.107)

In this reaction, the elements of HCN appear to have added across the C=C bond. Yet this is not a reaction of ordinary double bonds:

$$CH_3CH{=}CH_2 + HCN \xrightarrow[\text{EtOH}]{Na^+\ ^-CN} \text{no reaction} \qquad (22.108)$$

Nucleophilic addition to the double bond in an α,β-unsaturated carbonyl compound occurs because it gives a resonance-stabilized enolate ion intermediate:

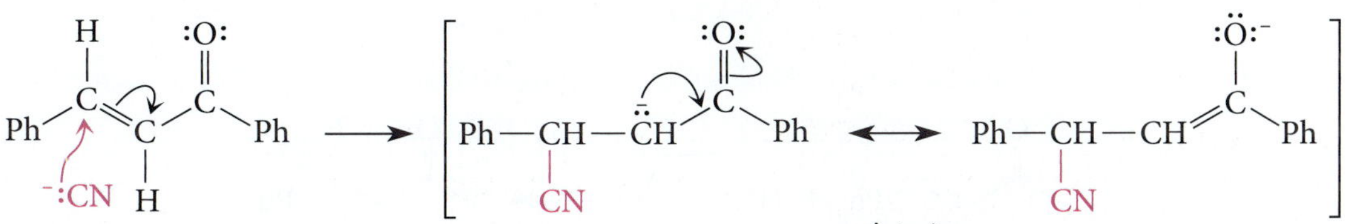

enolate ion

(22.109a)

(Nucleophilic addition to the alkene in Eq. 22.108, in contrast, would give a very unstable alkyl anion.) The enolate ion can be protonated on either oxygen or carbon. In either case, a carbonyl group is eventually regenerated because enols spontaneously form carbonyl compounds (Sec. 22.2). The *overall* result of the reaction is net addition to the double bond.

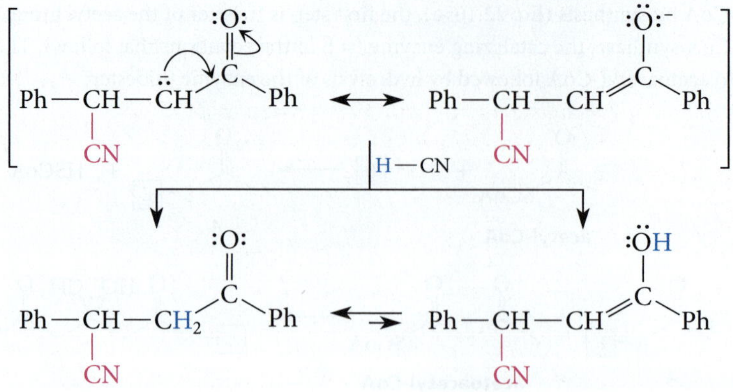

observed product　　enol form of product　　(22.109b)

Nucleophilic addition to the carbon–carbon double bonds of α,β-unsaturated aldehydes, ketones, esters, and nitriles is a rather general reaction called **nucleophilic conjugate addition**, which can be observed with a variety of nucleophiles. Some additional examples follow; try to write the mechanisms of these reactions.

Conjugate additions to α,β-unsaturated esters:

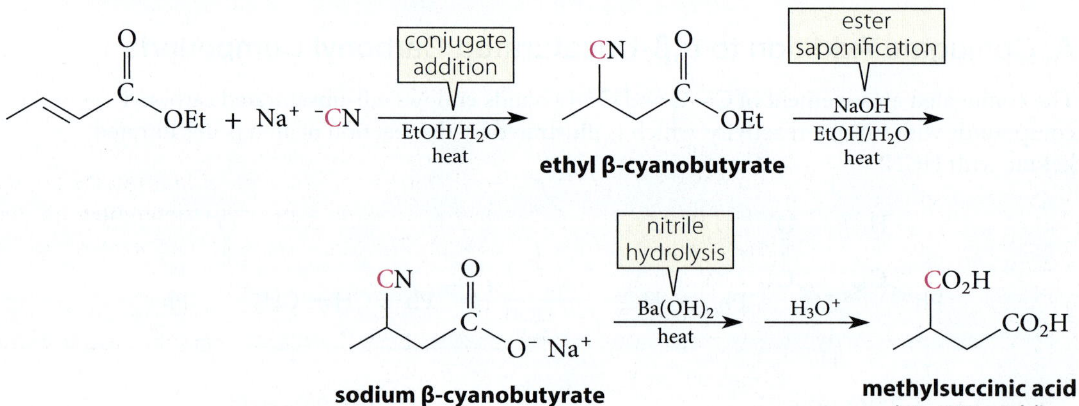

methylsuccinic acid
(66–70% yield)

(22.110)

$$\underset{\textbf{2-propanethiol}}{(CH_3)_2CH{-}SH} + \underset{\textbf{methyl acrylate}}{H_2C{=}CH{-}CO_2Me} \xrightarrow[\text{MeOH}]{\text{NaOMe}} \underset{\textbf{methyl 3-(isopropylthio)propanoate}}{(CH_3)_2CH{-}S{-}CH_2{-}CH_2{-}CO_2Me}$$

(97% yield)

(22.111)

Conjugate addition to an α,β-unsaturated ketone:

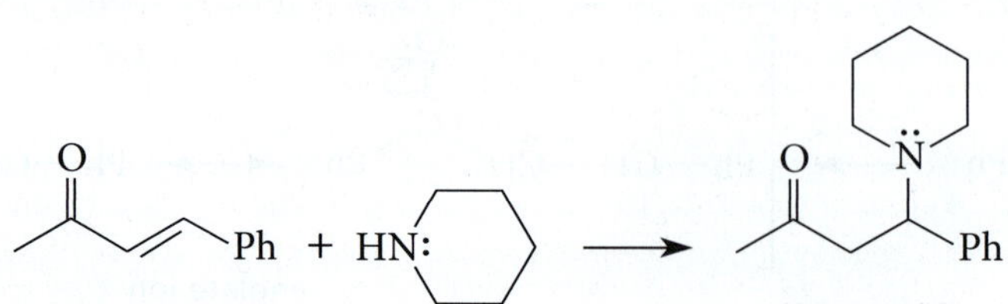

(85% yield)　　(22.112)

Conjugate addition to an α,β-unsaturated nitrile:

a cyanoethylation

$$\underset{\textbf{methanethiol}}{CH_3SH} + \underset{\textbf{acrylonitrile}}{H_2C{=}CH{-}CN} \xrightarrow[\text{MeOH}]{\text{NaOMe}} \underset{\textbf{3-(methylthio)propanenitrile}}{CH_3S{-}\overbrace{CH_2{-}CH_2{-}CN}^{\text{a }\beta\text{-cyanoethyl group}}}$$

(91% yield)　　(22.113)

Notice that the conjugate addition of cyanide (as in Eq. 22.110) forms a new carbon–carbon bond, and that the nitrile group can then be converted into a carboxylic acid group by hydrolysis. The addition of a nucleophile to acrylonitrile (as in Eq. 22.113) is a useful reaction called **cyanoethylation** because it introduces a β-cyanoethyl group to a nucleophilic center.

Because quinones (Sec. 18.8A) are α,β-unsaturated carbonyl compounds, they also undergo similar conjugate-addition reactions.

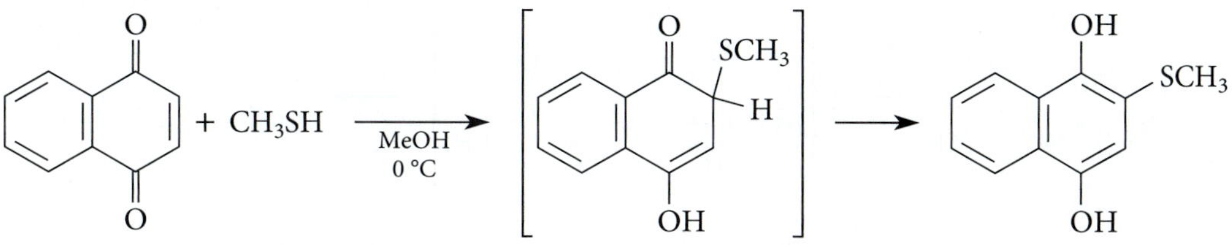

(22.114)

In this example, the reaction is driven to completion by enolization of the ketone in brackets to the phenol, which is aromatic. (See Eq. 22.19, Sec. 22.2.)

A Conjugate Addition Involved in Drug Toxicity

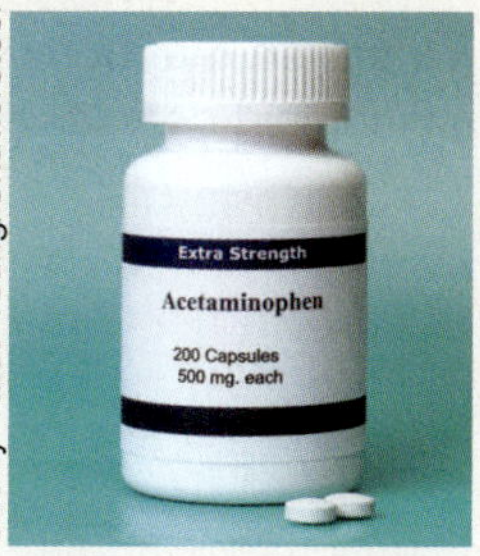

Sherry Yates Young/Shutterstock

Acetaminophen (sometimes called paracetamol) is a well-known drug that is used in a number of commercial analgesic medications for the relief of headaches and other minor aches and pains. Although it is generally regarded as safe at the recommended dosages, an overdose of acetaminophen can be very toxic. The basis of this toxicity involves a conjugate-addition reaction to a metabolite of acetaminophen.

In Sec. 17.5B, we showed that the liver enzyme cytochrome P450 (CyP450) is involved in phase I drug metabolism. CyP450 brings about the hydroxylation of acetaminophen at the nitrogen to give the *N*-hydroxy derivative.

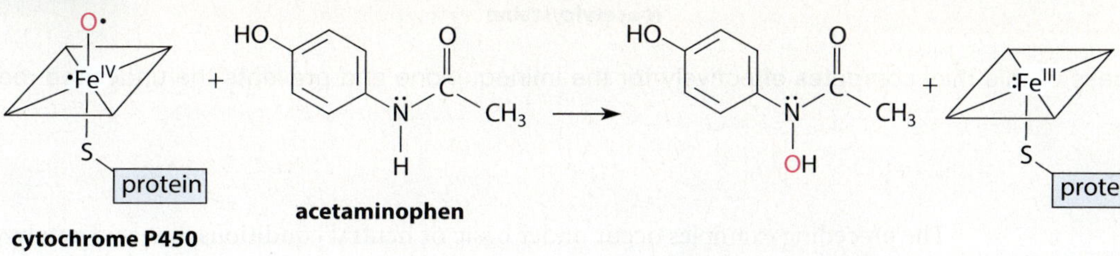

This derivative readily undergoes an elimination of water to form a quinone-like intermediate called an iminoquinone. Although aromaticity is lost in this reaction, the reaction is driven by loss of the very weak N—O bond.

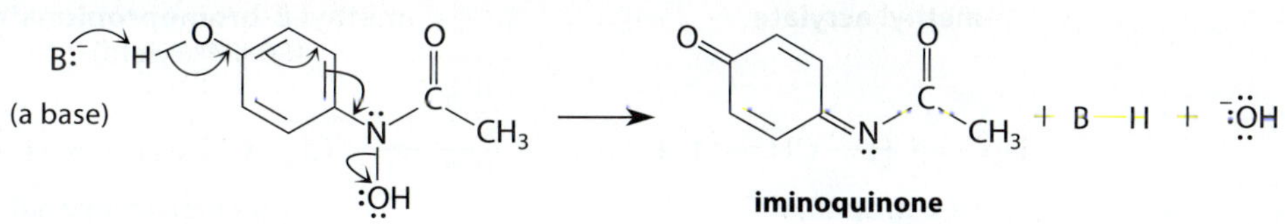

The resulting iminoquinone is usually scavenged by a conjugate-addition reaction with the thiol group of a small peptide, glutathione. The resulting product is converted into a water-soluble derivative that is carried into the urine and excreted. (A reaction with, and removal of, foreign electrophiles is a very important biological role of glutathione.)

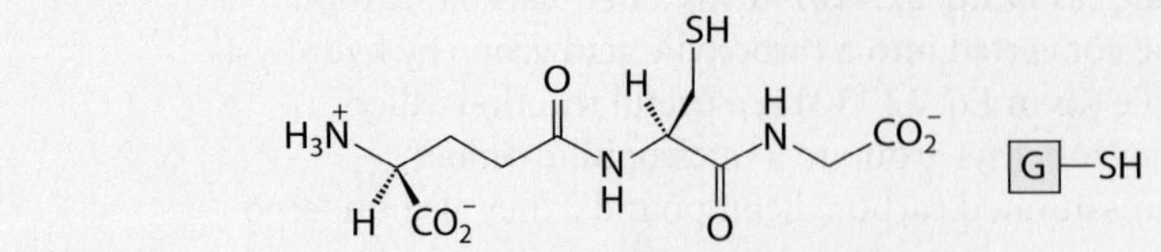

γ-glutamylcysteinylglycine (glutathione)

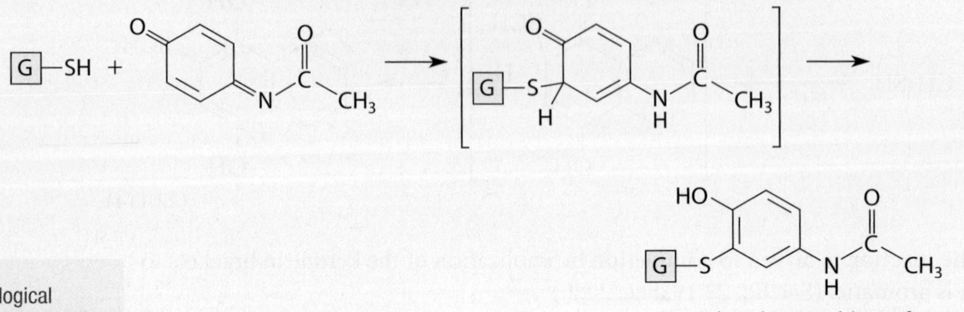

glutathione adduct of acetaminophen iminoquinine (converted into a derivative that is excreted in the urine)

For a related biological reaction of quinones, see the sidebar, "Poison Ivy and Itchy Quinones," in Sec. 18.8B.

This conjugate addition is highly favorable because it regenerates the aromatic ring system.

If too much iminoquinone forms because of a drug overdose, the glutathione supply in the liver is exhausted, and the iminoquinone can then react with the thiol groups of other proteins. These reactions of the iminoquinone are the source of the liver toxicity.

A conjugate-addition reaction of the iminoquinone is also used in the treatment of acetaminophen overdose. Treatments of acetaminophen overdose typically involve the administration of another thiol, the amino acid derivative *N*-acetylcysteine.

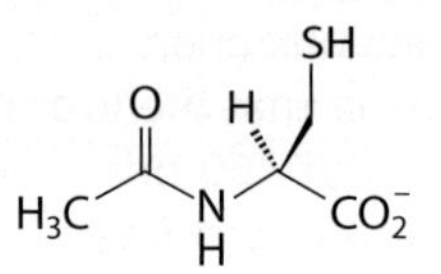

***N*-acetylcysteine**

The large excess of this thiol competes effectively for the iminoquinone and prevents the undesired reactions with protein thiols.

The preceding examples occur under basic or neutral conditions, but acid-catalyzed additions to the carbon–carbon double bonds of α,β-unsaturated carbonyl compounds are also known.

$$\underset{\textbf{methyl acrylate}}{H_2C{=}CH{-}CO_2Me} + HBr \xrightarrow[Et_2O]{} \underset{\substack{\textbf{methyl β-bromopropionate}\\ \text{(80–84\% yield)}}}{Br{-}CH_2CH_2{-}CO_2Me} \tag{22.115}$$

$$\underset{\substack{\textbf{2-propenal}\\ \textbf{(acrolein)}}}{H_2C{=}CH{-}CH{=}O} + HCl \xrightarrow[-15\ °C]{} \underset{\textbf{3-chloropropanal}}{Cl{-}CH_2CH_2{-}CH{=}O} \tag{22.116}$$

Although such reactions appear to be nothing more than simple additions to the carbon–carbon double bond, this is not the case. The more basic site of an α,β-unsaturated carbonyl compounds is not the double bond, but rather the carbonyl oxygen. Protonation on the carbonyl oxygen is followed by a reaction with the halide ion. The electrophilic oxygen can accept electrons as a result of a nucleophilic reaction of the halide ion either at the carbonyl carbon or, because of the conjugated arrangement of π bonds, at the β-carbon:

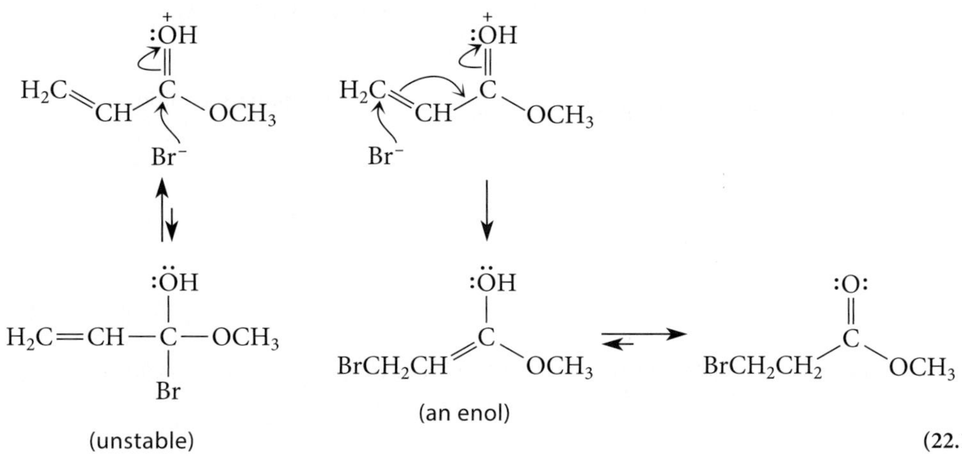

(22.117)

A reaction of Br^- at the carbonyl carbon yields a relatively unstable tetrahedral addition intermediate, which loses the Br^- leaving group and reverts back to the protonated ketone; a reaction at the β-carbon yields an enol, which rapidly reverts to the observed carbonyl product.

An acid-catalyzed addition to the double bond of an α,β-unsaturated carbonyl compound is another example of *electrophilic conjugate addition*. The mechanism of the conjugate addition of HBr shown in Eq. 22.117 is similar to the conjugate addition of HBr to 1,3-butadiene (Sec. 15.4A); both involve resonance-stabilized carbocation intermediates. However, the *nucleophilic conjugate addition*, such as the addition of cyanide in Eq. 22.110, has no parallel in the reactions of simple conjugated dienes.

We have in this section seen examples of nucleophilic conjugate addition, which involves carbanion intermediates, and electrophilic conjugate addition, which involves carbocation intermediates. We have previously discussed one other important conjugate addition of α,β-unsaturated carbonyl compounds—the *Diels–Alder reaction* (Sec. 15.3).

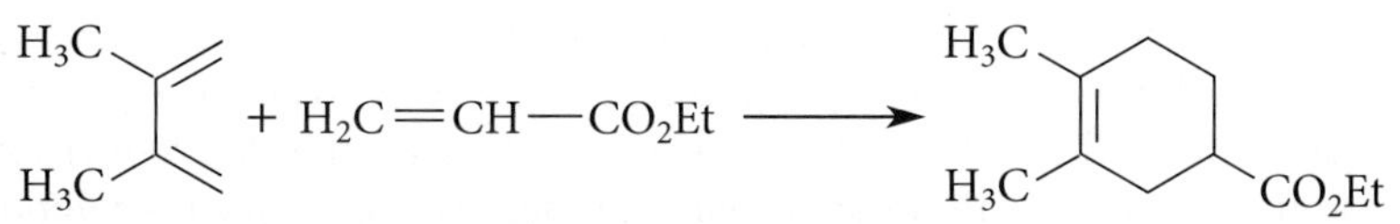

(22.118)

This is a *concerted conjugate* addition—that is, no ionic intermediates are involved. However, the most reactive dienophiles in the Diels–Alder reaction are alkenes with one or more electron-withdrawing substituents, an observation consistent with the idea that in the transition state, electron density is flowing from the diene into the dienophile. α,β-Unsaturated carbonyl compounds fit this description; they are among the most frequently used dienophiles in the Diels–Alder reaction, as Eq. 22.118 illustrates.

B. Conjugate-Addition Reactions versus Carbonyl-Group Reactions

Any conjugate-addition reaction *competes* with a carbonyl-group reaction. In the case of aldehydes and ketones, conjugate addition competes with addition to the carbonyl group. (Nuc = nucleophile; for example, in cyanide addition, H—Nuc = H—CN.)

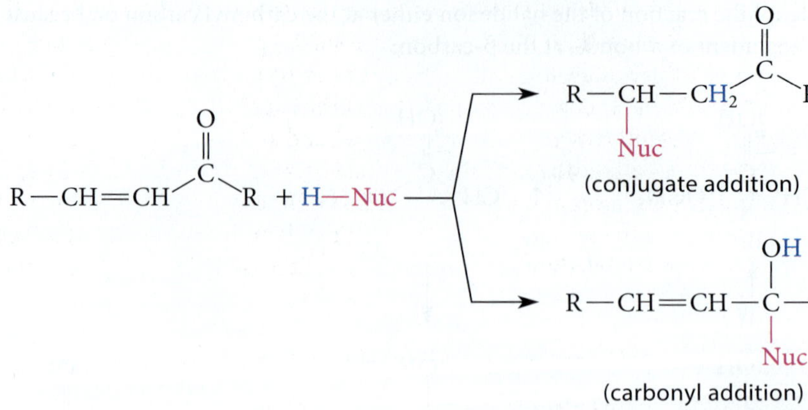

(22.119a)

In the case of esters, conjugate addition competes with *nucleophilic acyl substitution*.

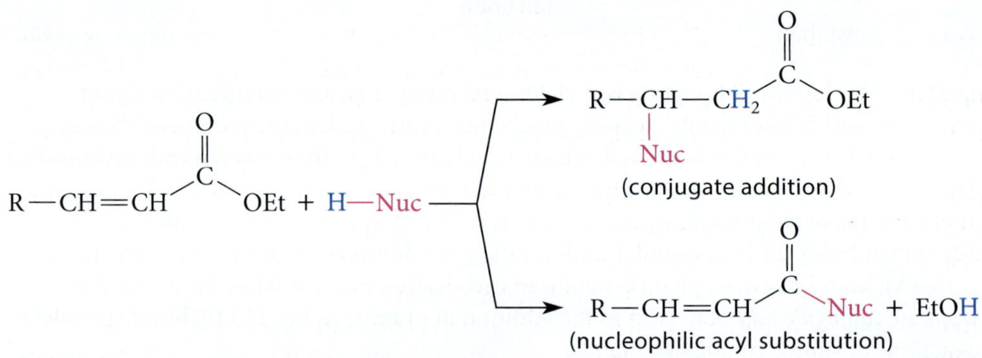

(22.119b)

When can we expect to observe conjugate addition, and when can we expect reactions at the carbonyl carbon?

Consider first the reactions of aldehydes and ketones. Relatively weak bases that give *reversible* carbonyl-addition reactions with ordinary aldehydes and ketones tend to give conjugate addition with α,β-unsaturated aldehydes and ketones. Among the relatively weak bases in this category are cyanide ion, amines, thiolate ions, and enolate ions derived from β-dicarbonyl compounds. Conjugate addition is observed with these nucleophiles because *the conjugate-addition products are more stable than the carbonyl-addition products*. If carbonyl addition is reversible—even if it occurs more rapidly—then conjugate addition can drain the carbonyl compound from the addition equilibrium, and the conjugate-addition product is formed ultimately.

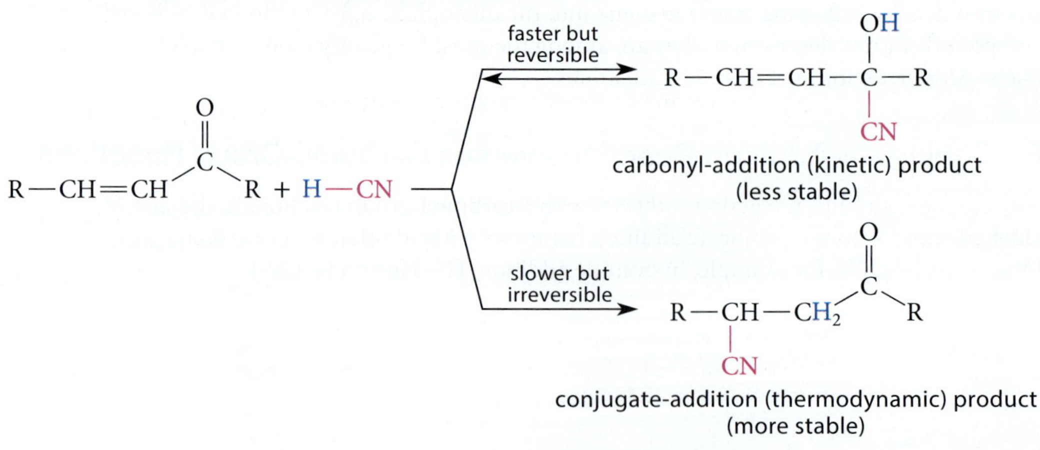

(22.120)

This, then, is another case of *kinetic versus thermodynamic control of a reaction* (Sec. 15.4B). The conjugate-addition product is the thermodynamic (more stable) product of the reaction.

The greater stability of the conjugate-addition product can be understood with a bond energy argument. Conjugate addition retains a carbonyl group at the expense of a carbon–carbon double bond. Carbonyl addition retains a carbon–carbon double bond at the expense of a carbonyl group. Because a C=O bond is considerably stronger than a C=C bond (see Table 10.2, Sec. 10.1E), conjugate addition gives a more stable product. (Other bonds are broken and formed as well, but the major effect is the relative strengths of the two kinds of double bonds.) These same factors are reflected in the relative heats of formation of the isomers allyl alcohol and propionaldehyde:

	$H_2C{=}CH{-}CH_2{-}OH$	$H_3C{-}CH_2{-}CH{=}O$	
	allyl alcohol	**propionaldehyde**	
ΔH_f°	-124 kJ mol^{-1}	-189 kJ mol^{-1}	
	(-29.6 kcal mol^{-1})	(-45.2 kcal mol^{-1})	(22.121)

As Eq. 22.120 suggests, carbonyl addition is in many cases the kinetically favored process—that is, it is faster than conjugate addition. When nucleophiles are used that undergo *irreversible* carbonyl additions, then the carbonyl-addition product is observed rather than the conjugate-addition product. This is exactly what happens with very powerful nucleophiles such as $LiAlH_4$ and organolithium reagents: These species add irreversibly to carbonyl groups and form carbonyl-addition products whether the reactant carbonyl compound is α,β-unsaturated or not. (These reactions are discussed further in Secs. 22.10 and 22.11A.)

Many of the same nucleophiles that undergo conjugate addition with aldehydes and ketones also undergo conjugate addition with esters. Stronger bases that react irreversibly to give carbonyl-addition products with aldehydes and ketones react with esters to give nucleophilic acyl substitution products. For example, hydroxide ion reacts with an α,β-unsaturated ester to give products of saponification, a nucleophilic acyl substitution reaction, because saponification is not reversible. Likewise, $LiAlH_4$ reduces α,β-unsaturated esters at the carbonyl group because the reaction of hydride ion at the carbonyl group is irreversible.

To summarize: Conjugate addition usually occurs with nucleophiles that are relatively weak bases. Stronger bases give irreversible carbonyl-addition or nucleophilic acyl substitution reactions.

Focused Problems

22.43 Give the product expected when methyl methacrylate (methyl 2-methylpropenoate) reacts with each of the following reagents.

(a) ^{-}CN and HCN in MeOH (b) C_2H_5SH and NaOMe catalyst in MeOH

(c) HBr (d) NaOH

22.44 Give a curved-arrow mechanism for each of the following reactions.

(a)

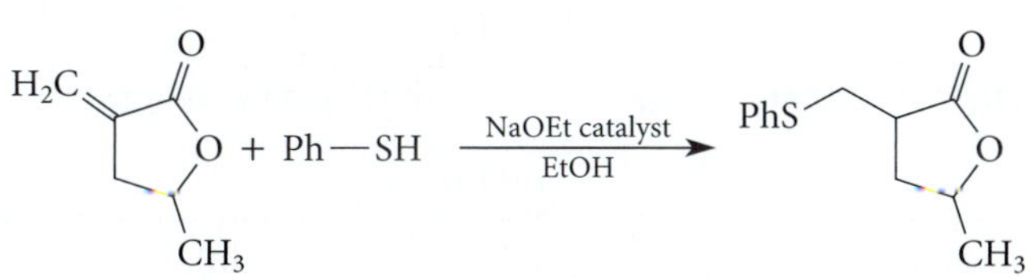

(mixture of stereoisomers; why?)

(b) $MeNH_2 + 2H_2C{=}CH{-}CO_2Me \longrightarrow MeO_2C{-}CH_2CH_2{-}\underset{\displaystyle Me}{\underset{|}{N}}{-}CH_2CH_2{-}CO_2Me$

(c)

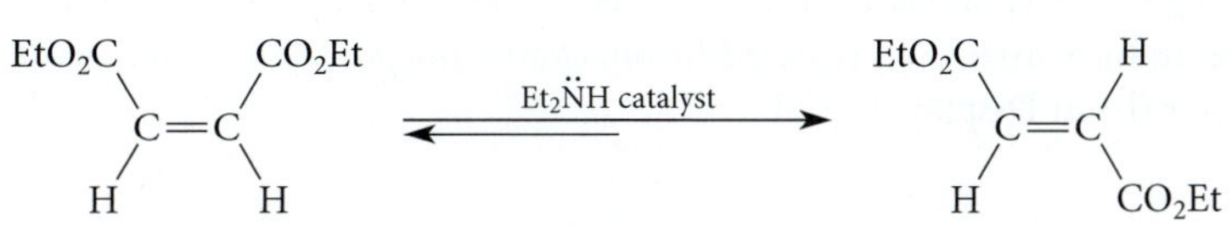

C. Conjugate Addition of Enolate Ions

Enolate ions, especially those derived from malonic ester derivatives, β-keto esters, and the like, undergo conjugate-addition reactions with α,β-unsaturated carbonyl compounds, as in the following example:

3-Buten-2-one, which finds frequent use in conjugate additions, is typically called "methyl vinyl ketone," or "MVK," by practicing organic chemists.

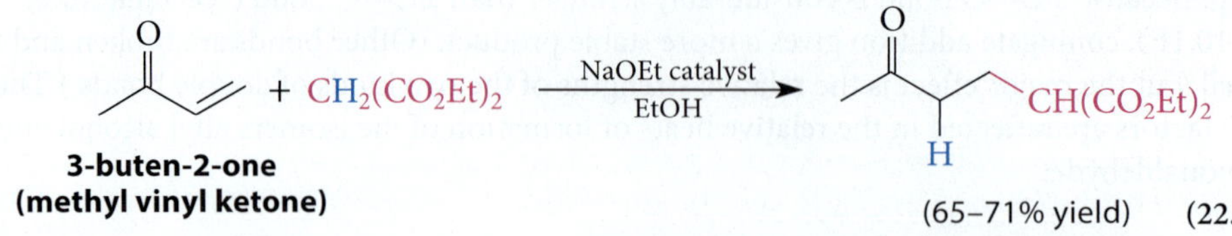

(65–71% yield) (22.122)

The mechanism of this reaction follows exactly the same pattern established for other nucleophilic conjugate additions; the nucleophile is the enolate ion formed in the reaction of ethoxide with diethyl malonate (Eq. 22.88a, Sec. 22.8A). In contrast to the Claisen ester condensation (Sec. 22.6A), this reaction requires only a catalytic amount of base. The reaction does *not* rely on ionization of the product to drive it to completion. It goes to completion because a carbon–carbon π bond in the starting α,β-unsaturated carbonyl compound is replaced by a stronger carbon–carbon σ bond.

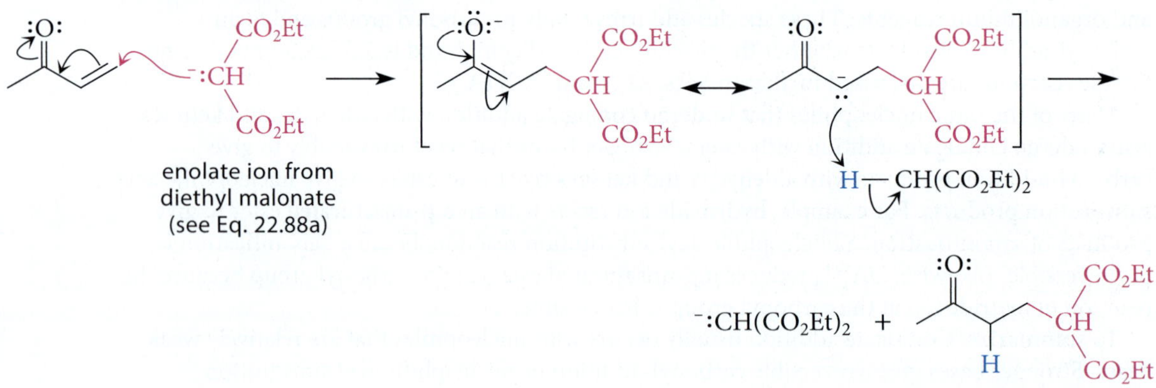

(22.123)

Some chemists refer to all nucleophilic conjugate additions as Michael additions. The justification for this view is that, from a mechanistic perspective, all nucleophilic conjugate additions are similar.

Conjugate additions of carbanions to α,β-unsaturated carbonyl compounds are called **Michael additions**, after Arthur Michael (1853–1942), a Harvard professor who investigated these reactions extensively.

Use of a Michael addition in a synthesis requires proper planning. The product of a given Michael addition might originate from two different pairs of reactants. For example, in the reaction shown in Eq. 22.123, the same product (in principle) might be obtained by the Michael addition reaction of either of the following pairs of reactants. (Convince yourself of this point.)

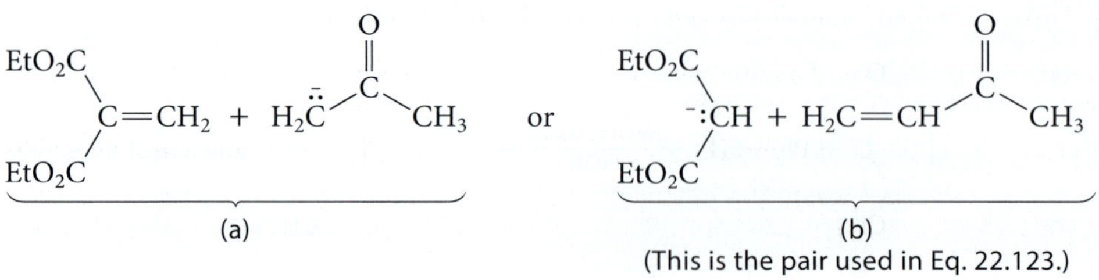

(22.124)

Which pair of reactants should be used? To answer this question, use the result in Sec. 22.9B: weaker bases tend to give conjugate addition, whereas stronger bases tend to give carbonyl-group reactions. Therefore, to maximize conjugate addition, *choose the pair of reactants with the less basic enolate ion*—pair (b) in Display 22.124.

In one useful application of the Michael addition, called the **Robinson annulation**, the immediate product of the conjugate addition can be subjected to an aldol condensation that closes a ring. (An annulation is a ring-forming reaction, from the Latin *annulus*, meaning "ring.")

The Robinson annulation was named for Sir Robert Robinson (1886–1975), a British chemist at Oxford University who pioneered its use. (Robinson received the 1947 Nobel Prize in Chemistry for his work in alkaloids, which are discussed in Sec. 23.12B.) Robinson is also credited with inventing the curved-arrow notation.

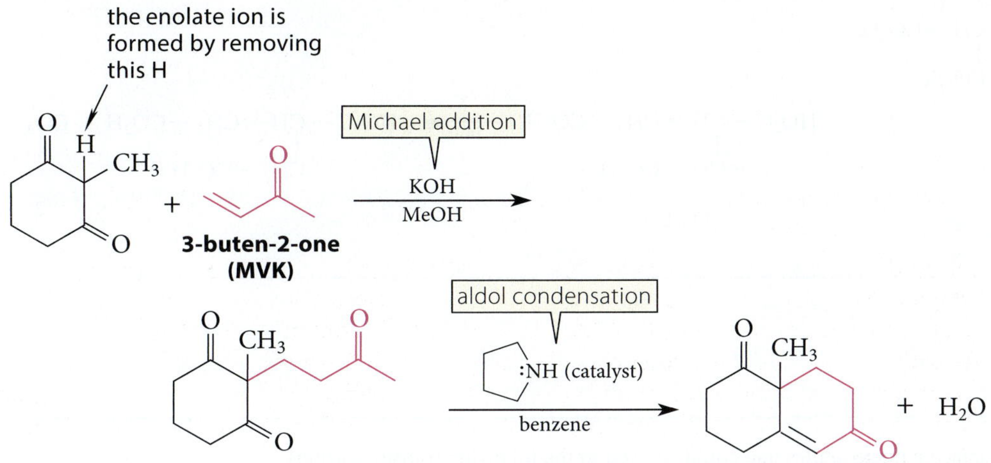

(63–65% yield) (22.125)

The Michael addition involves the enolate ion formed by ionization of the acidic proton of the β-diketone. (Write the curved-arrow mechanism of this addition.) The mechanism of the aldol condensation that follows is explored in Focused Problem 22.25 (Sec. 22.5).

Study Problem 22.6

Outline a synthesis of tricarballylic acid from diethyl fumarate and any other reagents.

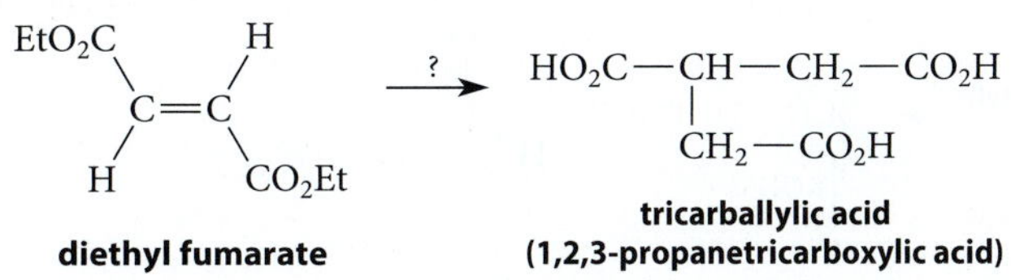

Solution Two of the carboxylic acid groups required in the target are already in place as the ester units in diethyl fumarate. A Michael addition of some species that could be converted into a $—CH_2CO_2H$ group is required. Notice that the desired product is conceptually a substituted acetic acid:

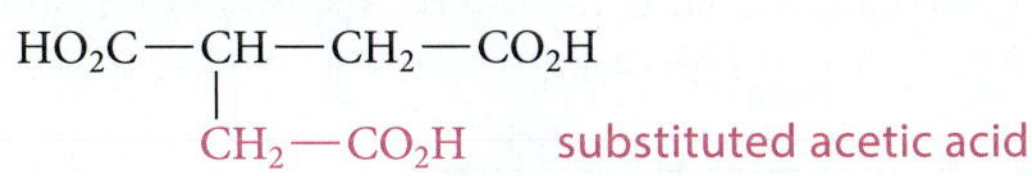

Recall that one way of preparing substituted acetic acids is the malonic ester synthesis (Sec. 22.8A). A variation of the malonic ester synthesis can be employed here in which alkylation of the conjugate-base anion of diethyl malonate is carried out by a Michael addition with diethyl fumarate instead of an S_N2 reaction with an alkyl halide.

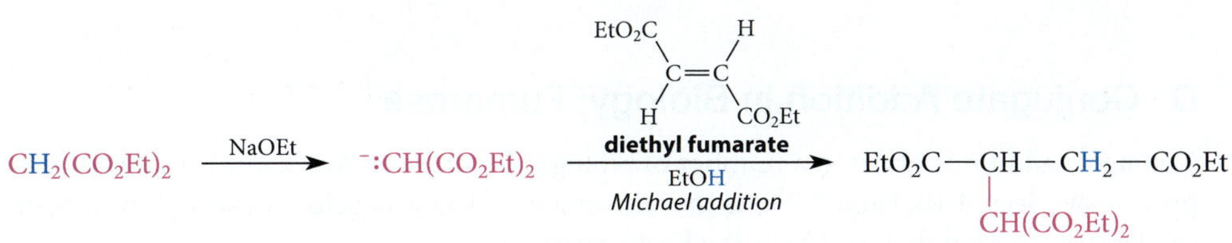

Saponification of all four ester groups, protonation, and decarboxylation yields the desired tricarboxylic acid:

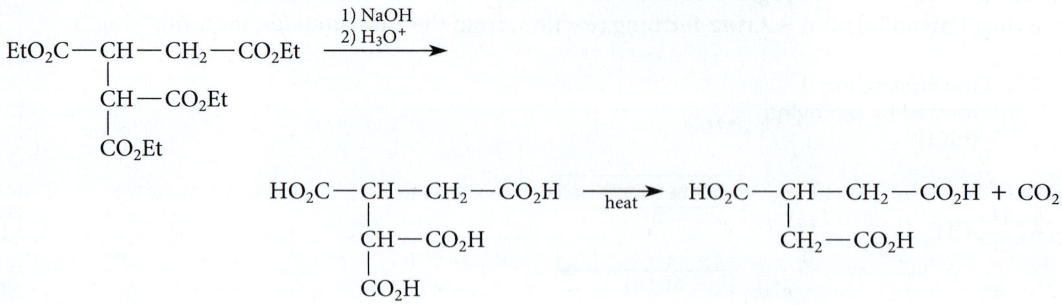

STUDY GUIDE LINK 22.7
Synthetic Equivalents in Conjugate Addition

Focused Problems

22.45 Provide structures for the missing nucleophiles that could be used in the following transformations.

(a)

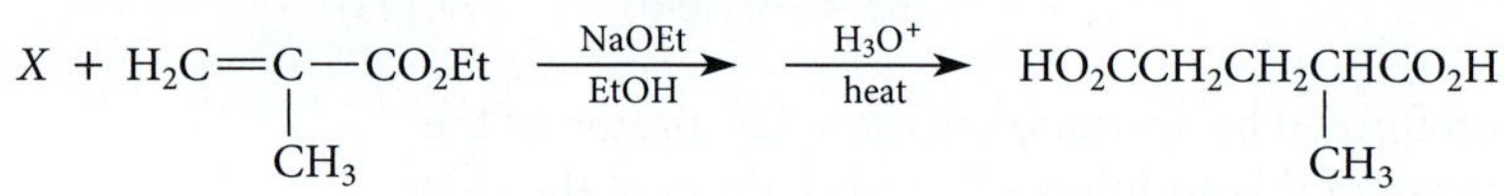

(b)

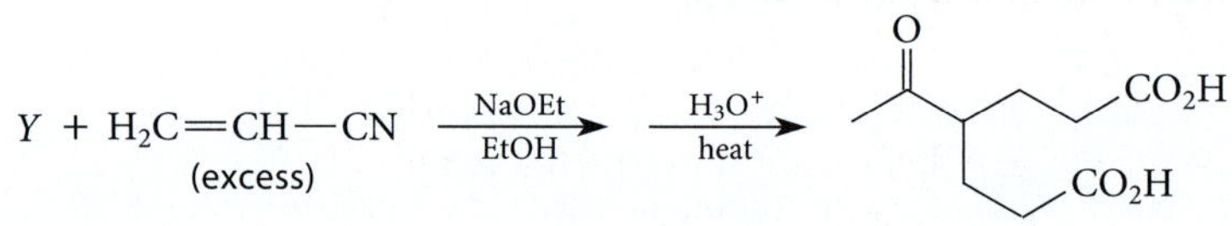

22.46 Give a curved-arrow mechanism for each of the following reactions. In each reaction identify the intermediate indicated by *A* or *B*.

(a)

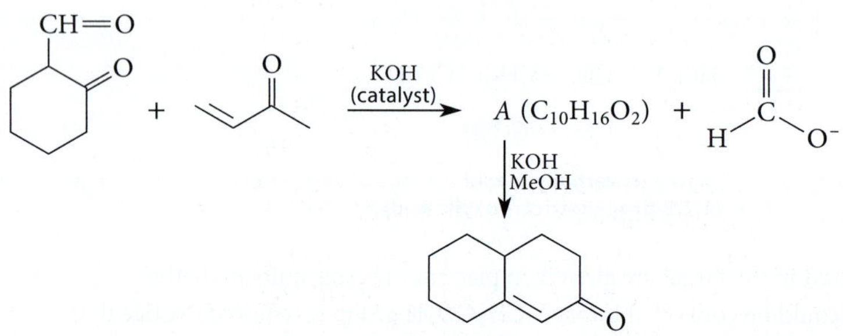

(b)

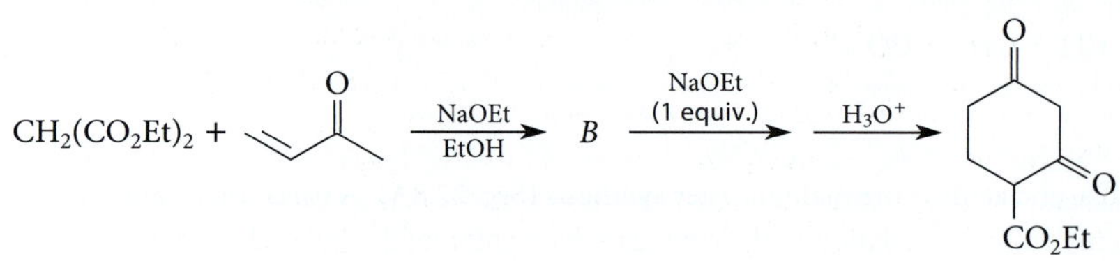

Chemical Biology Topic

D. Conjugate Addition in Biology: Fumarase

Conjugate additions occur in a number of biological pathways. An example that was presented previously (Secs. 4.10D and 7.7A) is the conversion of fumaric acid (fumarate) into (*S*)-malic acid (malate), a reaction that occurs in the Krebs cycle.

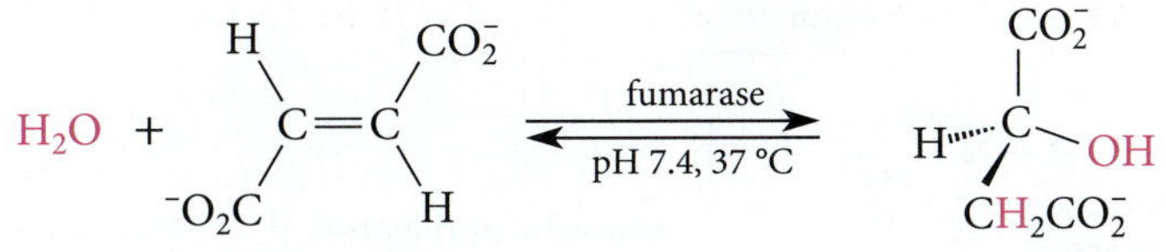

fumarate **(S)-malate** (22.126)

In Sec. 4.10D, this reaction was introduced as our first example of enzyme catalysis. In Sec. 7.7A, we considered the stereochemistry of this reaction.

This reaction is a conjugate addition of water to the double bond of fumarate. Let's consider first this reaction at physiological pH and why enzyme catalysis is necessary. If water is the nucleophile, the first step of the mechanism would be the nucleophilic reaction of water at the β-carbon of the double bond:

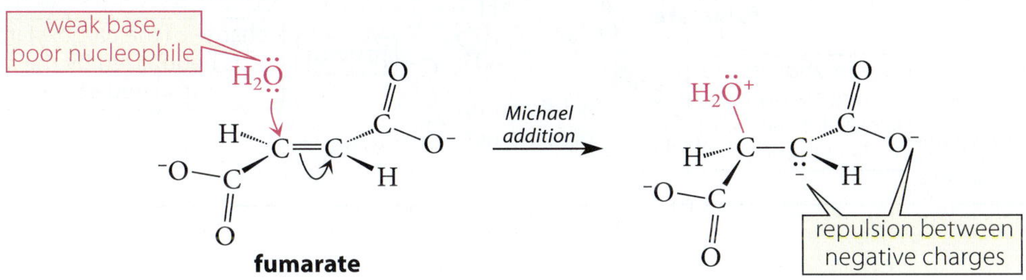

carbanion intermediate (22.127)

Water is a very weak base (conjugate acid $pK_a = 0$) and therefore is a poor nucleophile. Hydroxide would be a much better nucleophile, but the concentration of hydroxide ion at neutral pH is minuscule. Therefore, catalysis has to overcome the "poor nucleophile" problem.

Notice in Eq. 22.127 that the carbanion formed on addition of the nucleophile is destabilized by repulsion with the negative charge of the carboxylate ion. In fact, α,β-unsaturated carboxylates normally do not undergo conjugate additions under normal conditions for exactly this reason. We have shown (Eqs. 22.110 and 22.111) that esters are good conjugate-addition electrophiles; so, we might imagine that *un-ionized* carboxylic acids would also be good conjugate-addition electrophiles. However, under the usual basic conditions of the nucleophilic conjugate addition, carboxylic acids are fully ionized and cannot take on another negative charge without excessive charge repulsion. Therefore, catalysis has to overcome the "poor electrophile" problem.

The mechanism of fumarase catalysis is shown in **Fig. 22.4**. The electron pair from the basic nitrogen of a histidine in the enzyme active site is "relayed" to a proton of the nucleophilic water (*red*) by way of a second, immobilized, water molecule (*blue*), partially converting this water into more nucleophilic hydroxide. The acceptor carboxylate ion interacts strongly by both hydrogen bonding and electrostatic attraction with the protonated amino group of a lysine. This interaction not only helps to bind the fumarate ion in the active site but also neutralizes the negative charge on one carboxylate group; this carboxylate, then, is more like a carboxylic acid. Furthermore, delivery of a proton to the double bond by an acidic group in the enzyme active site avoids the formation of a highly basic carbanion and further reduces charge repulsion. Finally, the alignment of all of these elements within the active site provides a very large rate acceleration by the proximity effect (Sec. 12.8E). All of the proton transfers are believed to occur in a concerted manner. The enzyme active site is therefore constructed to convert what would be a very unfavorable conjugate-addition reaction into a very rapid one.

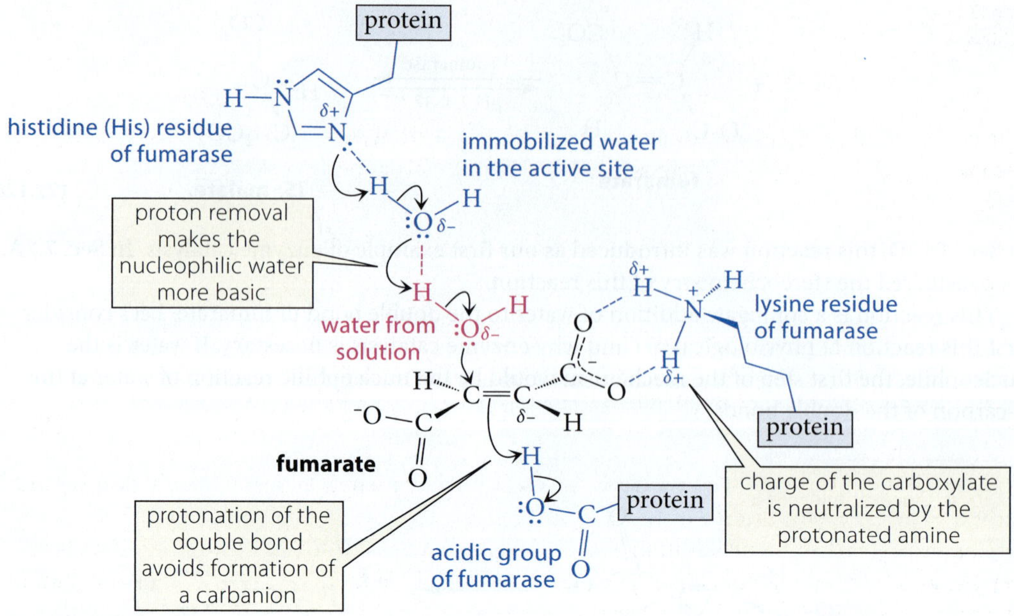

FIGURE 22.4 The mechanism of fumarate hydration, a conjugate addition, catalyzed by the enzyme fumarase. Enzyme active-site residues are shown in blue. The water shown in blue is an immobilized, permanent part of the active site. The pointers indicate important aspects of fumarase catalysis discussed in the text.

Focused Problem

22.47 The reducing agent in the double-bond reduction step of fatty-acid biosynthesis (the last step of Eq. 22.85, repeated here) is NADPH (Sec. 19.9B). (ACP = acyl carrier protein.)

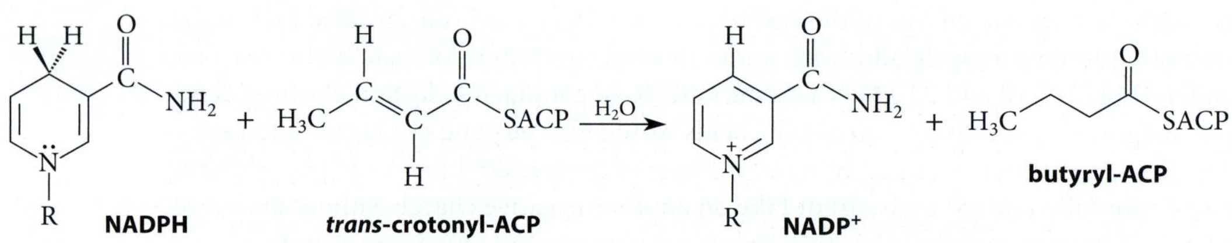

(a) Using the abbreviated structure for NADPH, and assuming that acids and bases are available as needed in the enzyme active site, draw a curved-arrow mechanism for this reaction.

(b) Explain why this reaction is an example of a conjugate addition.

(c) When NADPD (a deuterium-labeled NADPH, shown here) and D_2O are used in the enzyme-catalyzed reduction of *trans*-crotonyl-ACP, the product is the (2*S*,3*R*) stereoisomer of the doubly deuterated butyryl-ACP. Is this conjugate-addition reaction a syn or an anti addition? Give your reasoning.

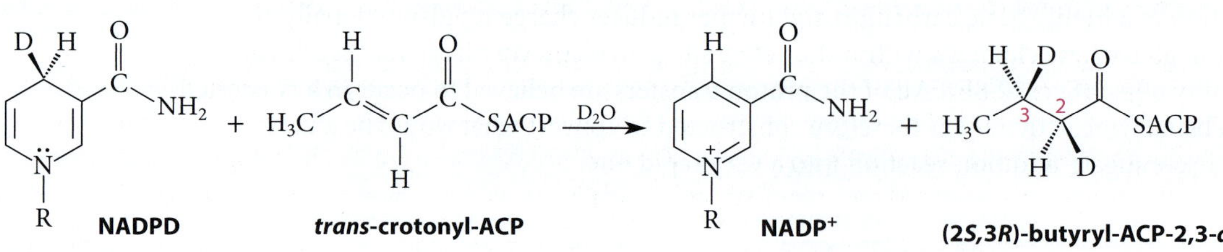

22.10 REDUCTION OF α,β-UNSATURATED CARBONYL COMPOUNDS

The carbonyl group of an α,β-unsaturated aldehyde or ketone, like that of an ordinary aldehyde or ketone (Sec. 19.9), is reduced to an alcohol with lithium aluminum hydride.

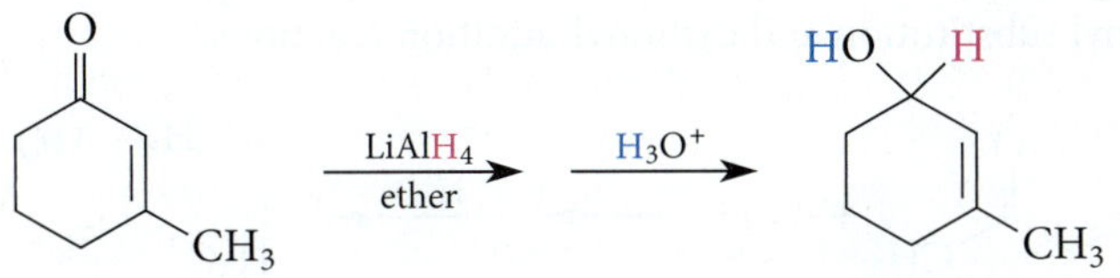

3-methyl-2-cyclohexenone **3-methyl-2-cyclohexenol** (98% yield) (22.128)

This reaction, like other $LiAlH_4$ reductions, involves the nucleophilic reaction of hydride at the carbonyl carbon and is therefore a carbonyl addition.

The reason that carbonyl addition, rather than conjugate addition, is observed follows from the discussion in Sec. 22.9B. Carbonyl addition is not only *faster* than conjugate addition but, in this case, is also *irreversible*. It is irreversible because hydride is a poor leaving group. Because carbonyl addition of $LiAlH_4$ is irreversible, conjugate addition never has a chance to occur and is therefore not observed.

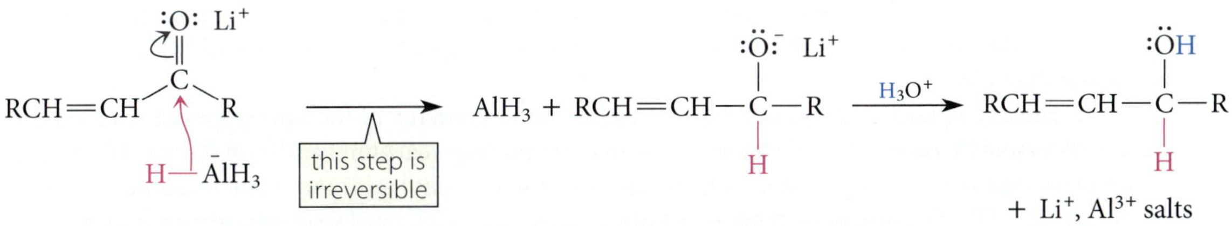

(22.129)

In other words, reduction of the carbonyl group with $LiAlH_4$ is a *kinetically controlled* reaction.

Many α,β-unsaturated aldehydes and ketones are reduced by $NaBH_4$ to give mixtures of both carbonyl-addition products and conjugate-addition products. Because mixtures are obtained, $NaBH_4$ reductions of α,β-unsaturated ketones are not useful. Why conjugate addition is observed with $NaBH_4$ is not well understood. Although some cases of conjugate addition with $LiAlH_4$ are known, this reagent usually reduces carbonyl groups, including the carbonyl groups of esters, without affecting double bonds.

The carbon–carbon double bond of an α,β-unsaturated carbonyl compound can in most cases be reduced selectively by catalytic hydrogenation. (See also Eq. 19.53, Sec. 19.9C.)

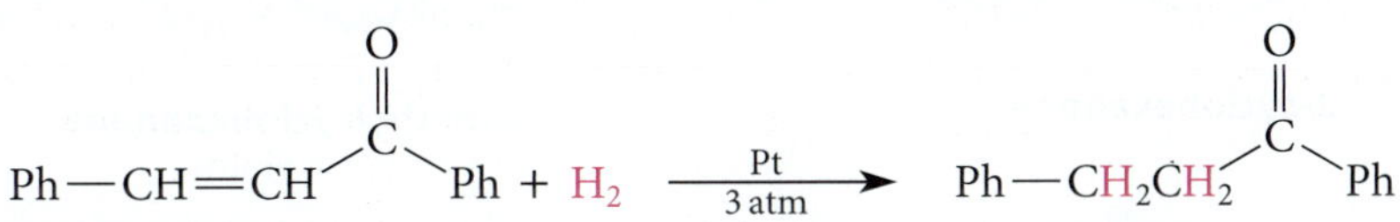

ethyl acetate (solvent)

1,3-diphenyl-2-propen-1-one (benzalacetophenone) **1,3-diphenyl-1-propanone** (81–95% yield) (22.130)

The reduction of double bonds in α,β-unsaturated carbonyl compounds occurs in biological pathways. (An example of such a reduction by NADPH is explored in Focused Problem 22.47c, Sec. 22.9D.)

Focused Problem

22.48 Show how ethyl 2-butenoate can be used as a starting material to prepare (a) ethyl butanoate and (b) 2-buten-1-ol.

22.11 REACTIONS OF α,β-UNSATURATED CARBONYL COMPOUNDS WITH ORGANOMETALLIC REAGENTS

A. Addition of Organolithium Reagents to the Carbonyl Group

Organolithium reagents react with α,β-unsaturated carbonyl compounds to yield products of consecutive carbonyl substitution and carbonyl-addition reactions.

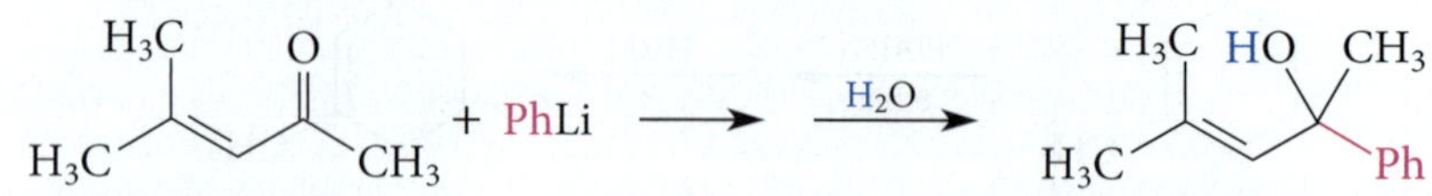

4-methyl-3-penten-2-one (mesityl oxide) → **4-methyl-2-phenyl-3-penten-2-ol** (67% yield) (22.131)

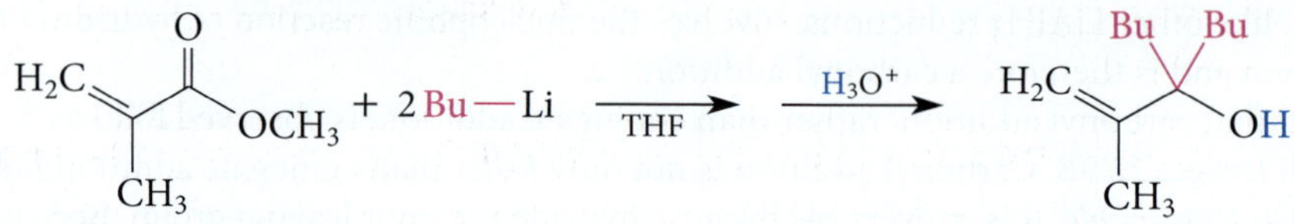

methyl 2-methylpropenoate → **3-butyl-2-methyl-1-hepten-3-ol** (89% yield) (22.132)

The reason carbonyl addition occurs rather than conjugate addition is the same as in the case of $LiAlH_4$ reduction (Sec. 22.10): carbonyl addition is more rapid than conjugate addition and it is also irreversible.

Because Grignard and organolithium reagents undergo many of the same types of reactions, it is reasonable to ask whether Grignard reagents also undergo carbonyl addition. Grignard reagents in many cases give mixtures of conjugate addition and carbonyl addition. (The reason is discussed in Sec. 22.11B.) Because both types of addition occur with Grignard reagents, organolithium reagents are used with α,β-unsaturated carbonyl compounds when only carbonyl addition is the desired reaction.

B. Conjugate Addition of Lithium Dialkylcuprate Reagents

Lithium dialkylcuprate reagents (Secs. 12.5C and 21.10B) give exclusively products of *conjugate addition* when they react with α,β-unsaturated esters and ketones.

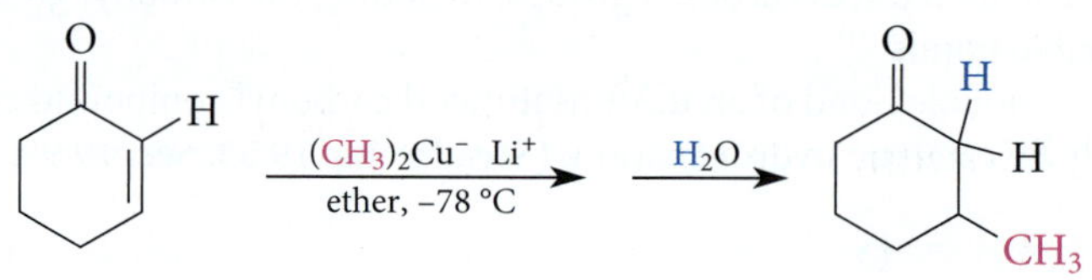

2-cyclohexenone → **3-methylcyclohexanone** (97% yield) (22.133)

Even α,β-unsaturated aldehydes, which are normally very reactive at the carbonyl group, give all or mostly products of conjugate addition, especially at low temperature.

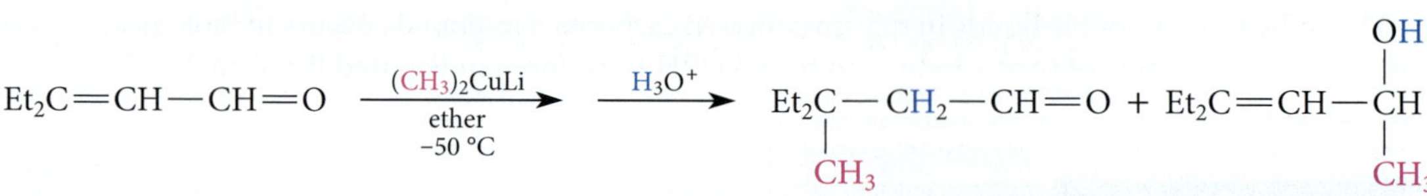

(95% of product) (5% of product)

70% total yield

(22.134)

FURTHER EXPLORATION 22.3
Conjugate Addition of Organocuprate Reagents

The fact that lithium dialkylcuprate reagents undergo conjugate addition might seem to contradict the notion that strong bases undergo carbonyl addition. However, there is good evidence that conjugate addition of lithium dialkylcuprate reagents proceeds by a special mechanism promoted by the presence of copper, and that this mechanism is particularly favorable for conjugate addition. (See Further Exploration 22.3.) For our purposes, however, the reaction can be envisioned mechanistically to be similar to other conjugate additions. The nucleophilic reaction of an anion—in this case the "alkyl anion" of the dialkylcuprate reagent—at the double bond gives a resonance-stabilized enolate ion.

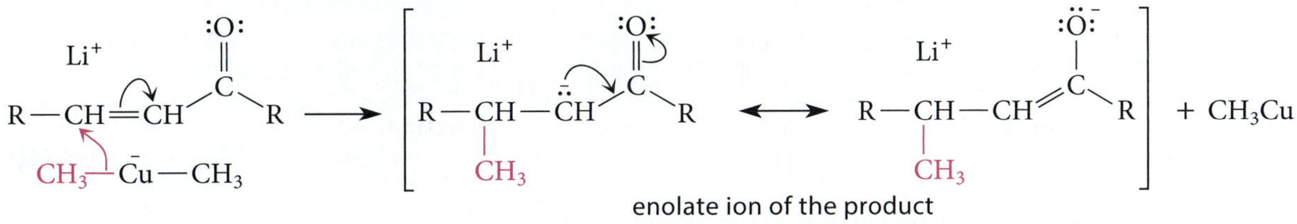

(22.135)

When water is added to the reaction mixture, protonation of the enolate ion gives the conjugate-addition product.

We noted in Sec. 22.10 that Grignard reagents react with α,β-unsaturated carbonyl compounds to give mixtures of carbonyl-addition and conjugate-addition products. Some chemists have theorized that the conjugate-addition products are due to small amounts of transition metals known to be present in commercial magnesium. Indeed, certain transition metals are known to promote conjugate addition of Grignard reagents. In fact, if a Grignard reagent is treated with CuCl, magnesium organocuprate reagents are formed, and these give exclusively conjugate addition like their lithium organocuprate counterparts.

To summarize: To carry out a *carbonyl-addition* reaction with an organometallic reagent, use an organolithium reagent. To carry out a *conjugate-addition* reaction, use a lithium organocuprate (or a Grignard reagent with added CuCl).

Focused Problems

22.49 Outline a synthesis of each of the following compounds from mesityl oxide (4-methyl-3-penten-2-one). Use an organometallic reagent in at least one step of each synthesis.

(a) 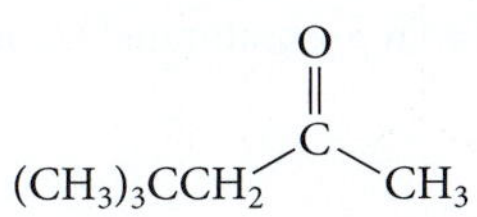

(b) 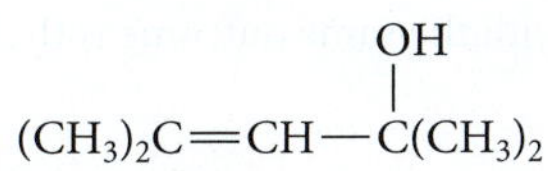

(c) 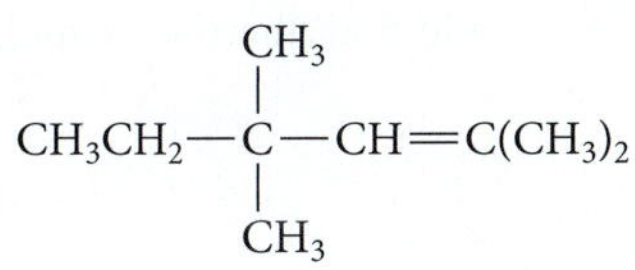

22.50 Complete the following reactions, and explain your reasoning.

(a) 

(b) $H_3C-C\equiv C-CO_2Me + Me_2CuLi \longrightarrow \xrightarrow{H_3O^+}$
(1 equiv.)

22.12 ORGANIC SYNTHESIS WITH CONJUGATE-ADDITION REACTIONS

When is a conjugate-addition reaction useful in an organic synthesis? One way to think of this problem is that any group at the β-position of a carbonyl compound (or nitrile) can *in principle* be delivered as a nucleophile in a conjugate addition. In principle, then, a conjugate addition can be mentally reversed by subtracting a nucleophilic group from the β-position of the target molecule and a positive fragment (usually a proton) from the α-position:

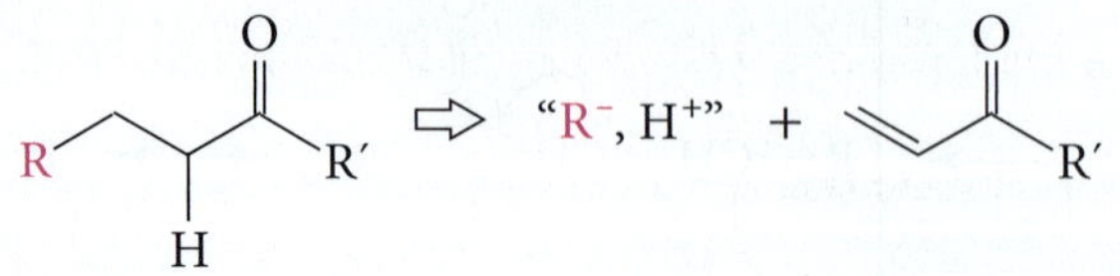

(22.136)

This approach is explored in the Study Problem 22.7.

Study Problem 22.7

Outline a preparation of 2-octanone by a conjugate-addition reaction.

Solution Two groups are attached to the β-carbon of 2-octanone: a hydrogen (*blue*) and a butyl group (*red*).

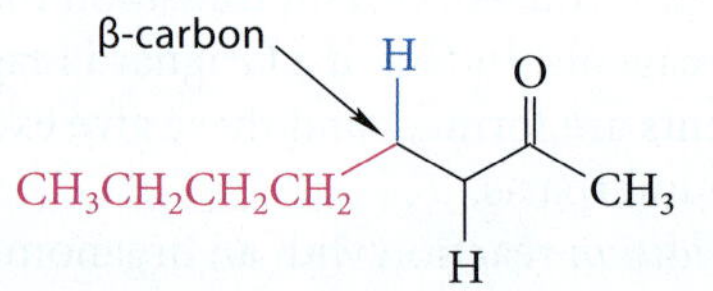

2-octanone

One choice for the "R⁻" group in Eq. 22.136 is a butyl group, which can be introduced as a "butyl anion" by the reaction of lithium dibutylcuprate with 3-buten-3-one (MVK); the α-proton is provided in the subsequent protonolysis step:

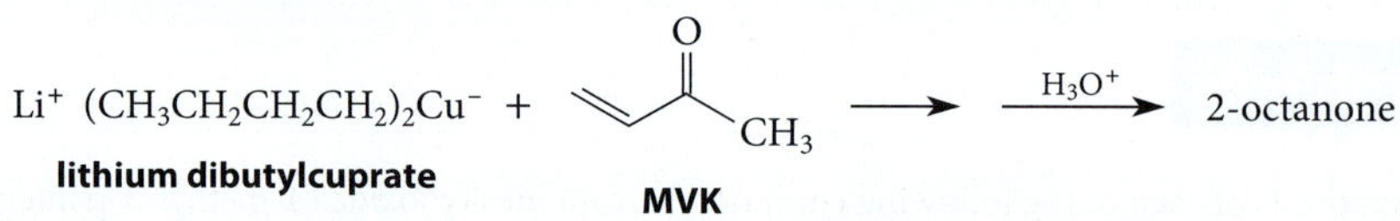

(22.137)

Another choice for "R⁻" is the hydrogen. Although we've not considered any ways for adding "H⁻" in a conjugate addition (there are some), a process with the same outcome is the hydrogenation of an α,β-unsaturated ketone:

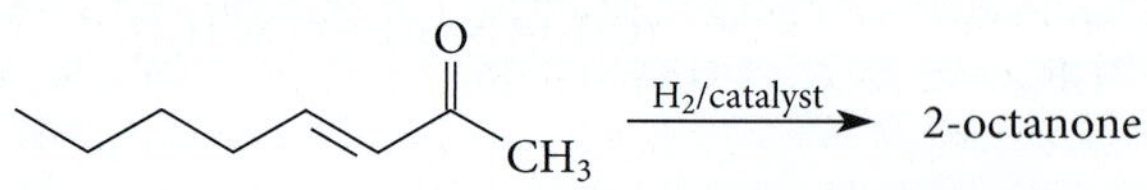

(22.138)

The type of analysis illustrated here will be even more useful if you keep in mind the notion of "synthetic equivalents" in Study Guide Link 22.7.

Focused Problem

22.51 Show how a conjugate addition can be used to prepare each of the following compounds.

(a) 3,4-Dimethyl-2-hexanone

(b) 3,4-Dimethyl-2-hexanone (another way)

(c)

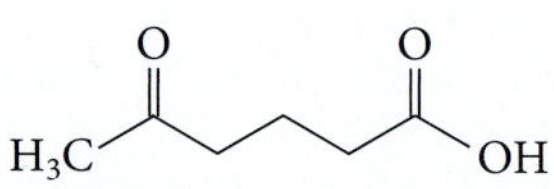

(d)

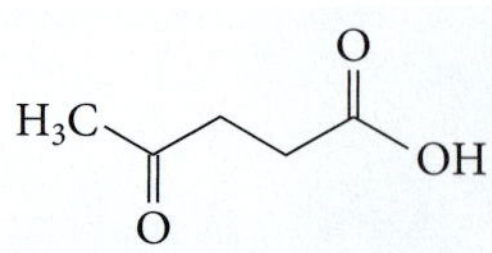

levulinic acid

Many of the reactions discussed in this chapter can be used to form carbon–carbon bonds:

1. Aldol addition and condensation reactions (Sec. 22.4)
2. Claisen and Dieckmann condensations (Sec. 22.6)
3. Malonic ester synthesis (Sec. 22.8A)
4. Acetoacetic ester synthesis (Sec. 22.8B)
5. Alkylation of ester enolates with amide bases and alkyl halides or tosylates (Sec. 22.8C)
6. Aldol addition of ester enolates (Sec. 22.8D)
7. Conjugate addition of cyanide ions (Sec. 22.9A) and enolate ions (Sec. 22.9C) to α,β-unsaturated carbonyl compounds
8. Reaction of lithium dialkylcuprates with α,β-unsaturated carbonyl compounds (Sec. 22.11B)

(A complete list of methods for forming carbon–carbon bonds is given in Appendix VI.) Their utility for carbon–carbon bond formation accounts in large measure for the importance of these reactions in organic chemistry.

CHAPTER SUMMARY *For a summary of the chapter, see Chapter 22 in the* Study Guide and Solutions Manual.

REACTION REVIEW *For a summary of reactions discussed in this chapter, see the* Reaction Review *section of Chapter 22 in the* Study Guide and Solutions Manual.

SKILLS OBJECTIVES WITH PROBLEMS

- **Identify acidic α-hydrogens in carbonyl and related compounds, give reasons for their enhanced acidity, and show with curved-arrow mechanisms how α-hydrogen ionization can lead to racemization or deuterium exchange. 22.1A**

22.52 Indicate which hydrogens are replaced by deuterium when each of the following compounds is treated with dilute NaOD in a large excess of CH_3OD.

(a)

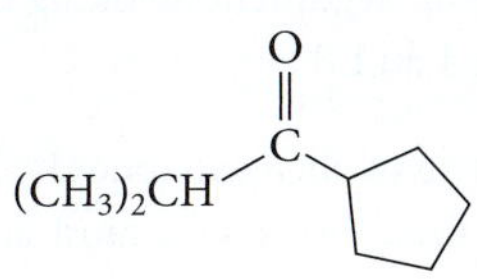

(b)

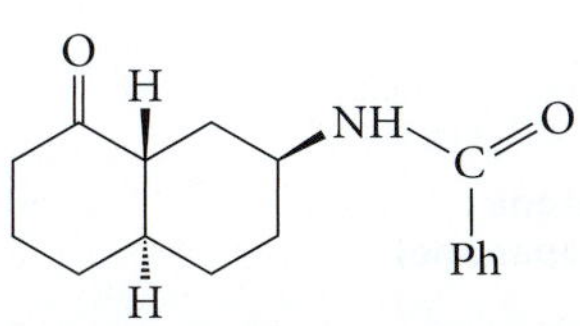

22.53 Although one enantiomer of the drug thalidomide is a sedative, the other is teratogenic (causes birth defects). Unfortunately, thalidomide is racemized in the body and, for that reason, both enantiomers are teratogenic. (See the sidebar, "Racemates in the Pharmaceutical Industry," Sec. 6.4B.) Assuming the availability of acids and bases as needed, give a curved-arrow mechanism for the base-catalyzed racemization of thalidomide in water.

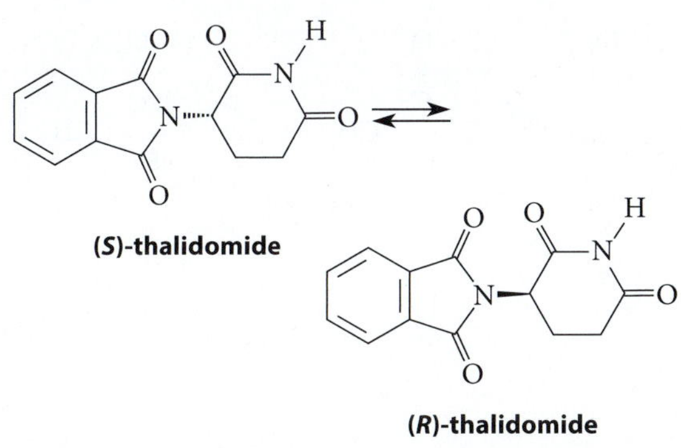

22.54 When compound *A* is treated with $NaOCH_3$ in CH_3OH, isomerization to compound *B* occurs.

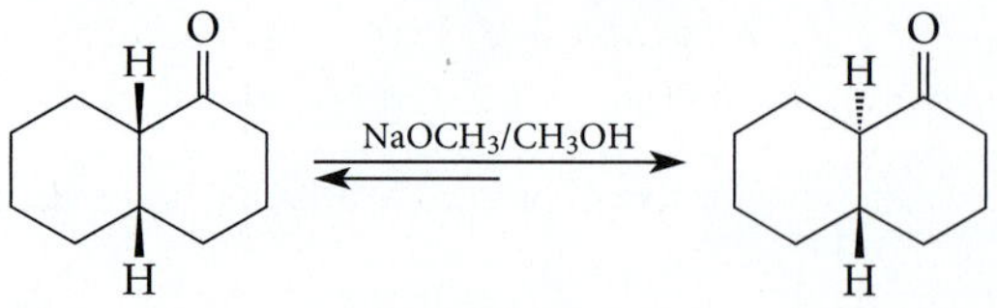

A *B*

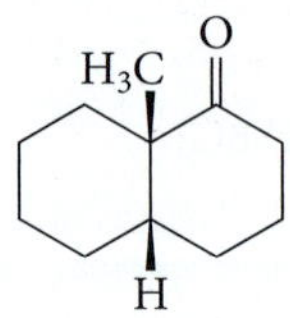

C

(a) Give a curved-arrow mechanism for the reaction, and explain why the equilibrium favors compound *B*.

(b) Explain why, when compound *C* is subjected to the same conditions, no isomerization occurs.

22.55 The pK_a of 2-nitropropane is 10. Give the structure of its conjugate base, and suggest reason(s) why 2-nitropropane has a particularly acidic C—H bond.

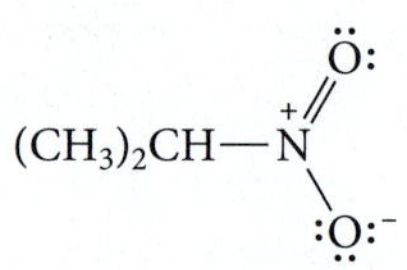

2-nitropropane

- **Compare the α-hydrogen acidities of esters and ketones, and give the reasons for the difference. 22.1A**

22.56 Which compound in each of the sets that follow is most acidic? Explain.

(a)

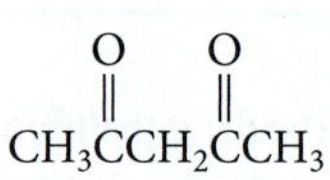

A

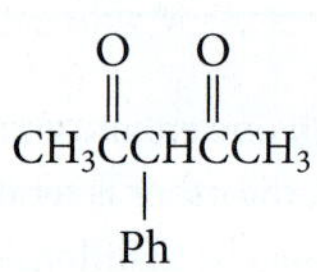

B

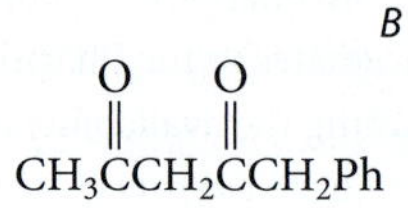

C

(b)

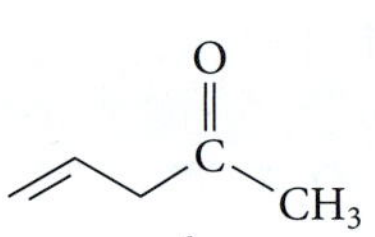

A

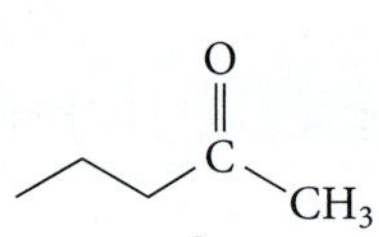

B

(c)

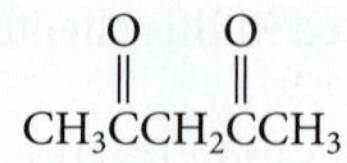

A

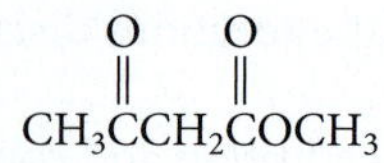

B

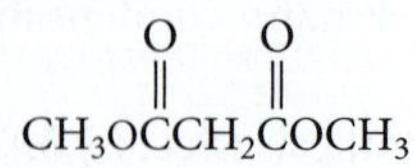

C

- **Draw the possible enol forms of any carbonyl compound with α-hydrogens. Given an enol structure, show the carbonyl compound from which it is formed. 22.2**

22.57 Each of the following compounds is unstable and either exists as an isomer or spontaneously decomposes to other compounds. In each case, give the more stable isomer or decomposition product, and explain.

(a) $H_3C—C≡C—OH$

(b)

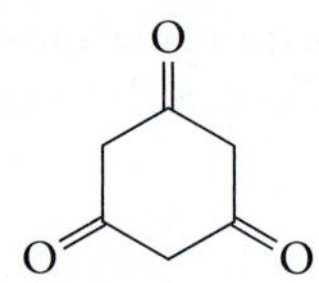

22.58 (a) Draw the structure of the conjugate base of each of the following compounds. What is the relationship between the two conjugate bases?

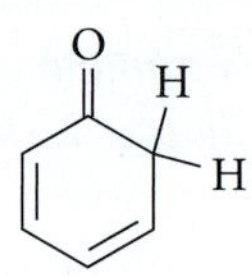

A

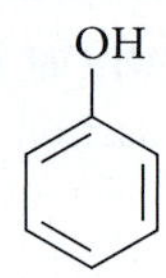

B

(b) Which compound is more acidic? Use an energy diagram to explain your reasoning.

22.59 (a) Show that the two following compounds have the same conjugate base.

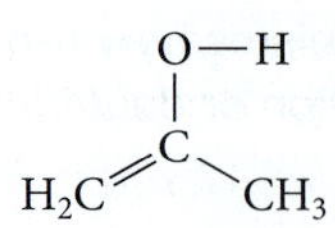

A

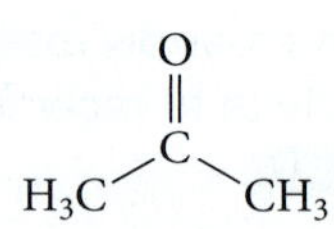

B

(b) Which compound is more acidic? Explain your reasoning using an argument invoking the relative free energies of *A* and *B*.

22.60 Draw all of the enol forms of phenylacetone (1-phenyl-2-propanone). Which of these forms is the most stable and why?

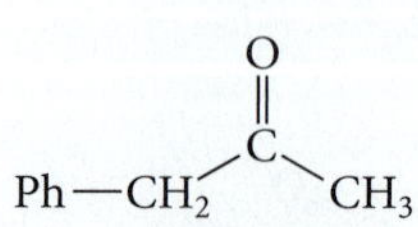

phenylacetone (1-phenyl-2-propanone)

- **Draw the curved-arrow mechanisms for acid- and base-catalyzed enolization of a carbonyl compound.** 22.2

22.61 For the ketone in Problem 22.60, give a curved-arrow mechanism for (a) acid-catalyzed enolization and (b) base-catalyzed enolization to give the most stable enol.

- **For a condensation reaction, analyze possible competing reactions, and determine whether they will interfere.** 22.6C, 22.8D

22.62 In each of the following reactions, give the major product expected, and outline the other possible reactions and why they do not occur.

(a)

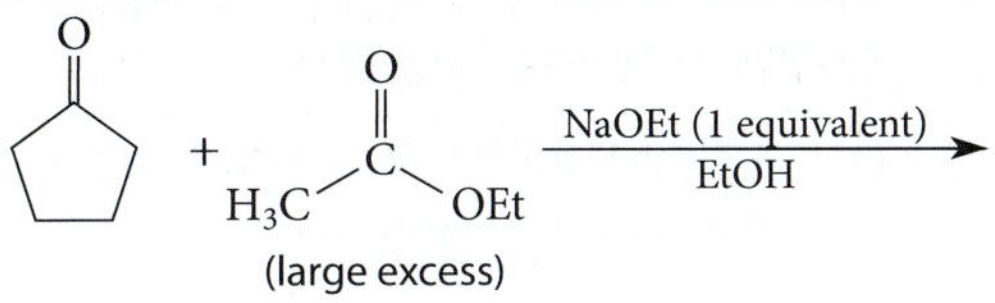

(b)

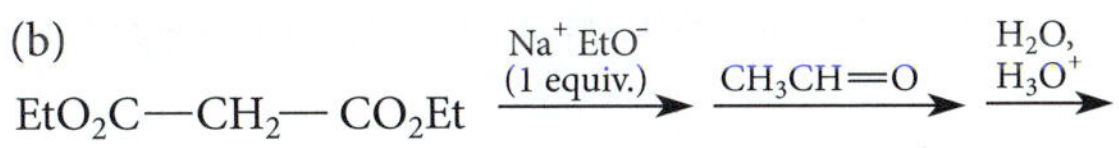

- **Provide the structures of missing reactants or products in the following reactions:**

(1) α-Halogenation of ketones and carboxylic acids 22.3

(2) Aldol additions and condensations 22.4

(3) Claisen condensations and crossed Claisen condensations 22.6

(4) Malonic ester synthesis 22.8A

(5) Acetoacetic ester synthesis 22.8B

(6) Alkylation reactions of ester enolates 22.8A,B,C

(7) Aldol reactions of ester enolates, including the Reformatsky reaction 22.8D

(8) Conjugate-addition reactions of α,β-unsaturated carbonyl compounds and nitriles 22.9A

(9) Selective carbonyl- and double-bond reductions of α,β-unsaturated carbonyl compounds 22.10

(10) Organometallic reactions of α,β-unsaturated carbonyl compounds 22.11A,B

22.63 Give the structure of the products produced in each of the following transformations.

(a) The reaction of 4-methylpentanoic acid with Br_2 and a catalytic amount of PBr_3

(b) The reaction of 4-methylpentanoic acid with Br_2 and a full equivalent of PBr_3

(c) The product of part (a) with one equivalent of trimethylamine $(CH_3)_3N$:

(d) The product of part (b) with one equivalent of 1-butanol

(e) The product of part (a) with potassium iodide in acetone

22.64 Give the principal organic product expected when 3-buten-2-one (methyl vinyl ketone, or MVK) reacts with each of the following reagents.

(a) HBr

(b) H_2, Pt (cat.)

(c) $LiAlH_4$, then H_2O

(d) HCN in water, pH 10

(e) Et_2CuLi, then H_3O^+

(f) Diethyl malonate and NaOEt, then H_3O^+

(g) Ethylene glycol, HCl (cat.)

(h) 1,3-Butadiene (*Hint:* See Sec. 15.3.)

22.65 Give the principal organic products expected when isobutyraldehyde (2-methylpropanal) reacts with each of the following reagents.

(a) The lithium enolate of acetone followed by H_2O/H_3O^+

(b) The lithium enolate of ethyl 2-methylpropanoate followed by dilute H_3O^+

(c) Ethyl α-bromoacetate + Zn, then H_3O^+

(d) Diethyl malonate + a secondary amine (R_2NH)/carboxylic acid catalyst

(e) Ethyl acetoacetate + a secondary amine/carboxylic acid catalyst

22.66 Give the principal organic products expected when ethyl *trans*-2-butenoate (ethyl crotonate) reacts with each of the following reagents.

(a) ^-CN in ethanol, then H_2O/H_3O^+, heat

(b) Me_2NH, room temperature

(c) NaOH, H_2O, heat

(d) CH_3Li (excess), then H_3O^+

(e) H_2, catalyst

(f) 1,3-Cyclopentadiene

22.67 Provide the missing products or reactants in the following transformations.

(a)

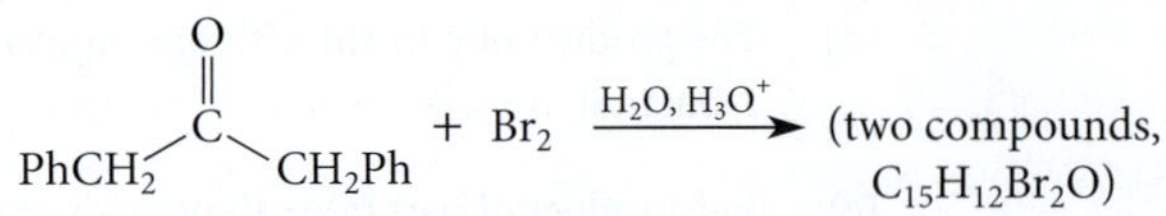

(b)

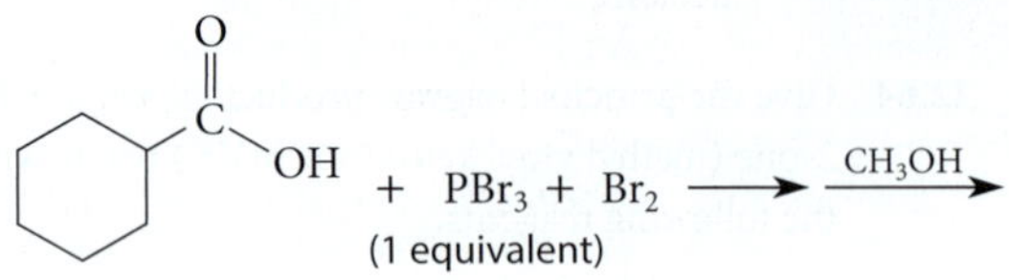

(c)

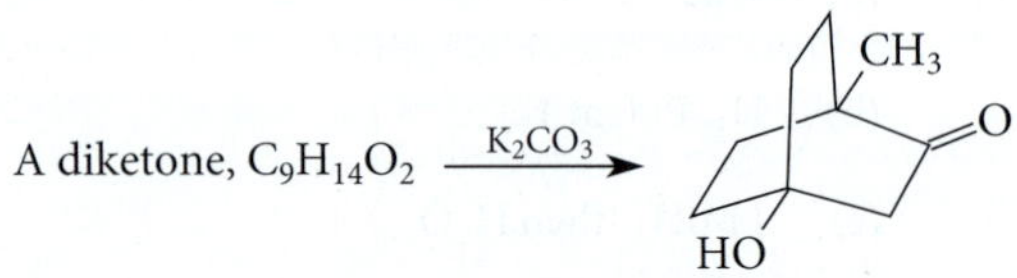

(d)

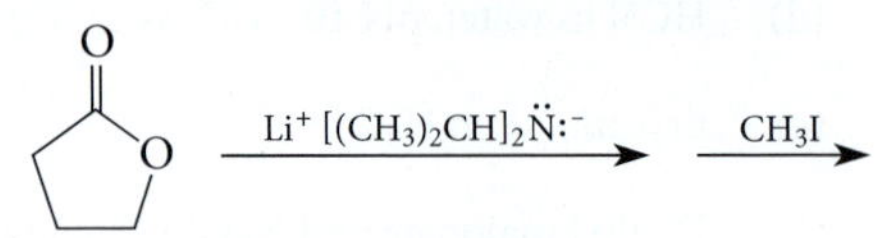

γ-butyrolactone

(e)

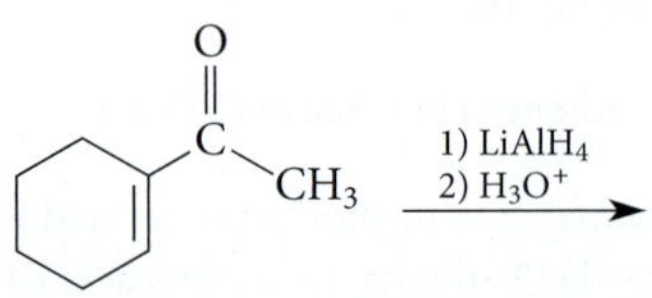

(f)

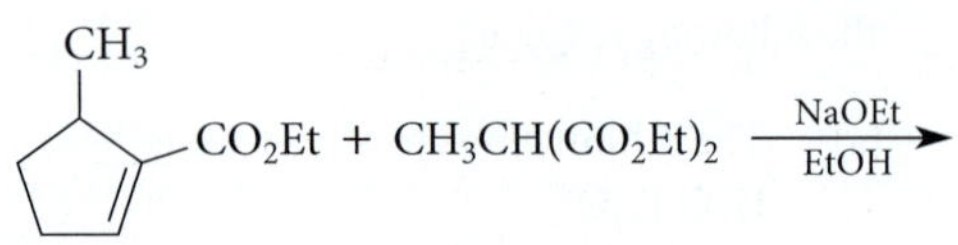

(g)

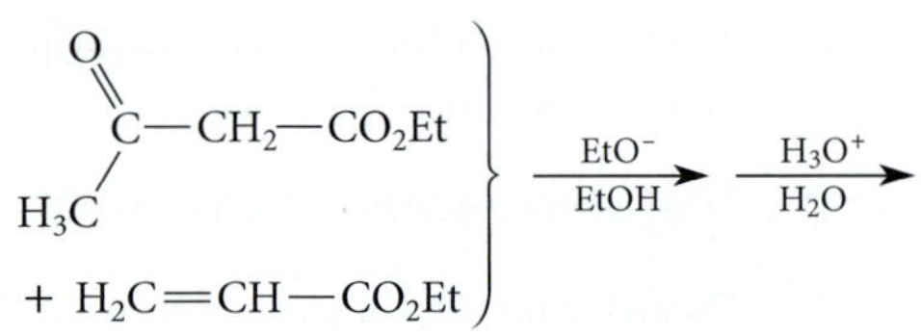

(h)

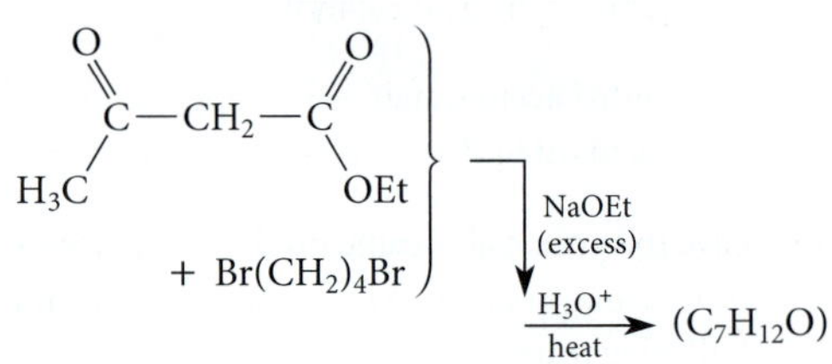

(i)

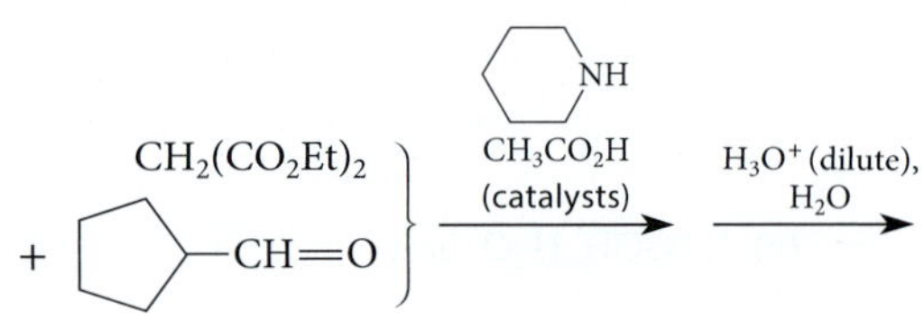

(j)

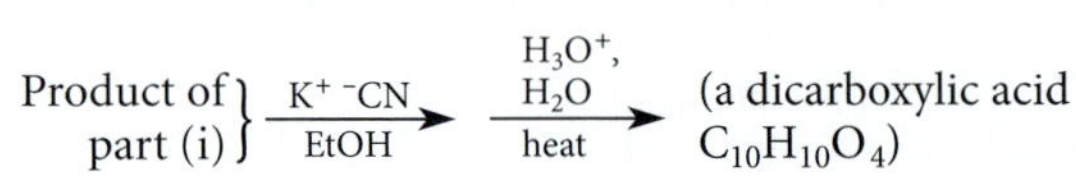

(k)

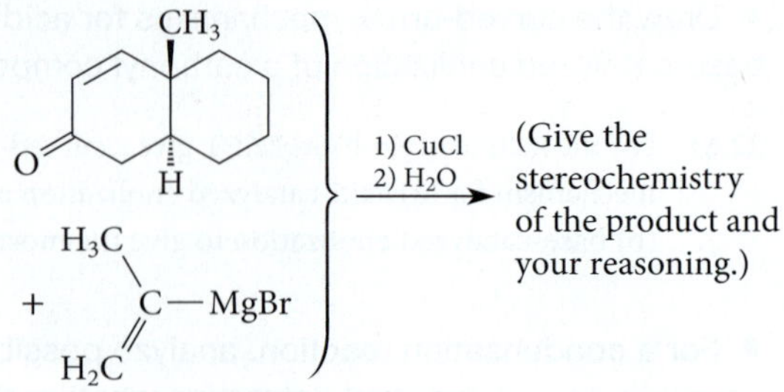

(*Hint:* Grignard reagents with added CuCl react like lithium organocuprate reagents.)

22.68 When 2,4-pentanedione in ether is treated with one equivalent of sodium hydride (NaH), a gas is evolved and an ionic compound *A* is formed.

(a) Give the structure of *A*. Which atoms of *A* should be nucleophilic? Explain.

(b) When *A* reacts with CH_3I, three isomeric compounds, *B*, *C*, and *D* ($C_6H_{10}O_2$), are formed. Suggest structures for these compounds.

● **Draw curved-arrow mechanisms for condensation and related reactions 22.1, 22.4, 22.6, 22.8**

22.69 Provide curved-arrow mechanisms for each of the following reactions.

(a)

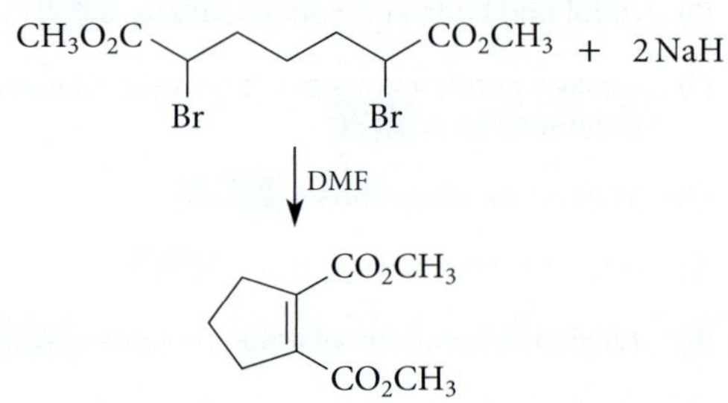

(b)

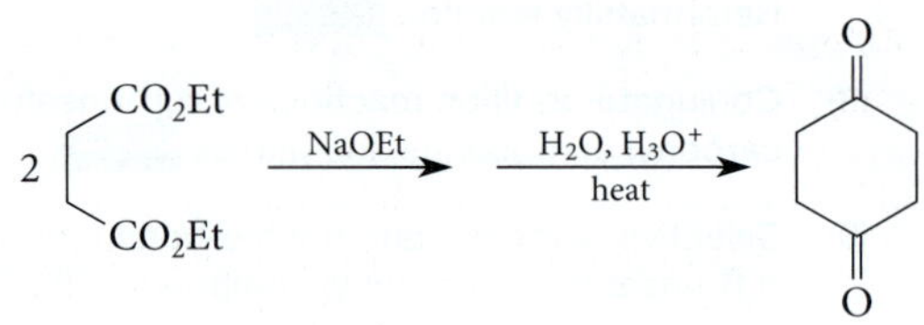

(c)

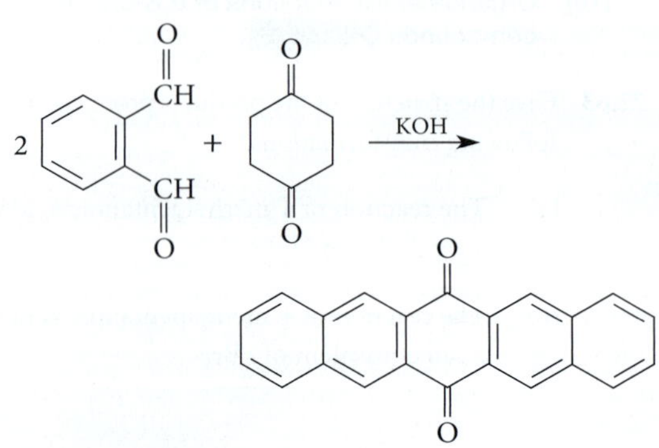

(d)

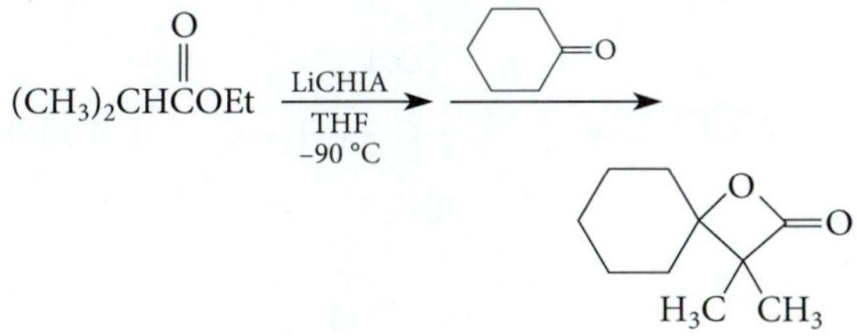

(e)

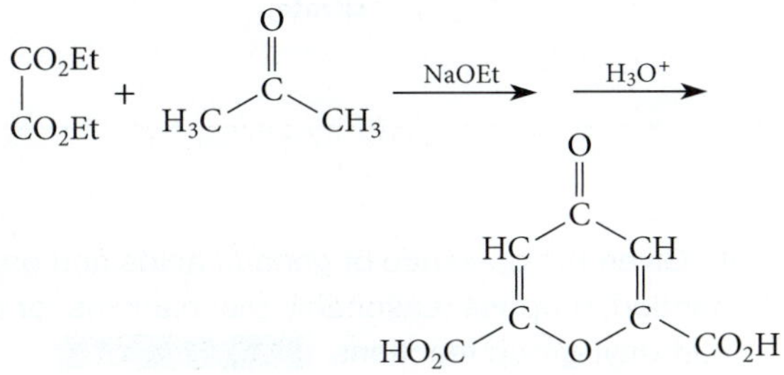

chelidonic acid

22.70 The reversibility of the aldol addition reaction is a major factor in each of the following reactions. Provide a curved-arrow mechanism for each transformation.

(a)

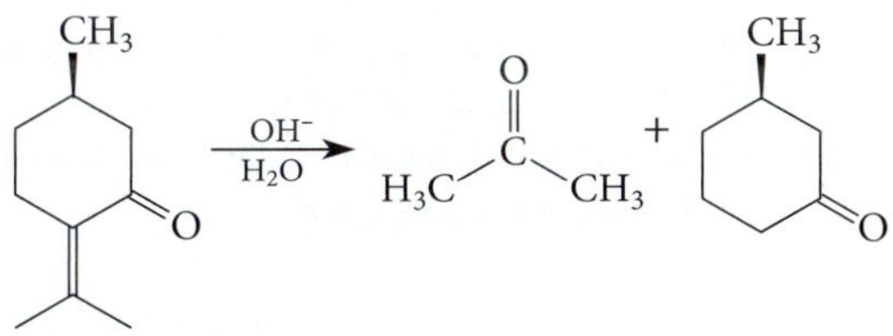

(*R*)-pulegone

(b)

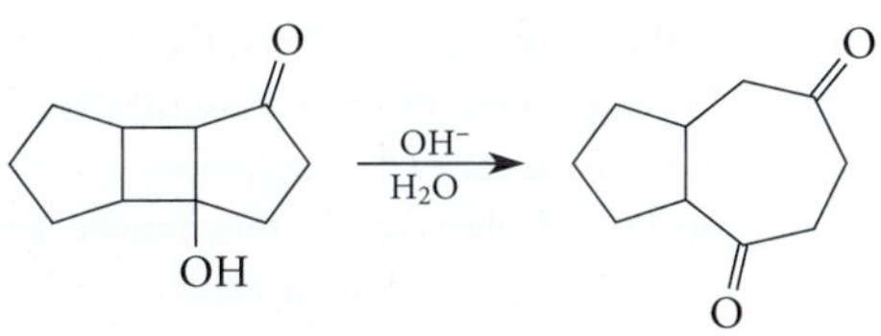

- **Incorporate the condensation reactions and conjugate additions discussed in this chapter into multistep syntheses. 22.3D, 22.4D, 22.6D, 22.12**

22.71 A useful diketone, *dimedone*, can be prepared in high yield by the synthesis shown in the following scheme. Provide structures for both the intermediate *A* (a Michael-addition product) and dimedone, and give a curved-arrow mechanism for each step up to compound *B*.

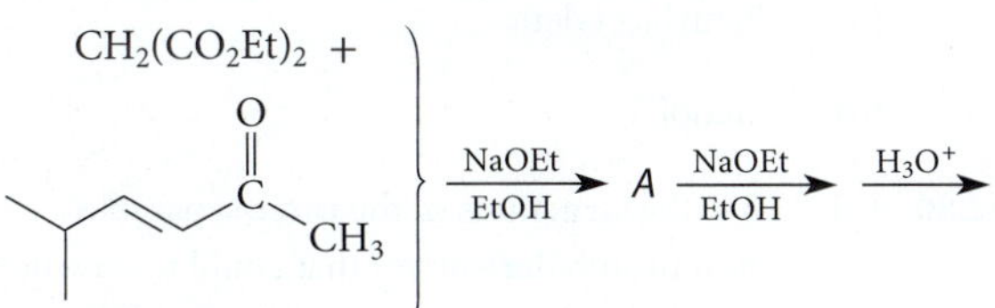

mesityl oxide

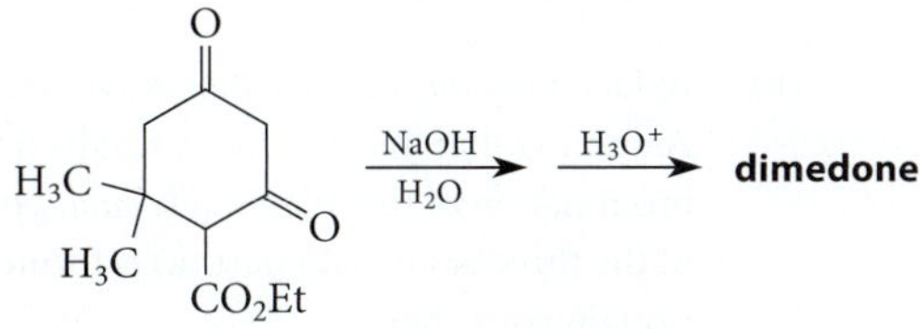

B

22.72 Propose syntheses of each of the following compounds from the indicated starting materials and any other reagents.

(a) 3-Ethylcyclopentanol from 2-cyclopentenone

(b) 1,3,3-Trimethylcyclohexanol from 3-methyl-2-cyclohexenone

(c) 2-Benzylcyclohexanone from $EtO_2C(CH_2)_5CO_2Et$

(d) 2,2-Dimethyl-1,3-propanediol from diethyl malonate

(e) $H_2NCH_2CH_2CH_2CH_2OH$ from ethyl acrylate (ethyl 2-propenoate)

(f) 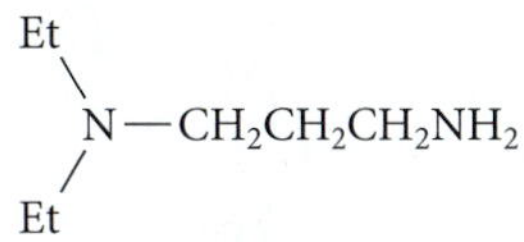 from acrylonitrile (propenenitrile)

(g) 2-Phenylbutanoic acid from phenylacetic acid. (Do not use branched amide bases; see Eq. 22.75 in Sec. 22.6C.)

(h) 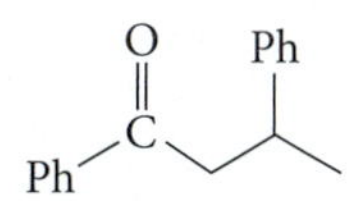 from 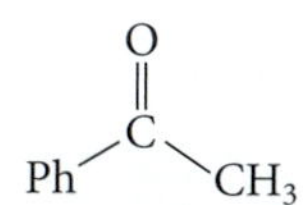

1,3-diphenyl-1-butanone **acetophenone**

(i) $Ph-CH(OH)-CD_2-Ph$ from phenylacetic acid

(j) 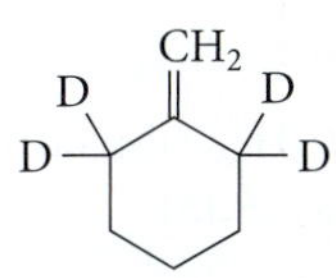 from cyclohexanone

(k) 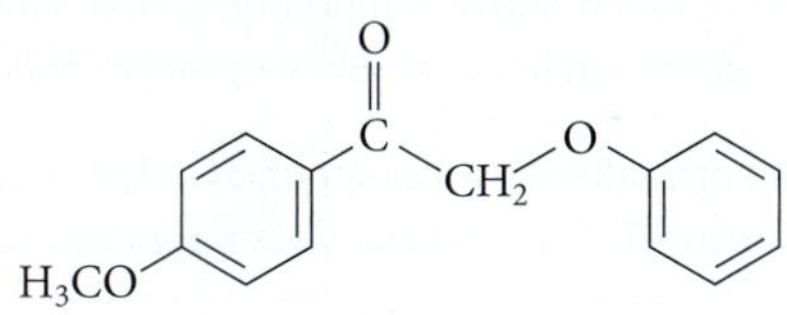 from acetyl chloride

(l) 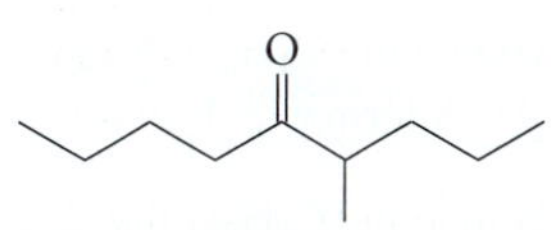 from diethyl malonate

(m) 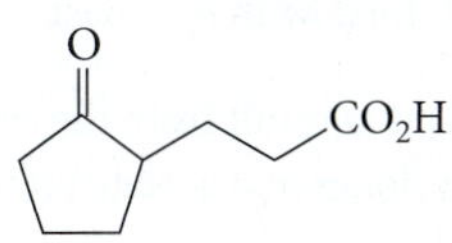 from 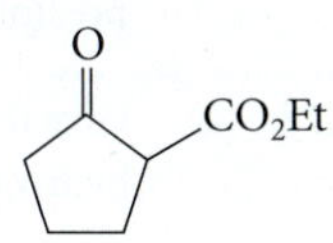

22.73 A biochemist, Sal Monella, has come to you to ask your assistance in testing a promising biosynthetic hypothesis. She wishes to have two samples of methylsuccinic acid specifically labeled with ^{14}C as shown in the following structures. The source of the isotope, for financial reasons,

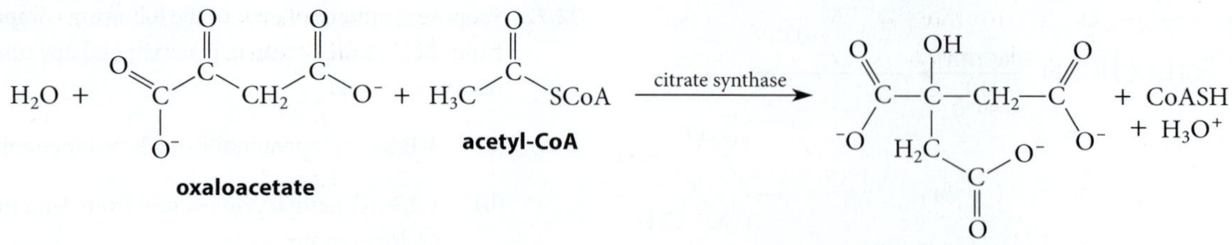

oxaloacetate acetyl-CoA citrate

FIGURE P22.75

is to be the salt $Na^{14}CN$. Outline syntheses that will accomplish the desired objective. (*C = ^{14}C.)

(a)

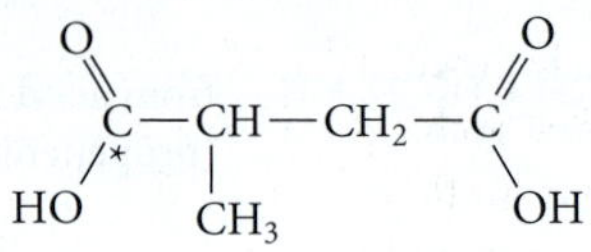

(b)

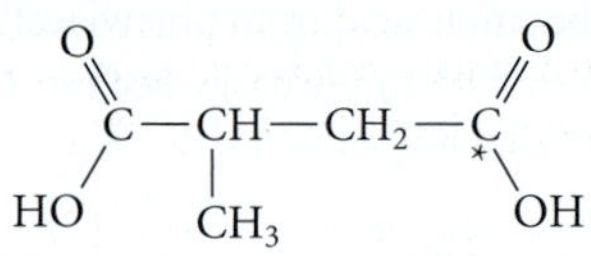

• **Given the presence of generic acids and bases as needed, propose reasonable mechanisms for biological carbonyl-group reactions. 22.5, 22.7, 22.9**

22.74 Using abbreviated structures like the ones used in Sec. 22.7, outline the steps that convert hexanoyl-ACP into octanoyl-ACP during fatty-acid biosynthesis.

22.75 In the Krebs cycle (tricarboxylic acid, or citric acid, cycle), the enzyme *citrate synthase* catalyzes the biosynthesis of citrate from oxaloacetate and acetyl-CoA as shown in **Fig. P22.75**. Assuming that acids and bases are provided as needed in the enzyme active site, propose a curved-arrow mechanism for this reaction.

INTEGRATED PROBLEMS

22.76 Give the structure of a compound that meets each criterion.

(a) An optically active compound $C_6H_{12}O$ that racemizes in base

(b) An achiral compound $C_6H_{12}O$ that does not give a positive Tollens test (Sec. 19.15)

(c) An optically active compound $C_6H_{12}O$ that neither racemizes in base nor gives a positive Tollens test

(d) An optically active compound $C_6H_{12}O$ that gives a positive Tollens test and does not racemize in base

22.77 Give the structure of a compound that meets each criterion.

(a) A carbonyl compound C_4H_8O that gives a precipitate of iodoform with I_2 in base

(b) A carbonyl compound C_4H_8O that does not give a precipitate of iodoform with I_2 in base

22.78 (a) Give the products that result from the ester hydrolysis of 1-cyclohexenyl acetate (compound *A*).

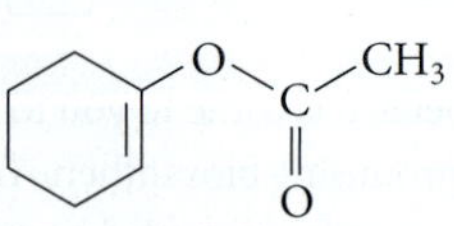

A

$\Delta G°$ (hydrolysis) = −67.3 kJ mol^{-1} (−16.1 kcal mol^{-1})

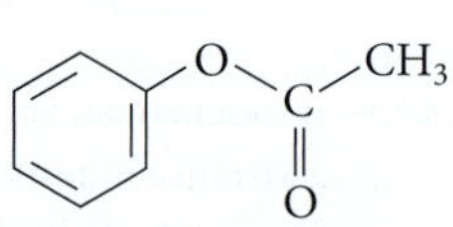

B

$\Delta G°$ (hydrolysis) = −30.9 kJ mol^{-1} (−7.39 kcal mol^{-1})

(b) As the preceding data show, the $\Delta G°$ for hydrolysis of *A* is *much* more negative than the $\Delta G°$ for hydrolysis of phenyl acetate (*B*). Explain why the equilibrium for the hydrolysis of *A* is much more favorable than the equilibrium for the hydrolysis of *B*.

22.79 Arrange the following compounds in order of increasing acidity and explain, including any point of uncertainty.

(1) Isobutyramide

(2) Octanoic acid

(3) Toluene

(4) Ethyl acetate

(5) Phenylacetylene

(6) Phenol

22.80 (a) Give the structures of the three separable monobromo derivatives that could form when 2-methylcyclohexanone is treated with Br_2 in the presence of HBr.

(b) In fact, only *one* of these derivatives is formed. Assuming that this derivative results from bromination of the *more stable enol*, predict which of the three isomers in part (a) is formed, and explain your choice.

22.81 When acetoacetic acid is decarboxylated in the presence of bromine, bromoacetone is isolated.

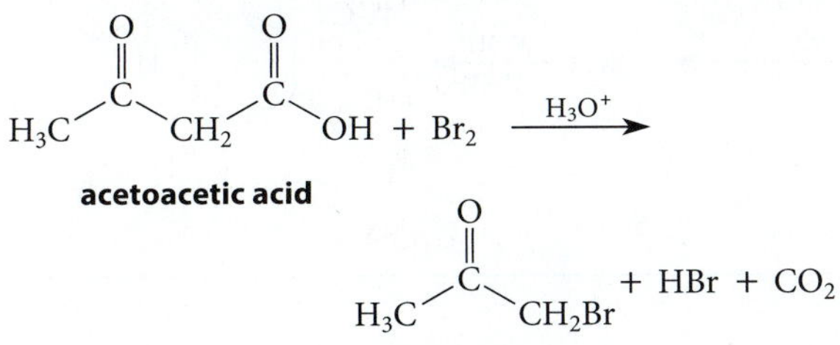

acetoacetic acid

bromoacetone

The rate of appearance of bromoacetone is described by the following rate law:

$$\text{rate} = k[\text{acetoacetic acid}]$$

(The reaction rate is zero order in bromine.) Suggest a mechanism for the reaction that is consistent with this rate law.

22.82 Account for the fact that treatment of 1,3-diphenyl-1,3-propanedione with I_2 and NaOH gives a precipitate of iodoform even though it is not a methyl ketone. Besides iodoform, the other product of the reaction, after acidification, is two equivalents of benzoic acid.

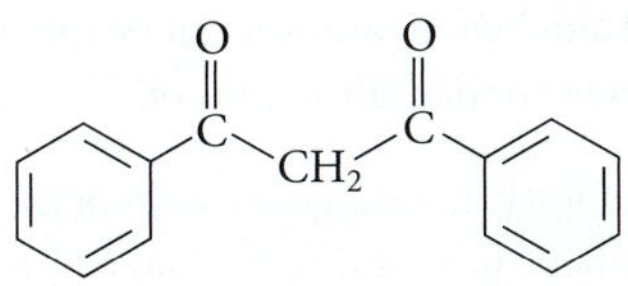

1,3-diphenyl-1,3-propanedione

22.83 In either acid or base, 3-cyclohexenone comes to equilibrium with 2-cyclohexenone:

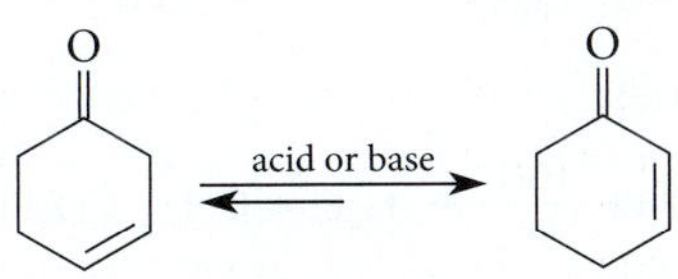

(a) Explain why the equilibrium favors the α,β-unsaturated ketone over its β,γ-unsaturated isomer.

(b) Give a mechanism for this reaction in aqueous NaOH.

(c) Give a mechanism for the same reaction in dilute aqueous H_2SO_4. (*Hint:* The enol 1,3-cyclohexadienol is an intermediate in the acid-catalyzed reaction.)

(d) Is the equilibrium constant for the analogous reaction of 4-methyl-3-cyclohexenone expected to be greater or smaller? Explain.

22.84 In 3-methyl-2-cyclohexenone, the eight hydrogens H^a, H^b, H^c, and H^d can be exchanged for deuterium in CH_3O^-/CH_3OD.

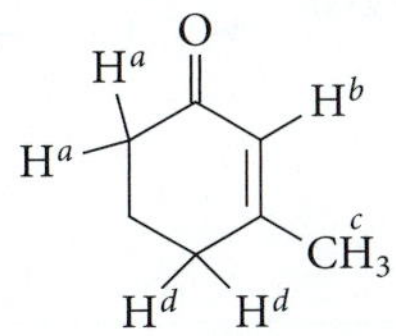

(a) Write curved-arrow mechanisms for the base-catalyzed exchange of hydrogens H^a, H^c, and H^d.

(b) Explain why hydrogen H^b is much less acidic than hydrogens H^a, H^c, and H^d, even though it is an α-hydrogen.

(c) Although hydrogen H^b is not unusually acidic, it nevertheless exchanges readily in base. Write a mechanism for the exchange of H^b. (*Hint:* Consider the equilibrium in Problem 22.83.)

22.85 Which hydrogens of the sex hormone testosterone would be exchanged for deuterium in CH_3O^-/CH_3OD?

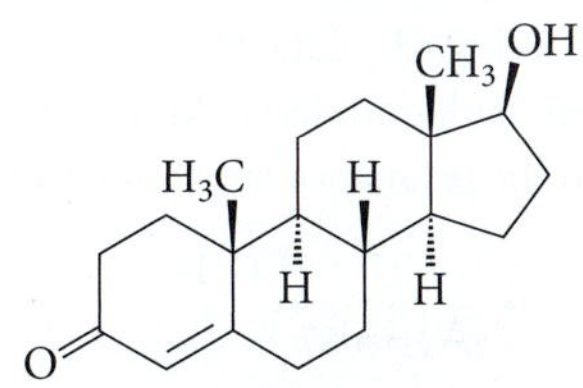

testosterone

22.86 Compound *A*, γ-pyrone, is an usually basic ketone with a conjugate-acid pK_a of −0.4. The pK_a of the conjugate acid of compound *B*, in contrast, is about −3.

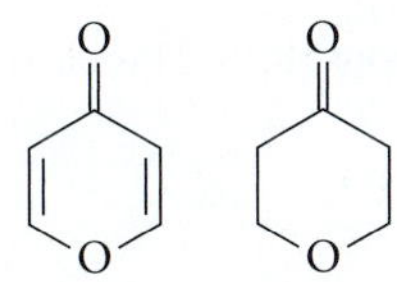

A *B*

γ-pyrone

(a) Draw the structures of the conjugate acids of both molecules, and explain why *A* is more basic than *B*.

(b) Tropone reacts with one equivalent of HBr to give a stable crystalline conjugate acid salt with a pK_a of −0.6, which is considerably greater (that is, less negative) than the pK_a values of most protonated α,β-unsaturated ketones. Give the structure of the conjugate acid of tropone, and explain why tropone is unusually basic.

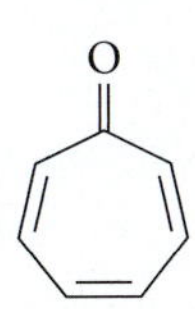

tropone

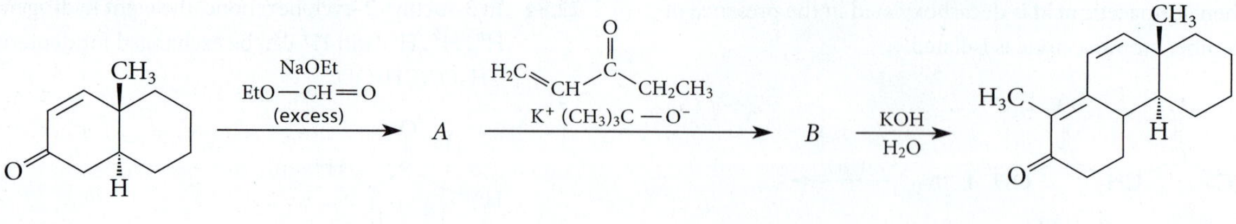

FIGURE P22.89

22.87 (a) An α,β-unsaturated carboxylic acid has the following resonance structures. Would this type of resonance interaction increase or diminish the acidity of an α,β-unsaturated carboxylic acid relative to that of an ordinary carboxylic acid? Explain.

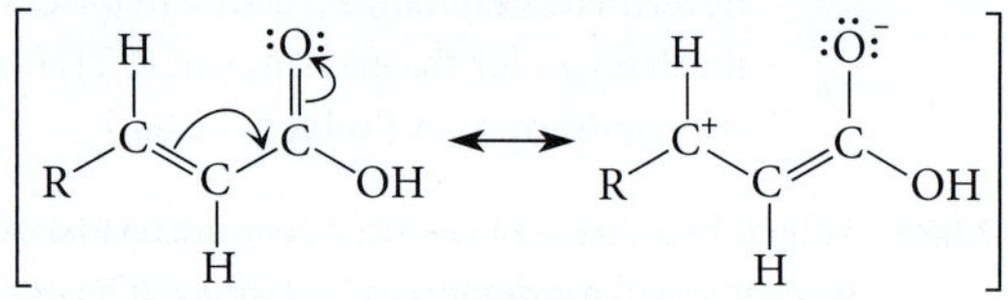

(b) Consider the following pK_a data. Show how these data are consistent with both the polar effect of the double bond and the resonance effect in part (a).

$CH_3CH_2CH_2—CO_2H$
pK_a = 4.87

$H_2C{=}CH—CH_2—CO_2H$
pK_a = 4.37

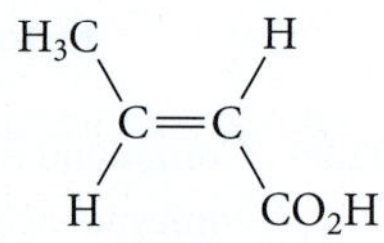
pK_a = 4.70

22.88 The pK_a of 2-nitropropane is about 10. (See Problem 22.55.)

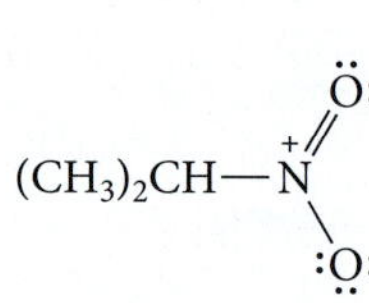

2-nitropropane

(a) When the conjugate base of 2-nitropropane is protonated, an isomer of 2-nitropropane is formed, which, on standing, is slowly converted into 2-nitropropane itself. Give the structure of this isomer.

(b) What product forms when 2-nitropropane reacts with ethyl acrylate ($H_2C{=}CH—CO_2Et$) in the presence of NaOEt in EtOH?

22.89 Give the structures of the intermediates *A* and *B* in the transformation shown in **Fig. P22.89**, and show how they are formed.

22.90 Explain the following findings.

(a) One full equivalent of base must be used in the Claisen or Dieckmann condensation.

(b) Ethyl acetate readily undergoes a Claisen condensation in the presence of one equivalent of sodium ethoxide, but phenyl acetate does *not* undergo a Claisen condensation in the presence of one equivalent of sodium phenoxide.

(c) Although the aldol condensation can be catalyzed by acid, the Claisen condensation cannot.

22.91 A student, Cringe Labrack, has suggested each of the following faulty synthetic procedures. Explain why each one cannot be expected to work as shown.

(a) $CH_3CH_2CO_2Et \xrightarrow[\text{EtOH}]{\text{NaOEt}} \xrightarrow{CH_3I} (CH_3)_2CHCO_2Et$

(b) $CH_2(CO_2Et)_2 \xrightarrow[\text{EtOH}]{\text{NaOEt}} \xrightarrow{\text{PhBr}} Ph—CH(CO_2Et)_2$

(c)

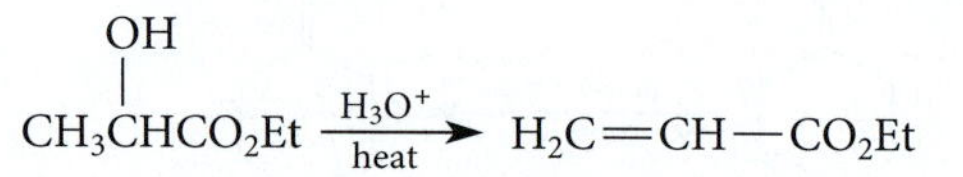

(d)

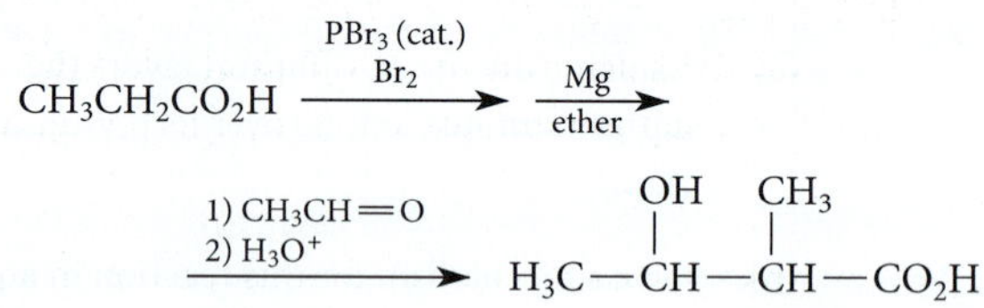

(e)

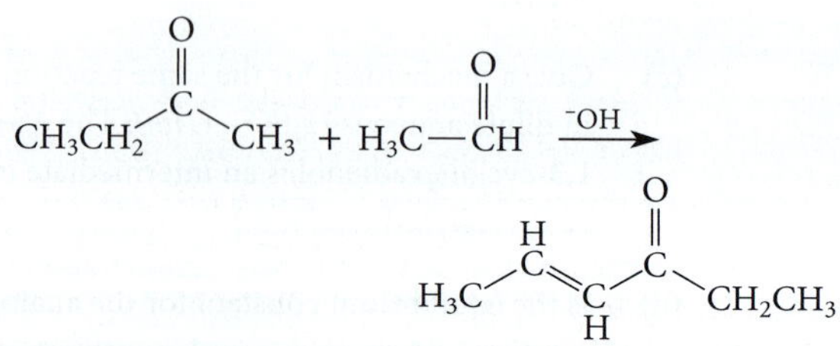

(f)

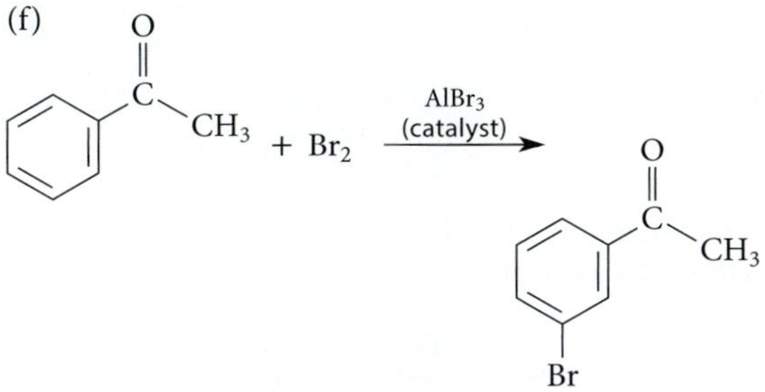

22.92 When the diethyl ester of a substituted malonic acid is treated with sodium ethoxide and urea, Veronal, a *barbiturate*, is formed. (Barbiturates are hypnotic drugs; some are actively used in modern anesthesia.) Using the curved-arrow notation, give a mechanism for the Veronal synthesis.

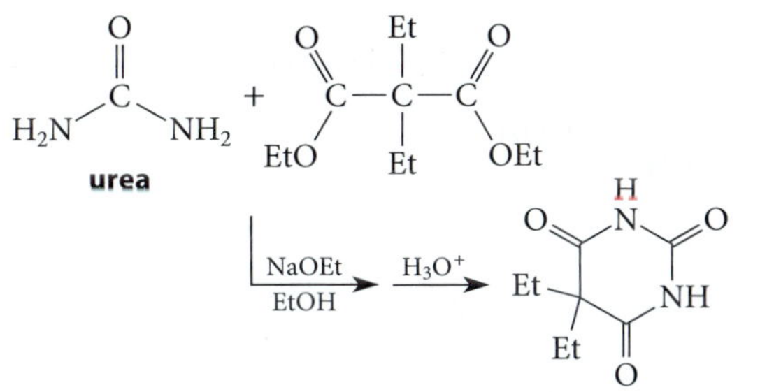

Veronal (barbital)
(a barbiturate)

22.93 Using a reaction similar to the one in Problem 22.92, outline a synthesis of pentothal from diethyl malonate and any other reagents. (The sodium salt of pentothal is a widely used injectable anesthetic.)

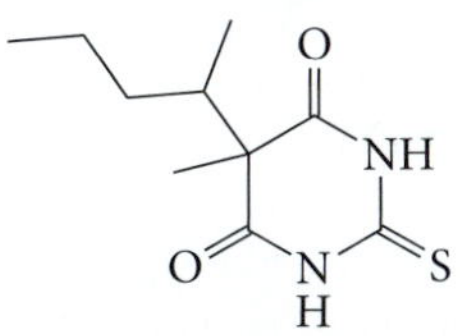

pentothal

22.94 When the epoxide 2-vinyloxirane reacts with lithium dibutylcuprate, followed by protonolysis, a compound *A* is the major product formed.

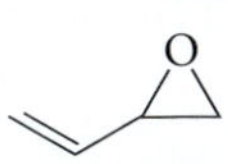

2-vinyloxirane

Oxidation of *A* with PCC yields *B*, a compound that gives a positive Tollens test and has an intense UV absorption around 215 nm. Treatment of *B* with Ag_2O, followed by catalytic hydrogenation, gives octanoic acid. Identify *A* and *B*, and outline a mechanism for the formation of *A*.

22.95 Using the curved-arrow notation, provide mechanisms for each of the reactions given in **Fig. P22.95**.

22.96 Outline curved-arrow mechanisms for each of the following known transformations that can be used to form three-membered rings. [Part (a) is an example of the *Darzens glycidic ester condensation*.]

(a)

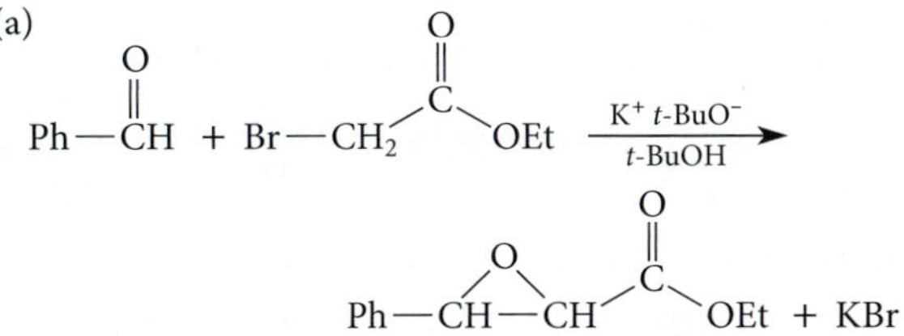

(mixture of stereoisomers)

(b)

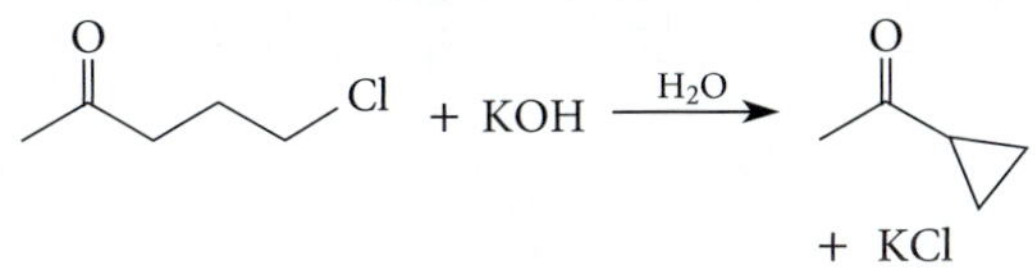

22.97 Ethyl vinyl ether, EtO—CH=CH_2, hydrolyzes in weakly acidic water to acetaldehyde and ethanol. Under the same conditions, diethyl ether does not hydrolyze. Quantitative comparisons of the hydrolysis rates of the two ethers under comparable conditions show that ethyl vinyl ether hydrolyzes about 10^{13} times faster than diethyl ether. The rapid hydrolysis of ethyl vinyl ether suggests an unusual mechanism for this reaction. The acetaldehyde formed when the hydrolysis of ethyl vinyl ether is carried out in D_2O/D_3O^+ contains one deuterium in its methyl group (that is, DCH_2—CH=O). Suggest a hydrolysis mechanism for ethyl vinyl ether consistent with these facts. (*Hint:* Vinylic ethers are also called *enol ethers*. Where do enols protonate? See Eq. 22.21b, Sec. 22.2.)

22.98 Bearing in mind the hydrolysis reaction discussed in Problem 22.97, identify the synthetic intermediates *A–C* and the final product *D*.

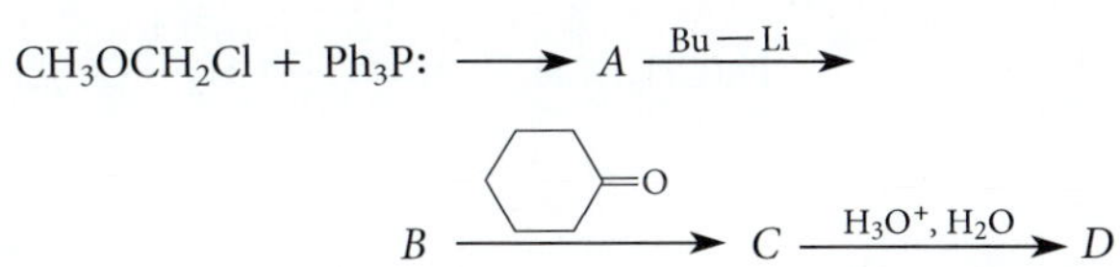

22.99 Complete each of the following reactions by providing the principal organic product. Explain your reasoning.

(a)

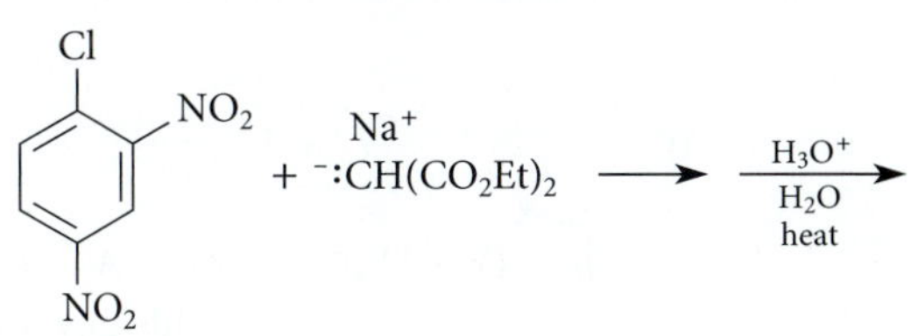

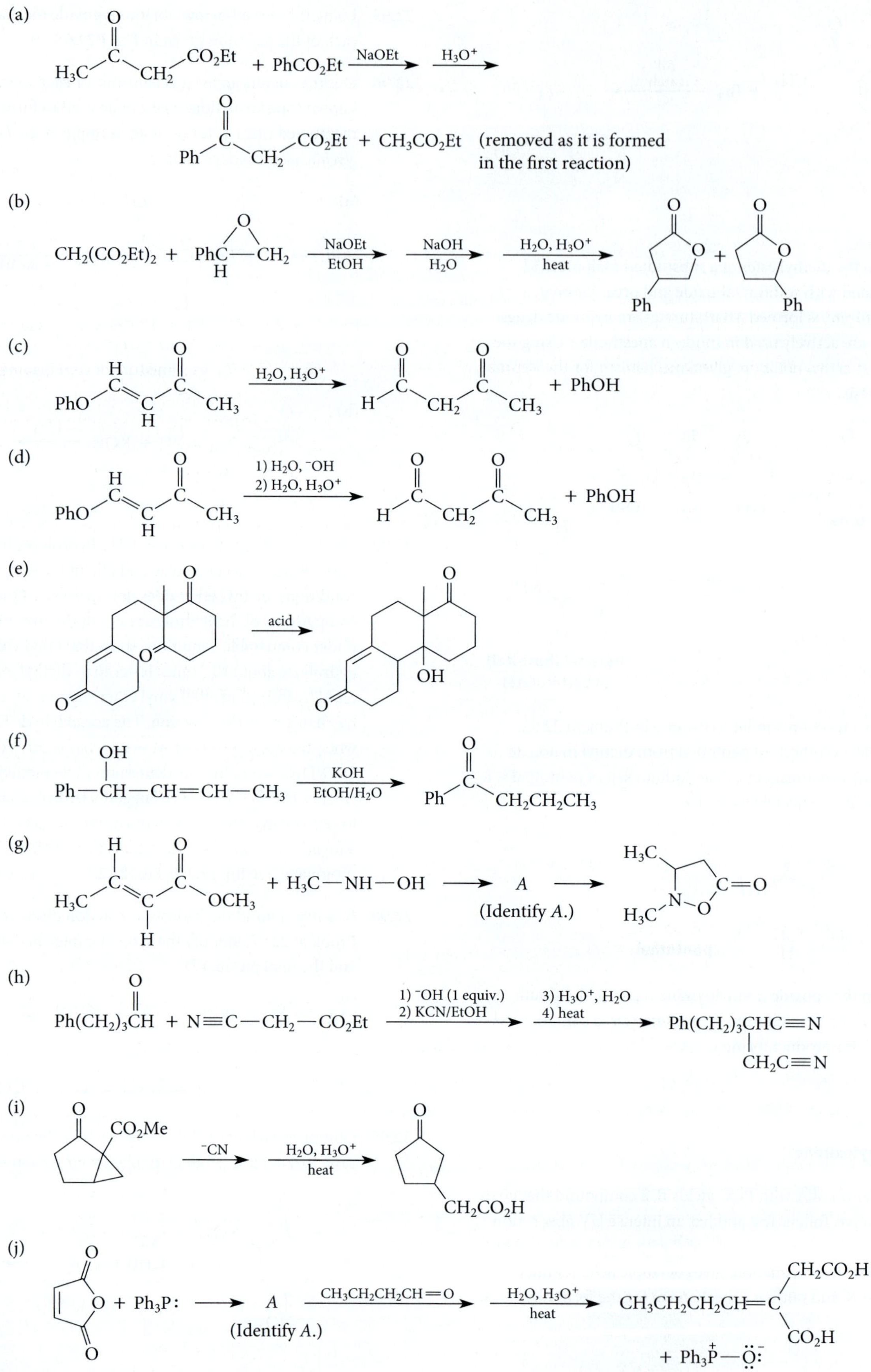

FIGURE P22.95

(b) 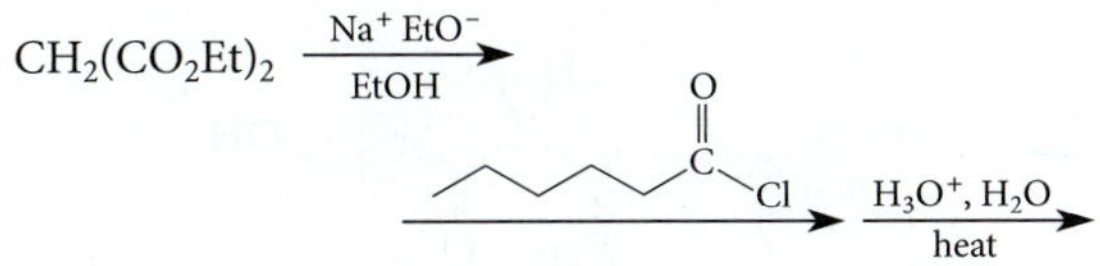

(c) 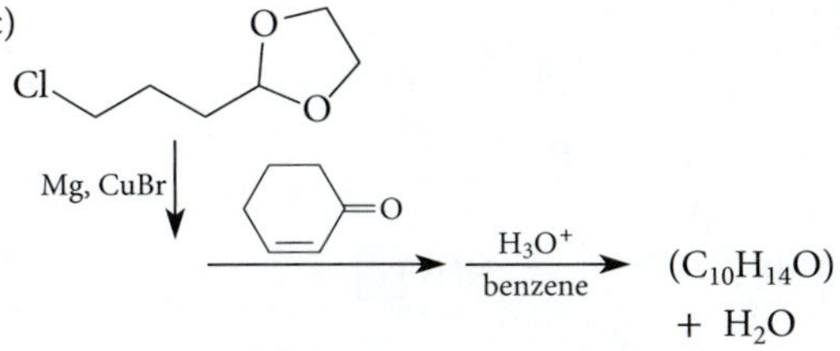

(*Hint:* Adding CuBr or CuCl to a Grignard reagent causes it to react like a lithium organocuprate.)

(d) Provide structures for the missing reactant *A* and the product *B*. [See the hint in part (c).]

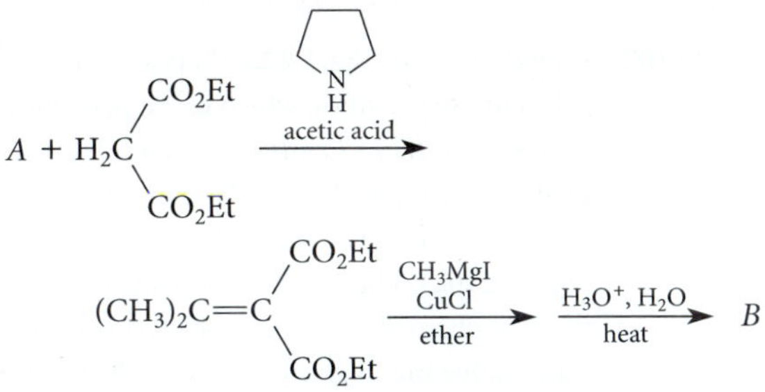

(e) 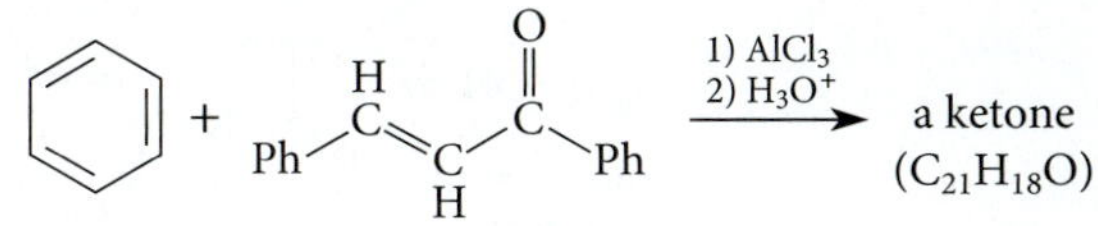

(f) 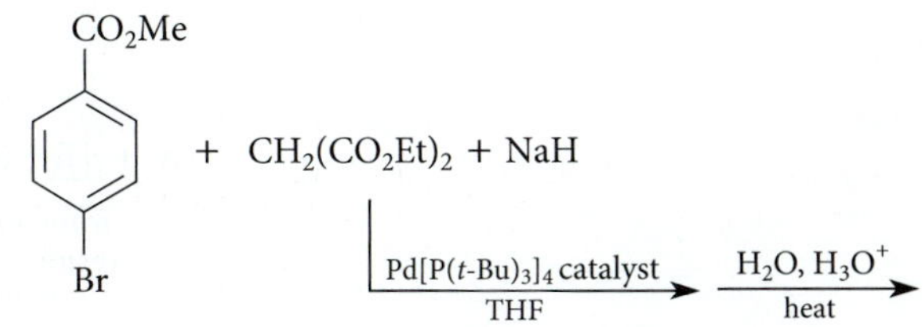

22.100 Treatment of (*S*)-(+)-5-methyl-2-cyclohexenone with lithium dimethylcuprate gives, after protonolysis, a good yield of a mixture containing mostly a dextrorotatory ketone *A* and a trace of an optically inactive isomer *B*. Treatment of *A* with zinc amalgam and HCl affords an optically active, dextrorotatory hydrocarbon *C*. Identify *A*, *B*, and *C*, and give the absolute stereochemical configurations of *A* and *C*.

22.101 The enzyme *dehydroquinate synthase* catalyzes reactions (1) and (3) in the sequence shown in **Fig. P22.101**, which is part of the pathway for the biosynthesis of aromatic amino acids such as tryptophan; reactions (2), (4), and (5) occur spontaneously.

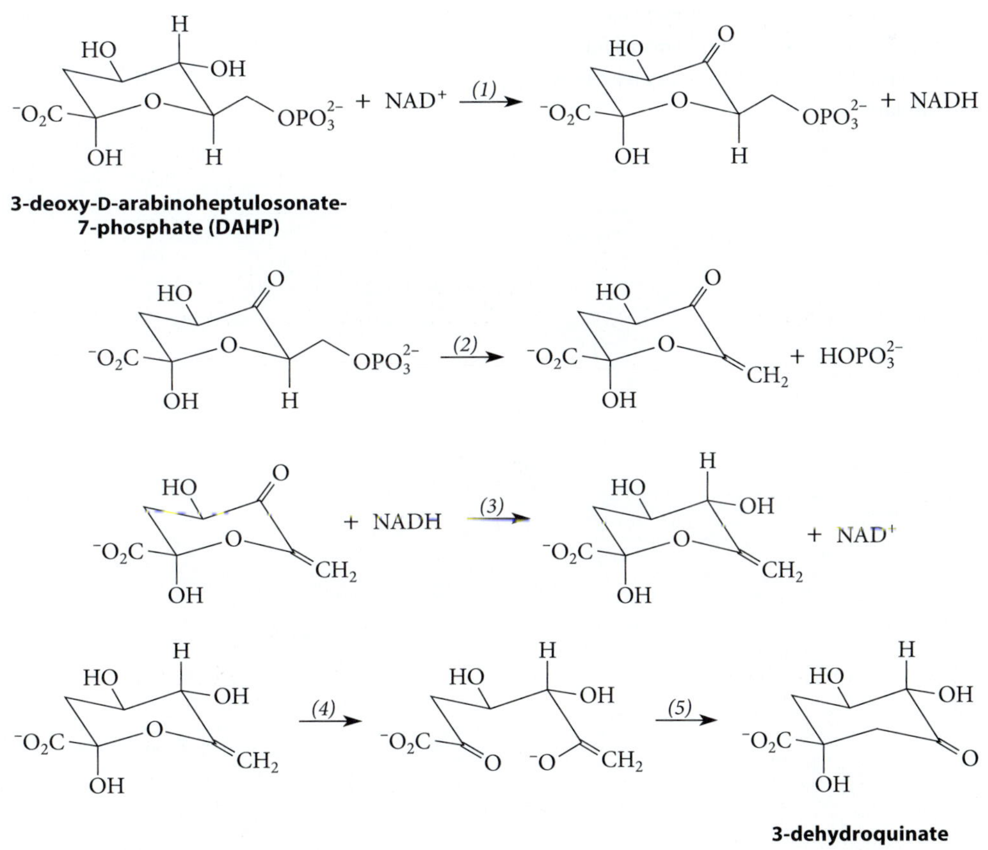

FIGURE P22.101

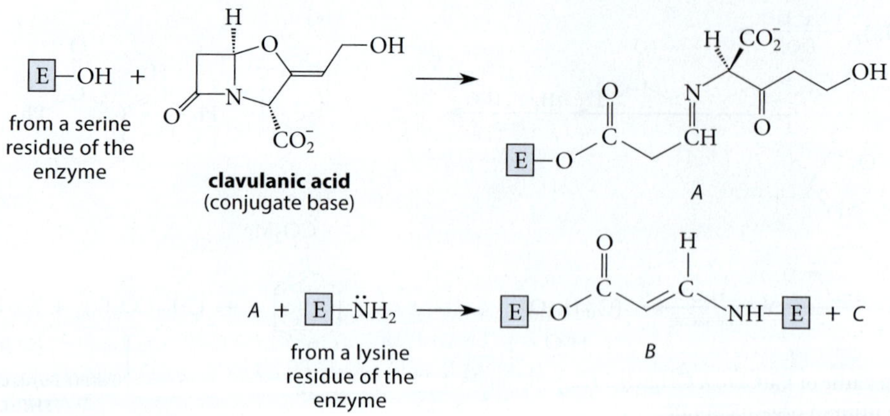

FIGURE P22.102

(a) Assuming that acids and bases are present as needed, propose a curved-arrow mechanism for reaction (2). To what laboratory reaction is this reaction analogous? [*Hint:* Phosphate is a leaving group in reaction (2).]

(b) The two enzyme-catalyzed steps (1) and (3) are, respectively, an oxidation and a reduction that consume and then regenerate an NAD^+ and result in no *net* change at the carbon involved. What is the chemical rationale for these two reactions? [*Hint:* Why is reaction (1) necessary for the success of reaction (2)?]

(c) Assuming that acids and bases are present as needed, propose curved-arrow mechanisms for reactions (4) and (5). To what laboratory reactions are these similar?

22.102 *Clavulanic acid* (**Fig. P22.102**) is an inhibitor of β-lactamase enzymes, which cause penicillin resistance. (See Sec. 21.8D.) Clavulanic acid inhibits β-lactamases by first reacting with the side-chain hydroxy group of an active-site serine residue (abbreviated as shown in Fig. P22.102). The resulting derivative *A* then reacts with the side-chain amino group of a lysine residue in the same active site to give the product *B*, which permanently blocks the active site, along with a by-product *C*, which is released into solution. Assuming acids and bases are present as needed, provide curved-arrow mechanisms for these reactions, and draw a structure of by-product *C*.

23 THE CHEMISTRY OF AMINES

Amines are organic derivatives of ammonia in which the ammonia hydrogens are replaced by alkyl or aryl groups. Amines are classified by the number of alkyl or aryl substituents (R groups) on the amine nitrogen. A **primary amine** has one alkyl substituent, a **secondary amine** has two, and a **tertiary amine** has three.

$\ddot{N}H_3$	$R-\ddot{N}H_2$	$R-\ddot{N}H-R$	$R-\underset{\cdot\cdot}{N}(R)-R$	
ammonia	primary amine	secondary amine	tertiary amine	**(23.1)**

Examples:

$CH_3CH_2-\ddot{N}H_2$	$CH_3CH_2-\ddot{N}H-CH_2CH_3$	$CH_3CH_2-\underset{\cdot\cdot}{N}(CH_2CH_3)-CH_2CH_3$	
ethylamine (a primary amine)	**diethylamine** (a secondary amine)	**triethylamine** (a tertiary amine)	**(23.2)**

This classification is like that of amides (see Sec. 21.1E). That is, amines and amides are classified according to the number of alkyl or aryl groups *on nitrogen*. Alcohols, on the other hand, are classified according to the number of alkyl or aryl groups on the α-carbon (Sec. 8.1).

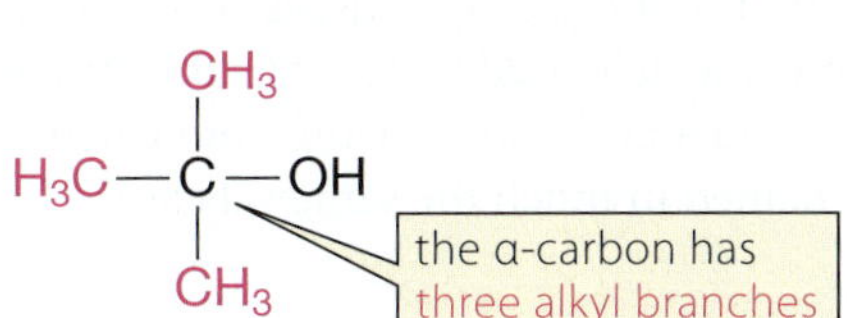

a tertiary alcohol

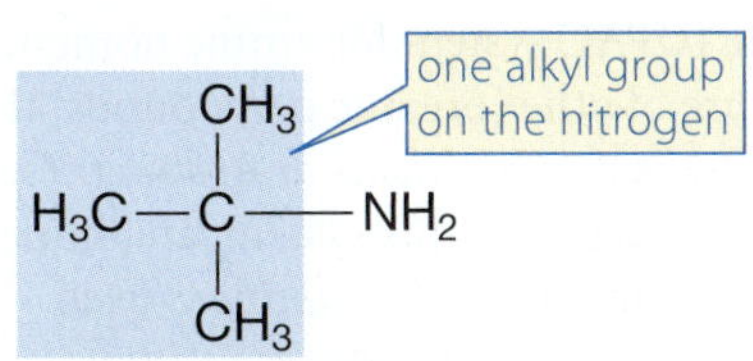

a primary amine (even though the α-carbon is tertiary)

The reason for the difference in the classification of amines and alcohols is that the substitution of the OH hydrogen of an alcohol by an alkyl or aryl group results in a different functional group—an ether. In contrast, substitutions of the hydrogens of amines give other amines, and a way is needed to classify amines by their degree of nitrogen substitution.

Besides amines, this chapter also considers briefly some other nitrogen-containing compounds that are formed from, or converted into, amines—namely, quaternary ammonium salts (Sec. 23.6), azobenzenes (Sec. 23.10B), diazonium salts (Sec. 23.10A), acyl azides (Sec. 23.11D), and nitro compounds (Sec. 23.11B).

23.1 NOMENCLATURE OF AMINES

A. Common Nomenclature

In common nomenclature, an amine is named by appending the suffix *amine* to the name of the alkyl group; the name of the amine is written as one word.

$CH_3CH_2NH_2$ — **ethylamine**

$(CH_3)_3N$ — **trimethylamine**

When two or more alkyl groups in a secondary or tertiary amine are different, the compound is named as an *N*-substituted derivative of the larger group.

$(CH_3)_2N—CH_2CH_2CH_2CH_3$

***N,N*-dimethylbutylamine**

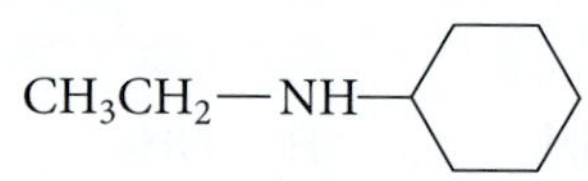

***N*-ethylcyclohexylamine**

This type of notation is required to indicate that the substituents are on the amine nitrogen and not on an alkyl group carbon.

Aromatic amines are named as derivatives of aniline.

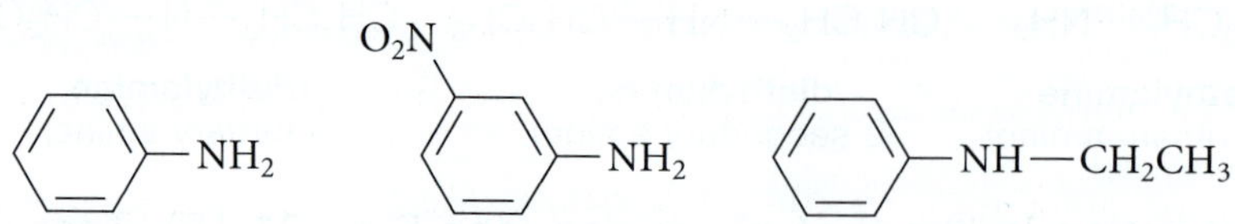

aniline **3-nitroaniline (*m*-nitroaniline)** ***N*-ethylaniline**

B. Substitutive Nomenclature

Because the IUPAC system for amine nomenclature is not logically consistent with IUPAC nomenclature of other organic compounds, the most widely used system of substitutive amine nomenclature is that of *Chemical Abstracts* (see the sidebar, "Nomenclature and Chemical Indexing," Sec. 2.4F). In this system, an amine is named in much the same way as the analogous alcohol, except that the suffix *amine* is used.

$CH_3CH(OH)CH_2CH_2CH_3$

2-pentanol

$CH_3CH(NH_2)CH_2CH_2CH_3$

2-pentanamine

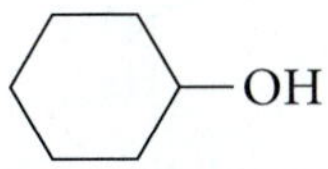

cyclohexanol

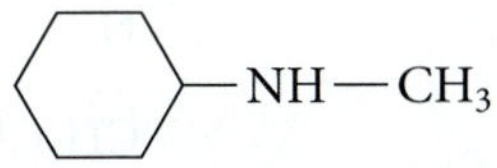

***N*-methylcyclohexanamine**

In diamine nomenclature, as in diol nomenclature, the final *e* of the hydrocarbon name is retained. In the second of the following examples, the prime is used to indicate that the ethyl and methyl groups are on different nitrogens.

$H_2N—CH_2CH_2CH_2CH_2CH_2—NH_2$

1,5-pentanediamine

$H_3C—NH—CH_2CH_2CH_2—NH—CH_2CH_3$

***N*-ethyl-*N*′-methyl-1,3-propanediamine**

The priority of citation of amine groups as principal groups is just below that of alcohols:

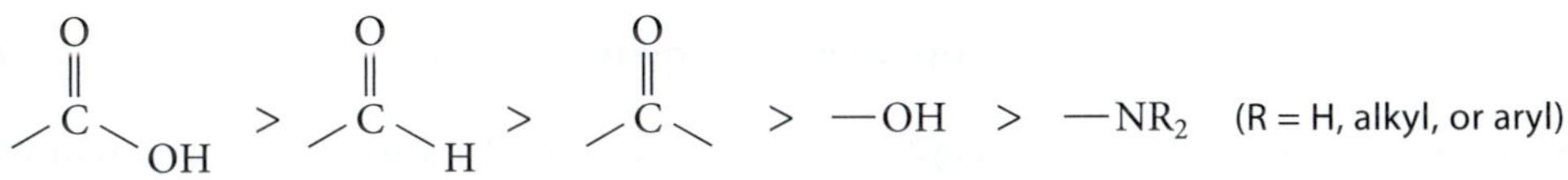

(carboxylic acid and derivatives) > (aldehyde) > (ketone) > (alcohol) > (amine) (R = H, alkyl, or aryl)

(23.3)

(A complete list of group priorities is given in Appendix I.)

When cited as a substituent, the —NH_2 group is called the **amino** group.

$H_2N—CH_2CH_2—OH$

2-aminoethanol
(OH has priority)

$H_2N—\overset{1}{C}H_2—\overset{2}{C}H=\overset{3}{C}H_2$

2-propen-1-amine
(NH_2 has priority)

$CH_3CH_2N(CH_3)CH_2CH_2CH(CH_3)CH_2CH_2Cl$ [the two CH_3 groups are drawn on vertical bonds: one below the N, one above the third CH]

5-chloro-*N*-ethyl-*N*,3-dimethyl-1-pentanamine

$(CH_3)_2N—CH_2CH_2CH_2—OH$

3-(dimethylamino)-1-propanol

An *N* designation in the last example is unnecessary because the position of the methyl groups is clear from the parentheses.

Although *Chemical Abstracts* calls aniline *benzenamine*, the usual practice is to use the common name *aniline* in substitutive nomenclature.

The nomenclature of *heterocyclic compounds* was introduced in Sec. 8.2C in the discussion of ether nomenclature. Many important nitrogen-containing heterocyclic compounds are known by specific names that should be learned. Some important saturated heterocyclic amines are the following:

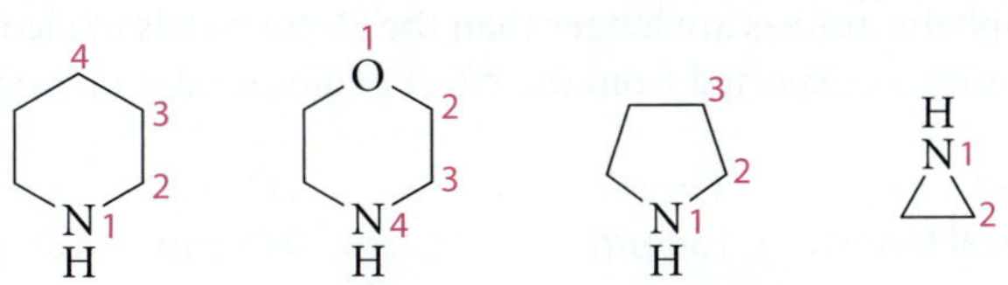

piperidine **morpholine** **pyrrolidine** **aziridine**

(23.4)

As in the oxygen heterocyclics, numbering generally begins with the heteroatom. The following are examples of substituted derivatives:

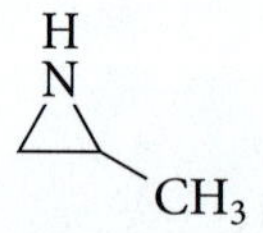

2-methylaziridine

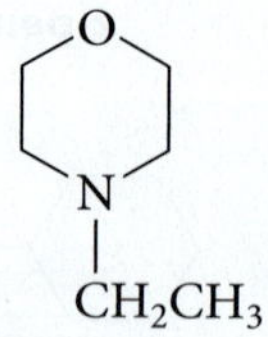

***N*-ethylmorpholine (4-ethylmorpholine)**

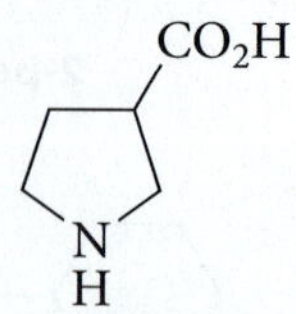

3-pyrrolidinecarboxylic acid

To a useful approximation, much of the chemistry of the saturated heterocyclic amines parallels the chemistry of the corresponding acyclic amines. There are also a number of unsaturated aromatic heterocyclic amines. Among these are pyridine and pyrrole, which were considered briefly in Sec. 15.7D.

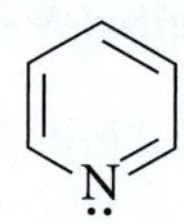

pyridine

pyrrole (23.5)

The chemistry of the aromatic heterocycles, which is quite different from that of their saturated counterparts, is considered in Chapter 26.

Focused Problems

23.1 Draw the structure of each of the following compounds.

(a) *N*-Isopropylaniline

(b) 3-Methoxypiperidine

(c) Ethyl-2-(diethylamino)pentanoate

(d) *N*,*N*-Diethyl-3-heptanamine

23.2 Give an acceptable name for each of the following compounds.

(a)

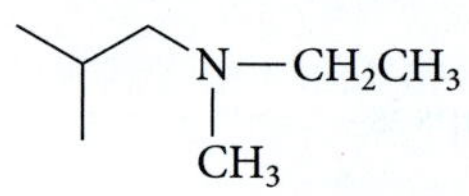

(b)

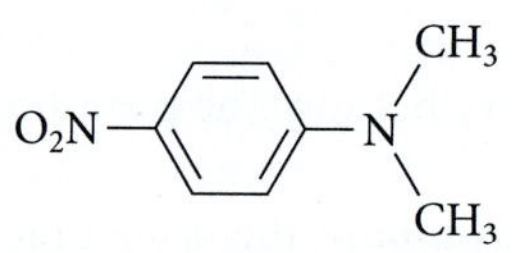

23.2 STRUCTURE OF AMINES

The C—N bonds of aliphatic amines are longer than the C—O bonds of alcohols, but shorter than the C—C bonds of alkanes, as expected from the effect of atomic size on bond length (Sec. 1.6B).

bond:	C—C	C—N	C—O	C—F
typical length:	154 pm	147 pm	143 pm	139 pm

(23.6)

Aliphatic amines have a trigonal pyramidal shape. Most amines undergo rapid *inversion* at nitrogen, which occurs through a planar transition state and converts a chiral amine into its mirror image. (See Fig. 6.9, Sec. 6.9B.)

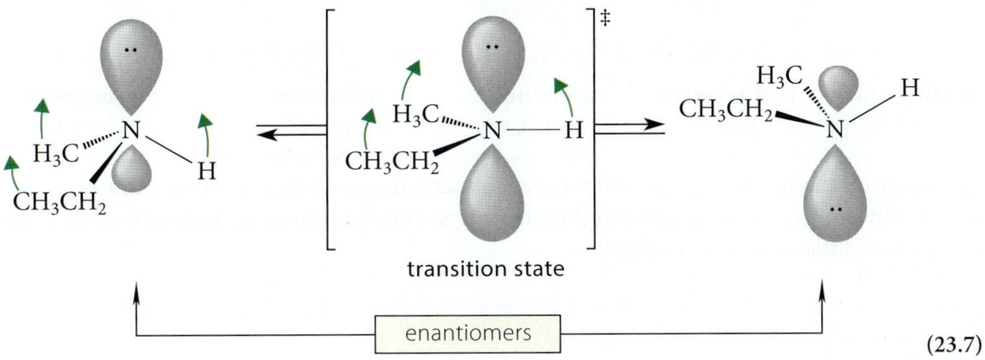

(23.7)

Because of this rapid inversion, amines in which the only asymmetric atom is the amine nitrogen cannot be resolved into enantiomers.

The C—N bond in aniline, which is 140 pm long, is shorter than the C—N bond in aliphatic amines. This reflects both the sp^2 hybridization of the adjacent carbon and the overlap of the nonbonding electrons on nitrogen with the π-electron system of the ring. This overlap, shown by the following resonance structures, gives some double-bond character to the C—N bond.

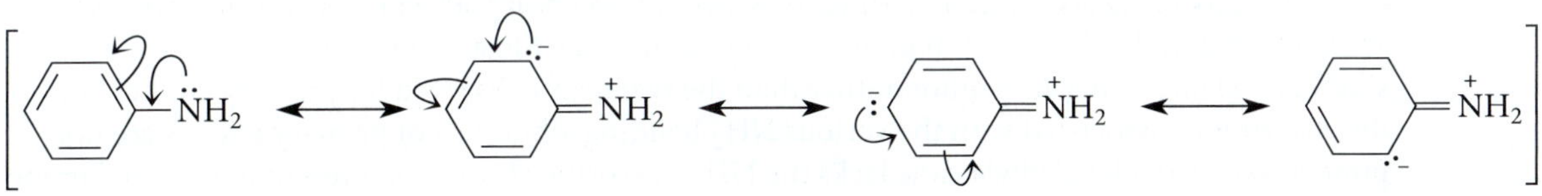

(23.8)

Focused Problem

23.3 Within each set, arrange the compounds in order of increasing C—N bond length. Explain your answers.

(a) *p*-Nitroaniline, aniline, cyclohexylamine

(b) $HN{=}CH_2$ (*A*) $H_3C{-}NH_2$ (*B*) $H_2C{=}CH{-}NH_2$ (*C*)

23.3 PHYSICAL PROPERTIES OF AMINES

Most amines are somewhat polar liquids with unpleasant odors that range from fishy to putrid. Primary and secondary amines, which can both donate and accept hydrogen bonds, have higher boiling points than isomeric tertiary amines, which can accept but cannot donate hydrogen bonds.

	$CH_3CH_2CH(CH_3)CH_3$	$CH_3CH_2N(CH_3)_2$	$(CH_3CH_2)_2NH$	$CH_3CH_2CH_2CH_2NH_2$
	isopentane	***N,N*-dimethylethylamine**	**diethylamine**	**butylamine**
boiling point:	27.8 °C	37.5 °C	56.3 °C	77.8 °C
dipole moment:	0 D	0.6 D	1.2–1.3 D	1.4 D

Increasing hydrogen bonding and polarity →

Because primary and secondary amines can both donate and accept hydrogen bonds, they have higher boiling points than ethers. Alcohols, on the other hand, are better hydrogen-bond donors than amines because alcohols are more acidic than amines. (See Sec. 23.5D.) Therefore, alcohols have higher boiling points than amines.

	$(CH_3CH_2)_2NH$	$(CH_3CH_2)_2O$	$(CH_3CH_2)_2CH_2$	$CH_3CH_2CH_2CH_2OH$	$CH_3CH_2CH_2CH_2NH_2$
	diethylamine	**diethyl ether**	**pentane**	**1-butanol**	**1-butanamine**
boiling point:	56.3 °C	37.5 °C	36.1 °C	117.3 °C	77.8 °C

The water miscibility of most primary and secondary amines with four or fewer carbons, as well as trimethylamine, is consistent with their hydrogen-bonding abilities. Amines with large carbon groups have little or no water solubility.

23.4 SPECTROSCOPY OF AMINES

A. IR Spectroscopy

The most important absorptions in the infrared spectra of primary amines are the N—H stretching absorptions, which usually occur as two peaks at 3200–3375 cm^{-1} corresponding to the NH_2 symmetric and asymmetric stretching vibrations. Also characteristic of primary amines is an NH_2 scissoring absorption (see Fig. 13.8, Sec. 13.3A) near 1600 cm^{-1}. These absorptions are illustrated in the IR spectrum of butylamine (**Fig. 23.1**). Most secondary amines show a single N—H stretching absorption rather than the two peaks observed for primary amines, and the absorptions associated with the various NH_2 bending vibrations of primary amines are not present. For example, diethylamine lacks the NH_2 scissoring absorption present in the butylamine spectrum. Tertiary amines show no absorptions associated with N—H vibrations. The C—N stretching absorptions of amines, which occur in the same general part of the spectrum as C—O stretching absorptions (1050–1225 cm^{-1}), are not very useful.

B. NMR Spectroscopy

The characteristic resonances in the proton NMR spectra of amines are those of the protons adjacent to the nitrogen (the α-protons) and the N—H protons. In alkylamines, the α-protons are observed in the δ 2.5–3.0 region of the spectrum. In aromatic amines, the α-protons of *N*-alkyl

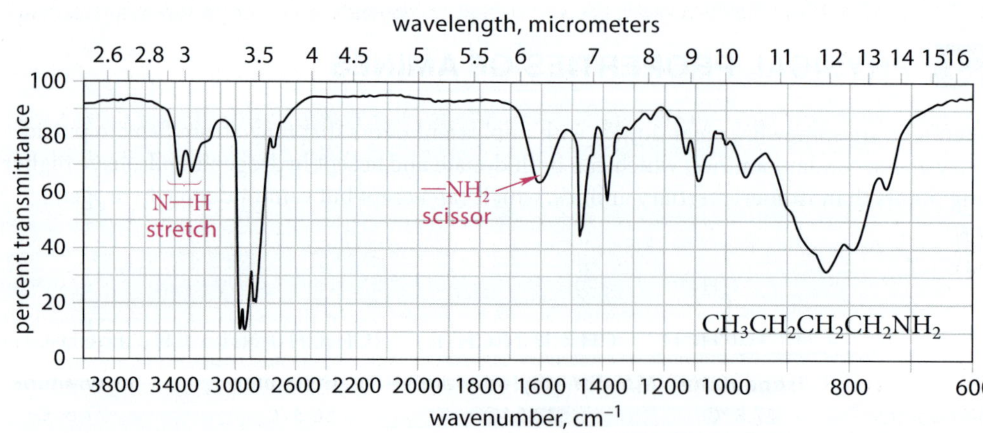

FIGURE 23.1 The IR spectrum of butylamine, a typical primary amine. The N—H stretching and NH_2 bending (scissoring) absorptions are the most important amine absorptions.

groups have somewhat greater chemical shifts, near δ 3. (Why? See Sec 14.3C.) The following chemical shifts are typical:

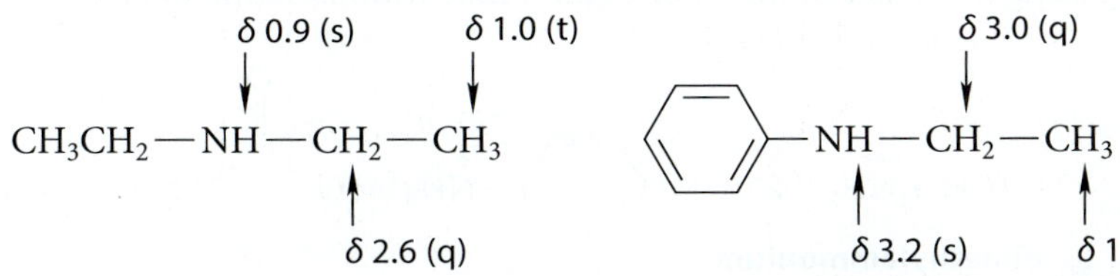

(23.9)

The chemical shift of the N—H proton, like that of the O—H proton in an alcohol, depends on the concentration of the amine and on other conditions of the NMR experiment. In alkylamines, this resonance occurs at rather small chemical shift—typically around δ 1. In aromatic amines, this resonance is at greater chemical shift, as in the second of the examples in Display 23.9.

Like the OH protons of alcohols, phenols, and carboxylic acids, the NH protons of amines under most conditions undergo rapid exchange (Secs. 14.6 and 14.7E). For this reason, splitting between the amine N—H and adjacent C—H groups is usually not observed. Thus, in the NMR spectrum of diethylamine, the N—H resonance is a singlet rather than the triplet expected from splitting by the adjacent —CH_2— protons. In some amine samples, the N—H resonance is broadened and, like the O—H proton of alcohols, it can be obliterated from the spectrum by exchange with D_2O (the "D_2O shake"; Eq. 14.23, Sec. 14.6).

The characteristic ^{13}C NMR absorptions of amines are those of the α-carbons—the carbons attached directly to the nitrogen. These absorptions occur in the δ 30–50 chemical-shift range. As expected from the relative electronegativities of oxygen and nitrogen, these shifts are somewhat less than the α-carbon shifts of ethers.

Focused Problems

23.4 Identify the compound that has an M + 1 ion at $m/z = 136$ in its CI mass spectrum, an IR absorption at 3279 cm^{-1}, and the following NMR spectrum: δ 0.91 (1*H*, s), δ 1.07 (3*H*, t, $J = 7$ Hz), δ 2.60 (2*H*, q, $J = 7$ Hz), δ 3.70 (2*H*, s), δ 7.18 (5*H*, apparent s).

23.5 A compound has IR absorptions at 3400–3500 cm^{-1} and the following NMR spectrum: δ 2.07 (6*H*, s), δ 2.16 (3*H*, s), δ 3.19 (broad, exchanges with D_2O), δ 6.63 (2*H*, s). To which one of the following compounds do these spectra belong? Explain.

(1) 2,4-Dimethylbenzylamine
(2) 2,4,6-Trimethylaniline
(3) *N,N*-Dimethyl-*p*-methylaniline
(4) 3,5-Dimethyl-*N*-methylaniline
(5) 4-Ethyl-2,6-dimethylaniline

23.6 Explain how you could distinguish between the two compounds in each of the following sets using *only* ^{13}C NMR spectroscopy.

(a) 2,2-Dimethyl-1-propanamine and 2-methyl-2-butanamine
(b) *trans*-1,2-Cyclohexanediamine and *trans*-1,4-cyclohexanediamine

23.5 BASICITY AND ACIDITY OF AMINES

A. Basicity of Amines

Amines, like ammonia, are strong enough bases that they are completely protonated in dilute acid solutions.

$$H_3C—\ddot{N}H_2 + H—Cl \rightleftharpoons H_3C—\overset{+}{N}H_3\ Cl^- + H_2O \tag{23.10}$$

methylamine **methylammonium chloride**

The salts of protonated amines are called **ammonium salts**. The ammonium salts of simple alkylamines are named as substituted derivatives of the ammonium ion. Other ammonium salts are named by replacing the final *e* in the name of the amine with the suffix *ium*.

$(CH_3)_2\overset{+}{N}H_2\ Cl^-$

dimethylammonium chloride

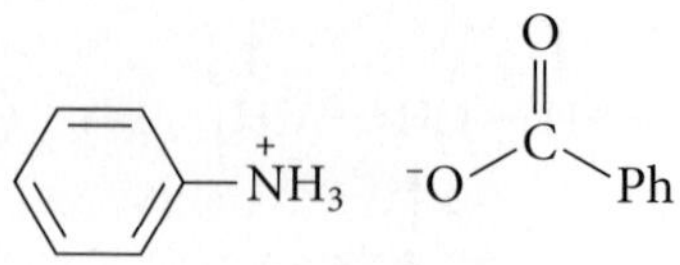

anilinium benzoate
(aniline + *ium*)

Always remember that ammonium salts are fully ionic compounds. Although ammonium chloride is often written as NH_4Cl, the structure is more properly represented as $^+NH_4\ Cl^-$. Although the N—H bonds are covalent, there is no covalent bond between the nitrogen and the chlorine. (A covalent bond would violate the octet rule.)

Many pharmaceuticals contain one or more amine functional groups and are marketed as salts. The basic amine nitrogen is protonated and positively charged, and the drug is accompanied by the pharmacologically inert anion. As noted in Sec. 8.6F, salts are much more soluble in water than many neutral organic compounds. The choice of counter ion is important because its identity affects the desired physical properties of the compound, including its melting point and rate of solubilization. In addition, protonating the nitrogen makes it resistant to oxidation, which can lead to degradation of the compound and a decrease in its efficacy.

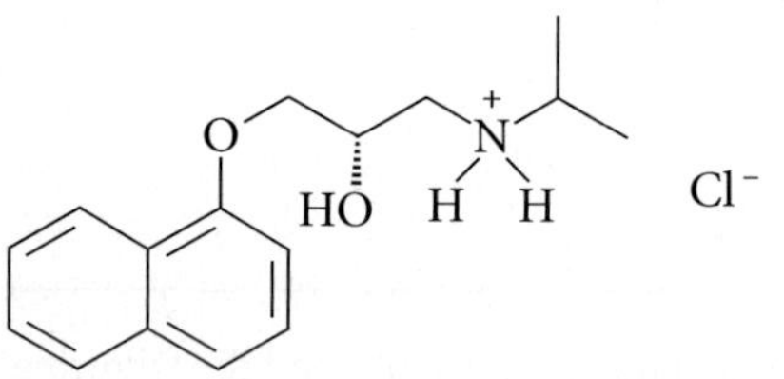

(*S*)-propranolol hydrochloride (Inderal)
(used to treat hypertension and anxiety)

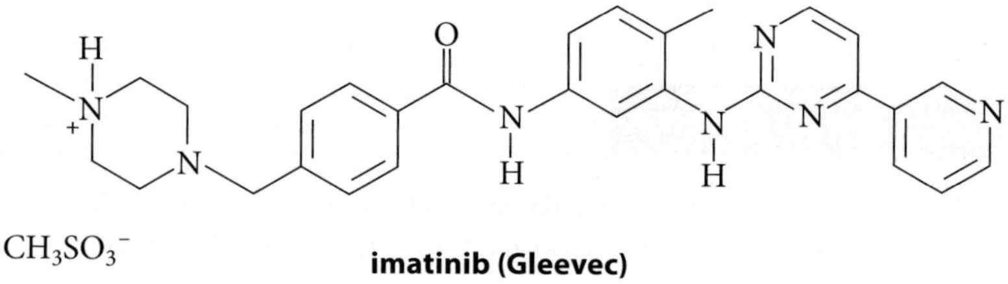

imatinib (Gleevec)
(leukemia treatment)

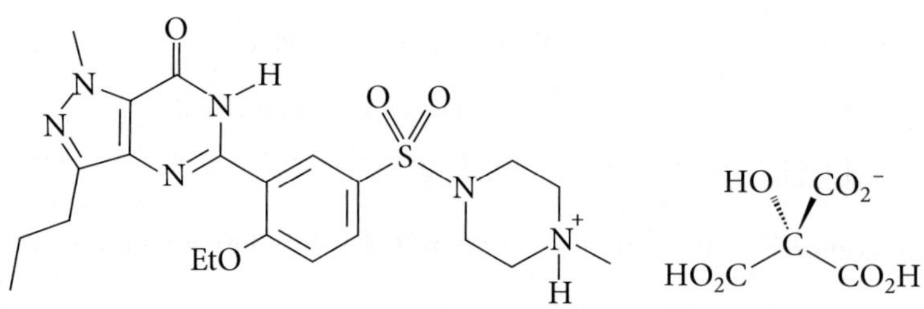

sildenafil citrate (Viagra)
(a phosphodiesterase type 5 inhibitor)

Recall that the basicity of any compound, including an amine, is expressed in terms of the pK_a of its conjugate acid (Sec. 3.6B). The higher the pK_a of an ammonium ion, the more basic is its conjugate-base amine.

B. Substituent Effects on Amine Basicity

The pK_a values for the conjugate acids of some representative amines are given in **Table 23.1**. As this table shows, the exact basicity of an amine depends on its structure. The following four factors influence the basicity of amines (these are the same effects that influence the acid–base properties of other compounds):

TABLE 23.1 Basicities of Some Amines
(Each pK_a value is for the dissociation of the corresponding conjugate-acid ammonium ion.)

Amine	pK_a	Amine	pK_a	Amine	pK_a
CH_3NH_2	10.62	$(CH_3)_2NH$	10.64	$(CH_3)_3N$	9.76
$CH_3CH_2NH_2$	10.63	$(CH_3CH_2)_2NH$	10.98	$(CH_3CH_2)_3N$	10.65
$PhCH_2NH_2$	9.34				
$PhNH_2$	4.62	$PhNHCH_3$	4.85	$PhN(CH_3)_2$	5.06
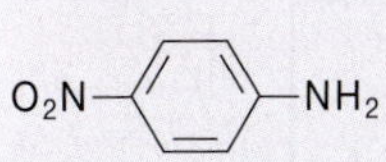	≈1.0	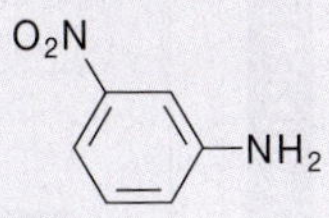	2.45	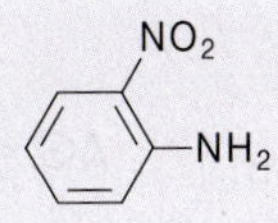	–0.26
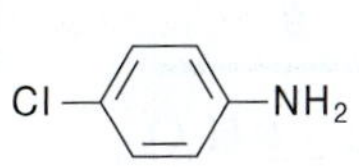	3.81	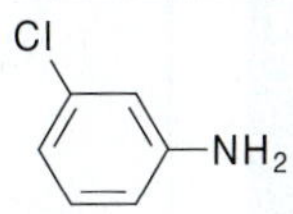	3.32	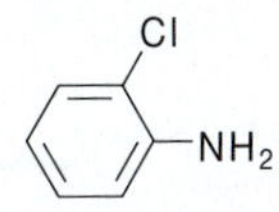	2.62
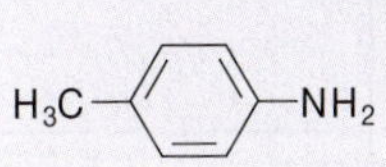	5.07	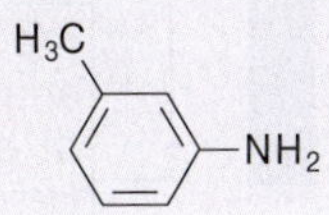	4.67	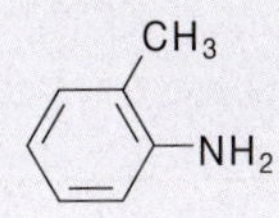	4.38

1. The effect of alkyl substitution
2. The polar effect
3. The resonance effect
4. The effect of charge

Recall that the pK_a of an ammonium ion, like that of any other acid, is directly related to the standard free-energy difference ΔG_a° between it and its conjugate base by Eq. 23.11 (this is a repeat of Eq. 3.61b, Sec. 3.6D).

$$\Delta G_a^\circ = 2.3RT(pK_a) \tag{23.11}$$

The effect of a substituent group on pK_a can be analyzed in terms of how it affects the energy of either an ammonium ion or its conjugate-base amine, as shown in **Fig. 23.2**. For example, if a substituent stabilizes an amine more than it stabilizes the conjugate-acid ammonium ion (Fig. 23.2a), the standard free energy of the amine is lowered, ΔG_a° is decreased, and the pK_a of the ammonium ion is reduced—that is, the amine is less basic than the amine without the substituent. If a substituent stabilizes the ammonium ion more than its conjugate-base amine (Fig. 23.2b), the opposite effect is observed—namely, the pK_a and the amine basicity are both increased.

Consider first the effect of alkyl substitution. Most common alkylamines are somewhat more basic than ammonia in aqueous solution:

	$\overset{+}{N}H_4$	$Me\overset{+}{N}H_3$	$Me_2\overset{+}{N}H_2$	$Me_3\overset{+}{N}H$	
pK_a in aqueous solution:	9.21	10.62	10.64	9.76	(23.12)

However, the increase in basicity that results from substitution of one hydrogen of ammonia by a methyl group is reversed as the number of alkyl substituents is increased to three. How can we explain this "turnaround" in amine basicity?

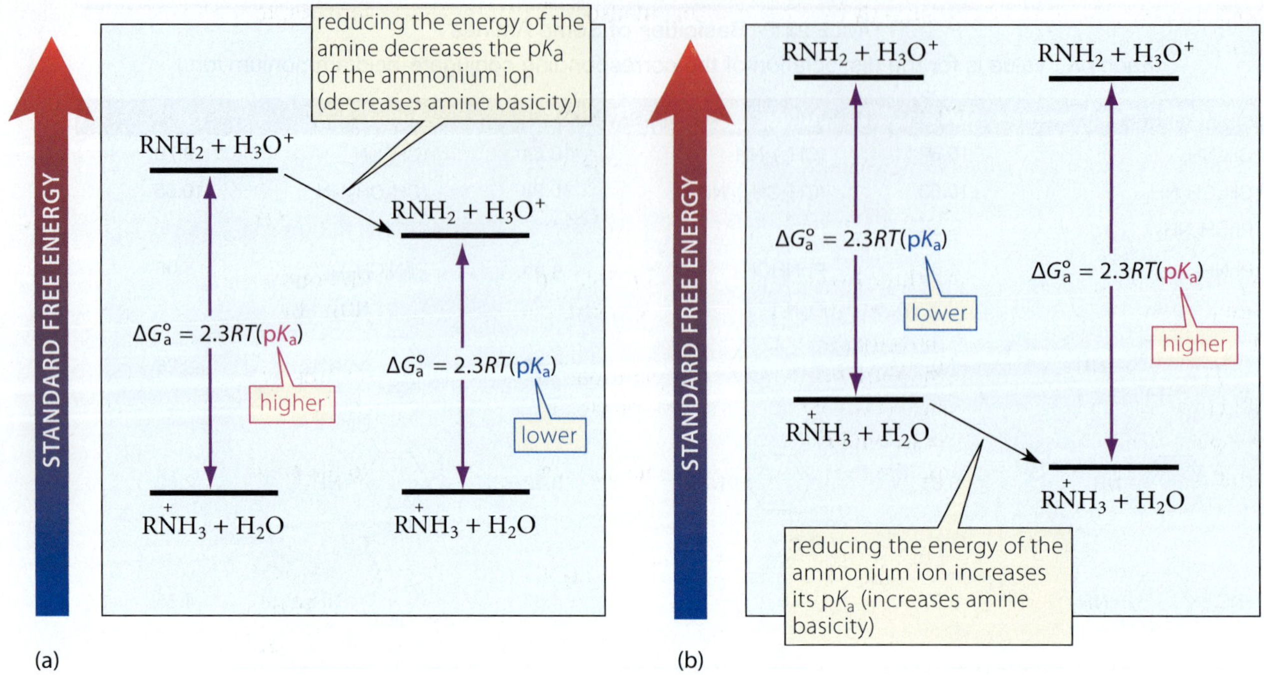

FIGURE 23.2 Effect of the relative free energies of ammonium ions and amines on the pK_a values of the ammonium ions. (a) Reducing the energy of an amine decreases the pK_a of its conjugate-acid ammonium ion and thus reduces the basicity of the amine. (b) Reducing the energy of an ammonium ion increases its pK_a and thus increases the basicity of its conjugate-base amine.

Two opposing factors are actually at work here. The first is the tendency of alkyl groups to stabilize charge through a *polarization* effect. The electron clouds of the alkyl groups distort so as to create a net attraction between them and the positive charge of the ammonium ion:

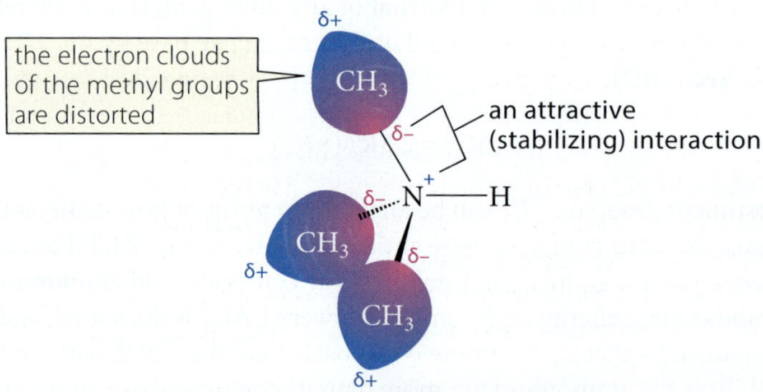

(23.13)

Because the ammonium ion is stabilized by this effect, its pK_a is increased (see Fig. 23.2b). This effect is evident in the *gas-phase* basicities of amines. In the gas phase, the acidity of ammonium ions decreases regularly with increasing alkyl substitution:

$$\text{Gas-phase acidity:}\quad \overset{+}{N}H_4 > Me\overset{+}{N}H_3 > Me_2\overset{+}{N}H_2 > Me_3\overset{+}{N}H \qquad (23.14)$$

FURTHER EXPLORATION 23.1 Alkyl Group Polarization in Ionization Reactions

(See Further Exploration 23.1 for a more extensive discussion.)

The second factor involved in the effect of alkyl substitution on amine basicity must be a *solvent effect*, because the basicity order of amines in the gas phase (Eq. 23.14) is different from that in aqueous solution (Display 23.12). In other words, the solvent water must play an important role in the solution basicity of amines. This solvent effect occurs because ammonium ions in solution are stabilized not only by alkyl groups, but also by hydrogen-bond donation to the solvent:

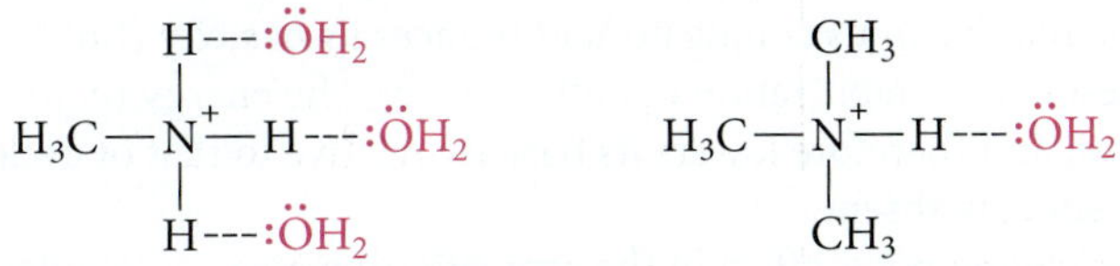

(23.15)

Primary ammonium salts have three hydrogens that can be donated to form hydrogen bonds, but a tertiary ammonium salt has only one. Thus, primary ammonium ions are stabilized by hydrogen bonding more than tertiary ones.

The pK_a values of alkylammonium salts reflect the operation of both hydrogen bonding and alkyl-group polarization. Because these effects work in opposite directions, the basicity in Display 23.12 maximizes at the secondary amine.

Ammonium-ion pK_a values, like the pK_a values of other acids, are also sensitive to the *polar effects* of substituents.

	$Et_2\overset{+}{N}HCH_2C{\equiv}N$	$Et_2\overset{+}{N}HCH_2CH_2C{\equiv}N$	$Et_2\overset{+}{N}H(CH_2)_4C{\equiv}N$	$Et_3\overset{+}{N}H$
pK_a:	4.55	7.65	10.08	10.65

(23.16)

An electronegative (electron-withdrawing) group such as halogen or cyano destabilizes an ammonium ion because of a repulsive electrostatic interaction between the positive charge on the ammonium ion and the positive end of the substituent bond dipole.

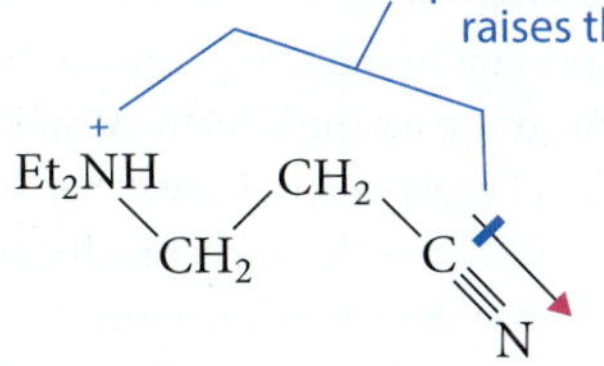

(23.17)

The polar effects of substituent groups operate largely on the *conjugate acid* of the amine—the alkylammonium ion—because this cation is the charged species in the acid–base equilibrium. (As explained in Sec. 3.6E, polar effects on stability are greatest on the charged species in a chemical equilibrium.) In the case of a carboxylic acid, alcohol, or phenol, the *conjugate-base anion* is the charged species; consequently, the polar effects of substituents are reflected primarily in their effects on the stabilities of these anions.

The data in Display 23.16 show that the base-weakening effect of electron-withdrawing substituents, like all polar effects, decreases significantly with distance between the substituent and the charged atom.

Resonance effects on amine basicity are illustrated by the difference between the pK_a values of the conjugate acids of aniline and cyclohexylamine, two primary amines of almost the same shape and molecular mass.

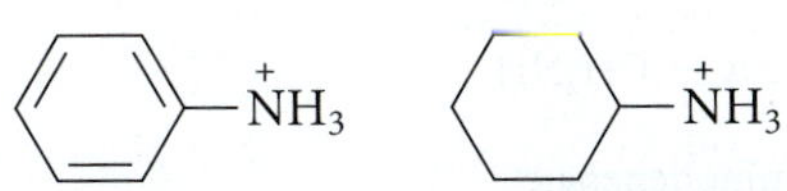

	conjugate acid of aniline	**conjugate acid of cyclohexylamine**
pK_a:	4.62	10.64

(23.18)

(Notice that the pK_a values of the substituted anilinium ions in Table 23.1 are considerably lower than the pK_a values of the alkylammonium ions.) Aniline is stabilized by a resonance interaction of the nonbonding electron pair on nitrogen with the aromatic ring. (This resonance interaction is shown in Eq. 23.8.) When aniline is protonated, this resonance stabilization is no longer present, because the nonbonding pair is bound to a proton and is "out of circulation." The stabilization of aniline relative to its conjugate acid reduces its basicity (Eq. 23.11 and Fig. 23.2). In other words, the resonance stabilization of aniline lowers the energy required for its formation from its conjugate acid, and therefore lowers its basicity relative to that of cyclohexylamine, in which the resonance effect is absent.

The electron-withdrawing polar effect of the aromatic ring also contributes significantly to the reduced basicity of aromatic amines. Recall that a similar polar effect is responsible for the increased acidities of phenols relative to alcohols (Sec. 18.7A).

Study Problem 23.1

Arrange the following three amines in order of increasing basicity.

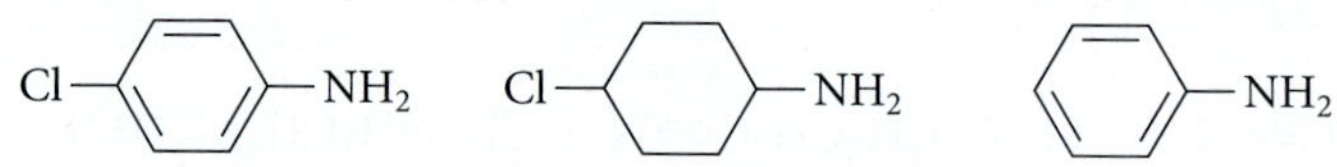

***p*-chloroaniline** **4-chlorocyclohexanamine** **aniline**

Solution Because the chloro substituent is an electron-withdrawing group, it reduces the basicity of an aniline. (The slight electron-donating resonance effect of a chloro substituent is much less important than its electron-withdrawing polar effect.) Hence, *p*-chloroaniline is less basic than aniline. Because of the resonance interaction of the amine nitrogen with the ring π-electron system, *p*-chloroaniline is also less basic than 4-chlorocyclohexanamine. But what about the relative basicity of aniline and 4-chlorocyclohexanamine? The resonance and polar effects of the aromatic ring and the polar effect of the chlorine are all base-weakening effects. The problem is to decide whether the effect of the aromatic ring or the effect of the chlorine is more important in reducing basicity. To make this decision, reason *by analogy.* Examine the effect of an electron-withdrawing group on the basicity of an amine in which the polar effect is the only effect that can operate. For example, the series in Display 23.16 shows that an electronegative group four carbons away from the amine nitrogen has a very modest effect. On the other hand, the comparison in Eq. 23.18 shows that the resonance and polar effects of an aromatic ring change the pK_a of an amine by about 6 units. Consequently, the base-weakening effect of "changing" a cyclohexane ring to a phenyl ring is much more important by many orders of magnitude than the base-weakening effect of "replacing" a hydrogen with a 4-chloro group in cyclohexanamine. Hence, the basicity order is

p-chloroaniline < aniline << 4-chlorocyclohexanamine

Focused Problems

23.7 Arrange the amines within each set in order of increasing basicity in aqueous solution, least basic first.

(a) Propylamine, ammonia, dipropylamine

(b) Methyl 3-aminopropanoate, *sec*-butylamine, $H_3\overset{+}{N}CH_2CH_2NH_2$

(c) Aniline, methyl *m*-aminobenzoate, methyl *p*-aminobenzoate

(d) Benzylamine, *p*-nitrobenzylamine, cyclohexylamine, aniline

23.8 Explain the basicity order of the following three amines: *p*-nitroaniline (*A*), *m*-nitroaniline (*B*), and aniline (*C*). The structures and pK_a data are shown in Table 23.1.

C. Separations Using Amine Basicity

Because ammonium salts are ionic compounds, many have appreciable water solubilities. Hence, when a water-insoluble amine is treated with dilute aqueous acid, such as 5% HCl solution, *the amine dissolves as its ammonium salt.* Upon treatment with base, the ammonium salt is converted back into the amine. These observations can be used to design separations of amines from other compounds, as Study Problem 23.2 illustrates.

Study Problem 23.2

A chemist has treated *p*-chloroaniline with acetic anhydride and wants to separate the amide product from any unreacted amine. Design a separation based on the basicities of the two compounds.

Solution If the mixture is treated with 5% aqueous HCl, the amine will form the hydrochloride salt and dissolve in the aqueous solution. The amide, however, is not basic enough to be protonated in 5% HCl and therefore does not dissolve.

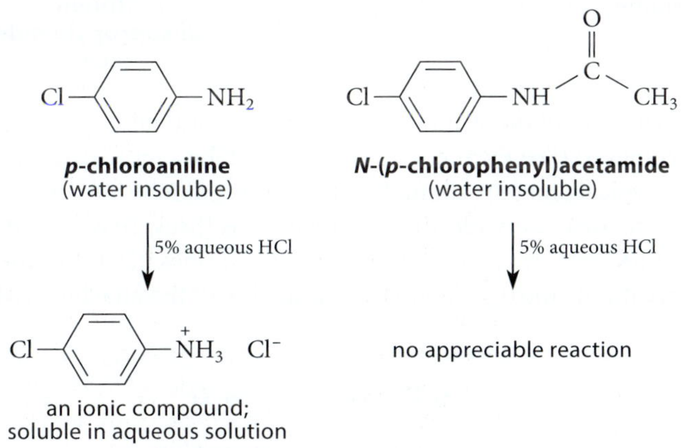

(23.19)

The water-insoluble amide can be filtered or extracted away from the aqueous solution of the ammonium salt. Then the aqueous solution of the ammonium salt can be treated with NaOH to liberate the free amine.

Amine basicities can play a key role in the design of enantiomeric resolutions. The enantiomeric resolution discussed in Sec. 6.10B should be reviewed with this idea in mind.

Focused Problems

23.9 Using their solubilities in acidic or basic solution, design a separation of *p*-chlorobenzoic acid, *p*-chloroaniline, and *p*-chlorotoluene from a mixture containing all three compounds.

23.10 Design an enantiomeric resolution of racemic 2-phenylpropanoic acid using a pure enantiomer of 1-phenylethanamine as resolving agent. (Assume that a solvent can be found in which the diasteromeric salts have different solubilities.) Discuss the importance of amine basicity to the success of this scheme. (See Sec. 6.10B.)

D. Acidity of Amines

Although amines are normally considered to be bases, primary and secondary amines are also weakly acidic. In other words, amines are *amphoteric compounds* (Sec. 3.4B). The conjugate base of an amine is called an **amide** (not to be confused with amide derivatives of carboxylic acids).

The amide conjugate base of ammonia itself is usually prepared by dissolving an alkali metal such as sodium in liquid ammonia in the presence of a trace of ferric ion. When sodium is used, the resulting base is called *sodium amide* (or *sodamide*).

$$2\,Na + 2\,\ddot{N}H_3 \xrightarrow{Fe^{3+}} 2\,(Na^+\ ^-\!:\!\ddot{N}H_2) + H_2 \qquad (23.20)$$

sodium amide (sodamide)

The conjugate bases of alkylamines are prepared by treating the amine with butyllithium in an ether solvent such as THF.

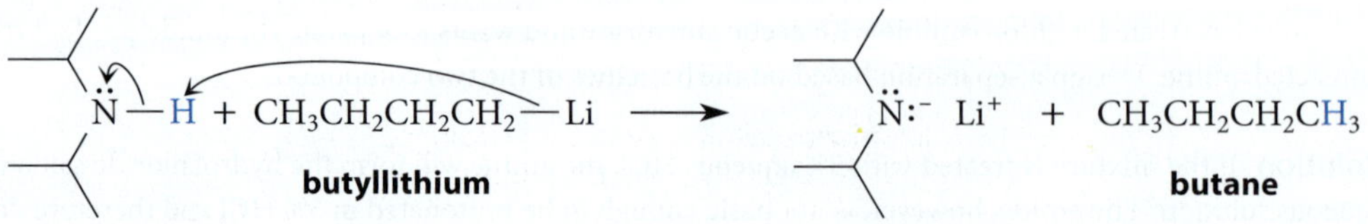

diisopropylamine **butyllithium** **lithium diisopropylamide (LDA)** **butane** (23.21)

Although the structures of these amide bases are conventionally drawn as in Eq. 23.21, they are actually more complex. (See Further Exploration 23.2 for a more extensive discussion.)

FURTHER EXPLORATION 23.2
Structures of Amide Bases

The pK_a of a typical amine is about 35. Thus, amide anions are *very* strong bases. This is why they can be used to form acetylide ions or enolate ions (Secs. 10.6A, 22.4C, and 22.8B). The difference in the pK_a values of ammonium ions and amines illustrates the effect of charge on pK_a. A positive charge on the nitrogen increases the acidity of the attached hydrogen by more than 20 pK_a units.

$$R_2\ddot{N}{-}H \qquad R_3\overset{+}{N}{-}H \qquad (23.22)$$

amines p$K_a \approx$ 32–35 ammonium ions p$K_a \approx$ 9–11

E. Summary of Acidity and Basicity

We have now surveyed the acidity and basicity of the most important organic functional groups. This information is summarized in the tables in Appendix VII. Although the values given in these tables are typical, remember that acidity and basicity are affected by alkyl substitution, polar effects, and resonance effects. The acid–base properties of organic compounds are important not only in predicting many of their chemical properties, but also in their industrial and medicinal applications.

23.6 QUATERNARY AMMONIUM AND PHOSPHONIUM SALTS

Closely related to ammonium salts are **quaternary ammonium salts**, in which all four hydrogens of $^+NH_4$ are replaced by alkyl or aryl groups. The following compounds are examples:

$(CH_3)_4N^+\ Cl^-$ **tetramethylammonium chloride**

$Ph\overset{+}{N}Et_3\ Br^-$ ***N,N,N*-triethylanilinium bromide**

$PhCH_2\overset{+}{N}(CH_3)_3\ ^-OH$ **benzyltrimethylammonium hydroxide (Triton B)**

$CH_3(CH_2)_{15}{-}\overset{+}{N}(CH_3)_2{-}CH_2{-}Ph\ \ Cl^-$

"benzalkonium chloride"
a cationic surfactant (Sec. 20.5)

Like the corresponding ammonium ions, quaternary ammonium salts are fully ionic compounds. Many quaternary ammonium salts containing large alkyl or aryl groups are soluble in nonaqueous solvents. Triton B, for example, is used as a source of hydroxide ion that is soluble in organic solvents. Benzalkonium chloride, a common antiseptic, acts as a surfactant in water (Sec. 20.5) and is also soluble in several organic solvents. The quaternary ammonium ions in such compounds can be conceptualized as "positive charges surrounded by greasy groups." (The term *greasy* is used to refer to materials soluble in hydrocarbon solvents.)

Quaternary phosphonium salts are the phosphorus analogs of quaternary ammonium salts.

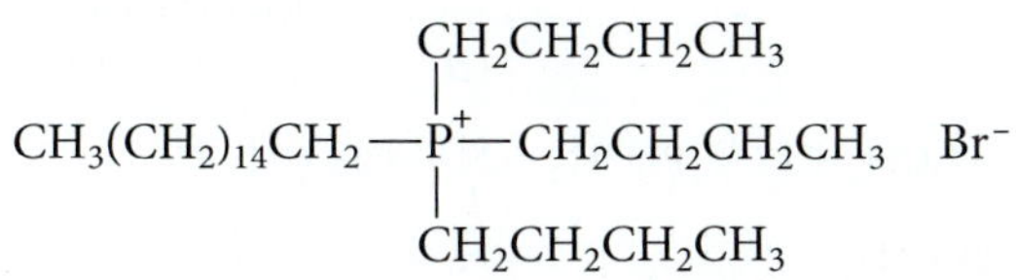

cetyltributylphosphonium bromide
(a quaternary phosphonium salt) (23.23)

Recall (Eq. 19.95a, Sec. 19.14) that phosphonium salts are the starting materials for the preparation of Wittig reagents.

An interesting and important use of both ammonium and phosphonium salts with large alkyl substituents is to catalyze organic reactions between an ionic reactant that is soluble in water and an organic reactant that is water insoluble. Such reactions typically involve two mutually insoluble layers: an aqueous phase and an organic phase. For example, if 1-bromooctane (alone or dissolved in a water-insoluble solvent) is treated with aqueous sodium cyanide, no reaction takes place, even with rapid stirring, because sodium cyanide, an ionic compound, is soluble only in the aqueous layer and cannot come into contact with the alkyl halide, which is insoluble in water. But when only a few percent of the quaternary phosphonium salt shown in Display 23.23 is added to the reaction mixture and the solution is stirred vigorously, the S_N2 reaction between the cyanide ion and the alkyl halide takes place readily.

$$\underset{\textbf{1-bromooctane}}{CH_3(CH_2)_6CH_2—Br} + Na^+ \ ^-CN \xrightarrow[H_2O]{\text{quaternary phosphonium salt}} \underset{\textbf{nonanenitrile}}{CH_3(CH_2)_6CH_2—CN} + Na^+ Br^- \quad (23.24)$$

When used in this way, the quaternary phosphonium salt is called a **phase-transfer catalyst**. The name describes the mechanism of action (**Fig. 23.3**). Because of its large alkyl groups, the quaternary phosphonium salt is soluble in the organic phase, which consists of a water-insoluble solvent and a mixture of the organic reactant and, ultimately, the product. The bromide ion in the organic phase readily exchanges with the cyanide ion from the aqueous phase, thus bringing the cyanide ion nucleophile into the organic phase, where it can react with the alkyl halide.

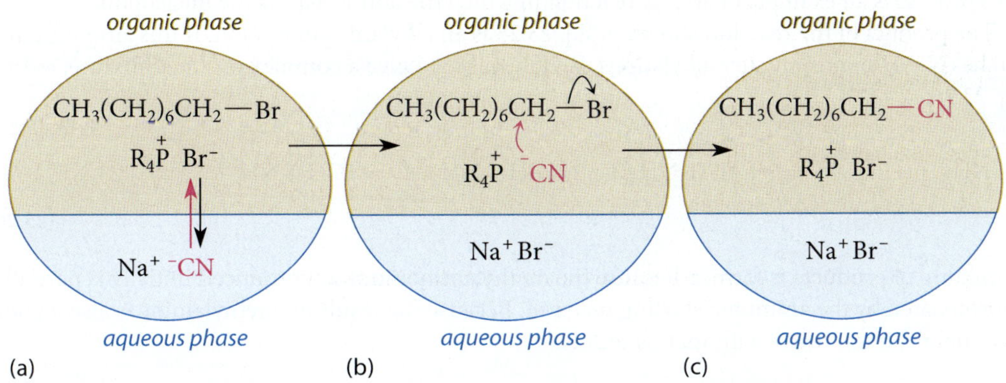

FIGURE 23.3 Phase-transfer catalysis by a quaternary phosphonium salt. (a) At the beginning of the reaction, the ionic nucleophile (red) is soluble in the aqueous layer. (b) Rapid equilibration of the nucleophile with the counterion of the quaternary phosphonium salt brings the nucleophile into the organic phase. (c) The nucleophile, now in the organic phase, can come into contact with the organic reactant, and a reaction occurs, forming the product and regenerating the phase-transfer catalyst.

Focused Problems

23.11 Draw the structure of each of the following quaternary ammonium salts.

(a) Tetraethylammonium fluoride

(b) Dibenzyldimethylammonium bromide

23.12 Explain why the quaternary ammonium salt *A* can be isolated in optically active form, but the trialkylammonium salt *B* cannot.

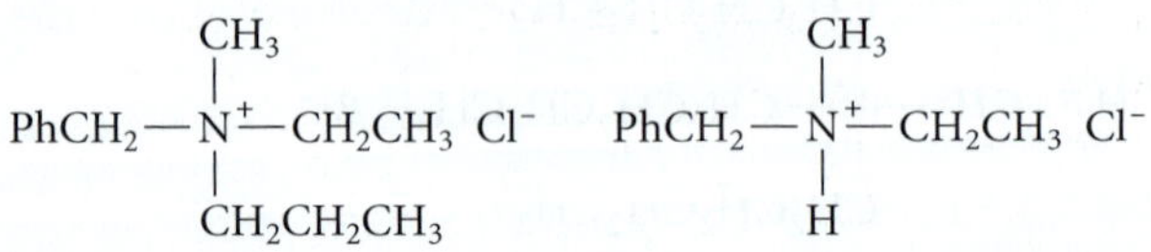

A *B*

23.7 ALKYLATION AND ACYLATION REACTIONS OF AMINES

Section 23.5 showed that amines are *Brønsted bases*. Amines, like many other Brønsted bases, are also *nucleophiles* (Lewis bases). Three reactions of nucleophiles are the following:

1. S_N2 reaction with alkyl halides, sulfonate esters, or epoxides (Secs. 9.1, 9.4, 11.4, and 12.5)
2. Addition to aldehydes, ketones, and α,β-unsaturated carbonyl compounds (Secs. 19.8, 19.12, and 22.9A)
3. Nucleophilic acyl substitution at the carbonyl carbons of carboxylic acid derivatives (Sec. 21.8)

This section covers or reviews reactions of amines that fit into each of these categories.

A. Direct Alkylation of Amines

Treatment of ammonia or an amine with an alkyl halide or other alkylating agent results in alkylation of the nitrogen.

$$R_3N\!: + H_3C{-}I \longrightarrow R_3\overset{+}{N}{-}CH_3\ I^- \qquad (23.25)$$

This process is an example of an S_N2 reaction in which the amine acts as the nucleophile.

The product of the reaction shown in Eq. 23.25 is an alkylammonium ion. If this ammonium ion has N—H bonds, further alkylations can take place to give a complex product mixture, as in Eq. 23.26:

$$\ddot{N}H_3 + CH_3I \longrightarrow CH_3\overset{+}{N}H_3\ I^- + (CH_3)_2\overset{+}{N}H_2\ I^- + (CH_3)_3\overset{+}{N}H\ I^- + (CH_3)_4N^+\ I^- \qquad (23.26)$$

A mixture of products is formed because the methylammonium ion produced initially is partially deprotonated by the ammonia starting material. Because the resulting methylamine is also a good nucleophile, it too reacts with methyl iodide.

$$\ddot{N}H_3 + H_3C{-}I \longrightarrow H_3\overset{+}{N}{-}CH_3\ I^- \qquad (23.27a)$$

(23.27b)

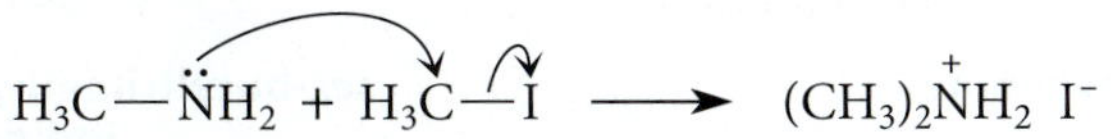

(23.27c)

Analogous deprotonation–alkylation reactions give the other products of the mixture shown in Eq. 23.26 (see Focused Problem 23.21, Sec. 23.7C).

Epoxides, as well as α,β-unsaturated carbonyl compounds and α,β-unsaturated nitriles, also react with amines and ammonia. As the following results show, multiple alkylations can occur with these alkylating agents as well.

(23.28)

$$NH_3\ \text{(excess)} + H_2C{=}CH{-}CN \longrightarrow \underset{\text{(32\% yield)}}{H_2N{-}CH_2CH_2CN} + \underset{\text{(57\% yield)}}{HN(CH_2CH_2CN)_2}$$

(23.29)

In an alkylation reaction, the exact amount of each product obtained depends on the precise reaction conditions and on the relative amounts of starting amine and alkyl halide. Because a mixture of products results, the utility of alkylation as a preparative method for amines is limited. In specific cases, however, conditions have been worked out to favor particular products. Section 23.11 discusses other methods that are more useful for the preparation of amines.

Focused Problem

23.13 Provide reaction mechanisms for the following equations.

(a) Equation 23.28 (*Hint:* Refer to Sec. 11.5A.)

(b) Equation 23.29 (*Hint:* Refer to Eqs. 22.101 and 22.102, Sec. 22.9A.)

(c) Suggest a reason why the reaction in Eq. 23.29 stops after two additions and a third doesn't occur in high yield. How might you change the reaction conditions to get a third addition?

Quaternization of Amines Amines can be converted into quaternary ammonium salts with excess alkyl halide. This process, called **quaternization**, is one of the most important synthetic applications of amine alkylation. The reaction is particularly useful when especially reactive alkyl halides, such as methyl iodide or benzylic halides, are used.

$$\underset{\textbf{benzyldimethylamine}}{PhCH_2\ddot{N}Me_2} + MeI \xrightarrow[\text{EtOH}]{} \underset{\substack{\textbf{benzyltrimethylammonium}\\ \textbf{iodide}\\ \text{(94–99\% yield)}}}{PhCH_2\overset{+}{N}Me_3\ I^-}$$

(23.30)

$$\underset{\textbf{\textit{N,N}-dimethyl-1-hexadecanamine}}{CH_3(CH_2)_{15}NMe_2} + \underset{\textbf{benzyl chloride}}{PhCH_2{-}Cl} \xrightarrow[\text{acetone}]{} \underset{\substack{\textbf{benzylhexadecyldimethylammonium}\\ \textbf{chloride}}}{CH_3(CH_2)_{15}\overset{+}{N}Me_2(CH_2Ph)\ Cl^-}$$

(23.31)

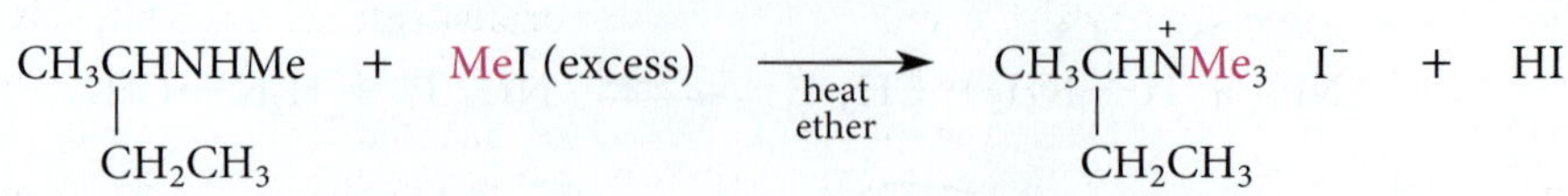

sec-butylmethylamine **sec-butyltrimethylammonium iodide** (23.32)

Conversion of an amine into a quaternary ammonium salt with excess methyl iodide (as in Eqs. 23.30 and 23.32) is called **exhaustive methylation**.

B. Reductive Amination

When primary and secondary amines react with either aldehydes or ketones, they form imines and enamines, respectively (Sec. 19.12). In the presence of a reducing agent, imines and enamines are reduced to amines.

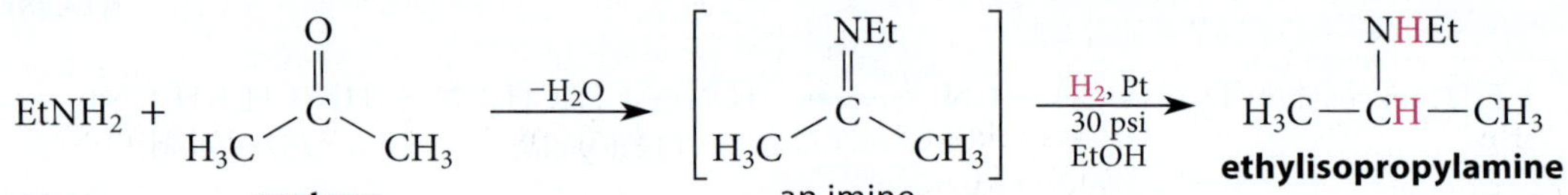

acetone an imine (not isolated) **ethylisopropylamine** (23.33)

Reduction of the C═N double bond is analogous to reduction of the C═O double bond (Sec. 19.9C). The imine or enamine does not have to be isolated, but is reduced within the reaction mixture as it forms. *Because imines and enamines are reduced more rapidly than carbonyl compounds*, reduction of the carbonyl compound is not a competing reaction.

Focused Problem

23.14 Provide a reaction mechanism for step 1, formation of the imine, in Eq. 23.33. (*Hint:* Refer to Sec. 19.12A.)

The formation of an amine from the reaction of an aldehyde or ketone with another amine and a reducing agent is called **reductive amination**. Two hydride reducing agents, sodium triacetoxyborohydride, $NaBH(OAc)_3$, and sodium cyanoborohydride, $NaBH_3CN$, find frequent use in reductive amination.

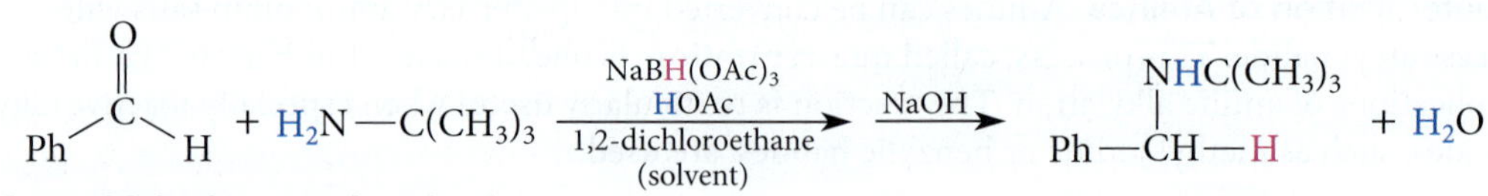

benzaldehyde ***tert*-butylamine** ***N*-*tert*-butylaniline** (95% yield) (23.34)

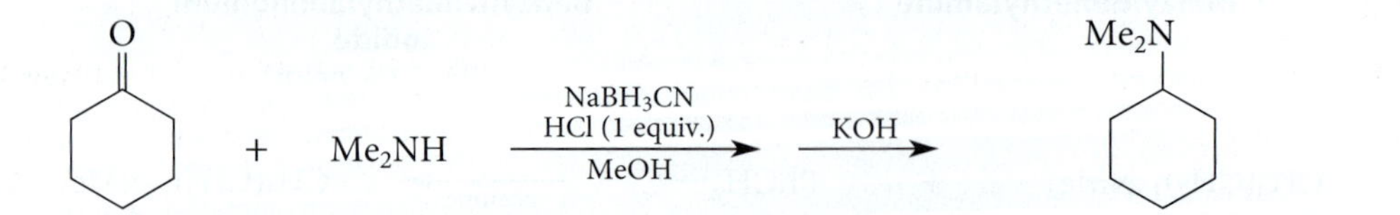

cyclohexanone **dimethylamine** ***N,N*-dimethylcyclohexanamine** (71% yield) (23.35)

Both sodium triacetoxyborohydride and sodium cyanoborohydride are commercially available, easily handled solids, and sodium cyanoborohydride can even be used in aqueous solutions above pH > 3. Reductive amination with $NaBH_3CN$ is known as the **Borch reaction**. Like $NaBH_4$ reductions, the Borch reaction requires a protic solvent or one equivalent of acid. A proton source is also required for reduction with sodium triacetoxyborohydride. In some cases, the water generated in the reaction is adequate for this purpose, and in other situations, a weak acid can be added. (Acetic acid serves this role in Eq. 23.34.)

The Borch reaction (or Borch reduction) is named for Richard F. Borch, who discovered and developed the reaction while he was a professor of chemistry at the University of Minnesota in 1971.

Focused Problem

23.15 Why is base (NaOH or KOH) added as a second step in each of the reactions shown in Eqs. 23.34 and 23.35? (*Hint:* What is the form of the product after the reduction step? Refer to Sec. 23.5A.)

Reductive amination, like catalytic hydrogenation, typically involves the imines or enamines and their conjugate acids as intermediates.

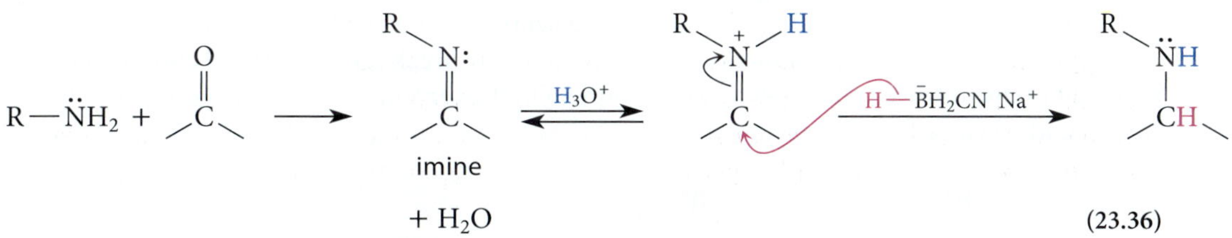

(23.36)

The success of reductive amination depends on the discrimination by the reducing agents between the imine intermediate and the carbonyl group of the aldehyde or ketone starting material. Each reagent is a sodium borohydride ($NaBH_4$) derivative in which one or more of the hydrides have been substituted with electron-withdrawing groups (—OAc or —CN). The polar effect of these groups reduces the effective negative charge on the hydride and, as a result, each reagent is less reactive than sodium borohydride itself. Each reagent is effectively "tuned" to be just reactive enough to reduce imines, but not reactive enough to reduce aldehydes or ketones. When $NaCNBH_3$ is used in protic solvents, hydrogen-bond donation by the solvent to the imine nitrogen (which is more basic than a carbonyl oxygen) catalyzes the reduction.

Formaldehyde can be reductively aminated with primary and secondary amines using the Borch reaction. This provides a way to introduce methyl groups to the level of a tertiary amine:

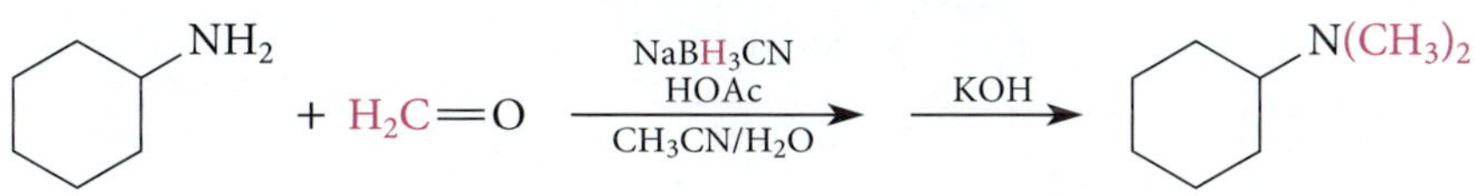

(84% yield) (23.37)

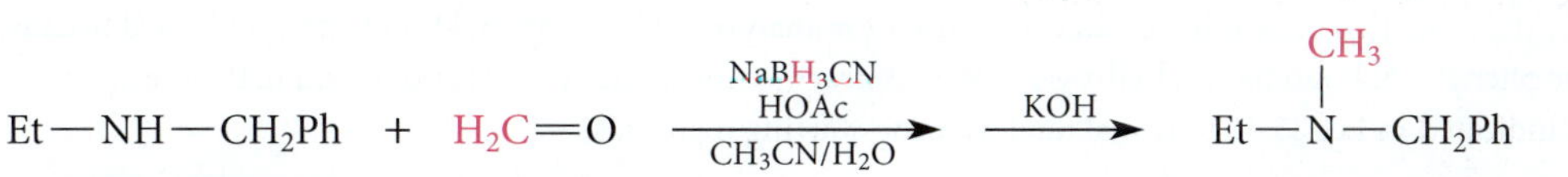

benzylethylamine **formaldehyde** **benzylethylmethylamine**

(80% yield)

(23.38)

Focused Problem

23.16 Quaternization (Sec. 23.6) does not occur in the reactions shown in Eqs. 23.37 and 23.38. Explain.

Neither an imine nor an enamine can be an intermediate in the reaction of a secondary amine with formaldehyde (Eq. 23.38). (Why?) In this case a small amount of a cationic intermediate, an *iminium ion*, is formed in solution by protonation of a carbinolamine intermediate and loss of water. The iminium ion, which is also a carbocation, is rapidly and irreversibly reduced by its reaction with hydride.

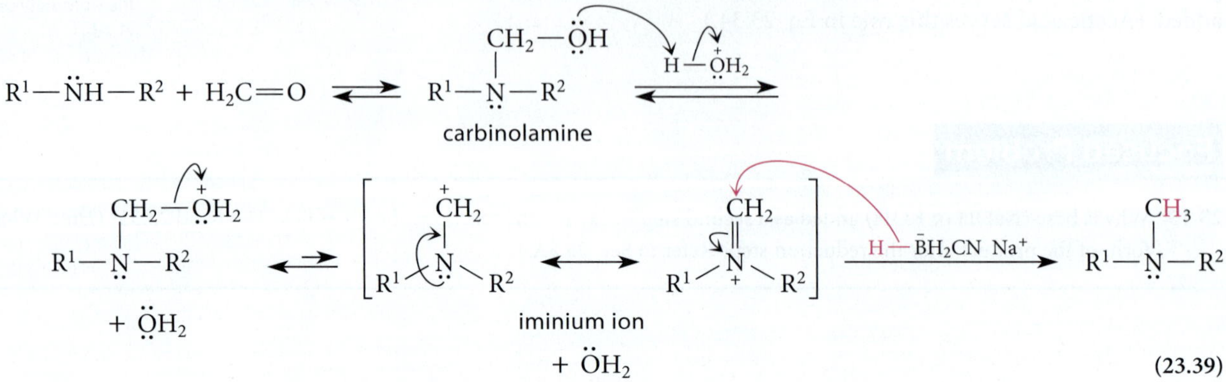

(23.39)

Suppose you want to prepare a given amine and want to determine whether reductive amination would be a suitable preparative method. How do you determine the required starting materials? Adopt the usual strategy for analyzing a synthesis: Start with the target molecule and mentally reverse the reductive amination process. Mentally break one of the C—N bonds and replace it on the nitrogen side with an N—H bond. On the carbon side, drop a hydrogen from the carbon and add a carbonyl oxygen.

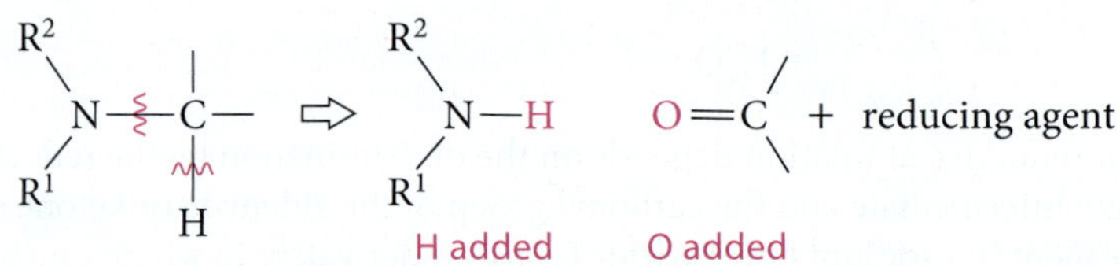

(23.40)

As this analysis shows, the target amine must have at least one hydrogen on the "disconnected" carbon. This process is applied in Study Problem 23.3.

Study Problem 23.3

Outline a preparation of *N*-ethyl-*N*-methylaniline from suitable starting materials using a reductive amination sequence.

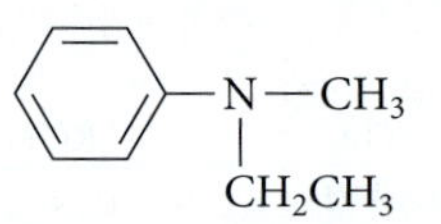

***N*-ethyl-*N*-methylaniline**

Solution Either the *N*-methyl or *N*-ethyl bond can be used for analysis. (The *N*-phenyl bond cannot be used because the carbon in the phenyl—N bond has no hydrogen.) We arbitrarily choose the N—CH_3 bond and make the appropriate replacements as indicated in Eq. 23.40 to reveal the following starting materials:

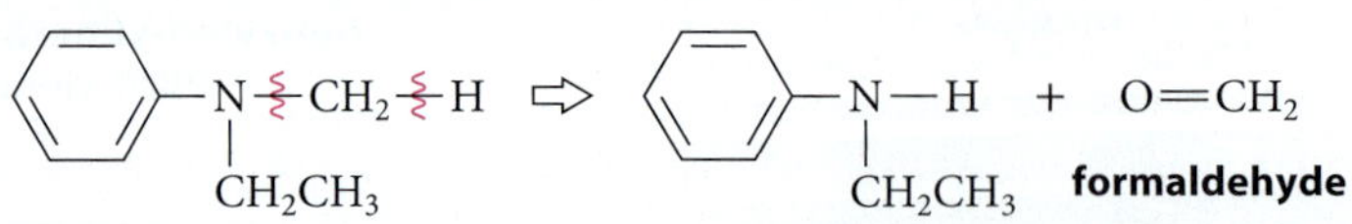

***N*-ethylaniline**

(23.41)

Thus, treatment of *N*-ethylaniline with formaldehyde and $NaBH_3CN$ should give the desired amine. (See Focused Problem 23.18.)

C. Acylation of Amines

Amines can be converted into amides by reaction with acid chlorides, anhydrides, or esters. These reactions are covered in Sec. 21.8.

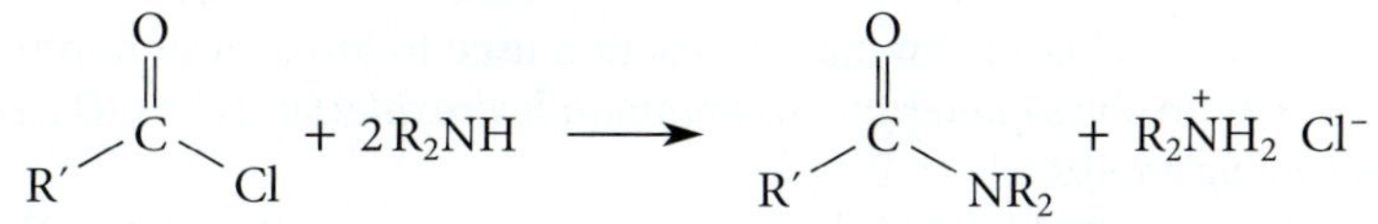

(23.42)

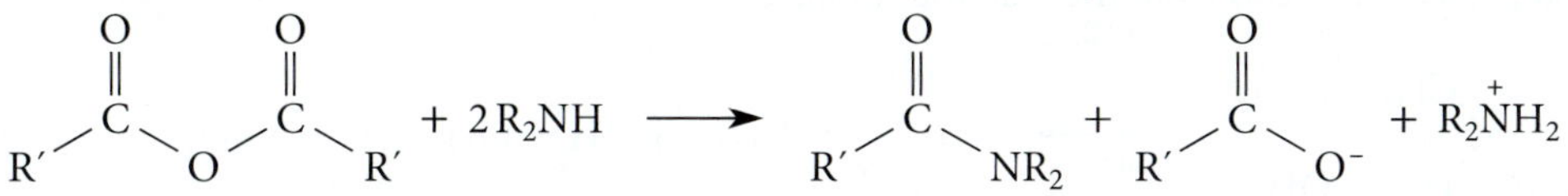

(23.43)

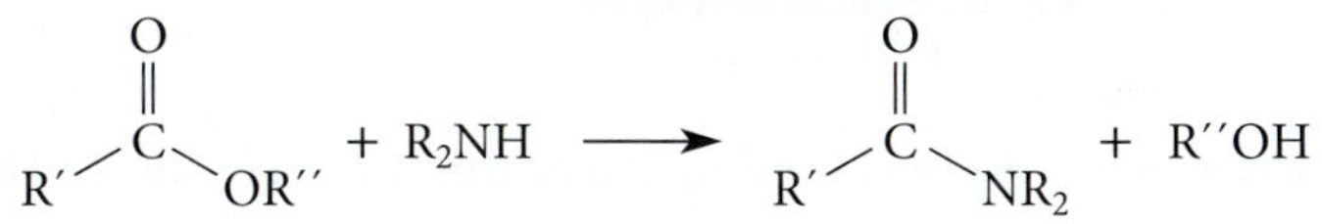

(23.44)

In this type of reaction, a bond is formed between the amine and a carbonyl carbon. These are all examples of *acylation*: a reaction involving the transfer of an *acyl group*.

Recall that the reaction of an amine with an acid chloride or an anhydride requires either *two equivalents* of the amine or one equivalent of the amine and an additional equivalent of another base such as a tertiary amine or hydroxide ion. These and other aspects of amine acylation can be reviewed in Sec. 21.8.

Focused Problems

23.17 Suggest two syntheses of *N*-ethylcyclohexanamine by reductive amination.

23.18 Outline a second synthesis of *N*-ethyl-*N*-methylaniline (the target molecule in Study Problem 23.3) by reductive amination.

23.19 Outline a synthesis of the quaternary ammonium salt $(CH_3)_3\overset{+}{N}CH_2Ph\ \overset{-}{Br}$ from each of the following combinations of starting materials.

(a) Dimethylamine and any other reagents

(b) Benzylamine and any other reagents

23.20 (a) A chemist, Caleb J. Cookbook, heated ammonia with bromobenzene expecting to form tetraphenylammonium bromide. Can Caleb expect this reaction to succeed? Explain.

(b) What type of catalyst might be used to bring about this reaction under relatively mild conditions? (See Sec. 23.11C.)

23.21 Continue the sequence of reactions in Eqs. 23.27a–c to show how trimethylammonium iodide is formed as one of the products in Eq. 23.26.

23.22 Outline a preparation of each of the following from an amine and an acid chloride.

(a) *N*-Phenylbenzamide

(b) *N*-Benzyl-*N*-ethylpropanamide

23.8 HOFMANN ELIMINATION OF QUATERNARY AMMONIUM HYDROXIDES

Section 23.7 discussed ways to *make* carbon–nitrogen bonds. In these reactions, amines react as *nucleophiles*. The subject of Sec. 23.8 is an elimination reaction used to *break* carbon–nitrogen bonds. In this reaction, which involves *quaternary ammonium hydroxides* ($R_4N^+ \ ^-OH$) as starting materials, amines act as *leaving groups*.

When a quaternary ammonium hydroxide is heated, a β-elimination reaction takes place to give an alkene, which distills from the reaction mixture.

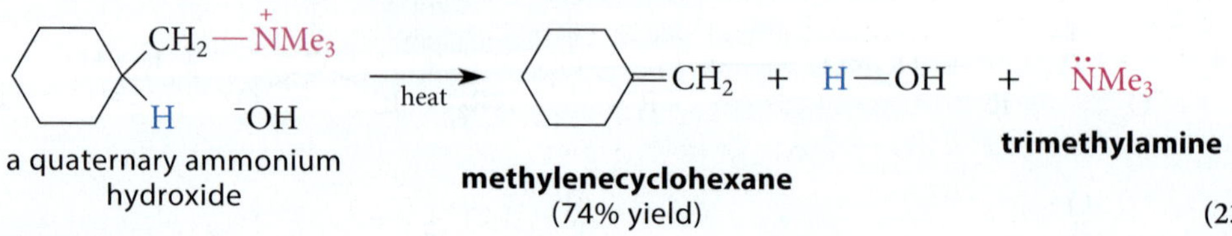

(23.45)

The Hofmann elimination is named for August Wilhelm Hofmann (1818–1892), a German chemist who became a professor at the Royal College of Chemistry in London and, later, at the University of Berlin. Hofmann was particularly noted for his work on amines.

This type of elimination reaction, when the leaving group is a neutral amine, is called a **Hofmann elimination**.

A quaternary ammonium hydroxide used as the starting material in Hofmann eliminations is formed by treating a quaternary ammonium salt with silver hydroxide (AgOH), which is essentially a hydrated form of silver(I) oxide (Ag_2O).

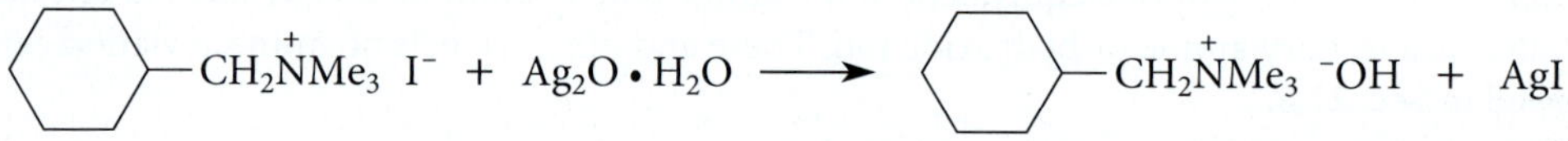

(23.46)

Alkenes, then, can be formed from amines by a three-step process: exhaustive methylation (see Eq. 23.30), conversion of the ammonium salt to the hydroxide (Eq. 23.46), and Hofmann elimination (Eq. 23.45).

The Hofmann elimination is conceptually analogous to the E2 reaction of alkyl halides (Sec. 9.5), in which a β-proton and a halide ion are eliminated; in the Hofmann elimination, a proton and a tertiary amine are eliminated. Because the amine leaving group is very basic, and therefore a relatively poor leaving group, the conditions of the Hofmann elimination are typically harsh.

Like the analogous E2 reaction of alkyl halides, the Hofmann elimination generally occurs as an anti elimination (Sec. 9.5E).

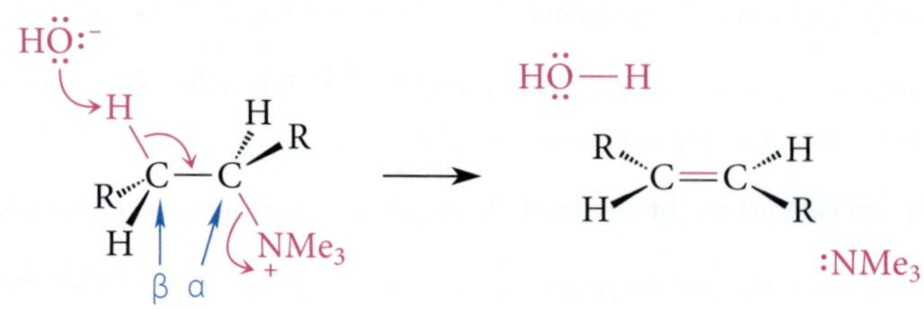

(23.47)

Focused Problems

23.23 What product (including its stereochemistry) is expected from the Hofmann elimination of each of the following stereoisomers?

(a)

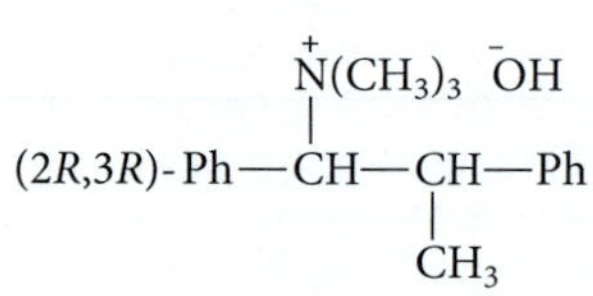

(b)

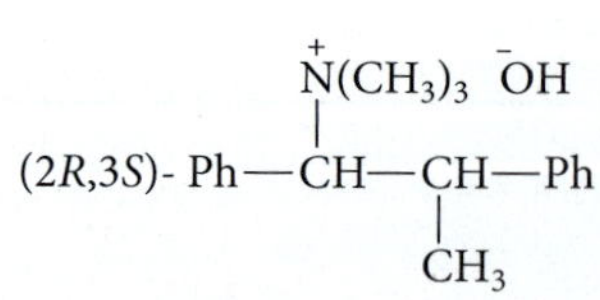

23.24 Give the major product formed when each of the following amines is treated exhaustively with methyl iodide and then heated with Ag_2O. Explain your reasoning.

(a) 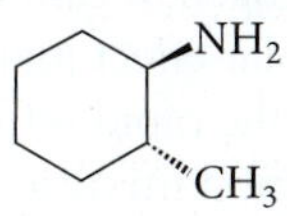 (b) 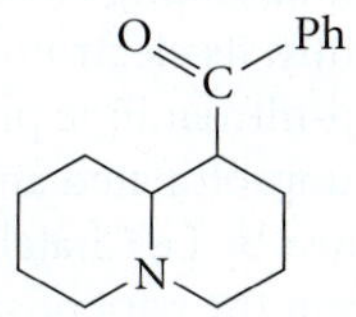

23.25 (+)-*Coniine* is the toxic component of the poison hemlock, the plant believed to have killed Socrates. Coniine has the molecular formula $C_8H_{17}N$. When coniine is exhaustively methylated, and the resulting product is then heated with Ag_2O, the mixture of compounds *A–C* is formed. [Compounds *A* and *B* are the (*E*) stereoisomers.]

$Me_2NCH_2CH_2CH_2CH_2CH{=}CHCH_2CH_3$ (*A*)

$Me_2NCH_2CH_2CH_2CH{=}CHCH_2CH_2CH_3$ (*B*)

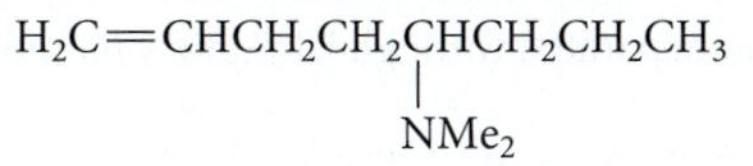

C

Propose a structure for coniine. (The absolute configuration of coniine cannot be determined from the data.)

23.9 AROMATIC SUBSTITUTION REACTIONS OF ANILINE DERIVATIVES

Aromatic amines can undergo *electrophilic aromatic substitution* reactions on the ring (Sec. 16.4). The amino group is one of the most powerful ortho, para-directing groups in electrophilic substitution. If the conditions of the reaction are not too acidic, aniline and its derivatives undergo rapid ring substitution. For example, aniline, like phenol, brominates three times under mild conditions.

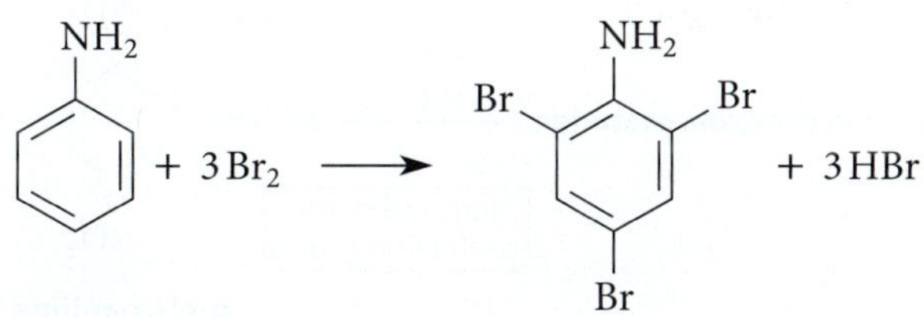

(23.48)

Focused Problem

23.26 Provide a reaction mechanism for the reaction shown in Eq. 23.48. (*Hint:* Refer to Sec. 16.4.)

When aniline is nitrated with a mixture of nitric and sulfuric acid, a mixture of meta and para nitration products is obtained.

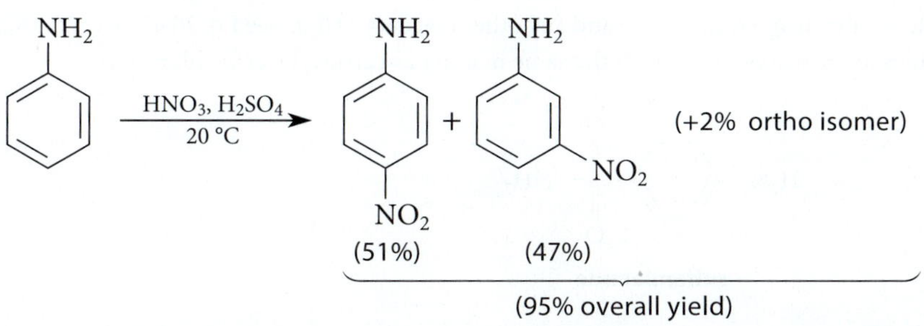

(23.49)

The formation of *m*-nitroaniline as one of the products is understandable because the major form of the starting material in the strongly acidic solution is the conjugate-acid ammonium ion. Because the ammonium ion has no nonbonding electron pair on nitrogen, it cannot donate electrons by a resonance effect; consequently, its electron-withdrawing polar effect makes it a meta-directing group. It is likely that the *p*-nitroaniline product arises by the rapid nitration of the minuscule amount of highly reactive unprotonated aniline in the reaction mixture. As the unprotonated aniline reacts, it is replenished by Le Chatelier's principle.

STUDY GUIDE LINK 23.1
Nitration of Aniline

Aniline can be nitrated regioselectively at the para position if the nitrogen is first *protected* from protonation. (The general idea of a *protecting group* was introduced in Sec. 19.11B.) This strategy is used in the solution to Study Problem 23.4.

Study Problem 23.4

Outline a preparation of *p*-nitroaniline from aniline and any other reagents. You can assume that ortho and para products, if formed together, can be separated through laboratory methods.

Solution An amide group is much less basic than an amino group. Hence, acylation of the amino group of aniline with acetyl chloride to give *N*-phenylacetamide (acetanilide) will protect the nitrogen from protonation. The acetamido group, although much less activating than a free amino group, is nevertheless an activating, ortho, para-directing group in aromatic substitution (Table 16.2 in Sec. 16.5A). Following nitration of acetanilide, the acetyl group is removed to give *p*-nitroaniline, the target compound.

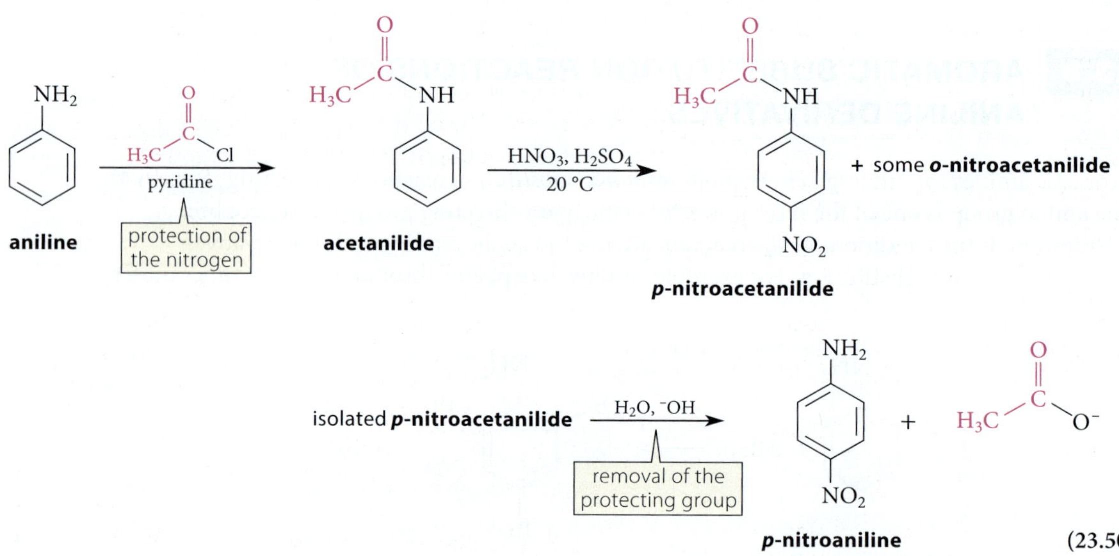

(23.50)

Because the acetamido group is considerably less basic than an amino group, it is only partially protonated under the acidic reaction conditions of nitration. Because the acetamido group is less activating than a free amino group (why?), nitration occurs only once.

Focused Problems

23.27 Outline a preparation of sulfanilamide, a sulfa drug, from aniline and any other reagents. (*Hint:* See Eq. 20.45, Sec. 20.9A; also note that sulfonyl chlorides react with amines to form amides in much the same manner as carboxylic acid chlorides.)

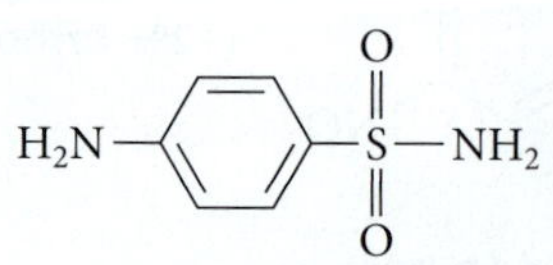

sulfanilamide

23.28 Outline a preparation of each of the following compounds from aniline and any other reagents.

(a) 2,4-Dinitroaniline

(b) Sulfathiazole, a sulfa drug (*Hint:* 2-Aminothiazole is a readily available amine.)

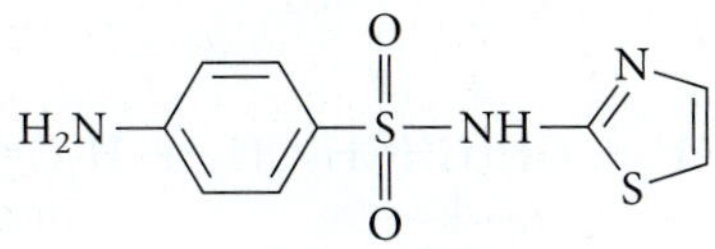

sulfathiazole

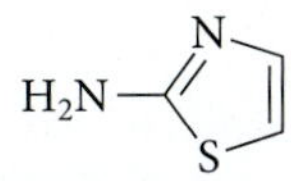

2-aminothiazole

23.10 DIAZOTIZATION; REACTIONS OF DIAZONIUM IONS

A. Formation and Substitution Reactions of Diazonium Salts

The reactions considered in previous sections show that the chemistry of the amino group vaguely resembles that of the hydroxy group. Amino groups, like hydroxy groups, can both donate and accept hydrogen bonds. Amino groups, like hydroxy groups, are basic and nucleophilic (only more so). Amino groups, like hydroxy groups (with suitable activation), can serve as leaving groups. And amino groups, like hydroxy groups, activate aromatic rings toward electrophilic aromatic substitution.

In contrast, when it comes to oxidation reactions, there is no parallel between amines and alcohols or phenols. Oxidation of amines generally occurs at the amino nitrogen, whereas oxidation of alcohols and phenols occurs at the α-carbon. An important oxidation reaction of amines that illustrates this point is called **diazotization**: the reaction of primary amines with nitrous acid (HNO_2) to form diazonium salts. A **diazonium salt** is a compound of the form $R—\overset{+}{N}{\equiv}N\colon\ X^-$, in which X^- is a typical anion (chloride, bromide, sulfate, and so on). Because nitrous acid is unstable, it is usually generated as needed by the reaction of sodium nitrite ($NaNO_2$) with a strong acid such as HCl or H_2SO_4. Both aliphatic and aromatic primary amines are readily diazotized:

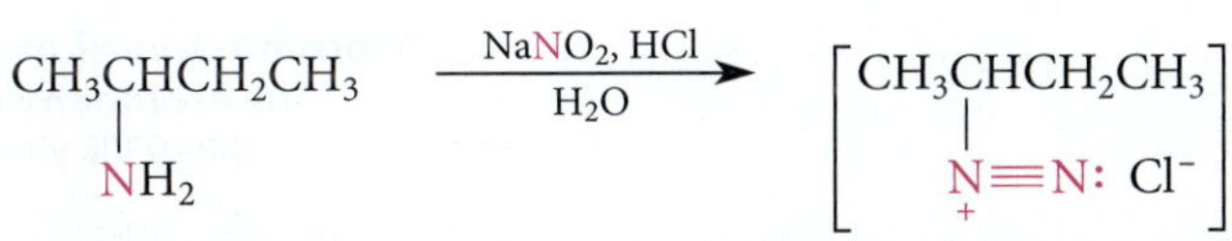

2-butanamine (*sec*-butylamine)

2-butanediazonium chloride (an aliphatic diazonium salt) unstable; cannot be isolated (23.51)

FURTHER EXPLORATION 23.1 Mechanism of Diazotization

$$\text{Ph—NH}_2 \xrightarrow[\text{H}_2\text{O}]{\text{NaNO}_2,\ \text{HCl}} \text{Ph—}\overset{+}{\text{N}}{\equiv}\text{N: Cl}^- \tag{23.52}$$

aniline → **benzenediazonium chloride** (an aromatic diazonium salt) can be isolated

Diazonium salts incorporate one of the very best leaving groups—molecular nitrogen (*red* in Eq. 23.51). For this reason, *aliphatic* diazonium salts react immediately as they are formed by S_N1, E1, and/or S_N2 mechanisms to give substitution and elimination products along with dinitrogen, a gas.

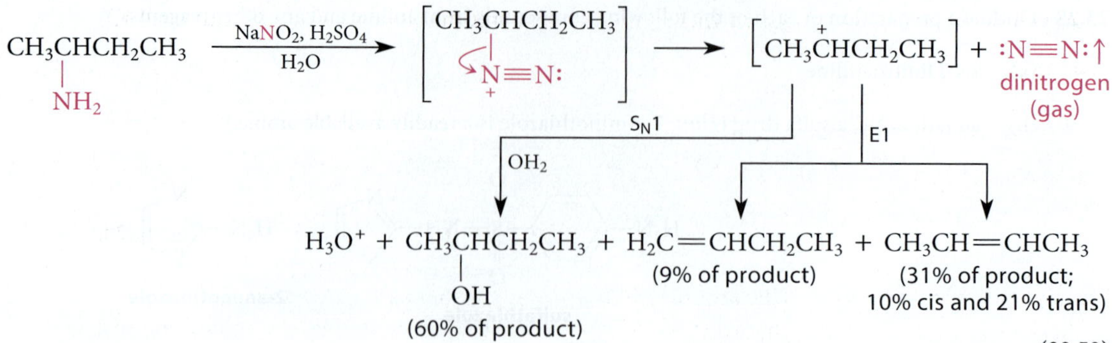

(23.53)

(The rapid liberation of dinitrogen on treatment with nitrous acid is a qualitative test for primary alkylamines.) Because of the complex mixture of solvolysis and elimination products that results, the reactions of aliphatic diazonium salts are not generally useful in organic synthesis.

Recall that benzene rings bearing good leaving groups do not readily undergo S_N1 or S_N2 reactions (Secs. 18.1 and 18.3). For this reason, *aryl*diazonium salts can be isolated and used in a variety of reactions. In practice, though, they are usually prepared in solution at 0–5 °C and used without isolation, because they lose nitrogen on heating and because they are explosive in the dry state.

Among the most important reactions of aryldiazonium salts are substitution reactions with cuprous halides; in these reactions, the diazonium group is replaced by a halogen.

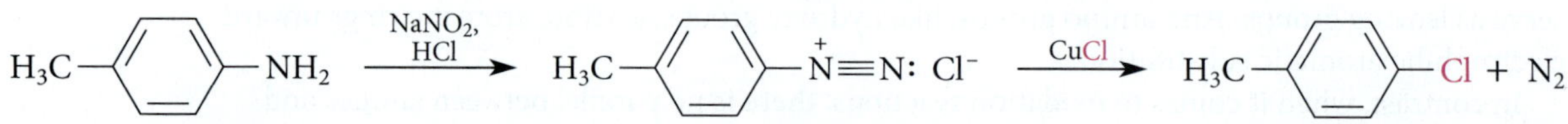

p-methylaniline (p-toluidine) **p-methylbenzenediazonium chloride** **p-chlorotoluene** (70–71% yield) (23.54)

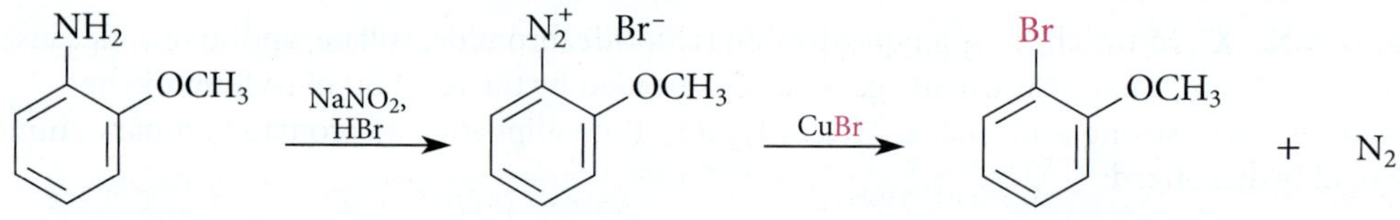

2-methoxyaniline **1-bromo-2-methoxybenzene (o-bromoanisole)** (88–93% yield) (23.55)

An analogous reaction occurs with cuprous cyanide, CuCN.

4-methylbenzonitrile (67% yield) (23.56)

This reaction is another way of forming a carbon–carbon bond, in this case to an aromatic ring (see Appendix VI). The resulting nitrile can be converted by hydrolysis into a carboxylic acid, which can, in turn, serve as the starting material for a variety of other types of compounds. The reaction of an aryldiazonium ion with a cuprous salt is called the **Sandmeyer reaction**. This reaction is an important method for the synthesis of aryl halides and nitriles.

Aryl iodides can also be made by the reaction of diazonium salts with the potassium salt KI.

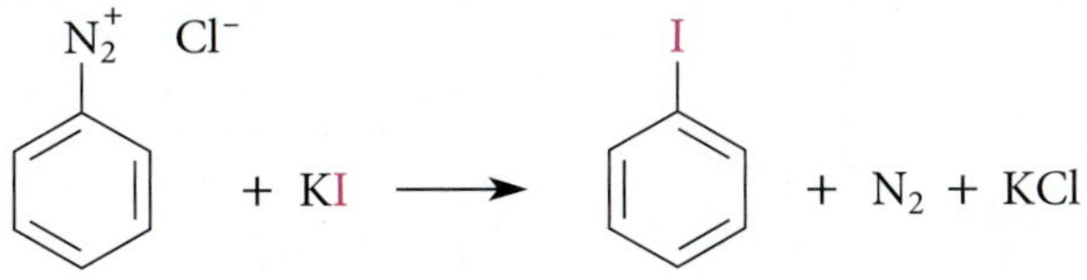

(74–76% yield) (23.57)

The analogous reactions with KBr and KCl do not work; cuprous salts are required.

Aryldiazonium salts can be hydrolyzed to phenols by heating them in water. A variation of this reaction reminiscent of the Sandmeyer reaction is the use of cuprous oxide (Cu_2O) and an excess of aqueous cupric nitrate [$Cu(NO_3)_2$] at room temperature.

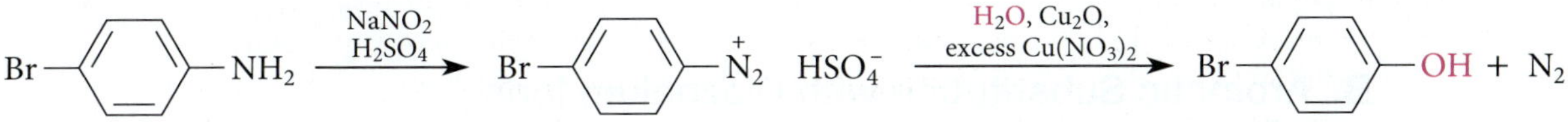

p-bromoaniline **p-bromophenol** (95% yield) (23.58)

Finally, the diazonium group is replaced by hydrogen when the diazonium salt is treated with hypophosphorous acid, H_3PO_2.

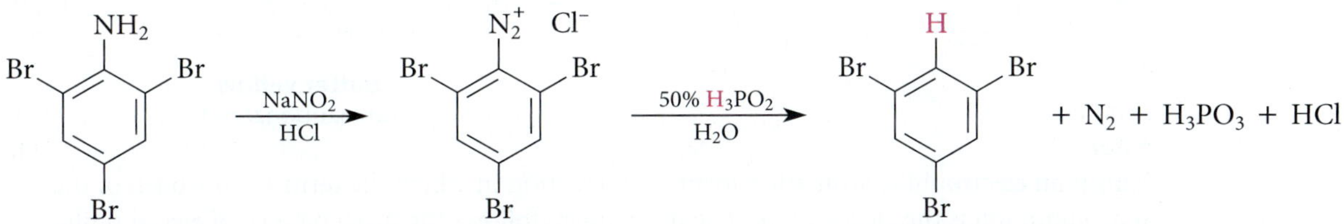

2,4,6-tribromoaniline **2,4,6-tribromobenzenediazonium chloride** **1,3,5-tribromobenzene** (70% yield) (23.59)

The product of Eq. 23.59, 1,3,5-tribromobenzene, cannot be prepared by the bromination of benzene itself. (Why?) Recall that the starting material, 2,4,6-tribromoaniline, is prepared by the bromination of aniline (Eq. 23.48). In this bromination reaction, the positions of the bromines are determined by the powerful directing effect of the amino nitrogen. Once the amino group has fulfilled its role as an activating and directing group, it can be removed using the reaction in Eq. 23.59.

The diazonium salt reactions shown in Eqs. 23.54–23.59 are all substitution reactions, but none are S_N2 or S_N1 reactions because aromatic rings do *not* undergo substitution by these mechanisms (Secs. 18.1 and 18.3). It turns out that the Sandmeyer and related reactions occur by radical mechanisms involving the copper. The reaction of diazonium salts with KI (Eq. 23.57), although not involving copper, probably occurs by a similar mechanism. The reaction of diazonium salts with H_3PO_2 (Eq. 23.59) has been shown definitively to be a free-radical chain reaction.

Focused Problems

23.29 Outline a synthesis for each of the following compounds from the indicated starting materials using a reaction sequence involving a diazonium salt.

(a) 2-Bromobenzoic acid from *o*-toluidine (*o*-methylaniline)

(b) 2,4,6-Tribromobenzoic acid from aniline

23.30 As shown in the following equation, when (*R*)-1-deuterio-1-butanamine is diazotized with nitrous acid in water, the alcohol product formed has the *S* configuration (D = ^{2}H).

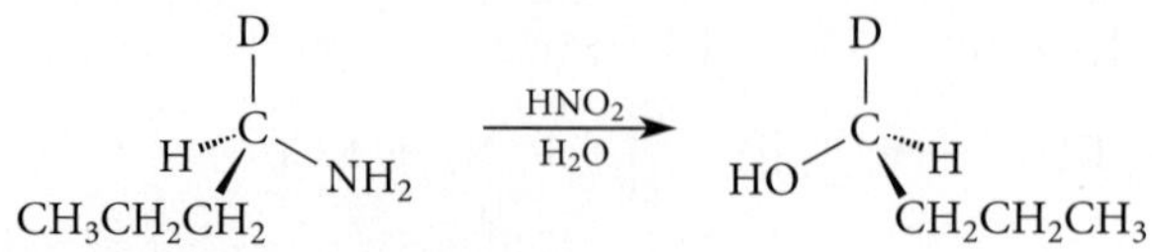

(*R*)-1-deuterio-1-butanamine

(a) Give the stereochemical configuration of the diazonium ion formed as an intermediate in this reaction. Draw its structure.

(b) What mechanism for reaction of the diazonium ion with water is consistent with the stereochemical result in the preceding equation?

B. Aromatic Substitution with Diazonium Ions

Aryldiazonium ions react with aromatic compounds containing strongly activating substituent groups, such as amines and phenols, to give substituted *azobenzenes*. (Azobenzene itself is Ph—N=N—Ph.)

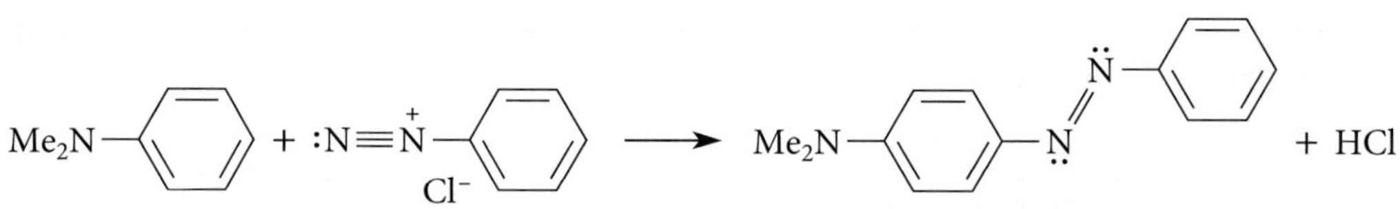

butter yellow
(an azobenzene) **(23.60)**

This is an electrophilic aromatic substitution reaction in which the terminal nitrogen of the diazonium ion is the electrophile. The mechanism follows the usual pattern of electrophilic aromatic substitution (Sec. 16.4B). First, the π electrons of the aromatic compound are donated to the electrophilic nitrogen to give a resonance-stabilized carbocation:

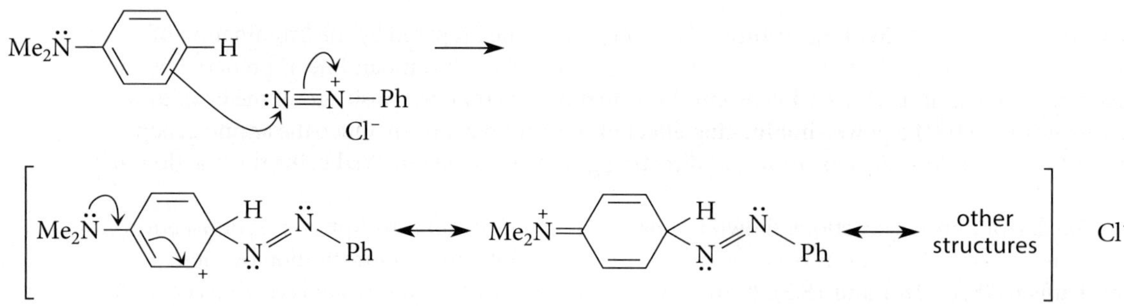

(23.61a)

This carbocation then loses a proton to give the substitution product.

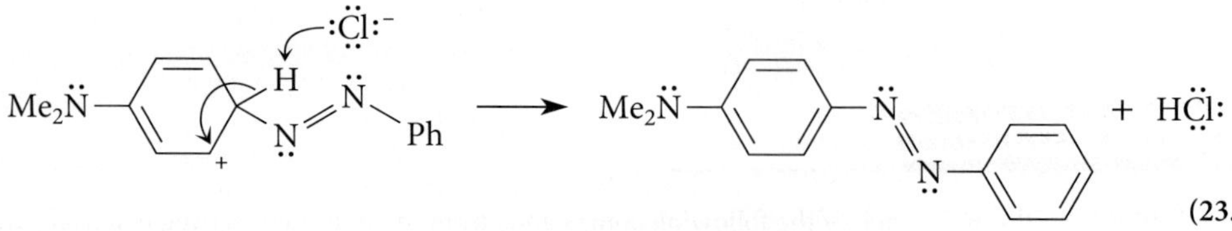

(23.61b)

(Why does substitution occur at the para position? See Sec. 16.5A.)

The azobenzene derivatives formed in these reactions have extensive conjugated π-electron systems, and most of them are colored (Sec. 15.2C). Some of these compounds are used as dyes and indicators; as a class, they are known as **azo dyes.** (An azo dye is a colored derivative of azobenzene.) For example, the azo dye methyl orange is a common acid–base indicator.

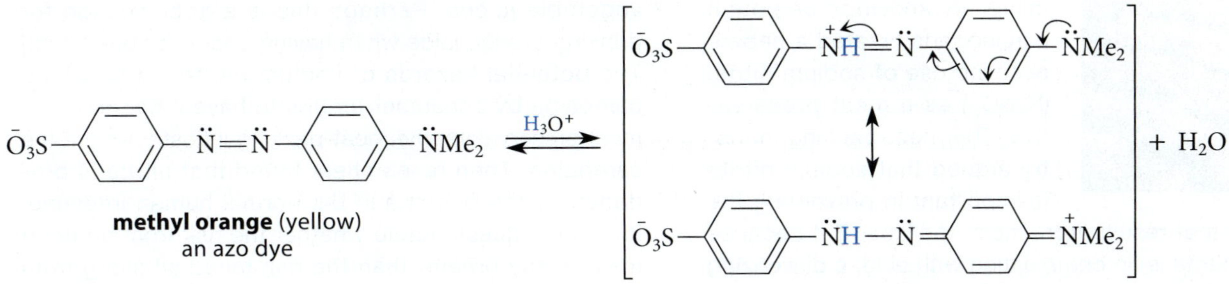

methyl orange (yellow)
an azo dye

protonated methyl orange (red)
pK_a = 3.5

(23.62)

Because methyl orange changes color when it is protonated, it can be used as an acid–base indicator at pH values near its pK_a of 3.5. Some azo dyes are used in dyeing fabrics, foodstuffs, and cosmetics. For example, FD & C Yellow No. 6 (FD & C = food, drug, and cosmetic) is a compound used to color gelatin desserts, ice cream, beverages, candy, and so on.

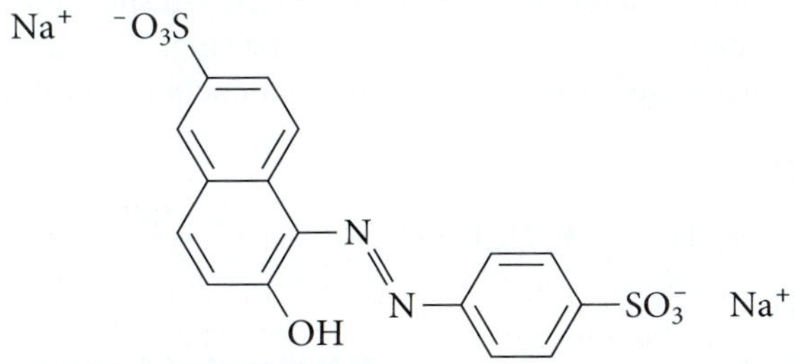

FD & C Yellow No. 6 ("Sunset Yellow")

C. Reactions of Secondary and Tertiary Amines with Nitrous Acid

Secondary amines react with nitrous acid to give *N*-nitrosoamines, compounds of the form R_2N—N═O, usually called simply **nitrosamines.**

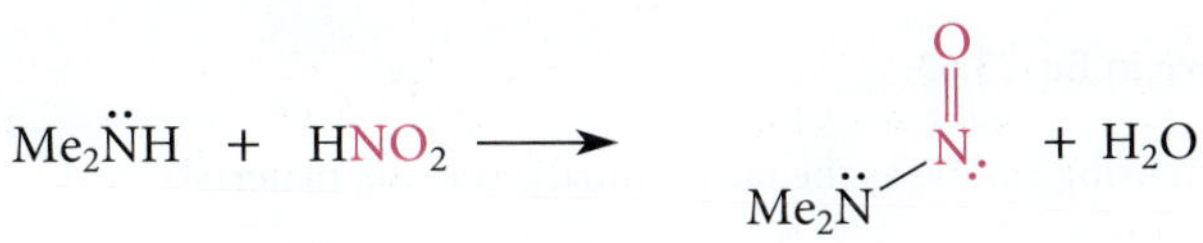

***N,N*-dimethylnitrosamine**
(89–90% yield)

(23.63)

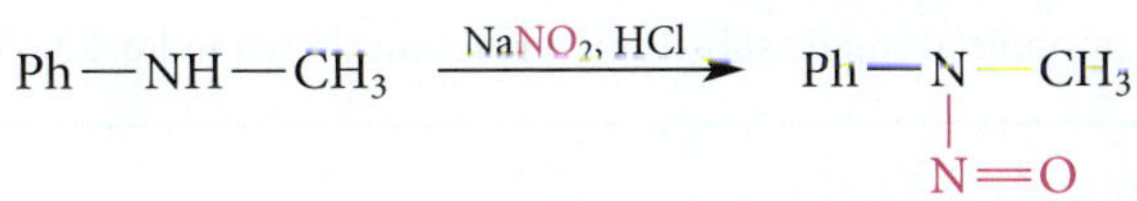

***N*-methyl-*N*-nitrosoaniline**

(23.64)

Nitrosamines, Cancer, and Breakfast Bacon

Joe Gough/Shutterstock

The fact that many nitrosamines are known to be potent carcinogens created a debate over the use of sodium nitrite ($NaNO_2$) as a meat preservative. The meat-packing industry argued that sodium nitrite is important in preventing the botulism that results from meat spoilage. But because sodium nitrite is, in combination with acid, a diazotizing reagent, it has the capacity for producing nitrosamines in the acidic environment of the stomach. For example, the frying of bacon generates nitrosamines that concentrate in the fat. [Well-drained bacon contains fewer nitrosamines, and nitrosamines are destroyed by ascorbic acid (vitamin C), which is present in fruit and vegetable juices. Perhaps this is a good reason for drinking orange juice when having bacon for breakfast!] The potential hazards of sodium nitrite led to a long campaign by consumer groups to have it banned as a meat preservative; the meat-packing industry fought the campaign. Then researchers found that nitrite is produced by the bacteria in the normal human intestine. It became questionable whether the risk from nitrite in meat is any greater than the risk faced all along from normal intestinal flora. These findings caused the Food and Drug Administration in 1980 to back away from banning sodium nitrite as a preservative, recommending only that it be kept to a minimum.

A tertiary amine cannot form a nitrosamine. However, *N,N*-disubstituted aromatic amines undergo electrophilic aromatic substitution on the benzene ring. The electrophile is the nitrosyl cation, $^{+}\ddot{N}{=}\ddot{O}$, which is generated from nitrous acid under acidic conditions.

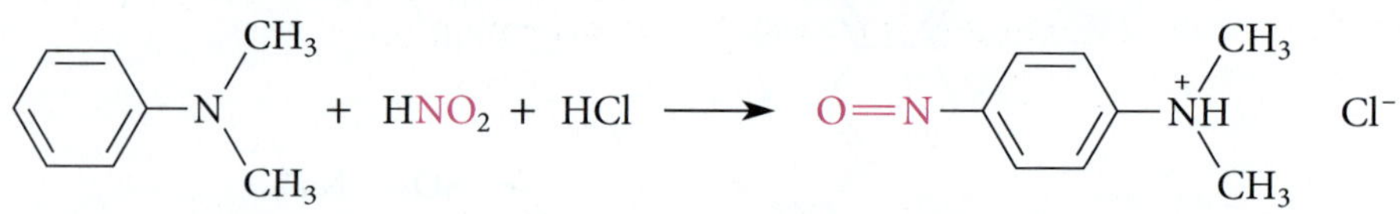

***N,N*-dimethylaniline**

***N,N*-dimethyl-4-nitrosoanilinium chloride**
(89–90% yield) (23.65)

Focused Problems

23.31 (a) Write a Lewis structure for HNO_2 in Eq. 23.63.

(b) Write a mechanism for the reaction shown in Eq. 23.63.

23.32 Design a synthesis of methyl orange (Eq. 23.62) using aniline as the only aromatic starting material.

23.33 What two compounds would react in a diazo coupling reaction to form FD & C Yellow No. 6?

23.34 (a) Using the curved-arrow notation, show how the nitrosyl cation, $^{+}\ddot{N}{=}\ddot{O}$, is generated from HNO_2 under acidic conditions.

(b) Draw a curved-arrow mechanism for the electrophilic aromatic substitution reaction shown in Eq. 23.65.

23.11 SYNTHESIS OF AMINES

Several reactions discussed in previous sections can be used for the synthesis of amines. In this section, five additional methods are presented, and, in Sec. 23.11E, all of the methods for preparing amines are summarized.

A. Synthesis of Primary Amines: The Gabriel Synthesis and the Staudinger Reaction

Recall that direct alkylation of ammonia is generally not a good synthetic method for the preparation of amines because multiple alkylation takes place (Sec. 23.7A). This problem can be avoided by protecting the amine nitrogen so that it can react only once with alkylating reagents. One approach of this sort begins with the imide *phthalimide.* Because the pK_a of phthalimide is 8.3, its conjugate-base anion is easily formed with KOH or NaOH. This anion is a good nucleophile, and it is alkylated by alkyl halides or sulfonate esters in S_N2 reactions.

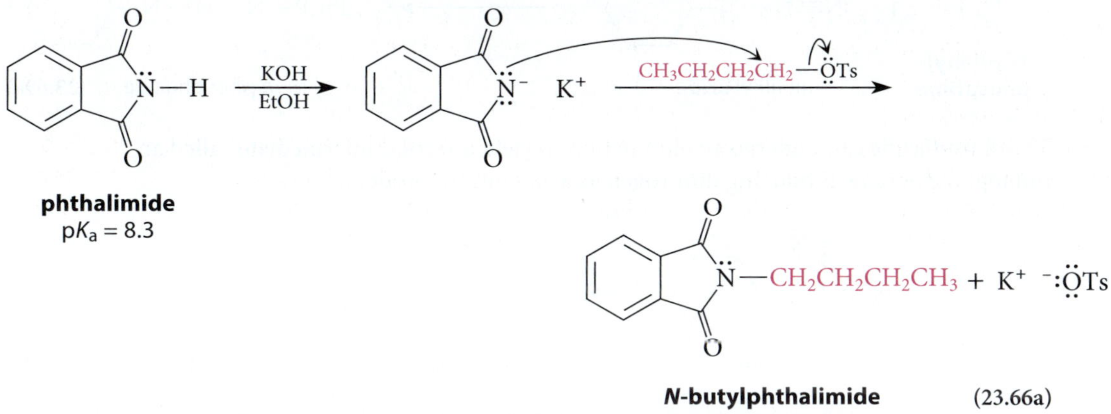

(23.66a)

The alkyl halides and sulfonates used in this reaction are primary or unbranched secondary. Because the *N*-alkylated phthalimide formed in this reaction is really a double amide, it can be converted into the free amine by amide hydrolysis in either strong acid or base.

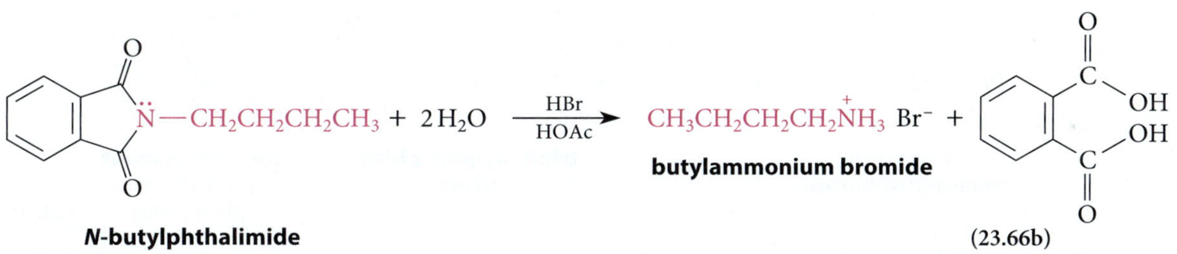

(23.66b)

In this example, acidic hydrolysis gives the ammonium salt, which can be converted into the free amine by neutralization with base.

The alkylation of phthalimide anion followed by hydrolysis of the alkylated derivative to the primary amine is called the **Gabriel synthesis**. Because the nitrogen in phthalimide has only one acidic hydrogen, it can be alkylated only once. Although *N*-alkylphthalimides also have a pair of nonbonding electrons on nitrogen, they do not alkylate further, because neutral imides are *much* less basic (why?), and therefore less nucleophilic, than the phthalimide anion. Hence, multiple alkylation, which occurs in the direct alkylation of ammonia, is avoided in the Gabriel synthesis.

The Gabriel synthesis is named for Siegmund Gabriel (1851–1924), a professor at the University of Berlin, who developed the reaction in 1887.

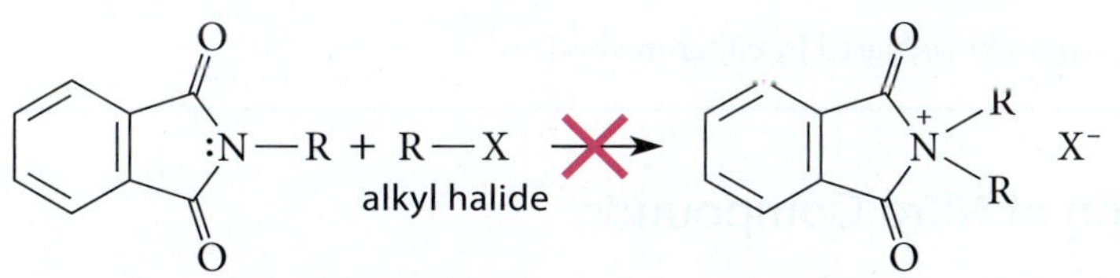

(23.67)

Another method for preparing primary amines while avoiding the multiple alkylation problem utilizes alkyl azides, and is called the **Staudinger reaction**. Alkyl azides can be prepared from an alkyl halide (or a sulfonate ester, such as a tosylate) and a source of azide ion, N_3^-, such as sodium azide (NaN_3). (See Table 9.1, Sec. 9.1A.) Like other S_N2 reactions, the reaction takes place most readily with relatively unhindered alkyl halides.

The Staudinger reaction is named for another German chemist, Hermann Staudinger (1881–1965), who developed the method in 1919 as a professor at the Swiss Federal Institute of Technology in Zürich, Switzerland. In 1953, he received the Nobel Prize in Chemistry for other pioneering work in the field of polymers, work that ultimately led to everyday products such as nylon and polyesters.

$$CH_3CH_2CH_2—Br + Na^+ N_3^- \xrightarrow{H_2O,\ THF} CH_3CH_2CH_2—N_3 + Na^+ Br^- \quad (23.68)$$

propyl azide
(88% yield)

Like the phthalimide in the Gabriel synthesis, the alkyl azide can be thought of as a protected amine nitrogen. The alkyl azide reacts with triphenylphosphine to form a phosphazide intermediate.

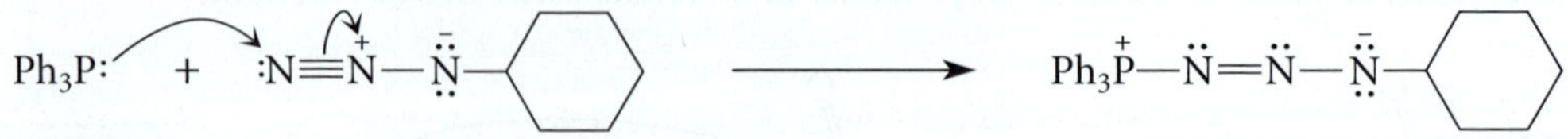

triphenyl-phosphine an alkyl azide a phosphazide (23.69a)

The phosphazide can undergo an elimination to yield a second intermediate called an iminophosphorane, producing dinitrogen as a gaseous by-product.

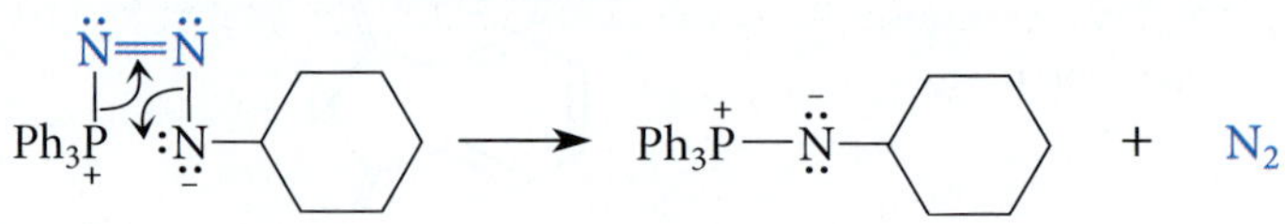

phosphazide an iminophosphorane **nitrogen gas** (23.69b)

Notice the similarity between the rearrangement of the phosphazide in Eq. 23.69b and the oxaphosphetane in the Wittig alkene synthesis (Eq. 19.94a, Sec. 19.14).

In the presence of water, the iminophosphorane is hydrolyzed to yield triphenylphosphine oxide and a primary amine. This reaction is analogous to imine hydrolysis, discussed in Sec. 19.12A.

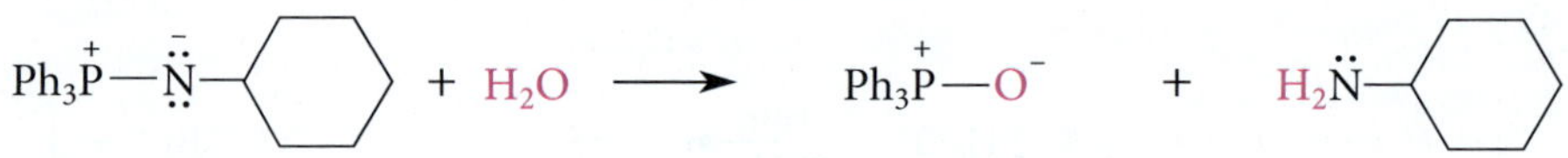

cyclohexyl-iminophosphorane **triphenylphosphine oxide** **cyclohexylamine** a primary amine (95% yield) (23.70)

Focused Problem

23.35 (a) Which one of the following three amines can be prepared by either the Gabriel synthesis or the Staudinger reaction: 2,2-dimethyl-1-propanamine, 3-methyl-1-pentanamine, or *N*-butylaniline?

(b) Starting with an alkyl halide, propose a synthesis for the compound you chose in part (a), using the Gabriel synthesis.

(c) Propose an alternative synthesis for the same compound using the Staudinger reaction and the same alkyl halide starting material.

(d) Explain why the other two amines cannot be prepared by either method.

B. Reduction of Nitro Compounds

Nitro compounds can be reduced easily to amines by catalytic hydrogenation:

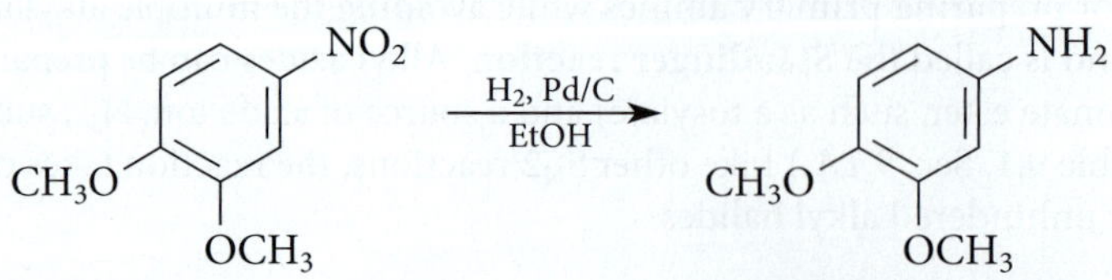

1,2-dimethoxy-4-nitrobenzene **3,4-dimethoxyaniline** (97% yield) (23.71)

In an older, but nevertheless effective, method, finely divided tin or iron powders and HCl can be used to convert aromatic nitro compounds into aniline derivatives.

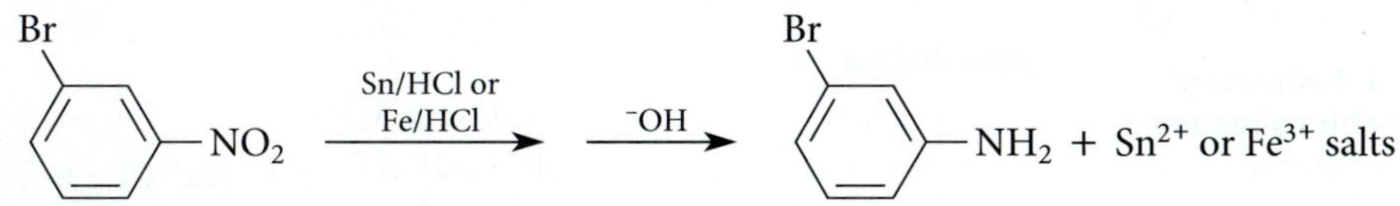

1-bromo-3-nitrobenzene **m-bromoaniline** (80% yield) (23.72)

In this reaction, the nitro compound is reduced *at nitrogen*, and the metal, which is oxidized to a metal salt, is the reducing agent. Although the methods shown in both Eqs. 23.71 and 23.72 also work with aliphatic nitro compounds, they are particularly important with aromatic nitro compounds as methods for introducing an amino group into an aromatic ring.

In view of the utility of lithium aluminum hydride ($LiAlH_4$) and sodium borohydride ($NaBH_4$) as reducing agents for other compounds, what happens when nitro compounds are treated with these reagents? Aromatic nitro compounds do react with $LiAlH_4$, but the reduction products are azobenzenes (Sec. 23.10B), not amines:

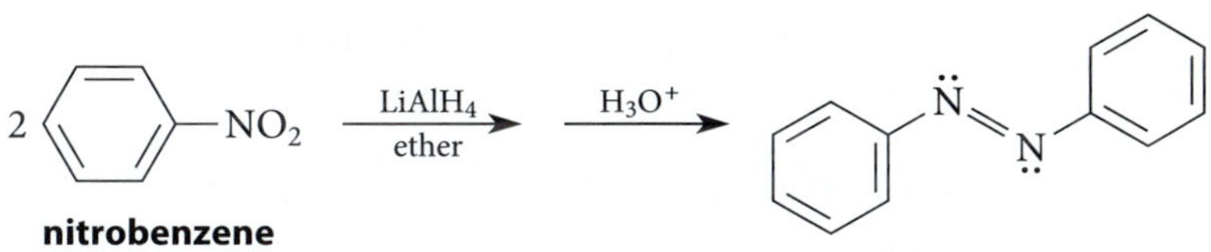

nitrobenzene

azobenzene (23.73)

Nitro groups do not react at all with sodium borohydride under the usual conditions.

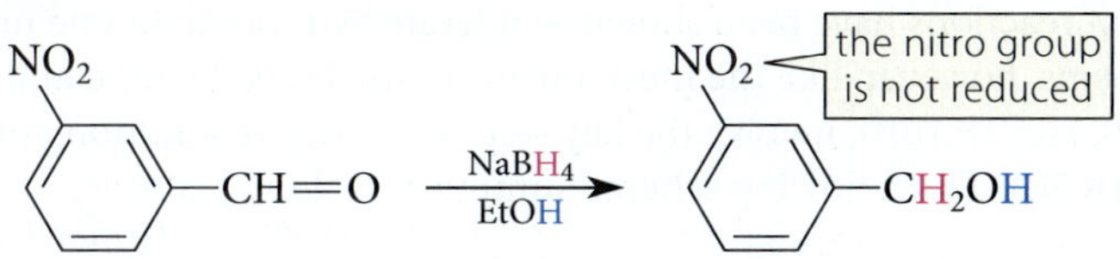

m-nitrobenzaldehyde **m-nitrobenzyl alcohol** (23.74)

Therefore, $LiAlH_4$ and $NaBH_4$ are *not* useful in forming aromatic amines from nitro compounds.

Focused Problem

23.36 Outline syntheses of the following compounds from the indicated starting materials.

(a) *p*-Iodoanisole from phenol and any other reagents

(b) *m*-Bromoiodobenzene from nitrobenzene

C. Amination of Aryl Halides and Aryl Triflates

Arylamines can be prepared by the direct amination of aryl chlorides and aryl bromides in the presence of a base and a Pd(0) catalyst.

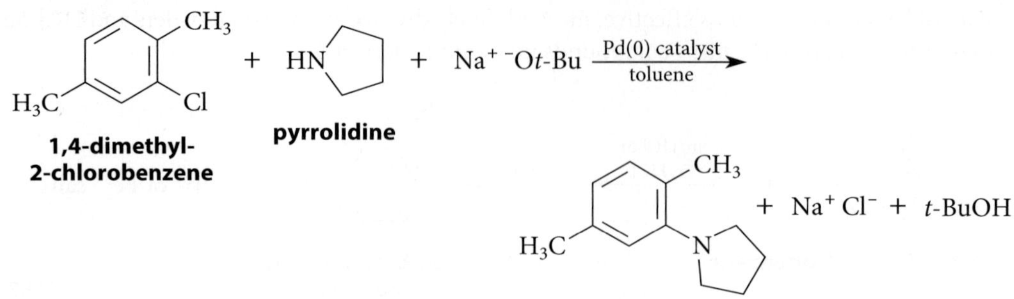

N-(2,5-dimethylphenyl)pyrrolidine
(98% yield) (23.75)

The direct amination of aryl halides is sometimes called **Buchwald–Hartwig amination** to recognize the two chemistry professors who led the research groups that developed these reactions: Stephen L. Buchwald of MIT and John F. Hartwig of the University of California, Berkeley.

A number of different catalysts have been explored for direct amination. These are typically of the form PdL_2, where L is a sterically demanding ligand such as the following:

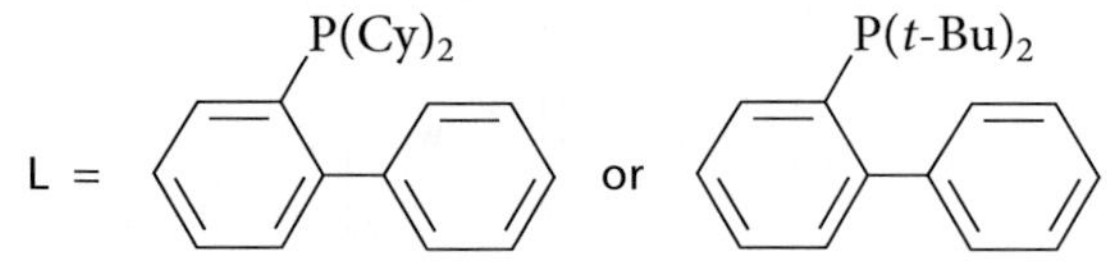

(Cy = cyclohexyl) (23.76)

These catalysts are formed by mixing palladium(II) acetate or other Pd precursors and two equivalents of the ligands.

These amination reactions have been shown to operate by more than one mechanism. All of the mechanisms, however, like the mechanisms of the Heck, Suzuki, and Stille reactions (Secs. 18.6A, 18.6B, and 18.10B), involve the key steps of oxidative addition and reductive elimination (Sec. 18.5E). The following scheme summarizes these features.

$$PdL_2 \rightleftharpoons PdL + L$$

$$\underset{\substack{\text{a } 12e^- \text{ complex} \\ \text{Pd(0)}}}{PdL + Ar—Cl} \xrightarrow[\text{addition}]{\text{oxidative}} \underset{\substack{\text{a } 14e^- \text{ complex} \\ \text{Pd(II)}}}{L \rightarrow \underset{\underset{Cl}{|}}{Pd}—Ar} \xrightarrow[\substack{\text{ligand} \\ \text{substitution}}]{\substack{\text{Base:}^- \\ R_2NH}} \underset{\substack{+\ \text{Base—H} \\ +\ Cl^-}}{L \rightarrow \underset{\underset{NR_2}{|}}{Pd}—Ar} \xrightarrow[\text{elimination}]{\text{reductive}} PdL + Ar—NR_2 \quad (23.77)$$

The amine used as the starting material in the amination reaction must lose a hydrogen in the reaction. Consequently, when a tertiary amine is the amination product, it cannot react further. However, when a primary amine is used as the starting material, the product is a secondary amine. It can in principle serve as the starting material in a competing second amination.

$$Ar—Cl \xrightarrow[\text{catalyst}]{RNH_2,\text{ base,}} Ar—NHR \xrightarrow[\text{catalyst}]{Ar—Cl,\text{ base,}} Ar—\overset{\overset{Ar}{|}}{N}R \quad (23.78)$$

The product of this second amination becomes an unwanted by-product. Nevertheless, amination with primary amines is practical if the primary amine is itself an arylamine, or if it has a large or highly branched alkyl group. In such cases, steric hindrance is used to advantage. The catalyst complex leading to the tertiary amine has significant steric repulsions; as a result, the undesired second amination is relatively slow and does not occur to a significant extent. This is one reason that the catalysts involve sterically demanding ligands.

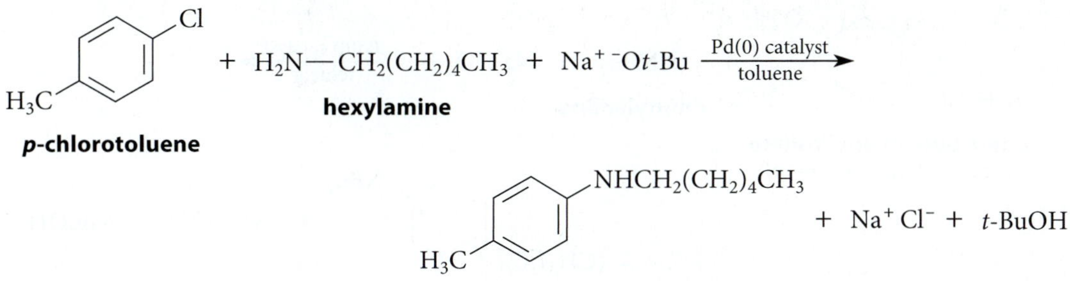

p-chlorotoluene

***N*-hexyl-4-methylaniline**
(85% yield) (23.79)

We showed in Sec. 23.7B that reductive amination is another way to prepare tertiary arylamines. Some tertiary arylamines, however, such as those containing nitrogen heterocycles (Eq. 23.75), would be difficult to prepare by reductive amination. Direct amination provides a straightforward route to these amines. Another attractive aspect of direct amination is that it, like other Pd-catalyzed coupling reactions, tolerates a wide variety of other functional groups, as Study Problem 23.5 illustrates.

Study Problem 23.5

Outline a synthesis of *p*-dipropylaminoacetophenone from chlorobenzene.

Solution Considering the problem retrosynthetically gives the following synthetic pathway, starting with the target molecule:

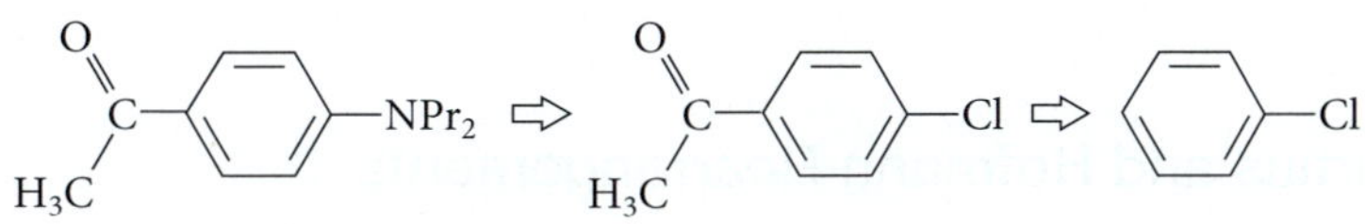

***p*-dipropylaminoacetophenone**

Direct amination of *p*-chloroacetophenone with dipropylamine and an appropriate Pd(0) catalyst gives the target:

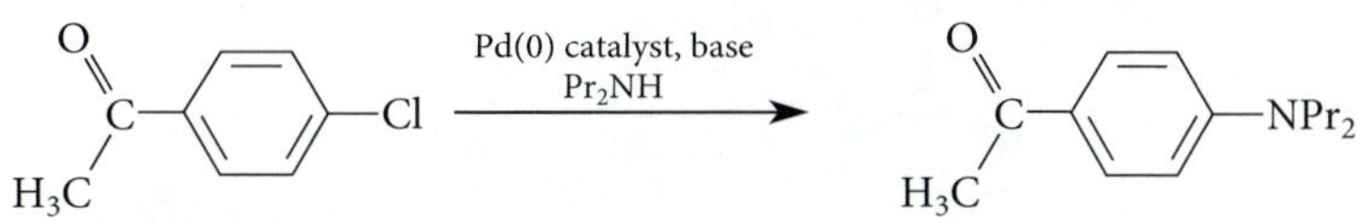

Reductive amination would not have worked, because the acetyl group would have been reduced under conditions of reductive amination.

The starting material for the amination, *p*-chloroacetophenone, can in turn be prepared by a Friedel–Crafts acylation reaction.

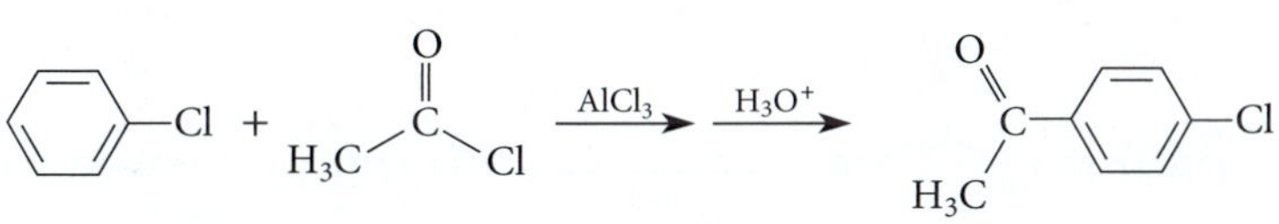

Aryl triflates can also be used as starting materials in amination. Because aryl triflates can be prepared from the corresponding phenols (Sec. 18.10B), this reaction provides a synthetic path from phenols to arylamines.

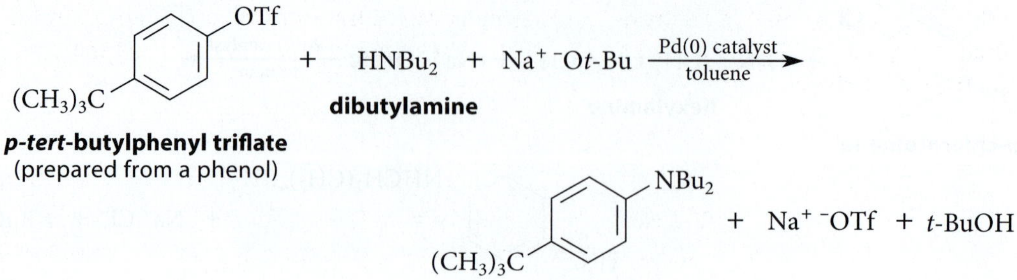

p-*tert*-butylphenyl triflate
(prepared from a phenol)

dibutylamine

***N,N*-dibutyl-4-*tert*-butylaniline**
(73% yield)

(23.80)

Focused Problem

23.37 Outline a synthesis of each of the following compounds from the indicated starting material and any other reagents.

(a) *N*-(*sec*-Butyl)-*N*-ethylaniline from chlorobenzene

(b)

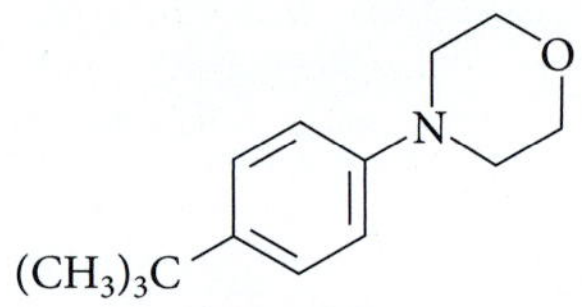

from phenol

(c)

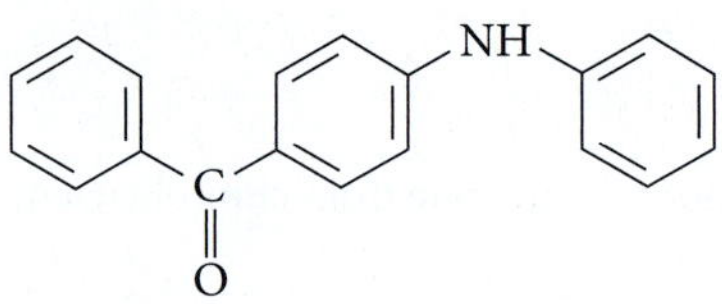

from chlorobenzene

D. Curtius and Hofmann Rearrangements

A useful synthesis of amines starts with a class of compounds called *acyl azides.* An **acyl azide** has the following general structure:

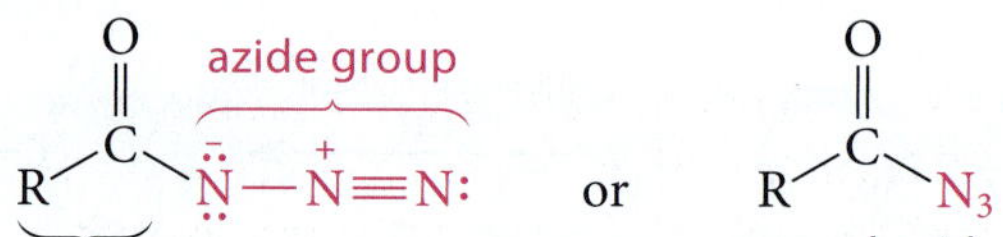

acyl group

an acyl azide

(23.81)

(The synthesis of acyl azides is discussed next—see Eqs. 23.88–23.90.) When an acyl azide is heated in an inert solvent such as benzene or toluene, it is transformed with loss of nitrogen into an **isocyanate**, a compound of the general structure R—N═C═O.

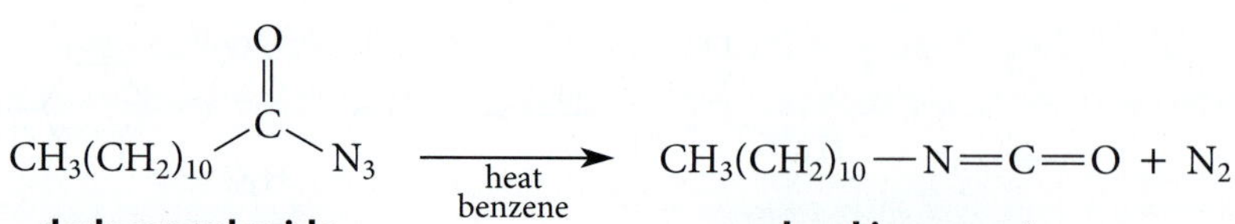

dodecanoyl azide

undecyl isocyanate
(81–86% yield)

(23.82)

This reaction, called the **Curtius rearrangement**, is a concerted reaction that can be represented as follows:

The rearrangement is named for its discoverer, Theodor Curtius (1857–1928), who was a professor of chemistry at Heidelberg University.

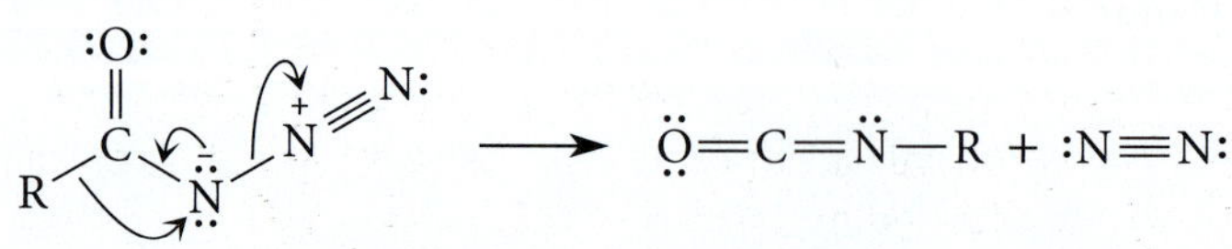

(23.83)

The isocyanate product of a Curtius rearrangement can be transformed into an amine by hydration in either acid or base. Hydration involves, first, the addition of water across the C=N bond to give a carbamic acid:

STUDY GUIDE LINK 23.2
Mechanism of the Curtius Rearrangement

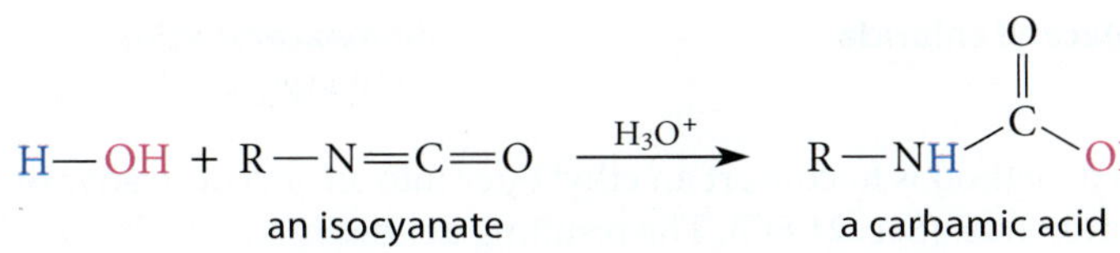

(23.84)

Carbamic acids are among those types of carboxylic acids that spontaneously decarboxylate (see Eq. 20.63b, Sec. 20.11A). Decarboxylation gives the amine, which is protonated under the acidic conditions of the reaction. The free amine is obtained by neutralization:

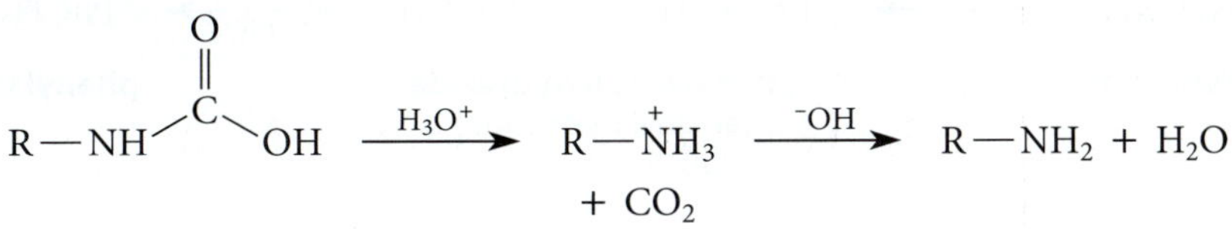

(23.85)

The overall transformation that occurs as a result of the Curtius rearrangement followed by hydration is the loss of the carbonyl carbon of the acyl azide as CO_2.

STUDY GUIDE LINK 23.3
Formation and Decarboxylation of Carbamic Acids

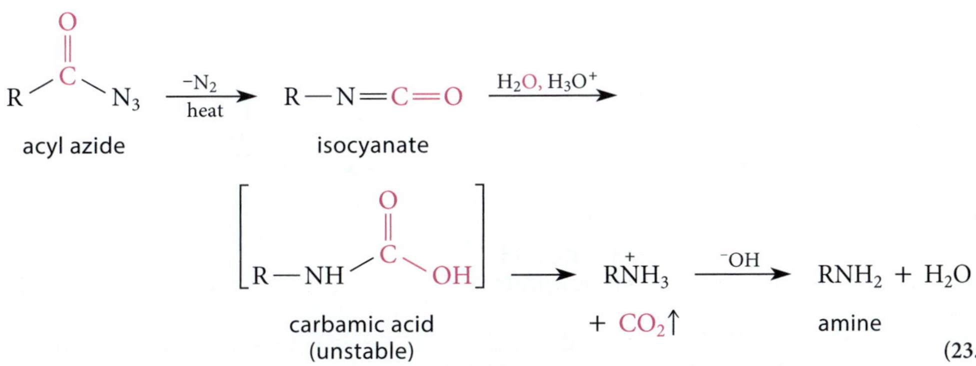

(23.86)

An important use of the Curtius rearrangement is for the preparation of carbamic acid derivatives (see Sec. 21.1G). Such derivatives are produced by allowing the isocyanate products to react with nucleophiles other than water. The reaction of isocyanates with alcohols or phenols yields carbamate esters, whereas the reaction with amines yields ureas.

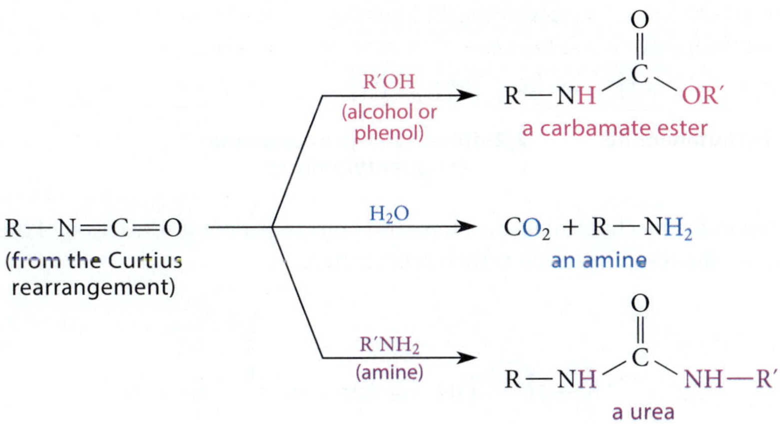

(23.87)

The key to the preparation of acyl azides used in the Curtius rearrangement is to recognize that these compounds are carboxylic acid derivatives. The most straightforward preparation is the reaction of an acid chloride with sodium azide.

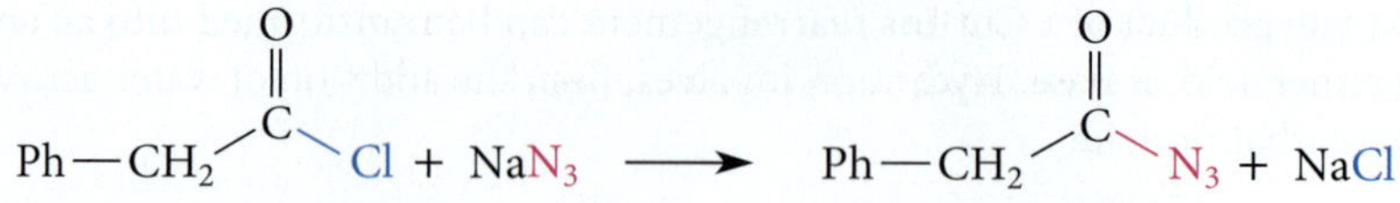

phenylacetyl chloride → **phenylacetyl azide** (an acyl azide) (23.88)

Another widely used method is to convert an ethyl ester into an acyl derivative of hydrazine ($H_2N—NH_2$) by aminolysis (Sec. 21.8C). The resulting amide, an *acyl hydrazide*, is then diazotized with nitrous acid to give the acyl azide.

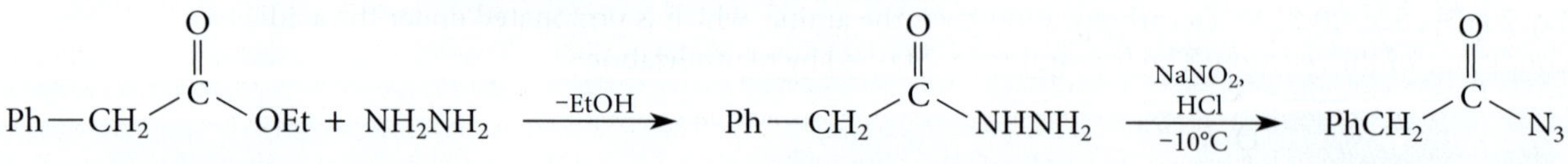

ethyl phenylacetate + **hydrazine** → **phenylacetyl hydrazide** (an acyl hydrazide) (80–100% yield) → **phenylacetyl azide** (23.89)

Notice the similarity of this diazotization to the diazotization of alkylamines:

Compare:

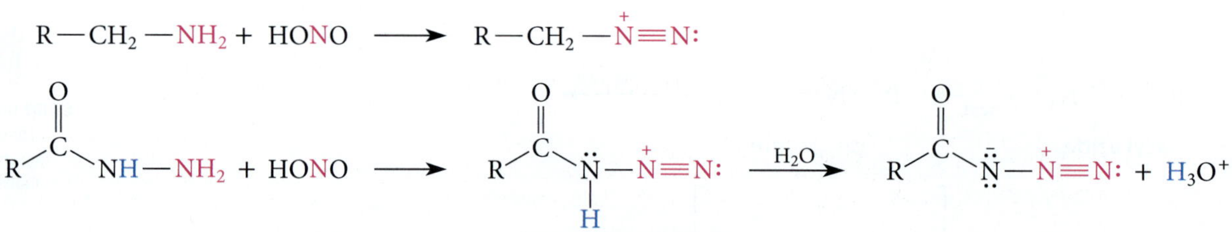

conjugate acid of the acyl azide (23.90)

Because the conjugate acid of the acyl azide is quite acidic (why?), it loses a proton from the adjacent nitrogen to the give the neutral acyl azide.

A reaction closely related to the Curtius rearrangement is the **Hofmann rearrangement** (also called the **Hofmann hypobromite reaction**). The starting material for this reaction is a primary amide rather than an acyl azide. Treatment of an amide with bromine in base gives rise to a rearrangement.

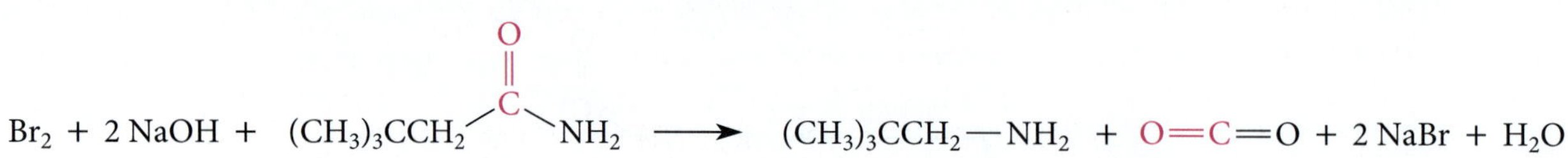

3,3-dimethylbutanamide → **2,2-dimethyl-1-propanamine (neopentylamine)** (23.91)

The first step in the mechanism of the Hofmann rearrangement is ionization of the amide N—H (Sec. 22.5); the resulting anion is then brominated.

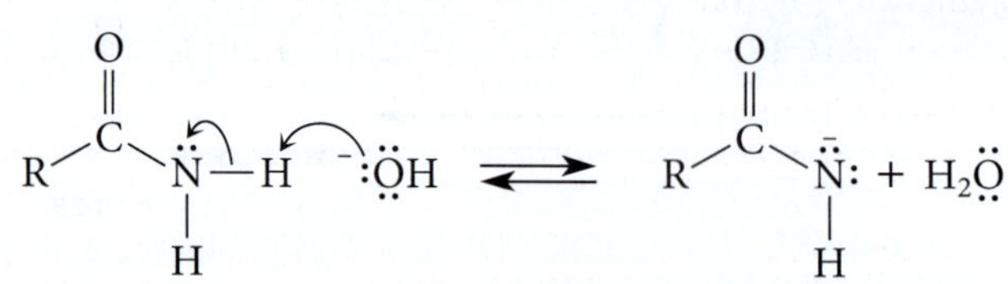

(23.92a)

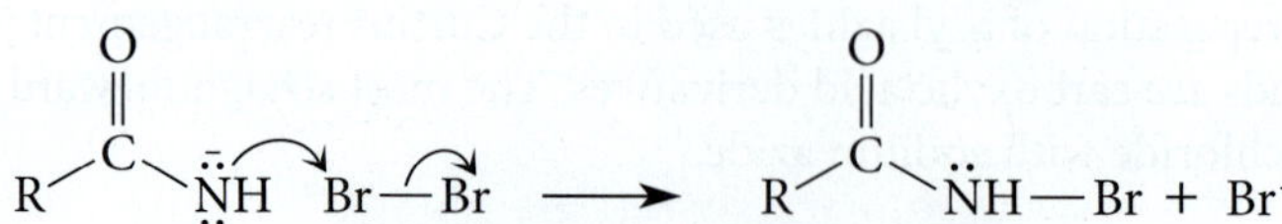

an *N*-bromoamide (23.92b)

(This reaction is analogous to the α-bromination of a ketone in base; Sec. 22.3B.) The *N*-bromoamide product is even more acidic than the amide starting material, and it too ionizes.

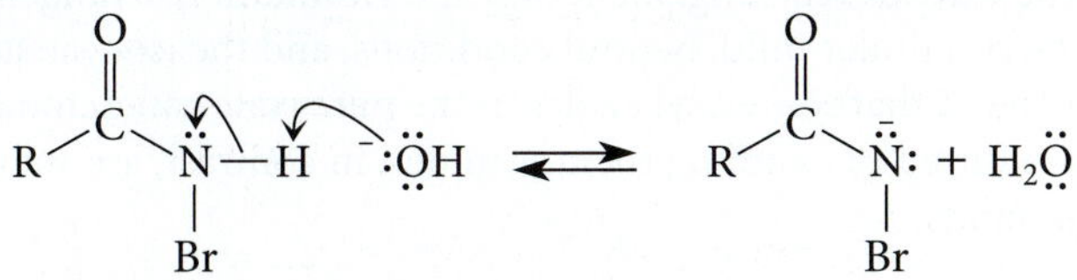

(23.92c)

The *N*-bromo anion then rearranges to an isocyanate.

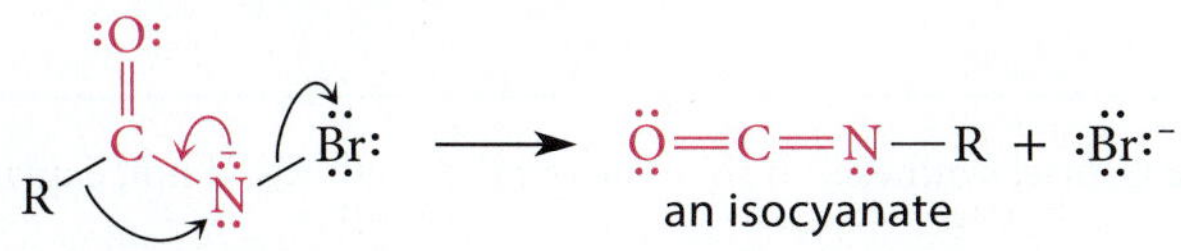

(23.92d)

Notice that the rearrangement steps of the Hofmann and Curtius reactions are conceptually identical; the only difference is the leaving group.

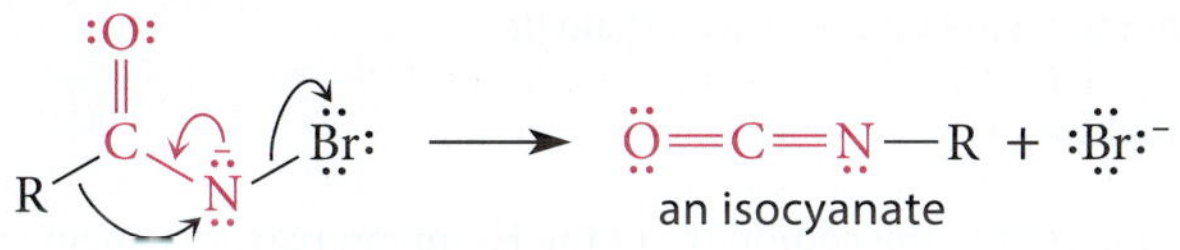

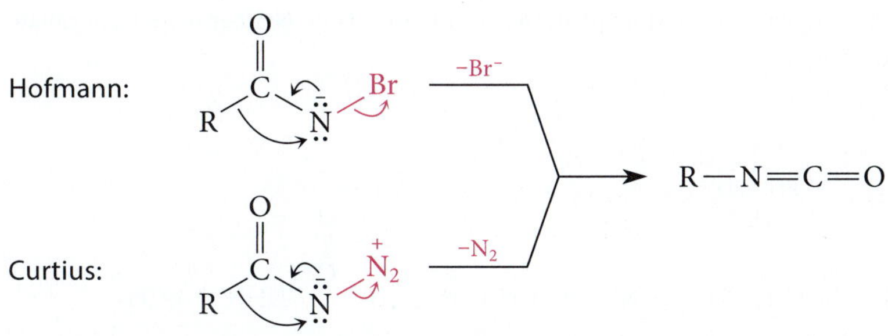

(23.93)

Because the Hofmann rearrangement is carried out in aqueous base, the isocyanate cannot be isolated as it is in the Curtius rearrangement. It spontaneously hydrates to form a carbamate ion, which then decarboxylates to the amine product under the strongly basic reaction conditions. (See Study Guide Link 23.3.)

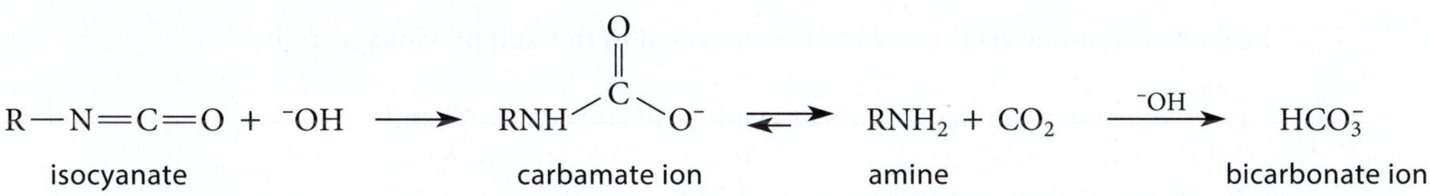

(23.94)

Although the reaction of amines with CO_2 is reversible, formation of the amine in the Hofmann rearrangement is driven to completion by the reaction of hydroxide ion with CO_2 to form bicarbonate ion (or carbonate ion) under the strongly basic conditions of the reaction.

An interesting and useful aspect of both the Hofmann and Curtius rearrangements is that they take place with complete *retention of stereochemical configuration* in the migrating alkyl group:

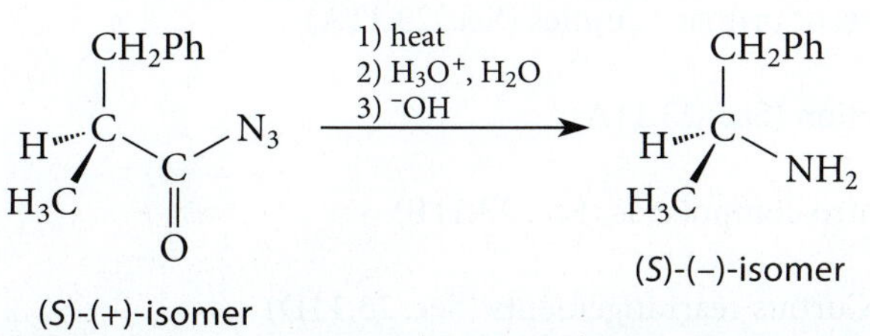

(23.95)

Equation 23.95 is a further illustration of the fact that there is no simple correlation between a compound's absolute configuration and the sign of its optical rotation.

Consequently, optically active carboxylic acid derivatives can be used to prepare optically active amines of known stereochemical configuration.

The advantage of the Curtius rearrangement over the Hofmann rearrangement is that the Curtius reaction can be run under mild, neutral conditions, and the isocyanate can be isolated if desired. The disadvantage is that some acyl azides in the pure state can detonate without warning, so extreme caution is required in handling them. Amides, in contrast, are stable and easily handled organic compounds.

Focused Problems

23.38 (a) Could *tert*-butylamine be prepared by the Gabriel synthesis? If so, write out the synthesis. If not, explain why.

(b) Propose a synthesis of *tert*-butylamine by another route.

23.39 Write a curved-arrow mechanism for each of the following reactions.

(a) Ethyl isocyanate ($CH_3CH_2—N{=}C{=}O$) with ethanol to yield ethyl *N*-ethylcarbamate

(b) Ethyl isocyanate with ethylamine to yield *N,N′*-diethylurea

23.40 What product is formed when 2-methylpropanamide is subjected to the conditions of the Hofmann rearrangement (a) in ethanol solvent? (b) in aqueous NaOH?

23.41 When hexanamide is subjected to the conditions of the Hofmann rearrangement, pentanamine (*A*) is obtained as expected. However, a significant by-product is *N,N′*-dipentylurea (*B*). Explain the origin of *B*. (*Hint:* Neither pentyl isocyanate nor pentanamine has appreciable solubility in aqueous base.)

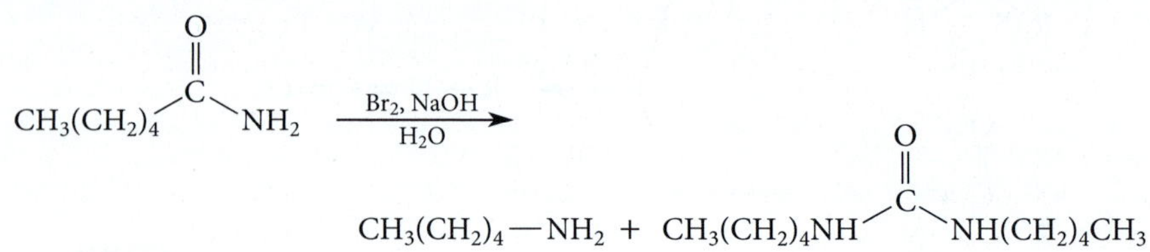

A *B*

E. Synthesis of Amines: Summary

The following amine syntheses have been covered in this and previous sections:

1. Reduction of amides and nitriles with $LiAlH_4$ (Secs. 21.9B and C)
2. Direct alkylation of amines (Sec. 23.7A)
3. Reductive amination (Sec. 23.7B)
4. Aromatic substitution reactions of anilines (Sec. 23.9)
5. Direct amination of aryl halides (Sec. 23.11C)
6. Gabriel synthesis of primary amines (Sec. 23.11A)
7. Staudinger reaction (Sec. 23.11A)
8. Reduction of nitro compounds (Sec. 23.11B)
9. Hofmann and Curtius rearrangements (Sec. 23.11D)

Methods 2, 3, 4, and 5 are used to prepare amines from other amines, and method 2 is really only useful for preparing quaternary ammonium salts. When an amide used in method 1 is prepared from an amine, this method, too, is one for obtaining one amine from another. Methods 6–9, as well as nitrile reduction in method 1, are limited to the preparation of primary amines, and methods 1, 3, 8, and 9 can be used for obtaining amines from other functional groups.

Focused Problem

23.42 Show how 2-cyclopentyl-*N*,*N*-dimethylethanamine could be synthesized from each of the following starting materials.

(a) 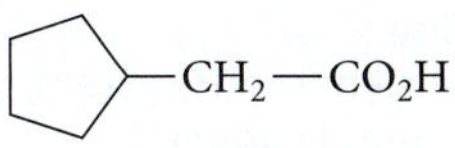

(b) 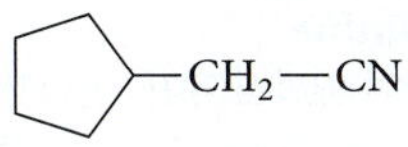

(c) 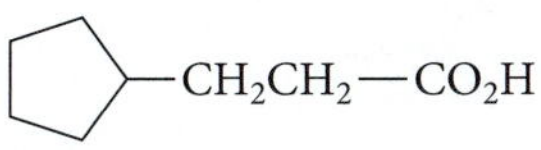

(d) 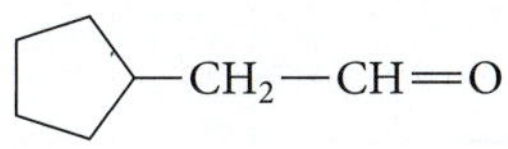 (two ways)

23.12 USE AND OCCURRENCE OF AMINES

A. Industrial Use of Amines and Ammonia

Chemistry in the Real World

Among the relatively few industrially important amines is hexamethylenediamine, $H_2N(CH_2)_6NH_2$, used in the synthesis of nylon-6,6 (Sec. 21.12A). Ammonia is also an important "amine" and is a key source of nitrogen in a number of manufacturing processes. In agricultural chemistry, for example, liquid ammonia itself and urea, which is made from ammonia and CO_2, are important nitrogen fertilizers. Ammonia is manufactured by the hydrogenation of N_2. Although it might not seem that the industrial synthesis of ammonia has anything to do with organic chemistry, the hydrogen used in its manufacture in fact comes from the cracking of alkanes (Sec. 5.10). Thus, the availability of ammonia is presently tied to the availability of hydrocarbons. However, there is significant interest in the development of methods for utilizing solar energy for *water splitting*—the conversion of water into H_2 and O_2. Should water splitting become practical, the production of ammonia would be completely uncoupled from the availability of petroleum.

B. Naturally Occurring Amines

Chemical Biology Topic

Alkaloids Among the many types of naturally occurring amines are the **alkaloids**—nitrogen-containing bases that occur naturally in plants. This simple definition encompasses a highly diverse group of compounds; the structures of a few alkaloids are shown in **Fig. 23.4**. Because amines are the most common organic bases, it is not surprising that most alkaloids are amines, including heterocyclic amines. It is believed that the first alkaloid ever isolated and studied was morphine, discovered in 1805. Many alkaloids have biological activity (see Fig. 23.4); others have no known activity, and their functions within the plants from which they come are, in many cases, obscure. Investigations dealing with the isolation, structure, and medicinal properties of alkaloids continue to be major research activities in organic chemistry.

Focused Problem

23.43 Illustrate the Brønsted basicity of (a) morphine and (b) mescaline (see Fig. 23.4) by giving the structures of their conjugate acids.

FIGURE 23.4 Structures of some alkaloids. Each compound has at least one basic amine group.

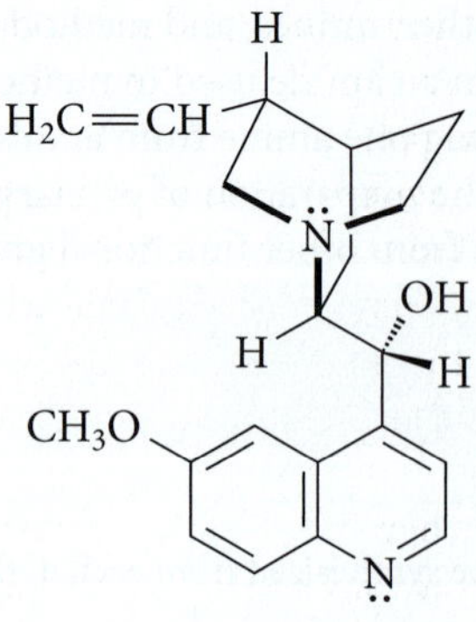

quinine
(an antimalarial drug)

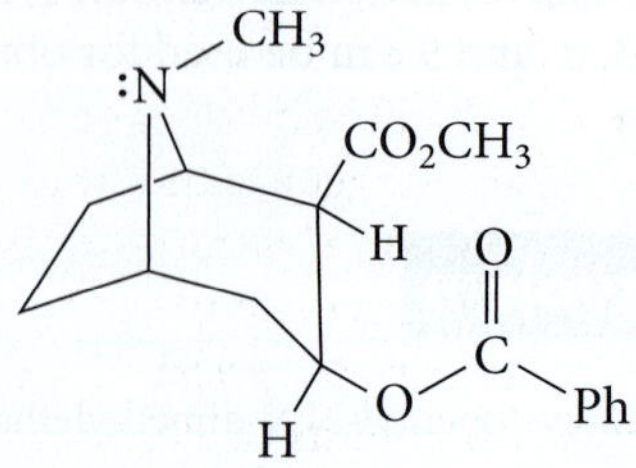

cocaine
(a stimulant of the central nervous system; induces euphoria; widely abused)

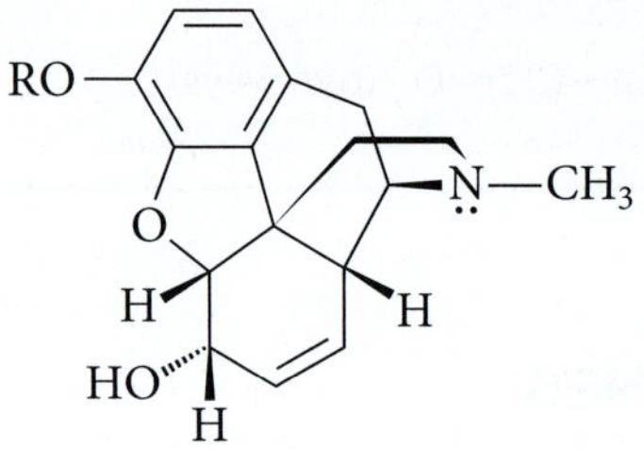

morphine (R = H)
codeine (R = CH_3)
(medically important analgesics)

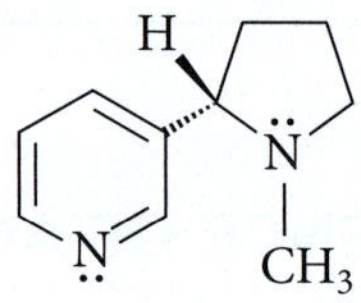

nicotine
(the principal alkaloid from tobacco)

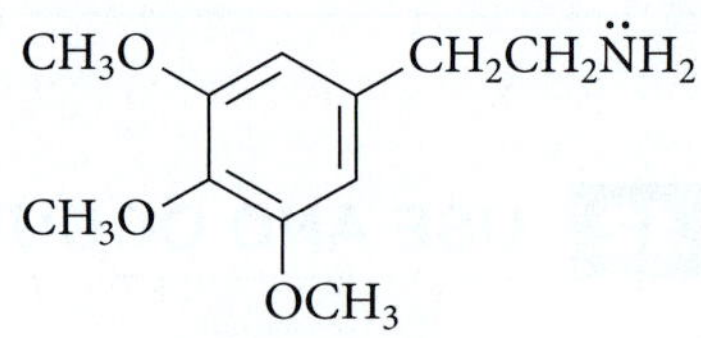

mescaline
(a hallucinogen from peyote cactus)

Hormones and Neurotransmitters Epinephrine (adrenaline) is an amine secreted by both the adrenal medulla and sympathetic nerve endings; it is an example of a **hormone**—a compound that regulates the biochemistry of multicellular organisms, particularly vertebrates.

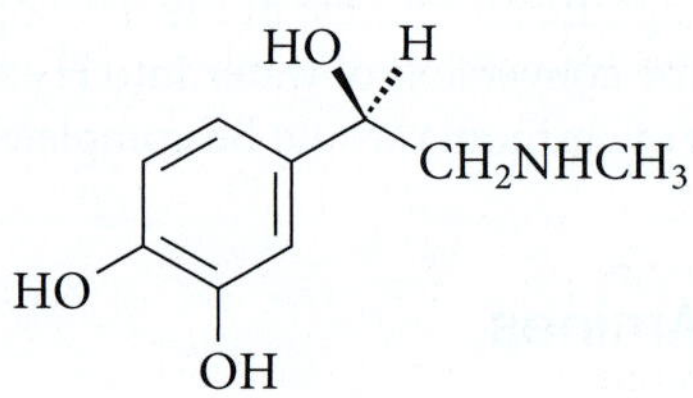

epinephrine

Epinephrine, for example, is associated with the "fight-or-flight" response to external stimuli; you might feel the effects of epinephrine secretion when you walk unprepared into your organic chemistry class and your instructor says, "Pop quiz today." The mechanisms by which hormones exert their effects are important research areas in contemporary biochemistry.

Norepinephrine, another amine, and acetylcholine, a quaternary ammonium ion, are examples of *neurotransmitters.*

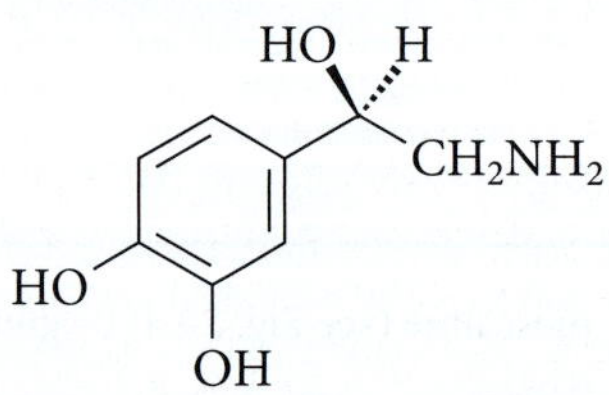

norepinephrine

acetylcholine

Neurotransmitters are molecules that are involved in the communication between nerve cells or between nerve cells and their target organs. This communication occurs at cellular junctions called *synapses.* A nerve impulse is transmitted when a neurotransmitter is released from a nerve cell on one side of the synapse, moves by diffusion across the synapse, and binds to a protein receptor molecule of another nerve cell or a target organ on the other side. (The involvement of pi–cation interactions in the binding of acetylcholine to its receptor protein is discussed in Sec. 15.8C.) This binding triggers either the transmission of the impulse down the nerve cell to the next synapse or a response by the target organ. Different neurotransmitters are involved in different parts of the nervous system.

Significant advances have occurred in understanding the chemistry that takes place in the human brain (*neurochemistry*). These advances are being made by teams of molecular biologists, biochemists, and organic chemists. It is conceivable that a deeper understanding of neurochemistry will lead to treatments for such widespread and tragic afflictions as Parkinson's disease and Alzheimer's disease. Sigmund Freud perhaps anticipated these developments when he wrote in 1930, "The hope of the future lies in organic chemistry."

CHAPTER SUMMARY *For a summary of the chapter, see Chapter 23 in the* Study Guide and Solutions Manual.

REACTION REVIEW *For a summary of reactions discussed in this chapter, see the* Reaction Review *section of Chapter 23 in the* Study Guide and Solutions Manual.

SKILLS OBJECTIVES WITH PROBLEMS

- **Given the name of an amine, draw its structure. Given the structure of an amine, provide its name using IUPAC substitutive nomenclature or, when appropriate, common nomenclature. 23.1**

23.44 Draw the structure of each of the following compounds.

(a) *tert*-Butylamine

(b) 2,2-Dimethyl-3-hexanamine

(c) *N,N*-Dimethyl-2-hexanamine

(d) *N*-Ethyl-*N*-methylisobutylamine

23.45 Give an acceptable name for each of the following compounds.

(a)

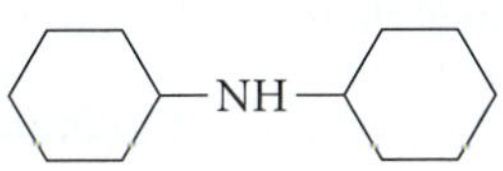

(b) $CH_3NHCHCH_2CH_2OH$ with CH_2CH_3 on the CH

(c)

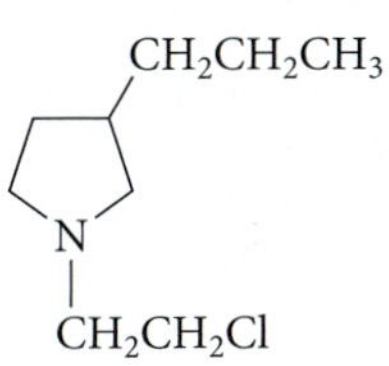

- **In a general way, show how the physical properties of amine compare with those of other functional classes. 23.3**

23.46 In each group, which compound would you expect to have the higher boiling point?

(a) CH_3NHCH_3 or $CH_3CH_2CH_3$

(b) $CH_3CH_2CH_2OH$ or $CH_3CH_2CH_2NH_2$

(c) $CH_3—\ddot{N}(CH_3)—CH_3$ or $CH_3CH_2—\ddot{N}(H)—CH_3$

- **Know the unique characteristics of the spectra of amines. Use spectra to determine the structures of unknown amines. 23.4**

23.47 Imagine that you have samples of the following four isomeric amines, but you don't know which is which. Explain how you could use proton NMR to distinguish among them.

A: $PhCH_2CH(NH_2)CH_3$

B: $PhCH_2N(CH_3)_2$

C: $PhCH(NH_2)CH_2CH_3$

D: $PhCH_2CH_2NHCH_3$

23.48 Propose a structure for the compound A ($C_6H_{15}O_2N$) that is unstable in aqueous acid and has the following NMR spectra:

Proton NMR: δ 2.30 (6H, s); δ 2.45 (2H, d, $J = 6$ Hz); δ 3.27 (6H, s); δ 4.50 (1H, t, $J = 6$ Hz)

^{13}C NMR: δ 46.3, δ 53.2, δ 68.8, δ 102.4

(*Hint:* Upon exposure to aqueous acid, compound A yields a product that shows an IR C=O stretch in the aldehyde region.)

- **Know the approximate pK_a values for amines and the conjugate acids of amines. Predict how the acidity and basicity of amines is affected by alkyl substitution, polar effects, and resonance effects. 23.5**

23.49 Arrange the amines within each set in order of increasing basicity in aqueous solution, least basic first.

(a) Ammonia, ethylamine, ethylmethylamine

(b) $H_3\overset{+}{N}CH_2CH_2NH_2$, $H_3\overset{+}{N}CH_2CH_2\overset{+}{N}H_3$, $H_2NCH_2CH_2NH_2$

(c) Aniline, 3-nitroaniline, 2,6-dinitroaniline, 4-nitroaniline

23.50 Explain the basicity order of the following three amines: p-chloroaniline (A), o-chloroaniline (B), and aniline (C). The structures and pK_a data are shown in Table 23.1.

23.51 Is the basicity of trifluralin, a widely used herbicide, much greater, about the same, or much less than the basicity of N,N-diethylaniline? Explain.

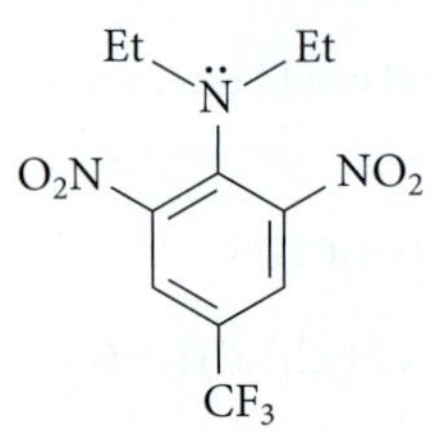

trifluralin

23.52 The following structure is given on the package insert for the drug *labetalol* (used in the control of blood pressure and hypertension):

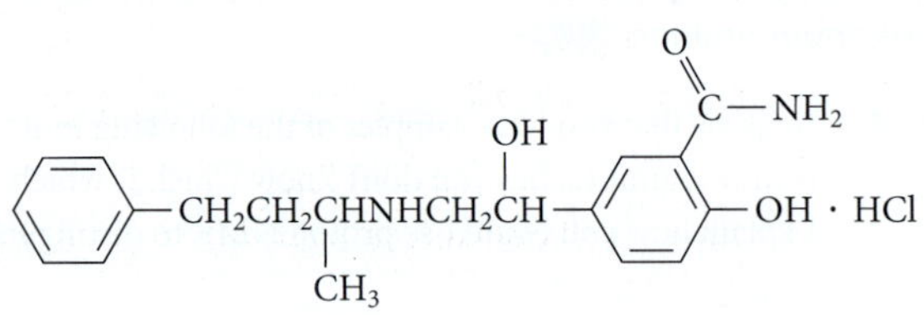

labetalol hydrochloride

(a) Labetalol is claimed to be a salt. Explain by giving a more detailed structure.

(b) What happens to labetalol•HCl when it is treated with one equivalent of NaOH at room temperature?

(c) What happens to labetalol when it is treated with an excess of aqueous NaOH and heat?

(d) What are the products formed when labetalol is treated with 6 M aqueous HCl and heat?

- **Predict the properties of quaternary ammonium salts on the basis of their structure. 23.6**

23.53 One of the following amines is optically active, whereas the other is not. Explain.

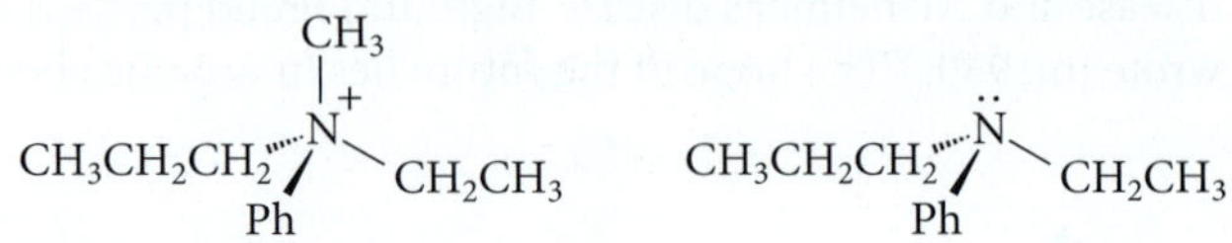

- **Provide missing starting materials or products in the reactions of amines.**

(1) Direct alkylation and quaternization of amines **23.7A**

(2) Reductive amination **23.7B**

(3) Acylation of amines **23.7C**

(4) Hofmann elimination of quaternary ammonium hydroxides **23.8**

(5) Aromatic substitutions with aniline derivatives and with diazonium ions **23.9, 23.10B**

(6) Formation and substitution reactions of diazonium salts **23.10A**

(7) Reactions of amines with nitrous acid **23.10C**

23.54 Give the principal organic product(s) expected when p-chloroaniline or other compound indicated reacts with each of the following reagents.

(a) Dilute HBr

(b) CH_3CH_2MgBr in ether

(c) $NaNO_2$, HCl, 0 °C

(d) p-Toluenesulfonyl chloride

(e) Product of part (c) with H_2O, Cu_2O, and excess $Cu(NO_3)_2$

(f) Product of part (c) with CuBr

(g) Product of part (c) with H_3PO_2

(h) Product of part (c) with CuCN

(i) Product of part (d) + NaOH, 25 °C

23.55 Give the principal organic product(s) expected when *N*-methylaniline reacts with each of the following reagents.

(a) Br_2

(b) Benzoyl chloride and pyridine

(c) Benzyl chloride (excess), then dilute NaOH

(d) *p*-Toluenesulfonic acid

(e) $NaNO_2$, HCl

(f) Excess CH_3I, heat, then Ag_2O

(g) $CH_3CH{=}O$, $NaBH(OAc)_3$, and HOAc in $ClCH_2CH_2Cl$, then KOH

(h) Chlorobenzene, K^+ *t*-BuO^-, and a Pd(0) catalyst

23.56 Give the principal organic product(s), if any, expected when isopropylamine or other compound indicated reacts with each of the following reagents.

(a) Dilute H_2SO_4

(b) Dilute NaOH solution

(c) Butyllithium in THF, −78 °C

(d) Acetyl chloride, pyridine

(e) $NaNO_2$, aqueous HBr, 0 °C

(f) Acetone, H_2, Pd/C

(g) Excess CH_3I, heat

(h) Benzoic acid, 25 °C

(i) Formaldehyde, $NaBH_3CN$, EtOH

(j) 2,4-Dimethylchlorobenzene, K^+ *t*-BuO^-, and a Pd(0) catalyst

(k) Product of part (g) + Ag_2O, then heat

(l) Product of part (d) with $LiAlH_4$, then H_3O^+, then ^-OH

• Provide missing starting materials or products in the synthesis of amines. 23.11A

23.57 Propose a synthesis for benzylamine, $PhCH_2NH_2$, using both the Gabriel synthesis and the Staudinger reaction. Provide all starting materials and reagents necessary.

23.58 Provide two other syntheses of benzylamine, $PhCH_2NH_2$, that start with different carboxylic acid derivatives.

• Provide reaction mechanisms for Curtius and Hoffman rearrangements to form amines. 23.11D

23.59 Draw a curved-arrow mechanism for each of the rearrangement reactions shown.

(a)

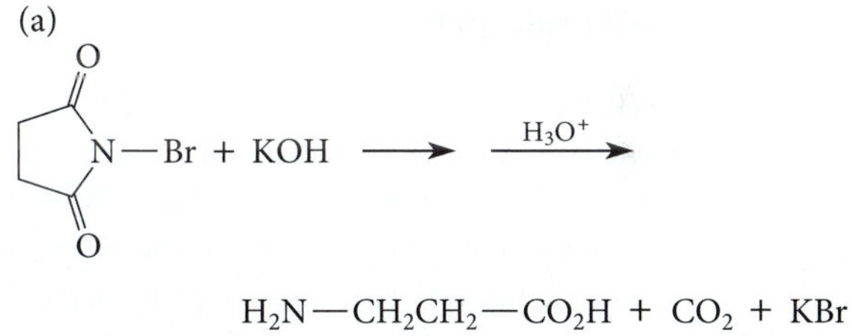

(b)

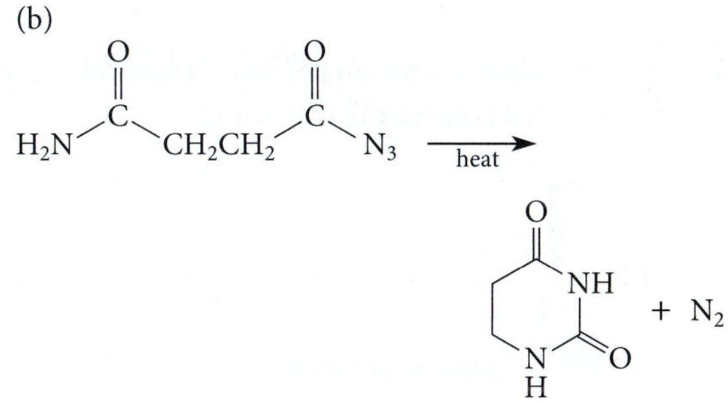

INTEGRATED PROBLEMS

23.60 Give the structure of a compound that fits each description. (There may be more than one correct answer for each.)

(a) A chiral primary amine C_4H_7N with no triple bonds

(b) A chiral primary amine $C_4H_{11}N$

(c) Two secondary amines, which, when treated with CH_3I, then Ag_2O and heat, give propene and *N,N*-dimethylaniline

(d) A compound C_4H_9N that reacts with $NaBH(OAc)_3$ and 1 equivalent of HOAc, then KOH, to give *N*-methyl-2-propanamine

23.61 Explain how you would distinguish the compounds within each set by a simple chemical test with readily observable results, such as solubility in acid or base, evolution of a gas, and so forth.

(a) *N*-Methylhexanamide; 1-octanamine; *N,N*-dimethyl-1-hexanamine

(b) *p*-Methylaniline, benzylamine, *p*-cresol, anisole

23.62 (a) Give the structure of cocaine (see Fig. 23.4) as it would exist in 1 *M* aqueous HCl solution.

(b) What products would form if cocaine were treated with an excess of aqueous NaOH and heat?

(c) What products would form if cocaine were treated with an excess of concentrated aqueous HCl and heat?

23.63 Design a separation of a mixture containing the following four compounds into its pure components. Describe exactly what you would do and what you would expect to observe.

nitrobenzene, aniline, *p*-chlorophenol, and *p*-nitrobenzoic acid

23.64 (a) When anthranilic acid is treated with $NaNO_2$ in aqueous HCl solution, and the resulting solution is treated with *N,N*-dimethylaniline, a dye called *methyl red* is formed. Give the structure of methyl red.

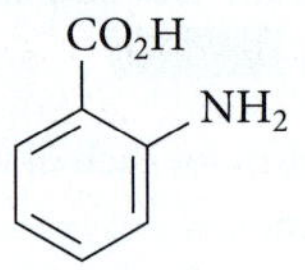

anthranilic acid

(b) When an acidic solution of methyl red is titrated with base, the dye behaves as a diprotic acid with pK_a values of 2.3 and 5.0. The color of the methyl red solution changes very little as the pH is raised past 2.3, but, as the pH is raised past 5.0, the color of the solution changes dramatically from red to yellow. Explain.

23.65 Alizarin yellow R is an azo dye that changes color from yellow to red between pH 10.2 and 12.2.

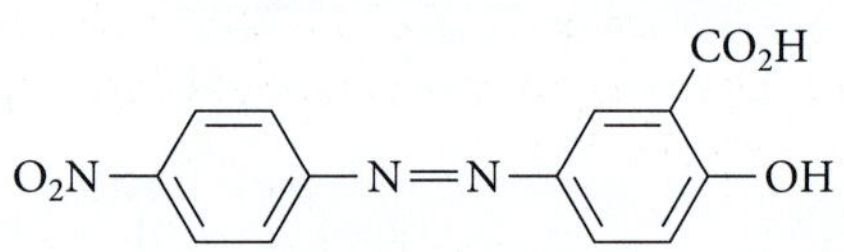

alizarin yellow R

(a) Outline a synthesis of alizarin yellow R from aniline, salicylic acid (*o*-hydroxybenzoic acid), and any other reagents.

(b) Draw the structure of alizarin yellow R as it exists in its yellow form at pH = 9. Note that the conjugate acid of a diazo group has a pK_a near 5.

(c) Draw the structure of alizarin yellow R as it exists at pH > 12. Why does it change color?

23.66 Amanda Amine, an organic chemistry student, has proposed the reactions given here. Indicate in each case why the reaction would not succeed as written.

(a)

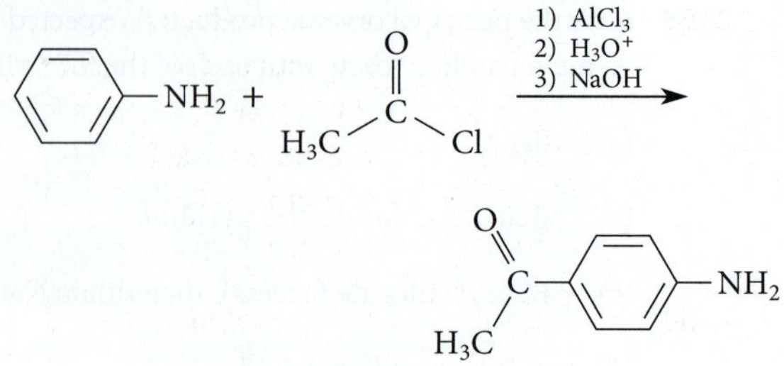

(b)

$$Me_3C{-}NH_2 + CH_3I \text{ (excess)} \longrightarrow \xrightarrow{^-OH} Me_3C{-}NH{-}CH_3 + I^-$$

(c)

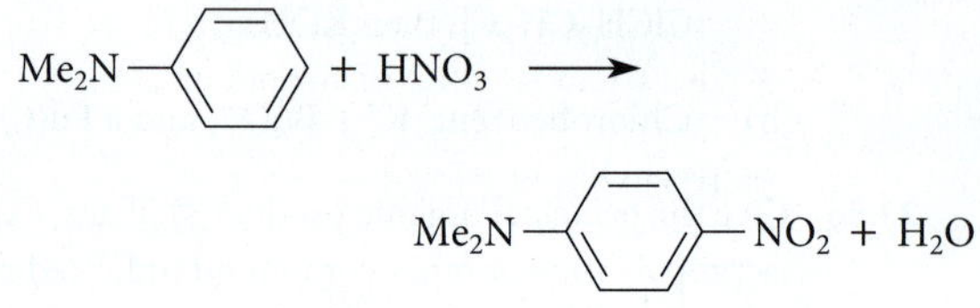

(d)

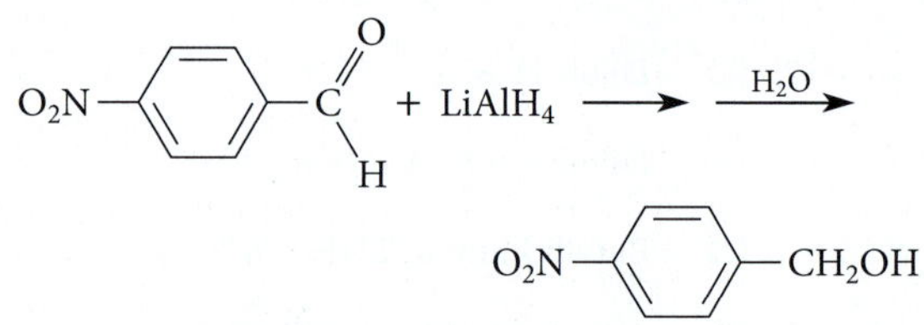

(e)

$$(CH_3)_2NH + CH_3CH_2CH_2OH \xrightarrow{H_2SO_4} CH_3CH_2CH_2N(CH_3)_2 + H_2O$$

(f)

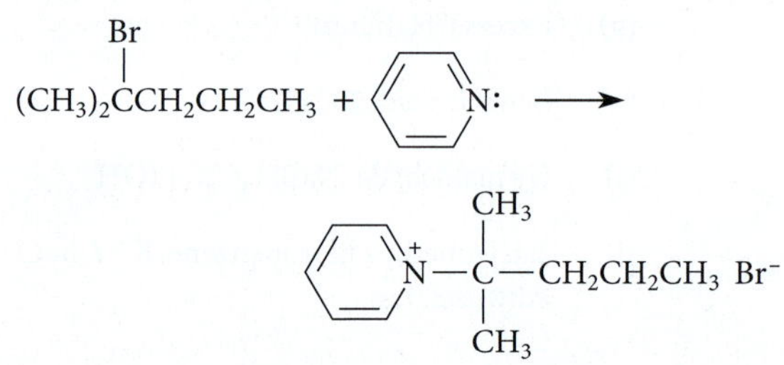

23.67 Outline a sequence of reactions that would bring about the conversion of aniline into each of the following compounds.

(a) Benzylamine

(b) Benzyl alcohol

(c) 2-Phenylethanamine

(d) *N*-Phenyl-2-butanamine

(e) *p*-Chlorobenzoic acid

(f) Diphenylamine

23.68 When *p*-aminophenol reacts with one molar equivalent of acetic anhydride, a compound acetaminophen (A, $C_8H_9NO_2$) is formed that dissolves in dilute NaOH. When A is treated with one equivalent of NaOH followed by ethyl iodide, an ethyl ether B is formed. What is the structure of acetaminophen? Explain your reasoning.

23.69 When 1,5-dibromopentane reacts with ammonia, among several products isolated is a water-soluble compound A that rapidly gives a precipitate of AgBr with acidic $AgNO_3$ solution. Compound A is unchanged when treated with dilute base, but treatment of A with concentrated NaOH and heat gives a new compound B ($C_{10}H_{19}N$) that decolorizes Br_2 in CCl_4. Compound B is identical to the product obtained from the reaction sequence shown. Identify A and B, and explain your reasoning.

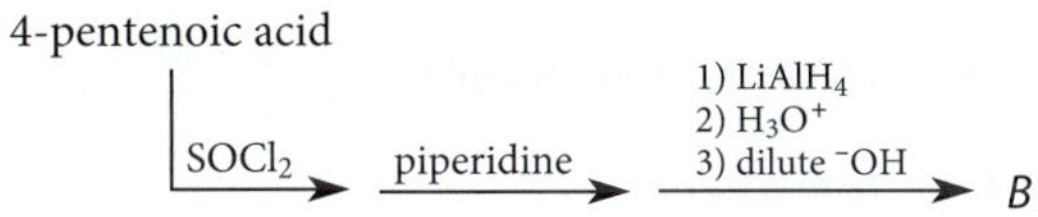

23.70 Give an explanation for each of the following facts.

(a) The barrier to internal rotation about the *N*-phenyl bond in *N*-methyl-*p*-nitroaniline is considerably higher (42–46 kJ mol^{-1}, or 10–11 kcal mol^{-1}) than that in *N*-methylaniline itself (about 25 kJ mol^{-1}, or 6 kcal mol^{-1}).

(b) *Cis*- and *trans*-1,3-dimethylpyrrolidine rapidly interconvert.

(c) $CH_3NH—CH_2—NHCH_3$ is unstable in aqueous solution.

(d) The following compound exists as the enamine isomer shown rather than as an imine:

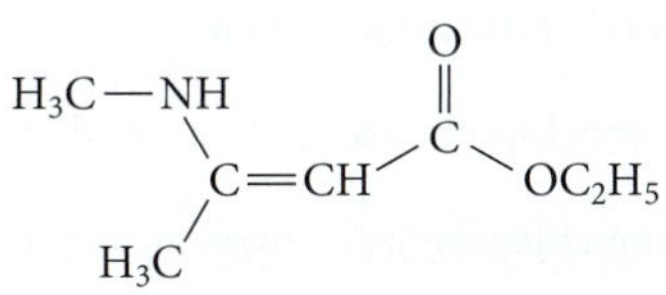

(e) Diazotization of 2,4-cyclopentadien-1-amine gives a diazonium salt, which, unlike most aliphatic diazonium ions, is relatively stable and does not decompose to a carbocation.

23.71 Imagine that you have been given a sample of racemic 2-phenylbutanoic acid. Outline steps that would allow you to obtain pure samples of each of the following compounds from this starting material and any other reagents. (Enantiomeric resolutions are time-consuming, so one resolution that would serve all five syntheses would be most efficient.)

(a)

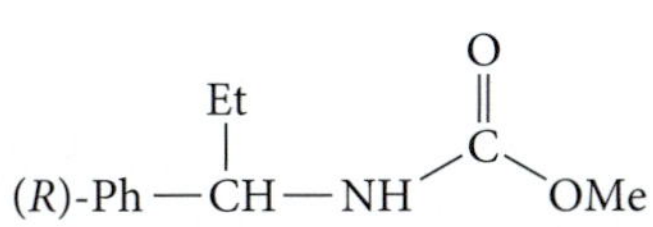

(b)

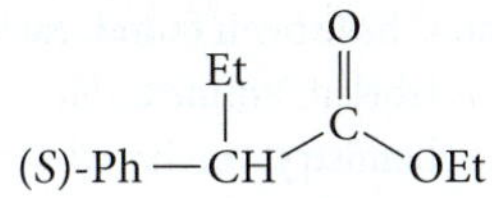

(c)

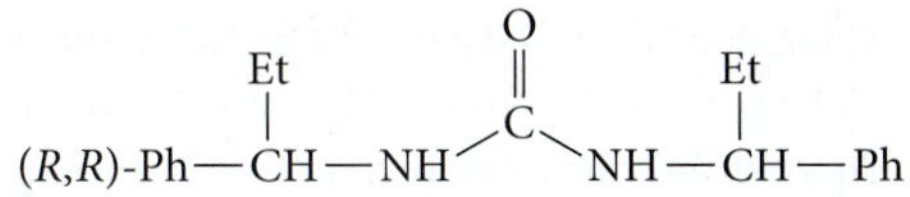

(d)

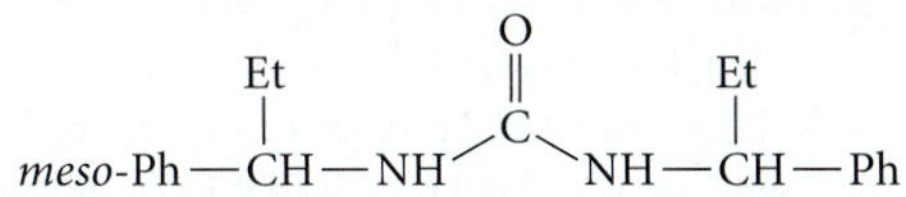

(e)

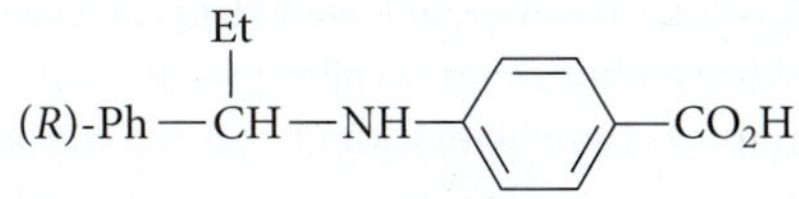

23.72 Offer an explanation for each of the following observations, including the structure of each product and the role of the quaternary ammonium salt.

(a) When sodium benzenethiolate, $Na^+ PhS^-$, is mixed with 1-bromooctane in water, no reaction takes place. However, when 1–2 mol % tetrabutylammonium bromide, $(CH_3CH_2CH_2CH_2)_4N^+ Br^-$, is included in the reaction mixture, a product is formed readily.

(b) When phenylacetonitrile, Ph—CH_2—C≡N, is mixed with aqueous sodium hydroxide and 1,4-dibromobutane, a separate organic layer forms and no reaction takes place. However, when the three components were rapidly stirred in dichloromethane solvent with a few mole percent of tetrabutylammonium bromide [structure in part (a)], a compound with the formula $C_{12}H_{13}N$ was formed in high yield.

(c) When morpholine (Display 23.4) and bromobenzene are allowed to react in toluene solvent in the presence of the catalyst Pd[P(*t*-Bu)$_3$]$_2$, a reaction takes place when aqueous NaOH is used as the base and 1 mol % of cetyltrimethylammonium bromide, $CH_3(CH_2)_{14}CH_2\overset{+}{N}(CH_3)_3\ \overset{-}{Br}$, is added to the reaction mixture.

23.73 A compound A ($C_{22}H_{27}NO$) is insoluble in acid and base but reacts with concentrated aqueous HCl and heat to give a clear aqueous solution from which, on cooling, benzoic acid precipitates. When the supernatant solution is made basic, a liquid B separates. Compound B is achiral. Treatment of B with benzoyl chloride in pyridine gives back A. Evolution of gas is not observed when B is treated with an aqueous solution of $NaNO_2$ and HCl. Treatment of B with excess CH_3I, then Ag_2O and heat, gives a compound C, $C_9H_{19}N$, plus styrene, Ph—CH=CH_2. Compound C, when treated with excess CH_3I, then Ag_2O and heat, gives a *single* alkene D that is identical to the compound obtained when cyclohexanone is treated with the ylid $:\overset{-}{C}H_2—\overset{+}{P}Ph_3$. Give the structure of A, and explain your reasoning.

23.74 Three bottles *A*, *B*, and *C* have been found, each of which contains a liquid and is labeled "amine $C_8H_{11}N$." As an expert in amine chemistry, you have been hired as a consultant and asked to identify each compound. Compounds *A* and *B* give off a gas when they react with $NaNO_2$ and HCl at 0 °C; *C* does not. However, when the aqueous reaction mixture from the diazotization of *C* is warmed, a gas is evolved. Compound *A* is optically inactive, but when it reacts with (+)-tartaric acid, two isomeric salts with different physical properties are obtained. Titration of *C* with aqueous HCl reveals that its conjugate acid has a $pK_a = 5.1$. Oxidation of *C* with H_2O_2 (a reagent known to oxidize amino groups to nitro groups), followed by vigorous oxidation with $KMnO_4$, gives *p*-nitrobenzoic acid. Oxidation of *B* in a similar manner yields 1,4-benzenedicarboxylic acid (terephthalic acid), and oxidation of *A* yields benzoic acid. Identify compounds *A*, *B*, and *C*.

23.75 Complete the reactions that follow by giving the structure(s) of the major product(s). Explain how you arrived at your answers.

(a) $NH_3 + HNO_2 \longrightarrow$

(b)

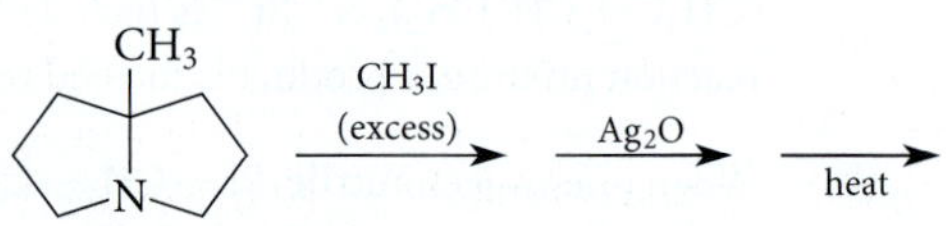

(c)

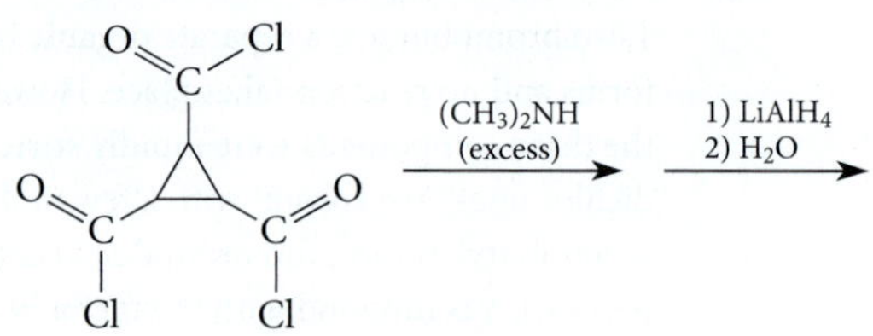

(d) Product of part (c) $\xrightarrow{\text{MeI (excess)}}$ $\xrightarrow{Ag_2O}$ $\xrightarrow[\text{heat}]{}$ (an isomer of benzene)

(e)

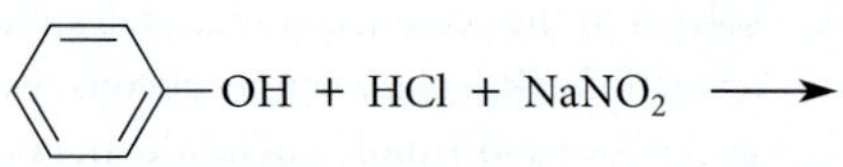

(f)

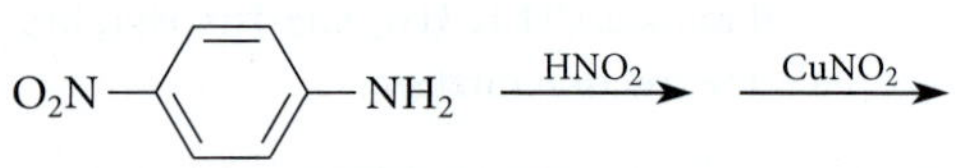

(g)

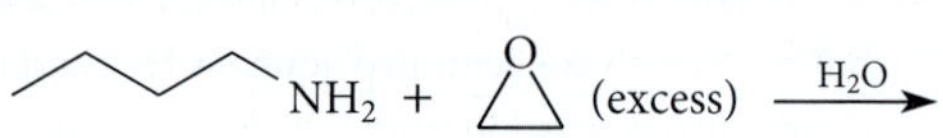

(h)

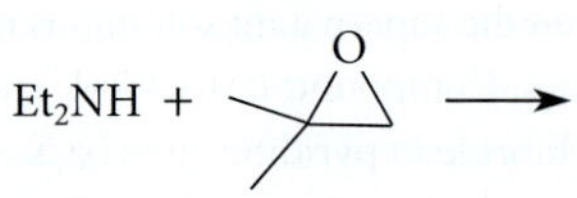

(i)

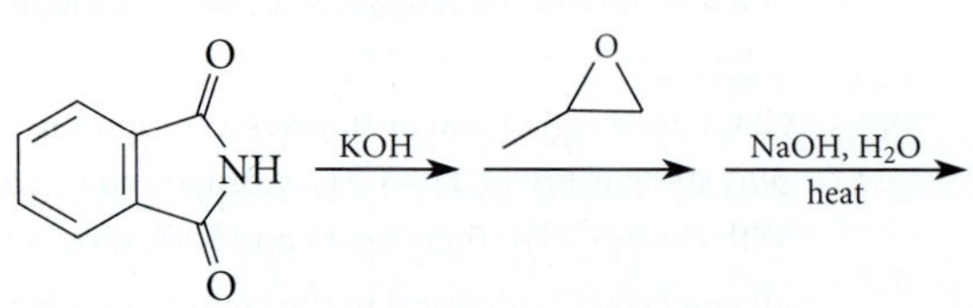

(j)

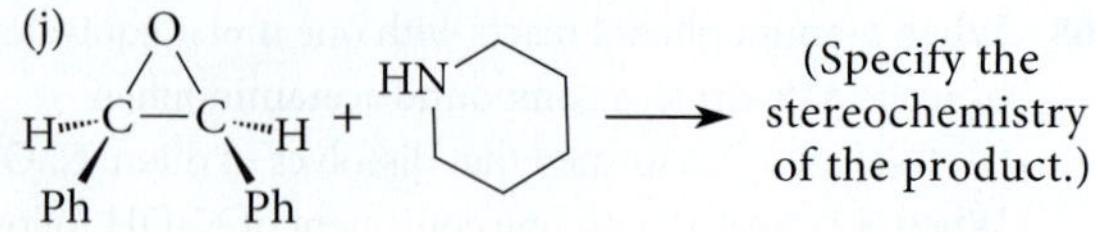

(k)

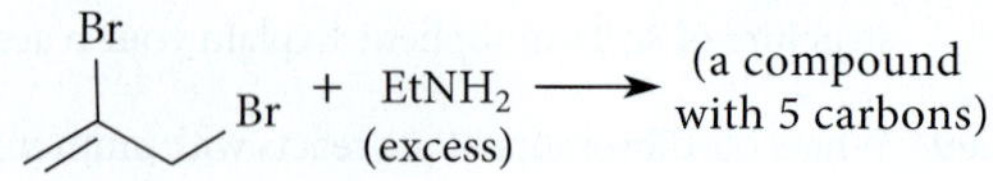

(l)

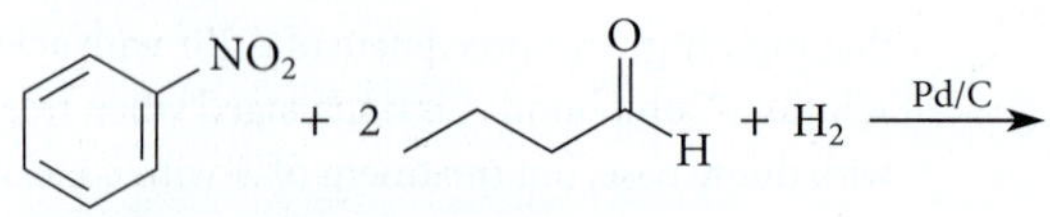

23.76 Outline a synthesis for each of the following compounds from the indicated starting materials and any other reagents. The starting material for the compounds in parts (a) through (e) is pentanoic acid.

(a) *N*-Methyl-1-hexanamine

(b) Pentylamine

(c) *N*,*N*-Dimethyl-1-pentanamine

(d) Butylamine

(e) Hexylamine

(f) $PhCH_2{-}\overset{+}{N}(CH_3)_2{-}CH_2CH_2CH_2CH_3\ Br^-$ from butyraldehyde

(g) *N*-Ethyl-3-phenyl-1-propanamine from toluene

(h) 2-Pentanamine from diethyl malonate

(i) Isobutylamine from acetone

(j) Isopentylamine from acetone

(k) *m*-Bromochlorobenzene from nitrobenzene

(l) *p*-Bromochlorobenzene from nitrobenzene

(m) *p*-Methoxybenzonitrile from phenol

(n) (*S*)-$CH_3CH(D)CH_2NH_2$ from (*R*)-$CH_3CH(D)OH$

(o)

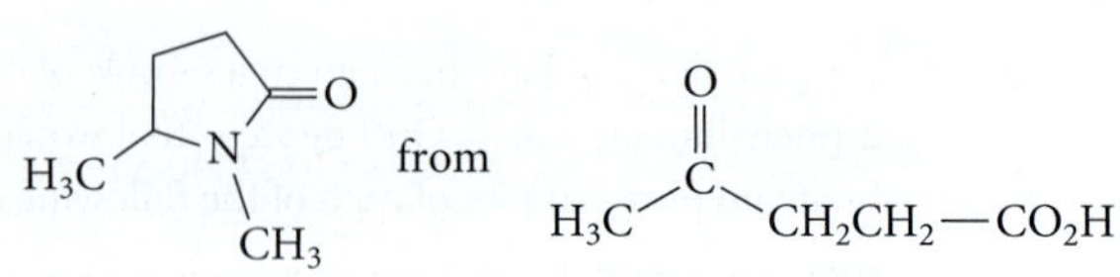

levulinic acid

(p) 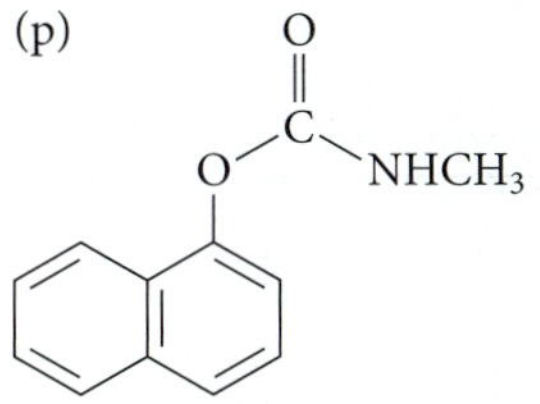 from $H_3C—N{=}C{=}O$

carbaryl (an insecticide) — **methyl isocyanate**

23.77 (a) An amine *A* has an EI mass spectrum with a base peak at $m/z = 72$. An amine *B* has an EI mass spectrum with a base peak at $m/z = 58$. One of the amines is 2-methyl-2-heptanamine, and the other is *N*-ethyl-4-methyl-2-pentanamine. Which is which? (*Hint:* A major fragmentation mechanism of amines in EI and CI mass spectra is α-cleavage; see Eq. 13.30, Sec. 13.6D.)

(b) Tributylamine has a CI mass spectrum with a strong M + 1 peak and one other major peak resulting from α-cleavage. At what m/z value does this peak occur?

23.78 In the NMR spectrum of a concentrated (4.5 *M*) aqueous solution of methylamine, the methyl group appears as a quartet when the solution pH = 1. At intermediate pH, the methyl group appears as a broad line. At pH = 9, the methyl group is observed as a single sharp line. Explain these observations.

23.79 Aniline has a UV spectrum with peaks at $\lambda_{max} = 230$ nm ($\varepsilon = 8600$) and 280 nm ($\varepsilon = 1430$). In the presence of dilute HCl, the spectrum of aniline changes dramatically: $\lambda_{max} = 203$ ($\varepsilon = 7500$) and 254 ($\varepsilon = 160$). This spectrum is nearly identical to the UV spectrum of benzene. Account for the effect of acid on the UV spectrum of aniline.

23.80 Three unidentified compounds have been found in the warehouse of the company Tumany Amines, Inc. The president of the company, Wotta Stench, has hired you to identify them from their spectra.

(a) Compound *A* ($C_9H_{13}NO$): IR spectrum: 3360, 3280 cm^{-1} (doublet); 1611 cm^{-1}; no carbonyl absorption. NMR spectrum shown in **Fig. P23.80a**.

(b) Compound *B* ($C_6H_{13}N$): IR spectrum: 3280, 1653, 898 cm^{-1}. NMR spectrum in **Fig. P23.80b**.

(c) Compound *C* ($C_6H_{16}N_2$): IR spectrum: 3281 cm^{-1}. NMR spectrum: δ 1.1 (8*H*, t, $J = 7$ Hz), δ 2.66 (4*H*, q, $J = 7$ Hz), δ 2.83 (4*H*, s). (*Hint:* The triplet at δ 1.1 conceals another broad resonance that contributes to the integral.)

23.81 (a) Propose a structure for an amine *A* (C_4H_9N), which liberates a gas when treated with $NaNO_2$ and HCl. The ^{13}C NMR spectrum of *A* is as follows, with attached protons in parentheses: δ 14(2), δ 34.3(2), δ 50.0(1).

(b) Propose a structure for an amine *B* (C_4H_9N), which does *not* liberate a gas when treated with $NaNO_2$ and HCl, and has IR absorptions at 917 cm^{-1}, 990 cm^{-1}, and 1640 cm^{-1}, as well as N—H absorption at 3300 cm^{-1}. The ^{13}C NMR spectrum of *B* is as follows: δ 36.0, δ 54.4, δ 115.8, δ 136.7.

23.82 Provide a curved-arrow mechanism for the example of the *Bayliss–Hilman reaction* shown here. Be sure that the role of the triethylamine catalyst is clearly indicated. (*Hint:* The role of the catalyst is *not* to remove the α-proton of the ester; this proton is not acidic. Why?)

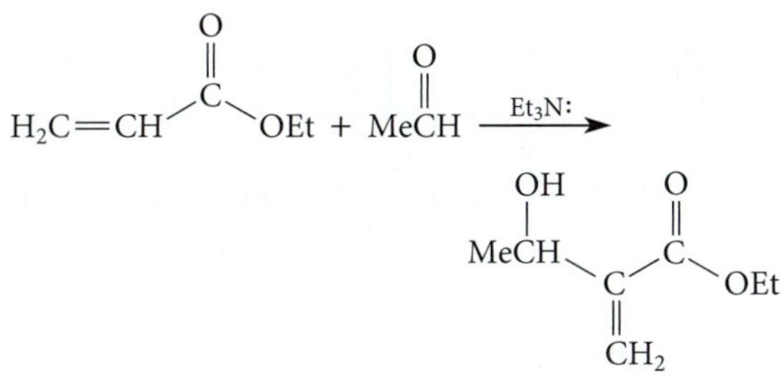

23.83 A chemist, Mada Meens, treated ammonia with pentanal in the presence of hydrogen gas and a catalyst in the expectation of obtaining 1-pentanamine by reductive amination. In addition to 1-pentanamine, however, she also obtained dipentylamine and tripentylamine. Explain how the by-products are formed.

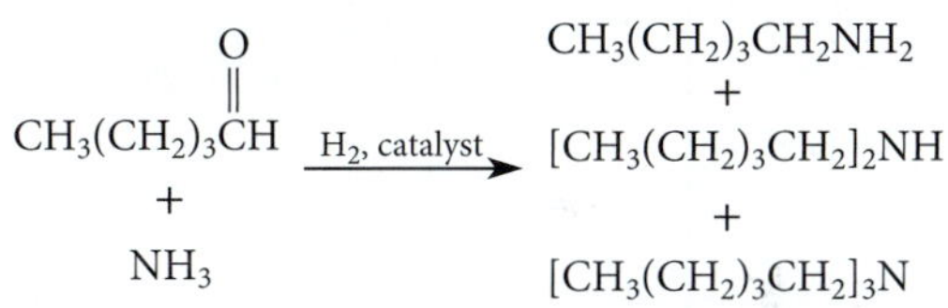

23.84 Around 1912, Swiss chemist Richard Willstätter (who subsequently was awarded the 1915 Nobel Prize in Chemistry) treated diamine *A* with methyl iodide and then with Ag_2O and heat, whereupon a hydrocarbon *B*, C_8H_8, distilled from the reaction mixture. Compound *B* reacted rapidly with Br_2 under mild conditions. Treatment of compound *C* in the same way gave a hydrocarbon *D*, C_6H_6, which did not react with Br_2.

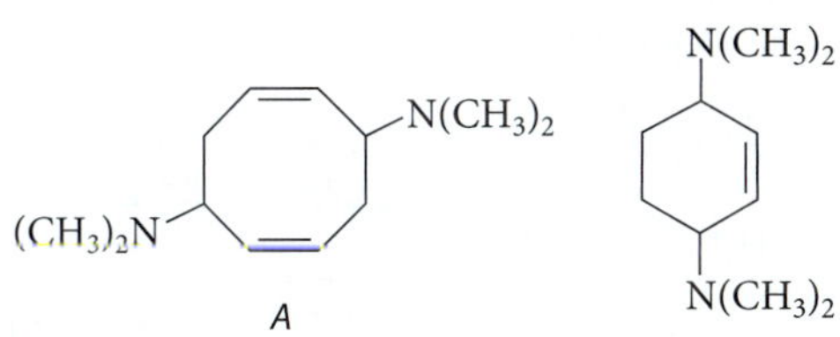

A *C*

Identify the two hydrocarbons *B* and *D*, and explain their very different behavior toward Br_2. (Willstätter concluded from these observations that compound *D* could not be an alkene.)

23.85 Explain the transformations shown here by showing relevant intermediates, providing analogies to known reactions and, where appropriate, giving curved-arrow mechanisms.

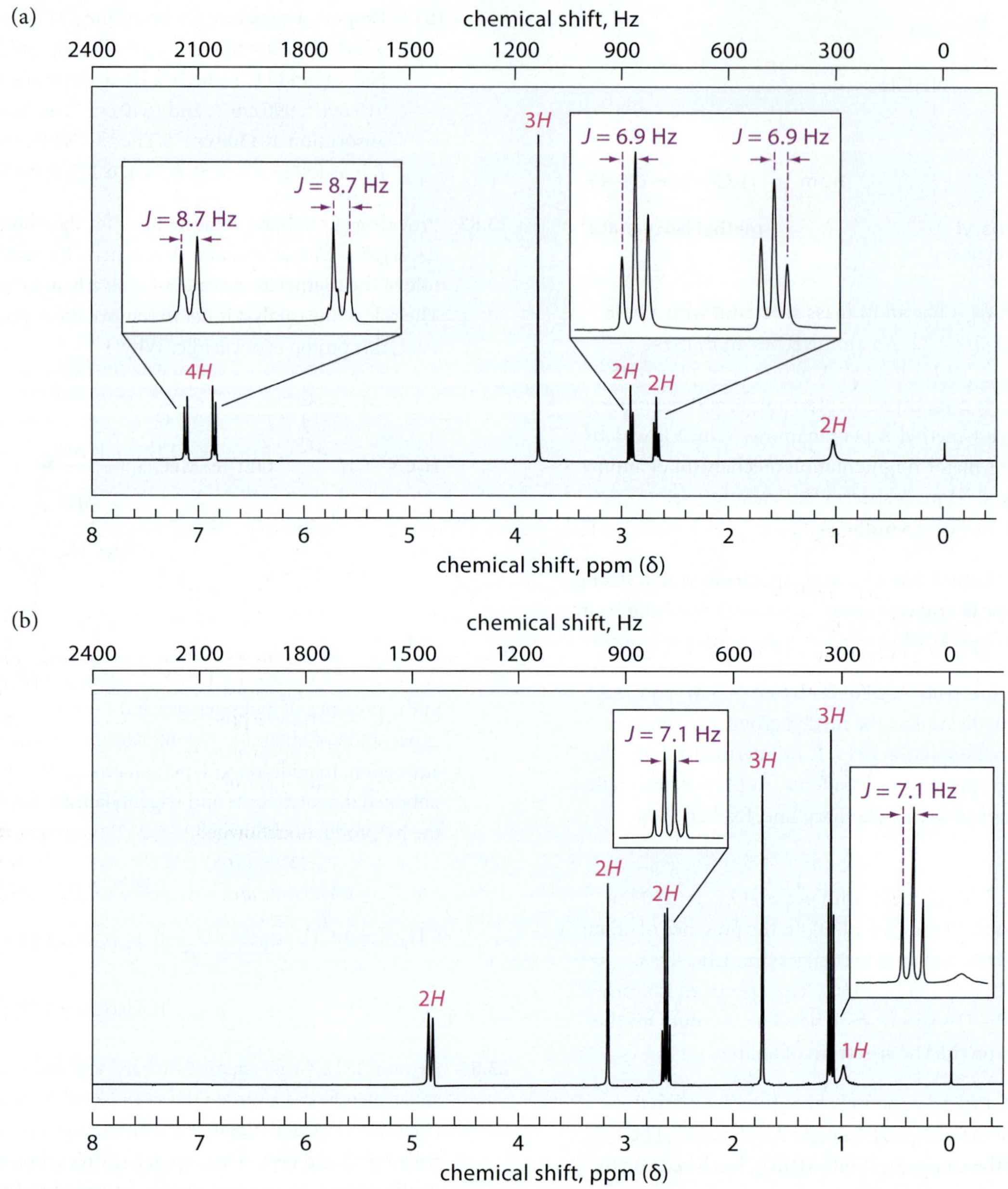

FIGURE P23.80 (a) The NMR spectrum for compound *A*, Problem 23.80(a). (b) The NMR spectrum for compound *B*, Problem 23.80(b). Integrals are shown in red above their respective resonances.

Problem 23.85, *continued*

(a)

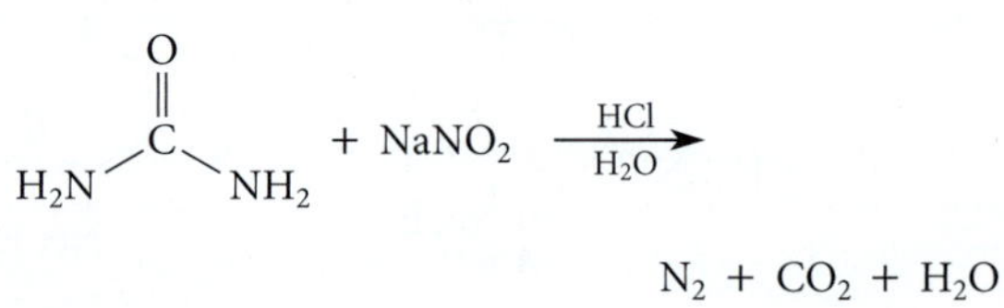

(This reaction can be used to scavenge unwanted nitrous acid.)

(b)

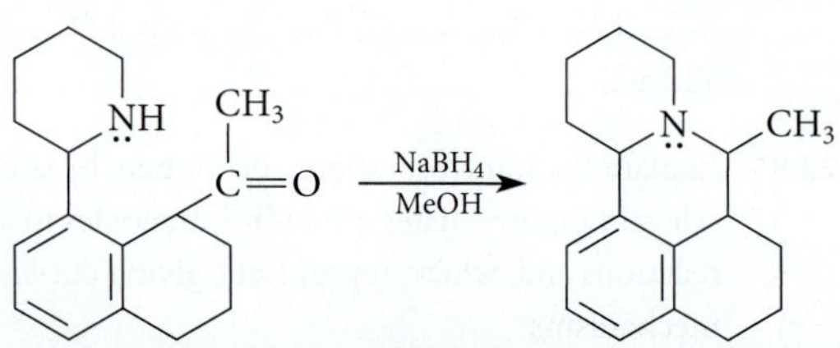

(c)

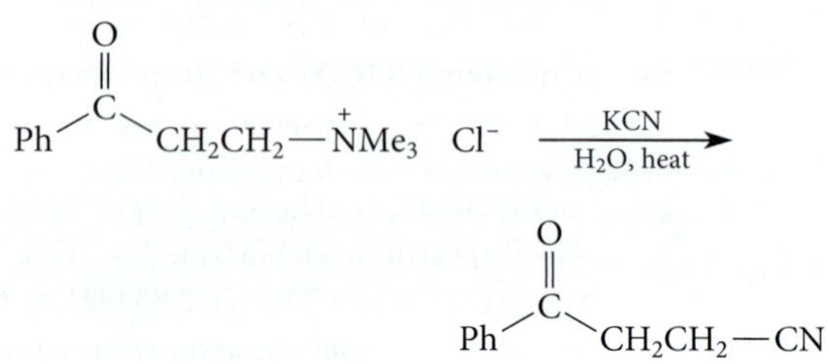

(d)

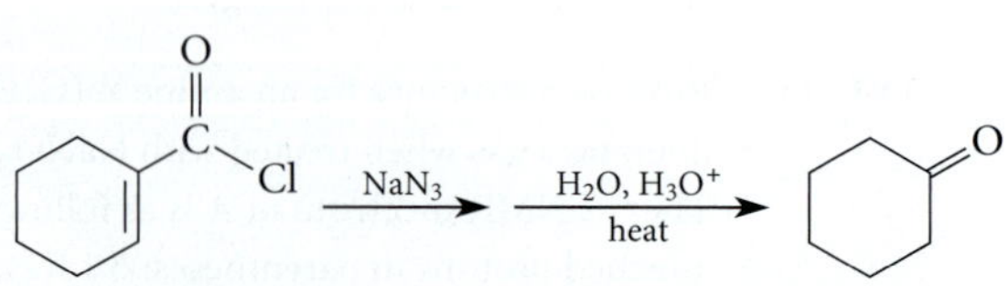

(e)

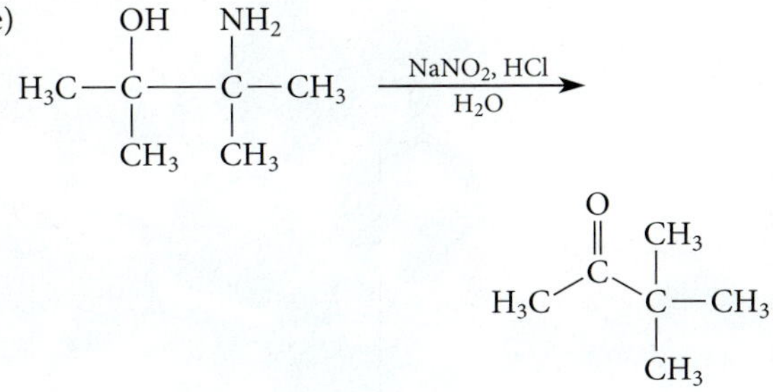

(f)

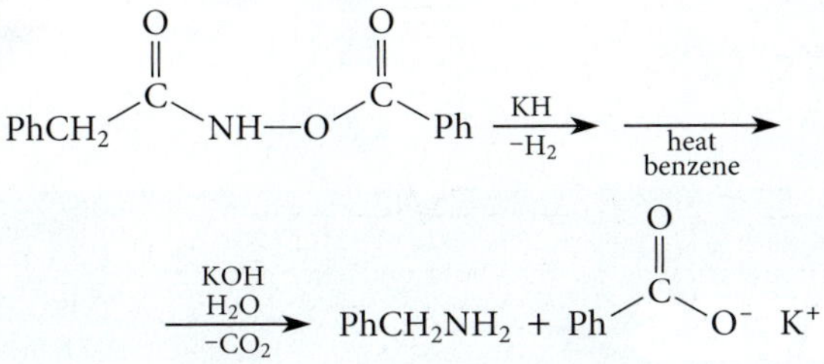

23.86 Explain why amines *A*–*C* have considerably different basicities, despite their similarities in structure. (*Hint:* Make a model of compound *B*. Look at the relationship of the nitrogen nonbonding electron pair to the p orbitals of the benzene ring.)

		conjugate-acid pK_a:
A	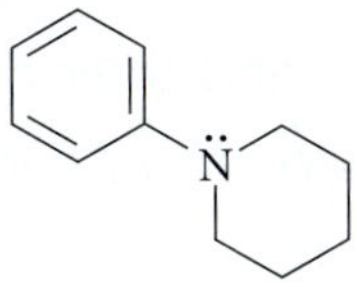	5.20
B	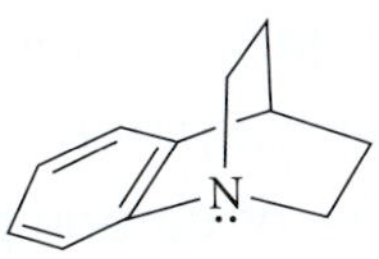	7.79
C	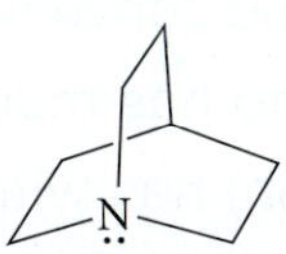	10.95

23.87 Amide *A*, δ-valerolactam, is a typical amide with a conjugate-acid pK_a of 0.8. The two cyclic tertiary amines *B* and *C* also have typical conjugate-acid pK_a values. In contrast, the conjugate-acid pK_a of amide *D* is unusually high for an amide, and it hydrolyzes much more rapidly than other amides. Draw the structure of the conjugate acid of amide *D*, and suggest a reason for both its unusual pK_a and its rapid hydrolysis.

		conjugate-acid pK_a:
A	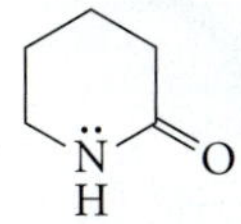	0.8
B	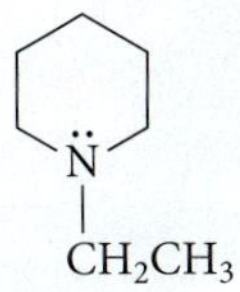	10.65
C	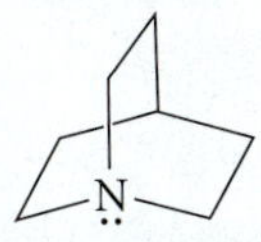	10.95
D	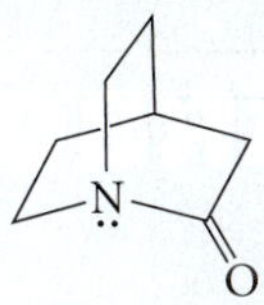	5.33

23.88 The Staudinger *ligation* is a method that has found increasing use in the growing field of chemical biology for labeling biomolecules, such as proteins, with fluorescent probes. Proteins possessing an azide functional group can be *ligated* onto a phosphine bearing the probe. (See **Fig. P23.88.**)

(a) Using your knowledge of the Staudinger *reaction* (Sec. 23.11A), propose a mechanism that accounts for the formation of the following key intermediate in the Staudinger *ligation*.

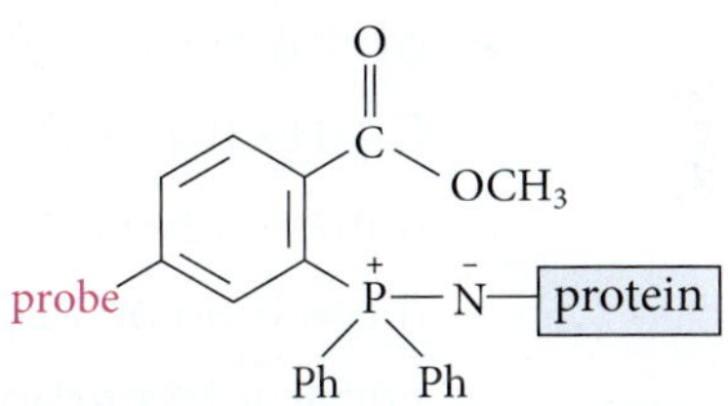

(b) Instead of reacting with water to release the free amine, as in the Staudinger reaction (Eq. 23.70), the probe-labeled protein product is formed. Draw a curved-arrow mechanism for the conversion of the intermediate in part (a) to the probe-labeled protein product in Fig. P23.88.

(c) In the absence of the methyl ester functional group, what would be the product of the reaction in the presence of water?

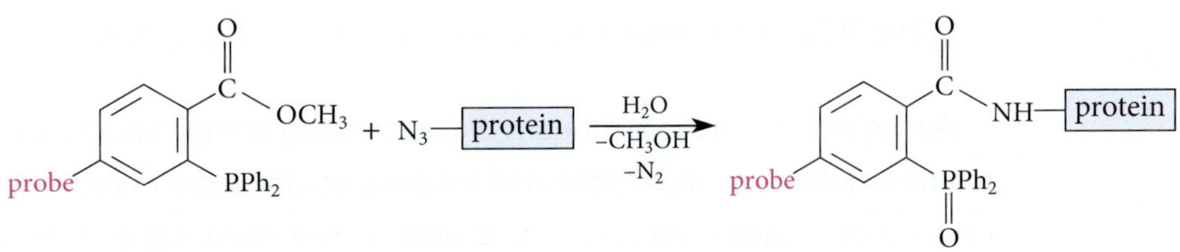

FIGURE P23.88

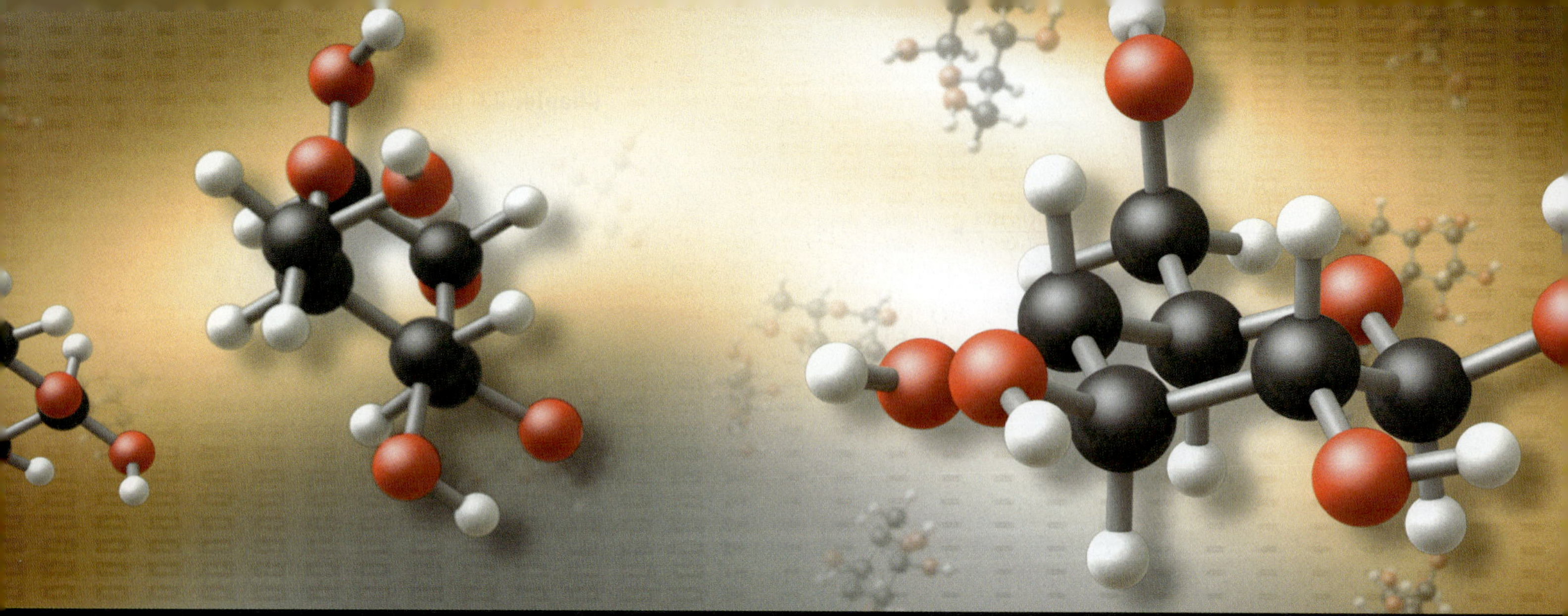

24 CARBOHYDRATES

Because of their abundance in the natural world and their importance to living things, sugars have been the subject of intense investigation since the earliest days of scientific inquiry. Scientists refer to sugars and their derivatives as carbohydrates. As this name implies, most of the common sugars have molecular formulas that fit a "hydrate-of-carbon" pattern—that is, a formula of the form $C_n(H_2O)_m$. For example, sucrose (table sugar) has the formula $C_{12}(H_2O)_{11}$ or $C_{12}H_{22}O_{11}$, and both glucose and fructose (sugars prevalent in honey) have the formula $C_6(H_2O)_6$ or $C_6H_{12}O_6$. This hydrate-of-carbon pattern is more than an apparent relationship. Anyone familiar with the conversion of table sugar into carbon by concentrated H_2SO_4 (or anyone who has made caramel sauce, a less extreme example of the same phenomenon) has witnessed the dehydration of carbohydrates:

> A quarter pound of nice white lump sugar put into a breakfast cup with the smallest possible dash of boiling water and then the addition of plenty of oil of vitriol [H_2SO_4] is a truly wonderful spectacle, and more instructive than much reading, to see the white sugar turn black, then boil spontaneously, and now, rising out of the cup in solemn black, it heaves and throbs as the oil of vitriol continues its work in the lower part of the cup, emitting volumes of steam. . . . [J.W. Pepper, *Scientific Amusements for Young People,* 1863]

As the result of a more modern understanding of their structures, **carbohydrates** are now defined as aldehydes and ketones containing a number of hydroxy groups on an unbranched carbon chain, as well as their chemical derivatives.

Two common carbohydrate structures:

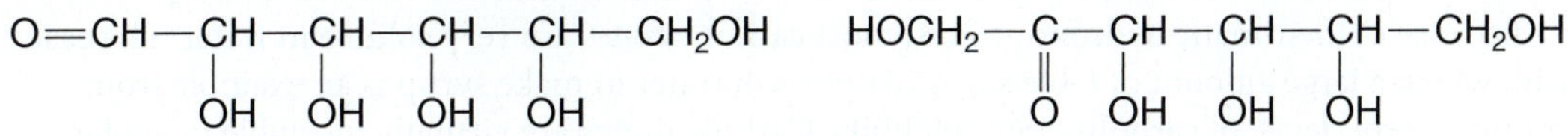

Less precisely, but more descriptively, carbohydrate chemistry can be regarded as the chemistry of sugars and their derivatives.

Carbohydrates are among the most abundant organic compounds on Earth. In polymerized form as cellulose, carbohydrates account for 50–80% of the dry weight of plants. Carbohydrates are a major source of food; sucrose (table sugar) and lactose (milk sugar) are examples. Even the shells of arthropods such as lobsters consist largely of carbohydrate.

The study of carbohydrates relies heavily on the principles of stereochemistry (Chapter 6) and on the conformational aspects of cyclohexane rings (Chapter 7). Therefore, molecular models should be very helpful as you study the material in this chapter.

24.1 CLASSIFICATION AND PROPERTIES OF CARBOHYDRATES

Carbohydrates can be classified in several ways based on structure. One type of classification is based on the type of carbonyl group in the carbohydrate. A carbohydrate with an aldehyde carbonyl group is called an **aldose**, and a carbohydrate with a ketone carbonyl group is called a **ketose**. Carbohydrates can also be classified by the number of carbon atoms they contain. A six-carbon carbohydrate is called a **hexose**, and a five-carbon carbohydrate is called a **pentose**. These two classifications can be combined: an **aldohexose** is an aldose containing six carbon atoms, and a **ketopentose** is a ketose containing five carbon atoms. A ketose can also be indicated with the suffix *ulose*; thus, a five-carbon ketose is also called a **pentulose**. These classifications are illustrated by the examples in Display 24.1.

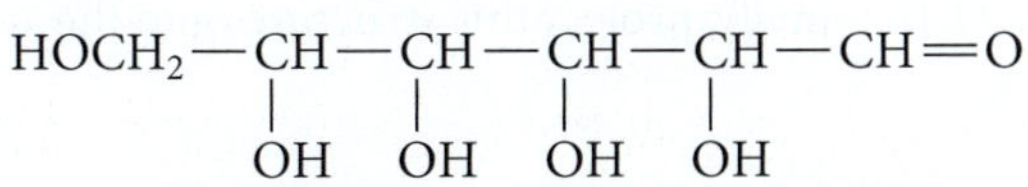

an *aldose* (aldehyde carbonyl group)
a *hexose* (six carbon atoms)
an *aldohexose* (combination of the above classifications)

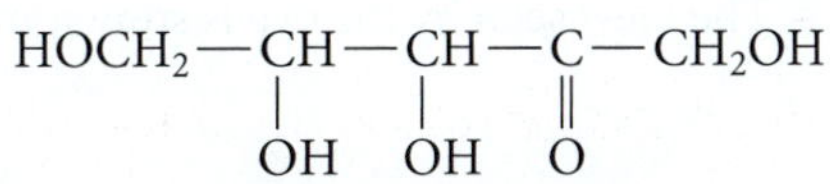

a *ketose* (ketone carbonyl group)
a *pentose* (five carbon atoms)
a *ketopentose* or *pentulose* (combination of the above classifications)

(24.1)

Another type of classification scheme is based on the hydrolysis of certain carbohydrates to simpler carbohydrates. **Monosaccharides** cannot be converted into simpler carbohydrates by hydrolysis. Glucose and fructose are examples of monosaccharides. Sucrose, however, is a **disaccharide**—a compound that can be converted by hydrolysis into two monosaccharides.

$$\text{sucrose } (C_{12}H_{22}O_{11}) + H_2O \xrightarrow{\text{acid or enzymes}} \text{glucose } (C_6H_{12}O_6) + \text{fructose } (C_6H_{12}O_6) \quad (24.2)$$

a disaccharide — monosaccharides (glucose and fructose)

Likewise, **trisaccharides** can be hydrolyzed to three monosaccharides, **oligosaccharides** to a "few" monosaccharides, and **polysaccharides** to a very large number of monosaccharides.

Because of their many hydroxy groups, most carbohydrates are very soluble in water. The ease with which a large amount of table sugar dissolves in water to make syrup is an example from common experience of carbohydrate solubility. Carbohydrates are virtually insoluble in apolar aprotic solvents.

24.2 FISCHER PROJECTIONS

Almost all carbohydrates are chiral molecules, and most have more than one asymmetric carbon. Many carbohydrates have several contiguous asymmetric carbons in an unbranched chain. For example, the aldoses have four such carbons, which are indicated by asterisks in the following structure:

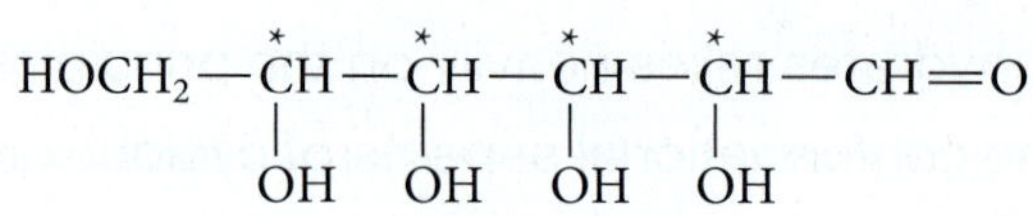

aldohexoses
four asymmetric carbons (*) (24.3)

To show the stereochemistry of such molecules, we could use line-and-wedge structures. However, a simpler system of showing stereochemistry was developed by the German chemist Emil Fischer, whose landmark work on the structure of glucose we discuss in Sec. 24.10. Fischer developed a way to represent three-dimensional structures on a two-dimensional surface (paper or blackboard) that does not require the use of wedges and dashed wedges. Such structures are called **Fischer projections**. We use Fischer projections extensively in this chapter. In this section, we explain how to draw and manipulate Fischer projections.

To illustrate the process of drawing a Fischer projection, we'll use the 2*R*,3*R* enantiomer of erythrose, an aldotetrose with two asymmetric carbons (**Fig. 24.1**). *You should follow this discussion with a molecular model.* To represent this molecule in a Fischer projection, arrange the molecule in an *all-eclipsed conformation* about the C2–C3 bond—the bond connecting the two asymmetric carbons. View the molecule as shown by the eye in Fig. 24.1a. Next, impose a reference plane containing the C2–C3 bond on the molecule. (This plane will ultimately be the plane of the page.) The plane should be oriented so that the other two carbon–carbon bonds are *receding behind this plane*, and the bonds to the OH and H groups *are emerging in front of this plane.* The view seen by the eye is shown in Fig. 24.1b. Finally, project this structure onto the

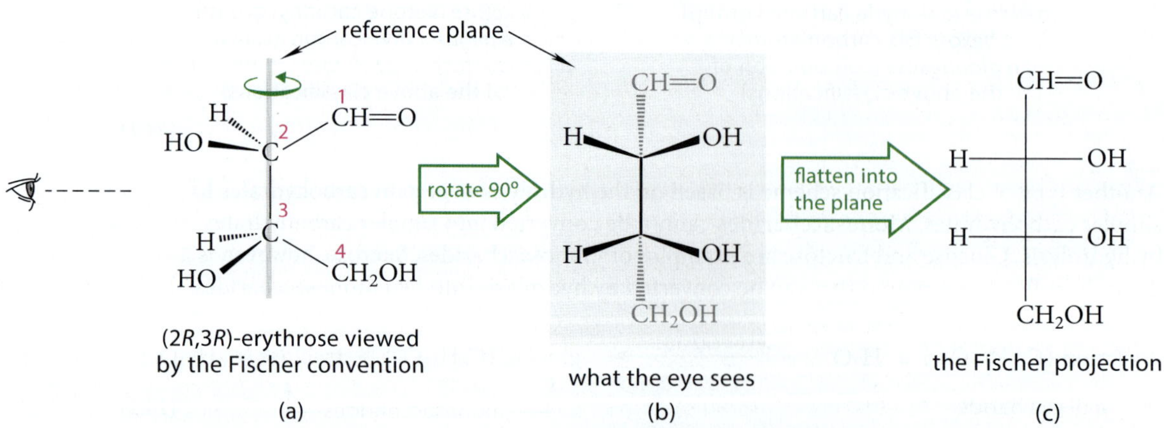

FIGURE 24.1 How to derive a Fischer projection for an aldotetrose. (a) The eclipsed conformation used to derive the projection, with the reference plane perpendicular to the page. (b) The view of the conformation in (a) as seen by the eye. The reference plane is now the plane of the page. The groups behind the plane are shown in gray. (c) The Fischer projection. The asymmetric carbons are located at the intersection of vertical and horizontal lines.

plane—that is, flatten it into the page—to give the Fischer projection. *The asymmetric carbons themselves are not drawn*, but are assumed to be located at the intersections of vertical and horizontal bonds (Fig. 24.1c). (As one student pointed out, the Fischer projection is the way that the molecule would look if we were to put it on the floor and step on it!)

The following five rules summarize the conventions used in the construction of Fischer projections:

1. A Fischer projection is based on an eclipsed molecular conformation.
2. The bonds connecting the asymmetric carbons are arranged in a vertical line.
3. The asymmetric carbons are located at the intersections of vertical and horizontal bonds and are not drawn explicitly.
4. Vertical bonds to the asymmetric carbons recede behind the page, away from the observer in the three-dimensional model.
5. Horizontal bonds to the asymmetric carbons emerge from the page, toward the observer in the three-dimensional model.

In other words, a flat Fischer projection can convey three-dimensional information if it requires a specific viewing mode (rules 1 and 2) and a strict adherence to a convention about the relationship of the horizontal and vertical bonds to the plane of the page (rules 4 and 5). Rule 3 alerts us to the fact that we are (or might be) dealing with a Fischer projection and not an "ordinary" Lewis structure.

To derive the Fischer projection of a molecule with more than two asymmetric carbons, a molecule is first placed (or imagined) in an eclipsed conformation in which the chain of asymmetric carbons is vertical and curving away from the observer, as if this chain were drawn on a convex surface such as a paper cylinder. This conformation is illustrated for the naturally occurring enantiomer of glucose, an aldohexose, in **Fig. 24.2a**. The horizontal bonds project toward the observer from this surface. All bonds are then projected onto this surface (**Fig. 24.2b**). Mentally cutting the cylinder and flattening it gives the Fischer projection (**Fig. 24.2c**).

By convention, Fischer projections of carbohydrates are typically drawn with the most oxidized carbon closest to the top. In the case of glucose the most oxidized carbon is C1, the aldehyde carbon.

The use of an eclipsed conformation to derive a Fischer projection does *not* mean that the molecule actually has such a conformation. As you've learned, most molecules actually exist in staggered conformations (Sec. 2.5). Fischer projections convey *no* information about molecular conformations. Their only purpose is to show the *absolute configuration* of each asymmetric carbon.

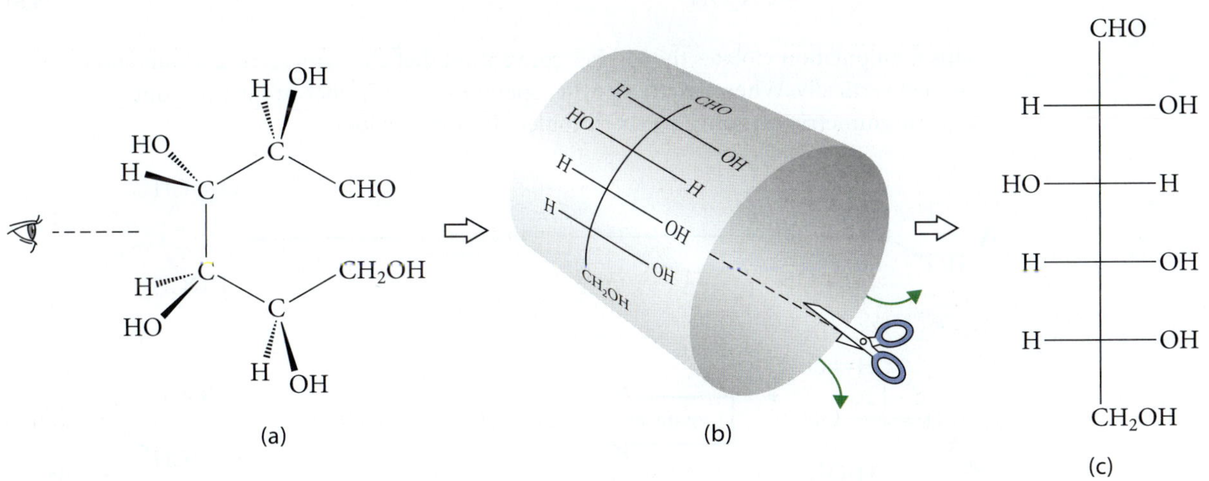

FIGURE 24.2 How to derive a Fischer projection for D-glucose, the naturally occurring enantiomer. (a) The eclipsed conformation is viewed with the chain of carbons oriented vertically and curving away from the observer, and the horizontal bonds projecting toward the observer. (b) This view is projected onto an imaginary curved cylinder. (c) Mentally cutting the cylinder and flattening it gives the Fischer projection.

To derive a three-dimensional model of a molecule from its Fischer projection, reverse the process just described. Always remember that the vertical bonds in the Fischer projection extend *away from* the observer, and the horizontal bonds extend *toward* the observer.

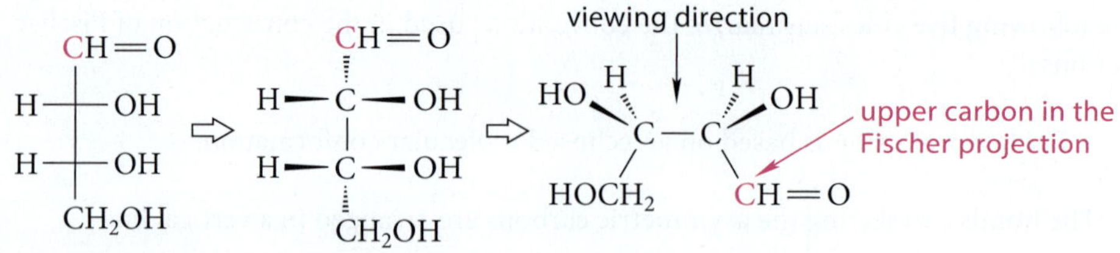

Fischer projection the corresponding line-and-wedge structures (24.4)

For any given molecule, *several valid Fischer projections can be drawn.* It is useful to be able to draw the different Fischer projections of a molecule without going back and forth to a three-dimensional model. For this purpose, some rules for manipulation of Fischer projections are helpful. Be sure to use models to convince yourself that these rules are valid:

1. *A Fischer projection may be turned 180° in the plane of the paper.*
   By this rule, the following two Fischer projections represent the same stereoisomer.

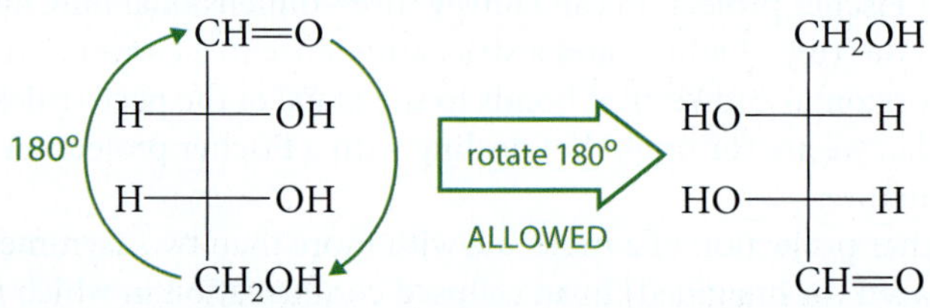

(24.5)

   This manipulation is allowed because it leaves horizontal bonds horizontal and vertical bonds vertical; therefore, it does not alter the meaning of the Fischer projection.

2. *A Fischer projection may not be turned 90° in the plane of the page.*

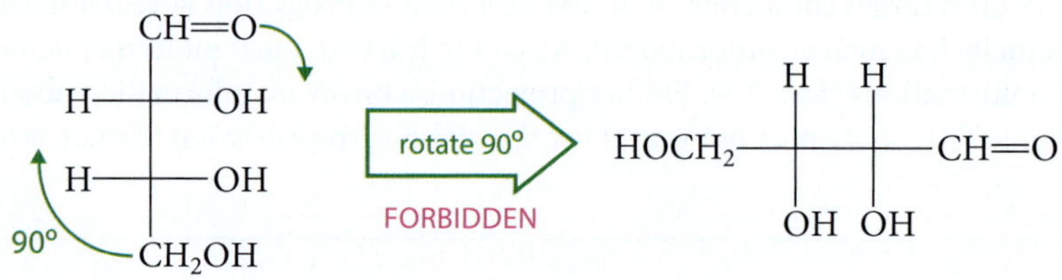

(24.6)

   This manipulation violates the Fischer convention that all asymmetric carbons should be aligned vertically. When we attempt this operation on a Fischer projection containing a *single* asymmetric carbon, a further problem becomes evident:

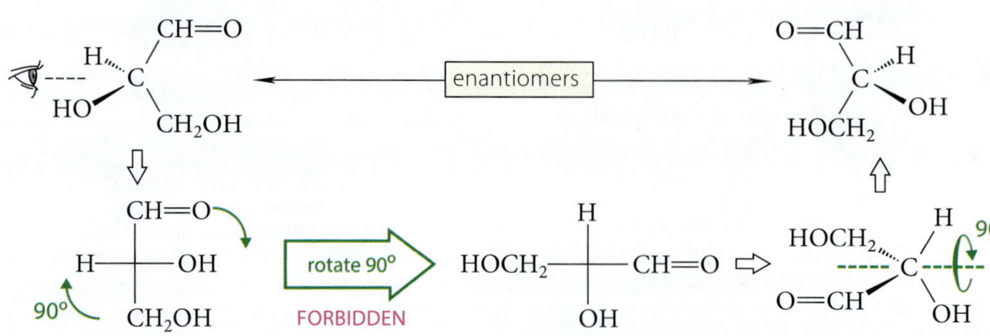

(24.7)

The 90° rotation exchanges horizontal and vertical groups and, in the process, *interconverts the original structure into its enantiomer.* This is disastrous, because the whole idea of Fischer projections is to convey stereochemical information.

3. *A Fischer projection may NOT be lifted from the plane of the paper and turned over.*

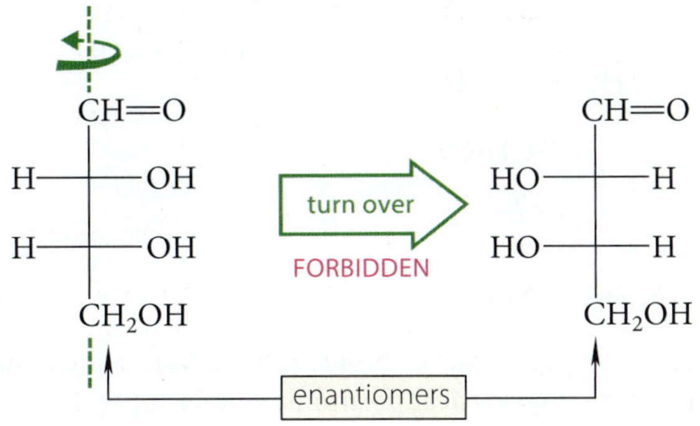

(24.8)

This rule has a similar rationale to rule 2. By flipping the structure over as shown, the stereochemistry of each asymmetric carbon is reversed.

4. *The three groups at either end of a Fischer projection may be interchanged in a* cyclic permutation. *That is, all three groups can be moved at the same time in a closed loop so that each occupies an adjacent position.*

(24.9)

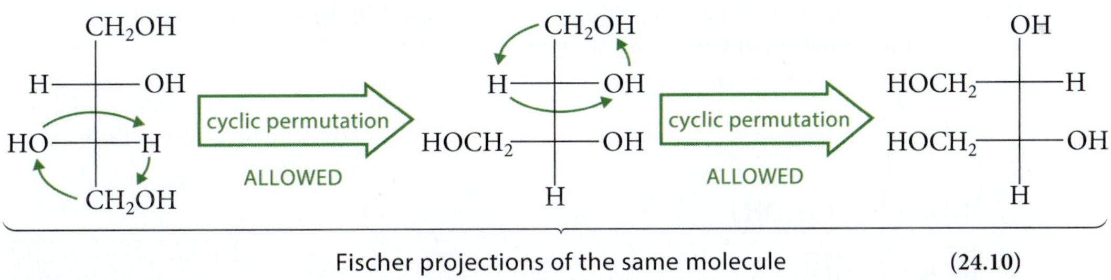

(24.10)

This operation is equivalent to an internal rotation. This point should become clear if you convert any one of the structures in Eq. 24.10 into a model. Leaving the model in an eclipsed conformation, carry out an internal rotation of 120° about the central carbon–carbon bond as shown by the colored arrows in Eq. 24.10, and then form a new Fischer projection from the resulting structure. Each 120° internal rotation is equivalent to one cyclic permutation described by rule 4. A different Fischer projection of the same molecule results from each different eclipsed conformation.

5. *An interchange of any two of the groups bonded to an asymmetric carbon changes the stereochemical configuration of that carbon.*
(Verify this rule with models.) This rule applies not only to Fischer projections, but also to three-dimensional models as well. It follows that a *pair* of interchanges leaves the configuration of the carbon unaffected; the first interchange changes the configuration, and the second interchange changes the configuration back to the original. In fact, the cyclic permutation in rule 4 is equivalent to a pair of interchanges.

It is particularly easy to recognize enantiomers and meso compounds from the appropriate Fischer projections, because planes of symmetry in the actual molecules reduce to lines of symmetry in their projections.

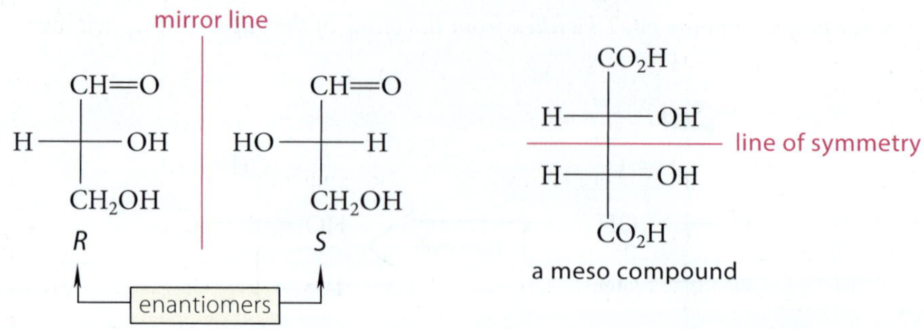

(24.11)

The *R*,*S* system can be applied to a Fischer projection without using a model. If the group of lowest priority is in either of the two vertical positions, simply apply the *R*,*S* priority rules to the remaining three groups.

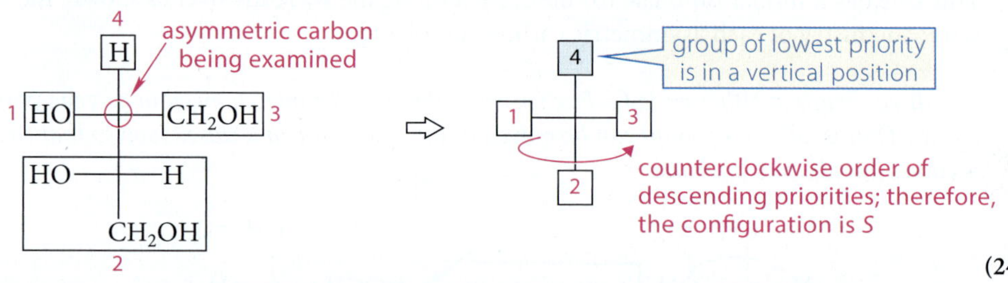

(24.12)

This method works because, if the lowest-priority group is in a vertical position in the Fischer projection, it is oriented away from the observer as required for application of the priority rules.

If the lowest-priority group is in a horizontal position, proceed in the same manner; but, since the molecule is being viewed incorrectly for assigning configuration, reverse the assignment. (This is a rare situation in which two wrongs make a right!)

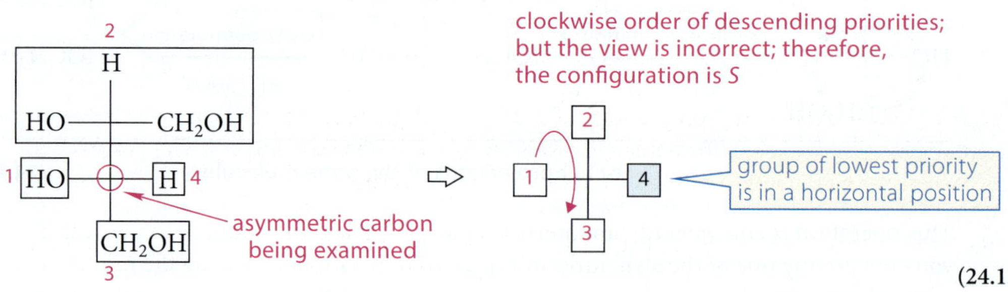

(24.13)

Focused Problems

24.1 Draw at least two Fischer projections for each of the following molecules.

(a)

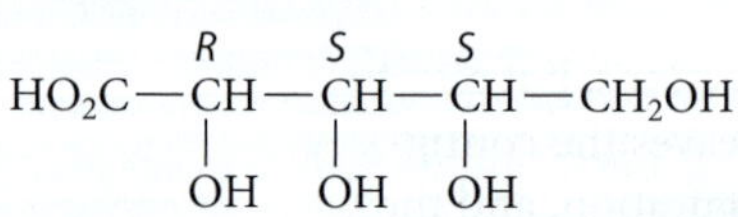

(b) (*S*)-2-Butanol

24.2 Indicate whether the structures in each of the following pairs are enantiomers, diastereomers, or identical molecules.

(a)

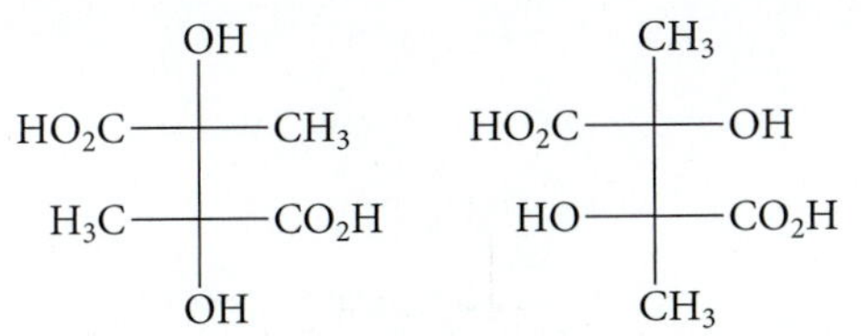

(b)

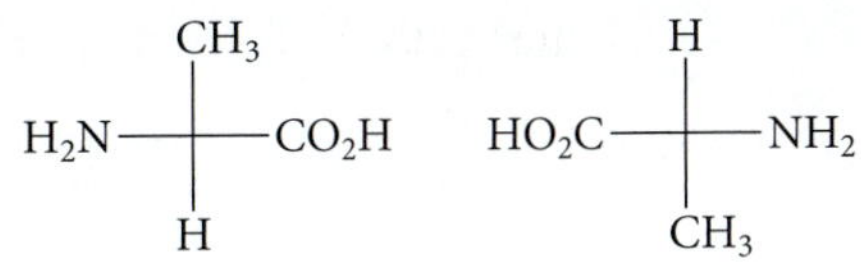

24.3 Which of the following are Fischer projections of a meso compound?

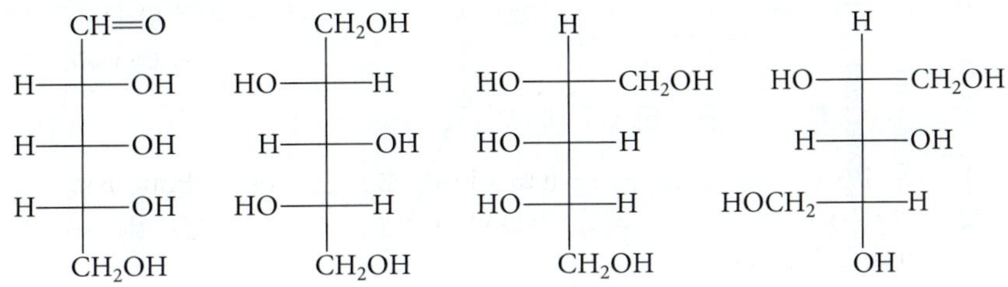

A *B* *C* *D*

24.3 STRUCTURES OF THE MONOSACCHARIDES

A. Stereochemistry and Configuration

We'll explore the stereochemistry of carbohydrates by focusing largely on the aldoses with six or fewer carbons. The aldohexoses have four asymmetric carbons and therefore exist as 2^4 or 16 possible stereoisomers. These can be divided into two enantiomeric sets of eight diastereomers.

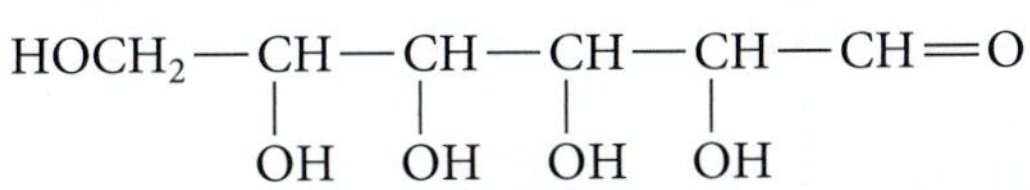

aldohexoses
$2^4 = 16$ stereoisomers (24.14)

Similarly, there are two enantiomeric sets of four diastereomers (eight stereoisomers total) in the aldopentose series. Each diastereomer is a *different carbohydrate* with *different properties*, known by a *different name*. The aldoses with six or fewer carbons are given in **Fig. 24.3** as Fischer projections.

Each of the monosaccharides in Fig. 24.3 has an enantiomer. For example, the two enantiomers of glucose have the following structures:

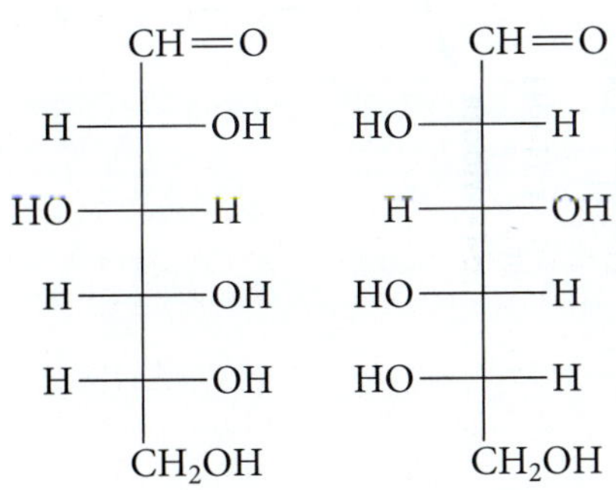

enantiomers of glucose (24.15)

It is important to specify the enantiomers of carbohydrates in a simple way. Suppose you have a model of one of these glucose enantiomers in your hand; how would you explain to someone who cannot see the model (for example, over the telephone) which enantiomer you are holding?

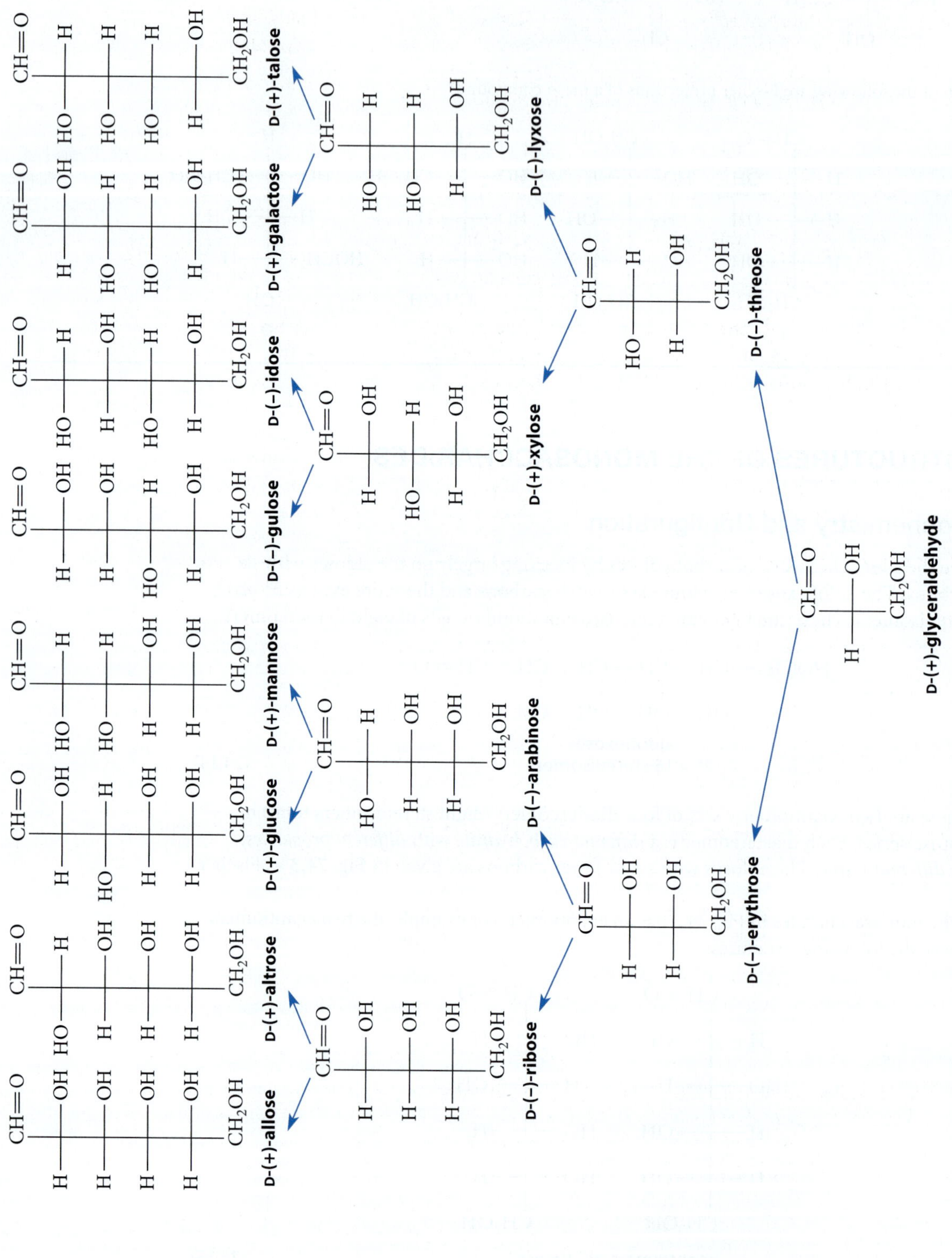

FIGURE 24.3 The D family of aldoses. Each compound shown here has an enantiomer in the L family. By convention, the aldehyde carbon is at the top of each structure. The blue arrows show how the aldoses are related by the Kiliani–Fischer synthesis (Sec. 24.9).

You could use the *R,S* system to describe the configuration of one or more of the asymmetric carbon atoms. A different system, however, was in use long before the *R,S* system was established. The **D,L system**, which came from proposals made in 1906 by M. A. Rosanoff, a New York University chemist, is still used today for this purpose. As this system is applied to carbohydrates, the configuration of a carbohydrate enantiomer is specified by applying the following conventions:

1. The configuration of the naturally occurring triose (+)-glyceraldehyde is designated as D, and the configuration of its enantiomer, (−)-glyceraldehyde, is designated as L. The OH group in the D stereoisomer is on the right when the CH=O group is in the upper vertical position and the CH_2OH group in the lower vertical position.

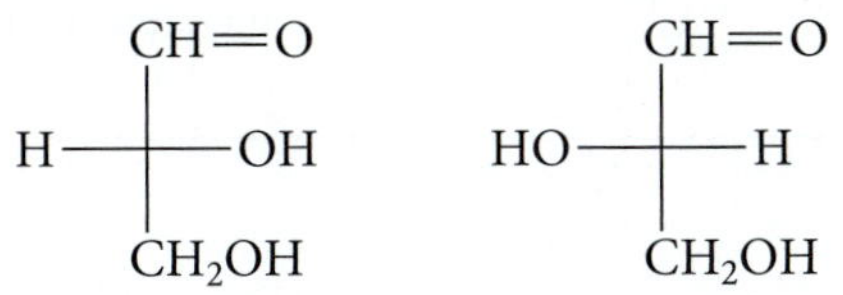

D-(+)-glyceraldehyde   L-(−)-glyceraldehyde   (24.16)

The basis for the use of the letters D and L was simply the fact that the D stereoisomer of glyceraldehyde is dextrorotatory and the L stereoisomer is levorotatory. As with the *R,S* system, however, there is no *general* correlation between configuration and the sign of optical rotation.

2. The other aldoses or ketoses are written in a Fischer projection with their carbon atoms in a straight vertical line, and the carbons are numbered consecutively as they would be in systematic nomenclature, so the carbonyl carbon receives the lowest possible number.

3. The *asymmetric carbon of highest number* is designated as the *configurational carbon*. If this carbon has the H, OH, and CH_2OH groups in the same relative configuration as the same three groups of D-glyceraldehyde, the carbohydrate is said to have the D configuration. If this carbon has the same configuration as L-glyceraldehyde, then the carbohydrate is said to have the L configuration.

The application of these conventions is illustrated in Study Problem 24.1.

Study Problem 24.1

Determine whether the following carbohydrate derivative, shown in Fischer projection, has the D or L configuration.

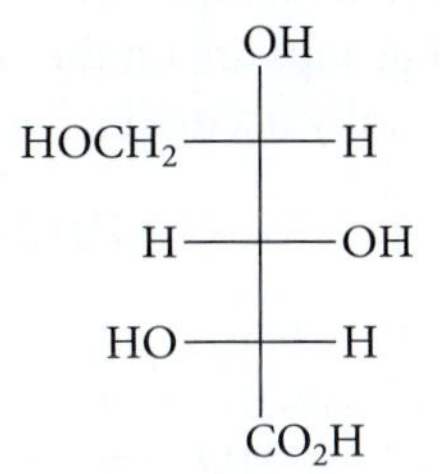

Solution First redraw the structure so that the carbon with the lowest number in substitutive nomenclature—the carbon of the carboxylic acid group—is at the top. This can be done by rotating the structure 180° in the plane of the page. Then carry out a cyclic permutation of the three groups at the bottom so that all carbons lie in a vertical line. As explained in Sec. 24.2, these are allowed manipulations of Fischer projections.

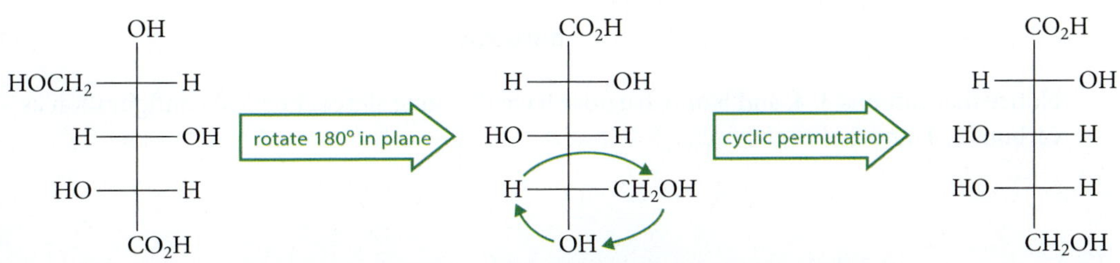

Finally, compare the configuration of the highest-numbered asymmetric carbon with that of D-glyceraldehyde. Because the configuration is different, the molecule has the L configuration.

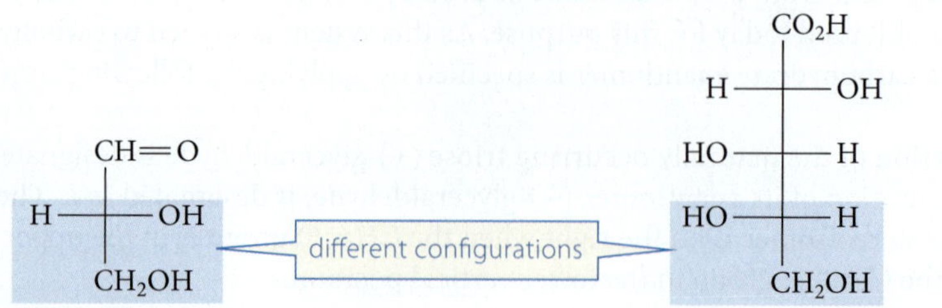

D-glyceraldehyde therefore, L configuration

The monosaccharides shown in Fig. 24.3 constitute the D family of enantiomers. Each of the compounds in this figure has an enantiomer with the L configuration. This figure illustrates the point that *there is no general correspondence between configuration and the sign of the optical rotation* (see Sec. 6.3C). For example, some D-aldoses have positive rotations, but others have negative rotations. Also, *there is no simple relationship between the D,L system and the R,S system*. The *R,S* system is used to specify the configuration of *each* asymmetric carbon atom in a molecule, but *the D,L system specifies a particular enantiomer of a molecule that might contain many asymmetric carbons.*

An annoying aspect of the D,L system is that each diastereomer is given a different nonsystematic name. This is one reason that the D,L system has been generally replaced with the *R,S* system, which can be used with systematic nomenclature. Nevertheless, the common names of many carbohydrates are so well entrenched that they remain important. Although use of the D,L system is fairly straightforward for carbohydrates and amino acids, it has been virtually abandoned for other compounds.

A few of the aldoses in Fig. 24.3 are particularly important, and you will find it helpful to learn their structures. D-Glucose, D-mannose, and D-galactose are the most important aldohexoses because of their wide natural occurrence. Start by learning the structure of D-glucose, then notice that the configurations of D-mannose and D-galactose differ in a simple way from the configuration of D-glucose. Specifically, D-glucose and D-mannose differ in configuration only at carbon-2, whereas D-glucose and D-galactose differ only at carbon-4. Diastereomers that differ in configuration at only one of several asymmetric carbons are called **epimers**. Thus, D-glucose and D-mannose are *epimeric* at carbon-2, whereas D-glucose and D-galactose are *epimeric* at carbon-4.

D-Ribose is a particularly important aldopentose. Its structure is easy to remember because, as shown in Fig. 24.3, all of its —OH groups are on the right in the standard Fischer projection.

D-Fructose is an important naturally occurring ketose:

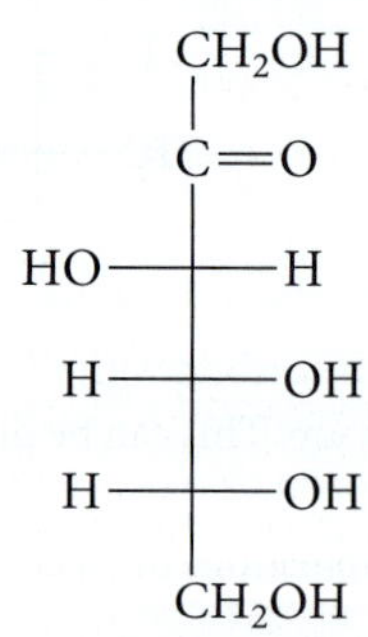

D-fructose (24.17)

Notice that carbons 3, 4, and 5 of D-fructose have the same stereochemical configuration as carbons 3, 4, and 5 of D-glucose.

Focused Problems

24.4 Classify each of the following aldoses as D or L.

(a)

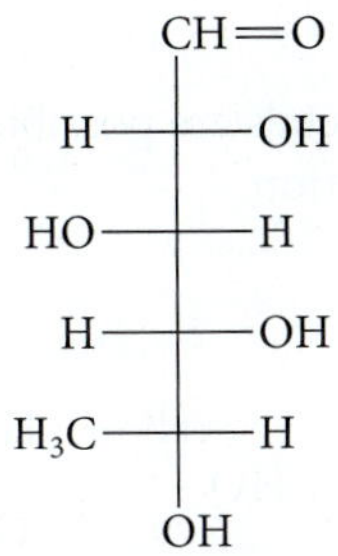

6-deoxyglucose

(b) The glucose enantiomer with the *R* configuration at carbon-3.

24.5 Which pair of the following aldoses are epimers, and which pair are enantiomers?

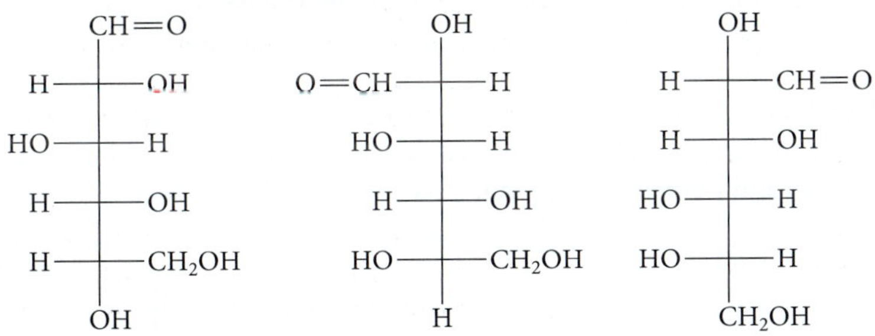

A *B* *C*

B. Cyclic Structures of the Monosaccharides

As noted in Sec. 19.11A, γ- or δ-hydroxy aldehydes exist predominantly as *cyclic hemiacetals.*

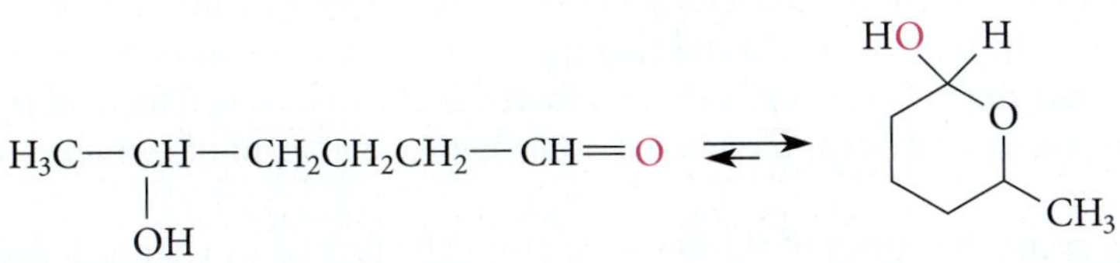

(24.18a)

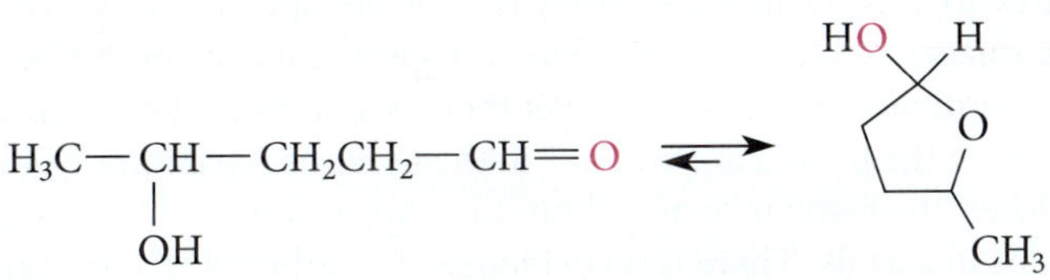

(24.18b)

The same is true of aldoses and ketoses. Although monosaccharides are often written by convention as acyclic carbonyl compounds, they exist predominantly as cyclic acetals. For example, in aqueous solution, glucose consists of about 0.003% aldehyde and a trace of the hydrate; the rest—more than 99.99%—is cyclic hemiacetals.

Focused Problem

24.6 Provide curved-arrow mechanisms for (a) Eq. 24.18a; (b) Eq. 24.18b.

In many carbohydrates, both five- and six-membered cyclic hemiacetals are possible, depending on which hydroxy group participates in the cyclization reaction.

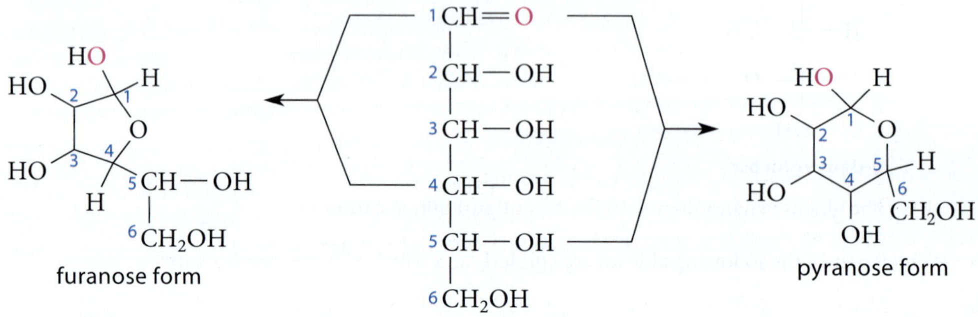

(24.19)

A five-membered cyclic acetal form of a carbohydrate is called a **furanose** (after furan, a five-membered oxygen heterocycle), whereas a six-membered cyclic acetal form of a carbohydrate is called a **pyranose** (after pyran, a six-membered oxygen heterocycle).

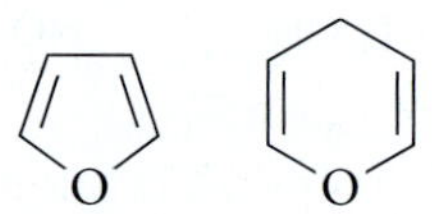

(24.20)

The aldohexoses and aldopentoses exist predominantly as pyranoses in aqueous solution, but the furanose forms of some carbohydrates are important.

A name such as *glucose* is used when referring to any or all forms of the carbohydrate. To name a cyclic hemiacetal form of a carbohydrate, start with a prefix derived from the name of the carbohydrate (for example, *gluco* for glucose or *manno* for mannose) followed by a suffix that indicates the type of hemiacetal ring (*pyranose* for a six-membered ring; *furanose* for a five-membered ring). Therefore, a six-membered cyclic hemiacetal form of D-glucose is called D-glucopyranose, and a five-membered cyclic hemiacetal form of D-mannose is called D-mannofuranose.

Although the cyclic structures of aldoses were originally proved by chemical degradations, these cyclic structures are readily apparent today from NMR spectroscopy. For example, the aldehydic proton resonance of glucose in the δ 9–10 region of its proton NMR spectrum is too weak to detect under ordinary circumstances, yet there is a doublet at δ 5.2 corresponding to a proton α to two oxygens: the proton at carbon-1 of the pyranose structure. Similarly, in the ^{13}C NMR spectra of aldoses, the resonance of carbon-1 occurs at about δ 92, which is very similar to the analogous carbons in acetals. There is no evidence of a carbonyl absorption near δ 200.

Anomers *The furanose or pyranose form of a carbohydrate has one more asymmetric carbon than the open-chain form*—carbon-1, in the case of the aldoses. Consequently, there are two possible diastereomers of D-glucopyranose.

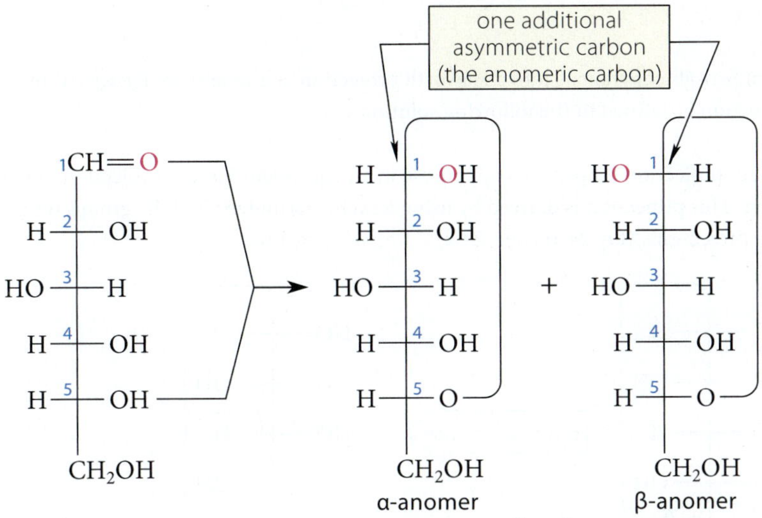

anomers of D-glucopyranose (24.21)

(The rings in the Fischer projections of these cyclic compounds are closed with a rather strange-looking long bond. We'll learn shortly how to draw more conventional representations of these cyclic structures.) Both of these compounds are forms of D-glucopyranose, and, in fact, glucose in solution exists as a mixture of both. They are diastereomers and are therefore separable compounds with different properties. When two cyclic forms of a carbohydrate differ in configuration only at their hemiacetal carbons, they are said to be **anomers**. In other words, anomers are cyclic forms of carbohydrates that are epimeric at the hemiacetal carbon. Thus, the two forms of D-glucopyranose are anomers of glucose. The hemiacetal carbon (carbon-1 of an aldose) is sometimes called the **anomeric carbon**.

As the structures in Eq. 24.21 illustrate, anomers are named with the Greek letters α and β. This nomenclature refers to the Fischer projection of the cyclic form of a carbohydrate, written with all carbon atoms in a straight vertical line. *In the α-anomer, the hemiacetal —OH group is on the same side of the Fischer projection as the oxygen at the configurational carbon.* (The configurational carbon is the one used for specifying the D,L designation; for example, carbon-5 for the aldohexoses.) Conversely, in the β-anomer, the hemiacetal —OH group is on the side of the Fischer projection opposite the oxygen at the configurational carbon. The application of these definitions to the nomenclature of the D-glucopyranose anomers is as follows:

FURTHER EXPLORATION 24.1
Nomenclature of Anomers

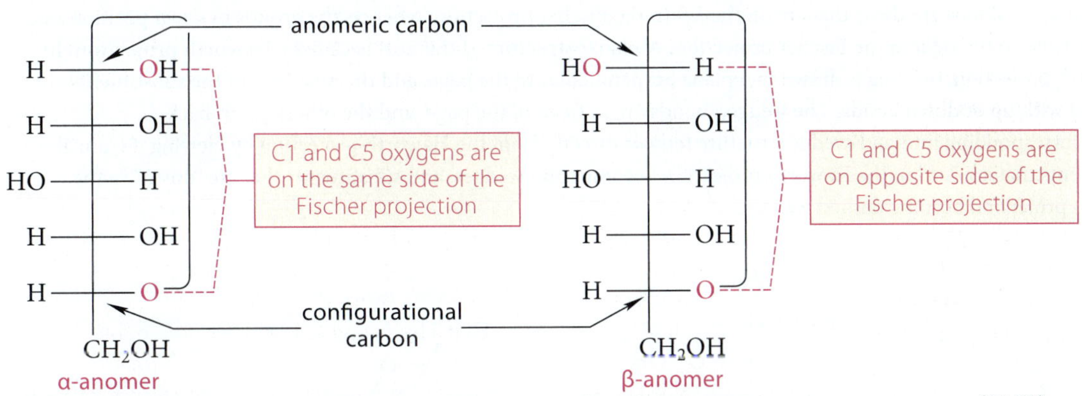

α-D-glucopyranose **β-D-glucopyranose** (24.22)

Haworth Projections, Line-and-Wedge Structures, and Conformational Representations of Pyranoses and Furanoses The goal of this section is to convert the Fischer projections of the pyranoses and furanoses into other representations of these cyclic carbohydrate structures. Study Problem 24.2 shows how to make these conversions in a systematic manner for the pyranoses.

Study Problem 24.2

Convert the Fischer projection of β-D-glucopyranose into a Haworth projection, a line-and-wedge structure, and a chair conformation. (A Haworth projection is defined in the following solution.)

Solution First redraw the Fischer projection for β-D-glucopyranose in an equivalent Fischer projection in which the ring oxygen is in a down position. This projection is derived by using a cyclic permutation of the groups on carbon-5, an allowed manipulation of Fischer projections (Eq. 24.10, Sec. 24.2).

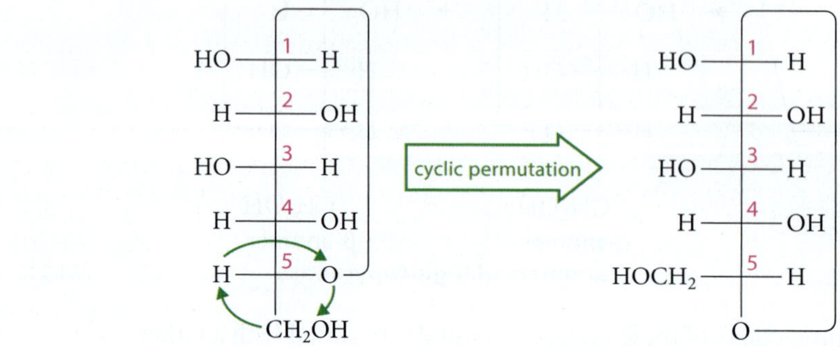

Recall that the carbon backbone of such a Fischer projection is imagined to be folded around a barrel or drum (see Fig. 24.2b, Sec. 24.2). Such an interpretation of the Fischer projection of β-D-glucopyranose yields the following structure, in which the ring lies in a plane that emerges from the page. (The ring hydrogens are not shown.)

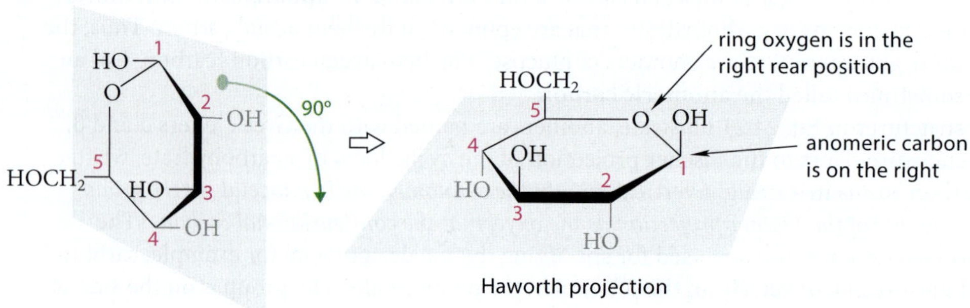

When the plane of the ring is turned 90° so that the *anomeric carbon is on the right and the ring oxygen is in the rear*, the groups in *up* positions are those that are on the *left* in the Fischer projection, whereas the groups in *down* positions are those that are on the *right* in the Fischer projection. A planar structure of this sort is called a **Haworth projection**. In a Haworth projection, the ring is drawn in a plane perpendicular to the page, and the positions of the substituents are indicated with up or down bonds. The wedged bonds are in front of the page, and the others are in back.

The corresponding line-and-wedge structure follows directly from the Haworth projection by viewing it from above. Substituents that are "up" in the Haworth projection become solid wedges, and substituents that are "down" in the Haworth projection become dashed wedges.

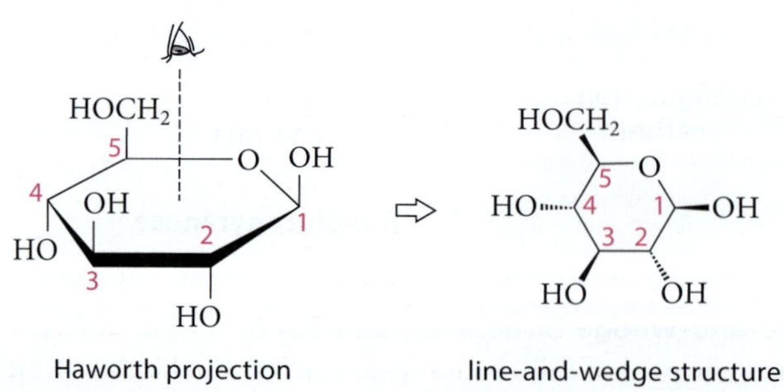

Neither a Haworth projection nor a line-and-wedge structure indicates the conformation of the ring. Six-membered carbohydrate rings resemble substituted cyclohexanes, and, like substituted cyclohexanes, they exist in chair conformations. So, to complete the conformational representation of β-D-glucopyranose, we apply the process given in Sec 7.4A for substituted cyclohexanes. Draw either one of the two chair conformations in which the anomeric carbon and the ring oxygen are in the same relative positions as they are in the preceding Haworth projection or line-and-wedge structure. Then place *up* and *down* groups in axial or equatorial positions, as appropriate.

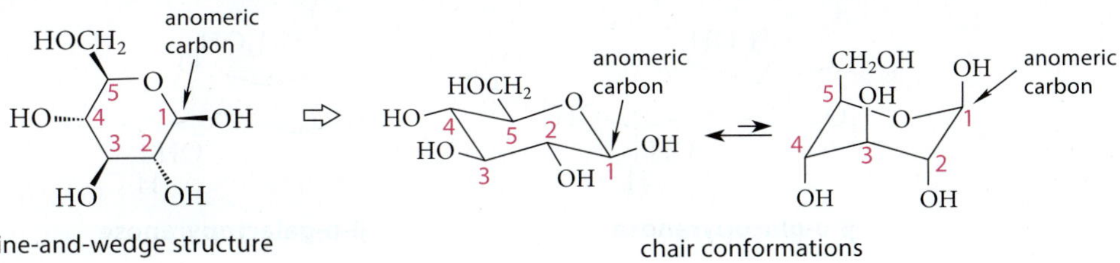

Remember: Although the chair interconversion changes equatorial groups to axial, and vice versa, it does not change whether a group is up or down. Consequently, it doesn't matter which of the two possible chair conformations you draw first. Once you have drawn one chair conformation and have converted it into the other, you can then decide which is the more stable conformation. For β-D-glucopyranose, the more stable conformation is the one on the left, because all of the OH groups are equatorial.

To summarize the conclusions of Study Problem 24.2: When a carbohydrate ring is drawn with the anomeric carbon on the right and the ring oxygen in the rear, substituents that are on the left in the Fischer projection are *up* in all of the other representations, whereas groups that are on the right in the Fischer projection are *down* in the other representations.

Although the five-membered rings of furanoses are nonplanar, they are close enough to planarity that Haworth projections are reasonable approximations to their actual structures. Haworth projections are frequently used for furanoses for this reason. Thus, a Haworth projection of β-D-ribofuranose is derived as follows:

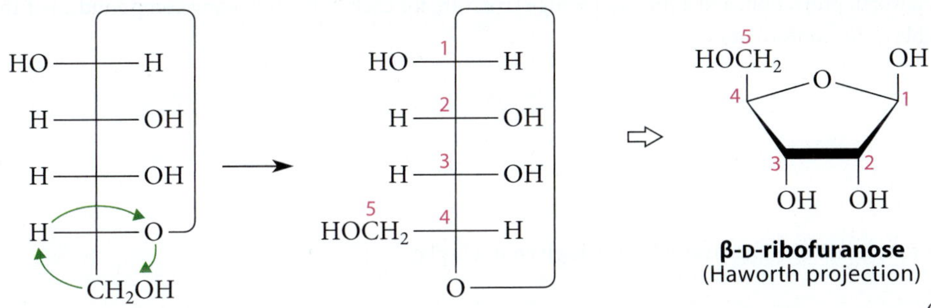

β-D-ribofuranose
(Haworth projection)

(24.23)

The corresponding line-and-wedge structure follows from the Haworth projection.

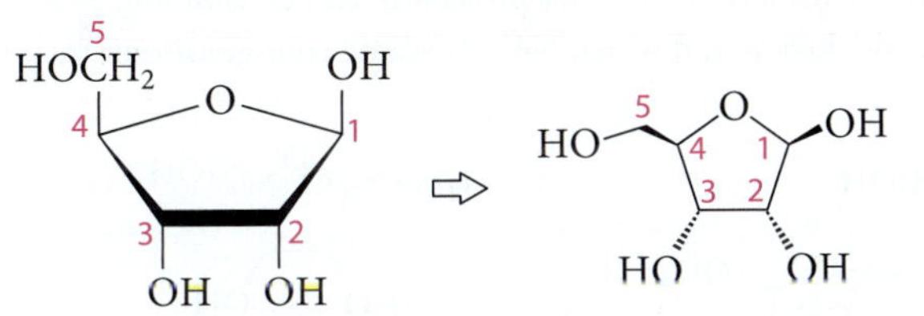

β-D-ribofuranose (24.24)

The Haworth projection is named for Sir Walter Norman Haworth (1883–1950), a noted British carbohydrate chemist who carried out important research on the cyclic structures of carbohydrates. Haworth received the Nobel Prize in Chemistry in 1937 and was knighted in 1947.

Although the procedure in Study Problem 24.2 can be used for any carbohydrate, in some cases it is simpler to derive a cyclic structure from its relationship to another cyclic structure. First, notice that the structure of β-D-glucopyranose is easy to remember because all ring substituents are equatorial in its most stable chair conformation (Study Problem 24.2). Suppose, now, that we want to draw the conformation of β-D-galactopyranose. Because D-galactose and

D-glucose are epimers at carbon-4, the conformational representation of β-D-galactopyranose can be derived easily by interchanging the —H and —OH groups at carbon-4 of β-D-glucopyranose.

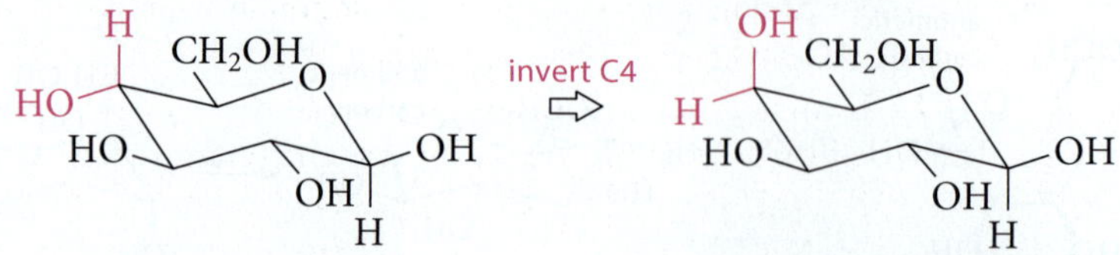

β-D-glucopyranose **β-D-galactopyranose** (24.25)

Likewise, because mannose and glucose are epimeric at carbon-2, the structure of a D-mannopyranose can be derived by interchanging the —H and —OH groups at carbon-2 of the corresponding D-glucopyranose structure.

Sometimes it is necessary to draw the conformation of a carbohydrate that is either a mixture of anomers or of uncertain anomeric composition. In such cases, the configuration at the anomeric carbon is represented by a "squiggly bond."

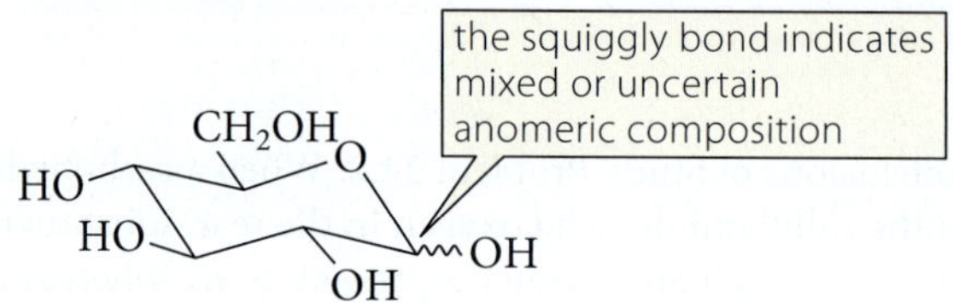

(24.26)

Focused Problems

24.7 Draw a Fischer projection, a Haworth projection, and a line-and wedge structure for each of the following compounds. For the pyranoses, draw the two possible chain conformations.

(a) α-D-Glucopyranose

(b) β-D-Mannopyranose

(c) β-D-Xylofuranose

(d) α-D-Fructopyranose (The structure of D-fructose, a ketose, is given in Display 24.17.)

(e) α-L-Glucopyranose

(f) A mixture of the α- and β-anomers of L-glucopyranose

24.8 Name each of the following aldoses. In part (a), work back to the Fischer projection and consult Fig. 24.3. In part (b), decide which carbons have configurations different from those of glucose and which have the same configurations; then use Fig. 24.3.

(a) 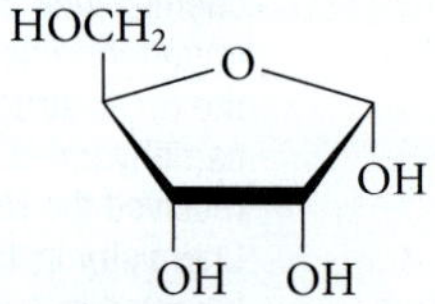

(b) 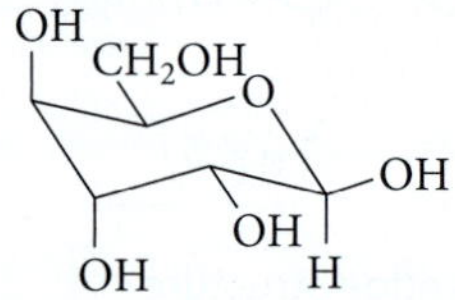

(c) 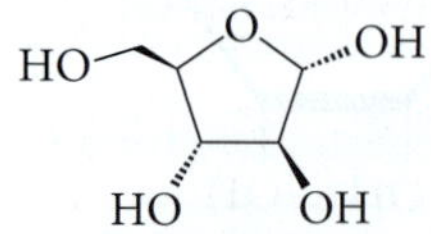

24.4 MUTAROTATION OF CARBOHYDRATES

When pure α-D-glucopyranose is dissolved in water, its specific rotation is found to be +112° mL g^{-1} dm^{-1}. With time, however, the specific rotation of the solution decreases, ultimately reaching a stable value of +52.7° mL g^{-1} dm^{-1}. When pure β-D-glucopyranose is dissolved in water, it has a specific rotation of +18.7° mL g^{-1} dm^{-1}. The specific rotation of this solution increases with time, also to +52.7° mL g^{-1} dm^{-1}. This change of optical rotation with time is called **mutarotation** (*muta*, meaning "change"). Mutarotation also occurs when pure anomers of other carbohydrates are dissolved in aqueous solution.

The mutarotation of glucose is caused by the conversion of the α- and β-glucopyranose anomers into an equilibrium mixture of both. The same equilibrium mixture is formed, as it must be, from either pure α-D-glucopyranose or β-D-glucopyranose. Mutarotation is catalyzed by both acid and base, but it also occurs slowly in pure water.

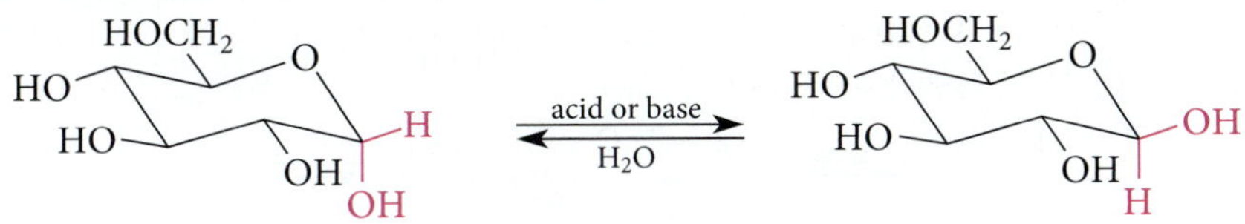

α-anomer $[\alpha]_D = +112°$ mL g^{-1} dm^{-1} β-anomer $[\alpha]_D = +18.7°$ mL g^{-1} dm^{-1}

equilibrium mixture: $[\alpha]_D = +52.7°$ mL g^{-1} dm^{-1} (24.27)

Mutarotation is characteristic of the *cyclic hemiacetal* forms of glucose; an aldehyde cannot undergo mutarotation, because an aldehyde carbon is not an asymmetric carbon. Mutarotation was one of the phenomena that suggested to early carbohydrate chemists that aldoses might exist as cyclic hemiacetals.

Mutarotation occurs, first, by opening of the pyranose ring to the free aldehyde form. This is nothing more than the reverse of hemiacetal formation (Sec. 19.11A). Then a 180° rotation about the carbon–carbon bond to the carbonyl group permits reclosure of the hemiacetal ring by the reaction of the hydroxy group at the opposite face of the carbonyl carbon.

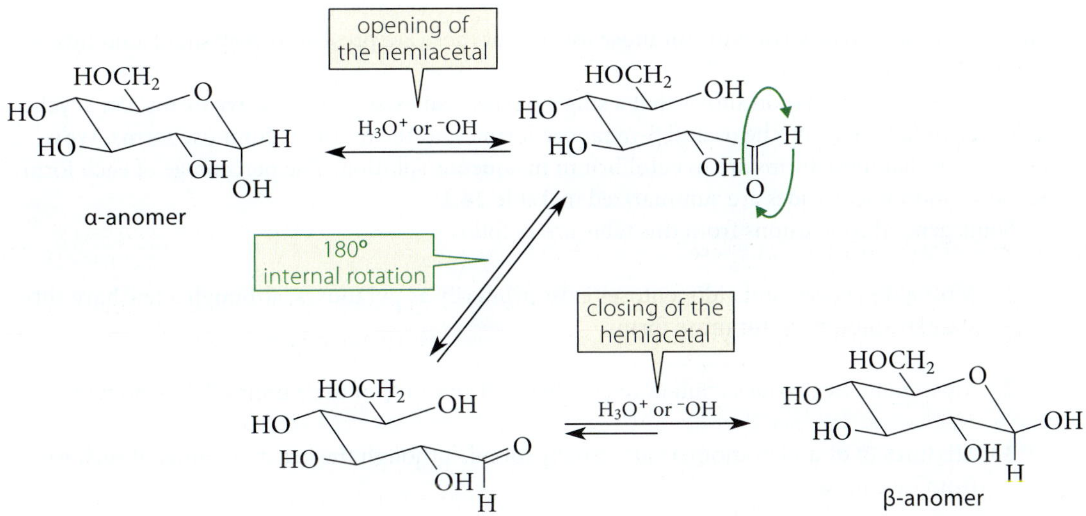

(24.28)

The mutarotation of glucose is due almost entirely to the interconversion of its two pyranose forms. Other carbohydrates, such as D-fructose, a 2-ketohexose, undergo more complex mutarotations. The structures of the cyclic hemiacetal forms of D-fructose can be derived from its carbonyl (ketone) form using the methods described in Sec. 24.3B (see also Focused Problem 24.7d):

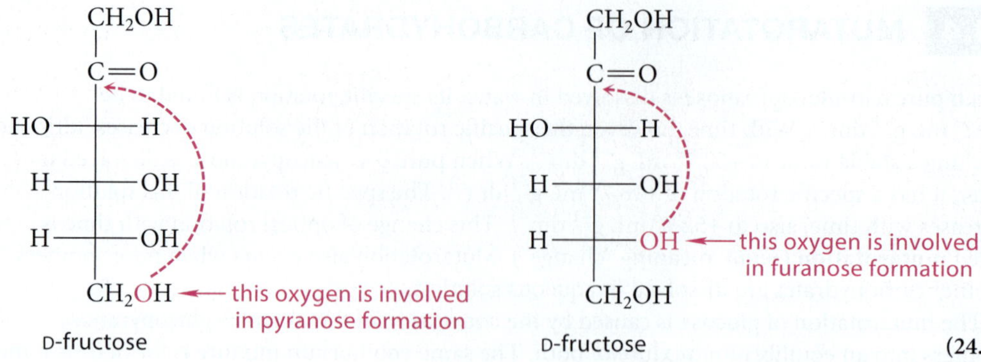

(24.29)

It happens that the crystalline form of D-fructose is β-D-fructopyranose. When crystals of this form are dissolved in water, it equilibrates to both pyranose and furanose forms.

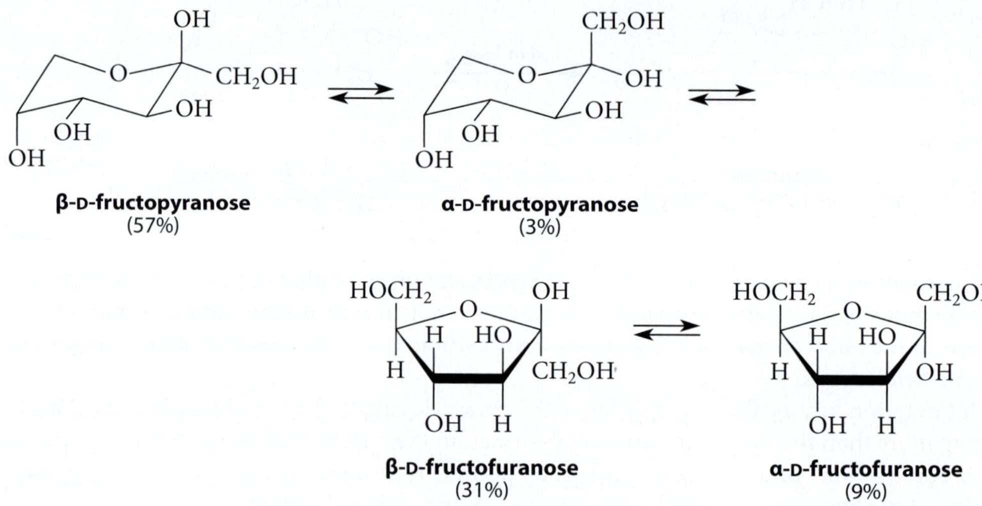

(24.30)

Glucose in solution also contains furanose forms, but these are present in very small amounts—about 0.2% each.

The foregoing discussion shows that a single hexose can exist in at least five forms: the acyclic aldehyde or ketone form, the α- and β-pyranose forms, and the α- and β-furanose forms. We've also shown that these forms are in equilibrium in aqueous solution. The percentage of each form for some monosaccharides are summarized in **Table 24.1**.

Some general conclusions from this table are as follows:

1. Most aldohexoses and aldopentoses exist primarily as pyranoses, although a few have substantial amounts of furanose forms.

2. Most monosaccharides contain relatively small amounts of their noncyclic carbonyl forms.

3. Mixtures of α- and β-anomers are usually found, although the exact amounts of each vary from case to case.

NMR spectroscopy has been one of the primary techniques that chemists have used to determine the ratios of the various forms of monosaccharides in solution. Using D-glucose as an example, Table 24.1 shows that at equilibrium there is nearly two times as much of the β-pyranose form (64%) as there is of the α-pyranose form (36%). Because the two pyranose forms are diastereomers (specifically, anomers), we would expect them to have different proton NMR spectra. However, because they are in equilibrium with each other, an NMR sample of D-glucose will contain both forms, and peaks from each will be seen.

TABLE 24.1 Compositions of Monosaccharides at Equilibrium in Aqueous Solution at 40 °C

	Percent at equilibrium				
	pyranose		furanose		aldehyde or ketone
Sugar	α	β	α	β	
D-glucose	36	64	trace*		0.003
D-galactose[†]	27–36	64–73	trace		trace
D-mannose	68	32	trace	0	trace
D-allose	18	70	5	7	
D-altrose	27	40	20	13	
D-idose[‡]	39	36	11	14	
D-talose	40	29	20	11	
D-arabinose[§]	63	34	3		
D-xylose	37	63			
D-ribose	20	56	6	18	0.02
D-fructose	0–3	57–75	4–9	21–31	0.25

*In 10% aqueous dioxane, glucose contains 0.1–0.2% of each furanose.
[†]At 25 °C, galactose contains 29% α-pyranose, 64% β-pyranose, 3% α-furanose, and 4% β-furanose.
[‡]25 °C
[§]At 25 °C, arabinose contains 60% α-pyranose, 35% β-pyranose, 3% α-furanose, and 2% β-furanose.

Let's consider the proton attached to carbon-1 (the anomeric carbon) in each pyranose form. In the β-form, this proton has a chemical shift of 4.63 ppm, and in the α-form it has a shift of 5.21 ppm. But how do we know which peak corresponds to which anomer? Both signals will be split into a doublet by the neighboring proton on carbon-2. However, the coupling constants between that proton and the carbon-1 protons will be different, depending on whether the anomeric proton is axial or equatorial (recall the discussion of the Karplus curve in Sec. 14.5A). Because the two protons are approximately anti to one another in the β-form, we expect a larger coupling constant ($J_{1,2} = 7.9$ Hz) than when the two protons are nearly gauche, as in the α-form ($J_{1,2} = 3.8$ Hz).

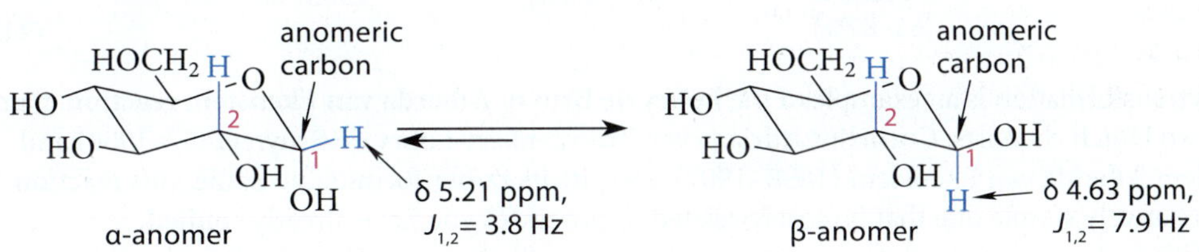

(24.31)

These two peaks integrate in a 64:36 ratio, so the amounts of each form present at equilibrium can be extrapolated. (The other peaks for protons attached to other carbons also integrate at the same ratio, but the anomeric protons are usually the most easily identified and are thus most often used for this purpose of determining ratios of anomers.)

The fraction of any form in solution at equilibrium is determined by its stability relative to that of all other forms. To predict the data in Table 24.1 for a given monosaccharide would require an understanding of all the factors that contribute to the stability or instability of *every* one of its isomeric forms in aqueous solution. In some cases, though, the principles of cyclohexane conformational analysis (Secs. 7.3 and 7.4) can be applied, as Focused Problem 24.11 illustrates.

Focused Problems

24.9 Using the curved-arrow notation, fill in the details for acid-catalyzed mutarotation of glucopyranose shown in Eq. 24.28. Begin by protonating the ring oxygen.

24.10 Using the curved-arrow notation, fill in the details for base-catalyzed mutarotation of glucopyranose. Begin by removing a proton from the hydroxy group at carbon-1.

24.11 Consider the β-D-pyranose forms of glucose and talose (see Fig. 24.3). Suggest one reason why talose contains a smaller fraction of β-pyranose form than glucose.

24.12 Draw a conformational representation of (a) β-D-allopyranose; (b) α-D-idofuranose.

24.13 From the specific rotations shown in Eq. 24.27, calculate the percentages of α- and β-D-glucopyranose present at equilibrium. (Assume that the amounts of aldehyde and furanose forms are negligible.) Compare your answer to the data given in Table 24.1.

24.5 BASE-CATALYZED ISOMERIZATION OF ALDOSES AND KETOSES

In base, aldoses and ketoses rapidly equilibrate to mixtures of other aldoses and ketoses.

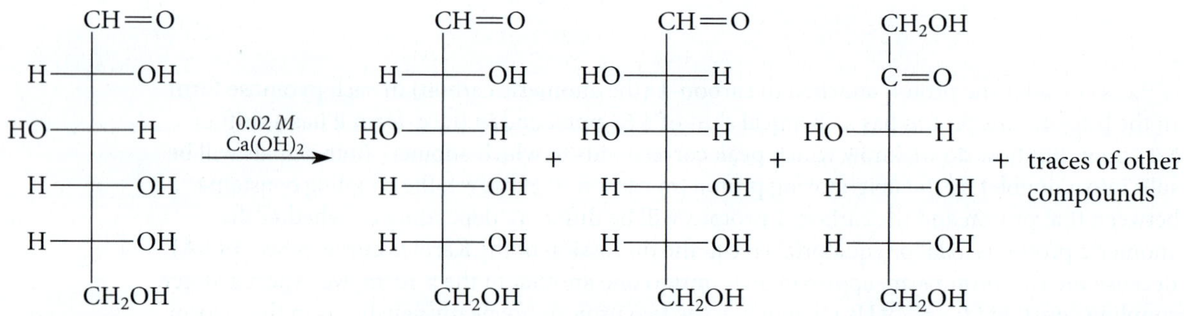

D-glucose **recovered D-glucose** (63–67%) **D-mannose** (0.8–2.4%) **D-fructose** (29–31%)

(24.32)

This transformation is an example of the **Lobry de Bruyn–Alberda van Ekenstein reaction**, named for two Dutch chemists, Cornelius Adriaan van Troostenbery Lobry de Bruyn (1857–1904) and Willem Alberda van Ekenstein (1858–1907). Despite its rather formidable name, this reaction is a relatively simple one that is closely related to processes you have already studied.

Although glucose in solution exists mostly in its cyclic hemiacetal forms, it is also in equilibrium with a small amount of its acyclic aldehyde form. This aldehyde, like other carbonyl compounds with α-hydrogens, ionizes to give small amounts of its enolate ion in base. Protonation of this enolate ion at one face of the double bond gives back glucose; protonation at the other face gives mannose. This is much like the process shown in Eqs. 22.13a and b, Sec. 22.1B.

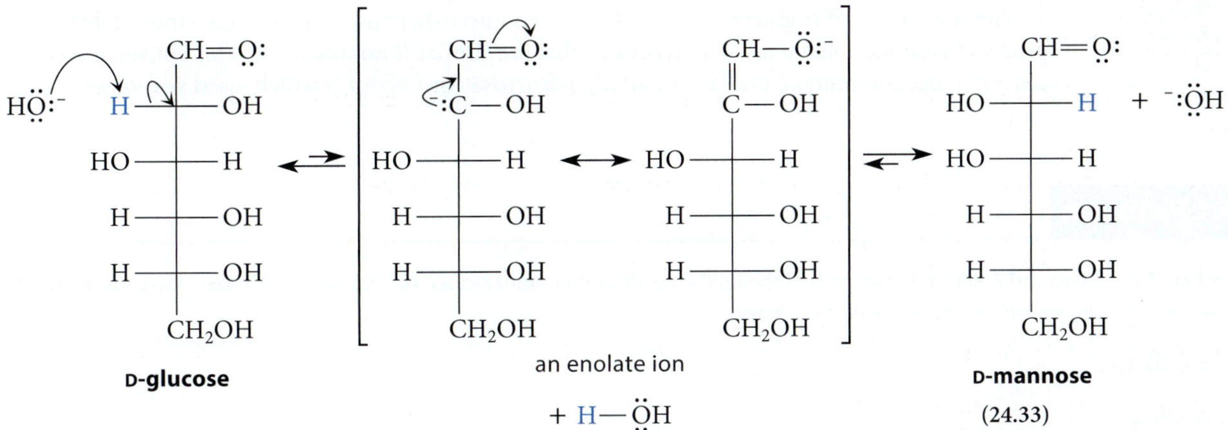

(24.33)

The enolate ion can also be protonated on oxygen to give a new enol, called an **enediol**. An enediol contains a hydroxy group at each end of a double bond. The enediol derived from glucose is simultaneously the enol of not only the aldoses glucose and mannose but also the ketose fructose.

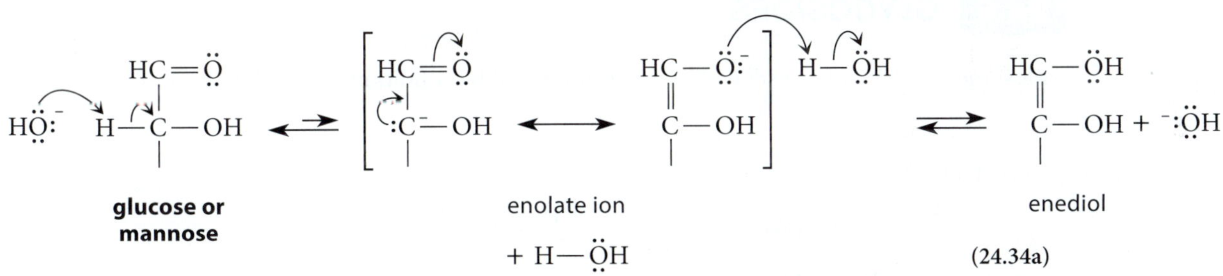

(24.34a)

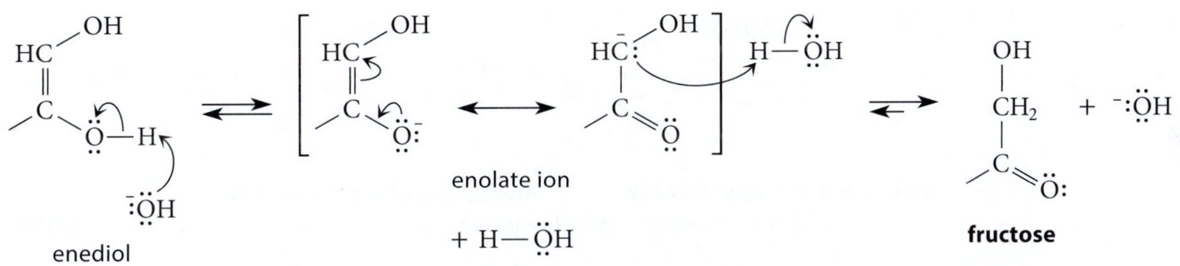

(24.34b)

Such base-catalyzed epimerizations and aldose–ketose equilibria need not stop at carbon-2. For example, D-fructose epimerizes at carbon-3 on prolonged treatment with base. (Why?)

Several transformations of this type are important in metabolism. One such reaction, the conversion of D-glucose-6-phosphate into D-fructose-6-phosphate, occurs in the breakdown of D-glucose (glycolysis), the series of reactions by which D-glucose is utilized as an energy (food) source. Because biochemical reactions occur near pH 7, too little hydroxide ion is present to catalyze the reaction. Instead, the reaction is catalyzed by an enzyme, D-glucose-6-phosphate isomerase, and the enzyme-catalyzed reaction also involves an enediol intermediate.

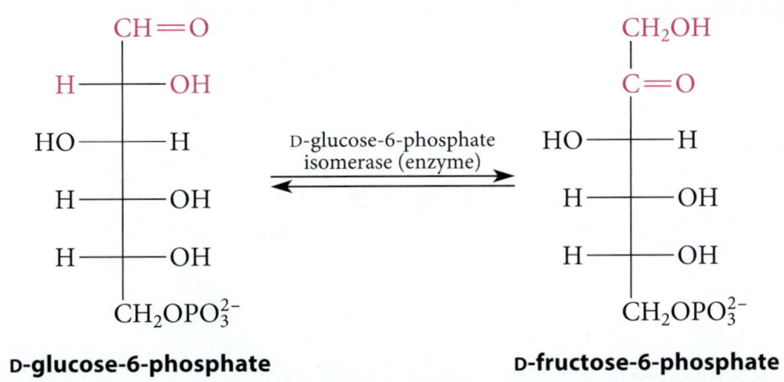

(24.35)

The conversion of D-glucose derived from corn into D-fructose is also an enzyme-catalyzed process that undoubtedly involves reactions that are similar if not identical. This conversion is central to the commercial production of high-fructose corn syrup, a widely used sweetener.

Focused Problem

24.14 Into what other aldose and 2-ketose would each of the following aldoses be transformed on treatment with base? Give the structure and name of the aldose, and the structure of the 2-ketose.

(a) D-Galactose

(b) D-Allose

24.6 GLYCOSIDES

Most monosaccharides react with alcohols under acidic conditions to yield cyclic acetals.

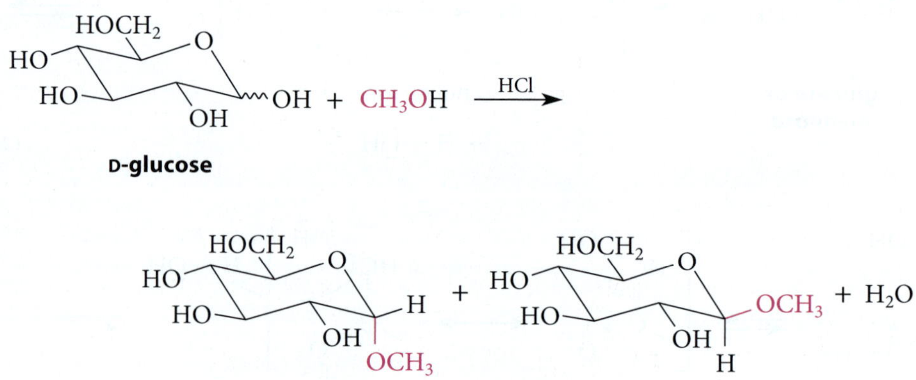

methyl α-D-glucopyranoside **methyl β-D-glucopyranoside**
(83–85% yield; separated by fractional crystallization) (24.36)

Such compounds are called **glycosides**. They are special types of acetals in which one of the oxygens of the acetal group is the ring oxygen of the pyranose or furanose.

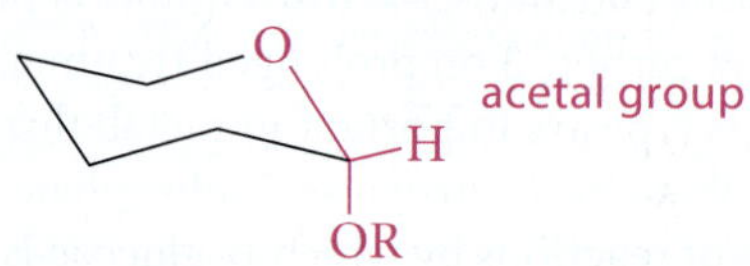

(24.37)

Contrast the reaction of a cyclic hemiacetal (such as glucopyranose) with the corresponding reaction of an ordinary aldehyde under the same conditions:

Glycoside formation:

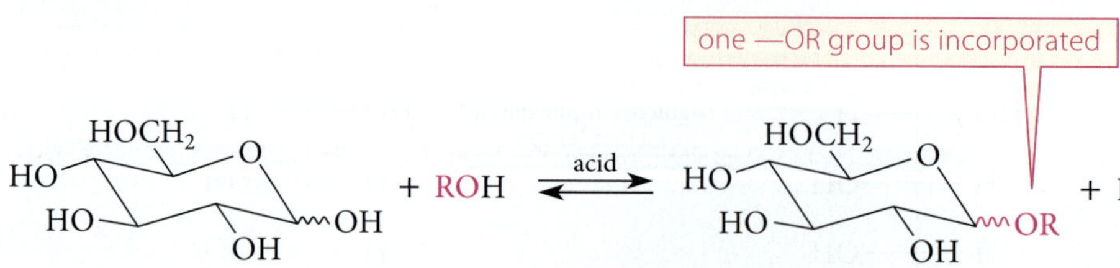

(24.38a)

Formation of an aldehyde acetal:

two —OR groups are incorporated

$$-CH{=}O + 2\,ROH \underset{}{\overset{\text{acid}}{\rightleftharpoons}} -\underset{}{\overset{OR}{\overset{|}{C}H}}-OR + H_2O \qquad (24.38b)$$

As Eqs. 24.38a and b show, a cyclic hemiacetal such as glucopyranose incorporates one alcohol —OR group, whereas an ordinary aldehyde incorporates two —OR groups. This difference between aldoses and ordinary aldehydes is another reason that early carbohydrate chemists suspected that aldoses exist as cyclic hemiacetals.

As illustrated in Eq. 24.36, glycosides are named as derivatives of the parent carbohydrate. The term *pyranoside* indicates that the glycoside ring is six-membered. The term *furanoside* is used for a five-membered ring.

Glycoside formation, like acetal formation, is catalyzed by acid and involves an α-alkoxy carbocation intermediate (Eq. 19.27, Sec. 19.7).

STUDY GUIDE LINK 24.1 Acid Catalysis of Carbohydrate Reactions

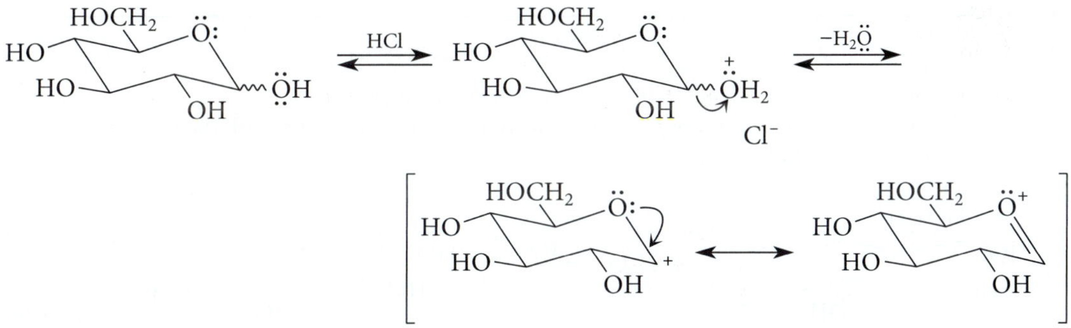

(24.39a)

There are then two stereochemically different fates for α-alkoxycarbocation:

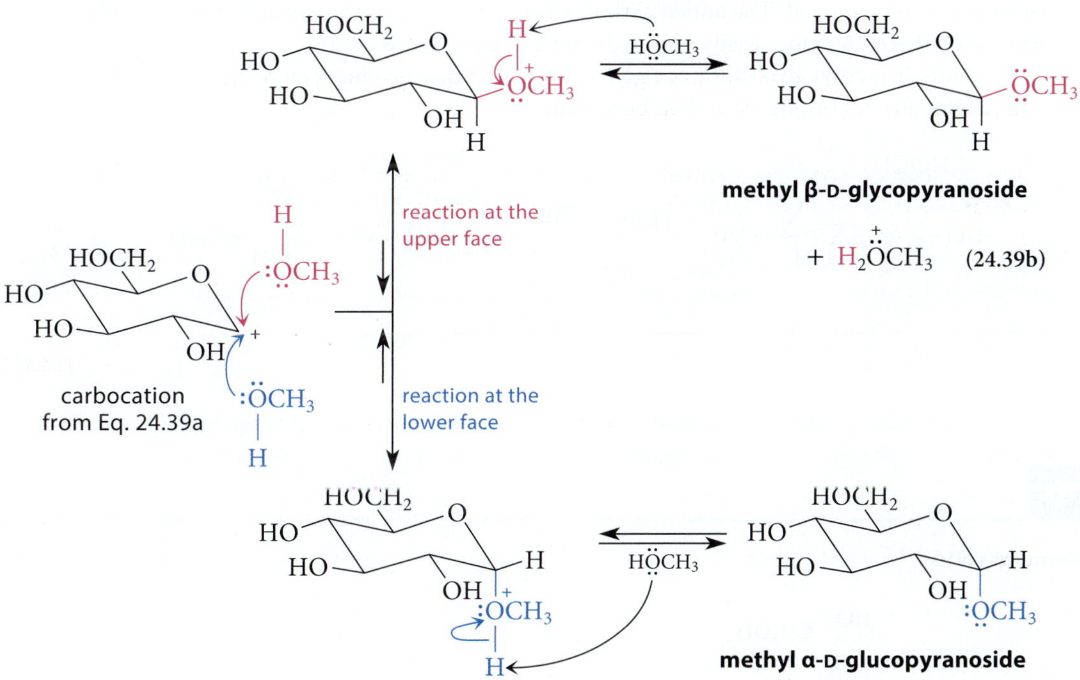

$+ H_2\overset{+}{O}CH_3$ (24.39b)

$+ H_2\overset{+}{O}CH_3$ (24.39c)

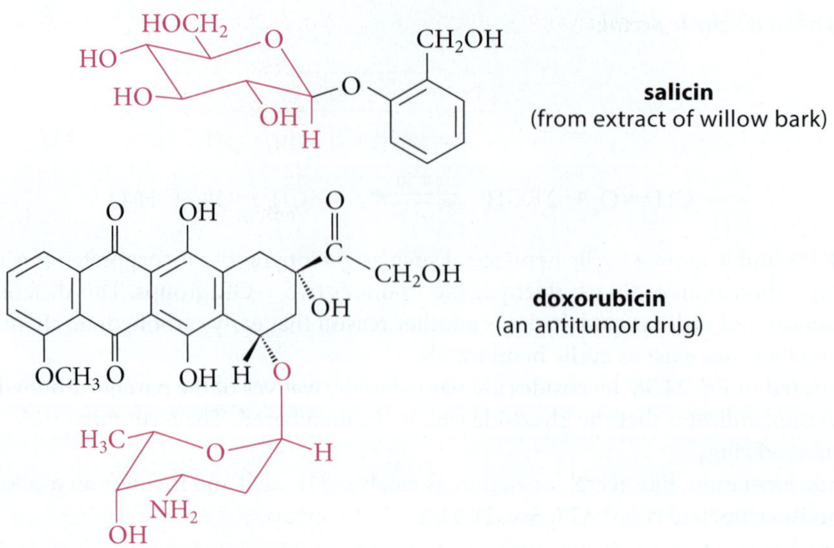

FIGURE 24.4 Two naturally occurring glycosides of medicinal interest. The carbohydrate part of each glycoside is shown in red.

Like other acetals, glycosides are *stable to base*, but, in dilute aqueous acid, they are hydrolyzed back to their parent carbohydrates.

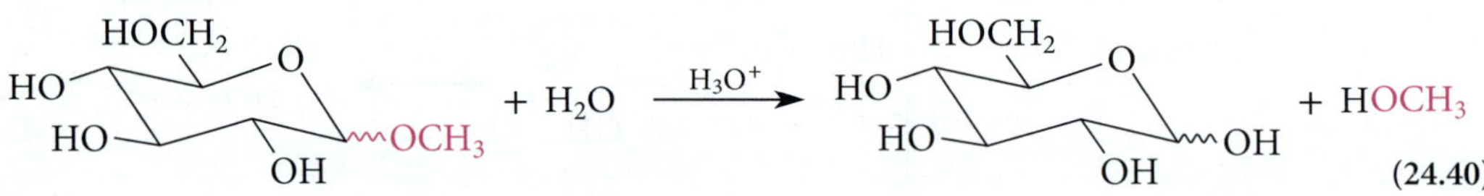

(24.40)

Many compounds occur naturally as glycosides; two examples are shown in **Fig. 24.4**. In addition, glycoside formation plays an important role in the removal of some chemicals from the body in phase II metabolism. In this process, a carbohydrate is joined to an —OH group of the substance to be removed. The added carbohydrate group makes the substance more soluble in water and, therefore, more easily excreted. (See Eq. 8.64, Sec. 8.6C.)

Like simple methyl glycosides, the glycoside of a natural product can be hydrolyzed to its component alcohol or phenol and carbohydrate.

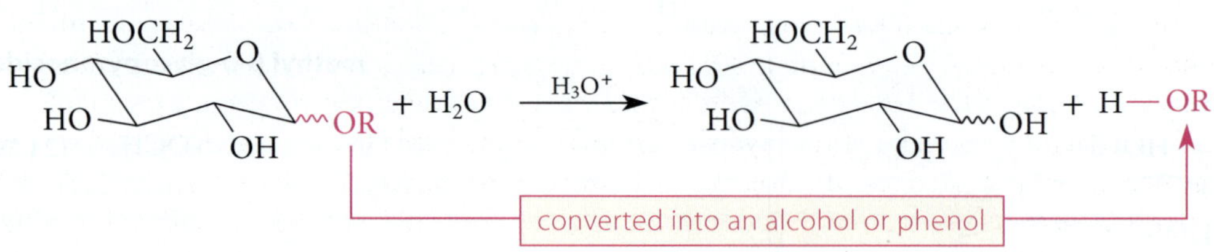

(24.41)

Focused Problems

24.15 (a) Name the following glycoside.

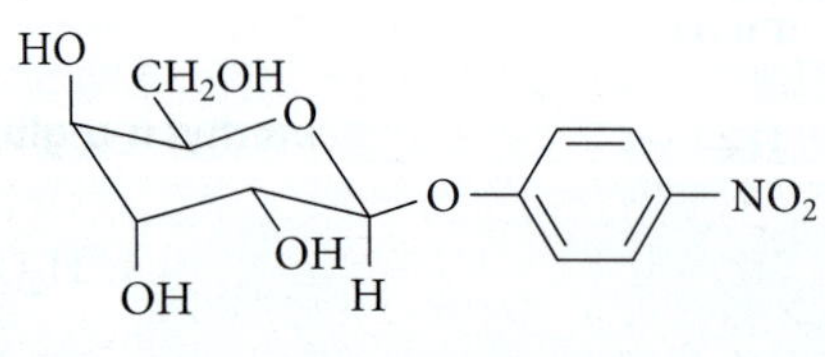

(b) Into what products will this glycoside be hydrolyzed in aqueous acid?

24.16 Vanillin (the natural vanilla flavoring) occurs in nature as a β-glycoside of D-glucose. Suggest a structure for this glycoside.

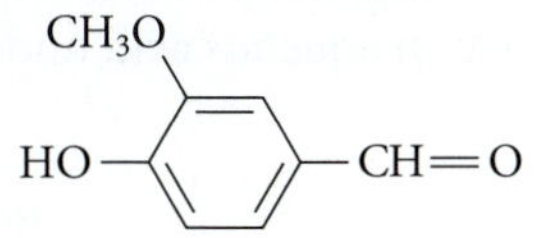

vanillin

24.17 Draw structures for (a) methyl β-D-fructofuranoside; (b) isopropyl α-D-galactopyranoside.

24.7 ETHER AND ESTER DERIVATIVES OF CARBOHYDRATES

Because carbohydrates contain many —OH groups, carbohydrates undergo many of the reactions of alcohols. One such reaction is ether formation. In the presence of concentrated base, carbohydrates are converted into ethers by reactive alkylating agents such as dimethyl sulfate (Eq. 11.40, Sec. 11.4C), methyl iodide, or benzyl chloride.

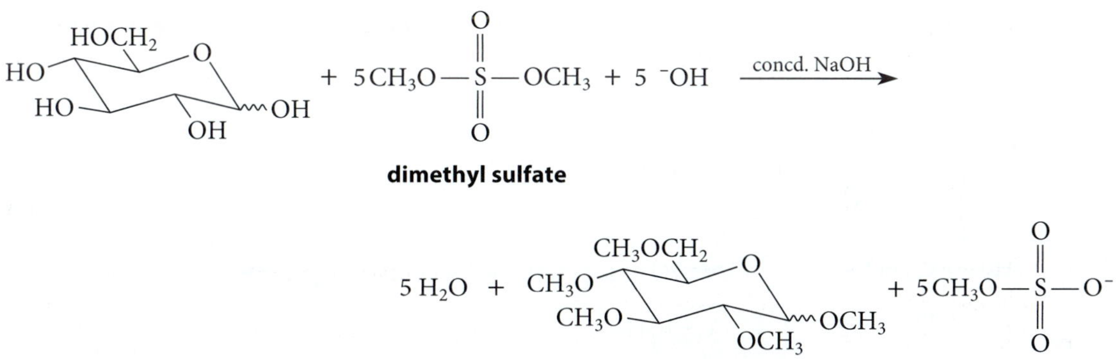

dimethyl sulfate

methyl 2,3,4,6-tetra-O-methyl-D-glucopyranoside (24.42)

(Notice that the ethers are named as *O*-alkyl derivatives of the carbohydrates.) These reactions are examples of the Williamson ether synthesis (Sec. 12.2A). The Williamson synthesis with most alcohols requires a base stronger than ^-OH to form the conjugate-base alkoxide. The hydroxy groups of carbohydrates, however, are more acidic ($pK_a \approx 12$) than those of ordinary alcohols. (The higher acidity of carbohydrate hydroxy groups is attributable to the polar effect of the many neighboring oxygens in the molecule.) Consequently, substantial concentrations of their conjugate-base alkoxide ions are formed in concentrated NaOH. A large excess of the alkylating reagent is used because hydroxide itself, present in large excess, also reacts with alkylating agents. Little or no base-catalyzed epimerization (Sec. 24.5) is observed in this reaction despite the strongly basic conditions used. Evidently, alkylation of the hydroxy group at the anomeric carbon is much faster than epimerization. Once this oxygen is alkylated, epimerization can no longer occur. (Why?)

In addition to the method described earlier, various other reagents are used to form methyl ethers of carbohydrates, including CH_3I/Ag_2O and the strongly basic $NaNH_2$ (sodium amide) in liquid NH_3 followed by CH_3I.

Remember that the alkoxy group at the anomeric carbon is different from the other alkoxy groups in an alkylated carbohydrate because it is part of the glycosidic linkage. Because it is an acetal, it can be hydrolyzed in aqueous acid under mild conditions:

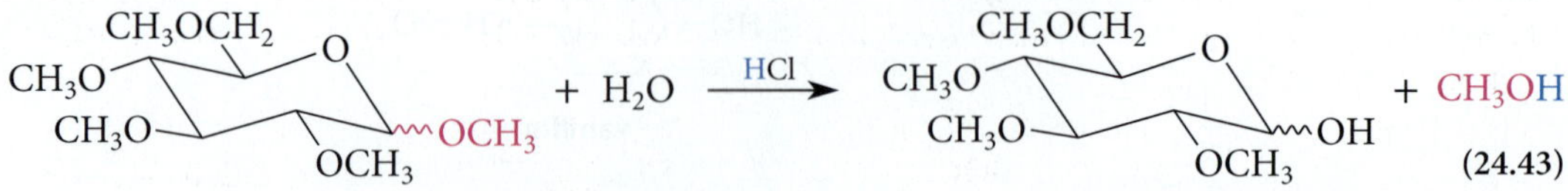

(24.43)

The other alkoxy groups are ordinary ethers and do not hydrolyze under these conditions. They require *much* harsher conditions for cleavage (Sec. 12.4).

Another reaction of alcohols is esterification; indeed, the hydroxy groups of carbohydrates, like those of other alcohols, can be esterified.

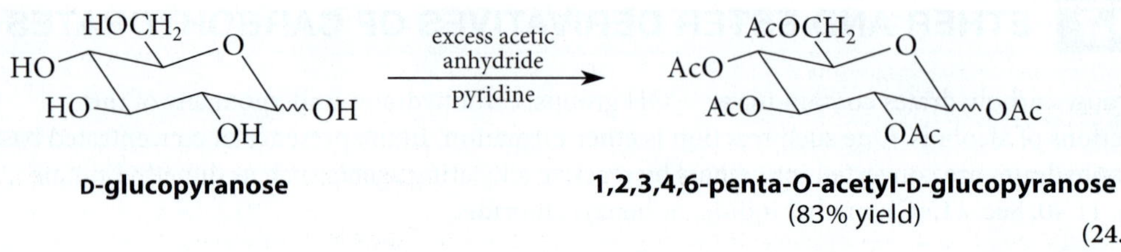

(24.44)

Ester derivatives of carbohydrates can be saponified in base or removed by transesterification with an alkoxide such as methoxide:

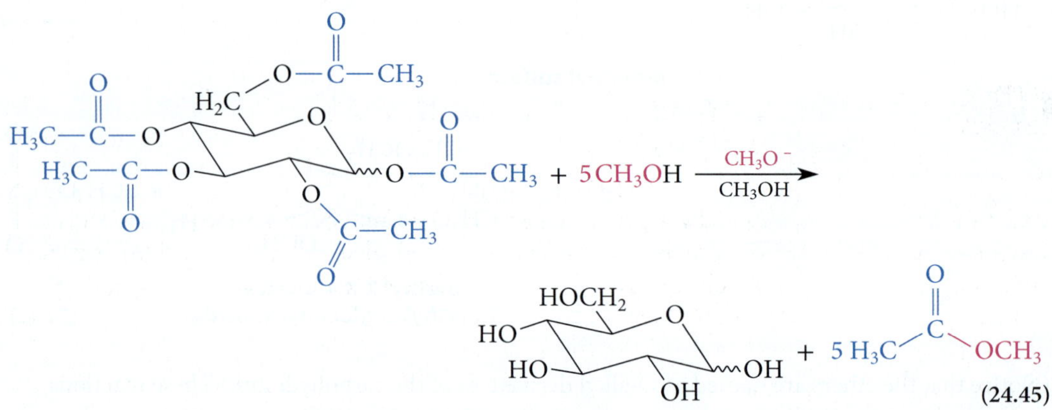

(24.45)

Ethers and esters are used as protecting groups in reactions involving carbohydrates. For example, acetate esters are used as protecting groups in the synthesis of ^{18}F-fluoro-D-glucose (FDG), an important imaging agent in positron emission tomography. (See the sidebar "The S_N2 Solvent Effect in Cancer Diagnosis" in Sec. 9.4E.) Because ethers and esters of carbohydrates have broader solubility characteristics and greater volatility than the carbohydrates themselves, they also find use in the characterization of carbohydrates by chromatography and mass spectrometry.

Study Problem 24.3

Outline a sequence of reactions by which D-glucose can be converted into methyl 2,3,4,6-tetra-*O*-acetyl-D-glucopyranoside.

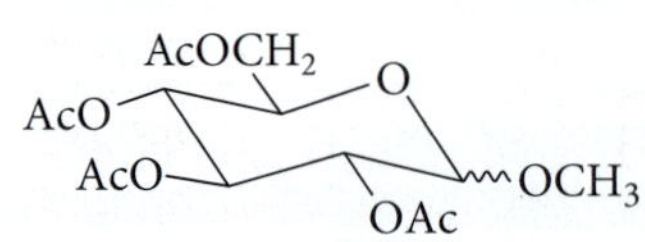

methyl 2,3,4,6-tetra-*O*-acetyl-D-glucopyranoside (24.46)

Solution In solving problems of this sort, in which apparently similar hydroxy groups are converted into different derivatives, the key is to recognize that hydroxy groups and alkoxy groups at the anomeric position (carbon-1 in aldoses) behave quite differently from these groups at other positions. As previously discussed, the hydroxy group at carbon-1 of glucose is part of a *hemiacetal* group, and alkoxy groups at the same carbon are part of an *acetal* group. Acetals are formed and hydrolyzed under much milder conditions than ordinary ethers. Consequently, the methyl "ether" (actually, a methyl acetal) at carbon-1 can be formed by treating D-glucose with methanol and acid. The remaining hydroxy groups can then be esterified with excess acetic anhydride (as in Eq. 24.44) to give the desired product:

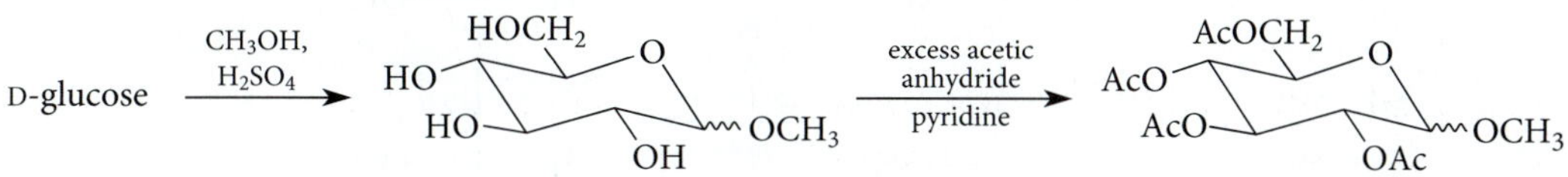

(24.47)

Focused Problems

24.18 Explain why acetals hydrolyze more rapidly than ordinary ethers. (*Hint:* Consider hydrolysis by a carbocation mechanism.)

24.19 Outline a sequence of reactions that will bring about each of the following conversions.

(a) D-Galactopyranose to ethyl 2,3,4,6-tetra-*O*-methyl-D-galactopyranoside

(b) D-Glucopyranose to 2,3,4,6-tetra-*O*-benzyl-D-glucopyranose

24.8 OXIDATION AND REDUCTION REACTIONS OF CARBOHYDRATES

Like simpler aldehydes, the aldehyde group of aldoses can be both oxidized and reduced. It is also possible to selectively oxidize the primary alcohol and aldehyde groups of an aldose without oxidizing the secondary alcohols. The structures and names of the common oxidation and reduction products of aldoses are summarized in **Table 24.2**.

Focused Problem

24.20 Using Table 24.2 to assist you, draw a Fischer projection for the structure of (a) galac-turonic acid, the uronic acid derived from galactose; (b) ribitol, the alditol derived from ribose.

TABLE 24.2 Structures of Common Oxidation and Reduction Products of Aldoses

General structure: $X\left(\begin{array}{c}CH\\|\\OH\end{array}\right)_n Y$

Derivative structure			
X—=	—Y=	General name	Example derived from glucose
$HOCH_2$—	—CH=O	aldose	glucose
$HOCH_2$—	—CO_2H	aldonic acid	gluconic acid
HO_2C—	—CO_2H	aldaric acid	glucaric acid
$HOCH_2$—	—CH_2OH	alditol	glucitol
HO_2C—	—CH=O	uronic acid	glucuronic acid

A. Oxidation to Aldonic Acids

Treatment of an aldose with bromine water oxidizes the aldehyde group to a carboxylic acid. The oxidation product is an *aldonic acid* (see Table 24.2).

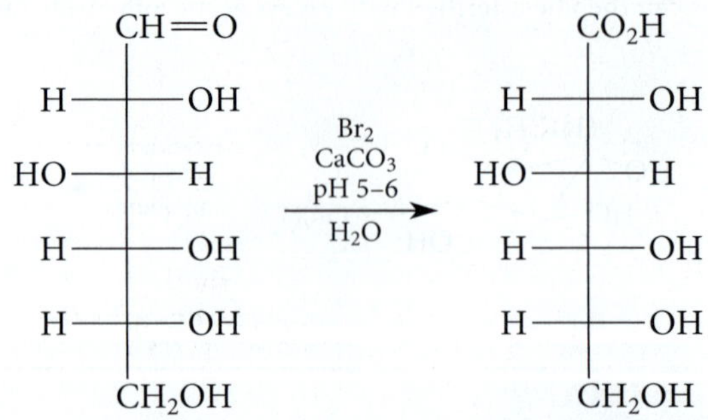

D-glucose **D-gluconic acid** (an aldonic acid; 77–96% yield as Ca^{2+} salt) (24.48)

Although it is customary to represent aldonic acids in the free carboxylic acid form, they, like other γ- and δ-hydroxy acids (Sec. 20.8A), exist in acidic solution as lactones called *aldonolactones.* The lactones with five-membered rings are somewhat more stable than those with six-membered rings.

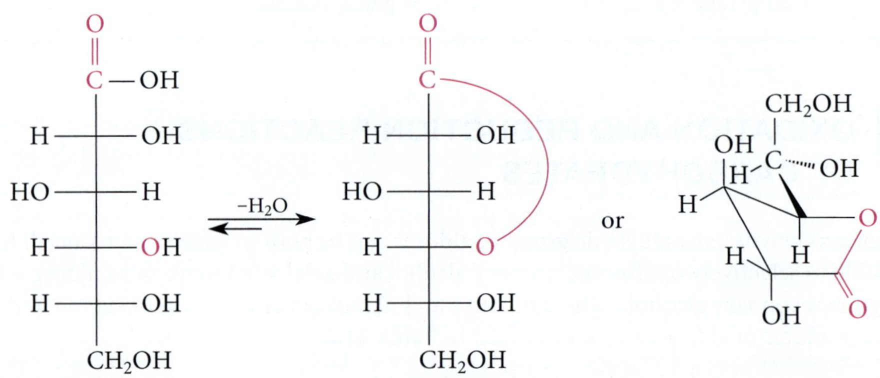

D-gluconic acid **D-γ-gluconolactone** (an aldonolactone) (24.49)

Oxidation with bromine water is a useful test for aldoses. Aldoses can also be oxidized with other reagents, such as the Tollens reagent [$Ag^+(NH_3)_2$; Sec. 19.15]. However, because the Tollens reagent is alkaline and causes base-catalyzed epimerization of aldoses (Sec. 24.5), it is less useful synthetically. Because the alkaline conditions of the Tollens test also promote the equilibration of aldoses and ketoses, ketoses also give positive Tollens tests. Glycosides are *not* oxidized by bromine water, because the aldehyde carbonyl group is protected as an acetal.

B. Oxidation to Aldaric Acids

Dilute nitric acid oxidizes aldehydes and primary alcohols to carboxylic acids *without affecting secondary alcohols.* Consequently, this is a very useful reagent for converting aldoses (or aldonic acids) into aldaric acids (Table 24.2).

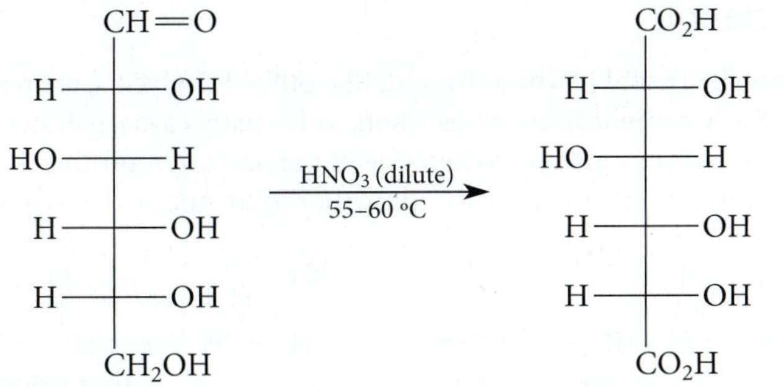

D-glucose

D-glucaric acid
(an aldaric acid)
(41% yield, isolated as Ca^{2+} salt) (24.50)

Like aldonic acids, aldaric acids form lactones in acidic solution. Two different five-membered lactones are possible, depending on which carboxylic acid group undergoes lactonization. Furthermore, under certain conditions, some aldaric acids can be isolated as dilactones, in which both carboxylic acid groups are lactonized.

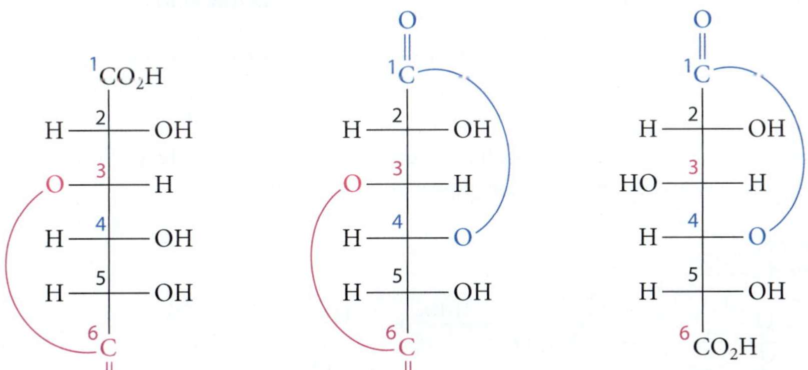

D-glucaric acid 3,6-lactone **D-glucaric acid 1,4 : 3,6-dilactone** **D-glucaric acid 1,4-lactone**

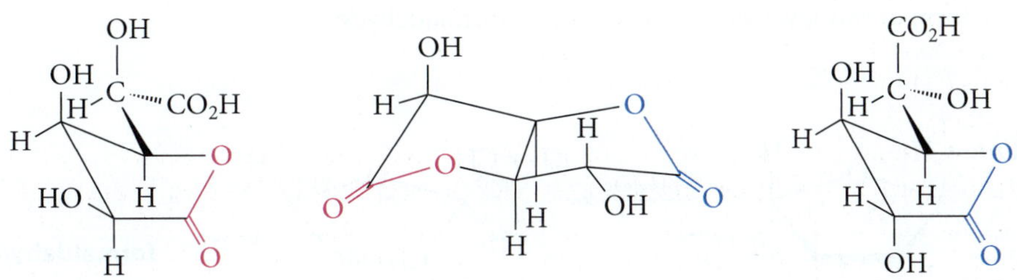

(24.51)

STUDY GUIDE LINK 24.2
Configurations of Aldaric Acids

Focused Problems

24.21 Give Fischer projections for the aldaric acids derived from both D-glucose and L-gulose. What is the relationship between these structures?

24.22 Draw a Fischer projection for the aldaric acid, and a structure of the 1,4-lactone, derived from the oxidation of (a) D-galactose; (b) D-mannose.

24.23 Give the product formed when each of the following alcohols is oxidized by dilute HNO_3.

(a) 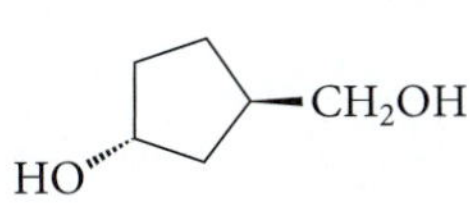

(b) $HOCH_2CH_2CH_2CH_2OH$

C. Periodate Oxidation

Many carbohydrates contain vicinal glycol units and, like other 1,2-glycols, are oxidized by periodic acid (Sec. 12.6B). A complication arises when, as in many carbohydrates, more than two adjacent carbons bear hydroxy groups. When one of the oxidation products is an α-hydroxy aldehyde, as in the following example, it is oxidized further to formic acid and another aldehyde.

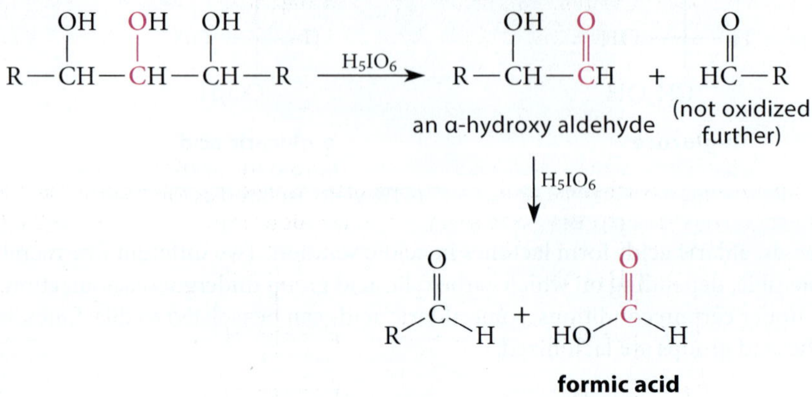

(24.52)

By analogy, an α-hydroxy ketone is oxidized to an aldehyde and a carboxylic acid.

Because it is possible to determine accurately both the amount of periodate consumed and the amount of formic acid produced, periodate oxidation can be used to differentiate between pyranose and furanose structures of saccharide derivatives. For example, periodate oxidation of methyl α-D-glucopyranoside liberates one equivalent of formic acid:

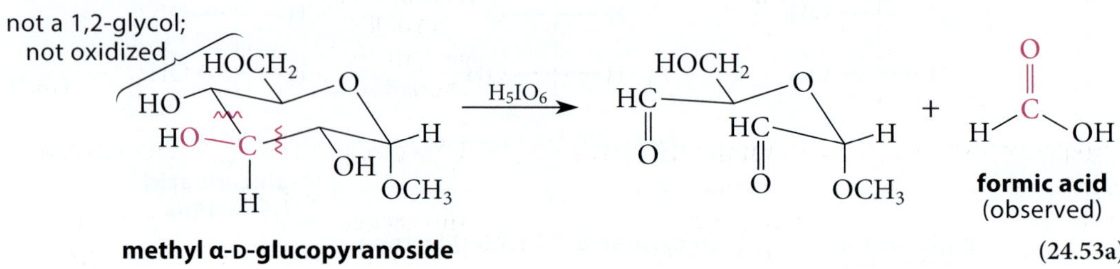

(24.53a)

A furanose form of this glycoside, however, gives formaldehyde:

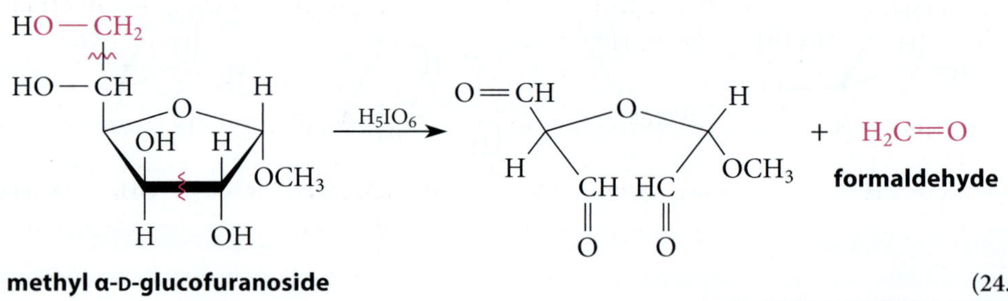

(24.53b)

The periodate oxidation of carbohydrates was developed by Claude S. Hudson (1881–1952), a noted American carbohydrate chemist at the National Institutes of Health. It was used extensively to relate the anomeric configurations of many carbohydrate derivatives. How this was done is suggested by Focused Problems 24.24 and 24.25.

Focused Problems

24.24 Explain why the methyl α-D-pyranosides of *all* D-aldohexoses give, in addition to formic acid, the same compound when oxidized by periodate.

24.25 Assuming you knew the properties of the compound obtained in Focused Problem 24.24, including its optical rotation, show how you could use periodate oxidation to distinguish methyl α-D-galactopyranoside from methyl β-D-galactopyranoside.

D. Reduction to Alditols

Aldohexoses, like ordinary aldehydes, undergo many of the usual carbonyl reductions. For example, sodium borohydride ($NaBH_4$, Sec. 19.9A) reduces aldoses to alditols (Table 24.2).

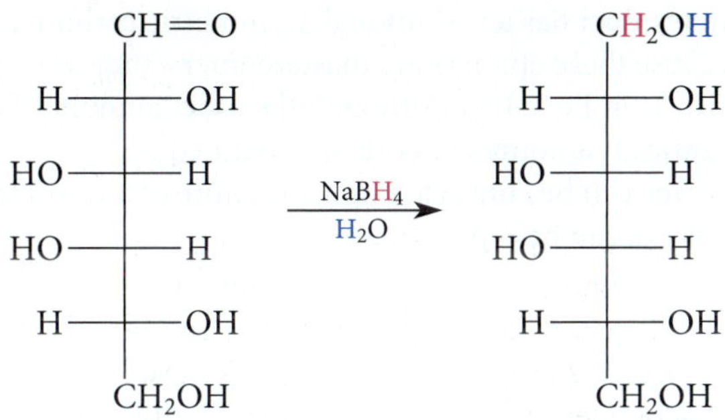

D-galactose

galactitol (dulcitol)
an alditol
(90% yield; neither D nor L because it is meso) (24.54)

Catalytic hydrogenation (for example, H_2 with a Raney nickel catalyst in aqueous ethanol) can also be used for the same transformation.

In the oxidation and reduction reactions discussed in this section, aldoses have been depicted in their carbonyl forms rather than in their cyclic hemiacetal forms. Do not lose sight of the fact that all forms are present at equilibrium, and the aldehyde form can react even though it is present in a very small amount. Once it reacts, it is immediately replenished (Le Chatelier's principle). Thus, when the aldehyde group reacts with $NaBH_4$ to give alditol, the equilibrium provides more of the aldehyde form:

$$\text{cyclic (hemiacetal) forms} \rightleftarrows \text{aldehyde form} \xrightarrow[H_2O]{NaBH_4} \text{alditol} \qquad (24.55)$$

Furthermore, the acidic or basic conditions of many aldehyde reactions (basic conditions in the case of $NaBH_4$ reduction) catalyze this equilibrium. Not only is more aldehyde formed, but it is also formed *rapidly*.

24.9 KILIANI–FISCHER SYNTHESIS

Aldoses, like other aldehydes, undergo addition of hydrogen cyanide to give cyanohydrins (Sec. 19.8). This reaction, like several others that have been discussed, involves the aldehyde form of the sugar.

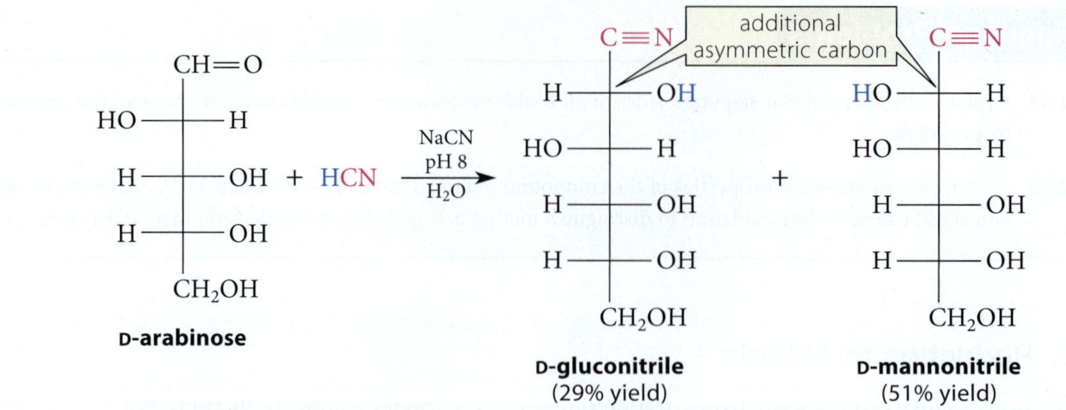

(24.56)

Because the cyanohydrin product has an additional asymmetric carbon, it is formed as a mixture of two epimers. Because these epimers are diastereomers, they are typically formed in different amounts (Sec. 7.7B), as in Eq. 24.56. Although the exact amount of each is not easily predicted, in most cases significant amounts of both are obtained.

The mixture of cyanohydrins can be converted into a mixture of aldoses by catalytic hydrogenation, and these aldoses can be separated.

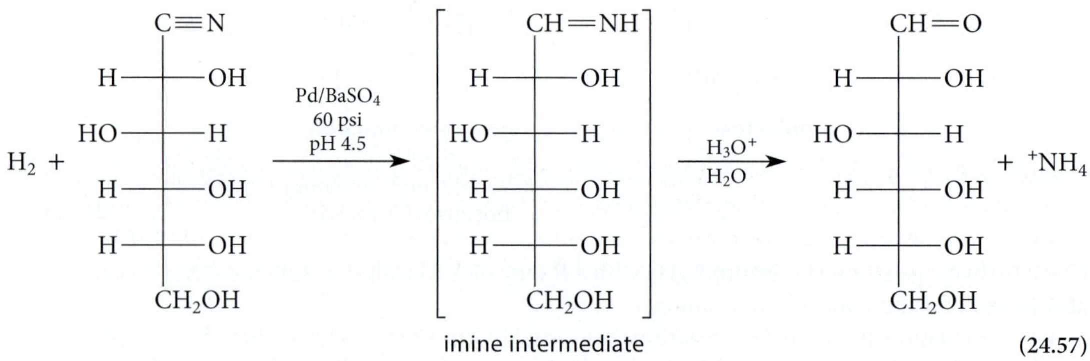

(24.57)

As Eq. 24.57 shows, the hydrogenation reaction involves reduction of the nitrile to an imine (or a cyclic carbinolamine derivative of the imine). Under the reaction conditions, the imine hydrolyzes readily to the aldose and ammonium ion.

This example shows that cyanohydrin formation followed by reduction converts an aldose into two epimeric aldoses with one additional carbon. That is, two aldohexoses, epimeric at carbon-2, are formed from an aldopentose. Notice particularly that this synthesis does *not* affect the stereochemistry of carbons 2, 3, and 4 in the starting material.

The formation of cyanohydrins from aldoses was developed by Heinrich Kiliani (1855–1945), head of the medicinal chemistry laboratory at the University of Freiburg. Kiliani also showed that the cyanohydrins could be hydrolyzed to aldonic acids. Emil Fischer, whose remarkable accomplishments in carbohydrate chemistry are described in Sec 24.10, developed a method to reduce the aldonic acids (as their lactones) to aldoses. The three processes—cyanohydrin formation, hydrolysis, and reduction—provided a way to convert an aldose into two other aldoses with one additional carbon. The overall transformation came to be known as the **Kiliani–Fischer synthesis**. The chemistry shown in Eqs. 24.56 and 24.57, which was developed in the 1970s, is a modern variation of the Kiliani–Fischer synthesis.

Kiliani, a noted authority on carbohydrates, also proved the structures of several monosaccharides, including the 2-ketose structure of fructose.

Focused Problem

24.26 Assuming the D configuration, identify *A* and *B*.

an aldopentose *A* —(dilute HNO_3)→ an aldaric acid, optically inactive

↓ Kiliani–Fischer synthesis

an aldohexose *B* (one of two formed) —(dilute HNO_3)→ an aldaric acid, optically inactive

24.10 THE PROOF OF GLUCOSE STEREOCHEMISTRY

The aldohexose structure of (+)-glucose (that is, the structure without any stereochemical details) was established around 1870. The van't Hoff–LeBel theory of the tetrahedral carbon atom, published in 1874 (Sec. 6.11), suggested the possibility that glucose and the other aldohexoses could be stereoisomers. Which one of the 2^4 possible stereoisomers, then, is (+)-glucose? This problem was solved in two stages.

A. Which Diastereomer? The Fischer Proof

The first (and major) part of the solution to the problem of glucose stereochemistry was published in 1891 by Emil Fischer. (See Fischer's biography in the sidebar later in this section.) It would be reason enough to study Fischer's proof as one of the most brilliant pieces of reasoning in the history of chemistry. However, it also will serve to sharpen your understanding of stereochemical relationships.

In Fischer's day, the tools for determining the absolute stereochemical configurations of chemical compounds had not yet been developed. Consequently, Fischer adopted the *arbitrary convention* that, at carbon-5 of (+)-glucose (the configurational carbon in the D,L system), the —OH is on the right in the standard Fischer projection—that is, Fischer assumed that (+)-glucose has what we now call the D configuration. Fischer, then, proved the stereochemistry of (+)-glucose *relative* to an assumed configuration at carbon-5. In other words, Fischer, of necessity, reduced the problem of 2^4 stereoisomers—two enantiomeric sets of eight diastereomers—to a problem of eight diastereomers.

Fischer had, as a starting point, pure samples of the naturally occurring aldopentose (−)-arabinose and the two aldohexoses, (+)-glucose and (+)-mannose. (+)-Glucose and (+)-mannose were known to be stereoisomers. The remarkable thing about Fischer's proof is that it allowed him to assign relative configurations in space for each asymmetric carbon in these compounds using only chemical reactions and optical activity. The logic involved is direct, simple, and elegant, and it can be summarized in four steps:

Step 1. **(−)-Arabinose, an aldopentose, is converted into both (+)-glucose and (+)-mannose by a Kiliani–Fischer synthesis.** From this observation (see Sec. 24.9), Fischer deduced that (+)-glucose and (+)-mannose are epimeric at carbon-2, and that the configuration of (−)-arabinose at carbons 2, 3, and 4 is the same as that of (+)-glucose and (+)-mannose at carbons 3, 4, and 5, respectively.

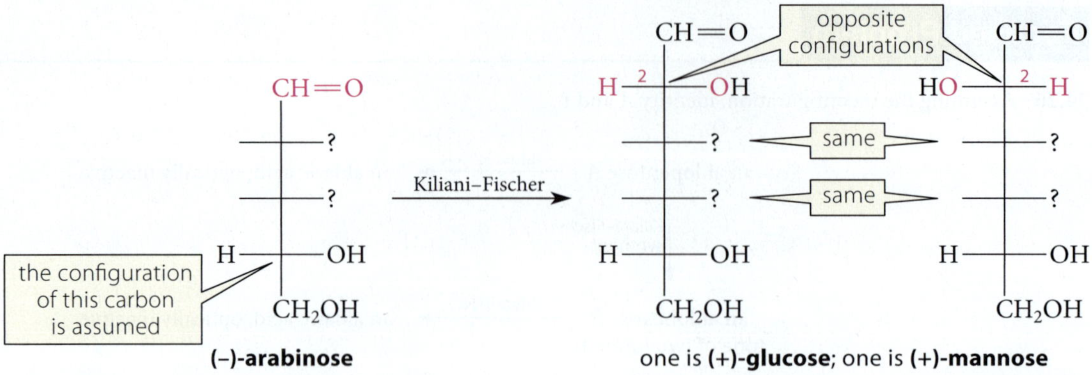

(24.58)

Step 2. **(–)-Arabinose can be oxidized by dilute HNO_3** (Sec. 24.8B) **to an optically active aldaric acid.** From this observation, Fischer concluded that the —OH group at carbon-2 of arabinose must be on the left. If this —OH group were on the right, then the aldaric acid of arabinose would have to be meso, and thus optically inactive, *regardless of the configuration of the —OH group at carbon-3.* [Be sure you see why this is so; if necessary, draw both possible structures for (–)-arabinose to verify this deduction.]

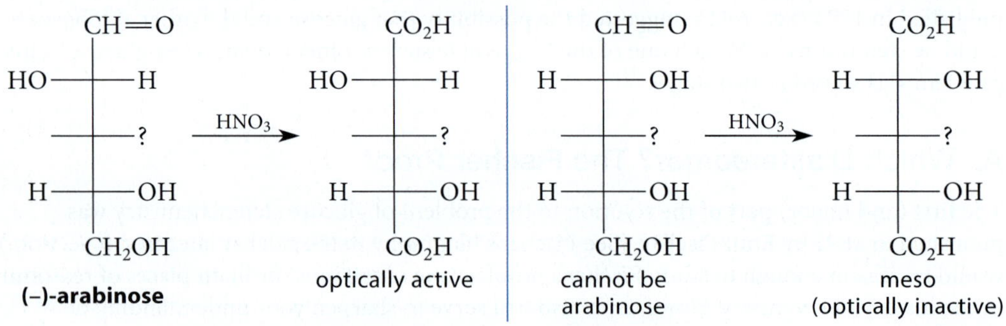

(24.59)

The relationships among arabinose, glucose, and mannose established in steps 1 and 2 require the following partial structures for (+)-glucose and (+)-mannose.

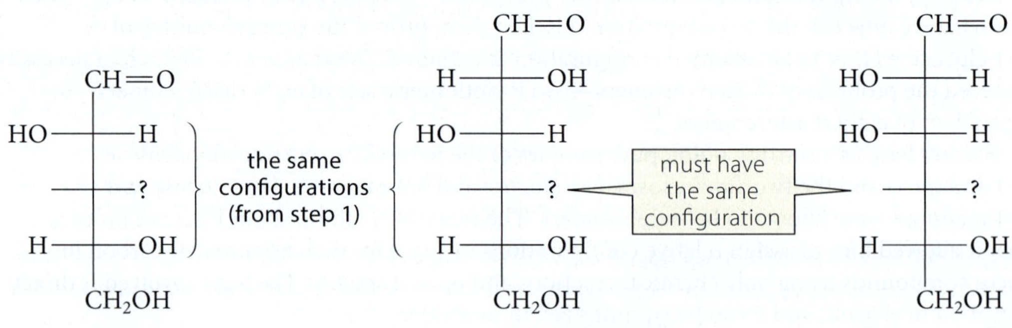

(–)-arabinose one is **(+)-glucose**; one is **(+)-mannose** (24.60)

Step 3. **Oxidations of both (+)-glucose and (+)-mannose with HNO_3 give optically active aldaric acids.** From this observation, Fischer deduced that the —OH group at carbon-4 is on the right in both (+)-glucose and (+)-mannose. Recall that, whatever the configuration at carbon-4 in these two aldohexoses, it must be the same in both. Only if the —OH is on the right will *both* structures yield, on oxidation, optically active aldaric acids. If the —OH were on the left, *one* of the two aldohexoses would have given a meso (and therefore, an optically inactive) aldaric acid.

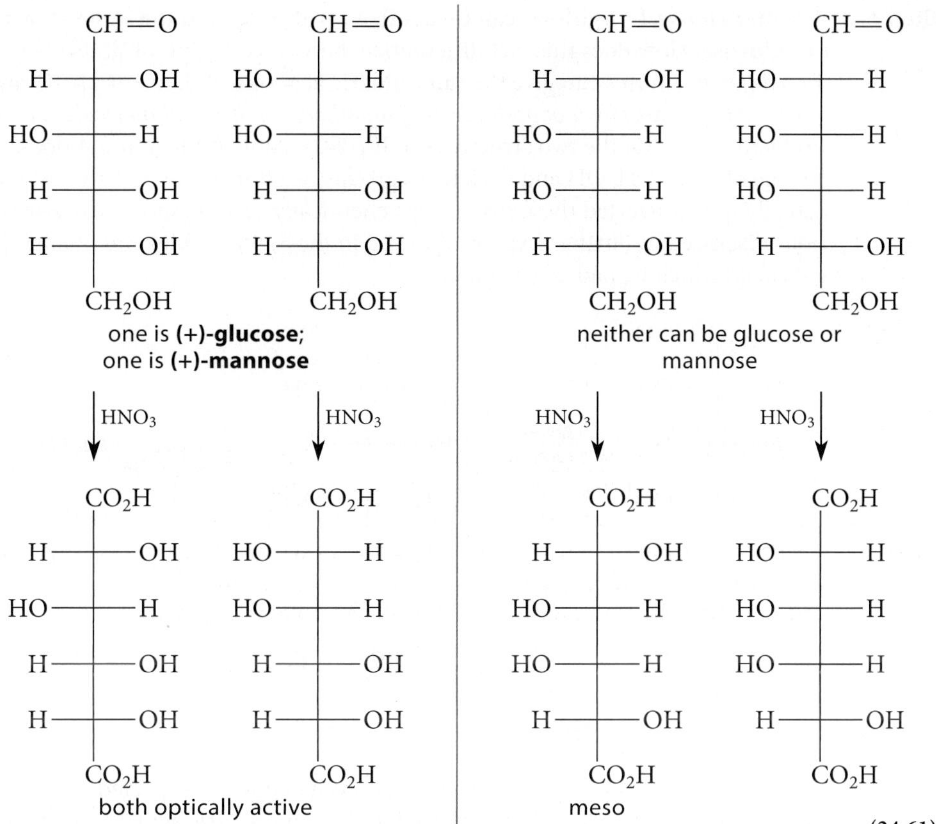

(24.61)

Because the configuration at carbon-4 of (+)-glucose and (+)-mannose is the same as that at carbon-3 of (−)-arabinose (step 1), at this point Fischer could deduce the complete structure of (−)-arabinose.

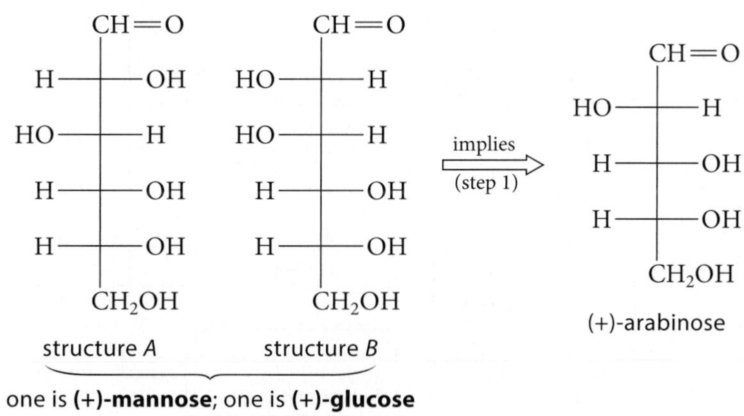

(24.62)

The previous steps had established that (+)-glucose had one of the two structures in Eq. 24.62 and (+)-mannose had the other, but Fischer did not yet know which structure goes with which sugar. This point can be confusing. Fischer's situation was like that of a woman who has just met two brothers, but she doesn't know their names. So she asks a friend, "What are their names?" The friend says, "Oh, they are Mannose and Glucose; only I don't know which is which!" Just because the woman knows *both* names doesn't mean that she can associate *each* name with *each* face. Similarly, although Fischer knew the structures associated with both (+)-glucose and (+)-mannose, he did not yet know how to correlate *each* aldose with *each* structure. This last problem was solved in step 4.

FURTHER EXPLORATION 24.2
More on the Fischer Proof

Step 4. **Another aldose, (+)-gulose, can be oxidized with HNO_3 to the same aldaric acid as (+)-glucose.** How does this fact differentiate between (+)-glucose and (+)-mannose? Two *different* aldoses can give the same aldaric acid only if their —CH=O and —CH_2OH groups are *at opposite ends of an otherwise identical molecule* (Focused Problem 24.21). Of the two structures in Eq. 24.62, only in structure *A* does an interchange of the —CH_2OH and —CH=O groups result in a different aldohexose. Fischer actually interconverted these two groups chemically on (+)-glucose (by a series of reactions discussed in Further Exploration 24.2 in the Study Guide) and obtained a different sugar, which he named (+)-gulose.

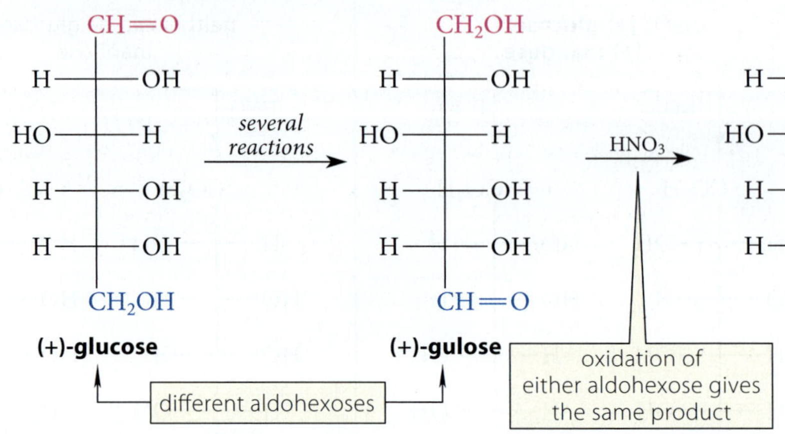

(24.63)

To be sure that the CH_2OH and CH=O interconversions had gone as expected, Fischer verified that both (+)-glucose and (+)-gulose were oxidized to the same aldaric acid, as shown in Eq. 24.63. The inescapable conclusion, then, is that the structure of (+)-glucose in Eq. 24.62 is *A*.

Completion of the proof requires that the same interconversions, when carried out on structure *B*, give the *same* aldohexose. (Verify this point by rotating either structure 180° in the plane of the page.)

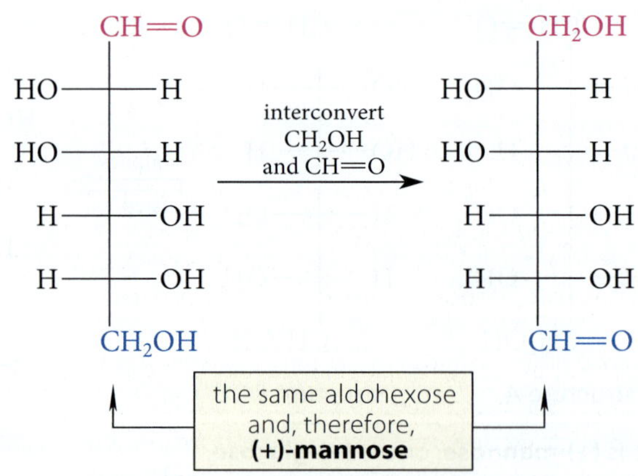

(24.64)

Therefore, structure *B* cannot be (+)-glucose. Because the only other possibility for *B* was (+)-mannose, the structure of this aldohexose was proved as well.

Emil Fischer

World History Archive/Superstock

Emil Fischer (1852–1919) studied with Adolph von Baeyer and ultimately became a professor at Berlin University in 1892. Fischer carried out important research on sugars, proteins, and heterocycles. Fischer was a technical advisor to Kaiser Wilhelm II. The following story gives some indication of the authority that Fischer commanded in Germany. It is said that one day he and the Kaiser were arguing questions of science policy, and the Kaiser sought to end debate by pounding his fist on the table, shouting, "Ich bin der Kaiser!" (I am the Emperor!) Fischer, not to be silenced, responded in kind: "Ich bin Fischer!" Another story, perhaps apocryphal, attributes an important laboratory function to Fischer's beard. It was said that when a student had difficulty crystallizing a sugar derivative (some of which are notoriously difficult to crystallize), Fischer would shake his beard over the flask containing the recalcitrant compound. The accumulated seed crystals in his beard would fall into the flask and bring about the desired crystallization. Fischer was awarded the Nobel Prize in Chemistry in 1902.

Focused Problems

24.27 An aldopentose *A* can be oxidized with dilute HNO_3 to an optically active aldaric acid. A Kiliani–Fischer synthesis starting with *A* gives two new aldoses: *B* and *C*. Aldose *B* can be oxidized to an achiral, and therefore optically inactive, aldaric acid, but aldose *C* is oxidized to an optically active aldaric acid. Assuming the D configuration, give the structures of *A*, *B*, and *C*.

24.28 An aldohexose *A* is either D-idose or D-gulose (see Fig. 24.3). It is found that a different aldohexose, L-(−)-glucose, gives the same aldaric acid as *A*. What is the identity of *A*?

B. Which Enantiomer? The Absolute Configuration of D-(+)-Glucose

Fischer never learned whether his arbitrary assignment of the absolute configuration of (+)-glucose was correct—that is, whether the —OH at carbon-5 of (+)-glucose was really on the right in its Fischer projection (as assumed) or on the left. The groundwork for solving this problem was laid when the configuration of (+)-glucose was correlated to that of (−)-tartaric acid. (Stereochemical correlation was introduced in Sec. 6.5.) This correlation was carried out in the following way.

(+)-Glucose was converted into (−)-arabinose by a reaction called the **Ruff degradation**. In this reaction sequence, an aldose is oxidized to its aldonic acid (Sec. 24.8), and the calcium salt of the aldonic acid is treated with ferric ion and hydrogen peroxide. This treatment decarboxylates the calcium salt and simultaneously oxidizes carbon-2 to an aldehyde.

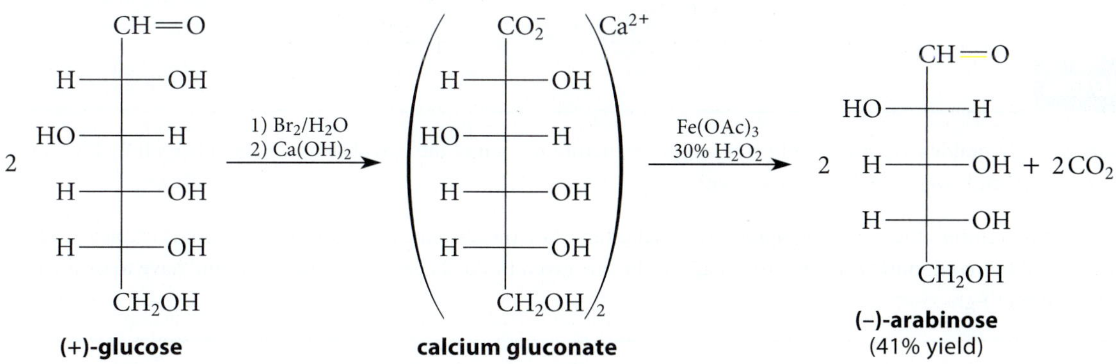

(24.65)

In other words, an aldose is degraded to another aldose with one fewer carbon atom, *its stereochemistry otherwise remaining the same.* Because the relationship between (+)-glucose and (−)-arabinose was already known from the Kiliani–Fischer synthesis (step 1 of the Fischer proof in the previous section), this reaction served to establish the course of the Ruff degradation. Next, (−)-arabinose was converted into (−)-erythrose by another cycle of the Ruff degradation.

$$(-)\text{-arabinose} \xrightarrow{\text{Ruff degradation}} (-)\text{-erythrose} \quad (24.66)$$

D-(+)-Glyceraldehyde, in turn, was related to D-(−)-erythrose by a Kiliani–Fischer synthesis:

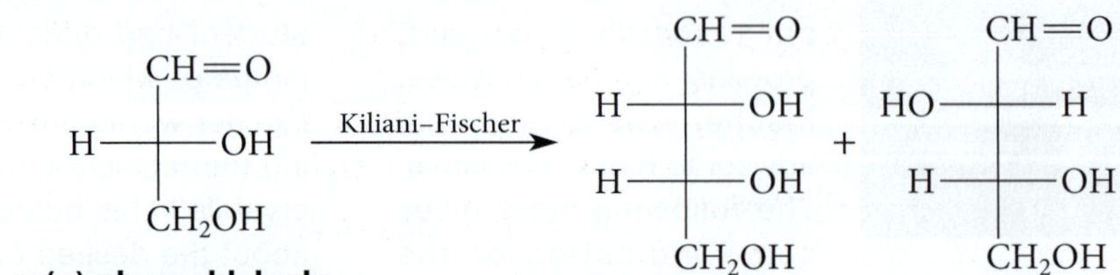

D-(+)-glyceraldehyde (absolute configuration assumed by convention)

D-(−)-erythrose **D-(−)-threose** (configurations at carbon-3 assumed by convention) (24.67)

This sequence of reactions showed that (+)-glucose, (−)-erythrose, (−)-threose, and (+)-glyceraldehyde were all of the same stereochemical series: the D series. Oxidation of D-(−)-threose with dilute HNO_3 gave D-(−)-tartaric acid.

In 1950, the absolute configuration of naturally occurring (+)-tartaric acid (as its potassium rubidium double salt) was determined by a special technique of X-ray crystallography called *anomalous dispersion.* This determination was made by J. M. Bijvoet, A. F. Peerdeman, and A. J. van Bommel, Dutch chemists who worked, appropriately enough, at the van't Hoff laboratory in Utrecht. If Fischer had made the right choice for the configuration at carbon-5 of (+)-glucose—what we now call the D configuration—the assumed structure for D-(−)-tartaric acid and the experimentally determined structure of (+)-tartaric acid determined by the Dutch crystallographers would be enantiomers. If Fischer had guessed incorrectly, the assumed structure for (−)-tartaric acid would be the same as the experimentally determined structure of (+)-tartaric acid, and would have to be reversed. To quote Bijvoet and his colleagues: *"The result is that Emil Fischer's convention* [for the D configuration] *appears to answer to reality."*

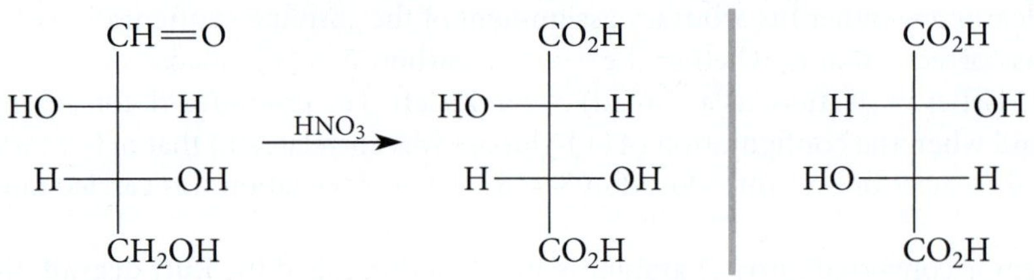

D-(−)-threose **D-(−)-tartaric acid** **L-(+)-tartaric acid** (by X-ray crystallography)

enantiomers

(24.68)

Focused Problems

24.29 Given the structure of D-glyceraldehyde, how would you assign a structure to each of the two aldoses obtained from it by Eq. 24.67, assuming that these compounds were previously unknown?

24.30 Imagine that a scientist reexamines the crystallographic work that established the absolute configuration of (+)-tartaric acid and finds that the structure of this compound is the mirror image of the one given in Eq. 24.68. What changes would have to be made in Fischer's structure of D-(+)-glucose?

24.11 DISACCHARIDES AND POLYSACCHARIDES

A. Disaccharides

Disaccharides consist of two monosaccharides connected by a glycosidic linkage. (+)-**Lactose** is a disaccharide. [(+)-Lactose is present to the extent of about 4.5% in cow's milk and 6–7% in human milk.]

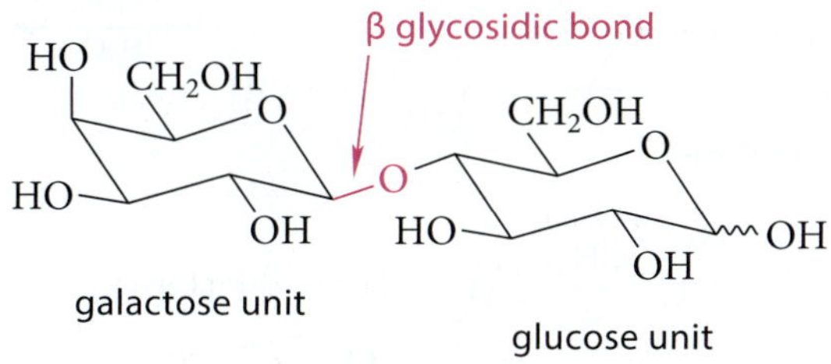

(+)-lactose
or **4-*O*-(β-D-galactopyranosyl)-D-glucopyranose** (24.69)

In (+)-lactose, a D-glucopyranose molecule is linked by its oxygen at carbon-4 to carbon-1 of D-galactopyranose. In effect, (+)-lactose is a glycoside in which galactose is the carbohydrate and glucose is the "alcohol." Recall that the glycosidic linkage is an acetal, and acetals hydrolyze under acidic conditions (Sec. 24.6). Therefore, (+)-lactose can be hydrolyzed in acidic solution to give one equivalent each of D-glucose and D-galactose, in the same sense that a methyl glycoside can be hydrolyzed to give methanol and a carbohydrate.

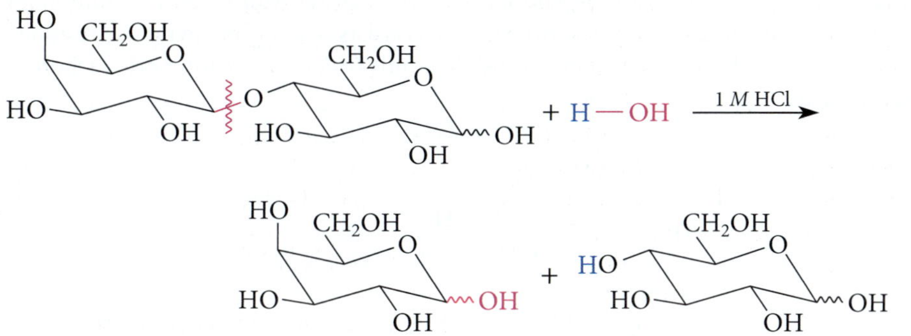

D-galactose **D-glucose** (24.70a)

Compare:

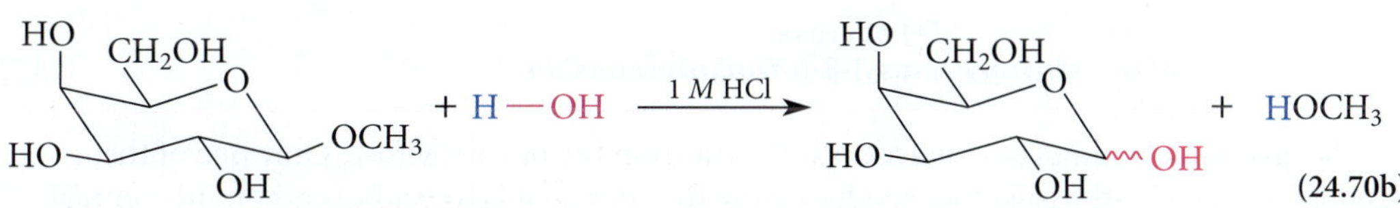

(24.70b)

Equation 24.70a demonstrates the structural basis for the definition of a disaccharide (Sec. 24.1) as a carbohydrate that can be hydrolyzed to two monosaccharides. Hydrolysis occurs at the glycosidic bond between the two monosaccharide residues.

The stereochemistry of the glycosidic bond in (+)-lactose is β. That is, the stereochemistry of the oxygen linking the two monosaccharide residues in the glycosidic bond corresponds to that in the β-anomer of D-galactopyranose. This stereochemistry is very important in biology, because higher animals possess an enzyme, β-galactosidase, that catalyzes the hydrolysis of this β-glycosidic linkage near neutral pH; this hydrolysis allows lactose to act as a source of glucose. α-Glycosides of galactose are inert to the action of this enzyme. (People who suffer from lactose intolerance must take a form of this enzyme orally to digest lactose-containing foods.)

Because carbon-1 of the galactose residue in (+)-lactose is involved in a glycosidic linkage, it cannot be oxidized. However, carbon-1 of the glucose residue is part of a hemiacetal group, which, like the hemiacetal group of monosaccharides, is in equilibrium with the free aldehyde and can undergo characteristic aldehyde reactions. Thus, treatment of (+)-lactose with bromine water (Sec. 24.7A) oxidizes the glucose residue:

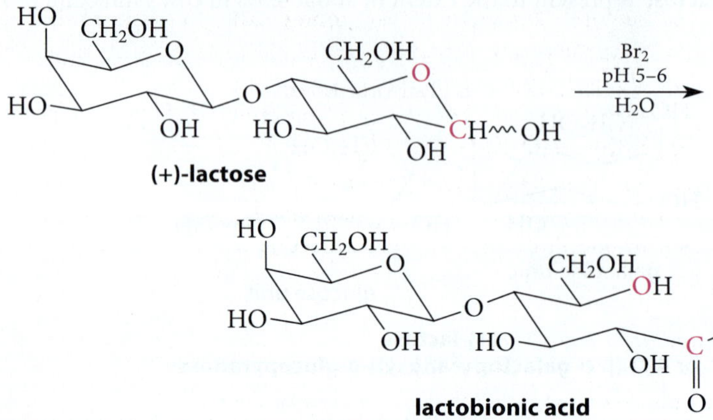

(24.71)

Carbohydrates such as (+)-lactose that can be oxidized in this way are called **reducing sugars** because, in being oxidized, they reduce the oxidizing agents. The glucose residue is said to be at the *reducing end* of the disaccharide, and the galactose residue at the *nonreducing end*. Because of its hemiacetal group, (+)-lactose also undergoes many other reactions of aldose hemiacetals, such as mutarotation.

(+)-**Sucrose** (table sugar) is another important disaccharide. More than 120 million tons of sucrose is produced annually in the world. Sucrose consists of a D-glucopyranose residue and a D-fructofuranose residue connected by glycosidic bonds (*color*) at the anomeric carbons of *both* monosaccharides.

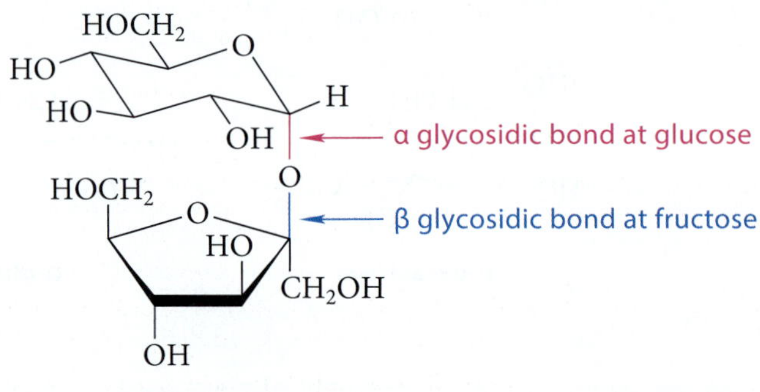

(+)-**sucrose**
or **α-D-glucopyranosyl-β-D-fructofuranoside** (24.72)

The glycosidic bond in (+)-sucrose is different from the one in lactose. Only one of the residues of lactose—the galactose residue—contains an acetal (glycosidic) carbon. In contrast, *both* residues of (+)-sucrose have an acetal carbon. The glycosidic bond in (+)-sucrose bridges carbon-2 of the fructofuranose residue and carbon-1 of the glucopyranose residue. These are the carbonyl carbons in the noncyclic forms of the individual monosaccharides; remember that the carbonyl carbons become the acetal or hemiacetal carbons in the cyclic forms.

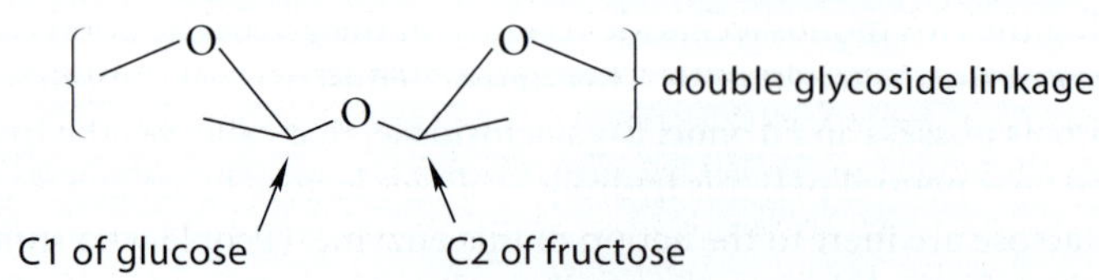

Thus, neither the fructose nor the glucose part of sucrose has a free hemiacetal group. Therefore, (+)-sucrose cannot be oxidized by bromine water, nor does it undergo mutarotation. Carbohydrates such as (+)-sucrose that cannot be oxidized by bromine water are classified as **nonreducing sugars**.

Like other glycosides, (+)-sucrose can be hydrolyzed to its component monosaccharides. Sucrose is hydrolyzed by aqueous acid or by enzymes (called *invertases*) to an equimolar mixture of D-glucose and D-fructose. This mixture is sometimes called *invert sugar* because, as hydrolysis of sucrose proceeds, the positive rotation of the solution changes to a negative rotation characteristic of the glucose–fructose mixture. This rotation is negative because the strongly negative rotation (–92° mL g^{-1} dm^{-1}) of fructose (sometimes called *levulose*) has a greater magnitude than the positive rotation (+52.7° mL g^{-1} dm^{-1}) of glucose (sometimes called *dextrose*). Fructose, which is the sweetest of the common sugars (about twice as sweet as sucrose), accounts for the intense sweetness of honey, which is mostly invert sugar.

The biosynthesis of disaccharides is discussed briefly in Sec. 25.7D.

Focused Problems

24.31 What products are expected from each of the following reactions?

(a) Lactobionic acid (Eq. 24.71) + 1 *M* aqueous HCl

(b) (+)-Lactose + dimethyl sulfate, NaOH

(c) Product of part (b) + 1 *M* aqueous H_2SO_4

24.32 Consider the structure of cellobiose, a disaccharide obtained from the hydrolysis of the polysaccharide cellulose. Into what monosaccharide(s) is cellobiose hydrolyzed by aqueous HCl?

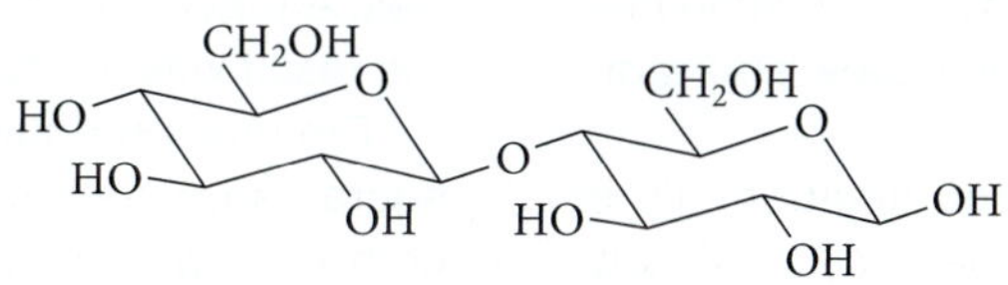

cellobiose

Tales of Serendipitous Sweet Discovery: Artificial Sweeteners

Steve Hamblin/Alamy

Artificial sweeteners are synthetic substitutes for sucrose and other natural sweeteners. Such compounds are in high demand because they allow consumers to enjoy a sweet taste without the calories of natural sweeteners. The annual worldwide market in artificial sweeteners is more than \$5 billion. It has been estimated that well over 100 million Americans use artificial sweeteners. Artificial sweeteners have been sought since the ancient Romans, who used lead acetate ("lead sugar") as an alternative to sugar, and many Romans suffered severe lead toxicity as a result.

To qualify as an artificial sweetener, a compound must have sweetness many times that of sugar so that very little sweetener has to be used to achieve the desired effect. Because so little is used, almost no calories are consumed. However, finding sweet compounds is not the major hurdle to the development of an artificial sweetener. Rather, a candidate compound must undergo rigorous testing to be sure that it is not toxic and does not have undesired side effects. Almost every sweetener that has been introduced has attracted its share of public suspicion despite the large amount of biological testing involved in getting it to market. The most widely used sweeteners in modern times have been sodium saccharin, sodium cyclamate, aspartame, and, most recently, sucralose.

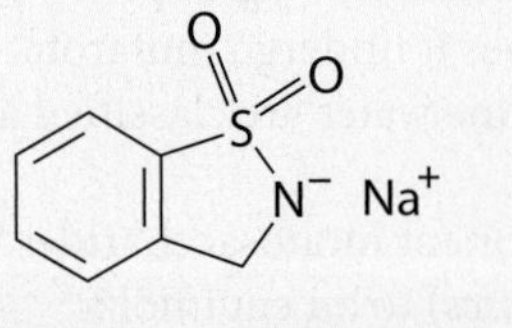

sodium saccharin (1879)
300 times as sweet as sucrose

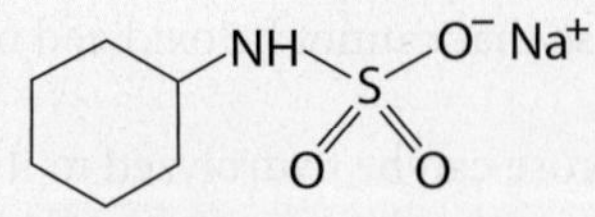

sodium cyclamate (1937)
30–50 times as sweet as sucrose

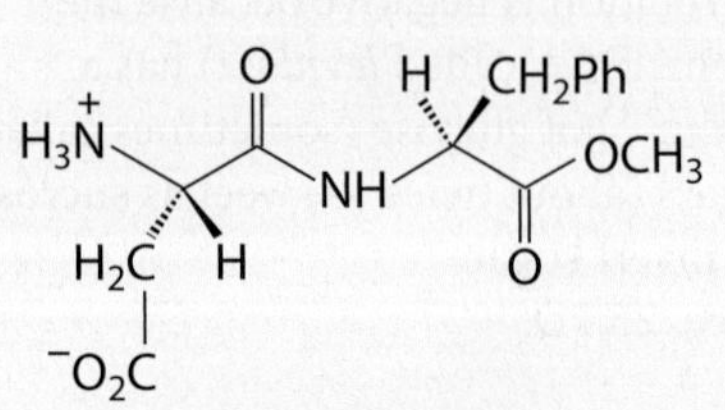

aspartame (1965)
180 times as sweet as sucrose

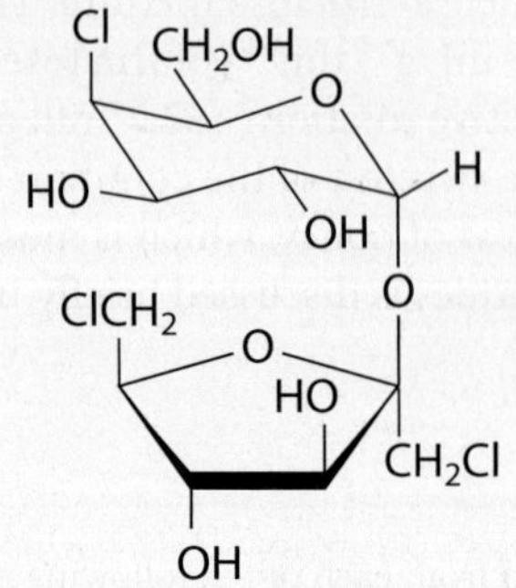

sucralose (1989)
600 times as sweet as sucrose

The discoveries of all of these sweeteners were serendipitous, and all resulted from the tasting of laboratory samples by the researchers involved. Tasting new compounds was actually once a common laboratory practice, and the tastes of new compounds were routinely reported in the chemical literature. However, tasting new compounds is no longer condoned as a safe laboratory practice.

Sodium saccharin was discovered accidentally in 1879 by Constantin Fahlberg, a student in the laboratory of Professor Ira Remsen at The Johns Hopkins University. At dinner, Fahlberg noticed a sweet taste on his fingers and concluded that it must have come from a compound he was working with. A "taste test" of many of his compounds led to sodium saccharin as the culprit. Fahlberg became wealthy from commercialization of the discovery, and when Professor Remsen did not receive any financial benefit, he became quite bitter toward his former student.

Aspartame was discovered accidentally in 1965 by Jim Schlatter, a chemist at G. D. Searle and Company, while he was working on the discovery of drugs to treat stomach ulcers. He noticed a sweet taste on his fingers after handling the compound. His supervisor convinced Searle that the compound was worth development, and the result was the NutraSweet brand of aspartame.

The most intriguing story surrounds the discovery of sucralose. Scientists from Tate & Lyle, a British sugar company, were working in 1989 with researcher Leslie Hough and his student, Shashikant Phadnis, at Queen Elizabeth College (now part of King's College) in London on a project that involved testing chlorinated sugars as chemical intermediates for the synthesis of other compounds. Hough asked Phadnis to "test" sucralose, but Phadnis misunderstood the word "test," thinking that he had been asked to "taste" the compound. The rest is history.

B. Polysaccharides

In principle, any number of monosaccharide residues can be linked with glycosidic bonds to form chains. When such chains are long, the sugars are called **polysaccharides**. This section surveys a few important polysaccharides.

Cellulose Cellulose, the principal structural component of plants, is the most abundant organic compound on Earth. Cotton is almost pure cellulose; wood is cellulose combined with a polymer called *lignin*. About 5×10^{14} kg of cellulose is biosynthesized and degraded annually on Earth.

Cellulose is a regular polymer of D-glucopyranose units connected by β-1,4-glycosidic linkages.

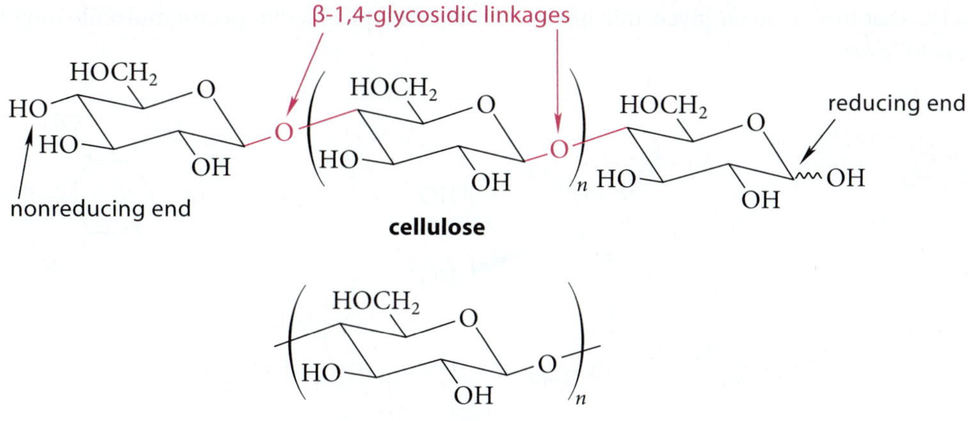

(24.73)

Like disaccharides, polysaccharides can be hydrolyzed to their constituent monosaccharides. Thus, cellulose can be hydrolyzed to D-glucose residues. Mammals lack the enzymes that catalyze the hydrolysis of the β-glycosidic linkages of cellulose; this is why humans cannot digest grasses, which are principally cellulose. Cattle, though, can derive nourishment from grasses, but this is because the bacteria in their rumens provide the appropriate enzymes that break down plant cellulose to glucose.

Processed cellulose (cellulose that has been specially treated) has many other uses. It can be spun into fibers (rayon) or made into wraps (cellophane). The paper on which this book is printed is largely processed cellulose. Nitration of the cellulose hydroxy groups gives nitrocellulose, a powerful explosive. Cellulose acetate, in which the hydroxy groups of cellulose are esterified with acetic acid, is known by the trade names Celanese, Arnel, and so on and is used in knitting yarn and decorative household articles.

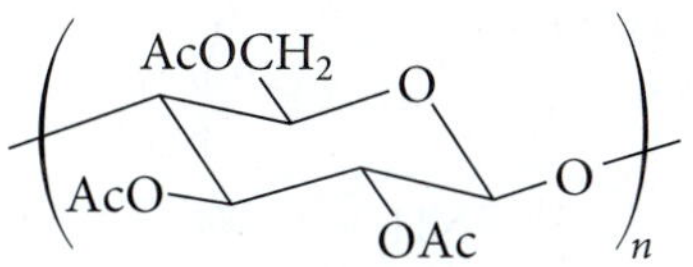

cellulose acetate (24.74)

Cellulose is potentially important as an alternative energy source. Biomass is largely cellulose, and cellulose is merely polymerized glucose. The glucose derived from the hydrolysis of cellulose can be fermented to ethanol, which can be used as a fuel (as in gasohol); and plants obtain the energy to manufacture cellulose from the Sun. Thus, the cellulose in plants—the most abundant source of carbon on Earth—can be regarded as a storehouse of solar energy. An important research problem is how to convert the more abundant sources of cellulose, such as wild grasses, into glucose without expending a large amount of energy. A solution to this problem would reduce or eliminate the need to use cultivated crops, such as corn, as a source of ethanol.

Starch Starch, like cellulose, is also a polymer of glucose. In fact, starch is a mixture of two different types of glucose polymer. In one, *amylose*, the glucose units are connected by α-1,4-glycosidic linkages. Conceptually, the only chemical difference between amylose and cellulose is the stereochemistry of the glycosidic bond.

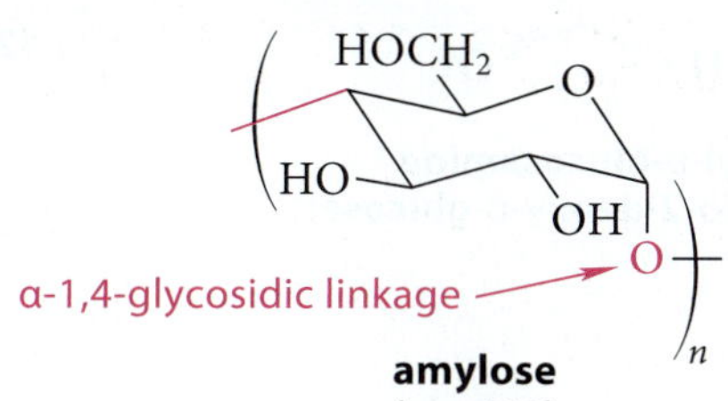

amylose
($n \approx 400$) (24.75)

The other constituent of starch is *amylopectin*, a branched polysaccharide. Amylopectin contains relatively short chains of glucose units in α-1,4-linkages. In addition, it contains branches that involve α-1,6-glycosidic linkages. Part of a typical amylopectin molecule might look as follows:

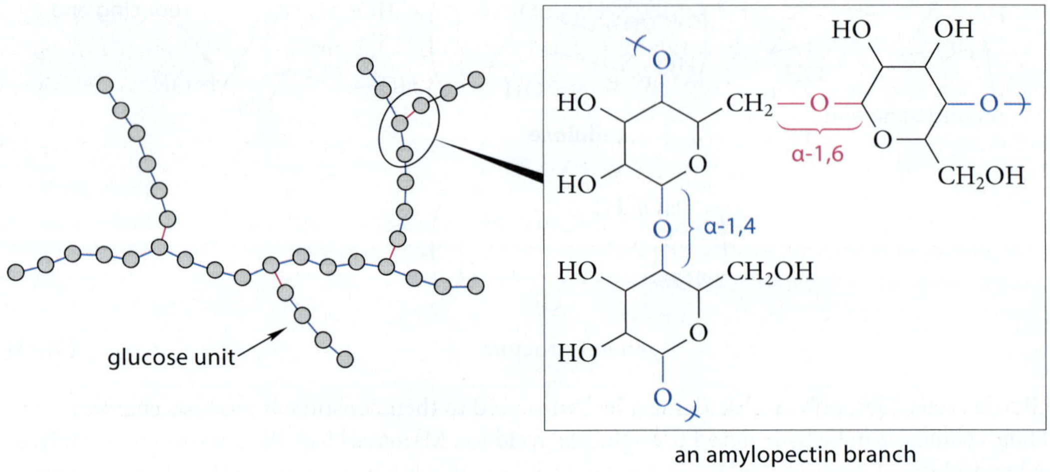

(24.76)

Starch is the important storage polysaccharide in corn, potatoes, and other starchy vegetables. Humans have enzymes that catalyze the hydrolysis of the α-glycosidic bonds in starch and can therefore use starch as a source of glucose.

Chitin *Chitin* is a polysaccharide that also occurs widely in nature—notably, in the shells of arthropods (for example, lobsters and crabs). Crab shell is an excellent source of nearly pure chitin.

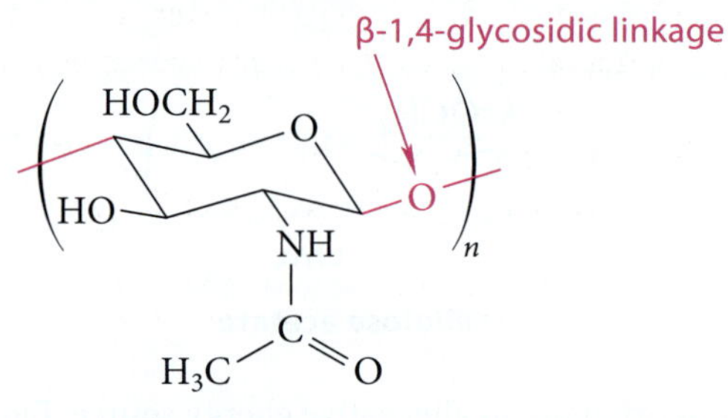

chitin (24.77)

Chitin is a polymer of *N*-acetyl-D-glucosamine (or, as it is named systematically, 2-acetamido-2-deoxy-D-glucose). Residues of this carbohydrate are connected by β-1,4-glycosidic linkages within the chitin polymer. *N*-Acetyl-D-glucosamine is liberated when chitin is hydrolyzed in aqueous acid. Stronger acid brings about hydrolysis of the amide bond to give D-glucosamine hydrochloride and acetic acid.

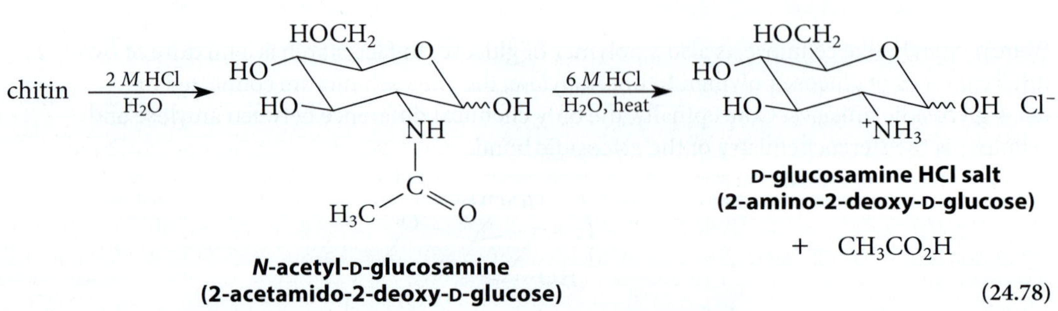

(24.78)

Discovery of D-Glucosamine

ex0rzist/Shutterstock

In 1876 Georg Ledderhose was a premedical student working in the laboratory of his uncle, Friedrich Wöhler (the same chemist who first synthesized urea; see Eq. 1.1, Sec. 1.1C). One day, Wöhler had lobster for lunch and returned to the laboratory carrying the lobster shell. "Find out what this is," he told his nephew. History does not record Ledderhose's thoughts on receiving the refuse from his uncle's lunch, but he proceeded to do what all chemists did with unknown material—he boiled it in concentrated HCl. After hydrolysis of the shell, crystals of the previously unknown D-glucosamine hydrochloride precipitated from the cooled solution (see Eq. 24.78).

Glucosamine and *N*-acetylglucosamine are the best-known examples of the **amino sugars**. A number of amino sugars occur widely in nature. Amino sugars linked to proteins (glycoproteins) are found at the outer surfaces of cell membranes, and some of these are responsible for blood-group specificity.

Principles of Polysaccharide Structure Studies of many polysaccharides have revealed the following generalizations about polysaccharide structure:

1. Polysaccharides are mostly long chains with some branches; there are no highly cross-linked, three-dimensional networks. Some cyclic oligosaccharides are known.

2. The linkages between monosaccharide units are in every case glycosidic linkages; thus, monosaccharides can be liberated from all polysaccharides by acid hydrolysis.

3. A given polysaccharide contains only one stereochemical type of glycoside linkage. Thus, the glycoside linkages in cellulose are all β; those in starch are all α.

Focused Problem

24.33 What product(s) would be obtained when cellulose is treated first exhaustively with dimethyl sulfate/NaOH, then with 1 *M* aqueous HCl?

CHAPTER SUMMARY *For a summary of the chapter, see Chapter 24 in the* Study Guide and Solutions Manual.

REACTION REVIEW *For a summary of reactions discussed in this chapter, see the* Reaction Review *section of Chapter 24 in the* Study Guide and Solutions Manual.

SKILLS OBJECTIVES WITH PROBLEMS

- **Interconvert line-and-wedge structures and Fischer projections. Properly manipulate Fischer projections. Determine stereochemical relationships between compounds represented by Fischer projections. 24.2, 24.3A**

24.34 Convert the following line-and-wedge structures into Fischer projections, then name each using Fig. 24.3 (Sec. 24.3A) as a reference. In each, place the aldehyde carbon (—CHO) at the top of the projection.

(a)

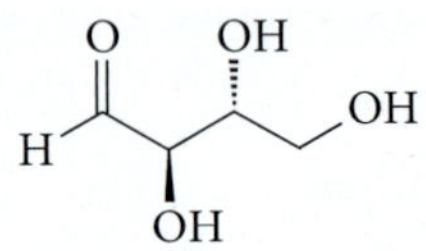

(b)

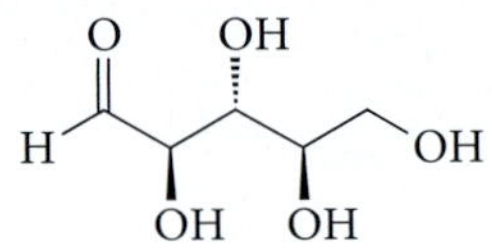

(c)

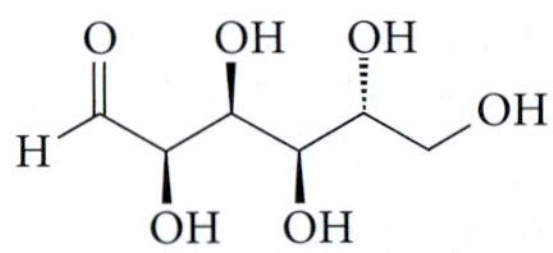

(d)

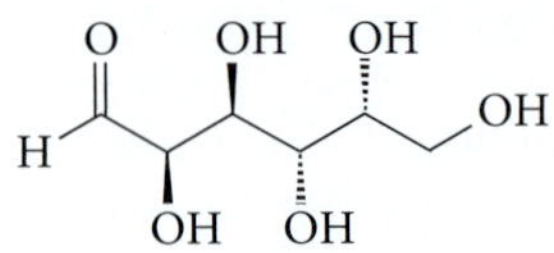

24.35 Assign *R* or *S* configurations to each asymmetric carbon in parts (a)–(d) in Problem 24.34.

24.36 Convert the following Fischer projections into line-and-wedge structures.

(a)

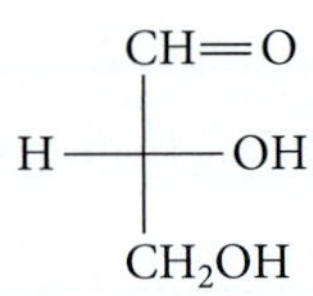

D-(+)-glyceraldehyde

(b)

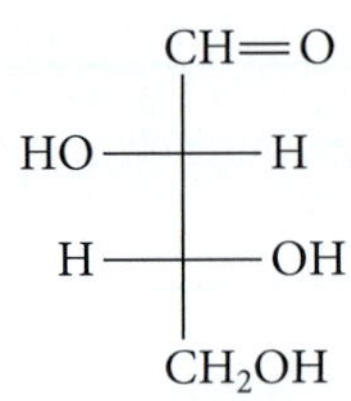

D-(−)-threose

(c)

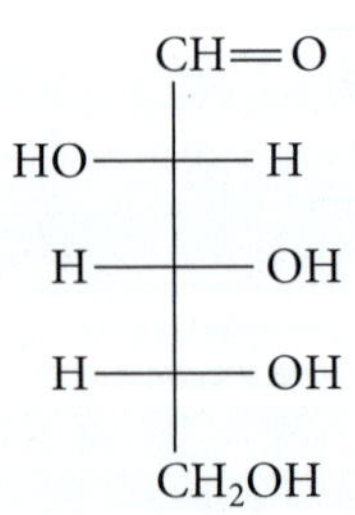

D-(−)-arabinose

(d)

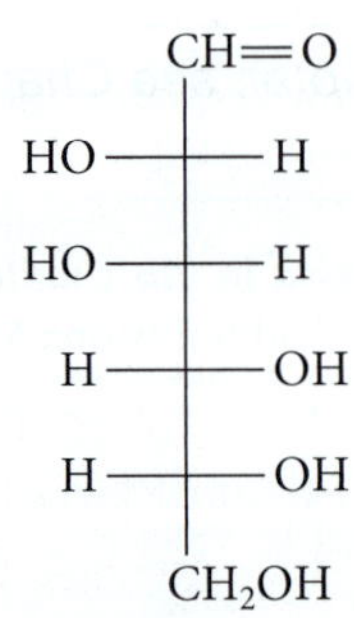

D-(+)-mannose

24.37 Assign *R* or *S* configurations to each asymmetric carbon in parts (a)–(d) in Problem 24.36. (*Hint:* Remember to use the shortcuts detailed in Eqs. 24.12 and 24.13 in Sec. 24.2 to quickly assign configurations.)

24.38 Use *R* and *S* configuration assignments to prove why "turning over" Fischer projections is forbidden if you want to preserve the stereochemistry of a structure, as shown in Eq. 24.8 (reprinted here).

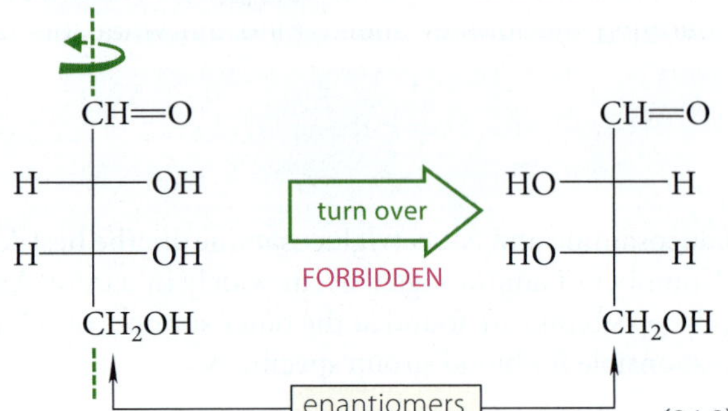

(24.8)

- **Draw line-and-wedge structures, Haworth projections, and chair conformations of cyclic monosaccharides. 24.3B**

24.39 Using the chair conformation of β-D-glucopyranose (Eq. 24.25, Sec. 24.3B) as a starting point, draw the most stable chair conformations of the seven other β-D-aldohexoses, shown on the top line in Fig. 24.3. In each, identify the anomeric carbon and the carbons that are epimeric relative to β-D-glucopyranose.

24.40 Draw line-and-wedge "planar" structures (where the five carbons and one oxygen of the ring are in the plane of the page) for each of the eight β-D-aldohexoses in Problem 24.39.

24.41 Using the Haworth projection of β-D-ribofuranose (Eq. 24.23, Sec. 24.3B) as a starting point, draw the β-anomers of the three other D-aldopentoses, shown on the second line in Fig. 24.3 (Sec. 24.3A).

24.42 Draw line-and-wedge "planar" structures (where the four carbons and one oxygen of the ring are in the plane of the page) for each of the four D-aldopentoses in Problem 24.41.

- **Justify mechanistically the mutarotation of carbohydrates. 24.4**

24.43 Using the curved-arrow notation, write a mechanism for acid-catalyzed mutarotation of β-D-galactopyranose (right side of Eq. 24.25, Sec. 24.3B) to α-D-galactopyranose. Begin by protonating the ring oxygen.

24.44 Using the curved-arrow notation, fill in the details for base-catalyzed mutarotation of galactopyranose. Begin by removing a proton from the hydroxy group at carbon-1.

24.45 Draw the Haworth projections of the two furanose forms of glucose that exist (in very small amounts) in aqueous solution, and write mechanisms for their formation. Use Eq. 24.28 (Sec. 24.4) as a starting point.

• Provide missing starting materials or products in the reactions of carbohydrates. Write reaction mechanisms when appropriate.

(1) Base-catalyzed isomerization of aldoses and ketoses 24.5

(2) Formation of glycosides 24.6

(3) Formation of ester and ether derivatives 24.7

(4) Oxidation and reduction reactions of carbohydrates 24.8

(5) Kiliani–Fischer synthesis 24.9

(6) Ruff degradation 24.10B

(7) Cleavage of disaccharides 24.11

24.46 Into what other aldose and 2-ketose would each of the following aldoses be transformed on treatment with base? Give the structure and name of the aldose, and the structure of the 2-ketose.

(a) D-Gulose

(b) D-Ribose

24.47 Give the product(s) expected when D-mannose (or other compound indicated) reacts with each of the following reagents. (Assume that cyclic mannose derivatives are pyranoses.)

(a) Dilute NaOH

(b) Br_2, $CaCO_3/H_2O$, then H_3O^+

(c) CH_3OH, HCl

(d) Acetic anhydride/pyridine

(e) Product of part (b) + $Ca(OH)_2$, then $Fe(OAc)_3$, H_2O_2

(f) Product of part (c) + $PhCH_2Cl$ (excess) and NaOH

24.48 Give the products expected when D-ribose (or other compound indicated) reacts with each of the following reagents.

(a) Dilute HNO_3

(b) NaCN, H_2O

(c) Product of part (b) + $H_2/Pd/BaSO_4$ + H_3O^+/H_2O

(d) CH_3OH, HCl (four isomeric compounds; two pyranosides and two furanosides)

(e) Products of part (d) + $(CH_3)_2SO_4$ (excess) and NaOH

• Label the types of glycosidic linkages in polysaccharides.

24.49 Draw the structure of 3-*O*-β-D-glucopyranosyl-α-D-arabinofuranose, a disaccharide that is the β-glycoside formed between D-glucopyranose at the nonreducing end and the —OH group at carbon-3 of α-D-arabinofuranose at the reducing end.

INTEGRATED PROBLEMS

24.50 Draw the indicated type of structure for each of the following compounds.

(a) α-D-Talopyranose (chair)

(b) Propyl β-L-arabinopyranoside (chair)

24.51 Name the specific form of each aldose shown here.

(a)

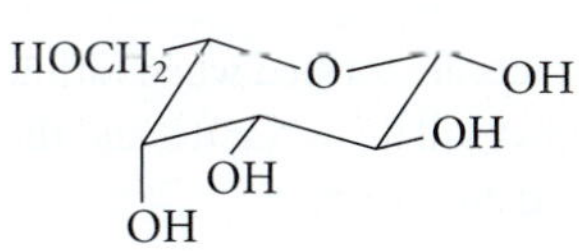

(b)

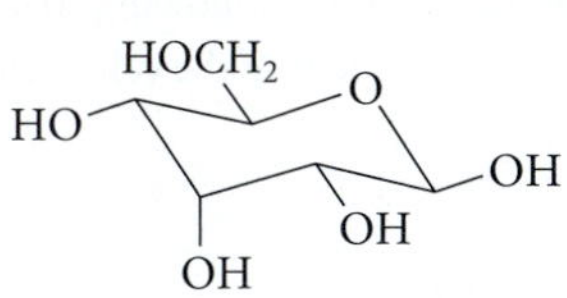

(c)

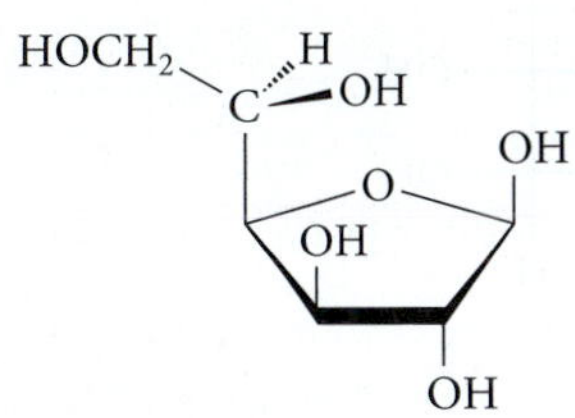

24.52 Draw the structure(s) of

(a) All of the 2-ketohexoses

(b) An achiral ketopentose $C_5H_{10}O_5$

(c) α-D-Galactofuranose

(d) β-D-Idofuranose

24.53 Specify the relationship(s) of the compounds in each of the following sets. Choose among the following terms: identical compounds, epimers, anomers, enantiomers, diastereomers, constitutional isomers, none of the above. (More than one answer may be correct.)

(a) α-D-Glucopyranose and β-D-glucopyranose

(b) α-D-Glucopyranose and α-D-mannopyranose

(c) β-D-Mannopyranose and β-L-mannopyranose

(d) α-D-Ribofuranose and α-D-ribopyranose

(e) Aldehyde form of D-glucose and α-D-glucopyranose

(f) Methyl α-D-fructofuranoside and 2-O-methyl-α-D-fructofuranose

(g) 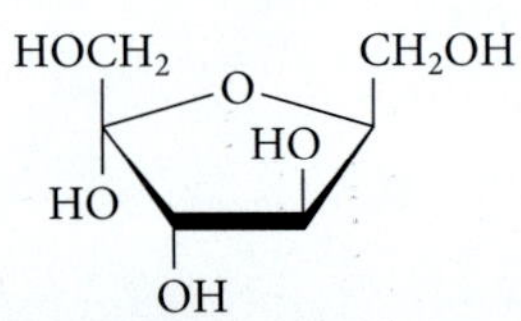

(h) 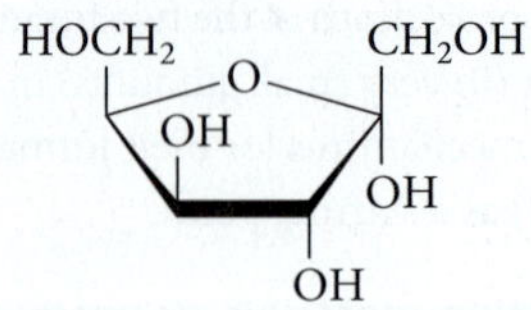

(i) 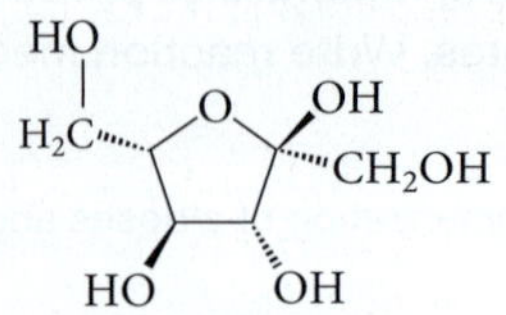

(j) 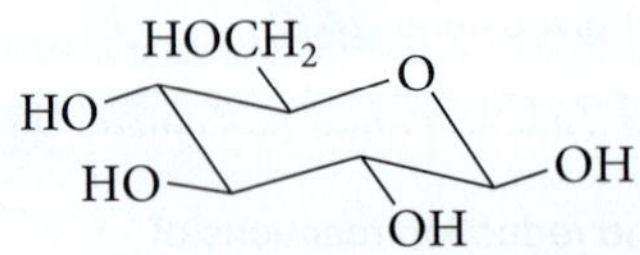

(k) 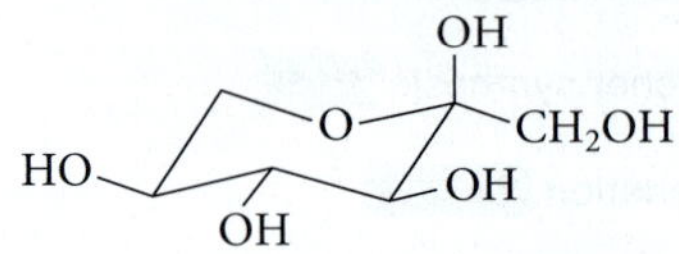

24.54 Tell whether each structure or term is a correct description of the L-sorbose structure shown here or a form with which it is in equilibrium.

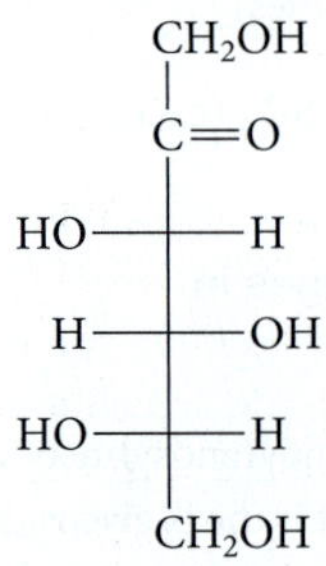

L-sorbose

(a) A hexose

(b) A ketohexose

(c) A glycoside

(d) An aldohexose

(e) 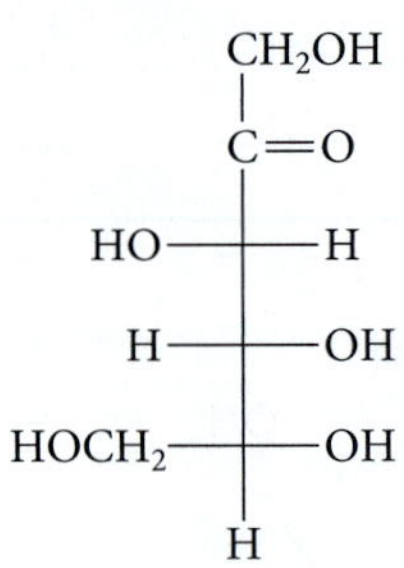

(f) 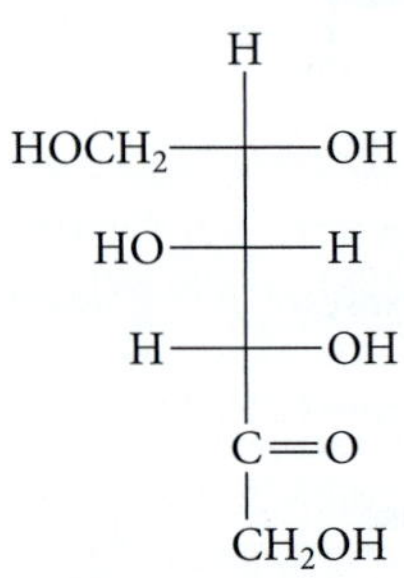

24.55 Consider the structure of *raffinose*, a trisaccharide found in sugar beets and a number of higher plants.

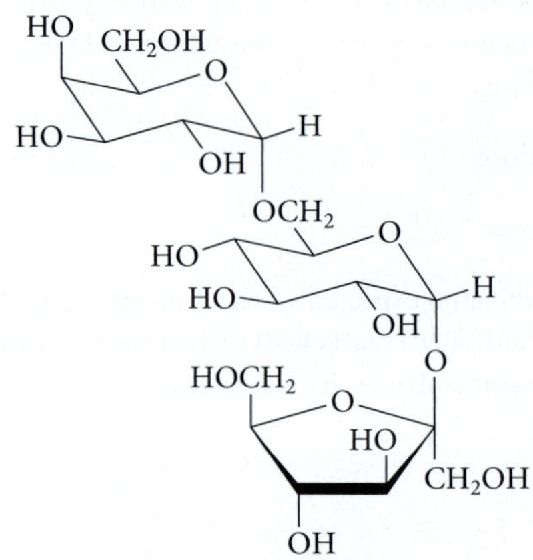

raffinose

(a) Classify raffinose as a reducing or nonreducing sugar, and tell how you know.

(b) Identify the glycoside linkages in raffinose, and classify each as either α or β.

(c) Name the monosaccharides formed when raffinose is hydrolyzed in aqueous acid.

(d) What products are formed when raffinose is treated with dimethyl sulfate in NaOH, and then with aqueous acid and heat?

24.56 Fucose, a carbohydrate with the following structure, has been identified as a component of the cell-surface antigens of certain tumor cells.

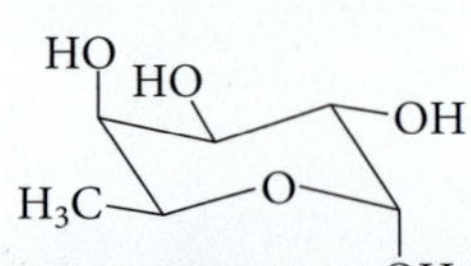

fucose

(a) Is this the D- or L-enantiomer of fucose? Explain.

(b) Is this the α- or β-anomer?

(c) Is fucose an aldose, a ketose, or neither? Explain.

(d) Draw a Fischer projection of the carbonyl form of this carbohydrate.

24.57 An important reaction used by Emil Fischer in his research on carbohydrate chemistry was the reaction of aldoses and ketoses with phenylhydrazine to give *osazones*.

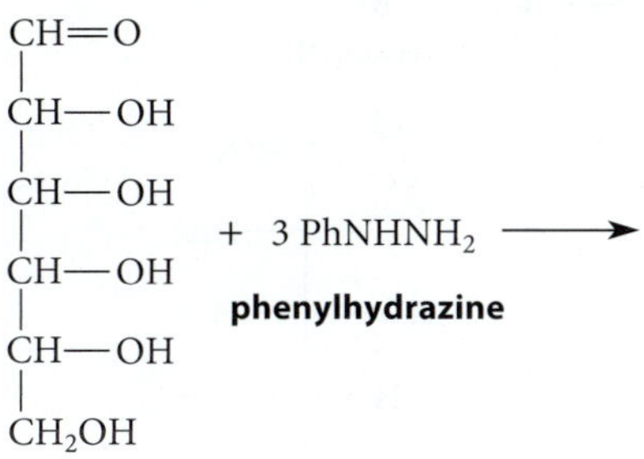

an aldose

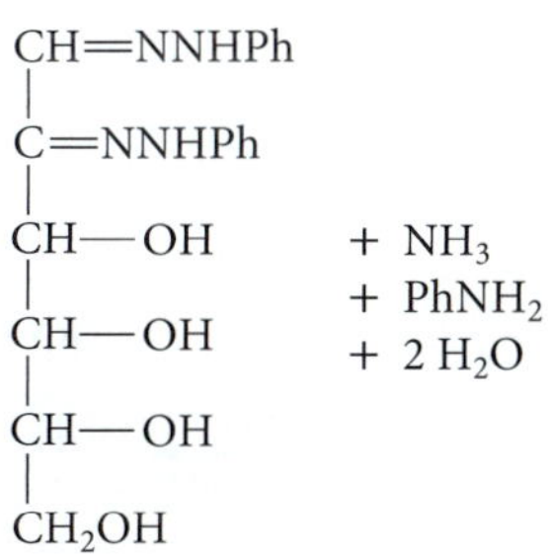

an osazone

Osazones, unlike many carbohydrates, form crystalline solids that are useful in characterizing carbohydrates.

(a) Glucose and mannose give the same osazone. Given that these two compounds are aldohexoses, what could a scientist who knows nothing about the stereochemistry of these carbohydrates deduce about their stereochemical relationship from this fact?

(b) What aldopentose gives the same osazone as D-arabinose?

24.58 Complete the reactions shown by giving the major organic product(s).

(a) Phenyl β-D-glucopyranoside + CH_3OH (solvent) $\xrightarrow{H_2SO_4}$

(b) $HOCH_2CH_2CH_2CH_2CH{=}O$ + CH_3OH (solvent) $\xrightarrow{HCl}$

(c)

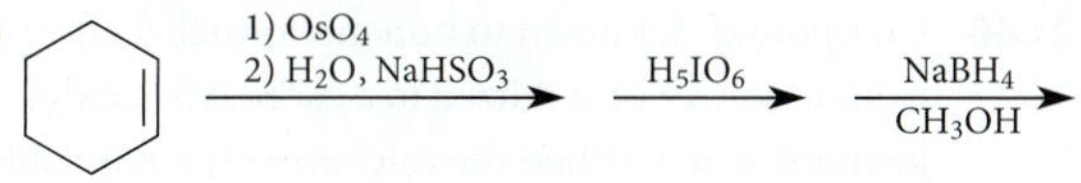

(d)

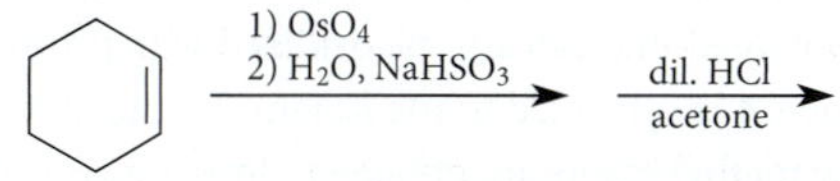

(e) (+)-Sucrose + $(CH_3)_2SO_4$ (excess) $\xrightarrow{\text{concd. NaOH}}$

(f) (+)-Lactose + CH_3CH_2OH $\xrightarrow[\text{heat}]{H_2SO_4}$

(g)

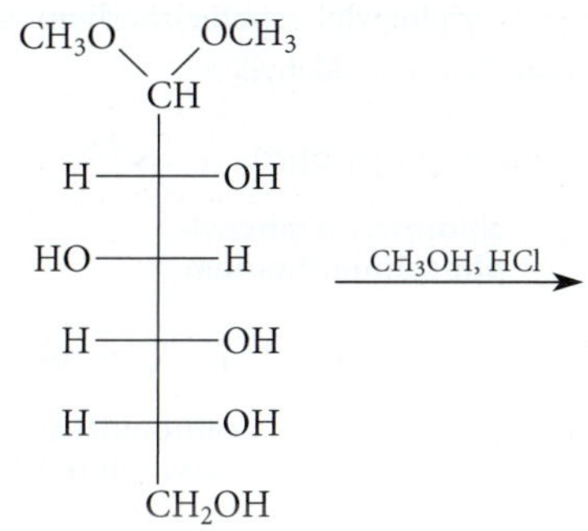

D-glucose dimethyl acetal

24.59 A biologist, Simone Spore, needs the following isotopically labeled aldoses for some feeding experiments. Realizing your expertise in the saccharide field, she has come to you to ask whether you will synthesize these compounds for her. She has agreed to provide an adequate supply of D-(−)-arabinose as a starting material. (* = ^{14}C, T = ^{3}H = tritium)

(a)

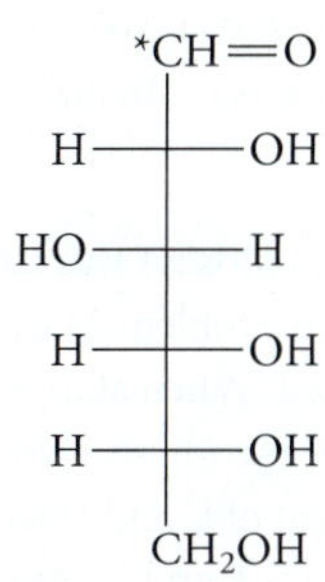

(b)

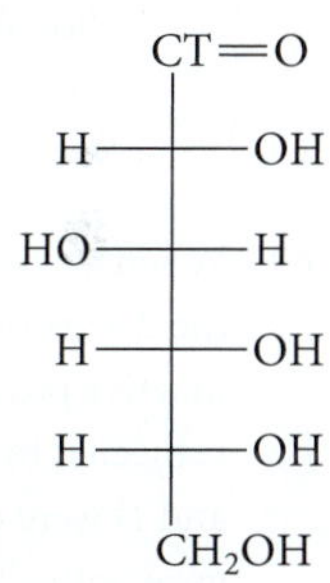

(c)

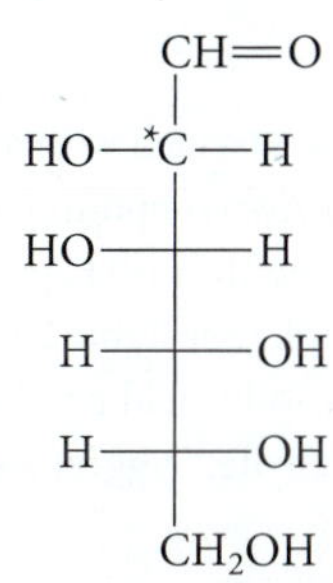

Available commercial sources of isotopes include $Na_2{}^*CO_3$, Na^*CN, $^3H_2$, and ^3H_2O. Outline a synthesis of each isotopically labeled compound.

24.60 Compound *A*, known to be a monomethyl ether of D-glucose, can be oxidized to a carboxylic acid *B* with bromine water. When the calcium salt of *B* is subjected to the Ruff degradation, another aldose monomethyl ether is obtained that can also be oxidized with bromine water. When *A* is subjected to the Kiliani–Fischer synthesis, two new methyl ethers are obtained. Both are optically active, and one of them can be oxidized with dilute nitric acid to an optically inactive compound. Suggest a structure for *A*, including its stereochemistry.

24.61 Chlorotris(triphenylphosphine)rhodium brings about the decarbonylation of aldehydes:

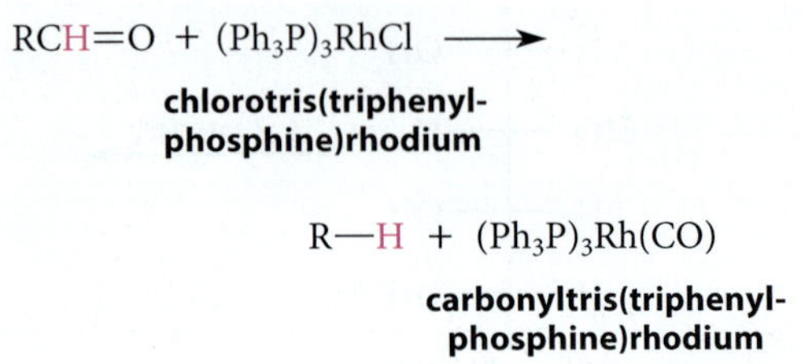

(a) What product is obtained when this reaction is applied to D-galactose?

(b) Suggest a reason why the reaction of an aldose requires a much higher temperature (130 °C) than the same reaction of an ordinary aldehyde (70 °C).

(c) Outline a mechanism for this reaction by showing the elementary steps involved and the important organometallic intermediates. (*Hint:* See Sec. 18.5E.)

24.62 When an optically active aldopentose *A* was subjected to the decarbonylation reaction in Problem 24.61, an optically inactive product *B* was obtained. When aldopentose *A* was subjected to the Kiliani–Fischer synthesis, two aldoses *C* and *D* were obtained. Treatment of *C* and *D* with HNO_3 gave optically active aldaric acids *E* and *F*, respectively. Identify compounds *A–F*.

24.63 When D-ribose-5-phosphate was treated with an extract of mouse spleen, an optically *inactive* compound *X*, $C_5H_{10}O_5$, was produced. Treatment of *X* with $NaBH_4$ gave a mixture of the alditols ribitol and xylitol. (See Table 24.2, Sec. 24.8.) Treatment of *X* with periodic acid produced two molar equivalents of formaldehyde. Suggest a structure for *X*.

24.64 The *Wohl degradation* can be used to convert an aldose into another aldose with one less carbon. Give the structure of the missing compounds as well as the curved-arrow mechanisms for the conversion of *B* to *C* and *C* to arabinose.

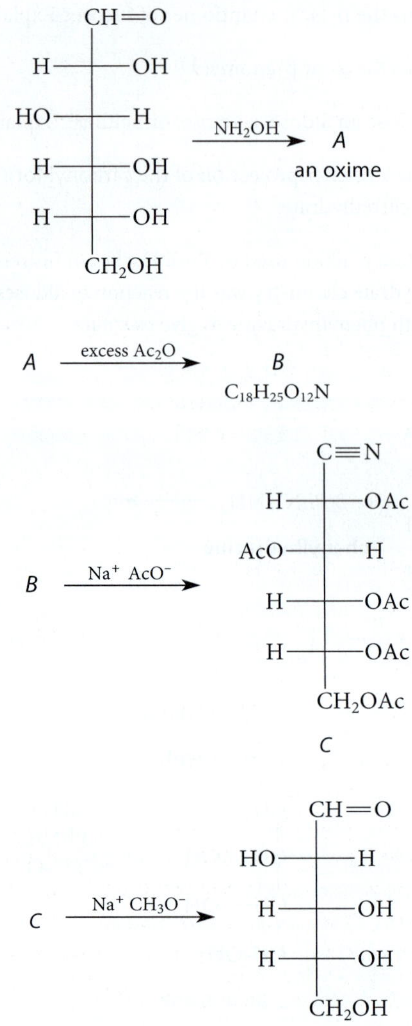

D-arabinose

24.65 A sequence of reactions called the *Weerman degradation* can be used to degrade an aldose to another aldose with one fewer carbon atom. Using glucose as the aldose, explain what is happening in each step of the sequence. Your explanation should include the identity of compounds *A* and *B*. (*Hint:* Compound *A* is a lactone, and a lactone is a type of ester.)

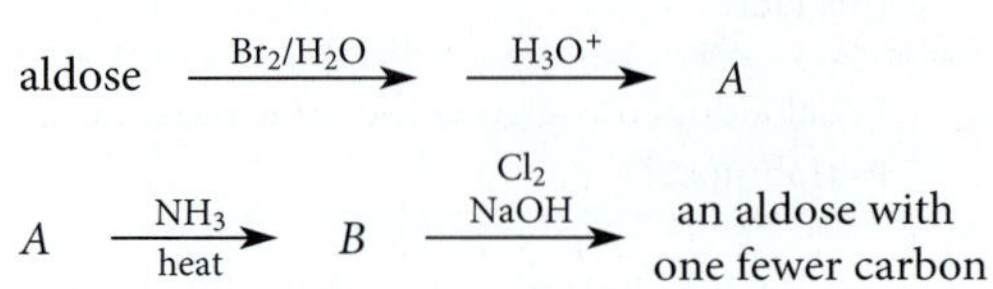

24.66 L-Rhamnose is a 6-deoxyaldose with the following structure. When a methyl glycoside of L-rhamnose, methyl α-L-rhamnopyranoside, was treated with periodic acid, compound *A*, $C_6H_{12}O_5$, was obtained that showed no evidence of a carbonyl group in its IR spectrum. Treatment of *A* with CH_3I/Ag_2O gave a derivative *B*, $C_8H_{16}O_5$. Treatment of *A* with H_2/Ni or $NaBH_4$ gave compound *C*, shown here in Fischer projection. Give the structure of *A*. Explain why *A* gives no detectable carbonyl absorption in its IR spectrum, yet reacts with $NaBH_4$.

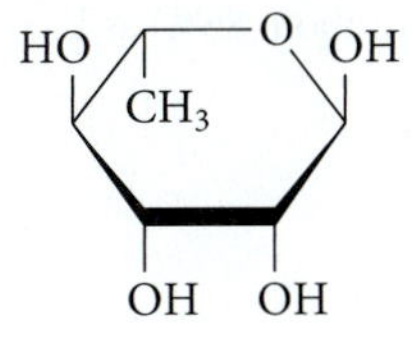

L-rhamnose
(Haworth projection)

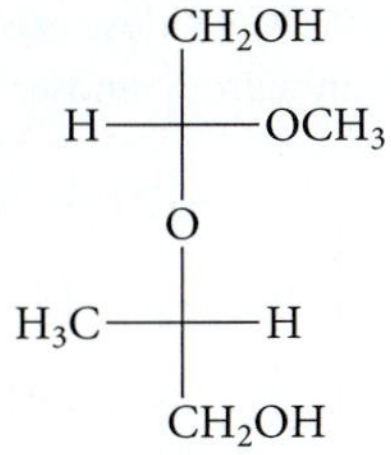

C

24.67 Oligosaccharides of the type shown in **Fig. P24.67** are obtained from the partial hydrolysis of starch amylopectin. What ratio of erythritol to glycerol would be obtained from successive treatment of a 12-unit oligosaccharide of the type shown with periodic acid, then $NaBH_4$, and then hydrolysis in aqueous acid?

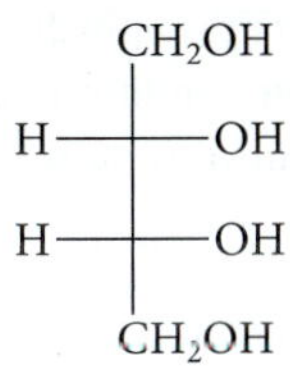

erythritol

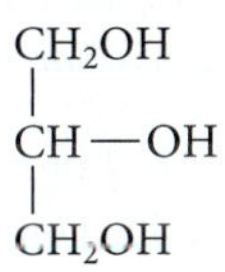

glycerol

24.68 Maltose is a disaccharide obtained from the hydrolysis of starch. Maltose can be hydrolyzed to two equivalents of glucose and can be oxidized to an acid, maltobionic acid, with bromine water. Treatment of maltose with dimethyl sulfate and sodium hydroxide, followed by hydrolysis of the product in aqueous acid, yields one equivalent each of 2,3,4,6-tetra-*O*-methyl-D-glucose and 2,3,6-tri-*O*-methyl-D-glucose. Hydrolysis of maltose is catalyzed by α-amylase, an enzyme known to affect only α-glycosidic linkages. Give *two* structures of maltose consistent with these data, and explain your answers.

Treatment of maltobionic acid with dimethyl sulfate and sodium hydroxide followed by hydrolysis of the product in aqueous acid gives 2,3,4,6-tetra-*O*-methyl-D-glucose and 2,3,5,6-tetra-*O*-methyl-D-gluconic acid. (See Eq. 24.48, Sec. 24.8A, for the structure of D-gluconic acid.) Give the structure of maltose.

24.69 Planteose, a carbohydrate isolated from tobacco seeds, can be hydrolyzed in dilute acid to yield one equivalent each of D-fructose, D-glucose, and D-galactose. Almond emulsin (an enzyme preparation that hydrolyzes α-galactosides) catalyzes the hydrolysis of planteose to D-galactose and sucrose. Planteose does not react with bromine water. Treatment of planteose with $(CH_3)_2SO_4/NaOH$, followed by dilute acid hydrolysis, yields, among other compounds, 1,3,4-tri-*O*-methyl-D-fructose. Suggest a structure for planteose.

24.70 A process called *sizing* chemically modifies the cellulose in paper. As a result, the paper resists wetting (which thus prevents inks from running). In addition, sizing leaves the paper in a slightly alkaline state. (Acid-free paper lasts much longer than paper that is not acid-free.) One sizing process involves treatment of cellulose with 2-alkylsuccinic anhydrides (where R and R′ are short alkyl groups—for example, ethyl or propyl groups):

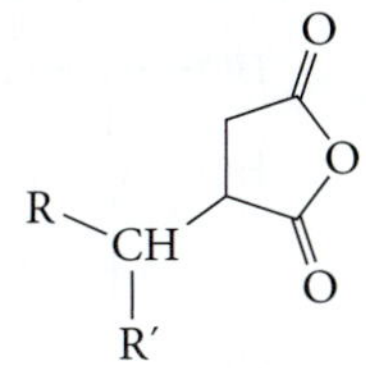

(a) What general reaction occurs when this sizing agent reacts with cellulose at pH 7?

(b) Why should this treatment cause the cellulose to become more resistant to wetting? (In answering this question, think of wetting as a solvation phenomenon.)

(c) Why does this treatment cause the paper to be slightly alkaline? That is, what basic group does this treatment introduce?

24.71 Write a curved-arrow mechanism for the following reaction, which is an example of the *Maillard reaction* followed by the *Amadori rearrangement.*

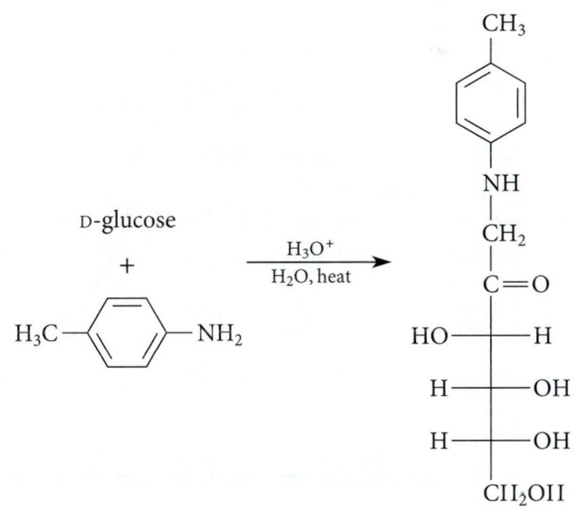

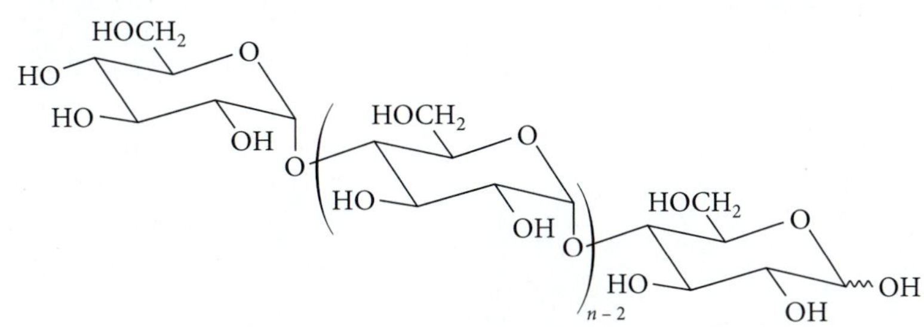

FIGURE P24.67

24.72 Explain with a curved-arrow mechanism why treatment of the 2-deoxy-2-amino derivative of D-glucose (D-glucosamine) with aqueous NaOH liberates ammonia.

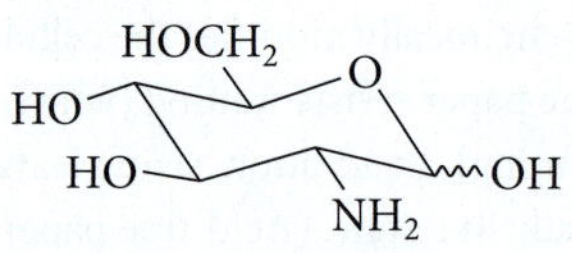

D-glucosamine

24.73 L-Ascorbic acid (vitamin C) has the following structure:

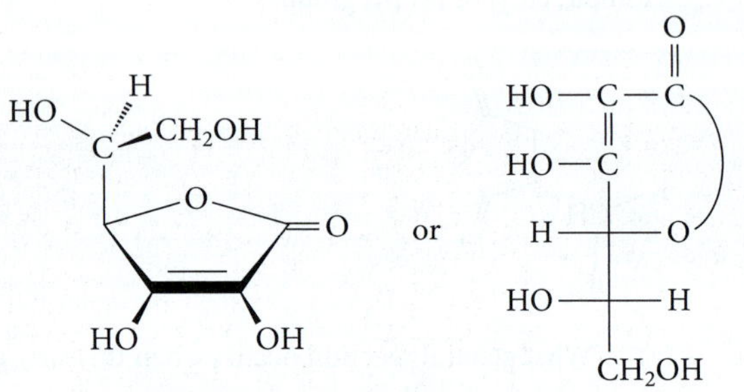

(a) Ascorbic acid has $pK_a = 4.21$, and is thus about as acidic as a typical carboxylic acid. Identify the acidic hydrogen, and explain.

(b) Annually, thousands of tons of ascorbic acid are made commercially from D-glucose by the process shown. Give the structures of the compounds *A–C*.

D-glucose $\xrightarrow{NaBH_4}$ *A* $\xrightarrow[\text{oxidation}]{\text{biological}}$ *B*

A = **D-glucitol (L-sorbitol)**; *B* = **L-sorbose** (a ketose)

B $\xrightarrow{C}$

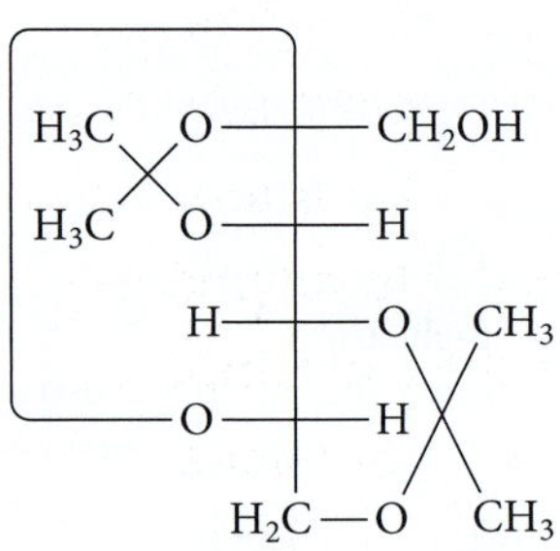

D

D $\xrightarrow{KMnO_4,\ OH^-}$ $\xrightarrow[H_2O,\ \text{heat}]{H_3O^+}$ L-ascorbic acid

24.74 At 100 °C, D-idose exists mostly (about 86%) as a 1,6-anhydropyranose:

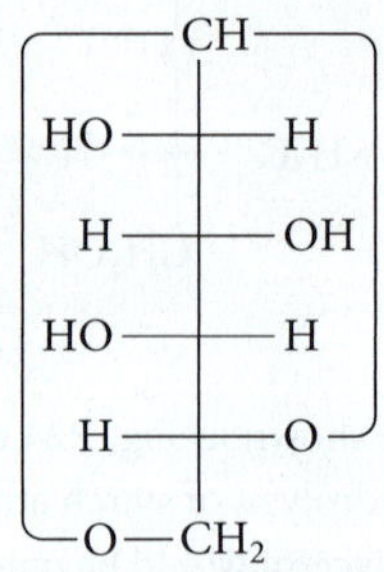

1,6-anhydro-D-idopyranose

(a) Draw the chair conformation of this compound.

(b) Explain why D-idose has more of the anhydro form than D-glucose. (Under the same conditions, glucose contains only 0.2% of the 1,6-anhydro form.)

24.75 The proton NMR of the C-1 proton region of D-glucopyranose shows that both anomers are present. (The large peak in the middle is residual HDO in the H_2O solvent.) The integrals are given in arbitrary units.

(a) Which is the resonance of the α-anomer, and which is the resonance of the β-anomer? How do you know?

(b) How much of each anomer is present in the mixture?

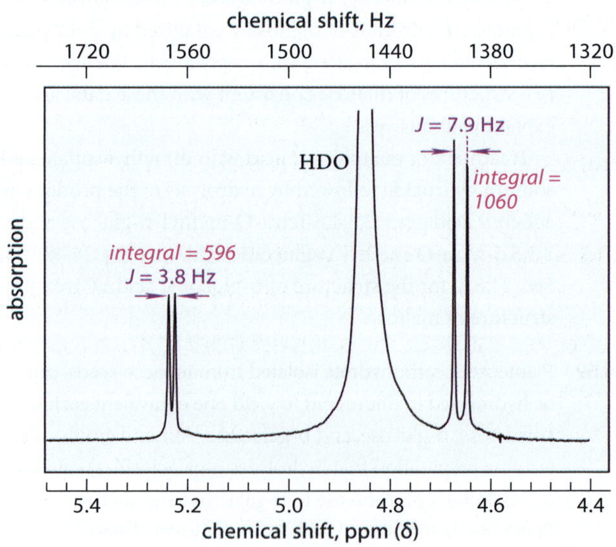

25 THE CHEMISTRY OF THIOESTERS, PHOSPHATE ESTERS, AND PHOSPHATE ANHYDRIDES

This chapter covers two compound classes. The first class is thioesters, which are the sulfur analogs of carboxylate esters. The second class is phosphate esters and phosphate anhydrides, which are derivatives of phosphoric acid. Unlike previous chapters, which focus on laboratory chemistry, the emphasis of this chapter is on reactions that are important in biology.

One goal of this chapter is to explore the strong analogy of thioester, phosphate ester, and phosphate anhydride chemistry to the chemistry of the analogous carboxylic acid derivatives that we considered in Chapters 20 and 21. A mastery of the principles in these chapters is important to an understanding of the material in this chapter. Despite this connection, there are important differences. Our second goal is to explain the unique aspects of thioester, phosphate ester, and phosphate anhydride chemistry that led to the evolution of these functional groups to their position of importance in biology.

25.1 THIOESTERS

Thioesters are the sulfur analogs of esters. There are actually two types of thioester. The more common type, the *S*-alkyl (or aryl) thioester, has a C═O and a sulfur-linked alkyl or aryl group. The much less common type, the *O*-alkyl (or aryl) thioester, has a C═S and an oxygen-linked alkyl or aryl group.

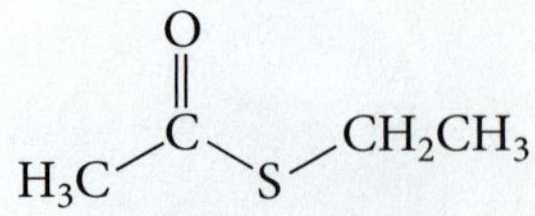

S-ethyl thioacetate
(an *S*-alkyl thioester)

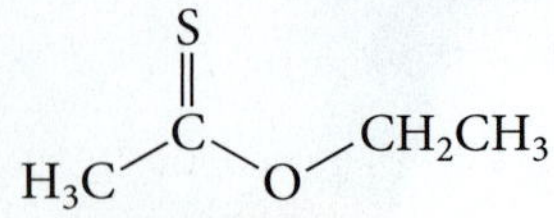

O-ethyl thioacetate
(an *O*-alkyl thioester)

(25.1)

The *S*-alkyl thioesters are the only ones we cover in this chapter, and we'll use the term *thioester* in this text to refer to them. In older literature, they are also called as *thiol esters* (that is, "esters of thiols").

The common nomenclature of thioesters uses the prefix *thio-* before the parent name of the acid and an *S* prefix before the name of the alkyl group to indicate that it is bonded to sulfur. This nomenclature is illustrated by the example in Display 25.1. *In common usage, the S prefix is understood.* Therefore, *S*-ethyl thioacetate is commonly called *ethyl thioacetate*.

In systematic IUPAC nomenclature, *thio* is inserted before the *ate* suffix in the name of the corresponding oxygen ester, as shown in Display 25.2. A final *e* must be added to the carboxylic acid stem (*pentan* in the second example, which becomes *pentane* because the stem name precedes a consonant). The last example shows the name of a thioester attached to a ring. In such cases the ring name is followed by the suffix *carbothioate*.

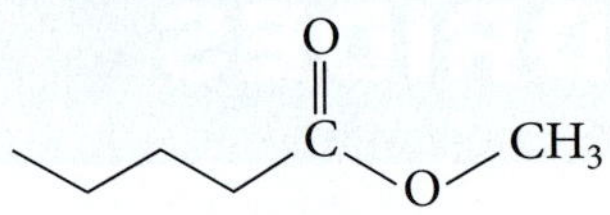

methyl pentanoate
(an oxygen ester)

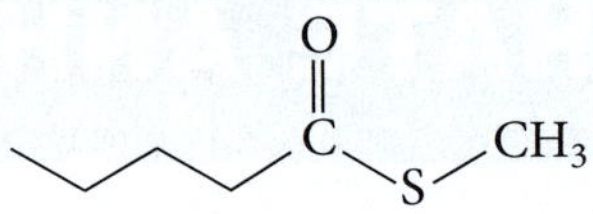

S-methyl pentanethioate
(a thioester)

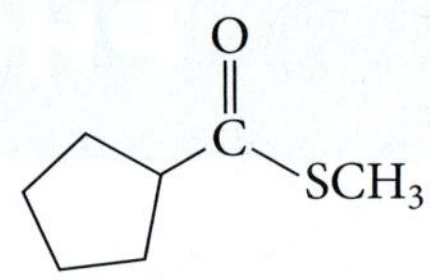

S-methyl cyclopentanecarbothioate

(25.2)

A few thioesters are biologically important; of these, the most important is **acetyl-CoA**, which is the thioester of acetic acid and a thiol called **coenzyme A**. The structure of acetyl-CoA and a few of its common abbreviations are shown in **Fig. 25.1**. As with many other biologically important compounds, a very small part of the molecule (the thioester in this case) is involved in its chemistry; the functionally rich remainder of the molecule is involved in noncovalent interactions when it binds to enzymes. We have considered some of the chemistry involving acetyl-CoA in Secs. 22.7 and 22.8D.

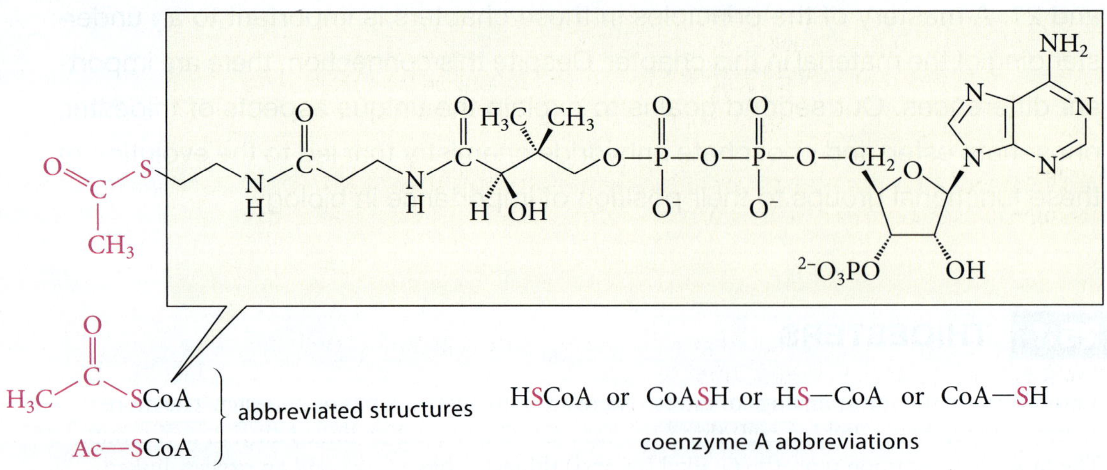

FIGURE 25.1 The structure of acetyl-coenzyme A (acetyl-CoA) and two abbreviated structures. Coenzyme A itself is the thiol HSCoA, shown at the lower right.

Focused Problems

25.1 Draw a structure for each of the following thioesters:

(a) Cyclohexyl thiobenzoate

(b) *S*-Isopropyl butanethioate

(c) *S*-Phenyl cyclohexanecarbothioate

25.2 Provide both common and IUPAC systematic names for the following thioester:

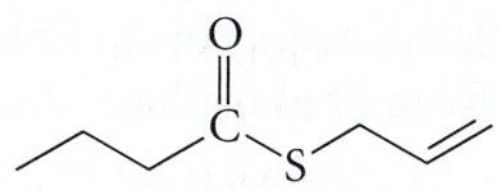

25.2 PHOSPHORIC ACID DERIVATIVES

The organic derivatives of phosphoric acid are conceptually similar to the corresponding carboxylic acid derivatives. Esters (phosphate esters), acid chlorides (phosphoryl chlorides), amides (phosphoramides), and anhydrides are well known.

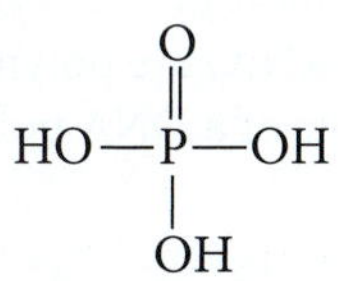

phosphoric acid
(un-ionized form)

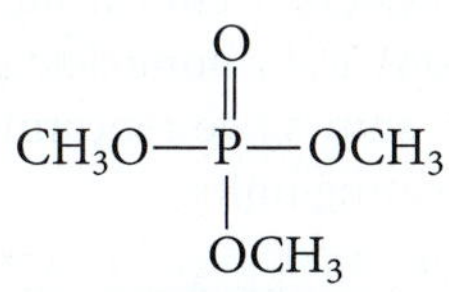

trimethyl phosphate
a phosphate ester

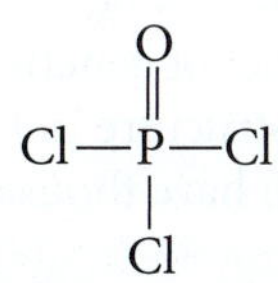

phosphoryl trichloride
(commonly known as **phosphorus oxychloride**)
a phosphoric acid chloride

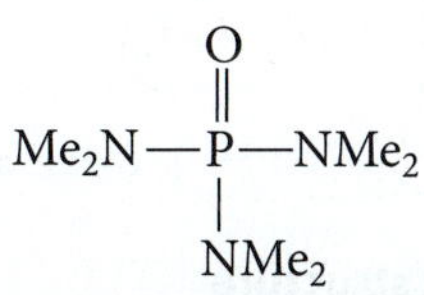

***N,N,N′,N′,N″,N″*-hexamethyl-phosphorotriamide (HMPT)**
a phosphoramide

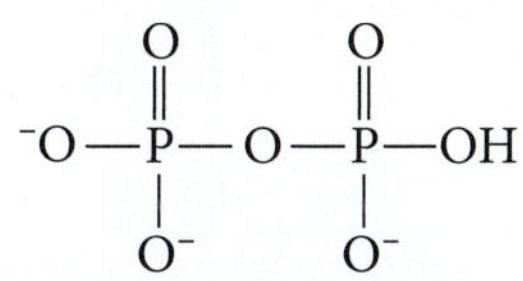

pyrophosphate
(ionized form at pH = 7.4)
a phosphate anhydride

(25.3)

We'll concern ourselves primarily with the two types of derivatives that are most important in biology: phosphate esters and anhydrides.

A. Phosphate Esters

One, two, or three of the acidic OH groups of phosphoric acid can be esterified to give monoesters, diesters, or triesters, respectively.

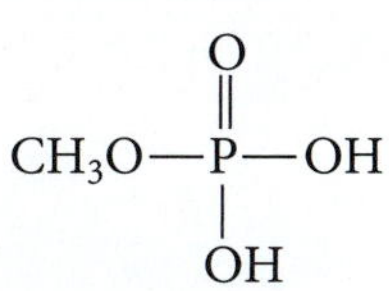

methyl phosphate (monomethyl phosphate)
a phosphate monoester

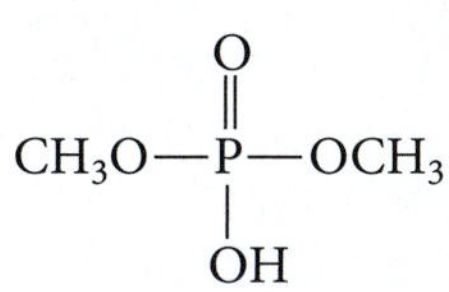

dimethyl phosphate
a phosphate diester

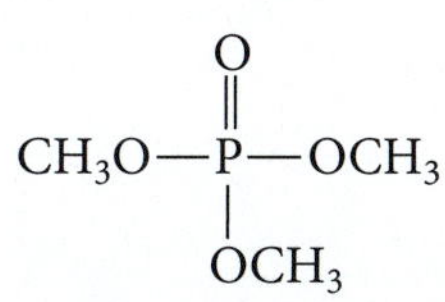

trimethyl phosphate
a phosphate triester

(25.4)

One of the most important differences between phosphate esters and carboxylate esters is that both phosphate diesters and phosphate monoesters have acidic hydrogens. Their pK_a values are very similar to the corresponding pK_a values of phosphoric acid.

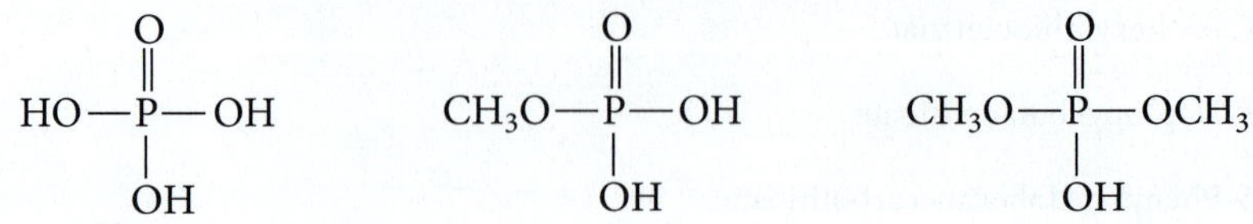

pK_a = 2.2, 7.2, 12.3 pK_a = 2.3, 6.7 pK_a = 2.3 (25.5)

At physiological pH (7.4), the major form of phosphate monoesters is the conjugate-base di-anion, and the major form of phosphate diesters is the conjugate-base anion:

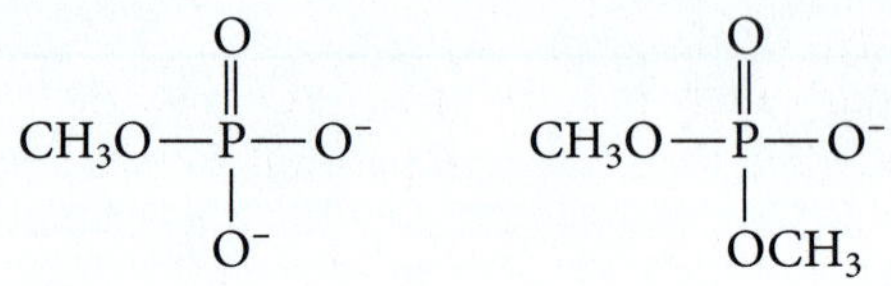

major forms at pH = 7.4 (25.6)

Therefore, at physiological pH, phosphate monoesters and phosphate diesters are *ionic compounds.* Because they are ionic, they have *significant water solubility.*

Chemical Biology Topic

Phosphate Esters in Biology Phosphate diesters occur throughout biology. Deoxyribonucleic acid (DNA, the carrier of genetic information) and ribonucleic acid (RNA) are polymeric phosphate diesters. The structure in Display 25.7 shows two repeating units of a DNA polymer chain. A DNA polymer can have thousands of repeating units.

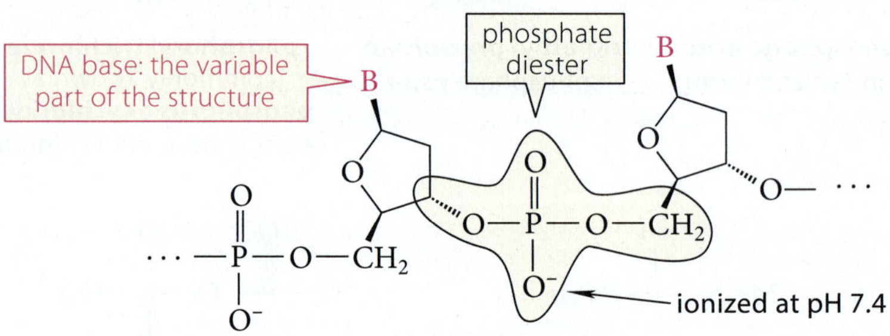

DNA covalent structure
(two units) (25.7)

The B groups shown in red are heterocyclic nitrogen-containing groups, called *DNA bases*, or *nucleobases*, which vary from unit to unit. These can have any of four different structures. (A detailed discussion of DNA structure is given in Sec. 26.5B.) What you should notice for now is the phosphate diester part of the molecule. Notice also that DNA at physiological pH is ionized; each diester unit has one negative charge. The negative charges of DNA are a very important aspect of the DNA three-dimensional structure (Sec. 26.5B).

An interesting and important phosphate diester that isn't a polymer is *cyclic AMP*, a very strained cyclic phosphate diester, which serves as an intracellular signaling molecule.

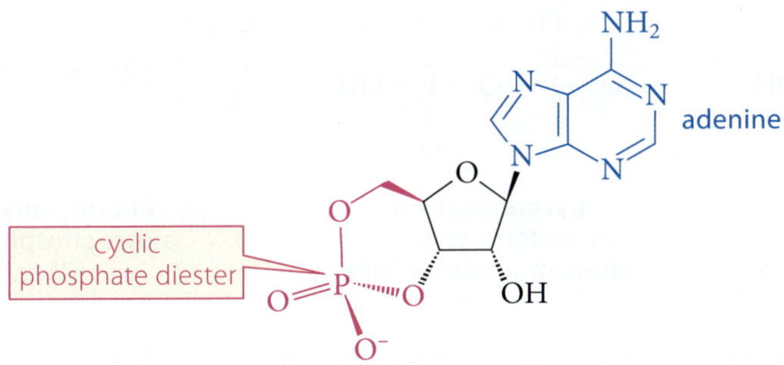

cyclic adenosine monophosphate (cyclic AMP) (25.8)

In this molecule, adenine (*blue*) is one of the nucleobases that occur in DNA and RNA. (We revisit the structures of the bases in Sec. 26.5A.)

Phosphate monoesters occur widely in biology. For example, glucose-6-phosphate is a metabolic intermediate in *glycolysis*, the biological pathway by which glucose is converted into pyruvate.

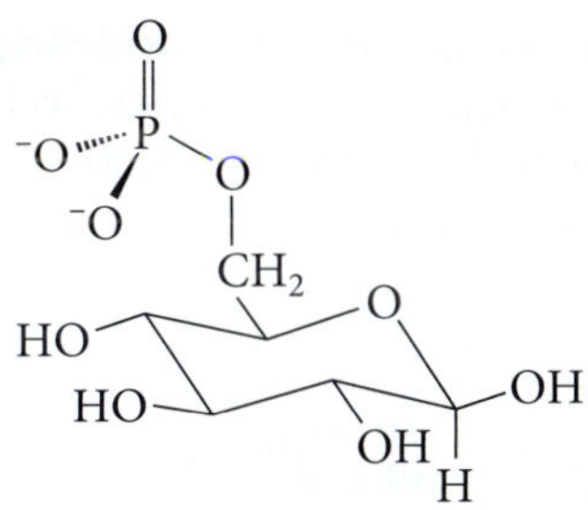

glucose-6-phosphate
a phosphate monoester (25.9)

A particularly interesting process involving phosphate monoesters is the biological conversion of tyrosine or serine residues of proteins, which are uncharged at physiological pH, into their phosphate monoesters. This conversion changes the charge of the tyrosine or serine from 0 to −2. As we show in Sec. 27.8E, this charge alteration can result in a profound change in the three-dimensional structure of a protein in which this conversion occurs.

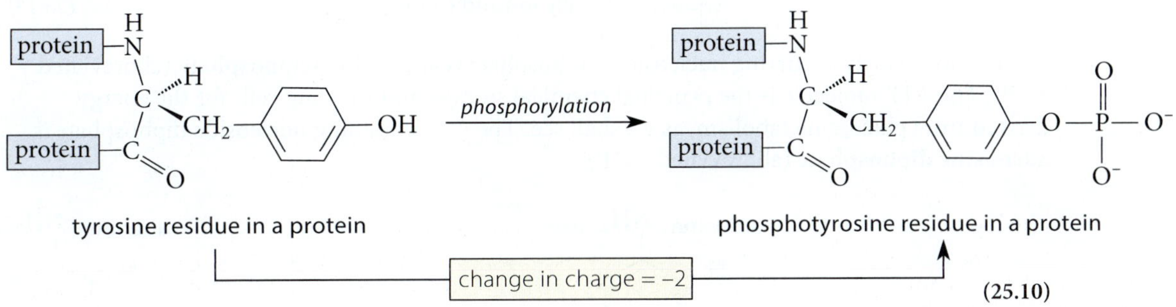

(25.10)

B. Phosphate Anhydrides

Chemical Biology Topic

A phosphate anhydride contains two (or more) linked phosphate groups. The simplest anhydride of phosphoric acid is pyrophosphoric acid, which exists at physiological pH as its conjugate-base tri-anion. Though not an organic compound, it plays an important role as a biological leaving group (Secs. 11.4E and 17.6B) when it is linked to an organic group as a pyrophosphate monoester.

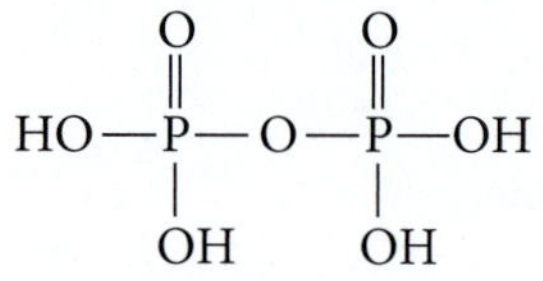

pyrophosphoric acid
(un-ionized form)
pK_a values = 0.9, 2.0, 6.6, 9.4

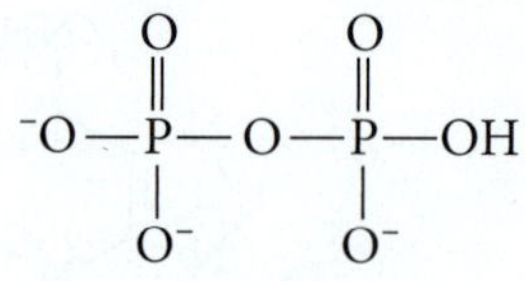

pyrophosphate
(ionized form at pH = 7.4)
often abbreviated $^{-}$**OPP**

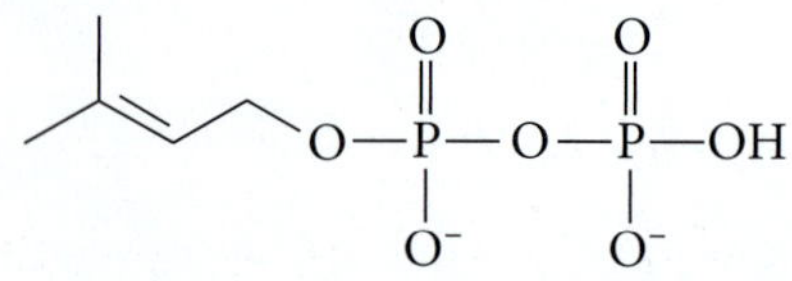

γ,γ-dimethylallyl pyrophosphate
a pyrophosphate monoester
(Sec. 17.6B)

(25.11)

The linkage of more than two phosphate groups is also possible. Such polyphosphates serve as phosphate reservoirs in biological systems.

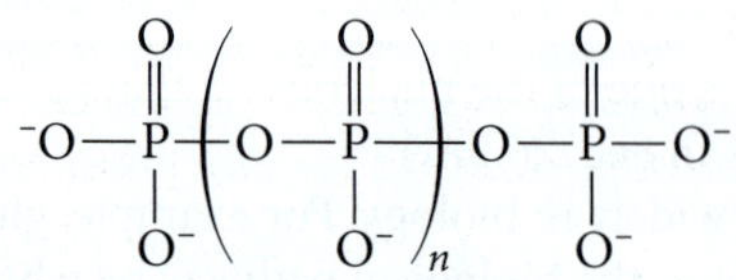

polyphosphate

(25.12)

Among the most important phosphate anhydrides are the **nucleoside triphosphates** and **nucleoside diphosphates**. These contain three structural units: a heterocyclic nucleobase, a sugar (ribose), and the phosphate anhydride group. In a nucleoside triphosphate, the phosphate anhydride actually contains two linked phosphate anhydrides.

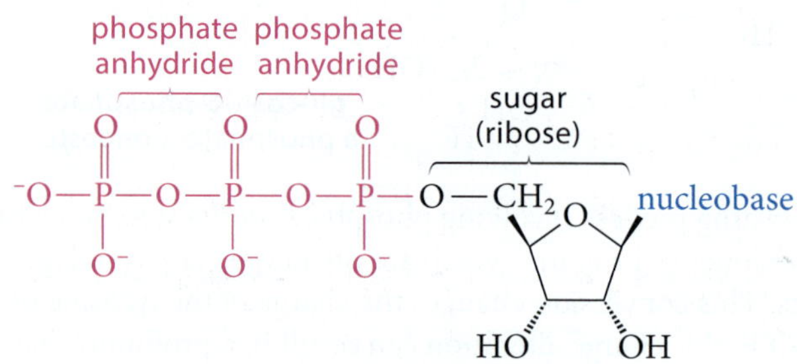

general structure of a nucleoside triphosphate
(shown in its fully ionized form)

(25.13)

The most widely occurring nucleoside triphosphate is **adenosine triphosphate** (abbreviated **ATP**). The ATP molecule is the principal chemical storage unit in living cells for the energy derived from glucose metabolism, as we shall see. The corresponding nucleoside diphosphate is **adenosine diphosphate** (abbreviated **ADP**).

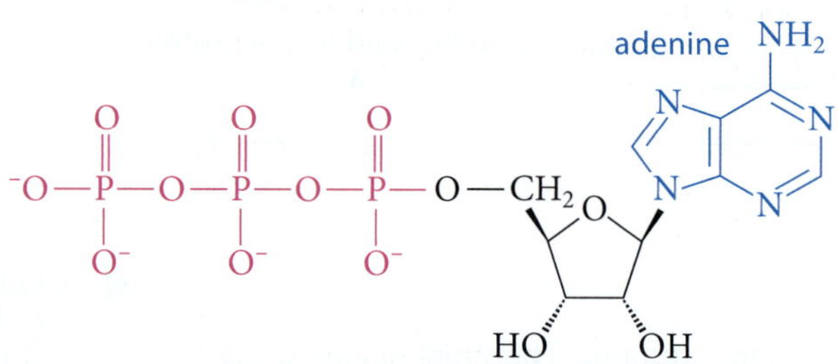

adenosine triphosphate (ATP)

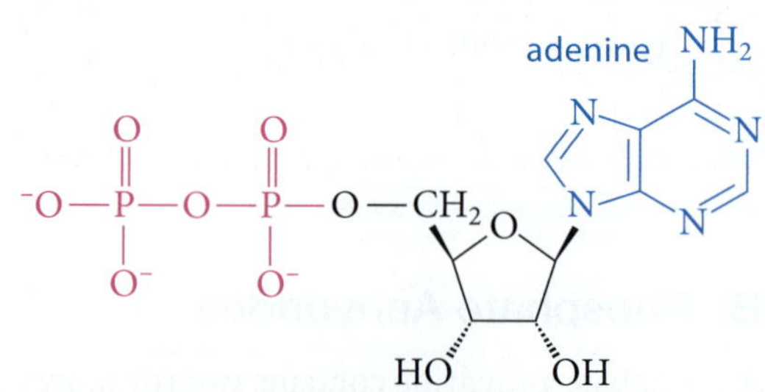

adenosine diphosphate (ADP)

(25.14a)

The heterocyclic nucleobase in ATP and ADP is adenine (*blue*), the same nucleobase that occurs in the structure of cyclic AMP.

Under physiological conditions, both ATP and ADP (and other nucleoside triphosphates and diphosphates) exist primarily as their complexes with Mg^{2+}, in which the Mg^{2+} is held, or chelated, by interaction with oxygen anions on the two terminal phosphate groups.

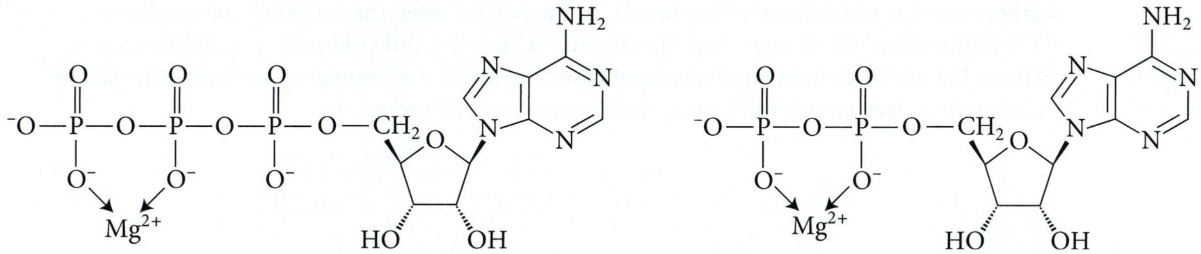

Mg^{2+} complex with ATP **Mg^{2+} complex with ADP**

(25.14b)

A few mixed anhydrides of phosphoric and carboxylic acids are important in biology. This type of compound is called an **acyl phosphate**. An example is carbamoyl phosphate, a mixed anhydride of carbamic acid (Eq. 20.63b, Sec. 20.11A) and phosphoric acid.

an acyl phosphate **carbamoyl phosphate** (25.15)

Focused Problems

25.3 Given the pK_a values of methyl phosphate shown in this section, calculate the percentage of the un-ionized form, the mono-anion form, and the di-anion form at pH 7.4.

25.4 The side chain —R of the amino acid serine is —CH_2OH. Draw the structure of the phosphate monoester of serine.

25.5 In the structure of acetyl-CoA (see Fig. 25.1), point out and identify the phosphorus-containing functional groups.

25.3 STRUCTURES OF THIOESTERS AND PHOSPHATE ESTERS

A. Structures of Thioesters

The structure of a simple thioester, methyl thioacetate, is shown in **Fig. 25.2** along with the structure of an oxygen ester, dimethyl carbonate. The longer carbon–sulfur bonds in comparison to the carbon–oxygen bonds are expected from the greater size of sulfur. The most important comparative aspect of the structures, however, is the relative lengths of the two types of

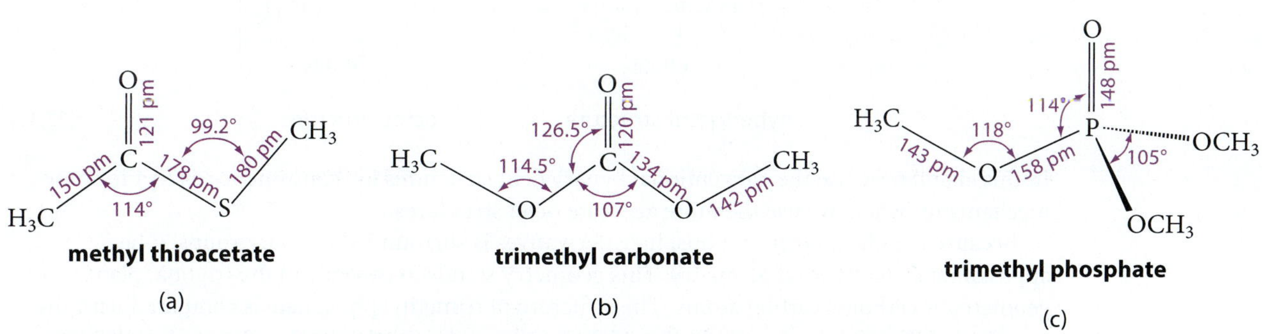

FIGURE 25.2 A comparison of the structures of (a) a thioester, (b) a carboxylic acid ester, and (c) a phosphate ester. Compare the thioester and the oxygen ester: the lengths of the two carbon–sulfur bonds of the thioester are nearly the same, whereas the lengths of the two carbon–oxygen bonds of the oxygen ester are significantly different. In the phosphate ester, notice the tetrahedral structure and the relative bond lengths of the phosphorus–oxygen double and single bonds.

carbon–oxygen and carbon–sulfur bonds. In the oxygen ester, the carbonyl–oxygen bond (134 pm) is about 6% shorter than the other C—O single bond (142 pm). This shortening reflects the partial double-bond character associated with the resonance overlap of the oxygen nonbonding electron pair with the π electrons of the carbonyl group.

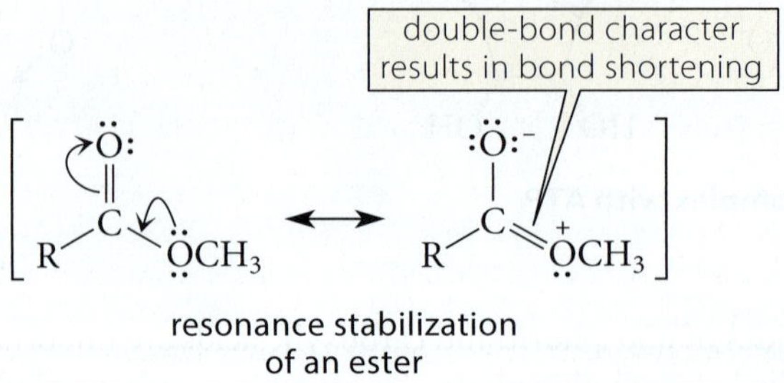

(25.16a)

In the thioester, however, the carbonyl–sulfur bond (178 pm Å) is only 1% shorter than the other C—S single bond (180 pm). Resonance overlap of a sulfur nonbonding pair with the π electrons of the carbonyl group is less important than it is in an oxygen ester; consequently, there is less double-bond character in the carbonyl–sulfur bond.

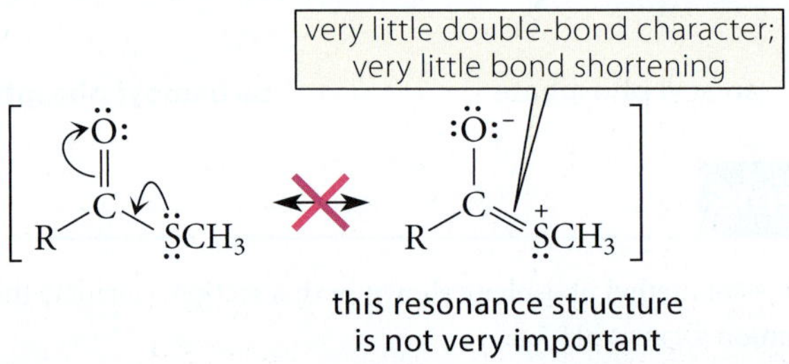

(25.16b)

The reason for the weaker resonance interaction in a thioester is that the sulfur resonance interaction involves 3p orbitals, which do not overlap well with the 2p orbitals of the carbonyl group. (See Fig. 16.7, Sec. 16.5B, for a similar situation.) *In other words, resonance does not stabilize a thioester as much as it stabilizes an oxygen ester.* This point will prove to be very important in understanding both the rates and equilibria for the reactions of thioesters.

B. Structures of Phosphate Esters

Phosphorus in phosphate esters and anhydrides is in the phosphorus(V) oxidation state. As usually drawn, the phosphorus atom in the structures of these compounds is hypervalent—that is, it has 10 shared electrons and therefore is involved in octet expansion. However, as we have pointed out in Sec. 11.10A and elsewhere, all of these compounds have very important octet structures in which there is charge separation (Sec. 11.10A).

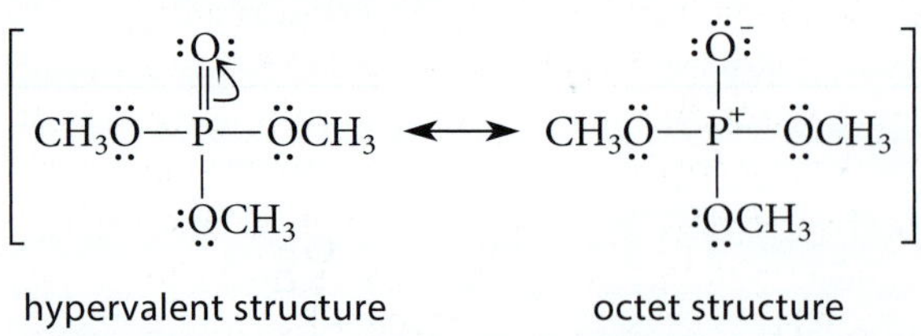

(25.17)

In this chapter, we use the conventional hypervalent structures for convenience except in some mechanisms, where we use the more accurate octet structures.

Because the phosphorus in phosphate derivatives is surrounded by four groups, it has approximately tetrahedral geometry. This geometry stands in contrast to the trigonal planar geometry at carbonyl carbon atoms. The structure of trimethyl phosphate is compared with the structure of dimethyl carbonate in Fig. 25.2. As expected from the greater size of phosphorus, bond lengths to phosphorus are greater than those to carbon. The P═O "double bond" is shorter than the P—O single bonds, but it is only 6% shorter. Compare this to the C═O bond in a carboxylate ester, which is more than 10% shorter than the C—O single bond. The shortening of the P═O "double bond" is equally consistent with the shortening of a P—O single bond in the octet structure caused by the electrostatic attraction of opposite charges on the two atoms.

Because of its tetrahedral geometry, the phosphorus of a phosphate ester can be an asymmetric atom when the four groups attached to phosphorus are different.

Focused Problems

25.6 What is the hybridization of the oxygen in the C—O bond of trimethyl phosphate? Give your reasoning.

25.7 The following chiral phosphate ester cannot be isolated in optically active form. Explain.

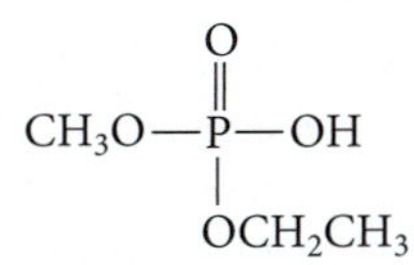

25.4 PROTON AND CARBON NMR SPECTROSCOPY OF PHOSPHORUS-CONTAINING MOLECULES

Phosphorus (^{31}P) has a nuclear spin of $\pm\frac{1}{2}$. Therefore, when protons or ^{13}C nuclei are being observed, a nearby phosphorus can cause splitting. Significant P–H splitting can occur over three bonds. This is analogous to the splitting of protons caused by ^{19}F (Sec. 14.7C). The splitting of the ^{13}C and ^{1}H resonances in trimethyl phosphate is illustrative.

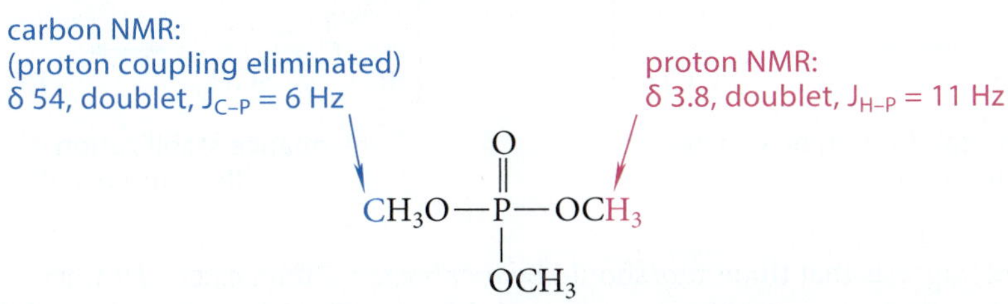

(25.18)

Comparison of the chemical shifts in this example with the chemical shifts in carboxylic esters given in Chapter 21 (Sec. 21.4B) shows that the chemical shifts of both carbons and protons are very similar in phosphate esters and carboxylic esters.

The ^{31}P NMR of the phosphorus atoms can be observed at an operating frequency different from the frequencies used to observe ^{13}C or ^{1}H (see Table 14.4, Sec. 14.9). In the ^{31}P NMR of trimethyl phosphate, for example, the phosphorus resonance is split into a 10-line pattern with $J = 11$ Hz by the nine methyl hydrogens. As with proton spectra, no splitting of the phosphorus resonance by ^{13}C is observed because of the low natural abundance of the ^{13}C isotope. Additional splitting by ^{13}C would be observed in a sample that is enriched with ^{13}C.

Focused Problem

25.8 Describe the splitting expected in the proton resonance of the —CH_2— group in triethyl phosphate. (The coupling constants for H–H splitting and P–H splitting happen to be identical in this case.) (*Hint:* See Fig. 14.11, Sec. 14.5A.)

25.5 REACTIONS OF THIOESTERS WITH NUCLEOPHILES

Thioesters, like esters, undergo nucleophilic substitution reactions. In these reactions, a thiol is the leaving group. In this section, we focus on the hydrolysis of thioesters and a few other reactions that have biological importance.

A. Hydrolysis of Thioesters

Thioesters, like esters, undergo saponification.

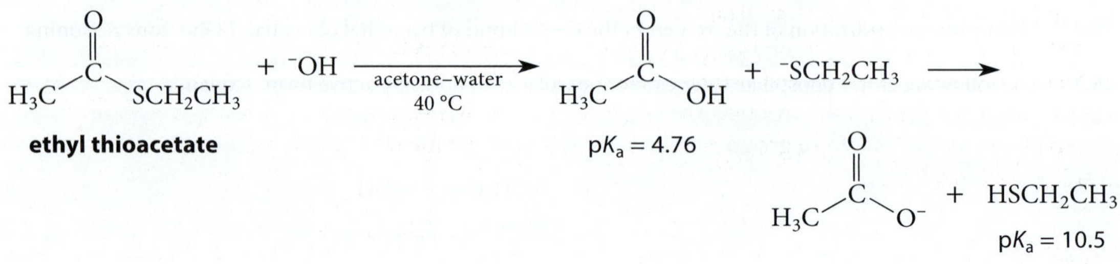

(25.19)

Like the saponification of esters, the saponification of thioesters is driven toward products by ionization of the carboxylic acid product. Thioesters have about the same reactivity in saponification as the corresponding oxygen esters. Let's consider why this similar reactivity is reasonable.

Recall (Sec. 21.7E) that relative reactivity in ester hydrolysis is governed by both the *stabilization of the ester*, which reduces reactivity, and the *stabilization of the tetrahedral intermediate*, which increases reactivity. As shown in Sec. 25.3A, the resonance stabilization of thioesters is much weaker than the resonance stabilization of esters.

resonance stabilization of an ester (important) — resonance stabilization of a thioester (less important)

(25.20)

This fact alone suggests that thioesters should be *more reactive* than esters. However, sulfur is considerably less electronegative than oxygen and chlorine. Therefore, the polar stabilization of the thioester transition state by sulfur is less than the polar stabilization of the ester transition state by oxygen. This fact alone suggests that thioesters should be *less reactive* than esters. Therefore, the two effects work in opposite directions and tend to cancel; in other words, the reactivities of oxygen esters and thioesters in saponification (and other nucleophilic carbonyl additions) are about the same.

The acid-catalyzed hydrolysis of thioesters, like that of esters, requires fairly strong acid.

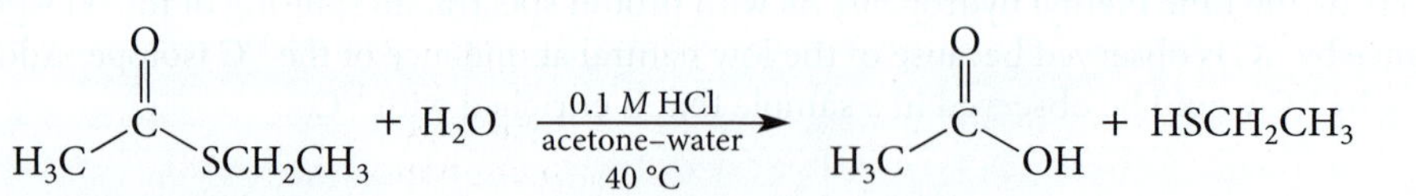

ethyl thioacetate

(25.21)

Thioesters are about 2% as reactive as the corresponding esters in acid-catalyzed hydrolysis, but this difference is not large in the overall reactivity spectrum of carboxylic acid derivatives. In summary, then, thioesters, depending on conditions, have comparable or somewhat lower reactivity toward hydrolysis than oxygen esters.

At pH 7.4 and 37 °C—physiological conditions—the hydrolysis rates of thioesters are negligible; their hydrolysis takes *years*. This means that thioesters such as acetyl-CoA can survive under cellular conditions until they are needed for enzyme-catalyzed reactions in biological processes.

The similar reactivity of oxygen esters and thioesters begs the question: Why did thioesters such as acetyl-CoA evolve in biology rather than the corresponding oxygen esters? Their relative reactivity is *not* the reason, because esters and thioesters have similar reactivities toward hydrolysis. We consider this interesting question in Sec. 25.8C.

B. Reactions of Thioesters with Other Nucleophiles

Chemical Biology Topic

Thioesters, like oxygen esters, can react with a variety of nucleophiles. Here we focus on the reaction types that are most important in biology.

One of the most common biological reactions of thioesters is the reaction with another thiol to form another thioester, called **transthioesterification**. A relatively common example of transthioesterification is the reaction of an ester of CoA—SH, such as malonyl-CoA, to a thiol group of a protein. This thiol group is invariably provided by the side chain of the amino acid residue cysteine.

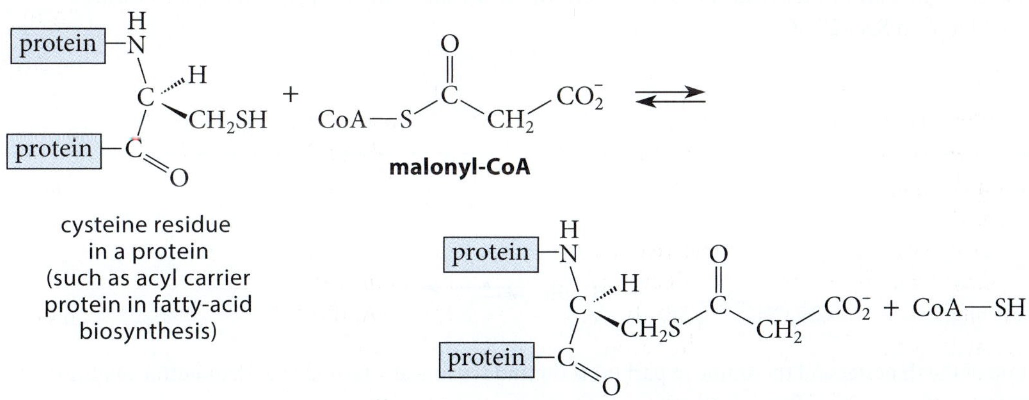

malonyl-CoA

cysteine residue in a protein (such as acyl carrier protein in fatty-acid biosynthesis)

a malonylated protein (25.22)

This reaction was illustrated in our discussion of fatty-acid biosynthesis (Eq. 22.84a, Sec. 22.7). We also showed the reactions of thioesters with nucleophilic enolate ions in the Claisen condensation steps of fatty-acid biosynthesis (Eq. 22.84b, Sec. 22.7).

The reaction of a thioester with an alcohol is illustrated by the biosynthesis of acetylcholine, an important neurotransmitter in the brain. This reaction involves the enzyme-catalyzed displacement of the thiol CoA—SH by the oxygen of choline.

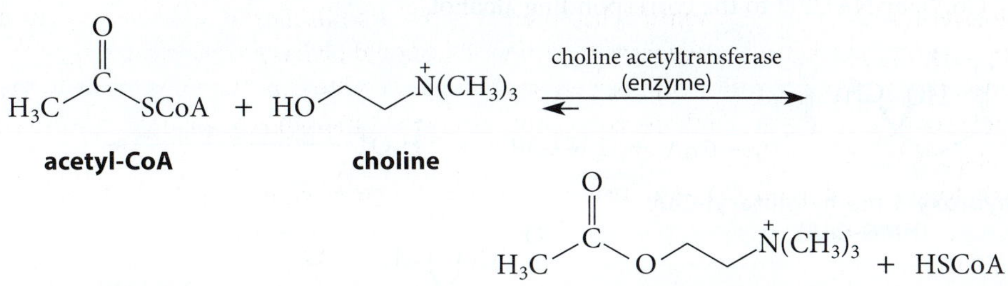

acetyl-CoA

choline

acetylcholine
a neurotransmitter (25.23)

The role of the enzyme is, first, to bind the acetyl-CoA and choline molecules in proximity (Sec. 12.8E). The charged nitrogen of choline and the many functional groups in the CoA part of acetyl-CoA (see Fig. 25.1) are molecular "handles" that provide sites for significant noncovalent attractions. Second, the enzyme provides a basic histidine residue that enhances the nucleophilicity of the choline hydroxy group by partial proton removal. The enzyme also provides a hydrogen-bond donor—the OH group of a serine residue—which enhances the electrophilicity of the carbonyl group of acetyl-CoA by forming a hydrogen bond to the carbonyl oxygen.

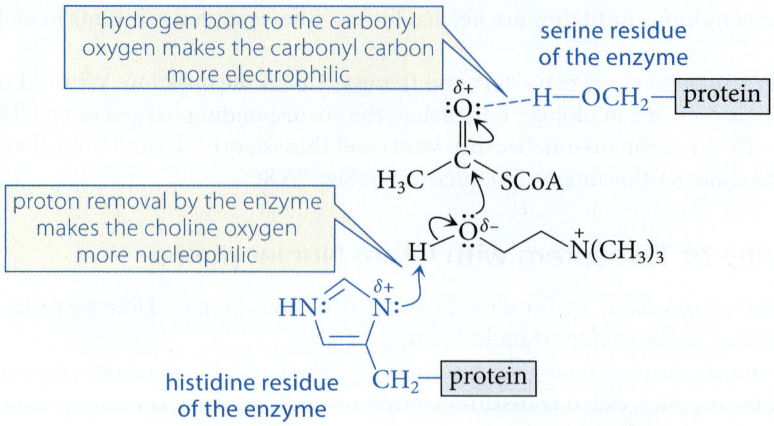

(25.24)

An important aspect of this reaction is its favorable equilibrium constant (K_{eq} = 12.3 at 38 °C). A favorable equilibrium constant (that is, a negative $\Delta G°$) is a characteristic of all reactions of acetyl-CoA and other thioesters with alcohols and water. The basis of this observation is considered in Sec. 25.8C.

Focused Problem

25.9 (a) Complete the following reaction.

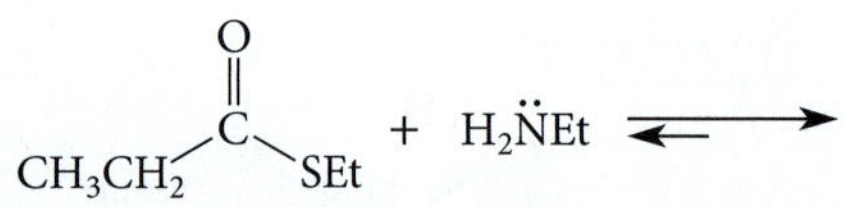

(b) Although the reaction of the thioester and the amine in part (a) is thermodynamically favorable under neutral conditions, the reaction is not observed in the presence of acid. Explain.

Chemical Biology Topic

C. Reduction of Thioesters: HMG-CoA Reductase

Thioesters undergo reduction with lithium aluminum hydride ($LiAlH_4$). (Recall that esters are also reduced by $LiAlH_4$; Sec. 21.9A.) The hydride reduction of esters has a medically important biological counterpart in the reduction of a thioester, (*S*)-3-hydroxy-3-methylglutaryl-CoA (HMG-CoA), by NADPH to the corresponding alcohol.

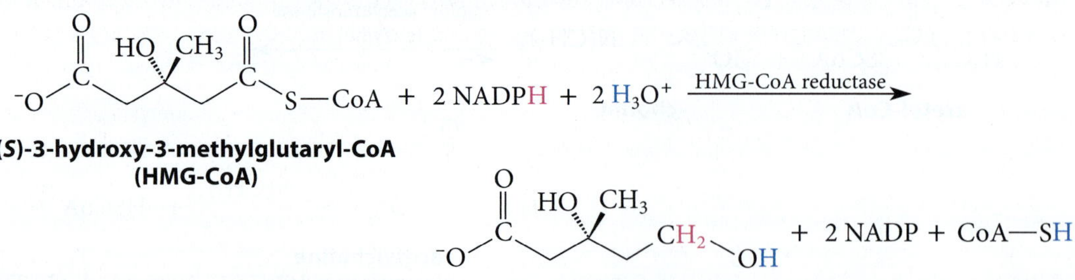

(25.25)

The reduction product, (*R*)-mevalonate, is the biological precursor of isopentenyl pyrophosphate, which is the starting material for the biosynthesis of isoprenoids and steroids, particularly cholesterol (Sec. 17.6C). This reaction, catalyzed by the enzyme HMG-CoA reductase, is the rate-limiting step in the biosynthesis of cholesterol. Inhibiting this enzyme has proven to be a highly effective strategy for lowering cholesterol. (See the sidebar, "The Statins: Blockbuster Drugs That Inhibit Cholesterol Biosynthesis," at the end of this section.)

We showed in Sec. 19.9B that one of the most common hydride donors for carbonyl-group reduction in biology is NADH or its phosphorylated variant, NADPH. In the reduction of HMG-CoA in humans, NADPH is the hydride donor. As in other enzyme-catalyzed hydride reductions, a hydrogen-bond donor (in this case a protonated amino group of a lysine residue) activates the carbonyl group by partial protonation. As the CoA—S⁻ leaving group is lost, it is protonated by another group on the enzyme, the conjugate acid of an imidazole group of a histidine residue. The first intermediate, as in laboratory ester reductions, is an aldehyde called *mevaldehyde*:

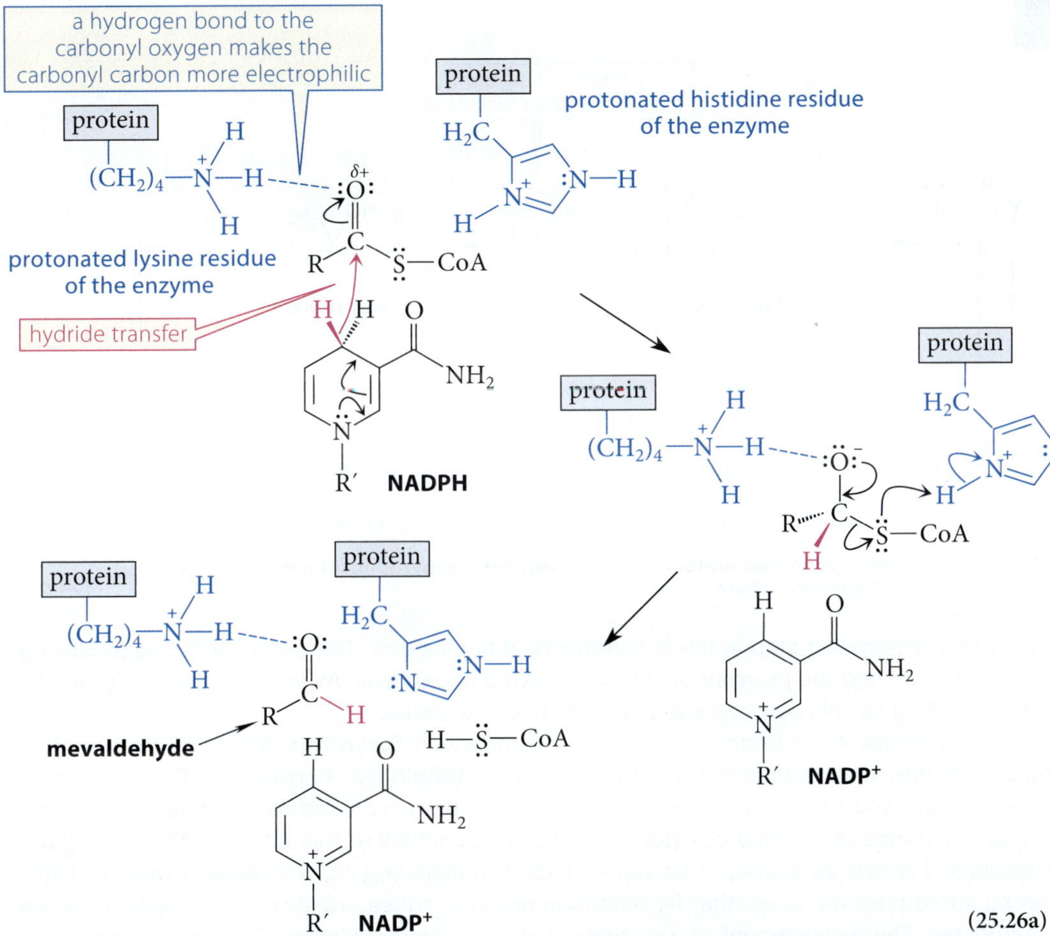

(25.26a)

Mevaldehyde is then reduced by a second molecule of NADPH to mevalonate. The mechanism of this second step is analogous to the one introduced in Sec. 19.9B. (See Focused Problem 25.10 at the end of this section.)

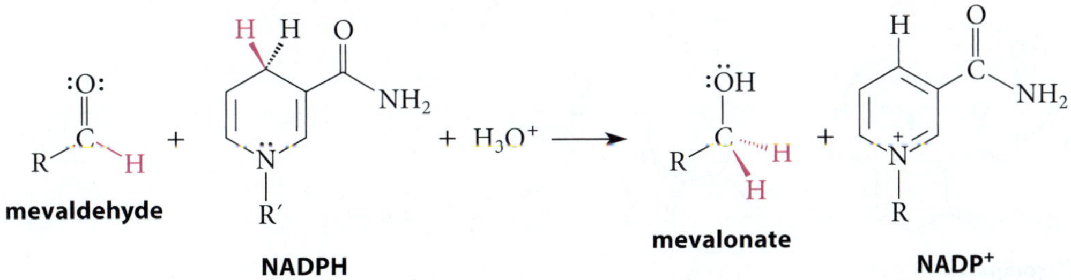

(25.26b)

The Statins: Blockbuster Drugs That Inhibit HMG-CoA Reductase

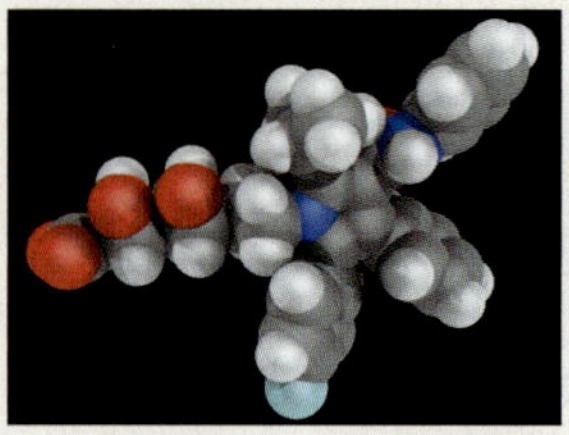

Laguna Design/Science Source

Coronary heart disease is caused by the formation of cholesterol-containing plaques in the coronary arteries, the blood vessels that supply blood to the heart. Therefore, inhibiting the biosynthesis of cholesterol should be a good strategy for eliminating or reducing these occlusive plaques. Key steps in the biosynthetic pathway that produces cholesterol can be summarized as follows:

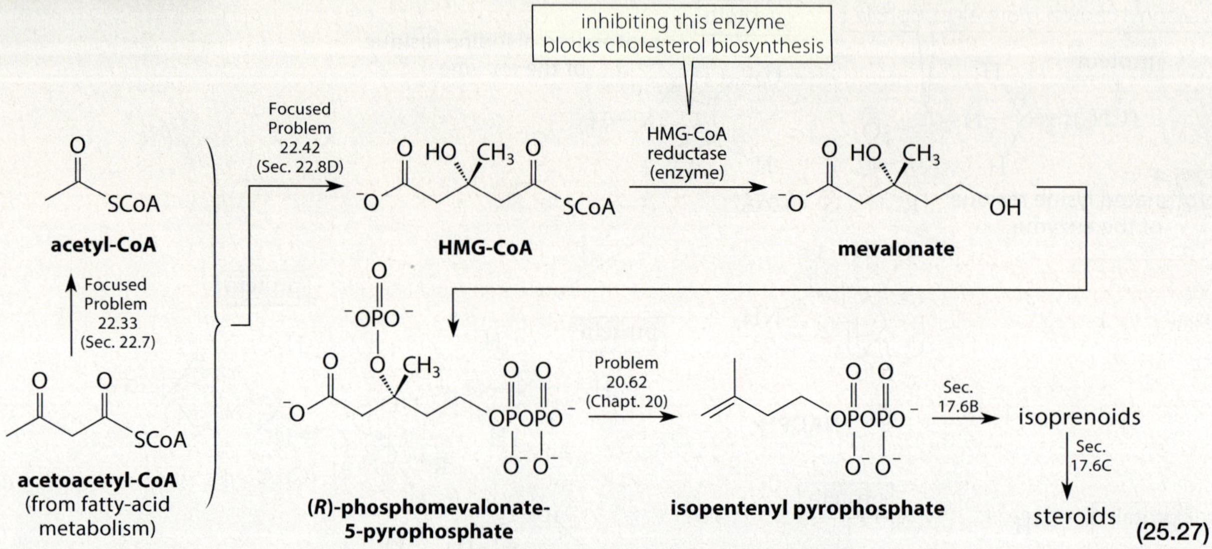

(25.27)

The reduction of HMG-CoA is the rate-limiting step in the biosynthesis of cholesterol. Therefore, inhibiting (blocking) the enzyme that catalyzes this step offered the promise of shutting down this pathway. As with many other scientific discoveries, the development of a drug for this purpose was somewhat serendipitous.

In the 1970s, Japanese microbiologist Akira Endo of the Sankyo Company in Tokyo was screening fermentation broths of *Penicillium citrinum* for new compounds that might be used as antibiotics. Fortunately, Endo was also interested in cholesterol metabolism, and he also screened the compounds he discovered for their ability to inhibit HMG-CoA reductase. He found a compound, called compactin, that strongly inhibited this enzyme. This turned out to be the first of many drugs, called *statins* as a group, that were effective in inhibiting HMG-CoA reductase. In 1978, scientists at Merck Research Laboratories found another fermentation product, subsequently called lovastatin, which also inhibited HMG-CoA reductase. The development of a number of statin analogs followed. The most widely used drug in this class in recent years has been atorvastatin (model above), marketed under the trade name Lipitor by the pharmaceutical company Pfizer.

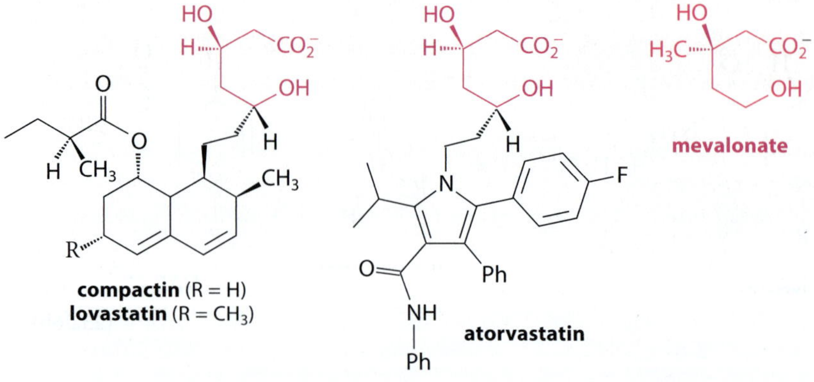

The sales of Lipitor reached $12 billion annually before patent protection for the drug expired in 2011. High-revenue drugs such as Lipitor that are taken indefinitely for chronic conditions are sometimes referred to as "blockbuster drugs."

The mode of action of the statins was further clarified by Michael S. Brown (b. 1941) and Joseph L. Goldstein (b. 1940), physician–scientists at the University of Texas Health Sciences Center in Dallas. Cholesterol is deposited into arterial plaques from transporter proteins called low-density lipoproteins (LDLs). Brown and Goldstein showed that when cholesterol biosynthesis is inhibited, ordinary cells, starved for needed cholesterol by the inhibitory action of statins, internalize LDL-cholesterol by expressing LDL receptors on the cell surface. These receptors bind the LDL and transport it into the cell. This takes cholesterol (as the offending LDL-cholesterol) out of circulation. For this work, Brown and Goldstein received the 1985 Nobel Prize for Physiology or Medicine.

Focused Problem

25.10 Using abbreviated structures like the ones in Eq. 25.26a, give a curved-arrow mechanism for the reduction of mevaldehyde shown in Eq. 25.26b.

25.6 HYDROLYSIS OF PHOSPHATE ESTERS AND ANHYDRIDES

In this section, we consider the hydrolysis reactions of phosphate esters and anhydrides with the goal of understanding the biologically important examples of these reactions. In Sec. 25.7, we discuss the reactions of these compounds with other nucleophiles.

A. Hydrolysis of Phosphate Esters

Hydrolysis of Phosphate Triesters Superficially, carboxylate ester hydrolysis and phosphate ester hydrolysis are similar reactions. For example, trialkyl phosphates undergo base-promoted hydrolysis (saponification) reactions just as carboxylic esters do.

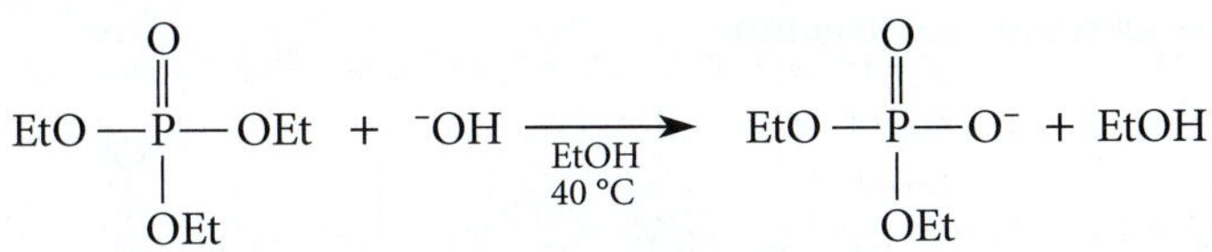

(25.28)

Notice that one ester group can be hydrolyzed without affecting the others, a significant point that we'll return to later. Triethyl phosphate saponification is about 3% as fast as ethyl acetate saponification; in other words, phosphate triesters are somewhat less reactive than carboxylate esters.

There are two possible pathways for the saponification reaction: C—O cleavage and P—O cleavage. These can be distinguished by using ^{-}OH containing the heavy oxygen isotope ^{18}O (*red*) and determining which product (diethyl phosphate or ethanol) contains the isotope.

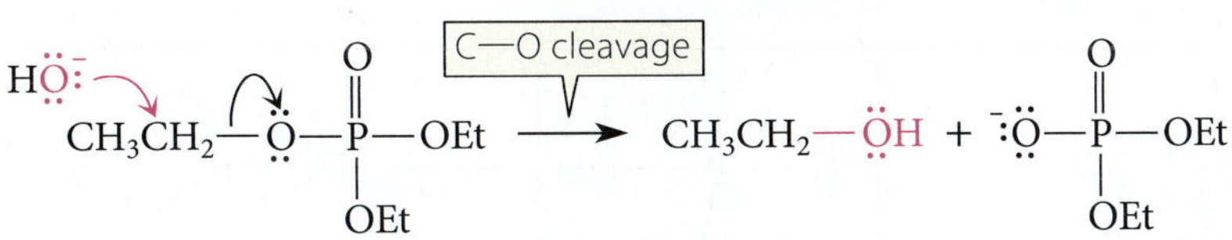

(25.29a)

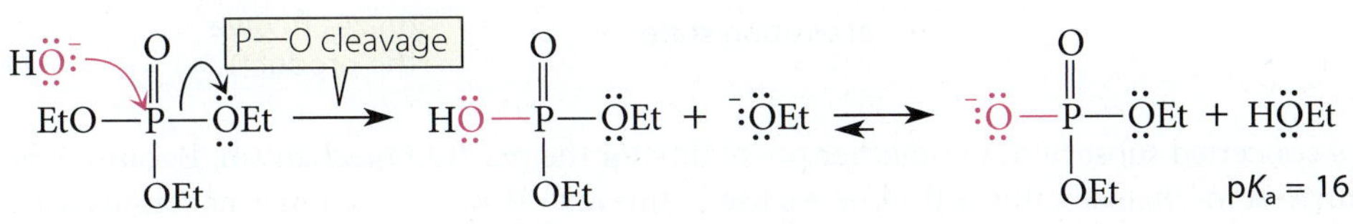

(25.29b)

In one pathway, ^{-}OH reacts as a nucleophile at the α-carbon of the alkyl group in an S_N2 reaction and the nucleophilic oxygen is incorporated into the alcohol product. This pathway is analogous to the S_N2 reaction of sulfonate esters (Sec. 11.4A). In the other, ^{-}OH reacts at the phosphorus and the nucleophilic oxygen is incorporated into the phosphate diester product. This pathway is analogous to the reaction of $^{-}$OH at the carbonyl group in ester saponification (Eq. 21.33a, Sec. 21.7A). For trimethyl phosphate, C—O cleavage accounts for about 90% of the product. However, ethyl derivatives are less reactive than methyl derivatives in S_N2 reactions (see Table 9.3, Sec. 9.4D). Therefore, C—O cleavage should be considerably slower and the P—O cleavage pathway should be observed for ethyl esters.

The observation of both C—O and P—O cleavage shows that the rates of these two processes are not very different. The P—O cleavage pathway is the most commonly observed one in biology with one exception, the cleavage of alkyl pyrophosphates, which occurs with C—O cleavage. (We consider that process in Sec. 25.7C.) For now, we focus on the mechanisms of P—O cleavage.

We can write phosphate ester saponification as a two-step addition–dissociation process. (We use the octet structure for phosphate esters in these mechanisms.)

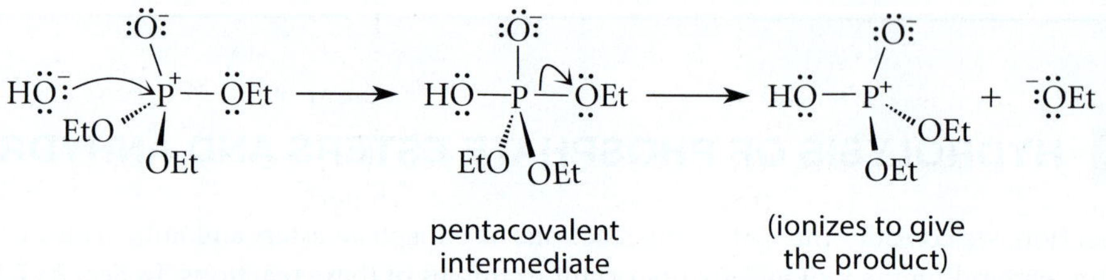

(25.30)

FURTHER EXPLORATION 25.1
Pentacovalent Intermediates in Phosphate Ester Hydrolysis

In this mechanism, the addition intermediate is *pentacovalent*, whereas the addition intermediate in carboxylic ester saponification is tetrahedral. However, phosphorus, because of octet expansion, can be pentacovalent. (For example, PF_5 is a well-known compound.) Although there is no direct evidence for such an intermediate in the saponification of simple phosphate esters (see Further Exploration 25.1), there is definitive evidence for pentacovalent addition intermediates in other cases. As discussed in Sec 3.3B (Eq. 3.24), however, the pentacovalent intermediate, if it exists, would be expected to have a high degree of ionic character because of the importance of an octet resonance structure:

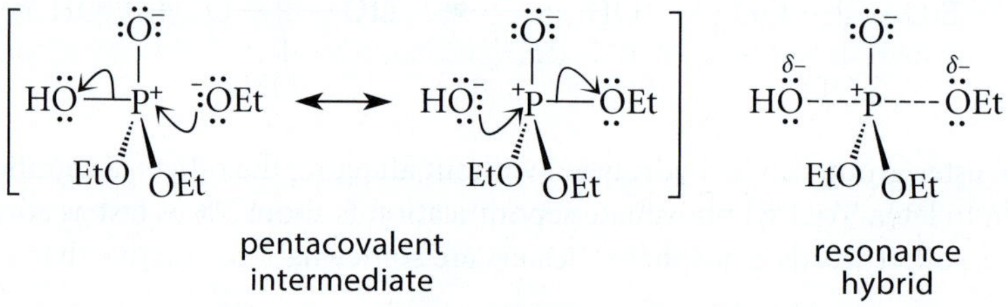

(25.31)

The resonance hybrid looks much like we would imagine for the transition state of a concerted, single-step substitution of hydroxide on phosphorus, which is analogous to an S_N2 reaction at carbon:

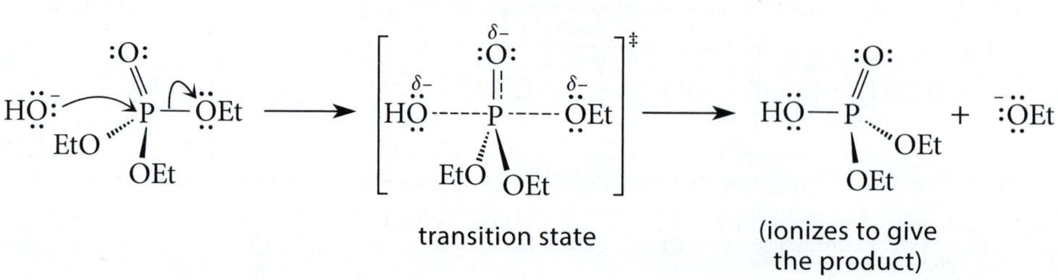

(25.32)

Until there is definitive evidence for a pentacovalent intermediate, the adoption of the one-step mechanism for phosphate-ester hydrolysis is justified by Occam's razor (William of Occam, ca. 1287–1347), a philosophical principle that says, "Entities (for example, reactive intermediates) should not be multiplied without necessity."

Such a concerted substitution is another possibility for the reaction mechanism. Because it is the simplest mechanism, this is the one we use in this text. However, you may encounter either mechanism if you study phosphate esters in biochemistry. As this analysis shows, the two mechanisms are only subtly different—the pentacovalent species is an intermediate in the two-step mechanism and a transition state in the concerted mechanism.

Hydrolysis of Phosphate Diesters As shown in Eq. 25.28, a phosphate triester can be hydrolyzed in base to a phosphate diester product, which can be isolated in high yield. The isolation of the diester shows that *phosphate diesters must hydrolyze much more slowly than phosphate triesters.* In fact, the saponification of phosphate diesters by the same P—O cleavage mechanism occurs at about 10^{-6} to 10^{-7} (that is, *one ten-millionth*) the rate of phosphate triester saponification. How do we account for this huge difference in rates?

In the P—O mechanism for saponification, the transition state for triester saponification contains one negative charge, whereas the transition state for diester saponification contains two negative charges:

triester saponification: one net negative charge in the transition state

diester saponification: two net negative charges in the transition state

(two δ− = one delocalized − charge) (25.33)

The repulsion between negative charges raises the energy of the transition state. This repulsion accounts for most of the difference in hydrolysis rates. To see that charge repulsion can have an effect of this magnitude, consider the first and second pK_a values of phosphoric acid:

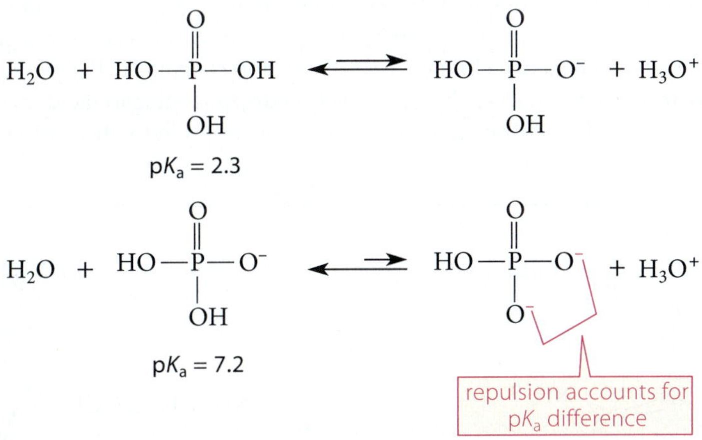

(25.34)

The introduction of one negative charge by the first ionization reduces the K_a for the second ionization by a factor of about 10^5. Charge repulsion has a similar effect on rate.

Now we are in a position to appreciate why phosphate diesters are ideally suited as the connections between the nucleic acid monomer units in DNA and RNA. First, these compounds are *polyanions* at pH 7.4. Anions do not readily cross the hydrophobic interior of membranes. This means that DNA and RNA molecules, once formed, do not "leak" from the cell. Second, because of the minuscule rate of phosphodiester hydrolysis, DNA and RNA do not undergo spontaneous hydrolysis. The integrity of the DNA and RNA polymers is crucial to the transmission of genetic information. It has been estimated that the spontaneous hydrolysis of a single phosphate ester bond in DNA under physiological conditions occurs with a half-life of greater than 100,000 years!

Half-life is a way of expressing a rate in terms of the lifetime of the reacting species rather than as a rate constant. The **half-life** is the time required for 50% of a compound to react. Therefore, in one half-life, 50% reacts; in two half-lives, 75%; in three half-lives, 87.5%; and so on. It takes about seven half-lives for 99% of a compound to react.

Chemical Biology Topic

Diester Hydrolysis in Biology; Nucleases Although phosphate diesters are very unreactive in solution, the hydrolysis reactions of specific phosphate diesters are important in biology and are catalyzed by enzymes. Let's examine one such reaction: the hydrolysis of DNA by *staphylococcal nuclease*, which we'll abbreviate as *StaphNuc*. A **nuclease** is an enzyme that breaks a phosphate diester bond in DNA or RNA. Staphylococcal nuclease is an enzyme secreted from the bacterium

Staphylococcus aureus, variations of which are responsible for "staph" infections. Other than the fact that StaphNuc is somehow involved in bacterial infection, its biological role is not well characterized. The following reaction is catalyzed by StaphNuc:

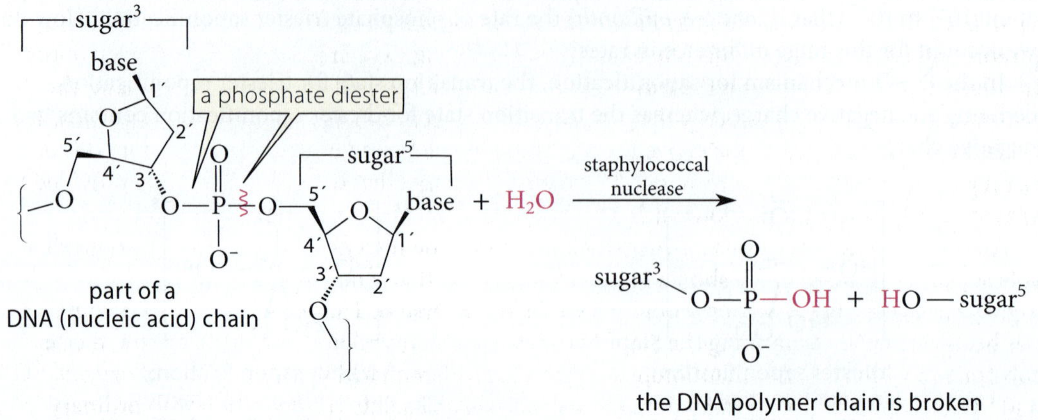

(25.35)

(Don't worry about the detailed structure of the DNA sugars; focus on the phosphate diester.)

Our goals in examining the catalytic mechanism of StaphNuc are (1) to see how enzymes overcome the intrinsic lack of reactivity of phosphate diesters and (2) to gain mechanistic insights that we might apply to the reactions of other phosphate esters. The mechanism of StaphNuc catalysis is shown in **Fig. 25.3**, along with the corresponding curved-arrow notation. Recall that the main reason for the low hydrolytic reactivity of phosphate diesters is their negative charge and the repulsion between negative charges in the transition state of the hydrolysis reaction. In StaphNuc, two arginine residues in their positively charged, conjugate-acid form—the form present at physiological pH—neutralize the two negative charges in the transition state for

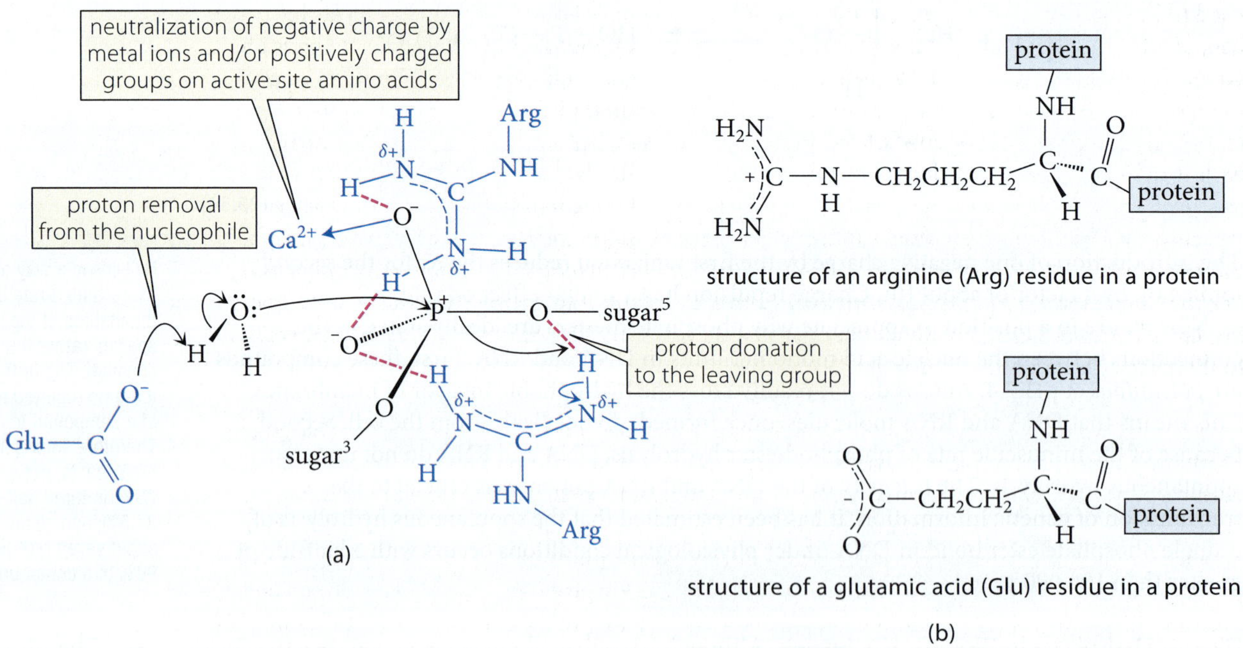

FIGURE 25.3 (a) The transition state for DNA (phosphate diester) cleavage catalyzed by the enzyme staphylococcal nuclease. The groups in blue are the metal ion and the catalytically important groups (derived from two arginine residues and a glutamic acid residue) in the active site of the enzyme. The dashed red lines indicate hydrogen bonds that are important for binding and/or catalysis. (b) The structures of the amino acid residues arginine and glutamic acid as they occur in proteins.

hydrolysis and, at the same time, form hydrogen bonds to the phosphate oxygens that contribute to phosphate binding to the active site. A calcium ion also interacts with an oxygen of the phosphate ester to neutralize charge. The carboxylate anion of a glutamic acid residue acts as a base to remove a proton from the water nucleophile as it reacts at the phosphorus. This water, then, is made more nucleophilic by its partial conversion into a stronger base, hydroxide. The protonated arginines also reduce the basicity of the leaving oxygen by donating hydrogen bonds. In other words, the basicity of the leaving group is reduced by partial protonation. Finally, this reaction is strongly accelerated by the *proximity effect* (Sec. 12.8), because all reactants, acids, and bases are pre-positioned in the active site and do not have to find each other by random diffusion as they would in an ordinary reaction. It has been estimated that the rate acceleration provided by enzyme catalysis in this reaction is a factor of more than 10^{17}.

An important aspect of this mechanism is the stereochemistry of the reaction. The opposite-side substitution mechanism shown in Fig. 25.3 predicts that if the phosphorus were an asymmetric center, the reaction would proceed with inversion. This stereochemical outcome has been verified by examining the StaphNuc-catalyzed conversion of a synthetic chiral diester substrate in which the two enantiotopic oxygens are differentiated by two isotopes of oxygen, ^{17}O and ^{18}O. Reaction of the optically pure diester of known absolute configuration with ordinary water (which contains ^{16}O) gives a chiral product:

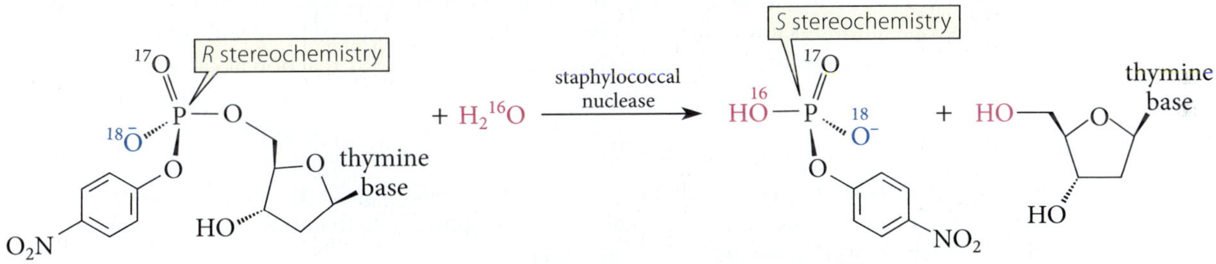

(25.36)

(Recall from the Cahn–Ingold–Prelog priority rules, discussed in Sec. 4.2B, that atomic mass is used to rank priorities for atoms of the same atomic number.) Some ingenious methods have been devised for determining the absolute configuration of phosphate diesters that are chiral by virtue of isotopic substitution. We won't concern ourselves with these methods; we only need to recognize that absolute configurations of isotopically chiral phosphate derivatives can be determined. Analysis of the absolute configuration shows whether the reaction has occurred with inversion, retention, or loss of configuration. The result is that *the reaction proceeds with inversion of stereochemistry*, as shown in Eq. 25.36. This stereochemistry, as well as the active-site structure shown in Fig. 25.3, fully supports the opposite-side substitution pathway for phosphate ester hydrolysis. In other words, *the stereochemistry of nucleophilic substitution at phosphorus is like the stereochemistry of the S_N2 reaction at carbon.* Chemists sometimes use the term *in-line displacement* to describe this stereochemistry at phosphorus. This term is completely equivalent to the term *opposite-side substitution.*

Hydrolysis of Phosphate Monoesters Phosphate monoesters contain *two* negative charges, and for that reason we might expect that these compounds would hydrolyze in base by P—O cleavage even more slowly than phosphate diesters. In fact, the nucleophilic substitution mechanism observed for phosphate triesters and phosphate diesters is so unfavorable that it does not occur; however, another P—O cleavage mechanism intervenes. In this mechanism, the cleavage is preceded by a very unfavorable acid–base reaction in which the proton is transferred from its normal position (at physiological pH) to the leaving group. Following this proton transfer, the monoester then undergoes a *unimolecular dissociation* to a very unstable species called **metaphosphate** (PO_3^-), which is the phosphorus analog of nitrate (NO_3^-).

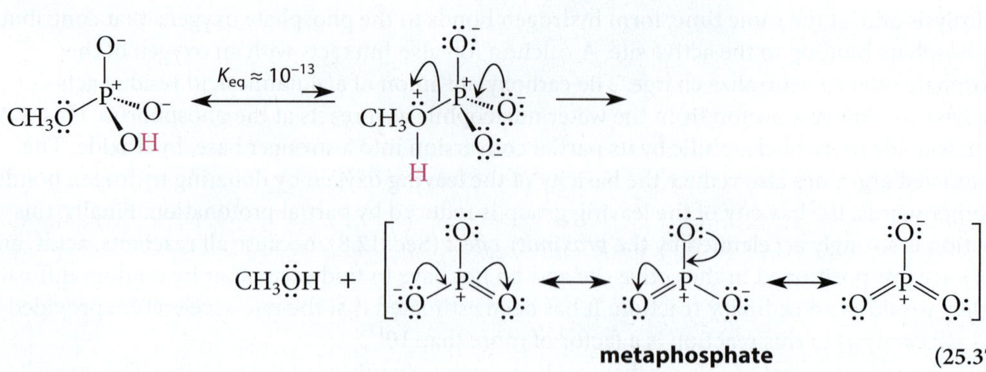

(25.37a)

This mechanism occurs because the leaving group, once protonated, is a weak base. Although this is the favored mechanism, a reaction that occurs by this mechanism is *very slow* because the initial proton transfer is very unfavorable and, consequently, the reactive species is present in very low concentration. This reaction is so slow that *phosphate monoesters are the least reactive phosphate esters.* An estimate of the hydrolysis rate at pH 7 and 37 °C suggests that the hydrolysis of an ordinary phosphate monoester di-anion has a half-life of 100 *billion years*—a time greater than the lifetime of our universe! Therefore, phosphate monoesters can survive indefinitely under physiological conditions until their hydrolysis reactions are catalyzed by specific enzymes.

This dissociative mechanism is somewhat analogous to the S_N1 reaction at carbon (Sec. 9.6), in which dissociation of a leaving group from carbon gives a carbocation. Metaphosphate, unlike its second-period analog nitrate, is a *very unstable* species. Its trigonal-planar geometry requires that the phosphorus is sp^2-hybridized. This means that oxygen 2p orbitals must overlap with phosphorus 3p orbitals, which have an additional node (**Fig. 25.4**). Therefore, resonance structures contribute much less to the stability of metaphosphate than they do to the stability of nitrate, in which overlap occurs among 2p orbitals.

Metaphosphate is instantly consumed by nucleophiles such as water to form phosphate:

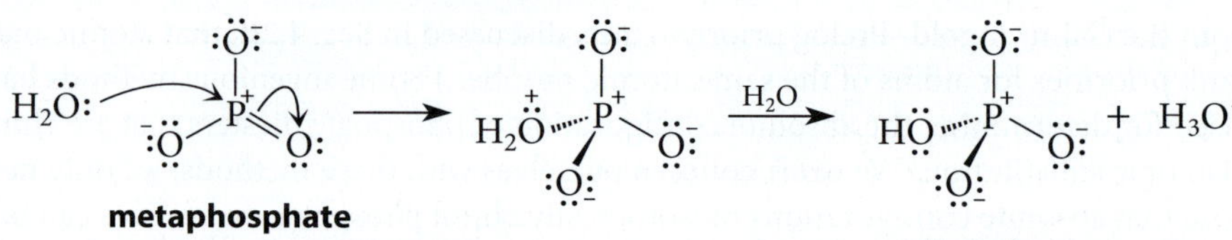

(25.37b)

Chemical Biology Topic

Monoester Hydrolysis in Biology; Phosphatases Enzymes that hydrolyze phosphate monoesters are called **phosphatases**. One example of a phosphatase is *fructose 1,6-bisphosphatase*, which we abbreviate F16BP. This enzyme catalyzes the hydrolysis of the 1-phosphate ester of fructose-1,6-bisphosphate to fructose-6-phosphate.

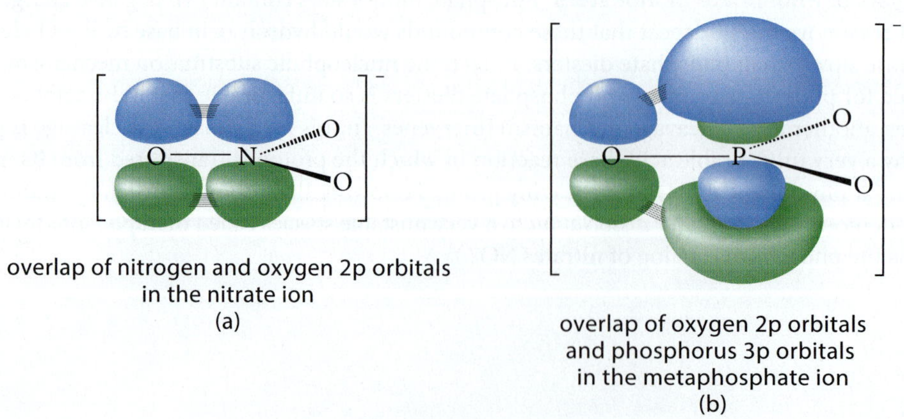

FIGURE 25.4 The overlap of p orbitals in (a) nitrate and (b) metaphosphate. In nitrate, the overlap between oxygen and nitrogen 2p orbitals, and therefore the corresponding resonance interaction, is strong. In metaphosphate, the overlap of the oxygen 2p orbitals with the phosphorus 3p orbitals, and therefore the resonance interaction, is relatively inefficient. The weakness of the resonance interaction accounts for the instability of metaphosphate.

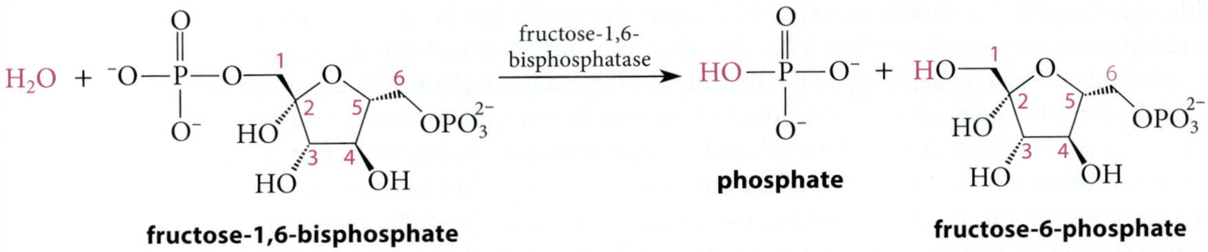

(25.38)

In this chapter we follow the traditional practice of referring to all conjugate-base forms of phosphoric acid (hydrogen phosphate, dihydrogen phosphate, and phosphate) simply as phosphate, because phosphate is a mixture of all three forms with the exact amount of each form depending on pH. At pH 7.4, hydrogen phosphate is typically the form present in greatest amount, as shown in Eq. 25.38.

The structure of this enzyme has been well characterized, and the transition state of the reaction hypothesized from the structure is shown in **Fig. 25.5**. The enzyme active site has evolved to overcome all the difficulties of phosphate monoester hydrolysis in solution. Recall that the major problem is the large amount of negative charge on the phosphate oxygens, which makes introduction of a third negative charge in the transition state very unfavorable. The active site contains possibly as many as *three* Mg^{2+} ions that are strongly associated with the phosphate oxygens of the ester that is hydrolyzed. This association in effect neutralizes the negative charge. The nucleophilic water molecule is deprotonated by a second water molecule, which in turn is deprotonated by a carboxylate ion within the active site. This deprotonation enhances the basicity of the nucleophilic water. The leaving group basicity is reduced by its interaction with a Mg^{2+} ion, which, for purposes of the mechanism, we can regard as a "large proton." Notice that the geometry at phosphorus is trigonal planar, which is consistent with either a trapped metaphosphate, a pentacovalent species, or a concerted substitution. The stereochemistry implied by this mechanism is *inversion*, as in the stereochemistry of phosphate diester hydrolysis. (This stereochemistry has been verified independently with an analog of fructose-1,6-bisphosphate; see Focused Problem 25.11.)

The catalysis of phosphate monoester hydrolysis by some phosphatases, unlike F16BP catalysis, shows overall *retention of stereochemistry.* For example, catalysis of phosphate ester hydrolysis by the well-known enzyme *alkaline phosphatase* from the bacterium *Escherichia coli* is in this class.

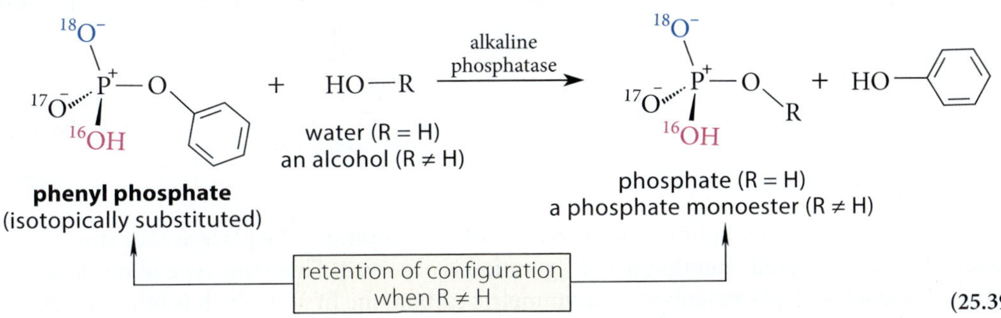

(25.39)

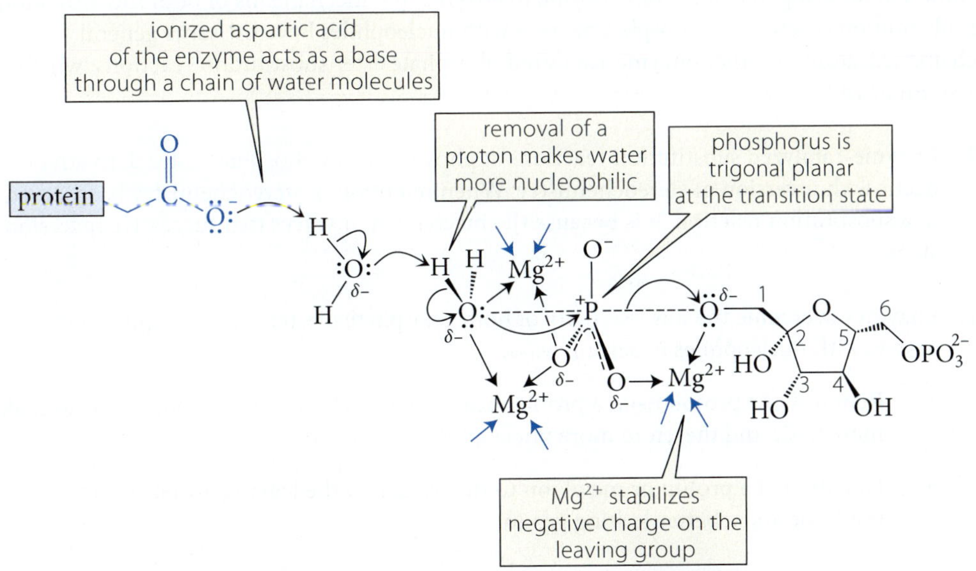

FIGURE 25.5 The transition state for hydrolysis of the 1-phosphate monoester in fructose-1,6-bisphosphate catalyzed by the enzyme fructose-1,6-bisphosphatase (F16BP). The carboxylate group in blue is part of the protein structure. The blue coordination arrows indicate that the magnesium ions are also coordinated to groups on the protein (typically carboxylate groups of glutamic and/or aspartic acid residues). Compare the catalytic features of this active site with those in Fig. 25.3.

(Although the proton is shown on a specific oxygen, the proton can be on any of the oxygens because proton exchange is very fast.) Because there are only three isotopes of oxygen, we can't study the stereochemistry of the hydrolysis reaction itself. (Notice that transfer to water—hydrolysis—gives an achiral product.) However, the enzyme is fairly promiscuous in the types of nucleophiles that can be used, and the conventional assumption is that using a nucleophile other than water will not affect the stereochemistry of the reaction. To study the stereochemistry of this reaction, chemists used an alcohol as a nucleophile, as shown in Eq. 25.39.

Phosphatases that give overall retention have, in all cases so far, been shown to involve a **phosphoenzyme intermediate**. That is, the *overall* substitution reaction occurs as a sequence of *two* substitution processes that involve a nucleophilic group in the enzyme active site. For example, the nucleophilic group in the active site of alkaline phosphatase is the alcohol group in the side chain of a serine residue:

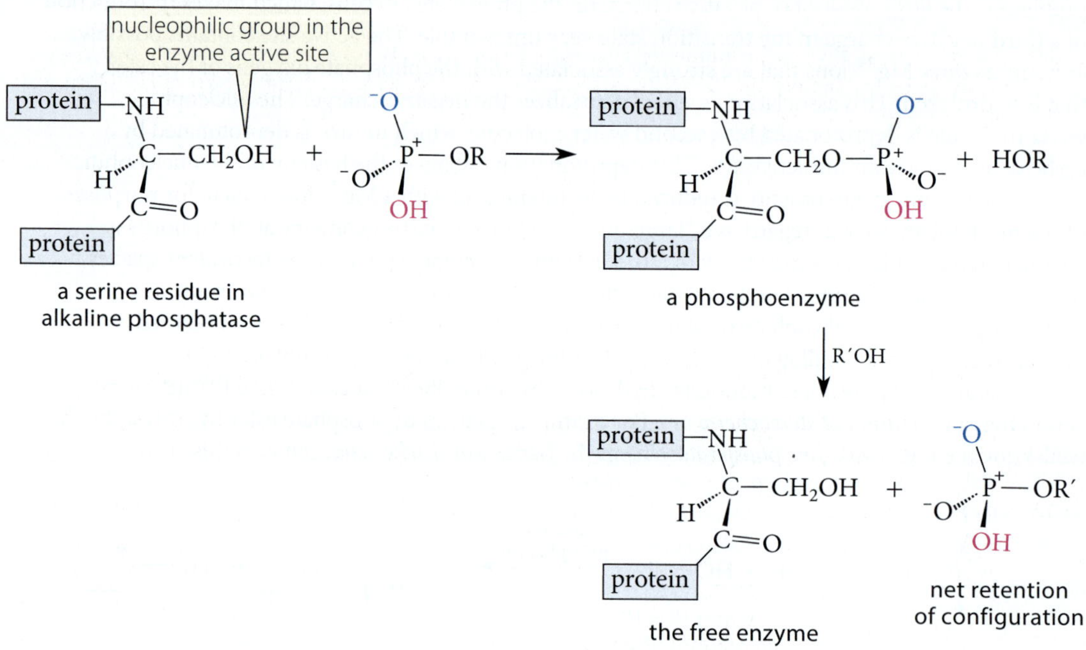

(25.40)

When two successive substitutions with inversion of configuration take place at the same asymmetric atom, overall retention of configuration is observed. This same type of result is obtained in double displacements at an asymmetric carbon, the first of which is intramolecular because of the proximity effect (Sec. 12.8D). In other words, each substitution at phosphorus occurs with inversion.

Although differing in detail from enzyme to enzyme, the mechanisms of both the formation and substitution reactions of phosphoenzymes with nucleophiles have the same general mechanistic features as other enzyme-catalyzed phosphate ester substitution reactions, which we now summarize:

We won't concern ourselves with why catalysis of phosphate monoester hydrolysis by some enzymes involves phosphoenzyme intermediates and why catalysis by others does not. We also won't discuss the biological purpose of phosphoenzyme mechanisms. We only need to recognize that when phosphoenzyme intermediates are involved, phosphate ester hydrolysis must involve two substitutions: formation of the phosphoenzyme followed by hydrolysis of the phosphoenzyme.

1. Enzyme-catalyzed substitution reactions at phosphorus in phosphate ester derivatives occur with inversion of stereochemistry. When retention of stereochemistry is observed in a substitution reaction, it is because the mechanism involves two successive inversion steps.

2. Enzymes overcome the low reactivity of both phosphate diesters and phosphate monoesters with nucleophiles in several ways:

   a. Removal of a proton from a protic nucleophile (such as an OH group) so as to make it more basic and therefore more nucleophilic

   b. Donation of a proton or metal ion to the oxygen of the leaving group so as to make it less basic and a better leaving group

c. Neutralization of charge on the phosphate oxygens either by hydrogen-bond donation from positively charged amino acid residues or by interactions with positively charged metal ions such as Mg^{2+} and Ca^{2+} within the active site, or both

d. The proximity effect: all reacting species—acids, bases, substrates, nucleophiles, and metal ions—are within the enzyme–substrate complex in ideal, or nearly ideal, arrangements for reaction

These principles will appear repeatedly as we study the enzyme-catalyzed reactions of other phosphate-containing molecules.

Focused Problems

25.11 The stereochemistry of substitutions in phosphate esters is sometimes studied with compounds in which sulfur is used in place of one oxygen. For example, the F16BP-catalyzed hydrolysis of fructose-1,6-bisphosphate in $H_2{}^{17}O$ was studied with the following sulfur-substituted analog:

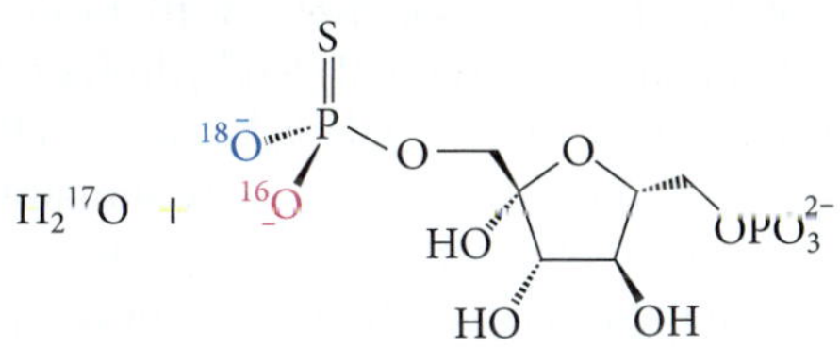

(a) Why is the substitution of sulfur for oxygen important when studying the stereochemistry of this reaction?

(b) Assuming that the substitution of sulfur for oxygen does not change the mechanism of the reaction, what is the product of this reaction and what is its stereochemistry?

25.12 The hydrolysis of phosphotyrosine esters in proteins is catalyzed by a family of enzymes called *protein phosphotyrosine phosphatases.*

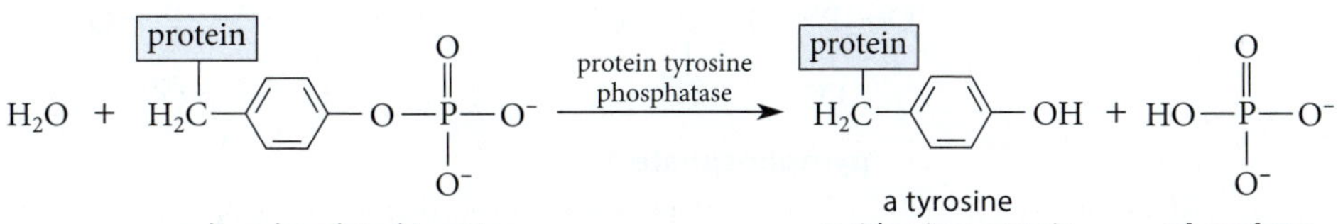

a phosphorylated tyrosine residue in a protein

a tyrosine residue in a protein

phosphate

These hydrolyses in many cases involve phosphoenzyme intermediates. The nucleophilic group of the phosphatase (the enzyme) is the thiol group of a cysteine residue in the active site.

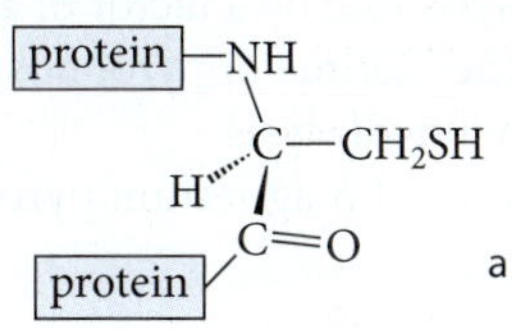

a cysteine residue in a protein

Using abbreviated structures like the ones in Eq. 25.40, write a curved-arrow mechanism for the hydrolysis that includes the structure of the phosphoenzyme intermediate. Assume that acids and bases are present in the enzyme active site as needed.

B. Hydrolysis of Phosphate Anhydrides

Chemical Biology Topic

Phosphate anhydrides, like carboxylic anhydrides, are much more reactive than the corresponding monoesters, but their hydrolysis rates are small enough that their spontaneous hydrolysis reactions do not compromise their survival under biological conditions. The reason for their low reactivity is the same as for the phosphate monoesters: the large amount of negative charge that causes repulsion when the phosphorus reacts with a nucleophile (see Display 25.33, Sec. 25.6A). For example, ATP at physiological pH bears four negative charges and ADP bears three.

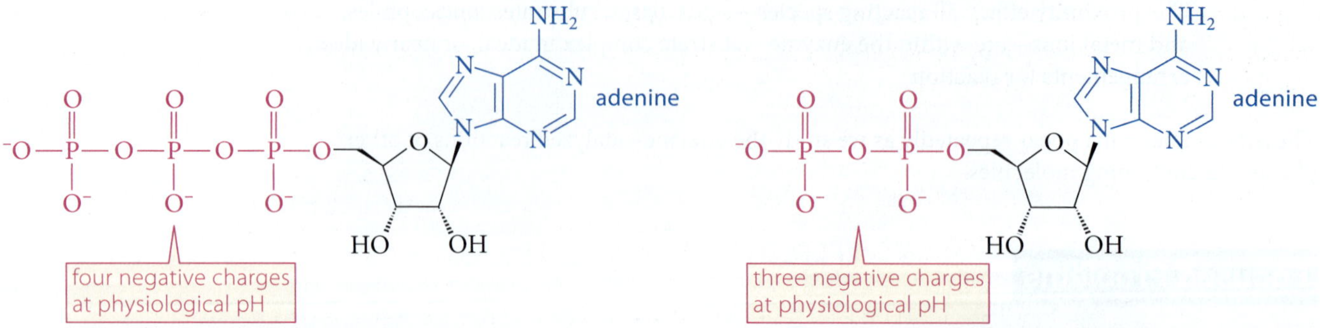

adenosine triphosphate (ATP) **adenosine triphosphate (ADP)** (25.41)

It has been estimated that the spontaneous hydrolysis of ATP at physiological pH and 25 °C has a half-life of 6.3 years. ATP forms a complex with divalent ions, such as Mg^{2+} (Sec. 25.2B); as we learned from studying the enzyme-catalyzed hydrolysis of phosphate esters, such ions accelerate hydrolysis, but only by a factor of about 10, because ATP forms a complex with only one Mg^{2+}. The half-lives of ADP and its Mg^{2+} complex are similar to those of ATP and its Mg^{2+} complex. These half-lives, although much smaller than phosphate ester half-lives, are nevertheless long enough that their spontaneous hydrolysis does not compromise their biological function.

The biologically important reactions of ATP involve its reactions with nucleophiles other than water. These reactions are considered in the discussion of the reactions of anhydrides with nucleophiles in Sec. 25.7A.

An anhydride that undergoes a biologically important hydrolysis reaction is the inorganic ion *pyrophosphate.*

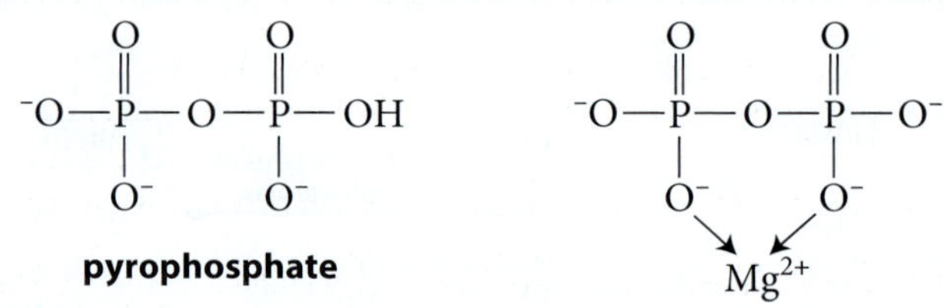

pyrophosphate

pyrophosphate
(complex with magnesium ion) (25.42)

Inorganic pyrophosphate bears three negative charges and is therefore resistant to reaction with nucleophiles. Its spontaneous hydrolysis at 37 °C has a half-life of about 35 years. Divalent ions such as Mg^{2+} increase the hydrolysis rate by a factor of about 4, but a half-life of 9 to 10 years is still very long in comparison to the lifetime of pyrophosphate in biological systems, in which it undergoes rapid enzyme-catalyzed hydrolysis.

The enzyme-catalyzed hydrolysis of magnesium pyrophosphate by water is a biologically significant process.

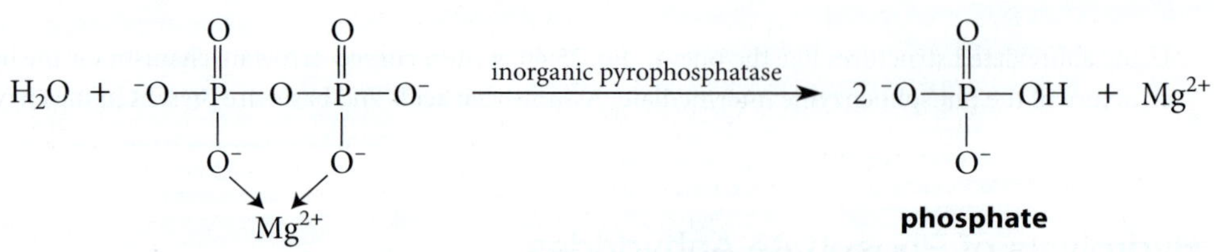

magnesium pyrophosphate **phosphate** (25.43)

In biology, this reaction is catalyzed by a family of enzymes called *inorganic pyrophosphatases*, which accelerate the pyrophosphate hydrolysis by a factor of about 10^{12}. Why is pyrophosphate hydrolysis important? Pyrophosphate serves as a leaving group in a number of biological substitution reactions. Two examples we have discussed are farnesylation (Eq. 11.48, Sec. 11.4E) and isoprenoid biosynthesis (Eq. 17.54, Sec. 17.6B). The enzyme-catalyzed hydrolysis of pyrophosphate, by removal of pyrophosphate as a reaction product, ensures that the substitution reactions are completely irreversible (Le Chatelier's principle).

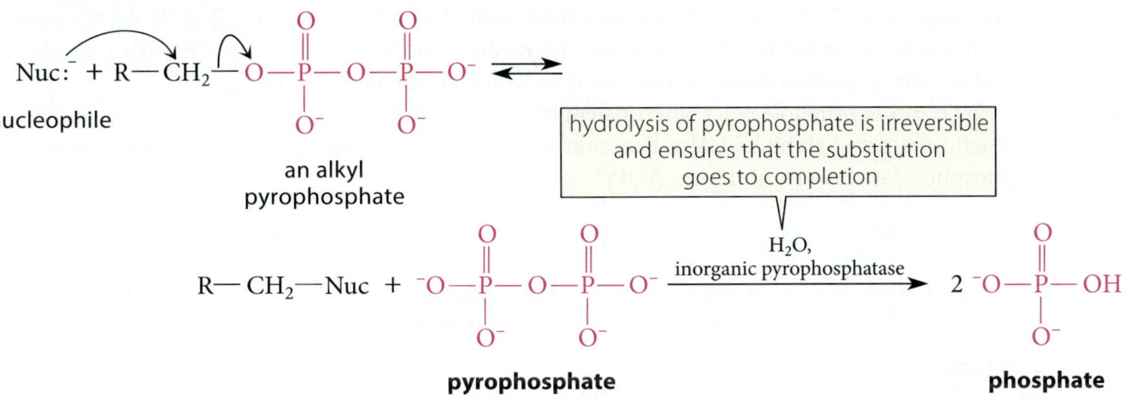

(25.44)

The active sites of pyrophosphatases show the same features we learned about in phosphate ester hydrolysis. You can examine the active site of one pyrophosphatase for yourself in Focused Problem 25.13 and in **Fig. 25.6**.

Focused Problem

25.13 The transition state for enzyme-catalyzed pyrophosphate hydrolysis, deduced from the structure of the enzyme inorganic pyrophosphatase, is shown in Fig. 25.6.

(a) Indicate the catalytic role of each numbered feature—that is, how each feature enhances catalysis.

(b) If the stereochemistry of the reaction could be determined by isotopic and sulfur substitution of the oxygens, would the result be retention or inversion of configuration at phosphorus?

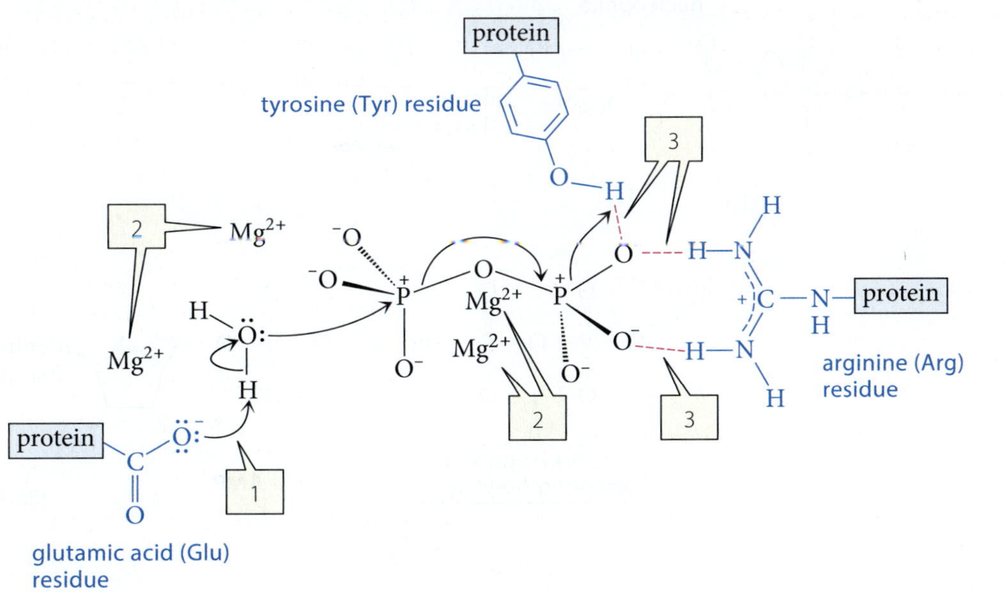

FIGURE 25.6 The active site for pyrophosphate hydrolysis catalyzed by yeast inorganic pyrophosphatase. (See Focused Problem 25.13.) The groups in blue are from amino acid residues (identified in parentheses) of the protein in the active site. The numbers refer to aspects of catalysis to be identified when working the problem.

Chemical Biology Topic

25.7 REACTIONS OF PHOSPHATE ANHYDRIDES WITH OTHER NUCLEOPHILES

A. Reactions of ATP with Nucleophiles

Phosphate anhydrides, like carboxylic acid anhydrides, react not only with water, but also with other nucleophiles, such as alcohols. Just as the hydrolysis reactions of phosphate anhydrides are very slow at pH 7, their spontaneous reactions with alcohols are also very slow. However, many such reactions are catalyzed by enzymes. Adenosine triphosphate (ATP) reacts with a number of biologically important nucleophiles, many of which are alcohols or phenols.

ATP can in principle react with nucleophiles at any of the three phosphorus atoms, and all reactions are known biologically. Reaction at the terminal phosphorus (γ-phosphate) gives a phosphorylated nucleophile and ADP:

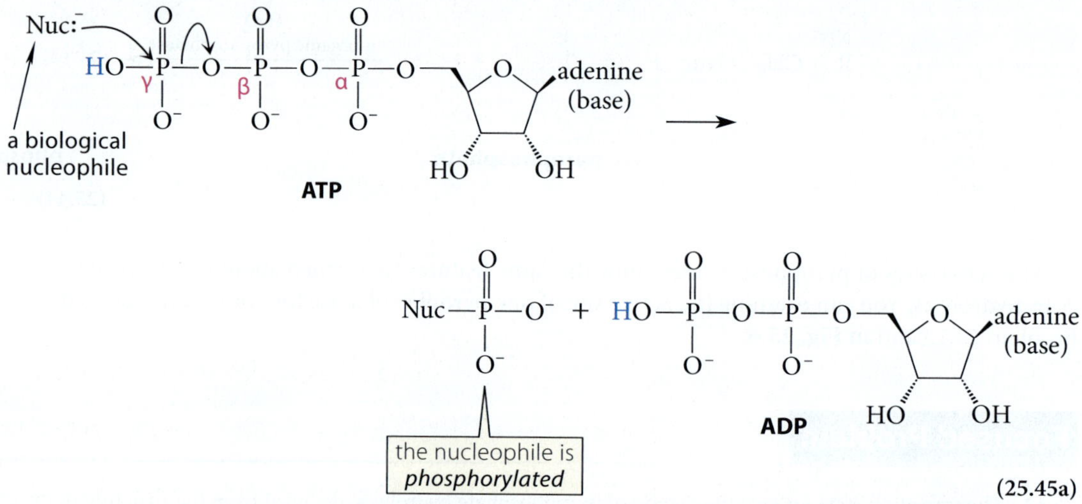

(25.45a)

In this reaction, the proton transfer steps are not shown explicitly. Moreover, the association of Mg^{2+} with ATP and ADP, which is important biologically, is not shown but is understood. Enzymes that catalyze this type of reaction are called **kinases**. (Contrast this reaction with the hydrolysis of phosphate monoesters, catalyzed by *phosphatases*, as in Eq. 25.38, Sec. 25.6A.)

Reaction at the middle phosphorus (β-phosphate) gives a nucleophile pyrophosphate and AMP:

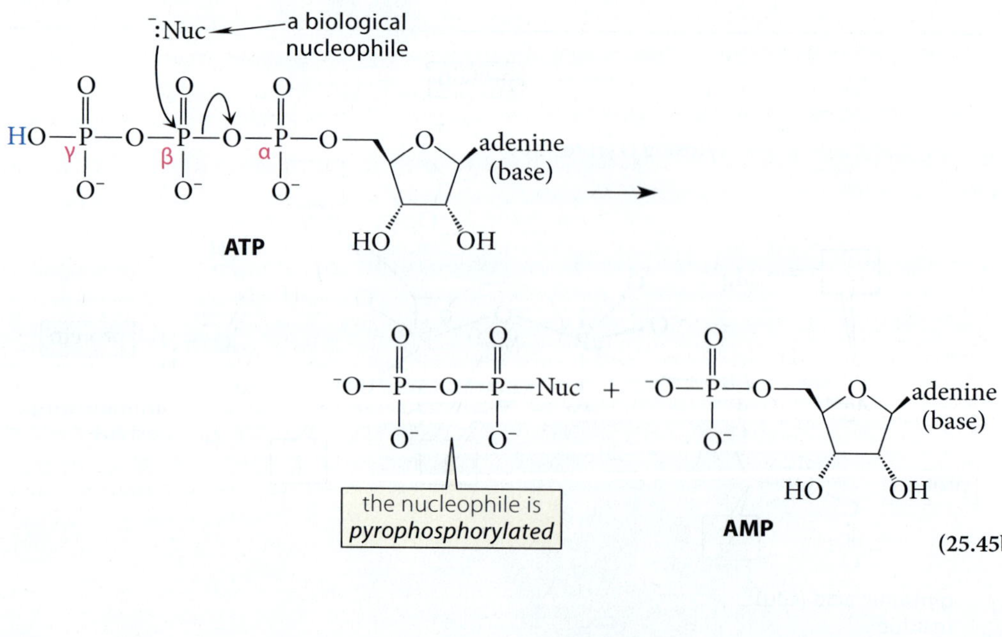

(25.45b)

In principle, there are two possible phosphate leaving groups in this reaction, but only the loss of AMP is biologically important. This leaving group is selectively activated by the catalyzing enzyme. Reaction at the β-phosphate is the least common of the three modes of reaction of ATP in biology.

Reaction at the phosphorus closest to the sugar (the α-phosphorus) could also in principle result in the loss of two possible leaving groups; but, of the two, the pyrophosphate leaving group is a much weaker base and a better leaving group. Loss of pyrophosphate results in *adenylation* of the nucleophile and the formation of pyrophosphate as a by-product. The pyrophosphate is hydrolyzed to phosphate in a reaction catalyzed by inorganic pyrophosphatase (Sec. 25.6B).

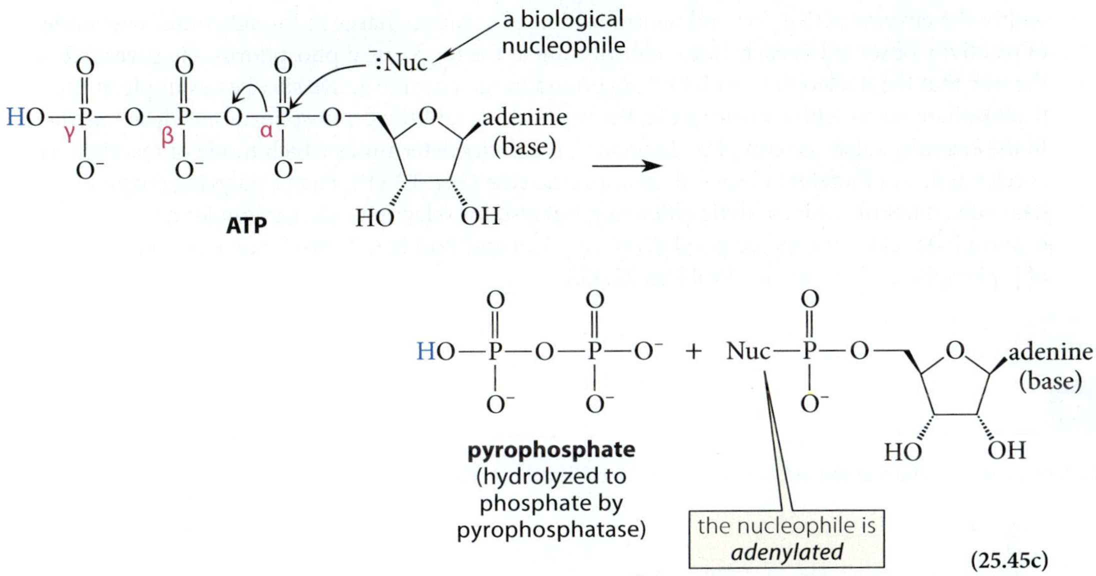

(25.45c)

Of the three pathways just shown, the most widely occurring reaction of ATP with nucleophiles in biology is the reaction of a nucleophile at the γ-phosphate to give a phosphorylated nucleophile and ADP (Eq. 25.45a). An example of this reaction is the phosphorylation of glucose in the first step of *glycolysis*, the reaction pathway by which glucose is broken down into smaller fragments and chemical energy is produced. The formation of glucose-6-phosphate is catalyzed by an enzyme called *hexokinase.*

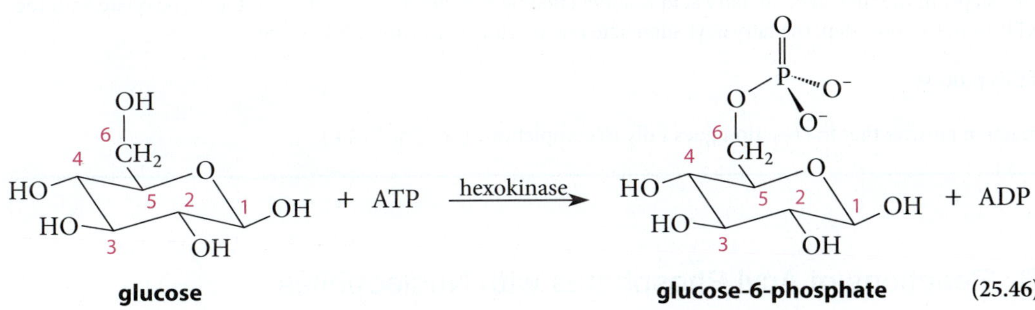

(25.46)

This enzyme-catalyzed reaction follows the now-familiar pattern for substitutions at phosphorus. It occurs with *inversion of configuration* at phosphorus, which was demonstrated with isotopically chiral ATP:

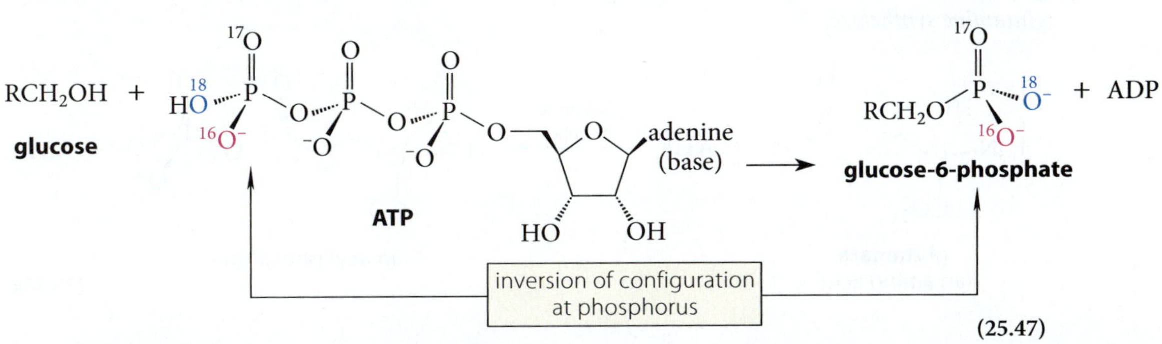

(25.47)

The active site of hexokinase contains features that we might expect from our consideration of phosphate-processing enzymes in Sec. 25.6A—namely, a base that removes a proton from the nucleophilic hydroxy group of glucose, a Mg^{2+} ion that bridges between the oxygens on the β- and γ-phosphate groups, and several groups that donate hydrogen bonds to the oxygens of the ADP leaving group.

Although we won't illustrate in the text the reactions of nucleophiles at the β- and γ-phosphates, these reactions fit the general pattern of nucleophilic substitutions at phosphorus—namely, opposite-side substitution, activation of the nucleophile by base catalysis within the enzyme active site, activation of the leaving group by proton donation or metal-ion activation within the enzyme active site, and neutralization of negative charge in the substrate. The mode of reactivity observed in each case—substitution at the α-, β-, or γ-phosphorus—is governed by the way that the nucleophile and ATP are bound in the enzyme active site. For example, if the β-phosphate reacts with a nucleophile, the nucleophile and the β-phosphorus are close together in the enzyme–substrate complex. Therefore, *proximity* determines which mode of reactivity is accelerated, and therefore observed, in a specific case (Sec. 12.8E). Each catalyzing enzyme has evolved not only for catalytic efficiency, but also for selectivity. (For examples of α-phosphate reactivity, see Focused Problem 25.14 and Problem 25.36; for an example of β-phosphate reactivity, see Problem 25.35.)

Focused Problem

25.14 The conversion of a fatty acid into a fatty acyl-CoA is one step in the utilization of fats:

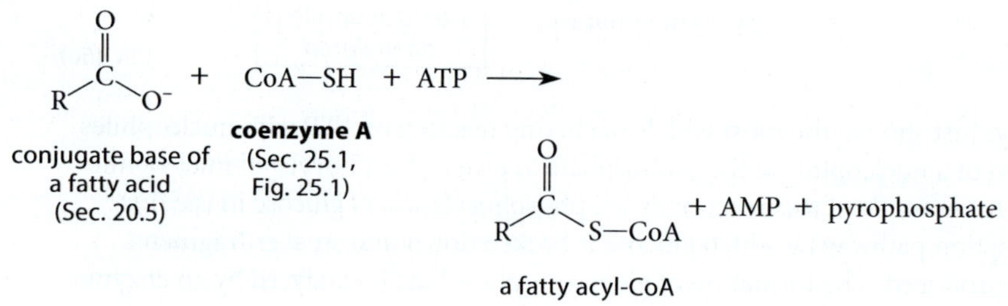

This process occurs in two steps. In the first step, the fatty acid is adenylated by a nucleophilic reaction of the carboxylate with the α-phosphate group of ATP. In the second step, the fatty acyl adenylate reacts with coenzyme A (CoA—SH).

(a) Show each step of this process.

(b) What additional reaction ensures that the reaction goes fully to completion? (See Eq. 25.44.)

B. Reactions of Acyl Phosphates with Nucleophiles

The reactions of a few mixed carboxylic phosphate anhydrides (acyl phosphates; Display 25.15, Sec. 25.2B) occur biologically. In one example, the side-chain carboxylic acid group of the amino acid glutamate is phosphorylated by ATP. The resulting acyl phosphate reacts with ammonia to give the amide form of glutamate, called *glutamine*. Both reactions are catalyzed by the enzyme *glutamine synthetase*.

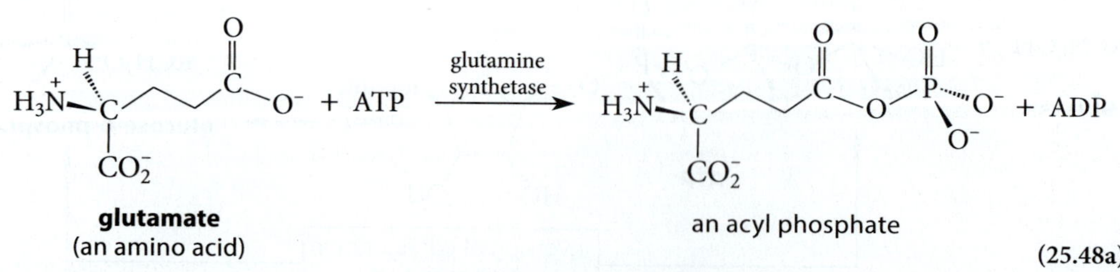

(25.48a)

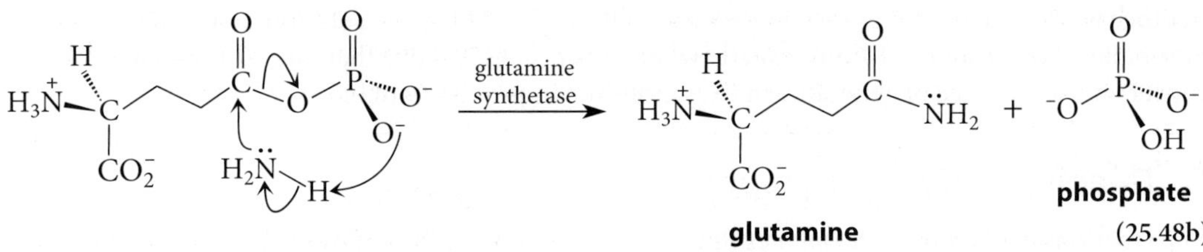

(25.48b)

Although this reaction is enzyme-catalyzed, the preferential reaction of the nucleophile ammonia at the carbonyl carbon rather than the phosphorus is consistent with the greater reactivity of carboxylic anhydrides.

Another interesting occurrence of an acyl phosphate is in the carboxylation of biotin to form *N*-carboxybiotin. (See Display 22.82, Sec. 22.7 for the structure of biotin and *N*-carboxybiotin.) Recall that biotin is nature's "carbon dioxide carrier," and that the purpose of *N*-carboxybiotin is to deliver a carboxy group to the α-carbon of acetyl-CoA to form malonyl-CoA in fatty-acid biosynthesis (Eq. 22.83, Sec. 22.7). The carboxylation of biotin is catalyzed by the enzyme *biotin carboxylase*. The source of the carboxyl group in *N*-carboxybiotin is *not* cellular CO_2, which is present in far too low a concentration to be useful. Rather the source is the bicarbonate ion, which is present in much higher cellular concentration than CO_2. Bicarbonate itself is not reactive enough to carboxylate biotin; so, bicarbonate is activated by an enzyme-catalyzed reaction with ATP to form carboxyphosphate, an acyl phosphate. This reaction is catalyzed by the enzyme biotin carboxylase.

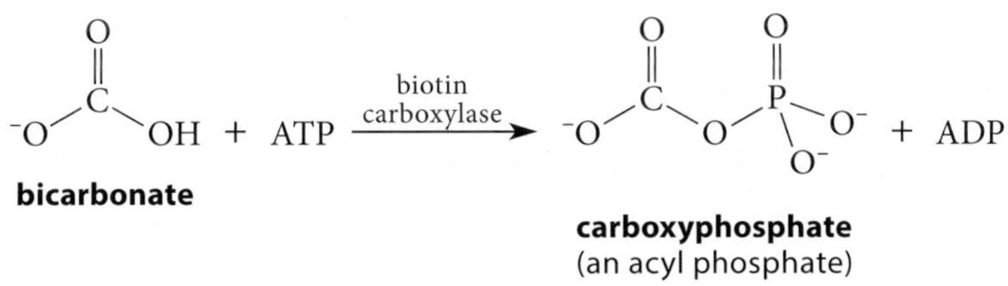

(25.49a)

The phosphate group of the carboxyphosphate provides an ionic group that is used to bind this molecule to the enzyme. This acyl phosphate is decomposed in a β-elimination reaction to CO_2. This reaction occurs very close to the biotin that is bound in the active site of the same enzyme.

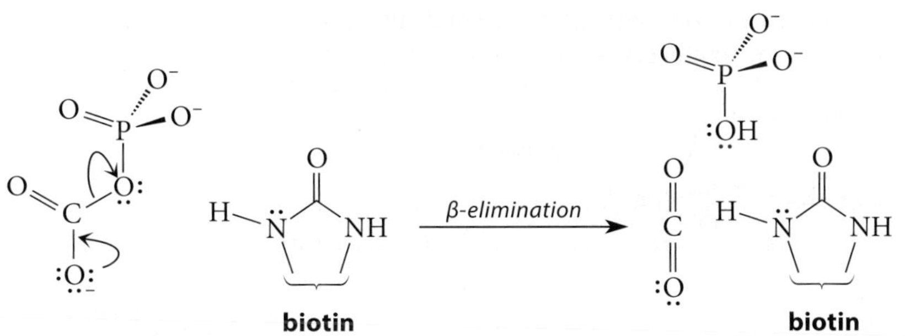

(25.49b)

The eliminated phosphate serves as a base to abstract a proton from the nearby biotin and convert it into its conjugate-base anion. Because of its proximity to the biotin anion, CO_2 reacts in a carbonyl addition to give *N*-carboxybiotin before it can escape from the active site into solution.

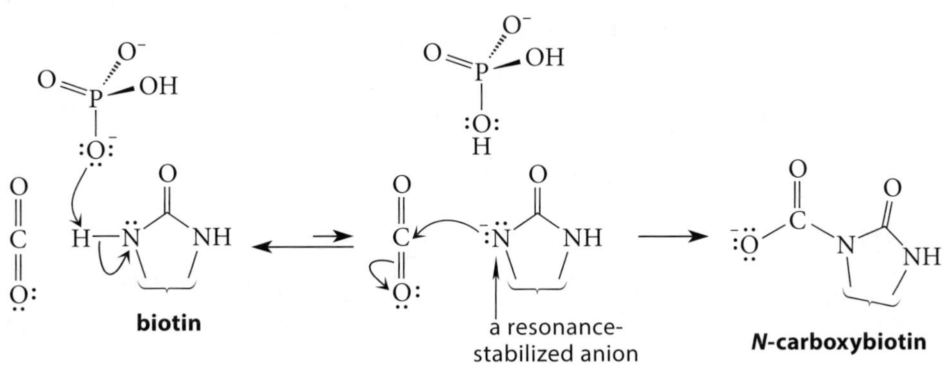

(25.49c)

In other words, the source of the carboxy group is not *free* CO_2 from solution, but rather a CO_2 molecule that is produced locally, exactly where it is needed. Notice that one ATP ⟶ ADP conversion is the price of generating a CO_2 molecule in proximity to biotin.

C. Reactions of Alkyl Pyrophosphates at Carbon

The reactions of alkyl pyrophosphates illustrate a completely different mode of reactivity of phosphate anhydrides. We've been considering the reactions of nucleophiles at phosphorus. However, nucleophiles can also react with alkyl pyrophosphates at *carbon* in S_N2 or S_N1 reactions to expel pyrophosphate as a leaving group (Nuc:$^-$ = a nucleophile):

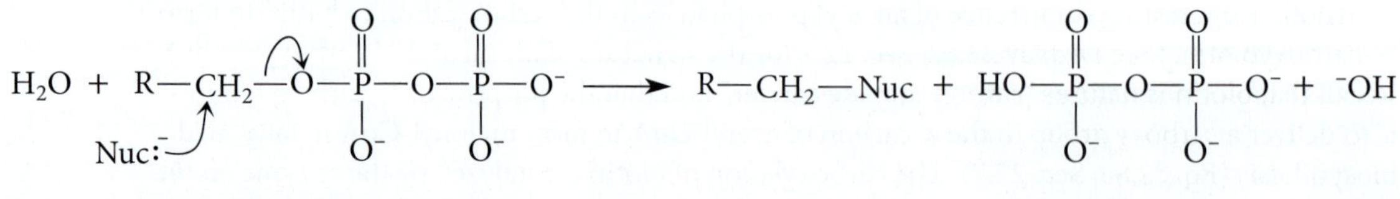

(25.50)

We might expect a nucleophilic reaction at carbon to be particularly favorable when the alkyl group has high S_N2 or S_N1 reactivity. In the two examples we considered in previous chapters, the alkyl groups are both allylic (farnesyl pyrophosphate, Sec. 11.4E; and dimethylallyl pyrophosphate, Sec. 17.6B). Recall (Secs. 17.1 and 17.4) that allylic systems are particularly reactive in S_N1 and S_N2 reactions. This reaction is conceptually similar to the S_N2 and S_N1 reactions of sulfonate esters (Sec. 11.4A). It appears that nature has evolved pyrophosphate as the preferred leaving group for nucleophilic reactions at carbon, and the reason is that these reactions can then be driven to completion by pyrophosphatase-catalyzed hydrolysis of pyrophosphate (Eq. 25.43, Sec. 25.6B).

D. Reactions of Other Phosphate Anhydrides at Carbon

Diphosphates other than pyrophosphate serve as leaving groups in the biosynthesis of disaccharides and polysaccharides, and in the biosynthesis of *glycoproteins*—biomolecules in which sugars are linked to proteins. In the biosynthesis of the disaccharide lactose, for example, UDP-galactose, a conjugate of the sugar galactose in its pyranose form and the nucleotide diphosphate UDP, reacts with glucose, also in its pyranose form:

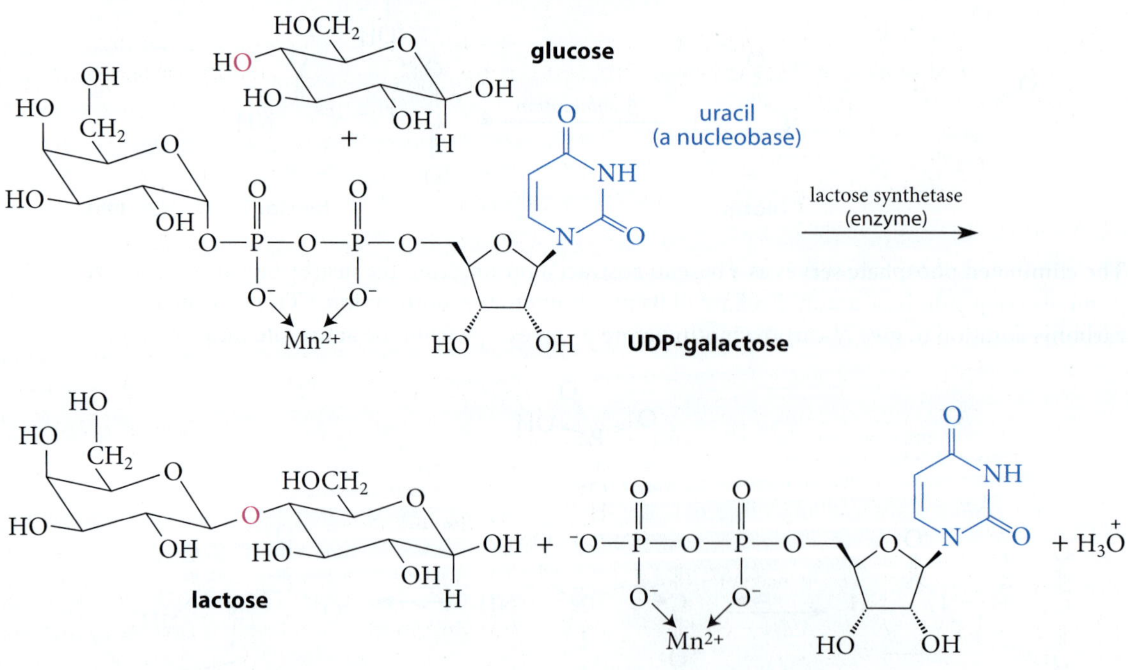

(25.51)

This is believed to be an S_N1-like reaction involving a trapped carbocation intermediate. The key features of enzyme catalysis are ones that we have encountered in previous examples—namely, a metal ion (in this case Mn^{2+}) that neutralizes charge on the diphosphate and makes it a better leaving group; a basic group on the enzyme, a carboxylate group of an ionized aspartic acid residue (*blue*) that removes a proton from the nucleophilic oxygen (*red*) of glucose, thereby making this oxygen more basic and more nucleophilic; and a binding site that holds all reactants in proximity, thus making all the reactions effectively intramolecular.

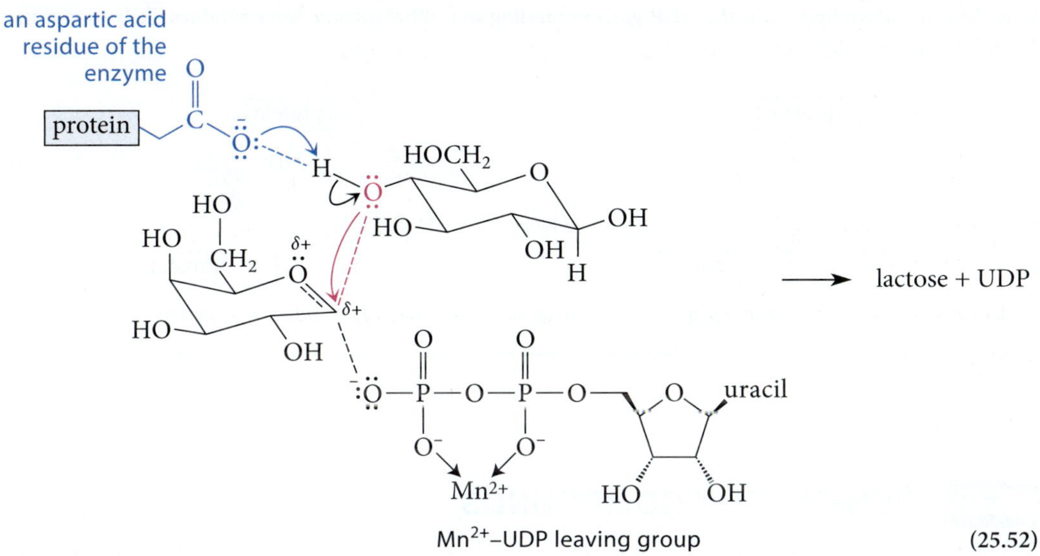

(25.52)

This reaction shows inversion of stereochemical configuration at the anomeric carbon of galactose, but other reactions that form disaccharides with retained (α) configuration at the anomeric carbon are well known. These reactions require either a double-displacement mechanism or a carbocation mechanism in which the face of the carbocation opposite to the leaving group is shielded by the active site (Focused Problem 25.16).

The glycosylation of proteins is discussed in Sec. 27.8E.

Focused Problems

25.15 In the biosynthesis of *S*-adenosylmethionine (SAM; Sec. 12.7B), the amino acid methionine undergoes an enzyme-catalyzed reaction with ATP to give SAM and triphosphate. (This is one of only two known biochemical reactions in which triphosphate serves as a leaving group.)

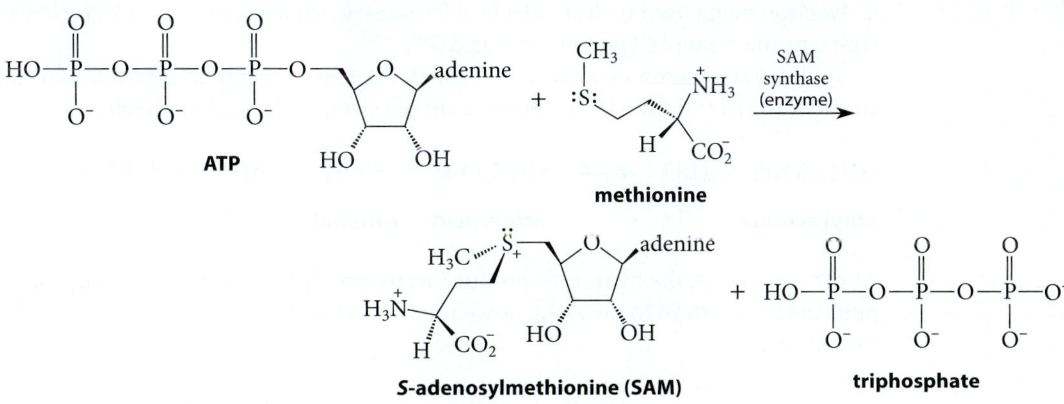

(a) Give a curved-arrow mechanism for this reaction, assuming it is an S_N2 process.

(b) The enzyme active site of the catalyzing enzyme, SAM synthase, contains two Mg^{2+} ions. What role might these ions have in the catalytic process?

25.16 *Sucrose synthase* is a plant enzyme that catalyzes the biosynthesis of the disaccharide sucrose (structure in Display 24.72, Sec. 24.11A) from UDP-glucose and β-D-fructofuranose.

(a) Draw the structure of UDP-glucose, assuming it has the same stereochemistry at the anomeric carbon as UDP-galactose.

(b) Draw out the reaction for sucrose biosynthesis. Does this reaction occur with retention or inversion of stereochemistry at the anomeric carbon of UDP-glucose?

(c) Sucrose synthase contains no metal cations in its active site. However, it does contain arginine and lysine residues in their cationic, conjugate-acid form (shown here) near the UDP-glucose binding site. What role are these residues likely to play in the synthase mechanism? (*Hint:* See Fig. 25.3, Sec. 25.6A.)

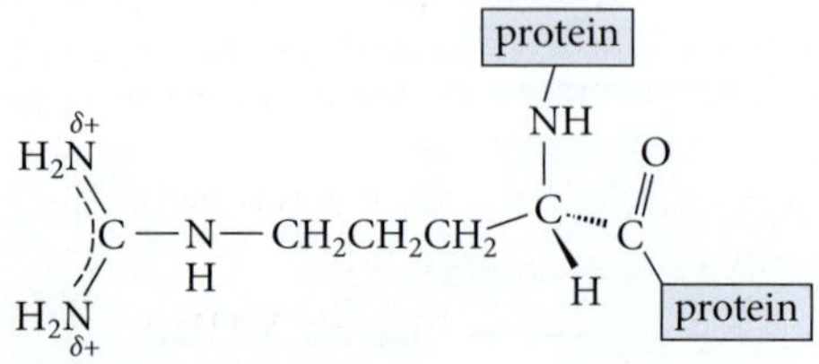

structure of an arginine (Arg) residue in a protein

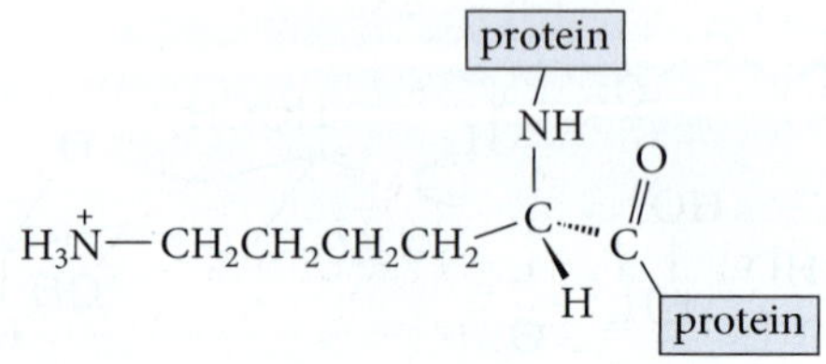

structure of a lysine (Lys) residue in a protein

Chemical Biology Topic

25.8 HIGH-ENERGY COMPOUNDS

A. The Concept of a High-Energy Compound

One of the important aspects of ATP and acetyl-CoA in biology is their utilization as a source of chemical energy. To understand this idea, we must first understand the convention that is commonly used for expressing free energy in biochemical systems. Recall (Sec. 3.5) that the definition of a standard free-energy change for a reaction, $\Delta G°$, is the free energy required to convert an ideal 1 M solution of reactants into an ideal 1 M solution of products at 298 K (25 °C). If protons are involved in the reaction, their concentration is also taken as 1 M (that is, pH 0). However, in biology, the standard state of the proton is taken as 10^{-7} M (that is, pH 7), because this is the whole-number pH closest to the physiological pH of 7.4. If the equilibrium for a reaction does not involve the uptake or release of protons, or the acid dissociation of a reactant or product, the $\Delta G°$ for a reaction in the two conventions is the same. However, as we now appreciate, many important biomolecules have dissociable groups with pK_a values within a few units of 7. In these cases, the $\Delta G°$ values in the two conventions are different. To be sure that the convention being used is clear, standard free-energy changes in the pH 7 convention are labeled with a prime symbol (′) and notated as $\mathbf{\Delta G°'}$.

To illustrate a common situation in which the two conventions give different $\Delta G°$ values, consider the $\Delta G°$ for the hydrolysis of a simple ester, such as ethyl acetate.

$$\underset{\textbf{ethyl acetate}}{CH_3CO_2Et} + H_2O \rightleftharpoons \underset{\textbf{acetic acid}}{CH_3CO_2H} + \underset{\textbf{ethanol}}{EtOH} \qquad \Delta G° = -6.95\ \text{kJ mol}^{-1}\ (-1.66\ \text{kcal mol}^{-1}) \tag{25.53a}$$

At pH 7, however, the hydrolysis product acetic acid ($pK_a = 4.76$) is ionized, and this ionization pulls the equilibrium to the right, as we learned in the discussion of ester saponification (Sec. 21.7A):

$$CH_3CO_2Et + H_2O \rightleftharpoons EtOH + \underset{pK_a = 4.76}{CH_3CO_2H} \xrightleftharpoons[]{H_2O,\ pH = 7} CH_3CO_2^- + H_3O^+ \tag{25.53b}$$

At pH 7, therefore, the free energy of hydrolysis is more favorable than it is at pH 0—that is, $\Delta G^{\circ\prime}$ is more negative than ΔG°. The value of $\Delta G^{\circ\prime}$ can be calculated from ΔG° by the general formula for free energy:

$$\Delta G^{\circ\prime} = \Delta G^{\circ} + 2.3RT \log \frac{(r_{\mathrm{CH_3CO_2H}})(r_{\mathrm{EtOH}})}{(r_{\mathrm{CH_3CO_2Et}})} \tag{25.54a}$$

where the r values are the ratios of concentrations of the various species in the equilibrium at pH 7 to their concentrations under standard conditions (pH 0). [Water is omitted because it has the unit activity standard state (Sec. 3.6A) in both conventions.] The only species in this reaction affected significantly by pH is acetic acid. Therefore, Eq. 25.54a becomes

$$\Delta G^{\circ\prime} = \Delta G^{\circ} + 2.3RT \log (r_{\mathrm{CH_3CO_2H}}) \tag{25.54b}$$

The ratio of CH_3CO_2H at pH 7 to its concentration at pH 0 can be calculated from the fractional dissociation formulas at each pH (Eqs. 3.67a and b, Sec. 3.6E). The fraction of un-ionized acetic acid at pH 0 is 1.0, and the fraction of un-ionized acetic acid at pH 7 is 0.0057. Therefore, $r = 0.0057$. Substituting this into Eq. 25.53b gives

$$\begin{aligned} \Delta G^{\circ\prime} &= \Delta G^{\circ} + 2.3RT \log (0.0057) \\ &= -6.95 + 5.71(-2.24) \\ &= -19.8 \text{ kJ mol}^{-1} \ (-4.73 \text{ kcal mol}^{-1}) \end{aligned} \tag{25.54c}$$

As Le Chatelier's principle suggests, the standard free energy in the pH 7 convention is indeed more favorable (more negative).

Having established the pH 7 convention for free energies, we're ready to consider high-energy compounds. Let's begin with a simple laboratory example. We showed that we can prepare ethyl acetate by letting acetic acid come to equilibrium with an excess of ethanol in the presence of an acid catalyst (Sec. 20.8A). However, we also showed that ethyl acetate can be prepared more rapidly by the reaction of ethanol with acetic anhydride or acetyl chloride (Sec. 21.8A). When we presented these reactions, our focus was on the *reactivity* of acetic anhydride and acetyl chloride. However, another reason that acetic anhydride or acetyl chloride is used is that *both of their esterification reactions are essentially irreversible because the equilibrium constants for these reactions are very large.* A large equilibrium constant means that the standard free-energy change for the reaction is very favorable (negative). The source of this large, negative standard free-energy change is *both* the high standard free energy of the reactants (acetic anhydride or acetyl chloride) and the low standard free energy of reaction products (an ester and either acetate ion or chloride ion). To take the reaction of an anhydride as an example, we showed in Eqs. 21.49 and 21.50 (Sec. 21.7E) that the resonance stabilization of acetic anhydride is less effective than the resonance stabilization of both an ester and the acetate ion. The gain in resonance stabilization releases significant energy (**Fig. 25.7**). Therefore, the relative free energies of *both* reactants and products contribute to the large negative $\Delta G^{\circ\prime}$ for ester formation. Biologists, however, tend to focus on the high-energy side of the equilibrium, and they would say that acetic anhydride and acetyl chloride are high-energy compounds.

In biology, a **high-energy compound** is one that has a large negative $\Delta G^{\circ\prime}$ for hydrolysis. (Typically, this concept is applied to carboxylic and phosphoric acid derivatives, all of which undergo hydrolysis to the corresponding acids.) *A compound is considered to be a high-energy compound if its $\Delta G^{\circ\prime}$ for hydrolysis is more negative than about -30 kJ mol^{-1} (-7.2 kcal mol^{-1}).* Let's see how this idea can be applied to the esterification of acetic acid. The $\Delta G^{\circ\prime}$ for the esterification of acetic acid is the negative of the $\Delta G^{\circ\prime}$ for ester hydrolysis [+19.8 kJ mol^{-1} (+4.73 kcal mol^{-1}); Eq. 25.53c]. Remember, this $\Delta G^{\circ\prime}$ value is actually the standard free energy for the esterification of acetic acid as it exists at pH 7—in its conjugate-base acetate form—and any protons present are at a concentration of 10^{-7} M (pH 7). The $\Delta G^{\circ\prime}$ for acetic anhydride hydrolysis has been experimentally determined to be -91.2 kJ mol^{-1} (-21.8 kcal mol^{-1}). This value puts acetic anhydride into the category of a high-energy compound. Remember, now, that *we can calculate the energy change for any reaction as the sum of the energy changes for other reactions* (Hess's law).

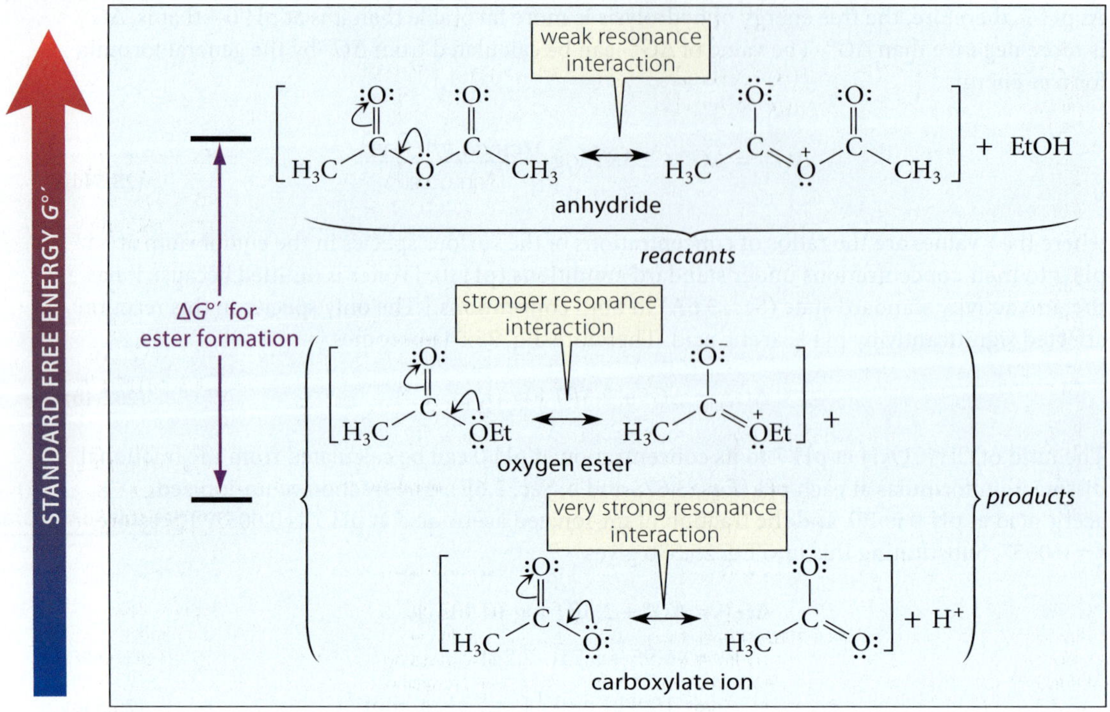

FIGURE 25.7 The reaction of ethanol with acetic anhydride releases considerable free energy because the resonance stabilization of both products is more effective than the resonance stabilization of the anhydride.

(We used this idea with heats of formation in Sec. 4.5A.) If we add the $\Delta G^{\circ\prime}$ for the esterification of acetic acid to the $\Delta G^{\circ\prime}$ for the hydrolysis of acetic anhydride, we can cancel one acetate ion, one water molecule, and one proton, and the result is the $\Delta G^{\circ\prime}$ for the formation of ethyl acetate from acetic anhydride.

$$H^+ + CH_3CO_2^- + EtOH \rightleftharpoons CH_3CO_2Et + H_2O \qquad \Delta G^{\circ\prime} = +19.8\ \text{kJ mol}^{-1}\ (+4.7\ \text{kcal mol}^{-1})$$

$$CH_3\overset{O}{\overset{\|}{C}}O\overset{O}{\overset{\|}{C}}CH_3 + H_2O \rightleftharpoons 2\,CH_3CO_2^- + 2\,H^+ \qquad \Delta G^{\circ\prime} = -91.2\ \text{kJ mol}^{-1}\ (-21.8\ \text{kcal mol}^{-1})$$

Sum:

$$CH_3\overset{O}{\overset{\|}{C}}O\overset{O}{\overset{\|}{C}}CH_3 + EtOH \rightleftharpoons CH_3CO_2Et + CH_3CO_2^- + H^+ \qquad \Delta G^{\circ\prime} = -71.4\ \text{kJ mol}^{-1}\ (-17.1\ \text{kcal mol}^{-1}) \tag{25.55}$$

The $\Delta G^{\circ\prime}$ for the formation of ethyl acetate from acetic acid at pH 7 and ethanol is actually unfavorable because of the ionization of acetic acid. The relatively large chemical free energy of acetic anhydride provides the free-energy "push" that makes esterification with the anhydride so favorable, and the $\Delta G^{\circ\prime}$ for the hydrolysis of acetic anhydride is our standard measure of this free energy (**Fig. 25.8**).

B. ATP as a High-Energy Compound

ATP is the most ubiquitous high-energy compound in biology. For example, Eq. 25.46 (Sec. 25.7A) shows an example of ATP as a phosphorylating reagent to form glucose-6-phosphate, a phosphate monoester.

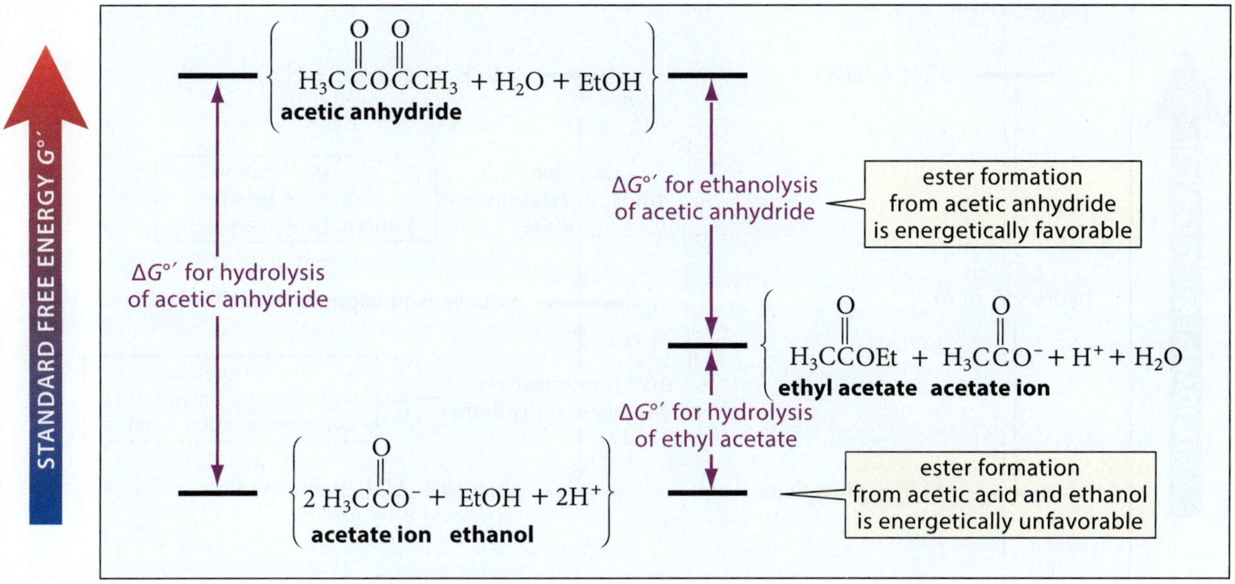

FIGURE 25.8 Diagram of Eq. 25.55. The species that cancel in the sum equation are included in the diagram. The relatively high free energy of acetic anhydride (see Fig. 25.7) makes its reaction with nucleophiles such as ethanol a very favorable process. The $\Delta G^{\circ\prime}$ for its hydrolysis is the standard way of expressing the high-energy character of acetic anhydride. Ester formation from acetic acid and ethanol is an unfavorable process because acetic acid is ionized to acetate at pH 7, the standard state for the energy calculation.

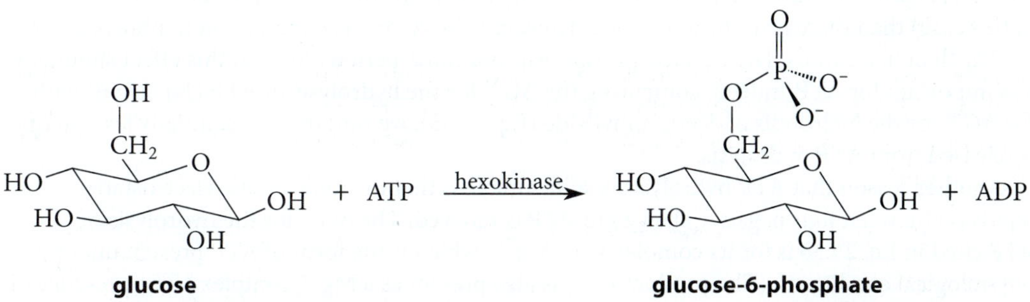

The $\Delta G^{\circ\prime}$ for the hydrolysis of ATP (to ADP and phosphate) at pH 7 in the presence of 10^{-3} *M* Mg^{2+} is –30.5 kJ mol^{-1} (–7.3 kcal mol^{-1}). The $\Delta G^{\circ\prime}$ for the formation of glucose-6-phosphate from phosphate and glucose is +13.8 kJ mol^{-1} (+3.3 kcal mol^{-1}). The positive $\Delta G^{\circ\prime}$ value shows that the formation of glucose-6-phosphate from phosphate and glucose is unfavorable. However, utilization of the phosphate anhydride ATP as the phosphorylating reagent makes phosphate monoester formation favorable:

$$\text{ATP} + \text{H}_2\text{O} \xrightleftharpoons{10\ \text{m}M\ \text{Mg}^{2+}} \text{ADP} + \text{phosphate} \qquad \Delta G^{\circ\prime} = -30.5\ \text{kJ mol}^{-1}\ (-7.3\ \text{kcal mol}^{-1})$$

$$\text{glucose} + \text{phosphate} \rightleftharpoons \text{glucose-6-phosphate} + \text{H}_2\text{O} \qquad \Delta G^{\circ\prime} = +13.8\ \text{kJ mol}^{-1}\ (+3.3\ \text{kcal mol}^{-1})$$

$$\textit{Sum:}\quad \text{glucose} + \text{ATP} \xrightleftharpoons{10\ \text{m}M\ \text{Mg}^{2+}} \text{glucose-6-phosphate} + \text{ADP} \qquad \Delta G^{\circ\prime} = -16.7\ \text{kJ mol}^{-1}\ (-4.0\ \text{kcal mol}^{-1}) \tag{25.56}$$

In reactions like this, it is sometimes said, "ATP hydrolysis is coupled to phosphate ester formation." As you can see, ATP is not really hydrolyzed. But a product of the reaction, ADP, is the same one that would form if ATP *were* hydrolyzed. This terminology is making use of the idea that ATP hydrolysis is a way of measuring the fact that it is a high-energy compound (**Fig. 25.9**).

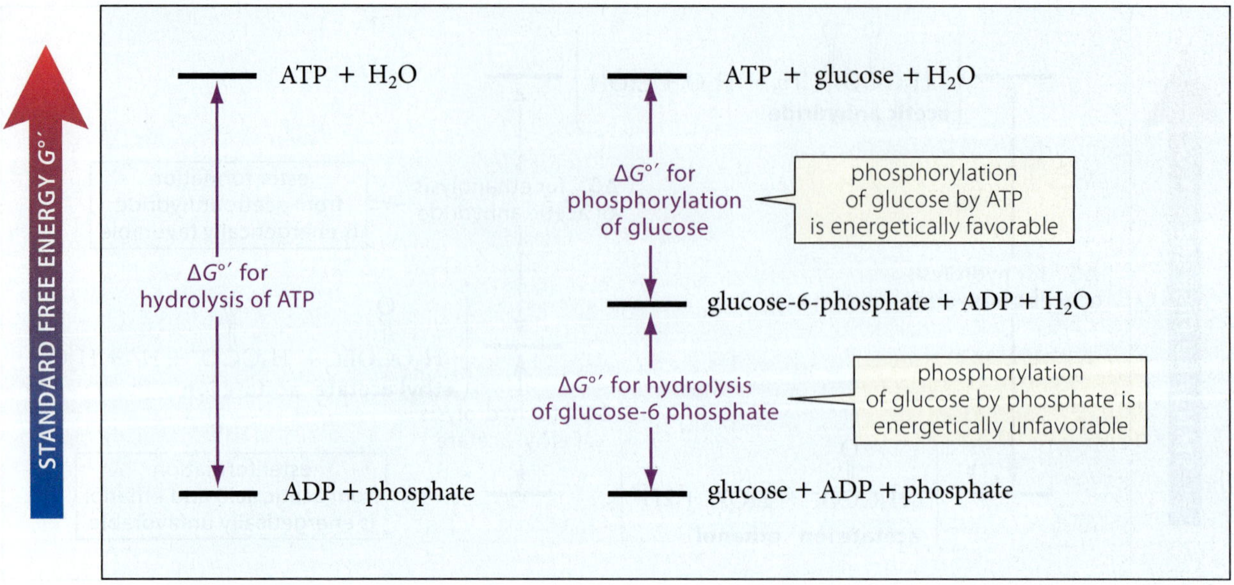

FIGURE 25.9 Diagram of Eq. 25.56. The $\Delta G°'$ for the phosphorylation of glucose by ATP is equal to the $\Delta G°'$ for the phosphorylation of glucose by phosphate plus the $\Delta G°'$ for ATP hydrolysis. The $\Delta G°'$ for ATP hydrolysis is the standard way of showing the high-energy character of ATP.

ATP hydrolysis releases a lot of free energy for the same reason that hydrolysis of carboxylic anhydrides does: because there is a greater resonance stabilization in the product (phosphate, in this case) than there is in the anhydride. However, the resonance interaction in phosphate is weaker than in a carboxylate because phosphorus is a third-period atom, so this effect should be less important for ATP. Indeed, comparing the $\Delta G°'$ for the hydrolysis of ATP (Eq. 25.56) with the $\Delta G°'$ for the hydrolysis of acetic anhydride (Eq. 25.55), we find that substantially less energy is released from ATP hydrolysis.

Another reason that ATP hydrolysis is so favorable is that the unfavorable electrostatic repulsion between four negative charges in ATP is relieved. The $\Delta G°'$ for the hydrolysis of ATP cited in Eq. 25.56 is for its complex with Mg^{2+}, which is the form of ATP present under physiological conditions. (The product ADP is also present as a Mg^{2+} complex.) The presence of Mg^{2+} partially offsets the charge repulsion in ATP and ADP. Indeed, when Mg^{2+} is not present, the $\Delta G°'$ for the hydrolysis of ATP is more negative by about 10 kJ mol^{-1} (2.4 kcal mol^{-1}).

C. Thioesters as High-Energy Compounds

Thioesters, and specifically acetyl-CoA, are also examples of high-energy compounds. The $\Delta G°'$ for the hydrolysis of acetyl-CoA is –31.4 kJ mol^{-1} (–7.5 kcal mol^{-1}). [The $\Delta G°'$ for the hydrolysis of ethyl acetate, an oxygen ester, is –19.8 kJ mol^{-1} (–4.73 kcal mol^{-1}).]

The reason that thioesters are high-energy compounds can be seen from an analysis of their resonance structures (Display 25.20, Sec. 25.5A). Resonance overlap between the sulfur nonbonding electron pair and the carbonyl 2p orbitals is weak, because the sulfur 3p orbitals have nodes that weaken this overlap. In the carboxylate hydrolysis product, resonance stabilization is very strong. It is the gain of resonance stabilization in the hydrolysis product that accounts for the highly negative $\Delta G°'$ for the hydrolysis of thioesters. Another way to express the same idea is that the high energy of thioesters is a reflection of their weak resonance stabilization. Resonance stabilization in oxygen esters is more effective. Therefore, less energy is released (the $\Delta G°'$ is "less negative") when oxygen esters undergo hydrolysis (**Fig. 25.10**).

The discussion in this section suggests a reason that thioesters evolved to become important in biology. As we showed in Sec. 25.5A, thioesters are not particularly reactive; under physiological conditions, acetyl-CoA and other thioesters are stable toward hydrolysis and other nucleophilic substitution reactions. Therefore, they will survive in the cell until they are needed for

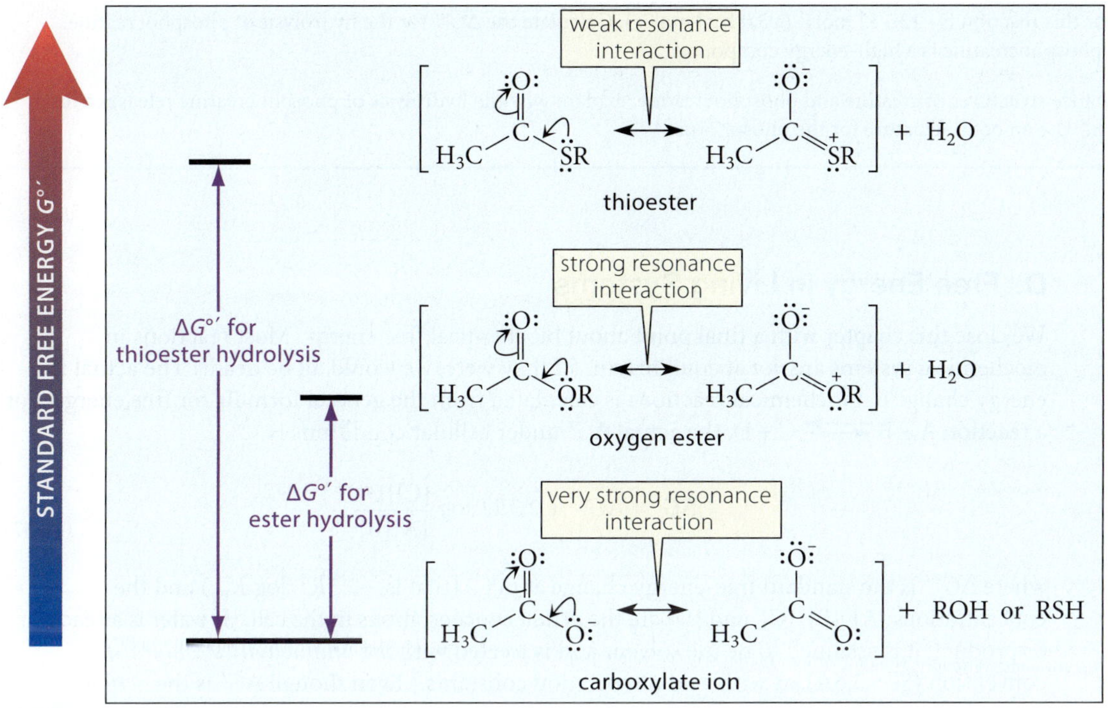

FIGURE 25.10 The weak resonance interaction between sulfur and the carbonyl π electrons results in a relatively high standard free energy for thioesters such as acetyl-CoA. Hydrolysis of the thioester or any other nucleophilic substitution resulting in replacement of the sulfur with nitrogen or oxygen releases significant free energy because a derivative is formed in which the resonance interaction is much stronger.

metabolically important enzyme-catalyzed reactions. However, their favorable $\Delta G^{\circ\prime}$ values for reactions with water and alcohols show that, when they do undergo enzyme-catalyzed reactions with these nucleophiles, their reactions will go essentially to completion.

Focused Problems

25.17 The $\Delta G^{\circ\prime}$ for ester hydrolysis of the neurotransmitter acetylcholine is –25.1 kJ mol^{-1} (–6.0 kcal mol^{-1}), and the $\Delta G^{\circ\prime}$ for the hydrolysis of acetyl-CoA is –31.4 kJ mol^{-1} (–7.5 kcal mol^{-1}).

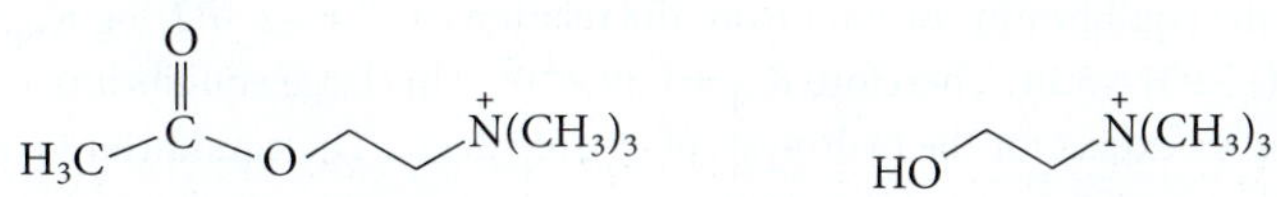

acetylcholine
a neurotransmitter

choline

(a) Draw a free-energy diagram that shows how the $\Delta G^{\circ\prime}$ for the biosynthesis of acetylcholine from acetyl-CoA and choline can be derived from the $\Delta G^{\circ\prime}$ for these two hydrolysis reactions.

(b) What is the $\Delta G^{\circ\prime}$ for this biosynthetic reaction?

(c) What is the equilibrium constant for this reaction at 37 °C? (2.3RT at 310 K = 5.92 kJ mol^{-1}.)

25.18 (a) Phosphocreatine (creatine phosphate) is a compound in muscle tissue that serves as a reservoir of high-energy phosphate. It reacts with ADP to replenish ATP stores in muscle:

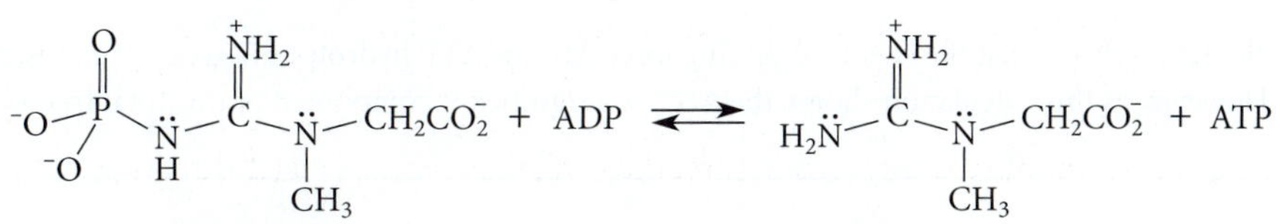

phosphocreatine **creatine**

The $\Delta G^{\circ\prime}$ for this reaction is -12.5 kJ mol^{-1} (-3.0 kcal mol^{-1}). Calculate the $\Delta G^{\circ\prime}$ for the hydrolysis of phosphocreatine. Verify that phosphocreatine is a high-energy compound.

(b) Using resonance structures of creatine and phosphocreatine, explain why the hydrolysis of phosphocreatine releases a lot of energy. (*Hint:* Use an octet structure for the phosphorus.)

D. Free Energy in Living Systems

We close this chapter with a final point about biochemical free energy: Most reactions in biochemical systems are *not* at equilibrium. (If they were, we would all be dead!) The actual free-energy change in biochemical reactions is calculated from the general formula for free energy. For a reaction $A + B \rightleftharpoons C + D$, the actual $\Delta G'$ under cellular conditions is

$$\Delta G' = \Delta G^{\circ\prime} + 2.3RT \log \frac{[C][D]}{[A][B]} \qquad (25.57)$$

where $\Delta G^{\circ\prime}$ is the standard free-energy change at pH 7 (that is, $-2.3RT \log K_{eq}$) and the concentrations [A], [B], [C], and [D] are the *actual* concentrations in the cell. [If water is a reactant or product, it is assumed to be the solvent and is treated with *the unit activity standard state* convention (Sec. 3.6A), as with acid dissociation constants.] Even though $\Delta G'$ is the *actual* chemical free-energy change under specific conditions, the $\Delta G^{\circ\prime}$ is a useful quantity because it provides the baseline, or standard, from which the actual $\Delta G'$ is calculated.

Study Problem 25.1

The $\Delta G^{\circ\prime}$ for ATP hydrolysis is -30.5 kJ mol^{-1} (-7.3 kcal mol^{-1}). In the red blood cell at 37 °C (310 K), the concentrations of ATP, ADP, and phosphate are 2.25, 0.25, and 1.65 mM, respectively. Show that these are not the equilibrium concentrations. Calculate the actual $\Delta G'$ under these conditions. [$2.3RT$ at 37 °C is 5.93 kJ mol^{-1} (1.42 kcal mol^{-1}).]

Solution The ATP hydrolysis reaction is

$$\text{ATP} + H_2O \xrightleftharpoons{10\ \text{m}M\ Mg^{2+}} \text{ADP} + \text{phosphate}$$

First, we calculate the equilibrium constant from the relation $\Delta G^{\circ\prime} = -2.3RT \log K_{eq}$, from which we find that $\log K_{eq} = (-30.5)/(-5.93) = 5.14$. Therefore $K_{eq} = 1.39 \times 10^5$. This large equilibrium constant means that ATP hydrolysis is very favorable, as we expect for the hydrolysis of an anhydride. The equilibrium constant is equal to the ratio [ADP][phosphate]/[ATP] *at equilibrium* at pH 7. At the concentrations given in the problem, this ratio is actually 1.83×10^{-4}. Therefore, under biological conditions, there is much more ATP than there would be at equilibrium. This is typically true in living systems, because ATP must be available to drive the many energy-requiring biological processes that depend on it.

To calculate the actual $\Delta G'$, we apply Eq. 25.57 and use the actual cellular concentrations of the three components of the reaction. The result is

$$\begin{aligned}\Delta G' &= \Delta G^{\circ\prime} + 2.3RT \log \frac{[\text{ADP}][\text{phosphate}]}{[\text{ATP}]}\\ &= -30.5 + (5.93)\log(1.83 \times 10^{-4})\\ &= -30.5 + (5.93)(-3.74) = -52.7 \text{ kJ mol}^{-1}\ (-12.6 \text{ kcal mol}^{-1})\end{aligned}$$

This calculation shows that the chemical driving force $\Delta G'$ for ATP hydrolysis is greater (that is, more negative) than the $\Delta G^{\circ\prime}$. However, as this calculation shows, the $\Delta G^{\circ\prime}$ is a significant component of this total free-energy change.

Focused Problems

25.19 What would the $\Delta G'$ be for ATP hydrolysis in a cell in which the concentrations of ATP, ADP, and phosphate were all 1 m*M*?

25.20 This question refers to the biosynthesis of *S*-adenosylmethionine (SAM) introduced in Focused Problem 25.15 (Sec. 25.7D). The hydrolysis of the triphosphate leaving group in this reaction is catalyzed by the enzyme SAM synthase, and the hydrolysis of the pyrophosphate product of this hydrolysis is hydrolyzed by pyrophosphatase.

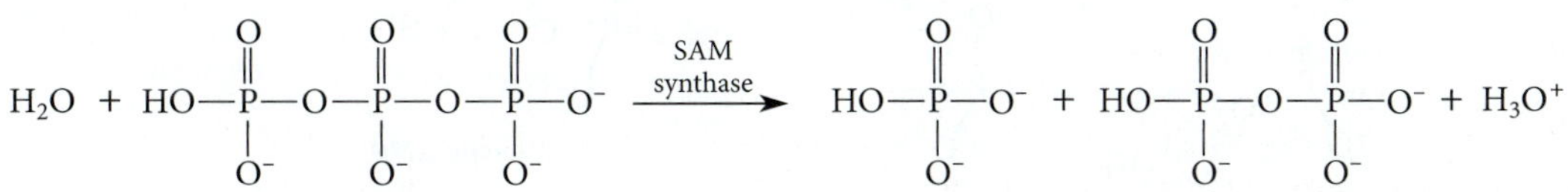

triphosphate **phosphate** **pyrophosphate** (hydrolysis is catalyzed by pyrophosphatase)

ADP rather than ATP could have evolved as an adenosyl donor, in which case pyrophosphate rather than triphosphate would have been the leaving group. What is the advantage of producing triphosphate as a leaving group rather than pyrophosphate? (*Hint:* What if methionine were present in very small concentration? Think about $\Delta G'$.)

CHAPTER SUMMARY *For a summary of the chapter, see Chapter 25 in the* Study Guide and Solutions Manual.

REACTION REVIEW *For a summary of reactions discussed in this chapter, see the* Reaction Review *section of Chapter 25 in the* Study Guide and Solutions Manual.

SKILLS OBJECTIVES WITH PROBLEMS

• **Given a structure for a thioester, provide a name; given a name, provide a structure. 25.1**

25.21 Draw the structure for each the following compounds.

(a) Isobutyl thiobenzoate

(b) *S*-Ethyl 3-oxohexanethioate

(c) Ethyl 3-ethylthiohexanoate

25.22 Name the following compounds.

(a)

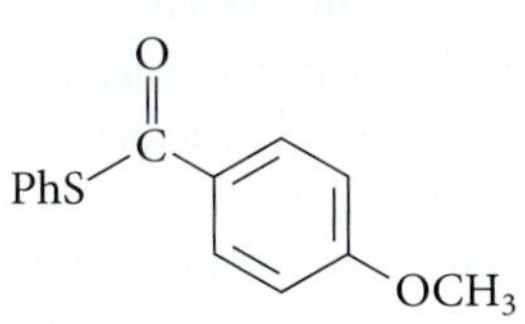

(b)

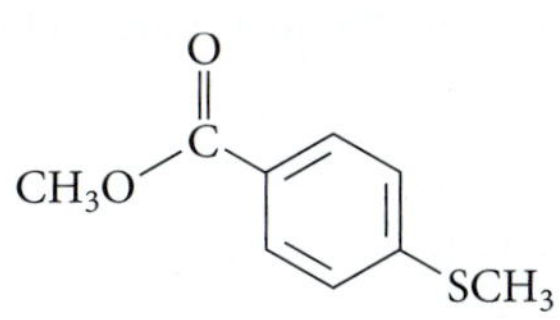

• **Point out the major structural differences between a thioester and the corresponding oxygen ester, and rationalize the differences using principles of structure. 25.3A**

25.23 Which of the following are the major structural differences between a thioester and an oxygen ester? Choose all that apply. For the items you select, list the specific difference you expect, and explain the reason of the difference.

(a) C═O bond lengths

(b) C—S and C—O bond lengths

(c) C—S—C and C—O—C bond angles

• **Recognize phosphate diester, monoester, pyrophosphate, and anhydride groups in biologically significant molecules. 25.2**

25.24 Identify the phosphorus-containing functional groups in each of the biologically occurring compounds shown in **Fig. P25.24.** Choose between phosphate monoester, phosphate diester, phosphate anhydride, and pyrophosphate monoester.

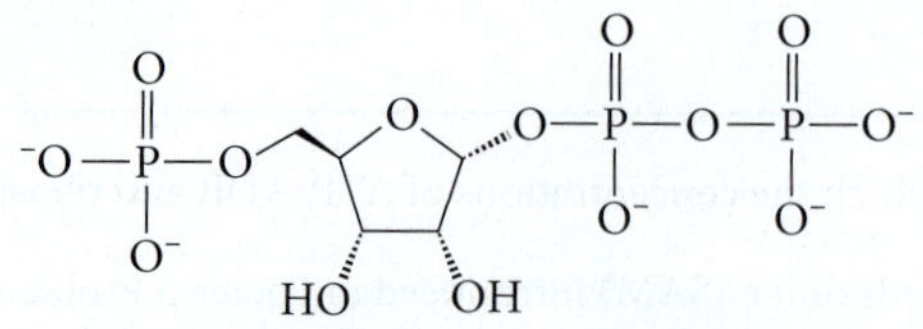

5-phosphoribosy-1-pyrophosphate

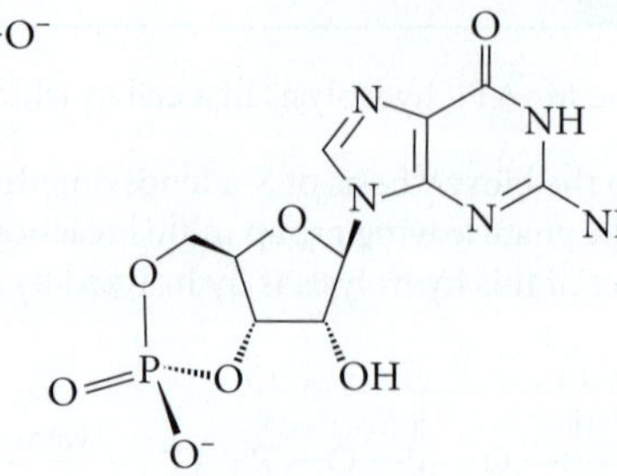

cyclic GMP

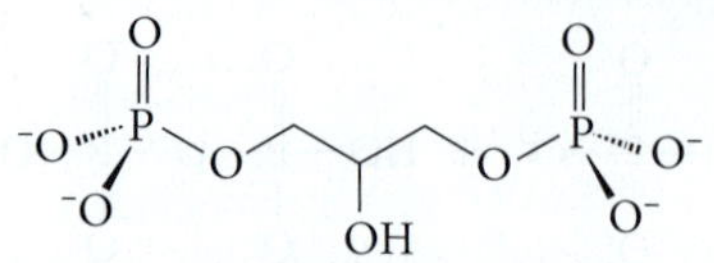

glycerol-1,3-diphosphate

FIGURE P25.24

- **Rank various phosphate-containing molecules according to their relative reactivity in hydrolysis without enzyme catalysis. 25.6A,B**

25.25 In each of the following sets, arrange the three compounds in order of increasing reactivity toward base-promoted hydrolysis, least reactive first, and explain your reasoning. (No enzymes are involved.)

(a)

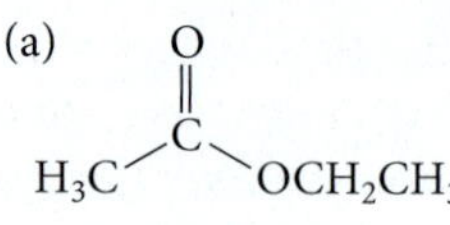

A

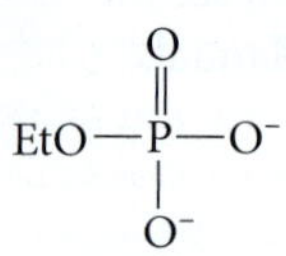

B

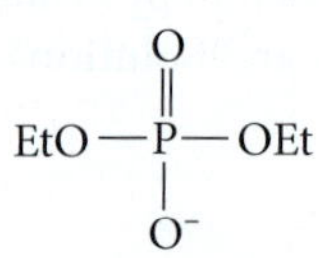

C

(b)

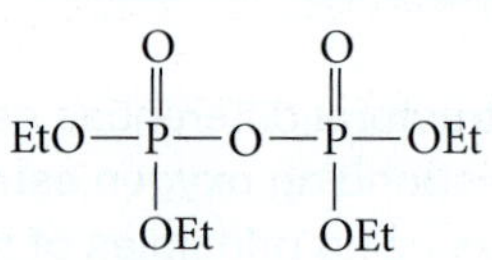

A

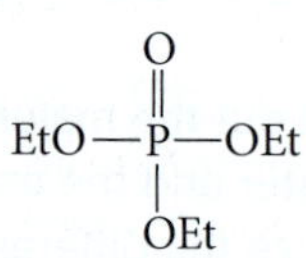

B

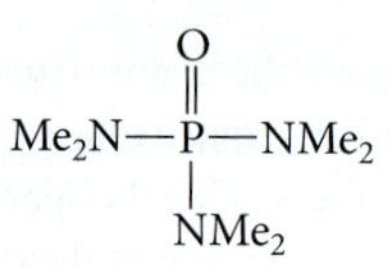

C

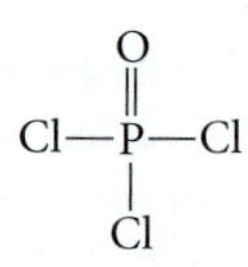

D

(c)

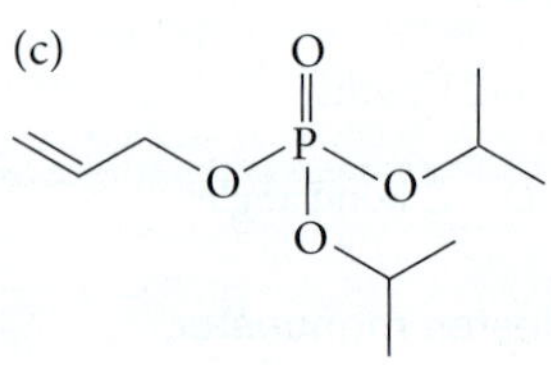

A

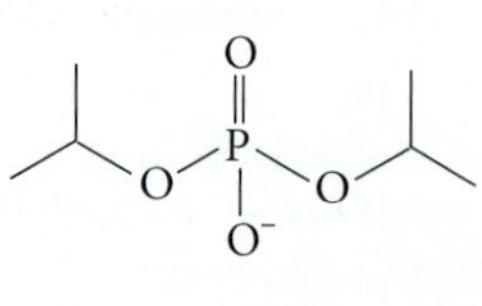

B

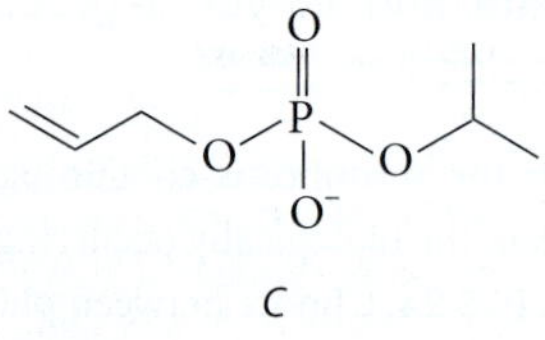

C

25.26 Why are the biologically important phosphate esters either diesters or monoesters? Why are there no triesters? Give two reasons.

- **Predict or explain the splittings in both ^{31}P and ^{1}H resonances in the NMR spectra of phosphorus-containing molecules. 25.4**

25.27 The anomeric proton (*red*) in the α-anomer of glucose-1-phosphate in the proton NMR spectrum appears at δ 5.45. It is split into a doublet of doublets—four lines of equal size—with coupling constants of 3.5 Hz and 7.3 Hz. Explain the origin of each splitting.

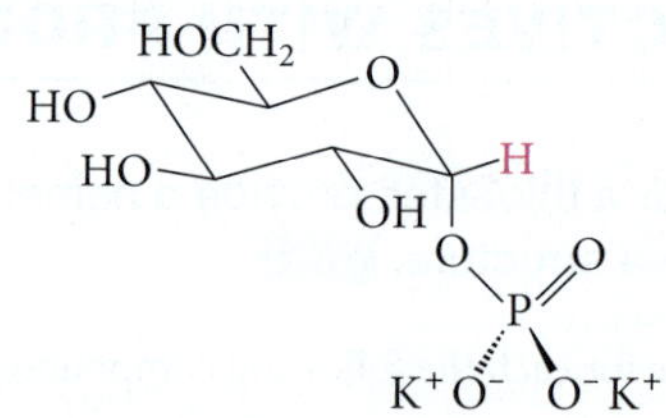

glucose-1-phosphate (dipotassium salt)

25.28 (a) Predict the proton splitting pattern in the proton NMR spectrum of trimethyl phosphite, $(CH_3O)_3P$.

(b) Predict the splitting pattern for phosphorus in the ^{31}P NMR spectrum of trimethyl phosphite.

25.29 (a) The ^{31}P NMR spectrum following compound has two resonances. The resonance at greater chemical shift is a seven-line pattern (septet) with a coupling constant of 2.6 Hz. The resonance at smaller chemical shift is also a septet with a coupling constant of 8.9 Hz. Explain the origin of each resonance and the appearance of the splitting patterns.

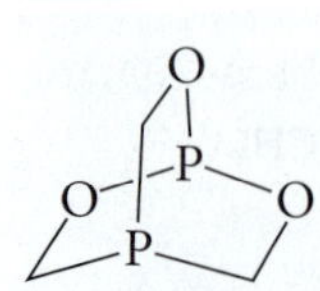

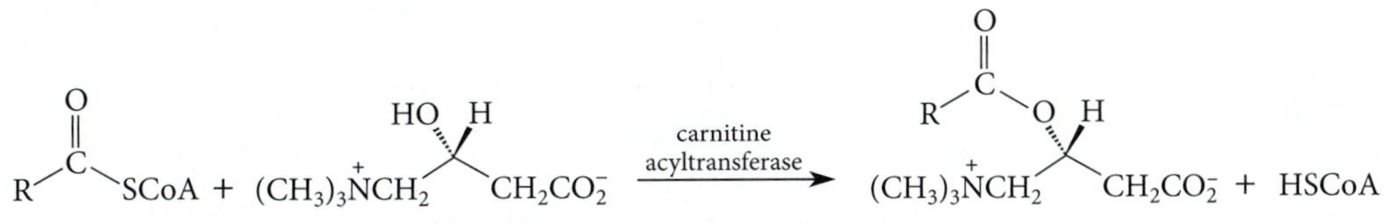

a fatty acyl-CoA — **carnitine** — a fatty acyl-carnitine — **coenzyme A**

FIGURE P25.30

(b) Predict the appearance of the proton NMR spectrum of the same compound. (*Hint:* All protons are chemically equivalent.)

- **Explain the advantage to biological systems in using thioesters instead of oxygen esters, and apply your reasoning to a specific case. 25.3A, 25.8**

25.30 The enzyme-catalyzed formation of fatty acyl-carnitine from fatty acyl-CoA (**Fig. P25.30**) is an early step in the metabolism of fatty acids. The equilibrium for this reaction strongly favors the products. Give a structural reason that the reaction is so favorable.

- **Give the mechanism for the reaction of a nucleophile with a phosphate monoester. 25.6A**

25.31 The tetrabutylammonium salt of isotopically chiral phenyl phosphate was heated in the polar aprotic solvent acetonitrile containing excess *tert*-butyl alcohol, and isotopically substituted *tert*-butyl phosphate was isolated. An analysis of its stereochemistry showed that it was completely racemic, as shown in the following equation. Explain this result with a mechanism. (*Hint:* Think about the reactive intermediate, and also consider the possible role of the solvent.)

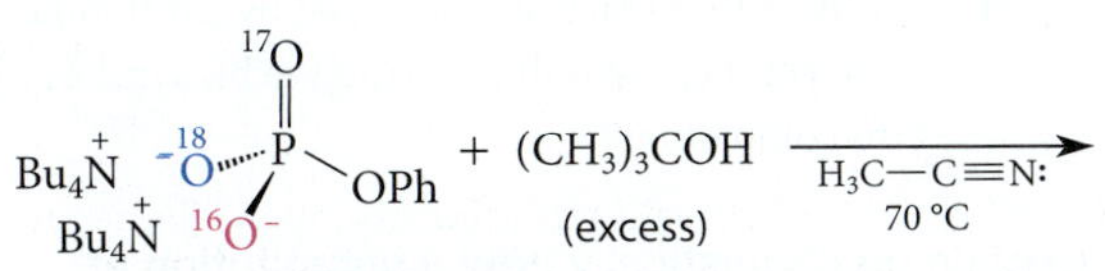

bis(tetrabutylammonium) phenyl phosphate
(*R* stereoisomer, isotopically chiral and enantiomerically pure)

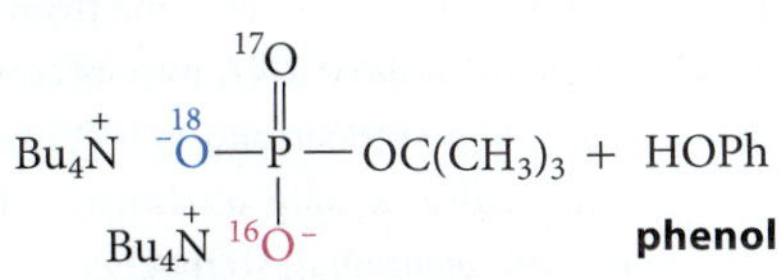

bis(tetrabutylammonium) *tert*-butyl phosphate
(isotopically chiral and racemic)

- **Outline the key structural features in the active sites of enzymes that catalyze phosphate-transfer reactions. Show how these features might apply to a specific case. 25.7**

25.32 List the key features that you would expect to find in the active site of an enzyme that catalyzes the reaction of a phosphate diester or monoester with a nucleophile such as an alcohol.

25.33 The conjugation of alcohols containing nonpolar groups and phenols to glucuronic acids is a key process by which these alcohols can be transformed into, and excreted as, water-soluble derivatives called *glucuronides* in phase II metabolism. (See Eq. 8.64, Sec. 8.6C.) These conjugation reactions involve the reaction of UDP-glucuronic acid and the alcohol or phenol catalyzed by a family of enzymes called *UDP-glucuronyl transferases.* (See **Fig. P25.33**.) What would you expect to find as key catalytic features in the active sites of these enzymes?

- **Show the three most common reaction types that can occur when ATP reacts as a phosphoryl donor. Analyze a particular case to show which of these reaction types is involved. 25.7A**

25.34 Show the three types of nucleophilic reactions that occur biologically at the phosphate groups of ATP.

25.35 The enzyme phosphoribosyl pyrophosphate synthetase (PRPP synthetase) catalyzes the conversion of ribose-5-phosphate into its 1-pyrophosphate derivative, as shown in the following equation. The pyrophosphate group in the product is provided by ATP (not shown).

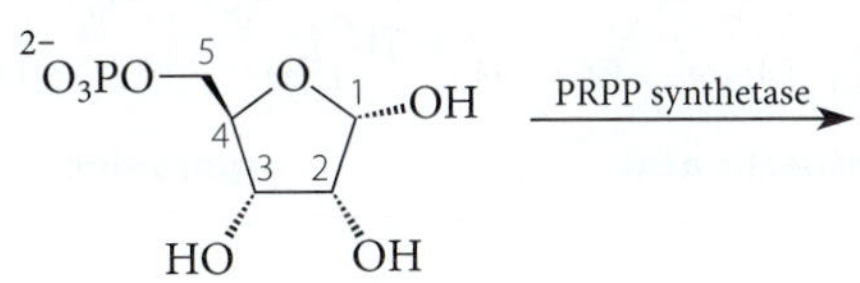

ribose-5-phosphate

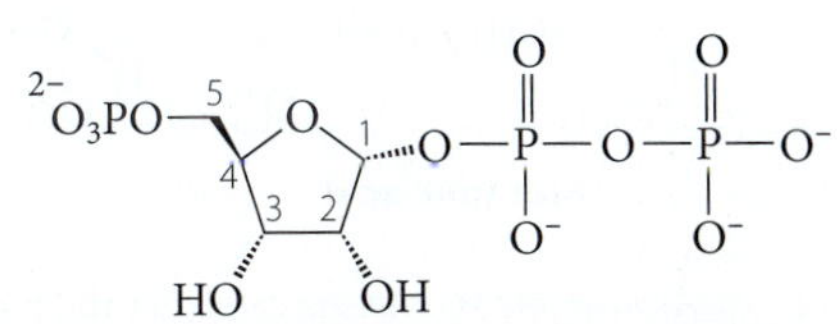

5-phosphoribosyl-1-pyrophosphate (PRPP)

(a) At which phosphate of ATP does the nucleophilic hydroxy group of ribose-5-phosphate react? What is the by-product of the reaction?

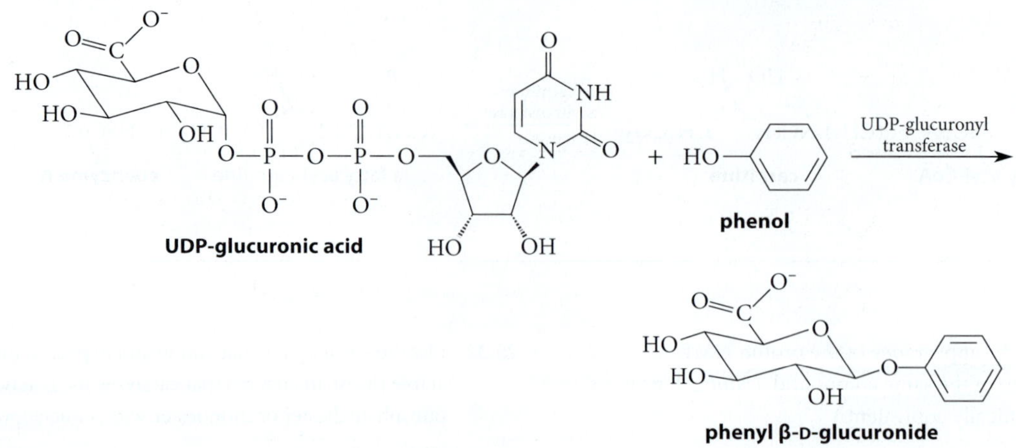

FIGURE P25.33

PRPP is an important intermediate in the biosynthesis of nucleic acids and amino acids. It reacts with nucleophiles at carbon-1 in typical S_N2-like processes.

(b) To illustrate this reactivity, show the reaction of ammonia with PRPP. Be sure to show the stereochemistry.

(c) A by-product of the reaction in part (b) is pyrophosphate. What other biochemical reaction ensures that the reaction you drew in part (b) runs to completion?

25.36 Aspartic acid and glutamic acid are the two naturally occurring α-amino acids that have carboxylate groups in their side chains. Asparagine and glutamine are the corresponding α-amino acids in which the side chains are primary amides. (All of these amino acids have the *S* configuration.)

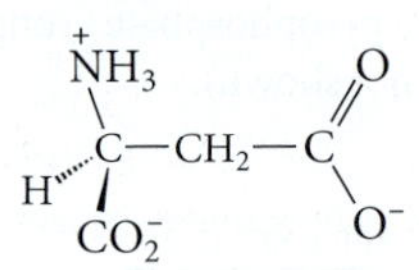

aspartic acid

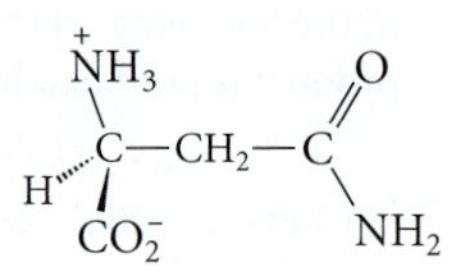

asparagine

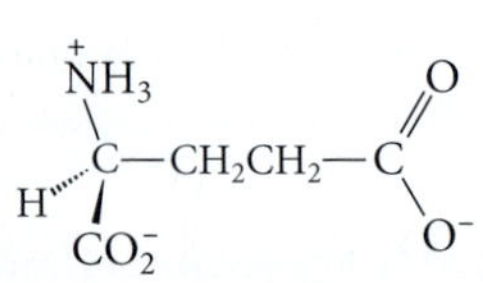

glutamic acid

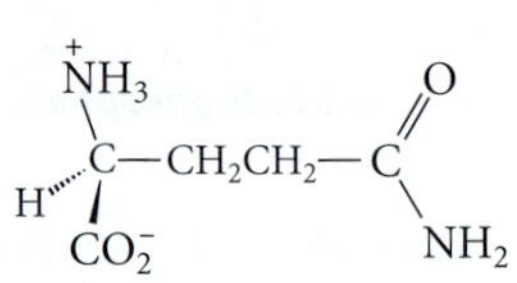

glutamine

Asparagine synthetase catalyzes the biosynthesis of asparagine from glutamine according to the following equation. Note that the amide —NH_2 group in asparagine comes from the amide —NH_2 group of glutamine.

$$\left.\begin{array}{l} H_2O + \text{glutamine} \\ + \text{ aspartic acid} + \text{ATP} \end{array}\right\} \xrightarrow{\text{asparagine synthetase}} \begin{array}{l} \text{glutamic acid} + \text{asparagine} \\ + \text{ AMP} + \text{pyrophosphate} \end{array}$$

(a) Suggest a sequence of steps for this reaction that involves the adenylation of aspartic acid.

(b) Glutamine and asparagine could in principle be equilibrated without ATP. Why is ATP utilized in this way?

(c) Calculate the $\Delta G°'$ for the biosynthesis of asparagine using the following $\Delta G°'$ values for hydrolysis (in kJ mol^{-1}): ATP to AMP + pyrophosphate, −45.6; asparagine to aspartic acid, −15.1; glutamine to glutamic acid, −14.2. (The corresponding values in kcal mol^{-1} are −10.9, −3.61, and −3.39, respectively.)

(d) It is sometimes said that asparagine biosynthesis is coupled to ATP hydrolysis (to AMP). What is meant by this statement? Is ATP actually hydrolyzed?

(e) What other reaction (catalyzed by a different enzyme) ensures that asparagine biosynthesis goes to completion?

• **Explain mechanistically how a substitution at phosphorus can occur with retention of configuration.** **25.6A**

25.37 Ribonuclease A, an enzyme from beef pancreas, catalyzes the hydrolysis of RNA into its component ribonucleic acids. When this reaction is studied with an isotopically chiral sulfur analog, as shown in **Fig. P25.37**, it occurs with overall *retention* of stereochemistry. No phosphoenzyme intermediate is formed. Give a curved-arrow mechanism that explains this stereochemical result. (Assume that acidic and basic groups are present in the enzyme active site as needed.)

25.38 Phosphoglycerate mutase catalyzes the interconversion of (*R*)-3-phosphoglycerate and (*R*)-2-phosphoglycerate, as shown in item (1) of **Fig. P25.38**. The enzyme contains a histidine residue in its active site, shown in (2); and the reaction in (1) involves a phosphohistidine intermediate (3) of Fig. P25.38.

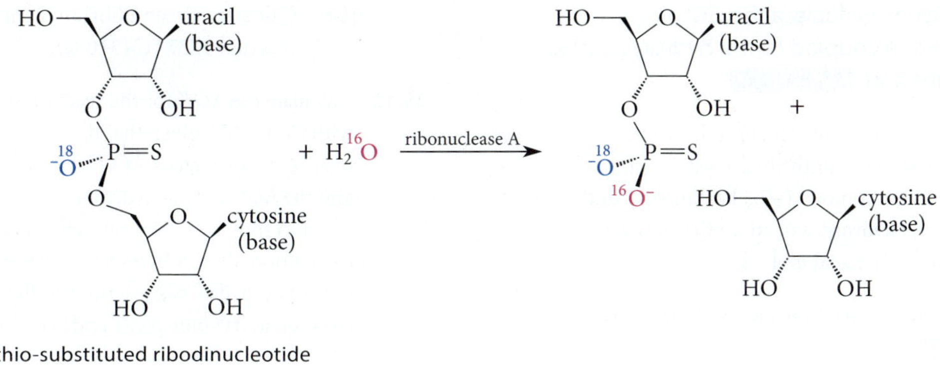

FIGURE P25.37

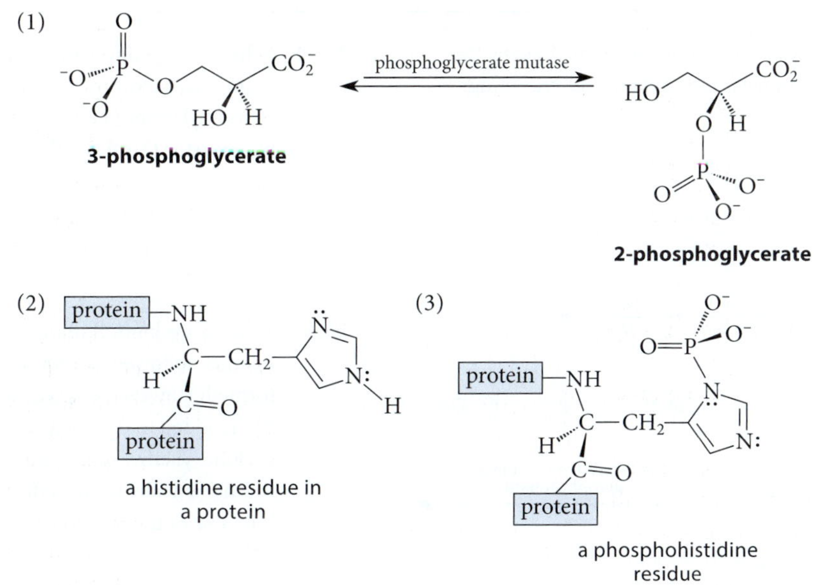

FIGURE P25.38

(a) Assuming the presence of acidic and basic groups in the active site of the enzyme as needed, draw a curved-arrow mechanism for the transformation of 3-phosphoglycerate to 2-phosphoglycerate that is consistent with the information given.

(b) Suppose that an enantiomerically pure 2-phosphoglycerate derivative containing an isotopically chiral phosphate group is subjected to this reaction. Would the isotopically chiral 3-phosphoglycerate formed have the *same* configuration at phosphorus, the *opposite configuration* at phosphorus, or equal amounts of the two configurations? Explain your reasoning.

- **Convert the $\Delta G°$ for a reaction into a $\Delta G°'$ for the same reaction. 25.8A**

25.39 The $\Delta G°$ for hydrolysis of the ester phenyl acetate is -30.9 kJ mol^{-1} (-7.39 kcal mol^{-1}). What is the $\Delta G°'$ for the same reaction? (The pK_a of acetic acid is 4.76, and the pK_a of phenol is 9.95.) Assume the unit activity standard state of water.

25.40 The $\Delta G°$ for hydrolysis of glycine ethyl ester to the amino acid glycine, shown in the following equation, is -8.24 kJ mol^{-1} (1.97 kcal mol^{-1}). Calculate the $\Delta G°'$ for this hydrolysis.

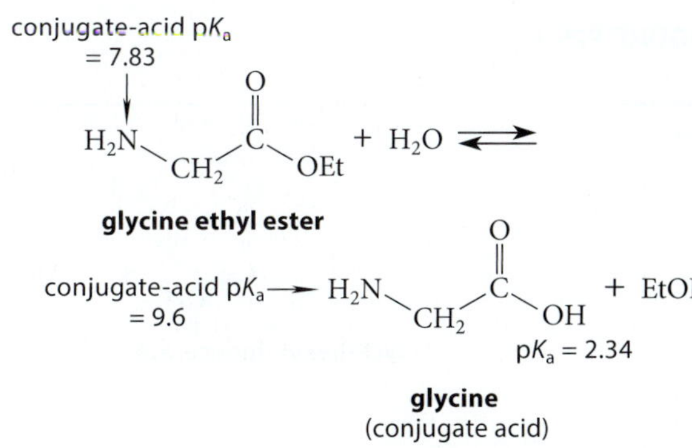

- **Apply Hess's law to calculate the $\Delta G^{\circ\prime}$ for a transformation that is coupled to the hydrolysis of a high-energy compound. 25.8A,B,C**

25.41 This problem relates to the reaction shown in Fig. P25.30. The $\Delta G^{\circ\prime}$ for the hydrolysis of an acyl-CoA is -31.4 kJ mol^{-1} (-7.5 kcal mol^{-1}), and the $\Delta G^{\circ\prime}$ for the hydrolysis of an acyl-carnitine is -17.4 kJ mol^{-1} (-4.2 kcal mol^{-1}).

(a) Calculate the $\Delta G^{\circ\prime}$ for the reaction shown in Fig. P25.30.

(b) Calculate the equilibrium constant (at pH 7) for this reaction at 37 °C (310 K).

25.42 Calculate the $\Delta G^{\circ\prime}$ for the reaction of acetyl phosphate with CoA—SH, given that the $\Delta G^{\circ\prime}$ for the hydrolysis of acetyl phosphate is -43.1 kJ mol^{-1} (-10.3 kcal mol^{-1}) and the $\Delta G^{\circ\prime}$ for the hydrolysis of acetyl-CoA is -31.4 kJ mol^{-1} (-7.52 kcal mol^{-1}). As part of your calculation, draw a free-energy diagram relating the $\Delta G^{\circ\prime}$ of the two hydrolysis reactions to the $\Delta G^{\circ\prime}$ of the reaction between acetyl phosphate and acetyl-CoA.

INTEGRATED PROBLEMS

25.43 Complete the following reactions shown by drawing the structures of the products, and explain your reasoning.

(a)

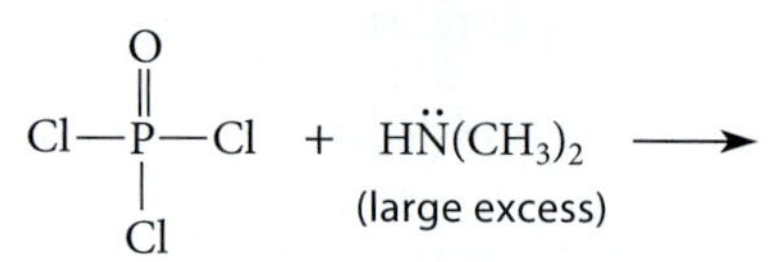

(b)

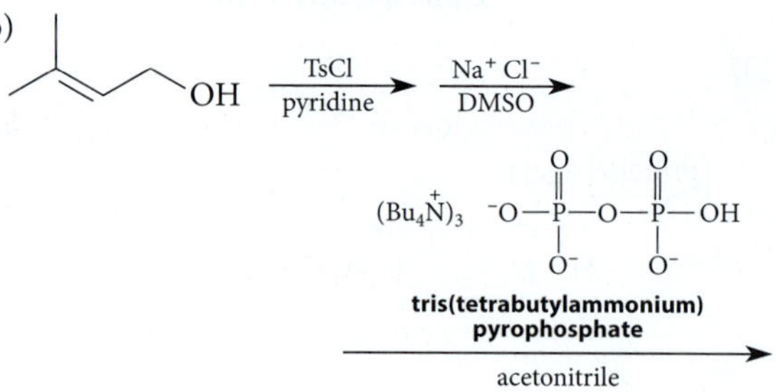

(c)

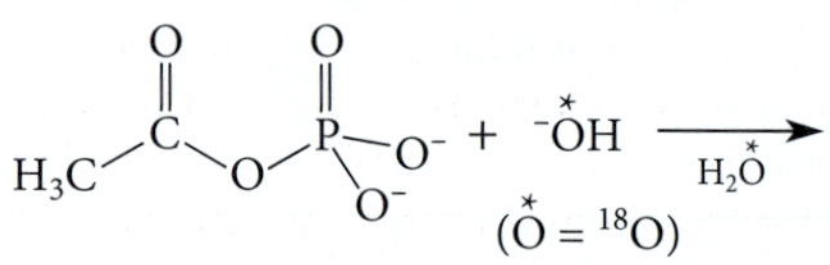

25.44 When *S*-ethyl thiobenzoate (*A*) is allowed to react with 2-aminoethanol (*B*), an amide *C* is formed if one equivalent of triethylamine (Et_3N:) is included in the reaction mixture, but an ester *D* is formed if one equivalent of a strong acid such as *p*-toluenesulfonic acid is added as a catalyst. (See **Fig. P25.44**.) Give the structures of *C* and *D*, and explain why different products are formed under different conditions.

25.45 When *S*-cyclohexyl thioacetate is reduced by $LiAlH_4$ in ether, followed by protonolysis, cyclohexanethiol is formed. However, when a large excess of the Lewis acid BF_3 is added to the reaction mixture before the reduction, cyclohexyl ethyl sulfide is formed (**Fig. P25.45**). Account mechanistically for the effect of BF_3 in changing the outcome of the reaction.

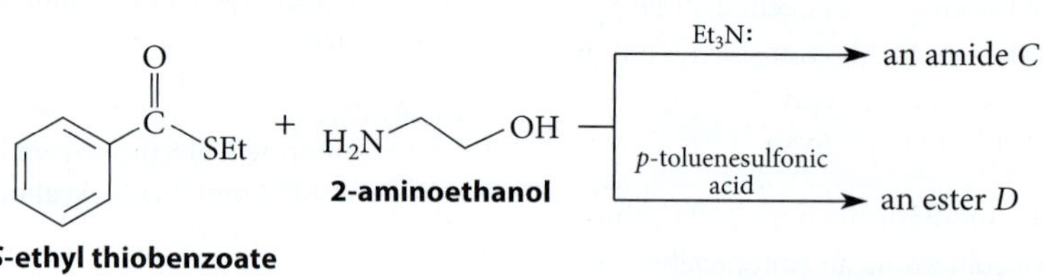

FIGURE P25.44

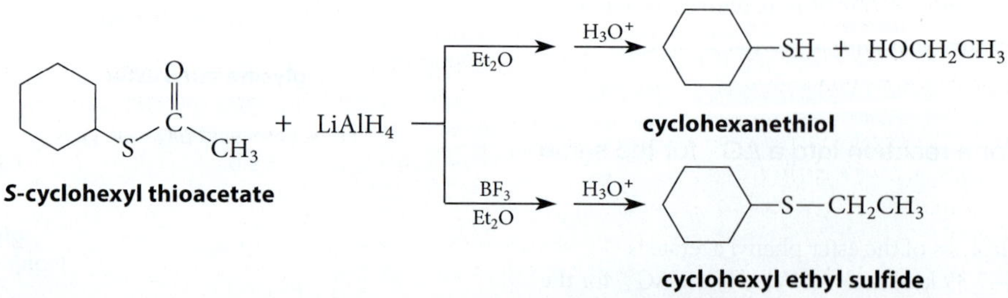

FIGURE P25.45

25.46 Sarin is an acid fluoride and ester of methylphosphonic acid.

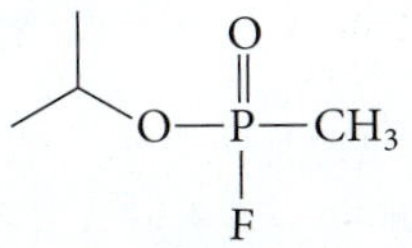

sarin
(isopropyl methylphosphonofluoridate)

Sarin is a deadly nerve gas that was outlawed by the Chemical Weapons Convention of 1993. However, it has been used at various times as a weapon of mass destruction by terrorists, including a 1995 attack in the Tokyo metro and a 2013 attack by the Syrian government on rebel forces during the Syrian civil war.

(a) Sarin is chiral; the *S* enantiomer is more active than the *R* enantiomer. Draw a line-and-wedge structure of the more active enantiomer.

(b) Sarin's biological effects are due to its reaction with a serine residue in the active site of acetylcholinesterase, an enzyme that is very important in nerve transmission.

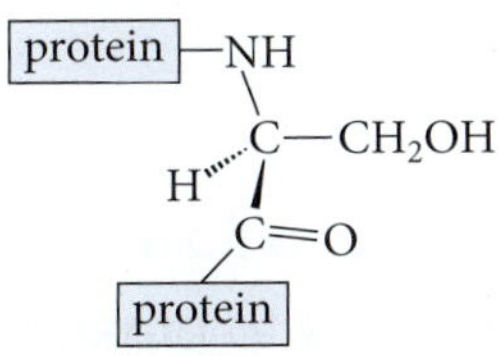

a serine residue in acetylcholinesterase

Give the product of the reaction of the more active enantiomer of sarin with serine. Include stereochemistry. (You can represent the serine as R—CH_2OH.)

25.47 (a) Give the products that result from the hydrolysis of the phosphate ester group of phosphoenolpyruvate (PEP), an important compound in the glycolysis pathway.

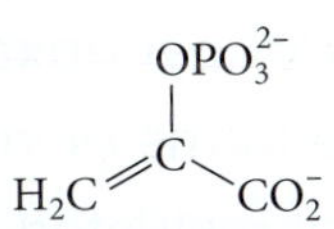

phosphoenolpyruvate (PEP)

(b) The $\Delta G^{\circ\prime}$ for the hydrolysis of PEP is −61.9 kJ mol^{-1} (14.8 kcal mol^{-1}), which puts it well into the category of a high-energy compound. Most phosphate esters have a $\Delta G^{\circ\prime}$ for hydrolysis of about −13.8 kJ mol^{-1} (3.3 kcal mol^{-1}). Explain why the hydrolysis of PEP releases much more energy than the hydrolysis of an ordinary phosphate ester.

25.48 Methanol containing the oxygen isotope ^{18}O is allowed to react, in separate reactions, with each of the acid chlorides shown in parts (a) through (c) shown here, and each of the resulting compounds *A*, *C*, and *E* is treated with one equivalent of sodium hydroxide. This treatment regenerates methanol in all three cases, along with hydrolysis products *B*, *D*, and *F*, respectively. Identify the compounds *A–F*, and account for the distribution of the isotope in all three cases. Specifically, why is isotopically substituted methanol formed in reaction (b) but not in reactions (a) and (c)?

(a)

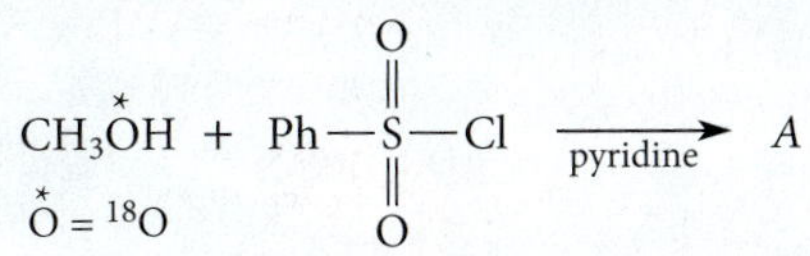

benzenesulfonyl chloride

$$A \xrightarrow{\text{NaOH}} CH_3OH + B$$

(contains no isotope)

(b)

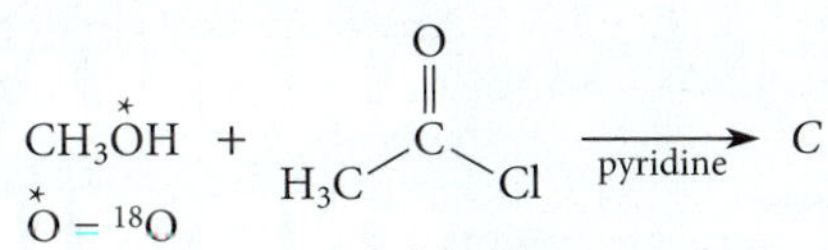

acetyl chloride

$$C \xrightarrow{\text{NaOH}} CH_3\overset{*}{O}H + D$$

(c)

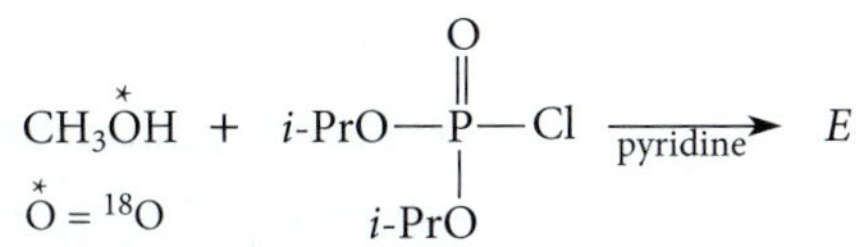

$$E \xrightarrow{\text{NaOH}} CH_3OH + F$$

(contains no ^{18}O isotope)

25.49 Account with a mechanism for the fact that the hydrolysis of trimethyl phosphate to dimethyl phosphate in acidic solution containing ^{18}O-labeled water gives methanol containing ^{18}O and dimethyl phosphate containing no isotope, as shown in the following equation.

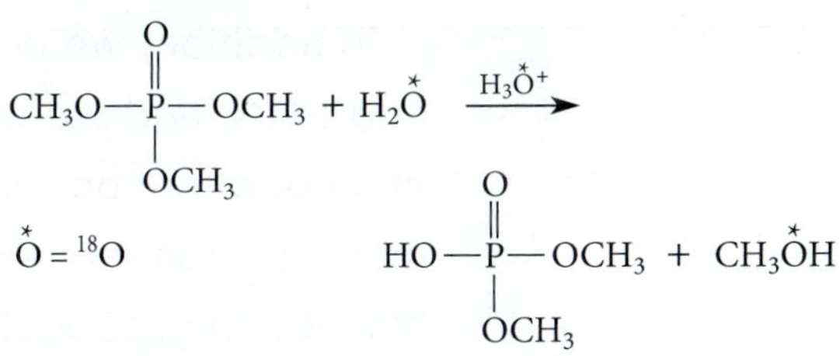

25.50 The reaction of water with metaphosphate ion is shown in Eq. 25.37b (Sec. 25.6A). Could nitrate ion undergo an analogous reaction with aqueous NaOH? If so, draw the structure of the product. If not, explain why.

25.51 The pK_a of the thiol group of CoASH is 9.6. How would the $\Delta G^{\circ\prime}$ for the hydrolysis of acetyl-CoA change if the hydrolysis were carried out at pH 11 instead of pH 7? Explain.

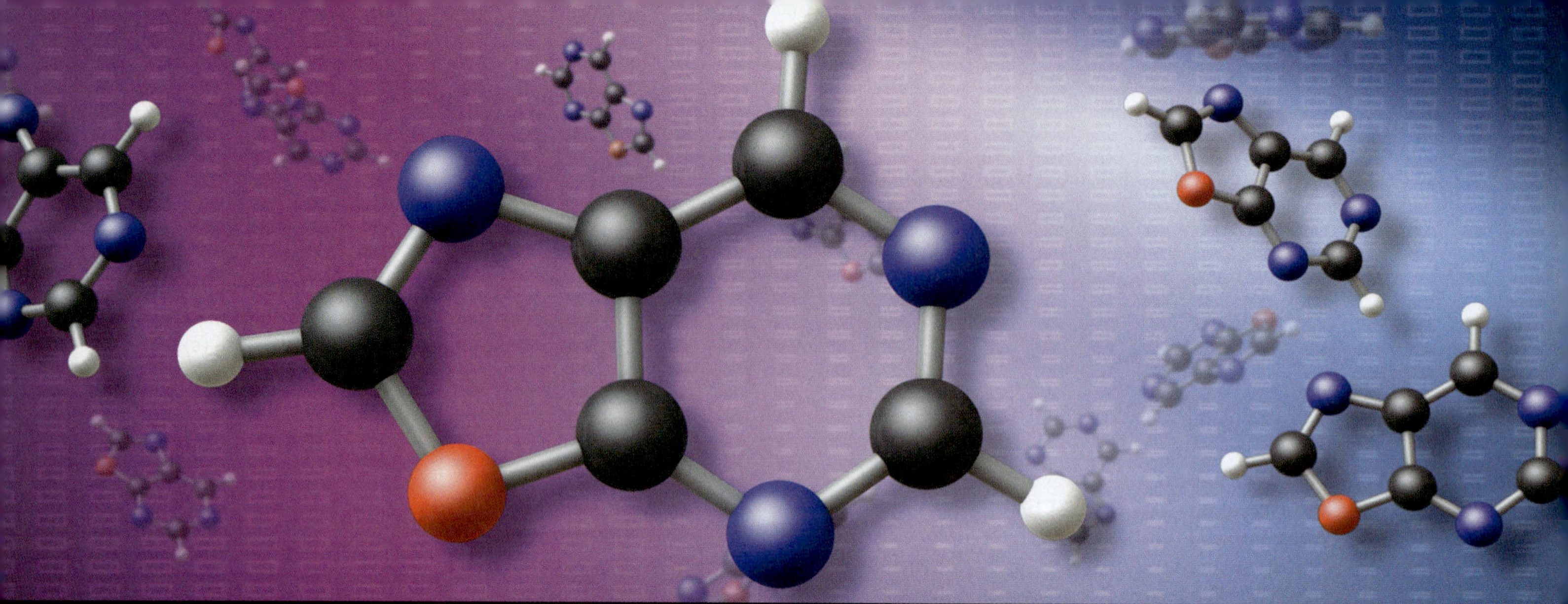

26 THE CHEMISTRY OF THE AROMATIC HETEROCYCLES AND NUCLEIC ACIDS

Heterocyclic compounds are compounds with rings that contain more than one element. The heterocyclic compounds of greatest interest to organic chemists have rings containing carbon and one or more **heteroatoms**—atoms other than carbon. Although the chemistry of many saturated heterocyclic compounds is analogous to that of their noncyclic counterparts, a significant number of unsaturated heterocyclic compounds exhibit aromatic behavior. This chapter focuses primarily on the unique chemistry of a few of these aromatic heterocycles. The principles that emerge should enable you to understand the chemistry and properties of other heterocyclic compounds that you may encounter.

In addition, we consider the structures of nucleosides, nucleotides, RNA, and DNA, and we examine some of the chemistry of pyridoxal phosphate, an important coenzyme in biology that is one of the forms of vitamin B_6. The medical importance of heterocyclic compounds is not restricted to their role in biochemistry; about 95% of small-molecule drugs contain heterocyclic groups.

26.1 NOMENCLATURE AND STRUCTURE OF THE AROMATIC HETEROCYCLES

A. Nomenclature

The names and structures of some important aromatic heterocyclic compounds are given in **Fig. 26.1**. This figure also shows how the rings are numbered in substitutive nomenclature. In all but a few cases, a heteroatom is given the number 1. (Isoquinoline is an exception.)

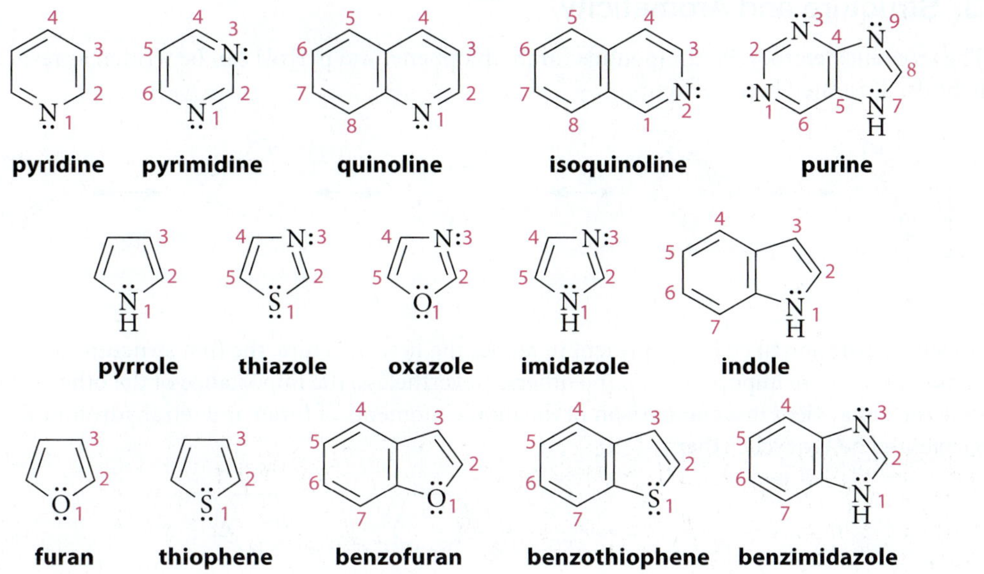

FIGURE 26.1 Common aromatic heterocyclic compounds. The numbers (red) are used in substitutive nomenclature.

The numbering convention illustrated by thiazole and oxazole is that oxygen and sulfur are given a lower number than nitrogen. Substituent groups are given the lowest number consistent with the ring numbering. (These are the same rules used in numbering and naming saturated heterocyclic compounds; see Secs. 8.2C and 23.1B.)

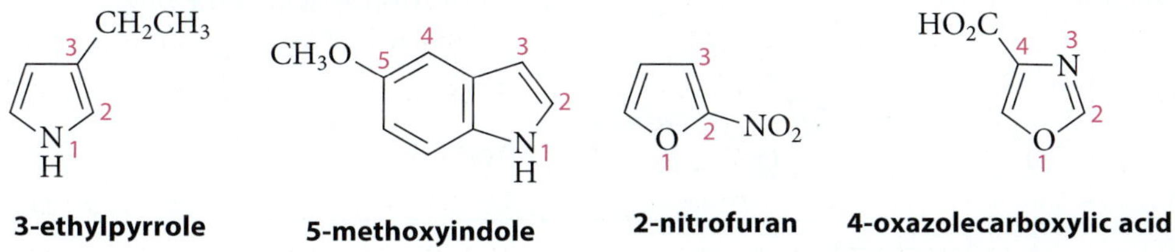

(26.1)

Focused Problems

26.1 Draw the structure of

(a) 4-(Dimethylamino)pyridine

(b) 4-Ethyl-2-nitroimidazole

26.2 Name the following compounds.

(a) 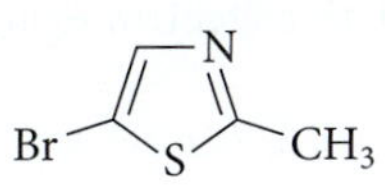

(b) 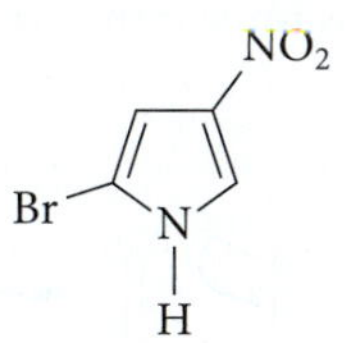

(c) 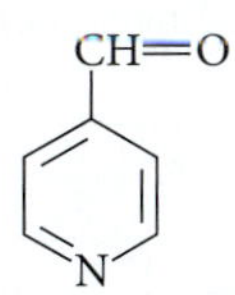

(d) 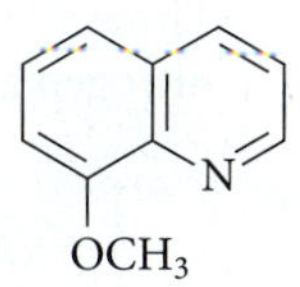

B. Structure and Aromaticity

The aromatic heterocyclic compounds furan, thiophene, and pyrrole can be written as resonance hybrids, illustrated here for furan.

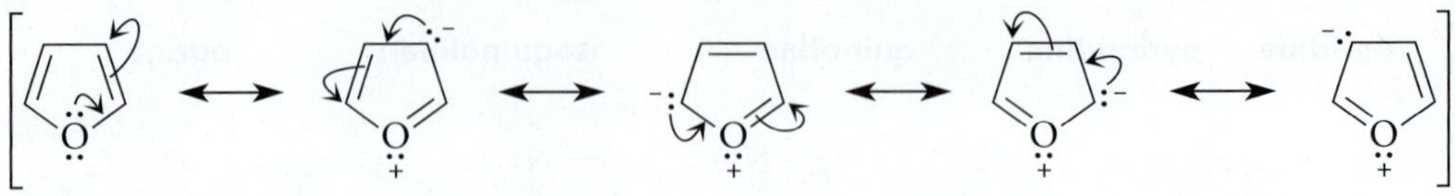

(26.2)

Because separation of charge is present in all but the first structure, the first structure is considerably more important than the others. Nevertheless, the importance of the other structures is evident in a comparison of the dipole moments of furan and tetrahydrofuran, a saturated heterocyclic ether.

	tetrahydrofuran	**furan**
dipole moment	1.7 D	0.7 D
boiling point	67 °C	31.4 °C

(26.3)

The dipole moment of tetrahydrofuran is attributable mostly to the bond dipoles of its polar C—O single bonds. That is, electrons in the σ bonds are pulled toward the oxygen because of its electronegativity. This same effect is present in furan, but in addition there is a second effect: the resonance delocalization of the oxygen nonbonding electrons into the ring shown in Eq. 26.2. This tends to push electrons away from oxygen into the π-electron system of the ring.

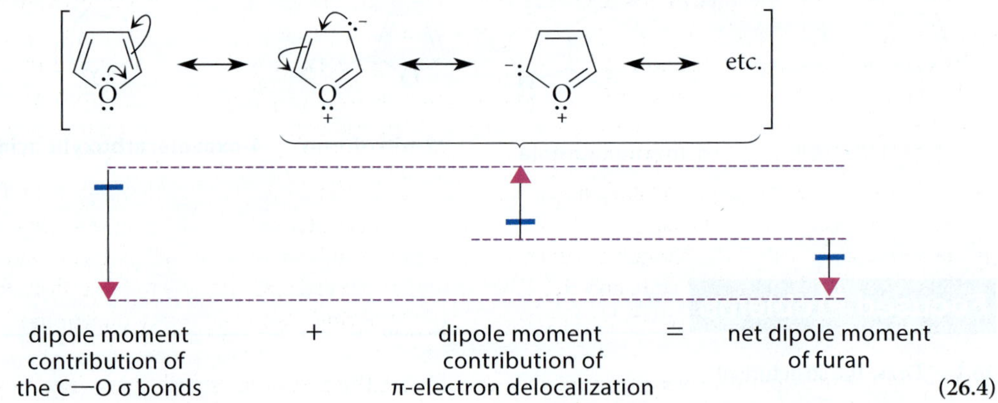

(26.4)

Because these two effects in furan nearly cancel, furan has a very small dipole moment. The relative boiling points of tetrahydrofuran and furan in Display 26.3 reflect the difference in their dipole moments (Sec. 8.5B).

Pyridine, like benzene, can be represented by two equivalent neutral resonance structures. Three additional pyridine structures of less importance reflect the relative electronegativities of nitrogen and carbon.

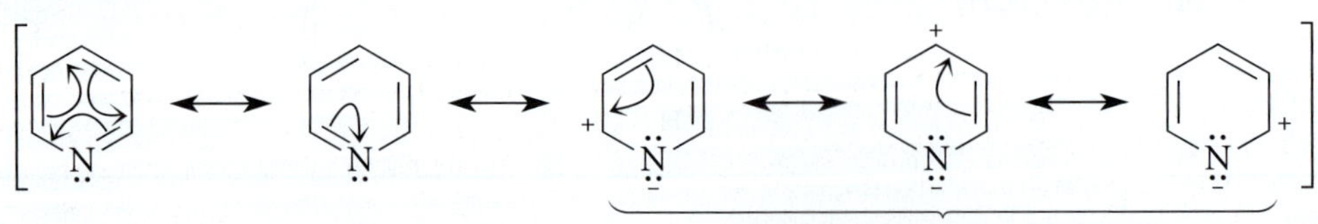

(26.5)

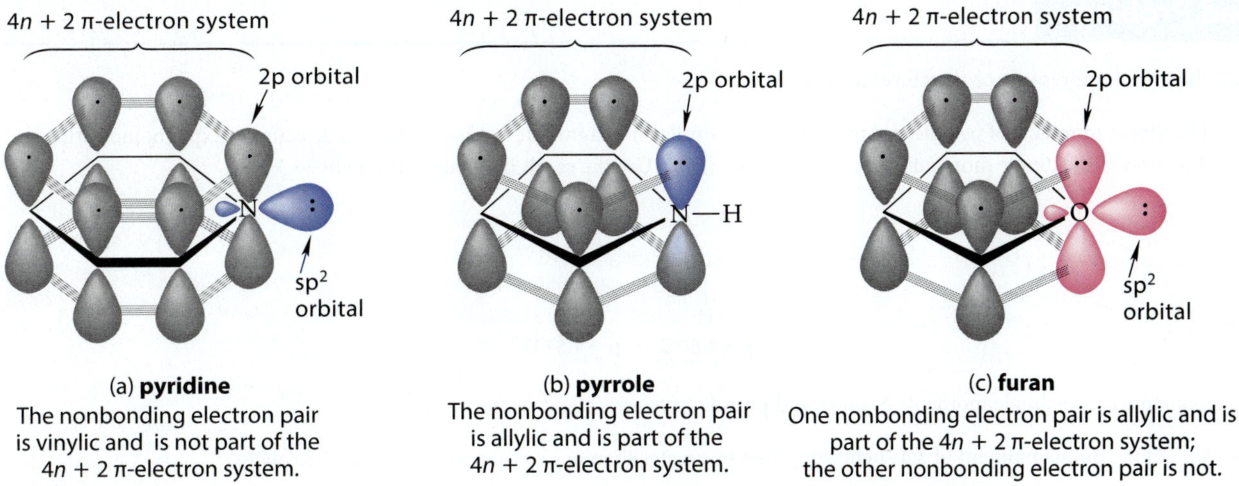

FIGURE 26.2 The orbital configurations in pyridine, pyrrole, and furan. (a) In pyridine, the nitrogen 2p orbital, like the carbon 2p orbitals, contributes one electron to the $4n + 2$ π-electron system, and the nitrogen nonbonding electron pair occupies an sp^2 orbital *(blue)*, which is not part of the π system. (b) In pyrrole, the nonbonding electron pair occupies a 2p orbital on nitrogen *(blue)*, and it contributes two electrons to the $4n + 2$ π-electron system. (c) Furan has two nonbonding electron pairs *(red)*. One nonbonding pair occupies a 2p orbital and contributes two electrons to the $4n + 2$ π-electron system, as in pyrrole. The other nonbonding pair occupies an sp^2 orbital. Like the nonbonding pair in pyridine, it does not contribute to the $4n + 2$ π-electron system.

The aromaticity of some heterocyclic compounds was considered in the discussion of the Hückel $4n + 2$ rule (Sec. 15.7D). It is important to understand which nonbonding electron pairs in a heterocyclic compound are part of the $4n + 2$ aromatic π-electron system, and which are not (**Fig. 26.2**). Vinylic nitrogens, such as the nitrogen of pyridine, contribute one π electron to the six π-electron aromatic system, just like each of the carbon atoms in the π system. The orbital containing the nonbonding electron pair of the pyridine nitrogen is perpendicular to the 2p orbitals of the ring and is therefore not involved in π bonding. In contrast, allylic heteroatoms, such as the nitrogen of pyrrole, contribute two electrons (a nonbonding pair) to the aromatic π-electron system. This nitrogen adopts sp^2 hybridization and trigonal planar geometry so that its nonbonding electron pair can occupy a 2p orbital, which has the optimum shape and orientation to overlap with the carbon 2p orbitals and thus to be part of the aromatic π-electron system. (See Display 1.79, Sec. 1.7E, and Display 3.18, Sec. 3.3B.) Consequently, the hydrogen of pyrrole lies in the plane of the ring. The oxygen of furan contributes one allylic nonbonding electron pair to the aromatic π-electron system, and the other nonbonding electron pair occupies a position analogous to the carbon–hydrogen bond of pyrrole—in the ring plane, perpendicular to the 2p orbitals of the ring.

The *empirical resonance energy* can be used to estimate the additional stability of a heterocyclic compound due to its aromaticity. (This concept was introduced in the discussion of the aromaticity of benzene in Sec. 15.7C.) The empirical resonance energies of benzene and some heterocyclic compounds are given in **Table 26.1**. To the extent that resonance energy is a measure of aromatic character, furan has the least aromatic character of the heterocyclic compounds in the table. The modest resonance energy of furan has significant consequences for its reactivity, as we show in Sec. 26.3B.

TABLE 26.1 Empirical Resonance Energies of Some Aromatic Compounds

	Resonance energy			Resonance energy	
Compound	kJ mol^{-1}	kcal mol^{-1}	Compound	kJ mol^{-1}	kcal mol^{-1}
benzene	138–151	33–36	thiophene	121	29
pyridine	96–117	23–28	pyrrole	89–92	21–22
			furan	67	16

Focused Problems

26.3 Draw the important resonance structures for pyrrole.

26.4 (a) The dipole moments of pyrrole and pyrrolidine are similar in magnitude but have opposite directions. Explain, indicating the direction of the dipole moment in each compound. (*Hint:* Use the result in Focused Problem 26.3.)

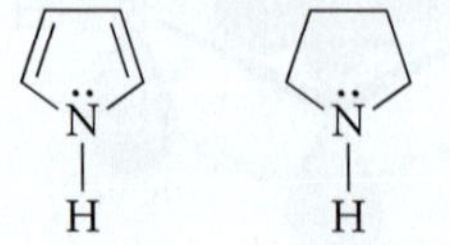

$\mu = 1.80$ D $\quad\mu = 1.57$ D

(b) Explain why the dipole moments of furan and pyrrole have opposite directions.

(c) Should the dipole moment of 3,4-dichloropyrrole be greater than or less than that of pyrrole? Explain.

26.5 Each of the following NMR chemical shifts goes with a proton at carbon-2 of pyridine, pyrrolidine, or pyrrole. Match each chemical shift with the appropriate heterocyclic compound, and explain your answer: δ 8.51, δ 6.41, and δ 2.82.

26.2 BASICITY AND ACIDITY OF THE NITROGEN HETEROCYCLES

A. Basicity of the Nitrogen Heterocycles

Pyridine and quinoline act as ordinary amine bases.

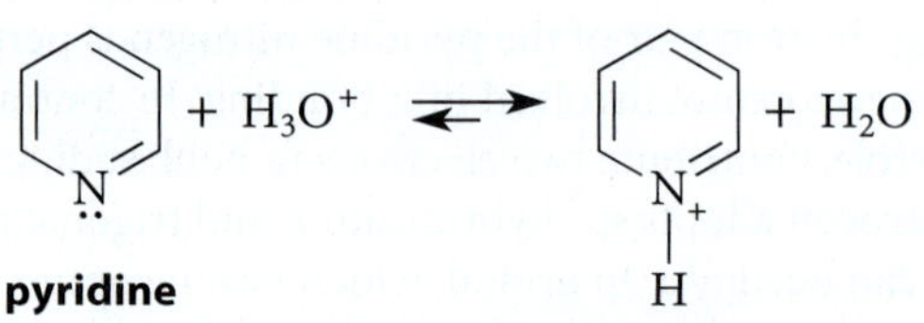

pyridine

$pK_a = 5.2$ (26.6)

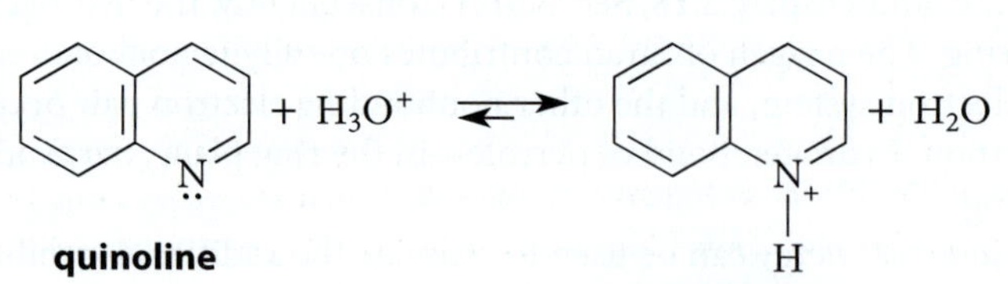

quinoline

$pK_a = 4.9$ (26.7)

Pyridine and quinoline are much less basic than aliphatic tertiary amines (Sec. 23.5A) because of the sp^2 hybridization of their nitrogen nonbonding electron pairs. (As noted in Sec. 3.7C, the basicity of a nonbonding electron pair decreases with increasing s character.)

Although pyrrole and indole look like amines, neither of these two heterocycles has appreciable basicity. These compounds are protonated only in strong acid, and protonation occurs on carbon, not nitrogen.

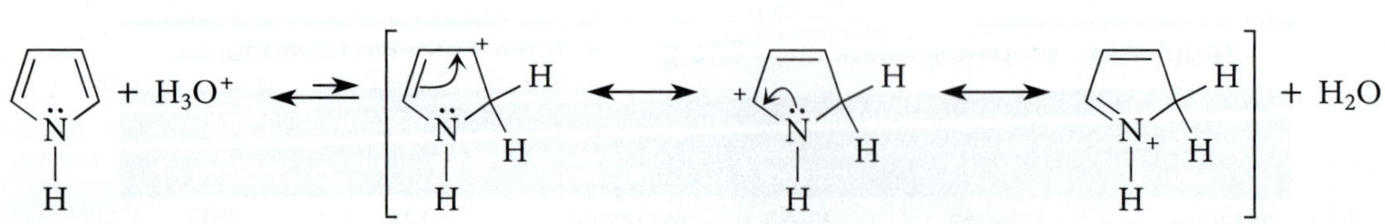

pyrrole

$pK_a \sim -4$ (26.8)

The marked contrast between the basicities of pyridine and pyrrole can be understood by considering the role of the nitrogen nonbonding electron pair in the aromaticity of each compound (see Fig. 26.2). Protonation of the pyrrole nitrogen would disrupt the aromatic system of six π electrons by taking the nitrogen's nonbonding pair "out of circulation."

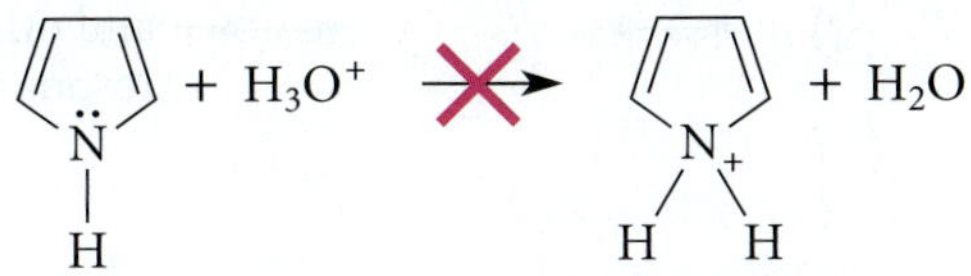

not aromatic;
is not formed (26.9)

Although protonation of the carbon of pyrrole (Eq. 26.8) also disrupts the aromatic π-electron system, at least the resulting cation is resonance-stabilized. On the other hand, protonation of the pyridine nonbonding electron pair occurs easily because this electron pair is not part of the π-electron system. Consequently, protonation of this electron pair does not disrupt pyridine's aromaticity.

The basicity of the nitrogen heterocycle imidazole, explored in Study Problem 26.1, is particularly relevant to biology because the amino acid histidine in proteins has an imidazole group in its side chain.

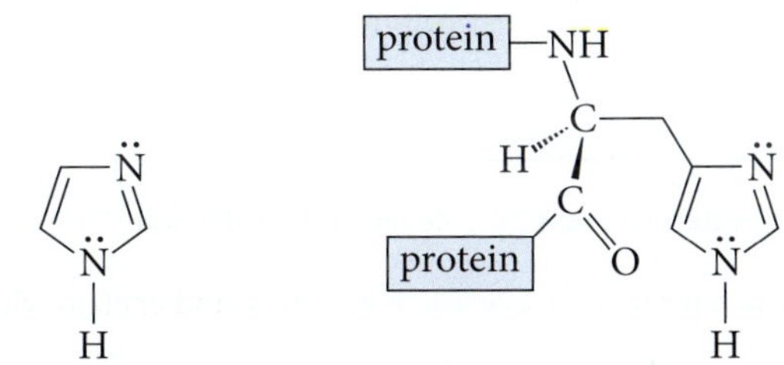

imidazole — a histidine residue in a protein (26.10)

In many cases this group serves as an acid–base catalyst or a nucleophile in enzyme-catalyzed reactions.

Study Problem 26.1

Imidazole is a base; the pK_a of its conjugate acid is 6.95. On which nitrogen is imidazole protonated?

Solution Imidazole has two nitrogen atoms: one has the electronic configuration of the nitrogen in pyridine—that is, it is vinylic—but the other is like the nitrogen of pyrrole—it is allylic. Consequently, protonation occurs on the pyridine-like nitrogen—the nitrogen whose nonbonding electron pair is *not* part of the aromatic sextet.

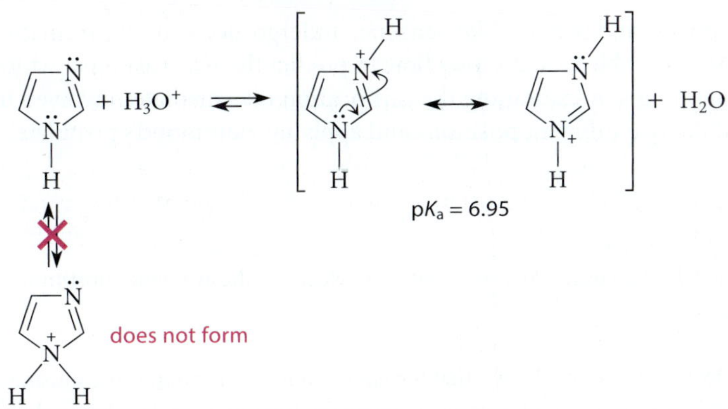

(26.11)

According to the resonance structures in Eq. 26.11, the two nitrogens of protonated imidazole are equivalent; consequently, deprotonation to give imidazole can occur at either nitrogen. Imidazole is more basic than pyridine because of the resonance stabilization of its conjugate-acid cation.

B. Acidity of Pyrrole and Indole

Pyrrole and indole are weak acids.

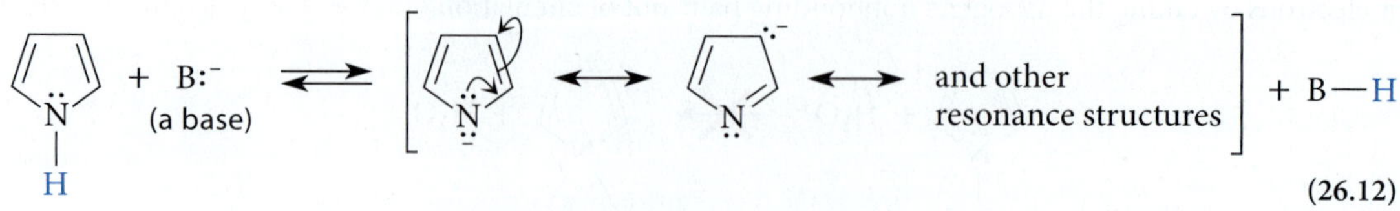

(26.12)

With pK_a values of about 17.5, pyrrole and indole are about as acidic as alcohols and about 15–17 pK_a units more acidic than primary and secondary amines (Sec. 23.5D). The greater acidity of pyrroles and indoles is a consequence of the resonance stabilization of their conjugate-base anions (Eq. 26.12; draw the three missing resonance structures in this equation). Pyrrole and indole are rapidly deprotonated by Grignard and organolithium reagents.

FURTHER EXPLORATION 26.1
Relative Acidities of 1,3-Cyclopentadiene and Pyrrole

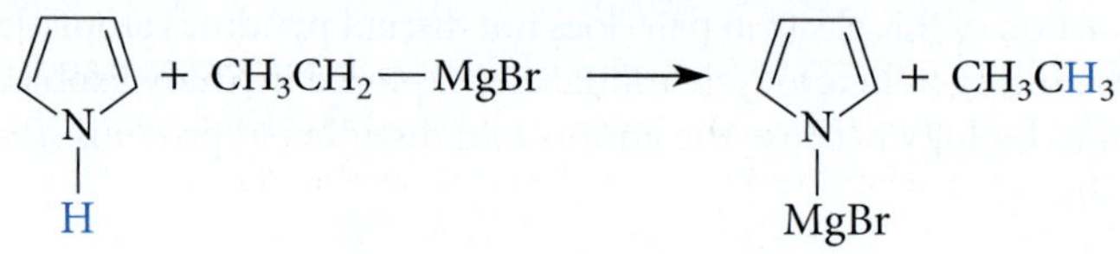

(26.13)

Focused Problems

26.6 (a) Suggest a reason why pyridine is miscible with water, whereas pyrrole has little water solubility.

(b) Indicate whether you would expect imidazole to have high or low water solubility, and explain why.

26.7 (a) The compound 4-(dimethylamino)pyridine protonates to give a conjugate acid with a pK_a value of 9.9. This compound is 4.7 pK_a units more basic than pyridine itself. Draw the structure of the conjugate acid of 4-(dimethylamino)pyridine, and, using resonance structures, explain why 4-(dimethylamino)pyridine is much more basic than pyridine.

(b) What product is expected when 4-(dimethylamino)pyridine reacts with CH_3I?

26.8 Protonation of aniline causes a dramatic shift of its UV spectrum to lower wavelengths, but protonation of pyridine has almost no effect on its UV spectrum. Explain the difference.

26.3 THE CHEMISTRY OF FURAN, PYRROLE, AND THIOPHENE

A. Electrophilic Aromatic Substitution

Furan, thiophene, and pyrrole, like benzene, undergo electrophilic aromatic substitution reactions. Study Problem 26.2 shows how to predict the ring carbon at which substitution occurs in these compounds by examining the carbocation intermediates involved in the substitution reactions at the two different positions and applying Hammond's postulate.

Study Problem 26.2

Using the nitration of pyrrole as an example, predict whether electrophilic aromatic substitution occurs predominantly at carbon-2 or carbon-3.

Solution Recall (Eqs. 16.19a and b, Sec. 16.4C) that the electrophile in nitration is the nitronium ion, $^+NO_2$. Substitution at the two different positions of pyrrole by the nitronium ion gives different carbocation intermediates:

Substitution at carbon-2:

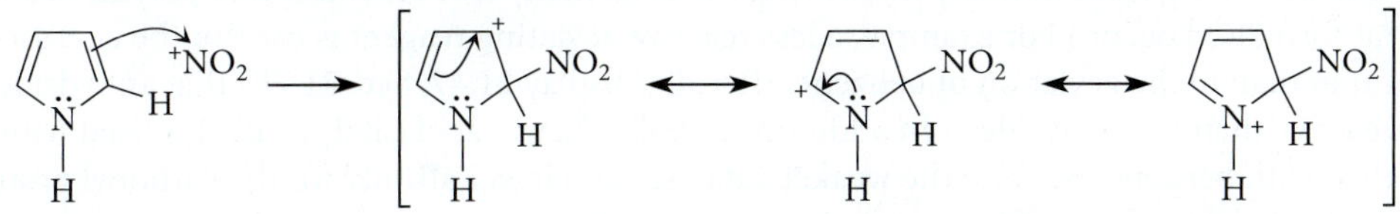

(26.14a)

Substitution at carbon-3:

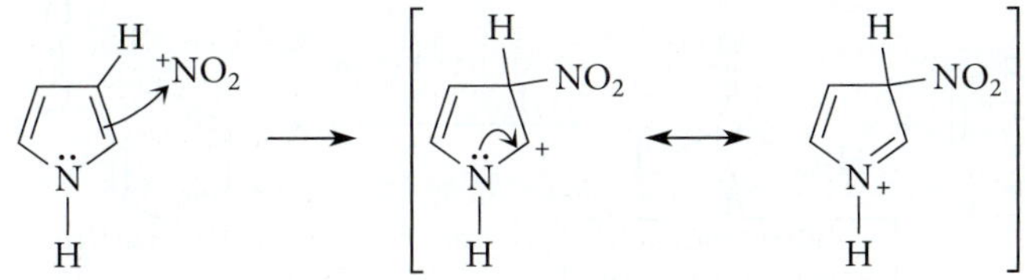

(26.14b)

The carbocation resulting from substitution at carbon-2 has more important resonance structures and is therefore more stable than the carbocation resulting from substitution at carbon-3. Applying Hammond's postulate, we predict that the reaction involving the more stable intermediate should be the faster reaction. Consequently, nitration should occur at carbon-2. The experimental facts are as follows:

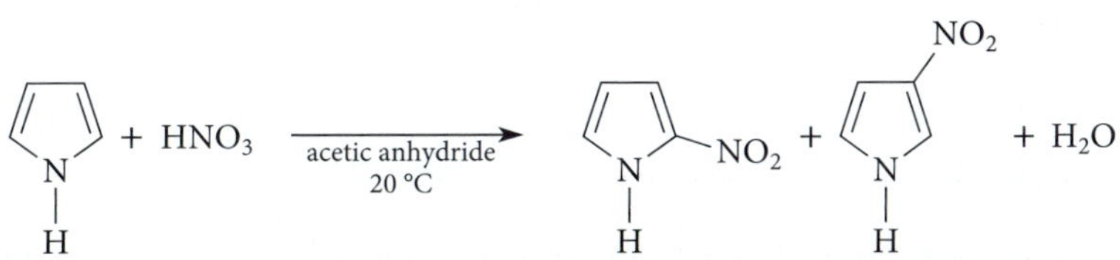

(50% yield) (15% yield) (26.15)

2-Nitropyrrole is the major nitration product of pyrrole, as predicted. Nothing is really wrong with substitution at carbon-3; substitution at carbon-2 is simply more favorable. As Eq. 26.15 shows, some 3-nitropyrrole is also obtained in the reaction.

As Study Problem 26.2 suggests, electrophilic substitution of pyrrole occurs predominantly at the 2-position. Similar results are observed with furan and thiophene:

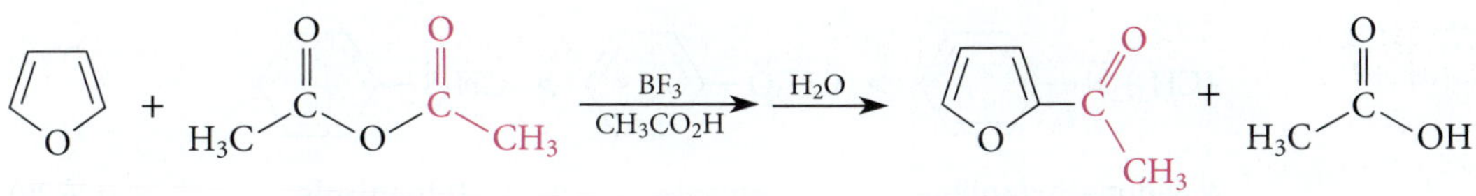

(a Friedel–Crafts acylation reaction) (75–92% yield) (26.16)

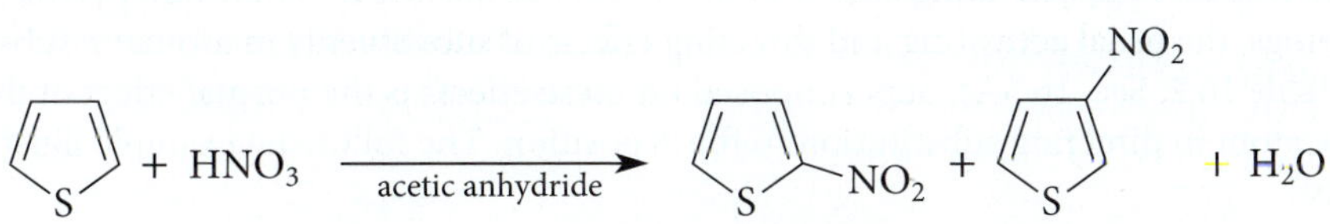

(70% yield) (5% yield) (26.17)

Pyrrole, furan, and thiophene are all much more reactive than benzene in electrophilic aromatic substitution. Although precise reactivity ratios depend on the particular reaction, the relative rates of bromination are typical:

pyrrole	>	furan	>	thiophene	>	benzene
3×10^{18}		6×10^{11}		5×10^{9}		1

(26.18)

Milder reaction conditions must be used with more reactive compounds. (Reaction conditions that are too vigorous in many cases bring about so many side reactions that polymerization and tar formation occur.) For example, a less reactive acylating reagent is used in the acylation of furan than in the acylation of benzene. [Recall (Display 21.47, Sec. 21.7E) that anhydrides are less reactive than acid chlorides.] In addition, a weaker Lewis-acid catalyst (BF_3) is used with furan than with benzene. Because the weaker catalyst has a lower affinity for the carbonyl group of the ketone product than $AlCl_3$, a catalytic amount of BF_3 can be used.

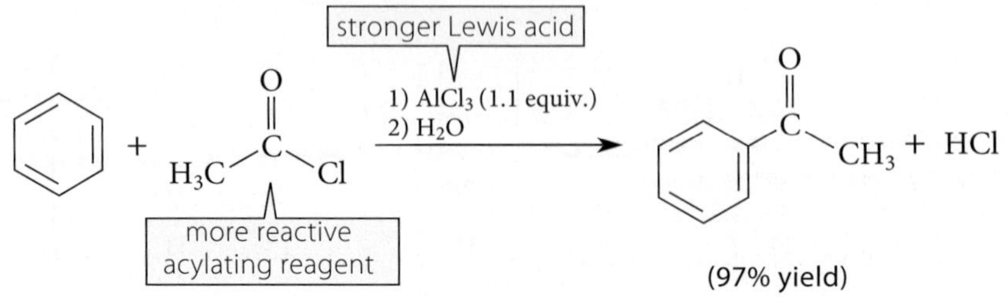

(97% yield) (26.19a)

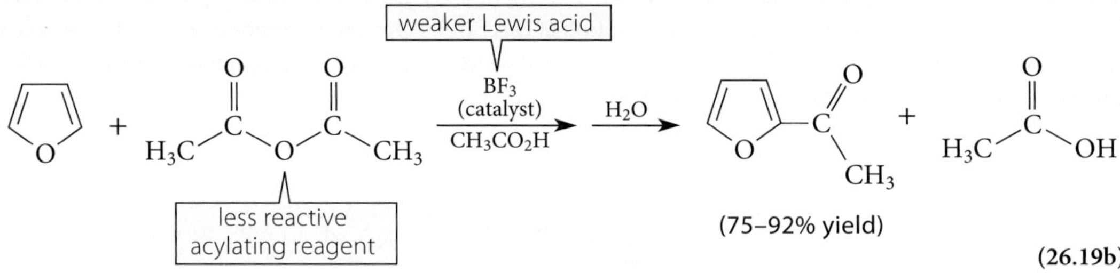

(75–92% yield) (26.19b)

The reactivity order of the heterocycles (Display 26.18) is a consequence of the relative abilities of the heteroatoms to stabilize positive charge in the intermediate carbocations (as in Eq. 26.14a, for example). Both pyrrole and furan have heteroatoms from the second period of the periodic table. Because nitrogen is better than oxygen at delocalizing positive charge—nitrogen is less electronegative—pyrrole is more reactive than furan. The sulfur of thiophene is a third-period element and, although it is less electronegative than oxygen, its 3p orbitals overlap less efficiently with the 2p orbitals of the aromatic π-electron system. (See Fig. 16.7, Sec. 16.5B, for the 3p–2p interaction.) In fact, the reactivity *order* of the heterocycles in aromatic substitution parallels the reactivity order of the correspondingly substituted benzene derivatives:

Relative reactivities in electrophilic aromatic substitution:

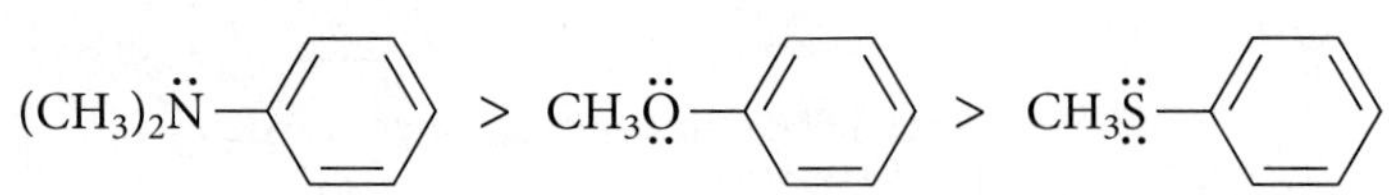

N,N-dimethylaniline **anisole** **thioanisole** (26.20)

When we consider the activating and directing effects of substituents in furan, pyrrole, and thiophene rings, the usual activating and directing effects of substituents in aromatic substitution apply (see Table 16.2, Sec. 16.5A). Superimposed on these effects is the normal effect of the heterocyclic atom in directing substitution to the 2-position. The following example illustrates these effects:

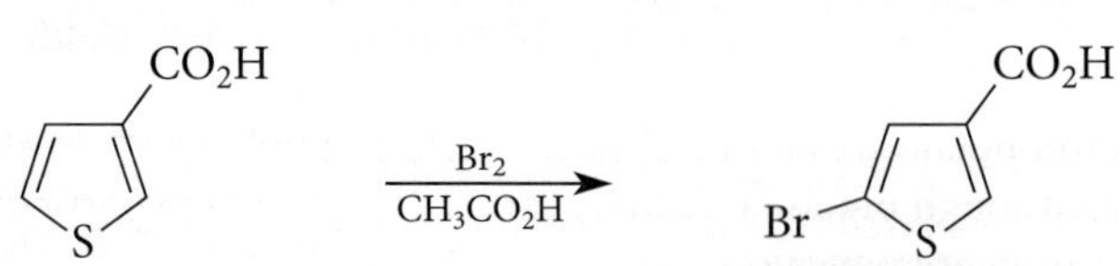

3-thiophenecarboxylic acid **5-bromo-3-thiophenecarboxylic acid**
(69% yield) (26.21)

The position of substitution follows from an analysis of the resonance and polar effects in the possible carbocation intermediates. This process, which is the same one used to understand the directing effects of substituents in benzene chemistry (Sec. 16.5), is illustrated in Study Problem 26.3.

Study Problem 26.3

Explain the position of substitution observed in the bromination of thiophene-3-carboxylic acid shown in Eq. 26.21. Draw the carbocation intermediates for all possible positions of substitution, and show that the intermediate in the observed substitution is the most stable one.

Solution The carbocation intermediates involved in the three possible positions of substitution are the following.

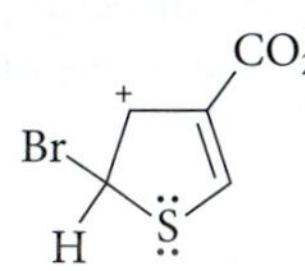

A
(substitution at C-5)

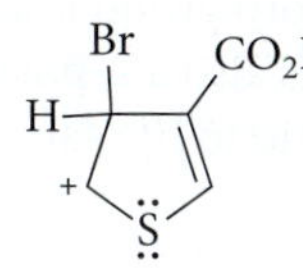

B
(substitution at C-4)

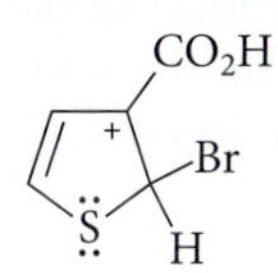

C
(substitution at C-2)

Carbocation *A* has three resonance structures:

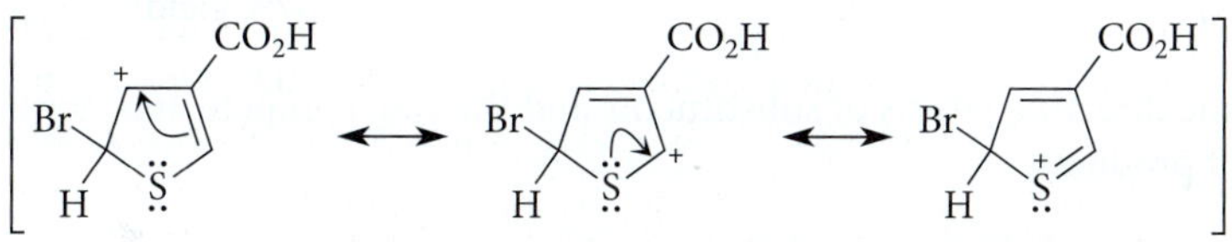

resonance structures for carbocation *A*

Carbocation *B* has only two resonance structures:

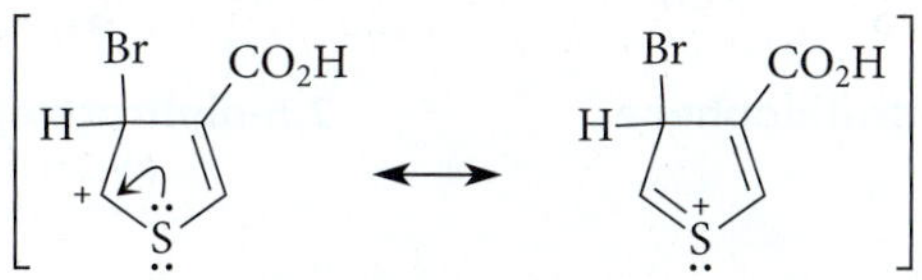

resonance structures for carbocation *B*

Carbocation *C*, like ion *A*, has three resonance structures. However, in one of the structures (*red*), the electron deficiency and positive charge are located on the carbon adjacent to an electron-withdrawing group, the carboxylic acid.

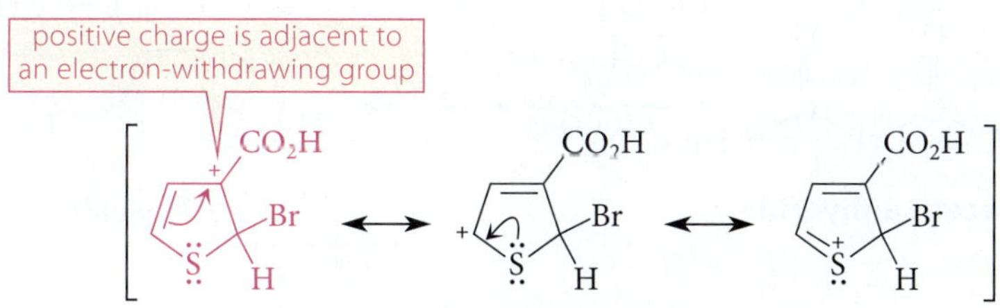

resonance structures for carbocation *C*

Therefore, carbocation *A* is the most stable carbocation intermediate because it has the greatest number of resonance structures without placing the positive charge next to the electron-withdrawing substituent. As a result, the most favorable substitution involves this intermediate—that is, substitution at C-5.

If we count around the carbon framework from the carboxylic acid substituent, we find that the substitution in Eq. 26.21 occurs at the carbon that is in a 1,3-relationship to the substituted carbon. In other words, substitution has occurred meta to the carboxylic acid group. Recall (see Table 16.2, Sec. 16.5A) that the carboxylic acid group is a "meta-directing" substituent.

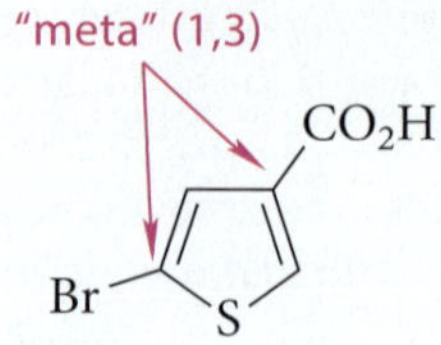

(26.22)

Thiophene itself substitutes at the carbon next to the sulfur. Therefore, the observed product satisfies the directing effects of both the heteroatom and the substitutions. The directing effects of substituents are exactly as they are in benzene substitution, provided that we view the ortho, para, or meta relationship *through the carbon framework and not through the heteroatom.*

In the following example, the chloro group is an ortho, para-directing group. Because the position "para" to the chloro group is also a 2-position, both the sulfur of the ring and the chloro group direct the incoming nitro group to the same position.

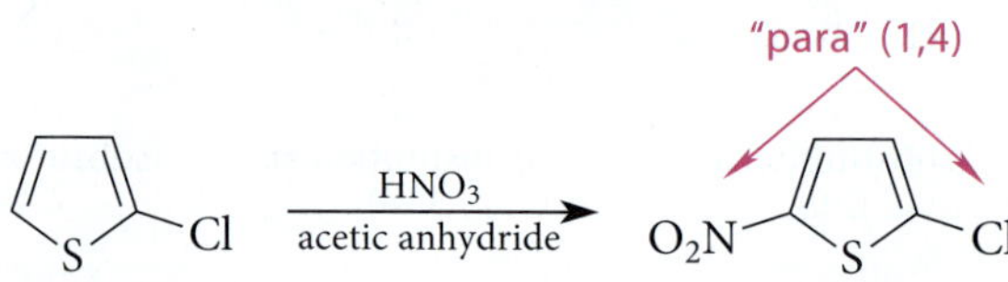

2-chlorothiophene **2-chloro-5-nitrothiophene** (57% yield) (26.23)

When the directing effects of substituents and the ring compete, it is not unusual to observe mixtures of products.

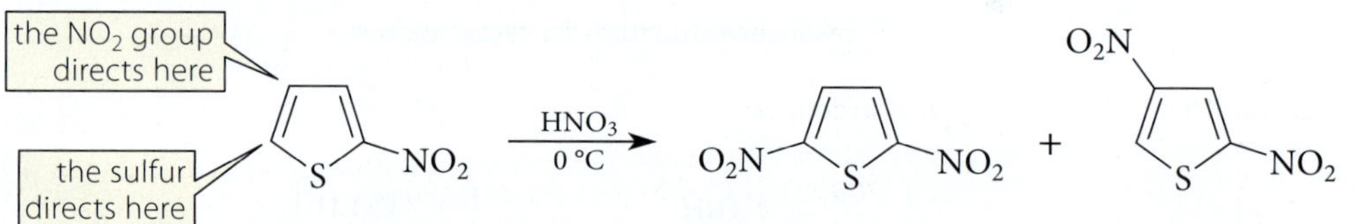

2-nitrothiophene **2,5-dinitrothiophene** (44%) **2,4-dinitrothiophene** (56%)

(60% yield) (26.24)

Finally, if both 2-positions are occupied, 3-substitution takes place.

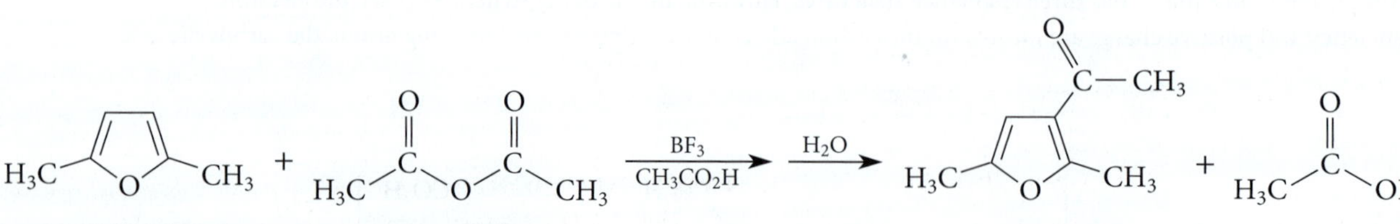

2,5-dimethylfuran **acetic anhydride** (65% yield) (26.25)

B. Addition Reactions of Furan

The previous sections focused on the aromatic character of furan, pyrrole, and thiophene. A furan, pyrrole, or thiophene could, however, be viewed as a 1,3-butadiene with its terminal carbons "tied down" by a heteroatom bridge.

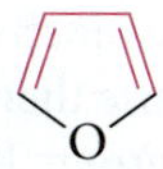

"butadiene" unit within furan

Do the heterocycles ever behave chemically as if they are conjugated dienes? Of the three heterocyclic compounds furan, pyrrole, and thiophene, furan has the least resonance energy (see Table 26.1, Sec. 26.1B) and, by implication, the least aromatic character. Consequently, of the three compounds, furan has the greatest tendency to behave like a conjugated diene.

One characteristic reaction of conjugated dienes is *conjugate addition* (Sec. 15.4A). Indeed, furan does undergo some conjugate addition reactions. For example, it undergoes conjugate addition of bromine and methanol in methanol solvent; the conjugate-addition product then undergoes an S_N1 reaction with the methanol. (Write mechanisms for both parts of this reaction; refer to Sec. 15.4A, if necessary.)

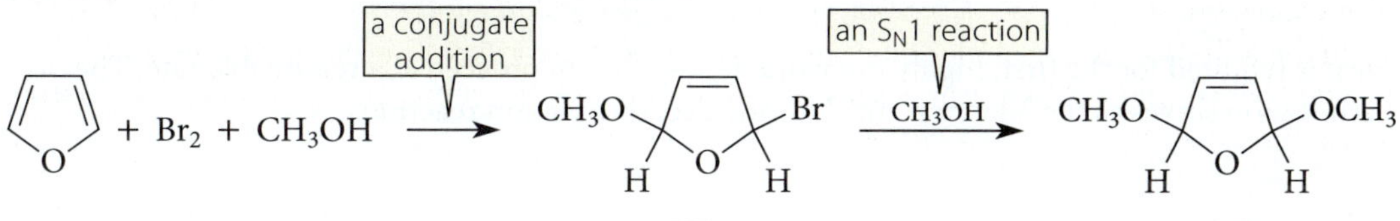

mixture of stereoisomers (72–76% yield)

(26.26)

Another manifestation of the conjugated-diene character of furan is that it undergoes Diels–Alder reactions (Sec. 15.3) with reactive dienophiles such as maleic anhydride.

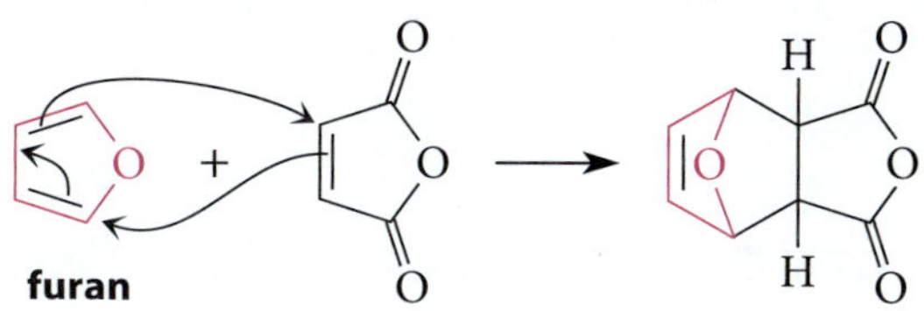

furan **maleic anhydride** (>90% yield) (26.27)

C. Side-Chain Reactions

Many reactions occur at the side chains of heterocyclic compounds without affecting the rings, just as some reactions occur at the side chain of a substituted benzene (Secs. 17.1–17.5).

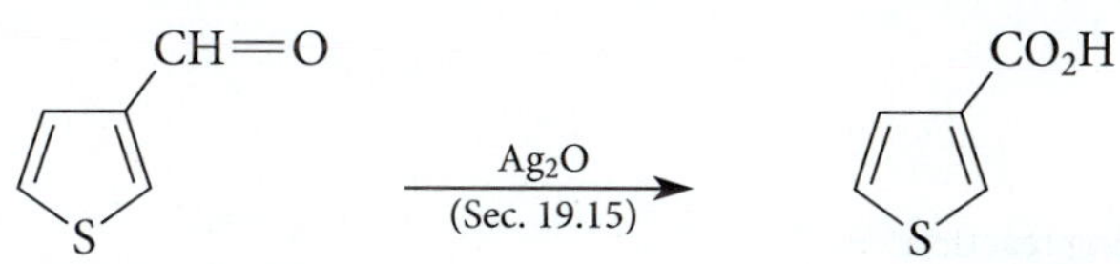

3-thiophenecarbaldehyde **3-thiophenecarboxylic acid** (95–97% yield) (26.28)

A particularly useful example of a side-chain reaction is removal of a carboxy group directly attached to the ring (*decarboxylation*). This reaction is brought about by strong heating, in some cases with catalysts.

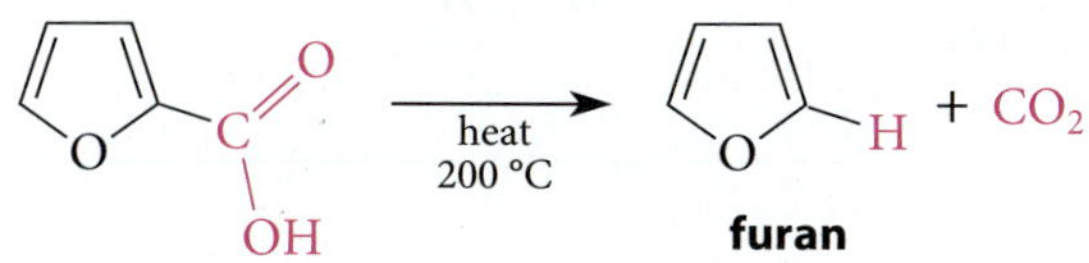

furan

2-furancarboxylic acid (26.29a)

The conditions of the uncatalyzed reaction are much harsher than the conditions of the decarboxylations discussed in Sec. 20.11 because there is no "electron sink" (Sec. 20.11A) to accept the electrons from the carboxylic acid group. It may be that this reaction proceeds by a very unfavorable internal protonation (or protonation by a second carboxylic acid molecule) to give a carbocation, which is actually the species that decarboxylates.

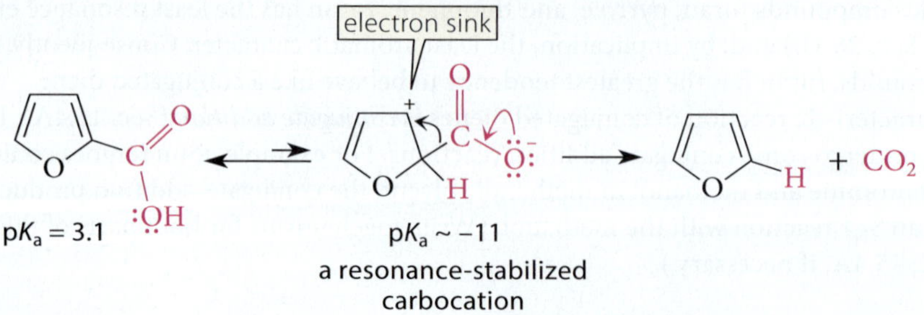

(26.29b)

Heat is required for the first, highly unfavorable, reaction to proceed at a reasonable rate. The carbocation serves as the "electron sink" for the decarboxylation reaction.

Focused Problems

26.9 Complete each of the following reactions by giving the principal organic product(s). For (b), write a curved-arrow mechanism that shows the carbocation intermediate and its resonance structures.

(a)

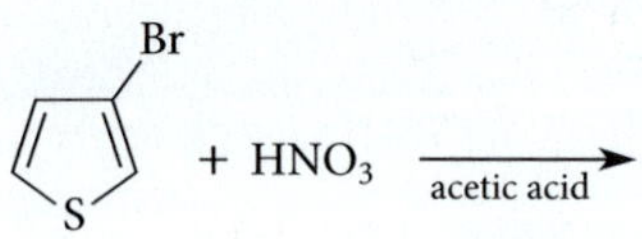

(b)

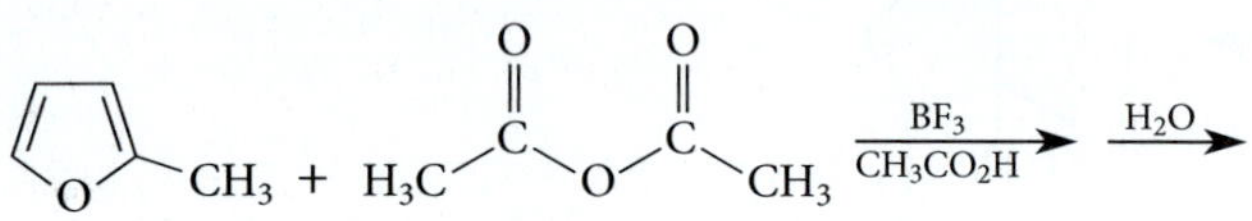

(c)

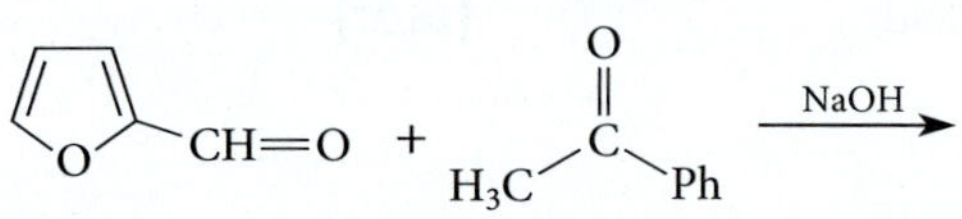

(d)

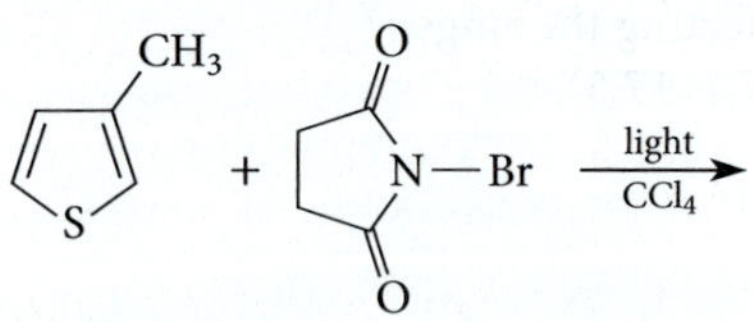

26.10 Write a curved-arrow mechanism for the following reaction.

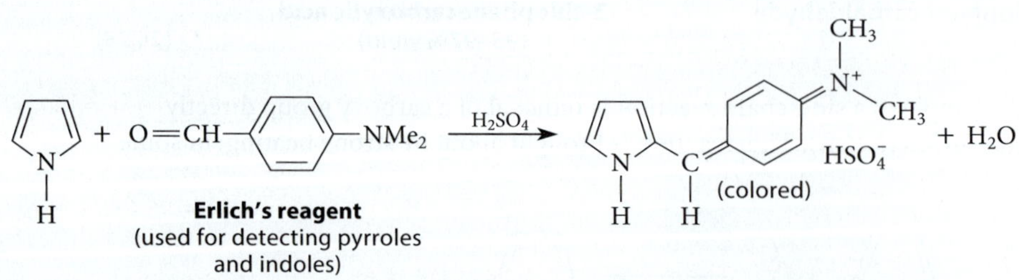

26.4 THE CHEMISTRY OF PYRIDINE

A. Electrophilic Aromatic Substitution

In general, it is difficult to prepare monosubstituted pyridines by electrophilic aromatic substitution because pyridine has a very low reactivity (much lower than benzene). An important reason for this low reactivity is that the nitrogen of pyridine is protonated under the very acidic conditions of most electrophilic aromatic substitution reactions (Eq. 26.6, Sec. 26.2A). The resulting positive charge on nitrogen makes it difficult to form a carbocation intermediate, which would place a second positive charge within the same ring.

Fortunately, a number of monosubstituted pyridines are available from natural sources. Among these are the methylpyridines, or *picolines*:

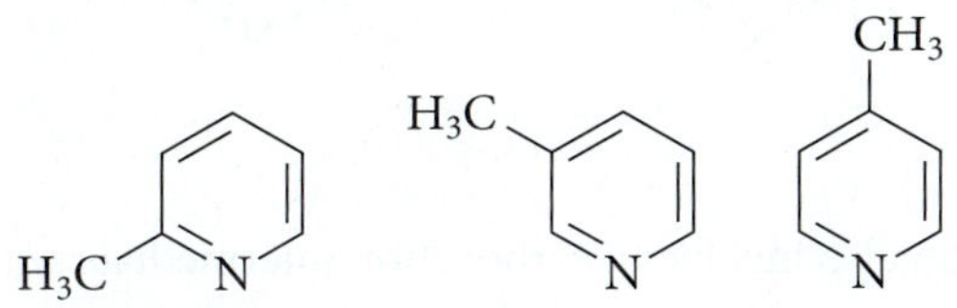

α-picoline **β-picoline** **γ-picoline** (26.30)

The picolines (and other methylated pyridines) are obtained from *coal tar* (Sec. 16.7). Another useful monosubstituted derivative of pyridine is *nicotinic acid* (pyridine-3-carboxylic acid), which is conveniently prepared in a number of ways, such as the side-chain oxidation of nicotine, an alkaloid present in tobacco (Fig. 23.4, Sec. 23.12B).

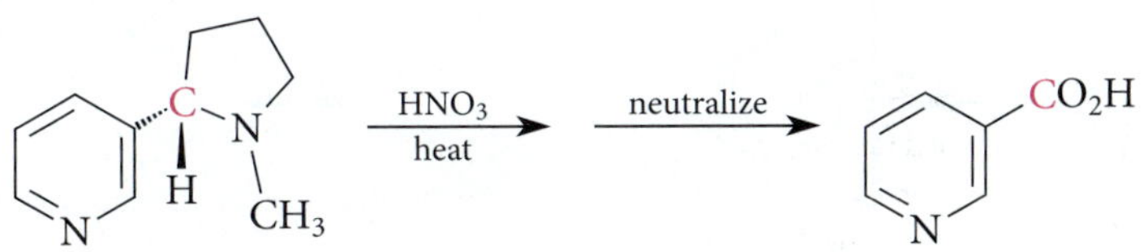

nicotine **nicotinic acid** (70% yield) (26.31)

(The nitric acid in this reaction is an oxidizing agent.)

Although electrophilic aromatic substitution reactions are not useful for introducing substituents into pyridine itself, pyridine rings substituted with activating groups such as methyl groups do undergo such reactions.

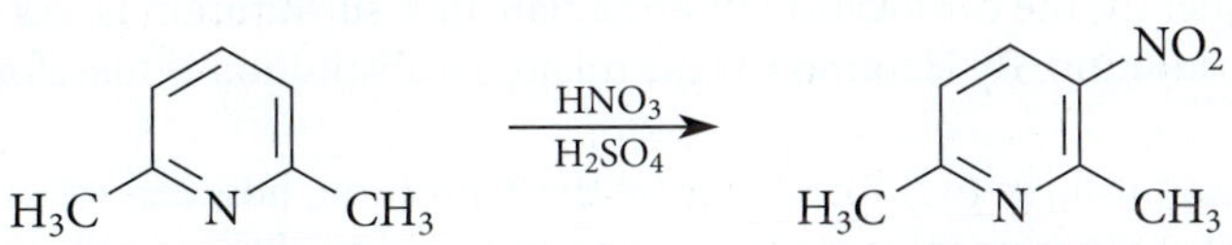

2,6-dimethylpyridine (2,6-lutidine) **2,6-dimethyl-3-nitropyridine** (81% yield) (26.32)

Reactions such as this, which occur in acidic solution, undoubtedly take place on the very small amount of the unprotonated pyridine that is in rapid equilibrium with the much larger amount of its conjugate acid. Because the reactive species—the unprotonated heterocycle—is present in very small concentration, the methyl-substituted pyridines are not very reactive, despite the presence of the activating methyl substituents.

As the example in Eq. 26.32 illustrates, substitution in pyridine generally takes place in the 3-position. Although the methyl groups in Eq. 26.32 also direct substitution to the 3-position, the tendency of pyridine to undergo 3-substitution is general even in the absence of such directing groups. As with other electrophilic substitutions, an understanding of this directing effect comes from an examination of the carbocation intermediates formed in substitution at different positions. Substitution in the 3-position gives a carbocation with three different resonance structures:

3-Substitution:

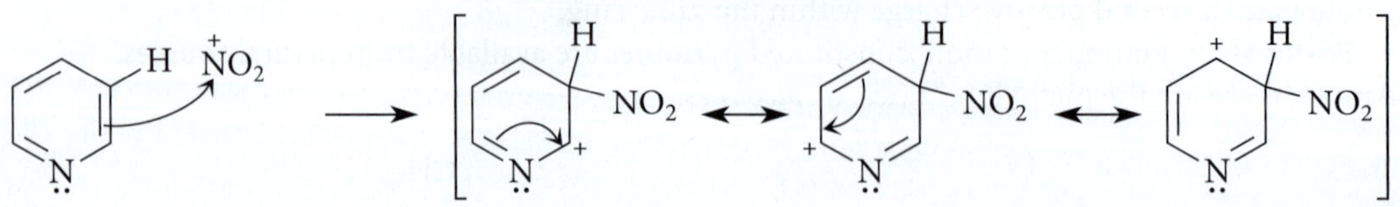

(26.33a)

Substitution at the 4-position also involves a carbocation intermediate with three resonance structures, but the one shown in red is particularly unstable and unimportant because *the nitrogen, an electronegative atom, is electron-deficient.* (See the rules for the relative importance of resonance structures in Sec. 3.3C.)

4-Substitution:

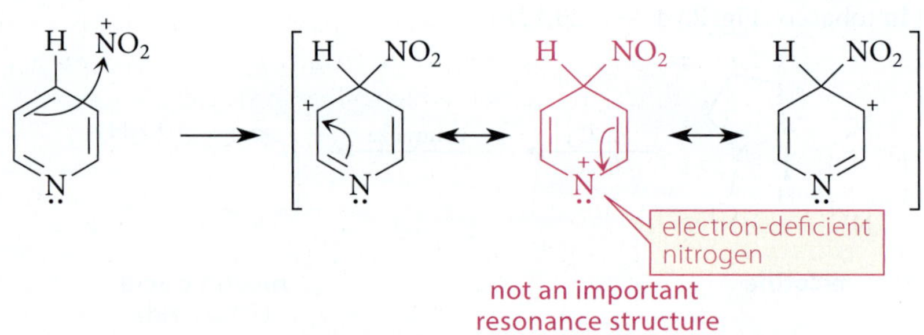

(26.33b)

Be sure to understand that the nitrogen in the red structure is *very* different from the nitrogen in pyrrole during electrophilic aromatic substitution (Eq. 26.14a, Sec. 26.3A). The pyrrole nitrogen is also positively charged, but it is not electron-deficient because it has a complete octet. In contrast, an *electron-deficient* electronegative atom such as the one in Eq. 26.33b is very unfavorable energetically. Consequently, the carbocation intermediate in 4-substitution is less stable than the intermediate in 3-substitution. By Hammond's postulate, 3-substitution is therefore the faster reaction.

If electrophilic substitution in pyridine occurs at the 3-position, how can we obtain pyridine derivatives substituted at other positions? One compound used to obtain 4-substituted pyridines is pyridine-*N*-oxide, formed by the oxidation of pyridine with 30% hydrogen peroxide.

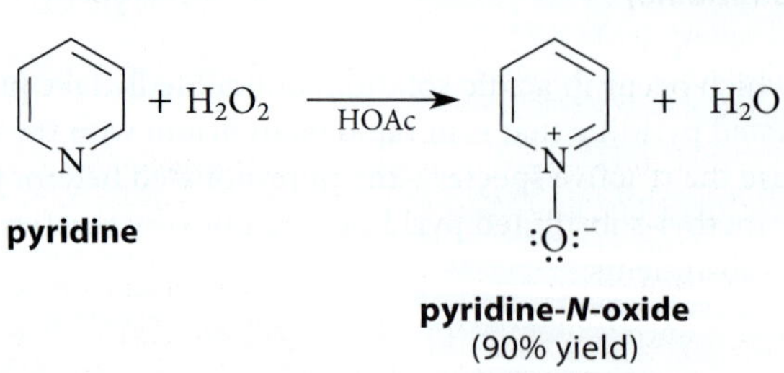

(26.34)

An analogy to pyridine-*N*-oxide from benzene chemistry is phenoxide, the conjugate base of phenol. Just as phenol or phenoxide is much more reactive in electrophilic aromatic substitution than benzene (Sec. 18.9), pyridine-*N*-oxide is much more reactive than pyridine. Because the nitrogen of pyridine-*N*-oxide has a positive charge, this compound is *much* less reactive than phenol or phenoxide. Nevertheless, pyridine-*N*-oxide undergoes useful aromatic substitution reactions, and substitution occurs in the 4-position.

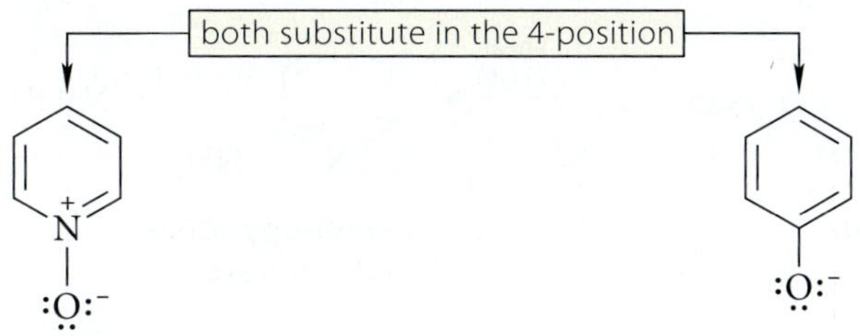

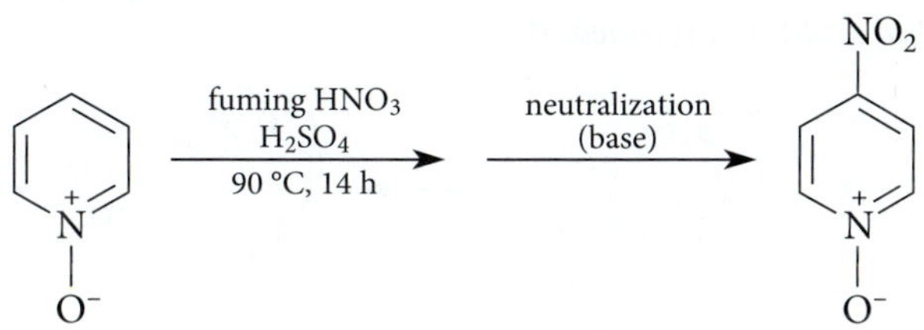

(90% yield) (26.35)

Once the *N*-oxide function is no longer needed, it can be removed by catalytic hydrogenation; this procedure also reduces the nitro group. A reaction with trivalent phosphorus compounds, such as PCl_3, removes the *N*-oxide function without reducing the nitro group.

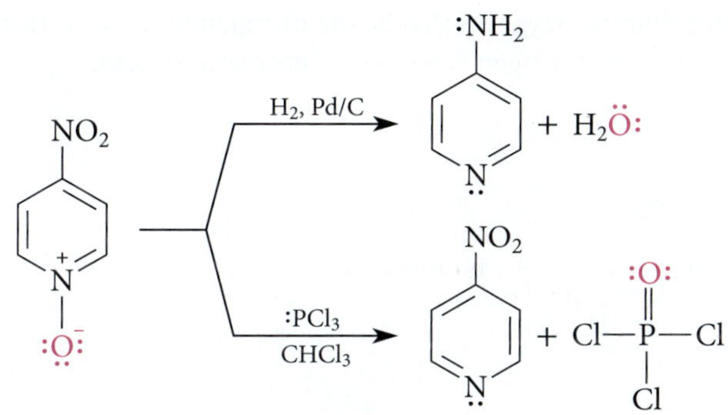

(26.36)

Focused Problems

26.11 Which should be more reactive in nitration: β-picoline or α-picoline? Explain using resonance structures, and give the major nitration product(s) in each case.

26.12 By drawing resonance structures for the carbocation intermediates, show why aromatic substitution in pyridine-*N*-oxide occurs at the 4-position rather than at the 3-position.

26.13 (a) Draw a curved-arrow mechanism for the reduction of pyridine-*N*-oxide by $:PCl_3$. (*Hint:* In the first step, the oxygen of the *N*-oxide reacts as a nucleophilic center with the phosphorus as an electrophilic center.)

(b) This is an oxidation–reduction reaction. What is oxidized and what is reduced? How many electrons are involved in this redox reaction?

The Chichibabin reaction is named for its discoverer, Aleksey Chichibábin (1871–1945). He became a professor at the Imperial College of Technology in Moscow in 1909 and remained there until 1929. After the death of his daughter, also a chemist, in a tragic chemical accident, he moved to Paris. (Chichibábin blamed the authorities for negligence that led to the accident.) In 1931 he began working at the Collège de France, remaining there until his death in 1945. He also served as a consultant to both French and American industry. He was stripped of his Soviet citizenship and his position in the Academy of Sciences in 1936, but his standing was restored posthumously in 1990.

B. Nucleophilic Aromatic Substitution

In contrast to its low reactivity in *electrophilic* aromatic substitution, the pyridine ring readily undergoes *nucleophilic* aromatic substitution. A rather unusual reaction of this type can be used to prepare 2-aminopyridine. In this reaction, called the **Chichibabin reaction**, treatment of a pyridine derivative with the strong base sodium amide, $Na^+ \ ^-NH_2$ (Sec. 10.6A) brings about the direct substitution of an amino group for a ring hydrogen.

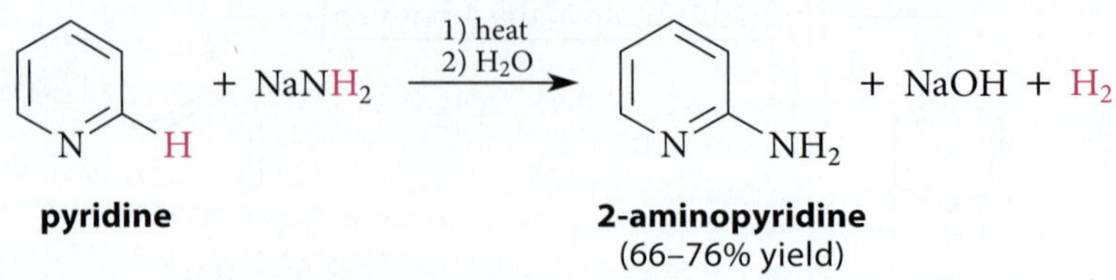

(26.37)

In the first step of the mechanism, the amide ion reacts as a nucleophile at the 2-position of the ring to form a *tetrahedral addition intermediate.*

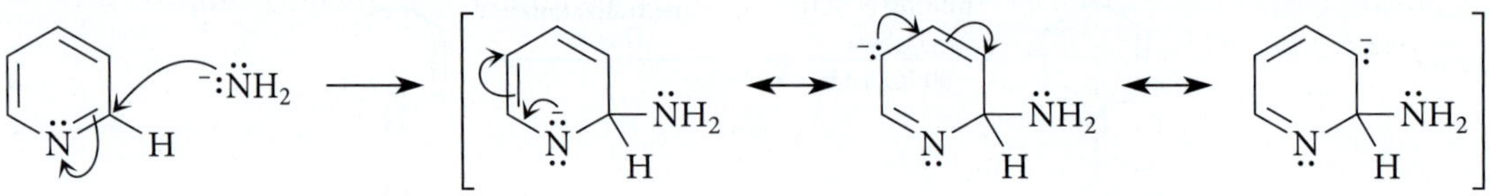

(26.38a)

This step of the mechanism can be understood by recognizing that *the C═N linkage of the pyridine ring is somewhat analogous to a carbonyl group*—that is, carbon at the 2-position has some of the character of a carbonyl carbon and can react with nucleophiles. The C═N group of pyridine, though, is *much* less reactive than a carbonyl group because a nitrogen is a considerably poorer electron acceptor than an oxygen—that is, the nitrogen anion is a stronger base than an oxygen anion—and because the nitrogen is part of an aromatic system.

Compare:

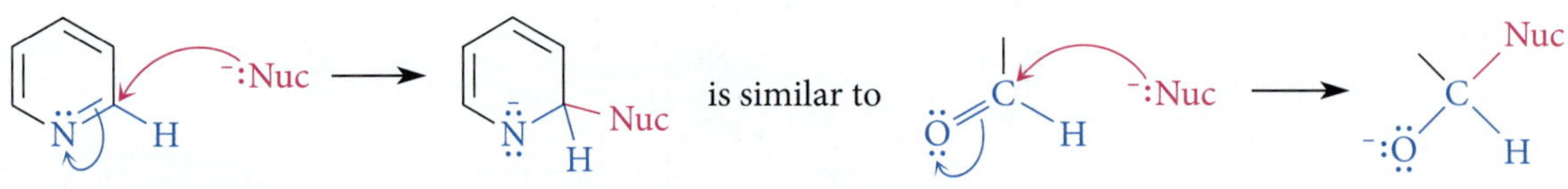

(⁻:Nuc = nucleophile)

(26.38b)

In the second step of the mechanism, the leaving group, a *hydride ion,* is lost.

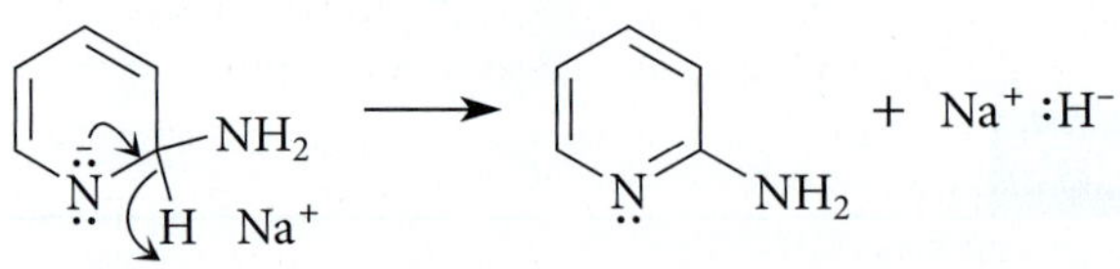

(26.38c)

Hydride ion is a poor, and therefore unusual, leaving group because it is very basic. This reaction occurs for two reasons. First, the aromatic pyridine ring is reformed; aromaticity lost in the formation of the tetrahedral addition intermediate is regained when the leaving group departs. Second, the basic hydride produced in the reaction reacts with the $—NH_2$ group irreversibly to form dihydrogen (a gas) and the resonance-stabilized conjugate-base anion of 2-aminopyridine.

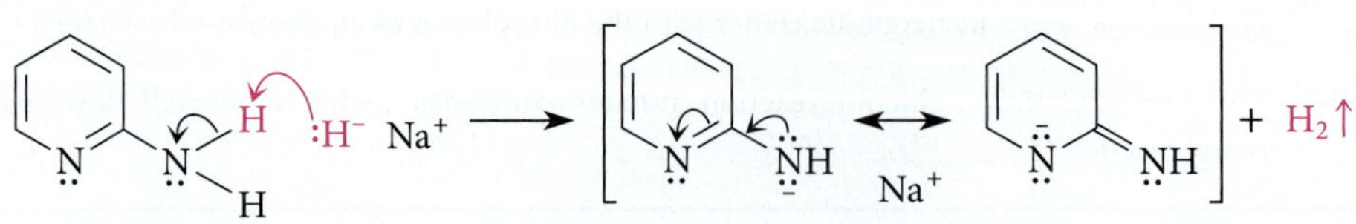

(two of several resonance structures)

(26.38d)

The neutral 2-aminopyridine is formed when water is added in a separate step.

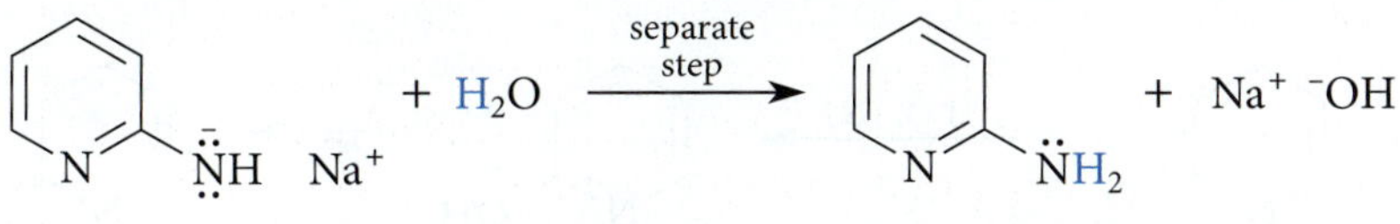

(26.38e)

A reaction similar to the Chichibabin reaction occurs with organolithium reagents.

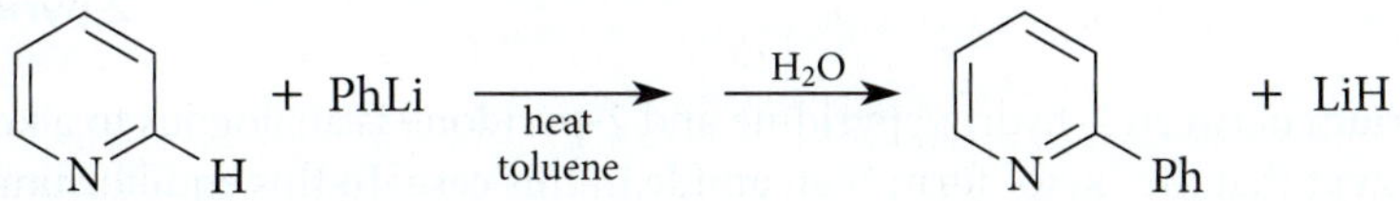

pyridine **2-phenylpyridine** (40–49% yield) (26.38f)

When pyridine is substituted with a much better leaving group than hydride at the 2-position, it reacts more rapidly with nucleophiles. The 2-halopyridines, for example, readily undergo substitution of the halogen by other nucleophiles under conditions that are much milder than those used in the Chichibabin reaction.

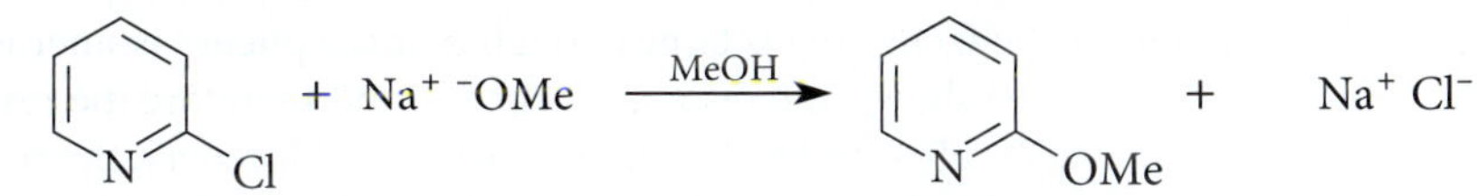

2-chloropyridine **2-methoxypyridine** (95% yield) (26.39)

This nucleophilic substitution can also be related to the analogous reaction of a carbonyl compound. This reaction of a 2-chloropyridine resembles the nucleophilic acyl substitution reaction of an acid chloride—except that acid chlorides are *much* more reactive than 2-halopyridines.

Compare:

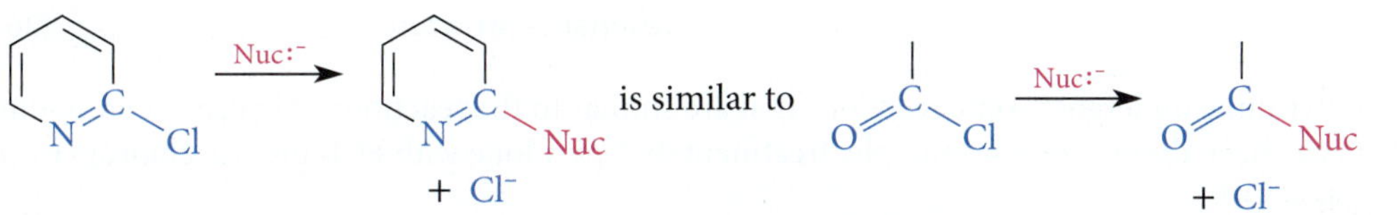

(26.40)

The nucleophilic substitution reactions of pyridines can be classified as *nucleophilic aromatic substitution* reactions. Recall that aryl halides undergo nucleophilic aromatic substitution when the benzene ring is substituted with electron-withdrawing groups (Sec. 18.4). The "electron-withdrawing group" in the reactions of pyridines is the pyridine nitrogen itself. The tetrahedral addition intermediate (Eq. 26.38a) is analogous to the Meisenheimer complex of nucleophilic aromatic substitution (Eqs. 18.20a–c, Sec. 18.4). This discussion shows that there is a mechanistic parallel between three types of reaction: (1) nucleophilic acyl substitution, a typical reaction of carboxylic acid derivatives; (2) nucleophilic aromatic substitution; and (3) nucleophilic substitution on the pyridine ring.

The 2-aminopyridines formed in the Chichibabin reaction (Eq. 26.37) serve as starting materials for a variety of other 2-substituted pyridines. For example, diazotization of 2-amino-pyridine gives a diazonium ion that can undergo substitution reactions (see Secs. 23.10A and B).

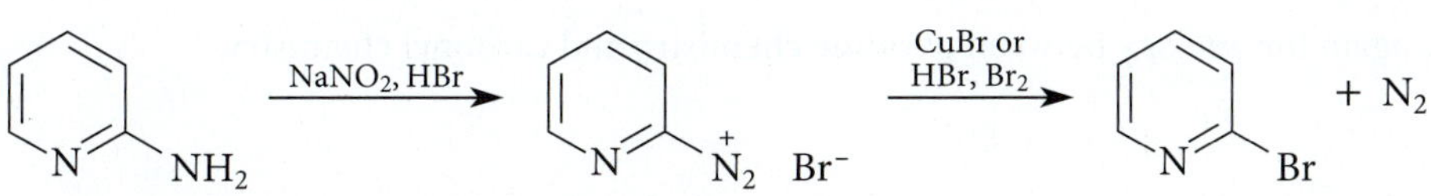

2-aminopyridine **2-pyridinediazonium bromide** **2-bromopyridine** (86–92% yield) (26.41)

When the diazonium salt reacts with water, it is hydrolyzed to 2-hydroxypyridine, which in most solvents exists in its carbonyl form, 2-pyridone.

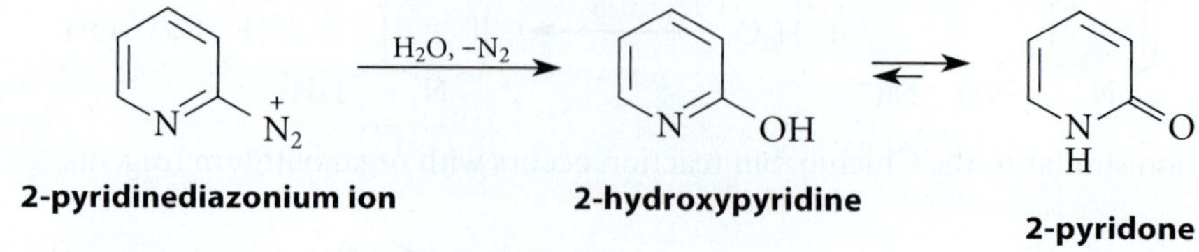

(26.42)

The equilibrium between 2-hydroxypyridine and 2-pyridone is analogous to a keto–enol equilibrium, except that the "keto" form is an amide in this case. In this equilibrium, the ratio of the hydroxy form to the carbonyl form is 1:910 in water, but the ratio varies with concentration and with solvent; in the vapor phase, the ratio is 1:0.4. The important points about this equilibrium, however, are (1) enough of each form is present so that either form can be involved in chemical reactions, and (2) a *much* greater proportion of carbonyl isomer is present than there is in phenol (Eq. 22.19, Sec. 22.2). Why should this be so? A major factor that determines whether an aromatic hydroxy compound exists as a carbonyl or hydroxy ("enol") form is whether the energetic advantage of aromaticity—that is, the *resonance stabilization* of the aromatic hydroxy isomer—outweighs the large carbonyl C═O bond energy. In the case of phenol itself, the resonance stabilization of the benzene ring is large enough that the phenol isomer is strongly preferred. As Table 26.1 (Sec. 26.1B) shows, the resonance energy, and therefore the resonance stabilization, of pyridine is considerably smaller than that of benzene. Moreover, the resonance interaction of the amide nitrogen with the carbonyl group further stabilizes 2-pyridone. As the following structure shows, the resulting resonance structure is aromatic. Consequently, 2-pyridone itself has a significant amount of aromatic character. The keto isomer of phenol has no stabilizing contribution of this sort.

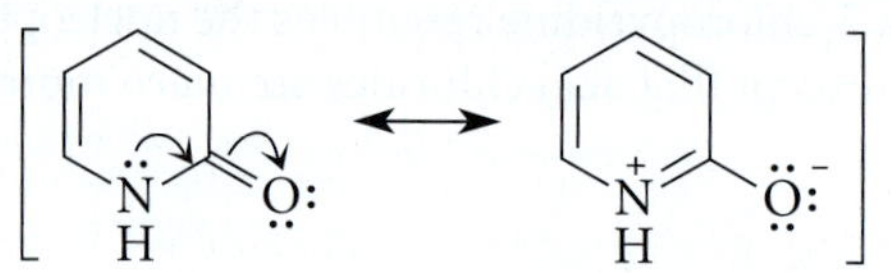

(26.43)

2-Pyridone undergoes some reactions that are similar to the reactions of hydroxy compounds that have been discussed. For example, treatment of 2-pyridone with PCl_5 gives 2-chloropyridine.

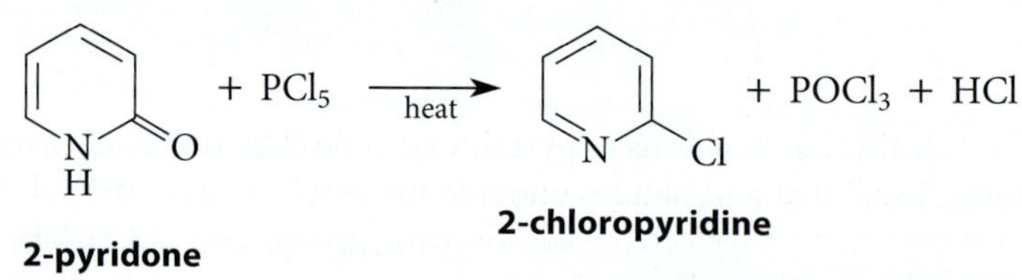

(26.44)

If we think of 2-pyridone in terms of its 2-hydroxypyridine isomer, this reaction is similar to the preparation of acid chlorides from carboxylic acids (Eq. 20.42, Sec. 20.9A).

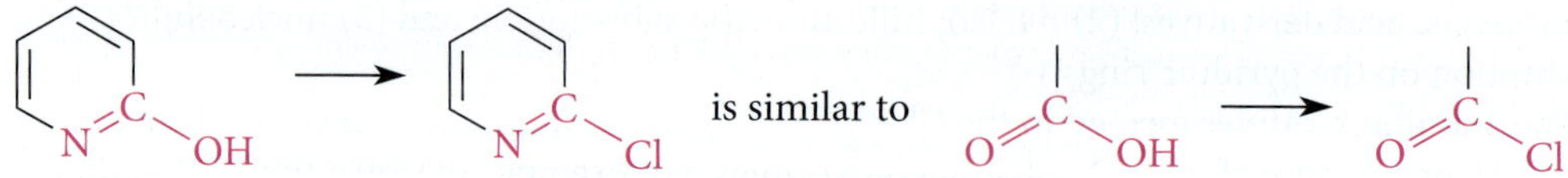

(26.45)

Notice again the analogy between pyridine chemistry and carbonyl chemistry.

Pyridines with leaving groups in the 4-position also undergo nucleophilic substitution reactions.

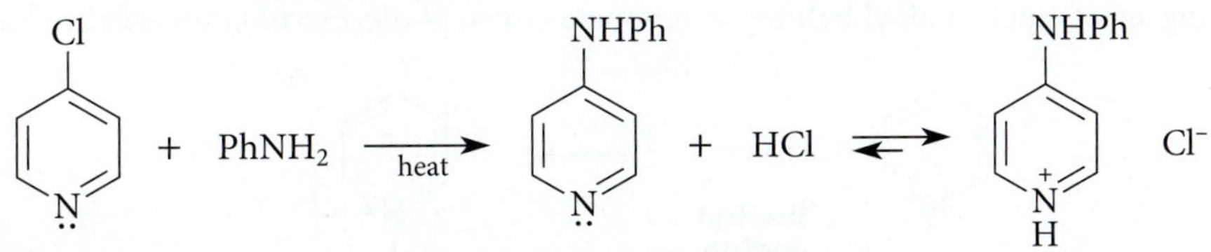

(26.46)

As the examples in this section suggest, nucleophilic substitution reactions at the 2- and 4-positions of a pyridine ring are particularly common. The reason follows from the mechanism of this type of reaction: Negative charge in the addition intermediate is delocalized onto the electronegative pyridine nitrogen.

Substitution at carbon-2: (Y = leaving group, $^{-}$:Nuc = nuclephile)

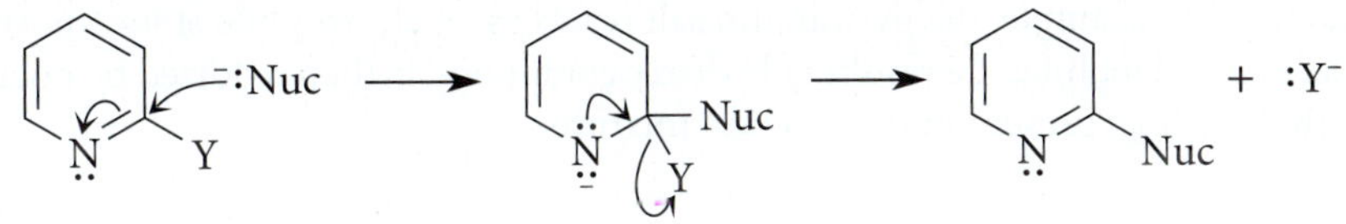

(26.47a)

Substitution at carbon-4:

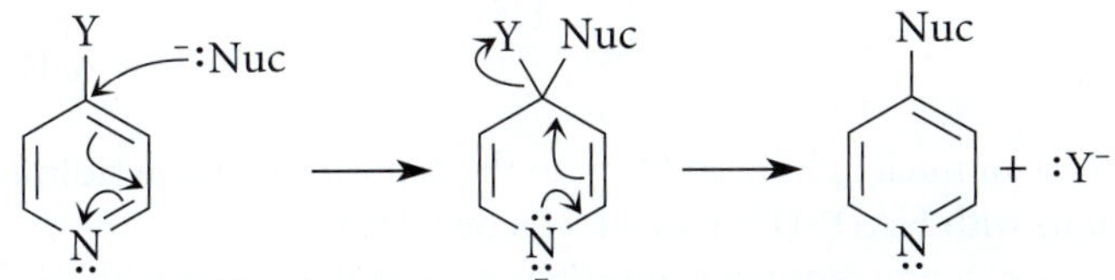

(26.47b)

What about substitution at carbon-3? 3-Substituted pyridines are *not* reactive in nucleophilic substitution because negative charge in the addition intermediate *cannot* be delocalized onto the electronegative nitrogen:

Substitution at carbon-3:

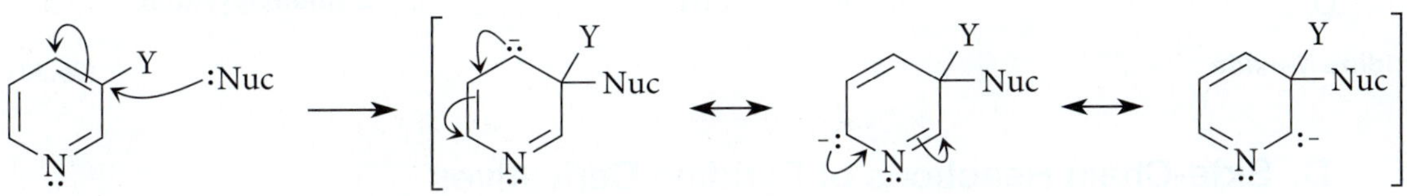

(26.47c)

Focused Problems

26.14 Give the structure of the product and a curved arrow mechanism for its formation in the reaction of 4-chloropyridine with sodium methoxide. Draw all important resonance structures for the addition intermediate.

26.15 Which compound should readily undergo substitution of the bromine by phenolate anion: 4-bromopyridine or 3-bromopyridine? Explain, and give the structure of the product.

C. *N*-Alkylpyridinium Salts and Their Reactions

Pyridine, like many other bases, is a nucleophile. When pyridines react in S_N2 reactions with alkylating agents such as alkyl halides or sulfonate esters, *N-alkylpyridinium salts* are formed.

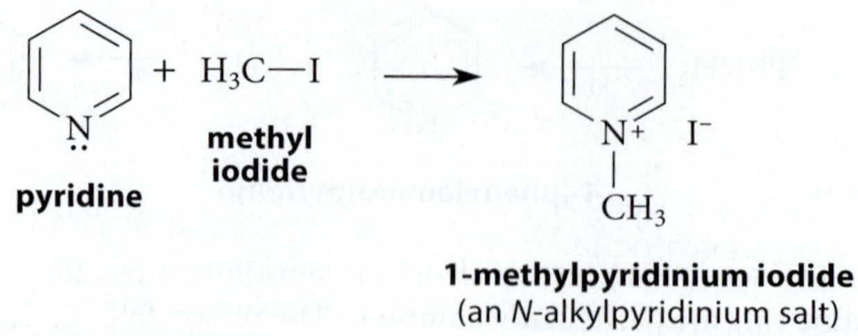

(26.48)

N-Alkylpyridinium salts are activated toward nucleophilic reactions at the 2- and 4-positions of the ring much more than pyridines themselves because the positively charged nitrogen is more electronegative, and is therefore a better electron acceptor, than the neutral nitrogen of a pyridine. When the nucleophiles in such displacement reactions are anions, charge is neutralized. In the following reaction, for example, the pyridinium salt reacts as an electrophile at its 2-position with the hydroxide ion nucleophile; the resulting hydroxy compound is then oxidized by potassium ferricyanide [$K_3Fe(CN)_6$] present in the reaction mixture.

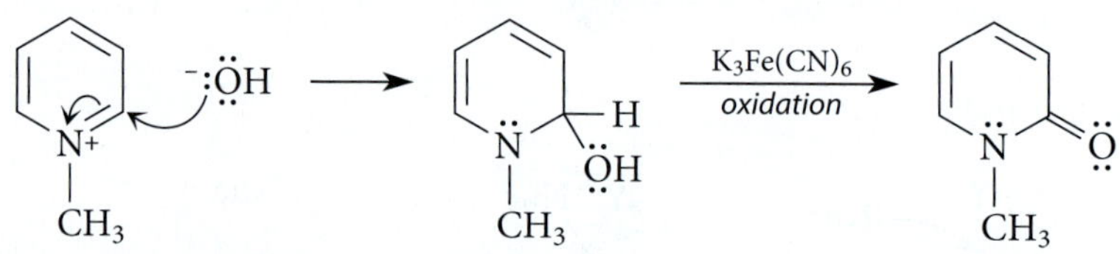

(26.49)

A biological example of nucleophilic addition to the 4-position of a pyridinium ring is found in biological oxidations with NAD^+ (Display 11.71a, Sec. 11.8).

Pyridine-*N*-oxides are in one sense pyridinium ions, and they react with nucleophiles in much the same way as *N*-alkylpyridinium salts:

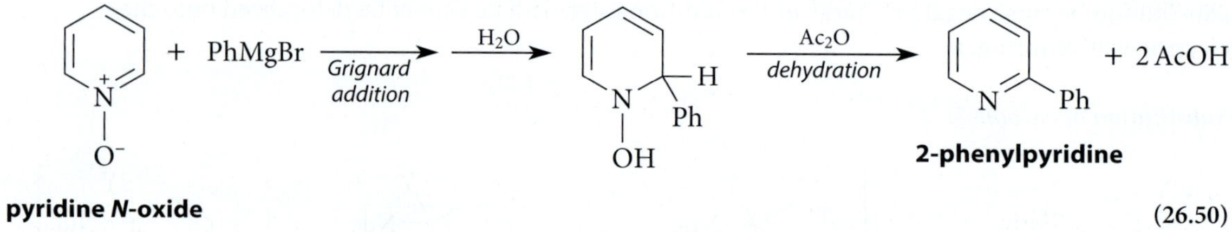

(26.50)

D. Side-Chain Reactions of Pyridine Derivatives

The "benzylic" hydrogens of an alkyl group at the 2- or 4-position of a pyridine ring are about 7 pK_a units more acidic than ordinary benzylic hydrogens because the electron pair (and charge) in the conjugate-base anion is delocalized onto the electronegative pyridine nitrogen.

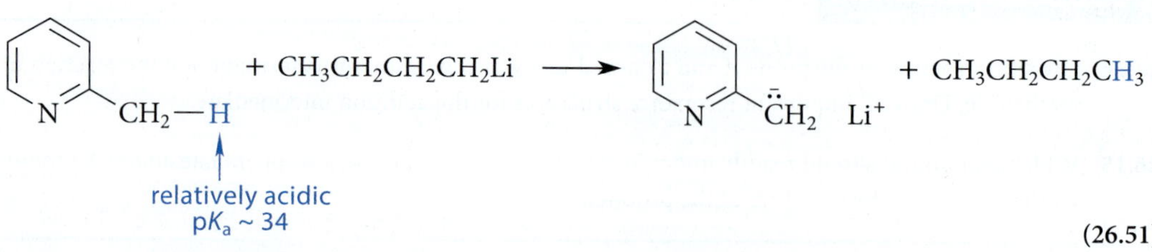

(26.51)

(Draw the resonance structures of this ion, and verify that charge is delocalized onto the pyridine nitrogen.) As the example in Eq. 26.51 illustrates, strongly basic reagents such as organolithium reagents or $NaNH_2$ abstract a "benzylic" hydrogen from 2- or 4-alkylpyridines. The anion formed in this way has a reactivity much like that of other organolithium reagents. In Eq. 26.52, for example, it adds to the carbonyl group of an aldehyde to give an alcohol (Sec. 19.10).

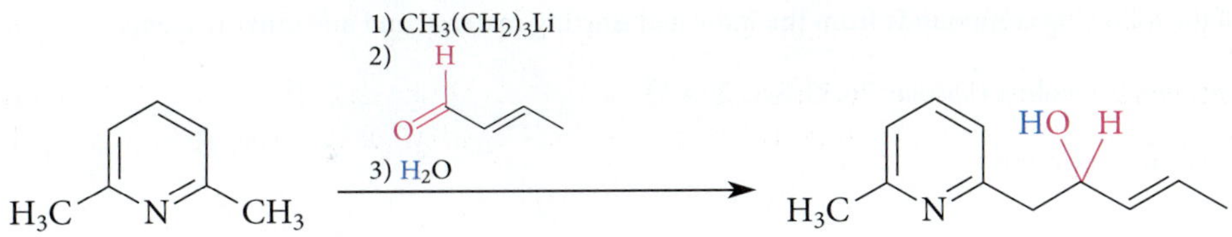

(26.52)

This reaction is another example of the analogy between pyridine chemistry and carbonyl chemistry. If the C═N linkage of a pyridine ring is analogous to a carbonyl group, then the "benzylic" anion is analogous to an enolate anion.

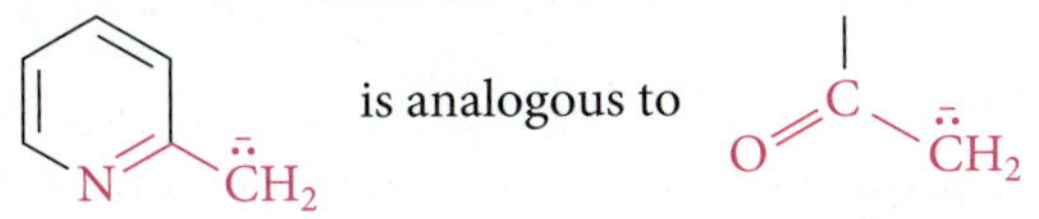

(26.53)

On the basis of this analogy, then, it is reasonable that these anions should undergo some of the reactions of enolate anions, such as the aldol-like addition in Eq. 26.52.

The "benzylic" hydrogens of 2- or 4-alkylpyridinium salts are much more acidic than those of the analogous pyridines because the conjugate-base "anion" is actually a neutral compound, as the following resonance structures show:

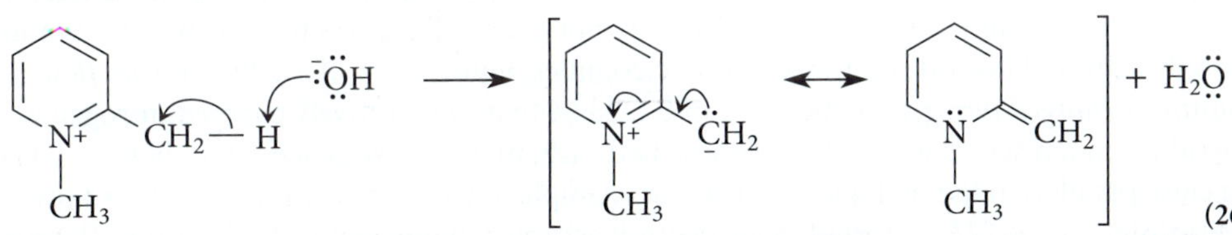

(26.54)

The "benzylic" hydrogens of 2- or 4-alkylpyridinium salts are acidic enough that the conjugate-base "anions" can be formed in useful concentrations by aqueous NaOH or amines. In the following reaction, which exploits this acidity, the conjugate base of a pyridinium salt is used as the "enolate" component in a variation of the Claisen–Schmidt condensation (Sec. 22.4C).

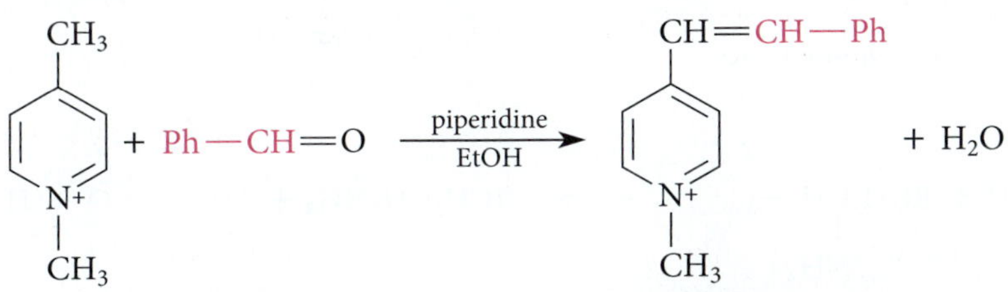

(probably the trans isomer; 85% yield)

(26.55)

Many side-chain reactions of pyridines are analogous to those of the corresponding benzene derivatives. For example, side-chain oxidation (Sec. 17.5C) is a useful reaction of both alkylbenzenes and alkylpyridines. The oxidation of nicotine to nicotinic acid (Eq. 26.31, Sec. 26.4A) is an example of such a reaction.

Focused Problems

26.16 Give the principal organic product in the reaction of quinoline (see Fig. 26.1, Sec. 26.1A) with each of the following reagents. (*Hint:* Consider the similar reactions of pyridine.)

(a) 30% H_2O_2

(b) $NaNH_2$, heat; then H_2O

(c) Product of part (a), then HNO_3, H_2SO_4

26.17 Outline a synthesis for each of the following compounds from the indicated starting material and any other reagents.

(a) 3-Methyl-4-nitropyridine from β-picoline (Display 26.30, Sec. 26.4A)

(b) 4-Methyl-3-nitropyridine from γ-picoline

(c) 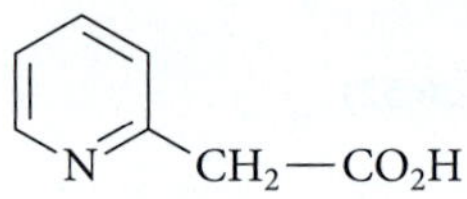

from α-picoline (*Hint:* See Sec. 20.6.)

(d) 3-Aminopyridine from β-picoline

26.18 Predict the predominant product in each of the following reactions. Explain your answer.

(a) 3,4-Dimethylpyridine + butyllithium (1 equiv.), then $CH_3I \longrightarrow (C_8H_{11}N)$

(b) 3,4-Dibromopyridine + NH_3, heat $\longrightarrow (C_5H_5BrN_2)$

Chemical Biology Topic

E. Pyridinium Ions in Biology: Pyridoxal Phosphate

The chemistry in the previous three sections occurs because the nitrogen of the pyridine ring can serve as an acceptor of electrons and because this electron-acceptor tendency is particularly enhanced in pyridinium ions. Review this idea by noticing in Eq. 26.49 that the pyridinium ion is strongly activated toward reactions with nucleophiles; *notice particularly the electron flow onto the positively charged nitrogen.* Notice also in Eq. 26.54 that the positively charged nitrogen of the pyridinium ion serves to stabilize the attached carbanion by resonance. This chemistry has some close parallels in the biological world. For example, reviewing Sec. 11.8 will show how the pyridinium ion of NAD^+ serves as an electron acceptor in biochemical reductions. (Notice particularly Eq. 11.71a, Sec. 11.8.) Another biologically important pyridine derivative, *pyridoxal phosphate*, fulfills a similar mechanistic role in other reactions. As shown in **Fig. 26.3**, pyridoxal phosphate is one of several forms of *vitamin B_6*.

Pyridoxol (Fig. 26.3) was the first form of vitamin B_6 discovered as a nutritional factor in 1934. In 1944, Esmond Snell (1914–2003), of the University of Texas, noticed that metabolites of pyridoxol secreted in the urine are more active. These metabolites turned out to be pyridoxal and pyridoxamine. Through the next decade, Snell and his co-workers elucidated the chemical role of these compounds.

Pyridoxal phosphate is an essential *coenzyme* (Eq. 11.69, Sec. 11.8) in several important biochemical transformations. Here are only three of many:

Decarboxylation of α-amino acids:

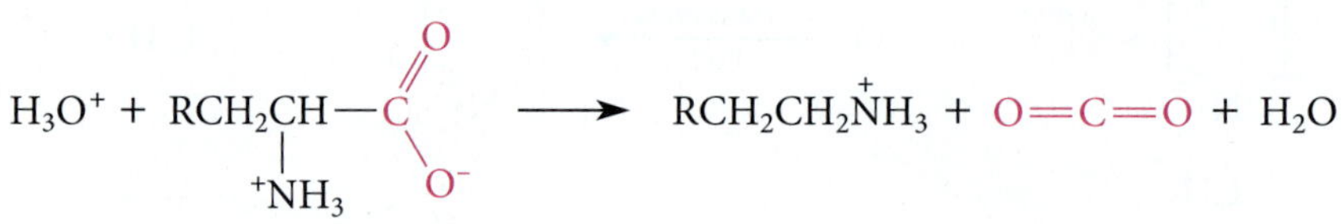

(26.56a)

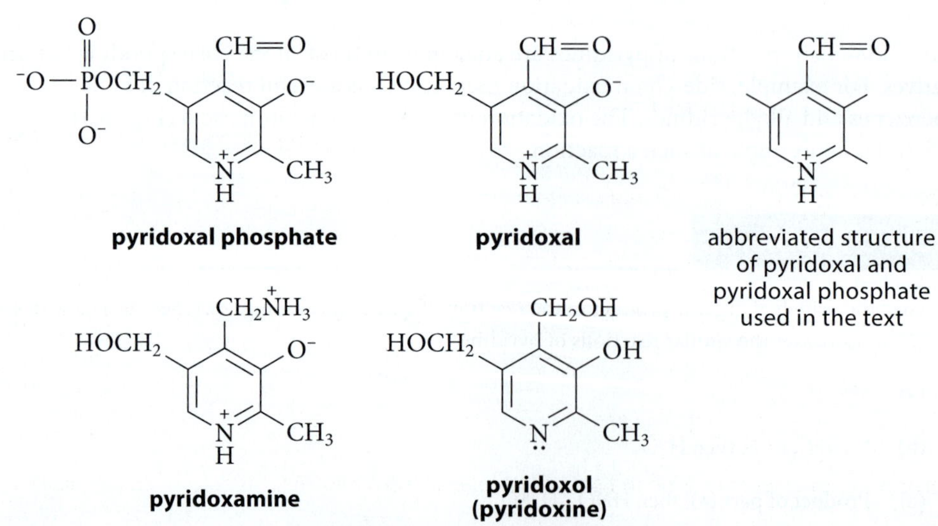

FIGURE 26.3 Various forms of vitamin B_6. Pyridoxol was the first form to be isolated, but any of the compounds shown can serve as a source of the vitamin. (For example, pyridoxol can be oxidized and phosphorylated to give pyridoxal phosphate.) Pyridoxal phosphate is the form of the vitamin involved in most biochemical transformations; pyridoxamine phosphate is an intermediate in some transformations. All compounds are shown in the ionization states in which they exist at physiological pH (7.4). An abbreviated structure of pyridoxal and pyridoxal phosphate, also shown, is used in the text.

This transformation is utilized for the production of biologically important amines, such as the neurotransmitters serotonin and dopamine in the brain, and the vasoconstrictor histamine. The biosynthesis of serotonin, for example, employs the α-amino acid L-tryptophan as a starting material. L-Tryptophan is oxidized, and the resulting α-amino acid, 5-hydroxy-L-tryptophan, is then decarboxylated.

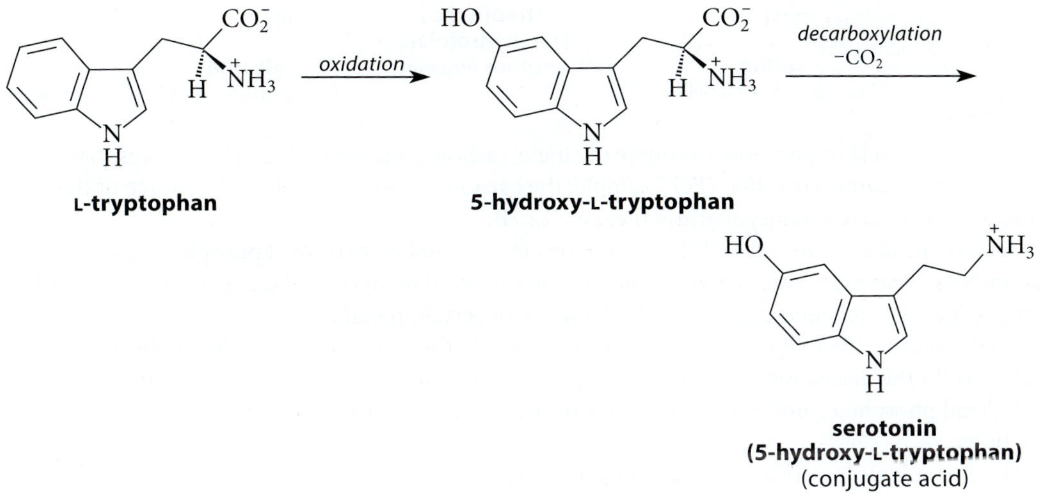

(26.56b)

The α-amino acid L-tyrosine is oxidized and decarboxylated to give dopamine, and the α-amino acid L-histidine is decarboxylated to give histamine.

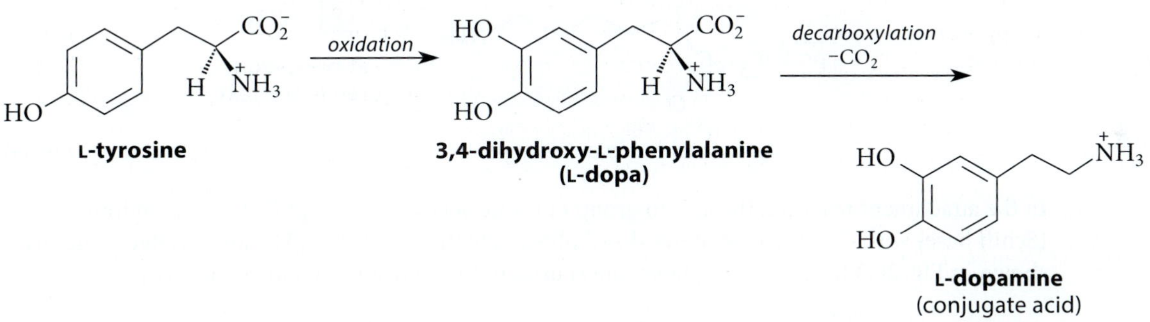

(26.56c)

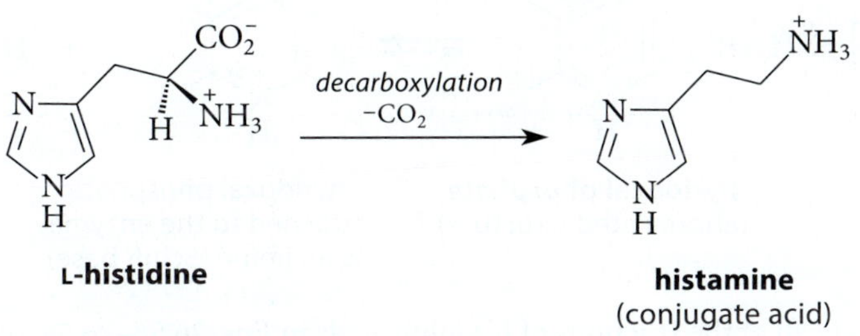

(26.56d)

Interconversion of α-amino acids and α-keto acids:

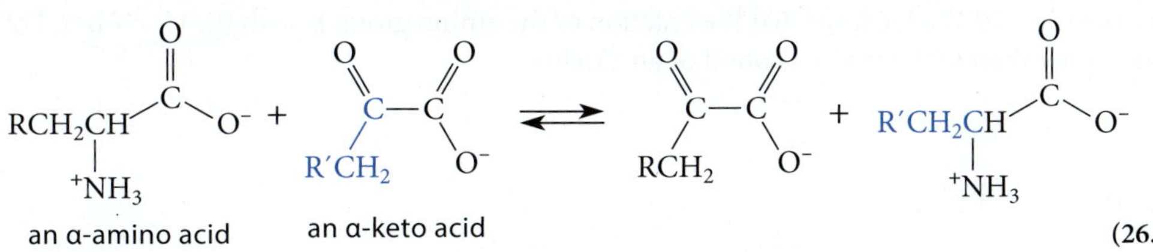

(26.57)

This process is an important one in the biological synthesis and degradation of amino acids.

Loss of formaldehyde from serine:

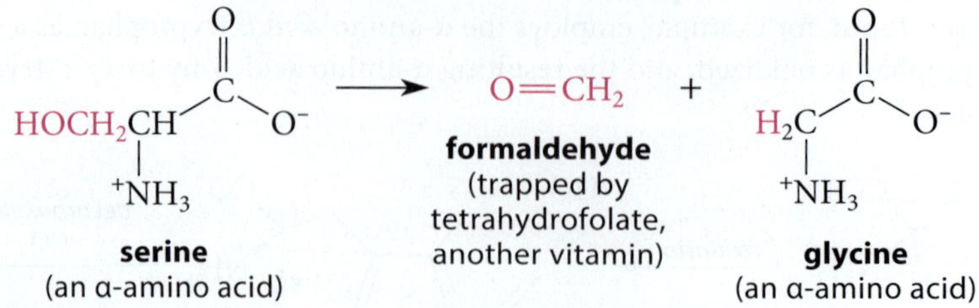

(26.58)

This conversion is important as a source of single-carbon units for biological processes that involve single-carbon transfer. (For example, the carbon of formaldehyde is the source of the methyl carbon in *S*-adenosylmethionine; Sec. 12.7B.)

In biological systems, each of these reactions is catalyzed by pyridoxal phosphate and an appropriate enzyme. These reactions can also be catalyzed by pyridoxal alone in the absence of enzymes at elevated temperatures in the presence of certain metal ions.

Let's examine the first of these reactions to illustrate the essentials of pyridoxal phosphate catalysis. In the discussion that follows, keep your eye on the protonated pyridine ring of pyridoxal phosphate, and relate the various transformations to the reactions of the previous sections.

In the biological world, pyridoxal phosphate typically is covalently attached as an imine (Schiff base) to various enzymes, usually through the side-chain amino group of the amino acid lysine within the protein structure:

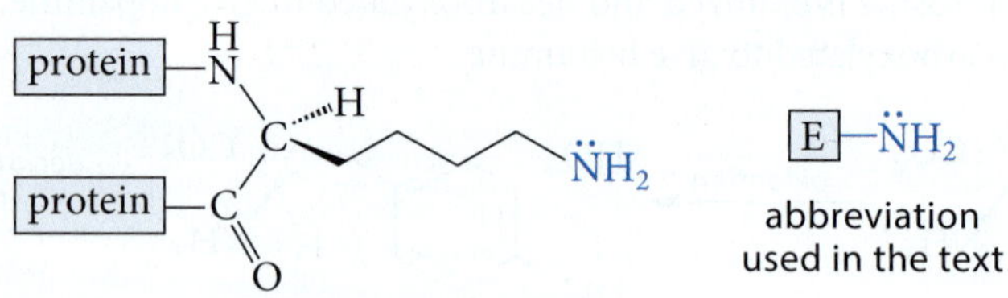

a lysine residue in an enzyme (26.59)

In the attachment reaction, the amino group of lysine acts as a nucleophile to form an imine (Schiff base) with the aldehyde of pyridoxal phosphate (Sec. 19.12A). (The abbreviated structure shown in Fig. 26.3 for pyridoxal phosphate is used in this and subsequent equations.)

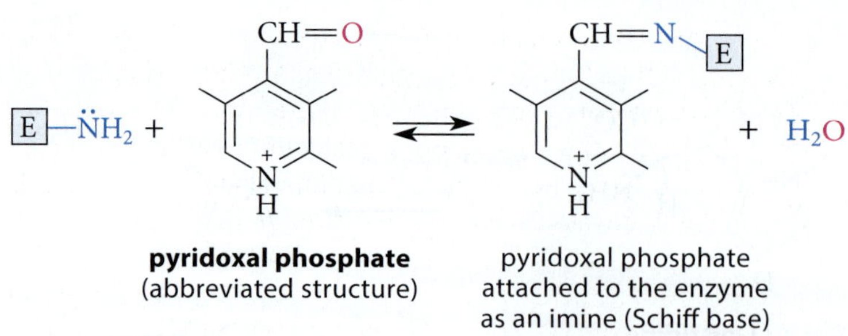

(26.60)

In the first step of all of the reactions of α-amino acids in Eqs. 26.56–26.58, the amino group of the α-amino acid, acting as a nucleophile in its unprotonated form, reacts with the imine product of Eq. 26.60 to form a new imine. This is exactly like imine formation from an amine and an aldehyde (Sec. 19.12A), except that the reaction of the amino group is with the C=N bond of an imine rather than with the C=O bond of an aldehyde.

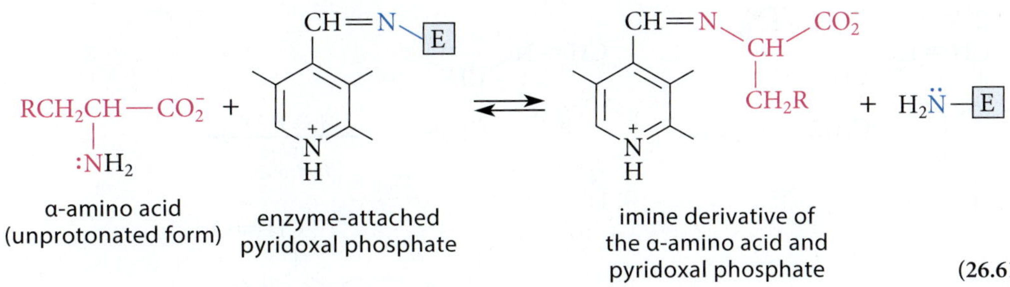

(26.61)

(Because imines are sometimes called *Schiff bases*, this reaction is sometimes whimsically called "trans-Schiffization.") Decarboxylation forms what appears to be a *carbanion intermediate.*

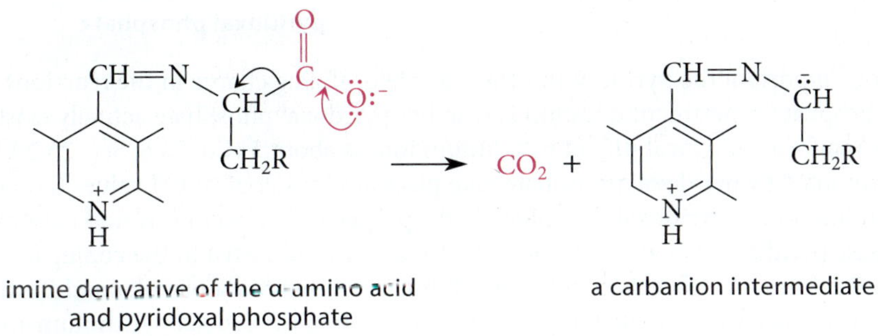

(26.62a)

But this is no ordinary carbanion. Most carbanions are such strong bases that they cannot exist under physiological conditions; however, this carbanion is a much weaker base because *it is stabilized by resonance*:

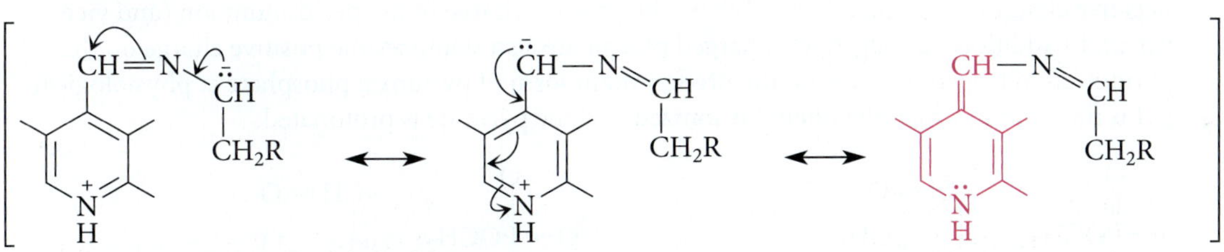

three of the many resonance structures for the carbanion intermediate (26.62b)

(Only three of the many possible resonance structures are shown; you should draw others.) The curved arrows in the middle structure, which result in the structure on the right, show how *the pyridinium ion stabilizes negative charge by accepting electrons.* In fact, the red part of the structure on the right shows that the "carbanion" is really not a carbanion at all—it is a neutral molecule. (Compare with Eq. 26.54, Sec. 26.4D.) The same type of "carbanion" is involved in all of the pyridoxal-catalyzed transformations discussed in this section. (See Focused Problem 26.19.)

As noted in Sec. 20.11A, decarboxylation reactions typically require an "electron sink" to accept the electrons from the carboxy group that is lost. Equations 26.62a and b show that the pyridinium ring of pyridoxal phosphate serves this role effectively.

Protonation of this anion and hydrolysis of the resulting imine gives pyridoxal phosphate and the product of Eq. 26.56a. ($^+$B—H and B: are acidic and basic groups in the enzyme active site.)

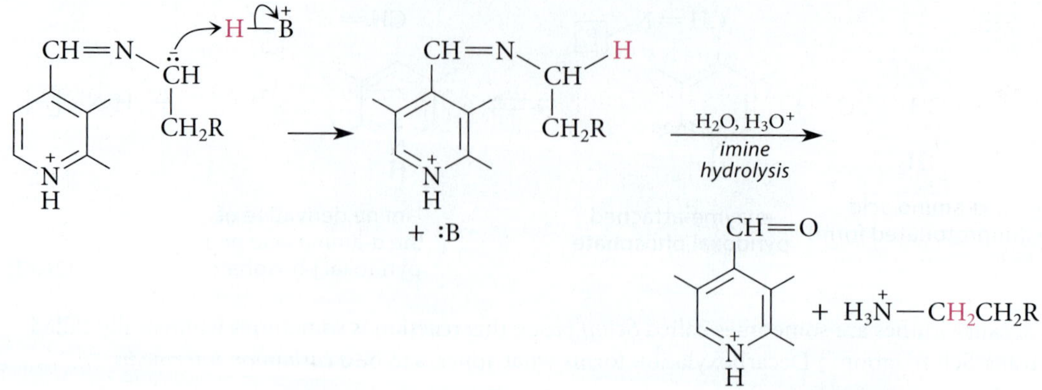

pyridoxal phosphate (26.62c)

Given how important the pyridinium ring is for delocalizing charge in the reactions of pyridoxal phosphate, a pertinent question is whether pyridoxal phosphate actually exists in the pyridinium-ion form. A typical pK_a of pyridinium ions is about 5 (Eq. 26.6, Sec. 26.2A). Yet the reactions promoted by pyridoxal phosphate take place at physiological pH values (about 7.4). If the pyridinium ion in pyridoxal phosphate had a pK_a near 5, most of it would exist as the conjugate-base pyridine form at pH 7.4; less than 1% of it would exist in the conjugate-acid pyridinium-ion form. (Verify this conclusion.) It turns out that the molecular architecture of pyridoxal phosphate ensures a much higher concentration of the crucial pyridinium-ion form. The key element in the structure is the —OH group in the 3-position and its ortho relationship to the aldehyde (see Eq. 26.63a). This ortho relationship makes the phenolic —OH group of pyridoxal phosphate *unusually acidic.* [Why? See Focused Problem 18.30(c), Sec. 18.7A.] Ionization of the phenolic —OH group, in turn, raises the pK_a of the pyridinium ion because the negative charge of the phenolate stabilizes the positive charge of the pyridinium ion (and vice versa). In addition, the negatively charged phosphate also stabilizes the positive charge on the pyridinium nitrogen. As a result, the predominant form of pyridoxal phosphate at physiological pH is the form in which the phenol is ionized and the pyridine is protonated:

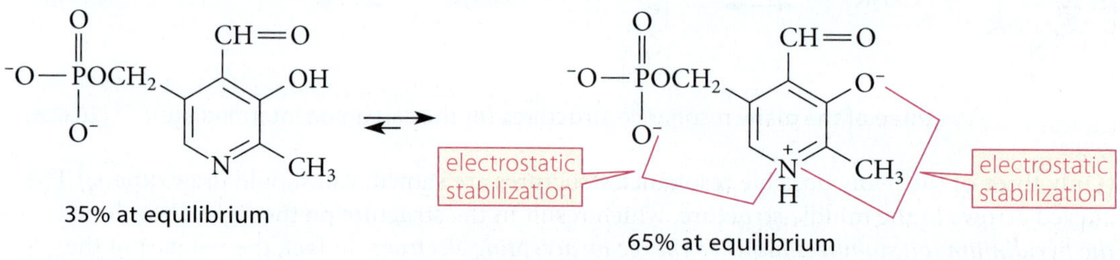

two forms of pyridoxal phosphate (26.63a)

But that's not all. When pyridoxal phosphate is bound to the enzymes that catalyze its reactions, the pyridinium form is further stabilized. In one well-studied case, this stabilization is the result of a favorable electrostatic interaction between an ionized carboxylate group and the positively charged nitrogen:

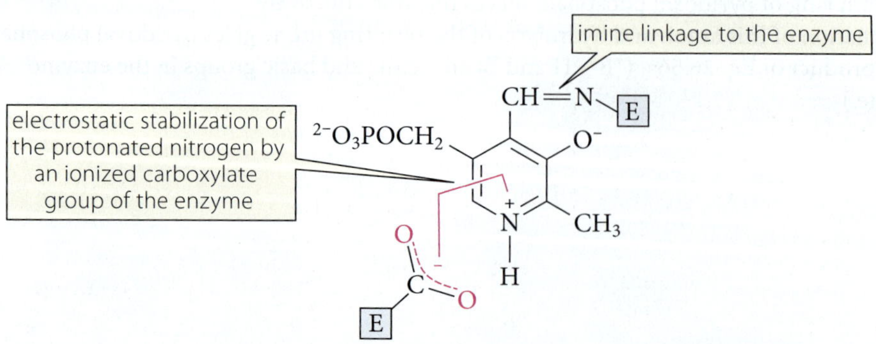

(26.63b)

As this discussion shows, everything conspires to ensure that the pyridinium nitrogen is protonated!

Focused Problems

26.19 (a) Pyridoxal catalysis of Eq. 26.57 involves the following transformations. (Running these reactions in the reverse direction with a different α-keto acid completes Eq. 26.57.)

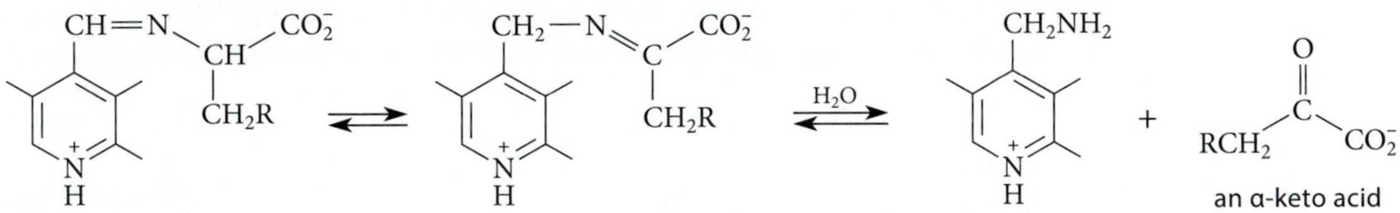

(from Eq. 26.61) **pyridoxamine phosphate**

Using bases (B:) and acids ($^{+}$BH) as needed, provide curved-arrow mechanisms for these reactions. Your mechanism should show the important intermediates. As part of your mechanism, explain the significance of the pyridinium ion in the catalysis of this reaction sequence.

(b) Using bases (B:) and acids ($^{+}$BH) as needed, provide a curved-arrow mechanism for the enzyme and pyridoxal phosphate-catalyzed conversion of the α-amino acid serine into formaldehyde and glycine (Eq. 26.58).

26.20 Isoniazid is an antituberculosis drug that operates by reacting with pyridoxal phosphate in the causative *Mycobacterium*. Show how isoniazid reacts with pyridoxal phosphate. (*Hint:* See Table 19.3, Sec. 19.12A.)

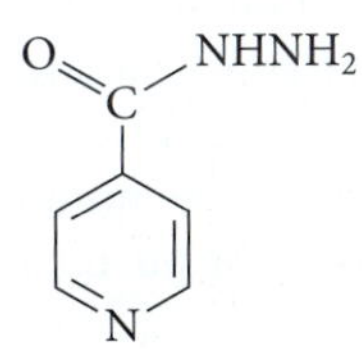

isoniazid

26.5 NUCLEOSIDES, NUCLEOTIDES, AND NUCLEIC ACIDS

Chemical Biology Topic

Heterocyclic compounds occur widely in living systems. Heterocyclic compounds, specifically derivatives of purine and pyrimidine (Fig. 26.1, Sec. 26.1A), play an important role in the structures of the nucleic acids DNA and RNA, polymers that are responsible for the storing and transmission of genetic information. This section introduces nucleic acids, and Sec. 26.6 introduces some of the other heterocyclic compounds that are important in biology.

A. Nucleosides and Nucleotides

A **ribonucleoside** is a compound formed between the furanose form of D-ribose and a heterocyclic compound. The heterocyclic group is commonly referred to as the **base**, and the ribose as the **sugar**. The stereochemistry of the bond between the base and the ribose is most commonly β (Sec. 24.3B). A **deoxyribonucleoside** is a similar derivative of D-2-deoxyribose and a heterocyclic base. The prefix *deoxy* means "without oxygen"; in other words, 2-deoxyribose is a ribose that has a second —H instead of an —OH group at carbon-2.

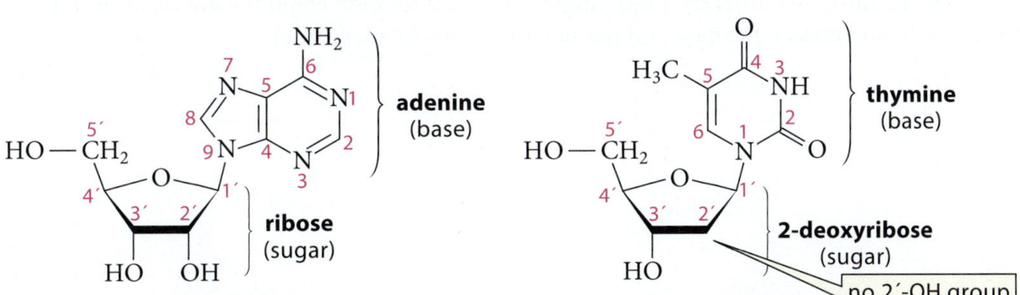

adenosine (a ribonucleoside) **2′-deoxythymidine** (a deoxyribonucleoside) (26.64)

In these structures the sugar ring and the heterocyclic ring are numbered separately. To differentiate the two sets of numbers, primes (′) are used in referring to the atoms of the sugar. For example, the 2′ (pronounced two-prime) carbon of adenosine is carbon-2 of the sugar ring.

The sugar (ribose or deoxyribose) is often represented, especially in biochemistry texts, in a *Haworth projection* (see Sec. 24.3B) in which the C2′–C3′ bond is shown as a heavy line, implying that it is in the front of the page. Previously in the text we've represented cyclic compounds with planar, line-and-wedge structures. The relationship between the two is as follows:

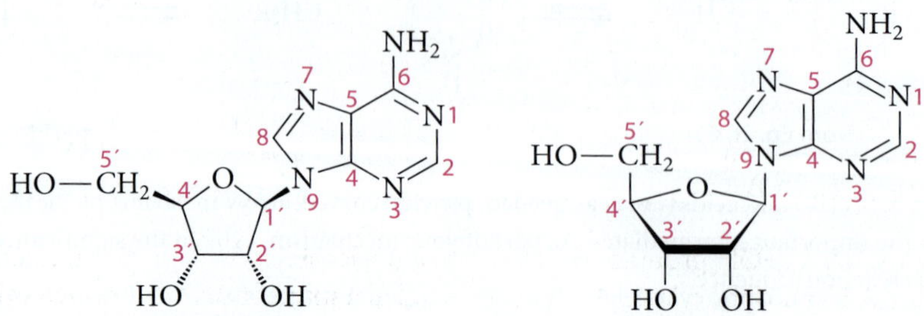

adenosine line-and-wedge structure for the sugar

adenosine Haworth projection for the sugar

(26.65)

Neither representation conveys the fact that the five-membered ring actually exists as an equilibrium mixture of several rapidly interconverting puckered conformations (see Sec. 7.5A). However, we won't need to be concerned with these conformations at this point.

The bases that occur most frequently in nucleosides are derived from two heterocyclic ring systems: *pyrimidine* and *purine.* (The numbering of these rings is shown in red.) Three bases of the pyrimidine type and two of the purine type occur most commonly.

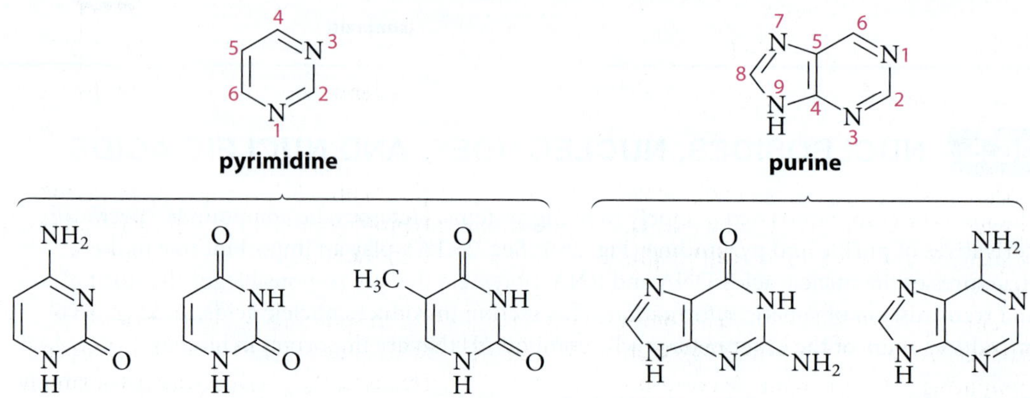

pyrimidine **purine**

cytosine (C) **uracil** (U; occurs in RNA) **thymine** (T; occurs in DNA) **guanine** (G) **adenine** (A)

(26.66)

The base is attached to the sugar at N-9 of the purines and N-1 of the pyrimidines, as in the preceding examples.

In a nucleoside or nucleotide, the base is conventionally shown in either of two conformations, called **syn** and **anti**, that differ by a 180° angle of internal rotation about the glycosidic bond—that is, the bond between the base and the sugar (C1′–N9 in adenosine).

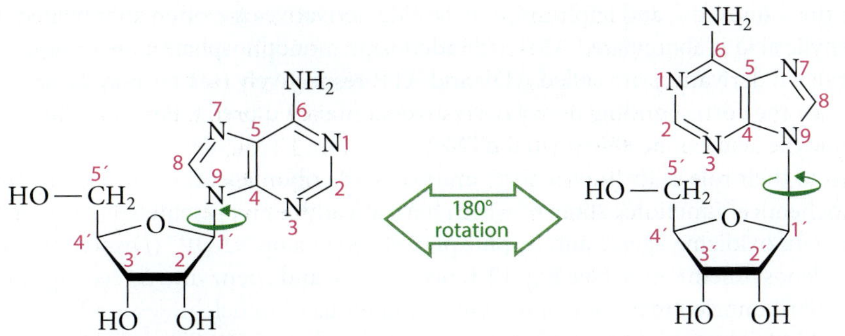

adenosine
(anti conformation about C1′–N9)

adenosine
(syn conformation about C1′–N9)

(26.67)

Drawing the syn conformation for a purine base requires drawing an unrealistically long C1′–N9 bond. This bond is not really long; it is simply an artifact of the Haworth projection. In this text we use the anti conformation in most cases, but be aware that you may see nucleosides and nucleotides drawn in the syn conformation in other places.

In living systems, the 5′—OH group of the ribose in a nucleoside is usually found esterified to a phosphate group. A 5′-phosphorylated nucleoside is called a **nucleotide**. A **ribonucleotide** is derived from the monosaccharide ribose; a **deoxyribonucleotide** is derived from 2′-deoxyribose. Some nucleotides contain a single phosphate group; others contain two or three phosphate groups condensed in phosphate anhydride linkages (See Sec. 25.2B).

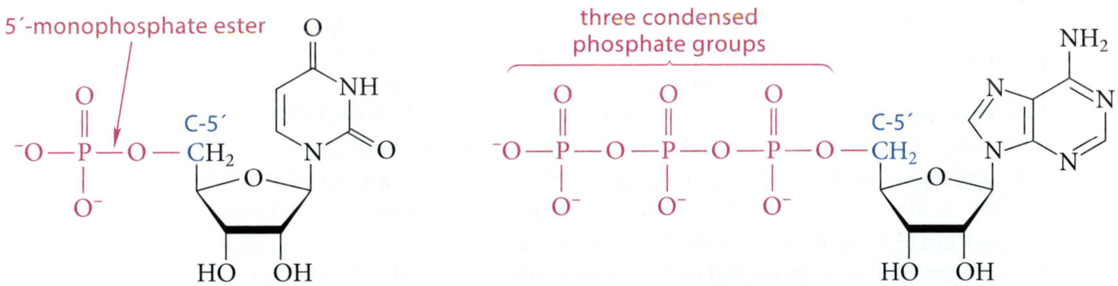

uridylic acid (UMP)

adenosine triphosphate (ATP)

(26.68)

Although the ionization state of the phosphate groups depends on pH, these groups are written conventionally in the ionized form. These structures are typically drawn with a hypervalent (that is, pentavalent) phosphorus, although, as we have stressed repeatedly, the octet structures are very important.

The nomenclature of the five common bases and their corresponding nucleosides and nucleotides is summarized in **Table 26.2**. This table gives the names of the ribonucleosides and ribonucleotides. To name the corresponding 2′-deoxy derivatives, the prefix *2′-deoxy* (or simply *deoxy*) is appended to the names of the corresponding ribose derivatives. For example, the 2′-deoxy analog of adenosine is called 2′-deoxyadenosine or simply deoxyadenosine. In addition,

TABLE 26.2 Nomenclature of Nucleic Acid Bases, Nucleosides, and Nucleotides*

Base	Nucleoside	Nucleotide (5′-monophosphate)	Abbreviation for the monophosphate
adenine (A)	adenosine	adenylic acid	AMP
uracil (U)	uridine	uridylic acid	UMP
thymine (T)	thymidine	thymidylic acid	TMP
cytosine (C)	cytidine	cytidylic acid	CMP
guanine (G)	guanosine	guanylic acid	GMP

*The deoxyribonucleosides and deoxyribonucleotides are named by appending the prefix *deoxy*, for example:

	deoxyadenosine	deoxyadenylic acid	dAMP

The prefix *deoxy* means 2′-deoxy unless stated otherwise.

the names of the mono-, di-, and triphosphonucleotide derivatives are often abbreviated. For example, adenylic acid is abbreviated AMP (for adenosine monophosphate); the di- and tri-phosphorylated derivatives are called ADP and ATP, respectively (see Display 26.68). The abbreviations for the corresponding deoxy derivatives contain a d prefix. For example, 2′-deoxythymidylic acid can be abbreviated dTMP.

In addition to their role as the monomeric units of RNA, ribonucleotides also have other important biochemical functions, some of which have already been presented. NAD^+, one of nature's important oxidizing agents and its phosphorylated analog $NADP^+$ (Display 11.68, Sec. 11.8), *S*-adenosylmethionine (see Fig. 12.1, Sec. 12.7B), and coenzyme A (see Fig. 25.1, Sec. 25.1) are all ribonucleotides. One of the most ubiquitous ribonucleotides is adenosine triphosphate (ATP), whose biological roles are discussed in Sec. 25.7A. The role of the ribonucleotide uridine diphosphate (UDP) in disaccharide biosynthesis is discussed in Sec. 27.8E.

Focused Problems

26.21 Draw the structure of (a) deoxythymidine monophosphate (dTMP); (b) GDP.

26.22 Draw and label the syn and anti conformations of the nucleoside guanosine about the C1′–N9 bond.

B. The Structures of DNA and RNA

Nucleic acids are polymers of nucleotides. **Deoxyribonucleic acid (DNA)** is a polymer of deoxyribonucleotides and is the storehouse of genetic information throughout all of nature (with the exception of certain viruses). The monomeric units of the DNA polymer are called **residues**. A three-residue segment of DNA is shown in **Fig. 26.4**. This figure shows that the nucleotide residues in DNA are interconnected by phosphate diester groups that are esterified both to the 3′—OH group of one ribose and the 5′—OH of another. The DNA polymer incorporates adenine, thymine, guanine, and cytosine as the nucleotide bases. Although only three residues are shown in Fig. 26.4, a typical strand of DNA might be thousands or even millions of nucleotides long.

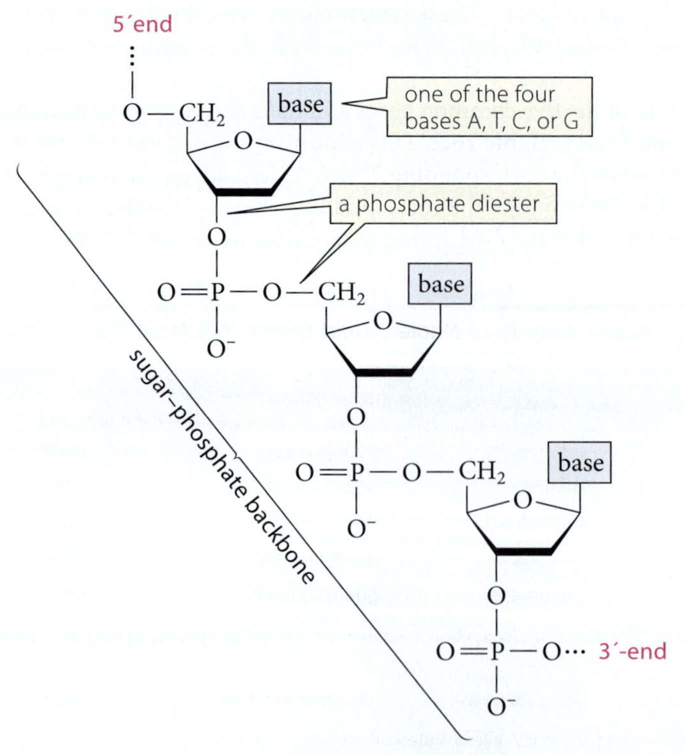

FIGURE 26.4 General structure of a single strand of DNA (base = A, T, G, or C; Table 26.2). Only three residues are shown here; a typical strand of DNA contains thousands or even millions of residues.

Each residue in a polynucleotide is differentiated by the identity of its base, and the sequence of bases encodes the genetic information in DNA. The DNA polymer, then, is a backbone of alternating phosphates and 2′-deoxyribose groups to which are connected bases that differ from residue to residue. The ends of the DNA polymer are labeled 3′ or 5′, corresponding to the deoxyribose carbon to which the terminal hydroxy group is attached.

Ribonucleic acid (RNA) polymers are conceptually much like DNA polymers, except that ribose, rather than 2′-deoxyribose, is the sugar. Three of the four bases in RNA are the same as in DNA; the fourth base, uracil, occurs in RNA instead of thymine (Display 26.66), and some rare bases (not considered here) are found in certain types of RNA. Although a number of biochemical processes utilize Watson–Crick base pairing of RNA with other RNA or DNA molecules, many types of RNA, unlike DNA, are single-stranded.

For many years before the detailed structure of DNA was determined, it was known that DNA carries genetic information and that it is *replicated*, or copied, during cell reproduction. In 1950, Erwin Chargaff (1905–2002) of Columbia University showed that the ratios of adenine to thymine, and guanosine to cytosine, in DNA are both 1.0; this observation has become known as *Chargaff's first parity rule*. How this rule relates to the storage and transmission of genetic information, however, remained a mystery. It became clear to a number of scientists that a knowledge of the three-dimensional structure of DNA would be essential to understand how DNA functions as it does. The importance of this problem was sufficiently obvious that several scientists worked feverishly to be the first to determine the three-dimensional structure of DNA. In 1953, James D. Watson (b. 1928) and Francis C. Crick (1916–2004), then at Cambridge University, proposed a structure for DNA. Their proposal was based on X-ray diffraction patterns of DNA fibers obtained by their colleagues at the Medical Research Council laboratory in England, Rosalind Franklin (1920–1958) and Maurice Wilkins (1916–2004).

For their work on the structure of DNA, Watson, Wilkins, and Crick were awarded the Nobel Prize in Medicine or Physiology in 1962. Rosalind Franklin did not share the prize posthumously because the terms of Nobel's bequest stipulate that the prize should go only to living scientists.

The Watson–Crick structure of DNA is shown in **Fig. 26.5**. The structure has the following important features:

1. The structure of DNA contains *two* right-handed helical polynucleotide chains coiled around a common axis forming a *double helix*. The helix makes a complete turn every 10 nucleotide residues. (Other helical conformations of DNA also occur.) The two polynucleotide chains run in opposite directions—that is, one chain runs in the 3′→5′ direction, and the other in the 5′→3′ direction.

2. The sugars and phosphates, which are rich in —OH groups and charges, are on the outside of the helix, where they can interact with solvent water or other hydrophilic compounds; the bases, which are hydrophobic, are buried in the interior of the double helix, away from water.

3. The chains are held together by hydrogen bonds between bases. Each adenine (A) in one chain forms hydrogen bonds to a thymine (T) in the other, and each guanosine (G) in one chain forms hydrogen bonds to a cytosine (C) in the other. That is, every purine in one chain is hydrogen-bonded to a pyrimidine in the other. For this reason, A is said to be *complementary* to T, and G is complementary to C. The hydrogen-bonded A–T and G–C pairs are often referred to as **Watson–Crick base pairs. Figure 26.6** provides a closer look at these Watson–Crick base pairs. Notice that the A–T pair has about the same spatial dimensions as the G–C pair.

4. The planes of successive complementary base pairs are stacked and somewhat offset, one on top of the other, and are perpendicular to the axis of the helix. The relative positions of the planes of the bases promotes a favorable offset-stacking interaction (Display 15.74, Sec. 15.8A) between the bases in adjacent planes. The distance between each successive base-pair plane is 340 pm (3.4 Å). The helix makes a complete turn every 10 residues, so the distance along the helix per complete turn is 10×340 pm = 3400 pm (34 Å).

5. The double-helical structure of DNA results in two grooves that wrap around the double helix along its periphery. The larger groove is called the *major groove*, and the smaller is called the *minor groove*. These are shown in Fig. 26.5a. These grooves, particularly the major groove, serve as the location of "docking" sites at which other macromolecules such as proteins interact with DNA.

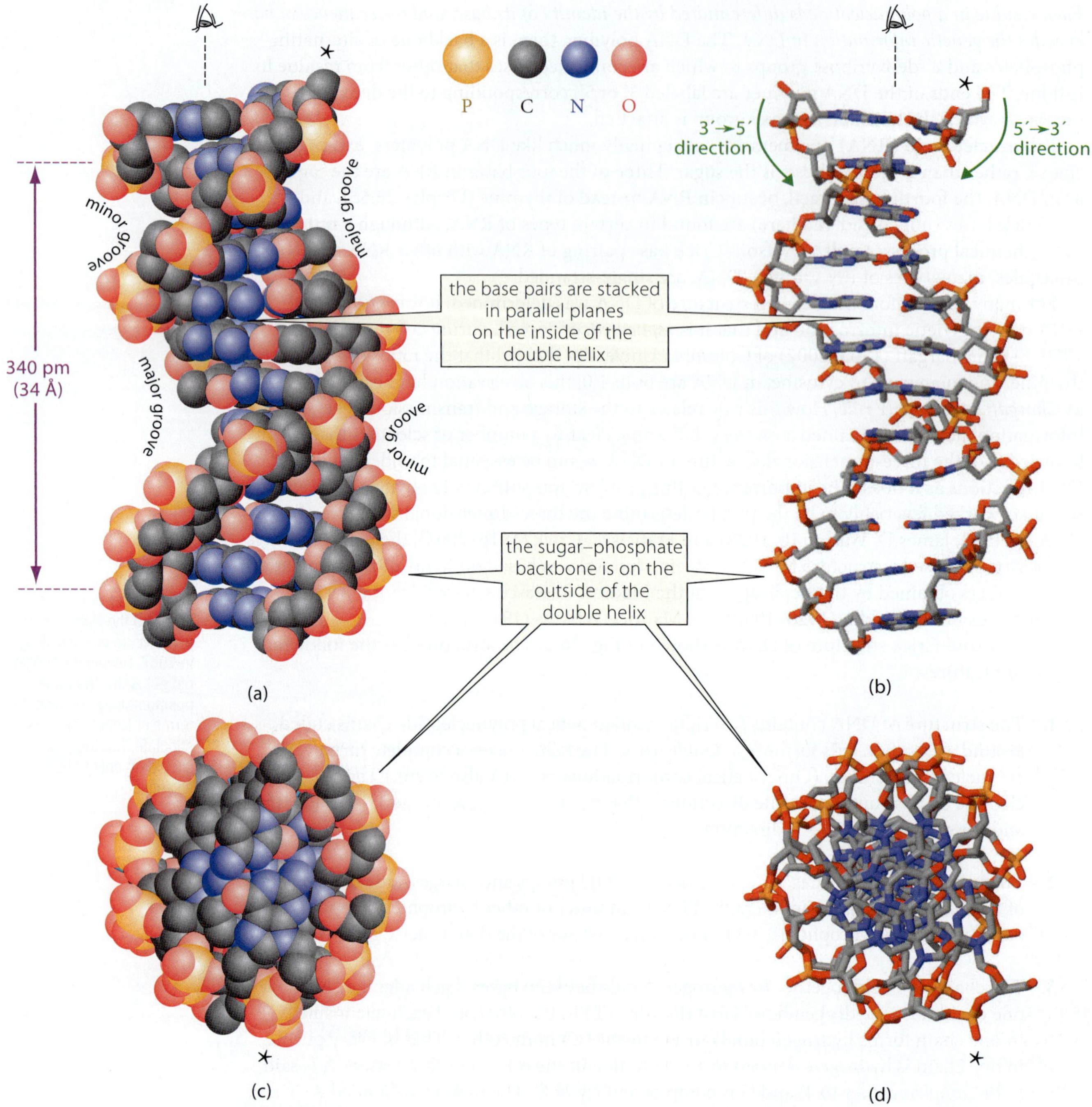

FIGURE 26.5 Molecular models of DNA from different perspectives. (a) A space-filling model in a side-on view, perpendicular to the helical axis. The major and minor grooves encircle the helix throughout its length. Many proteins that interact with DNA bind along the major groove. (b) A stick model in the same view as (a). This model shows more clearly the offset-stacking of the bases in parallel planes. The parallel-offset stacking of the rings provides the major stabilization of the double-helical structure. (c) and (d) Views of the same models from the top (along the helical axis; the direction is indicated by the eyeball, with a common point indicated by the star, *). The sugars and phosphates are on the outside of the double helix, where they can readily interact with water, and the base pairs are on the inside of the helix, where they can form complementary hydrogen bonds with each other.

6. There is no intrinsic restriction on the sequence of bases in a polynucleotide; however, because of the base pairing described in point 3, the sequence of one polynucleotide strand (the "Watson" strand) in the double helix is complementary to that in the other strand (the "Crick" strand). Consequently, everywhere there is an A in one strand, there is a T in the other; everywhere there is a G in one strand, there is a C in the other.

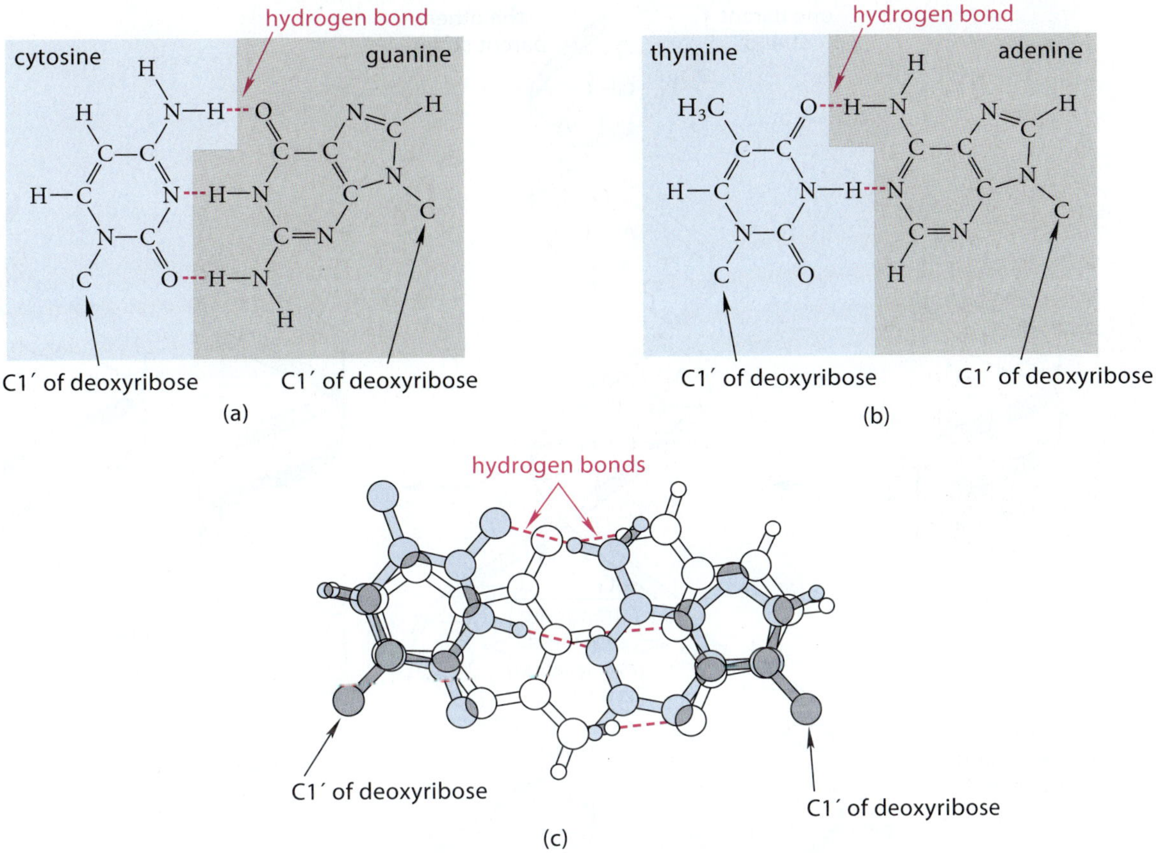

FIGURE 26.6 A closer look at the complementary base pairing in DNA. (a) A cytosine–guanine (C–G) base pair involves three hydrogen bonds. (b) A thymine–adenine (T–A) base pair involves two hydrogen bonds. (c) Superposition of the C–G *(white)* and T–A *(blue)* base pairs shows that the two occupy about the same space. (Regions of overlap are shown in gray.)

Hydrogen-bonding complementarity in DNA accounts for the Chargaff parity rule: if A always forms hydrogen bonds with T and G always forms hydrogen bonds with C, then the number of A residues must equal the number of T residues, and the number of G residues must equal the number of C residues. This structure also suggests a reasonable mechanism for the duplication of DNA during cell division—namely, the two strands come apart, and a new strand is grown as a complement of each original strand. In other words, *the proper sequence of each new DNA strand during cellular reproduction is ensured by hydrogen-bonding complementarity* (**Fig. 26.7**).

Although the complementary hydrogen bonds within the base pairs contribute to the stability of the double helix and account for the sequence fidelity of DNA replication, they are not the primary reason for the stability of double-helical DNA. There is good evidence that the offset-stacking interactions between the bases on successive planes provide the major part of the free-energy driving force for the formation of the double helix.

One of the most significant achievements in the history of biochemical analysis has been the development of methods for rapidly determining the sequential arrangement of individual bases in DNA. This type of analysis is called *DNA sequencing.* (These methods aren't discussed here.) In 1990, the National Institutes of Health and the Department of Energy co-sponsored the *Human Genome Project*, the primary goal of which was to identify the 20,000–25,000 human genes and to sequence all of the DNA in the human genome. Preliminary sequences were published in 2001, and high-quality sequences of almost all of the 24 human chromosomes were complete in 2003. The magnitude of this project can be appreciated from the fact that the human genome contains about 3 billion base pairs! The genome sequences of thousands of other organisms have been subsequently determined. These "sequenced organisms" come from every corner of the biological

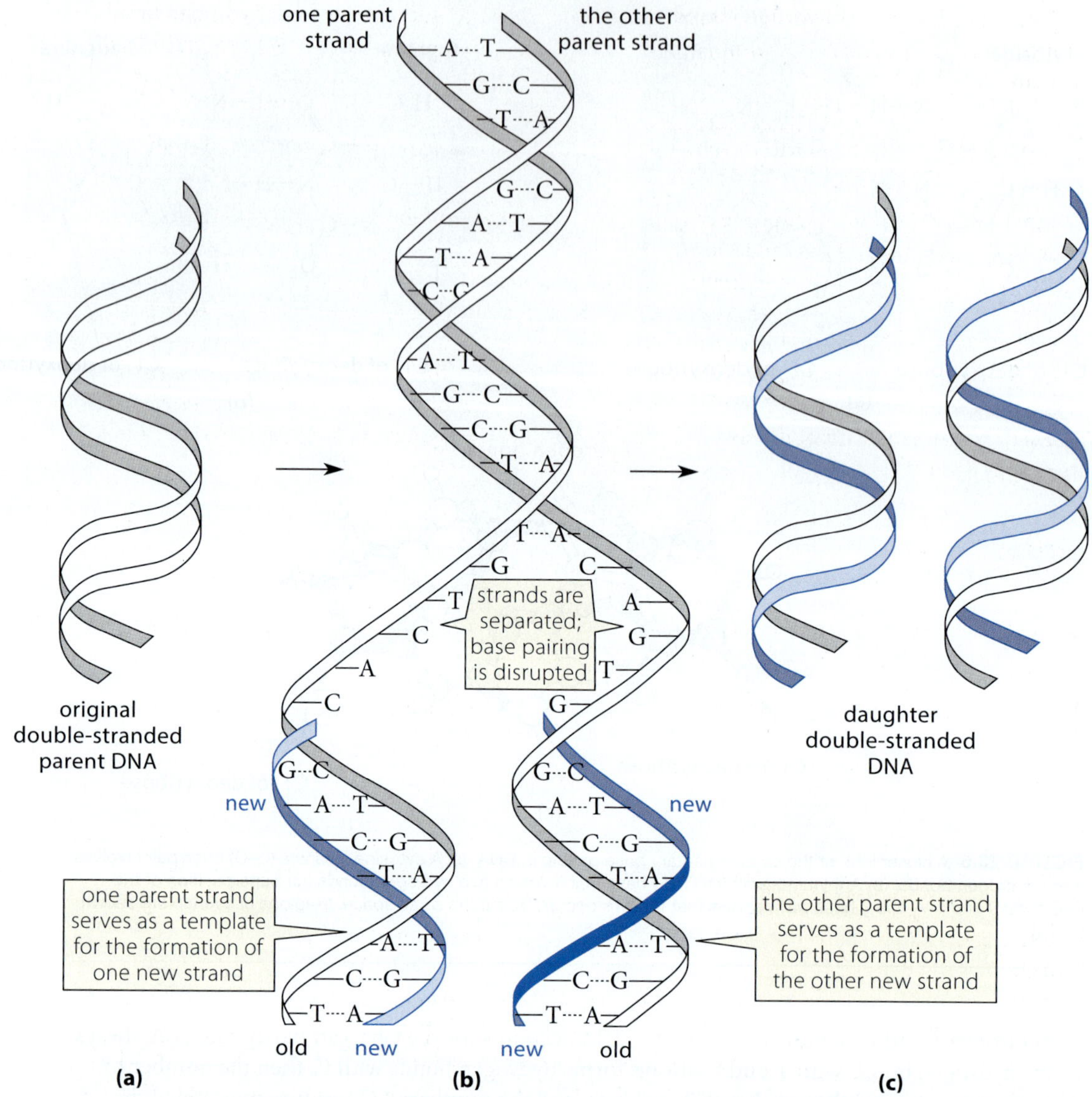

FIGURE 26.7 Complementary base pairing in DNA is crucial to its faithful replication. (a) A typical DNA double helix. (b) In the replicating DNA, a new strand grows on each of the original parent strands. (The synthesis of new DNA on a parent DNA template is catalyzed by several enzymes, which are not shown.) (c) The two new molecules of DNA each contain one parent strand and one new strand.

world—viruses [such as influenza A, human immunodeficiency virus (HIV), and the coronavirus responsible for the 2019 Covid-19 pandemic], bacteria (such as anthrax), plants (such as wheat and rice), insects (such as fruit flies), and higher animals (such as cows). Some viruses, called *retroviruses*, carry their genetic information in RNA rather than DNA; HIV and the Covid-19 virus are examples. When these viruses infect a cell, a viral enzyme, *reverse transcriptase*, translates the RNA code of the virus to viral DNA, forming an RNA–DNA hybrid. The RNA is then hydrolyzed back to individual ribonucleotides, and the remaining viral DNA is incorporated into the reproductive machinery of the cell, where eventual translation leads to the formation of many new virus particles and destruction of the host cell. Viral DNA can be produced in the laboratory from viral RNA with reverse transcriptase. Sequencing of this DNA provides the viral genome.

The central importance of DNA sequences is that they provide a linear code for biosynthesis of *messenger RNAs* (mRNAs), whose sequences, in turn, provide the code for the biosynthesis of proteins:

DNA sequence ⇨ mRNA sequence ⇨ linear sequence of amino acids in proteins

In other words, the linear sequence of amino acids in every protein is determined by a DNA code. (This process is discussed in more detail in Sec. 27.6B.) Therefore, it is possible to read the DNA "code" for a protein and thereby determine—without isolating the protein—the sequence of amino acids in the protein. These sequence data are unlocking the genetic basis of diseases as well as the significant genetic variations that occur among individuals. It is not unreasonable to imagine that, in the future, we will walk into our doctors' offices or pharmacies with a small magnetic card containing the complete sequence of our individual genome, which will be used to assess our individual risk of disease and to prescribe just the right medication and to determine its proper dose.

Focused Problems

26.23 Draw in detail the structure of a single-strand section of RNA four residues long that, from the 5′-end, has the following sequence of bases: A, U, C, G. Label the 3′ and 5′ ends.

26.24 Would you expect Chargaff's first parity rule to apply within an individual strand of DNA? Explain.

C. DNA Modification and Chemical Carcinogenesis

We've shown that the double-helical structure of DNA and DNA replication involves very specific Watson–Crick base-pairing complementarity. Other important processes, such as RNA biosynthesis and protein biosynthesis, also involve this type of complementarity. We showed that the molecular basis of this complementarity is the specific hydrogen bonding between a pyrimidine and a purine base. If this hydrogen bonding is disrupted, the Watson–Crick base-pairing complementarity can also be disrupted, and with it, some or all of the biological processes that rely on this phenomenon. There is strong evidence that chemical damage to DNA can interfere with this hydrogen-bonding complementarity and can in some cases trigger the state of uncontrolled cell division known as cancer.

One type of chemical damage to DNA is caused by *alkylating agents* (Sec. 11.4B). Certain types of alkylating agents react with DNA by alkylating one or more of the nucleotide bases. These same alkylating agents are also *carcinogens* (cancer-causing compounds). Examples of such compounds are the following:

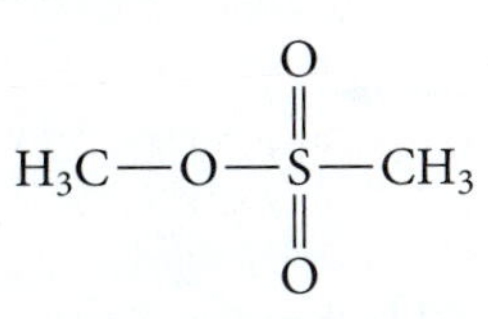

methyl methanesulfonate
(a weak carcinogen)

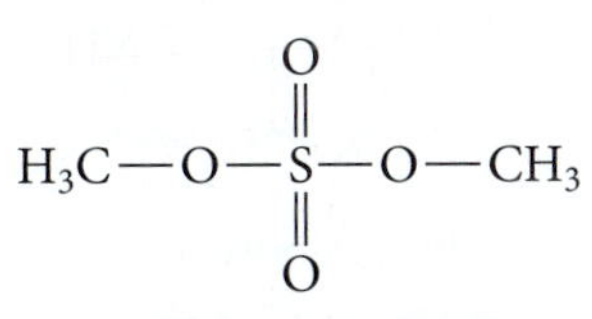

dimethyl sulfate
(a weak carcinogen)

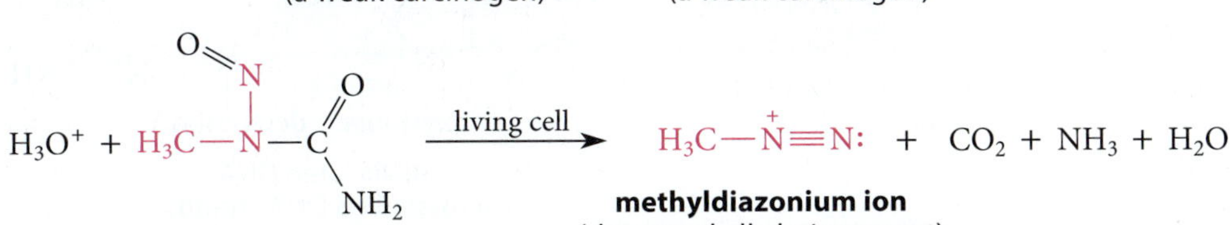

N-methyl-N-nitrosourea
(a potent carcinogen)

methyldiazonium ion
(the actual alkylating agent)

(26.69)

When such alkylating agents (abbreviated $H_3C—X$ in the following equations) react with DNA, alkylated guanosines are among the products. The major product is alkylated on N-7 of the guanine base, but an important minor product is alkylated on the oxygen at C-6 (called the O-6 position).

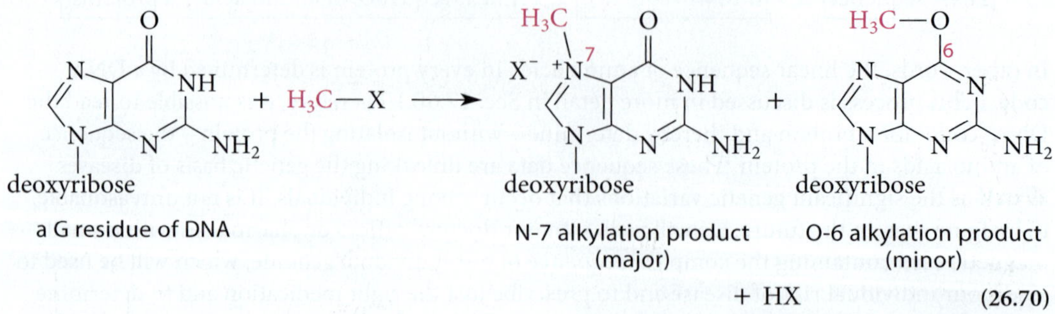

(26.70)

(An analogous alkylation occurs at O-4 of thymine; see Focused Problem 26.25.) Alkylation at O-6 prevents the N-1 nitrogen from acting as a hydrogen-bond donor in a Watson–Crick base pair (see Fig. 26.6) because the hydrogen is lost from this nitrogen as a result of alkylation. (B: = a base.)

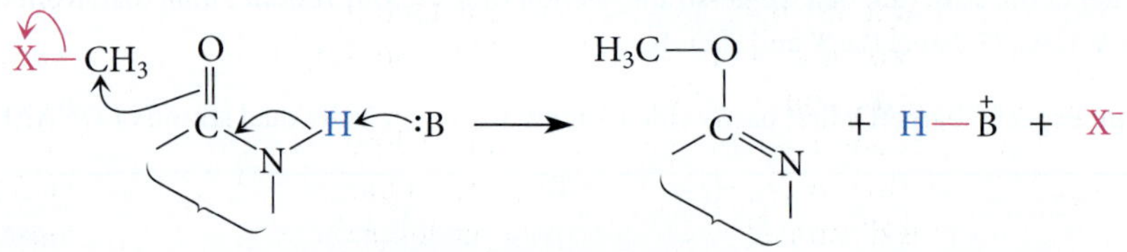

(26.71)

The N-7 alkylation, in contrast, does not directly affect any of the atoms involved in the hydrogen-bonding complementarity. It has been found that the alkylating agents that are the most potent carcinogens also yield the greatest number of the guanines alkylated at O-6 and thymines alkylated at O-4. This correlation suggests that these alkylations are primary events in carcinogenesis.

The nitrogen mustards used as antitumor drugs were described Sec. 12.8D. These drugs, as dialkylating agents, react at the N-7 positions on guanine bases in opposite strands of the DNA double helix. This reaction forms a crosslink between the two strands and prevents the strand separation that must accompany cell division (see Fig. 26.7).

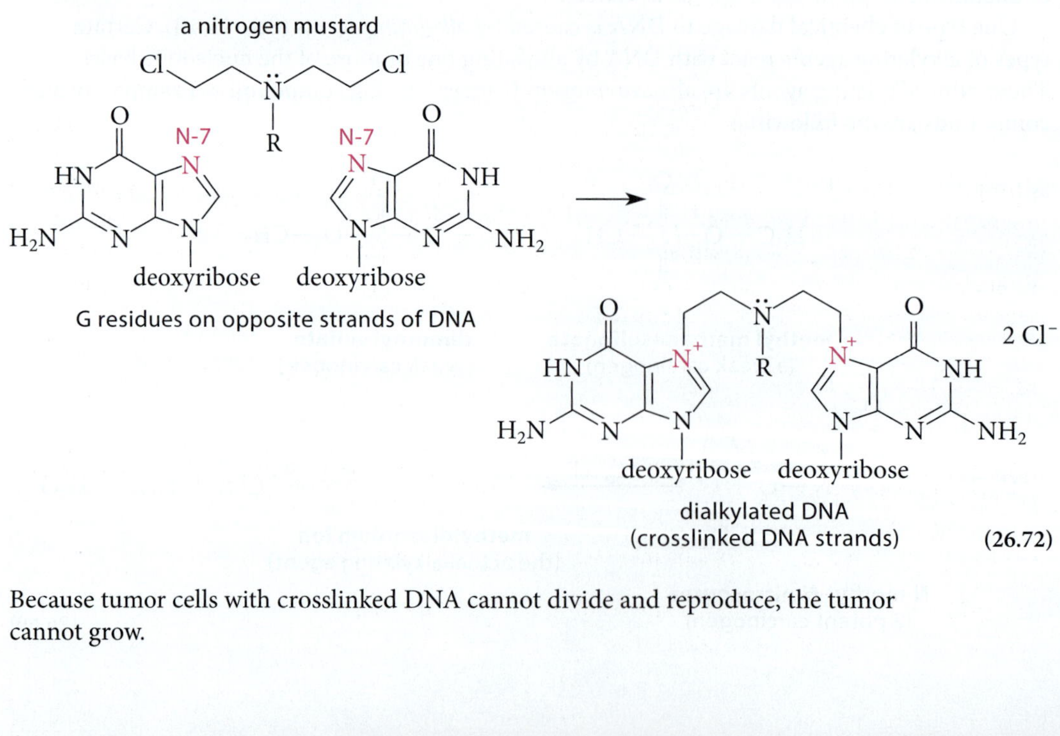

(26.72)

Because tumor cells with crosslinked DNA cannot divide and reproduce, the tumor cannot grow.

The conversion of aromatic hydrocarbons into carcinogenic diol epoxides by enzymes in living systems was discussed in Sec. 16.7. These epoxides react with G residues of DNA. Although the reaction can occur at N-7 (as with the mustards), the reaction that leads to carcinogenicity is the reaction with the amino group at N-2.

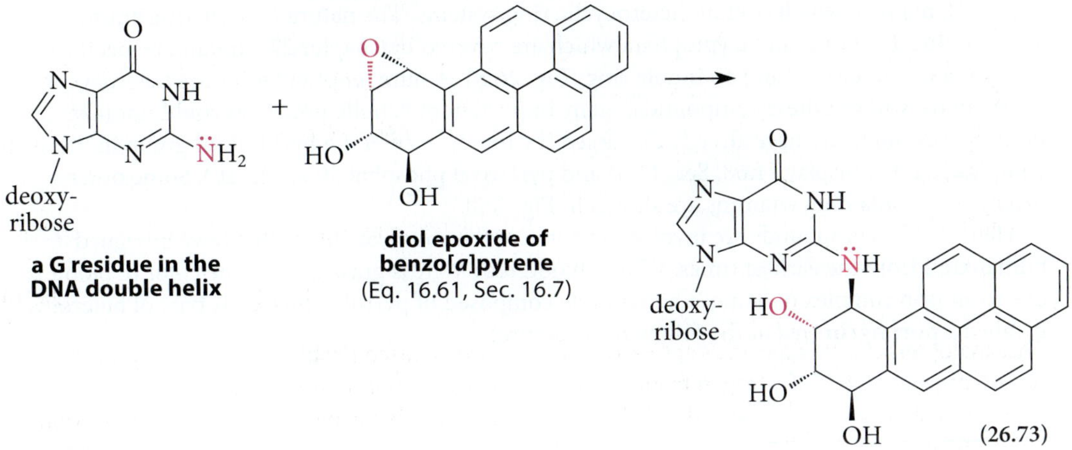

a G residue in the DNA double helix

diol epoxide of benzo[*a*]pyrene (Eq. 16.61, Sec. 16.7)

(26.73)

This nitrogen is also involved in the hydrogen-bonding interaction of G with C (see Fig. 26.6). It may be that alkylation by aromatic hydrocarbon epoxides also triggers the onset of cancer by interfering with the base-pairing complementarity.

DNA damage can also be caused by ultraviolet radiation. Ultraviolet light promotes a cycloaddition of two pyrimidines when they occur in adjacent positions on a strand of DNA. (This type of cycloaddition is discussed in Sec. 28.3.) In the following example, a thymine dimer is formed from two adjacent thymines.

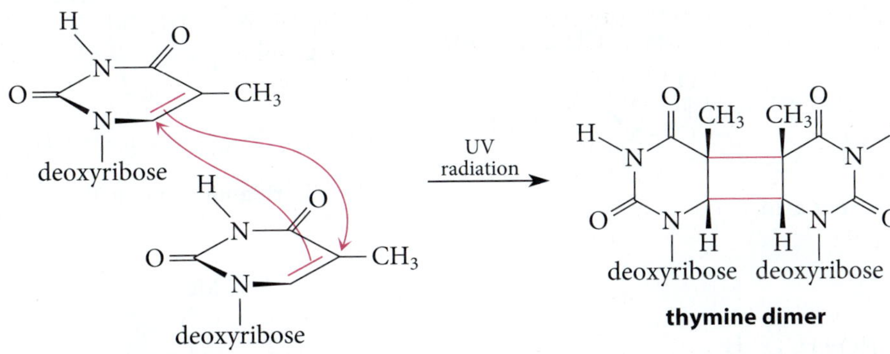

two thymines in DNA at adjacent positions on same strand

thymine dimer

(26.74)

Most people have a biological repair system that effects the removal of the modified pyrimidines and repairs the DNA. People with a rare skin disease, *xeroderma pigmentosum*, have a genetic deficiency in the enzyme that initiates this repair. Most of these people contract skin cancer and die at an early age. Here, then, is a situation in which the chemical modification of DNA has been clearly associated with the onset of cancer.

Focused Problem

26.25 There is evidence that alkylation at O-4 of thymine, like alkylation at O-6 of guanine, is another mutagenic event that can lead to cancer.

(a) Draw the structure of a thymine residue as it would exist after O-4 methylation.

(b) Explain why O-4 alkylation at thymine would disrupt Watson–Crick base pairing.

26.6 OTHER NATURALLY OCCURRING HETEROCYCLIC COMPOUNDS

Nitrogen heterocycles occur widely in nature. Section 23.12B introduced the *alkaloids* (see Fig. 23.4), many of which contain heterocyclic ring systems. The naturally occurring amino acids proline, histidine, and tryptophan, which are covered in Chapter 27, contain, respectively, a pyrrolidine, imidazole, and an indole ring (**Fig. 26.8**). A number of vitamins are heterocyclic compounds; without these compounds, many important metabolic processes could not take place. For example, we have already discussed the importance of the pyridinium group in the vitamins NAD^+ (Display 11.68, Sec. 11.8) and pyridoxal phosphate (Sec. 26.4E). Some other heterocycle-containing vitamins are shown in Fig. 26.8.

Heterocyclic compounds are involved in some of the colors of nature that have intrigued humankind from the earliest times. Why is blood red? Why is grass green? The color of blood is due to an iron complex of heme, a heterocycle composed of pyrrole units. This type of heterocycle is called a **porphyrin** (*red* in the following structure).

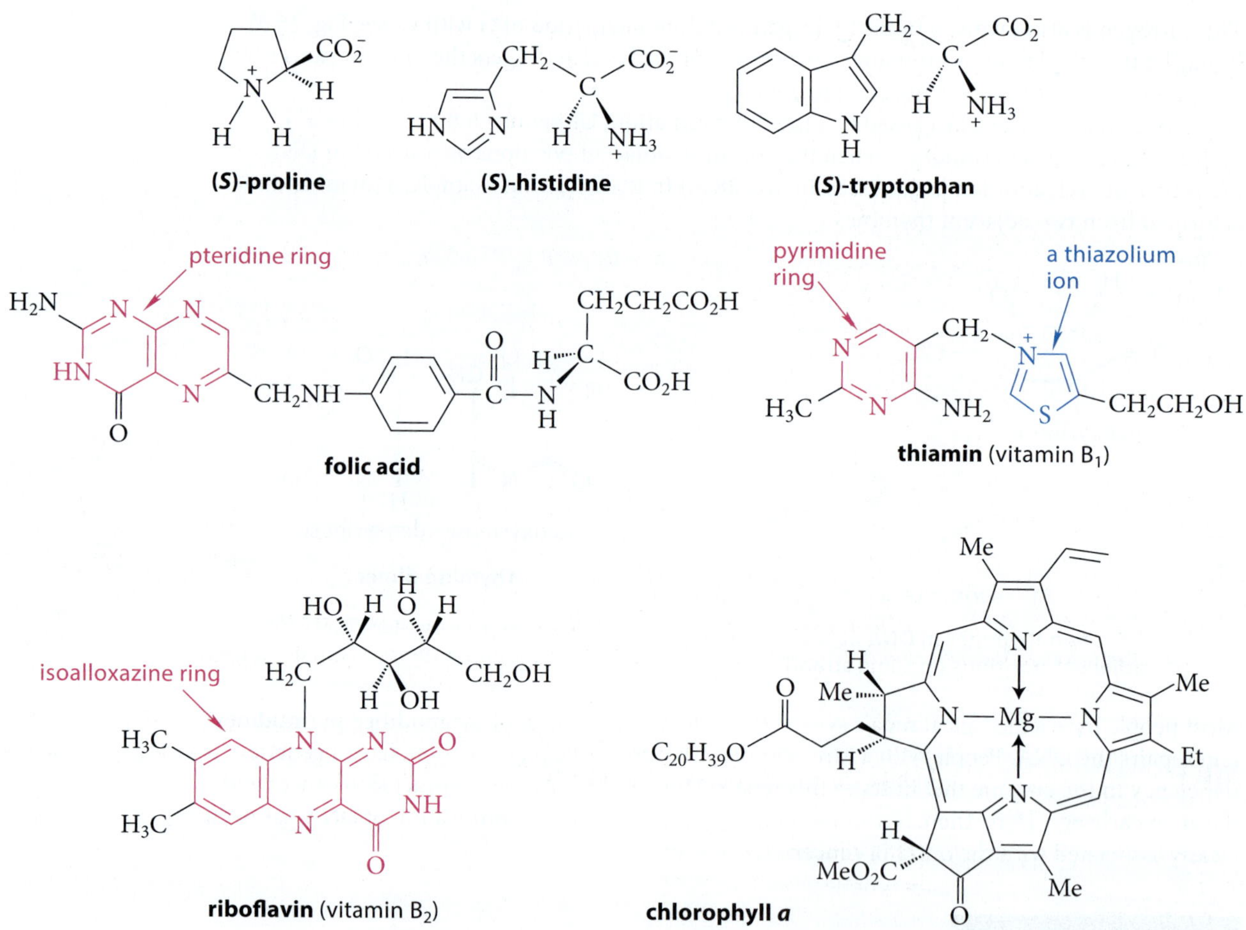

FIGURE 26.8 A few of the many naturally occurring heterocyclic compounds. The *S* enantiomers of proline, histidine, and tryptophan are α-amino acids. Folic acid, thiamin, and riboflavin are vitamins. The chlorophylls are the pigments responsible for the green color of plants. The $C_{20}H_{39}$ group is an isoprenoid side chain; see Sec. 17.6A. NAD^+ (Display 11.68, Sec. 11.8) and pyridoxal phosphate (see Fig. 26.3, Sec. 26.4E) are examples of important naturally occurring pyridine derivatives.

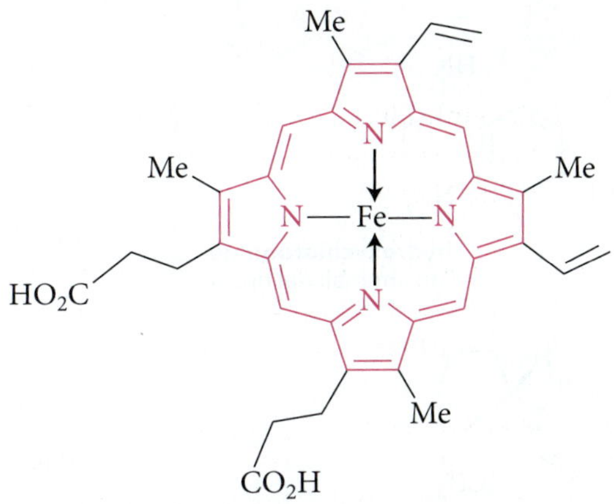

heme
(occurs in the protein complex hemoglobin)

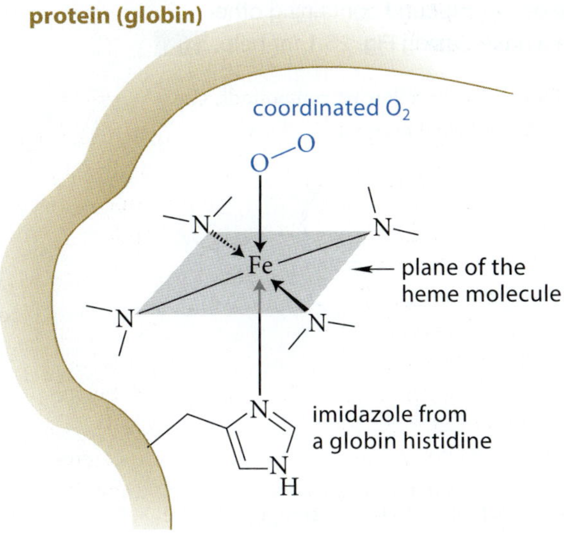

a schematic view of oxygenated heme in hemoblobin

(26.75)

Heme is the Fe(II) complex of an aromatic heterocycle that is found in red blood cells tightly bound to a protein called *globin*; the complex is called *hemoglobin*. The iron, held in position by coordination with the nitrogen atoms of heme and an imidazole of globin, reversibly forms a complex with oxygen. Hemoglobin is the oxygen carrier of blood, and the red color of blood is due to oxygenated hemoglobin. Carbon monoxide and cyanide, two well-known respiratory poisons, also form complexes with the iron in hemoglobin, as well as with iron in the heme groups of other respiratory proteins.

Protoporphyrin IX, the iron carrier in cytochrome P450 (see Fig. 17.4, Sec. 17.5B) is essentially a heme containing an iron in a high oxidation state.

The green color of plants is caused by *chlorophyll*, a class of compounds closely related to the porphyrins (see Fig. 26.8). The absorption of sunlight by chlorophylls is the first step in the conversion of sunlight into usable energy by plants. The chlorophyll molecules are nature's "solar energy collectors."

CHAPTER SUMMARY *For a summary of the chapter, see Chapter 26 in the* Study Guide and Solutions Manual.

REACTION REVIEW *For a summary of reactions discussed in this chapter, see the* Reaction Review *section of Chapter 26 in the* Study Guide and Solutions Manual.

SKILLS OBJECTIVES WITH PROBLEMS

- **Given the structure of a furan, thiophene, pyrrole, or pyridine derivative, provide a name. Given a name, provide a structure. 26.1**

26.26 Provide a name for each of the following compounds.

(a) 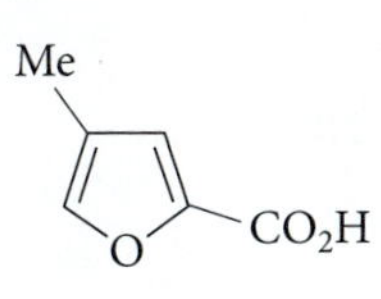

(b) 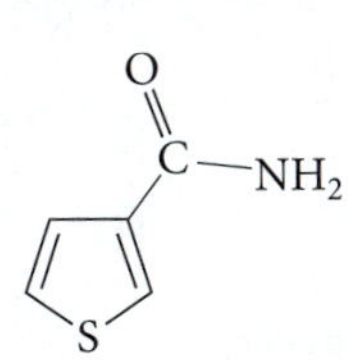

(c) 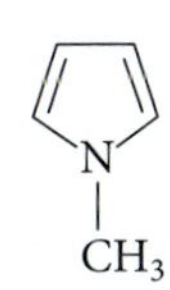

(d) 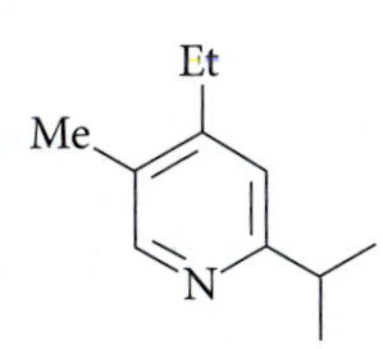

26.27 Draw a structure for each of the following compounds.

(a) 3,4-Dibromothiophene

(b) 2,5-Furandicarboxylic acid

(c) 1-(2-Furyl)-1-propanone

- **Given the structure of a compound containing other heterocycles, provide a name, using Fig. 26.1 for help.** **26.1**

26.28 Provide a name for each of the following compounds. Use Fig. 26.1 (Sec. 26.1A) for help if necessary.

(a)

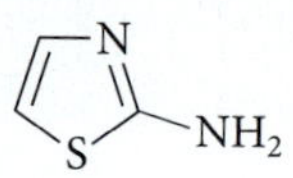

(b)

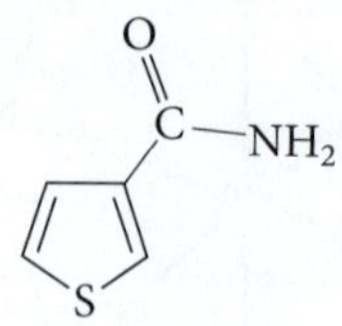

(c)

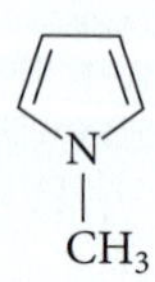

26.29 Draw a structure for each of the following compounds. Use Fig. 26.1 for help if necessary.

(a) 4-Oxazolecarbaldehyde

(b) 5-Methoxyindole

(c) 3-Quinolinecarboxylic acid

- **Use resonance structures and other structural criteria to predict relative basicities.** **26.2A**

  (1) **Identify the most basic atom in a compound, especially heterocyclic compounds with several nitrogen atoms, and explain your reasoning.**

  (2) **Rank heterocyclic compounds in a series according to increasing basicity, and explain your reasoning.**

26.30 Draw the structure of the major form of each of the following compounds present in an aqueous solution containing initially one molar equivalent of 1 *M* HCl. Explain your reasoning.

(a)

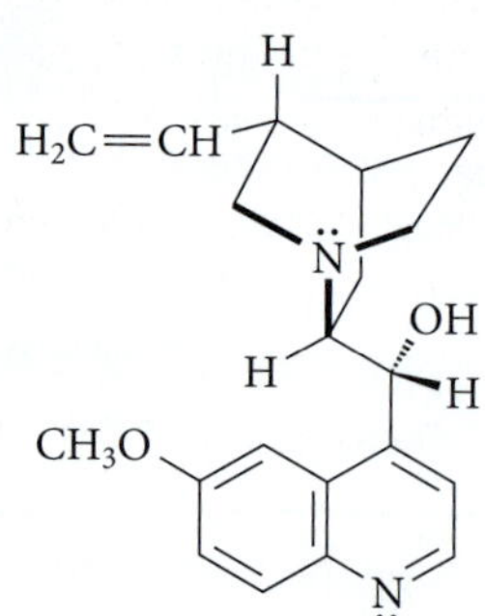

quinine

(b)

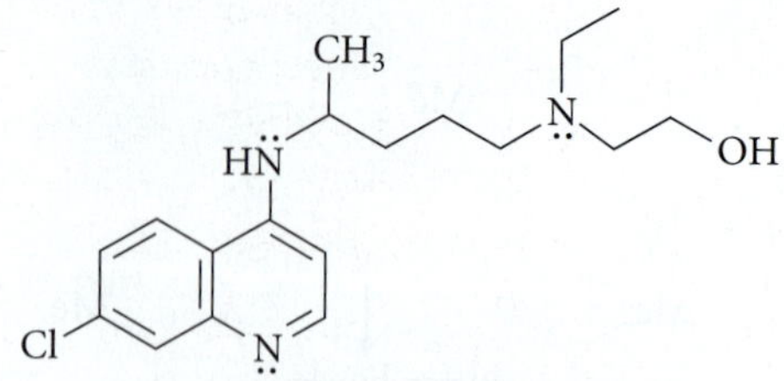

hydroxychloroquine
(an antimalarial drug)

(c)

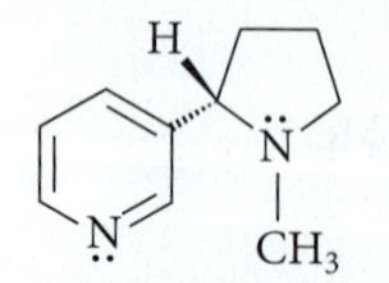

nicotine

(d) 3,4-Diaminopyridine

(e)

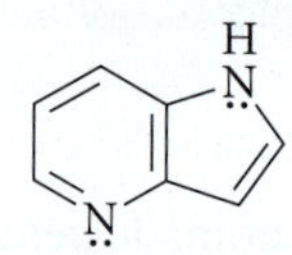

1,4-diazaindene

(f)

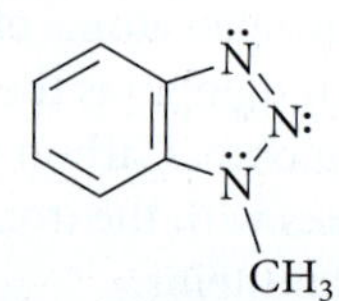

1-methyl-1,2,3-benzotriazole

(g) The compound in **Fig. P26.30.**

26.31 Rank the compounds within each of the following sets in order of increasing basicity, and explain your reasoning.

(a) Pyridine, 4-methoxypyridine, 5-methoxyindole, 3-methoxypyridine

(b) Pyridine, 3-nitropyridine, 3-chloropyridine

(c)

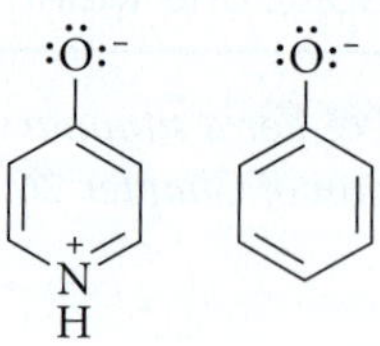

(d) Imidazole and oxazole

(e) Imidazole and thiazole

(h)

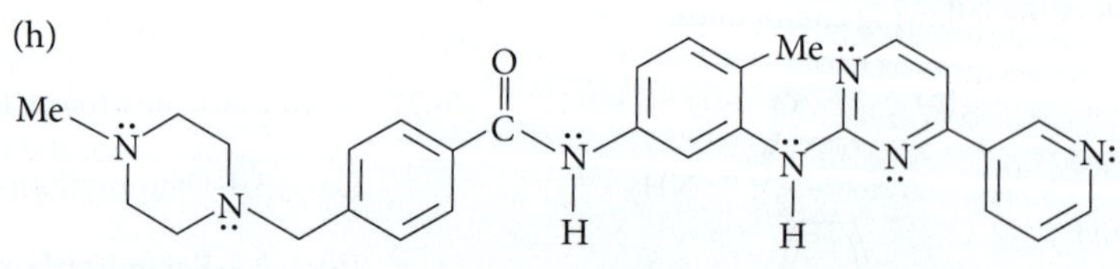

imatinib (Gleevec)
(an anticancer drug; see Sec. 1.1B)

FIGURE P26.30

- **Contrast the typical reactivities of pyridines with the reactivities of furans, thiophenes, and pyrroles in (a) electrophilic aromatic substitution reactions and (b) nucleophilic aromatic substitution reactions, and explain the difference. 26.3, 26.4A,B**

26.32 Explain the following two generalizations:

(a) Furan, pyrrole, and thiophene derivatives are typically very reactive in electrophilic aromatic substitution reactions, whereas pyridine derivatives are much less reactive.

(b) Pyridine derivatives with halogens in the 2- and/or 4-positions readily undergo nucleophilic aromatic substitution reactions, whereas halogen-substituted furan, pyrrole, and thiophene derivatives do not.

- **Rank derivatives of furan, pyrrole, thiophene, and pyridine in order of reactivity in electrophilic aromatic substitution reactions, and explain your choices. 26.3, 26.4A**

26.33 Rank the following compounds in order of increasing reactivity toward nitration with HNO_3, and explain your choices: thiophene (*A*), benzene (*B*), 3-methylpyridine (*C*), 3-methylthiophene (*D*), and 2-methylfuran (*E*).

- **Complete electrophilic aromatic substitution reactions in furan, pyrrole, thiophene, or pyridine derivatives. Predict the ring position at which substitution occurs, and support your prediction with resonance arguments. 26.3, 26.4A,B**

26.34 Predict the product in each of the following reactions, and explain your choices with resonance arguments.

(a)

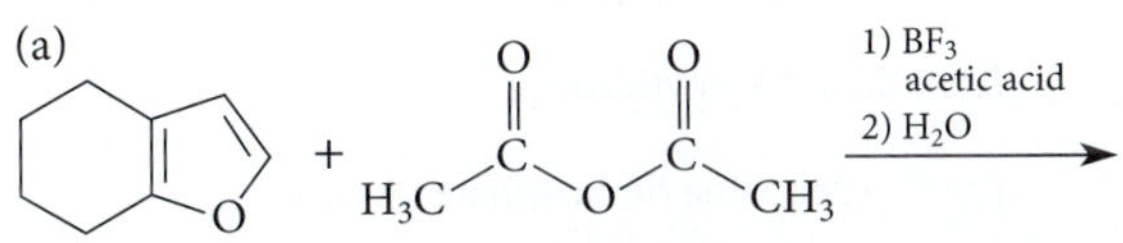

acetic anhydride

(b)

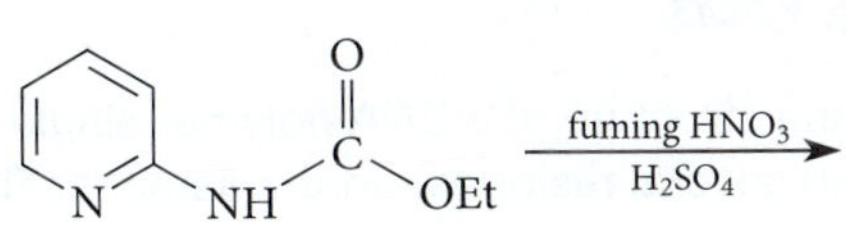

(c)

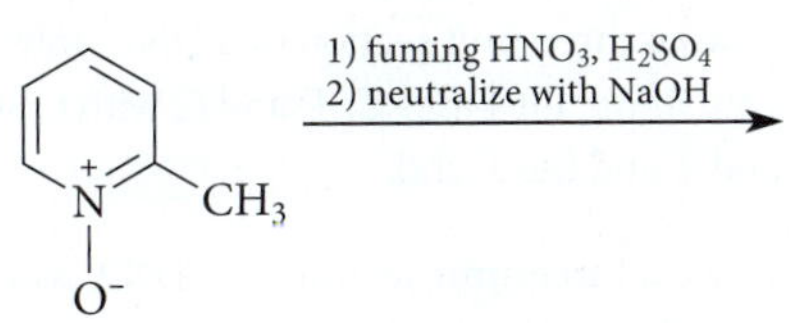

(d)

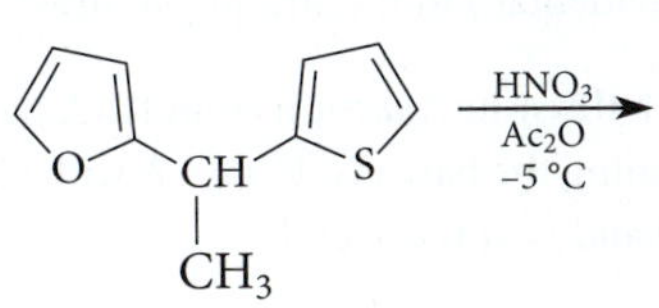

(a compound containing one nitro group)

(e)

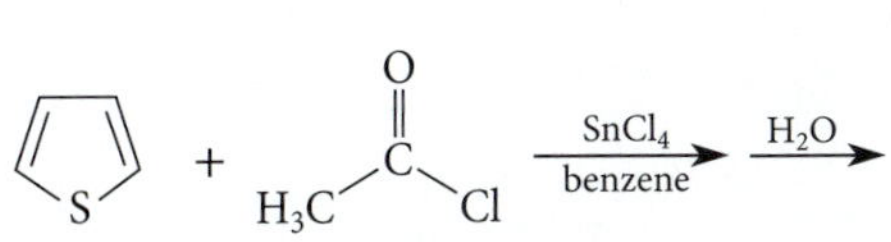

(f)

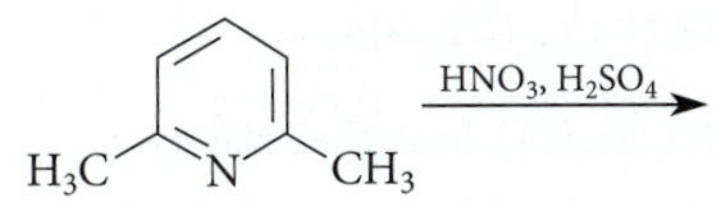

- **Use the mechanism of nucleophilic aromatic substitution to predict the relative reactivity of different compounds, or of different positions in the same compound. 26.4B**

26.35 Arrange the following compounds in order of their increasing rate of nucleophilic aromatic substitution with ethoxide ion in ethanol, and give reasons for your answer.

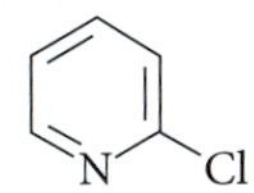

A

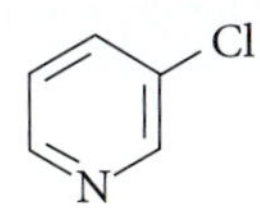

B

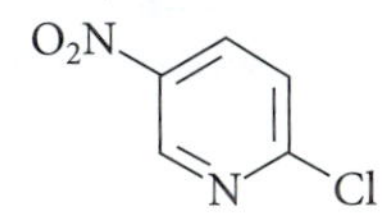

C

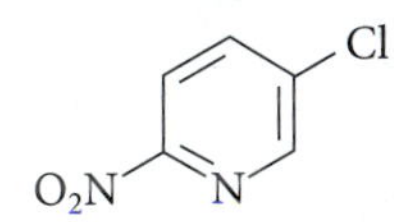

D

26.36 Give the product of the following reaction, and explain your reasoning.

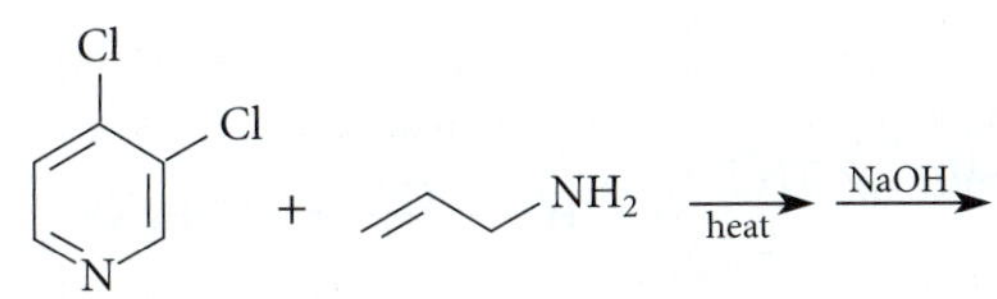

- **Complete examples of the reactions introduced in his chapter, including side-chain reactions, and give a curved-arrow mechanism where appropriate. 26.2, 26.3, 26.4A–D**

26.37 Give the principal organic product(s) expected when 2-methylthiophene or the other compound indicated reacts with each of the following reagents.

(a) Acetic anhydride, BF_3 catalyst, acid, then H_2O

(b) HNO_3

(c) *N*-Bromosuccinimide, CCl_4, light

(d) Dilute aqueous HCl

(e) Dilute aqueous NaOH

(f) Product of part (c) + Mg/ether, then CO_2, then H_3O^+

(g) Product of part (a) + Ph—CH=O and NaOH

26.38 Give the principal organic product(s) expected when 2-methylpyridine or the other compound indicated reacts with each of the following reagents.

(a) Dilute aqueous HCl

(b) Dilute aqueous NaOH

(c) $CH_3CH_2CH_2CH_2—Li$

(d) HNO_3, H_2SO_4, heat; then NaOH

(e) 30% H_2O_2

(f) CH_3I

(g) Product of part (c) + PhCH=O, then H_2O

(h) Product of part (e) + H_2, catalyst

26.39 Complete each of the following reactions by providing the structure of the missing organic products. Give the curved-arrow mechanism for the reactions indicated with an asterisk (*).

(a)*

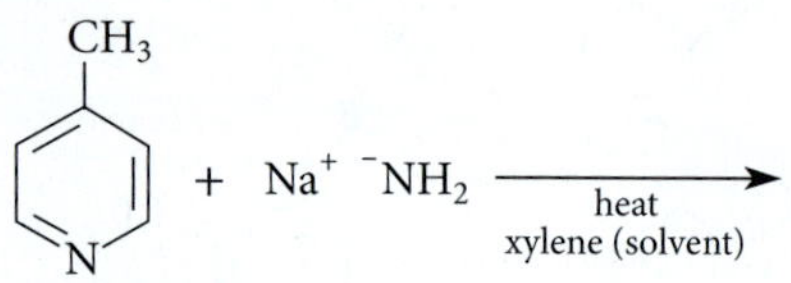

(b)*

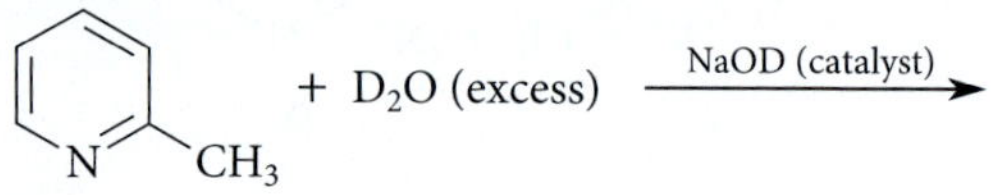

(c)*

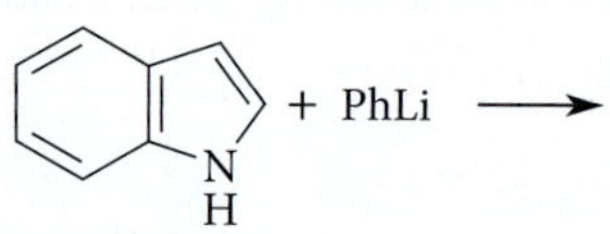

(d)

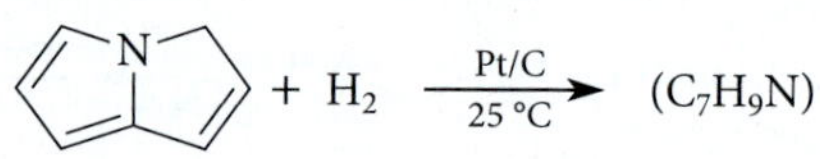

(e)*

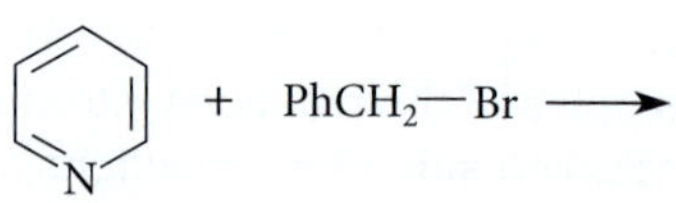

(f)

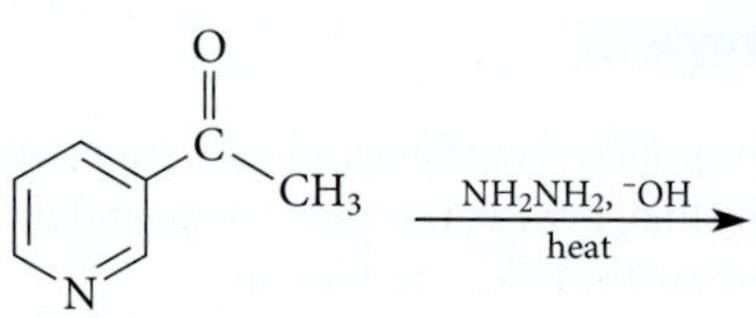

(g)*

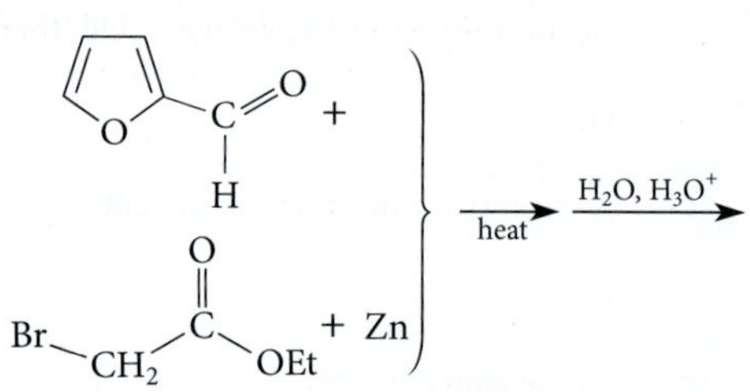

(h)

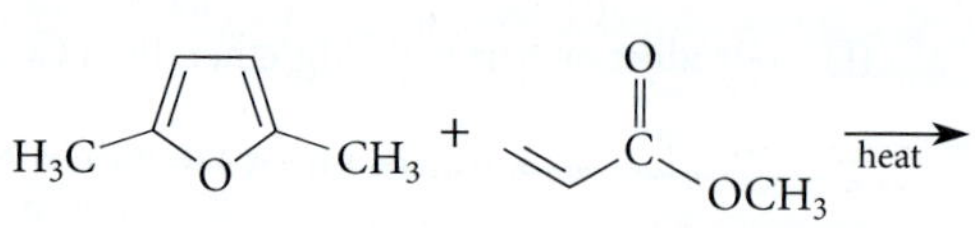

● **Write curved-arrow mechanisms for enzyme-catalyzed reactions involving pyridoxal phosphate as a coenzyme. 26.4E**

26.40 The *racemization* of amino acids is an important reaction in a number of bacteria.

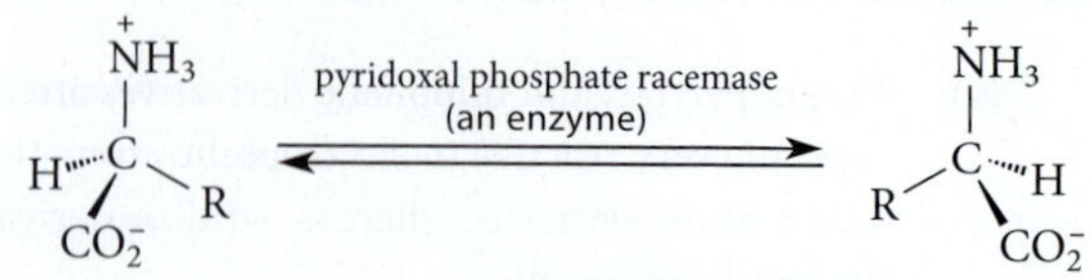

This is a pyridoxal-phosphate-catalyzed reaction. Outline a curved-arrow mechanism for this reaction showing clearly the role of pyridoxal phosphate. Assume that bases (B:) and acids ($^+$BH) are available as needed.

● **Give the structures of the four bases typically found in DNA; do the same for RNA. Identify whether each base is a purine or a pyrimidine. 26.5A,B**

26.41 (a) Give the structures and names of the four bases found in DNA. Classify each as a purine or pyrimidine.

(b) Give the structures and names of the four bases found in RNA. Classify each as a purine or pyrimidine.

● **Given the name or abbreviation for a nucleoside or nucleotide, draw its structure. Given a structure, provide the name or standard 3-letter abbreviation (as in Table 26.2). 26.5A**

26.42 Draw a structure for each of the following compounds.

(a) Uridine monophosphate (UMP, uridylic acid)

(b) 2′-Deoxycytidine

(c) Adenosine diphosphate (ADP)

26.43 Name each of the nucleotides or nucleosides shown in **Fig. P26.43.**

● **Draw the structure of a DNA polymer single strand containing several residues; do the same for RNA. 26.5B**

26.44 (a) Draw a three-unit section of a DNA polymer containing the bases C, T, and G with C at the 5′ end and T and the 3′ end.

(b) Draw a three-unit section of a DNA strand that is complementary to the strand you drew in part (a).

(c) Draw a three-unit section of an RNA polymer containing the bases G, U, and A with G at the 3′ end and U at the 5′ end.

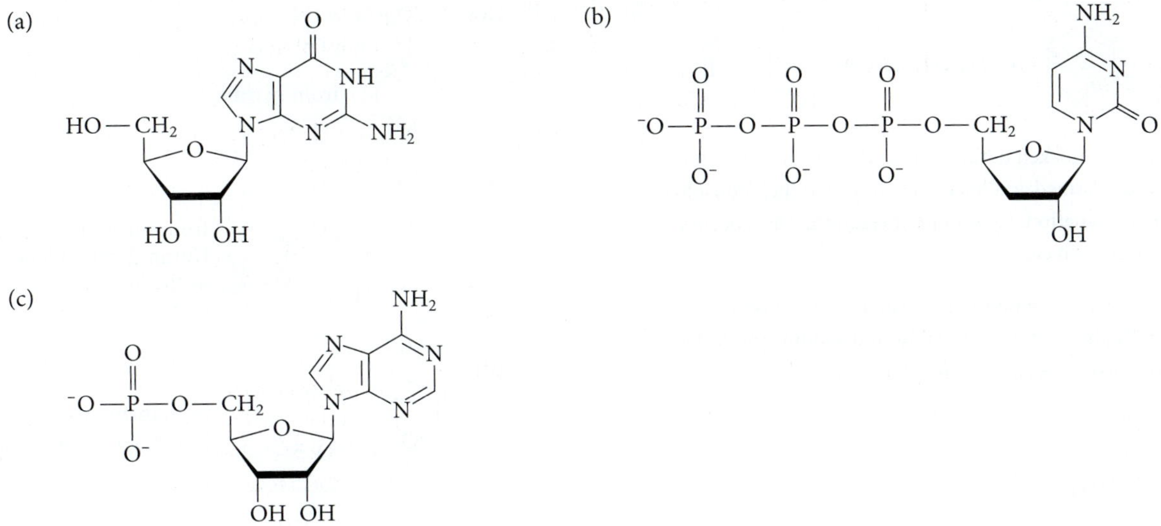

FIGURE 26.43

INTEGRATED PROBLEMS

26.45 The following compound is a very strong base; its conjugate acid has a pK_a of about 13.5. Give the structure of its conjugate acid, and show that it is stabilized by resonance.

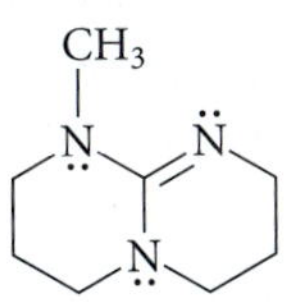

26.46 Which of the following two compounds should be the stronger base? Explain your reasoning.

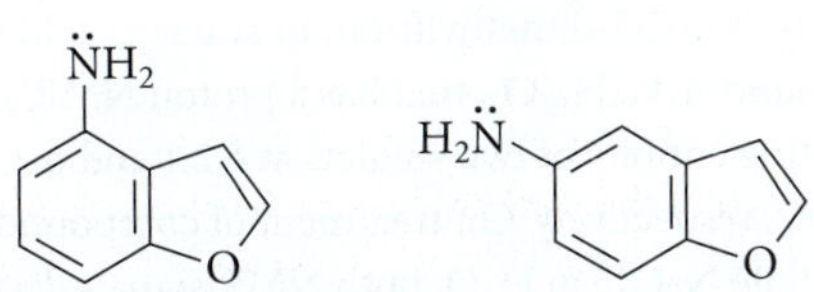

4-aminobenzofuran **5-aminobenzofuran**

26.47 Tetrahydrofuran (THF), like many ethers, protonates on oxygen; its conjugate-acid pK_a is about −2.1. Furan, in contrast, is a much weaker base and protonates on carbon; its conjugate-acid pK_a has been estimated to be −13. Account for the greatly different basicities of the two compounds.

tetrahydrofuran pK_a ~ −2.1 **furan** pK_a ~ −13

26.48 Rank each of the compounds within each part in order of increasing S_N1 solvolysis reactivity in ethanol, and explain your choices by drawing suitable structures.

(a)

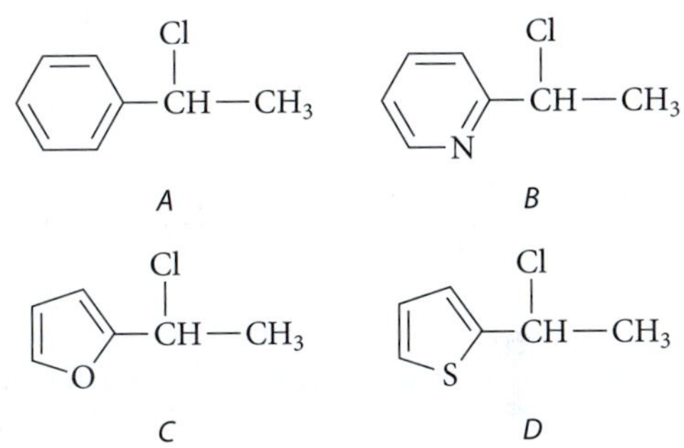

(b)

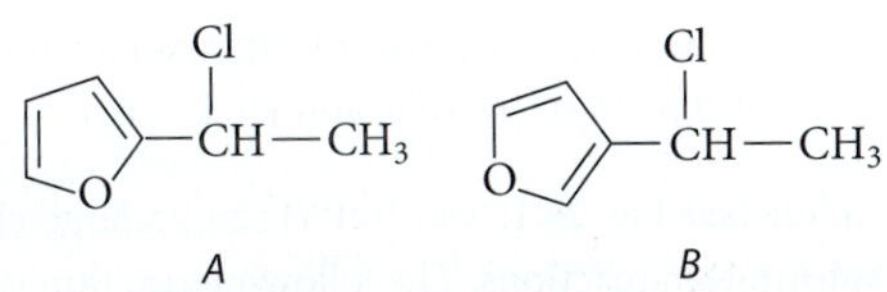

26.49 Think of the compounds in the following sets as enols. Then draw their carbonyl isomers. Which compound within each set contains the greatest percentage of carbonyl isomer? Explain.

(a) 2-Hydroxyfuran or 2-hydroxypyrrole

(b) Phenol or 4-hydroxypyridine

26.50 Chemists working for the MakeAmines corporation have proposed the following variations on the Chichibabin reaction:

(a)

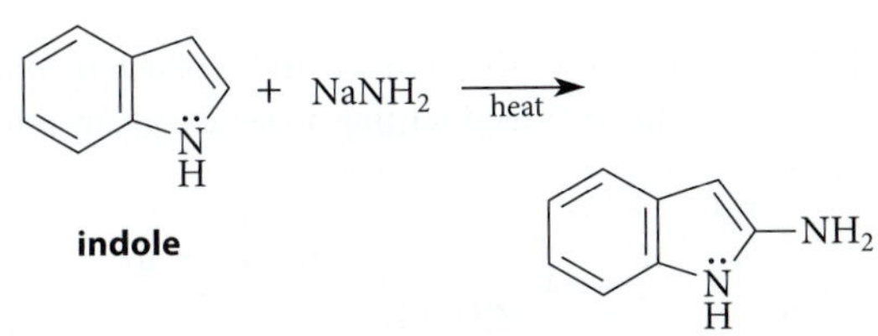

(b)

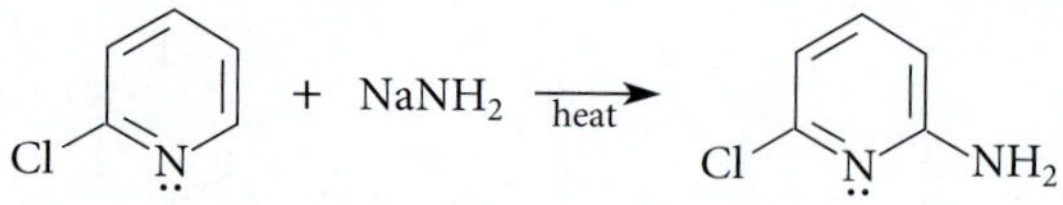

They were shocked to find that neither of these reactions works as planned and have come to you as their consultant for an explanation. Explain what reaction, if any, occurs instead in each case.

26.51 The following compound is isolated as a by-product in the Chichibabin reaction of pyridine and sodium amide. Give a curved-arrow mechanism for its formation.

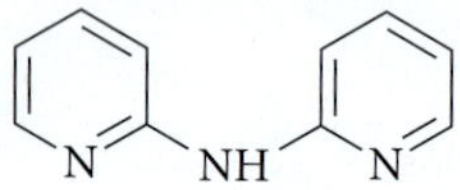

26.52 When pyrrole is treated with 5.5 *M* HCl at 0 °C for 30 s, a crystalline product *B* is obtained. A likely intermediate in this reaction is compound *A*.

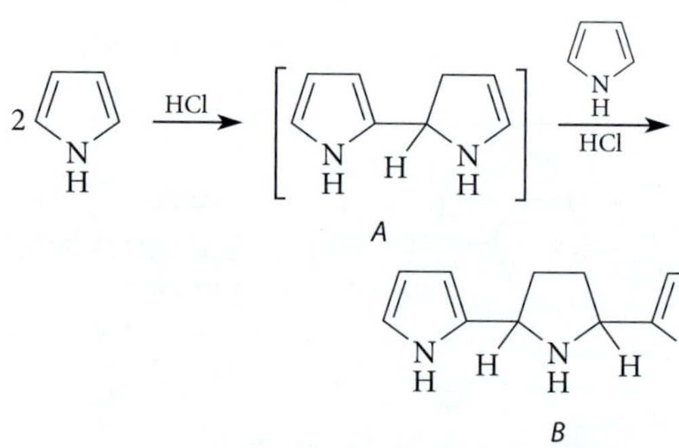

(a) Draw a curved-arrow mechanism for the formation of *A*.

(b) Draw a curved-arrow mechanism for the formation of *B* from *A*, pyrrole, and HCl.

26.53 Indole (see Fig. 26.1, Sec. 26.1A) can undergo electrophilic substitution reactions. The following reaction is an example:

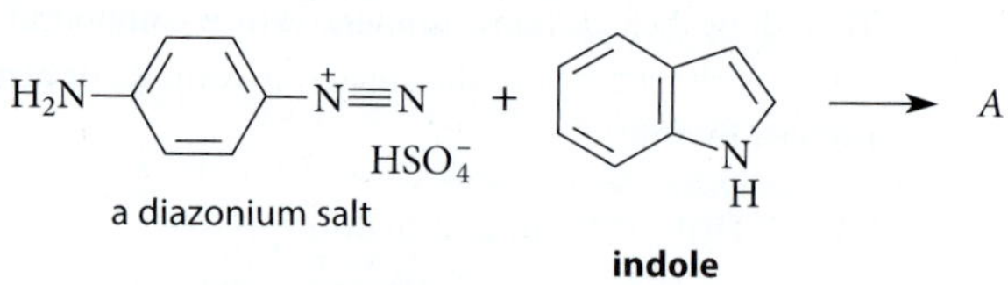

(See Eq. 23.61a, Sec. 23.10B.) Draw a curved-arrow mechanism for this electrophilic aromatic substitution reaction, and use it to derive the structure of the product *A*, an azo dye. (*Hint:* Indole undergoes electrophilic substitution at carbon-3.)

26.54 Outline a synthesis for each of the following compounds from the indicated starting material and any other reagents:

(a)

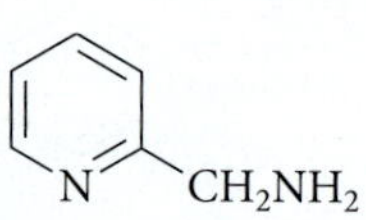

from pyridine

(b)

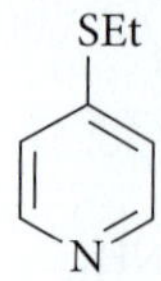

from pyridine

(c)

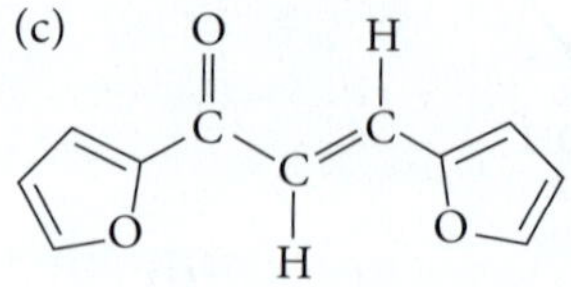

from furfural (furan-2-carbaldehyde) as the only source of furan rings

(d)

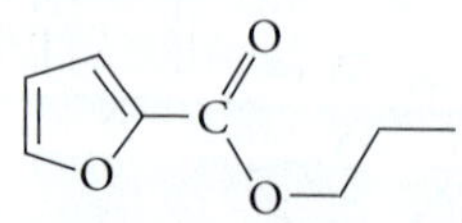

from furfural (furan-2-carbaldehyde)

(e)

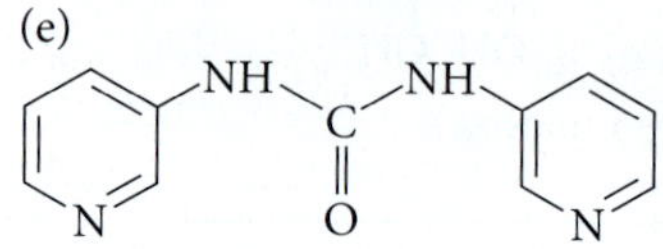

from 3-methylpyridine

(f)

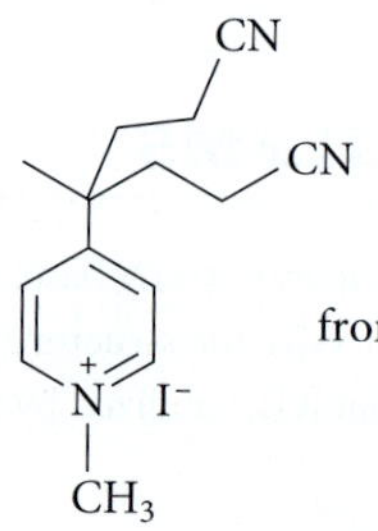

from 4-ethylpyridine

(g)

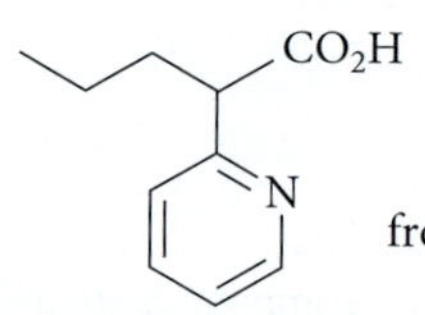

from 2-methylpyridine

26.55 Many furan derivatives are unstable in strong acid. Hydrolysis of 2,5-dimethylfuran in aqueous acid gives a compound *A*, $C_6H_{10}O_2$, that has a proton NMR spectrum consisting entirely of two singlets at δ 2.1 and δ 2.6 in the ratio 3:2, respectively. On treatment of compound *A* with very dilute NaOD in D_2O, both NMR signals disappear. Treatment of *A* with zinc amalgam and HCl gives hexane. Propose a structure for *A*, and then give a curved-arrow mechanism for its formation from 2,5-dimethylfuran.

26.56 Compound *A*, $C_8H_{11}NO$, which smells like dirty socks, can be resolved into enantiomers and it dissolves in 5% aqueous HCl. Oxidation of *A* with concentrated HNO_3 and heat gives nicotinic acid (3-pyridinecarboxylic acid; see Eq. 26.31, Sec. 26.4A). When *A* reacts with CrO_3 in pyridine, compound *B* (C_8H_9NO) is obtained. Compound *B*, when treated with dilute NaOD in D_2O, incorporates five deuterium atoms per molecule. Identify *A*, and explain your reasoning.

26.57 Aromatic sulfonation can be reversed by heating an arylsulfonic acid in steam:

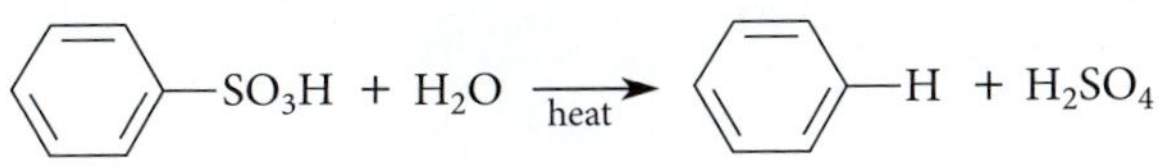

(a) Draw a curved-arrow mechanism for this transformation.

(b) Identify *A*, *B*, and *C* in the following scheme:

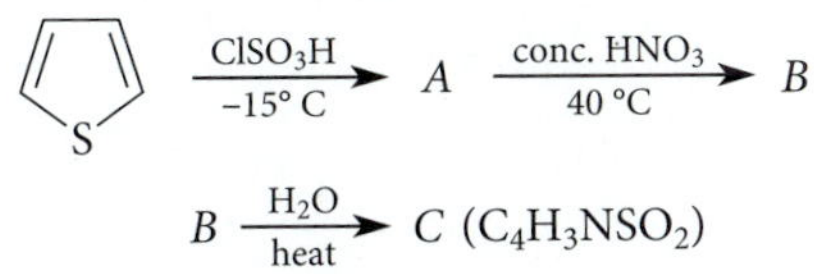

(c) Explain why compound *C* cannot be synthesized in one step from thiophene.

26.58 Provide curved-arrow mechanisms for each of the reactions given in **Fig. P26.58**. Give the structure for the intermediates *A* and *B* in parts (d) and (f).

26.59 Rank the following three compounds in order of their reactivity toward amide hydrolysis in aqueous NaOH, least reactive first, and explain your reasoning.

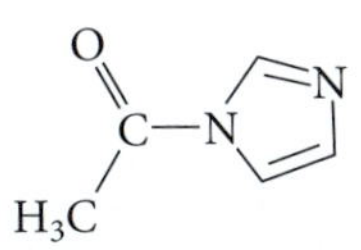

1-acetylimidazole
A

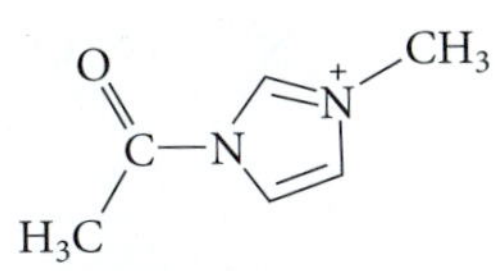

1-acetyl-3-methyl-imidazolium ion
B

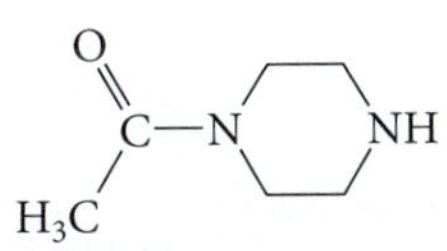

***N*-acetylpiperazine**
C

26.60 Decarboxylation of the amino acid L-histidine in *Lactobacillus* species involves an enzyme-attached amide (*blue*) of pyruvic acid, as shown in **Fig. P26.60**. (E = enzyme.)

(a) Assuming that bases (B:) and acids ($^+$BH) are available as needed, suggest a curved-arrow mechanism for this transformation. (*Hint:* An imine intermediate is involved.)

(b) In your mechanism, what group serves as the "electron sink" for the decarboxylation?

26.61 Pyridoxal phosphate and an enzyme, *tryptophan synthetase*, catalyze the last step in the biosynthesis of the amino acid *tryptophan* (see **Fig. P26.61**).

(a) The first part of this reaction involves the reaction of pyridoxal phosphate with serine to form species *A*, an imine of the unstable amino acid *dehydroserine*. [The red carbon is for part (b).]

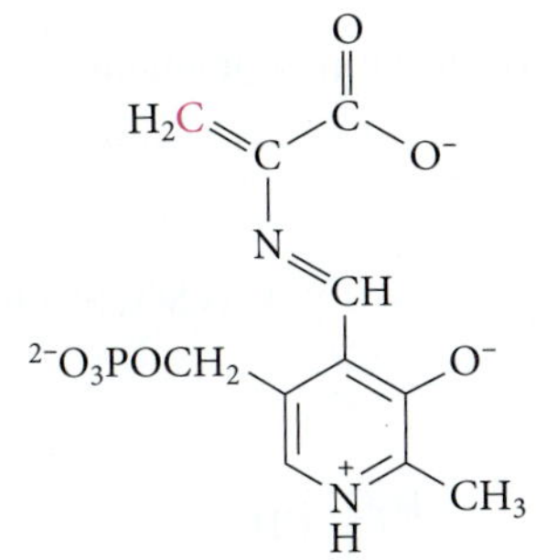

A
dehydroserine imine of pyridoxal phosphate

Assuming that bases (B:) and acids ($^+$BH) are available as needed, give a curved-arrow mechanism for the formation of *A*.

(b) Show with appropriate resonance structures that the carbon shown in red in the structure *A* has carbocation character.

(c) Recognizing that indole derivatives readily undergo electrophilic aromatic substitution at carbon-3, complete a curved-arrow mechanism for the biosynthesis of tryptophan. Assume that bases (B:) and acids ($^+$BH) are available as needed.

26.62 Explain each of the following facts.

(a) In the following compound, the hydrogens shown in red are readily exchanged for deuterium by dilute NaOD in D_2O, but those of the other methyl group are not.

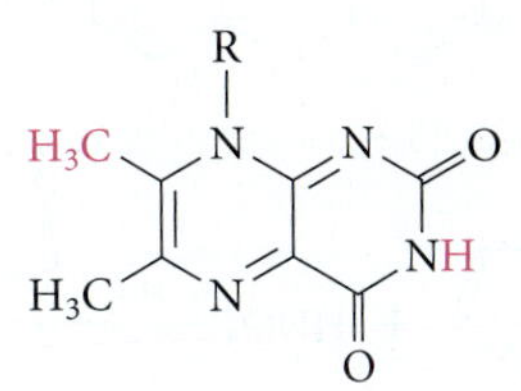

(b) In the following ion, the hydrogens of the methyl group shown in red are most acidic, even though the other methyl group is directly attached to the positively charged nitrogen.

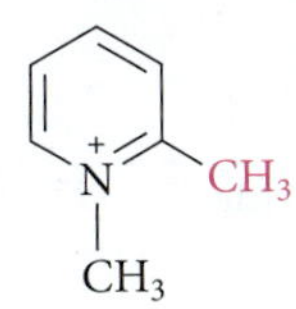

(a)

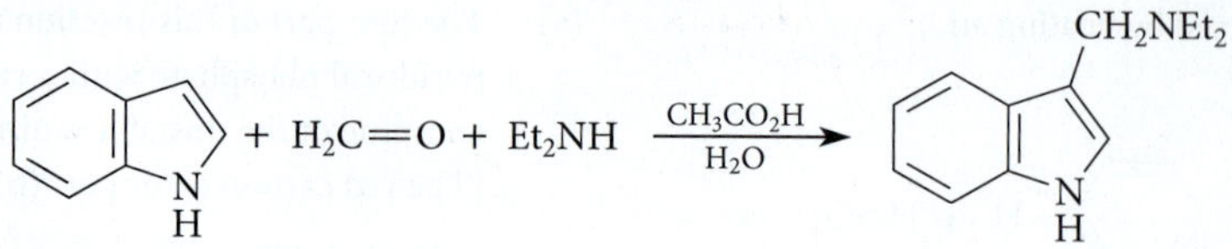

(*Hint*: Form the iminium ion $H_2C{=}\overset{+}{N}Et_2$ first.)

(b)

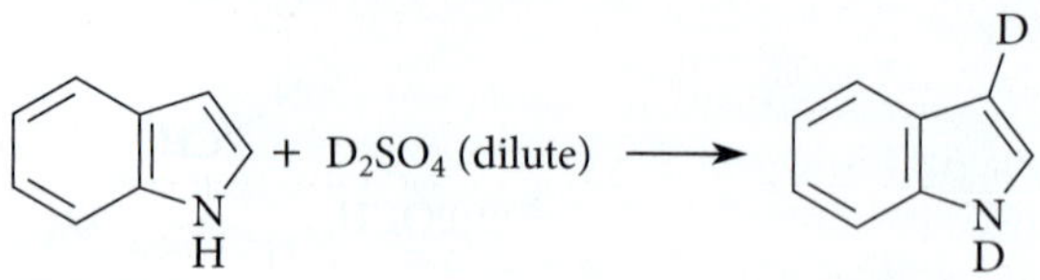

(c)

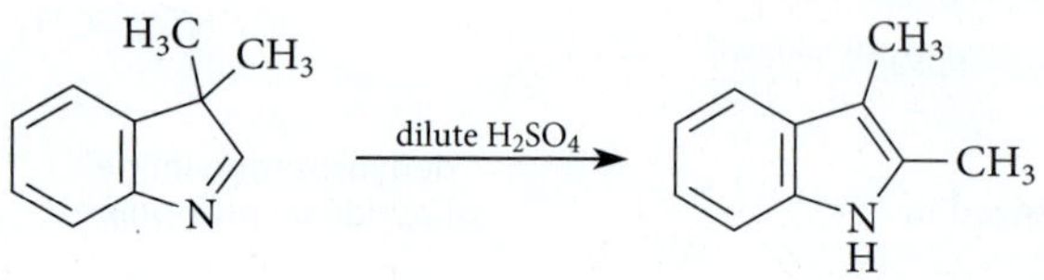

(d)

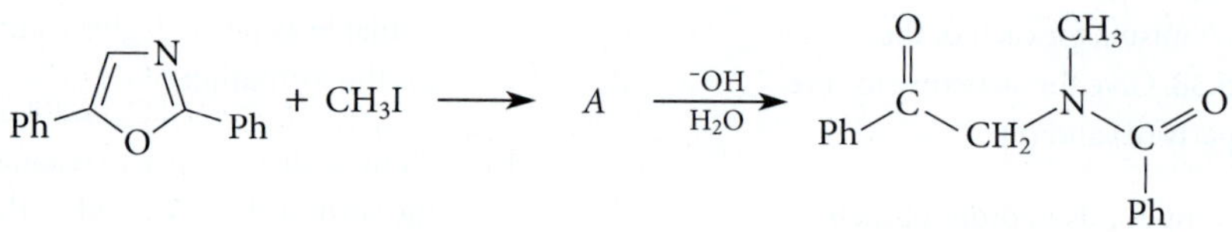

(e)

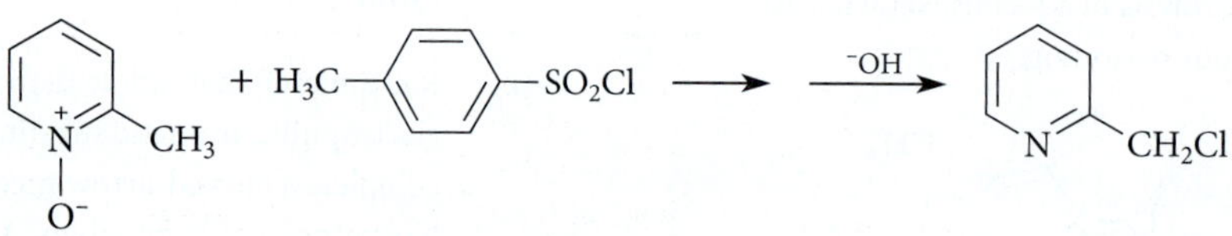

(f)

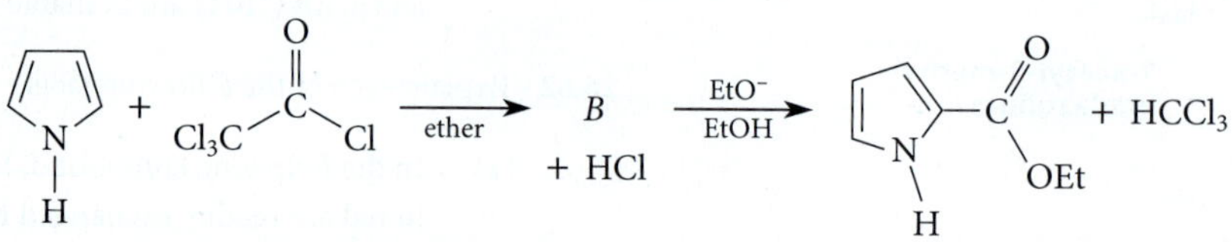

(g)

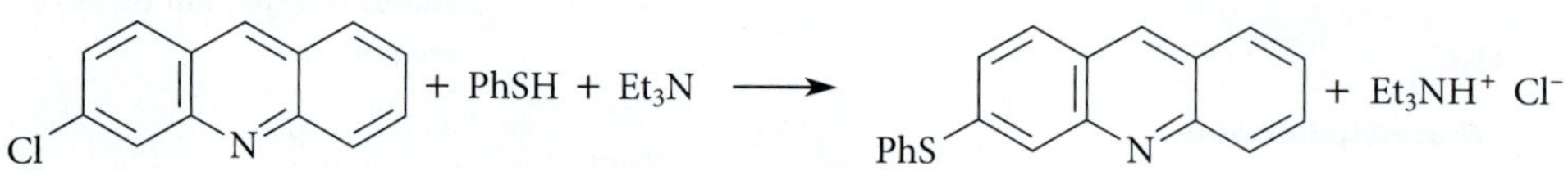

(h)

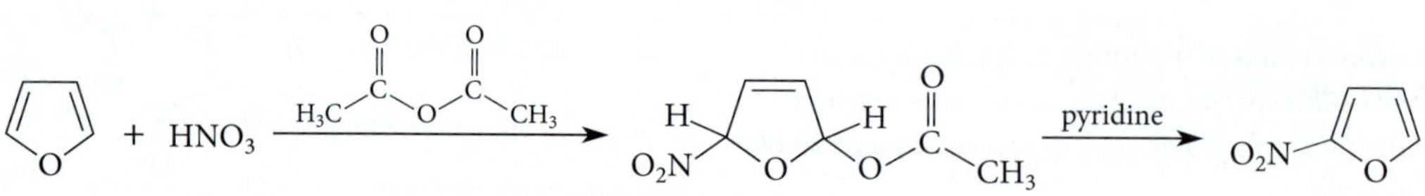

(*Hint:* HNO_3 is a source of the nitronium ion, $\overset{+}{N}O_2$.)

(i) Give the structure of compound *A* as well as the mechanisms of both reactions. (*Hint:* Think of the oxazole as a substituted furan.)

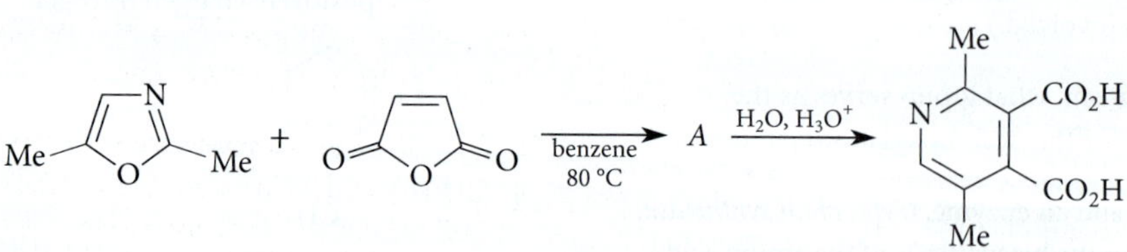

FIGURE P26.58

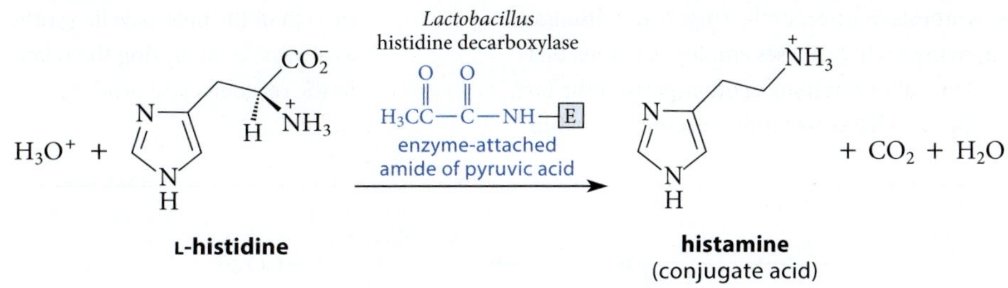

FIGURE P26.60

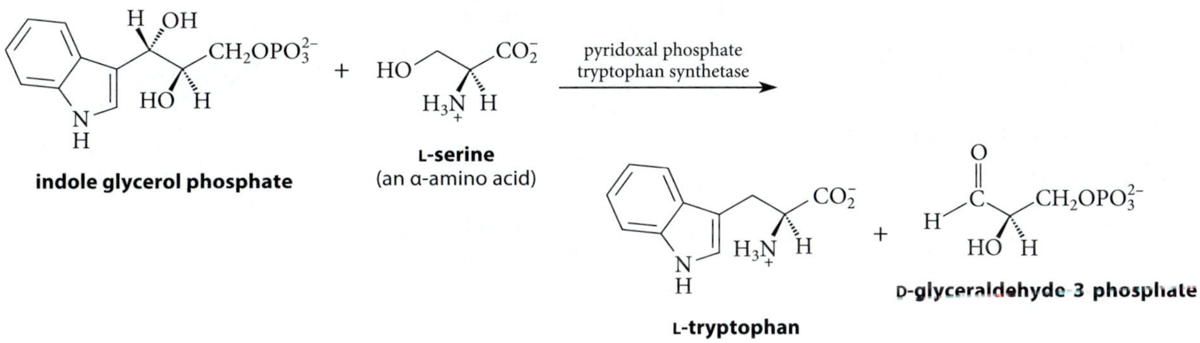

FIGURE P26.61

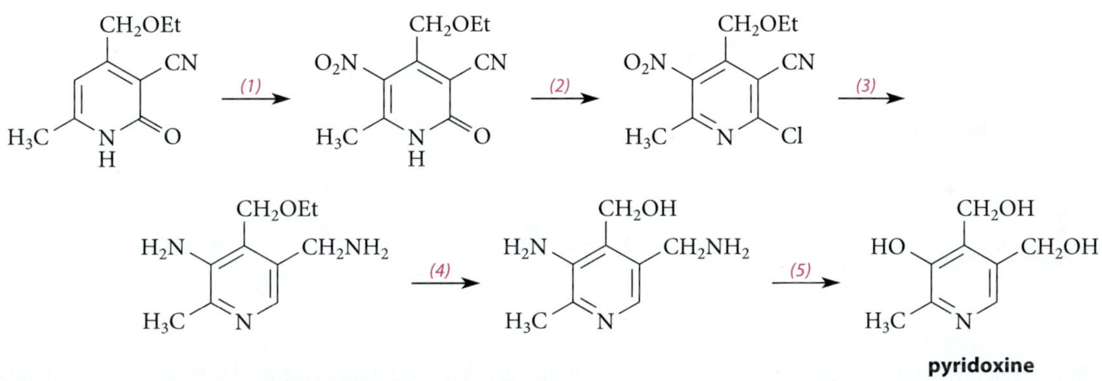

FIGURE P26.63

Problem 26.62, *continued*

(c) The following reaction takes place in aqueous base.

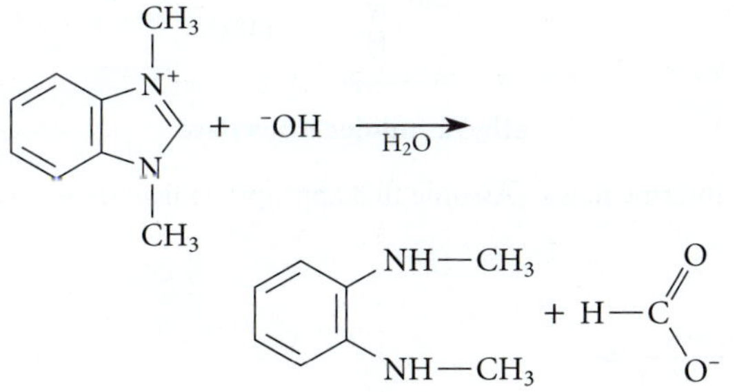

(d) The compound 2-pyridone does not hydrolyze in aqueous NaOH using conditions that bring about the rapid hydrolysis of δ-butyrolactam.

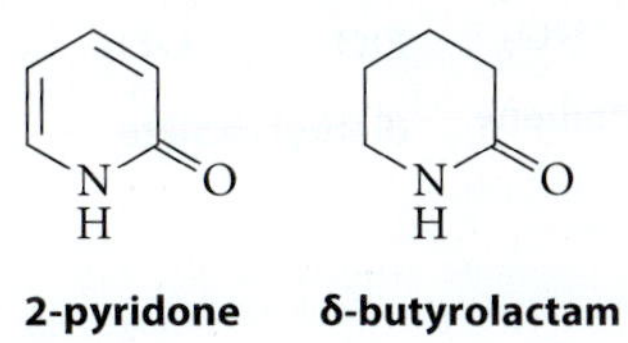

(e) Treatment of 4-chloropyridine with ammonia gives 4-aminopyridine, but treatment of 3-chloropyridine under the same conditions gives no reaction.

26.63 You work for a pharmaceutical company whose management has decided to produce synthetic vitamin B_6. The company is in possession of some fragmentary notes from Strong E. Nuff, one of their early chemists, that outline the synthesis of pyridoxine (a form of vitamin B_6) shown in **Fig. P26.63**. Unfortunately, reagents for each of the numbered steps have been omitted. The managers have hired you to serve as a consultant. Suggest the reagents that would accomplish each step.

26.64 Although the synthesis of heterocyclic rings is not discussed in this chapter, many such syntheses employ reactions that are similar or identical to reactions in other parts of the text. Give curved-arrow mechanisms for the reactions involved in each of the heterocyclic syntheses in **Fig. P26.64.** Begin, as always, by analyzing the relationship between the atoms in the reactants and products.

(a)

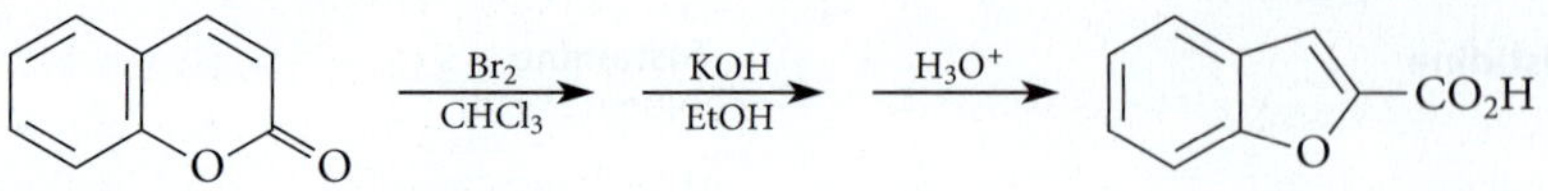

(b) Hinsberg thiophene synthesis:

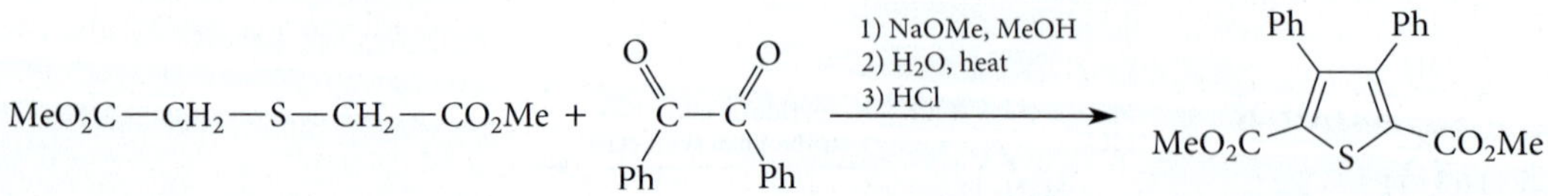

(c) Friedlander quinoline synthesis:

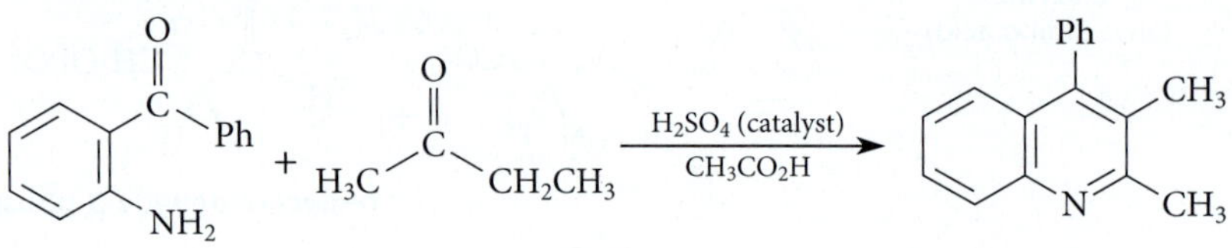

(d) Combes quinoline synthesis:

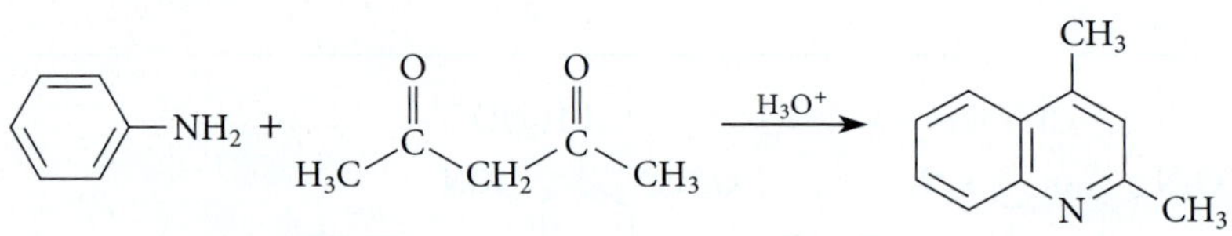

(e) Hantzsch dihydropyridine synthesis:

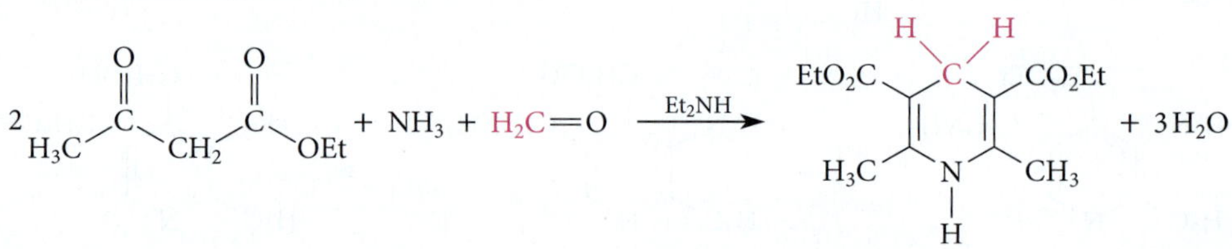

(f) Reissert indole synthesis. Identify compounds *A* and *B*, and give the mechanisms for the formation of compounds *A* and *C*.

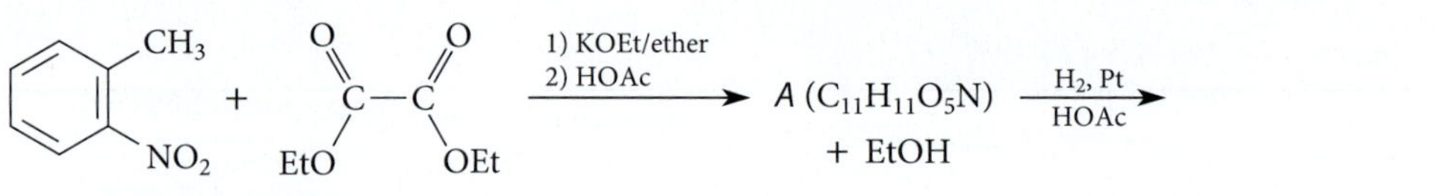

o-nitrotoluene **diethyl oxalate**

C

ethyl 2-indolecarboxylate

(g) Larsen–Chen indole synthesis. Show the different catalytic intermediates. (Assume that appropriate ligands are available for the catalyst; abbreviate these as "L.")

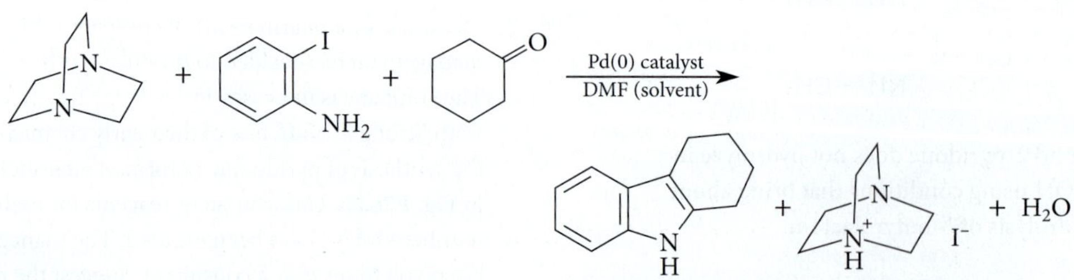

FIGURE P26.64

26.65 One theory of genetic mutation postulates that some mutations arise as the result of mispairing of bases in DNA caused by the existence of relatively rare isomeric forms of the bases. Show the hydrogen-bonding complementarity that can result from (a) the pairing of an imine isomer of C with A; (b) the pairing of an enol isomer of T with G.

26.66 When RNA is treated with periodic acid, and the product of that reaction is treated with base, only the nucleotide residue at the 3′-end is removed.

(a) Explain this transformation by showing its chemistry.

(b) Would the same reactions occur with DNA? Explain.

26.67 When DNA is treated with 0.5 *M* NaOH at 26 °C, no reaction takes place, but when RNA is subjected to the same conditions, it is rapidly cleaved into mononucleotide 2- and 3-phosphates. Explain. (*Hint:* What is the only structural difference between RNA and DNA? How can this difference promote the observed behavior? See Sec. 12.8.)

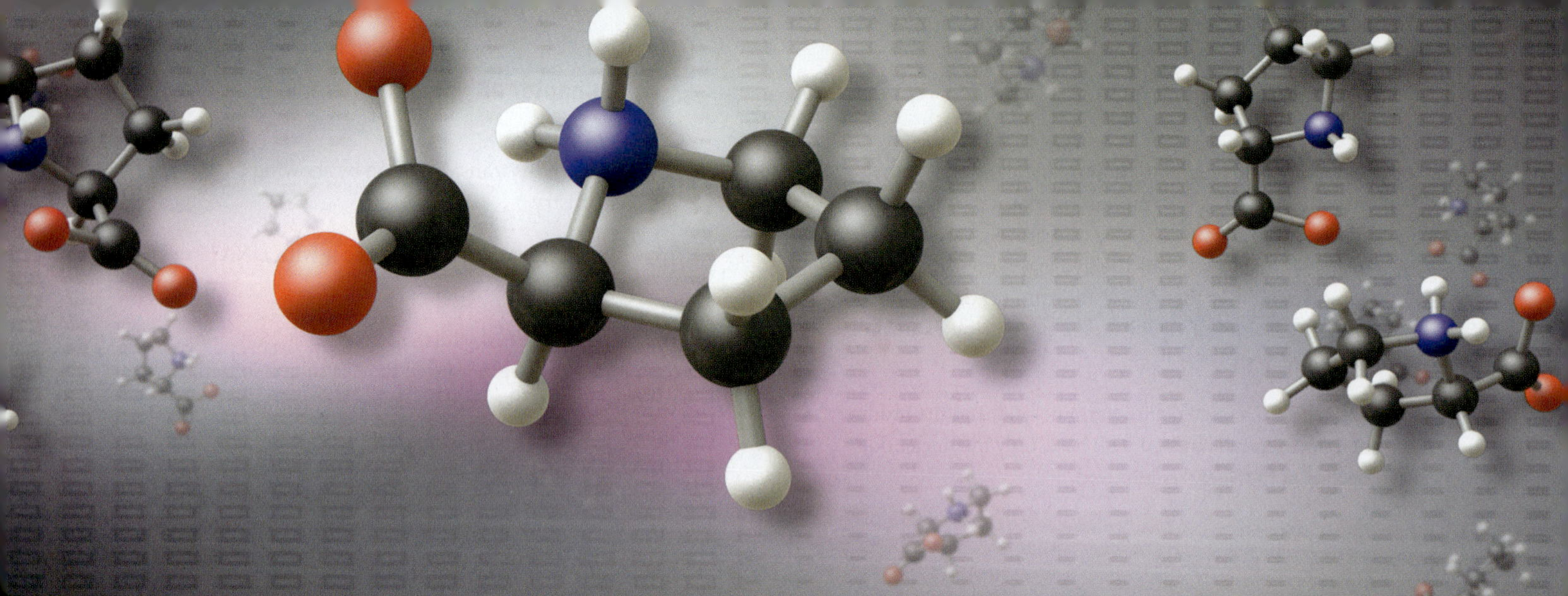

27 AMINO ACIDS, PEPTIDES, AND PROTEINS

Amino acids, as the name implies, are compounds that contain both an amino group and a carboxylic acid group.

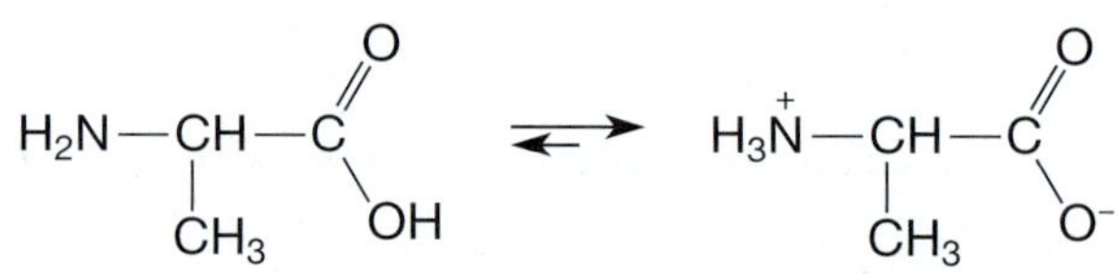

alanine
(an α-amino acid)

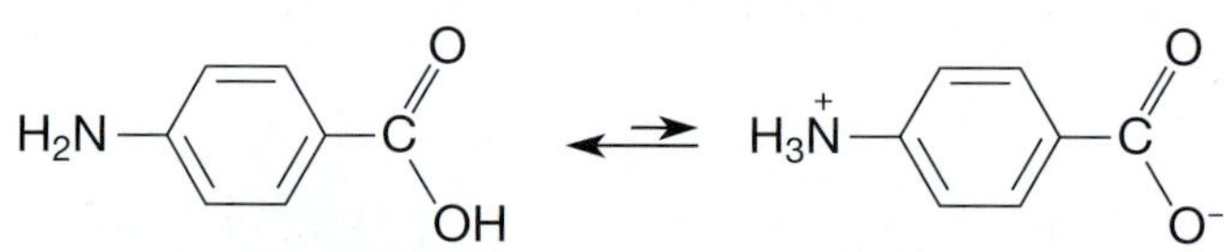

***p*-aminobenzoic acid**
(PABA, a component of folic acid, a vitamin)

(27.1)

As these structures show, a neutral amino acid—an amino acid with an *overall* charge of zero—can contain within the same molecule two groups of opposite charge. Molecules containing oppositely charged groups are called **zwitterions** (German, meaning "hybrid ion"). A zwitterionic structure is possible because the basic amino group can accept a proton and the acidic carboxylic acid group can lose a proton. Each of the **α-amino acids**, of which alanine is an example, has an amino group on the α-carbon—the carbon adjacent to the carboxylic acid group.

Peptides are biologically important polymers in which α-amino acids are joined into chains through amide bonds, called **peptide bonds**. A peptide bond is derived from the amino group of one amino acid and the carboxylic acid group of another. Peptides are *heteropolymers* because the R-groups in the repeating units can differ.

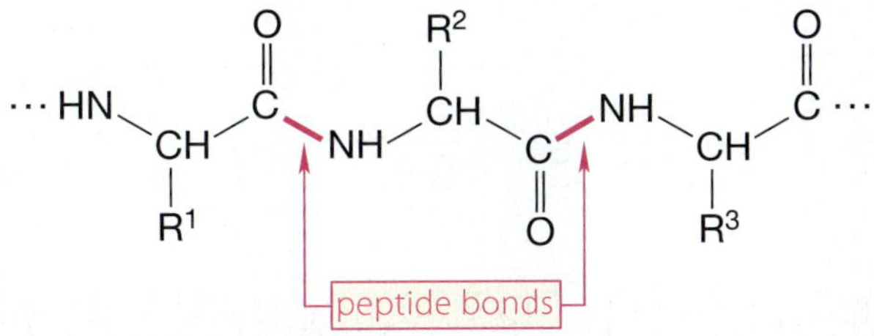

general peptide structure (27.2)

Proteins are very large peptides, and some proteins are aggregates of more than one peptide. The name *protein* (from a Greek word meaning "of first rank") is particularly apt because peptides and proteins serve many important roles in biology. For example, almost all enzymes (biological catalysts) and some hormones are peptides or proteins.

27.1 NOMENCLATURE OF AMINO ACIDS AND PEPTIDES

A. Nomenclature of Amino Acids

Some amino acids are named substitutively as carboxylic acids with amino substituents.

$$H_2N—CH_2CH_2CH_2—CO_2H \rightleftharpoons H_3\overset{+}{N}—CH_2CH_2CH_2—CO_2^-$$

4-aminobutanoic acid
(γ-aminobutyric acid)

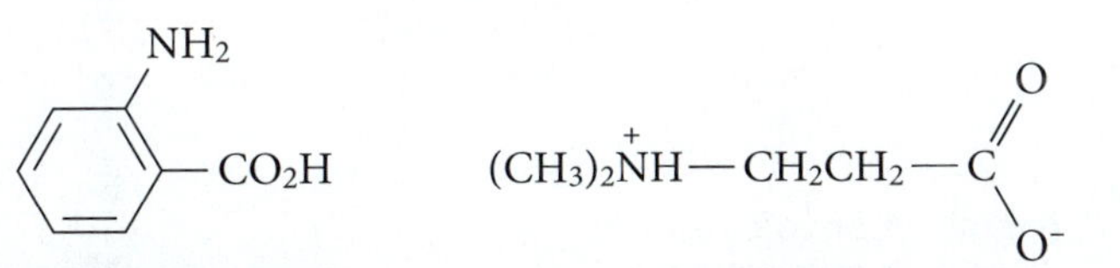

2-aminobenzoic acid
(*o*-aminobenzoic acid, anthranilic acid)

3-(dimethylamino)propanoic acid

(27.3)

Even if they exist as zwitterions, amino acids are named as uncharged compounds.

Twenty α-amino acids are known by widely accepted traditional names. These are the amino acids that occur commonly as constituents of proteins. The names and structures of these amino acids are given in **Table 27.1**.

Two points about the structures of the α-amino acids will help you to remember them. First, with the exception of proline, all α-amino acids have the same general structure, differing only in the identity of the side chain R.

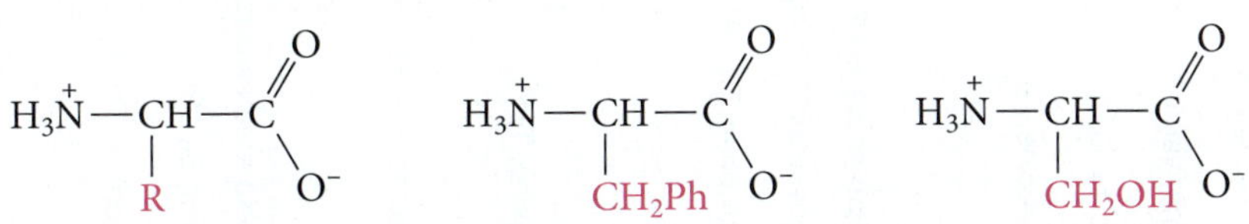

general structure

phenylalanine
($R = CH_2Ph$)

serine
($R = CH_2OH$)

(27.4)

TABLE 27.1 Names, Structures, Abbreviations, and Properties of the Twenty Common Naturally Occurring Amino Acids

General structure: $H_3\overset{+}{N}-\underset{\underset{R}{|}}{CH}-C(=O)O^-$

Name and abbreviations	R*	Optical rotation of L enantiomer in H_2O (sign of $[\alpha]_D$)	pK_{a1}	pK_{a2}	pK_{a3}	Isoelectric point, p*I*
Amino acids with simple aliphatic side chains						
glycine, Gly, G	$-H$		2.34	9.60	—	5.97
alanine, Ala, A	$-CH_3$	(+)	2.35	9.69	—	6.02
valine, Val, V	$-CH(CH_3)_2$	(+)	2.32	9.62	—	5.97
leucine, Leu, L	$-CH_2CH(CH_3)_2$	(–)	2.36	9.60	—	5.98
isoleucine, Ile, I	$-\underset{\underset{CH_3}{\vert}}{CH}-CH_2CH_3$ [*S* configuration]	(+)	2.36	9.68	—	6.02
Amino acids with aromatic side chains						
phenylalanine, Phe, F	$-CH_2Ph$	(–)	1.83	9.13	—	5.48
tryptophan, Trp, W	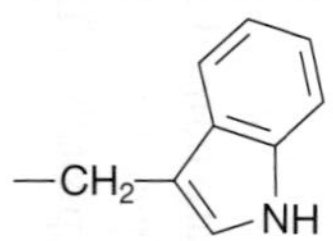	(–)	2.38	9.39	—	5.88
tyrosine, Tyr, Y	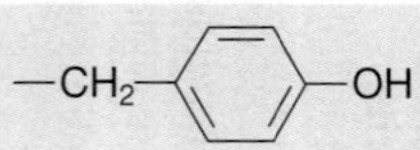	(–)	2.20	9.11	10.07	5.65
histidine, His, H[†]	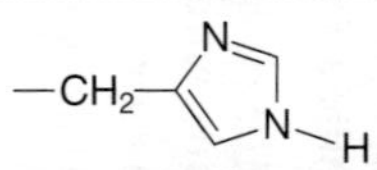	(–)	1.82	6.00	9.17	7.58
Amino acids with aliphatic side chains containing **—OH, —SH,** ***and*** **—SCH_3** ***groups***						
serine, Ser, S	$-CH_2-OH$	(–)	2.21	9.15	—	5.68
threonine, Thr, T	$-\underset{\underset{CH_3}{\vert}}{CH}-OH$ [*R* configuration]	(–)	2.71	9.62	—	5.16
methionine, Met, M	$-CH_2CH_2-SCH_3$	(–)	2.28	9.21	—	5.75
cysteine, Cys, C[‡]	$-CH_2-SH$	(–)	1.71	8.18	10.28	5.02

Name and abbreviations	R*	Optical rotation of L enantiomer in H_2O (sign of $[\alpha]_D$)	pK_{a1}	pK_{a2}	pK_{a3}	Isoelectric point, pI
Amino acids with side chains containing carboxylic acid or amide groups						
aspartic acid, Asp, D	$-CH_2-C(=O)-OH$	(+)	1.88	3.65	9.60	2.76
glutamic acid, Glu, E	$-CH_2CH_2-C(=O)-OH$	(–)	2.16	4.32	9.67	3.24
asparagine, Asn, N	$-CH_2-C(=O)-NH_2$	(–)	2.02	8.80	–	5.41
glutamine, Gln, Q	$-CH_2CH_2-C(=O)-NH_2$	(+)	2.17	9.13	–	5.65
Amino acids with side chains containing strongly basic groups						
lysine, Lys, K	$-(CH_2)_4-NH_2$	(+)	2.18	9.12	10.53	9.82
arginine, Arg, R	$-(CH_2)_3-C(=NH)-NH_2$	(+)	2.17	9.04	12.48	10.76
Cyclic (secondary) amino acid						
proline, Pro, P	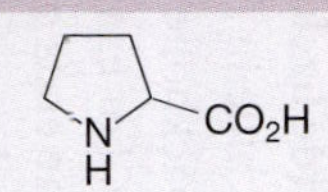	(–)	1.99	10.60	–	6.10

*Side chains are shown in their uncharged form

†Histidine is a weakly basic amino acid.

‡Cysteine often occurs in proteins as a disulfide dimer, called cystine. $H_2N-CH(CO_2H)-CH_2-S-S-CH_2-CH(CO_2H)-NH_2$. For this reason, cysteine is sometimes called **half-cystine** and abbreviated Cys/2.

Proline is the only naturally occurring amino acid with a secondary amino group. In proline the —NH— and the side chain are "tied together" in a ring.

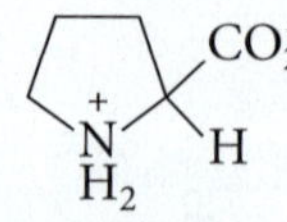

proline (27.5)

Second, as Table 27.1 shows, the amino acids can be organized into six groups according to the nature of their side chains.

- Amino acids with —H or aliphatic hydrocarbon side chains
- Amino acids with side chains containing aromatic groups
- Amino acids with aliphatic side chains containing —SH, —SCH_3, or —OH groups
- Amino acids with side chains containing carboxylic acid or amide groups
- Amino acids with basic side chains
- Proline

The α-amino acids are often designated by either three-letter or single-letter abbreviations, which are given in Table 27.1.

B. Nomenclature of Peptides

The terminology and nomenclature associated with peptides are best illustrated by an example. Consider the following peptide formed from the three amino acids alanine, valine, and lysine.

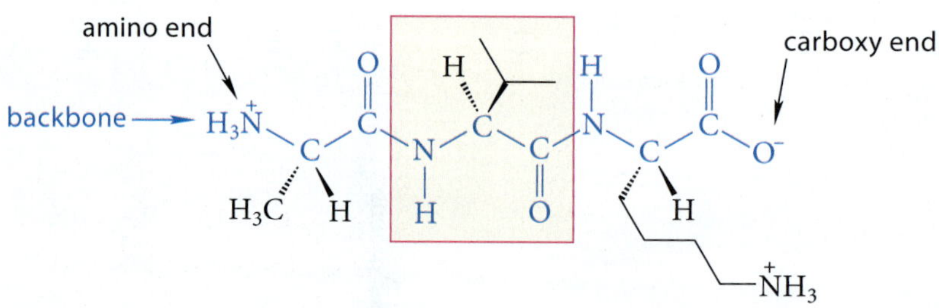

abbreviated **Ala-Val-Lys** or **A-V-K** (27.6)

The **peptide backbone** is the repeating sequence of nitrogen, α-carbon, and carbonyl groups shown in blue in Display 27.6. The characteristic amino acid side chains are attached to the peptide backbone at the respective α-carbon atoms. Each amino acid unit in the peptide is called a **residue**. For example, the part of the peptide derived from valine, the *valine residue*, is outlined in red. The ends of a peptide are labeled as the **amino end** or **amino terminus** and the **carboxy end** or **carboxy terminus**. A peptide can be characterized by the number of residues it contains. For example, the preceding peptide is a **tripeptide** because it contains three amino acid residues. A peptide containing two, three, or five amino acids is a **dipeptide**, **tripeptide**, or **pentapeptide**, respectively. A relatively short peptide of unspecified length containing a few amino acids is sometimes referred to as an **oligopeptide** (from a Greek root meaning "scant" or "few").

A peptide is conventionally named by giving successively the names of the amino acid residues, *starting at the amino end*. The names of all but the carboxy-terminal residue are formed by dropping the final suffix (*ine*, *ic*, or *an*) and replacing it with *yl*. For example, the peptide in Display 27.6 is named alanylvalyllysine. In practice, this type of nomenclature is cumbersome for all but the smallest peptides. A simpler way of naming peptides is to connect with hyphens the three-letter (or one-letter) abbreviations of the component amino acid residues beginning with the amino-terminal residue. For example, the preceding peptide is also written as Ala-Val-Lys or A-V-K.

Large peptides of biological importance are known by their common names. For example, *insulin* is an important peptide hormone that contains 51 amino acid residues; *ribonuclease*, an enzyme, is a protein containing 124 amino acid residues (and a rather small protein at that!).

Sometimes we want to focus on a single residue in a large protein. For example, suppose we want to focus on the role of a lysine residue in a protein such as an enzyme. We can employ the following condensed notations, which we have used throughout this text:

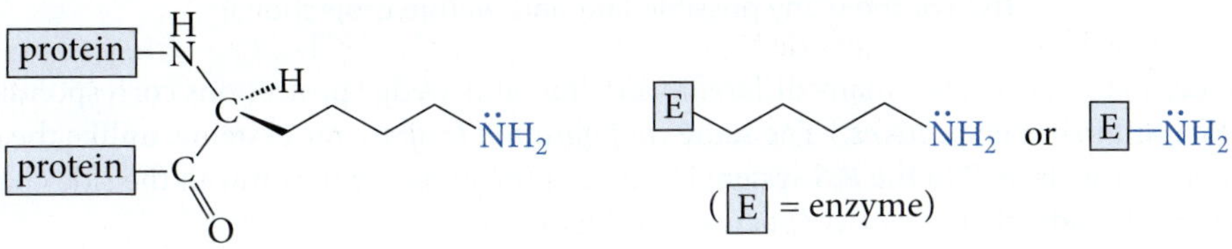

(27.7)

The "E" indicates that the protein is an enzyme.

Focused Problems

27.1 Draw the structures of the following peptides.

(a) Tryptophylglycylisoleucylaspartic acid

(b) Glu-Gln-Phe-Arg (or E-Q-F-R)

27.2 Using three-letter abbreviations for the amino acid residues, name the following peptide.

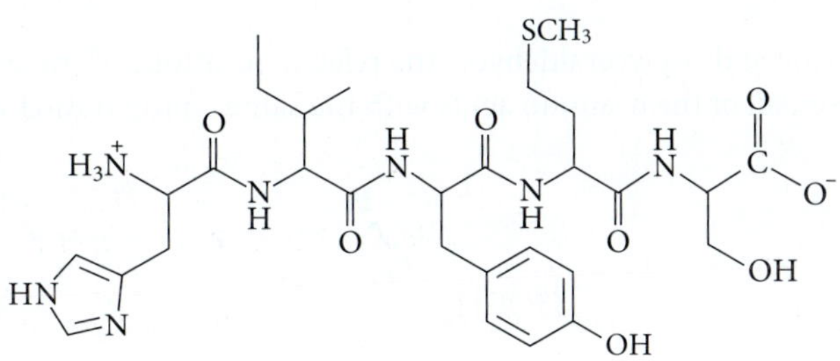

27.3 Draw condensed structures for each of the following:

(a) A phenylalanine residue in a large protein

(b) A serine residue in an enzyme

27.2 STEREOCHEMISTRY OF THE α-AMINO ACIDS

With the exception of glycine, all common naturally occurring α-amino acids have an asymmetric α-carbon atom and are chiral molecules. The chiral amino acids in Table 27.1 are found within naturally occurring proteins in only one enantiomeric form, which has the following configuration:

stereochemical configuration of the naturally occurring α-amino acids
(*S* for all chiral α-amino acids except cysteine)

(two of the many possible line-and-wedge projections) (27.8)

(Remember that we can draw many different valid line-and-wedge projections corresponding to views from different perspectives.) The same *configuration in space* for cysteine, unlike the other chiral α-amino acids, is *R* in the *R,S* system because of the presence of sulfur in the side chain, which raises the side-chain priority (Focused Problem 27.6).

The stereochemistry of α-amino acids is often specified with an older system, the **D,L system**, which was introduced for carbohydrates in Sec. 24.3A. The D,L system relates the configuration of the α-carbon of an amino acid to that of the 3-carbon aldose *glyceraldehyde* (see Sec. 24.3A). Serine is the reference α-amino acid that makes the comparison to glyceraldehyde easiest to see. Corresponding groups are shown in the same color.

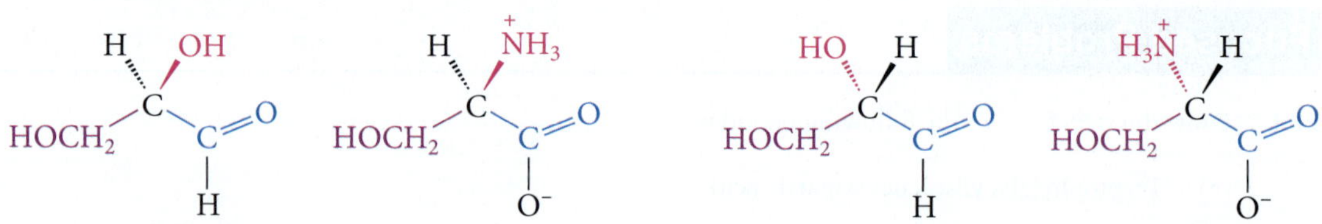

D-glyceraldehyde **D-serine [(*R*)-serine]** **L-glyceraldehyde** **L-serine [(*S*)-serine]**

(27.9)

In D-serine and D-glyceraldehyde, the relative positions of corresponding groups are the same. By extension, all of the α-amino acids with the same configuration as L-serine are L-amino acids.

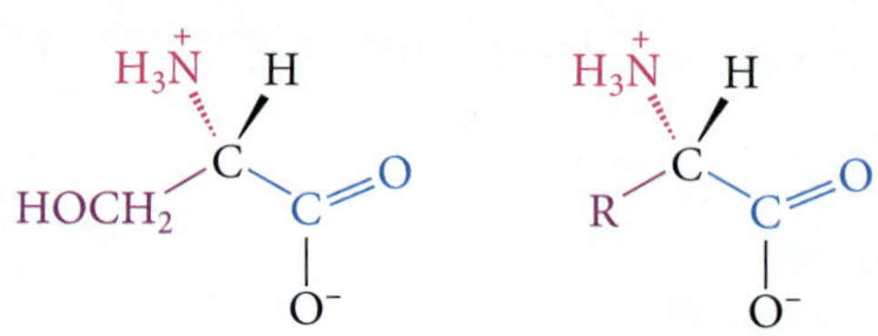

L-serine general structure of an L-amino acid (27.10)

Both L-serine and L-glyceraldehyde have the *S* configuration in the *R,S* system; both D-serine and D-glyceraldehyde have the *R* configuration. However, *this correspondence is not general.* For example, L-cysteine has the *R* configuration. (Why? See Focused Problem 27.6.) All of the naturally occurring amino acids have the L configuration, with a few rare exceptions. (For example, D-alanine occurs in bacterial cell walls; and certain antibiotics contain D-amino acids.)

As with the carbohydrates (Sec. 24.3A), the D and L designations specify the configuration of a *single reference carbon*. For α-amino acids, the reference carbon is the α-carbon. When an α-amino acid contains more than one asymmetric carbon, as in threonine and isoleucine, the diastereomers are given different names. For example, L-threonine has the 2*S*,3*R* configuration;

its diastereomer with the 2*S*,3*S* configuration is L-allothreonine. (The prefix *allo* comes from the Greek root *allos*, meaning "other"; L-allothreonine is the "other" threonine.)

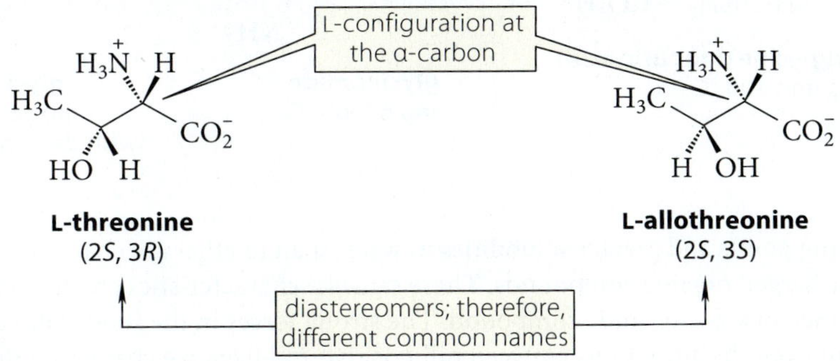

(27.11)

Although the use of different names for diastereomers undoubtedly makes organic nomenclature more colorful, this aspect of the D,L system is a major annoyance, especially for the beginner, because it is not systematic. Except for carbohydrates and amino acids, where it is widely used, the D,L system has been largely abandoned.

Focused Problems

27.4 Draw three valid line-and-wedge general structures for naturally occurring α-amino acids in addition to the ones shown in this section.

27.5 (a) L-Isoleucine has two asymmetric carbons and has the 2*S*,3*S* configuration. Complete the following line-and-wedge structure for L-isoleucine.

(b) Alloisoleucine is the diastereomer of isoleucine. Complete the line-and-wedge structure in part (a) for D-alloisoleucine.

27.6 (a) What is the α-carbon configuration of L-cysteine in the *R*,*S* system?

(b) Explain why L-cysteine and L-serine have different configurations in the *R*,*S* system.

27.3 ACID–BASE PROPERTIES OF AMINO ACIDS AND PEPTIDES

A. Zwitterionic Structures of Amino Acids and Peptides

As indicated in the introduction to this chapter, the neutral forms of the α-amino acids are *zwitterions*. Some of the evidence for zwitterionic structures is as follows:

1. Amino acids are insoluble in apolar aprotic solvents such as ether. On the other hand, most unprotonated amines and un-ionized carboxylic acids dissolve in ether.

2. Amino acids have very high melting points. For example, glycine melts at 262 °C (with decomposition), and tyrosine melts at 310 °C (also with decomposition). Hippuric acid, a much larger molecule than glycine, and glycinamide, the amide of glycine, have much lower melting points. The former compound lacks the amino group, the latter lacks the carboxylic acid group, and neither can exist as a zwitterion.

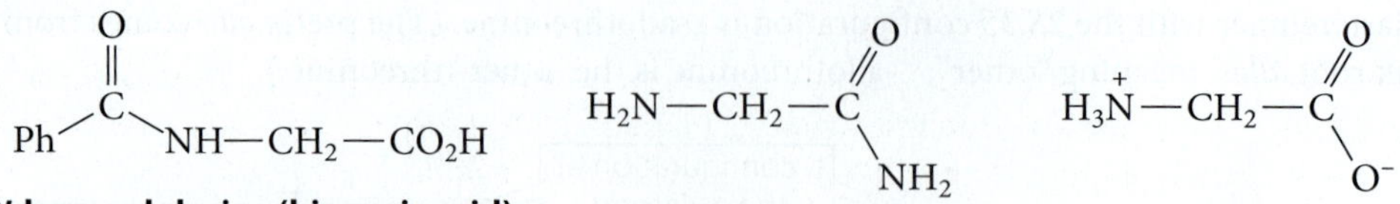

N-benzoylglycine (hippuric acid)
mp 190 °C

glycinamide
mp 67–68 °C

glycine
mp 262 °C
(with decomposition)

(27.12)

The high melting points and greater solubilities in water than in ether are characteristics expected of salts, not uncharged organic compounds. These salt-like characteristics are, however, what would be expected of a zwitterionic compound. The strong forces in the solid states of the amino acids that result from the attractions between full positive and negative charges on different molecules are much like those between the ions in a salt. These attractions stabilize the solid state and resist conversion of the solid into a liquid—whether a pure liquid melt or a solution. Water is the best solvent for most amino acids because it solvates ionic groups much as it solvates the ions of a salt (Sec. 8.6F).

3. The dipole moments of the amino acids are very large—much larger than those of similar-sized molecules with only one amine or carboxylic acid group.

$$\overset{+}{H_3N}—CH_2—CO_2^- \qquad H_3C—CH_2—CO_2H \qquad CH_3CH_2CH_2CH_2—NH_2$$

glycine $\mu = 14$ D — **propanoic acid** $\mu = 1.7$ D — **butylamine** $\mu = 1.4$ D (27.13)

A large dipole moment is expected for molecules that contain a great deal of separated charge (Sec. 1.9B and Further Exploration 1.3).

4. The pK_a values for amino acids are what would be expected for the zwitterionic forms of the neutral molecules.

Suppose a neutral amino acid is titrated with acid. When one equivalent of acid is added, the *basic* group of the amino acid will have been protonated. When this experiment is carried out with glycine, the pK_a of the basic group is found to be 2.3. If glycine is indeed a zwitterion, this basic group can only be the carboxylate ion. If glycine is not a zwitterion, this basic group has to be the amine.

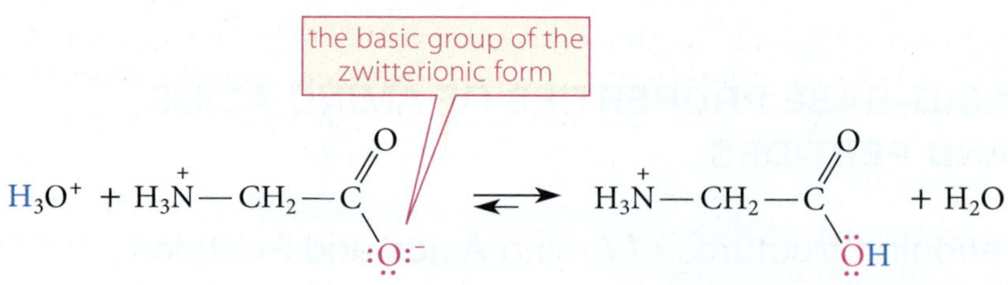

(27.14a)

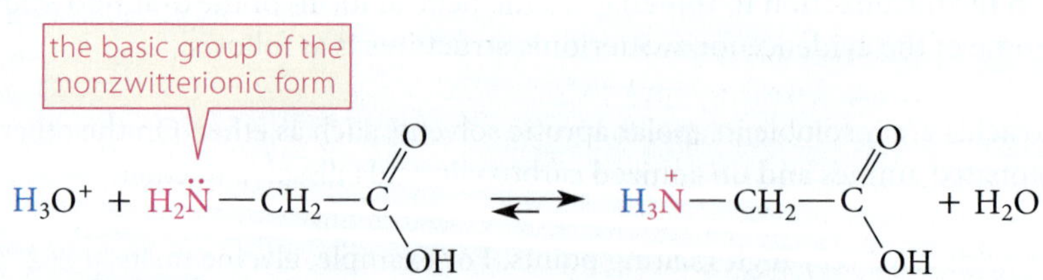

(27.14b)

Which is the correct description of the titration? The pK_a of 2.3 is that expected of a carboxylic acid in a molecule containing a nearby electron-withdrawing group (in this case, the H_3N^+— group). In contrast, the conjugate acids of amines have pK_a values in the 8–10 range. This analysis suggests that the zwitterion, not the uncharged form, is being titrated.

Along the same line, if NaOH is added to neutral glycine, a group is titrated with $pK_a = 9.6$. This is a reasonable pK_a value for an alkylammonium ion, but would be very unusual for a carboxylic acid. This comparison also suggests that the neutral form of glycine is a zwitterion.

The acid–base equilibria for glycine can be summarized as follows:

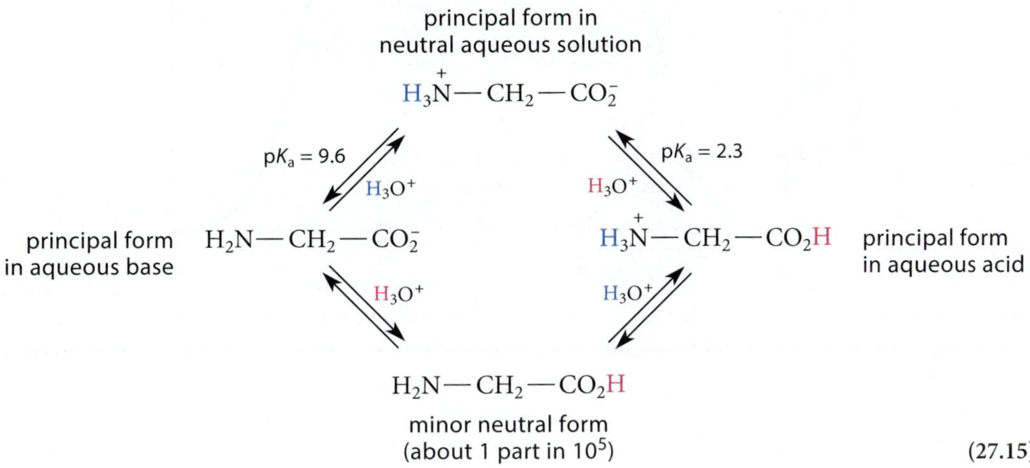

(27.15)

The major neutral form of any α-amino acid is the zwitterion. The ratio of the uncharged form of an α-amino acid to the zwitterion form is about 1 part in 10^5, as shown for glycine in Eq. 27.15.

Peptides also exist as zwitterions—that is, at pH values near 7, amino groups are protonated and carboxylic acid groups are ionized.

B. Acid–Base Equilibria of Amino Acids

The goal of this section is to explain how the various forms of the amino acids vary with pH. Consider the acid–base equilibria of the amino acid alanine.

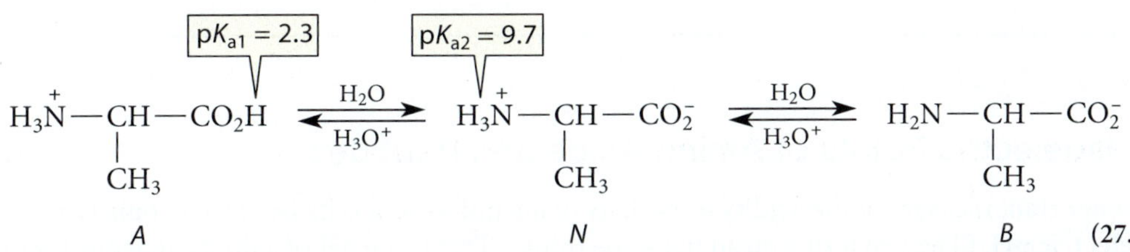

(27.16)

In very acidic solution alanine is fully protonated, and *A* is the only species present. As the pH is raised, the most acidic proton dissociates; this is the carboxylic acid proton, which has the lower pK_a. In this first dissociation equilibrium, the acid *A* and its conjugate base *N* are the only species involved. *When a molecule can undergo two (or more) proton dissociations, we can treat the two equilibria independently if the* pK_a *values are separated by 2 or more units.* Therefore, we can apply Eqs. 3.67a and b (Sec. 3.6E) with *A* as the acid and *N* as the base. When we plot these equations, we get a plot exactly like Fig. 3.1 (Sec. 3.6E), in which the fraction of *A* decreases and the fraction of *N* increases with increasing pH; the fractions of *A* and *N* are equal when $pH = pK_{a1} = 2.3$. This plot is shown on the left half of **Fig. 27.1**.

As the pH is raised further, the second dissociation becomes important. In this case, *N* is now the acid and *B* is its conjugate base. When we plot Eqs. 3.67a and b for this acid–base pair, we obtain the plot on the right half of Fig. 27.1. The fractions of *N* and *B* are equal with $pH = pK_{a2} = 9.7$. As the pH is raised further, *B* becomes the only species present.

The process we have just outlined can be applied to any two-dissociation system (not just amino acids), and it is readily extended to any number of dissociations. If two of the pK_a values involved differ by less than 2 units, the process is conceptually the same, except that the intermediate form *N* never becomes 100%.

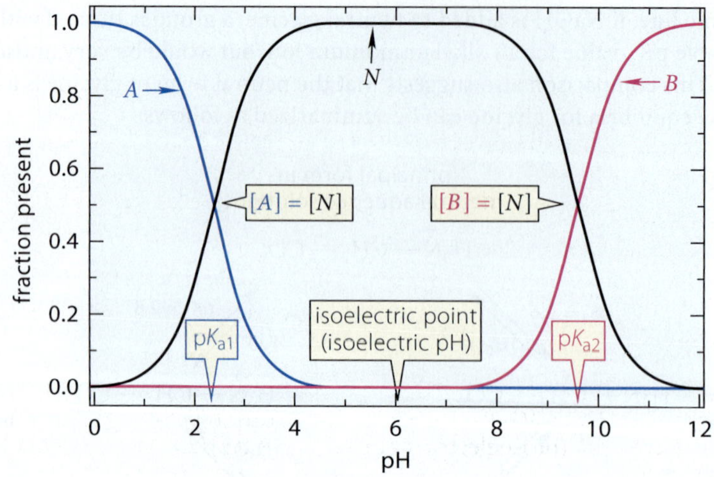

FIGURE 27.1 Variation in the concentrations of the three forms of alanine shown in Eq. 27.16 as a function of pH. The blue line represents the concentration of the conjugate acid A, the red line the concentration of the conjugate base B, and the black line the concentration of the neutral zwitterionic form N. The dissociation constant of A is K_{a1}, and that of N is K_{a2}. The concentration of the neutral form N is a maximum at a pH that is halfway between pK_{a1} and pK_{a2}; this pH is the isoelectric point.

Focused Problem

27.7 Obtain the three pK_a values of the α-amino acid *histidine* from Table 27.1. Because there are three pK_a values, the dissociation equilibria contain four species.

(a) With the species present at low pH on the left, and the fully dissociated species on the right, draw and label by letter all of the species involved in the acid–base dissociation equilibria of histidine. (*Hint:* The side-chain group protonates on the vinylic nitrogen of the imidazole ring.)

(b) Sketch a graph similar to Fig. 27.1 in which the fraction of all of the species are shown as a function of pH. Label each curve with a letter corresponding to one of the forms you drew in part (b). Indicate the pH values at which the fractions of two species are equal.

C. Isoelectric Points of Amino Acids and Peptides

An important measure of the acidity or basicity of an amino acid is its **isoelectric point** or **isoelectric pH**. (The two terms mean the same thing.) This is the pH of a dilute aqueous solution of the amino acid at which the total charge on all molecules of the amino acid is zero. We'll use the acid–base equilibria of alanine (Eq. 27.16) to illustrate the isoelectric point. Two conditions are met at the isoelectric point. First, the concentration of conjugate-acid molecules A equals the concentration of conjugate-base molecules B. Because the conjugate acid A is positively charged and the conjugate base B is negatively charged, the equality of the two concentrations means that the total charge on all molecules of the amino acid is zero. (The amount of N present doesn't matter because its net charge is zero.) Second, at the isoelectric point, the relative concentration of the zwitterion form N is greater than at any other pH. The first part of the definition—the equality of the B and A concentrations—is sufficient to calculate the isoelectric pH.

If the concentrations of A and B are equal at the isoelectric pH, their fractions are also equal. We use the following definitions: isoelectric pH = pH_i, $\Delta_1 = pH_i - pK_{a1}$, and $\Delta_2 = pH_i - pK_{a2}$. Applying Eqs. 3.67a and b (Sec. 3.6E) for the fractions of each species, and setting these fractions equal, we have

$$f_A = \frac{1}{1 + 10^{-\Delta_1}} = f_B = \frac{1}{1 + 10^{-\Delta_2}} \tag{27.17}$$

Simplifying,

$$1 + 10^{-\Delta_1} = 1 + 10^{-\Delta_2}$$

or

$$10^{-\Delta_1} = 10^{-\Delta_2}$$

Taking logs and applying the definitions of Δ_1 and Δ_2, we have

$$\mathrm{pH_i} - \mathrm{p}K_{\mathrm{a1}} = -(\mathrm{pH_i} - \mathrm{p}K_{\mathrm{a2}})$$

or

$$\mathrm{pH_i} = \frac{(\mathrm{p}K_{\mathrm{a1}} + \mathrm{p}K_{\mathrm{a2}})}{2} \qquad (27.18)$$

This result shows that *the isoelectric pH occurs at the average of the two pK_a values.* Therefore, alanine has an isoelectric pH (or isoelectric point) of $(2.3 + 9.7)/2 = 6.0$. This point is marked on the pH axis of Fig. 27.1.

The isoelectric point is significant because it indicates not only the pH value at which a solution of the amino acid contains the greatest amount of zwitterion form *N*, but also the sign of the net charge on the amino acid at *any* pH. For example, at a pH value lower (more acidic) than the isoelectric point, more molecules of an amino acid are in form *A* than in form *B*; in this situation, the amino acid has a net *positive charge*. At a pH value greater (more basic) than the isoelectric point, more molecules of an amino acid are in form *B* than in form *A*. In this situation, the amino acid has a net *negative charge*. To summarize:

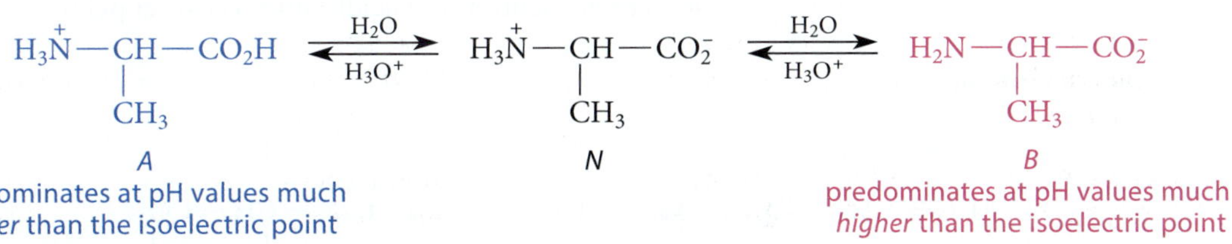

(27.19)

Section 27.3D discusses a separation technique that hinges on a knowledge of the net charge.

When an amino acid has a side chain containing an acidic or basic group, the isoelectric point is markedly changed. The amino acid lysine (Lys), for example, has a basic side-chain amino group as well as its α-amino and carboxy groups.

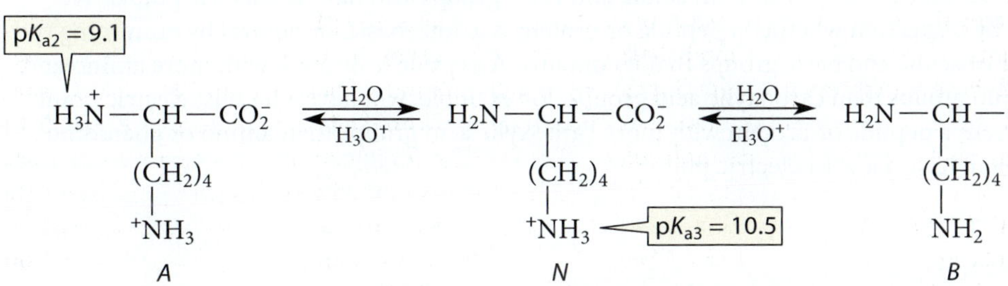

(27.20)

The isoelectric point of lysine is 9.82, which is the average of its two *highest* pK_a values of 9.12 and 10.53. At the isoelectric point of lysine, equal amounts of forms *A* and *B* of lysine are present, and form *N* has its maximum concentration. The lowest pK_a of lysine—the pK_a of the carboxylic acid group, 2.2—doesn't enter the picture because neither of the equilibria involving the neutral form *N* involves ionization of the carboxylic acid group.

Let's compare the charge state of alanine and lysine at pH 6. Because pH 6 is the isoelectric point of alanine, its charge is zero. Because pH 6 is much lower than the isoelectric point of lysine, the net charge on lysine molecules is positive at this pH. Amino acids with high isoelectric points are classified as *basic amino acids.* Lysine and arginine are the two most basic of the common naturally occurring amino acids (see Table 27.1, Sec. 27.1A). As indicated by its isoelectric point, arginine is the more basic of the two. Its side chain carries the basic guanidino group, the conjugate acid of which has a pK_a of 12.5. The basicity of this group results from the resonance stabilization of its conjugate acid.

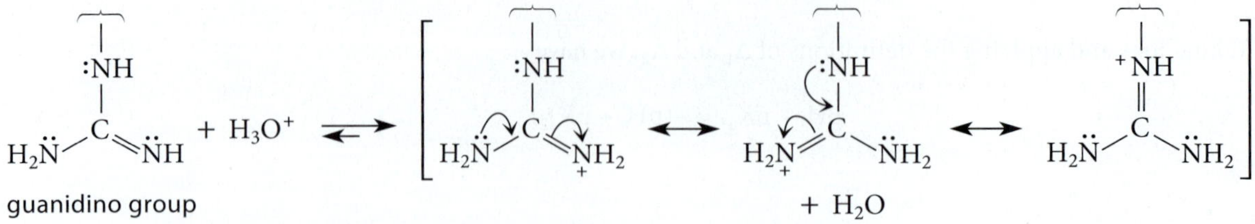

(27.21)

The amino acids aspartic acid (Asp) and glutamic acid (Glu) have carboxylic acid groups on their side chains and have low isoelectric points. The isoelectric point of aspartic acid, for example, is 2.76, the average of its two *lowest* pK_a values. (You should show why this is reasonable.) Amino acids with low isoelectric points are classified as *acidic amino acids.* Molecules of an acidic amino acid carry a net negative charge at pH 6. Aspartic acid and glutamic acid are the two most acidic of the common naturally occurring amino acids.

Amino acids with isoelectric points near 6, such as glycine or alanine, are classified as *neutral amino acids.* A pH of 6 is not *exactly* neutral; "neutral" amino acids are actually slightly acidic because the carboxylic acid group is somewhat more acidic than the amino group is basic. However, as Fig. 27.1 shows, even at pH 7 (neutral pH), the neutral amino acids are almost completely in form *N.*

Let's summarize the charge situation in basic, neutral, and acidic amino acids at pH 6:

Principal forms at pH 6:

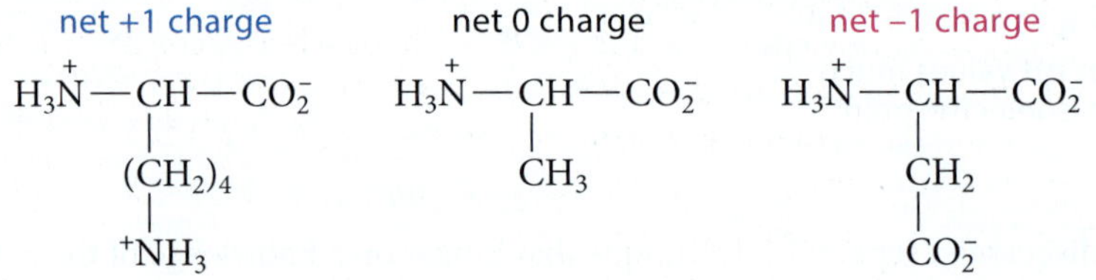

lysine (a *basic* amino acid) **alanine** (a *neutral* amino acid) **aspartic acid** (an *acidic* amino acid) (27.22)

Peptides and proteins with both acidic and basic groups also have isoelectric points. We can tell by inspection whether a peptide or protein is acidic, basic, or neutral by examining the number of acidic and basic groups that it contains. A peptide or protein with more amino and guanidino groups than carboxylic acid groups, for example, will have a high isoelectric point. Conversely, a peptide or protein with more carboxylic acid groups than amino or guanidino groups will have a low isoelectric point.

Focused Problems

27.8 (a) Point out the ionizable groups of the amino acid *tyrosine* (see Table 27.1).

(b) What is the net charge on tyrosine at pH 6? How do you know?

(c) Draw the structure of the major form(s) of tyrosine present at this pH.

27.9 (a) Estimate the isoelectric point of each of the following peptides.

A-K-V-I-M G-D-G-L-F

(b) Draw the structures of these peptides, indicating the predominant ionization state of each at its isoelectric point.

27.10 Classify the following peptides as acidic, basic, or neutral. What is the net charge on each peptide at pH = 6?

(a) Gly-Leu-Val

(b) Leu-Trp-Lys-Gly-Lys

(c) *N*-acetyl-Asp-Val-Ser-Arg-Arg (*N*-acetyl means that the terminal amino group of the peptide is acetylated.)

(d) Glu-Lys-Asp-Ala-Phe-Ile

D. Separations of Amino Acids and Peptides Using Acid–Base Properties

Isoelectric points are often used to design separations of amino acids and peptides. Consider, for example, the water solubilities of amino acids and peptides. Most peptides and amino acids, like carboxylic acids and amines, are most soluble when they carry a net charge and are least soluble in their neutral forms. Therefore, some peptides, proteins, and amino acids precipitate from water when the pH is adjusted to their isoelectric points. These same compounds are more soluble in water at pH values far from their isoelectric points, because they carry a net charge at these pH values.

A separation technique used a great deal in amino acid, peptide, and protein chemistry is *ion-exchange chromatography*. This method, too, depends on the isoelectric points of amino acids and peptides.

As noted in Sec. 6.10A, *chromatography* is a separation technique based on the relative adsorptions of compounds to a material called a *stationary phase*. If the stationary phase is contained in a column, compounds that are adsorbed weakly move through the column most rapidly and are eluted early; compounds that are more strongly adsorbed move through the column more slowly and emerge later. In other words, compounds are separated by their *differential adsorption* to the stationary phase. (Review Fig. 6.10, Sec. 6.10A.)

The type of chromatography that is used depends on the mechanism used to effect differential adsorption to the stationary phase. As explained in Sec. 6.10A, a chiral stationary phase is used to separate enantiomers. In **ion-exchange chromatography**, the column is filled with a buffer solution, and the stationary phase is a polymer called an *ion-exchange resin*. This resin bears charged groups. One popular resin, for example, is a sulfonated polystyrene—a polystyrene in which the phenyl rings contain strongly acidic sulfonic acid groups. If the pH of the buffer is such that the sulfonic acid groups are ionized, the resin bears a negative charge. This charge is the key to ion-exchange separations.

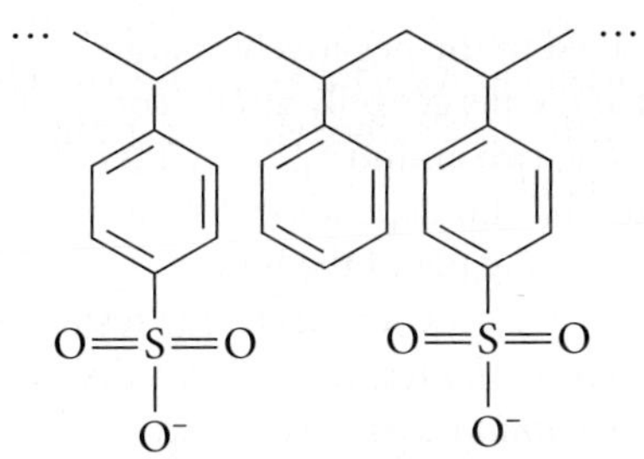

structure of sulfonated polystyrene with ionized sulfonic acid groups (27.23)

A separation using ion-exchange chromatography is illustrated in **Fig. 27.2**. Suppose the buffer in the column has a pH of 6. Because the pK_a of the sulfonic acid groups on the resin is about 1, these groups are ionized; therefore, at pH = 6, the resin is *anionic*. A solution containing a mixture of the two amino acids Val and Lys in the same buffer is added to the top of the column. Buffer is then allowed to flow through the column; a frit (a porous glass plate) keeps the resin from washing out. Because valine has zero charge at this pH, it is not attracted by the ionic groups on the column and is washed through the column with a relatively small volume of buffer. Lysine, on the other hand, has an isoelectric point of 9.8 and therefore bears a net positive charge at pH 6.

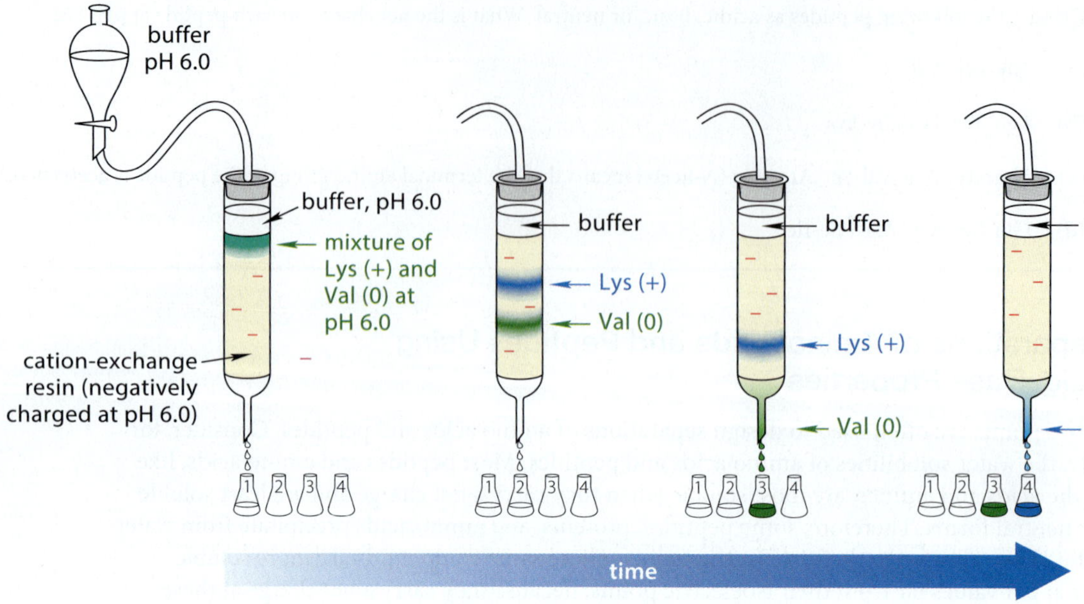

(a)

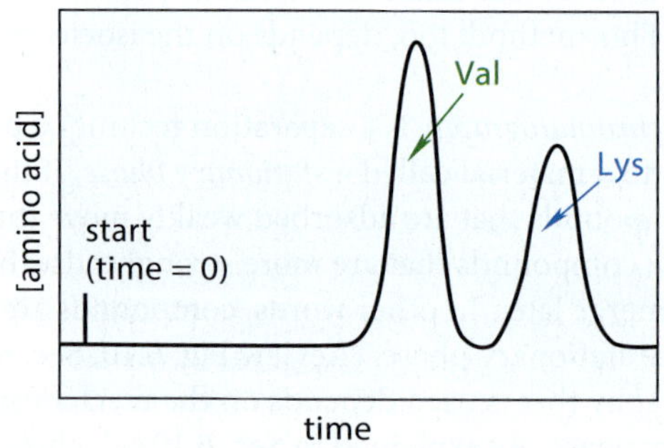

(b)

FIGURE 27.2 (a) The cation-exchange separation of valine (Val) and lysine (Lys). (b) The amino acid concentration in the eluent (the buffer emerging from the column) as a function of time. Lysine, which carries a positive charge at the pH of the buffer, is attracted to the negatively charged resin and moves through the column more slowly than valine, which carries zero charge. (The colors are for emphasis; Val and Lys are colorless.)

Therefore, lysine is strongly attracted to the negatively charged resin. Because of this attraction, lysine is retained on the column and emerges only after a considerably larger amount of buffer has passed through the column. The two amino acids are then separated. So, whether an amino acid or peptide is adsorbed by the column depends on its charge—which, in turn, depends on the relationship of its isoelectric point to the pH of the buffer.

In the experiment shown in Fig. 27.2, the ion-exchange resin is negatively charged and adsorbs cations; it is therefore called a *cation-exchange resin*. Resins that bear positively charged pendant groups adsorb anions and are called *anion-exchange resins*.

Ion Exchange and Water Softeners

BanksPhotos/Getty Images

Ion exchange has very important commercial applications, such as water treatment. Commercial water softeners contain cation-exchange resins much like the one used in this example, which adsorb the more highly charged calcium and magnesium ions in hard water and replace them with sodium ions with which the column is supplied. When the supply of sodium ions is exhausted, the column has to be flushed extensively, or regenerated, with concentrated NaCl solution to replace the adsorbed calcium and magnesium ions with sodium ions.

Focused Problem

27.11 (a) How might the structure of the resin in Eq. 27.23 be altered to make the resin an anion exchanger (that is, an anion-binding resin)? (*Hint:* What type of organic functional group carries a positive charge at neutral pH?)

(b) Predict the order of elution of the following peptides from an anion-exchange resin at pH 6: A-V-G, R-K-N-G, D-N-E-G. Explain your reasoning.

27.4 SYNTHESIS AND ENANTIOMERIC RESOLUTION OF α-AMINO ACIDS

A. Alkylation of Ammonia

Some α-amino acids can be prepared by the alkylation of ammonia with α-bromo carboxylic acids.

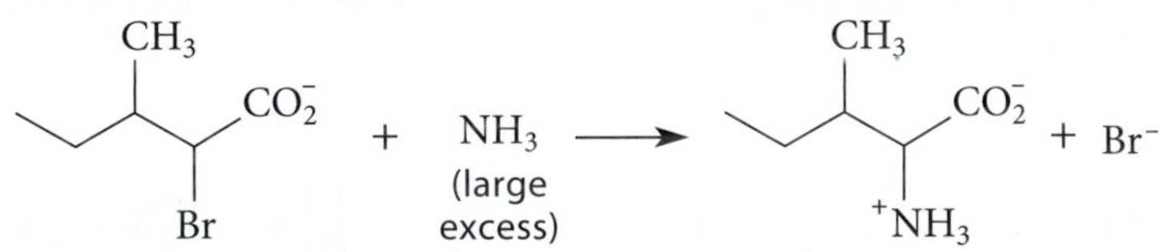

3-bromo-2-methylbutanoate **isoleucine** (49% yield) (27.24)

This is an S_N2 reaction in which ammonia acts as the nucleophile. (As noted in Sec. 22.3D, α-halo carbonyl compounds are very reactive in S_N2 reactions.) Alkylation of ammonia is usually not a good method for preparing primary amines because ammonia can be alkylated more than once to give complex mixtures (Sec. 23.7A). However, the use of a large excess of ammonia in this synthesis favors monoalkylation. Furthermore, amino acids are less reactive toward alkylating agents than simple alkylamines for two reasons: (1) the amino groups of amino acids are less basic, and therefore less nucleophilic, than ammonia and simple alkylamines; and (2) branching in amino acids provides a steric impediment to further alkylation.

B. Alkylation of Aminomalonate Derivatives

One of the most widely used methods for preparing α-amino acids is a variation of the malonic ester synthesis (Sec. 22.8A). The malonic ester derivative used is one in which a protected amino group is already in place: diethyl α-acetamidomalonate. This derivative is treated with sodium ethoxide in ethanol to form the enolate ion, which is then alkylated with an alkyl halide (benzyl chloride in the following example):

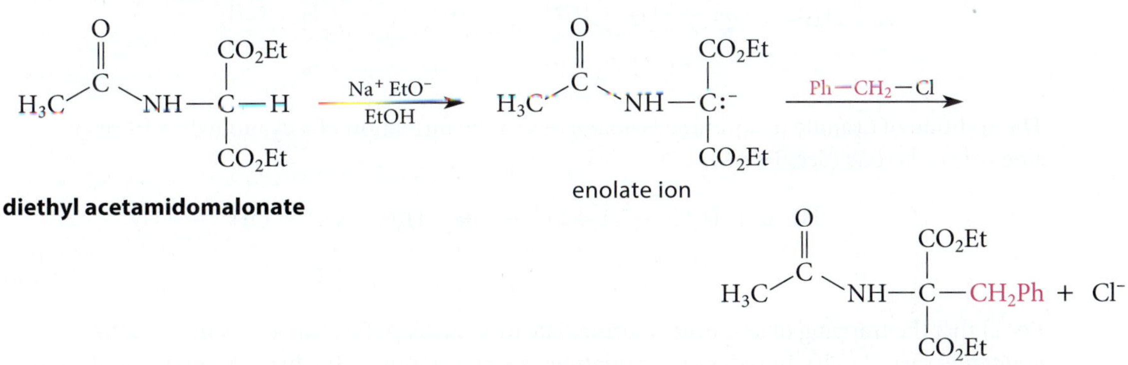

diethyl acetamidomalonate enolate ion

(27.25a)

The resulting compound is then treated with hot aqueous HCl or HBr. This acid treatment accomplishes three things: First, the ester groups are hydrolyzed to carboxylic acids (Sec. 21.7A) to yield a disubstituted malonic acid. Second, the malonic acid derivative decarboxylates under the reaction conditions (Sec. 20.11A). Third, the acetamido group, an amide, is also hydrolyzed (Sec. 21.7B). Neutralization gives the α-amino acid.

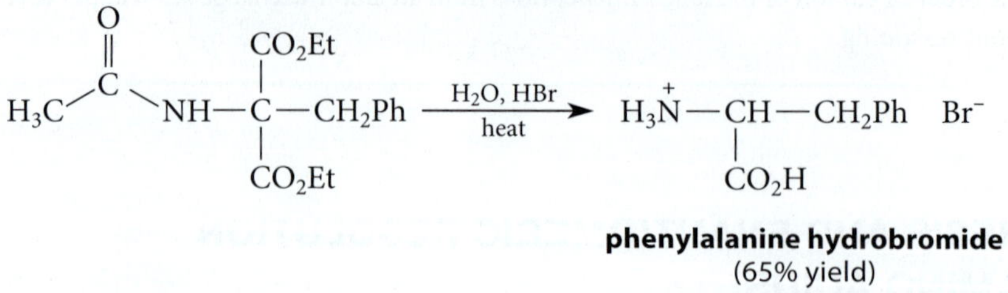

(27.25b)

C. Strecker Synthesis

An important method for synthesizing carboxylic acids is the hydrolysis of nitriles (Secs. 21.7C and 21.11). This reaction can be extended to the hydrolysis of α-amino nitriles to give α-amino acids. α-Amino nitriles, in turn, are prepared by the treatment of aldehydes with ammonia in the presence of cyanide ion.

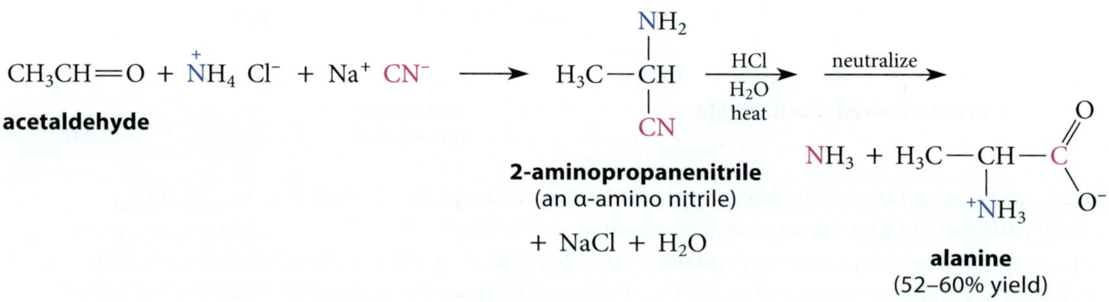

(27.26)

This preparation of α-amino acids is called the **Strecker synthesis.**

The mechanism of α-amino nitrile formation probably involves an imine intermediate.

(27.27a)

The conjugate acid of the imine reacts with cyanide under the conditions of the reaction to give the α-amino nitrile.

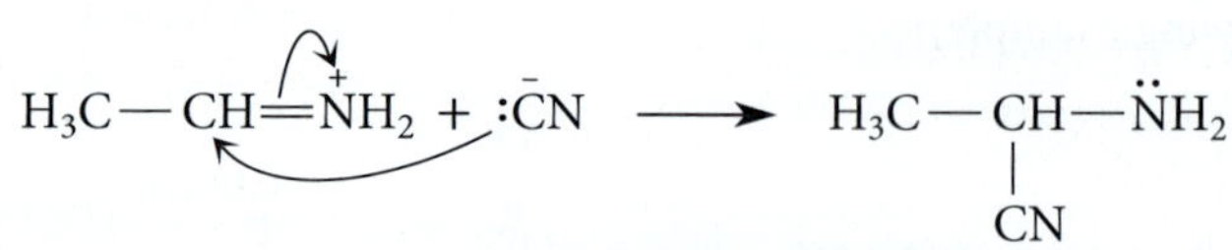

(27.27b)

The addition of cyanide to an imine is analogous to the formation of a cyanohydrin from an aldehyde or ketone (Sec. 19.8A).

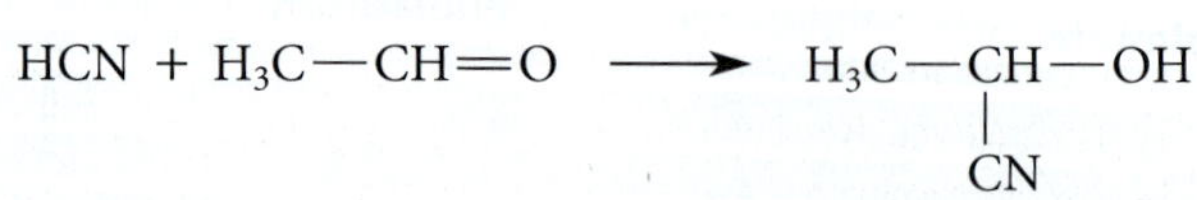

(27.28)

Recall that the trapping of an imine intermediate by a nucleophile also occurs in *reductive amination* (Sec. 23.7B). In reductive amination, the nucleophile is the hydride ion derived from $Na^{+}\,{}^{-}BH_3CN$ or $Na^{+}\,{}^{-}BH(OAc)_3$. In the Strecker synthesis, the nucleophile is cyanide ion.

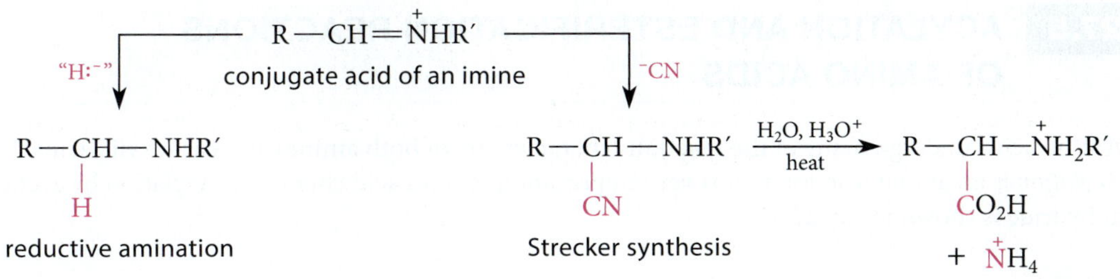

(27.29)

Focused Problem

27.12 Indicate which of the methods in this section can be used to prepare each of the following amino acids. For each method that can be used, give an equation. For each case in which a method would not work, give a reason.

(a) α-Phenylglycine

(b) Leucine

D. Enantiomeric Resolution of α-Amino Acids

Amino acids synthesized by common laboratory methods, such as the ones discussed in Secs. 27.4A–C, are *racemic*. Because enantiomerically pure amino acids are often needed, the racemic mixtures must be resolved. As useful as the diastereomeric salt method is (Sec. 6.10B), it can be tedious and time-consuming. An alternative approach to the preparation of enantiomerically pure amino acids, and one that is used industrially, is the synthesis of amino acids by microbiological fermentation. Some cultures of microorganisms can be used to produce industrial quantities of certain amino acids as the natural L enantiomer.

Certain enzymes can be used to resolve racemic amino acids into enantiomers by catalyzing the enantioselective hydrolysis of an *N*-amidated amino acid derivative. For example, a preparation of the enzyme *acylase* from hog kidney selectively catalyzes the hydrolysis of *N*-acetyl-L-amino acids and leaves the corresponding D isomers unaffected. Consequently, treatment of the *N*-acetylated racemate with this enzyme affords the free L-amino acid only:

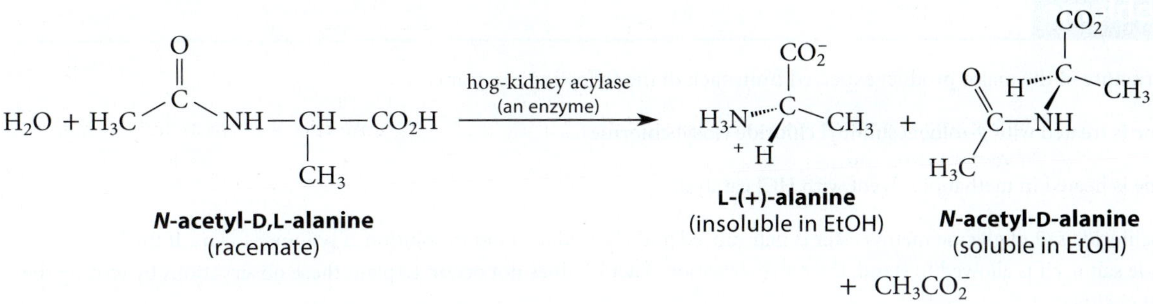

(27.30)

In this example, the liberated L-alanine is precipitated from ethanol; the *N*-acetyl-D-alanine remains in solution, from which it can be recovered and hydrolyzed in aqueous acid to D-alanine.

The enzyme differentiates between the two enantiomers of *N*-acetylalanine because it is an *enantiomerically pure chiral catalyst*. Recall that enantiomers have different reactivities with chiral reagents (Sec. 7.7A). The use of enzymes in this way is another practical example of the *principle of enantiomeric differentiation* (Sec. 6.10).

27.5 ACYLATION AND ESTERIFICATION REACTIONS OF AMINO ACIDS

Amino acids undergo many of the reactions characteristic of both amines and carboxylic acids. *Acylation* is an amine reaction that is very important in amino acid chemistry. Acylation by acetic anhydride is shown in Eq. 27.31.

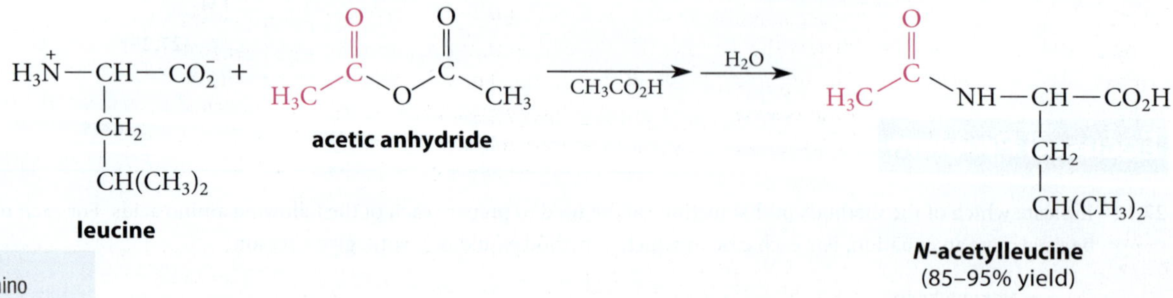

(27.31)

In Eq. 27.31, the amino group is protonated, yet the neutral form of the amine is required to serve as a nucleophile in the acylation reaction. Even in acidic solution, a very small amount of neutral amine is present. When this form reacts, the acid–base equilibrium shifts rapidly to replenish this form. More generally, a very minor component of an equilibrium can serve as a reactant in a reaction provided that (a) this component is sufficiently reactive and (b) the equilibrium can shift quickly enough to replenish the minor form once it reacts.

Acylation by acid chlorides is also a useful reaction (Sec. 21.8A).

Amino acids, like ordinary carboxylic acids, are easily esterified by heating with an alcohol and a strong acid catalyst (acid-catalyzed esterification; Sec. 20.8A).

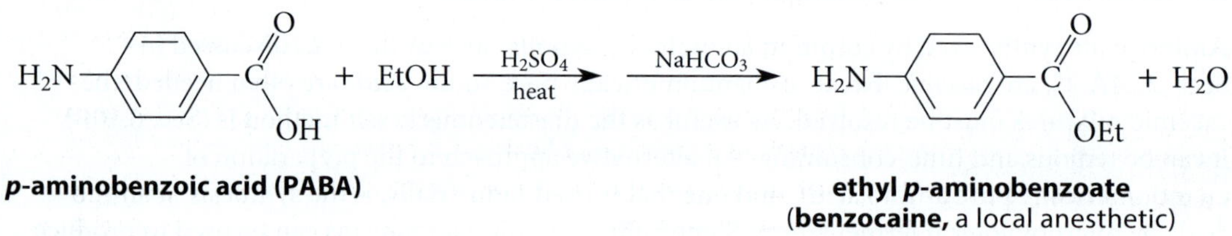

(27.32)

Focused Problems

27.13 Draw the structure of the major product expected from each of the following reactions.

(a) Leucine is treated with *p*-toluenesulfonyl chloride (tosyl chloride).

(b) Alanine is heated in methanol solvent with HCl catalyst.

27.14 If the hydrochloride salt of glycine methyl ester is neutralized and allowed to stand in solution, a polymer forms. If the hydrochloride salt itself is allowed to stand, the polymerization reaction does not occur. Explain these observations by writing the reaction that occurs.

27.6 PEPTIDE AND PROTEIN SYNTHESIS

This section addresses the synthesis of peptides and proteins from individual α-amino acids. In Sec. 27.6A, we discuss the laboratory synthesis of peptides. In Sec. 27.6B, we consider the biosynthesis of proteins: how living systems utilize the genetic code to assemble proteins from α-amino acids.

A. Solid-Phase Peptide Synthesis

A number of methods have been developed for peptide synthesis, but the most widely used are variations of an ingenious method called **solid-phase peptide synthesis**, which is universally known among its practitioners by the abbreviation **SPPS**. In this method, the carboxy-terminal amino acid is covalently anchored to an *insoluble* polymer, and the peptide is "grown" by adding one amino acid residue at a time to this polymer. Solutions containing the appropriate reagents are shaken with the polymer. At the conclusion of each step, the polymer containing the peptide is simply filtered away from the solution, which contains soluble by-products and impurities. The completed peptide is removed from the polymer by a reaction that breaks its bond to the resin, just as a plant is harvested by cutting it away from the ground. The advantage of this method is the ease with which the peptide is separated from soluble by-products of the reaction. The reactions used in SPPS also illustrate some important amino acid and peptide chemistry.

Solid-Phase Peptide Synthesis

Editorial/Alamy

Solid-phase peptide synthesis was devised by R. Bruce Merrifield (1921–2006) of The Rockefeller University and was first reported in the early 1960s. A particularly impressive achievement of the method was the synthesis of an active enzyme by Merrifield's research group in 1969 using a homemade machine in which the various steps of the method were preprogrammed. (Modern instruments for automated solid-phase peptide synthesis are available commercially.) The enzyme that was synthesized, ribonuclease, contains 124 amino acid residues; the synthesis required 369 separate reactions and 11,931 individual operations, yet it was carried out in 17% overall yield. (Several other proteins have since been prepared by solid-phase peptide synthesis.) For his invention and development of the solid-phase method, Merrifield was awarded the 1984 Nobel Prize in Chemistry.

Before considering an actual solid-phase synthesis of a peptide, we need to understand two particularly important aspects of peptide synthesis: the use of *protecting groups* and the use of *active esters.*

Protecting groups were introduced in Sec. 19.11B. Peptide synthesis involves a number of protecting groups for amino groups, carboxy groups, and side-chain functional groups, but we consider only the use of an amino protecting group in our synthesis. The purpose of an amino protecting group is to block an amine from reacting as a nucleophile at some point in the synthesis, as we show subsequently. One of the most widely used amino protecting groups is the (9-fluorenyl)methyloxycarbonyl group, known throughout the peptide-chemistry world as the **Fmoc group** (pronounced "eff-mock"). For example, an Fmoc-protected alanine has the following structure:

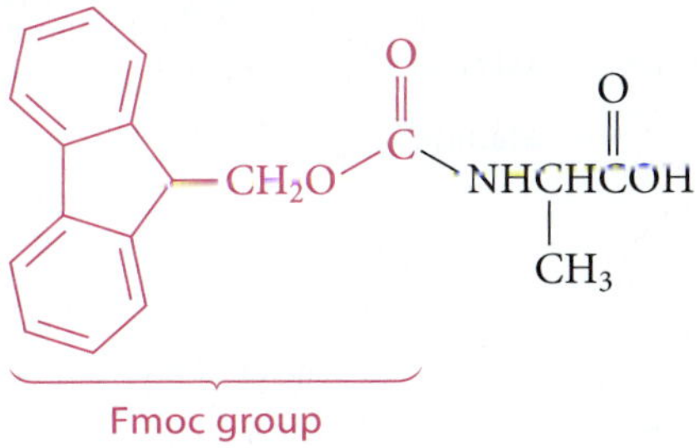

Fmoc-Ala

(27.33)

The Fmoc group was developed in 1972 by Professor Louis A. Carpino (1927–2019) of the University of Massachusetts. Professor Carpino, one of the giants of peptide synthesis, also developed a number of other useful techniques in this field.

The rationale behind the design of this group will become evident in Eq. 27.38.

Active esters are more reactive toward nucleophiles than ordinary alkyl esters. Two widely used active esters are based on two unusual "alcohols," *N*-hydroxysuccinimide and 1-hydroxy-1,2,3-benzotriazole. Both compounds have N—O bonds. These "alcohols" have very low pK_a values that are closer to the pK_a values of carboxylic acids than ordinary alcohols.

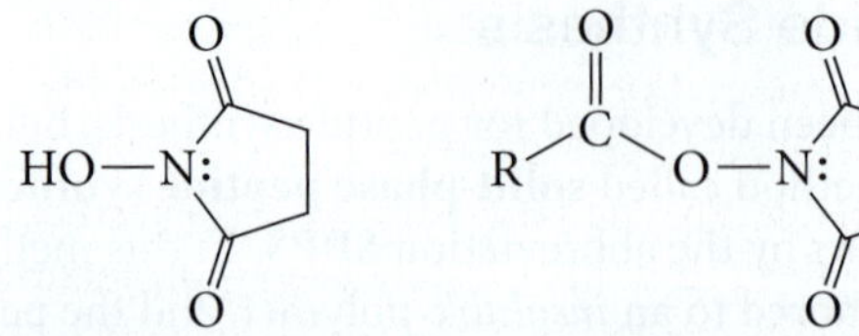

***N*-hydroxysuccinimide** (abbreviated **NHS**) $pK_a = 6.0$ — an NHS ester (27.34a)

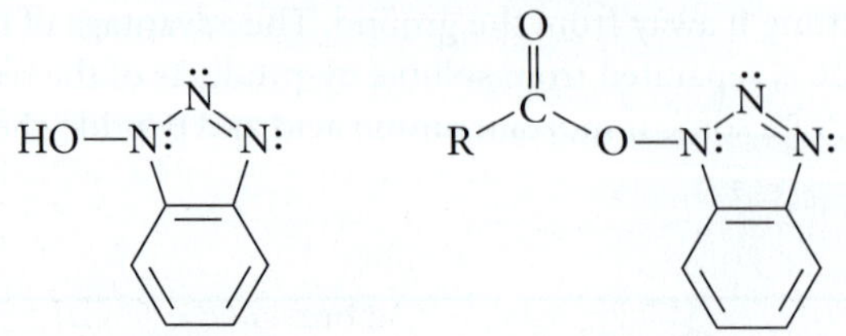

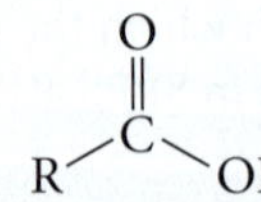

1-hydroxy-1,2,3-benzotriazole (abbreviated **HOBt**) $pK_a = 4.6$ — an HOBt ester — abbreviation for an HOBt ester (27.34b)

> Don't be confused by the fact that the active esters are nicknamed after the parent hydroxy compound, even though the H is removed as a result of the esterification. In other words, HOBt esters have an OBt group bonded to the carbonyl group. NHS esters have an *N*-oxysuccinimide group bonded to the carbonyl; the H of HOBt or NHS is lost to a base in the esterification, just as the H of ethyl alcohol is lost to solvent when an ethyl ester is formed from a carboxylic acid and ethanol. These esters might have been more systematically nicknamed OBt esters or NOS (*N*-oxysuccinimide) esters, but they weren't, and we just have to adapt.

Weak bases generally are good leaving groups in carbonyl substitution reactions, just as they are in S_N2 reactions (Sec. 9.4F), because the same property that makes them weak bases—their electron-withdrawing character—stabilizes the transition state for substitution (Sec. 21.7E). NHS and HOBt esters are much more reactive in nucleophilic substitution reactions than alkyl esters; HOBt esters verge on anhydride-like reactivity. Other rationales for use of these unusual leaving groups will become apparent in the discussion of Eq. 27.43.

We illustrate SPPS with the preparation of a tripeptide, Phe-Gly-Ala (F-G-A). The synthesis begins with the preparation of Fmoc-Ala, the protected alanine derivative shown in Display 27.33. The conjugate base of alanine is allowed to react with Fmoc-NHS, which undergoes an ester aminolysis (Sec. 21.8C).

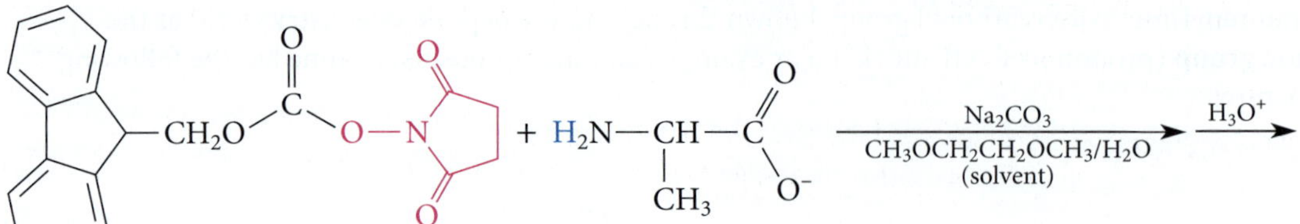

Fmoc-NHS — alanine

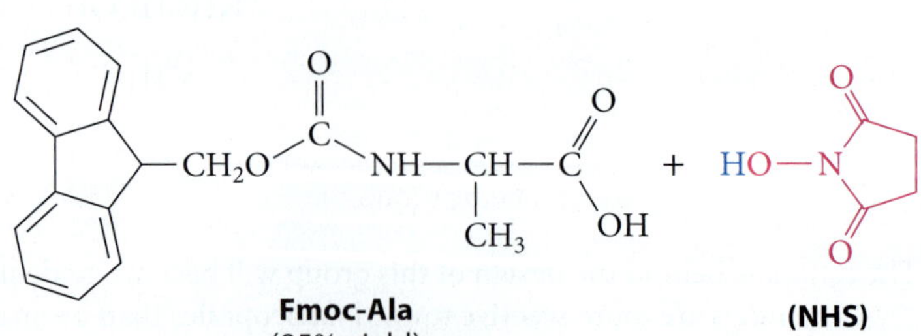

Fmoc-Ala (85% yield) (a carbamate ester) — **(NHS)** (27.35)

Fmoc-amino acids are available commercially from chemical supply firms that specialize in peptide-synthesis reagents.

This reaction illustrates the superior leaving-group properties of the NHS group. Fmoc-NHS is a carbonate ester; potentially, either of two alcohols could be a leaving group. Because of its low pK_a, NHS is a much better leaving group (as its conjugate base) than the conjugate base of (9-fluorenyl)methanol. Consequently, the NHS group is displaced and the (9-fluorenyl)methyloxy group remains intact as part of the carbamate-ester (Fmoc) protecting group.

The next step of the synthesis involves the solid phase for which the method is named. The Fmoc-Ala formed in Eq. 27.35 is anchored onto an insoluble solid polymeric support, called a *resin*, using the reactivity of its free carboxylate group. A variety of such resins are available commercially, and a popular one is the following:

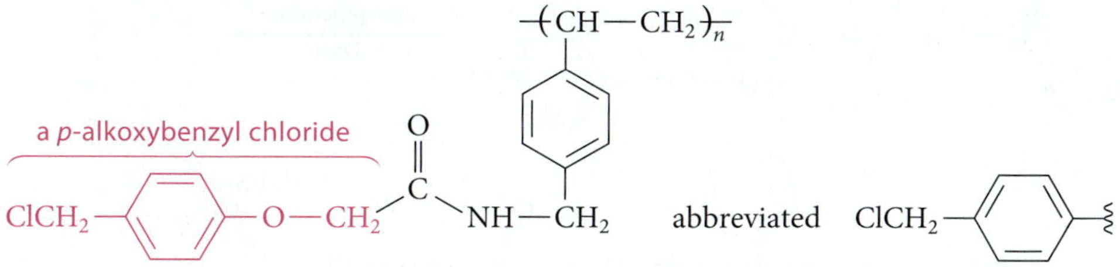

(27.36)

This is, in effect, an insoluble *p*-alkoxybenzyl chloride, and it has the enhanced S_N2 reactivity generally associated with benzylic halides (Sec. 17.4). (The reason for the para —OCH_2— group becomes apparent in Eq. 27.45.) The protected amino acid Fmoc-Ala is converted into its conjugate base. In an older version of SPPS, the cesium salt is used. In more recent versions, the salt is formed with a hindered tertiary amine such as diisopropylethylamine.

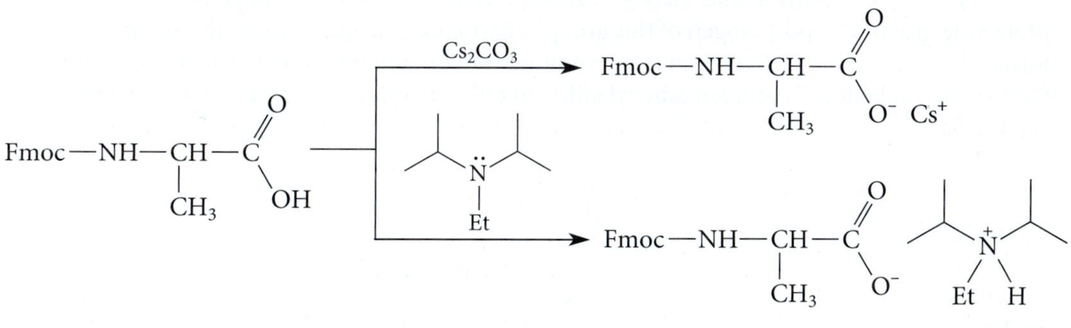

(27.37)

An S_N2 reaction between the conjugate base of Fmoc-Ala and the chloromethyl group of the resin results in the formation of an ester linkage to the resin by alkylation of the carboxylate ion. (See Sec. 20.8B for related chemistry.)

FURTHER EXPLORATION 27.1
Solid-Phase Peptide Synthesis

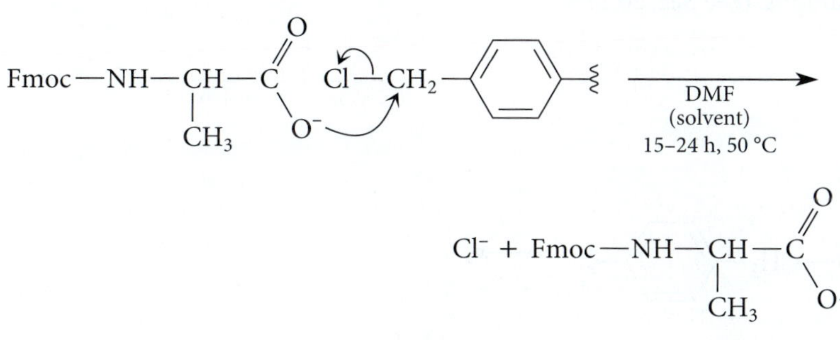

Fmoc-Ala linked to the resin

(27.38)

This equation shows why the Fmoc group is required—namely, to keep the amino group from competing with the carboxylate group as a nucleophile for the benzylic halide group on the resin.

The resin is supplied as a powder consisting of tiny spherical beads. Although the preceding equations show only one peptide on the resin, many peptide chains are anchored to each polymer bead, and many polymer beads are used in each synthesis.

Once the Fmoc-amino acid is anchored to the resin, the Fmoc protecting group is removed by treatment with piperidine, an amine base.

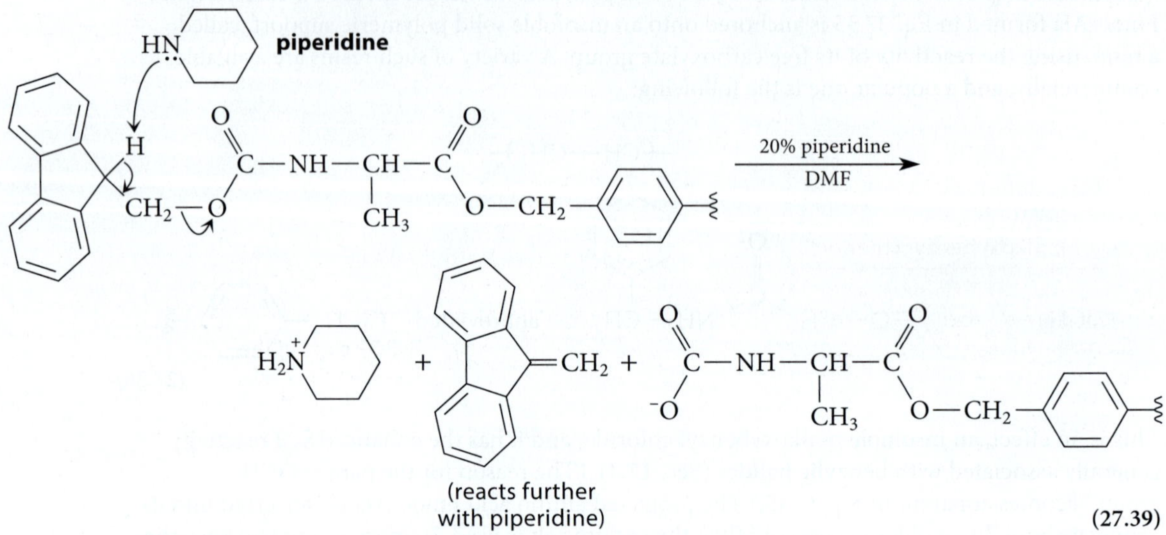

(27.39)

This is an E2 reaction. As explained in Sec. 17.3B, E2 reactions are particularly fast when the β-hydrogen is particularly acidic. Here we can appreciate the ingenious design of the Fmoc protecting group. The β-hydrogen of this group is particularly acidic because the anion that is formed by removal of this hydrogen as a proton is *aromatic* and therefore particularly stable. (Notice the "imbedded" cyclopentadienyl anion in red in Display 27.40; see Sec. 15.7D and Eq. 15.65a.)

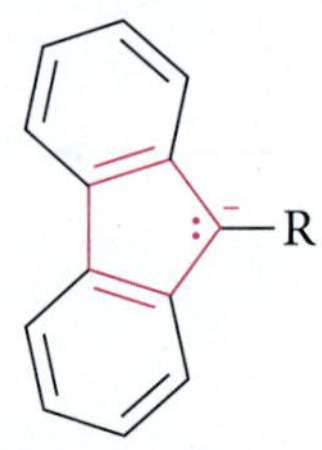

(27.40)

Because the product of the β-elimination in Eq. 27.39 is a carbamate anion, it decarboxylates under the reaction conditions. (See Sec. 20.11A.)

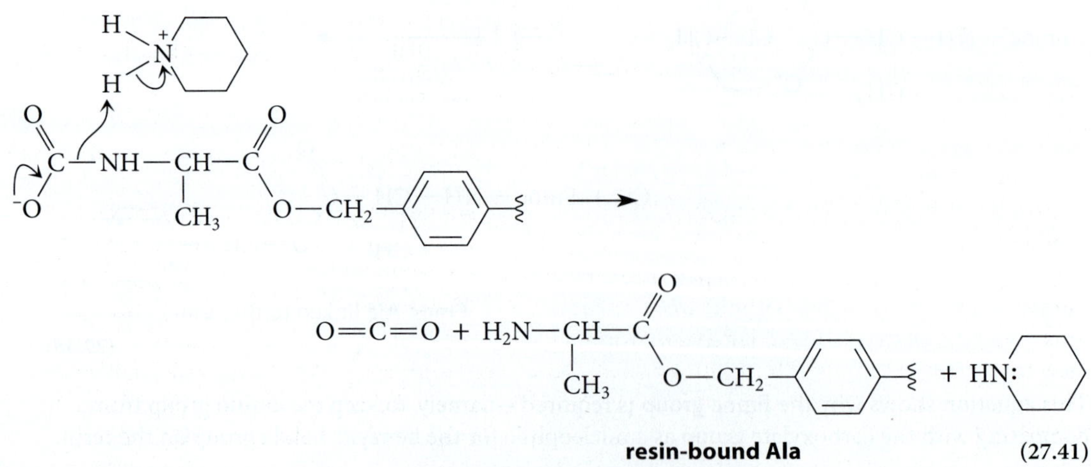

(27.41)

This reaction exposes the amino group of the resin-bound amino acid. This amino group serves as a nucleophile in the next reaction.

Next comes the formation of the first peptide bond. Fmoc-Gly is first converted into its conjugate base with Cs_2CO_3 or diisopropylethylamine, and then it is converted into its HOBt ester with a reagent nicknamed PyBOP. The carboxylate ion of the Fmoc amino acid reacts at the phosphorus of PyBOP to displace the conjugate base of HOBt, which instantly reacts at the carbonyl group of the amino acid to give the HOBt ester.

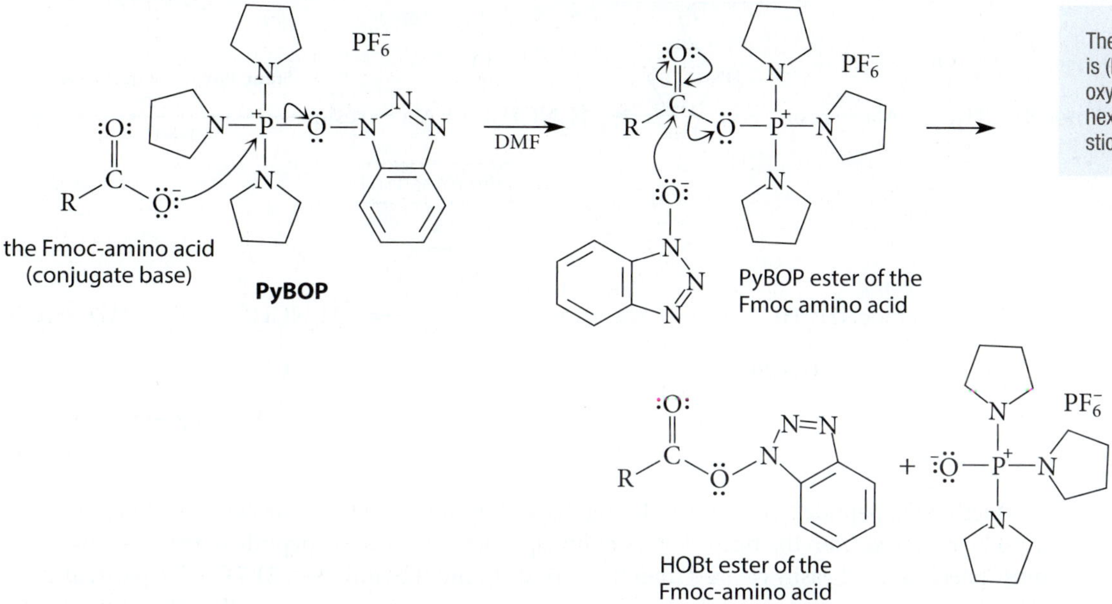

(27.42)

The IUPAC name of PyBOP is (benzotriazol-1-yl)oxytripyrrolidinophosphonium hexafluorophosphate. We'll stick with the nickname.

Because the PyBOP ester of the Fmoc-amino acid is so reactive, it reacts instantly with the nearby $^{-}$OBt group before it can diffuse away.

The HOBt ester reacts with the amino group of the resin-bound Ala in an ester aminolysis to form the peptide bond—that is, an amide linkage.

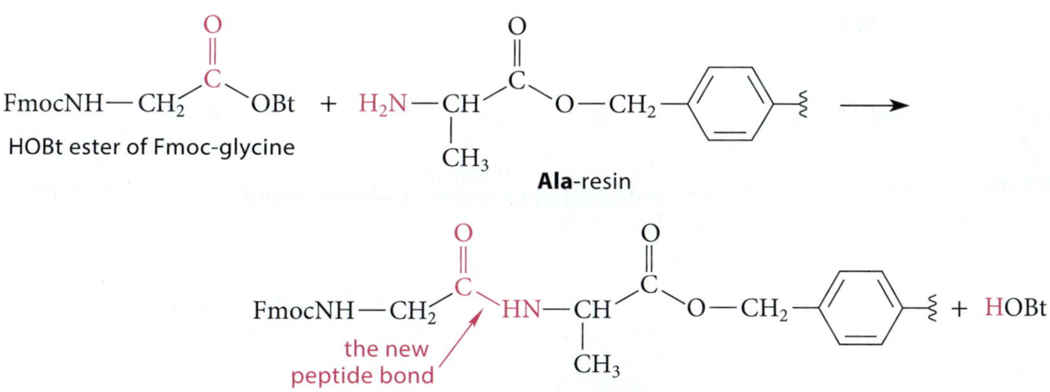

Fmoc-Gly-Ala-resin (27.43)

At this point it is reasonable to ask why we use the rather unusual reagent PyBOP. Why not simply make the acid chloride of the protected amino acid and let it react with HOBt? Or we could even ask why we need to prepare an HOBt ester at all. Why not let the acid chloride of the protected amino acid react with the free amino group to form the peptide bond? The answer is that acid chlorides are so reactive that they undergo a number of side reactions, some of which lead to racemization of the Fmoc-amino acid. These side reactions are disastrous if we are trying to prepare a biologically active peptide or protein, which has to be enantiomerically pure. It was found that these side reactions can be avoided by using the somewhat less reactive HOBt esters. This is another illustration of the *reactivity–selectivity principle* (Sec. 10.4B; see also the sidebar at the end of that section). The HOBt esters are reactive enough to undergo rapid aminolysis, but not so reactive that they undergo competing side reactions.

Completion of the peptide synthesis requires deprotection of the dipeptide-resin as in Eq. 27.39, a final coupling step with Fmoc-Phe (conjugate base) and PyBOP, and deprotection:

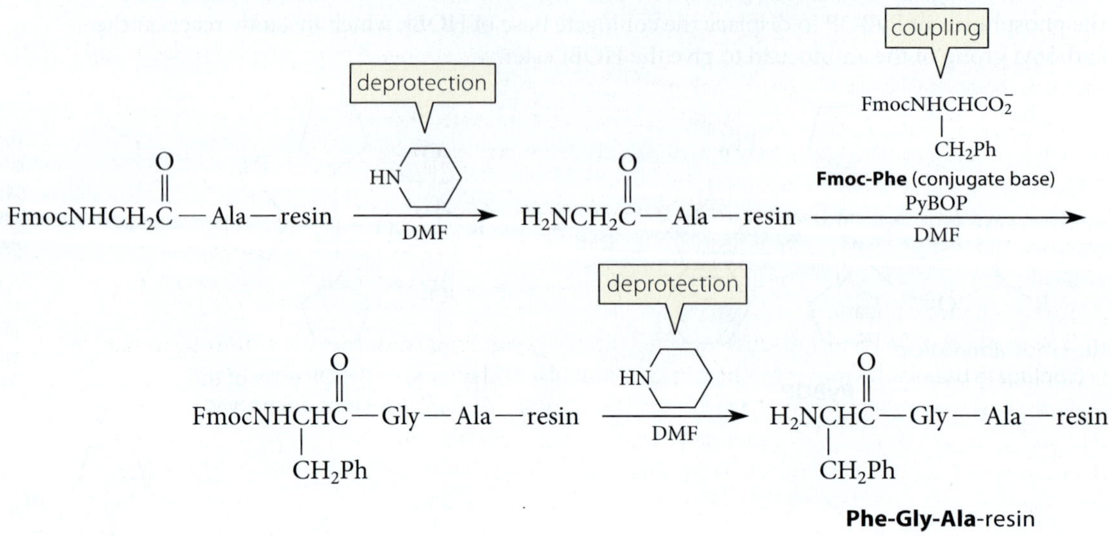

(27.44)

Once all the peptide bonds in the desired tripeptide are assembled, the completed peptide must be removed from the resin. The ester linkage that connects the peptide to the resin, like most esters, is more easily cleaved than the peptide (amide) bonds (Sec. 21.7E). The particular ester linkage used in this case is broken by a carbocation mechanism using 50–60% trifluoroacetic acid (TFA) in dichloromethane. (In this equation, the amino-terminal part of the peptide is abbreviated Pep^N.)

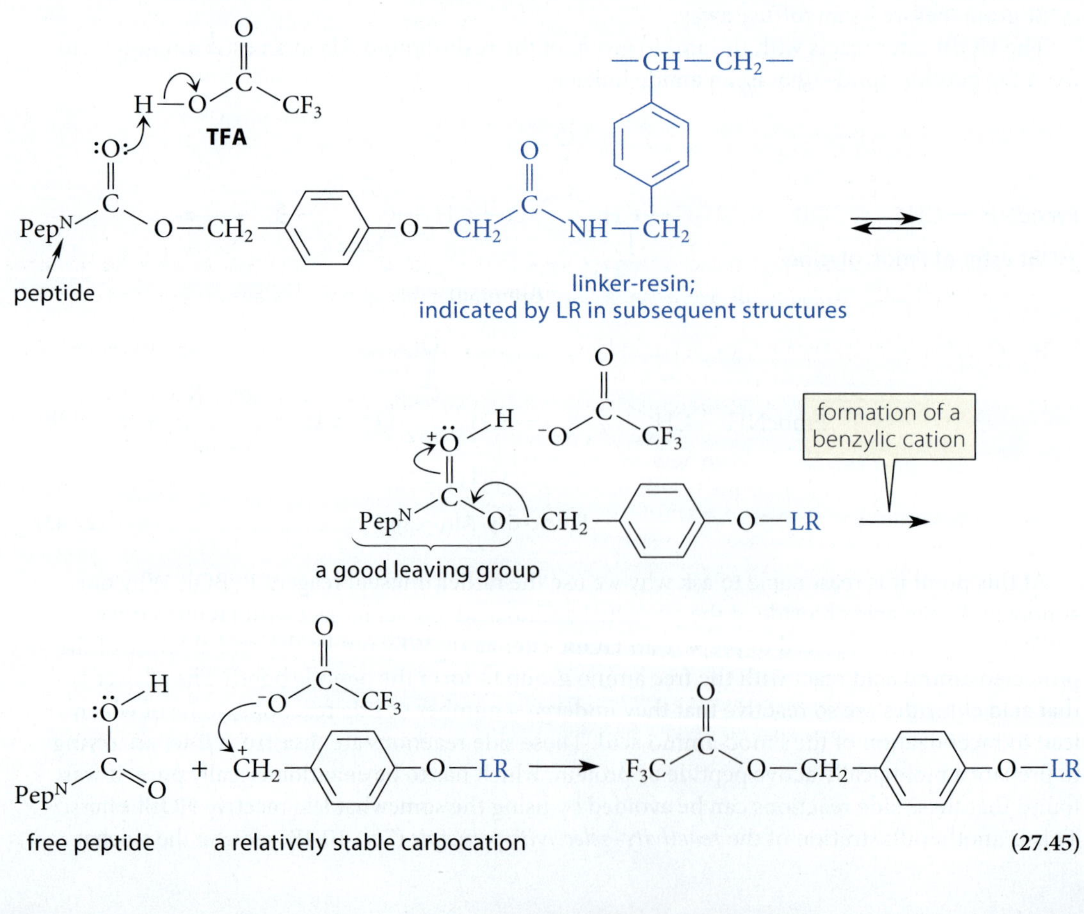

(27.45)

The acidic conditions promote breaking of the ester linkage by an S_N1 mechanism. Protonation of the peptide carbonyl converts this group into a good leaving group because it is the conjugate acid of a very weak base. The S_N1 cleavage yields a benzylic carbocation that is resonance-stabilized, not only by the benzene ring but also by the para oxygen. (Draw resonance structures that show this stabilization if this isn't clear.) The reason for inclusion of the para —OCH_2— group in the design of the resin, then, is that it stabilizes the carbocation intermediate in the cleavage reaction and, by Hammond's postulate, accelerates the release of the peptide. As a result of this reaction, the peptide is liberated into solution, from which it can be readily isolated.

Notice that the conditions of peptide synthesis and deprotection do *not* affect the ester group by which the peptide is linked to the resin. Benzylic esters undergo aminolysis very sluggishly with secondary amines such as piperidine because of steric hindrance between the phenyl hydrogens and the amine. Furthermore, the piperidine treatment required for removal of the Fmoc group takes only 1 minute. This is too brief a time for aminolysis of the ester to occur. However, the ester is cleaved by acidic conditions because of the ease with which it forms a relatively stable carbocation.

The method of SPPS just discussed, which employs the Fmoc group as the amino-terminal protecting group, is the major method in common use today.

Despite its advantages, SPPS has one unique problem. Suppose, for example, that a coupling reaction is incomplete, or that other side reactions take place to give impurities that remain covalently bonded to the resin. These are then carried along to the end of the synthesis, when they are also removed from the resin and must be separated (in some cases tediously) from the desired peptide product. To avoid impurities, then, each step in the solid-phase synthesis must occur in very high yield. Remarkably, this ideal is often approached closely in practice. (See Focused Problem 27.15.) Nevertheless, because of the accumulation of small amounts of impurities, the practical size limit for peptides prepared with "conventional" SPPS is about 30–50 residues, although there are exceptions. (See the sidebar at the beginning of this section.) Larger peptides can be prepared by *peptide-ligation reactions*, which are chemical methods for joining shorter fragments into longer ones. (See Problem 27.95.) In recent years, however, new methods for SPPS have been developed that involve flow techniques. These techniques employ the same fundamental chemistry but involve higher temperatures and higher concentrations of reagents that result in reaction times of only minutes per residue as well as greater product purity. Laboratories using these automated techniques are reporting peptide lengths as high as 160 residues with the expectation that higher limits are possible.

Given the modern capability of cloning genes and recruiting bacteria such as *Escherichia coli* to produce large quantities of medically important proteins (see the sidebar, "Human Proteins from Bacterial Factories," at the end of Sec. 27.6B), it is reasonable to ask why we would want to synthesize proteins chemically. One reason is that with peptide synthesis chemists can synthesize proteins that incorporate *unnatural* amino acids—that is, amino acids that are not among the 20 coded by the DNA–messenger RNA system discussed in the following section. Unnatural amino acids offer the possibility of producing *designer proteins*—proteins that can catalyze reactions that have industrial importance.

Focused Problems

27.15 Calculate the average yield of each of the 369 steps in the synthesis of ribonuclease by the solid-phase method discussed in the sidebar, "Solid-Phase Peptide Synthesis," at the beginning of this section. Assume the reported overall yield of 17%.

27.16 What average yield per amino acid would be required to synthesize a protein containing 100 amino acids in 50% overall yield?

27.17 (a) An aspiring peptide chemist, Mo Bonds, has decided to attempt the synthesis of the peptide Gly-Lys-Ala using the solid-phase method. To the Ala-resin he couples the following derivative of lysine:

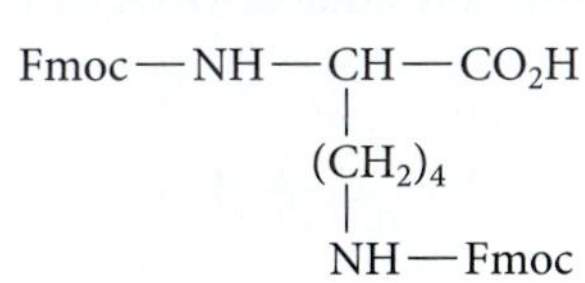

α,ε-di-Fmoc-lysine

Why are *two* protecting groups necessary for lysine?

(b) After the coupling, he deprotects his resin-bound peptide with 20% piperidine in DMF, and then he completes the synthesis in the usual way by coupling Fmoc-Gly, deprotecting the peptide and removing it from the resin. He is shocked to find a mixture of several peptide products. Two of them contain one residue of each of the amino acids Ala, Gly, and Lys, and one contains two residues of Gly, one residue of Ala, and one residue of Lys. Suggest a structure for each product, and explain how each is formed.

27.18 Another protecting group used in peptide synthesis is the *tert*-butoxycarbonyl group (nicknamed the *Boc* group).

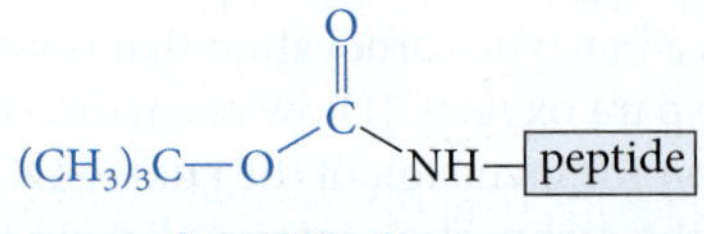

***tert*-butoxycarbonyl (Boc) group**

This group is removed by treatment of the peptide with CF_3CO_2H by an S_N1 mechanism. Using Eq. 27.45 as your model, show this deprotection reaction, and indicate why this group is susceptible to the S_N1 cleavage mechanism.

27.19 Consider the following solid-phase peptide synthesis:

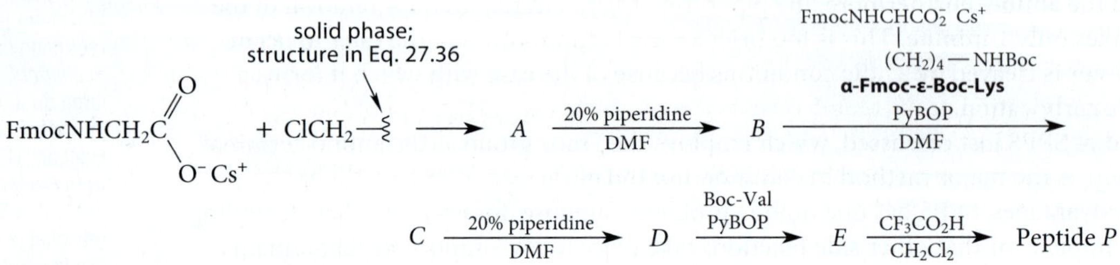

(a) Give the structure of each compound *A*–*E* and *P*.

(b) Explain the reason for the Boc group on the side chain of the Lys group in the reaction *B* ⟶ *C*.

(c) Explain why Boc-Val rather than Fmoc-Val is used in the *D* ⟶ *E* step of the synthesis.

Focused Problem 27.17 shows that certain amino acid side chains can also react under the conditions of peptide synthesis. Special protecting groups must be installed on these side chains to prevent undesired reactions. (Focused Problems 27.18 and 27.19 illustrate this situation.) These protecting groups must survive the entire synthesis, including the removal of the amino-protecting group at each stage, yet they must be removable at the end of the synthesis. A variety of side-chain protecting groups have been developed. The choice of protecting groups that can meet these exacting requirements is an important aspect of any peptide synthesis.

Chemical Biology Topic

B. The Biosynthesis of Proteins

Proteins are biosynthesized in nature by a system that translates the sequence of nucleotides in DNA into a sequence of amino acids in proteins. First, we outline the translation process, and then we consider the chemical aspects of the synthesis itself.

As discussed in Sec. 26.5B, the sequence of nucleotides in DNA forms a linear code for every protein and RNA molecule. To understand this point, let's see how the following strand of DNA, which might be imagined as part of a gene, could be used biologically to direct the synthesis of a specific protein. If this were the DNA from a cell, it would be one of the two strands of the double helix; each letter identifies a residue of DNA by its particular base.

⟵ 3′ end 5′ end ⟶

· · · A–A–A–G–A–T–T–C–A–C–C–C–C–T–C–A–T–C · · · (27.46a)

First, a strand of DNA directs the biosynthesis of a complementary strand of RNA. This process is called **transcription**. The sequence of the RNA transcript is complementary to the DNA strand from which it was transcribed—that is, each base of the DNA has hydrogen-bonding complementarity with a base of RNA (see Fig. 26.6, Sec. 26.5B). For example, everywhere there is a G (guanine) in DNA, there is a C (cytosine) in mRNA, and everywhere there is an A (adenine) in DNA, there is a U (uracil) in mRNA. (Transcribed RNA contains U instead of T; U is simply a T without the methyl group.) For example, the foregoing gene fragment would be transcribed as follows:

← 3′ end	DNA template	5′ end →
··· A–A–A–G–A–T–T–C–A–C–C–C–C–T–C–A–T–C ···		
··· U–U–U–C–U–A–A–G–U–G–G–G–G–A–G–U–A–G ···		
← 5′ end	RNA transcript	3′ end →

(27.46b)

Notice that the complementary sequence of the RNA transcript runs in the direction opposite to that of its parent DNA—the 5′-end of RNA matches the 3′-end of DNA, and vice versa.

DNA in the genes for most proteins contains interspersed noncoding sequences, called *introns*. Consequently, the RNA initially produced by transcription of DNA (called a *primary transcript*) contains the complementary RNA intron sequences. These introns are excised from the primary transcript (by a self-splicing reaction) to give the RNA that actually codes for the protein, called **messenger RNA (mRNA)**, which the cell uses to direct the synthesis of a specific protein from its component amino acids. This process is called **translation**. Each successive three-residue triplet, called a **codon**, in the sequence of mRNA is translated as a specific amino acid in the sequence of a protein according to the **genetic code** given in **Table 27.2**. For example, the particular stretch of RNA shown in Display 27.46b (assuming it is part of the ultimate mRNA) would be translated into a protein sequence as follows:

protein translation product

← amino end | carboxy end

··· Phe – Leu – Ser – Gly – Glu – STOP

··· U–U–U–C–U–A–A–G–U–G–G–G–G–A–G–U–A–G ···

← 5′ end | mRNA transcription product | 3′ end →

(27.46c)

Just as a sequence of dots and dashes in Morse code can be used to form words, the precise sequence of bases in DNA (by way of its complementary mRNA transcription product) codes for the successive amino acids of a protein. Morse code has two coding units—the dot and the dash. In DNA or mRNA, there are four: A, T (U in mRNA), G, and C, the four nucleotide bases. The sequences of DNA and mRNA contain no "commas." The protein-synthesizing system of the cell knows where one amino acid code ends and another starts because, as Table 27.2 shows, there

TABLE 27.2 The Genetic Code

5′-Terminal base of mRNA	Middle base of mRNA				3′-Terminal base of mRNA
	U	C	A	G	
U	Phe	Ser	Tyr	Cys	U
	Phe	Ser	Tyr	Cys	C
	Leu	Ser	(Stop)	(Stop)	A
	Leu	Ser	(Stop)	Trp	G
C	Leu	Pro	His	Arg	U
	Leu	Pro	His	Arg	C
	Leu	Pro	Gln	Arg	A
	Leu	Pro	Gln	Arg	G
A	Ile	Thr	Asn	Ser	U
	Ile	Thr	Asn	Ser	C
	Ile	Thr	Lys	Arg	A
	Met*	Thr	Lys	Arg	G
G	Val	Ala	Asp	Gly	U
	Val	Ala	Asp	Gly	C
	Val	Ala	Glu	Gly	A
	Val*	Ala	Glu	Gly	G

*Sometimes used as "start" codons.

is a specific "start" signal—either of the nucleotide sequences AUG or GUG—at the appropriate point in the mRNA. Because mRNA also contains "stop" signals (UAA, UGA, or UAG), protein synthesis is also terminated at the right place.

Some amino acids have multiple codes. For example, Table 27.2 shows that glycine, the most abundant amino acid in proteins, is coded by GGU, GGC, GGA, and GGG.

It is possible for the change of only one base in the DNA (and consequently in the mRNA) of an organism to cause the change of an amino acid in the corresponding protein. A dramatic example of such a change is the genetic disease *sickle-cell anemia*. In this painful disease, the red blood cells take on a peculiar sickle shape that causes them to clog capillaries. The molecular basis for this disease is a single amino acid substitution in hemoglobin, the protein that transports oxygen in the blood. In sickle-cell hemoglobin, glutamic acid at position 6 in one of the protein chains of normal hemoglobin is changed to valine. That is, sickle-cell disease results from a change in but 1 of the 141 amino acids in this hemoglobin chain! The mRNA genetic code for Glu is GAA and GAG, whereas the code for Val is GUA and GUG (among others). In other words, a change of only one nucleotide (A $\longrightarrow$ U) of the $(3 \times 141) = 423$ nucleotides that code for this chain of hemoglobin is responsible for the disease.

Now we consider the chemical aspects of peptide-bond formation in the biosynthesis of proteins. A different type of RNA, called **transfer RNA (tRNA)**, serves as the bridge, or adaptor, between the mRNA code and each amino acid. Each tRNA molecule contains between 73 and 94 nucleotides, depending on the amino acid to be encoded. Each amino acid has its own tRNA. The structure of a typical tRNA and the abbreviation that we'll use for tRNA in chemical equations are shown in **Fig. 27.3**. At one end of each tRNA molecule is a three-residue segment, called the **anticodon**, which forms a complementary hydrogen-bonding interaction with the codon of mRNA. *The codon–anticodon complementarity is the basis for the faithful translation of an mRNA sequence to a protein sequence.* For example, one of the codons for the amino acid alanine in mRNA (see Table 27.2) is GCU. The anticodon in alanine-tRNA that would "read" this code is CGA. (There is more than one alanine-tRNA; why?) At the other end of alanine-tRNA is the **acceptor stem**; this is where the amino acid is covalently attached.

The amino acid must be activated to react chemically with the acceptor stem of tRNA. The amino acid is converted into an *amino acid adenylate* by a reaction with ATP. (Adenylation was discussed in Sec. 25.7A, Eq. 25.45c.) In this reaction, the α-amino acid is converted into an *acyl phosphate.*

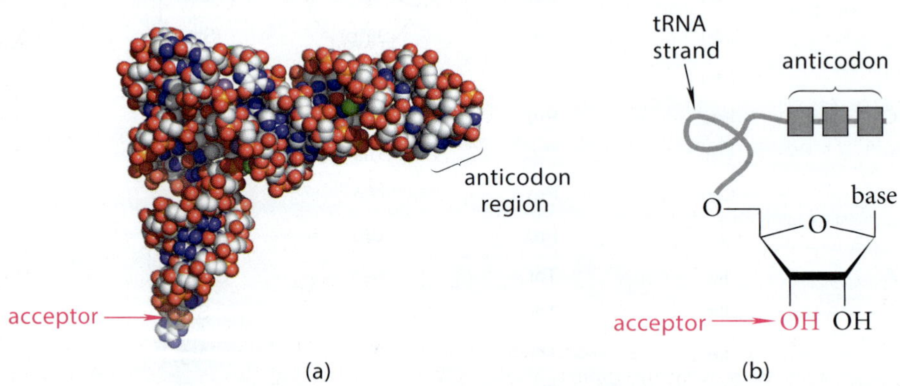

FIGURE 27.3 Transfer RNA (tRNA). (a) A space-filling model of a phenylalanine tRNA from yeast. (b) The abbreviated structure of tRNA used in the text.

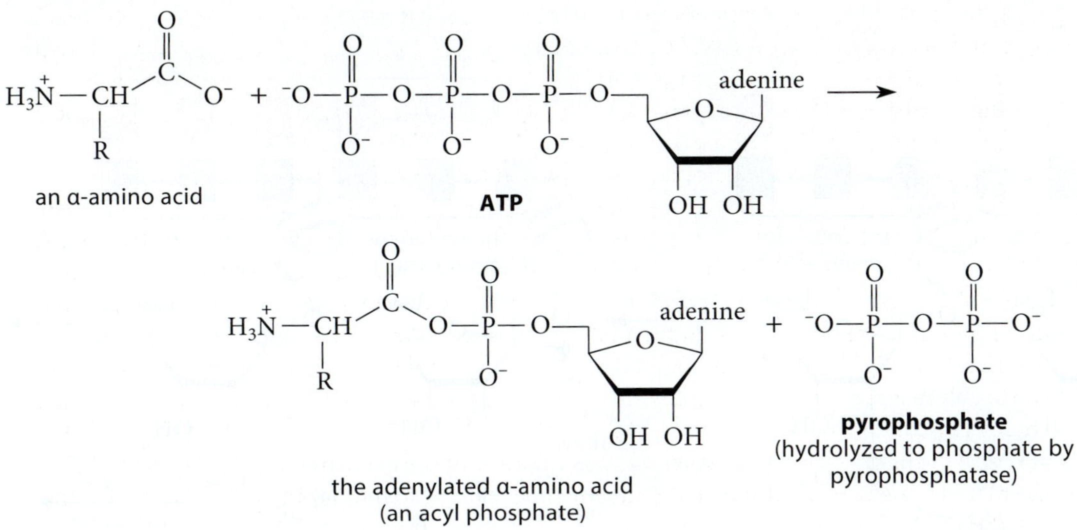

(27.47a)

In this reaction, pyrophosphate is lost as a leaving group; the reaction is driven to the right by the hydrolysis reaction of pyrophosphate with water, catalyzed by the enzyme pyrophosphatase, to give two molecules of phosphate (Eq. 25.44, Sec. 25.6B). Recall that acyl phosphates are excellent acylating agents (Sec. 25.7B). The aminoacyl phosphate reacts with the 3′-hydroxy group of the ribose at the 3′-end of the tRNA to give the aminoacyl-tRNA.

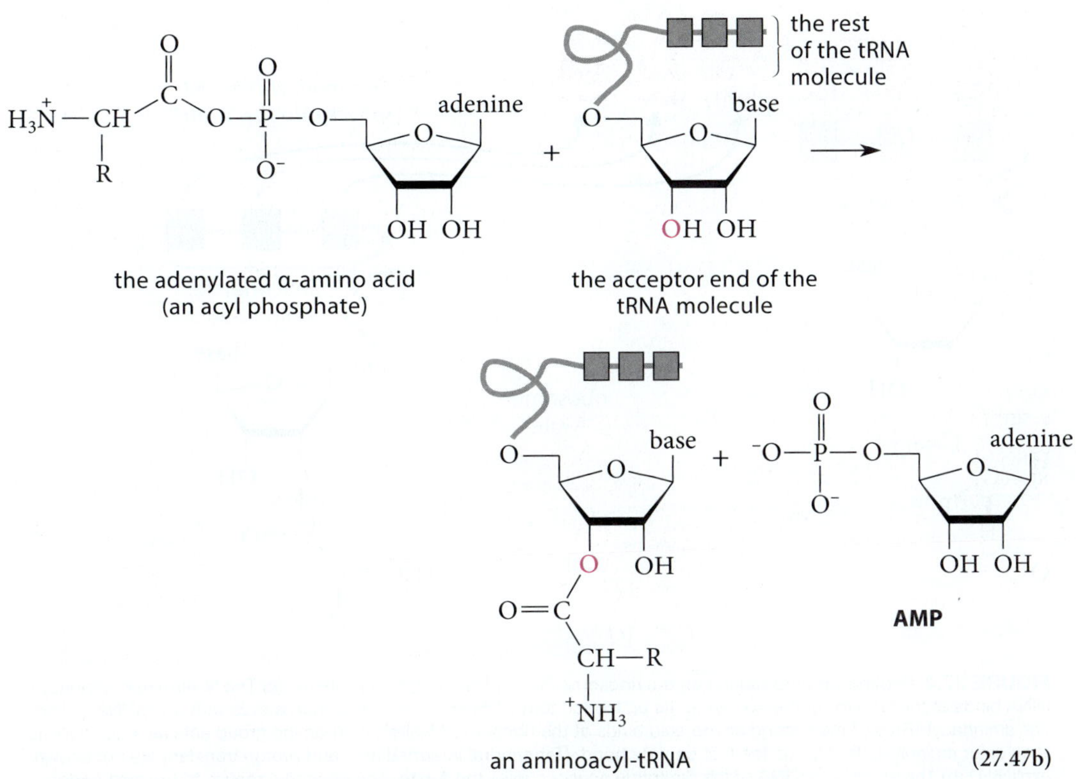

(27.47b)

(In some cases, the 2′-hydroxy group reacts, but in those cases the resulting 2′-aminoacyl-tRNA derivative is converted intramolecularly into the 3′-aminoacyl-tRNA.) Both reactions (Eqs. 27.47a and b) are catalyzed by the same enzyme, an aminoacyl-tRNA synthetase that is specific for the particular amino acid.

Proteins are synthesized from the amino end to the carboxy end. A large protein–RNA complex called the **ribosome** controls the reading of the mRNA and catalyzes the peptide bond formation. The RNA component of the ribosome (rRNA) actually catalyzes the peptide bond

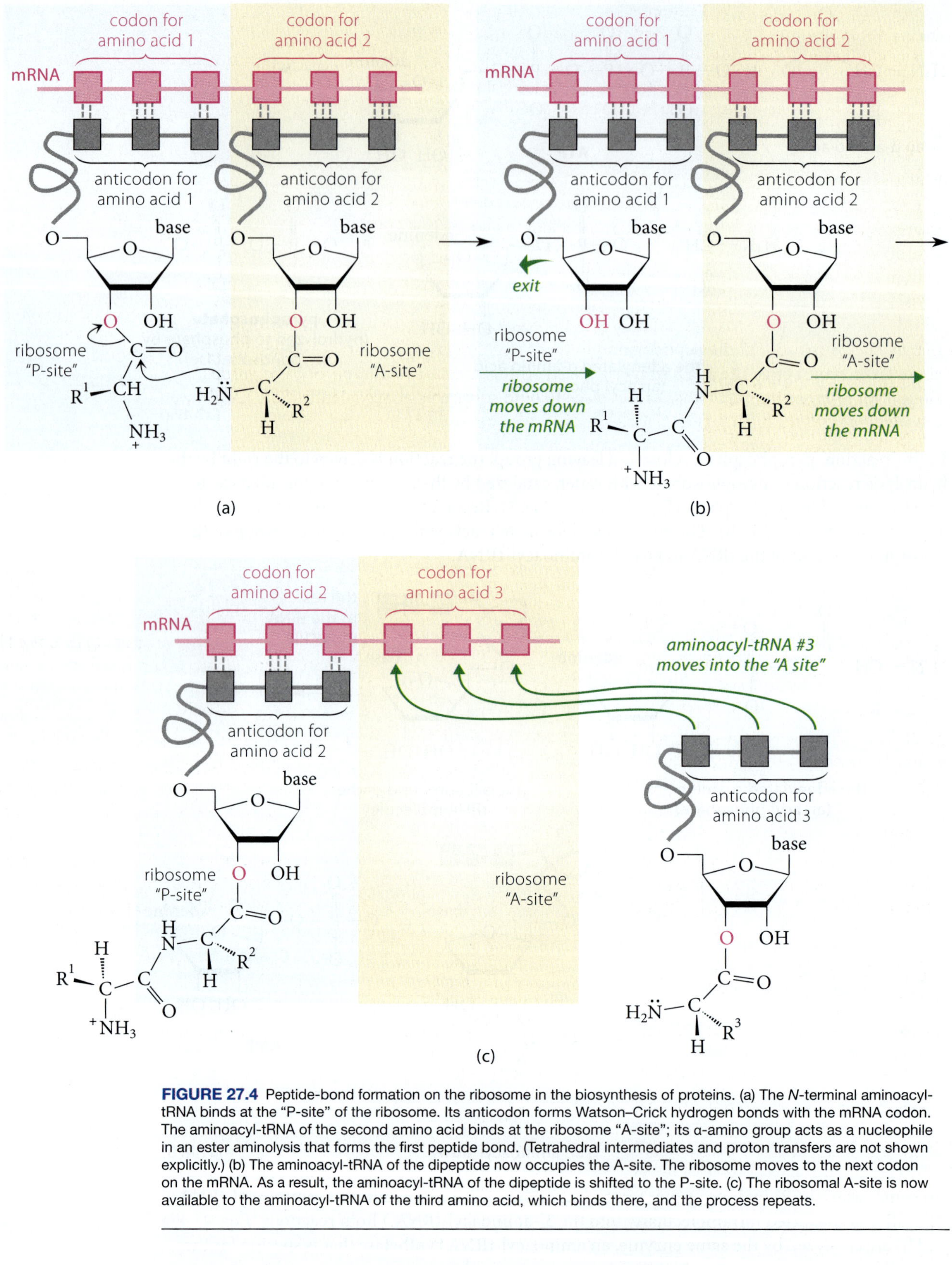

FIGURE 27.4 Peptide-bond formation on the ribosome in the biosynthesis of proteins. (a) The *N*-terminal aminoacyl-tRNA binds at the "P-site" of the ribosome. Its anticodon forms Watson–Crick hydrogen bonds with the mRNA codon. The aminoacyl-tRNA of the second amino acid binds at the ribosome "A-site"; its α-amino group acts as a nucleophile in an ester aminolysis that forms the first peptide bond. (Tetrahedral intermediates and proton transfers are not shown explicitly.) (b) The aminoacyl-tRNA of the dipeptide now occupies the A-site. The ribosome moves to the next codon on the mRNA. As a result, the aminoacyl-tRNA of the dipeptide is shifted to the P-site. (c) The ribosomal A-site is now available to the aminoacyl-tRNA of the third amino acid, which binds there, and the process repeats.

formation; it is therefore an unusual example of a nonprotein enzyme (called colloquially a "ribozyme"). The formation of the peptide bond is shown in **Fig. 27.4**. At the first codon of mRNA (which codes for the first amino acid), the aminoacyl-tRNA with the complementary anticodon forms a complex within the ribosome at a site called the "P-site." At an adjacent site on the ribosome, called the "A-site," the second aminoacyl-tRNA binds; it has an anticodon that is complementary to the second codon of mRNA. The binding of the two aminoacyl-tRNAs places the amino group of the second aminoacyl group in proximity to the carbonyl group of the first, and an ester aminolysis (Sec. 21.8C) occurs to produce the first peptide bond (Figs. 27.4a and b).

The empty tRNA formed exits the P-site. The ribosome moves down the mRNA so that the dipeptidyl-tRNA now occupies the P-site, a new aminoacyl-tRNA enters the A-site, and the process repeats (Fig. 27.4c).

To stress the chemical aspects of protein biosynthesis, we have omitted any discussion of other protein factors, called *elongation factors* and *release factors*, involved in this process. Hydrolysis of two guanosine triphosphate (GTP) molecules is involved in each elongation step, and part of the elongation process is a "proofreading" mechanism that ensures fidelity of translation. This hydrolysis, along with the two high-energy phosphates of ATP utilized during the formation of aminoacyl-tRNA molecules, results in an "energy cost" per peptide bond of four high-energy phosphate bonds—roughly 122 kJ mol^{-1} (29.2 kcal mol^{-1}) for the formation of each peptide bond. This high energetic cost ensures not only that the equilibrium for peptide-bond formation favors the product, but also that the peptide bond to the *correct* amino acid is formed. When we think about an enzyme active site as a highly organized arrangement of acidic and/or basic groups, along with residues in exactly the correct place to bind substrates noncovalently (Sec. 12.8E), we can now understand the high energetic price that must be paid for all of this organization.

Human Proteins from Bacterial Factories

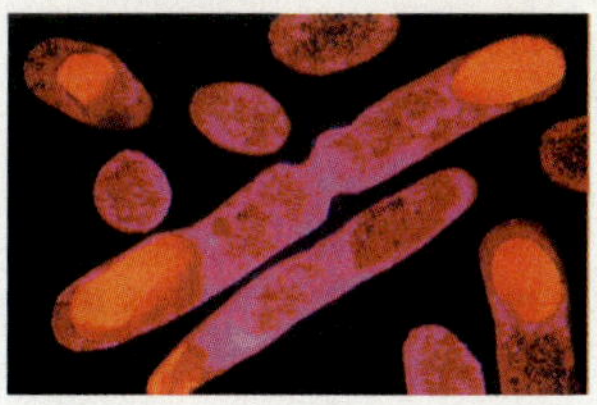

Volker Steger/Science Source

The technology for rapid gene sequencing and the ability to incorporate the genes from one organism into another [*recombinant DNA*, recognized with the 1980 Nobel Prize in Chemistry to Paul Berg of Stanford University (b. 1926), Walter Gilbert of Harvard University (b. 1932), and Frederick Sanger of the MRC Laboratory of Molecular Biology at Cambridge (1980–2013)], signaled the beginning of a race to produce biologically active peptides and proteins for human medicine in bacteria such as *Escherichia coli* by recombinant-DNA techniques. In these techniques, the protein-biosynthesis DNA code from the *human* genome is incorporated into the genome of *E. coli*, which can then be used to produce large amounts of the human protein of interest. (The photo shows an insulin-producing *E. coli.*) Herbert Boyer (b. 1936) and his colleagues at the University of California, San Francisco, in 1978 were the first to perfect this technology for the production of human insulin. Their discovery led to the founding of the biotechnology company Genentech and the licensing of this technology to Eli Lilly and Company for the commercial production of human insulin. This technology has been extended to the production of many other proteins, such as interferon, human growth hormone, and hepatitis-B vaccine in bacterial "factories." It is also possible to use genetically modified plants for the same purpose.

Focused Problems

27.20 (a) Give all of the mRNA codes for the peptide Phe-Arg-Gly-His-Trp.

(b) What are the DNA codes for the same peptide?

(c) What is the anticodon sequence in the tRNA for the Trp residue? (*Hint:* Be sure to specify the direction.)

27.21 In some tRNAs the anticodon contains an *inosine*. The heterocyclic base in inosine is *hypoxanthine.*

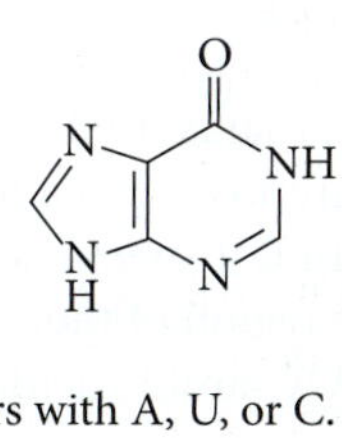

hypoxanthine

Inosine can form hydrogen-bonded base pairs with A, U, or C. This means that the inosine in tRNA can pair any of these bases in mRNA. Show the hydrogen-bonded base pair of hypoxanthine (a) with adenine; (b) with uracil.

27.7 HYDROLYSIS OF PEPTIDES AND PROTEINS

A. Complete Hydrolysis and Amino Acid Analysis

One reaction that all peptides and proteins have in common is amide hydrolysis. When peptides or proteins are heated with moderately concentrated aqueous acid or base, they are hydrolyzed to their component amino acids. This hydrolysis is typically carried to completion in 6 *M* aqueous HCl at 110 °C for 20–24 hours.

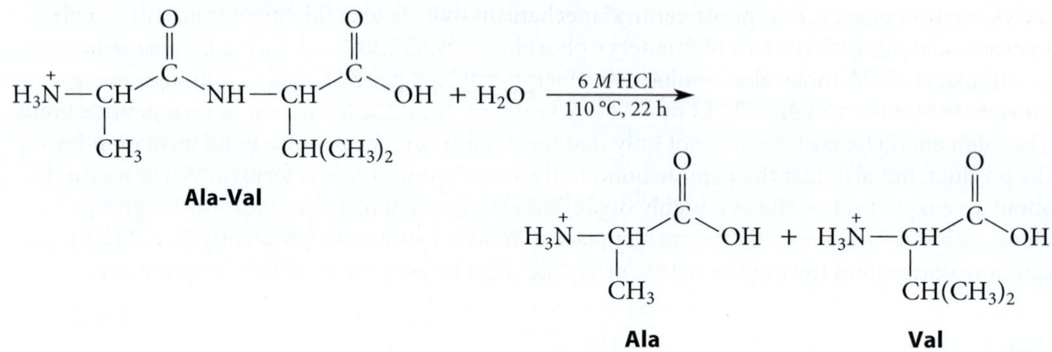

(27.48)

An important reason for hydrolyzing peptides or proteins of unknown structure is that the amino acid products that result from this hydrolysis can be separated and quantified. The determination of the identities and relative amounts of amino acids in a peptide or protein is called **amino acid analysis.**

Several reliable techniques are available for carrying out amino acid analysis. The methods in most common use today involve the conversion of the mixture of amino acids formed in the hydrolysis of a peptide into derivatives that are readily detected by spectroscopy. For example, in one method, the mixture of amino acids resulting from hydrolysis is allowed to react with 1-{[(6-quinolylamino)carbonyl]oxy}-2,5-pyrrolidinedione, a compound whose name in common usage is mercifully shortened to the acronym "AQC-NHS."

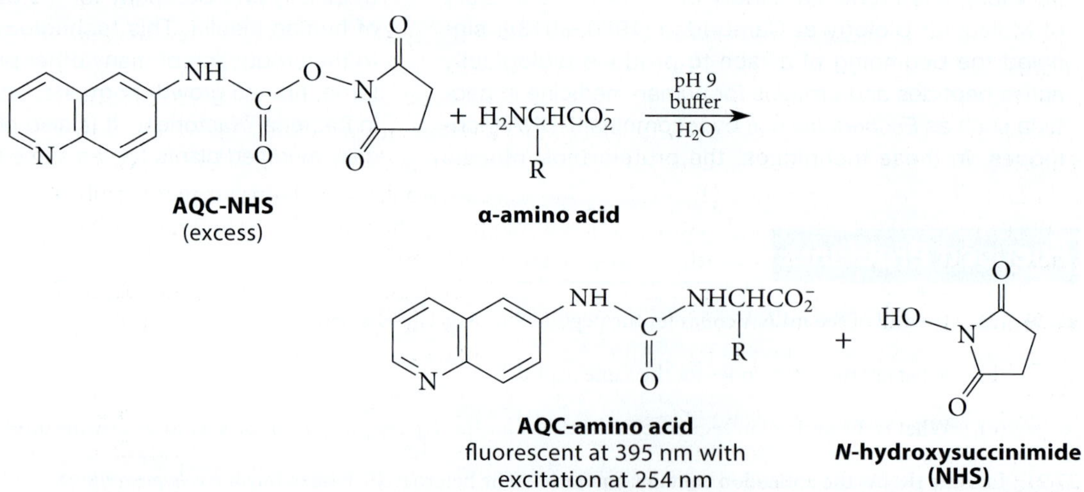

(27.49)

This is the same type of reaction that is used in peptide synthesis (Eq. 27.35, Sec. 27.6A). It results in the "tagging" of each amino acid in a hydrolysis mixture with the AQC group, which absorbs strongly at 254 nm in UV spectroscopy. This group is also *fluorescent*, emitting fluorescence at 395 nm, in the blue region of the visible spectrum. (Fluorescence was discussed in Sec. 15.2D.) After the various AQC-amino acids are separated, they can be quantified by measuring either UV absorption at 254 nm or fluorescence at 395 nm, because both techniques depend on the concentration of the absorbing or fluorescing species. (Fluorescence is more sensitive—that is, smaller quantities can be detected with fluorescence.)

Before the relative amounts of AQC-amino acids in a hydrolysis mixture can be determined, they must be separated. The separation of nearly 20 compounds of rather closely related structure might seem to be a daunting task, but conditions have been carefully worked out so that this separation is a routine matter. Again, liquid chromatography (Sec. 27.3C) is used; this type of liquid chromatography is called *C18 high-performance liquid chromatography*, or C18–HPLC.

Recall that, in chromatography, compounds are separated by their differential adsorptions on a stationary phase (Secs. 6.10A and 27.3D). In C18–HPLC, the stationary phase is a powder that consists of microscopic glass beads to which 18-carbon unbranched alkyl groups (that is, octadecyl groups) have been covalently bonded. We can represent the stationary phase schematically as follows:

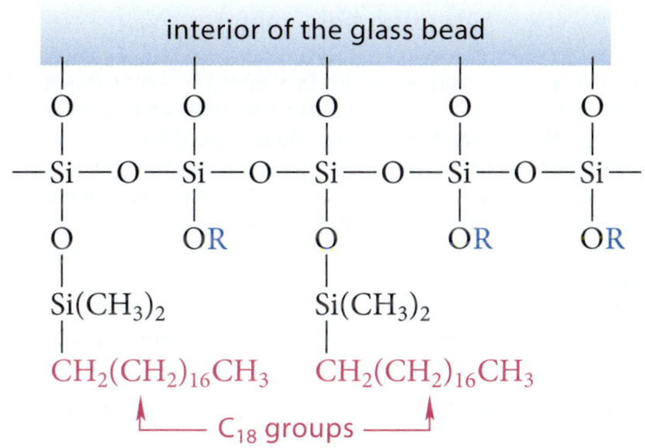

[R = some other group, for example., $(CH_3)_3Si$—] (27.50)

We can think of this stationary phase as glass with a hydrocarbon coat, and we can regard adsorption simply as a solubility phenomenon, with adsorption governed by the same noncovalent interactions that govern solubility. Compounds that are more soluble in hydrocarbons are adsorbed more strongly by the column. If we consider the structures of the various AQC-amino acids in this light, we would expect that the derivatives of amino acids with hydrocarbon side chains, such as leucine, isoleucine, and phenylalanine, would be adsorbed more strongly on the stationary phase. We would expect the AQC derivatives of amino acids with polar side chains, such as serine and aspartic acid, to be adsorbed less strongly for the same reasons that alcohols and carboxylic acids are not very soluble in hydrocarbons. This is exactly what happens. The C18–HPLC separation of a mixture of AQC-amino acids is shown in **Fig. 27.5**. The C18 column is first eluted with water. The AQC-amino acids with polar side chains are more soluble in the solvent and are less strongly attracted to the column, so they elute first. The AQC-amino acids with less polar, more hydrocarbon-like side chains are adsorbed by the column. They are eluted by changing the solvent composition gradually to about 20% acetonitrile; the adsorbed compounds are more soluble in acetonitrile than they are in water and are removed from the column by acetonitrile. As they emerge from the column, the various AQC-amino acids are detected by their fluorescence.

Once a standard mixture of AQC-amino acids has been through the C18 column and the relative fluorescences of the different compounds have been determined, the hydrolysate of a peptide of unknown structure can be "tagged" with AQC and treated in exactly the same way. The relative amounts of each amino acid are then calculated from the data.

As an example of amino acid analysis, imagine that a hypothetical peptide *P* has been hydrolyzed, tagged with AQC, and subjected to C18–HPLC, and that the results are as follows:

$$P: \quad \text{(Asp or Asn),Gly}_2\text{His,NH}_3\text{,Arg,Ala}_3\text{,Pro,Tyr,Val,Met,Lys,Ile,Leu,Phe,Trp}$$

According to this analysis, the peptide contains three times as much Ala and twice as much Gly as Arg, His, Lys, or the other amino acids present. The absolute number of each amino acid residue is unknown unless the molecular mass of the peptide is known. The relative order of the amino acid residues within the peptide is also unknown. In this sense, amino acid analysis is to the amino acid composition of a peptide as elemental analysis is to the molecular formula of an organic compound.

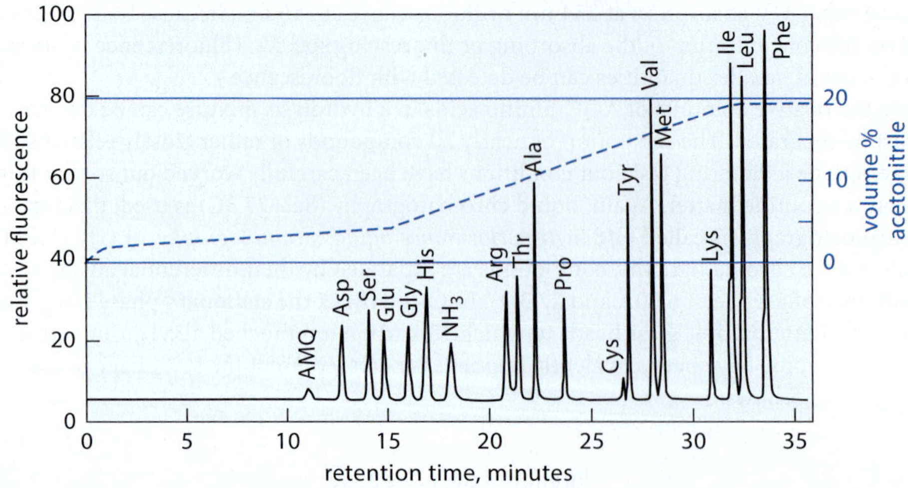

FIGURE 27.5 Separation of a mixture containing 50 picomoles (50×10^{-12} mol) of each AQC-amino acid by C18–HPLC in an aqueous buffer at pH 5.0 containing increasing percentages of acetonitrile. The percentage of acetonitrile in the eluting solvent is plotted in the blue overlay. Detection of the AQC-amino acids is by fluorescence. AQC-tryptophan (Trp) is not shown because tryptophan is destroyed by the strongly acidic conditions of peptide hydrolysis, but special base-hydrolysis methods can be used to detect Trp. AQC-glutamine (Gln) and AQC-asparagine (Asn) are also not shown because the side-chain amide groups of Asn and Gln are hydrolyzed under the conditions of amide hydrolysis; see Focused Problem 27.22. Cysteine (Cys) and lysine (Lys) are present at half the concentration of the other amino acids. The compound that elutes first, AMQ, is an ester-hydrolysis product of AQC-NHS that is formed in a side reaction during derivatization. The AQC-amino acids with the greatest hydrocarbon character are eluted last. Fluorescence intensity grows to the right because it is solvent-dependent and is greater in the solvents with a higher percentage of acetonitrile. (Data from Mary Bower, Purdue Biotechnology Facility)

Focused Problems

27.22 (a) Notice in peptide *P* (see previous discussion) that Asn and Asp are not distinguished by amino acid analysis. Explain. (*Hint:* Why is ammonia present in the amino acid analysis of peptide *P*?)

(b) What other pair of amino acids are not differentiated by amino acid analysis?

27.23 AQC-tryptophan is not shown in Fig. 27.5 because the indole ring does not survive the acid hydrolysis. In what general region of the chromatogram would you expect to find AQC-Trp if it were present? Explain.

27.24 The amino acids Lys and Cys, after "tagging" with AQC, are each found to contain *two* AQC groups. Explain; your explanation should involve the structures of the AQC-amino acids.

B. Enzyme-Catalyzed Peptide Hydrolysis

Peptides can be hydrolyzed at specific amino acid residues by treating them with certain enzymes, called **proteases**, **peptidases**, or **proteolytic enzymes**. These enzyme-catalyzed hydrolysis reactions are very useful in determining the structures of peptides, as we show in Sec. 27.8C.

One of the most widely used proteases is the enzyme *trypsin*. This is a digestive enzyme (obtained commercially from cattle), the biological role of which is to catalyze the hydrolysis of dietary proteins in the intestinal tract. Trypsin catalyzes the hydrolysis of peptides or proteins at the carbonyl group of arginine or lysine residues, provided that these residues are (a) not at the amino end of the protein, and (b) not followed by a proline residue. (In the following equations, Pep^{N} stands for the amino-terminal part of the peptide—the part attached by an amide bond to the α-nitrogen of arginine or lysine, in this case—and Pep^{C} stands for the carboxy-terminal part of the peptide.)

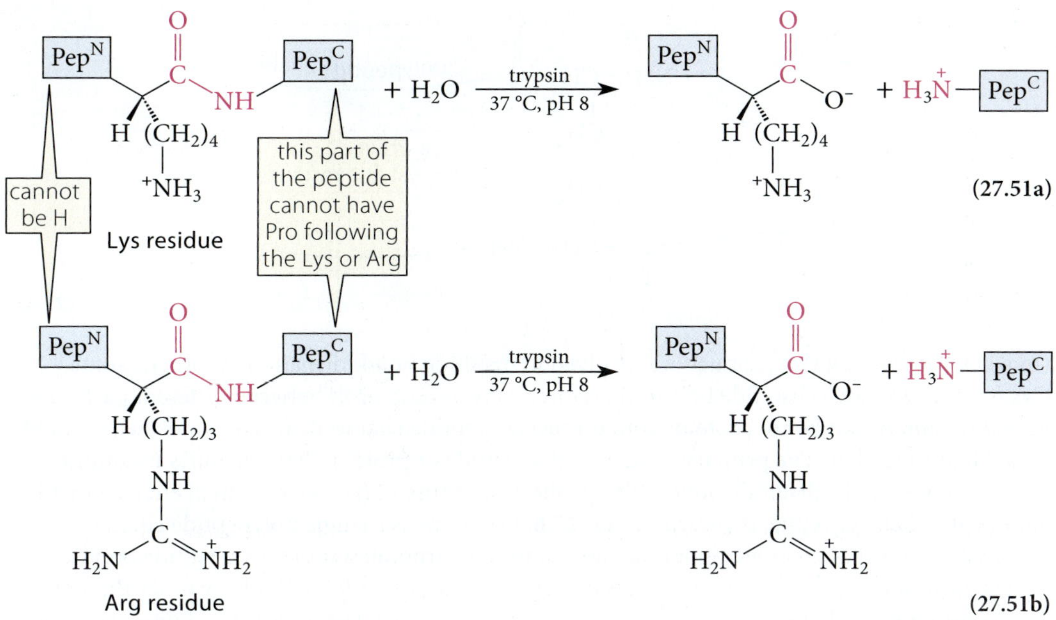

(27.51a)

(27.51b)

(The mechanism of trypsin-catalyzed hydrolysis and the reason for its lysine and arginine specificity are discussed in Sec. 27.10.) Because trypsin catalyzes the hydrolysis of peptides at internal rather than terminal residues, it is called an **endopeptidase**. (Enzymes that cleave peptides only at terminal residues are called **exopeptidases**. The prefixes *endo* and *exo* come from Greek roots meaning "inside" and "outside," respectively.)

Biochemists have developed an arsenal of different proteolytic enzymes that are used for the hydrolysis of peptides at specific sites. For example, chymotrypsin, another mammalian digestive protein related to trypsin, is used to catalyze the hydrolysis of peptides at amino acid residues with aromatic side chains and, to a lesser extent, at residues with large hydrocarbon side chains. Therefore, chymotrypsin cleaves peptides at Phe, Trp, Tyr, and, occasionally, at Leu and Ile residues. An important endopeptidase from the microorganism *Staphylococcus aureus* catalyzes the hydrolysis of peptides at glutamic acid residues.

Focused Problems

27.25 A peptide *P* has the sequence of amino acids E-R-G-A-N-I-K-K-H-E-M. What products would be formed if this peptide were subjected to trypsin-catalyzed hydrolysis?

27.26 When a peptide *Q* with the amino acid analysis (A,F,G_2,I,K,N,P,R,S,Y) is treated with trypsin, three new peptides are formed (amino acid analysis in parentheses): *T1*(A,F,R,S), *T2*(G,I,K), and *T3*(G,N,P,Y). When peptide *Q* is treated with chymotrypsin, four peptides are formed: *C1*(A,N,R,S,Y), *C2*(F,K), *C3*(G,I), and *C4*(G,P). What can you deduce about the order in which the amino acids in *P* are connected? Explain. What are the points of uncertainty?

27.8 PRIMARY STRUCTURE OF PEPTIDES AND PROTEINS

A. The Elements of Primary Structure

The structures of molecules as large as peptides and proteins can be described at different levels of complexity. The simplest description of a peptide or protein structure is its covalent structure, or **primary structure**. The most important aspect of any primary structure is the **amino acid sequence**, which is the order in which the amino acid residues are connected.

Peptide bonds are not the only covalent bonds that can connect amino acid residues. **Disulfide bonds** (Sec. 11.10B) link cysteine residues in different parts of a sequence.

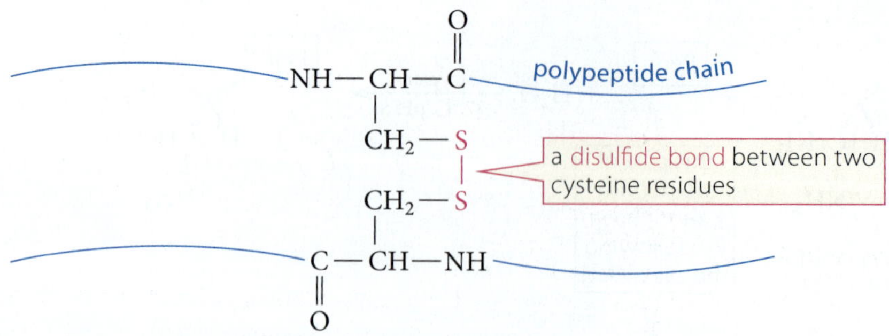

(27.52)

(In some proteins, all Cys residues are involved in disulfide bond formation; in others, some Cys residues are not.) Disulfide bonds, therefore, serve as crosslinks between different parts of a peptide chain. A number of proteins contain several peptide chains; disulfide bonds help to hold these chains together. The primary structure of a peptide or protein, then, includes its amino acid sequence and its disulfide bonds. The primary structure of *lysozyme*, a small enzyme that is abundant in hen egg white, is shown in **Fig. 27.6**. Lysozyme is a single polypeptide chain of 129 amino acids that includes eight cysteine residues incorporated into four disulfide bonds.

The disulfide bonds of a protein are readily reduced to free cysteine thiols by other thiols. Two commonly used thiol reagents are 2-mercaptoethanol ($HSCH_2CH_2OH$) and dithiothreitol (known to biochemists as DTT, or Cleland's reagent).

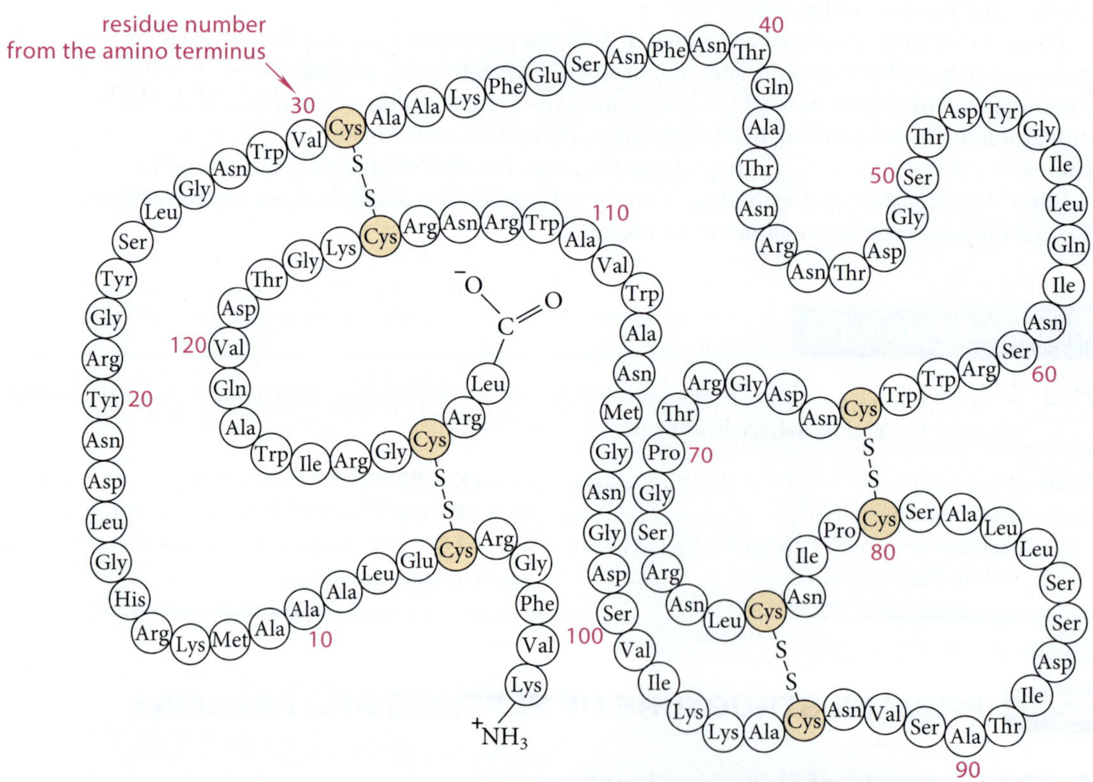

FIGURE 27.6 The primary structure of the enzyme lysozyme from hen egg white. Physiologically, lysozyme catalyzes the hydrolysis of polysaccharides in bacterial-cell walls. Different variants of this enzyme are found in tears, nasal mucus, and even viruses—anywhere antibacterial action is important. Lysozyme is one of the smallest known enzymes. Individual amino acid residues, connected by peptide bonds, are numbered from the amino terminus. The cysteine residues involved in the disulfide bonds are shown in orange.

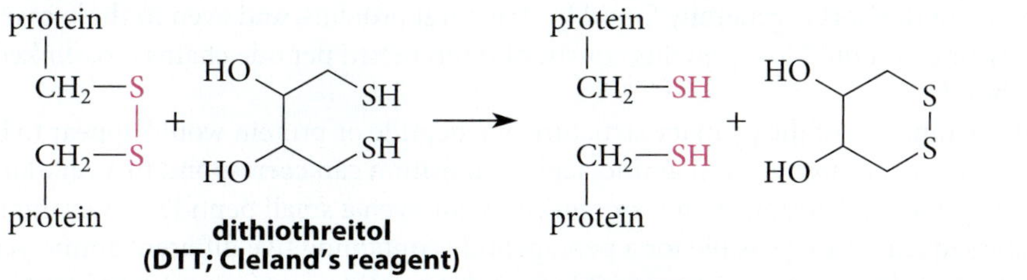

(27.53)

The reaction in Eq. 27.53 is a very clever application of the proximity effect (Sec. 12.8). The equilibrium constant between one thiol–disulfide pair and another would typically be close to 1.0. However, the formation of six-membered rings is favored entropically. As a result, the equilibrium strongly favors the right side of the equation.

This reaction is a biological example of the thiol–sulfide equilibrium shown in Eqs. 11.89a and b, Sec. 11.10B. Typically, when the extraneous thiols are removed, the thiols of the protein spontaneously reoxidize in air back to disulfides.

A Practical Example of Disulfide-Bond Reduction

Radu Razvan/Shutterstock

An interesting example of the biological effects of disulfide-bond reduction occurs in the use of permanent-wave preparations to curl the hair. Hair (protein) is treated with a thiol solution; this solution is responsible for the unpleasant smell of permanents. The thiol solution reduces the disulfide bonds in the hair. With the hair in curlers, the disulfides are allowed to reoxidize. The hair is then set by disulfide-bond reformation into the shape dictated by the curlers. Only after a long time do the disulfide bonds re-scramble to their normal configuration, when another permanent becomes necessary.

An industrial example of the use of disulfide bonds is the process of vulcanization (Sec. 15.5), which introduces disulfide bonds into both natural rubber and synthetic polymers. Vulcanization provides a polymer with greater strength and rigidity.

Disulfide bonds are the most common type of crosslink between peptide chains, but other types of crosslinks are possible. For example, the side-chain amino group of a Lys residue, or the side-chain carboxy groups of Asp or Glu residues, could form amide bonds that would create branches in peptide chains. One of the best-known examples of peptide crosslinking occurs in the bacterial cell wall. A rigid, two-dimensional network is formed by an all-glycine pentapeptide connected by amide bonds to the side-chain amino group of a Lys residue in one peptide chain and the carboxy group of a D-alanine residue in another.

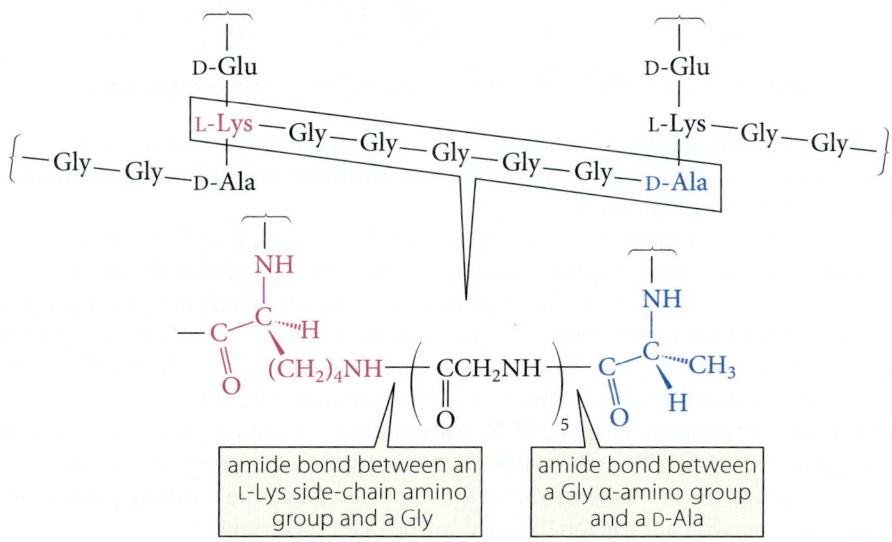

(27.54)

Crosslinking of this sort is generally found in structural proteins, and even in these cases it is relatively uncommon. Most proteins consist of unbranched peptide chains crosslinked by disulfide bonds.

The determination of the primary structure of a peptide or protein would appear to be a complex problem, because a given amino acid composition can correspond to a very large number of amino-acid arrangements, or *sequences*, for even a small peptide. For example, 120 unique sequences are possible for a pentapeptide containing five different amino acids, and more than 3.6 *million* sequences are possible for a decapeptide containing 10 different amino acids! Despite this apparent complexity, there are well-established methods for determining the primary structures of peptides. The determination of a primary structure is called **peptide sequencing**. In Sec. 27.8B, we discuss the use of mass spectrometry (Sec. 13.6) for peptide sequencing, and in Sec. 27.8C, we discuss a well-established chemical method. Finally, in Sec. 27.8D, we show how the primary amino acid sequences of large proteins are determined.

B. Peptide Sequencing by Mass Spectrometry

A typical peptide can be sequenced in the mass spectrometer using electrospray ionization (ESI), which we can regard as a chemical ionization (CI) technique (Sec. 13.6D). The peptide in the gas phase is protonated to give an $M+1$ ion, from which the molecular mass M of the peptide is determined. In a special type of mass spectrometer, the $M+1$ ion is subjected to a technique called **tandem mass spectrometry**, or simply **MS–MS**. In this technique, the ion is first energized in some way. Increasing the energy of this ion causes it to undergo fragmentation. For example, in one method, the $M+1$ ion is routed into a collision cell into which a noble gas such as argon or xenon is admitted. The collision of the ion with the gas molecules causes it to gain energy and undergo fragmentation. (This is something like what happens to a window if you throw baseballs at it.) Recall that, in fragmentation, a cationic fragment and a neutral fragment are produced, and the mass spectrometer detects the cation. Peptide fragmentation can in principle occur at every bond along the peptide backbone. When fragmentation at a particular bond takes place, the charge can remain with the amino-terminal fragment, or it can remain with the carboxy-terminal fragment. Using a four-residue peptide as an example, the possible fragments are categorized as follows:

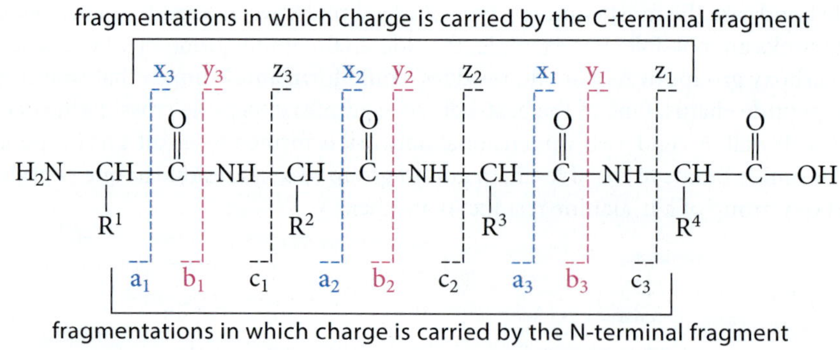

(27.55)

Each fragment shown in Display 27.55 arises from a *single* fragmentation of the $M+1$ ion. In other words, some $M+1$ ions undergo b_2 fragmentation, others undergo y_2 fragmentation, others undergo c_3 fragmentation, and so on.

In many cases, the b-type and y-type fragmentations occur most frequently. That is, *fragmentation often occurs at the peptide bond.* To illustrate the type of data obtained from a peptide mass spectrum, let's imagine, for simplicity, a case in which only b-type fragmentation is significant. An $M+1$ ion can be formed by protonation of any one of the carbonyl groups of the peptide as well as protonation of basic side-chain groups. In other words, the $M+1$ ion is actually a mixture of ions that differ in their sites of protonation. Consider, for example, the b_2 fragmentation of the peptide in Display. 27.55. Most of the protonation probably occurs at the carbonyl oxygen (Sec. 21.5). The carbonyl-protonated peptide can undergo a cyclization in which the carbonyl oxygen of the previous residue acts as a nucleophile. The resulting product is a protonated *oxazolone*, which is the ion detected by the mass spectrometer.

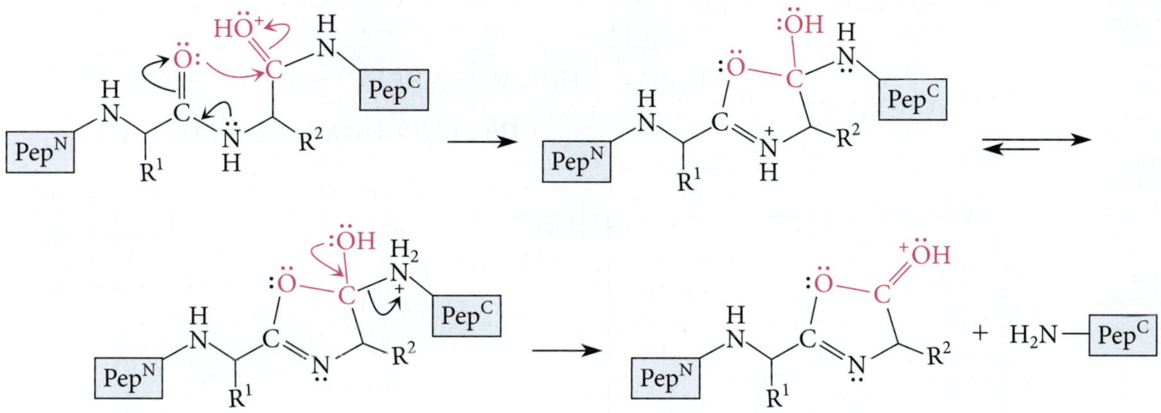

a protonated oxazolone (27.56a)

Another possible mechanism for b-type cleavage involves protonation of an amide nitrogen. Amide nitrogens are *much* less basic than amide carbonyl oxygens because all amide resonance is obliterated in the nitrogen-protonated species. However, once protonated, this nitrogen can serve as a good leaving group in the formation of an *acylium ion* (Eq. 16.31, Sec. 16.4F). The acylium ion has the same molecular mass as the protonated oxazolone.

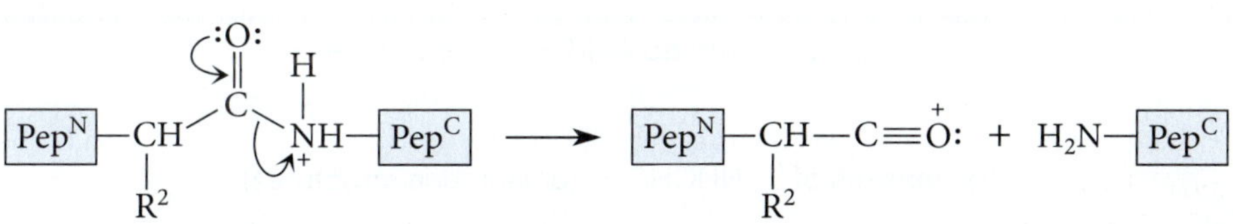

an *N*-protonated amide an acylium ion (27.56b)

Now imagine that b-type fragments are produced from other M + 1 ions that are protonated at other amide carbonyls or nitrogens, and further imagine that each of these fragmentations occurs in a measurable amount. As a result, fragmentation of the M + 1 ions results in a family of b-type ions—b_1, b_2, and b_3—the masses of which *differ by the atomic masses of the intervening residues*:

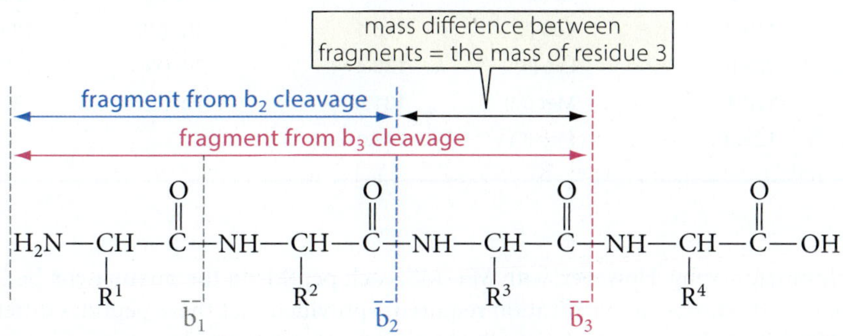

(27.57)

The sequence from the C-terminus can then be read directly from the mass spectrum, right-to-left, using the mass differences between the peaks. Such a sequence determination is illustrated in **Fig. 27.7** for a synthetic *N*-acylated tripeptide with the structure shown in the figure. The masses of the amino acid residues used to make the identification are given in **Table 27.3**.

When several types of fragmentation occur, each type of fragmentation gives a family of peaks, but each peak in a given family differs from the next by a residue mass. If the mass spectra are complex because of the occurrence of multiple fragmentation modes, computer programs are available that can assist in the analysis.

In MS–MS, all of the sequence information is obtained in one experiment. But MS–MS has another, even more powerful, capability. Suppose a large peptide is subjected to trypsin-catalyzed hydrolysis, and four or five smaller peptides are obtained. Traditionally, the sequencing of peptides required pure samples, and purification of a complex peptide mixture typically required

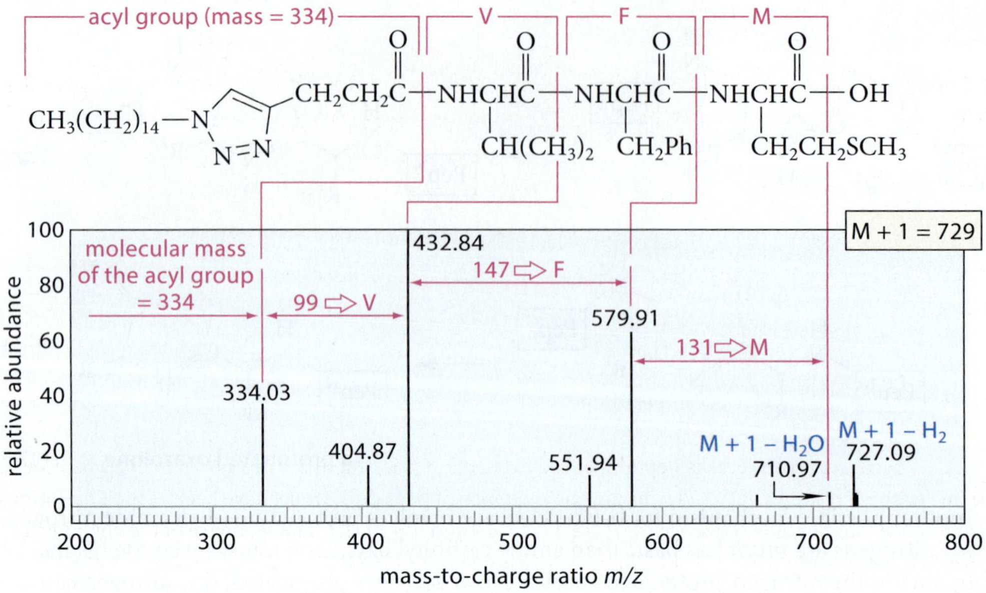

FIGURE 27.7 The MS–MS of an *N*-acylated tripeptide, V-F-M. The M + 1 ion was observed before it was subjected to fragmentation and found to have a mass of 729. The mass differences between the major peaks correspond to the residues between successive b-type fragmentation points. The fragmentation points are marked with dashed lines. The mass differences correspond to the residue masses in Table 27.3.

TABLE 27.3 Amino Acid Residue Masses

the masses of —NHCHC(=O)— (with R on the CH) (or isomeric structures),

where R = the residue side chains shown in Table 27.1, Sec. 27.1A

Residue	Mass	Residue	Mass	Residue	Mass
Ala (A)	71.0	Gly (G)	57.0	Pro (P)	97.1
Arg (R)	156.1	His (H)	137.1	Ser (S)	87.0
Asn (N)	114.0	Ile (I)	113.1	Thr (T)	101.0
Asp (D)	115.0	Leu (L)	113.1	Trp (W)	186.1
Cys (C)	103.0	Lys (K)	128.1	Tyr (Y)	163.1
Gln (Q)	128.1	Met (M)	131.0	Val (V)	99.1
Glu (E)	129.0	Phe (F)	147.1		

laborious chromatography. However, with MS–MS, each peptide in the mixture can be sequenced directly from the mixture—no purification required—provided that these peptides differ in mass. The MS–MS analyzer can sequester the M + 1 ions of each peptide in turn, energize them, and so produce separate mass spectra for all of them in the same experiment. In effect, the mass spectrometer performs the separation on the basis of the differing masses of the M + 1 ions. This high-throughput capability of MS–MS has revolutionized peptide sequence analysis.

The number of residues that can be sequenced by MS–MS depends on the specific case. Sequencing 7–10 residues is often possible, and 20 residues can be sequenced in favorable cases. Longer peptides, however, can be hydrolyzed with trypsin or other proteolytic enzymes to produce shorter peptides. The order of the shorter peptides in the overall sequence can be established by comparing the results of two or more digests resulting from the use of proteolytic enzymes with differing specificities. This process is called the *method of overlapping peptides.* This strategy is illustrated by Study Problem 27.1.

Study Problem 27.1

A peptide *P* with the amino acid composition (A,E,F,G_2,H,K,L,M,P,R,V,Y) did not yield to MS–MS sequencing, so it was hydrolyzed with trypsin to three peptides, *T1*, *T2*, and *T3*, which were found by MS–MS to have the following structures:

T1: A-H-K *T2:* E-M-V *T3:* (L,P)-F-G-G-Y-R

(The parentheses in *T3* mean that the order of L and P could not be determined.) Hydrolysis of *P* catalyzed by chymotrypsin yielded three peptides, *C1*, *C2*, and *C3*, which were sequenced by MS–MS and found to have the following structures:

C1: A-H-K-L-P-F *C2:* G-G-Y *C3:* R-E-M-V

What is the amino acid sequence of peptide *P*?

Solution Because trypsin breaks peptides at the C-terminal side of K and R, and because only peptide *T2* does not have one of these residues at its C-terminus, the sequence of peptide *T2* must have been at the C-terminus of peptide *P*. Therefore, the two possible sequences of *P* are *T1*-*T3*-*T2* and *T3*-*T1*-*T2*. The sequence of the chymotryptic peptide *C3* overlaps parts of the sequences of *T3* and *T2*. This overlap shows that, in the sequence of peptide *P*, an R residue precedes the E residue and that the sequence of *T3* precedes the sequence of *T2* in peptide *P*. The chymotryptic peptide *C1* resolves the ambiguity in the positions of L and P in peptide *T3*. Because the sequence of *C1* overlaps parts of the sequences of *T1* and *T3*, this sequence also confirms the sequence order *T1*-*T3*. Therefore, the sequence of peptide *P* is *T1*-*T3*-*T2*, or

P: A-H-K-L-P-F-G-G-Y-R-E-M-V

Focused Problems

27.27 (a) Give the m/z values of the fragment ions expected from b-type fragmentation of an M + 1 ion of the peptide N-F-E-S-G-K.

(b) Give the m/z values of the fragment ions expected from y-type fragmentation of an M + 1 ion of the peptide in part (a). All y-type fragments contain a protonated terminal amino group—that is, H_3N^+—.

27.28 Give the curved-arrow mechanism for the formation of each of the following fragment ions in Fig. 27.7 from an M + 1 ion.

(a) The fragment at $m/z = 551.94$. [*Hint:* This fragment results from an a-type cleavage (Display 27.55).]

(b) The fragment ion at $m/z = 710.97$

(c) The fragment ion at 727.09. Show how this mechanism might be tested with a deuterium-labeled peptide.

C. Peptide Sequencing by the Edman Degradation

Prior to the advent of peptide sequencing by MS–MS, the standard sequencing method was a chemical process called the **Edman degradation**, named after Pehr Victor Edman (1916–1977), a Swedish biochemist who devised the method in 1952. In an Edman degradation, the peptide is treated with *phenyl isothiocyanate* (often called the **Edman reagent**). The peptide reacts with the Edman reagent at its amino groups to give thiourea derivatives. Although reaction with the Edman reagent also occurs at the side-chain amino groups of lysine residues (see Problem 27.72), only the reaction at the terminal amino group is relevant to the degradation. (As before, the abbreviation Pep^C is used for the carboxy-terminal part of a peptide.)

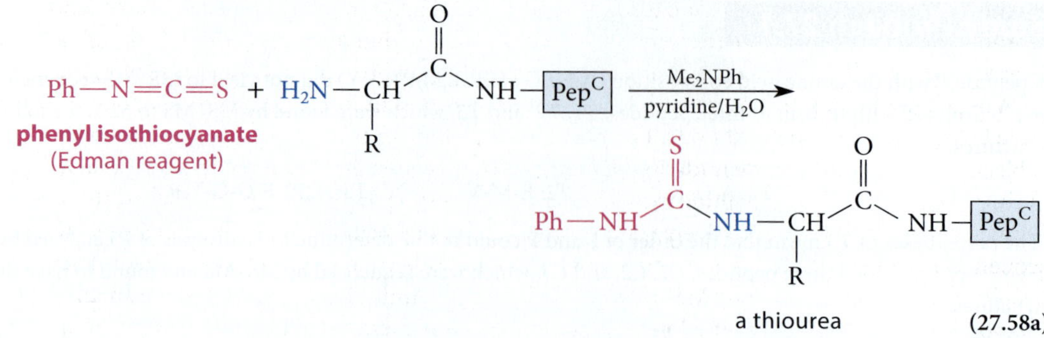

(27.58a)

This reaction is exactly analogous to the reaction of amines with *isocyanates*, the oxygen analogs of *isothiocyanates* (Eq. 23.87, Sec. 23.11D). Any remaining phenyl isothiocyanate is removed, and the modified peptide is then treated with anhydrous trifluoroacetic acid. As a result of this treatment, the sulfur of the thiourea, which is nucleophilic, displaces the amino group of the adjacent residue to yield a five-membered heterocycle called a *thiazolinone*; the other product of the reaction is *a peptide that is one residue shorter*.

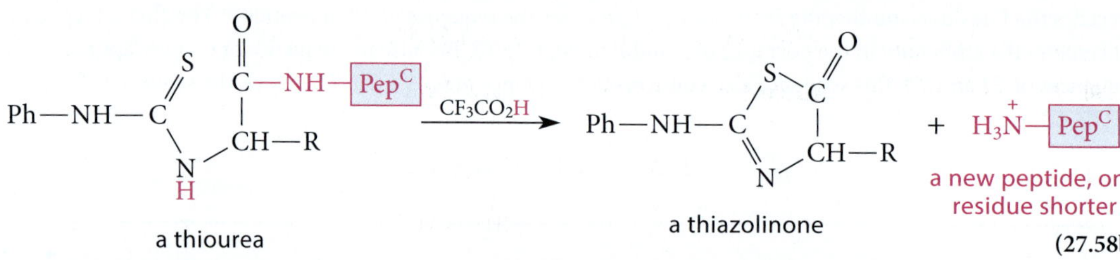

(27.58b)

When treated subsequently with aqueous acid, the thiazolinone derivative forms an isomer called a **phenylthiohydantoin**, or **PTH**. This probably occurs by reopening of the thiazolinone to the thiourea, followed by ring formation involving the thiourea nitrogen. Notice in this and the previous equation the formation of five-membered rings by intramolecular reactions.

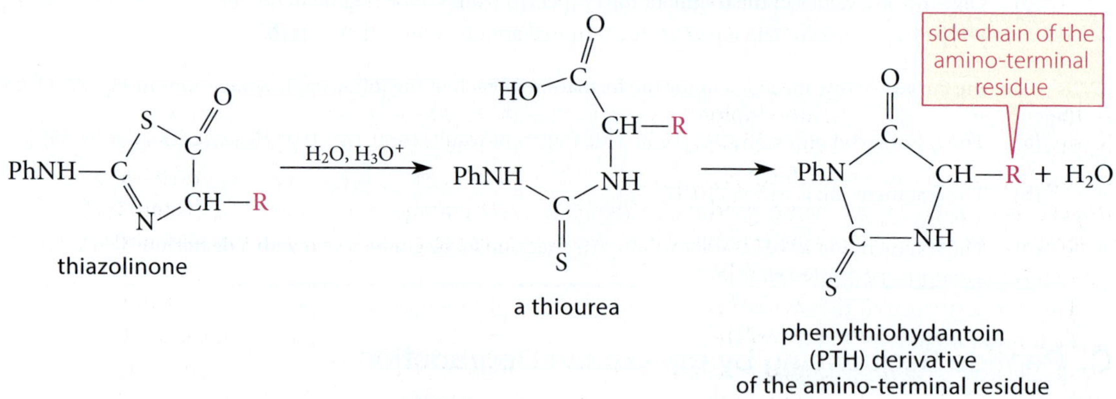

(27.58c)

Because the PTH derivative carries the characteristic side chain of the amino-terminal residue, identification of the PTH identifies the amino acid residue that was removed. Methods for identifying PTH derivatives by liquid chromatography are well established. The peptide liberated in Eq. 27.58b can be subjected in turn to the Edman degradation again to yield the PTH derivative of the next amino acid and a new peptide that is shorter by yet another residue.

In principle, the Edman degradation can be continued indefinitely for as many residues as necessary to define completely the sequence of a peptide. In practice, because the yields at each step are not perfectly quantitative, an increasingly complex mixture of peptides is formed with each successive step in the cleavage, and, after a number of such steps, the results become ambiguous. Therefore, the number of residues in a peptide sequence that can be determined by the Edman method is limited. Nevertheless, instruments are now in use that can apply Edman chemistry to the structure determination of peptides in a highly standardized, automated, and reproducible form. In such instruments, sequence determination of 20 residues is common, and the sequence determination of as many as 60 or 70 amino acid residues is sometimes possible.

When using the Edman degradation, a researcher has to wait for the completion of one cycle before initiating the next cycle. Furthermore, a peptide or protein must be purified before subjecting it to the Edman degradation. Unless multiple sequencing instruments are available in the laboratory, only one peptide can be sequenced at a time. Because of its chemistry, the Edman method works progressively from the amino end of a peptide, and, if a carboxy-terminal sequence is needed, the Edman method is of no use. (Another limitation is the subject of Focused Problem 27.30.) Sequencing by MS–MS has none of these limitations. However, the Edman method is extremely reliable, and longer peptides can be sequenced with the Edman technique than with MS–MS. Instruments for Edman sequencing are also much less expensive than MS–MS instruments. Although the Edman method is still used, it is used much less than it once was.

Focused Problems

27.29 Using the curved-arrow notation, write in detail the mechanisms for the reactions in the following equations:

(a) Eq. 27.58a (b) Eq. 27.58b (c) Eq. 27.58c

27.30 Some peptides found in nature have an amino-terminal acetyl group (*red*):

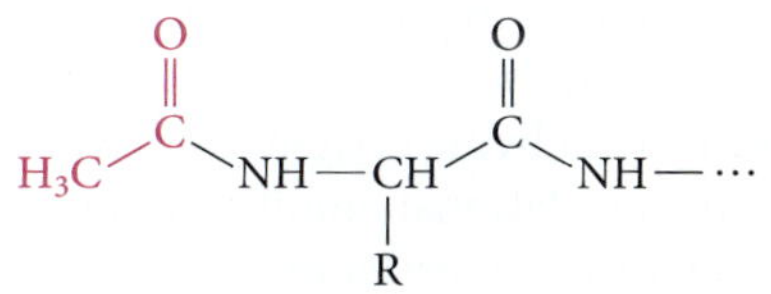

(a) Can these peptides undergo the Edman degradation? Explain.

(b) Does *N*-acylation have any adverse effect on sequencing by MS–MS? Explain.

D. Protein Sequencing

In the early history of structural biology, the complete primary sequence of a protein was determined by isolation of the individual peptide chains of the protein followed by tryptic and chymotryptic hydrolysis to afford overlapping peptides of manageable size. These peptides were then sequenced by the Edman degradation. Determining the complete sequence of a protein took several years of work. Nevertheless, many protein sequences were determined in this way, and these have proved to be quite reliable.

The development, in the late 1970s, of methods for rapidly sequencing the nucleotides in DNA made it possible to read the complete sequences of proteins directly from DNA sequences. The biosynthesis of proteins is coded within DNA by contiguous sequences of nucleotides in which each amino acid is represented by a three-base code. (This process was explained in Sec. 27.6B.) In principle, then, identification of the gene for a protein in a DNA sequence results automatically in the knowledge of the protein sequence. The one fly in the ointment, however, is that genetic DNA contains not only the code for protein sequences but also interspersed noncoding regions (introns). What is needed for reading the protein primary sequence is a DNA from which the introns have been excised—that is, DNA that contains only the protein-sequence information. Using a viral enzyme called *reverse transcriptase*, which produces DNA from an RNA template (Sec 26.5B), a **complementary DNA**, or **cDNA**, is produced from mRNA. (Recall that intron sequences have been excised from mRNA.) The sequence of a cDNA, then, contains all of the information necessary to read the sequence of a protein for which it codes. Almost all protein sequences are now determined by "translating" their cDNA codes.

As part of the effort to sequence the DNA of the complete genomes of many different species (including humans; see Sec. 26.5B), databases have been developed that show the protein-coding regions. When a researcher isolates an unknown protein from one of these species, a partial sequence is in many cases sufficient to obtain a unique match to a DNA coding region, from which the entire sequence of the protein can then be read. Standard computer software for matching partial sequences to genomic DNA sequences is available. This is one of the many reasons for the ascendancy of peptide sequencing by MS–MS. Tryptic digestion of a protein followed by the *simultaneous* partial sequencing, without purification, of the peptide products is often sufficient to establish a cDNA match, from which the sequence of the entire protein can then be determined.

Chemical Biology Topic

E. Posttranslational Modification of Proteins

Once proteins are produced within a living cell, the structures of many of them are modified by subsequent reactions. These reactions are called collectively **posttranslational modifications** because they occur after *translation*—that is, after the biosynthesis of the protein itself. More than 200 posttranslational modifications have been documented. The DNA-coding sequence for a protein carries no information about any posttranslational modifications that might occur. Currently, the only way to elucidate the posttranslational modifications of a protein is by studying the protein itself.

Most posttranslational modification reactions involve the chemical alteration of side-chain functional groups. These types of modifications serve many roles. They can provide unique structures so that another molecule can recognize the protein. They can serve as "molecular switches," turning on or off enzyme activity. They can control the lifetime of a protein within the cell, and they can be involved in *protein trafficking*—that is, in directing a protein to the proper destination within a cell. In short, posttranslational modifications increase the diversity of amino acid side chains available in living systems. MS–MS is widely used in studying these alterations. We here illustrate posttranslational modification with two of the most common types: *protein phosphorylation* and *protein glycosylation.*

Protein Phosphorylation Proteins are phosphorylated by the transfer of a γ-phosphate group from ATP to a serine, tyrosine, cysteine, or (more rarely) a histidine residue of proteins. These reactions are catalyzed by *protein kinases.* (The mechanism of phosphate transfer and the action of kinases was discussed in Sec. 25.7A; see Eq. 25.45a.) An analysis of the human genome predicts that there are more than 500 different protein kinases. Phosphorylation of serine is illustrated in the following equation.

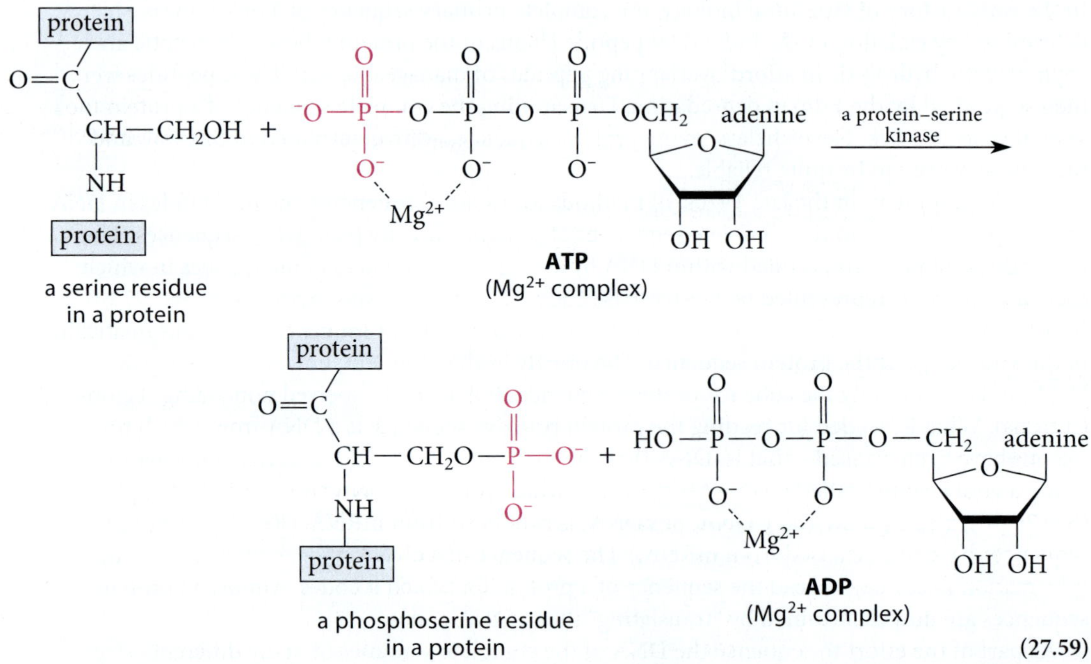

(27.59)

This reaction has an equilibrium constant $K'_{eq} \sim 100$ at pH = 7, the favorable equilibrium resulting from the anhydride character of ATP (see Sec. 25.8B).

Dephosphorylations of phosphoserine, phosphotyrosine, or phosphothreonine occur by an enzyme-catalyzed reaction of the phosphorylated amino acid residue with water. The enzymes that catalyze this reaction are called *phosphatases*; more than 100 phosphatases occur in humans. Phosphatases were discussed on Sec. 25.6A.

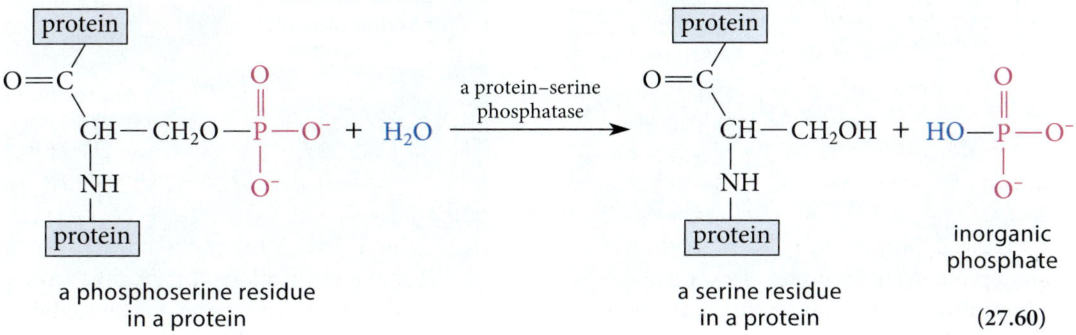

(27.60)

Phosphorylation in biology has several roles. One role is a regulatory function. Phosphorylation converts a neutral Ser, Tyr, Cys, or His residue into a residue with two negative charges. This charged state can result in a conformational change in the protein driven by the electrostatic and/or hydrogen-bonding interaction of the phosphate with a positively charged residue, such as a lysine or an arginine residue elsewhere in the protein, as shown schematically in **Fig. 27.8**.

An impressive example of a phosphorylation-induced conformational change in a protein occurs in the enzyme *glycogen phosphorylase*. The biological role of this enzyme is to initiate the

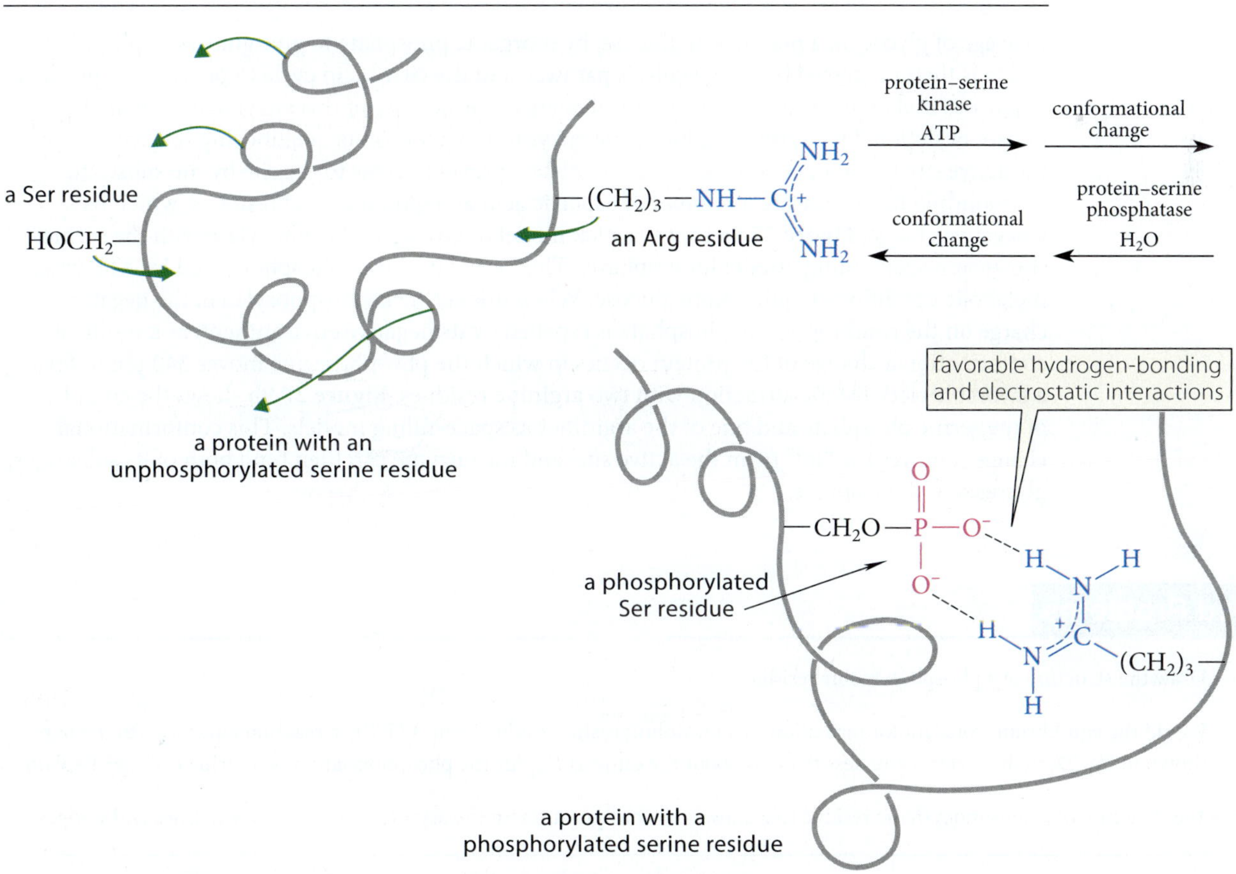

FIGURE 27.8 A diagram showing how phosphorylation of a serine residue can induce a conformational change in a protein. The gray strand represents the protein chain, and the green arrows show how the chain moves in the conformational change. Phosphorylation causes the change, and dephosphorylation reverses it.

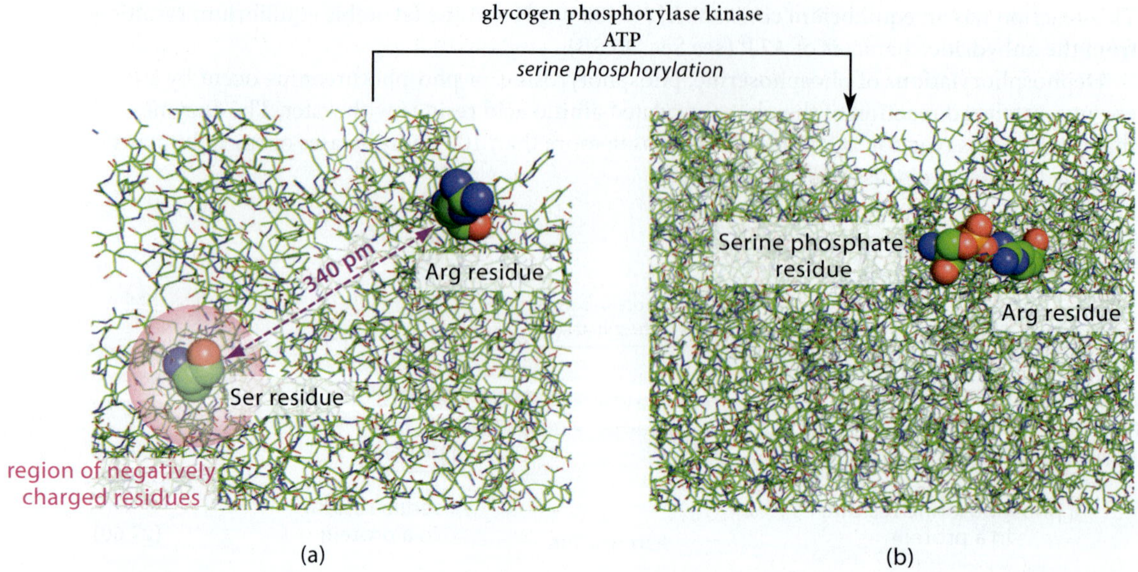

FIGURE 27.9 Part of the structure of the enzyme glycogen phosphorylase showing a phosphorylation-induced conformational change. The bonds of the enzyme are shown as green stick models except for one serine and one arginine residue, which are shown as space-filling models. (a) The enzyme in the unphosphorylated, inactive, state. The pink halo indicates a region of negatively charged residues surrounding the serine. When the serine is phosphorylated, the resulting serine phosphate is repelled by the surrounding negative charge. A conformational change occurs in the enzyme that brings the serine phosphate close to the arginine, with which it has a favorable electrostatic interaction (see Fig. 27.8). (A second arginine in the same region is not shown.) This conformational change unblocks the active site and allows the enzyme to become active.

cleavage of glycogen, a polymer of glucose, by inorganic phosphate to give glucose-1-phosphate, which is then processed by the glycolysis pathway and the citric acid cycle to provide the products of glucose metabolism along with chemical energy. The activity of this enzyme is regulated by phosphorylation. One serine residue of the enzyme and a few of its neighboring residues cover the active site like a lid on a bucket and completely prevent access to the site by the substrates. Surrounding this serine are a number of aspartic acid and glutamic acid residues, which have a negative charge. **Figure 27.9a** shows a stick model of glycogen phosphorylase with the serine shown as a space-filling model for emphasis. This serine becomes phosphorylated by ATP when metabolic conditions require more glucose. When this serine is phosphorylated, the negative charge on the resulting serine phosphate is repelled by its negative environment; as a result, a conformational change of the protein occurs in which the phosphoserine moves 340 pm to form a favorable electrostatic attraction with two arginine residues. **Figure 27.9b** shows the complex of the serine phosphate and one of the arginines as space-filling models. This conformational change removes the "lid" from the active site, and the enzyme can then bind both of its substrates, glycogen and phosphate.

Focused Problems

27.31 (a) Draw the structure of a phosphotyrosine residue.

(b) Would the equilibrium constant for formation of a phosphotyrosine residue from ATP (by a reaction analogous to the one shown in Eq. 27.59) be greater than, less than, or about the same as K_{eq} for the phosphorylation of a serine residue? Explain.

27.32 Draw the structure of a phosphocysteine residue that shows the configuration on the asymmetric carbons with lines and wedges.

Protein Glycosylation The most prevalent posttranslational modification of proteins is the attachment of oligosaccharides (Sec. 24.11B) to give **glycoproteins.** Glycoproteins serve a variety of functions. Perhaps the glycoproteins in most common experience are the mucins, proteins that are constituents of mucus. Glycosylation is responsible for the slimy feel of mucus. Glycoproteins are important in cell–cell recognition, in the immune response, and in connective tissue. Blood type depends on the oligosaccharides attached to proteins and lipids of red-blood-cell membranes. A few enzymes and peptide hormones are glycoproteins. The determination of the structure of a glycoprotein generally involves application of a combination of specific glycoside-hydrolyzing enzymes and mass spectrometry.

The attachment of a saccharide to a protein is called **glycosylation**. There are two broad types of glycosylation, ***N*-glycosylation** and ***O*-glycosylation**. In *N*-glycosylation, an oligosaccharide is attached by the reducing end of an *N*-acetylglucosamine residue to the side-chain amide nitrogen of an asparagine (Asn) residue that is part of a sequence Asn-X-Ser/Thr; X is any amino acid, and the residue on the carboxy-terminal side of residue X must be either serine or threonine.

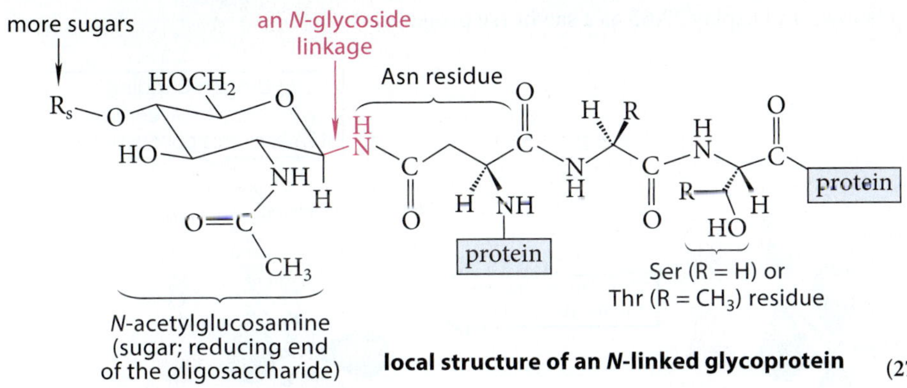

local structure of an *N*-linked glycoprotein (27.61)

The oligosaccharide, prior to its attachment to a protein, is built up, residue-by-residue, as a *dolichol diphosphate* derivative.

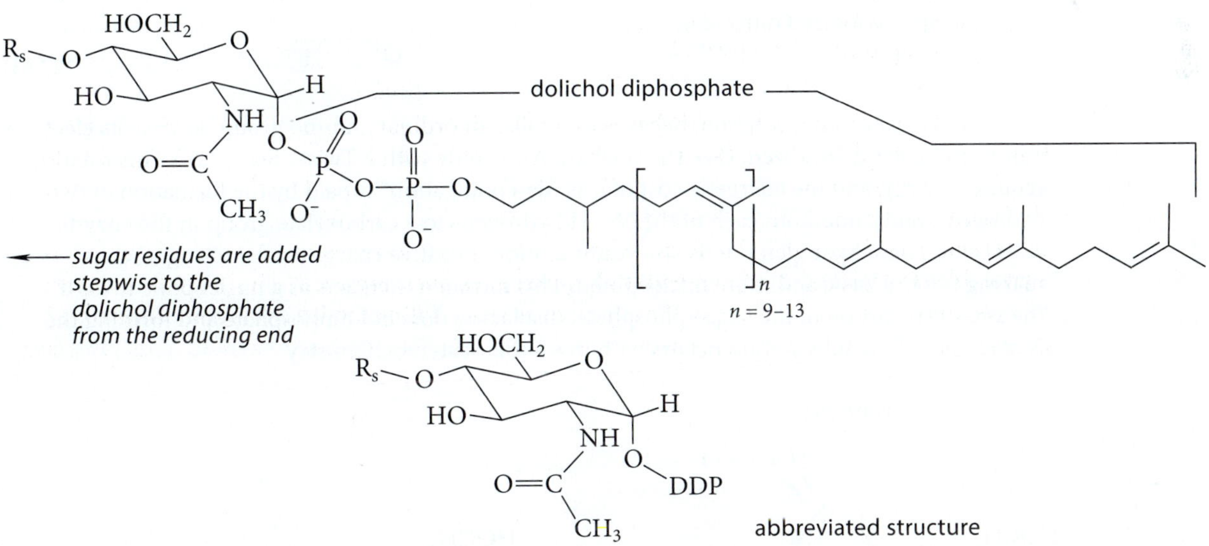

(27.62)

Dolichol is a long terpene hydrocarbon (typically 75–95 carbons). As we might expect from its hydrocarbon character, dolichol is anchored within the lipid bilayer of a membrane, in this case a membrane-like cellular structure called the *endoplasmic reticulum* (ER). Proteins are synthesized at the external surface of the ER, and the attachment of the oligosaccharide to the protein takes place as the protein chain is being elongated. The hydrocarbon chain of the dolichol ensures that the oligosaccharide-transfer process is localized at the surface of the ER, where the polypeptide chain is being formed—another example of a proximity effect (Sec. 12.8).

An apparent chemical problem in the process of *N*-glycosylation is that the amide $—NH_2$ group of the asparagine is normally not nucleophilic, because the nitrogen electron pair is delocalized into the carbonyl group.

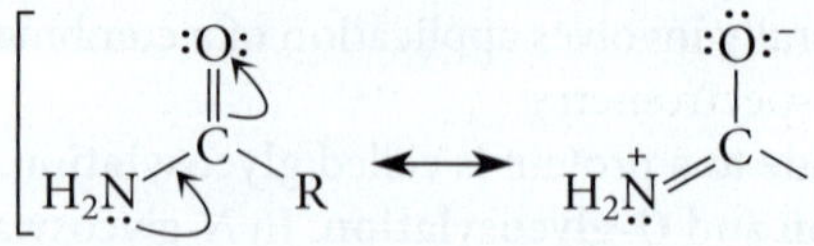

Also, an amide $—NH_2$ group is not acidic enough ($pK_a \sim 15$) to form significant amounts of its conjugate base at physiological pH. The catalyzing enzyme, a *glycosyl transferase*, overcomes this problem by forcing a 90° rotation of the amide nitrogen out of conjugation with the carbonyl group (shown in Display 27.63 as a sawhorse projection):

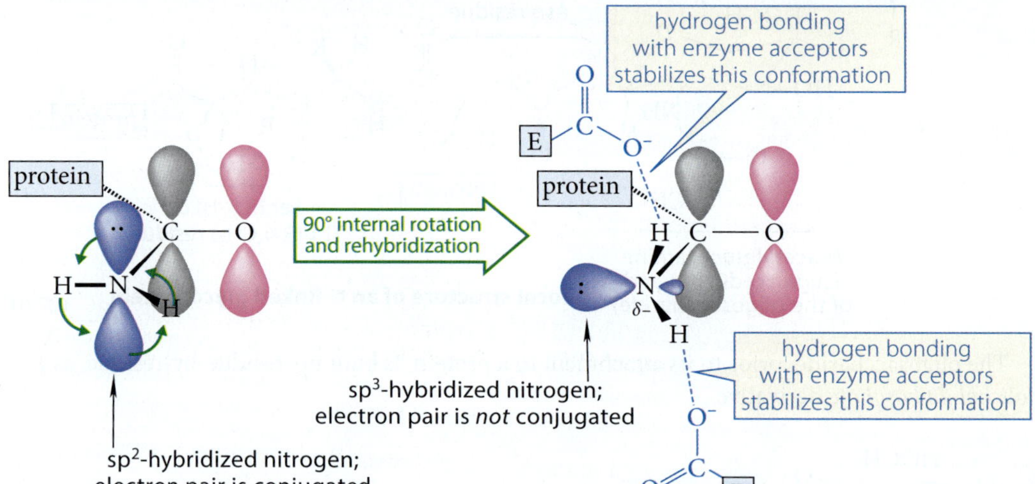

(27.63)

The $—NH_2$ group, after rotation, behaves more like an ordinary amino group because its electron pair is no longer delocalized. (See the sidebar, "An Amide with a Twist," Sec. 21.5.) This rotation requires energy, and the energetic cost of this "de-conjugation" is paid by the formation of two hydrogen bonds, one from each of the N—H hydrogens to a carboxylate group in the enzyme active site. These hydrogen bonds also result in more negative charge on the nitrogen, thus making it more basic and more nucleophilic. This nitrogen then acts as a nucleophile toward the anomeric carbon of the sugar-phosphate, displacing dolichol diphosphate and forming the *N*-glycoside. This substitution occurs with inversion of stereochemistry.

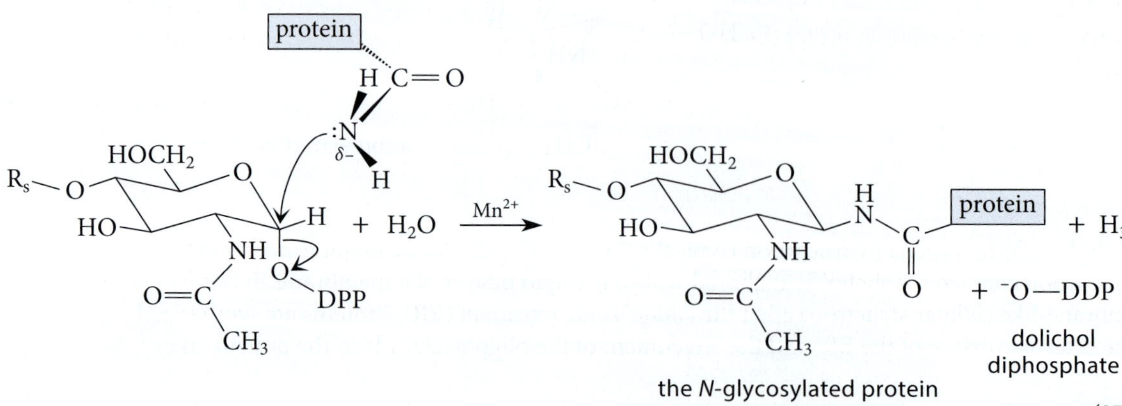

(27.64)

(The proton transfers are not shown.) This reaction is another example of the C—O cleavage of a pyrophosphate, a pattern that we have observed in the biosynthesis of terpenes (Secs. 17.6B and 25.7C).

In most cases of *O*-glycosylation, one sugar residue is attached to a serine or threonine residue of the protein acceptor, and subsequent sugars are added, one at a time, to complete the oligosaccharide. (Contrast this process with *N*-glycosylation, in which the pre-formed oligosaccharide is transferred to the protein.) The formation of the *O*-glycosyl bond between *N*-acetylgalactosamine and the oxygen of Ser or Thr in the proteins of mucin is typical. The sugar is transferred as an α-UDP-*N*-acetylgalactosamine derivative. The UDP serves as a leaving group.

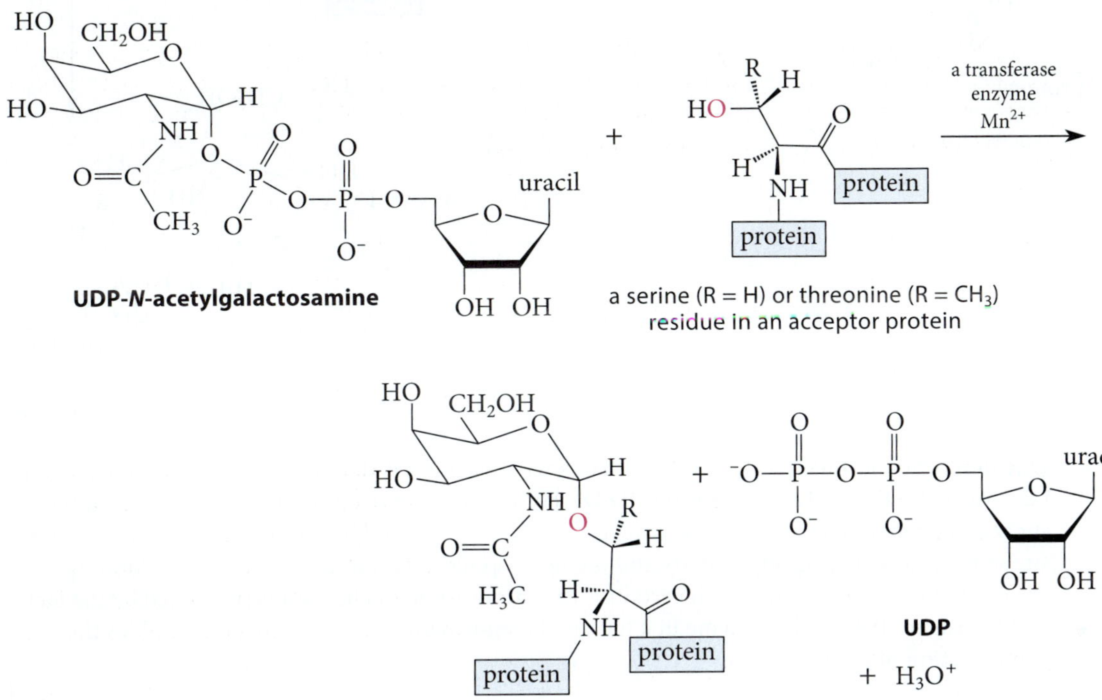

the *O*-glycosylated protein (27.65)

Once again we have an example of a pyrophosphate as a leaving group in a C—O cleavage. (The enzyme-catalyzed transfers of most saccharide groups, whether to proteins or to other sugars, always involve a nucleoside-diphosphate derivative of the sugar, and UDP-sugars are very common.)

Stereochemically, some *O*-glycosylations occur with inversion of configuration, and some occur with retention of configuration. The particular substitution considered here occurs with *retention of configuration*, as shown in Eq. 27.65. If the mechanism of substitution is strictly S_N2, then retention implies that two substitutions must occur at carbon-1 of the sugar (see Sec. 12.8C). However, another possibility is that the substitution mechanism resembles an S_N1 reaction with a carbocation intermediate.

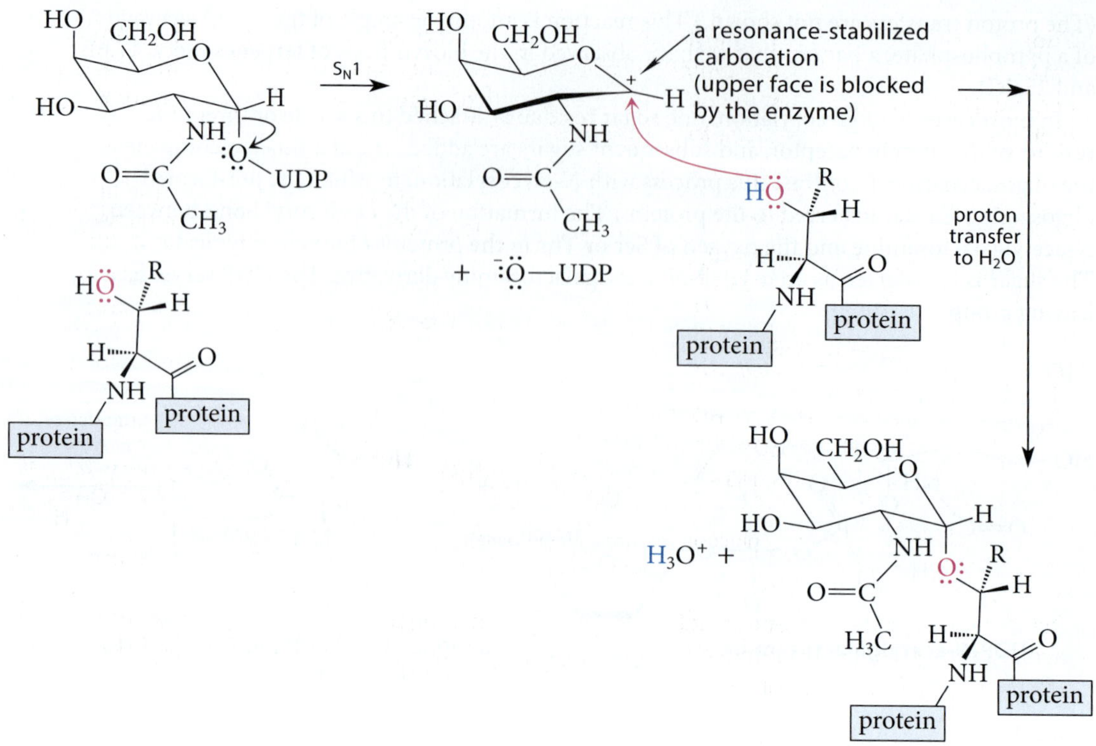

(27.66)

One stereochemical hallmark of S_N1 mechanisms in solution is that both retention and inversion take place (see Figs. 9.12 and 9.13, Sec. 9.6D). For retention to occur exclusively, the face of the sugar carbocation opposite to the leaving group must be blocked by groups in the enzyme active site so that the leaving group departs and the nucleophile enters from the same side. Although the mechanism is not known with certainty, the carbocation mechanism is supported by the fact that the sugar binds to the enzyme in a flattened conformation that very much resembles the conformation of a carbocation.

Focused Problems

27.33 Using abbreviated structures, draw the structure of an *N*-acetylgalactosamine conjugate with a threonine residue of an acceptor protein formed with *inversion* of configuration at the anomeric carbon.

27.34 (a) Draw a resonance structure for the carbocation intermediate in Eq. 27.66.

(b) All of the enzyme-catalyzed glycosylations require Mn^{2+}, a divalent cation. Suggest a role for the metal ion in the glycosylation mechanism.

Glycosylation in Clinical Diagnostics: The Hemoglobin A1c Test

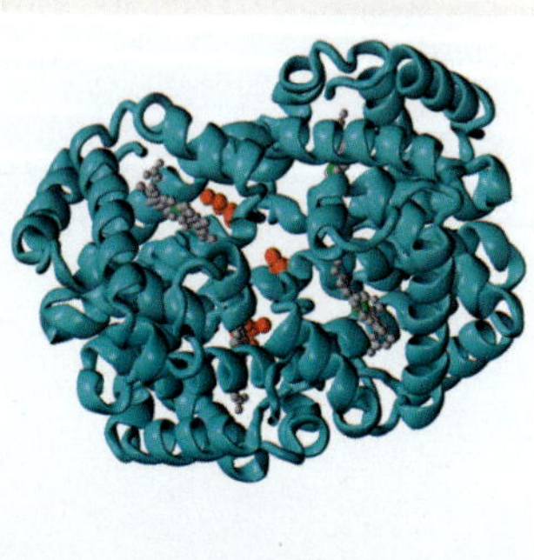

Kallista Images/Superstock

Glucose is present in the blood of all humans, and chronically elevated blood glucose is an indicator of diabetes. Physicians traditionally order "fasting blood glucose" tests to detect such elevated levels. This test determines the blood glucose level after a 12-hour fast (so that temporary elevations in glucose resulting from a recent meal don't skew the results). In the mid-1970s, Dr. Anthony Cerami (b. 1940), a biochemistry professor at The Rockefeller University, and his students found in blood an unusual variant of hemoglobin, the iron-containing protein of red blood cells, and they observed that the level of this unusual protein correlated with fasting blood glucose levels.

They called this variant hemoglobin A1c, abbreviated HbA1c (photo). They subsequently established that HbA1c results from the nonenzymatic glycosylation by glucose of the amino-terminal amino group in the β-chain of hemoglobin. In this reaction, the aldehyde form of glucose reacts with the amino group of hemoglobin to form an imine (Schiff base), which can re-close to the *N*-glycoside.

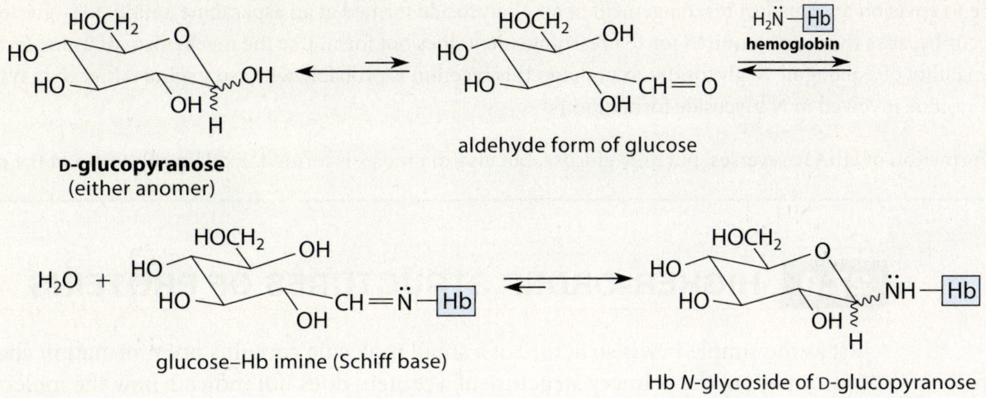

(27.67)

The imine undergoes a reaction called an Amadori rearrangement to give a new set of products. In this reaction, the imine carbon (carbon-1) is converted into a methylene group and the carbon next to it (carbon-2) is converted into a ketone. (See Focused Problem 27.35.)

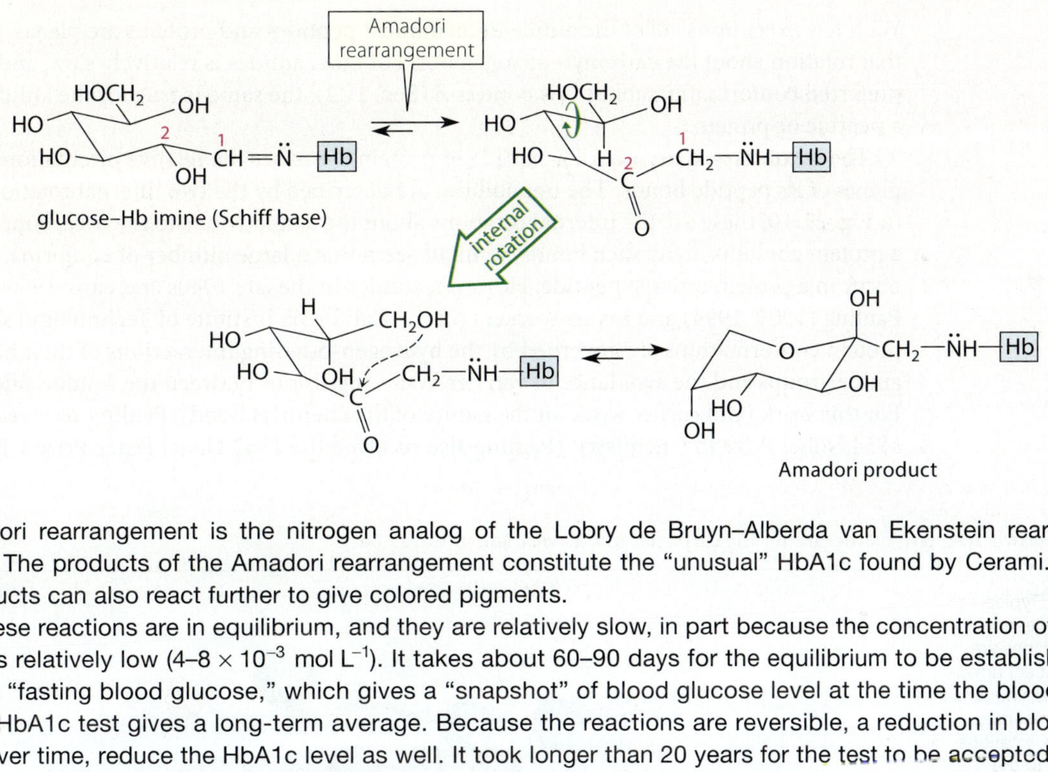

(27.68)

(The Amadori rearrangement is the nitrogen analog of the Lobry de Bruyn–Alberda van Ekenstein rearrangement; Sec. 24.5.) The products of the Amadori rearrangement constitute the "unusual" HbA1c found by Cerami. Over time, these products can also react further to give colored pigments.

All of these reactions are in equilibrium, and they are relatively slow, in part because the concentration of glucose in the blood is relatively low ($4–8 \times 10^{-3}$ mol L^{-1}). It takes about 60–90 days for the equilibrium to be established. Therefore, unlike "fasting blood glucose," which gives a "snapshot" of blood glucose level at the time the blood sample is taken, the HbA1c test gives a long-term average. Because the reactions are reversible, a reduction in blood glucose level will, over time, reduce the HbA1c level as well. It took longer than 20 years for the test to be accepted, but now it is a staple in the battery of tests for diabetes and prediabetes. In a nondiabetic person, less than 6% of the hemoglobin is present as Hb1Ac. An HbA1c level of greater than 6.5% is an indicator of diabetes.

Why does this one amino group of hemoglobin react, and not the many amino groups in the side chains of lysine residues? It turns out that the terminal amino group has a lower conjugate-acid pK_a than the lysine amino groups, and it is the conjugate-base form of the amine that reacts as a nucleophile with the aldehyde group of glucose.

If glucose can react with hemoglobin, then it can also react with other proteins. For example, glucose can react with the lens proteins of the eye, and the glycosylated proteins and their by-products can be one cause of cataracts. This is the reason that cataracts can result from uncontrolled diabetes.

Focused Problems

27.35 Draw a curved-arrow mechanism for the Amadori rearrangement in Eq. 27.68. Let H_3O^+ and H_2O be the acid–base pair, present as necessary.

27.36 It is possible to envision an Amadori rearrangement of an *N*-glycoside formed at an asparagine amide-NH_2 group. In fact, this does *not* occur because the imine required for the rearrangement does not form. Use the mechanism of imine formation to explain why. (The inability of asparagine *N*-glycosides to undergo this reaction is probably why asparagine rather than lysine evolved as the amino acid residue involved in *N*-glycoside formation.)

27.37 When the formation of HbA1c reverses, not only glucose, but also mannose, is formed. Explain the origin of the mannose.

27.9 HIGHER-ORDER STRUCTURES OF PROTEINS

Just as the simple Lewis structure of a small molecule contains no information about its conformation, the primary structure of a protein does not indicate how the molecule actually looks in three dimensions. The three-dimensional aspects of protein structure are described at three levels, which are called *secondary structure*, *tertiary structure*, and *quaternary structure*.

A. Secondary Structure

With few exceptions, all of the amide units in most peptides and proteins are planar. Recall that rotation about the carbonyl–nitrogen bond of most amides is relatively slow, and that the preferred conformation about this bond is *Z* (Sec. 21.2); the same is true for the amide bonds in a peptide or protein.

The **secondary structure** of a peptide or protein describes the relative orientations of the planes of its peptide bonds. The possibilities are described by the two internal rotations shown in **Fig. 27.10**; these are the internal rotations about the single bonds to the α-carbon. Because a protein contains many such bonds, it might seem that a large number of conformations could occur in a protein or large peptide. However, studies in the late 1940s and early 1950s by Linus Pauling (1901–1994) and his co-workers (of the California Institute of Technology) showed that protein conformations are governed by the hydrogen-bonding interactions of their backbone amide groups and the avoidance of van der Waals repulsions between the residue side chains. For this work (and earlier work on the nature of the chemical bond), Pauling received the 1954 Nobel Prize in Chemistry. (Pauling also received the 1962 Nobel Peace Prize.) Two major

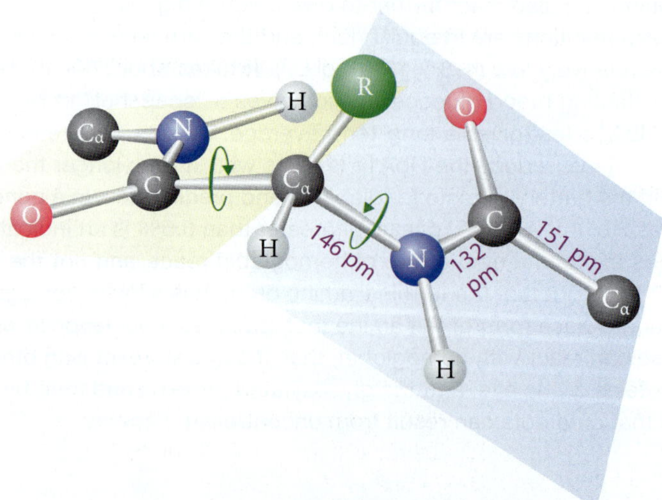

FIGURE 27.10 Typical dimensions of a peptide bond. The two planes are those of the adjacent amide groups, and the amino acid side chain is represented by R. In principle, rotations about the bonds to the α-carbons (marked with green arrows) are possible.

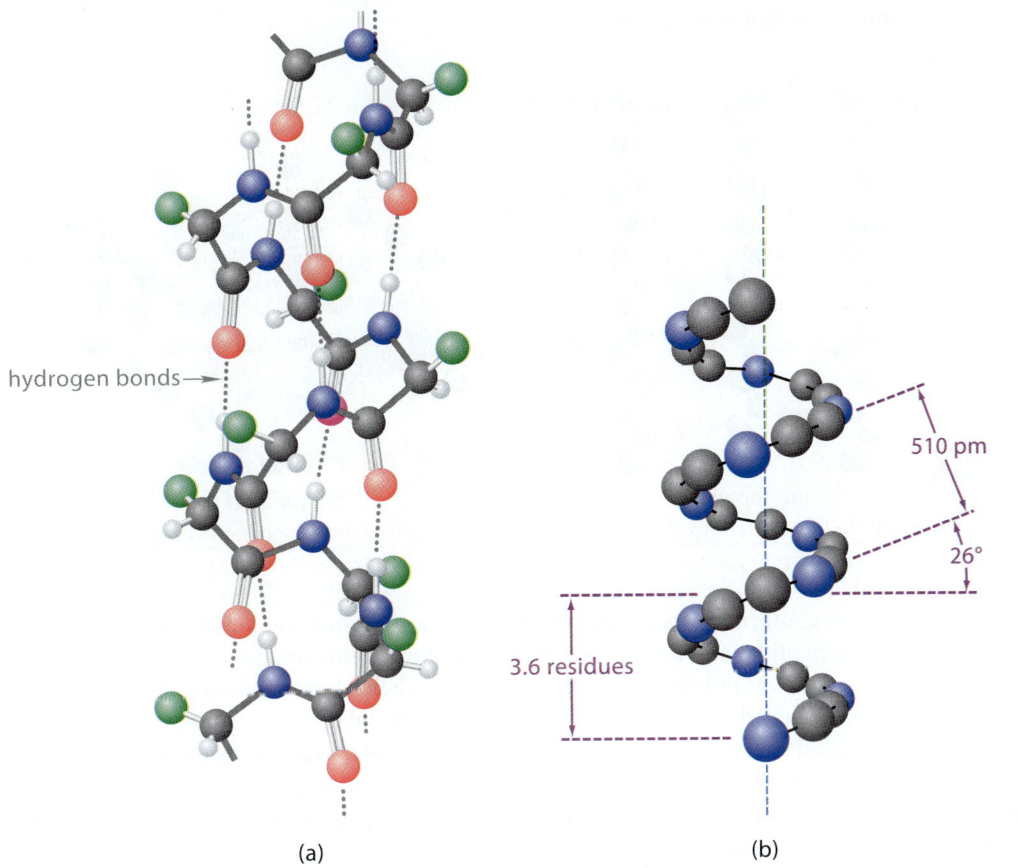

FIGURE 27.11 A peptide α-helix. (a) Hydrogen atoms are shown, and the side chains R are represented by green spheres. The side chains extend away from the helix on the outside. (b) Backbone atoms only, which form the α-helix itself, are shown. The typical α-helix has a pitch of 26°, a distance of 510 pm between turns, and 3.6 residues per turn.

conformations, the *α-helix* and the *β-sheet*, are common in proteins; as we show subsequently, hydrogen bonding plays a key role in maintaining these conformations. Within proteins are also found some regions of disorder, called *random coil.*

In the **right-handed α-helix**, shown in **Fig. 27.11**, the peptide chain adopts a conformation in which it turns in a clockwise manner along a helical axis. In this conformation, the side-chain groups are positioned on the outside of the helix, and the helix is stabilized by *hydrogen bonds* between the carbonyl oxygen of one residue and the amide N—H four residues further along the helix. The alpha (α) terminology refers to a characteristic X-ray diffraction pattern that was observed for certain proteins before chemists fully understood their structures. The α type of pattern was eventually shown to be associated with the right-handed helix—thus the name α-helix.

Another commonly occurring X-ray diffraction pattern, called a β pattern, was eventually found to be characteristic of a second peptide conformation, called **β-structure** or **pleated sheet**. In this type of structure, a peptide chain adopts an open, zigzag conformation and is engaged in hydrogen bonding with another peptide chain (or a different part of the same chain) in a similar conformation. The successive hydrogen-bonded chains can run (in the amino-terminal to carboxy-terminal sense) in the same direction (**parallel pleated sheet**) or, more commonly, in opposite directions (**antiparallel pleated sheet**). The antiparallel pleated sheet structure is shown in **Fig. 27.12**. The name "pleated sheet" is derived from the pleated surface described by the aggregate of several hydrogen-bonded chains (Fig. 27.12b). Notice that the side-chain R-groups alternate between positions above and below the sheet. When two parts of the same peptide chain interact in a β-sheet, they are typically connected by a very short turn, called a *β-turn*.

Peptides and proteins can contain regions of local disorder, which are called **random coil**. As the name implies, peptides and proteins that adopt a random coil show no discernible pattern in their conformations. An apt analogy for the random coil is the appearance of a tangled ball of yarn after an hour's encounter with a playful house cat.

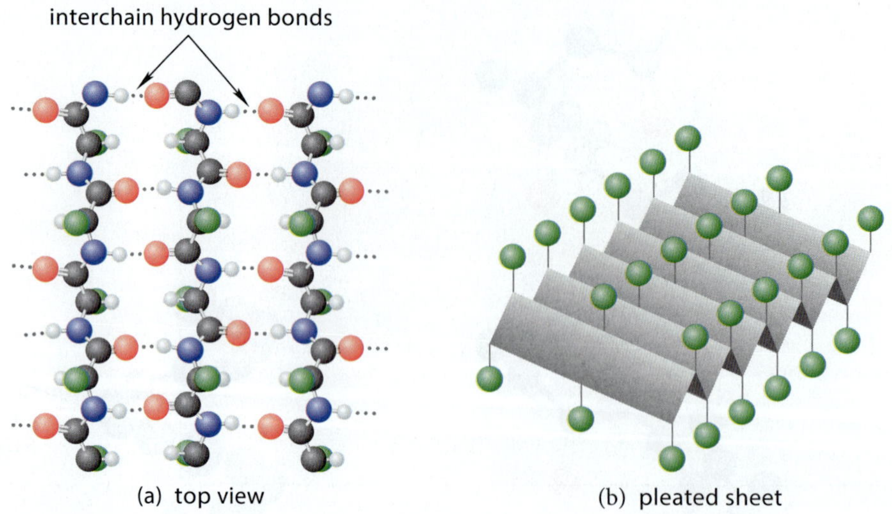

FIGURE 27.12 The β-antiparallel pleated-sheet structure of proteins. The amino acid side chains are shown as green spheres. (a) A top view of the antiparallel peptide chains. Notice the hydrogen bonds between chains (gray dotted lines). (b) The imaginary pleated-sheet surface formed by the backbone atoms.

Although other conformations are known in peptides, the α-helix and β-sheet are the major ones. Some peptides and proteins exist entirely in one conformation. For example, the α-keratins, major proteins of hair and wool, exist in the α-helical conformation. In these proteins, several α-helices are coiled about one another to form "molecular ropes." These structures have considerable physical strength. In contrast, silk fibroin, the fiber secreted by the silkworm, adopts the β-antiparallel pleated sheet conformation. Despite these examples, proteins that contain a single type of conformation are relatively rare. Rather, most proteins consist of regions of α-helix and β-sheet separated by short regions of random coil.

B. Tertiary Structure

The complete three-dimensional description of protein structure at the atomic level is called **tertiary structure**. The tertiary structures of proteins are determined by X-ray crystallography; each crystallographic structure analysis requires significant effort. Nevertheless, since the first protein crystallographic structure was determined in 1960, more than 130,000 protein structures have been elucidated, and more structures are continually appearing. High-field NMR is also being used with greater frequency for determining protein structures.

The tertiary structure of any given protein is an aggregate of α-helix, β-sheet, random coil, and other structural elements. It has become evident that, in many proteins, certain higher-order structural motifs are common. For example, one common motif is a bundle of four helices (called a *four-helix bundle*), each running approximately antiparallel to the next and separated by short turns in the peptide chain. Another common structural motif is the *β-barrel*, a "bag" consisting of β-sheets connected by short turns. Several such motifs can occur within a given protein, so a protein might consist of several smaller, relatively ordered structures connected by short turns. These ordered substructures are sometimes called **domains**.

A useful way to portray secondary structure as part of an overall protein structure is a *ribbon structure*, illustrated for hen-egg lysozyme in **Fig. 27.13**. In a **ribbon structure**, the *peptide backbone* (Sec. 27.1B) is portrayed as a ribbon. As noted in Sec. 27.9A, the amide bond is planar—that is, the carbonyl group, its attached N—H, and the two α-carbons attached to these groups lie in a common plane. The face of the ribbon defines the orientation of the planes of the peptide bonds. Twists and turns in the ribbon are defined by the two angles shown by arrows in Fig. 27.10. The ribbon structure of lysozyme clearly shows regions of α-helix and β-sheet. In this structure, the two domains of the protein are also clearly discernible.

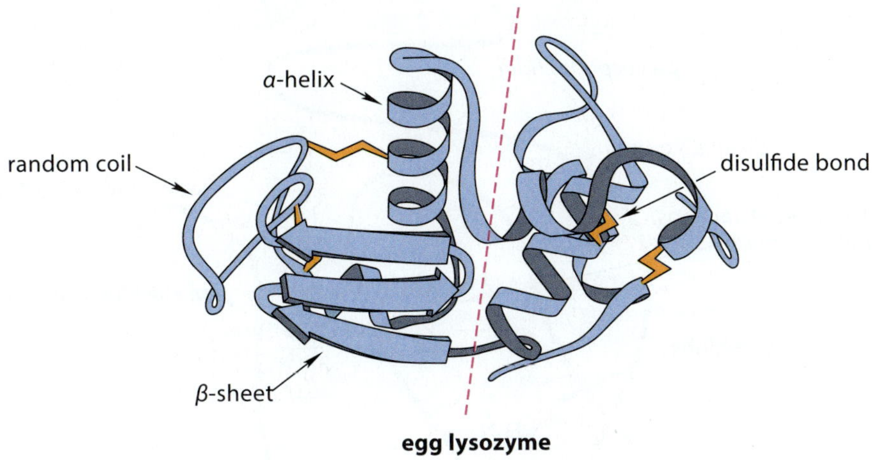

FIGURE 27.13 The three-dimensional structure of lysozyme from hens' eggs shown as a ribbon structure. The face of the ribbon shows the relative orientations of the planes of the peptide bond. Lysozyme contains regions of α-helix, β-sheet, and random coil. The two domains of lysozyme occur on either side of the plane indicated by the dashed line. (The primary structure of lysozyme is shown in Fig. 27.6.)

Richardson Ribbons: The Origin of Ribbon Structures

Courtesy of Duke University Photo

Ribbon structures were devised in 1980 by Jane Richardson (b. 1941), a crystallographer and biochemistry professor at Duke University, as a way to see the secondary structural elements within proteins at a glance. (One reviewer has said, "[These diagrams provide] protein structure with a sense of substance and design.") These diagrams (sometimes called Richardson diagrams) are now universally used to show secondary structure. Professor Richardson's first ribbon structures were hand-drawn. Now, all computer programs that render protein structures have a "ribbon structure" option that provides such structures automatically.

In general, the tertiary structures of proteins are determined by the preferred geometries of the peptide bonds, by the presence and location of disulfide bonds, and by *noncovalent interactions.* The effect of noncovalent interactions is a *balance* between interactions of groups in the protein with each other and interactions with the solvent water. Let's review the noncovalent interactions we have learned about and consider how these might affect protein structure. (These interactions are shown diagrammatically in **Fig. 27.14.**)

1. *Hydrogen bonds.* We have already seen the key role of hydrogen bonds within the protein backbone in the protein α-helix and β-sheet. In addition, the carbonyls and N—H groups of the backbone can serve as acceptors and donors, respectively, toward amino acid side chains that have hydrogen-bonding capability: Asp, Glu, His, Lys, Arg, Ser, Thr, Asn, Gln, Tyr, and the N—H proton in the indole ring of Trp; in some cases these amino acids can form hydrogen bonds with each other. A number of cases are known in which one or more water molecules are immobilized within the structure of a protein. These waters can serve as hydrogen-bonded anchors for a three-dimensional structure. Likewise, the amino acids just listed can form hydrogen bonds with the external solvent water.

2. *Interactions between hydrocarbon groups.* This type of interaction includes van der Waals attractions (Sec. 8.5A), pi-offset stacking (Sec. 15.8A), and associations caused by hydrophobic bonding—that is, the association of hydrocarbon groups driven by the unfavorable entropy of solvation by water (Sec. 8.6D). The hydrocarbon-like amino acids are Ala, Leu, Ile, Val, Phe, Trp, and the aromatic ring of Tyr.

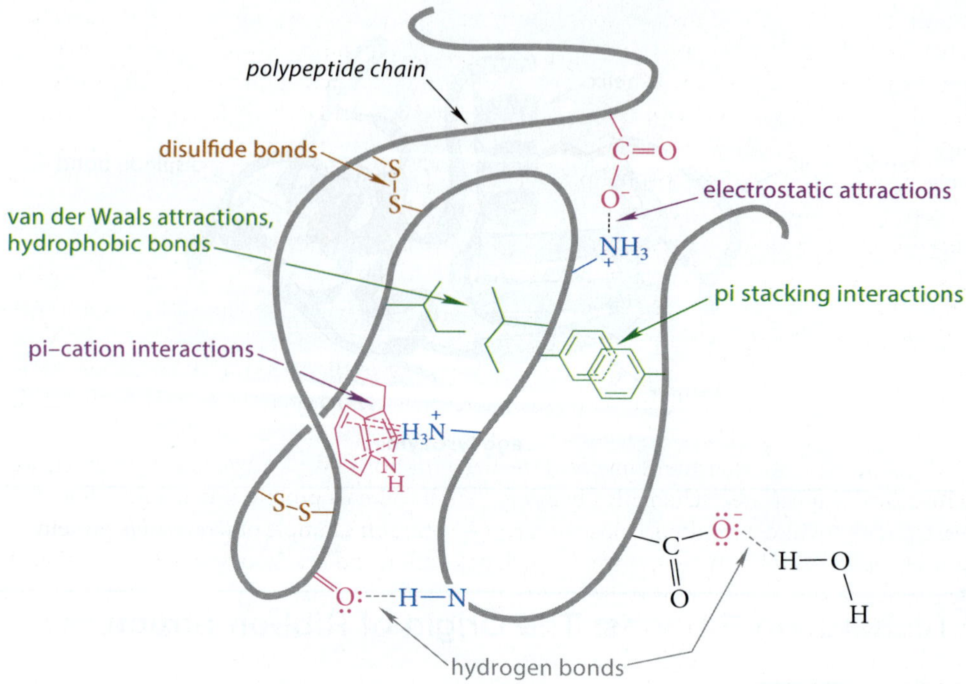

FIGURE 27.14 A diagram showing the various types of interaction responsible for the three-dimensional structures of proteins. The noncovalent interactions on the interior of the protein are balanced against the noncovalent interactions with the surrounding milieu.

3. *Electrostatic interactions.* These interactions include attractions between oppositely charged groups, pi–cation interactions, ion–dipole attractions, and dipole–dipole attractions. The effect of an attraction between oppositely charged groups was illustrated in Fig. 27.9 with the conformational change caused by the attraction between a phosphorylated serine and the conjugate base of an arginine. Pi–cation interactions were discussed in Sec. 15.8B. An important example of dipole–dipole attractions is illustrated by the antiparallel arrangement of α-helices when they are adjacent in a protein structure. The α-helix has a net dipole moment that is aligned along the helical axis from the amino end to the carboxy end (**Fig. 27.15**). This dipole moment results from the dipole moments of individual residues, which are oriented nearly parallel to the helical axis and oriented in the N-to-C direction. When α-helices are oriented in opposite directions, oppositely charged ends of the helices are adjacent and have attractive interactions.

The overall conformation of a protein is the balance of all these forces acting within the protein against the attractions with the surrounding aqueous solvent or other environment, such as the interior of a membrane. In water-soluble proteins, enough charged residues and residues that can form hydrogen bonds or favorable ion–dipole interactions with water are located at the protein surface to provide adequate solvation for the protein. However, even water-soluble proteins have a significant number of hydrocarbon groups on their surfaces. As a result, many water-soluble proteins are *roughly* globular in shape so as to minimize the water-exposed

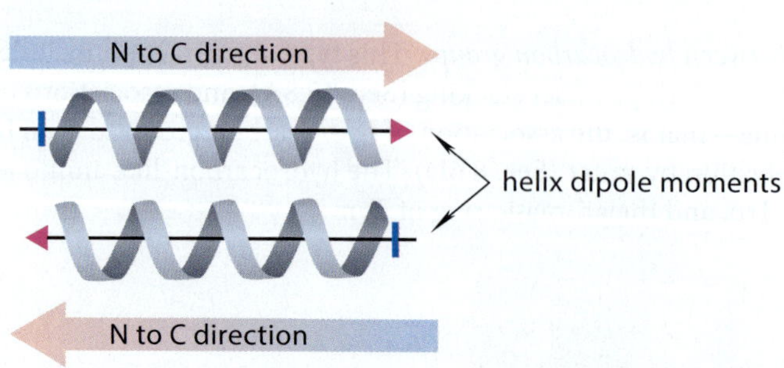

FIGURE 27.15 The antiparallel alignment of α-helices in proteins is favored because of the favorable interactions of the helical dipoles.

hydrocarbon surface area of hydrocarbon-containing residues. A number of proteins, such as ion channels (Sec. 8.7B) and many receptors, span the cell membrane or other membrane-like structures within the cell. Many of these proteins contain bundled antiparallel α-helices as a significant part of their structures. In the α-helix, the amino acid side chains point outward along the periphery of the helix (see Fig. 27.11). In membrane-embedded proteins, the surfaces of the helices exposed to the phospholipid bilayer, predictably, contain a high percentage of the hydrocarbon-like amino acid residues.

Suppose we were to synthesize a protein. Would the finished protein automatically "know" what conformation to assume, or is some external agent required to direct the protein into its naturally occurring conformation? This question was first answered by two elegant experiments with the enzyme ribonuclease. First of all, synthetic ribonuclease was prepared by the solid-phase method (Sec. 27.6A) and found to be an active enzyme. Because the enzyme must have its natural, or native, conformation to be active, it follows that ribonuclease, once synthesized, spontaneously folds into this conformation.

The second type of experiment involved the denaturation and renaturation of ribonuclease. **Denaturation** is illustrated schematically in **Fig. 27.16**. When a protein is denatured, it is converted entirely into a random-coil structure. (A common example of *irreversible* protein denaturation occurs when an egg is fried; the denaturation and precipitation of the proteins in egg white are responsible for the change in appearance of the white as it is cooked.) Some proteins, including ribonuclease, can be denatured *reversibly* by chemical agents. Typically, a protein is denatured by breaking its disulfide bonds with thiols, such as DTT or 2-mercaptoethanol (Eq. 27.53, Sec. 27.8A), and then by treating it with 8 *M* urea, detergents, or heat. Ribonuclease was denatured by treatment with 2-mercaptoethanol and 8 *M* urea. After the urea was removed and the cysteine —SH groups were allowed to reoxidize back to disulfides, the protein spontaneously reassumed its original, or *native*, conformation. This process is called **renaturation**. This experiment showed that *the amino acid sequence of ribonuclease specifies its conformation—that is, the native structure is the most stable structure.* If this were not so, another, more stable structure would have formed when the protein was allowed to refold after the urea was removed.

This renaturation experiment [for which Christian B. Anfinsen (1916–1995) of the U.S. National Institutes of Health shared the 1972 Nobel Prize in Chemistry] indicates that proteins spontaneously assume their native conformations at the time of their biosynthesis. In other words, *primary structure dictates tertiary structure.*

In some cases, protein folding is assisted by other proteins called *chaperones*. It seems likely that these molecules speed the folding process by helping proteins to avoid conformations other than the most stable ones. This is an active area of current research.

Protein and peptide misfolding in living organisms can be pathogenic. For example, Alzheimer's disease is known to correlate with the aggregation in the brain of abnormally folded peptides called β-amyloids into plaques. (See the sidebar, "Sulfide Oxidation and Protein-Aggregation Diseases" at the end of Sec. 12.9B.) The misfolding of a neural protein, α-synuclein, occurs in Parkinson's disease. Mad cow disease (bovine spongiform encephalopathy) and the related Creuzfeldt–Jakob disease in humans also result from the formation of protein plaques in the brain caused by infectious misfolded proteins called *prions* that somehow gain entry into the brain.

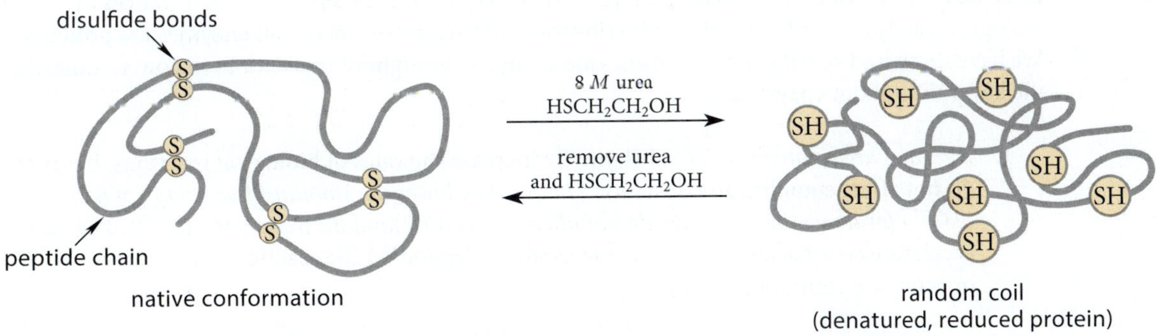

FIGURE 27.16 When a protein is denatured, its disulfide bonds are broken and it is converted entirely into a random coil. When the denaturing agents are removed, the protein renatures.

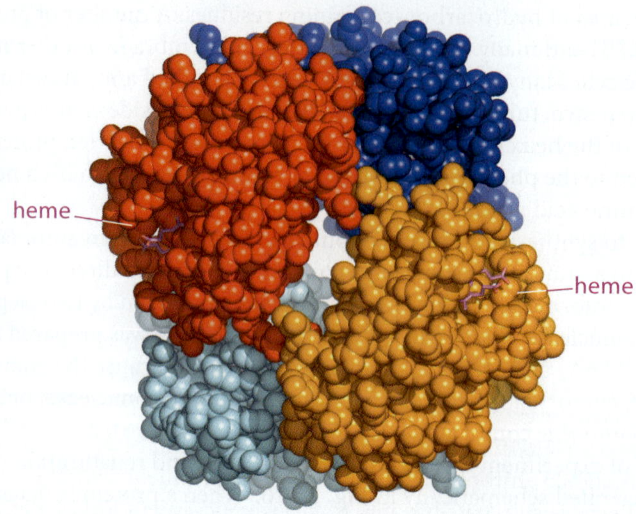

FIGURE 27.17 The tertiary and quaternary structure of hemoglobin. The hydrogen atoms are not shown. The two α chains are shown in red and orange, and the two β chains are shown in blue and turquoise. The heme groups are shown in magenta as framework models. The subunits are located at the corners of a regular tetrahedron.

C. Quaternary Structure

Many proteins are aggregates of other proteins. The best-known example of such proteins is *hemoglobin*, which transports oxygen in the bloodstream. Hemoglobin (**Fig. 27.17**) is an aggregate of four smaller proteins, or *subunits*, two of one type (called α subunits) and two of another (called β subunits). (This terminology has nothing to do with α- and β-structure.) The α and β subunits are similar, but differ somewhat in their primary structures. These subunits are held together solely by noncovalent forces. Notice in Fig. 27.17 that the individual subunits lie more or less at the vertices of a regular tetrahedron. This shape is the most compact arrangement that can be assumed by four objects. Many important proteins are aggregates of individual polypeptide subunits. In some proteins, the subunits are identical; in other cases, they are different. The description of the subunit arrangement in a protein is called **quaternary structure**.

Focused Problem

27.38 What would you expect to happen to the hemoglobin quaternary structure when hemoglobin is treated with a denaturant such as 8 *M* urea? Explain.

Chemical Biology Topic

27.10 ENZYMES: BIOLOGICAL CATALYSTS

A. The Catalytic Action of Enzymes

Enzymes are the catalysts for biological reactions (Sec. 12.8E). Except for a few instances of biological catalysis by ribonucleic acids (ribosomal RNA, "ribozymes"), all enzymes are proteins. We have discussed specific aspects of enzyme catalysis throughout this text. Let's now summarize the key points about enzymes:

1. *Enzymes are catalysts.* They substantially increase the rates of biological reactions, but they do not affect equilibrium constants. That is, *they lower the standard free energy of activation* ($\Delta G^{\circ\ddagger}$) *for a reaction, but they do not affect its overall standard free energy* (ΔG°). The rate accelerations associated with enzyme catalysis depend on the reaction, but factors of 10^{12} or more are common.

2. *The binding of substrates at the enzyme active site is an obligatory part of enzyme catalysis.* The compounds on which enzymes act are called **substrates**. A substrate is bound into a specific region of the enzyme called the **active site** prior to the actual reaction that converts bound substrate to bound product. The *active site* is the part of the enzyme in which the chemical transformations take place.

$$\mathrm{E} + \mathrm{S} \rightleftharpoons \mathrm{E}\bullet\mathrm{S} \rightleftharpoons \mathrm{E}\bullet\mathrm{P} \rightleftharpoons \mathrm{E} + \mathrm{P} \qquad (27.69)$$

E	S	E • S	E • P	E	P
enzyme	substrate	noncovalent enzyme–substrate complex	noncovalent enzyme–product complex	enzyme	product

3. *The binding of substrates to enzymes is noncovalent.* Noncovalent attractions, such as van der Waals attractions, hydrogen bonding, and electrostatic attractions of various types are used to pay the entropic price of substrate binding. In addition, solvating water can be "stripped off" of a substrate as part of substrate binding; the return of low-entropy solvation water to higher-entropy solvent water can contribute a favorable entropic component to binding. Because substrate binding involves noncovalent forces, binding is typically very fast; in many cases it occurs at the rate that the enzyme and substrate can diffuse together (10^8–10^9 M^{-1} s^{-1}).

4. *The rate accelerations observed in enzyme catalysis are largely due to the fact that the reactions within the active site are intramolecular* (Sec. 12.8E). Because the enzyme structure and the active site gather several reacting groups into proximity, concerted reaction mechanisms are observed that would be entropically impossible in the absence of enzymes. In addition, there may be a "solvent effect" associated with the active-site environment in some cases. When it contains no water, a protein active site may have an effective dielectric constant (Eq. 8.55 and discussion, Sec. 8.6B) that closely resembles that of a dipolar aprotic solvent ($\mu = 15$–20). Recall that substitution reactions are strongly accelerated in such solvents (Sec. 9.4E). Moreover, there is evidence in some cases that the noncovalent binding forces are capable of distorting a bound substrate toward its transition state by bending or stretching some of the bonds involved in the enzyme-catalyzed reaction.

5. *Enzymes are structurally specific for their substrates.* Each biological reaction has its own unique enzyme. In most cases, enzymes have very little tolerance for variations in substrate structure. For cases in which structural variations are tolerated, *certain parts of the structure* cannot be varied. The reason for this structural specificity is that each enzyme has evolved so that its specific substrate "fits" its active site. This idea, first enunciated by Emil Fischer in 1894, came to be called the **lock-and-key hypothesis**. This hypothesis implies that the substrate (the "key") fits into a rigid enzyme active site (the "lock"). However, enzymes are flexible and can undergo conformational changes, and such changes have been observed in many cases. In many situations an enzyme changes its conformation during the binding event to match the structure of its substrate. This idea, called **induced fit**, was proposed in 1958 by Daniel E. Koshland, Jr. (1920–2007), then a scientist at Brookhaven National Laboratory and subsequently a professor of biochemistry at the University of California, Berkeley. In either case, there is a unique fit between an enzyme and its substrates.

6. *Enzymes are stereochemically specific for their substrates.* As we have shown, enzymes are linear polymers of L-amino acids. They are therefore *chiral catalysts.* With very few exceptions, each enzyme is specific for a single stereoisomer of its substrate (Sec. 7.7A). Enzymes provide a powerful illustration of the *principle of enantiomeric differentiation.* (See margin note to right.)

If an enzyme consisting of all L-amino acid residues catalyzes a reaction of a chiral substrate, then the enantiomeric enzyme consisting of all D-amino acid residues should catalyze the same reaction of the *enantiomeric* substrate. Nature does not afford enantiomeric enzymes that can be used to test this prediction. However, an all-D enzyme was prepared in 1993 by chemists at The Scripps Research Institute in La Jolla, California, using solid-phase peptide synthesis. The natural enzyme that they prepared, HIV protease, is a small (and therefore synthetically accessible) peptide-hydrolyzing enzyme produced by the viral agent of the disease AIDS. (This enzyme, as a drug target, is discussed in Sec. 27.10B.) This enzyme catalyzes the hydrolysis of certain peptides consisting of amino acids with the L configuration. The synthetic all-D enzyme was found to be inactive with the natural substrates, but—as predicted—it hydrolyzes the *enantiomers* of the natural substrates with exactly the same catalytic efficiency as the natural enzyme hydrolyzes its all-L substrates.

7. *Enzymes are large molecules.* The smallest enzymes have molecular masses around 9000 g mol^{-1}, but many enzymes are much larger, and some have molecular masses exceeding 100,000 g mol^{-1}. The biosynthesis of an enzyme represents a substantial investment of biochemical energy (Sec. 27.6B). For that reason, we assume that the very specific organization of an active site that provides the necessary noncovalent interactions for

substrate binding and the precise placement of catalytic groups would not be possible in a much smaller protein. In addition, a large protein cannot escape from a cell or cellular compartment by passive diffusion as some small molecules can. Some enzymes consist of single polypeptide chains, such as lysozyme (see Fig. 27.6), whereas others such as trypsin consist of several polypeptide chains held together by disulfide bonds. (See the discussion of trypsin that follows.) Other enzymes consist of several identical subunits, each containing the same active site. Still others merge several different types of subunits that catalyze sequential reactions in a pathway; fatty acid synthase (Sec. 22.7) is one such enzyme.

8. *Coenzymes and other cofactors are involved in some enzyme-catalyzed reactions.* A **coenzyme** is a nonprotein biomolecule that is involved covalently in an enzyme-catalyzed reaction, but is itself either unchanged or later "recycled." Examples of coenzymes are NAD^+/NADH (Secs. 11.8 and 19.9B), *S*-adenosylmethionine (Sec. 12.7), thiamin pyrophosphate (Sec. 20.11B), biotin (Sec. 22.7), and pyridoxal phosphate (Sec. 26.4E). Metal ions (such as Mg^{2+}) are obligatory cofactors in some enzyme-catalyzed reactions. As we have shown, they are sometimes part of the catalytic mechanism.

To illustrate some of these points in context, let's consider the mechanism by which the enzyme *trypsin* catalyzes the hydrolysis of peptide bonds. Trypsin is the mammalian digestive enzyme used in the sequencing of proteins (Sec. 27.7B). With a molecular weight of about 24,000, trypsin is an enzyme of modest size. It is a soluble, globular protein containing three polypeptide chains held together by disulfide bonds. The following comparison provides some idea of the catalytic effectiveness of trypsin. Peptides in the presence of trypsin are rapidly hydrolyzed at 37 °C and pH = 8. In the absence of trypsin, peptide hydrolysis under the same conditions requires hundreds of years. As we showed in Sec. 27.7A, to hydrolyze proteins at a reasonable rate requires boiling them in 6 *M* HCl for several hours. Trypsin, in contrast to hot HCl solution, does not catalyze hydrolysis at just *any* peptide bonds; it is specific for the hydrolysis of the peptide bonds at lysine and arginine residues (Eqs. 27.51a and b, Sec. 27.7B). Moreover, it is specific only for substrates with the L stereochemical configuration.

The active site of trypsin consists of a cavity, or "pocket," that perfectly accommodates the amino acid side chain of a lysine or arginine residue from the substrate (**Fig 27.18**, first panel). Several hydrophobic residues line this cavity. At the bottom of the cavity is the side-chain carboxylic acid group of an aspartic acid residue (Asp-189 in the trypsin sequence). This group is ionized, and therefore *negatively charged*, at neutral pH. The amino group of a lysine side chain and the guanidino group of an arginine side chain are both in their conjugate-acid form, and therefore *positively charged*, at neutral pH. The favorable electrostatic attraction between the ionized Asp-189 side chain of the enzyme and the positively charged side chain of the substrate helps to stabilize the enzyme–substrate complex. This complex is also stabilized by the van der Waals interactions between the —CH_2— groups of the substrate side chain and the hydrophobic residues that line the cavity. Finally, a hydrogen bond from a backbone amide —NH— to the carbonyl oxygen of the Arg or Lys residue of the substrate (dashed line) also stabilizes the complex. These, then, are some of the reasons for the *specificity* of trypsin. The active site just "fits" the substrate (and vice versa), and it contains groups that are noncovalently attracted to groups on the substrate.

Near the mouth of the active site are two amino acid residues, Ser-195 and His-57, that serve a critical catalytic function. The way that these residues act to catalyze peptide bond hydrolysis at an Arg residue is also shown in Fig. 27.18. The —OH group of the serine side chain acts as a nucleophile to displace the peptide leaving group from the carbonyl group of the substrate. The hydrogen bond from the enzyme to the carbonyl oxygen activates the carbonyl carbon toward reaction with the nucleophile. The resulting product is an *acyl-enzyme*; in this transient covalent complex, the residual peptide substrate is actually esterified to the enzyme. The imidazole group of His-57 serves as a base catalyst to remove the proton from the nucleophilic serine hydroxy group. When water enters the active site, it too is deprotonated by the histidine as it reacts as a nucleophile at the carbonyl carbon of the acyl-enzyme, to give the free carboxy group of the substrate and regenerate the enzyme. After the product leaves the active site, the enzyme is ready for a new substrate molecule.

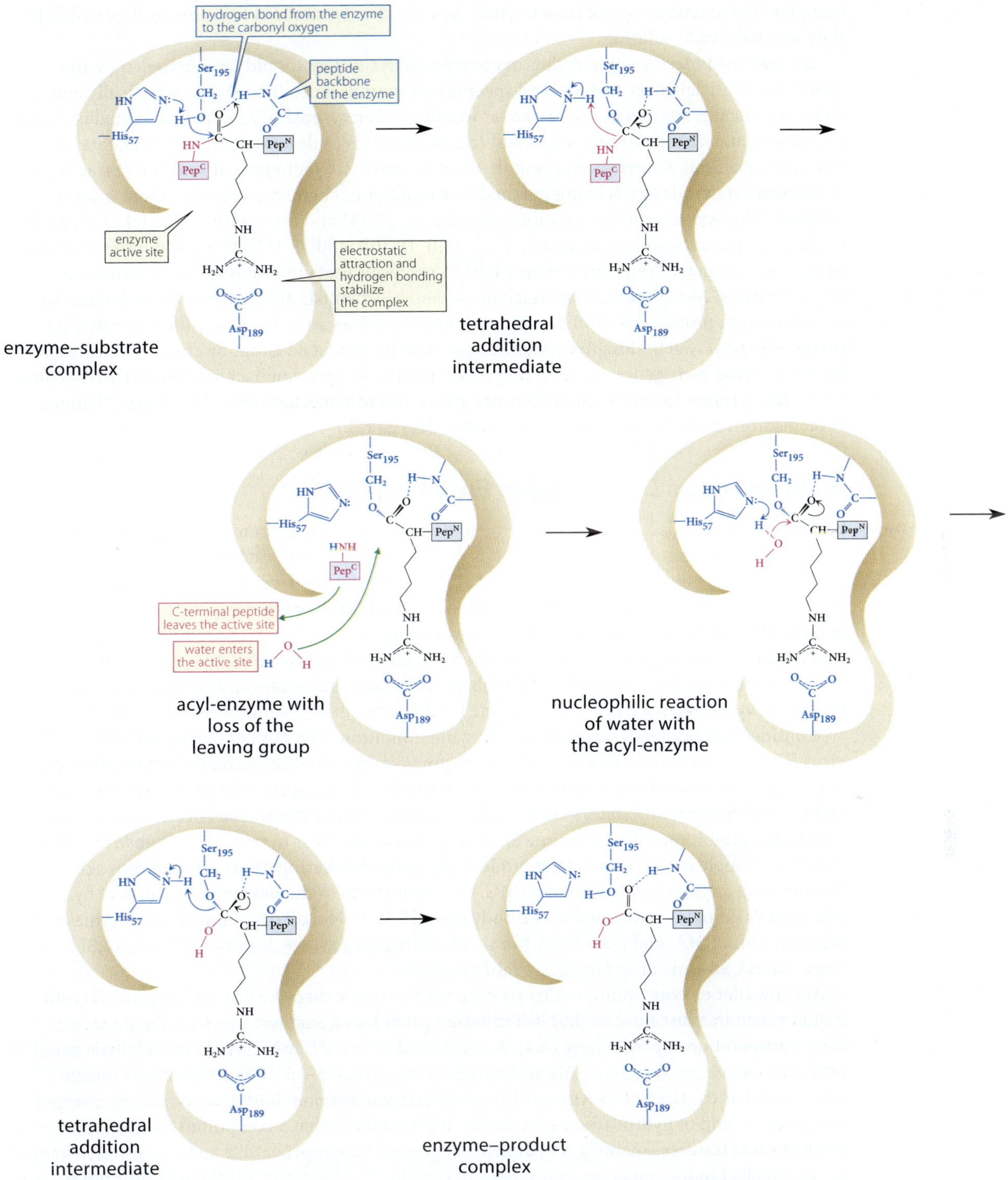

FIGURE 27.18 A diagrammatic representation of the trypsin active site showing the mechanism of the trypsin-catalyzed hydrolysis of an arginyl–peptide bond, beginning with the enzyme–substrate complex and ending with the enzyme–product complex. The imidazole in the side chain of the His-57 residue acts alternately as a base and, in its protonated form, as an acid to bring about the necessary proton transfers. Pep^N and Pep^C denote the amino-terminal and carboxy-terminal parts of the peptide substrate, respectively. Groups of the enzyme are colored blue.

The *catalytic efficiency* of trypsin, as well as that of other enzymes, is attributable mostly to the *proximity effect* (Sec. 12.8E). That is, all of the necessary reactive groups—the substrate carbonyl, a nucleophile (the serine —OH group), an acid–base catalyst (the imidazole of the histidine), and the hydrogen bond to the carbonyl oxygen—are positioned in proximity within the enzyme–substrate complex. These groups do not have to "find" one another by random collision, as they would if they were all free in solution.

If scientists understand the details of enzyme catalysis, they should be able to design and synthesize artificial enzymes that bind specific compounds and act on them catalytically and stereospecifically. Success in this endeavor would yield an arsenal of rationally designed molecules that could catalyze industrially important transformations under mild and environmentally friendly conditions. General success in this sort of activity is yet to be realized, and research in rational catalyst design is being pursued by a number of chemists. However, there is some progress. The asymmetric epoxidation catalyst (Sec. 12.11) and the transition-metal catalysts for aryl and vinylic substitution reactions (Secs. 18.6, 18.10B, and 23.11C) bring about the assembly of reactants on metal "templates" that results in reactions that are impossible in the absence of the catalysts. Along the same lines, chemists and biochemists are beginning to learn how to custom-design protein enzymes by altering known enzymes so as to change their specificities in predictable ways. Such "designer enzymes" can then be produced either by chemical synthesis or by synthesizing their genes and inserting them into the genomes of bacteria, which then become miniature "protein factories" when they are grown in the laboratory. (See the sidebar, "Human Proteins from Bacterial Factories," at the end of Sec. 27.6B.)

B. Enzymes as Drug Targets: Enzyme Inhibition

Suppose an enzyme is known to be an essential factor in the development of a certain disease state. By "knocking out" the enzyme (that is, by preventing it from catalyzing a reaction), we could prevent the disease. This strategy lies at the heart of drug development. One way to inactivate an enzyme is to treat it with an *inhibitor.* An **inhibitor** is a compound that prevents an enzyme from fulfilling its catalytic role. The most common inhibitors are *competitive inhibitors.* A compound is a **competitive inhibitor** when it binds to the enzyme's active site so tightly that the enzyme can no longer bind its usual substrate. Because substrate binding must precede catalysis (Eq. 27.69), prevention of substrate binding *always* results in the loss of catalysis.

Consider two impressive examples of enzyme inhibition. The human immunodeficiency virus (HIV, the virus responsible for AIDS) requires several unique enzymes for its replication and cellular infection; these enzymes have been used as drug targets. The anti-AIDS drug AZT and the HIV-protease inhibitors are two classes of anti-AIDS drug whose effectiveness is based on enzyme inhibition. (We'll examine an HIV-protease inhibitor in more detail subsequently.) Another example is the family of the modern cholesterol-lowering drugs. These drugs act by inhibiting an important enzyme (HMG-CoA reductase) in the biochemical pathway by which cholesterol is synthesized in the body (Sec. 25.5C). (Not all drug targets are enzymes; other proteins, RNA, and even DNA can serve as drug targets; see, for example, Sec. 12.8D. Nevertheless, enzyme drug targets are fairly common.)

We consider enzyme inhibition by starting with a simple case: the enzyme trypsin. Trypsin isn't involved in a disease state, but its inhibition provides an example in which the molecular basis for inhibition is particularly clear. As explained in Sec. 27.10A, trypsin specificity is based on a *hydrophobic pocket* containing an *ionized carboxy group* (Asp-189). Trypsin acts mostly on peptide bonds adjacent to Arg and Lys residues; these residues both have positively charged groups at the end of hydrocarbon side chains. It seems likely that a compound that has some or all of these same features—namely, a hydrocarbon group of appropriate size and a positively charged group attached to it—might be an inhibitor of trypsin. The idea is that if the inhibitor binds tightly enough, it should clog the active site and make the active site inaccessible to substrate.

Many such inhibitors for trypsin are known. One is the benzamidinium ion.

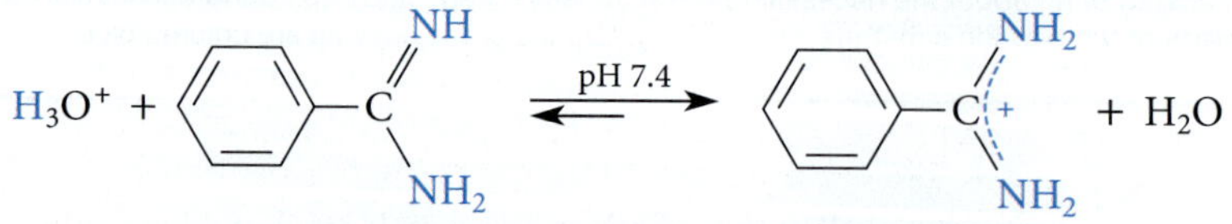

benzamidine **benzamidinium ion**

$pK_a = 11.6$ (27.70)

(This cation, the conjugate acid of benzamidine, has a pK_a value of 11.6, so it is fully protonated at the physiological pH of 7.4.) The design of this inhibitor utilizes its resemblance to the arginine residue at the trypsin cleavage site of peptides (Eq. 27.51a, Sec. 27.7B).

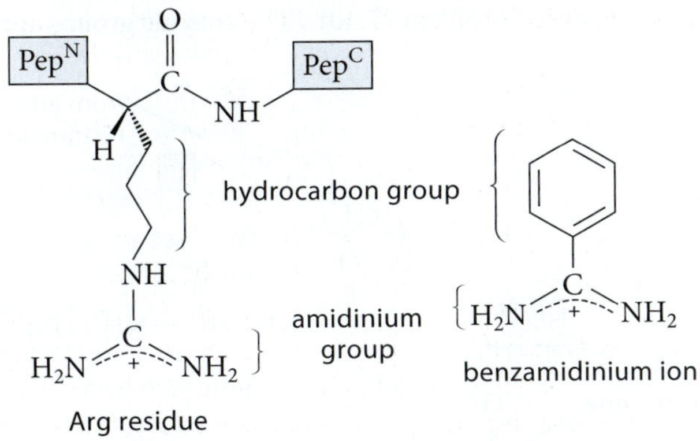

(27.71)

In the presence of 10^{-3} *M* benzamidine at pH 7.4, trypsin is inactive as a catalyst because the benzamidinium ion binds to the active site of trypsin, thus blocking the access of substrates to the active site. A model of trypsin containing a bound benzamidinium ion is shown in **Fig. 27.19a**. (This model comes from X-ray crystallography.) The benzamidinium ion is "stuck" in the active site like a counterfeit coin stuck in the slot of a vending machine. **Figure 27.19b** shows that the benzamidinium ion binds just as we would expect. The positively charged group of the cation is very near the ionized, and therefore anionic, side-chain carboxylic acid group of Asp-189. The benzamidinium ion is probably involved in significant hydrogen bonding with the carboxylate ion as well.

We illustrate the modern design of enzyme inhibitors as drugs with inhibitors for HIV protease, one of the enzymes of human immunodeficiency virus (HIV) involved in the virus's

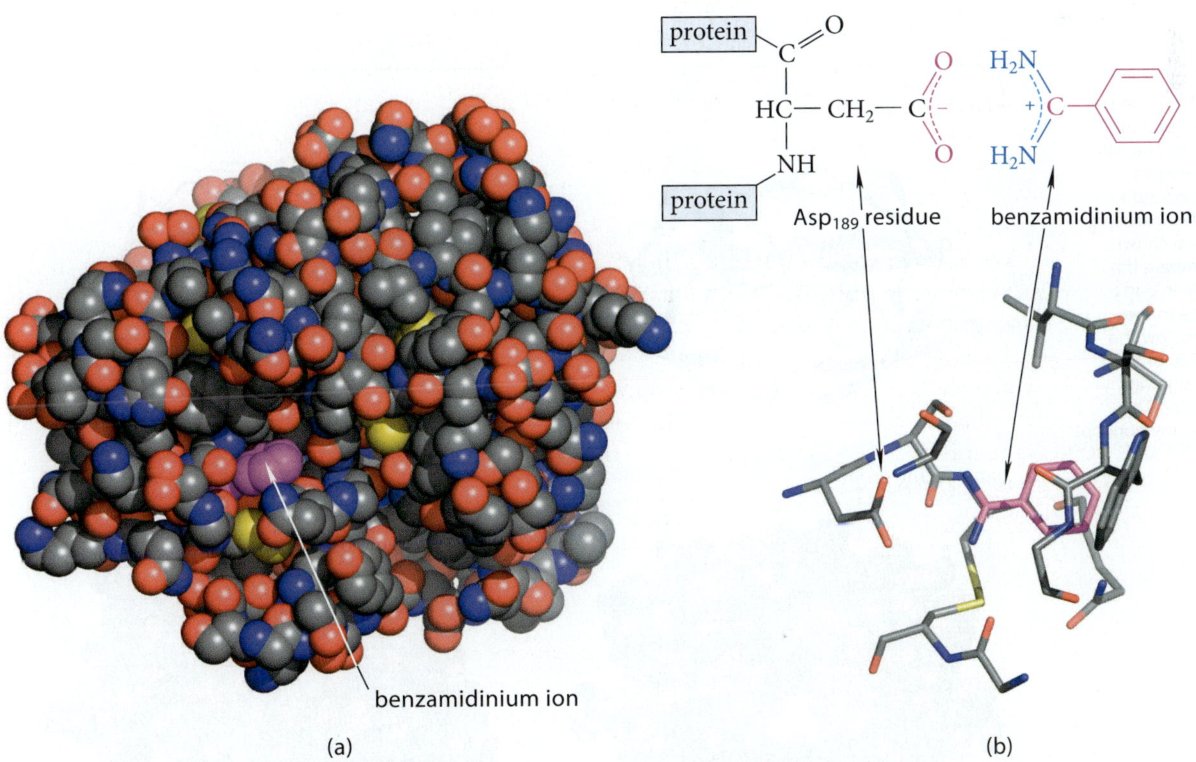

FIGURE 27.19 The trypsin–benzamidinium ion complex. (a) A space-filling model of trypsin with the bound benzamidinium ion shown in magenta. (Hydrogen atoms are not shown.) (b) A detailed diagram of the active site, showing the crucial Asp-189 and its spatial relationship to the positively charged group of the benzamidinium ion.

replication and infectious activity. Before the crystal structure of this enzyme was known, chemists had recognized that this enzyme resembled the digestive enzyme pepsin in its mechanism of action. Both pepsin and HIV protease catalyze peptide hydrolysis by a mechanism involving the side-chain carboxylic acid groups of two active-site Asp residues and a tightly bound water molecule (see Focused Problem 27.40). (The enzyme groups are shown in blue.)

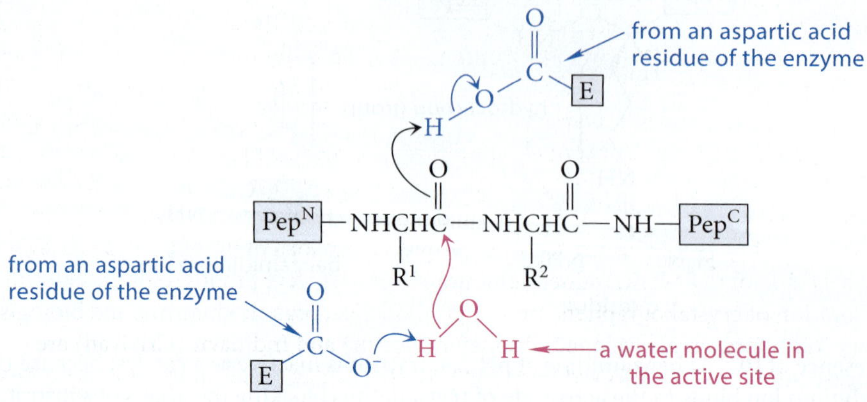

(27.72)

(The enzymes in this protease family are called *aspartyl proteases* for this reason.) A number of pepsin inhibitors were known, and these compounds were also found to be inhibitors of HIV protease. These would not make good drugs, however, because, ideally, a suitable drug should discriminate between the enzyme we want to inhibit—HIV protease—and enzymes such as pepsin, whose functions we do not want to compromise.

The determination of the X-ray crystal structure of HIV protease was a very important development in the design of inhibitors for this enzyme. **Figure 27.20a** shows the HIV protease structure.

The active site of HIV protease was easily identified by the presence of the two active-site Asp residues and the bound water. (The active site is exactly where it was expected—within the large "hole" in the enzyme structure.) Inhibitors for HIV protease were designed by

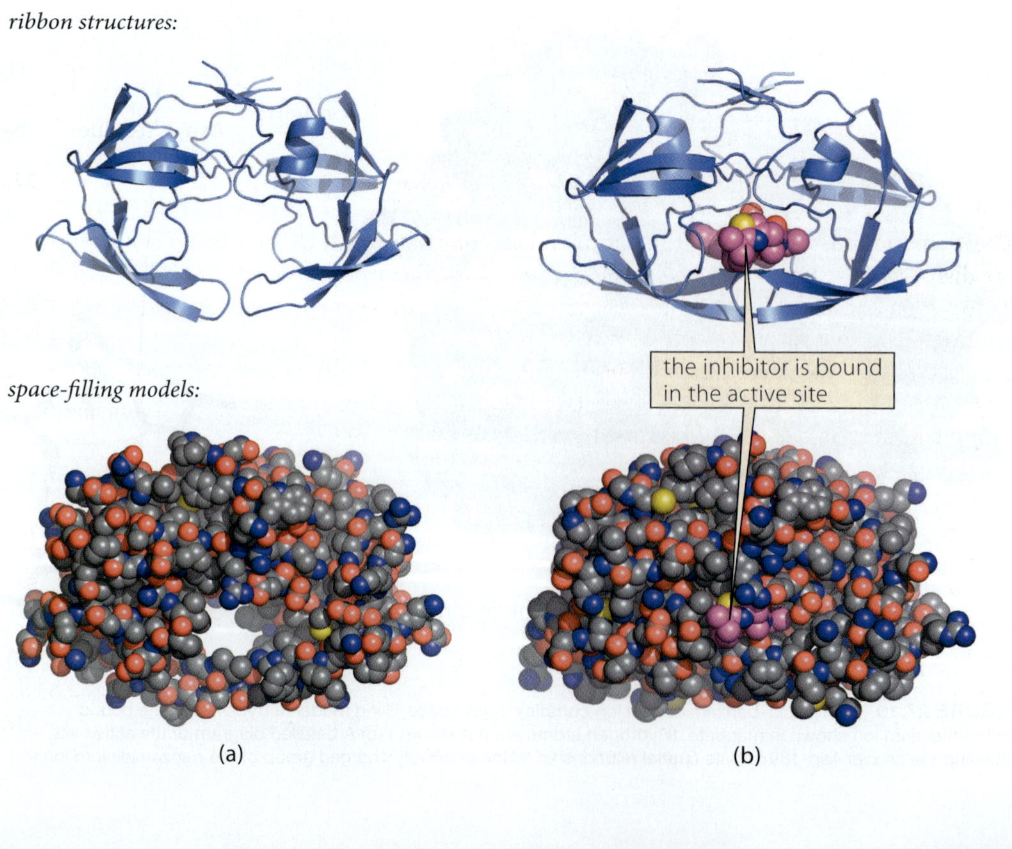

FIGURE 27.20 (a) The HIV protease. (Hydrogen atoms are not shown.) (b) The HIV protease containing an inhibitor (Norvir) bound to the active site. The carbons of the inhibitor are shown in magenta. Compare the ribbon structures in parts (a) and (b). Notice in the ribbon structures how the "arms" at the base of the protease come together around the inhibitor. This motion illustrates the concept of *induced fit*.

molecular modeling. Molecular modeling is conceptually like building models with model sets, with two important exceptions. First, it is done on a computer, so that *really large* molecules such as proteins can be handled easily and sophisticated graphics tools can be employed; second, the interaction energies between molecules (or between groups within molecules) can be calculated. These capabilities enable chemists to determine which inhibitors are likely to bind strongly to the enzyme. Inhibitors bind to HIV protease principally by an extensive network of hydrogen-bonding interactions, as well as by van der Waals attractions between nonpolar groups. (An electrostatic interaction like the one in trypsin is not involved.) The best inhibitors have hydrogen-bonding donor and acceptor sites that complement acceptor and donor sites on the protease, and they have nonpolar groups of appropriate size that interact well with similar groups on the enzyme. Molecular modeling studies suggested some possible unique structures for inhibitors. Organic chemists then prepared compounds with those and related structures, and they were tested as inhibitors. Crystallographers in some cases examined the complexes of these inhibitors with the protease to see whether the predictions of molecular modeling were correct. As a result of this work, refined inhibitor structures were proposed. Eventually, through the collaboration of crystallographers, molecular modelers, organic chemists, and biologists, satisfactory inhibitors were developed. Ritonavir (Norvir) and Indinavir (Crixivan) are HIV-protease inhibitors that evolved from such studies. Today these compounds are actively used as anti-HIV drugs. Darunavir was subsequently developed as a second-generation drug for the treatment of HIV infections that had become resistant to the other drugs. (Viruses, like bacteria, mutate into drug-resistant strains; see sidebar, Sec. 21.8D.)

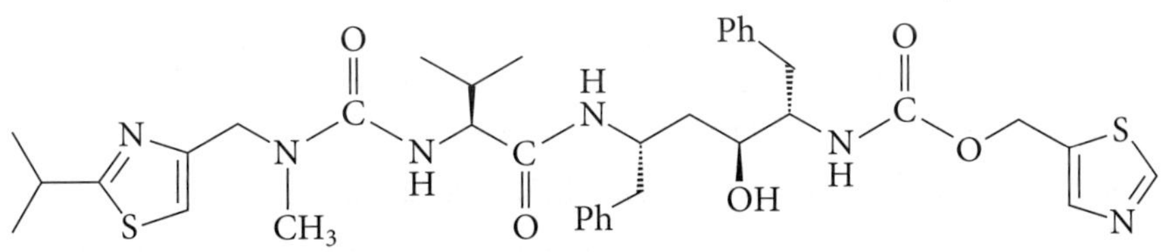

Norvir

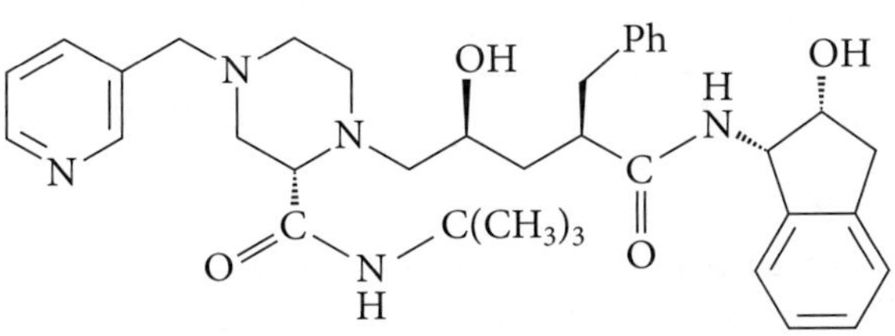

Crixivan

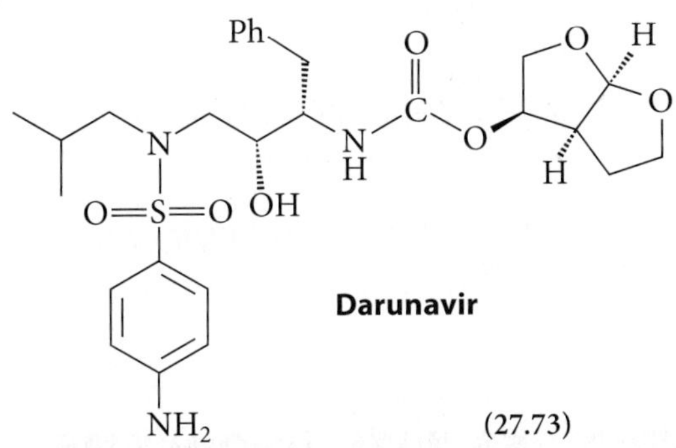

Darunavir

(27.73)

These compounds form very strong noncovalent complexes with HIV protease. For example, the dissociation constant for the complex of Norvir with the protease is 15×10^{-12} *M*, and the dissociation constant for the protease–Darunavir complex is 4.5×10^{-12} *M*. These are *very* strong bindings! The complex of Norvir with HIV protease is shown in **Fig. 27.20b**. In comparing HIV protease with and without inhibitor, notice how the "arms" of the protease at the base of the structure come together when the inhibitor is bound. (This has been called a "fireman's grip.") This change in the enzyme on binding the inhibitor illustrates the *induced fit* concept (point 5, Sec. 27.10A).

The strong binding of an inhibitor is a necessary but insufficient condition for it to be an effective drug. A good drug must also be nontoxic. It must not be metabolized (destroyed by the body) at too great a rate. It must have just the right water solubility. It must penetrate the appropriate tissues (that is, it must be *bioavailable*). Considerations such as these were very important in developing the final drugs, and a number of candidate compounds with excellent binding properties were tested before the final candidates were chosen. These and other HIV-protease drugs have led to a dramatic improvement in the life expectancy of patients with HIV infections. These cases illustrate how organic chemistry, when teamed with areas of the life sciences, can be used for the improvement of human health.

Focused Problems

27.39 Aeruginosin-B is a recently discovered natural product that inhibits trypsin.

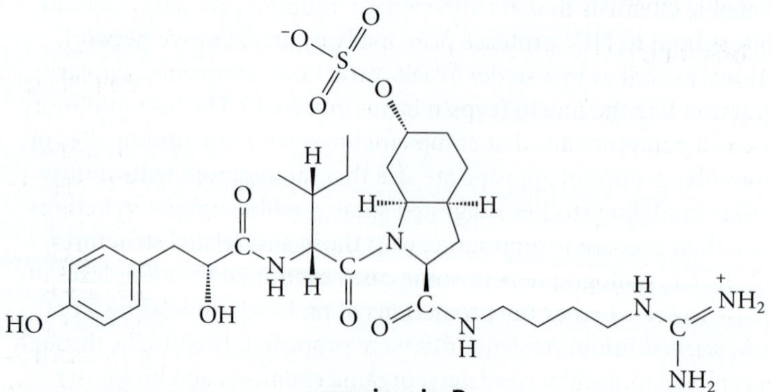

aeruginosin-B

Postulate one structural reason that aeruginosin-B binds to trypsin. Explain.

27.40 Display 27.72 shows the first step in the curved-arrow mechanism of peptide hydrolysis catalyzed by HIV protease. Complete the mechanism, using the two aspartic acid residues as catalytic groups.

27.41 The scientists who developed Norvir stated that one of the significant interactions of the inhibitor with the enzyme is hydrogen bonding of a thiazole nitrogen (the thiazole on the right side of the structure in Display 27.73) with a backbone N—H of a nearby peptide bond on the enzyme. Draw a thiazole such as the one in Norvir and, using it, show such a hydrogen-bonding interaction.

27.42 Show all of the potential hydrogen-bonding sites of Darunavir; explain whether they are acceptor or donor sites.

CHAPTER SUMMARY *For a summary of the chapter, see Chapter 27 in the* Study Guide and Solutions Manual.

REACTION REVIEW *For a summary of reactions discussed in this chapter, see the* Reaction Review *section of Chapter 27 in the* Study Guide and Solutions Manual.

SKILLS OBJECTIVES WITH PROBLEMS

- **Know the structures, names, three-letter abbreviations, and one-letter abbreviations of the 20 naturally occurring amino acids, and the conventions for using these abbreviations in peptide structures. 27.1A,B**

27.43 A peptide called *S-peptide* is obtained from pig liver ribonuclease by digestion with a proteolytic enzyme subtilisin. The primary sequence of S-peptide is as follows:

Lys-Glu-Ser-Ala-Ala-Ala-Lys-Phe-Glu-Arg-Gln-His-Met-Asp-Ser-Ser-Thr-Ser-Ser-Ala

(a) Give the sequence of S-peptide as single-letter abbreviations for the amino acids.

(b) Give the structures of residues 1, 5, 8, 9, 11, and 13 from the amino end.

- **Relate the D,L-system to the Cahn–Ingold–Prelog (*R*,*S*) stereochemical system for α-amino acids. 27.2**

27.44 You have just prepared a new α-amino acid, *isomethionine*, with the following structure and stereochemistry.

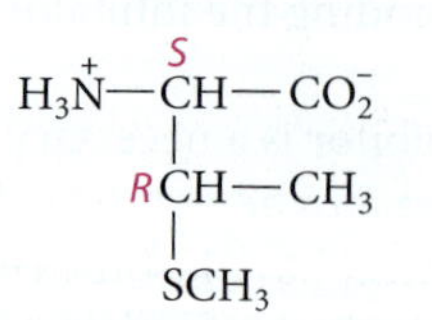

(2*S*,3*R*)-isomethionine

(a) Draw a line-and-wedge structure for this amino acid.

(b) Classify it as a D or L amino acid, and explain your reasoning.

(c) You decide to prepare the (2*S*,3*S*)-diastereomer of this compound. Is this a D or an L amino acid?

(d) How do you differentiate the isomer in part (c) from isomethionine in the D,L-system?

- **Show the acid–base equilibria for α-amino acids. 27.3A,B**

27.45 *Ornithine* is a naturally occurring amino acid that is *not* one of the 20 found in proteins.

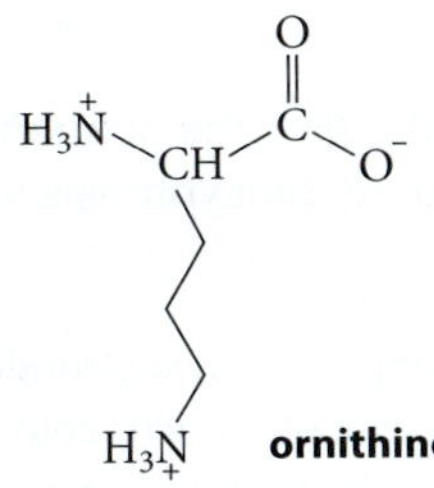

ornithine

The pK_a values for the conjugate acid of ornithine are 2.1, 8.5, and 10.5.

(a) Starting with the form of ornithine present in 1 *M* HCl solution, show the acid–base dissociation equilibria of ornithine, and label each dissociation with the appropriate pK_a value.

(b) Sketch a plot of the fraction of each form as a function of pH from pH 1 to pH 12.

- **Decide whether an α-amino acid or peptide is classified as acidic or basic; calculate its isoelectric point, and use the isoelectric points of amino acids and peptides to predict their separation behavior using techniques that separate peptides on the basis of charge. 27.3C,D**

27.46 Which of the following statements would correctly describe the isoelectric point of *cysteic acid*, an oxidation product of cysteine? Explain your answer.

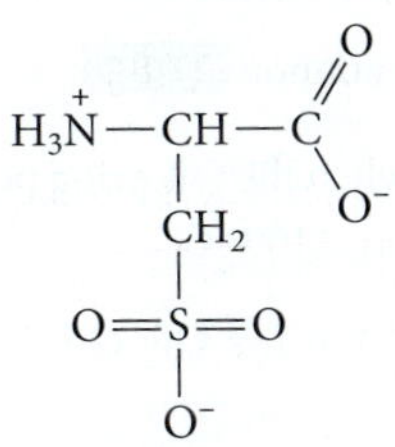

cysteic acid

(1) Lower than aspartic acid

(2) About the same as aspartic acid

(3) About the same as cysteine

(4) About the same as lysine

(5) Higher than lysine

27.47 According to its amino acid composition (see Fig. 27.6, Sec. 27.8A), lysozyme has an isoelectric point that is (choose one and explain):

(1) <<6 (2) about 6 (3) >>6

27.48 In *paper electrophoresis*, amino acids and peptides can be separated by their differential migration in an electric field. To the center of a strip of paper, a mixture of the following three peptides is applied in a single small spot: Gly-Lys, Gly-Asp, and Gly-Ala. The paper is soaked in a pH = 6 buffer, a positively charged electrode (anode) is attached to the left side of the paper, and a negatively charged electrode (cathode) is attached to the right side. A voltage is applied across the ends of the paper for a time, after which the peptides have separated into three spots: one near the cathode, one near the anode, and one in the center, at the location of the original spot. Which peptide is in each spot? Explain.

27.49 Consider the following three peptides, which may have been isolated from the floor of a Big Ten basketball arena circa 1990:

A: K-E-A-D-Y *B*: K-N-I-G-H-T *C*: P-A-I-N-T-E-R

(a) Which peptide is most basic (has the higher isoelectric point)? Explain.

(b) In what order are the three peptides predicted to emerge from an anion-exchange column developed with a pH 6.0 buffer?

- **Propose syntheses of amino acids and their derivatives using the methods discussed in this chapter. 27.4**

(1) Alkylation of ammonia

(2) Alkylation of aminomalonate derivatives

(3) The Strecker synthesis

27.50 Propose syntheses for each of the following compounds.

(a) 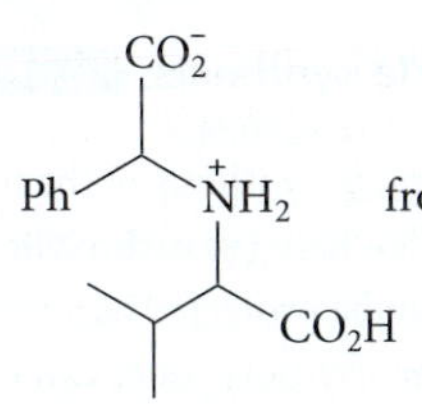 from valine

(*Hint*: Use a Strecker synthesis.)

(b) Norleucine (2-aminohexanoic acid) from 1-bromopentane.

(c) Glutamic acid from diethyl α-acetamidomalonate. (*Hint:* See Eq. 22.113 in Sec. 22.9A, and Sec. 22.9C.)

27.51 Show how the acetamidomalonate method (along with other steps) can be used to prepare the following unusual amino acids from the indicated starting material and any other reagents.

(a) $(CH_3)_2CDCH_2—CH(^{+}NH_3)—CO_2^-$ from isobutylene (methylpropene)

(b) $Ph—CHD—CH(^{+}NH_3)—CO_2^-$ from benzaldehyde

(c) 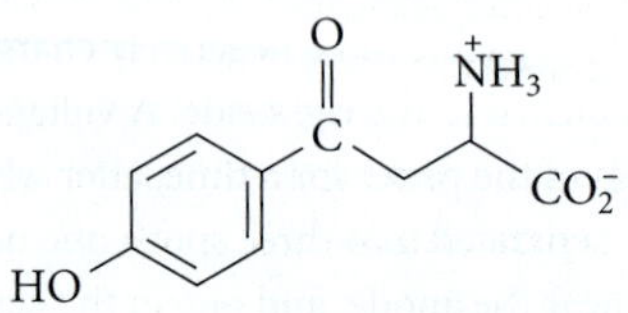 from anisole (methoxybenzene)

γ-oxohomotyrosine

• Complete esterification or acylation reactions of amino acids. 27.5

27.52 Complete the following reactions by giving the structure of the product.

(a) 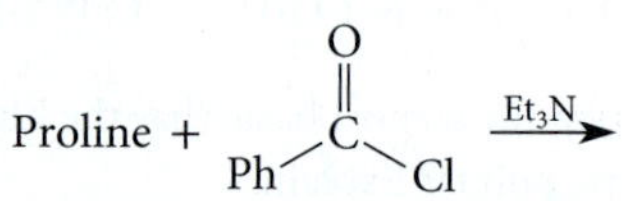

(b) Product of part (a) + $PhCH_2—Cl \xrightarrow[\text{acetone}]{K_2CO_3}$

(c) 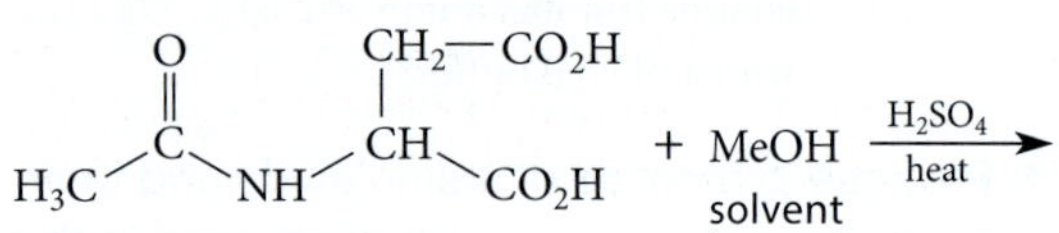

N-acetylaspartic acid

(d) 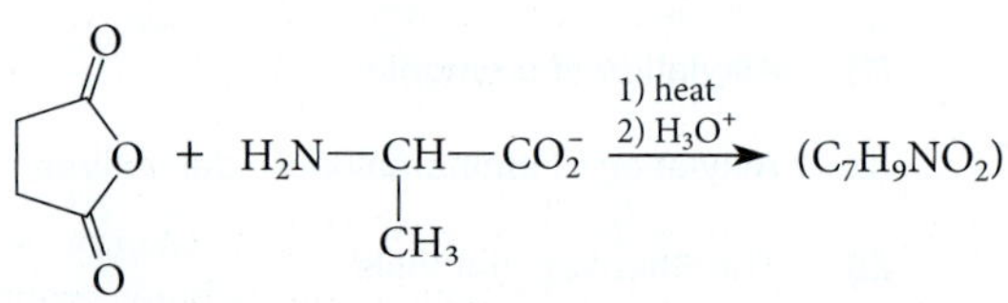

succinic anhydride

• Plan a solid-phase peptide synthesis. 27.6A

27.53 Outline a solid-phase peptide synthesis of the tetrapeptide Pro-Val-Glu-Ala. The side-chain (γ) carboxylic acid group of glutamic acid can be protected as a *tert*-butyl ester. Like Boc groups, the *tert*-butyl ester can be removed with trifluoroacetic acid. Assume for purposes of this synthesis that glutamic acid γ-*tert*-butyl ester is available commercially.

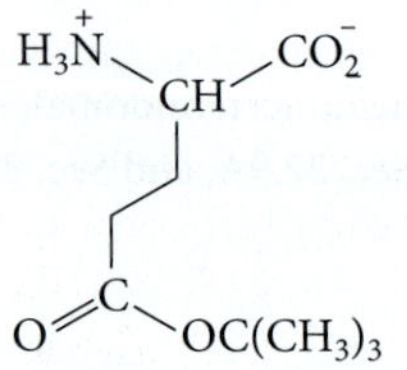

glutamic acid γ-*tert*-butyl ester

27.54 In repeated attempts to synthesize the dipeptide Val-Leu, a peptide chemist uses the following procedure.

> The cesium salt of leucine is allowed to react with chloromethyl resin (as in Eq. 27.38, Sec. 27.6A). The resulting derivative is treated with Fmoc-Val and PyBOP, then with trifluoroacetic acid.

The chemist obtains very little if any of the desired peptide at each attempt. Give a reason for this failure, and correct the procedure so that a good yield of Val-Leu should be obtained.

• Given the structure of a peptide, give the sequence of the DNA and mRNA templates for its biosynthesis with the aid of Table 27.2. 27.6B

27.55 (a) Which one of the following 2′-deoxynucleotide nucleotide sequences (3′-end on the left) could be the DNA sequence that codes for the biosynthesis of peptide *A* in Problem 27.49 (amino acid sequence K-E-A D-Y)? Explain your reasoning.

Sequence 1: CGUGAAGCCGACUACUAA

Sequence 2: TTCCTTCGGCTGATA

Sequence 3: ATTATGGTGCGGGATGCA

Sequence 4: GCACUUCGGCUGAUGAUU

(b) Using abbreviated structures like those used in Fig. 27.4 (Sec. 27.6B), give the structure of the aminoacyl-tRNA that codes for the first amino acid in the biosynthesis of peptide *A*. Include the sequence of residues in the anticodon region and the structure of the 3′ (acceptor) end.

• Give the products of the following reactions of peptides.

(1) Hydrolysis in concentrated acid 27.7A

(2) Trypsin-catalyzed hydrolysis 27.7B

(3) One cycle of the Edman degradation 27.8C

27.56 What are the products formed when the following peptide is hydrolyzed in 6 *N* HCl solution at 100 °C?

His-Ala-Gln-Val-Leu-Leu-Arg-Asn-Lys-Gly-Gly-Lys

27.57 What products are formed when the S-peptide (structure in Problem 27.43) is subjected to trypsin-catalyzed hydrolysis at pH 8? Explain your reasoning.

27.58 What products are formed when the S-peptide (structure in Problem 27.43) is subjected to one cycle of the Edman degradation? Give the structure of the PTH derivative, and use abbreviations for the peptide co-product.

27.59 The peptide hormone glucagon has the following amino acid sequence:

H-S-Q-G-T-F-T-S-D-Y-S-K-Y-L-D-S-R-R-A-Q-D-F-V-Q-W-L-M-N-T

Give the products that would be obtained when this peptide is treated with each of the following.

(a) Trypsin at pH 8

(b) Ph—N=C=S, then CF_3CO_2H, then aqueous acid

- **Know the difference between ordinary chemical-ionization mass spectroscopy and MS–MS. Interpret the MS–MS fragmentation pattern of a small peptide in terms of the peptide structure. 27.8B**

27.60 Both conventional chemical-ionization mass spectrometry (CI) and tandem mass spectrometry (MS–MS) provide protonated (that is, M + 1) fragment ions. How do the two techniques differ, and why is this difference important in peptide sequencing?

27.61 A pentapeptide obtained from tryptic digestion of a larger peptide is subjected to MS–MS, and the fragments are obtained at the following *m/z* values: 614.29, 598.28, 470.18, 371.08, 300.08, 113.98. One cycle of Edman degradation gave the PTH derivative of Leu. What is the sequence of the peptide?

- **Given the sequence of several smaller peptide fragments of a larger peptide, use the method of overlapping peptides to derive the sequence of the larger peptide. 27.8B**

27.62 A peptide *Q* has the following composition by amino acid analysis:

Q: Ala,Arg,Asp,Gly_2,Glu,Leu,Val_2,NH_3

Treatment of *Q* once with the Edman reagent followed by anhydrous acid gives a new peptide *R* with the following composition by amino acid analysis:

R: Ala,Arg,Asp,Gly_2,Glu,Val_2,NH_3

Treatment of *Q* and *R* with the enzyme *dipeptidylaminopeptidase* (DPAP) yields a mixture of the following peptides:

$Q \xrightarrow{\text{DPAP}}$ Arg-Gly, Gln-Ala, Leu-Val, Val-Asp, Gly

$R \xrightarrow{\text{DPAP}}$ Ala-Gly, Asp-Gln, Gly-Val, Val-Arg

What is the amino acid sequence of *Q*?

- **Know the two types of protein glycosylation and how they differ. 27.8E**

27.63 The text discussed two types of protein glycosylation: *O*-glycosylation and *N*-glycosylation. Tell which type of glycosylation fits each of the following descriptions. (Some statements may fit both, and some may fit neither.)

(a) The presence of Ser and Thr at or near the glycosylation site is required.

(b) Glycosylation occurs with inversion of configuration at the sugar anomeric carbon.

(c) The amide NH_2 of glutamine can be a site of glycosylation.

(d) The completed polysaccharide is delivered to the protein as a dolichol diphosphate derivative.

(e) The polysaccharide is built up one residue at a time while attached to the protein.

(f) The sugar residue directly attached to the protein is an *N*-acetylglucosamine.

- **Know the general reaction for protein phosphorylation and the protein residues that are commonly phosphorylated. 27.8E**

27.64 (a) Show the reaction for phosphorylation of a histidine residue in a protein. Explain your choice of the nitrogen on which phosphorylation occurs.

(b) What type of enzyme catalyzes the dephosphorylation of phosphohistidine?

(c) Phosphorylation of histidine is relatively rare. What are the more common protein residues at which phosphorylation occurs?

- **Describe the two types of secondary structure commonly found in proteins. 27.9A**

27.65 What are the two major types of secondary structure most commonly found in proteins? What is the principal type of noncovalent interaction responsible for the stability of these structures?

27.66 (a) In the following peptide, which residue donates a hydrogen bond to the carbonyl group of the valine residue when the peptide is in an α-helical conformation?

Ala-Leu-Arg-Met-His-Ala-Gln-Val-Leu-Ala-Glu-Trp-Glu-Arg-Glu-Asn-Lys-Met

(b) In the same peptide, to which residue does the N—H group of the valine residue donate a hydrogen bond?

(c) How far apart (as a vertical distance, in pm) are the two residues involved in hydrogen bonding with the valine?

27.67 Poly-L-lysine (a peptide containing only lysine residues) exists entirely in an α-helical conformation at pH > 11. Below pH 10, however, the peptide becomes a random coil. Poly-L-glutamic acid, on the other hand, exists in the α-helical conformation at pH < 4, but above pH 5 it becomes a random coil. Explain the effect of pH on the secondary structure of both polymers. That is, explain why low pH destroys the helical conformation of one peptide, whereas high pH destroys the helical conformation of the other. (*Hint:* Look carefully at the location of the amino acid side chains in Fig. 27.11, Sec. 27.9A.)

- **Explain in broad terms each of the following aspects of enzyme catalysis. 27.10A**

  (1) How enzymes are able to bring about large rate enhancements

  (2) Why enzymes show substrate specificity

  (3) Why the catalytic mechanisms of enzymes are difficult to duplicate in the laboratory with nonenzyme catalysts

27.68 (a) In general terms, why do enzymes bring about such large rate enhancements in the reactions that they catalyze?

(b) Why are these large rate enhancements difficult to duplicate with synthetic catalysts?

27.69 Use the mechanism of trypsin catalysis to explain the specificity of trypsin for peptide hydrolysis at Arg and Lys residues.

27.70 The *V8 protease* from the organism *Staphylococcus aureus* has found wide use in the specific hydrolysis of proteins because it catalyzes peptide hydrolysis specifically at the carbonyl group of Glu and Asp residues. What type of amino acid residue would you expect to find in the active site of this protease that would cause this type of specificity?

INTEGRATED PROBLEMS

27.71 Referring to Table 27.1, Sec. 27.1A, identify the amino acid(s) that satisfy each of the following criteria.

(a) The most acidic amino acid

(b) The most basic amino acid

(c) The amino acids that can exist as diastereomers

(d) The amino acid that has zero optical rotation under all conditions

(e) The amino acids that are converted into other amino acids on treatment with concentrated hot aqueous NaOH solution followed by neutralization

27.72 A peptide was subjected to one cycle of the Edman degradation, and the following compound was obtained. What is the amino-terminal residue of the peptide?

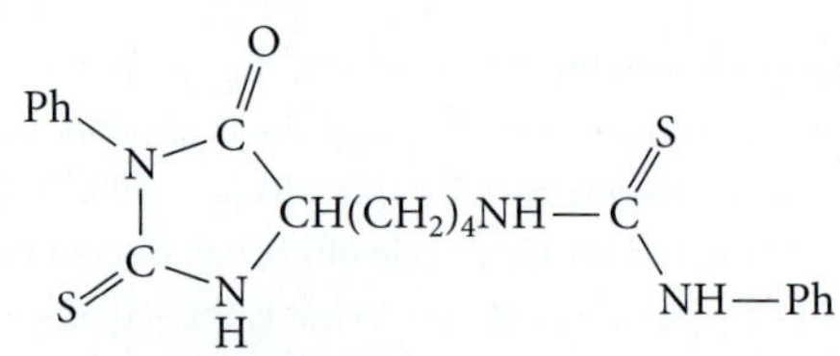

27.73 *Dansyl chloride* (5-dimethylamino-1-naphthalenesulfonyl chloride) reacts with amino groups to give a fluorescent derivative. After a peptide *P* with the composition (Arg,Asp,Gly,Leu_2,Thr,Val) reacts with dansyl chloride at pH 9, it is hydrolyzed in 6 *M* aqueous HCl. The derivative shown in the equation given in **Fig. P27.73**, detected by its fluorescence, is isolated after neutralization, along with the free amino acids Arg, Asp, Gly, Leu, and Thr. What conclusion can be drawn about the structure of the peptide from this result?

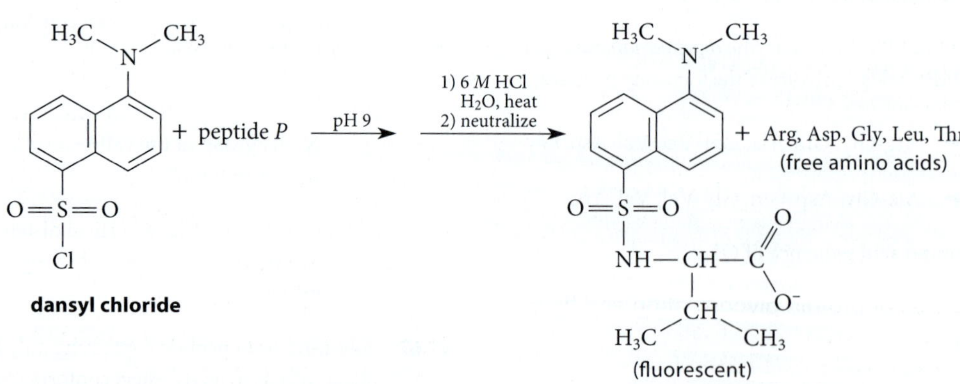

FIGURE P27.73

27.74 A peptide C was found to have a molecular mass of about 1000. Amino acid analysis of C revealed its composition to be (Ala$_2$,Arg,Gly,Ile). The peptide was unchanged on treatment with the Edman reagent, then CF_3CO_2H. Treatment of C with trypsin gave a *single* peptide D with an amino acid analysis identical to that of C. Three cycles of the Edman degradation applied to D revealed the partial sequence Ala-Ile-Gly.

(a) Suggest a structure for peptide C, and explain how you arrived at that structure.

(b) Describe what you would expect to see for the b-type fragmentation of the M + 1 ion of both peptides C and D in MS–MS.

27.75 When bovine insulin is treated with the Edman reagent followed by anhydrous CF_3CO_2H, then by aqueous acid, the PTH derivatives of *both* glycine and phenylalanine are obtained in nearly equal amounts. What can be deduced about the structure of insulin from this information?

27.76 A previously unknown amino acid, γ-carboxyglutamic acid (Gla), was discovered to be a posttranslational modification in the amino acid sequence of the blood-clotting protein prothrombin.

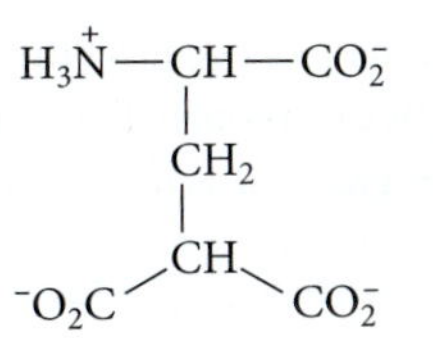

γ-carboxyglutamic acid (Gla)

This amino acid escaped detection for many years because, on acid hydrolysis, it is converted into another common amino acid. Explain this observation.

27.77 When a mixture of the amino acids Phe and Gly is subjected to chromatography in a pH 6 buffer on the ion-exchange resin shown in Display 27.23 (Sec. 27.3D), the Phe emerges from the column much later than the Gly, even though the two amino acids have the same isoelectric point. Explain.

27.78 (a) What reagent would be used to convert the corresponding chloromethyl polystyrene resin into the following resin?

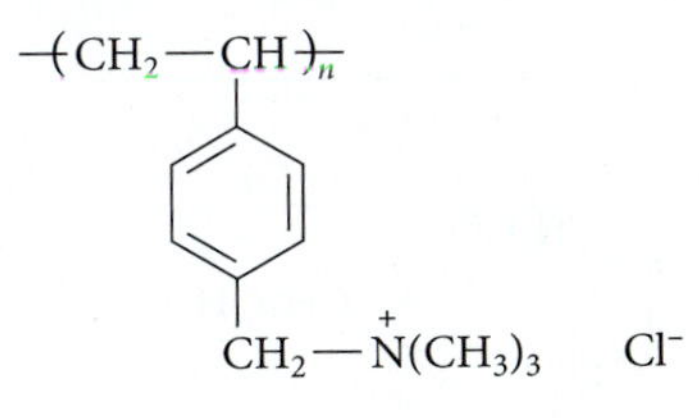

(b) A mixture of the amino acids Arg, Glu, and Leu is added to a column containing this resin suspended in a pH 6 buffer. The column is then eluted with the same buffer. In what order will the amino acids emerge from the column? Explain.

27.79 Explain each of the following observations.

(a) The optical rotation of alanine is different in water, 1 *M* HCl, and 1 *M* NaOH.

(b) Two mono-*N*-acetyl derivatives of lysine are known.

(c) The peptide Gly-Ala-Arg-Ala-Glu is readily hydrolyzed by trypsin in water at pH = 8, but it is inert to trypsin in 8 *M* urea at the same pH.

(d) After peptides containing cysteine are treated with $HSCH_2CH_2OH$, then with aziridine, they can be cleaved by trypsin at their (modified) cysteine residues.

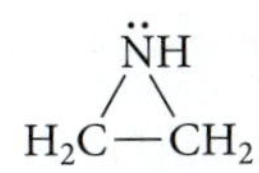

aziridine

(*Hint:* The conjugate acid of aziridine reacts like ethylene oxide.)

(e) When L-methionine is oxidized with H_2O_2, *two* separable methionine sulfoxides with the following structure are formed:

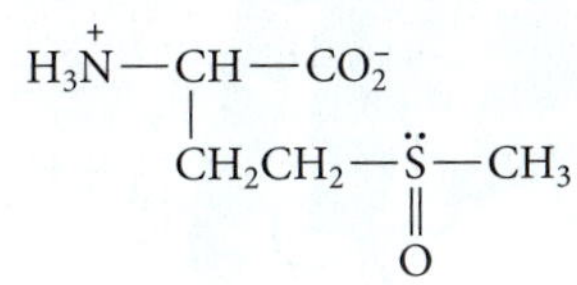

27.80 An amino acid A, isolated from the acid-catalyzed hydrolysis of a peptide antibiotic, incorporated two AQC groups when treated with AQC-NHS and had a specific optical rotation (HCl solution) of +37.5° mL g^{-1}dm^{-1}. Compound A was not identical to any of the amino acids in Table 27.1 (Sec. 27.1A). The isoelectric point of compound A was found to be 9.4. Compound A could be prepared by the reaction of L-glutamine with Br_2 in NaOH, followed by neutralization. (See Sec. 23.11D.) Suggest a structure for A.

27.81 (a) When proteins are prepared for sequencing, they are treated with DTT (Eq. 27.53, Sec. 27.8A) and then with an excess of iodoacetic acid, at pH = 8–9. Explain how iodoacetic acid reacts with the side-chain thiol group of a cysteine residue and why this reaction is a necessary prelude to sequencing.

(b) A different reaction that serves the same purpose is oxidation of the disulfide bonds with H_2O_2. What is the product of this oxidation?

27.82 One posttranslational side-chain modification of proteins is the methylation of aspartic acid residues to give a side-chain methyl ester. Draw the structure of this residue, and indicate what coenzyme is involved in this reaction. (*Hint:* See Sec. 12.7B.)

27.83 Sometimes preparations of the protease *chymotrypsin* (Sec. 27.7B) are contaminated with small amounts of trypsin. This can be a problem if the specific hydrolysis of peptides with *only* chymotrypsin is desired. How could trypsin-catalyzed hydrolysis with this chymotrypsin–trypsin mixture be avoided without having to separate the two enzymes? (*Hint:* See Fig. 27.19, Sec. 27.10B.)

27.84 Following is a ribbon diagram for one type of *opioid receptor*, a family of proteins that bind morphine and other opioids and initiate their physiological effects. The opioid receptor consists mostly of α-helices; the receptor spans the cell membrane. Parts of the helix (R-groups in the following diagram) are adjacent to the lipid bilayer of the membrane.

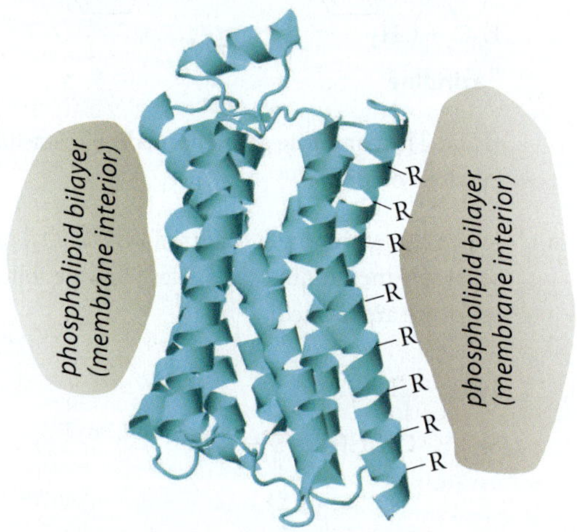

Which group of amino acid residues are likely to be found in greatest proportion near the phospholipid bilayer (choose one)? Explain why.

A: Asp and Glu | *B:* Lys, Arg, and His
C: Gly | *D:* Pro
E: Phe, Leu, Ile, and Val | *F:* Ser, Thr, Gln, and Asn

27.85 Complete the following reactions. Assume in each case that the amino acid residue is part of a peptide in aqueous solution and is at neither the amino nor the carboxy terminus.

(a) Lysine residue + $H_2C{=}O$ + $NaCNBH_3$ $\xrightarrow{\text{pH 9}}$
(excess of both)

(b)

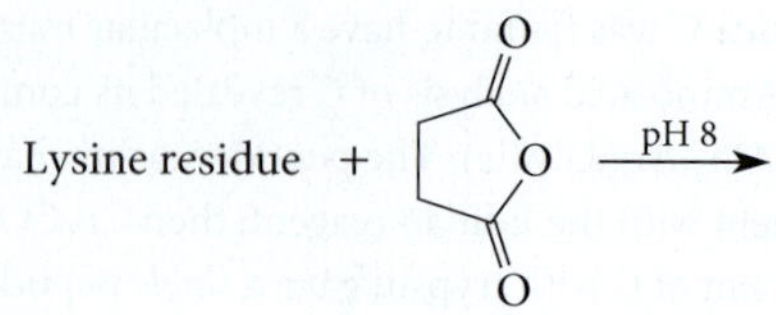

succinic anhydride

(c)

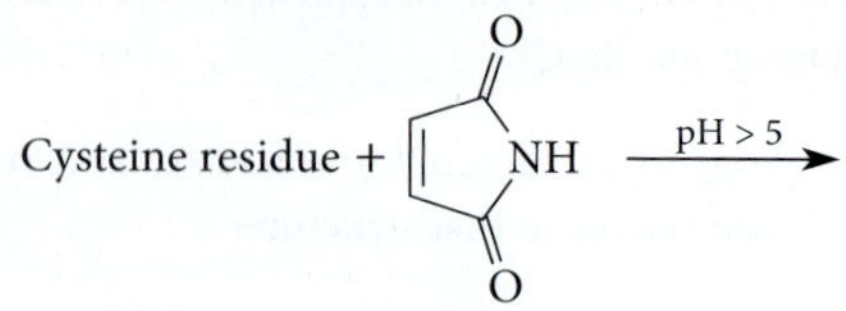

maleimide

(*Hint*: See Sec. 22.9A.)

(d) Tyrosine residue + $Ph\overset{+}{N}{\equiv}N$ Cl^- $\xrightarrow{\text{pH 9}}$

27.86 The artificial sweetener *aspartame* (sidebar, Sec. 24.11A) was withheld from the market for several years because, on storage for extended periods of time, particularly in carbonated soft drinks, it forms a *diketopiperazine* (see **Fig. P27.86**). (Extensive biological testing was required to show that this by-product was safe for consumers.) Give a curved-arrow mechanism for the formation of the diketopiperazine.

27.87 When glutamic acid or glutamine occurs at the amino-terminal residue of a peptide or protein, they often spontaneously form a pyroglutamic acid residue:

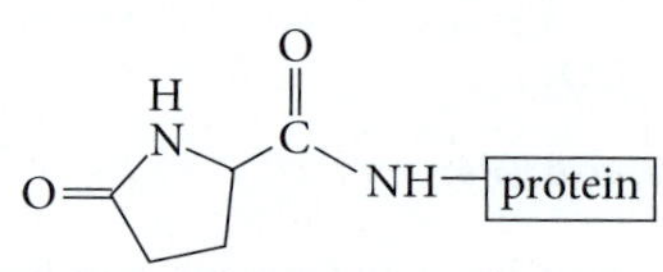

a pyroglutamic acid residue

(a) Assuming the presence of H_3O^+ and H_2O, propose a curved-arrow mechanism for the formation of a pyroglutamic acid residue from an *N*-terminal glutamic acid residue in a peptide.

(b) How does this residue affect the sequencing of this peptide by the Edman method? By MS–MS?

(c) Why don't *N*-terminal aspartic acid and asparagine undergo the same transformation?

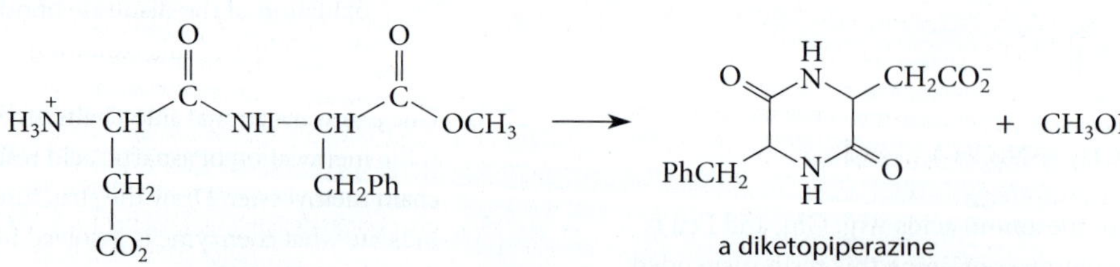

aspartame

a diketopiperazine

FIGURE P27.86

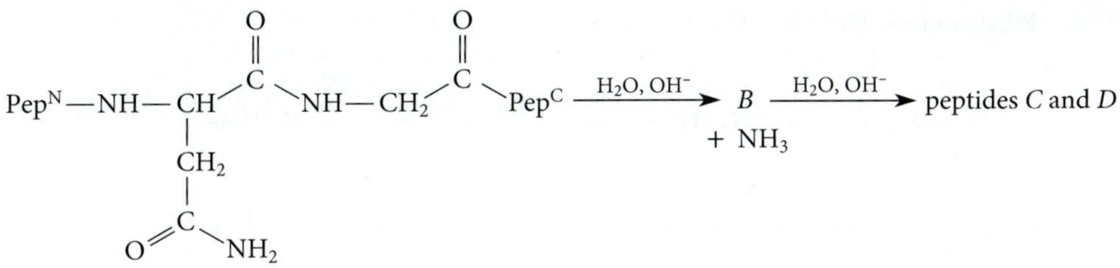

A

FIGURE P27.88

27.88 When peptides containing the Asn-Gly sequence, such as *A* in the equation given in **Fig. P27.88**, are stored in aqueous solution at neutral or slightly basic solution, ammonia is liberated and a derivative *B* is formed. On continued storage, species *B* reacts with water to give two new peptides: *C* and *D*. Peptide *C* is the same as peptide *A* except that Asn is replaced by Asp, and peptide *D* is an isomer of peptide *C*. Propose structures for *B* and for peptides *C* and *D*, and rationalize their formation using the curved-arrow notation. (These reactions are believed to be a major source of deterioration associated with aging in naturally occurring peptides and proteins.)

27.89 Outline a synthesis of each of the following compounds from the indicated starting material and any other reagents.

(a) 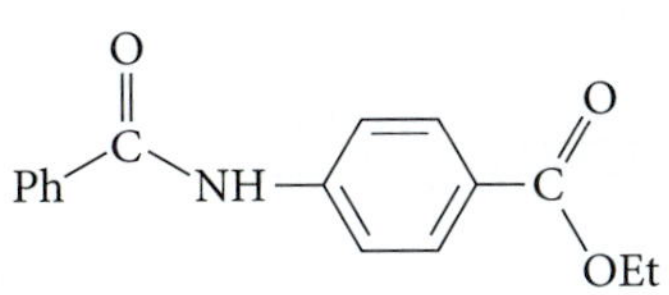 from *p*-aminobenzoic acid

(b) $Ph-CD(^{+}NH_3)-CO_2^-$ from benzoic acid

(c) $CD_3-CH(^{+}NH_3)-CO_2^-$ from $CD_3-CH{=}O$

(d) 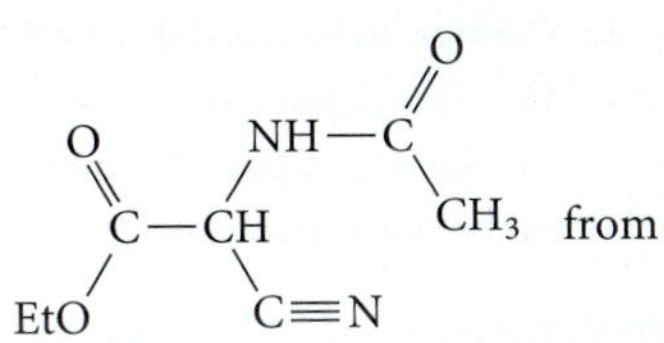 from 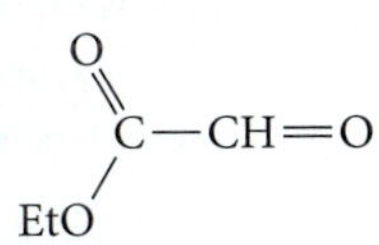

(e) 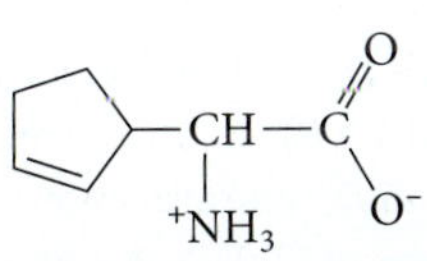 from the product of part (d) and cyclopentene

Cyclopentenyl amino acids are produced by certain plants. (*Hint:* Hydrogens α to a cyano group are about as acidic as those α to an ester group.)

(f) The polymer *p*-aramid from terephthalic acid (1,4-benzenedicarboxylic acid). (This polymer is used in tire cord and other applications that require rigidity and strength.)

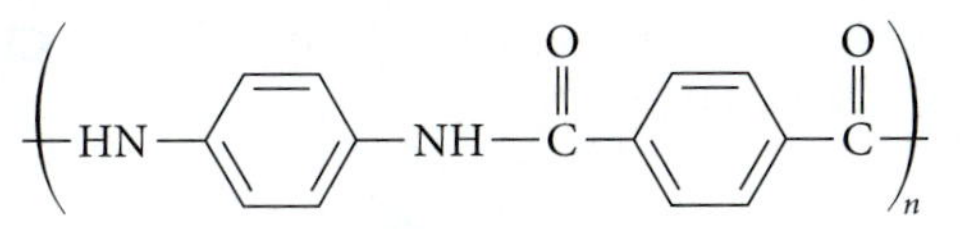

***p*-aramid**

27.90 Complete the reactions given in **Fig. P27.90** by giving the structure of the major organic product(s). Explain how you arrived at your answer.

27.91 Draw a curved-arrow mechanism for each of the reactions given in **Fig. P27.91**.

27.92 (a) In most peptides, the amide bonds have the *Z* conformation; explain why.

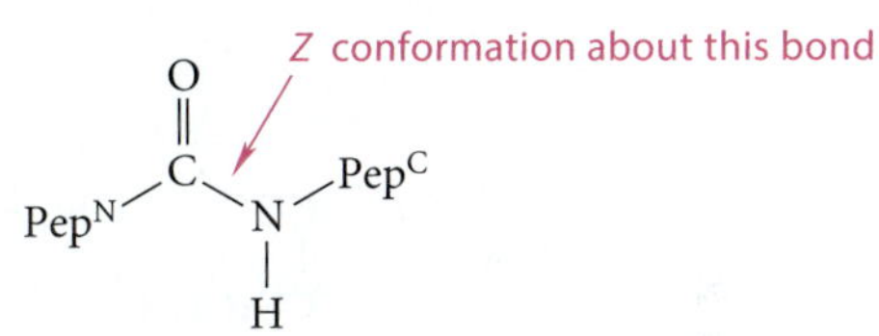

(b) One particular amino acid residue in the Pep^C position adopts the *E* conformation in some cases. Which amino acid residue should be most likely to assume an *E* conformation, and why?

27.93 (a) Explain why two monomethyl esters of *N*-acetyl-L-aspartic acid are known. Draw their structures.

(b) Explain why a mixture of these two compounds can be separated by cation-exchange chromatography at pH 3.0, but not at pH 7. (*Hint:* Use the pK_a values of aspartic acid in Table 27.1, Sec. 27.1A.) Your explanation should indicate which of the two compounds would emerge first from a cation-exchange column at pH 3.0.

27.94 When *N*-acetyl-L-aspartic acid is treated with acetic anhydride, an optically active compound *A*, $C_6H_7NO_4$, is formed. Treatment of *A* with the amino acid L-alanine yields two separable, isomeric peptides, *B* and *C*, that are both converted into a mixture of L-alanine and L-aspartic acid by prolonged acid hydrolysis. Suggest structures for *A*, *B*, and *C*.

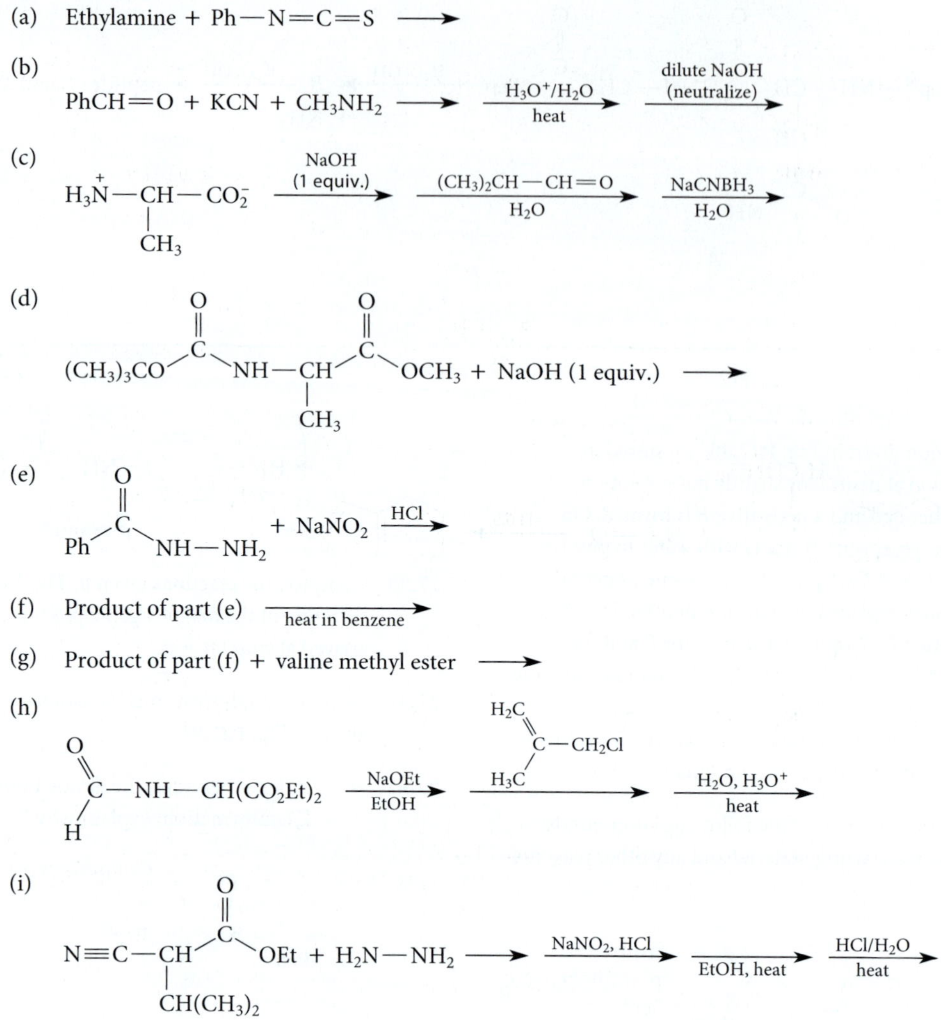

FIGURE P27.90

27.95 One technique of preparing large proteins by solid-phase peptide synthesis is to prepare fragments of the protein of manageable size and then form the amide bond between them to create a larger peptide. This process, called *peptide ligation*, was developed by chemist Stephen Kent and his colleagues at The Scripps Research Institute and reported in 1994. The process, which requires a cysteine residue at the amino end of one peptide and a special thiol ester at the carboxy end of the other, is shown in **Fig. P27.95**. (The thiol ester can be produced by solid-phase peptide synthesis.) The ligation is reported to work *only* with a cysteine residue at the amino terminus of peptide 2. Propose a curved-arrow mechanism for the ligation reaction that accounts for the role of the cysteine sulfur in the reaction. Explain the requirement for N-terminal cysteine in peptide 2. (*Hint:* Think about the effect of a protic solvent on the nucleophilic reactivities of an amine and a thiolate anion, which have similar basicities.)

27.96 In 2007, scientists in New Zealand isolated a peptide *P* from enzymatic digests of the pili of gram-positive bacteria. (A *pilus* is a fibrous appendage that a bacterium uses, among other things, to bind to cells, beginning the process of infection.) This peptide was sequenced by MS–MS, and the following two partial sequences were reconstructed from the mass spectrum:

A: L-T-V-T-K-N-L   *B*: N-S-L

The mass M of *P*, deduced from the m/z of its $M+1$ ion, equaled the mass of $(A+B)-17$ mass units. Two other fragment ions, *C* and *D*, were also observed. The mass of *C* equaled the mass of (T-K-N-L + N-S-L) − 17 mass units, and the mass of *D* equaled the mass of (V-T-K-N-L + N-S-L) − 17 mass units.

(a) The scientists used these data to propose a primary structure for *P* that contains an unusual peptide bond. What is this structure? Show how it is consistent with the 17 mass-unit differences.

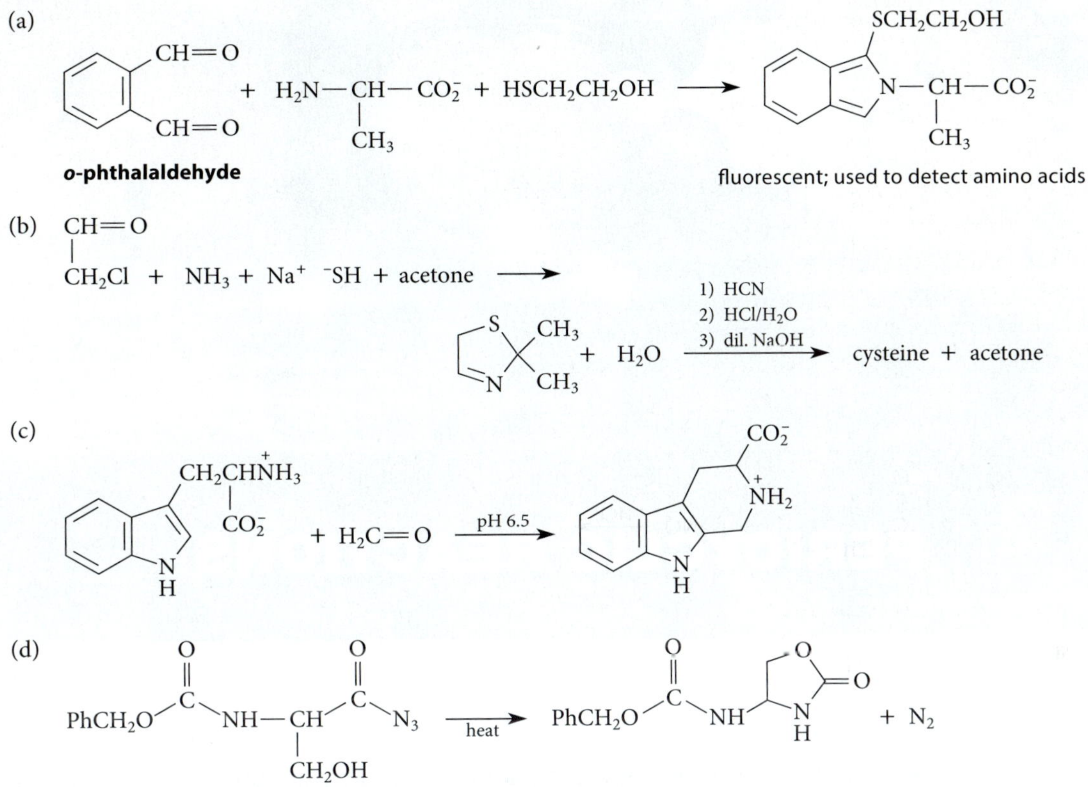

FIGURE P27.91

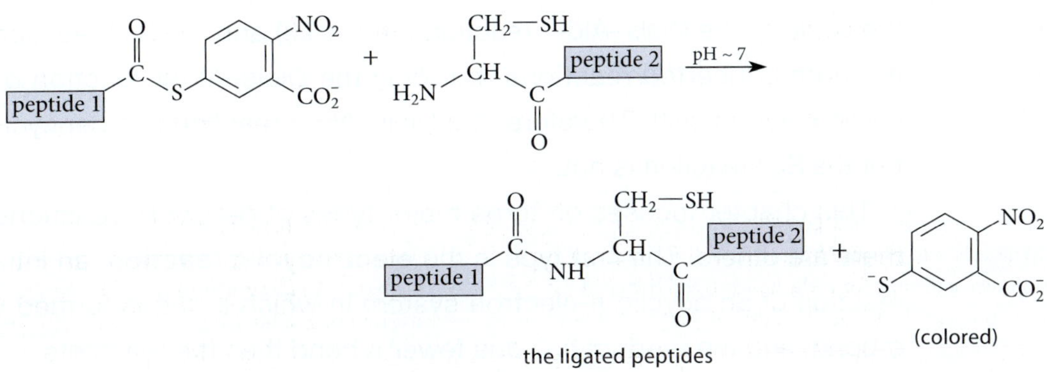

FIGURE P27.95

(b) The scientists also suggested that the unusual peptide bond forms spontaneously by a reaction between two peptide chains within the pilus. Show this reaction and its mechanism.

(c) No such reaction is observed when the peptides *A* and *B* are mixed in solution at physiological pH, and, evidently, no enzyme is involved. Why might this reaction nevertheless proceed at a rapid rate within the pilus structure? (*Hint:* See Sec. 12.8.)

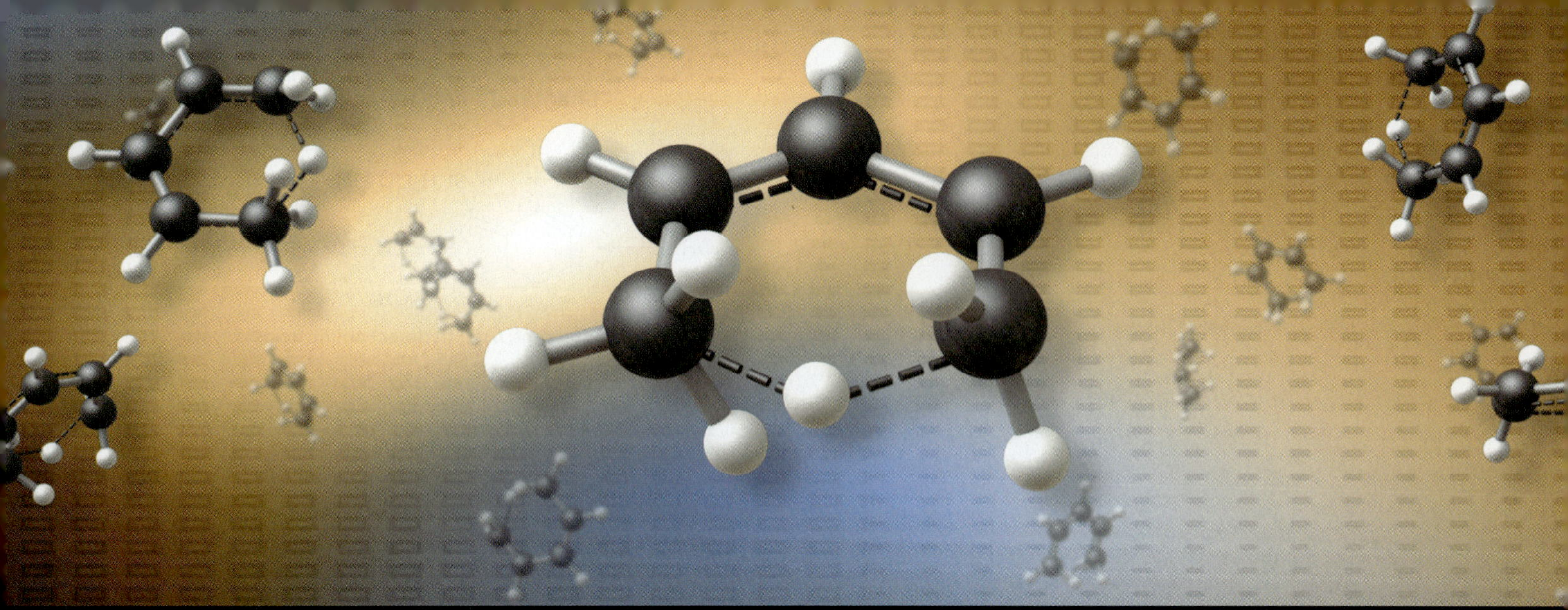

28 PERICYCLIC REACTIONS

Pericyclic reactions occur by a concerted cyclic shift of electrons. This definition states two key elements of pericyclic reactions. First, a pericyclic reaction is *concerted*. In a concerted reaction, reactant bonds are broken and product bonds are formed at the same time, without intermediates. Second, a pericyclic reaction involves a *cyclic shift of electrons*. (The word *pericyclic* means "around the circle.") The Diels–Alder reaction (Sec. 15.3) and the S_N2 reaction (Sec. 9.4) are both concerted reactions, but only the Diels–Alder reaction occurs by a cyclic electron shift. Therefore, the Diels–Alder reaction is a pericyclic reaction, but the S_N2 reaction is not.

This chapter focuses on three major types of pericyclic reactions, although there are others. The first type is the **electrocyclic reaction**: an intramolecular reaction of an acyclic π-electron system in which a ring is formed with a new σ bond, and the product has one fewer π bond than the reactants.

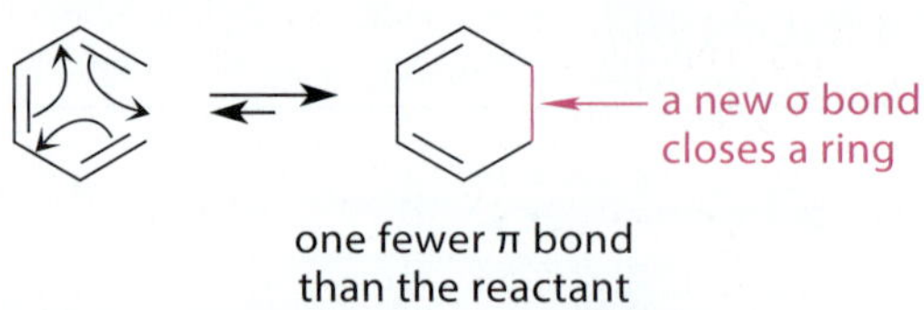

(28.1)

The second type of reaction is the **cycloaddition**: a reaction of two separate π-electron systems in which a ring is formed with two new σ bonds, and the product has two fewer π bonds than the reactants.

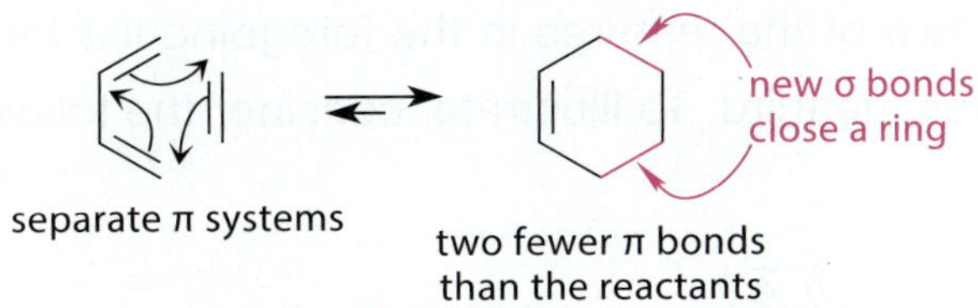

(28.2)

The third type of reaction is the **sigmatropic reaction**: a reaction in which an allylic σ bond at one end of a π-electron system appears to migrate to the other end of the π-electron system. The π bonds change positions in the process, but their total number is unchanged.

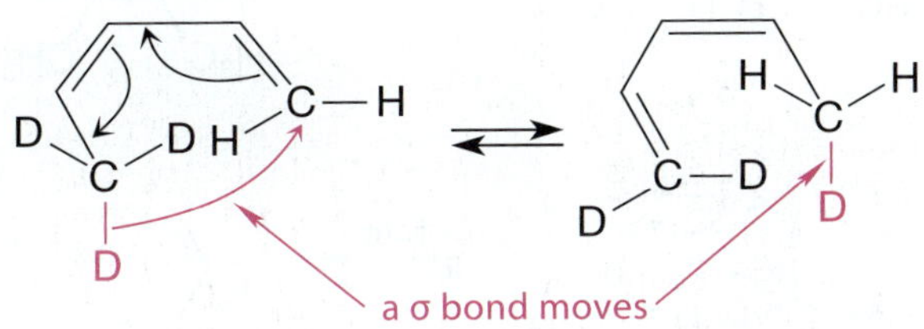

(28.3)

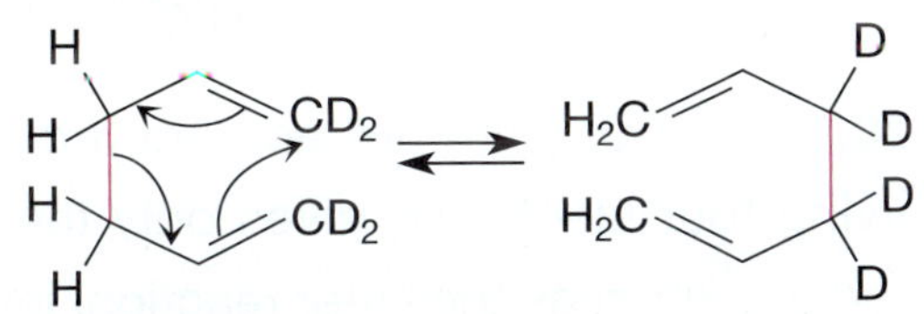

(28.4)

Three features of any given type of pericyclic reaction are intimately related:

1. The way the reaction is activated (heat or light)
2. The number of electrons involved in the reaction
3. The stereochemistry of the reaction

Before illustrating these points, let's clarify the first two terms in this list. Point 1 refers to the fact that many pericyclic reactions require no catalysts or reagents other than the reacting partners. Such reactions take place either on heating or on irradiation with ultraviolet light. Reactions that are activated by heat are not activated by light, and vice versa. Recall, for example, that Diels–Alder reactions occur merely on heating the diene and dienophile together (Sec. 15.3). These reactions are not activated by light.

The number of electrons involved in a pericyclic reaction (point 2), as in any heterolytic process, is twice the number of curved arrows required to write the reaction mechanism in the curved-arrow notation. For example:

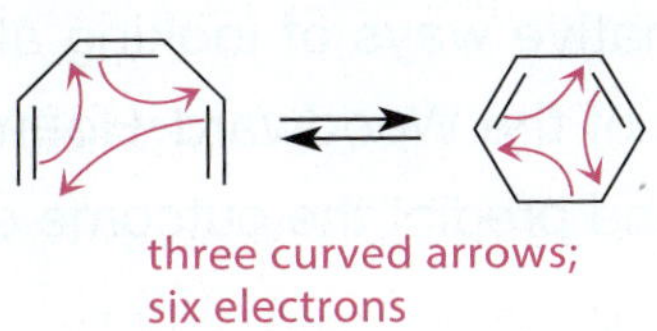

(28.5)

The direction of "electron flow" in many pericyclic reactions indicated by the curved arrows is arbitrary. Although it is clockwise in Eq. 28.5, it could be written counterclockwise and be equally correct.

Specifying any two of the features in the foregoing list for a particular type of reaction specifies the third. To illustrate, consider the following electrocyclic reactions:

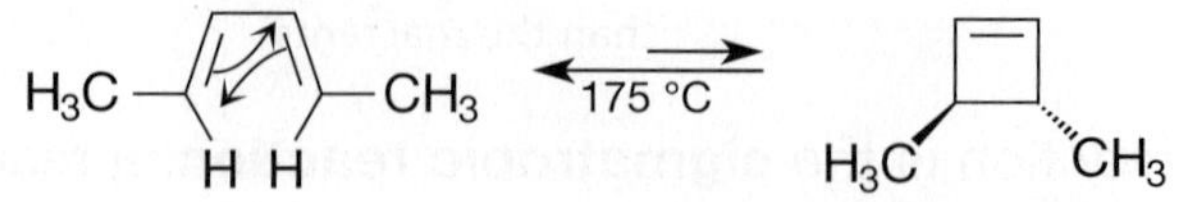

(28.6a)

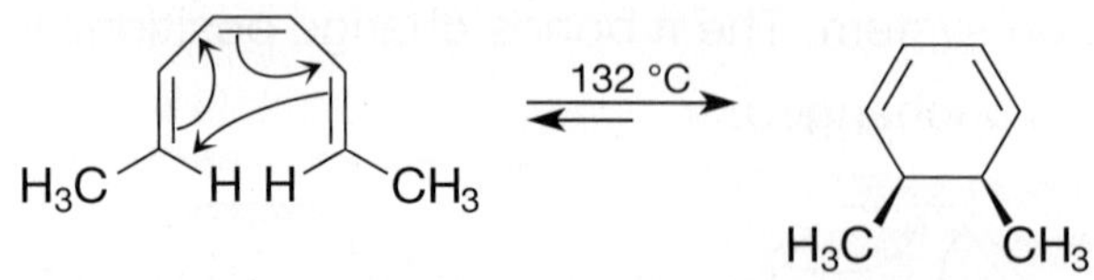

(28.6b)

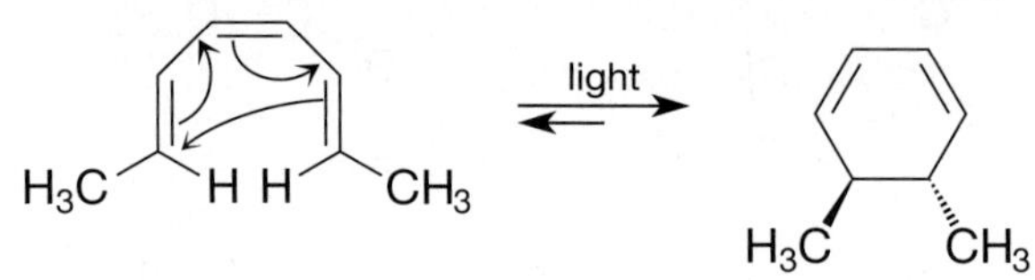

(28.6c)

First compare Eqs. 28.6a and b. Both are activated by heat; however, the former reaction, involving four electrons, gives only the trans-disubstituted isomer of the cyclic product, whereas the latter reaction, involving six electrons, gives only the cis-disubstituted isomer.

Next compare Eqs. 28.6b and c. Both reactions involve six electrons. When the starting material is heated, only the cis-disubstituted isomer of the cyclic product is obtained. When the starting material is irradiated with ultraviolet light, the only product obtained is the trans-disubstituted isomer.

Correlations such as these had been observed for many years, but the reasons for them were not understood until 1965. In that year, a theory that clearly explained these observations and successfully predicted many new ones was proposed by Robert B. Woodward (1917–1979), then a professor of chemistry at Harvard University, and Roald Hoffmann (b. 1937), at the time a junior fellow at Harvard and presently a professor emeritus at Cornell University. For this theory, called *conservation of orbital symmetry*, Hoffmann received the 1981 Nobel Prize in Chemistry. He shared the prize with Kenichi Fukui (1918–1998), a professor of chemistry at Kyoto University in Japan, who had advanced a related theory, called *frontier orbital theory*. (The two theories make the same predictions; they are alternative ways of looking at the same reactions.) This chapter presents elements of the Woodward–Hoffmann–Fukui theory that will enable you to understand and predict the outcome of pericyclic reactions.

Woodward undoubtedly would have also shared the 1981 Nobel Prize had he not died prior to its announcement. (The terms of Nobel's bequest require that the prize be awarded only to living scientists.) Woodward had, however, received an earlier Nobel Prize (1965) for his work in organic synthesis.

Focused Problem

28.1 Classify each of the following pericyclic reactions as an electrocyclic, cycloaddition, or sigmatropic reaction. Give the curved-arrow notation for each reaction, and tell how many electrons are involved.

(a)

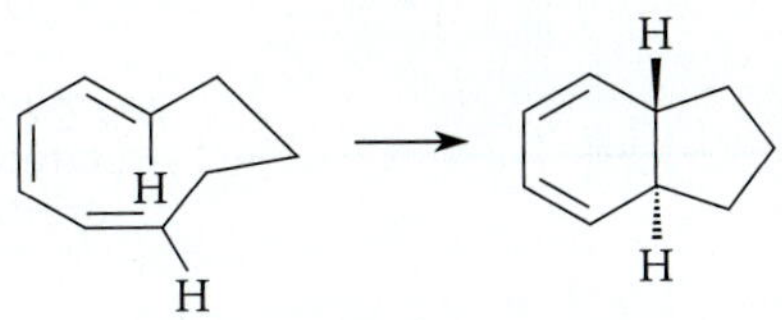

(b)

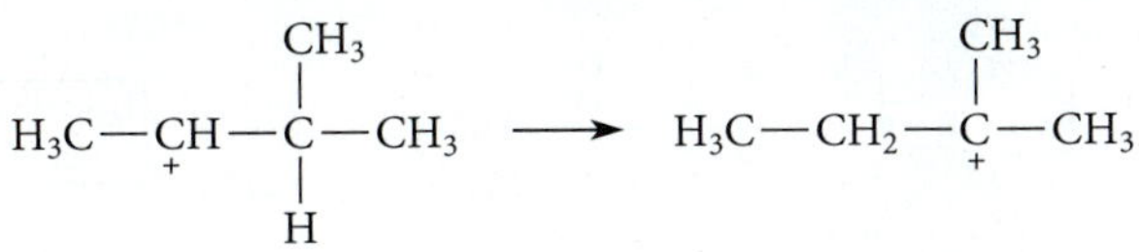

(c)

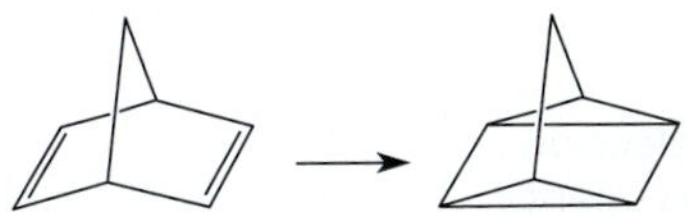

(d)

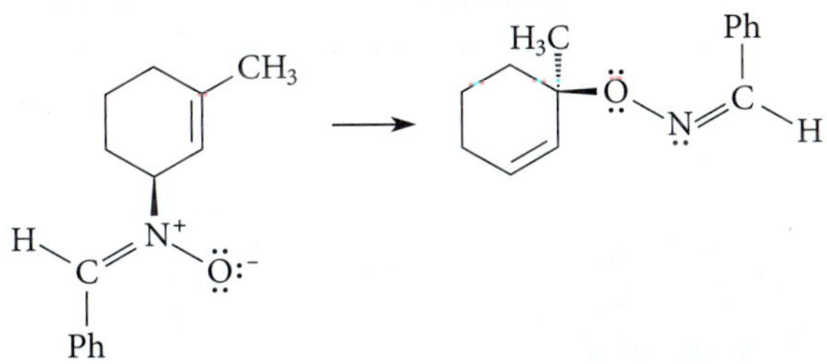

(e)

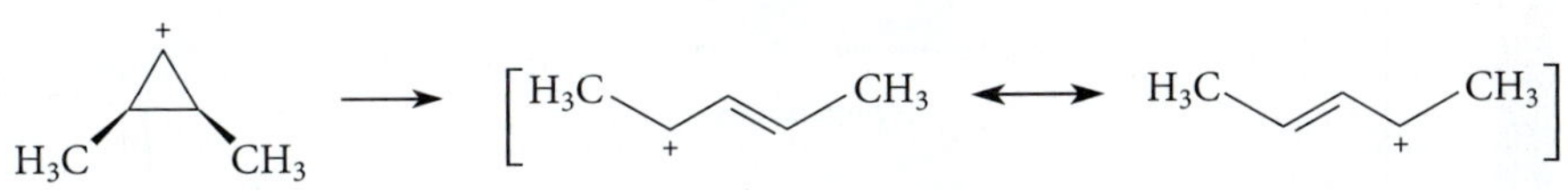

28.1 MOLECULAR ORBITALS OF CONJUGATED π-ELECTRON SYSTEMS

Understanding the theory of pericyclic reactions requires an understanding of some basics of *molecular orbital theory*, particularly as it applies to molecules containing π electrons. Molecular orbital theory was introduced in Secs. 1.8, 4.1C, 15.1A, and 15.6. The material that follows builds on those introductions.

A. Molecular Orbitals of Conjugated Alkenes

When 2p orbitals can overlap, pi (π) molecular orbitals (MOs) can form. The overlap of 2p orbitals to give π molecular orbitals is described by the mathematics of quantum theory. However, the mathematical aspects of this theory are not required to appreciate the results. This section considers the molecular orbital theory of ethylene and conjugated alkenes. The π molecular orbitals for such molecules can be constructed according to the following generalizations, which are applied to ethylene and 1,3-butadiene in **Figs. 28.1** and **28.2**. In these and subsequent figures, the carbons are flattened into the plane of the page and the hydrogens are not shown so that the nodes and symmetry relationships within the orbitals can be seen clearly. Perspective illustrations of these MOs can be found in Fig. 4.3 (Sec. 4.1C) for ethylene and in Fig. 15.1 (Sec. 15.1A) for 1,3-butadiene. Using Figs. 28.1 and 28.2, convince yourself that each generalization in the numbered list that follows applies to each example.

FIGURE 28.1 The π molecular orbitals (MOs) of ethylene. The carbons are shown as black dots, and the hydrogens are not shown. Wave peaks are shown in blue and wave troughs in green. The symmetry classification is described in Fig. 28.3 and the associated discussion. The node in π_2^*, shown with a heavy gray line, is perpendicular to the plane of the page. The nodal plane of the 2p orbitals is common to all π MOs. Only the nodes in addition to this one ("new nodes") are shown in this and subsequent figures.

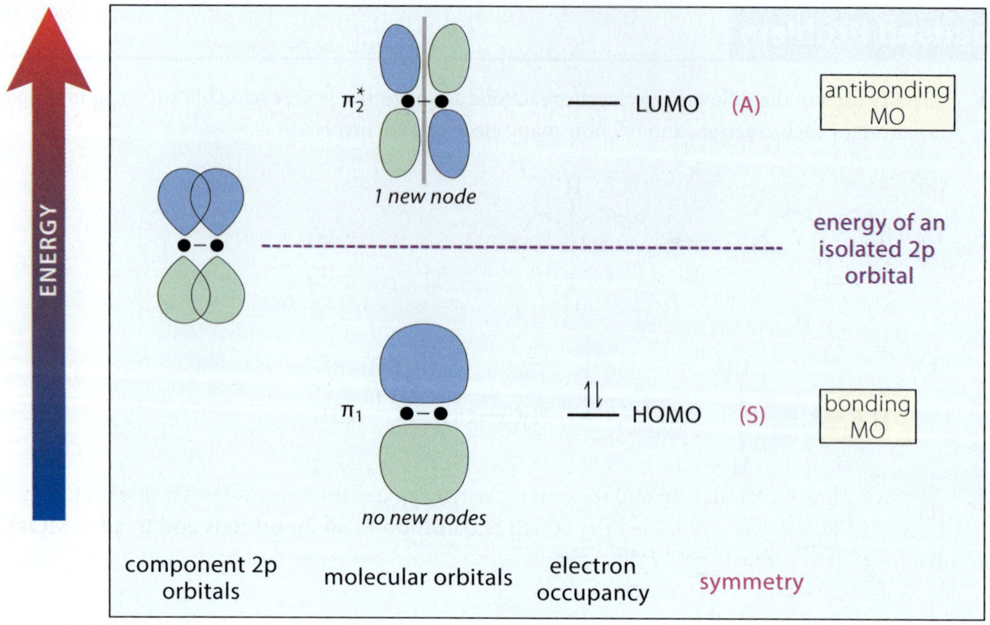

FIGURE 28.2 The π molecular orbitals of 1,3-butadiene. The carbons are flattened into the plane of the page and the hydrogens are not shown so that the nodes and symmetry relationships within the orbitals can be seen clearly. The symmetry classification is described in Fig. 28.3 and the associated discussion.

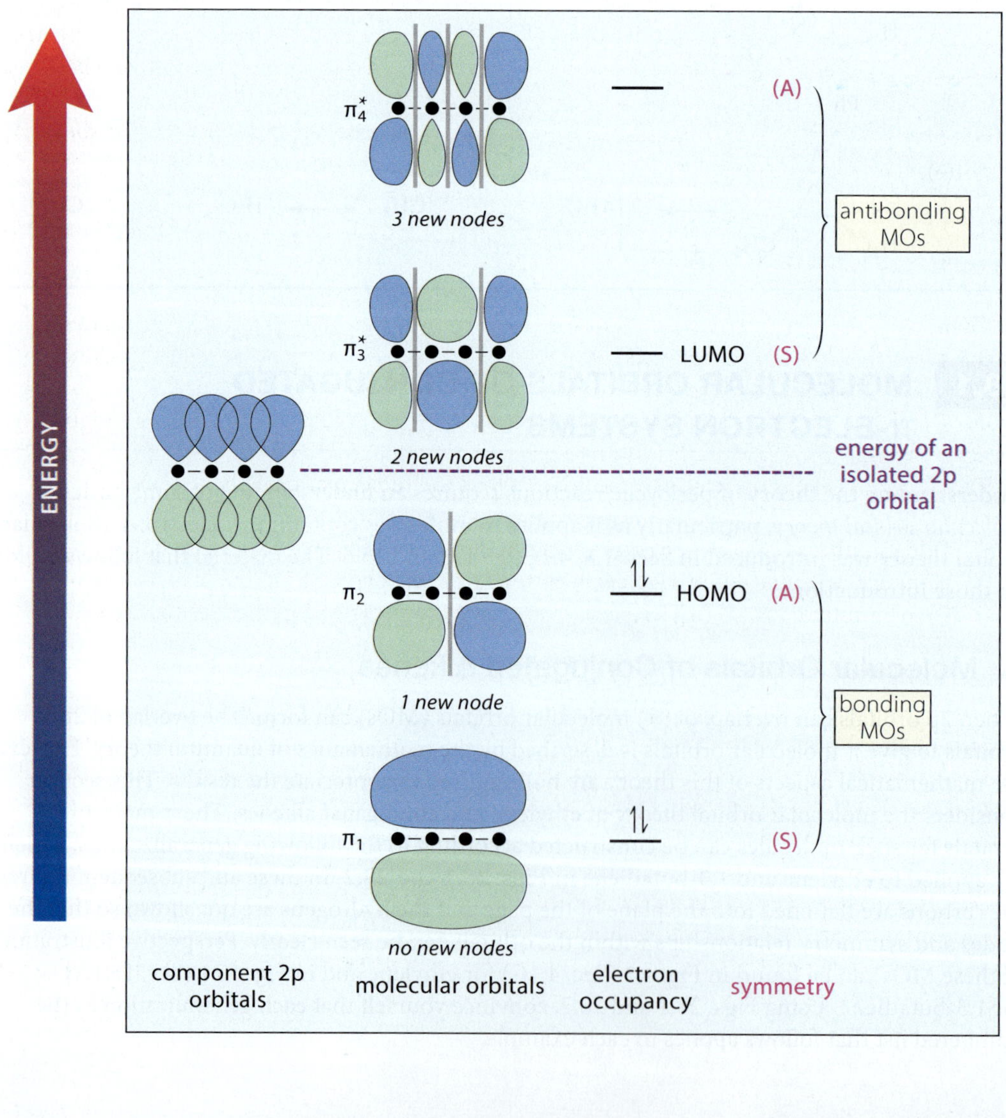

1. When a number *j* of atomic 2p orbitals interact, the resulting π-electron system contains the same number *j* of molecular orbitals (MOs), all with different energies.

Because two 2p orbitals contribute to the π-electron system of ethylene, this molecule has the same number—two—of π MOs, which are designated as π_1 and π_2. Similarly, the four 2p orbitals of 1,3-butadiene combine to form four MOs—π_1, π_2, π_3, and π_4.

2. Half of the molecular orbitals have lower energies than the isolated 2p orbitals. These are called **bonding molecular orbitals**. The other half have higher energies than the isolated 2p orbitals. These are called **antibonding molecular orbitals**.

To emphasize this distinction, antibonding MOs are indicated with asterisks. Thus, ethylene has one bonding MO (π_1) and one antibonding MO (π_2^*); 1,3-butadiene has two bonding MOs (π_1 and π_2) and two antibonding MOs (π_3^* and π_4^*).

3. The bonding molecular orbital of lowest energy, π_1, has no new nodes. (It does retain a node in the plane of the molecule, which is common to all 2p orbitals and to all π MOs). Each MO of increasingly higher energy has one additional node.

As explained in Sec. 1.2E, a *node* is a surface, in this case a plane, at which an electron wave (orbital) is zero; that is, when an electron is in a given MO, there is zero *probability of finding the electron*, or zero *electron density*, at the node. A particularly important feature of the node for understanding pericyclic reactions is that the electron wave has a *peak* on one side of a node (*blue*) and a *trough* on the other side (*green*). Because of the symmetry relationships within the MOs (discussed later in point 5), it doesn't matter whether a particular lobe of an MO is treated as a peak or a trough, *provided that a peak changes to a trough and a trough changes to a peak each time a node is crossed.*

For example, π_1 of ethylene has no new nodes, and π_2^* has one new node. In 1,3-butadiene, π_1 has no new nodes, π_2 has one new node, π_3^* has two, and π_4^* has three.

4. The nodes occur *between* atoms and are arranged symmetrically with respect to the center of the π-electron system.

The node in π_2^* of ethylene is between the two carbon atoms, in the center of the π system. The node in π_2 of 1,3-butadiene is also symmetrically placed in the center of the π system. The two nodes in π_3^* are placed between carbons 1 and 2, and between carbons 3 and 4, respectively—equidistant from the center of the π system. Each of the three nodes in π_4^*, the MO of highest energy, must occur between carbon atoms.

The next generalization relates to the *symmetry* of the molecular orbitals.

5. Odd-numbered MOs (π_1, π_3, π_5, . . .) are symmetric with respect to an imaginary *reference plane* at the center of the π-electron system and perpendicular to the plane of the molecule. Even-numbered MOs (π_2, π_4, π_6, . . .) are antisymmetric with respect to this plane.

The *reference plane* in this generalization is shown for the 1,3-butadiene molecule in **Fig. 28.3**. A **symmetric MO** is an MO in which peaks reflect across the reference plane into peaks and troughs reflect into troughs, as shown for π_3^* of 1,3-butadiene in Fig. 28.3. An **antisymmetric MO** is an MO in which peaks reflect into troughs, as shown for π_2 of 1,3-butadiene in Fig. 28.3. Of particular importance for the analysis of pericyclic reactions is the *relative phase* of each MO at its *terminal carbons*. The **relative phase** of an MO refers to the relative orientation of peaks and troughs at two different points. Within any symmetric MO, such as π_3^* of 1,3-butadiene, the relative phase at the two terminal carbons is *the same*. That is, the peaks on the two carbons are on the same side of the carbon chain (the upper side in Fig. 28.3), and the two troughs are on the same side. Within any antisymmetric MO, such as π_2 of 1,3-butadiene, the relative phase at the two terminal carbons is *different*. Be sure to verify for yourself that the other MOs in Figs. 28.1 and 28.2 fit this pattern.

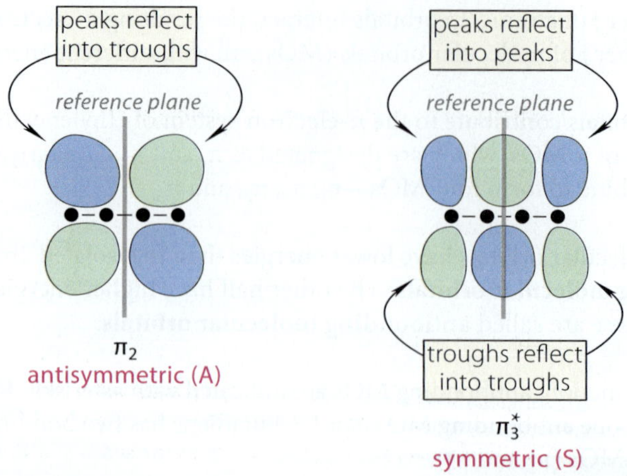

FIGURE 28.3 The antisymmetric and symmetric MO symmetry classifications illustrated for the π_2 and the π_3^* molecular orbitals of 1,3-butadiene. These classifications refer to the symmetry of an MO with respect to a reference plane (*gray*) through the center of the molecule and perpendicular to the plane of the page. (This plane is not the same thing as a node, although it may coincide with a node, as in the π_2 MO.) The symmetry classifications of the MOs in Figs. 28.1 and 28.2 are indicated by the abbreviations (A) for *antisymmetric* and (S) for *symmetric*. Although the molecules are flattened, the same symmetry relationships apply to their *s*-cis and *s*-trans conformations.

The last generalization deals with the distribution of the available π electrons within the MOs.

6. Electrons are placed pairwise into each molecular orbital, beginning with the orbital of lowest energy (aufbau principle).

This point is illustrated in Figs. 28.1 and 28.2 in the column labeled "electron occupancy." An alkene has the same number of π electrons as it has 2p orbitals. For example, ethylene, with two 2p atomic orbitals, has two π electrons. These are both placed (with opposite spin) into π_1 (see Fig. 28.1). 1,3-Butadiene, with four 2p atomic orbitals, has four π electrons. Two are placed in π_1 and two in π_2 (see Fig. 28.2). These examples show that the bonding MOs are fully filled in both simple and conjugated alkenes and that the antibonding MOs are empty.

Unconjugated substituents have small effects on the *energies* of the MOs. This is why, for example, the λ_{max} values in the UV–vis spectra of conjugated alkenes are increased slightly by alkyl substitution. (See Display 15.15, Sec. 15.2C.)

The presence of unconjugated substituents (for example, alkyl groups), to a useful approximation, does not alter the MO nodal properties of a conjugated alkene. For example, the π molecular orbital nodes in 1,3-butadiene and 1,3-pentadiene are essentially the same.

$$H_2C{=}CH{-}CH{=}CH_2 \qquad H_2C{=}CH{-}CH{=}CH{-}CH_3$$

1,3-butadiene **1,3-pentadiene**

the π MOs have the same nodes (28.7)

As noted in Sec. 15.1A, the π-electron contribution to the energy of a molecule is determined by the energies of its *occupied* MOs. Because bonding MOs have lower energies than isolated 2p orbitals, there is an energetic advantage to π molecular orbital formation; this is why π bonds exist.

Two MOs are of particular importance in understanding pericyclic reactions. One is the occupied molecular orbital of highest energy, called the **highest occupied molecular orbital (HOMO)**. The other is the unoccupied molecular orbital of lowest energy, called the **lowest unoccupied molecular orbital (LUMO)**. These are labeled in Figs. 28.1 and 28.2. In ethylene, π_1 is the HOMO and π_2^* is the LUMO; in 1,3-butadiene, π_2 is the HOMO and π_3^* is the LUMO. *The HOMO and LUMO of a conjugated alkene have opposite symmetries*. Also, *the HOMO has lower energy than the LUMO*.

The HOMO and LUMO are sometimes collectively referred to as the **frontier orbitals** because they are the molecular orbitals at the energy extremes: the HOMO is the occupied molecular orbital of *highest* energy, and the LUMO is the unoccupied molecular orbital of *lowest* energy. *The analysis of pericyclic reactions focuses heavily on the symmetries of frontier orbitals.*

Focused Problems

28.2 Answer the following questions for 1,3,5-hexatriene, the conjugated triene containing six carbons.

(a) How many π MOs are there?

(b) Classify each MO as symmetric or antisymmetric. (See Fig. 28.3.)

(c) Which MOs are bonding? Which are antibonding?

(d) Which MOs are the frontier molecular orbitals?

(e) Within the HOMO, is the phase at the terminal carbons the same or different?

(f) Within the LUMO, is the phase at the terminal carbons the same or different?

28.3 Without drawing the MOs, state whether the π molecular orbital π_6 in 1,3,5,7,9-decapentaene (a 10-carbon conjugated alkene) is symmetric or antisymmetric with respect to the reference plane; is bonding or antibonding; is a frontier MO; and, if so, is a HOMO or a LUMO.

B. Molecular Orbitals of Conjugated Ions and Radicals

Conjugated unbranched ions and radicals have an odd number of carbon atoms. For example, the allyl cation has three carbon atoms, three 2p orbitals, and three MOs.

$$\left[H_2C{=}CH{-}\overset{+}{C}H_2 \longleftrightarrow H_2\overset{+}{C}{-}CH{=}CH_2\right] \quad (28.8)$$

allyl cation

The MOs of such species follow many but not all of the same patterns as those of conjugated alkenes. The MOs for the allyl and 2,4-pentadienyl systems are shown in **Figs. 28.4** and **28.5**,

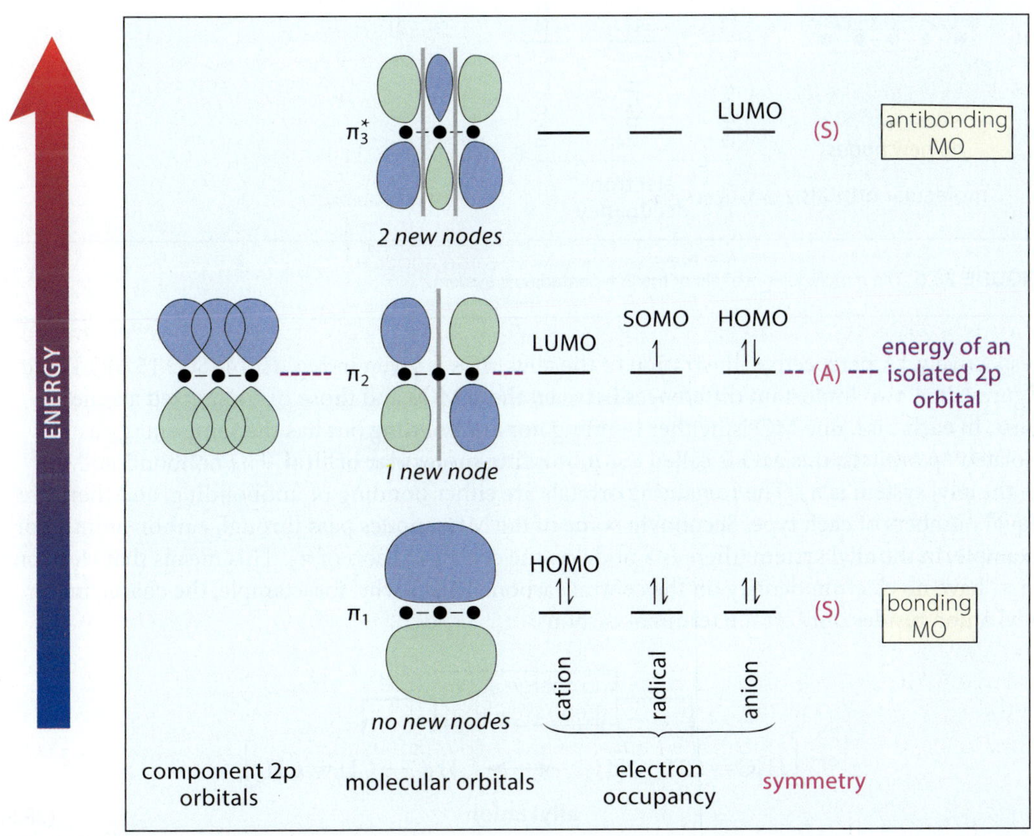

FIGURE 28.4 The π molecular orbitals of the allyl system. The nodal properties of the MOs are the same for the cation, the radical, and the anion. (The electron occupancy does affect the orbital energies somewhat, but this effect can be ignored.) In a radical, the molecular orbital containing the unpaired electron is called the singly occupied molecular orbital (SOMO). The designation of the HOMO and the LUMO depends on the electron occupancy.

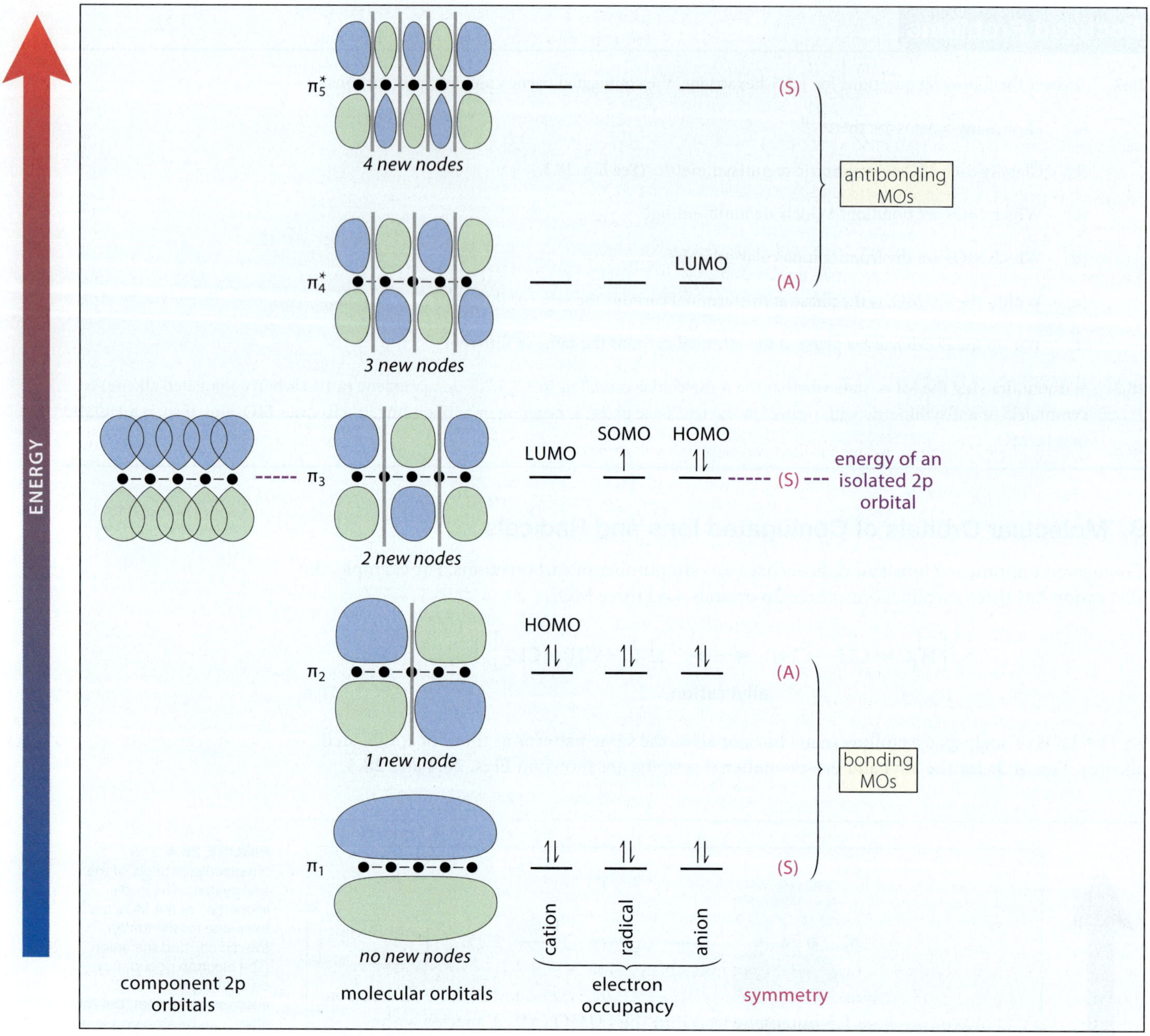

FIGURE 28.5 The π molecular orbitals of the 2,4-pentadienyl system.

respectively. [A perspective illustration of the allyl MOs is given in Fig. 15.14 (Sec. 15.6).] These figures show two important differences between these MOs and those of conjugated alkenes. First, in each case, one MO is neither bonding nor antibonding but has the same energy as the isolated 2p orbitals; this MO is called a **nonbonding molecular orbital**. The nonbonding MO in the allyl system is π_2. The remaining orbitals are either bonding or antibonding, and there are equal numbers of each type. Second, in some of the MOs, nodes pass through carbon atoms. For example, in the allyl system, there is a node on the central carbon of π_2. This means that electrons in π_2 have no electron density on the central carbon. This is why, for example, the charge in the allyl anion resides only on the terminal carbons:

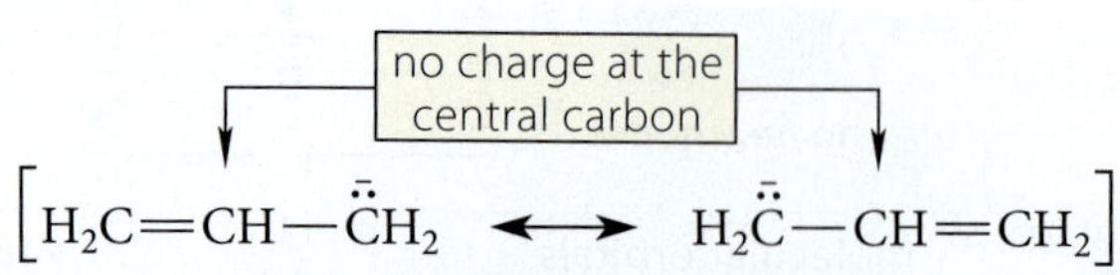

(28.9)

Just as the charge in an atomic anion is associated with an excess of valence electrons, the charge in a conjugated carbanion can be associated with the electrons in its HOMO. (See the discussion in Sec. 15.6 and Focused Problem 15.26 in that section.)

The MOs of a cation, a radical, and an anion involving the same π system have the same nodal properties. For example, the MOs of the allyl system apply equally well to the allyl cation, allyl radical, and allyl anion because all three species contain the same arrangement of 2p orbitals. These species differ only in the *number* of π electrons, as shown in the "electron occupancy" columns of Figs. 28.4 and 28.5. (The relative energies of the MOs differ somewhat, but we can ignore these differences.) Thus, the HOMO of the allyl cation is π_1, and the LUMO is π_2^*. In contrast, the HOMO of the allyl anion is π_2, and the LUMO is π_3^*. The HOMO of the allyl radical contains a single electron. Because this molecular orbital is "half-occupied," it is sometimes referred to as a **singly occupied molecular orbital (SOMO)**.

Focused Problems

28.4 Answer the following questions for the 2,4,6-heptatrienyl cation.

$$H_2C{=}CH{-}CH{=}CH{-}CH{=}CH{-}\overset{+}{C}H_2$$

2,4,6-heptatrienyl cation

(a) Which MO is nonbonding?

(b) Classify each MO as symmetric or antisymmetric.

(c) To which carbon atoms in this cation is the positive charge delocalized? Explain with both resonance structures and molecular orbital arguments.

28.5 Explain using (a) resonance arguments and (b) molecular orbital arguments why the unpaired electron in the allyl radical is delocalized to carbon-1 and carbon-3 but not to carbon-2.

C. Excited States

The molecules and ions we have been discussing can absorb energy from light of certain wavelengths. This process, which is also responsible for the UV spectra of these species (Sec. 15.2B), is shown in **Fig. 28.6** for 1,3-butadiene. The normal electronic configuration of any molecule is called the **ground state**. Energy from absorbed light is used to promote an electron from the HOMO of ground-state 1,3-butadiene (π_2) into the LUMO (π_3^*). A species with a promoted electron is said to be in an **excited state**. In the excited state of 1,3-butadiene, π_3^* is the HOMO, even though it contains only one electron. Notice that *the HOMOs of the ground state and the excited state have opposite symmetries.*

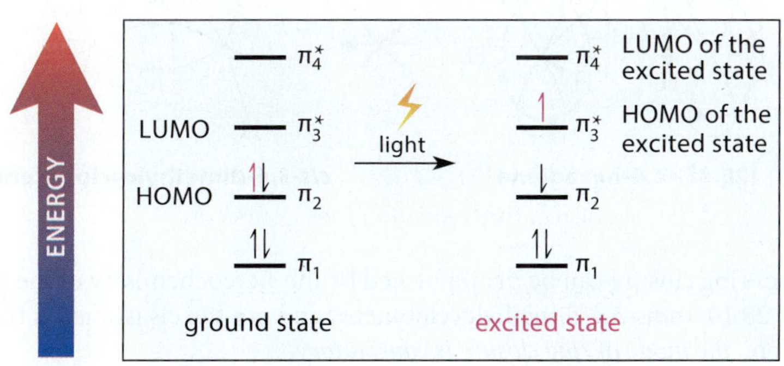

FIGURE 28.6 Light absorption by a conjugated species such as 1,3-butadiene (shown here) promotes an electron from the HOMO to the LUMO and produces an excited state.

In subsequent sections, we differentiate pericyclic reactions according to whether they are *thermal* or *photochemical*. A **thermal pericyclic reaction** is any pericyclic reaction *not* activated by light, whereas a **photochemical pericyclic reaction** is any pericyclic reaction activated by light. The fundamental distinction is that thermal reactions occur through molecular *ground states*, whereas photochemical reactions occur through molecular *excited states.*

The word *thermal* as used in this context may be a bit misleading. This term might suggest that thermal reactions require strong heating to occur. Although some thermal pericyclic reactions require high temperatures, others can occur at room temperature or below. The word *thermal* in this sense means "not activated by light." In other words, a thermal pericyclic reaction is any pericyclic reaction involving a ground state.

28.2 ELECTROCYCLIC REACTIONS

A. Ground-State (Thermal) Electrocyclic Reactions

This section begins the application of MO theory to pericyclic reactions with a discussion of *thermal* electrocyclic reactions—that is, electrocyclic reactions that proceed through ground states. Section 28.2B considers *photochemical* electrocyclic reactions—that is, reactions that proceed through excited states. (See Eq. 28.1 for the definition of electrocyclic reactions.)

When an electrocyclic reaction takes place, the carbons at each end of the conjugated π system must turn in a concerted fashion so that the 2p orbitals can overlap (and rehybridize) to form the σ bond that closes the ring. To illustrate, consider the reaction shown in Eq. 28.6a, the electrocyclic closure of (2*E*,4*E*)-2,4-hexadiene to give 3,4-dimethylcyclobutene. This turning can occur in two stereochemically distinct ways. In a **conrotatory** closure, the two carbon atoms turn in the same direction. (The green arrows show the direction of motion.)

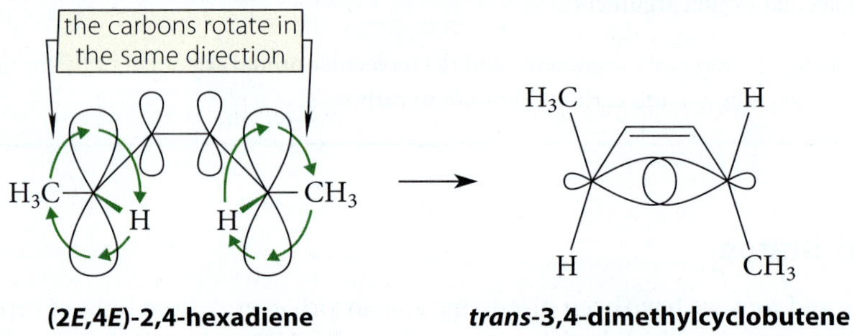

a conrotatory reaction (observed) (28.10)

(Of the two conrotatory modes, clockwise and counterclockwise, the clockwise mode is shown, but the counterclockwise mode in this case is equally probable.) In the second mode of ring closure, called **disrotatory**, the carbon atoms turn in opposite directions.

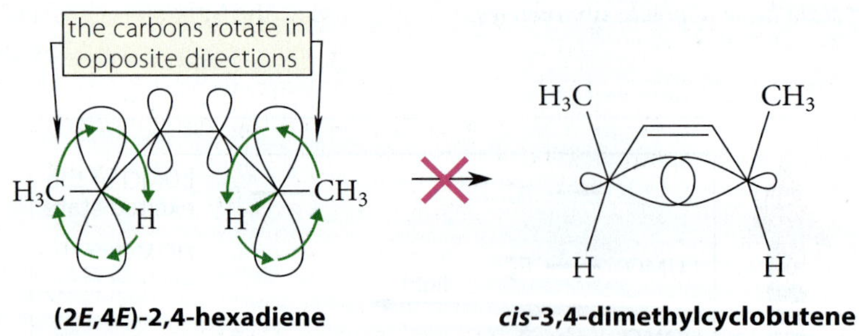

a disrotatory reaction (does not occur) (28.11)

The two modes of ring closure can be distinguished by the stereochemistry of the product. As shown in Eq. 28.10, *trans*-3,4-dimethylcyclobutene, and not the cis isomer, is the observed product. Therefore, *the mode of ring closure is conrotatory.*

Molecular orbital theory explains this result. A simple way to look at the reaction is to focus on the *HOMO of the diene*. This molecular orbital contains the π electrons of highest energy. *These π electrons are to a molecule as valence electrons are to an atom.* Just as the atomic valence electrons are involved in most chemical reactions, the electrons in the HOMO govern the course of pericyclic reactions. When the ring closure takes place, the two 2p orbitals on the ends of the π system must overlap. But simple overlap is not enough: they must overlap *in phase*. That is, the wave peak on one carbon must overlap with the wave peak on the other, or a wave trough must overlap with a wave trough. If a peak were to overlap with a trough, the electron waves would cancel and no bond would form.

Let's see what it takes to provide the required bonding overlap. First, the diene must assume the *s*-cis conformation; only in this conformation are the terminal carbons of the π-electron system close enough to each other that their 2p orbitals can overlap. (The *s*-cis and *s*-trans conformations of dienes are discussed in Sec. 15.1B.) Next, recall from Sec. 28.1A that alkyl substituents, to a useful approximation, do not affect the nodal properties of the MOs of a conjugated alkene. Consequently, the nodes of the π molecular orbitals of 2,4-hexadiene are the same as those of 1,3-butadiene (see Fig. 28.2). In other words, the methyl groups at each end of the molecule can be largely ignored when considering the MOs of the system. An examination of the HOMO of a conjugated diene (π_2 in Fig. 28.2) reveals that, because of the antisymmetric nature of π_2, conrotatory ring closure is required for in-phase, or bonding, overlap:

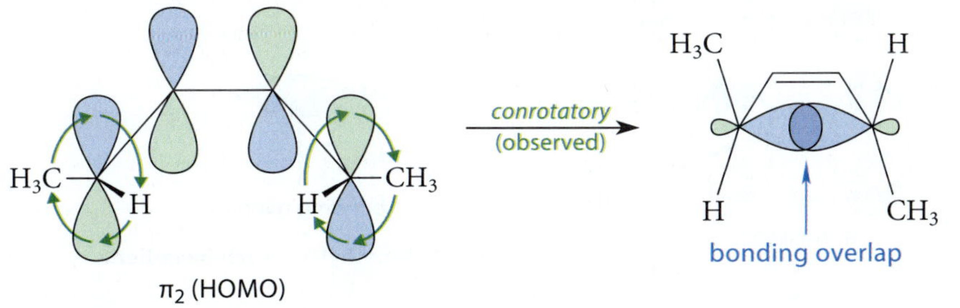

(28.12)

In contrast, disrotatory ring closure gives out-of-phase overlap, an antibonding (and therefore unstable) situation:

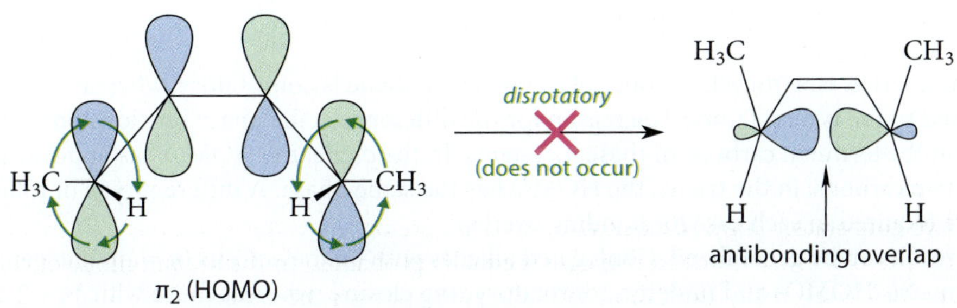

(28.13)

These examples show that it is the relative orbital phase at the terminal carbon atoms of the HOMO—the *orbital symmetry*—that determines whether the reaction is conrotatory or disrotatory. This observation suggests that *all* conjugated polyenes with *antisymmetric* HOMOs should undergo conrotatory ring closure, and indeed, such is the case. The electrocyclic reactions of conjugated alkenes with symmetric HOMOs can be predicted by a similar analysis, as Study Problem 28.1 illustrates.

Study Problem 28.1

Predict the stereochemistry of the thermal electrocyclic ring closure of (2*E*,4*Z*,6*E*)-2,4,6-octatriene to 5,6-dimethyl-1,3-cyclohexadiene.

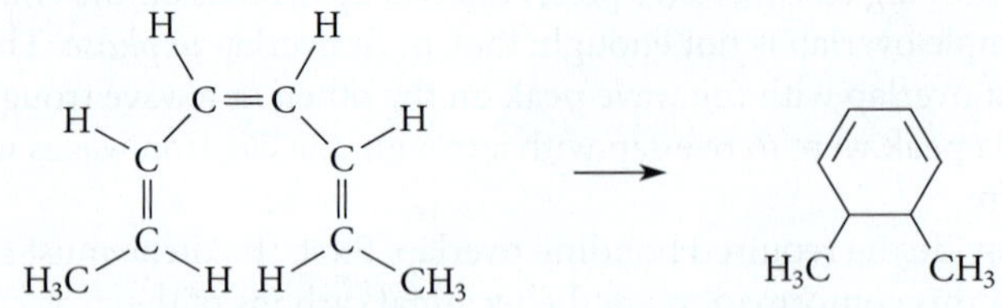

(2*E*,4*Z*,6*E*)-2,4,6-octatriene **5,6-dimethyl-1,3-cyclohexadiene** (28.14)

Solution First, examine the HOMO of the simplest conjugated triene, 1,3,5-hexatriene (Focused Problem 28.2). Because the HOMO of this triene (π_3) is *symmetric*, the HOMO has the *same phase* at each end of the π system. Therefore, bonding overlap can occur only if the ring closure is *disrotatory*.

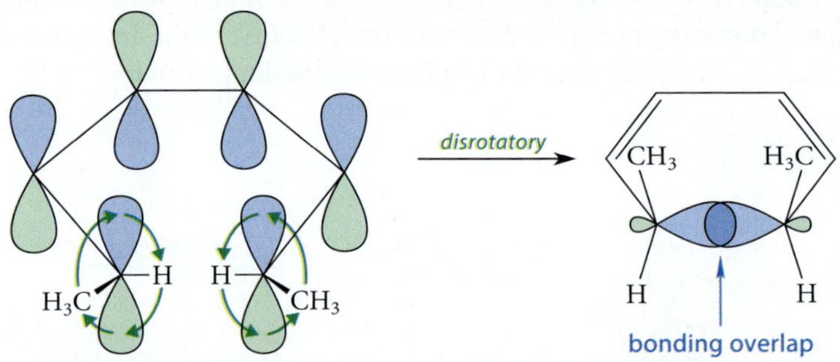

π_3 (HOMO) ***cis*-5,6-dimethyl-1,3-cyclohexadiene** (28.15)

The disrotatory motion, as Eq. 28.15 shows, requires that the methyl groups have a cis relationship in the product. And, as Eq. 28.6b shows, this is indeed the observed stereochemistry of the reaction.

To summarize, electrocyclic closure of a conjugated diene is conrotatory, whereas that of a conjugated triene is disrotatory. The reason for the difference is the phase relationships within the HOMO at the terminal carbons of these π systems. In the diene, the HOMO has opposite phase at these two carbons; in the triene, the HOMO has the same phase. A different type of rotation is therefore required in each case for bonding overlap.

This result can be generalized. Conjugated alkenes with $4n$ π electrons (n = any integer) have antisymmetric HOMOs and undergo conrotatory ring closure, whereas those with $4n + 2$ π electrons have symmetric HOMOs and undergo disrotatory ring closure. That is, conrotatory ring closure is *allowed* for systems with $4n$ π electrons, but it is *forbidden* for systems with $4n + 2$ π electrons. Conversely, disrotatory ring closure is *allowed* for systems with $4n + 2$ π electrons, but it is *forbidden* for systems with $4n$ π electrons.

B. Excited-State (Photochemical) Electrocyclic Reactions

When a molecule absorbs light, it reacts through its *excited state* (Sec. 28.1C). The HOMO of the excited state is different from the HOMO of the ground state, and it has different symmetry. For example, as Eq. 28.6c shows, the *photochemical* ring closure of (2*E*,4*Z*,6*E*)-2,4,6-octatriene is *conrotatory*. This is understandable in terms of the symmetry of π_4^*, the HOMO of the excited state.

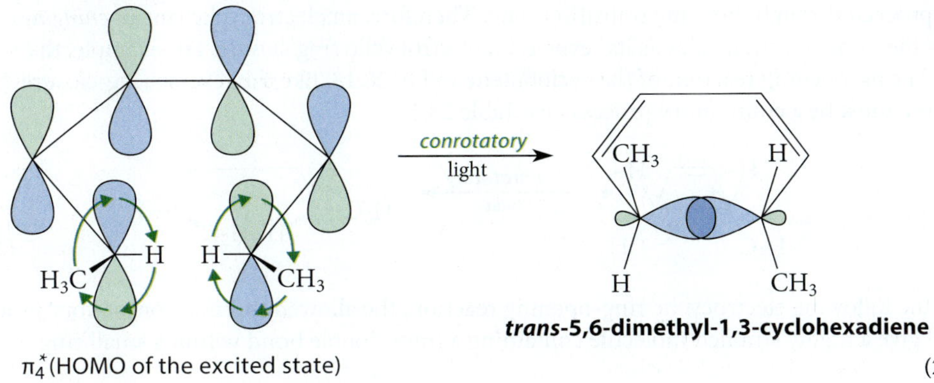

(28.16)

Contrast the stereochemistry of the product with that observed in the ground-state reaction of the same triene in Eq. 28.15. *The stereochemical result is different because the symmetry of the HOMO is different.*

To generalize this result: the mode of ring closure in *photochemical* electrocyclic reactions—reactions that occur through electronically excited states—differs from that of thermal electrocyclic reactions, which occur through electronic ground states. These results can be summarized with a series of *selection rules* for electrocyclic reactions, given in **Table 28.1**.

C. Selection Rules and Microscopic Reversibility

The orbital-symmetry selection rules in Table 28.1 (as well as others to be considered) refer to the *rates* of pericyclic reactions but have nothing to say about the *positions of the equilibria* involved. Therefore, the electrocyclic reaction of the diene in Eq. 28.6a to give a cyclobutene favors the diene at equilibrium because of the strain in the cyclobutene, but the electrocyclic reaction of the conjugated triene in Eq. 28.6b favors the cyclic compound because σ bonds are stronger than π bonds, and because six-membered rings are relatively stable.

It is also common for a photochemical reaction to favor the less stable isomer because the energy of light is harnessed to drive the reaction energetically "uphill." In the following reaction, for example, the conjugated alkene absorbs UV light, but the bicyclic compound does not; consequently, the photochemical reaction favors the latter.

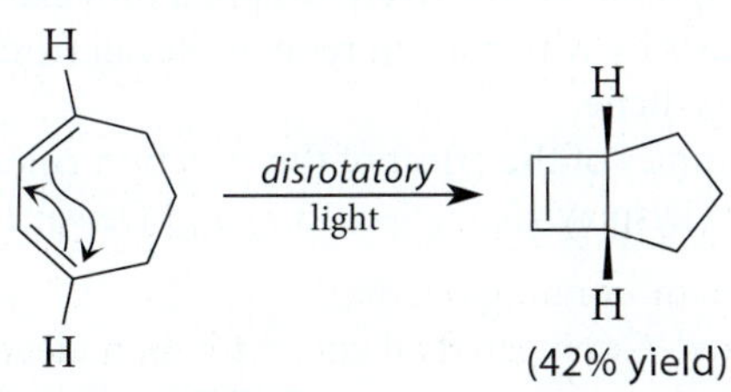

(28.17)

In summary, the selection rules do not indicate which component of an equilibrium will be favored—only whether the equilibrium will be established at a reasonable rate.

The *principle of microscopic reversibility* (Sec. 4.10C) ensures that selection rules apply equally well to the forward and reverse of any pericyclic reaction, because the reaction in both directions

TABLE 28.1 Selection Rules for Electrocyclic Reactions

Number of electrons*	Mode of activation	Allowed stereochemistry
$4n$	thermal	conrotatory
	photochemical	disrotatory
$4n + 2$	thermal	disrotatory
	photochemical	conrotatory

*n = an integer

must proceed through the same transition state. Therefore, an electrocyclic ring *opening* must follow the same selection rules as its reverse, an electrocyclic ring *closure*. For example, the thermal ring-opening reaction of the cyclobutene in Eq. 28.18, like the reverse ring-closure reaction, must be a conrotatory process (see Table 28.1).

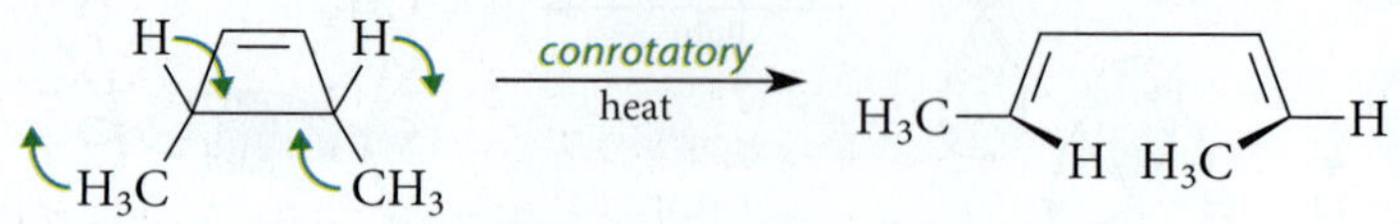

(28.18)

In the following electrocyclic ring-opening reaction, the allowed thermal conrotatory process would give a highly strained molecule containing a trans double bond within a small ring.

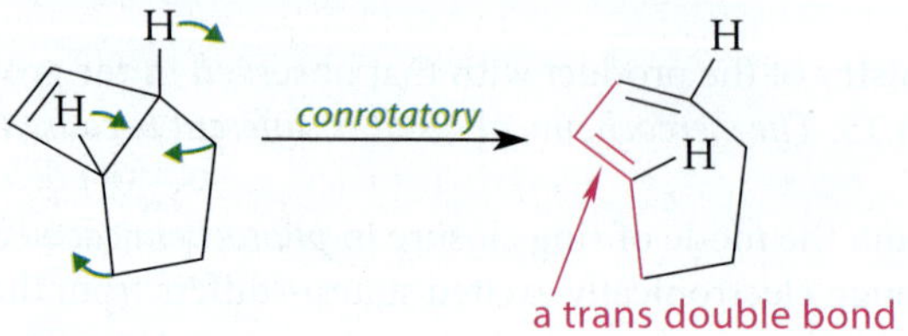

(28.19)

Although the selection rules suggest that the reaction could occur, it does not because of the strain in the product (Sec. 7.6C). In other words, "allowed" reactions are sometimes prevented from occurring for reasons having nothing to do with the selection rules. Furthermore, concerted ring opening to the relatively unstrained all-cis diene also does not occur, because this would be a disrotatory process—a process "forbidden" by the selection rules in Table 28.1 for a concerted $4n$-electron reaction.

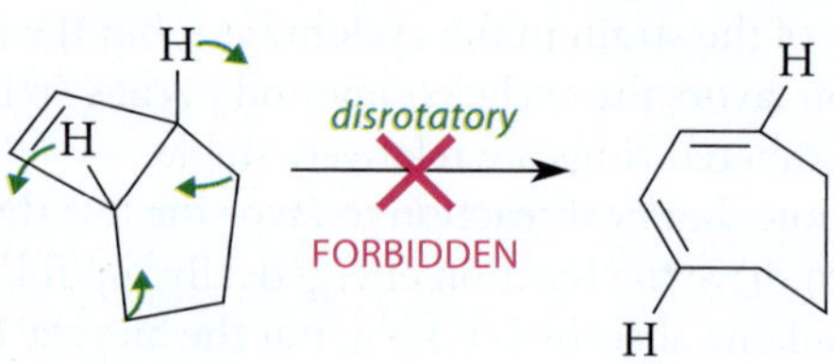

(28.20)

Consequently, the bicyclic compound is effectively "trapped into existence"—that is, no concerted thermal pathway exists by which it can reopen. Recall from Eq. 28.17 that it is formed *photochemically* from the all-cis diene.

Two rather spectacular examples of the effect of the selection rules for pericyclic reactions are the benzene isomers *prismane* (Display 15.57, Sec. 15.7) and *Dewar benzene*:

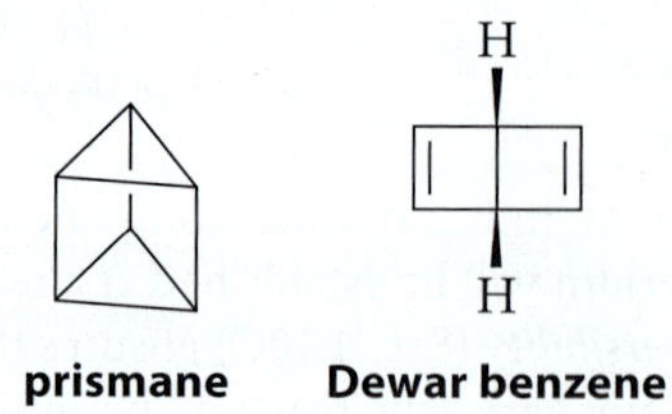

(28.21)

Despite the tremendous amount of strain in both molecules and the aromatic stability of benzene, neither prismane nor Dewar benzene is readily transformed into benzene because, in each case, such a transformation violates pericyclic selection rules (see Problem 28.61). Because no low-energy pathway exists for its conversion into the much more stable isomer benzene, prismane has been characterized in the chemical literature as a "caged tiger."

Focused Problems

28.6 Which one of the following electrocyclic reactions should occur readily by a concerted mechanism? Explain.

Reaction 1:

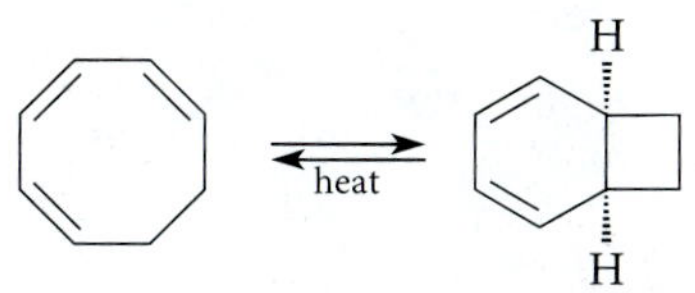

Reaction 2:

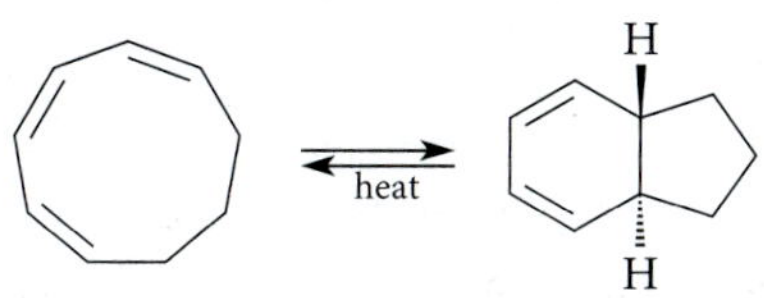

28.7 Show both conrotatory processes for the thermal electrocyclic conversion of (2*E*,4*E*)-2,4-hexadiene into 3,4-dimethylcyclobutene (Eq. 28.10). Explain why the two processes are equally likely.

28.8 In the thermal ring opening of *trans*-3,4-dimethylcyclobutene, *two* products could be formed by a conrotatory mechanism, but only one is observed. Give the two possible products. Which one is observed and why?

28.9 After heating to 200 °C, the following compound is converted in 95% yield into an isomer *A* that can be hydrogenated to cyclodecane. Give the structure of *A*, including its stereochemistry.

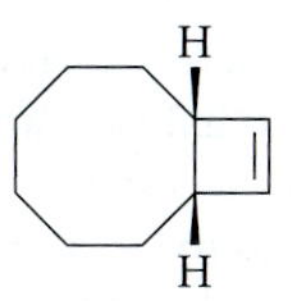

28.3 CYCLOADDITION REACTIONS

A *cycloaddition reaction* (Eq. 28.2) is classified, first, by the number of electrons involved in the reaction. The reaction in Eq. 28.22a is a [4 + 2] cycloaddition because the reaction involves four electrons from one reacting component and two electrons from the other. The reaction in Eq. 28.22b is a [2 + 2] cycloaddition.

[4 + 2] (28.22a)

[2 + 2] (28.22b)

As in electrocyclic reactions, the number of electrons involved is determined by writing the reaction mechanism in the curved-arrow notation. The number of electrons contributed by a given reactant is equal to twice the number of curved arrows originating from that component (two electrons per arrow).

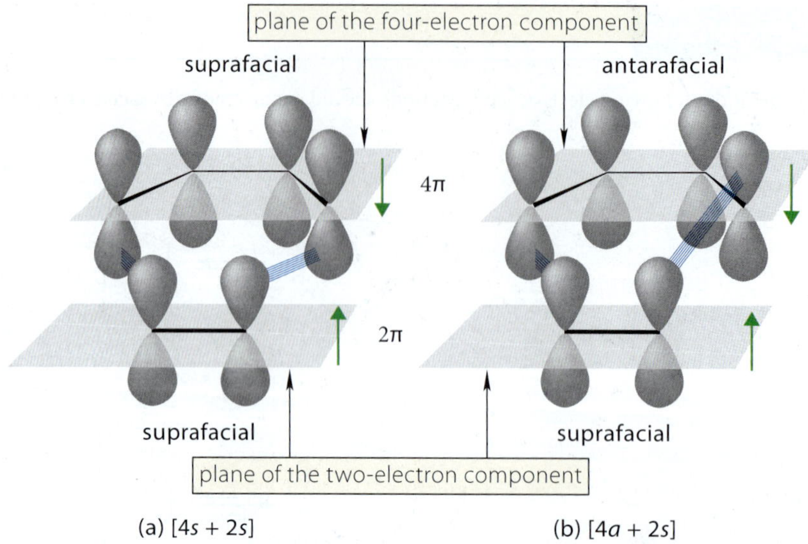

FIGURE 28.7 Classification of cycloaddition reactions, illustrated for (a) a [4*s* + 2*s*] cycloaddition and (b) a [4*a* + 2*s*] cycloaddition. The *s* and *a* designations refer to the stereochemistry of the cycloaddition (suprafacial or antarafacial) with respect to the planes of the reacting components. The two components approach each other in parallel planes (*green arrows*). In part (b), for example, the addition (*blue lines*) occurs below the plane of the 4π-electron component at one end and above the plane at the other and is therefore classified as 4*a*, or antarafacial on the 4π-electron component. The addition occurs on the 2π-electron component on the same side of its plane at both ends and is therefore classified as 2*s*, or suprafacial on the 2π-electron component. This reaction is therefore a [4*a* + 2*s*] cycloaddition.

A cycloaddition reaction is also classified by its stereochemistry with respect to *the plane of each reacting molecule*. (Recall that the carbons and their attached atoms in π-electron systems are coplanar.) This classification is shown for a [4 + 2] cycloaddition in **Fig. 28.7**. A cycloaddition may in principle occur either across the same face or across opposite faces of the planes in each reacting component. If the reaction occurs across the same face of a π system, the reaction is said to be **suprafacial** with respect to that π system. A suprafacial addition is the same thing as a syn addition (Sec. 7.8A). If the reaction occurs at opposite faces of a π system, it is said to be **antarafacial**. An antarafacial addition is an anti addition. Therefore, a [4*s* + 2*s*] cycloaddition is one that occurs suprafacially (or syn) on both the 4π component and the 2π component. A [4*a* + 2*s*] cycloaddition occurs antarafacially (or anti) on the 4π component, but suprafacially (or syn) on the 2π component.

For a cycloaddition to occur, bonding overlap must take place between the 2p orbitals at the terminal carbons of each π-electron system, because these are the carbons at which new bonds are formed. This bonding overlap begins when the HOMO of one component interacts with the LUMO of the other. The electrons in the HOMO of one component are analogous to the valence electrons in an atom: they are the reacting electrons. The LUMO of the other component is the empty orbital of lowest energy into which the electrons from the HOMO must flow. It doesn't matter whether we consider the HOMO from the 4π-electron component and the LUMO from the 2π-electron component, or vice versa. The important point is that the two frontier MOs involved in the interaction must have *matching phases* if bonding overlap is to be achieved.

STUDY GUIDE LINK 28.1
Frontier Orbitals

This phase match is achieved when a [4 + 2] cycloaddition occurs *suprafacially* on each component—that is, when the cycloaddition is a [4*s* + 2*s*] process. (A [4*a* + 2*a*] process is also theoretically allowed but is geometrically impossible.) Using the HOMO from the 4π-electron component and the LUMO from the 2π-electron component, this overlap can be represented as shown in Eq. 28.23:

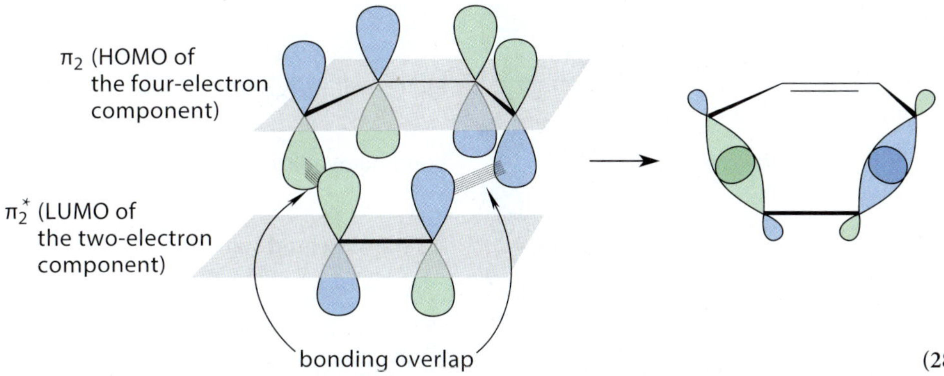

(28.23)

As explained in Sec. 15.3C, the Diels–Alder reaction, the most important example of a [4*s* + 2*s*] cycloaddition, occurs suprafacially on each component. This reaction mode is evident from the retention of stereochemistry observed in both the diene and dienophile in Eq. 28.24:

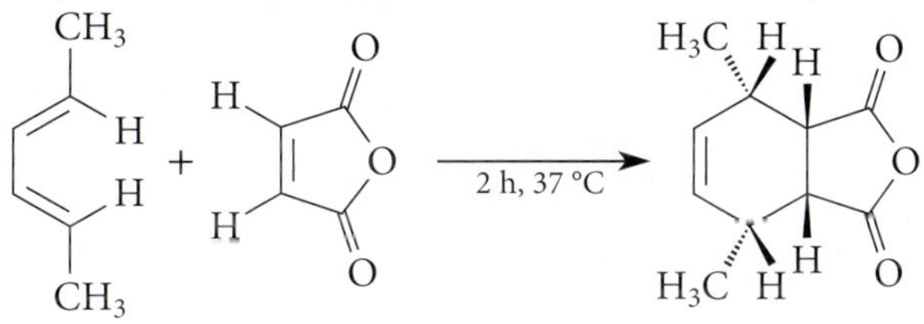

(28.24)

Convince yourself that the [4*s* + 2*a*] and [4*a* + 2*s*] modes of cycloaddition do *not* provide bonding overlap at both ends of the π-electron systems. You should also convince yourself that it doesn't matter which component provides the HOMO and which provides the LUMO (Focused Problem 28.10).

The situation is different in a [2 + 2] cycloaddition. Again, we use the HOMO of one component and the LUMO of the other. The orbital symmetries do *not* accommodate a cycloaddition that is suprafacial on both components:

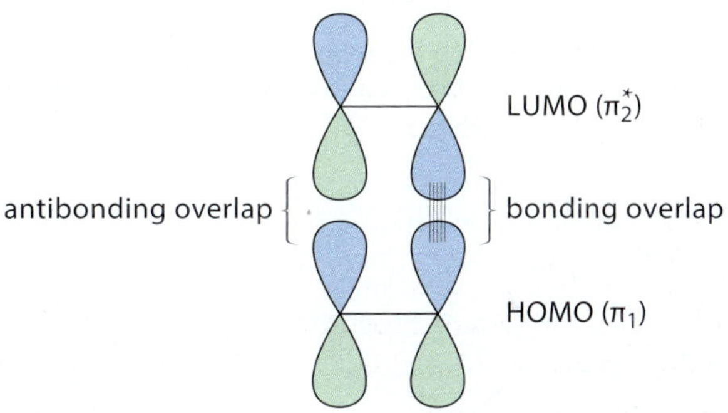

(28.25)

However, an addition that is suprafacial on one component but antarafacial on the other is allowed by orbital symmetry but is geometrically impossible. For this reason, the thermal [2 + 2] cycloaddition is a much less common reaction than the Diels–Alder reaction. All of the known thermal [2 + 2] additions occur by nonconcerted mechanisms and therefore do not fall under the purview of the rules for pericyclic reactions.

Although the [2*s* + 2*s*] cycloaddition is forbidden by orbital symmetry under *thermal* conditions, it is allowed under *photochemical* conditions. Under these conditions, the *excited state* of one alkene reacts with the other alkene, which is not photochemically excited. (Only a small fraction of the alkenes exist in an excited state under photochemical conditions.) The HOMO of the excited state has the proper symmetry to interact in a bonding way with the LUMO of the reacting partner:

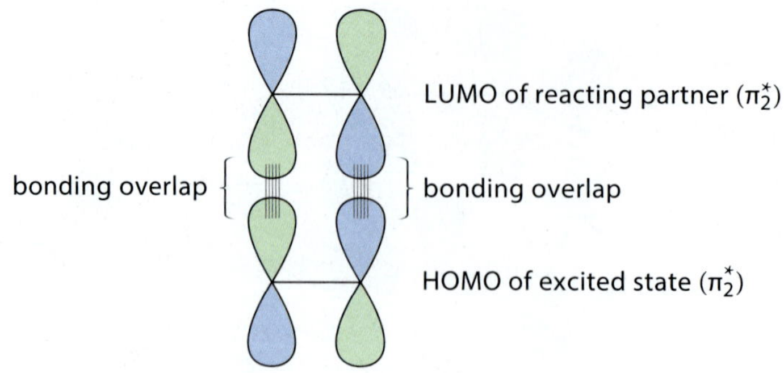

(28.26)

Indeed, many examples of photochemical [2s + 2s] cycloadditions are known. Such processes are widely used for making cyclobutanes.

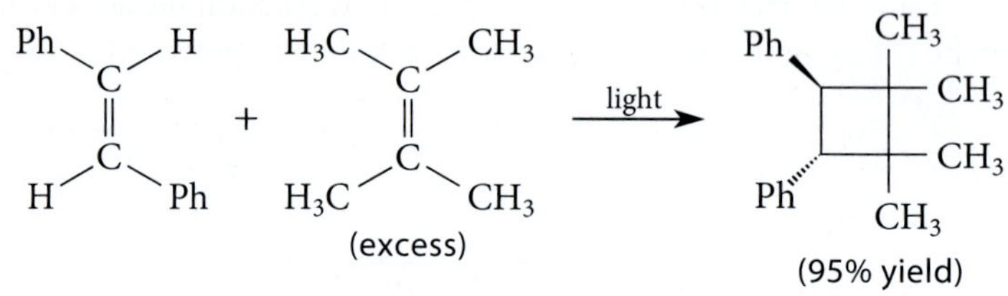

(28.27)

Notice the retention of alkene stereochemistry in this reaction: the phenyl groups, which are trans in the starting material, are trans in the product.

An important example of the allowed [2s + 2s] cycloaddition in biology is the photochemical reaction of two thymine bases in DNA to give a thymine dimer. (This reaction is shown in Eq. 26.74, Sec. 26.5C.) This reaction, which is one type of UV damage to DNA, is known to cause certain types of cancer and is probably a contributor to the onset of melanoma, a particularly virulent type of skin cancer.

The results of this section can be generalized to the cycloaddition selection rules given in **Table 28.2**. Notice that all-suprafacial cycloadditions are allowed thermally for systems in which the total number of reacting electrons is $4n + 2$, and they are allowed photochemically for systems in which the number is $4n$.

The following reaction is an example of a suprafacial cycloaddition involving more than six π electrons. This all-suprafacial cycloaddition is a [6s + 4s] process involving 10 electrons (five curved arrows). Notice that $4n + 2 = 10$ when $n = 2$.

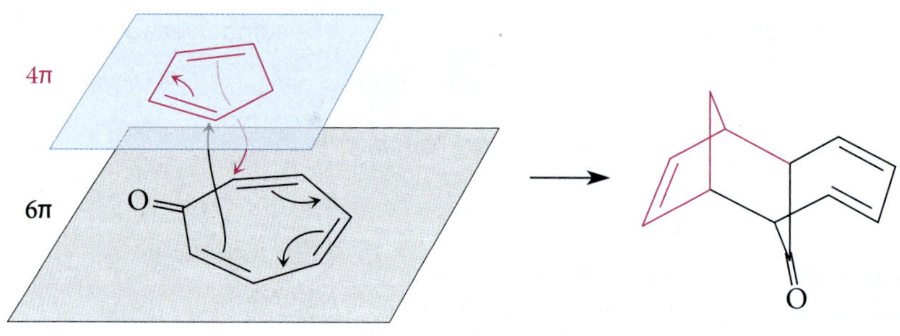

(28.28)

TABLE 28.2 Selection Rules for Cycloaddition Reactions

Number of electrons*	Mode of activation	Allowed stereochemistry†
$4n$	thermal	supra–antara
		antara–supra
	photochemical	supra–supra
		antara–antara
$4n+2$	thermal	supra–supra
		antara–antara
	photochemical	supra–antara
		antara–supra

*n = an integer

†supra = suprafacial; antara = antarafacial

Focused Problems

28.10 Show that using the HOMO from the 2π-electron component and the LUMO from the 4π-electron component also gives bonding overlap in a $[4s+2s]$ cycloaddition.

28.11 Show by a frontier orbital analysis that the $[4a+2s]$ and $[4s+2a]$ modes of cycloaddition are not allowed.

28.12 Give the product of the following reaction, which involves an $[8s+2s]$ cycloaddition:

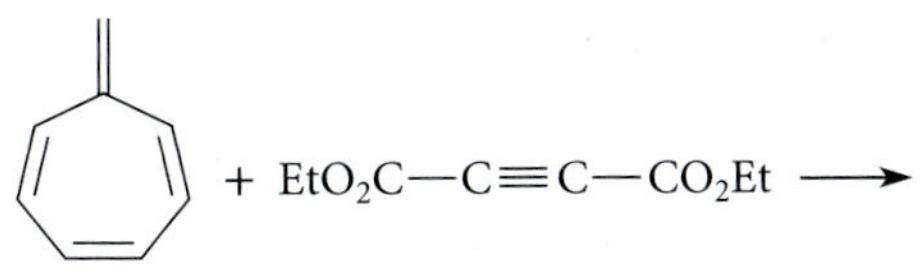

28.13 The photochemical cycloaddition of two molecules of *cis*-2-butene gives a mixture of two products: *A* and *B*. The analogous photochemical cycloaddition of *trans*-2-butene also gives a mixture of two products: *B* and *C*. The photochemical reaction of a *mixture* of *cis*- and *trans*-2-butene gives a mixture of *A*, *B*, and *C*, along with a fourth product, *D*. Propose structures for all four compounds.

28.4 THERMAL SIGMATROPIC REACTIONS

A. Classification and Stereochemistry

Sigmatropic reactions were defined in Eqs. 28.3 and 28.4 of the chapter introduction. In this section, we consider only thermal sigmatropic reactions—sigmatropic reactions that occur through ground states—because photochemical sigmatropic reactions are not common.

Sigmatropic reactions are classified by using bracketed numbers to indicate the number of atoms over which a σ bond appears to migrate. In some reactions, both ends of a σ bond migrate. In the following reaction, for example, each end of a σ bond migrates over three atoms. (Count the point of original attachment as atom #1.) The following reaction is therefore a [3,3] sigmatropic reaction.

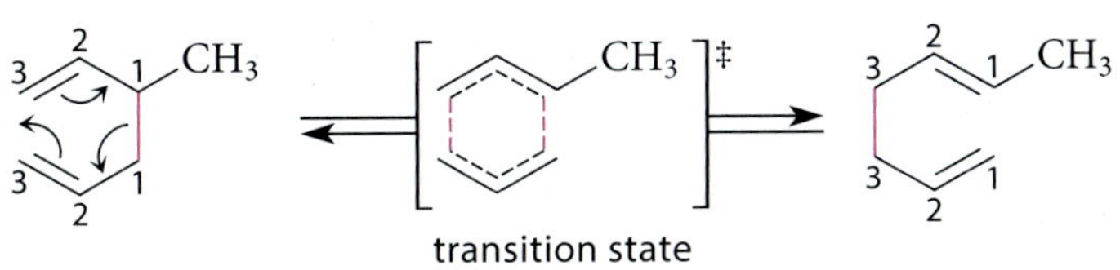

(28.29)

In other reactions, one end of a σ bond remains fixed to the same group and the other end migrates. For example, the following reaction is a [1,5] sigmatropic reaction because one end of the bond "moves" from atom #1 to atom #1 (that is, it doesn't move), and the other end moves over five atoms.

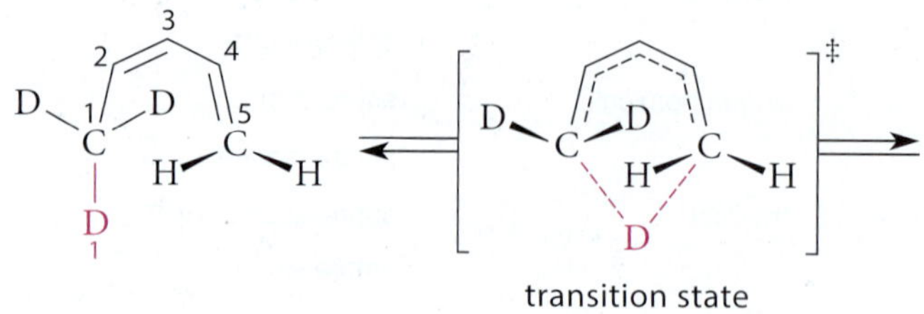

(28.30)

Sigmatropic reactions, like other pericyclic reactions, are classified by their stereochemistry. This classification is based on whether the migrating bond moves over the same face or between opposite faces of the π-electron system. If the migrating bond moves across one face of the π system, the reaction is said to be *suprafacial*. For example, if the [1,5] sigmatropic reaction of Eq. 28.30 were suprafacial, it would occur in the following manner:

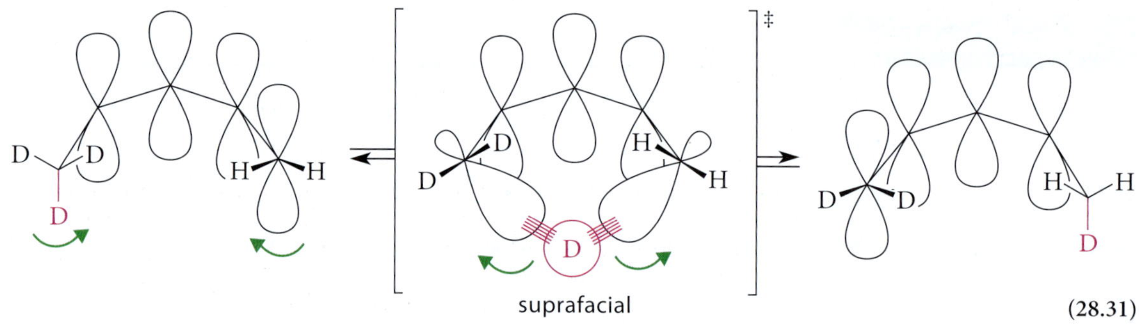

(28.31)

If the reaction were *antarafacial*, it would occur instead as shown in Eq. 28.32:

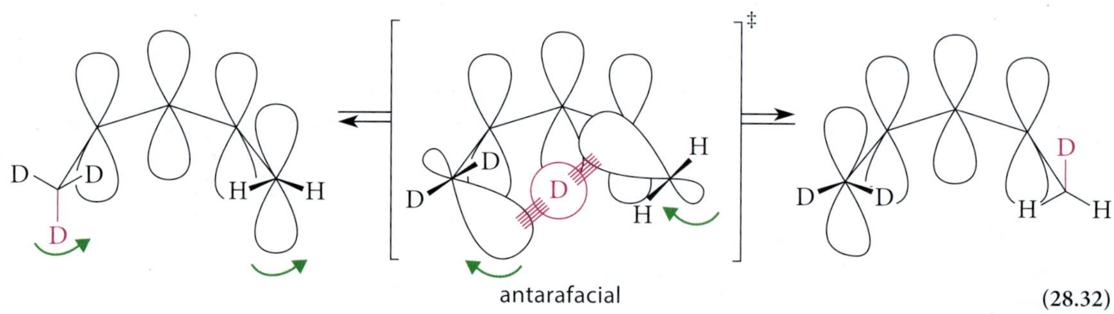

(28.32)

When both ends of a σ bond migrate, the reaction can be suprafacial or antarafacial with respect to either π system. For example, if the [3,3] sigmatropic reaction in Eq. 28.29 were suprafacial on both π systems, it could occur as follows:

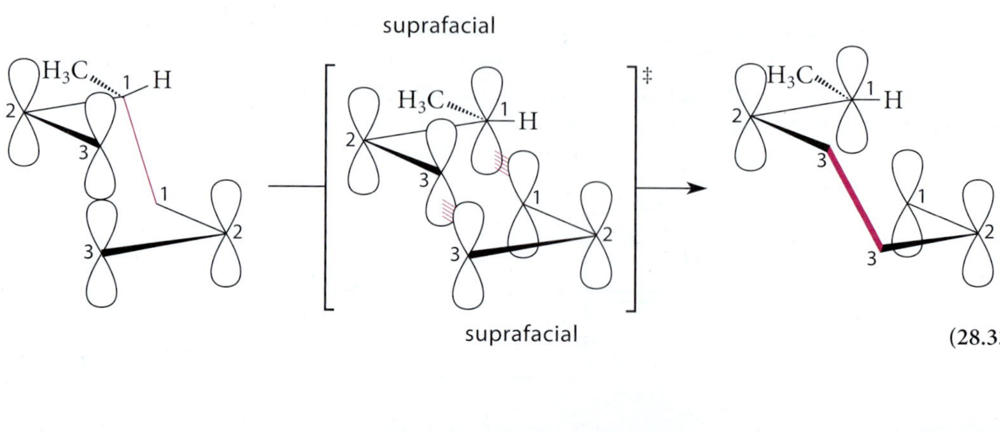

(28.33)

The stereochemistry of a sigmatropic reaction is revealed experimentally only if the molecules involved have stereocenters at the appropriate carbons. This point is illustrated in Study Problem 28.2.

Study Problem 28.2

Classify the following sigmatropic reaction by giving its bracketed-number designation and its stereochemistry with respect to the plane of the π-electron system.

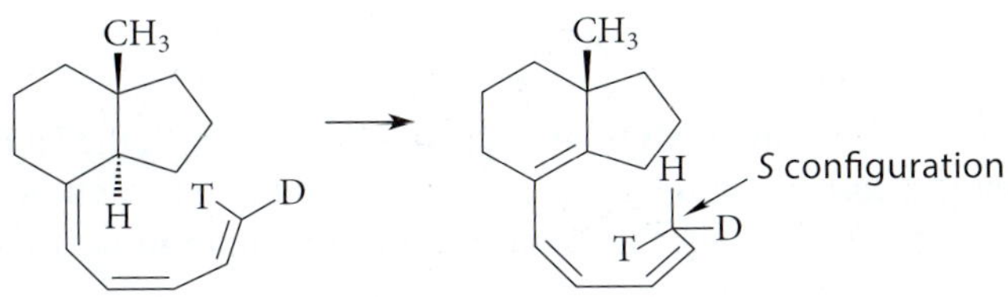

(28.34)

Solution First identify the sigma bond that is migrating. Because the hydrogen atom migrates, one end of the migrating bond remains fixed; in other words, this is a [1,?] sigmatropic reaction in which we have to determine "?" by counting the carbons over which the migration takes place:

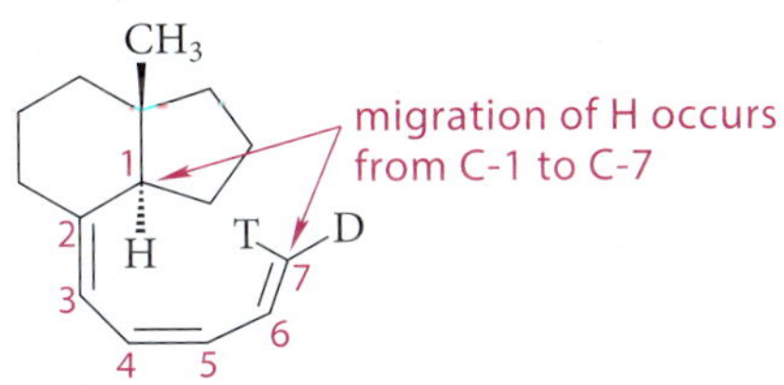

(28.35)

The original point of attachment is counted as carbon-1. Consequently, this is a [1,7] sigmatropic reaction. To determine the stereochemistry, imagine that carbons 1 through 7 all lie in the same plane. As the molecule is depicted, the migrating hydrogen is *below* that plane (*dashed wedge*). Because rotation about the C6–C7 bond cannot occur until after the reaction is over (it is a double bond), depict the T and D in the same relative orientations in the starting material and product (as they are depicted in the problem). This reveals that the hydrogen is *above* the plane of the π-electron system in the product. Consequently, the hydrogen has migrated from the lower to the upper face of the π-electron system. Therefore, the reaction is a [1,7] *antarafacial* sigmatropic reaction.

Focused Problems

28.14 (a) Refer to Study Problem 28.2 and, assuming an antarafacial migration, give the structure of a starting material that would give a stereoisomer of the product with the *R* configuration at the isotopically substituted carbon.

(b) Give the structure of *another* starting material that would give the same stereoisomer as in part (a).

28.15 Classify the following sigmatropic reactions with bracketed numbers.

(a)

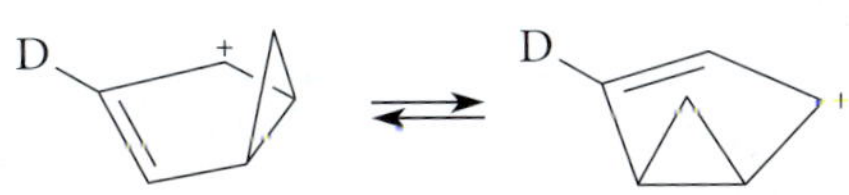

(b)

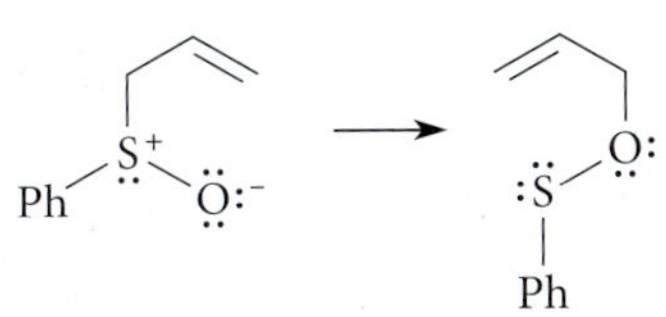

(c)

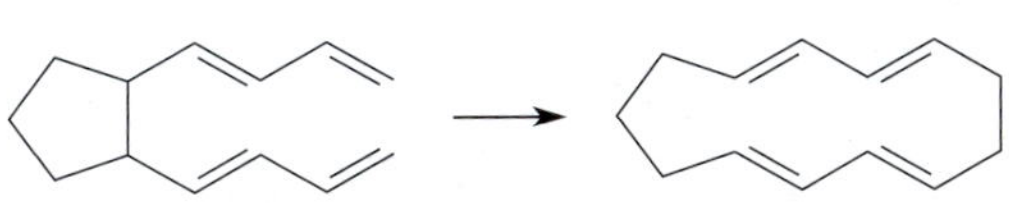

Orbital symmetry governs the connection between the type of sigmatropic reaction and its stereochemistry. Consider, for example, a [1,5] sigmatropic migration of hydrogen across a π-electron system. Think of this reaction as the migration of a proton from one end of the 2,4-pentadienyl anion (see Fig. 28.5) to the other:

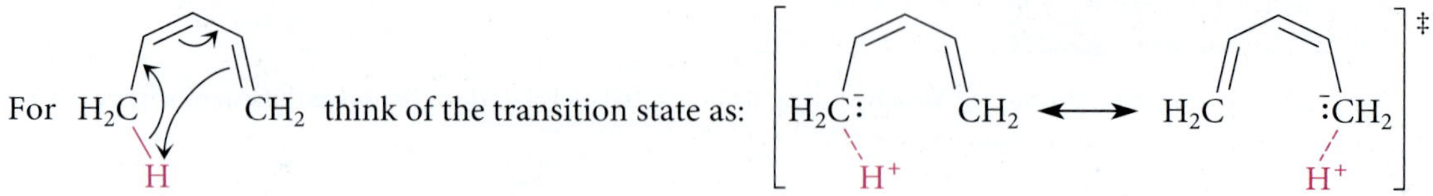

(28.36)

STUDY GUIDE LINK 28.2
Orbital Analysis of Sigmatropic Reactions

The interaction of the proton LUMO—an empty 1s orbital—with the HOMO of the π system controls the stereochemistry of the reaction. For the 2,4-pentadienyl anion (see Fig. 28.5), the HOMO is symmetric. This means that bonding overlap can occur only if the migration occurs suprafacially:

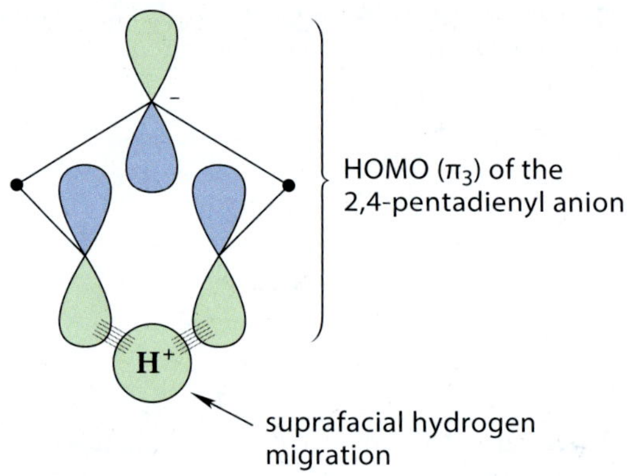

(28.37)

An ingenious experiment published in 1970 by Wolfgang Roth and his collaborators at the University of Cologne revealed the stereochemistry of the thermal [1,5] sigmatropic hydrogen shift. In the isotopically labeled, optically active alkene shown in Eq. 28.38, a suprafacial [1,5] hydrogen shift is possible from each of the conformations shown:

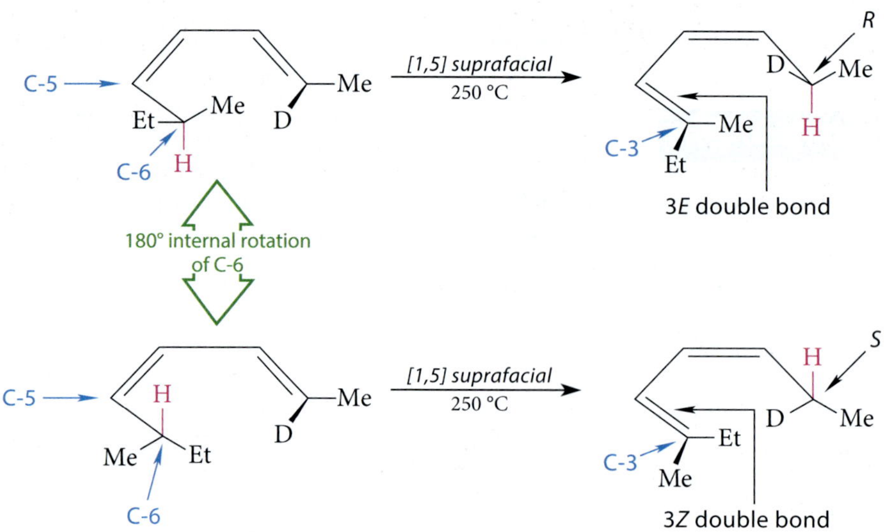

(28.38)

It follows from these equations that if the migration is suprafacial, the 3*E* isomer of the product must have the *R* configuration and the 3*Z* isomer of the product must have the *S* configuration. These were exactly the stereochemical results observed.

The [1,3] hydrogen shift involves the HOMO of the allyl anion, an antisymmetric orbital:

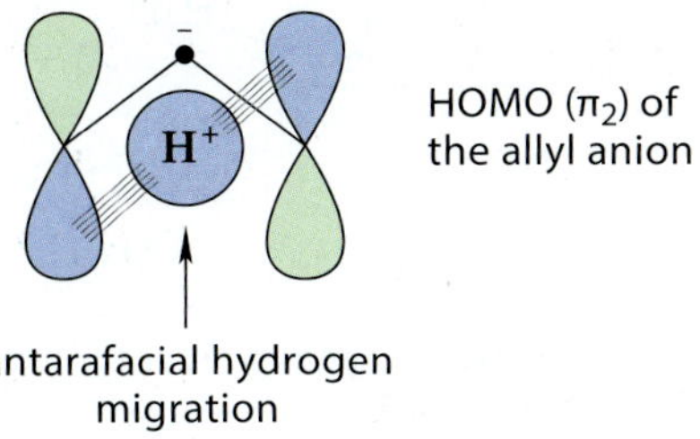

(28.39)

For a [1,3] hydrogen shift to occur, the migrating hydrogen must pass from one face of the allyl π system to the other. Despite the fact that this reaction is "allowed" by orbital symmetry, it requires that the migrating proton bridge too great a distance for adequate bonding. Alternatively, the terminal lobes of the allyl π system could twist; but then a new problem would arise: these lobes would not overlap with the 2p orbital of the central carbon. The resulting loss of orbital overlap would raise the energy of the transition state. As these arguments suggest, the concerted thermal sigmatropic [1,3] hydrogen shift is nonexistent in organic chemistry. (See Focused Problem 28.16.)

The interconversion of an enol and its isomeric carbonyl compound is a [1,3] hydrogen shift, which is disallowed as a concerted process by orbital symmetry.

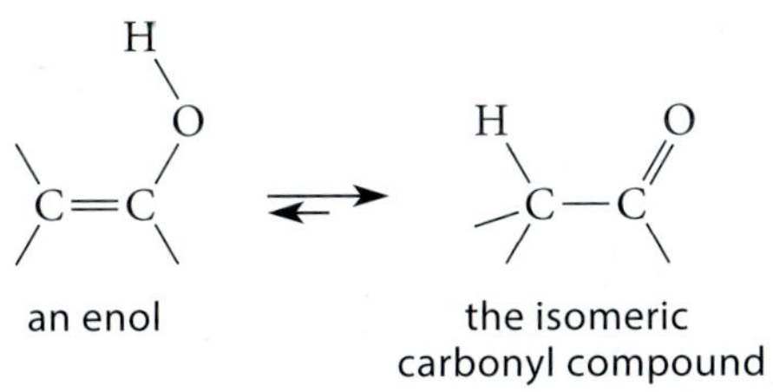

(28.40)

Yet, from Sec. 22.2, enols are rapidly converted into their corresponding carbonyl compounds. This is not a violation of orbital symmetry because all of the reactions by which enols and carbonyl compounds are interconverted involve *nonconcerted* pathways. (See, for example, Eqs. 22.21a and b.) In fact, chemists have succeeded in preparing enols in the absence of catalysts that promote their conversion into carbonyl compounds, and these enols have proved to be quite stable thermally, as the foregoing discussion of [1,3] hydrogen shifts suggests.

Several interesting experiments have been conducted in which a molecule could in principle undergo both [1,5] and [1,3] hydrogen shifts. In one particularly elegant experiment carried out in 1964, also by Roth, 1,3,5-cyclooctatriene was labeled at carbons 7 and 8 with deuterium and then allowed to undergo many hydrogen shifts for a long time. When the molecule undergoes [1,5] hydrogen shifts, the D should migrate part of the time, and the H part of the time. However, after a long time, the D should eventually scramble to all positions that have a 1,5-relationship. In such a case, only carbons 3, 4, 7, and 8 would be partially deuterated:

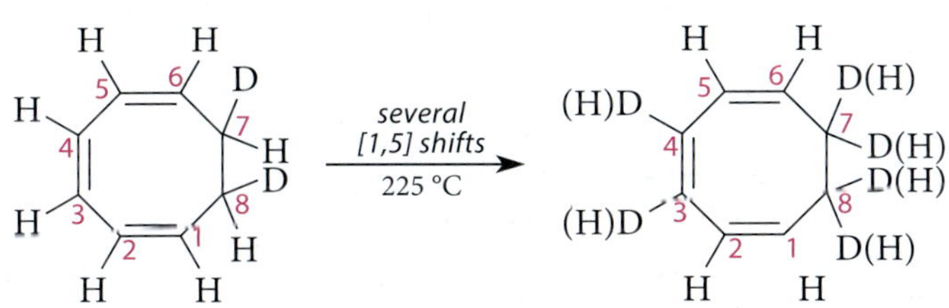

(28.41a)

(You should write a series of steps for this transformation to convince yourself that it is the predicted result.) On the other hand, if the molecule undergoes successive [1,3] hydrogen (or deuterium) shifts, the deuterium should be scrambled eventually to all positions.

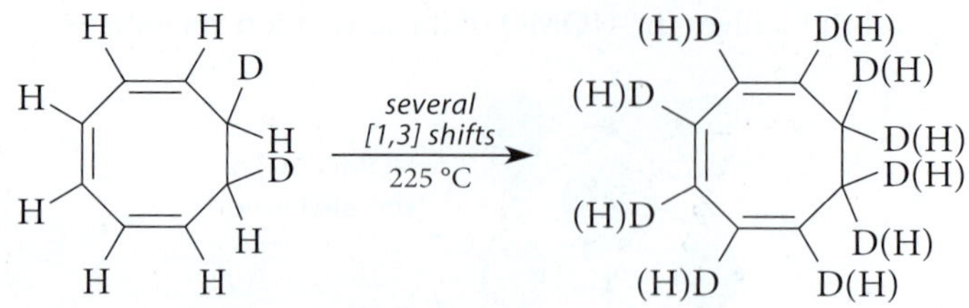

(28.41b)

The experimental result was that, even after very long reaction times, deuterium appeared only in the positions predicted by the [1,5] shift.

Although the suprafacial [1,3] shift of a hydrogen is *not* allowed, the corresponding shift of a carbon atom *is* allowed, provided that a stringent stereochemical condition is met. Suppose that an alkyl group (suitably substituted so that its stereochemistry can be traced) were to undergo a suprafacial [1,3] sigmatropic shift. This shift could occur in two stereochemically distinct ways. In the first way, the carbon migrates with *retention* of configuration.

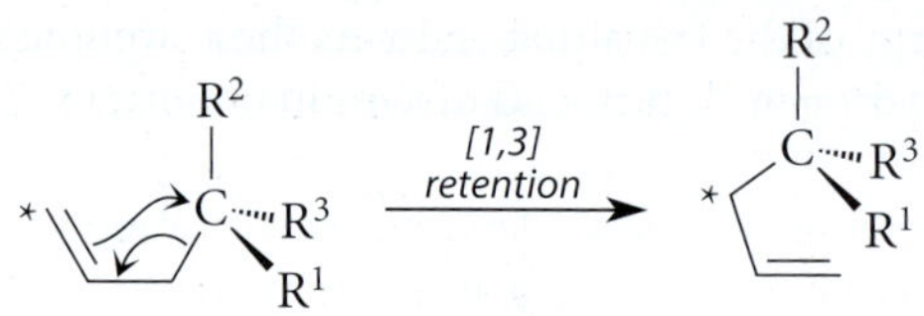

(28.42a)

(In this and the following equation, one carbon of the allyl group is marked with an asterisk so that its fate can be traced.) In the second way, the carbon migrates with *inversion* of configuration.

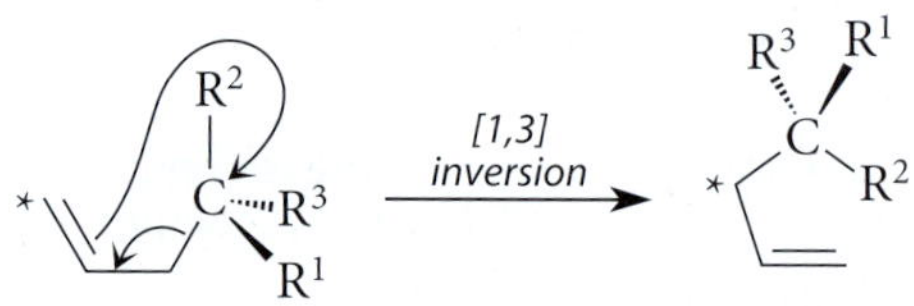

(28.42b)

Consider the orbital symmetry relationships in these two modes of reaction. Think of the migrating group as an alkyl cation migrating between the ends of an allyl anion. The LUMO of the alkyl cation—an empty 2p orbital—interacts with the HOMO of an allyl anion (π_2 in Fig. 28.4). In the case of migration with *retention*, the phase relationships between the orbitals involved lead to antibonding overlap:

Retention:

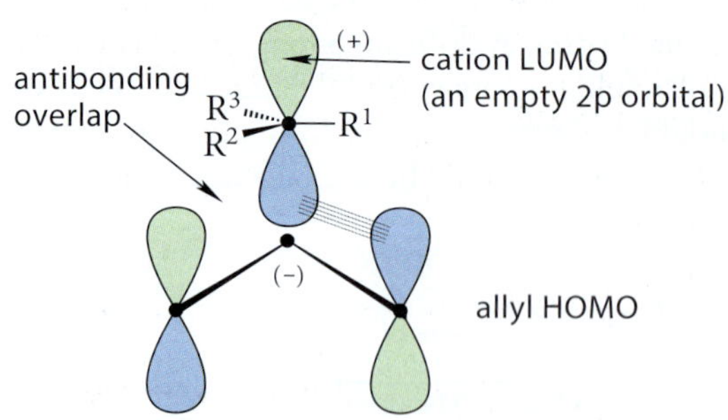

(28.43)

Therefore, suprafacial carbon migration with retention is forbidden by orbital symmetry, in the same sense that hydrogen migration is forbidden. If migration occurs with *inversion*, however, bonding overlap can occur in the transition state.

Inversion:

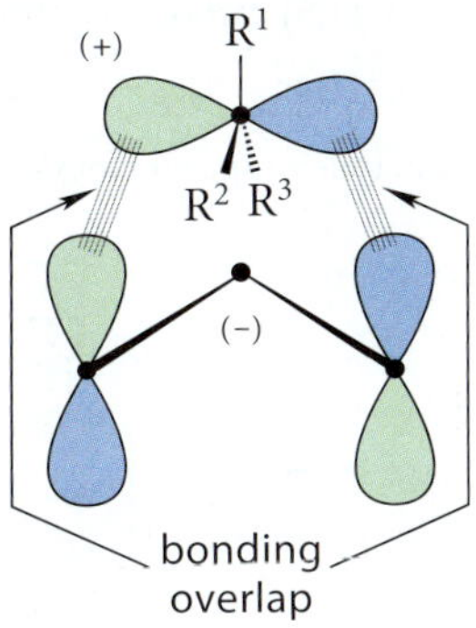

(28.44)

Thus, carbon migration with *inversion* is allowed by orbital symmetry.

This analysis shows that it is the node in the 2p orbital of the migrating carbon that makes the [1,3] suprafacial migration of this carbon possible; each of the two lobes of the 2p orbital, which have opposite phases, can overlap with each end of the allyl π system. Because a bond is broken at one side of the migrating carbon and formed at the other side, inversion of configuration is observed. In the migration of a hydrogen, the orbital involved is a 1s orbital, which has no nodes. As a result, [1,3] suprafacial migration of hydrogen is not allowed.

Orbital symmetry, then, makes a very straightforward prediction: The suprafacial [1,3] sigmatropic migration of carbon must occur with inversion of configuration. The following result confirms this prediction:

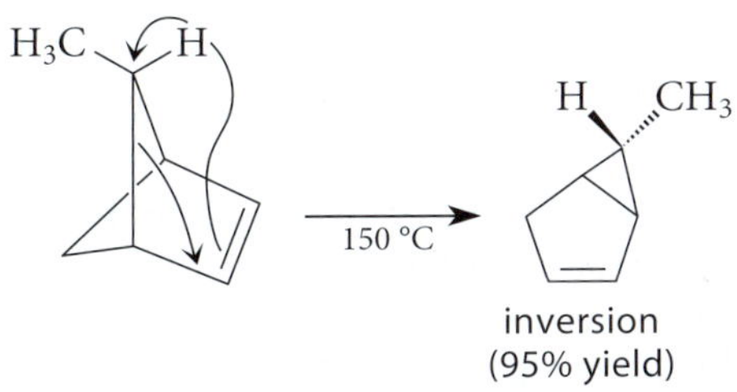

(28.45)

(See Problem 28.47 for another example.) Migration with retention of configuration might have been expected to be the most straightforward, least contorted pathway that the rearrangement could take; yet the theory of orbital symmetry predicts otherwise. One of the remarkable things about the theory is that it correctly predicts so many reactions that otherwise would have appeared unlikely.

As might be expected, orbital symmetry dictates the opposite stereochemistry for [1,5] migrations. Carbon, like hydrogen, undergoes suprafacial [1,5] migrations with retention of configuration (Focused Problem 28.18).

Focused Problems

28.16 Explain why the hydrogen migration shown in reaction (1) occurs readily and why the similar migration shown in (2) does *not* take place even under forcing conditions. (The asterisked carbons indicate a carbon isotope present so that the rearrangement can be detected.)

(1)

(2)

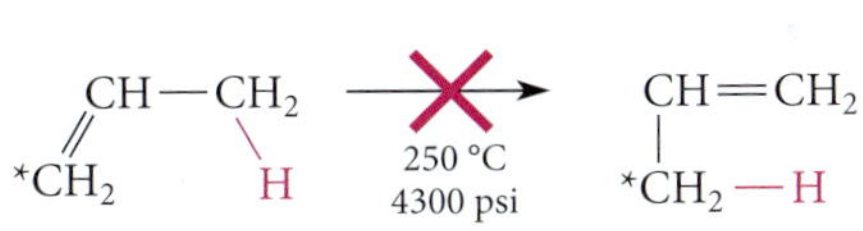

28.17 Predict the result that would have been expected in the experiment described by Eq. 28.38 for an antarafacial migration.

28.18 (a) Carry out an orbital symmetry analysis to show that suprafacial [1,5] carbon migrations should occur with retention of configuration in the migrating group.

(b) Indicate what type of sigmatropic reactions are involved in the following transformations. Is the stereochemistry of the first step in accord with the predictions of orbital symmetry?

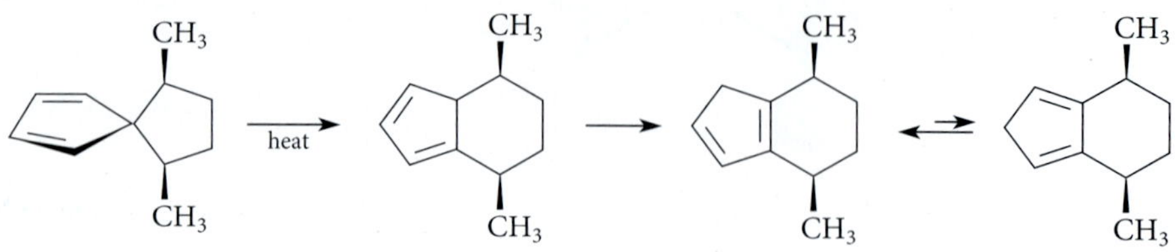

A *B*

B. Thermal [3,3] Sigmatropic Reactions

We now examine a sigmatropic reaction in which both ends of a σ bond change positions. One of the most common and useful examples of this type of reaction is the [3,3] sigmatropic rearrangement. Using the same logic as before, the transition state for this rearrangement can be visualized as the interaction of two allylic systems, one a cation and one an anion.

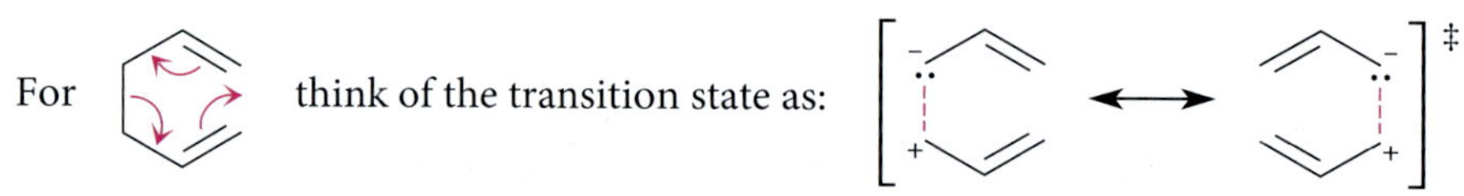

(28.46)

The frontier orbitals involved are the HOMO of the anion and the LUMO of the cation, which are the same orbital (π_2) of the allyl system (see Fig. 28.4).

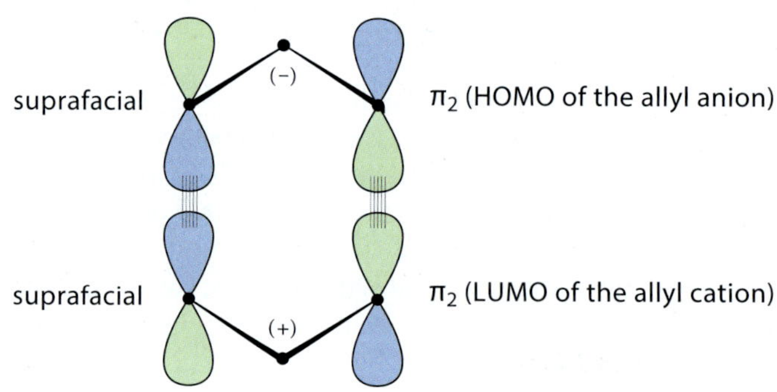

(28.47)

The two MOs involved achieve bonding overlap when the [3,3] sigmatropic rearrangement occurs suprafacially on both components. (You should convince yourself that a reaction that is antarafacial on both π systems is also allowed by orbital symmetry, but one that is suprafacial on one component and antarafacial on the other is forbidden.)

One of the best-known types of [3,3] sigmatropic rearrangement is the *Cope rearrangement*, which was extensively investigated by Professor Arthur C. Cope (1909–1966) of the Massachusetts Institute of Technology long before the principles of orbital symmetry were known. The **Cope rearrangement** is a 1,5-diene isomerization:

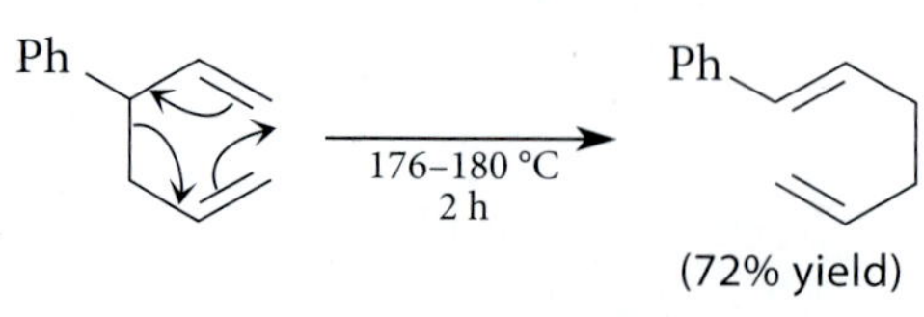

(28.48)

An interesting and useful variation of the Cope rearrangement is the **anionic oxy-Cope reaction**. In this reaction, the alkoxide ion derived from an alcohol undergoes a thermal [3,3] sigmatropic reaction. The alkoxide is generated with potassium hydride. The alkoxide is then heated in boiling THF. An enolate ion is formed initially; when it reacts with acid, the corresponding ketone is formed.

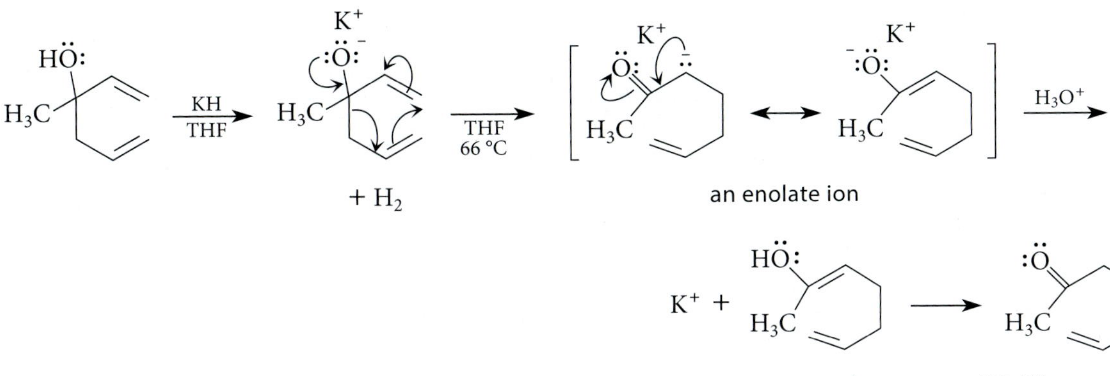

(28.49)

An oxy-Cope reaction can also take place without alkoxide formation, but the reaction then requires very strong heating. Alkoxide formation accelerates the reaction by a factor of at least 10^{10}. The reaction of the alkoxide calls to mind the breakdown of a tetrahedral addition intermediate in a carbonyl-addition reaction. The extra nonbonding electron pair on oxygen helps initiate the anionic oxy-Cope reaction by "pushing out" the bond to the neighboring carbon.

In the **Claisen rearrangement**, an ether that is both allylic and vinylic or an allylic aryl ether undergoes a [3,3] sigmatropic rearrangement. (The asterisk shows the fate of a carbon atom.)

The Claisen rearrangement is named for German chemist Rainer Ludwig Claisen (1851–1930), who investigated both this rearrangement and the Claisen condensation (Sec. 22.6A) extensively. He began his academic career at the University of Bonn in the Kekulé laboratory and, near the end of his career, collaborated with Emil Fischer (Secs. 24.9–24.10) at the University of Berlin.

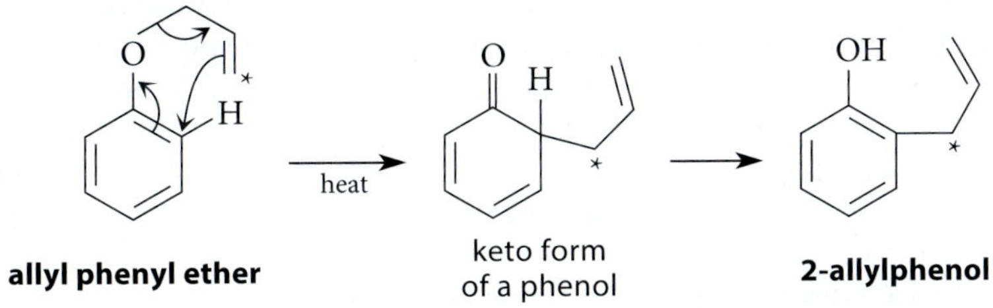

(28.50)

If both ortho positions are blocked by substituent groups, the para-substituted derivative is obtained:

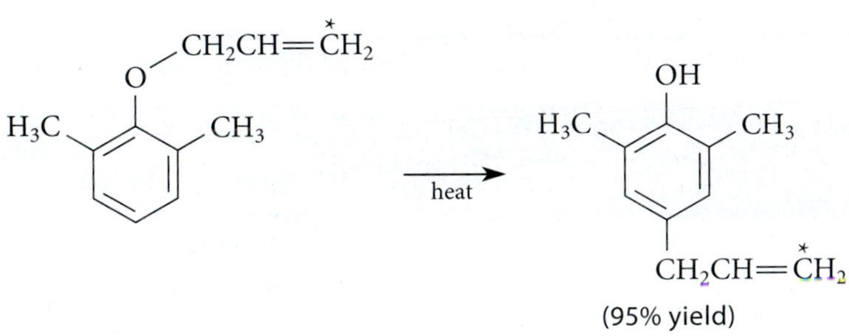

(28.51)

This reaction occurs by a sequence of two Claisen rearrangements, followed by isomerization of the product to the phenol:

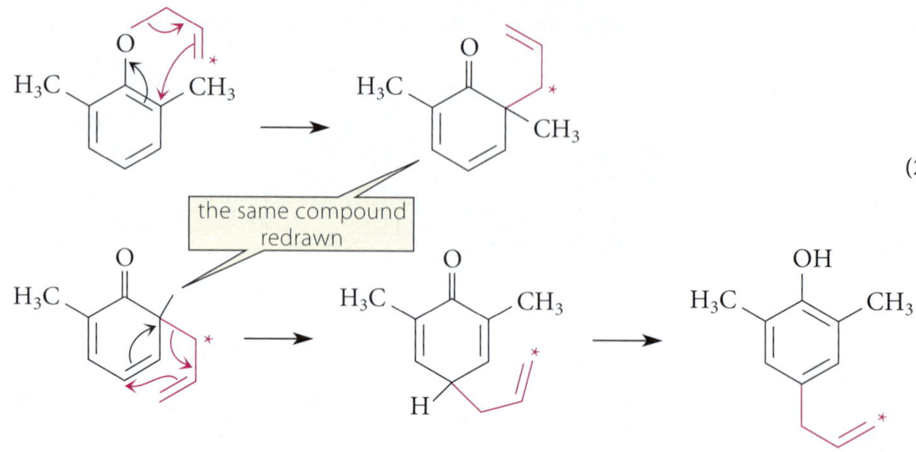

(28.52a)

(28.52b)

Focused Problem 28.20 considers the Claisen rearrangement of an aliphatic ether.

C. Summary: Selection Rules for Thermal Sigmatropic Reactions

The stereochemistry of sigmatropic reactions is a function of the number of electrons involved. As with other pericyclic reactions, the number of electrons involved is determined from the curved-arrow notation: Count the curved arrows and multiply by 2. All-suprafacial thermal sigmatropic reactions occur when $4n + 2$ electrons are involved in the reaction—that is, an odd number of electron pairs or curved arrows. In contrast, a thermal sigmatropic reaction must be antarafacial on one component and suprafacial on the other when $4n$ electrons (an even number of electron pairs or curved arrows) are involved. When a single carbon migrates, an additional issue is whether the carbon migrates with inversion of configuration or retention of configuration. With a system of $4n + 2$ electrons, a suprafacial carbon migration occurs with retention of configuration (Focused Problem 28.18), and with $4n$ electrons, a suprafacial carbon migration occurs with inversion (as in Displays 28.42b and 44). The stereochemistry at carbon is reversed for antarafacial migrations.

These generalizations are summarized in **Table 28.3** as selection rules for thermal sigmatropic reactions.

TABLE 28.3 Selection Rules for Thermal Sigmatropic Reactions

	Allowed stereochemistry*	
Number of electrons†	Generalized stereochemistry	Stereochemistry of single-atom migrations
$4n$	supra–antara	supra–inversion
	antara–supra	antara–retention
$4n + 2$	supra–supra	supra–retention
	antara–antara	antara–inversion

*supra = suprafacial; antara = antarafacial

†n = an integer

Focused Problems

28.19 (a) What allowed and reasonable sigmatropic reaction(s) can account for the following transformation?

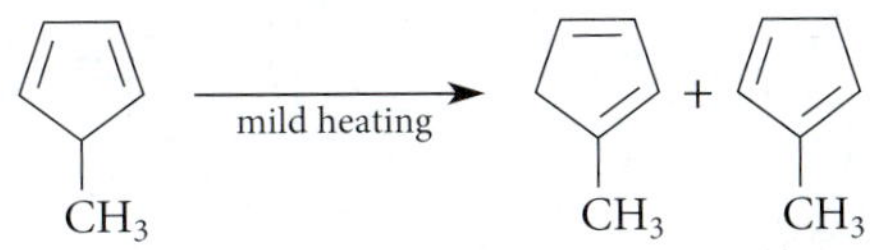

(b) What product(s) are expected from a similar reaction of 2,3-dimethyl-1,3-cyclopentadiene?

28.20 (a) Aliphatic allylic vinylic ethers undergo the Claisen rearrangement. Complete the following reaction:

$$(CH_3)_2C{=}CH{-}CH_2{-}O{-}CH{=}CH_2 \xrightarrow[\text{heat}]{}$$

(b) What starting material would give the following compound in an aliphatic Claisen rearrangement?

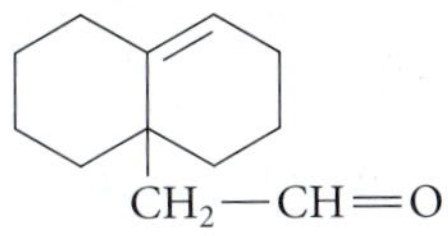

28.21 Show how the transition state for a [3,3] sigmatropic reaction can be analyzed as the interaction of two allylic radicals and that the same stereochemical outcome is predicted. (See Study Guide Link 28.2.)

28.22 (a) Give the curved-arrow mechanism for the anionic oxy-Cope reaction of compound *A*, and explain why the stereoisomer *B* does not react under the same conditions.

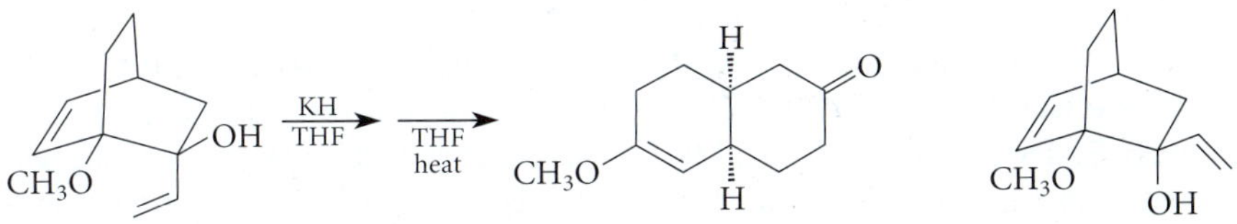

A

B
(unreactive under the same conditions)

(b) Give the structure of the product, including its stereochemistry, expected from the anion oxy-Cope reaction of the following compound.

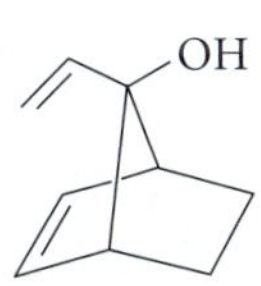

28.23 Show by an orbital symmetry analysis that a [3,3] sigmatropic reaction that is antarafacial on both components is allowed. Would you expect such a reaction to be very common? Why?

28.5 FLUXIONAL MOLECULES

A number of compounds continually undergo rapid sigmatropic rearrangements at room temperature. One such compound is *bullvalene*, which was first prepared in 1963.

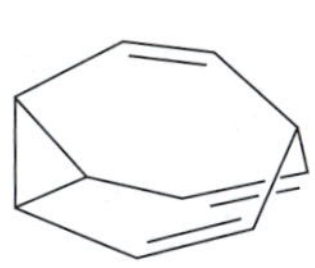

bullvalene (28.53)

The [3,3] sigmatropic rearrangements in bullvalene rapidly interconvert *identical forms of the molecule*:

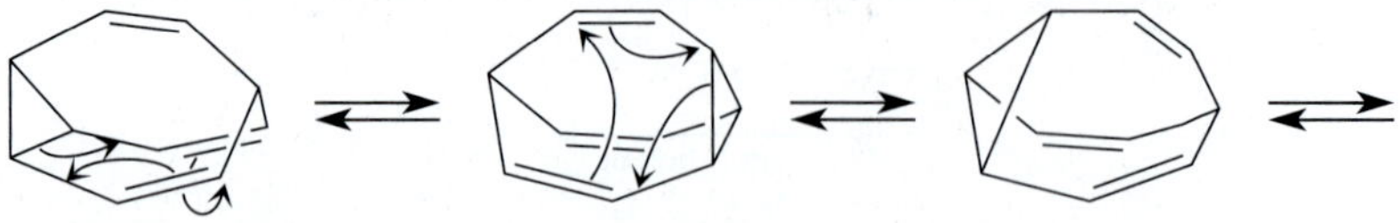

(28.54)

If the carbons could be individually labeled, there would be 1,209,600 equivalent structures of bullvalene in equilibrium! Each one of these forms is converted into another at a rate of about 2000 times per second at room temperature. (These reactions can be observed by the dynamic NMR methods discussed in Sec. 14.8.) At high temperature (above 100 °C), both the proton NMR spectrum and the ^{13}C NMR spectrum of bullvalene consist of a single line; the rapid fluxional isomerization averages the NMR absorptions so that all of the protons, as well as all of the carbons, become equivalent over time.

The different molecules related by rapid bond shifts are sometimes called **valence isomers.**

Molecules such as bullvalene that undergo rapid bond shifts are called **fluxional molecules**. Their atoms are in a continual state of motion associated with the rapid changes in bonding. Although the structures and the curved-arrow notation in Eq. 28.54 might suggest that they are resonance structures, fluxional molecules are *not* resonance structures because the nuclei actually move during their interconversion.

"The Bull" and Bullvalene

Odua Images/Shutterstock

The fluxional nature of bullvalene was predicted by William von E. Doering (1917–2011) while he was a professor at Yale University. The intriguing name of this compound seems to have originated from a seminar of his research group in late 1961 in which Doering first disclosed his prediction. Doering's research students had nicknamed him "The Bull," and Doering had been interested in a class of compounds known as the fulvalenes. When Doering wrote the structure on the board, one graduate student in the back of the room blurted out, "Bullvalene!" The name stuck. Bullvalene was first prepared serendipitously in 1963 by Gerhard Schröder, a chemist at Union Carbide Corporation in Brussels.

Focused Problem

28.24 Each of the following compounds exists as a fluxional molecule that is interconverted into one or more identical forms by the sigmatropic process indicated. Draw one structure in each case that demonstrates the process involved, and explain why each process is an allowed pericyclic reaction.

(a)

(b)

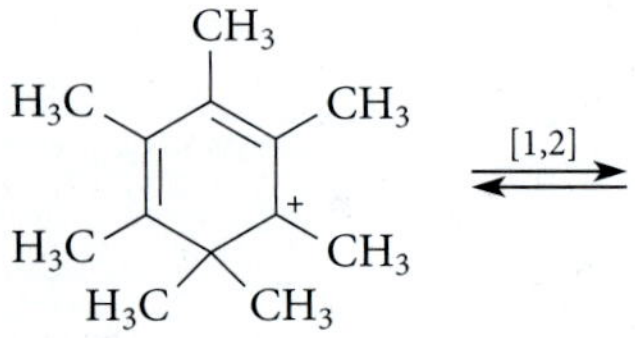

28.6 BIOLOGICAL PERICYCLIC REACTIONS. THE FORMATION OF VITAMIN D

Chemical Biology Topic

In areas of the world where winters are long and there is little sunlight, children have long suffered from a disease called *rickets* (from Old English, *wrickken*, meaning "to twist"). This disease is characterized by inadequate calcification of the bones. A similar disease in adults, *osteomalacia*, is particularly prominent among Bedouin Arab women who must remain completely covered when they are outdoors.

Rickets can be prevented by administration of any one of the forms of vitamin D, a hormone that controls calcium deposition in bone. The human body manufactures a chemical precursor to vitamin D called 7-dehydrocholesterol (structure in Eq. 28.55). This is converted into vitamin D_3, or *cholecalciferol* (structure in Eq. 28.56), only when the skin receives adequate ultraviolet radiation from the Sun or other source.

The reaction by which 7-dehydrocholesterol is converted into vitamin D_3 is a sequence of two pericyclic reactions. The first is a photochemical electrocyclic reaction:

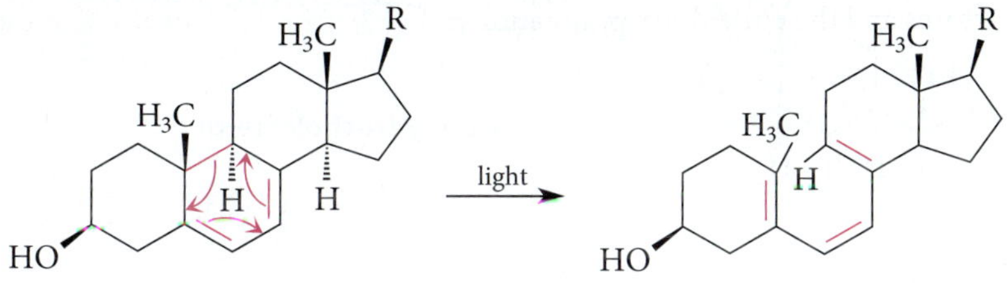

7-dehydrocholesterol **previtamin D_3**

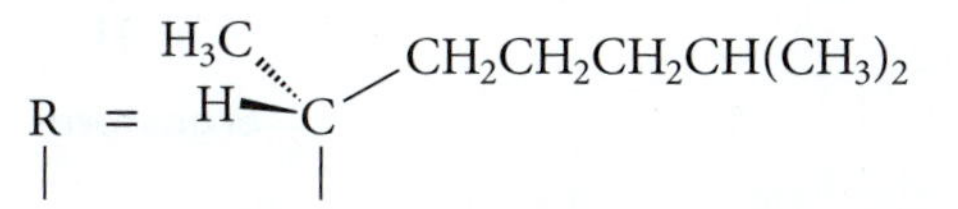

(28.55)

This reaction is a *conrotatory* process. (Be sure you understand why this is so; examine the reverse reaction if necessary.) This is precisely the stereochemistry required for a *photochemically allowed* electrocyclic reaction involving $4n + 2$ electrons. Sunlight ordinarily provides the UV radiation necessary for this reaction to occur in humans.

The final step in the formation of vitamin D_3 is a thermal [1,7] sigmatropic hydrogen shift:

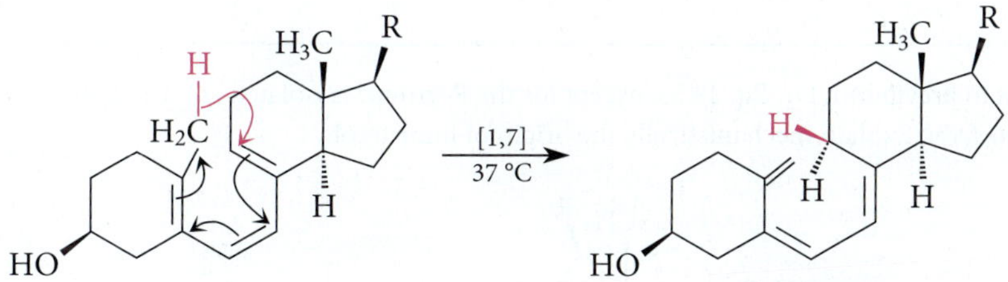

cholecalciferol (vitamin D_3) (28.56)

(The stereochemistry of this process is not defined by the structures of the reactant and product; see Focused Problem 28.28.) Vitamin D_3 exists in the more stable *s*-trans conformation, attained by internal rotation (*green arrow*):

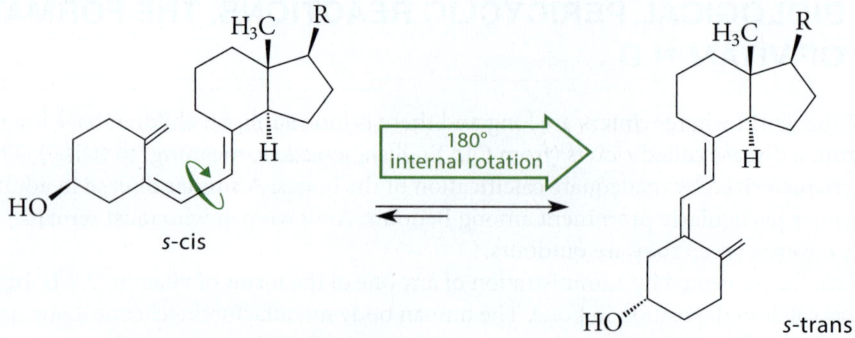

(28.57)

Vitamin D_2 (ergocalciferol or calciferol), a compound closely related to vitamin D_3, is produced commercially by irradiation of a steroid called *ergosterol.* Ergosterol is identical to 7-dehydrocholesterol (Eq. 28.55) except for the side-chain R:

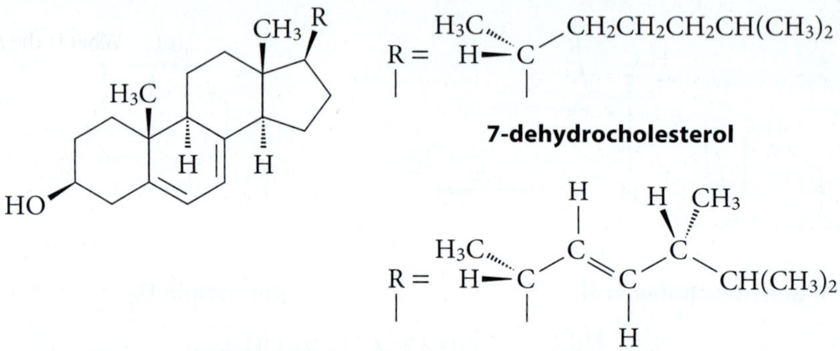

ergosterol (28.58)

Irradiation of ergosterol gives successively previtamin D_2 and vitamin D_2, which are identical to the products of Eqs. 28.55 and 28.56, respectively, except for the R-group. Vitamin D_2, sometimes called "irradiated ergosterol," is the form of vitamin D that is commonly added to milk and other foods as a dietary supplement.

Focused Problems

28.25 When previtamin D_2 (which is identical to previtamin D_3, Eq. 28.55, except for the R-group) is isolated and irradiated, ergosterol is obtained along with a stereoisomer, *lumisterol.* Explain mechanistically the origin of lumisterol.

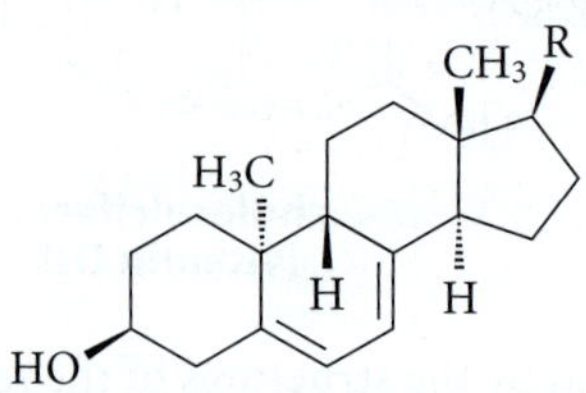

lumisterol

28.26 When previtamin D_2 is *heated*, two compounds, *A* and *B*, are obtained that are stereoisomers of both ergosterol and lumisterol. Suggest structures for these compounds, and explain mechanistically how they are formed.

28.27 When the compounds *A* and *B* in the previous problem are *irradiated*, two stereoisomeric compounds, *C* and *D*, respectively, are obtained, each of which contains a cyclobutene ring. Suggest structures for *C* and *D*, and explain mechanistically how they are formed. Explain why irradiation of either *A* or *B* does *not* give back previtamin D_2.

28.28 Although the stereochemistry of Eq. 28.56 cannot be determined from the reaction, what stereochemistry is expected from orbital symmetry considerations?

CHAPTER SUMMARY *For a summary of the chapter, see Chapter 28 in the* Study Guide and Solutions Manual.

REACTION REVIEW *For a summary of reactions discussed in this chapter, see the* Reaction Review *section of Chapter 28 in the* Study Guide and Solutions Manual.

SKILLS OBJECTIVES WITH PROBLEMS

- **Classify any pericyclic reaction as an electrocyclic reaction, a cycloaddition, or a sigmatropic reaction.** **Introduction**

28.29 Classify each of the following reactions as an electrocyclic reaction, a cycloaddition reaction, or a sigmatropic reaction. (*Hint:* Apply the principle of microscopic reversibility if necessary.)

(a)
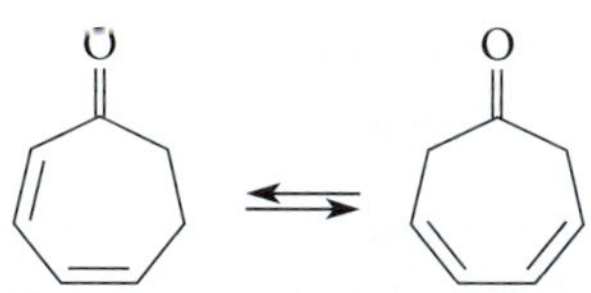

(b)
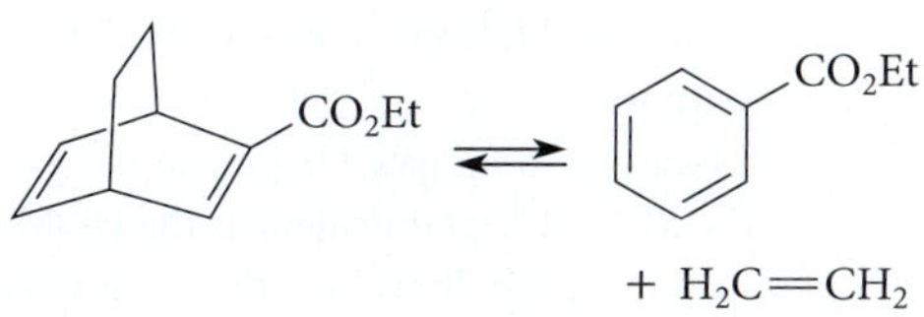

(c)

- **Sketch the π molecular orbitals of conjugated alkenes, classify them as symmetric or antisymmetric, and identify the HOMO and LUMO.** **28.1A**

28.30 (a) How many nodes does the HOMO of (3*Z*,5*Z*)-1,3,5,7-octatetraene have?

(b) Is this HOMO symmetric or antisymmetric?

(c) Represent the carbons as eight dots in a line, and sketch the position of the nodes (as lines) in the HOMO.

(d) Is the LUMO of this molecule symmetric or antisymmetric?

28.31 (a) How many π molecular orbitals does 3-methylene-1,4-pentadiene have?

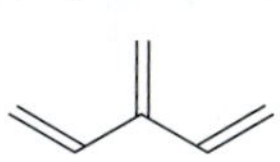
3-methylene-1,4-pentadiene

(b) Represent the carbons as dots in the plane of the page, and show the nodes for all occupied MOs.

(c) What is the symmetry of the HOMO?

- **Sketch the π molecular orbitals for conjugated cations, radicals, and anions; classify them as symmetric or antisymmetric; and identify the HOMO, LUMO (and if applicable) the SOMO.** **28.1B**

28.32 (a) How many π MOs are there in the 2,4,6,8-nonatetraen-1-yl radical?

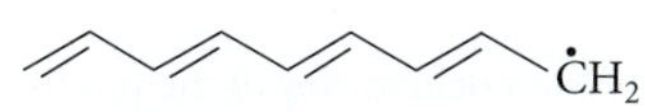
2,4,6,8-nonatetraen-1-yl radical

(b) Sketch an orbital energy diagram for this radical.

(c) Which MO is the singly occupied molecular orbital (SOMO)? How do you know?

(d) Where are the nodes in this MO? How do you know?

28.33 (a) How many π molecular orbitals does the benzyl cation have? How do you know?

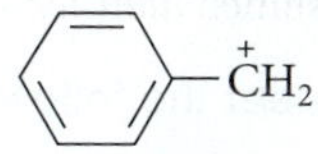
benzyl cation

(b) Number the MOs from lowest to highest energy, and identify the nonbonding MO (NBMO). Explain your reasoning.

(c) In the NBMO, which atoms are located at nodes? (*Hint:* Use resonance structures.)

- **Given the ground state of a molecule with π molecular orbitals, give the electronic configuration of the excited state and identify the HOMO and LUMO.** **28.1C**

28.34 (a) Draw an orbital energy diagram for 1,3,5-hexatriene (ignore stereochemistry), and populate the available MOs with electrons to form the ground state. Identify the HOMO.

(b) Draw another orbital energy diagram for the first excited state, and identify the HOMO.

(c) Contrast the symmetries of the HOMOs in the ground state and excited state. What is the significance of this symmetry difference for chemical reactions?

- **For electrocyclic reactions, meet the following objectives. 28.2**

  (1) **Know and apply the definition of conrotatory and disrotatory reaction stereochemistry.**

  (2) **Know the selection rules, and justify them with a frontier-orbital analysis.**

  (3) **Use the selection rules to determine the conditions (thermal or photochemical) for an electrocyclic reaction and its stereochemistry in either direction.**

28.35 Provide the following for the two reactions shown in **Fig. P28.35.**

(a) Show the stereochemistry of the product of the first reaction that would result from (1) conrotatory and (2) disrotatory reactions.

(b) For the second reaction, show the stereochemistry at the ring junction that would result from (1) conrotatory and (2) disrotatory reactions.

(c) For each reaction, apply the selection rules to determine the stereochemistry.

(d) For the second reaction, show that the predicted stereochemistry follows from the symmetry of the frontier molecular orbital involved.

- **For cycloaddition reactions, meet the following objectives. 28.3**

  (1) **Know and apply the numerical and stereochemical designations.**

  (2) **Know the selection rules, and justify them with a frontier-orbital analysis.**

  (3) **Use the selection rules to determine the conditions (thermal or photochemical) for an electrocyclic reaction and its stereochemistry.**

28.36 (a) For the following cycloaddition reaction, describe the possible modes of reaction with the numerical and stereochemical notation used for cycloaddition and the stereochemistry corresponding to each mode.

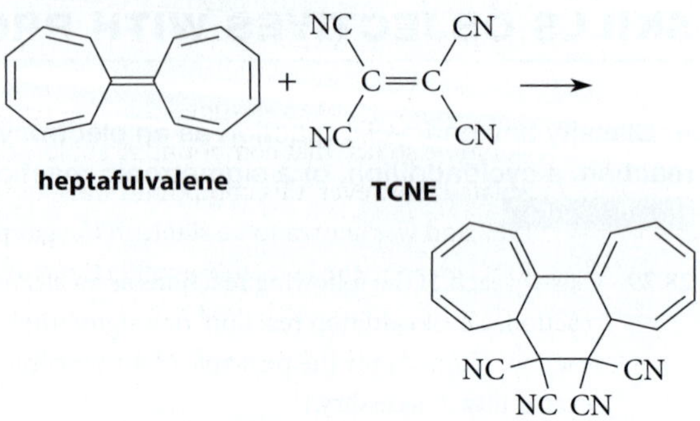

(b) What stereochemistry is predicted by the selection rules for cycloaddition?

(c) Although heptafulvalene is not an unbranched conjugated heptaene (a 14-carbon conjugated alkene with 7 double bonds), its HOMO has the same symmetry as such a heptaene. Using the MOs of the 14-carbon conjugated heptaene as a guide, which of the MOs of heptafulvalene is the HOMO? What is its symmetry? Show how this symmetry is consistent with the selection rule you invoked in part (b).

- **For thermal sigmatropic reactions, meet the following objectives. 28.4**

  (1) **Know and apply the numerical and stereochemical designations.**

  (2) **Know the selection rules, and justify them with a frontier-orbital analysis.**

  (3) **Use the selection rules to determine the stereochemistry of a thermal sigmatropic reaction.**

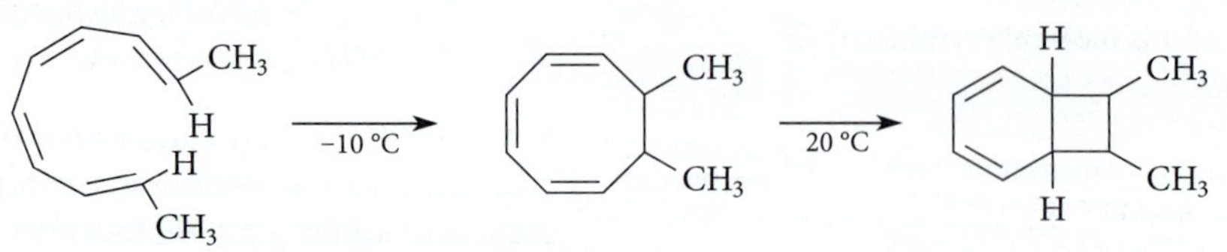

A *B* *C*

FIGURE P28.35

28.37 (a) Give the numerical description of the following sigmatropic reaction and, without considering the selection rules, show the two possible stereochemical paths for the reaction.

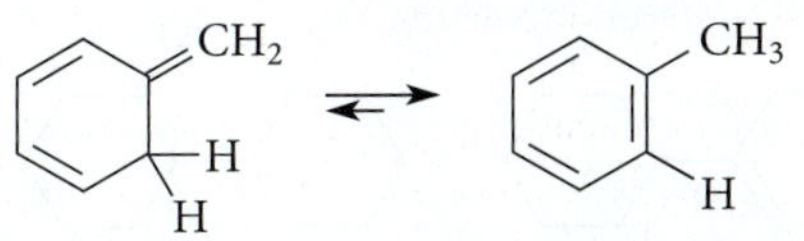

A **toluene** *(B)*

(b) Explain why the equilibrium for this reaction strongly favors the product toluene.

(c) Chemists had always assumed that this reaction would be so fast that compound *A* could never be isolated. However, this compound was prepared in 1962 and was shown to be stable in the gas phase at 70 °C, despite the favorable equilibrium constant for its transformation to *B*. Show why the conversion of *A* into *B* would *not* be expected to occur as a concerted reaction.

(d) Use the symmetry of the SOMO of compound *A* to justify your prediction.

(e) Would you expect a concerted mechanism for the following reaction to be equally slow? Why?

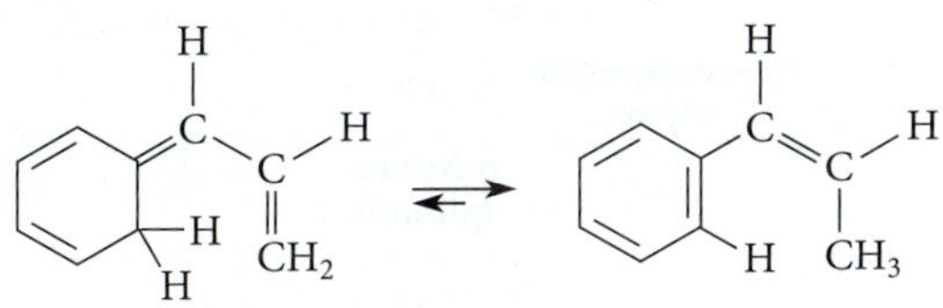

C *D*

(f) Why is the cis isomer of the alkene product rather than the more stable trans isomer formed in this reaction?

- **Complete examples of the following pericyclic reactions. 28.4B**

(1) Cope rearrangement

(2) Anionic oxy-Cope rearrangement

(3) Claisen rearrangement

(4) Diels–Alder reaction

28.38 Complete each of the following reactions by giving the product and the reasoning used to obtain it.

(a) Cope rearrangement:

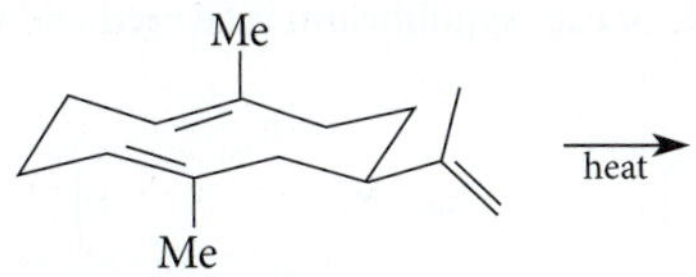

(b) Claisen rearrangement:

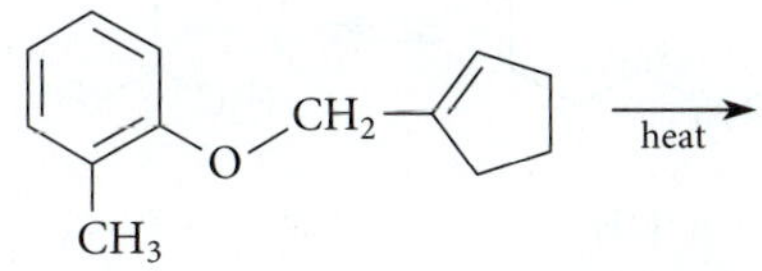

(c) Diels–Alder reaction:

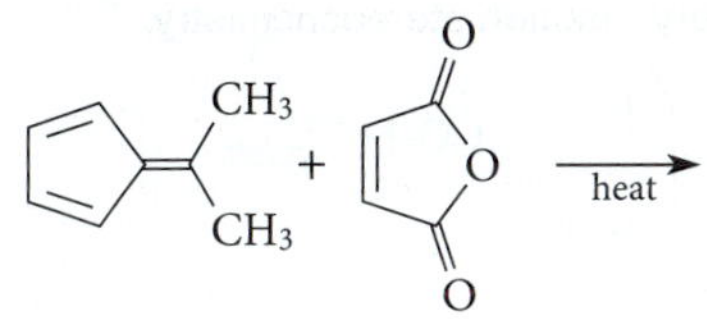

(d) Anionic oxy-Cope rearrangement:

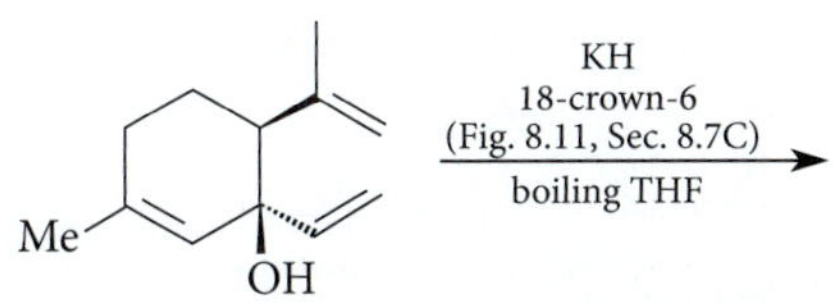

- **Know the difference between a fluxional molecule and resonance structures. 28.5**

28.39 Which of the following statements are true of a fluxional molecule? (The "movement of electrons" refers to electron shifts shown by the curved-arrow notation.)

(a) They are resonance structures that change with time.

(b) Electrons move but nuclei don't move in the fluxional interconversion.

(c) Electrons and nuclei both move in the fluxional interconversion.

(d) Nuclei move but electrons don't move in the fluxional interconversion.

(e) In principle, a fluxional molecule should have a temperature-dependent NMR spectrum if it contains chemically nonequivalent nuclei.

INTEGRATED PROBLEMS

28.40 Without consulting tables or figures, answer the following questions:

(a) Is a thermal disrotatory electrocyclic reaction involving 12 electrons allowed or forbidden?

(b) Is an $[8s + 4s]$ photochemical cycloaddition allowed?

(c) Is the HOMO of (*Z*)-3,4-dimethyl-1,3,5-hexatriene symmetric or antisymmetric?

28.41 What do the pericyclic selection rules have to say about the *position of equilibrium* in each of the following reactions? Which side of each equilibrium is favored and why?

(a)
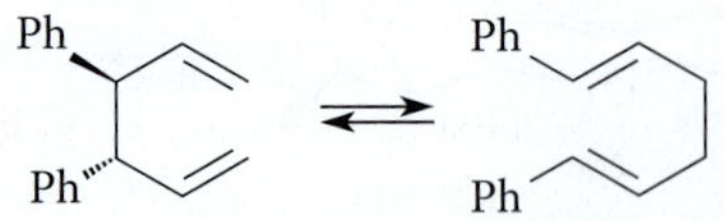

(b)
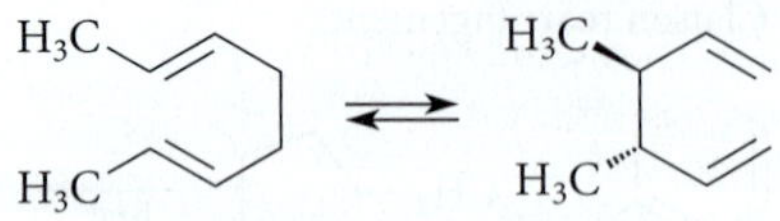

(c)
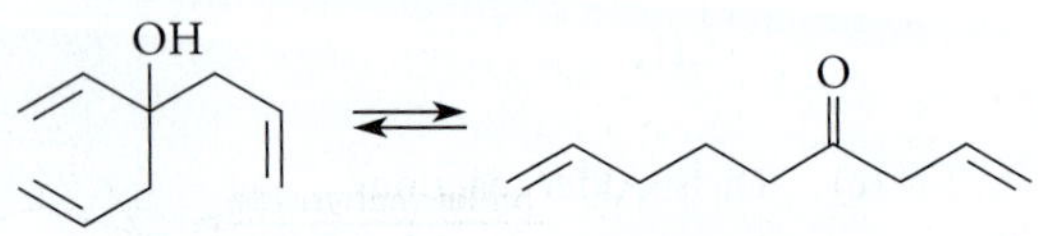

(d)
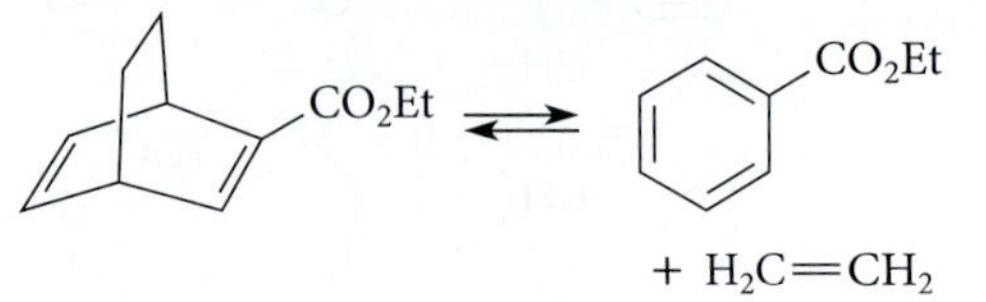

(e)
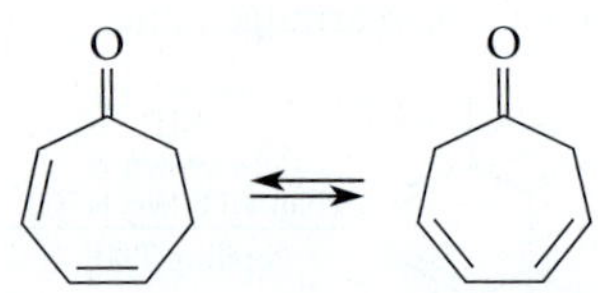

28.42 (a) Classify the following pericyclic reaction. (*Hint*: More than one classification is possible.)

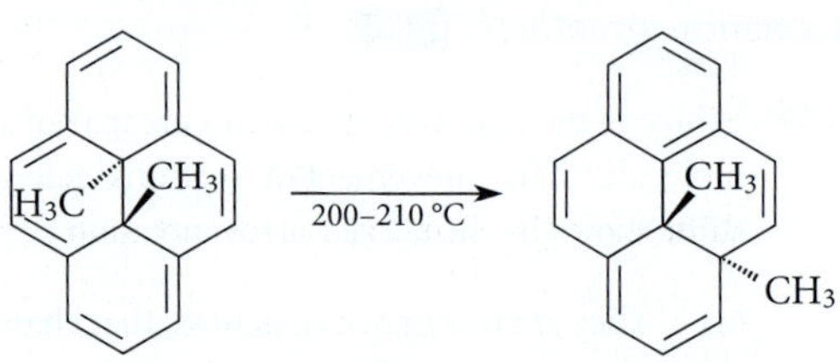

(80% yield)

(b) Suppose the migrating methyl group in part (a) were labeled with the hydrogen isotopes deuterium (D) and tritium (T) so that it is a —CHDT group with the *S* configuration. What would be the configuration of this group in the product? Explain your reasoning.

28.43 Complete the following reactions by giving the major organic product(s), including stereochemistry.

(a)
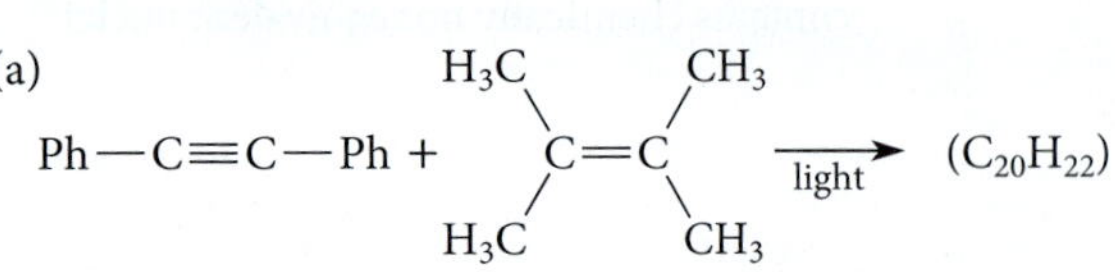

(b)
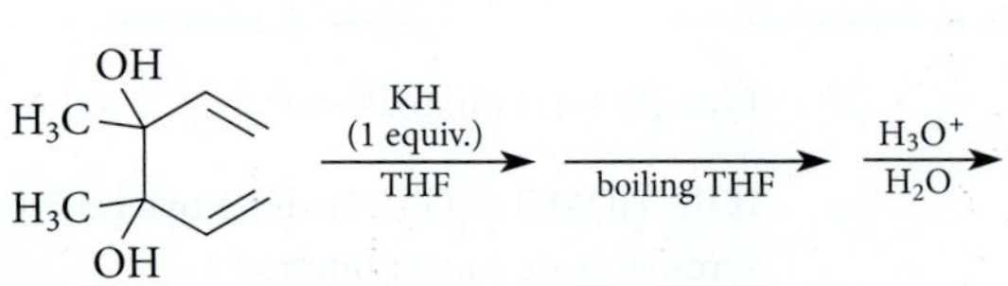

28.44 When compound *A* is irradiated with ultraviolet light for 115 hours in pentane, an isomeric compound *B* is obtained that decolorizes bromine in CH_2Cl_2 and reacts with ozone to give, after the usual workup, a compound *C*.

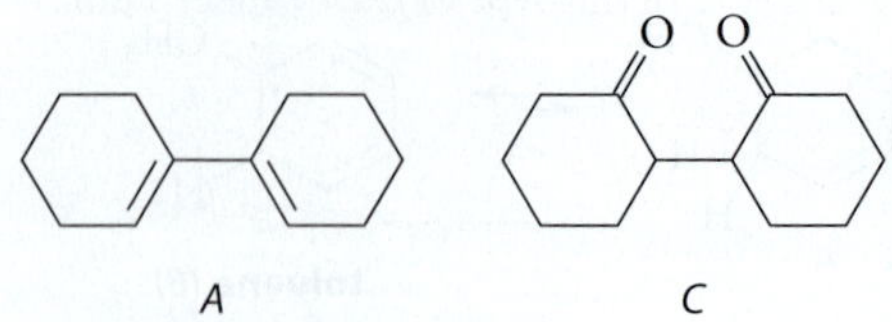

(a) Give the structure of *B* and the stereochemistry of both *B* and *C*.

(b) On heating to 90 °C, compound *D*, a stereoisomer of *B*, is converted into *A*, but compound *B* is virtually inert under the same conditions. Identify compound *D*, and account for these observations.

28.45 When 1,3-cyclopentadiene and *p*-benzoquinone are allowed to react at room temperature, a compound *X* is obtained. Irradiation of compound *X* gives compound *Y*. Give the structure and stereochemistry of *X*, and explain its conversion into *Y*.

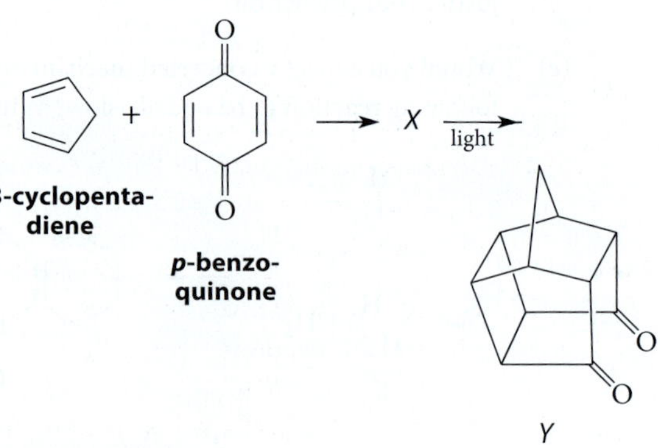

28.46 Suggest a mechanism for each of the following transformations. Some involve pericyclic reactions only; others involve pericyclic reactions as well as other steps. Invoke the appropriate selection rules to explain any stereochemical features observed.

(a)
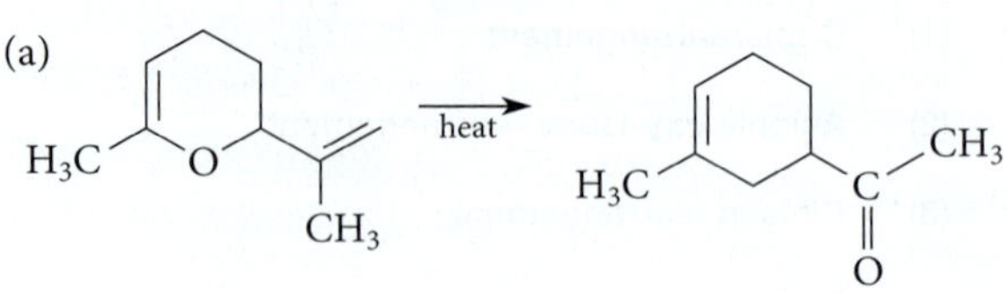

(b)
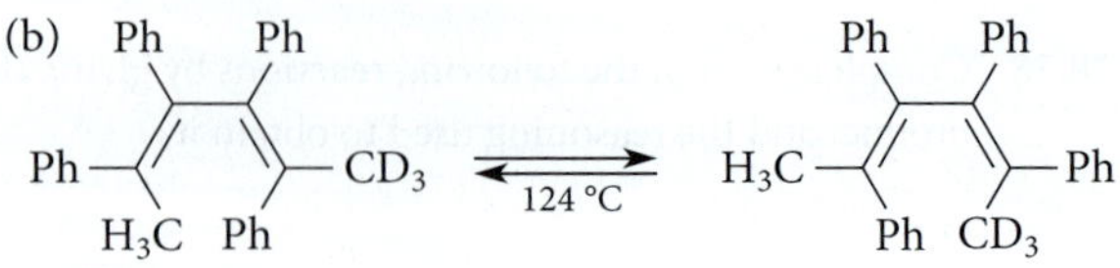

(c)
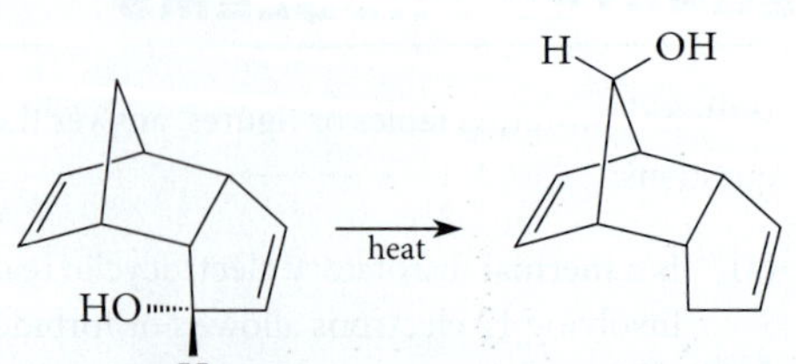

28.47 Classify the following sigmatropic reaction, give the curved-arrow notation, and show that the stereochemistry is that expected for a thermal concerted reaction. (This reaction, discovered by Professor J. A. Berson at Yale University, was the first example of this type of pericyclic reaction.)

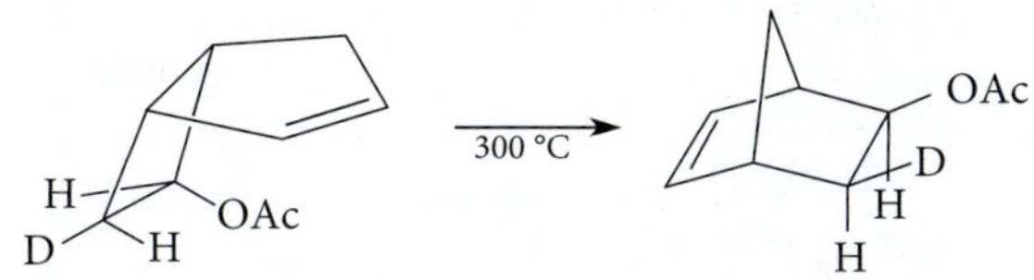

28.48 When 1,3,5-cyclooctatriene, *A*, is heated to 80–100 °C, it comes to equilibrium with an isomeric compound *B*. Treatment of the mixture of *A* and *B* with $MeO_2C—C{\equiv}C—CO_2Me$ gives a compound *C*, which, when heated to 200 °C for 20 minutes, gives dimethyl phthalate and cyclobutene. Identify compounds *B* and *C*, and explain what reactions have occurred.

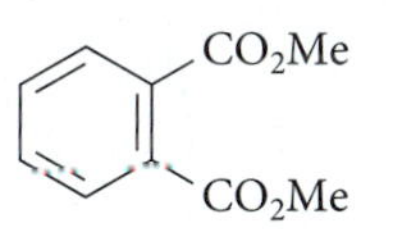

dimethyl phthalate

28.49 Each of the transformations in **Fig. P28.49** occurs as a sequence of two pericyclic reactions. Identify the intermediates *X*, *Y*, and *Z*, and describe the two reactions involved in each part.

28.50 Compound *A*, when heated, gives another compound *B*. When *B* is subjected to catalytic hydrogenation over a Pd catalyst, it gives a compound *C*, which finds use in the food industry as a fruity odor and flavor. Compound *C* has a strong IR absorption at 1740 cm^{-1}. Identify compounds *B* and *C*.

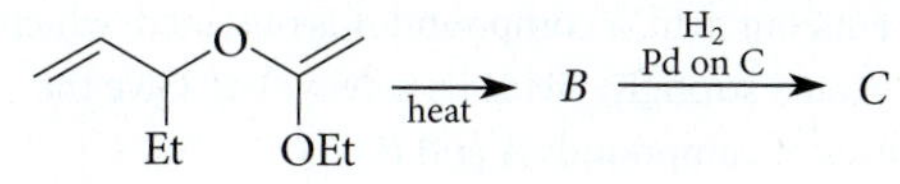

A

28.51 When each of the following compounds is heated in the presence of maleic anhydride, an intermediate is trapped as a Diels–Alder adduct. What is the intermediate formed in each reaction, and how is it formed from heating the starting material?

(a)

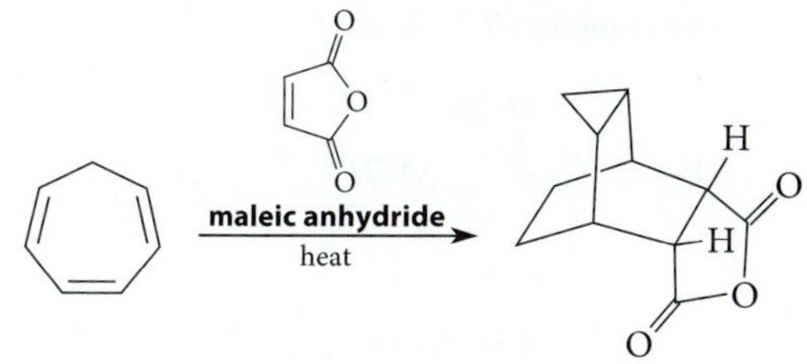

(b)

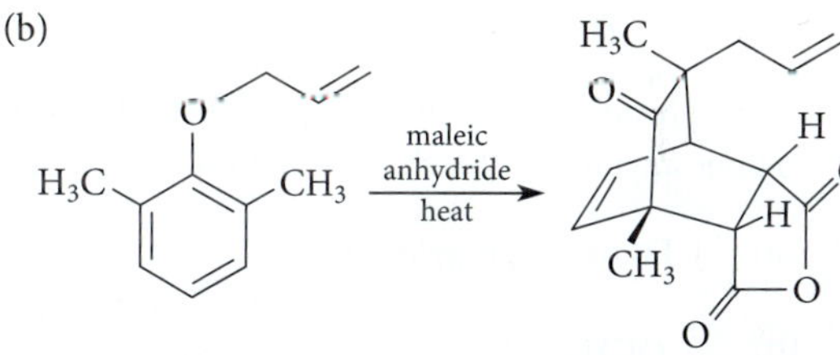

(mixture of stereoisomers)

(c)

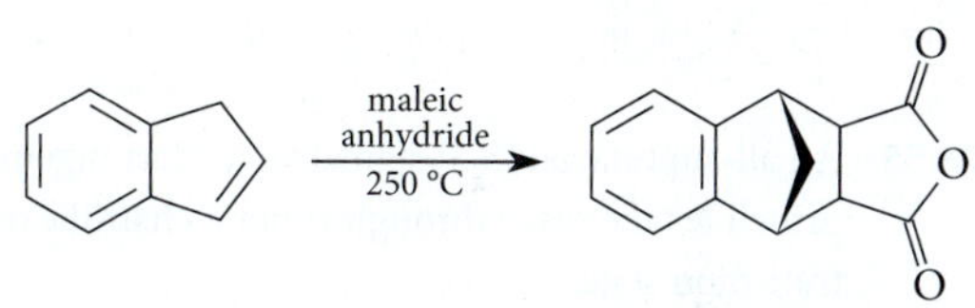

28.52 When 2-methyl-2-propenal is treated with allylmagnesium chloride ($H_2C{=}CH—CH_2—MgCl$) in ether, then with

---

(a)

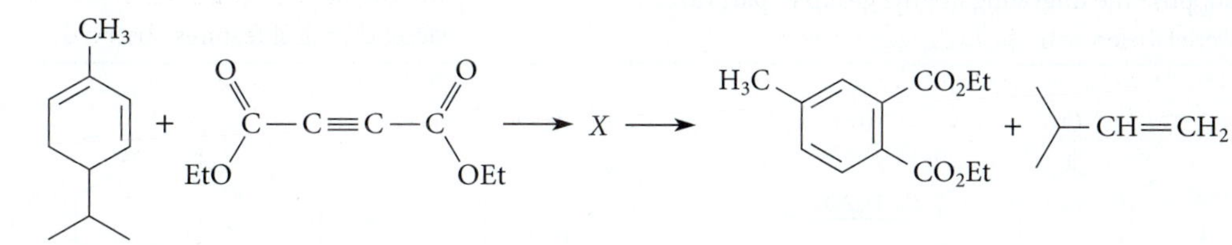

α-phellandrene

(b)

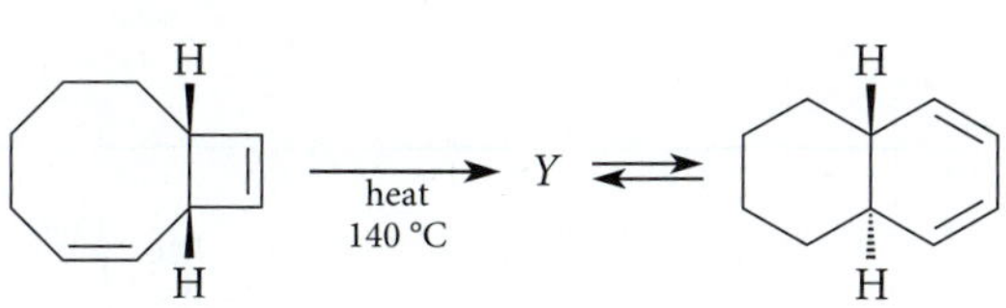

(c)

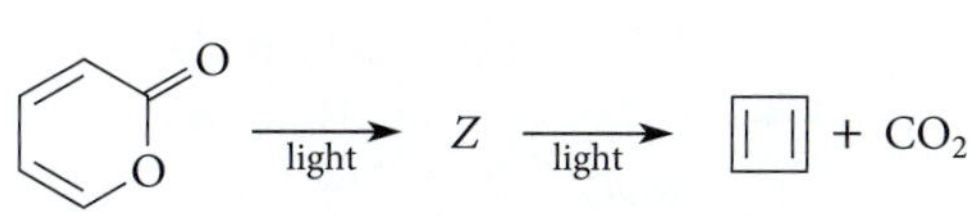

FIGURE P28.49

dilute aqueous acid, a compound *A* is obtained, which, when heated strongly, yields an aldehyde *B*. Give the structures of compounds *A* and *B*.

28.53 A compound *A* ($C_{11}H_{14}O_3$) is insoluble in base and gives an isomeric compound *B* when heated strongly. Compound *B* gives a sodium salt when treated with NaOH. Treatment of the sodium salt of *B* with dimethyl sulfate gives a new compound *C* ($C_{12}H_{16}O_3$) that is identical in all respects to a natural product *elemicin*. Ozonolysis of elemicin followed by oxidation gives the carboxylic acid *D*. Propose structures for compounds *A*, *B*, and *C*.

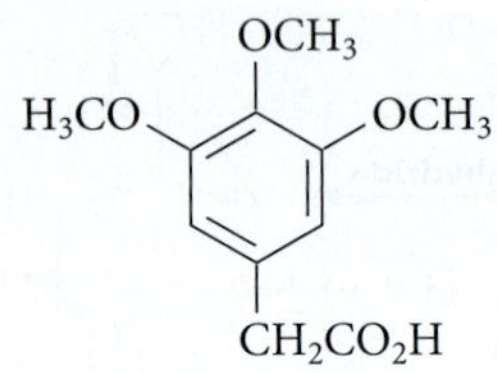

D

28.54 Using phenol and any other reagents as starting materials, outline a synthesis of each of the following compounds.

(a) 1-Ethoxy-2-propylbenzene

(b) 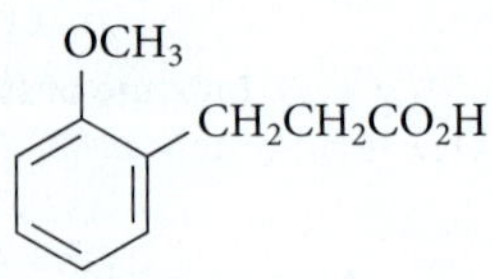

28.55 An all-suprafacial [3,3] sigmatropic rearrangement could in principle take place through either a chairlike or a boatlike transition state:

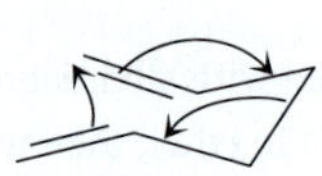

chairlike transition state

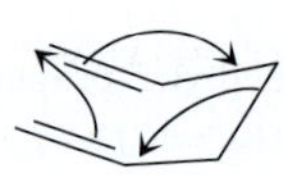

boatlike transition state

(a) According to the following results, which of these two transition states is preferred?

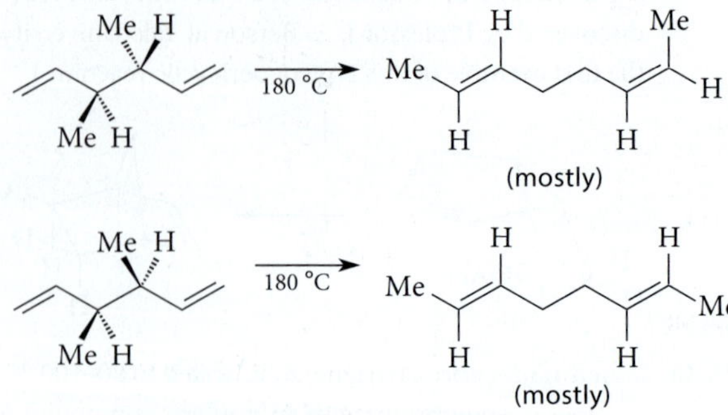

(b) When the terpene germacrone is distilled under reduced pressure at 165 °C, it is transformed to β-elemenone by a Cope rearrangement. Deduce the structure of germacrone, including its stereochemistry.

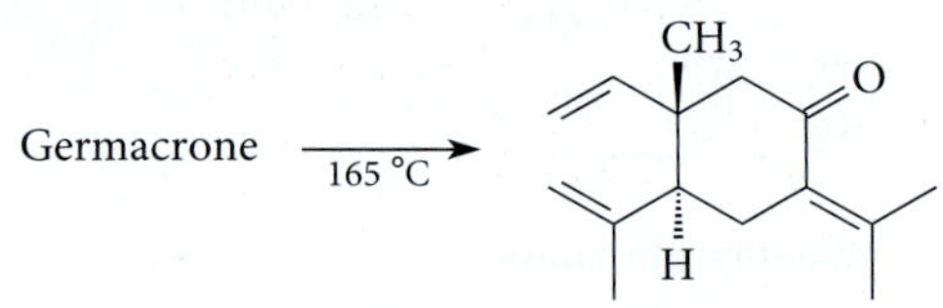

β-elemenone

28.56 Ions as well as neutral molecules undergo pericyclic reactions. Classify the pericyclic reactions of the cation involved in the transformation shown in **Fig. P28.56**. Tell whether the methyl groups are cis or trans and why.

28.57 (a) The transformation shown in **Fig. P28.57**, which involves a sequence of two pericyclic reactions, was used as a key step in a synthesis of the sex hormone estrone. Identify the unstable intermediate *A*, and give the mechanism for both its formation and its subsequent reaction.

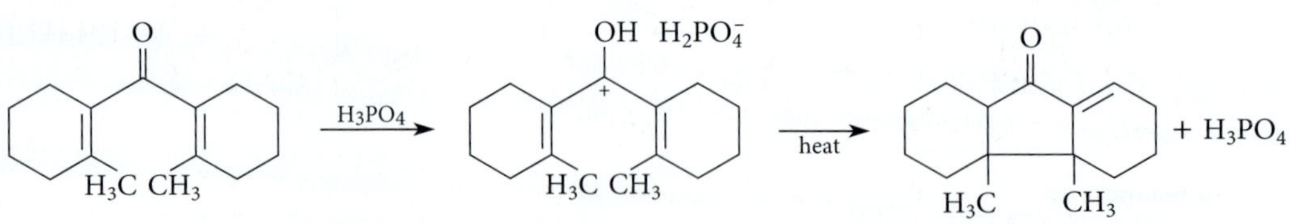

FIGURE P28.56

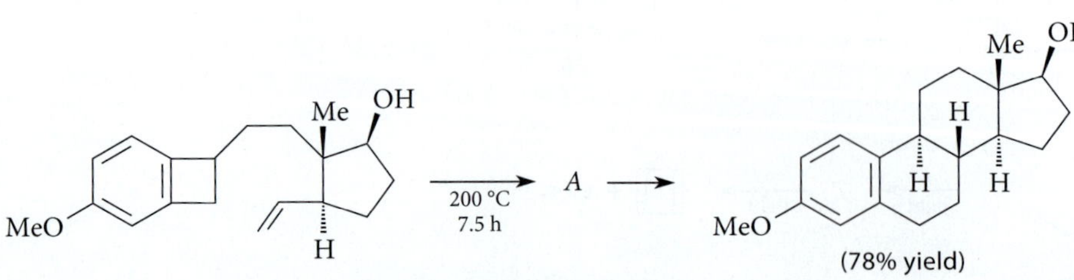

FIGURE P28.57

(b) Show how the product of part (a) can be converted into estrone.

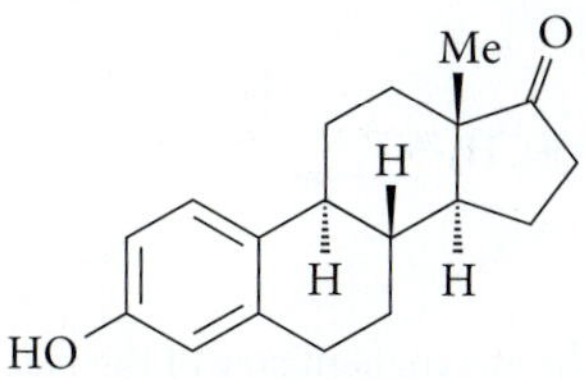

estrone

28.58 In 1985, two researchers at the University of California, Riverside, carried out the reaction given in **Fig. P28.58**. The equilibrium mixture contained compound *A* (22%), a *single* stereoisomer of *B* (47%), and a *single* stereoisomer of *C* (31%). Predict the stereochemistry of compounds *B* and *C* at the carbon marked with the asterisk (*). Explain how you arrived at your prediction.

28.59 An interesting heterocyclic compound *C* was prepared and trapped by the following sequence of reactions. Give the structure of all missing compounds, and explain what happens in each reaction.

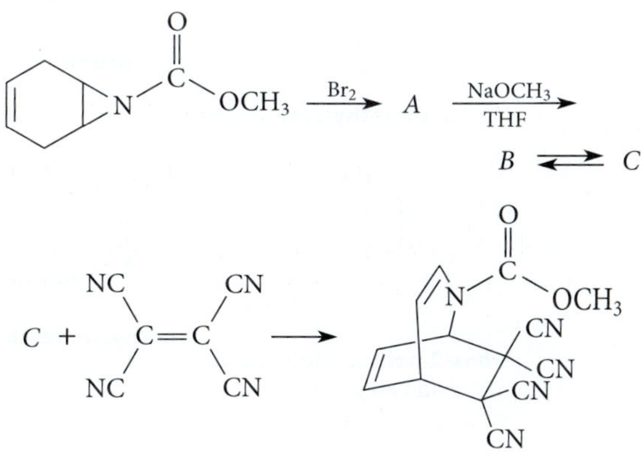

28.60 Anticipating the isolation of the potentially aromatic hydrocarbon *B*, a group of chemists irradiated compound *A* with ultraviolet light. Compound *C* was obtained as a product instead of *B*.

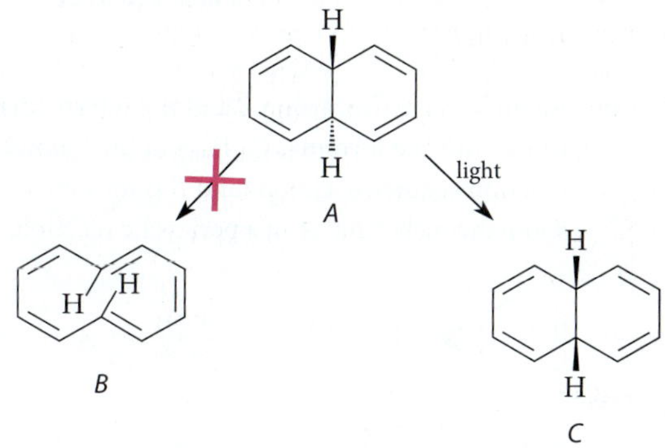

(a) Explain why compound *B* might be expected to be unstable in spite of its cyclic array of $4n + 2$ π electrons.

(b) Explain why the formation of compound *B* is allowed by the pericyclic selection rules.

(c) Account for the formation of the observed product *C*.

28.61 (a) What type of pericyclic reaction is required to form benzene from Dewar benzene?

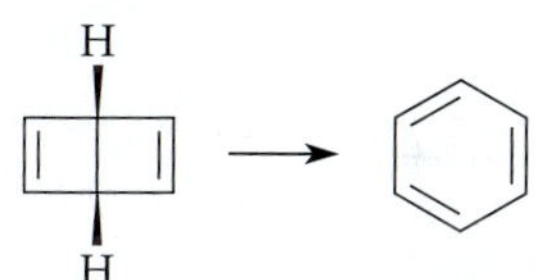

Dewar benzene **benzene**

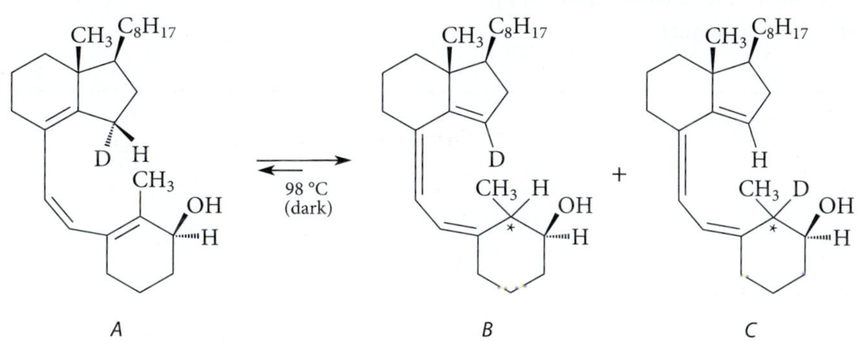

FIGURE P28.58

(b) Explain why Dewar benzene, although a very unstable molecule, is not readily transformed to benzene. (Although Dewar benzene forms benzene when heated, this reaction requires a surprisingly high temperature and is not believed to be concerted.)

28.62 (a) Identify the hydrocarbon *B* and the intermediate *A* (both with the formulas $C_{11}H_{10}$) in the following reaction sequence. Compound *B* is formed spontaneously from *A* in a pericyclic reaction.

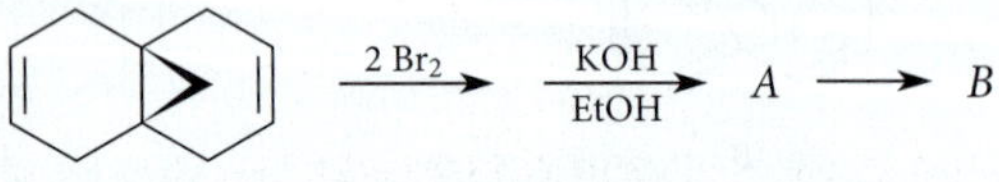

(b) The proton NMR spectrum of *B* consists of a complex absorption at δ 7.1 (8*H*) and a singlet at δ (−0.5) (2*H*). Account for the absorption at a negative chemical shift—that is, to which protons does this absorption correspond, and why do they have a *negative* chemical shift?

28.63 (a) Provide the curved-arrow notation for the following sigmatropic reaction, and classify it with the bracketed-number notation.

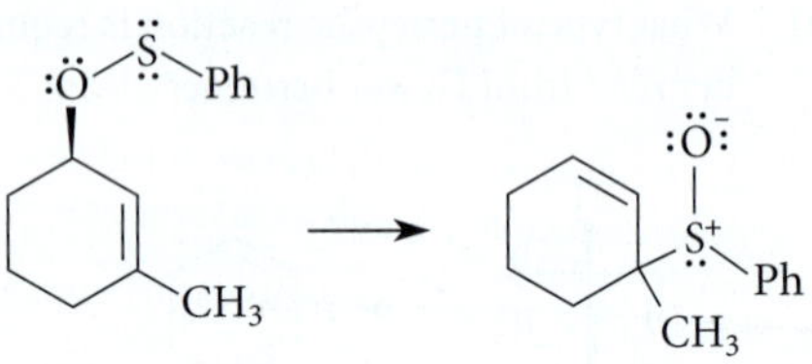

(b) How many electrons are involved in the reaction? Explain your reasoning.

(c) Use the thermal sigmatropic selection rules to specify whether the reaction is suprafacial or antarafacial, and predict the stereochemistry of the product. Explain your reasoning.

28.64 (a) Provide the curved-arrow notation for the following electrocyclic reaction.

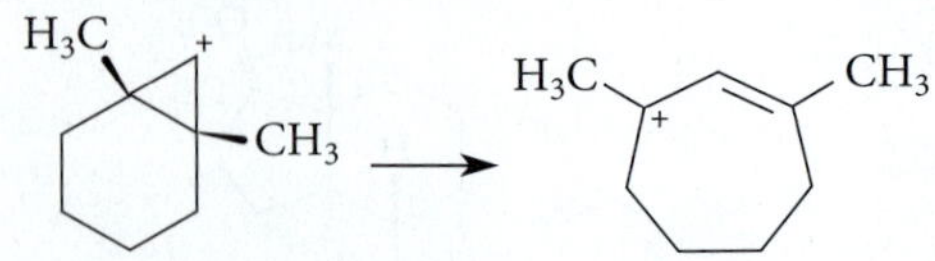

(b) Indicate the stereochemistry of the reaction required by the selection rules, and tell whether in fact that stereochemistry is observed.

(c) Why does the equilibrium for this reaction lie far to the right?

28.65 Each of the following reactions is a $[4s + 2s]$ cycloaddition. Both are examples of *dipolar cycloadditions*. Use the curved-arrow notation to help you complete each reaction, and provide the stereochemistry of the product. Explain your reasoning. (*Hint:* Both reactions form five-membered rings.)

(a)

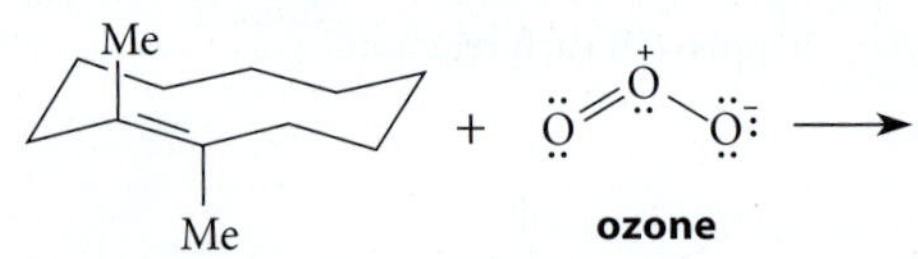

(*E*)-1,2-dimethylcyclodecene

(b)

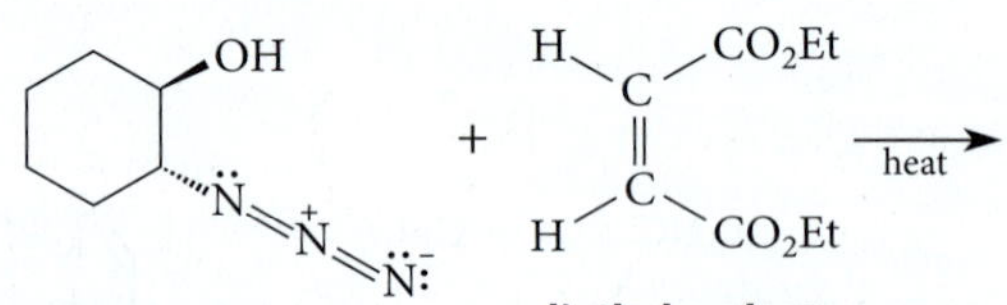

***trans*-2-azidocyclohexanol** (an alkyl azide) **diethyl maleate**

APPENDICES

APPENDIX I. SUBSTITUTIVE NOMENCLATURE OF ORGANIC COMPOUNDS

The substitutive name of an organic compound is based on its *principal group* and *principal chain*.

The *principal group* is assigned according to the following priorities:

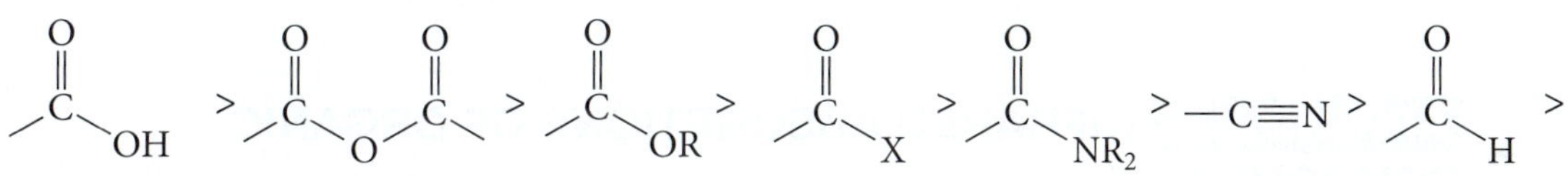

carboxylic acid > anhydride > ester > acid halide > amide > nitrile > aldehyde >

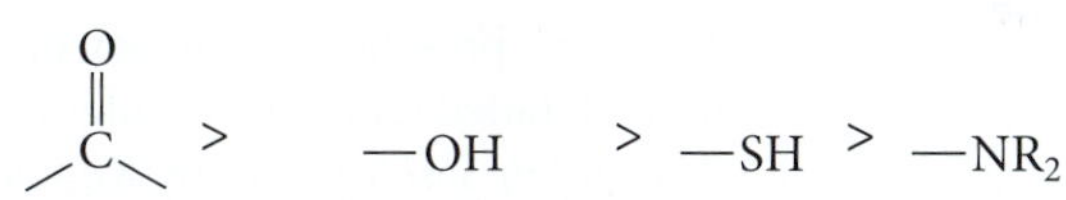

ketone > alcohol, phenol > thiol > amine

The *principal chain* is *identified* by applying the following criteria in order until a decision can be made:

1. Maximum number of substituents corresponding to the principal group
2. Maximum number of double and triple bonds considered together
3. Maximum length
4. Maximum number of substituents cited as prefixes

A *principal chain* is numbered by applying the following criteria in order until there is no ambiguity. Where multiple numbers are possible, comparisons are made at the first point of difference.

1. Lowest number for the principal group cited as a suffix—that is, the group on which the name is based
2. Lowest numbers for multiple bonds, with double bonds having priority over triple bonds in case of ambiguity
3. Lowest numbers for other substituents, taking into account the “first point of difference” rule (Sec. 2.4C, Rule 8)
4. Lowest number for the substituent named as a prefix that is cited first in the name

The *name* is *constructed* starting with the hydrocarbon corresponding to the principal chain:

1. Cite the principal group by its suffix and number; its number is the last one cited in the name.
2. If there is no principal group, name the compound as a substituted hydrocarbon.
3. Cite the names and numbers of the other substituents in alphabetical order at the beginning of the name.

These lists cover most of the cases cited in the text. (See Study Problems 8.1–8.3, Secs. 8.2B and C, for illustrations.) For a more complete discussion of nomenclature, see *Nomenclature of Organic Chemistry* (1979 Edition), by the International Union of Pure and Applied Chemistry (IUPAC), published by Pergamon Press.

In 1993, the IUPAC released *A Guide to IUPAC Nomenclature of Organic Compounds Recommendations 1993,* by R. Panico, W. H. Powell, and Jean-Claude Richer (senior editor), Blackwell Science. This publication advocated one major change that affects the nomenclature of relatively simple compounds. This change involves the way that principal groups are cited. The *1993 Recommendations* cite the principal group or multiple bond position with a number preceding the suffix itself, whereas the *1979 Recommendations* (followed in this text) cite the principal group or multiple bond position with a number preceding the hydrocarbon name. These differences are best illustrated by example.

	$H_2C{=}CHCH_2CH_2CH_3$	$HOCH_2CH_2CH_2CH_2CH_3$	$HOCH_2CH_2CH_2CH{=}CH_2$
1979 Recommendations:	**1-pentene**	**1-pentanol**	**4-penten-1-ol**
1993 Recommendations:	**pent-1-ene**	**pentan-1-ol**	**pent-4-en-1-ol**

The *1993 Recommendations* have not yet been generally adopted. Thus, names that adhere to either set of recommendations are acceptable.

APPENDIX II. INFRARED ABSORPTIONS OF ORGANIC COMPOUNDS

This table presents a summary of the important infrared absorptions discussed in this text. For more detailed tables, the reader may wish to consult more specialized texts, such as *Infrared Absorption Spectroscopy,* by Koji Nakanishi and Philippa H. Solomon, San Francisco: Holden-Day, 1977; or *Organic Structure Analysis,* by Philip Crews, Jaime Rodríguez, and Marcel Jaspars, 1998, Oxford University Press, Chapter 8.

Type of absorption	Frequency, cm^{-1} (Intensity)*	Comment
Alkanes		
C—H stretch	2850–3000 (m)	occurs in all compounds with aliphatic C—H bonds
Alkenes		
C═C stretch		
—CH═CH_2	1640 (m)	
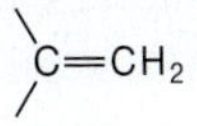	1655 (m)	
others	1660–1675 (w)	not observed if alkene is symmetrical
═C—H stretch	3000–3100 (m)	
═C—H bend		
—CH═CH_2	910, 990 (s)	two absorptions
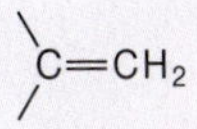	890 (s)	
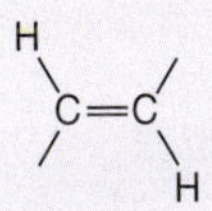	960–980 (s)	
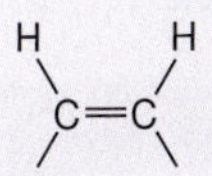	675–730 (s)	position is highly variable
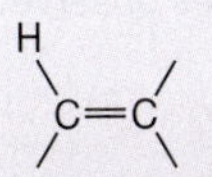	800–840 (s)	
Alcohols and Phenols		
O—H stretch	3200–3400 (s)	
C—O stretch	1050–1250 (s)	also present in other compounds with C—O bonds: ethers, esters, etc.
Alkynes		
C≡C stretch	2100–2200 (m)	not present or weak in many internal alkynes
≡C—H stretch	3300 (s)	present in 1-alkynes only
Aromatic Compounds		
C═C stretch	1500, 1600 (s)	two absorptions
C—H bend	650–750 (s)	
overtone	1660–2000 (w)	

*(s) = strong; (m) = medium; (w) = weak.

(*Table continues*)

Type of absorption	Frequency, cm^{-1} (Intensity)*	Comment
Aldehydes		
C═O stretch		
ordinary	1720–1725 (s)	
α,β-unsaturated	1680–1690 (s)	
benzaldehydes	1700 (s)	
C—H stretch	2720 (m)	
Ketones		
C═O stretch		
ordinary	1710–1715 (s)	increases with decreasing ring size (Table 21.3, Sec. 21.4A)
α,β-unsaturated	1670–1680 (s)	
aryl ketones	1680–1690 (s)	
Carboxylic Acids		
C═O stretch		
ordinary	1710 (s)	
benzoic acids	1680–1690 (s)	
O—H stretch	2400–3000 (s)	very broad
Esters and Lactones		
C═O stretch	1735–1745 (s)	increases with decreasing ring size (Table 21.3, Sec. 21.4A)
Acid Chlorides		
C═O stretch	1800 (s)	a second weaker band sometimes observed at 1700–1750
Anhydrides		
C═O stretch	1760, 1820 (s)	two bands; increases with decreasing ring size in cyclic anhydrides
Amides and Lactams		
C═O stretch	1650–1655 (s)	increases with decreasing ring size (Table 21.3, Sec. 21.4A)
N—H bend	1640 (s)	
N—H stretch	3200–3400 (m)	doublet absorption observed for some primary amides
Nitriles		
C≡N stretch	2200–2250 (m)	
Amines		
N—H stretch	3200–3375 (m)	several absorptions sometimes observed, especially for primary amines

*(s) = strong; (m) = medium; (w) = weak.

APPENDIX III. PROTON NMR CHEMICAL SHIFTS IN ORGANIC COMPOUNDS

This appendix is subdivided into a table of chemical shifts for protons that are *part of* functional groups and a table of chemical shifts for protons that are *adjacent to* functional groups.

A. Protons within Functional Groups

Group	Chemical shift, ppm
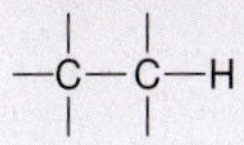	0.7–1.7
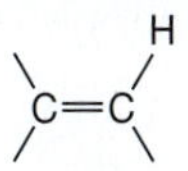	4.6–5.7
—O—H	varies with solvent and with acidity of O—H
—C≡C—H	1.7–2.5
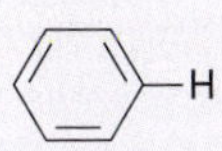	6.5–8.5

Group	Chemical shift, ppm
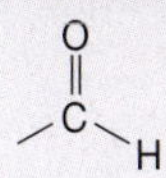	9–11
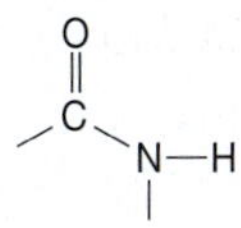	7.5–9.5
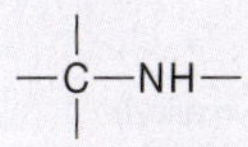	0.5–1.5
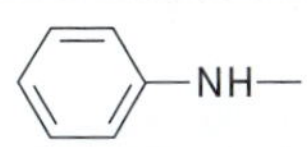	2.5–3.5

B. Protons Adjacent to Functional Groups

In this table, a range of chemical shifts is given for protons in the general environment

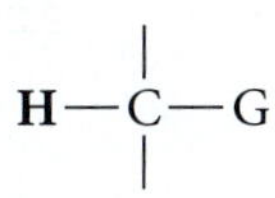

in which G is a group listed in column 1, and the two other bonds are to carbon or hydrogen. The remaining columns give the approximate chemical shifts for methyl protons (H_3C—G), methylene protons (—CH_2—G), and methine protons (—CH—G), respectively. The shifts in the following table are typical; some variation with structure of a few tenths of a ppm can be expected. The chemical shifts of methine protons are usually further downfield than those of methylene protons, which are further downfield than methyl protons. Each additional carbon substitution increases the chemical shift by 0.3–1.0 ppm.

Group, G	Chemical shift of H_3C—G, ppm	Chemical shift of —CH_2—G, ppm	Chemical shift of —CH—G, ppm
—H	0.2		
—CR_3	0.9	1.2	1.4
—F	4.3	4.5	4.8
—Cl	3.0	3.4	4.0
—Br	2.7	3.4	4.1
—I	2.2	3.2	4.2
—CR=CR_2 (R = alkyl, H)	1.8	2.0	2.3
—C≡CR (R = alkyl, H)	1.8	2.2	2.8
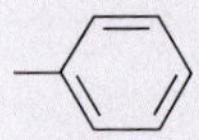	2.3	2.6	2.8
RO— (R = alkyl, H)	3.3 (R = alkyl) 3.5 (R = H)	3.4	3.6
RO— (R = aryl)	3.7	4.0	4.6
RS— (R = alkyl, H)	2.4	2.6	3.0
R—C(=O)—	2.1 (R = alkyl) 2.6 (R = aryl)	2.4 (R = alkyl) 2.7 (R = aryl)	2.6 (R = alkyl) 3.4 (R = aryl)
RO—C(=O)— (R = alkyl, H)	2.1	2.2	2.5
R—C(=O)—O— (R = alkyl, H)	3.6 (R = alkyl) 3.8 (R = aryl)	4.1 (R = alkyl, aryl)	5.0 (R = alkyl, aryl)
R_2N—C(=O)— (R = alkyl, H)	2.0	2.2	2.4
R—C(=O)—N(R)— (R = alkyl, H)	2.8	3.4	3.8
—NR_2 (R = alkyl, H)	2.2	2.4	2.8
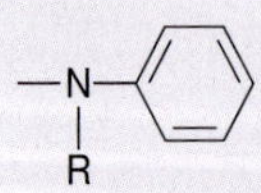 (R = alkyl, H)	2.6	3.1	3.6
N≡C—	2.0	2.4	2.9

APPENDIX IV. ^{13}C NMR CHEMICAL SHIFTS IN ORGANIC COMPOUNDS

This section is divided into a table of chemical shifts for carbons within functional groups and a table of chemical shifts for alkyl carbons adjacent to functional groups. A typical range of shifts is given for each case.

A. Chemical Shifts of Carbons within Functional Groups

Group	Chemical shift range, ppm
$-CH_3$	8–23
$-CH_2-$	20–30
—CH— (with one vertical bond)	21–33
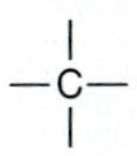	17–29
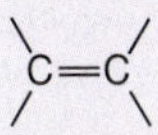	100–145*
—C≡C—	65–85*
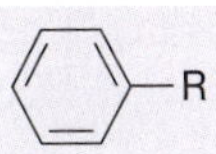	125–150*
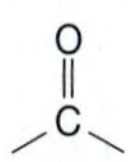	190–220
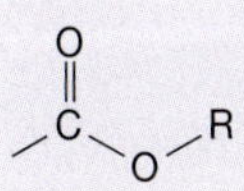 R = H, alkyl	160–190
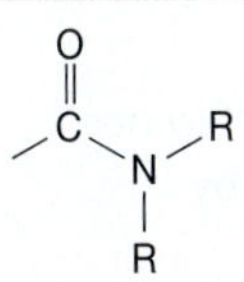 R = H, alkyl	160–180
—C≡N	110–130

*Alkyl substitution typically increases chemical shift.

B. Chemical Shifts of Carbons Adjacent to Functional Groups

In most cases, alkyl substitution on the carbon increases chemical shift. Methyl carbons will have shifts in the low end of the range; tertiary and quaternary carbons will have shifts in the upper end of the range.

Group G		Chemical shift of carbon in G—C—
$R_2C{=}CR-$		14–40
$HC{\equiv}C-$		18–28
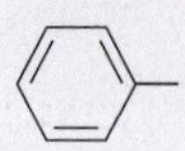		29–45
F—		83–91
Cl—		20–60
Br—		10–50
I—		−15 to 25
HO—		62–70
RO—	R = alkyl, H	70–79
R—C(=O)—	R = alkyl, H	25–52
RO—C(=O)—	R = alkyl, H	33–44
R_2N-	R = alkyl, H	41–51 (R = H) 53–60 (R = alkyl)
$N{\equiv}C-$		12–28

APPENDIX V. SUMMARY OF SYNTHETIC METHODS

The following methods are listed in order of their occurrence in the text; the section reference follows each reaction in parentheses. Thus, a review at any point in the text is possible by considering the methods listed for earlier sections.

Don't forget that in many cases, a method can be applied to compounds containing more than one functional group. Thus, catalytic hydrogenation can be used to convert phenols into alcohols, but it is listed under "Synthesis of Alkanes and Aromatic Hydrocarbons" because the actual transformation is the formation of $-CH_2-CH_2-$ groups from $-CH{=}CH-$ groups; the presence of the —OH group is incidental.

Reaction summaries for each chapter are found in the *Study Guide and Solutions Manual.*

A. Synthesis of Alkanes and Aromatic Hydrocarbons

1. Catalytic hydrogenation of alkenes (4.10A)
2. Catalytic hydrogenation of alkynes (4.10B)
3. Protonolysis of Grignard or related reagents (10.5C)
4. Cyclopropane formation by the addition of carbenoids to alkenes (Simmons–Smith reaction; 10.7B)
5. Friedel–Crafts alkylation of aromatic compounds (16.4E)
6. Catalytic hydrogenation of aromatic compounds (16.6)
7. Stille reaction of aryl triflates and aryl- or alkylstannanes to form substituted aromatic hydrocarbons (18.10B)
8. Wolff–Kishner or Clemmensen reductions of aldehydes or ketones (19.13)
9. Reaction of aryldiazonium salts with hypophosphorous acid (23.10A)

B. Synthesis of Alkenes

1. Catalytic hydrogenation of alkynes (gives cis-alkenes when used with internal alkynes; 4.10B)
2. β-Elimination reactions (E2) of alkyl halides or sulfonates (9.5, 11.4A, 17.3B)
3. Reduction of alkynes with alkali metals in liquid ammonia (gives trans-alkenes when used with internal alkenes; 10.2)
4. Acid-catalyzed dehydration of alcohols (11.2)
5. Diels–Alder reactions of dienes and alkenes to give cyclic alkenes (15.3, 28.3)
6. Heck reaction of aryl halides and alkenes to give aryl-substituted alkenes (18.6A)
7. Suzuki coupling of aryl or vinylic halides with aryl or vinylic boronic acids (18.6B)
8. Alkene metathesis (18.6C)
9. Wittig reaction of aldehydes and ketones (19.14)
10. Aldol condensation reactions of aldehydes or ketones to give α,β-unsaturated aldehydes or ketones (22.4)
11. Hofmann elimination of quaternary ammonium hydroxides (23.8)

C. Synthesis of Alkynes

1. Alkylation of acetylenic anions with alkyl halides or sulfonates (10.6B)
2. β-Elimination reactions of alkyl dihalides or vinylic halides (18.2)

D. Synthesis of Alkyl, Aryl, and Vinylic Halides

1. Addition of hydrogen halides to alkenes (4.7, 15.4A)
2. Addition of halogens or HBr to alkynes (4.9)
3. Addition of halogens to alkenes to give vicinal dihalides (5.2)
4. Peroxide-promoted addition of HBr to alkenes (10.1)
5. Synthesis of dihalocyclopropanes by the addition of dihalomethylene to alkenes (10.7A)
6. Reaction of alcohols with HBr, thionyl chloride, or triphenylphosphine dibromide (11.3, 11.4D)
7. Reaction of sulfonate esters or other alkyl halides with halide ions (11.4A, 17.4)
8. Halogenation of aromatic compounds (16.4A)
9. Allylic and benzylic bromination of alkenes or aromatic hydrocarbons (17.2)
10. α-Halogenation of aldehydes, ketones, or carboxylic acids (22.3A, C)
11. Synthesis of aryl halides by the reaction of cuprous chloride, cuprous bromide, or potassium iodide with aryldiazonium salts (Sandmeyer and related reactions; 23.10A)

E. Synthesis of Grignard Reagents and Related Organometallic Compounds

1. Reaction of alkyl or aryl halides with metals (10.5A)
2. Preparation of lithium dialkylcuprates by the reaction of alkyllithium reagents with cuprous halides (10.5B)
3. Preparation of acetylenic Grignard reagents by the metal–hydrogen exchange (10.6A)
4. Preparation of alkyl- and arylstannanes by the reaction of Grignard reagents with trialkylstannyl chlorides (18.10B)

F. Synthesis of Alcohols and Phenols

(Syntheses apply only to alcohols unless noted otherwise.)

1. Acid-catalyzed hydration of alkenes (used industrially, but generally not a good laboratory method; 4.10C)
2. Synthesis of halohydrins from alkenes (5.2B)
3. Oxymercuration–reduction of alkenes (5.4A)
4. Hydroboration–oxidation of alkenes (5.4B)
5. Ring-opening reactions of epoxides (12.5A, B)
6. Reaction of ethylene oxide with Grignard reagents (12.5C)
7. Reaction of epoxides with lithium organocuprates (12.5C)
8. Reduction of aldehydes or ketones (19.9, 22.10, 24.8D)
9. Reaction of aldehydes or ketones with Grignard or related reagents (19.10, 22.11A)
10. Reduction of carboxylic acids to primary alcohols (20.10)
11. Reduction of esters to primary alcohols (21.9A)
12. Reaction of esters with Grignard or related reagents (21.10A)
13. Aldol addition reactions of aldehydes or some ketones to give β-hydroxy aldehydes or ketones (22.4)
14. Reaction of diazonium salts with water to give phenols (23.10A)
15. Synthesis of phenols by the Claisen rearrangement of allylic aryl ethers (28.4B)

G. Synthesis of Glycols

1. Acid-catalyzed hydrolysis of epoxides (12.5B)
2. Reaction of alkenes with osmium tetroxide or alkaline potassium permanganate (12.6A)

H. Synthesis of Ethers, Acetals, and Sulfides

1. Alkylation of alkoxides, phenoxides, or thiolates with alkyl halides or alkyl sulfonates (Williamson synthesis; 12.2A, 18.7B, 24.7)
2. Alkoxymercuration–reduction of alkenes (12.2B)
3. Acid-catalyzed dehydration of alcohols (12.2C)
4. Acid-catalyzed addition of alcohols to alkenes (12.2C)
5. Reaction of epoxides with alkoxides and alcohols (12.5A, B)
6. Reaction of 2- and 4-nitroaryl halides with alkoxides (18.4)
7. Acetal formation by the acid-catalyzed reaction of alcohols with aldehydes or ketones (19.11A, 24.6)
8. Conjugate addition of thiols to α,β-unsaturated carbonyl compounds (22.9A)

I. Synthesis of Epoxides

1. Oxidation of alkenes with peroxycarboxylic acids (12.3A)
2. Cyclization of halohydrins (12.3B)
3. Asymmetric epoxidation of allylic alcohols (12.11)

J. Synthesis of Disulfides

1. Oxidation of thiols (11.10, 27.8)

K. Synthesis of Aldehydes

1. Hydroboration–oxidation of alkynes (5.6)
2. Ozonolysis of alkenes (of limited utility because carbon–carbon bonds are broken; 5.8)
3. Oxidation of primary alcohols (11.7A)
4. Oxidative cleavage of glycols (of limited utility because carbon–carbon bonds are broken; 12.6B, 24.8C)
5. Oxidation of allylic and benzylic alcohols with MnO_2 (17.5A)
6. Reduction of acid chlorides (21.9D)
7. Aldol addition reactions of aldehydes to give β-hydroxy aldehydes (22.4)
8. Aldol condensation reactions of aldehydes to give α,β-unsaturated aldehydes (22.4)
9. Synthesis of aldoses from other aldoses by the Kiliani–Fischer synthesis (24.9) and the Ruff degradation (24.10B)

L. Synthesis of Ketones

1. Mercuric-ion catalyzed hydration of alkynes (5.5)
2. Ozonolysis of alkenes (of limited utility because carbon–carbon bonds are broken; 5.8)
3. Oxidation of secondary alcohols (11.7A)
4. Oxidative cleavage of glycols (of limited utility because carbon–carbon bonds are broken; 12.6B)
5. Friedel–Crafts acylation of aromatic compounds (16.4F)
6. Oxidation of phenols to quinones (18.8A)
7. Reaction of acid chlorides with lithium dialkylcuprates (21.10B)
8. Aldol condensation reactions of ketones to give α,β-unsaturated ketones (22.4)
9. Claisen and Dieckmann condensation reactions of esters to give β-keto esters (22.6A, B)
10. Crossed Claisen condensation reactions of esters to give β-diketones (22.6C)
11. Acetoacetic ester synthesis (22.8C)
12. Conjugate addition reactions of α,β-unsaturated ketones (22.9), including the addition of lithium dialkylcuprate reagents (22.11B)

M. Synthesis of Sulfoxides and Sulfones

1. Oxidation of sulfides (12.9)

N. Synthesis of Carboxylic and Sulfonic Acids

(Syntheses apply only to carboxylic acids unless noted otherwise.)

1. Ozonolysis of alkenes (of limited utility because carbon–carbon bonds are broken; 5.8)
2. Oxidation of primary alcohols (11.7A)
3. Oxidation of thiols to sulfonic acids (11.10)
4. Sulfonation of aromatic compounds to give arylsulfonic acids (16.4D)
5. Side-chain oxidation of alkylbenzenes (17.5C)
6. Oxidation of aldehydes (19.15)
7. Reaction of Grignard or related reagents with carbon dioxide (20.6)
8. Hydrolysis of carboxylic acid derivatives, especially nitriles (21.7, 21.11, 25.5A, 27.4C)
9. Haloform reaction of methyl ketones (of limited utility because carbon–carbon bonds are broken; 22.3B)
10. Malonic ester synthesis (22.8A, 27.4B)
11. Strecker synthesis of α-amino acids (27.4C)

O. Synthesis of Esters

1. Reaction of alcohols and phenols with sulfonyl chlorides (for sulfonate esters; 11.4A, 18.10B)
2. Acid-catalyzed esterification of carboxylic acids with primary or secondary alcohols (20.8A, 24.7, 27.5)
3. Alkylation of carboxylic acids with diazomethane (20.8B)
4. Alkylation of carboxylate salts with alkyl halides (20.8B)
5. Reaction of acid chlorides, anhydrides, or esters with alcohols and phenols (21.8, 24.7)
6. Claisen and Dieckmann condensation reactions of esters to give β-keto esters (22.6A, B)
7. Alkylation of ester enolate ions; includes malonic ester synthesis, acetoacetic ester synthesis, and direct alkylation (22.8)
8. Conjugate addition reactions of α,β-unsaturated esters (22.9, 22.11B)

P. Synthesis of Anhydrides

1. Reaction of carboxylic acids with dehydrating agents (20.9B)
2. Reaction of acid chlorides with carboxylate salts (21.8A)

Q. Synthesis of Acid Chlorides

1. Reaction of carboxylic or sulfonic acids with thionyl chloride, phosphorus pentachloride, or related reagents (20.9A)
2. Synthesis of sulfonyl halides by chlorosulfonation of aromatic compounds (20.9A)

R. Synthesis of Amides

1. Reaction of acid chlorides, anhydrides, or esters with amines (21.8, 23.7C, 27.5)
2. Condensation of amines and carboxylate salts with PyBOP (27.6A)

S. Synthesis of Nitriles

1. Formation of cyanohydrins from aldehydes and some ketones (19.8A, B, 24.9)
2. Reaction of alkyl halides or alkyl sulfonates with cyanide ion (21.11)
3. Conjugate addition of cyanide ion to α,β-unsaturated carbonyl compounds (22.9A)
4. Reaction of cuprous cyanide with aryldiazonium salts (23.10A)

T. Synthesis of Amines

1. Reduction of amides (21.9B)
2. Reduction of nitriles to primary amines (21.9C)
3. Direct alkylation of ammonia or amines (of limited utility because of the possibility of over-alkylation; 23.7A, 27.4A)
4. Reductive amination of aldehydes and ketones (23.7B)
5. Aromatic substitution reactions of aniline derivatives (23.9)

6. Gabriel and Staudinger syntheses of primary amines (23.11A)
7. Reduction of nitro compounds (23.11B)
8. Pd-catalyzed amination of aryl halides and triflates (23.11C)
9. Curtius and Hofmann rearrangements (23.11D)
10. Strecker synthesis of α-amino acids (27.4C)

U. Synthesis of Nitro Compounds

1. Nitration of aromatic compounds (16.4C, 18.9)

APPENDIX VI. REACTIONS USED TO FORM CARBON–CARBON BONDS

Reactions that form carbon–carbon bonds have central importance in organic chemistry, because these reactions can be used to form carbon chains or rings. These reactions are listed in the order in which they are discussed in the text. The section reference follows each reaction in parentheses.

1. Reaction of acetylenic anions with alkyl halides or sulfonates (10.6)
2. Cyclopropane formation by addition of carbenes or carbenoids to alkenes (10.7)
3. Reaction of Grignard reagents with ethylene oxide (12.5C)
4. Reaction of epoxides with lithium organocuprates (12.5C)
5. Diels–Alder reactions (15.3, 28.3)
6. Friedel–Crafts alkylation (16.4E) and acylation reactions (16.4F)
7. The Heck reaction of alkenes with aryl halides (18.6A)
8. Suzuki coupling of aryl or vinylic halides with aryl or vinylic boronic acids (18.6B)
9. Alkene metathesis (18.6C)
10. The Stille reaction of organostannanes with aryl triflates (18.10B)
11. Cyanohydrin formation (19.8, 24.9, 27.4C)
12. Reaction of Grignard and related reagents with aldehydes and ketones (19.10)
13. Wittig alkene synthesis (19.14)
14. Reaction of Grignard and related reagents with carbon dioxide (20.6)
15. Reaction of Grignard and related reagents with esters (21.10A)
16. Reaction of lithium dialkylcuprates with acid chlorides (21.10B)
17. Reaction of cyanide ion with alkyl halides or sulfonates (21.11)
18. Aldol addition and condensation reactions (22.4)
19. Claisen and related condensation reactions (22.6)
20. Malonic ester synthesis (22.8A, 27.4B)
21. Alkylation of ester enolate ions with alkyl halides or tosylates (22.8B)
22. Acetoacetic ester synthesis (22.8B)
23. Conjugate-addition reactions of cyanide ion (22.9A) or enolate ions (22.9C) to α,β-unsaturated carbonyl compounds
24. Conjugate addition of lithium dialkylcuprate reagents to α,β-unsaturated carbonyl compounds (22.11B)
25. Reaction of aryldiazonium salts with cuprous cyanide (23.10A)
26. Formation of rings by electrocyclic reactions (28.2)
27. Claisen rearrangement (28.4B)

APPENDIX VII. TYPICAL ACIDITIES AND BASICITIES OF ORGANIC FUNCTIONAL GROUPS

A. Acidities of Groups That Ionize to Give Anionic Conjugate Bases

Functional group	Structure*	Structure of conjugate base	Typical pK_a
sulfonic acid	$R-SO_2-O-H$	$R-SO_2-O^-$	–2.8 (strong acid)
carboxylic acid	$R-C(=O)-O-H$	$R-C(=O)-O^-$	4–5
phenol	$X-C_6H_4-O-H$†	$X-C_6H_4-O^-$†	9–11
thiol	$R-S-H$	$R-S^-$	10–12
sulfonamide	$R-SO_2-N(H)-R$	$R-SO_2-\bar{N}-R$	10
amide	$R-C(=O)-N(H)-R$	$R-C(=O)-\bar{N}-R$	15–17
alcohol	$R-O-H$	$R-O^-$	15–19
aldehyde, ketone	$R-C(=O)-C(H)R_2$	$R-C(=O)-\bar{C}R_2$	16–20
ester	$R_2C(H)-C(=O)-OR$	$R_2\bar{C}-C(=O)-OR$	25
alkyne	$R-C\equiv C-H$	$R-C\equiv C^-$	25
nitrile	$R_2C(H)-C\equiv N$	$R_2\bar{C}-C\equiv N$	25
amine	R_2N-H	R_2N^-	32
alkene	$R_2C=C(H)R$	$R_2C=C^-R$	42
benzylic alkyl group	$Ar-C(H)R_2$	$Ar-\bar{C}R_2$	42
alkane	R_3C-H	R_3C^-	55–60

*In the structures, R = alkyl or H. The acidic hydrogen is shown in red.

†X = a general ring substituent group.

B. Basicities of Groups That Protonate to Give Cationic Conjugate Acids

One should be careful to distinguish between the behavior of a particular functional group as an acid and that of the same functional group as a base. For example, when an alcohol acid acts as an acid, it loses the RO—H proton to form an alkoxide (see the table in part A of this section). When it acts as a base, it *gains* a proton to form $R\overset{+}{O}H_2$. These are very different processes with different pK_a values. When we discuss the *acidity* of an alcohol, the relevant pK_a is that for the alcohol itself (see the table in part A). This same pK_a describes the *basicity* of the alkoxide, RO^-, which is the conjugate base of the alcohol. When we discuss the *basicity* of the alcohol itself, the relevant pK_a is the value for the acidity of $R\overset{+}{O}H_2$, given in the following table.

Functional group	Structure*,‡	Structure of conjugate acid*,‡	Typical conjugate-acid pK_a
alkylamine	R_3N	$R_3\overset{+}{N}{-}H$	9–11
pyridine	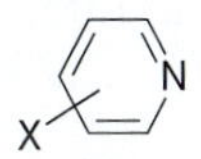	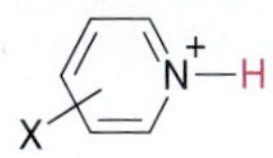	5
aromatic amine	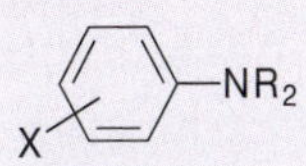	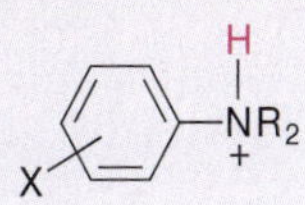	4–5
amide	$R{-}C({=}O){-}NR_2$	$R{-}C({=}\overset{+}{O}{-}H){-}NR_2$	–1
alcohol, ether	R—O—R	$R{-}\overset{+}{O}(H){-}R$	–2 to –3
ester, carboxylic acid	$R{-}C({=}O){-}OR$	$R{-}C({=}\overset{+}{O}{-}H){-}OR$	–6
phenol, aromatic ether†	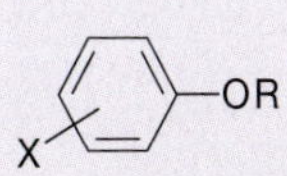	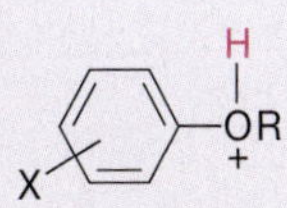	–6 to –7
thiol, sulfide	R—S—R	$R{-}\overset{+}{S}(H){-}R$	–6 to –7
aldehyde, ketone	$R{-}C({=}O){-}R$	$R{-}C({=}\overset{+}{O}{-}H){-}R$	–7

*In the structures, R = alkyl or H. In the conjugate acid, the acidic hydrogen is shown in red.
‡X = a general ring substituent group.
†A phenol or aromatic ether can be protonated on a ring carbon if the resulting carbocation can be strongly stabilized by the substituent groups.

(*Table continues*)

Functional group	Structure*	Structure of conjugate acid	Typical conjugate-acid pK_a
alkene	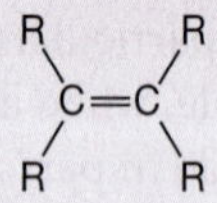	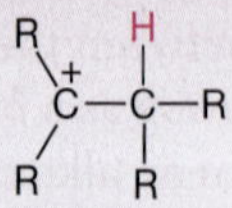	–8 to –10
nitrile	R—C≡N	R—C≡$\overset{+}{N}$—H	–10

*In the structures, R = alkyl or H.

INDEX

INDEX GUIDE

Leading Greek letters, numbers, stereochemical designations (e.g., *S*, *d*), and other locators (*e.g.*, *N*-) are ignored in alphabetizing except for cases in which they are the only elements that differentiate entries. Spaces and hyphens (except hyphens associated with nomenclature prefixes and numbers) take alphabetical precedence over letters and numbers. Spaces take precedence over hyphens. For example, Acetyl group comes before Acetyl-CoA, which comes before Acetylacetone. **Boldfaced** page numbers give the locations of boldfaced definitions in the text. The *italic* letters *f*, *p*, *sp*, *t*, and *v* following page numbers refer to entries found in figures, problems, study problems, tables, and vignettes (sidebars) or margin notes, respectively. The abbreviation *e.g.* means "for example."

A Periodic Table of the Elements

The shaded elements will be encountered most frequently in the text.

2 ← Group numbers recommended by the IUPAC are in gray.
2A ← Older group numbers. (These are used in the text for calculating formal charge for the A-group elements.)

(Atomic weights in parentheses are for the isotope of longest life.)

Period	1 1A	2 2A	3 3B	4 4B	5 5B	6 6B	7 7B	8 8B	9 8B	10 8B	11 1B	12 2B	13 3A	14 4A	15 5A	16 6A	17 7A	18 8A
1	1 **H** 1.008																	2 **He** 4.003
2	3 **Li** 6.941	4 **Be** 9.012											5 **B** 10.811	6 **C** 12.011	7 **N** 14.007	8 **O** 15.999	9 **F** 18.998	10 **Ne** 20.178
3	11 **Na** 22.99	12 **Mg** 24.305											13 **Al** 26.982	14 **Si** 28.085	15 **P** 30.974	16 **S** 32.06	17 **Cl** 35.452	18 **Ar** 39.948
4	19 **K** 39.098	20 **Ca** 40.078	21 **Sc** 44.956	22 **Ti** 47.867	23 **V** 50.942	24 **Cr** 51.996	25 **Mn** 54.938	26 **Fe** 55.845	27 **Co** 58.933	28 **Ni** 58.693	29 **Cu** 63.546	30 **Zn** 65.38	31 **Ga** 69.723	32 **Ge** 72.63	33 **As** 74.922	34 **Se** 78.971	35 **Br** 79.904	36 **Kr** 83.798
5	37 **Rb** 85.468	38 **Sr** 87.62	39 **Y** 88.906	40 **Zr** 91.224	41 **Nb** 92.906	42 **Mo** 95.95	43 **Tc** (97)	44 **Ru** 101.07	45 **Rh** 102.906	46 **Pd** 106.42	47 **Ag** 107.868	48 **Cd** 112.414	49 **In** 114.818	50 **Sn** 118.71	51 **Sb** 121.76	52 **Te** 127.6	53 **I** 126.904	54 **Xe** 131.293
6	55 **Cs** 132.905	56 **Ba** 137.327	57 **La** 138.905	72 **Hf** 178.49	73 **Ta** 180.948	74 **W** 183.84	75 **Re** 186.21	76 **Os** 190.23	77 **Ir** 192.217	78 **Pt** 195.084	79 **Au** 196.967	80 **Hg** 200.592	81 **Tl** 204.38	82 **Pb** 207.2	83 **Bi** 208.98	84 **Po** (209)	85 **At** (210)	86 **Rn** (222)
7	87 **Fr** (223)	88 **Ra** (226)	89 **Ac** (227)	104 **Rf** (267)	105 **Db** (270)	106 **Sg** (269)	107 **Bh** (270)	108 **Hs** (270)	109 **Mt** (278)	110 **Ds** (281)	111 **Rg** (281)	112 **Cn** (285)	113 **Nh** (286)	114 **Fl** (289)	115 **Mc** (289)	116 **Lv** (293)	117 **Ts** (293)	118 **Og** (294)

Transition elements: groups 3–12 (3B–2B)

Lanthanides	58 **Ce** 140.116	59 **Pr** 140.908	60 **Nd** 144.242	61 **Pm** (145)	62 **Sm** 150.36	63 **Eu** 151.964	64 **Gd** 157.25	65 **Tb** 158.925	66 **Dy** 162.5	67 **Ho** 164.930	68 **Er** 167.259	69 **Tm** 168.934	70 **Yb** 173.045	71 **Lu** 174.967
Actinides	90 **Th** 232.038	91 **Pa** 231.036	92 **U** 238.029	93 **Np** (237)	94 **Pu** (244)	95 **Am** (243)	96 **Cm** (247)	97 **Bk** (247)	98 **Cf** (251)	99 **Es** (252)	100 **Fm** (257)	101 **Md** (258)	102 **No** (259)	103 **Lr** (262)